Other available titles

Laxton's General Specification Volumes 1 & 2
(0 7506 3352 0)

Laxton's General Specification – electronic version
(0 7506 3693 9)

Laxton's Guide to Term Maintenance Contracts
(0 7506 2977 0)

Laxton's Measurement Rules for Contractors Quantities
(0 7506 2977 0)

Laxton's Trades Price Book: Small Works Repairs and Maintenance
(0 7506 2978 9)

Laxton's Guide to Budget Estimating
(0 7506 2967 3)

TO ORDER: CREDIT CARD HOTLINE - 01865 888180

BY MAIL: Technical Marketing Dept., Butterworth-Heinemann, FREEPOST, Oxford OX2 8BR

BY FAX: Heinemann Customer Services – 01865 314091

BY EMAIL: Send orders to: bhuk.orders@repp.co.uk

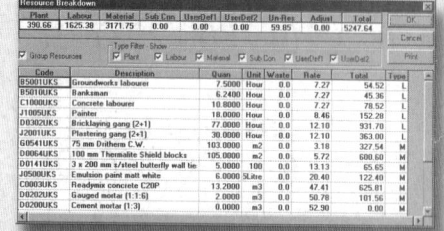

Laxton's
BUILDING PRICE BOOK
MAJOR & SMALL WORKS

Edited by **V.B. Johnson & Partners**

Chartered Quantity Surveyors • Construction Cost Consultants • Project Managers

ONE HUNDRED & SEVENTY SECOND EDITION

2000

Laxton's

An imprint of Butterworth-Heinemann

www.laxtonsprices.co.uk

Published by Laxton's, an imprint of Butterworth-Heinemann
Linacre House, Jordan Hill, Oxford OX2 8DP

A division of Reed Educational and Professional Publishing Ltd

A member of the Reed Elsevier plc group

OXFORD   AUCKLAND   BOSTON
JOHANNESBURG   MELBOURNE   NEW DELHI

First published 1999

British Library Cataloguing in Publication Data
A catalogue record for this book is available from the British Library

ISBN 0 7506 4599 7

Printed and bound in Great Britain

FOR EVERY TITLE THAT WE PUBLISH, BUTTERWORTH-HEINEMANN
WILL PAY FOR BTCV TO PLANT AND CARE FOR A TREE.

INTRODUCTION

Tender prices rose in the year to 1st quarter 1999 above the rate of inflation and are predicted to continue rising well into the Millennium.

The National agreed wage award this year came into force from 14th June 1999 and an additional days holiday for the Millennium has been awarded effectively adding 8.72% to the rate for craftsmen and 8% to the rate for Adult General Operatives, this is reflected within the prices herein.

As in previous years the "Build-up" of the "All-in" labour rates has been re-calculated; adjustment being made to changes in Labour Rates and in National Insurance contributions.

A section is included in Laxton's setting out the requirements of the Construction (Design and Management) Regulations 1994. This is now iaw and it is a criminal offence not to observe the requirements of these regulations. Clients, Designers (including Quantity Surveyors) and Contractors please note.

As with the previous edition, the maximum amount of pricing information is given in an easy to read format. Price rates are analysed under ten headings to show the complete estimating picture of materials, labour constants, labour rates and sundries all on one page.

The section headings are as follows:

Introduction
How to use
Essential Information
Contents
Index
Regional Variations
Preliminaries/General Conditions

Laxtons Building Price Book has been updated and structured generally in accordance with the Seventh Edition Revised 1998 of the Standard Method of Measurement of Building Works (SMM7)) which follows the format of the 'Common Arrangement of Work Sections for Building Works' 2nd edition which has been published to encourage Project Information to be co-ordinated throughout the industry.

The user is urged to read the relevant sections of SMM7 which requires certain measurement rules, definition rules and coverage rules to be observed. Certain labours are deemed to be included in the measured items (and prices) without mention and care should be taken to ascertain precisely the extent of these labours by reference to the Coverage Rules in SMM7.

The index on the front cover and the aligned markers on the edge of pages will assist the reader to find the sections in the book. The contents list and comprehensive index will further assist in locating specific items and subjects.

MAJOR WORKS

C	Existing Site/Buildings/Services
D	Groundwork
E	In-situ concrete/Large precast concrete
F	Masonry
G	Structural / Carcassing metal / timber
H	Cladding/Covering
J	Waterproofing
K	Linings/Sheathing/Dry partitioning
L	Windows/Doors/Stairs
M	Surface finishes
N	Furniture/Equipment
P	Building fabric sundries
Q	Paving/Planting/Fencing/Site furniture
R	Disposal systems
S	Piped supply systems
T	Mechanical heating/cooling/refrigeration systems
U	Ventilation/Air conditioning systems
V	Electrical supply/power/lighting systems
W	Communications/Security/Control systems
X	Transport systems
	Basic Prices of Materials
	Composite Prices for Approximate Estimating

SMALL WORKS

C	Existing Site/Buildings/Services
D	Groundwork
E	In-situ concrete/Large precast concrete
F	Masonry
G	Structural / Carcassing metal / timber
H	Cladding/Covering
J	Waterproofing
K	Linings/Sheathing/Dry partitioning
L	Windows/Doors/Stairs
M	Surface finishes
N	Furniture/Equipment
P	Building fabric sundries
Q	Paving/Planting/Fencing/Site furniture
R	Disposal systems
S	Piped supply systems
T	Mechanical heating/cooling/refrigeration
U	Ventilation/Air conditioning systems
V	Electrical supply/power/lighting systems
W	Communications/Security/Control systems
X	Transport systems
	Basic Prices of Materials
	Composite Prices for Approximate Estimating

General Information
Brands and Trade Names
Company Information
Products and Services
Index To Advertisers

The Landfill Tax was implemented during October 1996. The rate of tax currently stands at £2.00 per tonne for "inactive" waste and £10.00 per tonne for other waste. Users of Laxtons should note that the Landfill Tax has been included within Disposal items with alternatives for Inert, Active and Contaminated materials.

The Arbitration Act received Royal Assent on 17th June 1996 and came into operation on 31st January 1997. The Arbitration Act 1996 is printed in full in the General Information Section.

The Housing Grants Construction and Regeneration Act 1996 came into force on 1st May 1998, it includes mandatory provisions for disputes in construction contracts to be referred to adjudication, including any contracts which do not comply with these requirements. Also included are agreements made between clients and their consultants, further details are included at the end of the Arbitration Section.

Laxton's Approximate Estimating section has been rewritten, extended and costed to enable complete buildings to be priced following the format of the Standard Form of Cost Analysis as published by the Building Cost Information Service of the Royal Institution of Chartered Surveyors. The section gives examples of a detailed analysis for a detached house and an office block together with alternative composite prices enabling quick and comprehensive estimates to be produced and amended as detail design continues. This section has been tailored to the needs of the Quantity Surveyor for approximate estimating data in detail but in a simplified format than can be obtained from other sources.

A new Working Rule Agreement became effective on 29th June 1998 covering the building and civil engineering industry. The new Agreement includes a pay structure for a general operative, additional skilled rates, as well as the craft rate.

Basic Prices of Materials are included in both the Major and Small Work sections. The materials prices in the analysed rates are taken from these sections.

The pricing level of Laxton's is based on an average National level with indicative regional variations for overall pricing shown on the regional factors map.

All-in labour rates are shown at the head of each page in both the Major Works and Small Work sections. The build-up of these rates is illustrated in the Preliminaries/General Conditions pages.

Prices take account of the wage rates which became operative on 14th June 1999. Material prices used are those current during the second quarter of 1999.

Uniclass table J is based on the Common Arrangement and is used for organising information in specifications and bills of quantities and for classifying information on particular types of construction operations

Every endeavour has been made to ensure the accuracy of the information printed in the book but the Publishers and Editors cannot accept liability for any loss occasioned by the use of the information given.

The Editor wishes to thank the Professional bodies, Trade Organisations and firms who have kindly given help and provided information for this edition.

ACO Technologies PLC
Aerial Plastics
Air Diffusion Ltd.
Alfred McAlpine Slate
Ancon CCL
Angle Ring Co. Ltd.
Bar Fab Reinforcements
Beta Naco Ltd. (Louvre Windows)
Birtley Building Products Ltd.
Boddingtons Ltd.
Boulton & Paul
BRC
British Decorators Association
British Gypsum Ltd.
British Patent Glazing Ltd.
British Sisalkraft Ltd.
British Steel
Burlington Slate
CAMAS Building Materials
Cape Boards Ltd.
Caradon Celuform Ltd.
Caradon Terrain Ltd.
Carter Concrete Ltd.
Charcon Hard Landscaping
Construction Employers Federation Ltd.
Civil Engineering Contractors Association.
Civil Engineering Developments Ltd.
Clay Pipe Development Association Ltd.
Concrete Utilities
Crescent of Cambridge Ltd.
Crittall Windows Ltd.
Crosby Sarek Ltd.
Crowthorne Fencing
Deecrete Floors Ltd.
Del Piling Contractors
Dow Construction Products
Drainage Systems
Durable Berkeley.
E.C.C. Quarries Ltd.
Electrical Contractors' Association
Expanded Metal Co. Ltd.
Eternit
Filon
Fixatrad Ltd.
Fosroc Ltd.
Four Tees Engineering
Philip Grahame
Grace Construction Products Ltd.
Grass Concrete International Ltd.
Halfen
Hanson Bricks
Harris & Edgar
Heating and Ventilating Contractors Association
P.C Henderson Ltd.
Hepworth Building Products Ltd.
Hepworth Iron Co. Ltd.
Housing Corporation
H.S.S. Hire Shops
Ibstock Bricks
Jacksons Fencing
Johnston Pipes Ltd.
Joint Council for the Building and Civil Engineering Industry (N. Ireland)
Joint Industry Board for Plumbing
Marlflex
Mechanical Engineering Services in England and Wales
Kascade Drains Ltd.
Kerner Greenwood & Co
Kee Klamps Ltd.
Kingston Craftsmen Structural Timber
Kirkstone Quarries
Klargester Environmental Engineers Ltd

Kufa Plastics Ltd.
Kvaerner Cementation Ltd.
Light Alloy Ltd.
Madeley Paints (Distributors) Ltd.
Magnet Trade
Mandor Engineering Ltd.
Marley Buildings Ltd.
Marley Floors Ltd.
Marshalls Mono Ltd.
Masterbill Micro Systems
Mastic Asphalt Council & Employers Federation Ltd.
Mawrob Co (Engineers) Ltd.
Meshlite Ltd.
Metal Sections Ltd.
Microfloor Ltd.
Milton Pipes Ltd.
National Federation of Terrazzo, Marble and Mosaic Specialists
National Joint Council for the Building Industry
National Joint Council for the Laying Side of the Mastic Asphalt Industry
Natural Stone Products Ltd.
Nuway Manufacturing Co Ltd.
Oswestry Reinforced Plastics Ltd.
Owens Corning Building Products
Pendock Profiles Ltd.
Pre-Formed Components Ltd.
Promat Fire Protection
Ramsay & Sons(Forfar) Ltd.
Ranalah Gates Ltd.
Rawlplug Co Ltd.
R.C. Cutting & Co
Redland Bricks
Redland Roof Tiles Ltd.
Rentokil Ltd.
R.M.C
Rock & Alluvium Ltd.
Rockwool
Royal Institute of British Architects
Royal Institution of Chartered Surveyors
Ruberoid Building Products Ltd.
Sadolin(UK) Ltd.
Sealmaster Ltd.
Servicised Ltd.
Simplex
Solaglas Ltd.
Southern Evans Ltd.
Spit Fixings
B.M. Stainless Steel Drains Ltd.
Stoakes Systems Ltd.
Stowell Concrete Ltd.
Stressline Ltd.
Swish Products
Syston Rolling Shutters Ltd.
Tarmac Heavy Building Materials (UK) Ltd.
Tarmac Quarry Products(Southern Ltd).
Tarmac Topblock Ltd.
Thermalite Ltd.
Thomas Ness Ltd.
Thorn Lighting Ltd.
Timloc Building Products
Townscape Products Ltd.
Tremco Ltd.
Velux Co.Ltd.
Wards Cladding
Wavin Building Products Ltd.
Welconstruct Co Ltd.
Willan Building Services Ltd.

V.B.JOHNSON & PARTNERS,
Chartered Quantity Surveyors,
St John's House
304-310 St Albans Road,
Watford
Herts   WD2 5PE
Telephone : 01923 227236
Facsimile : 01923 231134
E Mail : office@ vbjw.demon.co.uk July 1999.

HOW TO USE

Laxton's Building Price Book is divided into 5 main sections; the Small Works section is printed on coloured paper. The sections are:

SECTIONS

1. MAJOR WORKS

Measured rates are given for all descriptions of building work generally in accordance with SMM7 Revised 1998. A complete breakdown is shown for all measured items, under 10 headings. This approach provides the reader with the complete price build-up at a glance and enables adjustments to be made easily. Net and gross rates are given for each item.

Prices reflect national average costs and the Major Works section is intended to apply to a building contract for new work or renovation within the range £250,000 to £1,000,000, with adjustment factors for tender values to £5,000,000.

2. SMALL WORKS (coloured paper)

This section follows an identical pattern to Major Works but is intended to apply to a contract in the range £25,000 to £75,000, with adjustment factors for tender values from £5,000 to £250,000.

The type of contract envisaged would involve one of the following :- (a) a small new build project, (b) a conversion and repair of existing two or three storey dwelling house with small extensions for additional accommodation, (c) the conversion, alteration and repair of existing office, factory or other type of building with a certain amount of internal reconstruction or renovation.

3. GENERAL INFORMATION

A useful reference section covering: Standard Rates of Wages, Builders' and Contractors' Plant, Guide Prices to Building Types, Cost Allowances, National Working Rule Agreement, Daywork Charges, Construction (Design and Management) Regulations 1994, Fees, Arbitration, Tables and Memoranda and Metric System.

4. BRANDS AND TRADE NAMES AND COMPANY INFORMATION

A unique list of Brands and Trade Names with the name and address of the manufacturer given in the adjacent Company Information Section. This list is useful for locating a particular branded item.

RATES GENERALLY

Rates given are within the context of a complete contract of construction where all trades work is involved. When pricing sub-contract work careful consideration must be given to the special circumstances pertaining to the pricing of such work and the very different labour outputs that can be achieved by specialists.

In total the Major and Small Works sections of Laxton's contain over 250,000 price elements and constants. The comprehensive index will take the reader straight to any particular measured item.

SCHEDULES OF RATES

In the event that Laxton's is to be used as a Schedule of Rates the employer should at tender stage clearly indicate whether rates are to be adjusted for regional factors, tender values and for overheads and profit.

It is suggested that the employer should state that works are to be valued in accordance with Laxton's Building Price Book 2000 Major Works or Small Works sections as appropriate. Net Rate column with the addition of a percentage to be stated by the contractor to include for regional factor adjustment, tender value adjustment, overheads, profit and also preliminary items if not covered elsewhere. Any proceedure for adjustment of fluctuation should be stated (eg either by using future editions of Laxton's price book or by the use of indices) and dates applicable.

Contractors must recognise that the net rates indicated within Laxtons Major Works and Small Works sections are national average costs for average projects in the range of £250,000 - £1,000,000 and £25,000 - £75,000 respectively. Minor items of maintenance works may require considerable adjustment.

ESSENTIAL INFORMATION

PRICING AND FORMAT

Materials

The delivered to site price is given together with a percentage allowance for waste on site, where appropriate, the resultant materials cost being shown in the shaded column.

Labour

Hours for craft operative and labourer are given, these being applied to the labour rates shown in the top left hand shaded panel, the resultant total labour cost being shown in the shaded column.

Sundries

These are incidental costs not included in the main materials or labour cost columns, e.g. mortar for brickwork, fixings for woodwork, plant specifically required for this item etc.

Mechanical Plant

Items of Mechanical Plant not included in Preliminaries /General conditions have been allowed for in the measured rates.

Plant and Transport costs are indicated separately in the measured rates for items where these are a major factor and material costs are insignificant as in the case of some groundwork items.

Rates

Net rates shown in the shaded column are exclusive of overheads and profit.

Gross rates are inclusive of 7.5% Main Contractor's overheads and profit in the Major Works section and 12.5% in the Small Works section. Readers can, of course, adjust the rates for any other percentages that they may wish to apply.

PERCENTAGE ADJUSTMENTS FOR TENDER VALUES

MAJOR WORKS

The measured rates in the Major Works section are based on new work contracts valued in the range of £250,000 to £1,000,000. As a guide to pricing works of larger value and for cost planning or budgetary purposes, the following adjustments may be applied to overall contract values

Contract value
£1,000,000 to £2,000,000	deduct 2.5%
£2,000,000 to £3,000,000	deduct 5.0%
£3,000,000 to £5,000,000	deduct 7.5%

SMALL WORKS

The measured rates in the Small Works section are based on new work contracts valued in the range of £25,000 to £75,000. As a guide to pricing works of lower or larger values and for cost planning or budgetary purposes, the following adjustments may be applied to overall contract values

Contract value
£ 5,000 to £ 15,000	add 20.0%
£ 15,000 to £ 25,000	add 10.0%
£ 25,000 to £ 75,000	rates as shown
£ 75,000 to £100,000	deduct 5.0%
£100,000 to £150,000	deduct 7.5%
£150,000 to £250,000	deduct 10.0%

RENOVATION WORKS AND WORK IN EXISTING UNOCCUPIED PREMISES

The rates in the Major Works or Small Works sections as appropriate may be used for this type of work however care must be taken to ensure adequate allowance is made for preparation works, difficulty of access and any double handling.

WORKING IN OCCUPIED PREMISES

Nearly all work of this nature is more costly in execution in that it has to be organised to suit the specific working conditions due to occupation and allowances should be made for the following:
(a) reduction in output arising therefrom.
(b) moving tradesmen on and off site as the progress of the contract demands.
(c) suppression of noise and dust.
As a guide, the extra labour involved in carrying out work in occupied premises could add between 50% and 100% to the labour cost of the same work in unoccupied premises.

REGIONAL VARIATIONS

Laxton's Building Price Book is based upon national average prices. The Regional Variations section and maps indicate factors that may be applied to adjust overall pricing to regional tender levels from data provided by the B.C.I.S.

USE OF FACTORS AND ADJUSTMENTS

Regional factors and tender value adjustments should only be applied to adjust overall pricing or total contract values. They should not be applied to individual rates or trades.

VALUE ADDED TAX

Prices in all sections exclude Value Added Tax.

Faber and Kell's
Heating and Air Conditioning of Buildings
Eighth Edition

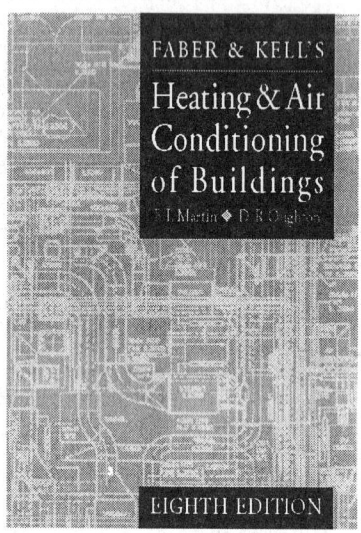

'**Faber and Kells**' has, for over fifty years, been accepted as the most practical and comprehensive book on heating and air conditioning design. It is regarded as the standard reference book for both students and practitioners.

In order to provide up-to-date information, this eight edition has been revised to include the latest changes to system design and covers many aspects in greater depth, whilst still retaining the character of previous editions.

Building services engineers, architects and others in the construction industry will find no better place for accessible and easily assimilated information on all aspects of the heating and air conditioning of buildings.

Paperback Edition, 1997, £35.00, isbn 0 7506 3778 1, 696pp

CONTENTS

The greatest care has been taken to ensure accuracy but the publishers can accept no responsibility for errors or omission.

INDEX

Regional Variations

The maps of Regional Factors are adapted from the BCIS Quarterly Review of Building Prices and are current as at the second quarter of 1999. A fuller analysis giving county factors is given in the "BCIS Quarterly Review of Building Prices" The information gives an approximate guide to the levels of overall pricing for the various regions of the United Kingdom. Prices in this book are based on a national average (1.00) and by multiplying by the factors indicated, comparative prices for regions may be obtained.

Considerable variations may also occur as a result of specific local factors, in particular within inner city areas and the reader should make allowance for these, further detail may be obtained from BCIS.

Introduction

The cost of a building is affected by its location. Many localised variables combine to produce a unique cost, including market factors such as demand and supply of labour and materials, workload, taxation and grants. The physical characteristics of a particular site, its size, accessibility and topography also contribute. Not even identical buildings built at the same time but in different localities will obtain identical tenders.

While all these factors are particular to a time and place, certain areas of the country tend to have different tender levels than others. The location factors given on the map are an attempt to identify some of these general differences using information derived from the BCIS Tender Price Index. The regions chosen are administrative areas and are not significant cost boundaries as far as the building industry is concerned.

It should be stressed that even within counties or large conurbations great variations in tender levels are evident and that in many cases these will outweigh the effect of general regional factors.

Calculation of Regional/County Factors

The location factors are based upon analysis of all the tender price indices calculated from mid 1982 onwards.

In order to convert the information to a common base, each individual index figure has been expressed as a factor of the relevant quarterly average index figure.

The individual job factors have been grouped into regions and counties. The Regional/County Factors have been calculated from the mean of the groups. The factors have been subject to statistical analysis which adjusts the individual job factors to reflect the regional trend.

Regions

The information has been analysed by county in England and Wales and by region in Scotland with the exception that Greater London has been further sub-divided into Outer London and London Postal Districts. These locations have been grouped into larger regions to give greater statistical reliability. The larger regions used are the Department of the Environment's Standard Statistical Regions except for the South East Region for which smaller groups are used and with Northern Ireland added.

The regions are defined as follows:

ENGLAND

Northern:
Cleveland, Cumbria, Durham, Northumberland, Tyne and Wear.

Yorkshire and Humberside:
Humberside, North Yorkshire, South Yorkshire, West Yorkshire.

East Midlands:
Derbyshire, Leicestershire, Lincolnshire, Northamptonshire, Nottinghamshire.

East Anglia:
Cambridgeshire, Norfolk, Suffolk.

South East:
Bedfordshire, Essex, Hertfordshire, Kent, Surrey, East Sussex and West Sussex, Berkshire, Buckinghamshire, Hampshire, Isle of Wight, Oxfordshire.

Greater London:
The area of the former Greater London Council.

South West:
Avon, Cornwall, Devon, Dorset, Gloucestershire, Somerset, Wiltshire.

West Midlands:
Hereford and Worcester, Shropshire, Staffordshire, Warwickshire, West Midlands.

North West:
Cheshire, Greater Manchester, Lancashire, Merseyside.

WALES

SCOTLAND

NORTHERN IRELAND

The map shows the regions and the regional factors at a glance.

Note:
The Regional/County Factors have been shown to represent fairly reliable average differences between locations. They are however averages and any individual project is unlikely to coincide exactly with the average result: the factors provide useful general guidance but on no account should they be used to adjust individual item prices.

This information is abstracted from a more detailed study of location factors which is included in the BCIS Quarterly Review of Building Prices available from BCIS, 12, Great George Street, Parliament Square, London, SW1P 3AD

Regional Factors

Scotland
0.95

Northern Ireland
0.75

Northern
0.94

Yorks & Humber
0.94

North West
1.00

East Midlands
0.93

West Midlands
0.94

East Anglia
0.96

Wales
0.94

South East
1.06

G.L 1.18

South West
0.99

Laxton's Building Price Book is based upon national average prices (=1.00). Indicative levels of tender pricing in regions as at the second quarter 1999 are given on this map and the factors shown may be applied to adjust overall pricing.

MAJOR WORKS – TENDER VALUES	SMALL WORKS – TENDER VALUES
The measured rates are based on contracts valued in the range of £250,000 to £1,000.00. As a guide to pricing works of larger value and for cost planning or budgetary purposes, the following adjustments may be applied to overall contract values. Contract Value £1,000,000 to £2,000,000.........deduct 2.5% £2,000,000 to £3,000,000.........deduct 5.0% £3,000,000 to £5,000,000.........deduct 7.5%	The measured rates are based on contracts valued in the range of £25,000 to £75,000. As a guide to pricing works of smaller or larger value and for cost planning or budgetary purposes, the following adjustments may be applied to overall contract values. Contract Value £5,000 to £15,000...............................add 20.0% £15,000 to £25,000............................add 10.0% £25,000 to £75,000........................rate as shown £75,000 to £100,000........................deduct 5.0% £100,000 to £150,000......................deduct7.5% £150,000 to £250,000......................deduct10.%

Preliminaries/General conditions

GENERALLY

The prices throughout this section offer in a convenient form
the means of arriving at approximate rates for the various operations
commonly met in connection with normal types of buildings. The
basis on which the rates are estimated is given below and should
be examined and understood before making any adjustments
necessary to adapt the prices to a specific contract or to rates of
wages other than those given.

Effective Dates of rates used:
Wages...June 28th, 1999
National Insurances................................. April 6th, 1999
Annual Holiday Stamps............................August 2nd, 1999
For earlier effective dates of the above see Standard Rates
of Wages Section
Includes additional Millenium day holiday

BASIC RATES OF WAGES
ALL-IN LABOUR RATES

The basic rates of wages used throughout are those which came
into force on June 29th 1998.

BUILDING		Craft Operatives £		General Operative £	Building
Guaranteed minimum weekly earnings..		235.95		177.45	
- 39 hours...		6.05		4.55	
2,015 hours (include inclement weather allowance)..	at £6.05	12190.75	at £ 4.55	9168.25	
Productivity Payments		1219.08		916.83	
Non-Productive Overtime 131 hours	at £6.05	792.55	at £ 4.55	596.05	
Sick Pay as WR.20 (per week)...		60.50		60.50	
Public Holidays 9 days x 8 hours = 72 hours	at £6.05	435.60	at £ 4.55	327.60	
		14698.48		11069.23	
National Insurance (Earnings threshold £83.00 per week)................... 12.2%	10382.48	1266.66	6753.23	823.89	
Holidays with Pay 47 weeks at £21.30..		1001.10		1001.10	
Training 0.38% of Payroll (CITB Levy)..		55.85		42.06	
		17022.09		12936.28	
Severance Pay 1.5% (Including loss of production during notice period absenteeism and turnover of labour)		255.33		194.04	
		17277.42		13130.32	
Employers Liability and Third Party Insurance	2%	345.55	2%	262.61	
Trade Supervision 3%..		518.32		393.91	
Cost per annum...		18141.29		13786.84	
Cost per hour - 1,965 working hours...		£ 9.23		£ 7.02	

Cost per hour - 1,965 working hours..
Skilled Operative Rate:-   4   £7.53   2   £8.49
                           3   £7.98   1   £8.80

Calculation of Hours Worked used above.

Summer: based on average 45 hours per week working		Hours	Hours
40 weeks			
Less Holidays			
Summer 2 weeks			
Easter 1 week	3 weeks		
37 weeks at 45 hours.... 1665			

Winter: based on average 43 hours per week working		Hours	Hours
12 weeks			
Less Holidays			
Winter 2 weeks	2 weeks		
10 weeks at 43 hours.....		430	
Less Bank Holiday 1 day at 8 hours...........		-8	
Less Sick leave 5 days at 8 hours.........		-40	382

ALL IN LABOUR RATES - cont'd

Less Bank Holidays-	4 days at 8 hours..........	32	1663

Add Summer Hours......................................	1633
Total Hours	2015
(Less Inclement Weather time)	50

Total Actual Hours Worked 1965

Calculation of Non-Productive Overtime hours included above.

Based on a 45 hour working week in Summer and a 43 hour working week in Winter, the calculation is as follows:-

				Hours
Summer				
6 hours overtime per week at time and a half	= 3 hours x 37 weeks	=		111
Winter				
4 hours overtime per week at time and a half	= 2 hours x 10 weeks	=		20
Non-Productive hours per annum				131

PLUMBING

The rates of wages used in the Plumbing and Mechanical Engineering Installations Section are those approved by The Joint Industry Board for Plumbing and Mechanical Engineering Services in England and Wales.

The All-in Labour rate calculations are as follows:

Effective dates of rates used:

Wages................................. August 24th 1998
National Insurance............... April 6th 1998

			Trained Plumber £		Advanced Plumber £		Technical Plumber £
47.8 weeks at 37.5 hours = 1792.5 hours	at £6.17	11059.73	at £7.20	12906.00	at £8.00	14340.00
Welding Supplement 1725 hours (Gas or Arc)....................		-	at £0.27	465.75	at £0.27	465.75
Travel time... .		464.60		464.60		464.60
		11524.33		13836.35		15270.35
Allowance for Incentive Pay	15%	1728.65	15%	2075.45	15%	2290.55
		13252.98		15911.80		17560.90
National Insurance (Employee Contracted Out).....................		854.40		1097.28		1247.52
Pension Contribution	6.5%	861.44	6.5%	1034.27	6.5%	1141.46
Redundancy Payments	1.5%	198.79	1.5%	238.68	1.5%	263.41
Employer's Liability Insurance	2%	265.06	2%	318.24	2%	351.22
Fares..		1414.50		1414.50		1414.50
Benefit stamps 52 weeks.................................... at £25.40		1320.80	at £28.55	1484.60	at £30.95	1609.40
Cost per annum.. ..		18167.97		21499.37		23588.41
Cost per hour - 1725 hours..		10.53		12.46		13.67
Inclement Weather time ....…….......	1%	0.11	1%	0.12	1%	0.14
	£	10.64	£	12.58	£	13.81

The Plumber's all-in wage rate used throughout the book is an average of 1 Trained, 3 Advanced and 1 Technical Plumbers rates giving an average rate of £12.44 per hour.

Note : 1. Total Weekly Stamp Value is the sum of Holiday Contribution and Welfare Credit.

2. Daily Travel Allowances based upon daily rate specified for each grade on an incremental scale of 5 miles for distances over 5 miles but not exceeding 50 miles.

ELECTRICAL

The rates of wages used in the Electrical Engineering Installation Section are those approved by the Joint Industry Board for the Electrical Contracting Industry.

The All-in Labour rate calculations are as follows:

Effective dates of rates used:

Wages.. April 1st 1999.

National Insurance........................ April 6th 1999.

	days	weeks	hours	APPROVED + 50p rate		APPROVED rate		ELEC. rate		TECH. rate	
a HOURS WORKED (less 1 week sick)	45.20	45.00		7.92	16109.28	7.42	15092.28	6.76	13749.84	8.48	17248.32
STATUTORY SICK PAY 2.00 (NOT PAYABLE FOR THE FIRST THREE DAYS)				11.54	23.08	11.54	23.08	11.54	23.08	11.54	23.08
b NPOT	45.20	3.50	7.92	1252.94	7.42	1173.84	6.76	1069.43	8.48	1341.54	
c INCENTIVE PAYMENT 17.15%				17.15%	2762.74	17.15%	2588.33	17.15%	2358.10	17.15%	2958.09
d TRAVEL TIME 45.20 (All taxable and Subject to NI)			45.00	0.55	1118.70	0.55	1118.70	0.55	1118.70	0.55	1118.70
e TRAVEL ALLNCE. (TA part taxable subject to NI)					0.00		0.00		0.00·		0.00
STAT HOLS 9.00			7.50	7.92	534.60	7.42	500.85	6.76	456.30	8.48	572.40
SUBTOTAL					21801.35		20497.08		18775.45		23262.12
NATIONAL INS				12.20%	2659.76	12.20%	2500.64	12.20%	2290.60	12.20%	2837.98
SUBTOTAL					24461.11		22997.72		21066.05		26100.10
TRAINING 2.50%				2.50%	545.03	2.50%	512.43	2.50%	469.39	2.50%	581.55
e TRAVEL ALLNCE. (TA Tax and NI exempt element)					0.00		0.00		0.00		0.00
JIB STAMP 52.00				29.46	1531.92	29.46	1531.92	27.46	1427.92	32.66	1698.32
SUBTOTAL					26538.06		25042.07		22963.36		28379.97
SEVERANCE PAY 2.00%				2.00%	530.76	2.00%	500.84	2.00%	459.27	2.00%	567.60
SUBTOTAL					27068.82		25542.91		23422.63		28947.57
EMP. LIAB & 3rd PARTY				2.00%	541.38	2 00%	510.86	2.00%	468.45	2.00%	578.95
COST PER YEAR					27610.20		26053.77		23891.08		29526.53

GANG RATE   **INDIVIDUAL RATES**

TECH	29526.53 x 1.00 Non productive	29526.53		£14.52
APPROVED ELEC.+	27610.20 x 1.00 100% productive	27610.20	£13.57	
APPROVED ELEC.	26053.77 x 2.00 100% productive	52107.54		£12.81
ELEC.	23891.08 x 4.00 100% productive	95564.32		£11.75
		204808.59		

Average man hours	45.20 x 45.00	2034 hrs
Average cost per man	7.00 men	**£14.38**

Based on National JIB rates

a.   basic week = 37.5 hrs.
b.   Hours to be worked before overtime paid = 38
c.   Rounded down to 15% will give a rate of £14.17, rounded up to 20% will give a rate of £14.67.
d.   Agreement which pays a supplementary rate on the worked hours
e.   This element to be phased out, starting with 0 – 10 mile band (which the above is based on)
    The above JIB hourly rates are from the 4/1/99. No further agreements have been announced.
    Hours remain at 45 as this seems to be the minimum paid in the industry at the moment
    Extra statutory holiday included
    National insurance increased to 12.2%

SITE PRELIMINARIES

GENERAL NOTES

The calculation of the costs to be included for Site Preliminaries requires careful consideration of the specific conditions relating to the particular project. No pre-conceived percentage addition can be adopted with any degree of accuracy. The cost of Site Preliminaries is dependent on the type of contract under consideration, its value, its planned programme time, its location and also on the anticipated requirements for the administration and management of the actual site.

Accurate assessment of the contractor's site costs are equally important in small works contracts as in major works.

Two factors which must be established prior to commencement of the detailed calculations are (a) the value of the main contractors work (from which is derived the value of the contractors labour force) and (b) the programme time for completion of the Works. If the latter is not stated in the contract conditions, it must be decided by the contractor. If it is stated in the contract conditions, it must be checked and confirmed by the contractor as being reasonable.

NOTES ON SITE PRELIMINARIES

Section A40 - Management and Staff

The numbers, types and grades of Site Staff to be included are matters for judgement by the Contractor.

A selection from the following list will provide the range of staff required to administer a normal building project:

Site Agent or Manager	General Foreman
Engineer	Checker
Storeman	Cashier/Wages Clerk
Materials Purchasing Clerk	Plant Clerk
Quantity Surveyor	Resources Manager
Productivity Controller	Safety Officer
Plant Fitter	Administrative Manager
Typist/Telephonist	

Allowance for Trades Supervision has been made in the Gross Wage build up.

On small contracts individual staff members may combine the duties of several of the foregoing grades. On larger contracts more than one member may be required for certain duties, for example two or more Engineers may be required in the initial stages for setting out the Works and for checking the frame construction.

From 2nd September 1996 new regulations will apply to the health, safety and welfare of workers on construction sites the associated costs of which are not specifically identified in this section and aquaintance with these regulations is recommended.

Section A41-1 - Site Accommodation

The size of the office required to accommodate the contractors site staff will depend on the number of staff envisaged. An allowance of 10m2 per staff member plus 25% for circulation space and toilets is normal.

The overall cost will vary according to the contract period if the hutting is to be hired. Otherwise a use and waste allowance can be made if the hutting is supplied from the contractors own resources. The cost should include for erection and eventual dismantling and, in the case of contracts extending over several years, for labour and materials in maintenance and redecoration and any ancillary accommodation which may be required for the Architect, Clerk of Works or Resident Engineer.

In works of renovation and rehabilitation, office accommodation and space for storage of materials is often available in the existing premises. Otherwise a simple lock up hut is all that is normally required for small works contracts.

Allowance should be made for heating, lighting, furniture and any attendance required. Accommodation priced in this Section could include Offices, Stores, Site Latrines, Canteen, Drying Rooms and Sanitary Facilities, First Aid Cabin, Plant Fitters Sheds, Reinforcement Bending Sheds.

Section A42-2 - Lighting

This item provides for any lighting required inside the buildings during construction or for external floodlighting installed during the winter months to maintain progress. To the cost of wiring, lights and other equipment, should be added the charge for electricity and electrician's time for the installation and removal of the system.

On small works contracts the contractor, by agreement with the client, may be permitted to use existing services. A nominal amount should be included for the use of electricity.

Section A42-4 - Water

Charges made by various Water Companies for the provision of water for the works differ considerably throughout the country and the actual rate to be paid should be ascertained regionally. Most water companies meter the water required for building purposes, a typical charge for which is currently 53.62 p/m3 (1m3 = 220 gallons) with a standing charge for the building water supplied of :

House / Bungalow (Meter size 25mm)	£76
Flat / Maisonette (Meter size 20mm)	£43
Light Industrial / Commercial (Meter size 25mm)	£76
Other Industrial / Commercial (Meter size 50mm)	£304

If a non-metered supply is already in existence on site then a charge may be applied based upon 0.17% of the contract value (minimum charge £53.50) and a yearly charge of £57.00.

A meter may also be fitted to an existing water supply for building purposes (prior to commencement of works) and charged at the rates stated previously.

To this must be added the cost of all temporary plumbing required on the site such as distribution piping, standpipes, hoses and tanks.

Typical clients connection charges relating to water supply are as follows :

	Av.cost. £
1. Connection charge	348.00
2. Pipelaying charge (per metre)	82.50
3. Infrastructure charges	211.00

Network charges are also levied on customers who use water for non-domestic purposes. The charge for this is based upon the diameter of the meter used.

Section A42-6 - Safety, Health and Welfare

Protective helmets, boots, goggles and protective clothing for operatives; wet weather clothing, drying sheds; provision of site latrines and items of a similar nature.

Section A42-7 - Storage of Materials

Consideration should be given to the provision of the following items:

Cement storage shed.
Aggregate storage bins.
Use and waste of tarpaulins and plastic sheeting.
Lockups and toolboxes for tradesmen's tools.

Section A42-8 - Rubbish disposal

Removal of rubbish during construction including clearing out after subcontract trades.

NOTES ON SITE PRELIMINARIES (Contd)

Section A42-9 - Cleaning

This Section includes keeping the site tidy throughout the construction period, clearing out after subcontract trades and the final cleaning up on completion of the Works prior to handover.

Section A42-10 - Drying Out

Allowance should be made in this Section for drying out the building by the use of blow heaters or, if permitted, by the provision of fuel to the heating system as installed.

Section A42-11 - Protection of Work

The protection of completed work during the contract period is usually priced in the relevant bill rates. Types of work requiring special protection include expensive stonework or precast concrete features, stair treads, floor finishes, wall linings, high class joinery and glazing to windows.

Section A42-12 - Security

Security of Site - Allow for the provision of a watchman if required or for periodic visits by a security organisation. Some allowance may also be required in particular localities for loss of materials due to theft.

Section A42-14 - Small plant and tools

Small plant and tools comprise ladders, wheelbarrows, trestles, picks, shovels, spades, sledgehammers, cold chisels and similar items. The cost of these minor items is usually allowed for by an addition to the overall cost of labour of between 1% and 2%.

Section A42-15 - Others

a) Insurance of Works

The Contract Bills of Quantities will normally state whether Fire Insurance is the responsibility of the Contractor or of the Employer. Rates of insurance vary according to the nature of the work but are usually in the range £0.10% to £0.30% of the total value to be insured. The total insurable value should include for full reinstatement of any damage caused, all cost escalation, any demolition of damaged work required and an appropriate percentage addition for professional fees.

Note: Allowance should be made for Terrorism insurance if required.

b) Travelling Time and Fares

Payments for travelling time and fares can be a major expense and since all contracts and particularly the JCT Conditions of Contract place on the Contractor the onus of including in the tender sum the cost of importing labour, investigation should be made into the prevailing conditions regarding the availability of local labour. When all enquiries have been made, the calculations of anticipated expenditure must be based on the Contractor's judgement of the labour position. Rates of payment for travelling time and fares are laid down in the relevant national and local Working Rule Agreements.

c) Miscellaneous

(1) Rates and Taxes on Temporary Buildings. The charges made by the Local Authority for rates on site accommodation should be ascertained locally.

(2) Signboards and notices.

(3) Glass breakage during unloading or after fixing.

(4) Sample panels.

(5) Attendance on site - provision of chainman and any special attendance required by the Clerk of Works.

(6) Setting out equipment including the use of theodolites, levels or other items and provision of pegs, paint etc.

(7) Provision for winter working. This can range from heating aggregates in the stockpiles to the provision of polythene sheeting to protect the operatives and/or the Works during construction.

(8) Allowance for the cost of labour and materials in rectifying minor defects prior to handover of the Works and during the Defects Liability Period.

(9) Sureties or bonds. The cost of any surety or performance bond for due completion of the Works required by the Contract Documents. Rates for this vary depending on the financial status of the Contractor.

(10) Special Conditions of Contract. These can refer to any particular point required by the Employer and a careful perusal of the Conditions and the Preambles to the Bills of Quantities or Specification is necessary to ensure that no requirement is overlooked. Typical items are the suppression of noisy mechanical equipment, regulations dealing with the restricted use of tower cranes or limitations to ground disturbance during piling operations.

Section A42-16 - General Attendance on Nominated Subcontractors

General Attendance on Nominated Subcontractors by the Main Contractor is deemed to include the use of the Contractors temporary roads, pavings and paths, standing scaffolding not required to be altered or retained, standing power-operated hoisting plant, the provision of temporary lighting and water supplies, clearing away rubbish, provision of space for the subcontractors own offices, the storage of his plant and materials and the use of messrooms, sanitary accommodation and welfare facilities provided by the Contractor for his own use.

The estimated cost of such attendance is generally added by the Contractor in the Bills of Quantities as a percentage of the Prime Cost Sum shown for each Nominated Subcontractor.

Alternatively it can be included as a lump sum according to the Contractors experience and judgement. The actual percentage can vary from as little as 0.1%, up to 2.5% or more, depending on the anticipated or known requirements of the Nominated Subcontractor.

Section A43 - Mechanical Plant

This section comprises the cost of all mechanical plant which has not been priced in the Bills of Quantities. It includes Tower Cranes, Mobile Cranes, Goods and Passenger hoists and Pumping plant for dewatering.

The apportioning of the cost of tower cranes and hoists to separate items of the Bills of Quantities is made difficult and unreliable by the fact that very often they deal with various materials in differing trades and the correct time allocation to the individual trades is virtually impossible. Moreover, although not always working to capacity they cannot be removed from site entirely and a certain proportion of their time is wasted in standing idle. It is therefore normal practice to view the contract as a whole, decide the overall period for which the particular tower crane or hoist will be required, and include the whole cost in the Preliminaries.

Tower Cranes - There is a large variety of cranes available in the market and the most suitable crane for each particular contract must be selected. The transport of the crane from the plant yard to the site can be a costly item and the installation of track, the erection, testing and final dismantling of the crane can, on occasion, constitute a greater expenditure than the hire charge for the crane.

NOTES ON PRELIMINARIES - cont'd

Section A43 - Mechanical Plant - cont'd

Moblie Tower Crane - A typical build up of cost of mobile tower crane is as follows:

Installation		£
(1)	Prepare ground and lay crane track......................	1600
(2)	Supply of track materials...........................	600
(3)	Transport crane to and from site...........................	2000
(4)	Erect and test................................	3000
(5)	Electricity to crane................................	850
(6)	Dismantle crane and track................................	2300
		10350

Running Costs		£
(1)	Weekly hire rate..........................	750
(2)	Running costs, electricity, oil and grease.....	135
(3)	Driver....................................	345
(4)	Attendant................................	264
		1494
	x 26 weeks...........................	38844
		£49194

Static Tower Crane - A typical build up of cost of small static tower crane is as follows:

Installation		£
(1)	Erection and testing..................................	850
(2)	Transport to and from site...........................	550
(3)	Electrician..........................	320
(4)	Tying-in materials use and waste................	215
(5)	Dismantle and remove................................	550
		2485

Running Costs		£
(1)	Weekly hire rate..............................	425
(2)	Running costs, electricity, oil and grease........	55
(3)	Driver....................................	345
(4)	Attendant................................	264
		1089
	x 6 weeks.................................	6534
		£ 9019

Mobile Cranes - A mobile crane may be required during the frame construction.for lifting heavy items such as precast concrete floors or cladding units. The weekly hire rates will vary according to the lifting capacity of the crane required but a typical build up of cost is as follows:

	£
Hire of mobile crane per week................................	320
Fuel, oil and grease................................	65
Driver and attendant..	590
	975
x 4 weeks.................................	3900
Transporting to and from site................................	300
	£ 4200

Section A44-3 - Access scaffolding

The cost of scaffolding varies not only with the superficial area of the structure but also with the length of time the scaffolding is required to stand and the nature of the building. When calculating the cost, the girth of the building should be multiplied by the height and the area priced at the appropriate rate as shown below, to which should be added the additional charge for standing time exceeding the basic 4 weeks as shown. Provision must also be made for any internal

Pumping of ground water - Provision for the pumping of ground water (as opposed to water from normal rainfall) is entirely dependent upon the conditions applying to the particular site in question. Under the JCT Standard Form of Contract this is usually covered by a Provisional or Prime Cost sum but in contracts where the onus is placed on the contractor it is most important that a visit should be paid, during the tendering period, to the site of the proposed works to ascertain, as far as possible, the nature of the ground and the likelihood of encountering water. When, as a result of such an inspection, it is decided that pumping will be necessary, say, during excavation of foundations and until the foundation walls have reached a certain height, the calculations will be based upon the length of time it is thought necessary to keep pumps working, a typical build up of cost is as follows:

Assuming one 75mm pump will be kept running continuously for five weeks and for three weeks a second 75mm pump will be required during working hours with a standby pump of 50mm diameter available for emergencies

		£
(a)	Hire of 75mm pump 5 weeks at £90.00...	450.00
	Operator's time attending, fuelling, etc., 28 hours at £7.53 per hour = £210.84 x 5 weeks................................	1054.20
	Hire of hoses (2 lengths x 6m) at £5.20 5 weeks x 2 hoses....................................	52.00
(b)	Hire of 75mm pump 3 weeks at £90.00....	270.00
	Operator's time 12 hours at £7.53 per hour = £90.36 x 3 weeks................................	271.08
	Hire of hoses 3 weeks x 2 hoses at £5.20..	31.20
(c)	Standby pump 5 weeks at £57.20..............	286.00
	Total cost of pumping £	2414.48

Section A43-4 - Site Transport

This section includes all requisite site transport not priced in the rates, such as site lorries, dumpers, vans, forklift trucks, tractors and trailers. The cost of transporting mechanical plant, scaffolding and temporary accommodation, is also included in this section. The cost of double handling materials during unloading may also be considered for inclusion in this section.

Section A44-1 - Temporary Roads

Temporary Roads - This section includes providing access to the site for heavy excavating equipment and for site roads and hardstandings constructed in ash, hardcore or timber sleepers. Allowance should be made for maintenance over the contract period and for eventual breaking up and removal. Crossovers, planked footways and any automatic stop/go lights required should be included. If the base courses of permanent roads can be utilised as temporary roads, an allowance should be made for making good the base prior to final surfacing.

Allow also for any cleaning and maintenance which may be required on Public or Private roads, especially if surplus excavated material has to be carted off site.

scaffolding required to lift and stair wells and also for fixing suspended ceilings.

NOTES ON PRELIMINARIES - cont'd

Section A44-3 - Access scaffolding - cont'd

	Small Works		Major Works	
	Putlog £	Independent £	Putlog £	Independent £
Scaffolding for brickwork average 6m high...............m2	3.62	5.31	3.15	4.62
Scaffolding for brickwork average 12m high.............m2	3.99	5.80	3.46	5.04
Scaffolding for brickwork average 18m high.............m2	4.35	6.34	3.78	5.51
Add for each additional week....................................m2	0.19	0.27	0.16	0.23

		Small Works				Major Works			
Height	2.00m £	4.00m £	6.00m £	8.00m £		2.00m £	4.00m £	6.00m £	8.00m £
Mobile towers...........for one week	28.00	42.00	82.90	121.65		22.28	33.60	66.30	97.35
For each additional week..........add	22.35	33.60	66.30	97.35		17.90	26.85	53.05	77.85

Section A44-4 - Support Scaffolding

	Small Works £	Major Works £
Bandstand scaffolding for suspended ceilings.. m3	2.00	1.50
Add for each additional week.......................... m3	1.50	1.00
Internal birdcage scaffolding to lift wells, etc.,...m3	4.52	3.94
Add for each additional week.......................... m3	0.19	0.17

Section A44-5 - Hoardings, fans and fencing

The nature of these items is so dependent upon the site, type of contract, and the frontage of the building area on the public highway that it is difficult to offer any general method of calculating the anticipated cost. Consideration must be given to the means of stabilizing hoardings on sites where basements or other deep excavations come right up to the line of the hoarding, and gates must be sited to give easy access from the road. Gantries may require artificial lighting throughout their length in certain conditions, particularly in narrow streets in city centres. Any police requirements for offloading materials must be ascertained. Allowance must also be made for giving notices and paying any licence fees required by Local Authorities.

SPECIMEN SITE PRELIMINARIES

The examples of the build up of Site Preliminaries which follow are based on building contract values of £60,000 and £1,000,000 (excluding Preliminaries) for Small Works and Major Works respectively made up as follows:

	Small Works £	Major Works £
Nominated Subcontractors.................	-	275000
Nominated Suppliers...........	-	50000
Provisional Sums.................	10000	50000
Main Contractors Labour.....	20000	275000
Main Contractors Materials..	20000	275000
Subcontractors...................	10000	75000
	£60,000	£1,000,000

The Contract Periods have been programmed as 26 weeks (Small Works) and 52 weeks (Major Works).
The Contract Document is the Standard Form of Building Contract 1998 Edition.

Section A40 - Management and Staff	Small Works Per Week £	No. of Weeks		Major Works Per Week £	No. of Weeks	
Project Manager/Site Agent............................	-	-	-	490	26	12740
General Foreman...	340 x 35%	26	3094	340	52	17680
Site Engineer...	-	-	-	420	6	2520
Wages Clerk/Productivity Controller................	-	-	-	260	26	6760
Typist/Telephonist..	-	-	-	165	26	4290
Carried to Summary			£ 3094			£ 43990

Section A41 - Site Accommodation						
Contractors Office: 50m2 at £26.00 per m2.....................			-			1300
Office furniture and equipment................................			225			425
Clerk of Works Office: 15m2 at £26.00 per m2................			-			390
Stores: 30m2 at £16.00 per m2.....................................			-			480
Canteen/Drying Room: 50m2 at £26.00 per m2.................			-			1300
Canteen equipment...			-			365
Cabins or site huts...			-			225
Lockups and toolboxes.......................................			-			130
Office: Use of existing accommodation. Allow for making good on completion...			150			-
Stores: Use existing. Allow for shelving and lockups for materials storage...			225			-
Carried to Summary			600			4615

SPECIMEN SITE PRELIMINARIES - cont'd

	Small Works £	£	Major Works £	£
Section A42 - 1 and 2 - Power and Lighting				
Connection to mains...	-		225	
Distribution unit, cables, floodlights, lamps, plugs and sundry materials..	-		1250	
Electrician in attendance......................................	-		1350	
Cost of electricity: 52 weeks at £21 per week..................	-		1092	
Heating and lighting offices: 52 weeks at £18.00 per week	-		936	
Power and lighting: 26 weeks at £13.00 per week...........	338	338	-	4853
Section A42 - 4 - Water				
Site Water connection.......................................	76		500	
Water Company charges...	-		348	
Service piping: 100m at £6.00 per m........................	-		600	
Stand pipes: 3 no. at £30....................................	-		90	
Hoses tanks and sundry materials....................................	100	176	200	1738
Section A42-5 - Telephone and Administration				
Installation...	-		180	
Rental...	-		225	
Calls: 52 weeks at £16 per week......................................	-		832	
Telephones: 26 weeks at £16.00 per week.....................	416	416	-	1237
Section A42-6 - Safety, Health and Welfare				
Site Latrines: 2 no. at £275.00....................................	-		550	
Latrine materials: 52 weeks at £4.00 per week..............	-		208	
Sewer connection...	-		250	
Drainage; installation and removal................................	-		600	
First Aid..	-		200	
Protective clothing....................................	120		750	
CDM...	100	220	500	3058
Section A42-8&9 - Rubbish Disposal and Cleaning Site				
Site clearance during construction: 26 weeks at average £20 per week	520		-	
Site clearance during construction: 50 weeks at average £24 per week	-		1200	
Final cleaning...	200	720	400	1600
Section A42-10 - Drying Out				
Heaters and Fuel..	100	100	500	500
Section A42-12 - Security				
Weekend site supervision: 50 weeks at average £30 per week...	-		1500	
Security: allow for loss of materials...............................	200	200		1500
Section A42-14 - Small Plant and Tools				
Labour value £275,000 x 1.5%................................	-		4125	
Labour value £ 20,000 x 1.5%................................	300	300	-	4125
Carried forward		2470		18611

SPECIMEN SITE PRELIMINARIES - cont'd

	Small Works		Major Works	
	£	£	£	£
Brought forward		2470		18611

Section A42-15 - Insurance of works

	Small Works		Major Works	
Contract Sum...	60000		1000000	
Allow for part demolition and clearing up........................	1600		25000	
	61600		1025000	
Increased costs over contract period: say 6%..............	3696	say 10%	102500	
	65296		1127500	
Increased costs over reconstruction period: say 6%......	3918	say 10%	112750	
	69214		1240250	
Professional fees: 16%..	11074		198440	
	£ 80288		£ 1438690	
Cost of Insurance: £0.30%		240	£0.20%	2877

Section A42-15 - Travelling Time and Fares

Main Contractors Labour.......	250000	
Labour in Preliminaries.......	25000	
	£ 275000	

- Average £275 per week = 1000 man/weeks in contract.

	Small Works		Major Works	
1000 man/weeks x 40% of men local..	-		-	
x 20% of men receive £5.00 per week............	-		1000	
x 20% of men receive £7.50 per week............	-		1500	
x 20% of men receive £9.50 per week............	-		1900	

Travelling time and fares:

	Small Works		Major Works	
Allow 90 man weeks at average £6.50 per week...........................	585	585	-	4400

Section A42-15 - Miscellaneous

	Small Works		Major Works	
Rates and Taxes..	225		1000	
Contractors signboards...	-		200	
Chainman attending Engineer......................................	-		550	
Winter working..	-		1250	
Sample panels..	-		300	
Testing concrete cubes..	-		300	
Labour attendance in offices.......................................	-		1000	
Defects Liability period				
Allow 2 tradesmen and 2 labourers for 1 week.............	-		975	
Materials..	-		380	
Handover and defects liability				
Labour..	315		-	
Materials...	105	645	-	5955
Carried to Summary	£	3940	£	31843

SPECIMEN PRELIMINARIES - cont'd

	Small Works £	£	Major Works £	£
Section A43-1 - Mechanical Plant				
(a) Hoists: Assuming a 500-750 kg hoist is required for 12 weeks...............				
Hire of hoist per week...........................	-		100	
Fuel, oil and grease.............................	-		30	
Operator (part-time).............................	-		100	
			£ 230	
			x 12 hoist weeks	2760
(b) Pumping and Dewatering. It is assumed that a Provisional Sum is included in the Bills of Quantities - A typical build up is shown in the notes.	-		-	
(c) Scaffold Hoist: Allow 4 weeks at £185 per week................	340		-	
(d) Pumps: Allow 50mm pump at £70 per week x 2 weeks.....	140	480	-	
Section A43-4 - Transport				
(a) Site tractor and trailer..	-		720	
Site van and driver..	-		960	
(b) Transport Plant to Site				
Machine excavator..	-		250	
Mechanical equipment including concrete mixers, hoists and pumps..	-		280	
Site Offices and storage hutting............................	-		250	
Scaffolding..	-		250	
Sundries..	-		275	
(c) Double handling materials on site...........................	215		-	
(d) Transport scaffolding and small plant to site...........................	260	475	-	2985
Carried to summary		£ 985		£ 5745
Section A44-1 and 6 - Temporary Roads and Hardstandings				
Temporary Roads and Hardstandings				
Hardcore road: 100m2 at £9.45 per m2...............................	-		945	
Hardstandings: 100m2 at £9.45 per m2...............................	-		945	
Temporary roads - Allow for access to site...	175	175	-	1890
Section A44-3 and 4 - Scaffolding				
Assuming a three storey office building 60m x 10.0m x 10.0m high in traditional brick construction				
Access scaffolding				
External putlog scaffolding				
1400m2 at £3.15 per m2.....................................	-		4410	
Additional weeks: 10 weeks at 1400m2 at £0.16 per m2 per week...............	-		2240	
Mobile tower 8m high for 12 weeks...........................	-		908	
Support scaffolding				
Internal scaffolding for suspended ceilings				
1800m3 at £1.50 per m3.....................................	-		2700	
Internal scaffolding for staircases				
300m3 at £3.94 per m3.....................................	-		1182	
Access scaffolding - Allow for mobile towers externally:				
4 Nr at £160 each including erection and dismantling............	640	640	-	11440
Section A44-5 - Hoardings, fans, fencing etc				
Hoarding 2m high: 100m at £20 per m...		-	2000	
Gate..	-		200	2200
Carried to summary		£ 815		£ 15530

SITE PRELIMINARIES

SUMMARY

SMM7 Section		Small Works £	£	Major Works £	£
A40	Management and Staff..	-	3094		43990
A41	Site Accommodation...	-	600		4615
A42	Power and Lighting...	338		4853	
	Water...	176		1738	
	Telephone...	416		1237	
	Safety, Health and Welfare..............................	220		3058	
	Rubbish Disposal and Cleaning Site...................	720		1600	
	Drying out building..	100		500	
	Security..	200		1500	
	Small Plant and Tools....................................	300		4125	
	Insurance of Works.......................................	240		2877	
	Travelling Time and Fares...............................	585		4400	
	Miscellaneous...	645	3940	5955	31843
A43	Mechanical Plant...	510		2760	
	Transport...	475	985	2985	5745
A44	Temporary Roads...	175		1890	
	Scaffolding...	640		11440	
	Hoardings..	-	815	2200	15530
			£ 9434		£ 101723
	Overheads and Profit 12.5%		£ 1179	7.5%	£ 7629
	Total of Site Preliminaries		£ 10523		£ 109352

FIRM PRICE ADDITION (Major Works)

The amount to be added for a firm price tender is dependent on the overall contract period of the project, the known or anticipated dates of wage awards and the estimated percentage amount of the award. Separate calculations must be carried out for anticipated labour increased costs, materials increases, plant increases, staff salary awards and the increased costs required by direct subcontractors, if any. The total of these represents the addition to be made for a firm price tender.

A typical calculation is as follows:

	£	£
Labour: Assume a 3.5% wage award half way through the 52 week contract period		
The value of labour affected by the award is calculated as £150000		
The firm price addition is 3.5% on £150000..		5250
Materials:The value of materials affected by the award is estimated as £150000		3750
Firm price addition for materials: allow average 2.5% on £150000..............		500
Staff: Allow 4% increase on £25000..		1000
Subcontractors: Net value affected by increased costs, say £65000		
Allow labour 3.5% on £30000..	1050	
Allow materials 2.5% on £35000...	875	1925
Net addition for firm price:		£ 12425

FIRM PRICE ADDITION (Small Works)

Firm Price: Most small works contracts are based on firm price tenders. This is due in the main to the relatively short contract periods normally involved. It is therefore the usual practice to make a lump sum allowance in the Preliminaries for any anticipated cost increases in labour and materials.

Allowance for firm price	200
Net adddition for firm price:	£ 200

LANDFILL TAX (Major and Small Works)

The Landfill Tax is included as indicated within the Disposal items of the Groundwork items within Laxton's rates in this edition.

There are two rates of Tax:-   £2.00 per tonne for inactive waste.
                                       £10.00 per tonne for all other waste.

Additional allowances should also be added for contaminated waste and specific quotations obtained as required.

The cost of excavation will be affected by the nature of the soil and it's subsequent bulking factor in handling and disposal.

A typical conversion rate is 2.10 tonnes per cubic metre. This would result in an addition of £4.20 per cubic metre for inactive waste and £21.00 per cubic metre for active waste to the disposal charges included within this edition.

Allowances for disposal and Land fill tax must be added to all items of demolitions, alterations and rubbish disposal as appropriate.

Regional Factors

Laxton's Building Price Book is based upon national average prices (=1.00). Indicative levels of tender pricing in regions as at the second quarter 1999 are given on this map and the factors shown may be applied to adjust overall pricing.

MAJOR WORKS – TENDER VALUES	SMALL WORKS – TENDER VALUES
The measured rates are based on contracts valued in the range of £250,000 to £1,000.00. As a guide to pricing works of larger value and for cost planning or budgetary purposes, the following adjustments may be applied to overall contract values.	The measured rates are based on contracts valued in the range of £25,000 to £75,000. As a guide to pricing works of smaller or larger value and for cost planning or budgetary purposes, the following adjustments may be applied to overall contract values.
Contract Value £1,000,000 to £2,000,000.........deduct 2.5% £2,000,000 to £3,000,000.........deduct 5.0% £3,000,000 to £5,000,000.........deduct 7.5%	Contract Value £5,000 to £15,000.............................add 20.0% £15,000 to £25,000............................add 10.0% £25,000 to £75,000........................rate as shown £75,000 to £100,000.......................deduct 5.0% £100,000 to £150,000......................deduct7.5% £150,000 to £250,000......................deduct10.%

MAJOR WORKS

The measured rates in this section are intended to apply to Contracts within the price range £250,000 to £1,000,000 based on National Average prices.

Adjustment to overall contract values for Regional Variations and contract values up to £5,000,000 can be made as noted in the "Essential Information" section.

Laxton's Guide to Budget Estimating

- **Covers all aspects of construction, including new build, refurbishment, civil engineering, landscaping and M&E systems**

- **Clear presentation and easy-to-follow elemental format**

- **Essential in the early stages of project assessment**

'Many projects will start on a firmer footing when the decision to go ahead is based on the sort of estimate which this book will help managers to prepare.' Martin Barnes, Martin Barnes Project Management

'The publisher is to be complemented on this ... addition to Laxton's building price books.'

300pp, £38.50, 1996, paperback, isbn 0 7506 2967 3

Labour hourly rates: (except Specialists) Craft Operatives £9.23 Labour £7.02 Rates are national average prices. Refer to REGIONAL VARIATIONS for indicative levels of overall pricing in regions	MATERIALS			LABOUR				RATES		
	Del to Site	Waste	Material Cost	Craft Optve	Lab	Labour Cost	Sunds	Nett Rate	Unit	Gross Rate (+7.5%)
	£	%	£	Hrs	Hrs	£	£	£		£

C20: DEMOLITION

Pulling down outbuildings

Demolishing individual structures

timber outbuilding 2.50 x 2.00 x 3.00m maximum high	-	-	-	-	11.00	77.22	59.50	136.72	nr	146.97
outbuilding with half brick walls, one brick piers and felt covered timber roof, 2.50 x 2.00 x 3.00m maximum high	-	-	-	-	18.00	126.36	119.00	245.36	nr	263.76
greenhouse with half brick dwarf walls and timber framing to upper walls and roof, 3.00 x 2.00 x 3.00m maximum high	-	-	-	-	18.00	126.36	102.00	228.36	nr	245.49

Demolishing individual structures; setting aside materials for re-use

metal framed greenhouse, 3.00 x 2.00 x 3.00m maximum high	-	-	-	4.00	7.00	86.06	-	86.06	nr	92.51
prefabricated concrete garage, 5.40 x 2.60 x 2.40m maximum high	-	-	-	6.00	11.00	132.60	-	132.60	nr	142.54

Demolishing individual structures; making good structures

timber framed lean-to outbuilding and remove flashings and make good facing brick wall at ridge and vertical abutments, 3.00 x 1.90 x 2.40m maximum high	-	-	-	1.50	12.00	98.08	63.00	161.09	nr	173.17

Unit rates for pricing the above and similar work

pull down building 3.00 x 1.90 x 2.40m maximum high	-	-	-	-	10.00	70.20	59.50	129.70	nr	139.43
remove flashing	-	-	-	-	0.12	0.84	0.34	1.18	m	1.27
make good facing brick wall at ridge	-	-	-	0.20	0.20	3.25	0.31	3.56	m	3.83
make good facing brick wall at vertical abutment	-	-	-	0.17	0.17	2.76	0.31	3.07	m	3.30
make good rendered wall at ridge	-	-	-	0.20	0.20	3.25	0.72	3.97	m	4.27
make good rendered wall at vertical abutment	-	-	-	0.20	0.20	3.25	0.72	3.97	m	4.27

Demolishing individual structures; making good structures

lean-to outbuilding with half brick walls and slate covered timber roof and remove flashings, hack off plaster to house wall and make good with rendering to match existing, 1.50 x 1.00 x 2.00m maximum high	-	-	-	4.00	12.50	124.67	40.04	164.71	nr	177.06

Unit rates for pricing the above and similar work

pull down building with half brick walls, 1.50 x 1.00 x 2.00m maximum high	-	-	-	-	5.30	37.21	27.20	64.41	nr	69.24
pull down building with one brick walls, 1.50 x 1.00 x 2.00m maximum high	-	-	-	-	10.00	70.20	51.00	121.20	nr	130.29
hack off plaster or rendering and make good to match existing brick facings	-	-	-	1.00	2.00	23.27	1.02	24.29	m2	26.11
hack off plaster or rendering and make good to match existing rendering	-	-	-	1.30	2.30	28.15	4.11	32.26	m2	34.67

Removal of old work

Demolishing parts of structures

in-situ plain concrete bed										
75mm thick	-	-	-	-	0.90	6.32	2.55	8.87	m2	9.53
100mm thick	-	-	-	-	1.20	8.42	3.40	11.82	m2	12.71
150mm thick	-	-	-	-	1.80	12.64	5.10	17.74	m2	19.07
200mm thick	-	-	-	-	2.40	16.85	6.80	23.65	m2	25.42
in-situ reinforced concrete flat roof										
100mm thick	-	-	-	-	1.90	13.34	3.40	16.74	m2	17.99
150mm thick	-	-	-	-	2.85	20.01	5.10	25.11	m2	26.99
200mm thick	-	-	-	-	3.80	26.68	6.80	33.48	m2	35.99
225mm thick	-	-	-	-	4.28	30.05	7.65	37.70	m2	40.52
in-situ reinforced concrete upper floor										
100mm thick	-	-	-	-	1.90	13.34	3.40	16.74	m2	17.99
150mm thick	-	-	-	-	2.85	20.01	5.10	25.11	m2	26.99

Labour hourly rates: (except Specialists) Craft Operatives £9.23 Labour £7.02 Rates are national average prices. Refer to REGIONAL VARIATIONS for indicative levels of overall pricing in regions	MATERIALS			LABOUR				RATES		
	Del to Site	Waste	Material Cost	Craft Optve	Lab	Labour Cost	Sunds	Nett Rate	Unit	Gross Rate (+7.5%)
	£	%	£	Hrs	Hrs	£	£	£		£
C20: DEMOLITION Cont.										
Removal of old work Cont.										
Demolishing parts of structures Cont.										
in-situ reinforced concrete upper floor Cont.										
200mm thick..................................	-	-	-	-	3.80	26.68	6.80	33.48	m2	35.99
225mm thick..................................	-	-	-	-	4.28	30.05	7.65	37.70	m2	40.52
in-situ reinforced concrete beam............	-	-	-	-	19.00	133.38	34.00	167.38	m3	179.93
in-situ reinforced concrete column..........	-	-	-	-	18.00	126.36	34.00	160.36	m3	172.39
in-situ reinforced concrete wall										
100mm thick..................................	-	-	-	-	1.80	12.64	3.40	16.04	m2	17.24
150mm thick..................................	-	-	-	-	2.70	18.95	5.10	24.05	m2	25.86
200mm thick..................................	-	-	-	-	3.60	25.27	6.80	32.07	m2	34.48
225mm thick..................................	-	-	-	-	4.05	28.43	7.65	36.08	m2	38.79
in-situ reinforced concrete casing to beam....	-	-	-	-	16.00	112.32	34.00	146.32	m3	157.29
in-situ reinforced concrete casing to column..	-	-	-	-	15.25	107.06	34.00	141.06	m3	151.63
brick internal walls in lime mortar										
102mm thick..................................	-	-	-	-	0.63	4.42	3.47	7.89	m2	8.48
215mm thick..................................	-	-	-	-	1.30	9.13	7.31	16.44	m2	17.67
327mm thick..................................	-	-	-	-	2.00	14.04	11.12	25.16	m2	27.05
brick internal walls in cement mortar										
102mm thick..................................	-	-	-	-	0.95	6.67	3.47	10.14	m2	10.90
215mm thick..................................	-	-	-	-	1.94	13.62	7.31	20.93	m2	22.50
327mm thick..................................	-	-	-	-	2.98	20.92	11.12	32.04	m2	34.44
reinforced brick internal walls in cement lime mortar										
102mm thick..................................	-	-	-	-	0.95	6.67	3.47	10.14	m2	10.90
215mm thick..................................	-	-	-	-	1.94	13.62	7.31	20.93	m2	22.50
hollow clay block internal walls in cement-lime mortar										
50mm thick...................................	-	-	-	-	0.30	2.11	1.70	3.81	m2	4.09
75mm thick...................................	-	-	-	-	0.40	2.81	2.55	5.36	m2	5.76
100mm thick..................................	-	-	-	-	0.52	3.65	3.40	7.05	m2	7.58
hollow clay block internal walls in cement mortar										
50mm thick...................................	-	-	-	-	0.32	2.25	1.70	3.95	m2	4.24
75mm thick...................................	-	-	-	-	0.43	3.02	2.55	5.57	m2	5.99
100mm thick..................................	-	-	-	-	0.57	4.00	3.40	7.40	m2	7.96
concrete block internal walls in cement lime mortar										
75mm thick...................................	-	-	-	-	0.35	2.46	2.55	5.01	m2	5.38
100mm thick..................................	-	-	-	-	0.53	3.72	3.40	7.12	m2	7.65
190mm thick..................................	-	-	-	-	0.70	4.91	6.46	11.37	m2	12.23
215mm thick..................................	-	-	-	-	1.41	9.90	7.31	17.21	m2	18.50
concrete block internal walls in cement mortar										
75mm thick...................................	-	-	-	-	0.56	3.93	2.55	6.48	m2	6.97
100mm thick..................................	-	-	-	-	0.73	5.12	3.40	8.52	m2	9.16
190mm thick..................................	-	-	-	-	0.91	6.39	6.46	12.85	m2	13.81
215mm thick..................................	-	-	-	-	1.67	11.72	7.31	19.03	m2	20.46
if internal walls plastered, add per side.....	-	-	-	-	0.12	0.84	0.68	1.52	m2	1.64
if internal walls rendered, add per side......	-	-	-	-	0.18	1.26	0.68	1.94	m2	2.09
brick external walls in lime mortar										
102mm thick..................................	-	-	-	-	0.48	3.37	3.47	6.84	m2	7.35
215mm thick..................................	-	-	-	-	0.98	6.88	7.31	14.19	m2	15.25
327mm thick..................................	-	-	-	-	1.51	10.60	11.12	21.72	m2	23.35
brick external walls in cement mortar										
102mm thick..................................	-	-	-	-	0.63	4.42	3.47	7.89	m2	8.48
215mm thick..................................	-	-	-	-	1.30	9.13	7.31	16.44	m2	17.67
327mm thick..................................	-	-	-	-	2.00	14.04	11.12	25.16	m2	27.05
reinforced brick external walls in cement lime mortar										
102mm thick..................................	-	-	-	-	0.63	4.42	3.47	7.89	m2	8.48
215mm thick..................................	-	-	-	-	1.30	9.13	7.31	16.44	m2	17.67
if external walls plastered, add per side.....	-	-	-	-	0.09	0.63	0.68	1.31	m2	1.41
if external walls rendered or rough cast, add per side.......................................	-	-	-	-	0.12	0.84	0.68	1.52	m2	1.64
clean old bricks in lime mortar and stack for re-use, per thousand...........................	-	-	-	-	15.00	105.30	-	105.30	nr	113.20
clean old bricks in cement mortar and stack for re-use, per thousand...........................	-	-	-	-	25.00	175.50	-	175.50	nr	188.66
rough rubble walling 600mm thick, in lime mortar......................................	-	-	-	-	3.75	26.32	20.40	46.73	m2	50.23
random rubble walling 350mm thick, in lime mortar......................................	-	-	-	-	2.20	15.44	11.90	27.34	m2	29.39
random rubble walling 500mm thick, in lime mortar......................................	-	-	-	-	3.10	21.76	17.00	38.76	m2	41.67
dressed stone walling 100mm thick, in gauged mortar......................................	-	-	-	2.00	2.30	34.61	3.40	38.01	m2	40.86
dressed stone walling 200mm thick, in gauged mortar......................................	-	-	-	3.90	4.50	67.59	6.80	74.39	m2	79.97
stone copings 300 x 50mm.....................	-	-	-	0.20	0.26	3.67	0.51	4.18	m	4.49
stone copings 375 x 100mm....................	-	-	-	0.40	0.52	7.34	1.28	8.62	m	9.27
stone staircases.............................	-	-	-	-	15.00	105.30	34.00	139.30	m3	149.75
stone steps..................................	-	-	-	-	14.75	103.55	34.00	137.54	m3	147.86
structural timbers										
50 x 75mm....................................	-	-	-	-	0.13	0.91	0.13	1.04	m	1.12
50 x 100mm...................................	-	-	-	-	0.15	1.05	0.17	1.22	m	1.31
50 x 225mm...................................	-	-	-	-	0.24	1.68	0.38	2.06	m	2.22
75 x 100mm...................................	-	-	-	-	0.18	1.26	0.26	1.52	m	1.64
75 x 150mm...................................	-	-	-	-	0.20	1.40	0.38	1.78	m	1.92
75 x 225mm...................................	-	-	-	-	0.31	2.18	0.57	2.75	m	2.95
steelwork										
steel beams, joists and lintels not exceeding 10 kg/m......................................	-	-	-	50.00	50.00	812.50	34.00	846.50	t	909.99
steel beams, joists and lintels 10 - 20 kg/m..	-	-	-	46.00	46.00	747.50	34.00	781.50	t	840.11
steel beams, joists and lintels 20 - 50 kg/m..	-	-	-	40.00	40.00	650.00	34.00	684.00	t	735.30

Labour hourly rates: (except Specialists) Craft Operatives £9.23 Labour £7.02 Rates are national average prices. Refer to REGIONAL VARIATIONS for indicative levels of overall pricing in regions	MATERIALS			LABOUR				RATES		
	Del to Site £	Waste %	Material Cost £	Craft Optve Hrs	Lab Hrs	Labour Cost £	Sunds £	Nett Rate £	Unit	Gross Rate (+7.5%) £
C20: DEMOLITION Cont.										
Removal of old work Cont.										
Demolishing parts of structures Cont. steelwork Cont.										
steel columns and stanchions not exceeding 10 kg/m	-	-	-	60.00	60.00	975.00	34.00	1009.00	t	1084.68
steel columns and stanchions 10 - 20 kg/m.....	-	-	-	56.00	56.00	910.00	34.00	944.00	t	1014.80
steel columns and stanchions 20 - 50 kg/m.....	-	-	-	50.00	50.00	812.50	34.00	846.50	t	909.99
steel purlins and rails not exceeding 10 kg/m.	-	-	-	50.00	50.00	812.50	34.00	846.50	t	909.99
steel purlins and rails 10 - 20 kg/m..........	-	-	-	50.00	50.00	812.50	34.00	846.50	t	909.99
C30: SHORING/FACADE RETENTION										
Support of structures not to be demolished										
Provide and erect timber raking shore complete with sole piece, cleats, needles and 25mm brace boarding with										
two 150 x 150mm rakers (total length 8.00m) and 50 x 175mm wall piece	122.00	10.00	134.20	4.00	4.00	65.00	19.52	218.72	nr	235.12
weekly cost of maintaining last	-	-	-	1.00	1.00	16.25	0.98	17.23	nr	18.52
three 225 x 225mm rakers (total length 21.00m) and 50 x 250mm wall piece	461.00	10.00	507.10	23.20	23.20	377.00	84.87	968.97	nr	1041.64
weekly cost of maintaining last	-	-	-	1.50	1.50	24.38	4.19	28.57	nr	30.71
Provide and erect timber flying shore of 100 x 150mm main member, 50 x 175mm wall pieces and 50 x 100mm straining pieces and 75 x 100mm struts, distance between wall faces										
4.00m	124.00	10.00	136.40	15.00	15.00	243.75	4.19	384.34	nr	413.17
weekly cost of maintaining last	-	-	-	2.50	2.50	40.63	0.98	41.60	nr	44.73
5.00m	144.00	10.00	158.40	17.80	17.80	289.25	4.88	452.53	nr	486.47
weekly cost of maintaining last	-	-	-	3.00	2.00	41.73	1.12	42.85	nr	46.06
6.00m	157.00	10.00	172.70	5.10	20.60	191.69	5.41	369.80	nr	397.53
weekly cost of maintaining last	-	-	-	3.50	3.50	56.88	1.28	58.16	nr	62.52
C40: CLEANING MASONRY/CONCRETE										
Cleaning surfaces										
Thoroughly clean existing concrete surfaces prior to applying damp proof membrane										
floors	-	-	-	-	0.19	1.33	0.02	1.35	m2	1.46
walls	-	-	-	-	0.21	1.47	0.02	1.49	m2	1.61
Thoroughly clean existing concrete surfaces, fill in nail holes and small surface imperfections and leave smooth										
walls	-	-	-	0.35	0.20	4.63	0.04	4.67	m2	5.03
soffits	-	-	-	0.50	0.29	6.65	0.04	6.69	m2	7.19
Thoroughly clean existing brick or block surfaces prior to applying damp proof membrane										
walls	-	-	-	-	0.25	1.75	-	1.75	m2	1.89
Clean out air brick 225 x 150mm	-	-	-	-	0.20	1.40	-	1.40	nr	1.51
C41: REPAIRING/RENOVATING/CONSERVING MASONRY										
Cut out decayed bricks and replace with new bricks in gauged mortar (1:2:9) and make good surrounding work										
Fair faced common bricks										
singly.......................................	0.20	10.00	0.22	0.17	0.17	2.76	0.14	3.12	nr	3.36
small patches	12.00	10.00	13.20	5.80	5.80	94.25	5.98	113.43	m2	121.94
Picked stock facing bricks										
singly.......................................	0.68	10.00	0.75	0.17	0.17	2.76	0.14	3.65	nr	3.92
small patches	40.80	10.00	44.88	5.80	5.80	94.25	5.98	145.11	m2	155.99
Cut out defective wall and re-build in gauged mortar (1:2:9) and tooth and bond to surrounding work (25% new bricks allowed)										
Half brick wall; stretcher bond										
common bricks	3.00	10.00	3.30	4.00	4.00	65.00	3.43	71.73	m2	77.11
One brick wall; English bond										
common bricks	6.00	10.00	6.60	7.90	7.90	128.38	7.37	142.35	m2	153.02
common bricks faced one side with picked stock facings	17.48	10.00	19.23	8.30	8.30	134.88	7.37	161.47	m2	173.58
picked stock facing bricks faced both sides	20.40	10.00	22.44	8.80	8.80	143.00	7.37	172.81	m2	185.77
One brick wall; Flemish bond										
common bricks	6.00	10.00	6.60	7.90	7.90	128.38	3.43	138.41	m2	148.79
common bricks faced one side with picked stock facings	17.48	10.00	19.23	8.30	8.30	134.88	7.37	161.47	m2	173.58
picked stock facing bricks faced both sides	20.40	10.00	22.44	8.80	8.80	143.00	7.37	172.81	m2	185.77
275mm hollow wall; skins in stretcher bond										
two half brick skins in common bricks	6.00	10.00	6.60	8.16	8.16	132.60	7.51	146.71	m2	157.71
one half brick skin in common bricks and one half brick skin in picked stock facings	14.55	10.00	16.00	8.60	8.60	139.75	7.51	163.26	m2	175.51

Labour hourly rates: (except Specialists) Craft Operatives £9.23 Labour £7.02 Rates are national average prices. Refer to REGIONAL VARIATIONS for indicative levels of overall pricing in regions	MATERIALS			LABOUR				RATES		
	Del to Site £	Waste %	Material Cost £	Craft Optve Hrs	Lab Hrs	Labour Cost £	Sunds £	Nett Rate £	Unit	Gross Rate (+7.5%) £
C41: REPAIRING/RENOVATING/CONSERVING MASONRY Cont.										
Cut out defective wall and re-build in gauged mortar (1:2:9) and tooth and bond to surrounding work (25% new bricks allowed) Cont.										
275mm hollow wall; skins in stretcher bond Cont. one 100mm skin in concrete blocks and one half brick skin in picked stock facings	12.58	10.00	**13.84**	8.26	8.26	**134.22**	7.02	**155.08**	m2	**166.71**
Cut out crack in brickwork and stitch across with new bricks in gauged mortar(1:2:9); average 450mm wide										
Half brick wall										
common bricks..........................	5.20	10.00	**5.72**	1.90	1.90	**30.88**	3.33	**39.92**	m	**42.92**
common bricks fair faced one side	5.20	10.00	**5.72**	2.30	2.30	**37.38**	3.33	**46.42**	m	**49.91**
picked stock facing bricks faced both sides	17.68	10.00	**19.45**	2.40	2.40	**39.00**	3.33	**61.78**	m	**66.41**
One brick wall										
common bricks	10.40	10.00	**11.44**	3.90	3.90	**63.38**	7.17	**81.98**	m	**88.13**
common bricks fair faced one side	10.40	10.00	**11.44**	4.40	4.40	**71.50**	7.17	**90.11**	m	**96.87**
common bricks faced one side with picked stock facings	29.12	10.00	**32.03**	4.60	4.60	**74.75**	7.17	**113.95**	m	**122.50**
picked stock facing bricks faced both sides	35.36	10.00	**38.90**	4.70	4.70	**76.38**	7.17	**122.44**	m	**131.62**
Cut out defective arch; re-build in picked stock facing bricks in gauged mortar (1:2:9) including centering										
Brick-on-edge flat arch										
102mm on soffit	4.76	10.00	**5.24**	1.00	1.00	**16.25**	0.84	**22.33**	m	**24.00**
215mm on soffit	9.06	10.00	**9.97**	0.95	0.95	**15.44**	1.47	**26.87**	m	**28.89**
Brick-on-end flat arch										
102mm on soffit	9.06	10.00	**9.97**	0.95	0.95	**15.44**	1.47	**26.87**	m	**28.89**
215mm on soffit	18.13	10.00	**19.94**	1.90	1.90	**30.88**	2.94	**53.76**	m	**57.79**
Segmental arch in two half brick rings										
102mm on soffit	9.06	10.00	**9.97**	2.90	2.90	**47.13**	1.47	**58.56**	m	**62.95**
215mm on soffit	18.13	10.00	**19.94**	2.90	2.90	**47.13**	2.94	**70.01**	m	**75.26**
Semi-circular arch in two half brick rings										
102mm on soffit	9.06	10.00	**9.97**	2.90	2.90	**47.13**	1.47	**58.56**	m	**62.95**
215mm on soffit	18.13	10.00	**19.94**	2.90	2.90	**47.13**	2.94	**70.01**	m	**75.26**
Cut out defective arch; replace with precast concrete, B.S.5328, designed mix C20, 20mm aggregate lintel; wedge and pin up to brickwork over with slates in cement mortar (1:3) and extend external rendering over										
Remove defective brick-on-end flat arch replace with 102 x 215mm precast concrete lintel reinforced with one 12mm mild steel bar	15.18	–	**15.18**	1.90	1.90	**30.88**	2.31	**48.37**	m	**51.99**
Cut out defective external sill and re-build										
Shaped brick-on-edge sill in picked stock facing bricks in gauged mortar (1:2:9) 225mm wide	9.06	10.00	**9.97**	1.00	1.00	**16.25**	2.05	**28.27**	m	**30.39**
Roofing tile sill set weathering and projecting in cement mortar (1:3) two courses	5.93	10.00	**6.52**	1.20	1.20	**19.50**	1.06	**27.08**	m	**29.11**
Take down defective brick-on-edge coping and re-build in picked stock facing bricks in gauged mortar (1:2:9)										
Coping										
to one brick wall	9.06	10.00	**9.97**	0.60	0.60	**9.75**	2.05	**21.77**	m	**23.40**
Coping with cement fillets both sides										
to one brick wall; with oversailing course	13.60	10.00	**14.96**	0.67	0.67	**10.89**	3.08	**28.93**	m	**31.10**
to one brick wall; with single course tile creasing	12.03	10.00	**13.23**	1.09	1.09	**17.71**	2.58	**33.53**	m	**36.04**
to one brick wall; with double course tile creasing	14.99	10.00	**16.49**	1.38	1.38	**22.43**	3.64	**42.55**	m	**45.75**
Cut out defective air brick and replace with new; bed and point in gauged mortar (1:2:9)										
Cast iron, light, square hole										
225 x 75mm	3.55	–	**3.55**	0.45	0.45	**7.31**	0.16	**11.02**	nr	**11.85**
225 x 150mm	6.51	–	**6.51**	0.45	0.45	**7.31**	0.26	**14.08**	nr	**15.14**
Terra cotta										
215 x 65mm	1.67	–	**1.67**	0.45	0.45	**7.31**	0.16	**9.14**	nr	**9.83**
215 x 140mm	2.31	–	**2.31**	0.45	0.45	**7.31**	0.26	**9.88**	nr	**10.62**
Rake out joints of old brickwork 20mm deep and re-point in cement mortar (1:3)										
For turned in edge of lead flashing										
flush pointing; horizontal	–	–	**–**	0.50	0.50	**8.13**	0.24	**8.37**	m	**8.99**
flush pointing; stepped	–	–	**–**	0.85	0.85	**13.81**	0.34	**14.15**	m	**15.21**

Labour hourly rates: (except Specialists) Craft Operatives £9.23 Labour £7.02 Rates are national average prices. Refer to REGIONAL VARIATIONS for indicative levels of overall pricing in regions	MATERIALS			LABOUR				RATES		
	Del to Site	Waste	Material Cost	Craft Optve	Lab	Labour Cost	Sunds	Nett Rate	Unit	Gross Rate (+7.5%)
	£	%	£	Hrs	Hrs	£	£	£		£
C41: REPAIRING/RENOVATING/CONSERVING MASONRY Cont.										
Rake out joints of old brickwork 20mm deep and re-point in cement mortar (1:3) Cont.										
For turned in edge of lead flashing Cont.										
weathered pointing; horizontal	-	-	-	0.55	0.55	8.94	0.24	9.18	m	9.87
weathered pointing; stepped	-	-	-	0.90	0.90	14.63	0.34	14.97	m	16.09
ironed in pointing; horizontal	-	-	-	0.55	0.55	8.94	0.24	9.18	m	9.87
ironed in pointing; stepped	-	-	-	0.90	0.90	14.63	0.34	14.97	m	16.09
Rake out joints of old brickwork 20mm deep and re-point in cement-lime mortar(1:1:6)										
Generally; English bond										
flush pointing	-	-	-	1.00	0.66	13.86	0.69	14.55	m2	15.64
weathered pointing	-	-	-	1.00	0.66	13.86	0.69	14.55	m2	15.64
ironed in pointing	-	-	-	1.20	0.80	16.69	0.69	17.38	m2	18.69
Generally; Flemish bond										
flush pointing	-	-	-	1.00	0.66	13.86	0.69	14.55	m2	15.64
weathered pointing	-	-	-	1.00	0.66	13.86	0.69	14.55	m2	15.64
ironed in pointing	-	-	-	1.20	0.80	16.69	0.69	17.38	m2	18.69
Isolated areas not exceeding 1.00m2; English bond										
flush pointing	-	-	-	1.50	1.00	20.86	0.69	21.56	m2	23.17
weathered pointing	-	-	-	1.50	1.00	20.86	0.69	21.56	m2	23.17
ironed in pointing	-	-	-	1.80	1.20	25.04	0.69	25.73	m2	27.66
Isolated areas not exceeding 1.00m2; Flemish bond										
flush pointing	-	-	-	1.50	1.00	20.86	0.69	21.56	m2	23.17
weathered pointing	-	-	-	1.50	1.00	20.86	0.69	21.56	m2	23.17
ironed in pointing	-	-	-	1.80	1.20	25.04	0.69	25.73	m2	27.66
For using coloured mortar										
add	-	-	-	-	-	-	0.50	0.50	m2	0.54
Rake out joints of old brickwork 20mm deep and re-point in cement mortar (1:3)										
Generally; English bond										
flush pointing	-	-	-	1.20	0.80	16.69	0.69	17.38	m2	18.69
weathered pointing	-	-	-	1.20	0.80	16.69	0.69	17.38	m2	18.69
Generally; Flemish bond										
flush pointing	-	-	-	1.20	0.80	16.69	0.69	17.38	m2	18.69
weathered pointing	-	-	-	1.20	0.80	16.69	0.69	17.38	m2	18.69
Isolated areas not exceeding 1.00m2; English bond										
flush pointing	-	-	-	1.80	1.20	25.04	0.69	25.73	m2	27.66
weathered pointing	-	-	-	1.80	1.20	25.04	0.69	25.73	m2	27.66
Isolated areas not exceeding 1.00m2; Flemish bond										
flush pointing	-	-	-	1.80	1.20	25.04	0.69	25.73	m2	27.66
weathered pointing	-	-	-	1.80	1.20	25.04	0.69	25.73	m2	27.66
Repairing/Renovating stone										
Note restoration and repair of stonework is work for a specialist; prices being obtained for specific projects. Some firms that specialise in this class of work are included within the list at the end of section F-Masonry										
Repairs in Portland stonework and set in gauged mortar (1:2:9) and point to match existing										
Cut out decayed stones in facings of walls, piers or the like, prepare for and set new 75mm thick stone facings										
single stone	388.00	-	388.00	15.00	15.00	243.75	6.75	638.50	m2	686.39
areas of two or more adjacent stones	388.00	-	388.00	12.00	12.00	195.00	6.75	589.75	m2	633.98
Take out decayed stone sill, prepare for and set new sunk weathered and throated sill										
175 x 100mm	211.00	-	211.00	1.80	1.80	29.25	0.91	241.16	m	259.25
250 x 125mm	268.00	-	268.00	2.50	2.50	40.63	1.54	310.17	m	333.43
Take out sections of decayed stone coping, prepare for and set new weathered and twice throated coping										
300 x 75mm	274.00	-	274.00	1.40	1.40	22.75	1.14	297.89	m	320.23
300 x 100mm	332.00	-	332.00	1.70	1.70	27.63	1.51	361.13	m	388.22
Repairs in York stonework and set in gauged mortar (1:2:9) and point to match existing										
Cut out decayed stones in facings of walls, piers or the like, prepare for and set new 75mm thick stone facings										
single stone	266.00	-	266.00	17.00	17.00	276.25	6.75	549.00	m2	590.17
areas of two or more adjacent stones	266.00	-	266.00	13.75	13.75	223.44	6.75	496.19	m2	533.40
Take out decayed stone sill, prepare for and set new sunk weathered and throated sill										
175 x 100mm	120.00	-	120.00	1.80	1.80	29.25	1.14	150.39	m	161.67
250 x 125mm	161.00	-	161.00	2.50	2.50	40.63	1.51	203.13	m	218.37

Labour hourly rates: (except Specialists) Craft Operatives £9.23 Labour £7.02 Rates are national average prices. Refer to REGIONAL VARIATIONS for indicative levels of overall pricing in regions	MATERIALS			LABOUR				RATES		
	Del to Site	Waste	Material Cost	Craft Optve	Lab	Labour Cost	Sunds	Nett Rate	Unit	Gross Rate (+7.5%)
	£	%	£	Hrs	Hrs	£	£	£		£
C41: REPAIRING/RENOVATING/CONSERVING MASONRY Cont.										
Repairs in York stonework and set in gauged mortar (1:2:9) and point to match existing Cont.										
Take out sections of decayed stone coping, prepare for and set new weathered and twice throated coping										
300 x 75mm	110.00	-	110.00	1.40	1.40	22.75	1.14	133.89	m	143.93
300 x 100mm	139.00	-	139.00	1.70	1.70	27.63	1.51	168.13	m	180.75
Rake out decayed mortar joints and re-point in gauged mortar (1:2:9)										
Re-point rubble walling										
coursed	-	-	-	1.25	1.25	20.31	1.12	21.43	m2	23.04
uncoursed	-	-	-	1.32	1.32	21.45	1.12	22.57	m2	24.26
squared	-	-	-	1.25	1.25	20.31	1.12	21.43	m2	23.04
Re-point stonework										
ashlar	-	-	-	1.20	1.20	19.50	1.12	20.62	m2	22.17
coursed block	-	-	-	1.20	1.20	19.50	1.12	20.62	m2	22.17
C42: REPAIRING/ RENOVATING/CONSERVING CONCRETE										
Re-bed loose copings										
Take off loose concrete coping and clean and remove old bedding mortar										
re-bed, joint and point in cement mortar (1:3)	-	-	-	1.00	1.00	16.25	1.00	17.25	m	18.54
Repairing cracks in concrete										
Repair cracks; clean out all dust and debris and fill with mortar mixed with a bonding agent										
up to 5mm wide	-	-	-	0.50	0.50	8.13	0.48	8.61	m	9.25
Repair cracks; cutting out to form groove, treat with bonding agent and fill with fine concrete mixed with a bonding agent										
25 x 25mm deep	-	-	-	0.50	0.50	8.13	0.87	8.99	m	9.67
40 x 40mm deep	-	-	-	0.60	0.60	9.75	2.06	11.81	m	12.70
C45: DAMP PROOF COURSE RENEWAL/INSERTION										
Insert damp proof course in old wall by cutting out one course of brickwork in alternate lengths not exceeding 1.00m and replace old bricks with 50mm bricks in cement mortar (1:3)										
B.S.6398 Class D, bitumen with hessian base laminated with lead in										
one brick wall	4.56	10.00	5.02	3.00	3.00	48.75	5.53	59.30	m	63.74
one and a half brick wall	6.83	10.00	7.51	4.00	4.00	65.00	8.31	80.82	m	86.88
254mm hollow wall	4.56	10.00	5.02	3.50	3.50	56.88	5.53	67.42	m	72.48
Hyload, pitch polymer in										
one brick wall	1.85	10.00	2.04	3.00	3.00	48.75	5.53	56.31	m	60.54
one and a half brick wall	2.77	10.00	3.05	4.00	4.00	65.00	8.31	76.36	m	82.08
254mm hollow wall	1.85	10.00	2.04	3.50	3.50	56.88	5.53	64.44	m	69.27
Two courses slates in cement mortar (1:3) in										
one brick wall	6.11	10.00	6.72	3.00	3.00	48.75	5.53	61.00	m	65.58
one and a half brick wall	9.17	10.00	10.09	4.00	4.00	65.00	8.31	83.40	m	89.65
254mm hollow wall	6.11	10.00	6.72	3.50	3.50	56.88	5.53	69.13	m	74.31
0.6mm bitumen coated copper in										
one brick wall	10.75	10.00	11.82	3.00	3.00	48.75	5.53	66.11	m	71.06
one and a half brick wall	16.13	10.00	17.74	4.00	4.00	65.00	8.31	91.05	m	97.88
254mm hollow wall	10.75	10.00	11.82	3.50	3.50	56.88	5.53	74.23	m	79.80
Insert damp proof course in old wall by hand sawing in 600mm lengths										
B.S.6398 Class D, bitumen with hessian base laminated with lead in										
one brick wall	4.56	10.00	5.02	2.50	2.50	40.63	8.27	53.91	m	57.95
one and a half brick wall	6.83	10.00	7.51	3.50	3.50	56.88	12.41	76.80	m	82.56
254mm hollow wall	4.56	10.00	5.02	3.00	3.00	48.75	8.27	62.04	m	66.69
Hyload, pitch polymer in										
one brick wall	1.85	10.00	2.04	2.50	2.50	40.63	8.27	50.93	m	54.75
one and a half brick wall	2.77	10.00	3.05	3.50	3.50	56.88	12.41	72.33	m	77.76
254mm hollow wall	1.85	10.00	2.04	3.00	3.00	48.75	8.27	59.06	m	63.48
Two courses slates in cement mortar (1:3) in										
one brick wall	6.11	10.00	6.72	2.50	2.50	40.63	8.27	55.62	m	59.79
one and a half brick wall	9.17	10.00	10.09	3.50	3.50	56.88	12.41	79.37	m	85.32
254mm hollow wall	6.11	10.00	6.72	3.00	3.00	48.75	8.27	63.74	m	68.52
0.6mm bitumen coated copper in										
one brick wall	10.75	10.00	11.82	2.50	2.50	40.63	8.27	60.72	m	65.27
one and a half brick wall	16.13	10.00	17.74	3.50	3.50	56.88	12.41	87.03	m	93.56
254mm hollow wall	10.75	10.00	11.82	3.00	3.00	48.75	8.27	68.84	m	74.01

Labour hourly rates: (except Specialists) Craft Operatives £9.23 Labour £7.02 Rates are national average prices. Refer to REGIONAL VARIATIONS for indicative levels of overall pricing in regions	MATERIALS			LABOUR				RATES		
	Del to Site £	Waste %	Material Cost £	Craft Optve Hrs	Lab Hrs	Labour Cost £	Sunds £	Nett Rate £	Unit	Gross Rate (+7.5%) £
C45: DAMP PROOF COURSE RENEWAL/INSERTION Cont.										
Insert damp proof course in old wall by machine sawing in 600mm lengths										
B.S.6398 Class D, bitumen with hessian base laminated with lead in										
one brick wall	4.56	10.00	5.02	2.00	2.00	32.50	10.98	48.50	m	52.13
one and a half brick wall	6.83	10.00	7.51	3.00	3.00	48.75	16.46	72.72	m	78.18
254mm hollow wall	4.56	10.00	5.02	2.50	2.50	40.63	10.98	56.62	m	60.87
Hyload, pitch polymer in										
one brick wall	1.85	10.00	2.04	2.00	2.00	32.50	10.98	45.52	m	48.93
one and a half brick wall	2.77	10.00	3.05	3.00	3.00	48.75	16.46	68.26	m	73.38
254mm hollow wall	1.85	10.00	2.04	2.50	2.50	40.63	10.98	53.64	m	57.66
Two courses slates in cement mortar (1:3) in										
one brick wall	6.11	10.00	6.72	2.00	2.00	32.50	10.98	50.20	m	53.97
one and a half brick wall	9.17	10.00	10.09	3.00	3.00	48.75	16.46	75.30	m	80.94
254mm hollow wall	6.11	10.00	6.72	2.50	2.50	40.63	10.98	58.33	m	62.70
0.6mm bitumen coated copper in										
one brick wall	10.75	10.00	11.82	2.00	2.00	32.50	10.98	55.31	m	59.45
one and a half brick wall	16.13	10.00	17.74	3.00	3.00	48.75	16.46	82.95	m	89.17
254mm hollow wall	10.75	10.00	11.82	2.50	2.50	40.63	10.98	63.43	m	68.19
Insert cavity tray in old wall by cutting out by hand in short lengths; insert individual trays and make good with picked stock facing bricks in gauged mortar (1:2:9) including pinning up with slates										
Type E Cavitray by Cavity Trays Ltd in										
half brick skin of cavity wall	7.53	10.00	8.28	1.75	1.75	28.44	7.02	43.74	m	47.02
external angle	5.30	10.00	5.83	1.00	1.00	16.25	6.45	28.53	nr	30.67
internal angle	5.30	10.00	5.83	1.25	1.25	20.31	6.45	32.59	nr	35.04
Type X Cavity Trays Ltd to suit 40 degree pitched roof complete with attached code 4 lead flashing and dress over tiles in										
half brick skin of cavity wall	34.17	10.00	37.59	4.25	4.25	69.06	14.05	120.70	m	129.75
ridge tray	8.48	10.00	9.33	1.00	1.00	16.25	3.27	28.85	nr	31.01
catchment tray	4.02	10.00	4.42	0.50	0.50	8.13	1.72	14.27	nr	15.34
corner catchment tray	8.87	10.00	9.76	1.00	1.00	16.25	3.27	29.28	nr	31.47
Damp proofing old brick wall by silicone injection method (excluding removal and reinstatement of plaster, etc.)										
Damp proofing										
one brick wall	7.10	-	7.10	1.00	1.00	16.25	0.75	24.10	m	25.91
one and a half brick wall	10.60	-	10.60	1.50	1.50	24.38	1.12	36.09	m	38.80
254mm hollow wall	7.10	-	7.10	1.35	1.35	21.94	0.75	29.79	m	32.02
C50: REPAIRING/RENOVATING/CONSERVING METAL										
Replace defective arch bar										
Take out defective arch bar including cutting away brickwork as necessary, replace with new mild steel arch bar primed all round and make good all work disturbed										
32 x 6mm flat arch bar	1.47	-	1.47	1.03	1.03	16.74	0.71	18.92	m	20.34
50 x 6mm flat arch bar	2.28	-	2.28	1.12	1.12	18.20	1.04	21.52	m	23.13
50 x 50 x 6mm angle arch bar	4.26	-	4.26	1.27	1.27	20.64	1.85	26.75	m	28.75
75 x 50 x 6mm angle arch bar	5.33	-	5.33	1.44	1.44	23.40	2.33	31.06	m	33.39
Replace defective baluster										
Cut out defective baluster, clean core and bottom rails, provide new mild steel baluster and weld on										
13mm diameter x 686mm long	0.86	-	0.86	0.92	0.92	14.95	1.48	17.29	nr	18.59
13mm diameter x 762mm long	0.86	-	0.86	0.94	0.94	15.28	1.48	17.61	nr	18.94
13 x 13 x 686mm long	1.17	-	1.17	0.94	0.94	15.28	1.75	18.20	nr	19.56
13 x 13 x 762mm long	1.17	-	1.17	0.95	0.95	15.44	1.75	18.36	nr	19.73
Repair damaged weld										
Clean off damaged weld including removing remains of weld and cleaning off surrounding paint back to bright metal and re-weld connection between 10mm rail and										
13mm diameter rod	-	-	-	0.50	0.50	8.13	2.18	10.31	nr	11.08
13 x 13mm bar	-	-	-	0.53	0.53	8.61	2.59	11.20	nr	12.04
25 x 25mm bar	-	-	-	0.60	0.60	9.75	3.25	13.00	nr	13.98
C51: REPAIRING/RENOVATING/CONSERVING TIMBER										
Cut out rotten or infected structural timber members including cutting away wall, etc. and shoring up adjacent work and replace with new pressure impregnated timber										
Plates and bed in cement mortar (1:3)										
50 x 75mm	1.26	10.00	1.39	0.30	0.30	4.88	0.17	6.43	m	6.91
75 x 100mm	2.70	10.00	2.97	0.50	0.50	8.13	0.33	11.43	m	12.28

Labour hourly rates: (except Specialists) Craft Operatives £9.23 Labour £7.02 Rates are national average prices. Refer to REGIONAL VARIATIONS for indicative levels of overall pricing in regions	MATERIALS			LABOUR				RATES		
	Del to Site £	Waste %	Material Cost £	Craft Optve Hrs	Lab Hrs	Labour Cost £	Sunds £	Nett Rate £	Unit	Gross Rate (+7.5%) £
C51: REPAIRING/RENOVATING/CONSERVING TIMBER Cont.										
Cut out rotten or infected structural timber members including cutting away wall, etc. and shoring up adjacent work and replace with new pressure impregnated timber Cont.										
Floor or roof joists										
50 x 175mm	2.93	10.00	3.22	0.57	0.57	9.26	0.38	12.87	m	13.83
50 x 225mm	3.77	10.00	4.15	0.67	0.67	10.89	0.49	15.52	m	16.69
75 x 175mm	4.74	10.00	5.21	0.79	0.79	12.84	0.58	18.63	m	20.03
Rafters										
38 x 100mm	1.64	10.00	1.80	0.35	0.35	5.69	0.17	7.66	m	8.24
50 x 100mm	1.67	10.00	1.84	0.38	0.38	6.17	0.22	8.23	m	8.85
Ceiling joists and collars										
50 x 100mm	1.67	10.00	1.84	0.38	0.38	6.17	0.22	8.23	m	8.85
50 x 150mm	2.51	10.00	2.76	0.52	0.52	8.45	0.33	11.54	m	12.41
Purlins, ceiling beams and struts										
75 x 100mm	2.70	10.00	2.97	0.55	0.55	8.94	0.33	12.24	m	13.16
100 x 225mm	8.52	10.00	9.37	1.23	1.23	19.99	0.99	30.35	m	32.63
Roof trusses										
75 x 100mm	2.70	10.00	2.97	0.92	0.92	14.95	0.33	18.25	m	19.62
75 x 150mm	4.05	10.00	4.46	1.37	1.37	22.26	0.49	27.21	m	29.25
Hangers in roof										
25 x 75mm	0.78	10.00	0.86	0.30	0.30	4.88	0.09	5.82	m	6.26
38 x 100mm	1.64	10.00	1.80	0.45	0.45	7.31	0.17	9.29	m	9.98
Cut out rotten or infected timber and prepare timber under to receive new										
Roof boarding; provide new 25mm pressure impregnated softwood boarding										
flat boarding; areas not exceeding 0.50m2	11.00	10.00	12.10	1.96	1.96	31.85	1.09	45.04	m2	48.42
flat boarding; areas 0.50 - 5.00m2	11.00	10.00	12.10	1.63	1.63	26.49	1.09	39.68	m2	42.65
sloping boarding; areas not exceeding 0.50m2	11.00	10.00	12.10	2.34	2.34	38.02	1.09	51.22	m2	55.06
sloping boarding; areas 0.50 - 5.00m2	11.00	10.00	12.10	1.89	1.89	30.71	1.09	43.90	m2	47.20
Gutter boarding and bearers; provide new 25mm pressure impregnated softwood boarding and 50 x 50mm bearers										
over 300mm wide	14.36	10.00	15.80	2.06	2.06	33.48	1.31	50.58	m2	54.37
200mm	3.88	15.00	4.46	0.55	0.55	8.94	0.44	13.84	m	14.88
225mm	4.38	15.00	5.04	0.62	0.62	10.07	0.48	15.59	m	16.76
Gutter sides; provide new 19mm pressure impregnated softwood boarding										
150mm	1.27	15.00	1.46	0.46	0.46	7.47	0.12	9.06	m	9.73
200mm	1.76	15.00	2.02	0.50	0.50	8.13	0.17	10.32	m	11.09
225mm	2.15	15.00	2.47	0.57	0.57	9.26	0.19	11.93	m	12.82
Eaves or verge soffit and bearers and provide new 25mm pressure impregnated softwood boarding and 25 x 50mm bearers										
over 300mm wide	14.31	10.00	15.74	2.10	2.10	34.13	1.31	51.18	m2	55.01
225mm	4.37	10.00	4.81	0.67	0.67	10.89	0.48	16.17	m	17.39
Fascia; provide new pressure impregnated softwood										
25 x 150mm	2.28	10.00	2.51	0.46	0.46	7.47	0.17	10.15	m	10.91
25 x 200mm	3.00	10.00	3.30	0.51	0.51	8.29	0.22	11.81	m	12.69
Barge board; provide new pressure impregnated softwood										
25 x 225mm	3.41	10.00	3.75	0.55	0.55	8.94	0.25	12.94	m	13.91
25 x 250mm	3.75	10.00	4.13	0.57	0.57	9.26	0.28	13.67	m	14.69
Take up floor boarding, remove all nails and clean joists under, re-lay boarding, clear away and replace damaged boards and sand surface (20% new softwood boards allowed)										
Square edged boarding										
25mm	2.70	15.00	3.11	1.73	1.73	28.11	0.44	31.66	m2	34.03
32mm	3.58	15.00	4.12	1.76	1.76	28.60	0.55	33.27	m2	35.76
Tongued and grooved boarding										
25mm	2.86	20.00	3.43	1.87	1.87	30.39	0.44	34.26	m2	36.83
32mm	3.79	20.00	4.55	1.89	1.89	30.71	0.55	35.81	m2	38.50
Take up floor boarding, remove all nails and clean joists under, re-lay boarding, destroy and replace infected boards with new pressure impregnated boards and sand surface (20% new softwood boards allowed)										
Square edged boarding										
25mm	2.97	15.00	3.42	1.81	1.81	29.41	0.44	33.27	m2	35.76
32mm	3.94	15.00	4.53	1.83	1.83	29.74	0.55	34.82	m2	37.43
Tongued and grooved boarding										
25mm	3.15	20.00	3.78	1.94	1.94	31.52	0.44	35.74	m2	38.43
32mm	4.17	20.00	5.00	1.95	1.95	31.69	0.55	37.24	m2	40.03

Labour hourly rates: (except Specialists) Craft Operatives £9.23 Labour £7.02 Rates are national average prices. Refer to REGIONAL VARIATIONS for indicative levels of overall pricing in regions	MATERIALS			LABOUR				RATES		
	Del to Site	Waste	Material Cost	Craft Optve	Lab	Labour Cost	Sunds	Nett Rate	Unit	Gross Rate (+7.5%)
	£	%	£	Hrs	Hrs	£	£	£		£
C51: REPAIRING/RENOVATING/CONSERVING TIMBER Cont.										
Take up worn or damaged floor boarding, remove all nails and clean joists under and replace with new softwood boarding										
Square edged boarding										
25mm in areas not exceeding 0.50m2	13.50	15.00	15.53	1.90	1.90	30.88	1.09	47.49	m2	51.05
32mm in areas not exceeding 0.50m2	17.90	15.00	20.59	1.94	1.94	31.52	1.39	53.50	m2	57.51
25mm in areas 0.50 - 3.00m2	13.50	15.00	15.53	1.57	1.57	25.51	1.09	42.13	m2	45.29
32mm in areas 0.50 - 3.00m2	17.90	15.00	20.59	1.61	1.61	26.16	1.39	48.14	m2	51.75
Tongued and grooved boarding										
25mm in areas not exceeding 0.50m2	14.85	20.00	17.82	2.00	2.00	32.50	1.09	51.41	m2	55.27
32mm in areas not exceeding 0.50m2	19.70	20.00	23.64	2.06	2.06	33.48	1.39	58.51	m2	62.89
25mm in areas 0.50 - 3.00m2	14.85	20.00	17.82	1.82	1.82	29.57	1.09	48.48	m2	52.12
32mm in areas 0.50 - 3.00m2	19.70	20.00	23.64	1.87	1.87	30.39	1.39	55.42	m2	59.57
Punch down all nails, remove all tacks, etc. and fill nail holes										
Surface of existing flooring										
generally	-	-	-	0.50	0.10	5.32	0.03	5.35	m2	5.75
generally and machine sand in addition	-	-	-	1.00	0.30	11.34	1.18	12.52	m2	13.45
Note										
removal of old floor coverings included elsewhere										
Easing and overhauling doors										
Ease and adjust door and oil ironmongery										
762 x 1981mm	-	-	-	0.33	0.06	3.47	0.02	3.49	nr	3.75
914 x 2134mm	-	-	-	0.39	0.06	4.02	0.02	4.04	nr	4.34
Take off door, shave 12mm off bottom edge and re-hang										
762 x 1981mm	-	-	-	0.66	0.11	6.86	-	6.86	nr	7.38
914 x 2134mm	-	-	-	0.79	0.13	8.20	-	8.20	nr	8.82
Adapting doors										
Take off door and re-hang on opposite hand, piece out as necessary and oil ironmongery										
762 x 1981mm	-	-	-	2.50	0.25	24.83	0.10	24.93	nr	26.80
914 x 2134mm	-	-	-	2.65	0.44	27.55	0.10	27.65	nr	29.72
Refixing removed doors and frames										
Doors										
762 x 1981mm	-	-	-	2.00	0.34	20.85	0.14	20.99	nr	22.56
914 x 2134mm	-	-	-	2.25	0.38	23.44	0.14	23.58	nr	25.34
Doors and frames; oil existing ironmongery										
762 x 1981mm	-	-	-	4.00	0.66	41.55	0.35	41.90	nr	45.05
914 x 2134mm	-	-	-	4.50	0.75	46.80	0.35	47.15	nr	50.69
Repairing doors and frames										
Piece in door where ironmongery removed										
rim lock	-	-	-	1.00	0.15	10.28	0.16	10.44	nr	11.23
rim lock or latch and furniture	-	-	-	1.50	0.25	15.60	0.39	15.99	nr	17.19
mortice lock	-	-	-	1.50	0.25	15.60	0.39	15.99	nr	17.19
mortice lock or latch and furniture	-	-	-	2.00	0.32	20.71	0.46	21.17	nr	22.75
butt hinge	-	-	-	0.50	0.08	5.18	0.22	5.40	nr	5.80
Piece in frame where ironmongery removed										
lock or latch keep	-	-	-	0.50	0.08	5.18	0.22	5.40	nr	5.80
butt hinge	-	-	-	0.50	0.08	5.18	0.22	5.40	nr	5.80
Take off door, dismantle as necessary, cramp and re-wedge with new glued wedges, pin and re-hang										
762 x 1981mm, panelled	-	-	-	5.00	0.84	52.05	0.16	52.21	nr	56.12
914 x 2134mm, panelled	-	-	-	6.00	1.00	62.40	0.16	62.56	nr	67.25
762 x 1981mm, glazed panelled	-	-	-	5.50	0.90	57.08	0.16	57.24	nr	61.54
914 x 2134mm, glazed panelled	-	-	-	6.50	1.08	67.58	0.16	67.74	nr	72.82
Take off softwood panelled door 762 x 1981mm, dismantle as necessary and replace the following damaged members and re-hang										
38 x 125mm top rail	5.25	10.00	5.78	2.15	0.36	22.37	0.45	28.60	nr	30.74
38 x 125mm hanging stile	13.60	10.00	14.96	2.36	0.39	24.52	1.14	40.62	nr	43.67
38 x 125mm locking stile	13.60	10.00	14.96	2.36	0.39	24.52	1.14	40.62	nr	43.67
38 x 200mm middle rail	8.40	10.00	9.24	4.23	0.71	44.03	0.74	54.01	nr	58.06
38 x 200mm bottom rail	8.40	10.00	9.24	2.20	0.37	22.90	0.74	32.88	nr	35.35
Take off afromosia panelled door 762 x 1981mm, dismantle as necessary and replace the following damaged members and re-hang										
38 x 125mm top rail	24.00	10.00	26.40	5.20	0.87	54.10	0.45	80.95	nr	87.02
38 x 125mm hanging stile	61.00	10.00	67.10	5.36	0.89	55.72	1.14	123.96	nr	133.26
38 x 125mm locking stile	61.00	10.00	67.10	5.36	0.89	55.72	1.14	123.96	nr	133.26
38 x 200mm middle rail	38.00	10.00	41.80	10.40	1.73	108.14	0.74	150.68	nr	161.98
38 x 200mm bottom rail	38.00	10.00	41.80	5.30	0.88	55.10	0.74	97.64	nr	104.96
Take off damaged softwood stop and replace with new										
12 x 38mm	0.46	10.00	0.51	0.35	0.06	3.65	0.10	4.26	m	4.58
25 (fin) x 50mm glue and screw on	0.81	10.00	0.89	0.45	0.08	4.72	0.12	5.73	m	6.16

Labour hourly rates: (except Specialists) Craft Operatives £9.23 Labour £7.02 Rates are national average prices. Refer to REGIONAL VARIATIONS for indicative levels of overall pricing in regions	MATERIALS			LABOUR				RATES		
	Del to Site £	Waste %	Material Cost £	Craft Optve Hrs	Lab Hrs	Labour Cost £	Sunds £	Nett Rate £	Unit	Gross Rate (+7.5%) £
C51: REPAIRING/RENOVATING/CONSERVING TIMBER Cont.										
Repairing doors and frames Cont.										
Take off damaged afromosia stop and replace with new										
12 x 38mm	1.60	10.00	1.76	0.45	0.08	4.72	0.10	6.58	m	7.07
25 (fin) x 50mm glue and screw on	3.20	10.00	3.52	0.69	0.12	7.21	0.12	10.85	m	11.66
Flush facing old doors										
Take off panelled door, remove ironmongery, cover both sides with 3.2mm hardboard; refix ironmongery, adjust stops and re-hang										
762 x 1981mm	5.40	10.00	5.94	3.10	0.50	32.12	0.25	38.31	nr	41.19
838 x 1981mm	5.40	10.00	5.94	3.15	0.51	32.65	0.25	38.84	nr	41.76
Take off panelled door, remove ironmongery, cover one side with 3.2mm hardboard and other side with 6mm fire resistant sheeting screwed on; refix ironmongery, adjust stops and re-hang										
762 x 1981mm	41.04	10.00	45.14	4.00	0.67	41.62	0.82	87.59	nr	94.16
838 x 1981mm	41.04	10.00	45.14	4.05	0.68	42.16	0.82	88.12	nr	94.73
Take off panelled door, remove ironmongery, cover one side with 3.2mm hardboard, fill panels on other side with 6mm fire resistant sheeting and cover with 6mm fire resistant sheeting screwed on; refix ironmongery, adjust stops and re-hang										
762 x 1981mm	60.21	10.00	66.23	5.00	0.83	51.98	1.05	119.26	nr	128.20
838 x 1981mm	60.21	10.00	66.23	5.05	0.84	52.51	1.05	119.79	nr	128.77
Take off flush door, remove ironmongery, cover one side with 6mm fire resistant sheeting screwed on; refix ironmongery and re-hang										
762 x 1981mm	38.34	10.00	42.17	2.50	0.42	26.02	0.53	68.73	nr	73.88
838 x 1981mm	38.34	10.00	42.17	2.50	0.43	26.09	0.53	68.80	nr	73.96
Easing and overhauling windows										
Ease and adjust and oil ironmongery										
casement opening light	-	-	-	0.41	0.07	4.28	0.02	4.30	nr	4.62
double hung sash and renew sash lines with best quality flax cord	0.99	10.00	1.09	1.64	0.27	17.03	0.05	18.17	nr	19.53
Refixing removed windows										
Windows										
casement opening light and oil ironmongery	-	-	-	1.03	0.17	10.70	0.04	10.74	nr	11.55
double hung sash and provide new best quality flax sash cord	0.99	10.00	1.09	1.70	0.28	17.66	0.05	18.80	nr	20.21
Single or multi-light timber casements and frames										
440 x 920mm	-	-	-	4.00	0.66	41.55	0.04	41.59	nr	44.71
1225 x 1070mm	-	-	-	6.00	1.00	62.40	0.05	62.45	nr	67.13
2395 x 1225mm	-	-	-	8.00	1.33	83.18	0.06	83.24	nr	89.48
Repairing windows										
Remove damaged members of cased frame and replace with new softwood										
parting bead	0.47	10.00	0.52	1.16	0.19	12.04	0.05	12.61	m	13.55
pulley stile	1.79	10.00	1.97	1.68	0.28	17.47	0.14	19.58	m	21.05
stop bead	0.68	10.00	0.75	0.62	0.10	6.42	0.05	7.22	m	7.76
Take out sash not exceeding 50mm thick, dismantle as necessary, replace the following damaged members in softwood and re-hang with new best quality flax sash cord										
glazing bar	2.38	10.00	2.62	3.14	0.52	32.63	0.05	35.30	m	37.95
stile	2.66	10.00	2.93	3.03	0.50	31.48	0.09	34.49	m	37.08
rail	4.17	10.00	4.59	2.80	0.45	29.00	0.26	33.85	m	36.39
Take out casement opening light not exceeding 50mm thick, dismantle as necessary, replace the following damaged members in softwood and re-hang										
glazing bar	1.32	10.00	1.45	1.97	0.32	20.43	0.05	21.93	m	23.58
hanging stile	1.60	10.00	1.76	1.85	0.30	19.18	0.10	21.04	m	22.62
shutting stile	1.60	10.00	1.76	1.85	0.30	19.18	0.10	21.04	m	22.62
rail	3.13	10.00	3.44	1.61	0.27	16.76	0.26	20.46	m	21.99
Take out sash, cramp and re-wedge with new glued wedges, pin and re-hang										
sliding sash	-	-	-	5.65	0.95	58.82	0.87	59.69	nr	64.17
opening casement light	-	-	-	5.00	0.85	52.12	0.87	52.99	nr	56.96
Repair corner of sash with 100 x 100mm angle repair plate										
let in flush	0.38	-	0.38	1.00	0.17	10.42	0.17	10.97	nr	11.80
Carefully rake out and dry crack in sill and fill with two part wood repair system	-	-	-	0.50	0.08	5.18	1.50	6.68	m	7.18

Labour hourly rates: (except Specialists) Craft Operatives £9.23 Labour £7.02 Rates are national average prices. Refer to REGIONAL VARIATIONS for indicative levels of overall pricing in regions	MATERIALS			LABOUR				RATES		
	Del to Site £	Waste %	Material Cost £	Craft Optve Hrs	Lab Hrs	Labour Cost £	Sunds £	Nett Rate £	Unit	Gross Rate (+7.5%) £
C51: REPAIRING/RENOVATING/CONSERVING TIMBER Cont.										
Repairs to softwood skirtings, picture rails, architraves, etc.										
Take off and refix										
skirting	-	-	-	0.40	0.07	4.18	0.04	4.22	m	4.54
dado or picture rail	-	-	-	0.35	0.06	3.65	0.03	3.68	m	3.96
architrave	-	-	-	0.26	0.05	2.75	0.03	2.78	m	2.99
Cut out damaged section and piece in new										
19 x 100mm square skirting	1.04	10.00	1.14	0.69	0.18	7.63	0.20	8.98	m	9.65
25 x 150mm square skirting	2.03	10.00	2.23	0.71	0.12	7.40	0.24	9.87	m	10.61
19 x 125mm moulded skirting	3.45	10.00	3.79	0.70	0.11	7.23	0.23	11.26	m	12.10
19 x 50mm moulded dado or picture rail	1.85	10.00	2.04	0.74	0.12	7.67	0.10	9.81	m	10.54
25 x 50mm moulded or splayed architrave	2.50	10.00	2.75	0.51	0.09	5.34	0.12	8.21	m	8.82
38 x 75mm moulded or splayed architrave	3.59	10.00	3.95	0.52	0.09	5.43	0.23	9.61	m	10.33
Cut out rotten or infected skirting and grounds and replace with new treated softwood										
19 x 100mm square skirting	1.48	10.00	1.63	1.00	0.17	10.42	0.27	12.32	m	13.25
25 x 150mm square skirting	2.47	10.00	2.72	1.07	0.18	11.14	0.37	14.23	m	15.29
19 x 125mm moulded skirting	3.89	10.00	4.28	1.03	0.17	10.70	0.31	15.29	m	16.44
Repairs to staircase members in softwood										
Cut out damaged members of staircase and piece in new										
32 x 275mm tread	5.80	10.00	6.38	4.33	0.72	45.02	0.46	51.86	m	55.75
38 x 275mm tread	6.88	10.00	7.57	4.85	0.81	50.45	0.53	58.55	m	62.94
19 x 200mm riser	2.82	10.00	3.10	2.25	0.38	23.44	0.18	26.72	m	28.72
25 x 200mm riser	3.63	10.00	3.99	3.03	0.51	31.55	0.27	35.81	m	38.50
Cut back old tread for a width of 125mm and provide and fix new nosing with glued and dowelled joints to tread										
32mm with rounded nosing	3.98	10.00	4.38	2.16	0.36	22.46	0.76	27.60	m	29.67
38mm with rounded nosing	4.60	10.00	5.06	2.55	0.43	26.56	0.85	32.47	m	34.90
32mm with moulded nosing	4.73	10.00	5.20	2.54	0.42	26.39	0.76	32.36	m	34.78
38mm with moulded nosing	5.37	10.00	5.91	3.06	0.51	31.82	0.90	38.63	m	41.53
Replace missing square bar baluster 914mm high; clean out old mortices										
25 x 25mm	0.55	-	0.55	0.56	0.09	5.80	-	6.35	nr	6.83
38 x 38mm	1.00	-	1.00	0.75	0.13	7.84	-	8.84	nr	9.50
Cut out damaged square bar baluster 914mm high and piece in new										
25 x 25mm	0.55	-	0.55	0.75	0.03	7.13	0.03	7.71	nr	8.29
38 x 38mm	1.00	-	1.00	0.91	0.15	9.45	0.05	10.50	nr	11.29
Cut out damaged handrail and piece in new										
50mm mopstick	20.30	10.00	22.33	1.95	0.33	20.32	3.20	45.85	m	49.28
75 x 100mm moulded	33.20	10.00	36.52	3.97	0.66	41.28	3.47	81.27	m	87.36
Refixing removed handrail										
to new balustrade	-	-	-	0.69	0.12	7.21	-	7.21	m	7.75
Repairs to staircase members in oak										
Cut out damaged members of staircase and piece in new										
32 x 275mm tread	39.00	10.00	42.90	6.09	1.02	63.37	0.46	106.73	m	114.74
38 x 275mm tread	47.00	10.00	51.70	7.18	1.20	74.70	0.53	126.93	m	136.44
19 x 200mm riser	21.00	10.00	23.10	3.34	0.56	34.76	0.18	58.04	m	62.39
25 x 200mm riser	26.00	10.00	28.60	4.56	0.76	47.42	0.29	76.31	m	82.04
Cut back old tread for a width of 125mm and provide and fix new nosing with glued and dowelled joints to tread										
32mm with rounded nosing	23.00	10.00	25.30	2.23	0.37	23.18	0.76	49.24	m	52.93
38mm with rounded nosing	26.00	10.00	28.60	2.92	0.49	30.39	0.90	59.89	m	64.38
32mm with moulded nosing	24.00	10.00	26.40	3.20	0.53	33.26	0.76	60.42	m	64.95
38mm with moulded nosing	27.00	10.00	29.70	3.95	0.66	41.09	0.90	71.69	m	77.07
Replace missing square bar baluster 914mm high; clean out old mortices										
25 x 25mm	3.40	-	3.40	0.94	0.16	9.80	-	13.20	nr	14.19
38 x 38mm	6.90	-	6.90	0.99	0.17	10.33	-	17.23	nr	18.52
Cut out damaged square bar baluster 914mm high and piece in new										
25 x 25mm	3.40	-	3.40	1.21	0.20	12.57	0.03	16.00	nr	17.20
38 x 38mm	6.00	-	6.00	1.32	0.22	13.73	0.05	19.78	nr	21.26
Cut out damaged handrail and piece in new										
50mm mopstick	43.00	10.00	47.30	3.64	0.61	37.88	3.20	88.38	m	95.01
75 x 100mm moulded	68.00	10.00	74.80	5.50	0.92	57.22	3.47	135.49	m	145.66
Refixing removed handrail										
to new balustrade	-	-	-	1.35	0.23	14.08	-	14.08	m	15.13
Remove broken ironmongery and fix only new to softwood										
Door ironmongery										
75mm butt hinges	-	-	-	0.23	0.04	2.40	0.01	2.41	nr	2.59
100mm butt hinges	-	-	-	0.23	0.04	2.40	0.01	2.41	nr	2.59

Labour hourly rates: (except Specialists) Craft Operatives £9.23 Labour £7.02 Rates are national average prices. Refer to REGIONAL VARIATIONS for indicative levels of overall pricing in regions	MATERIALS			LABOUR				RATES		
	Del to Site	Waste	Material Cost	Craft Optve	Lab	Labour Cost	Sunds	Nett Rate	Unit	Gross Rate (+7.5%)
	£	%	£	Hrs	Hrs	£	£	£		£
C51: REPAIRING/RENOVATING/CONSERVING TIMBER Cont.										
Remove broken ironmongery and fix only new to softwood Cont.										
Door ironmongery Cont.										
tee hinges up to 300mm	-	-	-	0.62	0.10	6.42	0.01	6.43	nr	6.92
single action floor spring hinges	-	-	-	3.00	0.50	31.20	1.30	32.50	nr	34.94
barrel bolt up to 300mm	-	-	-	0.41	0.07	4.28	0.01	4.29	nr	4.61
lock or latch furniture	-	-	-	0.44	0.07	4.55	0.01	4.56	nr	4.90
rim lock	-	-	-	0.80	0.13	8.30	0.01	8.31	nr	8.93
rim latch	-	-	-	0.80	0.13	8.30	0.01	8.31	nr	8.93
mortice lock	-	-	-	1.22	0.20	12.66	0.01	12.67	nr	13.63
mortice latch	-	-	-	1.08	0.18	11.23	0.01	11.24	nr	12.09
pull handle	-	-	-	0.40	0.07	4.18	0.03	4.21	nr	4.53
Window ironmongery										
sash fastener	-	-	-	1.09	0.18	11.32	0.01	11.33	nr	12.18
sash lift	-	-	-	0.23	0.04	2.40	0.01	2.41	nr	2.59
casement stay	-	-	-	0.40	0.07	4.18	0.01	4.19	nr	4.51
casement fastener	-	-	-	0.40	0.07	4.18	0.01	4.19	nr	4.51
Sundry ironmongery										
hat and coat hook	-	-	-	0.29	0.05	3.03	0.01	3.04	nr	3.27
toilet roll holder	-	-	-	0.29	0.05	3.03	0.01	3.04	nr	3.27
Remove broken ironmongery and fix only new to hardwood										
Door ironmongery										
75mm butt hinges	-	-	-	0.30	0.05	3.12	0.01	3.13	nr	3.36
100mm butt hinges	-	-	-	0.30	0.05	3.12	0.01	3.13	nr	3.36
tee hinges up to 300mm	-	-	-	0.86	0.14	8.92	0.01	8.93	nr	9.60
single action floor spring hinges	-	-	-	4.20	0.70	43.68	1.30	44.98	nr	48.35
barrel bolt up to 300mm	-	-	-	0.60	0.10	6.24	0.01	6.25	nr	6.72
lock or latch furniture	-	-	-	0.61	0.10	6.33	0.01	6.34	nr	6.82
rim lock	-	-	-	1.15	0.19	11.95	0.01	11.96	nr	12.86
rim latch	-	-	-	1.15	0.19	11.95	0.01	11.96	nr	12.86
mortice lock	-	-	-	1.77	0.30	18.44	0.01	18.45	nr	19.84
mortice latch	-	-	-	1.58	0.26	16.41	0.01	16.42	nr	17.65
pull handle	-	-	-	0.57	0.10	5.96	0.03	5.99	nr	6.44
Window ironmongery										
sash fastener	-	-	-	1.57	0.26	16.32	0.01	16.33	nr	17.55
sash lift	-	-	-	0.31	0.05	3.21	0.01	3.22	nr	3.46
casement stay	-	-	-	0.56	0.09	5.80	0.01	5.81	nr	6.25
casement fastener	-	-	-	0.56	0.09	5.80	0.01	5.81	nr	6.25
Sundry ironmongery										
hat and coat hook	-	-	-	0.42	0.07	4.37	0.01	4.38	nr	4.71
toilet roll holder	-	-	-	0.42	0.07	4.37	0.01	4.38	nr	4.71
C52: FUNGUS/BEETLE ERADICATION										
Protective treatment of existing timbers										
Treat with two coats of spray applied preservative										
boarding	0.56	10.00	0.62	-	0.15	1.05	0.04	1.71	m2	1.84
structural timbers	0.84	10.00	0.92	-	0.20	1.40	0.05	2.38	m2	2.56
Treat with two coats of brush applied preservative										
boarding	0.84	10.00	0.92	-	0.26	1.83	0.03	2.78	m2	2.99
structural timbers	1.26	10.00	1.39	-	0.80	5.62	0.05	7.05	m2	7.58
Insecticide treatment of existing timbers										
Treat worm infected timbers with spray applied proprietary insecticide										
boarding	0.84	10.00	0.92	-	0.24	1.68	0.05	2.66	m2	2.86
structural timbers	1.12	10.00	1.23	-	0.28	1.97	0.06	3.26	m2	3.50
joinery timbers	1.12	10.00	1.23	-	0.32	2.25	0.06	3.54	m2	3.80
Treat worm infected timbers with brush applied proprietary insecticide										
boarding	1.12	10.00	1.23	-	0.36	2.53	0.03	3.79	m2	4.07
structural timbers	1.31	10.00	1.44	-	0.42	2.95	0.05	4.44	m2	4.77
joinery timbers	1.31	10.00	1.44	-	0.48	3.37	0.05	4.86	m2	5.23
Treatment of wall surfaces										
Treat surfaces of concrete and brickwork adjoining areas where infected timbers removed										
with a blow lamp	-	-	-	-	0.25	1.75	0.03	1.78	m2	1.92
C90: ALTERATIONS - SPOT ITEMS										
Removal of fittings and fixtures										
Removing fittings and fixtures										
cupboard with doors, frames, architraves, etc	-	-	-	-	0.59	4.14	1.04	5.18	m2	5.57
worktop with legs and bearers	-	-	-	-	0.56	3.93	1.39	5.32	m2	5.72
shelving exceeding 300mm wide with bearers and brackets	-	-	-	-	0.45	3.16	0.88	4.04	m2	4.34
shelving not exceeding 300mm wide with bearers and brackets	-	-	-	-	0.20	1.40	0.28	1.68	m	1.81

Labour hourly rates: (except Specialists) Craft Operatives £9.23 Labour £7.02 Rates are national average prices. Refer to REGIONAL VARIATIONS for indicative levels of overall pricing in regions	MATERIALS			LABOUR				RATES		
	Del to Site	Waste	Material Cost	Craft Optve	Lab	Labour Cost	Sunds	Nett Rate	Unit	Gross Rate (+7.5%)
	£	%	£	Hrs	Hrs	£	£	£		£
C90: ALTERATIONS - SPOT ITEMS Cont.										
Removal of fittings and fixtures Cont.										
Removing fittings and fixtures Cont.										
draining board and bearers										
500 x 600mm..................................	-	-	-	-	0.28	1.97	0.53	2.50	nr	2.68
600 x 900mm..................................	-	-	-	-	0.37	2.60	0.94	3.54	nr	3.80
wall cupboard unit										
510 x 305 x 305mm...........................	-	-	-	-	0.47	3.30	0.84	4.14	nr	4.45
510 x 305 x 610mm...........................	-	-	-	-	0.65	4.56	1.63	6.19	nr	6.66
1020 x 305 x 610mm..........................	-	-	-	-	0.94	6.60	3.31	9.91	nr	10.65
1220 x 305 x 610mm..........................	-	-	-	-	1.15	8.07	3.94	12.01	nr	12.91
floor cupboard unit										
510 x 510 x 915mm...........................	-	-	-	-	0.26	1.83	4.14	5.97	nr	6.41
610 x 510 x 915mm...........................	-	-	-	-	1.21	8.49	4.94	13.43	nr	14.44
1020 x 510 x 915mm..........................	-	-	-	-	1.65	11.58	8.28	19.86	nr	21.35
1220 x 510 x 915mm..........................	-	-	-	-	1.93	13.55	9.92	23.47	nr	25.23
sink base unit										
585 x 510 x 895mm...........................	-	-	-	-	0.92	6.46	4.67	11.13	nr	11.96
1070 x 510 x 895mm..........................	-	-	-	-	1.65	11.58	8.50	20.08	nr	21.59
tall cupboard unit										
510 x 510 x 1505mm..........................	-	-	-	-	1.56	10.95	6.82	17.77	nr	19.10
610 x 510 x 1505mm..........................	-	-	-	-	1.77	12.43	8.15	20.58	nr	22.12
bath panel and bearers......................	-	-	-	-	0.47	3.30	0.88	4.18	nr	4.49
steel or concrete clothes line post with concrete base.............................	-	-	-	-	0.50	3.51	1.74	5.25	nr	5.64

This section continues on the next page

EXISTING SITE/BUILDINGS/SERVICES

Labour hourly rates: (except Specialists) Craft Operatives £12.44 Labour £7.02 Rates are national average prices. Refer to REGIONAL VARIATIONS for indicative levels of overall pricing in regions	MATERIALS			LABOUR				RATES		
	Del to Site	Waste	Material Cost	Craft Optve	Lab	Labour Cost	Sunds	Nett Rate	Unit	Gross Rate (+7.5%)
	£	%	£	Hrs	Hrs	£	£	£		£
C90: ALTERATIONS - SPOT ITEMS Cont.										
Removal of plumbing and electrical installations										
Removing plumbing and electrical installations										
100mm cast iron rainwater gutters	-	-	-	0.20	-	2.49	0.17	2.66	m	2.86
75mm cast iron rainwater pipes	-	-	-	0.25	-	3.11	0.19	3.30	m	3.55
100mm cast iron rainwater pipes	-	-	-	0.25	-	3.11	0.32	3.43	m	3.69
cast iron rainwater head	-	-	-	0.30	-	3.73	0.53	4.26	nr	4.58
cast iron soil and vent pipe with caulked joints										
50mm	-	-	-	0.35	-	4.35	0.10	4.45	m	4.79
75mm	-	-	-	0.35	-	4.35	0.19	4.54	m	4.88
100mm	-	-	-	0.35	-	4.35	0.32	4.67	m	5.02
150mm	-	-	-	0.50	-	6.22	0.76	6.98	m	7.50
100mm asbestos cement soil and vent pipe	-	-	-	0.25	-	3.11	0.32	3.43	m	3.69
150mm asbestos cement soil and vent pipe	-	-	-	0.25	-	3.11	0.76	3.87	m	4.16
100mm p.v.c. soil and vent pipe	-	-	-	0.25	-	3.11	0.32	3.43	m	3.69
150mm p.v.c. soil and vent pipe	-	-	-	0.25	-	3.11	0.76	3.87	m	4.16
100mm lead soil and vent pipe	-	-	-	0.35	-	4.35	0.32	4.67	m	5.02
150mm lead soil and vent pipe	-	-	-	0.35	-	4.35	0.76	5.11	m	5.50
w.c. suites	-	-	-	1.55	-	19.28	3.26	22.54	nr	24.23
lavatory basins	-	-	-	1.35	-	16.79	1.96	18.75	nr	20.16
baths	-	-	-	1.55	-	19.28	6.53	25.81	nr	27.75
glazed ware sinks	-	-	-	1.35	-	16.79	3.26	20.05	nr	21.56
stainless steel sinks	-	-	-	1.35	-	16.79	1.96	18.75	nr	20.16
stainless steel sinks with single drainer	-	-	-	1.35	-	16.79	2.62	19.41	nr	20.87
stainless steel sinks with double drainer	-	-	-	1.35	-	16.79	3.26	20.05	nr	21.56
hot and cold water, waste pipes, etc. 8 - 25mm diameter										
copper	-	-	-	0.15	-	1.87	0.04	1.91	m	2.05
lead	-	-	-	0.15	-	1.87	0.04	1.91	m	2.05
polythene or p.v.c.	-	-	-	0.15	-	1.87	0.04	1.91	m	2.05
stainless steel	-	-	-	0.25	-	3.11	0.04	3.15	m	3.39
steel	-	-	-	0.25	-	3.11	0.04	3.15	m	3.39
hot and cold water, waste pipes, etc. 32 - 50mm diameter										
copper	-	-	-	0.20	-	2.49	0.10	2.59	m	2.78
lead	-	-	-	0.20	-	2.49	0.10	2.59	m	2.78
polythene or p.v.c.	-	-	-	0.20	-	2.49	0.10	2.59	m	2.78
stainless steel	-	-	-	0.30	-	3.73	0.10	3.83	m	4.12
steel	-	-	-	0.30	-	3.73	0.10	3.83	m	4.12
cold water cisterns up to 454 litres capacity	-	-	-	2.00	-	24.88	14.82	39.70	nr	42.68
hot water tanks or cylinders up to 227 litre capacity	-	-	-	2.00	-	24.88	7.40	32.28	nr	34.70
wall type radiators up to 914mm long, 610 - 914mm high	-	-	-	1.35	-	16.79	1.31	18.10	nr	19.46
wall type radiators 914 - 1829mm long, 610 - 914mm high	-	-	-	1.95	-	24.26	2.62	26.88	nr	28.89
column type radiators up to 914mm long, 610 - 914mm high	-	-	-	1.35	-	16.79	6.85	23.64	nr	25.42
column type radiators 914 - 1829mm long, 610 - 914mm high	-	-	-	1.95	-	24.26	13.70	37.96	nr	40.80
pipe insulation from pipes 8 - 25mm diameter	-	-	-	0.10	-	1.24	0.10	1.34	m	1.44
pipe insulation from pipes 32 - 50mm diameter	-	-	-	0.13	-	1.62	0.19	1.81	m	1.94
water tank or calorifier insulation from tanks up to 227 litres	-	-	-	0.50	-	6.22	7.40	13.62	nr	14.64
water tank or calorifier insulation from tanks 227 - 454 litres	-	-	-	0.75	-	9.33	14.82	24.15	nr	25.96
Removing plumbing and electrical installations; setting aside for re-use										
cast iron soil and vent pipe with caulked joints										
50mm	-	-	-	0.55	-	6.84	-	6.84	m	7.36
75mm	-	-	-	0.60	-	7.46	-	7.46	m	8.02
100mm	-	-	-	0.80	-	9.95	-	9.95	m	10.70
150mm	-	-	-	1.00	-	12.44	-	12.44	m	13.37

This section continues on the next page

Labour hourly rates: (except Specialists) Craft Operatives £14.38 Labour £7.02 Rates are national average prices. Refer to REGIONAL VARIATIONS for indicative levels of overall pricing in regions	MATERIALS			LABOUR				RATES		
	Del to Site	Waste	Material Cost	Craft Optve	Lab	Labour Cost	Sunds	Nett Rate	Unit	Gross Rate (+7.5%)
	£	%	£	Hrs	Hrs	£	£	£		£
C90: ALTERATIONS - SPOT ITEMS Cont.										
Removal of plumbing and electrical installations Cont.										
Removing plumbing and electrical installations										
wall mounted electric fire	-	-	-	2.00	-	28.76	1.74	30.50	nr	32.79
1.5 kW night storage heater	-	-	-	2.50	-	35.95	5.22	41.17	nr	44.26
2 KW night storage heater	-	-	-	2.50	-	35.95	6.96	42.91	nr	46.13
3 KW night storage heater	-	-	-	2.50	-	35.95	10.44	46.39	nr	49.87
Removing plumbing and electrical installations; extending and making good finishings										
lighting points	-	-	-	1.25	-	17.98	1.42	19.40	nr	20.85
flush type switches	-	-	-	1.25	-	17.98	1.42	19.40	nr	20.85
surface mounted type switches	-	-	-	1.00	-	14.38	1.14	15.52	nr	16.68
flush type socket outlets	-	-	-	1.25	-	17.98	1.42	19.40	nr	20.85
surface mounted type socket outlets	-	-	-	1.00	-	14.38	1.14	15.52	nr	16.68
flush type fitting points	-	-	-	1.25	-	17.98	1.42	19.40	nr	20.85
surface mounted type fitting points	-	-	-	1.00	-	14.38	1.14	15.52	nr	16.68
surface mounted p.v.c. insulated and sheathed cables	-	-	-	0.04	-	0.58	0.06	0.64	m	0.68
surface mounted mineral insulated copper sheathed cables	-	-	-	0.10	-	1.44	0.12	1.56	m	1.67
conduits up to 25mm diameter with junction boxes .	-	-	-	0.15	-	2.16	0.17	2.33	m	2.50
surface mounted cable trunking up to 100 x 100mm .	-	-	-	0.55	-	7.91	0.62	8.53	m	9.17

This section continues
on the next page

Labour hourly rates: (except Specialists) Craft Operatives £9.23 Labour £7.02 Rates are national average prices. Refer to REGIONAL VARIATIONS for indicative levels of overall pricing in regions	MATERIALS			LABOUR				RATES		
	Del to Site £	Waste %	Material Cost £	Craft Optve Hrs	Lab Hrs	Labour Cost £	Sunds £	Nett Rate £	Unit	Gross Rate (+7.5%) £
C90: ALTERATIONS - SPOT ITEMS Cont.										
Removal of floor, wall and ceiling finishings										
Removing finishings										
cement and sand to floors	-	-	-	-	0.86	6.04	1.74	7.78	m2	8.36
granolithic to floors	-	-	-	-	0.92	6.46	1.74	8.20	m2	8.81
granolithic to treads and risers	-	-	-	-	1.43	10.04	1.74	11.78	m2	12.66
plastic or similar tiles to floors	-	-	-	-	0.20	1.40	0.18	1.58	m2	1.70
plastic or similar tiles to floors; cleaning off for new	-	-	-	-	0.47	3.30	0.18	3.48	m2	3.74
plastic or similar tiles to floors and screed under	-	-	-	-	1.05	7.37	1.92	9.29	m2	9.99
linoleum and underlay to floors	-	-	-	-	0.07	0.49	1.74	2.23	m2	2.40
ceramic or quarry tiles to floors	-	-	-	-	1.21	8.49	1.04	9.53	m2	10.25
ceramic or quarry tiles to floors and screed under	-	-	-	-	2.39	16.78	2.78	19.56	m2	21.02
extra; cleaning and keying surface of concrete under	-	-	-	-	0.53	3.72	0.04	3.76	m2	4.04
asphalt to floors on loose underlay	-	-	-	-	0.35	2.46	0.70	3.16	m2	3.39
asphalt to floors keyed to concrete or screed	-	-	-	-	0.76	5.34	0.70	6.04	m2	6.49
granolithic skirtings	-	-	-	-	0.33	2.32	0.18	2.50	m	2.68
plastic or similar skirtings; cleaning off for new	-	-	-	-	0.13	0.91	0.04	0.95	m	1.02
timber skirtings	-	-	-	-	0.12	0.84	0.18	1.02	m	1.10
ceramic or quarry tile skirtings	-	-	-	-	0.43	3.02	0.20	3.22	m	3.46
plaster to walls	-	-	-	-	0.60	4.21	0.70	4.91	m2	5.28
rendering to walls	-	-	-	-	0.87	6.11	0.70	6.81	m2	7.32
rough cast to walls	-	-	-	-	0.90	6.32	0.88	7.20	m2	7.74
match boarding linings to walls with battens	-	-	-	-	0.38	2.67	1.04	3.71	m2	3.99
plywood or similar sheet linings to walls with battens	-	-	-	-	0.31	2.18	1.04	3.22	m2	3.46
insulating board linings to walls with battens	-	-	-	-	0.31	2.18	1.04	3.22	m2	3.46
plasterboard to walls	-	-	-	-	0.31	2.18	0.53	2.71	m2	2.91
plasterboard dry linings to walls	-	-	-	-	0.50	3.51	0.53	4.04	m2	4.34
plasterboard and skim to walls	-	-	-	-	0.60	4.21	0.70	4.91	m2	5.28
lath and plaster to walls	-	-	-	-	0.44	3.09	0.88	3.97	m2	4.27
metal lath and plaster to walls	-	-	-	-	0.33	2.32	0.88	3.20	m2	3.44
ceramic tiles to walls	-	-	-	-	0.75	5.26	0.53	5.79	m2	6.23
ceramic tiles to walls and backing under	-	-	-	-	1.11	7.79	0.88	8.67	m2	9.32
asphalt coverings to walls keyed to concrete or brickwork	-	-	-	-	0.66	4.63	0.70	5.33	m2	5.73
plaster to ceilings	-	-	-	-	0.74	5.19	0.70	5.89	m2	6.34
match boarding linings to ceilings with battens	-	-	-	-	0.54	3.79	1.04	4.83	m2	5.19
plywood or similar sheet linings to ceilings with battens	-	-	-	-	0.38	2.67	1.04	3.71	m2	3.99
insulating board linings to ceilings with battens	-	-	-	-	0.26	1.83	1.04	2.87	m2	3.08
plasterboard to ceilings	-	-	-	-	0.33	2.32	0.53	2.85	m2	3.06
plasterboard and skim to ceilings	-	-	-	-	0.40	2.81	0.70	3.51	m2	3.77
lath and plaster to ceilings	-	-	-	-	0.59	4.14	0.88	5.02	m2	5.40
metal lath and plaster to ceilings	-	-	-	-	0.47	3.30	0.88	4.18	m2	4.49
hacking surfaces of concrete as key for new finishings										
ceiling	-	-	-	-	0.67	4.70	0.04	4.74	m2	5.10
floor	-	-	-	-	0.53	3.72	0.04	3.76	m2	4.04
wall	-	-	-	-	0.53	3.72	0.04	3.76	m2	4.04
hacking surfaces of brick wall and raking out joints as key for new finishings	-	-	-	-	0.60	4.21	0.04	4.25	m2	4.57
Removing finishings; extending and making good finishings										
plaster cornices up to 100mm girth on face	0.28	10.00	0.31	0.78	0.78	12.68	0.07	13.05	m	14.03
plaster cornices 100 - 200mm girth on face	0.46	10.00	0.51	0.88	0.88	14.30	0.18	14.99	m	16.11
plaster cornices 200 - 300mm girth on face	0.81	10.00	0.89	0.97	0.97	15.76	0.53	17.18	m	18.47
plaster ceiling roses up to 300mm diameter	0.14	10.00	0.15	0.41	0.41	6.66	0.14	6.96	nr	7.48
plaster ceiling roses 300 - 450mm diameter	0.28	10.00	0.31	0.64	0.64	10.40	0.28	10.99	nr	11.81
plaster ceiling roses 450 - 600mm diameter	0.55	10.00	0.60	0.90	0.90	14.63	0.55	15.78	nr	16.96
Removing finishings; carefully handling and disposing toxic or other special waste by approved method										
asbestos cement sheet linings to walls	-	-	-	2.00	2.00	32.50	2.78	35.28	m2	37.93
asbestos cement sheet linings to ceilings	-	-	-	3.00	3.00	48.75	2.78	51.53	m2	55.39
Removal of roof coverings										
Removing coverings										
felt to roofs	-	-	-	-	0.28	1.97	0.70	2.67	m2	2.87
felt skirtings to roofs	-	-	-	-	0.33	2.32	0.07	2.39	m	2.57
asphalt to roofs	-	-	-	-	0.35	2.46	0.88	3.34	m2	3.59
asphalt skirtings to roofs; on expanded metal reinforcement	-	-	-	-	0.20	1.40	0.14	1.54	m	1.66
asphalt skirtings to roofs; keyed to concrete, brickwork, etc.	-	-	-	-	0.30	2.11	0.14	2.25	m	2.41
asphalt coverings to roofs; on expanded metal reinforcement; per 100mm of width	-	-	-	-	0.13	0.91	0.11	1.02	m	1.10
asphalt coverings to roofs; keyed to concrete, brickwork, etc.; per 100mm of width	-	-	-	-	0.20	1.40	0.11	1.51	m	1.63
slate to roofs	-	-	-	-	0.34	2.39	0.70	3.09	m2	3.32
tiles to roofs	-	-	-	-	0.31	2.18	1.04	3.22	m2	3.46
extra; removing battens	-	-	-	-	0.28	1.97	0.70	2.67	m2	2.87
extra; removing counter battens	-	-	-	-	0.10	0.70	0.14	0.84	m2	0.91
extra; removing underfelt	-	-	-	-	0.20	1.40	0.28	1.68	m2	1.81
lead to roofs	-	-	-	-	0.48	3.37	0.70	4.07	m2	4.37
lead flashings to roofs; per 25mm of girth	-	-	-	-	0.05	0.35	0.04	0.39	m	0.42
zinc to roofs	-	-	-	-	0.34	2.39	0.70	3.09	m2	3.32
zinc flashings to roofs; per 25mm of girth	-	-	-	-	0.03	0.21	0.04	0.25	m	0.27
copper to roofs	-	-	-	-	0.34	2.39	0.70	3.09	m2	3.32
copper flashings to roofs; per 25mm of girth	-	-	-	-	0.03	0.21	0.04	0.25	m	0.27

Labour hourly rates: (except Specialists) Craft Operatives £9.23 Labour £7.02 Rates are national average prices. Refer to REGIONAL VARIATIONS for indicative levels of overall pricing in regions	MATERIALS			LABOUR				RATES		
	Del to Site	Waste	Material Cost	Craft Optve	Lab	Labour Cost	Sunds	Nett Rate	Unit	Gross Rate (+7.5%)
	£	%	£	Hrs	Hrs	£	£	£		£
C90: ALTERATIONS - SPOT ITEMS Cont.										
Removal of roof coverings Cont.										
Removing coverings Cont.										
corrugated metal sheeting to roofs	-	-	-	-	0.31	2.18	1.04	3.22	m2	3.46
corrugated translucent sheeting to roofs	-	-	-	-	0.31	2.18	1.04	3.22	m2	3.46
board roof decking	-	-	-	-	0.31	2.18	1.04	3.22	m2	3.46
woodwool roof decking	-	-	-	-	0.39	2.74	1.04	3.78	m2	4.06
Removing coverings; setting aside for re-use										
slate to roofs	-	-	-	-	0.60	4.21	-	4.21	m2	4.53
tiles to roofs	-	-	-	-	0.60	4.21	-	4.21	m2	4.53
clean and stack 405 x 205mm slates; per 100	-	-	-	-	1.50	10.53	-	10.53	nr	11.32
clean and stack 510 x 255mm slates; per 100	-	-	-	-	1.89	13.27	-	13.27	nr	14.26
clean and stack 610 x 305mm slates; per 100	-	-	-	-	2.24	15.72	-	15.72	nr	16.90
clean and stack concrete tiles; per 100	-	-	-	-	2.24	15.72	-	15.72	nr	16.90
clean and stack plain tiles; per 100	-	-	-	-	1.50	10.53	-	10.53	nr	11.32
Removing coverings; carefully handling and disposing toxic or other special waste by approved method										
asbestos cement sheeting to roofs	-	-	-	2.00	2.00	32.50	2.78	35.28	m2	37.93
asbestos cement roof decking	-	-	-	2.00	2.00	32.50	2.78	35.28	m2	37.93
Removal of woodwork										
Removing										
stud partitions plastered both sides	-	-	-	-	0.66	4.63	2.62	7.25	m2	7.80
roof boarding	-	-	-	-	0.19	1.33	0.88	2.21	m2	2.38
roof boarding; prepare joists for new	-	-	-	-	0.45	3.16	0.88	4.04	m2	4.34
gutter boarding and the like	-	-	-	-	0.19	1.33	0.88	2.21	m2	2.38
weather boarding and battens	-	-	-	-	0.24	1.68	1.04	2.72	m2	2.93
tilting fillets, angle fillets and the like	-	-	-	-	0.09	0.63	0.07	0.70	m	0.75
fascia boards 150 mm wide	-	-	-	-	0.10	0.70	0.14	0.84	m	0.91
barge boards 200mm wide	-	-	-	-	0.10	0.70	0.18	0.88	m	0.95
soffit boards 300mm wide	-	-	-	-	0.10	0.70	0.28	0.98	m	1.06
floor boarding	-	-	-	-	0.33	2.32	0.88	3.20	m2	3.44
handrails and brackets	-	-	-	-	0.10	0.70	0.55	1.25	m	1.35
balustrades complete down to and with cappings to aprons or strings	-	-	-	0.58	0.10	6.06	1.39	7.45	m	8.00
newel posts; cut off flush with landing or string	-	-	-	0.48	0.08	4.99	0.35	5.34	nr	5.74
ends of treads projecting beyond face of cut outer string including scotia under; make good treads where balusters removed	-	-	-	0.90	0.15	9.36	1.04	10.40	m	11.18
dado or picture rails with grounds	-	-	-	-	0.10	0.70	0.14	0.84	m	0.91
architrave	-	-	-	-	0.06	0.42	0.14	0.56	m	0.60
window boards and bearers	-	-	-	-	0.12	0.84	0.28	1.12	m	1.21
Removing; setting aside for reuse										
handrails and brackets	-	-	-	0.25	0.04	2.59	-	2.59	m	2.78
Removal of windows and doors										
Removing										
metal windows with internal and external sills; in conjunction with demolition										
997 x 923mm......................	-	-	-	-	0.40	2.81	1.67	4.48	nr	4.81
1486 x 923mm......................	-	-	-	-	0.50	3.51	2.40	5.91	nr	6.35
1486 x 1513mm.....................	-	-	-	-	0.60	4.21	3.94	8.15	nr	8.76
1994 x 1513mm.....................	-	-	-	-	0.70	4.91	5.26	10.17	nr	10.94
metal windows with internal and external sills; preparatory to filling openings; cut out lugs										
997 x 923mm......................	-	-	-	-	1.30	9.13	1.67	10.80	nr	11.61
1486 x 923mm......................	-	-	-	-	1.60	11.23	2.40	13.63	nr	14.65
1486 x 1513mm.....................	-	-	-	-	2.00	14.04	3.94	17.98	nr	19.33
1994 x 1513mm.....................	-	-	-	-	2.30	16.15	5.26	21.41	nr	23.01
wood single or multi-light casements and frames with internal and external sills; in conjunction with demolition										
440 x 920mm......................	-	-	-	-	0.28	1.97	1.08	3.05	nr	3.27
1225 x 1070mm.....................	-	-	-	-	0.29	2.04	3.41	5.45	nr	5.85
2395 x 1225mm.....................	-	-	-	-	0.30	2.11	7.66	9.77	nr	10.50
wood single or multi-light casements and frames with internal and external sills; preparatory to filling openings; cut out fixing cramps										
440 x 920mm......................	-	-	-	-	1.25	8.78	1.08	9.86	nr	10.59
1225 x 1070mm.....................	-	-	-	-	1.70	11.93	3.41	15.34	nr	16.49
2395 x 1225mm.....................	-	-	-	-	2.52	17.69	7.66	25.35	nr	27.25
wood cased frames and sashes with internal and external sills complete with accessories and weights; in conjunction with demolition										
610 x 1225mm.....................	-	-	-	-	0.61	4.28	2.62	6.90	nr	7.42
915 x 1525mm.....................	-	-	-	-	0.66	4.63	4.87	9.50	nr	10.22
1370 x 1525mm....................	-	-	-	-	0.75	5.26	7.27	12.54	nr	13.48
wood cased frames and sashes with internal and external sills complete with accessories and weights; preparatory to filling openings; cut out fixing cramps										
610 x 1225mm.....................	-	-	-	-	1.79	12.57	2.62	15.19	nr	16.32
915 x 1525mm.....................	-	-	-	-	2.56	17.97	4.87	22.84	nr	24.55
1370 x 1525mm....................	-	-	-	-	4.11	28.85	7.27	36.12	nr	38.83
single internal doors	-	-	-	-	0.23	1.61	2.09	3.70	nr	3.98
single internal doors and frames; in conjunction with demolition...................	-	-	-	-	0.18	1.26	4.70	5.96	nr	6.41
single internal doors and frames; preparatory to filling openings; cut out fixing cramps	-	-	-	-	1.00	7.02	4.70	11.72	nr	12.60

Labour hourly rates: (except Specialists) Craft Operatives £9.23 Labour £7.02 Rates are national average prices. Refer to REGIONAL VARIATIONS for indicative levels of overall pricing in regions	MATERIALS			LABOUR				RATES		
	Del to Site	Waste	Material Cost	Craft Optve	Lab	Labour Cost	Sunds	Nett Rate	Unit	Gross Rate (+7.5%)
	£	%	£	Hrs	Hrs	£	£	£		£
C90: ALTERATIONS - SPOT ITEMS Cont.										
Removal of windows and doors Cont.										
Removing Cont.										
wood cased frames and sashes with internal and external sills complete with accessories and weights; preparatory to filling openings; cut out fixing cramps Cont.										
double internal doors....................	-	-	-	-	0.47	3.30	4.18	7.48	nr	8.04
double internal doors and frames; in conjunction with demolition	-	-	-	-	0.19	1.33	7.14	8.47	nr	9.11
double internal doors and frames; preparatory to filling openings; cut out fixing cramps	-	-	-	-	1.20	8.42	7.14	15.56	nr	16.73
single external doors....................	-	-	-	-	0.25	1.75	3.48	5.24	nr	5.63
single external doors and frames; in conjunction with demolition..................	-	-	-	-	0.18	1.26	6.26	7.52	nr	8.09
single external doors and frames; preparatory to filling openings; cut out fixing cramps	-	-	-	-	1.00	7.02	6.26	13.28	nr	14.28
double external doors....................	-	-	-	-	0.47	3.30	4.18	7.48	nr	8.04
double external doors and frames; in conjunction with demolition..................	-	-	-	-	0.19	1.33	7.14	8.47	nr	9.11
double external doors and frames; preparatory to filling openings; cut out fixing cramps	-	-	-	-	1.20	8.42	7.14	15.56	nr	16.73
single door frames; in conjunction with demolition..................	-	-	-	-	0.17	1.19	2.62	3.81	nr	4.10
single door frames; preparatory to filling openings; cut out fixing cramps	-	-	-	-	0.77	5.41	2.62	8.03	nr	8.63
double door frames; in conjunction with demolition..................	-	-	-	-	0.18	1.26	2.96	4.22	nr	4.54
double door frames; preparatory to filling openings; cut out fixing cramps	-	-	-	-	0.82	5.76	2.96	8.72	nr	9.37
Removing; setting aside for reuse										
metal windows with internal and external sills; in conjunction with demolition										
997 x 923mm...............................	-	-	-	0.70	0.50	9.97	-	9.97	nr	10.72
1486 x 923mm..............................	-	-	-	0.85	0.65	12.41	-	12.41	nr	13.34
1486 x 1513mm.............................	-	-	-	0.90	0.75	13.57	-	13.57	nr	14.59
1994 x 1513mm.............................	-	-	-	1.00	0.85	15.20	-	15.20	nr	16.34
metal windows with internal and external sills; preparatory to filling openings; cut out lugs										
997 x 923mm...............................	-	-	-	0.70	1.40	16.29	-	16.29	nr	17.51
1486 x 923mm..............................	-	-	-	0.85	1.75	20.13	-	20.13	nr	21.64
1486 x 1513mm.............................	-	-	-	0.90	2.15	23.40	-	23.40	nr	25.16
1994 x 1513mm.............................	-	-	-	1.00	2.45	26.43	-	26.43	nr	28.41
wood single or multi-light casements and frames with internal and external sills; in conjunction with demolition										
440 x 920mm...............................	-	-	-	0.50	0.36	7.14	-	7.14	nr	7.68
1225 x 1070mm.............................	-	-	-	0.83	0.43	10.68	-	10.68	nr	11.48
2395 x 1225mm.............................	-	-	-	1.00	0.46	12.46	-	12.46	nr	13.39
wood single or multi-light casements and frames with internal and external sills; preparatory to filling openings; cut out fixing cramps										
440 x 920mm...............................	-	-	-	0.50	1.33	13.95	-	13.95	nr	15.00
1225 x 1070mm.............................	-	-	-	0.83	1.84	20.58	-	20.58	nr	22.12
2395 x 1225mm.............................	-	-	-	1.00	2.68	28.04	-	28.04	nr	30.15
wood cased frames and sashes with internal and external sills complete with accessories and weights; in conjunction with demolition										
610 x 1225mm.............................	-	-	-	0.97	0.78	14.43	-	14.43	nr	15.51
915 x 1525mm.............................	-	-	-	1.63	0.93	21.57	-	21.57	nr	23.19
1370 x 1525mm............................	-	-	-	1.94	1.07	25.42	-	25.42	nr	27.32
wood cased frames and sashes with internal and external sills complete with accessories and weights; preparatory to filling openings; cut out fixing cramps										
610 x 1225mm.............................	-	-	-	0.97	1.96	22.71	-	22.71	nr	24.42
915 x 1525mm.............................	-	-	-	1.63	2.83	34.91	-	34.91	nr	37.53
1370 x 1525mm............................	-	-	-	1.94	4.43	49.00	-	49.00	nr	52.68
note: the above rates assume that windows and sashes will be re-glazed when re-used										
single internal doors.........................	-	-	-	0.50	0.31	6.79	-	6.79	nr	7.30
single internal doors and frames; in conjunction with demolition....................	-	-	-	1.00	0.34	11.62	-	11.62	nr	12.49
single internal doors and frames; preparatory to filling openings; cut out fixing cramps	-	-	-	1.00	1.16	17.37	-	17.37	nr	18.68
double internal doors.........................	-	-	-	1.00	0.63	13.65	-	13.65	nr	14.68
double internal doors and frames; in conjunction with demolition....................	-	-	-	1.47	0.43	16.59	-	16.59	nr	17.83
double internal doors and frames; preparatory to filling openings; cut out fixing cramps	-	-	-	1.47	1.45	23.75	-	23.75	nr	25.53
single external doors.........................	-	-	-	0.50	0.33	6.93	-	6.93	nr	7.45
single external doors and frames; in conjunction with demolition....................	-	-	-	1.00	0.34	11.62	-	11.62	nr	12.49
single external doors and frames; preparatory to filling openings; cut out fixing cramps	-	-	-	1.00	1.16	17.37	-	17.37	nr	18.68
double external doors.........................	-	-	-	1.00	0.63	13.65	-	13.65	nr	14.68
double external doors and frames; in conjunction with demolition....................	-	-	-	1.47	0.44	16.66	-	16.66	nr	17.91
double external doors and frames; preparatory to filling openings; cut out fixing cramps	-	-	-	1.47	1.45	23.75	-	23.75	nr	25.53
single door frames; in conjunction with demolition....................	-	-	-	0.50	0.25	6.37	-	6.37	nr	6.85

Labour hourly rates: (except Specialists) Craft Operatives £9.23 Labour £7.02 Rates are national average prices. Refer to REGIONAL VARIATIONS for indicative levels of overall pricing in regions	MATERIALS			LABOUR				RATES		
	Del to Site	Waste	Material Cost	Craft Optve	Lab	Labour Cost	Sunds	Nett Rate	Unit	Gross Rate (+7.5%)
	£	%	£	Hrs	Hrs	£	£	£		£
C90: ALTERATIONS - SPOT ITEMS Cont.										
Removal of windows and doors Cont.										
Removing; setting aside for reuse Cont.										
note: the above rates assume that windows and sashes will be re-glazed when re-used Cont.										
single door frames; preparatory to filling openings; cut out fixing cramps	-	-	-	0.50	0.85	10.58	-	10.58	nr	11.38
double door frames; in conjunction with demolition..................................	-	-	-	0.50	0.26	6.44	-	6.44	nr	6.92
double door frames; preparatory to filling openings; cut out fixing cramps	-	-	-	0.50	0.90	10.93	-	10.93	nr	11.75
Removal of ironmongery										
Removing door ironmongery (piecing in doors and frames included elsewhere)										
butt hinges	-	-	-	-	0.06	0.42	0.01	0.43	nr	0.46
tee hinges up to 300mm	-	-	-	-	0.20	1.40	0.01	1.41	nr	1.52
floor spring hinges and top centres ...	-	-	-	-	0.92	6.46	0.20	6.66	nr	7.16
barrel bolts up to 200mm	-	-	-	-	0.13	0.91	0.01	0.92	nr	0.99
flush bolts up to 200mm	-	-	-	-	0.17	1.19	0.01	1.20	nr	1.29
indicating bolts	-	-	-	-	0.23	1.61	0.01	1.62	nr	1.75
double panic bolts	-	-	-	-	0.69	4.84	0.11	4.95	nr	5.33
single panic bolts	-	-	-	-	0.58	4.07	0.07	4.14	nr	4.45
Norfolk or Suffolk latches	-	-	-	-	0.23	1.61	0.01	1.62	nr	1.75
cylinder rim night latches	-	-	-	-	0.23	1.61	0.01	1.62	nr	1.75
rim locks or latches and furniture	-	-	-	-	0.23	1.61	0.01	1.62	nr	1.75
mortice dead locks	-	-	-	-	0.17	1.19	0.01	1.20	nr	1.29
mortice locks or latches and furniture .	-	-	-	-	0.29	2.04	0.01	2.05	nr	2.20
Bales catches	-	-	-	-	0.12	0.84	0.01	0.85	nr	0.92
overhead door closers, surface fixed ...	-	-	-	-	0.29	2.04	0.11	2.15	nr	2.31
pull handles	-	-	-	-	0.12	0.84	0.01	0.85	nr	0.92
push plates	-	-	-	-	0.12	0.84	0.01	0.85	nr	0.92
kicking plates	-	-	-	-	0.17	1.19	0.04	1.23	nr	1.33
letter plates	-	-	-	-	0.12	0.84	0.01	0.85	nr	0.92
Removing window ironmongery										
sash centres	-	-	-	-	0.12	0.84	0.01	0.85	nr	0.92
sash fasteners	-	-	-	-	0.17	1.19	0.01	1.20	nr	1.29
sash lifts	-	-	-	-	0.07	0.49	0.01	0.50	nr	0.54
sash screws	-	-	-	-	0.17	1.19	0.01	1.20	nr	1.29
casement fasteners	-	-	-	-	0.12	0.84	0.01	0.85	nr	0.92
casement stays	-	-	-	-	0.12	0.84	0.01	0.85	nr	0.92
quadrant stays	-	-	-	-	0.12	0.84	0.01	0.85	nr	0.92
fanlight catches	-	-	-	-	0.13	0.91	0.01	0.92	nr	0.99
curtain tracks	-	-	-	-	0.15	1.05	0.07	1.12	m	1.21
Removing sundry ironmongery										
hat and coat hooks	-	-	-	-	0.07	0.49	0.01	0.50	nr	0.54
cabin hooks and eyes	-	-	-	-	0.12	0.84	0.01	0.85	nr	0.92
shelf brackets	-	-	-	-	0.07	0.49	0.04	0.53	nr	0.57
toilet roll holders	-	-	-	-	0.09	0.63	0.04	0.67	nr	0.72
towel rollers	-	-	-	-	0.12	0.84	0.07	0.91	nr	0.98
Removal of metalwork										
Removing										
balustrades 1067mm high	-	-	-	-	0.81	5.69	0.90	6.59	m	7.08
wire mesh screens with timber or metal beads screwed on										
generally............................	-	-	-	-	0.62	4.35	0.88	5.23	m2	5.62
225 x 225mm	-	-	-	-	0.20	1.40	0.07	1.47	nr	1.58
305 x 305mm..........................	-	-	-	-	0.22	1.54	0.11	1.65	nr	1.78
guard bars of vertical bars at 100mm centres welded to horizontal fixing bars at 450mm centres, fixed with bolts	-	-	-	-	0.62	4.35	0.88	5.23	m2	5.62
guard bars of vertical bars at 100mm centres welded to horizontal fixing bars at 450mm centres, fixed with screws	-	-	-	-	0.39	2.74	0.88	3.62	m2	3.89
extra; piecing in softwood after removal of bolts	-	-	-	0.25	0.04	2.59	0.09	2.68	nr	2.88
small bracket........................	-	-	-	-	0.18	1.26	0.07	1.33	nr	1.43
Removing; making good finishings										
balustrades 1067mm high; making good mortices in treads................................	-	-	-	0.97	0.97	15.76	1.14	16.90	m	18.17
guard bars of vertical bars at 100mm centres welded to horizontal fixing bars at 450mm centres fixed in mortices; making good concrete or brickwork and plaster	-	-	-	0.60	1.42	15.51	1.19	16.70	m2	17.95
small bracket built or cast in; making good concrete or brickwork and plaster	-	-	-	0.25	0.25	4.06	0.47	4.53	nr	4.87
Removal of manholes										
Removing										
manholes overall size; remove cover and frame; break up brick sides and concrete bottom; fill in void with hardcore										
overall size 914 x 1067mm and 600mm deep to invert	1.92	30.00	2.50	-	10.00	70.20	3.48	76.18	nr	81.89
overall size 914 x 1219mm and 900mm deep to invert	3.84	30.00	4.99	-	11.50	80.73	3.83	89.55	nr	96.27
manhole covers and frames; clean off brickwork	-	-	-	-	0.50	3.51	0.62	4.13	nr	4.44

Labour hourly rates: (except Specialists) Craft Operatives £9.23 Labour £7.02 Rates are national average prices. Refer to REGIONAL VARIATIONS for indicative levels of overall pricing in regions	MATERIALS			LABOUR				RATES		
	Del to Site £	Waste %	Material Cost £	Craft Optve Hrs	Lab Hrs	Labour Cost £	Sunds £	Nett Rate £	Unit	Gross Rate (+7.5%) £
C90: ALTERATIONS – SPOT ITEMS Cont.										
Removal of manholes Cont.										
Removing Cont. manholes overall size; remove cover and frame; break up brick sides and concrete bottom; fill in void with hardcore Cont.										
fresh air inlets; clean out pipe socket.......	-	-	-	-	0.50	3.51	0.20	3.71	nr	3.99
Removal of fencing										
Removing chestnut pale fencing with posts										
610mm high......................................	-	-	-	-	0.30	2.11	0.55	2.66	m	2.86
914mm high......................................	-	-	-	-	0.34	2.39	0.84	3.23	m	3.47
1219mm high.....................................	-	-	-	-	0.38	2.67	1.12	3.79	m	4.07
close boarded fencing with posts										
1219mm high.....................................	-	-	-	-	0.68	4.77	1.28	6.05	m	6.51
1524mm high.....................................	-	-	-	-	0.81	5.69	1.60	7.29	m	7.83
1829mm high.....................................	-	-	-	-	0.91	6.39	1.92	8.31	m	8.93
chain link fencing with posts										
914mm high......................................	-	-	-	-	0.29	2.04	0.62	2.66	m	2.85
1219mm high.....................................	-	-	-	-	0.34	2.39	2.58	4.97	m	5.34
1524mm high.....................................	-	-	-	-	0.38	2.67	1.04	3.71	m	3.99
1829mm high.....................................	-	-	-	-	0.42	2.95	1.28	4.23	m	4.55
timber gate and posts..........................	-	-	-	-	0.98	6.88	2.62	9.50	nr	10.21
Strutting for forming openings										
Strutting generally the cost of strutting in connection with forming openings is to be added to the cost of forming openings; the following are examples of costs for strutting for various sizes and locations of openings. Costs are based on three uses of timber										
Strutting for forming small openings in internal load bearing walls										
opening size 900 x 2000mm	8.91	5.00	9.36	2.00	0.33	20.78	0.66	30.79	nr	33.10
opening size 2000 x 2300mm	19.80	5.00	20.79	5.00	0.83	51.98	1.43	74.20	nr	79.76
Strutting for forming openings in external one brick walls										
opening size 900 x 2000mm	13.37	5.00	14.04	3.00	0.50	31.20	1.14	46.38	nr	49.86
opening size 1500 x 1200mm	13.37	5.00	14.04	3.00	0.50	31.20	1.14	46.38	nr	49.86
Strutting for forming openings in external 280mm brick cavity walls										
opening size 900 x 2000mm	18.32	5.00	19.24	4.00	0.66	41.55	1.28	62.07	nr	66.72
opening size 1500 x 1200mm	18.32	5.00	19.24	4.00	0.66	41.55	1.28	62.07	nr	66.72
Strutting for forming large openings in internal load bearing walls on the ground floor of an average two storey building with timber floors and pitched roof; with load bearing surface 450mm below ground floor; including cutting holes and making good										
opening size 3700 x 2300mm	123.75	5.00	129.94	20.00	20.00	325.00	15.16	470.10	nr	505.35
Unit rates for pricing the above and similar work plates, struts, braces and wedges in supports to floor and roof...................................	1.73	5.00	1.82	0.28	0.28	4.55	0.22	6.59	m	7.08
dead shore and needle, sole plates, braces and wedges including cutting holes and making good ...	62.37	5.00	65.49	10.00	10.00	162.50	7.55	235.54	nr	253.20
Strutting for forming large openings in external one brick walls on the ground floor of an average two storey building as described above										
opening size 6000 x 2500mm	250.47	5.00	262.99	40.00	40.00	650.00	30.07	943.06	nr	1013.79
Unit rates for pricing the above and similar work strutting to window opening over new opening	12.47	5.00	13.09	2.00	2.00	32.50	1.49	47.08	nr	50.61
plates, struts, braces and wedges in supports to floor and roof...................................	1.68	5.00	1.76	0.28	0.28	4.55	0.22	6.53	m	7.02
dead shore and needle, sole plates, braces and wedges including cutting holes and making good ...	98.11	5.00	103.02	15.70	15.70	255.13	11.85	369.99	nr	397.74
set of two raking shores with 50mm wall piece, wedges and dogs	25.05	5.00	26.30	4.00	4.00	65.00	2.99	94.29	nr	101.36
Strutting for forming large openings in external 280mm brick cavity walls on the ground floor of an average two storey building as above described										
opening size 6000 x 2500mm	274.53	5.00	288.26	43.90	43.90	713.38	33.06	1034.69	nr	1112.29
Unit rates for pricing the above and similar work strutting to window opening over new opening	24.95	5.00	26.20	4.00	4.00	65.00	2.99	94.19	nr	101.25
plates, struts, braces and wedges in supports to floor and roof...................................	1.68	5.00	1.76	0.28	0.28	4.55	0.22	6.53	m	7.02
dead shore and needle, sole plates, braces and wedges including cutting holes and making good ...	103.95	5.00	109.15	16.70	16.70	271.38	12.63	393.15	nr	422.64
set of two raking shores with 50mm wall piece, wedges and dogs	25.34	5.00	26.61	4.00	4.00	65.00	2.99	94.60	nr	101.69

Labour hourly rates: (except Specialists) Craft Operatives £9.23 Labour £7.02 Rates are national average prices. Refer to REGIONAL VARIATIONS for indicative levels of overall pricing in regions	MATERIALS			LABOUR				RATES		
	Del to Site £	Waste %	Material Cost £	Craft Optve Hrs	Lab Hrs	Labour Cost £	Sunds £	Nett Rate £	Unit	Gross Rate (+7.5%) £
C90: ALTERATIONS - SPOT ITEMS Cont.										
Spot items										
Spot items generally the following items are usually specified as spot items where all work in all trades is included in one item; to assist in adapting these items for varying circumstances, unit rates have been given for the component operation where possible while these items are not in accordance with SMM7 either by description or measurement, it is felt that the pricing information given will be of value to the reader										
Breaking up and reinstatement of concrete floors for excavations										
Break up concrete bed and 150mm hardcore bed under for a width of 760mm for excavation of trench and reinstate with new hardcore and in-situ concrete, B.S.5328, DESIGNED mix C20, 20mm aggregate; make good up to new wall and existing concrete bed both side										
with 100mm plain concrete bed	9.20	10.00	10.12	–	2.70	18.95	6.46	35.53	m	38.20
with 150mm plain concrete bed	13.11	10.00	14.42	–	5.00	35.10	7.82	57.34	m	61.64
with 100mm concrete bed reinforced with steel fabric to B.S.4483, Reference A193	11.00	10.00	12.10	–	4.05	28.43	6.46	46.99	m	50.52
with 150mm concrete bed reinforced with steel fabric to B.S.4483, Reference A193	14.91	10.00	16.40	–	7.33	51.46	7.82	75.68	m	81.35
Openings through concrete										
Form opening for door frame through 150mm reinforced concrete wall plastered both sides and with skirting both sides; take off skirtings, cut opening through wall, square up reveals, make good existing plaster up to new frame both sides and form fitted ends on existing skirtings up to new frame; extend cement and sand floor screed and vinyl floor covering through opening and make good up to existing										
838 x 2032mm	15.80	–	15.80	22.50	28.00	404.24	30.20	450.24	nr	484.00
Unit rates for pricing the above and similar work										
cut opening through wall										
100mm thick.......................................	–	–	–	–	5.30	37.21	7.11	44.32	m2	47.64
150mm thick.......................................	–	–	–	–	7.90	55.46	10.63	66.09	m2	71.04
200mm thick.......................................	–	–	–	–	10.50	73.71	14.15	87.86	m2	94.45
make good fair face around opening...............	–	–	–	–	0.80	5.62	0.50	6.12	m	6.57
square up reveals to opening										
100mm wide.......................................	–	–	–	0.75	0.75	12.19	1.24	13.43	m	14.43
150mm wide.......................................	–	–	–	1.07	1.07	17.39	1.75	19.14	m	20.57
200mm wide.......................................	–	–	–	1.38	1.38	22.43	2.78	25.20	m	27.10
make good existing plaster up to new frame....	0.27	10.00	0.30	0.55	0.55	8.94	–	9.23	m	9.93
13mm two coat hardwall plaster to reveal not exceeding 300mm wide............................	0.52	10.00	0.57	0.64	0.64	10.40	–	10.97	m	11.79
take off old skirting............................	–	–	–	0.22	–	2.03	0.10	2.13	m	2.29
form fitted end on existing skirting up to new frame	–	–	–	0.38	–	3.51	–	3.51	nr	3.77
short length of old skirting up to 100mm long with mitre with existing one end	–	–	–	0.58	–	5.35	0.04	5.39	nr	5.80
short length of old skirting up to 200mm long with mitres with existing both ends	–	–	–	1.15	–	10.61	0.06	10.67	nr	11.48
38mm cement and sand (1:3) screeded bed in opening...	3.91	10.00	4.30	1.54	1.54	25.02	–	29.33	m2	31.53

vinyl or similar floor covering to match
    existing, fixed with adhesive, not exceeding
    300mm wide......................................
	5.40	10.00	5.94	0.15	0.15	2.44	0.19	8.57	m	9.21

iroko threshold; twice oiled; plugged and screwed										
15 x 100mm......................................	3.90	10.00	4.29	0.90	–	8.31	0.05	12.65	m	13.60
15 x 150mm......................................	5.70	10.00	6.27	1.20	–	11.08	0.06	17.41	m	18.71
15 x 200mm......................................	6.90	10.00	7.59	1.55	–	14.31	0.09	21.99	m	23.64
Form opening for staircase through 150mm reinforced concrete suspended floor plastered on soffit and with screed and vinyl floor covering on top; take up floor covering and screed, cut opening through floor, square up edges of slab, make good ceiling plaster up to lining and make good screed and floor covering to lining										
900 x 2134mm	59.60	–	59.60	19.00	39.00	449.15	39.00	547.75	nr	588.83
Unit rates for pricing the above and similar work										
take up vinyl or similar floor covering	–	–	–	–	0.55	3.86	0.14	4.00	m2	4.30
hack up screed	–	–	–	–	1.50	10.53	1.29	11.82	m2	12.71
cut opening through 150mm thick slab	–	–	–	–	7.90	55.46	10.63	66.09	m2	71.04
cut opening through 200mm thick slab	–	–	–	–	10.50	73.71	14.15	87.86	m2	94.45
make good fair face around opening	–	–	–	–	0.80	5.62	0.47	6.09	m	6.54
square up edges of 150mm thick slab	–	–	–	1.07	1.07	17.39	1.05	19.24	m	20.60
square up edges of 200mm thick slab	–	–	–	1.38	1.38	22.43	2.47	24.90	m	26.76
make good existing plaster up to new lining	0.27	10.00	0.30	0.55	0.55	8.94	–	9.23	m	9.93
make good existing 38mm floor screed up to new lining...	0.62	10.00	0.68	0.45	0.45	7.31	–	7.99	m	8.59
make good existing vinyl or similar floor covering up to new lining	7.20	10.00	7.92	0.67	0.67	10.89	0.30	19.11	m	20.54

Labour hourly rates: (except Specialists) Craft Operatives £9.23 Labour £7.02 Rates are national average prices. Refer to REGIONAL VARIATIONS for indicative levels of overall pricing in regions	MATERIALS			LABOUR				RATES		
	Del to Site £	Waste %	Material Cost £	Craft Optve Hrs	Lab Hrs	Labour Cost £	Sunds £	Nett Rate £	Unit	Gross Rate (+7.5%) £

C90: ALTERATIONS - SPOT ITEMS Cont.

Filling holes in concrete

Fill holes in concrete structure where pipes removed with concrete to B.S.5328, designed mix C20, 20mm aggregate; formwork
wall; thickness 100mm; pipe diameter

50mm	-	-	-	0.60	0.30	7.64	1.53	9.17	nr	9.86
100mm	-	-	-	0.78	0.39	9.94	2.02	11.96	nr	12.85
150mm	-	-	-	0.98	0.49	12.49	2.43	14.92	nr	16.03

wall; thickness 200mm; pipe diameter

50mm	-	-	-	0.78	0.39	9.94	2.02	11.96	nr	12.85
100mm	-	-	-	0.98	0.49	12.49	2.43	14.92	nr	16.03
150mm	-	-	-	1.00	0.77	14.64	3.12	17.76	nr	19.09

floor; thickness 200mm; pipe diameter

50mm	-	-	-	0.74	0.37	9.43	1.83	11.26	nr	12.10
100mm	-	-	-	0.93	0.46	11.81	2.34	14.15	nr	15.21
150mm	-	-	-	0.95	0.48	12.14	2.96	15.10	nr	16.23

Make good fair face to filling and surrounding work
to one side of wall; pipe diameter

50mm	-	-	-	0.10	0.10	1.63	0.21	1.84	nr	1.97
100mm	-	-	-	0.20	0.20	3.25	0.32	3.57	nr	3.84
150mm	-	-	-	0.25	0.25	4.06	0.39	4.45	nr	4.79

to soffit of floor; pipe diameter

50mm	-	-	-	0.10	0.10	1.63	0.21	1.84	nr	1.97
100mm	-	-	-	0.20	0.20	3.25	0.32	3.57	nr	3.84
150mm	-	-	-	0.25	0.25	4.06	0.39	4.45	nr	4.79

Make good plaster to filling and surrounding work
to one side of wall; pipe diameter

50mm	-	-	-	0.19	0.19	3.09	0.21	3.30	nr	3.54
100mm	-	-	-	0.36	0.36	5.85	0.32	6.17	nr	6.63
150mm	-	-	-	0.45	0.45	7.31	0.39	7.70	nr	8.28

to soffit of floor; pipe diameter

50mm	-	-	-	0.20	0.20	3.25	0.21	3.46	nr	3.72
100mm	-	-	-	0.40	0.40	6.50	0.32	6.82	nr	7.33
150mm	-	-	-	0.50	0.50	8.13	0.39	8.52	nr	9.15

Make good floor screed and surrounding work

pipe diameter 50mm	-	-	-	0.19	0.19	3.09	0.09	3.18	nr	3.42
pipe diameter 100mm	-	-	-	0.36	0.36	5.85	0.19	6.04	nr	6.49
pipe diameter 150mm	-	-	-	0.45	0.45	7.31	0.26	7.57	nr	8.14

Make good vinyl or similar floor covering and surrounding work

pipe diameter 50mm	-	-	-	0.50	0.50	8.13	1.56	9.69	nr	10.41
pipe diameter 100mm	-	-	-	1.00	1.00	16.25	2.09	18.34	nr	19.72
pipe diameter 150mm	-	-	-	1.20	1.20	19.50	2.93	22.43	nr	24.11

Fill holes in concrete structure where metal sections removed with concrete to B.S.5328, designed mix C20, 20mm aggregate; formwork

wall; thickness 100mm; section not exceeding 250mm deep	-	-	-	0.60	0.60	9.75	1.86	11.61	nr	12.48
wall; thickness 200mm; section not exceeding 250mm deep	-	-	-	0.75	0.75	12.19	2.71	14.90	nr	16.01
wall; thickness 100mm; section 250 - 500mm deep	-	-	-	0.75	0.75	12.19	2.71	14.90	nr	16.01
wall; thickness 200mm; section 250 - 500mm deep	-	-	-	0.90	0.90	14.63	3.27	17.90	nr	19.24

Make good fair face to filling and surrounding work

to one side of wall; section not exceeding 250mm deep	-	-	-	0.22	0.22	3.58	0.18	3.75	nr	4.04
to one side of wall; section 250 - 500mm deep	-	-	-	0.26	0.26	4.22	0.28	4.50	nr	4.84

Make good plaster to filling and surrounding work

to one side of wall; section not exceeding 250mm deep	-	-	-	0.35	0.35	5.69	0.49	6.18	nr	6.64
to one side of wall; section 250 - 500mm deep	-	-	-	0.43	0.43	6.99	0.74	7.73	nr	8.31

Filling openings in concrete

Fill opening where door removed in 150mm reinforced concrete wall with concrete reinforced with 12mm mild steel bars at 300mm centres both ways tied to existing structure; hack off plaster to reveals, hack up floor covering and screed in opening, prepare edges of opening to form joint with filling, drill edges of opening and tie in new reinforcement, plaster filling both sides and extended skirting both sides; making good junction of new and existing plaster and skirtings and make good floor screed and vinyl covering up to filling both sides

838 x 2032mm	100.10	-	100.10	22.50	34.50	449.87	9.63	559.60	nr	601.56

Unit rates for pricing the above and similar work
hack off plaster to reveal not exceeding 100mm

wide	-	-	-	-	0.24	1.68	0.07	1.75	m	1.89
hack off plaster to reveal 100 - 200mm wide	-	-	-	-	0.33	2.32	0.14	2.46	m	2.64
take up vinyl or similar floor covering and screed	-	-	-	-	2.00	14.04	1.43	15.47	m2	16.63
prepare edge of opening 100mm wide to form joint with filling	-	-	-	0.25	0.25	4.06	-	4.06	m	4.37
prepare edge of opening 200mm wide to form joint with filling	-	-	-	0.35	0.35	5.69	-	5.69	m	6.11

EXISTING SITE/BUILDINGS/SERVICES

Labour hourly rates: (except Specialists) Craft Operatives £9.23 Labour £7.02 Rates are national average prices. Refer to REGIONAL VARIATIONS for indicative levels of overall pricing in regions	MATERIALS			LABOUR				RATES		
	Del to Site £	Waste %	Material Cost £	Craft Optve Hrs	Lab Hrs	Labour Cost £	Sunds £	Nett Rate £	Unit	Gross Rate (+7.5%) £

C90: ALTERATIONS - SPOT ITEMS Cont.

Filling openings in concrete Cont.

Unit rates for pricing the above and similar work Cont.

	Del to Site	Waste	Material Cost	Craft Optve	Lab	Labour Cost	Sunds	Nett Rate	Unit	Gross Rate
concrete to B.S.5328, designed mix C20, 20mm aggregate in walls 150 - 450m thick	90.60	5.00	95.13	-	12.00	84.24	0.41	179.78	m3	193.26
concrete to B.S.5328, designed mix C20, 20mm aggregate in walls not exceeding 150mm thick	90.60	5.00	95.13	-	12.00	84.24	0.41	179.78	m3	193.26
12mm mild steel bar reinforcement; straight	752.00	10.00	827.20	85.00	-	784.55	4.51	1616.26	t	1737.48
drill edge of opening and tie in new reinforcement	-	-	-	-	1.00	7.02	0.41	7.43	m	7.99
formwork and basic finish to wall, per side	6.76	10.00	7.44	2.50	2.50	40.63	0.98	49.04	m2	52.72
formwork and fine formed finish to wall, per side	8.06	10.00	8.87	2.75	2.75	44.69	1.10	54.65	m2	58.75
junction between new and existing fair face	-	-	-	-	0.30	2.11	-	2.11	m	2.26
13mm two coat hardwall plaster over 300mm wide	1.76	10.00	1.94	1.30	1.30	21.13	-	23.06	m2	24.79
25 x 100mm softwood chamfered skirting, primed all round	1.46	10.00	1.61	0.56	0.09	5.80	0.11	7.52	m	8.08
junction with existing	-	-	-	0.50	0.08	5.18	-	5.18	nr	5.56
make good floor screed and vinyl or similar floor covering up to filling not exceeding 300mm wide	7.00	10.00	7.70	1.12	1.12	18.20	0.32	26.22	m	28.19

Fill opening where staircase removed in 150mm reinforced concrete floor with concrete reinforced with 12mm mild steel bars at 225mm centres both ways tied to existing structure; remove timber lining, prepare edges of opening to form joint with filling, drill edges of opening tie in new reinforcement, plaster filling on soffit and extend cement and sand screed and vinyl floor covering to match existing; make good junction of new and existing plaster and vinyl floor covering

	Del to Site	Waste	Material Cost	Craft Optve	Lab	Labour Cost	Sunds	Nett Rate	Unit	Gross Rate
900 x 2134mm	114.00	-	114.00	14.75	22.30	292.69	8.00	414.69	nr	445.79

Unit rates for pricing the above and similar work

	Del to Site	Waste	Material Cost	Craft Optve	Lab	Labour Cost	Sunds	Nett Rate	Unit	Gross Rate
take off timber lining	-	-	-	-	0.24	1.68	0.10	1.78	m	1.92
prepare edge of opening 150mm wide to form joint with filling	-	-	-	0.30	0.30	4.88	-	4.88	m	5.24
prepare edge of opening 200mm wide to form joint with filling	-	-	-	0.35	0.35	5.69	-	5.69	m	6.11
concrete to B.S.5328, designed mix C20, 20mm aggregate in slabs 150 - 450mm thick	90.60	5.00	95.13	-	12.00	84.24	0.41	179.78	m3	193.26
concrete to B.S.5328, designed mix C20, 20mm aggregate in slabs not exceeding 150mm thick	90.60	5.00	95.13	-	12.00	84.24	0.41	179.78	m3	193.26
12mm mild steel bar reinforcement; straight	752.00	10.00	827.20	85.00	-	784.55	4.51	1616.26	t	1737.48
drill edge of opening and tie in new reinforcement	-	-	-	-	1.00	7.02	0.41	7.43	m	7.99
formwork and basic finish to soffit, slab thickness not exceeding 200mm; height to soffit 1.50 - 3.00m	6.72	10.00	7.39	2.25	2.25	36.56	0.97	44.92	m2	48.29
formwork and basic finish to soffit, slab thickness 200 - 300mm; height to soffit 1.50 - 3.00m	8.06	10.00	8.87	2.50	2.50	40.63	1.10	50.59	m2	54.39
formwork and fine formed finish to soffit, slab thickness not exceeding 200mm; height to soffit 1.50 - 3.00m	6.72	10.00	7.39	2.50	2.50	40.63	1.12	49.14	m2	52.82
formwork and fine formed finish to soffit, slab thickness 200 - 300mm; height to soffit 1.50 - 3.00m	8.06	10.00	8.87	2.75	2.75	44.69	1.16	54.71	m2	58.82
junction between new and existing fair face	-	-	-	-	0.30	2.11	-	2.11	m	2.26
10mm two coat lightweight plaster over 300mm wide	1.73	10.00	1.90	1.30	1.30	21.13	-	23.03	m2	24.76
38mm cement and sand (1:3) trowelled bed	3.91	10.00	4.30	0.63	0.63	10.24	-	14.54	m2	15.63
vinyl or similar floor covering to match existing fixed with adhesive exceeding 300mm wide	18.00	10.00	19.80	1.25	0.50	15.05	0.56	35.41	m2	38.06

Openings through brickwork or blockwork

Form openings for door frames through 100mm block partition plastered both sides and with skirting both sides; take off skirtings, cut opening through partition, insert 100 x 150mm precast concrete lintel and wedge and pin up over, quoin up jambs, extend plaster to faces of lintel and make good junction with existing plaster; make good existing plaster up to new frame both sides and form fitted ends on existing skirtings up to new frame; extend softwood board flooring through opening on and with bearers and make good up to existing

	Del to Site	Waste	Material Cost	Craft Optve	Lab	Labour Cost	Sunds	Nett Rate	Unit	Gross Rate
838 x 2032mm	19.14	-	19.14	20.00	17.20	305.34	12.70	337.18	nr	362.47

Unit rates for pricing the above and similar work

	Del to Site	Waste	Material Cost	Craft Optve	Lab	Labour Cost	Sunds	Nett Rate	Unit	Gross Rate
cut opening through 75mm block partition plastered both sides	-	-	-	1.10	1.30	19.28	3.91	23.19	m2	24.93
cut opening through 100mm block partition plastered both sides	-	-	-	1.30	1.56	22.95	4.59	27.54	m2	29.61
make good fair face around opening	-	-	-	0.25	0.25	4.06	0.10	4.16	m	4.47
precast concrete, B.S.5328, designed mix C20, 20mm aggregate lintel 75 x 150 x 1200mm, reinforced with 1.01 kg of 12mm mild steel bars	6.98	10.00	7.68	0.80	0.80	13.00	0.25	20.93	nr	22.50
precast concrete, B.S.5328, designed mix C20, 20mm aggregate lintel 100 x 150 x 1200mm, reinforced with 1.01 kg of 12mm mild steel bars	9.63	10.00	10.59	1.00	1.00	16.25	0.33	27.17	nr	29.21
precast concrete, B.S.5328 designed mix C20, 20mm aggregate lintel 75 x 150 x 1200mm, reinforced with 1.01 kg of 12mm mild steel bars, fair finish all faces	8.33	10.00	9.16	0.80	0.80	13.00	0.25	22.41	nr	24.09

Labour hourly rates: (except Specialists) Craft Operatives £9.23 Labour £7.02 Rates are national average prices. Refer to REGIONAL VARIATIONS for indicative levels of overall pricing in regions	MATERIALS			LABOUR				RATES		
	Del to Site £	Waste %	Material Cost £	Craft Optve Hrs	Lab Hrs	Labour Cost £	Sunds £	Nett Rate £	Unit	Gross Rate (+7.5%) £
C90: ALTERATIONS – SPOT ITEMS Cont.										
Openings through brickwork or blockwork Cont.										
Unit rates for pricing the above and similar work Cont.										
precast concrete, B.S.5328 designed mix C20, 20mm aggregate lintel 100 x 150 x 1200mm, reinforced with 1.01 kg of 12mm mild steel bars, fair finish all faces	12.05	10.00	13.26	1.00	1.00	16.25	0.33	29.84	nr	32.07
wedge and pin up over lintel 75mm wide	-	-	-	0.50	0.50	8.13	0.62	8.74	m	9.40
wedge and pin up over lintel 100mm wide	-	-	-	0.66	0.66	10.73	0.82	11.55	m	12.41
quoin up 75mm wide jambs	-	-	-	0.52	0.52	8.45	0.62	9.07	m	9.75
quoin up 100mm wide jambs	-	-	-	0.68	0.68	11.05	0.82	11.87	m	12.76
extend 13mm hardwall plaster to face of lintel	1.76	10.00	1.94	2.13	2.13	34.61	0.06	36.61	m2	39.35
make good existing plaster up to new frame	0.27	10.00	0.30	0.55	0.55	8.94	-	9.23	m	9.93
take off old skirting	-	-	-	0.22	-	2.03	0.10	2.13	m	2.29
form fitted end on existing skirting up to new frame	-	-	-	3.91	-	36.09	-	36.09	nr	38.80
38mm cement and sand (1:3) screeded bed in opening	3.91	10.00	4.30	1.54	1.54	25.02	-	29.33	m2	31.53
extend 25mm softwood board flooring through opening on and with bearers	28.40	10.00	31.24	16.20	2.70	168.48	0.54	200.26	m2	215.28
iroko threshold; twice oiled; plugged and screwed										
19 x 75mm	3.45	10.00	3.79	0.80	-	7.38	0.07	11.25	m	12.09
19 x 100mm	3.90	10.00	4.29	0.90	-	8.31	0.08	12.68	m	13.63
Form openings for door frames through one brick wall plastered one side and with external rendering other side; take off skirting, cut opening through wall, insert 215 x 150mm precast concrete lintel and wedge and pin up over; quoin up jambs; extend plaster to face of lintel and make good junction with existing plaster; make good existing plaster up to new frame one side; extend external rendering to face of lintel and to reveals and make good up to existing rendering and new frame the other side; form fitted ends on existing skirting up to new frame										
914 x 2082mm	33.70	-	33.70	28.00	25.40	436.75	28.10	498.55	nr	535.94
Unit rates for pricing the above and similar work										
cut opening through half brick wall	-	-	-	1.95	2.40	34.85	4.76	39.61	m2	42.58
cut opening through one brick wall	-	-	-	3.80	4.70	68.07	9.01	77.08	m2	82.86
make good fair face around opening	-	-	-	0.50	0.50	8.13	0.62	8.74	m	9.40
make good facings to match existing around opening	2.38	10.00	2.62	0.50	0.50	8.13	0.62	11.36	m	12.22
precast concrete, B.S.5328, designed mix C20, 20mm aggregate lintel 102 x 150 x 1200mm, reinforced with 1.07 kg of 12mm mild steel bars	9.63	10.00	10.59	1.00	1.00	16.25	0.36	27.20	nr	29.24
precast concrete, B.S.5328, designed mix C20, 20mm aggregate lintel 215 x 150 x 1200mm, reinforced with 2.13 kg of 12mm mild steel bars	21.34	10.00	23.47	1.50	1.50	24.38	0.74	48.59	nr	52.23
precast concrete, B.S.5328, designed mix C20, 20mm aggregate lintel 102 x 150 x 1200mm, reinforced with 1.07 kg of 12mm mild steel bars, fair finish all faces	12.05	10.00	13.26	1.00	1.00	16.25	0.36	29.86	nr	32.10
precast concrete, B.S.5328, designed mix C20, 20mm aggregate lintel 215 x 150 x 1200mm, reinforced with 2.13 kg of 12mm mild steel bars, fair finish all faces	23.09	10.00	25.40	1.50	1.50	24.38	0.72	50.49	nr	54.28
wedge and pin up over lintel 102mm wide	-	-	-	0.66	0.66	10.73	0.82	11.55	m	12.41
wedge and pin up over lintel 215mm wide	-	-	-	1.00	1.00	16.25	1.75	18.00	m	19.35
quoin up half brick jambs	-	-	-	0.70	0.70	11.38	0.82	12.20	m	13.11
quoin up one brick jambs	-	-	-	1.27	1.27	20.64	1.75	22.39	m	24.07
facings to match existing to margin	1.02	10.00	1.12	0.67	0.67	10.89	0.82	12.83	m	13.79
make good existing plaster up to new frame	0.27	10.00	0.30	0.55	0.55	8.94	-	9.23	m	9.93
13mm two coat hardwall plaster to reveal not exceeding 300mm wide	0.53	10.00	0.58	0.78	0.78	12.68	0.02	13.28	m	14.27
make good existing external rendering up to new frame	0.21	10.00	0.23	0.47	0.47	7.64	-	7.87	m	8.46
13mm two coat cement and sand (1:3) external rendering to reveal not exceeding 300mm wide	0.42	10.00	0.46	0.74	0.74	12.03	0.02	12.51	m	13.45
take off old skirting	-	-	-	0.23	-	2.12	0.09	2.21	m	2.38
short length of old skirting up to 100mm long with mitre with existing one end	-	-	-	0.58	-	5.35	0.04	5.39	nr	5.80
short length of old skirting up to 300mm long with mitres with existing both ends	-	-	-	1.15	-	10.61	0.05	10.66	nr	11.46
38mm cement and sand (1:3) screeded bed in opening	3.91	10.00	4.30	1.54	1.54	25.02	-	29.33	m2	31.53
extend 25mm softwood board flooring through opening on and with bearers	5.36	10.00	5.90	3.65	0.61	37.97	0.14	44.01	nr	47.31
iroko threshold, twice oiled, plugged and screwed										
15 x 113mm	4.00	10.00	4.40	0.90	-	8.31	0.05	12.76	m	13.71
15 x 225mm	7.40	10.00	8.14	1.65	-	15.23	0.09	23.46	m	25.22
Form opening for window through 275mm hollow wall with two half brick skins, plastered one side and faced with picked stock facings the other; cut opening through wall, insert 275 x 225mm precast concrete boot lintel and wedge and pin up over; insert damp proof course and cavity gutter; quoin up jambs, close cavity at jambs with brickwork bonded to existing and with vertical damp proof course and close cavity at sill with one course of slates; face margin externally to match existing and extend plaster to face of lintel and to reveals and make good up to existing plaster and new frame										
900 x 900mm	51.30	-	51.30	15.90	16.80	264.69	14.60	330.59	nr	355.39

Labour hourly rates: (except Specialists) Craft Operatives £9.23 Labour £7.02 Rates are national average prices. Refer to REGIONAL VARIATIONS for indicative levels of overall pricing in regions	MATERIALS			LABOUR				RATES		
	Del to Site £	Waste %	Material Cost £	Craft Optve Hrs	Lab Hrs	Labour Cost £	Sunds £	Nett Rate £	Unit	Gross Rate (+7.5%) £
C90: ALTERATIONS - SPOT ITEMS Cont.										
Openings through brickwork or blockwork Cont.										
Unit rates for pricing the above and similar work										
cut opening through wall with two half brick skins	-	-	-	3.90	4.80	69.69	8.16	77.85	m2	83.69
cut opening through wall with one half brick and one 100mm block skins	-	-	-	3.25	3.96	57.80	8.16	65.96	m2	70.90
precast concrete, B.S.5328, designed mix C20, 20mm aggregate boot lintel 275 x 225 x 1200mm long reinforced with 3.60 kg of 12mm mild steel bars and 0.67 kg of 6mm mild steel links	31.99	10.00	35.19	2.30	2.30	37.38	0.82	73.38	nr	78.89
close cavity at jambs with blockwork bonded to existing and with 112.5mm wide bituminous hessian based damp proof course, B.S.6398, Class A	2.34	-	2.34	0.50	0.50	8.13	0.21	10.68	m	11.48
close cavity at jambs with brickwork bonded to existing and with 112.5mm wide bituminous hessian based damp proof course, B.S.6398, Class A	2.09	-	2.09	0.70	0.70	11.38	0.82	14.29	m	15.36
close cavity at jambs with slates in cement mortar (1:3) set vertically	3.84	10.00	4.22	0.44	0.44	7.15	0.21	11.58	m	12.45
close cavity at sill with one course of slates in cement mortar (1:3)	3.84	10.00	4.22	0.44	0.44	7.15	0.21	11.58	m	12.45
lead cored bituminous hessian based damp proof course and cavity tray, B.S.6398, Class F, width exceeding 225mm	20.28	10.00	22.31	0.80	0.80	13.00	0.21	35.52	m2	38.18
precast concrete, B.S.5328, designed mix C20, 20mm aggregate lintel 160 x 225 x 1200mm long reinforced with 2.13 kg of 12mm mild steel bars ..	23.34	10.00	25.67	1.50	1.50	24.38	0.62	50.67	nr	54.47
76 x 76 x 6mm mild steel angle arch bar, primed ..	6.39	5.00	6.71	0.50	0.50	8.13	0.62	15.45	m	16.61
take out three courses of facing bricks and build brick-on-end flat arch in picked stock facing bricks	9.06	5.00	9.51	1.00	1.00	16.25	1.81	27.57	m	29.64
For unit rates for finishes, quoining up jambs, etc see previous items Form opening for window where door removed, with window head at same level as old door head, through one brick wall plastered one side and faced the other with picked stock facings; remove old lintel and flat arch, cut away for and insert 102 x 225mm precast concrete lintel and wedge and pin up over; build brick-on-end flat arch in facing bricks on 76 x 76 x 6mm mild steel angle arch bar and point to match existing; cut away jambs of old door opening to extend width; quoin up jambs and face margins externally to match existing; fill in lower part of old opening with brickwork in gauged mortar (1:2:9), bond to existing and face externally and point to match existing										
1200 x 900mm; old door opening 914 x 2082mm; extend plaster to face of lintel, reveals and filling and make good up to existing plaster and new frame; extend skirting one side and make good junction with existing	110.60	-	110.60	22.00	21.50	353.99	25.00	489.59	nr	526.31
Unit rates for pricing the above and similar work										
take out old lintel	-	-	-	1.00	1.30	18.36	3.40	21.76	m	23.39
cut away one brick wall to increase width of opening	-	-	-	3.80	4.70	68.07	8.50	76.57	m2	82.31
cut away 275mm hollow wall of two half brick skins to increase width of opening	-	-	-	3.90	4.80	69.69	8.50	78.19	m2	84.06
For other unit rates for pricing the above and similar work see previous items and items in 'Filling openings in Brickwork and Blockwork' section Form opening for door frame where old window removed, with door head at same level as old window head, through one brick wall plastered one side and faced the other with picked stock facings; remove old lintel and flat arch, cut away for and insert 102 x 225mm precast concrete lintel and wedge and pin up over; build brick-on-end flat arch in facing bricks on 76 x 76 x 6mm mild steel angle arch bar and point, fill old opening at sides of new opening to reduce width with brickwork in gauged mortar (1:2:9), bond to existing, wedge and pin up at soffit, face externally and to margins and point to match existing; take off skirtings, cut away wall below sill of old opening, quion up jambs and face margins externally										
914 x 2082mm, old window opening 1200 x 900mm; extend plaster to face of lintel and filling and make good up to existing plaster and new frame; form fitted ends on existing skirting up to new frame...	66.80	-	66.80	24.00	23.80	388.60	28.50	483.90	nr	520.19

Labour hourly rates: (except Specialists) Craft Operatives £9.23 Labour £7.02 Rates are national average prices. Refer to REGIONAL VARIATIONS for indicative levels of overall pricing in regions	MATERIALS			LABOUR				RATES		
	Del to Site £	Waste %	Material Cost £	Craft Optve Hrs	Lab Hrs	Labour Cost £	Sunds £	Nett Rate £	Unit	Gross Rate (+7.5%) £

C90: ALTERATIONS - SPOT ITEMS Cont.

Openings through brickwork or blockwork Cont.

For unit rates for pricing the above and similar work see previous items and items in 'Filling openings in Brickwork and Blockwork' section

Air bricks in old walls

Cut opening through old one brick wall, render all round in cement mortar (1:3); build in clay air brick externally and fibrous plaster ventilator internally and make good facings and plaster

	Del to Site £	Waste %	Material Cost £	Craft Optve Hrs	Lab Hrs	Labour Cost £	Sunds £	Nett Rate £	Unit	Gross Rate £
225 x 150mm	3.82	-	3.82	1.75	1.75	28.44	0.55	32.81	nr	35.27
225 x 225mm	8.21	-	8.21	2.33	2.33	37.86	0.69	46.76	nr	50.27

Cut opening through old 275mm hollow brick wall and seal cavity with slates; build in clay air brick externally and fibrous plaster ventilator internally and make good facings and plaster

225 x 150mm	7.34	-	7.34	2.14	2.14	34.77	0.55	42.66	nr	45.86
225 x 225mm	13.13	-	13.13	2.81	2.81	45.66	0.69	59.48	nr	63.94

Openings through rubble walling

Form opening for door frame through 600mm rough rubble wall faced both sides; cut opening through wall, insert 600 x 150mm precast concrete lintel finished fair on all exposed faces and wedge and pin up over; quoin and face up jambs and extend softwood board flooring through opening on and with bearers and make good up to existing

838 x 2032mm	78.30	-	78.30	41.30	43.70	687.97	61.10	827.37	nr	889.43

Unit rates for pricing the above and similar work

cut opening through 350mm random rubble wall	-	-	-	6.00	7.00	104.52	11.90	116.42	m2	125.15
cut opening through 600mm rough rubble wall	-	-	-	10.00	12.00	176.54	20.40	196.94	m2	211.71
precast concrete, B.S.5328, designed mix C20, 20mm aggregate lintel 350 x 150 x 1140mm reinforced with 3.04 kg of 12mm mild steel bars; fair finish all faces	34.20	10.00	37.62	2.25	2.25	36.56	0.72	74.90	nr	80.52
precast concrete, B.S.5328, designed mix C20, 20mm aggregate lintel 600 x 150 x 1140mm reinforced with 5.00 kg of 12mm mild steel bars; fair finish all faces	57.80	10.00	63.58	3.65	3.65	59.31	1.13	124.02	nr	133.32
wedge and pin up over lintel 350mm wide	-	-	-	2.00	2.00	32.50	2.06	34.56	m	37.15
wedge and pin up over lintel 600mm wide	-	-	-	3.00	3.00	48.75	3.61	52.36	m	56.29
quoin up 350mm wide jambs	-	-	-	2.00	2.00	32.50	2.06	34.56	m	37.15
quoin up 600mm wide jambs	-	-	-	3.00	3.00	48.75	3.61	52.36	m	56.29
extend 25mm softwood board flooring through opening on and with bearers	13.40	10.00	14.74	9.70	1.60	100.76	0.39	115.89	nr	124.58

Openings through stone faced walls

Form opening for window through wall of half brick backing, plastered, and 100mm stone facing; cut opening through wall, insert 102 x 150mm precast concrete lintel and wedge and pin up over and build in 76 x 76 x 6mm mild steel angle arch bar to support stone facing, clean soffit of stone and point; quoin up jambs and face up and point externally; extend plaster to face of lintel and to reveals and make good up to existing plaster and new frame

900 x 900mm	24.40	-	24.40	19.20	19.80	316.21	18.90	359.51	nr	386.48

Unit rates for pricing the above and similar work

cut opening through wall	-	-	-	5.15	5.85	88.60	8.16	96.76	m2	104.02
precast concrete, B.S.5328, designed mix C20, 20mm aggregate lintel 102 x 150 x 1200mm reinforced with 1.07 kg of 12mm mild steel bars	9.63	10.00	10.59	1.05	1.05	17.06	0.31	27.97	nr	30.06
76 x 76 x 6mm mild steel angle arch bar, primed	6.39	5.00	6.71	0.50	0.50	8.13	0.62	15.45	m	16.61
quoin up jambs	-	-	-	0.70	0.70	11.38	0.93	12.31	m	13.23
face stone jambs and point	-	-	-	2.10	2.10	34.13	2.58	36.70	m	39.46
clean stone head and point	-	-	-	2.10	2.10	34.13	0.31	34.44	m	37.02

For unit rates for finishes, etc. see previous items

Filling openings in brickwork or blockwork

Fill opening where door removed in 100mm block wall with concrete blocks in gauged mortar (1:2:9); provide 50mm preservative treated softwood sole plate on timber floor in opening, bond to blockwork at jambs and wedge and pin up at head; plaster filling both sides and extend skirting both sides; make good junction of new and existing plaster and skirtings

838 x 2032mm	48.30	-	48.30	14.10	8.15	187.36	8.48	244.14	nr	262.45

Unit rates for pricing the above and similar work

75mm blockwork in filling	7.21	10.00	7.93	1.00	1.25	18.00	1.13	27.07	m2	29.10
100mm blockwork in filling	8.65	10.00	9.52	1.20	1.50	21.61	1.55	32.67	m2	35.12
75mm blockwork in filling and fair face and flush smooth pointing one side	7.21	15.00	8.29	1.25	1.50	22.07	1.34	31.70	m2	34.08
100mm blockwork in filling and fair face and flush smooth pointing one side	8.65	15.00	9.95	1.45	1.75	25.67	1.65	37.27	m2	40.06

Labour hourly rates: (except Specialists) Craft Operatives £9.23 Labour £7.02 Rates are national average prices. Refer to REGIONAL VARIATIONS for indicative levels of overall pricing in regions	MATERIALS			LABOUR				RATES		
	Del to Site £	Waste %	Material Cost £	Craft Optve Hrs	Lab Hrs	Labour Cost £	Sunds £	Nett Rate £	Unit	Gross Rate (+7.5%) £

C90: ALTERATIONS - SPOT ITEMS Cont.

Filling openings in brickwork or blockwork Cont.

Unit rates for pricing the above and similar work Cont.

	Del to Site £	Waste %	Material Cost £	Craft Optve Hrs	Lab Hrs	Labour Cost £	Sunds £	Nett Rate £	Unit	Gross Rate £
preservative treated softwood sole plate										
50 x 75mm	1.10	10.00	1.21	0.30	0.05	3.12	0.21	4.54	m	4.88
50 x 100mm	1.46	10.00	1.61	0.33	0.06	3.47	0.28	5.35	m	5.75
lead cored bituminous hessian based damp proof course, B.S.6398, Class F, width not exceeding 225mm	20.28	10.00	22.31	0.80	0.80	13.00	0.41	35.72	m2	38.40
wedge and pin up at head 75mm wide	-	-	-	0.10	0.10	1.63	0.41	2.04	m	2.19
wedge and pin up at head 100mm wide	-	-	-	0.12	0.12	1.95	0.62	2.57	m	2.76
cut pockets and bond 75mm filling to existing blockwork at jambs	-	-	-	0.40	0.40	6.50	0.72	7.22	m	7.76
cut pockets and bond 100mm filling to existing blockwork at jambs	-	-	-	0.50	0.50	8.13	1.02	9.14	m	9.83
12mm (average) cement and sand dubbing over 300mm wide on filling	1.24	10.00	1.36	0.54	0.54	8.78	-	10.14	m2	10.90
13mm two coat lightweight plaster over 300mm wide on filling	1.80	10.00	1.98	1.27	1.27	20.64	0.02	22.64	m2	24.34
19 x 100mm softwood chamfered skirting, primed all round	1.13	10.00	1.24	0.50	0.08	5.18	0.11	6.53	m	7.02
junction with existing	-	-	-	0.50	0.08	5.18	-	5.18	nr	5.56
make good vinyl or similar floor covering up to filling	6.00	10.00	6.60	0.50	0.50	8.13	0.14	14.87	m	15.98

Fill opening where door removed in half brick wall with brickwork in gauged mortar (1:2:9); provide lead cored damp proof course in opening, lapped with existing, bond to existing brickwork at jambs and wedge and pin up at head; plaster filling both sides and extend skirting both sides; make good junction of new and existing plaster and skirtings

	Del to Site £	Waste %	Material Cost £	Craft Optve Hrs	Lab Hrs	Labour Cost £	Sunds £	Nett Rate £	Unit	Gross Rate £
838 x 2032mm	47.30	-	47.30	15.70	12.75	234.42	10.40	292.12	nr	314.02

Unit rates for pricing the above and similar work

	Del to Site £	Waste %	Material Cost £	Craft Optve Hrs	Lab Hrs	Labour Cost £	Sunds £	Nett Rate £	Unit	Gross Rate £
half brick filling in common bricks	12.00	10.00	13.20	2.00	2.00	32.50	2.58	48.28	m2	51.90
half brick filling in common bricks fair faced and flush pointed one side	12.00	10.00	13.20	2.33	2.33	37.86	2.78	53.84	m2	57.88
50 x 102mm softwood sole plate	1.56	10.00	1.72	0.33	0.06	3.47	0.25	5.43	m	5.84
lead cored bituminous hessian based damp proof course, B.S.6398, Class F, width not exceeding 225mm	20.28	10.00	22.31	0.80	0.80	13.00	0.41	35.72	m2	38.40
wedge and pin up at head 102mm wide	-	-	-	0.12	0.12	1.95	0.62	2.57	m	2.76
cut pockets and bond half brick filling to existing brickwork at jambs	-	-	-	0.50	0.50	8.13	1.03	9.15	m	9.84

For unit rates for finishes, etc
see previous items

Fill opening where door removed in one brick external wall with brickwork in gauged mortar (1:2:9) faced externally with picked stock facings pointed to match existing; remove old lintel and arch, provide lead cored damp proof course in opening lapped with existing, bond to existing brickwork at jambs and wedge and pin up at head; plaster filling one side and extend skirting; make good junction of new and existing plaster and skirting

	Del to Site £	Waste %	Material Cost £	Craft Optve Hrs	Lab Hrs	Labour Cost £	Sunds £	Nett Rate £	Unit	Gross Rate £
914 x 2082mm	171.50	-	171.50	19.65	18.85	313.70	25.70	510.90	nr	549.21

Unit rates for pricing the above and similar work

	Del to Site £	Waste %	Material Cost £	Craft Optve Hrs	Lab Hrs	Labour Cost £	Sunds £	Nett Rate £	Unit	Gross Rate £
remove old lintel and arch	-	-	-	1.46	1.64	24.99	2.04	27.03	m	29.06
one brick filling in common bricks	24.00	10.00	26.40	4.00	4.00	65.00	6.18	97.58	m2	104.90
one brick filling in common bricks faced one side with picked stock facings and point to match existing	67.20	10.00	73.92	5.00	5.00	81.25	6.18	161.35	m2	173.45
one brick filling in common bricks fair faced and flush smooth pointing one side	24.00	10.00	26.40	4.33	4.33	70.36	6.18	102.94	m2	110.66
two course slate damp proof course, width not exceeding 225mm	25.60	10.00	28.16	1.02	1.02	16.57	0.72	45.45	m2	48.86
lead cored bituminous hessian based damp proof course, B.S.6398, Class F, width not exceeding 225mm	20.28	10.00	22.31	0.80	0.80	13.00	0.31	35.62	m2	38.29
wedge and pin up at head 215mm wide	-	-	-	0.20	0.20	3.25	1.44	4.69	m	5.04
cut pockets and bond one brick filling to existing brickwork at jambs	-	-	-	1.00	1.00	16.25	2.37	18.62	m	20.02
12mm cement and sand (1:3) two coat external rendering over 300mm wide on filling	1.24	10.00	1.36	1.25	1.25	20.31	0.04	21.72	m2	23.35

For unit rates for internal finishes, etc.
see previous items

Fill opening where door removed in 275mm external hollow wall with inner skin in common bricks and outer skin in picked stock facings pointed to match existing, in gauged mortar (1:2:9); remove old lintel and arch, provide lead cored combined damp proof course and cavity gutter lapped with existing; form cavity with ties; bond to existing brickwork at jambs and wedge and pin up at head; plaster filling one side and extend skirting; make good junction of new and existing plaster and skirting

	Del to Site £	Waste %	Material Cost £	Craft Optve Hrs	Lab Hrs	Labour Cost £	Sunds £	Nett Rate £	Unit	Gross Rate £
914 x 2082mm	121.80	-	121.80	19.90	19.10	317.76	26.25	465.81	nr	500.74

Labour hourly rates: (except Specialists) Craft Operatives £9.23 Labour £7.02 Rates are national average prices. Refer to REGIONAL VARIATIONS for indicative levels of overall pricing in regions	MATERIALS			LABOUR				RATES		
	Del to Site £	Waste %	Material Cost £	Craft Optve Hrs	Lab Hrs	Labour Cost £	Sunds £	Nett Rate £	Unit	Gross Rate (+7.5%) £

C90: ALTERATIONS - SPOT ITEMS Cont.

Filling openings in brickwork or blockwork Cont.

Unit rates for pricing the above and similar work

half brick filling in common bricks	12.00	10.00	13.20	2.00	2.00	32.50	2.58	48.28	m2	51.90
half brick filling in common bricks and fair face and flush smooth pointing one side	12.00	10.00	13.20	2.33	2.33	37.86	2.78	53.84	m2	57.88
half brick filling in picked stock facings and point to match existing	40.80	10.00	44.88	2.95	2.95	47.94	2.78	95.60	m2	102.77
form 50mm wide cavity with B.S.1243 Fig. 3 galvanised steel twisted wall ties built in	0.60	10.00	0.66	0.20	0.20	3.25	-	3.91	m2	4.20
lead cored bituminous hessian based damp proof course and cavity tray, B.S.6398, Class F, width exceeding 225mm	20.28	10.00	22.31	0.80	0.80	13.00	0.31	35.62	m2	38.29
wedge and pin up at head 102mm wide	-	-	-	0.12	0.12	1.95	0.62	2.57	m	2.76
cut pockets and bond half brick wall in common bricks to existing brickwork at jambs	-	-	-	0.50	0.50	8.13	1.03	9.15	m	9.84
cut pockets and bond half brick wall in facing bricks to existing brickwork at jambs and make good facings	-	-	-	0.60	0.60	9.75	2.16	11.91	m	12.80

For unit rates for internal finishes, etc. see previous items

Fillings openings in rubble walling

Fill opening where window removed in 600mm rough rubble wall with rubble walling in lime mortar (1:3) pointed to match existing; bond to existing walling at jambs and wedge and pin up at head; plaster filling one side and make good junction of new and existing plaster

900 x 900mm	119.90	-	119.90	12.00	12.00	195.00	15.90	330.80	nr	355.61

Unit rates for pricing the above and similar work

rough rubble walling in filling 600mm thick	124.00	10.00	136.40	6.00	6.00	97.50	12.36	246.26	m2	264.73
random rubble walling in filling 300mm thick...	68.20	10.00	75.02	5.80	5.80	94.25	6.18	175.45	m2	188.61
500mm thick...	113.00	10.00	124.30	10.90	10.90	177.13	10.30	311.73	m2	335.10
wedge and pin up at head 300mm wide.....................................	-	-	-	0.28	0.28	4.55	2.06	6.61	m	7.11
500mm wide.....................................	-	-	-	0.47	0.47	7.64	3.50	11.14	m	11.97
600mm wide.....................................	-	-	-	0.56	0.56	9.10	4.12	13.22	m	14.21
cut pockets and bond walling to existing at jambs 300mm	5.15	10.00	5.67	0.83	0.83	13.49	0.62	19.77	m	21.26
500mm	8.25	10.00	9.07	1.38	1.38	22.43	0.93	32.43	m	34.86
600mm	9.35	10.00	10.29	1.75	1.75	28.44	1.13	39.85	m	42.84

For unit rates for internal finishes, etc. see previous items

Work to chimney stacks

SEAL AND VENTILATE TOP OF FLUE - remove chimney pot and flaunching and provide 50mm precast concrete, B.S.5328, designed mix C20, 20mm aggregate, weathered and throated capping and bed in cement mortar (1:3) to seal top of stack; cut opening through brick side of stack and build in 225 x 225mm clay air brick to ventilate flue and make good facings

stack size 600 x 600mm with one flue	17.11	10.00	18.82	4.90	4.90	79.63	4.32	102.77	nr	110.47
stack size 825 x 600mm with two flues	28.84	10.00	31.72	9.00	9.00	146.25	7.96	185.93	nr	199.88
stack size 1050 x 600mm with three flues	40.57	10.00	44.63	13.10	13.10	212.88	10.91	268.41	nr	288.54

RENEW DEFECTIVE CHIMNEY POT - remove chimney pot and provide new clay chimney pot set and flaunched in cement mortar (1:3)

150mm diameter x 600mm high	28.97	-	28.97	2.50	2.50	40.63	3.91	73.50	nr	79.02

REBUILD DEFECTIVE CHIMNEY STACK - pull down defective stack to below roof level for a height of 2.00m; prepare for raising and rebuild in common brickwork faced with picked stock facings in gauged mortar (1:2:9) pointed to match existing; parge and core flues; provide No. 4 lead flashings and soakers; provide 150mm diameter clay chimney pots 600mm high set and flaunched in cement mortar (1:3); make good roof tiling and all other work disturbed

600 x 600mm with one flue	370.00	-	370.00	35.30	37.55	589.42	59.00	1018.42	nr	1094.80
825 x 600mm with two flues	476.00	-	476.00	47.15	47.90	771.45	82.00	1329.45	nr	1429.16

REMOVE DEFECTIVE CHIMNEY STACK - pull down stack to below roof level; remove all flashings and soakers; piece in 50 x 100mm rafters; extend roof tiling with machine made plain tiles to 100mm gauge nailed every fourth course with galvanised nails to and with 25 x 19mm battens; bond new tiling to existing

850 x 600mm for a height of 2.00m	34.90	-	34.90	26.00	22.30	396.53	36.00	467.43	nr	502.48

Labour hourly rates: (except Specialists) Craft Operatives £9.23 Labour £7.02 Rates are national average prices. Refer to REGIONAL VARIATIONS for indicative levels of overall pricing in regions	MATERIALS			LABOUR				RATES		
	Del to Site £	Waste %	Material Cost £	Craft Optve Hrs	Lab Hrs	Labour Cost £	Sunds £	Nett Rate £	Unit	Gross Rate (+7.5%) £

C90: ALTERATIONS - SPOT ITEMS Cont.

Work to chimney breasts and fireplaces

TAKE OUT FIREPLACE AND FILL IN OPENING - remove fire surround, fire back and hearth; hack up screed to hearth and extend 25mm tongued and grooved softwood floor boarding over hearth on and with bearers; fill in opening where fireplace removed with 75mm concrete blocks in gauged mortar (1:2:9) bonded to existing at jambs and wedged and pinned up to soffits; form 225 x 225mm opening and provide and build in 225 x 225mm fibrous plaster louvered air vent; plaster filling with 12mm two coat lightweight plaster on 12mm (average) dubbing plaster; extend 19 x 100mm softwood chamfered skirting over filling and join with existing; make good all new finishings to existing

	Del to Site	Waste	Material Cost	Craft Optve	Lab	Labour Cost	Sunds	Nett Rate	Unit	Gross Rate
fireplace opening 570 x 685mm, hearth 1030 x 405mm	35.00	-	35.00	8.00	7.45	126.14	10.30	171.44	nr	184.30
fireplace opening 800 x 760mm, hearth 1260 x 405mm	42.00	-	42.00	9.65	8.95	151.90	12.00	205.90	nr	221.34

REMOVE CHIMNEY BREAST - pull down chimney breast for full height from ground to roof level (two storeys) including removing two fire surrounds and hearths complete; make out brickwork where flues removed and make out brickwork where breasts removed; extend 50 x 175mm floor joists and 25mm tongued and grooved softwood floor boarding (ground and first floors); extend 50 x 100mm ceiling joists (first floor); plaster walls with 12mm two coat lightweight plaster on 12mm (average) plaster dubbing; extend ceiling plaster with expanded metal lathing and 19mm three coat lightweight plaster; run 19 x 100mm softwood chamfered skirting to walls to join with existing; make good all new finishings to existing

	Del to Site	Waste	Material Cost	Craft Optve	Lab	Labour Cost	Sunds	Nett Rate	Unit	Gross Rate
chimney breast 2.00m wide and 7.50m high including gathering in roof space	145.00	-	145.00	62.00	84.00	1161.94	144.00	1450.94	nr	1559.76

Form trap door in ceiling

Form new trap door in ceiling with 100mm deep joists and lath and plaster finish; cut away and trim ceiling joists around opening and insert new trimming and trimmer joists; provide 32x 127mm softwood lining with 15 x 25mm stop and 20 x 70mm architrave; provide and place in position 18mm blockboard trap door, lipped all round; make good existing ceiling plaster around openings

	Del to Site	Waste	Material Cost	Craft Optve	Lab	Labour Cost	Sunds	Nett Rate	Unit	Gross Rate
600 x 900mm	39.40	-	39.40	21.85	5.60	240.99	9.10	289.49	nr	311.20
Unit rates for pricing the above and similar work										
cut away plastered plasterboard ceiling	-	-	-	1.50	0.33	16.16	0.85	17.01	m2	18.29
cut away lath and plaster ceiling	-	-	-	2.00	0.41	21.34	1.02	22.36	m2	24.04
cut away and trim existing 100mm deep joists around opening										
664 x 964mm..........	-	-	-	0.95	0.16	9.89	0.71	10.60	nr	11.40
964 x 1444mm..........	-	-	-	1.90	0.33	19.85	1.43	21.28	nr	22.88
cut away and trim existing 150mm deep joists around opening										
664 x 964mm..........	-	-	-	1.43	0.24	14.88	1.09	15.97	nr	17.17
964 x 1444mm..........	-	-	-	2.86	0.48	29.77	2.14	31.91	nr	34.30
softwood trimming or trimmer joists										
75 x 100mm..........	2.46	10.00	2.71	0.95	0.16	9.89	0.36	12.96	m	13.93
75 x 150mm..........	3.68	10.00	4.05	1.00	0.17	10.42	0.39	14.86	m	15.98
softwood lining										
32 x 127mm..........	2.10	10.00	2.31	1.50	0.25	15.60	0.11	18.02	m	19.37
32 x 150mm..........	2.18	10.00	2.40	1.60	0.27	16.66	0.08	19.14	m	20.58
15 x 25mm softwood stop..........	0.45	10.00	0.50	0.33	0.05	3.40	0.05	3.94	m	4.24
19 x 50mm softwood twice rounded architrave to Patt. 18..........	0.59	10.00	0.65	0.33	0.05	3.40	0.05	4.10	m	4.40
18mm blockboard trap door lipped all round										
600 x 900mm..........	11.03	10.00	12.13	3.00	0.50	31.20	0.17	43.50	nr	46.77
900 x 1280mm..........	25.24	10.00	27.76	6.00	1.00	62.40	0.30	90.46	nr	97.25
75mm steel butt hinge to softwood..........	0.60	-	0.60	0.66	0.11	6.86	0.17	7.63	nr	8.21
make good plastered plasterboard ceiling around opening..........	-	-	-	0.75	0.75	12.19	1.23	13.42	m	14.42
make good lath and plaster ceiling around opening..........	-	-	-	0.83	0.83	13.49	1.51	15.00	m	16.12

Openings through stud partitions

Form opening for door frame through lath and plaster finished 50 x 100mm stud partition; take off skirtings, cut away plaster, cut away and trim studding around opening and insert new studding at head and jambs of opening; make good existing lath and plaster up to new frame both sides and form fitted ends on existing skirtings up to new frame; extend softwood board flooring through opening on and with bearers and make good up to existing

	Del to Site	Waste	Material Cost	Craft Optve	Lab	Labour Cost	Sunds	Nett Rate	Unit	Gross Rate
838 x 2032mm..........	11.80	-	11.80	18.10	6.10	209.88	16.10	237.79	nr	255.62
Unit rates for pricing the above and similar work										
cut away plastered plasterboard	-	-	-	1.00	0.25	10.98	0.85	11.84	m2	12.72
cut away lath and plaster	-	-	-	1.10	0.26	11.98	1.02	13.00	m2	13.97

Labour hourly rates: (except Specialists) Craft Operatives £9.23 Labour £7.02 Rates are national average prices. Refer to REGIONAL VARIATIONS for indicative levels of overall pricing in regions	MATERIALS			LABOUR				RATES		
	Del to Site	Waste	Material Cost	Craft Optve	Lab	Labour Cost	Sunds	Nett Rate	Unit	Gross Rate (+7.5%)
	£	%	£	Hrs	Hrs	£	£	£		£
C90: ALTERATIONS - SPOT ITEMS Cont.										
Openings through stud partitions Cont.										
Unit rates for pricing the above and similar work Cont.										
cut away and trim existing 75mm studding around opening										
838 x 2032mm..........................	-	-	-	2.05	0.36	21.45	2.18	23.63	nr	25.40
914 x 2032mm..........................	-	-	-	2.05	0.36	21.45	2.18	23.63	nr	25.40
cut away and trim existing 100mm studding around opening										
838 x 2032mm..........................	-	-	-	2.55	0.45	26.70	2.85	29.55	nr	31.76
914 x 2032mm..........................	-	-	-	2.55	0.45	26.70	2.85	29.55	nr	31.76
50 x 75mm softwood studding..............	1.10	10.00	1.21	0.50	0.08	5.18	0.11	6.50	m	6.98
50 x 100mm softwood studding.............	1.46	10.00	1.61	0.60	0.10	6.24	0.12	7.97	m	8.56
make good plastered plasterboard around opening	-	-	-	0.75	0.75	12.19	1.23	13.42	m	14.42
make good lath and plaster around opening	-	-	-	0.83	0.83	13.49	1.51	15.00	m	16.12
take off skirting.......................	-	-	-	0.21	-	1.94	0.18	2.12	m	2.28
form fitted end on existing skirting..........	-	-	-	0.57	-	5.26	-	5.26	nr	5.66
extend 25mm softwood tongued and grooved board flooring through opening on and with bearers ..	3.42	10.00	3.76	1.50	0.25	15.60	0.06	19.42	nr	20.88
Filling openings in stud partitions										
Fill opening where door removed in stud partition with 50 x 100mm studding and sole plate, covered on both sides with 9.5mm plasterboard baseboard with 5mm one coat board finish plaster; extend skirting on both sides; make good junction of new and existing plaster and skirtings										
838 x 2032mm..........................	31.10	-	31.10	16.00	5.36	185.31	3.00	219.41	nr	235.86
Unit rates for pricing the above and similar work										
50 x 75mm softwood sole plate	1.10	10.00	1.21	0.50	0.08	5.18	0.11	6.50	m	6.98
50 x 100mm softwood sole plate	1.46	10.00	1.61	0.60	0.10	6.24	0.14	7.99	m	8.58
50 x 75mm softwood studding	1.10	10.00	1.21	0.50	0.08	5.18	0.11	6.50	m	6.98
50 x 100mm softwood studding	1.46	10.00	1.61	0.60	0.10	6.24	0.14	7.99	m	8.58
9.5mm plasterboard baseboard to filling	1.83	10.00	2.01	0.50	0.10	5.32	0.28	7.61	m2	8.18
9.5mm plasterboard baseboard to filling including packing out not exceeding 10mm	1.83	10.00	2.01	1.00	0.17	10.42	0.37	12.81	m2	13.77
5mm one coat board finish plaster on plasterboard to filling	0.96	10.00	1.06	0.91	0.91	14.79	0.03	15.87	m2	17.06
10mm two coat hardwall plaster on plasterboard to filling	1.46	10.00	1.61	1.27	1.27	20.64	0.03	22.27	m2	23.94
19 x 100mm softwood chamfered skirting, primed all round..	1.24	10.00	1.36	0.54	0.09	5.62	0.18	7.16	m	7.70
junction with existing	-	-	-	0.57	-	5.26	-	5.26	nr	5.66
25 x 150mm softwood chamfered skirting, primed all round..	2.29	10.00	2.52	0.55	0.09	5.71	0.24	8.47	m	9.10
junction with existing	-	-	-	0.57	-	5.26	-	5.26	nr	5.66
Work to existing drains										
BREAK UP CONCRETE PAVING FOR EXCAVATING DRAIN TRENCH AND REINSTATE - break up 100mm concrete paving and 150mm hardcore bed under for excavation of drain trench and reinstate with new hardcore and concrete, B.S.5328, designed mix C20, 20mm aggregate and make good up to existing paving both sides										
400mm wide	4.90	10.00	5.39	-	3.27	22.96	3.40	31.75	m	34.13
TAKE UP FLAG PAVING FOR EXCAVATING DRAIN TRENCH AND REINSTATE - carefully take up 50mm precast concrete flag paving and set aside for re-use and break up 150mm hardcore bed under for excavation of drain trench and reinstate with new hardcore and re-lay salvaged flag paving on and with 25mm sand bed and grout in lime sand (1:3) and make good up to existing paving on both sides										
400mm wide	1.95	10.00	2.15	1.47	1.47	23.89	3.13	29.16	m	31.35
INSERT NEW JUNCTION IN EXISTING DRAIN -excavate for and trace and expose existing drain, break into glazed vitrified clay drain and insert junction with 100mm branch, short length of new pipe and double collar and joint to existing drain; support earthwork, make good concrete bed and haunching and backfill										
existing 100mm diameter drain, invert depth										
600mm	21.52	10.00	23.67	3.00	12.00	111.93	8.65	144.25	nr	155.07
750mm	21.52	10.00	23.67	3.00	14.25	127.72	9.35	160.75	nr	172.80
900mm	21.52	10.00	23.67	3.00	16.50	143.52	10.25	177.44	nr	190.75
existing 150mm diameter drain invert depth										
750mm	35.05	10.00	38.56	4.50	15.75	152.10	10.80	201.46	nr	216.56
900mm	35.05	10.00	38.56	4.50	18.00	167.90	12.25	218.70	nr	235.10
1050mm	35.05	10.00	38.56	4.50	20.25	183.69	13.40	235.65	nr	253.32
1200mm	35.05	10.00	38.56	4.50	22.50	199.49	14.65	252.69	nr	271.64

Labour hourly rates: (except Specialists) Craft Operatives £9.23 Labour £7.02 Rates are national average prices. Refer to REGIONAL VARIATIONS for indicative levels of overall pricing in regions	MATERIALS			LABOUR				RATES		
	Del to Site £	Waste %	Material Cost £	Craft Optve Hrs	Lab Hrs	Labour Cost £	Sunds £	Nett Rate £	Unit	Gross Rate (+7.5%) £

C90: ALTERATIONS - SPOT ITEMS Cont.

Work to existing drains Cont.

REPAIR DEFECTIVE DRAIN - excavate for and trace and expose existing drain, break out fractured glazed vitrified clay drain pipe, replace with new pipe and double collars and joint to existing drain; support earthwork, make good concrete bed and haunching and backfill
 existing 100mm diameter drain; single pipe length; invert depth

450mm	21.03	10.00	23.13	3.00	11.50	108.42	6.55	138.10	nr	148.46
600mm	21.03	10.00	23.13	3.00	13.50	122.46	7.90	153.49	nr	165.00
750mm	21.03	10.00	23.13	3.00	16.00	140.01	9.65	172.79	nr	185.75
add for each additional pipe length; invert depth										
450mm	11.74	10.00	12.91	2.00	6.75	65.84	6.25	85.01	nr	91.38
600mm	11.74	10.00	12.91	2.00	7.60	71.81	7.55	92.28	nr	99.20
750mm	11.74	10.00	12.91	2.00	8.80	80.24	9.05	102.20	nr	109.86

Work to existing manholes

RAISE TOP OF EXISTING MANHOLE - take off cover and frame and set aside for re-use, prepare level bed on existing one brick sides of manhole internal size 610 x 457mm and raise with Class B engineering brickwork in cement mortar (1:3) finished with a fair face and flush pointed; refix salvaged cover and frame, bed frame in cement mortar and cover in grease and sand

raising 150mm high	16.74	10.00	18.41	5.30	5.30	86.13	3.17	107.71	nr	115.79
raising 225mm high	25.11	10.00	27.62	6.75	6.75	109.69	4.70	142.01	nr	152.66
raising 300mm high	33.48	10.00	36.83	8.50	8.50	138.13	6.02	180.97	nr	194.55

INSERT NEW BRANCH BEND IN BOTTOM OF EXISTING MANHOLE - break into bottom and one brick side of manhole, insert new glazed vitrified clay three quarter section branch bend to discharge over existing main channel, build in end of new drain and make good benching and side of manhole to match existing

100mm diameter branch bend	7.66	10.00	8.43	5.75	5.75	93.44	2.27	104.13	nr	111.94

EXTEND EXISTING MANHOLE - take off cover and frame and set aside; break up one brick end wall of 457mm wide (internal) manhole and excavate for and extend manhole 225mm; support earthwork, level and compact bottom, part backfill and remove surplus spoil from site; extend bottom with 150mm concrete, BS 5328, ordinary prescribed mix C15P, 20mmaggregate bed; extend one brick sides and end with Class B engineering brickwork in cement mortar (1:3) fair faced and flush pointed and bond to existing; extend 100mm main channel and insert 100mm three quarter section branch channel bend; extend benching and build in end of new 100mm drain to one brick side, corbel over end of manhole
 manholes of the following invert depths; refix salvaged cover and frame, bed frame in cement mortar and cover in grease and sand

450mm	42.26	10.00	46.49	17.60	17.90	288.11	8.76	343.35	nr	369.10
600mm	47.37	10.00	52.11	19.80	20.20	324.56	11.26	387.93	nr	417.02
750mm	52.49	10.00	57.74	22.00	22.50	361.01	13.76	432.51	nr	464.95
900mm	57.60	10.00	63.36	24.20	24.80	397.46	16.26	477.08	nr	512.86

Temporary weatherproof coverings

Providing and erecting and clearing away on completion; for scaffold tube framework see preliminaries

corrugated iron sheeting	7.40	10.00	8.14	0.15	0.50	4.89	0.79	13.82	m2	14.86
flexible reinforced plastic sheeting	2.73	10.00	3.00	0.10	0.25	2.68	0.64	6.32	m2	6.80
tarpaulins	4.10	10.00	4.51	0.10	0.25	2.68	0.64	7.83	m2	8.42

Temporary internal screens

Providing and erecting and clearing away on completion; temporary screen of 50 x 50mm softwood framing and cover with

reinforced building paper	2.54	10.00	2.79	0.43	0.30	6.07	0.22	9.09	m2	9.77
heavy duty polythene sheeting	1.78	10.00	1.96	0.43	0.30	6.07	0.20	8.23	m2	8.85
3mm hardboard	2.65	10.00	2.92	0.53	0.32	7.14	0.33	10.38	m2	11.16

Providing and erecting and clearing away on completion; temporary dustproof screen of 50 x 75mm softwood framing securely fixed to walls, floor and ceiling, the joints and edges of lining sealed with masking tape and cover with

3mm hardboard lining one side	3.45	10.00	3.79	1.10	0.18	11.42	0.44	15.65	m2	16.83
3mm hardboard lining both sides	4.70	10.00	5.17	1.30	0.22	13.54	0.55	19.26	m2	20.71
6mm plywood lining one side	7.29	10.00	8.02	1.15	0.19	11.95	0.49	20.46	m2	21.99
6mm plywood lining both sides	12.38	10.00	13.62	1.40	0.23	14.54	0.77	28.92	m2	31.09
providing 35mm hardboard faced flush door size 838 x 1981mm in dustproof screen with 25 x 87mm softwood frame, 18 x 25mm softwood stop, pair of 100mm butts, pull handle and ball catch	135.00	10.00	148.50	1.50	0.25	15.60	2.78	166.88	nr	179.40

Labour hourly rates: (except Specialists) Craft Operatives £9.23 Labour £7.02 Rates are national average prices. Refer to REGIONAL VARIATIONS for indicative levels of overall pricing in regions	MATERIALS			LABOUR				RATES		
	Del to Site	Waste	Material Cost	Craft Optve	Lab	Labour Cost	Sunds	Nett Rate	Unit	Gross Rate (+7.5%)
	£	%	£	Hrs	Hrs	£	£	£		£
C90: ALTERATIONS - SPOT ITEMS Cont.										
Temporary dustproof corridors										

Note
    for prices of walls and doors, see dustproof
    screens above
Providing and erecting and clearing away on
completion; ceiling of 50 x 75mm softwood joists at
450mm centres, the joints and edges sealed with
masking tape and cover with										

3mm hardboard lining to soffit	3.69	10.00	4.06	1.10	0.18	11.42	0.44	15.92	m2	17.11
6mm plywood lining to soffit	7.53	10.00	8.28	1.15	0.19	11.95	0.55	20.78	m2	22.34
Temporary timber balustrades										
Providing and erecting and clearing away on completion; temporary softwood balustrade consisting of 50 x 75mm plate, 50 x 50mm standards at 900mm centres, four 25 x 150mm intermediate rails and 50 x 50mm handrail										
1150mm high	6.83	10.00	7.51	1.00	0.26	11.06	1.26	19.83	m	21.32
Temporary steel balustrades										
Providing and erecting and clearing away on completion; temporary steel balustrade constructed of 50mm diameter galvanised scaffold tubing and fittings with standards at 900mm centres intermediate rail, handrail, plastic mesh infill and with plates on ends of standards fixed to floor										
1150mm high	6.50	-	6.50	0.18	0.90	7.98	0.30	14.78	m	15.89
weekly cost of hire and maintenance	0.65	-	0.65	0.02	0.10	0.89	0.04	1.58	m	1.69
Temporary fillings to openings in external walls										
Providing and erecting and clearing away on completion; temporary filling to window or door opening of 50 x 100mm softwood framing and cover with										
corrugated iron sheeting	16.05	15.00	18.46	0.45	0.45	7.31	1.31	27.08	m2	29.11
25mm softwood boarding	12.42	12.50	13.97	0.60	0.60	9.75	1.53	25.25	m2	27.15
12mm external quality plywood	12.40	10.00	13.64	0.50	0.50	8.13	0.99	22.75	m2	24.46
REPAIRING/RENOVATING ASPHALT WORK										
Cut out crack in old covering and make good with new material to match existing										
Floor or tanking										
20mm thick in two coats, horizontal	-	-	-	0.66	0.66	10.73	0.50	11.23	m	12.07
30mm thick in three coats, horizontal	-	-	-	0.90	0.90	14.63	0.75	15.38	m	16.53
30mm thick in three coats, vertical	-	-	-	1.19	1.19	19.34	0.75	20.09	m	21.59
Cut out detached blister in old covering and make good with new material to match existing										
Floor or tanking										
20mm thick in two coats, horizontal	-	-	-	0.30	0.30	4.88	0.25	5.13	nr	5.51
30mm thick in three coats, horizontal	-	-	-	0.35	0.35	5.69	0.35	6.04	nr	6.49
30mm thick in three coats, vertical	-	-	-	0.48	0.48	7.80	0.35	8.15	nr	8.76
Cut out crack in old covering and make good with new material to match existing										
Roof covering										
20mm thick in two coats	-	-	-	0.66	0.66	10.73	0.50	11.23	m	12.07
Cut out detached blister in old covering and make good with new material to match existing										
20mm thick in two coats	-	-	-	0.30	0.30	4.88	0.25	5.13	nr	5.51
REPAIRING/RENOVATING ROOF COVERINGS										
Slate roofing repairs to roofs covered with 610 x 305mm slates										
Remove damaged slates and replace with new; sloping or vertical										
one slate	6.32	5.00	6.64	0.32	0.32	5.20	0.09	11.93	nr	12.82
patch of 10 slates	63.20	5.00	66.36	1.10	1.10	17.88	0.70	84.94	nr	91.31
patch of slates 1.00 - 3.00m2	77.42	5.00	81.29	1.25	1.25	20.31	0.76	102.36	m2	110.04
Examine battens, remove defective and provide 20% new										
19 x 38mm..	0.21	10.00	0.23	0.09	0.09	1.46	0.05	1.74	m2	1.87
25 x 50mm..	0.38	10.00	0.42	0.09	0.09	1.46	0.07	1.95	m2	2.10
Re-cover roof with slates previously removed and stacked and fix with slate nails										
with 75mm lap	-	-	-	0.77	0.77	12.51	0.09	12.60	m2	13.55
extra for providing 20% new slates	15.48	5.00	16.25	-	-	-	0.18	16.43	m2	17.67
Remove double course at eaves and										
refix...	-	-	-	0.24	0.24	3.90	0.31	4.21	m	4.53
extra for providing 20% new slates	5.25	5.00	5.51	-	-	-	0.09	5.60	m	6.02

Labour hourly rates: (except Specialists) Craft Operatives £9.23 Labour £7.02 Rates are national average prices. Refer to REGIONAL VARIATIONS for indicative levels of overall pricing in regions	MATERIALS			LABOUR				RATES		
	Del to Site £	Waste %	Material Cost £	Craft Optve Hrs	Lab Hrs	Labour Cost £	Sunds £	Nett Rate £	Unit	Gross Rate (+7.5%) £
REPAIRING/RENOVATING ROOF COVERINGS Cont.										
Slate roofing repairs to roofs covered with 610 x 305mm slates Cont.										
Remove double course at verge and replace with new slates and bed and point in mortar	23.64	5.00	24.82	0.29	0.29	4.71	0.58	30.11	m	32.37
Slate roofing repairs for roofs covered with 510 x 255mm slates										
Remove damaged slates and replace with new; sloping or vertical										
one slate	3.52	5.00	3.70	0.32	0.32	5.20	0.09	8.99	nr	9.66
patch of 10 slates	35.20	5.00	36.96	1.00	1.00	16.25	0.59	53.80	nr	57.84
patch of slates 1.00 - 3.00m2	63.36	5.00	66.53	1.49	1.49	24.21	0.75	91.49	m2	98.35
Examine battens, remove defective and provide 20% new										
19 x 38mm	0.24	10.00	0.26	0.11	0.11	1.79	0.07	2.12	m2	2.28
25 x 50mm	0.45	10.00	0.50	0.11	0.11	1.79	0.09	2.37	m2	2.55
Re-cover roof with slates previously removed and stacked and fix with slate nails										
with 75mm lap	-	-	-	1.00	1.00	16.25	0.13	16.38	m2	17.61
extra for providing 20% new slates	12.67	5.00	13.30	-	-	-	0.18	13.48	m2	14.49
Remove double course at eaves and										
refix	-	-	-	0.25	0.25	4.06	0.53	4.59	m	4.94
extra for providing 20% new slates	3.94	5.00	4.14	-	-	-	0.08	4.22	m	4.53
Remove double course at verge and replace with new slates and bed and point in mortar	16.29	5.00	17.10	0.29	0.29	4.71	0.57	22.39	m	24.07
Slate roofing repairs for roofs covered with 405 x 205mm slates										
Remove damaged slates and replace with new; sloping or vertical										
one slate	1.76	5.00	1.85	0.32	0.32	5.20	0.08	7.13	nr	7.66
patch of 10 slates	17.60	5.00	18.48	0.90	0.90	14.63	0.51	33.62	nr	36.14
patch of slates 1.00 - 3.00m2	52.80	5.00	55.44	2.07	2.07	33.64	0.83	89.91	m2	96.65
Examine battens, remove defective and provide 20% new										
19 x 38mm	0.32	10.00	0.35	0.13	0.13	2.11	0.08	2.54	m2	2.74
25 x 50mm	0.60	10.00	0.66	0.13	0.13	2.11	0.09	2.86	m2	3.08
Re-cover roof with slates previously removed and stacked and fix with slate nails										
with 75mm lap	-	-	-	1.14	1.14	18.52	0.19	18.72	m2	20.12
extra for providing 20% new slates	10.56	5.00	11.09	-	-	-	0.18	11.27	m2	12.11
Remove double course at eaves and										
refix	-	-	-	0.26	0.26	4.22	0.52	4.75	m	5.10
extra for providing 20% new slates	2.39	5.00	2.51	-	-	-	0.08	2.59	m	2.78
Remove double course at verge and replace with new slates and bed and point in mortar	10.26	5.00	10.77	0.30	0.30	4.88	0.63	16.28	m	17.50
Tile roofing repairs for roofs covered with machine made clay plain tiles laid to 100mm gauge										
Remove damaged tiles and replace with new, sloping or vertical										
one tile	0.49	10.00	0.54	0.32	0.32	5.20	0.03	5.77	nr	6.20
patch of 10 tiles	4.94	10.00	5.43	0.90	0.90	14.63	0.25	20.31	nr	21.83
patch of tiles 1.00 - 3.00m2	31.12	10.00	34.23	1.72	1.72	27.95	1.60	63.78	m2	68.57
Examine battens, remove defective and provide 20% new										
19 x 25mm	0.38	10.00	0.42	0.13	0.13	2.11	0.08	2.61	m2	2.81
19 x 38mm	0.51	10.00	0.56	0.14	0.14	2.27	0.09	2.93	m2	3.15
19 x 50mm	0.83	10.00	0.91	0.14	0.14	2.27	0.11	3.30	m2	3.55
Re-cover roof with tiles previously removed and stacked										
nail every fourth course with galvanised nails	-	-	-	1.00	1.00	16.25	0.17	16.42	m2	17.65
extra for nailing every course	-	-	-	0.20	0.20	3.25	0.45	3.70	m2	3.98
extra for providing 20% new tiles	6.22	10.00	6.84	-	-	-	0.18	7.02	m2	7.55
Tile roofing repairs to roofs covered with machine made clay plain tiles laid to 90mm gauge										
Remove damaged tiles and replace with new, sloping or vertical										
one tile	0.49	10.00	0.54	0.32	0.32	5.20	0.03	5.77	nr	6.20
patch of 10 tiles	4.94	10.00	5.43	0.90	0.90	14.63	0.25	20.31	nr	21.83
patch of tiles 1.00 - 3.00m2	34.58	10.00	38.04	2.05	2.05	33.31	1.78	73.13	m2	78.62
Examine battens, remove defective and provide 20% new										
19 x 25mm	0.44	10.00	0.48	0.14	0.14	2.27	0.09	2.85	m2	3.06
19 x 38mm	0.54	10.00	0.59	0.15	0.15	2.44	0.11	3.14	m2	3.38
19 x 50mm	0.93	10.00	1.02	0.15	0.15	2.44	0.12	3.58	m2	3.85

Labour hourly rates: (except Specialists) Craft Operatives £9.23 Labour £7.02 Rates are national average prices. Refer to REGIONAL VARIATIONS for indicative levels of overall pricing in regions	MATERIALS			LABOUR				RATES		
	Del to Site £	Waste %	Material Cost £	Craft Optve Hrs	Lab Hrs	Labour Cost £	Sunds £	Nett Rate £	Unit	Gross Rate (+7.5%) £
REPAIRING/RENOVATING ROOF COVERINGS Cont.										
Tile roofing repairs to roofs covered with machine made clay plain tiles laid to 90mm gauge Cont.										
Re-cover roof with tiles previously removed and stacked										
nail every fourth course with galvanised nails ...	-	-	-	1.15	1.15	18.69	0.19	18.88	m2	20.29
extra for nailing every course	-	-	-	0.23	0.23	3.74	0.47	4.21	m2	4.52
extra for providing 20% new tiles	6.92	10.00	7.61	-	-	-	0.18	7.79	m2	8.38
Tile roofing repairs for roofs covered with 381 x 227mm concrete interlocking tiles laid to 75mm laps										
Remove damaged tiles and replace with new, sloping or vertical										
one tile......................................	0.51	10.00	0.56	0.32	0.32	5.20	0.07	5.83	nr	6.27
patch of 10 tiles	5.12	10.00	5.63	0.90	0.90	14.63	0.73	20.99	nr	22.56
patch of tiles 1.00 - 3.00m2	7.17	10.00	7.89	0.85	0.85	13.81	1.16	22.86	m2	24.57
Examine battens, remove defective and provide 20% new										
19 x 25mm	0.13	10.00	0.14	0.05	0.05	0.81	0.07	1.03	m2	1.10
19 x 38mm	0.18	10.00	0.20	0.06	0.06	0.97	0.07	1.24	m2	1.34
19 x 50mm	0.25	10.00	0.28	0.07	0.07	1.14	0.08	1.49	m2	1.60
Re-cover roof with tiles previously removed and stacked										
nail every fourth course with galvanised nails ...	-	-	-	0.50	0.50	8.13	0.02	8.14	m2	8.76
extra for nailing every course	-	-	-	0.05	0.05	0.81	0.06	0.87	m2	0.94
extra for providing 20% new tiles	1.44	10.00	1.58	-	-	-	0.22	1.80	m2	1.94
Tile roofing repairs generally										
Remove damaged clay tiles and provide new										
half round ridge or hip and bed in mortar	11.00	10.00	12.10	0.60	0.60	9.75	1.09	22.94	m	24.66
bonnet hip.....................................	15.50	10.00	17.05	0.75	0.75	12.19	0.30	29.54	m	31.75
trough valley.................................	15.50	10.00	17.05	0.75	0.75	12.19	0.30	29.54	m	31.75
vertical angle................................	15.50	10.00	17.05	0.75	0.75	12.19	0.30	29.54	m	31.75
Remove double course and replace with new clay tiles and bed and point in mortar										
at eaves......................................	5.93	10.00	6.52	0.23	0.23	3.74	0.45	10.71	m	11.51
at verge......................................	12.35	10.00	13.59	0.38	0.38	6.17	1.39	21.15	m	22.74
Rake out defective pointing and re-point in cement mortar (1:3)										
ridge or hip tiles	-	-	-	0.34	0.34	5.53	0.17	5.70	m	6.12
Hack off defective cement mortar fillet and renew in cement mortar (1:3)	-	-	-	0.34	0.34	5.53	0.34	5.87	m	6.30
Remove defective hip hook and replace with new	1.14	-	1.14	0.34	0.34	5.53	0.07	6.74	nr	7.24
Corrugated fibre cement roofing repairs										
Remove damaged sheets and provide and fix new with screws and washers to timber purlins and ease and adjust edges of adjoining sheets as necessary										
one sheet; 1825mm long Profile 3 natural grey	14.07	10.00	15.48	1.23	1.23	19.99	1.99	37.45	nr	40.26
one sheet; 1825mm long Profile 3 standard colour .	16.88	10.00	18.57	1.23	1.23	19.99	1.99	40.55	nr	43.59
one sheet; 1825mm long Profile 6 natural grey	20.59	10.00	22.65	1.78	1.78	28.93	2.94	54.51	nr	58.60
one sheet; 1825mm long Profile 6 standard colour .	24.71	10.00	27.18	1.78	1.78	28.93	2.94	59.05	nr	63.47
over one sheet; Profile 3 natural grey	13.27	10.00	14.60	0.89	0.89	14.46	1.88	30.94	m2	33.26
over one sheet; Profile 3 standard colour	15.92	10.00	17.51	0.89	0.89	14.46	1.88	33.85	m2	36.39
over one sheet; Profile 6 natural grey	19.42	10.00	21.36	1.29	1.29	20.96	2.77	45.09	m2	48.48
over one sheet; Profile 6 standard colour	23.30	10.00	25.63	1.29	1.29	20.96	2.77	49.36	m2	53.06
Remove damaged sheets and provide and fix new with hook bolts and washers to steel purlins and ease and adjust edges of adjoining sheets as necessary										
one sheet; 1825mm long Profile 3 natural grey	14.07	10.00	15.48	1.37	1.37	22.26	1.99	39.73	nr	42.71
one sheet; 1825mm long Profile 3 standard colour .	16.88	10.00	18.57	1.37	1.37	22.26	1.99	42.82	nr	46.03
one sheet; 1825mm long Profile 6 natural grey	20.59	10.00	22.65	1.98	1.98	32.17	2.94	57.76	nr	62.10
one sheet; 1825mm long Profile 6 standard colour .	24.71	10.00	27.18	1.98	1.98	32.17	2.94	62.30	nr	66.97
over one sheet; Profile 3 natural grey	13.27	10.00	14.60	1.03	1.03	16.74	1.88	33.21	m2	35.71
over one sheet; Profile 3 standard colour	15.92	10.00	17.51	1.03	1.03	16.74	1.88	36.13	m2	38.84
over one sheet; Profile 6 natural grey	19.42	10.00	21.36	1.49	1.49	24.21	2.77	48.34	m2	51.97
over one sheet; Profile 6 standard colour	23.30	10.00	25.63	1.49	1.49	24.21	2.77	52.61	m2	56.56
Felt roofing repairs for roofs covered with bituminous felt roofing, B.S.747										
Sweep clean, cut out defective layer, re-bond adjacent felt and cover with one layer of felt bonded in hot bitumen; patches not exceeding 1.00m2										
fibre based roofing felt type 1B	1.42	15.00	1.63	0.88	0.88	14.30	2.38	18.31	m2	19.69
for each defective underlayer cut out and replaced, add	1.42	15.00	1.63	0.88	0.88	14.30	2.38	18.31	m2	19.69
fibre based mineral surfaced roofing felt type 1E	1.72	15.00	1.98	1.05	1.05	17.06	2.38	21.42	m2	23.03
high performance polyester based roofing felt type 5B	3.38	15.00	3.89	1.02	1.02	16.57	2.38	22.84	m2	24.56
for each defective underlayer cut out and replaced, add	3.38	15.00	3.89	1.02	1.02	16.57	2.38	22.84	m2	24.56
high performance polyester based mineral surfaced roofing felt type 5E	3.90	15.00	4.49	1.13	1.13	18.36	2.38	25.23	m2	27.12

Labour hourly rates: (except Specialists) Craft Operatives £9.23 Labour £7.02 Rates are national average prices. Refer to REGIONAL VARIATIONS for indicative levels of overall pricing in regions	MATERIALS			LABOUR				RATES		
	Del to Site £	Waste %	Material Cost £	Craft Optve Hrs	Lab Hrs	Labour Cost £	Sunds £	Nett Rate £	Unit	Gross Rate (+7.5%) £
REPAIRING/RENOVATING ROOF COVERINGS Cont.										
Felt roofing repairs for roofs covered with bituminous felt roofing, B.S.747 Cont.										
Sweep clean, cut out defective layer, re-bond adjacent felt and cover with one layer of felt bonded in hot bitumen; patches 1.00 - 3.00m2										
fibre based roofing felt type 1B	1.42	15.00	**1.63**	0.43	0.43	**6.99**	2.38	**11.00**	m2	**11.83**
for each defective underlayer cut out and										
replaced, add	1.42	15.00	**1.63**	0.43	0.43	**6.99**	2.38	**11.00**	m2	**11.83**
fibre based mineral surfaced roofing felt type 1E	1.72	15.00	**1.98**	0.56	0.56	**9.10**	2.38	**13.46**	m2	**14.47**
high performance polyester based roofing felt type 5B	3.38	15.00	**3.89**	0.53	0.53	**8.61**	2.38	**14.88**	m2	**16.00**
for each defective underlayer cut out and										
replaced, add	3.38	15.00	**3.89**	0.53	0.53	**8.61**	2.38	**14.88**	m2	**16.00**
high performance polyester based mineral surfaced roofing felt type 5E	3.90	15.00	**4.49**	0.65	0.65	**10.56**	2.38	**17.43**	m2	**18.73**
Sweep clean surface of existing roof, prime with hot bitumen and dress with										
13mm layer of limestone or granite chippings	1.80	10.00	**1.98**	-	0.23	**1.61**	1.33	**4.92**	m2	**5.29**
13mm layer of pea shingle	1.25	10.00	**1.38**	-	0.36	**2.53**	1.33	**5.23**	m2	**5.62**
extra; removing existing chippings or shingle	-	-	**-**	-	0.24	**1.68**	0.40	**2.08**	m2	**2.24**
Liquid bitumen proofing (black) brush applied on old roof covering including cleaning old covering										
Corrugated asbestos roofing										
one coat...	1.94	10.00	**2.13**	-	0.21	**1.47**	-	**3.61**	m2	**3.88**
two coats..	3.63	10.00	**3.99**	-	0.25	**1.75**	-	**5.75**	m2	**6.18**
Felt roofing										
one coat...	1.68	10.00	**1.85**	-	0.17	**1.19**	-	**3.04**	m2	**3.27**
two coats..	3.11	10.00	**3.42**	-	0.21	**1.47**	-	**4.90**	m2	**5.26**
For top coat in green in lieu black										
add..	0.52	10.00	**0.57**	-	-	**-**	-	**0.57**	m2	**0.61**

This section continues
on the next page

Labour hourly rates: (except Specialists) Craft Operatives £12.44 Labour £7.02 Rates are national average prices. Refer to REGIONAL VARIATIONS for indicative levels of overall pricing in regions	MATERIALS			LABOUR				RATES		
	Del to Site £	Waste %	Material Cost £	Craft Optve Hrs	Lab Hrs	Labour Cost £	Sunds £	Nett Rate £	Unit	Gross Rate (+7.5%) £
REPAIRING/RENOVATING ROOF COVERINGS Cont.										
Repairs to lead roofing										
Repair crack										
clean out and fill with copper bit solder	-	-	-	0.80	-	9.95	0.30	10.25	m	11.02
Turn back flashing										
and re-dress	-	-	-	0.48	-	5.97	-	5.97	m	6.42
Repairs to lead flashings, etc.; replace with new lead										
Remove flashing and provide new code Nr4 lead; wedge into groove										
150mm girth	2.10	5.00	2.21	1.06	-	13.19	0.26	15.65	m	16.83
240mm girth	3.36	5.00	3.53	1.49	-	18.54	0.26	22.32	m	24.00
300mm girth	4.20	5.00	4.41	1.82	-	22.64	0.26	27.31	m	29.36
Remove stepped flashing and provide new code Nr 4 lead; wedge into groove										
180mm girth	2.52	5.00	2.65	1.54	-	19.16	0.26	22.06	m	23.72
240mm girth	3.36	5.00	3.53	1.94	-	24.13	0.26	27.92	m	30.02
300mm girth	4.20	5.00	4.41	2.34	-	29.11	0.26	33.78	m	36.31
Remove apron flashing and provide new code Nr 5 lead; wedge into groove										
150mm girth	2.61	5.00	2.74	1.06	-	13.19	0.26	16.19	m	17.40
300mm girth	5.23	5.00	5.49	1.82	-	22.64	0.26	28.39	m	30.52
450mm girth	7.84	5.00	8.23	2.58	-	32.10	0.40	40.73	m	43.78
Remove lining to valley gutter and provide new code Nr 5 lead										
240mm girth	4.18	5.00	4.39	1.49	-	18.54	-	22.92	m	24.64
300mm girth	5.23	5.00	5.49	1.82	-	22.64	-	28.13	m	30.24
450mm girth	7.84	5.00	8.23	2.58	-	32.10	-	40.33	m	43.35
Remove gutter lining and provide new code Nr 6 lead										
360mm girth	7.42	5.00	7.79	2.12	-	26.37	-	34.16	m	36.73
420mm girth	8.66	5.00	9.09	2.43	-	30.23	-	39.32	m	42.27
450mm girth	9.27	5.00	9.73	2.58	-	32.10	-	41.83	m	44.97
Repairs to zinc flashings, etc.; replace with 0.8mm zinc										
Remove flashing and provide new; wedge into groove										
150mm girth	2.15	5.00	2.26	1.12	-	13.93	0.18	16.37	m	17.60
240mm girth	3.44	5.00	3.61	1.60	-	19.90	0.18	23.70	m	25.47
300mm girth	4.30	5.00	4.51	1.94	-	24.13	0.26	28.91	m	31.08
Remove stepped flashing and provide new; wedge into groove										
180mm girth	2.58	5.00	2.71	1.65	-	20.53	0.18	23.41	m	25.17
240mm girth	3.44	5.00	3.61	2.08	-	25.88	0.18	29.67	m	31.89
300mm girth	4.30	5.00	4.51	2.52	-	31.35	0.26	36.12	m	38.83
Remove apron flashing and provide new; wedge into groove										
150mm girth	2.15	5.00	2.26	1.12	-	13.93	0.18	16.37	m	17.60
300mm girth	4.30	5.00	4.51	1.94	-	24.13	0.26	28.91	m	31.08
450mm girth	6.45	5.00	6.77	2.76	-	34.33	0.32	41.43	m	44.53
Remove lining to valley gutter and provide new										
240mm girth	3.44	5.00	3.61	1.60	-	19.90	-	23.52	m	25.28
300mm girth	4.30	5.00	4.51	1.94	-	24.13	-	28.65	m	30.80
450mm girth	6.45	5.00	6.77	2.76	-	34.33	-	41.11	m	44.19
Remove gutter lining and provide new										
360mm girth	5.16	5.00	5.42	2.27	-	28.24	-	33.66	m	36.18
420mm girth	6.02	5.00	6.32	2.61	-	32.47	-	38.79	m	41.70
450mm girth	6.45	5.00	6.77	2.76	-	34.33	-	41.11	m	44.19
Repairs to copper flashings, etc.; replace with 0.6mm copper										
Remove flashing and provide new; wedge into groove										
150mm girth	2.49	5.00	2.61	1.06	-	13.19	0.36	16.16	m	17.37
240mm girth	3.99	5.00	4.19	1.49	-	18.54	0.36	23.09	m	24.82
300mm girth	4.98	5.00	5.23	1.82	-	22.64	0.36	28.23	m	30.35
Remove stepped flashing and provide new; wedge into groove										
180mm girth	2.99	5.00	3.14	1.54	-	19.16	0.36	22.66	m	24.36
240mm girth	3.99	5.00	4.19	1.94	-	24.13	0.36	28.68	m	30.83
300mm girth	4.98	5.00	5.23	2.34	-	29.11	0.36	34.70	m	37.30
Remove apron flashing and provide new; wedge into groove										
150mm girth	2.49	5.00	2.61	1.06	-	13.19	0.36	16.16	m	17.37
300mm girth	4.98	5.00	5.23	1.82	-	22.64	0.36	28.23	m	30.35
450mm girth	7.47	5.00	7.84	2.58	-	32.10	0.54	40.48	m	43.51
Remove lining to valley gutter and provide new										
240mm girth	3.99	5.00	4.19	1.49	-	18.54	-	22.73	m	24.43
300mm girth	4.98	5.00	5.23	1.82	-	22.64	-	27.87	m	29.96
450mm girth	7.47	5.00	7.84	2.58	-	32.10	-	39.94	m	42.93

Labour hourly rates: (except Specialists) Craft Operatives £12.44 Labour £7.02 Rates are national average prices. Refer to REGIONAL VARIATIONS for indicative levels of overall pricing in regions	MATERIALS			LABOUR				RATES		
	Del to Site £	Waste %	Material Cost £	Craft Optve Hrs	Lab Hrs	Labour Cost £	Sunds £	Nett Rate £	Unit	Gross Rate (+7.5%) £

REPAIRING/RENOVATING ROOF COVERINGS Cont.

Repairs to copper flashings, etc.; replace with 0.6mm copper Cont.

Remove gutter lining and provide new										
360mm girth	5.98	5.00	6.28	2.12	-	26.37	-	32.65	m	35.10
420mm girth	6.97	5.00	7.32	2.43	-	30.23	-	37.55	m	40.36
450mm girth	7.47	5.00	7.84	2.58	-	32.10	-	39.94	m	42.93

Repairs to aluminium flashings etc; replace with 0.60mm aluminium

Remove flashings and provide new; wedge into groove										
150mm girth	1.46	5.00	1.53	1.06	-	13.19	0.16	14.88	m	16.00
240mm girth	2.73	5.00	2.87	1.49	-	18.54	0.16	21.56	m	23.18
300mm girth	2.73	5.00	2.87	1.82	-	22.64	0.16	25.67	m	27.59
Remove stepped flashing and provide new; wedge into groove										
180mm girth	2.01	5.00	2.11	1.54	-	19.16	0.16	21.43	m	23.04
240mm girth	2.73	5.00	2.87	1.94	-	24.13	0.16	27.16	m	29.20
300mm girth	2.73	5.00	2.87	2.34	-	29.11	0.16	32.14	m	34.55
Remove apron flashing and provide new; wedge into groove										
150mm girth	1.46	5.00	1.53	1.06	-	13.19	0.16	14.88	m	16.00
300mm girth	2.73	5.00	2.87	1.82	-	22.64	0.16	25.67	m	27.59
450mm girth	3.80	5.00	3.99	2.56	-	31.85	3.00	38.84	m	41.75
Remove lining to valley gutter and provide new										
240mm girth	2.73	5.00	2.87	1.49	-	18.54	-	21.40	m	23.01
300mm girth	2.73	5.00	2.87	1.82	-	22.64	-	25.51	m	27.42
450mm girth	3.80	5.00	3.99	2.58	-	32.10	-	36.09	m	38.79
Remove gutter lining and provide new										
360mm girth	3.80	5.00	3.99	2.12	-	26.37	-	30.36	m	32.64
420mm girth	3.80	5.00	3.99	2.43	-	30.23	-	34.22	m	36.79
450mm girth	3.80	5.00	3.99	2.58	-	32.10	-	36.09	m	38.79

REPAIRING/RENOVATING PLUMBING

Repairs to rainwater goods

Clean out and make good loose fixings										
eaves gutter	-	-	-	0.40	-	4.98	-	4.98	m	5.35
rainwater pipe	-	-	-	0.40	-	4.98	-	4.98	m	5.35
rainwater head	-	-	-	0.75	-	9.33	-	9.33	nr	10.03
Clean out defective joint to existing eaves gutter and re-make joint										
with jointing compound and new bolt	1.00	-	1.00	0.75	-	9.33	-	10.33	nr	11.10
Remove remains of old balloon grating and provide new plastic grating in outlet or end of pipe										
50mm diameter	1.80	2.00	1.84	0.50	-	6.22	-	8.06	nr	8.66
75mm diameter	1.85	2.00	1.89	0.50	-	6.22	-	8.11	nr	8.72
100mm diameter	1.92	2.00	1.96	0.50	-	6.22	-	8.18	nr	8.79

Repair or replacement of stopcocks, valves, etc.

Turn off water supply drain down as necessary and renew washer to the following up to 19mm										
main stopcock	0.75	-	0.75	2.00	-	24.88	-	25.63	nr	27.55
service stopcock	0.75	-	0.75	2.00	-	24.88	-	25.63	nr	27.55
bib or pillar cock	0.75	-	0.75	2.00	-	24.88	-	25.63	nr	27.55
supatap	0.75	-	0.75	2.00	-	24.88	-	25.63	nr	27.55
draining tap	0.75	-	0.75	2.00	-	24.88	-	25.63	nr	27.55
Turn off water supply, drain down as necessary, take out old valves and re-new with joints to copper										
13mm brass stopcock (B.S.1010)	2.73	2.00	2.78	2.00	-	24.88	-	27.66	nr	29.74
13mm chromium plated sink pillar cock (B.S.1010)	12.15	2.00	12.39	2.00	-	24.88	-	37.27	nr	40.07
13mm brass bib cock (B.S.1010)	6.60	2.00	6.73	2.00	-	24.88	-	31.61	nr	33.98
13mm drain tap (B.S.2879) type 2	1.62	2.00	1.65	2.00	-	24.88	-	26.53	nr	28.52
19mm drain tap (B.S.2879) type 2	11.70	2.00	11.93	2.00	-	24.88	-	36.81	nr	39.58
13mm high pressure ball valve (B.S.1212) piston type with copper ball	7.78	2.00	7.94	2.00	-	24.88	-	32.82	nr	35.28
13mm high pressure ball valve (B.S.1212) piston type with plastic ball	4.27	2.00	4.36	2.00	-	24.88	-	29.24	nr	31.43
13mm low pressure ball valve (B.S.1212) piston type with copper ball	9.06	2.00	9.24	2.00	-	24.88	-	34.12	nr	36.68
13mm low pressure ball valve (B.S.1212) piston type with plastic ball	5.55	2.00	5.66	2.00	-	24.88	-	30.54	nr	32.83
19mm high pressure ball valve (B.S.1212) piston type with copper ball	16.29	2.00	16.62	2.50	-	31.10	-	47.72	nr	51.29
19mm high pressure ball valve (B.S.1212) piston type with plastic ball	11.50	2.00	11.73	2.50	-	31.10	-	42.83	nr	46.04
19mm low pressure ball valve (B.S.1212) piston type with copper ball	16.45	2.00	16.78	2.50	Lab	31.10	Sunds	47.88	nr	51.47
19mm low pressure ball valve (B.S.1212) piston type with plastic ball	11.68	2.00	11.91	2.50	-	31.10	-	43.01	nr	46.24

EXISTING SITE/BUILDINGS/SERVICES – MAJOR WORKS

Labour hourly rates: (except Specialists) Craft Operatives £12.44 Labour £7.02 Rates are national average prices. Refer to REGIONAL VARIATIONS for indicative levels of overall pricing in regions	MATERIALS			LABOUR				RATES		
	Del to Site £	Waste %	Material Cost £	Craft Optve Hrs	Lab Hrs	Labour Cost £	Sunds £	Nett Rate £	Unit	Gross Rate (+7.5%) £
REPAIRING/RENOVATING PLUMBING Cont.										
Replacement of traps										
Take out old plastic trap and replace with new 76mm seal trap (B.S.3943) with '0'' ring joint outlet										
36mm diameter P trap	2.37	2.00	2.42	1.25	-	15.55	-	17.97	nr	19.31
42mm diameter P trap	2.73	2.00	2.78	1.50	-	18.66	-	21.44	nr	23.05
36mm diameter S trap	3.00	2.00	3.06	1.25	-	15.55	-	18.61	nr	20.01
42mm diameter S trap	3.52	2.00	3.59	1.50	-	18.66	-	22.25	nr	23.92
42mm diameter bath trap with overflow connection	5.08	2.00	5.18	1.75	-	21.77	-	26.95	nr	28.97
Take out old copper trap and replace with new 76mm seal two piece trap (B.S.1184) with compression joint										
35mm diameter P trap	12.74	2.00	12.99	1.25	-	15.55	-	28.54	nr	30.69
42mm diameter P trap	15.02	2.00	15.32	1.50	-	18.66	-	33.98	nr	36.53
35mm diameter S trap	13.51	2.00	13.78	1.25	-	15.55	-	29.33	nr	31.53
42mm diameter S trap	16.24	2.00	16.56	1.50	-	18.66	-	35.22	nr	37.87
42mm diameter bath trap with male iron overflow connection	22.17	2.00	22.61	1.75	-	21.77	-	44.38	nr	47.71
Preparation of pipe for insertion of new pipe or fitting										
Cut into cast iron pipe and take out a length up to 1.00m for insertion of new pipe or fitting										
50mm diameter	-	-	-	3.70	-	46.03	-	46.03	nr	49.48
64 and 76mm diameter	-	-	-	4.30	-	53.49	-	53.49	nr	57.50
89 and 102mm diameter	-	-	-	4.90	-	60.96	-	60.96	nr	65.53
Cut into cast iron pipe with caulked joint and take out a length up to 1.00m for insertion of new pipe or fitting										
50mm diameter	-	-	-	3.70	-	46.03	-	46.03	nr	49.48
64 and 76mm diameter	-	-	-	4.30	-	53.49	-	53.49	nr	57.50
89 and 102mm diameter	-	-	-	4.90	-	60.96	-	60.96	nr	65.53
Cut into asbestos cement pipe and take out a length up to 1.00m for insertion of new pipe or fitting										
50mm diameter	-	-	-	3.15	-	39.19	-	39.19	nr	42.12
64 and 76mm diameter	-	-	-	3.75	-	46.65	-	46.65	nr	50.15
89 and 102mm diameter	-	-	-	4.35	-	54.11	-	54.11	nr	58.17
Cut into polythene or p.v.c. pipe and take out a length up to 1.00m for insertion of new pipe or fitting										
up to 25mm diameter	-	-	-	2.50	-	31.10	-	31.10	nr	33.43
32 - 50mm diameter	-	-	-	3.15	-	39.19	-	39.19	nr	42.12
64 and 76mm diameter	-	-	-	3.75	-	46.65	-	46.65	nr	50.15
89 and 102mm diameter	-	-	-	4.40	-	54.74	-	54.74	nr	58.84
Cut into steel pipe and take out a length up to 1.00m for insertion of new pipe or fitting										
up to 25mm diameter	-	-	-	3.05	-	37.94	-	37.94	nr	40.79
32 - 50mm diameter	-	-	-	3.70	-	46.03	-	46.03	nr	49.48
64 and 76mm diameter	-	-	-	4.30	-	53.49	-	53.49	nr	57.50
89 and 102mm diameter	-	-	-	4.90	-	60.96	-	60.96	nr	65.53
Cut into stainless steel pipe and take out a length up to 1.00m for insertion of new pipe or fitting										
up to 25mm diameter	-	-	-	2.90	-	36.08	-	36.08	nr	38.78
32 - 50mm diameter	-	-	-	3.50	-	43.54	-	43.54	nr	46.81
64 and 76mm diameter	-	-	-	4.10	-	51.00	-	51.00	nr	54.83
89 and 102mm diameter	-	-	-	4.75	-	59.09	-	59.09	nr	63.52
Cut into copper pipe and take out a length up to 1.00m for insertion of new pipe or fitting										
up to 25mm diameter	-	-	-	2.90	-	36.08	-	36.08	nr	38.78
32 - 50mm diameter	-	-	-	3.50	-	43.54	-	43.54	nr	46.81
64 and 76mm diameter	-	-	-	4.10	-	51.00	-	51.00	nr	54.83
89 and 102mm diameter	-	-	-	4.75	-	59.09	-	59.09	nr	63.52
Cut into lead pipe and take out a length up to 1.00m for insertion of new pipe or fitting										
up to 25mm diameter	-	-	-	3.05	-	37.94	-	37.94	nr	40.79
32 - 50mm diameter	-	-	-	3.70	-	46.03	-	46.03	nr	49.48
64 and 76mm diameter	-	-	-	4.30	-	53.49	-	53.49	nr	57.50
89 and 102mm diameter	-	-	-	4.90	-	60.96	-	60.96	nr	65.53
Removal of traps, valves, etc., for re-use										
Take out the following, including unscrewing or uncoupling, for re-use and including draining down as necessary										
trap up to 25mm diameter	-	-	-	0.70	-	8.71	-	8.71	nr	9.36
trap 32 - 50mm diameter	-	-	-	0.85	-	10.57	-	10.57	nr	11.37
stop valve up to 25mm diameter	-	-	-	2.00	-	24.88	-	24.88	nr	26.75
stop valve 32 - 50mm diameter	-	-	-	2.25	-	27.99	-	27.99	nr	30.09
radiator valve up to 25mm diameter	-	-	-	2.00	-	24.88	-	24.88	nr	26.75
radiator valve 32 - 50mm diameter	-	-	-	2.25	-	27.99	-	27.99	nr	30.09
tap up to 25mm diameter	-	-	-	2.00	-	24.88	-	24.88	nr	26.75
tap 32 - 50mm diameter	-	-	-	2.25	-	27.99	-	27.99	nr	30.09
extra; burning out one soldered joint up to 25mm diameter	-	-	-	0.50	-	6.22	-	6.22	nr	6.69
extra; burning out one soldered joint 32 - 50mm diameter	-	-	-	0.80	-	9.95	-	9.95	nr	10.70

38

Labour hourly rates: (except Specialists) Craft Operatives £12.44 Labour £7.02 Rates are national average prices. Refer to REGIONAL VARIATIONS for indicative levels of overall pricing in regions	MATERIALS			LABOUR				RATES		
	Del to Site £	Waste %	Material Cost £	Craft Optve Hrs	Lab Hrs	Labour Cost £	Sunds £	Nett Rate £	Unit	Gross Rate (+7.5%) £

REPAIRING/RENOVATING PLUMBING Cont.

Replacement of old sanitary fittings

Disconnect trap, valves and services pipes, remove old fitting and provide and fix new including re-connecting trap, valves and services pipes

560 x 405mm vitreous china lavatory basin with 32mm chromium plated waste outlet, plug, chain and stay and pair of cast iron brackets	21.00	5.00	22.05	6.00	–	74.64	–	96.69	nr	103.94
600 x 1000mm stainless steel sink with single drainer with 38mm chromium plated waste outlet, plug, chain and stay	48.25	5.00	50.66	6.00	–	74.64	–	125.30	nr	134.70
600 x 1500mm stainless steel sink with double drainer with 38mm chromium plated waste outlet, overflow plug, chain and stay	65.25	5.00	68.51	6.50	–	80.86	–	149.37	nr	160.58
1500mm pressed steel vitreous enamelled rectangular top bath with cradles and with 38mm chromium plated waste outlet, plug, chain and stay and 32mm chromium plated overflow and including removing old cast iron bath	82.50	5.00	86.63	8.00	–	99.52	–	186.15	nr	200.11
1700mm pressed steel vitreous enamelled rectangular top bath with cradles and with 38mm chromium plated waste outlet, plug, chain and stay and 32mm chromium plated overflow and including removing old cast iron bath	85.50	5.00	89.78	8.00	–	99.52	–	189.29	nr	203.49

Take off old W.C. seat and provide and fix new

ring pattern black plastic seat and cover	14.50	5.00	15.23	1.50	–	18.66	–	33.88	nr	36.43

Disconnect supply pipe, overflow and flush pipe and take out old W.C. cistern and provide and fix new 9 litre black plastic W.C. cistern with cover and connect pipe and ball valve

low level	49.75	5.00	52.24	5.00	–	62.20	–	114.44	nr	123.02
high level	42.00	5.00	44.10	5.00	–	62.20	–	106.30	nr	114.27

Disconnect flush pipe, take off seat and remove old W.C. pan, provide and fix new white china pan, connect flush pipe, form cement joint with drain and refix old seat

S or P trap pan	33.50	5.00	35.17	5.00	–	62.20	–	97.38	nr	104.68

Disconnect and remove old high level W.C. suite complete and provide and fix new suite with P or S trap pan, plastic ring seat and cover, plastic flush pipe or bend, 9 litre plastic W.C. cistern with cover, chain and pull or lever handle and 13mm low pressure ball valve; adapt supply pipe and overflow as necessary and connect to new suite and cement joint to drain

new high level suite	86.00	5.00	90.30	8.50	–	105.74	–	196.04	nr	210.74
new low level suite (vitreous china cistern)	87.25	5.00	91.61	9.00	–	111.96	–	203.57	nr	218.84

Disconnect and remove old low level W.C. suite complete and provide and fix new suite with P or S trap pan, plastic ring seat and cover, plastic flush bend, 9 litre vitreous china W.C. cistern with cover, lever handle and 13mm low pressure ball valve with plastic ball; re-connect supply pipe and overflow and cement joint to drain

new low level suite	87.25	5.00	91.61	8.50	–	105.74	–	197.35	nr	212.15

Replacement of old water tanks

Disconnect all pipework, set aside ball valve and take out cold water storage cistern in roof space; replace with new galvanised water storage cistern (B.S.417 Grade A); reconnect all pipework and ball valve; remove metal filings and clean inside of tank

type SC40 size 690 x 510 x 510mm (114 litre)	73.50	2.00	74.97	9.00	–	111.96	0.30	187.23	nr	201.27
type SC70 size 910 x 610 x 580mm (227 litre)	126.00	2.00	128.52	9.00	–	111.96	0.50	240.98	nr	259.05

Disconnect all pipework and take out hot water tank from first floor level; replace with new galvanised hot water tank (B.S.417 Grade A); reconnect all pipework

type T25/1 size 610 x 430 x 430mm (95 litre)	126.50	2.00	129.03	9.00	–	111.96	–	240.99	nr	259.06
type T30/1 size 410 x 460 x 480mm (114 litre)	136.00	2.00	138.72	9.00	–	111.96	–	250.68	nr	269.48

This section continues
on the next page

EXISTING SITE/BUILDINGS/SERVICES

abour hourly rates: (except Specialists) Craft Operatives £14.38 Labour £7.02 Rates are national average prices. Refer to REGIONAL VARIATIONS for indicative levels of overall pricing in regions	MATERIALS			LABOUR			RATES		
	Del to Site £	Waste %	Material Cost £	Craft Optve £	Lab £	Labour Cost £	Sunds £	Nett Rate £	Unit	Gross Rate (+7.5%) £

REPAIRING/RENOVATING ELECTRICAL INSTALLATIONS

Generally

Note
   the following approximate estimates of costs are
   dependent on the number and disposition of the
   points. Lamps and fittings together with cutting
   and making good are excluded
Strip out and re-wire complete

Provide new lamp holders, flush switches, flush
socket outlets, fitting outlets and consumer unit
with circuit breakers; a three bedroom house with ten
lighting points and associated switches, sixteen 13
amp socket outlets, three 13 amp fitting outlets, one
30 amp cooker control panel, three earth bonding
points and one eight way consumer unit

	Del to Site	Waste	Material Cost	Craft Optve	Lab	Labour Cost	Sunds	Nett Rate	Unit	Gross Rate
PVC cables in existing conduit	851.69	7.50	915.57	52.85	–	759.98	30.00	1705.55	nr	1833.47
PVC cables and conduit	941.97	7.50	1012.62	99.00	–	1423.62	30.00	2466.24	nr	2651.21

**Strip out and re-wire with PVC insulated and sheathed
cable in existing conduit**

Point to junction box, re-using existing lamp holder,
socket outlet etc

5 amp socket outlet	6.00	7.50	6.45	0.85	–	12.22	0.45	19.12	nr	20.56
13 amp socket or fitting outlet	8.00	7.50	8.60	0.85	–	12.22	0.45	21.27	nr	22.87
lighting point or switch	4.80	7.50	5.16	0.98	–	14.09	0.45	19.70	nr	21.18

**Strip out and re-wire with PVC insulated and sheathed
cable and new conduit**

Point to junction box, re-using existing lamp holder,
socket outlet etc

5 amp socket outlet	6.97	7.50	7.49	1.20	–	17.26	0.50	25.25	nr	27.14
13 amp socket outlet	9.35	7.50	10.05	1.20	–	17.26	0.50	27.81	nr	29.89
lighting point or switch	6.63	7.50	7.13	1.30	–	18.69	0.30	26.12	nr	28.08

Luminaires and accessories

Take out damaged and renew

lamp holder, batten type	4.75	7.50	5.11	0.56	–	8.05	0.25	13.41	nr	14.41
lamp holder, pendant type with rose	1.94	7.50	2.09	1.06	–	15.24	0.25	17.58	nr	18.90
switch, flush type	3.21	7.50	3.45	0.75	–	10.79	0.45	14.69	nr	15.79
switch, surface mounted type	3.01	7.50	3.24	0.75	–	10.79	0.45	14.47	nr	15.56
13 amp socket outlet, flush type	4.72	7.50	5.07	0.75	–	10.79	0.45	16.31	nr	17.53
13 amp socket outlet, surface mounted type	4.58	7.50	4.92	0.88	–	12.65	0.45	18.03	nr	19.38
5 amp socket outlet, flush type	7.89	7.50	8.48	0.70	–	10.07	0.45	19.00	nr	20.42
5 amp socket outlet, surface mounted type	7.67	7.50	8.25	0.88	–	12.65	0.45	21.35	nr	22.95

**This section continues
on the next page**

Labour hourly rates: (except Specialists) Craft Operatives £9.23 Labour £7.02 Rates are national average prices. Refer to REGIONAL VARIATIONS for indicative levels of overall pricing in regions	MATERIALS			LABOUR				RATES		
	Del to Site £	Waste %	Material Cost £	Craft Optve Hrs	Lab Hrs	Labour Cost £	Sunds £	Nett Rate £	Unit	Gross Rate (+7.5%) £

REPAIRING/RENOVATING FINISHINGS

Repairs to old wall and ceiling finishings

Cut out damaged two coat plaster to wall in patches and make out in new plaster to match existing										
under 1.00m2	2.26	10.00	**2.49**	2.13	2.13	**34.61**	0.52	**37.62**	m2	40.44
1.00 - 2.00m2	2.26	10.00	**2.49**	1.76	1.76	**28.60**	0.52	**31.61**	m2	33.98
2.00 - 4.00m2	2.26	10.00	**2.49**	1.39	1.39	**22.59**	0.52	**25.59**	m2	27.51
Cut out damaged three coat plaster to wall in patches and make out in new plaster to match existing										
under 1.00m2	3.00	10.00	**3.30**	2.50	2.50	**40.63**	0.80	**44.73**	m2	48.08
1.00 - 2.00m2	3.00	10.00	**3.30**	2.06	2.06	**33.48**	0.80	**37.58**	m2	40.39
2.00 - 4.00m2	3.00	10.00	**3.30**	1.62	1.62	**26.32**	0.80	**30.43**	m2	32.71
Cut out damaged two coat rendering to wall in patches and make out in new rendering to match existing										
under 1.00m2	1.34	10.00	**1.47**	1.93	1.93	**31.36**	0.52	**33.36**	m2	35.86
1.00 - 2.00m2	1.34	10.00	**1.47**	1.60	1.60	**26.00**	0.52	**27.99**	m2	30.09
2.00 - 4.00m2	1.34	10.00	**1.47**	1.26	1.26	**20.48**	0.52	**22.47**	m2	24.15
Cut out damaged three coat rendering to wall in patches and make out in new rendering to match existing										
under 1.00m2	2.06	10.00	**2.27**	2.28	2.28	**37.05**	0.80	**40.12**	m2	43.12
1.00 - 2.00m2	2.06	10.00	**2.27**	1.87	1.87	**30.39**	0.80	**33.45**	m2	35.96
2.00 - 4.00m2	2.06	10.00	**2.27**	1.47	1.47	**23.89**	0.80	**26.95**	m2	28.98
Cut out damaged two coat plaster to ceiling in patches and make out in new plaster to match existing										
under 1.00m2	2.26	10.00	**2.49**	2.38	2.38	**38.67**	0.52	**41.68**	m2	44.81
1.00 - 2.00m2	2.26	10.00	**2.49**	1.96	1.96	**31.85**	0.52	**34.86**	m2	37.47
2.00 - 4.00m2	2.26	10.00	**2.49**	1.54	1.54	**25.02**	0.52	**28.03**	m2	30.13
Cut out damaged three coat plaster to ceiling in patches and make out in new plaster to match existing										
under 1.00m2	3.00	10.00	**3.30**	2.79	2.79	**45.34**	0.80	**49.44**	m2	53.15
1.00 - 2.00m2	3.00	10.00	**3.30**	2.30	2.30	**37.38**	0.80	**41.48**	m2	44.59
2.00 - 4.00m2	3.00	10.00	**3.30**	1.81	1.81	**29.41**	0.80	**33.51**	m2	36.03
Cut out damaged 9.5mm plasterboard and two coat plaster to wall in patches and make out in new plasterboard and plaster to match existing										
under 1.00m2	4.07	10.00	**4.48**	2.23	2.23	**36.24**	0.89	**41.60**	m2	44.72
1.00 - 2.00m2	4.07	10.00	**4.48**	1.78	1.78	**28.93**	0.89	**34.29**	m2	36.86
2.00 - 4.00m2	4.07	10.00	**4.48**	1.36	1.36	**22.10**	0.89	**27.47**	m2	29.53
Cut out damaged 12.5mm plasterboard and two coat plaster to wall in patches and make out in new plasterboard and plaster to match existing										
under 1.00m2	4.40	10.00	**4.84**	2.28	2.28	**37.05**	0.99	**42.88**	m2	46.10
1.00 - 2.00m2	4.40	10.00	**4.84**	1.83	1.83	**29.74**	0.99	**35.57**	m2	38.24
2.00 - 4.00m2	4.40	10.00	**4.84**	1.38	1.38	**22.43**	0.99	**28.25**	m2	30.37
Cut out damaged 9.5mm plasterboard and two coat plaster to ceiling in patches and make out in new plasterboard and plaster to match existing										
under 1.00m2	4.10	10.00	**4.51**	2.54	2.54	**41.27**	0.89	**46.67**	m2	50.18
1.00 - 2.00m2	4.10	10.00	**4.51**	2.03	2.03	**32.99**	0.89	**38.39**	m2	41.27
2.00 - 4.00m2	4.10	10.00	**4.51**	1.55	1.55	**25.19**	0.89	**30.59**	m2	32.88
Cut out damaged 12.5mm plasterboard and two coat plaster to ceiling in patches and make out in new plasterboard and plaster to match existing										
under 1.00m2	4.40	10.00	**4.84**	2.56	2.56	**41.60**	0.99	**47.43**	m2	50.99
1.00 - 2.00m2	4.40	10.00	**4.84**	2.06	2.06	**33.48**	0.99	**39.31**	m2	42.25
2.00 - 4.00m2	4.40	10.00	**4.84**	1.57	1.57	**25.51**	0.99	**31.34**	m2	33.69
Cut out damaged lath and plaster to wall in patches and make out with new metal lathing and three coat plaster to match existing										
under 1.00m2	8.67	10.00	**9.54**	2.53	2.53	**41.11**	0.99	**51.64**	m2	55.51
1.00 - 2.00m2	8.67	10.00	**9.54**	2.14	2.14	**34.77**	0.99	**45.30**	m2	48.70
2.00 - 4.00m2	8.67	10.00	**9.54**	1.73	1.73	**28.11**	0.99	**38.64**	m2	41.54
Cut out damaged lath and plaster to ceiling in patches and make out with new metal lathing and three coat plaster to match existing										
under 1.00m2	8.67	10.00	**9.54**	2.87	2.87	**46.64**	0.99	**57.16**	m2	61.45
1.00 - 2.00m2	8.67	10.00	**9.54**	2.43	2.43	**39.49**	0.99	**50.01**	m2	53.77
2.00 - 4.00m2	8.67	10.00	**9.54**	1.95	1.95	**31.69**	0.99	**42.21**	m2	45.38
Cut out crack in plaster, form dovetailed key and make good with plaster										
not exceeding 50mm wide	0.11	10.00	**0.12**	0.20	0.20	**3.25**	0.03	**3.40**	m	3.66
Make good plaster where the following removed										
small pipe	-	-	**-**	0.16	0.16	**2.60**	0.03	**2.63**	nr	2.83
small steel section	-	-	**-**	0.21	0.21	**3.41**	0.04	**3.45**	nr	3.71
Cut out damaged plaster moulding or cornice and make out with new to match exisiting										
up to 100mm girth on face	0.69	10.00	**0.76**	1.25	0.63	**15.96**	0.21	**16.93**	m	18.20
100 - 150mm girth on face	1.14	10.00	**1.25**	1.86	0.93	**23.70**	0.41	**25.36**	m	27.26

Labour hourly rates: (except Specialists) Craft Operatives £9.23 Labour £7.02 Rates are national average prices. Refer to REGIONAL VARIATIONS for indicative levels of overall pricing in regions	MATERIALS			LABOUR				RATES		
	Del to Site £	Waste %	Material Cost £	Craft Optve Hrs	Lab Hrs	Labour Cost £	Sunds £	Nett Rate £	Unit	Gross Rate (+7.5%) £
REPAIRING/RENOVATING FINISHINGS Cont.										
Repairs to old wall and ceiling finishings Cont.										
Remove damaged fibrous plaster louvered ventilator, replace with new and make good										
225 x 75mm	1.26	10.00	1.39	0.50	0.50	8.13	0.10	9.61	nr	10.33
225 x 150mm	1.51	10.00	1.66	0.60	0.60	9.75	0.17	11.58	nr	12.45
Take out damaged 152 x 152 x 6mm white glazed wall tiles and renew to match existing										
isolated tile	0.45	5.00	0.47	0.32	0.32	5.20	0.06	5.73	nr	6.16
patch of 5 tiles	2.25	5.00	2.36	0.60	0.60	9.75	0.30	12.41	nr	13.34
patch 0.50 - 1.00m2	19.80	5.00	20.79	2.32	2.32	37.70	2.35	60.84	m2	65.40
patch 1.00 - 2.00m2	19.80	5.00	20.79	1.66	1.66	26.98	2.35	50.12	m2	53.87
Repairs to old floor finishings										
Take out damaged 150 x 150 x 12.5mm clay floor tiles and renew to match existing										
isolated tile	0.29	5.00	0.30	0.32	0.21	4.43	0.06	4.79	nr	5.15
patch of 5 tiles	1.45	5.00	1.52	0.65	0.45	9.16	0.30	10.98	nr	11.80
patch 0.50 - 1.00m2	13.05	5.00	13.70	2.40	1.61	33.45	2.35	49.51	m2	53.22
patch 1.00 - 2.00m2	13.05	5.00	13.70	1.90	1.27	26.45	2.35	42.50	m2	45.69
Cut out damaged 25mm granolithic paving in patches and make out in new granolithic to match existing										
under 1.00m2	6.26	5.00	6.57	1.50	1.50	24.38	1.10	32.05	m2	34.45
1.00 - 2.00m2	6.26	5.00	6.57	1.25	1.25	20.31	1.10	27.99	m2	30.08
2.00 - 4.00m2	6.26	5.00	6.57	1.07	1.07	17.39	1.10	25.06	m2	26.94
Cut out damaged 32mm granolithic paving in patches and make out in new granolithic to match existing										
under 1.00m2	9.02	5.00	9.47	1.68	1.68	27.30	1.29	38.06	m2	40.92
1.00 - 2.00m2	9.02	5.00	9.47	1.38	1.38	22.43	1.29	33.19	m2	35.67
2.00 - 4.00m2	9.02	5.00	9.47	1.15	1.15	18.69	1.29	29.45	m2	31.66
Cut out damaged 38mm granolithic paving in patches and make out in new granolithic to match existing										
under 1.00m2	9.50	5.00	9.97	1.80	1.80	29.25	1.52	40.74	m2	43.80
1.00 - 2.00m2	9.50	5.00	9.97	1.50	1.50	24.38	1.52	35.87	m2	38.56
2.00 - 4.00m2	9.50	5.00	9.97	1.25	1.25	20.31	1.52	31.81	m2	34.19
Cut out damaged 50mm granolithic paving in patches and make out in new granolithic to match existing										
under 1.00m2	12.51	5.00	13.14	2.12	2.12	34.45	1.85	49.44	m2	53.14
1.00 - 2.00m2	12.51	5.00	13.14	1.73	1.73	28.11	1.85	43.10	m2	46.33
2.00 - 4.00m2	12.51	5.00	13.14	1.45	1.45	23.56	1.85	38.55	m2	41.44
Cut out damaged granolithic paving to tread 275mm wide and make out in new granolithic to match existing										
25mm thick	1.74	5.00	1.83	0.44	0.44	7.15	0.31	9.29	m	9.98
32mm thick	2.19	5.00	2.30	0.48	0.48	7.80	0.35	10.45	m	11.23
38mm thick	2.63	5.00	2.76	0.50	0.50	8.13	0.41	11.30	m	12.14
50mm thick	3.50	5.00	3.67	0.59	0.59	9.59	0.51	13.77	m	14.81
Cut out damaged granolithic covering to plain riser 175mm wide and make out in new granolithic to match existing										
13mm thick	0.59	5.00	0.62	0.80	0.80	13.00	2.82	16.44	m	17.67
19mm thick	0.84	5.00	0.88	0.84	0.84	13.65	2.84	17.37	m	18.67
25mm thick	1.09	5.00	1.14	0.88	0.88	14.30	2.87	18.31	m	19.69
32mm thick	1.40	5.00	1.47	0.93	0.93	15.11	2.89	19.47	m	20.93
Cut out damaged granolithic covering to undercut riser 175mm wide and make out in new granolithic to match existing										
13mm thick	0.59	5.00	0.62	0.87	0.87	14.14	2.82	17.58	m	18.90
19mm thick	0.84	5.00	0.88	0.93	0.93	15.11	2.84	18.83	m	20.25
25mm thick	1.09	5.00	1.14	0.98	0.98	15.93	2.87	19.94	m	21.43
32mm thick	1.40	5.00	1.47	1.04	1.04	16.90	2.89	21.26	m	22.85
Cut out damaged granolithic skirting 150mm wide and make out in new granolithic to match existing										
13mm thick	0.50	5.00	0.53	0.72	0.72	11.70	2.41	14.64	m	15.73
19mm thick	0.72	5.00	0.76	0.76	0.76	12.35	2.44	15.55	m	16.71
25mm thick	0.95	5.00	1.00	0.80	0.80	13.00	2.46	16.46	m	17.69
32mm thick	1.19	5.00	1.25	0.82	0.82	13.32	2.49	17.06	m	18.34
Cut out crack in granolithic paving not exceeding 50mm wide, form dovetailed key and make good to match existing										
25mm thick	0.34	5.00	0.36	0.60	0.60	9.75	0.05	10.16	m	10.92
32mm thick	0.42	5.00	0.44	0.66	0.66	10.73	0.06	11.23	m	12.07
38mm thick	0.50	5.00	0.53	0.69	0.69	11.21	0.08	11.82	m	12.70
50mm thick	0.67	5.00	0.70	0.79	0.79	12.84	0.09	13.63	m	14.65
REPAIRING GLAZING										
Resecure glass										
Remove decayed putties, paint one coat on edge of rebate and re-putty										
wood window	0.13	10.00	0.14	0.69	-	6.37	0.02	6.53	m	7.02
metal window	0.13	10.00	0.14	0.76	-	7.01	0.02	7.18	m	7.72

Labour hourly rates: (except Specialists) Craft Operatives £9.23 Labour £7.02 Rates are national average prices. Refer to REGIONAL VARIATIONS for indicative levels of overall pricing in regions	MATERIALS			LABOUR				RATES		
	Del to Site £	Waste %	Material Cost £	Craft Optve Hrs	Lab Hrs	Labour Cost £	Sunds £	Nett Rate £	Unit	Gross Rate (+7.5%) £
REPAIRING GLAZING Cont.										
Resecure glass Cont.										
Remove beads, remove decayed bedding materials, paint one coat on edge of rebate, re-bed glass and re-fix beads										
wood window	0.13	10.00	0.14	1.22	-	11.26	0.02	11.42	m	12.28
metal window	0.13	10.00	0.14	1.32	-	12.18	0.02	12.35	m	13.27
REPAIRING/RENOVATING FENCING										
Repairs to softwood fencing										
Remove defective timber fence post and replace with new 127 x 102mm post; letting post 450mm into ground; backfilling around in concrete mix C20; securing ends of existing arris rails and gravel boards to new post										
to suit fencing 914mm high	11.33	-	11.33	2.04	2.04	33.15	0.41	44.89	nr	48.26
to suit fencing 1219mm high	12.17	-	12.17	2.06	2.06	33.48	0.47	46.12	nr	49.57
to suit fencing 1524mm high	13.01	-	13.01	2.08	2.08	33.80	0.55	47.36	nr	50.91
to suit fencing 1829mm high	13.86	-	13.86	2.10	2.10	34.13	0.64	48.63	nr	52.27
Remove defective pale from palisade fence and replace with new 19 x 75mm pointed pale; fixing with galvanised nails										
764mm long	0.72	-	0.72	0.55	0.09	5.71	0.18	6.61	nr	7.10
1069mm long	1.01	-	1.01	0.56	0.09	5.80	0.22	7.03	nr	7.56
1374mm long	1.30	-	1.30	0.58	0.10	6.06	0.25	7.61	nr	8.18
Remove defective pales from close boarded fence and replace with new 100mm wide feather edged pales; fixing with galvanised nails										
singly	1.63	-	1.63	0.25	0.04	2.59	0.21	4.43	m	4.76
areas up to 1.00m2	9.94	20.00	11.93	1.32	0.22	13.73	1.75	27.41	m2	29.46
areas over 1.00m2	9.94	20.00	11.93	0.98	0.16	10.17	1.75	23.85	m2	25.64
Remove defective gravel board and centre stump and replace with new 25 x 150mm gravel board and 50 x 50mm centre stump; fixing with galvanised nails										
securing to fence posts	2.44	5.00	2.56	0.53	0.09	5.52	0.21	8.30	m	8.92
Repairs to oak fencing										
Remove defective timber fence post and replace with new 127 x 102mm post; letting post 450mm into ground; backfilling around in concrete mix C20; securing ends of existing arris rails and gravel boards to new post										
to suit fencing 914mm high	20.70	-	20.70	2.17	2.17	35.26	0.41	56.37	nr	60.60
to suit fencing 1219mm high	24.16	-	24.16	2.21	2.21	35.91	0.47	60.54	nr	65.08
to suit fencing 1524mm high	27.62	-	27.62	2.25	2.25	36.56	0.55	64.73	nr	69.59
to suit fencing 1829mm high	31.08	-	31.08	2.29	2.29	37.21	0.64	68.93	nr	74.10
Remove defective pale from palisade fence and replace with new 19 x 75mm pointed pale; fixing with galvanised nails										
764mm long	1.73	-	1.73	0.70	0.12	7.30	0.18	9.21	nr	9.90
1069mm long	2.40	-	2.40	0.76	0.13	7.93	0.22	10.55	nr	11.34
1374mm long	3.10	-	3.10	0.83	0.14	8.64	0.25	11.99	nr	12.89
Remove defective pales from close boarded fence and replace with new 100mm wide feather edged pales; fixing with galvanised nails										
singly	3.88	-	3.88	0.27	0.05	2.84	0.21	6.93	m	7.45
areas up to 1.00m2	23.68	20.00	28.42	2.56	0.43	26.65	1.75	56.81	m2	61.07
areas over 1.00m2	23.68	20.00	28.42	2.07	0.35	21.56	1.75	51.73	m2	55.61
Remove defective gravel board and centre stump and replace with new 25 x 150mm gravel board and 50 x 50mm centre stump; fixing with galvanised nails										
securing to fence posts	5.78	5.00	6.07	1.10	0.18	11.42	0.21	17.70	m	19.02
Repairs to posts 100 x 100mm precast concrete spur 1219mm long; setting into ground; backfilling around in concrete mix C20										
bolting to existing timber post	12.44	-	12.44	0.76	0.76	12.35	1.66	26.45	nr	28.43
REPAIRING/RENOVATING EXTERNAL PAVINGS										
Repairs to precast concrete flag paving										
Take up uneven paving and set aside for re-use; level up and consolidate existing hardcore and relay salvaged paving										
bedding on 25mm sand bed; grouting in lime mortar (1:3)	0.75	10.00	0.82	0.67	0.93	12.71	0.12	13.66	m2	14.68
Repairs to York stone paving										
Take up old paving; re-square; relay random in random sizes										
bedding, jointing and pointing in cement mortar (1:3)	3.71	5.00	3.90	1.10	1.10	17.88	0.90	22.67	m2	24.37

EXISTING SITE/BUILDINGS/SERVICES – MAJOR WORKS

Labour hourly rates: (except Specialists) Craft Operatives £9.23 Labour £7.02 Rates are national average prices. Refer to REGIONAL VARIATIONS for indicative levels of overall pricing in regions	MATERIALS			LABOUR				RATES		
	Del to Site £	Waste %	Material Cost £	Craft Optve Hrs	Lab Hrs	Labour Cost £	Sunds £	Nett Rate £	Unit	Gross Rate (+7.5%) £
REPAIRING/RENOVATING EXTERNAL PAVINGS Cont.										
Repairs to granite sett paving										
Take up, clean and stack old paving	-	-	-	-	0.90	**6.32**	-	**6.32**	m2	**6.79**
Take up old paving, clean and re-lay bedding and grouting in cement mortar (1:3)	-	-	-	1.65	2.55	**33.13**	1.20	**34.33**	m2	**36.91**
Clean out joints of old paving and re-grout in cement mortar (1:3)	-	-	-	-	0.55	**3.86**	0.52	**4.38**	m2	**4.71**

Groundwork

Labour hourly rates: (except Specialists) Craft Operatives £9.23 Labour £7.02 Rates are national average prices. Refer to REGIONAL VARIATIONS for indicative levels of overall pricing in regions	PLANT & TRANSPORT			LABOUR				RATES		
	Plant Cost	Trans Cost	P & T Cost	Craft Optve	Lab	Labour Cost	Sunds	Nett Rate	Unit	Gross Rate (+7.5%)
	£	£	£	Hrs	Hrs	£	£	£		£
D20: EXCAVATING AND FILLING										
Site preparation										
Removing trees; filling voids left by removal of roots with selected material arising from excavation										
girth 0.60 - .50m	60.00	-	60.00	-	10.00	70.20	-	130.20	nr	139.97
girth 1.50 - 3.00m	110.00	-	110.00	-	20.00	140.40	-	250.40	nr	269.18
girth exceeding 3.00m	160.00	-	160.00	-	30.00	210.60	-	370.60	nr	398.39
Clearing site vegetation; filling voids left by removal of roots with selected material arising from excavation										
bushes, scrub, undergrowth, hedges, trees and tree stumps not exceeding 600mm girth.	0.55	-	0.55	-	-	-	-	0.55	m2	0.59
Lifting turf for preservation										
stacking on site average 100m distant for future use; watering	-	-	-	-	0.55	3.86	-	3.86	m2	4.15
Excavating - by machine										
Top soil for preservation										
average 150mm deep	0.25	-	0.25	-	-	-	-	0.25	m2	0.27
To reduce levels maximum depth not exceeding										
0.25m	1.90	-	1.90	-	-	-	-	1.90	m3	2.04
1.00m	1.00	-	1.00	-	-	-	-	1.00	m3	1.08
2.00m	1.15	-	1.15	-	-	-	-	1.15	m3	1.24
4.00m	1.30	-	1.30	-	-	-	-	1.30	m3	1.40
6.00m	1.60	-	1.60	-	-	-	-	1.60	m3	1.72
Basements and the like maximum depth not exceeding										
0.25m	1.90	-	1.90	-	0.39	2.74	-	4.64	m3	4.99
1.00m	1.20	-	1.20	-	0.41	2.88	-	4.08	m3	4.38
2.00m	1.40	-	1.40	-	0.47	3.30	-	4.70	m3	5.05
4.00m	2.00	-	2.00	-	0.60	4.21	-	6.21	m3	6.68
6.00m	2.80	-	2.80	-	0.74	5.19	-	7.99	m3	8.59
Trenches width not exceeding 0.30m maximum depth not exceeding										
0.25m	4.90	-	4.90	-	1.71	12.00	-	16.90	m3	18.17
1.00m	5.10	-	5.10	-	1.80	12.64	-	17.74	m3	19.07
Trenches width exceeding 0.30m maximum depth not exceeding										
0.25m	4.90	-	4.90	-	0.57	4.00	-	8.90	m3	9.57
1.00m	4.20	-	4.20	-	0.60	4.21	-	8.41	m3	9.04
2.00m	4.20	-	4.20	-	0.70	4.91	-	9.11	m3	9.80
4.00m	5.00	-	5.00	-	0.87	6.11	-	11.11	m3	11.94
6.00m	5.00	-	5.00	-	1.05	7.37	-	12.37	m3	13.30
commencing 1.50m below existing ground level, maximum depth not exceeding										
0.25m	4.20	-	4.20	-	0.85	5.97	-	10.17	m3	10.93
1.00m	4.20	-	4.20	-	0.90	6.32	-	10.52	m3	11.31
2.00m	4.20	-	4.20	-	1.13	7.93	-	12.13	m3	13.04
commencing 3.00m below existing ground level, maximum depth not exceeding										
0.25m	4.20	-	4.20	-	1.27	8.92	-	13.12	m3	14.10
1.00m	5.00	-	5.00	-	1.33	9.34	-	14.34	m3	15.41
2.00m	5.00	-	5.00	-	1.51	10.60	-	15.60	m3	16.77
For pile caps and ground beams between piles maximum depth not exceeding										
0.25m	5.10	-	5.10	-	0.84	5.90	-	11.00	m3	11.82
1.00m	5.00	-	5.00	-	0.88	6.18	-	11.18	m3	12.02
2.00m	5.00	-	5.00	-	1.00	7.02	-	12.02	m3	12.92

Labour hourly rates: (except Specialists) Craft Operatives £9.23 Labour £7.02 Rates are national average prices. Refer to REGIONAL VARIATIONS for indicative levels of overall pricing in regions	PLANT & TRANSPORT			LABOUR				RATES		
	Plant Cost	Trans Cost	P & T Cost	Craft Optve	Lab	Labour Cost	Sunds	Nett Rate	Unit	Gross Rate (+7.5%)
	£	£	£	Hrs	Hrs	£	£	£		£
D20: EXCAVATING AND FILLING Cont.										
Excavating - by machine Cont.										
Pits maximum depth not exceeding										
0.25m	4.30	-	4.30	-	0.84	5.90	-	10.20	m3	10.96
1.00m	5.30	-	5.30	-	0.88	6.18	-	11.48	m3	12.34
2.00m.	5.30	-	5.30	-	1.00	7.02	-	12.32	m3	13.24
4.00m	6.00	-	6.00	-	1.27	8.92	-	14.92	m3	16.03
Extra over any types of excavating irrespective of depth										
excavating below ground water level	4.32	-	4.32	-	-	-	-	4.32	m3	4.64
excavating in running silt, running sand or liquid mud	7.42	-	7.42	-	-	-	-	7.42	m3	7.98
excavating below ground water level in running silt, running sand or liquid mud	10.23	-	10.23	-	-	-	-	10.23	m3	11.00
excavating in heavy soil or clay	0.94	-	0.94	-	-	-	-	0.94	m3	1.01
excavating in gravel	1.48	-	1.48	-	-	-	-	1.48	m3	1.59
excavating in brash, loose rock or chalk	1.87	-	1.87	-	-	-	-	1.87	m3	2.01
Breaking out existing materials; extra over any types of excavating irrespective of depth										
sandstone	14.64	-	14.64	-	-	-	-	14.64	m3	15.74
hard rock	22.05	-	22.05	-	-	-	-	22.05	m3	23.70
concrete	18.31	-	18.31	-	-	-	-	18.31	m3	19.68
reinforced concrete	27.50	-	27.50	-	-	-	-	27.50	m3	29.56
brickwork, blockwork or stonework	12.22	-	12.22	-	-	-	-	12.22	m3	13.14
drain and concrete bed under										
100mm diameter	1.72	-	1.72	-	-	-	-	1.72	m	1.85
150mm diameter	2.06	-	2.06	-	-	-	-	2.06	m	2.21
225mm diameter	2.40	-	2.40	-	-	-	-	2.40	m	2.58
Breaking out existing hard pavings; extra over any types of excavating irrespective of depth										
concrete 150mm thick	2.89	-	2.89	-	-	-	-	2.89	m2	3.11
reinforced concrete 200mm thick	5.79	-	5.79	-	-	-	-	5.79	m2	6.22
coated macadam or asphalt 75mm thick	0.50	-	0.50	-	-	-	-	0.50	m2	0.54
Working space allowance to excavation; including additional earthwork support, disposal (excluding landfill tax) and backfilling										
reduce levels, basements and the like	4.40	-	4.40	-	0.28	1.97	-	6.37	m2	6.84
pits	5.00	-	5.00	-	0.60	4.21	-	9.21	m2	9.90
trenches	4.40	-	4.40	-	0.42	2.95	-	7.35	m2	7.90
Compacting bottoms of excavations	0.15	-	0.15	-	-	-	-	0.15	m2	0.16
Compacting with 680 kg vibratory roller ground	0.17	-	0.17	-	-	-	-	0.17	m2	0.18
Compacting with 6-8 tonnes smooth wheeled roller ground	0.34	-	0.34	-	-	-	-	0.34	m2	0.37
Trimming										
sides of cuttings; vertical or battered	0.41	-	0.41	-	-	-	-	0.41	m2	0.44
sides of embankments; vertical or battered	0.41	-	0.41	-	-	-	-	0.41	m2	0.44
Disposal of excavated material depositing on site in temporary spoil heaps where directed										
25m	-	1.61	1.61	-	0.10	0.70	-	2.31	m3	2.49
50m	-	1.69	1.69	-	0.10	0.70	-	2.39	m3	2.57
100m	-	1.84	1.84	-	0.10	0.70	-	2.54	m3	2.73
400m	-	2.18	2.18	-	0.10	0.70	-	2.88	m3	3.10
800m	-	2.84	2.84	-	0.10	0.70	-	3.54	m3	3.81
1200m	-	3.02	3.02	-	0.10	0.70	-	3.72	m3	4.00
1600m	-	3.21	3.21	-	0.10	0.70	-	3.91	m3	4.21
extra for each additional 1600m	-	0.50	0.50	-	-	-	-	0.50	m3	0.54
removing from site to tip a) Inert	-	13.00	13.00	-	-	-	-	13.00	m3	13.98
removing from site to tip b) Active	-	22.00	22.00	-	-	-	-	22.00	m3	23.65
removing from site to tip c) Contaminated (Guide price - always seek a quotation for specialist disposal costs.)	-	50.00	50.00	-	-	-	-	50.00	m3	53.75
Disposal of preserved top soil depositing on site in temporary spoil heaps where directed										
25m	-	1.61	1.61	-	0.10	0.70	-	2.31	m3	2.49
50m	-	1.69	1.69	-	0.10	0.70	-	2.39	m3	2.57
100m	-	1.84	1.84	-	0.10	0.70	-	2.54	m3	2.73
Excavating - by hand										
Top soil for preservation average 150mm deep	-	-	-	-	0.30	2.11	-	2.11	m2	2.26
To reduce levels maximum depth not exceeding										
0.25m	-	-	-	-	2.50	17.55	-	17.55	m3	18.87
1.00m	-	-	-	-	2.65	18.60	-	18.60	m3	20.00
2.00m	-	-	-	-	3.00	21.06	-	21.06	m3	22.64
4.00m	-	-	-	-	3.90	27.38	-	27.38	m3	29.43
6.00m	-	-	-	-	4.85	34.05	-	34.05	m3	36.60

Labour hourly rates: (except Specialists) Craft Operatives £9.23 Labour £7.02 Rates are national average prices. Refer to REGIONAL VARIATIONS for indicative levels of overall pricing in regions	PLANT & TRANSPORT			LABOUR				RATES		
	Plant' Cost	Trans Cost	P & T Cost	Craft Optve	Lab	Labour Cost	Sunds	Nett Rate	Unit	Gross Rate (+7.5%)
	£	£	£	Hrs	Hrs	£	£	£		£
D20: EXCAVATING AND FILLING Cont.										
Excavating - by hand Cont.										
Basements and the like										
maximum depth not exceeding										
0.25m.	-	-	-	-	2.60	18.25	-	18.25	m3	19.62
1.00m	-	-	-	-	2.75	19.31	-	19.31	m3	20.75
2.00m	-	-	-	-	3.10	21.76	-	21.76	m3	23.39
4.00m	-	-	-	-	4.00	28.08	-	28.08	m3	30.19
6.00m	-	-	-	-	4.85	34.05	-	34.05	m3	36.60
Trenches width not exceeding 0.30m										
maximum depth not exceeding										
0.25m	-	-	-	-	4.25	29.84	-	29.84	m3	32.07
1.00m	-	-	-	-	4.50	31.59	-	31.59	m3	33.96
Trenches width exceeding 0.30m										
maximum depth not exceeding										
0.25m	-	-	-	-	2.85	20.01	-	20.01	m3	21.51
1.00m	-	-	-	-	3.00	21.06	-	21.06	m3	22.64
2.00m	-	-	-	-	3.50	24.57	-	24.57	m3	26.41
4.00m	-	-	-	-	4.35	30.54	-	30.54	m3	32.83
6.00m	-	-	-	-	5.50	38.61	-	38.61	m3	41.51
commencing 1.50m below existing ground level,										
maximum depth not exceeding										
0.25m	-	-	-	-	4.25	29.84	-	29.84	m3	32.07
1.00m	-	-	-	-	4.50	31.59	-	31.59	m3	33.96
2.00m	-	-	-	-	5.65	39.66	-	39.66	m3	42.64
commencing 3.00m below existing ground level,										
maximum depth not exceeding										
0.25m	-	-	-	-	6.35	44.58	-	44.58	m3	47.92
1.00m.	-	-	-	-	6.65	46.68	-	46.68	m3	50.18
2.00m	-	-	-	-	7.65	53.70	-	53.70	m3	57.73
For pile caps and ground beams between piles										
maximum depth not exceeding										
0.25m.	-	-	-	-	3.35	23.52	-	23.52	m3	25.28
1.00m	-	-	-	-	3.50	24.57	-	24.57	m3	26.41
2.00m	-	-	-	-	4.00	28.08	-	28.08	m3	30.19
Pits										
maximum depth not exceeding										
0.25m	-	-	-	-	3.35	23.52	-	23.52	m3	25.28
1.00m	-	-	-	-	3.50	24.57	-	24.57	m3	26.41
2.00m	-	-	-	-	4.00	28.08	-	28.08	m3	30.19
4.00m	-	-	-	-	5.10	35.80	-	35.80	m3	38.49
Extra over any types of excavating irrespective of depth										
excavating below ground water level	-	-	-	-	1.50	10.53	-	10.53	m3	11.32
excavating in running silt, running sand or liquid mud	-	-	-	-	4.50	31.59	-	31.59	m3	33.96
excavating below ground water level in running silt, running sand or liquid mud	-	-	-	-	6.00	42.12	-	42.12	m3	45.28
excavating in heavy soil or clay	-	-	-	-	0.65	4.56	-	4.56	m3	4.91
excavating in gravel	-	-	-	-	1.65	11.58	-	11.58	m3	12.45
excavating in brash, loose rock or chalk	-	-	-	-	3.35	23.52	-	23.52	m3	25.28
Breaking out existing materials; extra over any types of excavating irrespective of depth										
sandstone	-	-	-	-	6.65	46.68	-	46.68	m3	50.18
hard rock	-	-	-	-	13.35	93.72	-	93.72	m3	100.75
concrete	-	-	-	-	10.00	70.20	-	70.20	m3	75.47
reinforced concrete	-	-	-	-	15.00	105.30	-	105.30	m3	113.20
brickwork, blockwork or stonework	-	-	-	-	7.00	49.14	-	49.14	m3	52.83
drain and concrete bed under										
100mm diameter	-	-	-	-	0.80	5.62	-	5.62	m	6.04
150mm diameter	-	-	-	-	1.00	7.02	-	7.02	m	7.55
225mm diameter	-	-	-	-	1.15	8.07	-	8.07	m	8.68
Breaking out existing hard pavings; extra over any types of excavating irrespective of depth										
concrete 150mm thick	-	-	-	-	1.50	10.53	-	10.53	m2	11.32
reinforced concrete 200mm thick	-	-	-	-	3.00	21.06	-	21.06	m2	22.64
coated macadam or asphalt 75mm thick	-	-	-	-	0.25	1.75	-	1.75	m2	1.89
Working space allowance to excavation; including additional earthwork support, disposal (excluding landfill tax) and backfilling										
reduce levels, basements and the like	-	-	-	-	2.60	18.25	-	18.25	m2	19.62
pits	-	-	-	-	3.12	21.90	-	21.90	m2	23.55
trenches	-	-	-	-	2.80	19.66	-	19.66	m2	21.13
Compacting										
bottoms of excavations	-	-	-	-	0.12	0.84	-	0.84	m2	0.91
Trimming										
sides of cuttings; vertical or battered	Plant' Cost	Trans Cost	P & T Cost	Craft Optve	0.20	1.40	-	1.40	m2	1.51
sides of embankments; vertical or battered	-	-	-	-	0.20	1.40	Sunds	1.40	m2	1.51
Disposal of excavated material										
depositing on site in temporary spoil heaps where directed										
25m	-	-	-	-	1.25	8.78	-	8.78	m3	9.43
50m	-	-	-	-	1.50	10.53	-	10.53	m3	11.32

GROUNDWORK

Labour hourly rates: (except Specialists) Craft Operatives £9.23 Labour £7.02 Rates are national average prices. Refer to REGIONAL VARIATIONS for indicative levels of overall pricing in regions	PLANT & TRANSPORT			LABOUR				RATES		
	Plant Cost £	Trans Cost £	P & T Cost £	Craft Optve Hrs	Lab Hrs	Labour Cost £	Sunds £	Nett Rate £	Unit	Gross Rate (+7.5%) £
D20: EXCAVATING AND FILLING Cont.										
Excavating - by hand Cont.										
Disposal of excavated material Cont.										
depositing on site in temporary spoil heaps where directed Cont.										
100m	-	-	-	-	2.00	14.04	-	14.04	m3	15.09
400m	-	2.18	2.18	-	1.50	10.53	-	12.71	m3	13.66
800m	-	2.84	2.84	-	1.50	10.53	-	13.37	m3	14.37
1200m	-	3.02	3.02	-	1.50	10.53	-	13.55	m3	14.57
1600m	-	3.21	3.21	-	1.50	10.53	-	13.74	m3	14.77
extra for each additional 1600m										
25m	-	0.50	0.50	-	-	-	-	0.50	m3	0.54
removing from site to tip a) Inert	-	13.00	13.00	-	1.35	9.48	-	22.48	m3	24.16
removing from site to tip b) Active	-	22.00	22.00	-	1.35	9.48	-	31.48	m3	33.84
removing from site to tip c) Contaminated (Guide price - always seek a quotation for specialist disposal costs.)	-	50.00	50.00	-	1.35	9.48	-	59.48	m3	63.94
Disposal of preserved top soil										
depositing on site in temporary spoil heaps where directed										
25m	-	-	-	-	1.25	8.78	-	8.78	m3	9.43
50m	-	-	-	-	1.50	10.53	-	10.53	m3	11.32
100m	-	-	-	-	2.00	14.04	-	14.04	m3	15.09
Excavating inside existing building - by hand										
To reduce levels										
maximum depth not exceeding										
0.25m	-	-	-	-	3.50	24.57	-	24.57	m3	26.41
1.00m	-	-	-	-	3.75	26.32	-	26.32	m3	28.30
2.00m	-	-	-	-	4.20	29.48	-	29.48	m3	31.70
4.00m	-	-	-	-	5.40	37.91	-	37.91	m3	40.75
Trenches width not exceeding 0.30m										
maximum depth not exceeding										
0.25m	-	-	-	-	6.00	42.12	-	42.12	m3	45.28
1.00m	-	-	-	-	6.25	43.88	-	43.88	m3	47.17
Trenches width exceeding 0.30m										
maximum depth not exceeding										
0.25m	-	-	-	-	4.00	28.08	-	28.08	m3	30.19
1.00m	-	-	-	-	4.25	29.84	-	29.84	m3	32.07
2.00m	-	-	-	-	5.00	35.10	-	35.10	m3	37.73
4.00m	-	-	-	-	6.15	43.17	-	43.17	m3	46.41
Pits										
maximum depth not exceeding										
0.25m	-	-	-	-	4.75	33.34	-	33.34	m3	35.85
1.00m	-	-	-	-	4.95	34.75	-	34.75	m3	37.36
2.00m	-	-	-	-	5.65	39.66	-	39.66	m3	42.64
4.00m	-	-	-	-	7.15	50.19	-	50.19	m3	53.96
Extra over any types of excavating irrespective of depth										
excavating below ground water level	-	-	-	-	2.15	15.09	-	15.09	m3	16.22
Breaking out existing materials; extra over any types of excavating irrespective of depth										
concrete	-	-	-	-	11.00	77.22	-	77.22	m3	83.01
reinforced concrete	-	-	-	-	15.50	108.81	-	108.81	m3	116.97
brickwork, blockwork or stonework	-	-	-	-	7.00	49.14	-	49.14	m3	52.83
drain and concrete bed under										
100mm diameter	-	-	-	-	1.15	8.07	-	8.07	m	8.68
150mm diameter	-	-	-	-	1.35	9.48	-	9.48	m	10.19
Breaking out existing hard pavings; extra over any types of excavating irrespective of depth										
concrete 150mm thick	-	-	-	-	1.50	10.53	-	10.53	m2	11.32
reinforced concrete 200mm thick	-	-	-	-	3.00	21.06	-	21.06	m2	22.64
Working space allowance to excavation; including additional earthwork support, disposal (excluding landfill tax) and backfilling										
pits	-	-	-	-	4.00	28.08	-	28.08	m2	30.19
trenches	-	-	-	-	3.70	25.97	-	25.97	m2	27.92
Compacting										
bottoms of excavations	-	-	-	-	0.14	0.98	-	0.98	m2	1.06
Disposal of excavated material										
depositing on site in temporary spoil heaps where directed										
25m	-	-	-	-	1.75	12.29	-	12.29	m3	13.21
50m	-	-	-	-	1.95	13.69	-	13.69	m3	14.72
removing from site to tip a) Inert	-	13.00	13.00	-	2.25	15.80	-	28.80	m3	30.95
removing from site to tip b) Active	-	22.00	22.00	-	2.25	15.80	-	37.80	m3	40.63
removing from site to tip c) Contaminated (Guide price - always seek a quotation for specialist disposal costs.)	-	50.00	50.00	-	2.25	15.80	-	65.80	m3	70.73

Labour hourly rates: (except Specialists) Craft Operatives £9.23 Labour £7.02 Rates are national average prices. Refer to REGIONAL VARIATIONS for indicative levels of overall pricing in regions	MATERIALS			LABOUR				RATES		
	Del to Site £	Waste %	Material Cost £	Craft Optve Hrs	Lab Hrs	Labour Cost £	Sunds £	Nett Rate £	Unit	Gross Rate (+7.5%) £

D20: EXCAVATING AND FILLING Cont.

Earthwork support

	Del to Site	Waste	Material Cost	Craft Optve	Lab	Labour Cost	Sunds	Nett Rate	Unit	Gross Rate
Earthwork support distance between opposing faces not exceeding 2.00m; maximum depth not exceeding										
1.00m	0.58	-	0.58	-	0.20	1.40	0.02	2.00	m2	2.15
2.00m	0.76	-	0.76	-	0.25	1.75	0.03	2.54	m2	2.74
4.00m	0.90	-	0.90	-	0.30	2.11	0.03	3.04	m2	3.26
6.00m	0.99	-	0.99	-	0.35	2.46	0.04	3.49	m2	3.75
distance between opposing faces 2.00 - 4.00m; maximum depth not exceeding										
1.00m.	0.68	-	0.68	-	0.21	1.47	0.02	2.17	m2	2.34
2.00m	0.84	-	0.84	-	0.26	1.83	0.03	2.70	m2	2.90
4.00m	0.96	-	0.96	-	0.32	2.25	0.03	3.24	m2	3.48
6.00m	1.12	-	1.12	-	0.37	2.60	0.04	3.76	m2	4.04
distance between opposing faces exceeding 4.00m; maximum depth not exceeding										
1.00m	0.76	-	0.76	-	0.22	1.54	0.03	2.33	m2	2.51
2.00m	0.91	-	0.91	-	0.28	1.97	0.03	2.91	m2	3.12
4.00m	1.05	-	1.05	-	0.33	2.32	0.04	3.41	m2	3.66
6.00m	1.21	-	1.21	-	0.39	2.74	0.04	3.99	m2	4.29
Earthwork support; unstable ground distance between opposing faces not exceeding 2.00m; maximum depth not exceeding										
1.00m.	1.05	-	1.05	-	0.33	2.32	0.04	3.41	m2	3.66
2.00m	1.27	-	1.27	-	0.42	2.95	0.04	4.26	m2	4.58
4.00m	1.49	-	1.49	-	0.50	3.51	0.05	5.05	m2	5.43
6.00m	1.67	-	1.67	-	0.59	4.14	0.06	5.87	m2	6.31
distance between opposing faces 2.00 - 4.00m; maximum depth not exceeding										
1.00m	1.18	-	1.18	-	0.35	2.46	0.04	3.68	m2	3.95
2.00m	1.40	-	1.40	-	0.43	3.02	0.05	4.47	m2	4.80
4.00m	1.61	-	1.61	-	0.53	3.72	0.06	5.39	m2	5.79
6.00m	1.86	-	1.86	-	0.62	4.35	0.06	6.27	m2	6.74
distance between opposing faces exceeding 4.00m; maximum depth not exceeding										
1.00m	1.27	-	1.27	-	0.37	2.60	0.04	3.91	m2	4.20
2.00m	1.52	-	1.52	-	0.47	3.30	0.05	4.87	m2	5.23
4.00m	1.78	-	1.78	-	0.55	3.86	0.06	5.70	m2	6.13
6.00m	2.04	-	2.04	-	0.65	4.56	0.07	6.67	m2	7.17
Earthwork support; next to roadways distance between opposing faces not exceeding 2.00m; maximum depth not exceeding										
1.00m	1.24	-	1.24	-	0.40	2.81	0.04	4.09	m2	4.39
2.00m	1.52	-	1.52	-	0.50	3.51	0.05	5.08	m2	5.46
4.00m	1.78	-	1.78	-	0.60	4.21	0.06	6.05	m2	6.51
6.00m	2.04	-	2.04	-	0.70	4.91	0.07	7.02	m2	7.55
distance between opposing faces 2.00 - 4.00m; maximum depth not exceeding										
1.00m	1.40	-	1.40	-	0.42	2.95	0.05	4.40	m2	4.73
2.00m	1.66	-	1.66	-	0.52	3.65	0.06	5.37	m2	5.77
4.00m	1.95	-	1.95	-	0.64	4.49	0.07	6.51	m2	7.00
6.00m.	2.22	-	2.22	-	0.74	5.19	0.08	7.49	m2	8.06
distance between opposing faces exceeding 4.00m; maximum depth not exceeding										
1.00m	1.49	-	1.49	-	0.44	3.09	0.05	4.63	m2	4.98
2.00m	1.84	-	1.84	-	0.56	3.93	0.06	5.83	m2	6.27
4.00m	2.14	-	2.14	-	0.66	4.63	0.07	6.84	m2	7.36
6.00m	2.44	-	2.44	-	0.78	5.48	0.08	8.00	m2	8.60
Earthwork support; left in distance between opposing faces not exceeding 2.00m; maximum depth not exceeding										
1.00m	2.97	-	2.97	-	0.16	1.12	0.04	4.13	m2	4.44
2.00m	2.97	-	2.97	-	0.20	1.40	0.05	4.42	m2	4.76
4.00m	2.97	-	2.97	-	0.24	1.68	0.05	4.70	m2	5.06
6.00m	2.97	-	2.97	-	0.28	1.97	0.06	5.00	m2	5.37
distance between opposing faces 2.00 - 4.00m; maximum depth not exceeding										
1.00m	2.97	-	2.97	-	0.17	1.19	0.04	4.20	m2	4.52
2.00m	2.97	-	2.97	-	0.21	1.47	0.05	4.49	m2	4.83
4.00m	2.97	-	2.97	-	0.26	1.83	0.06	4.86	m2	5.22
6.00m	2.97	-	2.97	-	0.30	2.11	0.07	5.15	m2	5.53
distance between opposing faces exceeding 4.00m; maximum depth not exceeding										
1.00m	3.96	-	3.96	-	0.18	1.26	0.05	5.27	m2	5.67
2.00m	3.96	-	3.96	-	0.22	1.54	0.05	5.55	m2	5.97
4.00m	3.96	-	3.96	-	0.26	1.83	0.06	5.85	m2	6.28
6.00m	3.96	-	3.96	-	0.31	2.18	0.07	6.21	m2	6.67
Earthwork support; next to roadways; left in distance between opposing faces not exceeding 2.00m; maximum depth not exceeding										
1.00m	9.90	-	9.90	-	0.32	2.25	0.08	12.23	m2	13.14
2.00m	9.90	-	9.90	-	0.40	2.81	0.09	12.80	m2	13.76
4.00m	9.90	-	9.90	Optve	0.49	3.44	0.11	13.45	m2	14.46
6.00m	9.90	%	9.90	-	0.56	3.93	0.12	13.95	m2	(+7.5%) 15.00
distance between opposing faces 2.00 - 4.00m; maximum depth not exceeding										
1.00m	11.88	-	11.88	-	0.34	2.39	0.08	14.35	m2	15.42
2.00m	11.88	-	11.88	-	0.42	2.95	0.10	14.93	m2	16.05
4.00m	11.88	-	11.88	-	0.51	3.58	0.12	15.58	m2	16.75

Labour hourly rates: (except Specialists) Craft Operatives £9.23 Labour £7.02 Rates are national average prices. Refer to REGIONAL VARIATIONS for indicative levels of overall pricing in regions	MATERIALS			LABOUR				RATES		
	Del to Site £	Waste %	Material Cost £	Craft Optve Hrs	Lab Hrs	Labour Cost £	Sunds £	Nett Rate £	Unit	Gross Rate (+7.5%) £
D20: EXCAVATING AND FILLING Cont.										
Earthwork support Cont.										
Earthwork support; next to roadways; left in Cont. distance between opposing faces 2.00 - 4.00m; maximum depth not exceeding Cont.										
6.00m	11.88	-	11.88	-	0.59	4.14	0.13	16.15	m2	17.36
distance between opposing faces exceeding 4.00m; maximum depth not exceeding										
1.00m	14.85	-	14.85	-	0.35	2.46	0.09	17.40	m2	18.70
2.00m	14.85	-	14.85	-	0.45	3.16	0.11	18.12	m2	19.48
4.00m	14.85	-	14.85	-	0.53	3.72	0.13	18.70	m2	20.10
6.00m	14.85	-	14.85	-	0.62	4.35	0.15	19.35	m2	20.80
Earthwork support; unstable ground; left in distance between opposing faces not exceeding 2.00m; maximum depth not exceeding										
1.00m	6.93	-	6.93	-	0.26	1.83	0.06	8.82	m2	9.48
2.00m	6.93	-	6.93	-	0.34	2.39	0.08	9.40	m2	10.10
4.00m	6.93	-	6.93	-	0.40	2.81	0.09	9.83	m2	10.57
6.00m	6.93	-	6.93	-	0.47	3.30	0.10	10.33	m2	11.10
distance between opposing faces 2.00 - 4.00m; maximum depth not exceeding										
1.00m	7.92	-	7.92	-	0.28	1.97	0.07	9.96	m2	10.70
2.00m	7.92	-	7.92	-	0.34	2.39	0.08	10.39	m2	11.17
4.00m	7.92	-	7.92	-	0.42	2.95	0.09	10.96	m2	11.78
6.00m	7.92	-	7.92	-	0.50	3.51	0.11	11.54	m2	12.41
distance between opposing faces exceeding 4.00m; maximum depth not exceeding										
1.00m	9.90	-	9.90	-	0.30	2.11	0.08	12.09	m2	12.99
2.00m	9.90	-	9.90	-	0.38	2.67	0.09	12.66	m2	13.61
4.00m	9.90	-	9.90	-	0.44	3.09	0.11	13.10	m2	14.08
6.00m	9.90	-	9.90	-	0.52	3.65	0.12	13.67	m2	14.70
Earthwork support; inside existing building distance between opposing faces not exceeding 2.00m; maximum depth not exceeding										
1.00m	0.65	-	0.65	-	0.22	1.54	0.02	2.21	m2	2.38
2.00m	0.82	-	0.82	-	0.28	1.97	0.03	2.82	m2	3.03
4.00m	0.93	-	0.93	-	0.30	2.11	0.03	3.07	m2	3.30
distance between opposing faces 2.00 - 4.00m; maximum depth not exceeding										
1.00m	0.93	-	0.93	-	0.23	1.61	0.02	2.56	m2	2.76
2.00m	0.90	-	0.90	-	0.29	2.04	0.03	2.97	m2	3.19
4.00m	0.99	-	0.99	-	0.35	2.46	0.04	3.49	m2	3.75
distance between opposing faces exceeding 4.00m; maximum depth not exceeding										
1.00m	0.81	-	0.81	-	0.24	1.68	0.03	2.52	m2	2.71
2.00m	0.94	-	0.94	-	0.31	2.18	0.03	3.15	m2	3.38
4.00m	1.13	-	1.13	-	0.36	2.53	0.04	3.70	m2	3.97
Earthwork support; unstable ground; inside existing building distance between opposing faces not exceeding 2.00m; maximum depth not exceeding										
1.00m.	1.13	-	1.13	-	0.36	2.53	0.04	3.70	m2	3.97
2.00m	1.34	-	1.34	-	0.46	3.23	0.05	4.62	m2	4.97
4.00m	1.56	-	1.56	-	0.55	3.86	0.05	5.47	m2	5.88
distance between opposing faces 2.00 - 4.00m; maximum depth not exceeding										
1.00m	1.22	-	1.22	-	0.39	2.74	0.04	4.00	m2	4.30
2.00m	1.48	-	1.48	-	0.47	3.30	0.05	4.83	m2	5.19
4.00m	1.73	-	1.73	-	0.58	4.07	0.06	5.86	m2	6.30
distance between opposing faces exceeding 4.00m; maximum depth not exceeding										
1.00m	1.34	-	1.34	-	0.41	2.88	0.05	4.27	m2	4.59
2.00m	1.59	-	1.59	-	0.52	3.65	0.05	5.29	m2	5.69
4.00m	1.85	-	1.85	-	0.61	4.28	0.06	6.19	m2	6.66
Excavated material arising from excavation - by machine										
Filling to excavation average thickness exceeding 0.25m	-	-	-	-	-	-	2.00	2.00	m3	2.15
Compacting filling	-	-	-	-	-	-	0.20	0.20	m2	0.22
Selected excavated material obtained from on site spoil heaps 25m distant - by machine										
Filling to make up levels; depositing in layers 150mm maximum thickness										
average thickness not exceeding 0.25m	-	-	-	-	-	-	2.90	2.90	m3	3.12
average thickness exceeding 0.25m	-	-	-	-	-	-	2.40	2.40	m3	2.58
Compacting with 680 kg vibratory roller filling	-	-	-	-	-	-	0.18	0.18	m2	0.19
Compacting with 6 - 8 tonnes smooth wheeled roller filling	-	%	-	Hrs	Hrs	-	0.35	0.35	m2	0.38
Trimming sides of cuttings; vertical or battered	-	-	-	-	-	-	0.90	0.90	m2	0.97
sides of embankments; vertical or battered	-	-	-	-	-	-	0.90	0.90	m2	0.97

Labour hourly rates: (except Specialists) Craft Operatives £9.23 Labour £7.02 Rates are national average prices. Refer to REGIONAL VARIATIONS for indicative levels of overall pricing in regions	MATERIALS			LABOUR				RATES		
	Del to Site £	Waste %	Material Cost £	Craft Optve Hrs	Lab Hrs	Labour Cost £	Sunds £	Nett Rate £	Unit	Gross Rate (+7.5%) £
D20: EXCAVATING AND FILLING Cont.										
Selected excavated material obtained from on site spoil heaps 50m distant – by machine										
Filling to make up levels; depositing in layers 150mm maximum thickness										
average thickness not exceeding 0.25m	-	-	-	-	-	-	3.70	**3.70**	m3	3.98
average thickness exceeding 0.25m	-	-	-	-	-	-	2.90	**2.90**	m3	3.12
Compacting with 680 kg vibratory roller										
filling	-	-	-	-	-	-	0.18	**0.18**	m2	0.19
Compacting with 6 – 8 tonnes smooth wheeled roller										
filling	-	-	-	-	-	-	0.35	**0.35**	m2	0.38
Trimming										
sides of cuttings; vertical or battered	-	-	-	-	-	-	0.90	**0.90**	m2	0.97
sides of embankments; vertical or battered	-	-	-	-	-	-	0.90	**0.90**	m2	0.97
Preserved topsoil obtained from on site spoil heaps not exceeding 100m distant – by machine										
Filling to make up levels										
average thickness not exceeding 0.25m	-	-	-	-	-	-	5.60	**5.60**	m3	6.02
Imported topsoil; wheeling not exceeding 50m – by machine										
Filling to make up levels										
average thickness not exceeding 0.25m	9.50	10.00	**10.45**	-	-	-	4.80	**15.25**	m3	16.39
Hard, dry, broken brick or stone to be obtained off site; wheeling not exceeding 25m – by machine										
Filling to make up levels; depositing in layers 150mm maximum thickness										
average thickness not exceeding 0.25m	12.40	25.00	**15.50**	-	-	-	2.90	**18.40**	m3	19.78
average thickness exceeding 0.25m	12.40	25.00	**15.50**	-	-	-	2.40	**17.90**	m3	19.24
Surface packing to filling										
to vertical or battered faces	-	-	-	-	-	-	0.50	**0.50**	m2	0.54
Compacting with 680 kg vibratory roller										
filling; blinding with sand, ashes or similar fine material	0.18	40.00	**0.25**	-	0.05	0.35	0.18	**0.78**	m2	0.84
Compacting with 6 – 8 tonnes smooth wheeled roller										
filling; blinding with sand, ashes or similar fine material	0.18	40.00	**0.25**	-	0.05	0.35	0.35	**0.95**	m2	1.02
Hard, dry, broken brick or stone to be obtained off site; wheeling not exceeding 50m – by machine										
Filling to make up levels; depositing in layers 150mm maximum thickness										
average thickness not exceeding 0.25m	12.40	25.00	**15.50**	-	-	-	3.70	**19.20**	m3	20.64
average thickness exceeding 0.25m	12.40	25.00	**15.50**	-	-	-	2.90	**18.40**	m3	19.78
Surface packing to filling										
to vertical or battered faces	-	-	-	-	-	-	0.50	**0.50**	m2	0.54
Compacting with 680 kg vibratory roller										
filling; blinding with sand, ashes or similar fine material	0.18	40.00	**0.25**	-	0.05	0.35	0.18	**0.78**	m2	0.84
Compacting with 6 – 8 tonnes smooth wheeled roller										
filling; blinding with sand, ashes or similar fine material	0.18	40.00	**0.25**	-	0.05	0.35	0.32	**0.92**	m2	0.99
MOT Type 1 to be obtained off site; wheeling not exceeding 25m – by machine										
Filling to make up levels; depositing in layers 150mm maximum thickness										
average thickness not exceeding 0.25m	14.50	25.00	**18.13**	-	-	-	2.90	**21.02**	m3	22.60
average thickness exceeding 0.25m	14.50	25.00	**18.13**	-	-	-	2.40	**20.52**	m3	22.06
Surface packing to filling										
to vertical or battered faces	-	-	-	-	-	-	0.50	**0.50**	m2	0.54
Compacting with 680 kg vibratory roller										
filling; blinding with sand, ashes or similar fine material	0.18	40.00	**0.25**	-	0.05	0.35	0.18	**0.78**	m2	0.84
Compacting with 6 – 8 tonnes smooth wheeled roller										
filling; blinding with sand, ashes or similar fine material	0.18	40.00	**0.25**	-	0.05	0.35	0.35	**0.95**	m2	1.02
MOT Type 1 to be obtained off site; wheeling not exceeding 50m – by machine										
Filling to make up levels; depositing in layers 150mm maximum thickness										
average thickness not exceeding 0.25m	14.50	25.00	**18.13**	-	-	-	3.70	**21.82**	m3	23.46
average thickness exceeding 0.25m	14.50	25.00	**18.13**	-	-	-	2.90	**21.02**	m3	22.60

GROUNDWORK

Labour hourly rates: (except Specialists) Craft Operatives £9.23 Labour £7.02 Rates are national average prices. Refer to REGIONAL VARIATIONS for indicative levels of overall pricing in regions	MATERIALS			LABOUR				RATES		
	Del to Site £	Waste %	Material Cost £	Craft Optve Hrs	Lab Hrs	Labour Cost £	Sunds £	Nett Rate £	Unit	Gross Rate (+7.5%) £
D20: EXCAVATING AND FILLING Cont.										
MOT Type 1 to be obtained off site; wheeling not exceeding 50m - by machine Cont.										
Surface packing to filling to vertical or battered faces	-	-	-	-	-	-	0.50	0.50	m2	0.54
Compacting with 680 kg vibratory roller filling; blinding with sand, ashes or similar fine material	0.18	40.00	0.25	-	0.05	0.35	0.18	0.78	m2	0.84
Compacting with 6 - 8 tonnes smooth wheeled roller filling; blinding with sand, ashes or similar fine material	0.18	40.00	0.25	-	0.05	0.35	0.35	0.95	m2	1.02
MOT Type 2 to be obtained off site; wheeling not exceeding 25m - by machine										
Filling to make up levels; depositing in layers 150mm maximum thickness										
average thickness not exceeding 0.25m	14.20	25.00	17.75	-	-	-	2.90	20.65	m3	22.20
average thickness exceeding 0.25m	14.20	25.00	17.75	-	-	-	2.40	20.15	m3	21.66
Surface packing to filling to vertical or battered faces	-	-	-	-	-	-	0.50	0.50	m2	0.54
Compacting with 680 kg vibratory roller filling; blinding with sand, ashes or similar fine material	0.18	40.00	0.25	-	0.05	0.35	0.18	0.78	m2	0.84
Compacting with 6 - 8 tonnes smooth wheeled roller filling; blinding with sand, ashes or similar fine material	0.18	40.00	0.25	-	0.05	0.35	0.35	0.95	m2	1.02
MOT Type 2 to be obtained off site; wheeling not exceeding 50m - by machine										
Filling to make up levels; depositing in layers 150mm maximum thickness										
average thickness not exceeding 0.25m	14.20	25.00	17.75	-	-	-	3.70	21.45	m3	23.06
average thickness exceeding 0.25m	14.20	25.00	17.75	-	-	-	2.90	20.65	m3	22.20
Surface packing to filling to vertical or battered faces	-	-	-	-	-	-	0.50	0.50	m2	0.54
Compacting with 680 kg vibratory roller filling; blinding with sand, ashes or similar fine material	0.18	40.00	0.25	-	0.05	0.35	0.18	0.78	m2	0.84
Compacting with 6 - 8 tonnes smooth wheeled roller filling; blinding with sand, ashes or similar fine material	0.18	40.00	0.25	-	0.05	0.35	0.35	0.95	m2	1.02
Ashes to be obtained off site; wheeling not exceeding 25m - by machine										
Filling to make up levels; depositing in layers 150mm maximum thickness										
average thickness not exceeding 0.25m	12.50	40.00	17.50	-	-	-	2.90	20.40	m3	21.93
Compacting with 680 kg vibratory roller filling...	-	-	-	-	-	-	0.18	0.18	m2	0.19
Compacting with 6 - 8 tonnes smooth wheeled roller filling...	-	-	-	-	-	-	0.35	0.35	m2	0.38
Ashes to be obtained off site; wheeling not exceeding 50m - by machine										
Filling to make up levels; depositing in layers 150mm maximum thickness										
average thickness not exceeding 0.25m	12.50	40.00	17.50	-	-	-	3.70	21.20	m3	22.79
Compacting with 680 kg vibratory roller filling...	-	-	-	-	-	-	0.18	0.18	m2	0.19
Compacting with 6 - 8 tonnes smooth wheeled roller filling...	-	-	-	-	-	-	0.35	0.35	m2	0.38
Sand to be obtained off site; wheeling not exceeding 25m - by machine										
Filling to make up levels; depositing in layers 150mm maximum thickness										
average thickness not exceeding 0.25m	9.25	33.00	12.30	-	-	-	2.90	15.20	m3	16.34
average thickness exceeding 0.25m	9.25	33.00	12.30	-	-	-	2.40	14.70	m3	15.81
Compacting with 680 kg vibratory roller filling...	-	-	-	-	-	-	0.18	0.18	m2	0.19
Compacting with 6 - 8 tonnes smooth wheeled roller filling...	-	-	-	-	-	-	0.35	0.35	m2	0.38

Labour hourly rates: (except Specialists) Craft Operatives £9.23 Labour £7.02. Rates are national average prices. Refer to REGIONAL VARIATIONS for indicative levels of overall pricing in regions	MATERIALS			LABOUR				RATES		
	Del to Site £	Waste %	Material Cost £	Craft Optve Hrs	Lab Hrs	Labour Cost £	Sunds £	Nett Rate £	Unit	Gross Rate (+7.5%) £
D20: EXCAVATING AND FILLING Cont.										
Sand to be obtained off site; wheeling not exceeding 50m - by machine										
Filling to make up levels; depositing in layers 150mm maximum thickness										
average thickness not exceeding 0.25m	9.25	33.00	12.30	-	-	-	3.70	16.00	m3	17.20
average thickness exceeding 0.25m	9.25	33.00	12.30	-	-	-	2.90	15.20	m3	16.34
Compacting with 680 kg vibratory roller										
filling	-	-	-	-	-	-	0.18	0.18	m2	0.19
Compacting with 6 - 8 tonnes smooth wheeled roller										
filling	-	-	-	-	-	-	0.35	0.35	m2	0.38
Hoggin to be obtained off site; wheeling not exceeding 25m - by machine										
Filling to make up levels; depositing in layers 150mm maximum thickness										
average thickness not exceeding 0.25m	8.00	33.00	10.64	-	-	-	2.90	13.54	m3	14.56
average thickness exceeding 0.25m	8.00	33.00	10.64	-	-	-	2.40	13.04	m3	14.02
Compacting with 680 kg vibratory roller										
filling	-	-	-	-	-	-	0.18	0.18	m2	0.19
Compacting with 6 - 8 tonnes smooth wheeled roller										
filling	-	-	-	-	-	-	0.35	0.35	m2	0.38
Hoggin to be obtained off site; wheeling not exceeding 50m - by machine										
Filling to make up levels; depositing in layers 150mm maximum thickness										
average thickness not exceeding 0.25m	8.00	33.00	10.64	-	-	-	3.70	14.34	m3	15.42
average thickness exceeding 0.25m	8.00	33.00	10.64	-	-	-	2.90	13.54	m3	14.56
Compacting with 680 kg vibratory roller										
filling	-	-	-	-	-	-	0.18	0.18	m2	0.19
Compacting with 6 - 8 tonnes smooth wheeled roller										
filling	-	-	-	-	-	-	0.35	0.35	m2	0.38
Excavated material arising from excavation - by hand										
Filling to excavation										
average thickness exceeding 0.25m	-	-	-	-	1.33	9.34	-	9.34	m3	10.04
Compacting										
filling	-	-	-	-	-	-	0.45	0.45	m2	0.48
Selected excavated material obtained from on site spoil heaps 25m distant - by hand										
Filling to make up levels; depositing in layers 150mm maximum thickness										
average thickness not exceeding 0.25m	-	-	-	-	1.75	12.29	-	12.29	m3	13.21
average thickness exceeding 0.25m	-	-	-	-	1.50	10.53	-	10.53	m3	11.32
Compacting with 680 kg vibratory roller										
filling	-	-	-	-	-	-	0.18	0.18	m2	0.19
Compacting with 6 - 8 tonnes smooth wheeled roller										
filling	-	-	-	-	-	-	0.35	0.35	m2	0.38
Trimming										
sides of cuttings; vertical or battered	-	-	-	-	0.20	1.40	-	1.40	m2	1.51
sides of embankments; vertical or battered	-	-	-	-	0.20	1.40	-	1.40	m2	1.51
Selected excavated material obtained from on site spoil heaps 50m distant - by hand										
Filling to make up levels; depositing in layers 150mm maximum thickness										
average thickness not exceeding 0.25m	-	-	-	-	2.00	14.04	-	14.04	m3	15.09
average thickness exceeding 0.25m	-	-	-	-	1.85	12.99	-	12.99	m3	13.96
Compacting with 680 kg vibratory roller										
filling	-	-	-	-	-	-	0.18	0.18	m2	0.19
Compacting with 6 - 8 tonnes smooth wheeled roller										
filling	-	-	-	-	-	-	0.35	0.35	m2	0.38
Trimming										
sides of cuttings; vertical or battered	-	-	-	-	0.20	1.40	-	1.40	m2	1.51
sides of embankments; vertical or battered	-	-	-	-	0.20	1.40	-	1.40	m2	1.51
Preserved topsoil obtained from on site spoil heaps not exceeding 100m distant - by hand										
Filling to make up levels										
average thickness not exceeding 0.25m	-	-	-	-	4.00	28.08	-	28.08	m3	30.19
Filling to external planters										
average thickness not exceeding 0.25m	-	-	-	-	5.00	35.10	-	35.10	m3	37.73

GROUNDWORK

Labour hourly rates: (except Specialists) Craft Operatives £9.23 Labour £7.02 Rates are national average prices. Refer to REGIONAL VARIATIONS for indicative levels of overall pricing in regions	MATERIALS			LABOUR				RATES		
	Del to Site £	Waste %	Material Cost £	Craft Optve Hrs	Lab Hrs	Labour Cost £	Sunds £	Nett Rate £	Unit	Gross Rate (+7.5%) £
D20: EXCAVATING AND FILLING Cont.										
Imported topsoil; wheeling not exceeding 50m - by hand										
Filling to make up levels										
average thickness not exceeding 0.25m	9.50	10.00	**10.45**	-	3.50	**24.57**	-	**35.02**	m3	37.65
Filling to external planters										
average thickness exceeding 0.25m	9.50	10.00	**10.45**	-	4.50	**31.59**	-	**42.04**	m3	45.19
Hard, dry, broken brick or stone to be obtained off site; wheeling not exceeding 25m - by hand										
Filling to make up levels; depositing in layers 150mm maximum thickness										
average thickness not exceeding 0.25m	12.40	25.00	**15.50**	-	1.50	**10.53**	-	**26.03**	m3	27.98
average thickness exceeding 0.25m	12.40	25.00	**15.50**	-	1.25	**8.78**	-	**24.27**	m3	26.10
Surface packing to filling										
to vertical or battered faces	-	-	-	-	0.20	**1.40**	-	**1.40**	m2	1.51
Compacting with 680 kg vibratory roller										
filling; blinding with sand, ashes or similar fine material	0.18	40.00	**0.25**	-	0.05	**0.35**	0.18	**0.78**	m2	0.84
Compacting with 6 - 8 tonnes smooth wheeled roller										
filling; blinding with sand, ashes or similar fine material	0.18	40.00	**0.25**	-	0.05	**0.35**	0.35	**0.95**	m2	1.02
Hard, dry, broken brick or stone to be obtained off site; wheeling not exceeding 50m - by hand										
Filling to make up levels; depositing in layers 150mm maximum thickness										
average thickness not exceeding 0.25m	12.40	25.00	**15.50**	-	1.75	**12.29**	-	**27.79**	m3	29.87
average thickness exceeding 0.25m	12.40	25.00	**15.50**	-	1.50	**10.53**	-	**26.03**	m3	27.98
Surface packing to filling										
to vertical or battered faces	-	-	-	-	0.20	**1.40**	-	**1.40**	m2	1.51
Compacting with 680 kg vibratory roller										
filling; blinding with sand, ashes or similar fine material	0.18	40.00	**0.25**	-	0.05	**0.35**	0.18	**0.78**	m2	0.84
Compacting with 6 - 8 tonnes smooth wheeled roller										
filling; blinding with sand, ashes or similar fine material	0.18	40.00	**0.25**	-	0.05	**0.35**	0.35	**0.95**	m2	1.02
MOT Type 1 to be obtained off site; wheeling not exceeding 25m - by hand										
Filling to make up levels; depositing in layers 150mm maximum thickness										
average thickness not exceeding 0.25m	14.50	25.00	**18.13**	-	1.50	**10.53**	-	**28.66**	m3	30.80
average thickness exceeding 0.25m	14.50	25.00	**18.13**	-	1.25	**8.78**	-	**26.90**	m3	28.92
Surface packing to filling										
to vertical or battered faces	-	-	-	-	0.20	**1.40**	-	**1.40**	m2	1.51
Compacting with 680 kg vibratory roller										
filling; blinding with sand, ashes or similar fine material	0.18	40.00	**0.25**	-	0.05	**0.35**	0.18	**0.78**	m2	0.84
Compacting with 6 - 8 tonnes smooth wheeled roller										
filling; blinding with sand, ashes or similar fine material	0.18	40.00	**0.25**	-	0.05	**0.35**	0.35	**0.95**	m2	1.02
MOT Type 1 to be obtained off site; wheeling not exceeding 50m - by hand										
Filling to make up levels; depositing in layers 150mm maximum thickness										
average thickness not exceeding 0.25m	14.50	25.00	**18.13**	-	1.75	**12.29**	-	**30.41**	m3	32.69
average thickness exceeding 0.25m	14.50	25.00	**18.13**	-	1.50	**10.53**	-	**28.66**	m3	30.80
Surface packing to filling										
to vertical or battered faces	-	-	-	-	0.20	**1.40**	-	**1.40**	m2	1.51
Compacting with 680 kg vibratory roller										
filling; blinding with sand, ashes or similar fine material	0.18	40.00	**0.25**	-	0.05	**0.35**	0.18	**0.78**	m2	0.84
Compacting with 6 - 8 tonnes smooth wheeled roller										
filling; blinding with sand, ashes or similar fine material	0.18	40.00	**0.25**	-	0.05	**0.35**	0.35	**0.95**	m2	1.02
MOT Type 2 to be obtained off site; wheeling not exceeding 25m - by hand										
Filling to make up levels; depositing in layers 150mm maximum thickness										
average thickness not exceeding 0.25m	14.20	25.00	**17.75**	-	1.50	**10.53**	-	**28.28**	m3	30.40
average thickness exceeding 0.25m	14.20	25.00	**17.75**	-	1.25	**8.78**	-	**26.52**	m3	28.51

Labour hourly rates: (except Specialists) Craft Operatives £9.23 Labour £7.02 Rates are national average prices. Refer to REGIONAL VARIATIONS for indicative levels of overall pricing in regions	MATERIALS			LABOUR				RATES		
	Del to Site £	Waste %	Material Cost £	Craft Optve Hrs	Lab Hrs	Labour Cost £	Sunds £	Nett Rate £	Unit	Gross Rate (+7.5%) £
D20: EXCAVATING AND FILLING Cont.										
MOT Type 2 to be obtained off site; wheeling not exceeding 25m - by hand Cont.										
Surface packing to filling to vertical or battered faces	-	-	-	-	0.20	1.40	-	1.40	m2	1.51
Compacting with 680 kg vibratory roller filling; blinding with sand, ashes or similar fine material	0.18	40.00	0.25	-	0.05	0.35	0.18	0.78	m2	0.84
Compacting with 6 - 8 tonnes smooth wheeled roller filling; blinding with sand, ashes or similar fine material	0.18	40.00	0.25	-	0.05	0.35	0.35	0.95	m2	1.02
MOT Type 2 to be obtained off site; wheeling not exceeding 50m - by hand										
Filling to make up levels; depositing in layers 150mm maximum thickness average thickness not exceeding 0.25m .... average thickness exceeding 0.25m	14.20 14.20	25.00 25.00	17.75 17.75	- -	1.75 1.50	12.29 10.53	- -	30.04 28.28	m3 m3	32.29 30.40
Surface packing to filling to vertical or battered faces	-	-	-	-	0.20	1.40	-	1.40	m2	1.51
Compacting with 680 kg vibratory roller filling; blinding with sand, ashes or similar fine material	0.18	40.00	0.25	-	0.05	0.35	0.18	0.78	m2	0.84
Compacting with 6 - 8 tonnes smooth wheeled roller filling; blinding with sand, ashes or similar fine material	0.18	40.00	0.25	-	0.05	0.35	0.35	0.95	m2	1.02
Ashes to be obtained off site; wheeling not exceeding 25m - by hand										
Filling to make up levels; depositing in layers 150mm maximum thickness average thickness not exceeding 0.25m	12.50	40.00	17.50	-	1.60	11.23	-	28.73	m3	30.89
Compacting with 680 kg vibratory roller filling.............................	-	-	-	-	-	-	0.18	0.18	m2	0.19
Compacting with 6 - 8 tonnes smooth wheeled roller filling...........................	-	-	-	-	-	-	0.35	0.35	m2	0.38
Ashes to be obtained off site; wheeling not exceeding 50m - by hand										
Filling to make up levels; depositing in layers 150mm maximum thickness average thickness not exceeding 0.25m	12.50	40.00	17.50	-	1.85	12.99	-	30.49	m3	32.77
Compacting with 680 kg vibratory roller filling...........................	-	-	-	-	-	-	0.18	0.18	m2	0.19
Compacting with 6 - 8 tonnes smooth wheeled roller filling...........................	-	-	-	-	-	-	0.35	0.35	m2	0.38
Sand to be obtained off site; wheeling not exceeding 25m - by hand										
Filling to make up levels; depositing in layers 150mm maximum thickness average thickness not exceeding 0.25m .... average thickness exceeding 0.25m	9.25 9.25	33.00 33.00	12.30 12.30	- -	1.75 1.50	12.29 10.53	- -	24.59 22.83	m3 m3	26.43 24.54
Compacting with 680 kg vibratory roller filling...........................	-	-	-	-	-	-	0.18	0.18	m2	0.19
Compacting with 6 - 8 tonnes smooth wheeled roller filling...........................	-	-	-	-	-	-	0.35	0.35	m2	0.38
Sand to be obtained off site; wheeling not exceeding 50m - by hand										
Filling to make up levels; depositing in layers 150mm maximum thickness average thickness not exceeding 0.25m .... average thickness exceeding 0.25m	9.25 9.25	33.00 33.00	12.30 12.30	- -	2.00 1.75	14.04 12.29	- -	26.34 24.59	m3 m3	28.32 26.43
Compacting with 680 kg vibratory roller filling..........................	-	-	-	-	-	-	0.18	0.18	m2	0.19
Compacting with 6 - 8 tonnes smooth wheeled roller filling.........................	-	-	-	-	-	-	0.35	0.35	m2	0.38
Hoggin to be obtained off site; wheeling not exceeding 25m - by hand										
Filling to make up levels; depositing in layers 150mm maximum thickness average thickness not exceeding 0.25m .... average thickness exceeding 0.25m	8.00 8.00	33.00 33.00	10.64 10.64	- -	1.50 1.25	10.53 8.78	- -	21.17 19.41	m3 m3	22.76 20.87

GROUNDWORK

Labour hourly rates: (except Specialists) Craft Operatives £9.23 Labour £7.02 Rates are national average prices. Refer to REGIONAL VARIATIONS for indicative levels of overall pricing in regions	MATERIALS			LABOUR				RATES		
	Del to Site £	Waste %	Material Cost £	Craft Optve Hrs	Lab Hrs	Labour Cost £	Sunds £	Nett Rate £	Unit	Gross Rate (+7.5%) £
D20: EXCAVATING AND FILLING Cont.										
Hoggin to be obtained off site; wheeling not exceeding 25m - by hand Cont.										
Compacting with 680 kg vibratory roller filling	-	-	-	-	-	-	0.18	0.18	m2	0.19
Compacting with 6 - 8 tonnes smooth wheeled roller filling	-	-	-	-	-	-	0.35	0.35	m2	0.38
Hoggin to be obtained off site; wheeling not exceeding 50m - by hand										
Filling to make up levels; depositing in layers 150mm maximum thickness										
average thickness not exceeding 0.25m	8.00	33.00	10.64	-	1.75	12.29	-	22.93	m3	24.64
average thickness exceeding 0.25m	8.00	33.00	10.64	-	1.50	10.53	-	21.17	m3	22.76
Compacting with 680 kg vibratory roller filling	-	-	-	-	-	-	0.18	0.18	m2	0.19
Compacting with 6 - 8 tonnes smooth wheeled roller filling	-	-	-	-	-	-	0.35	0.35	m2	0.38
Excavated material arising from excavation; work inside existing building - by hand										
Filling to excavation average thickness exceeding 0.25m	-	-	-	-	2.67	18.74	-	18.74	m3	20.15
Compacting filling	-	-	-	-	-	-	0.18	0.18	m2	0.19
Selected excavated material obtained from on site spoil heaps 25m distant; work inside existing building - by hand										
Filling to make up levels; depositing in layers 150mm maximum thickness										
average thickness not exceeding 0.25m	-	-	-	-	3.10	21.76	-	21.76	m3	23.39
average thickness exceeding 0.25m	-	-	-	-	2.93	20.57	-	20.57	m3	22.11
Compacting filling	-	-	-	-	-	-	0.18	0.18	m2	0.19
Selected excavated material obtained from on site spoil heaps 50m distant; work inside existing building - by hand										
Filling to make up levels; depositing in layers 150mm maximum thickness										
average thickness not exceeding 0.25m	-	-	-	-	3.35	23.52	-	23.52	m3	25.28
average thickness exceeding 0.25m	-	-	-	-	3.20	22.46	-	22.46	m3	24.15
Compacting filling	-	-	-	-	-	-	0.18	0.18	m2	0.19
Hard, dry, broken brick or stone to be obtained off site; wheeling not exceeding 25m; work inside existing building - by hand										
Filling to make up levels; depositing in layers 150mm maximum thickness										
average thickness not exceeding 0.25m	12.40	25.00	15.50	-	2.85	20.01	-	35.51	m3	38.17
average thickness exceeding 0.25m	12.40	25.00	15.50	-	2.68	18.81	-	34.31	m3	36.89
Surface packing to filling to vertical or battered faces	-	-	-	-	0.20	1.40	-	1.40	m2	1.51
Compacting filling; blinding with sand, ashes or similar fine material	0.18	40.00	0.25	-	0.05	0.35	0.18	0.78	m2	0.84
Hard, dry, broken brick or stone to be obtained off site; wheeling not exceeding 50m; work inside existing building - by hand										
Filling to make up levels; depositing in layers 150mm maximum thickness										
average thickness not exceeding 0.25m	12.40	25.00	15.50	-	3.10	21.76	-	37.26	m3	40.06
average thickness exceeding 0.25m	12.40	25.00	15.50	-	2.85	20.01	-	35.51	m3	38.17
Surface packing to filling to vertical or battered faces	-	-	-	-	0.20	1.40	-	1.40	m2	1.51
Compacting filling; blinding with sand, ashes or similar fine material	0.18	40.00	0.25	-	0.05	0.35	0.18	0.78	m2	0.84
Ashes to be obtained off site; wheeling not exceeding 25m; work inside existing building - by hand										
Filling to make up levels; depositing in layers 150mm maximum thickness										
average thickness not exceeding 0.25m	12.50	40.00	17.50	-	2.95	20.71	-	38.21	m3	41.07

Labour hourly rates: (except Specialists) Craft Operatives £9.23 Labour £7.02 Rates are national average prices. Refer to REGIONAL VARIATIONS for indicative levels of overall pricing in regions	MATERIALS			LABOUR				RATES		
	Del to Site £	Waste %	Material Cost £	Craft Optve Hrs	Lab Hrs	Labour Cost £	Sunds £	Nett Rate £	Unit	Gross Rate (+7.5%) £
D20: EXCAVATING AND FILLING Cont.										
Ashes to be obtained off site; wheeling not exceeding 25m; work inside existing building - by hand Cont.										
Compacting filling..	-	-	-	-	-	-	0.18	0.18	m2	0.19
Ashes to be obtained off site; wheeling not exceeding 50m; work inside existing building - by hand										
Filling to make up levels; depositing in layers 150mm maximum thickness average thickness not exceeding 0.25m	12.50	40.00	17.50	-	3.20	22.46	-	39.96	m3	42.96
Compacting filling..	-	-	-	-	-	-	0.18	0.18	m2	0.19
Sand to be obtained off site; wheeling not exceeding 25m; work inside existing building - by hand										

Filling to make up levels; depositing in layers 150mm maximum thickness
    average thickness not exceeding 0.25m ....
    average thickness exceeding 0.25m ....
	9.25
	9.25
		33.00
		33.00
			12.30
			12.30
				-
				-
					3.10
					2.85
						21.76
						20.01
							-
							-
								34.06
								32.31
									m3
									m3
										36.62
										34.73

Compacting filling..	-	-	-	-	-	-	0.18	0.18	m2	0.19
Sand to be obtained off site; wheeling not exceeding 50m; work inside existing building - by hand										

Filling to make up levels; depositing in layers 150mm maximum thickness
    average thickness not exceeding 0.25m ....
    average thickness exceeding 0.25m ....
	9.25
	9.25
		33.00
		33.00
			12.30
			12.30
				-
				-
					3.35
					3.10
						23.52
						21.76
							-
							-
								35.82
								34.06
									m3
									m3
										38.51
										36.62

Compacting filling..	-	-	-	-	-	-	0.18	0.18	m2	0.19
D30: CAST IN PLACE PILING										
Prices include for a 21 N/mm2 concrete mix, nominal reinforcement and a minimum number of 50 piles on any one contract. The working loads sizes and lengths given below will depend on the nature of the soils in which the piles will be founded as well as structure to be supported (SIMPLEX PILING LTD.)										

On/Off site charge
    in addition to the following prices add
    approximately ....
	-	-	Spclist	-	-	Spclist	-	3800.00	Sm	4085.00

Short auger piles
    up to 6m long; 450mm nominal diameter; 8 tonnes
    normal working load ....
    up to 6m long; 610mm nominal diameter; 30 tonnes
    normal working load ....
	-
	
	-
		-
		
		-
			Spclist
			
			Spclist
				-
				
				-
					-
					
					-
						Spclist
						
						Spclist
							-
							
							-
								22.26
								
								33.66
									m
									
									m
										23.93
										
										36.18

Bored cast in-situ piles
    up to 15m long; 450mm nominal diameter; 40 tonnes
    normal working load ....
    up to 15m long; 610mm nominal diameter; 120 tonnes
    normal working load ....
	-
	
	-
		-
		
		-
			Spclist
			
			Spclist
				-
				
				-
					-
					
					-
						Spclist
						
						Spclist
							-
							
							-
								20.54
								
								31.95
									m
									
									m
										22.08
										
										34.35

Auger piles
    up to 20m long; 450mm nominal diameter ....
    up to 30m long; 1200mm nominal diameter ....
	-
	-
		-
		-
			Spclist
			Spclist
				-
				-
					-
					-
						Spclist
						Spclist
							-
							-
								20.83
								103.28
									m
									m
										22.39
										111.03

Large diameter auger piles (plain shaft)
    up to 20m long; 610mm nominal diameter; 150 tonnes
    normal working load ....
    up to 30m long; 1525mm nominal diameter; 600
    tonnes normal working load ....
	-
	
	-
		-
		
		-
			Spclist
			
			Spclist
				-
				
				-
					-
					
					-
						Spclist
						
						Spclist
							-
							
							-
								31.68
								
								160.66
									m
									
									m
										34.06
										
										172.71

Large diameter auger piles (belled base)
    up to 20m long; 610mm nominal diameter; 150 tonnes
    normal working load ....
    up to 30m long; 1525mm nominal diameter; 1000
    tonnes normal working load ....
	-
	
	-
		-
		
		-
			Spclist
			
			Spclist
				-
				
				-
					-
					
					-
						Spclist
						
						Spclist
							-
							
							-
								43.87
								
								206.61
									m
									
									m
										47.16
										
										222.11

Boring through obstructions, rock like formations, etc
    undertaken on a time basis at a rate per piling
    rig per hour										

Cutting off tops of piles; including preparation and integration of reinforcement into pile cap or ground beam and disposal
    450-610mm nominal diameter ....
    610-1525mm nominal diameter ....
	-
	-
		-
		-
			Spclist
			Spclist
				-
				-
					-
					-
						Spclist
						Spclist
							-
							-
								7.62
								28.84
									m
									m
										8.19
										31.00

Labour hourly rates: (except Specialists) Craft Operatives £9.23 Labour £7.02. Rates are national average prices. Refer to REGIONAL VARIATIONS for indicative levels of overall pricing in regions	MATERIALS			LABOUR				RATES		
	Del to Site £	Waste %	Material Cost £	Craft Optve Hrs	Lab Hrs	Labour Cost £	Sunds £	Nett Rate £	Unit	Gross Rate (+7.5%) £
D32: STEEL PILING										
Mild steel Universal Bearing Piles 'H' section, to B.S.EN10025 Grade S275 and high yield steel Grade S275JR in lengths 9 - 15m supplied, handled, pitched and driven vertically with landbased plant (SIMPLEX PILING LTD.)										
Note the following prices are based on quantities of 25 - 150 tonnes										
On/Off site charge										
in addition to the following prices, add approximately	-	-	Spclist	-	-	Spclist	-	3150.00	Sm	3386.25
Mild steel piles										
203 x 203 x 45 kg/m, SWL 40 tonnes	-	-	Spclist	-	-	Spclist	-	30.43	m	32.71
203 x 203 x 54 kg/m, SWL 50 tonnes	-	-	Spclist	-	-	Spclist	-	34.31	m	36.88
254 x 254 x 63 kg/m, SWL 60 tonnes	-	-	Spclist	-	-	Spclist	-	38.78	m	41.69
254 x 254 x 71 kg/m, SWL 70 tonnes	-	-	Spclist	-	-	Spclist	-	41.77	m	44.90
305 x 305 x 79 kg/m, SWL 75 tonnes	-	-	Spclist	-	-	Spclist	-	45.99	m	49.44
254 x 254 x 85 kg/m, SWL 80 tonnes	-	-	Spclist	-	-	Spclist	-	48.81	m	52.47
305 x 305 x 88 kg/m, SWL 85 tonnes	-	-	Spclist	-	-	Spclist	-	53.32	m	57.32
305 x 305 x 95 kg/m, SWL 90 tonnes	-	-	Spclist	-	-	Spclist	-	56.69	m	60.94
356 x 368 x 109 kg/m, SWL 105 tonnes	-	-	Spclist	-	-	Spclist	-	67.94	m	73.04
305 x 305 x 110 kg/m, SWL 105 tonnes	-	-	Spclist	-	-	Spclist	-	63.91	m	68.70
305 x 305 x 126 kg/m, SWL 120 tonnes	-	-	Spclist	-	-	Spclist	-	72.90	m	78.37
305 x 305 x 149 kg/m, SWL 140 tonnes	-	-	Spclist	-	-	Spclist	-	80.46	m	86.49
356 x 368 x 133 kg/m, SWL 130 tonnes	-	-	Spclist	-	-	Spclist	-	84.21	m	90.53
356 x 368 x 152 kg/m, SWL 140 tonnes	-	-	Spclist	-	-	Spclist	-	89.38	m	96.08
356 x 368 x 174 kg/m, SWL 165 tonnes	-	-	Spclist	-	-	Spclist	-	102.79	m	110.50
305 x 305 x 186 kg/m, SWL 175 tonnes	-	-	Spclist	-	-	Spclist	-	109.02	m	117.20
305 x 305 x 223 kg/m, SWL 210 tonnes	-	-	Spclist	-	-	Spclist	-	122.32	m	131.49
Extra for high yield steel										
add	-	-	Spclist	-	-	Spclist	-	33.00	T	35.48
For quiet piling, add										
to terminal charge, lump sum	-	-	Spclist	-	-	Spclist	-	660.00	Sm	709.50
plus for all piles	-	-	Spclist	-	-	Spclist	-	1.63	m	1.75
Mild steel sheet piling to B.S.EN10025 Grade S275 or high yield steel Grade S275JR in lengths 4.5-15m supplied, handled, pitched and driven by land based plant in one visit (SIMPLEX PILING LTD.)										
On/Off site charge										
in addition to the following prices, add approximately	-	-	Spclist	-	-	Spclist	-	2880.00	Sm	3096.00
Mild steel sheet piling										
Larssen Section 6W	-	-	Spclist	-	-	Spclist	-	65.92	m2	70.86
extra for high yield steel	-	-	Spclist	-	-	Spclist	-	3.61	m2	3.88
extra for one coat L.B.V.(Lowca Varnish to B.S.1070 Type 2) before driving	-	-	Spclist	-	-	Spclist	-	2.64	m2	2.84
extra for corners	-	-	Spclist	-	-	Spclist	-	26.38	m	28.36
extra for junctions	-	-	Spclist	-	-	Spclist	-	35.97	m	38.67
Frodingham Section 1N and Larssen Section 9W	-	-	Spclist	-	-	Spclist	-	72.88	m2	78.35
extra for high yield steel	-	-	Spclist	-	-	Spclist	-	3.00	m2	3.23
extra for one coat L.B.V.(Lowca Varnish to B.S.1070 Type 2) before driving	-	-	Spclist	-	-	Spclist	-	2.64	m2	2.84
extra for corners	-	-	Spclist	-	-	Spclist	-	26.38	m	28.36
extra for junctions	-	-	Spclist	-	-	Spclist	-	35.97	m	38.67
Frodingham Section 2N and Larssen Section 12W	-	-	Spclist	-	-	Spclist	-	81.13	m2	87.21
extra for high yield steel	-	-	Spclist	-	-	Spclist	-	3.84	m2	4.13
extra for one coat L.B.V.(Lowca Varnish to B.S.1070 Type 2) before driving	-	-	Spclist	-	-	Spclist	-	2.64	m2	2.84
extra for corners	-	-	Spclist	-	-	Spclist	-	26.38	m	28.36
extra for junctions	-	-	Spclist	-	-	Spclist	-	35.97	m	38.67
Frodingham Section 1BXN	-	-	Spclist	-	-	Spclist	-	89.92	m2	96.66
extra for high yield steel	-	-	Spclist	-	-	Spclist	-	6.12	m2	6.58
extra for one coat L.B.V.(Lowca Varnish to B.S.1070 Type 2) before driving	-	-	Spclist	-	-	Spclist	-	2.40	m2	2.58
extra for corners	-	-	Spclist	-	-	Spclist	-	28.78	m	30.94
extra for junctions	-	-	Spclist	-	-	Spclist	-	35.97	m	38.67
Frodingham Section 3N and Larssen Section 16W	-	-	Spclist	-	-	Spclist	-	90.72	m2	97.52
extra for high yield steel	-	-	Spclist	-	-	Spclist	-	4.91	m2	5.28
extra for one coat L.B.V.(Lowca Varnish to B.S.1070 Type 2) before driving	-	-	Spclist	-	-	Spclist	-	2.40	m2	2.58
extra for corners	-	-	Spclist	-	-	Spclist	-	28.78	m	30.94
extra for junctions	-	-	Spclist	-	-	Spclist	-	35.97	m	38.67
Larssen Section 20W	-	-	Spclist	-	-	Spclist	-	102.65	m2	110.35
extra for high yield steel	-	-	Spclist	-	-	Spclist	-	5.40	m2	5.80
extra for one coat L.B.V.(Lowca Varnish to B.S.1070 Type 2) before driving	-	-	Spclist	-	-	Spclist	-	2.70	m2	2.90
extra for corners	-	-	Spclist	-	-	Spclist	-	31.17	m	33.51
extra for junctions	-	-	Spclist	-	-	Spclist	-	41.96	m	45.11
Frodingham Section 4N and Larssen Section 25W	-	-	Spclist	-	-	Spclist	-	113.67	m2	122.20
extra for high yield steel	-	-	Spclist	-	-	Spclist	-	5.88	m2	6.32
extra for one coat L.B.V.(Lowca Varnish to B.S.1070 Type 2) before driving	-	-	Spclist	-	-	Spclist	-	3.00	m2	3.23
extra for corners	-	-	Spclist	-	-	Spclist	-	33.57	m	36.09
extra for junctions	-	-	Spclist	-	-	Spclist	-	41.96	m	45.11
Larssen Section 32W	-	-	Spclist	-	-	Spclist	-	134.20	m2	144.26
extra for high yield steel	-	-	Spclist	-	-	Spclist	-	6.53	m2	7.02
extra for one coat L.B.V.(Lowca Varnish to B.S.1070 Type 2) before driving	-	-	Spclist	-	-	Spclist	-	3.24	m2	3.48

Labour hourly rates: (except Specialists) Craft Operatives £9.23 Labour £7.02 Rates are national average prices. Refer to REGIONAL VARIATIONS for indicative levels of overall pricing in regions	MATERIALS			LABOUR			RATES		
	Del to Site	Waste	Material Cost	Craft Optve	Lab	Labour Cost	Sunds	Nett Rate	Unit	Gross Rate (+7.5%)
	£	%	£	Hrs	Hrs	£	£	£		£
D32: STEEL PILING Cont.										
Mild steel sheet piling to B.S.EN10025 Grade S275 or high yield steel Grade S275JR in lengths 4.5-15m supplied, handled, pitched and driven by land based plant in one visit (SIMPLEX PILING LTD.) Cont.										
Mild steel sheet piling Cont.										
extra for corners	-	-	Spclist	-	-	Spclist	-	35.97	m	38.67
extra for junctions	-	-	Spclist	-	-	Spclist	-	44.36	m	47.69
Frodingham Section 5	-	-	Spclist	-	-	Spclist	-	151.44	m2	162.80
extra for high yield steel	-	-	Spclist	-	-	Spclist	-	8.63	m2	9.28
extra for one coat L.B.V.(Lowca Varnish to B.S.1070 Type 2) before driving	-	-	Spclist	-	-	Spclist	-	5.63	m2	6.05
extra for corners	-	-	Spclist	-	-	Spclist	-	35.97	m	38.67
extra for junctions	-	-	Spclist	-	-	Spclist	-	47.95	m	51.55
For quiet piling, add										
lump sum.......................................	-	-	Spclist	-	-	Spclist	-	600.00	Sm	645.00
plus for all sections, from	-	-	Spclist	-	-	Spclist	-	2.80	m2	3.01
to	-	-	Spclist	-	-	Spclist	-	4.65	m2	5.00
D40: EMBEDDED RETAINING WALLS										
Embedded retaining walls; contiguous panel construction; panel lengths not exceeding 5m. Note:- the following prices are indicative only: firm quotations should always be obtained (KVAERNER CEMENTATION FOUNDATIONS)										
On/Off site charge										
in addition to the following prices, add for bringing plant to site, erecting and dismantling, maintaining and removing from site, approximately.	-	-	Spclist	-	-	Spclist	-	60000.00	Sm	64500.00
Excavation and Bentonite slurry and disposal										
600mm thick wall; maximum depth										
5m	-	-	Spclist	-	-	Spclist	-	180.00	m3	193.50
10m	-	-	Spclist	-	-	Spclist	-	180.00	m3	193.50
15m	-	-	Spclist	-	-	Spclist	-	180.00	m3	193.50
20m	-	-	Spclist	-	-	Spclist	-	180.00	m3	193.50
800mm thick wall; maximum depth										
5m	-	-	Spclist	-	-	Spclist	-	150.00	m3	161.25
10m	-	-	Spclist	-	-	Spclist	-	150.00	m3	161.25
15m	-	-	Spclist	-	-	Spclist	-	150.00	m3	161.25
20m	-	-	Spclist	-	-	Spclist	-	150.00	m3	161.25
1000mm thick wall; maximum depth										
5m	-	-	Spclist	-	-	Spclist	-	120.00	m3	129.00
10m	-	-	Spclist	-	-	Spclist	-	120.00	m3	129.00
15m	-	-	Spclist	-	-	Spclist	-	120.00	m3	129.00
20m	-	-	Spclist	-	-	Spclist	-	120.00	m3	129.00
Excavating through obstructions, rock like formations, etc										
undertaken on a time basis at a rate per rig per hour	-	-	Spclist	-	-	Spclist	-	350.00	hr	376.25
Reinforced concrete; B.S.5328, designed mix C25, 20mm aggregate, minimum cement content 400 kg/m3										
600mm thick wall	-	-	Spclist	-	-	Spclist	-	75.00	m3	80.63
800mm thick wall	-	-	Spclist	-	-	Spclist	-	75.00	m3	80.63
1000mm thick wall	-	-	Spclist	-	-	Spclist	-	75.00	m3	80.63
Reinforced in-situ concrete; sulphate resisting; B.S.5328, designed mix C25, 20mm aggregate, minimum cement content 400 kg/m3										
600mm thick wall	-	-	Spclist	-	-	Spclist	-	85.00	m3	91.38
800mm thick wall	-	-	Spclist	-	-	Spclist	-	85.00	m3	91.38
1000mm thick wall	-	-	Spclist	-	-	Spclist	-	85.00	m3	91.38
Reinforcement bars; B.S.4449 hot rolled plain round mild steel; including hooks tying wire, and spacers and chairs which are at the discretion of the Contractor										
16mm; straight	-	-	Spclist	-	-	Spclist	-	535.00	t	575.13
20mm; straight	-	-	Spclist	-	-	Spclist	-	535.00	t	575.13
25mm; straight	-	-	Spclist	-	-	Spclist	-	535.00	t	575.13
32mm; straight	-	-	Spclist	-	-	Spclist	-	535.00	t	575.13
40mm; straight	-	-	Spclist	-	-	Spclist	-	535.00	t	575.13
16mm; bent	-	-	Spclist	-	-	Spclist	-	535.00	t	575.13
20mm; bent	-	-	Spclist	-	-	Spclist	-	535.00	t	575.13
25mm; bent	-	-	Spclist	-	-	Spclist	-	535.00	t	575.13
32mm; bent	-	-	Spclist	-	-	Spclist	-	535.00	t	575.13
40mm; bent	-	-	Spclist	-	-	Spclist	-	535.00	t	575.13
Reinforcement bars; B.S.4449 hot rolled deformed high yield steel; including hooks tying wire, and spacers and chairs which are at the discretion of the Contractor										
16mm; straight	-	-	Spclist	-	-	Spclist	-	535.00	t	575.13
20mm; straight	-	-	Spclist	-	-	Spclist	-	535.00	t	575.13
25mm; straight	-	-	Spclist	-	-	Spclist	-	535.00	t	575.13
32mm; straight	-	-	Spclist	-	-	Spclist	-	535.00	t	575.13
40mm; straight	-	-	Spclist	-	-	Spclist	-	535.00	t	575.13
16mm; bent	-	-	Spclist	-	-	Spclist	-	535.00	t	575.13
20mm; bent	-	-	Spclist	-	-	Spclist	-	535.00	t	575.13

Labour hourly rates: (except Specialists) Craft Operatives £9.23 Labour £7.02 Rates are national average prices. Refer to REGIONAL VARIATIONS for indicative levels of overall pricing in regions	MATERIALS			LABOUR				RATES		
	Del to Site £	Waste %	Material Cost £	Craft Optve Hrs	Lab Hrs	Labour Cost £	Sunds £	Nett Rate £	Unit	Gross Rate (+7.5%) £
D40: EMBEDDED RETAINING WALLS Cont.										
Embedded retaining walls; contiguous panel construction; panel lengths not exceeding 5m. Note:- the following prices are indicative only: firm quotations should always be obtained (KVAERNER CEMENTATION FOUNDATIONS) Cont.										
Reinforcement bars; B.S.4449 hot rolled deformed high yield steel; including hooks tying wire, and spacers and chairs which are at the discretion of the Contractor Cont.										
25mm; bent	-	-	Spclist	-	-	Spclist	-	535.00	t	575.13
32mm; bent	-	-	Spclist	-	-	Spclist	-	535.00	t	575.13
40mm; bent	-	-	Spclist	-	-	Spclist	-	535.00	t	575.13
Guide walls; excavation, disposal and support; reinforced in-situ concrete; B.S.5328, designed mix C25, 20mm aggregate, minimum cement content 290 kg/m3; reinforced with one layer fabric B.S.4483 reference A252, 3.95 kg/m2 including laps, tying wire, all cutting and bending, and spacers and chairs which are at the discretion of the Contractor; formwork both sides										
both sides; 1000mm high, 600mm apart; propped top and bottom at 2000mm centres	-	-	Spclist	-	-	Spclist	-	185.00	m	198.88
both sides; 1000mm high, 800mm apart; propped top and bottom at 2000mm centres	-	-	Spclist	-	-	Spclist	-	185.00	m	198.88
both sides; 1000mm high, 1000mm apart; propped top and bottom at 2000mm centres	-	-	Spclist	-	-	Spclist	-	185.00	m	198.88
both sides; 1500mm high, 600mm apart; propped top and bottom at 2000mm centres	-	-	Spclist	-	-	Spclist	-	285.00	m	306.38
both sides; 1500mm high, 800mm apart; propped top and bottom at 2000mm centres	-	-	Spclist	-	-	Spclist	-	285.00	m	306.38
both sides; 1500mm high, 1000mm apart; propped top and bottom at 2000mm centres	-	-	Spclist	-	-	Spclist	-	285.00	m	306.38
D41: CRIB WALLS/GABIONS/REINFORCED EARTH										
Retaining Walls										
Betoflor precast concrete landscape retaining walls including soil filling to pockets but excluding excavation, concrete foundations, stone backfill to rear of wall and planting which are all deemed measured separately										
Betoflor interlocking units 500mm long x 250mm wide x 200mm modular deep in wall 250mm wide	41.61	5.00	43.69	-	3.50	24.57	-	68.26	m2	73.38
Extra over for colours	1.50	5.00	1.58	-	0.50	3.51	-	5.09	m2	5.47
Betoatlas interlocking units 250mm long x 500mm wide x 200mm modular deep in wall 500mm wide	53.37	5.00	56.04	-	5.00	35.10	-	91.14	m2	97.97
Extra over for colours	3.00	5.00	3.15	-	0.10	0.70	-	3.85	m2	4.14
Betonap 150/50 woven mesh geotextile as reinforcement to backfill	3.26	5.00	3.42	-	0.20	1.40	-	4.83	m2	5.19
graded stone filling behind Betoflor or Betoatlas.	14.50	5.00	15.23	-	3.00	21.06	-	36.28	m3	39.01
concrete haunching to base of blocks	3.86	5.00	4.05	-	0.36	2.53	-	6.58	m	7.07
Betoflor range; Betojard precast concrete vertical acoustic wall including soil filling to pockets but excluding excavation, concrete foundations, stone backfill to rear of wall and planting which are all deemed measured separately										
Betojard standard 500mm long x 250mm wide x 200mm deep units in wall 250mm wide	49.57	5.00	52.05	-	4.00	28.08	-	80.13	m2	86.14
Extra over for colours	1.50	5.00	1.58	-	0.05	0.35	-	1.93	m2	2.07
Betojard pivot 290mm x 250mm wide x 200mm deep ...	49.57	5.00	52.05	-	5.00	35.10	-	87.15	m2	93.68
Extra over for colours	1.50	5.00	1.58	-	0.05	0.35	-	1.93	m2	2.07
Extra over fill core to pivot block with reinforced concrete with 4 No 10mm diameter vertical mild steel bars and 6mm links	4.67	5.00	4.90	-	1.50	10.53	-	15.43	m	16.59
Extra over dowelled connection between pivot and standard block with concrete infill to standard block and 1 No 10mm diameter x 200mm long galvanised dowel debonded one end at 300mm centres vertically.......................................	5.01	5.00	5.26	-	0.75	5.26	-	10.53	m	11.31
concrete haunching to base of blocks	3.86	5.00	4.05	-	0.36	2.53	-	6.58	m	7.07
D50: UNDERPINNING										
Information										
The work of underpinning in this section comprises work to be carried out in short lengths; prices are exclusive of shoring and other temporary supports										
Excavating										
Preliminary trenches maximum depth not exceeding										
1.00m	-	-	-	-	4.00	28.08	-	28.08	m3	30.19
2.00m	-	-	-	-	5.00	35.10	-	35.10	m3	37.73
4.00m	-	%	-	-	6.00	42.12	-	42.12	m3	45.28
Underpinning pits maximum depth not exceeding										
1.00m	-	-	-	-	5.20	36.50	-	36.50	m3	39.24

Labour hourly rates: (except Specialists) Craft Operatives £9.23 Labour £7.02 Rates are national average prices. Refer to REGIONAL VARIATIONS for indicative levels of overall pricing in regions	MATERIALS			LABOUR				RATES		
	Del to Site	Waste	Material Cost	Craft Optve	Lab	Labour Cost	Sunds	Nett Rate	Unit	Gross Rate (+7.5%)
	£	%	£	Hrs	Hrs	£	£	£		£
D50: UNDERPINNING Cont.										
Excavating Cont.										
Underpinning pits Cont. maximum depth not exceeding Cont.										
2.00m	-	-	-	-	6.50	45.63	-	45.63	m3	49.05
4.00m	-	-	-	-	7.80	54.76	-	54.76	m3	58.86

Earthwork support
  preliminary trenches; distance between opposing
  faces not exceeding 2.00m; maximum depth not
  exceeding										

1.00m	1.18	-	1.18	-	0.44	3.09	0.02	4.29	m2	4.61
2.00m	1.47	-	1.47	-	0.56	3.93	0.03	5.43	m2	5.84
4.00m	1.70	-	1.70	-	0.60	4.21	0.03	5.94	m2	6.39
preliminary trenches; distance between opposing faces 2.00 - 4.00m; maximum depth not exceeding										
1.00m	1.29	-	1.29	-	0.46	3.23	0.02	4.54	m2	4.88
2.00m	1.62	-	1.62	-	0.58	4.07	0.03	5.72	m2	6.15
4.00m	1.77	-	1.77	-	0.70	4.91	0.04	6.72	m2	7.23
underpinning pits; distance between opposing faces not exceeding 2.00m; maximum depth not exceeding										
1.00m	1.37	-	1.37	-	0.51	3.58	0.02	4.97	m2	5.34
2.00m	1.72	-	1.72	-	0.64	4.49	0.03	6.24	m2	6.71
4.00m	1.92	-	1.92	-	0.69	4.84	0.04	6.80	m2	7.31
underpinning pits; distance between opposing faces 2.00 - 4.00m; maximum depth not exceeding										
1.00m	1.46	-	1.46	-	0.53	3.72	0.02	5.20	m2	5.59
2.00m	1.85	-	1.85	-	0.67	4.70	0.03	6.58	m2	7.08
4.00m	2.13	-	2.13	-	0.81	5.69	0.04	7.86	m2	8.45
Cutting away existing projecting foundations										
masonry; maximum width 103mm; maximum depth 150mm	-	-	-	-	0.80	5.62	-	5.62	m	6.04
masonry; maximum width 154mm; maximum depth 225mm	-	-	-	-	0.95	6.67	-	6.67	m	7.17
concrete; maximum width 253mm; maximum depth 190mm	-	-	-	-	1.15	8.07	-	8.07	m	8.68
concrete; maximum width 304mm; maximum depth 300mm	-	-	-	-	1.35	9.48	-	9.48	m	10.19
Preparing the underside of the existing work to receive the pinning up of the new work										
350mm wide	-	-	-	-	0.40	2.81	-	2.81	m	3.02
500mm wide	-	-	-	-	0.50	3.51	-	3.51	m	3.77
1000mm wide	-	-	-	-	1.00	7.02	-	7.02	m	7.55
Compacting bottoms of excavations	-	-	-	-	0.40	2.81	-	2.81	m2	3.02

Disposal of excavated material
  removing from site to tip (including tipping
  charges but excluding landfill tax).
	-	-	-	-	3.00	21.06	10.60	31.66	m3	34.03

Excavated material arising from excavations										
Filling to excavations average thickness exceeding 0.25m	-	-	-	-	2.00	14.04	-	14.04	m3	15.09
Compacting filling	-	-	-	-	0.20	1.40	-	1.40	m2	1.51
Plain in-situ concrete; B.S.5328, ordinary prescribed **mix ST3, 20mm aggregate**										
Foundations; poured on or against earth or unblinded hardcore										
generally	48.30	7.50	51.92	-	4.75	33.34	-	85.27	m3	91.66
Plain in-situ concrete; B.S.5328, ordinary prescribed **mix ST4, 20mm aggregate**										
Foundations; poured on or against earth or unblinded hardcore										
generally	49.60	7.50	53.32	-	4.75	33.34	-	86.67	m3	93.16
Formwork and basic finish										
Sides of foundations; plain vertical										
height exceeding 1.00m	5.97	10.00	6.57	3.50	0.70	37.22	1.53	45.32	m2	48.71
height not exceeding 250mm	1.80	10.00	1.98	1.06	0.22	11.33	0.47	13.78	m	14.81
height 250 - 500mm	3.28	10.00	3.61	1.92	0.38	20.39	0.83	24.83	m	26.69
height 0.50m - 1.00m	6.02	10.00	6.62	3.58	0.72	38.10	1.58	46.30	m	49.77
Common bricks, B.S.3921, Category M, 215 x 102.5 x **65mm, compressive strength not less than 5.2 N/mm2;** **in cement mortar (1:3)**										
Walls; vertical										
215mm thick; English bond	17.73	5.00	18.62	4.70	3.40	67.25	4.78	90.65	m2	97.44
327mm thick; English bond	26.67	5.00	28.00	5.70	4.20	82.09	7.56	117.66	m2	126.48
440mm thick; English bond	35.46	5.00	37.23	6.30	4.75	91.49	9.94	138.67	m2	149.07
Bonding to existing including extra material										
thickness of new work 215mm	1.94	-	1.94	1.32	0.88	18.36	0.16	20.46	m	22.00
thickness of new work 327mm	2.91	-	2.91	1.90	1.27	26.45	0.21	29.57	m	31.79
thickness of new work 440mm	3.87	-	3.87	2.50	1.67	34.80	0.25	38.92	m	41.84

GROUNDWORK

Labour hourly rates: (except Specialists) Craft Operatives £9.23 Labour £7.02 Rates are national average prices. Refer to REGIONAL VARIATIONS for indicative levels of overall pricing in regions	MATERIALS			LABOUR				RATES		
	Del to Site £	Waste %	Material Cost £	Craft Optve Hrs	Lab Hrs	Labour Cost £	Sunds £	Nett Rate £	Unit	Gross Rate (+7.5%) £
D50: UNDERPINNING Cont.										
Milton Hall' Second Hard Stock bricks, B.S.3921, Category M, 215 x 102.5 x 65mm, in cement mortar (1:3)										
Walls; vertical										
215mm thick; English bond	34.51	5.00	36.24	4.70	3.40	67.25	4.78	108.26	m2	116.38
327mm thick; English bond	51.91	5.00	54.51	5.70	4.20	82.09	7.56	144.16	m2	154.97
440mm thick; English bond	69.02	5.00	72.47	6.30	4.75	91.49	9.94	173.91	m2	186.95
Bonding to existing including extra material										
thickness of new work 215mm	3.78	-	3.78	1.32	0.88	18.36	0.16	22.30	m	23.97
thickness of new work 327mm	5.65	-	5.65	1.90	1.27	26.45	0.21	32.31	m	34.74
thickness of new work 440mm	7.53	-	7.53	2.50	1.67	34.80	0.25	42.58	m	45.77
Engineering bricks, B.S.3921, Category F, 215 x 102.5 x 65mm, class A; in cement mortar (1:3)										
Walls; vertical										
215mm thick; English bond	59.40	5.00	62.37	5.20	3.75	74.32	4.78	141.47	m2	152.08
327mm thick; English bond	89.34	5.00	93.81	6.30	4.60	90.44	7.56	191.81	m2	206.19
440mm thick; English bond	118.08	5.00	123.98	6.95	5.25	101.00	9.94	234.93	m2	252.55
Bonding to existing including extra material										
thickness of new work 215mm	6.47	-	6.47	1.45	0.97	20.19	0.16	26.82	m	28.83
thickness of new work 327mm	9.72	-	9.72	2.10	1.40	29.21	0.21	39.14	m	42.08
thickness of new work 440mm	12.97	-	12.97	2.75	1.83	38.23	0.25	51.45	m	55.31
Engineering bricks, B.S.3921, Category F, 215 x 102.5 x 65mm, Class B; in cement mortar (1:3)										
Walls; vertical										
215mm thick; English bond	35.36	5.00	37.13	4.33	3.75	66.29	4.78	108.20	m2	116.31
327mm thick; English bond	53.19	5.00	55.85	6.30	4.60	90.44	7.56	153.85	m2	165.39
440mm thick; English bond	70.72	5.00	74.26	6.95	5.25	101.00	9.95	185.21	m2	199.10
Bonding to existing including extra material										
thickness of new work 215mm	3.87	-	3.87	1.45	0.97	20.19	0.16	24.22	m	26.04
thickness of new work 327mm	5.78	-	5.78	2.10	1.40	29.21	0.21	35.20	m	37.84
thickness of new work 440mm	7.73	-	7.73	2.75	1.83	38.23	0.25	46.21	m	49.67
Sundry items										
Wedging and pinning up to underside of existing construction with two courses slates in cement mortar (1:3)										
215mm walls	6.67	15.00	7.67	0.60	0.40	8.35	0.13	16.15	m	17.36
327mm walls	9.29	15.00	10.68	0.87	0.58	12.10	0.16	22.95	m	24.67
440mm walls	12.08	15.00	13.89	1.14	0.76	15.86	0.18	29.93	m	32.17

In situ concrete/Large precast concrete

Labour hourly rates: (except Specialists) Craft Operatives £9.23 Labour £7.02 Rates are national average prices. Refer to REGIONAL VARIATIONS for indicative levels of overall pricing in regions	MATERIALS			LABOUR				RATES		
	Del to Site £	Waste %	Material Cost £	Craft Optve Hrs	Lab Hrs	Labour Cost £	Sunds £	Nett Rate £	Unit	Gross Rate (+7.5%) £
E10: MIXING/CASTING/CURING IN-SITU CONCRETE - (READY MIXED)										
Plain in-situ concrete; B.S.5328, ordinary prescribed mix ST3, 20mm aggregate										
Foundations poured on or against earth or unblinded hardcore										
generally	48.30	7.50	51.92	-	2.10	14.74	-	66.66	m3	71.66
Isolated foundations; poured on or against earth or unblinded hardcore										
generally	48.30	7.50	51.92	-	2.50	17.55	-	69.47	m3	74.68
Beds; poured on or against earth or unblinded hardcore										
thickness exceeding 450mm	48.30	7.50	51.92	-	2.00	14.04	-	65.96	m3	70.91
thickness 150 - 450mm	48.30	7.50	51.92	-	2.25	15.80	-	67.72	m3	72.80
thickness not exceeding 150mm	48.30	7.50	51.92	-	2.85	20.01	-	71.93	m3	77.32
Filling hollow walls										
thickness not exceeding 150mm	48.30	5.00	50.72	-	5.00	35.10	-	85.81	m3	92.25
Plain in-situ concrete; B.S.5328, ordinary prescribed mix ST4, 20mm aggregate										
Foundations poured on or against earth or unblinded hardcore										
generally	49.60	7.50	53.32	-	2.10	14.74	-	68.06	m3	73.17
Isolated foundations; poured on or against earth or unblinded hardcore										
generally	49.60	7.50	53.32	-	2.50	17.55	-	70.87	m3	76.19
Beds; poured on or against earth or unblinded hardcore										
thickness exceeding 450mm	49.60	7.50	53.32	-	2.00	14.04	-	67.36	m3	72.41
thickness 150 - 450mm	49.60	7.50	53.32	-	2.25	15.80	-	69.11	m3	74.30
thickness not exceeding 150mm	49.60	7.50	53.32	-	2.85	20.01	-	73.33	m3	78.83
Filling hollow walls										
thickness not exceeding 150mm	49.60	5.00	52.08	-	5.00	35.10	-	87.18	m3	93.72
Plain in-situ concrete; B.S.5328, ordinary prescribed mix ST5, 20mm aggregate										
Foundations poured on or against earth or unblinded hardcore										
generally	50.83	7.50	54.64	-	2.10	14.74	-	69.38	m3	74.59
Isolated foundations; poured on or against earth or unblinded hardcore										
generally	50.83	7.50	54.64	-	2.50	17.55	-	72.19	m3	77.61
Beds; poured on or against earth or unblinded hardcore										
thickness exceeding 450mm	50.83	7.50	54.64	-	2.00	14.04	-	68.68	m3	73.83
thickness 150 - 450mm	50.83	7.50	54.64	-	2.25	15.80	-	70.44	m3	75.72
thickness not exceeding 150mm	50.83	7.50	54.64	-	2.85	20.01	-	74.65	m3	80.25
Filling hollow walls										
thickness not exceeding 150mm	50.83	5.00	53.37	-	5.00	35.10	-	88.47	m3	95.11
Reinforced in-situ concrete; B.S.5328, designed mix C15, 20mm aggregate, minimum cement content 220 kg/m3; vibrated										
Foundations										
generally	48.60	5.00	51.03	-	2.35	16.50	0.50	68.03	m3	73.13

Labour hourly rates: (except Specialists) Craft Operatives £9.23 Labour £7.02 Rates are national average prices. Refer to REGIONAL VARIATIONS for indicative levels of overall pricing in regions	MATERIALS			LABOUR				RATES		
	Del to Site £	Waste %	Material Cost £	Craft Optve Hrs	Lab Hrs	Labour Cost £	Sunds £	Nett Rate £	Unit	Gross Rate (+7.5%) £
E10: MIXING/CASTING/CURING IN-SITU CONCRETE - (READY MIXED) Cont.										
Reinforced in-situ concrete; B.S.5328, designed mix C15, 20mm aggregate, minimum cement content 220 kg/m3; vibrated Cont.										
Ground beams										
generally...	48.60	5.00	**51.03**	-	3.25	22.82	0.50	74.34	m3	79.92
Isolated foundations										
generally...	48.60	5.00	**51.03**	-	2.75	19.31	0.50	70.83	m3	76.15
Beds										
thickness exceeding 450mm	48.60	5.00	**51.03**	-	2.25	15.80	0.50	67.33	m3	72.37
thickness 150 - 450mm	48.60	5.00	**51.03**	-	2.50	17.55	0.50	69.08	m3	74.26
thickness not exceeding 150mm	48.60	5.00	**51.03**	-	3.10	21.76	0.50	73.29	m3	78.79
Slabs										
thickness exceeding 450mm	48.60	2.50	**49.81**	-	2.90	20.36	0.50	70.67	m3	75.97
thickness 150 - 450mm	48.60	2.50	**49.81**	-	3.15	22.11	0.50	72.43	m3	77.86
thickness not exceeding 150mm	48.60	2.50	**49.81**	-	4.15	29.13	0.50	79.45	m3	85.41
Walls										
thickness exceeding 450mm	48.60	2.50	**49.81**	-	3.20	22.46	0.50	72.78	m3	78.24
thickness 150 - 450mm	48.60	2.50	**49.81**	-	3.50	24.57	0.50	74.89	m3	80.50
thickness not exceeding 150mm	48.60	2.50	**49.81**	-	4.50	31.59	0.50	81.91	m3	88.05
Beams										
isolated...	48.60	2.50	**49.81**	-	4.25	29.84	0.50	80.15	m3	86.16
isolated deep	48.60	2.50	**49.81**	-	4.40	30.89	0.50	81.20	m3	87.29
attached deep	48.60	2.50	**49.81**	-	4.25	29.84	0.50	80.15	m3	86.16
Beam casings										
isolated...	48.60	2.50	**49.81**	-	4.50	31.59	0.50	81.91	m3	88.05
isolated deep	48.60	2.50	**49.81**	-	4.65	32.64	0.50	82.96	m3	89.18
attached deep	48.60	2.50	**49.81**	-	4.50	31.59	0.50	81.91	m3	88.05
Columns										
generally...	48.60	2.50	**49.81**	-	5.75	40.37	0.50	90.68	m3	97.48
Column casings										
generally...	48.60	2.50	**49.81**	-	6.00	42.12	0.50	92.44	m3	99.37
Staircases										
generally...	48.60	2.50	**49.81**	-	5.00	35.10	0.50	85.42	m3	91.82
Upstands										
generally...	48.60	2.50	**49.81**	-	7.00	49.14	0.50	99.45	m3	106.91
Reinforced in-situ concrete; B.S.5328, designed mix C20, 20mm aggregate, minimum cement content 240 kg/m3; vibrated										
Foundations										
generally...	49.20	5.00	**51.66**	-	2.35	16.50	0.50	68.66	m3	73.81
Ground beams										
generally...	49.20	5.00	**51.66**	-	3.25	22.82	0.50	74.97	m3	80.60
Isolated foundations										
generally...	49.20	5.00	**51.66**	-	2.75	19.31	0.50	71.47	m3	76.82
Beds										
thickness exceeding 450mm	49.20	5.00	**51.66**	-	2.25	15.80	0.50	67.95	m3	73.05
thickness 150 - 450mm	49.20	5.00	**51.66**	-	2.50	17.55	0.50	69.71	m3	74.94
thickness not exceeding 150mm	49.20	5.00	**51.66**	-	3.10	21.76	0.50	73.92	m3	79.47
Slabs										
thickness exceeding 450mm	49.20	2.50	**50.43**	-	2.90	20.36	0.50	71.29	m3	76.63
thickness 150 - 450mm	49.20	2.50	**50.43**	-	3.15	22.11	0.50	73.04	m3	78.52
thickness not exceeding 150mm	49.20	2.50	**50.43**	-	4.15	29.13	0.50	80.06	m3	86.07
Walls										
thickness exceeding 450mm	49.20	2.50	**50.43**	-	3.20	22.46	0.50	73.39	m3	78.90
thickness 150 - 450mm	49.20	2.50	**50.43**	-	3.50	24.57	0.50	75.50	m3	81.16
thickness not exceeding 150mm	49.20	2.50	**50.43**	-	4.50	31.59	0.50	82.52	m3	88.71
Beams										
isolated...	49.20	2.50	**50.43**	-	4.25	29.84	0.50	80.77	m3	86.82
isolated deep	49.20	2.50	**50.43**	-	4.40	30.89	0.50	81.82	m3	87.95
attached deep	49.20	2.50	**50.43**	-	4.25	29.84	0.50	80.77	m3	86.82
Beam casings										
isolated...	49.20	2.50	**50.43**	-	4.50	31.59	0.50	82.52	m3	88.71
isolated deep	49.20	2.50	**50.43**	-	4.65	32.64	0.50	83.57	m3	89.84
attached deep	49.20	2.50	**50.43**	-	4.50	31.59	0.50	82.52	m3	88.71
Columns										
generally...	49.20	2.50	**50.43**	-	5.75	40.37	0.50	91.30	m3	98.14
Column casings										
generally...	49.20	2.50	**50.43**	-	6.00	42.12	0.50	93.05	m3	100.03
Staircases										
generally...	49.20	2.50	**50.43**	-	5.00	35.10	0.50	86.03	m3	92.48

Labour hourly rates: (except Specialists) Craft Operatives £9.23 Labour £7.02 Rates are national average prices. Refer to REGIONAL VARIATIONS for indicative levels of overall pricing in regions	MATERIALS			LABOUR				RATES		
	Del to Site £	Waste %	Material Cost £	Craft Optve Hrs	Lab Hrs	Labour Cost £	Sunds £	Nett Rate £	Unit	Gross Rate (+7.5%) £
E10: MIXING/CASTING/CURING IN-SITU CONCRETE - (READY MIXED) Cont.										
Reinforced in-situ concrete; B.S.5328, designed mix C20, 20mm aggregate, minimum cement content 240 kg/m3; vibrated Cont.										
Upstands										
generally	49.20	2.50	50.43	-	7.00	49.14	0.50	100.07	m3	107.58
Reinforced in-situ concrete; B.S.5328, designed mix C25, 20mm aggregate, minimum cement content 290 kg/m3; vibrated										
Foundations										
generally	49.90	5.00	52.40	-	2.35	16.50	0.50	69.39	m3	74.60
Ground beams										
generally	49.90	5.00	52.40	-	3.25	22.82	0.50	75.71	m3	81.39
Isolated foundations										
generally	49.90	5.00	52.40	-	2.75	19.31	0.50	72.20	m3	77.61
Beds										
thickness exceeding 450mm	49.90	5.00	52.40	-	2.25	15.80	0.50	68.69	m3	73.84
thickness 150 - 450mm	49.90	5.00	52.40	-	2.50	17.55	0.50	70.44	m3	75.73
thickness not exceeding 150mm	49.90	5.00	52.40	-	3.10	21.76	0.50	74.66	m3	80.26
Slabs										
thickness exceeding 450mm	49.90	2.50	51.15	-	2.90	20.36	0.50	72.01	m3	77.41
thickness 150 - 450mm	49.90	2.50	51.15	-	3.15	22.11	0.50	73.76	m3	79.29
thickness not exceeding 150mm	49.90	2.50	51.15	-	4.15	29.13	0.50	80.78	m3	86.84
Walls										
thickness exceeding 450mm	49.90	2.50	51.15	-	3.20	22.46	0.50	74.11	m3	79.67
thickness 150 - 450mm	49.90	2.50	51.15	-	3.50	24.57	0.50	76.22	m3	81.93
thickness not exceeding 150mm	49.90	2.50	51.15	-	4.50	31.59	0.50	83.24	m3	89.48
Beams										
isolated	49.90	2.50	51.15	-	4.25	29.84	0.50	81.48	m3	87.59
isolated deep	49.90	2.50	51.15	-	4.40	30.89	0.50	82.54	m3	88.73
attached deep	49.90	2.50	51.15	-	4.25	29.84	0.50	81.48	m3	87.59
Beam casings										
isolated	49.90	2.50	51.15	-	4.50	31.59	0.50	83.24	m3	89.48
isolated deep	49.90	2.50	51.15	-	4.65	32.64	0.50	84.29	m3	90.61
attached deep	49.90	2.50	51.15	-	4.50	31.59	0.50	83.24	m3	89.48
Columns										
generally	49.90	2.50	51.15	-	5.75	40.37	0.50	92.01	m3	98.91
Column casings										
generally	49.90	2.50	51.15	-	6.00	42.12	0.50	93.77	m3	100.80
Staircases										
generally	49.90	2.50	51.15	-	5.00	35.10	0.50	86.75	m3	93.25
Upstands										
generally	49.90	2.50	51.15	-	7.00	49.14	0.50	100.79	m3	108.35
Reinforced in-situ concrete; B.S.5328, designed mix C30, 20mm aggregate, minimum cement content 290 kg/m3; vibrated										
Foundations										
generally	51.10	5.00	53.66	-	2.35	16.50	0.50	70.65	m3	75.95
Ground beams										
generally	51.10	5.00	53.66	-	3.25	22.82	0.50	76.97	m3	82.74
Isolated foundations										
generally	51.10	5.00	53.66	-	2.75	19.31	0.50	73.46	m3	78.97
Beds										
thickness exceeding 450mm	51.10	5.00	53.66	-	2.25	15.80	0.50	69.95	m3	75.20
thickness 150 - 450mm	51.10	5.00	53.66	-	2.50	17.55	0.50	71.70	m3	77.08
thickness not exceeding 150mm	51.10	5.00	53.66	-	3.10	21.76	0.50	75.92	m3	81.61
Slabs										
thickness exceeding 450mm	51.10	2.50	52.38	-	2.90	20.36	0.50	73.24	m3	78.73
thickness 150 - 450mm	51.10	2.50	52.38	-	3.15	22.11	0.50	74.99	m3	80.61
thickness not exceeding 150mm	51.10	2.50	52.38	-	4.15	29.13	0.50	82.01	m3	88.16
Walls										
thickness exceeding 450mm	51.10	2.50	52.38	-	3.20	22.46	0.50	75.34	m3	80.99
thickness 150 - 450mm	51.10	2.50	52.38	-	3.50	24.57	0.50	77.45	m3	83.26
thickness not exceeding 150mm	51.10	2.50	52.38	-	4.50	31.59	0.50	84.47	m3	90.80
Beams										
isolated	51.10	2.50	52.38	-	4.25	29.84	0.50	82.71	m3	88.92
isolated deep	51.10	2.50	52.38	-	4.40	30.89	0.50	83.77	m3	90.05
attached deep	51.10	2.50	52.38	-	4.25	29.84	0.50	82.71	m3	88.92
Beam casings										
isolated	51.10	2.50	52.38	-	4.50	31.59	0.50	84.47	m3	90.80
isolated deep	51.10	2.50	52.38	-	4.65	32.64	0.50	85.52	m3	91.93

Labour hourly rates: (except Specialists) Craft Operatives £9.23 Labour £7.02 Rates are national average prices. Refer to REGIONAL VARIATIONS for indicative levels of overall pricing in regions	MATERIALS			LABOUR				RATES		
	Del to Site £	Waste %	Material Cost £	Craft Optve Hrs	Lab Hrs	Labour Cost £	Sunds £	Nett Rate £	Unit	Gross Rate (+7.5%) £
E10: MIXING/CASTING/CURING IN-SITU CONCRETE – (READY MIXED) Cont.										
Reinforced in-situ concrete; B.S.5328, designed mix C30, 20mm aggregate, minimum cement content 290 kg/m3; vibrated Cont.										
Beam casings Cont.										
attached deep	51.10	2.50	**52.38**	–	4.50	31.59	0.50	84.47	m3	90.80
Columns										
generally	51.10	2.50	**52.38**	–	5.75	40.37	0.50	93.24	m3	100.24
Column casings										
generally	51.10	2.50	**52.38**	–	6.00	42.12	0.50	95.00	m3	102.12
Upstands										
generally	51.10	2.50	**52.38**	–	7.00	49.14	0.50	102.02	m3	109.67
E10: MIXING/CASTING/CURING IN-SITU CONCRETE – (SITE MIXED)										
Plain in-situ concrete; mix 1:6, all in aggregate										
Foundations poured on or against earth or unblinded hardcore										
generally	63.35	7.50	**68.10**	–	3.10	21.76	1.50	91.36	m3	98.22
Isolated foundations; poured on or against earth or unblinded hardcore										
generally	63.35	7.50	**68.10**	–	3.50	24.57	1.50	94.17	m3	101.23
Beds; poured on or against earth or unblinded hardcore										
thickness 150 - 450mm	63.35	7.50	**68.10**	–	3.25	22.82	1.50	92.42	m3	99.35
thickness not exceeding 150mm	63.35	7.50	**68.10**	–	3.85	27.03	1.50	96.63	m3	103.88
Beds; sloping not exceeding 15 degrees; poured on or against earth or unblinded hardcore										
thickness 150 - 450mm	63.35	7.50	**68.10**	–	3.45	24.22	1.50	93.82	m3	100.86
thickness not exceeding 150mm	63.35	7.50	**68.10**	–	4.05	28.43	1.50	98.03	m3	105.38
Beds; sloping exceeding 15 degrees; poured on or against earth or unblinded hardcore										
thickness 150 - 450mm	63.35	7.50	**68.10**	–	3.65	25.62	1.50	95.22	m3	102.37
thickness not exceeding 150mm	63.35	7.50	**68.10**	–	4.25	29.84	1.50	99.44	m3	106.89
Plain in-situ concrete; mix 1:8, all in aggregate										
Foundations poured on or against earth or unblinded hardcore										
generally	56.91	7.50	**61.18**	–	3.10	21.76	1.50	84.44	m3	90.77
Isolated foundations; poured on or against earth or unblinded hardcore										
generally	56.91	7.50	**61.18**	–	3.50	24.57	1.50	87.25	m3	93.79
Beds; poured on or against earth or unblinded hardcore										
thickness 150 - 450mm	56.91	7.50	**61.18**	–	3.25	22.82	1.50	85.49	m3	91.91
thickness not exceeding 150mm	56.91	7.50	**61.18**	–	3.85	27.03	1.50	89.71	m3	96.43
Beds; sloping not exceeding 15 degrees; poured on or against earth or unblinded hardcore										
thickness 150 - 450mm	56.91	7.50	**61.18**	–	3.45	24.22	1.50	86.90	m3	93.41
thickness not exceeding 150mm	56.91	7.50	**61.18**	–	4.05	28.43	1.50	91.11	m3	97.94
Beds; sloping exceeding 15 degrees; poured on or against earth or unblinded hardcore										
thickness 150 - 450mm	56.91	7.50	**61.18**	–	3.65	25.62	1.50	88.30	m3	94.92
thickness not exceeding 150mm	56.91	7.50	**61.18**	–	4.25	29.84	1.50	92.51	m3	99.45
Plain in-situ concrete; mix 1:12, all in aggregate										
Foundations poured on or against earth or unblinded hardcore										
generally	50.45	7.50	**54.23**	–	3.10	21.76	1.50	77.50	m3	83.31
Isolated foundations; poured on or against earth or unblinded hardcore										
generally	50.45	7.50	**54.23**	–	3.50	24.57	1.50	80.30	m3	86.33
Beds; poured on or against earth or unblinded hardcore										
thickness 150 - 450mm	50.45	7.50	**54.23**	–	3.25	22.82	1.50	78.55	m3	84.44
thickness not exceeding 150mm	50.45	7.50	**54.23**	–	3.85	27.03	1.50	82.76	m3	88.97
Beds; sloping not exceeding 15 degrees; poured on or against earth or unblinded hardcore										
thickness 150 - 450mm	50.45	7.50	**54.23**	–	3.45	24.22	1.50	79.95	m3	85.95
thickness not exceeding 150mm	50.45	7.50	**54.23**	–	4.05	28.43	1.50	84.16	m3	90.48
Beds; sloping exceeding 15 degrees; poured on or against earth or unblinded hardcore										
thickness 150 - 450mm	50.45	7.50	**54.23**	–	3.65	25.62	1.50	81.36	m3	87.46
thickness not exceeding 150mm	50.45	7.50	**54.23**	–	4.25	29.84	1.50	85.57	m3	91.99

Labour hourly rates: (except Specialists) Craft Operatives £9.23 Labour £7.02 Rates are national average prices. Refer to REGIONAL VARIATIONS for indicative levels of overall pricing in regions	MATERIALS			LABOUR				RATES		
	Del to Site £	Waste %	Material Cost £	Craft Optve Hrs	Lab Hrs	Labour Cost £	Sunds £	Nett Rate £	Unit	Gross Rate (+7.5%) £
E10: MIXING/CASTING/CURING IN-SITU CONCRETE – (SITE MIXED) Cont.										
Plain in-situ concrete; B.S.5328, ordinary prescribed mix C15P, 20mm aggregate										
Foundations poured on or against earth or unblinded hardcore										
generally	64.51	7.50	**69.35**	–	3.10	**21.76**	1.50	**92.61**	m3	**99.56**
Isolated foundations; poured on or against earth or unblinded hardcore										
generally	64.51	7.50	**69.35**	–	3.50	**24.57**	1.50	**95.42**	m3	**102.57**
Beds; poured on or against earth or unblinded hardcore										
thickness exceeding 450mm	64.51	7.50	**69.35**	–	3.00	**21.06**	1.50	**91.91**	m3	**98.80**
thickness 150 - 450mm	64.51	7.50	**69.35**	–	3.25	**22.82**	1.50	**93.66**	m3	**100.69**
thickness not exceeding 150mm	64.51	7.50	**69.35**	–	3.85	**27.03**	1.50	**97.88**	m3	**105.22**
Filling hollow walls										
thickness not exceeding 150mm	64.51	5.00	**67.74**	–	6.00	**42.12**	1.50	**111.36**	m3	**119.71**
Plain in-situ concrete; B.S.5328, ordinary prescribed mix C20P, 20mm aggregate										
Foundations poured on or against earth or unblinded hardcore										
generally	66.40	7.50	**71.38**	–	3.10	**21.76**	1.50	**94.64**	m3	**101.74**
Isolated foundations; poured on or against earth or unblinded hardcore										
generally	66.40	7.50	**71.38**	–	3.50	**24.57**	1.50	**97.45**	m3	**104.76**
Beds; poured on or against earth or unblinded hardcore										
thickness exceeding 450mm	66.40	7.50	**71.38**	–	3.00	**21.06**	1.50	**93.94**	m3	**100.99**
thickness 150 - 450mm	66.40	7.50	**71.38**	–	3.25	**22.82**	1.50	**95.69**	m3	**102.87**
thickness not exceeding 150mm	66.40	7.50	**71.38**	–	3.85	**27.03**	1.50	**99.91**	m3	**107.40**
Filling hollow walls										
thickness not exceeding 150mm	66.40	5.00	**69.72**	–	6.00	**42.12**	1.50	**113.34**	m3	**121.84**
Plain in-situ concrete; B.S.5328, ordinary prescribed mix C25P, 20mm aggregate										
Foundations poured on or against earth or unblinded hardcore										
generally	69.16	7.50	**74.35**	–	3.10	**21.76**	1.50	**97.61**	m3	**104.93**
Isolated foundations; poured on or against earth or unblinded hardcore										
generally	69.16	7.50	**74.35**	–	3.50	**24.57**	1.50	**100.42**	m3	**107.95**
Beds; poured on or against earth or unblinded hardcore										
thickness exceeding 450mm	69.16	7.50	**74.35**	–	3.00	**21.06**	1.50	**96.91**	m3	**104.18**
thickness 150 - 450mm	69.16	7.50	**74.35**	–	3.25	**22.82**	1.50	**98.66**	m3	**106.06**
thickness not exceeding 150mm	69.16	7.50	**74.35**	–	3.85	**27.03**	1.50	**102.87**	m3	**110.59**
Filling hollow walls										
thickness not exceeding 150mm	69.16	5.00	**72.62**	–	6.00	**42.12**	1.50	**116.24**	m3	**124.96**
Reinforced in-situ concrete; B.S.5328, designed mix C15, 20mm aggregate, minimum cement content 220 kg/m3; vibrated										
Foundations										
generally	60.30	5.00	**63.31**	–	3.35	**23.52**	2.00	**88.83**	m3	**95.49**
Ground beams										
generally	60.30	5.00	**63.31**	–	4.25	**29.84**	2.00	**95.15**	m3	**102.29**
Isolated foundations										
generally	60.30	5.00	**63.31**	–	3.75	**26.32**	2.00	**91.64**	m3	**98.51**
Beds										
thickness exceeding 450mm	60.30	5.00	**63.31**	–	3.25	**22.82**	2.00	**88.13**	m3	**94.74**
thickness 150 - 450mm	60.30	5.00	**63.31**	–	3.50	**24.57**	2.00	**89.89**	m3	**96.63**
thickness not exceeding 150mm	60.30	5.00	**63.31**	–	4.10	**28.78**	2.00	**94.10**	m3	**101.15**
Reinforced in-situ concrete; B.S.5328, designed mix C20, 20mm aggregate, minimum cement content 240 kg/m3; vibrated										
Foundations										
generally	61.36	5.00	**64.43**	–	3.35	**23.52**	2.00	**89.94**	m3	**96.69**
Ground beams										
generally	61.36	5.00	**64.43**	–	4.25	**29.84**	2.00	**96.26**	m3	**103.48**
Isolated foundations										
generally	61.36	5.00	**64.43**	–	3.75	**26.32**	2.00	**92.75**	m3	**99.71**
Beds										
thickness exceeding 450mm	61.36	5.00	**64.43**	–	3.25	**22.82**	2.00	**89.24**	m3	**95.94**

IN SITU CONCRETE/LARGE PRECAST CONCRETE

Labour hourly rates: (except Specialists) Craft Operatives £9.23 Labour £7.02 Rates are national average prices. Refer to REGIONAL VARIATIONS for indicative levels of overall pricing in regions	MATERIALS			LABOUR				RATES		
	Del to Site £	Waste %	Material Cost £	Craft Optve Hrs	Lab Hrs	Labour Cost £	Sunds £	Nett Rate £	Unit	Gross Rate (+7.5%) £
E10: MIXING/CASTING/CURING IN-SITU CONCRETE - (SITE MIXED) Cont.										
Reinforced in-situ concrete; B.S.5328, designed mix C20, 20mm aggregate, minimum cement content 240 kg/m3; vibrated Cont.										
Beds Cont.										
thickness 150 - 450mm	61.36	5.00	64.43	-	3.50	24.57	2.00	91.00	m3	97.82
thickness not exceeding 150mm	61.36	5.00	64.43	-	4.10	28.78	2.00	95.21	m3	102.35
Reinforced in-situ concrete; B.S.5328, designed mix C25, 20mm aggregate, minimum cement content 290 kg/m3; vibrated										
Foundations										
generally	64.95	5.00	68.20	-	3.35	23.52	2.00	93.71	m3	100.74
Ground beams										
generally	64.95	5.00	68.20	-	4.25	29.84	2.00	100.03	m3	107.53
Isolated foundations										
generally	64.95	5.00	68.20	-	3.75	26.32	2.00	96.52	m3	103.76
Beds										
thickness exceeding 450mm	64.95	5.00	68.20	-	3.25	22.82	2.00	93.01	m3	99.99
thickness 150 - 450mm	64.95	5.00	68.20	-	3.50	24.57	2.00	94.77	m3	101.88
thickness not exceeding 150mm	64.95	5.00	68.20	-	4.10	28.78	2.00	98.98	m3	106.40
Cement and sand (1:3); for grouting and the like										
Grouting.										
stanchion bases.	1.80	10.00	1.98	-	0.40	2.81	0.90	5.69	nr	6.11
E10: MIXING/CASTING/CURING IN-SITU CONCRETE - (SUNDRIES)										
The foregoing concrete is based on the use of Portland cement. For other cements and waterproofers ADD as follows										
Concrete 1:6										
rapid hardening cement	3.47	-	3.47	-	-	-	-	3.47	m3	3.73
sulphate resisting cement	4.35	-	4.35	-	-	-	-	4.35	m3	4.68
waterproofing powder	10.03	-	10.03	-	-	-	-	10.03	m3	10.78
waterproofing liquid	5.11	-	5.11	-	-	-	-	5.11	m3	5.49
Concrete 1:8										
rapid hardening cement	2.61	-	2.61	-	-	-	-	2.61	m3	2.81
sulphate resisting cement	3.27	-	3.27	-	-	-	-	3.27	m3	3.52
waterproofing powder	7.53	-	7.53	-	-	-	-	7.53	m3	8.09
waterproofing liquid	5.11	-	5.11	-	-	-	-	5.11	m3	5.49
Concrete 1:12										
rapid hardening cement	1.74	-	1.74	-	-	-	-	1.74	m3	1.87
sulphate resisting cement	2.17	-	2.17	-	-	-	-	2.17	m3	2.33
waterproofing powder	5.01	-	5.01	-	-	-	-	5.01	m3	5.39
waterproofing liquid	5.11	-	5.11	-	-	-	-	5.11	m3	5.49
Concrete ST3 and C15										
rapid hardening cement	2.32	-	2.32	-	-	-	-	2.32	m3	2.49
sulphate resisting cement	2.89	-	2.89	-	-	-	-	2.89	m3	3.11
waterproofing powder	6.69	-	6.69	-	-	-	-	6.69	m3	7.19
waterproofing liquid	5.11	-	5.11	-	-	-	-	5.11	m3	5.49
Concrete ST4 and C20										
rapid hardening cement	3.43	-	3.43	-	-	-	-	3.43	m3	3.69
sulphate resisting cement	4.30	-	4.30	-	-	-	-	4.30	m3	4.62
waterproofing powder	9.89	-	9.89	-	-	-	-	9.89	m3	10.63
waterproofing liquid	5.11	-	5.11	-	-	-	-	5.11	m3	5.49
Concrete ST5 and C25										
rapid hardening cement	4.63	-	4.63	-	-	-	-	4.63	m3	4.98
sulphate resisting cement	5.79	-	5.79	-	-	-	-	5.79	m3	6.22
waterproofing powder	13.36	-	13.36	-	-	-	-	13.36	m3	14.36
waterproofing liquid	5.11	-	5.11	-	-	-	-	5.11	m3	5.49
Reinforced in-situ lightweight concrete; 20.5 N/mm2; vibrated										
Walls										
thickness exceeding 450mm	79.00	2.50	80.97	-	4.40	30.89	2.00	113.86	m3	122.40
thickness 150 - 450mm	79.00	2.50	80.97	-	4.65	32.64	2.00	115.62	m3	124.29
thickness not exceeding 150mm	79.00	2.50	80.97	-	5.10	35.80	2.00	118.78	m3	127.69
Beam casings										
isolated	79.00	2.50	80.97	-	5.50	38.61	2.00	121.58	m3	130.70
Reinforced in-situ lightweight concrete; 26.0 N/mm2; vibrated										
Slabs										
thickness exceeding 450mm	82.80	2.50	84.87	-	4.00	28.08	2.00	114.95	m3	123.57
thickness 150 - 450mm	82.80	2.50	84.87	-	4.10	28.78	2.00	115.65	m3	124.33
thickness not exceeding 150mm	82.80	2.50	84.87	-	4.65	32.64	2.00	119.51	m3	128.48

Labour hourly rates: (except Specialists) Craft Operatives £9.23 Labour £7.02 Rates are national average prices. Refer to REGIONAL VARIATIONS for indicative levels of overall pricing in regions	MATERIALS			LABOUR				RATES		
	Del to Site	Waste	Material Cost	Craft Optve	Lab	Labour Cost	Sunds	Nett Rate	Unit	Gross Rate (+7.5%)
	£	%	£	Hrs	Hrs	£	£	£		£
E10: MIXING/CASTING/CURING IN-SITU CONCRETE – (SUNDRIES) Cont.										
Reinforced in-situ lightweight concrete; 41.5 N/mm2; vibrated										
Beams										
isolated............................	90.40	2.50	92.66	-	5.00	35.10	2.00	129.76	m3	139.49
Polythene sheeting as temporary protection to surface of concrete (use and waste)										
Polythene sheeting and laying										
125mu..............................	0.25	10.00	0.28	-	0.03	0.21	-	0.49	m2	0.52
Waterproof building paper, B.S.1521; laying on hardcore to receive concrete										
Waterproof building paper and laying										
Grade B1F.........................	0.35	25.00	0.44	-	0.05	0.35	-	0.79	m2	0.85
Grade B2..........................	0.28	25.00	0.35	-	0.05	0.35	-	0.70	m2	0.75
E20: FORMWORK FOR IN-SITU CONCRETE										
Formwork and basic finish										
Sides of foundations; plain vertical										
height exceeding 1.00m............	2.88	12.50	3.24	1.75	0.35	18.61	0.71	22.56	m2	24.25
height not exceeding 250mm........	0.89	12.50	1.00	0.53	0.11	5.66	0.22	6.89	m	7.40
height 250 - 500mm...............	1.57	12.50	1.77	0.96	0.19	10.19	0.41	12.37	m	13.30
height 0.50m - 1.00m.............	2.96	12.50	3.33	1.79	0.36	19.05	0.71	23.09	m	24.82
Sides of foundations; plain vertical; left in										
height exceeding 1.00m...........	12.96	-	12.96	1.00	0.20	10.63	0.43	24.02	m2	25.83
height not exceeding 250mm.......	3.97	-	3.97	0.30	0.06	3.19	0.15	7.31	m	7.86
height 250 - 500mm..............	7.14	-	7.14	0.55	0.11	5.85	0.24	13.23	m	14.22
height 0.50m - 1.00m............	13.27	-	13.27	1.03	0.21	10.98	0.46	24.71	m	26.56
Sides of ground beams and edges of beds; plain vertical										
height exceeding 1.00m...........	2.88	12.50	3.24	1.75	0.35	18.61	0.71	22.56	m2	24.25
height not exceeding 250mm.......	0.89	12.50	1.00	0.53	0.11	5.66	0.22	6.89	m	7.40
height 250 - 500mm..............	1.57	12.50	1.77	0.96	0.19	10.19	0.41	12.37	m	13.30
height 0.50m - 1.00m............	2.96	12.50	3.33	1.79	0.36	19.05	0.71	23.09	m	24.82
Sides of ground beams and edges of beds; plain vertical; left in										
height exceeding 1.00m...........	12.96	-	12.96	1.00	0.20	10.63	0.41	24.00	m2	25.80
height not exceeding 250mm.......	3.97	-	3.97	0.30	0.06	3.19	0.15	7.31	m	7.86
height 250 - 500mm..............	7.14	-	7.14	0.55	0.11	5.85	0.24	13.23	m	14.22
height 0.50m - 1.00m............	13.27	-	13.27	1.03	0.21	10.98	0.46	24.71	m	26.56
Edges of suspended slabs; plain vertical										
height not exceeding 250mm.......	1.09	12.50	1.23	0.66	0.13	7.00	0.25	8.48	m	9.12
height 250 - 500mm..............	2.00	12.50	2.25	1.20	0.24	12.76	0.49	15.50	m	16.66
extra; recesses; plain rectangular										
12 x 25mm.....................	0.02	-	0.02	0.05	-	0.46	-	0.48	m	0.52
25 x 50mm.....................	0.04	-	0.04	0.07	-	0.65	-	0.69	m	0.74
50 x 75mm.....................	0.14	-	0.14	0.09	-	0.83	-	0.97	m	1.04
50 x 100mm....................	0.21	-	0.21	0.11	-	1.02	-	1.23	m	1.32
extra; rebates; plain rectangular										
12 x 25mm.....................	0.02	-	0.02	0.07	-	0.65	-	0.67	m	0.72
25 x 50mm.....................	0.04	-	0.04	0.08	-	0.74	-	0.78	m	0.84
50 x 75mm.....................	0.14	-	0.14	0.10	-	0.92	-	1.06	m	1.14
50 x 100mm....................	0.21	-	0.21	0.13	-	1.20	-	1.41	m	1.52
extra; chamfers										
60mm wide.....................	0.07	-	0.07	0.05	-	0.46	-	0.53	m	0.57
75mm wide.....................	0.09	-	0.09	0.07	-	0.65	-	0.74	m	0.79
100mm wide....................	0.18	-	0.18	0.11	-	1.02	-	1.20	m	1.28
Edges of suspended slabs; plain vertical; curved 10m radius										
height not exceeding 250mm	1.59	12.50	1.79	1.00	0.20	10.63	0.43	12.85	m	13.82
Edges of suspended slabs; plain vertical; curved 1m radius										
height not exceeding 250mm	2.71	12.50	3.05	1.65	0.33	17.55	0.68	21.27	m	22.87
Edges of suspended slabs; plain vertical; curved 20m radius										
height not exceeding 250mm	1.21	12.50	1.36	0.74	0.15	7.88	0.28	9.52	m	10.24
Sides of upstands; plain vertical										
height not exceeding 250mm.......	1.26	12.50	1.42	0.75	0.15	7.98	0.28	9.67	m	10.40
height 250 - 500mm..............	1.49	12.50	1.68	0.88	0.18	9.39	0.37	11.43	m	12.29
extra; recesses; plain rectangular										
12 x 25mm.....................	0.02	-	0.02	0.05	-	0.46	-	0.48	m	0.52
25 x 50mm.....................	0.03	-	0.03	0.07	-	0.65	-	0.68	m	0.73
50 x 75mm.....................	0.14	-	0.14	0.09	-	0.83	-	0.97	m	1.04
50 x 100mm....................	0.21	-	0.21	0.11	-	1.02	-	1.23	m	1.32
extra; rebates; plain rectangular										
12 x 25mm.....................	0.02	%	0.02	0.07	-	0.65	-	0.67	m	0.72
25 x 50mm.....................	0.03	-	0.03	0.08	-	0.74	-	0.77	m	0.83
50 x 75mm.....................	0.14	-	0.14	0.10	-	0.92	-	1.06	m	1.14
50 x 100mm....................	0.21	-	0.21	0.13	-	1.20	-	1.41	m	1.52

Labour hourly rates: (except Specialists) Craft Operatives £9.23 Labour £7.02 Rates are national average prices. Refer to REGIONAL VARIATIONS for indicative levels of overall pricing in regions	MATERIALS			LABOUR				RATES		
	Del to Site £	Waste %	Material Cost £	Craft Optve Hrs	Lab Hrs	Labour Cost £	Sunds £	Nett Rate £	Unit	Gross Rate (+7.5%) £
E20: FORMWORK FOR IN-SITU CONCRETE Cont.										
Formwork and basic finish Cont.										
Sides of upstands; plain vertical Cont.										
extra; chamfers										
60mm wide..........................	0.07	-	0.07	0.05	-	0.46	-	0.53	m	0.57
75mm wide..........................	0.09	-	0.09	0.07	-	0.65	-	0.74	m	0.79
100mm wide.........................	0.18	-	0.18	0.11	-	1.02	-	1.20	m	1.28
Machine bases and plinths; plain vertical										
height exceeding 1.00m.............	3.15	12.50	3.54	1.93	0.38	20.48	0.78	24.81	m2	26.67
height not exceeding 250mm.........	0.94	12.50	1.06	0.58	0.11	6.13	0.24	7.42	m	7.98
height 250 - 500mm.................	1.72	12.50	1.94	1.06	0.20	11.19	0.44	13.56	m	14.58
height 0.50m - 1.00m...............	3.21	12.50	3.61	1.98	0.39	21.01	0.84	25.46	m	27.37
Soffits of slabs; horizontal										
slab thickness not exceeding 200mm; height to soffit										
1.50 - 3.00m.....................	8.69	10.00	9.56	1.50	0.30	15.95	1.00	26.51	m2	28.50
3.00 - 4.50m.....................	8.69	10.00	9.56	1.72	0.35	18.33	1.37	29.26	m2	31.46
4.50 - 6.00m.....................	8.69	10.00	9.56	1.94	0.40	20.71	1.71	31.98	m2	34.38
slab thickness 200 - 300mm; height to soffit										
1.50 - 3.00m.....................	9.59	10.00	10.55	1.65	0.33	17.55	1.12	29.22	m2	31.41
3.00 - 4.50m.....................	9.59	10.00	10.55	1.87	0.38	19.93	1.47	31.95	m2	34.34
4.50 - 6.00m.....................	9.59	10.00	10.55	2.09	0.43	22.31	1.80	34.66	m2	37.26
Soffits of slabs; horizontal; left in										
slab thickness not exceeding 200mm; height to soffit										
not exceeding 1.50m..............	23.30	-	23.30	1.00	0.20	10.63	0.64	34.57	m2	37.17
1.50 - 3.00m.....................	23.30	-	23.30	1.00	0.20	10.63	0.71	34.64	m2	37.24
Soffits of slabs; horizontal; with frequent uses of prefabricated panels										
slab thickness not exceeding 200mm; height to soffit										
1.50 - 3.00m.....................	2.62	15.00	3.01	1.50	0.30	15.95	1.00	19.96	m2	21.46
3.00 - 4.50m.....................	2.62	15.00	3.01	1.72	0.35	18.33	1.37	22.72	m2	24.42
4.50 - 6.00m.....................	2.62	15.00	3.01	1.94	0.40	20.71	1.71	25.44	m2	27.34
slab thickness 200 - 300mm; height to soffit										
1.50 - 3.00m.....................	2.85	15.00	3.28	1.65	0.33	17.55	1.09	21.91	m2	23.56
3.00 - 4.50m.....................	2.85	15.00	3.28	1.87	0.38	19.93	1.47	24.68	m2	26.53
4.50 - 6.00m.....................	2.85	15.00	3.28	2.09	0.43	22.31	1.80	27.39	m2	29.44
Soffits of slabs; sloping not exceeding 15 degrees										
slab thickness not exceeding 200mm; height to soffit										
1.50 - 3.00m.....................	9.60	10.00	10.56	1.65	0.33	17.55	1.08	29.19	m2	31.38
3.00 - 4.50m.....................	9.60	10.00	10.56	1.87	0.38	19.93	1.44	31.93	m2	34.32
4.50 - 6.00m.....................	9.60	10.00	10.56	2.09	0.43	22.31	1.80	34.67	m2	37.27
Soffits of slabs; sloping exceeding 15 degrees										
slab thickness not exceeding 200mm; height to soffit										
1.50 - 3.00m.....................	10.00	10.00	11.00	1.73	0.35	18.42	1.27	30.69	m2	33.00
3.00 - 4.50m.....................	10.00	10.00	11.00	1.95	0.40	20.81	1.62	33.43	m2	35.93
4.50 - 6.00m.....................	10.00	10.00	11.00	2.17	0.45	23.19	1.98	36.17	m2	38.88
Soffits of landings; horizontal										
slab thickness not exceeding 200mm; height to soffit										
not exceeding 1.50m..............	8.69	10.00	9.56	1.80	0.36	19.14	1.00	29.70	m2	31.93
1.50 - 3.00m.....................	8.69	10.00	9.56	2.02	0.41	21.52	1.37	32.45	m2	34.89
3.00 - 4.50m.....................	8.69	10.00	9.56	2.24	0.46	23.90	1.71	35.17	m2	37.81
Top formwork										
sloping exceeding 15 degrees	3.27	12.50	3.68	1.50	0.30	15.95	0.57	20.20	m2	21.71
Walls; vertical										
plain.........................	5.00	15.00	5.75	1.60	0.32	17.01	0.92	23.68	m2	25.46
plain; height exceeding 3.00m above floor level ..	5.62	15.00	6.46	1.60	0.32	17.01	1.03	24.51	m2	26.35
interrupted......................	5.52	15.00	6.35	1.75	0.35	18.61	1.01	25.97	m2	27.92
interrupted; height exceeding 3.00m above floor level	5.90	15.00	6.79	1.75	0.35	18.61	1.12	26.51	m2	28.50
extra; recesses; plain rectangular										
12 x 25mm......................	0.02	-	0.02	0.05	-	0.46	-	0.48	m	0.52
25 x 50mm......................	0.04	-	0.04	0.07	-	0.65	-	0.69	m	0.74
50 x 75mm......................	0.14	-	0.14	0.09	-	0.83	-	0.97	m	1.04
50 x 100mm.....................	0.21	-	0.21	0.11	-	1.02	-	1.23	m	1.32
Walls; vertical; curved 10m radius										
plain.........................	7.50	15.00	8.63	2.40	0.48	25.52	1.40	35.55	m2	38.21
Walls; vertical; curved 1m radius										
plain.........................	12.48	15.00	14.35	4.00	0.80	42.54	2.29	59.18	m2	63.62
Walls; vertical; curved 20m radius										
plain.........................	5.61	15.00	6.45	1.80	0.36	19.14	1.03	26.62	m2	28.62
Walls; battered										
plain.........................	5.82	15.00	6.69	1.92	0.38	20.39	1.12	28.20	m2	30.32
Beams; attached to slabs										
regular shaped; rectangular; height to soffit										
1.50 - 3.00m.....................	5.24	10.00	5.76	2.25	0.45	23.93	0.88	30.57	m2	32.86

Labour hourly rates: (except Specialists) Craft Operatives £9.23 Labour £7.02 Rates are national average prices. Refer to REGIONAL VARIATIONS for indicative levels of overall pricing in regions	MATERIALS			LABOUR				RATES		
	Del to Site £	Waste %	Material Cost £	Craft Optve Hrs	Lab Hrs	Labour Cost £	Sunds £	Nett Rate £	Unit	Gross Rate (+7.5%) £
E20: FORMWORK FOR IN-SITU CONCRETE Cont.										
Formwork and basic finish Cont.										
Beams; attached to slabs Cont.										
regular shaped; rectangular; height to soffit Cont.										
3.00 - 4.50m....................	5.24	10.00	5.76	2.25	0.45	23.93	1.07	30.76	m2	33.07
regular shaped; rectangular; with 50mm wide chamfers; height to soffit										
1.50 - 3.00m....................	5.34	10.00	5.87	2.31	0.45	24.48	0.88	31.23	m2	33.58
3.00 - 4.50m....................	5.34	10.00	5.87	2.31	0.45	24.48	1.07	31.42	m2	33.78
Beams; isolated										
regular shaped; rectangular; height to soffit										
1.50 - 3.00m....................	5.41	10.00	5.95	2.35	0.47	24.99	0.88	31.82	m2	34.21
3.00 - 4.50m....................	5.41	10.00	5.95	2.35	0.47	24.99	1.07	32.01	m2	34.41
regular shaped; rectangular; with 50mm wide chamfers; height to soffit										
1.50 - 3.00m....................	5.53	10.00	6.08	2.41	0.47	25.54	0.88	32.51	m2	34.94
3.00 - 4.50m....................	5.53	10.00	6.08	2.41	0.47	25.54	1.07	32.70	m2	35.15
Columns; attached to walls										
regular shaped; rectangular	3.64	10.00	4.00	2.30	0.46	24.46	0.83	29.29	m2	31.49
regular shaped; rectangular, height exceeding 3.00m above floor level	3.40	10.00	3.74	2.40	0.46	25.38	1.00	30.12	m2	32.38
Columns; isolated										
regular shaped; rectangular	3.64	10.00	4.00	2.25	0.45	23.93	0.88	28.81	m2	30.97
regular shaped; rectangular, height exceeding 3.00m above floor level	3.64	10.00	4.00	2.35	0.45	24.85	1.00	29.85	m2	32.09
regular shaped; rectangular; with 50mm wide chamfers	3.93	10.00	4.32	2.39	0.45	25.22	0.88	30.42	m2	32.70
regular shaped; rectangular; with 50mm wide chamfers; height exceeding 3.00m above floor level	3.93	10.00	4.32	2.50	0.45	26.23	1.00	31.56	m2	33.92
Column casings; attached to walls										
regular shaped; rectangular	3.64	10.00	4.00	2.30	0.46	24.46	0.88	29.34	m2	31.54
regular shaped; rectangular, height exceeding 3.00m above floor level	3.64	10.00	4.00	2.40	0.45	25.31	1.00	30.32	m2	32.59
Column casings; isolated										
regular shaped; rectangular	3.64	10.00	4.00	2.25	0.45	23.93	0.83	28.76	m2	30.92
regular shaped; rectangular, height exceeding 3.00m above floor level	3.64	10.00	4.00	2.30	0.45	24.39	1.00	29.39	m2	31.60
regular shaped; rectangular; with 50mm wide chamfers	3.93	10.00	4.32	2.39	0.45	25.22	0.88	30.42	m2	32.70
regular shaped; rectangular; with 50mm wide chamfers; height exceeding 3.00m above floor level	3.93	10.00	4.32	2.50	0.45	26.23	1.00	31.56	m2	33.92
Extra over a basic finish for a fine formed finish										
slabs	-	-	-	-	0.35	2.46	-	2.46	m2	2.64
walls	-	-	-	-	0.35	2.46	-	2.46	m2	2.64
beams	-	-	-	-	0.35	2.46	-	2.46	m2	2.64
columns	-	-	-	-	0.35	2.46	-	2.46	m2	2.64
Wall kickers										
straight	0.79	15.00	0.91	0.25	0.05	2.66	0.18	3.75	m	4.03
curved 2m radius	2.04	15.00	2.35	0.66	0.15	7.14	0.36	9.85	m	10.59
curved 10m radius	1.26	15.00	1.45	0.40	0.08	4.25	0.23	5.93	m	6.38
curved 20m radius	0.94	15.00	1.08	0.30	0.06	3.19	0.20	4.47	m	4.81
Suspended wall kickers										
straight	0.88	15.00	1.01	0.28	0.06	3.01	0.19	4.21	m	4.52
curved 2m radius	2.25	15.00	2.59	0.73	0.17	7.93	0.39	10.91	m	11.73
curved 10m radius	1.37	15.00	1.58	0.44	0.09	4.69	0.25	6.52	m	7.01
curved 20m radius	1.05	15.00	1.21	0.33	0.07	3.54	0.22	4.96	m	5.34
Wall ends, soffits and steps in walls; plain										
width not exceeding 250mm	1.70	15.00	1.96	0.53	0.11	5.66	0.28	7.90	m	8.49
Openings in walls; plain										
width not exceeding 250mm	1.70	15.00	1.96	0.53	0.11	5.66	0.28	7.90	m	8.49
Stairflights; 1000mm wide; 155mm thick waist; 178mm risers; includes formwork to soffits, risers and strings										
strings 300mm wide	12.97	10.00	14.27	6.05	1.21	64.34	2.08	80.68	m	86.73
string 300mm wide; junction with wall	12.97	10.00	14.27	6.05	1.21	64.34	2.08	80.68	m	86.73
Stairflights; 1500mm wide; 180mm thick waist; 178mm risers; includes formwork to soffits, risers and strings										
strings 325mm wide	16.22	10.00	17.84	7.56	1.51	80.38	2.60	100.82	m	108.38
string 325mm wide; junction with wall	16.22	10.00	17.84	7.56	1.51	80.38	2.60	100.82	m	108.38
Stairflights; 1000mm wide; 155mm thick waist; 178mm undercut risers; includes formwork to soffits, risers and strings										
strings 300mm wide	12.97	10.00	14.27	6.05	1.21	64.34	2.08	80.68	m	86.73
string 300mm wide; junction with wall	12.97	10.00	14.27	6.05	1.21	64.34	2.08	80.68	m	(86.73)
Stairflights; 1500mm wide; 180mm thick waist; 178mm undercut risers; includes formwork to soffits, risers and strings										
strings 325mm wide	16.22	10.00	17.84	7.56	1.51	80.38	2.60	100.82	m	108.38

IN SITU CONCRETE/LARGE PRECAST CONCRETE

Labour hourly rates: (except Specialists) Craft Operatives £9.23 Labour £7.02 Rates are national average prices. Refer to REGIONAL VARIATIONS for indicative levels of overall pricing in regions	MATERIALS			LABOUR				RATES		
	Del to Site £	Waste %	Material Cost £	Craft Optve Hrs	Lab Hrs	Labour Cost £	Sunds £	Nett Rate £	Unit	Gross Rate (+7.5%) £
E20: FORMWORK FOR IN-SITU CONCRETE Cont.										
Formwork and basic finish Cont.										
Stairflights; 1500mm wide; 180mm thick waist; 178mm undercut risers; includes formwork to soffits, risers and strings Cont.										
string 325mm wide; junction with wall	16.22	10.00	17.84	7.56	1.51	80.38	2.60	100.82	m	108.38
Mortices; rectangular										
girth not exceeding 500mm; depth not exceeding 250mm	0.77	10.00	0.85	0.48	0.10	5.13	0.20	6.18	nr	6.64
girth 0.50 - 1.00m; depth 250 - 500mm	2.56	10.00	2.82	1.54	0.31	16.39	0.63	19.84	nr	21.32
girth 0.50 - 1.00m; depth 0.50 - 1.00m	4.95	10.00	5.45	3.08	0.62	32.78	1.24	39.47	nr	42.43
Mortices; circular										
girth not exceeding 500mm; depth not exceeding 250mm	1.57	10.00	1.73	0.96	0.20	10.26	0.39	12.38	nr	13.31
Holes; rectangular										
girth not exceeding 500mm; depth not exceeding 250mm	0.77	10.00	0.85	0.48	0.10	5.13	0.20	6.18	nr	6.64
girth not exceeding 500mm; depth 250 - 500mm	1.57	10.00	1.73	0.88	0.18	9.39	0.36	11.47	nr	12.33
girth 0.50 - 1.00m; depth not exceeding 250mm	1.35	10.00	1.49	0.85	0.17	9.04	0.34	10.86	nr	11.68
girth 0.50 - 1.00m; depth 250 - 500mm	2.48	10.00	2.73	1.54	0.31	16.39	0.63	19.75	nr	21.23
girth 1.00m - 2.00m; depth not exceeding 250mm ...	1.95	10.00	2.15	1.21	0.24	12.85	0.49	15.49	nr	16.65
girth 2.00m - 3.00m; depth not exceeding 250mm ...	2.54	10.00	2.79	1.57	0.31	16.67	0.64	20.10	nr	21.61
girth 3.00m - 4.00m; depth not exceeding 250mm ...	3.16	10.00	3.48	1.93	0.39	20.55	0.78	24.81	nr	26.67
girth 1.00m - 2.00m; depth 250 - 500mm	3.53	10.00	3.88	2.20	0.44	23.39	0.88	28.16	nr	30.27
girth 2.00m - 3.00m; depth 250 - 500mm	4.60	10.00	5.06	2.85	0.57	30.31	1.16	36.53	nr	39.27
girth 3.00m - 4.00m; depth 250 - 500mm	5.75	10.00	6.33	3.50	0.70	37.22	1.41	44.95	nr	48.33
Holes; circular										
girth not exceeding 500mm; depth not exceeding 250mm	1.57	10.00	1.73	0.96	0.20	10.26	0.39	12.38	nr	13.31
girth not exceeding 500mm; depth 250 - 500mm	2.82	10.00	3.10	1.76	0.36	18.77	0.70	22.57	nr	24.27
girth 0.50 - 1.00m; depth not exceeding 250mm	2.72	10.00	2.99	1.70	0.34	18.08	0.68	21.75	nr	23.38
girth 0.50 - 1.00m; depth 250 - 500mm	4.95	10.00	5.45	3.08	0.62	32.78	1.24	39.47	nr	42.43
girth 1.00m - 2.00m; depth not exceeding 250mm ...	3.88	10.00	4.27	2.42	0.48	25.71	0.98	30.95	nr	33.28
girth 2.00m - 3.00m; depth not exceeding 250mm ...	5.05	10.00	5.55	3.14	0.62	33.33	1.27	40.16	nr	43.17
girth 3.00m - 4.00m; depth not exceeding 250mm ...	6.32	10.00	6.95	3.86	0.78	41.10	1.54	49.60	nr	53.32
girth 1.00m - 2.00m; depth 250 - 500mm	7.06	10.00	7.77	4.40	0.88	46.79	1.77	56.33	nr	60.55
girth 2.00m - 3.00m; depth 250 - 500mm	9.21	10.00	10.13	5.70	1.14	60.61	2.29	73.03	nr	78.51
girth 3.00m - 4.00m; depth 250 - 500mm	11.52	10.00	12.67	7.00	1.40	74.44	2.82	89.93	nr	96.67
Formwork and basic finish; coating with retarding agent										
Sides of foundations; plain vertical										
height exceeding 1.00m	3.47	12.50	3.90	1.75	0.59	20.29	0.71	24.91	m2	26.78
height not exceeding 250mm	0.96	12.50	1.08	0.53	0.14	5.87	0.22	7.17	m	7.71
height 250 - 500mm	1.81	12.50	2.04	0.96	0.28	10.83	0.41	13.27	m	14.27
height 0.50m - 1.00m	3.39	12.50	3.81	1.79	0.54	20.31	0.71	24.84	m	26.70
Sides of ground beams and edges of beds; plain vertical										
height exceeding 1.00m	3.47	12.50	3.90	1.75	0.59	20.29	0.71	24.91	m2	26.78
height not exceeding 250mm	0.96	12.50	1.08	0.53	0.14	5.87	0.22	7.17	m	7.71
height 250 - 500mm	1.81	12.50	2.04	0.96	0.28	10.83	0.41	13.27	m	14.27
height 0.50m - 1.00m	3.39	12.50	3.81	1.79	0.54	20.31	0.71	24.84	m	26.70
Edges of suspended slabs; plain vertical										
height not exceeding 250mm	1.13	12.50	1.27	0.66	0.16	7.21	0.25	8.74	m	9.39
height 250 - 500mm	2.25	12.50	2.53	1.20	0.33	13.39	0.49	16.41	m	17.64
Edges of suspended slabs; plain vertical; curved 10m radius										
height not exceeding 250mm	2.76	12.50	3.11	1.65	0.36	17.76	0.68	21.54	m	23.16
Edges of suspended slabs; plain vertical; curved 1m radius										
height not exceeding 250mm	1.68	12.50	1.89	1.00	0.23	10.84	0.43	13.16	m	14.15
Edges of suspended slabs; plain vertical; curved 20m radius										
height not exceeding 250mm	1.28	12.50	1.44	0.74	0.18	8.09	0.27	9.80	m	10.54
Sides of upstands; plain vertical										
height not exceeding 250mm	1.34	12.50	1.51	0.75	0.18	8.19	0.27	9.96	m	10.71
height 250 - 500mm	1.71	12.50	1.92	0.88	0.27	10.02	0.36	12.30	m	13.22
Machine bases and plinths; plain vertical										
height exceeding 1.00m	3.77	12.50	4.24	1.93	0.62	22.17	0.78	27.19	m2	29.23
height not exceeding 250mm	1.00	12.50	1.13	0.58	0.14	6.34	0.24	7.70	m	8.28
height 250 - 500mm	1.96	12.50	2.21	1.06	0.29	11.82	0.44	14.46	m	15.55
height 0.50m - 1.00m	3.67	12.50	4.13	1.98	0.57	22.28	0.82	27.23	m	29.27
Soffits of slabs; horizontal										
slab thickness not exceeding 200mm; height to soffit										
1.50 - 3.00m....................	9.31	10.00	10.24	1.50	0.54	17.64	1.00	28.88	m2	31.04
3.00 - 4.50m....................	9.31	10.00	10.24	1.72	0.59	20.02	1.37	31.63	m2	34.00
4.50 - 6.00m....................	9.31	10.00	10.24	1.94	0.64	22.40	1.71	34.35	m2	36.93
slab thickness 200 - 300mm; height to soffit										
1.50 - 3.00m....................	10.18	10.00	11.20	1.65	0.57	19.23	1.09	31.52	m2	33.88
3.00 - 4.50m....................	10.18	10.00	11.20	1.87	0.62	21.61	1.47	34.28	m2	36.85

Labour hourly rates: (except Specialists) Craft Operatives £9.23 Labour £7.02 Rates are national average prices. Refer to REGIONAL VARIATIONS for indicative levels of overall pricing in regions	MATERIALS			LABOUR				RATES		
	Del to Site £	Waste %	Material Cost £	Craft Optve Hrs	Lab Hrs	Labour Cost £	Sunds £	Nett Rate £	Unit	Gross Rate (+7.5%) £
E20: FORMWORK FOR IN-SITU CONCRETE Cont.										
Formwork and basic finish; coating with retarding agent Cont.										
Soffits of slabs; horizontal Cont.										
slab thickness 200 - 300mm; height to soffit Cont.										
4.50 - 6.00m..............................	10.18	10.00	**11.20**	2.09	0.67	**23.99**	1.80	**36.99**	m2	**39.77**
Soffits of slabs; horizontal; with frequent uses of prefabricated panels										
slab thickness not exceeding 200mm; height to soffit										
1.50 - 3.00m..............................	3.20	15.00	**3.68**	1.50	0.54	**17.64**	1.00	**22.32**	m2	**23.99**
3.00 - 4.50m..............................	3.20	15.00	**3.68**	1.72	0.59	**20.02**	1.37	**25.07**	m2	**26.95**
4.50 - 6.00m..............................	3.20	15.00	**3.68**	1.94	0.64	**22.40**	1.71	**27.79**	m2	**29.87**
slab thickness 200 - 300mm; height to soffit										
1.50 - 3.00m..............................	3.80	15.00	**4.37**	1.65	0.78	**20.71**	1.09	**26.17**	m2	**28.13**
3.00 - 4.50m..............................	3.80	15.00	**4.37**	1.87	0.83	**23.09**	1.47	**28.93**	m2	**31.10**
4.50 - 6.00m..............................	3.80	15.00	**4.37**	2.09	0.88	**25.47**	1.80	**31.64**	m2	**34.01**
Soffits of slabs; sloping not exceeding 15 degrees										
slab thickness not exceeding 200mm; height to soffit										
1.50 - 3.00m..............................	10.18	10.00	**11.20**	1.65	0.57	**19.23**	1.09	**31.52**	m2	**33.88**
3.00 - 4.50m..............................	10.18	10.00	**11.20**	1.87	0.62	**21.61**	1.47	**34.28**	m2	**36.85**
4.50 - 6.00m..............................	10.18	10.00	**11.20**	2.09	0.67	**23.99**	1.80	**36.99**	m2	**39.77**
Soffits of slabs; sloping exceeding 15 degrees										
slab thickness not exceeding 200mm; height to soffit										
1.50 - 3.00m..............................	10.59	10.00	**11.65**	1.73	0.59	**20.11**	1.27	**33.03**	m2	**35.51**
3.00 - 4.50m..............................	10.59	10.00	**11.65**	1.95	0.64	**22.49**	1.62	**35.76**	m2	**38.44**
4.50 - 6.00m..............................	10.59	10.00	**11.65**	2.17	0.69	**24.87**	1.98	**38.50**	m2	**41.39**
Soffits of landings; horizontal										
slab thickness not exceeding 200mm; height to soffit										
not exceeding 1.50m......................	9.31	10.00	**10.24**	1.80	0.60	**20.83**	1.00	**32.07**	m2	**34.47**
1.50 - 3.00m..............................	9.31	10.00	**10.24**	2.02	0.65	**23.21**	1.37	**34.82**	m2	**37.43**
3.00 - 4.50m..............................	9.31	10.00	**10.24**	2.24	0.70	**25.59**	1.71	**37.54**	m2	**40.36**
Top formwork										
sloping exceeding 15 degrees	3.86	12.50	**4.34**	1.50	0.54	**17.64**	0.57	**22.55**	m2	**24.24**
Walls; vertical										
plain...........................	5.61	15.00	**6.45**	1.60	0.56	**18.70**	0.92	**26.07**	m2	**28.03**
plain; height exceeding 3.00m above floor level ..	6.21	15.00	**7.14**	1.60	0.56	**18.70**	1.03	**26.87**	m2	**28.89**
interrupted	6.09	15.00	**7.00**	1.75	0.59	**20.29**	1.01	**28.31**	m2	**30.43**
interrupted; height exceeding 3.00m above floor level	6.51	15.00	**7.49**	1.75	0.59	**20.29**	1.12	**28.90**	m2	**31.07**
Walls; vertical; curved 10m radius										
plain......................................	8.08	15.00	**9.29**	2.40	0.72	**27.21**	1.40	**37.90**	m2	**40.74**
Walls; vertical; curved 1m radius										
plain......................................	13.06	15.00	**15.02**	4.00	1.04	**44.22**	2.29	**61.53**	m2	**66.14**
Walls; vertical; curved 20m radius										
plain......................................	6.18	15.00	**7.11**	1.80	0.60	**20.83**	1.03	**28.96**	m2	**31.14**
Walls; battered										
plain......................................	6.59	15.00	**7.58**	1.92	0.62	**22.07**	1.12	**30.77**	m2	**33.08**
Beams; attached to slabs										
regular shaped; rectangular; height to soffit										
1.50 - 3.00m..............................	5.84	10.00	**6.42**	2.25	0.69	**25.61**	0.88	**32.92**	m2	**35.38**
3.00 - 4.50m..............................	5.84	10.00	**6.42**	2.25	0.69	**25.61**	1.07	**33.11**	m2	**35.59**
regular shaped; rectangular; with 50mm wide chamfers; height to soffit										
1.50 - 3.00m..............................	5.97	10.00	**6.57**	2.31	0.69	**26.17**	0.88	**33.61**	m2	**36.13**
3.00 - 4.50m..............................	5.97	10.00	**6.57**	2.31	0.69	**26.17**	1.07	**33.80**	m2	**36.34**
Beams; isolated										
regular shaped; rectangular; height to soffit										
1.50 - 3.00m..............................	6.10	10.00	**6.71**	2.35	0.71	**26.67**	0.88	**34.26**	m2	**36.83**
3.00 - 4.50m..............................	6.10	10.00	**6.71**	2.35	0.71	**26.67**	1.07	**34.45**	m2	**37.04**
regular shaped; rectangular; with 50mm wide chamfers; height to soffit										
1.50 - 3.00m..............................	6.29	10.00	**6.92**	2.41	0.71	**27.23**	0.88	**35.03**	m2	**37.65**
3.00 - 4.50m..............................	6.29	10.00	**6.92**	2.41	0.71	**27.23**	1.07	**35.22**	m2	**37.86**
Columns; attached to walls										
regular shaped; rectangular	4.24	10.00	**4.66**	2.30	0.70	**26.14**	0.88	**31.69**	m2	**34.06**
regular shaped; rectangular, height exceeding 3.00m above floor level	4.24	10.00	**4.66**	2.43	0.70	**27.34**	1.00	**33.01**	m2	**35.48**
Columns; isolated										
regular shaped; rectangular	4.24	10.00	**4.66**	2.25	0.69	**25.61**	0.88	**31.16**	m2	**33.49**
regular shaped; rectangular, height exceeding 3.00m above floor level	4.24	10.00	**4.66**	2.48	0.69	**27.73**	1.00	**33.40**	m2	**35.90**
regular shaped; rectangular; with 50mm wide chamfers	4.55	10.00	**5.00**	2.39	0.69	**26.90**	0.88	**32.79**	m2	**35.25**
regular shaped; rectangular; with 50mm wide chamfers; height exceeding 3.00m above floor level	4.55	10.00	**5.00**	2.63	0.69	**29.12**	1.00	**35.12**	m2	**37.76**

IN SITU CONCRETE/LARGE PRECAST CONCRETE

Labour hourly rates: (except Specialists) Craft Operatives £9.23 Labour £7.02 Rates are national average prices. Refer to REGIONAL VARIATIONS for indicative levels of overall pricing in regions	MATERIALS			LABOUR				RATES		
	Del to Site £	Waste %	Material Cost £	Craft Optve Hrs	Lab Hrs	Labour Cost £	Sunds £	Nett Rate £	Unit	Gross Rate (+7.5%) £
E20: FORMWORK FOR IN-SITU CONCRETE Cont.										
Formwork and basic finish; coating with retarding agent Cont.										
Column casings; attached to walls										
regular shaped; rectangular	4.24	10.00	4.66	2.30	0.70	26.14	0.88	31.69	m2	34.06
regular shaped; rectangular, height exceeding 3.00m above floor level	4.24	10.00	4.66	2.43	0.70	27.34	1.00	33.01	m2	35.48
Column casings; isolated										
regular shaped; rectangular	4.24	10.00	4.66	2.25	0.69	25.61	0.88	31.16	m2	33.49
regular shaped; rectangular, height exceeding 3.00m above floor level	4.24	10.00	4.66	2.48	0.69	27.73	1.00	33.40	m2	35.90
regular shaped; rectangular; with 50mm wide chamfers	4.55	10.00	5.00	2.39	0.69	26.90	0.88	32.79	m2	35.25
regular shaped; rectangular; with 50mm wide chamfers; height exceeding 3.00m above floor level	4.55	10.00	5.00	2.63	0.69	29.12	1.00	35.12	m2	37.76
Extra over a basic finish for a fine formed finish										
slabs	-	-	-	-	0.35	2.46	-	2.46	m2	2.64
walls	-	-	-	-	0.35	2.46	-	2.46	m2	2.64
beams	-	-	-	-	0.35	2.46	-	2.46	m2	2.64
columns	-	-	-	-	0.35	2.46	-	2.46	m2	2.64
Wall kickers										
straight	0.92	15.00	1.06	0.25	0.09	2.94	0.18	4.18	m	4.49
curved 2m radius	2.15	15.00	2.47	0.66	0.17	7.29	0.36	10.12	m	10.88
curved 10m radius	1.33	15.00	1.53	0.40	0.12	4.53	0.23	6.29	m	6.77
curved 20m radius	1.02	15.00	1.17	0.30	0.10	3.47	0.20	4.84	m	5.21
Suspended wall kickers										
straight	1.02	15.00	1.17	0.28	0.10	3.29	0.20	4.66	m	5.01
curved 2m radius	2.34	15.00	2.69	0.73	0.19	8.07	0.40	11.16	m	12.00
curved 10m radius	1.50	15.00	1.73	0.44	0.13	4.97	0.25	6.95	m	7.47
curved 20m radius	1.13	15.00	1.30	0.33	0.11	3.82	0.22	5.34	m	5.74
Wall ends, soffits and steps in walls; plain										
width not exceeding 250mm	1.86	15.00	2.14	0.53	0.17	6.09	0.28	8.50	m	9.14
Openings in walls; plain										
width not exceeding 250mm	1.86	15.00	2.14	0.53	0.17	6.09	0.28	8.50	m	9.14
Stairflights; 1000mm wide; 155mm thick waist; 178mm risers; includes formwork to soffits, risers and strings										
strings 300mm wide	14.32	10.00	15.75	6.05	1.74	68.06	2.08	85.89	m	92.33
string 300mm wide; junction with wall	14.32	10.00	15.75	6.05	1.74	68.06	2.08	85.89	m	92.33
Stairflights; 1500mm wide; 180mm thick waist; 178mm risers; includes formwork to soffits, risers and strings										
strings 325mm wide	17.89	10.00	19.68	7.56	2.17	85.01	2.60	107.29	m	115.34
string 325mm wide; junction with wall	17.89	10.00	19.68	7.56	2.17	85.01	2.60	107.29	m	115.34
Stairflights; 1000mm wide; 155mm thick waist; 178mm undercut risers; includes formwork to soffits, risers and strings										
strings 300mm wide	14.32	10.00	15.75	6.05	1.74	68.06	2.08	85.89	m	92.33
string 300mm wide; junction with wall	14.32	10.00	15.75	6.05	1.74	68.06	2.08	85.89	m	92.33
Stairflights; 1500mm wide; 180mm thick waist; 178mm undercut risers; includes formwork to soffits, risers and strings										
strings 325mm wide	17.89	10.00	19.68	7.56	2.17	85.01	2.60	107.29	m	115.34
string 325mm wide; junction with wall	17.89	10.00	19.68	7.56	2.17	85.01	2.60	107.29	m	115.34
Claymaster low density expanded polystyrene permanent formwork; fixing with Clayfix hooks - 3/m2										
Sides of foundations; plain vertical 75mm thick										
height exceeding 1.00m	7.33	5.00	7.70	-	0.06	0.42	0.63	8.75	m2	9.40
height not exceeding 250mm	1.83	5.00	1.92	-	0.02	0.14	0.21	2.27	m	2.44
height 250-500mm	3.67	5.00	3.85	-	0.04	0.28	0.21	4.34	m	4.67
height 0.50m-1.00m	5.50	5.00	5.78	-	0.07	0.49	0.42	6.69	m	7.19
Sides of foundations; plain vertical 100mm thick										
height exceeding 1.00m	9.78	5.00	10.27	-	0.08	0.56	0.63	11.46	m2	12.32
height not exceeding 250mm	2.45	5.00	2.57	-	0.03	0.21	0.21	2.99	m	3.22
height 250-500mm	4.89	5.00	5.13	-	0.05	0.35	0.21	5.70	m	6.12
height 0.50m-1.00m	7.34	5.00	7.71	-	0.09	0.63	0.42	8.76	m	9.42
Sides of foundations; plain vertical 150mm thick										
height exceeding 1.00m	12.22	5.00	12.83	-	0.10	0.70	0.63	14.16	m2	15.23
height not exceeding 250mm	3.06	5.00	3.21	-	0.03	0.21	0.21	3.63	m	3.91
height 250-500mm	6.11	5.00	6.42	-	0.06	0.42	0.21	7.05	m	7.58
height 0.50m-1.00m	9.17	5.00	9.63	-	0.11	0.77	0.42	10.82	m	11.63
To underside of foundations; laid on earth or hardcore; 75mm thick										
horizontal	7.33	5.00	7.70	-	0.05	0.35	-	8.05	m2	8.65
To underside of foundations; laid on earth or hardcore; 100mm thick										
horizontal	9.78	5.00	10.27	-	0.06	0.42	-	10.69	m2	11.49

Labour hourly rates: (except Specialists) Craft Operatives £9.23 Labour £7.02 Rates are national average prices. Refer to REGIONAL VARIATIONS for indicative levels of overall pricing in regions	MATERIALS			LABOUR				RATES		
	Del to Site £	Waste %	Material Cost £	Craft Optve Hrs	Lab Hrs	Labour Cost £	Sunds £	Nett Rate £	Unit	Gross Rate (+7.5%) £
E20: FORMWORK FOR IN-SITU CONCRETE Cont.										
Claymaster low density expanded polystyrene permanent formwork; fixing with Clayfix hooks - 3/m2 Cont.										
To underside of foundations; laid on earth or hardcore; 150mm thick										
horizontal	12.22	5.00	12.83	-	0.07	0.49	-	13.32	m2	14.32
Expamet Hy-rib permanent shuttering and reinforcement										
Reference 2611, 4.86 kg/m2 to soffits of slabs; horizontal; one rib side laps; 150mm end laps										
slab thickness 75mm; strutting and supports at 750mm centres; height to soffit										
1.50 - 3.00m........................	10.96	5.00	11.51	1.19	0.30	13.09	1.07	25.67	m2	27.59
3.00 - 4.50m........................	10.96	5.00	11.51	1.37	0.34	15.03	1.32	27.86	m2	29.95
slab thickness 100mm; strutting and supports at 650mm centres; height to soffit										
1.50 - 3.00m........................	10.96	5.00	11.51	1.21	0.30	13.27	1.21	25.99	m2	27.94
3.00 - 4.50m........................	10.96	5.00	11.51	1.39	0.35	15.29	1.63	28.42	m2	30.56
slab thickness 125mm; strutting and supports at 550mm centres; height to soffit										
1.50 - 3.00m........................	10.96	5.00	11.51	1.23	0.31	13.53	1.36	26.40	m2	28.38
3.00 - 4.50m........................	10.96	5.00	11.51	1.41	0.35	15.47	1.97	28.95	m2	31.12
slab thickness 150mm; strutting and supports at 450mm centres; height to soffit										
1.50 - 3.00m........................	10.96	5.00	11.51	1.25	0.31	13.71	1.53	26.75	m2	28.76
3.00 - 4.50m........................	10.96	5.00	11.51	1.44	0.36	15.82	2.29	29.62	m2	31.84
Reference 2411, 6.34 kg/m2 to soffits of slabs; horizontal; one rib side laps; 150mm end laps										
slab thickness 75mm; strutting and supports at 850mm centres; height to soffit										
1.50 - 3.00m........................	14.00	5.00	14.70	1.22	0.31	13.44	1.07	29.21	m2	31.40
3.00 - 4.50m........................	14.00	5.00	14.70	1.40	0.35	15.38	1.32	31.40	m2	33.75
slab thickness 100mm; strutting and supports at 750mm centres; height to soffit										
1.50 - 3.00m........................	14.00	5.00	14.70	1.24	0.31	13.62	1.21	29.53	m2	31.75
3.00 - 4.50m........................	14.00	5.00	14.70	1.43	0.36	15.73	1.63	32.06	m2	34.46
slab thickness 125mm; strutting and supports at 650mm centres; height to soffit										
1.50 - 3.00m........................	14.00	5.00	14.70	1.25	0.32	13.78	1.36	29.84	m2	32.08
3.00 - 4.50m........................	14.00	5.00	14.70	1.45	0.36	15.91	1.97	32.58	m2	35.02
slab thickness 150mm; strutting and supports at 550mm centres; height to soffit										
1.50 - 3.00m........................	14.00	5.00	14.70	1.28	0.32	14.06	1.53	30.29	m2	32.56
3.00 - 4.50m........................	14.00	5.00	14.70	1.47	0.37	16.17	2.29	33.16	m2	35.64
slab thickness 175mm; strutting and supports at 450mm centres; height to soffit										
1.50 - 3.00m........................	14.00	5.00	14.70	1.31	0.33	14.41	1.76	30.87	m2	33.18
3.00 - 4.50m........................	12.59	5.00	13.22	1.51	0.38	16.60	2.73	32.55	m2	35.00
slab thickness 200mm; strutting and supports at 350mm centres; height to soffit										
1.50 - 3.00m........................	14.00	5.00	14.70	1.34	0.34	14.76	1.98	31.43	m2	33.79
3.00 - 4.50m........................	14.00	5.00	14.70	1.54	0.39	16.95	3.18	34.83	m2	37.44
Reference 2611, 4.86 kg/m2 to soffits of arched slabs; one rib side laps										
900mm span; 75mm rise; height to soffit										
1.50 - 3.00m........................	10.96	5.00	11.51	1.31	0.33	14.41	0.79	26.71	m2	28.71
1200mm span; 75mm rise; one row strutting and supports per span; height to soffit										
1.50 - 3.00m........................	10.96	5.00	11.51	1.31	0.33	14.41	1.07	26.99	m2	29.01
3.00 - 4.50m........................	10.96	5.00	11.51	1.49	0.37	16.35	1.32	29.18	m2	31.37
Reference 2411, 6.34 kg/m2 to soffits of arched slabs; one rib side laps										
1500mm span; 100mm rise; two rows strutting and supports per span; height to soffit										
1.50 - 3.00m........................	14.00	5.00	14.70	1.62	0.40	17.76	2.28	34.74	m2	37.35
3.00 - 4.50m........................	14.00	5.00	14.70	1.84	0.46	20.21	3.05	37.96	m2	40.81
1800mm span; 150mm rise; two rows strutting and supports per span; height to soffit										
1.50 - 3.00m........................	14.00	5.00	14.70	1.62	0.40	17.76	2.28	34.74	m2	37.35
3.00 - 4.50m........................	14.00	5.00	14.70	1.84	0.46	20.21	3.05	37.96	m2	40.81
E30: REINFORCEMENT FOR IN-SITU CONCRETE										
Reinforcement bars; B.S.4449, hot rolled plain round mild steel including hooks and tying wire, and spacers and chairs which are at the discretion of the Contractor										
Straight										
6mm...............................	335.00	5.00	351.75	46.00	4.00	452.66	17.91	822.32	t	883.99
8mm...............................	300.00	5.00	315.00	40.00	4.00	397.28	15.90	728.18	t	782.79
10mm...............................	290.00	5.00	304.50	33.00	4.00	332.67	12.87	650.04	t	698.79
12mm...............................	290.00	5.00	304.50	28.00	4.00	286.52	10.97	601.99	t	647.14
16mm...............................	290.00	5.00	304.50	22.00	4.00	231.14	9.02	544.66	t	585.51
20mm...............................	290.00	5.00	304.50	18.00	4.00	194.22	4.99	503.71	t	541.49
25mm...............................	290.00	5.00	304.50	18.00	4.00	194.22	4.99	503.71	t	541.49
Bent										
6mm...............................	335.00	5.00	351.75	46.00	4.00	452.66	17.91	822.32	t	883.99
8mm...............................	300.00	5.00	315.00	40.00	4.00	397.28	15.90	728.18	t	782.79

IN SITU CONCRETE/LARGE PRECAST CONCRETE

Labour hourly rates: (except Specialists) Craft Operatives £9.23 Labour £7.02 Rates are national average prices. Refer to REGIONAL VARIATIONS for indicative levels of overall pricing in regions	MATERIALS			LABOUR				RATES		
	Del to Site £	Waste %	Material Cost £	Craft Optve Hrs	Lab Hrs	Labour Cost £	Sunds £	Nett Rate £	Unit	Gross Rate (+7.5%) £
E30: REINFORCEMENT FOR IN-SITU CONCRETE Cont.										
Reinforcement bars; B.S.4449, hot rolled plain round mild steel including hooks and tying wire, and spacers and chairs which are at the discretion of the Contractor Cont.										
Bent Cont.										
10mm	290.00	5.00	304.50	33.00	4.00	332.67	12.87	650.04	t	698.79
12mm	290.00	5.00	304.50	28.00	4.00	286.52	10.97	601.99	t	647.14
16mm	290.00	5.00	304.50	22.00	4.00	231.14	9.02	544.66	t	585.51
20mm	290.00	5.00	304.50	18.00	4.00	194.22	4.99	503.71	t	541.49
25mm	290.00	5.00	304.50	18.00	4.00	194.22	4.99	503.71	t	541.49
Links										
6mm	350.00	5.00	367.50	60.00	4.00	581.88	17.91	967.29	t	1039.84
8mm	315.00	5.00	330.75	60.00	4.00	581.88	15.90	928.53	t	998.17
Reinforcement bars; B.S.4449, hot rolled deformed high yield steel including hooks and tying wire, and spacers and chairs which are at the discretion of the Contractor										
Straight										
6mm	345.00	5.00	362.25	46.00	4.00	452.66	17.91	832.82	t	895.28
8mm	310.00	5.00	325.50	40.00	4.00	397.28	15.90	738.68	t	794.08
10mm	300.00	5.00	315.00	33.00	4.00	332.67	12.87	660.54	t	710.08
12mm	300.00	5.00	315.00	28.00	4.00	286.52	10.97	612.49	t	658.43
16mm	300.00	5.00	315.00	22.00	4.00	231.14	9.02	555.16	t	596.80
20mm	300.00	5.00	315.00	18.00	4.00	194.22	4.99	514.21	t	552.78
25mm	300.00	5.00	315.00	18.00	4.00	194.22	4.99	514.21	t	552.78
Bent										
6mm	345.00	5.00	362.25	46.00	4.00	452.66	17.91	832.82	t	895.28
8mm	310.00	5.00	325.50	40.00	4.00	397.28	15.90	738.68	t	794.08
10mm	300.00	5.00	315.00	33.00	4.00	332.67	12.87	660.54	t	710.08
12mm	300.00	5.00	315.00	28.00	4.00	286.52	10.97	612.49	t	658.43
16mm	300.00	5.00	315.00	22.00	4.00	231.14	9.02	555.16	t	596.80
20mm	300.00	5.00	315.00	18.00	4.00	194.22	4.99	514.21	t	552.78
25mm	300.00	5.00	315.00	18.00	4.00	194.22	4.99	514.21	t	552.78
Links										
6mm	360.00	5.00	378.00	60.00	4.00	581.88	17.91	977.79	t	1051.12
8mm	325.00	5.00	341.25	60.00	4.00	581.88	15.90	939.03	t	1009.46
Take delivery, cut, bend and fix reinforcing rods; including hooks and tying wire, and spacers and chairs which are at the discretion of the Contractor										
Straight										
6mm	-	-	-	50.00	4.00	489.58	17.91	507.49	t	545.55
8mm	-	-	-	-	44.00	308.88	15.90	324.78	t	349.14
10mm	-	-	-	36.00	4.00	360.36	12.87	373.23	t	401.22
12mm	-	-	-	31.00	4.00	314.21	10.97	325.18	t	349.57
16mm	-	-	-	24.00	4.00	249.60	9.02	258.62	t	278.02
20mm	-	-	-	20.00	4.00	212.68	4.99	217.67	t	234.00
25mm	-	-	-	20.00	4.00	212.68	4.99	217.67	t	234.00
Bent										
6mm	-	-	-	68.00	4.00	655.72	17.91	673.63	t	724.15
8mm	-	-	-	60.00	4.00	581.88	15.90	597.78	t	642.61
10mm	-	-	-	50.00	4.00	489.58	12.87	502.45	t	540.13
12mm	-	-	-	43.00	4.00	424.97	10.97	435.94	t	468.64
16mm	-	-	-	34.00	4.00	341.90	9.02	350.92	t	377.24
20mm	-	-	-	28.00	4.00	286.52	4.99	291.51	t	313.37
25mm	-	-	-	26.00	4.00	268.06	4.99	273.05	t	293.53
Links										
6mm	-	-	-	90.00	4.00	858.78	17.91	876.69	t	942.44
8mm	-	-	-	90.00	4.00	858.78	15.90	874.68	t	940.28
Take delivery and fix reinforcing rods supplied cut to length and bent; including tying wire, and spacers and chairs which are at the discretion of the Contractor										
Straight										
6mm	-	-	-	46.00	4.00	452.66	17.91	470.57	t	505.86
8mm	-	-	-	40.00	4.00	397.28	15.90	413.18	t	444.17
10mm	-	-	-	33.00	4.00	332.67	12.87	345.54	t	371.46
12mm	-	-	-	28.00	4.00	286.52	10.97	297.49	t	319.80
16mm	-	-	-	22.00	4.00	231.14	9.02	240.16	t	258.17
20mm	-	-	-	18.00	4.00	194.22	4.99	199.21	t	214.15
25mm	-	-	-	18.00	4.00	194.22	4.99	199.21	t	214.15
Bent										
6mm	-	-	-	46.00	4.00	452.66	17.91	470.57	t	505.86
8mm	-	-	-	40.00	4.00	397.28	15.90	413.18	t	444.17
10mm	-	-	-	33.00	4.00	332.67	12.87	345.54	t	371.46
12mm	-	-	-	28.00	4.00	286.52	10.97	297.49	Unit	319.80
16mm	-	-	-	22.00	4.00	231.14	9.02	240.16	t	258.17
20mm	-	-	-	18.00	4.00	194.22	4.99	199.21	t	214.15
25mm	-	-	-	18.00	4.00	194.22	4.99	199.21	t	214.15
Links										
6mm	-	-	-	60.00	4.00	581.88	17.91	599.79	t	644.77

Labour hourly rates: (except Specialists) Craft Operatives £9.23 Labour £7.02 Rates are national average prices. Refer to REGIONAL VARIATIONS for indicative levels of overall pricing in regions	MATERIALS			LABOUR				RATES		
	Del to Site £	Waste %	Material Cost £	Craft Optve Hrs	Lab Hrs	Labour Cost £	Sunds £	Nett Rate £	Unit	Gross Rate (+7.5%) £
E30: REINFORCEMENT FOR IN-SITU CONCRETE Cont.										
Take delivery and fix reinforcing rods supplied cut to length and bent; including tying wire, and spacers and chairs which are at the discretion of the Contractor Cont.										
Links Cont.										
8mm	-	-	-	60.00	4.00	581.88	15.90	597.78	t	642.61
Reinforcement bars; high yield stainless steel B.S.6744, Type 2 (minimum yield stress 460 N/mm2); including hooks and tying wire, and spacers and chairs which are at the discretion of the Contractor										
Straight										
8mm	2840.00	5.00	2982.00	75.00	9.00	755.43	66.00	3803.43	t	4088.69
10mm	2840.00	5.00	2982.00	63.00	9.00	644.67	55.00	3681.67	t	3957.80
12mm	2840.00	5.00	2982.00	54.00	9.00	561.60	46.20	3589.80	t	3859.03
16mm	2840.00	5.00	2982.00	45.00	9.00	478.53	38.50	3499.03	t	3761.46
20mm	3000.00	5.00	3150.00	37.50	9.00	409.31	30.80	3590.11	t	3859.36
25mm	3050.00	5.00	3202.50	30.00	9.00	340.08	21.18	3563.76	t	3831.04
Bent										
8mm	3350.00	5.00	3517.50	75.00	9.00	755.43	66.00	4338.93	t	4664.35
10mm	3350.00	5.00	3517.50	63.00	9.00	644.67	55.00	4217.17	t	4533.46
12mm	3350.00	5.00	3517.50	54.00	9.00	561.60	46.20	4125.30	t	4434.70
16mm	3350.00	5.00	3517.50	45.00	9.00	478.53	38.50	4034.53	t	4337.12
20mm	3350.00	5.00	3517.50	37.50	9.00	409.31	30.80	3957.61	t	4254.43
25mm	3350.00	5.00	3517.50	30.00	9.00	340.08	21.18	3878.76	t	4169.67
Links										
6mm	2840.00	5.00	2982.00	120.00	9.00	1170.78	74.80	4227.58	t	4544.65
8mm	2840.00	5.00	2982.00	105.00	9.00	1032.33	66.00	4080.33	t	4386.35
Reinforcement bars; Tor Bar grade 250 mild steel ribbed and twisted bars; including hooks and tying wire, and spacers and chairs which are at the discretion of the Contractor										
Straight										
6mm	335.00	5.00	351.75	46.00	4.00	452.66	17.91	822.32	t	883.99
8mm	300.00	5.00	315.00	40.00	4.00	397.28	15.90	728.18	t	782.79
10mm	290.00	5.00	304.50	33.00	4.00	332.67	12.87	650.04	t	698.79
12mm	290.00	5.00	304.50	28.00	4.00	286.52	10.97	601.99	t	647.14
16mm	290.00	5.00	304.50	22.00	4.00	231.14	9.02	544.66	t	585.51
20mm	290.00	5.00	304.50	18.00	4.00	194.22	4.99	503.71	t	541.49
25mm	290.00	5.00	304.50	18.00	4.00	194.22	4.99	503.71	t	541.49
Bent										
6mm	335.00	5.00	351.75	46.00	4.00	452.66	17.91	822.32	t	883.99
8mm	300.00	5.00	315.00	40.00	4.00	397.28	15.90	728.18	t	782.79
10mm	290.00	5.00	304.50	33.00	4.00	332.67	12.87	650.04	t	698.79
12mm	290.00	5.00	304.50	28.00	4.00	286.52	10.97	601.99	t	647.14
16mm	290.00	5.00	304.50	22.00	4.00	231.14	9.02	544.66	t	585.51
20mm	290.00	5.00	304.50	18.00	4.00	194.22	4.99	503.71	t	541.49
25mm	290.00	5.00	304.50	18.00	4.00	194.22	4.99	503.71	t	541.49
Links										
6mm	350.00	5.00	367.50	60.00	4.00	581.88	17.91	967.29	t	1039.84
8mm	315.00	5.00	330.75	60.00	4.00	581.88	15.90	928.53	t	998.17
Reinforcement bars; Tor Bar grade 460 high yield steel ribbed and twisted bars; including hooks and tying wire, and spacers and chairs which are at the discretion of the Contractor										
Straight										
6mm	345.00	5.00	362.25	46.00	4.00	452.66	17.91	832.82	t	895.28
8mm	310.00	5.00	325.50	40.00	4.00	397.28	15.90	738.68	t	794.08
10mm	300.00	5.00	315.00	33.00	4.00	332.67	12.87	660.54	t	710.08
12mm	300.00	5.00	315.00	28.00	4.00	286.52	10.97	612.49	t	658.43
16mm	300.00	5.00	315.00	22.00	4.00	231.14	9.02	555.16	t	596.80
20mm	300.00	5.00	315.00	18.00	4.00	194.22	4.99	514.21	t	552.78
25mm	300.00	5.00	315.00	18.00	4.00	194.22	4.99	514.21	t	552.78
Bent										
6mm	345.00	5.00	362.25	46.00	4.00	452.66	17.91	832.82	t	895.28
8mm	310.00	5.00	325.50	40.00	4.00	397.28	15.90	738.68	t	794.08
10mm	300.00	5.00	315.00	33.00	4.00	332.67	12.87	660.54	t	710.08
12mm	300.00	5.00	315.00	28.00	4.00	286.52	10.97	612.49	t	658.43
16mm	300.00	5.00	315.00	22.00	4.00	231.14	9.02	555.16	t	596.80
20mm	300.00	5.00	315.00	18.00	4.00	194.22	4.99	514.21	t	552.78
25mm	300.00	5.00	315.00	18.00	4.00	194.22	4.99	514.21	t	552.78
Links										
6mm	360.00	5.00	378.00	60.00	4.00	581.88	17.91	977.79	t	1051.12
8mm	325.00	5.00	341.25	60.00	4.00	581.88	15.90	939.03	t	1009.46

IN SITU CONCRETE/LARGE PRECAST CONCRETE

Labour hourly rates: (except Specialists) Craft Operatives £9.23 Labour £7.02 Rates are national average prices. Refer to REGIONAL VARIATIONS for indicative levels of overall pricing in regions	MATERIALS			LABOUR				RATES		
	Del to Site £	Waste %	Material Cost £	Craft Optve Hrs	Lab Hrs	Labour Cost £	Sunds £	Nett Rate £	Unit	Gross Rate (+7.5%) £
E30: REINFORCEMENT FOR IN-SITU CONCRETE Cont.										
Reinforcement fabric; B.S.4483, hard drawn plain round steel; welded; including laps, tying wire, all cutting and bending, and spacers and chairs which are at the discretion of the Contractor										
Reference A98, 1.54 kg/m2; 200mm side laps; 200mm end laps										
generally	0.55	15.00	**0.63**	0.06	0.01	**0.62**	0.04	**1.30**	m2	**1.39**
strips in one width										
750mm wide.................................	0.55	15.00	**0.63**	0.09	0.01	**0.90**	0.04	**1.57**	m2	**1.69**
900mm wide.................................	0.55	15.00	**0.63**	0.08	0.01	**0.81**	0.04	**1.48**	m2	**1.59**
1050mm wide................................	0.55	15.00	**0.63**	0.08	0.01	**0.81**	0.04	**1.48**	m2	**1.59**
1200mm wide................................	0.55	15.00	**0.63**	0.07	0.01	**0.72**	0.04	**1.39**	m2	**1.49**
Reference A142, 2.22kg/m2; 200mm side laps; 200mm end laps										
generally	0.78	15.00	**0.90**	0.07	0.02	**0.79**	0.06	**1.74**	m2	**1.87**
strips in one width										
750mm wide.................................	0.78	15.00	**0.90**	0.11	0.02	**1.16**	0.06	**2.11**	m2	**2.27**
900mm wide.................................	0.78	15.00	**0.90**	0.10	0.02	**1.06**	0.06	**2.02**	m2	**2.17**
1050mm wide................................	0.78	15.00	**0.90**	0.09	0.02	**0.97**	0.06	**1.93**	m2	**2.07**
1200mm wide................................	0.78	15.00	**0.90**	0.08	0.02	**0.88**	0.06	**1.84**	m2	**1.97**
Reference A193, 3.02kg/m2; 200mm side laps; 200mm end laps										
generally	1.06	15.00	**1.22**	0.07	0.02	**0.79**	0.08	**2.09**	m2	**2.24**
strips in one width										
750mm wide.................................	1.06	15.00	**1.22**	0.11	0.02	**1.16**	0.08	**2.45**	m2	**2.64**
900mm wide.................................	1.06	15.00	**1.22**	0.10	0.02	**1.06**	0.08	**2.36**	m2	**2.54**
1050mm wide................................	1.06	15.00	**1.22**	0.09	0.02	**0.97**	0.08	**2.27**	m2	**2.44**
1200mm wide................................	1.06	15.00	**1.22**	0.08	0.02	**0.88**	0.08	**2.18**	m2	**2.34**
Reference A252, 3.95kg/m2; 200mm side laps; 200mm end laps										
generally	1.38	15.00	**1.59**	0.08	0.02	**0.88**	0.10	**2.57**	m2	**2.76**
strips in one width										
750mm wide.................................	1.38	15.00	**1.59**	0.12	0.02	**1.25**	0.10	**2.94**	m2	**3.16**
900mm wide.................................	1.38	15.00	**1.59**	0.11	0.02	**1.16**	0.10	**2.84**	m2	**3.06**
1050mm wide................................	1.38	15.00	**1.59**	0.10	0.02	**1.06**	0.10	**2.75**	m2	**2.96**
1200mm wide................................	1.38	15.00	**1.59**	0.10	0.02	**1.06**	0.10	**2.75**	m2	**2.96**
Reference A393, 6.16 kg/m2; 200mm side laps; 200mm end laps										
generally	2.15	15.00	**2.47**	0.09	0.20	**2.23**	0.15	**4.86**	m2	**5.22**
strips in one width										
750mm wide.................................	2.15	15.00	**2.47**	0.14	0.02	**1.43**	0.15	**4.06**	m2	**4.36**
900mm wide.................................	2.15	15.00	**2.47**	0.13	0.02	**1.34**	0.15	**3.96**	m2	**4.26**
1050mm wide................................	2.15	15.00	**2.47**	0.12	0.02	**1.25**	0.15	**3.87**	m2	**4.16**
1200mm wide................................	2.15	15.00	**2.47**	0.11	0.02	**1.16**	0.15	**3.78**	m2	**4.06**
Reference B1131, 10.90 kg/m2; 100mm side laps; 200mm end laps										
generally	3.88	15.00	**4.46**	0.12	0.30	**3.21**	0.27	**7.95**	m2	**8.54**
strips in one width										
750mm wide.................................	3.88	15.00	**4.46**	0.18	0.03	**1.87**	0.27	**6.60**	m2	**7.10**
900mm wide.................................	3.88	15.00	**4.46**	0.17	0.03	**1.78**	0.27	**6.51**	m2	**7.00**
1050mm wide................................	3.88	15.00	**4.46**	0.16	0.03	**1.69**	0.27	**6.42**	m2	**6.90**
1200mm wide................................	3.88	15.00	**4.46**	0.14	0.03	**1.50**	0.27	**6.23**	m2	**6.70**
Reference B196, 3.05 kg/m2; 100mm side laps; 200mm end laps										
generally	1.09	15.00	**1.25**	0.07	0.02	**0.79**	0.15	**2.19**	m2	**2.35**
strips in one width										
750mm wide.................................	1.09	15.00	**1.25**	0.11	0.02	**1.16**	0.15	**2.56**	m2	**2.75**
900mm wide.................................	1.09	15.00	**1.25**	0.10	0.02	**1.06**	0.15	**2.47**	m2	**2.65**
1050mm wide................................	1.09	15.00	**1.25**	0.09	0.02	**0.97**	0.15	**2.37**	m2	**2.55**
1200mm wide................................	1.09	15.00	**1.25**	0.08	0.02	**0.88**	0.15	**2.28**	m2	**2.45**
Reference B283, 3.73 kg/m2; 100mm side laps; 200mm end laps										
generally	1.32	15.00	**1.52**	0.07	0.02	**0.79**	0.09	**2.39**	m2	**2.57**
strips in one width										
750mm wide.................................	1.32	15.00	**1.52**	0.11	0.02	**1.16**	0.09	**2.76**	m2	**2.97**
900mm wide.................................	1.32	15.00	**1.52**	0.10	0.02	**1.06**	0.09	**2.67**	m2	**2.87**
1050mm wide................................	1.32	15.00	**1.52**	0.09	0.02	**0.97**	0.09	**2.58**	m2	**2.77**
1200mm wide................................	1.32	15.00	**1.52**	0.08	0.02	**0.88**	0.09	**2.49**	m2	**2.67**
Reference B385, 4.53 kg/m2; 100mm side laps; 200mm end laps										
generally	1.61	15.00	**1.85**	0.08	0.02	**0.88**	0.11	**2.84**	m2	**3.05**
strips in one width										
750mm wide.................................	1.61	15.00	**1.85**	0.12	0.02	**1.25**	0.11	**3.21**	m2	**3.45**
900mm wide.................................	1.61	15.00	**1.85**	0.11	0.02	**1.16**	0.11	**3.12**	m2	**3.35**
1050mm wide................................	1.61	15.00	**1.85**	0.10	0.02	**1.06**	0.11	**3.02**	m2	**3.25**
1200mm wide................................	1.61	15.00	**1.85**	0.10	0.02	**1.06**	0.11	**3.02**	m2	**3.25**
Reference B503, 5.93 kg/m2; 100mm side laps; 200mm end laps										
generally	2.09	15.00	**2.40**	0.09	0.02	**0.97**	0.15	**3.52**	m2	**3.79**
strips in one width										
750mm wide.................................	2.09	15.00	**2.40**	0.14	0.02	**1.43**	0.15	**3.99**	m2	**4.29**
900mm wide.................................	2.09	15.00	**2.40**	0.13	0.02	**1.34**	0.15	**3.89**	m2	**4.19**
1050mm wide................................	2.09	15.00	**2.40**	0.12	0.02	**1.25**	0.15	**3.80**	m2	**4.09**
1200mm wide................................	2.09	15.00	**2.40**	0.11	0.02	**1.16**	0.15	**3.71**	m2	**3.99**

IN-SITU CONCRETE/LARGE PRECAST CONCRETE – MAJOR WORKS

Labour hourly rates: (except Specialists) Craft Operatives £9.23 Labour £7.02 Rates are national average prices. Refer to REGIONAL VARIATIONS for indicative levels of overall pricing in regions	MATERIALS			LABOUR				RATES		
	Del to Site £	Waste %	Material Cost £	Craft Optve Hrs	Lab Hrs	Labour Cost £	Sunds £	Nett Rate £	Unit	Gross Rate (+7.5%) £
E30: REINFORCEMENT FOR IN-SITU CONCRETE Cont.										
Reinforcement fabric; B.S.4483, hard drawn plain round steel; welded; including laps, tying wire, all cutting and bending, and spacers and chairs which are at the discretion of the Contractor Cont.										
Reference B785, 8.14 kg/m2; 100mm side laps; 200mm end laps										
generally	2.87	15.00	3.30	0.10	0.03	1.13	0.20	4.63	m2	4.98
strips in one width										
750mm wide	2.87	15.00	3.30	0.15	0.03	1.60	0.20	5.10	m2	5.48
900mm wide	2.87	15.00	3.30	0.14	0.03	1.50	0.20	5.00	m2	5.38
1050mm wide	2.87	15.00	3.30	0.13	0.03	1.41	0.20	4.91	m2	5.28
1200mm wide	2.87	15.00	3.30	0.12	0.03	1.32	0.20	4.82	m2	5.18
Reference C283, 2.61 kg/m2; 100mm side laps; 400mm end laps										
generally	0.92	15.00	1.06	0.07	0.02	0.79	0.07	1.91	m2	2.06
strips in one width										
750mm wide	0.92	15.00	1.06	0.11	0.02	1.16	0.07	2.28	m2	2.45
900mm wide	0.92	15.00	1.06	0.11	0.02	1.16	0.07	2.28	m2	2.45
1050mm wide	0.92	15.00	1.06	0.09	0.02	0.97	0.07	2.10	m2	2.26
1200mm wide	0.92	15.00	1.06	0.08	0.02	0.88	0.07	2.01	m2	2.16
Reference C385, 3.41 kg/m2; 100mm side laps; 400mm end laps										
generally	1.20	15.00	1.38	0.07	0.02	0.79	0.09	2.26	m2	2.43
strips in one width										
750mm wide	1.20	15.00	1.38	0.11	0.02	1.16	0.09	2.63	m2	2.82
900mm wide	1.20	15.00	1.38	0.10	0.02	1.06	0.09	2.53	m2	2.72
1050mm wide	1.20	15.00	1.38	0.09	0.02	0.97	0.09	2.44	m2	2.62
1200mm wide	1.20	15.00	1.38	0.08	0.02	0.88	0.09	2.35	m2	2.52
Reference C503, 4.34 kg/m2; 100mm side laps; 400mm end laps										
generally	1.48	15.00	1.70	0.08	0.02	0.88	0.11	2.69	m2	2.89
strips in one width										
750mm wide	1.48	15.00	1.70	0.12	0.02	1.25	0.11	3.06	m2	3.29
900mm wide	1.48	15.00	1.70	0.11	0.02	1.16	0.11	2.97	m2	3.19
1050mm wide	1.48	15.00	1.70	0.10	0.02	1.06	0.11	2.88	m2	3.09
1200mm wide	1.48	15.00	1.70	0.10	0.02	1.06	0.11	2.88	m2	3.09
Reference C636, 5.55 kg/m2; 100mm side laps; 400mm end laps										
generally	1.97	15.00	2.27	0.09	0.02	0.97	0.14	3.38	m2	3.63
strips in one width										
750mm wide	1.97	15.00	2.27	0.14	0.02	1.43	0.14	3.84	m2	4.13
900mm wide	1.97	15.00	2.27	0.13	0.02	1.34	0.14	3.75	m2	4.03
1050mm wide	1.97	15.00	2.27	0.12	0.02	1.25	0.14	3.65	m2	3.93
1200mm wide	1.97	15.00	2.27	0.11	0.02	1.16	0.14	3.56	m2	3.83
Reference C785, 6.72 kg/m2; 100mm side laps; 400mm end laps										
generally	2.38	15.00	2.74	0.09	0.02	0.97	0.17	3.88	m2	4.17
strips in one width										
750mm wide	2.38	15.00	2.74	0.14	0.02	1.43	0.17	4.34	m2	4.67
900mm wide	2.38	15.00	2.74	0.13	0.02	1.34	0.17	4.25	m2	4.57
1050mm wide	2.38	15.00	2.74	0.12	0.02	1.25	0.17	4.16	m2	4.47
1200mm wide	2.38	15.00	2.74	0.11	0.02	1.16	0.17	4.06	m2	4.37
Reference D49, 0.77 kg/m2; 100mm side laps; 100mm end laps										
bent	0.56	15.00	0.64	0.03	0.01	0.35	0.06	1.05	m2	1.13
Reference D98, 1.54 kg/m2; 200mm side laps; 200mm end laps										
bent	0.50	15.00	0.57	0.03	0.01	0.35	0.06	0.98	m2	1.06
Expamet 76mm mesh expanded steel reinforcement (uncoated); including laps, tying wire, all cutting and bending, and spacers and chairs which are at the discretion of the Contractor										
Reference 8, 3.80 kg/m2; 150mm side laps; 150mm end laps										
generally	5.81	15.00	6.68	0.07	0.02	0.79	0.42	7.89	m2	8.48
strips in one width										
750mm wide	5.81	15.00	6.68	0.11	0.02	1.16	0.42	8.26	m2	8.88
900mm wide	5.81	15.00	6.68	0.10	0.02	1.06	0.42	8.16	m2	8.78
1050mm wide	5.81	15.00	6.68	0.09	0.02	0.97	0.42	8.07	m2	8.68
1200mm wide	5.81	15.00	6.68	0.08	0.02	0.88	0.42	7.98	m2	8.58
Expamet 'Securilath' flattened security mesh expanded steel reinforcement (uncoated); including laps										
Reference 2073F; 150mm side laps; 150mm end laps										
generally	16.76	15.00	19.27	0.10	0.02	1.06	0.20	20.54	m2	22.08

Labour hourly rates: (except Specialists) Craft Operatives £9.23 Labour £7.02 Rates are national average prices. Refer to REGIONAL VARIATIONS for indicative levels of overall pricing in regions	MATERIALS			LABOUR				RATES		
	Del to Site	Waste	Material Cost	Craft Optve	Lab	Labour Cost	Sunds	Nett Rate	Unit	Gross Rate (+7.5%)
	£	%	£	Hrs	Hrs	£	£	£		£
E40: DESIGNED JOINTS IN IN-SITU CONCRETE										
Formed joints										
Incorporating 10mm thick Korkpak, Servicised Ltd; formwork; reinforcement laid continuously across joint										
in concrete, depth not exceeding 150mm; horizontal	3.27	5.00	3.43	-	0.07	0.49	4.06	7.98	m	8.58
in concrete, depth 150 - 300mm; horizontal	5.36	5.00	5.63	-	0.08	0.56	7.47	13.66	m	14.68
in concrete, depth 300 - 450mm; horizontal	9.38	5.00	9.85	-	0.09	0.63	10.87	21.35	m	22.95
Incorporating 13mm thick Korkpak, Servicised Ltd; formwork; reinforcement laid continuously across joint										
in concrete, depth not exceeding 150mm; horizontal	3.83	5.00	4.02	-	0.07	0.49	4.06	8.57	m	9.22
in concrete, depth 150 - 300mm; horizontal	5.51	5.00	5.79	-	0.08	0.56	7.47	13.82	m	14.85
in concrete, depth 300 - 450mm; horizontal	9.56	5.00	10.04	-	0.09	0.63	10.87	21.54	m	23.16
Incorporating 19mm thick Korkpak, Servicised Ltd; formwork; reinforcement laid continuously across joint										
in concrete, depth not exceeding 150mm; horizontal	4.07	5.00	4.27	-	0.07	0.49	4.06	8.82	m	9.49
in concrete, depth 150 - 300mm; horizontal	7.63	5.00	8.01	-	0.08	0.56	7.47	16.04	m	17.25
in concrete, depth 300 - 450mm; horizontal	13.51	5.00	14.19	-	0.09	0.63	10.87	25.69	m	27.61
Incorporating 195mm wide 'Serviseal STD 195', external face type p.v.c waterstop, Servicised Ltd; heat welded joints; formwork; reinforcement laid continuously across joint										
in concrete, depth not exceeding 150mm; horizontal	4.09	10.00	4.50	-	0.40	2.81	4.06	11.37	m	12.22
in concrete, depth 150 - 300mm; horizontal	4.09	10.00	4.50	-	0.40	2.81	7.47	14.78	m	15.89
in concrete, depth 300 - 450mm; horizontal	4.09	10.00	4.50	-	0.40	2.81	10.87	18.18	m	19.54
vertical L piece	9.91	10.00	10.90	-	0.20	1.40	-	12.31	nr	13.23
flat L piece	7.86	10.00	8.65	-	0.20	1.40	-	10.05	nr	10.80
flat T piece	11.49	10.00	12.64	-	0.25	1.75	-	14.39	nr	15.47
flat X piece	12.90	10.00	14.19	-	0.30	2.11	-	16.30	nr	17.52
in concrete, width not exceeding 150mm; vertical .	4.09	10.00	4.50	-	0.50	3.51	4.06	12.07	m	12.97
in concrete, width 150 - 300mm; vertical	4.09	10.00	4.50	-	0.50	3.51	7.47	15.48	m	16.64
in concrete, width 300 - 450mm; vertical	4.09	10.00	4.50	-	0.50	3.51	10.87	18.88	m	20.29
flat L piece	7.86	10.00	8.65	-	0.20	1.40	-	10.05	nr	10.80
flat T piece	11.49	10.00	12.64	-	0.25	1.75	-	14.39	nr	15.47
flat X piece	12.90	10.00	14.19	-	0.30	2.11	-	16.30	nr	17.52
Incorporating 195mm wide 'Serviseal EXP 195', external face type p.v.c waterstop, Servicised Ltd; heat welded joints; formwork; reinforcement laid continuously across joint										
in concrete, depth not exceeding 150mm; horizontal	4.31	10.00	4.74	-	0.40	2.81	4.06	11.61	m	12.48
in concrete, depth 150 - 300mm; horizontal	4.31	10.00	4.74	-	0.40	2.81	7.47	15.02	m	16.15
in concrete, depth 300 - 450mm; horizontal	4.31	10.00	4.74	-	0.40	2.81	10.87	18.42	m	19.80
vertical L piece	10.87	10.00	11.96	-	0.20	1.40	-	13.36	nr	14.36
flat L piece	8.02	10.00	8.82	-	0.20	1.40	-	10.23	nr	10.99
flat T piece	11.66	10.00	12.83	-	0.25	1.75	-	14.58	nr	15.67
flat X piece	13.22	10.00	14.54	-	0.30	2.11	-	16.65	nr	17.90
in concrete, width not exceeding 150mm; vertical .	4.31	10.00	4.74	-	0.50	3.51	4.06	12.31	m	13.23
in concrete, width 150 - 300mm; vertical	4.31	10.00	4.74	-	0.50	3.51	7.47	15.72	m	16.90
in concrete, width 300 - 450mm; vertical	4.31	10.00	4.74	-	0.50	3.51	10.87	19.12	m	20.56
flat L piece	8.02	10.00	8.82	-	0.20	1.40	-	10.23	nr	10.99
flat T piece	11.66	10.00	12.83	-	0.25	1.75	-	14.58	nr	15.67
flat X piece	13.22	10.00	14.54	-	0.30	2.11	-	16.65	nr	17.90
Incorporating 240mm wide 'Serviseal HD240', heavy duty section external face type p.v.c. waterstop, Servicised Ltd; heat welded joints; formwork; reinforcement laid continuously across joint										
in concrete, depth not exceeding 150mm; horizontal	5.36	10.00	5.90	-	0.45	3.16	4.06	13.12	m	14.10
in concrete, depth 150 - 300mm; horizontal	5.36	10.00	5.90	-	0.45	3.16	7.47	16.52	m	17.76
in concrete, depth 300 - 450mm; horizontal	5.36	10.00	5.90	-	0.45	3.16	10.87	19.93	m	21.42
vertical L piece	10.16	10.00	11.18	-	0.20	1.40	-	12.58	nr	13.52
flat L piece	8.98	10.00	9.88	-	0.20	1.40	-	11.28	nr	12.13
flat T piece	13.10	10.00	14.41	-	0.25	1.75	-	16.16	nr	17.38
flat X piece	15.15	10.00	16.66	-	0.30	2.11	-	18.77	nr	20.18
in concrete, width not exceeding 150mm; vertical .	5.36	10.00	5.90	-	0.55	3.86	4.06	13.82	m	14.85
in concrete, width 150 - 300mm; vertical	5.36	10.00	5.90	-	0.55	3.86	7.47	17.23	m	18.52
in concrete, width 300 - 450mm; vertical	5.36	10.00	5.90	-	0.55	3.86	10.87	20.63	m	22.17
flat L piece	8.98	10.00	9.88	-	0.20	1.40	-	11.28	nr	12.13
flat T piece	13.10	10.00	14.41	-	0.25	1.75	-	16.16	nr	17.38
flat X piece	15.15	10.00	16.66	-	0.30	2.11	-	18.77	nr	20.18
Incorporating 240mm wide 'Serviseal EXP240', external face type p.v.c waterstop, Servicised Ltd; heat welded joints; formwork; reinforcement laid continuously across joint										
in concrete, depth not exceeding 150mm; horizontal	5.59	10.00	6.15	-	0.45	3.16	4.06	13.37	m	14.37
in concrete, depth 150 - 300mm; horizontal	5.59	10.00	6.15	-	0.45	3.16	7.47	16.78	m	18.04
in concrete, depth 300 - 450mm; horizontal	5.59	10.00	6.15	-	0.45	3.16	10.87	20.18	m	21.69
vertical L piece	11.31	10.00	12.44	-	0.20	1.40	-	13.85	nr	14.88
flat L piece	9.36	10.00	10.30	-	0.20	1.40	-	11.70	nr	12.58
flat T piece	13.67	10.00	15.04	-	0.25	1.75	-	16.79	nr	18.05
flat X piece	15.91	10.00	17.50	-	0.30	2.11	-	19.61	nr	21.08
in concrete, width not exceeding 150mm; vertical .	5.59	10.00	6.15	-	0.55	3.86	4.06	14.07	m	15.13
in concrete, width 150 - 300mm; vertical	5.59	10.00	6.15	-	0.55	3.86	7.47	17.48	m	18.79
in concrete, width 300 - 450mm; vertical	5.59	10.00	6.15	-	0.55	3.86	10.87	20.88	m	22.45
flat L piece	9.36	10.00	10.30	-	0.20	1.40	-	11.70	nr	12.58
flat T piece	13.67	10.00	15.04	-	0.25	1.75	-	16.79	nr	18.05
flat X piece	15.91	10.00	17.50	-	0.30	2.11	-	19.61	nr	21.08

Labour hourly rates: (except Specialists) Craft Operatives £9.23 Labour £7.02 Rates are national average prices. Refer to REGIONAL VARIATIONS for indicative levels of overall pricing in regions	MATERIALS			LABOUR				RATES		
	Del to Site £	Waste %	Material Cost £	Craft Optve Hrs	Lab Hrs	Labour Cost £	Sunds £	Nett Rate £	Unit	Gross Rate (+7.5%) £
E40: DESIGNED JOINTS IN IN-SITU CONCRETE Cont.										
Formed joints Cont.										
Incorporating 320mm wide 'Kicker 320', external face type p.v.c waterstop, Servicised Ltd; heat welded joints; formwork; reinforcement laid continuously across joint										
in concrete, depth not exceeding 150mm; horizontal	7.50	10.00	8.25	–	0.60	4.21	4.06	16.52	m	17.76
in concrete, depth 150 - 300mm; horizontal	7.50	10.00	8.25	–	0.60	4.21	7.47	19.93	m	21.43
in concrete, depth 300 - 450mm; horizontal	7.50	10.00	8.25	–	0.60	4.21	10.87	23.33	m	25.08
vertical L piece	10.30	10.00	11.33	–	0.25	1.75	–	13.09	nr	14.07
flat L piece	16.30	10.00	17.93	–	0.25	1.75	–	19.68	nr	21.16
flat T piece	22.95	10.00	25.25	–	0.30	2.11	–	27.35	nr	29.40
flat X piece	30.79	10.00	33.87	–	0.35	2.46	–	36.33	nr	39.05
Incorporating 170mm wide 'FD170' flat dumbell internally placed p.v.c. waterstop, Servicised Ltd; heat welded joints; formwork; reinforcement laid continuously across joint										
in concrete, depth not exceeding 150mm; horizontal	2.77	10.00	3.05	–	0.35	2.46	4.06	9.56	m	10.28
in concrete, depth 150 - 300mm; horizontal	2.77	10.00	3.05	–	0.35	2.46	7.47	12.97	m	13.95
in concrete, depth 300 - 450mm; horizontal	2.77	10.00	3.05	–	0.35	2.46	10.87	16.37	m	17.60
flat L piece	5.38	10.00	5.92	–	0.15	1.05	–	6.97	nr	7.49
flat T piece	7.53	10.00	8.28	–	0.20	1.40	–	9.69	nr	10.41
flat X piece	8.46	10.00	9.31	–	0.25	1.75	–	11.06	nr	11.89
in concrete, width not exceeding 150mm; vertical .	2.77	10.00	3.05	–	0.44	3.09	4.06	10.20	m	10.96
in concrete, width 150 - 300mm; vertical	2.77	10.00	3.05	–	0.44	3.09	7.47	13.61	m	14.63
in concrete, width 300 - 450mm; vertical	2.77	10.00	3.05	–	0.44	3.09	10.87	17.01	m	18.28
vertical L piece	8.81	10.00	9.69	–	0.15	1.05	–	10.74	nr	11.55
flat L piece	5.38	10.00	5.92	–	0.15	1.05	–	6.97	nr	7.49
vertical T piece	11.67	10.00	12.84	–	0.20	1.40	–	14.24	nr	15.31
flat T piece	7.53	10.00	8.28	–	0.20	1.40	–	9.69	nr	10.41
flat X piece	8.46	10.00	9.31	–	0.25	1.75	–	11.06	nr	11.89
Incorporating 210mm wide 'CB210' flat dumbell internally placed p.v.c. waterstop, Servicised Ltd; heat welded joints; formwork; reinforcement laid continuously across joint										
in concrete, depth not exceeding 150mm; horizontal	4.30	10.00	4.73	–	0.40	2.81	4.06	11.60	m	12.47
in concrete, depth 150 - 300mm; horizontal	4.30	10.00	4.73	–	0.40	2.81	7.47	15.01	m	16.13
in concrete, depth 300 - 450mm; horizontal	4.30	10.00	4.73	–	0.40	2.81	10.87	18.41	m	19.79
flat L piece	6.46	10.00	7.11	–	0.20	1.40	–	8.51	nr	9.15
flat T piece	9.34	10.00	10.27	–	0.25	1.75	–	12.03	nr	12.93
flat X piece	10.56	10.00	11.62	–	0.30	2.11	–	13.72	nr	14.75
in concrete, width not exceeding 150mm; vertical .	4.30	10.00	4.73	–	0.50	3.51	4.06	12.30	m	13.22
in concrete, width 150 - 300mm; vertical	4.30	10.00	4.73	–	0.50	3.51	7.47	15.71	m	16.89
in concrete, width 300 - 450mm; vertical	4.30	10.00	4.73	–	0.50	3.51	10.87	19.11	m	20.54
vertical L piece	9.18	10.00	10.10	–	0.20	1.40	–	11.50	nr	12.36
flat L piece	6.46	10.00	7.11	–	0.20	1.40	–	8.51	nr	9.15
vertical T piece	8.62	10.00	9.48	–	0.25	1.75	–	11.24	nr	12.08
flat T piece	9.34	10.00	10.27	–	0.25	1.75	–	12.03	nr	12.93
flat X piece	10.56	10.00	11.62	–	0.30	2.11	–	13.72	nr	14.75
Incorporating 250mm wide 'FD250' flat dumbell internally placed p.v.c. waterstop, Servicised Ltd; heat welded joints; formwork; reinforcement laid continuously across joint										
in concrete, depth not exceeding 150mm; horizontal	4.79	10.00	5.27	–	0.45	3.16	4.06	12.49	m	13.42
in concrete, depth 150 - 300mm; horizontal	4.79	10.00	5.27	–	0.45	3.16	7.47	15.90	m	17.09
in concrete, depth 300 - 450mm; horizontal	4.79	10.00	5.27	–	0.45	3.16	10.87	19.30	m	20.75
flat L piece	7.45	10.00	8.20	–	0.20	1.40	–	9.60	nr	10.32
flat T piece	10.87	10.00	11.96	–	1.04	7.30	–	19.26	nr	20.70
flat X piece	12.53	10.00	13.78	–	1.25	8.78	–	22.56	nr	24.25
in concrete, width not exceeding 150mm; vertical .	4.79	10.00	5.27	–	0.55	3.86	4.06	13.19	m	14.18
in concrete, width 150 - 300mm; vertical	4.79	10.00	5.27	–	0.55	3.86	7.47	16.60	m	17.84
in concrete, width 300 - 450mm; vertical	4.79	10.00	5.27	–	0.55	3.86	10.87	20.00	m	21.50
vertical L piece	7.81	10.00	8.59	–	0.20	1.40	–	9.99	nr	10.74
flat L piece	7.45	10.00	8.20	–	0.20	1.40	–	9.60	nr	10.32
vertical T piece	7.26	10.00	7.99	–	0.25	1.75	–	9.74	nr	10.47
flat T piece	10.87	10.00	11.96	–	0.25	1.75	–	13.71	nr	14.74
flat X piece	12.53	10.00	13.78	–	0.30	2.11	–	15.89	nr	17.08
Incorporating 160mm wide 'CB160' centre bulb type internally placed p.v.c. waterstop, Servicised Ltd; heat welded joints; formwork; reinforcement laid continuously across joint										
in concrete, depth not exceeding 150mm; horizontal	3.00	10.00	3.30	–	0.35	2.46	4.06	9.82	m	10.55
in concrete, depth 150 - 300mm; horizontal	3.00	10.00	3.30	–	0.35	2.46	7.47	13.23	m	14.22
in concrete, depth 300 - 450mm; horizontal	3.00	10.00	3.30	–	0.35	2.46	10.87	16.63	m	17.87
flat L piece	5.42	10.00	5.96	–	0.15	1.05	–	7.01	nr	7.54
flat T piece	7.58	10.00	8.34	–	0.20	1.40	–	9.74	nr	10.47
flat X piece	8.53	10.00	9.38	–	0.25	1.75	–	11.14	nr	11.97
in concrete, width not exceeding 150mm; vertical .	3.00	10.00	3.30	–	0.44	3.09	4.06	10.45	m	11.23
in concrete, width 150 - 300mm; vertical	3.00	10.00	3.30	–	0.44	3.09	7.47	13.86	m	14.90
in concrete, width 300 - 450mm; vertical	3.00	10.00	3.30	–	0.44	3.09	10.87	17.26	m	18.55
vertical L piece	9.01	10.00	9.91	–	0.15	1.05	–	10.96	nr	11.79
flat L piece	5.42	10.00	5.96	–	0.15	1.05	–	7.01	nr	7.54
vertical T piece	8.53	10.00	9.38	–	0.20	1.40	–	10.79	nr	11.60
flat T piece	7.58	10.00	8.34	–	0.20	1.40	–	9.74	nr	10.47
flat X piece	8.53	10.00	9.38	–	0.25	1.75	–	11.14	nr	11.97

IN SITU CONCRETE/LARGE PRECAST CONCRETE

Labour hourly rates: (except Specialists) Craft Operatives £9.23 Labour £7.02 Rates are national average prices. Refer to REGIONAL VARIATIONS for indicative levels of overall pricing in regions	MATERIALS			LABOUR				RATES		
	Del to Site £	Waste %	Material Cost £	Craft Optve Hrs	Lab Hrs	Labour Cost £	Sunds £	Nett Rate £	Unit	Gross Rate (+7.5%) £
E40: DESIGNED JOINTS IN IN-SITU CONCRETE Cont.										
Formed joints Cont.										
Incorporating 210mm wide 'FD210' centre bulb type internally placed p.v.c. waterstop, Servicised Ltd; heat welded joints; formwork; reinforcement laid continuously across joint										
in concrete, depth not exceeding 150mm; horizontal	3.62	10.00	3.98	-	0.40	2.81	4.06	10.85	m	11.66
in concrete, depth 150 - 300mm; horizontal	3.62	10.00	3.98	-	0.40	2.81	7.47	14.26	m	15.33
in concrete, depth 300 - 450mm; horizontal	3.62	10.00	3.98	-	0.40	2.81	10.87	17.66	m	18.98
flat L piece	6.52	10.00	7.17	-	0.20	1.40	-	8.58	nr	9.22
flat T piece	9.49	10.00	10.44	-	0.25	1.75	-	12.19	nr	13.11
flat X piece	10.69	10.00	11.76	-	0.30	2.11	-	13.87	nr	14.90
in concrete, width not exceeding 150mm; vertical .	3.62	10.00	3.98	-	0.50	3.51	4.06	11.55	m	12.42
in concrete, width 150 - 300mm; vertical	3.62	10.00	3.98	-	0.50	3.51	7.47	14.96	m	16.08
in concrete, width 300 - 450mm; vertical	3.62	10.00	3.98	-	0.50	3.51	10.87	18.36	m	19.74
vertical L piece	7.52	10.00	8.27	-	0.20	1.40	-	9.68	nr	10.40
flat L piece	6.52	10.00	7.17	-	0.20	1.40	-	8.58	nr	9.22
vertical T piece	7.05	10.00	7.75	-	0.25	1.75	-	9.51	nr	10.22
flat T piece	9.49	10.00	10.44	-	0.25	1.75	-	12.19	nr	13.11
flat X piece	10.69	10.00	11.76	-	0.30	2.11	-	13.87	nr	14.90
Incorporating 260mm wide 'CB260' centre bulb type internally placed p.v.c. waterstop, Servicised Ltd; heat welded joints; formwork; reinforcement laid continuously across joint										
in concrete, depth not exceeding 150mm; horizontal	5.02	10.00	5.52	-	0.45	3.16	4.06	12.74	m	13.70
in concrete, depth 150 - 300mm; horizontal	5.02	10.00	5.52	-	0.45	3.16	7.47	16.15	m	17.36
in concrete, depth 300 - 450mm; horizontal	5.02	10.00	5.52	-	0.45	3.16	10.87	19.55	m	21.02
flat L piece	7.51	10.00	8.26	-	0.20	1.40	-	9.66	nr	10.39
flat T piece	10.97	10.00	12.07	-	1.04	7.30	-	19.37	nr	20.82
flat X piece	12.63	10.00	13.89	-	1.25	8.78	-	22.67	nr	24.37
in concrete, width not exceeding 150mm; vertical .	5.02	10.00	5.52	-	0.55	3.86	4.06	13.44	m	14.45
in concrete, width 150 - 300mm; vertical	5.02	10.00	5.52	-	0.55	3.86	7.47	16.85	m	18.12
in concrete, width 300 - 450mm; vertical	5.02	10.00	5.52	-	0.55	3.86	10.87	20.25	m	21.77
vertical L piece	7.77	10.00	8.55	-	0.20	1.40	-	9.95	nr	10.70
flat L piece	7.51	10.00	8.26	-	0.20	1.40	-	9.66	nr	10.39
vertical T piece	7.30	10.00	8.03	-	0.25	1.75	-	9.79	nr	10.52
flat T piece	10.97	10.00	12.07	-	0.25	1.75	-	13.82	nr	14.86
flat X piece	12.63	10.00	13.89	-	0.30	2.11	-	16.00	nr	17.20
Incorporating 325mm wide 'CB325' centre bulb type internally placed p.v.c. waterstop, Servicised Ltd; heat welded joints; formwork; reinforcement laid continuously across joint										
in concrete, depth not exceeding 150mm; horizontal	11.15	10.00	12.27	-	0.50	3.51	4.06	19.84	m	21.32
in concrete, depth 150 - 300mm; horizontal	11.15	10.00	12.27	-	0.50	3.51	7.47	23.25	m	24.99
in concrete, depth 300 - 450mm; horizontal	11.15	10.00	12.27	-	0.50	3.51	10.87	26.65	m	28.64
flat L piece	14.56	10.00	16.02	-	0.25	1.75	-	17.77	nr	19.10
flat T piece	17.73	10.00	19.50	-	0.30	2.11	-	21.61	nr	23.23
flat X piece	20.59	10.00	22.65	-	0.35	2.46	-	25.11	nr	26.99
in concrete, width not exceeding 150mm; vertical .	11.15	10.00	12.27	-	0.60	4.21	4.06	20.54	m	22.08
in concrete, width 150 - 300mm; vertical	11.15	10.00	12.27	-	0.60	4.21	7.47	23.95	m	25.74
in concrete, width 300 - 450mm; vertical	11.15	10.00	12.27	-	0.60	4.21	10.87	27.35	m	29.40
vertical L piece	8.26	10.00	9.09	-	0.25	1.75	-	10.84	nr	11.65
flat L piece	14.56	10.00	16.02	-	0.25	1.75	-	17.77	nr	19.10
vertical T piece	7.71	10.00	8.48	-	0.30	2.11	-	10.59	nr	11.38
flat T piece	17.73	10.00	19.50	-	0.30	2.11	-	21.61	nr	23.23
flat X piece	20.59	10.00	22.65	-	0.35	2.46	-	25.11	nr	26.99
Incorporating 10 x 150mm 'Servi-tite CJ' flat dumbell internally placed p.v.c waterstop, Servicised Ltd; heat welded joints; formwork; reinforcement laid continuously across joint										
in concrete, depth not exceeding 150mm; horizontal	6.62	10.00	7.28	-	0.35	2.46	4.06	13.80	m	14.83
in concrete, depth 150 - 300mm; horizontal	6.62	10.00	7.28	-	0.35	2.46	7.47	17.21	m	18.50
in concrete, depth 300 - 450mm; horizontal	6.62	10.00	7.28	-	0.35	2.46	10.87	20.61	m	22.15
flat L piece	8.09	10.00	8.90	-	0.15	1.05	-	9.95	nr	10.70
flat T piece	13.91	10.00	15.30	-	0.20	1.40	-	16.70	nr	17.96
flat X piece	15.60	10.00	17.16	-	0.25	1.75	-	18.91	nr	20.33
in concrete, width not exceeding 150mm; vertical .	6.62	10.00	7.28	-	0.44	3.09	4.06	14.43	m	15.51
in concrete, width 150 - 300mm; vertical	6.62	10.00	7.28	-	0.44	3.09	7.47	17.84	m	19.18
in concrete, width 300 - 450mm; vertical	6.62	10.00	7.28	-	0.44	3.09	10.87	21.24	m	22.83
vertical L piece	9.30	10.00	10.23	-	0.15	1.05	-	11.28	nr	12.13
flat L piece	8.09	10.00	8.90	-	0.15	1.05	-	9.95	nr	10.70
vertical T piece	8.82	10.00	9.70	-	0.20	1.40	-	11.11	nr	11.94
flat T piece	13.91	10.00	15.30	-	0.20	1.40	-	16.70	nr	17.96
flat X piece	15.60	10.00	17.16	-	0.25	1.75	-	18.91	nr	20.33
Incorporating 10 x 230mm 'Servi-tite CJ' flat dumbell internally placed p.v.c waterstop, Servicised Ltd; heat welded joints; formwork; reinforcement laid continuously across joint										
in concrete, depth not exceeding 150mm; horizontal	9.26	10.00	10.19	-	0.45	3.16	4.06	17.41	m	18.71
in concrete, depth 150 - 300mm; horizontal	9.26	10.00	10.19	-	0.45	3.16	7.47	20.82	m	22.38
in concrete, depth 300 - 450mm; horizontal	9.26	10.00	10.19	-	0.45	3.16	10.87	24.22	m	26.03
flat L piece	11.67	10.00	12.84	-	0.20	1.40	-	14.24	nr	15.31
flat T piece	20.45	10.00	22.50	-	1.04	7.30	-	29.80	nr	32.03
flat X piece	23.34	10.00	25.67	-	1.25	8.78	-	34.45	Unit	37.03
in concrete, width not exceeding 150mm; vertical .	9.26	10.00	10.19	-	0.55	3.86	4.06	18.11	m	19.47
in concrete, width 150 - 300mm; vertical	9.26	10.00	10.19	-	0.55	3.86	7.47	21.52	m	23.13
in concrete, width 300 - 450mm; vertical	9.26	10.00	10.19	-	0.55	3.86	10.87	24.92	m	26.79
vertical L piece	7.32	10.00	8.05	-	0.20	1.40	-	9.46	nr	10.17
flat L piece	11.67	10.00	12.84	-	0.20	1.40	-	14.24	nr	15.31
vertical T piece	9.14	10.00	10.05	-	0.25	1.75	-	11.81	nr	12.69

Labour hourly rates: (except Specialists) Craft Operatives £9.23 Labour £7.02 Rates are national average prices. Refer to REGIONAL VARIATIONS for indicative levels of overall pricing in regions	MATERIALS			LABOUR				RATES		
	Del to Site £	Waste %	Material Cost £	Craft Optve Hrs	Lab Hrs	Labour Cost £	Sunds £	Nett Rate £	Unit	Gross Rate (+7.5%) £

E40: DESIGNED JOINTS IN IN-SITU CONCRETE Cont.

Formed joints Cont.

Incorporating 10 x 230mm 'Servi-tite CJ' flat dumbell internally placed p.v.c waterstop, Servicised Ltd; heat welded joints; formwork; reinforcement laid continuously across joint Cont.

flat T piece	20.45	10.00	22.50	-	0.25	1.75	-	24.25	nr	26.07
flat X piece	23.34	10.00	25.67	-	0.30	2.11	-	27.78	nr	29.86

Incorporating 10 x 305mm 'Servi-tite CJ' flat dumbell internally placed p.v.c waterstop, Servicised Ltd; heat welded joints; formwork; reinforcement laid continuously across joint

in concrete, depth not exceeding 150mm; horizontal	14.90	10.00	16.39	-	0.50	3.51	4.06	23.96	m	25.76
in concrete, depth 150 - 300mm; horizontal	14.90	10.00	16.39	-	0.50	3.51	7.47	27.37	m	29.42
in concrete, depth 300 - 450mm; horizontal	14.90	10.00	16.39	-	0.50	3.51	10.87	30.77	m	33.08
flat L piece	26.14	10.00	28.75	-	0.25	1.75	-	30.51	nr	32.80
flat T piece	34.51	10.00	37.96	-	0.30	2.11	-	40.07	nr	43.07
flat X piece	44.05	10.00	48.45	-	0.35	2.46	-	50.91	nr	54.73
in concrete, width not exceeding 150mm; vertical	14.90	10.00	16.39	-	0.60	4.21	4.06	24.66	m	26.51
in concrete, width 150 - 300mm; vertical	14.90	10.00	16.39	-	0.60	4.21	7.47	28.07	m	30.18
in concrete, width 300 - 450mm; vertical	14.90	10.00	16.39	-	0.60	4.21	10.87	31.47	m	33.83
vertical L piece	24.86	10.00	27.35	-	0.25	1.75	-	29.10	nr	31.28
flat L piece	26.14	10.00	28.75	-	0.25	1.75	-	30.51	nr	32.80
vertical T piece	31.99	10.00	35.19	-	0.30	2.11	-	37.30	nr	40.09
flat T piece	34.51	10.00	37.96	-	0.30	2.11	-	40.07	nr	43.07
flat X piece	44.05	10.00	48.45	-	0.35	2.46	-	50.91	nr	54.73

Incorporating 10 x 150mm 'Servi-tite XJ' centre bulb type internally placed p.v.c. waterstop, Servicised Ltd; heat welded joints; formwork; reinforcement laid continuously across joint

in concrete, depth not exceeding 150mm; horizontal	6.86	10.00	7.55	-	0.35	2.46	4.06	14.06	m	15.12
in concrete, depth 150 - 300mm; horizontal	6.86	10.00	7.55	-	0.35	2.46	7.47	17.47	m	18.78
in concrete, depth 300 - 450mm; horizontal	6.86	10.00	7.55	-	0.35	2.46	10.87	20.87	m	22.44
flat L piece	12.33	10.00	13.56	-	0.15	1.05	-	14.62	nr	15.71
flat T piece	18.06	10.00	19.87	-	0.20	1.40	-	21.27	nr	22.87
flat X piece	20.23	10.00	22.25	-	0.25	1.75	-	24.01	nr	25.81
in concrete, width not exceeding 150mm; vertical	6.86	10.00	7.55	-	0.44	3.09	4.06	14.69	m	15.80
in concrete, width 150 - 300mm; vertical	6.86	10.00	7.55	-	0.44	3.09	7.47	18.10	m	19.46
in concrete, width 300 - 450mm; vertical	6.86	10.00	7.55	-	0.44	3.09	10.87	21.50	m	23.12
vertical L piece	9.46	10.00	10.41	-	0.15	1.05	-	11.46	nr	12.32
flat L piece	12.33	10.00	13.56	-	0.15	1.05	-	14.62	nr	15.71
vertical T piece	8.90	10.00	9.79	-	0.20	1.40	-	11.19	nr	12.03
flat T piece	18.06	10.00	19.87	-	0.20	1.40	-	21.27	nr	22.87
flat X piece	20.23	10.00	22.25	-	0.25	1.75	-	24.01	nr	25.81

Incorporating 10 x 230mm 'Servi-tite XJ' centre bulb type internally placed p.v.c. waterstop, Servicised Ltd; heat welded joints; formwork; reinforcement laid continuously across joint

in concrete, depth not exceeding 150mm; horizontal	9.63	10.00	10.59	-	0.45	3.16	4.06	17.81	m	19.15
in concrete, depth 150 - 300mm; horizontal	9.63	10.00	10.59	-	0.45	3.16	7.47	21.22	m	22.81
in concrete, depth 300 - 450mm; horizontal	9.63	10.00	10.59	-	0.45	3.16	10.87	24.62	m	26.47
flat L piece	15.14	10.00	16.65	-	0.20	1.40	-	18.06	nr	19.41
flat T piece	27.12	10.00	29.83	-	1.04	7.30	-	37.13	nr	39.92
flat X piece	30.27	10.00	33.30	-	1.25	8.78	-	42.07	nr	45.23
in concrete, width not exceeding 150mm; vertical	9.63	10.00	10.59	-	0.55	3.86	4.06	18.51	m	19.90
in concrete, width 150 - 300mm; vertical	9.63	10.00	10.59	-	0.55	3.86	7.47	21.92	m	23.57
in concrete, width 300 - 450mm; vertical	9.63	10.00	10.59	-	0.55	3.86	10.87	25.32	m	27.22
vertical L piece	8.50	10.00	9.35	-	0.20	1.40	-	10.75	nr	11.56
flat L piece	15.14	10.00	16.65	-	0.20	1.40	-	18.06	nr	19.41
vertical T piece	7.95	10.00	8.74	-	0.25	1.75	-	10.50	nr	11.29
flat T piece	27.12	10.00	29.83	-	0.25	1.75	-	31.59	nr	33.96
flat X piece	30.27	10.00	33.30	-	0.30	2.11	-	35.40	nr	38.06

Incorporating 10 x 305mm 'Servi-tite XJ' centre bulb type internally placed p.v.c. waterstop, Servicised Ltd; heat welded joints; formwork; reinforcement laid continuously across joint

in concrete, depth not exceeding 150mm; horizontal	16.56	10.00	18.22	-	0.50	3.51	4.06	25.79	m	27.72
in concrete, depth 150 - 300mm; horizontal	16.56	10.00	18.22	-	0.50	3.51	7.47	29.20	m	31.39
in concrete, depth 300 - 450mm; horizontal	16.56	10.00	18.22	-	0.50	3.51	10.87	32.60	m	35.04
flat L piece	27.45	10.00	30.20	-	0.25	1.75	-	31.95	nr	34.35
flat T piece	41.98	10.00	46.18	-	0.30	2.11	-	48.28	nr	51.91
flat X piece	55.32	10.00	60.85	-	0.35	2.46	-	63.31	nr	68.06
in concrete, width not exceeding 150mm; vertical	16.56	10.00	18.22	-	0.60	4.21	4.06	26.49	m	28.47
in concrete, width 150 - 300mm; vertical	16.56	10.00	18.22	-	0.60	4.21	7.47	29.90	m	32.14
in concrete, width 300 - 450mm; vertical	16.56	10.00	18.22	-	0.60	4.21	10.87	33.30	m	35.80
vertical L piece	25.30	10.00	27.83	-	0.25	1.75	-	29.59	nr	31.80
flat L piece	27.45	10.00	30.20	-	0.25	1.75	-	31.95	nr	34.35
vertical T piece	33.59	10.00	36.95	-	0.30	2.11	-	39.06	nr	41.98
flat T piece	41.98	10.00	46.18	-	0.30	2.11	-	48.28	nr	51.91
flat X piece	55.32	10.00	60.85	-	0.35	2.46	-	63.31	nr	68.06

Incorporating Reference 2411, 6.34 kg/m2 Expamet Hy-rib permanent shuttering and reinforcement; 150mm end laps; temporary supports; formwork laid continuously across joint

in concrete, depth not exceeding 150mm; horizontal	2.10	5.00	2.21	0.23	0.06	2.54	4.50	9.25	m	9.94
in concrete, depth 150 - 300mm; horizontal	4.20	5.00	4.41	0.23	0.06	2.54	7.85	14.80	m	15.91
in concrete, width not exceeding 150mm; vertical	2.10	5.00	2.21	0.35	0.09	3.86	4.50	10.57	m	11.36
in concrete, width 150 - 300mm; vertical	4.20	5.00	4.41	0.35	0.09	3.86	7.85	16.12	m	17.33

IN SITU CONCRETE/LARGE PRECAST CONCRETE

Labour hourly rates: (except Specialists) Craft Operatives £9.23 Labour £7.02 Rates are national average prices. Refer to REGIONAL VARIATIONS for indicative levels of overall pricing in regions	MATERIALS			LABOUR				RATES		
	Del to Site £	Waste %	Material Cost £	Craft Optve Hrs	Lab Hrs	Labour Cost £	Sunds £	Nett Rate £	Unit	Gross Rate (+7.5%) £
E40: DESIGNED JOINTS IN IN-SITU CONCRETE Cont.										
Formed joints Cont.										
Incorporating Reference 2611, 4.86 kg/m2 Expamet Hy- rib permanent shuttering and reinforcement; 150mm end laps; temporary supports; formwork laid continuously across joint										
in concrete, depth not exceeding 150mm; horizontal	1.64	5.00	1.72	0.23	0.06	2.54	4.50	8.77	m	9.42
in concrete, depth 150 - 300mm; horizontal	3.29	5.00	3.45	0.23	0.06	2.54	7.85	13.85	m	14.89
in concrete, width not exceeding 150mm; vertical .	1.64	5.00	1.72	0.35	0.09	3.86	4.50	10.08	m	10.84
in concrete, width 150 - 300mm; vertical	3.29	5.00	3.45	0.35	0.09	3.86	7.85	15.17	m	16.30
Incorporating Reference 2811, 3.39 kg/m2 Expamet Hy- rib permanent shuttering and reinforcement; 150mm end laps; temporary supports; formwork laid continuously across joint										
in concrete, depth not exceeding 150mm; horizontal	1.45	5.00	1.52	0.23	0.06	2.54	4.50	8.57	m	9.21
in concrete, depth 150 - 300mm; horizontal	2.90	5.00	3.04	0.23	0.06	2.54	7.85	13.44	m	14.45
in concrete, width not exceeding 150mm; vertical .	1.45	5.00	1.52	0.35	0.09	3.86	4.50	9.88	m	10.63
in concrete, width 150 - 300mm; vertical	2.90	5.00	3.04	0.35	0.09	3.86	7.85	14.76	m	15.86
Sealant to joint; Servicised Ltd. 'Servigard DW'; including preparation, cleaners, primers and sealers										
10 x 25mm; horizontal	2.47	5.00	2.59	-	0.15	1.05	-	3.65	m	3.92
13 x 25mm; horizontal	3.20	5.00	3.36	-	0.20	1.40	-	4.76	m	5.12
19 x 25mm; horizontal	4.68	5.00	4.91	-	0.25	1.75	-	6.67	m	7.17
Sealant to joint; Servicised Ltd. 'Servimastic 96'; including preparation, cleaners, primers and sealers										
10 x 25mm; horizontal	1.93	5.00	2.03	-	0.10	0.70	-	2.73	m	2.93
13 x 25mm; horizontal	2.50	5.00	2.63	-	0.11	0.77	-	3.40	m	3.65
19 x 25mm; horizontal	3.66	5.00	3.84	-	0.12	0.84	-	4.69	m	5.04
E41: WORKED FINISHES/ CUTTING ON IN-SITU CONCRETE										
Worked finishes on in-situ concrete										
Vacuum dewatering; power floating and power trowelling										
surfaces	-	-	-	-	0.20	1.40	0.22	1.62	m2	1.75
Tamping unset concrete										
surfaces	-	-	-	-	0.20	1.40	0.27	1.67	m2	1.80
surfaces to falls	-	-	-	-	0.40	2.81	0.34	3.15	m2	3.38
Trowelling										
surfaces	-	-	-	-	0.33	2.32	-	2.32	m2	2.49
surfaces to falls	-	-	-	-	0.35	2.46	-	2.46	m2	2.64
Hacking; by hand										
surfaces	-	-	-	-	0.50	3.51	-	3.51	m2	3.77
to soffits	-	-	-	-	0.65	4.56	-	4.56	m2	4.91
Hacking; by machine										
surfaces	-	-	-	-	0.20	1.40	0.22	1.62	m2	1.75
to soffits	-	-	-	-	0.25	1.75	0.22	1.98	m2	2.12
Bush hammering										
surfaces	-	-	-	1.30	-	12.00	0.62	12.62	m2	13.57
to soffits	-	-	-	2.00	-	18.46	0.82	19.28	m2	20.73
Cutting on in-situ concrete										
Cutting chases										
depth not exceeding 50mm	-	-	-	-	0.17	1.19	-	1.19	m	1.28
Cutting chases; making good										
depth not exceeding 50mm	-	-	-	-	0.20	1.40	0.20	1.60	m	1.72
Cutting mortices										
50 x 50mm; depth not exceeding 100mm	-	-	-	-	0.50	3.51	-	3.51	nr	3.77
75 x 75mm; depth not exceeding 100mm	-	-	-	-	0.50	3.51	-	3.51	nr	3.77
100 x 100mm; depth 100 - 200mm	-	-	-	-	0.60	4.21	-	4.21	nr	4.53
38mm diameter; depth not exceeding 100mm	-	-	-	-	0.35	2.46	-	2.46	nr	2.64
Cutting holes										
225 x 150mm; depth not exceeding 100mm	-	-	-	-	1.32	9.27	0.29	9.56	nr	10.27
300 x 150mm; depth not exceeding 100mm	-	-	-	-	1.45	10.18	0.32	10.50	nr	11.29
300 x 300mm; depth not exceeding 100mm	-	-	-	-	1.45	10.18	0.32	10.50	nr	11.29
225 x 150mm; depth 100 - 200mm	-	-	-	-	1.92	13.48	0.46	13.94	nr	14.98
300 x 150mm; depth 100 - 200mm	-	-	-	-	2.11	14.81	0.47	15.28	nr	16.43
300 x 300mm; depth 100 - 200mm	-	-	-	-	2.11	14.81	0.47	15.28	nr	16.43
225 x 150mm; depth 200 - 300mm	-	-	-	-	2.21	15.51	0.48	15.99	nr	17.19
300 x 150mm; depth 200 - 300mm	-	-	-	-	2.41	16.92	0.53	17.45	nr	18.76
300 x 300mm; depth 200 - 300mm	-	-	-	-	2.41	16.92	0.53	17.45	nr	18.76
50mm diameter; depth not exceeding 100mm	-	-	-	-	0.85	5.97	0.18	6.15	nr	6.61
100mm diameter; depth not exceeding 100mm	-	-	-	-	0.94	6.60	0.23	6.83	nr	7.34
150mm diameter; depth not exceeding 100mm	-	-	-	-	0.94	6.60	0.23	6.83	nr	7.34
50mm diameter; depth 100 - 200mm	-	-	-	-	1.37	9.62	0.30	9.92	nr	10.66
100mm diameter; depth 100 - 200mm	-	-	-	-	1.47	10.32	0.32	10.64	nr	11.44
150mm diameter; depth 100 - 200mm	-	-	-	-	1.47	10.32	0.32	10.64	nr	11.44
50mm diameter; depth 200 - 300mm	-	-	-	-	1.63	11.44	0.37	11.81	nr	12.70
100mm diameter; depth 200 - 300mm	-	-	-	-	1.79	12.57	0.41	12.98	nr	13.95
150mm diameter; depth 200 - 300mm	-	-	-	-	1.79	12.57	0.41	12.98	nr	13.95

Labour hourly rates: (except Specialists) Craft Operatives £9.23 Labour £7.02 Rates are national average prices. Refer to REGIONAL VARIATIONS for indicative levels of overall pricing in regions	MATERIALS			LABOUR				RATES		
	Del to Site £	Waste %	Material Cost £	Craft Optve Hrs	Lab Hrs	Labour Cost £	Sunds £	Nett Rate £	Unit	Gross Rate (+7.5%) £
E41: WORKED FINISHES/ CUTTING ON IN-SITU CONCRETE Cont.										
Cutting on in-situ concrete Cont.										
Grouting into mortices with cement mortar (1:1); around steel										
50 x 50mm; depth not exceeding 100mm	-	-	-	-	0.12	0.84	0.45	1.29	nr	1.39
75 x 75mm; depth not exceeding 100mm	-	-	-	-	0.12	0.84	0.45	1.29	nr	1.39
100 x 100mm; depth 100 - 200mm	-	-	-	-	0.15	1.05	0.58	1.63	nr	1.76
38mm diameter; depth not exceeding 100mm	-	-	-	-	0.08	0.56	0.32	0.88	nr	0.95
Cutting on reinforced in-situ concrete										
Cutting holes										
225 x 150mm; depth not exceeding 100mm	-	-	-	-	1.64	11.51	0.37	11.88	nr	12.77
300 x 150mm; depth not exceeding 100mm	-	-	-	-	1.81	12.71	0.42	13.13	nr	14.11
300 x 300mm; depth not exceeding 100mm	-	-	-	-	1.81	12.71	0.42	13.13	nr	14.11
225 x 150mm; depth 100 - 200mm	-	-	-	-	2.39	16.78	0.53	17.31	nr	18.61
300 x 150mm; depth 100 - 200mm	-	-	-	-	2.63	18.46	0.58	19.04	nr	20.47
300 x 300mm; depth 100 - 200mm	-	-	-	-	2.63	18.46	0.53	18.99	nr	20.42
225 x 150mm; depth 200 - 300mm	-	-	-	-	2.78	19.52	0.62	20.14	nr	21.65
300 x 150mm; depth 200 - 300mm	-	-	-	-	3.03	21.27	0.68	21.95	nr	23.60
300 x 300mm; depth 200 - 300mm	-	-	-	-	3.03	21.27	0.68	21.95	nr	23.60
50mm diameter; depth not exceeding 100mm	-	-	-	-	1.05	7.37	0.19	7.56	nr	8.13
100mm diameter; depth not exceeding 100mm	-	-	-	-	1.18	8.28	0.28	8.56	nr	9.21
150mm diameter; depth not exceeding 100mm	-	-	-	-	1.18	8.28	0.28	8.56	nr	9.21
50mm diameter; depth 100 - 200mm	-	-	-	-	1.72	12.07	0.39	12.46	nr	13.40
100mm diameter; depth 100 - 200mm	-	-	-	-	1.84	12.92	0.42	13.34	nr	14.34
150mm diameter; depth 100 - 200mm	-	-	-	-	1.84	12.92	0.42	13.34	nr	14.34
50mm diameter; depth 200 - 300mm	-	-	-	-	2.04	14.32	0.46	14.78	nr	15.89
100mm diameter; depth 200 - 300mm	-	-	-	-	2.24	15.72	0.49	16.21	nr	17.43
150mm diameter; depth 200 - 300mm	-	-	-	-	2.24	15.72	0.49	16.21	nr	17.43
E42: ACCESSORIES CAST INTO IN-SITU CONCRETE										
Cast in accessories										
Hardwood fillets										
38 x 25mm; dovetail	0.63	10.00	0.69	-	0.10	0.70	-	1.40	m	1.50
38 x 25mm; 150mm long	0.20	10.00	0.22	-	0.04	0.28	-	0.50	nr	0.54
38 x 25mm; 225mm long	0.23	10.00	0.25	-	0.05	0.35	-	0.60	nr	0.65
Galvanised steel dowels										
12mm diameter x 150mm long	0.44	10.00	0.48	-	0.08	0.56	-	1.05	nr	1.12
12mm diameter x 300mm long	0.65	10.00	0.71	-	0.08	0.56	-	1.28	nr	1.37
12mm diameter x 600mm long	0.98	10.00	1.08	-	0.12	0.84	-	1.92	nr	2.06
Galvanised expanded metal tie										
300 x 50mm; bending and temporarily fixing to formwork	0.46	10.00	0.51	-	0.12	0.84	-	1.35	nr	1.45
Galvanised steel dovetailed masonry slots with 3mm thick twisted tie to suit 50mm cavity										
100mm long; temporarily fixing to formwork	0.47	10.00	0.52	0.10	-	0.92	-	1.44	nr	1.55
Stainless steel channel with anchors at 250mm centres										
28 x 15mm; temporarily fixing to formwork	7.60	5.00	7.98	0.28	-	2.58	-	10.56	m	11.36
28 x 15 x 150mm long; temporarily fixing to formwork	1.32	5.00	1.39	0.12	-	1.11	-	2.49	nr	2.68
M10 bolt 50mm long with `T' head nut and washers .	2.60	5.00	2.73	0.05	-	0.46	-	3.19	nr	3.43
fishtailed tie to suit 50mm cavity	1.20	5.00	1.26	0.06	-	0.55	-	1.81	nr	1.95
Stainless steel angle drilled at 450mm centres										
70 x 90 x 5mm .:......................	20.10	5.00	21.11	0.80	-	7.38	-	28.49	m	30.63
Copper dovetailed masonry slots with 3mm thick twisted tie to suit 50mm cavity										
100mm long; temporarily fixing to formwork	5.50	10.00	6.05	0.10	-	0.92	-	6.97	nr	7.50
Bolt boxes; The Expanded Metal Co. Ltd										
reference 220 for use with poured concrete; 75mm diameter										
150mm long................................	1.08	10.00	1.19	0.25	-	2.31	0.56	4.06	nr	4.36
225mm long................................	1.30	10.00	1.43	0.30	-	2.77	0.56	4.76	nr	5.12
300mm long................................	1.53	10.00	1.68	0.35	-	3.23	0.56	5.47	nr	5.88
reference 220 for use with poured concrete; 100mm diameter										
375mm long................................	2.47	10.00	2.72	0.40	-	3.69	0.56	6.97	nr	7.49
450mm long................................	2.78	10.00	3.06	0.45	-	4.15	0.56	7.77	nr	8.35
600mm long................................	3.08	10.00	3.39	0.50	-	4.62	0.56	8.56	nr	9.21
reference 220 for use with vibrated concrete; wrap with single layer of thin polythene sheet; 75mm diameter										
150mm long................................	1.21	10.00	1.33	0.25	-	2.31	0.67	4.31	nr	4.63
225mm long................................	1.45	10.00	1.60	0.30	-	2.77	0.67	5.03	nr	5.41
300mm long................................	1.68	10.00	1.85	0.35	-	3.23	0.67	5.75	nr	6.18
reference 220 for use with vibrated concrete; wrap with single layer of thin polythene sheet; 100mm diameter										
375mm long................................	2.62	10.00	2.88	0.40	-	3.69	0.67	7.24	nr	7.79
450mm long................................	2.96	10.00	3.26	0.45	-	4.15	0.67	8.08	nr	8.69
600mm long................................	3.25	10.00	3.58	0.50	-	4.62	0.67	8.86	nr	9.52
Steel rag bolts										
M 10 x 100mm long......................	0.92	5.00	0.97	0.09	-	0.83	-	1.80	nr	1.93

Labour hourly rates: (except Specialists) Craft Operatives £9.23 Labour £7.02 Rates are national average prices. Refer to REGIONAL VARIATIONS for indicative levels of overall pricing in regions	MATERIALS			LABOUR				RATES		
	Del to Site	Waste	Material Cost	Craft Optve	Lab	Labour Cost	Sunds	Nett Rate	Unit	Gross Rate (+7.5%)
	£	%	£	Hrs	Hrs	£	£	£		£
E42: ACCESSORIES CAST INTO IN-SITU CONCRETE Cont.										
Cast in accessories Cont.										
Steel rag bolts Cont.										
M 10 x 160mm long	0.99	5.00	1.04	0.10	–	0.92	–	1.96	nr	2.11
M 12 x 100mm long	1.05	5.00	1.10	0.09	–	0.83	–	1.93	nr	2.08
M 12 x 160mm long	1.18	5.00	1.24	0.10	–	0.92	–	2.16	nr	2.32
M 12 x 200mm long	1.27	5.00	1.33	0.12	–	1.11	–	2.44	nr	2.62
M 16 x 120mm long	1.58	5.00	1.66	0.09	–	0.83	–	2.49	nr	2.68
M 16 x 160mm long	1.80	5.00	1.89	0.10	–	0.92	–	2.81	nr	3.02
M 16 x 200mm long	2.02	5.00	2.12	0.12	–	1.11	–	3.23	nr	3.47
E60: PRECAST/COMPOSITE CONCRETE DECKING										
Precast concrete floors and hoist, bed and grout (BIRCHWOOD CONCRETE LTD)										
150mm thick floors of solid lightweight concrete units 600mm wide										
span between supports										
3000mm	30.29	2.50	31.05	–	0.66	4.63	4.12	39.80	m2	42.79
3600mm	30.29	2.50	31.05	–	0.65	4.56	4.07	39.68	m2	42.66
4200mm	30.29	2.50	31.05	–	0.63	4.42	3.93	39.40	m2	42.35
200mm thick floors of solid lightweight concrete units 600mm wide										
span between supports										
3000mm	40.49	2.50	41.50	–	0.69	4.84	4.24	50.59	m2	54.38
3600mm	40.49	2.50	41.50	–	0.68	4.77	4.19	50.47	m2	54.25
4200mm	40.49	2.50	41.50	–	0.66	4.63	4.07	50.21	m2	53.97
125mm thick floors of hollow prestressed concrete units 750mm wide										
span between supports										
3000mm	20.20	2.50	20.70	–	0.64	4.49	4.07	29.27	m2	31.46
3600mm	20.20	2.50	20.70	–	0.63	4.42	3.93	29.06	m2	31.24
4200mm	20.20	2.50	20.70	–	0.61	4.28	3.81	28.80	m2	30.96
4800mm	20.20	2.50	20.70	–	0.60	4.21	3.67	28.59	m2	30.73
150mm thick floors of hollow prestressed concrete units 750mm wide										
span between supports										
3000mm	21.32	2.50	21.85	–	0.66	4.63	4.07	30.56	m2	32.85
3600mm	21.32	2.50	21.85	–	0.65	4.56	3.99	30.41	m2	32.69
4200mm	21.32	2.50	21.85	–	0.63	4.42	3.93	30.21	m2	32.47
4800mm	21.32	2.50	21.85	–	0.61	4.28	3.75	29.89	m2	32.13
200mm thick floors of hollow prestressed concrete units 750mm wide										
span between supports										
3000mm	24.68	2.50	25.30	–	0.69	4.84	4.24	34.38	m2	36.96
3600mm	24.68	2.50	25.30	–	0.68	4.77	4.19	34.26	m2	36.83
4200mm	24.68	2.50	25.30	–	0.66	4.63	4.12	34.05	m2	36.60
4800mm	24.68	2.50	25.30	–	0.65	4.56	4.07	33.93	m2	36.47
Prestressed concrete ground floors; prestressed concrete beams and 100mm building blocks; hoist bed and grout										
155mm thick floors, distributed load 1.5 kN/m2										
span not exceeding 3.9m	11.64	2.50	11.93	–	1.05	7.37	0.40	19.70	m2	21.18
span not exceeding 4.8m	11.71	2.50	12.00	–	0.95	6.67	0.40	19.07	m2	20.50
span not exceeding 5.3m	12.05	2.50	12.35	–	0.90	6.32	0.40	19.07	m2	20.50
155mm thick floors, distributed load 3.0 kN/m2										
span not exceeding 3.4m	11.93	2.50	12.23	–	1.05	7.37	0.40	20.00	m2	21.50
span not exceeding 4.2m	11.99	2.50	12.29	–	0.95	6.67	0.40	19.36	m2	20.81
span not exceeding 4.6m	12.05	2.50	12.35	–	0.90	6.32	0.40	19.07	m2	20.50
155mm thick floors, distributed load 4.5 kN/m2										
span not exceeding 2.8m										
	12.27	2.50	12.58	–	1.05	7.37	0.40	20.35	m2	21.87
span not exceeding 3.8m	12.32	2.50	12.63	–	0.95	6.67	0.40	19.70	m2	21.17
span not exceeding 4.1m	12.39	2.50	12.70	–	0.90	6.32	0.40	19.42	m2	20.87
Hollow block floor with precast reinforced concrete planks at 650mm centres, precast hollow infill blocks, in-situ concrete, designed mix C25, filling between blocks and temporary supports and bearers										
Floor as described; height to soffit 1.50-3.00m										
200mm thick	25.56	2.50	26.20	–	0.55	3.86	3.17	33.23	m2	35.72
250mm thick	26.84	2.50	27.51	–	0.60	4.21	3.84	35.56	m2	38.23
Floor as described; with 50mm in-situ concrete, designed mix C25, topping; height to soffit 1.50-3.00m										
200mm total thickness	25.25	2.50	25.88	–	0.50	3.51	5.56	34.95	m2	37.57
250mm total thickness	25.56	2.50	26.20	–	0.55	3.86	5.76	35.82	m2	38.51
300mm total thickness	26.84	2.50	27.51	–	0.60	4.21	6.61	38.33	m2	41.21

Masonry

Labour hourly rates: (except Specialists) Craft Operatives £9.23 Labour £7.02 Rates are national average prices. Refer to REGIONAL VARIATIONS for indicative levels of overall pricing in regions	MATERIALS			LABOUR				RATES		
	Del to Site	Waste	Material Cost	Craft Optve	Lab	Labour Cost	Sunds	Nett Rate	Unit	Gross Rate (+7.5%)
	£	%	£	Hrs	Hrs	£	£	£		£
F10: BRICK/BLOCK WALLING										
Common bricks, B.S.3921, Category M, 215 x 102.5 x 65mm, compressive strength 20.5 N/mm2; in cement-lime mortar (1:2:9)										
Walls; vertical										
102mm thick; stretcher bond	8.79	5.00	9.23	1.45	1.10	21.11	1.80	32.13	m2	34.55
215mm thick; English bond	17.73	5.00	18.62	2.35	1.85	34.68	4.32	57.61	m2	61.94
327mm thick; English bond	26.67	5.00	28.00	2.85	2.30	42.45	6.83	77.28	m2	83.08
extra; grooved bricks; 50% wall surface keyed	0.14	5.00	0.15	-	-	-	-	0.15	m2	0.16
extra; grooved bricks; 100% wall surface keyed ...	0.27	5.00	0.28	-	-	-	-	0.28	m2	0.30
Walls; building against concrete (ties measured separately); vertical										
102mm thick; stretcher bond	8.79	5.00	9.23	1.60	1.20	23.19	1.80	34.22	m2	36.79
Walls; building against old brickwork; tie new to old with 3mm diameter galvanised twisted wire ties - 6/m2; vertical										
102mm thick; stretcher bond	8.79	5.00	9.23	1.60	1.20	23.19	2.32	34.74	m2	37.35
215mm thick; English bond	17.73	5.00	18.62	2.65	2.05	38.85	4.84	62.31	m2	66.98
Walls; bonding to stonework; including extra material; vertical										
102mm thick; stretcher bond	8.79	5.00	9.23	1.60	1.20	23.19	1.80	34.22	m2	36.79
215mm thick; English bond	17.73	5.00	18.62	2.65	2.05	38.85	4.32	61.79	m2	66.42
327mm thick; English bond	26.67	5.00	28.00	3.15	2.50	46.62	6.83	81.46	m2	87.57
440mm thick; English bond	35.46	5.00	37.23	3.70	3.00	55.21	9.36	101.80	m2	109.44
Walls; bonding to old brickwork; cutting pockets; including extra material; vertical										
102mm thick; stretcher bond	13.83	5.00	14.52	3.65	2.55	51.59	4.88	70.99	m2	76.32
215mm thick; English bond	22.77	5.00	23.91	4.60	3.30	65.62	8.89	98.42	m2	105.80
Walls; curved 2m radius; including extra material; vertical										
102mm thick; stretcher bond	8.79	5.00	9.23	2.85	2.05	40.70	1.80	51.73	m2	55.61
215mm thick; English bond	17.73	5.00	18.62	5.20	3.75	74.32	4.32	97.26	m2	104.55
327mm thick; English bond	26.67	5.00	28.00	7.10	5.10	101.33	6.83	136.17	m2	146.38
Walls; curved 6m radius; including extra material; vertical										
102mm thick; stretcher bond	8.79	5.00	9.23	2.15	1.55	30.73	1.80	41.76	m2	44.89
215mm thick; English bond	17.73	5.00	18.62	3.75	2.80	54.27	4.32	77.20	m2	83.00
327mm thick; English bond	26.67	5.00	28.00	4.95	3.70	71.66	6.83	106.50	m2	114.48
Isolated piers; vertical										
215mm thick; English bond	17.73	5.00	18.62	4.70	3.40	67.25	4.32	90.19	m2	96.95
327mm thick; English bond	26.67	5.00	28.00	5.70	4.20	82.09	6.83	116.93	m2	125.70
440mm thick; English bond	35.46	5.00	37.23	6.60	4.95	95.67	9.36	142.26	m2	152.93
Chimney stacks; vertical										
440mm thick; English bond	35.46	5.00	37.23	6.60	4.95	95.67	9.36	142.26	m2	152.93
890mm thick; English bond	70.92	5.00	74.47	10.40	8.00	152.15	18.68	245.30	m2	263.70
Projections; vertical										
215mm wide x 112mm projection	1.99	5.00	2.09	0.30	0.23	4.38	0.44	6.91	m	7.43
215mm wide x 215mm projection	3.97	5.00	4.17	0.50	0.39	7.35	0.85	12.37	m	13.30
327mm wide x 112mm projection	2.98	5.00	3.13	0.45	0.35	6.61	0.64	10.38	m	11.16
327mm wide x 215mm projection	5.96	5.00	6.26	0.75	0.59	11.06	1.27	18.59	m	19.99
Projections; horizontal										
215mm wide x 112mm projection	1.99	5.00	2.09	0.30	0.23	4.38	0.44	6.91	m	7.43
215mm wide x 215mm projection	3.97	5.00	4.17	0.50	0.39	7.35	0.85	12.37	m	13.30
327mm wide x 112mm projection	2.98	5.00	3.13	0.45	0.35	6.61	0.64	10.38	m	11.16
327mm wide x 215mm projection	5.96	5.00	6.26	0.75	0.59	11.06	1.27	18.59	m	19.99
Projections; bonding to old brickwork; cutting pockets; including extra material; vertical										
215mm wide x 112mm projection	3.31	5.00	3.48	0.93	0.65	13.15	0.89	17.51	m	18.83

MASONRY – MAJOR WORKS

Labour hourly rates: (except Specialists) Craft Operatives £9.23 Labour £7.02 Rates are national average prices. Refer to REGIONAL VARIATIONS for indicative levels of overall pricing in regions	MATERIALS			LABOUR				RATES		
	Del to Site £	Waste %	Material Cost £	Craft Optve Hrs	Lab Hrs	Labour Cost £	Sunds £	Nett Rate £	Unit	Gross Rate (+7.5%) £
F10: BRICK/BLOCK WALLING Cont.										
Common bricks, B.S.3921, Category M, 215 x 102.5 x 65mm, compressive strength 20.5 N/mm2; in cement-lime mortar (1:2:9) Cont.										
Projections; bonding to old brickwork; cutting pockets; including extra material; vertical Cont.										
215mm wide x 215mm projection	5.29	5.00	5.55	1.13	0.81	16.12	1.30	22.97	m	24.69
327mm wide x 112mm projection	4.97	5.00	5.22	1.40	0.98	19.80	1.33	26.35	m	28.33
327mm wide x 215mm projection	7.95	5.00	8.35	1.70	1.22	24.26	1.94	34.54	m	37.13
Projections; bonding to old brickwork; cutting pockets; including extra material; horizontal										
215mm wide x 112mm projection	3.31	5.00	3.48	0.93	0.65	13.15	0.89	17.51	m	18.83
215mm wide x 215mm projection	5.29	5.00	5.55	1.13	0.81	16.12	1.30	22.97	m	24.69
327mm wide x 112mm projection	4.97	5.00	5.22	1.40	0.98	19.80	1.33	26.35	m	28.33
327mm wide x 215mm projection	7.95	5.00	8.35	1.70	1.22	24.26	1.94	34.54	m	37.13
Arches including centering; flat										
112mm high on face; 215mm thick; width of exposed soffit 215mm; bricks-on-edge	1.99	5.00	2.09	1.50	1.00	20.86	1.88	24.83	m	26.70
215mm high on face; 112mm thick; width of exposed soffit 112mm; bricks-on-end	1.99	5.00	2.09	1.45	1.00	20.40	1.10	23.59	m	25.36
Arches including centering; semi-circular										
112mm high on face; 215mm thick; width of exposed soffit 215mm; bricks-on-edge	1.99	5.00	2.09	2.15	1.35	29.32	3.37	34.78	m	37.39
215mm high on face; 215mm thick; width of exposed soffit 215mm; bricks-on-edge	3.97	5.00	4.17	2.95	1.95	40.92	3.37	48.46	m	52.09
Closing cavities; vertical										
50mm wide with brickwork 102mm thick	0.97	5.00	1.02	0.40	0.27	5.59	0.22	6.83	m	7.34
75mm wide with brickwork 102mm thick	0.97	5.00	1.02	0.40	0.27	5.59	0.22	6.83	m	7.34
Closing cavities; horizontal										
50mm wide with slates	3.24	5.00	3.40	0.66	0.44	9.18	0.18	12.76	m	13.72
75mm wide with slates	3.24	5.00	3.40	0.66	0.44	9.18	0.18	12.76	m	13.72
50mm wide with brickwork 102mm thick	0.97	5.00	1.02	0.40	0.27	5.59	0.22	6.83	m	7.34
75mm wide with brickwork 102mm thick	0.97	5.00	1.02	0.40	0.27	5.59	0.22	6.83	m	7.34
Bonding ends to existing common brickwork; cutting pockets; extra material walls; bonding every third course										
102mm thick	0.66	5.00	0.69	0.37	0.25	5.17	0.14	6.00	m	6.45
215mm	1.32	5.00	1.39	0.66	0.44	9.18	0.17	10.74	m	11.54
327mm thick	1.99	5.00	2.09	0.95	0.63	13.19	0.22	15.50	m	16.66
Second hard stock bricks, B.S.3921, Category M, 215 x 102.5 x 65mm; in cement-lime mortar (1:2:9)										
Walls; vertical										
102mm thick; stretcher bond	17.11	5.00	17.97	1.45	1.10	21.11	1.80	40.87	m2	43.94
215mm thick; English bond	34.51	5.00	36.24	2.35	1.85	34.68	4.32	75.23	m2	80.88
327mm thick; English bond	51.91	5.00	54.51	2.85	2.30	42.45	6.83	103.79	m2	111.57
Walls; building against concrete (ties measured separately); vertical										
102mm thick; stretcher bond	17.11	5.00	17.97	1.60	1.20	23.19	1.80	42.96	m2	46.18
Walls; building against old brickwork; tie new to old with 3mm diameter galvanised twisted wire ties - 6/m2; vertical										
102mm thick; stretcher bond	17.11	5.00	17.97	1.60	1.20	23.19	2.32	43.48	m2	46.74
215mm thick; English bond	34.51	5.00	36.24	2.65	2.05	38.85	4.84	79.93	m2	85.92
Walls; bonding to stonework; including extra material; vertical										
102mm thick; stretcher bond	17.11	5.00	17.97	1.60	1.20	23.19	1.80	42.96	m2	46.18
215mm thick; English bond	34.51	5.00	36.24	2.65	2.05	38.85	4.32	79.41	m2	85.36
327mm thick; English bond	51.91	5.00	54.51	3.15	2.50	46.62	6.83	107.96	m2	116.06
440mm thick; English bond	69.02	5.00	72.47	3.70	3.00	55.21	9.36	137.04	m2	147.32
Walls; bonding to old brickwork; cutting pockets; including extra material; vertical										
102mm thick; stretcher bond	25.31	5.00	26.58	3.65	2.55	51.59	4.88	83.05	m2	89.27
215mm thick; English bond	44.31	5.00	46.53	4.60	3.30	65.62	8.89	121.04	m2	130.12
Walls; curved 2m radius; including extra material; vertical										
102mm thick; stretcher bond	17.11	5.00	17.97	2.85	2.05	40.70	1.80	60.46	m2	65.00
215mm thick; English bond	34.51	5.00	36.24	5.20	3.75	74.32	4.32	114.88	m2	123.49
327mm thick; English bond	51.91	5.00	54.51	7.10	5.10	101.33	6.83	162.67	m2	174.87
Walls; curved 6m radius; including extra material; vertical										
102mm thick; stretcher bond	17.11	5.00	17.97	2.15	1.55	30.73	1.80	50.49	m2	54.28
215mm thick; English bond	34.51	5.00	36.24	3.75	2.80	54.27	4.32	94.82	m2	101.94
327mm thick; English bond	51.91	5.00	54.51	4.95	3.70	71.66	6.83	133.00	m2	142.97
Isolated piers; vertical										
215mm thick; English bond	34.51	5.00	36.24	4.70	3.40	67.25	4.32	107.80	m2	115.89
327mm thick; English bond	51.91	5.00	54.51	5.70	4.20	82.09	6.83	143.43	m2	154.19
440mm thick; English bond	69.02	5.00	72.47	6.60	4.95	95.67	9.36	177.50	m2	190.81

Labour hourly rates: (except Specialists) Craft Operatives £9.23 Labour £7.02 Rates are national average prices. Refer to REGIONAL VARIATIONS for indicative levels of overall pricing in regions	MATERIALS			LABOUR				RATES		
	Del to Site £	Waste %	Material Cost £	Craft Optve Hrs	Lab Hrs	Labour Cost £	Sunds £	Nett Rate £	Unit	Gross Rate (+7.5%) £

F10: BRICK/BLOCK WALLING Cont.

Second hard stock bricks, B.S.3921, Category M, 215 x 102.5 x 65mm; in cement-lime mortar (1:2:9) Cont.

Chimney stacks; vertical

440mm thick; English bond	69.02	5.00	72.47	6.60	4.95	95.67	9.36	177.50	m2	190.81
890mm thick; English bond	138.04	5.00	144.94	10.40	8.00	152.15	18.68	315.77	m2	339.46

Projections; vertical

215mm wide x 112mm projection	3.87	5.00	4.06	0.30	0.23	4.38	0.44	8.89	m	9.55
215mm wide x 215mm projection	7.73	5.00	8.12	0.50	0.39	7.35	0.85	16.32	m	17.54
327mm wide x 112mm projection	5.80	5.00	6.09	0.45	0.35	6.61	0.64	13.34	m	14.34
327mm wide x 215mm projection	11.60	5.00	12.18	0.75	0.59	11.06	1.27	24.51	m	26.35

Projections; horizontal

215mm wide x 112mm projection	3.87	5.00	4.06	0.30	0.23	4.38	0.44	8.89	m	9.55
215mm wide x 215mm projection	7.73	5.00	8.12	0.50	0.39	7.35	0.85	16.32	m	17.54
327mm wide x 112mm projection	5.80	5.00	6.09	0.45	0.35	6.61	0.64	13.34	m	14.34
327mm wide x 215mm projection	11.60	5.00	12.18	0.75	0.59	11.06	1.27	24.51	m	26.35

Projections; bonding to old brickwork; cutting
pockets; including extra material; vertical

215mm wide x 112mm projection	6.45	5.00	6.77	0.93	0.65	13.15	0.89	20.81	m	22.37
215mm wide x 215mm projection	10.31	5.00	10.83	1.13	0.81	16.12	1.30	28.24	m	30.36
327mm wide x 112mm projection	9.67	5.00	10.15	1.40	0.98	19.80	1.33	31.29	m	33.63
327mm wide x 215mm projection	15.47	5.00	16.24	1.70	1.22	24.26	1.94	42.44	m	45.62

Projections; bonding to old brickwork; cutting
pockets; including extra material; horizontal

215mm wide x 112mm projection	6.45	5.00	6.77	0.93	0.65	13.15	0.89	20.81	m	22.37
215mm wide x 215mm projection	10.31	5.00	10.83	1.13	0.81	16.12	1.30	28.24	m	30.36
327mm wide x 112mm projection	9.67	5.00	10.15	1.40	0.98	19.80	1.33	31.29	m	33.63
327mm wide x 215mm projection	15.47	5.00	16.24	1.70	1.22	24.26	1.94	42.44	m	45.62

Arches including centering; flat

112mm high on face; 215mm thick; width of exposed soffit 215mm; bricks-on-edge	3.87	5.00	4.06	1.50	1.00	20.86	1.88	26.81	m	28.82
215mm high on face; 112mm thick; width of exposed soffit 112mm; bricks-on-end	3.87	5.00	4.06	1.45	1.00	20.40	1.10	25.57	m	27.48

Arches including centering; semi-circular

112mm high on face; 215mm thick; width of exposed soffit 215mm; bricks-on-edge	3.87	5.00	4.06	2.15	1.35	29.32	3.37	36.76	m	39.51
215mm high on face; 215mm thick; width of exposed soffit 215mm; bricks-on-edge	7.73	5.00	8.12	2.95	1.95	40.92	3.37	52.40	m	56.33

Closing cavities; vertical

50mm wide with brickwork 102mm thick	1.89	5.00	1.98	0.40	0.27	5.59	0.22	7.79	m	8.38
75mm wide with brickwork 102mm thick	1.89	5.00	1.98	0.40	0.27	5.59	0.22	7.79	m	8.38

Closing cavities; horizontal

50mm wide with slates	3.39	5.00	3.56	0.66	0.44	9.18	0.18	12.92	m	13.89
75mm wide with slates	3.39	5.00	3.56	0.66	0.44	9.18	0.18	12.92	m	13.89
50mm wide with brickwork 102mm thick	1.89	5.00	1.98	0.40	0.27	5.59	0.22	7.79	m	8.38
75mm wide with brickwork 102mm thick	1.89	5.00	1.98	0.40	0.27	5.59	0.22	7.79	m	8.38

Bonding ends to existing common brickwork; cutting
pockets; extra material
  walls; bonding every third course

102mm thick	1.29	5.00	1.35	0.37	0.25	5.17	0.14	6.66	m	7.16
215mm	2.58	5.00	2.71	0.66	0.44	9.18	0.17	12.06	m	12.96
327mm thick	3.87	5.00	4.06	0.95	0.63	13.19	0.22	17.47	m	18.79

Engineering bricks, B.S.3921, Category F, 215 x 102.5 x 65, Class B; in cement-lime mortar (1:2:9)

Walls; vertical

102mm thick; stretcher bond	17.53	5.00	18.41	1.60	1.20	23.19	1.80	43.40	m2	46.65
215mm thick; English bond	35.36	5.00	37.13	2.60	2.05	38.39	4.32	79.84	m2	85.82
327mm thick; English bond	53.19	5.00	55.85	3.15	2.55	46.98	6.83	109.66	m2	117.88

Walls; building against concrete (ties measured
separately); vertical

102mm thick; stretcher bond	17.53	5.00	18.41	1.75	1.30	25.28	1.80	45.48	m2	48.90

Walls; building against old brickwork; tie new to old
with 3mm diameter galvanised twisted wire ties -
6/m2; vertical

102mm thick; stretcher bond	17.53	5.00	18.41	1.75	1.30	25.28	2.32	46.01	m2	49.46
215mm thick; English bond	35.36	5.00	37.13	2.90	2.25	42.56	4.84	84.53	m2	90.87

Walls; bonding to stonework; including extra
material; vertical

102mm thick; stretcher bond	17.53	5.00	18.41	1.75	1.30	25.28	1.80	45.48	m2	48.90
215mm thick; English bond	35.36	5.00	37.13	2.90	2.25	42.56	4.32	84.01	m2	90.31
327mm thick; English bond	53.19	5.00	55.85	3.45	2.75	51.15	6.83	113.03	m2	122.37
440mm thick; English bond	70.72	5.00	74.26	4.05	3.30	60.55	9.36	144.16	m2	154.98

Walls; bonding to old brickwork; cutting pockets;
including extra material; vertical

102mm thick; stretcher bond	27.53	5.00	28.91	4.00	2.80	56.58	4.88	90.36	m2	97.14
215mm thick; English bond	45.36	5.00	47.63	5.05	3.65	72.23	8.89	128.75	m2	138.41

MASONRY

Labour hourly rates: (except Specialists) Craft Operatives £9.23 Labour £7.02 Rates are national average prices. Refer to REGIONAL VARIATIONS for indicative levels of overall pricing in regions	MATERIALS			LABOUR				RATES		
	Del to Site £	Waste %	Material Cost £	Craft Optve Hrs	Lab Hrs	Labour Cost £	Sunds £	Nett Rate £	Unit	Gross Rate (+7.5%) £
F10: BRICK/BLOCK WALLING Cont.										
Engineering bricks, B.S.3921, Category F, 215 x 102.5 x 65, Class B; in cement-lime mortar (1:2:9) Cont.										
Walls; curved 2m radius; including extra material; vertical										
102mm thick; stretcher bond	17.53	5.00	18.41	3.15	2.25	44.87	1.80	65.08	m2	69.96
215mm thick; English bond	35.36	5.00	37.13	5.70	4.10	81.39	4.32	122.84	m2	132.05
327mm thick; English bond	53.19	5.00	55.85	7.80	5.60	111.31	6.83	173.99	m2	187.03
Walls; curved 6m radius; including extra material; vertical										
102mm thick; stretcher bond	17.53	5.00	18.41	2.35	1.70	33.62	1.80	53.83	m2	57.87
215mm thick; English bond	35.36	5.00	37.13	4.10	3.10	59.60	4.32	101.05	m2	108.63
327mm thick; English bond	53.19	5.00	55.85	5.45	4.05	78.73	6.83	141.41	m2	152.02
Isolated piers; vertical										
215mm thick; English bond	35.36	5.00	37.13	5.15	3.75	73.86	4.32	115.31	m2	123.96
327mm thick; English bond	53.19	5.00	55.85	6.25	4.60	89.98	6.83	152.66	m2	164.11
440mm thick; English bond	70.72	5.00	74.26	7.25	5.45	105.18	9.36	188.79	m2	202.95
Chimney stacks; vertical										
440mm thick; English bond	70.72	5.00	74.26	7.25	5.45	105.18	9.36	188.79	m2	202.95
890mm thick; English bond	141.43	5.00	148.50	16.94	8.80	218.13	18.68	385.31	m2	414.21
Projections; vertical										
215mm wide x 112mm projection	3.96	5.00	4.16	0.33	0.25	4.80	0.44	9.40	m	10.10
215mm wide x 215mm projection	7.93	5.00	8.33	0.55	0.42	8.02	0.85	17.20	m	18.49
327mm wide x 112mm projection	5.94	5.00	6.24	0.50	0.38	7.28	0.64	14.16	m	15.22
327mm wide x 215mm projection	11.89	5.00	12.48	0.83	0.63	12.08	1.27	25.84	m	27.78
Projections; horizontal										
215mm wide x 112mm projection	3.96	5.00	4.16	0.33	0.25	4.80	0.44	9.40	m	10.10
215mm wide x 215mm projection	7.93	5.00	8.33	0.55	0.42	8.02	0.85	17.20	m	18.49
327mm wide x 112mm projection	5.94	5.00	6.24	0.50	0.38	7.28	0.64	14.16	m	15.22
327mm wide x 215mm projection	11.89	5.00	12.48	0.83	0.63	12.08	1.27	25.84	m	27.78
Projections; bonding to old brickwork; cutting pockets; including extra material; vertical										
215mm wide x 112mm projection	7.73	5.00	8.12	1.02	0.72	14.47	0.89	23.48	m	25.24
215mm wide x 215mm projection	11.59	5.00	12.17	1.24	0.89	17.69	1.30	31.16	m	33.50
327mm wide x 112mm projection	9.81	5.00	10.30	1.54	1.08	21.80	1.33	33.43	m	35.93
327mm wide x 215mm projection	15.75	5.00	16.54	1.87	1.34	26.67	1.94	45.14	m	48.53
Projections; bonding to old brickwork; cutting pockets; including extra material; horizontal										
215mm wide x 112mm projection	7.73	5.00	8.12	1.02	0.72	14.47	0.89	23.48	m	25.24
215mm wide x 215mm projection	11.59	5.00	12.17	1.24	0.89	17.69	1.30	31.16	m	33.50
327mm wide x 112mm projection	9.81	5.00	10.30	1.54	1.08	21.80	1.33	33.43	m	35.93
327mm wide x 215mm projection	15.75	5.00	16.54	1.87	1.34	26.67	1.94	45.14	m	48.53
Arches including centering; flat										
112mm high on face; 215mm thick; width of exposed soffit 215mm; bricks-on-edge	3.96	5.00	4.16	1.60	1.05	22.14	1.88	28.18	m	30.29
215mm high on face; 112mm thick; width of exposed soffit 112mm; bricks-on-end	3.96	5.00	4.16	1.55	1.10	22.03	1.10	27.29	m	29.33
Arches including centering; semi-circular										
112mm high on face; 215mm thick; width of exposed soffit 215mm; bricks-on-edge	3.96	5.00	4.16	2.25	1.40	30.60	3.37	38.12	m	40.98
215mm high on face; 215mm thick; width of exposed soffit 215mm; bricks-on-edge	7.93	5.00	8.33	3.10	2.05	43.00	3.37	54.70	m	58.80
Closing cavities; vertical										
50mm wide with brickwork 102mm thick	1.93	5.00	2.03	0.44	0.30	6.17	0.22	8.41	m	9.04
75mm wide with brickwork 102mm thick	1.93	5.00	2.03	0.44	0.30	6.17	0.22	8.41	m	9.04
Closing cavities; horizontal										
50mm wide with slates	3.39	5.00	3.56	0.66	0.44	9.18	0.18	12.92	m	13.89
75mm wide with slates	3.39	5.00	3.56	0.66	0.44	9.18	0.18	12.92	m	13.89
50mm wide with brickwork 102mm thick	3.03	5.00	3.18	0.44	0.30	6.17	0.22	9.57	m	10.29
75mm wide with brickwork 102mm thick	3.03	5.00	3.18	0.44	0.30	6.17	0.22	9.57	m	10.29
Bonding ends to existing common brickwork; cutting pockets; extra material walls; bonding every third course										
102mm thick	1.19	5.00	1.25	0.41	0.28	5.75	0.14	7.14	m	7.67
215mm	2.38	5.00	2.50	0.73	0.48	10.11	0.17	12.78	m	13.73
327mm thick	3.57	5.00	3.75	1.05	0.69	14.54	0.22	18.50	m	19.89
Engineering bricks, B.S.3921, Category F, 215 x 102.5 x 65mm, Class A; in cement-lime mortar (1:2:9)										
Walls; vertical										
102mm thick; stretcher bond	29.45	5.00	30.92	1.60	1.20	23.19	1.80	55.91	m2	60.11
215mm thick; English bond	59.40	5.00	62.37	2.60	2.05	38.39	4.32	105.08	m2	112.96
327mm thick; English bond	89.34	5.00	93.81	3.15	2.55	46.98	6.83	147.61	m2	158.68
Walls; building against concrete (ties measured separately); vertical										
102mm thick; stretcher bond	29.45	5.00	30.92	1.75	1.30	25.28	1.80	58.00	m2	62.35

Labour hourly rates: (except Specialists) Craft Operatives £9.23 Labour £7.02 Rates are national average prices. Refer to REGIONAL VARIATIONS for indicative levels of overall pricing in regions	MATERIALS			LABOUR				RATES		
	Del to Site £	Waste %	Material Cost £	Craft Optve Hrs	Lab Hrs	Labour Cost £	Sunds £	Nett Rate £	Unit	Gross Rate (+7.5%) £
F10: BRICK/BLOCK WALLING Cont.										
Engineering bricks, B.S.3921, Category F, 215 x 102.5 x 65mm, Class A; in cement-lime mortar (1:2:9) Cont.										
Walls; building against old brickwork; tie new to old with 3mm diameter galvanised twisted wire ties - 6/m2; vertical										
102mm thick; stretcher bond	29.45	5.00	30.92	1.75	1.30	25.28	2.32	58.52	m2	62.91
215mm thick; English bond	59.40	5.00	62.37	2.90	2.25	42.56	4.84	109.77	m2	118.00
Walls; bonding to stonework; including extra material; vertical										
102mm thick; stretcher bond	29.45	5.00	30.92	1.75	1.30	25.28	1.80	58.00	m2	62.35
215mm thick; English bond	59.40	5.00	62.37	2.90	2.25	42.56	4.32	109.25	m2	117.45
327mm thick; English bond	89.34	5.00	93.81	3.45	2.75	51.15	6.83	151.79	m2	163.17
440mm thick; English bond	118.79	5.00	124.73	4.05	3.30	60.55	9.36	194.64	m2	209.23
Walls; bonding to old brickwork; cutting pockets; including extra material; vertical										
102mm thick; stretcher bond	46.23	5.00	48.54	4.00	2.80	56.58	4.88	110.00	m2	118.25
215mm thick; English bond	76.16	5.00	79.97	5.05	3.65	72.23	8.89	161.09	m2	173.17
Walls; curved 2m radius; including extra material; vertical										
102mm thick; stretcher bond	29.45	5.00	30.92	3.15	2.25	44.87	1.80	77.59	m2	83.41
215mm thick; English bond	59.40	5.00	62.37	5.70	4.10	81.39	4.32	148.08	m2	159.19
327mm thick; English bond	89.34	5.00	93.81	7.80	5.60	111.31	6.83	211.94	m2	227.84
Walls; curved 6m radius; including extra material; vertical										
102mm thick; stretcher bond	29.45	5.00	30.92	2.35	1.70	33.62	1.80	66.35	m2	71.32
215mm thick; English bond	59.40	5.00	62.37	4.10	3.10	59.60	4.32	126.30	m2	135.77
327mm thick; English bond	89.34	5.00	93.81	5.45	4.05	78.73	6.83	179.37	m2	192.82
Isolated piers; vertical										
215mm thick; English bond	59.40	5.00	62.37	5.15	3.75	73.86	4.32	140.55	m2	151.09
327mm thick; English bond	89.34	5.00	93.81	6.25	4.60	89.98	6.83	190.62	m2	204.91
440mm thick; English bond	118.79	5.00	124.73	7.25	5.45	105.18	9.36	239.27	m2	257.21
Chimney stacks; vertical										
440mm thick; English bond	118.79	5.00	124.73	7.25	5.45	105.18	9.36	239.27	m2	257.21
890mm thick; English bond	237.58	5.00	249.46	16.94	8.80	218.13	18.68	486.27	m2	522.74
Projections; vertical										
215mm wide x 112mm projection	6.66	5.00	6.99	0.33	0.25	4.80	0.44	12.23	m	13.15
215mm wide x 215mm projection	13.31	5.00	13.98	0.55	0.42	8.02	0.85	22.85	m	24.56
327mm wide x 112mm projection	9.98	5.00	10.48	0.50	0.38	7.28	0.64	18.40	m	19.78
327mm wide x 215mm projection	19.97	5.00	20.97	0.83	0.63	12.08	1.27	34.32	m	36.90
Projections; horizontal										
215mm wide x 112mm projection	6.66	5.00	6.99	0.33	0.25	4.80	0.44	12.23	m	13.15
215mm wide x 215mm projection	13.31	5.00	13.98	0.55	0.42	8.02	0.85	22.85	m	24.56
327mm wide x 112mm projection	9.98	5.00	10.48	0.50	0.38	7.28	0.64	18.40	m	19.78
327mm wide x 215mm projection	19.97	5.00	20.97	0.83	0.63	12.08	1.27	34.32	m	36.90
Projections; bonding to old brickwork; cutting pockets; including extra material; vertical										
215mm wide x 112mm projection	12.97	5.00	13.62	1.02	0.72	14.47	0.89	28.98	m	31.15
215mm wide x 215mm projection	19.46	5.00	20.43	1.24	0.89	17.69	1.30	39.43	m	42.38
327mm wide x 112mm projection	16.47	5.00	17.29	1.54	1.08	21.80	1.33	40.42	m	43.45
327mm wide x 215mm projection	15.62	5.00	16.40	1.87	1.34	26.67	1.94	45.01	m	48.38
Projections; bonding to old brickwork; cutting pockets; including extra material; horizontal										
215mm wide x 112mm projection	12.97	5.00	13.62	1.02	0.72	14.47	0.89	28.98	m	31.15
215mm wide x 215mm projection	19.46	5.00	20.43	1.24	0.89	17.69	1.30	39.43	m	42.38
327mm wide x 112mm projection	16.47	5.00	17.29	1.54	1.08	21.80	1.33	40.42	m	43.45
327mm wide x 215mm projection	15.62	5.00	16.40	1.87	1.34	26.67	1.94	45.01	m	48.38
Arches including centering; flat										
112mm high on face; 215mm thick; width of exposed soffit 215mm; bricks-on-edge	6.31	5.00	6.63	1.60	1.05	22.14	1.88	30.64	m	32.94
215mm high on face; 112mm thick; width of exposed soffit 112mm; bricks-on-end	6.31	5.00	6.63	1.55	1.10	22.03	1.10	29.75	m	31.99
Arches including centering; semi-circular										
112mm high on face; 215mm thick; width of exposed soffit 215mm; bricks-on-edge	6.31	5.00	6.63	2.25	1.40	30.60	3.37	40.59	m	43.64
215mm high on face; 215mm thick; width of exposed soffit 215mm; bricks-on-edge	12.62	5.00	13.25	3.10	2.05	43.00	3.37	59.63	m	64.10
Closing cavities; vertical										
50mm wide with brickwork 102mm thick	3.08	5.00	3.23	0.44	0.30	6.17	0.22	9.62	m	10.34
75mm wide with brickwork 102mm thick	3.08	5.00	3.23	0.44	0.30	6.17	0.22	9.62	m	10.34
Closing cavities; horizontal										
50mm wide with slates	3.39	5.00	3.56	0.66	0.44	9.18	0.18	12.92	m	13.89
75mm wide with slates	3.39	5.00	3.56	0.66	0.44	9.18	0.18	12.92	m	13.89
50mm wide with brickwork 102mm thick	3.24	5.00	3.40	0.44	0.30	6.17	0.22	9.79	m	10.52
75mm wide with brickwork 102mm thick	3.24	5.00	3.40	0.44	0.30	6.17	0.22	9.79	m	10.52
Bonding ends to existing common brickwork; cutting pockets; extra material										
walls; bonding every third course 102mm thick	1.89	5.00	1.98	0.41	0.28	5.75	0.14	7.87	m	8.46

MASONRY

91

Labour hourly rates: (except Specialists) Craft Operatives £9.23 Labour £7.02 Rates are national average prices. Refer to REGIONAL VARIATIONS for indicative levels of overall pricing in regions	MATERIALS			LABOUR				RATES		
	Del to Site	Waste	Material Cost	Craft Optve	Lab	Labour Cost	Sunds	Nett Rate	Unit	Gross Rate (+7.5%)
	£	%	£	Hrs	Hrs	£	£	£		£
F10: BRICK/BLOCK WALLING Cont.										
Engineering bricks, B.S.3921, Category F, 215 x 102.5 x 65mm, Class A; in cement-lime mortar (1:2:9) Cont.										
Bonding ends to existing common brickwork; cutting pockets; extra material Cont. walls; bonding every third course Cont.										
215mm	3.79	5.00	3.98	0.73	0.48	10.11	0.17	14.26	m	15.33
327mm thick................	5.68	5.00	5.96	1.05	0.69	14.54	0.22	20.72	m	22.27
For other mortar mixes, ADD as follows										
For each half brick thickness										
1:1:6 cement-lime mortar	-	-	-	-	-	-	0.04	0.04	m2	0.04
1:4 cement mortar	-	-	-	-	-	-	0.01	0.01	m2	0.01
1:3 cement mortar	-	-	-	-	-	-	0.16	0.16	m2	0.17
For using Sulphate Resisting cement in lieu of Portland cement, ADD as follows										
For each half brick thickness										
1:2:9 cement-lime mortar	-	-	-	-	-	-	0.15	0.15	m2	0.16
1:1:6 cement-lime mortar	-	-	-	-	-	-	0.21	0.21	m2	0.23
1:4 cement mortar	-	-	-	-	-	-	0.33	0.33	m2	0.35
1:3 cement mortar	-	-	-	-	-	-	0.42	0.42	m2	0.45
Staffordshire wirecut bricks, B.S.3921,Category F, 215 x 102.5 x 65mm; in cement mortar (1:3)										
Walls; vertical										
215mm thick; English bond	45.82	5.00	48.11	2.60	2.05	38.39	4.78	91.28	m2	98.13
Staffordshire pressed bricks, B.S.3921,Category F, 215 x 102.5 x 65mm; in cement mortar (1:3)										
Walls; vertical										
215mm thick; English bond	44.03	5.00	46.23	2.60	2.05	38.39	4.78	89.40	m2	96.11
Composite walling of Staffordshire bricks, B.S.3921, Category F, 215 x 102.5 x 65mm; in cement mortar (1:3); wirecut bricks backing; pressed bricks facing; weather struck pointing as work proceeds										
Walls; vertical										
English bond; facework one side	44.10	5.00	46.31	2.90	2.25	42.56	4.78	93.65	m2	100.67
Flemish bond; facework one side	44.25	5.00	46.46	2.90	2.25	42.56	4.78	93.80	m2	100.84
Common bricks, B.S.3921, Category M, 215 x 102.5 x 65mm, compressive strength 20.5 N/mm2; in cement-lime mortar (1:2:9); flush smooth pointing as work proceeds										
Walls; vertical										
stretcher bond; facework one side										
102mm thick.....................	8.79	5.00	9.23	1.75	1.30	25.28	1.80	36.31	m2	39.03
stretcher bond; facework both sides										
102mm thick.....................	8.79	5.00	9.23	2.05	1.50	29.45	1.80	40.48	m2	43.52
English bond; facework one side										
215mm thick.....................	17.73	5.00	18.62	2.65	2.05	38.85	4.32	61.79	m2	66.42
English bond; facework both sides										
215mm thick.....................	17.73	5.00	18.62	2.95	2.25	43.02	4.32	65.96	m2	70.91
extra; special bricks; vertical angles; single bullnose; B.S.4729, type BN.1.1....	3.90	5.00	4.09	0.25	0.15	3.36	-	7.46	m	8.01
extra; special bricks; vertical angles; squint; B.S.4729, type AN.1.1...	4.90	5.00	5.14	0.25	0.15	3.36	-	8.51	m	9.14
extra; special bricks; intersections; birdsmouth; B.S.4729, type AN.4.1............	17.75	5.00	18.64	0.25	0.15	3.36	-	22.00	m	23.65
Engineering bricks, B.S.3921, Category F, 215 x 102.5 x 65mm, Class B; in cement-lime mortar (1:2:9); flush smooth pointing as work proceeds										
Walls; vertical										
stretcher bond; facework one side										
102mm thick.....................	17.53	5.00	18.41	1.90	1.40	27.36	1.80	47.57	m2	51.14
stretcher bond; facework both sides										
102mm thick.....................	17.53	5.00	18.41	2.20	1.60	31.54	1.80	51.74	m2	55.63
English bond; facework one side										
215mm thick.....................	35.36	5.00	37.13	2.90	2.25	42.56	4.32	84.01	m2	90.31
English bond; facework both sides										
215mm thick.....................	35.36	5.00	37.13	3.20	2.45	46.73	4.32	88.18	m2	94.80
Engineering bricks, B.S.3921, Category F, 215 x 102.5 x 65mm, Class A; in cement-lime mortar (1:2:9); flush smooth pointing as work proceeds										
Walls; vertical										
stretcher bond; facework one side										
102mm thick.....................	29.45	5.00	30.92	1.90	1.40	27.36	1.80	60.09	m2	64.59
stretcher bond; facework both sides										
102mm thick.....................	29.45	5.00	30.92	2.20	1.60	31.54	1.80	64.26	m2	69.08
English bond; facework one side										
215mm thick.....................	59.40	5.00	62.37	2.90	2.25	42.56	4.32	109.25	m2	117.45
English bond; facework both sides										
215mm thick.....................	59.40	5.00	62.37	3.20	2.45	46.73	4.32	113.43	m2	121.93

Labour hourly rates: (except Specialists) Craft Operatives £9.23 Labour £7.02. Rates are national average prices. Refer to REGIONAL VARIATIONS for indicative levels of overall pricing in regions	MATERIALS			LABOUR				RATES		
	Del to Site £	Waste %	Material Cost £	Craft Optve Hrs	Lab Hrs	Labour Cost £	Sunds £	Nett Rate £	Unit	Gross Rate (+7.5%) £
F10: BRICK/BLOCK WALLING Cont.										
Facing bricks, second hard stocks, B.S.3921, Category M, 215 x 102.5 x 65mm; in cement-lime mortar (1:2:9); flush smooth pointing as work proceeds										
Walls; vertical										
stretcher bond; facework one side										
102mm thick...........................	31.12	5.00	**32.68**	1.75	1.30	**25.28**	1.80	**59.75**	m2	**64.24**
stretcher bond; facework both sides										
102mm thick...........................	31.12	5.00	**32.68**	2.05	1.50	**29.45**	1.80	**63.93**	m2	**68.72**
English bond; facework one side										
215mm thick...........................	62.76	5.00	**65.90**	2.65	2.05	**38.85**	4.32	**109.07**	m2	**117.25**
English bond; facework both sides										
215mm thick...........................	62.76	5.00	**65.90**	2.95	2.25	**43.02**	4.32	**113.24**	m2	**121.73**
Facing bricks p.c. £150.00 per 1000, 215 x 102.5 x 65mm; in cement-lime mortar (1:2:9); flush smooth pointing as work proceeds										
Walls; vertical										
stretcher bond; facework one side										
102mm thick...........................	8.85	5.00	**9.29**	1.75	1.30	**25.28**	1.80	**36.37**	m2	**39.10**
stretcher bond; facework both sides										
102mm thick...........................	8.85	5.00	**9.29**	2.05	1.50	**29.45**	1.80	**40.54**	m2	**43.58**
English bond; facework both sides										
215mm thick...........................	17.85	5.00	**18.74**	2.65	2.05	**38.85**	4.32	**61.91**	m2	**66.56**
Flemish bond; facework both sides										
215mm thick...........................	17.85	5.00	**18.74**	2.95	2.25	**43.02**	4.32	**66.09**	m2	**71.04**
extra; special bricks; vertical angles; squint; B.S.4729, type AN.1.1..........	17.16	5.00	**18.02**	0.25	0.15	**3.36**	-	**21.38**	m	**22.98**
extra; special bricks; intersections; birdsmouth; B.S.4729, type AN.4.1.........	53.20	5.00	**55.86**	0.25	0.15	**3.36**	-	**59.22**	m	**63.66**
Walls; building overhand; vertical										
stretcher bond; facework one side										
102mm thick...........................	8.85	5.00	**9.29**	2.15	1.55	**30.73**	1.80	**41.82**	m2	**44.95**
stretcher bond; facework both sides										
102mm thick...........................	8.85	5.00	**9.29**	2.45	1.75	**34.90**	1.80	**45.99**	m2	**49.44**
English bond; facework both sides										
215mm thick...........................	17.85	5.00	**18.74**	3.05	2.30	**44.30**	4.32	**67.36**	m2	**72.41**
Flemish bond; facework both sides										
215mm thick...........................	17.85	5.00	**18.74**	3.35	2.50	**48.47**	4.32	**71.53**	m2	**76.90**
Arches including centering; flat										
112mm high on face; 215mm thick; width of exposed soffit 112mm; bricks-on-edge	2.00	5.00	**2.10**	1.05	0.80	**15.31**	1.80	**19.21**	m	**20.65**
112mm high on face; 215mm thick; width of exposed soffit 215mm; bricks-on-edge	2.00	5.00	**2.10**	1.50	1.00	**20.86**	1.80	**24.77**	m	**26.62**
215mm high on face; 112mm thick; width of exposed soffit 112mm; bricks-on-end	2.00	5.00	**2.10**	1.45	1.00	**20.40**	1.10	**23.60**	m	**25.37**
Arches including centering; segmental										
215mm high on face; 215mm thick; width of exposed soffit 112mm; bricks-on-edge	4.00	5.00	**4.20**	2.40	1.65	**33.73**	3.37	**41.31**	m	**44.40**
215mm high on face; 215mm thick; width of exposed soffit 215mm; bricks-on-edge	4.00	5.00	**4.20**	2.95	1.95	**40.92**	3.37	**48.49**	m	**52.12**
Arches including centering; semi-circular										
215mm high on face; 215mm thick; width of exposed soffit 112mm; bricks-on-edge	4.00	5.00	**4.20**	2.40	1.65	**33.73**	3.37	**41.31**	m	**44.40**
215mm high on face; 215mm thick; width of exposed soffit 215mm; bricks-on-edge	4.00	5.00	**4.20**	2.95	1.95	**40.92**	3.37	**48.49**	m	**52.12**
Facework copings; bricks-on-edge; pointing top and each side										
215 x 102.5mm; horizontal	2.00	5.00	**2.10**	0.70	0.50	**9.97**	0.47	**12.54**	m	**13.48**
215 x 102.5mm; with oversailing course and cement fillets both sides; horizontal	2.00	5.00	**2.10**	1.10	0.75	**15.42**	0.82	**18.34**	m	**19.71**
327 x 102.5mm; horizontal	3.00	5.00	**3.15**	1.05	0.75	**14.96**	0.69	**18.80**	m	**20.21**
327 x 102.5mm; with oversailing course and cement fillets both sides; horizontal	3.00	5.00	**3.15**	1.50	1.05	**21.22**	1.04	**25.41**	m	**27.31**
extra; galvanised iron coping cramp at angle or end..	1.68	5.00	**1.76**	0.04	0.03	**0.58**	-	**2.34**	nr	**2.52**
Facing bricks p.c. £200.00 per 1000, 215 x 102.5 x 65mm; in cement-lime mortar (1:2:9); flush smooth pointing as work proceeds										
Walls; vertical										
stretcher bond; facework one side										
102mm thick...........................	11.80	5.00	**12.39**	1.75	1.30	**25.28**	1.80	**39.47**	m2	**42.43**
stretcher bond; facework both sides										
102mm thick...........................	11.80	5.00	**12.39**	2.05	1.50	**29.45**	1.80	**43.64**	m2	**46.91**
English bond; facework both sides										
215mm thick...........................	23.80	5.00	**24.99**	2.65	2.05	**38.85**	4.32	**68.16**	m2	**73.27**
Flemish bond; facework both sides										
215mm thick...........................	23.80	5.00	**24.99**	2.95	2.25	**43.02**	4.32	**72.33**	m2	**77.76**
extra; special bricks; vertical angles; squint; B.S.4729, type AN.1.1..........	16.31	5.00	**17.13**	0.25	0.15	**3.36**	-	**20.49**	m	**22.02**
extra; special bricks; intersections; birdsmouth; B.S.4729, type AN.4.1.........	52.40	5.00	**55.02**	0.25	0.15	**3.36**	-	**58.38**	m	**62.76**
Walls; building overhand; vertical										
stretcher bond; facework one side										
102mm thick...........................	11.80	5.00	**12.39**	2.15	1.55	**30.73**	1.80	**44.92**	m2	**48.28**

Labour hourly rates: (except Specialists) Craft Operatives £9.23 Labour £7.02 Rates are national average prices. Refer to REGIONAL VARIATIONS for indicative levels of overall pricing in regions	MATERIALS			LABOUR				RATES		
	Del to Site	Waste	Material Cost	Craft Optve	Lab	Labour Cost	Sunds	Nett Rate	Unit	Gross Rate (+7.5%)
	£	%	£	Hrs	Hrs	£	£	£		£
F10: BRICK/BLOCK WALLING Cont.										
Facing bricks p.c. £200.00 per 1000, 215 x 102.5 x 65mm; in cement-lime mortar (1:2:9); flush smooth pointing as work proceeds Cont.										
Walls; building overhand; vertical Cont.										
stretcher bond; facework both sides										
102mm thick................................	11.80	5.00	12.39	2.45	1.75	34.90	1.80	49.09	m2	52.77
English bond; facework both sides										
215mm thick................................	23.80	5.00	24.99	3.05	2.30	44.30	4.32	73.61	m2	79.13
Flemish bond; facework both sides										
215mm thick................................	23.80	5.00	24.99	3.35	2.50	48.47	4.32	77.78	m2	83.61
Arches including centering; flat										
112mm high on face; 215mm thick; width of exposed soffit 112mm; bricks-on-edge................	2.67	5.00	2.80	1.05	0.80	15.31	1.80	19.91	m	21.40
112mm high on face; 215mm thick; width of exposed soffit 215mm; bricks-on-edge................	2.67	5.00	2.80	1.50	1.00	20.86	1.80	25.47	m	27.38
215mm high on face; 112mm thick; width of exposed soffit 112mm; bricks-on-end.................	2.67	5.00	2.80	1.45	1.00	20.40	1.10	24.31	m	26.13
Arches including centering; segmental										
215mm high on face; 215mm thick; width of exposed soffit 112mm; bricks-on-edge................	5.33	5.00	5.60	2.40	1.65	33.73	3.37	42.70	m	45.90
215mm high on face; 215mm thick; width of exposed soffit 215mm; bricks-on-edge................	5.33	5.00	5.60	2.95	1.95	40.92	3.37	49.88	m	53.63
Arches including centering; semi-circular										
215mm high on face; 215mm thick; width of exposed soffit 112mm; bricks-on-edge................	5.33	5.00	5.60	2.40	1.65	33.73	3.37	42.70	m	45.90
215mm high on face; 215mm thick; width of exposed soffit 215mm; bricks-on-edge................	5.33	5.00	5.60	2.95	1.95	40.92	3.37	49.88	m	53.63
Facework copings; bricks-on-edge; pointing top and each side										
215 x 102.5mm; horizontal...................	2.67	5.00	2.80	0.70	0.50	9.97	0.47	13.24	m	14.24
215 x 102.5mm; with oversailing course and cement fillets both sides; horizontal................	2.67	5.00	2.80	1.10	0.75	15.42	0.82	19.04	m	20.47
327 x 102.5mm; horizontal...................	4.00	5.00	4.20	1.05	0.75	14.96	0.69	19.85	m	21.33
327 x 102.5mm; with oversailing course and cement fillets both sides; horizontal................	4.00	5.00	4.20	1.50	1.05	21.22	1.04	26.46	m	28.44
extra; galvanised iron coping cramp at angle or end.......................................	1.68	5.00	1.76	0.04	0.03	0.58	–	2.34	nr	2.52
Facing bricks p.c. £250.00 per 1000, 215 x 102.5 x 65mm; in cement-lime mortar (1:2:9); flush smooth pointing as work proceeds										
Walls; vertical										
stretcher bond; facework one side										
102mm thick................................	14.75	5.00	15.49	1.75	1.30	25.28	1.80	42.57	m2	45.76
stretcher bond; facework both sides										
102mm thick................................	14.75	5.00	15.49	2.05	1.50	29.45	1.80	46.74	m2	50.24
English bond; facework both sides										
215mm thick................................	29.75	5.00	31.24	2.65	2.05	38.85	4.32	74.41	m2	79.99
Flemish bond; facework both sides										
215mm thick................................	29.75	5.00	31.24	2.95	2.25	43.02	4.32	78.58	m2	84.47
extra; special bricks; vertical angles; squint; B.S.4729, type AN.1.1......................	15.33	5.00	16.10	0.25	0.15	3.36	–	19.46	m	20.92
extra; special bricks; intersections; birdsmouth; B.S.4729, type AN.4.1.............	51.48	5.00	54.05	0.25	0.15	3.36	–	57.41	m	61.72
Walls; building overhand; vertical										
stretcher bond; facework one side										
102mm thick................................	14.75	5.00	15.49	2.15	1.55	30.73	1.80	48.01	m2	51.61
stretcher bond; facework both sides										
102mm thick................................	14.75	5.00	15.49	2.45	1.75	34.90	1.80	52.19	m2	56.10
English bond; facework both sides										
215mm thick................................	29.75	5.00	31.24	3.05	2.30	44.30	4.32	79.86	m2	85.84
Flemish bond; facework both sides										
215mm thick................................	29.75	5.00	31.24	3.35	2.50	48.47	4.32	84.03	m2	90.33
Arches including centering; flat										
112mm high on face; 215mm thick; width of exposed soffit 112mm; bricks-on-edge................	3.33	5.00	3.50	1.05	0.80	15.31	1.80	20.60	m	22.15
112mm high on face; 215mm thick; width of exposed soffit 215mm; bricks-on-edge................	3.33	5.00	3.50	1.50	1.00	20.86	1.80	26.16	m	28.12
215mm high on face; 112mm thick; width of exposed soffit 112mm; bricks-on-end.................	3.33	5.00	3.50	1.45	1.00	20.40	1.10	25.00	m	26.88
Arches including centering; segmental										
215mm high on face; 215mm thick; width of exposed soffit 112mm; bricks-on-edge................	6.67	5.00	7.00	2.40	1.65	33.73	3.37	44.11	m	47.42
215mm high on face; 215mm thick; width of exposed soffit 215mm; bricks-on-edge................	6.67	5.00	7.00	2.95	1.95	40.92	3.37	51.29	m	55.14
Arches including centering; semi-circular										
215mm high on face; 215mm thick; width of exposed soffit 112mm; bricks-on-edge................	6.67	5.00	7.00	2.40	1.65	33.73	3.37	44.11	m	47.42
215mm high on face; 215mm thick; width of exposed soffit 215mm; bricks-on-edge................	6.67	5.00	7.00	2.95	1.95	40.92	3.37	51.29	m	55.14
Facework copings; bricks-on-edge; pointing top and each side										
215 x 102.5mm; horizontal...................	3.33	5.00	3.50	0.70	0.50	9.97	0.47	13.94	m	14.98

Labour hourly rates: (except Specialists) Craft Operatives £9.23 Labour £7.02 Rates are national average prices. Refer to REGIONAL VARIATIONS for indicative levels of overall pricing in regions	MATERIALS			LABOUR				RATES		
	Del to Site £	Waste %	Material Cost £	Craft Optve Hrs	Lab Hrs	Labour Cost £	Sunds £	Nett Rate £	Unit	Gross Rate (+7.5%) £
F10: BRICK/BLOCK WALLING Cont.										
Facing bricks p.c. £250.00 per 1000, 215 x 102.5 x 65mm; in cement-lime mortar (1:2:9); flush smooth pointing as work proceeds Cont.										
Facework copings; bricks-on-edge; pointing top and each side Cont.										
215 x 102.5mm; with oversailing course and cement fillets both sides; horizontal	3.33	5.00	3.50	1.10	0.75	15.42	0.82	19.73	m	21.21
327 x 102.5mm; horizontal	5.00	5.00	5.25	1.05	0.75	14.96	0.69	20.90	m	22.46
327 x 102.5mm; with oversailing course and cement fillets both sides; horizontal	5.00	5.00	5.25	1.50	1.05	21.22	1.04	27.51	m	29.57
extra; galvanised iron coping cramp at angle or end	1.68	5.00	1.76	0.04	0.03	0.58	–	2.34	nr	2.52
Facing bricks p.c. £300.00 per 1000, 215 x 102.5 x 65mm; in cement-lime mortar (1:2:9); flush smooth pointing as work proceeds										
Walls; vertical										
stretcher bond; facework one side										
102mm thick................	17.70	5.00	18.59	1.75	1.30	25.28	1.80	45.66	m2	49.09
stretcher bond; facework both sides										
102mm thick................	17.70	5.00	18.59	2.05	1.50	29.45	1.80	49.84	m2	53.57
English bond; facework both sides										
215mm thick................	35.70	5.00	37.48	2.65	2.05	38.85	4.32	80.66	m2	86.70
Flemish bond; facework both sides										
215mm thick................	35.70	5.00	37.48	2.95	2.25	43.02	4.32	84.83	m2	91.19
extra; special bricks; vertical angles; squint; B.S.4729, type AN.1.1...................	14.48	5.00	15.20	0.25	0.15	3.36	–	18.56	m	19.96
extra; special bricks; intersections; birdsmouth; B.S.4729, type AN.4.1.............	50.63	5.00	53.16	0.25	0.15	3.36	–	56.52	m	60.76
Walls; building overhand; vertical										
stretcher bond; facework one side										
102mm thick................	17.70	5.00	18.59	2.15	1.55	30.73	1.80	51.11	m2	54.94
stretcher bond; facework both sides										
102mm thick................	17.70	5.00	18.59	2.45	1.75	34.90	1.80	55.28	m2	59.43
English bond; facework both sides										
215mm thick................	35.70	5.00	37.48	3.05	2.30	44.30	4.32	86.10	m2	92.56
Flemish bond; facework both sides										
215mm thick................	35.70	5.00	37.48	3.35	2.50	48.47	4.32	90.28	m2	97.05
Arches including centering; flat										
112mm high on face; 215mm thick; width of exposed soffit 112mm; bricks-on-edge	4.00	5.00	4.20	1.05	0.80	15.31	1.80	21.31	m	22.91
112mm high on face; 215mm thick; width of exposed soffit 215mm; bricks-on-edge	4.00	5.00	4.20	1.50	1.00	20.86	1.80	26.86	m	28.88
215mm high on face; 112mm thick; width of exposed soffit 112mm; bricks-on-end	4.00	5.00	4.20	1.45	1.00	20.40	1.10	25.70	m	27.63
Arches including centering; segmental										
215mm high on face; 215mm thick; width of exposed soffit 112mm; bricks-on-edge	8.00	5.00	8.40	2.40	1.65	33.73	3.37	45.51	m	48.92
215mm high on face; 215mm thick; width of exposed soffit 215mm; bricks-on-edge	8.00	5.00	8.40	2.95	1.95	40.92	3.37	52.69	m	56.64
Arches including centering; semi-circular										
215mm high on face; 215mm thick; width of exposed soffit 112mm; bricks-on-edge	8.00	5.00	8.40	2.40	1.65	33.73	3.37	45.51	m	48.92
215mm high on face; 215mm thick; width of exposed soffit 215mm; bricks-on-edge	8.00	5.00	8.40	2.95	1.95	40.92	3.37	52.69	m	56.64
Facework copings; bricks-on-edge; pointing top and each side										
215 x 102.5mm; horizontal	4.00	5.00	4.20	0.70	0.50	9.97	0.47	14.64	m	15.74
215 x 102.5mm; with oversailing course and cement fillets both sides; horizontal	4.00	5.00	4.20	1.10	0.75	15.42	0.82	20.44	m	21.97
327 x 102.5mm; horizontal	6.00	5.00	6.30	1.05	0.75	14.96	0.69	21.95	m	23.59
327 x 102.5mm; with oversailing course and cement fillets both sides; horizontal	6.00	5.00	6.30	1.50	1.05	21.22	1.04	28.56	m	30.70
extra; galvanised iron coping cramp at angle or end	1.68	5.00	1.76	0.04	0.03	0.58	–	2.34	nr	2.52
Facing bricks p.c. £350.00 per 1000, 215 x 102.5 x 65mm; in cement-lime mortar (1:2:9); flush smooth pointing as work proceeds										
Walls; vertical										
stretcher bond; facework one side										
102mm thick................	20.65	5.00	21.68	1.75	1.30	25.28	1.80	48.76	m2	52.42
stretcher bond; facework both sides										
102mm thick................	20.65	5.00	21.68	2.05	1.50	29.45	1.80	52.93	m2	56.90
English bond; facework both sides										
215mm thick................	41.65	5.00	43.73	2.65	2.05	38.85	4.32	86.90	m2	93.42
Flemish bond; facework both sides										
215mm thick................	41.65	5.00	43.73	2.95	2.25	43.02	4.32	91.08	m2	97.91
extra; special bricks; vertical angles; squint; B.S.4729, type AN.1.1...................	13.56	5.00	14.24	0.25	0.15	3.36	–	17.60	m	18.92
extra; special bricks; intersections; birdsmouth; B.S.4729, type AN.4.1.............	49.76	5.00	52.25	0.25	0.15	3.36	–	55.61	m	59.78
Walls; building overhand; vertical										
stretcher bond; facework one side										
102mm thick................	20.65	5.00	21.68	2.15	1.55	30.73	1.80	54.21	m2	58.27

	MATERIALS			LABOUR				RATES		
Labour hourly rates: (except Specialists) Craft Operatives £9.23 Labour £7.02. Rates are national average prices. Refer to REGIONAL VARIATIONS for indicative levels of overall pricing in regions	Del to Site £	Waste %	Material Cost £	Craft Optve Hrs	Lab Hrs	Labour Cost £	Sunds £	Nett Rate £	Unit	Gross Rate (+7.5%) £
F10: BRICK/BLOCK WALLING Cont.										
Facing bricks p.c. £350.00 per 1000, 215 x 102.5 x 65mm; in cement-lime mortar (1:2:9); flush smooth pointing as work proceeds Cont.										
Walls; building overhand; vertical Cont.										
stretcher bond; facework both sides										
102mm thick	20.65	5.00	21.68	2.45	1.75	34.90	1.80	58.38	m2	62.76
English bond; facework both sides										
215mm thick	41.65	5.00	43.73	3.05	2.30	44.30	4.32	92.35	m2	99.28
Flemish bond; facework both sides										
215mm thick	41.65	5.00	43.73	3.35	2.50	48.47	4.32	96.52	m2	103.76
Arches including centering; flat										
112mm high on face; 215mm thick; width of exposed soffit 112mm; bricks-on-edge	4.67	5.00	4.90	1.05	0.80	15.31	1.80	22.01	m	23.66
112mm high on face; 215mm thick; width of exposed soffit 215mm; bricks-on-edge	4.67	5.00	4.90	1.50	1.00	20.86	1.80	27.57	m	29.64
215mm high on face; 112mm thick; width of exposed soffit 112mm; bricks-on-end	4.67	5.00	4.90	1.45	1.00	20.40	1.10	26.41	m	28.39
Arches including centering; segmental										
215mm high on face; 215mm thick; width of exposed soffit 112mm; bricks-on-edge	9.33	5.00	9.80	2.40	1.65	33.73	3.37	46.90	m	50.42
215mm high on face; 215mm thick; width of exposed soffit 215mm; bricks-on-edge	9.33	5.00	9.80	2.95	1.95	40.92	3.37	54.08	m	58.14
Arches including centering; semi-circular										
215mm high on face; 215mm thick; width of exposed soffit 112mm; bricks-on-edge	9.33	5.00	9.80	2.40	1.65	33.73	3.37	46.90	m	50.42
215mm high on face; 215mm thick; width of exposed soffit 215mm; bricks-on-edge	9.33	5.00	9.80	2.95	1.95	40.92	3.37	54.08	m	58.14
Facework copings; bricks-on-edge; pointing top and each side										
215 x 102.5mm; horizontal	4.67	5.00	4.90	0.70	0.50	9.97	0.47	15.34	m	16.50
215 x 102.5mm; with oversailing course and cement fillets both sides; horizontal	4.67	5.00	4.90	1.10	0.75	15.42	0.82	21.14	m	22.73
327 x 102.5mm; horizontal	7.00	5.00	7.35	1.05	0.75	14.96	0.69	23.00	m	24.72
327 x 102.5mm; with oversailing course and cement fillets both sides; horizontal	7.00	5.00	7.35	1.50	1.05	21.22	1.04	29.61	m	31.83
extra; galvanised iron coping cramp at angle or end	1.68	5.00	1.76	0.04	0.03	0.58	–	2.34	nr	2.52
Composite walling of bricks 215 x 102.5x 65mm; in cement-lime mortar (1:2:9); common bricks B.S.3921 Category M backing, compressive strength 20.5 N/mm2; facing bricks p.c. £150.00 per 1000; flush smooth pointing as work proceeds										
Walls; vertical										
English bond; facework one side										
215mm thick	17.67	5.00	18.55	2.65	2.05	38.85	4.32	61.72	m2	66.35
327mm thick	26.46	5.00	27.78	3.15	2.50	46.62	6.83	81.24	m2	87.33
Flemish bond; facework one side										
215mm thick	17.66	5.00	18.54	2.65	2.05	38.85	4.32	61.71	m2	66.34
327mm thick	26.45	5.00	27.77	3.15	2.50	46.62	6.83	81.23	m2	87.32
Walls; building overhand; vertical										
English bond; facework one side										
215mm thick	17.67	5.00	18.55	3.05	2.35	44.65	4.32	67.52	m2	72.59
327mm thick	26.46	5.00	27.78	3.55	2.80	52.42	6.83	87.04	m2	93.56
Flemish bond; facework one side										
215mm thick	17.66	5.00	18.54	3.05	2.35	44.65	4.32	67.51	m2	72.57
327mm thick	26.46	5.00	27.78	3.55	2.80	52.42	6.83	87.04	m2	93.56
Composite walling of bricks 215 x 102.5x 65mm; in cement-lime mortar (1:2:9); common bricks B.S.3921 Category M backing, compressive strength 20.5 N/mm2; facing bricks p.c. £200.00 per 1000; flush smooth pointing as work proceeds										
Walls; vertical										
English bond; facework one side										
215mm thick	22.12	5.00	23.23	2.65	2.05	38.85	4.32	66.40	m2	71.38
327mm thick	30.91	5.00	32.46	3.15	2.50	46.62	6.83	85.91	m2	92.35
Flemish bond; facework one side										
215mm thick	21.61	5.00	22.69	2.65	2.05	38.85	4.32	65.86	m2	70.80
327mm thick	30.40	5.00	31.92	3.15	2.50	46.62	6.83	85.37	m2	91.78
Walls; building overhand; vertical										
English bond; facework one side										
215mm thick	22.12	5.00	23.23	3.05	2.35	44.65	4.32	72.19	m2	77.61
327mm thick	30.91	5.00	32.46	3.55	2.80	52.42	6.83	91.71	m2	98.59
Flemish bond; facework one side										
215mm thick	21.61	5.00	22.69	3.05	2.35	44.65	4.32	71.66	m2	77.03
327mm thick	30.40	5.00	31.92	3.55	2.80	52.42	6.83	91.17	m2	98.01

Labour hourly rates: (except Specialists) Craft Operatives £9.23 Labour £7.02 Rates are national average prices. Refer to REGIONAL VARIATIONS for indicative levels of overall pricing in regions	MATERIALS			LABOUR				RATES		
	Del to Site £	Waste %	Material Cost £	Craft Optve Hrs	Lab Hrs	Labour Cost £	Sunds £	Nett Rate £	Unit	Gross Rate (+7.5%) £
F10: BRICK/BLOCK WALLING Cont.										
Composite walling of bricks 215 x 102.5 x 65mm; in cement-lime mortar (1:2:9); common bricks B.S.3921 Category M backing, compressive strength 20.5 N/mm2; facing bricks p.c. £250.00 per 1000; flush smooth pointing as work proceeds										
Walls; vertical										
English bond; facework one side										
215mm thick.....................	26.57	5.00	**27.90**	2.65	2.05	**38.85**	4.32	**71.07**	m2	**76.40**
327mm thick.....................	35.36	5.00	**37.13**	3.15	2.50	**46.62**	6.83	**90.58**	m2	**97.38**
Flemish bond; facework one side										
215mm thick.....................	25.56	5.00	**26.84**	2.65	2.05	**38.85**	4.32	**70.01**	m2	**75.26**
327mm thick.....................	34.35	5.00	**36.07**	3.15	2.50	**46.62**	6.83	**89.52**	m2	**96.24**
Walls; building overhand; vertical										
English bond; facework one side										
215mm thick.....................	26.57	5.00	**27.90**	3.05	2.35	**44.65**	4.32	**76.87**	m2	**82.63**
327mm thick.....................	35.36	5.00	**37.13**	3.55	2.80	**52.42**	6.83	**96.38**	m2	**103.61**
Flemish bond; facework one side										
215mm thick.....................	25.56	5.00	**26.84**	3.05	2.35	**44.65**	4.32	**75.81**	m2	**81.49**
327mm thick.....................	34.35	5.00	**36.07**	3.55	2.80	**52.42**	6.83	**95.32**	m2	**102.47**
Composite walling of bricks 215 x 102.5 x 65mm; in cement-lime mortar (1:2:9); common bricks B.S.3921 Category M backing, compressive strength 20.5 N/mm2; facing bricks p.c. £300.00 per 1000; flush smooth pointing as work proceeds										
Walls; vertical										
English bond; facework one side										
215mm thick.....................	31.02	5.00	**32.57**	2.65	2.05	**38.85**	4.32	**75.74**	m2	**81.42**
327mm thick.....................	39.81	5.00	**41.80**	3.15	2.50	**46.62**	6.83	**95.25**	m2	**102.40**
Flemish bond; facework one side										
215mm thick.....................	29.51	5.00	**30.99**	2.65	2.05	**38.85**	4.32	**74.16**	m2	**79.72**
327mm thick.....................	38.30	5.00	**40.22**	3.15	2.50	**46.62**	6.83	**93.67**	m2	**100.69**
Walls; building overhand; vertical										
English bond; facework one side										
215mm thick.....................	31.02	5.00	**32.57**	3.05	2.35	**44.65**	4.32	**81.54**	m2	**87.65**
327mm thick.....................	39.81	5.00	**41.80**	3.55	2.80	**52.42**	6.83	**101.05**	m2	**108.63**
Flemish bond; facework one side										
215mm thick.....................	29.51	5.00	**30.99**	3.05	2.35	**44.65**	4.32	**79.95**	m2	**85.95**
327mm thick.....................	38.30	5.00	**40.22**	3.55	2.80	**52.42**	6.83	**99.47**	m2	**106.93**
Composite walling of bricks 215 x 102.5 x 65mm; in cement-lime mortar (1:2:9); common bricks B.S.3921 Category M backing, compressive strength 20.5 N/mm2; facing bricks p.c. £350.00 per 1000; flush smooth pointing as work proceeds										
Walls; vertical										
English bond; facework one side										
215mm thick.....................	35.47	5.00	**37.24**	2.65	2.05	**38.85**	4.32	**80.41**	m2	**86.45**
327mm thick.....................	44.26	5.00	**46.47**	3.15	2.50	**46.62**	6.83	**99.93**	m2	**107.42**
Flemish bond; facework one side										
215mm thick.....................	33.46	5.00	**35.13**	2.65	2.05	**38.85**	4.32	**78.30**	m2	**84.18**
327mm thick.....................	42.25	5.00	**44.36**	3.15	2.50	**46.62**	6.83	**97.82**	m2	**105.15**
Walls; building overhand; vertical										
English bond; facework one side										
215mm thick.....................	35.47	5.00	**37.24**	3.05	2.35	**44.65**	4.32	**86.21**	m2	**92.68**
327mm thick.....................	44.26	5.00	**46.47**	3.55	2.80	**52.42**	6.83	**105.73**	m2	**113.65**
Flemish bond; facework one side										
215mm thick.....................	33.46	5.00	**35.13**	3.05	2.35	**44.65**	4.32	**84.10**	m2	**90.41**
327mm thick.....................	42.25	5.00	**44.36**	3.55	2.80	**52.42**	6.83	**103.61**	m2	**111.39**
For each £10.00 difference in cost of 1000 facing bricks, add or deduct the following										
Wall in facing bricks										
102mm thick; English bond; facework one side	0.89	5.00	**0.93**	-	-	-	-	**0.93**	m2	**1.00**
215mm thick; facework both sides	1.19	5.00	**1.25**	-	-	-	-	**1.25**	m2	**1.34**
Composite wall; common brick backing; facing bricks facework										
215mm thick; English bond; facework one side	0.89	5.00	**0.93**	-	-	-	-	**0.93**	m2	**1.00**
215mm thick; Flemish bond; facework one side	0.79	5.00	**0.83**	-	-	-	-	**0.83**	m2	**0.89**
Extra over flush smooth pointing as work proceeds for the following types of pointing										
Pointing as work proceeds										
ironing in joints	-	-	-	0.03	0.02	**0.42**	-	**0.42**	m2	**0.45**
weathered pointing	-	-	-	0.03	0.02	**0.42**	-	**0.42**	m2	**0.45**
recessed pointing	-	-	-	0.04	0.03	**0.58**	-	**0.58**	m2	**0.62**
Raking out joints and pointing on completion										
flush smooth pointing with cement-lime mortar (1:2:9)	-	-	-	0.75	0.50	**10.43**	0.15	**10.58**	m2	**11.38**
flush smooth pointing with coloured cement-lime mortar (1:2:9)	-	-	-	0.75	0.50	**10.43**	0.20	**10.63**	m2	**11.43**
weathered pointing with cement-lime mortar (1:2:9)	-	-	-	0.83	0.55	**11.52**	0.15	**11.67**	m2	**12.55**
weathered pointing with coloured cement-lime mortar (1:2:9)	-	-	-	0.83	0.55	**11.52**	0.20	**11.72**	m2	**12.60**

MASONRY

Labour hourly rates: (except Specialists) Craft Operatives £9.23 Labour £7.02 Rates are national average prices. Refer to REGIONAL VARIATIONS for indicative levels of overall pricing in regions	MATERIALS			LABOUR				RATES		
	Del to Site	Waste	Material Cost	Craft Optve	Lab	Labour Cost	Sunds	Nett Rate	Unit	Gross Rate (+7.5%)
	£	%	£	Hrs	Hrs	£	£	£		£
F10: BRICK/BLOCK WALLING Cont.										
Extra over cement-lime mortar (1:2:9) for bedding and pointing for using coloured cement-lime mortar (1:2:9) for the following										
Wall										
102mm thick; pointing both sides	-	-	-	-	-	-	1.60	1.60	m2	1.72
215mm thick; pointing both sides	-	-	-	-	-	-	3.18	3.18	m2	3.42
Aerated concrete blocks, Thermalite, 440 x 215mm, Shield blocks (4.0 N/mm2);in cement-lime mortar (1:1:6)										
Walls; vertical										
75mm thick; stretcher bond	5.69	5.00	5.97	0.60	0.50	9.05	0.60	15.62	m2	16.79
90mm thick; stretcher bond	6.82	5.00	7.16	0.80	0.65	11.95	0.74	19.85	m2	21.34
100mm thick; stretcher bond	7.58	5.00	7.96	0.80	0.65	11.95	0.74	20.65	m2	22.19
140mm thick; stretcher bond	10.61	5.00	11.14	0.90	0.75	13.57	1.10	25.81	m2	27.75
190mm thick; stretcher bond	14.40	5.00	15.12	0.95	0.80	14.38	1.48	30.98	m2	33.31
Closing cavities; vertical										
50mm wide with blockwork 100mm thick	1.90	5.00	2.00	0.20	0.10	2.55	0.18	4.72	m	5.08
75mm wide with blockwork 100mm thick	1.90	5.00	2.00	0.20	0.10	2.55	0.18	4.72	m	5.08
Closing cavities; horizontal										
50mm wide with blockwork 100mm thick	1.90	5.00	2.00	0.20	0.10	2.55	0.18	4.72	m	5.08
75mm wide with blockwork 100mm thick	1.90	5.00	2.00	0.20	0.10	2.55	0.18	4.72	m	5.08
Bonding ends to common brickwork; forming pockets; extra material										
walls; bonding every third course										
75mm	0.57	-	0.57	0.15	0.10	2.09	-	2.66	m	2.86
90mm	0.68	-	0.68	0.25	0.20	3.71	-	4.39	m	4.72
100mm	0.76	-	0.76	0.25	0.20	3.71	-	4.47	m	4.81
140mm	1.06	-	1.06	0.35	0.25	4.99	-	6.05	m	6.50
190mm	1.44	-	1.44	0.40	0.30	5.80	-	7.24	m	7.78
Bonding ends to existing common brickwork; cutting pockets; extra material										
walls; bonding every third course										
75mm	1.14	-	1.14	0.30	0.20	4.17	-	5.31	m	5.71
90mm	1.36	-	1.36	0.50	0.40	7.42	-	8.78	m	9.44
100mm	1.52	-	1.52	0.50	0.40	7.42	-	8.94	m	9.61
140mm	2.12	-	2.12	0.70	0.50	9.97	-	12.09	m	13.00
190mm	2.88	-	2.88	0.80	0.60	11.60	-	14.48	m	15.56
Bonding ends to existing concrete blockwork; cutting pockets; extra material										
walls; bonding every third course										
75mm	1.14	-	1.14	0.25	0.20	3.71	-	4.85	m	5.22
90mm	1.36	-	1.36	0.35	0.30	5.34	-	6.70	m	7.20
100mm	1.52	-	1.52	0.40	0.30	5.80	-	7.32	m	7.87
140mm	2.12	-	2.12	0.55	0.40	7.88	-	10.00	m	10.75
190mm	2.88	-	2.88	0.65	0.50	9.51	-	12.39	m	13.32
Aerated concrete blocks, Thermalite, 440 x 215mm, Turbo blocks (2.8 N/mm2); in cement-lime mortar (1:1:6)										
Walls; vertical										
100mm thick; stretcher bond	7.86	5.00	8.25	0.80	0.65	11.95	0.74	20.94	m2	22.51
115mm thick; stretcher bond	9.04	5.00	9.49	0.80	0.65	11.95	0.74	22.18	m2	23.84
125mm thick; stretcher bond	9.83	5.00	10.32	0.85	0.70	12.76	1.10	24.18	m2	25.99
130mm thick; stretcher bond	10.22	5.00	10.73	0.85	0.70	12.76	1.10	24.59	m2	26.43
150mm thick; stretcher bond	11.79	5.00	12.38	0.90	0.75	13.57	1.10	27.05	m2	29.08
190mm thick; stretcher bond	14.93	5.00	15.68	0.95	0.80	14.38	1.48	31.54	m2	33.91
200mm thick; stretcher bond	15.72	5.00	16.51	1.00	0.85	15.20	1.48	33.18	m2	35.67
215mm thick; stretcher bond	16.90	5.00	17.75	1.00	0.85	15.20	1.85	34.79	m2	37.40
Closing cavities; vertical										
50mm wide with blockwork 100mm thick	1.97	5.00	2.07	0.20	0.10	2.55	0.18	4.80	m	5.16
75mm wide with blockwork 100mm thick	1.97	5.00	2.07	0.20	0.10	2.55	0.18	4.80	m	5.16
Closing cavities; horizontal										
50mm wide with blockwork 100mm thick	1.97	5.00	2.07	0.20	0.10	2.55	0.18	4.80	m	5.16
75mm wide with blockwork 100mm thick	1.97	5.00	2.07	0.20	0.10	2.55	0.18	4.80	m	5.16
Aerated concrete blocks, Thermalite, 440 x 215mm, Party wall blocks (4.0 N/mm2); in cement-lime mortar (1:1:6)										
Walls; vertical										
215mm thick; stretcher bond	16.56	5.00	17.39	1.00	0.85	15.20	1.85	34.44	m2	37.02
Aerated concrete blocks, Thermalite, 440 x 215mm, Hi-Strength 7 blocks (7.0 N/mm2); in cement-lime mortar (1:1:6)										
Walls; vertical										
100mm thick; stretcher bond	9.52	5.00	10.00	0.80	0.65	11.95	0.74	22.68	m2	24.38
140mm thick; stretcher bond	13.33	5.00	14.00	0.90	0.75	13.57	1.10	28.67	m2	30.82
150mm thick; stretcher bond	14.28	5.00	14.99	0.90	0.75	13.57	1.10	29.67	m2	31.89
190mm thick; stretcher bond	18.09	5.00	18.99	0.95	0.80	14.38	1.48	34.86	m2	37.47
200mm thick; stretcher bond	19.04	5.00	19.99	1.00	0.85	15.20	1.48	36.67	m2	39.42
215mm thick; stretcher bond	20.47	5.00	21.49	1.00	0.85	15.20	1.85	38.54	m2	41.43

Labour hourly rates: (except Specialists) Craft Operatives £9.23 Labour £7.02 Rates are national average prices. Refer to REGIONAL VARIATIONS for indicative levels of overall pricing in regions	MATERIALS			LABOUR				RATES		
	Del to Site £	Waste %	Material Cost £	Craft Optve Hrs	Lab Hrs	Labour Cost £	Sunds £	Nett Rate £	Unit	Gross Rate (+7.5%) £
F10: BRICK/BLOCK WALLING Cont.										
Aerated concrete blocks, Thermalite, 440 x 215mm, Hi-Strength 7 blocks (7.0 N/mm2); in cement-lime mortar (1:1:6) Cont.										
Closing cavities; vertical										
50mm wide with blockwork 100mm thick	2.38	5.00	2.50	0.20	0.10	2.55	0.18	5.23	m	5.62
75mm wide with blockwork 100mm thick	2.38	5.00	2.50	0.20	0.10	2.55	0.18	5.23	m	5.62
Closing cavities; horizontal										
50mm wide with blockwork 100mm thick	2.38	5.00	2.50	0.20	0.10	2.55	0.18	5.23	m	5.62
75mm wide with blockwork 100mm thick	2.38	5.00	2.50	0.20	0.10	2.55	0.18	5.23	m	5.62
Aerated concrete blocks, Thermalite, 440 x 215mm, Trenchblocks (4.0 N/mm2); in cement mortar (1:4)										
Walls; vertical										
255mm thick; stretcher bond	19.64	5.00	20.62	1.30	1.05	19.37	1.85	41.84	m2	44.98
275mm thick; stretcher bond	21.18	5.00	22.24	1.45	1.15	21.46	2.03	45.73	m2	49.15
305mm thick; stretcher bond	23.49	5.00	24.66	1.65	1.25	24.00	2.25	50.92	m2	54.74
355mm thick; stretcher bond	27.34	5.00	28.71	1.95	1.50	28.53	2.64	59.88	m2	64.37
Aerated concrete blocks, Tarmac Toplite foundation blocks; 440 x 215mm, (3.5 N/mm2); in cement mortar (1:4)										
Walls; vertical										
260mm thick; stretcher bond	17.21	5.00	18.07	1.40	0.95	19.59	1.91	39.57	m2	42.54
275mm thick; stretcher bond	18.21	5.00	19.12	1.60	1.05	22.14	2.03	43.29	m2	46.54
300mm thick; stretcher bond	19.86	5.00	20.85	1.70	1.10	23.41	2.21	46.48	m2	49.96
Tarmac Topblock, medium density concrete block, Hemelite, 440 x 215mm, solid Standard blocks (3.5 N/mm2); in cement-lime mortar (1:1:6)										
Walls; vertical										
75mm thick; stretcher bond	4.50	5.00	4.72	0.75	0.65	11.49	0.60	16.81	m2	18.07
90mm thick; stretcher bond	5.92	5.00	6.22	0.90	0.75	13.57	0.74	20.53	m2	22.07
100mm thick; stretcher bond	5.22	5.00	5.48	0.90	0.75	13.57	0.74	19.79	m2	21.28
140mm thick; stretcher bond	7.63	5.00	8.01	1.00	0.85	15.20	1.10	24.31	m2	26.13
190mm thick; stretcher bond	11.00	5.00	11.55	1.05	0.90	16.01	1.48	29.04	m2	31.22
215mm thick; stretcher bond	11.92	5.00	12.52	1.10	0.95	16.82	1.85	31.19	m2	33.53
Closing cavities; vertical										
50mm wide with blockwork 100mm thick	1.31	5.00	1.38	0.20	0.10	2.55	0.18	4.10	m	4.41
75mm wide with blockwork 100mm thick	1.31	5.00	1.38	0.20	0.10	2.55	0.18	4.10	m	4.41
Closing cavities; horizontal										
50mm wide with blockwork 100mm thick	1.31	5.00	1.38	0.20	0.10	2.55	0.18	4.10	m	4.41
75mm wide with blockwork 100mm thick	1.31	5.00	1.38	0.20	0.10	2.55	0.18	4.10	m	4.41
Bonding ends to common brickwork; forming pockets; extra material										
walls; bonding every third course										
75mm	0.45	–	0.45	0.25	0.20	3.71	–	4.16	m	4.47
90mm	0.59	–	0.59	0.30	0.25	4.52	–	5.11	m	5.50
100mm	0.52	–	0.52	0.30	0.25	4.52	–	5.04	m	5.42
140mm	0.76	–	0.76	0.40	0.30	5.80	–	6.56	m	7.05
190mm	1.10	–	1.10	0.45	0.35	6.61	–	7.71	m	8.29
215mm	1.19	–	1.19	0.55	0.40	7.88	–	9.07	m	9.76
Bonding ends to existing common brickwork; cutting pockets; extra material										
walls; bonding every third course										
75mm	0.90	–	0.90	0.50	0.40	7.42	–	8.32	m	8.95
90mm	1.18	–	1.18	0.60	0.50	9.05	–	10.23	m	11.00
100mm	1.04	–	1.04	0.60	0.50	9.05	–	10.09	m	10.84
140mm	1.52	–	1.52	0.80	0.60	11.60	–	13.12	m	14.10
190mm	2.20	–	2.20	0.90	0.70	13.22	–	15.42	m	16.58
215mm	2.38	–	2.38	1.10	0.80	15.77	–	18.15	m	19.51
Bonding ends to existing concrete blockwork; cutting pockets; extra material										
walls; bonding every third course										
75mm	0.90	–	0.90	0.40	0.30	5.80	–	6.70	m	7.20
90mm	1.18	–	1.18	0.45	0.35	6.61	–	7.79	m	8.37
100mm	1.04	–	1.04	0.45	0.35	6.61	–	7.65	m	8.22
140mm	1.52	–	1.52	0.60	0.45	8.70	–	10.22	m	10.98
190mm	2.20	–	2.20	0.70	0.55	10.32	–	12.52	m	13.46
215mm	2.38	–	2.38	0.85	0.60	12.06	–	14.44	m	15.52
Tarmac Topblock, medium density concrete block, Hemelite, 440 x 215mm, solid Standard blocks (7.0 N/mm2); in cement-lime mortar (1:1:6)										
Walls; vertical										
90mm thick; stretcher bond	6.32	5.00	6.64	0.90	0.75	13.57	0.74	20.95	m2	22.52
100mm thick; stretcher bond	5.62	5.00	5.90	0.90	0.75	13.57	0.74	20.21	m2	21.73
140mm thick; stretcher bond	7.84	5.00	8.23	1.00	0.85	15.20	1.10	24.53	m2	26.37
190mm thick; stretcher bond	11.39	5.00	11.96	1.05	0.90	16.01	1.48	29.45	m2	31.66
215mm thick; stretcher bond	12.40	5.00	13.02	1.10	0.95	16.82	1.85	31.69	m2	34.07
Closing cavities; vertical										
50mm wide with blockwork 100mm thick	1.41	5.00	1.48	0.20	0.10	2.55	0.18	4.21	m	4.52
75mm wide with blockwork 100mm thick	1.41	5.00	1.48	0.20	0.10	2.55	0.18	4.21	m	4.52

MASONRY

Labour hourly rates: (except Specialists) Craft Operatives £9.23 Labour £7.02 Rates are national average prices. Refer to REGIONAL VARIATIONS for indicative levels of overall pricing in regions	MATERIALS			LABOUR				RATES		
	Del to Site £	Waste %	Material Cost £	Craft Optve Hrs	Lab Hrs	Labour Cost £	Sunds £	Nett Rate £	Unit	Gross Rate (+7.5%) £
F10: BRICK/BLOCK WALLING Cont.										
Tarmac Topblock, medium density concrete block, Hemelite, 440 x 215mm, solid Standard blocks (7.0 N/mm2); in cement-lime mortar (1:1:6) Cont.										
Closing cavities; horizontal										
50mm wide with blockwork 100mm thick	1.41	5.00	1.48	0.20	0.10	2.55	0.18	4.21	m	4.52
75mm wide with blockwork 100mm thick	1.41	5.00	1.48	0.20	0.10	2.55	0.18	4.21	m	4.52
Tarmac Topblock, fair face concrete blocks, Lignacite, 440 x 215mm, solid Standard blocks (7.0 N/mm2); in cement-lime mortar (1:1:6)										
Walls; vertical										
100mm thick; stretcher bond	9.65	5.00	10.13	0.95	0.80	14.38	0.74	25.26	m2	27.15
140mm thick; stretcher bond	13.41	5.00	14.08	1.10	0.90	16.47	1.10	31.65	m2	34.03
190mm thick; stretcher bond	17.15	5.00	18.01	1.15	0.95	17.28	1.48	36.77	m2	39.53
Closing cavities; vertical										
50mm wide with blockwork 100mm thick	2.41	5.00	2.53	0.20	0.10	2.55	0.18	5.26	m	5.65
75mm wide with blockwork 100mm thick	2.41	5.00	2.53	0.20	0.10	2.55	0.18	5.26	m	5.65
Closing cavities; horizontal										
50mm wide with blockwork 100mm thick	2.41	5.00	2.53	0.20	0.10	2.55	0.18	5.26	m	5.65
75mm wide with blockwork 100mm thick	2.41	5.00	2.53	0.20	0.10	2.55	0.18	5.26	m	5.65
Tarmac Topblock, dense concrete blocks, Topcrete, 440 x 215mm, solid Standard blocks (7.0 N/mm2); in cement-lime mortar (1:1:6)										
Walls; vertical										
75mm thick; stretcher bond	5.41	5.00	5.68	1.00	0.80	14.85	0.60	21.13	m2	22.71
90mm thick; stretcher bond	5.95	5.00	6.25	1.15	0.95	17.28	0.74	24.27	m2	26.09
100mm thick; stretcher bond	6.62	5.00	6.95	1.20	1.00	18.10	0.74	25.79	m2	27.72
140mm thick; stretcher bond	9.26	5.00	9.72	1.30	1.10	19.72	1.10	30.54	m2	32.83
190mm thick; stretcher bond	12.57	5.00	13.20	1.40	1.15	21.00	1.48	35.67	m2	38.35
215mm thick; stretcher bond	14.23	5.00	14.94	1.50	1.20	22.27	1.84	39.05	m2	41.98
Closing cavities; vertical										
50mm wide with blockwork 100mm thick	1.66	5.00	1.74	0.20	0.10	2.55	0.18	4.47	m	4.81
75mm wide with blockwork 100mm thick	1.66	5.00	1.74	0.20	0.10	2.55	0.18	4.47	m	4.81
Closing cavities; horizontal										
50mm wide with blockwork 100mm thick	1.66	5.00	1.74	0.20	0.10	2.55	0.18	4.47	m	4.81
75mm wide with blockwork 100mm thick	1.66	5.00	1.74	0.20	0.10	2.55	0.18	4.47	m	4.81
Tarmac Topblock, dense concrete blocks, Topcrete, 440 x 215mm cellular standard blocks (7.0 N/mm2); in cement-lime mortar (1:1:6)										
Walls; vertical										
100mm thick; stretcher bond	5.41	5.00	5.68	1.10	0.58	14.22	0.79	20.70	m2	22.25
Closing cavities; vertical										
50mm wide with blockwork 100mm thick	1.35	5.00	1.42	0.20	0.10	2.55	0.18	4.15	m	4.46
75mm wide with blockwork 100mm thick	1.35	5.00	1.42	0.20	0.10	2.55	0.18	4.15	m	4.46
Closing cavities; horizontal										
50mm wide with blockwork 100mm thick	1.35	5.00	1.42	0.20	0.10	2.55	0.18	4.15	m	4.46
75mm wide with blockwork 100mm thick	1.35	5.00	1.42	0.20	0.10	2.55	0.18	4.15	m	4.46
Tarmac Topblock, dense concrete blocks, Topcrete, 440 x 215mm hollow standard blocks (7.0 N/mm2); in cement-lime mortar (1:1:6)										
Walls; vertical										
215mm thick; stretcher bond	10.55	5.00	11.08	1.50	0.81	19.53	1.84	32.45	m2	34.88
Aerated concrete blocks, Thermalite, 440 x 215mm, Smooth Face blocks (4.0 N/mm2); in cement-lime mortar (1:1:6); flush smooth pointing as work proceeds										
Walls; vertical										
stretcher bond; facework one side										
100mm thick...................................	13.00	5.00	13.65	1.00	0.80	14.85	0.74	29.24	m2	31.43
140mm thick...................................	18.20	5.00	19.11	1.10	0.90	16.47	1.10	36.68	m2	39.43
190mm thick...................................	24.70	5.00	25.93	1.15	0.95	17.28	1.48	44.70	m2	48.05
215mm thick...................................	27.95	5.00	29.35	1.20	1.00	18.10	1.84	49.28	m2	52.98
stretcher bond; facework both sides										
100mm thick...................................	13.00	5.00	13.65	1.20	0.95	17.75	0.74	32.13	m2	34.55
140mm thick...................................	18.20	5.00	19.11	1.30	1.05	19.37	1.10	39.58	m2	42.55
190mm thick...................................	24.70	5.00	25.93	1.35	1.10	20.18	1.48	47.60	m2	51.17
215mm thick...................................	27.95	5.00	29.35	1.40	1.15	21.00	1.84	52.18	m2	56.10
Tarmac Topblock, fair face concrete blocks, Lignacite, 440 x 215mm, solid Standard blocks (7.0 N/mm2); in cement-lime mortar (1:1:6); flush smooth pointing as work proceeds										
Walls; vertical										
stretcher bond; facework one side										
100mm thick...................................	9.65	5.00	10.13	1.15	0.95	17.28	0.74	28.16	m2	30.27
140mm thick...................................	13.41	5.00	14.08	1.30	1.05	19.37	1.10	34.55	m2	37.14
190mm thick...................................	17.15	5.00	18.01	1.35	1.10	20.18	1.48	39.67	m2	42.65

Labour hourly rates: (except Specialists) Craft Operatives £9.23 Labour £7.02 Rates are national average prices. Refer to REGIONAL VARIATIONS for indicative levels of overall pricing in regions	MATERIALS			LABOUR				RATES		
	Del to Site £	Waste %	Material Cost £	Craft Optve Hrs	Lab Hrs	Labour Cost £	Sunds £	Nett Rate £	Unit	Gross Rate (+7.5%) £

F10: BRICK/BLOCK WALLING Cont.

Tarmac Topblock, fair face concrete blocks, Lignacite, 440 x 215mm, solid Standard blocks (7.0 N/mm2); in cement-lime mortar (1:1:6); flush smooth pointing as work proceeds Cont.

Walls; vertical Cont.
  stretcher bond; facework both sides

	Del to Site	Waste	Material Cost	Craft Optve	Lab	Labour Cost	Sunds	Nett Rate	Unit	Gross Rate
100mm thick	9.65	5.00	10.13	1.35	1.10	20.18	0.74	31.06	m2	33.38
140mm thick	13.41	5.00	14.08	1.50	1.20	22.27	1.10	37.45	m2	40.26
190mm thick	17.15	5.00	18.01	1.55	1.25	23.08	1.48	42.57	m2	45.76

Reconstructed stone blocks, Marshalls Mono Ltd. Cromwell coursed random length Split Face buff walling blocks in cement-lime mortar (1:1:6); flat recessed pointing as work proceeds

Walls; vertical

	Del to Site	Waste	Material Cost	Craft Optve	Lab	Labour Cost	Sunds	Nett Rate	Unit	Gross Rate
100mm thick; alternate courses 102mm and 140mm high blocks; stretcher bond; facework one side	23.90	5.00	25.09	1.90	1.40	27.36	0.66	53.12	m2	57.10
100mm thick; one course 102mm high blocks and two courses 140mm high blocks; stretcher bond; facework one side	23.90	5.00	25.09	1.90	1.40	27.36	0.66	53.12	m2	57.10

Reconstructed stone blocks, Marshalls Mono Ltd. Cromwell coursed random length Pitched Face buff walling blocks in cement-lime mortar (1:1:6); flat recessed pointing as work proceeds

Walls; vertical

	Del to Site	Waste	Material Cost	Craft Optve	Lab	Labour Cost	Sunds	Nett Rate	Unit	Gross Rate
90mm thick; alternate courses 102mm and 140mm high blocks; stretcher bond; facework one side	26.15	5.00	27.46	1.90	1.40	27.36	0.66	55.48	m2	59.64
90mm thick; one course 102mm high blocks and two courses 140mm high blocks; stretcher bond; facework one side	26.15	5.00	27.46	1.90	1.40	27.36	0.66	55.48	m2	59.64

Firebricks, 215 x 102.5 x 65mm; in fire cement

Flue linings; bonding to surrounding brickwork with headers -4/m2

	Del to Site	Waste	Material Cost	Craft Optve	Lab	Labour Cost	Sunds	Nett Rate	Unit	Gross Rate
112mm thick; stretcher bond	84.58	5.00	88.81	1.90	1.40	27.36	5.43	121.60	m2	130.72

Flue linings; built clear of main brickwork but with one header in each course set projecting to contact main work

	Del to Site	Waste	Material Cost	Craft Optve	Lab	Labour Cost	Sunds	Nett Rate	Unit	Gross Rate
112mm thick; stretcher bond	84.58	5.00	88.81	1.90	1.40	27.36	5.43	121.60	m2	130.72
112mm thick; stretcher bond; in segmental top to flue	84.58	5.00	88.81	2.50	1.80	35.71	5.43	129.95	m2	139.70

HR Supra Flue lining bricks, 230 x 114 x 76mm, Hepworth Refractories, Sheffield, MPK 21 mortar

Flue linings; bonding to surrounding brickwork with headers -4/m2

	Del to Site	Waste	Material Cost	Craft Optve	Lab	Labour Cost	Sunds	Nett Rate	Unit	Gross Rate
114mm thick; stretcher bond	34.19	5.00	35.90	1.90	1.40	27.36	3.29	66.55	m2	71.55
230mm thick; English bond	61.24	5.00	64.30	3.15	2.35	45.57	5.87	115.74	m2	124.42

F11: GLASS BLOCK WALLING

SCREENS AND PANELS

Hollow glass blocks, white, cross ribbed, in cement mortar (1:3); continuous joints; flat recessed pointing as work proceeds

	Del to Site	Waste	Material Cost	Craft Optve	Lab	Labour Cost	Sunds	Nett Rate	Unit	Gross Rate
Screens or panels; 190 x 190mm blocks; vertical 80mm thick; facework both sides	235.00	10.00	258.50	3.00	3.00	48.75	26.00	333.25	m2	358.24
Screens or panels; 240 x 240mm blocks; vertical 80mm thick; facework both sides	215.00	10.00	236.50	2.50	2.50	40.63	19.00	296.13	m2	318.33

F20: NATURAL STONE RUBBLE WALLING

Stone rubble work; random stones of Yorkshire limestone; bedding and jointing in lime mortar (1:3); uncoursed

Walls; tapering both sides; including extra material

	Del to Site	Waste	Material Cost	Craft Optve	Lab	Labour Cost	Sunds	Nett Rate	Unit	Gross Rate
500mm thick	50.00	15.00	57.50	3.75	3.75	60.94	8.25	126.69	m2	136.19
600mm thick	55.00	15.00	63.25	4.50	4.50	73.13	9.90	146.28	m2	157.25

Stone rubble work; squared rubble face stones of Yorkshire limestone; bedding and jointing in lime mortar (1:3); irregular coursed; courses average 150mm high

Walls; vertical; face stones 100 - 150mm on bed; bonding to brickwork; including extra material; scappled or axed face; weather struck pointing as work proceeds

	Del to Site	Waste	Material Cost	Craft Optve	Lab	Labour Cost	Sunds	Nett Rate	Unit	Gross Rate
150mm thick; faced one side	35.00	15.00	40.25	3.00	3.00	48.75	2.75	91.75	m2	98.63

Walls; vertical; face stones 100 - 150mm on bed; bonding to brickwork; including extra material; hammer dressed face; weather struck pointing as work proceeds

	Del to Site	Waste	Material Cost	Craft Optve	Lab	Labour Cost	Sunds	Nett Rate	Unit	Gross Rate
150mm thick; faced one side	35.00	15.00	40.25	3.30	3.30	53.63	2.75	96.63	m2	103.87

MASONRY

Labour hourly rates: (except Specialists) Craft Operatives £9.23 Labour £7.02 Rates are national average prices. Refer to REGIONAL VARIATIONS for indicative levels of overall pricing in regions	MATERIALS			LABOUR				RATES		
	Del to Site £	Waste %	Material Cost £	Craft Optve Hrs	Lab Hrs	Labour Cost £	Sunds £	Nett Rate £	Unit	Gross Rate (+7.5%) £

F20: NATURAL STONE RUBBLE WALLING Cont.

Stone rubble work; squared rubble face stones of Yorkshire limestone; bedding and jointing in lime mortar (1:3); irregular coursed; courses average 150mm high Cont.

Walls; vertical; face stones 100 - 150mm on bed; bonding to brickwork; including extra material; rock worked face; pointing with a parallel joint as work proceeds

150mm thick; faced one side	37.00	15.00	42.55	3.60	3.60	58.50	2.75	103.80	m2	111.58

Stone rubble work; squared rubble face stones of Yorkshire limestone; bedding and jointing in lime mortar (1:3); regular coursed; courses average 150mm high

Walls; vertical; face stones 100 - 150mm on bed; bonding to brickwork; including extra material; scappled or axed face; weather struck pointing as work proceeds

150mm thick; faced one side	55.00	15.00	63.25	3.20	3.20	52.00	2.75	118.00	m2	126.85

Walls; vertical; face stones 100 - 150mm on bed; bonding to brickwork; including extra material; hammer dressed face; weather struck pointing as work proceeds

150mm thick; faced one side	55.00	15.00	63.25	3.50	3.50	56.88	2.75	122.88	m2	132.09

Walls; vertical; face stones 100 - 150mm on bed; bonding to brickwork; including extra material; rock worked face; pointing with a parallel joint as work proceeds

150mm thick; faced one side	57.00	15.00	65.55	3.80	3.80	61.75	2.75	130.05	m2	139.80

Stone rubble work; random rubble backing and squared rubble face stones of Yorkshire limestone; bedding and jointing in lime mortar (1:3); irregular coursed; courses average 150mm high

Walls; vertical; face stones 100 - 150mm on bed; scappled or axed face; weather struck pointing as work proceeds

350mm thick; faced one side	62.00	15.00	71.30	4.00	4.00	65.00	5.75	142.05	m2	152.70
500mm thick; faced one side	79.00	15.00	90.85	4.75	4.75	77.19	8.25	176.29	m2	189.51

Walls; vertical; face stones 100 - 150mm on bed; hammer dressed face; weather struck pointing as work proceeds

350mm thick; faced one side	62.00	15.00	71.30	4.30	4.30	69.88	5.75	146.93	m2	157.94
500mm thick; faced one side	79.00	15.00	90.85	5.05	5.05	82.06	8.25	181.16	m2	194.75

Walls; vertical; face stones 100 - 150mm on bed; rock worked face; pointing with a parallel joint as work proceeds

350mm thick; faced one side	62.00	15.00	71.30	4.60	4.60	74.75	5.75	151.80	m2	163.19
500mm thick; faced one side	79.00	15.00	90.85	5.35	5.35	86.94	8.25	186.04	m2	199.99

Quoin stones; scappled or axed face; attached faced two adjacent faces

250 x 200 x 350mm	21.50	5.00	22.57	0.50	0.25	6.37	0.28	29.23	nr	31.42
250 x 200 x 500mm	30.75	5.00	32.29	0.80	0.40	10.19	0.40	42.88	nr	46.10
250 x 250 x 350mm	26.50	5.00	27.82	0.65	0.35	8.46	0.35	36.63	nr	39.38
250 x 250 x 500mm	38.00	5.00	39.90	0.90	0.45	11.47	0.50	51.87	nr	55.76
380 x 200 x 350mm	32.75	5.00	34.39	0.80	0.40	10.19	0.45	45.03	nr	48.41
380 x 200 x 500mm	47.00	5.00	49.35	1.00	0.50	12.74	0.60	62.69	nr	67.39

Quoin stones; rock worked face; attached faced two adjacent faces

250 x 200 x 350mm	21.50	5.00	22.57	0.55	0.25	6.83	0.28	29.69	nr	31.91
250 x 200 x 500mm	30.75	5.00	32.29	0.85	0.40	10.65	0.40	43.34	nr	46.59
250 x 250 x 350mm	26.50	5.00	27.82	0.70	0.35	8.92	0.35	37.09	nr	39.87
250 x 250 x 500mm	38.00	5.00	39.90	0.95	0.45	11.93	0.50	52.33	nr	56.25
380 x 200 x 350mm	32.75	5.00	34.39	0.85	0.40	10.65	0.45	45.49	nr	48.90
380 x 200 x 500mm	47.00	5.00	49.35	1.10	0.50	13.66	0.60	63.61	nr	68.38

Arches; relieving

225mm high on face, 180mm wide on soffit	85.00	5.00	89.25	1.50	0.75	19.11	0.65	109.01	m	117.19
225mm high on face, 250mm wide on soffit	117.50	5.00	123.38	2.00	1.00	25.48	0.90	149.76	m	160.99

Grooves

12 x 25mm	-	-	-	0.25	-	2.31	-	2.31	m	2.48
12 x 38mm	-	-	-	0.30	-	2.77	-	2.77	m	2.98

Stone rubble work; random rubble backing and squared rubble face stones of Yorkshire limestone; bedding and jointing in lime mortar (1:3); regular coursed; courses average 150mm high

Walls; vertical; face stones 100 - 150mm on bed; scappled or axed face; weather struck pointing as work proceeds

350mm thick; faced one side	79.00	15.00	90.85	4.20	4.20	68.25	5.75	164.85	m2	177.21

Labour hourly rates: (except Specialists) Craft Operatives £9.23 Labour £7.02 Rates are national average prices. Refer to REGIONAL VARIATIONS for indicative levels of overall pricing in regions	MATERIALS			LABOUR				RATES		
	Del to Site £	Waste %	Material Cost £	Craft Optve Hrs	Lab Hrs	Labour Cost £	Sunds £	Nett Rate £	Unit	Gross Rate (+7.5%) £
F20: NATURAL STONE RUBBLE WALLING Cont.										
Stone rubble work; random rubble backing and squared rubble face stones of Yorkshire limestone; bedding and jointing in lime mortar (1:3); regular coursed; courses average 150mm high Cont.										
Walls; vertical; face stones 125 - 200mm on bed; scappled or axed face; weather struck pointing as work proceeds										
500mm thick; faced one side	97.00	15.00	**111.55**	4.95	4.95	**80.44**	8.25	**200.24**	m2	**215.26**
Walls; vertical; face stones 100 - 150mm on bed; hammer dressed face; weather struck pointing as work proceeds										
350mm thick; faced one side	79.00	15.00	**90.85**	4.50	4.50	**73.13**	5.75	**169.72**	m2	**182.45**
500mm thick; faced one side	97.00	15.00	**111.55**	5.25	5.25	**85.31**	8.25	**205.11**	m2	**220.50**
Walls; vertical; face stones 100 - 150mm on bed; rock worked face; pointing with a parallel joint as work proceeds										
350mm thick; faced one side	79.00	15.00	**90.85**	4.80	4.80	**78.00**	5.75	**174.60**	m2	**187.69**
500mm thick; faced one side	97.00	15.00	**111.55**	5.55	5.55	**90.19**	8.25	**209.99**	m2	**225.74**
Natural stonework										
Note										
stonework is work for a specialist; prices being obtained for specific projects. Some firms that specialise in this class of work are included within the list at the end of this section										
Natural stonework dressings; natural Dorset limestone; Portland Whitbed; bedding and jointing in mason's mortar (1:3:12); flush smooth pointing as work proceeds; slurrying with weak lime mortar and cleaning down on completion										
Walls; vertical; building against brickwork; B.S.1243 Fig 1 specification 3.5 wall ties - 4/m2 built in										
50mm thick; plain and rubbed one side	110.00	2.50	**112.75**	4.00	2.60	**55.17**	5.50	**173.42**	m2	**186.43**
75mm thick; plain and rubbed one side	140.00	2.50	**143.50**	4.50	3.15	**63.65**	6.00	**213.15**	m2	**229.13**
100mm thick; plain and rubbed one side	185.00	2.50	**189.63**	5.00	3.70	**72.12**	6.50	**268.25**	m2	**288.37**
Natural stonework; natural Dorset limestone, Portland Whitbed; bedding and jointing in cement-lime mortar (1:2:9); flush smooth pointing as work proceeds										
Lintels										
plain and rubbed faces -3; splayed and rubbed faces -1										
200 x 100mm......................................	58.00	2.50	**59.45**	1.25	0.85	**17.50**	1.25	**78.20**	m	**84.07**
225 x 125mm......................................	75.00	2.50	**76.88**	1.50	1.10	**21.57**	1.60	**100.04**	m	**107.55**
Sills										
plain and rubbed faces -3; sunk weathered and rubbed faces -1; grooves -1; throats -1										
200 x 75mm.......................................	80.00	2.50	**82.00**	1.00	0.65	**13.79**	1.90	**97.69**	m	**105.02**
250 x 75mm.......................................	97.00	2.50	**99.42**	1.20	0.80	**16.69**	2.20	**118.32**	m	**127.19**
300 x 75mm.......................................	112.00	2.50	**114.80**	1.40	1.00	**19.94**	2.45	**137.19**	m	**147.48**
Jamb stones; attached										
plain and rubbed faces -3; splayed and rubbed faces -1; rebates -1; grooves -1										
175 x 75mm.......................................	62.00	2.50	**63.55**	1.00	0.65	**13.79**	1.65	**78.99**	m	**84.92**
200 x 100mm......................................	67.00	2.50	**68.67**	1.25	1.00	**18.56**	2.15	**89.38**	m	**96.09**
Band courses; moulded; horizontal										
225 x 125mm......................................	132.00	2.50	**135.30**	1.60	1.15	**22.84**	3.00	**161.14**	m	**173.23**
250 x 150mm......................................	164.00	2.50	**168.10**	1.80	1.35	**26.09**	3.50	**197.69**	m	**212.52**
300 x 150mm......................................	178.00	2.50	**182.45**	2.00	1.55	**29.34**	4.10	**215.89**	m	**232.08**
Copings; horizontal										
plain and rubbed faces -2; weathered and rubbed faces -1; throats -2; cramped joints with stainless steel cramps										
300 x 50mm.......................................	97.00	2.50	**99.42**	0.75	0.50	**10.43**	4.20	**114.06**	m	**122.61**
300 x 75mm.......................................	113.00	2.50	**115.83**	0.90	0.70	**13.22**	4.70	**133.75**	m	**143.78**
375 x 100mm......................................	156.00	2.50	**159.90**	1.35	1.00	**19.48**	5.65	**185.03**	m	**198.91**
Natural stonework, natural Yorkshire sandstone, Bolton Wood; bedding and jointing in cement-lime mortar (1:2:9); flush smooth pointing as work proceeds										
Sills										
plain and rubbed faces -3; sunk weathered and rubbed faces -1; throats -1										
175 x 100mm......................................	58.00	2.50	**59.45**	1.25	0.85	**17.50**	1.25	**78.20**	m	**84.07**
250 x 125mm......................................	90.00	2.50	**92.25**	1.40	1.00	**19.94**	1.65	**113.84**	m	**122.38**
Copings; horizontal										
plain and rubbed faces -2; weathered and rubbed faces -2; throats -2; cramped joints with stainless steel cramps										
300 x 75mm.......................................	57.50	2.50	**58.94**	0.90	0.70	**13.22**	3.00	**75.16**	m	**80.80**
300 x 100mm......................................	70.00	2.50	**71.75**	1.10	0.80	**15.77**	3.30	**90.82**	m	**97.63**

MASONRY

Labour hourly rates: (except Specialists) Craft Operatives £9.23 Labour £7.02 Rates are national average prices. Refer to REGIONAL VARIATIONS for indicative levels of overall pricing in regions	MATERIALS			LABOUR				RATES		
	Del to Site £	Waste %	Material Cost £	Craft Optve Hrs	Lab Hrs	Labour Cost £	Sunds £	Nett Rate £	Unit	Gross Rate (+7.5%) £
F20: NATURAL STONE RUBBLE WALLING Cont.										
Natural stonework, natural Yorkshire sandstone, **Bolton Wood; bedding and jointing in cement-lime** **mortar (1:2:9); flush smooth pointing as work** **proceeds Cont.**										
Kerbs; horizontal plain and sawn faces - 4										
150 x 150mm	34.65	2.50	35.52	0.85	0.65	12.41	1.10	49.02	m	52.70
225 x 150mm	52.00	2.50	53.30	1.20	1.00	18.10	1.65	73.05	m	78.52
Cover stones 75 x 300; rough edges -2; plain and sawn face -1	31.10	2.50	31.88	0.85	0.65	12.41	2.20	46.49	m	49.97
Templates 150 x 300; rough edges -2; plain and sawn face -1	57.75	2.50	59.19	1.50	1.20	22.27	2.75	84.21	m	90.53
Steps; plain plain and rubbed top and front										
225 x 75mm	25.85	2.50	26.50	0.75	0.55	10.78	2.20	39.48	m	42.44
225 x 150mm	52.00	2.50	53.30	1.30	1.00	19.02	2.75	75.07	m	80.70
300 x 150mm	69.25	2.50	70.98	1.60	1.30	23.89	3.30	98.18	m	105.54

Landings
  75 x 900 x 900mm; sawn edges -4; plain and rubbed
  face -1
	93.50	2.50	95.84	2.00	1.75	30.75	3.30	129.88	nr	139.62

F22: CAST STONE ASHLAR WALLING/DRESSINGS										
Cast stonework dressings; simulated Dorset limestone; **Portland Whitbed; bedding and jointing in cement-lime** **mortar (1:2:9); flush smooth pointing as work** **proceeds**										
Walls; vertical building against brickwork; B.S.1243 Fig 1 specification 3.5 wall ties -4/m2 built in 100mm thick; plain and rubbed one side	72.50	5.00	76.13	4.00	2.50	54.47	5.50	136.10	m2	146.30
Fair raking cutting 100 thick	15.00	-	15.00	1.00	0.50	12.74	-	27.74	m	29.82
Cast stonework; simulated Dorset limestone; Portland **Whitbed; bedding and jointing in cement-lime mortar** **(1:2:9); flush smooth pointing as work proceeds**										

Lintels
  plain and rubbed faces -3; splayed and rubbed
  faces -1										

200 x 100mm	23.00	5.00	24.15	1.00	0.60	13.44	0.80	38.39	m	41.27
225 x 125mm	34.00	5.00	35.70	1.25	0.75	16.80	1.10	53.60	m	57.62

Sills
  plain and rubbed faces -3; sunk weathered and
  rubbed faces -1; grooves -1; throats -1										

200 x 75mm	21.00	5.00	22.05	0.90	0.50	11.82	1.10	34.97	m	37.59
extra; stoolings	5.00	5.00	5.25	-	-	-	-	5.25	nr	5.64
250 x 75mm	23.00	5.00	24.15	1.00	0.60	13.44	1.35	38.94	m	41.86
extra; stoolings	5.00	5.00	5.25	-	-	-	-	5.25	nr	5.64
300 x 75mm	27.00	5.00	28.35	1.25	0.75	16.80	1.65	46.80	m	50.31
extra; stoolings	5.00	5.00	5.25	-	-	-	-	5.25	nr	5.64

Jamb stones; attached
  plain and rubbed faces -3; splayed and rubbed
  faces -1; rebates -1; grooves -1										

175 x 75mm	22.50	5.00	23.63	0.90	0.50	11.82	0.85	36.29	m	39.01
200 x 100mm	29.00	5.00	30.45	1.00	0.60	13.44	1.65	45.54	m	48.96
Band courses; plain; horizontal										
225 x 125mm	22.50	5.00	23.63	1.40	1.00	19.94	1.35	44.92	m	48.29
extra; external return	16.00	5.00	16.80	-	-	-	-	16.80	nr	18.06
250 x 150mm	27.50	5.00	28.88	1.50	1.10	21.57	1.65	52.09	m	56.00
extra; external return	16.00	5.00	16.80	-	-	-	-	16.80	nr	18.06
300 x 150mm	35.00	5.00	36.75	1.60	1.20	23.19	1.90	61.84	m	66.48
extra; external return	16.00	5.00	16.80	-	-	-	-	16.80	nr	18.06

Copings; horizontal
  plain and rubbed faces -2; weathered and rubbed
  faces -2; throats -2; cramped joints with
  stainless steel cramps										

300 x 50mm	17.50	5.00	18.38	0.50	0.30	6.72	2.50	27.60	m	29.67
extra; internal angles	16.00	5.00	16.80	-	-	-	-	16.80	nr	18.06
extra; external angles	16.00	5.00	16.80	-	-	-	-	16.80	nr	18.06
300 x 75mm	20.00	5.00	21.00	0.60	0.35	8.00	2.75	31.75	m	34.13
extra; internal angles	16.00	5.00	16.80	-	-	-	-	16.80	nr	18.06
extra; external angles	16.00	5.00	16.80	-	-	-	-	16.80	nr	18.06
375 x 100mm	28.00	5.00	29.40	0.75	0.50	10.43	3.30	43.13	m	46.37
extra; internal angles	16.00	5.00	16.80	-	-	-	-	16.80	nr	18.06
extra; external angles	16.00	5.00	16.80	-	-	-	-	16.80	nr	18.06
Steps; plain 300 x 150mm; plain and rubbed tread and riser	27.00	5.00	28.35	1.25	0.90	17.86	2.20	48.41	m	52.04

Labour hourly rates: (except Specialists) Craft Operatives £9.23 Labour £7.02 Rates are national average prices. Refer to REGIONAL VARIATIONS for indicative levels of overall pricing in regions	MATERIALS			LABOUR				RATES		
	Del to Site £	Waste %	Material Cost £	Craft Optve Hrs	Lab Hrs	Labour Cost £	Sunds £	Nett Rate £	Unit	Gross Rate (+7.5%) £
F22: CAST STONE ASHLAR WALLING/DRESSINGS Cont.										
Cast stonework; simulated Dorset limestone; Portland Whitbed; bedding and jointing in cement-lime mortar (1:2:9); flush smooth pointing as work proceeds Cont.										
Steps; spandril 250mm wide tread; 180mm high riser; plain and rubbed tread and riser; carborundum finish to tread	45.00	5.00	**47.25**	2.00	1.25	**27.23**	1.65	**76.14**	m	**81.85**
Landings 150 x 900 x 900mm; plain and rubbed top surface ..	72.50	5.00	**76.13**	1.50	1.00	**20.86**	2.20	**99.19**	nr	**106.63**
F30: ACCESSORIES/SUNDRY ITEMS FOR BRICK/BLOCK/STONE WALLING										
Forming cavities in hollow walls										
Cavity with B.S.1243 Fig 1 galvanised wire butterfly wall ties, -4/m2 built in width of cavity 50mm	0.31	-	**0.31**	0.10	0.07	**1.41**	-	**1.72**	m2	**1.85**
Cavity with B.S.1243 Fig 1 stainless steel wire butterfly wall ties, -4/m2 built in width of cavity 50mm	0.34	-	**0.34**	0.10	0.07	**1.41**	-	**1.75**	m2	**1.89**
Cavity with B.S.1243 Fig 3 galvanised steel twisted wall ties, -4/m2 built in width of cavity 50mm	0.96	-	**0.96**	0.10	0.07	**1.41**	-	**2.37**	m2	**2.55**
Cavity with B.S.1243 Fig 3 stainless steel twisted wall ties, -4/m2 built in width of cavity 50mm	1.55	-	**1.55**	0.10	0.07	**1.41**	-	**2.96**	m2	**3.19**
Cavity with B.S.1243 Fig 3 galvanised steel twisted wall ties -4/m2 built in; 50 fibreglass resin bonded slab cavity insulation width of cavity 50mm	3.10	10.00	**3.41**	0.35	0.25	**4.99**	1.08	**9.48**	m2	**10.19**
Cavity with B.S.1243 Fig 3 galvanised steel twisted wall ties -4/m2 built in; 75 fibreglass resin bonded slab cavity insulation width of cavity 75mm	4.24	10.00	**4.66**	0.40	0.25	**5.45**	1.08	**11.19**	m2	**12.03**
Damp proof courses										
B.S.6398 Class A, bitumen with hessian base; 100mm laps; bedding in cement mortar (1:3); no allowance made for laps										
width not exceeding 225mm; vertical	5.50	5.00	**5.78**	0.45	-	**4.15**	-	**9.93**	m2	**10.67**
width exceeding 225mm; vertical	5.50	5.00	**5.78**	0.40	-	**3.69**	-	**9.47**	m2	**10.18**
width not exceeding 225mm; horizontal	5.50	5.00	**5.78**	0.30	-	**2.77**	-	**8.54**	m2	**9.18**
width exceeding 225mm; horizontal	5.50	5.00	**5.78**	0.25	-	**2.31**	-	**8.08**	m2	**8.69**
cavity trays; width not exceeding 225mm; horizontal	5.50	5.00	**5.78**	0.50	-	**4.62**	-	**10.39**	m2	**11.17**
cavity trays; width exceeding 225mm; horizontal ..	5.50	5.00	**5.78**	0.45	-	**4.15**	-	**9.93**	m2	**10.67**
B.S.6398 Class B, bitumen with fibre base; 100mm laps; bedding in cement mortar (1:3); no allowance made for laps										
width not exceeding 225mm; vertical	3.70	5.00	**3.88**	0.45	-	**4.15**	-	**8.04**	m2	**8.64**
width exceeding 225mm; vertical	3.70	5.00	**3.88**	0.40	-	**3.69**	-	**7.58**	m2	**8.15**
width not exceeding 225mm; horizontal	3.70	5.00	**3.88**	0.30	-	**2.77**	-	**6.65**	m2	**7.15**
width exceeding 225mm; horizontal	3.70	5.00	**3.88**	0.25	-	**2.31**	-	**6.19**	m2	**6.66**
B.S.6398 Class D, bitumen with hessian base laminated with lead; 100mm laps; bedding in cement mortar (1:3); no allowance made for laps										
width not exceeding 225mm; vertical	13.80	5.00	**14.49**	0.45	-	**4.15**	-	**18.64**	m2	**20.04**
width exceeding 225mm; vertical	13.80	5.00	**14.49**	0.40	-	**3.69**	-	**18.18**	m2	**19.55**
width not exceeding 225mm; horizontal	13.80	5.00	**14.49**	0.30	-	**2.77**	-	**17.26**	m2	**18.55**
width exceeding 225mm; horizontal	13.80	5.00	**14.49**	0.25	-	**2.31**	-	**16.80**	m2	**18.06**
B.S.6398 Class E, bitumen with fibre base laminated with lead; 100mm laps; bedding in cement mortar (1:3); no allowance made for laps										
width not exceeding 225mm; vertical	13.20	5.00	**13.86**	0.45	-	**4.15**	-	**18.01**	m2	**19.36**
width exceeding 225mm; vertical	13.20	5.00	**13.86**	0.40	-	**3.69**	-	**17.55**	m2	**18.87**
width not exceeding 225mm; horizontal	13.20	5.00	**13.86**	0.30	-	**2.77**	-	**16.63**	m2	**17.88**
width exceeding 225mm; horizontal	13.20	5.00	**13.86**	0.25	-	**2.31**	-	**16.17**	m2	**17.38**
Hyload, pitch polymer; 100mm laps; bedding in cement lime mortar (1:1:6); no allowance made for laps										
width not exceeding 225mm; vertical	6.75	5.00	**7.09**	0.45	-	**4.15**	-	**11.24**	m2	**12.08**
width exceeding 225mm; vertical	6.75	5.00	**7.09**	0.40	-	**3.69**	-	**10.78**	m2	**11.59**
width not exceeding 225mm; horizontal	6.75	5.00	**7.09**	0.30	-	**2.77**	-	**9.86**	m2	**10.60**
width exceeding 225mm; horizontal	6.75	5.00	**7.09**	0.25	-	**2.31**	-	**9.39**	m2	**10.10**
Synthaprufe bituminous latex emulsion; two coats brushed on; blinded with sand										
width not exceeding 225mm; vertical	2.98	5.00	**3.13**	-	0.50	**3.51**	0.14	**6.78**	m2	**7.29**
width exceeding 225mm; vertical	2.98	5.00	**3.13**	-	0.30	**2.11**	0.14	**5.38**	m2	**5.78**

MASONRY

Labour hourly rates: (except Specialists) Craft Operatives £9.23 Labour £7.02 Rates are national average prices. Refer to REGIONAL VARIATIONS for indicative levels of overall pricing in regions	MATERIALS			LABOUR				RATES		
	Del to Site	Waste	Material Cost	Craft Optve	Lab	Labour Cost	Sunds	Nett Rate	Unit	Gross Rate (+7.5%)
	£	%	£	Hrs	Hrs	£	£	£		£
F30: ACCESSORIES/SUNDRY ITEMS FOR BRICK/BLOCK/STONE WALLING Cont.										
Damp proof courses Cont.										
Synthaprufe bituminous latex emulsion; three coats brushed on; blinded with sand										
width not exceeding 225mm; vertical	4.56	5.00	4.79	–	0.70	4.91	0.16	9.86	m2	10.60
width exceeding 225mm; vertical	4.56	5.00	4.79	–	0.42	2.95	0.16	7.90	m2	8.49
Bituminous emulsion; two coats brushed on										
width not exceeding 225mm; vertical	1.47	5.00	1.54	–	0.42	2.95	0.14	4.63	m2	4.98
width exceeding 225mm; vertical	1.47	5.00	1.54	–	0.26	1.83	0.14	3.51	m2	3.77
Bituminous emulsion; three coats brushed on										
width not exceeding 225mm; vertical	2.01	5.00	2.11	–	0.60	4.21	0.16	6.48	m2	6.97
width exceeding 225mm; vertical	2.01	5.00	2.11	–	0.36	2.53	0.16	4.80	m2	5.16
One course slates in cement mortar (1:3)										
width not exceeding 225mm; vertical	14.00	15.00	16.10	1.30	0.90	18.32	–	34.42	m2	37.00
width exceeding 225mm; vertical	14.00	15.00	16.10	1.20	0.80	16.69	–	32.79	m2	35.25
width not exceeding 225mm; horizontal	14.00	15.00	16.10	0.80	0.55	11.24	–	27.34	m2	29.40
width exceeding 225mm; horizontal	14.00	15.00	16.10	0.70	0.50	9.97	–	26.07	m2	28.03
Two courses slates in cement mortar (1:3)										
width not exceeding 225mm; vertical	28.00	15.00	32.20	2.20	1.50	30.84	–	63.04	m2	67.76
width exceeding 225mm; vertical	28.00	15.00	32.20	2.00	1.35	27.94	–	60.14	m2	64.65
width not exceeding 225mm; horizontal	28.00	15.00	32.20	1.30	0.90	18.32	–	50.52	m2	54.31
width exceeding 225mm; horizontal	28.00	15.00	32.20	1.20	0.80	16.69	–	48.89	m2	52.56
Cavity trays										
Type G Cavitray by Cavity Trays Ltd in										
half brick skin of cavity wall	7.07	5.00	7.42	0.20	–	1.85	–	9.27	m	9.96
external angle	6.36	5.00	6.68	0.10	–	0.92	–	7.60	nr	8.17
internal angle	6.36	5.00	6.68	0.10	–	0.92	–	7.60	nr	8.17
Type X Cavitray by Cavity Trays Ltd to suit 40 degree pitched roof complete with attached code 4 lead flashing and dress over tiles in										
half brick skin of cavity wall	16.03	5.00	16.83	0.20	–	1.85	–	18.68	m	20.08
ridge tray	3.71	5.00	3.90	0.10	–	0.92	–	4.82	nr	5.18
catchment tray	1.90	5.00	2.00	0.10	–	0.92	–	2.92	nr	3.14
corner catchment angle tray	3.71	5.00	3.90	0.10	–	0.92	–	4.82	nr	5.18
Type W Cavity weep/ventilator by Cavity Trays Ltd in										
half brick skin of cavity wall	0.41	5.00	0.43	0.10	–	0.92	–	1.35	nr	1.46
extension duct	0.52	5.00	0.55	0.10	–	0.92	–	1.47	nr	1.58
Joint reinforcement										
Expamet grade 304/S15 Exmet reinforcement, stainless steel; 150mm laps; no allowance made for laps										
65mm wide	0.54	5.00	0.57	0.04	0.02	0.51	0.03	1.11	m	1.19
115mm wide	0.92	5.00	0.97	0.05	0.03	0.67	0.03	1.67	m	1.79
175mm wide	1.49	5.00	1.56	0.06	0.03	0.76	0.04	2.37	m	2.55
225mm wide	2.02	5.00	2.12	0.07	0.04	0.93	0.05	3.10	m	3.33
Bed Joint Reinforcement (UK) Ltd grade 304/S15 Brickspan reinforcement, stainless steel; 150mm laps; no allowance made for laps										
50mm wide	1.21	5.00	1.27	0.04	0.02	0.51	0.03	1.81	m	1.95
60mm wide	1.21	5.00	1.27	0.04	0.02	0.51	0.03	1.81	m	1.95
80mm wide	1.29	5.00	1.35	0.05	0.03	0.67	0.03	2.06	m	2.21
100mm wide	1.29	5.00	1.35	0.05	0.03	0.67	0.03	2.06	m	2.21
150mm wide	1.51	5.00	1.59	0.05	0.03	0.67	0.03	2.29	m	2.46
160mm wide	1.51	5.00	1.59	0.06	0.03	0.76	0.04	2.39	m	2.57
Bed Joint Reinforcement (UK) Ltd grade 304/S15 Wallspan reinforcement, stainless steel; 150mm laps; no allowance made for laps										
210mm wide	5.35	5.00	5.62	0.06	0.03	0.76	0.04	6.42	m	6.90
222.5mm wide	5.35	5.00	5.62	0.06	0.03	0.76	0.04	6.42	m	6.90
235mm wide	5.35	5.00	5.62	0.07	0.04	0.93	0.05	6.59	m	7.09
250mm wide	5.35	5.00	5.62	0.07	0.04	0.93	0.05	6.59	m	7.09
275mm wide	5.35	5.00	5.62	0.07	0.04	0.93	0.05	6.59	m	7.09
Pointing in flashings										
Cement mortar (1:3)										
horizontal	–	–	–	0.40	–	3.69	0.19	3.88	m	4.17
horizontal; in old wall	–	–	–	0.45	–	4.15	0.19	4.34	m	4.67
stepped	–	–	–	0.65	–	6.00	0.29	6.29	m	6.76
stepped; in old wall	–	–	–	0.70	–	6.46	0.29	6.75	m	7.26
Wedging and pinning										
Two courses slates in cement mortar (1:3)										
width of wall 215mm	6.43	15.00	7.39	0.50	0.25	6.37	0.10	13.86	m	14.90

Labour hourly rates: (except Specialists) Craft Operatives £9.23 Labour £7.02 Rates are national average prices. Refer to REGIONAL VARIATIONS for indicative levels of overall pricing in regions	MATERIALS			LABOUR				RATES		
	Del to Site £	Waste %	Material Cost £	Craft Optve Hrs	Lab Hrs	Labour Cost £	Sunds £	Nett Rate £	Unit	Gross Rate (+7.5%) £
F30: ACCESSORIES/SUNDRY ITEMS FOR BRICK/BLOCK/STONE WALLING Cont.										
Joints										
Expansion joints in facing brickwork 13mm wide; vertical; filling with Servicised Ltd, Fibrepack filler, Vertiseal compound pointing one side; including preparation, cleaners, primers and sealers										
102mm thick wall	2.49	7.50	2.68	0.09	0.18	2.09	-	4.77	m	5.13
215mm thick wall	3.47	7.50	3.73	0.10	0.20	2.33	-	6.06	m	6.51
Expansion joints in blockwork 13mm wide; vertical; filling with Servicised Ltd, Fibrepack filler, Vertiseal compound pointing one side; including preparation, cleaners, primers and sealers										
100mm thick wall	2.49	7.50	2.68	0.09	0.18	2.09	-	4.77	m	5.13
Expansion joints in glass blockwork 10mm wide; vertical; in filling with compressible material, polysulphide sealant both sides including preparation, cleaners, primers and sealers										
80mm thick wall	2.37	7.50	2.55	0.09	0.18	2.09	-	4.64	m	4.99
Slates and tiles for creasing										
Nibless creasing tiles, red, machine made, 265 x 165 x 10mm; in cement-lime mortar (1:1:6)										
one course 253mm wide	3.94	5.00	4.14	0.50	0.35	7.07	0.19	11.40	m	12.25
one course 365mm wide	6.39	5.00	6.71	0.70	0.55	10.32	0.26	17.29	m	18.59
two courses 253mm wide	7.88	5.00	8.27	0.85	0.55	11.71	0.38	20.36	m	21.89
two courses 365mm wide	12.78	5.00	13.42	1.15	0.85	16.58	0.51	30.51	m	32.80
Slate and tile sills										
Clay plain roofing tiles, red, B.S.402, machine made, 265 x 165 x 13mm; in cement-lime mortar										
one course 150mm wide; set weathering	2.43	5.00	2.55	0.45	0.30	6.26	0.21	9.02	m	9.70
two courses 150mm wide; set weathering	4.85	5.00	5.09	0.75	0.50	10.43	0.36	15.89	m	17.08
Fires and fire parts										
Solid one piece or two piece firebacks; B.S.1251; bed and joint in fire cement; concrete filling at back										
fire size 400mm	27.30	2.50	27.98	1.00	0.67	13.93	0.95	42.87	nr	46.08
fire size 450mm	32.83	2.50	33.65	1.10	0.73	15.28	1.06	49.99	nr	53.74
Solid one piece or two piece firebacks with cut out for boiler; B.S.1251; bed and joint in fire cement; concrete filling at back										
fire size 400mm	33.14	2.50	33.97	1.00	0.67	13.93	0.95	48.85	nr	52.52
fire size 450mm	38.60	2.50	39.56	1.10	0.73	15.28	1.06	55.90	nr	60.10
Frets and stools; black vitreous enamelled; place in position										
fire size 400mm	23.08	-	23.08	0.33	0.22	4.59	-	27.67	nr	29.75
fire size 450mm	36.07	-	36.07	0.36	0.24	5.01	-	41.08	nr	44.16
Frets and stools; lustre finish; place in position										
fire size 400mm	37.57	-	37.57	0.33	0.22	4.59	-	42.16	nr	45.32
fire size 450mm	24.62	-	24.62	0.36	0.24	5.01	-	29.63	nr	31.85
Continuous burning open fire; vitreous finish; self contained open fire; plugging and screwing; sealing to opening										
fire size 400mm	115.50	-	115.50	3.50	2.33	48.66	1.95	166.11	nr	178.57
fire size 450mm	132.00	-	132.00	3.75	2.50	52.16	3.79	187.95	nr	202.05
Gas ignited smokeless fuel fires; assembling; plugging and screwing; bedding and jointing in fire cement; sealing to opening grate only; black										
fire size 400mm	75.00	-	75.00	2.50	1.67	34.80	-	109.80	nr	118.03
fire size 450mm	76.00	-	76.00	2.50	1.67	34.80	-	110.80	nr	119.11
grate only; colours										
fire size 400mm	76.00	-	76.00	2.50	1.67	34.80	-	110.80	nr	119.11
fire size 450mm	77.00	-	77.00	2.50	1.67	34.80	-	111.80	nr	120.18
boiler and self contained flue										
fire size 400mm	174.00	-	174.00	4.50	3.00	62.59	1.95	238.54	nr	256.44
fire size 450mm	185.00	-	185.00	4.50	3.00	62.59	3.79	251.38	nr	270.24
threefold brick sets										
fire size 400mm	23.10	-	23.10	1.00	0.67	13.93	0.95	37.98	nr	40.83
fire size 450mm	23.10	-	23.10	1.00	0.67	13.93	1.06	38.09	nr	40.95
extra; chrome finish to fire front; fire size 400 or 450mm	23.10	-	23.10	-	-	-	-	23.10	nr	24.83
extra; back boiler unit with Bower Barffed rustless boiler; fire size 400 or 450mm	180.00	-	180.00	-	-	-	-	180.00	nr	193.50
extra; back boiler unit with copper boiler; 400 or 450mm	275.00	%	275.00	-	Hrs	£	-	275.00	nr	295.63

Labour hourly rates: (except Specialists) Craft Operatives £9.23 Labour £7.02 Rates are national average prices. Refer to REGIONAL VARIATIONS for indicative levels of overall pricing in regions	MATERIALS			LABOUR				RATES		
	Del to Site £	Waste %	Material Cost £	Craft Optve Hrs	Lab Hrs	Labour Cost £	Sunds £	Nett Rate £	Unit	Gross Rate (+7.5%) £
F30: ACCESSORIES/SUNDRY ITEMS FOR BRICK/BLOCK/STONE WALLING Cont.										
Flue linings										
Clay flue linings, B.S.1181; rebated joints; jointed in cement mortar (1:3)										
150mm diameter, Type 2	15.94	5.00	**16.74**	0.45	0.35	**6.61**	0.27	**23.62**	m	**25.39**
terminal Type 6F	15.25	5.00	**16.01**	0.45	0.35	**6.61**	0.14	**22.76**	nr	**24.47**
185mm diameter, Type 2	20.44	5.00	**21.46**	0.50	0.40	**7.42**	0.35	**29.23**	m	**31.43**
terminal Type 6F	16.58	5.00	**17.41**	0.50	0.40	**7.42**	0.16	**24.99**	nr	**26.87**
225mm diameter, Type 2	34.07	5.00	**35.77**	0.55	0.45	**8.24**	0.38	**44.39**	m	**47.72**
terminal Type 6F	17.85	5.00	**18.74**	0.55	0.45	**8.24**	0.20	**27.18**	nr	**29.22**
185 x 185mm, Type 1	21.95	5.00	**23.05**	0.55	0.45	**8.24**	0.35	**31.63**	m	**34.01**
terminal Type 4D	24.83	5.00	**26.07**	0.55	0.45	**8.24**	0.16	**34.47**	nr	**37.05**
Chimney pots										
Clay chimney pots; set and flaunched in cement mortar (1:3)										
tapered roll top; 600mm high	28.97	5.00	**30.42**	0.75	0.63	**11.35**	0.35	**42.11**	nr	**45.27**
tapered roll top; 750mm high	39.43	5.00	**41.40**	0.88	0.75	**13.39**	0.45	**55.24**	nr	**59.38**
tapered roll top; 900mm high	52.80	5.00	**55.44**	1.00	0.93	**15.76**	0.53	**71.73**	nr	**77.11**
Air bricks										
Clay air bricks, B.S.493, square hole pattern; opening with slate lintel over										
common brick wall 102mm thick; opening size										
225 x 75mm	1.52	-	**1.52**	0.20	0.10	**2.55**	0.53	**4.60**	nr	**4.94**
225 x 150mm	2.11	-	**2.11**	0.25	0.13	**3.22**	0.56	**5.89**	nr	**6.33**
225 x 225mm	5.81	-	**5.81**	0.30	0.15	**3.82**	0.60	**10.23**	nr	**11.00**
common brick wall 215mm thick; opening size										
225 x 75mm	1.52	-	**1.52**	0.30	0.15	**3.82**	1.00	**6.34**	nr	**6.82**
225 x 150mm	2.11	-	**2.11**	0.38	0.19	**4.84**	1.05	**8.00**	nr	**8.60**
225 x 225mm	5.81	-	**5.81**	0.45	0.23	**5.77**	1.12	**12.70**	nr	**13.65**
common brick wall 215mm thick; facework one side; opening size										
225 x 75mm	1.52	-	**1.52**	0.40	0.20	**5.10**	1.00	**7.62**	nr	**8.19**
225 x 150mm	2.11	-	**2.11**	0.50	0.25	**6.37**	1.05	**9.53**	nr	**10.24**
225 x 225mm	5.81	-	**5.81**	0.60	0.30	**7.64**	1.12	**14.57**	nr	**15.67**
common brick wall 327mm thick; facework one side; opening size										
225 x 75mm	1.52	-	**1.52**	0.60	0.30	**7.64**	1.62	**10.78**	nr	**11.59**
225 x 150mm	2.11	-	**2.11**	0.70	0.35	**8.92**	1.70	**12.73**	nr	**13.68**
225 x 225mm	5.81	-	**5.81**	0.80	0.40	**10.19**	1.75	**17.75**	nr	**19.08**
common brick wall 440mm thick; facework one side; opening size										
225 x 75mm	1.52	-	**1.52**	0.80	0.40	**10.19**	2.15	**13.86**	nr	**14.90**
225 x 150mm	2.11	-	**2.11**	0.90	0.45	**11.47**	2.23	**15.81**	nr	**16.99**
225 x 225mm	5.81	-	**5.81**	1.00	0.50	**12.74**	2.30	**20.85**	nr	**22.41**
cavity wall 252mm thick with 102mm facing brick outer skin, 100mm block inner skin and 50mm cavity; sealing cavity with slates in cement mortar (1:3); rendering all round with cement mortar (1:3); opening size										
225 x 75mm	1.52	-	**1.52**	0.60	0.30	**7.64**	1.94	**11.10**	nr	**11.94**
225 x 150mm	2.11	-	**2.11**	0.70	0.35	**8.92**	2.39	**13.42**	nr	**14.42**
225 x 225mm	5.81	-	**5.81**	0.80	0.40	**10.19**	2.80	**18.80**	nr	**20.21**
Gas flue blocks										
Marflex HP system precast refractory concrete gas flue blocks; bedding and jointing in Flue joint refractory mortar										
recess block reference HP1, 405 x 140 x 222mm	4.39	5.00	**4.61**	0.26	-	**2.40**	0.53	**7.54**	nr	**8.10**
cover block reference HP2, 385 x 140 x 222mm	6.33	5.00	**6.65**	0.45	-	**4.15**	0.53	**11.33**	nr	**12.18**
Standard block reference HP3, 355 x 140 x 72mm	4.25	5.00	**4.46**	0.30	-	**2.77**	0.53	**7.76**	nr	**8.34**
Standard block reference HP3, 355 x 140 x 112mm	4.25	5.00	**4.46**	0.30	-	**2.77**	0.53	**7.76**	nr	**8.34**
Standard block reference HP3, 355 x 140 x 222mm	4.25	5.00	**4.46**	0.30	-	**2.77**	0.53	**7.76**	nr	**8.34**
Vent unit for bathroom/kitchen extract fans reference HP1 and HP2	10.72	5.00	**11.26**	0.30	-	**2.77**	0.53	**14.56**	nr	**15.65**
standard block reference HP4, 280 x 140 x 72mm	4.10	5.00	**4.30**	0.25	-	**2.31**	0.47	**7.08**	nr	**7.61**
standard block reference HP4, 280 x 140 x 112mm	4.10	5.00	**4.30**	0.25	-	**2.31**	0.47	**7.08**	nr	**7.61**
standard block reference HP4, 280 x 140 x 222mm	4.10	5.00	**4.30**	0.25	-	**2.31**	0.47	**7.08**	nr	**7.61**
120mm side offset block reference HP5, 400 x 140 x 222mm	5.17	5.00	**5.43**	0.37	-	**3.42**	0.68	**9.52**	nr	**10.24**
70mm back offset block reference HP6, 280 x 210 x 222mm	13.31	5.00	**13.98**	0.35	-	**3.23**	0.68	**17.89**	nr	**19.23**
vertical exit block reference HP7, 280 x 181 x 222mm	8.82	5.00	**9.26**	0.35	-	**3.23**	0.53	**13.02**	nr	**14.00**
angled entry/exit block reference HP8, 280 x 140 x 230mm	8.82	5.00	**9.26**	0.40	-	**3.69**	0.72	**13.67**	nr	**14.70**
double rebate block used when in conjunction with a boiler reference HP9, 280 x 140 x 222mm	6.52	5.00	**6.85**	0.40	-	**3.69**	0.47	**11.01**	nr	**11.83**
corbel block reference HP10	8.33	5.00	**8.75**	0.40	-	**3.69**	0.53	**12.97**	nr	**13.94**
conversion unit for back boiler recess units reference HP25, 280 x 262 x 222mm	10.18	5.00	**10.69**	0.40	-	**3.69**	0.47	**14.85**	nr	**15.96**
Arch bars										
Steel flat arch bar										
30 x 6mm	1.57	-	**1.57**	0.70	-	**6.46**	-	**8.03**	m	**8.63**
50 x 6mm	2.42	-	**2.42**	0.70	-	**6.46**	-	**8.88**	m	**9.55**

Labour hourly rates: (except Specialists) Craft Operatives £9.23 Labour £7.02 Rates are national average prices. Refer to REGIONAL VARIATIONS for indicative levels of overall pricing in regions	MATERIALS			LABOUR				RATES		
	Del to Site	Waste	Material Cost	Craft Optve	Lab	Labour Cost	Sunds	Nett Rate	Unit	Gross Rate (+7.5%)
	£	%	£	Hrs	Hrs	£	£	£		£
F30: ACCESSORIES/SUNDRY ITEMS FOR BRICK/BLOCK/STONE WALLING Cont.										
Arch bars Cont.										
Steel angle arch bar										
50 x 50 x 6mm......................................	4.52	–	4.52	0.70	–	6.46	–	10.98	m	11.80
75 x 50 x 6mm......................................	5.74	–	5.74	0.80	–	7.38	–	13.12	m	14.11
Building in										
Building in metal windows; building in lugs; bedding in cement mortar (1:3), pointing with Secomastic standard mastic one side										
200mm high; lugs to brick jambs; plugging and screwing to brick sill and concrete head										
500mm wide..	–	–	–	0.53	0.22	6.44	0.87	7.31	nr	7.85
600mm wide..	–	–	–	0.58	0.24	7.04	1.08	8.12	nr	8.73
900mm wide..	–	–	–	0.74	0.29	8.87	1.48	10.35	nr	11.12
1200mm wide.......................................	–	–	–	0.90	0.34	10.69	1.90	12.59	nr	13.54
1500mm wide.......................................	–	–	–	1.06	0.39	12.52	2.30	14.82	nr	15.93
1800mm wide.......................................	–	–	–	1.22	0.44	14.35	2.70	17.05	nr	18.33
500mm high; lugs to brick jambs; plugging and screwing to brick sill and concrete head										
500mm wide..	–	–	–	0.64	0.25	7.66	1.33	8.99	nr	9.67
600mm wide..	–	–	–	0.69	0.27	8.26	1.47	9.73	nr	10.46
900mm wide..	–	–	–	0.85	0.32	10.09	1.89	11.98	nr	12.88
1200mm wide.......................................	–	–	–	1.02	0.37	12.01	2.28	14.29	nr	15.36
1500mm wide.......................................	–	–	–	1.18	0.43	13.91	2.70	16.61	nr	17.86
1800mm wide.......................................	–	–	–	1.34	0.48	15.74	3.08	18.82	nr	20.23
700mm high; lugs to brick jambs; plugging and screwing to brick sill and concrete head										
500mm wide..	–	–	–	0.75	0.28	8.89	1.60	10.49	nr	11.27
600mm wide..	–	–	–	0.80	0.30	9.49	1.73	11.22	nr	12.06
900mm wide..	–	–	–	0.97	0.35	11.41	2.15	13.56	nr	14.58
1200mm wide.......................................	–	–	–	1.13	0.41	13.31	2.55	15.86	nr	17.05
1500mm wide.......................................	–	–	–	1.30	0.46	15.23	2.95	18.18	nr	19.54
1800mm wide.......................................	–	–	–	1.46	0.52	17.13	3.34	20.47	nr	22.00
900mm high; lugs to brick jambs; plugging and screwing to brick sill and concrete head										
500mm wide..	–	–	–	0.85	0.30	9.95	1.85	11.80	nr	12.69
600mm wide..	–	–	–	0.91	0.32	10.65	2.00	12.65	nr	13.59
900mm wide..	–	–	–	1.08	0.38	12.64	2.40	15.04	nr	16.16
1200mm wide.......................................	–	–	–	1.25	0.44	14.63	2.80	17.43	nr	18.73
1500mm wide.......................................	–	–	–	1.42	0.50	16.62	3.21	19.83	nr	21.31
1800mm wide.......................................	–	–	–	1.58	0.55	18.44	3.59	22.03	nr	23.69
1100mm high; lugs to brick jambs; plugging and screwing to brick sill and concrete head										
500mm wide..	–	–	–	0.96	0.33	11.18	2.14	13.32	nr	14.32
600mm wide..	–	–	–	1.02	0.35	11.87	2.24	14.11	nr	15.17
900mm wide..	–	–	–	1.19	0.41	13.86	2.66	16.52	nr	17.76
1200mm wide.......................................	–	–	–	1.36	0.47	15.85	3.07	18.92	nr	20.34
1500mm wide.......................................	–	–	–	1.54	0.53	17.93	3.48	21.41	nr	23.02
1800mm wide.......................................	–	–	–	1.71	0.59	19.93	3.89	23.82	nr	25.60
1300mm high; lugs to brick jambs; plugging and screwing to brick sill and concrete head										
500mm wide..	–	–	–	1.07	0.36	12.40	2.37	14.77	nr	15.88
600mm wide..	–	–	–	1.12	0.38	13.01	2.50	15.51	nr	16.67
900mm wide..	–	–	–	1.30	0.44	15.09	2.91	18.00	nr	19.35
1200mm wide.......................................	–	–	–	1.48	0.50	17.17	3.32	20.49	nr	22.03
1500mm wide.......................................	–	–	–	1.65	0.57	19.23	3.74	22.97	nr	24.69
1800mm wide.......................................	–	–	–	1.83	0.63	21.31	4.14	25.45	nr	27.36
1500mm high; lugs to brick jambs; plugging and screwing to brick sill and concrete head										
500mm wide..	–	–	–	1.18	0.39	13.63	2.65	16.28	nr	17.50
600mm wide..	–	–	–	1.23	0.41	14.23	2.76	16.99	nr	18.27
900mm wide..	–	–	–	1.42	0.47	16.41	3.18	19.59	nr	21.05
1200mm wide.......................................	–	–	–	1.59	0.54	18.47	3.58	22.05	nr	23.70
1500mm wide.......................................	–	–	–	1.77	0.60	20.55	3.97	24.52	nr	26.36
1800mm wide.......................................	–	–	–	1.95	0.67	22.70	4.40	27.10	nr	29.13
2100mm high; lugs to brick jambs; plugging and screwing to brick sill and concrete head										
500mm wide..	–	–	–	1.39	0.44	15.92	3.43	19.35	nr	20.80
600mm wide..	–	–	–	1.45	0.46	16.61	3.56	20.17	nr	21.69
900mm wide..	–	–	–	1.64	0.53	18.86	3.96	22.82	nr	24.53
1200mm wide.......................................	–	–	–	1.82	0.60	21.01	4.40	25.41	nr	27.32
1500mm wide.......................................	–	–	–	2.01	0.67	23.26	4.77	28.03	nr	30.13
1800mm wide.......................................	–	–	–	2.19	0.74	25.41	5.20	30.61	nr	32.90
Building in factory glazed metal windows; screwing with galvanised screws; bedding in cement mortar (1:3), pointing with Secomastic standard mastic one side										
300mm high; plugging and screwing lugs to brick jambs and sill and concrete head										
600mm wide..	–	–	–	0.70	0.30	8.57	1.20	9.77	nr	10.50
900mm wide..	–	–	–	0.84	0.34	10.14	1.62	11.76	nr	12.64
1200mm wide.......................................	–	–	–	0.98	0.38	11.71	2.02	13.73	nr	14.76
1500mm wide.......................................	–	–	–	1.12	0.41	13.22	2.43	15.65	nr	16.82
1800mm wide.......................................	–	–	–	1.26	0.45	14.79	2.82	17.61	nr	18.93
2400mm wide.......................................	–	–	–	1.54	0.53	17.93	3.64	21.57	nr	23.19
3000mm wide.......................................	–	%	–	1.82	0.60	21.01	4.44	25.45	nr	27.36
700mm high; plugging and screwing lugs to brick jambs and sill and concrete head										
600mm wide..	–	–	–	0.96	0.40	11.67	1.73	13.40	nr	14.40
900mm wide..	–	–	–	1.17	0.48	14.17	2.15	16.32	nr	17.54

MASONRY

Labour hourly rates: (except Specialists) Craft Operatives £9.23 Labour £7.02 Rates are national average prices. Refer to REGIONAL VARIATIONS for indicative levels of overall pricing in regions	MATERIALS			LABOUR				RATES		
	Del to Site	Waste	Material Cost	Craft Optve	Lab	Labour Cost	Sunds	Nett Rate	Unit	Gross Rate (+7.5%)
	£	%	£	Hrs	Hrs	£	£	£		£
F30: ACCESSORIES/SUNDRY ITEMS FOR BRICK/BLOCK/STONE WALLING Cont.										
Building in Cont.										
Building in factory glazed metal windows; screwing with galvanised screws; bedding in cement mortar (1:3), pointing with Secomastic standard mastic one side Cont.										
700mm high; plugging and screwing lugs to brick jambs and sill and concrete head Cont.										
1200mm wide....................................	-	-	-	1.37	0.57	16.65	2.55	19.20	nr	20.64
1500mm wide....................................	-	-	-	1.58	0.64	19.08	2.95	22.03	nr	23.68
1800mm wide....................................	-	-	-	1.79	0.73	21.65	3.34	24.99	nr	26.86
2400mm wide....................................	-	-	-	2.21	0.90	26.72	4.17	30.89	nr	33.20
3000mm wide....................................	-	-	-	2.62	1.06	31.62	5.00	36.62	nr	39.37
900mm high; plugging and screwing lugs to brick jambs and sill and concrete head										
600mm wide.....................................	-	-	-	1.09	0.45	13.22	2.00	15.22	nr	16.36
900mm wide.....................................	-	-	-	1.33	0.55	16.14	2.40	18.54	nr	19.93
1200mm wide....................................	-	-	-	1.57	0.66	19.12	2.80	21.92	nr	23.57
1500mm wide....................................	-	-	-	1.81	0.76	22.04	3.21	25.25	nr	27.15
1800mm wide....................................	-	-	-	2.06	0.87	25.12	3.59	28.71	nr	30.86
2400mm wide....................................	-	-	-	2.54	1.08	31.03	4.42	35.45	nr	38.10
3000mm wide....................................	-	-	-	3.02	1.29	36.93	5.24	42.17	nr	45.33
1100mm high; plugging and screwing lugs to brick jambs and sill and concrete head										
600mm wide.....................................	-	-	-	1.22	0.50	14.77	2.25	17.02	nr	18.30
900mm wide.....................................	-	-	-	1.49	0.62	18.11	2.66	20.77	nr	22.32
1200mm wide....................................	-	-	-	1.77	0.75	21.60	3.06	24.66	nr	26.51
1500mm wide....................................	-	-	-	2.04	0.88	25.01	3.48	28.49	nr	30.62
1800mm wide....................................	-	-	-	2.32	1.01	28.50	3.89	32.39	nr	34.82
2400mm wide....................................	-	-	-	2.87	1.27	35.41	4.69	40.10	nr	43.10
3000mm wide....................................	-	-	-	3.42	1.52	42.24	5.51	47.75	nr	51.33
1300mm high; plugging and screwing lugs to brick jambs and sill and concrete head										
600mm wide.....................................	-	-	-	1.34	0.54	16.16	2.66	18.82	nr	20.23
900mm wide.....................................	-	-	-	1.66	0.70	20.24	2.91	23.15	nr	24.88
1200mm wide....................................	-	-	-	1.96	0.85	24.06	3.33	27.39	nr	29.44
1500mm wide....................................	-	-	-	2.28	0.99	27.99	3.74	31.73	nr	34.11
1800mm wide....................................	-	-	-	2.59	1.15	31.98	4.14	36.12	nr	38.83
2400mm wide....................................	-	-	-	3.21	1.45	39.81	4.92	44.73	nr	48.08
3000mm wide....................................	-	-	-	3.83	1.75	47.64	5.73	53.37	nr	57.37
1500mm high; plugging and screwing lugs to brick jambs and sill and concrete head										
600mm wide.....................................	-	-	-	1.47	0.59	17.71	2.76	20.47	nr	22.01
900mm wide.....................................	-	-	-	1.82	0.77	22.20	3.18	25.38	nr	27.29
1200mm wide....................................	-	-	-	2.16	0.94	26.54	3.58	30.12	nr	32.37
1500mm wide....................................	-	-	-	2.51	1.11	30.96	4.00	34.96	nr	37.58
1800mm wide....................................	-	-	-	2.85	1.29	35.36	4.40	39.76	nr	42.74
2400mm wide....................................	-	-	-	3.54	1.64	44.19	5.21	49.40	nr	53.10
3000mm wide....................................	-	-	-	4.23	1.98	52.94	6.01	58.95	nr	63.37
2100mm high; plugging and screwing lugs to brick jambs and sill and concrete head										
600mm wide.....................................	-	-	-	1.86	0.74	22.36	3.56	25.92	nr	27.87
900mm wide.....................................	-	-	-	2.31	0.98	28.20	3.96	32.16	nr	34.57
1200mm wide....................................	-	-	-	2.75	1.22	33.95	4.39	38.34	nr	41.21
1500mm wide....................................	-	-	-	3.20	1.46	39.79	4.77	44.56	nr	47.90
1800mm wide....................................	-	-	-	3.65	1.71	45.69	5.17	50.86	nr	54.68
2400mm wide....................................	-	-	-	4.54	2.19	57.28	5.99	63.27	nr	68.01
3000mm wide....................................	-	-	-	5.43	2.67	68.86	6.79	75.65	nr	81.33
Building in wood windows; building in lugs; bedding in cement mortar (1:3), pointing with Secomastic standard mastic one side										
768mm high; lugs to brick jambs										
438mm wide.....................................	-	-	-	1.43	0.09	13.83	1.62	15.45	nr	16.61
641mm wide.....................................	-	-	-	1.64	0.09	15.77	1.91	17.68	nr	19.00
1225mm wide....................................	-	-	-	2.24	0.10	21.38	2.72	24.10	nr	25.90
1809mm wide....................................	-	-	-	2.86	0.11	27.17	3.54	30.71	nr	33.01
2394mm wide....................................	-	-	-	3.46	0.12	32.78	4.37	37.15	nr	39.93
920mm high; lugs to brick jambs										
438mm wide.....................................	-	-	-	1.61	0.09	15.49	1.80	17.29	nr	18.59
641mm wide.....................................	-	-	-	1.82	0.10	17.50	2.08	19.58	nr	21.05
1225mm wide....................................	-	-	-	2.45	0.11	23.39	2.91	26.30	nr	28.27
1809mm wide....................................	-	-	-	3.07	0.13	29.25	3.74	32.99	nr	35.46
2394mm wide....................................	-	-	-	3.70	0.14	35.13	4.55	39.68	nr	42.66
1073mm high; lugs to brick jambs										
438mm wide.....................................	-	-	-	1.80	0.10	17.32	2.02	19.34	nr	20.79
641mm wide.....................................	-	-	-	2.01	0.11	19.32	2.30	21.62	nr	23.25
1225mm wide....................................	-	-	-	2.66	0.12	25.39	3.13	28.52	nr	30.66
1809mm wide....................................	-	-	-	3.30	0.14	31.44	3.93	35.37	nr	38.02
2394mm wide....................................	-	-	-	3.95	0.16	37.58	4.76	42.34	nr	45.52
1225mm high; lugs to brick jambs										
438mm wide.....................................	-	-	-	1.98	0.10	18.98	2.23	21.21	nr	22.80
641mm wide.....................................	-	-	-	2.20	0.11	21.08	2.49	23.57	nr	25.34
1225mm wide....................................	-	-	-	2.87	0.13	27.40	3.32	30.72	nr	33.03
1809mm wide....................................	-	-	-	3.52	0.16	33.61	4.10	37.71	nr	40.54
2394mm wide....................................	-	-	-	4.19	0.18	39.94	4.94	44.88	nr	48.24
1378mm high; lugs to brick jambs										
438mm wide.....................................	-	-	-	2.16	0.10	20.64	2.42	23.06	nr	24.79
641mm wide.....................................	-	-	-	2.39	0.11	22.83	2.70	25.53	nr	27.45
1225mm wide....................................	-	-	-	3.07	0.14	29.32	3.53	32.85	nr	35.31
1809mm wide....................................	-	-	-	3.75	0.17	35.81	4.34	40.15	nr	43.16
2394mm wide....................................	-	-	-	4.43	0.20	42.29	5.16	47.45	nr	51.01

Labour hourly rates: (except Specialists) Craft Operatives £9.23 Labour £7.02 Rates are national average prices. Refer to REGIONAL VARIATIONS for indicative levels of overall pricing in regions	MATERIALS			LABOUR				RATES		
	Del to Site £	Waste %	Material Cost £	Craft Optve Hrs	Lab Hrs	Labour Cost £	Sunds £	Nett Rate £	Unit	Gross Rate (+7.5%) £

F30: ACCESSORIES/SUNDRY ITEMS FOR BRICK/BLOCK/STONE WALLING Cont.

Building in Cont.

Building in fireplace interior with slabbed tile surround and loose hearth tiles for 400mm opening; bed and joint interior in fire cement and fill with concrete at back; plug and screw on surround lugs; bed hearth tiles in cement mortar (1:3) and point in

with firebrick back	-	-	-	5.00	3.50	70.72	4.10	74.82	nr	80.43
with back boiler unit and self contained flue	-	-	-	6.00	4.00	83.46	4.94	88.40	nr	95.03

Proprietary items

Precast prestressed concrete lintels; Stressline Ltd; Roughcast; bedding in cement-lime mortar (1:1:6)

100 x 70mm
900mm long	2.99	2.50	3.06	0.24	0.12	3.06	0.22	6.34	nr	6.82
1050mm long	3.49	2.50	3.58	0.27	0.14	3.47	0.24	7.29	nr	7.84
1200mm long	3.99	2.50	4.09	0.31	0.16	3.98	0.29	8.36	nr	8.99
1500mm long	4.78	2.50	4.90	0.38	0.19	4.84	0.35	10.09	nr	10.85
1800mm long	5.97	2.50	6.12	0.44	0.22	5.61	0.38	12.10	nr	13.01
2100mm long	6.97	2.50	7.14	0.50	0.25	6.37	0.44	13.95	nr	15.00
2400mm long	7.97	2.50	8.17	0.55	0.28	7.04	0.50	15.71	nr	16.89
2700mm long	8.97	2.50	9.19	0.60	0.30	7.64	0.54	17.38	nr	18.68
3000mm long	9.95	2.50	10.20	0.66	0.33	8.41	0.60	19.21	nr	20.65

150 x 70mm
900mm long	4.06	2.50	4.16	0.25	0.13	3.22	0.22	7.60	nr	8.17
1050mm long	4.74	2.50	4.86	0.28	0.14	3.57	0.26	8.69	nr	9.34
1200mm long	5.42	2.50	5.56	0.32	0.16	4.08	0.29	9.92	nr	10.67
1500mm long	6.77	2.50	6.94	0.39	0.20	5.00	0.35	12.29	nr	13.21
1800mm long	8.12	2.50	8.32	0.45	0.23	5.77	0.39	14.48	nr	15.57
2100mm long	9.47	2.50	9.71	0.51	0.26	6.53	0.47	16.71	nr	17.96
2400mm long	10.83	2.50	11.10	0.57	0.29	7.30	0.50	18.90	nr	20.31
2700mm long	12.18	2.50	12.48	0.62	0.31	7.90	0.53	20.91	nr	22.48
3000mm long	13.53	2.50	13.87	0.68	0.34	8.66	0.62	23.15	nr	24.89

225 x 70mm
900mm long	6.26	2.50	6.42	0.26	0.13	3.31	0.22	9.95	nr	10.70
1050mm long	7.30	2.50	7.48	0.29	0.15	3.73	0.26	11.47	nr	12.33
1200mm long	8.35	2.50	8.56	0.33	0.17	4.24	0.29	13.09	nr	14.07
1500mm long	10.44	2.50	10.70	0.40	0.20	5.10	0.35	16.15	nr	17.36
1800mm long	12.53	2.50	12.84	0.46	0.23	5.86	0.39	19.09	nr	20.53
2100mm long	14.62	2.50	14.99	0.52	0.26	6.62	0.47	22.08	nr	23.74
2400mm long	16.70	2.50	17.12	0.58	0.29	7.39	0.53	25.04	nr	26.91
2700mm long	18.78	2.50	19.25	0.63	0.32	8.06	0.59	27.90	nr	29.99
3000mm long	20.86	2.50	21.38	0.69	0.35	8.83	0.62	30.83	nr	33.14

255 x 70mm
900mm long	6.66	2.50	6.83	0.26	0.13	3.31	0.25	10.39	nr	11.17
1050mm long	7.77	2.50	7.96	0.30	0.15	3.82	0.27	12.06	nr	12.96
1200mm long	8.89	2.50	9.11	0.33	0.17	4.24	0.30	13.65	nr	14.68
1500mm long	11.11	2.50	11.39	0.40	0.20	5.10	0.35	16.83	nr	18.10
1800mm long	13.32	2.50	13.65	0.47	0.24	6.02	0.42	20.10	nr	21.60
2100mm long	15.55	2.50	15.94	0.53	0.27	6.79	0.49	23.22	nr	24.96
2400mm long	17.77	2.50	18.21	0.60	0.30	7.64	0.53	26.39	nr	28.37
2700mm long	19.99	2.50	20.49	0.64	0.32	8.15	0.59	29.23	nr	31.43
3000mm long	22.21	2.50	22.77	0.70	0.35	8.92	0.62	32.30	nr	34.73

100 x 145mm
900mm long	6.05	2.50	6.20	0.26	0.13	3.31	0.22	9.73	nr	10.46
1050mm long	7.06	2.50	7.24	0.29	0.15	3.73	0.26	11.23	nr	12.07
1200mm long	8.06	2.50	8.26	0.33	0.17	4.24	0.30	12.80	nr	13.76
1500mm long	10.08	2.50	10.33	0.40	0.20	5.10	0.35	15.78	nr	16.96
1800mm long	12.09	2.50	12.39	0.46	0.23	5.86	0.42	18.67	nr	20.07
2100mm long	14.11	2.50	14.46	0.52	0.26	6.62	0.47	21.56	nr	23.17
2400mm long	16.12	2.50	16.52	0.58	0.29	7.39	0.53	24.44	nr	26.28
2700mm long	18.13	2.50	18.58	0.63	0.32	8.06	0.59	27.23	nr	29.28
3000mm long	20.15	2.50	20.65	0.69	0.35	8.83	0.62	30.10	nr	32.36

Galvanised steel lintels; SUPERGALV (BIRTLEY) lintels reference CB 50; bedding in cement-lime mortar (1:1:6)
140mm deep x 750mm long	13.42	2.50	13.76	0.28	0.28	4.55	-	18.31	nr	19.68
140mm deep x 1200mm long	21.43	2.50	21.97	0.38	0.38	6.17	-	28.14	nr	30.25
140mm deep x 1350mm long	24.98	2.50	25.60	0.41	0.41	6.66	-	32.27	nr	34.69
150mm deep x 1500mm long	27.87	2.50	28.57	0.44	0.44	7.15	-	35.72	nr	38.40
150mm deep x 1650mm long	31.46	2.50	32.25	0.47	0.47	7.64	-	39.88	nr	42.88
140mm deep x 1800mm long	34.31	2.50	35.17	0.50	0.50	8.13	-	43.29	nr	46.54
165mm deep x 1950mm long	37.34	2.50	38.27	0.53	0.53	8.61	-	46.89	nr	50.40
165mm deep x 2100mm long	39.42	2.50	40.41	0.56	0.56	9.10	-	49.51	nr	53.22
190mm deep x 2250mm long	44.37	2.50	45.48	0.60	0.60	9.75	-	55.23	nr	59.37
190mm deep x 2400mm long	47.33	2.50	48.51	0.63	0.63	10.24	-	58.75	nr	63.16
200mm deep x 2550mm long	54.28	2.50	55.64	0.67	0.67	10.89	-	66.52	nr	71.51
200mm deep x 2850mm long	72.08	2.50	73.88	0.71	0.71	11.54	-	85.42	nr	91.83
210mm deep x 3000mm long	77.41	2.50	79.35	0.75	0.75	12.19	-	91.53	nr	98.40
225mm deep x 3300mm long	85.22	2.50	87.35	0.81	0.81	13.16	-	100.51	nr	108.05
215mm deep x 3600mm long	96.79	2.50	99.21	0.88	0.88	14.30	-	113.51	nr	122.02
215mm deep x 3900mm long	115.02	2.50	117.90	0.94	0.94	15.28	-	133.17	nr	143.16

Galvanised steel lintels; SUPERGALV (BIRTLEY) lintels reference CB 50 H.D.; bedding in cement-lime mortar (1:1:6)
165mm deep x 750mm long	15.50	2.50	15.89	0.28	0.28	4.55	-	20.44	nr	21.97
165mm deep x 1050mm long	20.32	2.50	20.83	0.34	0.34	5.53	-	26.35	nr	28.33
165mm deep x 1200mm long	23.64	2.50	24.23	0.38	0.38	6.17	-	30.41	nr	32.69
165mm deep x 1350mm long	28.20	2.50	28.91	0.41	0.41	6.66	-	35.57	nr	38.24
165mm deep x 1500mm long	31.61	2.50	32.40	0.44	0.44	7.15	-	39.55	nr	42.52

MASONRY

Labour hourly rates: (except Specialists) Craft Operatives £9.23 Labour £7.02 Rates are national average prices. Refer to REGIONAL VARIATIONS for indicative levels of overall pricing in regions	MATERIALS			LABOUR				RATES		
	Del to Site £	Waste %	Material Cost £	Craft Optve Hrs	Lab Hrs	Labour Cost £	Sunds £	Nett Rate £	Unit	Gross Rate (+7.5%) £
F30: ACCESSORIES/SUNDRY ITEMS FOR BRICK/BLOCK/STONE WALLING Cont.										
Proprietary items Cont.										
Galvanised steel lintels; SUPERGALV (BIRTLEY) lintels reference CB 50 H.D.; bedding in cement-lime mortar (1:1:6) Cont.										
165mm deep x 1650mm long	36.78	2.50	37.70	0.47	0.47	7.64	–	45.34	nr	48.74
200mm deep x 1800mm long	41.29	2.50	42.32	0.50	0.50	8.13	–	50.45	nr	54.23
200mm deep x 2100mm long	48.71	2.50	49.93	0.56	0.56	9.10	–	59.03	nr	63.45
200mm deep x 2250mm long	60.16	2.50	61.66	0.60	0.60	9.75	–	71.41	nr	76.77
225mm deep x 2400mm long	71.17	2.50	72.95	0.63	0.63	10.24	–	83.19	nr	89.43
215mm deep x 2550mm long	77.33	2.50	79.26	0.67	0.67	10.89	–	90.15	nr	96.91
Galvanised steel lintels; SUPERGALV (BIRTLEY) lintels reference CB 70; bedding in cement-lime mortar (1:1:6)										
140mm deep x 750mm long	13.56	2.50	13.90	0.29	0.29	4.71	–	18.61	nr	20.01
140mm deep x 1200mm long	21.69	2.50	22.23	0.40	0.40	6.50	–	28.73	nr	30.89
140mm deep x 1350mm long	24.95	2.50	25.57	0.43	0.43	6.99	–	32.56	nr	35.00
150mm deep x 1500mm long	27.96	2.50	28.66	0.46	0.46	7.47	–	36.13	nr	38.84
150mm deep x 1650mm long	31.68	2.50	32.47	0.49	0.49	7.96	–	40.43	nr	43.47
140mm deep x 1800mm long	34.56	2.50	35.42	0.53	0.53	8.61	–	44.04	nr	47.34
165mm deep x 1950mm long	38.08	2.50	39.03	0.56	0.56	9.10	–	48.13	nr	51.74
165mm deep x 2100mm long	40.23	2.50	41.24	0.59	0.59	9.59	–	50.82	nr	54.63
190mm deep x 2250mm long	44.63	2.50	45.75	0.63	0.63	10.24	–	55.98	nr	60.18
190mm deep x 2400mm long	47.59	2.50	48.78	0.66	0.66	10.73	–	59.50	nr	63.97
200mm deep x 2550mm long	54.41	2.50	55.77	0.70	0.70	11.38	–	67.15	nr	72.18
200mm deep x 2850mm long	72.28	2.50	74.09	0.75	0.75	12.19	–	86.27	nr	92.75
210mm deep x 3000mm long	77.65	2.50	79.59	0.79	0.79	12.84	–	92.43	nr	99.36
225mm deep x 3300mm long	87.25	2.50	89.43	0.85	0.85	13.81	–	103.24	nr	110.99
215mm deep x 3600mm long	95.91	2.50	98.31	0.92	0.92	14.95	–	113.26	nr	121.75
215mm deep x 3900mm long	125.06	2.50	128.19	0.99	0.99	16.09	–	144.27	nr	155.09
Galvanised steel lintels; SUPERGALV (BIRTLEY) lintels reference CB 70 H.D.; bedding in cement-lime mortar (1:1:6)										
165mm deep x 750mm long	15.50	2.50	15.89	0.29	0.29	4.71	–	20.60	nr	22.15
165mm deep x 1050mm long	20.68	2.50	21.20	0.36	0.36	5.85	–	27.05	nr	29.08
165mm deep x 1200mm long	23.59	2.50	24.18	0.40	0.40	6.50	–	30.68	nr	32.98
165mm deep x 1350mm long	28.56	2.50	29.27	0.43	0.43	6.99	–	36.26	nr	38.98
165mm deep x 1500mm long	31.76	2.50	32.55	0.46	0.46	7.47	–	40.03	nr	43.03
165mm deep x 1650mm long	40.76	2.50	41.78	0.49	0.49	7.96	–	49.74	nr	53.47
200mm deep x 1800mm long	44.37	2.50	45.48	0.53	0.53	8.61	–	54.09	nr	58.15
200mm deep x 2100mm long	51.76	2.50	53.05	0.59	0.59	9.59	–	62.64	nr	67.34
200mm deep x 2250mm long	61.53	2.50	63.07	0.63	0.63	10.24	–	73.31	nr	78.80
225mm deep x 2400mm long	71.52	2.50	73.31	0.66	0.66	10.73	–	84.03	nr	90.34
225mm deep x 2550mm long	77.72	2.50	79.66	0.70	0.70	11.38	–	91.04	nr	97.87
Galvanised steel lintels; SUPERGALV (BIRTLEY) lintels reference CB 50/130; bedding in cement-lime mortar (1:1:6)										
140mm deep x 750mm long	13.46	2.50	13.80	0.31	0.31	5.04	–	18.83	nr	20.25
140mm deep x 1200mm long	21.48	2.50	22.02	0.42	0.42	6.83	–	28.84	nr	31.01
140mm deep x 1350mm long	24.59	2.50	25.20	0.45	0.45	7.31	–	32.52	nr	34.96
150mm deep x 1500mm long	27.55	2.50	28.24	0.48	0.48	7.80	–	36.04	nr	38.74
150mm deep x 1650mm long	31.28	2.50	32.06	0.52	0.52	8.45	–	40.51	nr	43.55
140mm deep x 1800mm long	34.13	2.50	34.98	0.55	0.55	8.94	–	43.92	nr	47.21
165mm deep x 1950mm long	37.16	2.50	38.09	0.58	0.58	9.43	–	47.51	nr	51.08
165mm deep x 2100mm long	40.14	2.50	41.14	0.62	0.62	10.07	–	51.22	nr	55.06
190mm deep x 2250mm long	46.70	2.50	47.87	0.66	0.66	10.73	–	58.59	nr	62.99
190mm deep x 2400mm long	49.81	2.50	51.06	0.69	0.69	11.21	–	62.27	nr	66.94
200mm deep x 2550mm long	54.18	2.50	55.53	0.74	0.74	12.03	–	67.56	nr	72.63
200mm deep x 2850mm long	76.35	2.50	78.26	0.78	0.78	12.68	–	90.93	nr	97.75
210mm deep x 3000mm long	80.37	2.50	82.38	0.83	0.83	13.49	–	95.87	nr	103.06
225mm deep x 3300mm long	88.14	2.50	90.34	0.89	0.89	14.46	–	104.81	nr	112.67
215mm deep x 3600mm long	97.07	2.50	99.50	0.97	0.97	15.76	–	115.26	nr	123.90
215mm deep x 3900mm long	139.40	2.50	142.88	1.03	1.03	16.74	–	159.62	nr	171.59
Galvanised steel lintels; SUPERGALV (BIRTLEY) lintels reference CB 50/130 H.D.; bedding in cement-lime mortar (1:1:6)										
165mm deep x 750mm long	18.83	2.50	19.30	0.31	0.31	5.04	–	24.34	nr	26.16
165mm deep x 1200mm long	28.18	2.50	28.88	0.42	0.42	6.83	–	35.71	nr	38.39
165mm deep x 1350mm long	35.27	2.50	36.15	0.45	0.45	7.31	–	43.46	nr	46.72
165mm deep x 1500mm long	38.85	2.50	39.82	0.48	0.48	7.80	–	47.62	nr	51.19
165mm deep x 1650mm long	44.51	2.50	45.62	0.52	0.52	8.45	–	54.07	nr	58.13
200mm deep x 1800mm long	48.17	2.50	49.37	0.55	0.55	8.94	–	58.31	nr	62.69
200mm deep x 1950mm long	59.99	2.50	61.49	0.58	0.58	9.43	–	70.91	nr	76.23
200mm deep x 2250mm long	69.01	2.50	70.74	0.66	0.66	10.73	–	81.46	nr	87.57
225mm deep x 2400mm long	74.59	2.50	76.45	0.69	0.69	11.21	–	87.67	nr	94.24
225mm deep x 2550mm long	81.78	2.50	83.82	0.74	0.74	12.03	–	95.85	nr	103.04
225mm deep x 2700mm long	86.39	2.50	88.55	0.80	0.80	13.00	–	101.55	nr	109.17
Galvanised steel lintels; SUPERGALV (BIRTLEY) lintels reference CB 50/150; bedding in cement-lime mortar (1:1:6)										
140mm deep x 750mm long	17.13	2.50	17.56	0.31	0.31	5.04	–	22.60	nr	24.29
140mm deep x 1200mm long	27.68	2.50	28.37	0.42	0.42	6.83	–	35.20	nr	37.84
140mm deep x 1350mm long	31.49	2.50	32.28	0.45	0.45	7.31	–	39.59	nr	42.56
150mm deep x 1500mm long	34.78	2.50	35.65	0.48	0.48	7.80	–	43.45	nr	46.71
150mm deep x 1650mm long	41.88	2.50	42.93	0.52	0.52	8.45	–	51.38	nr	55.23
140mm deep x 1800mm long	43.74	2.50	44.83	0.55	0.55	8.94	–	53.77	nr	57.80
165mm deep x 1950mm long	47.07	2.50	48.25	0.58	0.58	9.43	–	57.67	nr	62.00
165mm deep x 2100mm long	50.97	2.50	52.24	0.62	0.62	10.07	–	62.32	nr	66.99

Labour hourly rates: (except Specialists) Craft Operatives £9.23 Labour £7.02 Rates are national average prices. Refer to REGIONAL VARIATIONS for indicative levels of overall pricing in regions	MATERIALS			LABOUR				RATES		
	Del to Site £	Waste %	Material Cost £	Craft Optve Hrs	Lab Hrs	Labour Cost £	Sunds £	Nett Rate £	Unit	Gross Rate (+7.5%) £
F30: ACCESSORIES/SUNDRY ITEMS FOR BRICK/BLOCK/STONE WALLING Cont.										
Proprietary items Cont.										
Galvanised steel lintels; SUPERGALV (BIRTLEY) lintels reference CB 50/150; bedding in cement-lime mortar (1:1:6) Cont.										
190mm deep x 2250mm long	57.25	2.50	58.68	0.66	0.66	10.73	–	69.41	nr	74.61
190mm deep x 2400mm long	65.50	2.50	67.14	0.69	0.69	11.21	–	78.35	nr	84.23
200mm deep x 2550mm long	71.58	2.50	73.37	0.74	0.74	12.03	–	85.39	nr	91.80
200mm deep x 2850mm long	90.52	2.50	92.78	0.78	0.78	12.68	–	105.46	nr	113.37
210mm deep x 3000mm long	95.58	2.50	97.97	0.83	0.83	13.49	–	111.46	nr	119.82
225mm deep x 3300mm long	106.11	2.50	108.76	0.89	0.89	14.46	–	123.23	nr	132.47
Galvanised steel lintels; SUPERGALV (BIRTLEY) lintels reference Box 50; bedding in cement-lime mortar (1:1:6)										
215mm deep x 3900mm long	124.82	2.50	127.94	1.02	1.02	16.57	–	144.52	nr	155.35
215mm deep x 4200mm long	138.81	2.50	142.28	1.05	1.05	17.06	–	159.34	nr	171.29
215mm deep x 4500mm long	160.45	2.50	164.46	1.09	1.09	17.71	–	182.17	nr	195.84
215mm deep x 4800mm long	167.00	2.50	171.18	1.12	1.12	18.20	–	189.38	nr	203.58
Galvanised steel lintels; SUPERGALV (BIRTLEY) lintels reference Box 50 H.D.; bedding in cement-lime mortar (1:1:6)										
215mm deep x 3300mm long	124.09	2.50	127.19	0.85	0.85	13.81	–	141.00	nr	151.58
215mm deep x 3600mm long	135.37	2.50	138.75	0.92	0.92	14.95	–	153.70	nr	165.23
Galvanised steel lintels; SUPERGALV (BIRTLEY) lintels reference SB100; bedding in cement-lime mortar (1:1:6)										
75mm deep x 750mm long	9.08	2.50	9.31	0.45	0.45	7.31	–	16.62	nr	17.87
75mm deep x 900mm long	10.76	2.50	11.03	0.48	0.48	7.80	–	18.83	nr	20.24
75mm deep x 1050mm long	12.56	2.50	12.87	0.51	0.51	8.29	–	21.16	nr	22.75
75mm deep x 1200mm long	14.17	2.50	14.52	0.54	0.54	8.78	–	23.30	nr	25.05
75mm deep x 1500mm long	17.74	2.50	18.18	0.60	0.60	9.75	–	27.93	nr	30.03
150mm deep x 1800mm long	26.49	2.50	27.15	0.65	0.65	10.56	–	37.71	nr	40.54
150mm deep x 1950mm long	30.15	2.50	30.90	0.69	0.69	11.21	–	42.12	nr	45.27
150mm deep x 2100mm long	32.01	2.50	32.81	0.73	0.73	11.86	–	44.67	nr	48.02
150mm deep x 2250mm long	34.44	2.50	35.30	0.77	0.77	12.51	–	47.81	nr	51.40
150mm deep x 2400mm long	37.11	2.50	38.04	0.82	0.82	13.32	–	51.36	nr	55.21
150mm deep x 2550mm long	40.05	2.50	41.05	0.87	0.87	14.14	–	55.19	nr	59.33
150mm deep x 2700mm long	42.55	2.50	43.61	0.92	0.92	14.95	–	58.56	nr	62.96
225mm deep x 2850mm long	64.37	2.50	65.98	0.93	0.93	15.11	–	81.09	nr	87.17
225mm deep x 3000mm long	68.23	2.50	69.94	0.95	0.95	15.44	–	85.37	nr	91.78
Galvanised steel lintels; SUPERGALV (BIRTLEY) lintels reference SB140; bedding in cement-lime mortar (1:1:6)										
75mm deep x 750mm long	11.31	2.50	11.59	0.45	0.45	7.31	–	18.91	nr	20.32
75mm deep x 900mm long	13.46	2.50	13.80	0.48	0.48	7.80	–	21.60	nr	23.22
75mm deep x 1050mm long	15.86	2.50	16.26	0.51	0.51	8.29	–	24.54	nr	26.38
75mm deep x 1200mm long	18.40	2.50	18.86	0.54	0.54	8.78	–	27.64	nr	29.71
75mm deep x 1350mm long	20.58	2.50	21.09	0.57	0.57	9.26	–	30.36	nr	32.63
75mm deep x 1500mm long	23.71	2.50	24.30	0.60	0.60	9.75	–	34.05	nr	36.61
150mm deep x 1650mm long	28.07	2.50	28.77	0.62	0.62	10.07	–	38.85	nr	41.76
150mm deep x 1800mm long	30.96	2.50	31.73	0.65	0.65	10.56	–	42.30	nr	45.47
150mm deep x 1950mm long	34.17	2.50	35.02	0.69	0.69	11.21	–	46.24	nr	49.70
150mm deep x 2100mm long	37.08	2.50	38.01	0.73	0.73	11.86	–	49.87	nr	53.61
150mm deep x 2250mm long	41.32	2.50	42.35	0.77	0.77	12.51	–	54.87	nr	58.98
150mm deep x 2400mm long	43.54	2.50	44.63	0.82	0.82	13.32	–	57.95	nr	62.30
150mm deep x 2550mm long	45.97	2.50	47.12	0.87	0.87	14.14	–	61.26	nr	65.85
150mm deep x 2700mm long	48.67	2.50	49.89	0.92	0.92	14.95	–	64.84	nr	69.70
225mm deep x 2850mm long	68.57	2.50	70.28	0.95	0.95	15.44	–	85.72	nr	92.15
225mm deep x 3000mm long	71.38	2.50	73.16	0.97	0.97	15.76	–	88.93	nr	95.60
Galvanised steel lintels; SUPERGALV (BIRTLEY) lintels reference SBL200; bedding in cement-lime mortar (1:1:6)										
150mm deep x 750mm long	15.34	2.50	15.72	0.45	0.45	7.31	–	23.04	nr	24.76
150mm deep x 900mm long	18.40	2.50	18.86	0.48	0.48	7.80	–	26.66	nr	28.66
150mm deep x 1050mm long	21.38	2.50	21.91	0.51	0.51	8.29	–	30.20	nr	32.47
150mm deep x 1200mm long	24.35	2.50	24.96	0.54	0.54	8.78	–	33.73	nr	36.26
150mm deep x 1350mm long	27.31	2.50	27.99	0.60	0.60	9.75	–	37.74	nr	40.57
150mm deep x 1500mm long	31.39	2.50	32.17	0.65	0.65	10.56	–	42.74	nr	45.94
150mm deep x 1650mm long	35.27	2.50	36.15	0.67	0.67	10.89	–	47.04	nr	50.57
150mm deep x 1800mm long	38.48	2.50	39.44	0.72	0.72	11.70	–	51.14	nr	54.98
150mm deep x 1950mm long	41.68	2.50	42.72	0.76	0.76	12.35	–	55.07	nr	59.20
150mm deep x 2100mm long	45.06	2.50	46.19	0.80	0.80	13.00	–	59.19	nr	63.63
150mm deep x 2250mm long	51.66	2.50	52.95	0.77	0.77	12.51	–	65.46	nr	70.37
150mm deep x 2400mm long	54.79	2.50	56.16	0.82	0.82	13.32	–	69.48	nr	74.70
150mm deep x 2550mm long	58.19	2.50	59.64	0.87	0.87	14.14	–	73.78	nr	79.32
150mm deep x 2700mm long	60.85	2.50	62.37	0.92	0.92	14.95	–	77.32	nr	83.12
Galvanised steel lintels; SUPERGALV (BIRTLEY) internal door lintels reference INT100; bedding in cement-lime mortar (1:1:6)										
100mm wide, 900mm long	3.22	2.50	3.30	0.33	0.33	5.36	–	8.66	nr	9.31
100mm wide, 1050mm long	3.69	2.50	3.78	0.36	0.36	5.85	–	9.63	nr	10.35
100mm wide, 1200mm long	4.09	2.50	4.19	0.38	0.38	6.17	–	10.37	nr	11.14
Stainless steel Furfix wall extension profiles; single flange; plugging and screwing to brickwork and building ties in to joints of new walls										
100mm	10.95	5.00	11.50	0.30	–	2.77	–	14.27	m	15.34

MASONRY

MASONRY – MAJOR WORKS

Labour hourly rates: (except Specialists) Craft Operatives £9.23 Labour £7.02. Rates are national average prices. Refer to REGIONAL VARIATIONS for indicative levels of overall pricing in regions	MATERIALS			LABOUR				RATES		
	Del to Site £	Waste %	Material Cost £	Craft Optve Hrs	Lab Hrs	Labour Cost £	Sunds £	Nett Rate £	Unit	Gross Rate (+7.5%) £
F30: ACCESSORIES/SUNDRY ITEMS FOR BRICK/BLOCK/STONE WALLING Cont.										
Proprietary items Cont.										
Stainless steel Furfix wall extension profiles; single flange; plugging and screwing to brickwork and building ties in to joints of new walls Cont.										
150mm	13.78	5.00	14.47	0.35	–	3.23	–	17.70	m	19.03
215mm	19.81	5.00	20.80	0.40	–	3.69	–	24.49	m	26.33
Harris and Edgar Ltd; Hemax restraint channel ref. 36/8; fixing to steelwork with Hemax M8 screws complete with plate washer at 450mm centres; incorporating ties at 450mm centres										
100mm long ties reference BP36 fishtail	8.14	5.00	8.55	0.50	0.25	6.37	0.63	15.55	m	16.71
125mm long ties reference BP36 fishtail	8.23	5.00	8.64	0.50	0.25	6.37	0.74	15.75	m	16.93
150mm long ties reference BP36 fishtail	8.30	5.00	8.71	0.50	0.25	6.37	0.84	15.93	m	17.12
Thermabate insulated cavity closers; fixing to timber with nails										
Thermabate 50; vertical	4.66	5.00	4.89	0.33	0.16	4.17	–	9.06	m	9.74
Thermabate 75; vertical	5.05	5.00	5.30	0.33	0.16	4.17	–	9.47	m	10.18
Thermabate 50; horizontal	4.66	5.00	4.89	0.33	0.16	4.17	–	9.06	m	9.74
Thermabate 75; horizontal	5.05	5.00	5.30	0.33	0.16	4.17	–	9.47	m	10.18
Thermabate insulated cavity closers; fixing to masonry with PVC-U ties at 225mm centres										
Thermabate 50; vertical	5.28	5.00	5.54	0.33	0.16	4.17	–	9.71	m	10.44
Thermabate 75; vertical	5.67	5.00	5.95	0.33	0.16	4.17	–	10.12	m	10.88
Thermabate 50; horizontal	5.28	5.00	5.54	0.33	0.16	4.17	–	9.71	m	10.44
Thermabate 75; horizontal	5.67	5.00	5.95	0.33	0.16	4.17	–	10.12	m	10.88
F31: PRECAST CONCRETE SILLS/LINTELS/COPINGS/FEATURES										
Precast concrete sills; B.S.5642 Part 1; bedding in cement lime mortar (1:1:6)										
Sills; figure 2 or figure 4										
50 x 150 x 300mm splayed and grooved	1.07	2.50	1.10	0.15	0.15	2.44	0.11	3.64	nr	3.92
50 x 150 x 400mm splayed and grooved	1.40	2.50	1.44	0.20	0.20	3.25	0.13	4.82	nr	5.18
50 x 150 x 700mm splayed and grooved	2.47	2.50	2.53	0.30	0.30	4.88	0.19	7.60	nr	8.17
50 x 150 x 1300mm splayed and grooved	4.53	2.50	4.64	0.50	0.50	8.13	0.28	13.05	nr	14.03
extra for stooled end	1.27	–	1.27	–	–	–	–	1.27	nr	1.37
75 x 150 x 300mm splayed and grooved	1.41	2.50	1.45	0.17	0.17	2.76	0.11	4.32	nr	4.64
75 x 150 x 400mm splayed and grooved	1.88	2.50	1.93	0.22	0.22	3.58	0.13	5.63	nr	6.05
75 x 150 x 700mm splayed and grooved	3.24	2.50	3.32	0.35	0.35	5.69	0.19	9.20	nr	9.89
75 x 150 x 1300mm splayed and grooved	6.11	2.50	6.26	0.55	0.55	8.94	0.28	15.48	nr	16.64
extra for stooled end	1.38	–	1.38	–	–	–	–	1.38	nr	1.48
100 x 150 x 300mm splayed and grooved	1.84	2.50	1.89	0.20	0.20	3.25	0.11	5.25	nr	5.64
100 x 150 x 400mm splayed and grooved	2.39	2.50	2.45	0.25	0.25	4.06	0.13	6.64	nr	7.14
100 x 150 x 700mm splayed and grooved	3.31	2.50	3.39	0.40	0.40	6.50	0.19	10.08	nr	10.84
100 x 150 x 1300mm splayed and grooved	7.72	2.50	7.91	0.60	0.60	9.75	0.28	17.94	nr	19.29
extra for stooled end	1.48	–	1.48	–	–	–	–	1.48	nr	1.59
Precast concrete copings B.S.5642 Part 2; bedding in cement-lime mortar (1:1:6)										
Copings; figure 1; horizontal										
75 x 200mm; splayed; rebated joints	4.28	2.50	4.39	0.25	0.13	3.22	0.16	7.77	m	8.35
extra; stopped ends	1.16	–	1.16	–	–	–	–	1.16	nr	1.25
extra; internal angles	2.48	–	2.48	–	–	–	–	2.48	nr	2.67
100 x 300mm; splayed; rebated joints	8.12	2.50	8.32	0.50	0.25	6.37	0.25	14.94	m	16.06
extra; stopped ends	1.55	–	1.55	–	–	–	–	1.55	nr	1.67
extra; internal angles	3.08	–	3.08	–	–	–	–	3.08	nr	3.31
75 x 200mm; saddleback; rebated joints	4.28	2.50	4.39	0.25	0.13	3.22	0.16	7.77	m	8.35
extra; hipped ends	1.16	–	1.16	–	–	–	–	1.16	nr	1.25
extra; internal angles	2.48	–	2.48	–	–	–	–	2.48	nr	2.67
100 x 300mm; saddleback; rebated joints	8.25	2.50	8.46	0.50	0.25	6.37	0.25	15.08	m	16.21
extra; hipped ends	1.60	–	1.60	–	–	–	–	1.60	nr	1.72
extra; internal angles	3.25	–	3.25	–	–	–	–	3.25	nr	3.49
Precast concrete lintels; B.S.5328, designed mix C25, 20mm aggregate minimum cement content 360 kg/m3; vibrated; reinforcement bars B.S.4449 hot rolled plain round mild steel; bedding in cement-lime mortar (1:1:6)										
Lintels										
75 x 150 x 1200mm; reinforced with 1.20 kg of 12mm bars	4.26	2.50	4.37	0.18	0.09	2.29	0.11	6.77	nr	7.28
75 x 150 x 1650mm; reinforced with 1.60 kg of 12mm bars	5.83	2.50	5.98	0.25	0.13	3.22	0.14	9.34	nr	10.04
102 x 150 x 1200mm; reinforced with 1.20 kg of 12mm bars	5.46	2.50	5.60	0.30	0.15	3.82	0.12	9.54	nr	10.25
102 x 150 x 1650mm; reinforced with 1.60 kg of 12mm bars	7.43	2.50	7.62	0.41	0.20	5.19	0.14	12.94	nr	13.91
215 x 150 x 1200mm; reinforced with 2.40 kg of 12mm bars	11.31	2.50	11.59	0.60	0.30	7.64	0.18	19.42	nr	20.87
215 x 150 x 1650mm; reinforced with 3.20 kg of 12mm bars	14.90	2.50	15.27	0.83	0.41	10.54	0.24	26.05	nr	28.01
215 x 225 x 1200mm; reinforced with 2.40 kg of 12mm bars	15.17	2.50	15.55	0.78	0.39	9.94	0.19	25.68	nr	27.60
215 x 225 x 1650mm; reinforced with 3.20 kg of 12mm bars	20.70	2.50	21.22	1.07	0.53	13.60	0.25	35.06	nr	37.69

Labour hourly rates: (except Specialists) Craft Operatives £9.23 Labour £7.02 Rates are national average prices. Refer to REGIONAL VARIATIONS for indicative levels of overall pricing in regions	MATERIALS			LABOUR				RATES		
	Del to Site £	Waste %	Material Cost £	Craft Optve Hrs	Lab Hrs	Labour Cost £	Sunds £	Nett Rate £	Unit	Gross Rate (+7.5%) £
F31: PRECAST CONCRETE SILLS/LINTELS/COPINGS/FEATURES Cont.										
Precast concrete lintels; B.S.5328, designed mix C25, 20mm aggregate minimum cement content 360 kg/m3; vibrated; reinforcement bars B.S.4449 hot rolled plain round mild steel; bedding in cement-lime mortar (1:1:6) Cont.										
Lintels Cont.										
327 x 150 x 1200mm; reinforced with 3.60 kg of 12mm bars	16.78	2.50	17.20	0.78	0.39	9.94	0.19	27.33	nr	29.38
327 x 150 x 1650mm; reinforced with 4.80 kg of 12mm bars	22.92	2.50	23.49	1.07	0.54	13.67	0.25	37.41	nr	40.22
327 x 225 x 1200mm; reinforced with 3.60 kg of 12mm bars	24.10	2.50	24.70	1.20	0.60	15.29	0.23	40.22	nr	43.24
327 x 225 x 1650mm; reinforced with 4.80 kg of 12mm bars	32.91	2.50	33.73	1.65	0.83	21.06	0.29	55.08	nr	59.21
75 x 150 x 1200mm; reinforced with 1.20 kg of 12mm bars; fair finish two faces	5.40	2.50	5.54	0.18	0.04	1.94	0.11	7.59	nr	8.16
75 x 150 x 1650mm; reinforced with 1.60 kg of 12mm bars; fair finish two faces	7.37	2.50	7.55	0.25	0.13	3.22	0.14	10.91	nr	11.73
102 x 150 x 1200mm; reinforced with 1.20 kg of 12mm bars; fair finish two faces	6.58	2.50	6.74	0.30	0.15	3.82	0.11	10.68	nr	11.48
102 x 150 x 1650mm; reinforced with 1.60 kg of 12mm bars; fair finish two faces	8.97	2.50	9.19	0.41	0.20	5.19	0.14	14.52	nr	15.61
215 x 150 x 1200mm; reinforced with 2.40 kg of 12mm bars; fair finish two faces	12.43	2.50	12.74	0.60	0.30	7.64	0.18	20.56	nr	22.11
215 x 150 x 1650mm; reinforced with 3.20 kg of 12mm bars; fair finish two faces	16.45	2.50	16.86	0.83	0.41	10.54	0.24	27.64	nr	29.71
215 x 225 x 1200mm; reinforced with 2.40 kg of 12mm bars; fair finish two faces	17.24	2.50	17.67	0.78	0.39	9.94	0.19	27.80	nr	29.88
215 x 225 x 1650mm; reinforced with 3.20 kg of 12mm bars; fair finish two faces	22.95	2.50	23.52	1.07	0.53	13.60	0.25	37.37	nr	40.17
327 x 150 x 1200mm; reinforced with 3.60 kg of 12mm bars; fair finish two faces	18.32	2.50	18.78	0.78	0.39	9.94	0.19	28.91	nr	31.07
327 x 150 x 1650mm; reinforced with 4.80 kg of 12mm bars; fair finish two faces	24.42	2.50	25.03	1.07	0.54	13.67	0.25	38.95	nr	41.87
327 x 225 x 1200mm; reinforced with 3.60 kg of 12mm bars; fair finish two faces	26.40	2.50	27.06	1.20	0.60	15.29	0.22	42.57	nr	45.76
327 x 225 x 1650mm; reinforced with 4.80 kg of 12mm bars; fair finish two faces	35.18	2.50	36.06	1.65	0.83	21.06	0.33	57.45	nr	61.75
Boot lintels										
215 x 150 x 1200mm; reinforced with 3.60 kg of 12mm bars and 0.51 kg of 6mm links	10.67	2.50	10.94	0.60	0.30	7.64	0.17	18.75	nr	20.16
215 x 150 x 1800mm; reinforced with 5.19 kg of 12mm bars and 0.77 kg of 6mm links	15.96	2.50	16.36	0.90	0.30	10.41	0.25	27.02	nr	29.05
215 x 225 x 1200mm; reinforced with 3.60 kg of 12mm bars and 0.62 kg of 6mm links	14.89	2.50	15.26	0.78	0.45	10.36	0.18	25.80	nr	27.74
215 x 225 x 1800mm; reinforced with 5.19 kg of 12mm bars and 0.93 kg of 6mm links	22.39	2.50	22.95	1.18	0.39	13.63	0.26	36.84	nr	39.60
252 x 150 x 1200mm; reinforced with 3.60 kg of 12mm bars and 0.56 kg of 6mm links	11.35	2.50	11.63	0.72	0.59	10.79	0.22	22.64	nr	24.34
252 x 150 x 1800mm; reinforced with 5.19 kg of 12mm bars and 0.83 kg of 6mm links	18.37	2.50	18.83	1.08	0.36	12.50	0.30	31.62	nr	34.00
252 x 225 x 1200mm; reinforced with 3.60 kg of 12mm bars and 0.67 kg of 6mm links	17.70	2.50	18.14	0.90	0.54	12.10	0.22	30.46	nr	32.74
252 x 225 x 1800mm; reinforced with 5.19 kg of 12mm bars and 1.00 kg of 6mm links	25.15	2.50	25.78	1.35	0.39	15.20	0.31	41.29	nr	44.38
Precast concrete padstones; B.S.5328, designed mix C25, 20mm aggregate, minimum cement content 360 kg/m3; vibrated; reinforced at contractor's discretion; bedding in cement-lime mortar (1:1:6)										
Padstones										
215 x 215 x 75mm	2.13	2.50	2.18	0.06	0.03	0.76	0.11	3.06	nr	3.29
215 x 215 x 150mm	4.14	2.50	4.24	0.08	0.04	1.02	0.11	5.37	nr	5.78
327 x 215 x 150mm	6.41	2.50	6.57	0.11	0.05	1.37	0.12	8.06	nr	8.66
327 x 327 x 150mm	7.72	2.50	7.91	0.17	0.08	2.13	0.13	10.17	nr	10.94
440 x 215 x 150mm	6.84	2.50	7.01	0.17	0.08	2.13	0.14	9.28	nr	9.98
440 x 327 x 150mm	10.39	2.50	10.65	0.20	0.10	2.55	0.15	13.35	nr	14.35
440 x 440 x 150mm	11.31	2.50	11.59	0.20	0.10	2.55	0.16	14.30	nr	15.37

MASONRY

Spence Geddes'
Estimating for
Building and Civil Engineering Works

This is a fully updated edition of **Spence Geddes'** standard reference publication on estimating. The opportunity has been taken to make certain alterations whilst maintaining the essential features of earlier editions.

It deals in a practical and reasonable way with many of the estimating problems which can arise where building and civil engineering works are carried out and to include comprehensive estimating data within the guidelines of good practice.

£35.00, Hardback, 1996, 448pp, isbn 0 7506 2797 2

TO ORDER

CREDIT CARD HOTLINE - 01865 888180

BY MAIL: Technical Marketing Dept., Butterworth-Heinemann, FREEPOST, Oxford OX2 8BR

BY FAX: Heinemann Customer Services – 01865 314091

BY EMAIL: Send orders to: bhuk.orders@repp.co.uk

Structural/Carcassing metal/timber

Labour hourly rates: (except Specialists) Craft Operatives £9.23 Labourer £7.02 Rates are national average prices. Refer to REGIONAL VARIATIONS for indicative levels of overall pricing in regions	MATERIALS			LABOUR				RATES		
	Del to Site £	Waste %	Material Cost £	Craft Optve Hrs	Lab Hrs	Labour Cost £	Sunds £	Nett Rate £	Unit	Gross Rate (+7.5%) £
G10: STRUCTURAL STEEL FRAMING										
Prices generally for fabricated steelwork										

Note
    prices can vary considerably dependent upon the
    character of the work; the following are average
    prices for work fabricated and delivered to site
Framing, fabrication; including shop and site black
bolts, nuts and washers for structural framing to
structural framing connections										

Weldable steel, B.S.EN 10025 : 1993 Grade S275JR (formerly B.S.4360 Grade 43B), hot rolled sections B.S.4 Part 1 (Euronorm 54); welded fabrication in accordance with B.S.5950 Part 2										
Columns										
weight less than 40 Kg/m......................	850.00	-	850.00	-	-	-	-	850.00	t	913.75
weight 40-100 Kg/m.......................	850.00	-	850.00	-	-	-	-	850.00	t	913.75
weight exceeding 100 Kg/m..................	850.00	-	850.00	-	-	-	-	850.00	t	913.75
Beams										
weight less than 40 Kg/m......................	850.00	-	850.00	-	-	-	-	850.00	t	913.75
weight 40-100 Kg/m.......................	850.00	-	850.00	-	-	-	-	850.00	t	913.75
weight exceeding 100 Kg/m..................	850.00	-	850.00	-	-	-	-	850.00	t	913.75
Beams, castellated										
weight less than 40 Kg/m......................	1050.00	-	1050.00	-	-	-	-	1050.00	t	1128.75
weight 40-100 Kg/m.......................	1050.00	-	1050.00	-	-	-	-	1050.00	t	1128.75
weight exceeding 100 Kg/m..................	1050.00	-	1050.00	-	-	-	-	1050.00	t	1128.75
Beams, curved										
weight less than 40 Kg/m......................	1300.00	-	1300.00	-	-	-	-	1300.00	t	1397.50
weight 40-100 Kg/m.......................	1300.00	-	1300.00	-	-	-	-	1300.00	t	1397.50
weight exceeding 100 Kg/m..................	1300.00	-	1300.00	-	-	-	-	1300.00	t	1397.50
Bracings, tubular										
weight less than 40 Kg/m......................	900.00	-	900.00	-	-	-	-	900.00	t	967.50
weight 40-100 Kg/m.......................	900.00	-	900.00	-	-	-	-	900.00	t	967.50
weight exceeding 100 Kg/m..................	900.00	-	900.00	-	-	-	-	900.00	t	967.50
Purlins and cladding rails										
weight less than 40 Kg/m......................	870.00	-	870.00	-	-	-	-	870.00	t	935.25
weight 40-100 Kg/m.......................	870.00	-	870.00	-	-	-	-	870.00	t	935.25
weight exceeding 100 Kg/m..................	870.00	-	870.00	-	-	-	-	870.00	t	935.25
portal frames....................	800.00	-	800.00	-	-	-	-	800.00	t	860.00
trusses 12m to 18m span.....................	1350.00	-	1350.00	-	-	-	-	1350.00	t	1451.25
trusses 6m to 12m span.....................	1400.00	-	1400.00	-	-	-	-	1400.00	t	1505.00
trusses up to 6m span.....................	1500.00	-	1500.00	-	-	-	-	1500.00	t	1612.50
Weldable steel B.S.EN 10210 : 1994 Grade S275JOH (formerly B.S.4360 Grade 43C), hot rolled sections B.S.4848 Part2 (Euronorm 57); welded fabrication in accordance with B.S.5950 Part 2										
column, square hollow section										
weight less than 40 Kg/m......................	990.00	-	990.00	-	-	-	-	990.00	t	1064.25
weight 40-100 Kg/m.......................	990.00	-	990.00	-	-	-	-	990.00	t	1064.25
weight exceeding 100 Kg/m..................	990.00	-	990.00	-	-	-	-	990.00	t	1064.25
column, rectangular hollow section										
weight less than 40 Kg/m......................	1005.00	-	1005.00	-	-	-	-	1005.00	t	1080.38
weight 40-100 Kg/m.......................	1005.00	-	1005.00	-	-	-	-	1005.00	t	1080.38
weight exceeding 100 Kg/m..................	1005.00	-	1005.00	-	-	-	-	1005.00	t	1080.38
Holding down bolts or assemblies; mild steel										
rag or indented bolts; M 10 with nuts and washers										
100mm long.....................................	1.04	5.00	1.09	0.10	-	0.92	-	2.02	nr	2.17
160mm long.....................................	1.08	5.00	1.13	0.11	-	1.02	-	2.15	nr	2.31
rag or indented bolts; M 12 with nuts and washers										
100mm long.....................................	1.12	5.00	1.18	0.10	-	0.92	-	2.10	nr	2.26
160mm long.....................................	1.30	5.00	1.37	0.13	-	1.20	-	2.56	nr	2.76
200mm long.....................................	1.46	5.00	1.53	0.14	-	1.29	-	2.83	nr	3.04
rag or indented bolts; M 16 with nuts and washers										
120mm long.....................................	1.69	5.00	1.77	0.11	-	1.02	-	2.79	nr	3.00
160mm long.....................................	1.94	5.00	2.04	0.14	-	1.29	-	3.33	nr	3.58
200mm long.....................................	2.26	5.00	2.37	0.16	-	1.48	-	3.85	nr	4.14

Labour hourly rates: (except Specialists) Craft Operatives £9.23 Labourer £7.02 Rates are national average prices. Refer to REGIONAL VARIATIONS for indicative levels of overall pricing in regions	MATERIALS			LABOUR				RATES		
	Del to Site £	Waste %	Material Cost £	Craft Optve Hrs	Lab Hrs	Labour Cost £	Sunds £	Nett Rate £	Unit	Gross Rate (+7.5%) £
G10: STRUCTURAL STEEL FRAMING Cont.										
Framing, fabrication; including shop and site black bolts, nuts and washers for structural framing to structural framing connections Cont.										
Holding down bolts or assemblies; mild steel Cont. holding down bolt assembly; M 20 bolt with 100 x 100 x 10mm plate washer tack welded to head; with nuts and washers										
300mm long	3.90	5.00	4.09	0.28	–	2.58	–	6.68	nr	7.18
350mm long	4.45	5.00	4.67	0.32	–	2.95	–	7.63	nr	8.20
400mm long	5.05	5.00	5.30	0.35	–	3.23	–	8.53	nr	9.17
450mm long	5.78	5.00	6.07	0.55	–	5.08	–	11.15	nr	11.98
High strength friction grip bolts; B.S.4395 Part 1 - general grade										
M 16; with nuts and washers										
50mm long	0.63	5.00	0.66	0.08	–	0.74	–	1.40	nr	1.50
60mm long	0.69	5.00	0.72	0.08	–	0.74	–	1.46	nr	1.57
70mm long	0.74	5.00	0.78	0.08	–	0.74	–	1.52	nr	1.63
80mm long	0.80	5.00	0.84	0.09	–	0.83	–	1.67	nr	1.80
90mm long	0.85	5.00	0.89	0.09	–	0.83	–	1.72	nr	1.85
100mm long	0.88	5.00	0.92	0.10	–	0.92	–	1.85	nr	1.99
120mm long	0.97	5.00	1.02	0.10	–	0.92	–	1.94	nr	2.09
140mm long	1.10	5.00	1.16	0.11	–	1.02	–	2.17	nr	2.33
150mm long	1.11	5.00	1.17	0.12	–	1.11	–	2.27	nr	2.44
160mm long	1.23	5.00	1.29	0.13	–	1.20	–	2.49	nr	2.68
180mm long	1.39	5.00	1.46	0.13	–	1.20	–	2.66	nr	2.86
200mm long	3.53	5.00	3.71	0.14	–	1.29	–	5.00	nr	5.37
220mm long	3.75	5.00	3.94	0.14	–	1.29	–	5.23	nr	5.62
M 20; with nuts and washers										
60mm long	1.20	5.00	1.26	0.08	–	0.74	–	2.00	nr	2.15
70mm long	1.28	5.00	1.34	0.08	–	0.74	–	2.08	nr	2.24
80mm long	1.33	5.00	1.40	0.09	–	0.83	–	2.23	nr	2.39
90mm long	1.40	5.00	1.47	0.10	–	0.92	–	2.39	nr	2.57
100mm long	1.45	5.00	1.52	0.10	–	0.92	–	2.45	nr	2.63
120mm long	1.62	5.00	1.70	0.11	–	1.02	–	2.72	nr	2.92
140mm long	1.82	5.00	1.91	0.12	–	1.11	–	3.02	nr	3.24
150mm long	2.02	5.00	2.12	0.13	–	1.20	–	3.32	nr	3.57
160mm long	2.07	5.00	2.17	0.14	–	1.29	–	3.47	nr	3.73
180mm long	2.37	5.00	2.49	0.14	–	1.29	–	3.78	nr	4.06
200mm long	4.10	5.00	4.30	0.15	–	1.38	–	5.69	nr	6.12
220mm long	4.43	5.00	4.65	0.17	–	1.57	–	6.22	nr	6.69
240mm long	4.78	5.00	5.02	0.18	–	1.66	–	6.68	nr	7.18
260mm long	5.11	5.00	5.37	0.20	–	1.85	–	7.21	nr	7.75
280mm long	5.46	5.00	5.73	0.21	–	1.94	–	7.67	nr	8.25
300mm long	5.81	5.00	6.10	0.22	–	2.03	–	8.13	nr	8.74
M 24; with nuts and washers										
70mm long	2.14	5.00	2.25	0.09	–	0.83	–	3.08	nr	3.31
80mm long	2.42	5.00	2.54	0.11	–	1.02	–	3.56	nr	3.82
90mm long	2.46	5.00	2.58	0.12	–	1.11	–	3.69	nr	3.97
100mm long	2.58	5.00	2.71	0.12	–	1.11	–	3.82	nr	4.10
120mm long	2.91	5.00	3.06	0.13	–	1.20	–	4.26	nr	4.57
140mm long	3.30	5.00	3.46	0.15	–	1.38	–	4.85	nr	5.21
150mm long	3.40	5.00	3.57	0.15	–	1.38	–	4.95	nr	5.33
160mm long	3.65	5.00	3.83	0.16	–	1.48	–	5.31	nr	5.71
180mm long	4.15	5.00	4.36	0.17	–	1.57	–	5.93	nr	6.37
200mm long	5.95	5.00	6.25	0.18	–	1.66	–	7.91	nr	8.50
220mm long	6.48	5.00	6.80	0.20	–	1.85	–	8.65	nr	9.30
240mm long	6.98	5.00	7.33	0.21	–	1.94	–	9.27	nr	9.96
260mm long	7.50	5.00	7.88	0.22	–	2.03	–	9.91	nr	10.65
280mm long	8.00	5.00	8.40	0.24	–	2.22	–	10.62	nr	11.41
300mm long	8.54	5.00	8.97	0.25	–	2.31	–	11.27	nr	12.12
Framing, erection; in accordance with B.S.5950 Part 2										
Note the following prices for erection include for the site to be reasonably clear, ease of access and the erection carried out during normal working hours										
Permanent erection of fabricated steelwork on site with bolted connections										
framing	–	–	–	23.00	–	212.29	–	212.29	t	228.21
Cold rolled zed purlins and cladding rails; Metsec, galvanised steel										
Purlin sleeved system (section only); fixing to cleats on frame members at 6000mm centres; hoist and fix 3.00m above ground level; fixing with bolts										
reference 14214; 3.03 kg/m	7.31	2.50	7.49	0.20	0.03	2.06	–	9.55	m	10.27
reference 14216; 3.47 kg/m	8.27	2.50	8.48	0.20	0.03	2.06	–	10.53	m	11.32
reference 17214; 3.66 kg/m	8.67	2.50	8.89	0.25	0.03	2.52	–	11.40	m	12.26
reference 17216; 4.11 kg/m	9.84	2.50	10.09	0.25	0.03	2.52	–	12.60	m	13.55
reference 20216; 4.49 kg/m	10.77	2.50	11.04	0.30	0.03	3.05	–	14.09	m	15.15
reference 20218; 5.03 kg/m	12.12	2.50	12.42	0.30	0.04	3.05	–	15.47	m	16.63
reference 20220; 5.57 kg/m	13.33	2.50	13.66	0.30	0.04	3.05	–	16.71	m	17.97
reference 23218; 5.73 kg/m	13.78	2.50	14.12	0.33	0.04	3.33	–	17.45	m	18.76
reference 23220; 6.34 kg/m	15.26	2.50	15.64	0.33	0.04	3.33	–	18.97	m	20.39
Purlin sleeved system (section only); fixing to cleats on frame members at 6000mm centres; hoist and fix 6.00m above ground level; fixing with bolts										
reference 14214; 3.03 kg/m	7.31	2.50	7.49	0.22	0.03	2.24	–	9.73	m	10.46

Labour hourly rates: (except Specialists) Craft Operatives £9.23 Labourer £7.02 Rates are national average prices. Refer to REGIONAL VARIATIONS for indicative levels of overall pricing in regions	MATERIALS			LABOUR				RATES		
	Del to Site £	Waste %	Material Cost £	Craft Optve Hrs	Lab Hrs	Labour Cost £	Sunds £	Nett Rate £	Unit	Gross Rate (+7.5%) £
G10: STRUCTURAL STEEL FRAMING Cont.										
Cold rolled zed purlins and cladding rails; Metsec, galvanised steel Cont.										
Purlin sleeved system (section only); fixing to cleats on frame members at 6000mm centres; hoist and fix 6.00m above ground level; fixing with bolts Cont.										
reference 14216; 3.47 kg/m	8.27	2.50	8.48	0.22	0.03	2.24	-	10.72	m	11.52
reference 17214; 3.66 kg/m	8.67	2.50	8.89	0.28	0.04	2.87	-	11.75	m	12.63
reference 17216; 4.11 kg/m	9.84	2.50	10.09	0.28	0.04	2.87	-	12.95	m	13.92
reference 20216; 4.49 kg/m	10.77	2.50	11.04	0.33	0.04	3.33	-	14.37	m	15.44
reference 20218; 5.03 kg/m	12.12	2.50	12.42	0.33	0.04	3.33	-	15.75	m	16.93
reference 20220; 5.57 kg/m	13.33	2.50	13.66	0.33	0.04	3.33	-	16.99	m	18.26
reference 23218; 5.73 kg/m	13.78	2.50	14.12	0.36	0.05	3.67	-	17.80	m	19.13
reference 23220; 6.34 kg/m	15.26	2.50	15.64	0.36	0.05	3.67	-	19.32	m	20.76
Purlin overlap system (cleat only); fixing to cleats on frame members at 6000mm centres; hoist and fix 3.00m above ground level; fixing with bolts galvanised										
reference 142	3.48	2.50	3.57	0.20	0.03	2.06	-	5.62	m	6.05
reference 172	3.77	2.50	3.86	0.25	0.03	2.52	-	6.38	m	6.86
reference 202	5.25	2.50	5.38	0.30	0.04	3.05	-	8.43	m	9.06
reference 232	6.97	2.50	7.14	0.33	0.04	3.33	-	10.47	m	11.26
Purlin overlap system (cleat only); fixing to cleats on frame members at 6000mm centres; hoist and fix 6.00m above ground level; fixing with bolts galvanised										
reference 142	3.48	2.50	3.57	0.22	0.03	2.24	-	5.81	m	6.24
reference 172	3.77	2.50	3.86	0.28	0.04	2.87	-	6.73	m	7.23
reference 202	5.25	2.50	5.38	0.33	0.04	3.33	-	8.71	m	9.36
reference 232	6.97	2.50	7.14	0.36	0.05	3.67	-	10.82	m	11.63
Purlin Non-Continuous system (section only); fixing to cleats on frame members at 6000mm centres; hoist and fix 3.00m above ground level; fixing with bolts										
reference 17215; 3.85 kg/m	9.22	2.50	9.45	0.25	0.03	2.52	-	11.97	m	12.87
reference 17216; 4.11 kg/m	9.84	2.50	10.09	0.25	0.03	2.52	-	12.60	m	13.55
reference 20216; 4.49 kg/m	10.77	2.50	11.04	0.30	0.04	3.05	-	14.09	m	15.15
reference 20218; 5.03 kg/m	12.12	2.50	12.42	0.30	0.04	3.05	-	15.47	m	16.63
reference 20220; 5.57 kg/m	13.33	2.50	13.66	0.30	0.04	3.05	-	16.71	m	17.97
reference 23218; 5.73 kg/m	13.78	2.50	14.12	0.33	0.04	3.33	-	17.45	m	18.76
reference 23223; 7.26 kg/m	17.56	2.50	18.00	0.33	0.04	3.33	-	21.33	m	22.93
reference 26223; 7.92 kg/m	19.22	2.50	19.70	0.40	0.05	4.04	-	23.74	m	25.52
reference 26229; 9.88 kg/m	23.84	2.50	24.44	0.40	0.05	4.04	-	28.48	m	30.61
Purlin Non-Continuous system (section only); fixing to cleats on frame members at 6000mm centres; hoist and fix 6.00m above ground level; fixing with bolts										
reference 17215; 3.85 kg/m	9.22	2.50	9.45	0.28	0.04	2.87	-	12.32	m	13.24
reference 17216; 4.11 kg/m	9.84	2.50	10.09	0.28	0.04	2.87	-	12.95	m	13.92
reference 20216; 4.49 kg/m	10.77	2.50	11.04	0.33	0.04	3.33	-	14.37	m	15.44
reference 20218; 5.03 kg/m	12.12	2.50	12.42	0.33	0.04	3.33	-	15.75	m	16.93
reference 20220; 5.57 kg/m	13.33	2.50	13.66	0.33	0.04	3.33	-	16.99	m	18.26
reference 23218; 5.73 kg/m	13.78	2.50	14.12	0.36	0.05	3.67	-	17.80	m	19.13
reference 23223; 7.26 kg/m	17.56	2.50	18.00	0.36	0.05	3.67	-	21.67	m	23.30
reference 26223; 7.92 kg/m	19.22	2.50	19.70	0.44	0.06	4.48	-	24.18	m	26.00
reference 26229; 9.88 kg/m	23.84	2.50	24.44	0.44	0.06	4.48	-	28.92	m	31.09
Galvanised purlin cleats; weld on										
Reference 142 for 142mm deep purlins	4.99	2.50	5.11	-	-	-	-	5.11	nr	5.50
Reference 172 for 172mm deep purlins	5.95	2.50	6.10	-	-	-	-	6.10	nr	6.56
Reference 202 for 202mm deep purlins	7.78	2.50	7.97	-	-	-	-	7.97	nr	8.57
Reference 232 for 232mm deep purlins	10.19	2.50	10.44	-	-	-	-	10.44	nr	11.23
Reference 262 for 262mm deep purlins	10.88	2.50	11.15	-	-	-	-	11.15	nr	11.99
Round Lok anti-sag rods; push fit to purlins										
purlins at 1150mm centres	2.58	2.50	2.64	0.25	0.03	2.52	-	5.16	nr	5.55
purlins at 1350mm centres	3.00	2.50	3.08	0.25	0.03	2.52	-	5.59	nr	6.01
purlins at 1550mm centres	3.39	2.50	3.47	0.33	0.04	3.33	-	6.80	nr	7.31
purlins at 1700mm centres	3.79	2.50	3.88	0.33	0.04	3.33	-	7.21	nr	7.75
purlins at 1950mm centres	4.18	2.50	4.28	0.33	0.04	3.33	-	7.61	nr	8.18
Side rail sleeved system (section only); fixing to cleats on stanchions at 6000mm centres; hoist and fix 3.00m above ground level; fixing with bolts										
reference 14214; 3.03 kg/m	7.31	2.50	7.49	0.20	0.03	2.06	-	9.55	m	10.27
reference 14216; 3.47 kg/m	8.27	2.50	8.48	0.20	0.03	2.06	-	10.53	m	11.32
reference 17215; 3.85 kg/m	9.22	2.50	9.45	0.25	0.03	2.52	-	11.97	m	12.87
reference 17216; 4.11 kg/m	9.84	2.50	10.09	0.25	0.03	2.52	-	12.60	m	13.55
reference 20216; 4.49 kg/m	10.77	2.50	11.04	0.30	0.04	3.05	-	14.09	m	15.15
reference 20218; 5.03 kg/m	12.12	2.50	12.42	0.30	0.04	3.05	-	15.47	m	16.63
reference 20220; 5.57 kg/m	13.33	2.50	13.66	0.30	0.04	3.05	-	16.71	m	17.97
reference 23218; 5.73 kg/m	13.78	2.50	14.12	0.33	0.04	3.33	-	17.45	m	18.76
reference 23223; 7.26 kg/m	17.66	2.50	18.10	0.33	0.04	3.33	-	21.43	m	23.04
Side rail sleeved system (section only); fixing to cleats on stanchions at 6000mm centres; hoist and fix 6.00m above ground level; fixing with bolts										
reference 14214; 3.03 kg/m	7.31	2.50	7.49	0.22	0.03	2.24	-	9.73	m	10.46
reference 14216; 3.47 kg/m	8.27	2.50	8.48	0.22	0.03	2.24	-	10.72	m	11.52
reference 17215; 3.85 kg/m	9.22	2.50	9.45	0.28	0.04	2.87	-	12.32	m	13.24
reference 17216; 4.11 kg/m	9.84	2.50	10.09	0.28	0.04	2.87	-	12.95	m	13.92
reference 20216; 4.49 kg/m	10.77	2.50	11.04	0.33	0.04	3.33	-	14.37	m	15.44

STRUCTURAL/CARCASSING METAL/TIMBER

Labour hourly rates: (except Specialists) Craft Operatives £9.23 Labourer £7.02 Rates are national average prices. Refer to REGIONAL VARIATIONS for indicative levels of overall pricing in regions	MATERIALS			LABOUR				RATES		
	Del to Site £	Waste %	Material Cost £	Craft Optve Hrs	Lab Hrs	Labour Cost £	Sunds £	Nett Rate £	Unit	Gross Rate (+7.5%) £
G10: STRUCTURAL STEEL FRAMING Cont.										
Cold rolled zed purlins and cladding rails; Metsec, galvanised steel Cont.										
Side rail sleeved system (section only); fixing to cleats on stanchions at 6000mm centres; hoist and fix 6.00m above ground level; fixing with bolts Cont.										
reference 20218; 5.03 kg/m	12.12	2.50	12.42	0.33	0.04	3.33	-	15.75	m	16.93
reference 20220; 5.57 kg/m	13.13	2.50	13.46	0.33	0.04	3.33	-	16.78	m	18.04
reference 23218; 5.73 kg/m	13.78	2.50	14.12	0.36	0.05	3.67	-	17.80	m	19.13
reference 23223; 7.26 kg/m	17.66	2.50	18.10	0.36	0.05	3.67	-	21.78	m	23.41
Side rail single span system (section only) ; fixing to cleats on stanchions at 6000mm centres; hoist and fix 3.00m above ground level; fixing with bolts										
reference 14214; 3.03 kg/m	7.31	2.50	7.49	0.20	0.03	2.06	-	9.55	m	10.27
reference 14216; 3.47 kg/m	8.27	2.50	8.48	0.20	0.03	2.06	-	10.53	m	11.32
reference 17215; 3.85 kg/m	9.22	2.50	9.45	0.25	0.03	2.52	-	11.97	m	12.87
reference 17216; 4.11 kg/m	9.84	2.50	10.09	0.25	0.03	2.52	-	12.60	m	13.55
reference 20216; 4.49 kg/m	10.77	2.50	11.04	0.30	0.04	3.05	-	14.09	m	15.15
reference 20218; 5.03 kg/m	12.12	2.50	12.42	0.30	0.04	3.05	-	15.47	m	16.63
reference 20220; 5.57 kg/m	13.13	2.50	13.46	0.30	0.04	3.05	-	16.51	m	17.75
reference 23218; 5.73 kg/m	13.78	2.50	14.12	0.33	0.04	3.33	-	17.45	m	18.76
reference 23223; 7.26 kg/m	17.66	2.50	18.10	0.33	0.04	3.33	-	21.43	m	23.04
Side rail single span system (section only) ; fixing to cleats on stanchions at 6000mm centres; hoist and fix 6.00m above ground level; fixing with bolts										
reference 14214; 3.03 kg/m	7.31	2.50	7.49	0.22	0.03	2.24	-	9.73	m	10.46
reference 14216; 3.47 kg/m	8.27	2.50	8.48	0.22	0.03	2.24	-	10.72	m	11.52
reference 17215; 3.85 kg/m	9.22	2.50	9.45	0.28	0.04	2.87	-	12.32	m	13.24
reference 17216; 4.11 kg/m	9.84	2.50	10.09	0.28	0.04	2.87	-	12.95	m	13.92
reference 20216; 4.49 kg/m	10.77	2.50	11.04	0.33	0.04	3.33	-	14.37	m	15.44
reference 20218; 5.03 kg/m	12.12	2.50	12.42	0.33	0.04	3.33	-	15.75	m	16.93
reference 20220; 5.57 kg/m	13.13	2.50	13.46	0.33	0.04	3.33	-	16.78	m	18.04
reference 23218; 5.73 kg/m	13.78	2.50	14.12	0.36	0.05	3.67	-	17.80	m	19.13
reference 23223; 7.26 kg/m	17.66	2.50	18.10	0.36	0.05	3.67	-	21.78	m	23.41
Galvanised side rail cleats; weld on										
Reference 142 for 142mm wide rails	4.99	2.50	5.11	-	-	-	-	5.11	nr	5.50
Reference 172 for 172mm wide rails	5.95	2.50	6.10	-	-	-	-	6.10	nr	6.56
Reference 202 for 202mm wide rails	7.78	2.50	7.97	-	-	-	-	7.97	nr	8.57
Reference 232 for 232mm wide rails	10.19	2.50	10.44	-	-	-	-	10.44	nr	11.23
Galvanised side rail supports; 122-262 series; weld on										
rail 1000mm long	7.53	2.50	7.72	0.35	0.04	3.51	-	11.23	nr	12.07
rail 1400mm long	8.23	2.50	8.44	0.35	0.04	3.51	-	11.95	nr	12.84
rail 1600mm long	9.13	2.50	9.36	0.45	0.06	4.57	-	13.93	nr	14.98
rail 1800mm long	10.98	2.50	11.25	0.45	0.06	4.57	-	15.83	nr	17.02
Diagonal tie wire ropes (assembled with end brackets); bolt on										
1700mm long	14.70	2.50	15.07	0.30	0.04	3.05	-	18.12	nr	19.48
2200mm long	16.44	2.50	16.85	0.30	0.04	3.05	-	19.90	nr	21.39
2600mm long	18.20	2.50	18.66	0.40	0.05	4.04	-	22.70	nr	24.40
3600mm long	22.46	2.50	23.02	0.40	0.05	4.04	-	27.06	nr	29.09
Surface preparation at works										
Note notwithstanding the requirement of SMM7 to measure painting on structural steelwork in m2, painting off site has been given in tonnes of structural steelwork in accordance with normal steelwork contractors practice Blast cleaning										
surfaces of steelwork	-	-	-	-	-	-	-	70.00	t	75.25
Surface treatment at works										
One coat micaceous oxide primer, 75 microns										
surfaces of steelwork	-	-	-	-	-	-	-	75.00	t	80.63
Two coats micaceous oxide primer, 150 microns										
surfaces of steelwork	-	-	-	-	-	-	-	140.00	t	150.50
G12: ISOLATED STRUCTURAL METAL MEMBERS										
Isolated unfabricated structural members										
Weldable steel, B.S.EN 10025 : 1993 Grade S275 (formerly B.S.4360 Grade 43A), hot rolled sections plain member; beam										
weight less than 40 Kg/m	780.00	-	780.00	44.12	-	407.23	-	1187.23	t	1276.27
weight 40-100 Kg/m	780.00	-	780.00	41.18	-	380.09	-	1160.09	t	1247.10
weight exceeding 100 Kg/m	780.00	-	780.00	36.75	-	339.20	-	1119.20	t	1203.14
Steel short span lattice joists primed at works; treated timber inserts in top and bottom chords; full depth end seatings for bolting to supporting structure; hoist and fix 3.00m above ground level										
200mm deep, 8.2 kg/m	26.30	2.50	26.96	0.66	0.08	6.65	-	33.61	m	36.13
250mm deep, 8.5 kg/m	27.40	2.50	28.09	0.68	0.09	6.91	-	34.99	m	37.62
300mm deep, 10.7 kg/m	34.60	2.50	35.47	0.75	0.09	7.55	-	43.02	m	46.25

Labour hourly rates: (except Specialists) Craft Operatives £9.23 Labourer £7.02 Rates are national average prices. Refer to REGIONAL VARIATIONS for indicative levels of overall pricing in regions	MATERIALS			LABOUR				RATES		
	Del to Site £	Waste %	Material Cost £	Craft Optve Hrs	Lab Hrs	Labour Cost £	Sunds £	Nett Rate £	Unit	Gross Rate (+7.5%) £
G12: ISOLATED STRUCTURAL METAL MEMBERS Cont.										
Isolated unfabricated structural members Cont.										
Steel short span lattice joists primed at works; treated timber inserts in top and bottom chords; full depth end seatings for bolting to supporting structure; hoist and fix 3.00m above ground level Cont.										
350mm deep, 11.6 kg/m	37.20	2.50	38.13	0.81	0.10	8.18	–	46.31	m	49.78
350mm deep, 12.8 kg/m	40.65	2.50	41.67	0.90	0.11	9.08	–	50.75	m	54.55
Steel short span lattice joists primed at works; treated timber inserts in top and bottom chords; full depth end seatings for bolting to supporting structure; hoist and fix 6.00m above ground level										
200mm deep, 8.2 kg/m	26.30	2.50	26.96	0.72	0.09	7.28	–	34.23	m	36.80
250mm deep, 8.5 kg/m	27.40	2.50	28.09	0.75	0.09	7.55	–	35.64	m	38.31
300mm deep, 10.7 kg/m	34.60	2.50	35.47	0.82	0.10	8.27	–	43.74	m	47.02
350mm deep, 11.6 kg/m	37.20	2.50	38.13	0.89	0.11	8.99	–	47.12	m	50.65
350mm deep, 12.8 kg/m	40.65	2.50	41.67	0.98	0.12	9.89	–	51.55	m	55.42
Steel intermediate span lattice joists primed at works; treated timber inserts in top and bottom chords; full depth end seatings for bolting to supporting structure; hoist and fix 3.00m above ground level										
450mm deep, 12.2 kg/m	35.60	2.50	36.49	0.85	0.11	8.62	–	45.11	m	48.49
500mm deep, 15.8 kg/m	44.00	2.50	45.10	1.03	0.13	10.42	–	55.52	m	59.68
550mm deep, 19.4 kg/m	52.00	2.50	53.30	1.26	0.16	12.75	–	66.05	m	71.01
600mm deep, 22.5 kg/m	59.50	2.50	60.99	1.35	0.17	13.65	–	74.64	m	80.24
650mm deep, 29.7 kg/m	78.40	2.50	80.36	1.78	0.22	17.97	–	98.33	m	105.71
Steel intermediate span lattice joists primed at works; treated timber inserts in top and bottom chords; full depth end seatings for bolting to supporting structure; hoist and fix 6.00m above ground level										
450mm deep, 12.2 kg/m	35.60	2.50	36.49	0.94	0.12	9.52	–	46.01	m	49.46
500mm deep, 15.8 kg/m	44.00	2.50	45.10	1.13	0.14	11.41	–	56.51	m	60.75
550mm deep, 19.4 kg/m	52.00	2.50	53.30	1.39	0.17	14.02	–	67.32	m	72.37
600mm deep, 22.5 kg/m	59.50	2.50	60.99	1.49	0.19	15.09	–	76.07	m	81.78
650mm deep, 29.7 kg/m	78.40	2.50	80.36	1.96	0.25	19.85	–	100.21	m	107.72
Steel long span lattice joists primed at works; treated timber inserts in top and bottom chords; full depth end seatings for bolting to supporting structure; hoist and fix 3.00m above ground level										
700mm deep, 39.2 kg/m	101.50	2.50	104.04	1.96	0.25	19.85	–	123.88	m	133.17
800mm deep, 44.1 kg/m	108.60	2.50	111.32	2.21	0.28	22.36	–	133.68	m	143.70
900mm deep, 45.3 kg/m	109.20	2.50	111.93	2.27	0.28	22.92	–	134.85	m	144.96
1000mm deep, 46.1 kg/m	111.50	2.50	114.29	2.31	0.29	23.36	–	137.64	m	147.97
1500mm deep, 54.2 kg/m	132.00	2.50	135.30	2.40	0.30	24.26	–	159.56	m	171.52
Steel long span lattice joists primed at works; treated timber inserts in top and bottom chords; full depth end seatings for bolting to supporting structure; hoist and fix 6.00m above ground level										
700mm deep, 39.2 kg/m	101.50	2.50	104.04	2.16	0.27	21.83	–	125.87	m	135.31
800mm deep, 44.1 kg/m	108.60	2.50	111.32	2.43	0.30	24.53	–	135.85	m	146.04
900mm deep, 45.3 kg/m	109.20	2.50	111.93	2.50	0.31	25.25	–	137.18	m	147.47
1000mm deep, 46.1 kg/m	111.50	2.50	114.29	2.54	0.32	25.69	–	139.98	m	150.48
1500mm deep, 54.2 kg/m	132.00	2.50	135.30	2.64	0.33	26.68	–	161.98	m	174.13
Fixing bolts; steel										
Black bolts, B.S.4190 grade 4.6										
M 8; with nuts and washers										
30mm long	0.11	5.00	0.12	0.06	–	0.55	–	0.67	nr	0.72
40mm long	0.12	5.00	0.13	0.06	–	0.55	–	0.68	nr	0.73
50mm long	0.13	5.00	0.14	0.06	–	0.55	–	0.69	nr	0.74
60mm long	0.14	5.00	0.15	0.06	–	0.55	–	0.70	nr	0.75
70mm long	0.15	5.00	0.16	0.07	–	0.65	–	0.80	nr	0.86
80mm long	0.17	5.00	0.18	0.07	–	0.65	–	0.82	nr	0.89
90mm long	0.19	5.00	0.20	0.07	–	0.65	–	0.85	nr	0.91
100mm long	0.21	5.00	0.22	0.07	–	0.65	–	0.87	nr	0.93
120mm long	0.25	5.00	0.26	0.08	–	0.74	–	1.00	nr	1.08
M 10; with nuts and washers										
40mm long	0.17	5.00	0.18	0.06	–	0.55	–	0.73	nr	0.79
50mm long	0.18	5.00	0.19	0.06	–	0.55	–	0.74	nr	0.80
60mm long	0.20	5.00	0.21	0.06	–	0.55	–	0.76	nr	0.82
70mm long	0.21	5.00	0.22	0.07	–	0.65	–	0.87	nr	0.93
80mm long	0.26	5.00	0.27	0.07	–	0.65	–	0.92	nr	0.99
90mm long	0.28	5.00	0.29	0.07	–	0.65	–	0.94	nr	1.01
100mm long	0.32	5.00	0.34	0.07	–	0.65	–	0.98	nr	1.06
120mm long	0.34	5.00	0.36	0.08	–	0.74	–	1.10	nr	1.18
140mm long	0.44	5.00	0.46	0.09	–	0.83	–	1.29	nr	1.39
150mm long	0.50	5.00	0.53	0.10	–	0.92	–	1.45	nr	1.56
M 12; with nuts and washers										
40mm long	0.24	5.00	0.25	0.07	–	0.65	–	0.90	nr	0.97
50mm long	0.25	5.00	0.26	0.07	–	0.65	–	0.91	nr	0.98
60mm long	0.27	5.00	0.28	0.07	–	0.65	–	0.93	nr	1.00
70mm long	0.31	5.00	0.33	0.07	–	0.65	–	0.97	nr	1.04
80mm long	0.34	5.00	0.36	0.07	–	0.65	–	1.00	nr	1.08
90mm long	0.35	5.00	0.37	0.07	–	0.65	–	1.01	nr	1.09
100mm long	0.38	5.00	0.40	0.08	–	0.74	–	1.14	nr	1.22

Labour hourly rates: (except Specialists) Craft Operatives £9.23 Labourer £7.02 Rates are national average prices. Refer to REGIONAL VARIATIONS for indicative levels of overall pricing in regions	MATERIALS			LABOUR				RATES		
	Del to Site £	Waste %	Material Cost £	Craft Optve Hrs	Lab Hrs	Labour Cost £	Sunds £	Nett Rate £	Unit	Gross Rate (+7.5%) £
G12: ISOLATED STRUCTURAL METAL MEMBERS Cont.										
Fixing bolts; steel Cont.										
Black bolts, B.S.4190 grade 4.6 Cont.										
M 12; with nuts and washers Cont.										
120mm long.................................	0.46	5.00	0.48	0.08	-	0.74	-	1.22	nr	1.31
140mm long.................................	0.57	5.00	0.60	0.10	-	0.92	-	1.52	nr	1.64
150mm long.................................	0.72	5.00	0.76	0.10	-	0.92	-	1.68	nr	1.80
160mm long.................................	1.03	5.00	1.08	0.10	-	0.92	-	2.00	nr	2.15
180mm long.................................	1.08	5.00	1.13	0.11	-	1.02	-	2.15	nr	2.31
200mm long.................................	1.13	5.00	1.19	0.12	-	1.11	-	2.29	nr	2.47
220mm long.................................	1.19	5.00	1.25	0.13	-	1.20	-	2.45	nr	2.63
240mm long.................................	1.24	5.00	1.30	0.14	-	1.29	-	2.59	nr	2.79
260mm long.................................	1.30	5.00	1.37	0.15	-	1.38	-	2.75	nr	2.96
280mm long.................................	1.37	5.00	1.44	0.16	-	1.48	-	2.92	nr	3.13
300mm long.................................	1.42	5.00	1.49	0.17	-	1.57	-	3.06	nr	3.29
M 16; with nuts and washers										
50mm long.................................	0.44	5.00	0.46	0.07	-	0.65	-	1.11	nr	1.19
60mm long.................................	0.49	5.00	0.51	0.07	-	0.65	-	1.16	nr	1.25
70mm long.................................	0.54	5.00	0.57	0.07	-	0.65	-	1.21	nr	1.30
80mm long.................................	0.59	5.00	0.62	0.08	-	0.74	-	1.36	nr	1.46
90mm long.................................	0.63	5.00	0.66	0.08	-	0.74	-	1.40	nr	1.50
100mm long.................................	0.74	5.00	0.78	0.09	-	0.83	-	1.61	nr	1.73
120mm long.................................	0.86	5.00	0.90	0.09	-	0.83	-	1.73	nr	1.86
140mm long.................................	0.96	5.00	1.01	0.10	-	0.92	-	1.93	nr	2.08
150mm long.................................	1.01	5.00	1.06	0.11	-	1.02	-	2.08	nr	2.23
160mm long.................................	1.40	5.00	1.47	0.12	-	1.11	-	2.58	nr	2.77
180mm long.................................	1.46	5.00	1.53	0.12	-	1.11	-	2.64	nr	2.84
200mm long.................................	1.55	5.00	1.63	0.13	-	1.20	-	2.83	nr	3.04
220mm long.................................	1.64	5.00	1.72	0.13	-	1.20	-	2.92	nr	3.14
M 20; with nuts and washers										
60mm long.................................	0.83	5.00	0.87	0.07	-	0.65	-	1.52	nr	1.63
70mm long.................................	0.92	5.00	0.97	0.07	-	0.65	-	1.61	nr	1.73
80mm long.................................	1.04	5.00	1.09	0.08	-	0.74	-	1.83	nr	1.97
90mm long.................................	1.05	5.00	1.10	0.09	-	0.83	-	1.93	nr	2.08
100mm long.................................	1.20	5.00	1.26	0.09	-	0.83	-	2.09	nr	2.25
120mm long.................................	1.41	5.00	1.48	0.10	-	0.92	-	2.40	nr	2.58
140mm long.................................	1.64	5.00	1.72	0.11	-	1.02	-	2.74	nr	2.94
150mm long.................................	1.68	5.00	1.76	0.12	-	1.11	-	2.87	nr	3.09
160mm long.................................	2.07	5.00	2.17	0.13	-	1.20	-	3.37	nr	3.63
180mm long.................................	2.18	5.00	2.29	0.13	-	1.20	-	3.49	nr	3.75
200mm long.................................	2.33	5.00	2.45	0.14	-	1.29	-	3.74	nr	4.02
220mm long.................................	2.42	5.00	2.54	0.15	-	1.38	-	3.93	nr	4.22
240mm long.................................	2.56	5.00	2.69	0.16	-	1.48	-	4.16	nr	4.48
260mm long.................................	2.63	5.00	2.76	0.17	-	1.57	-	4.33	nr	4.66
280mm long.................................	2.70	5.00	2.84	0.18	-	1.66	-	4.50	nr	4.83
300mm long.................................	2.86	5.00	3.00	0.20	-	1.85	-	4.85	nr	5.21
M 24; with nuts and washers										
70mm long.................................	2.09	5.00	2.19	0.09	-	0.83	-	3.03	nr	3.25
80mm long.................................	2.18	5.00	2.29	0.09	-	0.83	-	3.12	nr	3.35
90mm long.................................	2.25	5.00	2.36	0.10	-	0.92	-	3.29	nr	3.53
100mm long.................................	2.33	5.00	2.45	0.10	-	0.92	-	3.37	nr	3.62
120mm long.................................	2.49	5.00	2.61	0.10	-	0.92	-	3.54	nr	3.80
140mm long.................................	2.80	5.00	2.94	0.11	-	1.02	-	3.96	nr	4.25
150mm long.................................	2.97	5.00	3.12	0.13	-	1.20	-	4.32	nr	4.64
160mm long.................................	3.07	5.00	3.22	0.13	-	1.20	-	4.42	nr	4.76
180mm long.................................	3.15	5.00	3.31	0.13	-	1.20	-	4.51	nr	4.85
200mm long.................................	3.30	5.00	3.46	0.14	-	1.29	-	4.76	nr	5.11
220mm long.................................	3.64	5.00	3.82	0.15	-	1.38	-	5.21	nr	5.60
240mm long.................................	3.74	5.00	3.93	0.16	-	1.48	-	5.40	nr	5.81
260mm long.................................	3.99	5.00	4.19	0.18	-	1.66	-	5.85	nr	6.29
280mm long.................................	4.12	5.00	4.33	0.20	-	1.85	-	6.17	nr	6.63
300mm long.................................	4.24	5.00	4.45	0.22	-	2.03	-	6.48	nr	6.97
Black bolts, B.S.4190 grade 4.6; galvanised										
M 8; with nuts and washers										
30mm long.................................	0.15	5.00	0.16	0.06	-	0.55	-	0.71	nr	0.76
40mm long.................................	0.17	5.00	0.18	0.06	-	0.55	-	0.73	nr	0.79
50mm long.................................	0.18	5.00	0.19	0.06	-	0.55	-	0.74	nr	0.80
60mm long.................................	0.20	5.00	0.21	0.06	-	0.55	-	0.76	nr	0.82
70mm long.................................	0.22	5.00	0.23	0.07	-	0.65	-	0.88	nr	0.94
80mm long.................................	0.23	5.00	0.24	0.07	-	0.65	-	0.89	nr	0.95
90mm long.................................	0.26	5.00	0.27	0.07	-	0.65	-	0.92	nr	0.99
100mm long.................................	0.32	5.00	0.34	0.07	-	0.65	-	0.98	nr	1.06
120mm long.................................	0.38	5.00	0.40	0.08	-	0.74	-	1.14	nr	1.22
M 10; with nuts and washers										
40mm long.................................	0.23	5.00	0.24	0.06	-	0.55	-	0.80	nr	0.85
50mm long.................................	0.26	5.00	0.27	0.06	-	0.55	-	0.83	nr	0.89
60mm long.................................	0.29	5.00	0.30	0.06	-	0.55	-	0.86	nr	0.92
70mm long.................................	0.32	5.00	0.34	0.07	-	0.65	-	0.98	nr	1.06
80mm long.................................	0.39	5.00	0.41	0.07	-	0.65	-	1.06	nr	1.13
90mm long.................................	0.43	5.00	0.45	0.07	-	0.65	-	1.10	nr	1.18
100mm long.................................	0.47	5.00	0.49	0.07	-	0.65	-	1.14	nr	1.23
120mm long.................................	0.50	5.00	0.53	0.08	-	0.74	-	1.26	nr	1.36
140mm long.................................	0.65	5.00	0.68	0.09	-	0.83	-	1.51	nr	1.63
150mm long.................................	0.75	5.00	0.79	0.10	-	0.92	-	1.71	nr	1.84
M 12; with nuts and washers										
40mm long.................................	0.39	5.00	0.41	0.07	-	0.65	-	1.06	nr	1.13
50mm long.................................	0.41	5.00	0.43	0.07	-	0.65	-	1.08	nr	1.16
60mm long.................................	0.44	5.00	0.46	0.07	-	0.65	-	1.11	nr	1.19
70mm long.................................	0.49	5.00	0.51	0.07	-	0.65	-	1.16	nr	1.25
80mm long.................................	0.53	5.00	0.56	0.07	-	0.65	-	1.20	nr	1.29
90mm long.................................	0.55	5.00	0.58	0.07	-	0.65	-	1.22	nr	1.32
100mm long.................................	0.60	5.00	0.63	0.08	-	0.74	-	1.37	nr	1.47
120mm long.................................	0.74	5.00	0.78	0.08	-	0.74	-	1.52	nr	1.63

Labour hourly rates: (except Specialists) Craft Operatives £9.23 Labourer £7.02 Rates are national average prices. Refer to REGIONAL VARIATIONS for indicative levels of overall pricing in regions	MATERIALS			LABOUR				RATES		
	Del to Site £	Waste %	Material Cost £	Craft Optve Hrs	Lab Hrs	Labour Cost £	Sunds £	Nett Rate £	Unit	Gross Rate (+7.5%) £

G12: ISOLATED STRUCTURAL METAL MEMBERS Cont.

Fixing bolts; steel Cont.

Black bolts, B.S.4190 grade 4.6; galvanised Cont.

	Del to Site	Waste	Material Cost	Craft Optve	Lab	Labour Cost	Sunds	Nett Rate	Unit	Gross Rate
M 12; with nuts and washers Cont.										
140mm long	0.87	5.00	0.91	0.10	–	0.92	–	1.84	nr	1.97
150mm long	1.10	5.00	1.16	0.10	–	0.92	–	2.08	nr	2.23
160mm long	1.55	5.00	1.63	0.10	–	0.92	–	2.55	nr	2.74
180mm long	1.63	5.00	1.71	0.11	–	1.02	–	2.73	nr	2.93
200mm long	1.70	5.00	1.78	0.12	–	1.11	–	2.89	nr	3.11
220mm long	1.79	5.00	1.88	0.13	–	1.20	–	3.08	nr	3.31
240mm long	1.86	5.00	1.95	0.14	–	1.29	–	3.25	nr	3.49
260mm long	1.97	5.00	2.07	0.15	–	1.38	–	3.45	nr	3.71
280mm long	2.03	5.00	2.13	0.16	–	1.48	–	3.61	nr	3.88
300mm long	2.12	5.00	2.23	0.17	–	1.57	–	3.80	nr	4.08
M 16; with nuts and washers										
50mm long	0.71	5.00	0.75	0.07	–	0.65	–	1.39	nr	1.50
60mm long	0.77	5.00	0.81	0.07	–	0.65	–	1.45	nr	1.56
70mm long	0.83	5.00	0.87	0.07	–	0.65	–	1.52	nr	1.63
80mm long	0.90	5.00	0.94	0.08	–	0.74	–	1.68	nr	1.81
90mm long	0.95	5.00	1.00	0.08	–	0.74	–	1.74	nr	1.87
100mm long	1.00	5.00	1.05	0.09	–	0.83	–	1.88	nr	2.02
120mm long	1.30	5.00	1.37	0.09	–	0.83	–	2.20	nr	2.36
140mm long	1.45	5.00	1.52	0.10	–	0.92	–	2.45	nr	2.63
150mm long	1.54	5.00	1.62	0.11	–	1.02	–	2.63	nr	2.83
160mm long	2.08	5.00	2.18	0.12	–	1.11	–	3.29	nr	3.54
180mm long	2.21	5.00	2.32	0.12	–	1.11	–	3.43	nr	3.69
200mm long	2.35	5.00	2.47	0.13	–	1.20	–	3.67	nr	3.94
220mm long	2.45	5.00	2.57	0.13	–	1.20	–	3.77	nr	4.06
M 20; with nuts and washers										
60mm long	1.28	5.00	1.34	0.07	–	0.65	–	1.99	nr	2.14
70mm long	1.41	5.00	1.48	0.07	–	0.65	–	2.13	nr	2.29
80mm long	1.59	5.00	1.67	0.08	–	0.74	–	2.41	nr	2.59
90mm long	1.60	5.00	1.68	0.09	–	0.83	–	2.51	nr	2.70
100mm long	1.83	5.00	1.92	0.09	–	0.83	–	2.75	nr	2.96
120mm long	2.12	5.00	2.23	0.10	–	0.92	–	3.15	nr	3.39
140mm long	2.47	5.00	2.59	0.11	–	1.02	–	3.61	nr	3.88
150mm long	2.53	5.00	2.66	0.12	–	1.11	–	3.76	nr	4.05
160mm long	3.13	5.00	3.29	0.13	–	1.20	–	4.49	nr	4.82
180mm long	3.23	5.00	3.39	0.13	–	1.20	–	4.59	nr	4.94
200mm long	3.44	5.00	3.61	0.14	–	1.29	–	4.90	nr	5.27
220mm long	3.58	5.00	3.76	0.15	–	1.38	–	5.14	nr	5.53
240mm long	3.80	5.00	3.99	0.16	–	1.48	–	5.47	nr	5.88
260mm long	3.89	5.00	4.08	0.17	–	1.57	–	5.65	nr	6.08
280mm long	3.99	5.00	4.19	0.18	–	1.66	–	5.85	nr	6.29
300mm long	4.23	5.00	4.44	0.20	–	1.85	–	6.29	nr	6.76
M 24; with nuts and washers										
70mm long	3.09	5.00	3.24	0.09	–	0.83	–	4.08	nr	4.38
80mm long	3.23	5.00	3.39	0.09	–	0.83	–	4.22	nr	4.54
90mm long	3.33	5.00	3.50	0.10	–	0.92	–	4.42	nr	4.75
100mm long	3.46	5.00	3.63	0.10	–	0.92	–	4.56	nr	4.90
120mm long	3.65	5.00	3.83	0.10	–	0.92	–	4.76	nr	5.11
140mm long	4.14	5.00	4.35	0.11	–	1.02	–	5.36	nr	5.76
150mm long	4.40	5.00	4.62	0.13	–	1.20	–	5.82	nr	6.26
160mm long	4.54	5.00	4.77	0.13	–	1.20	–	5.97	nr	6.41
180mm long	4.65	5.00	4.88	0.13	–	1.20	–	6.08	nr	6.54
200mm long	4.70	5.00	4.93	0.14	–	1.29	–	6.23	nr	6.69
220mm long	5.13	5.00	5.39	0.15	–	1.38	–	6.77	nr	7.28
240mm long	5.24	5.00	5.50	0.16	–	1.48	–	6.98	nr	7.50
260mm long	5.51	5.00	5.79	0.18	–	1.66	–	7.45	nr	8.01
280mm long	5.65	5.00	5.93	0.20	–	1.85	–	7.78	nr	8.36
300mm long	5.83	5.00	6.12	0.22	–	2.03	–	8.15	nr	8.76

G20: CARPENTRY/TIMBER FRAMING/FIRST FIXING

Sasco Trussed Rafters in softwood, sawn, B.S.4978, GS grade, impregnated; note:- prices are guide prices only, based on contracts within 30 mile radius of Widnes and are subject to adjustment for quantity: firm quotations should always be obtained

	Del to Site	Waste	Material Cost	Craft Optve	Lab	Labour Cost	Sunds	Nett Rate	Unit	Gross Rate
Trussed rafters; Hydro-Nail plated joints										
22.5 degree Standard duo pitch; 450mm overhangs; fixing with clips (included elsewhere); span over wall plates										
5000mm	22.56	2.50	23.12	1.10	0.14	11.14	5.50	39.76	nr	42.74
6000mm	25.96	2.50	26.61	1.20	0.15	12.13	6.00	44.74	nr	48.09
7000mm	29.05	2.50	29.78	1.30	0.16	13.12	6.50	49.40	nr	53.10
8000mm	34.39	2.50	35.25	1.40	0.17	14.12	7.00	56.37	nr	60.59
9000mm	39.81	2.50	40.81	1.50	0.19	15.18	7.50	63.48	nr	68.25
10000mm	48.20	2.50	49.41	1.60	0.20	16.17	8.00	73.58	nr	79.10
35 degree Standard duo pitch; 450mm overhangs; fixing with clips (included elsewhere); span over wall plates										
5000mm	23.45	2.50	24.04	1.10	0.14	11.14	5.50	40.67	nr	43.72
6000mm	27.10	2.50	27.78	1.20	0.15	12.13	6.00	45.91	nr	49.35
7000mm	30.20	2.50	30.95	1.30	0.16	13.12	6.50	50.58	nr	54.37
8000mm	36.25	2.50	37.16	1.40	0.17	14.12	7.00	58.27	nr	62.64
9000mm	40.35	2.50	41.36	1.50	0.19	15.18	7.50	64.04	nr	68.84
45 degree Standard duo pitch; 450mm overhangs; fixing with clips (included elsewhere); span over wall plates										
5000mm	36.81	2.50	37.73	1.10	0.14	11.14	5.50	54.33	nr	58.41
6000mm	44.05	2.50	45.15	1.20	0.15	12.13	6.00	63.28	nr	68.03
7000mm	51.37	2.50	52.65	1.30	0.16	13.12	6.50	72.28	nr	77.70

Labour hourly rates: (except Specialists) Craft Operatives £9.23 Labourer £7.02 Rates are national average prices. Refer to REGIONAL VARIATIONS for indicative levels of overall pricing in regions	MATERIALS			LABOUR				RATES		
	Del to Site £	Waste %	Material Cost £	Craft Optve Hrs	Lab Hrs	Labour Cost £	Sunds £	Nett Rate £	Unit	Gross Rate (+7.5%) £
G20: CARPENTRY/TIMBER FRAMING/FIRST FIXING Cont.										
Sasco Trussed Rafters in softwood, sawn, B.S.4978, GS grade, impregnated; note:- prices are guide prices only, based on contracts within 30 mile radius of Widnes and are subject to adjustment for quantity: firm quotations should always be obtained Cont.										
Trussed rafters; Hydro-Nail plated joints Cont.										
45 degree Standard duo pitch; 450mm overhangs; fixing with clips (included elsewhere); span over wall plates Cont.										
8000mm	58.75	2.50	60.22	1.40	0.17	14.12	7.00	81.33	nr	87.43
9000mm	66.20	2.50	67.86	1.50	0.19	15.18	7.50	90.53	nr	97.32
22.5 degree Bobtail duo pitch; 450mm overhangs; fixing with clips (included elsewhere); span over wall plates										
4000mm	24.80	2.50	25.42	1.00	0.13	10.14	5.00	40.56	nr	43.60
5000mm	30.91	2.50	31.68	1.10	0.14	11.14	5.50	48.32	nr	51.94
6000mm	37.13	2.50	38.06	1.20	0.15	12.13	6.00	56.19	nr	60.40
7000mm	43.37	2.50	44.45	1.30	0.16	13.12	6.50	64.08	nr	68.88
8000mm	49.60	2.50	50.84	1.40	0.17	14.12	7.00	71.96	nr	77.35
9000mm	55.89	2.50	57.29	1.50	0.19	15.18	7.50	79.97	nr	85.96
35 degree Bobtail duo pitch; 450mm overhangs; fixing with clips (included elsewhere); span over wall plates										
4000mm	25.87	2.50	26.52	1.00	0.13	10.14	5.00	41.66	nr	44.78
5000mm	32.24	2.50	33.05	1.10	0.14	11.14	5.50	49.68	nr	53.41
6000mm	38.79	2.50	39.76	1.20	0.15	12.13	6.00	57.89	nr	62.23
7000mm	45.08	2.50	46.21	1.30	0.16	13.12	6.50	65.83	nr	70.77
8000mm	51.56	2.50	52.85	1.40	0.17	14.12	7.00	73.96	nr	79.51
9000mm	57.94	2.50	59.39	1.50	0.19	15.18	7.50	82.07	nr	88.22
45 degree Bobtail duo pitch; 450mm overhangs; fixing with clips (included elsewhere); span over wall plates										
4000mm	40.13	2.50	41.13	1.00	0.13	10.14	5.00	56.28	nr	60.50
5000mm	50.12	2.50	51.37	1.10	0.14	11.14	5.50	68.01	nr	73.11
6000mm	60.22	2.50	61.73	1.20	0.15	12.13	6.00	79.85	nr	85.84
7000mm	70.26	2.50	72.02	1.30	0.16	13.12	6.50	91.64	nr	98.51
8000mm	86.66	2.50	88.83	1.40	0.17	14.12	7.00	109.94	nr	118.19
9000mm	90.36	2.50	92.62	1.50	0.19	15.18	7.50	115.30	nr	123.95
22.5 degree Monopitch; 450mm overhangs; fixing with clips (included elsewhere); span over wall plates										
2000mm	12.14	2.50	12.44	0.80	0.10	8.09	4.00	24.53	nr	26.37
3000mm	16.09	2.50	16.49	0.90	0.11	9.08	4.50	30.07	nr	32.33
4000mm	20.02	2.50	20.52	1.00	0.13	10.14	5.00	35.66	nr	38.34
5000mm	24.47	2.50	25.08	1.10	0.14	11.14	5.50	41.72	nr	44.85
6000mm	31.60	2.50	32.39	1.20	0.15	12.13	6.00	50.52	nr	54.31
35 degree Monopitch; 450mm overhangs; fixing with clips (included elsewhere); span over wall plates										
2000mm	12.98	2.50	13.30	0.80	0.10	8.09	4.00	25.39	nr	27.29
3000mm	17.10	2.50	17.53	0.90	0.11	9.08	4.50	31.11	nr	33.44
4000mm	21.30	2.50	21.83	1.00	0.13	10.14	5.00	36.98	nr	39.75
5000mm	26.26	2.50	26.92	1.10	0.14	11.14	5.50	43.55	nr	46.82
6000mm	33.84	2.50	34.69	1.20	0.15	12.13	6.00	52.81	nr	56.78
45 degree Monopitch; 450mm overhangs; fixing with clips (included elsewhere); span over wall plates										
2000mm	13.86	2.50	14.21	0.80	0.10	8.09	4.00	26.29	nr	28.26
3000mm	26.40	2.50	27.06	0.90	0.11	9.08	4.50	40.64	nr	43.69
4000mm	31.22	2.50	32.00	1.00	0.13	10.14	5.00	47.14	nr	50.68
22.5 degree duo pitch Girder Truss; fixing with clips (included elsewhere); span over wall plates										
5000mm	91.11	2.50	93.39	1.10	0.14	11.14	5.50	110.02	nr	118.28
6000mm	103.13	2.50	105.71	1.20	0.15	12.13	6.00	123.84	nr	133.13
7000mm	111.76	2.50	114.55	1.30	0.16	13.12	6.50	134.18	nr	144.24
8000mm	129.65	2.50	132.89	1.40	0.17	14.12	7.00	154.01	nr	165.56
9000mm	144.14	2.50	147.74	1.50	0.19	15.18	7.50	170.42	nr	183.20
10000mm	169.85	2.50	174.10	1.60	0.20	16.17	8.00	198.27	nr	213.14
35 degree duo pitch Girder Truss; fixing with clips (included elsewhere); span over wall plates										
5000mm	95.44	2.50	97.83	1.10	0.14	11.14	5.50	114.46	nr	123.05
6000mm	106.31	2.50	108.97	1.20	0.15	12.13	6.00	127.10	nr	136.63
7000mm	116.24	2.50	119.15	1.30	0.16	13.12	6.50	138.77	nr	149.18
8000mm	134.15	2.50	137.50	1.40	0.17	14.12	7.00	158.62	nr	170.52
9000mm	148.54	2.50	152.25	1.50	0.19	15.18	7.50	174.93	nr	188.05
10000mm	174.21	2.50	178.57	1.60	0.20	16.17	8.00	202.74	nr	217.94
45 degree duo pitch Girder Truss; fixing with clips (included elsewhere); span over wall plates										
5000mm	129.20	2.50	132.43	1.10	0.14	11.14	5.50	149.07	nr	160.25
6000mm	157.06	2.50	160.99	1.20	0.15	12.13	6.00	179.12	nr	192.55
7000mm	179.94	2.50	184.44	1.30	0.16	13.12	6.50	204.06	nr	219.37
8000mm	201.56	2.50	206.60	1.40	0.17	14.12	7.00	227.71	nr	244.79
9000mm	223.82	2.50	229.42	1.50	0.19	15.18	7.50	252.09	nr	271.00
22.5 degree pitch gable ladder; 450mm overhang; span over wall plate										
5000mm	12.02	2.50	12.32	1.10	0.14	11.14	5.50	28.96	nr	31.13
6000mm	13.73	2.50	14.07	1.20	0.15	12.13	6.00	32.20	nr	34.62
7000mm	15.39	2.50	15.77	1.30	0.16	13.12	6.50	35.40	nr	38.05
8000mm	18.45	2.50	18.91	1.40	0.17	14.12	7.00	40.03	Unit	43.03
9000mm	21.13	2.50	21.66	1.50	0.19	15.18	7.50	44.34	nr	47.66
10000mm.	25.63	2.50	26.27	1.60	0.20	16.17	8.00	50.44	nr	54.23
35 degree pitch gable ladder; 450mm overhang; span over wall plate										
5000mm	12.47	2.50	12.78	1.10	0.14	11.14	5.50	29.42	nr	31.62
6000mm	14.43	2.50	14.79	1.20	0.15	12.13	6.00	32.92	nr	35.39

Labour hourly rates: (except Specialists) Craft Operatives £9.23 Labourer £7.02 Rates are national average prices. Refer to REGIONAL VARIATIONS for indicative levels of overall pricing in regions	MATERIALS			LABOUR				RATES		
	Del to Site £	Waste %	Material Cost £	Craft Optve Hrs	Lab Hrs	Labour Cost £	Sunds £	Nett Rate £	Unit	Gross Rate (+7.5%) £

G20: CARPENTRY/TIMBER FRAMING/FIRST FIXING Cont.

Sasco Trussed Rafters in softwood, sawn, B.S.4978, GS grade, impregnated; note:- prices are guide prices only, based on contracts within 30 mile radius of Widnes and are subject to adjustment for quantity: firm quotations should always be obtained Cont.

Trussed rafters; Hydro-Nail plated joints Cont.
    35 degree pitch gable ladder; 450mm overhang; span over wall plate Cont.

	Del to Site £	Waste %	Material Cost £	Craft Optve Hrs	Lab Hrs	Labour Cost £	Sunds £	Nett Rate £	Unit	Gross Rate (+7.5%) £
7000mm	15.82	2.50	16.22	1.30	0.16	13.12	6.50	35.84	nr	38.53
8000mm	19.20	2.50	19.68	1.40	0.17	14.12	7.00	40.80	nr	43.86
9000mm	21.93	2.50	22.48	1.50	0.19	15.18	7.50	45.16	nr	48.54
10000mm	26.58	2.50	27.24	1.60	0.20	16.17	8.00	51.42	nr	55.27

45 degree pitch gable ladder; 450mm overhang; span over wall plate

	Del to Site £	Waste %	Material Cost £	Craft Optve Hrs	Lab Hrs	Labour Cost £	Sunds £	Nett Rate £	Unit	Gross Rate (+7.5%) £
5000mm	19.59	2.50	20.08	1.10	0.14	11.14	5.50	36.72	nr	39.47
6000mm	23.45	2.50	24.04	1.20	0.15	12.13	6.00	42.17	nr	45.33
7000mm	27.27	2.50	27.95	1.30	0.16	13.12	6.50	47.57	nr	51.14
8000mm	30.66	2.50	31.43	1.40	0.17	14.12	7.00	52.54	nr	56.48
9000mm	34.86	2.50	35.73	1.50	0.19	15.18	7.50	58.41	nr	62.79
10000mm	38.42	2.50	39.38	1.60	0.20	16.17	8.00	63.55	nr	68.32

Softwood, wrot, B.S.4978, GS grade - laminated beams

Glued laminated beams; B.S.4169

	Del to Site £	Waste %	Material Cost £	Craft Optve Hrs	Lab Hrs	Labour Cost £	Sunds £	Nett Rate £	Unit	Gross Rate (+7.5%) £
65 x 150 x 4000mm	41.17	2.50	42.20	0.50	0.06	5.04	3.20	50.44	nr	54.22
65 x 175 x 4000mm	47.95	2.50	49.15	0.60	0.07	6.03	3.75	58.93	nr	63.35
65 x 200 x 4000mm	54.72	2.50	56.09	0.60	0.07	6.03	4.25	66.37	nr	71.34
65 x 225 x 4000mm	61.81	2.50	63.36	0.80	0.10	8.09	4.75	76.19	nr	81.91
65 x 250 x 4000mm	68.91	2.50	70.63	0.90	0.11	9.08	5.30	85.01	nr	91.39
65 x 250 x 6000mm	103.31	2.50	105.89	1.10	0.14	11.14	5.90	122.93	nr	132.15
65 x 275 x 4000mm	75.63	2.50	77.52	1.00	0.13	10.14	5.90	93.56	nr	100.58
65 x 275 x 6000mm	114.92	2.50	117.79	1.20	0.15	12.13	6.40	136.32	nr	146.55
65 x 300 x 4000mm	81.55	2.50	83.59	1.20	0.15	12.13	6.40	102.12	nr	109.78
65 x 300 x 6000mm	123.03	2.50	126.11	1.30	0.16	13.12	6.90	146.13	nr	157.09
65 x 325 x 4000mm	90.58	2.50	92.84	1.30	0.16	13.12	6.90	112.87	nr	121.33
65 x 325 x 6000mm	133.19	2.50	136.52	1.40	0.18	14.19	7.45	158.16	nr	170.02
65 x 325 x 8000mm	178.17	2.50	182.62	1.50	0.19	15.18	8.00	205.80	nr	221.24
90 x 150 x 4000mm	52.20	2.50	53.51	0.70	0.09	7.09	4.25	64.85	nr	69.71
90 x 175 x 4000mm	61.52	2.50	63.06	0.80	0.10	8.09	4.75	75.89	nr	81.59
90 x 200 x 4000mm	69.98	2.50	71.73	0.90	0.11	9.08	5.30	86.11	nr	92.57
90 x 225 x 4000mm	78.16	2.50	80.11	1.00	0.13	10.14	5.90	96.16	nr	103.37
90 x 250 x 4000mm	88.26	2.50	90.47	1.10	0.14	11.14	6.40	108.00	nr	116.10
90 x 250 x 6000mm	130.38	2.50	133.64	1.30	0.16	13.12	6.90	153.66	nr	165.19
90 x 275 x 4000mm	96.51	2.50	98.92	1.20	0.15	12.13	6.90	117.95	nr	126.80
90 x 275 x 6000mm	145.33	2.50	148.96	1.40	0.18	14.19	7.45	170.60	nr	183.39
90 x 300 x 4000mm	104.41	2.50	107.02	1.40	0.18	14.19	7.45	128.66	nr	138.31
90 x 300 x 6000mm	159.15	2.50	163.13	1.50	0.19	15.18	8.05	186.36	nr	200.33
90 x 325 x 4000mm	115.41	2.50	118.30	1.50	0.19	15.18	8.05	141.52	nr	152.14
90 x 325 x 6000mm	171.57	2.50	175.86	1.60	0.20	16.17	8.55	200.58	nr	215.62
90 x 325 x 8000mm	227.16	2.50	232.84	1.70	0.21	17.17	9.10	259.10	nr	278.54
90 x 350 x 4000mm	121.91	2.50	124.96	1.60	0.20	16.17	8.55	149.68	nr	160.91
90 x 350 x 6000mm	185.12	2.50	189.75	1.70	0.21	17.17	9.10	216.01	nr	232.21
90 x 350 x 8000mm	247.20	2.50	253.38	1.80	0.22	18.16	9.60	281.14	nr	302.22
90 x 375 x 4000mm	130.38	2.50	133.64	1.70	0.21	17.17	9.10	159.90	nr	171.90
90 x 375 x 6000mm	196.40	2.50	201.31	1.80	0.22	18.16	9.60	229.07	nr	246.25
90 x 375 x 8000mm	263.28	2.50	269.86	1.90	0.24	19.22	10.20	299.28	nr	321.73
90 x 375 x 10000mm	329.04	2.50	337.27	2.00	0.25	20.22	10.65	368.13	nr	395.74
90 x 400 x 4000mm	141.66	2.50	145.20	1.80	0.22	18.16	9.60	172.96	nr	185.93
90 x 400 x 6000mm	209.95	2.50	215.20	1.90	0.24	19.22	10.20	244.62	nr	262.97
90 x 400 x 8000mm	280.50	2.50	287.51	2.00	0.25	20.22	10.65	318.38	nr	342.26
90 x 400 x 10000mm	350.19	2.50	358.94	2.10	0.26	21.21	11.20	391.35	nr	420.70
90 x 425 x 6000mm	234.78	2.50	240.65	2.00	0.25	20.22	10.65	271.51	nr	291.88
90 x 425 x 8000mm	295.73	2.50	303.12	2.10	0.26	21.21	11.20	335.53	nr	360.70
90 x 425 x 10000mm	368.54	2.50	377.75	2.20	0.28	22.27	11.70	411.73	nr	442.60
90 x 425 x 12000mm	442.47	2.50	453.53	2.30	0.29	23.26	12.25	489.05	nr	525.73
90 x 450 x 6000mm	237.04	2.50	242.97	2.10	0.26	21.21	11.20	275.37	nr	296.03
90 x 450 x 8000mm	315.49	2.50	323.38	2.20	0.28	22.27	11.70	357.35	nr	384.15
90 x 450 x 10000mm	393.37	2.50	403.20	2.30	0.29	23.26	12.25	438.72	nr	471.62
90 x 450 x 12000mm	472.95	2.50	484.77	2.40	0.30	24.26	12.80	521.83	nr	560.97
115 x 250 x 4000mm	111.47	2.50	114.26	1.30	0.16	13.12	7.45	134.83	nr	144.94
115 x 250 x 6000mm	164.86	2.50	168.98	1.50	0.19	15.18	8.05	192.21	nr	206.63
115 x 275 x 4000mm	121.06	2.50	124.09	1.40	0.18	14.19	8.05	146.32	nr	157.30
115 x 275 x 6000mm	185.12	2.50	189.75	1.60	0.20	16.17	8.55	214.47	nr	230.56
115 x 300 x 4000mm	134.04	2.50	137.39	1.60	0.20	16.17	8.55	162.11	nr	174.27
115 x 300 x 6000mm	201.20	2.50	206.23	1.70	0.21	17.17	9.10	232.50	nr	249.93
115 x 325 x 4000mm	145.33	2.50	148.96	1.70	0.21	17.17	9.10	175.23	nr	188.37
115 x 325 x 6000mm	218.42	2.50	223.88	1.80	0.22	18.16	9.60	251.64	nr	270.51
115 x 325 x 8000mm	291.77	2.50	299.06	1.90	0.24	19.22	10.20	328.49	nr	353.12
115 x 350 x 4000mm	158.87	2.50	162.84	1.80	0.22	18.16	9.60	190.60	nr	204.90
115 x 350 x 6000mm	237.04	2.50	242.97	1.90	0.24	19.22	10.20	272.39	nr	292.82
115 x 350 x 8000mm	315.49	2.50	323.38	2.00	0.25	20.22	10.65	354.24	nr	380.81
115 x 375 x 4000mm	170.16	2.50	174.41	1.90	0.24	19.22	10.20	203.84	nr	219.12
115 x 375 x 6000mm	253.41	2.50	259.75	2.00	0.25	20.22	10.65	290.61	nr	312.41
115 x 375 x 8000mm	337.76	2.50	346.20	2.10	0.26	21.21	11.20	378.61	nr	407.01
115 x 400 x 4000mm	180.04	2.50	184.54	2.00	0.25	20.22	10.65	215.41	nr	231.56
115 x 400 x 6000mm	270.90	2.50	277.67	2.10	0.26	21.21	11.20	310.08	nr	333.34
115 x 400 x 8000mm	361.20	2.50	370.23	2.20	0.28	22.27	11.70	404.20	nr	434.52
115 x 400 x 10000mm	452.07	2.50	463.37	2.30	0.29	23.26	12.25	498.89	nr	536.30
115 x 425 x 6000mm	280.50	2.50	287.51	2.20	0.28	22.27	11.70	321.48	nr	345.60
115 x 425 x 8000mm	375.02	2.50	384.40	2.30	0.29	23.26	12.25	419.91	nr	451.40
115 x 425 x 10000mm	466.74	2.50	478.41	2.40	0.30	24.26	12.80	515.47	nr	554.13
115 x 425 x 12000mm	558.73	2.50	572.70	2.50	0.31	25.25	13.35	611.30	nr	657.15
115 x 450 x 6000mm	296.86	2.50	304.28	2.30	0.29	23.26	12.25	339.80	nr	365.28

STRUCTURAL/CARCASSING METAL/TIMBER

Labour hourly rates: (except Specialists) Craft Operatives £9.23 Labourer £7.02 Rates are national average prices. Refer to REGIONAL VARIATIONS for indicative levels of overall pricing in regions	MATERIALS			LABOUR				RATES		
	Del to Site	Waste	Material Cost	Craft Optve	Lab	Labour Cost	Sunds	Nett Rate	Unit	Gross Rate (+7.5%)
	£	%	£	Hrs	Hrs	£	£	£		£

G20: CARPENTRY/TIMBER FRAMING/FIRST FIXING Cont.

Softwood, wrot, B.S.4978, GS grade - laminated beams Cont.

Glued laminated beams; B.S.4169 Cont.

	Del to Site £	Waste %	Material Cost £	Craft Optve Hrs	Lab Hrs	Labour Cost £	Sunds £	Nett Rate £	Unit	Gross Rate £
115 x 450 x 8000mm	396.19	2.50	406.09	2.40	0.30	24.26	12.80	443.15	nr	476.39
115 x 450 x 10000mm	496.65	2.50	509.07	2.50	0.31	25.25	13.35	547.67	nr	588.74
115 x 450 x 12000mm	595.98	2.50	610.88	2.60	0.33	26.31	13.85	651.04	nr	699.87
115 x 475 x 6000mm	315.49	2.50	323.38	2.40	0.30	24.26	12.80	360.44	nr	387.47
115 x 475 x 8000mm	421.02	2.50	431.55	2.50	0.31	25.25	13.35	470.15	nr	505.41
115 x 475 x 10000mm	526.67	2.50	539.84	2.60	0.33	26.31	13.85	580.00	nr	623.50
115 x 475 x 12000mm	629.28	2.50	645.01	2.70	0.34	27.31	14.45	686.77	nr	738.28
115 x 475 x 14000mm	733.69	2.50	752.03	2.80	0.35	28.30	14.95	795.28	nr	854.93
115 x 500 x 6000mm	332.98	2.50	341.30	2.50	0.31	25.25	13.35	379.91	nr	408.40
115 x 500 x 8000mm	441.91	2.50	452.96	2.60	0.33	26.31	13.85	493.12	nr	530.11
115 x 500 x 10000mm	555.07	2.50	568.95	2.70	0.34	27.31	14.45	610.70	nr	656.51
115 x 500 x 12000mm	665.37	2.50	682.00	2.80	0.35	28.30	14.95	725.26	nr	779.65
115 x 500 x 14000mm	774.89	2.50	794.26	2.90	0.36	29.29	15.50	839.06	nr	901.99
140 x 350 x 4000mm	191.04	2.50	195.82	2.00	0.25	20.22	10.65	226.68	nr	243.68
140 x 350 x 6000mm	289.24	2.50	296.47	2.10	0.26	21.21	11.20	328.88	nr	353.55
140 x 350 x 8000mm	384.90	2.50	394.52	2.20	0.28	22.27	11.70	428.49	nr	460.63
140 x 375 x 4000mm	203.45	2.50	208.54	2.10	0.26	21.21	11.20	240.94	nr	259.02
140 x 375 x 6000mm	319.16	2.50	327.14	2.20	0.28	22.27	11.70	361.11	nr	388.19
140 x 375 x 8000mm	408.61	2.50	418.83	2.30	0.29	23.26	12.25	454.34	nr	488.42
140 x 375 x 10000mm	510.20	2.50	522.96	2.40	0.30	24.26	12.80	560.01	nr	602.01
140 x 400 x 6000mm	322.82	2.50	330.89	2.30	0.29	23.26	12.25	366.41	nr	393.89
140 x 400 x 8000mm	433.44	2.50	444.28	2.40	0.30	24.26	12.80	481.33	nr	517.43
140 x 400 x 10000mm	540.11	2.50	553.61	2.50	0.31	25.25	13.35	592.21	nr	636.63
140 x 425 x 6000mm	348.78	2.50	357.50	2.40	0.30	24.26	12.80	394.56	nr	424.15
140 x 425 x 8000mm	466.74	2.50	478.41	2.50	0.31	25.25	13.35	517.01	nr	555.79
140 x 425 x 10000mm	583.56	2.50	598.15	2.60	0.33	26.31	13.85	638.31	nr	686.19
140 x 425 x 12000mm	700.39	2.50	717.90	2.70	0.34	27.31	14.45	759.66	nr	816.63
140 x 450 x 6000mm	367.41	2.50	376.60	2.50	0.31	25.25	13.35	415.20	nr	446.34
140 x 450 x 8000mm	490.45	2.50	502.71	2.60	0.33	26.31	13.85	542.88	nr	583.59
140 x 450 x 10000mm	611.78	2.50	627.07	2.70	0.34	27.31	14.45	668.83	nr	718.99
140 x 450 x 12000mm	733.69	2.50	752.03	2.80	0.35	28.30	14.95	795.28	nr	854.93
140 x 475 x 6000mm	384.90	2.50	394.52	2.60	0.33	26.31	13.85	434.69	nr	467.29
140 x 475 x 8000mm	512.73	2.50	525.55	2.70	0.34	27.31	14.45	567.31	nr	609.85
140 x 475 x 10000mm	641.98	2.50	658.03	2.80	0.35	28.30	14.95	701.28	nr	753.88
140 x 475 x 12000mm	770.94	2.50	790.21	2.90	0.36	29.29	15.50	835.01	nr	897.63
140 x 475 x 14000mm	895.10	2.50	917.48	3.00	0.38	30.36	16.00	963.84	nr	1036.12
165 x 425 x 6000mm	410.87	2.50	421.14	2.60	0.33	26.31	13.85	461.31	nr	495.90
165 x 425 x 8000mm	548.57	2.50	562.28	2.70	0.34	27.31	14.45	604.04	nr	649.35
165 x 425 x 10000mm	685.58	2.50	702.72	2.80	0.35	28.30	14.95	745.97	nr	801.92
165 x 425 x 12000mm	821.30	2.50	841.83	2.90	0.36	29.29	15.50	886.63	nr	953.12
165 x 450 x 6000mm	428.37	2.50	439.08	2.70	0.34	27.31	14.45	480.84	nr	516.90
165 x 450 x 8000mm	572.28	2.50	586.59	2.80	0.35	28.30	14.95	629.84	nr	677.08
165 x 450 x 10000mm	715.07	2.50	732.95	2.90	0.36	29.29	15.50	777.74	nr	836.07
165 x 450 x 12000mm	858.98	2.50	880.45	3.00	0.38	30.36	16.00	926.81	nr	996.32
165 x 475 x 6000mm	454.33	2.50	465.69	2.80	0.35	28.30	14.95	508.94	nr	547.11
165 x 475 x 8000mm	607.27	2.50	622.45	2.90	0.36	29.29	15.50	667.25	nr	717.29
165 x 475 x 10000mm	757.39	2.50	776.32	3.00	0.38	30.36	16.00	822.68	nr	884.38
165 x 475 x 12000mm	911.18	2.50	933.96	3.10	0.39	31.35	16.55	981.86	nr	1055.50
165 x 475 x 14000mm	1061.59	2.50	1088.13	3.20	0.40	32.34	17.10	1137.57	nr	1222.89
190 x 475 x 6000mm	526.28	2.50	539.44	3.00	0.38	30.36	16.00	585.79	nr	629.73
190 x 475 x 8000mm	700.10	2.50	717.60	3.10	0.39	31.35	16.55	765.50	nr	822.92
190 x 475 x 10000mm	875.35	2.50	897.23	3.20	0.40	32.34	17.10	946.68	nr	1017.68
190 x 475 x 12000mm	1050.31	2.50	1076.57	3.30	0.41	33.34	17.60	1127.50	nr	1212.07
190 x 475 x 14000mm	1225.27	2.50	1255.90	3.40	0.42	34.33	18.10	1308.33	nr	1406.46

Softwood, sawn, B.S.4978, GS grade - carcassing

Floor members

	Del to Site £	Waste %	Material Cost £	Craft Optve Hrs	Lab Hrs	Labour Cost £	Sunds £	Nett Rate £	Unit	Gross Rate £
38 x 75mm	0.57	10.00	0.63	0.12	0.01	1.18	0.02	1.82	m	1.96
38 x 100mm	0.76	10.00	0.84	0.14	0.02	1.43	0.02	2.29	m	2.46
50 x 75mm	0.68	10.00	0.75	0.14	0.02	1.43	0.02	2.20	m	2.37
50 x 100mm	0.90	10.00	0.99	0.16	0.02	1.62	0.03	2.64	m	2.83
50 x 175mm	1.58	10.00	1.74	0.24	0.03	2.43	0.04	4.20	m	4.52
50 x 200mm	1.80	10.00	1.98	0.27	0.04	2.77	0.04	4.79	m	5.15
50 x 225mm	2.03	10.00	2.23	0.29	0.04	2.96	0.05	5.24	m	5.63
50 x 250mm	2.25	10.00	2.48	0.32	0.04	3.23	0.05	5.76	m	6.19
75 x 150mm	2.14	10.00	2.35	0.29	0.04	2.96	0.05	5.36	m	5.76
75 x 175mm	2.49	10.00	2.74	0.33	0.04	3.33	0.05	6.12	m	6.57
75 x 200mm	2.85	10.00	3.13	0.35	0.05	3.58	0.06	6.78	m	7.28
75 x 225mm	3.21	10.00	3.53	0.40	0.05	4.04	0.06	7.63	m	8.21
75 x 250mm	3.56	10.00	3.92	0.45	0.06	4.57	0.07	8.56	m	9.20

Wall or partition members

	Del to Site £	Waste %	Material Cost £	Craft Optve Hrs	Lab Hrs	Labour Cost £	Sunds £	Nett Rate £	Unit	Gross Rate £
38 x 75mm	0.57	10.00	0.63	0.18	0.02	1.80	0.03	2.46	m	2.64
38 x 100mm	0.76	10.00	0.84	0.21	0.03	2.15	0.03	3.01	m	3.24
50 x 75mm	0.68	10.00	0.75	0.21	0.03	2.15	0.03	2.93	m	3.15
50 x 100mm	0.90	10.00	0.99	0.24	0.03	2.43	0.04	3.46	m	3.71
75 x 100mm	1.43	10.00	1.57	0.30	0.04	3.05	0.05	4.67	m	5.02
38 x 75mm; fixing to masonry	0.57	10.00	0.63	0.36	0.04	3.60	0.06	4.29	m	4.61
38 x 100mm; fixing to masonry	0.76	10.00	0.84	0.39	0.05	3.95	0.06	4.85	m	5.21
50 x 75mm; fixing to masonry	0.68	10.00	0.75	0.39	0.05	3.95	0.06	4.76	m	5.12
50 x 100mm; fixing to masonry	0.90	10.00	0.99	0.43	0.05	4.32	0.07	5.38	m	5.78
75 x 100mm; fixing to masonry	1.43	10.00	1.57	0.50	0.06	5.04	0.08	6.69	m	7.19

Plates

	Del to Site £	Waste %	Material Cost £	Craft Optve Hrs	Lab Hrs	Labour Cost £	Sunds £	Nett Rate £	Unit	Gross Rate £
38 x 75mm	0.57	10.00	0.63	0.18	0.02	1.80	0.03	2.46	m	2.64
38 x 100mm	0.76	10.00	0.84	0.21	0.03	2.15	0.03	3.01	m	3.24
50 x 75mm	0.68	10.00	0.75	0.21	0.03	2.15	0.03	2.93	m	3.15
50 x 100mm wrot	0.90	10.00	0.99	0.24	0.03	2.43	0.04	3.46	m	3.71
75 x 150mm	2.14	10.00	2.35	0.40	0.05	4.04	0.06	6.46	m	6.94

Labour hourly rates: (except Specialists) Craft Operatives £9.23 Labourer £7.02 Rates are national average prices. Refer to REGIONAL VARIATIONS for indicative levels of overall pricing in regions	MATERIALS			LABOUR				RATES		
	Del to Site £	Waste %	Material Cost £	Craft Optve Hrs	Lab Hrs	Labour Cost £	Sunds £	Nett Rate £	Unit	Gross Rate (+7.5%) £
G20: CARPENTRY/TIMBER FRAMING/FIRST FIXING Cont.										
Softwood, sawn, B.S.4978, GS grade - carcassing Cont.										
Plates Cont.										
38 x 75mm; fixing by bolting	0.57	10.00	0.63	0.22	0.03	2.24	-	2.87	m	3.08
38 x 100mm; fixing by bolting	0.76	10.00	0.84	0.25	0.03	2.52	-	3.35	m	3.61
50 x 75mm; fixing by bolting	0.68	10.00	0.75	0.25	0.03	2.52	-	3.26	m	3.50
50 x 100mm; fixing by bolting	0.90	10.00	0.99	0.29	0.04	2.96	-	3.95	m	4.24
75 x 150mm; fixing by bolting	2.14	10.00	2.35	0.48	0.06	4.85	-	7.21	m	7.75
Roof members; flat										
50 x 175mm	1.66	10.00	1.83	0.24	0.03	2.43	0.04	4.29	m	4.61
50 x 200mm	1.90	10.00	2.09	0.27	0.04	2.77	0.04	4.90	m	5.27
50 x 225mm	2.14	10.00	2.35	0.29	0.04	2.96	0.05	5.36	m	5.76
50 x 250mm	2.38	10.00	2.62	0.32	0.04	3.23	0.05	5.90	m	6.35
75 x 175mm	2.49	10.00	2.74	0.33	0.04	3.33	0.05	6.12	m	6.57
75 x 200mm	2.85	10.00	3.13	0.35	0.05	3.58	0.06	6.78	m	7.28
75 x 225mm	3.21	10.00	3.53	0.40	0.05	4.04	0.06	7.63	m	8.21
75 x 250mm	3.56	10.00	3.92	0.45	0.06	4.57	0.07	8.56	m	9.20
Roof members; pitched										
25 x 75mm	0.38	10.00	0.42	0.14	0.02	1.43	0.02	1.87	m	2.01
25 x 125mm	0.63	10.00	0.69	0.16	0.02	1.62	0.03	2.34	m	2.52
32 x 175mm	1.12	10.00	1.23	0.22	0.03	2.24	0.04	3.51	m	3.78
32 x 225mm	1.44	10.00	1.58	0.25	0.03	2.52	0.04	4.14	m	4.45
38 x 75mm	0.57	10.00	0.63	0.15	0.02	1.52	0.02	2.17	m	2.33
38 x 100mm	0.76	10.00	0.84	0.18	0.02	1.80	0.03	2.67	m	2.87
50 x 75mm	0.68	10.00	0.75	0.18	0.02	1.80	0.03	2.58	m	2.77
50 x 100mm	0.90	10.00	0.99	0.20	0.03	2.06	0.03	3.08	m	3.31
50 x 125mm	1.13	10.00	1.24	0.23	0.03	2.33	0.04	3.62	m	3.89
50 x 150mm	1.35	10.00	1.49	0.26	0.03	2.61	0.04	4.14	m	4.45
50 x 225mm	2.03	10.00	2.23	0.36	0.05	3.67	0.06	5.97	m	6.41
50 x 250mm	2.25	10.00	2.48	0.40	0.05	4.04	0.06	6.58	m	7.07
75 x 100mm	1.43	10.00	1.57	0.26	0.03	2.61	0.04	4.22	m	4.54
75 x 150mm	2.14	10.00	2.35	0.35	0.05	3.58	0.06	6.00	m	6.45
100 x 150mm	2.85	10.00	3.13	0.45	0.06	4.57	0.07	7.78	m	8.36
100 x 225mm	4.28	10.00	4.71	0.60	0.08	6.10	0.10	10.91	m	11.73
Joist strutting; herringbone; depth of joist 175mm										
50 x 50mm	0.99	10.00	1.09	0.65	0.08	6.56	0.11	7.76	m	8.34
Joist strutting; herringbone; depth of joist 200mm										
50 x 50mm	1.04	10.00	1.14	0.65	0.08	6.56	0.11	7.82	m	8.40
Joist strutting; herringbone; depth of joist 225mm										
50 x 50mm	1.08	10.00	1.19	0.65	0.08	6.56	0.11	7.86	m	8.45
Joist strutting; herringbone; depth of joist 250mm										
50 x 50mm	1.13	10.00	1.24	0.65	0.08	6.56	0.11	7.91	m	8.51
Joist strutting; block; depth of joist 150mm										
50 x 150mm	1.35	10.00	1.49	0.65	0.08	6.56	0.11	8.16	m	8.77
Joist strutting; block; depth of joist 175mm										
50 x 175mm	1.58	10.00	1.74	0.70	0.09	7.09	0.11	8.94	m	9.61
Joist strutting; block; depth of joist 200mm										
50 x 200mm	1.80	10.00	1.98	0.70	0.09	7.09	0.11	9.18	m	9.87
Joist strutting; block; depth of joist 225mm										
50 x 225mm	2.03	10.00	2.23	0.75	0.09	7.55	0.12	9.91	m	10.65
Joist strutting; block; depth of joist 250mm										
50 x 250mm	2.25	10.00	2.48	0.75	0.09	7.55	0.12	10.15	m	10.91
Noggings to joists										
50 x 50mm	0.45	10.00	0.50	0.28	0.04	2.87	0.05	3.41	m	3.67
50 x 75mm	0.68	10.00	0.75	0.33	0.04	3.33	0.05	4.12	m	4.43
Wrot surfaces										
plain; 50mm wide	-	-	-	0.15	-	1.38	-	1.38	m	1.49
plain; 75mm wide	-	-	-	0.20	-	1.85	-	1.85	m	1.98
plain; 100mm wide	-	-	-	0.25	-	2.31	-	2.31	m	2.48
plain; 150mm wide	-	-	-	0.30	-	2.77	-	2.77	m	2.98
plain; 200mm wide	-	-	-	0.35	-	3.23	-	3.23	m	3.47
Softwood, sawn, B.S.4978, GS grade, impregnated - carcassing										
Floor members										
38 x 75mm	0.64	10.00	0.70	0.12	0.01	1.18	0.02	1.90	m	2.04
38 x 100mm	0.86	10.00	0.95	0.14	0.02	1.43	0.02	2.40	m	2.58
50 x 75mm	0.77	10.00	0.85	0.14	0.02	1.43	0.02	2.30	m	2.47
50 x 100mm	1.03	10.00	1.13	0.16	0.02	1.62	0.03	2.78	m	2.99
50 x 175mm	1.79	10.00	1.97	0.24	0.03	2.43	0.04	4.43	m	4.77
50 x 200mm	2.05	10.00	2.25	0.27	0.04	2.77	0.04	5.07	m	5.45
50 x 225mm	2.31	10.00	2.54	0.29	0.04	2.96	0.05	5.55	m	5.96
50 x 250mm	2.56	10.00	2.82	0.32	0.04	3.23	0.05	6.10	m	6.56
75 x 150mm	2.42	10.00	2.66	0.29	0.04	2.96	0.05	5.67	m	6.09
75 x 175mm	2.82	10.00	3.10	0.33	0.04	3.33	0.05	6.48	m	6.96
75 x 200mm	3.23	10.00	3.55	0.35	0.05	3.58	0.06	7.19	m	7.73
75 x 225mm	3.63	10.00	3.99	0.40	0.05	4.04	0.06	8.10	m	8.70
75 x 250mm	4.03	10.00	4.43	0.45	0.06	4.57	0.07	9.08	m	9.76

Labour hourly rates: (except Specialists) Craft Operatives £9.23 Labourer £7.02 Rates are national average prices. Refer to REGIONAL VARIATIONS for indicative levels of overall pricing in regions	MATERIALS			LABOUR				RATES		
	Del to Site	Waste	Material Cost	Craft Optve	Lab	Labour Cost	Sunds	Nett Rate	Unit	Gross Rate (+7.5%)
	£	%	£	Hrs	Hrs	£	£	£		£
G20: CARPENTRY/TIMBER FRAMING/FIRST FIXING Cont.										
Softwood, sawn, B.S.4978, GS grade, impregnated – carcassing Cont.										
Wall or partition members										
38 x 75mm	0.64	10.00	0.70	0.18	0.02	1.80	0.03	2.54	m	2.73
38 x 100mm	0.86	10.00	0.95	0.21	0.03	2.15	0.03	3.12	m	3.36
50 x 75mm	0.77	10.00	0.85	0.21	0.03	2.15	0.03	3.03	m	3.25
50 x 100mm	1.03	10.00	1.13	0.24	0.03	2.43	0.04	3.60	m	3.87
75 x 100mm	1.61	10.00	1.77	0.30	0.04	3.05	0.05	4.87	m	5.24
38 x 75mm; fixing to masonry	0.64	10.00	0.70	0.36	0.04	3.60	0.06	4.37	m	4.70
38 x 100mm; fixing to masonry	0.86	10.00	0.95	0.39	0.05	3.95	0.06	4.96	m	5.33
50 x 75mm; fixing to masonry	0.77	10.00	0.85	0.39	0.05	3.95	0.06	4.86	m	5.22
50 x 100mm; fixing to masonry	1.03	10.00	1.13	0.43	0.05	4.32	0.07	5.52	m	5.94
75 x 100mm; fixing to masonry	1.61	10.00	1.77	0.50	0.06	5.04	0.08	6.89	m	7.40
Plates										
38 x 75mm	0.64	10.00	0.70	0.18	0.02	1.80	0.03	2.54	m	2.73
38 x 100mm	0.86	10.00	0.95	0.21	0.03	2.15	0.03	3.12	m	3.36
50 x 75mm	0.77	10.00	0.85	0.21	0.03	2.15	0.03	3.03	m	3.25
50 x 100mm	1.03	10.00	1.13	0.24	0.03	2.43	0.04	3.60	m	3.87
75 x 150mm	2.42	10.00	2.66	0.40	0.05	4.04	0.06	6.76	m	7.27
38 x 75mm; fixing by bolting	0.64	10.00	0.70	0.22	0.03	2.24	-	2.95	m	3.17
38 x 100mm; fixing by bolting	0.86	10.00	0.95	0.25	0.03	2.52	-	3.46	m	3.72
50 x 75mm; fixing by bolting	0.77	10.00	0.85	0.25	0.03	2.52	-	3.37	m	3.62
50 x 100mm; fixing by bolting	1.03	10.00	1.13	0.29	0.04	2.96	-	4.09	m	4.40
75 x 150mm; fixing by bolting	2.42	10.00	2.66	0.48	0.06	4.85	-	7.51	m	8.08
Roof members; flat										
50 x 175mm	1.79	10.00	1.97	0.24	0.03	2.43	0.04	4.43	m	4.77
50 x 200mm	2.05	10.00	2.25	0.27	0.04	2.77	0.04	5.07	m	5.45
50 x 225mm	2.31	10.00	2.54	0.29	0.04	2.96	0.05	5.55	m	5.96
50 x 250mm	2.56	10.00	2.82	0.32	0.04	3.23	0.05	6.10	m	6.56
75 x 175mm	2.82	10.00	3.10	0.33	0.04	3.33	0.05	6.48	m	6.96
75 x 200mm	3.23	10.00	3.55	0.35	0.05	3.58	0.06	7.19	m	7.73
75 x 225mm	3.63	10.00	3.99	0.40	0.05	4.04	0.06	8.10	m	8.70
75 x 250mm	4.03	10.00	4.43	0.45	0.06	4.57	0.07	9.08	m	9.76
Roof members; pitched										
25 x 75mm	0.42	10.00	0.46	0.14	0.02	1.43	0.02	1.91	m	2.06
25 x 125mm	0.70	10.00	0.77	0.16	0.02	1.62	0.03	2.42	m	2.60
32 x 175mm	1.26	10.00	1.39	0.22	0.03	2.24	0.04	3.67	m	3.94
32 x 225mm	1.62	10.00	1.78	0.25	0.03	2.52	0.04	4.34	m	4.67
38 x 75mm	0.64	10.00	0.70	0.15	0.02	1.52	0.02	2.25	m	2.42
38 x 100mm	0.86	10.00	0.95	0.18	0.02	1.80	0.03	2.78	m	2.99
50 x 75mm	0.77	10.00	0.85	0.18	0.02	1.80	0.03	2.68	m	2.88
50 x 100mm	1.03	10.00	1.13	0.20	0.03	2.06	0.03	3.22	m	3.46
50 x 125mm	1.28	10.00	1.41	0.23	0.03	2.33	0.04	3.78	m	4.07
50 x 150mm	1.54	10.00	1.69	0.26	0.03	2.61	0.04	4.34	m	4.67
50 x 225mm	2.31	10.00	2.54	0.36	0.05	3.67	0.06	6.27	m	6.75
50 x 250mm	2.56	10.00	2.82	0.40	0.05	4.04	0.06	6.92	m	7.44
75 x 100mm	1.61	10.00	1.77	0.26	0.03	2.61	0.04	4.42	m	4.75
75 x 150mm	2.42	10.00	2.66	0.35	0.05	3.58	0.06	6.30	m	6.78
100 x 150mm	3.23	10.00	3.55	0.45	0.06	4.57	0.07	8.20	m	8.81
100 x 225mm	4.61	10.00	5.07	0.60	0.08	6.10	0.10	11.27	m	12.12
Joist strutting; herringbone; depth of joist 175mm										
50 x 50mm	1.13	10.00	1.24	0.65	0.08	6.56	0.11	7.91	m	8.51
Joist strutting; herringbone; depth of joist 200mm										
50 x 50mm	1.18	10.00	1.30	0.65	0.08	6.56	0.11	7.97	m	8.57
Joist strutting; herringbone; depth of joist 225mm										
50 x 50mm	1.23	10.00	1.35	0.65	0.08	6.56	0.11	8.02	m	8.63
Joist strutting; herringbone; depth of joist 250mm										
50 x 50mm	1.28	10.00	1.41	0.65	0.08	6.56	0.11	8.08	m	8.69
Joist strutting; block; depth of joist 150mm										
50 x 150mm	1.54	10.00	1.69	0.65	0.08	6.56	0.11	8.37	m	8.99
Joist strutting; block; depth of joist 175mm										
50 x 175mm	1.79	10.00	1.97	0.70	0.09	7.09	0.11	9.17	m	9.86
Joist strutting; block; depth of joist 200mm										
50 x 200mm	2.05	10.00	2.25	0.70	0.09	7.09	0.11	9.46	m	10.17
Joist strutting; block; depth of joist 225mm										
50 x 225mm	2.31	10.00	2.54	0.75	0.09	7.55	0.12	10.22	m	10.98
Joist strutting; block; depth of joist 250mm										
50 x 250mm	2.56	10.00	2.82	0.75	0.09	7.55	0.12	10.49	m	11.28
Noggings to joists										
50 x 50mm	0.51	10.00	0.56	0.28	0.04	2.87	0.05	3.48	m	3.74
50 x 75mm	0.77	10.00	0.85	0.33	0.04	3.33	0.05	4.22	m	4.54
Wrot surfaces										
plain; 50mm wide	-	-	-	0.15	-	1.38	-	1.38	m	1.49
plain; 75mm wide	-	-	-	0.20	-	1.85	-	1.85	m	1.98
plain; 100mm wide	-	-	-	0.25	-	2.31	-	2.31	m	2.48
plain; 150mm wide	-	-	-	0.30	-	2.77	-	2.77	m	2.98
plain; 200mm wide	-	-	-	0.35	-	3.23	-	3.23	m	3.47

Labour hourly rates: (except Specialists) Craft Operatives £9.23 Labourer £7.02 Rates are national average prices. Refer to REGIONAL VARIATIONS for indicative levels of overall pricing in regions	MATERIALS			LABOUR				RATES		
	Del to Site	Waste	Material Cost	Craft Optve	Lab	Labour Cost	Sunds	Nett Rate	Unit	Gross Rate (+7.5%)
	£	%	£	Hrs	Hrs	£	£	£		£

G20: CARPENTRY/TIMBER FRAMING/FIRST FIXING Cont.

Oak, sawn - carcassing

Floor members										
38 x 75mm	3.70	10.00	4.07	0.21	0.03	2.15	0.03	6.25	m	6.72
38 x 100mm	4.94	10.00	5.43	0.24	0.03	2.43	0.04	7.90	m	8.49
50 x 75mm	4.87	10.00	5.36	0.24	0.03	2.43	0.04	7.82	m	8.41
50 x 100mm	6.50	10.00	7.15	0.28	0.04	2.87	0.05	10.07	m	10.82
50 x 175mm	11.37	10.00	12.51	0.42	0.05	4.23	0.07	16.80	m	18.06
50 x 200mm	13.00	10.00	14.30	0.47	0.06	4.76	0.08	19.14	m	20.57
50 x 225mm	14.62	10.00	16.08	0.51	0.06	5.13	0.08	21.29	m	22.89
50 x 250mm	16.25	10.00	17.88	0.56	0.07	5.66	0.09	23.63	m	25.40
75 x 150mm	14.62	10.00	16.08	0.51	0.06	5.13	0.08	21.29	m	22.89
75 x 175mm	17.07	10.00	18.78	0.58	0.07	5.84	0.09	24.71	m	26.57
75 x 200mm	19.50	10.00	21.45	0.61	0.08	6.19	0.10	27.74	m	29.82
75 x 225mm	21.95	10.00	24.15	0.70	0.09	7.09	0.11	31.35	m	33.70
75 x 250mm	24.37	10.00	26.81	0.79	0.10	7.99	0.13	34.93	m	37.55
Wall or partition members										
38 x 75mm	3.70	10.00	4.07	0.32	0.04	3.23	0.05	7.35	m	7.91
38 x 100mm	4.94	10.00	5.43	0.37	0.05	3.77	0.06	9.26	m	9.95
50 x 75mm	4.87	10.00	5.36	0.37	0.05	3.77	0.06	9.18	m	9.87
50 x 100mm	6.50	10.00	7.15	0.42	0.05	4.23	0.07	11.45	m	12.31
75 x 100mm	9.75	10.00	10.73	0.52	0.06	5.22	0.08	16.03	m	17.23
38 x 75mm; fixing to masonry	3.70	10.00	4.07	0.53	0.07	5.38	0.09	9.54	m	10.26
38 x 100mm; fixing to masonry	4.94	10.00	5.43	0.59	0.07	5.94	0.10	11.47	m	12.33
50 x 75mm; fixing to masonry	4.87	10.00	5.36	0.59	0.07	5.94	0.10	11.39	m	12.25
50 x 100mm; fixing to masonry	6.50	10.00	7.15	0.65	0.08	6.56	0.11	13.82	m	14.86
75 x 100mm; fixing to masonry	9.75	10.00	10.73	0.78	0.10	7.90	0.13	18.76	m	20.16
Plates										
38 x 75mm	3.70	10.00	4.07	0.32	0.04	3.23	0.05	7.35	m	7.91
38 x 100mm	4.94	10.00	5.43	0.37	0.05	3.77	0.06	9.26	m	9.95
50 x 75mm	4.87	10.00	5.36	0.37	0.05	3.77	0.06	9.18	m	9.87
50 x 100mm	6.50	10.00	7.15	0.42	0.05	4.23	0.07	11.45	m	12.31
75 x 150mm	14.62	10.00	16.08	0.52	0.06	5.22	0.08	21.38	m	22.99
38 x 75mm; fixing by bolting	3.70	10.00	4.07	0.38	0.05	3.86	-	7.93	m	8.52
38 x 100mm; fixing by bolting	4.94	10.00	5.43	0.44	0.05	4.41	-	9.85	m	10.58
50 x 75mm; fixing by bolting	4.87	10.00	5.36	0.44	0.06	4.48	-	9.84	m	10.58
50 x 100mm; fixing by bolting	6.50	10.00	7.15	0.50	0.06	5.04	-	12.19	m	13.10
75 x 150mm; fixing by bolting	14.62	10.00	16.08	0.62	0.07	6.21	-	22.30	m	23.97
Roof members; flat										
50 x 175mm	11.37	10.00	12.51	0.42	0.05	4.23	0.07	16.80	m	18.06
50 x 200mm	13.00	10.00	14.30	0.47	0.06	4.76	0.08	19.14	m	20.57
50 x 225mm	14.62	10.00	16.08	0.51	0.06	5.13	0.08	21.29	m	22.89
50 x 250mm	16.25	10.00	17.88	0.56	0.07	5.66	0.09	23.63	m	25.40
75 x 175mm	17.07	10.00	18.78	0.58	0.07	5.84	0.09	24.71	m	26.57
75 x 200mm	19.50	10.00	21.45	0.61	0.08	6.19	0.10	27.74	m	29.82
75 x 225mm	21.95	10.00	24.15	0.70	0.09	7.09	0.11	31.35	m	33.70
75 x 250mm	23.59	10.00	25.95	0.79	0.10	7.99	0.13	34.07	m	36.63
Roof members; pitched										
25 x 75mm	2.45	10.00	2.69	0.25	0.03	2.52	0.04	5.25	m	5.65
25 x 125mm	4.07	10.00	4.48	0.28	0.04	2.87	0.05	7.39	m	7.95
32 x 175mm	7.28	10.00	8.01	0.39	0.05	3.95	0.06	12.02	m	12.92
32 x 225mm	9.36	10.00	10.30	0.44	0.06	4.48	0.07	14.85	m	15.96
38 x 75mm	3.70	10.00	4.07	0.26	0.03	2.61	0.04	6.72	m	7.22
38 x 100mm	4.94	10.00	5.43	0.32	0.04	3.23	0.05	8.72	m	9.37
50 x 75mm	4.87	10.00	5.36	0.32	0.04	3.23	0.05	8.64	m	9.29
50 x 100mm	6.50	10.00	7.15	0.35	0.04	3.51	0.06	10.72	m	11.53
50 x 125mm	8.12	10.00	8.93	0.40	0.05	4.04	0.06	13.04	m	14.01
50 x 150mm	9.75	10.00	10.73	0.46	0.06	4.67	0.08	15.47	m	16.63
50 x 225mm	14.62	10.00	16.08	0.63	0.08	6.38	0.10	22.56	m	24.25
50 x 250mm	16.25	10.00	17.88	0.70	0.09	7.09	0.11	25.08	m	26.96
75 x 100mm	9.75	10.00	10.73	0.46	0.06	4.67	0.08	15.47	m	16.63
75 x 150mm	14.62	10.00	16.08	0.61	0.08	6.19	0.10	22.37	m	24.05
100 x 150mm	19.50	10.00	21.45	0.79	0.10	7.99	0.13	29.57	m	31.79
100 x 225mm	29.25	10.00	32.17	1.05	0.13	10.60	0.17	42.95	m	46.17
Joist strutting; herringbone; depth of joist 175mm										
50 x 50mm	7.15	10.00	7.87	1.10	0.14	11.14	0.18	19.18	m	20.62
Joist strutting; herringbone; depth of joist 200mm										
50 x 50mm	7.47	10.00	8.22	1.10	0.14	11.14	0.18	19.53	m	21.00
Joist strutting; herringbone; depth of joist 225mm										
50 x 50mm	7.80	10.00	8.58	1.10	0.14	11.14	0.18	19.90	m	21.39
Joist strutting; herringbone; depth of joist 250mm										
50 x 50mm	8.12	10.00	8.93	1.10	0.14	11.14	0.18	20.25	m	21.77
Joist strutting; block; depth of joist 150mm										
50 x 150mm	9.75	10.00	10.73	1.10	0.14	11.14	0.18	22.04	m	23.69
Joist strutting; block; depth of joist 175mm										
50 x 175mm	11.37	10.00	12.51	1.20	0.15	12.13	0.20	24.84	m	26.70
Joist strutting; block; depth of joist 200mm										
50 x 200mm	13.00	10.00	14.30	1.20	0.15	12.13	0.20	26.63	m	28.63
Joist strutting; block; depth of joist 225mm										
50 x 225mm	14.62	10.00	16.08	1.30	0.16	13.12	0.21	29.41	m	31.62

Labour hourly rates: (except Specialists) Craft Operatives £9.23 Labourer £7.02 Rates are national average prices. Refer to REGIONAL VARIATIONS for indicative levels of overall pricing in regions	MATERIALS			LABOUR				RATES		
	Del to Site £	Waste %	Material Cost £	Craft Optve Hrs	Lab Hrs	Labour Cost £	Sunds £	Nett Rate £	Unit	Gross Rate (+7.5%) £
G20: CARPENTRY/TIMBER FRAMING/FIRST FIXING Cont.										
Oak, sawn - carcassing Cont.										
Joist strutting; block; depth of joist 250mm										
50 x 250mm	16.25	10.00	17.88	1.30	0.16	13.12	0.21	31.21	m	33.55
Noggings to joists										
50 x 50mm	3.25	10.00	3.58	0.50	0.06	5.04	0.08	8.69	m	9.34
50 x 75mm	4.87	10.00	5.36	0.60	0.08	6.10	0.10	11.56	m	12.42
Wrot surfaces										
plain; 50mm wide	-	-	-	0.30	-	2.77	-	2.77	m	2.98
plain; 75mm wide	-	-	-	0.40	-	3.69	-	3.69	m	3.97
plain; 100mm wide	-	-	-	0.50	-	4.62	-	4.62	m	4.96
plain; 150mm wide	-	-	-	0.60	-	5.54	-	5.54	m	5.95
plain; 200mm wide	-	-	-	0.70	-	6.46	-	6.46	m	6.95
Stress grading softwood										
General Structural (SS) grade included in rates Special Structural (SS) grade add 10% to materials prices										
Flame proofing treatment to softwood										
For timbers treated with proofing process to Class 1 add 100% to materials prices										
Softwood, sawn - supports										
Butt jointed supports										
width exceeding 300mm; 19 x 38mm at 300mm centres; fixing to masonry	0.96	10.00	1.06	1.00	0.13	10.14	0.16	11.36	m2	12.21
width exceeding 300mm; 25 x 50mm at 300mm centres; fixing to masonry	1.68	10.00	1.85	1.10	0.14	11.14	0.18	13.16	m2	14.15
Framed supports										
width exceeding 300mm; 38 x 38mm members at 400mm centres one way; 38 x 38mm subsidiary members at 400mm centres one way	1.44	10.00	1.58	1.00	0.13	10.14	0.16	11.89	m2	12.78
width exceeding 300mm; 38 x 38mm members at 500mm centres one way; 38 x 38mm subsidiary members at 400mm centres one way	1.30	10.00	1.43	0.80	0.10	8.09	0.13	9.65	m2	10.37
width exceeding 300mm; 50 x 50mm members at 400mm centres one way; 50 x 50mm subsidiary members at 400mm centres one way	2.50	10.00	2.75	1.10	0.14	11.14	0.18	14.07	m2	15.12
width exceeding 300mm; 50 x 50mm members at 500mm centres one way; 50 x 50mm subsidiary members at 400mm centres one way	2.25	10.00	2.48	0.90	0.11	9.08	0.15	11.70	m2	12.58
width exceeding 300mm; 19 x 38mm members at 300mm centres one way; 19 x 38mm subsidiary members at 300mm centres one way; fixing to masonry	0.96	10.00	1.06	2.25	0.28	22.73	0.36	24.15	m2	25.96
width exceeding 300mm; 25 x 50mm members at 300mm centres one way; 25 x 50mm subsidiary members at 300mm centres one way; fixing to masonry	1.68	10.00	1.85	2.35	0.29	23.73	0.38	25.95	m2	27.90
width not exceeding 300mm; 38 x 38mm members longitudinally; 38 x 38mm subsidiary members at 400mm centres laterally	0.80	10.00	0.88	0.40	0.05	4.04	0.06	4.98	m	5.36
width not exceeding 300mm; 50 x 50mm members longitudinally; 50 x 50mm subsidiary members at 400mm centres laterally	1.40	10.00	1.54	0.45	0.06	4.57	0.07	6.18	m	6.65
Individual supports										
6 x 38mm	0.08	10.00	0.09	0.08	0.01	0.81	0.01	0.91	m	0.97
12 x 38mm	0.13	10.00	0.14	0.08	0.01	0.81	0.01	0.96	m	1.03
19 x 38mm	0.14	10.00	0.15	0.09	0.01	0.90	0.01	1.06	m	1.14
25 x 50mm	0.25	10.00	0.28	0.10	0.01	0.99	0.02	1.29	m	1.38
38 x 50mm	0.38	10.00	0.42	0.11	0.01	1.09	0.02	1.52	m	1.64
50 x 50mm	0.50	10.00	0.55	0.12	0.02	1.25	0.02	1.82	m	1.95
50 x 150mm	1.50	10.00	1.65	0.17	0.02	1.71	0.03	3.39	m	3.64
50 x 300mm	3.00	10.00	3.30	0.23	0.03	2.33	0.04	5.67	m	6.10
6 x 38mm; fixing to masonry	0.08	10.00	0.09	0.23	0.03	2.33	0.04	2.46	m	2.65
12 x 38mm; fixing to masonry	0.13	10.00	0.14	0.23	0.03	2.33	0.04	2.52	m	2.71
19 x 38mm; fixing to masonry	0.14	10.00	0.15	0.24	0.03	2.43	0.04	2.62	m	2.82
25 x 50mm; fixing to masonry	0.25	10.00	0.28	0.25	0.03	2.52	0.04	2.83	m	3.05
38 x 50mm; fixing to masonry	0.38	10.00	0.42	0.27	0.03	2.70	0.04	3.16	m	3.40
50 x 50mm; fixing to masonry	0.50	10.00	0.55	0.28	0.04	2.87	0.05	3.47	m	3.73
Softwood, sawn, impregnated - supports										
Butt jointed supports										
width exceeding 300mm; 50mm wide x 25mm average depth at 450mm centres	1.24	10.00	1.36	0.30	0.04	3.05	0.05	4.46	m2	4.80
width exceeding 300mm; 50mm wide x 25mm average depth at 600mm centres	0.95	10.00	1.04	0.23	0.03	2.33	0.04	3.42	m2	3.67
width exceeding 300mm; 50mm wide x 50mm average depth at 450mm centres	2.48	10.00	2.73	0.36	0.05	3.67	0.06	6.46	m2	6.95
width exceeding 300mm; 50mm wide x 50mm average depth at 600mm centres	1.80	10.00	1.98	0.27	0.03	2.70	0.04	4.72	m2	5.08
width exceeding 300mm; 50mm wide x 63mm average depth at 450mm centres	3.15	10.00	3.46	0.40	0.05	4.04	0.06	7.57	m2	8.14
width exceeding 300mm; 50mm wide x 63mm average depth at 600mm centres	2.48	10.00	2.73	0.30	0.04	3.05	0.05	5.83	m2	6.26
width exceeding 300mm; 50mm wide x 75mm average depth at 450mm centres	3.71	10.00	4.08	0.43	0.05	4.32	0.07	8.47	m2	9.11

Labour hourly rates: (except Specialists) Craft Operatives £9.23 Labourer £7.02 Rates are national average prices. Refer to REGIONAL VARIATIONS for indicative levels of overall pricing in regions	MATERIALS			LABOUR				RATES		
	Del to Site £	Waste %	Material Cost £	Craft Optve Hrs	Lab Hrs	Labour Cost £	Sunds £	Nett Rate £	Unit	Gross Rate (+7.5%) £
G20: CARPENTRY/TIMBER FRAMING/FIRST FIXING Cont.										
Softwood, sawn, impregnated - supports Cont.										
Butt jointed supports Cont.										
width exceeding 300mm; 50mm wide x 75mm average depth at 600mm centres	2.81	10.00	3.09	0.33	0.04	3.33	0.05	6.47	m2	6.95
Individual supports										
25 x 25mm; 1 labours	0.18	10.00	0.20	0.11	0.01	1.09	0.02	1.30	m	1.40
25 x 25mm; 2 labours	0.21	10.00	0.23	0.11	0.01	1.09	0.02	1.34	m	1.44
25 x 38mm; 1 labours	0.27	10.00	0.30	0.11	0.01	1.09	0.02	1.40	m	1.51
25 x 38mm; 2 labours	0.32	10.00	0.35	0.11	0.01	1.09	0.02	1.46	m	1.57
triangular; extreme dimensions										
25 x 25mm...........................	0.18	10.00	0.20	0.11	0.01	1.09	0.02	1.30	m	1.40
38 x 75mm...........................	0.80	10.00	0.88	0.13	0.02	1.34	0.02	2.24	m	2.41
50 x 75mm...........................	1.05	10.00	1.16	0.15	0.02	1.52	0.02	2.70	m	2.90
50 x 100mm..........................	1.41	10.00	1.55	0.17	0.02	1.71	0.03	3.29	m	3.54
75 x 100mm..........................	2.11	10.00	2.32	0.19	0.02	1.89	0.03	4.25	m	4.56
rounded roll for lead										
50 x 50mm...........................	1.69	10.00	1.86	0.22	0.03	2.24	0.04	4.14	m	4.45
50 x 75mm...........................	2.53	10.00	2.78	0.25	0.03	2.52	0.04	5.34	m	5.74
rounded roll for lead; birdsmouthed on to ridge or hip										
50 x 50mm...........................	1.69	10.00	1.86	0.38	0.05	3.86	0.06	5.78	m	6.21
50 x 75mm...........................	2.53	10.00	2.78	0.41	0.05	4.14	0.07	6.99	m	7.51
roll for zinc; 32 x 44mm...........	0.95	10.00	1.04	0.16	0.02	1.62	0.03	2.69	m	2.89
Softwood, wrot, impregnated - gutter boarding										
Gutter boards including sides, tongued and grooved joints										
width exceeding 300mm; 19mm thick	6.60	10.00	7.26	2.00	0.25	20.22	0.32	27.80	m2	29.88
width exceeding 300mm; 25mm thick	8.70	10.00	9.57	2.25	0.28	22.73	0.36	32.66	m2	35.11
width not exceeding 300mm; 19mm thick										
150mm wide...................................	0.99	10.00	1.09	0.30	0.04	3.05	0.05	4.19	m	4.50
225mm wide...................................	1.50	10.00	1.65	0.40	0.05	4.04	0.06	5.75	m	6.18
width not exceeding 300mm; 25mm thick										
150mm wide...................................	1.32	10.00	1.45	0.33	0.04	3.33	0.05	4.83	m	5.19
225mm wide...................................	1.95	10.00	2.15	0.44	0.05	4.41	0.07	6.63	m	7.12
cesspool 225 x 225 x 150mm deep	1.68	10.00	1.85	1.50	0.19	15.18	0.24	17.27	nr	18.56
Chimney gutter boards, butt joints										
width not exceeding 300mm; 25mm thick x 100mm average wide	1.13	10.00	1.24	0.50	0.06	5.04	0.08	6.36	m	6.84
gusset end	0.30	10.00	0.33	0.35	0.04	3.51	0.06	3.90	nr	4.19
width not exceeding 300mm; 25mm thick x 175mm average wide	1.68	10.00	1.85	0.65	0.08	6.56	0.11	8.52	m	9.16
gusset end	0.42	10.00	0.46	0.40	0.05	4.04	0.06	4.57	nr	4.91
Lier boards; tongued and grooved joints										
width exceeding 300mm; 19mm thick	6.60	10.00	7.26	0.90	0.11	9.08	0.15	16.49	m2	17.73
width exceeding 300mm; 25mm thick	8.70	10.00	9.57	1.00	0.12	10.07	0.16	19.80	m2	21.29
Valley sole boards; butt joints										
width not exceeding 300mm; 25mm thick										
100mm wide....................................	0.75	10.00	0.82	0.45	0.06	4.57	0.07	5.47	m	5.88
150mm wide....................................	1.13	10.00	1.24	0.50	0.06	5.04	0.08	6.36	m	6.84
Plywood B.S.6566, II/III grade, WBP bonded, butt joints - gutter boarding										
Gutter boards including sides, butt joints										
width exceeding 300mm; 18mm thick	10.55	10.00	11.61	1.50	0.19	15.18	0.24	27.02	m2	29.05
width exceeding 300mm; 25mm thick	14.66	10.00	16.13	1.65	0.21	16.70	0.27	33.10	m2	35.58
width not exceeding 300mm; 18mm thick										
150mm wide....................................	1.58	10.00	1.74	0.20	0.03	2.06	0.03	3.82	m	4.11
225mm wide....................................	2.37	10.00	2.61	0.22	0.03	2.24	0.04	4.89	m	5.25
width not exceeding 300mm; 25mm thick										
150mm wide....................................	2.20	10.00	2.42	0.23	0.03	2.33	0.04	4.79	m	5.15
225mm wide....................................	3.30	10.00	3.63	0.25	0.03	2.52	0.04	6.19	m	6.65
cesspool 225 x 225 x 150mm deep	2.93	10.00	3.22	1.50	0.19	15.18	0.24	18.64	nr	20.04
Softwood, wrot, impregnated - eaves and verge boarding										
Fascia and barge boards										
width not exceeding 300mm; 25mm thick										
150mm wide....................................	1.07	10.00	1.18	0.22	0.03	2.24	0.04	3.46	m	3.72
200mm wide....................................	1.43	10.00	1.57	0.23	0.03	2.33	0.04	3.95	m	4.24
Fascia and barge boards, tongued, grooved and veed joints										
width not exceeding 300mm; 25mm thick										
225mm wide....................................	1.85	10.00	2.04	0.30	0.04	3.05	0.05	5.13	m	5.52
250mm wide....................................	2.05	10.00	2.25	0.32	0.04	3.23	0.05	5.54	m	5.95
Eaves or verge soffit boards										
width exceeding 300mm; 25mm thick	7.13	10.00	7.84	1.00	0.12	10.07	0.16	18.08	m2	19.43
width not exceeding 300mm; 25mm thick										
225mm wide....................................	1.60	10.00	1.76	0.30	0.04	3.05	0.05	4.86	m	5.22

Labour hourly rates: (except Specialists) Craft Operatives £9.23 Labourer £7.02 Rates are national average prices. Refer to REGIONAL VARIATIONS for indicative levels of overall pricing in regions	MATERIALS			LABOUR				RATES		
	Del to Site £	Waste %	Material Cost £	Craft Optve Hrs	Lab Hrs	Labour Cost £	Sunds £	Nett Rate £	Unit	Gross Rate (+7.5%) £
G20: CARPENTRY/TIMBER FRAMING/FIRST FIXING Cont.										
Plywood B.S.6566, II/III grade, WBP bonded, butt joints - eaves and verge boarding										
Fascia and barge boards width not exceeding 300mm; 18mm thick										
150mm wide	1.58	10.00	1.74	0.22	0.03	2.24	0.04	4.02	m	4.32
225mm wide	2.37	10.00	2.61	0.23	0.03	2.33	0.04	4.98	m	5.35
250mm wide	2.63	10.00	2.89	0.24	0.03	2.43	0.04	5.36	m	5.76
Eaves or verge soffit boards width exceeding 300mm; 18mm thick	10.55	10.00	11.61	0.75	0.09	7.55	0.12	19.28	m2	20.73
width not exceeding 300mm; 18mm thick										
225mm wide	2.37	10.00	2.61	0.25	0.03	2.52	0.04	5.17	m	5.55
Fascia and bargeboards; Caradon Celuform Ltd										
Fascia reference 4247; butt joint with joint trims reference 4560; fixing to timber with Polytop screws										
150mm wide	8.66	5.00	9.09	0.20	0.03	2.06	0.48	11.63	m	12.50
200mm wide	10.43	5.00	10.95	0.20	0.03	2.06	0.48	13.49	m	14.50
250mm wide	13.71	5.00	14.40	0.20	0.03	2.06	0.96	17.41	m	18.72
Elite solid fascia board reference 4245; butt joint with joint trims reference 4560; fixing to timber with Polytop screws										
150mm wide	8.72	5.00	9.16	0.20	0.03	2.06	0.48	11.69	m	12.57
200mm wide	10.73	5.00	11.27	0.20	0.03	2.06	0.48	13.80	m	14.84
250mm wide	14.47	5.00	15.19	0.20	0.03	2.06	0.96	18.21	m	19.58
Supaliner fascia board reference 4270; butt joint with joint trims reference 4586; fixing to timber with Polytop screws										
150mm wide	7.02	5.00	7.37	0.30	0.04	3.05	0.48	10.90	m	11.72
170mm wide	8.08	5.00	8.48	0.30	0.04	3.05	0.48	12.01	m	12.91
225mm wide	10.26	5.00	10.77	0.30	0.04	3.05	0.48	14.30	m	15.38
250mm wide	11.33	5.00	11.90	0.35	0.04	3.51	0.96	16.37	m	17.60
Celuvent multi purpose/soffite board reference 4258; butt joint with panel joint trim reference 4570; fixing to timber with stainless steel nails; fixing soffite board channel reference 4572 to masonry with Polytop screws										
100mm wide	8.51	5.00	8.94	0.40	0.06	4.11	0.48	13.53	m	14.54
225mm wide	14.27	5.00	14.98	0.40	0.06	4.11	0.96	20.06	m	21.56
325mm wide	18.82	5.00	19.76	0.60	0.08	6.10	1.44	27.30	m	29.35
Multi purpose/soffite board reference 4558; butt joint with panel joint trim reference 4570; fixing to timber with stainless steel nails; fixing soffite board channel reference 4572 to masonry with Polytop screws										
125mm wide	9.55	5.00	10.03	0.40	0.06	4.11	0.48	14.62	m	15.72
175mm wide	11.62	5.00	12.20	0.40	0.06	4.11	0.48	16.79	m	18.05
200mm wide	12.10	5.00	12.71	0.40	0.06	4.11	0.96	17.78	m	19.11
300mm wide	17.62	5.00	18.50	0.60	0.08	6.10	1.44	26.04	m	27.99
Steel - tie rods and straps										
Tie rods										
19mm diameter; threaded	4.76	2.50	4.88	0.25	0.03	2.52	–	7.40	m	7.95
extra; with nut and washer	0.23	2.50	0.24	0.20	0.03	2.06	–	2.29	nr	2.46
25mm diameter; threaded	6.44	2.50	6.60	0.30	0.04	3.05	–	9.65	m	10.37
extra; with nut and washer	0.42	2.50	0.43	0.30	0.04	3.05	–	3.48	nr	3.74
Straps 3 x 38; holes -4; fixing with bolts (bolts included elsewhere)										
500mm long	0.95	2.50	0.97	0.13	0.02	1.34	–	2.31	nr	2.49
750mm long	1.40	2.50	1.44	0.19	0.02	1.89	–	3.33	nr	3.58
1000mm long	1.91	2.50	1.96	0.25	0.03	2.52	–	4.48	nr	4.81
6 x 50; holes -4; fixing with bolts (bolts included elsewhere)										
500mm long	2.08	2.50	2.13	0.16	0.02	1.62	–	3.75	nr	4.03
750mm long	3.08	2.50	3.16	0.23	0.03	2.33	–	5.49	nr	5.90
1000mm long	4.14	2.50	4.24	0.30	0.04	3.05	–	7.29	nr	7.84
Steel; galvanised - straps										
Straps and clips; BAT Building ProductsLtd. 30 x 2.5mm standard strapping; fixing with nails										
600mm long	0.64	5.00	0.67	0.15	0.02	1.52	0.02	2.22	nr	2.38
800mm long	0.94	5.00	0.99	0.17	0.02	1.71	0.03	2.73	nr	2.93
1000mm long	1.21	5.00	1.27	0.18	0.02	1.80	0.03	3.10	nr	3.33
1200mm long	1.46	5.00	1.53	0.19	0.02	1.89	0.03	3.46	nr	3.72
1600mm long	1.90	5.00	2.00	0.20	0.02	1.99	0.03	4.01	nr	4.31
30 x 2.5mm standard strapping; twists -1; fixing with nails										
600mm long	0.71	5.00	0.75	0.15	0.02	1.52	0.02	2.29	nr	2.46
800mm long	1.01	5.00	1.06	0.17	0.02	1.71	0.03	2.80	nr	3.01
1000mm long	1.27	5.00	1.33	0.18	0.02	1.80	0.03	3.17	nr	3.40
1200mm long	1.52	5.00	1.60	0.19	0.02	1.89	0.03	3.52	nr	3.78
1600mm long	1.96	5.00	2.06	0.20	0.02	1.99	0.03	4.07	nr	4.38

Labour hourly rates: (except Specialists) Craft Operatives £9.23 Labourer £7.02 Rates are national average prices. Refer to REGIONAL VARIATIONS for indicative levels of overall pricing in regions	MATERIALS			LABOUR				RATES		
	Del to Site £	Waste %	Material Cost £	Craft Optve Hrs	Lab Hrs	Labour Cost £	Sunds £	Nett Rate £	Unit	Gross Rate (+7.5%) £
G20: CARPENTRY/TIMBER FRAMING/FIRST FIXING Cont.										
Steel; galvanised - straps Cont.										
Straps and clips; BAT Building ProductsLtd. Cont. 30 x 5mm M 305 strapping; bends -1; nailing one end (building in included elsewhere)										
600mm long............................	1.36	5.00	1.43	0.08	0.02	0.88	0.02	2.33	nr	2.50
800mm long............................	1.74	5.00	1.83	0.08	0.02	0.88	0.02	2.73	nr	2.93
1000mm long...........................	2.21	5.00	2.32	0.09	0.02	0.97	0.02	3.31	nr	3.56
1200mm long...........................	2.66	5.00	2.79	0.09	0.02	0.97	0.02	3.78	nr	4.07
1600mm long...........................	3.51	5.00	3.69	0.10	0.02	1.06	0.02	4.77	nr	5.13
30 x 5mm M305 strapping; bends -1; twists -1; nailing one end (building in included elsewhere)										
600mm long............................	1.41	5.00	1.48	0.08	0.02	0.88	0.02	2.38	nr	2.56
800mm long............................	1.78	5.00	1.87	0.08	0.02	0.88	0.02	2.77	nr	2.98
1000mm long...........................	2.27	5.00	2.38	0.09	0.02	0.97	0.02	3.37	nr	3.63
1200mm long...........................	2.73	5.00	2.87	0.09	0.02	0.97	0.02	3.86	nr	4.15
1600mm long...........................	3.57	5.00	3.75	0.10	0.02	1.06	0.02	4.83	nr	5.19
truss clips; fixing with nails										
for 38mm thick members...............	0.25	5.00	0.26	0.20	0.02	1.99	0.03	2.28	nr	2.45
for 50mm thick members...............	0.27	5.00	0.28	0.20	0.02	1.99	0.03	2.30	nr	2.47
Steel; galvanised - joist hangers										
Joist hangers; BAT Building Products Ltd; (building in where required included elsewhere)										
SPH type S, for 50 x 100mm joist	1.35	2.50	1.38	-	-	-	-	1.38	nr	1.49
SPH type S, for 50 x 125mm joist	1.38	2.50	1.41	-	-	-	-	1.41	nr	1.52
SPH type S, for 50 x 150mm joist	1.38	2.50	1.41	-	-	-	-	1.41	nr	1.52
SPH type S, for 50 x 175mm joist	1.44	2.50	1.48	-	-	-	-	1.48	nr	1.59
SPH type S, for 50 x 200mm joist	1.60	2.50	1.64	-	-	-	-	1.64	nr	1.76
SPH type S, for 50 x 225mm joist	1.71	2.50	1.75	-	-	-	-	1.75	nr	1.88
SPH type S, for 50 x 250mm joist	2.19	2.50	2.24	-	-	-	-	2.24	nr	2.41
SPH type S, for 63 x 100mm joist	1.72	2.50	1.76	-	-	-	-	1.76	nr	1.90
SPH type S, for 63 x 125mm joist	1.75	2.50	1.79	-	-	-	-	1.79	nr	1.93
SPH type S, for 63 x 150mm joist	2.00	2.50	2.05	-	-	-	-	2.05	nr	2.20
SPH type S, for 63 x 175mm joist	1.88	2.50	1.93	-	-	-	-	1.93	nr	2.07
SPH type S, for 63 x 200mm joist	2.00	2.50	2.05	-	-	-	-	2.05	nr	2.20
SPH type S, for 63 x 225mm joist	2.14	2.50	2.19	-	-	-	-	2.19	nr	2.36
SPH type S, for 63 x 250mm joist	2.29	2.50	2.35	-	-	-	-	2.35	nr	2.52
SPH type R, for 50 x 100mm joist	2.04	2.50	2.09	-	-	-	-	2.09	nr	2.25
SPH type R, for 50 x 125mm joist	2.07	2.50	2.12	-	-	-	-	2.12	nr	2.28
SPH type R, for 50 x 150mm joist	2.14	2.50	2.19	-	-	-	-	2.19	nr	2.36
SPH type R, for 50 x 175mm joist	2.22	2.50	2.28	-	-	-	-	2.28	nr	2.45
SPH type R, for 50 x 200mm joist	2.36	2.50	2.42	-	-	-	-	2.42	nr	2.60
SPH type R, for 50 x 225mm joist	2.59	2.50	2.65	-	-	-	-	2.65	nr	2.85
SPH type R, for 50 x 250mm joist	3.07	2.50	3.15	-	-	-	-	3.15	nr	3.38
SPH type R, for 63 x 100mm joist	2.49	2.50	2.55	-	-	-	-	2.55	nr	2.74
SPH type R, for 63 x 125mm joist	2.53	2.50	2.59	-	-	-	-	2.59	nr	2.79
SPH type R, for 63 x 150mm joist	2.81	2.50	2.88	-	-	-	-	2.88	nr	3.10
SPH type R, for 63 x 175mm joist	2.64	2.50	2.71	-	-	-	-	2.71	nr	2.91
SPH type R, for 63 x 200mm joist	2.81	2.50	2.88	-	-	-	-	2.88	nr	3.10
SPH type R, for 63 x 225mm joist	3.00	2.50	3.08	-	-	-	-	3.08	nr	3.31
SPH type R, for 63 x 250mm joist	3.17	2.50	3.25	-	-	-	-	3.25	nr	3.49
SPH type ST for 50 x 100mm joist	3.57	2.50	3.66	-	-	-	-	3.66	nr	3.93
SPH type ST for 50 x 125mm joist	3.67	2.50	3.76	-	-	-	-	3.76	nr	4.04
SPH type ST for 50 x 150mm joist	3.50	2.50	3.59	-	-	-	-	3.59	nr	3.86
SPH type ST for 50 x 175mm joist	3.99	2.50	4.09	-	-	-	-	4.09	nr	4.40
SPH type ST for 50 x 200mm joist	4.19	2.50	4.29	-	-	-	-	4.29	nr	4.62
SPH type ST for 50 x 225mm joist	4.58	2.50	4.69	-	-	-	-	4.69	nr	5.05
SPH type ST for 50 x 250mm joist	5.55	2.50	5.69	-	-	-	-	5.69	nr	6.12
Speedy Minor type for the following size joists; fixing with nails										
38 x 100mm............................	0.38	2.50	0.39	0.15	0.02	1.52	0.02	1.93	nr	2.08
50 x 100mm............................	0.38	2.50	0.39	0.15	0.02	1.52	0.02	1.93	nr	2.08
Speedy short leg type for the following size joists fixing with nails										
38 x 100mm............................	0.70	2.50	0.72	0.15	0.02	1.52	0.02	2.26	nr	2.43
50 x 175mm............................	0.70	2.50	0.72	0.15	0.02	1.52	0.02	2.26	nr	2.43
Speedy Standard Leg type for the following size joists; fixing with nails										
38 x 100mm............................	0.76	2.50	0.78	0.15	0.02	1.52	0.02	2.32	nr	2.50
50 x 175mm............................	0.76	2.50	0.78	0.15	0.02	1.52	0.02	2.32	nr	2.50
63 x 225mm............................	0.79	2.50	0.81	0.15	0.02	1.52	0.02	2.35	nr	2.53
75 x 225mm............................	0.79	2.50	0.81	0.17	0.02	1.71	0.03	2.55	nr	2.74
100 x 225mm...........................	0.87	2.50	0.89	0.17	0.02	1.71	0.03	2.63	nr	2.83
Steel; galvanised - joist struts										
Herringbone joist struts; BAT Building Products Ltd. to suit joists at the following centres; fixing with nails										
400mm	0.27	5.00	0.28	0.25	0.03	2.52	0.04	2.84	m	3.05
450mm	0.30	5.00	0.32	0.25	0.03	2.52	0.04	2.87	m	3.09
600mm	0.34	5.00	0.36	0.20	0.03	2.06	0.04	2.45	m	2.64
Steel; galvanised - truss plates and framing anchors										
Truss plates; BAT Building Products Ltd. fixing with nails										
51 x 114mm............................	0.14	5.00	0.15	0.35	0.04	3.51	0.06	3.72	nr	4.00
76 x 254mm............................	0.52	5.00	0.55	1.00	0.13	10.14	0.16	10.85	nr	11.66
114 x 152mm...........................	0.42	5.00	0.44	1.00	0.13	10.14	0.16	10.74	nr	11.55
114 x 254mm...........................	0.72	5.00	0.76	1.50	0.19	15.18	0.24	16.17	nr	17.39
152 x 152mm...........................	0.58	5.00	0.61	1.60	0.20	16.17	0.26	17.04	nr	18.32

STRUCTURAL/CARCASSING METAL/TIMBER

Labour hourly rates: (except Specialists) Craft Operatives £9.23 Labourer £7.02 Rates are national average prices. Refer to REGIONAL VARIATIONS for indicative levels of overall pricing in regions	MATERIALS			LABOUR				RATES		
	Del to Site	Waste	Material Cost	Craft Optve	Lab	Labour Cost	Sunds	Nett Rate	Unit	Gross Rate (+7.5%)
	£	%	£	Hrs	Hrs	£	£	£		£
G20: CARPENTRY/TIMBER FRAMING/FIRST FIXING Cont.										
Steel; galvanised - truss plates and framing anchors Cont.										
Framing anchors; BAT Building Products Ltd.										
type A; fixing with nails	0.35	5.00	0.37	0.20	0.02	1.99	0.03	2.38	nr	2.56
type B; fixing with nails	0.35	5.00	0.37	0.20	0.02	1.99	0.03	2.38	nr	2.56
type C; fixing with nails	0.35	5.00	0.37	0.20	0.02	1.99	0.03	2.38	nr	2.56
Steel - timber connectors										
Connectors B.S.1579										
split ring connectors table 1; 64mm diameter	0.93	5.00	0.98	0.08	0.01	0.81	-	1.79	nr	1.92
split ring connectors table 1; 100 mm diameter ...	2.56	5.00	2.69	0.08	0.01	0.81	-	3.50	nr	3.76
shear plate connectors, table 2; 67mm diameter ...	1.53	5.00	1.61	0.08	0.01	0.81	-	2.42	nr	2.60
single sided round toothed-plate connectors, table 4										
38mm diameter.................................	0.16	5.00	0.17	0.08	0.01	0.81	-	0.98	nr	1.05
51mm diameter.................................	0.21	5.00	0.22	0.08	0.01	0.81	-	1.03	nr	1.11
64mm diameter.................................	0.29	5.00	0.30	0.08	0.01	0.81	-	1.11	nr	1.20
76mm diameter.................................	0.41	5.00	0.43	0.08	0.01	0.81	-	1.24	nr	1.33
double sided round toothed-plate connector, table 4										
38mm diameter.................................	0.21	5.00	0.22	0.08	0.01	0.81	-	1.03	nr	1.11
51mm diameter.................................	0.24	5.00	0.25	0.08	0.01	0.81	-	1.06	nr	1.14
64mm diameter.................................	0.31	5.00	0.33	0.08	0.01	0.81	-	1.13	nr	1.22
76mm diameter.................................	0.45	5.00	0.47	0.08	0.01	0.81	-	1.28	nr	1.38
Cast iron - connectors										
Connectors B.S.1579										
shear plate connectors, table 2; 102mm diameter ..	6.10	5.00	6.41	0.12	0.01	1.18	-	7.58	nr	8.15
Steel - bolts and nuts										
Black bolts, B.S.4190 grade 4.6, hexagon head										
M 12; with nuts and washers										
100mm long....................................	0.41	5.00	0.43	0.09	0.01	0.90	-	1.33	nr	1.43
120mm long....................................	0.50	5.00	0.53	0.11	0.01	1.09	-	1.61	nr	1.73
160mm long....................................	1.06	5.00	1.11	0.11	0.01	1.09	-	2.20	nr	2.36
180mm long....................................	1.11	5.00	1.17	0.12	0.01	1.18	-	2.34	nr	2.52
200mm long....................................	1.16	5.00	1.22	0.14	0.02	1.43	-	2.65	nr	2.85
M 16; with nuts and washers										
100mm long....................................	0.57	5.00	0.60	0.09	0.01	0.90	-	1.50	nr	1.61
120mm long....................................	0.88	5.00	0.92	0.11	0.01	1.09	-	2.01	nr	2.16
160mm long....................................	1.43	5.00	1.50	0.12	0.01	1.18	-	2.68	nr	2.88
180mm long....................................	1.51	5.00	1.59	0.13	0.02	1.34	-	2.93	nr	3.15
200mm long....................................	1.60	5.00	1.68	0.14	0.02	1.43	-	3.11	nr	3.35
M 20; with nuts and washers										
100mm long....................................	1.24	5.00	1.30	0.10	0.01	0.99	-	2.30	nr	2.47
120mm long....................................	1.69	5.00	1.77	0.11	0.01	1.09	-	2.86	nr	3.07
160mm long....................................	2.13	5.00	2.24	0.14	0.02	1.43	-	3.67	nr	3.94
180mm long....................................	2.15	5.00	2.26	0.14	0.02	1.43	-	3.69	nr	3.97
200mm long....................................	2.37	5.00	2.49	0.14	0.02	1.43	-	3.92	nr	4.22
M 12; with nuts and 38 x 38 x 3mm square plate washers										
100mm long....................................	0.42	5.00	0.44	0.09	0.01	0.90	-	1.34	nr	1.44
120mm long....................................	0.52	5.00	0.55	0.11	0.01	1.09	-	1.63	nr	1.75
160mm long....................................	1.07	5.00	1.12	0.11	0.01	1.09	-	2.21	nr	2.37
180mm long....................................	1.15	5.00	1.21	0.12	0.01	1.18	-	2.39	nr	2.56
200mm long....................................	1.18	5.00	1.24	0.14	0.02	1.43	-	2.67	nr	2.87
M 12; with nuts and 50 x 50 x 3mm square plate washers										
100mm long....................................	0.45	5.00	0.47	0.09	0.01	0.90	-	1.37	nr	1.48
120mm long....................................	0.53	5.00	0.56	0.11	0.01	1.09	-	1.64	nr	1.77
160mm long....................................	1.10	5.00	1.16	0.11	0.01	1.09	-	2.24	nr	2.41
180mm long....................................	1.15	5.00	1.21	0.12	0.01	1.18	-	2.39	nr	2.56
200mm long....................................	1.21	5.00	1.27	0.14	0.02	1.43	-	2.70	nr	2.91
Black bolts, B.S.4933 Grade 4.6, cup head, square neck										
M 6; with nuts and washers										
25mm long....................................	0.07	5.00	0.07	0.07	0.01	0.72	-	0.79	nr	0.85
50mm long....................................	0.08	5.00	0.08	0.07	0.01	0.72	-	0.80	nr	0.86
75mm long....................................	0.10	5.00	0.11	0.08	0.01	0.81	-	0.91	nr	0.98
100mm long....................................	0.14	5.00	0.15	0.08	0.01	0.81	-	0.96	nr	1.03
150mm long....................................	0.20	5.00	0.21	0.09	0.01	0.90	-	1.11	nr	1.19
M 8; with nuts and washers										
25mm long....................................	0.10	5.00	0.11	0.07	0.01	0.72	-	0.82	nr	0.88
50mm long....................................	0.11	5.00	0.12	0.07	0.01	0.72	-	0.83	nr	0.89
75mm long....................................	0.14	5.00	0.15	0.08	0.01	0.81	-	0.96	nr	1.03
100mm long....................................	0.21	5.00	0.22	0.08	0.01	0.81	-	1.03	nr	1.11
150mm long....................................	0.30	5.00	0.32	0.09	0.01	0.90	-	1.22	nr	1.31
M 10; with nuts and washers										
50mm long....................................	0.18	5.00	0.19	0.07	0.01	0.72	-	0.91	nr	0.97
75mm long....................................	0.20	5.00	0.21	0.08	0.01	0.81	-	1.02	nr	1.09
100mm long....................................	0.30	5.00	0.32	0.08	0.01	0.81	-	1.12	nr	1.21
150mm long....................................	0.43	5.00	0.45	0.10	0.01	0.99	-	1.44	nr	1.55
M 12; with nuts and washers										
50mm long....................................	0.26	5.00	0.27	0.08	0.01	0.81	-	1.08	nr	1.16
75mm long....................................	0.34	5.00	0.36	0.08	0.01	0.81	-	1.17	nr	1.25
100mm long....................................	0.42	5.00	0.44	0.09	0.01	0.90	-	1.34	nr	1.44
150mm long....................................	0.58	5.00	0.61	0.11	0.01	1.09	-	1.69	nr	1.82

Labour hourly rates: (except Specialists) Craft Operatives £9.23 Labourer £7.02 Rates are national average prices. Refer to REGIONAL VARIATIONS for indicative levels of overall pricing in regions	MATERIALS			LABOUR				RATES		
	Del to Site £	Waste %	Material Cost £	Craft Optve Hrs	Lab Hrs	Labour Cost £	Sunds £	Nett Rate £	Unit	Gross Rate (+7.5%) £
G20: CARPENTRY/TIMBER FRAMING/FIRST FIXING Cont.										
Steel; galvanised - bolts and nuts										
Black bolts, B.S.4190 grade 4.6, hexagon head										
M 12; with nuts and washers										
100mm long...............................	0.60	5.00	0.63	0.09	0.01	0.90	-	1.53	nr	1.65
120mm long...............................	0.74	5.00	0.78	0.11	0.01	1.09	-	1.86	nr	2.00
160mm long...............................	1.55	5.00	1.63	0.11	0.01	1.09	-	2.71	nr	2.92
180mm long...............................	1.63	5.00	1.71	0.12	0.01	1.18	-	2.89	nr	3.11
200mm long...............................	1.70	5.00	1.78	0.14	0.02	1.43	-	3.22	nr	3.46
M 16; with nuts and washers										
100mm long...............................	1.00	5.00	1.05	0.09	0.01	0.90	-	1.95	nr	2.10
120mm long...............................	1.30	5.00	1.37	0.11	0.01	1.09	-	2.45	nr	2.63
160mm long...............................	2.08	5.00	2.18	0.12	0.01	1.18	-	3.36	nr	3.61
180mm long...............................	2.21	5.00	2.32	0.13	0.02	1.34	-	3.66	nr	3.94
200mm long...............................	2.35	5.00	2.47	0.14	0.02	1.43	-	3.90	nr	4.19
M 20; with nuts and washers										
100mm long...............................	1.83	5.00	1.92	0.10	0.01	0.99	-	2.91	nr	3.13
120mm long...............................	2.47	5.00	2.59	0.11	0.01	1.09	-	3.68	nr	3.95
160mm long...............................	3.12	5.00	3.28	0.14	0.02	1.43	-	4.71	nr	5.06
180mm long...............................	3.23	5.00	3.39	0.14	0.02	1.43	-	4.82	nr	5.19
200mm long...............................	3.44	5.00	3.61	0.14	0.02	1.43	-	5.04	nr	5.42
M 12; with nuts and 38 x 38 x 3mm square plate washers										
100mm long...............................	0.66	5.00	0.69	0.09	0.01	0.90	-	1.59	nr	1.71
120mm long...............................	0.79	5.00	0.83	0.11	0.01	1.09	-	1.92	nr	2.06
160mm long...............................	1.67	5.00	1.75	0.11	0.01	1.09	-	2.84	nr	3.05
180mm long...............................	1.69	5.00	1.77	0.12	0.01	1.18	-	2.95	nr	3.17
200mm long...............................	1.77	5.00	1.86	0.14	0.02	1.43	-	3.29	nr	3.54
M 12; with nuts and 50 x 50 x 3mm square plate washers										
100mm long...............................	0.70	5.00	0.73	0.09	0.01	0.90	-	1.64	nr	1.76
120mm long...............................	0.83	5.00	0.87	0.11	0.01	1.09	-	1.96	nr	2.10
160mm long...............................	1.65	5.00	1.73	0.11	0.01	1.09	-	2.82	nr	3.03
180mm long...............................	1.74	5.00	1.83	0.12	0.01	1.18	-	3.00	nr	3.23
200mm long...............................	1.81	5.00	1.90	0.14	0.02	1.43	-	3.33	nr	3.58
Steel - expanding bolts										
Expanding bolts; bolt projecting Rawlbolts; drilling masonry										
with nuts and washers; reference										
44 - 505.............................	0.69	5.00	0.72	0.20	0.02	1.99	-	2.71	nr	2.91
44 - 510.............................	0.77	5.00	0.81	0.20	0.02	1.99	-	2.79	nr	3.00
44 - 515.............................	0.81	5.00	0.85	0.20	0.02	1.99	-	2.84	nr	3.05
44 - 555.............................	0.87	5.00	0.91	0.22	0.03	2.24	-	3.15	nr	3.39
44 - 560.............................	0.92	5.00	0.97	0.22	0.03	2.24	-	3.21	nr	3.45
44 - 565.............................	0.98	5.00	1.03	0.22	0.03	2.24	-	3.27	nr	3.52
44 - 605.............................	1.19	5.00	1.25	0.25	0.03	2.52	-	3.77	nr	4.05
44 - 610.............................	1.25	5.00	1.31	0.25	0.03	2.52	-	3.83	nr	4.12
44 - 615.............................	1.30	5.00	1.37	0.25	0.03	2.52	-	3.88	nr	4.17
44 - 655.............................	1.90	5.00	2.00	0.27	0.03	2.70	-	4.70	nr	5.05
44 - 660.............................	2.03	5.00	2.13	0.27	0.03	2.70	-	4.83	nr	5.20
44 - 665.............................	2.54	5.00	2.67	0.27	0.03	2.70	-	5.37	nr	5.77
44 - 705.............................	4.49	5.00	4.71	0.30	0.04	3.05	-	7.76	nr	8.35
44 - 710.............................	4.85	5.00	5.09	0.30	0.04	3.05	-	8.14	nr	8.75
44 - 715.............................	5.09	5.00	5.34	0.30	0.04	3.05	-	8.39	nr	9.02
44 - 755.............................	6.76	5.00	7.10	0.33	0.04	3.33	-	10.42	nr	11.21
44 - 760.............................	7.30	5.00	7.67	0.33	0.04	3.33	-	10.99	nr	11.82
44 - 765.............................	8.30	5.00	8.71	0.33	0.04	3.33	-	12.04	nr	12.94
Expanding bolts; Sleeve Anchor (Rawlok); bolt projecting type; drilling masonry										
with nuts and washers; reference										
69 - 504.............................	0.20	5.00	0.21	0.16	0.02	1.62	-	1.83	nr	1.96
69 - 506.............................	0.24	5.00	0.25	0.16	0.02	1.62	-	1.87	nr	2.01
69 - 508.............................	0.25	5.00	0.26	0.18	0.02	1.80	-	2.06	nr	2.22
69 - 510.............................	0.32	5.00	0.34	0.18	0.02	1.80	-	2.14	nr	2.30
69 - 512.............................	0.42	5.00	0.44	0.18	0.02	1.80	-	2.24	nr	2.41
69 - 514.............................	0.36	5.00	0.38	0.20	0.03	2.06	-	2.43	nr	2.62
69 - 516.............................	0.47	5.00	0.49	0.20	0.03	2.06	-	2.55	nr	2.74
69 - 518.............................	0.60	5.00	0.63	0.20	0.03	2.06	-	2.69	nr	2.89
69 - 520.............................	0.55	5.00	0.58	0.22	0.03	2.24	-	2.82	nr	3.03
69 - 522.............................	0.62	5.00	0.65	0.22	0.03	2.24	-	2.89	nr	3.11
69 - 524.............................	0.86	5.00	0.90	0.22	0.03	2.24	-	3.14	nr	3.38
69 - 525.............................	1.06	5.00	1.11	0.22	0.03	2.24	-	3.35	nr	3.61
69 - 526.............................	1.13	5.00	1.19	0.25	0.03	2.52	-	3.70	nr	3.98
69 - 528.............................	1.32	5.00	1.39	0.25	0.03	2.52	-	3.90	nr	4.20
69 - 530.............................	1.90	5.00	2.00	0.25	0.03	2.52	-	4.51	nr	4.85
69 - 533.............................	1.91	5.00	2.01	0.27	0.03	2.70	-	4.71	nr	5.06
69 - 534.............................	2.46	5.00	2.58	0.27	0.03	2.70	-	5.29	nr	5.68
69 - 536.............................	2.97	5.00	3.12	0.27	0.03	2.70	-	5.82	nr	6.26
Expanding bolts; loose bolt Rawlbolts; drilling masonry										
with washers; reference										
44 - 015.............................	0.69	5.00	0.72	0.20	0.02	1.99	-	2.71	nr	2.91
44 - 020.............................	0.73	5.00	0.77	0.20	0.02	1.99	-	2.75	nr	2.96
44 - 025.............................	0.74	5.00	0.78	0.20	0.02	1.99	-	2.76	nr	2.97
44 - 055.............................	0.87	5.00	0.91	0.22	0.03	2.24	-	3.15	nr	3.39
44 - 060.............................	0.90	5.00	0.94	0.22	0.03	2.24	-	3.19	nr	3.43
44 - 065.............................	0.95	5.00	1.00	0.22	0.03	2.24	-	3.24	nr	3.48
44 - 105.............................	1.15	5.00	1.21	0.25	0.03	2.52	-	3.73	nr	4.01
44 - 110.............................	1.19	5.00	1.25	0.25	0.03	2.52	-	3.77	nr	4.05
44 - 115.............................	1.25	5.00	1.31	0.25	0.03	2.52	-	3.83	nr	4.12

STRUCTURAL/CARCASSING METAL/TIMBER

Labour hourly rates: (except Specialists) Craft Operatives £9.23 Labourer £7.02 Rates are national average prices. Refer to REGIONAL VARIATIONS for indicative levels of overall pricing in regions	MATERIALS			LABOUR				RATES		
	Del to Site	Waste	Material Cost	Craft Optve	Lab	Labour Cost	Sunds	Nett Rate	Unit	Gross Rate (+7.5%)
	£	%	£	Hrs	Hrs	£	£	£		£

G20: CARPENTRY/TIMBER FRAMING/FIRST FIXING Cont.

Steel – expanding bolts Cont.

Expanding bolts; loose bolt Rawlbolts; drilling masonry Cont.
  with washers; reference Cont.

	Del to Site	Waste	Material Cost	Craft Optve	Lab	Labour Cost	Sunds	Nett Rate	Unit	Gross Rate
44 – 120	1.30	5.00	1.37	0.25	0.03	2.52	–	3.88	nr	4.17
44 – 155	1.72	5.00	1.81	0.27	0.03	2.70	–	4.51	nr	4.85
44 – 160	1.90	5.00	2.00	0.27	0.03	2.70	–	4.70	nr	5.05
44 – 165	1.98	5.00	2.08	0.27	0.03	2.70	–	4.78	nr	5.14
44 – 170	2.09	5.00	2.19	0.27	0.03	2.70	–	4.90	nr	5.26
44 – 205	3.99	5.00	4.19	0.30	0.04	3.05	–	7.24	nr	7.78
44 – 210	4.61	5.00	4.84	0.30	0.04	3.05	–	7.89	nr	8.48
44 – 215	4.99	5.00	5.24	0.30	0.04	3.05	–	8.29	nr	8.91
44 – 255	7.83	5.00	8.22	0.33	0.04	3.33	–	11.55	nr	12.41
44 – 260	8.13	5.00	8.54	0.33	0.04	3.33	–	11.86	nr	12.75

Expanding bolts; Sleeve Anchor (Rawlok); loose bolt type with countersunk bolt; drilling masonry
  reference

	Del to Site	Waste	Material Cost	Craft Optve	Lab	Labour Cost	Sunds	Nett Rate	Unit	Gross Rate
69 – 572	0.38	5.00	0.40	0.16	0.02	1.62	–	2.02	nr	2.17
69 – 574	0.40	5.00	0.42	0.16	0.02	1.62	–	2.04	nr	2.19
69 – 576	0.47	5.00	0.49	0.16	0.02	1.62	–	2.11	nr	2.27
59 – 578	0.55	5.00	0.58	0.18	0.02	1.80	–	2.38	nr	2.56
69 – 580	0.64	5.00	0.67	0.18	0.02	1.80	–	2.47	nr	2.66
69 – 582	0.78	5.00	0.82	0.20	0.02	1.99	–	2.81	nr	3.02
69 – 584	0.89	5.00	0.93	0.20	0.02	1.99	–	2.92	nr	3.14

Expanding bolts; Sleeve Anchor (Rawlok); loose bolt type with round head bolt; drilling masonry
  reference

	Del to Site	Waste	Material Cost	Craft Optve	Lab	Labour Cost	Sunds	Nett Rate	Unit	Gross Rate
69 – 604	0.40	5.00	0.42	0.16	0.02	1.62	–	2.04	nr	2.19
69 – 608	0.50	5.00	0.53	0.16	0.02	1.62	–	2.14	nr	2.30
69 – 610	0.56	5.00	0.59	0.18	0.02	1.80	–	2.39	nr	2.57
69 – 612	0.64	5.00	0.67	0.18	0.02	1.80	–	2.47	nr	2.66
69 – 614	0.79	5.00	0.83	0.20	0.02	1.99	–	2.82	nr	3.03
69 – 616	0.87	5.00	0.91	0.20	0.02	1.99	–	2.90	nr	3.12

Spit Mega high performance safety anchors; loose nut version; drilling masonry
  with nuts and washers, reference

	Del to Site	Waste	Material Cost	Craft Optve	Lab	Labour Cost	Sunds	Nett Rate	Unit	Gross Rate
E10 – 15/20	1.48	5.00	1.55	0.25	0.03	2.52	–	4.07	nr	4.38
E10 – 15/45	1.63	5.00	1.71	0.25	0.03	2.52	–	4.23	nr	4.55
E10 – 15/65	1.74	5.00	1.83	0.25	0.03	2.52	–	4.35	nr	4.67
E12 – 18/25	2.09	5.00	2.19	0.27	0.03	2.70	–	4.90	nr	5.26
E12 – 18/45	2.17	5.00	2.28	0.27	0.03	2.70	–	4.98	nr	5.35
E12 – 18/65	2.32	5.00	2.44	0.27	0.03	2.70	–	5.14	nr	5.52
E16 – 24/25	5.86	5.00	6.15	0.30	0.03	2.98	–	9.13	nr	9.82
E16 – 24/45	6.22	5.00	6.53	0.30	0.04	3.05	–	9.58	nr	10.30
E16 – 24/95	7.65	5.00	8.03	0.30	0.04	3.05	–	11.08	nr	11.91
E20 – 28/25	8.57	5.00	9.00	0.33	0.04	3.33	–	12.33	nr	13.25
E20 – 28/45	10.26	5.00	10.77	0.33	0.04	3.33	–	14.10	nr	15.16
E20 – 28/95	10.32	5.00	10.84	0.33	0.04	3.33	–	14.16	nr	15.22

Spit Mega high performance safety anchors; bolt head version; drilling masonry
  with washers, reference

	Del to Site	Waste	Material Cost	Craft Optve	Lab	Labour Cost	Sunds	Nett Rate	Unit	Gross Rate
V10 – 15/20	1.47	5.00	1.54	0.25	0.03	2.52	–	4.06	nr	4.37
V12 – 18/25	2.09	5.00	2.19	0.27	0.03	2.70	–	4.90	nr	5.26
V16 – 24/25	4.52	5.00	4.75	0.30	0.04	3.05	–	7.80	nr	8.38
V20 – 28/25	8.57	5.00	9.00	0.33	0.04	3.33	–	12.33	nr	13.25

Spit Fix high performance through bolt BZP anchors; drilling masonry
  with nuts and washers, reference

	Del to Site	Waste	Material Cost	Craft Optve	Lab	Labour Cost	Sunds	Nett Rate	Unit	Gross Rate
6/10	0.15	5.00	0.16	0.18	0.03	1.87	–	2.03	nr	2.18
6/20	0.16	5.00	0.17	0.18	0.03	1.87	–	2.04	nr	2.19
6/50	0.17	5.00	0.18	0.18	0.03	1.87	–	2.05	nr	2.20
8/10	0.18	5.00	0.19	0.20	0.03	2.06	–	2.25	nr	2.41
8/50	0.24	5.00	0.25	0.20	0.03	2.06	–	2.31	nr	2.48
8/90	0.27	5.00	0.28	0.20	0.03	2.06	–	2.34	nr	2.52
10/10	0.28	5.00	0.29	0.22	0.03	2.24	–	2.54	nr	2.73
10/25	0.29	5.00	0.30	0.22	0.03	2.24	–	2.55	nr	2.74
10/45	0.31	5.00	0.33	0.22	0.03	2.24	–	2.57	nr	2.76
12/10	0.37	5.00	0.39	0.25	0.03	2.52	–	2.91	nr	3.12
12/40	0.51	5.00	0.54	0.25	0.03	2.52	–	3.05	nr	3.28
12/80	0.59	5.00	0.62	0.25	0.03	2.52	–	3.14	nr	3.37
12/120	0.75	5.00	0.79	0.25	0.03	2.52	–	3.31	nr	3.55
12/160	1.27	5.00	1.33	0.25	0.03	2.52	–	3.85	nr	4.14
16/10	1.00	5.00	1.05	0.27	0.03	2.70	–	3.75	nr	4.03
16/45	1.02	5.00	1.07	0.27	0.03	2.70	–	3.77	nr	4.06
16/95	1.04	5.00	1.09	0.27	0.03	2.70	–	3.79	nr	4.08
20/20	1.09	5.00	1.14	0.30	0.03	2.98	–	4.12	nr	4.43
20/60	1.19	5.00	1.25	0.30	0.04	3.05	–	4.30	nr	4.62
20/115	1.32	5.00	1.39	0.30	0.04	3.05	–	4.44	nr	4.77

Steel; plated – expanding bolts

Expanding bolts; hook Rawlbolts; drilling masonry
  with nuts and washers; reference

	Del to Site	Waste	Material Cost	Craft Optve	Lab	Labour Cost	Sunds	Nett Rate	Unit	Gross Rate
44 – 401	0.93	5.00	0.98	0.20	0.03	2.06	–	3.03	nr	3.26
44 – 406	1.13	5.00	1.19	0.22	0.03	2.24	–	3.43	nr	3.68
44 – 411	1.83	5.00	1.92	0.25	0.03	2.52	–	4.44	nr	4.77
44 – 416	2.89	5.00	3.03	0.27	0.03	2.70	–	5.74	nr	6.17

	MATERIALS			LABOUR				RATES		
Labour hourly rates: (except Specialists) Craft Operatives £9.23 Labourer £7.02 Rates are national average prices. Refer to REGIONAL VARIATIONS for indicative levels of overall pricing in regions	Del to Site £	Waste %	Material Cost £	Craft Optve Hrs	Lab Hrs	Labour Cost £	Sunds £	Nett Rate £	Unit	Gross Rate (+7.5%) £
G20: CARPENTRY/TIMBER FRAMING/FIRST FIXING Cont.										
Steel; plated - expanding bolts Cont.										
Expanding bolts; eye Rawlbolts; drilling masonry with nuts and washers; reference										
44 - 432	2.05	5.00	2.15	0.20	0.03	2.06	-	4.21	nr	4.52
44 - 437	2.34	5.00	2.46	0.22	0.03	2.24	-	4.70	nr	5.05
44 - 442	3.14	5.00	3.30	0.25	0.03	2.52	-	5.82	nr	6.25
44 - 447	4.69	5.00	4.92	0.27	0.03	2.70	-	7.63	nr	8.20
Expanding bolts; bolt projecting Through Bolts; drilling masonry with nuts and washers; reference										
56 - 102	0.30	5.00	0.32	0.18	0.02	1.80	-	2.12	nr	2.28
56 - 104	0.31	5.00	0.33	0.18	0.02	1.80	-	2.13	nr	2.29
56 - 108	0.33	5.00	0.35	0.18	0.02	1.80	-	2.15	nr	2.31
56 - 112	0.36	5.00	0.38	0.18	0.02	1.80	-	2.18	nr	2.34
56 - 114	0.43	5.00	0.45	0.20	0.03	2.06	-	2.51	nr	2.70
56 - 116	0.45	5.00	0.47	0.20	0.03	2.06	-	2.53	nr	2.72
56 - 120	0.47	5.00	0.49	0.20	0.03	2.06	-	2.55	nr	2.74
56 - 124	0.50	5.00	0.53	0.20	0.03	2.06	-	2.58	nr	2.78
56 - 126	0.60	5.00	0.63	0.20	0.03	2.06	-	2.69	nr	2.89
56 - 128	0.58	5.00	0.61	0.22	0.03	2.24	-	2.85	nr	3.06
56 - 129	0.55	5.00	0.58	0.22	0.03	2.24	-	2.82	nr	3.03
56 - 132	0.61	5.00	0.64	0.22	0.03	2.24	-	2.88	nr	3.10
56 - 136	0.65	5.00	0.68	0.22	0.03	2.24	-	2.92	nr	3.14
56 - 138	0.76	5.00	0.80	0.22	0.03	2.24	-	3.04	nr	3.27
56 - 139	0.86	5.00	0.90	0.25	0.03	2.52	-	3.42	nr	3.68
56 - 140	0.87	5.00	0.91	0.25	0.03	2.52	-	3.43	nr	3.69
56 - 144	0.94	5.00	0.99	0.25	0.03	2.52	-	3.51	nr	3.77
56 - 148	1.01	5.00	1.06	0.25	0.03	2.52	-	3.58	nr	3.85
56 - 150	1.05	5.00	1.10	0.25	0.03	2.52	-	3.62	nr	3.89
56 - 152	1.58	5.00	1.66	0.27	0.03	2.70	-	4.36	nr	4.69
56 - 153	1.40	5.00	1.47	0.27	0.03	2.70	-	4.17	nr	4.49
56 - 156	1.91	5.00	2.01	0.27	0.03	2.70	-	4.71	nr	5.06
56 - 158	2.48	5.00	2.60	0.27	0.03	2.70	-	5.31	nr	5.70
56 - 159	2.48	5.00	2.60	0.30	0.04	3.05	-	5.65	nr	6.08
56 - 160	2.70	5.00	2.84	0.30	0.04	3.05	-	5.88	nr	6.33
56 - 164	3.45	5.00	3.62	0.30	0.04	3.05	-	6.67	nr	7.17
56 - 166	4.30	5.00	4.51	0.30	0.04	3.05	-	7.56	nr	8.13
56 - 168	4.93	5.00	5.18	0.33	0.04	3.33	-	8.50	nr	9.14
56 - 172	6.07	5.00	6.37	0.33	0.04	3.33	-	9.70	nr	10.43
Stainless steel grade 316 - expanding bolts										
Expanding bolts; bolt projecting Through Bolts; drilling masonry with nuts and washers; reference										
56 - 604	1.41	5.00	1.48	0.18	0.02	1.80	-	3.28	nr	3.53
56 - 616	1.72	5.00	1.81	0.20	0.02	1.99	-	3.79	nr	4.08
56 - 624	2.13	5.00	2.24	0.20	0.02	1.99	-	4.22	nr	4.54
56 - 628	2.22	5.00	2.33	0.22	0.03	2.24	-	4.57	nr	4.92
56 - 636	2.63	5.00	2.76	0.22	0.03	2.24	-	5.00	nr	5.38
56 - 638	2.93	5.00	3.08	0.22	0.03	2.24	-	5.32	nr	5.72
56 - 640	4.04	5.00	4.24	0.25	0.03	2.52	-	6.76	nr	7.27
56 - 648	4.86	5.00	5.10	0.25	0.03	2.52	-	7.62	nr	8.19
56 - 650	5.41	5.00	5.68	0.25	0.03	2.52	-	8.20	nr	8.81
56 - 652	6.95	5.00	7.30	0.27	0.03	2.70	-	10.00	nr	10.75
56 - 658	10.04	5.00	10.54	0.27	0.03	2.70	-	13.24	nr	14.24
56 - 660	13.61	5.00	14.29	0.30	0.04	3.05	-	17.34	nr	18.64
56 - 666	20.20	5.00	21.21	0.30	0.04	3.05	-	24.26	nr	26.08
56 - 672	27.86	5.00	29.25	0.33	0.04	3.33	-	32.58	nr	35.02
Chemical anchors										
Chemical anchors; Kemfix capsules and standard studs; drilling masonry										
capsule reference 60-428; stud reference 60-708; with nuts and washers	1.31	5.00	1.38	0.24	0.03	2.43	-	3.80	nr	4.09
capsule reference 60-430; stud reference 60-710; with nuts and washers	1.47	5.00	1.54	0.27	0.03	2.70	-	4.25	nr	4.56
capsule reference 60-432; stud reference 60-712; with nuts and washers	1.79	5.00	1.88	0.30	0.04	3.05	-	4.93	nr	5.30
capsule reference 60-603; stud reference 60-014; with nuts and washers	3.12	5.00	3.28	0.33	0.04	3.33	-	6.60	nr	7.10
Chemical anchors; Kemfix capsules and stainless steel studs; drilling masonry										
capsule reference 60-428; stud reference 60-906; with nuts and washers	2.31	5.00	2.43	0.24	0.03	2.43	-	4.85	nr	5.22
capsule reference 60-430; stud reference 60-911; with nuts and washers	3.08	5.00	3.23	0.27	0.03	2.70	-	5.94	nr	6.38
capsule reference 60-432; stud reference 60-916; with nuts and washers	4.24	5.00	4.45	0.30	0.04	3.05	-	7.50	nr	8.06
capsule reference 60-436; stud reference 60-921; with nuts and washers	7.21	5.00	7.57	0.33	0.04	3.33	-	10.90	nr	11.71
capsule reference 60-440; stud reference 60-926; with nuts and washers	11.13	5.00	11.69	0.35	0.04	3.51	-	15.20	nr	16.34
capsule reference 60-444; stud reference 60-931; with nuts and washers	17.99	5.00	18.89	0.38	0.05	3.86	-	22.75	nr	24.45
Chemical anchors; Kemfix capsules and standard internal threaded sockets; drilling masonry										
capsule reference 60-428; socket reference 60-622	1.80	5.00	1.89	0.30	0.04	3.05	-	4.94	nr	5.31
capsule reference 60-430; socket reference 60-624	2.08	5.00	2.18	0.33	0.04	3.33	-	5.51	nr	5.92

Labour hourly rates: (except Specialists) Craft Operatives £9.23 Labourer £7.02 Rates are national average prices. Refer to REGIONAL VARIATIONS for indicative levels of overall pricing in regions	MATERIALS			LABOUR				RATES		
	Del to Site £	Waste %	Material Cost £	Craft Optve Hrs	Lab Hrs	Labour Cost £	Sunds £	Nett Rate £	Unit	Gross Rate (+7.5%) £

G20: CARPENTRY/TIMBER FRAMING/FIRST FIXING Cont.

Chemical anchors Cont.

Chemical anchors; Kemfix capsules and standard
internal threaded sockets; drilling masonry Cont.

	Del to Site £	Waste %	Material Cost £	Craft Optve Hrs	Lab Hrs	Labour Cost £	Sunds £	Nett Rate £	Unit	Gross Rate £
capsule reference 60-432; socket reference 60-628	2.81	5.00	2.95	0.35	0.04	3.51	-	6.46	nr	6.95
capsule reference 60-436; socket reference 60-630	5.19	5.00	5.45	0.38	0.05	3.86	-	9.31	nr	10.01

Chemcial anchors; Kemfix capsules and stainless steel
internal threaded sockets; drilling masonry

capsule reference 60-428; socket reference 60-987	3.09	5.00	3.24	0.30	0.04	3.05	-	6.29	nr	6.77
capsule reference 60-430; socket reference 60-989	4.01	5.00	4.21	0.33	0.04	3.33	-	7.54	nr	8.10
capsule reference 60-432; socket reference 60-993	5.16	5.00	5.42	0.35	0.04	3.51	-	8.93	nr	9.60
capsule reference 60-436; socket reference 60-995	10.16	5.00	10.67	0.38	0.05	3.86	-	14.53	nr	15.62

Chemical anchors in low density material; Kemfix
capsules, perforated sleeves and standard studs;
drilling masonry

capsule reference 60-428; sleeve reference 60-105; stud reference 60-708; with nuts and washers	1.93	5.00	2.03	0.24	0.03	2.43	-	4.45	nr	4.79
capsule reference 60-430; sleeve reference 60-107; stud reference 60-710; with nuts and washers	2.12	5.00	2.23	0.27	0.03	2.70	-	4.93	nr	5.30
capsule reference 60-432; sleeve reference 60-113; stud reference 60-712; with nuts and washers	2.54	5.00	2.67	0.30	0.04	3.05	-	5.72	nr	6.15
capsule reference 60-436; sleeve reference 60-117; stud reference 60-014; with nuts and washers	3.95	5.00	4.15	0.35	0.04	3.51	-	7.66	nr	8.23

Chemical anchors in low density material; Kemfix
capsules, perforated sleeves and stainless steel
studs; drilling masonry

capsule reference 60-428; sleeve reference 60-105; stud reference 60-906; with nuts and washers	2.92	5.00	3.07	0.24	0.03	2.43	-	5.49	nr	5.90
capsule reference 60-430; sleeve reference 60-107; stud reference 60-917; with nuts and washers	2.73	5.00	2.87	0.27	0.03	2.70	-	5.57	nr	5.99
capsule reference 60-432; sleeve reference 60-113; stud reference 60-916; with nuts and washers	4.99	5.00	5.24	0.30	0.04	3.05	-	8.29	nr	8.91
capsule reference 60-436; sleeve reference 60-117; stud reference 60-921; with nuts and washers	8.13	5.00	8.54	0.35	0.04	3.51	-	12.05	nr	12.95

Chemical anchors in low density material; Kemfix
capsules, perforated sleeves and standard internal
threaded sockets; drilling masonry

capsule reference 60-428; sleeve reference 60-105; socket reference 60-622	2.42	5.00	2.54	0.24	0.03	2.43	-	4.97	nr	5.34
capsule reference 60-430; sleeve reference 60-107; socket reference 60-624	2.70	5.00	2.84	0.27	0.03	2.70	-	5.54	nr	5.95
capsule reference 60-432; sleeve reference 60-113; socket reference 60-628	3.56	5.00	3.74	0.30	0.04	3.05	-	6.79	nr	7.30
capsule reference 60-436; sleeve reference 60-117; socket reference 60-630	6.10	5.00	6.41	0.33	0.04	3.33	-	9.73	nr	10.46

Chemical anchors in low density material; Kemfix
capsules, perforated sleeves and stainless steel
internal threaded sockets; drilling masonry

capsule reference 60-428; sleeve reference 60-105; socket reference 60-687	3.71	5.00	3.90	0.24	0.03	2.43	-	6.32	nr	6.80
capsule reference 60-430; sleeve reference 60-107; socket reference 60-689	4.66	5.00	4.89	0.27	0.03	2.70	-	7.60	nr	8.17
capsule reference 60-432; sleeve reference 60-113; socket reference 60-993	5.91	5.00	6.21	0.30	0.04	3.05	-	9.26	nr	9.95
capsule reference 60-436; sleeve reference 60-117; socket reference 60-995	11.08	5.00	11.63	0.33	0.04	3.33	-	14.96	nr	16.08

Spit Maxi high performance chemical anchors; capsules
and zinc coated steelstuds; drilling masonry

capsule reference M8 stud reference SM8; with nuts and washers	1.10	5.00	1.16	0.25	0.03	2.52	-	3.67	nr	3.95
capsule reference M10; stud reference SM10; with nuts and washers	1.24	5.00	1.30	0.27	0.03	2.70	-	4.00	nr	4.31
capsule reference M12; stud reference SM12; with nuts and washers	1.50	5.00	1.58	0.25	0.03	2.52	-	4.09	nr	4.40
capsule reference M16; stud reference SM16; with nuts and washers	2.08	5.00	2.18	0.27	0.03	2.70	-	4.89	nr	5.25
capsule reference M20; stud reference SM20; with nuts and washers	3.72	5.00	3.91	0.30	0.04	3.05	-	6.96	nr	7.48
capsule reference M24; stud reference SM24; with nuts and washers	5.37	5.00	5.64	0.35	0.04	3.51	—	9.15	nr	9.84

Spit Maxi high performance chemical anchors; capsules
and stainless steel studs Grade 316 (A4); drilling
masonry

capsule reference M8 stud reference SM8i; with nuts and washers	1.70	5.00	1.78	0.25	0.03	2.52	-	4.30	nr	4.63
capsule reference M10; stud reference SM10i; with nuts and washers	2.33	5.00	2.45	0.27	0.03	2.70	-	5.15	nr	5.54
capsule reference M12; stud reference SM12i; with nuts and washers	3.37	5.00	3.54	0.25	0.03	2.52	-	6.06	nr	6.51
capsule reference M16; stud reference SM16i; with nuts and washers	5.52	5.00	5.80	0.27	0.03	2.70	-	8.50	nr	9.14
capsule reference M20; stud reference SM20i; with nuts and washers	10.06	5.00	10.56	0.30	0.04	3.05	-	13.61	nr	14.63
capsule referenceM24; stud reference SM24i; with nuts and washers	14.83	5.00	15.57	0.35	0.04	3.51	-	19.08	nr	20.51

Labour hourly rates: (except Specialists) Craft Operatives £9.23 Labourer £7.02 Rates are national average prices. Refer to REGIONAL VARIATIONS for indicative levels of overall pricing in regions	MATERIALS			LABOUR				RATES		
	Del to Site £	Waste %	Material Cost £	Craft Optve Hrs	Lab Hrs	Labour Cost £	Sunds £	Nett Rate £	Unit	Gross Rate (+7.5%) £
G30: METAL PROFILED SHEET DECKING										
Galvanised steel troughed decking; 0.7mm thick metal, 35mm overall depth; natural soffit and top surface; 150mm end laps and one corrugation side laps										
Decking; fixed to steel rails at 1200mm centres with self tapping screws; drilling holes										
sloping; 10 degrees pitch	-	-	Spclist	-	-	Spclist	-	21.30	m2	22.90
Extra over decking for										
raking cutting	-	-	Spclist	-	-	Spclist	-	5.40	m	5.80
holes 50mm diameter; formed on site	-	-	Spclist	-	-	Spclist	-	6.10	nr	6.56
Galvanised steel troughed decking; 0.7mm thick metal, 48mm overall depth; natural soffit and top surface; 150mm end laps and one corrugation side laps										
Decking; fixed to steel rails at 1500mm centres with self tapping screws; drilling holes										
sloping; 10 degrees pitch	-	-	Spclist	-	-	Spclist	-	22.20	m2	23.86
Extra over decking for										
raking cutting	-	-	Spclist	-	-	Spclist	-	5.75	m	6.18
holes 50mm diameter; formed on site	-	-	Spclist	-	-	Spclist	-	6.10	nr	6.56
Galvanised steel troughed decking; 0.7mm thick metal, 63mm overall depth; natural soffit and top surface; 150mm end laps and one corrugation side laps										
Decking; fixed to steel rails at 2000mm centres with self tapping screws; drilling holes										
sloping; 10 degrees pitch	-	-	Spclist	-	-	Spclist	-	23.85	m2	25.64
Extra over decking for										
raking cutting	-	-	Spclist	-	-	Spclist	-	6.10	m	6.56
holes 50mm diameter; formed on site	-	-	Spclist	-	-	Spclist	-	6.10	nr	6.56
Galvanised steel troughed decking; 0.7mm thick metal, 100mm overall depth;natural soffit and top surface; 150mm end laps and one corrugation side laps										
Decking; fixed to steel rails at 3500mm centres with self tapping screws; drilling holes										
sloping; 10 degrees pitch	-	-	Spclist	-	-	Spclist	-	26.95	m2	28.97
Extra over decking for										
raking cutting	-	-	Spclist	-	-	Spclist	-	6.85	m	7.36
holes 50mm diameter; formed on site	-	-	Spclist	-	-	Spclist	-	6.10	nr	6.56
Galvanised steel troughed decking; 0.9mm thick metal, 35mm overall depth; natural soffit and top surface; 150mm end laps and one corrugation side laps										
Decking; fixed to steel rails at 1500mm centres with self tapping screws; drilling holes										
sloping; 10 degrees pitch	-	-	Spclist	-	-	Spclist	-	22.30	m2	23.97
Extra over decking for										
raking cutting	-	-	Spclist	-	-	Spclist	-	5.65	m	6.07
holes 50mm diameter; formed on site	-	-	Spclist	-	-	Spclist	-	6.10	nr	6.56
Galvanised steel troughed decking; 0.9mm thick metal, 48mm overall depth; natural soffit and top surface; 150mm end laps and one corrugation side laps										
Decking; fixing to steel rails at 2000mm centres with self tapping screws; drilling holes										
sloping; 10 degrees pitch	-	-	Spclist	-	-	Spclist	-	23.35	m2	25.10
Extra over decking for										
raking cutting	-	-	Spclist	-	-	Spclist	-	6.00	m	6.45
holes 50mm diameter; formed on site	-	-	Spclist	-	-	Spclist	-	6.10	nr	6.56
Galvanised steel troughed decking; 0.9mm thick metal, 63mm overall depth; natural soffit and top surface; 150mm end laps and one corrugation side laps										
Decking; fixing to steel rails at 2500mm centres with self tapping screws; drilling holes										
sloping; 10 degrees pitch	-	-	Spclist	-	-	Spclist	-	25.05	m2	26.93
Extra over decking for										
raking cutting	-	-	Spclist	-	-	Spclist	-	6.20	m	6.67
holes 50mm diameter; formed on site	-	-	Spclist	-	-	Spclist	-	6.10	nr	6.56
Galvanised steel troughed decking; 0.9mm thick metal, 100mm overall depth;natural soffit and top surface; 150mm end laps and one corrugation side laps										
Decking; fixing to steel rails at 4000mm centres with self tapping screws; drilling holes										
sloping; 10 degrees pitch	-	-	Spclist	-	-	Spclist	-	28.05	m2	30.15
Extra over decking for										
raking cutting	-	-	Spclist	-	-	Spclist	-	7.15	m	7.69

Labour hourly rates: (except Specialists) Craft Operatives £9.23 Labourer £7.02 Rates are national average prices. Refer to REGIONAL VARIATIONS for indicative levels of overall pricing in regions	MATERIALS			LABOUR				RATES		
	Del to Site £	Waste %	Material Cost £	Craft Optve Hrs	Lab Hrs	Labour Cost £	Sunds £	Nett Rate £	Unit	Gross Rate (+7.5%) £
G30: METAL PROFILED SHEET DECKING Cont.										
Galvanised steel troughed decking; 0.9mm thick metal, 100mm overall depth;natural soffit and top surface; 150mm end laps and one corrugation side laps Cont.										
Extra over decking for Cont.										
holes 50mm diameter; formed on site	-	-	Spclist	-	-	Spclist	-	6.10	nr	6.56
Galvanised steel troughed decking; 1.2mm thick metal, 35mm overall depth; natural soffit and top surface; 150mm end laps and one corrugation side laps										
Decking; fixing to steel rails at 2000mm centres with self tapping screws; drilling holes										
sloping; 10 degrees pitch	-	-	Spclist	-	-	Spclist	-	25.25	m2	27.14
Extra over decking for										
raking cutting	-	-	Spclist	-	-	Spclist	-	6.10	m	6.56
holes 50mm diameter; formed on site	-	-	Spclist	-	-	Spclist	-	6.10	nr	6.56
Galvanised steel troughed decking; 1.2mm thick metal, 48mm overall depth; natural soffit and top surface; 150mm end laps and one corrugation side laps										
Decking; fixing to steel rails at 2500mm centres with self tapping screws; drilling holes										
sloping; 10 degrees pitch	-	-	Spclist	-		Spclist	-	25.15	m2	27.04
Extra over decking for										
raking cutting	-	-	Spclist	-	-	Spclist	-	6.30	m	6.77
holes 50mm diameter; formed on site	-	-	Spclist	-	-	Spclist	-	6.10	nr	6.56
Galvanised steel troughed decking; 1.2mm thick metal, 63mm overall depth; natural soffit and top surface; 150mm end laps and one corrugation side laps										
Decking; fixing to steel rails at 3000mm centres with self tapping screws; drilling holes										
sloping; 10 degrees pitch	-	-	Spclist	-	-	Spclist	-	26.95	m2	28.97
Extra over decking for										
raking cutting	-	-	Spclist	-	-	Spclist	-	6.75	m	7.26
holes 50mm diameter; formed on site	-	-	Spclist	-	-	Spclist	-	6.10	nr	6.56
Galvanised steel troughed decking; 1.2mm thick metal, 100mm overall depth;natural soffit and top surface; 150mm end laps and one corrugation side laps										
Decking; fixing to steel rails at 4500mm centres with self tapping screws; drilling holes										
sloping; 10 degrees pitch	-	-	Spclist	-	-	Spclist	-	30.95	m2	33.27
Extra over decking for										
raking cutting	-	-	Spclist	-	-	Spclist	-	7.20	m	7.74
holes 50mm diameter; formed on site	-	-	Spclist	-	-	Spclist	-	6.10	nr	6.56
Aluminium troughed decking; 0.9mm thick metal 35mm overall depth; natural soffit and top surface; 150mm end laps and one corrugation side laps										
Decking; fixing to steel rails at 900mm centres with self tapping screws; drilling holes										
sloping; 10 degrees pitch	-	-	Spclist	-	-	Spclist	-	25.50	m2	27.41
Extra over decking for										
raking cutting	-	-	Spclist	-	-	Spclist	-	6.50	m	6.99
holes 50mm diameter; formed on site	-	-	Spclist	-	-	Spclist	-	6.10	nr	6.56
Aluminium troughed decking; 0.9mm thick metal, 48mm overall depth; natural soffit and top surface; 150mm end laps and one corrugation side laps										
Decking; fixing to steel rails at 1200mm centres with self tapping screws; drilling holes										
sloping; 10 degrees pitch	-	-	Spclist	-	-	Spclist	-	27.30	m2	29.35
Extra over decking for										
raking cutting	-	-	Spclist	-	-	Spclist	-	6.85	m	7.36
holes 50mm diameter; formed on site	-	-	Spclist	-	-	Spclist	-	6.10	nr	6.56
Aluminium troughed decking; 0.9mm thick metal, 63mm overall depth; natural soffit and top surface; 150mm end laps and one corrugation side laps										
Decking; fixing to steel rails at 1500mm centres with self tapping screws; drilling holes										
sloping; 10 degrees pitch	-	-	Spclist	-	-	Spclist	-	28.90	m2	31.07
Extra over decking for										
raking cutting	-	-	Spclist	-	-	Spclist	-	7.40	m	7.96
holes 50mm diameter; formed on site	-	-	Spclist	-	-	Spclist	-	6.10	nr	6.56

Labour hourly rates: (except Specialists) Craft Operatives £9.23 Labourer £7.02 Rates are national average prices. Refer to REGIONAL VARIATIONS for indicative levels of overall pricing in regions	MATERIALS			LABOUR				RATES		
	Del to Site £	Waste %	Material Cost £	Craft Optve Hrs	Lab Hrs	Labour Cost £	Sunds £	Nett Rate £	Unit	Gross Rate (+7.5%) £
G30: METAL PROFILED SHEET DECKING Cont.										
Aluminium troughed decking; 0.9mm thick metal, 100mm overall depth; natural soffit and top surface; 150mm end laps and one corrugation side laps										
Decking; fixing to steel rails at 2900mm centres with self tapping screws; drilling holes										
sloping; 10 degrees pitch	-	-	Spclist	-	-	Spclist	-	32.30	m2	34.72
Extra over decking for										
raking cutting	-	-	Spclist	-	-	Spclist	-	8.25	m	8.87
holes 50mm diameter; formed on site	-	-	Spclist	-	-	Spclist	-	6.10	nr	6.56
Aluminium troughed decking; 1.2mm thick metal, 35mm overall depth; natural soffit and top surface; 150mm end laps and one corrugation side laps										
Decking; fixing to steel rails at 1200mm centres with self tapping screws; drilling holes										
sloping; 10 degrees pitch	-	-	Spclist	-	-	Spclist	-	27.40	m2	29.45
Extra over decking for										
raking cutting	-	-	Spclist	-	-	Spclist	-	6.85	m	7.36
holes 50mm diameter; formed on site	-	-	Spclist	-	-	Spclist	-	6.10	nr	6.56
Aluminium troughed decking; 1.2mm thick metal, 48mm overall depth; natural soffit and top surface; 150mm end laps and one corrugation side laps										
Decking; fixing to steel rails at 1700mm centres with self tapping screws; drilling holes										
sloping; 10 degrees pitch	-	-	Spclist	-	-	Spclist	-	29.10	m2	31.28
Extra over decking for										
raking cutting	-	-	Spclist	-	-	Spclist	-	7.45	m	8.01
holes 50mm diameter; formed on site	-	-	Spclist	-	-	Spclist	-	6.10	nr	6.56
Aluminium troughed decking; 1.2mm thick metal, 63mm overall depth; natural soffit and top surface; 150mm end laps and one corrugation side laps										
Decking; fixing to steel rails at 2000mm centres with self tapping screws; drilling holes										
sloping; 10 degrees pitch	-	-	Spclist	-	-	Spclist	-	30.85	m2	33.16
Extra over decking for										
raking cutting	-	-	Spclist	-	-	Spclist	-	8.00	m	8.60
holes 50mm diameter; formed on site	-	-	Spclist	-	-	Spclist	-	6.10	nr	6.56
Aluminium troughed decking; 1.2mm thick metal, 100mm overall depth; natural soffit and top surface; 150mm end laps and one corrugation side laps										
Decking; fixing to steel rails at 3000mm centres with self tapping screws; drilling holes										
sloping; 10 degrees pitch	-	-	Spclist	-	-	Spclist	-	34.10	m2	36.66
Extra over decking for										
raking cutting	-	-	Spclist	-	-	Spclist	-	8.55	m	9.19
holes 50mm diameter; formed on site	-	-	Spclist	-	-	Spclist	-	6.10	nr	6.56
G32: EDGE SUPPORTED/ REINFORCED WOODWOOL SLAB DECKING										
Channel reinforced woodwool slabs, B.S.1105 Type SB; nominal thickness 50mm; natural both sides; butt joints										
Decking; fixing to timber joists at 2000mm general spacing with galvanised mild steel nails and gripper plates										
sloping; 10 degrees pitch	16.08	5.00	16.88	0.44	0.06	4.48	0.22	21.59	m2	23.21
Decking; fixing to steel rails at 2000mm centres with sherardised clips										
sloping; 10 degrees pitch	16.08	5.00	16.88	0.33	0.04	3.33	0.43	20.64	m2	22.19
Extra over decking for										
raking cutting	3.22	-	3.22	0.26	0.03	2.61	-	5.83	m	6.27
holes 100mm diameter; formed on site	-	-	-	0.15	-	1.38	-	1.38	nr	1.49
Channel reinforced woodwool slabs, B.S.1105 Type SB; nominal thickness 75mm; natural both sides; butt joints										
Decking; fixing to timber joists at 3000mm general spacing with galvanised mild steel nails and gripper plates										
sloping; 10 degrees pitch	23.60	5.00	24.78	0.55	0.07	5.57	0.32	30.67	m2	32.97
Decking; fixing to steel rails at 3000mm centres with sherardised clips										
sloping; 10 degrees pitch	23.60	5.00	24.78	0.40	0.05	4.04	0.68	29.50	m2	31.72
Extra over decking for										
raking cutting	4.72	-	4.72	0.34	0.04	3.42	-	8.14	m	8.75

Labour hourly rates: (except Specialists) Craft Operatives £9.23 Labourer £7.02 Rates are national average prices. Refer to REGIONAL VARIATIONS for indicative levels of overall pricing in regions	MATERIALS			LABOUR				RATES		
	Del to Site £	Waste %	Material Cost £	Craft Optve Hrs	Lab Hrs	Labour Cost £	Sunds £	Nett Rate £	Unit	Gross Rate (+7.5%) £
G32: EDGE SUPPORTED/ REINFORCED WOODWOOL SLAB DECKING Cont.										
Channel reinforced woodwool slabs, B.S.1105 Type SB; nominal thickness 75mm; natural both sides; butt joints Cont.										
Extra over decking for Cont. holes 100mm diameter; formed on site	-	-	-	0.20	-	1.85	-	1.85	nr	1.98
Channel reinforced woodwool slabs, B.S.1105 Type SB; nominal thickness 100mm; natural both sides; butt joints										
Decking; fixing to timber joists at 4000mm general spacing with galvanised mild steel nails and gripper plates sloping; 10 degrees pitch	27.65	5.00	29.03	0.72	0.09	7.28	0.43	36.74	m2	39.50
Decking; fixing to steel rails at 4000mm centres with sherardised clips sloping; 10 degrees pitch	27.65	5.00	29.03	0.54	0.07	5.48	0.86	35.37	m2	38.02

Extra over decking for
    raking cutting.................................
    holes 100mm diameter; formed on site ....
	5.53
	-
		-
		-
			5.53
			-
				0.43
				0.25
					0.05
					-
						4.32
						2.31
							-
							-
								9.85
								2.31
									m
									nr
										10.59
										2.48

Channel reinforced woodwool slabs, B.S.1105 Type SB; nominal thickness 50mm; pre-textured soffit, natural top surface; butt joints										
Decking; fixing to timber joists at 2000mm general spacing with galvanised mild steel nails and gripper plates sloping; 10 degrees pitch	18.74	5.00	19.68	0.44	0.06	4.48	0.22	24.38	m2	26.21
Decking; fixing to steel rails at 2000mm centres with sherardised clips sloping; 10 degrees pitch	18.74	5.00	19.68	0.33	0.04	3.33	0.43	23.43	m2	25.19

Extra over decking for
    raking cutting.................................
    holes 100mm diameter; formed on site ....
	3.75
	-
		-
		-
			3.75
			-
				0.26
				0.15
					0.03
					-
						2.61
						1.38
							-
							-
								6.36
								1.38
									m
									nr
										6.84
										1.49

Channel reinforced woodwool slabs, B.S.1105 Type SB; nominal thickness 75mm; pre-textured soffit, natural top surface; butt joints										
Decking; fixing to timber joists at 3000mm general spacing with galvanised mild steel nails and gripper plates sloping; 10 degrees pitch	26.26	5.00	27.57	0.55	0.07	5.57	0.32	33.46	m2	35.97
Decking; fixing to steel rails at 3000mm centres with sherardised clips sloping; 10 degrees pitch	26.26	5.00	27.57	0.40	0.05	4.04	0.68	32.30	m2	34.72

Extra over decking for
    raking cutting.................................
    holes 100mm diameter; formed on site ....
	5.25
	-
		-
		-
			5.25
			-
				0.34
				0.20
					0.04
					-
						3.42
						1.85
							-
							-
								8.67
								1.85
									m
									nr
										9.32
										1.98

Channel reinforced woodwool slabs, B.S.1105 Type SB; nominal thickness 100mm; pre-textured soffit, natural top surface; butt joints										
Decking; fixing to timber joists at 4000mm general spacing with galvanised mild steel nails and gripper plates sloping; 10 degrees pitch	30.31	5.00	31.83	0.72	0.09	7.28	0.43	39.53	m2	42.50
Decking; fixing to steel rails at 4000mm centres with sherardised clips sloping; 10 degrees pitch	30.31	5.00	31.83	0.54	0.07	5.48	0.86	38.16	m2	41.02

Extra over decking for
    raking cutting.................................
    holes 100mm diameter; formed on site ....
	6.06
	-
		-
		-
			6.06
			-
				0.43
				0.25
					0.05
					-
						4.32
						2.31
							-
							-
								10.38
								2.31
									m
									nr
										11.16
										2.48

Cladding/Covering

Labour hourly rates: (except Specialists) Craft Operatives £9.23 Labour £7.02 Rates are national average prices. Refer to REGIONAL VARIATIONS for indicative levels of overall pricing in regions	MATERIALS			LABOUR				RATES		
	Del to Site £	Waste %	Material Cost £	Craft Optve Hrs	Lab Hrs	Labour Cost £	Sunds £	Nett Rate £	Unit	Gross Rate (+7.5%) £
H10: PATENT GLAZING										
Patent glazing with aluminium alloy bars 2000mm long with aluminium wings and seatings for glass, spaced at approximately 600mm centres, glazed with B.S.952 Georgian wired cast, 7mm thick										
Roof areas										
single tier	83.21	5.00	**87.37**	1.50	1.50	**24.38**	2.40	**114.15**	m2	122.71
multi-tier	89.24	5.00	**93.70**	1.70	1.70	**27.63**	2.40	**123.73**	m2	133.01
Vertical surfaces										
single tier	92.62	5.00	**97.25**	1.50	1.50	**24.38**	2.40	**124.03**	m2	133.33
multi-tier	98.74	5.00	**103.68**	1.70	1.70	**27.63**	2.40	**133.70**	m2	143.73
Patent glazing with aluminium alloy bars 2000mm long with aluminium wings and seatings for glass, spaced at approximately 600mm centres, glazed with B.S.952 Georgian wired polished, 6mm thick										
Roof areas										
single tier	130.03	5.00	**136.53**	1.65	1.65	**26.81**	2.40	**165.74**	m2	178.17
multi-tier	139.77	5.00	**146.76**	1.90	1.90	**30.88**	2.40	**180.03**	m2	193.54
Vertical surfaces										
single tier	139.54	5.00	**146.52**	1.65	1.65	**26.81**	2.40	**175.73**	m2	188.91
multi-tier	148.42	5.00	**155.84**	1.90	1.90	**30.88**	2.40	**189.12**	m2	203.30
Patent glazing with aluminium alloy bars 3000mm long with aluminium wings and seatings for glass, spaced at approximately 600mm centres, glazed with B.S.952 Georgian wired cast, 7mm thick										
Roof areas										
single tier	98.87	5.00	**103.81**	1.50	1.50	**24.38**	2.40	**130.59**	m2	140.38
multi-tier	105.23	5.00	**110.49**	1.70	1.70	**27.63**	2.40	**140.52**	m2	151.06
Vertical surfaces										
single tier	107.47	5.00	**112.84**	1.50	1.50	**24.38**	2.40	**139.62**	m2	150.09
multi-tier	114.74	5.00	**120.48**	1.70	1.70	**27.63**	2.40	**150.50**	m2	161.79
Patent glazing with aluminium alloy bars 3000mm long with aluminium wings and seatings for glass, spaced at approximately 600mm centres, glazed with B.S.952 Georgian wired polished, 6mm thick										
Roof areas										
single tier	144.95	5.00	**152.20**	1.65	1.65	**26.81**	2.40	**181.41**	m2	195.02
multi-tier	155.79	5.00	**163.58**	1.90	1.90	**30.88**	2.40	**196.85**	m2	211.62
Vertical surfaces										
single tier	154.35	5.00	**162.07**	1.65	1.65	**26.81**	2.40	**191.28**	m2	205.63
multi-tier	165.28	5.00	**173.54**	1.90	1.90	**30.88**	2.40	**206.82**	m2	222.33
Extra over patent glazing in roof areas with aluminium alloy bars 2000mm long with aluminium wings and seatings for glass, spaced at approximately 600mm centres, glazed with B.S.952 Georgian wired polished, 6mm thick										
Opening lights including opening gear										
600 x 600mm	199.67	5.00	**209.65**	4.40	4.40	**71.50**	-	**281.15**	nr	302.24
600 x 900mm	242.54	5.00	**254.67**	5.50	5.50	**89.38**	-	**344.04**	nr	369.85

Labour hourly rates: (except Specialists) Craft Operatives £9.23 Labour £7.02 Rates are national average prices. Refer to REGIONAL VARIATIONS for indicative levels of overall pricing in regions	MATERIALS			LABOUR				RATES		
	Del to Site £	Waste %	Material Cost £	Craft Optve Hrs	Lab Hrs	Labour Cost £	Sunds £	Nett Rate £	Unit	Gross Rate (+7.5%) £
H10: PATENT GLAZING Cont.										
Extra over patent glazing in roof areas with aluminium alloy bars 2000mm long with aluminium wings and seatings for glass, spaced at approximately 600mm centres, glazed with B.S.952 Georgian wired cast, 7mm thick										
Opening lights including opening gear										
600 x 600mm	199.67	5.00	209.65	4.00	4.00	65.00	–	274.65	nr	295.25
600 x 900mm	242.54	5.00	254.67	5.00	5.00	81.25	–	335.92	nr	361.11
Extra over patent glazing in vertical surfaces with aluminium alloy bars 3000mm long with aluminium wings and seatings for glass, spaced at approximately 600mm centres, glazed with B.S.952 Georgian wired cast, 7mm thick										
Opening lights including opening gear										
600 x 600mm	291.40	5.00	305.97	2.00	3.00	39.52	–	345.49	nr	371.40
600 x 900mm	354.33	5.00	372.05	2.00	3.00	39.52	–	411.57	nr	442.43
Extra over patent glazing to vertical surfaces with aluminium alloy bars 3000mm long with aluminium wings and seatings for glass, spaced at approximately 600mm centres, glazed with B.S.952 Georgian wired polished, 6mm thick										
Opening lights including opening gear										
600 x 600mm	291.40	5.00	305.97	2.20	2.20	35.75	–	341.72	nr	367.35
600 x 900mm	354.15	5.00	371.86	3.30	3.30	53.63	–	425.48	nr	457.39
H21: TIMBER WEATHERBOARDING										
Softwood, wrought, impregnated										
Boarding to walls, shiplapped joints; 19mm thick, 150mm wide boards										
width exceeding 300mm; fixing to timber	6.84	10.00	7.52	0.70	0.09	7.09	0.11	14.73	m2	15.83
Boarding to walls, shiplapped joints; 25mm thick, 150mm wide boards										
width exceeding 300mm; fixing to timber	9.00	10.00	9.90	0.75	0.10	7.62	0.12	17.64	m2	18.97
Western Red Cedar										
Boarding to walls, shiplapped joints; 19mm thick, 150mm wide boards										
width exceeding 300mm; fixing to timber	22.80	10.00	25.08	0.75	0.10	7.62	0.12	32.82	m2	35.29
Boarding to walls, shiplapped joints; 25mm thick, 150mm wide boards										
width exceeding 300mm; fixing to timber	30.00	10.00	33.00	0.80	0.10	8.09	0.13	41.22	m2	44.31
Abutments										
19mm thick softwood boarding										
raking cutting	0.68	–	0.68	0.06	–	0.55	–	1.23	m	1.33
curved cutting	1.03	–	1.03	0.09	–	0.83	–	1.86	m	2.00
25mm thick softwood boarding										
raking cutting	0.90	–	0.90	0.07	–	0.65	–	1.55	m	1.66
curved cutting	1.35	–	1.35	0.10	–	0.92	–	2.27	m	2.44
19mm thick Western Red Cedar boarding										
raking cutting	2.28	–	2.28	0.08	–	0.74	–	3.02	m	3.24
curved cutting	3.42	–	3.42	0.12	–	1.11	–	4.53	m	4.87
25mm thick Western Red Cedar boarding										
raking cutting	3.00	–	3.00	0.09	–	0.83	–	3.83	m	4.12
curved cutting	4.50	–	4.50	0.13	–	1.20	–	5.70	m	6.13
H30: FIBRE CEMENT PROFILED SHEET CLADDING/COVERING/SIDING										
Roof coverings; corrugated reinforced cement Profile 3 sheeting, standard grey colour; lapped one and a half corrugations at sides and 150mm at ends										
Coverings; fixing to timber joists at 900mm general spacing with galvanised mild steel drive screws and washers; drilling holes										
pitch 30 degrees from horizontal	9.08	15.00	10.44	0.25	0.25	4.06	0.41	14.91	m2	16.03
Coverings; fixing to steel purlins at 900mm general spacing with galvanised hook bolts and washers; drilling holes										
pitch 30 degrees from horizontal	9.08	15.00	10.44	0.33	0.33	5.36	0.82	16.62	m2	17.87
Eaves										
eaves filler pieces	6.70	10.00	7.37	0.10	0.10	1.63	–	8.99	m	9.67
Ridges										
two piece plain angular adjustable ridge tiles	9.70	10.00	10.67	0.20	0.20	3.25	–	13.92	m	14.96

Labour hourly rates: (except Specialists) Craft Operatives £9.23 Labour £7.02 Rates are national average prices. Refer to REGIONAL VARIATIONS for indicative levels of overall pricing in regions	MATERIALS			LABOUR				RATES		
	Del to Site £	Waste %	Material Cost £	Craft Optve Hrs	Lab Hrs	Labour Cost £	Sunds £	Nett Rate £	Unit	Gross Rate (+7.5%) £

H30: FIBRE CEMENT PROFILED SHEET CLADDING/COVERING/SIDING Cont.

Roof coverings; corrugated reinforced cement Profile 3 sheeting, standard grey colour; lapped one and a half corrugations at sides and 150mm at ends Cont.

	Del to Site £	Waste %	Material Cost £	Craft Optve Hrs	Lab Hrs	Labour Cost £	Sunds £	Nett Rate £	Unit	Gross Rate £
Barge boards standard barge boards	7.10	10.00	**7.81**	0.10	0.10	**1.63**	-	**9.44**	m	**10.14**
Aprons/sills apron flashings	6.74	10.00	**7.41**	0.13	0.13	**2.11**	-	**9.53**	m	**10.24**
Finials standard one piece ridge cap finials	9.34	10.00	**10.27**	0.15	0.15	**2.44**	-	**12.71**	nr	**13.66**
Cutting raking	1.82	-	**1.82**	0.25	0.25	**4.06**	-	**5.88**	m	**6.32**
Holes for pipes, standards or the like	-	-	-	0.50	-	**4.62**	-	**4.62**	nr	**4.96**

Roof coverings; corrugated reinforced cement Profile 3 sheeting, standard coloured; lapped one and a half corrugations at sides and 150mm at ends

	Del to Site £	Waste %	Material Cost £	Craft Optve Hrs	Lab Hrs	Labour Cost £	Sunds £	Nett Rate £	Unit	Gross Rate £
Coverings; fixing to timber joists at 900mm general spacing with galvanised mild steel drive screws and washers; drilling holes pitch 30 degrees from horizontal	10.90	15.00	**12.54**	0.25	0.25	**4.06**	0.54	**17.14**	m2	**18.42**
Coverings; fixing to steel purlins at 900mm general spacing with galvanised hook bolts and washers; drilling holes pitch 30 degrees from horizontal	10.90	15.00	**12.54**	0.33	0.33	**5.36**	1.06	**18.96**	m2	**20.38**
Eaves eaves filler pieces	8.38	10.00	**9.22**	0.10	0.10	**1.63**	-	**10.84**	m	**11.66**
Ridges two piece plain angular adjustable ridge tiles ...	11.69	10.00	**12.86**	0.20	0.20	**3.25**	-	**16.11**	m	**17.32**
Barge boards standard barge boards	8.85	10.00	**9.73**	0.10	0.10	**1.63**	-	**11.36**	m	**12.21**
Aprons/sills apron flashings	8.40	10.00	**9.24**	0.13	0.13	**2.11**	-	**11.35**	m	**12.20**
Finials standard one piece ridge cap finials	11.67	10.00	**12.84**	0.15	0.15	**2.44**	-	**15.27**	nr	**16.42**
Cutting raking	2.18	-	**2.18**	0.25	0.25	**4.06**	-	**6.24**	m	**6.71**
Holes for pipes, standards or the like	-	-	-	0.50	-	**4.62**	-	**4.62**	nr	**4.96**

Roof coverings; corrugated reinforced cement Profile 6 sheeting, standard grey colour; lapped half a corrugation at sides and 150mm at ends

	Del to Site £	Waste %	Material Cost £	Craft Optve Hrs	Lab Hrs	Labour Cost £	Sunds £	Nett Rate £	Unit	Gross Rate £
Coverings; fixing to timber joists at 900mm general spacing with galvanised mild steel drive screws and washers; drilling holes pitch 30 degrees from horizontal	7.94	15.00	**9.13**	0.25	0.25	**4.06**	0.43	**13.62**	m2	**14.65**
Coverings; fixing to steel purlins at 900mm general spacing with galvanised hook bolts and washers; drilling holes pitch 30 degrees from horizontal	7.94	15.00	**9.13**	0.33	0.33	**5.36**	0.86	**15.35**	m2	**16.51**
Eaves eaves filler pieces	5.28	10.00	**5.81**	0.10	0.10	**1.63**	-	**7.43**	m	**7.99**
Ridges two piece plain angular adjustable ridge tiles ...	9.71	10.00	**10.68**	0.20	0.20	**3.25**	-	**13.93**	m	**14.98**
Barge boards standard barge boards	6.52	10.00	**7.17**	0.10	0.10	**1.63**	-	**8.80**	m	**9.46**
Aprons/sills apron flashings	5.69	10.00	**6.26**	0.13	0.13	**2.11**	-	**8.37**	m	**9.00**
Finials standard one piece ridge cap finials	6.33	10.00	**6.96**	0.15	0.15	**2.44**	-	**9.40**	nr	**10.11**
Cutting raking	1.59	-	**1.59**	0.25	0.25	**4.06**	-	**5.65**	m	**6.08**
Holes for pipes, standards or the like	-	%	-	0.50	-	**4.62**	-	**4.62**	nr	**4.96**

CLADDING/COVERING

145

Labour hourly rates: (except Specialists) Craft Operatives £9.23 Labour £7.02 Rates are national average prices. Refer to REGIONAL VARIATIONS for indicative levels of overall pricing in regions	MATERIALS			LABOUR				RATES		
	Del to Site £	Waste %	Material Cost £	Craft Optve Hrs	Lab Hrs	Labour Cost £	Sunds £	Nett Rate £	Unit	Gross Rate (+7.5%) £
H30: FIBRE CEMENT PROFILED SHEET CLADDING/COVERING/SIDING Cont.										
Roof coverings; corrugated reinforced cement Profile 6 sheeting, standard coloured; lapped half a corrugation at sides and 150mm at ends										
Coverings; fixing to timber joists at 900mm general spacing with galvanised mild steel drive screws and washers; drilling holes										
pitch 30 degrees from horizontal	9.53	15.00	**10.96**	0.25	0.25	**4.06**	0.53	**15.55**	m2	**16.72**
Coverings; fixing to steel purlins at 900mm general spacing with galvanised hook bolts and washers; drilling holes										
pitch 30 degrees from horizontal	9.53	15.00	**10.96**	0.33	0.33	**5.36**	1.06	**17.38**	m2	**18.69**
Eaves										
eaves filler pieces	6.60	10.00	**7.26**	0.10	0.10	**1.63**	-	**8.88**	m	**9.55**
Ridges										
two piece plain angular adjustable ridge tiles ...	12.14	10.00	**13.35**	0.20	0.20	**3.25**	-	**16.60**	m	**17.85**
Barge boards										
standard barge boards	8.15	10.00	**8.96**	0.10	0.10	**1.63**	-	**10.59**	m	**11.38**
Aprons/sills										
apron flashings	7.12	10.00	**7.83**	0.13	0.13	**2.11**	-	**9.94**	m	**10.69**
Finials										
standard one piece ridge cap finials	7.90	10.00	**8.69**	0.15	0.15	**2.44**	-	**11.13**	nr	**11.96**
Cutting										
raking..	1.91	-	**1.91**	0.25	0.25	**4.06**	-	**5.97**	m	**6.42**
Holes										
for pipes, standards or the like	-	-	**-**	0.50	-	**4.62**	-	**4.62**	nr	**4.96**
H31: METAL PROFILED/FLAT SHEET CLADDING/COVERING/SIDING										
Wall cladding; PVC colour coated both sides galvanised steel profiled sheeting 0.70mm thick and with sheets secured at seams and laps										
Coverings; fixing to steel rails at 900mm general spacing with galvanised hook bolts and washers; drilling holes										
pitch 90 degrees from horizontal	9.00	5.00	**9.45**	0.35	0.35	**5.69**	1.02	**16.16**	m2	**17.37**
Vertical angles										
vertical corner flashings 500mm girth	6.00	5.00	**6.30**	0.15	0.15	**2.44**	-	**8.74**	m	**9.39**
Filler blocks										
polyethylene	1.37	5.00	**1.44**	0.10	0.10	**1.63**	-	**3.06**	m	**3.29**
PVC	1.71	5.00	**1.80**	0.10	0.10	**1.63**	-	**3.42**	m	**3.68**
black synthetic rubber	1.90	5.00	**2.00**	0.10	0.10	**1.63**	-	**3.62**	m	**3.89**
Cutting										
raking...	1.80	-	**1.80**	0.25	0.25	**4.06**	-	**5.86**	m	**6.30**
Roof coverings; PVC colour coated both sides galvanised steel profiled sheeting 0.70mm thick and with sheets secured at seams and laps										
Coverings; fixing to steel purlins at 900mm general spacing with galvanised hook bolts and washers; drilling holes										
pitch 30 degrees from horizontal	9.00	5.00	**9.45**	0.30	0.30	**4.88**	1.02	**15.35**	m2	**16.50**
Ridges										
ridge cappings 500mm girth	6.00	5.00	**6.30**	0.15	0.15	**2.44**	-	**8.74**	m	**9.39**
Flashings										
gable flashings 500mm girth	6.00	5.00	**6.30**	0.15	0.15	**2.44**	-	**8.74**	m	**9.39**
eaves flashings 500mm girth	6.00	5.00	**6.30**	0.15	0.15	**2.44**	-	**8.74**	m	**9.39**
Filler blocks										
polyethylene	1.37	5.00	**1.44**	0.10	0.10	**1.63**	-	**3.06**	m	**3.29**
PVC	1.71	5.00	**1.80**	0.10	0.10	**1.63**	-	**3.42**	m	**3.68**
black synthetic rubber	1.90	5.00	**2.00**	0.10	0.10	**1.63**	-	**3.62**	m	**3.89**
Cutting										
raking...	1.80	-	**1.80**	0.25	0.25	**4.06**	-	**5.86**	m	**6.30**
H32: PLASTICS PROFILED SHEET CLADDING/ COVERING/SIDING										
Roof coverings; standard corrugated glass fibre 1.3mm thick reinforced translucent sheeting; lapped one and a half corrugations at sides and 150mm at ends										
Coverings; fixing to timber joists at 900mm general spacing with galvanised mild steel drive screws and washers; drilling holes										
pitch 30 degrees from horizontal	12.41	5.00	**13.03**	0.25	0.25	**4.06**	0.28	**17.37**	m2	**18.68**

Labour hourly rates: (except Specialists) Craft Operatives £9.23 Labour £7.02 Rates are national average prices. Refer to REGIONAL VARIATIONS for indicative levels of overall pricing in regions	MATERIALS			LABOUR				RATES		
	Del to Site £	Waste %	Material Cost £	Craft Optve Hrs	Lab Hrs	Labour Cost £	Sunds £	Nett Rate £	Unit	Gross Rate (+7.5%) £
H32: PLASTICS PROFILED SHEET CLADDING/ COVERING/SIDING Cont.										
Roof coverings; standard corrugated glass fibre 1.3mm thick reinforced translucent sheeting; lapped one and a half corrugations at sides and 150mm at ends Cont.										
Coverings; fixing to steel purlins at 900mm general spacing with galvanised hook bolts and washers; drilling holes										
pitch 30 degrees from horizontal	12.41	5.00	13.03	0.30	0.30	4.88	0.44	18.35	m2	19.72
Cutting										
raking	2.48	-	2.48	0.25	0.25	4.06	-	6.54	m	7.03
Roof coverings; fire resisting corrugated glass fibre reinforced translucent sheeting; lapped one and a half corrugations at sides and 150mm at ends										
Coverings; fixing to timber joists at 900mm general spacing with galvanised mild steel drive screws and washers; drilling holes										
pitch 30 degrees from horizontal	17.38	5.00	18.25	0.25	0.25	4.06	0.28	22.59	m2	24.29
Coverings; fixing to steel purlins at 900mm general spacing with galvanised hook bolts and washers; drilling holes										
pitch 30 degrees from horizontal	17.38	5.00	18.25	0.30	0.30	4.88	0.44	23.56	m2	25.33
Cutting										
raking	3.47	-	3.47	0.25	0.25	4.06	-	7.53	m	8.10
Roof coverings; standard vinyl corrugated sheeting; lapped one and a half corrugations at sides and 150mm at ends; 1.3mm thick										
Coverings; fixing to timber joists at 900mm general spacing with galvanised mild steel drive screws and washers; drilling holes										
pitch 30 degrees from horizontal	15.44	5.00	16.21	0.25	0.25	4.06	0.28	20.55	m2	22.10
Coverings; fixing to steel purlins at 900mm general spacing with galvanised hook bolts and washers; drilling holes										
pitch 30 degrees from horizontal	15.44	5.00	16.21	0.30	0.30	4.88	0.44	21.53	m2	23.14
Cutting										
raking	3.09	-	3.09	0.25	0.25	4.06	-	7.15	m	7.69
Wall cladding; Swish Products high impact rigid uPVC profiled sections; colour white; secured with starter sections and clips										
Coverings with shiplap profile code C002 giving 150mm cover; fixing to timber vertical cladding										
sections applied horizontally..............	32.80	5.00	34.44	0.38	0.38	6.17	1.66	42.27	m2	45.45
sections applied vertically.................	32.80	5.00	34.44	0.38	0.38	6.17	1.66	42.27	m2	45.45
Coverings with open V profile code C003 giving 150mm cover; fixing to timber vertical cladding										
sections applied horizontally..............	29.04	5.00	30.49	0.40	0.40	6.50	2.37	39.36	m2	42.31
sections applied vertically.................	29.04	5.00	30.49	0.40	0.40	6.50	2.37	39.36	m2	42.31
Coverings with open V profile code C269 giving 100mm cover; fixing to timber vertical cladding										
sections applied horizontally..............	32.80	5.00	34.44	0.35	0.35	5.69	1.66	41.79	m2	44.92
sections applied vertically.................	32.80	5.00	34.44	0.35	0.35	5.69	1.66	41.79	m2	44.92
Vertical angles										
section C030 for vertically applied section	3.32	5.00	3.49	0.10	0.10	1.63	0.13	5.24	m	5.63
H51: NATURAL STONE SLAB CLADDING/FEATURES										
English blue/grey slate facings (Best Quality); natural riven finish; bedding, jointing and pointing in gauged mortar (1:2:9)										
450 x 600 x 20mm units to walls on brickwork or blockwork base										
plain, width exceeding 300mm	105.95	2.50	108.60	2.00	2.00	32.50	0.40	141.50	m2	152.11
450 x 600 x 30mm units to walls on brickwork or blockwork base										
plain, width exceeding 300mm	160.15	2.50	164.15	2.25	2.25	36.56	0.46	201.18	m2	216.26
450 x 600 x 40mm units to walls on brickwork or blockwork base										
plain, width exceeding 300mm	204.15	2.50	209.25	2.50	2.50	40.63	0.62	250.50	m2	269.29
450 x 600 x 50mm units to walls on brickwork or blockwork base										
plain, width exceeding 300mm	266.70	2.50	273.37	3.00	3.00	48.75	0.76	322.88	m2	347.09

CLADDING/COVERING

Labour hourly rates: (except Specialists) Craft Operatives £9.23 Labour £7.02 Rates are national average prices. Refer to REGIONAL VARIATIONS for indicative levels of overall pricing in regions	MATERIALS			LABOUR				RATES		
	Del to Site £	Waste %	Material Cost £	Craft Optve Hrs	Lab Hrs	Labour Cost £	Sunds £	Nett Rate £	Unit	Gross Rate (+7.5%) £
H51: NATURAL STONE SLAB CLADDING/FEATURES Cont.										
English blue/grey slate facings (Best quality); fine rubbed finish, one face; bedding, jointing and pointing in gauged mortar (1:2:9)										
750 x 1200 x 20mm units to walls on brickwork or blockwork base										
plain, width exceeding 300mm	115.00	2.50	117.88	2.00	2.00	32.50	1.16	151.54	m2	162.90
extra; rubbed square edges	8.35	-	8.35	-	-	-	-	8.35	m	8.98
extra; half rounded edges	10.40	-	10.40	-	-	-	-	10.40	m	11.18
extra; full rounded edges	15.50	-	15.50	-	-	-	-	15.50	m	16.66
extra; rebated joints	15.50	-	15.50	-	-	-	-	15.50	m	16.66
750 x 1200 x 30mm units to walls on brickwork or blockwork base										
plain, width exceeding 300mm	172.00	2.50	176.30	2.25	2.25	36.56	1.23	214.09	m2	230.15
extra; rubbed square edges	9.25	-	9.25	-	-	-	-	9.25	m	9.94
extra; half rounded edges	11.15	-	11.15	-	-	-	-	11.15	m	11.99
extra; full rounded edges	16.70	-	16.70	-	-	-	-	16.70	m	17.95
extra; rebated joints	16.70	-	16.70	-	-	-	-	16.70	m	17.95
750 x 1200 x 40mm units to walls on brickwork or blockwork base										
plain, width exceeding 300mm	212.75	2.50	218.07	2.50	2.50	40.63	1.38	260.07	m2	279.58
extra; rubbed square edges	10.71	-	10.71	-	-	-	-	10.71	m	11.51
extra; half rounded edges	12.20	-	12.20	-	-	-	-	12.20	m	13.12
extra; full rounded edges	18.00	-	18.00	-	-	-	-	18.00	m	19.35
extra; rebated joints	18.00	-	18.00	-	-	-	-	18.00	m	19.35
750 x 1200 x 50mm units to walls on brickwork or blockwork base										
plain, width exceeding 300mm	286.50	2.50	293.66	3.00	3.00	48.75	1.53	343.94	m2	369.74
extra; rubbed square edges	11.50	-	11.50	-	-	-	-	11.50	m	12.36
extra; half rounded edges	12.70	-	12.70	-	-	-	-	12.70	m	13.65
extra; full rounded edges	18.45	-	18.45	-	-	-	-	18.45	m	19.83
extra; rebated joints	18.45	-	18.45	-	-	-	-	18.45	m	19.83
English blue/grey slate facings (Best quality); fine rubbed finish, both faces; bedding, jointing and pointing in gauged mortar (1:2:9)										
750 x 1200 x 20mm units to walls on brickwork or blockwork base										
plain, width exceeding 300mm	146.50	2.50	150.16	2.50	2.50	40.63	1.92	192.71	m2	207.16
750 x 1200 x 30mm units to walls on brickwork or blockwork base										
plain, width exceeding 300mm	203.50	2.50	208.59	2.75	2.75	44.69	2.02	255.29	m2	274.44
750 x 1200 x 40mm units to walls on brickwork or blockwork base										
plain, width exceeding 300mm	244.25	2.50	250.36	3.00	3.00	48.75	2.16	301.27	m2	323.86
750 x 1200 x 50mm units to walls on brickwork or blockwork base										
plain, width exceeding 300mm	317.95	2.50	325.90	3.50	3.50	56.88	2.29	385.06	m2	413.94
Kirkstone green slate; bedding, jointing and pointing in gauged mortar (1:2:9)										
30mm thick units to sills on brickwork or blockwork base										
width 125mm; weathered, throated and grooved	48.81	2.50	50.03	0.55	0.55	8.94	0.50	59.47	m	63.93
width 190mm; weathered, throated and grooved	61.92	2.50	63.47	0.60	0.60	9.75	0.65	73.87	m	79.41
38mm thick units to sills on brickwork or blockwork base										
width 125mm	56.40	2.50	57.81	0.60	0.60	9.75	0.50	68.06	m	73.16
extra; stooling for jambs	26.00	-	26.00	-	-	-	-	26.00	nr	27.95
extra; notching for jambs and mullions	15.60	-	15.60	-	-	-	-	15.60	nr	16.77
width 190mm	72.79	2.50	74.61	0.75	0.75	12.19	0.65	87.45	m	94.01
extra; stooling for jambs	26.00	-	26.00	-	-	-	-	26.00	nr	27.95
extra; notching for jambs and mullions	15.60	-	15.60	-	-	-	-	15.60	nr	16.77
50mm thick units to sills on brickwork or blockwork base										
width 125mm	67.95	2.50	69.65	0.75	0.75	12.19	0.50	82.34	m	88.51
extra; stooling for jambs	26.00	-	26.00	-	-	-	-	26.00	nr	27.95
extra; notching for jambs and mullions	15.60	-	15.60	-	-	-	-	15.60	nr	16.77
width 190mm	91.92	2.50	94.22	1.05	1.05	17.06	0.65	111.93	m	120.33
extra; stooling for jambs	26.00	-	26.00	-	-	-	-	26.00	nr	27.95
extra; notching for jambs and mullions	15.60	-	15.60	-	-	-	-	15.60	nr	16.77
30mm units to window boards on brickwork or blockwork base										
width 230mm	58.99	2.50	60.46	0.75	0.75	12.19	0.83	73.48	m	78.99
25mm units to combined sills and window boards on brickwork or blockwork base										
width 360mm	85.30	2.50	87.43	1.00	1.00	16.25	1.16	104.84	m	112.71

Labour hourly rates: (except Specialists) Craft Operatives £9.23 Labour £7.02 Rates are national average prices. Refer to REGIONAL VARIATIONS for indicative levels of overall pricing in regions	MATERIALS			LABOUR				RATES		
	Del to Site £	Waste %	Material Cost £	Craft Optve Hrs	Lab Hrs	Labour Cost £	Sunds £	Nett Rate £	Unit	Gross Rate (+7.5%) £
H60: PLAIN ROOF TILING										
Quantities required for 1m2 of tiling										
Plain tiles to 65mm lap										
tiles, 60nr										
battens, 10m										
Plain tiles to 85mm lap										
tiles, 70nr										
battens, 11m										
Roof coverings; clayware machine made plain tiles B.S.402, red, 265 x 165mm; fixing every fourth course with two galvanised nails per tile to 65mm lap; 38 x 19mm pressure impregnated softwood battens fixed with galvanised nails; underlay B.S.747 type 1F reinforced bitumen felt; 150mm laps; fixing with galvanised steel clout nails										
Pitched 50 degrees from horizontal										
generally	21.30	5.00	**22.36**	0.62	0.31	**7.90**	5.09	**35.35**	m2	**38.01**
holes	-	-	-	0.50	0.25	**6.37**	-	**6.37**	nr	**6.85**
abutments; square	2.13	-	**2.13**	0.25	0.13	**3.22**	-	**5.35**	m	**5.75**
abutments; raking	3.04	-	**3.04**	0.38	0.19	**4.84**	-	**7.88**	m	**8.47**
double course at eaves; purpose made eaves tile ..	4.00	5.00	**4.20**	0.30	0.15	**3.82**	0.47	**8.49**	m	**9.13**
verges; bed and point in coloured cement-lime mortar (1:1:6)	2.00	5.00	**2.10**	0.35	0.18	**4.49**	0.34	**6.93**	m	**7.45**
verges; single extra undercloak course plain tiles; bed and point in coloured cement-lime mortar (1:1:6)	2.67	5.00	**2.80**	0.50	0.25	**6.37**	0.34	**9.51**	m	**10.23**
ridge tiles, half round; in 300mm effective lengths; butt jointed; bedding and pointing in coloured cement-lime mortar (1:1:6)	14.89	5.00	**15.63**	0.30	0.15	**3.82**	0.34	**19.80**	m	**21.28**
hip tiles, half round; in 300mm effective lengths; butt jointed; bedding and pointing in coloured cement-lime mortar (1:1:6)	14.80	5.00	**15.54**	0.30	0.15	**3.82**	0.34	**19.70**	m	**21.18**
hip tiles; angular	14.89	5.00	**15.63**	0.30	0.15	**3.82**	0.34	**19.80**	m	**21.28**
hip tiles; bonnet pattern; bedding and pointing in coloured cement-lime mortar (1:1:6)	14.89	5.00	**15.63**	0.30	0.15	**3.82**	0.34	**19.80**	m	**21.28**
valley tiles; angular	14.89	5.00	**15.63**	0.45	0.23	**5.77**	0.34	**21.74**	m	**23.37**
Roof coverings; clayware machine made plain tiles B.S.402, red, 265 x 165mm; fixing every fourth course with two galvanised nails per tile to 65mm lap; 50 x 19mm pressure impregnated softwood battens fixed with galvanised nails; underlay B.S.747 type 1F reinforced bitumen felt; 150mm laps; fixing with galvanised steel clout nails										
Pitched 50 degrees from horizontal										
generally	21.30	5.00	**22.36**	0.62	0.31	**7.90**	6.09	**36.35**	m2	**39.08**
Roof coverings; clayware machine made plain tiles B.S.402, red, 265 x 165mm; fixing every course with two galvanised nails per tile to 65mm lap; 38 x 19mm pressure impregnated softwood battens fixed with galvanised nails; underlay B.S.747 type 1F reinforced bitumen felt; 150mm laps; fixing with galvanised steel clout nails										
Pitched 50 degrees from horizontal										
generally	21.30	5.00	**22.36**	0.74	0.37	**9.43**	6.24	**38.03**	m2	**40.89**
Roof coverings; clayware machine made plain tiles B.S.402, red, 265 x 165mm; fixing every fourth course with two copper nails per tile to 65mm lap; 38 x19mm pressure impregnated softwood battens fixed with galvanised nails; underlay B.S.747 type 1F reinforced bitumen felt; 150mm laps; fixing with galvanised steel clout nails										
Pitched 50 degrees from horizontal										
generally	21.30	5.00	**22.36**	0.62	0.31	**7.90**	5.76	**36.02**	m2	**38.73**
Roof coverings; clayware machine made plain tiles B.S.402, red, 265 x 165mm; fixing every course with two copper nails per tile to 65mm lap; 38 x 19mm pressure impregnated softwood battens fixed with galvanised nails; underlay B.S.747 type 1F reinforced bitumen felt; 150mm laps; fixing with galvanised steel clout nails										
Pitched 50 degrees from horizontal										
generally	21.30	5.00	**22.36**	0.74	0.37	**9.43**	8.26	**40.05**	m2	**43.06**
Roof coverings; clayware machine made plain tiles B.S.402, red, 265 x 165mm; fixing every fourth course with two galvanised nails per tile to 85mm lap; 38 x 19mm pressure impregnated softwood battens fixed with galvanised nails; underlay B.S.747 type 1F reinforced bitumen felt; 150mm laps; fixing with galvanised steel clout nails										
Pitched 40 degrees from horizontal										
generally	24.85	5.00	**26.09**	0.70	0.35	**8.92**	5.34	**40.35**	m2	**43.38**

Labour hourly rates: (except Specialists) Craft Operatives £9.23 Labour £7.02 Rates are national average prices. Refer to REGIONAL VARIATIONS for indicative levels of overall pricing in regions	MATERIALS			LABOUR				RATES		
	Del to Site £	Waste %	Material Cost £	Craft Optve Hrs	Lab Hrs	Labour Cost £	Sunds £	Nett Rate £	Unit	Gross Rate (+7.5%) £
H60: PLAIN ROOF TILING Cont.										
Wall coverings; clayware machine made plain tiles B.S. 402, red, 265 x 165mm;fixing every course with two galvanised nails per tile to 38mm lap; 38 x 19mm pressure impregnated softwood battens fixed with galvanised nails; underlay B.S. 747 type 1F reinforced bitumen felt; 150mm laps; fixing with galvanised steel clout nails										
Vertical generally	22.76	5.00	23.90	0.75	0.38	9.59	5.66	39.15	m2	42.08
Roof coverings; clayware hand made plain tiles B.S.402, red, 265 x 165mm; fixing every fourth course with two galvanised nails per tile to 65mm lap; 38 x 19mm pressure impregnated softwood battens fixed with galvanised nails; underlay B.S.747 type 1F reinforced bitumen felt; 150mm laps; fixing with galvanised steel clout nails										
Pitched 50 degrees from horizontal										
generally	35.28	5.00	37.04	0.62	0.31	7.90	5.09	50.03	m2	53.79
holes	-	-	-	0.50	0.25	6.37	-	6.37	nr	6.85
abutments; square	3.53	-	3.53	0.25	0.13	3.22	-	6.75	m	7.26
abutments; raking	5.04	-	5.04	0.38	0.19	4.84	-	9.88	m	10.62
double course at eaves; purpose made eaves tile ..	7.75	5.00	8.14	0.30	0.15	3.82	0.47	12.43	m	13.36
verges; bed and point in coloured cement-lime mortar (1:1:6)	3.87	5.00	4.06	0.35	0.18	4.49	0.34	8.90	m	9.56
verges; single extra undercloak course plain tiles; bed and point in coloured cement-lime mortar (1:1:6)	5.17	5.00	5.43	0.50	0.25	6.37	0.34	12.14	m	13.05
ridge tiles, half round; in 300mm effective lengths; butt jointed; bedding and pointing in coloured cement-lime mortar (1:1:6)	14.89	5.00	15.63	0.30	0.15	3.82	0.34	19.80	m	21.28
hip tiles, half round; in 300mm effective lengths; butt jointed; bedding and pointing in coloured cement-lime mortar (1:1:6)	14.80	5.00	15.54	0.30	0.15	3.82	0.34	19.70	m	21.18
hip tiles; angular	14.89	5.00	15.63	0.30	0.15	3.82	0.34	19.80	m	21.28
hip tiles; bonnet pattern; bedding and pointing in coloured cement-lime mortar (1:1:6)	14.89	5.00	15.63	0.30	0.15	3.82	0.34	19.80	m	21.28
valley tiles; angular	14.89	5.00	15.63	0.45	0.23	5.77	0.34	21.74	m	23.37
Roof coverings; clayware hand made plain tiles B.S.402, red, 265 x 165mm; fixing every fourth course with two galvanised nails per tile to 85mm lap; 38 x 19mm pressure impregnated softwood battens fixed with galvanised nails; underlay B.S.747 type 1F reinforced bitumen felt; 150mm laps; fixing with galvanised steel clout nails										
Pitched 40 degrees from horizontal										
generally	41.16	5.00	43.22	0.70	0.35	8.92	5.34	57.48	m2	61.79
Fittings										
Hip irons galvanised mild steel; 32 x 3 x 380mm girth; scrolled; fixing with galvanised steel screws to timber	1.79	5.00	1.88	0.10	0.05	1.27	-	3.15	nr	3.39
Lead soakers fixing only	-	-	-	0.30	0.15	3.82	-	3.82	nr	4.11
Roof coverings; Redland Plain granular faced tiles, 268 x 165mm; fixing every fifth course with two aluminium nails to each tile to 65mm lap; 32 x 19mm pressure impregnated softwood battens fixed with galvanised nails; underlay B.S.747 type 1F bitumen reinforced felt; 150mm laps; fixing with galvanised steel clout nails										
Pitched 40 degrees from horizontal										
generally	15.93	5.00	16.73	0.62	0.31	7.90	4.25	28.88	m2	31.04
double course at eaves and nailing each tile with two aluminium nails	1.59	5.00	1.67	0.30	0.15	3.82	0.42	5.91	m	6.35
verges; Redland Dry Verge system with plain tiles and tile-and-a-half tiles in alternate courses, clips and aluminium nails	10.76	5.00	11.30	0.25	0.12	3.15	0.90	15.35	m	16.50
verges; plain tile undercloak; tiles and undercloak bedded and pointed in tinted cement mortar (1:3) and with plain tiles and tile-and-a-half tiles in alternate courses and aluminium nails	1.59	5.00	1.67	0.25	0.12	3.15	0.37	5.19	m	5.58
double course at top edges with clips and nailing each tile with two aluminium nails	1.61	5.00	1.69	0.30	0.15	3.82	1.82	7.33	m	7.88
ridge or hip tiles, Redland Plain angle; butt jointed; bedding and pointing in tinted cement mortar (1:3)	16.48	5.00	17.30	0.30	0.15	3.82	0.34	21.47	m	23.08
ridge or hip tiles, Redland half round; butt jointed; bedding and pointing in tinted cement mortar (1:3)	11.17	5.00	11.73	0.30	0.15	3.82	0.34	15.89	m	17.08
valley tiles , Redland Plain valley tiles	11.62	5.00	12.20	0.30	0.15	3.82	0.34	16.36	m	17.59

Labour hourly rates: (except Specialists) Craft Operatives £9.23 Labour £7.02 Rates are national average prices. Refer to REGIONAL VARIATIONS for indicative levels of overall pricing in regions	MATERIALS			LABOUR				RATES		
	Del to Site £	Waste %	Material Cost £	Craft Optve Hrs	Lab Hrs	Labour Cost £	Sunds £	Nett Rate £	Unit	Gross Rate (+7.5%) £
H60: PLAIN ROOF TILING Cont.										
Roof coverings; Redland Plain granular faced tiles, 268 x 165mm; fixing every fifth course with two aluminium nails to each tile to 65mm lap; 32 x 25mm pressure impregnated softwood battens fixed with galvanised nails; underlay; B.S.747 type 1F bitumen reinforced felt; 150mm laps; fixing with galvanised steel clout nails										
Pitched 40 degrees from horizontal										
generally	15.93	5.00	**16.73**	0.62	0.31	**7.90**	5.15	**29.78**	m2	32.01
abutments; square	1.59	-	**1.59**	0.25	0.13	**3.22**	-	**4.81**	m	5.17
abutments; raking	2.28	-	**2.28**	0.38	0.19	**4.84**	-	**7.12**	m	7.66
abutments; curved to 3000mm radius	3.19	-	**3.19**	0.50	0.25	**6.37**	-	**9.56**	m	10.28
Roof coverings; Redland Plain granular faced tiles, 268 x 165mm; fixing each tile with two aluminium nails to 65mm lap; 32 x 19mm pressure impregnated softwood battens fixed with galvanised nails; underlay; B.S.747 type 1F bitumen reinforced felt; 150mm laps; fixing with galvanised steel clout nails										
Pitched 40 degrees from horizontal										
generally	15.93	5.00	**16.73**	0.74	0.37	**9.43**	6.11	**32.26**	m2	34.68
Roof coverings; Plain through coloured tiles, 268 x 165mm; fixing every fifth course with two aluminium nails to each tile to 65mm lap; 32 x 19mm pressure impregnated softwood battens fixed with galvanised nails; underlay; B.S.747 type 1F bitumen reinforced felt; 150mm laps; fixing with galvanised steel clout nails										
Pitched 40 degrees from horizontal										
generally	15.93	5.00	**16.73**	0.62	0.31	**7.90**	4.25	**28.88**	m2	31.04
double course at eaves and nailing each tile with two aluminium nails	1.59	5.00	**1.67**	0.30	0.15	**3.82**	0.42	**5.91**	m	6.35
verges; Redland Dry Verge system with plain tiles and tile-and-a-half tiles in alternate courses, clips and aluminium nails	10.76	5.00	**11.30**	0.25	0.12	**3.15**	0.90	**15.35**	m	16.50
verges; plain tile undercloak; tiles and undercloak bedded and pointed in tinted cement mortar (1:3) and with plain tiles and tile-and-a-half tiles in alternate courses and aluminium nails	1.59	5.00	**1.67**	0.25	0.12	**3.15**	0.37	**5.19**	m	5.58
double course at top edges with clips and nailing each tile with two aluminium nails	1.61	5.00	**1.69**	0.30	0.15	**3.82**	1.82	**7.33**	m	7.88
ridge or hip tiles, Redland Plain angle; butt jointed; bedding and pointing in tinted cement mortar (1:3)	16.48	5.00	**17.30**	0.30	0.15	**3.82**	0.34	**21.47**	m	23.08
ridge or hip tiles, Redland half round; butt jointed; bedding and pointing in tinted cement mortar (1:3)	11.17	5.00	**11.73**	0.30	0.15	**3.82**	0.34	**15.89**	m	17.08
valley tiles, Redland Plain valley tiles	11.62	5.00	**12.20**	0.30	0.15	**3.82**	0.34	**16.36**	m	17.59
Roof coverings; Redland Downland plain granular faced tiles, 268 x 165mm; fixing every fifth course with two aluminium nails to each tile to 65mm lap; 32 x 19mm pressure impregnated softwood battens fixed with galvanised nails; underlay; B.S.747 type 1F bitumen reinforced felt; 150mm laps; fixing with galvanised steel clout nails										
Pitched 40 degrees from horizontal										
generally	19.12	5.00	**20.08**	0.62	0.31	**7.90**	4.25	**32.22**	m2	34.64
double course at eaves and nailing each tile with two aluminium nails	1.91	5.00	**2.01**	0.30	0.15	**3.82**	0.42	**6.25**	m	6.72
verges; Redland Dry Verge system with plain tiles and tile-and-a-half tiles in alternate courses, clips and aluminium nails	10.92	5.00	**11.47**	0.25	0.12	**3.15**	0.90	**15.52**	m	16.68
verges; plain tile undercloak; tiles and undercloak bedded and pointed in tinted cement mortar (1:3) and with plain tiles and tile-and-a-half tiles in alternate courses and aluminium nails	1.91	5.00	**2.01**	0.25	0.12	**3.15**	0.37	**5.53**	m	5.94
double course at top edges with clips and nailing each tile with two aluminium nails	1.93	5.00	**2.03**	0.30	0.15	**3.82**	1.82	**7.67**	m	8.24
ridge or hip tiles, Redland Plain angle; butt jointed; bedding and pointing in tinted cement mortar (1:3)	16.48	5.00	**17.30**	0.25	0.15	**3.36**	0.34	**21.00**	m	22.58
ridge or hip tiles, Redland half round; butt jointed; bedding and pointing in tinted cement mortar (1:3)	11.17	5.00	**11.73**	0.30	0.15	**3.82**	0.34	**15.89**	m	17.08
valley tiles, Redland Plain valley tiles	9.79	5.00	**10.28**	0.30	0.15	**3.82**	0.34	**14.44**	m	15.52
Roof coverings; Redland Downland plain through coloured tiles, 268 x 165mm; fixing every fifth course with two aluminium nails to each tile to 65mm lap; 32 x 19mm pressure impregnated softwood battens fixed with galvanised nails; underlay; B.S.747 type 1F bitumen reinforced felt; 150mm laps; fixing with galvanised steel clout nails										
Pitched 40 degrees from horizontal										
generally	19.12	5.00	**20.08**	0.62	0.31	**7.90**	4.25	**32.22**	m2	34.64

CLADDING/COVERING

Labour hourly rates: (except Specialists) Craft Operatives £9.23 Labour £7.02 Rates are national average prices. Refer to REGIONAL VARIATIONS for indicative levels of overall pricing in regions	MATERIALS			LABOUR				RATES		
	Del to Site	Waste	Material Cost	Craft Optve	Lab	Labour Cost	Sunds	Nett Rate	Unit	Gross Rate (+7.5%)
	£	%	£	Hrs	Hrs	£	£	£		£

H60: PLAIN ROOF TILING Cont.

Roof coverings; Redland Downland plain through coloured tiles, 268 x 165mm; fixing every fifth course with two aluminium nails to each tile to 65mm lap; 32 x 19mm pressure impregnated softwood battens fixed with galvanised nails; underlay; B.S.747 type 1F bitumen reinforced felt; 150mm laps; fixing with galvanised steel clout nails Cont.

Pitched 40 degrees from horizontal Cont.

	Del to Site £	Waste %	Material Cost £	Craft Optve Hrs	Lab Hrs	Labour Cost £	Sunds £	Nett Rate £	Unit	Gross Rate £
double course at eaves and nailing each tile with two aluminium nails	1.91	5.00	2.01	0.30	0.15	3.82	0.42	6.25	m	6.72
verges; Redland Dry Verge system with plain tiles and tile-and-a-half tiles in alternate courses, clips and aluminium nails	10.92	5.00	11.47	0.25	0.12	3.15	0.90	15.52	m	16.68
verges; plain tile undercloak; tiles and undercloak bedded and pointed in tinted cement mortar (1:3) and with plain tiles and tile-and-a-half tiles in alternate courses and aluminium nails	1.91	5.00	2.01	0.25	0.12	3.15	0.37	5.53	m	5.94
double course at top edges with clips and nailing each tile with two aluminium nails	1.93	5.00	2.03	0.30	0.15	3.82	1.82	7.67	m	8.24
ridge or hip tiles, Redland Plain angle; butt jointed; bedding and pointing in tinted cement mortar (1:3)	16.48	5.00	17.30	0.30	0.15	3.82	0.34	21.47	m	23.08
ridge or hip tiles, Redland half round; butt jointed; bedding and pointing in tinted cement mortar (1:3)	11.17	5.00	11.73	0.30	0.15	3.82	0.34	15.89	m	17.08
valley tiles, Redland Plain valley tiles	9.79	5.00	10.28	0.30	0.15	3.82	0.34	14.44	m	15.52

Fittings

Ventilator tiles

	Del to Site £	Waste %	Material Cost £	Craft Optve Hrs	Lab Hrs	Labour Cost £	Sunds £	Nett Rate £	Unit	Gross Rate £
Red Vent ridge ventilation terminal 450mm long for half round ridge	46.40	5.00	48.72	1.30	0.65	16.56	0.58	65.86	nr	70.80
universal Delta ridge	56.90	5.00	59.74	1.50	0.75	19.11	1.04	79.89	nr	85.89
universal angle ridge	41.30	5.00	43.37	1.30	0.65	16.56	1.04	60.97	nr	65.54
Redvent eaves ventilators fixing with aluminium nails	8.12	5.00	8.53	0.50	0.25	6.37	0.03	14.93	m	16.05
Red Vent Thruvent tiles for Stonewold slates	31.40	5.00	32.97	0.50	0.25	6.37	0.48	39.82	nr	42.81
Delta tiles	30.70	5.00	32.23	0.50	0.25	6.37	0.48	39.09	nr	42.02
Regent tiles	30.70	5.00	32.23	0.50	0.25	6.37	0.48	39.09	nr	42.02
Grovebury double pantiles	30.70	5.00	32.23	0.50	0.25	6.37	0.48	39.09	nr	42.02
Norfolk pantiles	30.70	5.00	32.23	0.50	0.25	6.37	0.48	39.09	nr	42.02
Redland 49 tiles	30.70	5.00	32.23	0.50	0.25	6.37	0.48	39.09	nr	42.02
Renown tiles	30.70	5.00	32.23	0.50	0.25	6.37	0.48	39.09	nr	42.02
Redland 50 double roman tiles	30.70	5.00	32.23	0.50	0.25	6.37	0.48	39.09	nr	42.02
Redland plain tiles	46.50	5.00	48.83	0.50	0.25	6.37	0.24	55.44	nr	59.59
Redland Downland plain tiles	42.30	5.00	44.41	0.50	0.25	6.37	0.24	51.02	nr	54.85
Red line ventilation tile complete with underlay seal and fixing clips for Stonewold slates	38.30	5.00	40.22	1.00	0.50	12.74	0.48	53.44	nr	57.44
Regent tiles	38.30	5.00	40.22	1.00	0.50	12.74	0.48	53.44	nr	57.44
Grovebury double pantiles	38.30	5.00	40.22	1.00	0.50	12.74	0.48	53.44	nr	57.44
Renown tiles	38.30	5.00	40.22	1.00	0.50	12.74	0.48	53.44	nr	57.44
Redland 50 double roman tiles	38.30	5.00	40.22	1.00	0.50	12.74	0.48	53.44	nr	57.44

Gas terminals

	Del to Site £	Waste %	Material Cost £	Craft Optve Hrs	Lab Hrs	Labour Cost £	Sunds £	Nett Rate £	Unit	Gross Rate £
Gas Flue ridge terminal Mark III, 450mm long with sealing gasket and fixing brackets half round ridge	60.40	5.00	63.42	1.50	0.75	19.11	0.58	83.11	nr	89.34
universal Delta ridge	69.90	5.00	73.39	1.70	0.85	21.66	2.72	97.77	nr	105.11
universal angle ridge	54.30	5.00	57.02	1.50	0.75	19.11	2.72	78.84	nr	84.76
Gas Flue ridge terminal Mark III, 450mm long with sealing gasket and fixing brackets with 150mm R type adaptor for half round ridge	60.40	5.00	63.42	1.50	0.75	19.11	0.58	83.11	nr	89.34
universal Delta ridge	70.90	5.00	74.44	1.70	0.85	21.66	2.72	98.82	nr	106.23
universal angle ridge	55.30	5.00	58.06	1.50	2.56	31.82	2.72	92.60	nr	99.55
extra for extension adaptor and gasket	27.10	5.00	28.45	0.30	0.15	3.82	1.07	33.35	nr	35.85

Hip irons

	Del to Site £	Waste %	Material Cost £	Craft Optve Hrs	Lab Hrs	Labour Cost £	Sunds £	Nett Rate £	Unit	Gross Rate £
galvanised mild steel; 32 x 3 x 380mm girth; scrolled; fixing with galvanised steel screws to timber	1.79	-	1.79	0.10	0.05	1.27	-	3.06	nr	3.29

Lead soakers

	Del to Site £	Waste %	Material Cost £	Craft Optve Hrs	Lab Hrs	Labour Cost £	Sunds £	Nett Rate £	Unit	Gross Rate £
fixing only	-	-	-	0.30	0.15	3.82	-	3.82	nr	4.11

This section continues
on the next page

Labour hourly rates: (except Specialists) Craft Operatives £9.23 Labour £7.02 Rates are national average prices. Refer to REGIONAL VARIATIONS for indicative levels of overall pricing in regions	MATERIALS			LABOUR				RATES		
	Del to Site £	Waste %	Material Cost £	Craft Optve Hrs	Lab Hrs	Labour Cost £	Sunds £	Nett Rate £	Unit	Gross Rate (+7.5%) £
H61: FIBRE CEMENT SLATING										
Quantities required for 1m2 of slating										

Slates to 70mm lap
  400 x 240mm, 25.3nr
Slates to 102mm lap
  600 x 300mm, 13nr
  500 x 250mm, 19.5nr
Roof coverings; asbestos-free cement slates, 600 x 300mm; centre fixing with copper nails and copper disc rivets to 102mm lap; 38 x 19mm pressure impregnated softwood battens fixed with galvanised nails; underlay; B.S.747 type 1F bitumen reinforced felt; 150mm laps; fixing with galvanised steel clout nails										

Pitched 30 degrees from horizontal										
generally	17.93	5.00	**18.83**	0.52	0.26	**6.62**	3.69	**29.14**	m2	31.33
holes	-	-	-	0.50	0.25	**6.37**	-	**6.37**	nr	6.85
abutments; square	1.71	5.00	**1.80**	0.15	0.08	**1.95**	-	**3.74**	m	4.02
abutments; raking	2.70	-	**2.70**	2.70	0.12	**25.76**	-	**28.46**	m	30.60
abutments; curved to 3000mm radius ..	3.59	-	**3.59**	0.30	0.15	**3.82**	-	**7.41**	m	7.97
double course at eaves	4.83	5.00	**5.07**	0.15	0.08	**1.95**	0.62	**7.64**	m	8.21
verges; slate undercloak and point in cement mortar (1:3)	2.76	5.00	**2.90**	0.08	0.04	**1.02**	0.50	**4.42**	m	4.75
ridges or hips; asbestos free cement; fixing with nails	17.87	5.00	**18.76**	0.20	0.10	**2.55**	0.34	**21.65**	m	23.28
Roof coverings; asbestos-free cement slates, 600 x 300mm; centre fixing with copper nails and copper disc rivets to 102mm lap; 50 x 25mm pressure impregnated softwood battens fixed with galvanised nails; underlay; B.S.747 type 1F bitumen reinforced felt; 150mm laps; fixing with galvanised steel clout nails										
Pitched 30 degrees from horizontal										
generally	17.93	5.00	**18.83**	0.56	0.28	**7.13**	4.41	**30.37**	m2	32.65
Roof coverings; asbestos-free cement slates, 600 x 300mm; centre fixing with copper nails and copper disc rivets to 102mm lap (close boarding on rafters included elsewhere); underlay; B.S.747 type 1B underslating felt; 150mm laps; fixing with galvanised steel clout nails										
Pitched 30 degrees from horizontal										
generally	17.93	5.00	**18.83**	0.42	0.21	**5.35**	2.33	**26.51**	m2	28.50
Roof coverings; asbestos-free cement slates, 500 x 250mm; centre fixing with copper nails and copper disc rivets to 102mm lap; 38 x 19mm pressure impregnated softwood battens fixed with galvanised nails; underlay; B.S.747 type 1F bitumen reinforced felt; 150mm laps; fixing with galvanised steel clout nails										
Pitched 30 degrees from horizontal										
generally	20.22	5.00	**21.23**	0.58	0.29	**7.39**	4.46	**33.08**	m2	35.56
holes	-	-	-	0.50	0.25	**6.37**	-	**6.37**	nr	6.85
abutments; square	2.03	-	**2.03**	0.20	0.10	**2.55**	-	**4.58**	m	4.92
abutments; raking	3.05	-	**3.05**	0.30	0.15	**3.82**	-	**6.87**	m	7.39
abutments; curved to 3000mm radius ..	4.05	-	**4.05**	0.40	0.20	**5.10**	-	**9.15**	m	9.83
double course at eaves	4.15	5.00	**4.36**	0.18	0.09	**2.29**	0.66	**7.31**	m	7.86
verges; slate undercloak and point in cement mortar (1:3)	2.59	5.00	**2.72**	0.10	0.05	**1.27**	0.54	**4.53**	m	4.87
ridges or hips; asbestos free cement; fixing with nails	17.87	5.00	**18.76**	0.20	0.10	**2.55**	0.34	**21.65**	m	23.28
Roof coverings; asbestos-free cement slates, 500 x 250mm; centre fixing with copper nails and copper disc rivets to 102mm lap; 50 x 25mm pressure impregnated softwood battens fixed with galvanised nails; underlay; B.S.747 type 1F bitumen reinforced felt; 150mm laps; fixing with galvanised steel clout nails										
Pitched 30 degrees from horizontal										
generally	20.22	5.00	**21.23**	0.63	0.32	**8.06**	5.36	**34.65**	m2	37.25
Roof coverings; asbestos-free cement slates, 400 x 200mm; centre fixing with copper nails and copper disc rivets to 102mm lap; 38 x 19mm pressure impregnated softwood battens fixed with galvanised nails; underlay; B.S.747 type 1F bitumen reinforced felt; 150mm laps; fixing with galvanised steel clout nails										
Pitched 30 degrees from horizontal										
generally	21.25	5.00	**22.31**	0.62	0.31	**7.90**	5.71	**35.92**	m2	38.62
holes	-	%	-	0.50	0.25	**6.37**	-	**6.37**	nr	6.85
abutments; square	2.12	-	**2.12**	0.25	0.13	**3.22**	-	**5.34**	m	5.74
abutments; raking	3.18	-	**3.18**	0.38	0.19	**4.84**	-	**8.02**	m	8.62
abutments; curved to 3000mm radius ..	4.24	-	**4.24**	0.50	0.25	**6.37**	-	**10.61**	m	11.41
double course at eaves	3.32	5.00	**3.49**	0.20	0.10	**2.55**	0.74	**6.77**	m	7.28

CLADDING/COVERING – MAJOR WORKS

Labour hourly rates: (except Specialists) Craft Operatives £9.23 Labour £7.02. Rates are national average prices. Refer to REGIONAL VARIATIONS for indicative levels of overall pricing in regions	MATERIALS			LABOUR				RATES		
	Del to Site £	Waste %	Material Cost £	Craft Optve Hrs	Lab Hrs	Labour Cost £	Sunds £	Nett Rate £	Unit	Gross Rate (+7.5%) £
H61: FIBRE CEMENT SLATING Cont.										
Roof coverings; asbestos-free cement slates, 400 x 200mm; centre fixing with copper nails and copper disc rivets to 102mm lap; 38 x 19mm pressure impregnated softwood battens fixed with galvanised nails; underlay; B.S.747 type 1F bitumen reinforced felt; 150mm laps; fixing with galvanised steel clout nails Cont.										
Pitched 30 degrees from horizontal Cont. verges; slate undercloak and point in cement mortar (1:3)	2.00	5.00	2.10	0.12	0.06	1.53	0.58	4.21	m	4.52
ridges or hips; asbestos free cement; fixing with nails	17.87	5.00	18.76	0.20	0.10	2.55	0.34	21.65	m	23.28
Roof coverings; asbestos-free cement slates, 400 x 200mm; centre fixing with copper nails and copper disc rivets to 102mm lap; 50 x 25mm pressure impregnated softwood battens fixed with galvanised nails; underlay; B.S.747 type 1F bitumen reinforced felt; 150mm laps; fixing with galvanised steel clout nails										
Pitched 30 degrees from horizontal generally	21.25	5.00	22.31	0.70	0.35	8.92	6.79	38.02	m2	40.87
Fittings										
Hip irons galvanised mild steel; 32 x 3 x 380mm girth; scrolled; fixing with galvanised steel screws to timber	1.79	5.00	1.88	0.10	0.05	1.27	-	3.15	nr	3.39
Lead soakers fixing only	-	-	-	0.30	0.15	3.82	-	3.82	nr	4.11
H62: NATURAL SLATING										
Quantities required for 1m2 of slating										
Slates to 75mm lap 610 x 305mm, 12.25nr 510 x 255mm, 18nr 405 x 205mm, 30nr										
Roof coverings; blue/grey slates, 610 x 305mm, 6.5mm thick; fixing with slate nails to 75mm lap; 38 x 19mm pressure impregnated softwood battens fixed with galvanised nails; underlay; B.S.747 type 1F bitumen reinforced felt; 150mm laps; fixing with galvanised steel clout nails										
Pitched 30 degrees from horizontal generally	51.21	5.00	53.77	0.50	0.25	6.37	2.77	62.91	m2	67.63
holes	-	-	-	0.50	0.25	6.37	-	6.37	nr	6.85
abutments; square	5.12	-	5.12	0.15	0.08	1.95	-	7.07	m	7.60
abutments; raking	7.68	-	7.68	0.23	0.12	2.97	-	10.65	m	11.44
double course at eaves	13.00	5.00	13.65	0.14	0.07	1.78	0.66	16.09	m	17.30
verges; slate undercloak and point in cement mortar (1:3)	8.36	5.00	8.78	0.08	0.04	1.02	0.38	10.18	m	10.94
Roof coverings; blue/grey slates, 610 x 305mm, 6.5mm thick; fixing with aluminium nails to 75mm lap; 38 x 19mm pressure impregnated softwood battens fixed with galvanised nails; underlay; B.S.747 type 1F bitumen reinforced felt; 150mm laps; fixing with galvanised steel clout nails										
Pitched 30 degrees from horizontal generally	51.21	5.00	53.77	0.50	0.25	6.37	3.26	63.40	m2	68.16
Roof coverings; blue/grey slates, 610 x 305mm, 6.5mm thick; fixing with copper nails to 75mm laps; 38 x 19mm pressure impregnated softwood battens fixed with galvanised nails; underlay; B.S.747 type 1F bitumen reinforced felt; 150mm laps; fixing with galvanised steel clout nails										
Pitched 30 degrees from horizontal generally	51.21	5.00	53.77	0.50	0.25	6.37	3.51	63.65	m2	68.42
Roof coverings; blue/grey slates, 610 x 305mm, 6.5mm thick; fixing with slate nails to 75mm lap; 50 x 25mm pressure impregnated softwood battens fixed with galvanised nails; underlay; B.S.747 type 1F bitumen reinforced felt; 150mm laps; fixing with galvanised steel clout nails										
Pitched 30 degrees from horizontal generally	51.21	5.00	53.77	0.54	0.27	6.88	3.49	64.14	m2	68.95

Labour hourly rates: (except Specialists) Craft Operatives £9.23 Labour £7.02 Rates are national average prices. Refer to REGIONAL VARIATIONS for indicative levels of overall pricing in regions	MATERIALS			LABOUR				RATES		
	Del to Site £	Waste %	Material Cost £	Craft Optve Hrs	Lab Hrs	Labour Cost £	Sunds £	Nett Rate £	Unit	Gross Rate (+7.5%) £

H62: NATURAL SLATING Cont.

Roof coverings; blue/grey slates, 610 x 305mm, 6.5mm thick; fixing with slate nails to 75mm lap; 50 x 25mm pressure impregnated softwood counterbattens at 1067mm centres and 38 x 19mm pressure impregnated softwood battens fixed with galvanised nails; underlay; B.S.747 type 1B underslating felt; 150mm laps; fixing with galvanised steel clout nails

Pitched 30 degrees from horizontal generally..	51.21	5.00	53.77	0.54	0.27	6.88	3.26	63.91	m2	68.70

Roof coverings; blue/grey slates, 510 x 255mm, 6.5mm thick; fixing with slate nails to 75mm lap; 38 x 19mm pressure impregnated softwood battens fixed with galvanised nails; underlay; B.S.747 type 1F bitumen reinforced felt; 150mm laps; fixing with galvanised steel clout nails

Pitched 30 degrees from horizontal										
generally..	33.85	5.00	35.54	0.58	0.29	7.39	3.26	46.19	m2	49.66
holes..	-	-	-	0.50	0.25	6.37	-	6.37	nr	6.85
abutments; square	3.39	-	3.39	0.20	0.10	2.55	-	5.94	m	6.38
abutments; raking	5.08	-	5.08	0.30	0.15	3.82	-	8.90	m	9.57
double course at eaves	7.44	5.00	7.81	0.17	0.09	2.20	0.67	10.68	m	11.48
verges; slate undercloak and point in cement mortar (1:3)	4.65	5.00	4.88	0.10	0.05	1.27	0.39	6.55	m	7.04

Roof coverings; blue/grey slates, 510 x 255mm, 6.5mm thick; fixing with aluminium nails to 75mm lap; 38 x 19mm pressure impregnated softwood battens fixed with galvanised nails; underlay; B.S.747 type 1F bitumen reinforced felt; 150mm laps; fixing with galvanised steel clout nails

Pitched 30 degrees from horizontal generally..	33.85	5.00	35.54	-	0.29	2.04	3.99	41.57	m2	44.69

Roof coverings; blue/grey slates, 510 x 255mm, 6.5mm thick; fixing with copper nails to 75mm lap; 38 x 19mm pressure impregnated softwood battens fixed with galvanised nails; underlay; B.S.747 type 1F bitumen reinforced felt; 150mm laps; fixing with galvanised steel clout nails

Pitched 30 degrees from horizontal generally..	33.85	5.00	35.54	-	0.29	2.04	4.35	41.93	m2	45.07

Roof coverings; blue/grey slates, 510 x 255mm, 6.5mm thick; fixing with slate nails to 75mm lap; 50 x 25mm pressure impregnated softwood battens fixed with galvanised nails; underlay; B.S.747 type 1F bitumen reinforced felt; 150mm laps; fixing with galvanised steel clout nails

Pitched 30 degrees from horizontal generally..	33.85	5.00	35.54	0.63	0.32	8.06	4.16	47.76	m2	51.35

Roof coverings; blue/grey slates, 405 x 205mm, 6.5mm thick; fixing with slate nails to 75mm lap; 38 x 19mm pressure impregnated softwood battens fixed with galvanised nails; underlay; B.S.747 type 1F bitumen reinforced felt; 150mm laps; fixing with galvanised steel clout nails

Pitched 30 degrees from horizontal										
generally..	26.10	5.00	27.41	0.70	0.35	8.92	3.99	40.31	m2	43.34
holes..	-	-	-	0.50	0.25	6.37	-	6.37	nr	6.85
abutments; square	2.61	-	2.61	0.25	0.13	3.22	-	5.83	m	6.27
abutments; raking	3.92	-	3.92	0.38	0.19	4.84	-	8.76	m	9.42
double course at eaves	4.35	5.00	4.57	0.20	0.10	2.55	0.35	7.47	m	8.03
verges; slate undercloak and point in cement mortar (1:3)	2.61	5.00	2.74	0.12	0.06	1.53	0.40	4.67	m	5.02

Roof coverings; blue/grey slates, 405 x 205mm, 6.5mm thick; fixing with aluminium nails to 75mm lap; 38 x 19mm pressure impregnated softwood battens fixed with galvanised nails; underlay; B.S.747 type 1F bitumen reinforced felt; 150mm laps; fixing with galvanised steel clout nails

Pitched 30 degrees from horizontal generally..	26.10	5.00	27.41	0.70	0.35	8.92	5.19	41.51	m2	44.63

Roof coverings; blue/grey slates, 405 x205mm, 6.5mm thick; fixing with copper nails to 75mmlap; 38 x 19mm pressure impregnated softwood battens fixed with galvanised nails; underlay; B.S.747 type 1F bitumen reinforced felt; 150mm laps; fixing with galvanised steel clout nails

Pitched 30 degrees from horizontal generally..	26.10	5.00	27.41	0.70	0.35	8.92	5.79	42.11	m2	45.27

CLADDING/COVERING

Labour hourly rates: (except Specialists) Craft Operatives £9.23 Labour £7.02 Rates are national average prices. Refer to REGIONAL VARIATIONS for indicative levels of overall pricing in regions	MATERIALS			LABOUR				RATES		
	Del to Site £	Waste %	Material Cost £	Craft Optve Hrs	Lab Hrs	Labour Cost £	Sunds £	Nett Rate £	Unit	Gross Rate (+7.5%) £

H62: NATURAL SLATING Cont.

Roof coverings; blue/grey slates, 405 x 205mm, 6.5mm thick; fixing with slate nails to 75mm lap; 50 x 25mm pressure impregnated softwood battens fixed with galvanised nails; underlay; B.S.747 type 1F bitumen reinforced felt; 150mm laps; fixing with galvanised steel clout nails

Pitched 30 degrees from horizontal										
generally	26.10	5.00	27.41	0.76	0.38	9.68	5.07	42.16	m2	45.32

Roof coverings; Burlington green slates, best random fixed in diminishing courses with alloy nails to 75mm lap; 38 x 19mm pressure impregnated softwood battens fixed with galvanised nails; underlay; B.S.747 type 1F bitumen reinforced felt; 150mm laps; fixing with galvanised steel clout nails

Pitched 30 degrees from horizontal										
generally	102.63	5.00	107.76	0.75	0.38	9.59	4.56	121.91	m2	131.05
abutments; square	10.26	-	10.26	0.20	0.10	2.55	-	12.81	m	13.77
abutments; raking	15.39	-	15.39	0.30	0.15	3.82	0.86	20.07	m	21.58
double course at eaves	22.42	5.00	23.54	0.20	0.10	2.55	0.52	26.61	m	28.60
verges; slate undercloak and point in cement mortar (1:3)	14.95	5.00	15.70	0.20	0.10	2.55	0.52	18.77	m	20.17
hips, valleys and angles; close mitred	17.44	5.00	18.31	0.35	0.18	4.49	0.55	23.36	m	25.11

Fittings

Hip irons galvanised mild steel; 32 x 3 x 380mm girth; scrolled; fixing with galvanised steel screws to timber	1.79	5.00	1.88	0.10	0.05	1.27	-	3.15	nr	3.39
Lead soakers fixing only	-	-	-	0.30	0.15	3.82	-	3.82	nr	4.11

H63: RECONSTRUCTED STONE SLATING/ TILING

Roof coverings; Redland Cambrian interlocking riven textured slates, 300 x 336mm; fixing every slate with two stainless steel nails and one stainless steel clip to 50mm laps; 38 x 25mm pressure impregnated softwood battens fixed with galvanised nails; underlay; B.S.747 type 1F bitumen reinforced felt; 150mm laps; fixing with galvanised steel clout nails

Pitched 40 degrees from horizontal										
generally	23.43	5.00	24.60	0.48	0.24	6.12	4.46	35.18	m2	37.81
abutments; square	2.34	-	2.34	0.10	0.05	1.27	-	3.61	m	3.89
abutments; raking	3.51	-	3.51	0.15	0.08	1.95	-	5.46	m	5.87
abutments; curved to 3000mm radius	4.69	-	4.69	0.20	0.10	2.55	-	7.24	m	7.78
supplementary fixing at eaves with stainless steel eaves clip to each slate	0.48	5.00	0.50	0.04	0.02	0.51	-	1.01	m	1.09
verges; 150 x 6mm fibre cement undercloak; slates and undercloak bedded and pointed in tinted cement mortar (1:3) with full slates and slate and a half slates in alternate courses and stainless steel verge clips	5.27	5.00	5.53	0.20	0.10	2.55	3.73	11.81	m	12.70
verges; 150 x 6mm fibre cement undercloak; slates and undercloak bedded and pointed in tinted cement mortar (1:3) with verge slates and slate and a half verge slates in alternate courses and stainless steel verge clips	5.27	5.00	5.53	0.20	0.10	2.55	3.73	11.81	m	12.70
ridge or hip tiles, Redland third round; butt jointed; bedding and pointing in tinted cement mortar (1:3)	7.63	5.00	8.01	0.20	0.10	2.55	0.34	10.90	m	11.72
ridge or hip tiles, Redland half round; butt jointed; bedding and pointing in tinted cement mortar (1:3)	7.63	5.00	8.01	0.20	0.10	2.55	0.34	10.90	m	11.72
ridge or hip tiles, Redland universal angle; butt jointed; bedding and pointing in tinted cement mortar (1:3)	8.14	5.00	8.55	0.20	0.10	2.55	0.34	11.44	m	12.29
ridge tiles, Redland Dry Ridge system with half round ridge tiles; fixing with stainless steel batten straps and ring shanked fixing nails with neoprene washers and sleeves; polypropylene ridge seals and uPVC profile filler units	24.03	5.00	25.23	0.40	0.20	5.10	-	30.33	m	32.60
ridge tiles, Redland Dry Ridge system with universal angle ridge tiles; fixing with stainless steel batten straps and ring shanked fixing nails with neoprene washers and sleeves; polypropylene ridge seals and uPVC profile filler units	24.54	5.00	25.77	0.40	0.20	5.10	-	30.86	m	33.18
ridge tiles, Redland Dry Vent Ridge system with half round ridge tiles; fixing with stainless steel batten straps and ring shanked nails with neoprene washers and sleeves; polypropylene ridge seals, PVC air flow control units and uPVC ventilated profile filler units	24.03	5.00	25.23	0.40	0.20	5.10	-	30.33	m	32.60
ridge tiles, Redland Dry Vent Ridge system with Universal angle ridge tiles; fixing with stainless steel batten straps and ring shanked nails with neoprene washers and sleeves; polypropylene ridge seals, PVC air flow control units and uPVC ventilated profile filler units	24.54	5.00	25.77	0.40	0.20	5.10	-	30.86	m	33.18

Labour hourly rates: (except Specialists) Craft Operatives £9.23 Labour £7.02 Rates are national average prices. Refer to REGIONAL VARIATIONS for indicative levels of overall pricing in regions	MATERIALS			LABOUR				RATES		
	Del to Site £	Waste %	Material Cost £	Craft Optve Hrs	Lab Hrs	Labour Cost £	Sunds £	Nett Rate £	Unit	Gross Rate (+7.5%) £
H63: RECONSTRUCTED STONE SLATING/ TILING Cont.										
Roof coverings; Redland Cambrian interlocking riven textured slates, 300 x 336mm; fixing every slate with two stainless steel nails and one stainless steel clip to 50mm laps; 38 x 25mm pressure impregnated softwood battens fixed with galvanised nails; underlay; B.S.747 type 1F bitumen reinforced felt; 150mm laps; fixing with galvanised steel clout nails Cont.										
Pitched 40 degrees from horizontal Cont. Monoridge filler units in conjunction with top tile, bed tile in nonsetting mastic sealant and screw on through filler unit with screws, washers and caps	34.28	5.00	35.99	0.20	0.10	2.55	-	38.54	m	41.43
valley tiles, Redland Universal valley troughs; laid with 100mm laps	21.66	5.00	22.74	0.50	0.25	6.37	0.34	29.45	m	31.66
Roof coverings; Redland Cambrian interlocking riven textured slates, 300 x 336mm; fixing every slate with two stainless steel nails and one stainless steel clip to 90mm lap; 38 x 25mm pressure impregnated softwood battens fixed with galvanised nails; underlay; B.S.747 type 1F bitumen reinforced felt; 150mm laps; fixing with galvanised steel clout nails										
Pitched 40 degrees from horizontal generally	27.95	5.00	29.35	0.55	0.28	7.04	5.35	41.74	m2	44.87
abutments; square	2.79	-	2.79	0.10	0.05	1.27	-	4.06	m	4.37
abutments; raking	4.19	-	4.19	0.15	0.08	1.95	-	6.14	m	6.60
abutments; curved to 3000mm radius	5.59	-	5.59	0.20	0.10	2.55	-	8.14	m	8.75
Fittings										
Ventilator tiles Red Vent ridge ventilation terminal 450mm long for half round ridge	46.40	5.00	48.72	1.30	0.65	16.56	0.58	65.86	nr	70.80
universal angle ridge...................	46.30	5.00	48.62	1.30	0.65	16.56	0.58	65.76	nr	70.69
Red Vent eaves ventilators; fixing with stainless steel nails...................	11.27	5.00	11.83	0.40	0.20	5.10	0.58	17.51	m	18.82
Redland Cambrian Thruvent interlocking slate complete with weather cap, underlay seal and fixing clips...................	37.50	5.00	39.38	0.50	0.25	6.37	0.58	46.33	nr	49.80
Gas terminals Gas Flue ridge terminal Mark III, 450mm long with sealing gasket and fixing brackets half round ridge...................	60.40	5.00	63.42	1.50	0.75	19.11	0.58	83.11	nr	89.34
Gas Flue ridge terminal Mark III, 450mm long with sealing gasket and fixing brackets with 150mm R type adaptor for half round ridge...................	60.40	5.00	63.42	1.50	0.75	19.11	27.68	110.21	nr	118.48
Hip irons galvanised mild steel; 32 x 3 x 380mm girth; scrolled; fixing with galvanised steel screws to timber	1.79	5.00	1.88	0.10	0.05	1.27	-	3.15	nr	3.39
Lead soakers fixing only	-	-	-	0.30	0.15	3.82	-	3.82	nr	4.11
Fittings										
Timloc uPVC Mark 3 eaves ventilators; reference 1126/40 fixing with nails to timber at 400mm centres 330mm wide...................	2.80	5.00	2.94	0.25	0.12	3.15	0.26	6.35	m	6.83
Timloc uPVC Mark 3 eaves ventilators; reference 1123 fixing with nails to timber at 600mm centres 330mm wide...................	2.08	5.00	2.18	0.20	0.10	2.55	0.26	4.99	m	5.37
Timlock uPVC soffit ventilators; reference 1137, 10mm airflow; fixing with screws to timber type C...................	1.66	5.00	1.74	0.15	0.05	1.74	0.10	3.58	m	3.85
Timlock uPVC Mark 2 eaves ventilators; reference 1122; fitting between trusses at 600mm centres 300mm girth...................	2.17	5.00	2.28	0.20	0.10	2.55	0.26	5.09	m	5.47
Timlock polypropylene over-fascia ventilators; reference 3011; fixing with screws to timber to top of fascia...................	2.97	5.00	3.12	0.20	0.10	2.55	0.16	5.83	m	6.26
H65: SINGLE LAP ROOF TILING										
Roof coverings; Redland Stonewold through coloured slates, 430 x 380mm; fixing every fourth course with galvanised nails to 75mm lap; 38 x 22mm pressure impregnated softwood battens fixed with galvanised nails; underlay; B.S.747 type 1F reinforced bitumen felt; 150mm laps; fixing with galvanised steel clout nails										
Pitched 40 degrees from horizontal generally...........................	10.07	5.00	10.57	0.42	0.21	5.35	2.90	18.82	m2	20.24

Labour hourly rates: (except Specialists) Craft Operatives £9.23 Labour £7.02 Rates are national average prices. Refer to REGIONAL VARIATIONS for indicative levels of overall pricing in regions	MATERIALS			LABOUR				RATES		
	Del to Site £	Waste %	Material Cost £	Craft Optve Hrs	Lab Hrs	Labour Cost £	Sunds £	Nett Rate £	Unit	Gross Rate (+7.5%) £
H65: SINGLE LAP ROOF TILING Cont.										
Roof coverings; Redland Stonewold through coloured slates, 430 x 380mm; fixing every fourth course with galvanised nails to 75mm lap; 38 x 22mm pressure impregnated softwood battens fixed with galvanised nails; underlay; B.S.747 type 1F reinforced bitumen felt; 150mm laps; fixing with galvanised steel clout nails Cont.										
Pitched 40 degrees from horizontal Cont.										
abutments; square	1.01	-	1.01	0.10	0.05	1.27	-	2.28	m	2.46
abutments; raking	1.51	-	1.51	0.15	0.08	1.95	-	3.46	m	3.72
abutments; curved to 3000mm radius	2.01	-	2.01	0.20	0.10	2.55	-	4.56	m	4.90
supplementary fixing eaves course with one clip per slate	0.36	-	0.36	0.03	0.02	0.42	-	0.78	m	0.84
verges; Redland Dry Verge system with half slates, full slates and clips	13.65	5.00	14.33	0.25	0.12	3.15	1.80	19.28	m	20.73
verges; Redland Dry Verge system with half slates, verge slates and clips	13.65	5.00	14.33	0.25	0.12	3.15	1.80	19.28	m	20.73
verges; 150 x 6mm fibre cement undercloak, slates and undercloak bedded and pointed in tinted cement mortar (1:3) and with half slates, full slates and clips	3.69	5.00	3.87	0.20	0.10	2.55	3.73	10.15	m	10.91
verges; 150 x 6mm fibre cement undercloak, slates and undercloak bedded and pointed in tinted cement mortar (1:3) and with half slates, verge slates and clips	3.69	5.00	3.87	0.20	0.10	2.55	3.73	10.15	m	10.91
ridge or hip tiles, Redland third round; butt jointed; bedding and pointing in tinted cement mortar (1:3)	7.63	5.00	8.01	0.20	0.10	2.55	0.34	10.90	m	11.72
ridge or hip tiles, Redland half round; butt jointed; bedding and pointing in tinted cement mortar (1:3)	7.63	5.00	8.01	0.20	0.10	2.55	0.34	10.90	m	11.72
ridge tiles, Redland universal angle; butt jointed; bedding and pointing in tinted cement mortar (1:3)	8.14	5.00	8.55	0.20	0.10	2.55	0.34	11.44	m	12.29
ridge tiles, Redland universal Stonewold type Monopitch; butt jointed; fixing with aluminium nails and bedding and pointing in tinted cement mortar (1:3)	19.12	5.00	20.08	0.20	0.10	2.55	0.34	22.96	m	24.69
ridge tiles, Redland universal half round type Monopitch; butt jointed; fixing with aluminium nails and bedding and pointing in tinted cement mortar (1:3)	19.12	5.00	20.08	0.20	0.10	2.55	0.34	22.96	m	24.69
ridge tiles, Redland Dry Ridge system with half round ridge tiles; fixing with stainless steel batten straps and ring shanked fixing nails with neoprene washers and sleeves; polypropylene ridge seals and uPVC profile filler units	24.03	5.00	25.23	0.40	0.20	5.10	-	30.33	m	32.60
ridge tiles, Redland Dry Ridge system with universal angle ridge tiles; fixing with stainless steel batten straps and ring shanked fixing nails with neoprene washers and sleeves; polypropylene ridge seals and uPVC profile filler units	24.54	5.00	25.77	0.40	0.20	5.10	-	30.86	m	33.18
ridge tiles, Redland Dry Vent Ridge system with half round ridge tiles; fixing with stainless steel batten straps and ring shanked nails with neoprene washers and sleeves; polypropylene ridge seals, PVC air flow control units and uPVC ventilated profile filler units	24.03	5.00	25.23	0.40	0.20	5.10	-	30.33	m	32.60
ridge tiles, Redland Dry Vent Ridge system with Universal angle ridge tiles; fixing with stainless steel batten straps and ring shanked nails with neoprene washers and sleeves; polypropylene ridge seals, PVC air flow control units and uPVC ventilated profile filler units	24.54	5.00	25.77	0.40	0.20	5.10	-	30.86	m	33.18
Monoridge filler units in conjunction with top tile, bed tile in nonsetting mastic sealant and screw on through filler unit with screws, washers and caps......................	34.28	5.00	35.99	0.20	0.10	2.55	0.34	38.88	m	41.80
valley tiles, Redland Universal valley troughs; laid with 100mm laps........................	21.66	5.00	22.74	0.50	0.25	6.37	0.34	29.45	m	31.66
Roof coverings; Redland Stonewold through coloured slates, 430 x 380mm; fixing every fourth course with galvanised nails to 75mm lap; 38 x 25mm pressured impregnated softwood battens fixed with galvanised nails; underlay; B.S.747 type 1F reinforced bitumen felt; 150mm laps; fixing with galvanised steel clout nails										
Pitched 40 degrees from horizontal										
generally..	10.07	5.00	10.57	0.42	0.21	5.35	2.03	17.95	m2	19.30
Roof coverings; Redland Stonewold through coloured slates, 430 x 380mm; fixing each course with galvanised nails to 75mm; 38 x 22mm pressure impregnated softwood battens fixed with galvanised nails; underlay; B.S 747 type 1F reinforced bitumen felt; 150mm laps; fixing with galvanised steel clout nails										
Pitched 40 degrees from horizontal										
generally..	10.07	5.00	10.57	0.46	0.23	5.86	4.38	20.81	m2	22.37

Labour hourly rates: (except Specialists) Craft Operatives £9.23 Labour £7.02 Rates are national average prices. Refer to REGIONAL VARIATIONS for indicative levels of overall pricing in regions	MATERIALS			LABOUR				RATES		
	Del to Site £	Waste %	Material Cost £	Craft Optve Hrs	Lab Hrs	Labour Cost £	Sunds £	Nett Rate £	Unit	Gross Rate (+7.5%) £

H65: SINGLE LAP ROOF TILING Cont.

Roof coverings; Redland Stonewold through coloured slates, 430 x 380mm; fixing every fourth course with galvanised nails to 75mm lap; 38 x 22mm pressure impregnated softwood battens fixed with galvanised nails; underlay, B.S.747 type 1F aluminium foil surfaced reinforced bitumen felt; 150mm laps; fixing with galvanised steel clout nails

| Pitched 40 degrees from horizontal generally...... | 10.07 | 5.00 | 10.57 | 0.43 | 0.21 | 5.44 | 5.26 | 21.28 | m2 | 22.87 |

Roof coverings; Redland Stonewold through coloured slates, 430 x 380mm; fixing every fourth course with galvanised nails to 75mm lap; 38 x 22mm pressure impregnated softwood battens fixed with galvanised nails; underlay, B.S.747 type 1F reinforced bitumen felt with 50mm glass fibre insulation bonded on; 150mm laps; fixing with galvanised steel clout nails

| Pitched 40 degrees from horizontal generally...... | 10.07 | 5.00 | 10.57 | 0.52 | 0.26 | 6.62 | 6.90 | 24.10 | m2 | 25.91 |

Roof coverings; Redland Delta through coloured tiles, 430 x 380mm; fixing every fourth course with galvanised nails to 75mm lap; 38 x 22mm pressure impregnated softwood battens fixed with galvanised nails; underlay; B.S.747 type 1F reinforced bitumen felt; 150mm laps; fixing with galvanised steel clout nails

Pitched 40 degrees from horizontal generally......	13.74	5.00	14.43	0.42	0.21	5.35	2.90	22.68	m2	24.38
Reform eaves filler unit and eaves clip to each tile......	0.72	5.00	0.76	0.06	0.03	0.76	0.24	1.76	m	1.89
verges; 150 x 6mm fibre cement undercloak; tiles and undercloak bedded and pointed in tinted cement mortar (1:3) and with standard tiles and clips ...	5.03	5.00	5.28	0.20	0.10	2.55	3.73	11.56	m	12.43
verges; 150 x 6mm fibre cement undercloak; tiles and undercloak bedded and pointed in tinted cement mortar (1:3) and with verge tiles and clips	5.03	5.00	5.28	0.20	0.10	2.55	3.73	11.56	m	12.43
ridge tiles, Redland Delta cut ridge tiles; butt jointed; bedding and pointing in tinted cement mortar (1:3)......	11.09	5.00	11.64	0.20	0.10	2.55	0.34	14.53	m	15.62
ridge tiles, Redland Delta type Monopitch ridge tiles; butt jointed; fixing with aluminium nails and bedding and pointing in tinted cement mortar (1:3)......	16.95	5.00	17.80	0.20	0.10	2.55	0.39	20.74	m	22.29

Roof coverings; Redland Stonewold through coloured slates and Delta through coloured tiles, 430 x 380mm; alternating one tile and one slate in each course; fixing every fourth course with galvanised nails to 75mm lap; 38 x 22mm pressure impregnated softwood battens fixed with galvanised nails; underlay; B.S.747 type 1F reinforced bitumen felt; 150mm laps; fixing with galvanised steel clout nails

Pitched 40 degrees from horizontal generally......	11.91	5.00	12.51	0.42	0.21	5.35	2.90	20.76	m2	22.31
Reform eaves filler unit to Delta tiles and eaves clip to each slate or tile	0.72	5.00	0.76	0.06	0.03	0.76	0.24	1.76	m	1.89

Roof coverings; Redland Stonewold through coloured slates and Delta through coloured tiles, 430 x 380mm; alternating two tiles and one slate in each course; fixing every fourth course with galvanised nails to 75mm lap; 38 x 22mm pressure impregnated softwood battens fixed with galvanised nails; underlay; B.S.747 type 1F reinforced bitumen felt; 150mm laps; fixing with galvanised steel clout nails

Pitched 40 degrees from horizontal generally......	12.52	5.00	13.15	0.42	0.21	5.35	2.90	21.40	m2	23.00
abutments; square	1.25	–	1.25	0.10	0.05	1.27	–	2.52	m	2.71
abutments; raking	1.88	–	1.88	0.15	0.08	1.95	–	3.83	m	4.11
abutments; curved to 3000mm radius	2.50	–	2.50	0.20	0.10	2.55	–	5.05	m	5.43

Roof coverings; Redland Stonewold through coloured slates and Delta through coloured tiles, 430 x 380mm; alternating three tiles and one slate in each course; fixing every fourth course with galvanised nails to 75mm lap; 38 x 22mm pressure impregnated softwood battens fixed with galvanised nails; underlay; B.S.747 type 1F bitumen reinforced felt; 150mm laps; fixing with galvanised steel clout nails

Pitched 40 degrees from horizontal generally......	12.82	5.00	13.46	0.42	0.21	5.35	2.90	21.71	m2	23.34

CLADDING/COVERING

	MATERIALS			LABOUR				RATES		
Labour hourly rates: (except Specialists) Craft Operatives £9.23 Labour £7.02 Rates are national average prices. Refer to REGIONAL VARIATIONS for indicative levels of overall pricing in regions	Del to Site £	Waste %	Material Cost £	Craft Optve Hrs	Lab Hrs	Labour Cost £	Sunds £	Nett Rate £	Unit	Gross Rate (+7.5%) £

H65: SINGLE LAP ROOF TILING Cont.

Roof coverings; Redland Stonewold through coloured slates and Delta through coloured tiles, 430 x 380mm; alternating one tile and two slates in each course; fixing every fourth course with galvanised nails to 75mm lap; 38 x 22mm pressure impregnated softwood battens fixed with galvanised nails; underlay; B.S.747 type 1F bitumen reinforced felt; 150mm laps; fixing with galvanised steel clout nails

Pitched 40 degrees from horizontal

Description	Del to Site £	Waste %	Material Cost £	Craft Optve Hrs	Lab Hrs	Labour Cost £	Sunds £	Nett Rate £	Unit	Gross Rate £
generally	11.30	5.00	11.87	0.42	0.21	5.35	2.90	20.12	m2	21.62

Roof coverings; Redland Regent granular faced tiles, 418 x 332mm; fixing every fourth course with galvanised nails to 75mm lap; 38 x 22mm pressure impregnated softwood battens fixed with galvanised nails; underlay; B.S.747 type 1F bitumen reinforced felt; 150mm laps; fixing with galvanised steel clout nails

Pitched 40 degrees from horizontal

Description	Del to Site £	Waste %	Material Cost £	Craft Optve Hrs	Lab Hrs	Labour Cost £	Sunds £	Nett Rate £	Unit	Gross Rate £
generally	6.06	5.00	6.36	0.42	0.21	5.35	2.75	14.46	m2	15.55
reform eaves filler unit and eaves clip to each eaves tile	0.72	5.00	0.76	0.06	0.03	0.76	0.24	1.76	m	1.89
verges; cloaked verge tiles and aluminium nails	8.85	5.00	9.29	0.15	0.08	1.95	0.06	11.30	m	12.15
verges; half tiles, cloaked verge tiles and aluminium nails	10.10	5.00	10.61	0.20	0.10	2.55	0.10	13.25	m	14.25
verges; 150 x 6mm fibre cement undercloak; tiles and undercloak bedded and pointed in tinted cement mortar (1:3) and with standard tiles and clips	1.87	5.00	1.96	0.20	0.10	2.55	3.73	8.24	m	8.86
verges; 150 x 6mm fibre cement undercloak; tiles and undercloak bedded and pointed in tinted cement mortar (1:3) and with verge tiles and clips	1.87	5.00	1.96	0.20	0.10	2.55	3.73	8.24	m	8.86
verges; 150 x 6mm fibre cement undercloak; tiles and undercloak bedded and pointed in tinted cement mortar (1:3) and with half tiles, standard tiles and clips	5.23	5.00	5.49	0.20	0.10	2.55	3.73	11.77	m	12.65
verges; 150 x 6mm fibre cement undercloak; tiles and undercloak bedded and pointed in tinted cement mortar (1:3) and with half tiles, verge tiles and clips	5.23	5.00	5.49	0.20	0.10	2.55	3.73	11.77	m	12.65
ridge or hip tiles, Redland third round; butt jointed; bedding and pointing in tinted cement mortar (1:3); dentil slips in pan of each tile set in bedding	7.63	5.00	8.01	0.30	0.15	3.82	1.48	13.31	m	14.31
ridge or hip tiles, Redland half round; butt jointed; bedding and pointing in tinted cement mortar (1:3); dentil slips in pan of each tile set in bedding	7.63	5.00	8.01	0.30	0.15	3.82	1.48	13.31	m	14.31
ridge tiles, Redland universal half round type Monopitch; butt jointed; fixing with aluminium nails and bedding and pointing in tinted cement mortar (1:3); dentil slips in pan of each tile set in bedding	8.14	5.00	8.55	0.30	0.15	3.82	1.48	13.85	m	14.89
valley tiles, Redland universal valley troughs; laid with 100mm laps	21.66	5.00	22.74	0.30	0.15	3.82	0.34	26.91	m	28.92

Roof coverings; Redland Regent granular faced tiles, 418 x 332mm; fixing every fourth course with aluminium nails to 75mm lap; 38 x 22mm pressure impregnated softwood battens fixed with galvanised nails; underlay; B.S.747 type 1F bitumen reinforced felt; 150mm laps; fixing with galvanised steel clout nails

Pitched 40 degrees from horizontal

Description	Del to Site £	Waste %	Material Cost £	Craft Optve Hrs	Lab Hrs	Labour Cost £	Sunds £	Nett Rate £	Unit	Gross Rate £
generally	6.06	5.00	6.36	0.42	0.21	5.35	2.90	14.61	m2	15.71

Roof coverings; Redland Regent granular faced tiles, 418 x 332mm; fixing every fourth course with galvanised nails to 100mm lap; 38 x 22mm pressure impregnated softwood battens fixed with galvanised nails; underlay; B.S.747 type 1F bitumen reinforced felt; 150mm laps; fixing with galvanised steel clout nails

Pitched 40 degrees from horizontal

Description	Del to Site £	Waste %	Material Cost £	Craft Optve Hrs	Lab Hrs	Labour Cost £	Sunds £	Nett Rate £	Unit	Gross Rate £
generally	6.56	5.00	6.89	0.42	0.21	5.35	2.53	14.77	m2	15.88
abutments; square	0.66	-	0.66	0.10	0.05	1.27	-	1.93	m	2.08
abutments; raking	0.98	-	0.98	0.15	0.08	1.95	-	2.93	m	3.15
abutments; curved to 3000mm radius	1.31	-	1.31	0.20	0.10	2.55	-	3.86	m	4.15

Roof coverings; Redland Regent through coloured tiles, 418 x 332mm; fixing every fourth course with galvanised nails to 75mm lap; 38 x 22mm pressure impregnated softwood battens fixed with galvanised nails; underlay; B.S.747 type 1F bitumen reinforced felt; 150mm laps; fixing with galvanised steel clout nails

Pitched 40 degrees from horizontal

Description	Del to Site £	Waste %	Material Cost £	Craft Optve Hrs	Lab Hrs	Labour Cost £	Sunds £	Nett Rate £	Unit	Gross Rate £
generally	6.06	5.00	6.36	0.42	0.21	5.35	2.75	14.46	m2	15.55
reform eaves filler unit and eaves clip to each eaves tile	0.72	5.00	0.76	0.06	0.03	0.76	0.24	1.76	m	1.89

Labour hourly rates: (except Specialists) Craft Operatives £9.23 Labour £7.02 Rates are national average prices. Refer to REGIONAL VARIATIONS for indicative levels of overall pricing in regions	MATERIALS			LABOUR				RATES		
	Del to Site £	Waste %	Material Cost £	Craft Optve Hrs	Lab Hrs	Labour Cost £	Sunds £	Nett Rate £	Unit	Gross Rate (+7.5%) £

H65: SINGLE LAP ROOF TILING Cont.

Roof coverings; Redland Regent through coloured tiles, 418 x 332mm; fixing every fourth course with galvanised nails to 75mm lap; 38 x 22mm pressure impregnated softwood battens fixed with galvanised nails; underlay; B.S.747 type 1F bitumen reinforced felt; 150mm laps; fixing with galvanised steel clout nails Cont.

Pitched 40 degrees from horizontal Cont.

Description	Del to Site £	Waste %	Material Cost £	Craft Optve Hrs	Lab Hrs	Labour Cost £	Sunds £	Nett Rate £	Unit	Gross Rate £
verges; cloaked verge tiles and aluminium nails ..	8.85	5.00	9.29	0.15	0.08	1.95	0.06	11.30	m	12.15
verges; 150 x 6mm fibre cement undercloak; tiles and undercloak bedded and pointed in tinted cement mortar (1:3) and with standard tiles and clips ...	1.87	5.00	1.96	0.20	0.10	2.55	3.73	8.24	m	8.86
verges; 150 x 6mm fibre cement undercloak; tiles and undercloak bedded and pointed in tinted cement mortar (1:3) and with verge tiles and clips	1.87	5.00	1.96	0.20	0.10	2.55	3.73	8.24	m	8.86
verges; 150 x 6mm fibre cement undercloak; tiles and undercloak bedded and pointed in tinted cement mortar (1:3) and with half tiles, standard tiles and clips	5.23	5.00	5.49	0.20	0.10	2.55	3.73	11.77	m	12.65
verges; 150 x 6mm fibre cement undercloak; tiles and undercloak bedded and pointed in tinted cement mortar (1:3) and with half tiles, verge tiles and clips	5.23	5.00	5.49	0.20	0.10	2.55	3.73	11.77	m	12.65
valley tiles, Redland universal valley troughs; laid with 100mm laps	21.66	5.00	22.74	0.30	0.15	3.82	0.34	26.91	m	28.92

Roof coverings; Redland Regent through coloured tiles, 418 x 332mm; fixing every fourth course with galvanised nails to 100mm lap; 38 x 22mm pressure impregnated softwood battens fixed with galvanised nails; underlay; B.S.747 type 1F bitumen reinforced felt; 150mm laps; fixing with galvanised steel clout nails

Pitched 40 degrees from horizontal

Description	Del to Site £	Waste %	Material Cost £	Craft Optve Hrs	Lab Hrs	Labour Cost £	Sunds £	Nett Rate £	Unit	Gross Rate £
generally	6.56	5.00	6.89	0.42	0.21	5.35	2.53	14.77	m2	15.88
abutments; square	0.66	-	0.66	0.10	0.05	1.27	-	1.93	m	2.08
abutments; raking	0.98	-	0.98	0.15	0.08	1.95	-	2.93	m	3.15
abutments; curved to 3000mm radius	1.31	-	1.31	0.20	0.10	2.55	-	3.86	m	4.15

Roof coverings; Redland Grovebury granular faced double pantiles, 418 x 332mm; fixing every fourth course with galvanised nails to 75mm lap; 38 x 22mm pressure impregnated softwood battens fixed with galvanised nails; underlay; B.S.747 type 1F bitumen reinforced felt; 150mm laps; fixing with galvanised steel clout nails

Pitched 40 degrees from horizontal

Description	Del to Site £	Waste %	Material Cost £	Craft Optve Hrs	Lab Hrs	Labour Cost £	Sunds £	Nett Rate £	Unit	Gross Rate £
generally	6.06	5.00	6.36	0.42	0.21	5.35	2.75	14.46	m2	15.55
reform eaves filler unit and eaves clip to each eaves tile	0.72	5.00	0.76	0.06	0.03	0.76	0.24	1.76	m	1.89
verges; cloaked verge tiles and aluminium nails ..	8.85	5.00	9.29	0.15	0.08	1.95	0.06	11.30	m	12.15
verges; half tiles, cloaked verge tiles and aluminium nails	10.10	5.00	10.61	0.20	0.10	2.55	0.10	13.25	m	14.25
verges; 150 x 6mm fibre cement undercloak; tiles and undercloak bedded and pointed in tinted cement mortar (1:3) and with standard tiles and clips ...	1.87	5.00	1.96	0.20	0.10	2.55	3.73	8.24	m	8.86
verges; 150 x 6mm fibre cement undercloak; tiles and undercloak bedded and pointed in tinted cement mortar (1:3) and with treble roll verge tiles and clips	1.87	5.00	1.96	0.20	0.10	2.55	3.73	8.24	m	8.86
ridge or hip tiles, Redland third round; butt jointed; bedding and pointing in tinted cement mortar (1:3)	7.63	5.00	8.01	0.30	0.15	3.82	1.48	13.31	m	14.31
ridge or hip tiles, Redland half round; butt jointed; bedding and pointing in tinted cement mortar (1:3)	7.63	5.00	8.01	0.30	0.15	3.82	1.48	13.31	m	14.31
valley tiles, Redland Universal valley troughs; laid with 100mm laps	21.66	5.00	22.74	0.30	0.15	3.82	0.34	26.91	m	28.92

Roof coverings; Redland Grovebury through coloured double pantiles, 418 x 332mm; fixing every fourth course with galvanised nails to 75mm lap; 38 x 22mm pressure impregnated softwood battens fixed with galvanised nails; underlay; B.S.747 type 1F bitumen reinforced felt; 150mm laps; fixing with galvanised steel clout nails

Pitched 40 degrees from horizontal

Description	Del to Site £	Waste %	Material Cost £	Craft Optve Hrs	Lab Hrs	Labour Cost £	Sunds £	Nett Rate £	Unit	Gross Rate £
generally	6.06	5.00	6.36	0.42	0.21	5.35	2.75	14.46	m2	15.55
reform eaves filler unit and eaves clip to each eaves tile	0.72	5.00	0.76	0.06	0.03	0.76	0.24	1.76	m	1.89
verges; cloaked verge tiles and aluminium nails ..	8.85	5.00	9.29	0.15	0.08	1.95	0.06	11.30	m	12.15
verges; half tiles, cloaked verge tiles and aluminium nails	10.10	5.00	10.61	0.20	0.10	2.55	0.10	13.25	m	14.25
verges; 150 x 6mm fibre cement undercloak; tiles and undercloak bedded and pointed in tinted cement mortar (1:3) and with standard tiles and clips ...	1.87	5.00	1.96	0.20	0.10	2.55	3.73	8.24	m	8.86
verges; 150 x 6mm fibre cement undercloak; tiles and undercloak bedded and pointed in tinted cement mortar (1:3) and with treble roll verge tiles and clips	1.87	5.00	1.96	0.20	0.10	2.55	3.73	8.24	m	8.86

Labour hourly rates: (except Specialists) Craft Operatives £9.23 Labour £7.02 Rates are national average prices. Refer to REGIONAL VARIATIONS for indicative levels of overall pricing in regions	MATERIALS			LABOUR				RATES		
	Del to Site £	Waste %	Material Cost £	Craft Optve Hrs	Lab Hrs	Labour Cost £	Sunds £	Nett Rate £	Unit	Gross Rate (+7.5%) £
H65: SINGLE LAP ROOF TILING Cont.										
Roof coverings; Redland Grovebury through coloured double pantiles, 418 x 332mm; fixing every fourth course with galvanised nails to 75mm lap; 38 x 22mm pressure impregnated softwood battens fixed with galvanised nails; underlay; B.S.747 type 1F bitumen reinforced felt; 150mm laps; fixing with galvanised steel clout nails Cont.										
Pitched 40 degrees from horizontal Cont.										
ridge or hip tiles, Redland third round; butt jointed; bedding and pointing in tinted cement mortar (1:3)	7.63	5.00	8.01	0.30	0.15	3.82	1.48	13.31	m	14.31
ridge or hip tiles, Redland half round; butt jointed ; bedding and pointing in tinted cement mortar (1:3)	7.63	5.00	8.01	0.30	0.15	3.82	1.48	13.31	m	14.31
valley tiles, Redland Universal valley troughs; laid with 100mm laps	21.66	5.00	22.74	0.30	0.15	3.82	0.34	26.91	m	28.92
Roof coverings; Redland Norfolk through coloured pantiles, 381 x 227mm; fixing every fourth course with galvanised nails to 75mm lap; 38 x 22mm pressure impregnated softwood battens fixed with galvanised nails; underlay; B.S.747 type 1F bitumen reinforced felt; 150mm laps; fixing with galvanised steel clout nails										
Pitched 40 degrees from horizontal										
generally	7.14	5.00	7.50	0.54	0.27	6.88	3.27	17.65	m2	18.97
abutments; square	0.71	-	0.71	0.10	0.05	1.27	-	1.98	m	2.13
abutments; raking	1.07	-	1.07	0.15	0.08	1.95	-	3.02	m	3.24
abutments; curved to 3000mm radius	1.43	-	1.43	0.20	0.10	2.55	-	3.98	m	4.28
reform eaves filler unit and aluminium nails to each eaves tile	0.57	5.00	0.60	0.06	0.03	0.76	0.24	1.60	m	1.72
verges; plain tile undercloak; tiles and undercloak bedded and pointed in tinted cement mortar (1:3) and with standard tiles and aluminium nails	1.31	5.00	1.38	0.10	0.05	1.27	0.37	3.02	m	3.25
verges; plain tile undercloak; tiles and undercloak bedded and pointed in tinted cement mortar (1:3) and with standard tiles and clips	1.31	5.00	1.38	0.10	0.05	1.27	3.43	6.08	m	6.54
verges; 150 x 6mm fibre cement undercloak; tiles and undercloak bedded and pointed in tinted cement mortar (1:3) and with standard tiles and aluminium nails	1.31	5.00	1.38	0.15	0.08	1.95	2.05	5.37	m	5.77
verges; 150 x 6mm fibre cement undercloak; tiles and undercloak bedded and pointed in tinted cement mortar (1:3) and with standard tiles and clips	1.31	5.00	1.38	0.15	0.08	1.95	5.08	8.40	m	9.03
ridge or hip tiles, Redland third round; butt jointed; bedding and pointing in tinted cement mortar (1:3)	7.63	5.00	8.01	0.30	0.15	3.82	0.34	12.17	m	13.09
ridge or hip tiles, Redland half round; butt jointed; bedding and pointing in tinted cement mortar (1:3)	7.63	5.00	8.01	0.30	0.15	3.82	0.34	12.17	m	13.09
valley tiles, Redland Universal valley troughs; laid with 100mm laps	21.66	5.00	22.74	0.30	0.15	3.82	0.34	26.91	m	28.92
Roof coverings; Redland Norfolk through coloured pantiles, 381 x 227mm; fixing every fourth course with galvanised nails to 75mm lap; 38 x 25mm pressure impregnated softwood battens fixed with galvanised nails; underlay; B.S.747 type 1F bitumen reinforced felt; 150mm laps; fixing with galvanised steel clout nails										
Pitched 40 degrees from horizontal										
generally	7.14	5.00	7.50	0.54	0.27	6.88	3.43	17.81	m2	19.14
Roof coverings; Redland Norfolk through coloured pantiles, 381 x 227mm; fixing every fourth course with galvanised nails to 100mm lap; 38 x 22mm pressure impregnated softwood battens fixed with galvanised nails; underlay; B.S.747 type 1F bitumen reinforced felt; 150mm laps; fixing with galvanised steel clout nails										
Pitched 40 degrees from horizontal										
generally	7.80	5.00	8.19	0.59	0.30	7.55	3.32	19.06	m2	20.49
Roof coverings; Redland 49 granular faced tiles, 381 x 227mm; fixing every fourth course with galvanised nails to 75mm lap; 38 x 22mm pressure impregnated softwood battens fixed with galvanised nails; underlay B.S.747 type 1F bitumen reinforced felt; 150mm laps; fixing with galvanised steel clout nails										
Pitched 40 degrees from horizontal										
generally	6.92	5.00	7.27	0.54	0.27	6.88	3.39	17.54	m2	18.85
supplementary fixing eaves course with one aluminium nail per tile	-	-	-	0.02	0.01	0.25	0.03	0.28	m	0.31
verges; Redland Dry Verge system with standard tiles and clips	11.23	5.00	11.79	0.25	0.12	3.15	1.80	16.74	m	18.00
verges; Redland Dry Verge system with verge tiles and clips	11.23	5.00	11.79	0.25	0.12	3.15	1.80	16.74	m	18.00

Labour hourly rates: (except Specialists) Craft Operatives £9.23 Labour £7.02 Rates are national average prices. Refer to REGIONAL VARIATIONS for indicative levels of overall pricing in regions	MATERIALS			LABOUR				RATES		
	Del to Site £	Waste %	Material Cost £	Craft Optve Hrs	Lab Hrs	Labour Cost £	Sunds £	Nett Rate £	Unit	Gross Rate (+7.5%) £
H65: SINGLE LAP ROOF TILING Cont.										
Roof coverings; Redland 49 granular faced tiles, 381 x 227mm; fixing every fourth course with galvanised nails to 75mm lap; 38 x 22mm pressure impregnated softwood battens fixed with galvanised nails; underlay B.S.747 type 1F bitumen reinforced felt; 150mm laps; fixing with galvanised steel clout nails Cont.										
Pitched 40 degrees from horizontal Cont.										
verges; plain tile undercloak; tiles and undercloak bedded and pointed in tinted cement mortar (1:3) and with standard tiles and aluminium nails	1.27	5.00	1.33	0.10	0.05	1.27	0.40	3.01	m	3.23
verges; plain tile undercloak; tiles and undercloak bedded and pointed in tinted cement mortar (1:3) and with verge tiles and aluminium nails	1.27	5.00	1.33	0.10	0.05	1.27	0.40	3.01	m	3.23
verges; 150 x 6mm fibre cement undercloak; tiles and undercloak bedded and pointed in tinted cement mortar (1:3) and with standard tiles and aluminium nails	1.27	5.00	1.33	0.15	0.08	1.95	2.05	5.33	m	5.73
verges; 150 x 6mm fibre cement undercloak; tiles and undercloak bedded and pointed in tinted cement mortar (1:3) and with verge tiles and aluminium nails	1.27	5.00	1.33	0.15	0.08	1.95	2.05	5.33	m	5.73
ridge or hip tiles, Redland third round; butt jointed; bedding and pointing in tinted cement mortar (1:3)	7.63	5.00	8.01	0.30	0.15	3.82	0.34	12.17	m	13.09
ridge or hip tiles, Redland half round; butt jointed; bedding and pointing in tinted cement mortar (1:3)	7.63	5.00	8.01	0.30	0.15	3.82	0.34	12.17	m	13.09
valley tiles, Redland Universal valley troughs; laid with 100mm laps	21.66	5.00	22.74	0.30	0.15	3.82	0.34	26.91	m	28.92
Roof coverings; Redland 49 granular faced tiles, 381 x 227mm; fixing every fourth course with galvanised nails to 100mm lap; 38 x 22mm pressure impregnated softwood battens fixed with galvanised nails; underlay; B.S.747 type 1F bitumen reinforced felt; 150mm laps; fixing with galvanised steel clout nails										
Pitched 40 degrees from horizontal										
generally	7.56	5.00	7.94	0.59	0.30	7.55	3.48	18.97	m2	20.39
abutments; square	0.76	-	0.76	0.10	0.05	1.27	-	2.03	m	2.19
abutments; raking	1.13	-	1.13	0.15	0.08	1.95	-	3.08	m	3.31
abutments; curved to 3000mm radius	1.51	-	1.51	0.20	0.10	2.55	-	4.06	m	4.36
Roof coverings; Redland 49 through coloured tiles, 381 x 227mm; fixing every fourth course with galvanised nails to 75mm lap; 38 x 22mm pressure impregnated softwood battens fixed with galvanised nails; underlay; B.S.747 type 1F bitumen reinforced felt; 150mm laps; fixing with galvanised steel clout nails										
Pitched 40 degrees from horizontal										
generally	6.92	5.00	7.27	0.54	0.27	6.88	3.39	17.54	m2	18.85
abutments; square	0.69	-	0.69	0.10	0.05	1.27	-	1.96	m	2.11
abutments; raking	1.04	-	1.04	0.15	0.08	1.95	-	2.99	m	3.21
abutments; curved to 3000mm radius	1.38	-	1.38	0.20	0.10	2.55	-	3.93	m	4.22
supplementary fixing eaves course with one aluminium nail per tile	-	-	-	0.02	0.01	0.25	0.03	0.28	m	0.31
verges; Redland Dry Verge system with standard tiles and clips	11.23	5.00	11.79	0.25	0.12	3.15	1.80	16.74	m	18.00
verges; Redland Dry Verge system with verge tiles and clips	11.23	5.00	11.79	0.25	0.12	3.15	1.80	16.74	m	18.00
verges; plain tile undercloak; tiles and undercloak bedded and pointed in tinted cement mortar (1:3) and with standard tiles and aluminium nails	1.27	5.00	1.33	0.10	0.05	1.27	0.40	3.01	m	3.23
verges; plain tile undercloak; tiles and undercloak bedded and pointed in tinted cement mortar (1:3) and with verge tiles and aluminium nails	1.27	5.00	1.33	0.10	0.05	1.27	0.40	3.01	m	3.23
verges; 150 x 6mm fibre cement undercloak; tiles and undercloak bedded and pointed in tinted cement mortar (1:3) and with standard tiles and aluminium nails	1.27	5.00	1.33	0.15	0.08	1.95	2.05	5.33	m	5.73
verges; 150 x 6mm fibre cement undercloak; tiles and undercloak bedded and pointed in tinted cement mortar (1:3) and with verge tiles and aluminium nails	1.27	5.00	1.33	0.15	0.08	1.95	2.05	5.33	m	5.73
ridge or hip tiles, Redland third round; butt jointed; bedding and pointing in tinted cement mortar (1:3)	7.63	5.00	8.01	0.30	0.15	3.82	0.34	12.17	m	13.09
ridge or hip tiles, Redland half round; butt jointed; bedding and pointing in tinted cement mortar (1:3)	7.63	5.00	8.01	0.30	0.15	3.82	0.34	12.17	m	13.09
valley tiles, Redland Universal valley troughs; laid with 100mm laps	21.66	5.00	22.74	0.30	0.15	3.82	0.34	26.91	m	28.92

Labour hourly rates: (except Specialists) Craft Operatives £9.23 Labour £7.02. Rates are national average prices. Refer to REGIONAL VARIATIONS for indicative levels of overall pricing in regions	MATERIALS			LABOUR				RATES		
	Del to Site £	Waste %	Material Cost £	Craft Optve Hrs	Lab Hrs	Labour Cost £	Sunds £	Nett Rate £	Unit	Gross Rate (+7.5%) £

H65: SINGLE LAP ROOF TILING Cont.

Roof coverings; Redland 49 through coloured tiles, 381 x 227mm; fixing every fourth course with galvanised nails to 100mm lap; 38 x 22mm pressure impregnated softwood battens fixed with galvanised nails; underlay; B.S.747 type 1F bitumen reinforced felt; 150mm laps; fixing with galvanised steel clout nails

Pitched 40 degrees from horizontal

	Del to Site	Waste	Material Cost	Craft Optve	Lab	Labour Cost	Sunds	Nett Rate	Unit	Gross Rate
generally	7.56	5.00	7.94	0.59	0.30	7.55	3.48	18.97	m2	20.39

Roof coverings; Redland Renown granular faced tiles, 418 x 330mm; fixing every fourth course with galvanised nails to 75mm lap; 38 x 22mm pressure impregnated softwood battens fixed with galvanised nails; underlay; B.S.747 type 1F bitumen reinforced felt; 150mm laps; fixing with galvanised steel clout nails

Pitched 40 degrees from horizontal

	Del to Site	Waste	Material Cost	Craft Optve	Lab	Labour Cost	Sunds	Nett Rate	Unit	Gross Rate
generally	5.74	5.00	6.03	0.42	0.21	5.35	2.75	14.13	m2	15.19
abutments; square	0.57	–	0.57	0.10	0.05	1.27	–	1.84	m	1.98
abutments; raking	0.86	–	0.86	0.15	0.08	1.95	–	2.81	m	3.02
abutments; curved to 3000mm radius	1.15	–	1.15	0.20	0.10	2.55	–	3.70	m	3.98
reform eaves filler unit and aluminium nails to each eaves tile	0.57	5.00	0.60	0.06	0.03	0.76	0.27	1.63	m	1.76
verges; cloaked verge tiles and aluminium nails	8.85	5.00	9.29	0.15	0.08	1.95	0.06	11.30	m	12.15
verges; half tiles, cloaked verge tiles and aluminium nails	10.03	5.00	10.53	0.20	0.10	2.55	0.10	13.18	m	14.17
verges; plain tile undercloak; tiles and undercloak bedded and pointed in tinted cement mortar (1:3) and with standard tiles and aluminium nails	1.77	5.00	1.86	0.10	0.05	1.27	0.40	3.53	m	3.80
verges; plain tile undercloak; tiles and undercloak bedded and pointed in tinted cement mortar (1:3) and with verge tiles and aluminium nails	1.77	5.00	1.86	0.10	0.05	1.27	0.40	3.53	m	3.80
verges; 150 x 6mm fibre cement undercloak; tiles and undercloak bedded and pointed in tinted cement mortar (1:3) and with standard tiles and aluminium nails	1.77	5.00	1.86	0.15	0.08	1.95	2.05	5.85	m	6.29
verges; 150 x 6mm fibre cement undercloak; tiles and undercloak bedded and pointed in tinted cement mortar (1:3) and with verge tiles and aluminium nails	1.77	5.00	1.86	0.15	0.08	1.95	2.05	5.85	m	6.29
ridge or hip tiles, Redland third round; butt jointed; bedding and pointing in tinted cement mortar (1:3)	7.63	5.00	8.01	0.30	0.15	3.82	0.34	12.17	m	13.09
ridge or hip tiles, Redland half round; butt jointed; bedding and pointing in tinted cement mortar (1:3)	7.63	5.00	8.01	0.30	0.15	3.82	0.34	12.17	m	13.09
valley tiles, Redland Universal valley troughs; laid with 100mm laps	21.66	5.00	22.74	0.30	0.15	3.82	0.34	26.91	m	28.92

Roof coverings; Redland 50 granular faced double roman tiles, 418 x 330mm; fixing every fourth course with galvanised nails to 75mm lap; 38 x 22mm pressure impregnated softwood battens fixed with galvanised nails; underlay; B.S.747 type 1F bitumen reinforced felt; 150mm laps; fixing with galvanised steel clout nails

Pitched 40 degrees from horizontal

	Del to Site	Waste	Material Cost	Craft Optve	Lab	Labour Cost	Sunds	Nett Rate	Unit	Gross Rate
generally	5.74	5.00	6.03	0.42	0.21	5.35	2.75	14.13	m2	15.19
reform eaves filler unit and aluminium nails to each eaves tile	0.57	5.00	0.60	0.06	0.03	0.76	0.24	1.60	m	1.72
verges; cloaked verge tiles and aluminium nails	8.85	5.00	9.29	0.15	0.08	1.95	0.06	11.30	m	12.15
verges; half tiles, cloaked verge tiles and aluminium nails	8.85	5.00	9.29	0.20	0.10	2.55	0.06	11.90	m	12.79
verges; plain tile undercloak; tiles and undercloak bedded and pointed in tinted cement mortar (1:3) and with standard tiles and aluminium nails	1.77	5.00	1.86	0.10	0.05	1.27	0.40	3.53	m	3.80
verges; plain tile undercloak; tiles and undercloak bedded and pointed in tinted cement mortar (1:3) and with treble roll verge tiles and aluminium nails	2.66	5.00	2.79	0.10	0.05	1.27	0.40	4.47	m	4.80
verges; 150 x 6mm fibre cement undercloak; tiles and undercloak bedded and pointed in tinted cement mortar (1:3) and with standard tiles and aluminium nails	1.77	5.00	1.86	0.15	0.08	1.95	2.05	5.85	m	6.29
verges; 150 x 6mm fibre cement undercloak; tiles and undercloak bedded and pointed in tinted cement mortar (1:3) and with treble roll verge tiles and aluminium nails	2.66	5.00	2.79	0.15	0.08	1.95	0.40	5.14	m	5.52
ridge or hip tiles, Redland third round; butt jointed; bedding and pointing in tinted cement mortar (1:3)	7.63	5.00	8.01	0.30	0.15	3.82	0.34	12.17	Um	13.09
ridge or hip tiles, Redland half round; butt jointed; bedding and pointing in tinted cement mortar (1:3)	7.63	5.00	8.01	0.30	0.15	3.82	0.34	12.17	m	13.09
valley tiles, Redland Universal valley troughs; laid with 100mm laps	21.66	5.00	22.74	0.30	0.15	3.82	0.34	26.91	m	28.92

Labour hourly rates: (except Specialists) Craft Operatives £9.23 Labour £7.02 Rates are national average prices. Refer to REGIONAL VARIATIONS for indicative levels of overall pricing in regions	MATERIALS			LABOUR				RATES		
	Del to Site £	Waste %	Material Cost £	Craft Optve Hrs	Lab Hrs	Labour Cost £	Sunds £	Nett Rate £	Unit	Gross Rate (+7.5%) £
H65: SINGLE LAP ROOF TILING Cont.										
Roof coverings; Redland 50 through coloured double roman tiles, 418 x 330mm; fixing every fourth course with galvanised nails to 75mm lap; 38 x 22mm pressure impregnated softwood battens fixed with galvanised nails; underlay; B.S.747 type 1F bitumen reinforced felt; 150mm laps; fixing with galvanised steel clout nails										
Pitched 40 degrees from horizontal										
generally	5.74	5.00	6.03	0.42	0.21	5.35	2.75	14.13	m2	15.19
reform eaves filler unit and aluminium nails to each eaves tile	0.57	5.00	0.60	0.06	0.03	0.76	0.24	1.60	m	1.72
verges; cloaked verge tiles and aluminium nails	8.85	5.00	9.29	0.15	0.08	1.95	0.06	11.30	m	12.15
verges; half tiles, cloaked verge tiles and aluminium nails	8.85	5.00	9.29	0.20	0.10	2.55	0.40	12.24	m	13.16
verges; plain tile undercloak; tiles and undercloak bedded and pointed in tinted cement mortar (1:3) and with standard tiles and aluminium nails	1.77	5.00	1.86	0.10	0.05	1.27	0.06	3.19	m	3.43
verges; plain tile undercloak; tiles and undercloak bedded and pointed in tinted cement mortar (1:3) and with treble roll verge tiles and aluminium nails	2.66	5.00	2.79	0.10	0.05	1.27	0.40	4.47	m	4.80
verges; 150 x 6mm fibre cement undercloak; tiles and undercloak bedded and pointed in tinted cement mortar (1:3) and with standard tiles and aluminium nails	1.77	5.00	1.86	0.15	0.08	1.95	2.05	5.85	m	6.29
verges; 150 x 6mm fibre cement undercloak; tiles and undercloak bedded and pointed in tinted cement mortar (1:3) and with treble roll verge tiles and aluminium nails	2.66	5.00	2.79	0.15	0.08	1.95	0.40	5.14	m	5.52
ridge or hip tiles, Redland third round; butt jointed; bedding and pointing in tinted cement mortar (1:3)	7.63	5.00	8.01	0.30	0.15	3.82	0.34	12.17	m	13.09
ridge or hip tiles, Redland half round; butt jointed; bedding and pointing in tinted cement mortar (1:3)	7.63	5.00	8.01	0.30	0.15	3.82	0.34	12.17	m	13.09
valley tiles, Redland Universal valley troughs; laid with 100mm laps	21.66	5.00	22.74	0.30	0.15	3.82	0.34	26.91	m	28.92
Fittings										
Ventilator tiles										
Red Vent ridge ventilation terminal 450mm long for half round ridge	46.40	5.00	48.72	1.30	0.65	16.56	0.58	65.86	nr	70.80
universal Delta ridge	64.00	5.00	67.20	1.50	0.75	19.11	0.58	86.89	nr	93.41
universal angle ridge	46.30	5.00	48.62	1.30	0.65	16.56	0.58	65.76	nr	70.69
Redvent eaves ventilators fixing with aluminium nails	8.12	5.00	8.53	0.50	0.25	6.37	0.03	14.93	m	16.05
Red Vent Thruvent tiles for										
Stonewold slates	35.30	5.00	37.06	0.50	0.25	6.37	0.24	43.67	nr	46.95
Delta tiles	35.30	5.00	37.06	0.50	0.25	6.37	0.24	43.67	nr	46.95
Regent tiles	34.50	5.00	36.23	0.50	0.25	6.37	0.24	42.84	nr	46.05
Grovebury double pantiles	34.50	5.00	36.23	0.50	0.25	6.37	0.24	42.84	nr	46.05
Norfolk pantiles	34.50	5.00	36.23	0.50	0.25	6.37	0.24	42.84	nr	46.05
Redland 49 tiles	34.50	5.00	36.23	0.50	0.25	6.37	0.24	42.84	nr	46.05
Renown tiles	34.50	5.00	36.23	0.50	0.25	6.37	0.24	42.84	nr	46.05
Redland 50 double roman tiles	-	-	-	0.50	0.25	6.37	0.24	6.61	nr	7.11
Red line ventilation tile complete with underlay seal and fixing clips for										
Stonewold slates	42.80	5.00	44.94	1.00	0.50	12.74	0.24	57.92	nr	62.26
Regent tiles	42.80	5.00	44.94	1.00	0.50	12.74	0.24	57.92	nr	62.26
Grovebury double pantiles	42.80	5.00	44.94	1.00	0.50	12.74	0.24	57.92	nr	62.26
Renown tiles	42.80	5.00	44.94	1.00	0.50	12.74	0.24	57.92	nr	62.26
Redland 50 double roman tiles	42.80	5.00	44.94	1.00	0.50	12.74	0.24	57.92	nr	62.26
Gas terminals										
Gas Flue ridge terminal Mark III, 450mm long with sealing gasket and fixing brackets										
half round ridge	-	-	-	1.50	0.75	19.11	0.34	19.45	nr	20.91
universal Delta ridge	-	-	-	1.70	0.85	21.66	0.34	22.00	nr	23.65
extra for 150mm extension adaptor	-	-	-	0.30	0.15	3.82	0.24	4.06	nr	4.37
Hip irons										
galvanised mild steel; 32 x 3 x 380mm girth; scrolled; fixing with galvanised steel screws to timber	1.79	5.00	1.88	0.10	0.05	1.27	-	3.15	nr	3.39
Lead soakers										
fixing only	-	-	-	0.30	0.15	3.82	-	3.82	nr	4.11

This section continues
on the next page

CLADDING/COVERING

Labour hourly rates: (except Specialists) Craft Operatives £12.44 Labourer £7.02 Rates are national average prices. Refer to REGIONAL VARIATIONS for indicative levels of overall pricing in regions	MATERIALS			LABOUR				RATES		
	Del to Site	Waste	Material Cost	Craft Optve	Lab	Labour Cost	Sunds	Nett Rate	Unit	Gross Rate (+7.5%)
	£	%	£	Hrs	Hrs	£	£	£		£
H71: LEAD SHEET COVERINGS/FLASHINGS										
Sheet lead										
Technical data										
Code 3, 1.32mm thick, 14.97 kg/m2, colour code green										
Code 4, 1.80mm thick, 20.41 kg/m2, colour code blue										
Code 5, 2.24mm thick, 25.40 kg/m2, colour code red										
Code 6, 2.65mm thick, 30.05 kg/m2, colour code black										
Code 7, 3.15mm thick, 35.72 kg/m2, colour code white										
Code 8, 3.55mm thick, 40.26 kg/m2, colour code orange										
Milled lead sheet, B.S.1178										
Nr 3 roof coverings; fixing to timber with milled lead cleats and galvanised screws										
pitch 7.5 degrees from horizontal	10.27	5.00	10.78	3.85	-	47.89	0.20	58.88	m2	63.29
pitch 40 degrees from horizontal	10.27	5.00	10.78	4.40	-	54.74	0.20	65.72	m2	70.65
pitch 75 degrees from horizontal	10.27	5.00	10.78	4.95	-	61.58	0.20	72.56	m2	78.00
Nr 4 roof coverings; fixing to timber with milled lead cleats and galvanised screws										
pitch 7.5 degrees from horizontal	14.00	5.00	14.70	4.40	-	54.74	0.26	69.70	m2	74.92
pitch 40 degrees from horizontal	14.00	5.00	14.70	4.95	-	61.58	0.26	76.54	m2	82.28
pitch 75 degrees from horizontal	14.00	5.00	14.70	5.50	-	68.42	0.26	83.38	m2	89.63
Nr 5 roof coverings; fixing to timber with milled lead cleats and galvanised screws										
pitch 7.5 degrees from horizontal	17.42	5.00	18.29	4.95	-	61.58	0.32	80.19	m2	86.20
pitch 40 degrees from horizontal	17.42	5.00	18.29	5.50	-	68.42	0.32	87.03	m2	93.56
pitch 75 degrees from horizontal	17.42	5.00	18.29	6.05	-	75.26	0.32	93.87	m2	100.91
extra; oil patination to surfaces	1.40	10.00	1.54	0.20	-	2.49	-	4.03	m2	4.33
Nr 3 wall coverings; fixing to timber with milled lead cleats and galvanised screws										
vertical	10.27	5.00	10.78	6.95	-	86.46	0.20	97.44	m2	104.75
Nr 4 wall coverings; fixing to timber with milled lead cleats and galvanised screws										
vertical	14.00	5.00	14.70	7.50	-	93.30	0.26	108.26	m2	116.38
Nr 5 wall coverings; fixing to timber with milled lead cleats and galvanised screws										
vertical	17.42	5.00	18.29	8.05	-	100.14	0.32	118.75	m2	127.66
Flashings; horizontal										
Nr 3; 150mm lapped joints; fixing to masonry with milled lead clips and lead wedges										
150mm girth	1.54	5.00	1.62	0.40	-	4.98	0.20	6.79	m	7.30
240mm girth	2.46	5.00	2.58	0.64	-	7.96	0.20	10.74	m	11.55
300mm girth	3.08	5.00	3.23	0.80	-	9.95	0.20	13.39	m	14.39
Nr 4; 150mm lapped joints; fixing to masonry with milled lead clips and lead wedges										
150mm girth	2.10	5.00	2.21	0.44	-	5.47	0.26	7.94	m	8.53
240mm girth	3.36	5.00	3.53	0.71	-	8.83	0.26	12.62	m	13.57
300mm girth	4.20	5.00	4.41	0.89	-	11.07	0.26	15.74	m	16.92
Nr 5; 150mm lapped joints; fixing to masonry with milled lead clips and lead wedges										
150mm girth	2.61	5.00	2.74	0.50	-	6.22	0.32	9.28	m	9.98
240mm girth	4.18	5.00	4.39	0.80	-	9.95	0.32	14.66	m	15.76
300mm girth	5.23	5.00	5.49	1.00	-	12.44	0.32	18.25	m	19.62
Nr 3; 150mm lapped joints; fixing to timber with copper nails										
150mm girth	1.54	5.00	1.62	0.40	-	4.98	0.25	6.84	m	7.36
240mm girth	2.46	5.00	2.58	0.64	-	7.96	0.25	10.79	m	11.60
300mm girth	3.08	5.00	3.23	0.80	-	9.95	0.25	13.44	m	14.44
Nr 4; 150mm lapped joints; fixing to timber with copper nails										
150mm girth	2.10	5.00	2.21	0.44	-	5.47	0.25	7.93	m	8.52
240mm girth	3.36	5.00	3.53	0.71	-	8.83	0.25	12.61	m	13.56
300mm girth	4.20	5.00	4.41	0.89	-	11.07	0.25	15.73	m	16.91
Nr 5; 150mm lapped joints; fixing to timber with copper nails										
150mm girth	2.61	5.00	2.74	0.50	-	6.22	0.25	9.21	m	9.90
240mm girth	4.18	5.00	4.39	0.80	-	9.95	0.25	14.59	m	15.69
300mm girth	5.23	5.00	5.49	1.00	-	12.44	0.25	18.18	m	19.55
Flashings; stepped										
Nr 3; 150mm lapped joints; fixing to masonry with milled lead clips and lead wedges										
180mm girth	1.85	5.00	1.94	0.64	-	7.96	0.24	10.14	m	10.90
240mm girth	2.46	5.00	2.58	0.86	-	10.70	0.24	13.52	m	14.54
300mm girth	3.08	5.00	3.23	1.08	-	13.44	0.24	16.91	m	18.18

Labour hourly rates: (except Specialists) Craft Operatives £12.44 Labourer £7.02 Rates are national average prices. Refer to REGIONAL VARIATIONS for indicative levels of overall pricing in regions	MATERIALS			LABOUR				RATES		
	Del to Site £	Waste %	Material Cost £	Craft Optve Hrs	Lab Hrs	Labour Cost £	Sunds £	Nett Rate £	Unit	Gross Rate (+7.5%) £

H71: LEAD SHEET COVERINGS/FLASHINGS Cont.

Flashings; stepped Cont.

Nr 4; 150mm lapped joints; fixing to masonry with
milled lead clips and lead wedges

180mm girth	2.52	5.00	2.65	0.71	–	8.83	0.40	11.88	m	12.77
240mm girth	3.36	5.00	3.53	0.95	–	11.82	0.40	15.75	m	16.93
300mm girth	4.20	5.00	4.41	1.19	–	14.80	0.40	19.61	m	21.08

Nr 5; 150mm lapped joints; fixing to masonry with
milled lead clips and lead wedges

180mm girth	3.14	5.00	3.30	0.80	–	9.95	0.48	13.73	m	14.76
240mm girth	4.18	5.00	4.39	1.07	–	13.31	0.48	18.18	m	19.54
300mm girth	5.23	5.00	5.49	1.33	–	16.55	0.48	22.52	m	24.21

Aprons; horizontal

Nr 3; 150mm lapped joints; fixing to masonry with
milled lead clips and lead wedges

150mm girth	1.54	5.00	1.62	0.40	–	4.98	0.20	6.79	m	7.30
240mm girth	2.46	5.00	2.58	0.64	–	7.96	0.20	10.74	m	11.55
300mm girth	3.08	5.00	3.23	0.80	–	9.95	0.20	13.39	m	14.39
450mm girth	4.62	5.00	4.85	1.20	–	14.93	0.30	20.08	m	21.58

Nr 4; 150mm lapped joints; fixing to masonry with
milled lead clips and lead wedges

150mm girth	2.10	5.00	2.21	0.44	–	5.47	0.26	7.94	m	8.53
240mm girth	3.36	5.00	3.53	0.71	–	8.83	0.26	12.62	m	13.57
300mm girth	4.20	5.00	4.41	0.89	–	11.07	0.26	15.74	m	16.92
450mm girth	6.30	5.00	6.62	1.33	–	16.55	0.40	23.56	m	25.33

Nr 5; 150mm lapped joints; fixing to masonry with
milled lead clips and lead wedges

150mm girth	2.61	5.00	2.74	0.50	–	6.22	0.32	9.28	m	9.98
240mm girth	4.18	5.00	4.39	0.80	–	9.95	0.32	14.66	m	15.76
300mm girth	5.23	5.00	5.49	1.00	–	12.44	0.32	18.25	m	19.62
450mm girth	7.84	5.00	8.23	1.50	–	18.66	0.48	27.37	m	29.42

Hips; sloping; dressing over slating and tiling

Nr 3; 150mm lapped joints; fixing to timber with
milled lead clips

240mm girth	2.46	5.00	2.58	0.64	–	7.96	0.20	10.74	m	11.55
300mm girth	3.08	5.00	3.23	0.80	–	9.95	0.20	13.39	m	14.39
450mm girth	4.62	5.00	4.85	1.20	–	14.93	0.30	20.08	m	21.58

Nr 4; 150mm lapped joints; fixing to timber with
milled lead clips

240mm girth	3.36	5.00	3.53	0.71	–	8.83	0.26	12.62	m	13.57
300mm girth	4.20	5.00	4.41	0.89	–	11.07	0.26	15.74	m	16.92
450mm girth	6.30	5.00	6.62	1.33	–	16.55	0.40	23.56	m	25.33

Nr 5; 150mm lapped joints; fixing to timber with
milled lead clips

240mm girth	4.18	5.00	4.39	0.80	–	9.95	0.32	14.66	m	15.76
300mm girth	5.23	5.00	5.49	1.00	–	12.44	0.32	18.25	m	19.62
450mm girth	7.84	5.00	8.23	1.50	–	18.66	0.48	27.37	m	29.42

Kerbs; horizontal

Nr 3; 150mm lapped joints; fixing to timber with
copper nails

240mm girth	2.46	5.00	2.58	0.64	–	7.96	0.20	10.74	m	11.55
300mm girth	3.08	5.00	3.23	0.80	–	9.95	0.20	13.39	m	14.39
450mm girth	4.62	5.00	4.85	1.20	–	14.93	0.30	20.08	m	21.58

Nr 4; 150mm lapped joints; fixing to timber with
copper nails

240mm girth	3.36	5.00	3.53	0.71	–	8.83	0.26	12.62	m	13.57
300mm girth	4.20	5.00	4.41	0.89	–	11.07	0.26	15.74	m	16.92
450mm girth	6.30	5.00	6.62	1.33	–	16.55	0.40	23.56	m	25.33

Nr 5; 150mm lapped joints; fixing to timber with
copper nails

240mm girth	4.18	5.00	4.39	0.80	–	9.95	0.32	14.66	m	15.76
300mm girth	5.23	5.00	5.49	1.00	–	12.44	0.32	18.25	m	19.62
450mm girth	7.84	5.00	8.23	1.50	–	18.66	0.48	27.37	m	29.42

Ridges; horizontal; dressing over slating and tiling

Nr 3; 150mm lapped joints; fixing to timber with
milled lead clips

240mm girth	2.46	5.00	2.58	0.64	–	7.96	0.20	10.74	m	11.55
300mm girth	3.08	5.00	3.23	0.80	–	9.95	0.20	13.39	m	14.39
450mm girth	4.62	5.00	4.85	1.20	–	14.93	0.30	20.08	m	21.58

Nr 4; 150mm lapped joints; fixing to timber with
milled lead clips

240mm girth	3.36	5.00	3.53	0.71	–	8.83	0.26	12.62	m	13.57
300mm girth	4.20	5.00	4.41	0.89	–	11.07	0.26	15.74	m	16.92
450mm girth	6.30	5.00	6.62	1.33	–	16.55	0.40	23.56	m	25.33

Nr 5; 150mm lapped joints; fixing to timber with
milled lead clips

240mm girth	4.18	5.00	4.39	0.80	–	9.95	0.32	14.66	m	15.76

Labour hourly rates: (except Specialists) Craft Operatives £12.44 Labourer £7.02 Rates are national average prices. Refer to REGIONAL VARIATIONS for indicative levels of overall pricing in regions	MATERIALS			LABOUR				RATES		
	Del to Site £	Waste %	Material Cost £	Craft Optve Hrs	Lab Hrs	Labour Cost £	Sunds £	Nett Rate £	Unit	Gross Rate (+7.5%) £
H71: LEAD SHEET COVERINGS/FLASHINGS Cont.										
Ridges; horizontal; dressing over slating and tiling Cont.										
Nr 5; 150mm lapped joints; fixing to timber with milled lead clips Cont.										
300mm girth	5.23	5.00	5.49	1.00	–	12.44	0.32	18.25	m	19.62
450mm girth	7.84	5.00	8.23	1.50	–	18.66	0.48	27.37	m	29.42
Valleys; sloping										
Nr 3; dressing over tilting fillets -1; 150mm lapped joints; fixing to timber with copper nails										
240mm girth	2.46	5.00	2.58	0.64	–	7.96	0.20	10.74	m	11.55
300mm girth	3.08	5.00	3.23	0.80	–	9.95	0.20	13.39	m	14.39
450mm girth	4.62	5.00	4.85	1.20	–	14.93	0.30	20.08	m	21.58
Nr 4; dressing over tilting fillets -1; 150mm lapped joints; fixing to timber with copper nails										
240mm girth	3.36	5.00	3.53	0.71	–	8.83	0.26	12.62	m	13.57
300mm girth	4.20	5.00	4.41	0.89	–	11.07	0.26	15.74	m	16.92
450mm girth	6.30	5.00	6.62	1.33	–	16.55	0.40	23.56	m	25.33
Nr 5; dressing over tilting fillets -1; 150mm lapped joints; fixing to timber with copper nails										
240mm girth	4.18	5.00	4.39	0.80	–	9.95	0.32	14.66	m	15.76
300mm girth	5.23	5.00	5.49	1.00	–	12.44	0.32	18.25	m	19.62
450mm girth	7.84	5.00	8.23	1.50	–	18.66	0.48	27.37	m	29.42
Cavity gutters										
Nr 3; 150mm lapped joints; bedding in cement mortar (1:3)										
225mm girth	2.31	5.00	2.43	0.64	–	7.96	0.40	10.79	m	11.60
345mm girth	3.34	5.00	3.51	0.96	–	11.94	0.60	16.05	m	17.25
Nr 3 edges										
welted	–	–	–	0.17	–	2.11	–	2.11	m	2.27
beaded	–	–	–	0.17	–	2.11	–	2.11	m	2.27
Nr 4 edges										
welted	–	–	–	0.17	–	2.11	–	2.11	m	2.27
beaded	–	–	–	0.17	–	2.11	–	2.11	m	2.27
Nr 5 edges										
welted	–	–	–	0.17	–	2.11	–	2.11	m	2.27
beaded	–	–	–	0.17	–	2.11	–	2.11	m	2.27
Nr 3 seams										
leadburned	0.55	–	0.55	0.20	–	2.49	–	3.04	m	3.27
Nr 4 seams										
leadburned	0.70	–	0.70	0.20	–	2.49	–	3.19	m	3.43
Nr 5 seams										
leadburned	0.86	–	0.86	0.20	–	2.49	–	3.35	m	3.60
Nr 3 dressings										
corrugated roofing; fibre cement; down corrugations	–	–	–	0.25	–	3.11	–	3.11	m	3.34
corrugated roofing; fibre cement; across corrugations	–	–	–	0.25	–	3.11	–	3.11	m	3.34
glass and glazing bars; timber	–	–	–	0.25	–	3.11	–	3.11	m	3.34
Nr 4 dressings										
corrugated roofing; fibre cement; down corrugations	–	–	–	0.25	–	3.11	–	3.11	m	3.34
corrugated roofing; fibre cement; across corrugations	–	–	–	0.25	–	3.11	–	3.11	m	3.34
glass and glazing bars; timber	–	–	–	0.25	–	3.11	–	3.11	m	3.34
Nr 5 dressings										
corrugated roofing; fibre cement; down corrugations	–	–	–	0.25	–	3.11	–	3.11	m	3.34
corrugated roofing; fibre cement; across corrugations	–	–	–	0.25	–	3.11	–	3.11	m	3.34
glass and glazing bars; timber	–	–	–	0.25	–	3.11	–	3.11	m	3.34
Nr 3 soakers and slates handed to others for fixing										
180 x 180mm	0.33	5.00	0.35	0.20	–	2.49	–	2.83	nr	3.05
180 x 300mm	0.56	5.00	0.59	0.20	–	2.49	–	3.08	nr	3.31
450 x 450mm	2.08	5.00	2.18	0.20	–	2.49	–	4.67	nr	5.02
Nr 4 soakers and slates handed to others for fixing										
180 x 180mm	0.45	5.00	0.47	0.20	–	2.49	–	2.96	nr	3.18
180 x 300mm	0.76	5.00	0.80	0.20	–	2.49	–	3.29	nr	3.53
450 x 450mm	2.84	5.00	2.98	0.20	–	2.49	–	5.47	nr	5.88
Nr 5 soakers and slates handed to others for fixing										
180 x 180mm	0.57	5.00	0.60	0.20	–	2.49	–	3.09	nr	3.32
180 x 300mm	0.94	5.00	0.99	0.20	–	2.49	–	3.48	nr	3.74
450 x 450mm	3.53	5.00	3.71	0.20	–	2.49	–	6.19	nr	6.66

Labour hourly rates: (except Specialists) Craft Operatives £12.44 Labourer £7.02 Rates are national average prices. Refer to REGIONAL VARIATIONS for indicative levels of overall pricing in regions	MATERIALS			LABOUR				RATES		
	Del to Site £	Waste %	Material Cost £	Craft Optve Hrs	Lab Hrs	Labour Cost £	Sunds £	Nett Rate £	Unit	Gross Rate (+7.5%) £
H71: LEAD SHEET COVERINGS/FLASHINGS Cont.										
Cavity gutters Cont.										
Nr 3 collars around pipes, standards and the like 150mm long; soldered joints to metal covering										
50mm diameter..................................	2.65	5.00	**2.78**	1.00	–	12.44	–	15.22	nr	16.36
100mm diameter.................................	2.98	5.00	**3.13**	1.00	–	12.44	–	15.57	nr	16.74
Nr 4 collars around pipes, standards and the like 150mm long; soldered joints to metal covering										
50mm diameter..................................	3.42	5.00	**3.59**	1.00	–	12.44	–	16.03	nr	17.23
100mm diameter.................................	3.82	5.00	**4.01**	1.00	–	12.44	–	16.45	nr	17.68
Nr 5 collars around pipes, standards and the like 150mm long; soldered joints to metal covering										
50mm diameter..................................	4.27	5.00	**4.48**	1.00	–	12.44	–	16.92	nr	18.19
100mm diameter.................................	4.76	5.00	**5.00**	1.00	–	12.44	–	17.44	nr	18.75
Nr 3 dots										
cast lead	0.46	5.00	**0.48**	0.67	–	8.33	–	8.82	nr	9.48
soldered	5.40	5.00	**5.67**	0.75	–	9.33	–	15.00	nr	16.13
Nr 4 dots										
cast lead	0.60	5.00	**0.63**	0.67	–	8.33	–	8.96	nr	9.64
soldered	5.40	5.00	**5.67**	0.75	–	9.33	–	15.00	nr	16.13
Nr 5 dots										
cast lead	0.72	5.00	**0.76**	0.67	–	8.33	–	9.09	nr	9.77
soldered	5.40	5.00	**5.67**	0.75	–	9.33	–	15.00	nr	16.13
H72: ALUMINIUM SHEET STRIP COVERINGS/FLASHINGS										
Aluminium sheet, B.S.1470 grade S1BO, commercial purity										
0.60mm thick roof coverings; fixing to timber with aluminium cleats and aluminium alloy screws										
pitch 7.5 degrees from horizontal	8.06	5.00	**8.46**	4.40	–	54.74	0.12	63.32	m2	68.07
pitch 40 degrees from horizontal	8.06	5.00	**8.46**	4.95	–	61.58	0.12	70.16	m2	75.42
pitch 75 degrees from horizontal	8.06	5.00	**8.46**	5.50	–	68.42	0.12	77.00	m2	82.78
Flashings; horizontal 0.60mm thick; 150mm lapped joints; fixing to masonry with aluminium clips and wedges										
150mm girth...................................	1.46	5.00	**1.53**	0.44	–	5.47	0.12	7.13	m	7.66
240mm girth...................................	2.73	5.00	**2.87**	0.71	–	8.83	0.12	11.82	m	12.71
300mm girth...................................	2.73	5.00	**2.87**	0.89	–	11.07	0.12	14.06	m	15.11
Flashings; stepped 0.60mm thick; 150mm lapped joints; fixing to masonry with aluminium clips and wedges										
180mm girth...................................	2.01	5.00	**2.11**	0.71	–	8.83	0.18	11.12	m	11.95
240mm girth...................................	2.73	5.00	**2.87**	0.95	–	11.82	0.18	14.86	m	15.98
300mm girth...................................	2.73	5.00	**2.87**	1.19	–	14.80	0.18	17.85	m	19.19
Aprons; horizontal 0.60mm thick; 150mm lapped joints; fixing to masonry with aluminium clips and wedges										
150mm girth...................................	1.46	5.00	**1.53**	0.44	–	5.47	0.12	7.13	m	7.66
180mm girth...................................	2.01	5.00	**2.11**	0.71	–	8.83	0.12	11.06	m	11.89
240mm girth...................................	2.73	5.00	**2.87**	0.89	–	11.07	0.12	14.06	m	15.11
450mm girth...................................	3.80	5.00	**3.99**	1.33	–	16.55	0.18	20.72	m	22.27
Hips; sloping 0.60mm thick; 150mm lapped joints; fixing to timber with aluminium clips										
240mm girth...................................	2.73	5.00	**2.87**	0.71	–	8.83	0.12	11.82	m	12.71
300mm girth...................................	2.73	5.00	**2.87**	0.89	–	11.07	0.12	14.06	m	15.11
450mm girth...................................	3.80	5.00	**3.99**	1.33	–	16.55	0.18	20.72	m	22.27
Kerbs; horizontal 0.60mm thick; 150mm lapped joints; fixing to timber with aluminium clips										
240mm girth...................................	2.73	5.00	**2.87**	0.71	–	8.83	0.12	11.82	m	12.71
300mm girth...................................	2.73	5.00	**2.87**	0.89	–	11.07	0.12	14.06	m	15.11
450mm girth...................................	3.80	5.00	**3.99**	1.33	–	16.55	0.18	20.72	m	22.27
Ridges; horizontal 0.60mm thick; 150mm lapped joints; fixing to timber with aluminium clips										
240mm girth...................................	2.73	5.00	**2.87**	0.71	–	8.83	0.12	11.82	m	12.71
300mm girth...................................	2.73	5.00	**2.87**	0.89	–	11.07	0.12	14.06	m	15.11
450mm girth...................................	3.80	5.00	**3.99**	1.33	–	16.55	0.18	20.72	m	22.27
Valleys; sloping 0.60mm thick; 150mm lapped joints; fixing to timber with aluminium clips										
240mm girth...................................	2.73	5.00	**2.87**	0.71	–	8.83	0.12	11.82	m	12.71
300mm girth...................................	2.73	5.00	**2.87**	0.89	–	11.07	0.12	14.06	m	15.11
450mm girth...................................	3.80	5.00	**3.99**	1.33	–	16.55	0.18	20.72	m	22.27
0.60mm thick edges										
welted	–	–	**–**	0.17	–	2.11	–	2.11	m	2.27
beaded	–	–	**–**	0.17	–	2.11	–	2.11	m	2.27

CLADDING/COVERING

Labour hourly rates: (except Specialists) Craft Operatives £12.44 Labourer £7.02. Rates are national average prices. Refer to REGIONAL VARIATIONS for indicative levels of overall pricing in regions	MATERIALS			LABOUR				RATES		
	Del to Site £	Waste %	Material Cost £	Craft Optve Hrs	Lab Hrs	Labour Cost £	Sunds £	Nett Rate £	Unit	Gross Rate (+7.5%) £
H72: ALUMINIUM SHEET STRIP COVERINGS/FLASHINGS Cont.										
Aluminium sheet, B.S.1470 grade S1BO, commercial purity Cont.										
0.60mm thick soakers and slates handed to others for fixing										
180 x 180mm	0.36	5.00	0.38	0.20	–	2.49	–	2.87	nr	3.08
180 x 300mm	0.60	5.00	0.63	0.20	–	2.49	–	3.12	nr	3.35
H73: COPPER STRIP SHEET COVERINGS/FLASHINGS										
Copper sheet; B.S.2870										
0.60mm thick roof coverings; fixing to timber with copper cleats and copper nails										
pitch 7.5 degrees from horizontal	16.60	5.00	17.43	4.40	–	54.74	0.36	72.53	m2	77.97
pitch 40 degrees from horizontal	16.60	5.00	17.43	4.95	–	61.58	0.36	79.37	m2	85.32
pitch 75 degrees from horizontal	16.60	5.00	17.43	5.50	–	68.42	0.36	86.21	m2	92.68
0.70mm thick roof coverings; fixing to timber with copper cleats and copper nails										
pitch 7.5 degrees from horizontal	20.84	5.00	21.88	4.95	–	61.58	0.45	83.91	m2	90.20
pitch 40 degrees from horizontal	20.84	5.00	21.88	5.50	–	68.42	0.45	90.75	m2	97.56
pitch 75 degrees from horizontal	20.84	5.00	21.88	6.05	–	75.26	0.45	97.59	m2	104.91
Flashings; horizontal 0.60mm thick; 150mm lapped joints; fixing to masonry with copper clips and wedges										
150mm girth	2.49	5.00	2.61	0.44	–	5.47	0.36	8.45	m	9.08
240mm girth	3.99	5.00	4.19	0.71	–	8.83	0.36	13.38	m	14.39
300mm girth	4.98	5.00	5.23	0.89	–	11.07	0.36	16.66	m	17.91
0.70mm thick; 150mm lapped joints; fixing to masonry with copper clips and wedges										
150mm girth	3.13	5.00	3.29	0.50	–	6.22	0.45	9.96	m	10.70
240mm girth	5.00	5.00	5.25	0.80	–	9.95	0.45	15.65	m	16.83
300mm girth	6.26	5.00	6.57	1.00	–	12.44	0.45	19.46	m	20.92
Flashings; stepped 0.60mm thick; 150mm lapped joints; fixing to masonry with copper clips and wedges										
180mm girth	2.99	5.00	3.14	0.71	–	8.83	0.54	12.51	m	13.45
240mm girth	3.99	5.00	4.19	0.95	–	11.82	0.54	16.55	m	17.79
300mm girth	4.98	5.00	5.23	1.19	–	14.80	0.54	20.57	m	22.12
0.70mm thick; 150mm lapped joints; fixing to masonry with copper clips and wedges										
180mm girth	3.75	5.00	3.94	0.80	–	9.95	0.68	14.57	m	15.66
240mm girth	5.00	5.00	5.25	1.07	–	13.31	0.68	19.24	m	20.68
300mm girth	6.26	5.00	6.57	1.33	–	16.55	0.68	23.80	m	25.58
Aprons; horizontal 0.60mm thick; 150mm lapped joints; fixing to timber with copper clips										
150mm girth	2.49	5.00	2.61	0.44	–	5.47	0.36	8.45	m	9.08
240mm girth	3.99	5.00	4.19	0.71	–	8.83	0.36	13.38	m	14.39
300mm girth	4.98	5.00	5.23	0.89	–	11.07	0.36	16.66	m	17.91
450mm girth	7.47	5.00	7.84	1.33	–	16.55	0.54	24.93	m	26.80
0.70mm thick; 150mm lapped joints; fixing to timber with copper clips										
150mm girth	3.13	5.00	3.29	0.50	–	6.22	0.45	9.96	m	10.70
240mm girth	5.00	5.00	5.25	0.80	–	9.95	0.45	15.65	m	16.83
300mm girth	6.25	5.00	6.56	1.00	–	12.44	0.45	19.45	m	20.91
450mm girth	9.38	5.00	9.85	1.50	–	18.66	0.68	29.19	m	31.38
Hips; sloping 0.60mm thick; 150mm lapped joints; fixing to timber with copper clips										
240mm girth	3.99	5.00	4.19	0.71	–	8.83	0.36	13.38	m	14.39
300mm girth	4.98	5.00	5.23	0.89	–	11.07	0.36	16.66	m	17.91
450mm girth	7.47	5.00	7.84	1.33	–	16.55	0.54	24.93	m	26.80
0.70mm thick; 150mm lapped joints; fixing to timber with copper clips										
240mm girth	5.00	5.00	5.25	0.80	–	9.95	0.45	15.65	m	16.83
300mm girth	6.27	5.00	6.58	1.00	–	12.44	0.45	19.47	m	20.93
450mm girth	9.38	5.00	9.85	1.50	–	18.66	0.68	29.19	m	31.38
Kerbs; horizontal 0.60mm thick; 150mm lapped joints; fixing to timber with copper clips										
240mm girth	3.99	5.00	4.19	0.71	–	8.83	0.36	13.38	m	14.39
300mm girth	4.98	5.00	5.23	0.89	–	11.07	0.36	16.66	m	17.91
450mm girth	7.47	5.00	7.84	1.33	–	16.55	0.54	24.93	m	26.80
0.70mm thick; 150mm lapped joints; fixing to timber with copper clips										
240mm girth	5.00	5.00	5.25	0.80	–	9.95	0.45	15.65	m	16.83
300mm girth	6.25	5.00	6.56	1.00	–	12.44	0.45	19.45	m	20.91
450mm girth	9.38	5.00	9.85	1.50	–	18.66	0.68	29.19	m	31.38
Ridges; horizontal 0.60mm thick; 150mm lapped joints; fixing to timber with copper clips										
240mm girth	3.99	5.00	4.19	0.71	–	8.83	0.36	13.38	m	14.39
300mm girth	4.98	5.00	5.23	0.89	–	11.07	0.36	16.66	m	17.91
450mm girth	7.47	5.00	7.84	1.33	–	16.55	0.54	24.93	m	26.80

Labour hourly rates: (except Specialists) Craft Operatives £12.44 Labourer £7.02 Rates are national average prices. Refer to REGIONAL VARIATIONS for indicative levels of overall pricing in regions	MATERIALS			LABOUR				RATES		
	Del to Site £	Waste %	Material Cost £	Craft Optve Hrs	Lab Hrs	Labour Cost £	Sunds £	Nett Rate £	Unit	Gross Rate (+7.5%) £
H73: COPPER STRIP SHEET COVERINGS/FLASHINGS Cont.										
Copper sheet; B.S.2870 Cont.										
Ridges; horizontal Cont. 0.70mm thick; 150mm lapped joints; fixing to timber with copper clips										
240mm girth	5.00	5.00	**5.25**	0.80	-	**9.95**	0.45	**15.65**	m	16.83
300mm girth	6.26	5.00	**6.57**	1.00	-	**12.44**	0.45	**19.46**	m	20.92
450mm girth	9.38	5.00	**9.85**	1.50	-	**18.66**	0.68	**29.19**	m	31.38
Valleys; sloping 0.60mm thick; 150mm lapped joints; fixing to timber with copper clips										
240mm girth	3.99	5.00	**4.19**	0.71	-	**8.83**	0.36	**13.38**	m	14.39
300mm girth	4.98	5.00	**5.23**	0.89	-	**11.07**	0.36	**16.66**	m	17.91
450mm girth	7.47	5.00	**7.84**	1.33	-	**16.55**	0.54	**24.93**	m	26.80
0.70mm thick; 150mm lapped joints; fixing to timber with copper clips										
240mm girth	5.00	5.00	**5.25**	0.80	-	**9.95**	0.45	**15.65**	m	16.83
300mm girth	6.26	5.00	**6.57**	1.07	-	**13.31**	0.45	**20.33**	m	21.86
450mm girth	9.38	5.00	**9.85**	1.33	-	**16.55**	0.68	**27.07**	m	29.10
0.60mm thick edges										
welted	-	-	**-**	0.17	-	**2.11**	-	**2.11**	m	2.27
beaded	-	-	**-**	0.17	-	**2.11**	-	**2.11**	m	2.27
0.70mm thick edges										
welted	-	-	**-**	0.17	-	**2.11**	-	**2.11**	m	2.27
beaded	-	-	**-**	0.17	-	**2.11**	-	**2.11**	m	2.27
H74: ZINC STRIP SHEET COVERINGS/FLASHINGS										
Zinc alloy sheet; B.S.6561										
0.65mm thick roof coverings; 25mm standing seam; fixing to timber with zinc clips										
pitch 7.5 degrees from horizontal	11.67	5.00	**12.25**	4.40	-	**54.74**	0.13	**67.12**	m2	72.15
pitch 40 degrees from horizontal	11.67	5.00	**12.25**	4.95	-	**61.58**	0.13	**73.96**	m2	79.51
pitch 75 degrees from horizontal	11.67	5.00	**12.25**	5.50	-	**68.42**	0.13	**80.80**	m2	86.86
0.8mm thick roof coverings; roll cap; fixing to timber with zinc clips										
pitch 7.5 degrees from horizontal	14.33	5.00	**15.05**	4.95	-	**61.58**	0.18	**76.80**	m2	82.56
pitch 40 degrees from horizontal	14.33	5.00	**15.05**	5.50	-	**68.42**	0.18	**83.65**	m2	89.92
pitch 75 degrees from horizontal	14.33	5.00	**15.05**	6.05	-	**75.26**	0.20	**90.51**	m2	97.30
Flashings; horizontal 0.8mm thick; 150mm lapped joints; fixing to masonry with zinc clips and wedges										
150mm girth	2.15	5.00	**2.26**	0.44	-	**5.47**	0.13	**7.86**	m	8.45
240mm girth	3.44	5.00	**3.61**	0.71	-	**8.83**	0.13	**12.57**	m	13.52
300mm girth	4.30	5.00	**4.51**	0.89	-	**11.07**	0.13	**15.72**	m	16.90
Flashings; stepped 0.8mm thick; 150mm lapped joints; fixing to masonry with zinc clips and wedges										
180mm girth	2.58	5.00	**2.71**	0.58	-	**7.22**	0.20	**10.12**	m	10.88
240mm girth	3.44	5.00	**3.61**	0.95	-	**11.82**	0.20	**15.63**	m	16.80
300mm girth	4.30	5.00	**4.51**	1.19	-	**14.80**	0.20	**19.52**	m	20.98
Aprons; horizontal 0.8mm thick; 150mm lapped joints; fixing to masonry with zinc clips and wedges										
150mm girth	2.15	5.00	**2.26**	0.44	-	**5.47**	0.13	**7.86**	m	8.45
240mm girth	3.44	5.00	**3.61**	0.71	-	**8.83**	0.13	**12.57**	m	13.52
300mm girth	4.30	5.00	**4.51**	0.89	-	**11.07**	0.13	**15.72**	m	16.90
450mm girth	6.45	5.00	**6.77**	1.33	-	**16.55**	0.20	**23.52**	m	25.28
Hips; sloping 0.8mm thick; 150mm lapped joints; fixing to timber with zinc clips										
240mm girth	3.44	5.00	**3.61**	0.71	-	**8.83**	0.13	**12.57**	m	13.52
300mm girth	4.30	5.00	**4.51**	0.89	-	**11.07**	0.13	**15.72**	m	16.90
450mm girth	6.45	5.00	**6.77**	1.33	-	**16.55**	0.20	**23.52**	m	25.28
Kerbs; horizontal 0.8mm thick; 150mm lapped joints; fixing to timber with zinc clips										
240mm girth	3.44	5.00	**3.61**	0.71	-	**8.83**	0.13	**12.57**	m	13.52
300mm girth	4.30	5.00	**4.51**	0.89	-	**11.07**	0.13	**15.72**	m	16.90
450mm girth	6.45	5.00	**6.77**	1.33	-	**16.55**	0.20	**23.52**	m	25.28
Ridges; horizontal 0.8mm thick; 150mm lapped joints; fixing to timber with zinc clips										
240mm girth	3.44	5.00	**3.61**	0.71	-	**8.83**	0.13	**12.57**	m	13.52
300mm girth	4.30	5.00	**4.51**	0.89	-	**11.07**	0.13	**15.72**	m	16.90
450mm girth	6.45	5.00	**6.77**	1.33	-	**16.55**	0.20	**23.52**	m	25.28
Valleys; sloping 0.8mm thick; 150mm lapped joints; fixing to timber with zinc clips										
240mm girth	3.44	5.00	**3.61**	0.71	-	**8.83**	0.13	**12.57**	m	13.52
300mm girth	4.30	5.00	**4.51**	0.89	-	**11.07**	0.13	**15.72**	m	16.90
450mm girth	6.45	5.00	**6.77**	1.33	-	**16.55**	0.20	**23.52**	m	25.28

Labour hourly rates: (except Specialists) Craft Operatives £12.44 Labourer £7.02 Rates are national average prices. Refer to REGIONAL VARIATIONS for indicative levels of overall pricing in regions	MATERIALS			LABOUR				RATES		
	Del·to Site £	Waste %	Material Cost £	Craft Optve Hrs	Lab Hrs	Labour Cost £	Sunds £	Nett Rate £	Unit	Gross Rate (+7.5%) £
H74: ZINC STRIP SHEET COVERINGS/FLASHINGS Cont.										
Zinc alloy sheet; B.S.6561 Cont.										
0.65mm thick edges										
welted.............................	-	-	-	0.17	-	2.11	-	2.11	m	2.27
beaded.............................	-	-	-	0.17	-	2.11	-	2.11	m	2.27
0.8mm thick edges										
welted.............................	-	-	-	0.17	-	2.11	-	2.11	m	2.27
beaded.............................	-	-	-	0.17	-	2.11	-	2.11	m	2.27
0.65mm thick soakers and slates handed to others for fixing										
180 x 180mm......................	0.38	5.00	0.40	0.20	-	2.49	-	2.89	nr	3.10
180 x 300mm......................	0.63	5.00	0.66	0.20	-	2.49	-	3.15	nr	3.39
450 x 450mm......................	2.35	5.00	2.47	0.20	-	2.49	-	4.96	nr	5.33
0.8mm thick soakers and slates - handed to others for fixing										
180 x 180mm......................	0.46	5.00	0.48	0.20	-	2.49	-	2.97	nr	3.19
180 x 300mm......................	0.77	5.00	0.81	0.20	-	2.49	-	3.30	nr	3.54
450 x 450mm......................	2.90	5.00	3.04	0.20	-	2.49	-	5.53	nr	5.95
0.65mm thick collars around pipes, standards and the like										
150mm long; soldered joints to metal covering										
50mm diameter....................	2.34	5.00	2.46	1.00	-	12.44	-	14.90	nr	16.01
100mm diameter...................	2.64	5.00	2.77	1.00	-	12.44	-	15.21	nr	16.35
0.8mm thick collars around pipes, standards and the like										
150mm long; soldered joints to metal covering										
50mm diameter....................	2.89	5.00	3.03	1.00	-	12.44	-	15.47	nr	16.64
100mm diameter...................	3.23	5.00	3.39	1.00	-	12.44	-	15.83	nr	17.02

Waterproofing

Labour hourly rates: (except Specialists) Craft Operatives £9.23 Labourer £7.02 Rates are national average prices. Refer to REGIONAL VARIATIONS for indicative levels of overall pricing in regions	MATERIALS			LABOUR				RATES		
	Del to Site £	Waste %	Material Cost £	Craft Optve Hrs	Lab Hrs	Labour Cost £	Sunds £	Nett Rate £	Unit	Gross Rate (+7.5%) £
J20: MASTIC ASPHALT TANKING/ DAMP PROOF MEMBRANES										
Tanking and damp proofing B.S.6925 (limestone aggregate)										
Tanking and damp proofing 13mm thick; coats of asphalt -1; to concrete base; horizontal work subsequently covered										
width not exceeding 150mm.....................	-	-	Spclist	-	-	Spclist	-	30.80	m2	33.11
width 150 - 225mm...........................	-	-	Spclist	-	-	Spclist	-	25.80	m2	27.73
width 225 - 300mm...........................	-	-	Spclist	-	-	Spclist	-	20.65	m2	22.20
width exceeding 300mm.......................	-	-	Spclist	-	-	Spclist	-	15.45	m2	16.61
Tanking and damp proofing 20mm thick; coats of asphalt -2; to concrete base; horizontal work subsequently covered										
width not exceeding 150mm.....................	-	-	Spclist	-	-	Spclist	-	37.20	m2	39.99
width 150 - 225mm...........................	-	-	Spclist	-	-	Spclist	-	31.20	m2	33.54
width 225 - 300mm...........................	-	-	Spclist	-	-	Spclist	-	24.80	m2	26.66
width exceeding 300mm.......................	-	-	Spclist	-	-	Spclist	-	18.80	m2	20.21
Tanking and damp proofing 30mm thick; coats of asphalt -3; to concrete base; horizontal work subsequently covered										
width not exceeding 150mm.....................	-	-	Spclist	-	-	Spclist	-	54.10	m2	58.16
width 150 - 225mm...........................	-	-	Spclist	-	-	Spclist	-	45.30	m2	48.70
width 225 - 300mm...........................	-	-	Spclist	-	-	Spclist	-	36.05	m2	38.75
width exceeding 300mm.......................	-	-	Spclist	-	-	Spclist	-	27.05	m2	29.08
Tanking and damp proofing 20mm thick; coats of asphalt -2; to concrete base; vertical work subsequently covered										
width not exceeding 150mm.....................	-	-	Spclist	-	-	Spclist	-	64.00	m2	68.80
width 150 - 225mm...........................	-	-	Spclist	-	-	Spclist	-	53.40	m2	57.41
width 225 - 300mm...........................	-	-	Spclist	-	-	Spclist	-	42.65	m2	45.85
width exceeding 300mm.......................	-	-	Spclist	-	-	Spclist	-	31.95	m2	34.35
Tanking and damp proofing 20mm thick; coats of asphalt -3; to concrete base; vertical work subsequently covered										
width not exceeding 150mm.....................	-	-	Spclist	-	-	Spclist	-	88.55	m2	95.19
width 150 - 225mm...........................	-	-	Spclist	-	-	Spclist	-	74.00	m2	79.55
width 225 - 300mm...........................	-	-	Spclist	-	-	Spclist	-	59.05	m2	63.48
width exceeding 300mm.......................	-	-	Spclist	-	-	Spclist	-	44.25	m2	47.57
Internal angle fillets to concrete base; priming base with bitumen										
two coats; work subsequently covered	-	-	Spclist	-	-	Spclist	-	4.60	m	4.95
J21: MASTIC ASPHALT ROOFING/INSULATION/FINISHES										
Roofing; B.S.6925 (limestone aggregate); sand rubbing; B.S.747 type 4A sheathing felt isolating membrane, butt joints										
Roofing 20mm thick; coats of asphalt -2; to concrete base; pitch not exceeding 6 degrees from horizontal										
width exceeding 300mm	-	-	Spclist	-	-	Spclist	-	16.70	m2	17.95
extra; solid water check roll	-	-	Spclist	-	-	Spclist	-	11.45	m	12.31
Roofing 20mm thick; coats of asphalt -2; to concrete base; pitch 30 degrees from horizontal										
width exceeding 300mm	-	-	Spclist	-	-	Spclist	-	30.25	m2	32.52
Roofing 20mm thick; coats of asphalt -2; to concrete base; pitch 45 degrees from horizontal										
width exceeding 300mm	-	-	Spclist	-	-	Spclist	-	30.25	m2	32.52
Paint two coats of Solar reflective roof paint on surfaces of asphalt										
width exceeding 300mm	-	-	Spclist	-	-	Spclist	-	3.45	m2	3.71

Labour hourly rates: (except Specialists) Craft Operatives £9.23 Labourer £7.02 Rates are national average prices. Refer to REGIONAL VARIATIONS for indicative levels of overall pricing in regions	MATERIALS			LABOUR				RATES		
	Del to Site £	Waste %	Material Cost £	Craft Optve Hrs	Lab Hrs	Labour Cost £	Sunds £	Nett Rate £	Unit	Gross Rate (+7.5%) £
J21: MASTIC ASPHALT ROOFING/INSULATION/FINISHES Cont.										
Roofing; B.S.6925 (limestone aggregate); sand rubbing; B.S.747 type 4A sheathing felt isolating membrane, butt joints Cont.										
Skirtings 13mm thick; coats of asphalt -2; to brickwork base										
girth not exceeding 150mm	-	-	Spclist	-	-	Spclist	-	11.35	m	12.20
Coverings to kerbs 20mm thick; coats of asphalt -2; to concrete base										
girth 600mm...........................	-	-	Spclist	-	-	Spclist	-	31.00	m	33.33
Jointing new roofing to existing										
20mm thick	-	-	Spclist	-	-	Spclist	-	8.50	m	9.14
Edge trim; aluminium; silver anodised priming with bituminous primer; butt joints with internal jointing sleeves; fixing with aluminium alloy screws to timber; working two coat asphalt into grooves -1										
63.5mm wide x 44.4mm face depth..............	5.57	5.00	5.85	0.36	-	3.32	1.25	10.42	m	11.20
63.5mm wide x 76.2mm face depth..............	6.92	5.00	7.27	0.36	-	3.32	1.31	11.90	m	12.79
104.8mm wide x 44.4mm face depth.............	7.03	5.00	7.38	0.36	-	3.32	1.31	12.01	m	12.92
104.8mm wide x 76.2mm face depth.............	10.18	5.00	10.69	0.44	-	4.06	1.49	16.24	m	17.46
76.2mm fixing arm at 10 degrees x 38.1mm face depth	7.37	5.00	7.74	0.36	-	3.32	1.31	12.37	m	13.30
Edge trim; glass fibre reinforced butt joints; fixing with stainless steel screws to timber; working two coat asphalt into grooves -1 .	8.33	5.00	8.75	0.30	-	2.77	1.49	13.01	m	13.98
Roof Ventilators										
plastic; setting in position	6.08	2.50	6.23	0.75	-	6.92	0.49	13.64	nr	14.67
aluminium; setting in position	6.75	2.50	6.92	0.75	-	6.92	0.49	14.33	nr	15.41
Roofing; B.S.6925 (limestone aggregate); covering with 13mm white spar chippings in hot bitumen; B.S.747 type 4A sheathing felt isolating membrane; butt joints										
Roofing 20mm thick; coats of asphalt -2; to concrete base; pitch not exceeding 6 degrees from horizontal										
width exceeding 300mm	-	-	Spclist	-	-	Spclist	-	19.40	m2	20.86
Roofing 20mm thick; coats of asphalt -2; to concrete base; pitch 30 degrees from horizontal										
width exceeding 300mm	-	-	Spclist	-	-	Spclist	-	31.00	m2	33.33
J30: LIQUID APPLIED TANKING/DAMP PROOF MEMBRANES										
Synthaprufe bituminous emulsion; blinding with sand										
Coverings, coats -2; to concrete base; horizontal										
width not exceeding 150mm	3.05	10.00	3.36	-	0.40	2.81	0.13	6.29	m2	6.76
width 150 - 225mm	3.05	10.00	3.36	-	0.33	2.32	0.13	5.80	m2	6.24
width 225 - 300mm	3.05	10.00	3.36	-	0.26	1.83	0.13	5.31	m2	5.71
width exceeding 300mm	3.05	10.00	3.36	-	0.20	1.40	0.13	4.89	m2	5.26
Coverings, coats -3; to concrete base; horizontal										
width not exceeding 150mm	4.57	10.00	5.03	-	0.56	3.93	0.15	9.11	m2	9.79
width 150 - 225mm	4.57	10.00	5.03	-	0.46	3.23	0.15	8.41	m2	9.04
width 225 - 300mm	4.57	10.00	5.03	-	0.36	2.53	0.15	7.70	m2	8.28
width exceeding 300mm	4.57	10.00	5.03	-	0.28	1.97	0.15	7.14	m2	7.68
Coverings; coats -2; to concrete base; vertical										
width not exceeding 150mm	3.05	10.00	3.36	-	0.60	4.21	0.13	7.70	m2	8.27
width 150 - 225mm	3.05	10.00	3.36	-	0.50	3.51	0.13	7.00	m2	7.52
width 225 - 300mm	3.05	10.00	3.36	-	0.40	2.81	0.13	6.29	m2	6.76
width exceeding 300mm	3.05	10.00	3.36	-	0.30	2.11	0.13	5.59	m2	6.01
Coverings; coats -3; to concrete base; vertical										
width not exceeding 150mm	4.57	10.00	5.03	-	0.84	5.90	0.15	11.07	m2	11.90
width 150 - 225mm	4.57	10.00	5.03	-	0.70	4.91	0.15	10.09	m2	10.85
width 225 - 300mm	4.57	10.00	5.03	-	0.56	3.93	0.15	9.11	m2	9.79
width exceeding 300mm	4.57	10.00	5.03	-	0.42	2.95	0.15	8.13	m2	8.73
Bituminous emulsion										
Coverings, coats -2; to concrete base; horizontal										
width not exceeding 150mm	1.44	10.00	1.58	-	0.34	2.39	0.13	4.10	m2	4.41
width 150 - 225mm	1.44	10.00	1.58	-	0.28	1.97	0.13	3.68	m2	3.96
width 225 - 300mm	1.44	10.00	1.58	-	0.22	1.54	0.13	3.26	m2	3.50
width exceeding 300mm	1.44	10.00	1.58	-	0.17	1.19	0.13	2.91	m2	3.13
Coverings, coats -3; to concrete base; horizontal										
width not exceeding 150mm	1.99	10.00	2.19	-	0.48	3.37	0.15	5.71	m2	6.14
width 150 - 225mm	1.99	10.00	2.19	-	0.40	2.81	0.15	5.15	m2	5.53
width 225 - 300mm	1.99	10.00	2.19	-	0.32	2.25	0.15	4.59	m2	4.93
width exceeding 300mm	1.99	10.00	2.19	-	0.24	1.68	0.15	4.02	m2	4.33
Coverings; coats -2; to concrete base; vertical										
width not exceeding 150mm	1.44	10.00	1.58	-	0.51	3.58	0.13	5.29	m2	5.69
width 150 - 225mm	1.44	10.00	1.58	-	0.42	2.95	0.13	4.66	m2	5.01
width 225 - 300mm	1.44	10.00	1.58	-	0.33	2.32	0.13	4.03	m2	4.33

Labour hourly rates: (except Specialists) Craft Operatives £9.23 Labourer £7.02 Rates are national average prices. Refer to REGIONAL VARIATIONS for indicative levels of overall pricing in regions	MATERIALS			LABOUR				RATES		
	Del to Site £	Waste %	Material Cost £	Craft Optve Hrs	Lab Hrs	Labour Cost £	Sunds £	Nett Rate £	Unit	Gross Rate (+7.5%) £
J30: LIQUID APPLIED TANKING/DAMP PROOF MEMBRANES **Cont.**										
Bituminous emulsion Cont.										
Coverings; coats -2; to concrete base; vertical Cont.										
width exceeding 300mm	1.44	10.00	1.58	-	0.26	1.83	0.13	3.54	m2	3.80
Coverings; coats -3; to concrete base; vertical										
width not exceeding 150mm	1.99	10.00	2.19	-	0.72	5.05	0.15	7.39	m2	7.95
width 150 - 225mm	1.99	10.00	2.19	-	0.60	4.21	0.15	6.55	m2	7.04
width 225 - 300mm	1.99	10.00	2.19	-	0.48	3.37	0.15	5.71	m2	6.14
width exceeding 300mm	1.99	10.00	2.19	-	0.36	2.53	0.15	4.87	m2	5.23
R.I.W. liquid asphaltic composition										
Coverings, coats -2; to concrete base; horizontal										
width not exceeding 150mm	3.05	10.00	3.36	-	0.40	2.81	0.13	6.29	m2	6.76
width 150 - 225mm	3.05	10.00	3.36	-	0.33	2.32	0.13	5.80	m2	6.24
width 225 - 300mm	3.05	10.00	3.36	-	0.26	1.83	0.13	5.31	m2	5.71
width exceeding 300mm	3.05	10.00	3.36	-	0.20	1.40	0.13	4.89	m2	5.26
Coverings, coats -3; to concrete base; horizontal										
width not exceeding 150mm	4.57	10.00	5.03	-	0.56	3.93	0.15	9.11	m2	9.79
width 150 - 225mm	4.57	10.00	5.03	-	0.46	3.23	0.15	8.41	m2	9.04
width 225 - 300mm	4.57	10.00	5.03	-	0.36	2.53	0.15	7.70	m2	8.28
width exceeding 300mm	4.57	10.00	5.03	-	0.28	1.97	0.15	7.14	m2	7.68
Coverings; coats -2; to concrete base; vertical										
width not exceeding 150mm	3.05	10.00	3.36	-	0.60	4.21	0.13	7.70	m2	8.27
width 150 - 225mm	3.05	10.00	3.36	-	0.50	3.51	0.13	7.00	m2	7.52
width 225 - 300mm	3.05	10.00	3.36	-	0.40	2.81	0.13	6.29	m2	6.76
width exceeding 300mm	3.05	10.00	3.36	-	0.30	2.11	0.13	5.59	m2	6.01
Coverings; coats -3; to concrete base; vertical										
width not exceeding 150mm	4.57	10.00	5.03	-	0.84	5.90	0.15	11.07	m2	11.90
width 150 - 225mm	4.57	10.00	5.03	-	0.70	4.91	0.15	10.09	m2	10.85
width 225 - 300mm	4.57	10.00	5.03	-	0.56	3.93	0.15	9.11	m2	9.79
width exceeding 300mm	4.57	10.00	5.03	-	0.42	2.95	0.15	8.13	m2	8.73
J40: FLEXIBLE SHEET TANKING/DAMP PROOFING MEMBRANES										
Polythene sheeting; 100mm welted laps										
Tanking and damp proofing; 125mu										
horizontal; on concrete base	0.30	20.00	0.36	-	0.05	0.35	-	0.71	m2	0.76
vertical; on concrete base	0.30	20.00	0.36	-	0.08	0.56	-	0.92	m2	0.99
Tanking and damp proofing; 250mu										
horizontal; on concrete base	0.36	20.00	0.43	-	0.08	0.56	-	0.99	m2	1.07
vertical; on concrete base	0.36	20.00	0.43	-	0.12	0.84	-	1.27	m2	1.37
Tanking and damp proofing; 300mu										
horizontal; on concrete base	0.49	20.00	0.59	-	0.09	0.63	-	1.22	m2	1.31
vertical; on concrete base	0.49	20.00	0.59	-	0.14	0.98	-	1.57	m2	1.69
Bituthene 500X self adhesive damp proof membrane; **25mm lapped joints**										
Tanking and damp proofing										
horizontal; on concrete base	3.54	20.00	4.25	-	0.15	1.05	-	5.30	m2	5.70
vertical; on concrete base	3.54	20.00	4.25	-	0.25	1.75	-	6.00	m2	6.45
Bituthene 500X self adhesive damp proof membrane; **75mm lapped joints**										
Tanking and damp proofing										
horizontal; on concrete base	4.94	20.00	5.93	-	0.20	1.40	-	7.33	m2	7.88
vertical; on concrete base; priming with										
Servicised B primer	4.94	20.00	5.93	-	0.30	2.11	0.24	8.27	m2	8.89
extra; Bituthene internal angle fillet; 40 x 40mm	4.65	10.00	5.11	-	0.10	0.70	-	5.81	m	6.25
British Sisalkraft Ltd 728 damp proof membrane; 150mm **laps sealed with tape**										
Tanking and damp proofing										
horizontal; on concrete base	1.15	20.00	1.38	-	0.04	0.28	0.10	1.76	m2	1.89
Visqueen 1200 super damp proof membrane; 150mm laps **sealed with tape and mastic**										
Tanking and damp proofing										
horizontal; on concrete base	0.49	20.00	0.59	-	0.04	0.28	0.10	0.97	m2	1.04
J41: BUILT UP FELT ROOF COVERINGS										
Roofing; felt B.S.747, comprising 3G Rubervent **underlay 2.6 kg/m2, 1 layer Ruberfort HP180 underlay** **3.4 kg/m2, 1 layer Ruberfort HP180 mineral surface** **top sheet 3.6 kg/m2**										
Roof coverings; layers of felt -3; bonding with hot bitumen compound to timber base										
pitch 7.5 degrees from horizontal	5.94	15.00	6.83	0.40	0.05	4.04	2.98	13.85	m2	14.89
pitch 40 degrees from horizontal	5.94	15.00	6.83	0.67	0.08	6.75	2.98	16.56	m2	17.80
pitch 75 degrees from horizontal	5.94	15.00	6.83	0.80	0.10	8.09	2.98	17.90	m2	19.24

WATERPROOFING

Labour hourly rates: (except Specialists) Craft Operatives £9.23 Labourer £7.02 Rates are national average prices. Refer to REGIONAL VARIATIONS for indicative levels of overall pricing in regions	MATERIALS			LABOUR				RATES		
	Del to Site £	Waste %	Material Cost £	Craft Optve Hrs	Lab Hrs	Labour Cost £	Sunds £	Nett Rate £	Unit	Gross Rate (+7.5%) £
J41: BUILT UP FELT ROOF COVERINGS Cont.										
Roofing; felt B.S.747, comprising 3G Rubervent underlay 2.6 kg/m2, 1 layer Ruberfort HP180 underlay 3.4 kg/m2, 1 layer Ruberfort HP180 mineral surface top sheet 3.6 kg/m2 Cont.										
Roof coverings; layers of felt -3; bonding with hot bitumen compound to cement and sand or concrete base										
pitch 7.5 degrees from horizontal	5.94	15.00	6.83	0.40	0.05	4.04	2.98	13.85	m2	14.89
pitch 40 degrees from horizontal	5.94	15.00	6.83	0.67	0.08	6.75	2.98	16.56	m2	17.80
pitch 75 degrees from horizontal	5.94	15.00	6.83	0.80	0.10	8.09	2.98	17.90	m2	19.24
Skirtings; layers of felt - 3; bonding with hot bitumen compound to brickwork base										
girth 0 - 200mm	1.19	15.00	1.37	0.25	0.03	2.52	0.59	4.48	m	4.81
girth 200 - 400mm	2.38	15.00	2.74	0.40	0.05	4.04	1.18	7.96	m	8.56
Collars around pipes; standard and the like; layers of felt -1; bonding with hot bitumen compound to metal base										
50mm diameter x 150mm long; hole in 3 layer covering	0.23	15.00	0.26	0.15	0.02	1.52	0.15	1.94	nr	2.08
100mm diameter x 150mm long; hole in 3 layer covering	0.36	15.00	0.41	0.20	0.03	2.06	0.23	2.70	nr	2.90
Eaves trim; aluminium; silver anodised butt joints over matching sleeve pieces 200mm long; fixing to timber base with aluminium alloy screws; bedding in mastic										
63.5mm wide x 44.4mm face depth	5.99	5.00	6.29	0.36	-	3.32	1.37	10.98	m	11.81
63.5mm wide x 47.6mm face depth	6.03	5.00	6.33	0.36	-	3.32	1.37	11.02	m	11.85
63.5mm wide x 76.2mm face depth	7.37	5.00	7.74	0.36	-	3.32	1.44	12.50	m	13.44
104.8mm wide x 44.4mm face depth	7.58	5.00	7.96	0.36	-	3.32	1.44	12.72	m	13.68
104.8mm wide x 76.2mm face depth	10.95	5.00	11.50	0.44	-	4.06	1.61	17.17	m	18.46
Eaves trim; glass fibre reinforced butt joints; fixing to timber base with stainless steel screws; bedding in mastic	8.97	5.00	9.42	0.30	-	2.77	1.61	13.80	m	14.83
Roof ventilators										
plastic; setting in position	6.53	2.50	6.69	0.75	-	6.92	0.50	14.12	nr	15.17
aluminium; setting in position	7.28	2.50	7.46	0.75	-	6.92	0.50	14.88	nr	16.00
Roofing; B.S.747, all layers type 3B -1.8 kg/m2, 75mm laps; fully bonding layers with hot bitumen bonding compound										
Roof coverings; layers of felt -2; bonding with hot bitumen compound to timber base										
pitch 7.5 degrees from horizontal	1.38	15.00	1.59	0.33	0.04	3.33	2.08	6.99	m2	7.52
pitch 40 degrees from horizontal	1.38	15.00	1.59	0.55	0.07	5.57	2.08	9.23	m2	9.93
pitch 75 degrees from horizontal	1.38	15.00	1.59	0.66	0.08	6.65	2.08	10.32	m2	11.09
Roof coverings; layers of felt -2; bonding with hot bitumen compound to cement and sand or concrete base										
pitch 7.5 degrees from horizontal	1.38	15.00	1.59	0.33	0.04	3.33	2.08	6.99	m2	7.52
pitch 40 degrees from horizontal	1.38	15.00	1.59	0.55	0.07	5.57	2.08	9.23	m2	9.93
pitch 75 degrees from horizontal	1.38	15.00	1.59	0.66	0.08	6.65	2.08	10.32	m2	11.09
Roof coverings; layers of felt -3; bonding with hot bitumen compound to timber base										
pitch 7.5 degrees from horizontal	2.07	15.00	2.38	0.48	0.06	4.85	3.11	10.34	m2	11.12
pitch 40 degrees from horizontal	2.07	15.00	2.38	0.80	0.10	8.09	3.11	13.58	m2	14.59
pitch 75 degrees from horizontal	2.07	15.00	2.38	0.96	0.12	9.70	3.11	15.19	m2	16.33
Roof coverings; layers of felt -3; bonding with hot bitumen compound to cement and sand or concrete base										
pitch 7.5 degrees from horizontal	2.07	15.00	2.38	0.48	0.06	4.85	3.11	10.34	m2	11.12
pitch 40 degrees from horizontal	2.07	15.00	2.38	0.80	0.10	8.09	3.11	13.58	m2	14.59
pitch 75 degrees from horizontal	2.07	15.00	2.38	0.96	0.12	9.70	3.11	15.19	m2	16.33
Roofing; felt B.S.747, bottom and intermediate layers type 3B -1.8 kg/m2, top layer type 3E -2.8 kg/m2, 75mm laps; fully bonding layers with hot bitumen bonding compound										
Roof coverings; layers of felt -2; bonding with hot bitumen compound to timber base										
pitch 7.5 degrees from horizontal	2.21	15.00	2.54	0.33	0.04	3.33	2.08	7.95	m2	8.54
pitch 40 degrees from horizontal	2.21	15.00	2.54	0.55	0.07	5.57	2.08	10.19	m2	10.95
pitch 75 degrees from horizontal	2.21	15.00	2.54	0.66	0.08	6.65	2.08	11.27	m2	12.12
Roof coverings; layers of felt -2; bonding with hot bitumen compound to cement and sand or concrete base										
pitch 7.5 degrees from horizontal	2.21	15.00	2.54	0.33	0.04	3.33	2.08	7.95	m2	8.54
pitch 40 degrees from horizontal	2.21	15.00	2.54	0.55	0.07	5.57	2.08	10.19	m2	10.95
pitch 75 degrees from horizontal	2.21	15.00	2.54	0.66	0.08	6.65	2.08	11.27	m2	12.12
Roof coverings; layers of felt -3; bonding with hot bitumen compound to timber base										
pitch 7.5 degrees from horizontal	2.90	15.00	3.34	0.48	0.06	4.85	3.11	11.30	m2	12.14
pitch 40 degrees from horizontal	2.90	15.00	3.34	0.80	0.10	8.09	3.11	14.53	m2	15.62
pitch 75 degrees from horizontal	2.90	15.00	3.34	0.96	0.12	9.70	3.11	16.15	m2	17.36

Labour hourly rates: (except Specialists) Craft Operatives £9.23 Labourer £7.02 Rates are national average prices. Refer to REGIONAL VARIATIONS for indicative levels of overall pricing in regions	MATERIALS			LABOUR				RATES		
	Del to Site	Waste	Material Cost	Craft Optve	Lab	Labour Cost	Sunds	Nett Rate	Unit	Gross Rate (+7.5%)
	£	%	£	Hrs	Hrs	£	£	£		£
J41: BUILT UP FELT ROOF COVERINGS Cont.										
Roofing; felt B.S.747, bottom and intermediate layers type 3B -1.8 kg/m2, top layer type 3E -2.8 kg/m2, 75mm laps; fully bonding layers with hot bitumen bonding compound Cont.										
Roof coverings; layers of felt -3; bonding with hot bitumen compound to cement and sand or concrete base										
pitch 7.5 degrees from horizontal	2.90	15.00	3.34	0.48	0.06	4.85	3.11	11.30	m2	12.14
pitch 40 degrees from horizontal	2.90	15.00	3.34	0.80	0.10	8.09	3.11	14.53	m2	15.62
pitch 75 degrees from horizontal	2.90	15.00	3.34	0.96	0.12	9.70	3.11	16.15	m2	17.36
Skirtings; layers of felt - 3; bonding with hot bitumen compound to brickwork base										
girth 0 - 200mm	0.60	15.00	0.69	0.25	0.03	2.52	0.62	3.83	m	4.12
girth 200 - 400mm	1.20	15.00	1.38	0.40	0.05	4.04	1.24	6.66	m	7.16
Collars around pipes; standard and the like; layers of felt -1; bonding with hot bitumen compound to metal base										
50mm diameter x 150mm long; hole in 3 layer covering	0.23	15.00	0.26	0.15	0.02	1.52	0.15	1.94	nr	2.08
100mm diameter x 150mm long; hole in 3 layer covering	0.36	15.00	0.41	0.20	0.03	2.06	0.23	2.70	nr	2.90
Roofing; sheet, Ruberoid Building Products Superflex Ultrabond system, bottom layer Rubervent B.S. 747 type 3G; intermediate layer Superbase Ultrabond; top layer Superflex Ultrabond slate surfaced cap sheet, 50mm side laps, 75mm end laps; partially bonding bottom layer, fully bonding other layers in Ruberoid 95/25 hot bonding bitumen										
Roof coverings; layers of sheet -3; bonding with hot bitumen compound to timber base										
pitch not exceeding 5.0 degrees from horizontal ..	14.47	15.00	16.64	0.48	0.24	6.12	0.96	23.72	m2	25.49
Roof coverings; layers of sheet -3; bonding with hot bitumen compound to cement and sand or concrete base										
pitch not exceeding 5.0 degrees from horizontal ..	14.47	15.00	16.64	0.48	0.24	6.12	0.96	23.72	m2	25.49
Roof coverings; layers of sheet -3; bonding with hot bitumen compound to metal base										
pitch not exceeding 5.0 degrees from horizontal ..	14.47	15.00	16.64	0.52	0.24	6.48	0.96	24.08	m2	25.89
Skirtings; layers of sheet -2; bonding with hot bitumen compound to brickwork base										
girth 0 - 200mm	2.89	15.00	3.32	0.25	0.03	2.52	0.44	6.28	m	6.75
girth 200 - 400mm	5.78	15.00	6.65	0.40	0.05	4.04	0.87	11.56	m	12.43
Roofing; sheet, Ruberoid Building Products Superflex Ultrabond system, bottom layer Rubervent B.S. 747 type 3G; intermediate layer Superbase Ultrabond; top layer Superflex Ultrabond slate surfaced cap sheet, 50mm side laps, 75mm end laps; partially bonding bottom layer, fully bonding other layers in Ruberoid 95/25 hot bonding bitumen										
Roof coverings; layers of sheet -3; bonding with hot bitumen compound to timber base										
pitch not exceeding 5.0 degrees from horizontal ..	19.84	15.00	22.82	0.96	0.31	11.04	0.96	34.81	m2	37.42
Roof coverings; layers of sheet -3; bonding with hot bitumen compound to cement and sand or concrete base										
pitch not exceeding 5.0 degrees from horizontal ..	19.84	15.00	22.82	0.96	0.31	11.04	0.96	34.81	m2	37.42
Roof coverings; layers of sheet -3; bonding with hot bitumen compound to metal base										
pitch not exceeding 5.0 degrees from horizontal ..	19.84	15.00	22.82	1.00	0.31	11.41	0.96	35.18	m2	37.82
Roofing; sheet, Ruberoid Building Products Superflex Ultrabond system, bottom layer Rubervent B.S. 747 type 3G; intermediate layer Superbase Ultrabond; top layer Superflex Ultrabond slate surfaced cap sheet, 50mm side laps, 75mm end laps; partially bonding bottom layer, fully bonding other layers in Ruberoid 95/25 hot bonding bitumen										
Roof coverings; layers of sheet -3; bonding with hot bitumen compound to timber base										
pitch not exceeding 5.0 degrees from horizontal ..	19.84	15.00	22.82	0.98	0.31	11.22	0.96	35.00	m2	37.62
Roofing; sheet, Ruberoid Building Products Superflex Ultrabond system, bottom layer Rubervent B.S. 747 type 3G; intermediate layer Superbase Ultrabond; top layer Superflex Ultrabond slate surfaced cap sheet, 50mm side laps, 75mm end laps; partially bonding bottom layer, fully bonding other layers in Ruberoid 95/25 hot bonding bitumen										
Roof coverings; layers of sheet -3; bonding with hot bitumen compound to timber base										
pitch 7.5 degrees from horizontal	20.56	15.00	23.64	1.00	0.31	11.41	0.96	36.01	m2	38.71

WATERPROOFING

Labour hourly rates: (except Specialists) .. Craft Operatives £9.23 Labourer £7.02 Rates are national average prices. Refer to REGIONAL VARIATIONS for indicative levels of overall pricing in regions	MATERIALS			LABOUR				RATES		
	Del to Site £	Waste %	Material Cost £	Craft Optve Hrs	Lab Hrs	Labour Cost £	Sunds £	Nett Rate £	Unit	Gross Rate (+7.5%) £
J41: BUILT UP FELT ROOF COVERINGS Cont.										
Roofing; sheet, Ruberoid Building Products Superflex Ultrabond system, bottom layer Rubervent B.S. 747 type 3G; intermediate layer Superbase Ultrabond; top layer Superflex Ultrabond slate surfaced cap sheet, 50mm side laps, 75mm end laps; partially bonding bottom layer, fully bonding other layers in Ruberoid 95/25 hot bonding bitumen										
Roof coverings; layers of sheet -3; bonding with hot bitumen compound to timber base										
pitch 7.5 degrees from horizontal	20.96	15.00	**24.10**	1.02	0.31	**11.59**	0.96	**36.65**	m2	**39.40**
Roofing; sheet, Ruberoid Building Products Superflex Ultrabond system, bottom layer Rubervent B.S. 747 type 3G; intermediate layer Superbase Ultrabond; top layer Superflex Ultrabond slate surfaced cap sheet, 50mm side laps, 75mm end laps; partially bonding bottom layer, fully bonding other layers in Ruberoid 95/25 hot bonding bitumen										
Roof coverings; layers of sheet -3; bonding with hot bitumen compound to timber base										
pitch not exceeding 5.0 degrees from horizontal ..	21.80	15.00	**25.07**	1.05	0.31	**11.87**	0.96	**37.90**	m2	**40.74**
Roofing; sheet, Ruberoid Building Products Superflex Ultratorch system, bottom layer Superbase Ultratorch; top layer Superflex Ultratorch slate surfaced cap sheet, 75mm side and end laps; fully bonding both layers (by torching); SuperRock Torch, mineral wool insulation with bitumen primed surface; Superbar vapour control layer										
Roof coverings; layers of sheet -3; bonding to timber base										
pitch not exceeding 5.0 degrees from horizontal ..	14.56	15.00	**16.74**	0.48	0.24	**6.12**	1.88	**24.74**	m2	**26.59**
Roof coverings; layers of sheet -3; bonding to cement and sand or concrete base										
pitch not exceeding 5.0 degrees from horizontal ..	14.56	15.00	**16.74**	0.48	0.24	**6.12**	1.88	**24.74**	m2	**26.59**
Roof coverings; layers of sheet -3; bonding to metal base										
pitch not exceeding 5.0 degrees from horizontal ..	14.56	15.00	**16.74**	0.52	0.24	**6.48**	1.88	**25.11**	m2	**26.99**
Skirtings; layers of sheet -2; bonding with hot bitumen compound to brickwork base										
girth 0 - 200mm	2.91	15.00	**3.35**	0.25	0.03	**2.52**	0.44	**6.30**	m	**6.78**
girth 200 - 400mm	5.82	15.00	**6.69**	0.40	0.05	**4.04**	0.87	**11.61**	m	**12.48**
Roofing; sheet, Ruberoid Building Products Superflex Ultratorch system, bottom layer Superbase Ultratorch; top layer Superflex Ultratorch slate surfaced cap sheet, 75mm side and end laps; fully bonding both layers (by torching); SuperRock Torch, mineral wool insulation with bitumen primed surface; Superbar vapour control layer										
Roof coverings; layers of sheet -3; bonding to timber base										
pitch not exceeding 5.0 degrees from horizontal ..	19.93	15.00	**22.92**	0.96	0.31	**11.04**	1.88	**35.84**	m2	**38.52**
Roof coverings; layers of sheet -3; bonding to cement and sand or concrete base										
pitch not exceeding 5.0 degrees from horizontal ..	19.93	15.00	**22.92**	1.02	0.31	**11.59**	1.88	**36.39**	m2	**39.12**
Roof coverings; layers of sheet -3; bonding to metal base										
pitch not exceeding 5.0 degrees from horizontal ..	19.93	15.00	**22.92**	1.00	0.31	**11.41**	1.88	**36.21**	m2	**38.92**
Roofing; sheet, Ruberoid Building Products Superflex Ultratorch system, bottom layer Superbase Ultratorch; top layer Superflex Ultratorch slate surfaced cap sheet, 75mm side and end laps; fully bonding both layers (by torching); SuperRock Torch, mineral wool insulation with bitumen primed surface; Superbar vapour control layer										
Roof coverings; layers of sheet -3; bonding to timber base										
pitch not exceeding 5.0 degrees from horizontal ..	19.93	15.00	**22.92**	0.98	0.31	**11.22**	1.88	**36.02**	m2	**38.72**
Roofing; sheet, Ruberoid Building Products Superflex Ultratorch system, bottom layer Superbase Ultratorch; top layer Superflex Ultratorch slate surfaced cap sheet, 75mm side and end laps; fully bonding both layers (by torching); SuperRock Torch, mineral wool insulation with bitumen primed surface; Superbar vapour control layer										
Roof coverings; layers of sheet -3; bonding to timber base										
pitch not exceeding 5.0 degrees from horizontal ..	20.65	15.00	**23.75**	1.00	0.31	**11.41**	1.88	**37.03**	m2	**39.81**

Labour hourly rates: (except Specialists) Craft Operatives £9.23 Labourer £7.02 Rates are national average prices. Refer to REGIONAL VARIATIONS for indicative levels of overall pricing in regions	MATERIALS			LABOUR				RATES		
	Del to Site £	Waste %	Material Cost £	Craft Optve Hrs	Lab Hrs	Labour Cost £	Sunds £	Nett Rate £	Unit	Gross Rate (+7.5%) £
J41: BUILT UP FELT ROOF COVERINGS Cont.										
Roofing; sheet, Ruberoid Building Products Superflex Ultratorch system, bottom layer Superbase Ultratorch; top layer Superflex Ultratorch slate surfaced cap sheet, 75mm side and end laps; fully bonding both layers (by torching); SuperRock Torch, mineral wool insulation with bitumen primed surface; Superbar vapour control layer										
Roof coverings; layers of sheet -3; bonding to timber base										
pitch not exceeding 5.0 degrees from horizontal ..	21.05	15.00	24.21	1.02	0.31	11.59	1.88	37.68	m2	40.50
Roofing; sheet, Ruberoid Building Products Superflex Ultratorch system, bottom layer Superbase Ultratorch; top layer Superflex Ultratorch slate surfaced cap sheet, 75mm side and end laps; fully bonding both layers (by torching); SuperRock Torch, mineral wool insulation with bitumen primed surface; Superbar vapour control layer										
Roof coverings; layers of sheet -3; bonding to timber base										
pitch not exceeding 5.0 degrees from horizontal ..	21.89	15.00	25.17	1.05	0.31	11.87	1.88	38.92	m2	41.84

WATERPROOFING

Linings/Sheathing/Dry partitioning

Labour hourly rates: (except Specialists) Craft Operatives £9.23 Labourer £7.02 Rates are national average prices. Refer to REGIONAL VARIATIONS for indicative levels of overall pricing in regions	MATERIALS			LABOUR				RATES		
	Del to Site £	Waste %	Material Cost £	Craft Optve Hrs	Lab Hrs	Labour Cost £	Sunds £	Nett Rate £	Unit	Gross Rate (+7.5%) £
K10: PLASTERBOARD DRY LINING/PARTITIONS/CEILINGS										
PLASTERBOARD FOR DIRECT DECORATION - Linings; tapered edge sheeting of one layer 9.5mm thick Gyproc wall board for direct decoration, B.S.1230, butt joints; fixing with galvanised nails to timber base; butt joints filled with joint filler tape and joint finish, spot filling										
Walls										
height 2100 - 2400mm	3.86	5.00	4.05	0.84	0.42	10.70	1.28	16.03	m	17.24
height 2400 - 2700mm	4.35	5.00	4.57	0.95	0.48	12.14	1.44	18.15	m	19.51
height 2700 - 3000mm	4.83	5.00	5.07	1.05	0.53	13.41	1.60	20.08	m	21.59
Beams; 3nr faces										
total girth 600 - 1200mm	1.93	10.00	2.12	0.88	0.21	9.60	0.64	12.36	m	13.29
Columns; 4nr faces										
total girth 600 - 1200mm	1.93	10.00	2.12	0.88	0.21	9.60	0.64	12.36	m	13.29
total girth 1200 - 1800mm	2.90	10.00	3.19	1.31	0.32	14.34	0.96	18.49	m	19.87
Reveals and soffits of openings and recesses										
width not exceeding 300mm	0.48	10.00	0.53	0.21	0.05	2.29	0.16	2.98	m	3.20
Ceilings										
generally	1.61	5.00	1.69	0.40	0.18	4.96	0.53	7.18	m2	7.71
PLASTERBOARD FOR DIRECT DECORATION - Linings; tapered edge sheeting of one layer 12.5mm thick Gyproc wallboard for direct decoration, B.S.1230 butt joints; fixing with galvanised nails to timber base; butt joints filled with joint filler tape and joint finish, spot filling										
Walls										
height 2100 - 2400mm	4.49	5.00	4.71	0.96	0.48	12.23	1.29	18.23	m	19.60
height 2400 - 2700mm	5.05	5.00	5.30	1.08	0.54	13.76	1.45	20.51	m	22.05
height 2700 - 3000mm	5.61	5.00	5.89	1.20	0.60	15.29	1.61	22.79	m	24.50
Beams; 3nr faces										
total girth 600 - 1200mm	2.24	10.00	2.46	1.02	0.24	11.10	0.64	14.20	m	15.27
Columns; 4nr faces										
total girth 600 - 1200mm	2.24	10.00	2.46	1.02	0.24	11.10	0.64	14.20	m	15.27
total girth 1200 - 1800mm	3.37	10.00	3.71	1.53	0.36	16.65	0.97	21.33	m	22.93
Reveals and soffits of openings and recesses										
width not exceeding 300mm	0.56	10.00	0.62	0.21	0.06	2.36	0.16	3.14	m	3.37
Ceilings										
generally	1.87	5.00	1.96	0.46	0.20	5.65	0.54	8.15	m2	8.76
PLASTERBOARD FOR DIRECT DECORATION - Linings; tapered edge sheeting of one layer 15mm thick Gyproc wall board for direct decoration, B.S.1230, butt joints; fixing with galvanised nails to timber base; butt joints filled with joint filler tape and joint finish, spot filling										
Walls										
height 2100 - 2400mm	4.94	5.00	5.19	1.08	1.23	18.60	1.29	25.08	m	26.96
height 2400 - 2700mm	5.56	5.00	5.84	1.21	0.61	15.45	1.45	22.74	m	24.44
height 2700 - 3000mm	6.18	5.00	6.49	1.35	0.68	17.23	1.61	25.33	m	27.23
Beams; 3nr faces										
total girth 600 - 1200mm	2.47	10.00	2.72	1.16	0.27	12.60	0.64	15.96	m	17.16
Columns; 4nr faces										
total girth 600 - 1200mm	2.47	10.00	2.72	1.16	0.27	12.60	0.64	15.96	m	17.16
total girth 1200 - 1800mm	3.71	10.00	4.08	1.74	0.41	18.94	0.97	23.99	m	25.79

Labour hourly rates: (except Specialists) .. Craft Operatives £9.23 Labourer £7.02 Rates are national average prices. Refer to REGIONAL VARIATIONS for indicative levels of overall pricing in regions	MATERIALS			LABOUR				RATES		
	Del to Site £	Waste %	Material Cost £	Craft Optve Hrs	Lab Hrs	Labour Cost £	Sunds £	Nett Rate £	Unit	Gross Rate (+7.5%) £

K10: PLASTERBOARD DRY LINING/PARTITIONS/CEILINGS Cont.

PLASTERBOARD FOR DIRECT DECORATION – Linings; tapered edge sheeting of one layer 15mm thick Gyproc wall board for direct decoration, B.S.1230, butt joints; fixing with galvanised nails to timber base; butt joints filled with joint filler tape and joint finish, spot filling Cont.

	Del to Site £	Waste %	Material Cost £	Craft Optve Hrs	Lab Hrs	Labour Cost £	Sunds £	Nett Rate £	Unit	Gross Rate £
Reveals and soffits of openings and recesses										
width not exceeding 300mm	0.62	10.00	0.68	0.22	0.06	2.45	0.16	3.29	m	3.54
Ceilings										
generally	2.06	5.00	2.16	0.52	0.22	6.34	0.54	9.05	m2	9.73

PLASTERBOARD FOR DIRECT DECORATION – Linings; tapered edge sheeting of one layer 9.5mm thick Gyproc Duplex wallboard for direct decoration, B.S.1230 butt joints; fixing with galvanised nails to timber base; butt joints filled with joint filler tape and joint finish, spot filling

	Del to Site £	Waste %	Material Cost £	Craft Optve Hrs	Lab Hrs	Labour Cost £	Sunds £	Nett Rate £	Unit	Gross Rate £
Walls										
height 2100 - 2400mm	5.26	5.00	5.52	0.84	0.42	10.70	1.28	17.50	m	18.82
height 2400 - 2700mm	5.91	5.00	6.21	0.95	0.48	12.14	1.44	19.78	m	21.27
height 2700 - 3000mm	6.57	5.00	6.90	1.05	0.53	13.41	1.60	21.91	m	23.55
Beams; 3nr faces										
total girth 600 - 1200mm	2.63	10.00	2.89	0.88	0.21	9.60	0.64	13.13	m	14.11
Columns; 4nr faces										
total girth 600 - 1200mm	2.63	10.00	2.89	0.88	0.21	9.60	0.64	13.13	m	14.11
total girth 1200 - 1800mm	3.94	10.00	4.33	1.31	0.32	14.34	0.96	19.63	m	21.10
Reveals and soffits of openings and recesses										
width not exceeding 300mm	0.66	10.00	0.73	0.21	0.05	2.29	0.16	3.18	m	3.41
Ceilings										
generally	2.19	5.00	2.30	0.40	0.18	4.96	0.53	7.79	m2	8.37

PLASTERBOARD FOR DIRECT DECORATION – Linings; tapered edge sheeting of one layer 12.5mm thick Gyproc Duplex wallboard for direct decoration, B.S.1230, butt joints; fixing with galvanised nails to timber base; butt joints filled with joint filler tape and joint finish, spot filling

	Del to Site £	Waste %	Material Cost £	Craft Optve Hrs	Lab Hrs	Labour Cost £	Sunds £	Nett Rate £	Unit	Gross Rate £
Walls										
height 2100 - 2400mm	5.88	5.00	6.17	0.96	0.48	12.23	1.29	19.69	m	21.17
height 2400 - 2700mm	6.62	5.00	6.95	1.08	0.54	13.76	1.45	22.16	m	23.82
height 2700 - 3000mm	7.35	5.00	7.72	1.20	0.60	15.29	1.61	24.62	m	26.46
Beams; 3nr faces										
total girth 600 - 1200mm	2.94	10.00	3.23	1.02	0.24	11.10	0.64	14.97	m	16.10
Columns; 4nr faces										
total girth 600 - 1200mm	2.94	10.00	3.23	1.02	0.24	11.10	0.64	14.97	m	16.10
total girth 1200 - 1800mm	4.41	10.00	4.85	1.53	0.36	16.65	0.97	22.47	m	24.16
Reveals and soffits of openings and recesses										
width not exceeding 300mm	0.74	10.00	0.81	0.21	0.06	2.36	0.16	3.33	m	3.58
Ceilings										
generally	2.45	5.00	2.57	0.46	0.20	5.65	0.54	8.76	m2	9.42

PLASTERBOARD FOR DIRECT DECORATION – Linings; tapered edge sheeting of one layer 15mm thick Gyproc Duplex wallboard for direct decoration, B.S.1230, butt joints; fixing with galvanised nails to timber base; butt joints filled with joint filler tape and joint finish, spot filling

	Del to Site £	Waste %	Material Cost £	Craft Optve Hrs	Lab Hrs	Labour Cost £	Sunds £	Nett Rate £	Unit	Gross Rate £
Walls										
height 2100 - 2400mm	6.46	5.00	6.78	1.08	1.23	18.60	1.29	26.68	m	28.68
height 2400 - 2700mm	7.26	5.00	7.62	1.21	0.61	15.45	1.45	24.52	m	26.36
height 2700 - 3000mm	8.07	5.00	8.47	1.35	0.68	17.23	1.61	27.32	m	29.37
Beams; 3nr faces										
total girth 600 - 1200mm	3.23	10.00	3.55	1.16	0.27	12.60	0.64	16.80	m	18.05
Columns; 4nr faces										
total girth 600 - 1200mm	3.23	10.00	3.55	1.16	0.27	12.60	0.64	16.80	m	18.05
total girth 1200 - 1800mm	4.84	10.00	5.32	1.74	0.41	18.94	0.97	25.23	m	27.12
Reveals and soffits of openings and recesses										
width not exceeding 300mm	0.81	10.00	0.89	0.22	0.06	2.45	0.16	3.50	m	3.77
Ceilings										
generally	2.69	5.00	2.82	0.52	0.22	6.34	0.54	9.71	m2	10.44

Labour hourly rates: (except Specialists) Craft Operatives £9.23 Labourer £7.02 Rates are national average prices. Refer to REGIONAL VARIATIONS for indicative levels of overall pricing in regions	MATERIALS			LABOUR				RATES		
	Del to Site	Waste	Material Cost	Craft Optve	Lab	Labour Cost	Sunds	Nett Rate	Unit	Gross Rate (+7.5%)
	£	%	£	Hrs	Hrs	£	£	£		£
K10: PLASTERBOARD DRY LINING/PARTITIONS/CEILINGS Cont.										
PLASTERBOARD FOR DIRECT DECORATION - Linings; tapered edge sheeting of one layer 19mm thick Gyproc plank for direct decoration, B.S.1230, butt joints; fixing with galvanised nails to timber base; butt joints filled with joint filler tape and joint finish, spot filling										
Walls										
height 2100 - 2400mm	7.34	5.00	7.71	1.20	0.60	15.29	1.20	24.20	m	26.01
height 2400 - 2700mm	8.26	5.00	8.67	1.35	0.68	17.23	1.34	27.25	m	29.29
height 2700 - 3000mm	9.18	5.00	9.64	1.50	0.75	19.11	1.49	30.24	m	32.51
Beams; 3nr faces										
total girth 600 - 1200mm	3.67	10.00	4.04	1.50	0.30	15.95	0.60	20.59	m	22.13
Columns; 4nr faces										
total girth 600 - 1200mm	3.67	10.00	4.04	1.50	0.30	15.95	0.60	20.59	m	22.13
total girth 1200 - 1800mm	5.51	10.00	6.06	2.25	0.45	23.93	0.90	30.89	m	33.20
Reveals and soffits of openings and recesses										
width not exceeding 300mm	0.92	10.00	1.01	0.30	0.08	3.33	0.15	4.49	m	4.83
Ceilings										
generally	3.06	5.00	3.21	0.60	0.25	7.29	0.50	11.01	m2	11.83
PLASTERBOARD FOR DIRECT DECORATION - Linings; two layers of Gypsum wallboard, first layer square edge sheeting 9.5mm thick, second layer tapered edge sheeting 9.5mm thick for direct decoration, B.S.1230, butt joints; fixing with galvanised nails to timber base; butt joints of second layer filled with joint filler tape and joint finish, spot filling										
Walls										
height 2100 - 2400mm	7.73	5.00	8.12	1.44	0.72	18.35	1.60	28.06	m	30.17
height 2400 - 2700mm	8.69	5.00	9.12	1.62	0.81	20.64	1.80	31.56	m	33.93
height 2700 - 3000mm	9.66	5.00	10.14	1.80	0.90	22.93	2.00	35.08	m	37.71
Beams; 3nr faces										
total girth 600 - 1200mm	3.86	10.00	4.25	1.62	0.36	17.48	0.80	22.53	m	24.22
Columns; 4nr faces										
total girth 600 - 1200mm	3.86	10.00	4.25	1.62	0.36	17.48	0.80	22.53	m	24.22
total girth 1200 - 1800mm	5.80	10.00	6.38	2.43	0.54	26.22	1.20	33.80	m	36.33
Reveals and soffits of openings and recesses										
width not exceeding 300mm	0.97	10.00	1.07	0.33	0.09	3.68	0.20	4.94	m	5.32
Ceilings										
generally	3.22	5.00	3.38	0.70	0.30	8.57	0.67	12.62	m2	13.56
PLASTERBOARD FOR DIRECT DECORATION - Linings; two layers of Gypsum wallboard, first layer square edge sheeting 9.5mm thick, second layer tapered edge sheeting 12.5mm thick for direct decoration, B.S.1230, butt joints; fixing with galvanised nails to timber base; butt joints of second layer filled with joint filler tape and joint finish, spot filling										
Walls										
height 2100 - 2400mm	8.35	5.00	8.77	1.56	0.78	19.87	1.60	30.24	m	32.51
height 2400 - 2700mm	9.40	5.00	9.87	1.76	0.88	22.42	1.80	34.09	m	36.65
height 2700 - 3000mm	10.44	5.00	10.96	1.95	0.98	24.88	2.00	37.84	m	40.68
Beams; 3nr faces										
total girth 600 - 1200mm	4.18	10.00	4.60	1.78	0.39	19.17	0.80	24.57	m	26.41
Columns; 4nr faces										
total girth 600 - 1200mm	4.18	10.00	4.60	1.78	0.39	19.17	0.80	24.57	m	26.41
total girth 1200 - 1800mm	6.26	10.00	6.89	2.66	0.59	28.69	1.20	36.78	m	39.54
Reveals and soffits of openings and recesses										
width not exceeding 300mm	1.04	10.00	1.14	0.36	0.10	4.02	0.20	5.37	m	5.77
Ceilings										
generally	3.48	5.00	3.65	0.76	0.33	9.33	0.67	13.66	m2	14.68
PLASTERBOARD FOR DIRECT DECORATION - Linings; two layers of Gypsum wallboard, first layer square edge sheeting 12.5mm thick, second layer tapered edge sheeting 12.5mm thick for direct decoration, B.S.1230, butt joints; fixing with galvanised nails to timber base; butt joints of second layer filled with joint filler tape and joint finish, spot filling										
Walls										
height 2100 - 2400mm	8.98	5.00	9.43	1.68	0.84	21.40	1.61	32.44	m	34.88
height 2400 - 2700mm	10.10	5.00	10.61	1.89	0.95	24.11	1.81	36.53	m	39.27
height 2700 - 3000mm	11.22	5.00	11.78	2.10	1.05	26.75	2.01	40.55	m	43.59
Beams; 3nr faces										
total girth 600 - 1200mm	4.49	10.00	4.94	1.92	0.42	20.67	0.81	26.42	m	28.40

Labour hourly rates: (except Specialists) Craft Operatives £9.23 Labourer £7.02 Rates are national average prices. Refer to REGIONAL VARIATIONS for indicative levels of overall pricing in regions	MATERIALS			LABOUR				RATES		
	Del to Site £	Waste %	Material Cost £	Craft Optve Hrs	Lab Hrs	Labour Cost £	Sunds £	Nett Rate £	Unit	Gross Rate (+7.5%) £

K10: PLASTERBOARD DRY LINING/PARTITIONS/CEILINGS Cont.

PLASTERBOARD FOR DIRECT DECORATION – Linings; two layers of Gypsum wallboard, first layer square edge sheeting 12.5mm thick, second layer tapered edge sheeting 12.5mm thick for direct decoration, B.S.1230, butt joints; fixing with galvanised nails to timber base; butt joints of second layer filled with joint filler tape and joint finish, spot filling Cont.

	Del to Site £	Waste %	Material Cost £	Craft Optve Hrs	Lab Hrs	Labour Cost £	Sunds £	Nett Rate £	Unit	Gross Rate £
Columns; 4nr faces										
total girth 600 - 1200mm	4.49	10.00	4.94	1.92	0.42	20.67	0.81	26.42	m	28.40
total girth 1200 - 1800mm	6.73	10.00	7.40	2.88	0.63	31.00	1.21	39.62	m	42.59
Reveals and soffits of openings and recesses										
width not exceeding 300mm	1.12	10.00	1.23	0.39	0.11	4.37	0.20	5.80	m	6.24
Ceilings										
generally	3.74	5.00	3.93	0.82	0.35	10.03	0.67	14.62	m2	15.72

PLASTERBOARD FOR DIRECT DECORATION – Linings; two layers of Gypsum wallboard, first layer square edge sheeting 15mm thick, second layer tapered edge sheeting 15mm thick for direct decoration, B.S.1230, butt joints; fixing with galvanised nails to timber base; butt joints of second layer filled with joint filler tape and joint finish, spot filling

	Del to Site £	Waste %	Material Cost £	Craft Optve Hrs	Lab Hrs	Labour Cost £	Sunds £	Nett Rate £	Unit	Gross Rate £
Walls										
height 2100 - 2400mm	9.89	5.00	10.38	1.80	0.90	22.93	1.62	34.94	m	37.56
height 2400 - 2700mm	11.12	5.00	11.68	2.02	1.01	25.73	1.82	39.23	m	42.17
height 2700 - 3000mm	12.36	5.00	12.98	2.25	1.13	28.70	2.03	43.71	m	46.99
Beams; 3nr faces										
total girth 600 - 1200mm	4.94	10.00	5.43	1.68	0.45	18.67	0.81	24.91	m	26.78
Columns; 4nr faces										
total girth 600 - 1200mm	4.94	10.00	5.43	1.68	0.45	18.67	0.81	24.91	m	26.78
total girth 1200 - 1800mm	7.42	10.00	8.16	2.52	0.68	28.03	1.22	37.42	m	40.22
Reveals and soffits of openings and recesses										
width not exceeding 300mm	1.24	10.00	1.36	0.45	0.11	4.93	0.20	6.49	m	6.98
Ceilings										
generally	4.12	5.00	4.33	0.86	0.38	10.61	0.68	15.61	m2	16.78

PLASTERBOARD FOR DIRECT DECORATION – Linings; two layers of Gypsum wallboard, first layer Duplex square edge sheeting 9.5mm thick, second layer tapered edge sheeting 9.5mm thick for direct decoration, B.S.1230, butt joints; fixing with galvanised nails to timber base; butt joints of second layer filled with joint filler tape and joint finish, spot filling

	Del to Site £	Waste %	Material Cost £	Craft Optve Hrs	Lab Hrs	Labour Cost £	Sunds £	Nett Rate £	Unit	Gross Rate £
Walls										
height 2100 - 2400mm	9.12	5.00	9.58	1.44	0.72	18.35	1.60	29.52	m	31.74
height 2400 - 2700mm	10.26	5.00	10.77	1.62	0.81	20.64	1.80	33.21	m	35.70
height 2700 - 3000mm	11.40	5.00	11.97	1.80	0.90	22.93	2.00	36.90	m	39.67
Beams; 3nr faces										
total girth 600 - 1200mm	4.56	10.00	5.02	1.62	0.36	17.48	0.80	23.30	m	25.04
Columns; 4nr faces										
total girth 600 - 1200mm	4.56	10.00	5.02	1.62	0.36	17.48	0.80	23.30	m	25.04
total girth 1200 - 1800mm	6.84	10.00	7.52	2.43	0.54	26.22	1.20	34.94	m	37.56
Reveals and soffits of openings and recesses										
width not exceeding 300mm	1.14	10.00	1.25	0.33	0.09	3.68	0.20	5.13	m	5.52
Ceilings										
generally	3.80	5.00	3.99	0.70	0.30	8.57	0.67	13.23	m2	14.22

PLASTERBOARD FOR DIRECT DECORATION – Linings; two layers of Gypsum wallboard, first layer Duplex square edge sheeting 9.5mm thick, second layer tapered edge sheeting 12.5mm thick for direct decoration, B.S.1230, butt joints; fixing with galvanised nails to timber base; butt joints of second layer filled with joint filler tape and joint finish, spot filling

	Del to Site £	Waste %	Material Cost £	Craft Optve Hrs	Lab Hrs	Labour Cost £	Sunds £	Nett Rate £	Unit	Gross Rate £
Walls										
height 2100 - 2400mm	9.74	5.00	10.23	1.56	0.78	19.87	1.60	31.70	m	34.08
height 2400 - 2700mm	10.96	5.00	11.51	1.76	0.88	22.42	1.80	35.73	m	38.41
height 2700 - 3000mm	12.18	5.00	12.79	1.95	0.98	24.88	1.98	39.65	m	42.62
Beams; 3nr faces										
total girth 600 - 1200mm	4.87	10.00	5.36	1.78	0.39	19.17	0.80	25.32	m	27.22
Columns; 4nr faces										
total girth 600 - 1200mm	4.87	10.00	5.36	1.78	0.39	19.17	0.80	25.32	m	27.22
total girth 1200 - 1800mm	7.31	10.00	8.04	2.66	0.59	28.69	1.20	37.93	m	40.78
Reveals and soffits of openings and recesses										
width not exceeding 300mm	1.22	10.00	1.34	0.36	0.10	4.02	0.20	5.57	m	5.98

Labour hourly rates: (except Specialists) Craft Operatives £9.23 Labourer £7.02 Rates are national average prices. Refer to REGIONAL VARIATIONS for indicative levels of overall pricing in regions	MATERIALS			LABOUR				RATES		
	Del to Site £	Waste %	Material Cost £	Craft Optve Hrs	Lab Hrs	Labour Cost £	Sunds £	Nett Rate £	Unit	Gross Rate (+7.5%) £
K10: PLASTERBOARD DRY LINING/PARTITIONS/CEILINGS Cont.										
PLASTERBOARD FOR DIRECT DECORATION - Linings; two layers of Gypsum wallboard, first layer Duplex square edge sheeting 9.5mm thick, second layer tapered edge sheeting 12.5mm thick for direct decoration, B.S.1230, butt joints; fixing with galvanised nails to timber base; butt joints of second layer filled with joint filler tape and joint finish, spot filling Cont.										
Ceilings										
generally...	4.06	5.00	**4.26**	0.76	0.33	**9.33**	0.67	**14.26**	m2	**15.33**
PLASTERBOARD FOR DIRECT DECORATION - Linings; two layers of Gypsum wallboard, first layer Duplex square edge sheeting 12.5mm thick, second layer tapered edge sheeting 12.5mm thick for direct decoration, B.S.1230, butt joints; fixing with galvanised nails to timber base; butt joints of second layer filled with joint filler tape and joint finish, spot filling										
Walls										
height 2100 - 2400mm	10.37	5.00	**10.89**	1.68	0.84	**21.40**	1.61	**33.90**	m	**36.44**
height 2400 - 2700mm	11.66	5.00	**12.24**	1.89	0.95	**24.11**	1.81	**38.17**	m	**41.03**
height 2700 - 3000mm	12.96	5.00	**13.61**	2.10	1.05	**26.75**	2.01	**42.37**	m	**45.55**
Beams; 3nr faces										
total girth 600 - 1200mm	5.18	10.00	**5.70**	1.92	0.42	**20.67**	0.81	**27.18**	m	**29.22**
Columns; 4nr faces										
total girth 600 - 1200mm	5.18	10.00	**5.70**	1.92	0.42	**20.67**	0.81	**27.18**	m	**29.22**
total girth 1200 - 1800mm	7.78	10.00	**8.56**	2.88	0.63	**31.00**	1.21	**40.77**	m	**43.83**
Reveals and soffits of openings and recesses										
width not exceeding 300mm	1.30	10.00	**1.43**	0.39	0.11	**4.37**	0.20	**6.00**	m	**6.45**
Ceilings										
generally...	4.32	5.00	**4.54**	0.82	0.35	**10.03**	0.67	**15.23**	m2	**16.37**
PLASTERBOARD FOR DIRECT DECORATION - Linings; two layers of Gypsum wallboard, first layer Duplex square edge sheeting 15mm thick, second layer tapered edge sheeting 15mm thick for direct decoration, B.S.1230, butt joints; fixing with galvanised nails to timber base; butt joints of second layer filled with joint filler tape and joint finish, spot filling										
Walls										
height 2100 - 2400mm	11.40	5.00	**11.97**	1.80	0.90	**22.93**	1.62	**36.52**	m	**39.26**
height 2400 - 2700mm	12.83	5.00	**13.47**	2.02	1.01	**25.73**	1.82	**41.03**	m	**44.10**
height 2700 - 3000mm	14.25	5.00	**14.96**	2.25	1.13	**28.70**	2.03	**45.69**	m	**49.12**
Beams; 3nr faces										
total girth 600 - 1200mm	5.70	10.00	**6.27**	1.68	0.45	**18.67**	0.81	**25.75**	m	**27.68**
Columns; 4nr faces										
total girth 600 - 1200mm	5.70	10.00	**6.27**	1.68	0.45	**18.67**	0.81	**25.75**	m	**27.68**
total girth 1200 - 1800mm	8.55	10.00	**9.40**	2.52	0.68	**28.03**	1.22	**38.66**	m	**41.56**
Reveals and soffits of openings and recesses										
width not exceeding 300mm	1.43	10.00	**1.57**	0.45	0.11	**4.93**	0.20	**6.70**	m	**7.20**
Ceilings										
generally...	4.75	5.00	**4.99**	0.86	0.38	**10.61**	0.68	**16.27**	m2	**17.49**
PLASTIC FACED PLASTERBOARD - Linings; square edge sheeting of one layer 9.5mm thick Gyproc industrial grade plastic faced plasterboard, B.S.1230, butt joints; fixing with galvanised nails with plastic coated nail caps to timber base; butt joints covered with Gyproc batten section and trim										
Walls										
height 2100 - 2400mm	10.85	5.00	**11.39**	0.96	0.48	**12.23**	9.06	**32.68**	m	**35.13**
height 2400 - 2700mm	12.20	5.00	**12.81**	0.68	0.34	**8.66**	10.19	**31.66**	m	**34.04**
height 2700 - 3000mm	13.56	5.00	**14.24**	0.75	0.38	**9.59**	11.32	**35.15**	m	**37.78**
Beams; 3nr faces										
total girth 600 - 1200mm	5.42	10.00	**5.96**	0.76	0.15	**8.07**	4.53	**18.56**	m	**19.95**
Columns; 4nr faces										
total girth 600 - 1200mm	5.42	10.00	**5.96**	0.76	0.15	**8.07**	4.53	**18.56**	m	**19.95**
total girth 1200 - 1800mm	8.14	10.00	**8.95**	1.13	0.23	**12.04**	6.79	**27.79**	m	**29.87**
Reveals and soffits of openings and recesses										
width not exceeding 300mm	1.36	10.00	**1.50**	0.15	0.04	**1.67**	1.13	**4.29**	m	**4.61**
Ceilings										
generally...	4.52	5.00	**4.75**	0.30	0.13	**3.68**	3.77	**12.20**	m2	**13.11**
3.50 - 5.00m above floor	4.52	5.00	**4.75**	0.33	0.13	**3.96**	3.77	**12.47**	m2	**13.41**

LININGS/SHEATHING/DRY PARTITIONING

Labour hourly rates: (except Specialists) Craft Operatives £9.23 Labourer £7.02 Rates are national average prices. Refer to REGIONAL VARIATIONS for indicative levels of overall pricing in regions	MATERIALS			LABOUR				RATES		
	Del to Site £	Waste %	Material Cost £	Craft Optve Hrs	Lab Hrs	Labour Cost £	Sunds £	Nett Rate £	Unit	Gross Rate (+7.5%) £

K10: PLASTERBOARD DRY LINING/PARTITIONS/CEILINGS Cont.

PLASTIC FACED PLASTERBOARD - Linings; square edge sheeting of one layer 12.5mm thick Gyproc industrial grade plastic faced plasterboard, B.S.1230, butt joints; fixing with galvanised nails with plastic coated nail caps to timber base; butt joints covered with Gyproc batten section and trim

Walls

height 2100 - 2400mm	11.69	5.00	12.27	0.70	0.35	8.92	9.07	30.26	m	32.53
height 2400 - 2700mm	13.15	5.00	13.81	0.78	0.39	9.94	10.20	33.94	m	36.49
height 2700 - 3000mm	14.61	5.00	15.34	0.86	0.43	10.96	11.33	37.63	m	40.45

Beams; 3nr faces

total girth 600 - 1200mm	5.84	10.00	6.42	0.87	0.17	9.22	4.53	20.18	m	21.69

Columns; 4nr faces

total girth 600 - 1200mm	5.84	10.00	6.42	0.87	0.17	9.22	4.53	20.18	m	21.69
total girth 1200 - 1800mm	8.77	10.00	9.65	1.30	0.26	13.82	6.80	30.27	m	32.54

Reveals and soffits of openings and recesses

width not exceeding 300mm	1.46	10.00	1.61	0.17	0.05	1.92	1.13	4.66	m	5.01

Ceilings

generally	4.87	5.00	5.11	0.35	0.15	4.28	3.78	13.18	m2	14.17
3.50 - 5.00m above floor	4.87	5.00	5.11	0.38	0.15	4.56	3.78	13.45	m2	14.46

THERMAL BOARD - Linings; tapered edged sheeting of one layer 22mm thick Gyproc Thermal board vapour check grade for direct decoration, butt joints; fixing with galvanised nails to timber base; butt joints filled with joint filler tape and joint finish, spot filling

Walls

height 2100 - 2400mm	10.58	5.00	11.11	1.44	0.72	18.35	1.30	30.75	m	33.06
height 2400 - 2700mm	11.91	5.00	12.51	1.62	0.81	20.64	1.46	34.60	m	37.20
height 2700 - 3000mm	13.23	5.00	13.89	1.80	0.90	22.93	1.62	38.44	m	41.33

Reveals and soffits of openings and recesses

width not exceeding 300mm	1.32	10.00	1.45	0.36	0.09	3.95	0.16	5.57	m	5.98

Ceilings

generally	4.41	5.00	4.63	0.72	0.30	8.75	0.54	13.92	m2	14.97

THERMAL BOARD - Linings; tapered edged sheeting of one layer 30mm thick Gyproc thermal board vapour check grade for direct decoration, butt joints; fixing with galvanised nails to timber base; butt joints filled with joint filler tape and joint finish, spot filling

Walls

height 2100 - 2400mm	12.05	5.00	12.65	1.68	0.84	21.40	1.40	35.46	m	38.11
height 2400 - 2700mm	13.55	5.00	14.23	1.89	0.95	24.11	1.57	39.91	m	42.90
height 2700 - 3000mm	15.06	5.00	15.81	2.10	1.05	26.75	1.75	44.32	m	47.64

Reveals and soffits of openings and recesses

width not exceeding 300mm	1.51	10.00	1.66	0.42	0.11	4.65	0.17	6.48	m	6.97

Ceilings

generally	5.02	5.00	5.27	0.84	0.36	10.28	0.58	16.13	m2	17.34

THERMAL BOARD - Linings; tapered edged sheeting of one layer 40mm thick Gyproc Thermal board vapour check grade for direct decoration, butt joints; fixing with galvanised nails to timber base; butt joints filled with joint filler tape and joint finish, spot filling

Walls

height 2100 - 2400mm	18.70	5.00	19.64	1.92	0.96	24.46	1.40	45.50	m	48.91
height 2400 - 2700mm	21.03	5.00	22.08	2.16	1.08	27.52	1.57	51.17	m	55.01
height 2700 - 3000mm	23.37	5.00	24.54	2.40	1.20	30.58	1.75	56.86	m	61.13

Reveals and soffits of openings and recesses

width not exceeding 300mm	2.34	10.00	2.57	0.48	0.12	5.27	0.35	8.20	m	8.81

Ceilings

generally	7.79	5.00	8.18	0.96	0.42	11.81	0.58	20.57	m2	22.11

THERMAL BOARD - Linings; tapered edged sheeting of one layer 50mm thick Gyproc Thermal board vapour check grade for direct decoration, butt joints; fixing with galvanised nails to timber base; butt joints filled with joint filler tape and joint finish, spot filling

Walls

height 2100 - 2400mm	20.16	5.00	21.17	2.16	1.08	27.52	1.46	50.15	m	53.91
height 2400 - 2700mm	22.68	5.00	23.81	2.43	1.22	30.99	1.65	56.46	m	60.69
height 2700 - 3000mm	25.20	5.00	26.46	2.70	1.35	34.40	1.83	62.69	m	67.39

Labour hourly rates: (except Specialists) Craft Operatives £9.23 Labourer £7.02 Rates are national average prices. Refer to REGIONAL VARIATIONS for indicative levels of overall pricing in regions	MATERIALS			LABOUR				RATES		
	Del to Site £	Waste %	Material Cost £	Craft Optve Hrs	Lab Hrs	Labour Cost £	Sunds £	Nett Rate £	Unit	Gross Rate (+7.5%) £
K10: PLASTERBOARD DRY LINING/PARTITIONS/CEILINGS Cont.										
THERMAL BOARD - Linings; tapered edged sheeting of one layer 50mm thick Gyproc Thermal board vapour check grade for direct decoration, butt joints; fixing with galvanised nails to timber base; butt joints filled with joint filler tape and joint finish, spot filling Cont.										
Reveals and soffits of openings and recesses width not exceeding 300mm	2.52	10.00	**2.77**	0.54	0.14	**5.97**	0.18	**8.92**	m	**9.59**
Ceilings generally	8.40	5.00	**8.82**	1.08	0.48	**13.34**	0.61	**22.77**	m2	**24.48**
CELLULAR PARTITIONS - Proprietary partitions; Paramount single leaf, tapered edge panels for direct decoration, butt joints; fixing with galvanised nails to 19 x 57mm (f sizes) impregnated wrot softwood sole plate with 19 x 37 x 300mm (f sizes) locating blocks at 900mm centres and 19x 37mm (f sizes) impregnated wrought softwood wall and ceiling battens; applying continuous beads of Gyproc acoustical sealant to perimeters of construction both sides; butt joints filled with joint filler tape and joint finish, spot filling; fixing to masonry with cartridge fired nails										
Height 2100 - 2400mm, 57mm thick boarded both sides	15.94	5.00	**16.74**	2.60	1.30	**33.12**	8.71	**58.57**	m	**62.96**
Height 2400 - 2700mm, 57mm thick boarded both sides	17.93	5.00	**18.83**	2.85	1.43	**36.34**	9.42	**64.59**	m	**69.43**
Height 2100 - 2400mm, 63mm thick boarded both sides	21.65	5.00	**22.73**	2.90	1.45	**36.95**	8.75	**68.43**	m	**73.56**
Height 2400 - 2700mm, 63mm thick boarded both sides	24.35	5.00	**25.57**	3.20	1.60	**40.77**	9.46	**75.80**	m	**81.48**
CELLULAR PARTITIONS - Angles to partitions; 37 x 19mm softwood batten, 37 x 37mm softwood batten, fixing together through partitions; one panel rebated by cutting back one face by full panel thickness; joints filled with joint filler tape and joint finish										
Plain 57mm thick partitions	0.62	10.00	**0.68**	0.25	0.13	**3.22**	0.85	**4.75**	m	**5.11**
CELLULAR PARTITIONS - Tee junctions to partitions; 37 x 19mm softwood batten, 37 x 37mm softwood batten, fixing together through partitions; joints filled with joint filler tape and joint finish										
Plain 57mm thick partitions	0.62	10.00	**0.68**	0.15	0.08	**1.95**	0.85	**3.48**	m	**3.74**
CELLULAR PARTITIONS - Fair ends to partitions										

57 mm thick partitions
    facing with 9.5mm Gyproc wallboard; fixing to 37 x
    37mm softwood batten with galvanised nails ....
	0.41	10.00	**0.45**	0.10	0.05	**1.27**	0.96	**2.69**	m	**2.89**

CELLULAR PARTITIONS - Fixings for heavy fittings										

Sinks
    insert 37 x 37mm softwood fixing block into core
    prior to erection of panels; not exceeding 150mm
    long ....
	0.06	10.00	**0.07**	0.10	-	**0.92**	-	**0.99**	nr	**1.06**

insert 37 x 37mm softwood fixing block into core prior to erection panels; 150 - 300mm long	0.12	10.00	**0.13**	0.10	-	**0.92**	-	**1.05**	nr	**1.13**

Radiators
    insert 37 x 37mm softwood fixing block into core
    prior to erection of panels; not exceeding 150mm
    long ....
	0.06	10.00	**0.07**	0.10	-	**0.92**	-	**0.99**	nr	**1.06**

insert 37 x 37mm softwood fixing block into core prior to erection panels; 150 - 300mm long	0.12	10.00	**0.13**	0.10	-	**0.92**	-	**1.05**	nr	**1.13**
LAMINATED PARTITIONS - Proprietary partitions; Gyproc laminated partition; tapered edge sheeting of one layer 12.5mm thick Gyproc wall board each side for direct decoration, B.S.1230, butt joints; applying continuous beads of Gyproc acoustical sealant to perimeters of construction on both sides, fixing with nails to timber frame comprising 25 x 38mm (f sizes) perimeter members with one layer 19mm thick Gyproc plank infill; butt joints filled with joint filler tape and joint finish, spot filling; fixing to masonry with cartridge fired nails										
Height 2100 - 2400mm, 50mm thick boarded both sides	16.32	5.00	**17.14**	2.90	1.45	**36.95**	7.25	**61.33**	m	**65.93**
Height 2400 - 2700mm, 50mm thick boarded both sides	18.36	5.00	**19.28**	3.20	1.60	**40.77**	7.72	**67.77**	m	**72.85**

LININGS/SHEATHING/DRY PARTITIONING

Labour hourly rates: (except Specialists) Craft Operatives £9.23 Labourer £7.02 Rates are national average prices. Refer to REGIONAL VARIATIONS for indicative levels of overall pricing in regions	MATERIALS			LABOUR				RATES		
	Del to Site £	Waste %	Material Cost £	Craft Optve Hrs	Lab Hrs	Labour Cost £	Sunds £	Nett Rate £	Unit	Gross Rate (+7.5%) £
K10: PLASTERBOARD DRY LINING/PARTITIONS/CEILINGS Cont.										
LAMINATED PARTITIONS - Proprietary partitions; Gyproc laminated partition; tapered edge sheeting of one layer 19mm thick Gyproc plank for direct decoration, B.S 1230, butt joints; applying continuous beads of Gyproc acoustical sealant to perimeters of construction on both sides, fixing with nails to timber frame comprising 25 x 38mm (f sizes) perimeter members with one layer 19mm thick Gyproc plank infill; butt joints filled with joint filler tape and joint finish, spot filling; fixing to masonry with cartridge fired nails										
Height 2100 - 2400mm, 65mm thick boarded both sides	22.03	5.00	**23.13**	3.15	1.58	**40.17**	7.25	**70.55**	m	**75.84**
Height 2400 - 2700mm, 65mm thick boarded both sides	24.79	5.00	**26.03**	3.45	1.73	**43.99**	7.72	**77.74**	m	**83.57**
Height 2700 - 3000mm, 65mm thick boarded both sides	27.54	5.00	**28.92**	3.80	1.90	**48.41**	8.17	**85.50**	m	**91.91**
LAMINATED PARTITIONS - Angles to partitions; 38 x 25mm softwood batten, fixing together through partitions; each panel rebated by cutting back core and inner layers; joints filled with joint filler tape and joint finish										
Plain										
50mm thick partitions	0.29	10.00	**0.32**	0.25	-	**2.31**	0.85	**3.48**	m	**3.74**
65mm thick partitions	0.29	10.00	**0.32**	0.25	-	**2.31**	0.84	**3.47**	m	**3.73**
LAMINATED PARTITIONS - Tee junctions to partitions; forming groove by cutting back one face layer; bedding end in groove in bonding compound										
Plain										
50mm thick partitions	-	-	**-**	0.20	-	**1.85**	0.99	**2.84**	m	**3.05**
65mm thick partitions	-	-	**-**	0.20	-	**1.85**	0.99	**2.84**	m	**3.05**
METAL STUD PARTITIONS - Proprietary partitions; Gyproc metal stud; tapered edge sheeting of one layer 12.5mm thick Gyproc wallboard for direct decoration, B.S.1230, butt joints; applying continuous beads of Gyproc acoustical sealant to perimeters of construction both sides, fixing with Pozidriv head screws to steel frame comprising 50mm head and floor channels, 48mm vertical studs at 600mm centres; butt joints filled with joint filler tape and joint finish, spot filling; fixing to masonry with cartridge fired nails										
Height 2100 - 2400mm, 75mm thick boarded both sides	14.86	5.00	**15.60**	2.50	1.25	**31.85**	6.07	**53.52**	m	**57.54**
METAL STUD PARTITIONS - Proprietary partitions; Gyproc metal stud; tapered edge sheeting of two layers 12.5mm thick Gyproc wallboard for direct decoration, B.S.1230, butt joints; applying continuous beads of Gyproc acoustical sealant to perimeters of construction both sides, fixing with Pozidriv head screws to steel frame comprising 50mm head and floor channels, 48mm vertical studs at 600mm centres; butt joints filled with joint filler tape and joint finish, spot filling; fixing to masonry with cartridge fired nails										
Height 2100 - 2400mm, 100mm thick boarded both sides	24.03	5.00	**25.23**	2.80	1.40	**35.67**	6.07	**66.97**	m	**72.00**
Height 2400 - 2700mm, 100mm thick boarded both sides	26.82	5.00	**28.16**	3.15	1.58	**40.17**	6.49	**74.82**	m	**80.43**
Height 2700 - 3000mm, 100mm thick boarded both sides	29.63	5.00	**31.11**	3.50	1.75	**44.59**	6.90	**82.60**	m	**88.80**
METAL STUD PARTITIONS - Angles to partitions; fixing end studs together through partition; extending sheeting to one side across end of partition; joints filled with joint filler tape and joint finish										
Plain										
75mm thick partitions	0.17	10.00	**0.19**	0.10	-	**0.92**	0.85	**1.96**	m	**2.11**
100mm thick partitions	0.46	10.00	**0.51**	0.10	-	**0.92**	0.85	**2.28**	m	**2.45**

Labour hourly rates: (except Specialists) Craft Operatives £9.23 Labourer £7.02 Rates are national average prices. Refer to REGIONAL VARIATIONS for indicative levels of overall pricing in regions	MATERIALS			LABOUR				RATES		
	Del to Site £	Waste %	Material Cost £	Craft Optve Hrs	Lab Hrs	Labour Cost £	Sunds £	Nett Rate £	Unit	Gross Rate (+7.5%) £
K10: PLASTERBOARD DRY LINING/PARTITIONS/CEILINGS: STAIRCASE AREAS										
PLASTERBOARD FOR DIRECT DECORATION - Linings; tapered edge sheeting of one layer 9.5mm thick Gyproc wall board for direct decoration, B.S.1230, butt joints; fixing with galvanised nails to timber base; butt joints filled with joint filler tape and joint finish, spot filling										
Walls										
height 2100 - 2400mm	3.86	5.00	4.05	0.98	0.42	11.99	1.28	17.33	m	18.63
height 2400 - 2700mm	4.35	5.00	4.57	1.11	0.48	13.61	1.44	19.62	m	21.09
height 2700 - 3000mm	4.83	5.00	5.07	1.23	0.53	15.07	1.60	21.75	m	23.38
Reveals and soffits of openings and recesses										
width not exceeding 300mm	0.48	10.00	0.53	0.20	0.05	2.20	0.16	2.88	m	3.10
Ceilings										
generally	1.61	5.00	1.69	0.44	0.18	5.32	0.53	7.55	m2	8.11
PLASTERBOARD FOR DIRECT DECORATION - Linings; tapered edge sheeting of one layer 12.5mm thick Gyproc wallboard for direct decoration, B.S.1230 butt joints; fixing with galvanised nails to timber base; butt joints filled with joint filler tape and joint finish, spot filling										
Walls										
height 2100 - 2400mm	4.49	5.00	4.71	1.15	0.48	13.98	1.29	19.99	m	21.49
height 2400 - 2700mm	5.05	5.00	5.30	1.30	0.54	15.79	1.45	22.54	m	24.23
height 2700 - 3000mm	5.61	5.00	5.89	1.44	0.60	17.50	1.61	25.00	m	26.88
Reveals and soffits of openings and recesses										
width not exceeding 300mm	0.56	10.00	0.62	0.23	0.06	2.54	0.16	3.32	m	3.57
Ceilings										
generally	1.87	5.00	1.96	0.51	0.20	6.11	0.54	8.61	m2	9.26
PLASTERBOARD FOR DIRECT DECORATION - Linings; tapered edge sheeting of one layer 15mm thick Gyproc wallboard for direct decoration, B.S.1230 butt joints; fixing with galvanised nails to timber base; butt joints filled with joint filler tape and joint finish, spot filling										
Walls										
height 2100 - 2400mm	4.94	5.00	5.19	1.24	0.54	15.24	1.29	21.71	m	23.34
height 2400 - 2700mm	5.56	5.00	5.84	1.39	0.61	17.11	1.45	24.40	m	26.23
height 2700 - 3000mm	6.18	5.00	6.49	1.55	0.68	19.08	1.61	27.18	m	29.22
Reveals and soffits of openings and recesses										
width not exceeding 300mm	0.62	10.00	0.68	0.25	0.06	2.73	0.16	3.57	m	3.84
Ceilings										
generally	2.06	5.00	2.16	0.57	0.22	6.81	0.54	9.51	m2	10.22
PLASTERBOARD FOR DIRECT DECORATION - Linings; tapered edge sheeting of one layer 9.5mm thick Gyproc Duplex wallboard for direct decoration, B.S.1230 butt joints; fixing with galvanised nails to timber base; butt joints filled with joint filler tape and joint finish, spot filling										
Walls										
height 2100 - 2400mm	5.26	5.00	5.52	0.98	0.42	11.99	1.28	18.80	m	20.21
height 2400 - 2700mm	5.91	5.00	6.21	1.11	0.48	13.61	1.44	21.26	m	22.85
height 2700 - 3000mm	6.57	5.00	6.90	1.23	0.53	15.07	1.60	23.57	m	25.34
Reveals and soffits of openings and recesses										
width not exceeding 300mm	0.66	10.00	0.73	0.20	0.05	2.20	0.16	3.08	m	3.31
Ceilings										
generally	2.19	5.00	2.30	0.44	0.18	5.32	0.53	8.15	m2	8.77
PLASTERBOARD FOR DIRECT DECORATION - Linings; tapered edge sheeting of one layer 12.5mm thick Gyproc Duplex wallboard for direct decoration, B.S.1230, butt joints; fixing with galvanised nails to timber base; butt joints filled with joint filler tape and joint finish, spot filling										
Walls										
height 2100 - 2400mm	5.88	5.00	6.17	1.15	0.48	13.98	1.29	21.45	m	23.06
height 2400 - 2700mm	6.62	5.00	6.95	1.30	0.54	15.79	1.45	24.19	m	26.01
height 2700 - 3000mm	7.35	5.00	7.72	1.44	0.60	17.50	1.61	26.83	m	28.84
Reveals and soffits of openings and recesses										
width not exceeding 300mm	0.74	10.00	0.81	0.23	0.06	2.54	0.16	3.52	m	3.78
Ceilings										
generally	2.45	5.00	2.57	0.51	0.20	6.11	0.54	9.22	m2	9.92

LININGS/SHEATHING/DRY PARTITIONING

Labour hourly rates: (except Specialists) Craft Operatives £9.23 Labourer £7.02 Rates are national average prices. Refer to REGIONAL VARIATIONS for indicative levels of overall pricing in regions	MATERIALS			LABOUR				RATES		
	Del to Site £	Waste %	Material Cost £	Craft Optve Hrs	Lab Hrs	Labour Cost £	Sunds £	Nett Rate £	Unit	Gross Rate (+7.5%) £
K10: PLASTERBOARD DRY LINING/PARTITIONS/CEILINGS: STAIRCASE AREAS Cont.										
PLASTERBOARD FOR DIRECT DECORATION - Linings; tapered edge sheeting of one layer 15mm thick Gyproc Duplex wallboard for direct decoration, B.S.1230 butt joints; fixing with galvanised nails to timber base; butt joints filled with joint filler tape and joint finish, spot filling										
Walls										
height 2100 - 2400mm	6.46	5.00	6.78	1.24	0.54	15.24	1.29	23.31	m	25.06
height 2400 - 2700mm	7.26	5.00	7.62	1.39	0.61	17.11	1.45	26.18	m	28.15
height 2700 - 3000mm	8.07	5.00	8.47	1.55	0.68	19.08	1.61	29.16	m	31.35
Reveals and soffits of openings and recesses										
width not exceeding 300mm	0.81	10.00	0.89	0.25	0.06	2.73	0.16	3.78	m	4.06
Ceilings										
generally	2.69	5.00	2.82	0.57	0.22	6.81	0.54	10.17	m2	10.93
PLASTERBOARD FOR DIRECT DECORATION - Linings; tapered edge sheeting of one layer 19mm thick Gyproc plank for direct decoration, B.S.1230, butt joints; fixing with galvanised nails to timber base; butt joints filled with joint filler tape and joint finish, spot filling										
Walls										
height 2100 - 2400mm	7.34	5.00	7.71	1.51	0.60	18.15	1.20	27.06	m	29.09
height 2400 - 2700mm	8.26	5.00	8.67	1.70	0.68	20.46	1.34	30.48	m	32.76
height 2700 - 3000mm	9.18	5.00	9.64	1.89	0.75	22.71	1.49	33.84	m	36.38
Reveals and soffits of openings and recesses										
width not exceeding 300mm	0.92	10.00	1.01	0.34	0.08	3.70	0.15	4.86	m	5.23
Ceilings										
generally	3.06	5.00	3.21	0.68	0.25	8.03	0.50	11.74	m2	12.63
PLASTERBOARD FOR DIRECT DECORATION - Linings; two layers of Gypsum wallboard, first layer square edge sheeting 9.5mm thick, second layer tapered edge sheeting 9.5mm thick for direct decoration, B.S.1230, butt joints; fixing with galvanised nails to timber base; butt joints of second layer filled with joint filler tape and joint finish, spot filling										
Walls										
height 2100 - 2400mm	7.73	5.00	8.12	1.75	0.72	21.21	1.60	30.92	m	33.24
height 2400 - 2700mm	8.69	5.00	9.12	1.97	0.81	23.87	1.80	34.79	m	37.40
height 2700 - 3000mm	9.66	5.00	10.14	2.19	0.90	26.53	2.00	38.67	m	41.58
Reveals and soffits of openings and recesses										
width not exceeding 300mm	0.97	10.00	1.07	0.37	0.09	4.05	0.20	5.31	m	5.71
Ceilings										
generally	3.22	5.00	3.38	0.78	0.30	9.31	0.67	13.36	m2	14.36
PLASTERBOARD FOR DIRECT DECORATION - Linings; two layers of Gypsum wallboard, first layer square edge sheeting 9.5mm thick, second layer tapered edge sheeting 12.5mm thick for direct decoration, B.S.1230, butt joints; fixing with galvanised nails to timber base; butt joints of second layer filled with joint filler tape and joint finish, spot filling										
Walls										
height 2100 - 2400mm	8.35	5.00	8.77	1.90	0.78	23.01	1.60	33.38	m	35.88
height 2400 - 2700mm	9.40	5.00	9.87	2.13	0.88	25.84	1.80	37.51	m	40.32
height 2700 - 3000mm	10.44	5.00	10.96	2.37	0.98	28.75	2.00	41.72	m	44.85
Reveals and soffits of openings and recesses										
width not exceeding 300mm	1.04	10.00	1.14	0.40	0.10	4.39	0.20	5.74	m	6.17
Ceilings										
generally	3.48	5.00	3.65	0.84	0.33	10.07	0.67	14.39	m2	15.47
PLASTERBOARD FOR DIRECT DECORATION - Linings; two layers of Gypsum wallboard, first layer square edge sheeting 12.5mm thick, second layer tapered edge sheeting 12.5mm thick for direct decoration, B.S.1230, butt joints; fixing with galvanised nails to timber base; butt joints of second layer filled with joint filler tape and joint finish, spot filling										
Walls										
height 2100 - 2400mm	8.98	5.00	9.43	2.04	0.84	24.73	1.61	35.77	m	38.45
height 2400 - 2700mm	10.10	5.00	10.61	2.30	0.95	27.90	1.81	40.31	m	43.34
height 2700 - 3000mm	11.22	5.00	11.78	2.55	1.05	30.91	2.01	44.70	m	48.05
Reveals and soffits of openings and recesses										
width not exceeding 300mm	1.12	10.00	1.23	0.44	0.11	4.83	0.20	6.27	m	6.74
Ceilings										
generally	3.74	5.00	3.93	0.91	0.35	10.86	0.67	15.45	m2	16.61

Labour hourly rates: (except Specialists) Craft Operatives £9.23 Labourer £7.02 Rates are national average prices. Refer to REGIONAL VARIATIONS for indicative levels of overall pricing in regions	MATERIALS			LABOUR				RATES		
	Del to Site	Waste	Material Cost	Craft Optve	Lab	Labour Cost	Sunds	Nett Rate	Unit	Gross Rate (+7.5%)
	£	%	£	Hrs	Hrs	£	£	£		£
K10: PLASTERBOARD DRY LINING/PARTITIONS/CEILINGS: STAIRCASE AREAS Cont.										
PLASTERBOARD FOR DIRECT DECORATION - Linings; two layers of Gypsum wallboard, first layer square edge sheeting 15mm thick, second layer tapered edge sheeting 15mm thick for direct decoration, B.S.1230, butt joints; fixing with galvanised nails to timber base; butt joints of second layer filled with joint filler tape and joint finish, spot filling										
Walls										
height 2100 - 2400mm	9.89	5.00	**10.38**	2.16	0.90	**26.25**	1.62	**38.26**	m	41.13
height 2400 - 2700mm	11.12	5.00	**11.68**	2.43	1.01	**29.52**	1.82	**43.02**	m	46.24
height 2700 - 3000mm	12.36	5.00	**12.98**	2.70	1.13	**32.85**	2.03	**47.86**	m	51.45
Reveals and soffits of openings and recesses										
width not exceeding 300mm	1.24	10.00	**1.36**	0.50	0.11	**5.39**	0.20	**6.95**	m	7.47
Ceilings										
generally......................................	4.12	5.00	**4.33**	0.95	0.38	**11.44**	0.68	**16.44**	m2	17.68
PLASTERBOARD FOR DIRECT DECORATION - Linings; two layers of Gypsum wallboard, first layer Duplex square edge sheeting 9.5mm thick, second layer tapered edge sheeting 9.5mm thick for direct decoration, B.S.1230, butt joints; finish with galvanised nails to timber base; butt joint on second layer filled with joint filler tape and joint finish, spot filling										
Walls										
height 2100 - 2400mm	9.12	5.00	**9.58**	1.75	0.72	**21.21**	1.60	**32.38**	m	34.81
height 2400 - 2700mm	14.06	5.00	**14.76**	1.97	0.81	**23.87**	-	**38.63**	m	41.53
height 2700 - 3000mm	11.40	5.00	**11.97**	2.19	0.90	**26.53**	2.00	**40.50**	m	43.54
Reveals and soffits of openings and recesses										
width not exceeding 300mm	1.14	10.00	**1.25**	0.37	0.09	**4.05**	0.20	**5.50**	m	5.91
Ceilings										
generally......................................	3.80	5.00	**3.99**	0.78	0.30	**9.31**	0.67	**13.97**	m2	15.01
PLASTERBOARD FOR DIRECT DECORATION - Linings; two layers of Gypsum wallboard, first layer Duplex square edge sheeting 9.5mm thick, second layer tapered edge sheeting 12.5mm thick for direct decoration, B.S.1230, butt joints; fixing with galvanised nails to timber base; butt joints of second layer filled with joint filler tape and joint finish, spot filling										
Walls										
height 2100 - 2400mm	9.74	5.00	**10.23**	1.90	0.78	**23.01**	1.60	**34.84**	m	37.45
height 2400 - 2700mm	10.96	5.00	**11.51**	2.13	0.88	**25.84**	1.80	**39.15**	m	42.08
height 2700 - 3000mm	12.18	5.00	**12.79**	2.37	0.98	**28.75**	2.00	**43.54**	m	46.81
Reveals and soffits of openings and recesses										
width not exceeding 300mm	1.22	10.00	**1.34**	0.40	0.10	**4.39**	0.20	**5.94**	m	6.38
Ceilings										
generally......................................	4.06	5.00	**4.26**	0.84	0.33	**10.07**	0.67	**15.00**	m2	16.13
PLASTERBOARD FOR DIRECT DECORATION - Linings; two layers of Gypsum wallboard, first layer Duplex square edge sheeting 12.5mm thick, second layer tapered edge sheeting 12.5mm thick for direct decoration, B.S.1230, butt joints; fixing with galvanised nails to timber base; butt joints of second layer filled with joint filler tape and joint finish, spot filling										
Walls										
height 2100 - 2400mm	10.37	5.00	**10.89**	2.04	0.84	**24.73**	1.61	**37.22**	m	40.02
height 2400 - 2700mm	11.66	5.00	**12.24**	2.30	0.95	**27.90**	1.81	**41.95**	m	45.10
height 2700 - 3000mm	12.96	5.00	**13.61**	2.55	1.05	**30.91**	2.01	**46.53**	m	50.01
Reveals and soffits of openings and recesses										
width not exceeding 300mm	1.30	10.00	**1.43**	0.44	0.11	**4.83**	0.20	**6.46**	m	6.95
Ceilings										
generally......................................	4.32	5.00	**4.54**	0.91	0.35	**10.86**	0.67	**16.06**	m2	17.27
PLASTERBOARD FOR DIRECT DECORATION - Linings; two layers of Gypsum wallboard, first layer Duplex square edge sheeting 15mm thick, second layer tapered edge sheeting 15mm thick for direct decoration, B.S.1230, butt joints; fixing with galvanised nails to timber base; butt joints of second layer filled with joint filler tape and joint finish, spot filling										
Walls										
height 2100 - 2400mm	11.40	5.00	**11.97**	2.16	0.90	**26.25**	1.62	**39.84**	m	42.83
height 2400 - 2700mm	12.83	5.00	**13.47**	2.43	1.01	**29.52**	1.82	**44.81**	m	48.17
height 2700 - 3000mm	14.25	5.00	**14.96**	2.70	1.13	**32.85**	2.03	**49.85**	m	53.58
Reveals and soffits of openings and recesses										
width not exceeding 300mm	1.43	10.00	**1.57**	0.50	0.11	**5.39**	0.20	**7.16**	m	7.70

Labour hourly rates: (except Specialists) Craft Operatives £9.23 Labourer £7.02 Rates are national average prices. Refer to REGIONAL VARIATIONS for indicative levels of overall pricing in regions	MATERIALS			LABOUR				RATES		
	Del to Site £	Waste %	Material Cost £	Craft Optve Hrs	Lab Hrs	Labour Cost £	Sunds £	Nett Rate £	Unit	Gross Rate (+7.5%) £
K10: PLASTERBOARD DRY LINING/PARTITIONS/CEILINGS: STAIRCASE AREAS Cont.										
PLASTERBOARD FOR DIRECT DECORATION - Linings; two layers of Gypsum wallboard, first layer Duplex square edge sheeting 15mm thick, second layer tapered edge sheeting 15mm thick for direct decoration, B.S.1230, butt joints; fixing with galvanised nails to timber base; butt joints of second layer filled with joint filler tape and joint finish, spot filling Cont.										
Ceilings										
generally	4.75	5.00	4.99	0.95	0.38	11.44	0.68	17.10	m2	18.39
PLASTIC FACED - Linings; square edge sheeting of one layer 9.5mm thick Gyproc industrial grade plastic faced plasterboard, B.S.1230, butt joints; fixing with galvanised nails with plastic coated nail caps to timber base; butt joints covered with Gyproc batten section and trim										
Walls										
height 2100 - 2400mm	10.85	5.00	11.39	0.74	0.30	8.94	9.06	29.39	m	31.59
height 2400 - 2700mm	12.20	5.00	12.81	0.84	0.34	10.14	10.19	33.14	m	35.63
height 2700 - 3000mm	13.56	5.00	14.24	0.93	0.38	11.25	11.32	36.81	m	39.57
Reveals and soffits of openings and recesses										
width not exceeding 300mm	1.36	10.00	1.50	0.18	0.04	1.94	1.13	4.57	m	4.91
Ceilings										
generally	4.52	5.00	4.75	0.34	0.13	4.05	3.77	12.57	m2	13.51
PLASTIC FACED - Linings; square edge sheeting of one layer 12.5mm thick Gyproc industrial grade plastic faced plasterboard, B.S.1230, butt joints; fixing with galvanised nails with plastic coated nail caps to timber base; butt joints covered with Gyproc batten section and trim										
Walls										
height 2100 - 2400mm	11.69	5.00	12.27	0.85	0.35	10.30	9.07	31.65	m	34.02
height 2400 - 2700mm	13.15	5.00	13.81	0.97	0.39	11.69	10.20	35.70	m	38.38
height 2700 - 3000mm	14.61	5.00	15.34	1.07	0.44	12.96	11.33	39.64	m	42.61
Reveals and soffits of openings and recesses										
width not exceeding 300mm	1.46	10.00	1.61	0.21	0.05	2.29	1.13	5.03	m	5.40
Ceilings										
generally	4.87	5.00	5.11	0.39	0.15	4.65	3.78	13.55	m2	14.56
K11: RIGID SHEET FLOORING/SHEATHING/LININGS/CASINGS										
Douglas Fir plywood, unsanded select sheathing quality, WBP bonded; tongued and grooved joints long edges										
15mm thick sheeting to floors										
width exceeding 300mm	5.76	7.50	6.19	0.29	0.04	2.96	0.10	9.25	m2	9.94
18mm thick sheeting to floors										
width exceeding 300mm	7.30	7.50	7.85	0.30	0.04	3.05	0.10	11.00	m2	11.82
Birch faced plywood, BB quality, WBP bonded; tongued and grooved joints all edges										
12mm thick sheeting to floors										
width exceeding 300mm	17.30	7.50	18.60	0.27	0.04	2.77	0.10	21.47	m2	23.08
15mm thick sheeting to floors										
width exceeding 300mm	20.77	7.50	22.33	0.29	0.04	2.96	0.10	25.39	m2	27.29
18mm thick sheeting to floors										
width exceeding 300mm	25.19	7.50	27.08	0.30	0.04	3.05	0.10	30.23	m2	32.50
Wood chipboard, B.S.5669, type II flooring, square edges; butt joints										
18mm thick sheeting to floors										
width exceeding 300mm	5.13	7.50	5.51	0.30	0.04	3.05	0.10	8.66	m2	9.31
22mm thick sheeting to floors										
width exceeding 300mm	6.95	7.50	7.47	0.30	0.04	3.05	0.10	10.62	m2	11.42
Wood chipboard, B.S.5669, type II flooring; tongued and grooved joints										
18mm thick sheeting to floors										
width exceeding 300mm	5.42	7.50	5.83	0.30	0.04	3.05	0.10	8.98	m2	9.65
22mm thick sheeting to floors										
width exceeding 300mm	7.21	7.50	7.75	0.30	0.04	3.05	0.10	10.90	m2	11.72
Abutments										
12mm thick plywood linings										
raking cutting	1.73	-	1.73	0.03	-	0.28	-	2.01	m	2.16

Labour hourly rates: (except Specialists) Craft Operatives £9.23 Labourer £7.02 Rates are national average prices. Refer to REGIONAL VARIATIONS for indicative levels of overall pricing in regions	MATERIALS			LABOUR				RATES		
	Del to Site £	Waste %	Material Cost £	Craft Optve Hrs	Lab Hrs	Labour Cost £	Sunds £	Nett Rate £	Unit	Gross Rate (+7.5%) £
K11: RIGID SHEET FLOORING/SHEATHING/LININGS/CASINGS Cont.										
Abutments Cont.										
12mm thick plywood linings Cont.										
curved cutting	2.60	–	2.60	0.05	–	0.46	–	3.06	m	3.29
15mm thick plywood linings										
raking cutting	2.08	–	2.08	0.04	–	0.37	–	2.45	m	2.63
curved cutting	3.12	–	3.12	0.06	–	0.55	–	3.67	m	3.95
18mm thick plywood linings										
raking cutting	2.52	–	2.52	0.04	–	0.37	–	2.89	m	3.11
curved cutting	3.78	–	3.78	0.06	–	0.55	–	4.33	m	4.66
18mm thick wood chipboard linings										
raking cutting	0.51	–	0.51	0.03	–	0.28	–	0.79	m	0.85
curved cutting	0.77	–	0.77	0.05	–	0.46	–	1.23	m	1.32
22mm thick wood chipboard linings										
raking cutting	0.70	–	0.70	0.04	–	0.37	–	1.07	m	1.15
curved cutting	1.04	–	1.04	0.06	–	0.55	–	1.59	m	1.71
SOUND INSULATED FLOORS - Foam backed softwood battens covered with tongued and grooved chipboard or plywood panels										
19mm chipboard panels to level sub-floors										
width exceeding 300mm; battens at 450mm centres; elevation										
50mm	16.16	10.00	17.78	0.27	0.03	2.70	0.26	20.74	m2	22.29
75mm	18.57	10.00	20.43	0.27	0.03	2.70	0.28	23.41	m2	25.17
width exceeding 300mm; battens at 600mm centres; elevation										
50mm	14.73	10.00	16.20	0.27	0.03	2.70	0.22	19.13	m2	20.56
75mm	17.14	10.00	18.85	0.27	0.03	2.70	0.23	21.79	m2	23.42
19mm chipboard panels with improved moisture resistance to level sub-floors										
width exceeding 300mm; battens at 450mm centres; elevation										
50mm	17.59	10.00	19.35	0.27	0.03	2.70	0.26	22.31	m2	23.99
75mm	20.00	10.00	22.00	0.27	0.03	2.70	0.28	24.98	m2	26.86
width exceeding 300mm; battens at 600mm centres; elevation										
50mm	16.16	10.00	17.78	0.27	0.03	2.70	0.22	20.70	m2	22.25
75mm	18.57	10.00	20.43	0.27	0.03	2.70	0.23	23.36	m2	25.11
19mm plywood panels to level sub-floors										
width exceeding 300mm; battens at 450mm centres; elevation										
50mm	24.71	10.00	27.18	0.27	0.03	2.70	0.26	30.14	m2	32.40
75mm	27.12	10.00	29.83	0.27	0.03	2.70	0.28	32.81	m2	35.28
FIBREBOARDS AND HARDBOARDS - Standard hardboard, B.S.1142, type SHA; butt joints										
3.2mm thick linings to walls										
width exceeding 300mm	0.90	5.00	0.94	0.33	0.04	3.33	0.06	4.33	m2	4.66
width not exceeding 300mm	0.27	10.00	0.30	0.20	0.01	1.92	0.05	2.26	m	2.43
4.8mm thick linings to walls										
width exceeding 300mm	1.50	5.00	1.58	0.34	0.04	3.42	0.08	5.07	m2	5.45
width not exceeding 300mm	0.45	10.00	0.50	0.20	0.01	1.92	0.05	2.46	m	2.65
6.4mm thick linings to walls										
width exceeding 300mm	1.95	5.00	2.05	0.35	0.04	3.51	0.08	5.64	m2	6.06
width not exceeding 300mm	0.59	10.00	0.65	0.21	0.01	2.01	0.05	2.71	m	2.91
3.2mm thick sheeting to floors										
width exceeding 300mm	0.90	5.00	0.94	0.17	0.02	1.71	0.04	2.69	m2	2.90
3.2mm thick linings to ceilings										
width exceeding 300mm	0.90	5.00	0.94	0.41	0.05	4.14	0.10	5.18	m2	5.57
width not exceeding 300mm	0.27	10.00	0.30	0.25	0.02	2.45	0.06	2.80	m	3.02
4.8mm thick linings to ceilings										
width exceeding 300mm	1.50	5.00	1.58	0.43	0.05	4.32	0.10	5.99	m2	6.44
width not exceeding 300mm	0.45	10.00	0.50	0.26	0.02	2.54	0.06	3.10	m	3.33
6.4mm thick linings to ceilings										
width exceeding 300mm	1.95	5.00	2.05	0.44	0.05	4.41	0.10	6.56	m2	7.05
width not exceeding 300mm	0.59	10.00	0.65	0.26	0.02	2.54	0.06	3.25	m	3.49
FIBREBOARDS AND HARDBOARDS - Tempered hardboard, B.S.1142, Type THE; butt joints										
3.2mm thick linings to walls										
width exceeding 300mm	1.50	5.00	1.58	0.33	0.04	3.33	0.06	4.96	m2	5.33
width not exceeding 300mm	0.45	10.00	0.50	0.20	0.01	1.92	0.05	2.46	m	2.65
4.8mm thick linings to walls										
width exceeding 300mm	2.11	5.00	2.22	0.34	0.04	3.42	0.08	5.71	m2	6.14
width not exceeding 300mm	0.63	10.00	0.69	0.20	0.01	1.92	0.05	2.66	m	2.86

Labour hourly rates: (except Specialists) Craft Operatives £9.23 Labourer £7.02 Rates are national average prices. Refer to REGIONAL VARIATIONS for indicative levels of overall pricing in regions	MATERIALS			LABOUR				RATES		
	Del to Site £	Waste %	Material Cost £	Craft Optve Hrs	Lab Hrs	Labour Cost £	Sunds £	Nett Rate £	Unit	Gross Rate (+7.5%) £
K11: RIGID SHEET FLOORING/SHEATHING/LININGS/CASINGS **Cont.**										
FIBREBOARDS AND HARDBOARDS - Tempered hardboard, **B.S.1142, Type THE; butt joints Cont.**										
6.4mm thick linings to walls										
width exceeding 300mm	2.77	5.00	**2.91**	0.35	0.04	**3.51**	0.08	**6.50**	m2	**6.99**
width not exceeding 300mm	0.83	10.00	**0.91**	0.21	0.01	**2.01**	0.05	**2.97**	m	**3.19**
3.2mm thick linings to ceilings										
width exceeding 300mm	1.50	5.00	**1.58**	0.41	0.05	**4.14**	0.10	**5.81**	m2	**6.25**
width not exceeding 300mm	0.45	10.00	**0.50**	0.25	0.02	**2.45**	0.06	**3.00**	m	**3.23**
4.8mm thick linings to ceilings										
width exceeding 300mm	2.11	5.00	**2.22**	0.43	0.05	**4.32**	0.10	**6.64**	m2	**7.13**
width not exceeding 300mm	0.63	10.00	**0.69**	0.26	0.02	**2.54**	0.06	**3.29**	m	**3.54**
6.4mm thick linings to ceilings										
width exceeding 300mm	2.77	5.00	**2.91**	0.44	0.05	**4.41**	0.10	**7.42**	m2	**7.98**
width not exceeding 300mm	0.83	10.00	**0.91**	0.26	0.02	**2.54**	0.06	**3.51**	m	**3.78**
FIBREBOARDS AND HARDBOARDS - Flame retardant **hardboard, B.S.1142 tested to B.S.476 Part 7 Class 1;** **butt joints**										
3.2mm thick linings to walls										
width exceeding 300mm	5.25	5.00	**5.51**	0.33	0.04	**3.33**	0.06	**8.90**	m2	**9.57**
width not exceeding 300mm	1.58	10.00	**1.74**	0.20	0.01	**1.92**	0.05	**3.70**	m	**3.98**
4.8mm thick linings to walls										
width exceeding 300mm	6.05	5.00	**6.35**	0.34	0.04	**3.42**	0.08	**9.85**	m2	**10.59**
width not exceeding 300mm	1.82	10.00	**2.00**	0.20	0.01	**1.92**	0.05	**3.97**	m	**4.27**
6.4mm thick linings to walls										
width exceeding 300mm	6.85	5.00	**7.19**	0.35	0.04	**3.51**	0.08	**10.78**	m2	**11.59**
width not exceeding 300mm	2.06	10.00	**2.27**	0.21	0.01	**2.01**	0.05	**4.32**	m	**4.65**
3.2mm thick linings to ceilings										
width exceeding 300mm	5.25	5.00	**5.51**	0.41	0.05	**4.14**	0.10	**9.75**	m2	**10.48**
width not exceeding 300mm	1.58	10.00	**1.74**	0.25	0.02	**2.45**	0.06	**4.25**	m	**4.56**
4.8mm thick linings to ceilings										
width exceeding 300mm	6.05	5.00	**6.35**	0.43	0.05	**4.32**	0.10	**10.77**	m2	**11.58**
width not exceeding 300mm	1.82	10.00	**2.00**	0.26	0.02	**2.54**	0.06	**4.60**	m	**4.95**
6.4mm thick linings to ceilings										
width exceeding 300mm	6.85	5.00	**7.19**	0.44	0.05	**4.41**	0.10	**11.70**	m2	**12.58**
width not exceeding 300mm	2.06	10.00	**2.27**	0.26	0.02	**2.54**	0.06	**4.87**	m	**5.23**
FIBREBOARDS AND HARDBOARDS - Low density medium **board, B.S.1142, type LMN; butt joints**										
6mm thick linings to walls										
width exceeding 300mm	6.44	5.00	**6.76**	0.35	0.04	**3.51**	0.08	**10.35**	m2	**11.13**
width not exceeding 300mm	1.93	10.00	**2.12**	0.21	0.01	**2.01**	0.05	**4.18**	m	**4.50**
9mm thick linings to walls										
width exceeding 300mm	8.24	5.00	**8.65**	0.36	0.05	**3.67**	0.09	**12.42**	m2	**13.35**
width not exceeding 300mm	2.47	10.00	**2.72**	0.22	0.02	**2.17**	0.05	**4.94**	m	**5.31**
12mm thick linings to walls										
width exceeding 300mm	10.51	5.00	**11.04**	0.38	0.05	**3.86**	0.09	**14.98**	m2	**16.11**
width not exceeding 300mm	3.15	10.00	**3.46**	0.23	0.02	**2.26**	0.05	**5.78**	m	**6.21**
6mm thick linings to ceilings										
width exceeding 300mm	6.44	5.00	**6.76**	0.44	0.06	**4.48**	0.11	**11.35**	m2	**12.21**
width not exceeding 300mm	1.93	10.00	**2.12**	0.26	0.02	**2.54**	0.06	**4.72**	m	**5.08**
9mm thick linings to ceilings										
width exceeding 300mm	8.24	5.00	**8.65**	0.45	0.06	**4.57**	0.11	**13.34**	m2	**14.34**
width not exceeding 300mm	2.47	10.00	**2.72**	0.27	0.02	**2.63**	0.06	**5.41**	m	**5.82**
12mm thick linings to ceilings										
width exceeding 300mm	10.51	5.00	**11.04**	0.48	0.06	**4.85**	0.11	**16.00**	m2	**17.20**
width not exceeding 300mm	3.15	10.00	**3.46**	0.29	0.02	**2.82**	0.07	**6.35**	m	**6.83**
FIBREBOARDS AND HARDBOARDS - Low density medium **board, B.S.1142, type LMN; butt joints**										
6mm thick linings to walls										
width exceeding 300mm	5.51	5.00	**5.79**	0.35	0.04	**3.51**	0.08	**9.38**	m2	**10.08**
width not exceeding 300mm	1.65	10.00	**1.81**	0.21	0.01	**2.01**	0.05	**3.87**	m	**4.16**
9mm thick linings to walls										
width exceeding 300mm	6.85	5.00	**7.19**	0.36	0.05	**3.67**	0.09	**10.96**	m2	**11.78**
width not exceeding 300mm	2.06	10.00	**2.27**	0.22	0.02	**2.17**	0.05	**4.49**	m	**4.82**
6mm thick linings to ceilings										
width exceeding REGIOANL 300mm	5.51	5.00	**5.79**	0.44	0.06	**4.48**	0.11	**10.38**	m2	**11.16**
width not exceeding 300mm	1.65	10.00	**1.81**	0.26	0.02	**2.54**	0.06	**4.42**	m	**4.75**
9mm thick linings to ceilings										
width exceeding 300mm	6.85	5.00	**7.19**	0.45	0.06	**4.57**	0.11	**11.88**	m2	**12.77**
width not exceeding 300mm	2.06	10.00	**2.27**	0.27	0.02	**2.63**	0.06	**4.96**	m	**5.33**

Labour hourly rates: (except Specialists) Craft Operatives £9.23 Labourer £7.02 Rates are national average prices. Refer to REGIONAL VARIATIONS for indicative levels of overall pricing in regions	MATERIALS			LABOUR				RATES		
	Del to Site	Waste	Material Cost	Craft Optve	Lab	Labour Cost	Sunds	Nett Rate	Unit	Gross Rate (+7.5%)
	£	%	£	Hrs	Hrs	£	£	£		£
K11: RIGID SHEET FLOORING/SHEATHING/LININGS/CASINGS Cont.										
FIBREBOARDS AND HARDBOARDS - Medium density fibreboard, B.S.1142, type MDF; butt joints										
9mm thick linings to walls										
width exceeding 300mm	3.40	5.00	3.57	0.35	0.04	3.51	0.08	7.16	m2	7.70
width not exceeding 300mm	1.02	10.00	1.12	0.21	0.01	2.01	0.05	3.18	m	3.42
12mm thick linings to walls										
width exceeding 300mm	4.33	5.00	4.55	0.38	0.05	3.86	0.09	8.49	m2	9.13
width not exceeding 300mm	1.30	10.00	1.43	0.23	0.02	2.26	0.06	3.75	m	4.03
18mm thick linings to walls										
width exceeding 300mm	5.25	5.00	5.51	0.41	0.05	4.14	0.10	9.75	m2	10.48
width not exceeding 300mm	1.58	10.00	1.74	0.24	0.02	2.36	0.06	4.15	m	4.47
25mm thick linings to walls										
width exceeding 300mm	7.93	5.00	8.33	0.44	0.06	4.48	0.11	12.92	m2	13.89
width not exceeding 300mm	2.38	10.00	2.62	0.26	0.02	2.54	0.06	5.22	m	5.61
9mm thick linings to ceilings										
width exceeding 300mm	3.40	5.00	3.57	0.44	0.06	4.48	0.11	8.16	m2	8.77
width not exceeding 300mm	1.02	10.00	1.12	0.26	0.02	2.54	0.06	3.72	m	4.00
12mm thick linings to ceilings										
width exceeding 300mm	4.33	5.00	4.55	0.48	0.06	4.85	0.11	9.51	m2	10.22
width not exceeding 300mm	1.30	10.00	1.43	0.29	0.02	2.82	0.07	4.32	m	4.64
18mm thick linings to ceilings										
width exceeding 300mm	5.25	5.00	5.51	0.52	0.07	5.29	0.13	10.93	m2	11.75
width not exceeding 300mm	1.58	10.00	1.74	0.31	0.02	3.00	0.07	4.81	m	5.17
25mm thick linings to ceilings										
width exceeding 300mm	7.93	5.00	8.33	0.56	0.07	5.66	0.13	14.12	m2	15.18
width not exceeding 300mm	2.38	10.00	2.62	0.34	0.02	3.28	0.08	5.98	m	6.42
FIBREBOARDS AND HARDBOARDS - Insulating softboard, B.S.1142; type SBN butt joints										
13mm thick linings to walls										
width exceeding 300mm	1.86	5.00	1.95	0.38	0.05	3.86	0.09	5.90	m2	6.34
width not exceeding 300mm	0.56	10.00	0.62	0.23	0.02	2.26	0.05	2.93	m	3.15
13mm thick linings to ceilings										
width exceeding 300mm	1.86	5.00	1.95	0.48	0.06	4.85	0.11	6.91	m2	7.43
width not exceeding 300mm	0.56	10.00	0.62	0.29	0.02	2.82	0.07	3.50	m	3.77
FIBREBOARDS AND HARDBOARDS - Bitumen impregnated insulating board, B.S.1142, type SBI; butt joints										
13mm thick linings to walls										
width exceeding 300mm	2.20	5.00	2.31	0.38	0.05	3.86	0.09	6.26	m2	6.73
width not exceeding 300mm	0.66	10.00	0.73	0.23	0.02	2.26	0.05	3.04	m	3.27
13mm thick linings to ceilings										
width exceeding 300mm	2.20	5.00	2.31	0.48	0.06	4.85	0.11	7.27	m2	7.82
width not exceeding 300mm	0.66	10.00	0.73	0.29	0.02	2.82	0.07	3.61	m	3.88
ASBESTOS FREE BOARD - Masterboard Class 0 fire resisting boards; butt joints										
6mm thick linings to walls										
width exceeding 300mm	6.85	5.00	7.19	0.45	0.06	4.57	0.11	11.88	m2	12.77
width not exceeding 300mm	2.06	10.00	2.27	0.27	0.02	2.63	0.06	4.96	m	5.33
9mm thick linings to walls										
width exceeding 300mm	13.85	5.00	14.54	0.49	0.06	4.94	0.12	19.61	m2	21.08
width not exceeding 300mm	4.16	10.00	4.58	0.29	0.02	2.82	0.07	7.46	m	8.02
12mm thick linings to walls										
width exceeding 300mm	18.30	5.00	19.22	0.53	0.07	5.38	0.13	24.73	m2	26.58
width not exceeding 300mm	5.49	10.00	6.04	0.32	0.02	3.09	0.07	9.20	m	9.89
6mm thick linings to ceilings										
width exceeding 300mm	6.85	5.00	7.19	0.57	0.07	5.75	0.14	13.09	m2	14.07
width not exceeding 300mm	2.06	10.00	2.27	0.34	0.02	3.28	0.08	5.62	m	6.05
9mm thick linings to ceilings										
width exceeding 300mm	13.85	5.00	14.54	0.61	0.08	6.19	0.15	20.88	m2	22.45
width not exceeding 300mm	4.16	10.00	4.58	0.37	0.02	3.56	0.08	8.21	m	8.83
12mm thick linings to ceilings										
width exceeding 300mm	18.30	5.00	19.22	0.67	0.08	6.75	0.16	26.12	m2	28.08
width not exceeding 300mm	5.49	10.00	6.04	0.40	0.02	3.83	0.09	9.96	m	10.71
ASBESTOS FREE BOARD - Supalux fire resisting boards, Sanded; butt joints; fixing to timber with screws, countersinking										
6mm thick linings to walls										
width exceeding 300mm	10.10	5.00	10.61	0.45	0.06	4.57	0.11	15.29	m2	16.44
width not exceeding 300mm	3.03	10.00	3.33	0.27	0.02	2.63	0.06	6.03	m	6.48

Labour hourly rates: (except Specialists) Craft Operatives £9.23 Labourer £7.02 Rates are national average prices. Refer to REGIONAL VARIATIONS for indicative levels of overall pricing in regions	MATERIALS			LABOUR				RATES		
	Del to Site £	Waste %	Material Cost £	Craft Optve Hrs	Lab Hrs	Labour Cost £	Sunds £	Nett Rate £	Unit	Gross Rate (+7.5%) £
K11: RIGID SHEET FLOORING/SHEATHING/LININGS/CASINGS Cont.										
ASBESTOS FREE BOARD - Supalux fire resisting boards, Sanded; butt joints; fixing to timber with screws, countersinking Cont.										
9mm thick linings to walls										
width exceeding 300mm	15.00	5.00	15.75	0.49	0.06	4.94	0.11	20.80	m2	22.36
width not exceeding 300mm	4.50	10.00	4.95	0.29	0.02	2.82	0.07	7.84	m	8.42
12mm thick linings to walls										
width exceeding 300mm	21.80	5.00	22.89	0.53	0.07	5.38	0.13	28.40	m2	30.53
width not exceeding 300mm	6.54	10.00	7.19	0.32	0.02	3.09	0.07	10.36	m	11.13
6mm thick linings to ceilings										
width exceeding 300mm	10.10	5.00	10.61	0.57	0.07	5.75	0.14	16.50	m2	17.73
width not exceeding 300mm	3.03	10.00	3.33	0.34	0.02	3.28	0.08	6.69	m	7.19
9mm thick linings to ceilings										
width exceeding 300mm	15.00	5.00	15.75	0.61	0.08	6.19	0.15	22.09	m2	23.75
width not exceeding 300mm	4.50	10.00	4.95	0.37	0.02	3.56	0.08	8.59	m	9.23
12mm thick linings to ceilings										
width exceeding 300mm	21.80	5.00	22.89	0.67	0.08	6.75	0.16	29.80	m2	32.03
width not exceeding 300mm	6.54	10.00	7.19	0.40	0.02	3.83	0.09	11.12	m	11.95
6mm thick casings to isolated beams or the like										
total girth not exceeding 600mm	10.10	10.00	11.11	1.13	0.14	11.41	0.27	22.79	m2	24.50
total girth 600 - 1200mm	10.10	5.00	10.61	0.68	0.09	6.91	0.16	17.67	m2	19.00
9mm thick casings to isolated beams or the like										
total girth not exceeding 600mm	15.00	10.00	16.50	1.21	0.15	12.22	0.29	29.01	m2	31.19
total girth 600 - 1200mm	15.00	5.00	15.75	0.74	0.09	7.46	0.18	23.39	m2	25.15
12mm thick casings to isolated beams or the like										
total girth not exceeding 600mm	21.80	10.00	23.98	1.33	0.17	13.47	0.32	37.77	m2	40.60
total girth 600 - 1200mm	21.80	5.00	22.89	0.80	0.10	8.09	0.19	31.17	m2	33.50
6mm thick casings to isolated columns or the like										
total girth not exceeding 600mm	10.10	10.00	11.11	1.13	0.14	11.41	0.27	22.79	m2	24.50
total girth 600 - 1200mm	10.10	5.00	10.61	0.68	0.09	6.91	0.16	17.67	m2	19.00
9mm thick casings to isolated columns or the like										
total girth not exceeding 600mm	15.00	10.00	16.50	1.21	0.15	12.22	0.29	29.01	m2	31.19
total girth 600 - 1200mm	15.00	5.00	15.75	0.74	0.09	7.46	0.18	23.39	m2	25.15
12mm thick casings to isolated columns or the like										
total girth not exceeding 600mm	21.80	10.00	23.98	1.33	0.17	13.47	0.32	37.77	m2	40.60
total girth 600 - 1200mm	21.80	5.00	22.89	0.80	0.10	8.09	0.19	31.17	m2	33.50
ASBESTOS FREE BOARD - Vermiculux fire resisting boards; butt joints; fixing to steel with self tapping screws, countersinking										
20mm thick casings to isolated beams or the like										
total girth not exceeding 600mm	16.60	10.00	18.26	1.50	0.19	15.18	0.36	33.80	m2	36.33
total girth 600 - 1200mm	16.60	5.00	17.43	0.90	0.11	9.08	0.22	26.73	m2	28.73
30mm thick casings to isolated beams or the like										
total girth not exceeding 600mm	27.15	10.00	29.86	1.65	0.21	16.70	0.40	46.97	m2	50.49
total girth 600 - 1200mm	27.15	5.00	28.51	1.00	0.13	10.14	0.24	38.89	m2	41.81
40mm thick casings to isolated beams or the like										
total girth not exceeding 600mm	41.20	10.00	45.32	1.80	0.23	18.23	0.43	63.98	m2	68.78
total girth 600 - 1200mm	41.20	5.00	43.26	1.08	0.14	10.95	0.26	54.47	m2	58.56
50mm thick casings to isolated beams or the like										
total girth not exceeding 600mm	54.85	10.00	60.34	1.95	0.24	19.68	0.47	80.49	m2	86.52
total girth 600 - 1200mm	54.85	5.00	57.59	1.17	0.15	11.85	0.28	69.72	m2	74.95
60mm thick casings to isolated beams or the like										
total girth not exceeding 600mm	65.40	10.00	71.94	2.10	0.26	21.21	0.50	93.65	m2	100.67
total girth 600 - 1200mm	65.40	5.00	68.67	1.26	0.16	12.75	0.30	81.72	m2	87.85
20mm thick casings to isolated columns or the like										
total girth 600 - 1200mm	16.60	10.00	18.26	1.50	0.19	15.18	0.36	33.80	m2	36.33
total girth 1200 - 1800mm	16.60	5.00	17.43	0.90	0.11	9.08	0.22	26.73	m2	28.73
30mm thick casings to isolated columns or the like										
total girth 600 - 1200mm	27.15	10.00	29.86	1.65	0.21	16.70	0.40	46.97	m2	50.49
total girth 1200 - 1800mm	27.15	5.00	28.51	1.00	0.13	10.14	0.24	38.89	m2	41.81
40mm thick casings to isolated columns or the like										
total girth 600 - 1200mm	41.20	10.00	45.32	1.80	0.23	18.23	0.43	63.98	m2	68.78
total girth 1200 - 1800mm	41.20	5.00	43.26	1.08	0.14	10.95	0.26	54.47	m2	58.56
50mm thick casings to isolated columns or the like										
total girth 600 - 1200mm	54.85	10.00	60.34	1.95	0.24	19.68	0.47	80.49	m2	86.52
total girth 1200 - 1800mm	54.85	5.00	57.59	1.17	0.15	11.85	0.28	69.72	m2	74.95
60mm thick casings to isolated columns or the like										
total girth 600 - 1200mm	65.40	10.00	71.94	2.10	0.26	21.21	0.50	93.65	m2	100.67
total girth 1200 - 1800mm	65.40	5.00	68.67	1.26	0.16	12.75	0.30	81.72	m2	87.85

Labour hourly rates: (except Specialists) Craft Operatives £9.23 Labourer £7.02 Rates are national average prices. Refer to REGIONAL VARIATIONS for indicative levels of overall pricing in regions	MATERIALS			LABOUR				RATES		
	Del to Site	Waste	Material Cost	Craft Optve	Lab	Labour Cost	Sunds	Nett Rate	Unit	Gross Rate (+7.5%)
	£	%	£	Hrs	Hrs	£	£	£		£
K11: RIGID SHEET FLOORING/SHEATHING/LININGS/CASINGS Cont.										
ASBESTOS FREE BOARD - Vermiculux fire resisting boards; butt joints; fixing to steel with self tapping screws, countersinking Cont.										
50 x 25mm x 20g mild steel fixing angle										
fixing to masonry	1.29	5.00	**1.35**	0.40	0.05	**4.04**	0.10	**5.50**	m	**5.91**
fixing to steel	1.29	5.00	**1.35**	0.40	0.05	**4.04**	0.10	**5.50**	m	**5.91**
ASBESTOS FREE BOARD - Tacboard Class 0 fire resisting boards; butt joints										
6mm thick linings to walls										
width exceeding 300mm	6.70	5.00	**7.04**	0.45	0.06	**4.57**	0.11	**11.72**	m2	**12.60**
width not exceeding 300mm	2.01	10.00	**2.21**	0.27	0.02	**2.63**	0.06	**4.90**	m	**5.27**
9mm thick linings to walls										
width exceeding 300mm	12.34	5.00	**12.96**	0.49	0.06	**4.94**	0.12	**18.02**	m2	**19.37**
width not exceeding 300mm	3.70	10.00	**4.07**	0.29	0.02	**2.82**	0.07	**6.96**	m	**7.48**
12mm thick linings to walls										
width exceeding 300mm	16.06	5.00	**16.86**	0.53	0.07	**5.38**	0.13	**22.38**	m2	**24.05**
width not exceeding 300mm	4.82	10.00	**5.30**	0.32	0.02	**3.09**	0.07	**8.47**	m	**9.10**
6mm thick linings to ceilings										
width exceeding 300mm	6.70	5.00	**7.04**	0.57	0.07	**5.75**	0.14	**12.93**	m2	**13.90**
width not exceeding 300mm	2.01	10.00	**2.21**	0.34	0.02	**3.28**	0.08	**5.57**	m	**5.99**
ASBESTOS FREE BOARD - New Tacfire Class 1 fire resisting boards; butt joints										
6mm thick linings to walls										
width exceeding 300mm	8.95	5.00	**9.40**	0.45	0.06	**4.57**	0.11	**14.08**	m2	**15.14**
width not exceeding 300mm	2.69	10.00	**2.96**	0.27	0.02	**2.63**	0.06	**5.65**	m	**6.08**
9mm thick linings to walls										
width exceeding 300mm	13.71	5.00	**14.40**	0.49	0.06	**4.94**	0.12	**19.46**	m2	**20.92**
width not exceeding 300mm	4.11	10.00	**4.52**	0.29	0.02	**2.82**	0.07	**7.41**	m	**7.96**
12mm thick linings to walls										
width exceeding 300mm	18.08	5.00	**18.98**	0.53	0.07	**5.38**	0.13	**24.50**	m2	**26.33**
width not exceeding 300mm	5.42	10.00	**5.96**	0.32	0.02	**3.09**	0.07	**9.13**	m	**9.81**
6mm thick linings to ceilings; 6mm cover fillets										
width exceeding 300mm	8.95	5.00	**9.40**	0.57	0.07	**5.75**	1.13	**16.28**	m2	**17.50**
width not exceeding 300mm	2.69	10.00	**2.96**	0.34	0.02	**3.28**	0.37	**6.61**	m	**7.10**
9mm thick linings to ceilings; 9mm cover fillets										
width exceeding 300mm	13.71	5.00	**14.40**	0.61	0.08	**6.19**	2.04	**22.63**	m2	**24.32**
width not exceeding 300mm	4.11	10.00	**4.52**	0.37	0.02	**3.56**	0.65	**8.73**	m	**9.38**
12mm thick linings to ceilings; 9mm cover fillets										
width exceeding 300mm	18.08	5.00	**18.98**	0.67	0.08	**6.75**	2.06	**27.79**	m2	**29.87**
width not exceeding 300mm	5.42	10.00	**5.96**	0.40	0.02	**3.83**	0.66	**10.45**	m	**11.24**
ASBESTOS FREE BOARD - Vicuclad 900 fire resisting boards; butt joints; joints filled with c/v cement; fixing with screws; countersinking										
18mm thick casings to isolated beams or the like; including noggins										
total girth not exceeding 600mm	13.52	10.00	**14.87**	1.42	0.18	**14.37**	2.54	**31.78**	m2	**34.17**
total girth 600 - 1200mm	13.52	5.00	**14.20**	0.85	0.11	**8.62**	3.54	**26.35**	m2	**28.33**
20mm thick casings to isolated beams or the like; including noggins										
total girth not exceeding 600mm	14.08	10.00	**15.49**	1.50	0.19	**15.18**	2.69	**33.36**	m2	**35.86**
total girth 600 - 1200mm	14.08	5.00	**14.78**	0.90	0.11	**9.08**	3.51	**27.37**	m2	**29.43**
25mm thick casings to isolated beams or the like; including noggins										
total girth not exceeding 600mm	15.34	10.00	**16.87**	1.58	0.20	**15.99**	3.06	**35.92**	m2	**38.62**
total girth 600 - 1200mm	15.34	5.00	**16.11**	0.95	0.12	**9.61**	3.87	**29.59**	m2	**31.81**
30mm thick casings to isolated beams or the like; including noggins										
total girth not exceeding 600mm	18.60	10.00	**20.46**	1.65	0.21	**16.70**	3.02	**40.18**	m2	**43.20**
total girth 600 - 1200mm	18.60	5.00	**19.53**	1.00	0.13	**10.14**	4.63	**34.30**	m2	**36.88**
18mm thick casings to isolated columns or the like										
total girth not exceeding 600mm	13.52	10.00	**14.87**	1.42	0.18	**14.37**	1.57	**30.81**	m2	**33.12**
total girth 600 - 1200mm	13.52	5.00	**14.20**	0.85	0.11	**8.62**	1.43	**24.24**	m2	**26.06**
20mm thick casings to isolated columns or the like										
total girth not exceeding 600mm	14.08	10.00	**15.49**	1.50	0.19	**15.18**	1.72	**32.39**	m2	**34.82**
total girth 600 - 1200mm	14.08	5.00	**14.78**	0.90	0.11	**9.08**	1.58	**25.44**	m2	**27.35**
25mm thick casings to isolated columns or the like										
total girth not exceeding 600mm	15.34	10.00	**16.87**	1.58	0.20	**15.99**	2.09	**34.95**	m2	**37.57**
total girth 600 - 1200mm	15.34	5.00	**16.11**	0.95	0.12	**9.61**	1.94	**27.66**	m2	**29.73**
30mm thick casings to isolated columns or the like										
total girth not exceeding 600mm	18.60	10.00	**20.46**	1.65	0.21	**16.70**	2.45	**39.61**	m2	**42.58**
total girth 600 - 1200mm	18.60	5.00	**19.53**	1.00	0.13	**10.14**	2.29	**31.96**	m2	**34.36**

LININGS/SHEATHING/DRY PARTITIONING

Labour hourly rates: (except Specialists) Craft Operatives £9.23 Labourer £7.02 Rates are national average prices. Refer to REGIONAL VARIATIONS for indicative levels of overall pricing in regions	MATERIALS			LABOUR				RATES		
	Del to Site	Waste	Material Cost	Craft Optve	Lab	Labour Cost	Sunds	Nett Rate	Unit	Gross Rate (+7.5%)
	£	%	£	Hrs	Hrs	£	£	£		£
K11: RIGID SHEET FLOORING/SHEATHING/LININGS/CASINGS Cont.										
DUCT FRONTS - Plywood B.S.6566, II/III grade, MR bonded; butt joints; fixing to timber with screws, countersinking - duct fronts										
12mm thick linings to walls										
width exceeding 300mm	6.72	20.00	8.06	1.00	0.13	10.14	0.16	18.37	m2	19.74
width exceeding 300mm; fixing to timber with brass screws and cups	6.72	20.00	8.06	1.50	0.19	15.18	0.58	23.82	m2	25.61
DUCT FRONTS - Blockboard, B.S.3444, 2/2 grade, MR bonded; butt joints - duct fronts										
18mm thick linings to walls										
width exceeding 300mm	9.62	20.00	11.54	1.20	0.15	12.13	0.20	23.87	m2	25.66
width exceeding 300mm; fixing to timber with brass screws and cups	9.62	20.00	11.54	1.75	0.22	17.70	0.62	29.86	m2	32.10
DUCT FRONTS - Extra over plywood B.S.6566, II/III grade, MR bonded; butt joints; 12mm thick linings to walls; width exceeding 300mm										
Access panels										
300 x 600mm	-	-	-	0.50	0.05	4.97	0.08	5.05	nr	5.42
600 x 600mm	-	-	-	0.65	0.07	6.49	0.10	6.59	nr	7.09
600 x 900mm	-	-	-	0.80	0.08	7.95	0.13	8.08	nr	8.68
DUCT FRONTS - Extra over blockboard, B.S.3444, 2/2 grade, MR bonded, butt joints; 18mm thick linings to walls; width exceeding 300mm										
Access panels										
300 x 600mm	-	-	-	0.60	0.07	6.03	0.10	6.13	nr	6.59
600 x 600mm	-	-	-	0.80	0.08	7.95	0.13	8.08	nr	8.68
600 x 900mm	-	-	-	1.10	0.11	10.93	0.18	11.11	nr	11.94
PIPE CASINGS - Standard hardboard, B.S.1142, type SHA; butt joints - pipe casings										
3mm thick linings to pipe ducts and casings										
width exceeding 300mm	1.05	20.00	1.26	0.80	0.10	8.09	0.13	9.48	m2	10.19
width not exceeding 300mm	0.31	20.00	0.37	0.25	0.03	2.52	0.04	2.93	m	3.15
PIPE CASINGS - Plywood B.S.6566, II/III grade, MR bonded; butt joints; fixing to timber with screws, countersinking - pipe casings										
12mm thick linings to pipe ducts and casings										
width exceeding 300mm	6.72	20.00	8.06	1.50	0.19	15.18	0.24	23.48	m2	25.24
width not exceeding 300mm	2.02	20.00	2.42	0.50	0.06	5.04	0.08	7.54	m	8.11
PIPE CASINGS - Supalux Class 0 fire resisting boards, sanded; butt joints; fixing to timber with screws, countersinking - pipe casings										
9mm thick linings to pipe ducts and casings										
width exceeding 300mm	16.35	20.00	19.62	1.50	0.19	15.18	0.24	35.04	m2	37.67
width not exceeding 300mm	4.91	20.00	5.89	0.50	0.06	5.04	0.08	11.01	m	11.83
PIPE CASINGS - Pendock Profiles Ltd preformed plywood casings; white melamine finish; butt joints; fixing to timber with polytop white screws, countersinking										
5mm thick casings; horizontal										
reference TK110, 45 x 110mm	5.38	20.00	6.46	0.40	0.05	4.04	0.20	10.70	m	11.50
extra; internal corner	5.18	2.50	5.31	0.25	0.03	2.52	0.20	8.03	nr	8.63
extra; external corner	6.75	2.50	6.92	0.25	0.03	2.52	0.20	9.64	nr	10.36
extra; stop end	4.68	2.50	4.80	0.15	0.02	1.52	0.10	6.42	nr	6.90
reference TK150, 45 x 150mm	6.80	20.00	8.16	0.42	0.05	4.23	0.20	12.59	m	13.53
extra; internal corner	6.12	2.50	6.27	0.25	0.03	2.52	0.20	8.99	nr	9.67
extra; external corner	7.76	2.50	7.95	0.25	0.03	2.52	0.20	10.67	nr	11.47
extra; stop end	6.55	2.50	6.71	0.15	0.02	1.52	0.10	8.34	nr	8.96
reference TK190, 45 x 190mm	8.93	20.00	10.72	0.44	0.05	4.41	0.20	15.33	m	16.48
extra; internal corner	6.16	2.50	6.31	0.25	0.03	2.52	0.20	9.03	nr	9.71
extra; external corner	7.80	2.50	8.00	0.25	0.03	2.52	0.20	10.71	nr	11.52
extra; stop end	7.80	2.50	8.00	0.15	0.02	1.52	0.10	9.62	nr	10.34
reference TK125, 25 x 125mm	5.37	20.00	6.44	0.40	0.05	4.04	0.20	10.69	m	11.49
extra; internal corner	5.19	2.50	5.32	0.25	0.03	2.52	0.20	8.04	nr	8.64
extra; external corner	6.75	2.50	6.92	0.25	0.03	2.52	0.20	9.64	nr	10.36
extra; stop end	4.68	2.50	4.80	0.15	0.02	1.52	0.10	6.42	nr	6.90
5mm thick casings; vertical										
reference TK110, 45 x 110mm	4.90	20.00	5.88	0.40	0.05	4.04	0.20	10.12	m	10.88
reference TK135, 45 x 135mm	7.15	20.00	8.58	0.41	0.05	4.14	0.20	12.92	m	13.88
reference TK150, 45 x 150mm	6.27	20.00	7.52	0.42	0.05	4.23	0.20	11.95	m	12.85
reference TK190, 45 x 190mm	8.31	20.00	9.97	0.44	0.05	4.41	0.20	14.58	m	15.68
5mm thick casings; horizontal										
reference MX 100/100, 100 x 100mm	8.97	20.00	10.76	0.42	0.05	4.23	0.20	15.19	m	16.33
extra; stop end	7.78	2.50	7.97	0.15	0.02	1.52	0.10	9.60	nr	10.32
8mm thick casings; horizontal										
reference MX 150/150, 150 x 150mm	13.58	20.00	16.30	0.45	0.06	4.57	0.20	21.07	m	22.65
extra; internal corner	9.67	2.50	9.91	0.25	0.03	2.52	0.20	12.63	nr	13.58

Labour hourly rates: (except Specialists) Craft Operatives £9.23 Labourer £7.02 Rates are national average prices. Refer to REGIONAL VARIATIONS for indicative levels of overall pricing in regions	MATERIALS			LABOUR				RATES		
	Del to Site £	Waste %	Material Cost £	Craft Optve Hrs	Lab Hrs	Labour Cost £	Sunds £	Nett Rate £	Unit	Gross Rate (+7.5%) £
K11: RIGID SHEET FLOORING/SHEATHING/LININGS/CASINGS Cont.										
PIPE CASINGS - Pendock Profiles Ltd preformed plywood casings; white melamine finish; butt joints; fixing to timber with polytop white screws, countersinking Cont.										
8mm thick casings; horizontal Cont.										
extra; external corner	14.58	2.50	14.94	0.25	0.03	2.52	0.20	17.66	nr	18.99
extra; stop end	7.78	2.50	7.97	0.15	0.02	1.52	0.10	9.60	nr	10.32
reference MX 200/150, 200 x 150mm	15.05	20.00	18.06	0.46	0.06	4.67	0.20	22.93	m	24.65
extra; stop end	7.78	2.50	7.97	0.15	0.02	1.52	0.10	9.60	nr	10.32
reference MX 200/200, 200 x 200mm	16.41	20.00	19.69	0.47	0.06	4.76	0.20	24.65	m	26.50
extra; internal corner	9.75	2.50	9.99	0.25	0.03	2.52	0.20	12.71	nr	13.67
extra; external corner	14.74	2.50	15.11	0.25	0.03	2.52	0.20	17.83	nr	19.16
extra; stop end	7.78	2.50	7.97	0.15	0.02	1.52	0.10	9.60	nr	10.32
reference MX 200/300, 200 x 300mm	19.26	20.00	23.11	0.48	0.06	4.85	0.20	28.16	m	30.28
extra; stop end	8.54	2.50	8.75	0.15	0.02	1.52	0.10	10.38	nr	11.16
reference MX 300/150, 300 x 150mm	18.21	20.00	21.85	0.48	0.06	4.85	0.20	26.90	m	28.92
extra; internal corner	10.35	2.50	10.61	0.25	0.03	2.52	0.20	13.33	nr	14.33
extra; external corner	15.38	2.50	15.76	0.25	0.03	2.52	0.20	18.48	nr	19.87
extra; stop end	8.54	2.50	8.75	0.15	0.02	1.52	0.10	10.38	nr	11.16
reference MX 300/300, 300 x 300mm	21.96	20.00	26.35	0.50	0.06	5.04	0.20	31.59	m	33.96
extra; internal corner	11.04	2.50	11.32	0.25	0.03	2.52	0.20	14.03	nr	15.09
extra; external corner	16.08	2.50	16.48	0.25	0.03	2.52	0.20	19.20	nr	20.64
extra; stop end	10.61	2.50	10.88	0.15	0.02	1.52	0.10	12.50	nr	13.44
5mm thick casings; vertical										
reference MX 100/100, 100 x 100mm	7.51	20.00	9.01	0.42	0.05	4.23	0.20	13.44	m	14.45
8mm thick casings; vertical										
reference MX 150/150, 150 x 150mm	12.10	20.00	14.52	0.45	0.06	4.57	0.20	19.29	m	20.74
reference MX 200/150, 200 x 150mm	13.59	20.00	16.31	0.46	0.06	4.67	0.20	21.18	m	22.76
reference MX 200/200, 200 x 200mm	14.96	20.00	17.95	0.47	0.06	4.76	0.20	22.91	m	24.63
reference MX 200/300, 200 x 300mm	17.80	20.00	21.36	0.48	0.06	4.85	0.20	26.41	m	28.39
reference MX 300/150, 300 x 150mm	16.75	20.00	20.10	0.49	0.06	4.94	0.20	25.24	m	27.14
reference MX 300/300, 300 x 300mm	20.49	20.00	24.59	0.50	0.06	5.04	0.20	29.82	m	32.06
ROOF BOARDING - Plywood B.S.6566, II/III grade, WBP bonded; butt joints										
18mm thick sheeting to roofs; external										
width exceeding 300mm	10.55	10.00	11.61	0.30	0.04	3.05	0.05	14.70	m2	15.81
extra; cesspool 225 x 225 x 150mm	2.11	10.00	2.32	1.40	0.18	14.19	0.23	16.74	nr	17.99
extra; cesspool 300 x 300 x 150mm	2.90	10.00	3.19	1.50	0.19	15.18	0.24	18.61	nr	20.00
width exceeding 300mm; sloping	10.55	10.00	11.61	0.40	0.05	4.04	0.06	15.71	m2	16.89
24mm thick sheeting to roofs; external										
width exceeding 300mm	14.66	10.00	16.13	0.35	0.04	3.51	0.05	19.69	m2	21.16
extra; cesspool 225 x 225 x 150mm	2.93	10.00	3.22	1.50	0.19	15.18	0.24	18.64	nr	20.04
extra; cesspool 300 x 300 x 150mm	4.03	10.00	4.43	1.60	0.20	16.17	0.26	20.86	nr	22.43
width exceeding 300mm; sloping	14.66	10.00	16.13	0.45	0.06	4.57	0.07	20.77	m2	22.33
ROOF BOARDING - Wood chipboard, B.S.5669, type 1 standard; butt joints										
12mm thick sheeting to roofs; external										
width exceeding 300mm	2.15	10.00	2.37	0.25	0.03	2.52	0.04	4.92	m2	5.29
extra; cesspool 225 x 225 x 150mm	0.43	10.00	0.47	1.30	0.16	13.12	0.21	13.81	nr	14.84
extra; cesspool 300 x 300 x 150mm	0.59	10.00	0.65	1.40	0.18	14.19	0.23	15.06	nr	16.19
width exceeding 300mm; sloping	2.15	10.00	2.37	0.35	0.04	3.51	0.06	5.94	m2	6.38
18mm thick sheeting to roofs; external										
width exceeding 300mm	2.88	10.00	3.17	0.30	0.04	3.05	0.05	6.27	m2	6.74
extra; cesspool 225 x 225 x 150mm	0.58	10.00	0.64	1.40	0.18	14.19	0.23	15.05	nr	16.18
extra; cesspool 300 x 300 x 150mm	0.79	10.00	0.87	1.50	0.19	15.18	0.24	16.29	nr	17.51
width exceeding 300mm; sloping	2.88	10.00	3.17	0.40	0.05	4.04	0.06	7.27	m2	7.82
ROOF BOARDING - Abutments										
18mm thick plywood linings										
raking cutting	1.06	-	1.06	0.30	-	2.77	-	3.83	m	4.12
curved cutting	1.58	-	1.58	0.45	-	4.15	-	5.73	m	6.16
rebates	-	-	-	0.18	-	1.66	-	1.66	m	1.79
grooves	-	-	-	0.18	-	1.66	-	1.66	m	1.79
chamfers	-	-	-	0.13	-	1.20	-	1.20	m	1.29
24mm thick plywood linings										
raking cutting	1.47	-	1.47	0.40	-	3.69	-	5.16	m	5.55
curved cutting	2.20	-	2.20	0.60	-	5.54	-	7.74	m	8.32
rebates	-	-	-	0.18	-	1.66	-	1.66	m	1.79
grooves	-	-	-	0.18	-	1.66	-	1.66	m	1.79
chamfers	-	-	-	0.13	-	1.20	-	1.20	m	1.29
12mm thick wood chipboard linings										
raking cutting	0.21	-	0.21	0.25	-	2.31	-	2.52	m	2.71
curved cutting	0.32	-	0.32	0.35	-	3.23	-	3.55	m	3.82
rebates	-	-	-	0.18	-	1.66	-	1.66	m	1.79
grooves	-	-	-	0.18	-	1.66	-	1.66	m	1.79
chamfers	-	-	-	0.13	-	1.20	-	1.20	m	1.29
18mm thick wood chipboard linings										
raking cutting	0.29	-	0.29	0.30	-	2.77	-	3.06	m	3.29
curved cutting	0.43	-	0.43	0.45	-	4.15	-	4.58	m	4.93
rebates	-	-	-	0.18	-	1.66	-	1.66	m	1.79

LININGS/SHEATHING/DRY PARTITIONING

Labour hourly rates: (except Specialists) Craft Operatives £9.23 Labourer £7.02 Rates are national average prices. Refer to REGIONAL VARIATIONS for indicative levels of overall pricing in regions	MATERIALS			LABOUR				RATES		
	Del to Site £	Waste %	Material Cost £	Craft Optve Hrs	Lab Hrs	Labour Cost £	Sunds £	Nett Rate £	Unit	Gross Rate (+7.5%) £
K11: RIGID SHEET FLOORING/SHEATHING/LININGS/CASINGS Cont.										
ROOF BOARDING - Abutments Cont.										
18mm thick wood chipboard linings Cont.										
grooves	-	-	-	0.18	-	1.66	-	1.66	m	1.79
chamfers	-	-	-	0.13	-	1.20	-	1.20	m	1.29
WOOD WOOL BUILDING SLABS - Wood wool slabs, B.S.1105 Type A; natural both sides; butt joints										
25mm thick linings to walls										
width exceeding 300mm	4.79	5.00	5.03	0.31	0.04	3.14	0.05	8.22	m2	8.84
38mm thick linings to walls										
width exceeding 300mm	5.34	5.00	5.61	0.37	0.05	3.77	0.06	9.43	m2	10.14
50mm thick linings to walls										
width exceeding 300mm	5.38	5.00	5.65	0.44	0.06	4.48	0.07	10.20	m2	10.97
25mm thick linings to ceilings										
width exceeding 300mm	4.79	5.00	5.03	0.34	0.04	3.42	0.05	8.50	m2	9.14
38mm thick linings to ceilings										
width exceeding 300mm	5.34	5.00	5.61	0.41	0.05	4.14	0.06	9.80	m2	10.54
50mm thick linings to ceilings										
width exceeding 300mm	5.38	5.00	5.65	0.48	0.06	4.85	0.07	10.57	m2	11.36
WOOD WOOL BUILDING SLABS - Wood wool slabs, B.S.1105 Type B; natural both sides; butt joints										
50mm thick sheeting to roofs; external										
width exceeding 300mm; fixing to timber with galvanised nails and gripper plates	5.50	5.00	5.78	0.44	0.06	4.48	0.12	10.38	m2	11.16
width exceeding 300mm; fixing to steel rails with sherardised clips	5.50	5.00	5.78	0.33	0.04	3.33	0.24	9.34	m2	10.04
75mm thick sheeting to roofs; external										
width exceeding 300mm; fixing to timber with galvanised nails and gripper plates	8.84	5.00	9.28	0.56	0.07	5.66	0.15	15.09	m2	16.22
width exceeding 300mm; fixing to steel rails with sherardised clips	8.84	5.00	9.28	0.42	0.05	4.23	0.30	13.81	m2	14.85
100mm thick sheeting to roofs; external										
width exceeding 300mm; fixing to timber with galvanised nails and gripper plates	12.42	5.00	13.04	0.72	0.09	7.28	0.19	20.51	m2	22.05
width exceeding 300mm; fixing to steel rails with sherardised clips	12.42	5.00	13.04	0.54	0.07	5.48	0.38	18.90	m2	20.31
WOOD WOOL BUILDING SLABS - Wood wool slabs, B.S.1105 Type B; natural soffit, pre-felted top surface; butt joints										
50mm thick sheeting to roofs; external										
width exceeding 300mm; fixing to timber with galvanised nails and gripper plates	9.90	5.00	10.40	0.44	0.06	4.48	0.12	15.00	m2	16.12
width exceeding 300mm; fixing to steel rails with sherardised clips	9.90	5.00	10.40	0.33	0.04	3.33	0.24	13.96	m2	15.01
75mm thick sheeting to roofs; external										
width exceeding 300mm; fixing to timber with galvanised nails and gripper plates	13.23	5.00	13.89	0.56	0.07	5.66	0.15	19.70	m2	21.18
width exceeding 300mm; fixing to steel rails with sherardised clips	13.23	5.00	13.89	0.42	0.05	4.23	0.30	18.42	m2	19.80
100mm thick sheeting to roofs; external										
width exceeding 300mm; fixing to timber with galvanised nails and gripper plates	16.80	5.00	17.64	0.72	0.09	7.28	0.19	25.11	m2	26.99
width exceeding 300mm; fixing to steel rails with sherardised clips	16.80	5.00	17.64	0.54	0.07	5.48	0.38	23.50	m2	25.26
WOOD WOOL BUILDING SLABS - Wood wool slabs, B.S.1105 Type B; natural soffit, pre-screeded top surface; butt joints										
50mm thick sheeting to roofs; external										
width exceeding 300mm; fixing to timber with galvanised nails and gripper plates	8.37	5.00	8.79	0.44	0.06	4.48	0.12	13.39	m2	14.40
width exceeding 300mm; fixing to steel rails with sherardised clips	8.37	5.00	8.79	0.33	0.04	3.33	0.24	12.36	m2	13.28
75mm thick sheeting to roofs; external										
width exceeding 300mm; fixing to timber with galvanised nails and gripper plates	11.71	5.00	12.30	0.56	0.07	5.66	0.15	18.11	m2	19.46
width exceeding 300mm; fixing to steel rails with sherardised clips	11.71	5.00	12.30	0.42	0.05	4.23	0.30	16.82	m2	18.08
100mm thick sheeting to roofs; external										
width exceeding 300mm; fixing to timber with galvanised nails and gripper plates	15.30	5.00	16.07	0.72	0.09	7.28	0.19	23.53	m2	25.30
width exceeding 300mm; fixing to steel rails with sherardised clips	15.30	5.00	16.07	0.54	0.07	5.48	0.38	21.92	m2	23.56

Labour hourly rates: (except Specialists) Craft Operatives £9.23 Labourer £7.02 Rates are national average prices. Refer to REGIONAL VARIATIONS for indicative levels of overall pricing in regions	MATERIALS			LABOUR				RATES		
	Del to Site £	Waste %	Material Cost £	Craft Optve Hrs	Lab Hrs	Labour Cost £	Sunds £	Nett Rate £	Unit	Gross Rate (+7.5%) £
K11: RIGID SHEET FLOORING/SHEATHING/ LININGS/CASINGS; STAIRCASE AREAS										
ASBESTOS FREE BOARD - Masterboard Class 0 fire resisting boards; butt joints										
6mm thick linings to walls										
width exceeding 300mm	6.85	5.00	7.19	0.57	0.07	5.75	0.12	13.07	m2	14.04
width not exceeding 300mm	2.06	10.00	2.27	0.34	0.02	3.28	0.07	5.61	m	6.04
9mm thick linings to walls										
width exceeding 300mm	13.85	5.00	14.54	0.61	0.08	6.19	0.13	20.86	m2	22.43
width not exceeding 300mm	4.16	10.00	4.58	0.37	0.02	3.56	0.08	8.21	m	8.83
12mm thick linings to walls										
width exceeding 300mm	18.30	5.00	19.22	0.67	0.08	6.75	0.15	26.11	m2	28.07
width not exceeding 300mm	5.49	10.00	6.04	0.40	0.02	3.83	0.08	9.95	m	10.70
6mm thick linings to ceilings										
width exceeding 300mm	6.85	5.00	7.19	0.61	0.08	6.19	0.13	13.51	m2	14.53
width not exceeding 300mm	2.06	10.00	2.27	0.37	0.02	3.56	0.08	5.90	m	6.34
9mm thick linings to ceilings										
width exceeding 300mm	13.85	5.00	14.54	0.67	0.08	6.75	0.15	21.44	m2	23.05
width not exceeding 300mm	4.16	10.00	4.58	0.40	0.02	3.83	0.08	8.49	m	9.13
12mm thick linings to ceilings										
width exceeding 300mm	18.30	5.00	19.22	0.72	0.09	7.28	0.16	26.65	m2	28.65
width not exceeding 300mm	5.49	10.00	6.04	0.43	0.03	4.18	0.09	10.31	m	11.08
ASBESTOS FREE BOARD - Supalux fire resisting boards, sanded; butt joints; fixing to timber with screws, countersinking										
6mm thick linings to walls										
width exceeding 300mm	10.10	5.00	10.61	0.57	0.07	5.75	0.12	16.48	m2	17.71
width not exceeding 300mm	3.03	10.00	3.33	0.34	0.02	3.28	0.07	6.68	m	7.18
9mm thick linings to walls										
width exceeding 300mm	15.00	5.00	15.75	0.61	0.08	6.19	0.13	22.07	m2	23.73
width not exceeding 300mm	4.50	10.00	4.95	0.37	0.02	3.56	0.08	8.59	m	9.23
12mm thick linings to walls										
width exceeding 300mm	21.80	5.00	22.89	0.67	0.08	6.75	0.15	29.79	m2	32.02
width not exceeding 300mm	6.54	10.00	7.19	0.40	0.02	3.83	0.08	11.11	m	11.94
6mm thick linings to ceilings										
width exceeding 300mm	10.10	5.00	10.61	0.61	0.08	6.19	0.13	16.93	m2	18.20
width not exceeding 300mm	3.03	10.00	3.33	0.37	0.02	3.56	0.08	6.97	m	7.49
9mm thick linings to ceilings										
width exceeding 300mm	15.00	5.00	15.75	0.67	0.08	6.75	0.15	22.65	m2	24.34
width not exceeding 300mm	4.50	10.00	4.95	0.40	0.02	3.83	0.08	8.86	m	9.53
12mm thick linings to ceilings										
width exceeding 300mm	21.80	5.00	22.89	0.72	0.09	7.28	0.16	30.33	m2	32.60
width not exceeding 300mm	6.54	10.00	7.19	0.43	0.03	4.18	0.09	11.46	m	12.32
K12: UNDER PURLIN/INSIDE RAIL PANEL LININGS										
PLASTIC FACED PLASTERBOARD - Paraclip metal grid fixing system; square edge sheeting of one layer 9.5mm thick Gyproc industrial grade board, B.S.1230, butt joints; laying in position in zinc coated mild steel grid; fixing to metal with self-tapping screws										
9.5mm thick linings to walls										
width exceeding 300mm	4.52	5.00	4.75	0.80	0.40	10.19	2.37	17.31	m2	18.61
width not exceeding 300mm	1.36	10.00	1.50	0.48	0.12	5.27	1.42	8.19	m	8.80
9.5mm thick linings to ceilings										
width exceeding 300mm	4.52	5.00	4.75	1.00	0.50	12.74	2.37	19.86	m2	21.35
width not exceeding 300mm	1.36	10.00	1.50	0.60	0.15	6.59	1.42	9.51	m	10.22
K13: RIGID SHEET FINE LININGS/ PANELLING										
Softwood, wrot - wall panelling										
19mm thick panelled linings to walls; square framed; including grounds										
width exceeding 300mm	66.60	2.50	68.27	1.60	0.20	16.17	0.27	84.71	m2	91.06
width exceeding 300mm; fixing to masonry with screws	66.60	2.50	68.27	2.60	0.32	26.24	0.44	94.95	m2	102.07
25mm thick panelled linings to walls; square framed; including grounds										
width exceeding 300mm	67.50	2.50	69.19	1.70	0.21	17.17	0.29	86.64	m2	93.14
width exceeding 300mm; fixing to masonry with screws	67.50	2.50	69.19	2.70	0.34	27.31	0.47	96.97	m2	104.24
19mm thick panelled linings to walls; square framed; obstructed by integral services; including grounds										
width exceeding 300mm	66.60	2.50	68.27	2.10	0.26	21.21	0.36	89.83	m2	96.57
width exceeding 300mm; fixing to masonry with screws	66.60	2.50	68.27	3.10	0.39	31.35	0.53	100.15	m2	107.66

Labour hourly rates: (except Specialists) Craft Operatives £9.23 Labourer £7.02. Rates are national average prices. Refer to REGIONAL VARIATIONS for indicative levels of overall pricing in regions	MATERIALS			LABOUR				RATES		
	Del to Site £	Waste %	Material Cost £	Craft Optve Hrs	Lab Hrs	Labour Cost £	Sunds £	Nett Rate £	Unit	Gross Rate (+7.5%) £
K13: RIGID SHEET FINE LININGS/ PANELLING Cont.										
Softwood, wrot - wall panelling Cont.										
25mm thick panelled linings to walls; square framed; obstructed by integral services; including grounds										
width exceeding 300mm	67.50	2.50	69.19	2.20	0.27	22.20	0.38	91.77	m2	98.65
width exceeding 300mm; fixing to masonry with screws	67.50	2.50	69.19	3.20	0.40	32.34	0.55	102.08	m2	109.74
19mm thick panelled linings to walls; moulded; including grounds										
width exceeding 300mm	72.90	2.50	74.72	1.70	0.21	17.17	0.29	92.18	m2	99.09
width exceeding 300mm; fixing to masonry with screws	72.90	2.50	74.72	2.70	0.34	27.31	0.47	102.50	m2	110.19
25mm thick panelled linings to walls; moulded; including grounds										
width exceeding 300mm	75.60	2.50	77.49	1.80	0.22	18.16	0.30	95.95	m2	103.14
width exceeding 300mm; fixing to masonry with screws	75.60	2.50	77.49	2.80	0.35	28.30	0.48	106.27	m2	114.24
19mm thick panelled linings to walls; moulded; obstructed by integral services; including grounds										
width exceeding 300mm	72.90	2.50	74.72	2.20	0.27	22.20	0.38	97.30	m2	104.60
width exceeding 300mm; fixing to masonry with screws	72.90	2.50	74.72	3.20	0.40	32.34	0.55	107.62	m2	115.69
25mm thick panelled linings to walls; moulded; obstructed by integral services; including grounds										
width exceeding 300mm	75.60	2.50	77.49	2.30	0.29	23.26	0.39	101.14	m2	108.73
width exceeding 300mm; fixing to masonry with screws	75.60	2.50	77.49	3.30	0.41	33.34	0.57	111.40	m2	119.75
Afromosia, wrot, selected for transparent finish - wall panelling										
19mm thick panelled linings to walls; square framed; including grounds										
width exceeding 300mm	194.40	2.50	199.26	2.25	0.28	22.73	0.39	222.38	m2	239.06
width exceeding 300mm; fixing to masonry with screws	194.40	2.50	199.26	3.25	0.41	32.88	0.56	232.70	m2	250.15
25mm thick panelled linings to walls; square framed; including grounds										
width exceeding 300mm	199.80	2.50	204.79	2.35	0.29	23.73	0.40	228.92	m2	246.09
width exceeding 300mm; fixing to masonry with screws	199.80	2.50	204.79	3.35	0.42	33.87	0.58	239.24	m2	257.19
19mm thick panelled linings to walls; square framed; obstructed by integral services; including grounds										
width exceeding 300mm	194.40	2.50	199.26	2.75	0.34	27.77	0.47	227.50	m2	244.56
width exceeding 300mm; fixing to masonry with screws	194.40	2.50	199.26	3.75	0.47	37.91	0.64	237.81	m2	255.65
25mm thick panelled linings to walls; square framed; obstructed by integral services; including grounds										
width exceeding 300mm	199.80	2.50	204.79	2.85	0.36	28.83	0.48	234.11	m2	251.67
width exceeding 300mm; fixing to masonry with screws	199.80	2.50	204.79	3.85	0.48	38.91	0.66	244.36	m2	262.69
19mm thick panelled linings to walls; moulded; including grounds										
width exceeding 300mm	208.80	2.50	214.02	2.40	0.30	24.26	0.41	238.69	m2	256.59
width exceeding 300mm; fixing to masonry with screws	208.80	2.50	214.02	3.40	0.42	34.33	0.58	248.93	m2	267.60
25mm thick panelled linings to walls; moulded; including grounds										
width exceeding 300mm	222.30	2.50	227.86	2.50	0.31	25.25	0.43	253.54	m2	272.55
width exceeding 300mm; fixing to masonry with screws	222.30	2.50	227.86	3.50	0.44	35.39	0.60	263.85	m2	283.64
19mm thick panelled linings to walls; moulded; obstructed by integral services; including grounds										
width exceeding 300mm	200.80	2.50	205.82	2.90	0.36	29.29	0.49	235.60	m2	253.27
width exceeding 300mm; fixing to masonry with screws	200.80	2.50	205.82	3.90	0.49	39.44	0.66	245.92	m2	264.36
25mm thick panelled linings to walls; moulded; obstructed by integral services; including grounds										
width exceeding 300mm	222.30	2.50	227.86	3.00	0.37	30.29	0.51	258.65	m2	278.05
width exceeding 300mm; fixing to masonry with screws	222.30	2.50	227.86	4.00	0.50	40.43	0.68	268.97	m2	289.14
Sapele, wrot, selected for transparent finish - wall panelling										
19mm thick panelled linings to walls; square framed; including grounds										
width exceeding 300mm	130.50	2.50	133.76	2.05	0.26	20.75	0.35	154.86	m2	166.47
width exceeding 300mm; fixing to masonry with screws	130.50	2.50	133.76	3.05	0.38	30.82	0.53	165.11	m2	177.49

Labour hourly rates: (except Specialists) Craft Operatives £9.23 Labourer £7.02 Rates are national average prices. Refer to REGIONAL VARIATIONS for indicative levels of overall pricing in regions	MATERIALS			LABOUR				RATES		
	Del to Site £	Waste %	Material Cost £	Craft Optve Hrs	Lab Hrs	Labour Cost £	Sunds £	Nett Rate £	Unit	Gross Rate (+7.5%) £
K13: RIGID SHEET FINE LININGS/ PANELLING Cont.										
Sapele, wrot, selected for transparent finish - wall panelling Cont.										
25mm thick panelled linings to walls; square framed; including grounds										
width exceeding 300mm	136.80	2.50	**140.22**	2.15	0.27	**21.74**	0.37	**162.33**	m2	**174.50**
width exceeding 300mm; fixing to masonry with screws	136.80	2.50	**140.22**	3.15	0.39	**31.81**	0.54	**172.57**	m2	**185.52**
19mm thick panelled linings to walls; square framed; obstructed by integral services; including grounds										
width exceeding 300mm	130.50	2.50	**133.76**	2.55	0.32	**25.78**	0.44	**159.99**	m2	**171.98**
width exceeding 300mm; fixing to masonry with screws	130.50	2.50	**133.76**	3.55	0.44	**35.86**	0.61	**170.23**	m2	**182.99**
25mm thick panelled linings to walls; square framed; obstructed by integral services; including grounds										
width exceeding 300mm	136.80	2.50	**140.22**	2.65	0.33	**26.78**	0.45	**167.45**	m2	**180.00**
width exceeding 300mm; fixing to masonry with screws	136.80	2.50	**140.22**	3.65	0.46	**36.92**	0.62	**177.76**	m2	**191.09**
19mm thick panelled linings to walls; moulded; including grounds										
width exceeding 300mm	144.00	2.50	**147.60**	2.15	0.27	**21.74**	0.37	**169.71**	m2	**182.44**
width exceeding 300mm; fixing to masonry with screws	144.00	2.50	**147.60**	3.15	0.39	**31.81**	0.54	**179.95**	m2	**193.45**
25mm thick panelled linings to walls; moulded; including grounds										
width exceeding 300mm	149.40	2.50	**153.13**	2.30	0.29	**23.26**	0.39	**176.79**	m2	**190.05**
width exceeding 300mm; fixing to masonry with screws	149.40	2.50	**153.13**	3.30	0.41	**33.34**	0.57	**187.04**	m2	**201.07**
19mm thick panelled linings to walls; moulded; obstructed by integral services; including grounds										
width exceeding 300mm	144.00	2.50	**147.60**	2.65	0.33	**26.78**	0.45	**174.83**	m2	**187.94**
width exceeding 300mm; fixing to masonry with screws	144.00	2.50	**147.60**	3.65	0.46	**36.92**	0.62	**185.14**	m2	**199.02**
25mm thick panelled linings to walls; moulded; obstructed by integral services; including grounds										
width exceeding 300mm	149.40	2.50	**153.13**	2.80	0.35	**28.30**	0.48	**181.92**	m2	**195.56**
width exceeding 300mm; fixing to masonry with screws	149.40	2.50	**153.13**	3.80	0.47	**38.37**	0.65	**192.16**	m2	**206.57**
VENEERED PLYWOOD PANELLING - Plywood, pre-finished, decorative veneers; butt joints - wall panelling Note. prices are for panelling with Afromosia, Ash, Beech, Cherry, Elm, Oak, Knotted Pine, Sapele or Teak faced veneers										
4mm thick linings to walls										
width exceeding 300mm	9.56	15.00	**10.99**	0.80	0.10	**8.09**	0.13	**19.21**	m2	**20.65**
width exceeding 300 mm; fixing to timber with adhesive	9.56	15.00	**10.99**	0.95	0.12	**9.61**	2.00	**22.60**	m2	**24.30**
width exceeding 300mm; fixing to plaster with adhesive	9.56	15.00	**10.99**	0.90	0.11	**9.08**	2.00	**22.07**	m2	**23.73**
VENEERED PLYWOOD PANELLING - Plywood, flame retardant, pre-finished, decorative, veneers not matched; random V-grooves on face; butt joints - wall panelling. Note. prices are for panelling with Afromosia, Ash, Beech, Cherry, Elm, Oak, Knotted Pine, Sapele or Teak faced veneers										
4mm thick linings to walls										
width exceeding 300mm	25.65	15.00	**29.50**	0.80	0.10	**8.09**	0.13	**37.71**	m2	**40.54**
width exceeding 300 mm; fixing to timber with adhesive	25.65	15.00	**29.50**	0.95	0.12	**9.61**	2.00	**41.11**	m2	**44.19**
width exceeding 300mm; fixing to plaster with adhesive	25.65	15.00	**29.50**	0.90	0.11	**9.08**	2.00	**40.58**	m2	**43.62**
K20: TIMBER BOARD FLOORING/SHEATHING/ LININGS/CASINGS										
Softwood, sawn; fixing to timber - boarded flooring										
Boarding to floors, square edges; 19mm thick, 75mm wide boards										
width exceeding 300mm	3.40	7.50	**3.65**	0.60	0.08	**6.10**	0.10	**9.85**	m2	**10.59**
Boarding to floors, square edges; 25mm thick, 125mm wide boards										
width exceeding 300mm	4.25	7.50	**4.57**	0.55	0.07	**5.57**	0.09	**10.23**	m2	**10.99**
Boarding to floors, square edges; 32mm thick, 150mm wide boards										
width exceeding 300mm	5.41	7.50	**5.82**	0.55	0.07	**5.57**	0.09	**11.47**	m2	**12.33**
Softwood, wrought; fixing to timber - boarded flooring										
Boarding to floors, square edges; 19mm thick, 75mm wide boards										
width exceeding 300mm	5.67	7.50	**6.10**	0.70	0.09	**7.09**	0.11	**13.30**	m2	**14.30**

Labour hourly rates: (except Specialists) Craft Operatives £9.23 Labourer £7.02 Rates are national average prices. Refer to REGIONAL VARIATIONS for indicative levels of overall pricing in regions	MATERIALS			LABOUR				RATES		
	Del to Site £	Waste %	Material Cost £	Craft Optve Hrs	Lab Hrs	Labour Cost £	Sunds £	Nett Rate £	Unit	Gross Rate (+7.5%) £
K20: TIMBER BOARD FLOORING/SHEATHING/ LININGS/CASINGS Cont.										
Softwood, wrought; fixing to timber - boarded flooring Cont.										
Boarding to floors, square edges; 25mm thick, 125mm wide boards										
width exceeding 300mm	7.09	7.50	7.62	0.65	0.08	6.56	0.11	14.29	m2	15.36
Boarding to floors, square edges; 32mm thick, 150mm wide boards										
width exceeding 300mm	9.02	7.50	9.70	0.65	0.08	6.56	0.11	16.37	m2	17.60
Boarding to floors, tongued and grooved joints; 19mm thick, 75mm wide boards										
width exceeding 300mm	6.48	7.50	6.97	0.80	0.10	8.09	0.13	15.18	m2	16.32
Boarding to floors, tongued and grooved joints; 25mm thick, 125mm wide boards										
width exceeding 300mm	7.76	7.50	8.34	0.75	0.10	7.62	0.12	16.09	m2	17.29
Boarding to floors, tongued and grooved joints; 32mm thick, 150mm wide boards										
width exceeding 300mm	9.72	7.50	10.45	0.75	0.10	7.62	0.12	18.19	m2	19.56
Abutments										
19mm thick softwood boarding										
raking cutting	0.57	-	0.57	0.06	-	0.55	-	1.12	m	1.21
curved cutting	0.85	-	0.85	0.09	-	0.83	-	1.68	m	1.81
25mm thick softwood boarding										
raking cutting	0.71	-	0.71	0.06	-	0.55	-	1.26	m	1.36
curved cutting	1.06	-	1.06	0.09	-	0.83	-	1.89	m	2.03
32mm thick softwood boarding										
raking cutting	0.90	-	0.90	0.06	-	0.55	-	1.45	m	1.56
curved cutting	1.35	-	1.35	0.09	-	0.83	-	2.18	m	2.34
SURFACE TREATMENT OF EXISTING WOOD FLOORING - Sanding and sealing existing wood flooring										
Machine sanding										
width exceeding 300mm	-	-	-	0.50	-	4.62	2.00	6.62	m2	7.11
Prepare, one priming coat and one finish coat of seal										
width exceeding 300mm	-	-	-	0.16	-	1.48	1.00	2.48	m2	2.66
WALL AND CEILING BOARDING - Softwood, wrought; fixing to timber - wall and ceiling boarding										
Boarding to walls, tongued, grooved and veed joints; 19mm thick, 100mm wide boards										
width exceeding 300mm	9.48	10.00	10.43	1.00	0.13	10.14	0.16	20.73	m2	22.29
Boarding to walls, tongued, grooved and veed joints; 19mm thick, 150mm wide boards										
width exceeding 300mm	8.65	10.00	9.52	0.70	0.09	7.09	0.11	16.72	m2	17.97
Boarding to walls, tongued, grooved and veed joints; 25mm thick, 100mm wide boards										
width exceeding 300mm	12.36	10.00	13.60	1.05	0.13	10.60	0.17	24.37	m2	26.20
Boarding to walls, tongued, grooved and veed joints; 25mm thick, 150mm wide boards										
width exceeding 300mm	11.54	10.00	12.69	0.75	0.09	7.55	0.12	20.37	m2	21.90
Boarding to ceilings, tongued, grooved and veed joints; 19mm thick, 100mm wide boards										
width exceeding 300mm	9.48	10.00	10.43	1.50	0.19	15.18	0.24	25.85	m2	27.79
Boarding to ceilings, tongued, grooved and veed joints; 19mm thick, 150mm wide boards										
width exceeding 300mm	8.65	10.00	9.52	1.00	0.13	10.14	0.16	19.82	m2	21.30
Boarding to ceilings, tongued, grooved and veed joints; 25mm thick, 100mm wide boards										
width exceeding 300mm	12.36	10.00	13.60	1.55	0.19	15.64	0.25	29.49	m2	31.70
Boarding to ceilings, tongued, grooved and veed joints; 25mm thick, 150mm wide boards										
width exceeding 300mm	11.54	10.00	12.69	1.05	0.13	10.60	0.17	23.47	m2	25.23
WALL AND CEILING BOARDING - Western Red Cedar, wrought; fixing to timber; wall and ceiling boarding										
Boarding to walls, tongued, grooved and veed joints; 19mm thick, 100mm wide boards										
width exceeding 300mm	38.80	10.00	42.68	1.05	0.13	10.60	0.17	53.45	m2	57.46
Boarding to walls, tongued, grooved and veed joints; 19mm thick, 150mm wide boards										
width exceeding 300mm	33.00	10.00	36.30	0.75	0.10	7.62	0.12	44.04	m2	47.35

Labour hourly rates: (except Specialists) Craft Operatives £9.23 Labourer £7.02 Rates are national average prices. Refer to REGIONAL VARIATIONS for indicative levels of overall pricing in regions	MATERIALS			LABOUR				RATES		
	Del to Site £	Waste %	Material Cost £	Craft Optve Hrs	Lab Hrs	Labour Cost £	Sunds £	Nett Rate £	Unit	Gross Rate (+7.5%) £
K20: TIMBER BOARD FLOORING/SHEATHING/ LININGS/CASINGS Cont.										
WALL AND CEILING BOARDING - Western Red Cedar, wrought; fixing to timber; wall and ceiling boarding Cont.										
Boarding to walls, tongued, grooved and veed joints; 25mm thick, 100mm wide boards										
width exceeding 300mm	48.75	10.00	53.63	1.10	0.14	11.14	0.18	64.94	m2	69.81
Boarding to walls, tongued, grooved and veed joints; 25mm thick, 150mm wide boards										
width exceeding 300mm	43.90	10.00	48.29	0.80	0.10	8.09	0.13	56.51	m2	60.74
WALL AND CEILING BOARDING - Knotty Pine, wrought, selected for transparent finish; fixing to timber - wall and ceiling boarding										
Boarding to walls, tongued, grooved and veed joints; 19mm thick, 100mm wide boards										
width exceeding 300mm	32.80	10.00	36.08	1.30	0.16	13.12	0.21	49.41	m2	53.12
Boarding to walls, tongued, grooved and veed joints; 19mm thick, 150mm wide boards										
width exceeding 300mm	28.00	10.00	30.80	0.90	0.11	9.08	0.15	40.03	m2	43.03
Boarding to walls, tongued, grooved and veed joints; 25mm thick, 100mm wide boards										
width exceeding 300mm	41.25	10.00	45.38	1.35	0.17	13.65	0.22	59.25	m2	63.69
Boarding to walls, tongued, grooved and veed joints; 25mm thick, 150mm wide boards										
width exceeding 300mm	37.15	10.00	40.87	0.95	0.12	9.61	0.15	50.63	m2	54.42
Boarding to ceilings, tongued, grooved and veed joints; 19mm thick, 100mm wide boards										
width exceeding 300mm	32.80	10.00	36.08	1.85	0.23	18.69	0.30	55.07	m2	59.20
Boarding to ceilings, tongued, grooved and veed joints; 19mm thick, 150mm wide boards										
width exceeding 300mm	28.00	10.00	30.80	1.35	0.17	13.65	0.22	44.67	m2	48.02
Boarding to ceilings, tongued, grooved and veed joints; 25mm thick, 100mm wide boards										
width exceeding 300mm	41.25	10.00	45.38	1.90	0.24	19.22	0.31	64.91	m2	69.77
Boarding to ceilings, tongued, grooved and veed joints; 25mm thick, 150mm wide boards										
width exceeding 300mm	37.15	10.00	40.87	1.40	0.18	14.19	0.23	55.28	m2	59.43
WALL AND CEILING BOARDING - Sapele, wrought, selected for transparent finish; fixing to timber - wall and ceiling boarding										
Boarding to walls, tongued, grooved and veed joints; 19mm thick, 100mm wide boards										
width exceeding 300mm	44.35	10.00	48.78	1.45	0.18	14.65	0.24	63.67	m2	68.45
Boarding to walls, tongued, grooved and veed joints; 19mm thick, 150mm wide boards										
width exceeding 300mm	38.60	10.00	42.46	1.00	0.12	10.07	0.16	52.69	m2	56.64
Boarding to walls, tongued, grooved and veed joints; 25mm thick, 100mm wide boards										
width exceeding 300mm	56.20	10.00	61.82	1.50	0.19	15.18	0.24	77.24	m2	83.03
Boarding to walls, tongued, grooved and veed joints; 25mm thick, 150mm wide boards										
width exceeding 300mm	50.50	10.00	55.55	1.05	0.13	10.60	0.17	66.32	m2	71.30
Boarding to ceilings, tongued, grooved and veed joints; 19mm thick, 100mm wide boards										
width exceeding 300mm	44.35	10.00	48.78	2.00	0.25	20.22	0.33	69.33	m2	74.53
Boarding to ceilings, tongued, grooved and veed joints; 19mm thick, 150mm wide boards										
width exceeding 300mm	38.60	10.00	42.46	1.50	0.19	15.18	0.24	57.88	m2	62.22
Boarding to ceilings, tongued, grooved and veed joints; 25mm thick, 100mm wide boards										
width exceeding 300mm	56.20	10.00	61.82	2.05	0.26	20.75	0.33	82.90	m2	89.11
Boarding to ceilings, tongued, grooved and veed joints; 25mm thick, 150mm wide boards										
width exceeding 300mm	50.50	10.00	55.55	1.55	0.19	15.64	0.25	71.44	m2	76.80
WALL AND CEILING BOARDING - Abutments										
19mm thick softwood boarding										
raking cutting	0.95	-	0.95	0.06	-	0.55	-	1.50	m	1.62
curved cutting	1.42	-	1.42	0.09	-	0.83	-	2.25	m	2.42
25mm thick softwood boarding										
raking cutting	1.24	-	1.24	0.07	-	0.65	-	1.89	m	2.03
curved cutting	1.85	-	1.85	0.10	-	0.92	-	2.77	m	2.98

Labour hourly rates: (except Specialists) Craft Operatives £9.23 Labourer £7.02 Rates are national average prices. Refer to REGIONAL VARIATIONS for indicative levels of overall pricing in regions	MATERIALS			LABOUR				RATES		
	Del to Site £	Waste %	Material Cost £	Craft Optve Hrs	Lab Hrs	Labour Cost £	Sunds £	Nett Rate £	Unit	Gross Rate (+7.5%) £
K20: TIMBER BOARD FLOORING/SHEATHING/ LININGS/CASINGS Cont.										
WALL AND CEILING BOARDING - Abutments Cont.										
19mm thick Western Red Cedar boarding										
raking cutting	3.88	-	3.88	0.07	-	0.65	-	4.53	m	4.87
curved cutting	5.82	-	5.82	0.10	-	0.92	-	6.74	m	7.25
25mm thick Western Red Cedar boarding										
raking cutting	4.88	-	4.88	0.08	-	0.74	-	5.62	m	6.04
curved cutting	7.31	-	7.31	0.11	-	1.02	-	8.33	m	8.95
19 mm thick hardwood boarding										
raking cutting	4.44	-	4.44	0.14	-	1.29	-	5.73	m	6.16
curved cutting	6.65	-	6.65	0.21	-	1.94	-	8.59	m	9.23
25 mm thick hardwood boarding										
raking cutting	5.62	-	5.62	0.16	-	1.48	-	7.10	m	7.63
curved cutting	8.43	-	8.43	0.24	-	2.22	-	10.65	m	11.44
WALL AND CEILING BOARDING - Finished angles										
External; 19mm thick softwood boarding										
tongued and mitred	-	-	-	0.67	-	6.18	-	6.18	m	6.65
External; 25mm thick softwood boarding										
tongued and mitred	-	-	-	0.70	-	6.46	-	6.46	m	6.95
External; 19mm thick hardwood boarding										
tongued and mitred	-	-	-	0.95	-	8.77	-	8.77	m	9.43
External; 25mm thick hardwood boarding										
tongued and mitred	-	-	-	1.00	-	9.23	-	9.23	m	9.92
ROOF BOARDING - Softwood, sawn, impregnated; fixing to timber - roof boarding										
Boarding to roofs, butt joints; 19mm thick, 75mm wide boards; external										
width exceeding 300mm	3.90	7.50	4.19	0.60	0.07	6.03	0.10	10.32	m2	11.10
width exceeding 300mm; sloping	3.90	7.50	4.19	0.70	0.09	7.09	0.11	11.40	m2	12.25
Boarding to roofs, butt joints; 25mm thick, 125mm wide boards; external										
width exceeding 300mm	5.13	7.50	5.51	0.55	0.07	5.57	0.09	11.17	m2	12.01
width exceeding 300mm; sloping	5.13	7.50	5.51	0.65	0.08	6.56	0.11	12.19	m2	13.10
ROOF BOARDING - Softwood, wrought, impregnated; fixing to timber - roof boarding										
Boarding to roofs, tongued and grooved joints; 19mm thick, 75mm wide boards external										
width exceeding 300mm	7.80	7.50	8.38	0.80	0.10	8.09	0.13	16.60	m2	17.85
extra; cross rebated and rounded drip; 50mm wide .	0.81	10.00	0.89	0.20	0.03	2.06	0.03	2.98	m	3.20
extra; dovetailed cesspool 225 x 225 x 150mm	1.30	10.00	1.43	1.40	0.17	14.12	0.23	15.78	nr	16.96
extra; dovetailed cesspool 300 x 300 x 150mm	1.63	10.00	1.79	1.60	0.20	16.17	0.26	18.23	nr	19.59
width exceeding 300mm; sloping	7.80	7.50	8.38	0.90	0.11	9.08	0.15	17.61	m2	18.94
Boarding to roofs, tongued and grooved joints; 25mm thick, 125mm wide boards external										
width exceeding 300mm	9.34	7.50	10.04	0.75	0.09	7.55	0.12	17.71	m2	19.04
extra; cross rebated and rounded drip; 50mm wide .	0.81	10.00	0.89	0.20	0.03	2.06	0.03	2.98	m	3.20
extra; dovetailed cesspool 225 x 225 x 150mm	1.63	10.00	1.79	1.50	0.19	15.18	0.24	17.21	nr	18.50
extra; dovetailed cesspool 300 x 300 x 150mm	2.19	10.00	2.41	1.70	0.21	17.17	0.28	19.85	nr	21.34
width exceeding 300mm; sloping	9.34	7.50	10.04	0.85	0.11	8.62	0.14	18.80	m2	20.21
ROOF BOARDING - Abutments										
19mm thick softwood boarding										
raking cutting	0.78	-	0.78	0.35	-	3.23	-	4.01	m	4.31
curved cutting	1.17	-	1.17	0.55	-	5.08	-	6.25	m	6.71
rebates	-	-	-	0.18	-	1.66	-	1.66	m	1.79
grooves	-	-	-	0.18	-	1.66	-	1.66	m	1.79
chamfers	-	-	-	0.13	-	1.20	-	1.20	m	1.29
25mm thick softwood boarding										
raking cutting	0.93	-	0.93	0.45	-	4.15	-	5.08	m	5.46
curved cutting	1.40	-	1.40	0.65	-	6.00	-	7.40	m	7.95
rebates	-	-	-	0.18	-	1.66	-	1.66	m	1.79
grooves	-	-	-	0.18	-	1.66	-	1.66	m	1.79
chamfers	-	-	-	0.13	-	1.20	-	1.20	m	1.29
K21: TIMBER STRIP/ BOARD FINE FLOORING/ LININGS										
STRIP FLOORING - Hardwood, wrought, selected for transparent finish; sanded, two coats sealer finish										
Boarding to floors, tongued and grooved joints; 22mm thick, 75mm wide boards; fixing on and with 25 x 50mm impregnated softwood battens to masonry										
width exceeding 300 mm; secret fixing to timber ..	34.07	7.50	36.63	1.25	0.15	12.59	0.20	49.42	m2	53.12
Boarding to floors, tongued and grooved joints; 12mm thick, 75mm wide boards; overlay										
width exceeding 300 mm; secret fixing to timber ..	28.35	7.50	30.48	1.00	0.13	10.14	0.16	40.78	m2	43.84

Labour hourly rates: (except Specialists) Craft Operatives £9.23 Labourer £7.02 Rates are national average prices. Refer to REGIONAL VARIATIONS for indicative levels of overall pricing in regions	MATERIALS			LABOUR				RATES		
	Del to Site £	Waste %	Material Cost £	Craft Optve Hrs	Lab Hrs	Labour Cost £	Sunds £	Nett Rate £	Unit	Gross Rate (+7.5%) £
K21: TIMBER STRIP/ BOARD FINE FLOORING/ LININGS Cont.										
SEMI-SPRUNG FLOORS - Foam backed softwood battens covered with hardwood tongued and grooved strip flooring										

22mm Standard Beech strip flooring to level sub-floors
  width exceeding 300mm; battens at approximately
  400mm centres; elevation
    75mm ....
	-	-	Spclist	-	-	Spclist	-	49.10	m2	52.78

22mm 'Sylva Squash' Beech strip flooring to level sub-floors
  width exceeding 300mm; battens at approximately
  430mm centres; elevation
    75mm ....
	-	-	Spclist	-	-	Spclist	-	52.50	m2	56.44

20mm Prime Maple strip flooring to level sub-floors
  width exceeding 300mm; battens at approximately
  300mm centres; elevation
    75mm ....
	-	-	Spclist	-	-	Spclist	-	39.95	m2	42.95

20mm First Grade Maple strip flooring to level sub-floors
  width exceeding 300mm; battens at approximately
  300mm centres; elevation
    75mm ....
	-	-	Spclist	-	-	Spclist	-	43.37	m2	46.62

K40: DEMOUNTABLE SUSPENDED CEILINGS										
Suspended ceilings; 300 x 300mm bevelled, grooved and rebated asbestos-free fire resisting tiles; laying in position in metal suspension system of main channel members on wire or rod hangers										
Depth of suspension 150 - 500mm 9mm thick linings; fixing hangers to masonry	14.90	5.00	15.65	0.75	0.38	9.59	4.35	29.59	m2	31.80
9mm thick linings not exceeding 300mm wide; fixing hangers to masonry	4.50	10.00	4.95	0.45	0.11	4.93	2.60	12.48	m	13.41
Suspended ceilings; 600 x 600mm bevelled, grooved and rebated asbestos-free fire resisting tiles; laying in position in metal suspension system of main channel members on wire or rod hangers										
Depth of suspension 150 - 500mm 9mm thick linings; fixing hangers to masonry	14.65	5.00	15.38	0.60	0.30	7.64	2.65	25.68	m2	27.60
9mm thick linings not exceeding 300mm wide; fixing hangers to masonry	4.40	10.00	4.84	0.36	0.09	3.95	1.60	10.39	m	11.17
Suspended ceilings; 300 x 300mm bevelled, grooved and rebated plain mineral fibre tiles; laying in position in metal suspension system of main channel members on wire or rod hangers										
Depth of suspension 150 - 500mm 15.8mm thick linings; fixing hangers to masonry ..	9.10	5.00	9.55	0.75	0.38	9.59	4.35	23.50	m2	25.26
15.8mm thick linings not exceeding 300mm wide; fixing hangers to masonry	2.70	10.00	2.97	0.45	0.11	4.93	2.60	10.50	m	11.28
Suspended ceilings; 600 x 600mm bevelled, grooved and rebated plain mineral fibre tiles; laying in position in metal suspension system of main channel members on wire or rod hangers										
Depth of suspension 150 - 500mm 15.8mm thick linings; fixing hangers to masonry ..	9.10	5.00	9.55	0.60	0.30	7.64	2.65	19.85	m2	21.34
15.8mm thick linings not exceeding 300mm wide; fixing hangers to masonry	2.70	10.00	2.97	0.36	0.09	3.95	2.60	9.52	m	10.24
Suspended ceilings; 300 x 300mm bevelled, grooved and rebated textured mineral fibre tiles; laying in position in metal suspension system of main channel members on wire or rod hangers										
Depth of suspension 150 - 500mm 15.8mm thick linings; fixing hangers to masonry ..	11.30	5.00	11.87	0.75	0.38	9.59	4.35	25.81	m2	27.74
15.8mm thick linings not exceeding 300mm wide; fixing hangers to masonry	3.40	10.00	3.74	0.36	0.09	3.95	2.60	10.29	m	11.07
Suspended ceilings; 600 x 600mm bevelled, grooved and rebated regular drilled mineral fibre tiles; laying in position in metal suspension system of main channel members on wire or rod hangers										
Depth of suspension 150 - 500mm 15.8mm thick linings; fixing hangers to masonry ..	11.30	5.00	11.87	0.60	0.30	7.64	2.65	22.16	m2	23.82
15.8mm thick linings not exceeding 300mm wide; fixing hangers to masonry	3.40	10.00	3.74	0.36	0.09	3.95	1.60	9.29	m	9.99

Labour hourly rates: (except Specialists) Craft Operatives £9.23 Labourer £7.02 Rates are national average prices. Refer to REGIONAL VARIATIONS for indicative levels of overall pricing in regions	MATERIALS			LABOUR				RATES		
	Del to Site £	Waste %	Material Cost £	Craft Optve Hrs	Lab Hrs	Labour Cost £	Sunds £	Nett Rate £	Unit	Gross Rate (+7.5%) £
K40: DEMOUNTABLE SUSPENDED CEILINGS Cont.										
Suspended ceilings; 300 x 300mm bevelled, grooved and rebated patterned tiles p.c. £15.00/m2; laying in position in metal suspension system of main channel members on wire or rod hangers										
Depth of suspension 150 - 500mm										
15.8mm thick linings; fixing hangers to masonry ..	14.75	5.00	**15.49**	0.75	0.38	**9.59**	4.35	**29.43**	m2	**31.63**
15.8mm thick linings not exceeding 300mm wide; fixing hangers to masonry	4.45	10.00	**4.89**	0.36	0.09	**3.95**	2.60	**11.45**	m	**12.31**
Suspended ceilings; 600 x 600mm bevelled, grooved and rebated patterned tiles p.c. £15.00/m2; laying in position in metal suspension system of main channel members on wire or rod hangers										
Depth of suspension 150 - 500mm										
15.8mm thick linings; fixing hangers to masonry ..	14.75	5.00	**15.49**	0.60	0.30	**7.64**	2.65	**25.78**	m2	**27.72**
15.8mm thick linings not exceeding 300mm wide; fixing hangers to masonry	4.45	10.00	**4.89**	0.36	0.09	**3.95**	1.60	**10.45**	m	**11.23**
Suspended ceiling; British Gypsum Ltd, 1200 x 600mm Glasroc GRG tiles; clip fastening into Quicklock fire rated grid system with Q417 clips on suspended wire or rod hangers										
Depth of suspension 150-500mm										
10mm thick linings; fixing hangers to masonry	4.80	10.00	**5.28**	0.36	0.09	**3.95**	2.65	**11.88**	m2	**12.78**
10mm thick linings not exceeding 300mm wide; fixing hangers to masonry	4.80	10.00	**5.28**	0.36	0.09	**3.95**	1.60	**10.83**	m	**11.65**
Edge trims										
Plain										
angle section; white stove enamelled aluminium; fixing to masonry with screws at 450mm centres ...	0.90	5.00	**0.94**	0.40	0.20	**5.10**	-	**6.04**	m	**6.49**
K41: RAISED ACCESS FLOORS										
Flooring; Microfloor 'Bonded 600' light grade full access system; 600 x 600mm high density particle board panels, B.S.5669 and DIN 68761; 100 x 100mm precast lightweight concrete pedestals at 600mm centres fixed to sub-floor with epoxy resin adhesive										
Thickness of panel 30mm										
finished floor height										
50mm	-	-	Spclist	-	-	Spclist	-	16.73	m2	17.98
75mm	-	-	Spclist	-	-	Spclist	-	16.85	m2	18.11
100mm	-	-	Spclist	-	-	Spclist	-	17.06	m2	18.34
125mm	-	-	Spclist	-	-	Spclist	-	17.24	m2	18.53
150mm	-	-	Spclist	-	-	Spclist	-	17.59	m2	18.91
175mm	-	-	Spclist	-	-	Spclist	-	17.87	m2	19.21
200mm	-	-	Spclist	-	-	Spclist	-	18.03	m2	19.38
Flooring; Microfloor 'Bonded 600' medium grade full access system; 600 x 600mm high density particle board panels, B.S.5669 and DIN 68761; 100 x 100mm precast lightweight concrete pedestals at 600mm centres fixed to sub-floor with epoxy resin adhesive										
Thickness of panel 38mm										
finished floor height										
50mm	-	-	Spclist	-	-	Spclist	-	19.03	m2	20.46
75mm	-	-	Spclist	-	-	Spclist	-	19.03	m2	20.46
100mm	-	-	Spclist	-	-	Spclist	-	19.39	m2	20.84
125mm	-	-	Spclist	-	-	Spclist	-	19.56	m2	21.03
150mm	-	-	Spclist	-	-	Spclist	-	20.19	m2	21.70
175mm	-	-	Spclist	-	-	Spclist	-	20.36	m2	21.89
200mm	-	-	Spclist	-	-	Spclist	-	20.51	m2	22.05
Flooring; Microfloor 'Bonded 600' office loadings grade full access system; 600 x 600mm high density particle board panels, B.S. 5669 and DIN 68761; 100 x 100mm precast lightweight concrete pedestals at 600mm centres fixed to sub-floor with epoxy resin adhesive										
Thickness of panel 30mm										
finished floor height										
50mm	-	-	Spclist	-	-	Spclist	-	13.03	m2	14.01
75mm	-	-	Spclist	-	-	Spclist	-	13.10	m2	14.08
100mm	-	-	Spclist	-	-	Spclist	-	13.39	m2	14.39
125mm	-	-	Spclist	-	-	Spclist	-	13.56	m2	14.58
150mm	-	-	Spclist	-	-	Spclist	-	13.89	m2	14.93
175mm	-	-	Spclist	-	-	Spclist	-	14.20	m2	15.27
200mm	-	-	Spclist	-	-	Spclist	-	14.35	m2	15.43

Labour hourly rates: (except Specialists) Craft Operatives £9.23 Labourer £7.02 Rates are national average prices. Refer to REGIONAL VARIATIONS for indicative levels of overall pricing in regions	MATERIALS			LABOUR				RATES		
	Del to Site	Waste	Material Cost	Craft Optve	Lab	Labour Cost	Sunds	Nett Rate	Unit	Gross Rate (+7.5%)
	£	%	£	Hrs	Hrs	£	£	£		£

L10: WINDOWS/ROOFLIGHTS/SCREENS/LOUVRES

TIMBER WINDOWS - Casements in softwood, wrought; knotting and priming by manufacturer

Magnet Trade; without bars; softwood sub-sills; hinges; fasteners; fixing to masonry with galvanised steel cramps -4 nr, 25 x 3 x 150mm girth, flat section, holes -2

Description	Del to Site	Waste	Material Cost	Craft Optve	Lab	Labour Cost	Sunds	Nett Rate	Unit	Gross Rate
488 x 900mm overall; N09V	57.95	2.50	59.40	1.20	0.60	15.29	1.40	76.09	nr	81.79
631 x 750mm overall; 107V	62.09	2.50	63.64	1.20	0.60	15.29	1.40	80.33	nr	86.36
631 x 900mm overall; 109V	62.54	2.50	64.10	1.30	0.65	16.56	1.40	82.07	nr	88.22
631 x 1050mm overall; 110V	64.16	2.50	65.76	1.50	0.75	19.11	1.40	86.27	nr	92.74
631 x 1200mm overall; 112v	66.23	2.50	67.89	1.60	0.80	20.38	1.40	89.67	nr	96.39
915 x 900mm overall; 2N09W	76.58	2.50	78.49	1.70	0.85	21.66	1.40	101.55	nr	109.17
915 x 1050mm overall; 2N10W	77.57	2.50	79.51	1.70	0.85	21.66	1.40	102.57	nr	110.26
915 x 1200mm overall; 2N12W	79.64	2.50	81.63	1.80	0.90	22.93	1.40	105.96	nr	113.91
1200 x 900mm overall; 209W	96.20	2.50	98.61	1.70	0.85	21.66	1.40	121.66	nr	130.79
1200 x 1050mm overall; 210C	94.13	2.50	96.48	1.90	0.95	24.21	1.40	122.09	nr	131.25
1200 x 1050mm overall; 210W	97.73	2.50	100.17	1.90	0.95	24.21	1.40	125.78	nr	135.21
1200 x 1050mm overall; 210CV	119.51	2.50	122.50	1.90	0.95	24.21	1.40	148.10	nr	159.21
1200 x 1200mm overall; 212C	98.27	2.50	100.73	2.00	1.00	25.48	1.40	127.61	nr	137.18
1200 x 1200mm overall; 212W	99.80	2.50	102.30	2.00	1.00	25.48	1.40	129.18	nr	138.86
1200 x 1200mm overall; 212CV	124.19	2.50	127.29	2.00	1.00	25.48	1.40	154.17	nr	165.74
1524 x 1050mm overall; 3NN10WW	134.54	2.50	137.90	2.10	1.05	26.75	1.40	166.06	nr	178.51
1769 x 1050mm overall; 310C	112.76	2.50	115.58	2.40	1.20	30.58	1.40	147.56	nr	158.62
1769 x 1050mm overall; 310WW	179.54	2.50	184.03	2.40	1.20	30.58	1.40	216.00	nr	232.20
1769 x 1050mm overall; 310CVC	163.52	2.50	167.61	2.40	1.20	30.58	1.40	199.58	nr	214.55
1769 x 1200mm overall; 312C	116.90	2.50	119.82	2.50	1.25	31.85	1.40	153.07	nr	164.55
1769 x 1200mm overall; 312WW	183.14	2.50	187.72	2.50	1.25	31.85	1.40	220.97	nr	237.54
1769 x 1200mm overall; 312CVC	169.73	2.50	173.97	2.50	1.25	31.85	1.40	207.22	nr	222.76

TIMBER WINDOWS - Fully reversible windows in softwood, wrought; base coat stain by manufacturer

Boulton and Paul Hi-Profile; horizontal tilt; safety and reverse locking catch; locking fastener with high security espagnolette bolt; fully weatherstripped; fixing to masonry with galvanised steel cramps -4 nr, 25 x 3 x 150mm girth, holes

Description	Del to Site	Waste	Material Cost	Craft Optve	Lab	Labour Cost	Sunds	Nett Rate	Unit	Gross Rate
600 x 900mm overall; R0906	207.68	2.50	212.87	1.35	0.67	17.16	1.40	231.44	nr	248.79
900 x 900mm overall; R0909	216.37	2.50	221.78	1.60	0.80	20.38	1.40	243.56	nr	261.83
900 x 1050mm overall; R0910	220.56	2.50	226.07	1.70	0.85	21.66	1.40	249.13	nr	267.82
900 x 1200mm overall; R0912	226.76	2.50	232.43	1.80	0.90	22.93	1.40	256.76	nr	276.02
1200 x 900mm overall; R1209	234.96	2.50	240.83	1.80	0.90	22.93	1.40	265.17	nr	285.05
1200 x 1050mm overall; R1210	240.45	2.50	246.46	1.90	0.95	24.21	1.40	272.07	nr	292.47
1200 x 1200mm overall; R1212	246.97	2.50	253.14	2.00	1.00	25.48	1.40	280.02	nr	301.03

TIMBER WINDOWS - Windows in hardwood, wrought, preservative treated; one coat Redwood stain by manufacturer

Crosby Sarek Ltd. Alpha Energy Saving windows; glazing beads; weatherstripping, hinges; fasteners; fixing to masonry with galvanised steel cramps -4 nr, 25 x 3 x 150mm girth; flat section, holes -2; side hung-non bar type

Description	Del to Site	Waste	Material Cost	Craft Optve	Lab	Labour Cost	Sunds	Nett Rate	Unit	Gross Rate
630 x 750mm overall; X107C	133.76	2.50	137.10	1.60	0.80	20.38	1.40	158.89	nr	170.80
630 x 900mm overall; X109C	140.69	2.50	144.21	1.85	0.92	23.53	1.40	169.14	nr	181.83
630 x 1050mm overall; X110C	147.72	2.50	151.41	2.05	1.02	26.08	1.40	178.89	nr	192.31
630 x 1200mm overall; X112C	154.83	2.50	158.70	2.20	1.10	28.03	1.40	188.13	nr	202.24
630 x 1350mm overall; X113C	167.46	2.50	171.65	2.40	1.20	30.58	1.40	203.62	nr	218.89
1200 x 750mm overall; X207C	178.75	2.50	183.22	2.15	1.07	27.36	1.40	211.97	nr	227.87
1200 x 900mm overall; X209C	188.80	2.50	193.52	2.40	1.20	30.58	1.40	225.50	nr	242.41
1200 x 1050mm overall; X210C	195.78	2.50	200.67	2.55	1.27	32.45	1.40	234.53	nr	252.12
1200 x 1200mm overall; X212C	205.61	2.50	210.75	2.75	1.37	35.00	1.40	247.15	nr	265.69
1200 x 1350mm overall; X213C	220.92	2.50	226.44	2.95	1.47	37.55	1.40	265.39	nr	285.30
1770 x 750mm overall; X307C	227.18	2.50	232.86	2.75	1.37	35.00	1.40	269.26	nr	289.45
1770 x 750mm overall; X307CC	282.66	2.50	289.73	2.75	1.37	35.00	1.40	326.13	nr	350.59
1770 x 900mm overall; X309C	230.04	2.50	235.79	2.95	1.47	37.55	1.40	274.74	nr	295.34
1770 x 900mm overall; X309CC	297.31	2.50	304.74	2.95	1.47	37.55	1.40	343.69	nr	369.47

Labour hourly rates: (except Specialists) Craft Operatives £9.23 Labourer £7.02 Rates are national average prices. Refer to REGIONAL VARIATIONS for indicative levels of overall pricing in regions	MATERIALS			LABOUR				RATES		
	Del to Site £	Waste %	Material Cost £	Craft Optve Hrs	Lab Hrs	Labour Cost £	Sunds £	Nett Rate £	Unit	Gross Rate (+7.5%) £

L10: WINDOWS/ROOFLIGHTS/SCREENS/LOUVRES Cont.

TIMBER WINDOWS - Windows in hardwood, wrought, preservative treated; one coat Redwood stain by manufacturer Cont.

Crosby Sarek Ltd. Alpha Energy Saving windows; glazing beads; weatherstripping, hinges; fasteners; fixing to masonry with galvanised steel cramps -4 nr, 25 x 3 x 150mm girth; flat section, holes -2; side hung-non bar type Cont.

	Del to Site £	Waste %	Material Cost £	Craft Optve Hrs	Lab Hrs	Labour Cost £	Sunds £	Nett Rate £	Unit	Gross Rate £
1770 x 1050mm overall; X310C	237.86	2.50	243.81	3.10	1.55	39.49	1.40	284.70	nr	306.05
1770 x 1050mm overall; X310CC	310.10	2.50	317.85	3.10	1.55	39.49	1.40	358.75	nr	385.65
1770 x 1200mm overall; X312C	244.50	2.50	250.61	3.35	1.67	42.64	1.40	294.66	nr	316.76
1770 x 1200mm overall; X312CC	323.23	2.50	331.31	3.35	1.67	42.64	1.40	375.35	nr	403.51
1770 x 1350mm overall; X313C	265.68	2.50	272.32	3.60	1.80	45.86	1.40	319.59	nr	343.55
1770 x 1350mm overall; X313CC	347.76	2.50	356.45	3.60	1.80	45.86	1.40	403.72	nr	434.00
2339 x 900mm overall; X409CMC	372.90	2.50	382.22	3.60	1.80	45.86	1.40	429.49	nr	461.70
2339 x 1050mm overall; X410CMC	390.73	2.50	400.50	3.85	1.92	49.01	1.40	450.91	nr	484.73
2339 x 1200mm overall; X412CMC	409.16	2.50	419.39	4.05	2.02	51.56	1.40	472.35	nr	507.78
2339 x 1350mm overall; X413CMC	438.00	2.50	448.95	4.25	2.12	54.11	1.40	504.46	nr	542.29

Crosby Sarek Ltd; Alpha Energy Saving windows; glazing beads; weatherstripping; hinges; fasteners; fixing to masonry with galvanised steel cramps -4nr, 25 x 3 x 150mm girth, flat section, holes -2; non bar side hung/vent type

	Del to Site £	Waste %	Material Cost £	Craft Optve Hrs	Lab Hrs	Labour Cost £	Sunds £	Nett Rate £	Unit	Gross Rate £
630 x 1050mm overall; X110T	211.04	2.50	216.32	2.05	1.02	26.08	1.40	243.80	nr	262.08
630 x 1200mm overall; X112T	218.24	2.50	223.70	2.20	1.10	28.03	1.40	253.12	nr	272.11
1200 x 750mm overall; X207CV	233.73	2.50	239.57	2.15	1.07	27.36	1.40	268.33	nr	288.45
1200 x 900mm overall; X209CV	243.89	2.50	249.99	2.40	1.20	30.58	1.40	281.96	nr	303.11
1200 x 1050mm overall; X210CV	252.11	2.50	258.41	2.55	1.27	32.45	1.40	292.26	nr	314.18
1200 x 1050mm overall; X210T	253.55	2.50	259.89	2.55	1.27	32.45	1.40	293.74	nr	315.77
1200 x 1200mm overall; X212CV	261.71	2.50	268.25	2.75	1.37	35.00	1.40	304.65	nr	327.50
1200 x 1200mm overall; X212T	262.33	2.50	268.89	2.75	1.37	35.00	1.40	305.29	nr	328.18
1200 x 1350mm overall; X213CV	276.41	2.50	283.32	2.95	1.47	37.55	1.40	322.27	nr	346.44
1200 x 1350mm overall; X213T	272.04	2.50	278.84	2.95	1.47	37.55	1.40	317.79	nr	341.62
1200 x 1500mm overall; X215T	279.48	2.50	286.47	3.15	1.57	40.10	1.40	327.96	nr	352.56
1770 x 900mm overall; X309CVC	351.90	2.50	360.70	2.95	1.47	37.55	1.40	399.65	nr	429.62
1770 x 1050mm overall; X310CVC	365.47	2.50	374.61	3.10	1.55	39.49	1.40	415.50	nr	446.66
1770 x 1050mm overall; X310CW	329.61	2.50	337.85	3.10	1.55	39.49	1.40	378.74	nr	407.15
1770 x 1050mm overall; X310WW	356.54	2.50	365.45	3.10	1.55	39.49	1.40	406.35	nr	436.82
1770 x 1200mm overall; X312CVC	379.22	2.50	388.70	3.35	1.67	42.64	1.40	432.74	nr	465.20
1770 x 1200mm overall; X312CW	339.37	2.50	347.85	3.35	1.67	42.64	1.40	391.90	nr	421.29
1770 x 1200mm overall; X312WW	366.48	2.50	375.64	3.35	1.67	42.64	1.40	419.69	nr	451.16
1770 x 1350mm overall; X313CVC	403.91	2.50	414.01	3.60	1.80	45.86	1.40	461.27	nr	495.87
1770 x 1350mm overall; X313CW	354.42	2.50	363.28	3.60	1.80	45.86	1.40	410.54	nr	441.34
2339 x 1050mm overall; X410CWC	451.96	2.50	463.26	3.85	1.92	49.01	1.40	513.67	nr	552.20
2339 x 1050mm overall; X410TT	482.46	2.50	494.52	3.85	1.92	49.01	1.40	544.94	nr	585.81
2339 x 1200mm overall; X412CWC	467.88	2.50	479.58	4.05	2.02	51.56	1.40	532.54	nr	572.48
2339 x 1200mm overall; X412TT	495.77	2.50	508.16	4.05	2.02	51.56	1.40	561.13	nr	603.21
2339 x 1350mm overall; X413CWC	495.77	2.50	508.16	4.25	2.12	54.11	1.40	563.67	nr	605.95
2339 x 1350mm overall; X413TT	510.12	2.50	522.87	4.25	2.12	54.11	1.40	578.38	nr	621.76
2339 x 1500mm overall; X415TT	525.09	2.50	538.22	4.40	2.20	56.06	1.40	595.67	nr	640.35

Crosby Sarek Ltd; Alpha Energy Saving windows; glazing beads; weatherstripping; hinges; fasteners; fixing to masonry with galvanised steel cramps -4nr, 25 x 3 x 150mm girth, flat section, holes -2; top hung type

	Del to Site £	Waste %	Material Cost £	Craft Optve Hrs	Lab Hrs	Labour Cost £	Sunds £	Nett Rate £	Unit	Gross Rate £
630 x 600mm overall; X106A	138.62	2.50	142.09	1.50	0.75	19.11	1.40	162.60	nr	174.79
630 x 750mm overall; X107A	146.22	2.50	149.88	1.60	0.80	20.38	1.40	171.66	nr	184.53
630 x 900mm overall; X109A	152.07	2.50	155.87	1.85	0.92	23.53	1.40	180.81	nr	194.37
630 x 1050mm overall; X110A	159.69	2.50	163.68	2.05	1.02	26.08	1.40	191.16	nr	205.50
630 x 1200mm overall; X112A	166.85	2.50	171.02	2.20	1.10	28.03	1.40	200.45	nr	215.48
915 x 600mm overall; X2N06A	174.62	2.50	178.99	1.80	0.90	22.93	1.40	203.32	nr	218.57
915 x 750mm overall; X2N07A	182.48	2.50	187.04	1.95	0.97	24.81	1.40	213.25	nr	229.24
915 x 900mm overall; X2N09A	198.23	2.50	203.19	2.15	1.07	27.36	1.40	231.94	nr	249.34
915 x 1050mm overall; X2N10A	206.84	2.50	212.01	2.35	1.17	29.90	1.40	243.31	nr	261.56
915 x 1200mm overall; X2N12A	213.92	2.50	219.27	2.50	1.25	31.85	1.40	252.52	nr	271.46
1200 x 600mm overall; X206A	201.93	2.50	206.98	2.05	1.02	26.08	1.40	234.46	nr	252.04
1200 x 750mm overall; X207A	209.75	2.50	214.99	2.15	1.07	27.36	1.40	243.75	nr	262.03
1200 x 900mm overall; X209A	224.50	2.50	230.11	2.40	1.20	30.58	1.40	262.09	nr	281.75
1200 x 1050mm overall; X210A	232.72	2.50	238.54	2.55	1.27	32.45	1.40	272.39	nr	292.82
1200 x 1200mm overall; X212A	240.02	2.50	246.02	2.75	1.37	35.00	1.40	282.42	nr	303.60
1770 x 600mm overall; X306AE	242.82	2.50	248.89	2.65	1.32	33.73	1.40	284.02	nr	305.32
1770 x 750mm overall; X307AE	252.56	2.50	258.87	2.75	1.37	35.00	1.40	295.27	nr	317.42
1770 x 900mm overall; X309AE	269.76	2.50	276.50	2.95	1.47	37.55	1.40	315.45	nr	339.11
1770 x 1050mm overall; X310AE	280.65	2.50	287.67	3.10	1.55	39.49	1.40	328.56	nr	353.20
1770 x 1200mm overall; X312AE	290.09	2.50	297.34	3.35	1.67	42.64	1.40	341.39	nr	366.99

Crosby Sarek Ltd; Alpha Energy Saving windows; glazing beads; weatherstripping; hinges; fasteners; fixing to masonry with galvanised steel cramps -4nr, 25 x 3 x 150mm girth, flat section, holes -2; design guide - horizontal glazing bar and fanlight type

	Del to Site £	Waste %	Material Cost £	Craft Optve Hrs	Lab Hrs	Labour Cost £	Sunds £	Nett Rate £	Unit	Gross Rate £
488 x 900mm overall; XHN09D	180.42	2.50	184.93	1.60	0.80	20.38	1.40	206.71	nr	222.22
488 x 1050mm overall; XHN10D	188.62	2.50	193.34	1.80	0.90	22.93	1.40	217.67	nr	233.99
488 x 1200mm overall; XHN12D	195.54	2.50	200.43	2.00	1.00	25.48	1.40	227.31	nr	244.36
488 x 1350mm overall; XHN13D	208.52	2.50	213.73	2.10	1.05	26.75	1.40	241.89	nr	260.03
630 x 900mm overall; XH109D	190.70	2.50	195.47	1.85	0.92	23.53	1.40	220.40	nr	236.93
630 x 1050mm overall; XH110D	197.23	2.50	202.16	2.05	1.02	26.08	1.40	229.64	nr	246.87
630 x 1200mm overall; XH112D	203.61	2.50	208.70	2.20	1.10	28.03	1.40	238.13	nr	255.99
630 x 1350mm overall; XH113D	209.37	2.50	214.60	2.40	1.20	30.58	1.40	246.58	nr	265.07
915 x 900mm overall; H2N09D	218.98	2.50	224.45	2.15	1.07	27.36	1.40	253.21	nr	272.20
915 x 1050mm overall; H2N10D	214.39	2.50	219.75	2.35	1.17	29.90	1.40	251.05	nr	269.88

Labour hourly rates: (except Specialists) Craft Operatives £9.23 Labourer £7.02 Rates are national average prices. Refer to REGIONAL VARIATIONS for indicative levels of overall pricing in regions	MATERIALS			LABOUR				RATES		
	Del to Site £	Waste %	Material Cost £	Craft Optve Hrs	Lab Hrs	Labour Cost £	Sunds £	Nett Rate £	Unit	Gross Rate (+7.5%) £

L10: WINDOWS/ROOFLIGHTS/SCREENS/LOUVRES Cont.

TIMBER WINDOWS - Windows in hardwood, wrought, preservative treated; one coat Redwood stain by manufacturer Cont.

Crosby Sarek Ltd; Alpha Energy Saving windows; glazing beads; weatherstripping; hinges; fasteners; fixing to masonry with galvanised steel cramps -4nr, 25 x 3 x 150mm girth, flat section, holes -2; design guide - horizontal glazing bar and fanlight type Cont.

	Del to Site £	Waste %	Material Cost £	Craft Optve Hrs	Lab Hrs	Labour Cost £	Sunds £	Nett Rate £	Unit	Gross Rate £
915 x 1200mm overall; H2N12D	220.42	2.50	225.93	2.50	1.25	31.85	1.40	259.18	nr	278.62
915 x 1350mm overall; XH2N13D	229.32	2.50	235.05	2.70	1.35	34.40	1.40	270.85	nr	291.16
1200 x 1050mm overall; XH210CD	348.99	2.50	357.71	2.55	1.27	32.45	1.40	391.57	nr	420.93
1200 x 1200mm overall; XH212CD	362.79	2.50	371.86	2.75	1.37	35.00	1.40	408.26	nr	438.88
1200 x 1350mm overall; XH213CD	377.22	2.50	386.65	3.00	1.50	38.22	1.40	426.27	nr	458.24
1770 x 1050mm overall; XH310CDC	491.31	2.50	503.59	3.10	1.55	39.49	1.40	544.49	nr	585.32
1770 x 1200mm overall; XH312CDC	511.91	2.50	524.71	3.40	1.70	43.32	1.40	569.42	nr	612.13
1770 x 1350mm overall; XH313CDC	536.11	2.50	549.51	3.60	1.80	45.86	1.40	596.78	nr	641.54

Crosby Sarek Ltd; Alpha Energy Saving windows; glazing beads; weatherstripping; hinges; fasteners; fixing to masonry with galvanised steel cramps -4nr, 25 x 3 x 150mm girth, flat section, holes -2; design guide - horizontal glazing bar type

	Del to Site £	Waste %	Material Cost £	Craft Optve Hrs	Lab Hrs	Labour Cost £	Sunds £	Nett Rate £	Unit	Gross Rate £
630 x 900mm overall; XH109C	149.64	2.50	153.38	1.85	0.92	23.53	1.40	178.31	nr	191.69
630 x 1050mm overall; XH110C	155.66	2.50	159.55	2.05	1.02	26.08	1.40	187.03	nr	201.06
630 x 1200mm overall; XH112C	162.11	2.50	166.16	2.20	1.10	28.03	1.40	195.59	nr	210.26
630 x 1200mm overall; XH113C	173.70	2.50	178.04	2.40	1.20	30.58	1.40	210.02	nr	225.77
1200 x 900mm overall; XH209C	210.42	2.50	215.68	2.40	1.20	30.58	1.40	247.66	nr	266.23
1200 x 1050mm overall; XH210C	218.75	2.50	224.22	2.55	1.27	32.45	1.40	258.07	nr	277.43
1200 x 1200mm overall; XH212C	228.24	2.50	233.95	2.75	1.37	35.00	1.40	270.35	nr	290.62
1200 x 1350mm overall; XH213C	241.26	2.50	247.29	3.00	1.50	38.22	1.40	286.91	nr	308.43
1770 x 1050mm overall; XH310CC	342.23	2.50	350.79	3.10	1.55	39.49	1.40	391.68	nr	421.06
1770 x 1200mm overall; XH312CC	354.96	2.50	363.83	3.40	1.70	43.32	1.40	408.55	nr	439.19
1770 x 1350mm overall; XH313CC	378.77	2.50	388.24	3.60	1.80	45.86	1.40	435.50	nr	468.17
2339 x 1050mm overall; XH410CMC	436.33	2.50	447.24	3.85	1.77	47.96	1.40	496.60	nr	533.84
2339 x 1200mm overall; XH412CMC	452.85	2.50	464.17	4.05	2.02	51.56	1.40	517.13	nr	555.92
2339 x 1350mm overall; XH413CMC	481.18	2.50	493.21	4.30	2.15	54.78	1.40	549.39	nr	590.60

Crosby Sarek Ltd; Alpha Energy Saving windows; glazing beads; weatherstripping; hinges; fasteners; fixing to masonry with galvanised steel cramps -4nr, 25 x 3 x 150mm girth, flat section, holes -2; curved head sash -side hung type

	Del to Site £	Waste %	Material Cost £	Craft Optve Hrs	Lab Hrs	Labour Cost £	Sunds £	Nett Rate £	Unit	Gross Rate £
630 x 750mm overall; XS107C	156.62	2.50	160.54	1.60	0.80	20.38	1.40	182.32	nr	195.99
630 x 900mm overall; XS109C	163.55	2.50	167.64	1.85	0.92	23.53	1.40	192.57	nr	207.02
630 x 1050mm overall; XS110C	170.58	2.50	174.84	2.05	1.02	26.08	1.40	202.33	nr	217.50
630 x 1200mm overall; XS112C	177.67	2.50	182.11	2.20	1.10	28.03	1.40	211.54	nr	227.41
630 x 1350mm overall; XS113C	190.31	2.50	195.07	2.35	1.17	29.90	1.40	226.37	nr	243.35
915 x 750mm overall; XS2N07C	222.48	2.50	228.04	1.95	0.97	24.81	1.40	254.25	nr	273.32
915 x 900mm overall; XS2N09C	230.76	2.50	236.53	2.15	1.07	27.36	1.40	265.28	nr	285.18
915 x 1050mm overall; XS2N10C	238.76	2.50	244.73	2.35	1.17	29.90	1.40	276.03	nr	296.74
915 x 1200mm overall; XS2N12C	246.39	2.50	252.55	2.50	1.25	31.85	1.40	285.80	nr	307.23
915 x 1350mm overall; XS2N13C	253.44	2.50	259.78	2.70	1.35	34.40	1.40	295.57	nr	317.74
1200 x 750mm overall; XS207C	232.61	2.50	238.43	2.15	1.07	27.36	1.40	267.18	nr	287.22
1200 x 750mm overall; XS207CC	304.57	2.50	312.18	2.15	1.07	27.36	1.40	340.94	nr	366.51
1200 x 900mm overall; XS209C	242.67	2.50	248.74	2.40	1.20	30.58	1.40	280.71	nr	301.77
1200 x 900mm overall; XS209CC	318.12	2.50	326.07	2.40	1.20	30.58	1.40	358.05	nr	384.90
1200 x 1050mm overall; XS210C	249.64	2.50	255.88	2.55	1.27	32.45	1.40	289.73	nr	311.46
1200 x 1050mm overall; XS210CC	333.23	2.50	341.56	2.55	1.27	32.45	1.40	375.41	nr	403.57
1200 x 1200mm overall; XS212C	259.48	2.50	265.97	2.75	1.37	35.00	1.40	302.37	nr	325.04
1200 x 1200mm overall; XS212CC	346.26	2.50	354.92	2.75	1.37	35.00	1.40	391.32	nr	420.67
1200 x 1350mm overall; XS213C	274.79	2.50	281.66	3.00	1.50	38.22	1.40	321.28	nr	345.38
1200 x 1350mm overall; XS213CC	373.11	2.50	382.44	3.00	1.50	38.22	1.40	422.06	nr	453.71

TIMBER WINDOWS - Purpose made windows in softwood, wrought

	Del to Site £	Waste %	Material Cost £	Craft Optve Hrs	Lab Hrs	Labour Cost £	Sunds £	Nett Rate £	Unit	Gross Rate £
38mm moulded casements or fanlights										
in one pane	71.00	2.50	72.78	1.50	0.19	15.18	0.24	88.19	m2	94.81
divided into panes 0.10 - 0.50m2	81.65	2.50	83.69	1.50	0.19	15.18	0.24	99.11	m2	106.54
divided into panes not exceeding 0.10m2	92.30	2.50	94.61	1.50	0.19	15.18	0.24	110.03	m2	118.28
50mm moulded casements or fanlights										
in one pane	81.65	2.50	83.69	1.60	0.20	16.17	0.26	100.12	m2	107.63
divided into panes 0.10 - 0.50m2	92.30	2.50	94.61	1.60	0.20	16.17	0.26	111.04	m2	119.37
divided into panes not exceeding 0.10m2	102.95	2.50	105.52	1.60	0.20	16.17	0.26	121.96	m2	131.10
38mm moulded casements with semi-circular heads (measured square)										
in one pane	142.00	2.50	145.55	1.65	0.21	16.70	0.27	162.52	m2	174.71
divided into panes 0.10 - 0.50m2	152.65	2.50	156.47	1.65	0.21	16.70	0.27	173.44	m2	186.45
divided into panes not exceeding 0.10m2	163.30	2.50	167.38	1.65	0.21	16.70	0.27	184.36	m2	198.18
50mm moulded casements with semi-circular heads (measured square)										
in one pane	163.30	2.50	167.38	1.75	0.22	17.70	0.28	185.36	m2	199.26
divided into panes 0.10 - 0.50m2	173.95	2.50	178.30	1.75	0.22	17.70	0.28	196.28	m2	211.00
divided into panes not exceeding 0.10m2	184.60	2.50	189.22	1.75	0.22	17.70	0.28	207.19	m2	222.73
38mm bullseye casements										
457mm diameter in one pane	106.50	2.50	109.16	1.00	0.12	10.07	0.16	119.39	nr	128.35
762mm diameter in one pane	142.00	2.50	145.55	1.50	0.19	15.18	0.24	160.97	nr	173.04

WINDOWS/DOORS/STAIRS

Labour hourly rates: (except Specialists) Craft Operatives £9.23 Labourer £7.02 Rates are national average prices. Refer to REGIONAL VARIATIONS for indicative levels of overall pricing in regions	MATERIALS			LABOUR				RATES		
	Del to Site	Waste	Material Cost	Craft Optve	Lab	Labour Cost	Sunds	Nett Rate	Unit	Gross Rate (+7.5%)
	£	%	£	Hrs	Hrs	£	£	£		£
L10: WINDOWS/ROOFLIGHTS/SCREENS/LOUVRES Cont.										
TIMBER WINDOWS - Purpose made windows in softwood, wrought Cont.										
50mm bullseye casements										
457mm diameter in one pane	113.60	2.50	116.44	1.10	0.14	11.14	0.18	127.76	nr	137.34
762mm diameter in one pane	152.65	2.50	156.47	1.65	0.21	16.70	0.27	173.44	nr	186.45
Labours										
check throated edge	0.36	2.50	0.37	-	-	-	-	0.37	m	0.40
rebated and splayed bottom rail	0.36	2.50	0.37	-	-	-	-	0.37	m	0.40
rebated and beaded meeting stile	0.50	2.50	0.51	-	-	-	-	0.51	m	0.55
fitting and hanging casement or fanlight on butts (included elsewhere)										
38mm	-	-	-	0.67	0.08	6.75	-	6.75	nr	7.25
50mm	-	-	-	0.74	0.09	7.46	-	7.46	nr	8.02
fitting and hanging casement or fanlight on sash centres (included elsewhere)										
38mm	-	-	-	1.50	0.19	15.18	-	15.18	nr	16.32
50mm	-	-	-	1.65	0.21	16.70	-	16.70	nr	17.96
TIMBER WINDOWS - Purpose made windows in Afrormosia, wrought										
38mm moulded casements or fanlights										
in one pane	178.00	2.50	182.45	2.25	0.28	22.73	0.37	205.55	m2	220.97
divided into panes 0.10 - 0.50m2	204.70	2.50	209.82	2.25	0.28	22.73	0.37	232.92	m2	250.39
divided into panes not exceeding 0.10m2	231.40	2.50	237.19	2.25	0.28	22.73	0.37	260.29	m2	279.81
50mm moulded casements or fanlights										
in one pane	204.70	2.50	209.82	2.40	0.30	24.26	0.39	234.47	m2	252.05
divided into panes 0.10 - 0.50m2	231.40	2.50	237.19	2.40	0.30	24.26	0.39	261.83	m2	281.47
divided into panes not exceeding 0.10m2	258.10	2.50	264.55	2.40	0.30	24.26	0.39	289.20	m2	310.89
38mm moulded casements with semi-circular heads (measured square)										
in one pane	356.00	2.50	364.90	2.45	0.31	24.79	0.40	390.09	m2	419.35
divided into panes 0.10 - 0.50m2	382.70	2.50	392.27	2.45	0.31	24.79	0.40	417.46	m2	448.77
divided into panes not exceeding 0.10m2	409.40	2.50	419.63	2.45	0.31	24.79	0.40	444.82	m2	478.19
50mm moulded casements with semi-circular heads (measured square)										
in one pane	409.40	2.50	419.63	2.60	0.32	26.24	0.42	446.30	m2	479.77
divided into panes 0.10 - 0.50m2	436.10	2.50	447.00	2.60	0.32	26.24	0.42	473.67	m2	509.19
divided into panes not exceeding 0.10m2	462.80	2.50	474.37	2.60	0.32	26.24	0.42	501.03	m2	538.61
38mm bullseye casements										
457mm diameter in one pane	267.00	2.50	273.68	1.50	0.19	15.18	0.24	289.09	nr	310.78
762mm diameter in one pane	356.00	2.50	364.90	2.25	0.28	22.73	0.37	388.00	nr	417.10
50mm bullseye casements										
457mm diameter in one pane	284.80	2.50	291.92	1.65	0.21	16.70	0.27	308.89	nr	332.06
762mm diameter in one pane	382.70	2.50	392.27	2.45	0.31	24.79	0.40	417.46	nr	448.77
Labours										
check throated edge	0.89	2.50	0.91	-	-	-	-	0.91	m	0.98
rebated and splayed bottom rail	0.89	2.50	0.91	-	-	-	-	0.91	m	0.98
rebated and beaded meeting stile	1.25	2.50	1.28	-	-	-	-	1.28	m	1.38
fitting and hanging casement or fanlight on butts (included elsewhere)										
38mm	-	-	-	1.00	0.12	10.07	-	10.07	nr	10.83
50mm	-	-	-	1.10	0.14	11.14	-	11.14	nr	11.97
fitting and hanging casement or fanlight on sash centres (included elsewhere)										
38mm	-	-	-	2.25	0.28	22.73	-	22.73	nr	24.44
50mm	-	-	-	2.45	0.31	24.79	-	24.79	nr	26.65
TIMBER WINDOWS - Purpose made windows in European Oak, wrought										
38mm moulded casements or fanlights										
in one pane	213.00	2.50	218.32	3.00	0.37	30.29	0.49	249.10	m2	267.79
divided into panes 0.10 - 0.50m2	244.95	2.50	251.07	3.00	0.37	30.29	0.49	281.85	m2	302.99
divided into panes not exceeding 0.10m2	276.90	2.50	283.82	3.00	0.37	30.29	0.49	314.60	m2	338.19
50mm moulded casements or fanlights										
in one pane	244.95	2.50	251.07	3.20	0.40	32.34	0.52	283.94	m2	305.23
divided into panes 0.10 - 0.50m2	276.90	2.50	283.82	3.20	0.40	32.34	0.52	316.69	m2	340.44
divided into panes not exceeding 0.10m2	308.85	2.50	316.57	3.20	0.40	32.34	0.52	349.44	m2	375.64
38mm moulded casements with semi-circular heads (measured square)										
in one pane	426.00	2.50	436.65	3.30	0.41	33.34	0.54	470.53	m2	505.82
divided into panes 0.10 - 0.50m2	457.95	2.50	469.40	3.30	0.41	33.34	0.54	503.28	m2	541.02
divided into panes not exceeding 0.10m2	489.90	2.50	502.15	3.30	0.41	33.34	0.54	536.02	m2	576.23
50mm moulded casements with semi-circular heads (measured square)										
in one pane	489.90	2.50	502.15	3.50	0.44	35.39	0.57	538.11	m2	578.47
divided into panes 0.10 - 0.50m2	521.85	2.50	534.90	3.50	0.44	35.39	0.57	570.86	m2	613.67
divided into panes not exceeding 0.10m2	553.80	2.50	567.64	3.50	0.44	35.39	0.57	603.61	m2	648.88
38mm bullseye casements										
457mm diameter in one pane	319.50	2.50	327.49	2.00	0.25	20.22	0.33	348.03	nr	374.13
762mm diameter in one pane	426.00	2.50	436.65	3.00	0.37	30.29	0.49	467.43	nr	502.48

Labour hourly rates: (except Specialists) Craft Operatives £9.23 Labourer £7.02 Rates are national average prices. Refer to REGIONAL VARIATIONS for indicative levels of overall pricing in regions	MATERIALS			LABOUR				RATES		
	Del to Site £	Waste %	Material Cost £	Craft Optve Hrs	Lab Hrs	Labour Cost £	Sunds £	Nett Rate £	Unit	Gross Rate (+7.5%) £
L10: WINDOWS/ROOFLIGHTS/SCREENS/LOUVRES Cont.										
TIMBER WINDOWS - Purpose made windows in European Oak, wrought Cont.										
50mm bullseye casements										
457mm diameter in one pane	340.80	2.50	349.32	2.20	0.27	22.20	0.36	371.88	nr	399.77
762mm diameter in one pane	457.95	2.50	469.40	3.30	0.41	33.34	0.54	503.28	nr	541.02
Labours										
check throated edge	1.07	2.50	1.10	-	-	-	-	1.10	m	1.18
rebated and splayed bottom rail	1.07	2.50	1.10	-	-	-	-	1.10	m	1.18
rebated and beaded meeting stile	1.49	2.50	1.53	-	-	-	-	1.53	m	1.64
fitting and hanging casement or fanlight on butts (included elsewhere)										
38mm	-	-	-	1.30	0.16	13.12	-	13.12	nr	14.11
50mm	-	-	-	1.45	0.18	14.65	-	14.65	nr	15.75
fitting and hanging casement or fanlight on sash centres (included elsewhere)										
38mm	-	-	-	3.00	0.37	30.29	-	30.29	nr	32.56
50mm	-	-	-	3.30	0.41	33.34	-	33.34	nr	35.84
TIMBER WINDOWS - Sash windows in softwood, wrought, preservative treated; priming or staining by manufacturer										
Magnet Trade; weather stripping; spiral balances; brass plated fittings; fixing to masonry with galvanised steel cramps -4 nr, 25 x 3 x 150mm girth, flat section holes -2										
825 x 1094mm overall; SS3	221.93	2.50	227.48	3.30	1.65	42.04	1.40	270.92	nr	291.24
825 x 1394mm overall; SS4	235.79	2.50	241.68	3.60	1.80	45.86	1.40	288.95	nr	310.62
920 x 1225mm overall; SS2	247.85	2.50	254.05	3.60	1.80	45.86	1.40	301.31	nr	323.91
1051 x 1094mm overall; SSW3	235.97	2.50	241.87	3.60	1.80	45.86	1.40	289.13	nr	310.82
1051 x 1394mm overall; SSW4	252.53	2.50	258.84	3.90	1.95	49.69	1.40	309.93	nr	333.17
1051 x 1694mm overall; SSW5	278.36	2.50	285.32	4.20	2.10	53.51	1.40	340.23	nr	365.74
Magnet Trade; Georgian, open rebated for glass; weatherstripping; spiral balances; brass plated fittings; fixing to masonry with galvanised steel cramps -4 nr, 25 x 3 x 150mm girth, flat section, holes -2										
825 x 1094mm overall; GS3	246.77	2.50	252.94	2.25	1.12	28.63	1.40	282.97	nr	304.19
825 x 1394mm overall; GS4	268.01	2.50	274.71	2.60	1.30	33.12	1.40	309.23	nr	332.43
1051 x 1394mm overall; GSW4	296.00	2.50	303.40	2.30	0.29	23.26	0.37	327.03	nr	351.56
TIMBER WINDOWS - Sash windows mainly in softwood, wrought										
Cased frames; 25mm inside and outside linings; 32mm pulley stiles and head; 10mm beads, back linings, etc; 76mm oak sunk weathered and throated sills; 38mm moulded sashes										
in one pane	140.00	2.50	143.50	2.30	0.29	23.26	0.37	167.13	m2	179.67
divided into panes 0.50 - 1.00m2	147.00	2.50	150.68	2.30	0.29	23.26	0.37	174.31	m2	187.38
divided into panes 0.10 - 0.50m2	161.00	2.50	165.03	2.30	0.29	23.26	0.37	188.66	m2	202.81
divided into panes not exceeding 0.10m2	182.00	2.50	186.55	2.30	0.29	23.26	0.37	210.18	m2	225.95
Cased frames; 25mm inside and outside linings; 32mm pulley stiles and head; 10mm beads, back linings, etc; 76mm oak sunk weathered and throated sills; 50mm moulded sashes										
in one pane	161.00	2.50	165.03	2.50	0.31	25.25	0.41	190.69	m2	204.99
divided into panes 0.50 - 1.00m2	168.00	2.50	172.20	2.50	0.31	25.25	0.41	197.86	m2	212.70
divided into panes 0.10 - 0.50m2	182.00	2.50	186.55	2.50	0.31	25.25	0.41	212.21	m2	228.13
divided into panes not exceeding 0.10m2	203.00	2.50	208.07	2.50	0.31	25.25	0.41	233.74	m2	251.27
extra; windows in three lights with boxed mullions	28.00	2.50	28.70	0.70	0.09	7.09	0.11	35.90	nr	38.60
extra; windows with moulded horns	35.00	2.50	35.88	-	-	-	-	35.88	nr	38.57
extra; deep bottom rails and draught beads	3.50	2.50	3.59	-	-	-	-	3.59	m	3.86
Labours										
throats	0.35	2.50	0.36	-	-	-	-	0.36	m	0.39
grooves for jamb linings	0.42	2.50	0.43	-	-	-	-	0.43	m	0.46
fitting and hanging sashes in double hung windows; providing brass faced iron pulleys, best flax cords and iron weights										
38mm thick weighing 14 lbs per sash	35.00	2.50	35.88	1.40	0.17	14.12	-	49.99	nr	53.74
50mm thick weighing 20 lbs per sash	40.00	2.50	41.00	1.40	0.17	14.12	-	55.12	nr	59.25
Note for small frames and sashes, i.e. 1.25m2 and under, add 20% to the foregoing prices										
TIMBER WINDOWS - Sash windows in Afrormosia, wrought										
Cased frames; 25mm inside and outside linings; 32mm pulley stiles and head; 10mm beads, back linings, etc; 76mm oak sunk weathered and throated sills; 38mm moulded sashes										
in one pane	350.00	2.50	358.75	4.00	0.50	40.43	0.65	399.83	m2	429.82
divided into panes 0.50 - 1.00m2	367.50	2.50	376.69	4.00	0.50	40.43	0.65	417.77	m2	449.10
divided into panes 0.10 - 0.50m2	402.50	2.50	412.56	4.00	0.50	40.43	0.65	453.64	m2	487.67
divided into panes not exceeding 0.10m2	455.00	2.50	466.38	4.00	0.50	40.43	0.65	507.45	m2	545.51

WINDOWS/DOORS/STAIRS

Labour hourly rates: (except Specialists) Craft Operatives £9.23 Labourer £7.02 Rates are national average prices. Refer to REGIONAL VARIATIONS for indicative levels of overall pricing in regions	MATERIALS			LABOUR				RATES		
	Del to Site	Waste	Material Cost	Craft Optve	Lab	Labour Cost	Sunds	Nett Rate	Unit	Gross Rate (+7.5%)
	£	%	£	Hrs	Hrs	£	£	£		£
L10: WINDOWS/ROOFLIGHTS/SCREENS/LOUVRES Cont.										
TIMBER WINDOWS – Sash windows in Afrormosia, wrought Cont.										
Cased frames; 25mm inside and outside linings; 32mm pulley stiles and head; 10mm beads, back linings, etc; 76mm oak sunk weathered and throated sills; 50mm moulded sashes										
in one pane	402.50	2.50	412.56	4.40	0.55	44.47	0.72	457.76	m2	492.09
divided into panes 0.50 - 1.00m2	420.00	2.50	430.50	4.40	0.55	44.47	0.72	475.69	m2	511.37
divided into panes 0.10 - 0.50m2	455.00	2.50	466.38	4.40	0.55	44.47	0.72	511.57	m2	549.94
divided into panes not exceeding 0.10m2	507.50	2.50	520.19	4.40	0.55	44.47	0.72	565.38	m2	607.78
extra; windows in three lights with boxed mullions	70.00	2.50	71.75	1.20	0.15	12.13	0.20	84.08	nr	90.38
extra; windows with moulded horns	87.50	2.50	89.69	-	-	-	-	89.69	nr	96.41
extra; deep bottom rails and draught beads	8.75	2.50	8.97	-	-	-	-	8.97	m	9.64
Labours										
throats	0.88	2.50	0.90	-	-	-	-	0.90	m	0.97
grooves for jamb linings	1.05	2.50	1.08	-	-	-	-	1.08	m	1.16
fitting and hanging sashes in double hung windows; providing brass faced iron pulleys, best flax cords and iron weights										
38mm thick weighing 14 lbs per sash..........	35.00	2.50	35.88	2.50	0.31	25.25	-	61.13	nr	65.71
50mm thick weighing 20 lbs per sash..........	40.00	2.50	41.00	2.50	0.31	25.25	-	66.25	nr	71.22
Note for small frames and sashes, i.e. 1.25m2 and under, add 20% to the foregoing prices										
TIMBER WINDOWS – Sash windows in European Oak, wrought										
Cased frames; 25mm inside and outside linings; 32mm pulley stiles and head; 10mm beads, back linings, etc; 76mm oak sunk weathered and throated sills; 38mm moulded sashes										
in one pane	420.00	2.50	430.50	4.60	0.57	46.46	0.75	477.71	m2	513.54
divided into panes 0.50 - 1.00m2	441.00	2.50	452.02	4.60	0.57	46.46	0.75	499.23	m2	536.68
divided into panes 0.10 - 0.50m2	483.00	2.50	495.07	4.60	0.57	46.46	0.75	542.28	m2	582.96
divided into panes not exceeding 0.10m2	546.00	2.50	559.65	4.60	0.57	46.46	0.75	606.86	m2	652.37
Cased frames; 25mm inside and outside linings; 32mm pulley stiles and head; 10mm beads, back linings, etc; 76mm oak sunk weathered and throated sills; 50mm moulded sashes										
in one pane	483.00	2.50	495.07	5.00	0.62	50.50	0.81	546.39	m2	587.37
divided into panes 0.50 - 1.00m2	504.00	2.50	516.60	5.00	0.62	50.50	0.81	567.91	m2	610.51
divided into panes 0.10 - 0.50m2	546.00	2.50	559.65	5.00	0.62	50.50	0.81	610.96	m2	656.78
divided into panes not exceeding 0.10m2	609.00	2.50	624.23	5.00	0.62	50.50	0.81	675.54	m2	726.20
extra; windows in three lights with boxed mullions	84.00	2.50	86.10	1.40	0.17	14.12	0.23	100.45	nr	107.98
extra; windows with moulded horns	105.00	2.50	107.63	-	-	-	-	107.63	nr	115.70
extra; deep bottom rails and draught beads	10.50	2.50	10.76	-	-	-	-	10.76	m	11.57
Labours										
throats	1.05	2.50	1.08	-	-	-	-	1.08	m	1.16
grooves for jamb linings	1.26	2.50	1.29	-	-	-	-	1.29	m	1.39
fitting and hanging sashes in double hung windows; providing brass faced iron pulleys, best flax cords and iron weights										
38mm thick weighing 14 lbs per sash..........	35.00	2.50	35.88	2.80	0.35	28.30	-	64.18	nr	68.99
50mm thick weighing 20 lbs per sash..........	40.00	2.50	41.00	2.80	0.35	28.30	-	69.30	nr	74.50
Note for small frames and sashes, i.e. 1.25m2 and under, add 20% to the foregoing prices										
TIMBER ROOFLIGHTS – Skylights in softwood, wrought										
Chamfered, straight bar										
38mm......................................	67.45	2.50	69.14	1.30	0.16	13.12	0.21	82.47	m2	88.65
50mm......................................	78.10	2.50	80.05	1.50	0.19	15.18	0.24	95.47	m2	102.63
63mm......................................	88.75	2.50	90.97	1.80	0.22	18.16	0.29	109.42	m2	117.62
Moulded, straight bar										
38mm......................................	71.00	2.50	72.78	1.30	0.16	13.12	0.21	86.11	m2	92.57
50mm......................................	81.65	2.50	83.69	1.50	0.19	15.18	0.24	99.11	m2	106.54
63mm......................................	92.30	2.50	94.61	1.80	0.22	18.16	0.29	113.06	m2	121.54
TIMBER ROOFLIGHTS – Skylights in Oak, wrought										
Chamfered, straight bar										
38mm......................................	202.35	2.50	207.41	2.00	0.25	20.22	0.33	227.95	m2	245.05
50mm......................................	234.30	2.50	240.16	2.25	0.28	22.73	0.37	263.26	m2	283.01
63mm......................................	266.25	2.50	272.91	2.70	0.34	27.31	0.44	300.65	m2	323.20
Moulded, straight bar										
38mm......................................	213.00	2.50	218.32	2.00	0.25	20.22	0.33	238.87	m2	256.79
50mm......................................	244.95	2.50	251.07	2.25	0.28	22.73	0.37	274.18	m2	294.74
63mm......................................	276.90	2.50	283.82	2.70	0.34	27.31	0.44	311.57	m2	334.94
TIMBER ROOFLIGHTS – Skylight kerbs in softwood, wrought										
Kerbs; dovetailed at angles										
38 x 225mm..	9.20	2.50	9.43	0.27	0.03	2.70	0.04	12.17	m	13.09
50 x 225mm..	11.03	2.50	11.31	0.30	0.04	3.05	0.05	14.41	m	15.49
38 x 225mm; chamfers -1 nr	9.65	2.50	9.89	0.27	0.03	2.70	0.04	12.63	m	13.58

Labour hourly rates: (except Specialists) Craft Operatives £9.23 Labourer £7.02 Rates are national average prices. Refer to REGIONAL VARIATIONS for indicative levels of overall pricing in regions	MATERIALS			LABOUR				RATES		
	Del to Site	Waste	Material Cost	Craft Optve	Lab	Labour Cost	Sunds	Nett Rate	Unit	Gross Rate (+7.5%)
	£	%	£	Hrs	Hrs	£	£	£		£

L10: WINDOWS/ROOFLIGHTS/SCREENS/LOUVRES Cont.

TIMBER ROOFLIGHTS - Skylight kerbs in softwood, wrought Cont.

Kerbs; dovetailed at angles Cont.

50 x 225mm; chamfers -1 nr	11.49	2.50	11.78	0.30	0.04	3.05	0.05	14.88	m	15.99

Kerbs; in two thicknesses to circular skylights

38 x 225mm	27.56	2.50	28.25	0.41	0.05	4.14	0.07	32.45	m	34.89
50 x 225mm	33.04	2.50	33.87	0.46	0.06	4.67	0.08	38.61	m	41.51
38 x 225mm; chamfers -1 nr	28.49	2.50	29.20	0.41	0.05	4.14	0.07	33.41	m	35.91
50 x 225mm; chamfers -1 nr	34.00	2.50	34.85	0.46	0.06	4.67	0.08	39.60	m	42.57

TIMBER ROOFLIGHTS - Skylight kerbs in Oak, wrought

Kerbs; dovetailed at angles

38 x 225mm	43.00	2.50	44.08	0.54	0.07	5.48	0.09	49.64	m	53.36
50 x 225mm	51.60	2.50	52.89	0.60	0.08	6.10	0.10	59.09	m	63.52
38 x 225mm; chamfers -1 nr	45.15	2.50	46.28	0.54	0.07	5.48	0.09	51.84	m	55.73
50 x 225mm; chamfers -1 nr	53.75	2.50	55.09	0.60	0.08	6.10	0.10	61.29	m	65.89

Kerbs; in two thicknesses to circular skylights

38 x 225mm	129.00	2.50	132.22	0.82	0.10	8.27	0.13	140.63	m	151.17
50 x 225mm	154.80	2.50	158.67	0.92	0.12	9.33	0.15	168.15	m	180.77
38 x 225mm; chamfers -1 nr	133.30	2.50	136.63	0.82	0.10	8.27	0.13	145.03	m	155.91
50 x 225mm; chamfers -1 nr	159.10	2.50	163.08	0.92	0.12	9.33	0.15	172.56	m	185.50

TIMBER ROOFLIGHTS - Roof windows in Nordic red pine, wrought, treated

Velux roof windows, The Velux Company Ltd; aluminium clad externally; hinges; fittings; factory glazed clear float double glazed sealed unit; type EDZ 0000 flashings; fixing to timber with screws; dressing flashings

550 x 980mm overall, GGL104	186.83	2.50	191.50	4.50	2.25	57.33	-	248.83	nr	267.49
660 x 1180mm overall, GGL206	218.61	2.50	224.08	5.25	2.62	66.85	-	290.93	nr	312.74
660 x 1180mm overall, GHL206	284.51	2.50	291.62	5.25	2.62	66.85	-	358.47	nr	385.36
780 x 980mm overall, GGL304	216.86	2.50	222.28	5.00	2.50	63.70	-	285.98	nr	307.43
780 x 980mm overall, GHL304	278.48	2.50	285.44	5.00	2.50	63.70	-	349.14	nr	375.33
780 x 1400mm overall, GGL308	255.11	2.50	261.49	5.50	2.75	70.07	-	331.56	nr	356.42
780 x 1400mm overall, GHL308	346.65	2.50	355.32	5.50	2.75	70.07	-	425.39	nr	457.29
940 x 1600mm overall, GGL410	306.48	2.50	314.14	6.00	3.00	76.44	-	390.58	nr	419.88
1140 x 1180mm overall, GGL606	287.70	2.50	294.89	5.75	2.87	73.22	-	368.11	nr	395.72
1140 x 1180mm overall, GHL606	353.71	2.50	362.55	5.75	2.87	73.22	-	435.77	nr	468.46
1340 x 980mm overall, GGL804	292.22	2.50	299.53	5.75	2.87	73.22	-	372.75	nr	400.70
1340 x 980mm overall, GHL804	358.47	2.50	367.43	5.75	2.87	73.22	-	440.65	nr	473.70
1340 x 1400mm overall, GGL808	346.21	2.50	354.87	6.50	3.25	82.81	-	437.68	nr	470.50
1340 x 1400mm overall, GHL808	426.85	2.50	437.52	6.50	3.25	82.81	-	520.33	nr	559.36

Velux roof windows, The Velux Company Ltd; aluminium clad externally; hinges; fittings; factory glazed clear float double glazed sealed unit; type EDL 0000 flashings; fixing to timber with screws; dressing flashings

550 x 980mm overall, GGL104	182.50	2.50	187.06	4.50	2.25	57.33	-	244.39	nr	262.72
660 x 1180mm overall, GGL206	214.04	2.50	219.39	5.25	2.62	66.85	-	286.24	nr	307.71
660 x 1180mm overall, GHL206	279.94	2.50	286.94	5.25	2.62	66.85	-	353.79	nr	380.32
780 x 980mm overall, GGL304	211.72	2.50	217.01	5.00	2.50	63.70	-	280.71	nr	301.77
780 x 980mm overall, GHL304	273.35	2.50	280.18	5.00	2.50	63.70	-	343.88	nr	369.68
780 x 1400mm overall, GGL308	249.09	2.50	255.32	5.50	2.75	70.07	-	325.39	nr	349.79
780 x 1400mm overall, GHL308	340.63	2.50	349.15	5.50	2.75	70.07	-	419.22	nr	450.66
940 x 1600mm overall, GGL410	300.79	2.50	308.31	6.00	3.00	76.44	-	384.75	nr	413.61
1140 x 1180mm overall, GGL606	281.20	2.50	288.23	5.75	2.87	73.22	-	361.45	nr	388.56
1140 x 1180mm overall, GHL606	347.24	2.50	355.92	5.75	2.87	73.22	-	429.14	nr	461.33
1340 x 980mm overall, GGL804	284.17	2.50	291.27	5.75	2.87	73.22	-	364.49	nr	391.83
1340 x 980mm overall, GHL804	350.36	2.50	359.12	5.75	2.87	73.22	-	432.34	nr	464.76
1340 x 1400mm overall, GGL808	337.85	2.50	346.30	6.50	3.25	82.81	-	429.11	nr	461.29
1340 x 1400mm overall, GHL808	418.50	2.50	428.96	6.50	3.25	82.81	-	511.77	nr	550.16

Velux roof windows, The Velux Company Ltd; aluminium clad externally; hinges; fittings; factory glazed clear float/heat absorbing glass double glazed sealed unit; type EDZ 0000 flashings; fixing to timber with screws; dressing flashings

550 x 980mm overall, GGL104	222.48	2.50	228.04	4.50	2.25	57.33	-	285.37	nr	306.77
660 x 1180mm overall, GGL206	257.16	2.50	263.59	5.25	2.62	66.85	-	330.44	nr	355.22
660 x 1180mm overall, GHL206	322.07	2.50	330.12	5.25	2.62	66.85	-	396.97	nr	426.74
780 x 980mm overall, GGL304	254.23	2.50	260.59	5.00	2.50	63.70	-	324.29	nr	348.61
780 x 980mm overall, GHL304	314.90	2.50	322.77	5.00	2.50	63.70	-	386.47	nr	415.46
780 x 1400mm overall, GGL308	301.83	2.50	309.38	5.50	2.75	70.07	-	379.45	nr	407.90
780 x 1400mm overall, GHL308	391.88	2.50	401.68	5.50	2.75	70.07	-	471.75	nr	507.13
940 x 1600mm overall, GGL410	356.11	2.50	365.02	6.00	3.00	76.44	-	441.45	nr	474.56
1140 x 1180mm overall, GGL606	339.72	2.50	348.21	5.75	2.87	73.22	-	421.43	nr	453.04
1140 x 1180mm overall, GHL606	404.65	2.50	414.77	5.75	2.87	73.22	-	487.99	nr	524.59
1340 x 980mm overall, GGL804	343.91	2.50	352.51	5.75	2.87	73.22	-	425.73	nr	457.66
1340 x 980mm overall, GHL804	408.82	2.50	419.04	5.75	2.87	73.22	-	492.26	nr	529.18
1340 x 1400mm overall, GGL808	402.57	2.50	412.63	6.50	3.25	82.81	-	495.44	nr	532.60
1340 x 1400mm overall, GHL808	481.74	2.50	493.78	6.50	3.25	82.81	-	576.59	Unr	619.84

	MATERIALS			LABOUR				RATES		
Labour hourly rates: (except Specialists) Craft Operatives £9.23 Labourer £7.02 Rates are national average prices. Refer to REGIONAL VARIATIONS for indicative levels of overall pricing in regions	Del to Site £	Waste %	Material Cost £	Craft Optve Hrs	Lab Hrs	Labour Cost £	Sunds £	Nett Rate £	Unit	Gross Rate (+7.5%) £

L10: WINDOWS/ROOFLIGHTS/SCREENS/LOUVRES Cont.

TIMBER ROOFLIGHTS - Roof windows in Nordic red pine, wrought, treated Cont.

Velux roof windows, The Velux Company Ltd; aluminium clad externally; hinges; fittings; factory glazed clear float/heat absorbing glass double glazed sealed unit; type EDL 0000 flashings; fixing to timber with screws; dressing flashings

	Del to Site	Waste	Material Cost	Craft Optve	Lab	Labour Cost	Sunds	Nett Rate	Unit	Gross Rate
550 x 980mm overall, GGL104	218.14	2.50	223.59	4.50	2.25	57.33	–	280.92	nr	301.99
660 x 1180mm overall, GGL206	252.59	2.50	258.90	5.25	2.62	66.85	–	325.75	nr	350.19
660 x 1180mm overall, GHL206	317.50	2.50	325.44	5.25	2.62	66.85	–	392.29	nr	421.71
780 x 980mm overall, GGL304	249.10	2.50	255.33	5.00	2.50	63.70	–	319.03	nr	342.95
780 x 980mm overall, GHL304	309.78	2.50	317.52	5.00	2.50	63.70	–	381.22	nr	409.82
780 x 1400mm overall, GGL308	295.82	2.50	303.22	5.50	2.75	70.07	–	373.29	nr	401.28
780 x 1400mm overall, GHL308	385.86	2.50	395.51	5.50	2.75	70.07	–	465.58	nr	500.49
940 x 1600mm overall, GGL410	350.41	2.50	359.17	6.00	3.00	76.44	–	435.61	nr	468.28
1140 x 1180mm overall, GGL606	333.46	2.50	341.80	5.75	2.87	73.22	–	415.02	nr	446.14
1140 x 1180mm overall, GHL606	398.17	2.50	408.12	5.75	2.87	73.22	–	481.34	nr	517.44
1340 x 980mm overall, GGL804	335.79	2.50	344.18	5.75	2.87	73.22	–	417.40	nr	448.71
1340 x 980mm overall, GHL804	400.72	2.50	410.74	5.75	2.87	73.22	–	483.96	nr	520.25
1340 x 1400mm overall, GGL808	394.20	2.50	404.06	6.50	3.25	82.81	–	486.87	nr	523.38
1340 x 1400mm overall, GHL808	473.38	2.50	485.21	6.50	3.25	82.81	–	568.02	nr	610.63

Velux roof windows, The Velux Company Ltd; aluminium clad externally; hinges; fittings; factory glazed clear float/clear laminated glass double glazed sealed unit; type EDZ 0000 flashings; fixing to timber with screws; dressing flashings

	Del to Site	Waste	Material Cost	Craft Optve	Lab	Labour Cost	Sunds	Nett Rate	Unit	Gross Rate
550 x 980mm overall, GGL104	230.93	2.50	236.70	4.50	2.25	57.33	–	294.03	nr	316.09
660 x 1180mm overall, GGL206	274.96	2.50	281.83	5.25	2.62	66.85	–	348.68	nr	374.84
660 x 1180mm overall, GHL206	339.42	2.50	347.91	5.25	2.62	66.85	–	414.76	nr	445.86
780 x 980mm overall, GGL304	270.85	2.50	277.62	5.00	2.50	63.70	–	341.32	nr	366.92
780 x 980mm overall, GHL304	331.13	2.50	339.41	5.00	2.50	63.70	–	403.11	nr	433.34
780 x 1400mm overall, GGL308	329.02	2.50	337.25	5.50	2.75	70.07	–	407.32	nr	437.86
780 x 1400mm overall, GHL308	418.61	2.50	429.08	5.50	2.75	70.07	–	499.15	nr	536.58
940 x 1600mm overall, GGL410	399.03	2.50	409.01	6.00	3.00	76.44	–	485.45	nr	521.85
1140 x 1180mm overall, GGL606	377.34	2.50	386.77	5.75	2.87	73.22	–	459.99	nr	494.49
1140 x 1180mm overall, GHL606	440.81	2.50	451.83	5.75	2.87	73.22	–	525.05	nr	564.43
1340 x 980mm overall, GGL804	379.83	2.50	389.33	5.75	2.87	73.22	–	462.55	nr	497.24
1340 x 980mm overall, GHL804	443.84	2.50	454.94	5.75	2.87	73.22	–	528.16	nr	567.77
1340 x 1400mm overall, GGL808	459.23	2.50	470.71	6.50	3.25	82.81	–	553.52	nr	595.03
1340 x 1400mm overall, GHL808	536.79	2.50	550.21	6.50	3.25	82.81	–	633.02	nr	680.50

Velux roof windows, The Velux Company Ltd; aluminium clad externally; hinges; fittings; factory glazed clear float/clear laminated glass double glazed sealed unit; type EDL 0000 flashings; fixing to timber with screws; dressing flashings

	Del to Site	Waste	Material Cost	Craft Optve	Lab	Labour Cost	Sunds	Nett Rate	Unit	Gross Rate
550 x 980mm overall, GGL104	226.60	2.50	232.26	4.50	2.25	57.33	–	289.60	nr	311.31
660 x 1180mm overall, GGL206	270.40	2.50	277.16	5.25	2.62	66.85	–	344.01	nr	369.81
660 x 1180mm overall, GHL206	334.87	2.50	343.24	5.25	2.62	66.85	–	410.09	nr	440.85
780 x 980mm overall, GGL304	265.71	2.50	272.35	5.00	2.50	63.70	–	336.05	nr	361.26
780 x 980mm overall, GHL304	326.00	2.50	334.15	5.00	2.50	63.70	–	397.85	nr	427.69
780 x 1400mm overall, GGL308	323.52	2.50	331.61	5.50	2.75	70.07	–	401.68	nr	431.80
780 x 1400mm overall, GHL308	412.61	2.50	422.93	5.50	2.75	70.07	–	493.00	nr	529.97
940 x 1600mm overall, GGL410	393.34	2.50	403.17	6.00	3.00	76.44	–	479.61	nr	515.58
1140 x 1180mm overall, GGL606	370.85	2.50	380.12	5.75	2.87	73.22	–	453.34	nr	487.34
1140 x 1180mm overall, GHL606	434.31	2.50	445.17	5.75	2.87	73.22	–	518.39	nr	557.27
1340 x 980mm overall, GGL804	371.72	2.50	381.01	5.75	2.87	73.22	–	454.23	nr	488.30
1340 x 980mm overall, GHL804	435.72	2.50	446.61	5.75	2.87	73.22	–	519.83	nr	558.82
1340 x 1400mm overall, GGL808	459.23	2.50	470.71	6.50	3.25	82.81	–	553.52	nr	595.03
1340 x 1400mm overall, GHL808	528.60	2.50	541.82	6.50	3.25	82.81	–	624.63	nr	671.47

TIMBER ROOFLIGHTS - Bedding and pointing frames

Bedding in cement mortar (1:3); pointing with Secomastic standard mastic one side

	Del to Site	Waste	Material Cost	Craft Optve	Lab	Labour Cost	Sunds	Nett Rate	Unit	Gross Rate
wood frames	1.00	10.00	1.10	0.18	–	1.66	–	2.76	m	2.97

Bedding in cement mortar (1:3); pointing with coloured two part polysulphide mastic one side

	Del to Site	Waste	Material Cost	Craft Optve	Lab	Labour Cost	Sunds	Nett Rate	Unit	Gross Rate
wood frames	2.00	10.00	2.20	0.18	–	1.66	–	3.86	m	4.15

METAL WINDOWS - Windows in galvanised steel

Crittall Windows Ltd. Duralife Homelight; weatherstripping; fixing to masonry with lugs
fixed lights

	Del to Site	Waste	Material Cost	Craft Optve	Lab	Labour Cost	Sunds	Nett Rate	Unit	Gross Rate
508 x 292mm, NG5	15.07	2.50	15.45	0.53	0.22	6.44	–	21.88	nr	23.52
508 x 923mm, NC5	20.65	2.50	21.17	0.85	0.30	9.95	–	31.12	nr	33.45
508 x 1218mm, ND5	23.61	2.50	24.20	1.07	0.36	12.40	–	36.60	nr	39.35
997 x 628mm, NE13	27.28	2.50	27.96	0.97	0.35	11.41	–	39.37	nr	42.33
997 x 923mm, NC13	31.96	2.50	32.76	1.08	0.38	12.64	–	45.40	nr	48.80
997 x 1218mm, ND13	37.08	2.50	38.01	1.30	0.44	15.09	–	53.09	nr	57.08
1486 x 628mm, NE14	32.70	2.50	33.52	1.30	0.46	15.23	–	48.75	nr	52.40
1486 x 923mm, NC14	36.72	2.50	37.64	1.42	0.50	16.62	–	54.25	nr	58.32
1486 x 1218mm, ND14	41.58	2.50	42.62	1.65	0.57	19.23	–	61.85	nr	66.49
top hung lights										
508 x 292mm, NG1	50.68	2.50	51.95	0.53	0.22	6.44	–	58.38	nr	62.76
508 x 457mm, NH1	58.81	2.50	60.28	0.64	0.25	7.66	–	67.94	nr	73.04
508 x 628mm, NE1	62.38	2.50	63.94	0.75	0.28	8.89	–	72.83	nr	78.29
997 x 628mm, NE13E	83.31	2.50	85.39	0.97	0.35	11.41	–	96.80	nr	104.06
997 x 923mm, NC13C	89.66	2.50	91.90	1.08	0.38	12.64	–	104.54	nr	112.38

Labour hourly rates: (except Specialists) Craft Operatives £9.23 Labourer £7.02 Rates are national average prices. Refer to REGIONAL VARIATIONS for indicative levels of overall pricing in regions	MATERIALS			LABOUR				RATES		
	Del to Site £	Waste %	Material Cost £	Craft Optve Hrs	Lab Hrs	Labour Cost £	Sunds £	Nett Rate £	Unit	Gross Rate (+7.5%) £
L10: WINDOWS/ROOFLIGHTS/SCREENS/LOUVRES Cont.										
METAL WINDOWS - Windows in galvanised steel Cont.										
Crittall Windows Ltd. Duralife Homelight; weatherstripping; fixing to masonry with lugs Cont.										
bottom hung lights										
508 x 628mm, NL1....................	69.30	2.50	71.03	0.75	0.28	8.89	-	79.92	nr	85.91
side hung lights										
508 x 628mm, NES1...................	64.64	2.50	66.26	0.75	0.28	8.89	-	75.14	nr	80.78
508 x 923mm, NC1....................	69.45	2.50	71.19	0.85	0.30	9.95	-	81.14	nr	87.22
508 x 1067mm, NCO1..................	73.63	2.50	75.47	0.96	0.33	11.18	-	86.65	nr	93.15
508 x 1218mm, ND1...................	78.37	2.50	80.33	1.07	0.36	12.40	-	92.73	nr	99.69
mixed lights										
279 x 923mm, NC6F..................	55.99	2.50	57.39	0.85	0.30	9.95	-	67.34	nr	72.39
508 x 923mm, NC5F..................	61.38	2.50	62.91	0.85	0.30	9.95	-	72.87	nr	78.33
508 x 1067mm, NCO5F................	62.83	2.50	64.40	0.96	0.33	11.18	-	75.58	nr	81.25
997 x 292mm, NG2..................	69.06	2.50	70.79	0.74	0.29	8.87	-	79.65	nr	85.63
997 x 457mm, NH2..................	72.98	2.50	74.80	0.85	0.32	10.09	-	84.90	nr	91.26
997 x 628mm, NE2..................	76.90	2.50	78.82	0.97	0.35	11.41	-	90.23	nr	97.00
997 x 628mm, NES2.................	87.56	2.50	89.75	0.97	0.35	11.41	-	101.16	nr	108.75
997 x 923mm, NC2..................	95.76	2.50	98.15	1.08	0.38	12.64	-	110.79	nr	119.10
997 x 923mm, NC2F.................	127.12	2.50	130.30	1.08	0.38	12.64	-	142.93	nr	153.65
997 x 1067mm, NCO2................	104.31	2.50	106.92	1.19	0.41	13.86	-	120.78	nr	129.84
997 x 1067mm, NCO2F...............	138.68	2.50	142.15	1.19	0.41	13.86	-	156.01	nr	167.71
997 x 1218mm, ND2.................	115.68	2.50	118.57	1.30	0.44	15.09	-	133.66	nr	143.68
997 x 1218mm, ND2F................	140.15	2.50	143.65	1.30	0.44	15.09	-	158.74	nr	170.65
997 x 1513mm, NDV2FSB.............	161.54	2.50	165.58	1.42	0.47	16.41	-	181.98	nr	195.63
1486 x 628mm, NE3.................	104.27	2.50	106.88	1.30	0.46	15.23	-	122.10	nr	131.26
1486 x 923mm, NC4.................	158.38	2.50	162.34	1.42	0.50	16.62	-	178.96	nr	192.38
1486 x 923mm, NC4F................	193.85	2.50	198.70	1.42	0.50	16.62	-	215.31	nr	231.46
1486 x 1067mm, NCO4...............	188.81	2.50	193.53	1.54	0.53	17.93	-	211.47	nr	227.32
1486 x 1067mm, NCO4F..............	213.36	2.50	218.69	1.54	0.53	17.93	-	236.63	nr	254.38
1486 x 1218mm, ND4...............	203.34	2.50	208.42	1.65	0.57	19.23	-	227.65	nr	244.73
1486 x 1218mm, ND4F..............	216.43	2.50	221.84	1.65	0.57	19.23	-	241.07	nr	259.15
1486 x 1513mm, NDV4FSB...........	238.91	2.50	244.88	1.77	0.60	20.55	-	265.43	nr	285.32
1994 x 923mm, NC11F..............	213.21	2.50	218.54	1.75	0.60	20.36	-	238.90	nr	256.82
1994 x 1218mm, ND11F.............	241.75	2.50	247.79	2.00	0.70	23.37	-	271.17	nr	291.51
1994 x 1513mm, NDV11FSB..........	265.55	2.50	272.19	2.15	0.75	25.11	-	297.30	nr	319.60
Extra for										
mullions										
292mm high..........................	8.65	2.50	8.87	0.20	0.10	2.55	-	11.41	nr	12.27
457mm high..........................	11.20	2.50	11.48	0.30	0.15	3.82	-	15.30	nr	16.45
628mm high..........................	12.09	2.50	12.39	0.40	0.20	5.10	-	17.49	nr	18.80
923mm high..........................	18.19	2.50	18.64	0.50	0.25	6.37	-	25.01	nr	26.89
1067mm high.........................	20.01	2.50	20.51	0.60	0.30	7.64	-	28.15	nr	30.27
1218mm high.........................	21.20	2.50	21.73	0.80	0.40	10.19	-	31.92	nr	34.32
1513mm high.........................	25.99	2.50	26.64	1.00	0.50	12.74	-	39.38	nr	42.33
2056mm high.........................	31.09	2.50	31.87	1.20	0.60	15.29	-	47.16	nr	50.69
transoms										
279mm wide..........................	8.65	2.50	8.87	0.20	0.10	2.55	-	11.41	nr	12.27
508mm wide..........................	12.10	2.50	12.40	0.30	0.15	3.82	-	16.22	nr	17.44
997mm wide..........................	20.43	2.50	20.94	0.60	0.30	7.64	-	28.58	nr	30.73
1486mm wide.........................	28.48	2.50	29.19	1.00	0.50	12.74	-	41.93	nr	45.08
Controlair Ventilators, permanent, mill finish										
508mm wide..........................	23.81	2.50	24.41	-	-	-	-	24.41	nr	26.24
997mm wide..........................	36.68	2.50	37.60	-	-	-	-	37.60	nr	40.42
1486mm wide.........................	42.80	2.50	43.87	-	-	-	-	43.87	nr	47.16
Controlair Ventilators, permanent, mill finish, flyscreen										
508mm wide..........................	28.93	2.50	29.65	-	-	-	-	29.65	nr	31.88
997mm wide..........................	46.46	2.50	47.62	-	-	-	-	47.62	nr	51.19
1486mm wide.........................	54.78	2.50	56.15	-	-	-	-	56.15	nr	60.36
Controlair Ventilators, adjustable										
508mm wide 1-LT.....................	31.33	2.50	32.11	-	-	-	-	32.11	nr	34.52
997mm wide 2-LT.....................	45.72	2.50	46.86	-	-	-	-	46.86	nr	50.38
Controlair Ventilators, adjustable, flyscreen										
508mm wide 1-LT.....................	35.98	2.50	36.88	-	-	-	-	36.88	nr	39.65
997mm wide 2-LT.....................	55.67	2.50	57.06	-	-	-	-	57.06	nr	61.34
locks with Parkes locking handles for										
side hung lights...................	26.59	2.50	27.25	-	-	-	-	27.25	nr	29.30
horizontally pivoted lights........	26.59	2.50	27.25	-	-	-	-	27.25	nr	29.30
METAL WINDOWS - Windows in galvanised steel; **polyester powder coated; white matt**										
Crittall Windows Ltd. Duralife Homelight; weatherstripping; fixing to masonry with lugs										
fixed lights										
508 x 292mm, NG5....................	18.71	2.50	19.18	0.53	0.22	6.44	-	25.61	nr	27.54
508 x 923mm, NC5....................	25.62	2.50	26.26	0.85	0.30	9.95	-	36.21	nr	38.93
508 x 1218mm, ND5..................	29.30	2.50	30.03	1.07	0.36	12.40	-	42.44	nr	45.62
997 x 628mm, NE13.................	33.86	2.50	34.71	0.97	0.35	11.41	-	46.12	nr	49.58
997 x 923mm, NC13.................	39.65	2.50	40.64	1.08	0.38	12.64	-	53.28	nr	57.27
997 x 1218mm, ND13................	45.68	2.50	46.82	1.30	0.44	15.09	-	61.91	nr	66.55
1486 x 628mm, NE14................	40.66	2.50	41.68	1.30	0.46	16.23	-	56.90	nr	61.17
1486 x 923mm, NC14................	45.55	2.50	46.69	1.42	0.50	16.62	-	63.31	nr	68.05
1486 x 1218mm, ND14..............	51.56	2.50	52.85	1.65	0.57	19.23	-	72.08	nr	77.49
top hung lights										
508 x 292mm, NG1...................	59.57	2.50	61.06	0.53	0.22	6.44	-	67.50	nr	72.56
508 x 457mm, NH1...................	69.31	2.50	71.04	0.64	0.25	7.66	-	78.70	nr	84.61
508 x 628mm, NE1...................	71.27	2.50	73.05	0.75	0.28	8.89	-	81.94	nr	88.09
997 x 628mm, NE13E................	98.96	2.50	101.43	0.97	0.35	11.41	-	112.84	nr	121.31
997 x 923mm, NC13C................	109.56	2.50	112.30	1.08	0.38	12.64	-	124.94	nr	134.31

Labour hourly rates: (except Specialists) Craft Operatives £9.23 Labourer £7.02 Rates are national average prices. Refer to REGIONAL VARIATIONS for indicative levels of overall pricing in regions	MATERIALS			LABOUR				RATES		
	Del to Site £	Waste %	Material Cost £	Craft Optve Hrs	Lab Hrs	Labour Cost £	Sunds £	Nett Rate £	Unit	Gross Rate (+7.5%) £
L10: WINDOWS/ROOFLIGHTS/SCREENS/LOUVRES Cont.										
METAL WINDOWS - Windows in galvanised steel; **polyester powder coated; white matt Cont.**										
Crittall Windows Ltd. Duralife Homelight; weatherstripping; fixing to masonry with lugs Cont.										
bottom hung lights										
508 x 628mm, NL1...........................	85.99	2.50	88.14	0.75	0.28	8.89	-	97.03	nr	104.30
side hung lights										
508 x 628mm, NES1.........................	76.86	2.50	78.78	0.75	0.28	8.89	-	87.67	nr	94.24
508 x 923mm, NC1..........................	82.60	2.50	84.67	0.85	0.30	9.95	-	94.62	nr	101.71
508 x 1067mm, NCO1.......................	87.61	2.50	89.80	0.96	0.33	11.18	-	100.98	nr	108.55
508 x 1218mm, ND1........................	93.41	2.50	95.75	1.07	0.36	12.40	-	108.15	nr	116.26
mixed lights										
279 x 923mm, NC6F........................	64.97	2.50	66.59	0.85	0.30	9.95	-	76.55	nr	82.29
508 x 923mm, NC5F........................	72.89	2.50	74.71	0.85	0.30	9.95	-	84.66	nr	91.01
508 x 1067mm, NCO5F......................	74.80	2.50	76.67	0.96	0.33	11.18	-	87.85	nr	94.44
997 x 292mm, NG2.........................	82.15	2.50	84.20	0.74	0.29	8.87	-	93.07	nr	100.05
997 x 457mm, NH2.........................	86.45	2.50	88.61	0.85	0.32	10.09	-	98.70	nr	106.11
997 x 628mm, NE2.........................	91.13	2.50	93.41	0.97	0.35	11.41	-	104.82	nr	112.68
997 x 628mm, NES2........................	104.77	2.50	107.39	0.97	0.35	11.41	-	118.80	nr	127.71
997 x 923mm, NC2.........................	114.23	2.50	117.09	1.08	0.38	12.64	-	129.72	nr	139.45
997 x 923mm, NC2F........................	150.17	2.50	153.92	1.08	0.38	12.64	-	166.56	nr	179.05
997 x 1067mm, NCO2.......................	123.98	2.50	127.08	1.19	0.41	13.86	-	140.94	nr	151.51
997 x 1067mm, NCO2F......................	163.89	2.50	167.99	1.19	0.41	13.86	-	181.85	nr	195.49
997 x 1218mm, ND2........................	138.04	2.50	141.49	1.30	0.44	15.09	-	156.58	nr	168.32
997 x 1218mm, ND2F.......................	165.71	2.50	169.85	1.30	0.44	15.09	-	184.94	nr	198.81
997 x 1513mm, NDV2FSB....................	192.19	2.50	196.99	1.42	0.47	16.41	-	213.40	nr	229.41
1486 x 628mm, NE3........................	123.90	2.50	127.00	1.30	0.46	15.23	-	142.23	nr	152.89
1486 x 923mm, NC4........................	187.05	2.50	191.73	1.42	0.50	16.62	-	208.34	nr	223.97
1486 x 923mm, NC4F.......................	228.71	2.50	234.43	1.42	0.50	16.62	-	251.04	nr	269.87
1486 x 1067mm, NCO4......................	223.80	2.50	229.40	1.54	0.53	17.93	-	247.33	nr	265.88
1486 x 1067mm, NCO4F.....................	251.55	2.50	257.84	1.54	0.53	17.93	-	275.77	nr	296.46
1486 x 1218mm, ND4.......................	241.32	2.50	247.35	1.65	0.57	19.23	-	266.58	nr	286.58
1486 x 1218mm, ND4F......................	255.28	2.50	261.66	1.65	0.57	19.23	-	280.89	nr	301.96
1486 x 1513mm, NDV4FSB...................	295.93	2.50	303.33	1.77	0.60	20.55	-	323.88	nr	348.17
1994 x 923mm, NC11F......................	252.05	2.50	258.35	1.75	0.60	20.36	-	278.72	nr	299.62
1994 x 1218mm, ND11F.....................	285.89	2.50	293.04	2.00	0.70	23.37	-	316.41	nr	340.14
1994 x 1513mm, NDV11FSB..................	401.23	2.50	411.26	2.15	0.75	25.11	-	436.37	nr	469.10
Extra for										
mullions										
292mm high..............................	10.79	2.50	11.06	0.20	0.10	2.55	-	13.61	nr	14.63
457mm high..............................	14.06	2.50	14.41	0.30	0.15	3.82	-	18.23	nr	19.60
628mm high..............................	15.20	2.50	15.58	0.40	0.20	5.10	-	20.68	nr	22.23
923mm high..............................	23.06	2.50	23.64	0.50	0.25	6.37	-	30.01	nr	32.26
1067mm high.............................	25.15	2.50	25.78	0.60	0.30	7.64	-	33.42	nr	35.93
1218mm high.............................	27.68	2.50	28.37	0.80	0.40	10.19	-	38.56	nr	41.46
1513mm high.............................	26.48	2.50	27.14	1.00	0.50	12.74	-	39.88	nr	42.87
2056mm high.............................	38.83	2.50	39.80	1.20	0.60	15.29	-	55.09	nr	59.22
transoms										
279mm wide..............................	11.94	2.50	12.24	0.20	0.10	2.55	-	14.79	nr	15.90
508mm wide..............................	15.19	2.50	15.57	0.30	0.15	3.82	-	19.39	nr	20.85
997mm wide..............................	25.56	2.50	26.20	0.60	0.30	7.64	-	33.84	nr	36.38
1486mm wide.............................	35.55	2.50	36.44	1.00	0.50	12.74	-	49.18	nr	52.87
Controlair Ventilators, permanent, mill finish										
508mm wide..............................	29.85	2.50	30.60	-	-	-	-	30.60	nr	32.89
997mm wide..............................	45.81	2.50	46.96	-	-	-	-	46.96	nr	50.48
1486mm wide.............................	53.63	2.50	54.97	-	-	-	-	54.97	nr	59.09
Controlair Ventilators, permanent, mill finish, flyscreen										
508mm wide..............................	34.91	2.50	35.78	-	-	-	-	35.78	nr	38.47
997mm wide..............................	55.61	2.50	57.00	-	-	-	-	57.00	nr	61.28
1486mm wide.............................	65.48	2.50	67.12	-	-	-	-	67.12	nr	72.15
Controlair Ventilators, adjustable										
508mm wide 1-LT.........................	38.86	2.50	39.83	-	-	-	-	39.83	nr	42.82
997mm wide 2-LT.........................	57.20	2.50	58.63	-	-	-	-	58.63	nr	63.03
Controlair Ventilators, adjustable, flyscreen										
508mm wide 1-LT.........................	43.91	2.50	45.01	-	-	-	-	45.01	nr	48.38
997mm wide 2-LT.........................	66.94	2.50	68.61	-	-	-	-	68.61	nr	73.76
locks with Parkes locking handles for										
side hung lights.....................	26.38	2.50	27.04	-	-	-	-	27.04	nr	29.07
horizontally pivoted lights	26.38	2.50	27.04	-	-	-	-	27.04	nr	29.07
METAL WINDOWS - Windows in aluminium; polyester **powder finish; white matt**										
Kawneer Products 102 Casement range; factory glazed, 4mm, one pane clear glass; fixing to masonry with lugs										
fixed lights										
600 x 400mm.............................	53.00	2.50	54.33	0.70	0.30	8.57	-	62.89	nr	67.61
600 x 800mm.............................	65.00	2.50	66.63	1.09	0.45	13.22	-	79.84	nr	85.83
600 x 1000mm............................	72.00	2.50	73.80	1.22	0.50	14.77	-	88.57	nr	95.21
600 x 1200mm............................	78.00	2.50	79.95	1.47	0.59	17.71	-	97.66	nr	104.98
600 x 1600mm............................	93.00	2.50	95.33	1.47	0.59	17.71	-	113.03	nr	121.51
800 x 400mm.............................	57.00	2.50	58.42	0.84	0.34	10.14	-	68.56	nr	73.71
800 x 800mm.............................	72.00	2.50	73.80	1.33	0.55	16.14	-	89.94	nr	96.68
800 x 1000mm............................	80.00	2.50	82.00	1.49	0.62	18.11	-	100.11	nr	107.61
800 x 1200mm............................	87.00	2.50	89.17	1.82	0.77	22.20	£	111.38	nr	119.73
800 x 1600mm............................	104.00	2.50	106.60	1.82	0.77	22.20	£	128.80	nr	138.46
1200 x 400mm............................	69.00	2.50	70.72	0.98	0.38	11.71	-	82.44	nr	88.62
1200 x 800mm............................	82.00	2.50	84.05	1.57	0.66	19.12	-	103.17	nr	110.91
1200 x 1000mm...........................	98.00	2.50	100.45	1.77	0.75	21.60	-	122.05	nr	131.21
1200 x 1200mm...........................	107.00	2.50	109.68	2.16	0.94	26.54	-	136.21	nr	146.43

Labour hourly rates: (except Specialists) Craft Operatives £9.23 Labourer £7.02 Rates are national average prices. Refer to REGIONAL VARIATIONS for indicative levels of overall pricing in regions	MATERIALS			LABOUR				RATES		
	Del to Site £	Waste %	Material Cost £	Craft Optve Hrs	Lab Hrs	Labour Cost £	Sunds £	Nett Rate £	Unit	Gross Rate (+7.5%) £
L10: WINDOWS/ROOFLIGHTS/SCREENS/LOUVRES Cont.										
METAL WINDOWS - Windows in aluminium; polyester powder finish; white matt Cont.										
Kawneer Products 102 Casement range; factory glazed, 4mm, one pane clear glass; fixing to masonry with lugs Cont.										
fixed lights Cont.										
1200 x 1600mm.....................	127.00	2.50	130.18	2.16	0.94	26.54	–	156.71	nr	168.46
1400 x 800mm.....................	95.00	2.50	97.38	1.81	0.76	22.04	–	119.42	nr	128.37
1400 x 1000mm.....................	106.00	2.50	108.65	2.04	0.88	25.01	–	133.66	nr	143.68
1400 x 1400mm.....................	127.00	2.50	130.18	2.51	1.11	30.96	–	161.13	nr	173.22
1400 x 1600mm.....................	151.00	2.50	154.78	2.51	1.11	30.96	–	185.73	nr	199.66
top hung casement										
600 x 800mm.....................	139.00	2.50	142.47	1.09	0.45	13.22	–	155.69	nr	167.37
600 x 1000mm.....................	150.00	2.50	153.75	1.22	0.50	14.77	–	168.52	nr	181.16
600 x 1200mm.....................	160.00	2.50	164.00	1.47	0.59	17.71	–	181.71	nr	195.34
800 x 800mm.....................	150.00	2.50	153.75	1.33	0.55	16.14	–	169.89	nr	182.63
800 x 1000mm.....................	162.00	2.50	166.05	1.49	0.62	18.11	–	184.16	nr	197.97
800 x 1200mm.....................	173.00	2.50	177.32	1.82	0.77	22.20	–	199.53	nr	214.49
top hung fanlights										
600 x400mm.....................	138.00	2.50	141.45	0.70	0.30	8.57	–	150.02	nr	161.27
800 x 600mm.....................	154.00	2.50	157.85	0.84	0.34	10.14	–	167.99	nr	180.59
1200 x 600mm.....................	167.00	2.50	171.18	0.98	0.38	11.71	–	182.89	nr	196.60
Kawneer Products 102 Casement range; factory glazed, 4mm, one pane obscure glass; fixing to masonry with lugs										
fixed lights										
600 x 400mm.....................	53.00	2.50	54.33	0.70	0.30	8.57	–	62.89	nr	67.61
600 x 800mm.....................	66.00	2.50	67.65	1.09	0.45	13.22	–	80.87	nr	86.93
600 x 1000mm.....................	73.00	2.50	74.83	1.22	0.50	14.77	–	89.60	nr	96.32
600 x 1200mm.....................	80.00	2.50	82.00	1.47	0.59	17.71	–	99.71	nr	107.19
600 x 1600mm.....................	95.00	2.50	97.38	1.47	0.59	17.71	–	115.08	nr	123.72
800 x 400mm.....................	57.00	2.50	58.42	0.84	0.34	10.14	–	68.56	nr	73.71
800 x 800mm.....................	73.00	2.50	74.83	1.33	0.55	16.14	–	90.96	nr	97.78
800 x 1000mm.....................	82.00	2.50	84.05	1.49	0.62	18.11	–	102.16	nr	109.82
1200 x 400mm.....................	70.00	2.50	71.75	0.98	0.38	11.71	–	83.46	nr	89.72
1200 x 800mm.....................	89.00	2.50	91.22	1.57	0.66	19.12	–	110.35	nr	118.63
top hung casement										
600 x 800mm.....................	140.00	2.50	143.50	1.09	0.45	13.22	–	156.72	nr	168.47
600 x 1000mm.....................	151.00	2.50	154.78	1.22	0.50	14.77	–	169.55	nr	182.26
600 x 1200mm.....................	161.00	2.50	165.03	1.47	0.59	17.71	–	182.73	nr	196.44
800 x 800mm.....................	151.00	2.50	154.78	1.33	0.55	16.14	–	170.91	nr	183.73
800 x 1000mm.....................	163.00	2.50	167.07	1.49	0.62	18.11	–	185.18	nr	199.07
800 x 1200mm.....................	175.00	2.50	179.38	1.82	0.77	22.20	–	201.58	nr	216.70
top hung fanlights										
600 x400mm.....................	140.00	2.50	143.50	0.70	0.30	8.57	–	152.07	nr	163.47
800 x 600mm.....................	156.00	2.50	159.90	0.84	0.34	10.14	–	170.04	nr	182.79
1200 x 600mm.....................	171.00	2.50	175.28	0.98	0.38	11.71	–	186.99	nr	201.01
METAL WINDOWS - Sliding windows in aluminium; polyester powder finish; white matt										
Kawneer Products Kingsley Equal series; factory double glazed, 18mm - 4/12/4 units with clear glass; fixing to masonry with lugs										
vertical sliders										
600 x 800mm.....................	249.00	2.50	255.22	1.09	0.45	13.22	–	268.44	nr	288.58
600 x 1000mm.....................	292.00	2.50	299.30	1.22	0.50	14.77	–	314.07	nr	337.63
600 x 1200mm.....................	342.00	2.50	350.55	1.34	0.54	16.16	–	366.71	nr	394.21
800 x 800mm.....................	446.00	2.50	457.15	1.33	0.55	16.14	–	473.29	nr	508.78
800 x 1000mm.....................	494.00	2.50	506.35	1.49	0.62	18.11	–	524.46	nr	563.79
800 x 1200mm.....................	552.00	2.50	565.80	1.66	0.70	20.24	–	586.04	nr	629.99
1200 x 800mm.....................	885.00	2.50	907.13	1.77	0.75	21.60	–	928.73	nr	998.38
1200 x 1000mm.....................	942.00	2.50	965.55	1.96	0.85	24.06	–	989.61	nr	1063.83
1200 x 1200mm.....................	1022.00	2.50	1047.55	2.16	0.94	26.54	–	1074.09	nr	1154.64
1400 x 800mm.....................	1119.00	2.50	1146.97	2.28	1.11	28.84	–	1175.81	nr	1264.00
1400 x 1000mm.....................	1193.00	2.50	1222.83	2.51	1.11	30.96	–	1253.78	nr	1347.82
1400 x 1000mm.....................	1248.00	2.50	1279.20	2.51	1.11	30.96	–	1310.16	nr	1408.42
horizontal sliders										
1200 x 800mm.....................	233.00	2.50	238.82	1.57	0.66	19.12	–	257.95	nr	277.30
1200 x 1000mm.....................	246.00	2.50	252.15	1.77	0.75	21.60	–	273.75	nr	294.28
1200 x 1400mm.....................	272.00	2.50	278.80	1.96	0.85	24.06	–	302.86	nr	325.57
1400 x 800mm.....................	245.00	2.50	251.13	1.81	0.76	22.04	–	273.17	nr	293.65
1400 x 1000mm.....................	261.00	2.50	267.52	2.04	0.88	25.01	–	292.53	nr	314.47
1400 x 1400mm.....................	288.00	2.50	295.20	2.28	0.99	27.99	–	323.19	nr	347.43
1400 x 1600mm.....................	304.00	2.50	311.60	2.51	1.11	30.96	–	342.56	nr	368.25
1600 x 800mm.....................	260.00	2.50	266.50	2.06	0.87	25.12	–	291.62	nr	313.49
1600 x 1000mm.....................	276.00	2.50	282.90	2.32	1.01	28.50	–	311.40	nr	334.76
1600 x 1600mm.....................	322.00	2.50	330.05	2.85	1.29	35.36	–	365.41	nr	392.82
Window sills in galvanised pressed steel B.S.6510; fixing to metal										
for opening										
508mm wide; AWA.....................	11.52	2.50	11.81	0.50	0.25	6.37	–	18.18	nr	19.54
508mm wide; AWB.....................	22.49	2.50	23.05	0.50	0.25	6.37	–	29.42	nr	31.63
508mm wide; PS.....................	13.13	2.50	13.46	0.50	0.25	6.37	–	19.83	nr	21.32
508mm wide; RPS.....................	13.13	2.50	13.46	0.50	0.25	6.37	–	19.83	nr	21.32
628mm wide; AWA.....................	12.84	2.50	13.16	0.60	0.30	7.64	–	20.81	nr	22.37
628mm wide; AWB.....................	14.00	2.50	14.35	0.60	0.30	7.64	–	21.99	nr	23.64
628mm wide; PS.....................	15.77	2.50	16.16	0.60	0.30	7.64	–	23.81	nr	25.59
628mm wide; RPS.....................	15.77	2.50	16.16	0.60	0.30	7.64	–	23.81	nr	25.59
997mm wide; AWA.....................	18.03	2.50	18.48	0.70	0.35	8.92	–	27.40	nr	29.45
997mm wide; AWB.....................	20.29	2.50	20.80	0.70	0.35	8.92	–	29.72	nr	31.94

WINDOWS/DOORS/STAIRS

Labour hourly rates: (except Specialists) Craft Operatives £9.23 Labourer £7.02 Rates are national average prices. Refer to REGIONAL VARIATIONS for indicative levels of overall pricing in regions	MATERIALS			LABOUR				RATES		
	Del to Site £	Waste %	Material Cost £	Craft Optve Hrs	Lab Hrs	Labour Cost £	Sunds £	Nett Rate £	Unit	Gross Rate (+7.5%) £
L10: WINDOWS/ROOFLIGHTS/SCREENS/LOUVRES Cont.										
METAL WINDOWS - Sliding windows in aluminium; polyester powder finish; white matt Cont.										
Window sills in galvanised pressed steel B.S.6510; fixing to metal Cont.										
for opening Cont.										
997mm wide; PS...............................	21.17	2.50	21.70	0.70	0.35	8.92	-	30.62	nr	32.91
997mm wide; RPS..............................	21.17	2.50	21.70	0.70	0.35	8.92	-	30.62	nr	32.91
1237mm wide; AWA.............................	20.29	2.50	20.80	0.80	0.40	10.19	-	30.99	nr	33.31
1237mm wide; AWB.............................	21.88	2.50	22.43	0.80	0.40	10.19	-	32.62	nr	35.07
1237mm wide; PS..............................	26.36	2.50	27.02	0.80	0.40	10.19	-	37.21	nr	40.00
1237mm wide; RPS.............................	26.36	2.50	27.02	0.80	0.40	10.19	-	37.21	nr	40.00
1486mm wide; AWA.............................	23.57	2.50	24.16	0.90	0.45	11.47	-	35.63	nr	38.30
1486mm wide; AWB.............................	26.28	2.50	26.94	0.90	0.45	11.47	-	38.40	nr	41.28
1486mm wide; PS..............................	28.42	2.50	29.13	0.90	0.45	11.47	-	40.60	nr	43.64
1486mm wide; RPS.............................	28.42	2.50	29.13	0.90	0.45	11.47	-	40.60	nr	43.64
1846mm wide; AWA.............................	26.73	2.50	27.40	1.00	0.50	12.74	-	40.14	nr	43.15
1846mm wide; AWB.............................	30.87	2.50	31.64	1.00	0.50	12.74	-	44.38	nr	47.71
1846mm wide; PS..............................	34.34	2.50	35.20	1.00	0.50	12.74	-	47.94	nr	51.53
1846mm wide; RPS.............................	34.34	2.50	35.20	1.00	0.50	12.74	-	47.94	nr	51.53
METAL ROOFLIGHTS - Rooflights, mainly in mill finish aluminium										
Rooflights; factory glazed with one layer 6mm thick Georgian wired cast glass										
non-ventilating base frame; fixing to masonry with screws										
600 x 600mm.............................	83.50	2.50	85.59	1.80	0.90	22.93	-	108.52	nr	116.66
600 x 900mm.............................	101.50	2.50	104.04	2.00	1.00	25.48	-	129.52	nr	139.23
600 x 1200mm............................	118.50	2.50	121.46	2.00	1.00	25.48	-	146.94	nr	157.96
900 x 900mm.............................	125.00	2.50	128.13	2.00	1.00	25.48	-	153.60	nr	165.13
900 x 1200mm............................	149.00	2.50	152.72	2.20	1.10	28.03	-	180.75	nr	194.31
1200 x 1200mm...........................	180.00	2.50	184.50	2.50	1.25	31.85	-	216.35	nr	232.58
non-ventilating base frame and kerb; fixing to masonry with screws										
600 x 600mm.............................	160.00	2.50	164.00	1.80	0.90	22.93	-	186.93	nr	200.95
600 x 900mm.............................	220.00	2.50	225.50	2.00	1.00	25.48	-	250.98	nr	269.80
600 x 1200mm............................	233.00	2.50	238.82	2.00	1.00	25.48	-	264.31	nr	284.13
900 x 900mm.............................	240.00	2.50	246.00	2.00	1.00	25.48	-	271.48	nr	291.84
900 x 1200mm............................	280.00	2.50	287.00	2.20	1.10	28.03	-	315.03	nr	338.66
1200 x 1200mm...........................	334.00	2.50	342.35	2.50	1.25	31.85	-	374.20	nr	402.26
ventilating base frame; fixing to masonry with screws										
600 x 600mm.............................	180.00	2.50	184.50	1.80	0.90	22.93	-	207.43	nr	222.99
600 x 900mm.............................	203.00	2.50	208.07	2.00	1.00	25.48	-	233.56	nr	251.07
600 x 1200mm............................	227.00	2.50	232.68	2.00	1.00	25.48	-	258.15	nr	277.52
900 x 900mm.............................	233.00	2.50	238.82	2.00	1.00	25.48	-	264.31	nr	284.13
900 x 1200mm............................	261.00	2.50	267.52	2.20	1.10	28.03	-	295.55	nr	317.72
1200 x 1200mm...........................	303.00	2.50	310.57	2.50	1.25	31.85	-	342.43	nr	368.11
ventilating base frame and kerb; fixing to masonry with screws										
600 x 600mm.............................	257.00	2.50	263.43	1.80	0.90	22.93	-	286.36	nr	307.83
600 x 900mm.............................	298.00	2.50	305.45	2.00	1.00	25.48	-	330.93	nr	355.75
600 x 1200mm............................	339.00	2.50	347.48	2.00	1.00	25.48	-	372.95	nr	400.93
900 x 900mm.............................	347.00	2.50	355.68	2.00	1.00	25.48	-	381.15	nr	409.74
900 x 1200mm............................	393.00	2.50	402.82	2.20	1.10	28.03	-	430.85	nr	463.17
1200 x 1200mm...........................	459.00	2.50	470.48	2.50	1.25	31.85	-	502.32	nr	540.00
Rooflights; factory glazed with one layer 6mm thick rough cast glass, one layer 6mm thick Georgian wired cast glass, 6mm air space										
non-ventilating base frame; fixing to masonry with screws										
600 x 600mm.............................	108.00	2.50	110.70	2.20	1.10	28.03	-	138.73	nr	149.13
600 x 900mm.............................	138.00	2.50	141.45	2.50	1.25	31.85	-	173.30	nr	186.30
600 x 1200mm............................	167.00	2.50	171.18	2.85	1.45	36.48	-	207.66	nr	223.23
900 x 900mm.............................	180.00	2.50	184.50	2.85	1.45	36.48	-	220.98	nr	237.56
900 x 1200mm............................	221.00	2.50	226.53	3.25	1.65	41.58	-	268.11	nr	288.21
1200 x 1200mm...........................	274.00	2.50	280.85	3.60	1.80	45.86	-	326.71	nr	351.22
non-ventilating base frame and kerb; fixing to masonry with screws										
600 x 600mm.............................	185.00	2.50	189.63	2.20	1.10	28.03	-	217.65	nr	233.98
600 x 900mm.............................	233.00	2.50	238.82	2.50	1.25	31.85	-	270.68	nr	290.98
600 x 1200mm............................	279.00	2.50	285.98	2.85	1.45	36.48	-	322.46	nr	346.64
900 x 900mm.............................	293.00	2.50	300.32	2.85	1.45	36.48	-	336.81	nr	362.07
900 x 1200mm............................	352.00	2.50	360.80	3.25	1.65	41.58	-	402.38	nr	432.56
1200 x 1200mm...........................	428.00	2.50	438.70	3.60	1.80	45.86	-	484.56	nr	520.91
ventilating base frame; fixing to masonry with screws										
600 x 600mm.............................	197.00	2.50	201.93	2.20	1.10	28.03	-	229.95	nr	247.20
600 x 900mm.............................	233.00	2.50	238.82	2.50	1.25	31.85	-	270.68	nr	290.98
600 x 1200mm............................	268.00	2.50	274.70	2.85	1.45	36.48	-	311.18	nr	334.52
900 x 900mm.............................	293.00	2.50	300.32	2.85	1.45	36.48	-	336.81	nr	362.07
900 x 1200mm............................	321.00	2.50	329.02	3.25	1.65	41.58	-	370.61	nr	398.40
1200 x 1200mm...........................	382.00	2.50	391.55	3.60	1.80	45.86	-	437.41	nr	470.22
ventilating base frame and kerb; fixing to masonry with screws										
600 x 600mm.............................	274.00	2.50	280.85	2.20	1.10	28.03	-	308.88	nr	332.04
600 x 900mm.............................	329.00	2.50	337.23	2.50	1.25	31.85	£	369.07	nr	396.76
600 x 1200mm............................	382.00	2.50	391.55	2.85	1.45	36.48	-	428.03	nr	460.14
900 x 900mm.............................	406.00	2.50	416.15	2.85	1.45	36.48	-	452.63	nr	486.58
900 x 1200mm............................	453.00	2.50	464.32	3.25	1.65	41.58	-	505.91	nr	543.85
1200 x 1200mm...........................	537.00	2.50	550.42	3.60	1.80	45.86	-	596.29	nr	641.01

Labour hourly rates: (except Specialists) Craft Operatives £9.23 Labourer £7.02 Rates are national average prices. Refer to REGIONAL VARIATIONS for indicative levels of overall pricing in regions	MATERIALS			LABOUR				RATES		
	Del to Site £	Waste %	Material Cost £	Craft Optve Hrs	Lab Hrs	Labour Cost £	Sunds £	Nett Rate £	Unit	Gross Rate (+7.5%) £
L10: WINDOWS/ROOFLIGHTS/SCREENS/LOUVRES Cont.										
METAL SCREENS, BORROWED LIGHTS, FRAMES AND GRILLES - **Screens in mild steel**										
Screens; 6 x 50 x 50mm angle framing to perimeter; 75 x 75 x 3mm infill; welded connections										
2000 x 2000mm overall; fixing to masonry with screws	251.00	5.00	**263.55**	6.00	3.00	**76.44**	-	**339.99**	nr	365.49
3000 x 2000mm overall; fixing to masonry with screws	624.00	5.00	**655.20**	8.00	4.00	**101.92**	-	**757.12**	nr	813.90
Screens; 6 x 50 x 50mm angle framing to perimeter; 75 x 75 x 5mm mesh infill; welded connections										
2000 x 2000mm overall; fixing to masonry with screws	257.00	5.00	**269.85**	6.00	3.00	**76.44**	-	**346.29**	nr	372.26
Screens; 6 x 50 x 75mm angle framing to perimeter; 75 x 75 x 3mm mesh infill; welded connections										
3000 x 2000mm overall; fixing to masonry with screws	379.00	5.00	**397.95**	8.00	4.00	**101.92**	-	**499.87**	nr	537.36
Screens; 6 x 50 x 50mm angle framing to perimeter; 6 x 51 x 102mm tee mullion; 75 x 75 x 3mm mesh infill; welded connections										
2000 x 4000mm overall; mullions -1 nr; fixing to masonry with screws	649.00	5.00	**681.45**	8.00	4.00	**101.92**	-	**783.37**	nr	842.12
Screens; 38.1 x 38.1 x 3.2mm hollow section framing to perimeter; 75 x 75 x 3mm mesh infill; welded connections										
2000 x 2000mm overall; fixing to masonry with screws	337.00	5.00	**353.85**	6.00	3.00	**76.44**	-	**430.29**	nr	462.56
Screens; 38.1 x 38.1 x 3.2mm hollow section framing to perimeter; 75 x 75 x 5mm mesh infill; welded connections										
2000 x 2000mm overall; fixing to masonry with screws	355.00	5.00	**372.75**	6.00	3.00	**76.44**	-	**449.19**	nr	482.88
Screens; 76.2 x 38.1 x 4mm hollow section framing to perimeter; 75 x 75 x 3mm mesh infill; welded connections										
2000 x 2000mm overall; fixing to masonry with screws	368.00	5.00	**386.40**	6.00	3.00	**76.44**	-	**462.84**	nr	497.55
METAL SCREENS, BORROWED LIGHTS, FRAMES AND GRILLES - **Grilles in mild steel**										
Grilles; 13 x 51mm flat bar framing to perimeter; 13mm diameter vertical infill bars at 150mm centres; welded connections										
1000 x 1000mm overall; fixing to masonry with screws	251.00	5.00	**263.55**	6.00	3.00	**76.44**	-	**339.99**	nr	365.49
2000 x 1000mm overall; fixing to masonry with screws	337.00	5.00	**353.85**	8.00	4.00	**101.92**	-	**455.77**	nr	489.95
Grilles; 13 x 51mm flat bar framing to perimeter; 13mm diameter vertical infill bars at 300mm centres; welded connections										
1000 x 1000mm overall; fixing to masonry with screws	239.00	5.00	**250.95**	4.00	2.00	**50.96**	-	**301.91**	nr	324.55
Grilles; 13 x 51mm flat bar framing to perimeter; 18mm diameter vertical infill bars at 150mm centres; welded connections										
1000 x 1000mm overall; fixing to masonry with screws	245.00	5.00	**257.25**	4.00	2.00	**50.96**	-	**308.21**	nr	331.33
METAL SCREENS, BORROWED LIGHTS, FRAMES AND GRILLES - **Glazing frames in mild steel**										
Glazing frames; 13 x 38 x 3mm angle framing to perimeter; welded connections; 13 x 13mm glazing beads, fixed with screws										
500 x 500mm overall; fixing to timber with screws	147.00	5.00	**154.35**	3.00	1.50	**38.22**	-	**192.57**	nr	207.01
Glazing frames; 15 x 21 x 3mm angle framing to perimeter; welded connections; 13 x 13mm glazing beads, fixed with screws										
500 x 1000mm overall; fixing to timber with screws	239.00	5.00	**250.95**	3.00	1.50	**38.22**	-	**289.17**	nr	310.86
Glazing frames; 18 x 25 x 3mm angle framing to perimeter; welded connections; 9 x 25mm glazing beads, fixed with screws										
1000 x 1000mm overall; fixing to timber with screws	288.00	5.00	**302.40**	4.00	2.00	**50.96**	-	**353.36**	nr	379.86
Glazing frames; 18 x 25 x 3mm angle framing to perimeter; welded connections; 13 x 13mm glazing beads, fixed with screws										
1000 x 1000mm overall; fixing to timber with screws	307.00	5.00	**322.35**	4.00	2.00	**50.96**	-	**373.31**	nr	401.31

WINDOWS/DOORS/STAIRS

	MATERIALS			LABOUR				RATES		
Labour hourly rates: (except Specialists) Craft Operatives £9.23 Labourer £7.02 Rates are national average prices. Refer to REGIONAL VARIATIONS for indicative levels of overall pricing in regions	Del to Site £	Waste %	Material Cost £	Craft Optve Hrs	Lab Hrs	Labour Cost £	Sunds £	Nett Rate £	Unit	Gross Rate (+7.5%) £

L10: WINDOWS/ROOFLIGHTS/SCREENS/LOUVRES Cont.

METAL SCREENS, BORROWED LIGHTS, FRAMES AND GRILLES – Glazing frames in mild steel Cont.

	Del to Site	Waste	Material Cost	Craft Optve	Lab	Labour Cost	Sunds	Nett Rate	Unit	Gross Rate
Glazing frames; 18 x 38 x 3mm angle framing to perimeter; welded connections; 9 x 25mm glazing beads, fixed with screws 1000 x 1000mm overall; fixing to timber with screws	380.00	5.00	399.00	3.00	1.50	38.22	–	437.22	nr	470.01
Glazing frames; 18 x 25 x 3mm angle framing to perimeter; welded connections; 15 x 15 x 3mm channel glazing beads; fixed with screws 1000 x 1000mm overall; fixing to timber with screws	380.00	5.00	399.00	3.00	1.50	38.22	–	437.22	nr	470.01
Glazing frames; 18 x 38 x 3mm angle framing to perimeter; welded connections; 15 x 15 x 3mm channel glazing beads; fixed with screws 1000 x 1000mm overall; fixing to timber with screws	380.00	5.00	399.00	3.00	1.50	38.22	–	437.22	nr	470.01

METAL SCREENS, BORROWED LIGHTS, FRAMES AND GRILLES – Glazing frames in aluminium

	Del to Site	Waste	Material Cost	Craft Optve	Lab	Labour Cost	Sunds	Nett Rate	Unit	Gross Rate
Glazing frames; 13 x 38 x 3mm angle framing to perimeter; welded connections; 13 x 13mm glazing beads, fixed with screws 500 x 500mm overall; fixing to timber with screws	223.00	5.00	234.15	3.00	1.50	38.22	–	272.37	nr	292.80
Glazing frames; 15 x 21 x 3mm angle framing to perimeter; welded connections; 13 x 13mm glazing beads, fixed with screws 500 x 1000mm overall; fixing to timber with screws	245.00	5.00	257.25	3.00	1.50	38.22	–	295.47	nr	317.63
Glazing frames; 18 x 25 x 3mm angle framing to perimeter; welded connections; 9 x 25mm glazing beads, fixed with screws 1000 x 1000mm overall; fixing to timber with screws	258.00	5.00	270.90	4.00	2.00	50.96	–	321.86	nr	346.00
Glazing frames; 18 x 25 x 3mm angle framing to perimeter; welded connections; 13 x 13mm glazing beads, fixed with screws 1000 x 1000mm overall; fixing to timber with screws	265.00	5.00	278.25	4.00	2.00	50.96	–	329.21	nr	353.90
Glazing frames; 18 x 38 x 3mm angle framing to perimeter; welded connections; 9 x 25mm glazing beads, fixed with screws 1000 x 1000mm overall; fixing to timber with screws	307.00	5.00	322.35	4.00	2.00	50.96	–	373.31	nr	401.31
Glazing frames; 18 x 25 x 3mm angle framing to perimeter; welded connections; 15 x 15 x 3mm channel glazing beads; fixed with screws 1000 x 1000mm overall; fixing to timber with screws	307.00	5.00	322.35	4.00	2.00	50.96	–	373.31	nr	401.31
Glazing frames; 18 x 38 x 3mm angle framing to perimeter; welded connections; 15 x 15 x 3mm channel glazing beads; fixed with screws 1000 x 1000mm overall; fixing to timber with screws	342.00	5.00	359.10	4.00	2.00	50.96	–	410.06	nr	440.81

METAL SCREENS, BORROWED LIGHTS, FRAMES AND GRILLES – Bedding and pointing frames

	Del to Site	Waste	Material Cost	Craft Optve	Lab	Labour Cost	Sunds	Nett Rate	Unit	Gross Rate
Pointing with Secomastic standard metal frames one side	0.75	10.00	0.82	0.80	–	7.38	–	8.21	m	8.82
Bedding in cement mortar (1:3); pointing with Secomastic standard mastic one side metal frames	0.84	10.00	0.92	0.18	–	1.66	–	2.59	m	2.78

PLASTICS WINDOWS – Windows in uPVC; white

Windows; factory glazed 20mm double glazed units; hinges; fastenings
  fixed light; fixing to masonry with cleats and screws; overall size

	Del to Site	Waste	Material Cost	Craft Optve	Lab	Labour Cost	Sunds	Nett Rate	Unit	Gross Rate
600 x 600mm	95.34	2.50	97.72	2.40	1.20	30.58	1.40	129.70	nr	139.43
600 x 900mm	102.17	2.50	104.72	3.00	1.50	38.22	1.40	144.34	nr	155.17
600 x 1050mm	108.97	2.50	111.69	3.30	1.65	42.04	1.40	155.14	nr	166.77
600 x 1200mm	117.56	2.50	120.50	3.25	1.62	41.37	1.40	163.27	nr	175.51
600 x 1500mm	137.98	2.50	141.43	3.80	1.90	48.41	1.40	191.24	nr	205.58
750 x 600mm	98.74	2.50	101.21	2.45	1.22	31.18	1.40	133.79	nr	143.82
750 x 900mm	115.78	2.50	118.67	3.00	1.50	38.22	1.40	158.29	nr	170.17
750 x 1050mm	122.59	2.50	125.65	3.25	1.62	41.37	1.40	168.42	nr	181.06
750 x 1200mm	129.40	2.50	132.63	3.50	1.75	44.59	1.40	178.63	nr	192.02
750 x 1500mm	151.53	2.50	155.32	4.05	2.02	51.56	1.40	208.28	nr	223.90
900 x 600mm	107.27	2.50	109.95	2.70	1.35	34.40	1.40	145.75	nr	156.68
900 x 900mm	141.31	2.50	144.84	3.25	1.62	41.37	1.40	187.61	nr	201.68
900 x 1050mm	136.21	2.50	139.62	3.40	1.70	43.32	1.40	184.33	nr	198.16
900 x 1200mm	146.42	2.50	150.08	3.80	1.90	48.41	1.40	199.89	nr	214.88

	MATERIALS			LABOUR				RATES		
Labour hourly rates: (except Specialists) Craft Operatives £9.23 Labourer £7.02 Rates are national average prices. Refer to REGIONAL VARIATIONS for indicative levels of overall pricing in regions	Del to Site	Waste	Material Cost	Craft Optve	Lab	Labour Cost	Sunds	Nett Rate	Unit	Gross Rate (+7.5%)
	£	%	£	Hrs	Hrs	£	£	£		£

L10: WINDOWS/ROOFLIGHTS/SCREENS/LOUVRES Cont.

PLASTICS WINDOWS – Windows in uPVC; white Cont.

Windows; factory glazed 20mm double glazed units; hinges; fastenings Cont.

	Del to Site £	Waste %	Material Cost £	Craft Optve Hrs	Lab Hrs	Labour Cost £	Sunds £	Nett Rate £	Unit	Gross Rate (+7.5%) £
fixed light; fixing to masonry with cleats and screws; overall size Cont.										
900 x 1500mm	172.04	2.50	176.34	4.30	2.15	54.78	1.40	232.52	nr	249.96
1200 x 600mm	117.56	2.50	120.50	3.05	1.52	38.82	1.40	160.72	nr	172.77
1200 x 900mm	141.38	2.50	144.91	3.60	1.80	45.86	1.40	192.18	nr	206.59
1200 x 1050mm	155.00	2.50	158.88	3.85	1.92	49.01	1.40	209.29	nr	224.99
1200 x 1200mm	168.63	2.50	172.85	4.15	2.07	52.84	1.40	227.08	nr	244.11
1200 x 1500mm	200.91	2.50	205.93	4.75	2.37	60.48	1.40	267.81	nr	287.90
fixed light with fixed light; fixing to masonry with cleats and screws; overall size										
600 x 1350mm	170.27	2.50	174.53	3.50	1.75	44.59	1.40	220.52	nr	237.06
600 x 1500mm	183.89	2.50	188.49	3.80	1.90	48.41	1.40	238.30	nr	256.17
600 x 1800mm	202.68	2.50	207.75	4.05	2.02	51.56	1.40	260.71	nr	280.26
600 x 2100mm	221.33	2.50	226.86	4.30	2.15	54.78	1.40	283.05	nr	304.27
750 x 1350mm	183.89	2.50	188.49	3.80	1.90	48.41	1.40	238.30	nr	256.17
750 x 1500mm	195.87	2.50	200.77	4.05	2.02	51.56	1.40	253.73	nr	272.76
750 x 1800mm	221.33	2.50	226.86	4.30	2.15	54.78	1.40	283.05	nr	304.27
750 x 2100mm	248.58	2.50	254.79	4.60	2.30	58.60	1.40	314.80	nr	338.41
900 x 1350mm	199.28	2.50	204.26	4.05	2.02	51.56	1.40	257.22	nr	276.52
900 x 1500mm	211.12	2.50	216.40	4.30	2.15	54.78	1.40	272.58	nr	293.02
900 x 1800mm	236.73	2.50	242.65	4.50	2.25	57.33	1.40	301.38	nr	323.98
900 x 2100mm	262.06	2.50	268.61	4.85	2.42	61.75	1.40	331.77	nr	356.65
1200 x 1350mm	224.74	2.50	230.36	4.50	2.25	57.33	1.40	289.09	nr	310.77
1200 x 1500mm	240.15	2.50	246.15	4.75	2.37	60.48	1.40	308.03	nr	331.14
1200 x 1800mm	270.78	2.50	277.55	5.05	2.52	64.30	1.40	343.25	nr	369.00
1200 x 2100mm	299.65	2.50	307.14	5.30	2.65	67.52	1.40	376.06	nr	404.27
overall, tilt/turn; fixing to masonry with cleats and screws; overall size										
600 x 600mm	277.59	2.50	284.53	2.15	1.07	27.36	1.40	313.29	nr	336.78
600 x 900mm	306.47	2.50	314.13	2.70	1.35	34.40	1.40	349.93	nr	376.17
600 x 1050mm	330.16	2.50	338.41	3.00	1.50	38.22	1.40	378.03	nr	406.39
600 x 1200mm	340.52	2.50	349.03	3.25	1.62	41.37	1.40	391.80	nr	421.19
600 x 1350mm	364.20	2.50	373.31	3.50	1.75	44.59	1.40	419.30	nr	450.74
600 x 1500mm	383.16	2.50	392.74	3.80	1.90	48.41	1.40	442.55	nr	475.74
750 x 600mm	288.54	2.50	295.75	2.45	1.22	31.18	1.40	328.33	nr	352.96
750 x 900mm	321.87	2.50	329.92	3.00	1.50	38.22	1.40	369.54	nr	397.25
750 x 1050mm	340.83	2.50	349.35	3.25	1.62	41.37	1.40	392.12	nr	421.53
750 x 1200mm	359.31	2.50	368.29	3.50	1.75	44.59	1.40	414.28	nr	445.35
750 x 1350mm	377.98	2.50	387.43	3.80	1.90	48.41	1.40	437.24	nr	470.03
750 x 1500mm	395.00	2.50	404.88	4.05	2.02	51.56	1.40	457.84	nr	492.17
900 x 900mm	340.52	2.50	349.03	3.25	1.62	41.37	1.40	391.80	nr	421.19
900 x 1050mm	360.95	2.50	369.97	3.40	1.70	43.32	1.40	414.69	nr	445.79
900 x 1200mm	379.75	2.50	389.24	3.80	1.90	48.41	1.40	439.06	nr	471.98
900 x 1350mm	398.40	2.50	408.36	4.05	2.02	51.56	1.40	461.32	nr	495.92
900 x 1500mm	422.24	2.50	432.80	4.30	2.15	54.78	1.40	488.98	nr	525.65
1200 x 1050mm	405.22	2.50	415.35	3.85	1.92	49.01	1.40	465.76	nr	500.70
1200 x 1200mm	429.05	2.50	439.78	4.15	2.07	52.84	1.40	494.01	nr	531.06
1200 x 1350mm	447.86	2.50	459.06	4.50	2.25	57.33	1.40	517.79	nr	556.62
1200 x 1500mm	471.69	2.50	483.48	4.75	2.37	60.48	1.40	545.36	nr	586.26
tilt/turn with fixed lights; fixing to masonry with cleats and screws; overall size										
600 x 1350mm	355.91	2.50	364.81	3.50	1.75	44.59	1.40	410.80	nr	441.61
600 x 1500mm	374.58	2.50	383.94	3.80	1.90	48.41	1.40	433.76	nr	466.29
600 x 2100mm	432.45	2.50	443.26	4.30	2.15	54.78	1.40	499.44	nr	536.90
750 x 1350mm	379.75	2.50	389.24	3.80	1.90	48.41	1.40	439.06	nr	471.98
750 x 1500mm	396.77	2.50	406.69	4.05	2.02	51.56	1.40	459.65	nr	494.12
750 x 2100mm	453.03	2.50	464.36	4.60	2.30	58.60	1.40	524.36	nr	563.69
900 x 1350mm	407.15	2.50	417.33	4.05	2.02	51.56	1.40	470.29	nr	505.56
900 x 1500mm	422.24	2.50	432.80	4.30	2.15	54.78	1.40	488.98	nr	525.65
900 x 2100mm	483.54	2.50	495.63	4.85	2.42	61.75	1.40	558.78	nr	600.69
1200 x 1350mm	459.70	2.50	471.19	4.50	2.25	57.33	1.40	529.92	nr	569.67
1200 x 1500mm	478.50	2.50	490.46	4.75	2.37	60.48	1.40	552.34	nr	593.77
1200 x 2100mm	529.42	2.50	542.66	5.30	2.65	67.52	1.40	611.58	nr	657.45
tilt/turn sash with fixed side light; fixing to masonry with cleats and screws; overall size										
1800 x 900mm	453.03	2.50	464.36	4.40	2.20	56.06	1.40	521.81	nr	560.95
1800 x 1050mm	483.54	2.50	495.63	4.70	2.35	59.88	1.40	556.91	nr	598.67
1800 x 1200mm	514.19	2.50	527.04	5.05	2.52	64.30	1.40	592.75	nr	637.20
1800 x 1350mm	544.82	2.50	558.44	5.40	2.70	68.80	1.40	628.64	nr	675.78
1800 x 1500mm	575.46	2.50	589.85	5.65	2.82	71.95	1.40	663.19	nr	712.93
tilt/turn sash with centre fixed light; fixing to masonry with cleats and screws; overall size										
2400 x 900mm	715.09	2.50	732.97	4.40	2.20	56.06	1.40	790.42	nr	849.70
2400 x 1050mm	766.15	2.50	785.30	5.75	2.87	73.22	1.40	859.92	nr	924.42
2400 x 1200mm	817.24	2.50	837.67	6.10	3.05	77.71	1.40	916.78	nr	985.54
2400 x 1350mm	859.88	2.50	881.38	6.40	3.20	81.54	1.40	964.31	nr	1036.64
2400 x 1500mm	893.93	2.50	916.28	6.65	3.32	84.69	1.40	1002.36	nr	1077.54

PLASTICS ROOFLIGHTS – Roof domelights in glass reinforced plastics

	Del to Site £	Waste %	Material Cost £	Craft Optve Hrs	Lab Hrs	Labour Cost £	Sunds £	Nett Rate £	Unit	Gross Rate (+7.5%) £
Domelights, translucent										
plain; fixing to masonry with screws										
610 x 610mm	82.20	2.50	84.25	1.00	1.00	16.25	0.27	100.78	nr	108.33
914 x 914mm	105.61	2.50	108.25	2.00	2.00	32.50	0.54	141.29	nr	151.89
1219 x 1219mm	140.44	2.50	143.95	3.00	3.00	48.75	0.81	193.51	nr	208.02
fixed ventilator; fixing to masonry with screws										
610 x 610mm	146.15	2.50	149.80	1.00	1.00	16.25	0.27	166.32	nr	178.80
914 x 914mm	175.83	2.50	180.23	2.00	2.00	32.50	0.54	213.27	nr	229.26
1219 x 1219mm	216.95	2.50	222.37	3.00	3.00	48.75	0.81	271.93	nr	292.33

WINDOWS/DOORS/STAIRS

Labour hourly rates: (except Specialists) Craft Operatives £9.23 Labourer £7.02 Rates are national average prices. Refer to REGIONAL VARIATIONS for indicative levels of overall pricing in regions	MATERIALS			LABOUR				RATES		
	Del to Site £	Waste %	Material Cost £	Craft Optve Hrs	Lab Hrs	Labour Cost £	Sunds £	Nett Rate £	Unit	Gross Rate (+7.5%) £
L10: WINDOWS/ROOFLIGHTS/SCREENS/LOUVRES Cont.										
PLASTICS ROOFLIGHTS - Roof domelights in glass reinforced plastics Cont.										
Domelights, translucent Cont. controlled ventilator; fixing to masonry with screws										
610 x 610mm....................	163.85	2.50	167.95	1.00	1.00	16.25	0.27	184.47	nr	198.30
914 x 914mm....................	222.64	2.50	228.21	2.00	2.00	32.50	0.54	261.25	nr	280.84
1219 x 1219mm..................	298.64	2.50	306.11	3.00	3.00	48.75	0.81	355.67	nr	382.34
PLASTICS ROOFLIGHTS - Bedding and pointing frames										
Pointing with Secomastic standard mastic plastics frames one side	1.30	10.00	1.43	0.12	-	1.11	-	2.54	m	2.73
Bedding in cement mortar (1:3); pointing with Secomastic standard mastic one side plastics frames	1.70	10.00	1.87	0.18	-	1.66	-	3.53	m	3.80
L20: DOORS/SHUTTERS/HATCHES										
TIMBER DOORS - Doors in Scandinavian softwood, wrought										
Internal panelled doors; SA; Magnet Trade 762 x 1981 x 34mm (f sizes)	58.85	2.50	60.32	1.20	0.15	12.13	-	72.45	nr	77.88
Internal panelled doors; SA; glazed with bevelled glass; Magnet Trade 762 x 1981 x 34mm (f sizes)	164.85	2.50	168.97	4.20	0.53	42.49	-	211.46	nr	227.32
Internal panelled doors; Blenheim with timber; Magnet Trade										
686 x 1981 x 34mm (f sizes)	52.28	2.50	53.59	1.00	0.13	10.14	-	63.73	nr	68.51
762 x 1981 x 34mm (f sizes)	54.08	2.50	55.43	1.20	0.15	12.13	-	67.56	nr	72.63
Internal panelled doors; Blenheim with MDF panels; Magnet Trade										
686 x 1981 x 34mm (f sizes)	66.68	2.50	68.35	1.00	0.13	10.14	-	78.49	nr	84.38
762 x 1981 x 34mm (f sizes)	70.28	2.50	72.04	1.00	0.13	10.14	-	82.18	nr	88.34
Internal panelled doors; Victorian softwood panels; Magnet Trade										
686 x 1981 x 34mm (f sizes)	48.68	2.50	49.90	1.00	0.13	10.14	-	60.04	nr	64.54
762 x 1981 x 34mm (f sizes)	51.38	2.50	52.66	1.20	0.15	12.13	-	64.79	nr	69.65
TIMBER DOORS - Doors in Scandinavian softwood, wrought, preservative treated										
External ledged and braced doors; L & B Magnet Trade										
686 x 1981 x 44mm (f sizes)	68.75	2.50	70.47	1.30	0.16	13.12	-	83.59	nr	89.86
762 x 1981 x 44mm (f sizes)	74.78	2.50	76.65	1.30	0.16	13.12	-	89.77	nr	96.50
838 x 1981 x 44mm (f sizes)	74.78	2.50	76.65	1.50	0.19	15.18	-	91.83	nr	98.72
External framed, ledged and braced doors; YX; Magnet Trade										
686 x 1981 x 44mm (f sizes)	102.50	2.50	105.06	1.10	0.14	11.14	-	116.20	nr	124.91
762 x 1981 x 44mm (f sizes)	103.58	2.50	106.17	1.30	0.16	13.12	-	119.29	nr	128.24
813 x 1981 x 44mm (f sizes)	103.58	2.50	106.17	1.50	0.19	15.18	-	121.35	nr	130.45
838 x 1981 x 44mm (f sizes)	103.58	2.50	106.17	1.50	0.19	15.18	-	121.35	nr	130.45
External panelled doors; Cavendish; Magnet Trade										
762 x 1981 x 44mm (f sizes)	144.35	2.50	147.96	1.30	0.16	13.12	-	161.08	nr	173.16
838 x 1981 x 44mm (f sizes)	144.35	2.50	147.96	1.50	0.19	15.18	-	163.14	nr	175.37
External panelled doors; KXT; Magnet Trade										
762 x 1981 x 44mm (f sizes)	80.90	2.50	82.92	1.30	0.16	13.12	-	96.04	nr	103.25
External panelled doors; Kentucky; Magnet Trade										
838 x 1981 x 44mm (f sizes)	142.55	2.50	146.11	1.50	0.19	15.18	-	161.29	nr	173.39
External panelled doors; Pembroke; Magnet Trade										
838 x 1981 x 44mm (f sizes)	156.41	2.50	160.32	1.50	0.19	15.18	-	175.50	nr	188.66
External panelled doors; Stable; softwood; Magnet Trade										
762 x 1981 x 44mm (f sizes)	123.83	2.50	126.93	2.40	0.30	24.26	-	151.18	nr	162.52
838 x 1981 x 44mm (f sizes)	126.17	2.50	129.32	2.60	0.33	26.31	-	155.64	nr	167.31
External panelled doors; Stable upper door glazed with Geneva glazing panel, Magnet Trade										
762 x 1981 x 44mm (f sizes)	169.36	2.50	173.59	2.70	0.34	27.31	-	200.90	nr	215.97
838 x 1981 x 44mm (f sizes)	171.70	2.50	175.99	2.90	0.36	29.29	-	205.29	nr	220.68
External panelled doors; Stable upper door glazed with Salisbury glazing panel, Magnet Trade										
762 x 1981 x 44mm (f sizes)	178.09	2.50	182.54	2.70	0.34	27.31	-	209.85	nr	225.59
838 x 1981 x 44mm (f sizes)	180.43	2.50	184.94	2.90	0.36	29.29	-	214.23	nr	230.30
External panelled doors; 2XG; Magnet Trade										
762 x 1981 x 44mm (f sizes)	52.82	2.50	54.14	1.30	0.16	13.12	-	67.26	nr	72.31
813 x 2032 x 44mm (f sizes)	54.35	2.50	55.71	1.50	0.19	15.18	-	70.89	nr	76.20
838 x 1981 x 44mm (f sizes)	54.35	2.50	55.71	1.50	0.19	15.18	-	70.89	nr	76.20

Labour hourly rates: (except Specialists) Craft Operatives £9.23 Labourer £7.02 Rates are national average prices. Refer to REGIONAL VARIATIONS for indicative levels of overall pricing in regions	MATERIALS			LABOUR				RATES		
	Del to Site £	Waste %	Material Cost £	Craft Optve Hrs	Lab Hrs	Labour Cost £	Sunds £	Nett Rate £	Unit	Gross Rate (+7.5%) £
L20: DOORS/SHUTTERS/HATCHES Cont.										
TIMBER DOORS - Doors in Scandinavian softwood, wrought, preservative treated Cont.										
External panelled doors; 2XGG; Magnet Trade										
762 x 1981 x 44mm (f sizes)	48.77	2.50	49.99	1.30	0.16	13.12	-	63.11	nr	67.84
838 x 1981 x 44mm (f sizes)	50.21	2.50	51.47	1.50	0.19	15.18	-	66.64	nr	71.64
External panelled doors; 2XG, glazed with 4mm clear tempered safety glass; Magnet Trade										
762 x 1981 x 44mm (f sizes)	69.37	2.50	71.10	1.70	0.21	17.17	-	88.27	nr	94.89
813 x 2032 x 44mm (f sizes)	75.49	2.50	77.38	1.90	0.24	19.22	-	96.60	nr	103.84
838 x 1981 x 44mm (f sizes)	74.50	2.50	76.36	1.90	0.24	19.22	-	95.58	nr	102.75
External panelled doors; 2XG, glazed with 4mm obscure tempered safety glass; Magnet Trade										
762 x 1981 x 44mm (f sizes)	72.97	2.50	74.79	1.70	0.21	17.17	-	91.96	nr	98.86
813 x 2032 x 44mm (f sizes)	76.57	2.50	78.48	1.90	0.24	19.22	-	97.71	nr	105.03
838 x 1981 x 44mm (f sizes)	76.57	2.50	78.48	1.90	0.24	19.22	-	97.71	nr	105.03
External panelled doors; 2XGG, glazed with 4mm obscure tempered safety glass; Magnet Trade										
762 x 1981 x 44mm (f sizes)	81.33	2.50	83.36	2.00	0.25	20.22	-	103.58	nr	111.35
838 x 1981 x 44mm (f sizes)	85.83	2.50	87.98	2.20	0.28	22.27	-	110.25	nr	118.52
Garage doors; MFL, pair, side hung; Magnet Trade										
1981 x 2134 x 44mm (f sizes) overall	276.46	2.50	283.37	3.50	0.44	35.39	-	318.77	nr	342.67
2134 x 2134 x 44mm (f sizes) overall	283.66	2.50	290.75	3.65	0.46	36.92	-	327.67	nr	352.25
Garage doors; 301, pair, side hung; Magnet Trade										
1981 x 2134 x 44mm (f sizes) overall	313.36	2.50	321.19	3.50	0.44	35.39	-	356.59	nr	383.33
2134 x 2134 x 44mm (f sizes) overall	321.64	2.50	329.68	3.65	0.46	36.92	-	366.60	nr	394.09
TIMBER DOORS - Doors in Hemlock, wrought										
Internal panelled doors; 2G; Magnet Trade										
762 x 1981 x 34mm (f sizes)	71.54	2.50	73.33	1.20	0.15	12.13	-	85.46	nr	91.87
Internal panelled doors; 2GG; Magnet Trade										
762 x 1981 x 34mm (f sizes)	68.21	2.50	69.92	1.20	0.15	12.13	-	82.04	nr	88.20
Internal panelled doors; SA; Magnet Trade										
686 x 1981 x 34mm (f sizes)	92.60	2.50	94.92	1.00	0.13	10.14	-	105.06	nr	112.94
762 x 1981 x 34mm (f sizes)	93.50	2.50	95.84	1.20	0.15	12.13	-	107.97	nr	116.06
813 x 2032 x 34mm (f sizes)	101.15	2.50	103.68	1.40	0.18	14.19	-	117.86	nr	126.70
838 x 1981 x 34mm (f sizes)	101.15	2.50	103.68	1.40	0.18	14.19	-	117.86	nr	126.70
Internal panelled doors; 10; Magnet Trade										
762 x 1981 x 34mm (f sizes)	54.53	2.50	55.89	1.20	0.15	12.13	-	68.02	nr	73.12
Internal panelled doors; 2G, glazed with 4mm clear tempered safety glass; Magnet Trade										
762 x 1981 x 34mm (f sizes)	91.20	2.50	93.48	1.60	0.20	16.17	-	109.65	nr	117.88
Internal panelled doors; 2GG, glazed with 4mm clear tempered safety glass; Magnet Trade										
762 x 1981 x 34mm (f sizes)	102.30	2.50	104.86	1.90	0.24	19.22	-	124.08	nr	133.39
Internal panelled doors; 10, glazed with 4mm clear tempered safety glass; Magnet Trade										
762 x 1981 x 34mm (f sizes)	89.67	2.50	91.91	1.70	0.22	17.24	-	109.15	nr	117.33
Internal panelled doors; 2G, glazed with 4mm obscure tempered safety glass; Magnet Trade										
762 x 1981 x 34mm (f sizes)	94.80	2.50	97.17	1.60	0.20	16.17	-	113.34	nr	121.84
Internal panelled doors; 2GG, glazed with 4mm obscure tempered safety glass; Magnet Trade										
762 x 1981 x 34mm (f sizes)	103.87	2.50	106.47	1.90	0.24	19.22	-	125.69	nr	135.12
Internal panelled doors; 10, glazed with 4mm obscure tempered safety glass; Magnet Trade										
762 x 1981 x 34mm (f sizes)	91.74	2.50	94.03	1.70	0.22	17.24	-	111.27	nr	119.61
Internal panelled doors; SA, glazed with bevelled glass; Magnet Trade										
762 x 1981 x 34mm (f sizes)	199.50	2.50	204.49	4.20	0.53	42.49	-	246.97	nr	265.50
813 x 2032 x 34mm (f sizes)	213.90	2.50	219.25	4.40	0.55	44.47	-	263.72	nr	283.50
838 x 1981 x 34mm (f sizes)	212.91	2.50	218.23	4.40	0.55	44.47	-	262.71	nr	282.41
External panelled doors; 2XG; Magnet Trade										
762 x 1981 x 44mm (f sizes)	78.83	2.50	80.80	1.30	0.16	13.12	-	93.92	nr	100.97
813 x 2032 x 44mm (f sizes)	82.52	2.50	84.58	1.50	0.19	15.18	-	99.76	nr	107.24
838 x 1981 x 44mm (f sizes)	82.52	2.50	84.58	1.50	0.19	15.18	-	99.76	nr	107.24
External panelled doors; 2XGG; Magnet Trade										
762 x 1981 x 44mm (f sizes)	75.05	2.50	76.93	1.30	0.16	13.12	-	90.05	nr	96.80
813 x 2032 x 44mm (f sizes)	78.74	2.50	80.71	1.50	0.19	15.18	-	95.89	nr	103.08
838 x 1981 x 44mm (f sizes)	78.74	2.50	80.71	1.50	0.19	15.18	-	95.89	nr	103.08
External panelled doors; KXT; Magnet Trade										
762 x 1981 x 44mm (f sizes)	132.47	2.50	135.78	1.30	0.16	13.12	-	148.90	nr	160.07
838 x 1981 x 44mm (f sizes)	136.07	2.50	139.47	1.50	0.19	15.18	-	154.65	nr	166.25

WINDOWS/DOORS/STAIRS

Labour hourly rates: (except Specialists) Craft Operatives £9.23 Labourer £7.02 Rates are national average prices. Refer to REGIONAL VARIATIONS for indicative levels of overall pricing in regions	MATERIALS			LABOUR				RATES		
	Del to Site £	Waste %	Material Cost £	Craft Optve Hrs	Lab Hrs	Labour Cost £	Sunds £	Nett Rate £	Unit	Gross Rate (+7.5%) £
L20: DOORS/SHUTTERS/HATCHES Cont.										
TIMBER DOORS - Doors in Hemlock, wrought Cont.										
External panelled doors; SA; Magnet Trade										
762 x 1981 x 44mm (f sizes)	100.88	2.50	103.40	1.30	0.16	13.12	-	116.52	nr	125.26
838 x 1981 x 44mm (f sizes)	104.39	2.50	107.00	1.50	0.19	15.18	-	122.18	nr	131.34
External panelled doors; 10; Magnet Trade										
762 x 1981 x 44mm (f sizes)	74.06	2.50	75.91	1.30	0.16	13.12	-	89.03	nr	95.71
External panelled doors; 2XG, glazed with 4mm clear tempered safety glass; Magnet Trade										
762 x 1981 x 44mm (f sizes)	95.38	2.50	97.76	1.70	0.21	17.17	-	114.93	nr	123.55
813 x 2032 x 44mm (f sizes)	103.66	2.50	106.25	1.90	0.24	19.22	-	125.47	nr	134.88
838 x 1981 x 44mm (f sizes)	104.74	2.50	107.36	1.90	0.24	19.22	-	126.58	nr	136.07
External panelled doors; 2XGG, glazed with 4mm clear tempered safety glass; Magnet Trade										
762 x 1981 x 44mm (f sizes)	102.93	2.50	105.50	2.00	0.25	20.22	-	125.72	nr	135.15
813 x 2032 x 44mm (f sizes)	112.29	2.50	115.10	2.20	0.28	22.27	-	137.37	nr	147.67
838 x 1981 x 44mm (f sizes)	111.30	2.50	114.08	2.20	0.28	22.27	-	136.35	nr	146.58
External panelled doors; 10, glazed with 4mm clear tempered safety glass; Magnet Trade										
762 x 1981 x 44mm (f sizes)	104.02	2.50	106.62	1.80	0.23	18.23	-	124.85	nr	134.21
External panelled doors; 2XG, glazed with 4mm obscure tempered safety glass; Magnet Trade										
762 x 1981 x 44mm (f sizes)	98.98	2.50	101.45	1.70	0.21	17.17	-	118.62	nr	127.52
813 x 2032 x 44mm (f sizes)	104.74	2.50	107.36	1.90	0.24	19.22	-	126.58	nr	136.07
838 x 1981 x 44mm (f sizes)	104.74	2.50	107.36	1.90	0.24	19.22	-	126.58	nr	136.07
External panelled doors; 2XGG, glazed with 4mm obscure tempered safety glass; Magnet Trade										
762 x 1981 x 44mm (f sizes)	107.61	2.50	110.30	2.00	0.25	20.22	-	130.52	nr	140.30
813 x 2032 x 44mm (f sizes)	114.36	2.50	117.22	2.20	0.28	22.27	-	139.49	nr	149.95
838 x 1981 x 44mm (f sizes)	114.36	2.50	117.22	2.00	0.28	20.43	-	137.64	nr	147.97
External panelled doors; 10, glazed with 4mm obscure tempered safety glass; Magnet Trade										
762 x 1981 x 44mm (f sizes)	106.09	2.50	108.74	1.80	0.23	18.23	-	126.97	nr	136.49
External panelled doors; SA, glazed with bevelled glass; Magnet Trade										
762 x 1981 x 44mm (f sizes)	209.94	2.50	215.19	4.30	0.54	43.48	-	258.67	nr	278.07
838 x 1981 x 44mm (f sizes)	219.21	2.50	224.69	4.50	0.56	45.47	-	270.16	nr	290.42
TIMBER DOORS - Doors in Hardwood, wrought										
External panelled doors; Alicante; Magnet Trade										
813 x 2032 x 44mm (f sizes)	206.36	2.50	211.52	2.30	0.29	23.26	-	234.78	nr	252.39
838 x 1981 x 44mm (f sizes)	206.36	2.50	211.52	2.30	0.29	23.26	-	234.78	nr	252.39
External panelled doors; Carolina, glazed in clear glass; Magnet Trade										
813 x 2032 x 44mm (f sizes)	225.16	2.50	230.79	2.30	0.29	23.26	-	254.05	nr	273.11
838 x 1981 x 44mm (f sizes)	225.16	2.50	230.79	2.30	0.29	23.26	-	254.05	nr	273.11
External panelled doors; Alicante, glazed with bevelled glass; Magnet Trade										
813 x 2032 x 44mm (f sizes)	258.55	2.50	265.01	4.10	0.51	41.42	-	306.44	nr	329.42
838 x 1981 x 44mm (f sizes)	258.55	2.50	265.01	4.10	0.51	41.42	-	306.44	nr	329.42
TIMBER DOORS - Doors in Hardwood (solid, laminated or veneered), wrought										
External panelled doors, fire resisting; FD20 2XGG; Magnet Trade										
838 x 1981 x 44mm (f sizes)	158.85	2.50	162.82	2.20	0.28	22.27	-	185.09	nr	198.97
External panelled doors; Airedale; Magnet Trade										
838 x 1981 x 44mm (f sizes)	373.94	2.50	383.29	2.30	0.29	23.26	-	406.55	nr	437.04
External panelled doors; Belvoir; Magnet Trade										
762 x 1981 x 44mm (f sizes)	146.42	2.50	150.08	2.10	0.26	21.21	-	171.29	nr	184.14
838 x 1981 x 44mm (f sizes)	146.42	2.50	150.08	2.30	0.29	23.26	-	173.35	nr	186.35
External panelled doors; Cadiz; Magnet Trade										
838 x 1981 x 44mm (f sizes)	290.78	2.50	298.05	2.30	0.29	23.26	-	321.31	nr	345.41
External panelled doors; Conway; Magnet Trade										
762 x 1981 x 44mm (f sizes)	199.16	2.50	204.14	2.10	0.26	21.21	-	225.35	nr	242.25
813 x 2032 x 44mm (f sizes)	199.16	2.50	204.14	2.30	0.29	23.26	-	227.40	nr	244.46
838 x 1981 x 44mm (f sizes)	199.16	2.50	204.14	2.30	0.29	23.26	-	227.40	nr	244.46
External panelled doors; Elizabethan; Magnet Trade										
813 x 2032 x 44mm (f sizes)	222.74	2.50	228.31	2.30	0.29	23.26	-	251.57	nr	270.44
838 x 1981 x 44mm (f sizes)	222.74	2.50	228.31	2.30	0.29	23.26	-	251.57	nr	270.44
External panelled doors; Manilla; Magnet Trade										
762 x 1981 x 44mm (f sizes)	215.90	2.50	221.30	2.10	0.26	21.21	-	242.51	nr	260.69
813 x 2032 x 44mm (f sizes)	215.90	2.50	221.30	2.30	0.29	23.26	-	244.56	nr	262.90
838 x 1981 x 44mm (f sizes)	215.90	2.50	221.30	2.30	0.29	23.26	-	244.56	nr	262.90

Labour hourly rates: (except Specialists) Craft Operatives £9.23 Labourer £7.02 Rates are national average prices. Refer to REGIONAL VARIATIONS for indicative levels of overall pricing in regions	MATERIALS			LABOUR				RATES		
	Del to Site £	Waste %	Material Cost £	Craft Optve Hrs	Lab Hrs	Labour Cost £	Sunds £	Nett Rate £	Unit	Gross Rate (+7.5%) £

L20: DOORS/SHUTTERS/HATCHES Cont.

TIMBER DOORS - Doors in Hardwood (solid, laminated or veneered), wrought Cont.

	Del to Site £	Waste %	Material Cost £	Craft Optve Hrs	Lab Hrs	Labour Cost £	Sunds £	Nett Rate £	Unit	Gross Rate £
External panelled doors; Richmond; Magnet Trade										
838 x 1981 x 44mm (f sizes)	283.76	2.50	290.85	2.30	0.29	23.26	-	314.12	nr	337.68
External panelled doors; Rutland; Magnet Trade										
762 x 1981 x 44mm (f sizes)	144.90	2.50	148.52	2.10	0.26	21.21	-	169.73	nr	182.46
838 x 1981 x 44mm (f sizes)	144.90	2.50	148.52	2.30	0.29	23.26	-	171.79	nr	184.67
External panelled doors; Stable door; Magnet Trade										
762 x 1981 x 44mm (f sizes)	297.53	2.50	304.97	3.60	0.45	36.39	-	341.36	nr	366.96
838 x 1981 x 44mm (f sizes)	320.84	2.50	328.86	3.85	0.48	38.91	-	367.77	nr	395.35
External panelled doors; Stourbridge; Magnet Trade										
838 x 1981 x 44mm (f sizes)	287.72	2.50	294.91	2.30	0.29	23.26	-	318.18	nr	342.04
External panelled doors; Stuart; Magnet Trade										
762 x 1981 x 44mm (f sizes)	170.90	2.50	175.17	2.10	0.26	21.21	-	196.38	nr	211.11
838 x 1981 x 44mm (f sizes)	177.83	2.50	182.28	2.30	0.29	23.26	-	205.54	nr	220.96
External panelled doors; Belvoir, glazed with clear glass; Magnet Trade										
762 x 1981 x 44mm (f sizes)	162.97	2.50	167.04	2.50	0.31	25.25	-	192.30	nr	206.72
838 x 1981 x 44mm (f sizes)	166.57	2.50	170.73	2.70	0.34	27.31	-	198.04	nr	212.90
External panelled doors; Belvoir, glazed with obscure glass; Magnet Trade										
762 x 1981 x 44mm (f sizes)	166.57	2.50	170.73	2.50	0.31	25.25	-	195.99	nr	210.68
838 x 1981 x 44mm (f sizes)	168.64	2.50	172.86	2.70	0.34	27.31	-	200.16	nr	215.18
External panelled doors; Rutland, glazed with clear glass; Magnet Trade										
762 x 1981 x 44mm (f sizes)	172.78	2.50	177.10	2.80	0.35	28.30	-	205.40	nr	220.81
838 x 1981 x 44mm (f sizes)	177.46	2.50	181.90	3.00	0.38	30.36	-	212.25	nr	228.17
External panelled doors; Rutland, glazed with obscure glass; Magnet Trade										
762 x 1981 x 44mm (f sizes)	177.46	2.50	181.90	2.80	0.35	28.30	-	210.20	nr	225.96
838 x 1981 x 44mm (f sizes)	180.52	2.50	185.03	3.00	0.38	30.36	-	215.39	nr	231.54
External panelled doors; Stourbridge, glazed with clear glass; Magnet Trade										
838 x 1981 x 44mm (f sizes)	305.72	2.50	313.36	5.30	0.66	53.55	-	366.92	nr	394.43
External panelled doors; Stourbridge, glazed with obscure glass; Magnet Trade										
838 x 1981 x 44mm (f sizes)	314.72	2.50	322.59	5.30	0.66	53.55	-	376.14	nr	404.35
External panelled doors; Airedale, glazed with bevelled glass; Magnet Trade										
838 x 1981 x 44mm (f sizes)	423.44	2.50	434.03	4.10	0.51	41.42	-	475.45	nr	511.11
External panelled doors; Stuart, glazed with bevelled glass; Magnet Trade										
762 x 1981 x 44mm (f sizes)	250.54	2.50	256.80	5.10	0.64	51.57	-	308.37	nr	331.50
838 x 1981 x 44mm (f sizes)	257.47	2.50	263.91	5.30	0.66	53.55	-	317.46	nr	341.27
External panelled doors; Stable door, upper door glazed with bevelled glass; Magnet Trade										
762 x 1981 x 44mm (f sizes)	264.94	2.50	271.56	5.00	0.63	50.57	-	322.14	nr	346.30
838 x 1981 x 44mm (f sizes)	264.94	2.50	271.56	5.25	0.66	53.09	-	324.65	nr	349.00
External panelled doors; Cadiz, glazed with Geneva glazing panel; Magnet Trade										
838 x 1981 x 44mm (f sizes)	366.37	2.50	375.53	2.60	0.33	26.31	-	401.84	nr	431.98
External panelled doors; Cadiz, glazed with Salisbury glazing panel; Magnet Trade										
838 x 1981 x 44mm (f sizes)	375.10	2.50	384.48	2.60	0.33	26.31	-	410.79	nr	441.60
External panelled doors; Richmond, glazed with Malton glazing panel; Magnet Trade										
838 x 1981 x 44mm (f sizes)	421.36	2.50	431.89	2.90	0.36	29.29	-	461.19	nr	495.78
External panelled doors; Belvoir, glazed with clear glass double glazing unit; Magnet Trade										
762 x 1981 x 44mm (f sizes)	205.36	2.50	210.49	2.50	0.31	25.25	-	235.75	nr	253.43
838 x 1981 x 44mm (f sizes)	210.23	2.50	215.49	2.70	0.34	27.31	-	242.79	nr	261.00
External panelled doors; Belvoir, glazed with leaded glass double glazing unit; Magnet Trade										
762 x 1981 x 44mm (f sizes)	273.85	2.50	280.70	2.50	0.31	25.25	-	305.95	nr	328.89
838 x 1981 x 44mm (f sizes)	273.85	2.50	280.70	2.70	0.34	27.31	-	308.00	nr	331.10
External panelled doors; Belvoir, glazed with obscure glass double glazing unit; Magnet Trade										
762 x 1981 x 44mm (f sizes)	206.44	2.50	211.60	2.50	0.31	25.25	-	236.85	nr	254.62
838 x 1981 x 44mm (f sizes)	212.65	2.50	217.97	2.70	0.34	27.31	-	245.27	nr	263.67
External panelled doors; Rutland, glazed with clear glass double glazing unit; Magnet Trade										
762 x 1981 x 44mm (f sizes)	234.88	2.50	240.75	2.80	0.35	28.30	-	269.05	nr	289.23
838 x 1981 x 44mm (f sizes)	241.82	2.50	247.87	3.00	0.38	30.36	-	278.22	nr	299.09

WINDOWS/DOORS/STAIRS

Labour hourly rates: (except Specialists) Craft Operatives £9.23 Labourer £7.02 Rates are national average prices. Refer to REGIONAL VARIATIONS for indicative levels of overall pricing in regions	MATERIALS			LABOUR				RATES		
	Del to Site £	Waste %	Material Cost £	Craft Optve Hrs	Lab Hrs	Labour Cost £	Sunds £	Nett Rate £	Unit	Gross Rate (+7.5%) £
L20: DOORS/SHUTTERS/HATCHES Cont.										
TIMBER DOORS - Doors in Hardwood (solid, laminated or veneered), wrought Cont.										
External panelled doors; Rutland, glazed with leaded glass double glazing unit; Magnet Trade										
762 x 1981 x 44mm (f sizes)	324.86	2.50	332.98	2.80	0.35	28.30	-	361.28	nr	388.38
838 x 1981 x 44mm (f sizes)	340.90	2.50	349.42	3.00	0.38	30.36	-	379.78	nr	408.26
External panelled doors; Rutland, glazed with obscure glass double glazing unit; Magnet Trade										
762 x 1981 x 44mm (f sizes)	246.31	2.50	252.47	2.80	0.35	28.30	-	280.77	nr	301.83
838 x 1981 x 44mm (f sizes)	246.31	2.50	252.47	3.00	0.38	30.36	-	282.83	nr	304.04
Garage doors; Chevron; Magnet Trade										
2134 x 2134 x 44mm (f sizes)	772.00	2.50	791.30	4.75	0.59	47.98	-	839.28	nr	902.23
TIMBER DOORS - Doors in softwood and fibreboard core, Mahogany or Sapele facings										
Internal panelled doors; Granada; Magnet Trade										
686 x 1981 x 34mm (f sizes)	259.73	2.50	266.22	1.90	0.24	19.22	-	285.45	nr	306.85
762 x 1981 x 34mm (f sizes)	264.23	2.50	270.84	2.10	0.26	21.21	-	292.04	nr	313.95
Internal panelled doors; Malaga; Magnet Trade										
762 x 1981 x 34mm (f sizes)	219.41	2.50	224.90	2.10	0.26	21.21	-	246.10	nr	264.56
Internal panelled doors; Palma; Magnet Trade										
762 x 1981 x 34mm (f sizes)	199.80	2.50	204.79	2.10	0.26	21.21	-	226.00	nr	242.95
TIMBER DOORS - Doors in Hardwood (solid, laminated or veneered), wrought										
Internal panelled doors; SA; Magnet Trade										
686 x 1981 x 34mm (f sizes)	154.61	2.50	158.48	1.90	0.24	19.22	-	177.70	nr	191.02
762 x 1981 x 34mm (f sizes)	155.24	2.50	159.12	2.10	0.26	21.21	-	180.33	nr	193.85
Internal panelled doors; Victorian; Magnet Trade										
686 x 1981 x 34mm (f sizes)	183.14	2.50	187.72	1.90	0.24	19.22	-	206.94	nr	222.46
762 x 1981 x 34mm (f sizes)	186.29	2.50	190.95	2.10	0.26	21.21	-	212.16	nr	228.07
Internal panelled doors; Windsor; Magnet Trade										
686 x 1981 x 34mm (f sizes)	183.14	2.50	187.72	1.90	0.24	19.22	-	206.94	nr	222.46
762 x 1981 x 34mm (f sizes)	186.29	2.50	190.95	2.10	0.26	21.21	-	212.16	nr	228.07
Internal panelled doors; 10; Magnet Trade										
762 x 1981 x 34mm (f sizes)	72.22	2.50	74.03	2.10	0.26	21.21	-	95.23	nr	102.38
Internal panelled doors; 10, glazed with 4mm clear tempered safety glass; Magnet Trade										
762 x 1981 x 34mm (f sizes)	106.18	2.50	108.83	2.60	0.33	26.31	-	135.15	nr	145.29
Internal panelled doors; 10, glazed with 4mm obscure tempered safety glass; Magnet Trade										
762 x 1981 x 34mm (f sizes)	106.45	2.50	109.11	2.60	0.33	26.31	-	135.43	nr	145.58
Internal panelled doors; SA, glazed with bevelled glass; Magnet Trade										
762 x 1981 x 34mm (f sizes)	238.48	2.50	244.44	5.10	0.64	51.57	-	296.01	nr	318.21
TIMBER DOORS - Doors in solid or laminated construction, Oak facings										
Internal panelled doors; Louis; Magnet Trade										
762 x 1981 x 34mm (f sizes)	225.26	2.50	230.89	2.10	0.26	21.21	-	252.10	nr	271.01
TIMBER DOORS - Doors in hardboard, embossed										
Internal panelled doors; Colinist; Magnet Trade										
610 x 1981 x 34mm (f sizes)	54.89	2.50	56.26	1.60	0.20	16.17	-	72.43	nr	77.87
686 x 1981 x 34mm (f sizes)	55.16	2.50	56.54	1.70	0.21	17.17	-	73.70	nr	79.23
762 x 1981 x 34mm (f sizes)	55.34	2.50	56.72	1.90	0.24	19.22	-	75.95	nr	81.64
Internal panelled doors; Sentinel; Magnet Trade										
686 x 1981 x 34mm (f sizes)	58.22	2.50	59.68	1.70	0.21	17.17	-	76.84	nr	82.60
762 x 1981 x 34mm (f sizes)	58.49	2.50	59.95	1.90	0.24	19.22	-	79.17	nr	85.11
TIMBER DOORS - Flush doors										
Internal; hardboard facings; Magnet Trade										
305 x 1981 x 34mm	21.59	2.50	22.13	1.00	0.13	10.14	-	32.27	nr	34.69
381 x 1981 x 34mm	22.40	2.50	22.96	1.10	0.14	11.14	-	34.10	nr	36.65
457 x 1981 x 34mm	22.40	2.50	22.96	1.20	0.15	12.13	-	35.09	nr	37.72
533 x 1981 x 34mm	22.40	2.50	22.96	1.30	0.16	13.12	-	36.08	nr	38.79
610 x 1981 x 34mm	22.40	2.50	22.96	1.40	0.18	14.19	-	37.15	nr	39.93
686 x 1981 x 34mm	22.85	2.50	23.42	1.50	0.19	15.18	-	38.60	nr	41.50
711 x 1981 x 34mm	23.21	2.50	23.79	1.60	0.20	16.17	-	39.96	nr	42.96
762 x 1981 x 34mm	23.21	2.50	23.79	1.70	0.21	17.17	-	40.96	nr	44.03
813 x 2032 x 34mm	24.38	2.50	24.99	1.90	0.24	19.22	-	44.21	nr	47.53
838 x 1981 x 34mm	24.38	2.50	24.99	1.90	0.24	19.22	-	44.21	nr	47.53
Internal; plywood facings; Magnet Trade										
381 x 1981 x 34mm	30.32	2.50	31.08	1.10	0.14	11.14	-	42.21	nr	45.38
457 x 1981 x 34mm	30.32	2.50	31.08	1.20	0.15	12.13	-	43.21	nr	46.45
533 x 1981 x 34mm	30.32	2.50	31.08	1.30	0.16	13.12	-	44.20	nr	47.52

Labour hourly rates: (except Specialists) Craft Operatives £9.23 Labourer £7.02 Rates are national average prices. Refer to REGIONAL VARIATIONS for indicative levels of overall pricing in regions	MATERIALS			LABOUR				RATES		
	Del to Site	Waste	Material Cost	Craft Optve	Lab	Labour Cost	Sunds	Nett Rate	Unit	Gross Rate (+7.5%)
	£	%	£	Hrs	Hrs	£	£	£		£
L20: DOORS/SHUTTERS/HATCHES Cont.										
TIMBER DOORS - Flush doors Cont.										
Internal; plywood facings; Magnet Trade Cont.										
610 x 1981 x 34mm	30.32	2.50	**31.08**	1.40	0.18	**14.19**	–	**45.26**	nr	48.66
686 x 1981 x 34mm	30.68	2.50	**31.45**	1.50	0.19	**15.18**	–	**46.63**	nr	50.12
711 x 1981 x 34mm	31.13	2.50	**31.91**	1.60	0.20	**16.17**	–	**48.08**	nr	51.69
762 x 1981 x 34mm	31.13	2.50	**31.91**	1.70	0.21	**17.17**	–	**49.07**	nr	52.75
813 x 2032 x 34mm	32.57	2.50	**33.38**	1.90	0.24	**19.22**	–	**52.61**	nr	56.55
838 x 1981 x 34mm	32.57	2.50	**33.38**	1.90	0.24	**19.22**	–	**52.61**	nr	56.55
Internal; Sapele Showpiece; Magnet Trade										
381 x 1981 x 34mm	38.87	2.50	**39.84**	1.30	0.16	**13.12**	–	**52.96**	nr	56.94
457 x 1981 x 34mm	38.87	2.50	**39.84**	1.40	0.18	**14.19**	–	**54.03**	nr	58.08
533 x 1981 x 34mm	38.87	2.50	**39.84**	1.50	0.19	**15.18**	–	**55.02**	nr	59.15
610 x 1981 x 34mm	38.87	2.50	**39.84**	1.60	0.20	**16.17**	–	**56.01**	nr	60.21
686 x 1981 x 34mm	31.87	2.50	**32.67**	1.70	0.21	**17.17**	–	**49.83**	nr	53.57
762 x 1981 x 34mm	38.87	2.50	**39.84**	2.10	0.26	**21.21**	–	**61.05**	nr	65.63
838 x 1981 x 34mm	40.49	2.50	**41.50**	2.10	0.26	**21.21**	–	**62.71**	nr	67.41
External; MF1X, plywood facings; Magnet Trade										
762 x 1981 x 44mm	53.90	2.50	**55.25**	1.80	0.23	**18.23**	–	**73.48**	nr	78.99
838 x 1981 x 44mm	55.79	2.50	**57.18**	2.00	0.25	**20.22**	–	**77.40**	nr	83.20
External; MF2X, plywood facings; Magnet Trade										
762 x 1981 x 44mm; glazing aperture 457 x 457mm	67.76	2.50	**69.45**	1.80	0.23	**18.23**	–	**87.68**	nr	94.26
838 x 1981 x 44mm; glazing aperture 457 x 457mm	69.56	2.50	**71.30**	2.00	0.25	**20.22**	–	**91.51**	nr	98.38
External; MF4X, plywood facings; Magnet Trade										
762 x 1981 x 44mm; glazing aperture 559 x 864mm	67.76	2.50	**69.45**	1.80	0.23	**18.23**	–	**87.68**	nr	94.26
838 x 1981 x 44mm; glazing aperture 635 x 864mm	69.56	2.50	**71.30**	2.00	0.25	**20.22**	–	**91.51**	nr	98.38
Internal; fire resisting; Magnaseal; FD30; Magnet Trade										
610 x 1981 x 44mm	52.37	2.50	**53.68**	1.80	0.23	**18.23**	–	**71.91**	nr	77.30
686 x 1981 x 44mm	53.63	2.50	**54.97**	1.70	0.21	**17.17**	–	**72.14**	nr	77.55
762 x 1981 x 44mm	53.63	2.50	**54.97**	1.90	0.24	**19.22**	–	**74.19**	nr	79.76
813 x 2032 x 44mm	56.06	2.50	**57.46**	2.20	0.28	**22.27**	–	**79.73**	nr	85.71
838 x 1981 x 44mm	56.06	2.50	**57.46**	2.20	0.28	**22.27**	–	**79.73**	nr	85.71
Internal; fire resisting; plywood facings; FDG30; Magnet Trade										
762 x 1981 x 44mm; glazing aperture 508 x 508mm	67.49	2.50	**69.18**	1.90	0.24	**19.22**	–	**88.40**	nr	95.03
813 x 2032 x 44mm; glazing aperture 559 x 559mm	69.65	2.50	**71.39**	2.20	0.29	**22.34**	–	**93.73**	nr	100.76
838 x 1981 x 44mm; glazing aperture 584 x 584mm	69.65	2.50	**71.39**	2.20	0.29	**22.34**	–	**93.73**	nr	100.76
External; fire resisting; plywood facings; FD30; Magnet Trade										
762 x 1981 x 44mm	69.47	2.50	**71.21**	1.90	0.24	**19.22**	–	**90.43**	nr	97.21
838 x 1981 x 44mm	72.35	2.50	**74.16**	2.30	0.29	**23.26**	–	**97.42**	nr	104.73
Internal; fire resisting; Sapele facings; FD30; Magnet Trade										
762 x 1981 x 44mm	73.97	2.50	**75.82**	2.20	0.29	**22.34**	–	**98.16**	nr	105.52
838 x 1981 x 44mm	77.21	2.50	**79.14**	2.40	0.30	**24.26**	–	**103.40**	nr	111.15
TIMBER DOORS - Trap doors in softwood, wrought										
19mm matchboarding on 25 x 75mm ledges										
457 x 610mm	22.61	2.50	**23.18**	0.18	0.02	**1.80**	–	**24.98**	nr	26.85
610 x 610mm	27.96	2.50	**28.66**	0.19	0.02	**1.89**	–	**30.55**	nr	32.84
762 x 610mm	33.90	2.50	**34.75**	0.20	0.02	**1.99**	–	**36.73**	nr	39.49
TIMBER DOORS - Trap doors in B.C. Pine, wrought										
19mm matchboarding on 25 x 75mm ledges										
457 x 610mm	28.00	2.50	**28.70**	0.18	0.02	**1.80**	–	**30.50**	nr	32.79
610 x 610mm	39.59	2.50	**40.58**	0.19	0.02	**1.89**	–	**42.47**	nr	45.66
762 x 610mm	50.61	2.50	**51.88**	0.20	0.02	**1.99**	–	**53.86**	nr	57.90
TIMBER DOORS - Panelled doors										
Note the following prices for panelled doors are for doors to detail in moderate numbers: doors in large numbers to one pattern would cost considerably less										
TIMBER DOORS - Panelled doors in softwood, wrought										
38mm square framed (or chamfered or moulded one or both sides)										
two panel	63.95	2.50	**65.55**	0.85	0.11	**8.62**	–	**74.17**	m2	79.73
four panel	71.12	2.50	**72.90**	0.85	0.11	**8.62**	–	**81.52**	m2	87.63
six panel	78.55	2.50	**80.51**	0.85	0.11	**8.62**	–	**89.13**	m2	95.82
add if upper panels open moulded in small squares for glass	15.44	2.50	**15.83**	–	–	–	–	**15.83**	m2	17.01
50mm square framed (or chamfered or moulded one or both sides)										
two panel	71.12	2.50	**72.90**	1.00	0.13	**10.14**	–	**83.04**	m2	89.27
four panel	78.55	2.50	**80.51**	1.00	0.11	**10.00**	–	**90.52**	m2	97.30
six panel	86.00	2.50	**88.15**	1.00	0.11	**10.00**	–	**98.15**	m2	105.51
add if upper panels open moulded in small squares for glass	15.44	2.50	**15.83**	–	–	–	–	**15.83**	m2	17.01

Labour hourly rates: (except Specialists) Craft Operatives £9.23 Labourer £7.02 Rates are national average prices. Refer to REGIONAL VARIATIONS for indicative levels of overall pricing in regions	MATERIALS			LABOUR				RATES		
	Del to Site £	Waste %	Material Cost £	Craft Optve Hrs	Lab Hrs	Labour Cost £	Sunds £	Nett Rate £	Unit	Gross Rate (+7.5%) £
L20: DOORS/SHUTTERS/HATCHES Cont.										
TIMBER DOORS - Panelled doors in Afromosia, wrought										
38mm square framed (or chamfered or moulded one or both sides)										
two panel	189.10	2.50	193.83	1.50	0.19	15.18	-	209.01	m2	224.68
four panel	201.72	2.50	206.76	1.50	0.19	15.18	-	221.94	m2	238.59
six panel	214.33	2.50	219.69	1.50	0.19	15.18	-	234.87	m2	252.48
add if upper panels open moulded in small squares for glass	35.93	2.50	36.83	-	-	-	-	36.83	m2	39.59
50mm square framed (or chamfered or moulded one or both sides)										
two panel	208.03	2.50	213.23	1.75	0.22	17.70	-	230.93	m2	248.25
four panel	220.63	2.50	226.15	1.75	0.22	17.70	-	243.84	m2	262.13
six panel	233.24	2.50	239.07	1.75	0.22	17.70	-	256.77	m2	276.03
add if upper panels open moulded in small squares for glass	35.93	2.50	36.83	-	-	-	-	36.83	m2	39.59
TIMBER DOORS - Panelled doors in Sapele, wrought										
38mm square framed (or chamfered or moulded one or both sides)										
two panel	137.76	2.50	141.20	1.25	0.16	12.66	-	153.86	m2	165.40
four panel	148.52	2.50	152.23	1.25	0.16	12.66	-	164.89	m2	177.26
six panel	159.29	2.50	163.27	1.25	0.16	12.66	-	175.93	m2	189.13
add if upper panels open moulded in small squares for glass	25.83	2.50	26.48	-	-	-	-	26.48	m2	28.46
50mm square framed (or chamfered or moulded one or both sides)										
two panel	148.52	2.50	152.23	1.50	0.19	15.18	-	167.41	m2	179.97
four panel	159.29	2.50	163.27	1.50	0.19	15.18	-	178.45	m2	191.83
six panel	170.05	2.50	174.30	1.50	0.19	15.18	-	189.48	m2	203.69
add if upper panels open moulded in small squares for glass	25.83	2.50	26.48	-	-	-	-	26.48	m2	28.46
TIMBER DOORS - Garage doors in softwood, wrought										
Side hung; framed, tongued and grooved boarded										
2134 x 1981 x 44mm overall, pair	282.19	2.50	289.24	4.10	0.51	41.42	-	330.67	nr	355.47
2134 x 2134 x 44mm overall, pair	293.48	2.50	300.82	4.25	0.53	42.95	-	343.77	nr	369.55
Up and over; framed, tongued and grooved boarded; up and over door gear										
2134 x 1981 x 44mm	395.06	2.50	404.94	8.00	1.00	80.86	-	485.80	nr	522.23
TIMBER DOORS - Garage doors in Western Red Cedar, wrought										
Side hung; framed, tongued and grooved boarded										
2134 x 1981 x 44mm overall, pair	407.93	2.50	418.13	4.50	0.56	45.47	-	463.59	nr	498.36
2134 x 2134 x 44mm overall, pair	418.95	2.50	429.42	4.75	0.59	47.98	-	477.41	nr	513.21
Up and over; framed, tongued and grooved boarded; up and over door gear										
2134 x 1981 x 44mm	518.18	2.50	531.13	8.40	1.05	84.90	-	616.04	nr	662.24
TIMBER DOORS - Garage doors in Oak, wrought										
Side hung; framed, tongued and grooved boarded										
2134 x 1981 x 44mm overall, pair	1090.85	2.50	1118.12	7.15	0.90	72.31	-	1190.43	nr	1279.72
2134 x 2134 x 44mm overall, pair	1151.46	2.50	1180.25	7.40	0.95	74.97	-	1255.22	nr	1349.36
Up and over; framed, tongued and grooved boarded; up and over door gear										
2134 x 1981 x 44mm	1272.67	2.50	1304.49	12.00	1.50	121.29	-	1425.78	nr	1532.71
TIMBER DOORS - Patio doors in hard wood, wrought										
Double glazed clear toughened glass, Magnet Trade, one opening leaf, doors and frames treated with base coat stains, fixing frame to masonry with screws										
2073 x 1805mm overall, HP 6	680.00	2.50	697.00	12.00	1.50	121.29	-	818.29	nr	879.66
2073 X 2387mm overall, HP 8	893.62	2.50	915.96	12.00	1.50	121.29	-	1425.78	nr	1532.71
extra; ventilator head, for HP 6	46.81	2.50	47.98	-	-	-	-	47.98	nr	51.58
extra; ventilator head, for HP 8	52.77	2.50	54.09	-	-	-	-	54.09	nr	58.15
DOORSETS - Doorsets mainly in softwood, wrought										
Doorsets; 28mm thick jambs, head and transome with 12mm thick stop to suit 100mm thick wall; 14mm thick hardwood threshold; honeycomb core plywood faced flush door lipped two long edges; 6mm plywood transome panel fixed with pinned beads; 65mm snap in hinges; 57mm backset mortice latch; fixing frame to masonry with screws										
526 x 2040 x 40mm flush door -1 nr; basic dimensions										
600 x 2100mm.............................	120.70	2.50	123.72	1.25	0.16	12.66	-	136.38	nr	146.61
600 x 2400mm.............................	147.58	2.50	151.27	1.35	0.17	13.65	-	164.92	nr	177.29
626 x 2040 x 40mm flush door -1 nr; basic dimensions										
700 x 2100mm.............................	120.70	2.50	123.72	1.25	0.16	12.66	-	136.38	nr	146.61
700 x 2400mm.............................	147.58	2.50	151.27	1.35	0.17	13.65	-	164.92	nr	177.29

Labour hourly rates: (except Specialists) Craft Operatives £9.23 Labourer £7.02 Rates are national average prices. Refer to REGIONAL VARIATIONS for indicative levels of overall pricing in regions	Del to Site £	Waste %	Material Cost £	Craft Optve Hrs	Lab Hrs	Labour Cost £	Sunds £	Nett Rate £	Unit	Gross Rate (+7.5%) £
L20: DOORS/SHUTTERS/HATCHES Cont.										
DOORSETS - Doorsets mainly in softwood, wrought Cont.										
Doorsets; 28mm thick jambs, head and transome with 12mm thick stop to suit 100mm thick wall; 14mm thick hardwood threshold; honeycomb core plywood faced flush door lipped two long edges; 6mm plywood transome panel fixed with pinned beads; 65mm snap in hinges; 57mm backset mortice latch; fixing frame to masonry with screws Cont.										
726 x 2040 x 40mm flush door -1 nr; basic dimensions										
800 x 2100mm	120.70	2.50	123.72	1.25	0.16	12.66	-	136.38	nr	146.61
800 x 2400mm	147.58	2.50	151.27	1.35	0.17	13.65	-	164.92	nr	177.29
826 x 2040 x 40mm flush door -1 nr; basic dimensions										
900 x 2100mm	125.95	2.50	129.10	1.25	0.16	12.66	-	141.76	nr	152.39
900 x 2400mm	150.73	2.50	154.50	1.35	0.17	13.65	-	168.15	nr	180.76
Note actual frame size is 14mm narrower and 14mm shorter than basic size dimensions										
ROLLER SHUTTERS - Roller shutters in wood										
Note roller shutters are always purpose made to order and the following prices are indicative only. Firm quotations should always be obtained										
Pole and hook operation										
2134 x 2134mm	875.00	-	875.00	12.00	12.00	195.00	25.00	1095.00	nr	1177.13
LININGS - Door frames and door lining sets in softwood, wrought										
Sets										
linings; fixing to masonry with screws										
32 x 113mm	4.84	2.50	4.96	0.40	0.05	4.04	0.07	9.07	m	9.75
32 x 150mm	5.81	2.50	5.96	0.40	0.05	4.04	0.07	10.07	m	10.82
32 x 225mm	8.69	2.50	8.91	0.60	0.07	6.03	0.10	15.04	m	16.16
32 x 330mm	13.65	2.50	13.99	0.60	0.07	6.03	0.10	20.12	m	21.63
38 x 113mm	5.61	2.50	5.75	0.40	0.05	4.04	0.07	9.86	m	10.60
38 x 150mm	6.79	2.50	6.96	0.40	0.05	4.04	0.07	11.07	m	11.90
38 x 225mm	9.99	2.50	10.24	0.60	0.07	6.03	0.10	16.37	m	17.60
38 x 330mm	16.14	2.50	16.54	0.60	0.07	6.03	0.10	22.67	m	24.37
linings										
32 x 113mm; labours -1	4.97	2.50	5.09	0.20	0.02	1.99	0.04	7.12	m	7.65
32 x 113mm; labours -2	5.04	2.50	5.17	0.20	0.02	1.99	0.04	7.19	m	7.73
32 x 113mm; labours -3	5.09	2.50	5.22	0.20	0.02	1.99	0.04	7.24	m	7.79
32 x 113mm; labours -4	5.16	2.50	5.29	0.20	0.02	1.99	0.04	7.32	m	7.86
38 x 113mm; labours -1	5.75	2.50	5.89	0.20	0.02	1.99	0.04	7.92	m	8.51
38 x 113mm; labours -2	5.81	2.50	5.96	0.20	0.02	1.99	0.04	7.98	m	8.58
38 x 113mm; labours -3	5.88	2.50	6.03	0.20	0.02	1.99	0.04	8.05	m	8.66
38 x 113mm; labours -4	5.94	2.50	6.09	0.20	0.02	1.99	0.04	8.11	m	8.72
Internal door lining sets; supplied unassembled; fixing to masonry with screws										
32 x 115mm rebated linings; assembling										
for 610 x 1981mm doors	21.57	2.50	22.11	2.00	0.25	20.22	0.32	42.64	nr	45.84
for 686 x 1981mm doors	21.57	2.50	22.11	2.00	0.25	20.22	0.32	42.64	nr	45.84
for 762 x 1981mm doors	21.57	2.50	22.11	2.00	0.25	20.22	0.32	42.64	nr	45.84
32 x 140mm rebated linings; assembling										
for 610 x 1981mm doors	23.53	2.50	24.12	2.00	0.25	20.22	0.32	44.65	nr	48.00
for 686 x 1981mm doors	23.53	2.50	24.12	2.00	0.25	20.22	0.32	44.65	nr	48.00
for 762 x 1981mm doors	23.53	2.50	24.12	2.00	0.25	20.22	0.32	44.65	nr	48.00
LININGS - Door frames and door lining sets in Afromosia, wrought, selected for transparent finish										
Sets										
linings; fixing to masonry with screws										
32 x 113mm	12.03	2.50	12.33	0.60	0.08	6.10	0.10	18.53	m	19.92
32 x 150mm	14.56	2.50	14.92	0.60	0.08	6.10	0.10	21.12	m	22.71
32 x 225mm	21.57	2.50	22.11	0.90	0.11	9.08	0.15	31.34	m	33.69
32 x 330mm	33.59	2.50	34.43	0.90	0.11	9.08	0.15	43.66	m	46.93
38 x 113mm	14.03	2.50	14.38	0.60	0.08	6.10	0.10	20.58	m	22.12
38 x 150mm	16.80	2.50	17.22	0.60	0.08	6.10	0.10	23.42	m	25.18
38 x 225mm	24.63	2.50	25.25	0.90	0.11	9.08	0.15	34.47	m	37.06
38 x 330mm	39.19	2.50	40.17	0.90	0.11	9.08	0.15	49.40	m	53.10
linings										
32 x 113mm; labours -1	12.20	2.50	12.51	0.30	0.04	3.05	0.06	15.61	m	16.79
32 x 113mm; labours -2	12.32	2.50	12.63	0.30	0.04	3.05	0.06	15.74	m	16.92
32 x 113mm; labours -3	12.43	2.50	12.74	0.30	0.04	3.05	0.06	15.85	m	17.04
32 x 113mm; labours -4	12.56	2.50	12.87	0.30	0.04	3.05	0.06	15.98	m	17.18
38 x 113mm; labours -1	14.20	2.50	14.56	0.30	0.04	3.05	0.06	17.66	m	18.99
38 x 113mm; labours -2	14.32	2.50	14.68	0.30	0.04	3.05	0.06	17.79	m	19.12
38 x 113mm; labours -3	14.44	2.50	14.80	0.30	0.04	3.05	0.06	17.91	m	19.25
38 x 113mm; labours -4	14.56	2.50	14.92	0.30	0.04	3.05	0.06	18.03	m	19.39
LININGS - Door frames and door lining sets in Sapele, wrought, selected for transparent finish										
Sets										
linings; fixing to masonry with screws										
32 x 113mm	9.19	2.50	9.42	0.55	0.07	5.57	0.09	15.08	m	16.21
32 x 150mm	11.03	2.50	11.31	0.55	0.07	5.57	0.09	16.96	m	18.24

Labour hourly rates: (except Specialists) Craft Operatives £9.23 Labourer £7.02 Rates are national average prices. Refer to REGIONAL VARIATIONS for indicative levels of overall pricing in regions	Del to Site £	Waste %	Material Cost £	Craft Optve Hrs	Lab Hrs	Labour Cost £	Sunds £	Nett Rate £	Unit	Gross Rate (+7.5%) £
L20: DOORS/SHUTTERS/HATCHES Cont.										
LININGS - Door frames and door lining sets in Sapele, wrought, selected for transparent finish Cont.										
Sets Cont.										
linings; fixing to masonry with screws Cont.										
32 x 225mm	16.54	2.50	16.95	0.80	0.08	7.95	0.14	25.04	m	26.92
32 x 330mm	25.73	2.50	26.37	0.80	0.08	7.95	0.14	34.46	m	37.04
38 x 113mm	10.76	2.50	11.03	0.55	0.07	5.57	0.09	16.69	m	17.94
38 x 150mm	12.86	2.50	13.18	0.55	0.07	5.57	0.09	18.84	m	20.25
38 x 225mm	18.90	2.50	19.37	0.80	0.08	7.95	0.14	27.46	m	29.52
38 x 330mm	30.45	2.50	31.21	0.80	0.08	7.95	0.14	39.30	m	42.24
linings										
32 x 113mm; labours -1	9.35	2.50	9.58	0.27	0.03	2.70	0.04	12.33	m	13.25
32 x 113mm; labours -2	9.45	2.50	9.69	0.27	0.03	2.70	0.04	12.43	m	13.36
32 x 113mm; labours -3	9.56	2.50	9.80	0.27	0.03	2.70	0.04	12.54	m	13.48
32 x 113mm; labours -4	9.66	2.50	9.90	0.27	0.03	2.70	0.04	12.64	m	13.59
38 x 113mm; labours -1	10.92	2.50	11.19	0.27	0.03	2.70	0.04	13.94	m	14.98
38 x 113mm; labours -2	11.03	2.50	11.31	0.27	0.03	2.70	0.04	14.05	m	15.10
38 x 113mm; labours -3	11.13	2.50	11.41	0.27	0.03	2.70	0.04	14.15	m	15.21
38 x 113mm; labours -4	11.24	2.50	11.52	0.27	0.03	2.70	0.04	14.26	m	15.33
FRAMES - Door frames and door lining sets in softwood, wrought - frames										
Sets										
38 x 75mm jambs	4.53	2.50	4.64	0.18	0.02	1.80	0.03	6.48	m	6.96
38 x 100mm jambs	5.24	2.50	5.37	0.18	0.02	1.80	0.03	7.20	m	7.74
50 x 75mm jambs	5.24	2.50	5.37	0.20	0.03	2.06	0.03	7.46	m	8.02
50 x 100mm jambs	6.08	2.50	6.23	0.20	0.03	2.06	0.03	8.32	m	8.94
50 x 125mm jambs	6.79	2.50	6.96	0.20	0.03	2.06	0.03	9.05	m	9.72
63 x 100mm jambs	6.79	2.50	6.96	0.20	0.03	2.06	0.03	9.05	m	9.72
75 x 100mm jambs	8.69	2.50	8.91	0.22	0.03	2.24	0.04	11.19	m	12.03
75 x 113mm jambs	9.93	2.50	10.18	0.22	0.03	2.24	0.04	12.46	m	13.39
100 x 100mm jambs	10.97	2.50	11.24	0.25	0.03	2.52	0.04	13.80	m	14.84
100 x 113mm jambs	12.61	2.50	12.93	0.27	0.03	2.70	0.04	15.67	m	16.84
100 x 125mm jambs	14.25	2.50	14.61	0.27	0.03	2.70	0.04	17.35	m	18.65
113 x 113mm jambs	14.25	2.50	14.61	0.27	0.03	2.70	0.04	17.35	m	18.65
113 x 125mm jambs	15.80	2.50	16.20	0.30	0.04	3.05	0.05	19.29	m	20.74
113 x 150mm jambs	17.39	2.50	17.82	0.30	0.04	3.05	0.05	20.92	m	22.49
38 x 75mm jambs; labours -1	4.64	2.50	4.76	0.18	0.02	1.80	0.03	6.59	m	7.08
38 x 75mm jambs; labours -2	4.69	2.50	4.81	0.18	0.02	1.80	0.03	6.64	m	7.14
38 x 75mm jambs; labours -3	4.78	2.50	4.90	0.18	0.02	1.80	0.03	6.73	m	7.24
38 x 75mm jambs; labours -4	4.84	2.50	4.96	0.18	0.02	1.80	0.03	6.79	m	7.30
38 x 75mm heads	4.53	2.50	4.64	0.18	0.02	1.80	0.03	6.48	m	6.96
38 x 100mm heads	5.24	2.50	5.37	0.18	0.02	1.80	0.03	7.20	m	7.74
50 x 75mm heads	5.24	2.50	5.37	0.20	0.03	2.06	0.03	7.46	m	8.02
50 x 100mm heads	6.08	2.50	6.23	0.20	0.03	2.06	0.03	8.32	m	8.94
50 x 125mm heads	6.79	2.50	6.96	0.20	0.03	2.06	0.03	9.05	m	9.72
63 x 100mm heads	6.79	2.50	6.96	0.20	0.03	2.06	0.03	9.05	m	9.72
75 x 100mm heads	8.69	2.50	8.91	0.22	0.03	2.24	0.04	11.19	m	12.03
75 x 113mm heads	9.94	2.50	10.19	0.22	0.03	2.24	0.04	12.47	m	13.40
100 x 100mm heads	10.97	2.50	11.24	0.25	0.03	2.52	0.04	13.80	m	14.84
100 x 113mm heads	12.61	2.50	12.93	0.27	0.03	2.70	0.04	15.67	m	16.84
100 x 125mm heads	14.26	2.50	14.62	0.27	0.03	2.70	0.04	17.36	m	18.66
113 x 113mm heads	14.26	2.50	14.62	0.27	0.03	2.70	0.04	17.36	m	18.66
113 x 125mm heads	15.80	2.50	16.20	0.30	0.04	3.05	0.05	19.29	m	20.74
113 x 150mm heads	17.39	2.50	17.82	0.30	0.04	3.05	0.05	20.92	m	22.49
38 x 75mm heads; labours -1	4.64	2.50	4.76	0.18	0.02	1.80	0.03	6.59	m	7.08
38 x 75mm heads; labours -2	4.69	2.50	4.81	0.18	0.02	1.80	0.03	6.64	m	7.14
38 x 75mm heads; labours -3	4.78	2.50	4.90	0.18	0.02	1.80	0.03	6.73	m	7.24
38 x 75mm heads; labours -4	4.84	2.50	4.96	0.18	0.02	1.80	0.03	6.79	m	7.30
63 x 125mm sills	10.32	2.50	10.58	0.23	0.03	2.33	0.04	12.95	m	13.92
75 x 125mm sills	11.76	2.50	12.05	0.24	0.03	2.43	0.04	14.52	m	15.61
75 x 150mm sills	13.20	2.50	13.53	0.25	0.03	2.52	0.04	16.09	m	17.29
63 x 125mm sills; labours -1	10.45	2.50	10.71	0.23	0.03	2.33	0.04	13.08	m	14.07
63 x 125mm sills; labours -2	10.50	2.50	10.76	0.23	0.03	2.33	0.04	13.14	m	14.12
63 x 125mm sills; labours -3	10.59	2.50	10.85	0.23	0.03	2.33	0.04	13.23	m	14.22
63 x 125mm sills; labours -4	10.64	2.50	10.91	0.23	0.03	2.33	0.04	13.28	m	14.28
63 x 100mm mullions	6.79	2.50	6.96	0.05	0.01	0.53	-	7.49	m	8.05
75 x 100mm mullions	8.69	2.50	8.91	0.05	0.01	0.53	-	9.44	m	10.15
75 x 113mm mullions	9.93	2.50	10.18	0.05	0.01	0.53	-	10.71	m	11.51
100 x 100mm mullions	10.97	2.50	11.24	0.05	0.01	0.53	-	11.78	m	12.66
100 x 125mm mullions	14.26	2.50	14.62	0.05	0.01	0.53	-	15.15	m	16.28
63 x 100mm mullions; labours -1	6.93	2.50	7.10	0.05	0.01	0.53	-	7.63	m	8.21
63 x 100mm mullions; labours -2	6.99	2.50	7.16	0.05	0.01	0.53	-	7.70	m	8.27
63 x 100mm mullions; labours -3	7.07	2.50	7.25	0.05	0.01	0.53	-	7.78	m	8.36
63 x 100mm mullions; labours -4	7.12	2.50	7.30	0.05	0.01	0.53	-	7.83	m	8.42
63 x 100mm transoms	6.79	2.50	6.96	0.05	0.01	0.53	-	7.49	m	8.05
75 x 100mm transoms	8.69	2.50	8.91	0.05	0.01	0.53	-	9.44	m	10.15
75 x 113mm transoms	9.93	2.50	10.18	0.05	0.01	0.53	-	10.71	m	11.51
100 x 100mm transoms	10.97	2.50	11.24	0.05	0.01	0.53	-	11.78	m	12.66
100 x 125mm transoms	14.26	2.50	14.62	0.05	0.01	0.53	-	15.15	m	16.28
63 x 100mm transoms; labours -1	6.93	2.50	7.10	0.05	0.01	0.53	-	7.63	m	8.21
63 x 100mm transoms; labours -2	6.99	2.50	7.16	0.05	0.01	0.53	-	7.70	m	8.27
63 x 100mm transoms; labours -3	7.07	2.50	7.25	0.05	0.01	0.53	-	7.78	m	8.36
63 x 100mm transoms; labours -4	7.12	2.50	7.30	0.05	0.01	0.53	-	7.83	m	8.42
Internal door frame sets; supplied unassembled 38 x 63mm jambs and head; stops (supplied loose); assembling										
for 686 x 1981mm doors	19.61	2.50	20.10	1.00	0.13	10.14	0.16	30.40	nr	32.68
for 762 x 1981mm doors	19.61	2.50	20.10	1.00	0.13	10.14	0.16	30.40	nr	32.68

Labour hourly rates: (except Specialists) Craft Operatives £9.23 Labourer £7.02 Rates are national average prices. Refer to REGIONAL VARIATIONS for indicative levels of overall pricing in regions	MATERIALS			LABOUR				RATES		
	Del to Site £	Waste %	Material Cost £	Craft Optve Hrs	Lab Hrs	Labour Cost £	Sunds £	Nett Rate £	Unit	Gross Rate (+7.5%) £

L20: DOORS/SHUTTERS/HATCHES Cont.

FRAMES - Door frames and door lining sets in softwood, wrought - frames Cont.

Internal door frame sets; supplied unassembled Cont.
38 x 75mm jambs and head; stops (supplied loose); assembling

for 686 x 1981mm doors	20.91	2.50	**21.43**	1.00	0.13	**10.14**	0.16	**31.74**	nr	34.12
for 762 x 1981mm doors	20.91	2.50	**21.43**	1.00	0.13	**10.14**	0.16	**31.74**	nr	34.12

50 x 100mm jambs and head; stops (supplied loose); assembling

for 686 x 1981mm doors	32.67	2.50	**33.49**	1.00	0.13	**10.14**	0.16	**43.79**	nr	47.07
for 762 x 1981mm doors	32.67	2.50	**33.49**	1.00	0.13	**10.14**	0.16	**43.79**	nr	47.07

50 x 113mm jambs and head; stops (supplied loose); assembling

for 686 x 1981mm doors	40.52	2.50	**41.53**	1.00	0.13	**10.14**	0.16	**51.84**	nr	55.72
for 762 x 1981mm doors	40.52	2.50	**41.53**	1.00	0.13	**10.14**	0.16	**51.84**	nr	55.72

External door frame sets; one coat external primer before delivery to site
63 x 75mm jambs and head; rebates -1

for 762 x 1981 x 50mm doors	35.28	2.50	**36.16**	1.00	0.13	**10.14**	0.16	**46.46**	nr	49.95
for 838 x 1981 x 50mm doors	36.59	2.50	**37.50**	1.00	0.13	**10.14**	0.16	**47.81**	nr	51.39

63 x 88mm jambs and head; rebates -1

for 762 x 1981 x 50mm doors	36.59	2.50	**37.50**	1.00	0.13	**10.14**	0.16	**47.81**	nr	51.39
for 838 x 1981 x 50mm doors	37.91	2.50	**38.86**	1.00	0.13	**10.14**	0.16	**49.16**	nr	52.85

63 x 75mm jambs and head, rebates -1 nr; hardwood sill

for 762 x 1981 x 50mm doors	78.41	2.50	**80.37**	1.25	0.16	**12.66**	0.20	**93.23**	nr	100.22
for 838 x 1981 x 50mm doors	79.72	2.50	**81.71**	1.25	0.16	**12.66**	0.20	**94.57**	nr	101.67

63 x 88mm jambs and head, rebates -1 nr; hardwood sill

for 762 x 1981 x 50mm doors	81.00	2.50	**83.03**	1.25	0.16	**12.66**	0.20	**95.89**	nr	103.08
for 838 x 1981 x 50mm doors	82.33	2.50	**84.39**	1.25	0.16	**12.66**	0.20	**97.25**	nr	104.54
for 1168 x 1981 x 50mm doors	83.62	2.50	**85.71**	1.50	0.19	**15.18**	0.24	**101.13**	nr	108.71

External garage door frame sets; one coat external primer before delivery to site; supplied unassembled
75 x 100mm jambs and head; assembling

for 2134 x 1981mm side hung doors	57.68	2.50	**59.12**	2.25	0.28	**22.73**	0.37	**82.23**	nr	88.39
for 2134 x 2134mm side hung doors	61.51	2.50	**63.05**	2.25	0.28	**22.73**	0.37	**86.15**	nr	92.61

75 x 75mm jambs and head; assembling

for 2134 x 1981mm up and over doors	53.57	2.50	**54.91**	2.25	0.28	**22.73**	0.37	**78.01**	nr	83.86

FRAMES - Door frames and door lining sets in Afromosia, wrought, selected for transparent finish - frames

Sets

38 x 75mm jambs	11.20	2.50	**11.48**	0.31	0.04	**3.14**	0.05	**14.67**	m	15.77
38 x 100mm jambs	12.90	2.50	**13.22**	0.31	0.04	**3.14**	0.05	**16.41**	m	17.65
50 x 75mm jambs	12.90	2.50	**13.22**	0.35	0.04	**3.51**	0.06	**16.79**	m	18.05
50 x 100mm jambs	15.15	2.50	**15.53**	0.35	0.04	**3.51**	0.06	**19.10**	m	20.53
50 x 125mm jambs	16.80	2.50	**17.22**	0.35	0.04	**3.51**	Sunds	**20.79**	m	22.35
63 x 100mm jambs	16.80	2.50	**17.22**	0.35	0.04	**3.51**	0.06	**20.79**	m	22.35
75 x 100mm jambs	21.57	2.50	**22.11**	0.38	0.05	**3.86**	0.06	**26.03**	m	27.98
75 x 113mm jambs	24.63	2.50	**25.25**	0.38	0.05	**3.86**	0.06	**29.16**	m	31.35
100 x 100mm jambs	27.46	2.50	**28.15**	0.44	0.05	**4.41**	0.07	**32.63**	m	35.08
100 x 113mm jambs	31.36	2.50	**32.14**	0.47	0.06	**4.76**	0.08	**36.98**	m	39.76
100 x 125mm jambs	35.31	2.50	**36.19**	0.47	0.06	**4.76**	0.08	**41.03**	m	44.11
113 x 113mm jambs	35.31	2.50	**36.19**	0.47	0.06	**4.76**	0.08	**41.03**	m	44.11
113 x 125mm jambs	39.19	2.50	**40.17**	0.52	0.07	**5.29**	0.09	**45.55**	m	48.97
113 x 150mm jambs	43.15	2.50	**44.23**	0.52	0.07	**5.29**	0.09	**49.61**	m	53.33
38 x 75mm jambs; labours -1	11.37	2.50	**11.65**	0.31	0.04	**3.14**	0.05	**14.85**	m	15.96
38 x 75mm jambs; labours -2	11.49	2.50	**11.78**	0.31	0.04	**3.14**	0.05	**14.97**	m	16.09
38 x 75mm jambs; labours -3	11.67	2.50	**11.96**	0.31	0.04	**3.14**	0.05	**15.15**	m	16.29
38 x 75mm jambs; labours -4	11.85	2.50	**12.15**	0.31	0.04	**3.14**	0.05	**15.34**	m	16.49
38 x 75mm heads	11.20	2.50	**11.48**	0.31	0.04	**3.14**	0.05	**14.67**	m	15.77
38 x 100mm heads	12.90	2.50	**13.22**	0.31	0.04	**3.14**	0.05	**16.41**	m	17.65
50 x 75mm heads	12.90	2.50	**13.22**	0.31	0.04	**3.14**	0.05	**16.41**	m	17.65
50 x 100mm heads	15.15	2.50	**15.53**	0.35	0.04	**3.51**	0.06	**19.10**	m	20.53
50 x 125mm heads	16.80	2.50	**17.22**	0.35	0.04	**3.51**	0.06	**20.79**	m	22.35
63 x 100mm heads	16.80	2.50	**17.22**	0.35	0.04	**3.51**	0.06	**20.79**	m	22.35
75 x 100mm heads	21.57	2.50	**22.11**	0.38	0.05	**3.86**	0.06	**26.03**	m	27.98
75 x 113mm heads	24.63	2.50	**25.25**	0.38	0.05	**3.86**	0.06	**29.16**	m	31.35
100 x 100mm heads	27.46	2.50	**28.15**	0.44	0.05	**4.41**	0.07	**32.63**	m	35.08
100 x 113mm heads	31.36	2.50	**32.14**	0.47	0.06	**4.76**	0.08	**36.98**	m	39.76
100 x 125mm heads	35.31	2.50	**36.19**	0.47	0.06	**4.76**	0.08	**41.03**	m	44.11
113 x 113mm heads	35.31	2.50	**36.19**	0.47	0.06	**4.76**	0.08	**41.03**	m	44.11
113 x 125mm heads	39.19	2.50	**40.17**	0.52	0.07	**5.29**	0.09	**45.55**	m	48.97
113 x 150mm heads	43.15	2.50	**44.23**	0.52	0.07	**5.29**	0.09	**49.61**	m	53.33
38 x 75mm heads; labours -1	11.37	2.50	**11.65**	0.31	0.04	**3.14**	0.05	**14.85**	m	15.96
38 x 75mm heads; labours -2	11.49	2.50	**11.78**	0.31	0.04	**3.14**	0.05	**14.97**	m	16.09
38 x 75mm heads; labours -3	11.67	2.50	**11.96**	0.31	0.04	**3.14**	0.05	**15.15**	m	16.29
38 x 75mm heads; labours -4	11.85	2.50	**12.15**	0.31	0.04	**3.14**	0.05	**15.34**	m	16.49
63 x 125mm sills	25.46	2.50	**26.10**	0.40	0.05	**4.04**	0.06	**30.20**	m	32.46
75 x 125mm sills	29.12	2.50	**29.85**	0.44	0.06	**4.48**	0.07	**34.40**	m	36.98
75 x 150mm sills	32.78	2.50	**33.60**	0.45	0.06	**4.57**	0.07	**38.24**	m	41.11
63 x 125mm sills; labours -1	25.76	2.50	**26.40**	0.40	0.05	**4.04**	0.06	**30.51**	m	32.80
63 x 125mm sills; labours -2	25.88	2.50	**26.53**	0.40	0.05	**4.04**	0.06	**30.63**	m	32.93
63 x 125mm sills; labours -3	25.99	2.50	**26.64**	0.40	0.05	**4.04**	0.06	**30.74**	m	33.05
63 x 125mm sills; labours -4	26.11	2.50	**26.76**	0.40	0.05	**4.04**	0.06	**30.87**	m	33.18
63 x 100mm mullions	16.80	2.50	**17.22**	0.05	0.01	**0.53**	-	**17.75**	m	19.08
75 x 100mm mullions	21.57	2.50	**22.11**	0.05	0.01	**0.53**	-	**22.64**	m	24.34
75 x 113mm mullions	24.63	2.50	**25.25**	0.05	0.01	**0.53**	-	**25.78**	m	27.71

Labour hourly rates: (except Specialists) Craft Operatives £9.23 Labourer £7.02. Rates are national average prices. Refer to REGIONAL VARIATIONS for indicative levels of overall pricing in regions	MATERIALS			LABOUR				RATES		
	Del to Site £	Waste %	Material Cost £	Craft Optve Hrs	Lab Hrs	Labour Cost £	Sunds £	Nett Rate £	Unit	Gross Rate (+7.5%) £
L20: DOORS/SHUTTERS/HATCHES Cont.										
FRAMES - Door frames and door lining sets in Afromosia, wrought, selected for transparent finish - frames Cont.										
Sets Cont.										
100 x 100mm mullions	27.46	2.50	28.15	0.05	0.01	0.53	–	28.68	m	30.83
100 x 125mm mullions	35.31	2.50	36.19	0.05	0.01	0.53	–	36.72	m	39.48
63 x 100mm mullions; labours -1	16.97	2.50	17.39	0.05	0.01	0.53	–	17.93	m	19.27
63 x 100mm mullions; labours -2	17.09	2.50	17.52	0.05	0.01	0.53	–	18.05	m	19.40
63 x 100mm mullions; labours -3	17.21	2.50	17.64	0.05	0.01	0.53	–	18.17	m	19.53
63 x 100mm mullions; labours -4	17.33	2.50	17.76	0.05	0.01	0.53	–	18.29	m	19.67
63 x 100mm transoms	16.80	2.50	17.22	0.05	0.01	0.53	–	17.75	m	19.08
75 x 100mm transoms	21.57	2.50	22.11	0.05	0.01	0.53	–	22.64	m	24.34
75 x 113mm transoms	24.63	2.50	25.25	0.05	0.01	0.53	–	25.78	m	27.71
100 x 100mm transoms	27.46	2.50	28.15	0.05	0.01	0.53	–	28.68	m	30.83
100 x 125mm transoms	35.31	2.50	36.19	0.05	0.01	0.53	–	36.72	m	39.48
63 x 100mm transoms; labours -1	16.97	2.50	17.39	0.05	0.01	0.53	–	17.93	m	19.27
63 x 100mm transoms; labours -2	17.09	2.50	17.52	0.05	0.01	0.53	–	18.05	m	19.40
63 x 100mm transoms; labours -3	17.21	2.50	17.64	0.05	0.01	0.53	–	18.17	m	19.53
63 x 100mm transoms; labours -4	17.33	2.50	17.76	0.05	0.01	0.53	–	18.29	m	19.67
FRAMES - Door frames and door lining sets in Sapele, wrought, selected for transparent finish - frames										
Sets										
38 x 75mm jambs	8.61	2.50	8.83	0.27	0.03	2.70	0.04	11.57	m	12.44
38 x 100mm jambs	9.92	2.50	10.17	0.27	0.03	2.70	0.04	12.91	m	13.88
50 x 75mm jambs	9.92	2.50	10.17	0.30	0.04	3.05	0.05	13.27	m	14.26
50 x 100mm jambs	11.55	2.50	11.84	0.30	0.04	3.05	0.05	14.94	m	16.06
50 x 125mm jambs	12.86	2.50	13.18	0.30	0.04	3.05	0.05	16.28	m	17.50
63 x 100mm jambs	12.86	2.50	13.18	0.30	0.04	3.05	0.05	16.28	m	17.50
75 x 100mm jambs	16.54	2.50	16.95	0.33	0.04	3.33	0.05	20.33	m	21.85
75 x 113mm jambs	18.90	2.50	19.37	0.33	0.04	3.33	0.05	22.75	m	24.46
100 x 100mm jambs	20.90	2.50	21.42	0.37	0.05	3.77	0.06	25.25	m	27.14
100 x 113mm jambs	23.94	2.50	24.54	0.40	0.05	4.04	0.07	28.65	m	30.80
100 x 125mm jambs	27.04	2.50	27.72	0.40	0.05	4.04	0.07	31.83	m	34.22
113 x 113mm jambs	27.04	2.50	27.72	0.40	0.05	4.04	0.07	31.83	m	34.22
113 x 125mm jambs	30.14	2.50	30.89	0.45	0.06	4.57	0.07	35.54	m	38.20
113 x 150mm jambs	33.08	2.50	33.91	0.45	0.06	4.57	0.07	38.55	m	41.44
38 x 75mm jambs; labours -1	8.77	2.50	8.99	0.27	0.03	2.70	0.04	11.73	m	12.61
38 x 75mm jambs; labours -2	8.87	2.50	9.09	0.27	0.03	2.70	0.04	11.83	m	12.72
38 x 75mm jambs; labours -3	8.98	2.50	9.20	0.27	0.03	2.70	0.04	11.95	m	12.84
38 x 75mm jambs; labours -4	9.08	2.50	9.31	0.27	0.03	2.70	0.04	12.05	m	12.95
38 x 75mm heads	8.61	2.50	8.83	0.27	0.03	2.70	0.04	11.57	m	12.44
38 x 100mm heads	9.92	2.50	10.17	0.27	0.03	2.70	0.04	12.91	m	13.88
50 x 75mm heads	9.92	2.50	10.17	0.30	0.04	3.05	0.05	13.27	m	14.26
50 x 100mm heads	11.55	2.50	11.84	0.30	0.04	3.05	0.05	14.94	m	16.06
50 x 125mm heads	12.86	2.50	13.18	0.30	0.04	3.05	0.05	16.28	m	17.50
63 x 100mm heads	12.86	2.50	13.18	0.30	0.04	3.05	0.05	16.28	m	17.50
75 x 100mm heads	13.39	2.50	13.72	0.33	0.04	3.33	0.05	17.10	m	18.38
75 x 113mm heads	18.90	2.50	19.37	0.33	0.04	3.33	0.05	22.75	m	24.46
100 x 100mm heads	20.90	2.50	21.42	0.37	0.05	3.77	0.06	25.25	m	27.14
100 x 113mm heads	23.94	2.50	24.54	0.40	0.05	4.04	0.07	28.65	m	30.80
100 x 125mm heads	27.04	2.50	27.72	0.40	0.05	4.04	0.07	31.83	m	34.22
113 x 113mm heads	27.04	2.50	27.72	0.40	0.05	4.04	0.07	31.83	m	34.22
113 x 125mm heads	30.14	2.50	30.89	0.45	0.06	4.57	0.07	35.54	m	38.20
113 x 150mm heads	33.08	2.50	33.91	0.45	0.06	4.57	0.07	38.55	m	41.44
38 x 75mm heads; labours -1	8.77	2.50	8.99	0.27	0.03	2.70	0.04	11.73	m	12.61
38 x 75mm heads; labours -2	8.77	2.50	8.99	0.27	0.03	2.70	0.04	11.73	m	12.61
38 x 75mm heads; labours -3	8.98	2.50	9.20	0.27	0.03	2.70	0.04	11.95	m	12.84
38 x 75mm heads; labours -4	9.08	2.50	9.31	0.27	0.03	2.70	0.04	12.05	m	12.95
63 x 125mm sills	19.64	2.50	20.13	0.35	0.04	3.51	0.06	23.70	m	25.48
75 x 125mm sills	22.31	2.50	22.87	0.37	0.05	3.77	0.06	26.69	m	28.70
75 x 150mm sills	25.20	2.50	25.83	0.37	0.05	3.77	0.06	29.66	m	31.88
63 x 125mm sills; labours -1	19.79	2.50	20.28	0.35	0.04	3.51	0.06	23.86	m	25.65
63 x 125mm sills; labours -2	19.90	2.50	20.40	0.35	0.04	3.51	0.06	23.97	m	25.77
63 x 125mm sills; labours -3	20.00	2.50	20.50	0.35	0.04	3.51	0.06	24.07	m	25.88
63 x 125mm sills; labours -4	20.11	2.50	20.61	0.35	0.04	3.51	0.06	24.18	m	26.00
63 x 100mm mullions	12.86	2.50	13.18	0.05	0.01	0.53	–	13.71	m	14.74
75 x 100mm mullions	13.39	2.50	13.72	0.05	0.01	0.53	–	14.26	m	15.33
75 x 113mm mullions	18.90	2.50	19.37	0.05	0.01	0.53	–	19.90	m	21.40
100 x 100mm mullions	20.90	2.50	21.42	0.05	0.01	0.53	–	21.95	m	23.60
100 x 125mm mullions	27.04	2.50	27.72	0.05	0.01	0.53	–	28.25	m	30.37
63 x 100mm mullions; labours -1	13.02	2.50	13.35	0.05	0.01	0.53	–	13.88	m	14.92
63 x 100mm mullions; labours -2	13.13	2.50	13.46	0.05	0.01	0.53	–	13.99	m	15.04
63 x 100mm mullions; labours -3	13.23	2.50	13.56	0.05	0.01	0.53	–	14.09	m	15.15
63 x 100mm mullions; labours -4	13.34	2.50	13.67	0.05	0.01	0.53	–	14.21	m	15.27
63 x 100mm transoms	12.86	2.50	13.18	0.05	0.01	0.53	–	13.71	m	14.74
75 x 100mm transoms	13.39	2.50	13.72	0.05	0.01	0.53	–	14.26	m	15.33
75 x 113mm transoms	18.90	2.50	19.37	0.05	0.01	0.53	–	19.90	m	21.40
100 x 100mm transoms	20.90	2.50	21.42	0.05	0.01	0.53	–	21.95	m	23.60
100 x 125mm transoms	27.04	2.50	27.72	0.05	0.01	0.53	–	28.25	m	30.37
63 x 100mm transoms; labours -1	13.02	2.50	13.35	0.05	0.01	0.53	–	13.88	m	14.92
63 x 100mm transoms; labours -2	13.13	2.50	13.46	0.05	0.01	0.53	–	13.99	m	15.04
63 x 100mm transoms; labours -3	13.23	2.50	13.56	0.05	0.01	0.53	–	14.09	m	15.15
63 x 100mm transoms; labours -4	13.34	2.50	13.67	0.05	0.01	0.53	–	14.21	m	15.27
FRAMES - Intumescent strips and smoke seals										
Albi-Flex, white PVC sleeved self adhesive; Rentokil Ltd.										
12 x 4mm intumescent strip, half hour application; setting into groove in timber frame or door	2.95	10.00	3.25	0.15	–	1.38	–	4.63	m	4.98

Labour hourly rates: (except Specialists) Craft Operatives £9.23 Labourer £7.02 Rates are national average prices. Refer to REGIONAL VARIATIONS for indicative levels of overall pricing in regions	MATERIALS			LABOUR				RATES		
	Del to Site £	Waste %	Material Cost £	Craft Optve Hrs	Lab Hrs	Labour Cost £	Sunds £	Nett Rate £	Unit	Gross Rate (+7.5%) £
L20: DOORS/SHUTTERS/HATCHES Cont.										
FRAMES - Intumescent strips and smoke seals Cont.										
Albi-Flex, white PVC sleeved self adhesive; Rentokil Ltd. Cont.										
18 x 4mm intumescent strip, one hour application; setting into groove in timber frame or door	4.14	10.00	4.55	0.15	-	1.38	-	5.94	m	6.38
22 x 4mm intumescent strip with integral cold smoke seal, half hour application; setting into groove in timber frame or door	6.26	10.00	6.89	0.15	-	1.38	-	8.27	m	8.89
28 x 4mm intumescent strip with integral cold smoke seal, one hour application; setting into groove in timber frame or door	7.27	10.00	8.00	0.15	-	1.38	-	9.38	m	10.09
12 x 4mm intumescent strip, half hour application; fixing to both sides of glass behind glazing beads	5.91	10.00	6.50	0.30	-	2.77	-	9.27	m	9.97
25 x 4mm intumescent strip, one hour application; fixing to both sides of glass behind glazing beads	10.63	10.00	11.69	0.30	-	2.77	-	14.46	m	15.55
10 x 4mm cold smoke seal; setting into groove in timber frame or door	3.24	10.00	3.56	0.15	-	1.38	-	4.95	m	5.32
Lorient Polyproducts Ltd; System 36										
15 x 12mm glazing channel reference LG1512; fitting over the edge of the pane	1.38	10.00	1.52	0.15	-	1.38	-	2.90	m	3.12
Lorient Polyproducts Ltd; System 90										
intumescent lining reference LX4402 to suit 44mm thick doors; fixing to timber with adhesive	3.00	10.00	3.30	0.30	-	2.77	-	6.07	m	6.52
intumescent lining reference LX5402 to suit 54mm thick doors; fixing to timber with adhesive	3.36	10.00	3.70	0.30	-	2.77	-	6.46	m	6.95
Mann McGowan Fabrications Ltd										
Pyroglaze 30 half hour application; fixing to both sides of glass behind glazing beads	2.43	10.00	2.67	0.35	-	3.23	-	5.90	m	6.35
Pyroglaze 60 one hour application; fixing to both sides of glass behind glazing beads	3.80	10.00	4.18	0.35	-	3.23	-	7.41	m	7.97
FRAMES - Bedding and pointing frames										
Bedding in cement mortar (1:3); pointing one side										
wood frames	0.20	10.00	0.22	0.12	-	1.11	-	1.33	m	1.43
Bedding in cement mortar (1:3); pointing each side										
wood frames	0.20	10.00	0.22	0.18	-	1.66	-	1.88	m	2.02
Bedding in cement mortar (1:3); pointing with Secomastic standard mastic one side										
wood frames one side	1.10	10.00	1.21	0.18	-	1.66	-	2.87	m	3.09
METAL DOORS - Doors and sidelights in galvanised steel										
Crittall Windows Ltd. Duralife Homelight; weatherstripping; fixing to masonry with lugs										
761 x 2056mm doors NA15	413.07	2.50	423.40	2.80	1.40	35.67	-	459.07	nr	493.50
997 x 2056mm doors NA2	622.02	2.50	637.57	3.00	1.50	38.22	-	675.79	nr	726.47
1143 x 2056mm doors NA25	634.37	2.50	650.23	3.20	1.60	40.77	-	691.00	nr	742.82
279 x 2056mm sidelights, NA6	69.62	2.50	71.36	1.20	0.60	15.29	-	86.65	nr	93.15
508 x 2056mm sidelights, NA5	84.61	2.50	86.73	1.50	0.75	19.11	-	105.84	nr	113.77
997 x 2056mm sidelights, NA13F	160.97	2.50	164.99	1.80	0.90	22.93	-	187.93	nr	202.02
METAL DOORS - Doors and sidelights in galvanised steel; polyester powder coated; white matt										
Crittall Windows Ltd. Duralife Homelight; weatherstripping; fixing to masonry with lugs										
761 x 2056mm doors NA15	520.72	2.50	533.74	2.50	1.40	32.90	-	566.64	nr	609.14
997 x 2056mm doors NA2	783.58	2.50	803.17	3.00	1.50	38.22	-	841.39	nr	904.49
1143 x 2056mm doors NA25	798.68	2.50	818.65	3.20	1.60	40.77	-	859.41	nr	923.87
279 x 2056mm sidelights, NA6	86.29	2.50	88.45	1.20	0.60	15.29	-	103.74	nr	111.52
508 x 2056mm sidelights, NA5	104.93	2.50	107.55	1.50	0.75	19.11	-	126.66	nr	136.16
997 x 2056mm sidelights, NA13F	195.72	2.50	200.61	1.80	0.90	22.93	-	223.54	nr	240.31
METAL DOORS - Garage doors; Birtley Durham Type in galvanised steel; one coat primer by manufacturer										
Overhead garage doors; tensioning device fixing to timber with screws; for opening size										
2135 x 1980mm	230.00	2.50	235.75	4.00	4.00	65.00	-	300.75	nr	323.31
2135 x 2135mm	242.00	2.50	248.05	4.00	4.00	65.00	-	313.05	nr	336.53
Overhead garage doors; counterbalanced by springs										
4270 x 1980mm	690.00	2.50	707.25	6.00	6.00	97.50	-	804.75	nr	865.11
4270 x 2135mm	690.00	2.50	707.25	6.00	6.00	97.50	-	804.75	nr	865.11
METAL DOORS - Patio doors in aluminium, with wrought hardwood subframes										
Double glazed clear toughened glass, Magnet Trade Magnastar, one opening leaf, fixing subframe to masonry with screws										
2085 x 1501mm overall, Bp5 white	416.58	2.50	427.00	6.00	6.00	97.50	-	524.50	nr	563.84
2085 x 1805mm overall, Bp6 white	533.28	2.50	546.61	6.00	6.00	97.50	-	644.11	nr	692.42
2085 x 2086mm overall, Bp7 white	593.08	2.50	607.91	6.00	6.00	97.50	-	705.41	nr	758.32
2085 x 2387mm overall, Bp8 white	702.38	2.50	719.94	6.00	6.00	97.50	-	817.44	nr	878.75

WINDOWS/DOORS/STAIRS

Labour hourly rates: (except Specialists) Craft Operatives £9.23 Labourer £7.02 Rates are national average prices. Refer to REGIONAL VARIATIONS for indicative levels of overall pricing in regions	MATERIALS			LABOUR				RATES		
	Del to Site £	Waste %	Material Cost £	Craft Optve Hrs	Lab Hrs	Labour Cost £	Sunds £	Nett Rate £	Unit	Gross Rate (+7.5%) £
L20: DOORS/SHUTTERS/HATCHES Cont.										
METAL DOORS - Patio doors and frames in aluminium										
Double glazed clear toughened glass, Magnet Trade Magnaplus, one opening leaf, fixing subframe to masonry with screws										
2090 x 1501mm overall, THP5 white	500.69	2.50	513.21	6.00	6.00	97.50	-	610.71	nr	656.13
2090 x 1805mm overall, THP6 white	572.33	2.50	586.64	6.00	6.00	97.50	-	684.14	nr	735.45
2090 x 2086mm overall, THP7 white	619.31	2.50	634.79	6.00	6.00	97.50	-	732.29	nr	787.21
2090 X 2387 overall, THP8 white	681.95	2.50	699.00	6.00	6.00	97.50	-	796.50	nr	856.24
extra; former for THP5	80.25	2.50	82.26	-	-	-	-	82.26	nr	88.43
extra; former for THP6	82.20	2.50	84.26	-	-	-	-	84.26	nr	90.58
extra; former for THP7	85.10	2.50	87.23	-	-	-	-	87.23	nr	93.77
extra; former for THP8	91.48	2.50	93.77	-	-	-	-	93.77	nr	100.80
ROLLER SHUTTERS - Roller shutters in steel, galvanised										
Note roller shutters are always purpose made to order and the following prices are indicative only. Firm quotations should always be obtained										
Crank handle operation										
4572 x 3658mm overall; fixing to masonry with screws	1200.00	2.50	1230.00	16.00	16.00	260.00	-	1490.00	nr	1601.75
Endless hand chain operation										
5486 x 4267mm overall; fixing to masonry with screws	1405.00	2.50	1440.13	20.00	20.00	325.00	-	1765.13	nr	1897.51
Electric motor operation										
6096 x 6096mm overall; fixing to masonry with screws	2880.00	2.50	2952.00	30.00	30.00	487.50	-	3439.50	nr	3697.46
ROLLER SHUTTERS - Roller shutters in aluminium										
Endless hand chain operation										
3553 x 3048mm overall; fixing to masonry with screws	1135.00	2.50	1163.38	20.00	20.00	325.00	-	1488.38	nr	1600.00
ROLLER SHUTTERS - Bedding and pointing frames										
Pointing with Secomastic standard mastic										
metal frames one side	0.84	5.00	0.88	0.12	0.06	1.53	-	2.41	m	2.59
Bedding in cement mortar (1:3); pointing with Secomastic standard mastic one side										
metal frames one side	0.96	5.00	1.01	0.12	0.10	1.81	-	2.82	m	3.03
PLASTICS/RUBBER DOORS - Flexible doors mainly in plastics and rubber										
Mancuna standard doors; Mandor Engineering Ltd; 43mm diameter steel tube frame; top and bottom plates with pivoting arrangements and spring unit; 8mm rubber panel doors with triangular perspex windows in each leaf										
fixing to masonry; for opening size										
1800 x 2100mm..................................	836.00	2.50	856.90	4.00	4.00	65.00	-	921.90	nr	991.04
2440 x 2400mm..................................	972.00	2.50	996.30	5.00	5.00	81.25	-	1077.55	nr	1158.37
Mancuna standard doors; Mandor Engineering Ltd; 43mm diameter steel tube frame; top and bottom plates with pivoting arrangements and spring unit; 12mm rubber panel doors with triangular perspex windows in each leaf										
fixing to masonry; for opening size										
1800 x 2100mm..................................	972.00	2.50	996.30	4.50	4.50	73.13	-	1069.43	nr	1149.63
2440 x 2400mm..................................	1206.00	2.50	1236.15	5.50	5.50	89.38	-	1325.53	nr	1424.94
Mancuna standard doors; Mandor Engineering Ltd; 43mm diameter steel tube frame; top and bottom plates with pivoting arrangements and spring unit; 6mm clear flexible panel doors										
fixing to masonry; for opening size										
1800 x 2100mm..................................	634.00	2.50	649.85	3.50	3.50	56.88	-	706.73	nr	759.73
2440 x 2400mm..................................	697.00	2.50	714.42	4.00	4.00	65.00	-	779.42	nr	837.88
Mancuna standard doors; Mandor Engineering Ltd; 43mm diameter steel tube frame; top and bottom plates with pivoting arrangements and spring unit; 9mm clear flexible panel doors										
fixing to masonry; for opening size										
1800 x 2100mm..................................	728.00	2.50	746.20	3.75	3.75	60.94	-	807.14	nr	867.67
2440 x 2400mm..................................	822.00	2.50	842.55	4.25	4.25	69.06	-	911.61	nr	979.98
Mandor heavy duty doors; Mandor Engineering Ltd; 63mm diameter steel tube frame; top and bottom plates with pivoting arrangements vand spring unit; 12mm rubber panel doors with oval panel windows in each leaf										
fixing to masonry; for opening size										
2440 x 3050mm..................................	1529.00	2.50	1567.22	8.00	8.00	130.00	-	1697.22	nr	1824.52
2740 x 3350mm..................................	1851.00	2.50	1897.28	9.00	9.00	146.25	-	2043.53	nr	2196.79
3050 x 3660mm..................................	1960.00	2.50	2009.00	10.00	10.00	162.50	-	2171.50	nr	2334.36

Labour hourly rates: (except Specialists) Craft Operatives £9.23 Labourer £7.02 Rates are national average prices. Refer to REGIONAL VARIATIONS for indicative levels of overall pricing in regions	MATERIALS			LABOUR				RATES		
	Del to Site	Waste	Material Cost	Craft Optve	Lab	Labour Cost	Sunds	Nett Rate	Unit	Gross Rate (+7.5%)
	£	%	£	Hrs	Hrs	£	£	£		£
L20: DOORS/SHUTTERS/HATCHES Cont.										
PLASTICS/RUBBER DOORS - Flexible doors mainly in plastics and rubber Cont.										
Mandor heavy duty doors; Mandor Engineering Ltd; 63mm diameter steel tube frame; top and bottom plates with pivoting arrangements and spring unit; 9mm clear flexible panel doors										
fixing to masonry; for opening size										
2440 x 3050mm	1024.00	2.50	1049.60	6.00	6.00	97.50	-	1147.10	nr	1233.13
2740 x 3350mm	1170.00	2.50	1199.25	7.00	7.00	113.75	-	1313.00	nr	1411.47
3050 x 3660mm	1347.00	2.50	1380.68	8.00	8.00	130.00	-	1510.68	nr	1623.98
Manby aluminium doors; Mandor Engineering Ltd; extruded aluminium frame; pivot arrangement and spring unit; clear pvc panel doors										
fixing to masonry; for opening size										
1800 x 2100mm	692.00	2.50	709.30	3.75	3.75	60.94	-	770.24	nr	828.01
2440 x 2400mm	759.00	2.50	777.98	4.75	4.75	77.19	-	855.16	nr	919.30
Manby aluminium doors; Mandor Engineering Ltd; extruded aluminium frame; pivot arrangement and spring unit; half coloured half clear pvc panel doors										
fixing to masonry; for opening size										
1800 x 2100mm	759.00	2.50	777.98	3.75	3.75	60.94	-	838.91	nr	901.83
2440 x 2400mm	832.00	2.50	852.80	4.75	4.75	77.19	-	929.99	nr	999.74
PLASTICS/RUBBER DOORS - Flexible strip curtains mainly in pvc										
Mandor strip curtains; Mandor Engineering Ltd; 200mm wide x 2.0mm thick clear pvc strip curtains, suspended from 80 x 80 x 6mm steel angle; fixing angle to masonry with bolts										
double overlap; for opening size										
1800 x 2100mm	103.00	2.50	105.58	2.00	2.00	32.50	-	138.07	nr	148.43
2100 x 2440mm	141.00	2.50	144.53	2.30	2.30	37.38	-	181.90	nr	195.54
single overlap; for opening size										
1800 x 2100mm	83.00	2.50	85.08	1.80	1.80	29.25	-	114.33	nr	122.90
2100 x 2440mm	113.00	2.50	115.83	2.10	2.10	34.13	-	149.95	nr	161.20
Mandor strip curtains; Mandor Engineering Ltd; 300mm wide x 2.5mm thick clear pvc strip curtains, suspended from 80 x 80 x 6mm steel angle; fixing angle to masonry with bolts										
double overlap; for opening size										
2440 x 2740mm	200.00	2.50	205.00	2.64	2.64	42.90	-	247.90	nr	266.49
2740 x 2900mm	237.00	2.50	242.93	2.94	2.94	47.77	-	290.70	nr	312.50
single overlap; for opening size										
2440 x 2740mm	154.00	2.50	157.85	2.44	2.44	39.65	-	197.50	nr	212.31
2740 x 2900mm	184.00	2.50	188.60	2.74	2.74	44.52	-	233.13	nr	250.61
Wavespan 20T strip curtains; Mandor Engineering Ltd.; 305mm wide overlapping nylon reinforced clear pvc strips, stapled to 50 x 50mm softwood batten										
fixing batten to masonry with screws; for opening size										
1000 x 2440mm	67.00	2.50	68.67	1.25	1.25	20.31	-	88.99	nr	95.66
1000 x 2740mm	76.00	2.50	77.90	1.50	1.50	24.38	-	102.28	nr	109.95
1000 x 3050mm	84.00	2.50	86.10	1.75	1.75	28.44	-	114.54	nr	123.13
2000 x 2440mm	133.00	2.50	136.32	2.50	2.50	40.63	-	176.95	nr	190.22
2000 x 2740mm	151.00	2.50	154.78	2.75	2.75	44.69	-	199.46	nr	214.42
2000 x 3050mm	164.00	2.50	168.10	3.00	3.00	48.75	-	216.85	nr	233.11
3000 x 2440mm	202.00	2.50	207.05	4.00	4.00	65.00	-	272.05	nr	292.45
3000 x 2740mm	227.00	2.50	232.68	4.50	4.50	73.13	-	305.80	nr	328.74
3000 x 3050mm	248.00	2.50	254.20	4.00	4.00	65.00	-	319.20	nr	343.14
L30: STAIRS/WALKWAYS/BALUSTRADES										
TIMBER STAIRS - Stairs in softwood, wrought										
Straight flight staircase and balustrade; 25mm treads; 19mm risers, glued; wedged and blocked; 25mm wall string; 38mm outer string; 75 x 75mm newels; 25 x 25mm balusters; 38 x 75mm hardwood handrail										
864mm wide x 2600mm rise overall; balustrade to one side; fixing to masonry with screws	635.00	2.50	650.88	12.00	1.50	121.29	4.25	776.41	nr	834.65
864mm wide x 2600mm rise overall with 3 nr winders at bottom; balustrade to one side; fixing to masonry with screws	762.00	2.50	781.05	16.00	2.00	161.72	4.50	947.27	nr	1018.32
864mm wide x 2600mm rise overall with 3 nr winders at top; balustrade to one side; fixing to masonry with screws	762.00	2.50	781.05	16.00	2.00	161.72	4.50	947.27	nr	1018.32
TIMBER STAIRS - Balustrades in softwood, wrought										
Isolated balustrades; 25 x 25mm balusters at 150mm centres, housed construction (handrail included elsewhere)										
914mm high; fixing to timber with screws	33.35	2.50	34.18	0.65	0.08	6.56	0.11	40.85	m	43.92
Isolated balustrades; 38 x 38mm balusters at 150mm centres, housed construction (handrail included elsewhere)										
914mm high; fixing to timber with screws	48.88	2.50	50.10	0.95	0.12	9.61	0.15	59.86	m	64.35

WINDOWS/DOORS/STAIRS

Labour hourly rates: (except Specialists) Craft Operatives £9.23 Labourer £7.02 Rates are national average prices. Refer to REGIONAL VARIATIONS for indicative levels of overall pricing in regions	MATERIALS			LABOUR				RATES		
	Del to Site £	Waste %	Material Cost £	Craft Optve Hrs	Lab Hrs	Labour Cost £	Sunds £	Nett Rate £	Unit	Gross Rate (+7.5%) £
L30: STAIRS/WALKWAYS/BALUSTRADES Cont.										
TIMBER STAIRS - Balustrades in softwood, wrought Cont.										
Isolated balustrades; 50 x 50mm balusters at 150mm centres, housed construction (handrail included elsewhere)										
914mm high; fixing to timber with screws	64.22	2.50	65.83	1.10	0.14	11.14	0.18	77.14	m	82.93
Isolated balustrades; 25 x 25mm balusters at 150mm centres, 50mm mopstick handrail, housed construction										
914mm high; fixing to timber with screws	55.90	2.50	57.30	1.00	0.12	10.07	0.16	67.53	m	72.59
extra; ramps	34.31	2.50	35.17	0.90	0.11	9.08	0.15	44.40	nr	47.73
extra; wreaths	68.61	2.50	70.33	1.80	0.23	18.23	0.29	88.84	nr	95.51
extra; bends	34.31	2.50	35.17	0.60	0.08	6.10	0.10	41.37	nr	44.47
Isolated balustrades; 38 x 38mm balusters at 150mm centres, 50mm mopstick handrail, housed construction										
914mm high; fixing to timber with screws	71.15	2.50	72.93	1.30	0.16	13.12	0.21	86.26	m	92.73
extra; ramps	34.31	2.50	35.17	0.90	0.11	9.08	0.15	44.40	nr	47.73
extra; wreaths	68.61	2.50	70.33	1.80	0.23	18.23	0.29	88.84	nr	95.51
extra; bends	34.31	2.50	35.17	0.60	0.08	6.10	0.10	41.37	nr	44.47
Isolated balustrades; 50 x 50mm balusters at 150mm centres, 50mm mopstick handrail, housed construction										
914mm high; fixing to timber with screws	86.39	2.50	88.55	1.45	0.18	14.65	0.24	103.44	m	111.19
extra; ramps	34.31	2.50	35.17	0.90	0.11	9.08	0.15	44.40	nr	47.73
extra; wreaths	68.61	2.50	70.33	1.80	0.23	18.23	0.29	88.84	nr	95.51
extra; bends	34.31	2.50	35.17	0.60	0.08	6.10	0.10	41.37	nr	44.47
Isolated balustrades; 25 x 25mm balusters at 150mm centres, 75 x 100mm moulded handrail; housed construction										
914mm high; fixing to timber with screws	68.61	2.50	70.33	1.05	0.13	10.60	0.17	81.10	m	87.18
extra; ramps	50.82	2.50	52.09	2.00	0.25	20.22	0.33	72.64	nr	78.08
extra; wreaths	101.64	2.50	104.18	4.00	0.50	40.43	0.65	145.26	nr	156.16
extra; bends	50.82	2.50	52.09	1.20	0.15	12.13	0.20	64.42	nr	69.25
Isolated balustrades; 38 x 38mm balusters at 150mm centres, 75 x 100mm moulded handrail; housed construction										
914mm high; fixing to timber with screws	83.85	2.50	85.95	1.35	0.17	13.65	0.22	99.82	m	107.31
extra; ramps	50.82	2.50	52.09	2.00	0.25	20.22	0.33	72.64	nr	78.08
extra; wreaths	101.64	2.50	104.18	4.00	0.50	40.43	0.65	145.26	nr	156.16
extra; bends	50.82	2.50	52.09	1.20	0.15	12.13	0.20	64.42	nr	69.25
Isolated balustrades; 50 x 50mm balusters at 150mm centres, 75 x 100mm moulded handrail; housed construction										
914mm high; fixing to timber with screws	99.10	2.50	101.58	1.50	0.19	15.18	0.24	117.00	m	125.77
extra; ramps	50.82	2.50	52.09	2.00	0.25	20.22	0.33	72.64	nr	78.08
extra; wreaths	101.64	2.50	104.18	4.00	0.50	40.43	0.65	145.26	nr	156.16
extra; bends	50.82	2.50	52.09	1.20	0.15	12.13	0.20	64.42	nr	69.25
TIMBER STAIRS - Balustrades in European Oak, wrought, selected for transparent finish										
Isolated balustrades; 25 x 25mm balusters at 150mm centres, 50mm mopstick handrail, housed construction										
914mm high; fixing to timber with screws	111.80	2.50	114.60	1.70	0.21	17.17	0.28	132.04	m	141.94
extra; ramps	68.62	2.50	70.34	1.50	0.19	15.18	0.24	85.75	nr	92.19
extra; wreaths	137.21	2.50	140.64	3.00	0.38	30.36	0.49	171.49	nr	184.35
extra; bends	68.62	2.50	70.34	1.10	0.14	11.14	0.18	81.65	nr	87.78
Isolated balustrades; 38 x 38mm balusters at 150mm centres, 50mm mopstick handrail, housed construction										
914mm high; fixing to timber with screws	142.30	2.50	145.86	2.15	0.27	21.74	0.35	167.95	m	180.54
extra; ramps	68.62	2.50	70.34	1.50	0.19	15.18	0.24	85.75	nr	92.19
extra; wreaths	137.21	2.50	140.64	3.00	0.38	30.36	0.49	171.49	nr	184.35
extra; bends	68.62	2.50	70.34	1.10	0.14	11.14	0.18	81.65	nr	87.78
Isolated balustrades; 50 x 50mm balusters at 150mm centres, 50mm mopstick handrail, housed construction										
914mm high; fixing to timber with screws	172.79	2.50	177.11	2.50	0.31	25.25	0.41	202.77	m	217.98
extra; ramps	68.62	2.50	70.34	1.50	0.19	15.18	0.24	85.75	nr	92.19
extra; wreaths	137.21	2.50	140.64	3.00	0.38	30.36	0.49	171.49	nr	184.35
extra; bends	68.62	2.50	70.34	1.10	0.14	11.14	0.18	81.65	nr	87.78
Isolated balustrades; 25 x 25mm balusters at 150mm centres, 75 x 100mm moulded handrail; housed construction										
914mm high; fixing to timber with screws	137.21	2.50	140.64	1.80	0.23	18.23	0.29	159.16	m	171.10
extra; ramps	101.64	2.50	104.18	3.10	0.39	31.35	0.50	136.03	nr	146.23
extra; wreaths	203.28	2.50	208.36	6.20	0.78	62.70	1.00	272.06	nr	292.47
extra; bends	101.64	2.50	104.18	2.15	0.27	21.74	0.35	126.27	nr	135.74
Isolated balustrades; 38 x 38mm balusters at 150mm centres, 75 x 100mm moulded handrail; housed construction										
914mm high; fixing to timber with screws	167.71	2.50	171.90	2.25	0.28	22.73	0.37	195.01	m	209.63
extra; ramps	101.64	2.50	104.18	3.10	0.39	31.35	0.50	136.03	nr	146.23
extra; wreaths	203.28	2.50	208.36	6.20	0.78	62.70	1.00	272.06	nr	292.47
extra; bends	101.64	2.50	104.18	2.15	0.27	21.74	0.35	126.27	nr	135.74

Labour hourly rates: (except Specialists) Craft Operatives £9.23 Labourer £7.02 Rates are national average prices. Refer to REGIONAL VARIATIONS for indicative levels of overall pricing in regions	MATERIALS			LABOUR				RATES		
	Del to Site £	Waste %	Material Cost £	Craft Optve Hrs	Lab Hrs	Labour Cost £	Sunds £	Nett Rate £	Unit	Gross Rate (+7.5%) £
L30: STAIRS/WALKWAYS/BALUSTRADES Cont.										
TIMBER STAIRS - Balustrades in European Oak, wrought, selected for transparent finish Cont.										
Isolated balustrades; 50 x 50mm balusters at 150mm centres, 75 x 100mm moulded handrail; housed construction										
914mm high; fixing to timber with screws	198.20	2.50	203.16	2.60	0.33	26.31	0.42	229.89	m	247.13
extra; ramps	101.64	2.50	104.18	3.10	0.39	31.35	0.50	136.03	nr	146.23
extra; wreaths	203.28	2.50	208.36	6.20	0.78	62.70	1.00	272.06	nr	292.47
extra; bends	101.64	2.50	104.18	2.15	0.27	21.74	0.35	126.27	nr	135.74
TIMBER STAIRS - Handrails in softwood, wrought										
Associated handrails										
50mm mopstick; fixing through metal backgrounds with screws	19.06	5.00	20.01	0.35	0.04	3.51	0.06	23.58	m	25.35
extra; ramps	34.31	5.00	36.03	1.00	0.13	10.14	0.16	46.33	nr	49.80
extra; wreaths	68.61	5.00	72.04	2.00	0.25	20.22	0.33	92.59	nr	99.53
extra; bends	34.31	5.00	36.03	1.00	0.13	10.14	0.16	46.33	nr	49.80
75 x 100mm; moulded; fixing through metal backgrounds with screws	31.77	5.00	33.36	0.40	0.05	4.04	0.07	37.47	m	40.28
extra; ramps	50.82	5.00	53.36	1.20	0.15	12.13	0.20	65.69	nr	70.62
extra; wreaths	101.64	5.00	106.72	2.40	0.30	24.26	0.39	131.37	nr	141.22
extra; bends	50.82	5.00	53.36	1.20	0.15	12.13	0.20	65.69	nr	70.62
TIMBER STAIRS - Handrails in African Mahogany, wrought, selected for transparent finish										
Associated handrails										
50mm mopstick; fixing through metal backgrounds with screws	23.83	5.00	25.02	0.50	0.06	5.04	0.08	30.14	m	32.40
extra; ramps	42.89	5.00	45.03	1.50	0.19	15.18	0.24	60.45	nr	64.99
extra; wreaths	85.76	5.00	90.05	3.00	0.38	30.36	0.49	120.90	nr	129.96
extra; bends	42.89	5.00	45.03	1.50	0.19	15.18	0.24	60.45	nr	64.99
75 x 100mm; moulded; fixing through metal backgrounds with screws	39.71	5.00	41.70	0.60	0.08	6.10	0.10	47.90	m	51.49
extra; ramps	63.53	5.00	66.71	1.80	0.23	18.23	0.29	85.23	nr	91.62
extra; wreaths	127.05	5.00	133.40	3.60	0.45	36.39	0.59	170.38	nr	183.16
extra; bends	63.53	5.00	66.71	1.80	0.73	21.74	0.36	88.81	nr	95.47
TIMBER STAIRS - Handrails in European Oak, wrought, selected for transparent finish										
Associated handrails										
50mm mopstick; fixing through metal backgrounds with screws	38.13	5.00	40.04	0.50	0.06	5.04	0.08	45.15	m	48.54
extra; ramps	68.62	5.00	72.05	1.50	0.19	15.18	0.24	87.47	nr	94.03
extra; wreaths	137.21	5.00	144.07	3.00	0.38	30.36	0.49	174.92	nr	188.04
extra; bends	68.62	5.00	72.05	1.50	0.19	15.18	0.24	87.47	nr	94.03
75 x 100mm; moulded; fixing through metal backgrounds with screws	63.54	5.00	66.72	0.60	0.08	6.10	0.10	72.92	m	78.39
extra; ramps	101.64	5.00	106.72	1.80	0.23	18.23	0.29	125.24	nr	134.63
extra; wreaths	203.28	5.00	213.44	3.60	0.45	36.39	0.59	250.42	nr	269.20
extra; bends	101.64	5.00	106.72	1.80	0.23	18.23	0.36	125.31	nr	134.71
METAL STAIRS AND BALUSTRADES - Staircases in steel										
Straight flight staircases; 180 x 10mm flat stringers, shaped ends to top and bottom; 6 x 250mm on plain raised pattern plate treads, with 50 x 50 x 6mm shelf angles and 40 x 40 x 6mm stiffening bars, bolted to stringers; welded, cleated and bolted connections										
770mm wide x 3000mm going x 2600mm rise overall; fixing to masonry with 4 nr Rawlbolts	-	-	Spclist	-	-	Spclist	-	1495.00	nr	1607.13
920mm wide x 3000mm going x 2600mm rise overall; fixing to masonry with 4 nr Rawlbolts	-	-	Spclist	-	-	Spclist	-	1710.00	nr	1838.25
Straight flight staircases, 180 x 10mm flat stringers, shaped ends to top and bottom; 6 x 250mm tray treads, with three 6mm diameter reinforcing bars welded to inside, and 50 x 50 x 6mm shelf angles, bolted to stringers; welded, cleated and bolted connections										
770mm wide x 3000mm going x 2600mm rise overall; fixing to masonry with 4 nr Rawlbolts	-	-	Spclist	-	-	Spclist	-	1415.00	nr	1521.13
920mm wide x 3000mm going x 2600mm rise overall; fixing to masonry with 4 nr Rawlbolts	-	-	Spclist	-	-	Spclist	-	1460.00	nr	1569.50
Straight flight staircases; 178 x 76mm channel stringers, shaped ends to top and bottom; 6 x 250mm on plain raised pattern plate treads, with 50 x 50 x 6mm shelf angles and 40 x 40 x 6mm stiffening bars, bolted to stringers; welded, cleated and bolted connections										
770mm wide x 3000mm going x 2600mm rise overall; fixing to masonry with 4 nr Rawlbolts	-	-	Spclist	-	-	Spclist	-	1625.00	nr	1746.88
920mm wide x 3000mm going x 2600mm rise overall; fixing to masonry with 4 nr Rawlbolts	-	-	Spclist	-	-	Spclist	-	1850.00	nr	1988.75

WINDOWS/DOORS/STAIRS

239

Labour hourly rates: (except Specialists) Craft Operatives £9.23 Labourer £7.02 Rates are national average prices. Refer to REGIONAL VARIATIONS for indicative levels of overall pricing in regions	MATERIALS			LABOUR				RATES		
	Del to Site £	Waste %	Material Cost £	Craft Optve Hrs	Lab Hrs	Labour Cost £	Sunds £	Nett Rate £	Unit	Gross Rate (+7.5%) £
L30: STAIRS/WALKWAYS/BALUSTRADES Cont.										
METAL STAIRS AND BALUSTRADES - Staircases in steel Cont.										
Straight flight staircases and balustrades; 180 x 10mm flat stringers, shaped ends to top and bottom; 6 x 250mm on plain raised pattern plate treads with 50 x 50 x 6mm shelf angles and 40 x 40 x 6mm stiffening bars, bolted to stringers; 915mm high balustrade to both sides consisting of 25mm diameter solid bar handrail and 32mm diameter solid bar standards at 250mm centres with base plate welded on and bolted to face of stringer, and ball type joints at intersections; welded, cleated and bolted connections 770mm wide x 3000mm rise overall; fixing to masonry with 4 nr Rawlbolts	-	-	Spclist	-	-	Spclist	-	2235.00	nr	2402.63
Straight flight staircases and balustrades; 180 x 10mm flat stringers, shaped ends to top and bottom; 6 x 250mm on plain raised pattern plate treads with 50 x 50 x 6mm shelf angles and 40 x 40 x 6mm stiffening bars, bolted to stringers; 915mm high balustrade to both sides consisting of 25mm diameter solid bar handrail and 32mm diameter solid bar standards at 250mm centres with base plate welded on and bolted to top of stringer, and ball type joints at intersections; welded, cleated and bolted connections 770mm wide x 3000mm going x 2600mm rise overall; fixing to masonry with 4 nr Rawlbolts	-	-	Spclist	-	-	Spclist	-	2185.00	nr	2348.88
Straight flight staircases and balustrades; 180 x 10mm flat stringers, shaped ends to top and bottom; 6 x 250mm on plain raised pattern plate treads with 50 x 50 x 6mm shelf angles and 40 x 40 x 6mm stiffening bars, bolted to stringers; 915mm high balustrade to one side consisting of 25mm diameter solid bar handrail and 32mm diameter solid bar standards at 250mm centres with base plate welded on and bolted to face of stringer, and ball type joints at intersections; welded, cleated and bolted connections 770mm wide x 3000mm going x 2600mm rise overall; fixing to masonry with 4 nr Rawlbolts	-	-	Spclist	-	-	Spclist	-	2035.00	nr	2187.63
Straight flight staircases and balustrades; 180 x 10mm flat stringers, shaped ends to top and bottom; 6 x 250mm on plain raised pattern plate treads with 50 x 50 x 6mm shelf angles and 40 x 40 x 6mm stiffening bars, bolted to stringers; 915mm high balustrade to one side consisting of 25mm diameter solid bar handrail and 32mm diameter solid bar standards at 250mm centres with base plate welded on and bolted to top of stringer, and ball type joints at intersections; welded, cleated and bolted connections 770mm wide x 3000mm going x 2600mm rise overall; fixing to masonry with 4 nr Rawlbolts	-	-	Spclist	-	-	Spclist	-	2035.00	nr	2187.63
Straight flight staircases and balustrades; 180 x 10mm flat stringers, shaped ends to top and bottom; 6 x 250mm on plain raised pattern plate treads with 50 x 50 x 6mm shelf angles and 40 x 40 x 6mm stiffening bars, bolted to stringers; 1070mm high balustrade to both sides consisting of 25mm diameter solid bar handrail and 32mm diameter solid bar standards at 250mm centres with base plate welded on and bolted to top of stringer, and ball type joints at intersections; welded, cleated and bolted connections 770mm wide x 3000mm going x 2600mm rise overall; fixing to masonry with 4 nr Rawlbolts	-	-	Spclist	-	-	Spclist	-	2235.00	nr	2402.63
Straight flight staircases and balustrades; 180 x 10mm flat stringers, shaped ends to top and bottom; 6 x 250mm on plain raised pattern plate treads with 50 x 50 x 6mm shelf angles and 40 x 40 x 6mm stiffening bars, bolted to stringers; 1070mm high balustrade to one side consisting of 25mm diameter solid bar handrail and 32mm diameter solid bar standards at 250mm centres with base plate welded on and bolted to top of stringer, and ball type joints at intersections; welded, cleated and bolted connections 770mm wide x 3000mm going x 2600mm rise overall; fixing to masonry with 4 nr Rawlbolts	-	-	Spclist	-	-	Spclist	-	2090.00	nr	2246.75
Straight flight staircases and balustrades; 180 x 10mm flat stringers, shaped ends to top and bottom; 6 x 250mm on plain raised pattern plate treads with 50 x 50 x 6mm shelf angles and 40 x 40 x 6mm stiffening bars, bolted to stringers; 915mm high balustrade to one side consisting of 25mm diameter solid bar handrail and intermediate rail, and 32mm diameter solid bar standards at 250mm centres with base plate welded on and bolted to top of stringer, and ball type joints at intersections 770mm wide x 3000mm going x 2600mm rise overall; fixing to masonry with 4 nr Rawlbolts	-	-	Spclist	-	-	Spclist	-	2035.00	nr	2187.63

Labour hourly rates: (except Specialists) Craft Operatives £9.23 Labourer £7.02 Rates are national average prices. Refer to REGIONAL VARIATIONS for indicative levels of overall pricing in regions	MATERIALS			LABOUR				RATES		
	Del to Site £	Waste %	Material Cost £	Craft Optve Hrs	Lab Hrs	Labour Cost £	Sunds £	Nett Rate £	Unit	Gross Rate (+7.5%) £
L30: STAIRS/WALKWAYS/BALUSTRADES Cont.										
METAL STAIRS AND BALUSTRADES - Staircases in steel Cont.										
Straight flight staircases and balustrades; 180 x 10mm flat stringers, shaped ends to top and bottom; 6 x 250mm on plain raised pattern plate treads with 50 x 50 x 6mm shelf angles and 40 x 40 x 6mm stiffening bars, bolted to stringers; 1070mm high balustrade to one side consisting of 25mm diameter solid bar handrail and intermediate rail, and 32mm diameter solid bar standards at 250mm centres with base plate welded on and bolted to top of stringer, and ball type joints at intersections										
770mm wide x 3000mm going x 2600mm rise overall; fixing to masonry with 4 nr Rawlbolts	-	-	Spclist	-	-	Spclist	-	2095.00	nr	2252.13
Quarter landing staircases, in two flights; 180 x 10mm flat stringers, shaped ends to top and bottom; 6 x 250mm on plain raised pattern plate treads, with 50 x 50 x 6mm shelf angles and 40 x 40 x 6mm stiffening bars, bolted to stringers; 6mm on plain raised pattern plate landing, welded on; 100 x 8mm flat kicking plates welded on; welded, cleated and bolted connections										
770mm wide x 2000mm going first flight excluding landing x 1000mm going second flight excluding landing x 2600mm rise overall; 770 x 770mm landing overall; fixing to masonry with 4 nr Rawlbolts ...	-	-	Spclist	-	-	Spclist	-	3335.00	nr	3585.13
920mm wide x 2000mm going first flight excluding landing x 1000mm going second flight excluding landing x 2600mm rise overall; 920 x 920mm landing overall; fixing to masonry with 4 nr Rawlbolts ...	-	-	Spclist	-	-	Spclist	-	3505.00	nr	3767.88
Half landing staircases, in two flights; 180 x 10mm flat stringers, shaped ends to top and bottom; 6 x 250mm on plain raised pattern plate treads, with 50 x 50 x 6mm shelf angles and 40 x 40 x 6mm stiffening bars, bolted to stringers; 6mm on plain raised pattern plate landing, welded on; 100 x 8mm flat kicking plates welded on; welded, cleated and bolted connections										
770mm wide x 2000mm going first flight excluding landing x 1000mm going second flight excluding landing x 2600mm rise overall; 770 x 1640mm landing overall; fixing to masonry with 4 nr Rawlbolts	-	-	Spclist	-	-	Spclist	-	3750.00	nr	4031.25
920mm wide x 2000mm going first flight excluding landing x 1000mm going second flight excluding landing x 2600mm rise overall; 920 x 1940mm landing overall; fixing to masonry with 4 nr Rawlbolts	-	-	Spclist	-	-	Spclist	-	4245.00	nr	4563.38
METAL STAIRS AND BALUSTRADES - Balustrades in steel										
Isolated balustrades; 6 x 38mm flat core rail; 13mm diameter balusters at 250mm centres; welded fabrication ground to smooth finish; casting into mortices in concrete; wedging in position; temporary wedges										
838mm high; level	101.30	2.50	103.83	1.00	0.50	12.74	-	116.57	m	125.32
extra; ramps	17.27	2.50	17.70	1.00	0.50	12.74	-	30.44	nr	32.72
extra; wreaths	28.86	2.50	29.58	2.00	1.00	25.48	-	55.06	nr	59.19
extra; bends	15.69	2.50	16.08	1.00	0.50	12.74	-	28.82	nr	30.98
838mm high; raking	104.53	2.50	107.14	1.25	0.65	16.10	-	123.24	m	132.49
914mm high; level	111.35	2.50	114.13	1.00	0.50	12.74	-	126.87	m	136.39
extra; ramps	17.27	2.50	17.70	1.00	0.50	12.74	-	30.44	nr	32.72
extra; wreaths	28.86	2.50	29.58	2.00	1.00	25.48	-	55.06	nr	59.19
extra; bends	15.70	2.50	16.09	1.00	0.50	12.74	-	28.83	nr	30.99
914mm high; raking	114.71	2.50	117.58	1.25	0.65	16.10	-	133.68	m	143.70
Isolated balustrades; 6 x 38mm flat core rail; 13 x 13mm balusters at 250mm centres; welded fabrication ground to smooth finish; casting into mortices in concrete; wedging in position; temporary wedges										
838mm high; level	105.95	2.50	108.60	1.00	0.50	12.74	-	121.34	m	130.44
838mm high; raking	109.31	2.50	112.04	1.00	0.50	12.74	-	124.78	m	134.14
914mm high; level	114.71	2.50	117.58	1.00	0.50	12.74	-	130.32	m	140.09
914mm high; raking	121.85	2.50	124.90	1.00	0.50	12.74	-	137.64	m	147.96
Isolated balustrades; 13 x 51mm rounded handrail; 13mm diameter balusters at 250mm centres; welded fabrication ground to smooth finish; casting into mortices in concrete; wedging in position; temporary wedges										
838mm high; level	121.38	2.50	124.41	1.00	0.50	12.74	-	137.15	m	147.44
extra; ramps	18.84	2.50	19.31	1.00	0.50	12.74	-	32.05	nr	34.45
extra; wreaths	31.87	2.50	32.67	2.00	1.00	25.48	-	58.15	nr	62.51
extra; bends	17.27	2.50	17.70	1.00	0.50	12.74	-	30.44	nr	32.72
838mm high; raking	124.79	2.50	127.91	1.00	0.50	12.74	-	140.65	m	151.20
914mm high; level	129.94	2.50	133.19	1.00	0.50	12.74	-	145.93	m	156.87
extra; ramps	18.84	2.50	19.31	1.00	0.50	12.74	-	32.05	nr	34.45
extra; wreaths	31.88	2.50	32.68	2.00	1.00	25.48	-	58.16	nr	62.52
extra; bends	17.27	2.50	17.70	1.00	0.50	12.74	-	30.44	nr	32.72
914mm high; raking	135.50	2.50	138.89	1.00	0.50	12.74	-	151.63	m	163.00

WINDOWS/DOORS/STAIRS

Labour hourly rates: (except Specialists)
Craft Operatives £9.23 Labourer £7.02
Rates are national average prices.
Refer to REGIONAL VARIATIONS for indicative levels of overall pricing in regions

	MATERIALS			LABOUR				RATES		
	Del to Site £	Waste %	Material Cost £	Craft Optve Hrs	Lab Hrs	Labour Cost £	Sunds £	Nett Rate £	Unit	Gross Rate (+7.5%) £
L30: STAIRS/WALKWAYS/BALUSTRADES Cont.										
METAL STAIRS AND BALUSTRADES - Balustrades in steel Cont.										
Isolated balustrades; 6 x 38mm flat core rail; 10 x 51mm flat bottom rail; 13mm diameter infill balusters at 250mm centres; 25 x 25mm standards at 3000mm centres; welded fabrication ground to smooth finish; casting into mortices in concrete; wedging in position; temporary wedges										
838mm high; level	134.03	2.50	137.38	1.00	0.50	12.74	-	150.12	m	161.38
extra; ramps	32.55	2.50	33.36	1.00	0.50	12.74	-	46.10	nr	49.56
extra; wreaths	40.58	2.50	41.59	2.00	1.00	25.48	-	67.07	nr	72.11
extra; bends	28.80	2.50	29.52	1.00	0.50	12.74	-	42.26	nr	45.43
838mm high; raking	135.50	2.50	138.89	1.00	0.50	12.74	-	151.63	m	163.00
914mm high; level	147.32	2.50	151.00	1.00	0.50	12.74	-	163.74	m	176.02
extra; ramps	32.60	2.50	33.41	1.00	0.50	12.74	-	46.16	nr	49.62
extra; wreaths	40.58	2.50	41.59	2.00	1.00	25.48	-	67.07	nr	72.11
extra; bends	28.82	2.50	29.54	1.00	0.50	12.74	-	42.28	nr	45.45
914mm high; raking	152.15	2.50	155.95	1.00	0.50	12.74	-	168.69	m	181.35
METAL STAIRS AND BALUSTRADES - Balustrades in steel tubing, B.S.1387, medium grade; galvanised after fabrication										
Isolated balustrades; 40mm diameter handrails; 40mm diameter standards at 1500mm centres; welded fabrication ground to smooth finish; casting into mortices in concrete; wedging in position; temporary wedges										
838mm high; level	58.22	2.50	59.68	0.75	0.40	9.73	-	69.41	m	74.61
838mm high; raking	60.80	2.50	62.32	0.75	0.40	9.73	-	72.05	m	77.45
914mm high; level	66.41	2.50	68.07	0.75	0.40	9.73	-	77.80	m	83.64
914mm high; raking	70.40	2.50	72.16	0.75	0.40	9.73	-	81.89	m	88.03
Isolated balustrades; 40mm diameter handrails; 40mm diameter standards at 1500mm centres, with 6 x 75mm diameter fixing plates welded on to end, holes -3 nr; welded fabrication ground to smooth finish; fixing to masonry with Rawlbolts										
838mm high; level	71.24	2.50	73.02	1.00	0.50	12.74	-	85.76	m	92.19
838mm high; raking	74.87	2.50	76.74	1.00	0.50	12.74	-	89.48	m	96.19
914mm high; level	84.53	2.50	86.64	1.00	0.50	12.74	-	99.38	m	106.84
914mm high; raking	87.57	2.50	89.76	1.00	0.50	12.74	-	102.50	m	110.19
METAL STAIRS AND BALUSTRADES - Spiral staircases in steel; one coat red oxide primer before delivery to site										
Crescent 'H' range domestic spiral staircases; model 15 'H'; Crescent of Cambridge Ltd.; comprising tread modules complete with centre column, tread support, wooden tread, tread baluster, handrail section, PVC handrail cover, baluster wicket infill panel and tread riser bar, and with attachment brackets for fixing to floor and newel extending from centre column; assembling and bolting together										
1568mm diameter; for floor to floor height units between 2160mm and 2640mm; plywood treads with hardwood edgings; fixing base and ground plates to timber	981.00	2.50	1005.53	30.00	15.00	382.20	-	1387.72	nr	1491.80
extra; additional tread modules, 180 - 220mm per rise	72.10	2.50	73.90	1.50	0.75	19.11	-	93.01	nr	99.99
extra; mini landing to match treads, at upper floor level	180.00	2.50	184.50	2.00	1.00	25.48	-	209.98	nr	225.73
1568mm diameter; for floor to floor height units between 2160mm and 2640mm; plywood treads with carpet recess; fixing base and ground plates to timber	1120.00	2.50	1148.00	30.00	15.00	382.20	-	1530.20	nr	1644.96
extra; additional tread modules, 180 - 220mm per rise	85.00	2.50	87.13	2.00	1.00	25.48	-	112.61	nr	121.05
1568mm diameter; for floor to floor height units between 2160mm and 2640mm; hardwood treads, plain or with hardwood surround providing carpet recess; fixing base and ground plates to timber	1273.00	2.50	1304.83	30.00	15.00	382.20	-	1687.03	nr	1813.55
extra; additional tread modules, 180 - 220mm per rise	98.00	2.50	100.45	2.00	1.00	25.48	-	125.93	nr	135.37
extra; mini landing to match treads, at upper floor level	216.00	2.50	221.40	2.00	1.00	25.48	-	246.88	nr	265.40
stairwell balustrading with straight bar infill and PVC handrail cover	79.60	2.50	81.59	1.75	0.90	22.47	-	104.06	m	111.87
stairwell balustrading to match staircase including PVC handrail cover	131.50	2.50	134.79	1.75	0.90	22.47	-	157.26	m	169.05

Labour hourly rates: (except Specialists) Craft Operatives £9.23 Labourer £7.02 Rates are national average prices. Refer to REGIONAL VARIATIONS for indicative levels of overall pricing in regions	MATERIALS			LABOUR				RATES		
	Del to Site £	Waste %	Material Cost £	Craft Optve Hrs	Lab Hrs	Labour Cost £	Sunds £	Nett Rate £	Unit	Gross Rate (+7.5%) £
L30: STAIRS/WALKWAYS/BALUSTRADES Cont.										
METAL STAIRS AND BALUSTRADES – Spiral staircases in steel; one coat red oxide primer before delivery to site Cont.										
Crescent 'H' range domestic spiral staircases; model 18 'H'; Crescent of Cambridge Ltd.; comprising tread modules complete with centre column, tread support, wooden tread, tread baluster, handrail section, PVC handrail cover, baluster wicket infill panel and tread riser bar, and with attachment brackets for fixing to floor and newel extending from centre column; assembling and bolting together										
1956 diameter; for floor to floor height units between 2160mm and 2640mm; plywood treads with hardwood edgings; fixing base and ground plates to timber	1014.00	2.50	1039.35	30.00	15.00	382.20	–	1421.55	nr	1528.17
extra; additional tread modules. 180-220mm per rise	75.30	2.50	77.18	1.50	0.75	19.11	–	96.29	nr	103.51
extra; mini landing to match treads, at upper floor level	187.00	2.50	191.68	2.00	1.00	25.48	–	217.16	nr	233.44
1956 diameter; for floor to floor height units between 2160mm and 2640mm; plywood treads with carpet recess fixing base and ground plates to timber	1156.00	2.50	1184.90	30.00	15.00	382.20	–	1567.10	nr	1684.63
extra; additional tread modules, 180-220mm per rise	87.00	2.50	89.17	1.50	0.75	19.11	–	108.29	nr	116.41
1956 diameter; for floor to floor height units between 2160mm and 2640mm; hardwood treads, plain or with hardwood surround providing carpet recess; fixing base and ground plates to timber	1294.00	2.50	1326.35	30.00	15.00	382.20	–	1708.55	nr	1836.69
extra; additional tread modules, 180-220mm per rise	99.80	2.50	102.30	1.50	0.75	19.11	–	121.41	nr	130.51
extra; mini landing to match treads, at upper floor level	222.50	2.50	228.06	2.00	1.00	25.48	–	253.54	nr	272.56
stairwell balustrading with straight bar infill and PVC handrail cover	79.60	2.50	81.59	1.75	0.90	22.47	–	104.06	m	111.87
stairwell balustrading to match staircase including PVC handrail cover	131.50	2.50	134.79	1.75	0.90	22.47	–	157.26	nr	169.05
Crescent 'H' range domestic spiral staircases; model 20 'H'; Crescent of Cambridge Ltd.; comprising tread modules complete with centre column, tread support, wooden tread, tread baluster, handrail section, PVC handrail cover, baluster wicket infill panel and tread riser bar, and with attachment brackets for fixing to floor and newel extending from centre column; assembling and bolting together										
2092mm diameter; for floor to floor height units between 2160mm and 2640mm; plywood treads with hardwood edgings; fixing base and ground plates to timber	1051.00	2.50	1077.28	30.00	15.00	382.20	–	1459.47	nr	1568.94
extra; additional tread modules, 180 - 220mm per rise	77.60	2.50	79.54	1.50	0.75	19.11	–	98.65	nr	106.05
extra; mini landing to match treads, at upper floor level	193.00	2.50	197.82	2.00	1.00	25.48	–	223.31	nr	240.05
2092mm diameter; for floor to floor height units between 2160mm and 2640mm; plywood treads with carpet recess; fixing base and ground plates to timber	1188.00	2.50	1217.70	30.00	15.00	382.20	–	1599.90	nr	1719.89
extra; additional tread modules, 180 - 220mm per rise	88.00	2.50	90.20	1.50	0.75	19.11	–	109.31	nr	117.51
2092mm diameter; for floor to floor height units between 2160mm and 2640mm; hardwood treads, plain or with hardwood surround providing carpet recess; fixing base and ground plates to timber	1317.00	2.50	1349.93	31.50	16.00	403.07	–	1752.99	nr	1884.46
extra; additional tread modules, 180 - 220mm per rise	102.00	2.50	104.55	1.50	0.75	19.11	–	123.66	nr	132.93
extra; mini landing to match treads, at upper floor level	229.00	2.50	234.72	2.00	1.00	25.48	–	260.20	nr	279.72
stairwell balustrading with straight bar infill and PVC handrail cover	79.60	2.50	81.59	1.75	0.90	22.47	–	104.06	m	111.87
stairwell balustrading to match staircase including PVC handrail cover	131.50	2.50	134.79	1.75	0.90	22.47	–	157.26	m	169.05
Crescent 'S' range shop, office, factory and domestic spiral staircases; model 15 'S'; Crescent of Cambridge Ltd.; comprising tread modules complete with tread/collar section, baluster bar and section of handrail, and with upper floor attachment details with flanged column tube section, newel and top handrail baluster; assembling and bolting together										
1548mm diameter; for floor to floor height units between 2520mm and 3080mm; plain or ribbed treads; fixing base and ground plates and upper floor attachment to masonry with expanding bolts	1083.00	2.50	1110.08	35.00	17.50	446.90	–	1555.97	nr	1672.67
extra; additional tread modules, 180 - 220mm per rise	72.10	2.50	73.90	1.50	0.75	19.11	–	93.01	nr	99.99
1548mm diameter; for floor to floor height units between 2520mm and 3080mm; recessed treads; fixing base and ground plates and upper floor attachment to masonry with expanding bolts	1156.00	2.50	1184.90	36.50	19.00	470.27	–	1655.18	nr	1779.31
extra; additional tread modules, 180 - 220mm per rise	76.40	2.50	78.31	1.50	0.75	19.11	–	97.42	nr	104.73

Labour hourly rates: (except Specialists) Craft Operatives £9.23 Labourer £7.02 Rates are national average prices. Refer to REGIONAL VARIATIONS for indicative levels of overall pricing in regions	MATERIALS			LABOUR				RATES		
	Del to Site £	Waste %	Material Cost £	Craft Optve Hrs	Lab Hrs	Labour Cost £	Sunds £	Nett Rate £	Unit	Gross Rate (+7.5%) £

L30: STAIRS/WALKWAYS/BALUSTRADES Cont.

METAL STAIRS AND BALUSTRADES - Spiral staircases in steel; one coat red oxide primer before delivery to site Cont.

Crescent 'S' range shop, office, factory and domestic spiral staircases; model 15 'S'; Crescent of Cambridge Ltd.; comprising tread modules complete with tread/collar section, baluster bar and section of handrail, and with upper floor attachment details with flanged column tube section, newel and top handrail baluster; assembling and bolting together Cont.

	Del to Site	Waste	Material Cost	Craft Optve	Lab	Labour Cost	Sunds	Nett Rate	Unit	Gross Rate
1548mm diameter; for floor to floor height units between 2520mm and 3080mm; recessed treads with studded rubber tread profile covers; fixing base and ground plates and upper floor attachment to masonry with expanding bolts	1475.00	2.50	1511.88	36.50	19.00	470.27	-	1982.15	nr	2130.81
extra; additional tread modules, 180 - 220mm per rise	99.80	2.50	102.30	1.50	0.75	19.11	-	121.41	nr	130.51
extra; 825 x 825mm plain or Durbar plate landing at upper floor level, complete with column attachment section, column newel, top handrail, baluster and handrail section, and balustrading complete to one side	329.00	2.50	337.23	3.00	1.50	38.22	-	375.44	nr	403.60
extra; 825 x 825mm recessed landing at upper floor level complete with column attachment section, column newel, top handrail, baluster and handrail section, and balustrading complete to one side	358.00	2.50	366.95	3.00	1.50	38.22	-	405.17	nr	435.56
extra; stairwell balustrade including handrail balusters and ground rail	68.80	2.50	70.52	1.75	0.90	22.47	-	92.99	m	99.96
extra; additional tread balusters	7.45	2.50	7.64	0.80	0.40	10.19	-	17.83	nr	19.17
extra; riser bars	6.10	2.50	6.25	0.80	0.40	10.19	-	16.44	nr	17.68
extra; PVC handrail cover	3.70	2.50	3.79	0.30	0.15	3.82	-	7.61	m	8.19

Crescent 'S' range shop, office, factory and domestic spiral staircases; model 20 'S'; Crescent of Cambridge Ltd.; comprising tread modules complete with tread/collar section, baluster bar and section of handrail, and with upper floor attachment details with flanged column tube section, newel and top handrail baluster; assembling and bolting together

	Del to Site	Waste	Material Cost	Craft Optve	Lab	Labour Cost	Sunds	Nett Rate	Unit	Gross Rate
2072mm diameter; for floor to floor height units between 2520mm and 3080mm; plain or ribbed treads; fixing base and ground plates and upper floor attachment to masonry with expanding bolts	1203.00	2.50	1233.08	36.50	19.00	470.27	-	1703.35	nr	1831.10
extra; additional tread modules, 180 - 220mm per rise	79.50	2.50	81.49	1.50	0.75	19.11	-	100.60	nr	108.14
2072mm diameter; for floor to floor height units between 2520mm and 3080mm; recessed treads; fixing base and ground plates and upper floor attachment to masonry with expanding bolts	1265.00	2.50	1296.63	36.50	19.00	470.27	-	1766.90	nr	1899.42
extra; additional tread modules, 180 - 220mm per rise	85.00	2.50	87.13	1.50	0.75	19.11	-	106.24	nr	114.20
2072mm diameter; for floor to floor height units between 2520mm and 3080mm; recessed treads with studded rubber tread profile covers; fixing base and ground plates and upper floor attachment to masonry with expanding bolts	1687.00	2.50	1729.18	36.50	19.00	470.27	-	2199.45	nr	2364.41
extra; additional tread modules, 180 - 220mm per rise	114.00	2.50	116.85	1.50	0.75	19.11	-	135.96	nr	146.16
extra; 1100 x 1100mm plain or Durbar plate landing at upper floor level, complete with column attachment section, column newel, top handrail, baluster and handrail section, and balustrading complete to one side	392.00	2.50	401.80	3.00	1.50	38.22	-	440.02	nr	473.02
extra; 1100 x 1100mm recessed landing at upper floor level complete with column attachment section, column newel, top handrail, baluster and handrail section, and balustrading complete to one side	426.00	2.50	436.65	3.50	1.75	44.59	-	481.24	nr	517.33
extra; stairwell balustrade including handrail balusters and ground rail	66.89	2.50	68.56	1.75	0.90	22.47	-	91.03	m	97.86
extra; additional tread balusters	7.45	2.50	7.64	0.80	0.40	10.19	-	17.83	nr	19.17
extra; riser bars	6.10	2.50	6.25	0.80	0.40	10.19	-	16.44	nr	17.68
extra; PVC handrail cover	3.70	2.50	3.79	0.30	0.15	3.82	-	7.61	m	8.19

METAL STAIRS AND BALUSTRADES - Spiral staircases in steel; hot dip galvanised finish

Crescent 'S' range shop, office, factory and domestic spiral staircases; model 15 'S'; Crescent of Cambridge Ltd.; comprising tread modules complete with tread/collar section, baluster bar and section of handrail, and with upper floor attachment details with flanged column tube section, newel and top handrail baluster; assembling and bolting together

	Del to Site	Waste	Material Cost	Craft Optve	Lab	Labour Cost	Sunds	Nett Rate	Unit	Gross Rate
1548mm diameter; for floor to floor height units between 2520mm and 3080mm; plain or ribbed treads; fixing base and ground plates and upper floor attachment to masonry with expanding bolts	1156.00	2.50	1184.90	40.00	20.00	509.60	-	1694.50	nr	1821.59
extra; additional tread modules, 180 - 220mm per rise	77.60	2.50	79.54	1.50	0.75	19.11	-	98.65	nr	106.05

Labour hourly rates: (except Specialists) Craft Operatives £9.23 Labourer £7.02 Rates are national average prices. Refer to REGIONAL VARIATIONS for indicative levels of overall pricing in regions	MATERIALS			LABOUR				RATES		
	Del to Site £	Waste %	Material Cost £	Craft Optve Hrs	Lab Hrs	Labour Cost £	Sunds £	Nett Rate £	Unit	Gross Rate (+7.5%) £
L30: STAIRS/WALKWAYS/BALUSTRADES Cont.										
METAL STAIRS AND BALUSTRADES – Spiral staircases in steel; hot dip galvanised finish Cont.										
Crescent 'S' range shop, office, factory and domestic spiral staircases; model 15 'S'; Crescent of Cambridge Ltd.; comprising tread modules complete with tread/collar section, baluster bar and section of handrail, and with upper floor attachment details with flanged column tube section, newel and top handrail baluster; assembling and bolting together Cont.	..									
1548mm diameter; for floor to floor height units between 2520mm and 3080mm; recessed treads; fixing base and ground plates and upper floor attachment to masonry with expanding bolts	1231.00	2.50	1261.78	36.50	19.00	470.27	-	1732.05	nr	1861.95
extra; additional tread modules, 180 - 220mm per rise	83.00	2.50	85.08	1.50	0.75	19.11	-	104.19	nr	112.00
extra; 825 x 825mm plain or Durbar plate landing at upper floor level, complete with column attachment section, column newel, top handrail, baluster and handrail section, and balustrading complete to one side	366.00	2.50	375.15	3.00	1.50	38.22	-	413.37	nr	444.37
extra; 825 x 825mm recessed landing at upper floor level complete with column attachment section, column newel, top handrail, baluster and handrail section, and balustrading complete to one side ...	394.00	2.50	403.85	3.50	1.75	44.59	-	448.44	nr	482.07
extra; stairwell balustrade including handrail balusters and ground rail	74.50	2.50	76.36	1.75	0.90	22.47	-	98.83	m	106.25
extra; additional tread balusters	8.50	2.50	8.71	0.80	0.40	10.19	-	18.90	nr	20.32
extra; riser bars	7.45	2.50	7.64	0.80	0.40	10.19	-	17.83	nr	19.17
Crescent 'S' range shop, office, factory and domestic spiral staircases; model 20 'S'; Crescent of Cambridge Ltd.; comprising tread modules complete with tread/collar section, baluster bar and section of handrail, and with upper floor attachment details with flanged column tube section, newel and top handrail baluster; assembling and bolting together										
2072mm diameter; for floor to floor height units between 2520mm and 3080mm; plain or ribbed treads; fixing base and ground plates and upper floor attachment to masonry with expanding bolts	1273.00	2.50	1304.83	36.50	19.00	470.27	-	1775.10	nr	1908.23
extra; additional tread modules, 180 - 220mm per rise	87.00	2.50	89.17	1.50	0.75	19.11	-	108.29	nr	116.41
2072mm diameter; for floor to floor height units between 2520mm and 3080mm; recessed treads; fixing base and ground plates and upper floor attachment to masonry with expanding bolts	1332.00	2.50	1365.30	36.50	19.00	470.27	-	1835.58	nr	1973.24
extra; additional tread modules, 180 - 220mm per rise	91.00	2.50	93.28	1.50	0.75	19.11	-	112.39	nr	120.81
extra; 1100 x 1100mm plain or Durbar plate landing at upper floor level, complete with column attachment section, column newel, top handrail, baluster and handrail section, and balustrading complete to one side	446.00	2.50	457.15	3.00	1.50	38.22	-	495.37	nr	532.52
extra; 1100 x 1100mm recessed landing at upper floor level complete with column attachment section, column newel, top handrail, baluster and handrail section, and balustrading complete to one side	485.00	2.50	497.13	3.50	1.75	44.59	-	541.72	nr	582.34
extra; stairwell balustrade including handrail balusters and ground rail	74.50	2.50	76.36	1.75	0.90	22.47	-	98.83	m	106.25
extra; additional tread balusters	8.50	2.50	8.71	0.80	0.40	10.19	-	18.90	nr	20.32
extra; riser bars	7.45	2.50	7.64	0.80	0.40	10.19	-	17.83	nr	19.17
Crescent modular spiral fire escape staircases; model 15; Crescent of Cambridge Ltd.; comprising tread modules with plain or ribbed treads, and complete with centre core, spacers, baluster bar and section of handrail, tread riser bars and standard landing with balustrade one side, bars at 115mm centres; assembling and bolting together										
1548mm diameter; ground to first floor assembly, with base and ground plates; for floor to floor heights between 2520mm and 3080mm; fixing base plates and landing to masonry with expanding bolts	1888.00	2.50	1935.20	40.00	20.00	509.60	-	2444.80	nr	2628.16
extra; base extension	122.00	2.50	125.05	2.00	1.00	25.48	-	150.53	nr	161.82
1548mm diameter; intermediate floor assembly; for floor to floor heights between 2520mm and 3080mm; fixing landing to masonry with expanding bolts ...	1814.00	2.50	1859.35	40.00	20.00	509.60	-	2368.95	nr	2546.62
1548mm diameter; top floor assembly with two additional lengths of balustrade to landing; for floor to floor heights between 2520mm and 3080mm; fixing landing to masonry with expanding bolts ...	2106.00	2.50	2158.65	30.00	15.00	382.20	-	2540.85	nr	2731.41
extra; additional tread modules 180 - 220mm per rise	98.10	2.50	100.55	1.50	0.75	19.11	-	119.66	nr	128.64
extra; shield type infill panel in lieu of additional balusters to tread module	145.00	2.50	148.63	1.00	0.50	12.74	-	161.37	nr	173.47
extra; 'wicket bar' type infill panel in lieu of additional balusters to tread module	219.00	2.50	224.47	1.00	0.50	12.74	-	237.22	nr	255.01
extra; PVC handrail cover	3.73	2.50	3.82	0.30	0.15	3.82	-	7.65	m	8.22

WINDOWS/DOORS/STAIRS

Labour hourly rates: (except Specialists) Craft Operatives £9.23 Labourer £7.02 Rates are national average prices. Refer to REGIONAL VARIATIONS for indicative levels of overall pricing in regions	MATERIALS			LABOUR				RATES		
	Del to Site £	Waste %	Material Cost £	Craft Optve Hrs	Lab Hrs	Labour Cost £	Sunds £	Nett Rate £	Unit	Gross Rate (+7.5%) £

L30: STAIRS/WALKWAYS/BALUSTRADES Cont.

METAL STAIRS AND BALUSTRADES - Spiral staircases in steel; hot dip galvanised finish Cont.

Crescent modular spiral fire escape staircases; model 20; Crescent of Cambridge Ltd.; comprising tread modules with plain or ribbed treads, and complete with centre core, spacers, baluster bar and section of handrail, tread riser bars and standard landing with balustrade one side, bars at 115mm centres; assembling and bolting together

2072mm diameter; ground to first floor assembly, with base and ground plates; for floor to floor heights between 2520mm and 3080mm; fixing base plates and landing to masonry with expanding bolts	2244.00	2.50	2300.10	40.00	20.00	509.60	-	2809.70	nr	3020.43
extra; base extension	135.00	2.50	138.38	2.00	1.00	25.48	-	163.85	nr	176.14
2072mm diameter; intermediate floor assembly; for floor to floor heights between 2520mm and 3080mm; fixing landing to masonry with expanding bolts ...	2169.00	2.50	2223.22	30.00	15.00	382.20	-	2605.43	nr	2800.83
2072mm diameter; top floor assembly with two additional lengths of balustrade to landing; for floor to floor heights between 2520mm and 3080mm; fixing landing to masonry with expanding bolts ...	2461.00	2.50	2522.53	30.00	15.00	382.20	-	2904.72	nr	3122.58
extra; additional tread modules 180 - 220mm per rise	97.00	2.50	99.42	1.50	0.75	19.11	-	118.54	nr	127.43
extra; 'wicket bar' type infill panel in lieu of additional balusters to tread module	122.00	2.50	125.05	1.00	0.50	12.74	-	137.79	nr	148.12
extra; PVC handrail cover	3.73	2.50	3.82	0.30	0.15	3.82	-	7.65	m	8.22

METAL STAIRS AND BALUSTRADES - Ladders in steel

Vertical ladders; 65 x 10mm flat stringers; 65 x 10 x 250mm girth stringer brackets, bent once, bolted to stringers; 20mm diameter solid bar rungs welded to stringers; fixing to masonry with expanding bolts

395mm wide x 3000mm long overall; brackets -6 nr; rungs -7 nr..	395.00	2.50	404.88	10.00	10.00	162.50	-	567.38	nr	609.93
395mm wide x 5000mm long overall; brackets -8 nr; rungs -14 nr...	395.00	2.50	404.88	12.00	12.00	195.00	-	599.88	nr	644.87
470mm wide x 3000mm long overall; brackets -6 nr; rungs -7 nr..	335.00	2.50	343.38	10.00	10.00	162.50	-	505.88	nr	543.82
470mm wide x 5000mm long overall; brackets -8 nr; rungs -14 nr...	590.00	2.50	604.75	12.00	12.00	195.00	-	799.75	nr	859.73
395mm wide x 3000mm long overall, stringers rising and returning 900mm above top fixing points; brackets -6 nr; rungs -7 nr	335.00	2.50	343.38	10.00	10.00	162.50	-	505.88	nr	543.82
395mm wide x 5000mm long overall, stringers rising and returning 900mm above top fixing points; brackets -8 nr; rungs -14 nr	590.00	2.50	604.75	12.00	12.00	195.00	-	799.75	nr	859.73
470mm wide x 3000mm long overall, stringers rising and returning 900mm above top fixing points; brackets -6 nr; rungs -7 nr	590.00	2.50	604.75	10.00	10.00	162.50	-	767.25	nr	824.79
470mm wide x 5000mm long overall, stringers rising and returning 900mm above top fixing points; brackets -8 nr; rungs -14 nr	715.00	2.50	732.88	12.00	12.00	195.00	-	927.88	nr	997.47

Vertical ladders; 65 x 10mm flat stringers; 65 x 10 x 250mm girth stringer brackets, bent once, bolted to stringers; 20mm diameter solid bar rungs welded to stringers; 65 x 10mm flat back hoops 900mm diameter, welded to stringers

395mm wide x 3000mm long overall; brackets -6 nr; rungs -7 nr; hoops -7 nr	715.00	2.50	732.88	15.00	15.00	243.75	-	976.63	nr	1049.87
395mm wide x 5000mm long overall; brackets -8 nr; rungs -14 nr; hoops -14 nr	870.00	2.50	891.75	20.00	20.00	325.00	-	1216.75	nr	1308.01
470mm wide x 3000mm long overall; brackets -6 nr; rungs -7 nr; hoops -7 nr	815.00	2.50	835.38	15.00	15.00	243.75	-	1079.13	nr	1160.06
470mm wide x 5000mm long overall; brackets -8 nr; rungs -14 nr; hoops -14 nr	960.00	2.50	984.00	20.00	20.00	325.00	-	1309.00	nr	1407.18
395mm wide x 3000mm long overall, stringers rising and returning 900mm above top fixing points; brackets -6 nr; rungs -7 nr; hoops -7 nr	690.00	2.50	707.25	15.00	15.00	243.75	-	951.00	nr	1022.33
395mm wide x 5000mm long overall, stringers rising and returning 900mm above top fixing points; brackets -8 nr; rungs -14 nr; hoops -14 nr	855.00	2.50	876.38	20.00	20.00	325.00	-	1201.38	nr	1291.48
470mm wide x 3000mm long overall, stringers rising and returning 900mm above top fixing points; brackets -6 nr; rungs -7 nr; hoops -7 nr	780.00	2.50	799.50	15.00	15.00	243.75	-	1043.25	nr	1121.49
470mm wide x 5000mm long overall, stringers rising and returning 900mm above top fixing points; brackets -8 nr; rungs -14 nr; hoops -14 nr	1020.00	2.50	1045.50	20.00	20.00	325.00	-	1370.50	nr	1473.29

Labour hourly rates: (except Specialists) Craft Operatives £9.23 Labourer £7.02 Rates are national average prices. Refer to REGIONAL VARIATIONS for indicative levels of overall pricing in regions	MATERIALS			LABOUR				RATES		
	Del to Site	Waste	Material Cost	Craft Optve	Lab	Labour Cost	Sunds	Nett Rate	Unit	Gross Rate (+7.5%)
	£	%	£	Hrs	Hrs	£	£	£		£
L30: STAIRS/WALKWAYS/BALUSTRADES Cont.										
METAL STAIRS AND BALUSTRADES - Ladders in steel Cont.										
Ship type ladders; 75 x 10mm flat stringers, shaped and cleated ends to tops and bottoms; 65 x 10 x 350mm girth stringer brackets, bent once, bolted to stringers; 8 x 75mm on plain raised pattern plate treads with 50 x 50 x 6mm shelf angles, bolted to stringers; 25mm diameter solid bar handrails to both sides, with flattened ends bolted to stringers; 32mm diameter x 175mm long solid bar standards with three way ball type joint at intersection with handrail, and flattened ends bolted to stringers; fixing to masonry with expanding bolts										
420mm wide x 3000mm long overall, handrails rising and returning 900mm above top of stringers; brackets -4 nr; treads -12 nr; standards -6 nr ...	540.00	2.50	553.50	10.00	10.00	162.50	-	716.00	nr	769.70
420mm wide x 5000mm long overall, handrails rising and returning 900mm above top of stringers; brackets -4 nr; treads -21 nr; standards -10 nr ..	765.00	2.50	784.13	12.00	12.00	195.00	-	979.13	nr	1052.56
470mm wide x 3000mm long overall, handrails rising and returning 900mm above top of stringers; brackets -4 nr; treads -12 nr; standards -6 nr ...	630.00	2.50	645.75	10.00	10.00	162.50	-	808.25	nr	868.87
470mm wide x 5000mm long overall, handrails rising and returning 900mm above top of stringers; brackets -4 nr; treads -21 nr; standards -10 nr ..	855.00	2.50	876.38	12.00	12.00	195.00	-	1071.38	nr	1151.73
Ship type ladders; 75 x 10mm flat stringers, shaped and cleated ends to tops and bottoms; 65 x 10 x 350mm girth stringer brackets, bent once, bolted to stringers; 8 x 75mm on plain raised pattern plate treads with 50 x 50 x 6mm shelf angles, bolted to stringers; 25mm diameter solid bar handrails to both sides, with flattened ends bolted to stringers; 32mm diameter x 330mm long solid bar standards with three way ball type joint at intersection with handrail, and flattened ends bolted to stringers; fixing to masonry with expanding bolts										
420mm wide x 3000mm long overall, handrails rising and returning 900mm above top of stringers; brackets -4 nr; treads -12 nr; standards -6 nr ...	545.00	2.50	558.63	10.00	10.00	162.50	-	721.13	nr	775.21
420mm wide x 5000mm long overall, handrails rising and returning 900mm above top of stringers; brackets -4 nr; treads -21 nr; standards -10 nr ..	780.00	2.50	799.50	12.00	12.00	195.00	-	994.50	nr	1069.09
470mm wide x 3000mm long overall, handrails rising and returning 900mm above top of stringers; brackets -4 nr; treads -12 nr; standards -6 nr ...	675.00	2.50	691.88	10.00	10.00	162.50	-	854.38	nr	918.45
470mm wide x 5000mm long overall, handrails rising and returning 900mm above top of stringers; brackets -4 nr; treads -21 nr; standards -10 nr ..	870.00	2.50	891.75	12.00	12.00	195.00	-	1086.75	nr	1168.26
L40: GENERAL GLAZING										
Glass; B.S.952, clear float										
3mm thick to wood rebates with B.S.544 putty										
not exceeding 0.15m2	18.20	5.00	19.11	0.80	0.08	7.95	0.27	27.33	m2	29.38
0.15 - 4.00m2	18.20	5.00	19.11	0.40	0.04	3.97	0.13	23.21	m2	24.95
4mm thick to wood rebates with B.S.544 putty										
not exceeding 0.15m2	18.20	5.00	19.11	0.80	0.08	7.95	0.27	27.33	m2	29.38
0.15 - 4.00m2	18.20	5.00	19.11	0.40	0.04	3.97	0.13	23.21	m2	24.95
5mm thick to wood rebates with B.S.544 putty										
not exceeding 0.15m2	26.60	5.00	27.93	0.90	0.08	8.87	0.27	37.07	m2	39.85
0.15 - 4.00m2	26.60	5.00	27.93	0.45	0.04	4.43	0.13	32.49	m2	34.93
6mm thick to wood rebates with B.S.544 putty										
not exceeding 0.15m2	26.60	5.00	27.93	0.90	0.08	8.87	0.27	37.07	m2	39.85
0.15 - 4.00m2	26.60	5.00	27.93	0.45	0.04	4.43	0.13	32.49	m2	34.93
3mm thick to wood rebates with bradded wood beads (included elsewhere) and B.S.544 putty										
not exceeding 0.15m2	18.20	5.00	19.11	1.00	0.08	9.79	0.27	29.17	m2	31.36
0.15 - 4.00m2	18.20	5.00	19.11	0.50	0.04	4.90	0.13	24.14	m2	25.95
4mm thick to wood rebates with bradded wood beads (included elsewhere) and B.S.544 putty										
not exceeding 0.15m2	18.20	5.00	19.11	1.00	0.08	9.79	0.27	29.17	m2	31.36
0.15 - 4.00m2	18.20	5.00	19.11	0.50	0.04	4.90	0.13	24.14	m2	25.95
5mm thick to wood rebates with bradded wood beads (included elsewhere) and B.S.544 putty										
not exceeding 0.15m2	26.60	5.00	27.93	1.10	0.08	10.71	0.27	38.91	m2	41.83
0.15 - 4.00m2	26.60	5.00	27.93	0.55	0.04	5.36	0.13	33.42	m2	35.92
6mm thick to wood rebates with bradded wood beads (included elsewhere) and B.S.544 putty										
not exceeding 0.15m2	26.60	5.00	27.93	1.10	0.08	10.71	0.27	38.91	m2	41.83
0.15 - 4.00m2	26.60	5.00	27.93	0.55	0.04	5.36	0.13	33.42	m2	35.92
3mm thick to wood rebates with screwed wood beads and glazing strip (included elsewhere)										
not exceeding 0.15m2	18.20	5.00	19.11	1.40	0.08	13.48	0.27	32.86	m2	35.33
0.15 - 4.00m2	18.20	5.00	19.11	0.70	0.04	6.74	0.13	25.98	m2	27.93

Labour hourly rates: (except Specialists) Craft Operatives £9.23 Labourer £7.02 Rates are national average prices. Refer to REGIONAL VARIATIONS for indicative levels of overall pricing in regions	MATERIALS			LABOUR				RATES		
	Del to Site £	Waste %	Material Cost £	Craft Optve Hrs	Lab Hrs	Labour Cost £	Sunds £	Nett Rate £	Unit	Gross Rate (+7.5%) £
L40: GENERAL GLAZING Cont.										
Glass; B.S.952, clear float Cont.										
4mm thick to wood rebates with screwed wood beads and glazing strip (included elsewhere)										
not exceeding 0.15m2	18.20	5.00	19.11	1.40	0.08	13.48	0.27	32.86	m2	35.33
0.15 - 4.00m2	18.20	5.00	19.11	0.70	0.04	6.74	0.13	25.98	m2	27.93
5mm thick to wood rebates with screwed wood beads and glazing strip (included elsewhere)										
not exceeding 0.15m2	26.60	5.00	27.93	1.50	0.08	14.41	0.27	42.61	m2	45.80
0.15 - 4.00m2	26.60	5.00	27.93	0.75	0.04	7.20	0.13	35.26	m2	37.91
6mm thick to wood rebates with screwed wood beads and glazing strip (included elsewhere)										
not exceeding 0.15m2	26.60	5.00	27.93	1.50	0.08	14.41	0.27	42.61	m2	45.80
0.15 - 4.00m2	26.60	5.00	27.93	0.75	0.04	7.20	0.13	35.26	m2	37.91
3mm thick to metal rebates with metal casement glazing compound										
not exceeding 0.15m2	18.20	5.00	19.11	0.90	0.08	8.87	0.31	28.29	m2	30.41
0.15 - 4.00m2	18.20	5.00	19.11	0.45	0.04	4.43	0.15	23.69	m2	25.47
4mm thick to metal rebates with metal casement glazing compound										
not exceeding 0.15m2	18.20	5.00	19.11	0.90	0.08	8.87	0.31	28.29	m2	30.41
0.15 - 4.00m2	18.20	5.00	19.11	0.45	0.04	4.43	0.15	23.69	m2	25.47
5mm thick to metal rebates with metal casement glazing compound										
not exceeding 0.15m2	26.60	5.00	27.93	1.00	0.08	9.79	0.31	38.03	m2	40.88
0.15 - 4.00m2	26.60	5.00	27.93	0.50	0.04	4.90	0.15	32.98	m2	35.45
6mm thick to metal rebates with metal casement glazing compound										
not exceeding 0.15m2	26.60	5.00	27.93	1.00	0.08	9.79	0.31	38.03	m2	40.88
0.15 - 4.00m2	26.60	5.00	27.93	0.50	0.04	4.90	0.15	32.98	m2	35.45
3mm thick to metal rebates with clipped metal beads and gaskets (included elsewhere)										
not exceeding 0.15m2	18.20	5.00	19.11	1.20	0.08	11.64	0.27	31.02	m2	33.34
0.15 - 4.00m2	18.20	5.00	19.11	0.60	0.04	5.82	0.13	25.06	m2	26.94
4mm thick to metal rebates with clipped metal beads and gaskets (included elsewhere)										
not exceeding 0.15m2	18.20	5.00	19.11	1.20	0.08	11.64	0.27	31.02	m2	33.34
0.15 - 4.00m2	18.20	5.00	19.11	0.60	0.04	5.82	0.13	25.06	m2	26.94
5mm thick to metal rebates with clipped metal beads and gaskets (included elsewhere)										
not exceeding 0.15m2	26.60	5.00	27.93	1.30	0.08	12.56	0.27	40.76	m2	43.82
0.15 - 4.00m2	26.60	5.00	27.93	0.65	0.04	6.28	0.13	34.34	m2	36.92
6mm thick to metal rebates with clipped metal beads and gaskets (included elsewhere)										
not exceeding 0.15m2	26.60	5.00	27.93	1.30	0.08	12.56	0.27	40.76	m2	43.82
0.15 - 4.00m2	26.60	5.00	27.93	0.65	0.04	6.28	0.13	34.34	m2	36.92
Glass; B.S.952, rough cast										
6mm thick to wood rebates with B.S.544 putty										
not exceeding 0.15m2	34.03	5.00	35.73	0.90	0.08	8.87	0.27	44.87	m2	48.24
0.15 - 4.00m2	34.03	5.00	35.73	0.45	0.04	4.43	0.13	40.30	m2	43.32
6mm thick to wood rebates with bradded wood beads (included elsewhere) and B.S.544 putty										
not exceeding 0.15m2	34.03	5.00	35.73	1.10	0.08	10.71	0.27	46.72	m2	50.22
0.15 - 4.00m2	34.03	5.00	35.73	0.55	0.04	5.36	0.13	41.22	m2	44.31
6mm thick to wood rebates with screwed wood beads and glazing strip (included elsewhere)										
not exceeding 0.15m2	34.03	5.00	35.73	1.50	0.08	14.41	0.27	50.41	m2	54.19
0.15 - 4.00m2	34.03	5.00	35.73	0.75	0.04	7.20	0.13	43.06	m2	46.29
6mm thick to metal rebates with metal casement glazing compound										
not exceeding 0.15m2	34.03	5.00	35.73	1.00	0.08	9.79	0.31	45.83	m2	49.27
0.15 - 4.00m2	34.03	5.00	35.73	0.50	0.04	4.90	0.15	40.78	m2	43.84
6mm thick to metal rebates with clipped metal beads and gaskets (included elsewhere)										
not exceeding 0.15m2	34.03	5.00	35.73	1.30	0.08	12.56	0.31	48.60	m2	52.25
0.15 - 4.00m2	34.03	5.00	35.73	0.65	0.04	6.28	0.15	42.16	m2	45.32
Glass; B.S.952, Pyroshield clear										
6mm thick to wood rebates with screwed beads and intumescent glazing strip (included elsewhere)										
not exceeding 0.15m2	60.56	5.00	63.59	1.80	0.08	17.18	-	80.76	m2	86.82
not exceeding 0.15m2; aligning panes with adjacent panes	60.56	8.50	65.71	2.00	0.08	19.02	-	84.73	m2	91.08
0.15 - 4.00m2	60.56	5.00	63.59	0.90	0.04	8.59	-	72.18	m2	77.59
0.15 - 4.00m2; aligning panes with adjacent panes	60.56	8.50	65.71	1.00	0.04	9.51	-	75.22	m2	80.86

Labour hourly rates: (except Specialists) Craft Operatives £9.23 Labourer £7.02 Rates are national average prices. Refer to REGIONAL VARIATIONS for indicative levels of overall pricing in regions	MATERIALS			LABOUR				RATES		
	Del to Site	Waste	Material Cost	Craft Optve	Lab	Labour Cost	Sunds	Nett Rate	Unit	Gross Rate (+7.5%)
	£	%	£	Hrs	Hrs	£	£	£		£

L40: GENERAL GLAZING Cont.

Glass; B.S.952, Pyroshield clear Cont.

6mm thick to wood rebates with screwed beads and
Pyroglazing strip (included elsewhere)

not exceeding 0.15m2	60.56	5.00	63.59	1.80	0.08	17.18	–	80.76	m2	86.82
not exceeding 0.15m2; aligning panes with adjacent panes	60.56	8.50	65.71	2.00	0.08	19.02	–	84.73	m2	91.08
0.15 – 4.00m2	60.56	5.00	63.59	0.90	0.04	8.59	–	72.18	m2	77.59
0.15 – 4.00m2; aligning panes with adjacent panes	60.56	8.50	65.71	1.00	0.04	9.51	–	75.22	m2	80.86

6mm thick to wood rebates with screwed beads and
Lorient System 90 (included elsewhere)

not exceeding 0.15m2	60.56	5.00	63.59	2.00	0.08	19.02	–	82.61	m2	88.81
not exceeding 0.15m2; aligning panes with adjacent panes	60.56	8.50	65.71	2.20	0.08	20.87	–	86.58	m2	93.07
0.15 – 4.00m2	60.56	5.00	63.59	1.00	0.04	9.51	–	73.10	m2	78.58
0.15 – 4.00m2; aligning panes with adjacent panes	60.56	8.50	65.71	1.10	0.04	10.43	–	76.14	m2	81.85

6mm thick to metal rebates with screwed metal beads
and mild steel flat strips (included elsewhere)

not exceeding 0.15m2	60.56	5.00	63.59	1.80	0.08	17.18	–	80.76	m2	86.82
not exceeding 0.15m2; aligning panes with adjacent panes	60.56	8.50	65.71	2.00	0.08	19.02	–	84.73	m2	91.08
0.15 – 4.00m2	60.56	5.00	63.59	0.90	0.04	8.59	–	72.18	m2	77.59
0.15 – 4.00m2; aligning panes with adjacent panes	60.56	8.50	65.71	1.00	0.04	9.51	–	75.22	m2	80.86

Glass; B.S.952, Pyroshield safety clear

6mm thick to wood rebates with screwed beads and
intumescent glazing strip (included elsewhere)

not exceeding 0.15m2	71.31	5.00	74.88	1.80	0.08	17.18	–	92.05	m2	98.95
not exceeding 0.15m2; aligning panes with adjacent panes	71.31	8.50	77.37	2.00	0.08	19.02	–	96.39	m2	103.62
0.15 – 4.00m2	71.31	5.00	74.88	0.90	0.04	8.59	–	83.46	m2	89.72
0.15 – 4.00m2; aligning panes with adjacent panes	71.31	8.50	77.37	1.00	0.04	9.51	–	86.88	m2	93.40

6mm thick to wood rebates with screwed beads and
Pyroglazing strip (included elsewhere)

not exceeding 0.15m2	71.31	5.00	74.88	1.80	0.08	17.18	–	92.05	m2	98.95
not exceeding 0.15m2; aligning panes with adjacent panes	71.31	8.50	77.37	2.00	0.08	19.02	–	96.39	m2	103.62
0.15 – 4.00m2	71.31	5.00	74.88	0.90	0.04	8.59	–	83.46	m2	89.72
0.15 – 4.00m2; aligning panes with adjacent panes	71.31	8.50	77.37	1.00	0.04	9.51	–	86.88	m2	93.40

6mm thick to wood rebates with screwed beads and
Lorient System 90 (included elsewhere)

not exceeding 0.15m2	71.31	5.00	74.88	2.00	0.08	19.02	–	93.90	m2	100.94
not exceeding 0.15m2; aligning panes with adjacent panes	71.31	8.50	77.37	2.20	0.08	20.87	–	98.24	m2	105.61
0.15 – 4.00m2	71.31	5.00	74.88	1.00	0.04	9.51	–	84.39	m2	90.72
0.15 – 4.00m2; aligning panes with adjacent panes	71.31	8.50	77.37	1.10	0.04	10.43	–	87.81	m2	94.39

6mm thick to metal rebates with screwed metal beads
and mild steel flat strips (included elsewhere)

not exceeding 0.15m2	71.31	5.00	74.88	1.80	0.08	17.18	–	92.05	m2	98.95
not exceeding 0.15m2; aligning panes with adjacent panes	71.31	8.50	77.37	2.00	0.08	19.02	–	96.39	m2	103.62
0.15 – 4.00m2	71.31	5.00	74.88	0.90	0.04	8.59	–	83.46	m2	89.72
0.15 – 4.00m2; aligning panes with adjacent panes	71.31	8.50	77.37	1.00	0.04	9.51	–	86.88	m2	93.40

Glass; B.S.952, Pyroshield texture

7mm thick to wood rebates with screwed beads and
intumescent glazing strip (included elsewhere)

not exceeding 0.15m2	28.02	5.00	29.42	1.80	0.08	17.18	–	46.60	m2	50.09
not exceeding 0.15m2; aligning panes with adjacent panes	28.02	8.50	30.40	2.00	0.08	19.02	–	49.42	m2	53.13
0.15 – 4.00m2	28.02	5.00	29.42	0.90	0.04	8.59	–	38.01	m2	40.86
0.15 – 4.00m2; aligning panes with adjacent panes	28.02	8.50	30.40	1.00	0.04	9.51	–	39.91	m2	42.91

7mm thick to wood rebates with screwed beads and
Pyroglazing strip (included elsewhere)

not exceeding 0.15m2	28.02	5.00	29.42	1.80	0.08	17.18	–	46.60	m2	50.09
not exceeding 0.15m2; aligning panes with adjacent panes	28.02	8.50	30.40	2.00	0.08	19.02	–	49.42	m2	53.13
0.15 – 4.00m2	28.02	5.00	29.42	0.90	0.04	8.59	–	38.01	m2	40.86
0.15 – 4.00m2; aligning panes with adjacent panes	28.02	8.50	30.40	1.00	0.04	9.51	–	39.91	m2	42.91

7mm thick to wood rebates with screwed beads and
Lorient System 90 (included elsewhere)

not exceeding 0.15m2	28.02	5.00	29.42	2.00	0.08	19.02	–	48.44	m2	52.08
not exceeding 0.15m2; aligning panes with adjacent panes	28.02	8.50	30.40	2.20	0.08	20.87	–	51.27	m2	55.11
0.15 – 4.00m2	28.02	5.00	29.42	1.00	0.04	9.51	–	38.93	m2	41.85
0.15 – 4.00m2; aligning panes with adjacent panes	28.02	8.50	30.40	1.10	0.04	10.43	–	40.84	m2	43.90

7mm thick to metal rebates with screwed metal beads
and mild steel flat strips (included elsewhere)

not exceeding 0.15m2	28.02	5.00	29.42	1.80	0.08	17.18	–	46.60	m2	50.09
not exceeding 0.15m2; aligning panes with adjacent panes	28.02	8.50	30.40	2.00	0.08	19.02	–	49.42	m2	53.13
0.15 – 4.00m2	28.02	5.00	29.42	0.90	0.04	8.59	–	38.01	m2	40.86
0.15 – 4.00m2; aligning panes with adjacent panes	28.02	8.50	30.40	1.00	0.04	9.51	–	39.91	m2	42.91

WINDOWS/DOORS/STAIRS

Labour hourly rates: (except Specialists) Craft Operatives £9.23 Labourer £7.02 Rates are national average prices. Refer to REGIONAL VARIATIONS for indicative levels of overall pricing in regions	MATERIALS			LABOUR				RATES		
	Del to Site	Waste	Material Cost	Craft Optve	Lab	Labour Cost	Sunds	Nett Rate	Unit	Gross Rate (+7.5%)
	£	%	£	Hrs	Hrs	£	£	£		£

L40: GENERAL GLAZING Cont.

Glass; B.S.952, Pyroshield safety texture

7mm thick to wood rebates with screwed beads and intumescent glazing strip (included elsewhere)										
not exceeding 0.15m2	36.45	5.00	38.27	1.80	0.08	17.18	–	55.45	m2	59.61
not exceeding 0.15m2; aligning panes with adjacent panes	36.45	8.50	39.55	2.00	0.08	19.02	–	58.57	m2	62.96
0.15 - 4.00m2	36.45	5.00	38.27	0.90	0.04	8.59	–	46.86	m2	50.37
0.15 - 4.00m2; aligning panes with adjacent panes	36.45	8.50	39.55	1.00	0.04	9.51	–	49.06	m2	52.74
7mm thick to wood rebates with screwed beads and Pyroglazing strip (included elsewhere)										
not exceeding 0.15m2	36.45	5.00	38.27	1.80	0.08	17.18	–	55.45	m2	59.61
not exceeding 0.15m2; aligning panes with adjacent panes	36.45	8.50	39.55	2.00	0.08	19.02	–	58.57	m2	62.96
0.15 - 4.00m2	36.45	5.00	38.27	0.90	0.04	8.59	–	46.86	m2	50.37
0.15 - 4.00m2; aligning panes with adjacent panes	36.45	8.50	39.55	1.00	0.04	9.51	–	49.06	m2	52.74
7mm thick to wood rebates with screwed beads and Lorient System 90 (included elsewhere)										
not exceeding 0.15m2	36.45	5.00	38.27	2.00	0.08	19.02	–	57.29	m2	61.59
not exceeding 0.15m2; aligning panes with adjacent panes	36.45	8.50	39.55	2.20	0.08	20.87	–	60.42	m2	64.95
0.15 - 4.00m2	36.45	5.00	38.27	1.00	0.04	9.51	–	47.78	m2	51.37
0.15 - 4.00m2; aligning panes with adjacent panes	36.45	8.50	39.55	1.10	0.04	10.43	–	49.98	m2	53.73
7mm thick to metal rebates with screwed metal beads and mild steel flat strips (included elsewhere)										
not exceeding 0.15m2	36.45	5.00	38.27	1.80	0.08	17.18	–	55.45	m2	59.61
not exceeding 0.15m2; aligning panes with adjacent panes	36.45	8.50	39.55	2.00	0.08	19.02	–	58.57	m2	62.96
0.15 - 4.00m2	36.45	5.00	38.27	0.90	0.04	8.59	–	46.86	m2	50.37
0.15 - 4.00m2; aligning panes with adjacent panes	36.45	8.50	39.55	1.00	0.04	9.51	–	49.06	m2	52.74

Glass; B.S.952, white patterned

4mm thick to wood rebates with B.S.544 putty										
not exceeding 0.15m2	22.60	5.00	23.73	0.80	0.08	7.95	0.27	31.95	m2	34.34
0.15 - 4.00m2	22.60	5.00	23.73	0.40	0.04	3.97	0.13	27.83	m2	29.92
6mm thick to wood rebates with B.S.544 putty										
not exceeding 0.15m2	34.03	5.00	35.73	0.90	0.08	8.87	0.27	44.87	m2	48.24
0.15 - 4.00m2	34.03	5.00	35.73	0.45	0.04	4.43	0.13	40.30	m2	43.32
4mm thick to wood rebates with bradded wood beads (included elsewhere) and B.S.544 putty										
not exceeding 0.15m2	22.60	5.00	23.73	1.00	0.08	9.79	0.27	33.79	m2	36.33
0.15 - 4.00m2	22.60	5.00	23.73	0.50	0.04	4.90	0.13	28.76	m2	30.91
6mm thick to wood rebates with bradded wood beads (included elsewhere) and B.S.544 putty										
not exceeding 0.15m2	34.03	5.00	35.73	1.10	0.08	10.71	0.27	46.72	m2	50.22
0.15 - 4.00m2	34.03	5.00	35.73	0.55	0.04	5.36	0.13	41.22	m2	44.31
4mm thick to wood rebates with screwed wood beads and glazing strip (included elsewhere)										
not exceeding 0.15m2	22.60	5.00	23.73	1.40	0.08	13.48	0.27	37.48	m2	40.29
0.15 - 4.00m2	22.60	5.00	23.73	0.70	0.04	6.74	0.13	30.60	m2	32.90
6mm thick to wood rebates with screwed wood beads and glazing strip (included elsewhere)										
not exceeding 0.15m2	34.03	5.00	35.73	1.50	0.08	14.41	0.27	50.41	m2	54.19
0.15 - 4.00m2	34.03	5.00	35.73	0.75	0.04	7.20	0.13	43.06	m2	46.29
4mm thick to metal rebates with metal casement glazing compound										
not exceeding 0.15m2	22.60	5.00	23.73	0.90	0.08	8.87	0.31	32.91	m2	35.38
0.15 - 4.00m2	22.60	5.00	23.73	0.45	0.04	4.43	0.15	28.31	m2	30.44
6mm thick to metal rebates with metal casement glazing compound										
not exceeding 0.15m2	34.03	5.00	35.73	1.00	0.08	9.79	0.31	45.83	m2	49.27
0.15 - 4.00m2	34.03	5.00	35.73	0.50	0.04	4.90	0.15	40.78	m2	43.84
4mm thick to metal rebates with clipped metal beads and gaskets (included elsewhere)										
not exceeding 0.15m2	22.60	5.00	23.73	1.20	0.08	11.64	0.27	35.64	m2	38.31
0.15 - 4.00m2	22.60	5.00	23.73	0.60	0.04	5.82	0.13	29.68	m2	31.90
6mm thick to metal rebates with clipped metal beads and gaskets (included elsewhere)										
not exceeding 0.15m2	34.03	5.00	35.73	1.30	0.08	12.56	0.27	48.56	m2	52.20
0.15 - 4.00m2	34.03	5.00	35.73	0.65	0.04	6.28	0.13	42.14	m2	45.30

Glass; B.S.952, tinted patterned

4mm thick to wood rebates with B.S.544 putty										
not exceeding 0.15m2	33.38	5.00	35.05	0.80	0.08	7.95	0.27	43.26	m2	46.51
0.15 - 4.00m2	33.38	5.00	35.05	0.40	0.04	3.97	0.13	39.15	m2	42.09
6mm thick to wood rebates with B.S.544 putty										
not exceeding 0.15m2	37.77	5.00	39.66	0.90	0.08	8.87	0.27	48.80	m2	52.46
0.15 - 4.00m2	37.77	5.00	39.66	0.45	0.04	4.43	0.13	44.22	m2	47.54

Labour hourly rates: (except Specialists) Craft Operatives £9.23 Labourer £7.02 Rates are national average prices. Refer to REGIONAL VARIATIONS for indicative levels of overall pricing in regions	MATERIALS			LABOUR				RATES		
	Del to Site £	Waste %	Material Cost £	Craft Optve Hrs	Lab Hrs	Labour Cost £	Sunds £	Nett Rate £	Unit	Gross Rate (+7.5%) £

L40: GENERAL GLAZING Cont.

Glass; B.S.952, tinted patterned Cont.

4mm thick to wood rebates with bradded wood beads (included elsewhere) and B.S.544 putty

not exceeding 0.15m2	33.38	5.00	35.05	1.00	0.08	9.79	0.27	45.11	m2	48.49
0.15 - 4.00m2	33.38	5.00	35.05	0.50	0.04	4.90	0.13	40.07	m2	43.08

6mm thick to wood rebates with bradded wood beads (included elsewhere) and B.S.544 putty

not exceeding 0.15m2	37.77	5.00	39.66	1.10	0.08	10.71	0.27	50.64	m2	54.44
0.15 - 4.00m2	37.77	5.00	39.66	0.55	0.04	5.36	0.13	45.15	m2	48.53

4mm thick to wood rebates with screwed wood beads and glazing strip (included elsewhere)

not exceeding 0.15m2	33.38	5.00	35.05	1.40	0.08	13.48	0.27	48.80	m2	52.46
0.15 - 4.00m2	33.38	5.00	35.05	0.70	0.04	6.74	0.13	41.92	m2	45.06

6mm thick to wood rebates with screwed wood beads and glazing strip (included elsewhere)

not exceeding 0.15m2	37.77	5.00	39.66	1.50	0.08	14.41	0.27	54.34	m2	58.41
0.15 - 4.00m2	37.77	5.00	39.66	0.75	0.04	7.20	0.13	46.99	m2	50.52

4mm thick to metal rebates with metal casement glazing compound

not exceeding 0.15m2	33.38	5.00	35.05	0.90	0.08	8.87	0.31	44.23	m2	47.54
0.15 - 4.00m2	33.38	5.00	35.05	0.45	0.04	4.43	0.15	39.63	m2	42.61

6mm thick to metal rebates with metal casement glazing compound

not exceeding 0.15m2	37.77	5.00	39.66	1.00	0.08	9.79	0.31	49.76	m2	53.49
0.15 - 4.00m2	37.77	5.00	39.66	0.50	0.04	4.90	0.15	44.70	m2	48.06

4mm thick to metal rebates with clipped metal beads and gaskets (included elsewhere)

not exceeding 0.15m2	33.38	5.00	35.05	1.20	0.08	11.64	0.27	46.96	m2	50.48
0.15 - 4.00m2	33.38	5.00	35.05	0.60	0.04	5.82	0.13	41.00	m2	44.07

6mm thick to metal rebates with clipped metal beads and gaskets (included elsewhere)

not exceeding 0.15m2	37.77	5.00	39.66	1.30	0.08	12.56	0.27	52.49	m2	56.43
0.15 - 4.00m2	37.77	5.00	39.66	0.65	0.04	6.28	0.13	46.07	m2	49.52

Patterns available

4mm Bronze tint
    Autumn; Cotswold; Everglade; Sycamore
4mm white
    Arctic
6mm white
    Deep Flemish
4 and 6mm white
    Autumn; Cotswold; Driftwood; Everglade; Flemish;
    Linkon; Mayflower; Reeded; Stippolyte

Glass; B.S.952, antisun float; grey

4mm thick to wood rebates with B.S.544 putty

not exceeding 2400 x 1200mm	34.71	5.00	36.45	0.60	0.10	6.24	0.19	42.88	m2	46.09

6mm thick to wood rebates with B.S.544 putty

not exceeding 5950 x 3150mm	50.08	5.00	52.58	0.65	0.10	6.70	0.19	59.48	m2	63.94

10mm thick to wood rebates with B.S.544 putty

not exceeding 5950 x 3150mm	89.54	5.00	94.02	1.50	0.10	14.55	0.24	108.80	m2	116.96

12mm thick to wood rebates with B.S.544 putty

not exceeding 5950 x 3150mm	123.59	5.00	129.77	2.00	0.10	19.16	0.30	149.23	m2	160.42

4mm thick to wood rebates with bradded wood beads (included elsewhere) and B.S.544 putty

not exceeding 2400 x 1200mm	34.71	5.00	36.45	0.70	0.10	7.16	0.19	43.80	m2	47.08

6mm thick to wood rebates with bradded wood beads (included elsewhere) and B.S.544 putty

not exceeding 5950 x 3150mm	50.08	5.00	52.58	0.75	0.10	7.62	0.19	60.40	m2	64.93

10mm thick to wood rebates with bradded wood beads (included elsewhere) and B.S.544 putty

not exceeding 5950 x 3150mm	89.54	5.00	94.02	1.60	0.10	15.47	0.24	109.73	m2	117.96

12mm thick to wood rebates with bradded wood beads (included elsewhere) and B.S.544 putty

not exceeding 5950 x 3150mm	123.59	5.00	129.77	2.10	0.10	20.09	0.30	150.15	m2	161.42

4mm thick to wood rebates with screwed wood beads and glazing strip (included elsewhere)

not exceeding 2400 x 1200mm	34.71	5.00	36.45	0.90	0.10	9.01	-	45.45	m2	48.86

6mm thick to wood rebates with screwed wood beads and glazing strip (included elsewhere)

not exceeding 5950 x 3150mm	50.08	5.00	52.58	0.95	0.10	9.47	-	62.05	m2	66.71

10mm thick to wood rebates with screwed wood beads and glazing strip (included elsewhere)

not exceeding 5950 x 3150mm	89.54	5.00	94.02	1.70	0.10	16.39	-	110.41	m2	118.69

Labour hourly rates: (except Specialists) Craft Operatives £9.23 Labourer £7.02 Rates are national average prices. Refer to REGIONAL VARIATIONS for indicative levels of overall pricing in regions	MATERIALS			LABOUR				RATES		
	Del to Site £	Waste %	Material Cost £	Craft Optve Hrs	Lab Hrs	Labour Cost £	Sunds £	Nett Rate £	Unit	Gross Rate (+7.5%) £
L40: GENERAL GLAZING Cont.										
Glass; B.S.952, antisun float; grey Cont.										
12mm thick to wood rebates with screwed wood beads and glazing strip (included elsewhere) not exceeding 5950 x 3150mm	123.59	5.00	129.77	2.20	0.10	21.01	-	150.78	m2	162.09
Glass; B.S.952, antisun float; bronze										
4mm thick to wood rebates with B.S.544 putty not exceeding 2400 x 1200mm	34.71	5.00	36.45	0.60	0.10	6.24	0.19	42.88	m2	46.09
6mm thick to wood rebates with B.S.544 putty not exceeding 5950 x 3150mm	50.08	5.00	52.58	0.65	0.10	6.70	0.19	59.48	m2	63.94
10mm thick to wood rebates with B.S.544 putty not exceeding 5950 x 3150mm	89.54	5.00	94.02	1.50	0.10	14.55	0.24	108.80	m2	116.96
12mm thick to wood rebates with B.S.544 putty not exceeding 5950 x 3150mm	123.59	5.00	129.77	2.00	0.10	19.16	0.30	149.23	m2	160.42
4mm thick to wood rebates with bradded wood beads (included elsewhere) and B.S.544 putty not exceeding 2400 x 1200mm	34.71	5.00	36.45	0.70	0.10	7.16	0.19	43.80	m2	47.08
6mm thick to wood rebates with bradded wood beads (included elsewhere) and B.S.544 putty not exceeding 5950 x 3150mm	50.08	5.00	52.58	0.75	0.10	7.62	0.19	60.40	m2	64.93
10mm thick to wood rebates with bradded wood beads (included elsewhere) and B.S.544 putty not exceeding 5950 x 3150mm	89.54	5.00	94.02	1.60	0.10	15.47	0.24	109.73	m2	117.96
12mm thick to wood rebates with bradded wood beads (included elsewhere) and B.S.544 putty not exceeding 5950 x 3150mm	123.59	5.00	129.77	2.10	0.10	20.09	0.30	150.15	m2	161.42
4mm thick to wood rebates with screwed wood beads and glazing strip (included elsewhere) not exceeding 2400 x 1200mm	34.71	5.00	36.45	0.90	0.10	9.01	-	45.45	m2	48.86
6mm thick to wood rebates with screwed wood beads and glazing strip (included elsewhere) not exceeding 5950 x 3150mm	50.08	5.00	52.58	0.95	0.10	9.47	-	62.05	m2	66.71
10mm thick to wood rebates with screwed wood beads and glazing strip (included elsewhere) not exceeding 5950 x 3150mm	89.54	5.00	94.02	1.70	0.10	16.39	-	110.41	m2	118.69
12mm thick to wood rebates with screwed wood beads and glazing strip (included elsewhere) not exceeding 5950 x 3150mm	123.09	5.00	129.24	2.20	0.10	21.01	-	150.25	m2	161.52
Glass; B.S.952, antisun float; green										
6mm thick to wood rebates with B.S.544 putty not exceeding 3150 x 2050mm	64.20	5.00	67.41	0.65	0.10	6.70	0.19	74.30	m2	79.87
6mm thick to wood rebates with bradded wood beads (included elsewhere) and B.S.544 putty not exceeding 3150 x 2050mm	64.20	5.00	67.41	0.75	0.10	7.62	0.19	75.22	m2	80.87
6mm thick to wood rebates with screwed wood beads and glazing strip (included elsewhere) not exceeding 3150 x 2050mm	64.20	5.00	67.41	0.95	0.10	9.47	-	76.88	m2	82.65
Glass; B.S.952, clear float										
10mm thick to wood rebates with screwed wood beads and glazing strip (included elsewhere) not exceeding 5950 x 3150mm	53.88	5.00	56.57	2.25	0.10	21.47	-	78.04	m2	83.90
12mm thick to wood rebates with screwed wood beads and glazing strip (included elsewhere) not exceeding 5950 x 3150mm	62.06	5.00	65.16	2.75	0.10	26.08	-	91.25	m2	98.09
15mm thick to wood rebates with screwed wood beads and glazing strip (included elsewhere) not exceeding 2950 x 2000mm	103.26	5.00	108.42	3.50	0.10	33.01	-	141.43	m2	152.04
19mm thick to wood rebates with screwed wood beads and glazing strip (included elsewhere) not exceeding 2950 x 2000mm	146.27	5.00	153.58	4.00	0.10	37.62	-	191.21	m2	205.55
25mm thick to wood rebates with screwed wood beads and glazing strip (included elsewhere) not exceeding 2950 x 2000mm	232.19	5.00	243.80	4.50	0.10	42.24	-	286.04	m2	307.49
Glass; B.S.952, toughened clear float										
4mm thick to metal rebates with screwed metal beads and gaskets (included elsewhere) not exceeding 2400 x 1300mm	26.85	5.00	28.19	0.60	0.10	6.24	-	34.43	m2	37.01

Labour hourly rates: (except Specialists) Craft Operatives £9.23 Labourer £7.02 Rates are national average prices. Refer to REGIONAL VARIATIONS for indicative levels of overall pricing in regions	MATERIALS			LABOUR				RATES		
	Del to Site £	Waste %	Material Cost £	Craft Optve Hrs	Lab Hrs	Labour Cost £	Sunds £	Nett Rate £	Unit	Gross Rate (+7.5%) £

L40: GENERAL GLAZING Cont.

Glass; B.S.952, toughened clear float Cont.

5mm thick to metal rebates with screwed metal beads and gaskets (included elsewhere) not exceeding 2500 x 1520mm	40.27	5.00	42.28	0.60	0.10	6.24	-	48.52	m2	52.16
6mm thick to metal rebates with screwed metal beads and gaskets (included elsewhere) not exceeding 2500 x 1520mm	40.27	5.00	42.28	0.65	0.10	6.70	-	48.98	m2	52.66
10mm thick to metal rebates with screwed metal beads and gaskets (included elsewhere) not exceeding 2500 x 1520mm	63.89	5.00	67.08	1.50	0.10	14.55	-	81.63	m2	87.75
12mm thick to metal rebates with screwed metal beads and gaskets (included elsewhere) not exceeding 2500 x 1520mm	125.19	5.00	131.45	2.00	0.10	19.16	-	150.61	m2	161.91

Glass; B.S.952, toughened white patterned

4mm thick to metal rebates with screwed metal beads and gaskets (included elsewhere) not exceeding 2100 x 1300mm	54.12	5.00	56.83	0.60	0.10	6.24	-	63.07	m2	67.80
6mm thick to metal rebates with screwed metal beads and gaskets (included elsewhere) not exceeding 2100 x 1300mm	56.50	5.00	59.33	0.65	0.10	6.70	-	66.03	m2	70.98

Glass; B.S.952, toughened tinted patterned

4mm thick to metal rebates with screwed metal beads and gaskets (included elsewhere) not exceeding 2100 x 1300mm	54.12	5.00	56.83	0.60	0.10	6.24	-	63.07	m2	67.80
6mm thick to metal rebates with screwed metal beads and gaskets (included elsewhere) not exceeding 2100 x 1300mm	56.50	5.00	59.33	0.65	0.10	6.70	-	66.03	m2	70.98

Patterns available

4mm tinted
    Everglade
4mm white
    Reeded
4mm tinted
    Autumn; Cotswold
4 and 6mm white
    Autumn; Cotswold; Driftwood; Everglade; Flemish;
    Linkon; Mayflower; Stippolyte; Sycamore

Glass; B.S.952, toughened antisun float; grey

4mm thick to metal rebates with screwed metal beads and gaskets (included elsewhere) not exceeding 2100 x 1250mm	49.22	5.00	51.68	0.60	0.10	6.24	-	57.92	m2	62.27
6mm thick to metal rebates with screwed metal beads and gaskets (included elsewhere) not exceeding 2500 x 1520mm	51.36	5.00	53.93	0.65	0.10	6.70	-	60.63	m2	65.18
10mm thick to metal rebates with screwed metal beads and gaskets (included elsewhere) not exceeding 2500 x 1520mm	101.65	5.00	106.73	1.50	0.10	14.55	-	121.28	m2	130.38
12mm thick to metal rebates with screwed metal beads and gaskets (included elsewhere) not exceeding 2500 x 1520mm	133.75	5.00	140.44	2.00	0.10	19.16	-	159.60	m2	171.57

Glass; B.S.952, toughened antisun float; bronze

4mm thick to metal rebates with screwed metal beads and gaskets (included elsewhere) not exceeding 2100 x 1250mm	49.22	5.00	51.68	0.60	0.10	6.24	-	57.92	m2	62.27
6mm thick to metal rebates with screwed metal beads and gaskets (included elsewhere) not exceeding 2500 x 1520mm	51.36	5.00	53.93	0.65	0.10	6.70	-	60.63	m2	65.18
10mm thick to metal rebates with screwed metal beads and gaskets (included elsewhere) not exceeding 2500 x 1520mm	101.65	5.00	106.73	1.50	0.10	14.55	-	121.28	m2	130.38
12mm thick to metal rebates with screwed metal beads and gaskets (included elsewhere) not exceeding 2500 x 1520mm	133.75	5.00	140.44	2.00	0.10	19.16	-	159.60	m2	171.57

Glass; B.S.952, laminated safety, clear float

4.4mm thick to metal rebates with screwed metal beads and gaskets (included elsewhere) not exceeding 2100 x 1200mm	40.89	5.00	42.93	1.60	0.10	15.47	-	58.40	m2	62.78
6.4mm thick to metal rebates with screwed metal beads and gaskets (included elsewhere) not exceeding 3210 x 2000mm	34.28	5.00	35.99	2.00	0.10	19.16	-	55.16	m2	59.29

Labour hourly rates: (except Specialists) Craft Operatives £9.23 Labourer £7.02 Rates are national average prices. Refer to REGIONAL VARIATIONS for indicative levels of overall pricing in regions	MATERIALS			LABOUR				RATES		
	Del to Site £	Waste %	Material Cost £	Craft Optve Hrs	Lab Hrs	Labour Cost £	Sunds £	Nett Rate £	Unit	Gross Rate (+7.5%) £
L40: GENERAL GLAZING Cont.										
Glass; B.S.952, laminated anti-bandit, clear float										
7.5mm thick to metal rebates with screwed metal beads and gaskets (included elsewhere)										
not exceeding 3210 x 2000mm	69.42	5.00	72.89	2.50	0.10	23.78	–	96.67	m2	103.92
9.5mm thick to metal rebates with screwed metal beads and gaskets (included elsewhere)										
not exceeding 3600 x 2500mm	68.02	5.00	71.42	2.75	0.10	26.08	–	97.51	m2	104.82
11.5mm thick to metal rebates with screwed metal beads and gaskets (included elsewhere)										
not exceeding 3180 x 2000mm	73.83	5.00	77.52	3.25	0.10	30.70	–	108.22	m2	116.34
not exceeding 4500 x 2500mm	89.35	5.00	93.82	3.25	0.10	30.70	–	124.52	m2	133.86
Materials resembling glass; UVA stabilised polycarbonate sheet, latex paper masked both sides										
3mm thick to metal rebates with screwed metal beads and gaskets (included elsewhere)										
standard grade	58.93	10.00	64.82	1.00	0.10	9.93	–	74.75	m2	80.36
4mm thick to metal rebates with screwed metal beads and gaskets (included elsewhere)										
standard grade	78.81	10.00	86.69	1.00	0.10	9.93	–	96.62	m2	103.87
5mm thick to metal rebates with screwed metal beads and gaskets (included elsewhere)										
standard grade	98.60	10.00	108.46	1.00	0.10	9.93	–	118.39	m2	127.27
6mm thick to metal rebates with screwed metal beads and gaskets (included elsewhere)										
standard grade	118.05	10.00	129.85	1.15	0.10	11.32	–	141.17	m2	151.76
8mm thick to metal rebates with screwed metal beads and gaskets (included elsewhere)										
standard grade	157.40	10.00	173.14	1.35	0.10	13.16	–	186.30	m2	200.28
9.5mm thick to metal rebates with screwed metal beads and gaskets (included elsewhere)										
standard grade	187.01	10.00	205.71	1.50	0.10	14.55	–	220.26	m2	236.78
3mm thick to metal rebates with screwed metal beads and gaskets (included elsewhere)										
abrasion resistant hard coated grade	91.39	10.00	100.53	1.00	0.10	9.93	–	110.46	m2	118.75
4mm thick to metal rebates with screwed metal beads and gaskets (included elsewhere)										
abrasion resistant hard coated grade	108.84	10.00	119.72	1.00	0.10	9.93	–	129.66	m2	139.38
5mm thick to metal rebates with screwed metal beads and gaskets (included elsewhere)										
abrasion resistant hard coated grade	130.50	10.00	143.55	1.00	0.10	9.93	–	153.48	m2	164.99
6mm thick to metal rebates with screwed metal beads and gaskets (included elsewhere)										
abrasion resistant hard coated grade	145.85	10.00	160.44	1.15	0.10	11.32	–	171.75	m2	184.63
8mm thick to metal rebates with screwed metal beads and gaskets (included elsewhere)										
abrasion resistant hard coated grade	176.53	10.00	194.18	1.35	0.10	13.16	–	207.35	m2	222.90
9.5mm thick to metal rebates with screwed metal beads and gaskets (included elsewhere)										
abrasion resistant hard coated grade	199.44	10.00	219.38	1.50	0.10	14.55	–	233.93	m2	251.48
Materials resembling glass; Meshlite vandal resistant glazing; Georgian, clear, smooth both faces or crinkle one face										
3mm thick to wood rebates with screwed wood beads (included elsewhere) and B.S.544 putty										
general purpose grade	28.52	5.00	29.95	1.00	0.10	9.93	0.14	40.02	m2	43.02
fire retardant grade, class 2	34.40	5.00	36.12	1.00	0.10	9.93	0.14	46.19	m2	49.66
fire retardant grade, class 0	40.40	5.00	42.42	1.00	0.10	9.93	0.14	52.49	m2	56.43
4mm thick to wood rebates with screwed wood beads (included elsewhere) and B.S.544 putty										
general purpose grade	36.85	5.00	38.69	1.00	0.10	9.93	0.14	48.76	m2	52.42
fire retardant grade, class 2	44.49	5.00	46.71	1.00	0.10	9.93	0.14	56.79	m2	61.05
fire retardant grade, class 0	52.38	5.00	55.00	1.00	0.10	9.93	0.14	65.07	m2	69.95
6mm thick to wood rebates with screwed wood beads (included elsewhere) and B.S.544 putty										
general purpose grade	49.26	5.00	51.72	1.15	0.10	11.32	0.14	63.18	m2	67.92
fire retardant grade, class 2	62.42	5.00	65.54	1.15	0.10	11.32	0.14	77.00	m2	82.77
fire retardant grade, class 0	75.80	5.00	79.59	1.15	0.10	11.32	0.14	91.05	m2	97.87
3mm thick to metal rebates with screwed metal beads and gaskets (included elsewhere)										
general purpose grade	28.52	5.00	29.95	1.00	0.10	9.93	–	39.88	m2	42.87
fire retardant grade, class 2	34.40	5.00	36.12	1.00	0.10	9.93	–	46.05	m2	49.51
fire retardant grade, class 0	40.40	5.00	42.42	1.00	0.10	9.93	–	52.35	m2	56.28

Labour hourly rates: (except Specialists) Craft Operatives £9.23 Labourer £7.02 Rates are national average prices. Refer to REGIONAL VARIATIONS for indicative levels of overall pricing in regions	MATERIALS			LABOUR				RATES		
	Del to Site £	Waste %	Material Cost £	Craft Optve Hrs	Lab Hrs	Labour Cost £	Sunds £	Nett Rate £	Unit	Gross Rate (+7.5%) £

L40: GENERAL GLAZING Cont.

Materials resembling glass; Meshlite vandal resistant glazing; Georgian, clear, smooth both faces or crinkle one face Cont.

4mm thick to metal rebates with screwed metal beads and gaskets (included elsewhere)

general purpose grade	36.85	5.00	38.69	1.00	0.10	9.93	–	48.62	m2	52.27
fire retardant grade, class 2	44.49	5.00	46.71	1.00	0.10	9.93	–	56.65	m2	60.89
fire retardant grade, class 0	52.38	5.00	55.00	1.00	0.10	9.93	–	64.93	m2	69.80

6mm thick to metal rebates with screwed metal beads and gaskets (included elsewhere)

general purpose grade	49.26	5.00	51.72	1.15	0.10	11.32	–	63.04	m2	67.77
fire retardant grade, class 2	62.42	5.00	65.54	1.15	0.10	11.32	–	76.86	m2	82.62
fire retardant grade, class 0	75.80	5.00	79.59	1.15	0.10	11.32	–	90.91	m2	97.72

Materials resembling glass; Meshlite Vandal resistant glazing; plain, opaque colours or clear, smooth both faces or crinkle one face

2mm thick to wood rebates with screwed wood beads (included elsewhere) and B.S.544 putty

general purpose grade	21.60	5.00	22.68	1.00	0.10	9.93	0.14	32.75	m2	35.21

3mm thick to wood rebates with screwed wood beads (included elsewhere) and B.S.544 putty

general purpose grade	25.85	5.00	27.14	1.00	0.10	9.93	0.14	37.21	m2	40.01
fire retardant grade, class 2	32.51	5.00	34.14	1.00	0.10	9.93	0.14	44.21	m2	47.52
fire retardant grade, class 0	39.16	5.00	41.12	1.00	0.10	9.93	0.14	51.19	m2	55.03

4mm thick to wood rebates with screwed wood beads (included elsewhere) and B.S.544 putty

general purpose grade	34.40	5.00	36.12	1.00	0.10	9.93	0.14	46.19	m2	49.66
fire retardant grade, class 2	42.25	5.00	44.36	1.00	0.10	9.93	0.14	54.43	m2	58.52
fire retardant grade, class 0	51.10	5.00	53.66	1.00	0.10	9.93	0.14	63.73	m2	68.51

6mm thick to wood rebates with screwed wood beads (included elsewhere) and B.S.544 putty

general purpose grade	45.42	5.00	47.69	1.15	0.10	11.32	0.14	59.15	m2	63.58
fire retardant grade, class 2	59.89	5.00	62.88	1.15	0.10	11.32	0.14	74.34	m2	79.92
fire retardant grade, class 0	74.36	5.00	78.08	1.15	0.10	11.32	0.14	89.53	m2	96.25

2mm thick to metal rebates with screwed metal beads and gaskets (included elsewhere)

general purpose grade; plain	21.60	5.00	22.68	1.00	0.10	9.93	–	32.61	m2	35.06

3mm thick to metal rebates with screwed metal beads and gaskets (included elsewhere)

general purpose grade	25.85	5.00	27.14	1.00	0.10	9.93	–	37.07	m2	39.86
fire retardant grade, class 2	32.51	5.00	34.14	1.00	0.10	9.93	–	44.07	m2	47.37
fire retardant grade, class 0	39.16	5.00	41.12	1.00	0.10	9.93	–	51.05	m2	54.88

4mm thick to metal rebates with screwed metal beads and gaskets (included elsewhere)

general purpose grade	34.40	5.00	36.12	1.00	0.10	9.93	–	46.05	m2	49.51
fire retardant grade, class 2	42.25	5.00	44.36	1.00	0.10	9.93	–	54.29	m2	58.37
fire retardant grade, class 0	57.10	5.00	59.95	1.00	0.10	9.93	–	69.89	m2	75.13

6mm thick to metal rebates with screwed metal beads and gaskets (included elsewhere)

general purpose grade	45.42	5.00	47.69	1.15	0.10	11.32	–	59.01	m2	63.43
fire retardant grade, class 2	59.89	5.00	62.88	1.15	0.10	11.32	–	74.20	m2	79.77
fire retardant grade, class 0	74.36	5.00	78.08	1.15	0.10	11.32	–	89.39	m2	96.10

Materials resembling glass; Meshlite Vandal resistant glazing; diamond, clear, smooth both faces or crinkle one face

3mm thick to wood rebates with screwed wood beads (included elsewhere) and B.S.544 putty

general purpose grade	29.06	5.00	30.51	1.00	0.10	9.93	0.14	40.59	m2	43.63
fire retardant grade, class 2	34.37	5.00	36.09	1.00	0.10	9.93	0.14	46.16	m2	49.62
fire retardant grade, class 0	39.80	5.00	41.79	1.00	0.10	9.93	0.14	51.86	m2	55.75

4mm thick to wood rebates with screwed wood beads (included elsewhere) and B.S.544 putty

general purpose grade	38.04	5.00	39.94	1.00	0.10	9.93	0.14	50.01	m2	53.77
fire retardant grade, class 2	45.01	5.00	47.26	1.00	0.10	9.93	0.14	57.33	m2	61.63
fire retardant grade, class 0	51.98	5.00	54.58	1.00	0.10	9.93	0.14	64.65	m2	69.50

6mm thick to wood rebates with screwed wood beads (included elsewhere) and B.S.544 putty

general purpose grade	50.02	5.00	52.52	1.15	0.10	11.32	0.14	63.98	m2	68.78
fire retardant grade, class 2	62.79	5.00	65.93	1.15	0.10	11.32	0.14	77.39	m2	83.19
fire retardant grade, class 0	75.28	5.00	79.04	1.15	0.10	11.32	0.14	90.50	m2	97.29

3mm thick to metal rebates with screwed metal beads and gaskets (included elsewhere)

general purpose grade	29.06	5.00	30.51	1.00	0.10	9.93	–	40.45	m2	43.48
fire retardant grade, class 2	34.37	5.00	36.09	1.00	0.10	9.93	–	46.02	m2	49.47
fire retardant grade, class 0	39.80	5.00	41.79	1.00	0.10	9.93	–	51.72	m2	55.60

WINDOWS/DOORS/STAIRS

Labour hourly rates: (except Specialists) Craft Operatives £9.23 Labourer £7.02. Rates are national average prices. Refer to REGIONAL VARIATIONS for indicative levels of overall pricing in regions	MATERIALS			LABOUR				RATES		
	Del to Site £	Waste %	Material Cost £	Craft Optve Hrs	Lab Hrs	Labour Cost £	Sunds £	Nett Rate £	Unit	Gross Rate (+7.5%) £
L40: GENERAL GLAZING Cont.										
Materials resembling glass; Meshlite Vandal resistant glazing; diamond, clear, smooth both faces or crinkle one face Cont.										
4mm thick to metal rebates with screwed metal beads and gaskets (included elsewhere)										
general purpose grade	38.04	5.00	39.94	1.00	0.10	9.93	-	49.87	m2	53.61
fire retardant grade, class 2	45.01	5.00	47.26	1.00	0.10	9.93	-	57.19	m2	61.48
fire retardant grade, class 0	51.98	5.00	54.58	1.00	0.10	9.93	-	64.51	m2	69.35
6mm thick to metal rebates with screwed metal beads and gaskets (included elsewhere)										
general purpose grade	50.02	5.00	52.52	1.15	0.10	11.32	-	63.84	m2	68.63
fire retardant grade, class 2	62.29	5.00	65.40	1.15	0.10	11.32	-	76.72	m2	82.48
fire retardant grade, class 0	75.28	5.00	79.04	1.15	0.10	11.32	-	90.36	m2	97.14
Factory made double glazed hermetically sealed units										
Two panes B.S.952, clear float 3 or 4mm thick; to metal rebates with screwed metal beads and gaskets (included elsewhere)										
521mm wide x 421mm high	10.23	5.00	10.74	1.00	-	9.23	-	19.97	nr	21.47
521mm wide x 621mm high	14.88	5.00	15.62	1.25	-	11.54	-	27.16	nr	29.20
740mm wide x 740mm high	25.58	5.00	26.86	1.75	-	16.15	-	43.01	nr	46.24
848mm wide x 848mm high	33.49	5.00	35.16	2.00	-	18.46	-	53.62	nr	57.65
1048mm wide x 1048mm high	51.16	5.00	53.72	2.50	-	23.07	-	76.79	nr	82.55
1148mm wide x 1248mm high	66.51	5.00	69.84	3.00	-	27.69	-	97.53	nr	104.84
Two panes B.S.952, clear float 5 or 6mm thick; to metal rebates with screwed metal beads and gaskets (included elsewhere)										
521mm wide x 421mm high	15.35	5.00	16.12	1.10	-	10.15	-	26.27	nr	28.24
521mm wide x 621mm high	22.33	5.00	23.45	1.40	-	12.92	-	36.37	nr	39.10
740mm wide x 740mm high	38.37	5.00	40.29	1.90	-	17.54	-	57.83	nr	62.16
848mm wide x 848mm high	50.23	5.00	52.74	2.20	-	20.31	-	73.05	nr	78.53
1048mm wide x 1048mm high	76.74	5.00	80.58	2.75	-	25.38	-	105.96	nr	113.91
1148mm wide x 1248mm high	99.77	5.00	104.76	3.30	-	30.46	-	135.22	nr	145.36
Inner pane B.S.952, clear float 4mm thick; outer pane B.S.952, white patterned 4mm thick; to metal rebates with screwed metal beads and gaskets (included elsewhere)										
521mm wide x 421mm high	12.28	5.00	12.89	1.00	-	9.23	-	22.12	nr	23.78
521mm wide x 621mm high	17.86	5.00	18.75	1.25	-	11.54	-	30.29	nr	32.56
740mm wide x 740mm high	30.70	5.00	32.23	1.75	-	16.15	-	48.39	nr	52.02
848mm wide x 848mm high	40.19	5.00	42.20	2.00	-	18.46	-	60.66	nr	65.21
1048mm wide x 1048mm high	61.40	5.00	64.47	2.50	-	23.07	-	87.55	nr	94.11
1148mm wide x 1248mm high	79.81	5.00	83.80	3.00	-	27.69	-	111.49	nr	119.85
Inner pane B.S.952, clear float 4mm thick; outer pane B.S.952, white patterned 6mm thick; to metal rebates with screwed metal beads and gaskets (included elsewhere)										
521mm wide x 421mm high	18.42	5.00	19.34	1.10	-	10.15	-	29.49	nr	31.71
521mm wide x 621mm high	26.79	5.00	28.13	1.40	-	12.92	-	41.05	nr	44.13
740mm wide x 740mm high	46.05	5.00	48.35	1.90	-	17.54	-	65.89	nr	70.83
848mm wide x 848mm high	60.28	5.00	63.29	2.20	-	20.31	-	83.60	nr	89.87
1048mm wide x 1048mm high	92.09	5.00	96.69	2.75	-	25.38	-	122.08	nr	131.23
1148mm wide x 1248mm high	119.72	5.00	125.71	3.30	-	30.46	-	156.16	nr	167.88
Polyester window films; Durable Berkeley Company Ltd.,										
3M Scotchshield Safety films; applying to glass										
type SH4CLL, optically clear	-	-	Spclist	-	-	Spclist	-	26.90	m2	28.92
type SH4S1L, combination solar/safety	-	-	Spclist	-	-	Spclist	-	32.28	m2	34.70
3M Scotchtint Solar Control films; applying to glass										
type P18, silver	-	-	Spclist	-	-	Spclist	-	29.59	m2	31.81
type RE15S1X, external	-	-	Spclist	-	-	Spclist	-	37.66	m2	40.48
type RE35NEARL, neutral	-	-	Spclist	-	-	Spclist	-	29.59	m2	31.81
3M Scotchtint Plus All Seasons insulating films; applying to glass										
type LE20 S1AR (silver)	-	-	Spclist	-	-	Spclist	-	37.66	m2	40.48
type LE35 AMARL (bronze)	-	-	Spclist	-	-	Spclist	-	37.66	m2	40.48
type LE50 AMARL (bronze)	-	-	Spclist	-	-	Spclist	-	37.66	m2	40.48
Glass doors; glass, B.S.952, toughened clear, polished plate; fittings finish BMA, satin chrome or polished chrome; prices include the provision of floor springs but exclude handles										
12mm thick										
750mm wide x 2150mm high	1002.00	-	1002.00	8.00	8.00	130.00	-	1132.00	nr	1216.90
762mm wide x 2134mm high	946.00	-	946.00	8.00	8.00	130.00	-	1076.00	nr	1156.70
800mm wide x 2150mm high	1036.00	-	1036.00	8.00	8.00	130.00	-	1166.00	nr	1253.45
838mm wide x 2134mm high	994.00	-	994.00	8.00	8.00	130.00	-	1124.00	nr	1208.30
850mm wide x 2150mm high	1066.00	-	1066.00	8.00	8.00	130.00	-	1196.00	nr	1285.70
900mm wide x 2150mm high	1102.00	-	1102.00	8.00	8.00	130.00	-	1232.00	nr	1324.40
914mm wide x 2134mm high	1035.00	-	1035.00	8.00	8.00	130.00	-	1165.00	nr	1252.38
950mm wide x 2150mm high	1170.00	-	1170.00	9.00	9.00	146.25	-	1316.25	nr	1414.97
1000mm wide x 2150mm high	1163.00	-	1163.00	9.00	9.00	146.25	-	1309.25	nr	1407.44
1100mm wide x 2150mm high	1385.00	-	1385.00	10.00	10.00	162.50	-	1547.50	nr	1663.56

Labour hourly rates: (except Specialists) Craft Operatives £9.23 Labourer £7.02 Rates are national average prices. Refer to REGIONAL VARIATIONS for indicative levels of overall pricing in regions	MATERIALS			LABOUR				RATES		
	Del to Site £	Waste %	Material Cost £	Craft Optve Hrs	Lab Hrs	Labour Cost £	Sunds £	Nett Rate £	Unit	Gross Rate (+7.5%) £
L40: GENERAL GLAZING Cont.										
Glass doors; glass, B.S.952, toughened clear, polished plate; fittings finish BMA, satin chrome or polished chrome; prices include the provision of floor springs but exclude handles Cont.										
12mm thick Cont.										
1200mm wide x 2150mm high	1450.00	–	1450.00	10.00	10.00	162.50	–	1612.50	nr	1733.44
Drilling										
Hole through sheet or float glass; not exceeding 6mm thick										
6 - 15mm diameter	3.25	–	3.25	–	–	-	–	3.25	nr	3.49
16 - 38mm diameter	4.59	–	4.59	–	–	-	–	4.59	nr	4.93
exceeding 38mm diameter	9.26	–	9.26	–	–	-	–	9.26	nr	9.95
Hole through sheet or float glass; not exceeding 10mm thick										
6 - 15mm diameter	4.20	–	4.20	–	–	-	–	4.20	nr	4.51
16 - 38mm diameter	6.18	–	6.18	–	–	-	–	6.18	nr	6.64
exceeding 38mm diameter	11.24	–	11.24	–	–	-	–	11.24	nr	12.08
Hole through sheet or float glass; not exceeding 12mm thick										
6 - 15mm diameter	5.21	–	5.21	–	–	-	–	5.21	nr	5.60
16 - 38mm diameter	7.37	–	7.37	–	–	-	–	7.37	nr	7.92
exceeding 38mm diameter	13.33	–	13.33	–	–	-	–	13.33	nr	14.33
Hole through sheet or float glass; not exceeding 19mm thick										
6 - 15mm diameter	6.51	–	6.51	–	–	-	–	6.51	nr	7.00
16 - 38mm diameter	9.26	–	9.26	–	–	-	–	9.26	nr	9.95
exceeding 38mm diameter	16.41	–	16.41	–	–	-	–	16.41	nr	17.64
Hole through sheet or float glass; not exceeding 25mm thick										
6 - 15mm diameter	8.19	–	8.19	–	–	-	–	8.19	nr	8.80
16 - 38mm diameter	11.60	–	11.60	–	–	-	–	11.60	nr	12.47
exceeding 38mm diameter	20.45	–	20.45	–	–	-	–	20.45	nr	21.98
For wired and laminated glass add 50% For countersunk holes add 33 1/3%										
Bedding edges of panes										
Wash leather strips										
to edges of 3mm thick glass or the like	0.57	5.00	0.60	0.05	–	0.46	–	1.06	m	1.14
to edges of 6mm thick glass or the like	0.70	5.00	0.73	0.06	–	0.55	–	1.29	m	1.39
Rubber glazing strips										
to edges of 3mm thick glass or the like	0.53	5.00	0.56	0.05	–	0.46	–	1.02	m	1.09
to edges of 6mm thick glass or the like	0.63	5.00	0.66	0.06	–	0.55	–	1.22	m	1.31
Hacking out existing glass; preparing for re-glazing										
Float glass										
wood rebates	-	–	-	0.38	0.38	6.17	0.16	6.34	m	6.81
wood rebates and screwed wood beads; storing beads for re-use ...	-	–	-	0.45	0.45	7.31	0.16	7.47	m	8.03
metal rebates	-	–	-	0.40	0.40	6.50	0.16	6.66	m	7.16
metal rebates and screwed metal beads; storing beads for re-use	-	–	-	0.48	0.48	7.80	0.16	7.96	m	8.56
Float glass behind guard bars in position										
wood rebates and screwed wood beads; storing beads for re-use	-	–	-	0.75	0.75	12.19	0.16	12.35	m	13.27
metal rebates and screwed metal beads; storing beads for re-use	-	–	-	0.80	0.80	13.00	0.16	13.16	m	14.15

Estimating for Builders and Quantity Surveyors

R D Buchan, F W Fleming, J R Kelly

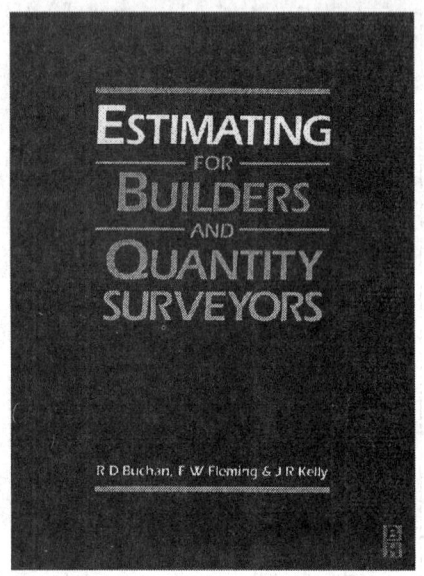

Written for students taking courses in building and surveying at HNC/D and BSc level, this textbook explains the calculation of rates for the items in bills of quantities, based on SMM7. For convenience of use in practice, the arrangement is by individual topic or trade with appropriate reference to subsections of SMM7. Care has been taken, in both discussion and examples; to emphasize the different approaches used when pricing the various types of work. The worked examples reflect both traditional and up-to-date technology.

For all the trades included, a core of the most useful examples is provided, with particular attention paid to those areas that require rather more detailed explanation than has been provided in some more summary texts.

A chapter on computerized estimating is complemented by the Appendix which shows how a spreadsheet program can be developed for the calculation of the cost of mortars etc. Chapters on tendering strategy are included to set estimating in its professional context.

£17.99, Paperback, 1991, 300pp, isbn 0 7506 0041 1

Surface finishes

Labour hourly rates: (except Specialists) Craft Operatives £9.23 Labourer £7.02 Rates are national average prices. Refer to REGIONAL VARIATIONS for indicative levels of overall pricing in regions	MATERIALS			LABOUR				RATES		
	Del to Site £	Waste %	Material Cost £	Craft Optve Hrs	Lab Hrs	Labour Cost £	Sunds £	Nett Rate £	Unit	Gross Rate (+7.5%) £
M10: CEMENT: SAND/CONCRETE SCREEDS/TOPPINGS										
Mortar, cement and sand (1:3) - screeds										
13mm work to walls on brickwork or blockwork base; one coat; screeded										
width exceeding 300mm	1.13	5.00	1.19	0.40	0.20	5.10	–	6.28	m2	6.75
width not exceeding 300mm	0.34	10.00	0.37	0.24	0.06	2.64	–	3.01	m	3.24
13mm work to walls on brickwork or blockwork base; one coat; trowelled										
width exceeding 300mm	1.13	5.00	1.19	0.52	0.20	6.20	–	7.39	m2	7.94
width not exceeding 300mm	0.34	10.00	0.37	0.31	0.06	3.28	–	3.66	m	3.93
13mm work to walls on brickwork or blockwork base; one coat; floated										
width exceeding 300mm	1.13	5.00	1.19	0.50	0.20	6.02	–	7.21	m2	7.75
width not exceeding 300mm	0.34	10.00	0.37	0.30	0.06	3.19	–	3.56	m	3.83
19mm work to floors on concrete base; one coat; screeded										
level and to falls only not exceeding 15 degrees from horizontal	1.66	5.00	1.74	0.24	0.15	3.27	0.18	5.19	m2	5.58
to falls and crossfalls and to slopes not exceeding 15 degrees from horizontal	1.66	5.00	1.74	0.34	0.15	4.19	0.18	6.11	m2	6.57
25mm work to floors on concrete base; one coat; screeded										
level and to falls only not exceeding 15 degrees from horizontal	2.18	5.00	2.29	0.26	0.16	3.52	0.18	5.99	m2	6.44
to falls and crossfalls and to slopes not exceeding 15 degrees from horizontal	2.18	5.00	2.29	0.36	0.16	4.45	0.18	6.92	m2	7.43
32mm work to floors on concrete base; one coat; screeded										
level and to falls only not exceeding 15 degrees from horizontal	2.79	5.00	2.93	0.29	0.18	3.94	0.18	7.05	m2	7.58
to falls and crossfalls and to slopes not exceeding 15 degrees from horizontal	2.79	5.00	2.93	0.39	0.18	4.86	0.18	7.97	m2	8.57
38mm work to floors on concrete base; one coat; screeded										
level and to falls only not exceeding 15 degrees from horizontal	3.32	5.00	3.49	0.31	0.19	4.20	0.18	7.86	m2	8.45
to falls and crossfalls and to slopes not exceeding 15 degrees from horizontal	3.32	5.00	3.49	0.41	0.19	5.12	0.18	8.78	m2	9.44
50mm work to floors on concrete base; one coat; screeded										
level and to falls only not exceeding 15 degrees from horizontal	4.36	5.00	4.58	0.36	0.21	4.80	0.18	9.55	m2	10.27
to falls and crossfalls and to slopes not exceeding 15 degrees from horizontal	4.36	5.00	4.58	0.46	0.21	5.72	0.18	10.48	m2	11.26
19mm work to floors on concrete base; one coat; trowelled										
level and to falls only not exceeding 15 degrees from horizontal	1.66	5.00	1.74	0.36	0.15	4.38	0.18	6.30	m2	6.77
to falls and crossfalls and to slopes not exceeding 15 degrees from horizontal	1.66	5.00	1.74	0.46	0.15	5.30	0.18	7.22	m2	7.76
25mm work to floors on concrete base; one coat; trowelled										
level and to falls only not exceeding 15 degrees from horizontal	2.18	5.00	2.29	0.38	0.16	4.63	0.18	7.10	m2	7.63
to falls and crossfalls and to slopes not exceeding 15 degrees from horizontal	2.18	5.00	2.29	0.48	0.16	5.55	0.18	8.02	m2	8.62

Labour hourly rates: (except Specialists) Craft Operatives £9.23 Labourer £7.02 Rates are national average prices. Refer to REGIONAL VARIATIONS for indicative levels of overall pricing in regions	MATERIALS			LABOUR				RATES		
	Del to Site £	Waste %	Material Cost £	Craft Optve Hrs	Lab Hrs	Labour Cost £	Sunds £	Nett Rate £	Unit	Gross Rate (+7.5%) £
M10: CEMENT: SAND/CONCRETE SCREEDS/TOPPINGS Cont.										
Mortar, cement and sand (1:3) - screeds Cont.										
32mm work to floors on concrete base; one coat; trowelled										
level and to falls only not exceeding 15 degrees from horizontal	2.79	5.00	2.93	0.41	0.18	5.05	0.18	8.16	m2	8.77
to falls and crossfalls and to slopes not exceeding 15 degrees from horizontal	2.79	5.00	2.93	0.51	0.18	5.97	0.18	9.08	m2	9.76
38mm work to floors on concrete base; one coat; trowelled										
level and to falls only not exceeding 15 degrees from horizontal	3.32	5.00	3.49	0.43	0.19	5.30	0.18	8.97	m2	9.64
to falls and crossfalls and to slopes not exceeding 15 degrees from horizontal	3.32	5.00	3.49	0.53	0.19	6.23	0.18	9.89	m2	10.63
50mm work to floors on concrete base; one coat; trowelled										
level and to falls only not exceeding 15 degrees from horizontal	4.36	5.00	4.58	0.48	0.21	5.90	0.18	10.66	m2	11.46
to falls and crossfalls and to slopes not exceeding 15 degrees from horizontal	4.36	5.00	4.58	0.58	0.21	6.83	0.18	11.59	m2	12.45
19mm work to floors on concrete base; one coat; floated										
level and to falls only not exceeding 15 degrees from horizontal	1.66	5.00	1.74	0.34	0.15	4.19	0.18	6.11	m2	6.57
to falls and crossfalls and to slopes not exceeding 15 degrees from horizontal	1.66	5.00	1.74	0.44	0.15	5.11	0.18	7.04	m2	7.56
25mm work to floors on concrete base; one coat; floated										
level and to falls only not exceeding 15 degrees from horizontal	2.18	5.00	2.29	0.36	0.16	4.45	0.18	6.92	m2	7.43
to falls and crossfalls and to slopes not exceeding 15 degrees from horizontal	2.18	5.00	2.29	0.46	0.16	5.37	0.18	7.84	m2	8.43
32mm work to floors on concrete base; one coat; floated										
level and to falls only not exceeding 15 degrees from horizontal	2.79	5.00	2.93	0.39	0.18	4.86	0.18	7.97	m2	8.57
to falls and crossfalls and to slopes not exceeding 15 degrees from horizontal	2.79	5.00	2.93	0.49	0.18	5.79	0.18	8.90	m2	9.56
38mm work to floors on concrete base; one coat; floated										
level and to falls only not exceeding 15 degrees from horizontal	3.32	5.00	3.49	0.41	0.19	5.12	0.18	8.78	m2	9.44
to falls and crossfalls and to slopes not exceeding 15 degrees from horizontal	3.32	5.00	3.49	0.51	0.19	6.04	0.18	9.71	m2	10.44
50mm work to floors on concrete base; one coat; floated										
level and to falls only not exceeding 15 degrees from horizontal	4.36	5.00	4.58	0.46	0.21	5.72	0.18	10.48	m2	11.26
to falls and crossfalls and to slopes not exceeding 15 degrees from horizontal	4.36	5.00	4.58	0.56	0.21	6.64	0.18	11.40	m2	12.26
Mortar, cement and sand (1:3) - paving										
25mm work to floors on concrete base; one coat; trowelled										
level and to falls only not exceeding 15 degrees from horizontal	2.18	5.00	2.29	0.38	0.16	4.63	0.18	7.10	m2	7.63
to falls and crossfalls and to slopes not exceeding 15 degrees from horizontal	2.18	5.00	2.29	0.48	0.16	5.55	0.18	8.02	m2	8.62
to slopes exceeding 15 degrees from horizontal	2.18	5.00	2.29	0.53	0.16	6.02	0.18	8.48	m2	9.12
32mm work to floors on concrete base; one coat; trowelled										
level and to falls only not exceeding 15 degrees from horizontal	2.79	5.00	2.93	0.41	0.18	5.05	0.18	8.16	m2	8.77
to falls and crossfalls and to slopes not exceeding 15 degrees from horizontal	2.79	5.00	2.93	0.51	0.18	5.97	0.18	9.08	m2	9.76
to slopes exceeding 15 degrees from horizontal	2.79	5.00	2.93	0.56	0.18	6.43	0.18	9.54	m2	10.26
38mm work to floors on concrete base; one coat; trowelled										
level and to falls only not exceeding 15 degrees from horizontal	3.32	5.00	3.49	0.43	0.19	5.30	0.18	8.97	m2	9.64
to falls and crossfalls and to slopes not exceeding 15 degrees from horizontal	3.32	5.00	3.49	0.53	0.19	6.23	0.18	9.89	m2	10.63
to slopes exceeding 15 degrees from horizontal	3.32	5.00	3.49	0.58	0.19	6.69	0.18	10.35	m2	11.13
50mm work to floors on concrete base; one coat; trowelled										
level and to falls only not exceeding 15 degrees from horizontal	4.36	5.00	4.58	0.48	0.21	5.90	0.18	10.66	m2	11.46
to falls and crossfalls and to slopes not exceeding 15 degrees from horizontal	4.36	5.00	4.58	0.58	0.21	6.83	0.18	11.59	m2	12.45
to slopes exceeding 15 degrees from horizontal	4.36	5.00	4.58	0.63	0.21	7.29	0.18	12.05	m2	12.95
If paving oil proofed, add										
25mm work	1.37	10.00	1.51	-	0.03	0.21	-	1.72	m2	1.85

Labour hourly rates: (except Specialists) Craft Operatives £9.23 Labourer £7.02 Rates are national average prices. Refer to REGIONAL VARIATIONS for indicative levels of overall pricing in regions	MATERIALS			LABOUR				RATES		
	Del to Site £	Waste %	Material Cost £	Craft Optve Hrs	Lab Hrs	Labour Cost £	Sunds £	Nett Rate £	Unit	Gross Rate (+7.5%) £
M10: CEMENT: SAND/CONCRETE SCREEDS/TOPPINGS Cont.										
Mortar, cement and sand (1:3) - paving Cont.										
If paving oil proofed, add Cont.										
32mm work	1.76	10.00	**1.94**	–	0.03	**0.21**	–	**2.15**	m2	2.31
38mm work	2.09	10.00	**2.30**	–	0.04	**0.28**	–	**2.58**	m2	2.77
50mm work	2.75	10.00	**3.02**	–	0.05	**0.35**	–	**3.38**	m2	3.63
19mm work to skirtings on brickwork or blockwork base										
height 100mm	0.25	10.00	**0.28**	0.70	0.05	**6.81**	–	**7.09**	m	7.62
height 150mm	0.37	10.00	**0.41**	0.80	0.05	**7.74**	–	**8.14**	m	8.75
25mm work to skirtings on brickwork or blockwork base										
height 100mm	0.33	10.00	**0.36**	0.75	0.05	**7.27**	–	**7.64**	m	8.21
height 150mm	0.49	10.00	**0.54**	0.85	0.05	**8.20**	–	**8.74**	m	9.39
Granolithic; cement and granite chippings (2:5); steel trowelled										
25mm work to floors on concrete base; one coat										
level and to falls only not exceeding 15 degrees from horizontal	4.24	5.00	**4.45**	0.42	0.18	**5.14**	0.36	**9.95**	m2	10.70
to falls and crossfalls and to slopes not exceeding 15 degrees from horizontal	4.24	5.00	**4.45**	0.52	0.18	**6.06**	0.36	**10.88**	m2	11.69
to slopes exceeding 15 degrees from horizontal	4.24	5.00	**4.45**	0.57	0.18	**6.52**	0.36	**11.34**	m2	12.19
32mm work to floors on concrete base; one coat										
level and to falls only not exceeding 15 degrees from horizontal	5.43	5.00	**5.70**	0.45	0.20	**5.56**	0.36	**11.62**	m2	12.49
to falls and crossfalls and to slopes not exceeding 15 degrees from horizontal	5.43	5.00	**5.70**	0.55	0.20	**6.48**	0.36	**12.54**	m2	13.48
to slopes exceeding 15 degrees from horizontal	5.43	5.00	**5.70**	0.61	0.20	**7.03**	0.36	**13.10**	m2	14.08
38mm work to floors on concrete base; one coat										
level and to falls only not exceeding 15 degrees from horizontal	6.45	5.00	**6.77**	0.48	0.21	**5.90**	0.36	**13.04**	m2	14.01
to falls and crossfalls and to slopes not exceeding 15 degrees from horizontal	6.45	5.00	**6.77**	0.58	0.21	**6.83**	0.36	**13.96**	m2	15.01
to slopes exceeding 15 degrees from horizontal	6.45	5.00	**6.77**	0.63	0.21	**7.29**	0.36	**14.42**	m2	15.50
50mm work to floors on concrete base; one coat										
level and to falls only not exceeding 15 degrees from horizontal	8.48	5.00	**8.90**	0.54	0.24	**6.67**	0.36	**15.93**	m2	17.13
to falls and crossfalls and to slopes not exceeding 15 degrees from horizontal	8.48	5.00	**8.90**	0.64	0.24	**7.59**	0.36	**16.86**	m2	18.12
to slopes exceeding 15 degrees from horizontal	8.48	5.00	**8.90**	0.69	0.24	**8.05**	0.36	**17.32**	m2	18.62
If paving tinted, add according to tint										
25mm work	3.82	10.00	**4.20**	–	0.03	**0.21**	–	**4.41**	m2	4.74
32mm work	4.89	10.00	**5.38**	–	0.03	**0.21**	–	**5.59**	m2	6.01
38mm work	5.80	10.00	**6.38**	–	0.04	**0.28**	–	**6.66**	m2	7.16
50mm work	7.63	10.00	**8.39**	–	0.05	**0.35**	–	**8.74**	m2	9.40
If carborundum trowelled into surface										
add	2.50	10.00	**2.75**	0.08	0.04	**1.02**	–	**3.77**	m2	4.05
25mm linings to channels on concrete base										
150mm girth on face; to falls	0.95	10.00	**1.04**	0.20	0.05	**2.20**	0.05	**3.29**	m	3.54
225mm girth on face; to falls	1.43	10.00	**1.57**	0.25	0.05	**2.66**	0.08	**4.31**	m	4.63
32mm linings to channels on concrete base										
150mm girth on face; to falls	1.22	10.00	**1.34**	0.23	0.05	**2.47**	0.05	**3.87**	m	4.16
225mm girth on face; to falls	1.83	10.00	**2.01**	0.28	0.05	**2.94**	0.08	**5.03**	m	5.41
38mm linings to channels on concrete base										
150mm girth on face; to falls	1.45	10.00	**1.60**	0.25	0.05	**2.66**	0.05	**4.30**	m	4.63
225mm girth on face; to falls	2.18	10.00	**2.40**	0.30	0.05	**3.12**	0.08	**5.60**	m	6.02
50mm linings to channels on concrete base										
150mm girth on face; to falls	1.91	10.00	**2.10**	0.30	0.05	**3.12**	0.05	**5.27**	m	5.67
225mm girth on face; to falls	2.86	10.00	**3.15**	0.35	0.05	**3.58**	0.08	**6.81**	m	7.32
13mm work to skirtings on concrete base										
height 150mm	0.50	10.00	**0.55**	0.75	0.05	**7.27**	–	**7.82**	m	8.41
19mm work to skirtings on concrete base										
height 150mm	0.73	10.00	**0.80**	0.80	0.05	**7.74**	–	**8.54**	m	9.18
25mm work to skirtings on concrete base										
height 150mm	0.95	10.00	**1.04**	0.85	0.05	**8.20**	–	**9.24**	m	9.93
32mm work to skirtings on concrete base										
height 150mm	1.22	10.00	**1.34**	0.90	0.05	**8.66**	–	**10.00**	m	10.75
Rounded angles and intersections										
10 100mm radius	–				0.18	**1.26**		**1.26**	m	1.36
Surface hardeners on paving										
Proprietary surface hardener (Sodium Silicate Base)										
two coats	0.70	5.00	**0.73**	–	0.20	**1.40**	–	**2.14**	m2	2.30
three coats	1.05	5.00	**1.10**	–	0.25	**1.75**	–	**2.86**	m2	3.07
Polyurethane floor sealer										
two coats	1.32	5.00	**1.39**	–	0.20	**1.40**	–	**2.79**	m2	3.00

Labour hourly rates: (except Specialists) Craft Operatives £9.23 Labourer £7.02 Rates are national average prices. Refer to REGIONAL VARIATIONS for indicative levels of overall pricing in regions	MATERIALS			LABOUR				RATES		
	Del to Site £	Waste %	Material Cost £	Craft Optve Hrs	Lab Hrs	Labour Cost £	Sunds £	Nett Rate £	Unit	Gross Rate (+7.5%) £
M10: CEMENT: SAND/CONCRETE SCREEDS/TOPPINGS Cont.										
Surface hardeners on paving Cont.										
Polyurethane floor sealer Cont.										
three coats	1.92	5.00	2.02	–	0.25	1.75	–	3.77	m2	4.05
Nitoflor Lithurin concrete surface dressing										
one coat	0.36	10.00	0.40	–	0.17	1.19	–	1.59	m2	1.71
two coats	0.47	10.00	0.52	–	0.24	1.68	–	2.20	m2	2.37
If on old floors										
add for cleaning and degreasing	0.50	5.00	0.53	–	0.35	2.46	–	2.98	m2	3.21
Vermiculite screed consisting of cement and Vermiculite aggregate (1:6) finished with 20mm cement and sand (1:4) screeded bed										
45mm work to roofs on concrete base; two coats to falls and crossfalls and to slopes not exceeding 15 degrees from horizontal	4.19	5.00	4.40	0.45	0.48	7.52	0.18	12.10	m2	13.01
60mm work to roofs on concrete base; two coats to falls and crossfalls and to slopes not exceeding 15 degrees from horizontal	5.77	5.00	6.06	0.48	0.51	8.01	0.18	14.25	m2	15.32
70mm work to roofs on concrete base; two coats to falls and crossfalls and to slopes not exceeding 15 degrees from horizontal	6.82	5.00	7.16	0.51	0.54	8.50	0.18	15.84	m2	17.03
80mm work to roofs on concrete base; two coats to falls and crossfalls and to slopes not exceeding 15 degrees from horizontal	7.86	5.00	8.25	0.53	0.56	8.82	0.18	17.26	m2	18.55
Lightweight concrete screed consisting of cement and lightweight aggregate, medium grade, 10 - 5 gauge, 799 kg/m3 (1:10) finished with 15mm cement and sand (1:4) trowelled bed										
50mm work to roofs on concrete base; two coats to falls and crossfalls and to slopes not exceeding 15 degrees from horizontal	3.80	5.00	3.99	0.60	0.51	9.12	0.18	13.29	m2	14.28
75mm work to roofs on concrete base; two coats to falls and crossfalls and to slopes not exceeding 15 degrees from horizontal	5.45	5.00	5.72	0.65	0.56	9.93	0.18	15.83	m2	17.02
100mm work to roofs on concrete base; two coats to falls and crossfalls and to slopes not exceeding 15 degrees from horizontal	7.40	5.00	7.77	0.70	0.61	10.74	0.18	18.69	m2	20.10
Screed reinforcement										
Reinforcement; galvanised wire netting, B.S.1485, 13mm mesh, 22 gauge wire; 150mm laps; placing in position										
floors	2.58	5.00	2.71	0.05	–	0.46	–	3.17	m2	3.41
Reinforcement; galvanised wire netting, B.S.1485, 25mm mesh, 22 gauge wire; 150mm laps; placing in position										
floors	1.29	5.00	1.35	0.05	–	0.46	–	1.82	m2	1.95
Reinforcement; galvanised wire netting, B.S.1485, 38mm mesh, 19 gauge wire; 150mm laps; placing in position										
floors	1.44	5.00	1.51	0.05	–	0.46	–	1.97	m2	2.12
Reinforcement; galvanised wire netting, B.S.1485, 50mm mesh, 19 gauge wire; 150mm laps; placing in position										
floors	1.08	5.00	1.13	0.05	–	0.46	–	1.60	m2	1.72
Division strips										
Aluminium dividing strips; setting in bed										
5 x 25mm; flat section	4.60	5.00	4.83	0.15	–	1.38	–	6.21	m	6.68
Brass dividing strips; setting in bed										
5 x 25mm; flat section	5.70	5.00	5.99	0.15	–	1.38	–	7.37	m	7.92
M10: CEMENT: SAND/CONCRETE SCREEDS/TOPPINGS; STAIRCASE AREAS										
Mortar, cement and sand (1:3) - paving										
13mm work to walls on brickwork or blockwork base; one coat; screeded										
width exceeding 300mm	1.13	5.00	1.19	0.60	0.20	6.94	–	8.13	m2	8.74
width not exceeding 300mm	0.34	10.00	0.37	0.36	0.06	3.74	–	4.12	m	4.43
13mm work to walls on brickwork or blockwork base; one coat; trowelled										
width exceeding 300mm	1.13	5.00	1.19	0.78	0.20	8.60	–	9.79	m2	10.52
width not exceeding 300mm	0.34	10.00	0.37	0.47	0.06	4.76	–	5.13	m	5.52

	MATERIALS			LABOUR				RATES		
Labour hourly rates: (except Specialists) Craft Operatives £9.23 Labourer £7.02 Rates are national average prices. Refer to REGIONAL VARIATIONS for indicative levels of overall pricing in regions	Del to Site £	Waste %	Material Cost £	Craft Optve Hrs	Lab Hrs	Labour Cost £	Sunds £	Nett Rate £	Unit	Gross Rate (+7.5%) £

M10: CEMENT: SAND/CONCRETE SCREEDS/TOPPINGS; STAIRCASE AREAS Cont.

Mortar, cement and sand (1:3) - paving Cont.

	Del to Site £	Waste %	Material Cost £	Craft Optve Hrs	Lab Hrs	Labour Cost £	Sunds £	Nett Rate £	Unit	Gross Rate (+7.5%) £
13mm work to walls on brickwork or blockwork base; one coat; floated										
width exceeding 300mm	1.13	5.00	1.19	0.75	0.20	8.33	-	9.51	m2	10.23
width not exceeding 300mm	0.34	10.00	0.37	0.45	0.06	4.57	-	4.95	m	5.32
19mm work to floors on concrete base; one coat; screeded										
level and to falls only not exceeding 15 degrees from horizontal	1.66	5.00	1.74	0.36	0.15	4.38	0.18	6.30	m2	6.77
25mm work to floors on concrete base; one coat; screeded										
level and to falls only not exceeding 15 degrees from horizontal	2.18	5.00	2.29	0.39	0.16	4.72	0.18	7.19	m2	7.73
32mm work to floors on concrete base; one coat; screeded										
level and to falls only not exceeding 15 degrees from horizontal	2.79	5.00	2.93	0.44	0.18	5.32	0.18	8.43	m2	9.07
38mm work to floors on concrete base; one coat; screeded										
level and to falls only not exceeding 15 degrees from horizontal	3.32	5.00	3.49	0.47	0.19	5.67	0.18	9.34	m2	10.04
50mm work to floors on concrete base; one coat; screeded										
level and to falls only not exceeding 15 degrees from horizontal	4.36	5.00	4.58	0.54	0.21	6.46	0.18	11.22	m2	12.06
19mm work to floors on concrete base; one coat; trowelled										
level and to falls only not exceeding 15 degrees from horizontal	1.66	5.00	1.74	0.54	0.15	6.04	0.18	7.96	m2	8.56
25mm work to floors on concrete base; one coat; trowelled										
level and to falls only not exceeding 15 degrees from horizontal	2.18	5.00	2.29	0.57	0.16	6.38	0.18	8.85	m2	9.52
32mm work to floors on concrete base; one coat; trowelled										
level and to falls only not exceeding 15 degrees from horizontal	2.79	5.00	2.93	0.62	0.18	6.99	0.18	10.10	m2	10.85
38mm work to floors on concrete base; one coat; trowelled										
level and to falls only not exceeding 15 degrees from horizontal	3.32	5.00	3.49	0.65	0.19	7.33	0.18	11.00	m2	11.82
50mm work to floors on concrete base; one coat; trowelled										
level and to falls only not exceeding 15 degrees from horizontal	4.36	5.00	4.58	0.72	0.21	8.12	0.18	12.88	m2	13.84
19mm work to floors on concrete base; one coat; floated										
level and to falls only not exceeding 15 degrees from horizontal	1.66	5.00	1.74	0.51	0.15	5.76	0.18	7.68	m2	8.26
25mm work to floors on concrete base; one coat; floated										
level and to falls only not exceeding 15 degrees from horizontal	2.18	5.00	2.29	0.54	0.16	6.11	0.18	8.58	m2	9.22
32mm work to floors on concrete base; one coat; floated										
level and to falls only not exceeding 15 degrees from horizontal	2.79	5.00	2.93	0.59	0.18	6.71	0.18	9.82	m2	10.56
38mm work to floors on concrete base; one coat; floated										
level and to falls only not exceeding 15 degrees from horizontal	3.32	5.00	3.49	0.62	0.19	7.06	0.18	10.72	m2	11.53
50mm work to floors on concrete base; one coat; floated										
level and to falls only not exceeding 15 degrees from horizontal	4.36	5.00	4.58	0.69	0.21	7.84	0.18	12.60	m2	13.55
19mm work to treads on concrete base; screeded										
width 275mm	0.68	10.00	0.75	0.15	0.04	1.67	0.05	2.46	m	2.65
25mm work to treads on concrete base; screeded										
width 275mm	0.90	10.00	0.99	0.16	0.04	1.76	0.05	2.80	m	3.01
32mm work to treads on concrete base; screeded										
width 275mm	1.15	10.00	1.26	0.18	0.05	2.01	0.05	3.33	m	3.58
38mm work to treads on concrete base; screeded										
width 275mm	1.37	10.00	1.51	0.20	0.05	2.20	0.05	3.75	m	4.04

SURFACE FINISHES

Labour hourly rates: (except Specialists) Craft Operatives £9.23 Labourer £7.02 Rates are national average prices. Refer to REGIONAL VARIATIONS for indicative levels of overall pricing in regions	MATERIALS			LABOUR				RATES		
	Del to Site £	Waste %	Material Cost £	Craft Optve Hrs	Lab Hrs	Labour Cost £	Sunds £	Nett Rate £	Unit	Gross Rate (+7.5%) £
M10: CEMENT: SAND/CONCRETE SCREEDS/TOPPINGS; STAIRCASE AREAS Cont.										
Mortar, cement and sand (1:3) - paving Cont.										
19mm work to treads on concrete base; floated width 275mm	0.68	10.00	0.75	0.20	0.04	2.13	0.05	2.92	m	3.14
25mm work to treads on concrete base; floated width 275mm	0.90	10.00	0.99	0.21	0.04	2.22	0.05	3.26	m	3.50
32mm work to treads on concrete base; floated width 275mm	1.15	10.00	1.26	0.23	0.05	2.47	0.05	3.79	m	4.07
38mm work to treads on concrete base; floated width 275mm	1.37	10.00	1.51	0.25	0.05	2.66	0.05	4.22	m	4.53
19mm work to treads on concrete base; trowelled width 275mm	0.68	10.00	0.75	0.25	0.04	2.59	0.05	3.39	m	3.64
25mm work to treads on concrete base; trowelled width 275mm	0.90	10.00	0.99	0.26	0.04	2.68	0.05	3.72	m	4.00
32mm work to treads on concrete base; trowelled width 275mm	1.15	10.00	1.26	0.28	0.05	2.94	0.05	4.25	m	4.57
38mm work to treads on concrete base; trowelled width 275mm	1.37	10.00	1.51	0.30	0.05	3.12	0.05	4.68	m	5.03
19mm work to plain risers on concrete base; keyed height 175mm	0.44	10.00	0.48	0.30	0.03	2.98	-	3.46	m	3.72
25mm work to plain risers on concrete base; keyed height 175mm	0.57	10.00	0.63	0.35	0.03	3.44	-	4.07	m	4.37
32mm work to plain risers on concrete base; keyed height 175mm	0.73	10.00	0.80	0.40	0.03	3.90	-	4.71	m	5.06
19mm work to plain risers on concrete base; trowelled height 175mm	0.44	10.00	0.48	0.50	0.03	4.83	-	5.31	m	5.71
25mm work to plain risers on concrete base; trowelled height 175mm	0.57	10.00	0.63	0.55	0.03	5.29	-	5.91	m	6.36
32mm work to plain risers on concrete base; trowelled height 175mm	0.73	10.00	0.80	0.60	0.03	5.75	-	6.55	m	7.04
19mm work to undercut risers on concrete base; keyed height 175mm	0.44	10.00	0.48	0.35	0.03	3.44	-	3.93	m	4.22
25mm work to undercut risers on concrete base; keyed height 175mm	0.57	10.00	0.63	0.40	0.03	3.90	-	4.53	m	4.87
32mm work to undercut risers on concrete base; keyed height 175mm	0.73	10.00	0.80	0.45	0.03	4.36	-	5.17	m	5.55
19mm work to undercut risers on concrete base; trowelled height 175mm	0.44	10.00	0.48	0.55	0.03	5.29	-	5.77	m	6.20
25mm work to undercut risers on concrete base; trowelled height 175mm	0.57	10.00	0.63	0.60	0.03	5.75	-	6.38	m	6.85
32mm work to undercut risers on concrete base; trowelled height 175mm	0.73	10.00	0.80	0.65	0.03	6.21	-	7.01	m	7.54
Granolithic; cement and granite chippings (2:5); steel trowelled - paving										
25mm work to floors on concrete base; one coat level and to falls only not exceeding 15 degrees from horizontal	4.24	5.00	4.45	0.63	0.18	7.08	0.36	11.89	m2	12.78
32mm work to floors on concrete base; one coat level and to falls only not exceeding 15 degrees from horizontal	5.43	5.00	5.70	0.68	0.20	7.68	0.36	13.74	m2	14.77
38mm work to floors on concrete base; one coat level and to falls only not exceeding 15 degrees from horizontal	6.45	5.00	6.77	0.72	0.21	8.12	0.36	15.25	m2	16.40
50mm work to floors on concrete base; one coat level and to falls only not exceeding 15 degrees from horizontal	8.48	5.00	8.90	0.81	0.24	9.16	0.36	18.43	m2	19.81
25mm work to treads on concrete base width 275mm	1.75	10.00	1.93	0.25	0.05	2.66	0.10	4.68	m	5.03
32mm work to treads on concrete base width 275mm	2.24	10.00	2.46	0.30	0.06	3.19	0.10	5.75	m	6.19
38mm work to treads on concrete base width 275mm	2.66	10.00	2.93	0.35	0.06	3.65	0.10	6.68	m	7.18

SURFACE FINISHES – MAJOR WORKS

Labour hourly rates: (except Specialists) Craft Operatives £9.23 Labourer £7.02 Rates are national average prices. Refer to REGIONAL VARIATIONS for indicative levels of overall pricing in regions	MATERIALS			LABOUR				RATES		
	Del to Site	Waste	Material Cost	Craft Optve	Lab	Labour Cost	Sunds	Nett Rate	Unit	Gross Rate (+7.5%)
	£	%	£	Hrs	Hrs	£	£	£		£
M10: CEMENT: SAND/CONCRETE SCREEDS/TOPPINGS; STAIRCASE AREAS Cont.										
Granolithic; cement and granite chippings (2:5); steel trowelled - paving Cont.										
50mm work to treads on concrete base width 275mm	3.50	10.00	**3.85**	0.40	0.07	**4.18**	0.10	**8.13**	m	**8.74**
13mm work to plain risers on concrete base height 175mm	0.58	10.00	**0.64**	0.50	0.05	**4.97**	-	**5.60**	m	**6.02**
19mm work to plain risers on concrete base height 175mm	0.85	10.00	**0.94**	0.55	0.05	**5.43**	-	**6.36**	m	**6.84**
25mm work to plain risers on concrete base height 175mm	1.11	10.00	**1.22**	0.60	0.05	**5.89**	-	**7.11**	m	**7.64**
32mm work to plain risers on concrete base height 175mm	1.43	10.00	**1.57**	0.65	0.05	**6.35**	-	**7.92**	m	**8.52**
13mm work to undercut risers on concrete base height 175mm	0.58	10.00	**0.64**	0.55	0.05	**5.43**	-	**6.07**	m	**6.52**
19mm work to undercut risers on concrete base height 175mm	0.85	10.00	**0.94**	0.60	0.05	**5.89**	-	**6.82**	m	**7.34**
25mm work to undercut risers on concrete base height 175mm	1.11	10.00	**1.22**	0.65	0.05	**6.35**	-	**7.57**	m	**8.14**
32mm work to undercut risers on concrete base height 175mm	1.43	10.00	**1.57**	0.70	0.05	**6.81**	-	**8.38**	m	**9.01**
13mm work to strings on concrete base height 300mm	0.99	10.00	**1.09**	0.70	0.05	**6.81**	-	**7.90**	m	**8.49**
19mm work to strings on concrete base height 300mm	1.45	10.00	**1.60**	0.75	0.05	**7.27**	-	**8.87**	m	**9.53**
25mm work to strings on concrete base height 300mm	1.91	10.00	**2.10**	0.80	0.05	**7.74**	-	**9.84**	m	**10.57**
32mm work to strings on concrete base height 300mm	2.44	10.00	**2.68**	0.85	0.06	**8.27**	-	**10.95**	m	**11.77**
13mm work to strings on brickwork or blockwork base height 275mm	0.91	10.00	**1.00**	0.65	0.05	**6.35**	-	**7.35**	m	**7.90**
19mm work to strings on brickwork or blockwork base height 275mm	1.33	10.00	**1.46**	0.70	0.05	**6.81**	-	**8.28**	m	**8.90**
25mm work to strings on brickwork or blockwork base height 275mm	1.75	10.00	**1.93**	0.75	0.05	**7.27**	-	**9.20**	m	**9.89**
32mm work to strings on brickwork or blockwork base height 275mm	2.24	10.00	**2.46**	0.80	0.06	**7.81**	-	**10.27**	m	**11.04**
13mm work to aprons on concrete base height 150mm	0.50	10.00	**0.55**	0.50	0.05	**4.97**	-	**5.52**	m	**5.93**
19mm work to aprons on concrete base height 150mm	0.73	10.00	**0.80**	0.55	0.05	**5.43**	-	**6.23**	m	**6.70**
25mm work to aprons on concrete base height 150mm	0.95	10.00	**1.04**	0.60	0.05	**5.89**	-	**6.93**	m	**7.45**
32mm work to aprons on concrete base height 150mm	1.22	10.00	**1.34**	0.65	0.05	**6.35**	-	**7.69**	m	**8.27**
M11: MASTIC ASPHALT FLOORING/FLOOR UNDERLAYS										
Paving; B.S.6925 (limestone aggregate) Grade III, brown, smooth floated finish										
Flooring and underlay 15mm thick; coats of asphalt - 1; to concrete base; flat width not exceeding 150mm	-	-	Spclist	-	-	Spclist	-	24.75	m2	26.61
width 150 - 225mm	-	-	Spclist	-	-	Spclist	-	20.25	m2	21.77
width 225 - 300mm	-	-	Spclist	-	-	Spclist	-	16.10	m2	17.31
width exceeding 300mm	-	-	Spclist	-	-	Spclist	-	12.20	m2	13.12
Skirtings 15mm thick; coats of asphalt - 1; no underlay; to brickwork base girth not exceeding 150mm	-	-	Spclist	-	-	Spclist	-	7.80	m	8.38
Paving; B.S.6925 (limestone aggregate) Grade III, red, smooth floated finish										
Flooring and underlay 15mm thick; coats of asphalt - 1; to concrete base; flat width not exceeding 150mm	-	-	Spclist	-	-	Spclist	-	26.25	m2	28.22
width 150 - 225mm	-	-	Spclist	-	-	Spclist	-	21.80	m2	23.43
width 225 - 300mm	-	-	Spclist	-	-	Spclist	-	17.50	m2	18.81
width exceeding 300mm	-	-	Spclist	-	-	Spclist	-	13.10	m2	14.08

Labour hourly rates: (except Specialists) Craft Operatives £9.23 Labourer £7.02 Rates are national average prices. Refer to REGIONAL VARIATIONS for indicative levels of overall pricing in regions	MATERIALS			LABOUR				RATES		
	Del to Site	Waste	Material Cost	Craft Optve	Lab	Labour Cost	Sunds	Nett Rate	Unit	Gross Rate (+7.5%)
	£	%	£	Hrs	Hrs	£	£	£		£
M11: MASTIC ASPHALT FLOORING/FLOOR UNDERLAYS Cont.	..									
Paving; B.S.6925 (limestone aggregate) Grade III, red, smooth floated finish Cont.										
Skirtings 15mm thick; coats of asphalt - 1; no underlay; to brickwork base										
girth not exceeding 150mm	-	-	Spclist	-	-	Spclist	-	8.95	m	9.62
M12: TROWELLED BITUMEN/RESIN/RUBBER LATEX FLOORING										
One coat levelling screeds										
3mm work to floors on concrete base; one coat level and to falls only not exceeding 15 degrees from horizontal	2.64	5.00	2.77	0.40	0.05	4.04	-	6.82	m2	7.33
3mm work to floors on existing timber boarded base; one coat level and to falls only not exceeding 15 degrees from horizontal	2.64	5.00	2.77	0.50	0.06	5.04	-	7.81	m2	8.39
M20: PLASTERED/RENDERED/ROUGHCAST COATINGS										
PLASTERBOARD - Baseboarding; gypsum baseboard, Thistle; 5mm joints, filled with plaster and scrimmed; fixing with galvanised nails; internal										
9.5mm work to walls on timber base										
width exceeding 300mm	1.61	5.00	1.69	0.25	0.13	3.22	0.17	5.08	m2	5.46
width not exceeding 300mm	0.48	10.00	0.53	0.15	0.04	1.67	0.10	2.29	m	2.47
9.5mm work to ceilings on timber base										
width exceeding 300mm	1.61	5.00	1.69	0.30	0.13	3.68	0.17	5.54	m2	5.96
width not exceeding 300mm	0.48	10.00	0.53	0.18	0.04	1.94	0.10	2.57	m	2.76
9.5mm work to isolated beams on timber base										
width exceeding 300mm	1.61	5.00	1.69	0.38	0.13	4.42	0.17	6.28	m2	6.75
width not exceeding 300mm	0.48	10.00	0.53	0.23	0.04	2.40	0.10	3.03	m	3.26
9.5mm work to isolated columns on timber base										
width exceeding 300mm	1.61	5.00	1.69	0.38	0.13	4.42	0.17	6.28	m2	6.75
width not exceeding 300mm	0.48	10.00	0.53	0.23	0.04	2.40	0.10	3.03	m	3.26
PLASTERBOARD - Baseboarding; gypsum baseboard, Thistle Duplex; 5mm joints filled with plaster and scrimmed; fixing with galvanised nails; internal										
9.5mm work to walls on timber base										
width exceeding 300mm	2.19	5.00	2.30	0.25	0.13	3.22	0.17	5.69	m2	6.12
width not exceeding 300mm	0.66	10.00	0.73	0.15	0.04	1.67	0.10	2.49	m	2.68
9.5mm work to ceilings on timber base										
width exceeding 300mm	2.19	5.00	2.30	0.30	0.13	3.68	0.17	6.15	m2	6.61
width not exceeding 300mm	0.66	10.00	0.73	0.18	0.04	1.94	0.10	2.77	m	2.98
9.5mm work to isolated beams on timber base										
width exceeding 300mm	2.19	5.00	2.30	0.38	0.13	4.42	0.17	6.89	m2	7.41
width not exceeding 300mm	0.66	10.00	0.73	0.23	0.04	2.40	0.10	3.23	m	3.47
9.5mm work to isolated columns on timber base										
width exceeding 300mm	2.19	5.00	2.30	0.38	0.13	4.42	0.17	6.89	m2	7.41
width not exceeding 300mm	0.66	10.00	0.73	0.23	0.04	2.40	0.10	3.23	m	3.47
PLASTERBOARD - Baseboarding; Gyproc lath; 3mm joints, filled with plaster; fixing with galvanised nails; internal										
9.5mm work to walls on timber base										
width exceeding 300mm	1.87	5.00	1.96	0.25	0.13	3.22	0.13	5.31	m2	5.71
width not exceeding 300mm	0.56	10.00	0.62	0.15	0.04	1.67	0.08	2.36	m	2.54
12.5mm work to walls on timber base										
width exceeding 300mm	2.22	5.00	2.33	0.30	0.15	3.82	0.13	6.28	m2	6.75
width not exceeding 300mm	0.67	10.00	0.74	0.18	0.05	2.01	0.08	2.83	m	3.04
9.5mm work to ceilings on timber base										
width exceeding 300mm	1.87	5.00	1.96	0.30	0.13	3.68	0.13	5.78	m2	6.21
width not exceeding 300mm	0.56	10.00	0.62	0.18	0.05	2.01	0.08	2.71	m	2.91
12.5mm work to ceilings on timber base										
width exceeding 300mm	2.22	5.00	2.33	0.36	0.15	4.38	0.13	6.84	m2	7.35
width not exceeding 300mm	0.67	10.00	0.74	0.22	0.05	2.38	0.08	3.20	m	3.44
9.5mm work to isolated beams on timber base										
width exceeding 300mm	1.87	5.00	1.96	0.38	0.13	4.42	0.13	6.51	m2	7.00
width not exceeding 300mm	0.56	10.00	0.62	0.23	0.04	2.40	0.08	3.10	m	3.33
12.5mm work to isolated beams on timber base										
width exceeding 300mm	2.22	5.00	2.33	0.45	0.15	5.21	0.13	7.67	m2	8.24
width not exceeding 300mm	0.67	10.00	0.74	0.27	0.05	2.84	0.08	3.66	m	3.93
9.5mm work to isolated columns on timber base										
width exceeding 300mm	1.87	5.00	1.96	0.38	0.13	4.42	0.13	6.51	m2	7.00
width not exceeding 300mm	0.56	10.00	0.62	0.23	0.04	2.40	0.08	3.10	m	3.33

Labour hourly rates: (except Specialists) Craft Operatives £9.23 Labourer £7.02 Rates are national average prices. Refer to REGIONAL VARIATIONS for indicative levels of overall pricing in regions	MATERIALS			LABOUR				RATES		
	Del to Site £	Waste %	Material Cost £	Craft Optve Hrs	Lab Hrs	Labour Cost £	Sunds £	Nett Rate £	Unit	Gross Rate (+7.5%) £
M20: PLASTERED/RENDERED/ROUGHCAST COATINGS Cont.										
PLASTERBOARD - Baseboarding; Gyproc lath; 3mm joints, filled with plaster; fixing with galvanised nails; internal Cont.										
12.5mm work to isolated columns on timber base										
width exceeding 300mm	2.22	5.00	**2.33**	0.45	0.15	**5.21**	0.13	**7.67**	m2	**8.24**
width not exceeding 300mm	0.67	10.00	**0.74**	0.27	0.05	**2.84**	0.08	**3.66**	m	**3.93**
PLASTERBOARD - Baseboarding; Gyproc plank, square edge; 5mm joints, filled with plaster and scrimmed; internal										
19mm work to walls on timber base										
width exceeding 300mm	3.06	5.00	**3.21**	0.40	0.20	**5.10**	0.16	**8.47**	m2	**9.10**
width not exceeding 300mm	0.92	10.00	**1.01**	0.24	0.06	**2.64**	0.10	**3.75**	m	**4.03**
19mm work to ceilings on timber base										
width exceeding 300mm	3.06	5.00	**3.21**	0.48	0.20	**5.83**	0.16	**9.21**	m2	**9.90**
width not exceeding 300mm	0.92	10.00	**1.01**	0.29	0.06	**3.10**	0.10	**4.21**	m	**4.53**
19mm work to isolated beams on timber base										
width exceeding 300mm	3.06	5.00	**3.21**	0.60	0.20	**6.94**	0.16	**10.32**	m2	**11.09**
width not exceeding 300mm	0.92	10.00	**1.01**	0.36	0.04	**3.60**	0.10	**4.72**	m	**5.07**
19mm work to isolated columns on timber base										
width exceeding 300mm	3.06	5.00	**3.21**	0.60	0.20	**6.94**	0.16	**10.32**	m2	**11.09**
width not exceeding 300mm	0.92	10.00	**1.01**	0.36	0.04	**3.60**	0.10	**4.72**	m	**5.07**
PLASTERBOARD - Baseboarding; Gyproc square edge wallboard; 3mm joints filled with plaster and scrimmed; fixing with galvanised nails; internal										
9.5mm work to walls on timber base										
width exceeding 300mm	1.61	5.00	**1.69**	0.25	0.13	**3.22**	0.17	**5.08**	m2	**5.46**
width not exceeding 300mm	0.48	10.00	**0.53**	0.15	0.04	**1.67**	0.10	**2.29**	m	**2.47**
12.5mm work to walls on timber base										
width exceeding 300mm	1.87	5.00	**1.96**	0.30	0.15	**3.82**	0.17	**5.96**	m2	**6.40**
width not exceeding 300mm	0.56	10.00	**0.62**	0.18	0.05	**2.01**	0.10	**2.73**	m	**2.93**
15mm work to walls on timber base										
width exceeding 300mm	2.06	5.00	**2.16**	0.35	0.18	**4.49**	0.17	**6.83**	m2	**7.34**
width not exceeding 300mm	0.62	10.00	**0.68**	0.21	0.05	**2.29**	0.10	**3.07**	m	**3.30**
9.5mm work to ceilings on timber base										
width exceeding 300mm	1.61	5.00	**1.69**	0.30	0.13	**3.68**	0.17	**5.54**	m2	**5.96**
width not exceeding 300mm	0.48	10.00	**0.53**	0.18	0.04	**1.94**	0.10	**2.57**	m	**2.76**
12.5mm work to ceilings on timber base										
width exceeding 300mm	1.87	5.00	**1.96**	0.36	0.15	**4.38**	0.17	**6.51**	m2	**7.00**
width not exceeding 300mm	0.56	10.00	**0.62**	0.22	0.05	**2.38**	0.10	**3.10**	m	**3.33**
15mm work to ceilings on timber base										
width exceeding 300mm	2.06	5.00	**2.16**	0.42	0.18	**5.14**	0.17	**7.47**	m2	**8.03**
width not exceeding 300mm	0.62	10.00	**0.68**	0.26	0.05	**2.75**	0.10	**3.53**	m	**3.80**
9.5mm work to isolated beams on timber base										
width exceeding 300mm	1.61	5.00	**1.69**	0.38	0.13	**4.42**	0.17	**6.28**	m2	**6.75**
width not exceeding 300mm	0.48	10.00	**0.53**	0.23	0.04	**2.40**	0.10	**3.03**	m	**3.26**
12.5mm work to isolated beams on timber base										
width exceeding 300mm	1.87	5.00	**1.96**	0.45	0.15	**5.21**	0.17	**7.34**	m2	**7.89**
width not exceeding 300mm	0.56	10.00	**0.62**	0.27	0.05	**2.84**	0.10	**3.56**	m	**3.83**
15mm work to isolated beams on timber base										
width exceeding 300mm	2.06	5.00	**2.16**	0.52	0.18	**6.06**	0.17	**8.40**	m2	**9.03**
width not exceeding 300mm	0.62	10.00	**0.68**	0.31	0.05	**3.21**	0.10	**3.99**	m	**4.29**
9.5mm work to isolated columns on timber base										
width exceeding 300mm	1.61	5.00	**1.69**	0.38	0.13	**4.42**	0.17	**6.28**	m2	**6.75**
width not exceeding 300mm	0.48	10.00	**0.53**	0.23	0.04	**2.40**	0.10	**3.03**	m	**3.26**
12.5mm work to isolated columns on timber base										
width exceeding 300mm	1.87	5.00	**1.96**	0.45	0.15	**5.21**	0.17	**7.34**	m2	**7.89**
width not exceeding 300mm	0.56	10.00	**0.62**	0.27	0.05	**2.84**	0.10	**3.56**	m	**3.83**
15mm work to isolated columns on timber base										
width exceeding 300mm	2.06	5.00	**2.16**	0.52	0.18	**6.06**	0.17	**8.40**	m2	**9.03**
width not exceeding 300mm	0.62	10.00	**0.68**	0.31	0.05	**3.21**	0.10	**3.99**	m	**4.29**
PLASTERBOARD - Baseboarding; Gyproc square edge Duplex wallboard; 3mm joints filled with plaster and scrimmed; fixing with galvanised nails; internal										
9.5mm work to walls on timber base										
width Oexceeding 300mm	2.19	5.00	**2.30**	0.25	0.13	**3.22**	0.17	**5.69**	m2	**6.12**
width not exceeding 300mm	0.66	10.00	**0.73**	0.15	0.04	**1.67**	0.10	**2.49**	m	**2.68**
12.5mm work to walls on timber base										
width exceeding 300mm	2.45	5.00	**2.57**	0.30	0.15	**3.82**	0.17	**6.56**	m2	**7.06**
width not exceeding 300mm	0.74	10.00	**0.81**	0.18	0.05	**2.01**	0.10	**2.93**	m	**3.15**
15mm work to walls on timber base										
width exceeding 300mm	2.69	5.00	**2.82**	0.35	0.18	**4.49**	0.17	**7.49**	m2	**8.05**

SURFACE FINISHES

Labour hourly rates: (except Specialists) Craft Operatives £9.23 Labourer £7.02 Rates are national average prices. Refer to REGIONAL VARIATIONS for indicative levels of overall pricing in regions	MATERIALS			LABOUR				RATES		
	Del to Site £	Waste %	Material Cost £	Craft Optve Hrs	Lab Hrs	Labour Cost £	Sunds £	Nett Rate £	Unit	Gross Rate (+7.5%) £
M20: PLASTERED/RENDERED/ROUGHCAST COATINGS Cont.										
PLASTERBOARD - Baseboarding; Gyproc square edge Duplex wallboard; 3mm joints filled with plaster and scrimmed; fixing with galvanised nails; internal Cont.										
15mm work to walls on timber base Cont.										
width not exceeding 300mm	0.81	10.00	0.89	0.21	0.05	2.29	0.10	3.28	m	3.53
9.5mm work to ceilings on timber base										
width exceeding 300mm	2.19	5.00	2.30	0.30	0.13	3.68	0.17	6.15	m2	6.61
width not exceeding 300mm	0.66	10.00	0.73	0.18	0.04	1.94	0.10	2.77	m	2.98
12.5mm work to ceilings on timber base										
width exceeding 300mm	2.45	5.00	2.57	0.36	0.15	4.38	0.17	7.12	m2	7.65
width not exceeding 300mm	0.74	10.00	0.81	0.22	0.05	2.38	0.10	3.30	m	3.54
15mm work to ceilings on timber base										
width exceeding 300mm	2.69	5.00	2.82	0.42	0.18	5.14	0.17	8.13	m2	8.74
width not exceeding 300mm	0.81	10.00	0.89	0.26	0.05	2.75	0.10	3.74	m	4.02
9.5mm work to isolated beams on timber base										
width exceeding 300mm	2.19	5.00	2.30	0.38	0.13	4.42	0.17	6.89	m2	7.41
width not exceeding 300mm ..:.............	0.66	10.00	0.73	0.23	0.04	2.40	0.10	3.23	m	3.47
12.5mm work to isolated beams on timber base										
width exceeding 300mm	2.45	5.00	2.57	0.45	0.15	5.21	0.17	7.95	m2	8.55
width not exceeding 300mm	0.74	10.00	0.81	0.27	0.05	2.84	0.10	3.76	m	4.04
15mm work to isolated beams on timber base										
width exceeding 300mm	2.69	5.00	2.82	0.52	0.18	6.06	0.17	9.06	m2	9.74
width not exceeding 300mm	0.81	10.00	0.89	0.31	0.05	3.21	0.10	4.20	m	4.52
9.5mm work to isolated columns on timber base										
width exceeding 300mm	2.19	5.00	2.30	0.38	0.13	4.42	0.17	6.89	m2	7.41
width not exceeding 300mm	0.66	10.00	0.73	0.23	0.04	2.40	0.10	3.23	m	3.47
12.5mm work to isolated columns on timber base										
width exceeding 300mm	2.45	5.00	2.57	0.45	0.15	5.21	0.17	7.95	m2	8.55
width not exceeding 300mm	0.74	10.00	0.81	0.27	0.05	2.84	0.10	3.76	m	4.04
15mm work to isolated columns on timber base										
width exceeding 300mm	2.69	5.00	2.82	0.52	0.18	6.06	0.17	9.06	m2	9.74
width not exceeding 300mm	0.81	10.00	0.89	0.31	0.05	3.21	0.10	4.20	m	4.52
PLASTERBOARD - Baseboarding; two layers 9.5mm Gyproc square edge wallboard; second layer with 3mm joints filled with plaster and scrimmed; fixing with galvanised nails; internal										
19mm work to walls on timber base										
width exceeding 300mm	3.22	5.00	3.38	0.50	0.25	6.37	0.25	10.00	m2	10.75
width not exceeding 300mm	0.97	10.00	1.07	0.30	0.08	3.33	0.15	4.55	m	4.89
19mm work to ceilings on timber base										
width exceeding 300mm	3.22	5.00	3.38	0.60	0.25	7.29	0.25	10.92	m2	11.74
width not exceeding 300mm	0.97	10.00	1.07	0.36	0.08	3.88	0.15	5.10	m	5.48
19mm work to isolated beams on timber base										
width exceeding 300mm	3.22	5.00	3.38	0.75	0.25	8.68	0.25	12.31	m2	13.23
width not exceeding 300mm	0.97	10.00	1.07	0.45	0.08	4.72	0.15	5.93	m	6.38
19mm work to isolated columns on timber base										
width exceeding 300mm	3.22	5.00	3.38	0.75	0.25	8.68	0.25	12.31	m2	13.23
width not exceeding 300mm	0.97	10.00	1.07	0.45	0.08	4.72	0.15	5.93	m	6.38
PLASTERBOARD - Baseboarding; one layer 9.5mm and one layer 12.5mm Gyproc square edge wallboard; second layer with 3mm joints filled with plaster and scrimmed; fixing with galvanised nails, internal										
22mm work to walls on timber base										
width exceeding 300mm	3.48	5.00	3.65	0.55	0.28	7.04	0.25	10.95	m2	11.77
width not exceeding 300mm	1.04	10.00	1.14	0.33	0.08	3.61	0.15	4.90	m	5.27
22mm work to ceilings on timber base										
width exceeding 300mm	3.48	5.00	3.65	0.66	0.28	8.06	0.25	11.96	m2	12.86
width not exceeding 300mm	1.04	10.00	1.14	0.40	0.08	4.25	0.15	5.55	m	5.96
22mm work to isolated beams on timber base										
width exceeding 300mm	3.48	5.00	3.65	0.83	0.28	9.63	0.25	13.53	m2	14.55
width not exceeding 300mm	1.04	10.00	1.14	0.50	0.08	5.18	0.15	6.47	m	6.96
22mm work to isolated columns on timber base										
width exceeding 300mm	3.48	5.00	3.65	0.83	0.28	9.63	0.25	13.53	m2	14.55
width not exceeding 300mm	1.04	10.00	1.14	0.50	0.08	5.18	0.15	6.47	m	6.96
PLASTERBOARD - Baseboarding; two layers 12.5mm Gyproc square edge wallboard; second layer with 3mm joints filled with plaster and scrimmed; fixing with galvanised nails; internal										
25mm work to walls on timber base										
width exceeding 300mm	3.74	5.00	3.93	0.60	0.30	7.64	0.25	11.82	m2	12.71
width not exceeding 300mm	1.12	10.00	1.23	0.36	0.09	3.95	0.15	5.34	m	5.74

Labour hourly rates: (except Specialists) Craft Operatives £9.23 Labourer £7.02 Rates are national average prices. Refer to REGIONAL VARIATIONS for indicative levels of overall pricing in regions	MATERIALS			LABOUR			RATES		
	Del to Site £	Waste %	Material Cost £	Craft Optve Hrs	Lab Hrs	Labour Cost £	Sunds £	Nett Rate £	Unit	Gross Rate (+7.5%) £

M20: PLASTERED/RENDERED/ROUGHCAST COATINGS Cont.

PLASTERBOARD - Baseboarding; two layers 12.5mm Gyproc square edge wallboard; second layer with 3mm joints filled with plaster and scrimmed; fixing with galvanised nails; internal Cont.

	Del to Site £	Waste %	Material Cost £	Craft Optve Hrs	Lab Hrs	Labour Cost £	Sunds £	Nett Rate £	Unit	Gross Rate (+7.5%) £
25mm work to ceilings on timber base										
width exceeding 300mm	3.74	5.00	3.93	0.72	0.30	8.75	0.25	12.93	m2	13.90
width not exceeding 300mm	1.12	10.00	1.23	0.43	0.09	4.60	0.15	5.98	m	6.43
25mm work to isolated beams on timber base										
width exceeding 300mm	3.74	5.00	3.93	0.90	0.30	10.41	0.25	14.59	m2	15.68
width not exceeding 300mm	1.12	10.00	1.23	0.54	0.09	5.62	0.15	7.00	m	7.52
25mm work to isolated columns on timber base										
width exceeding 300mm	3.74	5.00	3.93	0.90	0.30	10.41	0.25	14.59	m2	15.68
width not exceeding 300mm	1.12	10.00	1.23	0.54	0.09	5.62	0.15	7.00	m	7.52

PLASTERBOARD - Baseboarding; two layers 15mm Gyproc square edge wallboard; second layer with 3mm joints filled with plaster and scrimmed; fixing with galvanised nails; internal

	Del to Site £	Waste %	Material Cost £	Craft Optve Hrs	Lab Hrs	Labour Cost £	Sunds £	Nett Rate £	Unit	Gross Rate (+7.5%) £
30mm work to walls on timber base										
width exceeding 300mm	4.12	5.00	4.33	0.65	0.33	8.32	0.25	12.89	m2	13.86
width not exceeding 300mm	1.24	10.00	1.36	0.39	0.10	4.30	0.15	5.82	m	6.25
30mm work to ceilings on timber base										
width exceeding 300mm	4.12	5.00	4.33	0.78	0.33	9.52	0.25	14.09	m2	15.15
width not exceeding 300mm	1.24	10.00	1.36	0.47	0.10	5.04	0.15	6.55	m	7.05
30mm work to isolated beams on timber base										
width exceeding 300mm	4.12	5.00	4.33	0.98	0.33	11.36	0.25	15.94	m2	17.13
width not exceeding 300mm	1.24	10.00	1.36	0.59	0.10	6.15	0.15	7.66	m	8.24
30mm work to isolated columns on timber base										
width exceeding 300mm	4.12	5.00	4.33	0.98	0.33	11.36	0.25	15.94	m2	17.13
width not exceeding 300mm	1.24	10.00	1.36	0.59	0.10	6.15	0.15	7.66	m	8.24

PLASTERBOARD - Baseboarding; two layers; first layer 9.5mm Gyproc Duplex wallboard, second layer 9.5mm Gyproc wallboard both layers with square edge boards; second layer with 3mm joints filled with plaster and scrimmed; fixing with galvanised nails; internal

	Del to Site £	Waste %	Material Cost £	Craft Optve Hrs	Lab Hrs	Labour Cost £	Sunds £	Nett Rate £	Unit	Gross Rate (+7.5%) £
19mm work to walls on timber base										
width exceeding 300mm	3.80	5.00	3.99	0.50	0.25	6.37	0.25	10.61	m2	11.41
width not exceeding 300mm	1.14	10.00	1.25	0.30	0.08	3.33	0.15	4.73	m	5.09
19mm work to ceilings on timber base										
width exceeding 300mm	3.80	5.00	3.99	0.60	0.25	7.29	0.25	11.53	m2	12.40
width not exceeding 300mm	1.14	10.00	1.25	0.36	0.08	3.88	0.15	5.29	m	5.69
19mm work to isolated beams on timber base										
width exceeding 300mm	3.80	5.00	3.99	0.75	0.25	8.68	0.25	12.92	m2	13.89
width not exceeding 300mm	1.14	10.00	1.25	0.45	0.08	4.72	0.15	6.12	m	6.58
19mm work to isolated columns on timber base										
width exceeding 300mm	3.80	5.00	3.99	0.75	0.25	8.68	0.25	12.92	m2	13.89
width not exceeding 300mm	1.14	10.00	1.25	0.45	0.08	4.72	0.15	6.12	m	6.58

PLASTERBOARD - Baseboarding; two layers; first layer 12.5mm Gyproc Duplex wallboard, second layer 9.5mm Gyproc wallboard, both layers with square edge boards; second layer with 3mm joints filled with plaster and scrimmed; fixing with galvanised nails; internal

	Del to Site £	Waste %	Material Cost £	Craft Optve Hrs	Lab Hrs	Labour Cost £	Sunds £	Nett Rate £	Unit	Gross Rate (+7.5%) £
22mm work to walls on timber base										
width exceeding 300mm	4.06	5.00	4.26	0.55	0.28	7.04	0.25	11.56	m2	12.42
width not exceeding 300mm	1.22	10.00	1.34	0.33	0.08	3.61	0.15	5.10	m	5.48
22mm work to ceilings on timber base										
width exceeding 300mm	4.06	5.00	4.26	0.66	0.28	8.06	0.25	12.57	m2	13.51
width not exceeding 300mm	1.22	10.00	1.34	0.40	0.08	4.25	0.15	5.75	m	6.18
22mm work to isolated beams on timber base										
width exceeding 300mm	4.06	5.00	4.26	0.83	0.28	9.63	0.25	14.14	m2	15.20
width not exceeding 300mm	1.22	10.00	1.34	0.50	0.08	5.18	0.15	6.67	m	7.17
22mm work to isolated columns on timber base										
width exceeding 300mm	4.06	5.00	4.26	0.83	0.28	9.63	0.25	14.14	m2	15.20
width not exceeding 300mm	1.22	10.00	1.34	0.50	0.08	5.18	0.15	6.67	m	7.17

PLASTERBOARD - Baseboarding; two layers; first layer 12.5mm Gyproc Duplex wallboard, second layer 12.5mm Gyproc wallboard, both layers with square edge boards; second layer with 3mm joints filled with plaster and scrimmed; fixing with galvanised nails; internal

	Del to Site £	Waste %	Material Cost £	Craft Optve Hrs	Lab Hrs	Labour Cost £	Sunds £	Nett Rate £	Unit	Gross Rate (+7.5%) £
25mm work to walls on timber base										
width exceeding 300mm	4.32	5.00	4.54	0.60	0.30	7.64	0.25	12.43	m2	13.36
width not exceeding 300mm	1.30	10.00	1.43	0.36	0.09	3.95	0.15	5.53	m	5.95

	MATERIALS			LABOUR				RATES		
Labour hourly rates: (except Specialists) Craft Operatives £9.23 Labourer £7.02 Rates are national average prices. Refer to REGIONAL VARIATIONS for indicative levels of overall pricing in regions	Del to Site £	Waste %	Material Cost £	Craft Optve Hrs	Lab Hrs	Labour Cost £	Sunds £	Nett Rate £	Unit	Gross Rate (+7.5%) £

M20: PLASTERED/RENDERED/ROUGHCAST COATINGS Cont.

PLASTERBOARD - Baseboarding; two layers; first layer 12.5mm Gyproc Duplex wallboard, second layer 12.5mm Gyproc wallboard, both layers with square edge boards; second layer with 3mm joints filled with plaster and scrimmed; fixing with galvanised nails; internal Cont.

	Del to Site	Waste	Material Cost	Craft Optve	Lab	Labour Cost	Sunds	Nett Rate	Unit	Gross Rate
25mm work to ceilings on timber base										
width exceeding 300mm	4.32	5.00	4.54	0.72	0.30	8.75	0.25	13.54	m2	14.55
width not exceeding 300mm	1.30	10.00	1.43	0.43	0.09	4.60	0.15	6.18	m	6.64
25mm work to isolated beams on timber base										
width exceeding 300mm	4.32	5.00	4.54	0.90	0.30	10.41	0.25	15.20	m2	16.34
width not exceeding 300mm	1.30	10.00	1.43	0.54	0.09	5.62	0.15	7.20	m	7.74
25mm work to isolated columns on timber base										
width exceeding 300mm	4.32	5.00	4.54	0.90	0.30	10.41	0.25	15.20	m2	16.34
width not exceeding 300mm	1.30	10.00	1.43	0.54	0.09	5.62	0.15	7.20	m	7.74

PLASTERBOARD - Baseboarding; two layers; first layer 15mm Gyproc Duplex wallboard, second layer 15mm Gyproc wallboard, both layers with square edge boards; second layer with 3mm joints filled with plaster and scrimmed; fixing with galvanised nails; internal

	Del to Site	Waste	Material Cost	Craft Optve	Lab	Labour Cost	Sunds	Nett Rate	Unit	Gross Rate
30mm work to walls on timber base										
width exceeding 300mm	4.75	5.00	4.99	0.65	0.33	8.32	0.25	13.55	m2	14.57
width not exceeding 300mm	1.43	10.00	1.57	0.39	0.10	4.30	0.15	6.02	m	6.48
30mm work to ceilings on timber base										
width exceeding 300mm	4.75	5.00	4.99	0.78	0.33	9.52	0.25	14.75	m2	15.86
width not exceeding 300mm	1.43	10.00	1.57	0.47	0.10	5.04	0.15	6.76	m	7.27
30mm work to isolated beams on timber base										
width exceeding 300mm	4.75	5.00	4.99	0.98	0.33	11.36	0.25	16.60	m2	17.84
width not exceeding 300mm	1.43	10.00	1.57	0.59	0.10	6.15	0.15	7.87	m	8.46
30mm work to isolated columns on timber base										
width exceeding 300mm	4.75	5.00	4.99	0.98	0.33	11.36	0.25	16.60	m2	17.84
width not exceeding 300mm	1.43	10.00	1.57	0.59	0.10	6.15	0.15	7.87	m	8.46

DAMP WALL TREATMENT - Newlath; fixing with masonry nails; internal

	Del to Site	Waste	Material Cost	Craft Optve	Lab	Labour Cost	Sunds	Nett Rate	Unit	Gross Rate
Work to walls on brickwork or blockwork base										
width exceeding 300mm	5.40	5.00	5.67	0.36	0.18	4.59	1.35	11.61	m2	12.48
width not exceeding 300mm	1.62	10.00	1.78	0.22	0.05	2.38	0.44	4.60	m	4.95

PORTLAND CEMENT WORK - Plain face, first and finishing coats cement and sand (1:3), total 13mm thick; wood floated; external

	Del to Site	Waste	Material Cost	Craft Optve	Lab	Labour Cost	Sunds	Nett Rate	Unit	Gross Rate
13mm work to walls on brickwork or blockwork base										
width exceeding 300mm	1.13	5.00	1.19	0.45	0.23	5.77	-	6.95	m2	7.48
width exceeding 300mm; dubbing average 6mm thick	1.66	5.00	1.74	0.68	0.44	9.37	-	11.11	m2	11.94
width exceeding 300mm; dubbing average 12mm thick	2.18	5.00	2.29	0.68	0.44	9.37	-	11.65	m2	12.53
width exceeding 300mm; dubbing average 19mm thick	2.79	5.00	2.93	0.70	0.57	10.46	-	13.39	m2	14.40
width exceeding 300mm; curved to 3000mm radius	1.13	5.00	1.19	0.60	0.23	7.15	-	8.34	m2	8.96
width not exceeding 300mm	0.34	10.00	0.37	0.27	0.07	2.98	-	3.36	m	3.61
width not exceeding 300mm; dubbing average 6mm thick	0.50	10.00	0.55	0.41	0.13	4.70	-	5.25	m	5.64
width not exceeding 300mm; dubbing average 12mm thick	0.65	10.00	0.71	0.41	0.13	4.70	-	5.41	m	5.82
width not exceeding 300mm; dubbing average 19mm thick	0.84	10.00	0.92	0.42	0.17	5.07	-	5.99	m	6.44
width not exceeding 300mm; curved to 3000mm radius	0.34	10.00	0.37	0.36	0.07	3.81	-	4.19	m	4.50
If plain face waterproofed										
add	0.55	10.00	0.60	-	0.02	0.14	-	0.75	m2	0.80
Rounded angles										
radius 10 - 100mm	-	-	-	0.24	-	2.22	-	2.22	m	2.38
Work forming flush skirting; 13mm thick										
height 150mm	0.26	50.00	0.39	0.50	0.05	4.97	-	5.36	m	5.76
height 225mm	0.38	50.00	0.57	0.55	0.05	5.43	-	6.00	m	6.45
height 150mm; curved to 3000mm radius	0.26	50.00	0.39	0.67	0.05	6.54	-	6.93	m	7.44
height 225mm; curved to 3000mm radius	0.38	50.00	0.57	0.73	0.05	7.09	-	7.66	m	8.23
Work forming projecting skirting; 13mm projection										
height 150mm	0.26	50.00	0.39	0.75	0.05	7.27	-	7.66	m	8.24
height 225mm	0.38	50.00	0.57	0.80	0.05	7.74	-	8.30	m	8.93
height 150mm; curved to 3000mm radius	0.26	50.00	0.39	1.00	0.05	9.58	-	9.97	m	10.72
height 225mm; curved to 3000mm radius	0.38	50.00	0.57	1.07	0.05	10.23	-	10.80	m	11.61

ROUGH CAST - Render and dry dash; first and finishing coats cement and sand (1:3), total 15mm thick; wood floated; dry dash of pea shingle; external

	Del to Site	Waste	Material Cost	Craft Optve	Lab	Labour Cost	Sunds	Nett Rate	Unit	Gross Rate
15mm work to walls on brickwork or blockwork base										
width exceeding 300mm	1.46	5.00	1.53	0.75	0.38	9.59	-	11.12	m2	11.96
extra; spatterdash coat	0.10	5.00	0.11	0.20	0.10	2.55	-	2.65	m2	2.85
width exceeding 300mm; curved to 3000mm radius	1.46	5.00	1.53	1.00	0.38	11.90	-	13.43	m2	14.44

SURFACE FINISHES – MAJOR WORKS

Labour hourly rates: (except Specialists) Craft Operatives £9.23 Labourer £7.02 Rates are national average prices. Refer to REGIONAL VARIATIONS for indicative levels of overall pricing in regions	MATERIALS			LABOUR				RATES		
	Del to Site £	Waste %	Material Cost £	Craft Optve Hrs	Lab Hrs	Labour Cost £	Sunds £	Nett Rate £	Unit	Gross Rate (+7.5%) £
M20: PLASTERED/RENDERED/ROUGHCAST COATINGS Cont.										
ROUGH CAST - Render and dry dash; first and finishing coats cement and sand (1:3), total 15mm thick; wood floated; dry dash of pea shingle; external Cont.										
15mm work to walls on brickwork or blockwork base Cont.										
width not exceeding 300mm	0.44	10.00	0.48	0.45	0.11	4.93	–	5.41	m	5.82
extra; spatterdash coat	0.03	10.00	0.03	0.12	0.03	1.32	–	1.35	m	1.45
width not exceeding 300mm; curved to 3000mm radius	0.44	10.00	0.48	0.60	0.11	6.31	–	6.79	m	7.30
ROUGH CAST - Render and wet dash; first and finishing coats cement and sand (1:3), total 15mm thick; wood floated; wet dash of crushed stone or shingle and cement slurry										
15mm work to walls on brickwork or blockwork base										
width exceeding 300mm	1.99	5.00	2.09	0.85	0.43	10.86	–	12.95	m2	13.93
extra; spatterdash coat	0.45	5.00	0.47	0.25	0.13	3.22	–	3.69	m2	3.97
width exceeding 300mm; curved to 3000mm radius	1.99	5.00	2.09	1.13	0.43	13.45	–	15.54	m2	16.70
width not exceeding 300mm	0.60	10.00	0.66	0.51	0.13	5.62	–	6.28	m	6.75
extra; spatterdash coat	0.14	10.00	0.15	0.15	0.04	1.67	–	1.82	m	1.96
width not exceeding 300mm; curved to 3000mm radius	0.60	10.00	0.66	0.68	0.13	7.19	–	7.85	m	8.44
TYROLEAN FINISH - Render and Tyrolean finish; first and finishing coats cement and sand (1:3), total 15mm thick; wood floated; Tyrolean finish of 'Cullamix' mixture applied by machine										
15mm work to walls on brickwork or blockwork base										
width exceeding 300mm	3.25	5.00	3.41	0.90	0.25	10.06	0.55	14.02	m2	15.08
width not exceeding 300mm	1.00	10.00	1.10	0.54	0.08	5.55	0.33	6.98	m	7.50
HARDWALL PLASTERING - Plaster, B.S.1191, Part 1, Class B; finishing coat of board finish, 3mm thick; steel trowelled; internal										
3mm work to walls on concrete or plasterboard base										
width exceeding 300mm	0.42	5.00	0.44	0.25	0.13	3.22	0.01	3.67	m2	3.95
width exceeding 300mm; curved to 3000mm radius	0.42	5.00	0.44	0.33	0.13	3.96	0.01	4.41	m2	4.74
width not exceeding 300mm	0.13	10.00	0.14	0.15	0.04	1.67	–	1.81	m	1.94
width not exceeding 300mm; curved to 3000mm radius	0.13	10.00	0.14	0.20	0.04	2.13	–	2.27	m	2.44
3mm work to ceilings on concrete or plasterboard base										
width exceeding 300mm	0.42	5.00	0.44	0.30	0.13	3.68	0.01	4.13	m2	4.44
width exceeding 300mm; 3.50 - 5.00m above floor	0.42	5.00	0.44	0.33	0.13	3.96	0.01	4.41	m2	4.74
width not exceeding 300mm	0.13	10.00	0.14	0.18	0.04	1.94	–	2.09	m	2.24
width not exceeding 300mm; 3.50 - 5.00m above floor	0.13	10.00	0.14	0.20	0.04	2.13	–	2.27	m	2.44
3mm work to isolated beams on concrete or plasterboard base										
width exceeding 300mm	0.42	5.00	0.44	0.38	0.13	4.42	0.01	4.87	m2	5.24
width not exceeding 300mm	0.13	10.00	0.14	0.23	0.04	2.40	–	2.55	m	2.74
3mm work to isolated columns on concrete or plasterboard base										
width exceeding 300mm	0.42	5.00	0.44	0.38	0.13	4.42	0.01	4.87	m2	5.24
width not exceeding 300mm	0.13	10.00	0.14	0.23	0.04	2.40	–	2.55	m	2.74
HARDWALL PLASTERING - Plaster; first coat of hardwall, 11mm thick; finishing coat of multi-finish 2mm thick; steel trowelled; internal										
13mm work to walls on brickwork or blockwork base										
width exceeding 300mm	1.54	5.00	1.62	0.45	0.23	5.77	0.04	7.43	m2	7.98
width exceeding 300mm; curved to 3000mm radius	1.54	5.00	1.62	0.60	0.23	7.15	0.04	8.81	m2	9.47
width not exceeding 300mm	0.46	10.00	0.51	0.27	0.07	2.98	0.01	3.50	m	3.76
width not exceeding 300mm; curved to 3000mm radius	0.46	10.00	0.51	0.36	0.07	3.81	0.01	4.33	m	4.65
13mm work to ceilings on concrete base										
width exceeding 300mm	1.54	5.00	1.62	0.54	0.23	6.60	0.04	8.26	m2	8.87
width exceeding 300mm; 3.50 - 5.00m above floor	1.54	5.00	1.62	0.59	0.23	7.06	0.04	8.72	m2	9.37
width not exceeding 300mm	0.46	10.00	0.51	0.32	0.07	3.44	0.01	3.96	m	4.26
width not exceeding 300mm; 3.50 - 5.00m above floor	0.46	10.00	0.51	0.35	0.07	3.72	0.01	4.24	m	4.56
13mm work to isolated beams on concrete base										
width exceeding 300mm	1.54	5.00	1.62	0.68	0.23	7.89	0.04	9.55	m2	10.26
width not exceeding 300mm	0.46	10.00	0.51	0.41	0.07	4.28	0.01	4.79	m	5.15
13mm work to isolated columns on concrete base										
width exceeding 300mm	1.54	5.00	1.62	0.68	0.23	7.89	0.04	9.55	m2	10.26
width not exceeding 300mm	0.46	10.00	0.51	0.41	0.07	4.28	0.01	4.79	m	5.15
Rounded angles										
radius 10 - 100mm	–	–	–	0.24	–	2.22	–	2.22	m	2.38

SURFACE FINISHES

Labour hourly rates: (except Specialists) Craft Operatives £9.23 Labourer £7.02 Rates are national average prices. Refer to REGIONAL VARIATIONS for indicative levels of overall pricing in regions	MATERIALS			LABOUR				RATES		
	Del to Site £	Waste %	Material Cost £	Craft Optve Hrs	Lab Hrs	Labour Cost £	Sunds £	Nett Rate £	Unit	Gross Rate (+7.5%) £
M20: PLASTERED/RENDERED/ROUGHCAST COATINGS Cont.										
HARDWALL PLASTERING - Plaster; Thistle Universal one coat plaster, 13mm thick; steel trowelled; internal										
Note the thickness is from the face of the metal lathing										
13mm work to walls on metal lathing base										
width exceeding 300mm	2.00	5.00	2.10	0.35	0.18	4.49	0.05	6.64	m2	7.14
width exceeding 300mm; curved to 3000mm radius ...	2.00	5.00	2.10	0.47	0.18	5.60	0.05	7.75	m2	8.33
width not exceeding 300mm	0.60	10.00	0.66	0.21	0.05	2.29	0.01	2.96	m	3.18
width not exceeding 300mm; curved to 3000mm radius	0.60	10.00	0.66	0.28	0.05	2.94	0.01	3.61	m	3.88
13mm work to ceilings on metal lathing base										
width exceeding 300mm	2.00	5.00	2.10	0.42	0.18	5.14	0.05	7.29	m2	7.84
width exceeding 300mm; 3.50 - 5.00m above floor ..	2.00	5.00	2.10	0.46	0.18	5.51	0.05	7.66	m2	8.23
width not exceeding 300mm	0.60	10.00	0.66	0.25	0.05	2.66	0.01	3.33	m	3.58
width not exceeding 300mm; 3.50 - 5.00m above floor	0.60	10.00	0.66	0.28	0.05	2.94	0.01	3.61	m	3.88
13mm work to isolated beams on metal lathing base										
width exceeding 300mm	2.00	5.00	2.10	0.53	0.18	6.16	0.05	8.31	m2	8.93
width not exceeding 300mm	0.60	10.00	0.66	0.32	0.05	3.30	0.01	3.97	m	4.27
13mm work to isolated columns on metal lathing base										
width exceeding 300mm	2.00	5.00	2.10	0.53	0.18	6.16	0.05	8.31	m2	8.93
width not exceeding 300mm	0.60	10.00	0.66	0.32	0.05	3.30	0.01	3.97	m	4.27
Rounded angles										
radius 10 - 100mm	-	-	-	0.18	-	1.66	-	1.66	m	1.79
LIGHTWEIGHT PLASTERING - Plaster, Carlite; pre-mixed; floating coat of browning, 11mm thick; finishing coat of finish, 2mm thick; steel trowelled; internal										
13mm work to walls on brickwork or blockwork base										
width exceeding 300mm	1.31	5.00	1.38	0.40	0.20	5.10	0.03	6.50	m2	6.99
width exceeding 300mm; curved to 3000mm radius ...	1.31	5.00	1.38	0.53	0.20	6.30	0.03	7.70	m2	8.28
width exceeding 300mm; dubbing average 6mm thick .	1.89	5.00	1.98	0.63	0.38	8.48	0.05	10.52	m2	11.31
width exceeding 300mm; dubbing average 6mm thick; curved to 3000mm radius	1.89	5.00	1.98	0.84	0.38	10.42	0.05	12.46	m2	13.39
width exceeding 300mm; dubbing average 12mm thick	2.47	5.00	2.59	0.63	0.38	8.48	0.06	11.14	m2	11.97
width exceeding 300mm; dubbing average 12mm thick; curved to 3000mm radius	2.47	5.00	2.59	0.84	0.38	10.42	0.06	13.07	m2	14.05
width exceeding 300mm; dubbing average 19mm thick	3.14	5.00	3.30	0.65	0.49	9.44	0.08	12.82	m2	13.78
width exceeding 300mm; dubbing average 19mm thick; curved to 3000mm radius	3.14	5.00	3.30	0.87	0.49	11.47	0.08	14.85	m2	15.96
width not exceeding 300mm	0.39	10.00	0.43	0.24	0.06	2.64	0.01	3.08	m	3.31
width not exceeding 300mm; curved to 3000mm radius	0.39	10.00	0.43	0.32	0.06	3.37	0.01	3.81	m	4.10
width not exceeding 300mm; dubbing average 6mm thick	0.57	10.00	0.63	0.36	0.09	3.95	0.01	4.59	m	4.94
width not exceeding 300mm; dubbing average 6mm thick; curved to 3000mm radius	0.57	10.00	0.63	0.48	0.09	5.06	0.01	5.70	m	6.13
width not exceeding 300mm; dubbing average 12mm thick	0.74	10.00	0.81	0.38	0.09	4.14	0.02	4.97	m	5.35
width not exceeding 300mm; dubbing average 12mm thick; curved to 3000mm radius	0.74	10.00	0.81	0.51	0.09	5.34	0.02	6.17	m	6.64
width not exceeding 300mm; dubbing average 19mm thick	0.94	10.00	1.03	0.39	0.15	4.65	0.02	5.71	m	6.13
width not exceeding 300mm; dubbing average 19mm thick; curved to 3000mm radius	0.94	10.00	1.03	0.52	0.15	5.85	0.02	6.91	m	7.42
Rounded angles										
radius 10 - 100mm	-	-	-	0.24	-	2.22	-	2.22	m	2.38
LIGHTWEIGHT PLASTERING - Plaster, Carlite; pre-mixed; floating coat of bonding 8mm thick; finishing coat of finish, 2mm thick; steel trowelled; internal										
10mm work to walls on concrete or plasterboard base										
width exceeding 300mm	1.23	5.00	1.29	0.40	0.20	5.10	0.03	6.42	m2	6.90
width exceeding 300mm; curved to 3000mm radius ...	1.23	5.00	1.29	0.53	0.20	6.30	0.03	7.62	m2	8.19
width not exceeding 300mm	0.37	10.00	0.41	0.24	0.06	2.64	0.01	3.05	m	3.28
width not exceeding 300mm; curved to 3000mm radius	0.37	10.00	0.41	0.32	0.06	3.37	0.01	3.79	m	4.08
10mm work to ceilings on concrete or plasterboard base										
width exceeding 300mm	1.23	5.00	1.29	0.48	0.20	5.83	0.03	7.16	m2	7.69
width exceeding 300mm; 3.50 - 5.00m above floor ..	1.23	5.00	1.29	0.52	0.20	6.20	0.03	7.53	m2	8.09
width not exceeding 300mm	0.37	10.00	0.41	0.29	0.06	3.10	0.01	3.51	m	3.78
width not exceeding 300mm; 3.50 - 5.00m above floor	0.37	10.00	0.41	0.31	0.06	3.28	0.01	3.70	m	3.98
10mm work to isolated beams on concrete or plasterboard base										
width exceeding 300mm	1.23	5.00	1.29	0.60	0.20	6.94	0.03	8.26	m2	8.88
width not exceeding 300mm	0.37	10.00	0.41	0.36	0.06	3.74	0.01	4.16	m	4.47
10mm work to isolated columns on concrete or plasterboard base										
width exceeding 300mm	1.23	5.00	1.29	0.60	0.20	6.94	0.03	8.26	m2	8.88
width not exceeding 300mm	0.37	10.00	0.41	0.36	0.06	3.74	0.01	4.16	m	4.47
Rounded angles										
radius 10 - 100mm	-	-	-	0.24	-	2.22	-	2.22	m	2.38

Labour hourly rates: (except Specialists) Craft Operatives £9.23 Labourer £7.02 Rates are national average prices. Refer to REGIONAL VARIATIONS for indicative levels of overall pricing in regions	MATERIALS			LABOUR				RATES		
	Del to Site	Waste	Material Cost	Craft Optve	Lab	Labour Cost	Sunds	Nett Rate	Unit	Gross Rate (+7.5%)
	£	%	£	Hrs	Hrs	£	£	£		£
M20: PLASTERED/RENDERED/ROUGHCAST COATINGS Cont.										
LIGHTWEIGHT PLASTERING - Plaster, Carlite; pre-mixed; floating coat of bonding 11mm thick; finishing coat of finish, 2mm thick; steel trowelled; internal										
13mm work to ceilings on precast concrete beam and infill blocks base										
width exceeding 300mm	1.88	5.00	1.97	0.48	0.20	5.83	0.05	7.86	m2	8.45
width not exceeding 300mm	0.56	10.00	0.62	0.29	0.06	3.10	0.01	3.72	m	4.00
LIGHTWEIGHT PLASTERING - Plaster, Carlite; pre-mixed; pricking up and floating coats metal lathing, 11mm thick; finishing coat of finish, 2mm thick; steel trowelled; internal										
Note the thickness is from the face of the metal lathing										
13mm work to walls on metal lathing base										
width exceeding 300mm	2.45	5.00	2.57	0.55	0.28	7.04	0.06	9.67	m2	10.40
width exceeding 300mm; curved to 3000mm radius	2.45	5.00	2.57	0.73	0.28	8.70	0.06	11.34	m2	12.19
width not exceeding 300mm	0.74	10.00	0.81	0.33	0.08	3.61	0.02	4.44	m	4.77
width not exceeding 300mm; curved to 3000mm radius	0.74	10.00	0.81	0.44	0.08	4.62	0.02	5.46	m	5.87
13mm work to ceilings on metal lathing base										
width exceeding 300mm	2.45	5.00	2.57	0.66	0.28	8.06	0.06	10.69	m2	11.49
width exceeding 300mm; 3.50 - 5.00m above floor	2.45	5.00	2.57	0.72	0.28	8.61	0.06	11.24	m2	12.09
width not exceeding 300mm	0.74	10.00	0.81	0.40	0.08	4.25	0.02	5.09	m	5.47
width not exceeding 300mm; 3.50 - 5.00m above floor	0.74	10.00	0.81	0.43	0.08	4.53	0.02	5.36	m	5.77
13mm work to isolated beams on metal lathing base										
width exceeding 300mm	2.45	5.00	2.57	0.83	0.28	9.63	0.06	12.26	m2	13.18
width not exceeding 300mm	0.74	10.00	0.81	0.50	0.08	5.18	0.02	6.01	m	6.46
13mm work to isolated columns on metal lathing base										
width exceeding 300mm	2.45	5.00	2.57	0.83	0.28	9.63	0.06	12.26	m2	13.18
width not exceeding 300mm	0.74	10.00	0.81	0.50	0.08	5.18	0.02	6.01	m	6.46
Rounded angles										
radius 10 - 100mm	-	-	-	0.30	-	2.77	-	2.77	m	2.98
RENOVATING PLASTER - Renovating plaster, Thistle; undercoat 11mm thick; finishing coat 2mm thick; steel trowelled; internal										
13mm work to walls on existing brickwork or blockwork base										
width exceeding 300mm	1.63	5.00	1.71	0.50	0.25	6.37	0.04	8.12	m2	8.73
width exceeding 300mm; curved to 3000mm radius	1.63	5.00	1.71	0.67	0.25	7.94	0.04	9.69	m2	10.42
width exceeding 300mm; dubbing average 6mm thick	2.32	5.00	2.44	0.70	0.37	9.06	0.06	11.55	m2	12.42
width exceeding 300mm; dubbing average 6mm thick; curved to 3000mm radius	2.32	5.00	2.44	0.93	0.37	11.18	0.06	13.68	m2	14.70
width exceeding 300mm; dubbing average 12mm thick	3.01	5.00	3.16	0.73	0.48	10.11	0.07	13.34	m2	14.34
width exceeding 300mm; dubbing average 12mm thick; curved to 3000mm radius	3.01	5.00	3.16	0.97	0.48	12.32	0.07	15.55	m2	16.72
width not exceeding 300mm	0.49	10.00	0.54	0.30	0.08	3.33	0.01	3.88	m	4.17
width not exceeding 300mm; curved to 3000mm radius	0.49	10.00	0.54	0.40	0.08	4.25	0.01	4.80	m	5.16
width not exceeding 300mm; dubbing average 6mm thick	0.70	10.00	0.77	0.42	0.12	4.72	0.02	5.51	m	5.92
width not exceeding 300mm; dubbing average 6mm thick; curved to 3000mm radius	0.70	10.00	0.77	0.56	0.12	6.01	0.02	6.80	m	7.31
width not exceeding 300mm; dubbing average 12mm thick	0.90	10.00	0.99	0.44	0.15	5.11	0.02	6.12	m	6.58
width not exceeding 300mm; dubbing average 12mm thick; curved to 3000mm radius	0.90	10.00	0.99	0.59	0.15	6.50	0.02	7.51	m	8.07
Rounded angles										
radius 10 - 100mm	-	-	-	0.24	-	2.22	-	2.22	m	2.38
PLASTER BEADS AND STOPS - Accessories										
Galvanised steel angle beads Expamet fixing to brickwork or blockwork with masonry nails										
Reference 550	0.58	10.00	0.64	0.08	-	0.74	-	1.38	m	1.48
Reference 553	0.70	10.00	0.77	0.08	-	0.74	-	1.51	m	1.62
Reference 554	0.91	10.00	1.00	0.08	-	0.74	-	1.74	m	1.87
Galvanised steel stop beads Expamet fixing to brickwork or blockwork with masonry nails										
Reference 560	0.58	10.00	0.64	0.08	-	0.74	-	1.38	m	1.48
Reference 561	0.75	10.00	0.82	0.08	-	0.74	-	1.56	m	1.68
Reference 562	0.68	10.00	0.75	0.08	-	0.74	-	1.49	m	1.60
Reference 563	0.68	10.00	0.75	0.08	-	0.74	-	1.49	m	1.60
Reference 565	0.87	10.00	0.96	0.08	-	0.74	-	1.70	m	1.82
Reference 566	0.87	10.00	0.96	0.08	-	0.74	-	1.70	m	1.82
Galvanised steel plasterboard edging beads Expamet fixing to timber with galvanised nails										
Reference 567	0.99	10.00	1.09	0.08	-	0.74	-	1.83	m	1.96
Reference 568	0.99	10.00	1.09	0.08	-	0.74	-	1.83	m	1.96

SURFACE FINISHES

	MATERIALS			LABOUR				RATES		
Labour hourly rates: (except Specialists) Craft Operatives £9.23 Labourer £7.02 Rates are national average prices. Refer to REGIONAL VARIATIONS for indicative levels of overall pricing in regions	Del to Site £	Waste %	Material Cost £	Craft Optve Hrs	Lab Hrs	Labour Cost £	Sunds £	Nett Rate £	Unit	Gross Rate (+7.5%) £

M20: PLASTERED/RENDERED/ROUGHCAST COATINGS Cont.

PLASTER BEADS AND STOPS – Accessories Cont.

Description	Del to Site £	Waste %	Material Cost £	Craft Optve Hrs	Lab Hrs	Labour Cost £	Sunds £	Nett Rate £	Unit	Gross Rate £
Galvanised steel depth gauge beads Expamet fixing to brickwork or blockwork with masonry nails Reference 569	0.55	10.00	0.60	0.08	–	0.74	–	1.34	m	1.44

PLASTER BEADS AND STOPS – Accessories; external

Description	Del to Site £	Waste %	Material Cost £	Craft Optve Hrs	Lab Hrs	Labour Cost £	Sunds £	Nett Rate £	Unit	Gross Rate £
Galvanised steel stop beads Expamet fixing to brickwork or blockwork with masonry nails Reference 570	0.70	10.00	0.77	0.08	–	0.74	–	1.51	m	1.62
Stainless steel angle beads Expamet fixing to brickwork or blockwork with masonry nails Reference 545	1.69	10.00	1.86	0.08	–	0.74	–	2.60	m	2.79
Stainless steel stop beads Expamet fixing to brickwork or blockwork with masonry nails Reference 547	1.49	10.00	1.64	0.08	–	0.74	–	2.38	m	2.56

ANTI-CRACK STRIPS – Accessories – anti-crack strips

Lathing; BB galvanised expanded metal lath, 9mm mesh X 0.500mm thick x 1.11 kg/m2; butt joints fixing with nails

Description	Del to Site £	Waste %	Material Cost £	Craft Optve Hrs	Lab Hrs	Labour Cost £	Sunds £	Nett Rate £	Unit	Gross Rate £
100mm wide to walls; one edge to timber; one edge to brickwork or blockwork	0.40	10.00	0.44	0.08	–	0.74	0.02	1.20	m	1.29
100mm wide to walls; one edge to timber; one edge to concrete	0.40	10.00	0.44	0.10	–	0.92	0.02	1.38	m	1.49
100mm wide to walls; one edge to brickwork or blockwork; one edge to concrete	0.40	10.00	0.44	0.12	–	1.11	0.02	1.57	m	1.69

Lathing; BB galvanised expanded metal lath, 9mm mesh x 0.725mm thick x 1.61 kg/m2; butt joints; fixing with nails

Description	Del to Site £	Waste %	Material Cost £	Craft Optve Hrs	Lab Hrs	Labour Cost £	Sunds £	Nett Rate £	Unit	Gross Rate £
100mm wide to walls; one edge to timber; one edge to brickwork or blockwork	0.47	10.00	0.52	0.08	–	0.74	0.02	1.28	m	1.37
100mm wide to walls; one edge to timber; one edge to concrete	0.47	10.00	0.52	0.10	–	0.92	0.02	1.46	m	1.57
100mm wide to walls; one edge to brickwork or blockwork; one edge to concrete	0.47	10.00	0.52	0.12	–	1.11	0.02	1.64	m	1.77

BONDING FLUID – Prepare and apply bonding fluid to receive plaster or cement rendering

Description	Del to Site £	Waste %	Material Cost £	Craft Optve Hrs	Lab Hrs	Labour Cost £	Sunds £	Nett Rate £	Unit	Gross Rate £
Work to walls on existing cement and sand base width exceeding 300mm	0.57	10.00	0.63	0.20	–	1.85	–	2.47	m2	2.66
Work to walls on existing glazed tiling base width exceeding 300mm	0.38	10.00	0.42	0.15	–	1.38	–	1.80	m2	1.94
Work to walls on existing painted surface width exceeding 300mm	0.45	10.00	0.50	0.18	–	1.66	–	2.16	m2	2.32
Work to walls on existing concrete base width exceeding 300mm	0.50	10.00	0.55	0.20	–	1.85	–	2.40	m2	2.58
Work to ceilings on existing cement and sand base width exceeding 300mm	0.57	10.00	0.63	0.24	–	2.22	–	2.84	m2	3.06
Work to ceilings on existing painted surface width exceeding 300mm	0.45	10.00	0.50	0.22	–	2.03	–	2.53	m2	2.72
Work to ceilings on existing concrete base width exceeding 300mm	0.50	10.00	0.55	0.24	–	2.22	–	2.77	m2	2.97

M20: PLASTERED/RENDERED/ROUGHCAST COATINGS; STAIRCASE AREAS

PLASTERBOARD – Baseboarding; gypsum baseboard, Thistle; 5mm joints, filled with plaster and scrimmed; fixing with galvanised nails; internal

Description	Del to Site £	Waste %	Material Cost £	Craft Optve Hrs	Lab Hrs	Labour Cost £	Sunds £	Nett Rate £	Unit	Gross Rate £
9.5mm work to walls on timber base width exceeding 300mm	1.61	5.00	1.69	0.31	0.13	3.77	0.17	5.63	m2	6.06
width not exceeding 300mm	0.48	10.00	0.53	0.19	0.04	2.03	0.10	2.66	m	2.86
9.5mm work to ceilings on timber base width exceeding 300mm	1.61	5.00	1.69	0.38	0.13	4.42	0.17	6.28	m2	6.75
width not exceeding 300mm	0.48	10.00	0.53	0.23	0.04	2.40	0.10	3.03	m	3.26

PLASTERBOARD – Baseboarding; gypsum baseboard, Thistle Duplex; 5mm joints filled with plaster and scrimmed; fixing with galvanised nails; internal

Description	Del to Site £	Waste %	Material Cost £	Craft Optve Hrs	Lab Hrs	Labour Cost £	Sunds £	Nett Rate £	Unit	Gross Rate £
9.5mm work to walls on timber base width exceeding 300mm	2.19	5.00	2.30	0.31	0.13	3.77	0.17	6.24	m2	6.71
width not exceeding 300mm	0.66	10.00	0.73	0.19	0.04	2.03	0.10	2.86	m	3.08

Labour hourly rates: (except Specialists) Craft Operatives £9.23 Labourer £7.02 Rates are national average prices. Refer to REGIONAL VARIATIONS for indicative levels of overall pricing in regions	MATERIALS			LABOUR				RATES		
	Del to Site £	Waste %	Material Cost £	Craft Optve Hrs	Lab Hrs	Labour Cost £	Sunds £	Nett Rate £	Unit	Gross Rate (+7.5%) £
M20: PLASTERED/RENDERED/ROUGHCAST COATINGS; STAIRCASE AREAS Cont.										
PLASTERBOARD - Baseboarding; gypsum baseboard, Thistle Duplex; 5mm joints filled with plaster and scrimmed; fixing with galvanised nails; internal Cont.										
9.5mm work to ceilings on timber base										
width exceeding 300mm	2.19	5.00	2.30	0.38	0.13	4.42	0.17	6.89	m2	7.41
width not exceeding 300mm	0.66	10.00	0.73	0.23	0.04	2.40	0.10	3.23	m	3.47
PLASTERBOARD - Baseboarding; Gyproc lath; 3mm joints, filled with plaster; fixing with galvanised nails; internal										
9.5mm work to walls on timber base										
width exceeding 300mm	1.87	5.00	1.96	0.31	0.13	3.77	0.13	5.87	m2	6.31
width not exceeding 300mm	0.56	10.00	0.62	0.19	0.04	2.03	0.08	2.73	m	2.94
12.5mm work to walls on timber base										
width exceeding 300mm	2.22	5.00	2.33	0.38	0.15	4.56	0.13	7.02	m2	7.55
width not exceeding 300mm	0.67	10.00	0.74	0.23	0.05	2.47	0.08	3.29	m	3.54
9.5mm work to ceilings on timber base										
width exceeding 300mm	1.87	5.00	1.96	0.38	0.13	4.42	0.13	6.51	m2	7.00
width not exceeding 300mm	0.56	10.00	0.62	0.23	0.05	2.47	0.08	3.17	m	3.41
12.5mm work to ceilings on timber base										
width exceeding 300mm	2.22	5.00	2.33	0.45	0.15	5.21	0.13	7.67	m2	8.24
width not exceeding 300mm	0.67	10.00	0.74	0.27	0.05	2.84	0.08	3.66	m	3.93
PLASTERBOARD - Baseboarding; Gyproc plank, square edge; 5mm joints, filled with plaster and scrimmed; internal										
19mm work to walls on timber base										
width exceeding 300mm	3.06	5.00	3.21	0.50	0.20	6.02	0.16	9.39	m2	10.10
width not exceeding 300mm	0.92	10.00	1.01	0.30	0.06	3.19	0.10	4.30	m	4.62
19mm work to ceilings on timber base										
width exceeding 300mm	3.06	5.00	3.21	0.60	0.20	6.94	0.16	10.32	m2	11.09
width not exceeding 300mm	0.92	10.00	1.01	0.36	0.06	3.74	0.10	4.86	m	5.22
PLASTERBOARD - Baseboarding; Gyproc square edge wallboard; 3mm joints filled with plaster and scrimmed; fixing with galvanised nails; internal										
9.5mm work to walls on timber base										
width exceeding 300mm	1.61	5.00	1.69	0.31	0.13	3.77	0.17	5.63	m2	6.06
width not exceeding 300mm	0.48	10.00	0.53	0.19	0.04	2.03	0.10	2.66	m	2.86
12.5mm work to walls on timber base										
width exceeding 300mm	1.87	5.00	1.96	0.38	0.15	4.56	0.17	6.69	m2	7.20
width not exceeding 300mm	0.56	10.00	0.62	0.23	0.05	2.47	0.10	3.19	m	3.43
15mm work to walls on timber base										
width exceeding 300mm	2.06	5.00	2.16	0.45	0.18	5.42	0.17	7.75	m2	8.33
width not exceeding 300mm	0.62	10.00	0.68	0.27	0.05	2.84	0.10	3.63	m	3.90
9.5mm work to ceilings on timber base										
width exceeding 300mm	1.61	5.00	1.69	0.38	0.13	4.42	0.17	6.28	m2	6.75
width not exceeding 300mm	0.48	10.00	0.53	0.23	0.04	2.40	0.10	3.03	m	3.26
12.5mm work to ceilings on timber base										
width exceeding 300mm	1.87	5.00	1.96	0.45	0.15	5.21	0.17	7.34	m2	7.89
width not exceeding 300mm	0.56	10.00	0.62	0.27	0.05	2.84	0.10	3.56	m	3.83
15mm work to ceilings on timber base										
width exceeding 300mm	2.06	5.00	2.16	0.52	0.18	6.06	0.17	8.40	m2	9.03
width not exceeding 300mm	0.62	10.00	0.68	0.31	0.05	3.21	0.10	3.99	m	4.29
PLASTERBOARD - Baseboarding; Gyproc square edge Duplex wallboard; 3mm joints filled with plaster and scrimmed; fixing with galvanised nails; internal										
9.5mm work to walls on timber base										
width exceeding 300mm	2.19	5.00	2.30	0.31	0.13	3.77	0.17	6.24	m2	6.71
width not exceeding 300mm	0.66	10.00	0.73	0.19	0.04	2.03	0.10	2.86	m	3.08
12.5mm work to walls on timber base										
width exceeding 300mm	2.45	5.00	2.57	0.38	0.15	4.56	0.17	7.30	m2	7.85
width not exceeding 300mm	0.74	10.00	0.81	0.23	0.05	2.47	0.10	3.39	m	3.64
15mm work to walls on timber base										
width exceeding 300mm	2.69	5.00	2.82	0.45	0.18	5.42	0.17	8.41	m2	9.04
width not exceeding 300mm	0.81	10.00	0.89	0.27	0.05	2.84	0.10	3.83	m	4.12
9.5mm work to ceilings on timber base										
width exceeding 300mm	2.19	5.00	2.30	0.38	0.13	4.42	0.17	6.89	m2	7.41
width not exceeding 300mm	0.66	10.00	0.73	0.23	0.04	2.40	0.10	3.23	m	3.47
12.5mm work to ceilings on timber base										
width exceeding 300mm	2.45	5.00	2.57	0.45	0.15	5.21	0.17	7.95	m2	8.55
width not exceeding 300mm	0.74	10.00	0.81	0.27	0.05	2.84	0.10	3.76	m	4.04

SURFACE FINISHES

Labour hourly rates: (except Specialists) Craft Operatives £9.23 Labourer £7.02 Rates are national average prices. Refer to REGIONAL VARIATIONS for indicative levels of overall pricing in regions	MATERIALS			LABOUR				RATES		
	Del to Site £	Waste %	Material Cost £	Craft Optve Hrs	Lab Hrs	Labour Cost £	Sunds £	Nett Rate £	Unit	Gross Rate (+7.5%) £
M20: PLASTERED/RENDERED/ROUGHCAST COATINGS; STAIRCASE AREAS Cont.										
PLASTERBOARD - Baseboarding; Gyproc square edge Duplex wallboard; 3mm joints filled with plaster and scrimmed; fixing with galvanised nails; internal Cont.										
15mm work to ceilings on timber base										
width exceeding 300mm	2.69	5.00	2.82	0.52	0.18	6.06	0.17	9.06	m2	9.74
width not exceeding 300mm	0.81	10.00	0.89	0.31	0.05	3.21	0.10	4.20	m	4.52
PLASTERBOARD - Baseboarding; two layers 9.5mm Gyproc square edge wallboard; second layer with 3mm joints filled with plaster and scrimmed; fixing with galvanised nails; internal										
19mm work to walls on timber base										
width exceeding 300mm	3.22	5.00	3.38	0.63	0.25	7.57	0.25	11.20	m2	12.04
width not exceeding 300mm	0.97	10.00	1.07	0.38	0.08	4.07	0.15	5.29	m	5.68
19mm work to ceilings on timber base										
width exceeding 300mm	3.22	5.00	3.38	0.75	0.25	8.68	0.25	12.31	m2	13.23
width not exceeding 300mm	0.97	10.00	1.07	0.45	0.08	4.72	0.15	5.93	m	6.38
PLASTERBOARD - Baseboarding; one layer 9.5mm and one layer 12.5mm Gyproc square edge wallboard; second layer with 3mm joints filled with plaster and scrimmed; fixing with galvanised nails, internal										
22mm work to walls on timber base										
width exceeding 300mm	3.48	5.00	3.65	0.69	0.28	8.33	0.25	12.24	m2	13.16
width not exceeding 300mm	1.04	10.00	1.14	0.41	0.08	4.35	0.15	5.64	m	6.06
22mm work to ceilings on timber base										
width exceeding 300mm	3.48	5.00	3.65	0.83	0.28	9.63	0.25	13.53	m2	14.55
width not exceeding 300mm	1.04	10.00	1.14	0.50	0.08	5.18	0.15	6.47	m	6.96
PLASTERBOARD - Baseboarding; two layers 12.5mm Gyproc square edge wallboard; second layer with 3mm joints filled with plaster and scrimmed; fixing with galvanised nails; internal										
25mm work to walls on timber base										
width exceeding 300mm	3.74	5.00	3.93	0.75	0.30	9.03	0.25	13.21	m2	14.20
width not exceeding 300mm	1.12	10.00	1.23	0.45	0.09	4.79	0.15	6.17	m	6.63
25mm work to ceilings on timber base										
width exceeding 300mm	3.74	5.00	3.93	0.90	0.30	10.41	0.25	14.59	m2	15.68
width not exceeding 300mm	1.12	10.00	1.23	0.54	0.09	5.62	0.15	7.00	m	7.52
PLASTERBOARD - Baseboarding; two layers 15mm Gyproc square edge wallboard; second layer with 3mm joints filled with plaster and scrimmed; fixing with galvanised nails; internal										
30mm work to walls on timber base										
width exceeding 300mm	4.12	5.00	4.33	0.81	0.33	9.79	0.25	14.37	m2	15.45
width not exceeding 300mm	1.24	10.00	1.36	0.49	0.10	5.22	0.15	6.74	m	7.24
30mm work to ceilings on timber base										
width exceeding 300mm	4.12	5.00	4.33	0.97	0.33	11.27	0.25	15.85	m2	17.03
width not exceeding 300mm	1.24	10.00	1.36	0.58	0.10	6.06	0.15	7.57	m	8.14
PLASTERBOARD - Baseboarding; two layers; first layer 9.5mm Gyproc Duplex wallboard, second layer 9.5mm Gyproc wallboard both layers with square edge boards; second layer with 3mm joints filled with plaster and scrimmed; fixing with galvanised nails; internal										
19mm work to walls on timber base										
width exceeding 300mm	3.80	5.00	3.99	0.63	0.25	7.57	0.25	11.81	m2	12.70
width not exceeding 300mm	1.14	10.00	1.25	0.38	0.08	4.07	0.15	5.47	m	5.88
19mm work to ceilings on timber base										
width exceeding 300mm	3.80	5.00	3.99	0.75	0.25	8.68	0.25	12.92	m2	13.89
width not exceeding 300mm	1.14	10.00	1.25	0.45	0.08	4.72	0.15	6.12	m	6.58
PLASTERBOARD - Baseboarding; two layers; first layer 12.5mm Gyproc Duplex wallboard, second layer 9.5mm Gyproc wallboard, both layers with square edge boards; second layer with 3mm joints filled with plaster and scrimmed; fixing with galvanised nails; internal										
22mm work to walls on timber base										
width exceeding 300mm	4.06	5.00	4.26	0.69	0.28	8.33	0.25	12.85	m2	13.81
width not exceeding 300mm	1.22	10.00	1.34	0.41	0.08	4.35	0.15	5.84	m	6.28
22mm work to ceilings on timber base										
width exceeding 300mm	4.06	5.00	4.26	0.83	0.28	9.63	0.25	14.14	m2	15.20
width not exceeding 300mm	1.22	10.00	1.34	0.50	0.08	5.18	0.15	6.67	m	7.17

Labour hourly rates: (except Specialists) Craft Operatives £9.23 Labourer £7.02 Rates are national average prices. Refer to REGIONAL VARIATIONS for indicative levels of overall pricing in regions	MATERIALS			LABOUR				RATES		
	Del to Site £	Waste %	Material Cost £	Craft Optve Hrs	Lab Hrs	Labour Cost £	Sunds £	Nett Rate £	Unit	Gross Rate (+7.5%) £
M20: PLASTERED/RENDERED/ROUGHCAST COATINGS; STAIRCASE AREAS Cont.										
PLASTERBOARD - Baseboarding; two layers; first layer 12.5mm Gyproc Duplex wallboard, second layer 12.5mm Gyproc wallboard, both layers with square edge boards; second layer with 3mm joints filled with plaster and scrimmed; fixing with galvanised nails; internal										
25mm work to walls on timber base										
width exceeding 300mm	4.32	5.00	4.54	0.75	0.30	9.03	0.25	13.81	m2	14.85
width not exceeding 300mm	1.30	10.00	1.43	0.45	0.09	4.79	0.15	6.37	m	6.84
25mm work to ceilings on timber base										
width exceeding 300mm	4.32	5.00	4.54	0.90	0.30	10.41	0.25	15.20	m2	16.34
width not exceeding 300mm	1.30	10.00	1.43	0.54	0.09	5.62	0.15	7.20	m	7.74
PLASTERBOARD - Baseboarding; two layers; first layer 15mm Gyproc Duplex wallboard, second layer 15mm Gyproc wallboard, both layers with square edge boards; second layer with 3mm joints filled with plaster and scrimmed; fixing with galvanised nails; internal										
30mm work to walls on timber base										
width exceeding 300mm	4.75	5.00	4.99	0.81	0.33	9.79	0.25	15.03	m2	16.16
width not exceeding 300mm	1.43	10.00	1.57	0.49	0.10	5.22	0.15	6.95	m	7.47
30mm work to ceilings on timber base										
width exceeding 300mm	4.75	5.00	4.99	0.97	0.33	11.27	0.25	16.51	m2	17.75
width not exceeding 300mm	1.43	10.00	1.57	0.58	0.10	6.06	0.15	7.78	m	8.36
HARDWALL PLASTERING - Plaster, B.S.1191, Part 1, Class B; finishing coat of board finish, 3mm thick; steel trowelled; internal										
3mm work to walls on concrete or plasterboard base										
width exceeding 300mm	0.42	5.00	0.44	0.31	0.13	3.77	0.01	4.22	m2	4.54
width not exceeding 300mm	0.13	10.00	0.14	0.19	0.04	2.03	-	2.18	m	2.34
3mm work to ceilings on concrete or plasterboard base										
width exceeding 300mm	0.42	5.00	0.44	0.38	0.13	4.42	0.01	4.87	m2	5.24
width not exceeding 300mm	0.13	10.00	0.14	0.23	0.04	2.40	-	2.55	m	2.74
HARDWALL PLASTERING - Plaster; first coat of hardwall, 11mm thick; finishing coat of multi-finish 2mm thick; steel trowelled; internal										
13mm work to walls on brickwork or blockwork base										
width exceeding 300mm	1.54	5.00	1.62	0.56	0.23	6.78	0.04	8.44	m2	9.07
width not exceeding 300mm	0.46	10.00	0.51	0.34	0.07	3.63	0.01	4.15	m	4.46
13mm work to ceilings on concrete base										
width exceeding 300mm	1.54	5.00	1.62	0.68	0.23	7.89	0.04	9.55	m2	10.26
width not exceeding 300mm	0.46	10.00	0.51	0.41	0.07	4.28	0.01	4.79	m	5.15
HARDWALL PLASTERING - Plaster; Thistle Universal one coat plaster, 13mm thick; steel trowelled; internal										
Note the thickness is from the face of the metal lathing										
13mm work to walls on metal lathing base										
width exceeding 300mm	2.00	5.00	2.10	0.44	0.18	5.32	0.05	7.47	m2	8.04
width not exceeding 300mm	0.60	10.00	0.66	0.26	0.05	2.75	0.01	3.42	m	3.68
13mm work to ceilings on metal lathing base										
width exceeding 300mm	2.00	5.00	2.10	0.53	0.18	6.16	0.05	8.31	m2	8.93
width not exceeding 300mm	0.60	10.00	0.66	0.32	0.05	3.30	0.01	3.97	m	4.27
LIGHTWEIGHT PLASTERING										
LIGHTWEIGHT PLASTERING - Plaster, Carlite; pre-mixed; floating coat of browning, 11mm thick; finishing coat of finish, 2mm thick; steel trowelled; internal										
13mm work to walls on brickwork or blockwork base										
width exceeding 300mm	1.31	5.00	1.38	0.50	0.20	6.02	0.03	7.42	m2	7.98
width not exceeding 300mm	0.39	10.00	0.43	0.30	0.06	3.19	0.01	3.63	m	3.90
LIGHTWEIGHT PLASTERING - Plaster, Carlite; pre-mixed; floating coat of bonding 8mm thick; finishing coat of finish, 2mm thick; steel trowelled; internal										
10mm work to walls on concrete or plasterboard base										
width exceeding 300mm	1.23	5.00	1.29	0.50	0.20	6.02	0.03	7.34	m2	7.89
width not exceeding 300mm	0.37	10.00	0.41	0.30	0.06	3.19	0.01	3.61	m	3.88
10mm work to ceilings on concrete or plasterboard base										
width exceeding 300mm	1.23	5.00	1.29	0.60	0.20	6.94	0.03	8.26	m2	8.88
width not exceeding 300mm	0.37	10.00	0.41	0.36	0.06	3.74	0.01	4.16	m	4.47

SURFACE FINISHES

Labour hourly rates: (except Specialists) Craft Operatives £9.23 Labourer £7.02 Rates are national average prices. Refer to REGIONAL VARIATIONS for indicative levels of overall pricing in regions	MATERIALS			LABOUR				RATES		
	Del to Site £	Waste %	Material Cost £	Craft Optve Hrs	Lab Hrs	Labour Cost £	Sunds £	Nett Rate £	Unit	Gross Rate (+7.5%) £
M20: PLASTERED/RENDERED/ROUGHCAST COATINGS; STAIRCASE AREAS Cont.										
LIGHTWEIGHT PLASTERING - Plaster, Carlite; pre-mixed; pricking up and floating coats metal lathing, 11mm thick; finishing coat of finish, 2mm thick; steel trowelled; internal										
Note										
the thickness is from the face of the metal lathing										
13mm work to walls on metal lathing base										
width exceeding 300mm	2.45	5.00	2.57	0.69	0.28	8.33	0.06	10.97	m2	11.79
width not exceeding 300mm	0.74	10.00	0.81	0.41	0.08	4.35	0.02	5.18	m	5.57
13mm work to ceilings on metal lathing base										
width exceeding 300mm	2.45	5.00	2.57	0.83	0.28	9.63	0.06	12.26	m2	13.18
width not exceeding 300mm	0.74	10.00	0.81	0.50	0.08	5.18	0.02	6.01	m	6.46
M30: METAL MESH LATHING/ANCHORED REINFORCEMENT FOR PLASTERED COATINGS										
Lathing; BB galvanised expanded metal lath, 9mm mesh x 0.500mm thick x 1.11 kg/m2; butt joints; fixing with galvanised staples										
Work to walls										
width exceeding 300mm	3.79	5.00	3.98	0.15	0.08	1.95	0.05	5.98	m2	6.42
width not exceeding 300mm	1.33	10.00	1.46	0.09	0.02	0.97	0.03	2.46	m	2.65
Work to ceilings										
width exceeding 300mm	3.79	5.00	3.98	0.18	0.08	2.22	0.07	6.27	m2	6.74
width not exceeding 300mm	1.33	10.00	1.46	0.11	0.02	1.16	0.04	2.66	m	2.86
Work to isolated beams										
width exceeding 300mm	3.79	5.00	3.98	0.23	0.08	2.68	0.07	6.73	m2	7.24
width not exceeding 300mm	1.33	10.00	1.46	0.14	0.02	1.43	0.04	2.94	m	3.16
Work to isolated columns										
width exceeding 300mm	3.79	5.00	3.98	0.23	0.08	2.68	0.07	6.73	m2	7.24
width not exceeding 300mm	1.33	10.00	1.46	0.14	0.02	1.43	0.04	2.94	m	3.16
Lathing; BB galvanised expanded metal lath 9mm mesh x 0.725mm thick x 1.61 kg/m2; butt joints; fixing with galvanised staples										
Work to walls										
width exceeding 300mm	4.43	5.00	4.65	0.15	0.08	1.95	0.05	6.65	m2	7.15
width not exceeding 300mm	1.55	10.00	1.71	0.09	0.02	0.97	0.03	2.71	m	2.91
Work to ceilings										
width exceeding 300mm	4.43	5.00	4.65	0.18	0.08	2.22	0.07	6.94	m2	7.47
width not exceeding 300mm	1.55	10.00	1.71	0.11	0.02	1.16	0.04	2.90	m	3.12
Work to isolated beams										
width exceeding 300mm	4.43	5.00	4.65	0.23	0.08	2.68	0.07	7.41	m2	7.96
width not exceeding 300mm	1.55	10.00	1.71	0.14	0.02	1.43	0.04	3.18	m	3.42
Work to isolated columns										
width exceeding 300mm	4.43	5.00	4.65	0.23	0.08	2.68	0.07	7.41	m2	7.96
width not exceeding 300mm	1.55	10.00	1.71	0.14	0.02	1.43	0.04	3.18	m	3.42
Lathing; galvanised expanded metal lath 9mm mesh x 0.950mm thick x 2.50 kg/m2; butt joints; fixing with galvanised staples										
Work to walls										
width exceeding 300mm	7.20	5.00	7.56	0.15	0.08	1.95	0.05	9.56	m2	10.27
width not exceeding 300mm	2.52	10.00	2.77	0.09	0.02	0.97	0.03	3.77	m	4.06
Work to ceilings										
width exceeding 300mm	7.20	5.00	7.56	0.18	0.08	2.22	0.07	9.85	m2	10.59
width not exceeding 300mm	2.52	10.00	2.77	0.11	0.02	1.16	0.04	3.97	m	4.27
Work to isolated beams										
width exceeding 300mm	7.20	5.00	7.56	0.23	0.08	2.68	0.07	10.31	m2	11.09
width not exceeding 300mm	2.52	10.00	2.77	0.14	0.02	1.43	0.04	4.24	m	4.56
Work to isolated columns										
width exceeding 300mm	7.20	5.00	7.56	0.23	0.08	2.68	0.07	10.31	m2	11.09
width not exceeding 300mm	2.52	10.00	2.77	0.14	0.02	1.43	0.04	4.24	m	4.56
Lathing; Expamet galvanised Rib-Lath, 0.400mm thick x 1.78 kg/m2; butt joints; fixing with galvanised staples										
Work to walls										
width exceeding 300mm	4.76	5.00	5.00	0.15	0.08	1.95	0.05	6.99	m2	7.52
width not exceeding 300mm	1.67	10.00	1.84	0.09	0.02	0.97	0.03	2.84	m	3.05
Work to ceilings										
width exceeding 300mm	4.76	5.00	5.00	0.18	0.08	2.22	0.07	7.29	m2	7.84
width not exceeding 300mm	1.67	10.00	1.84	0.11	0.02	1.16	0.04	3.03	m	3.26
Work to isolated beams										
width exceeding 300mm	4.76	5.00	5.00	0.23	0.08	2.68	0.07	7.75	m2	8.33
width not exceeding 300mm	1.67	10.00	1.84	0.14	0.02	1.43	0.04	3.31	m	3.56

SURFACE FINISHES – MAJOR WORKS

Labour hourly rates: (except Specialists) Craft Operatives £9.23 Labourer £7.02 Rates are national average prices. Refer to REGIONAL VARIATIONS for indicative levels of overall pricing in regions	MATERIALS			LABOUR				RATES		
	Del to Site £	Waste %	Material Cost £	Craft Optve Hrs	Lab Hrs	Labour Cost £	Sunds £	Nett Rate £	Unit	Gross Rate (+7.5%) £
M30: METAL MESH LATHING/ANCHORED REINFORCEMENT FOR PLASTERED COATINGS Cont.										
Lathing; Expamet galvanised Rib-Lath, 0.400mm thick x 1.78 kg/m2; butt joints; fixing with galvanised staples Cont.										
Work to isolated columns										
width exceeding 300mm	4.76	5.00	5.00	0.23	0.08	2.68	0.07	7.75	m2	8.33
width not exceeding 300mm	1.67	10.00	1.84	0.14	0.02	1.43	0.04	3.31	m	3.56
Lathing; Expamet galvanised Rib-Lath, 0.500mm thick x 2.25 kg/m2; butt joints; fixing with galvanised staples										
Work to walls										
width exceeding 300mm	5.47	5.00	5.74	0.15	0.08	1.95	0.05	7.74	m2	8.32
width not exceeding 300mm	1.91	10.00	2.10	0.09	0.02	0.97	0.03	3.10	m	3.33
Work to ceilings										
width exceeding 300mm	5.47	5.00	5.74	0.18	0.08	2.22	0.07	8.04	m2	8.64
width not exceeding 300mm	1.91	10.00	2.10	0.11	0.02	1.16	0.04	3.30	m	3.54
Work to isolated beams										
width exceeding 300mm	5.47	5.00	5.74	0.23	0.08	2.68	0.07	8.50	m2	9.14
width not exceeding 300mm	1.91	10.00	2.10	0.14	0.02	1.43	0.04	3.57	m	3.84
Work to isolated columns										
width exceeding 300mm	5.47	5.00	5.74	0.23	0.08	2.68	0.07	8.50	m2	9.14
width not exceeding 300mm	1.91	10.00	2.10	0.14	0.02	1.43	0.04	3.57	m	3.84
Lathing; Expamet stainless steel Rib-Lath, 0.300mm thick x 1.52 kg/m2; butt joints; fixing with stainless steel staples										
Work to walls										
width exceeding 300mm	12.09	5.00	12.69	0.15	0.08	1.95	0.05	14.69	m2	15.79
width not exceeding 300mm	4.23	10.00	4.65	0.09	0.02	0.97	0.03	5.65	m	6.08
Work to ceilings										
width exceeding 300mm	12.09	5.00	12.69	0.18	0.08	2.22	0.07	14.99	m2	16.11
width not exceeding 300mm	4.23	10.00	4.65	0.11	0.02	1.16	0.04	5.85	m	6.29
Work to isolated beams										
width exceeding 300mm	12.09	5.00	12.69	0.23	0.08	2.68	0.07	15.45	m2	16.61
width not exceeding 300mm	4.23	10.00	4.65	0.14	0.02	1.43	0.04	6.13	m	6.59
Work to isolated columns										
width exceeding 300mm	12.09	5.00	12.69	0.23	0.08	2.68	0.07	15.45	m2	16.61
width not exceeding 300mm	4.23	10.00	4.65	0.14	0.02	1.43	0.04	6.13	m	6.59
Lathing; Expamet galvanised Spraylath, 0.500mm thick x 2.25 kg/m2; butt joints; fixing with galvanised staples										
Work to walls										
width exceeding 300mm	6.66	5.00	6.99	0.15	0.08	1.95	0.05	8.99	m2	9.66
width not exceeding 300mm	2.33	10.00	2.56	0.09	0.02	0.97	0.03	3.56	m	3.83
Work to ceilings										
width exceeding 300mm	6.66	5.00	6.99	0.18	0.08	2.22	0.07	9.29	m2	9.98
width not exceeding 300mm	2.33	10.00	2.56	0.11	0.02	1.16	0.04	3.76	m	4.04
Work to isolated beams										
width exceeding 300mm	6.66	5.00	6.99	0.23	0.08	2.68	0.07	9.75	m2	10.48
width not exceeding 300mm	2.33	10.00	2.56	0.14	0.02	1.43	0.04	4.04	m	4.34
Work to isolated columns										
width exceeding 300mm	6.66	5.00	6.99	0.23	0.08	2.68	0.07	9.75	m2	10.48
width not exceeding 300mm	2.33	10.00	2.56	0.14	0.02	1.43	0.04	4.04	m	4.34
Lathing; galvanised Red-rib lath, 0.400mm thick x 1.78 kg/m2, butt joints; fixing with galvanised staples										
Work to walls										
width exceeding 300mm	6.10	5.00	6.41	0.15	0.08	1.95	0.05	8.40	m2	9.03
width not exceeding 300mm	2.14	10.00	2.35	0.09	0.02	0.97	0.03	3.36	m	3.61
Work to ceilings										
width exceeding 300mm	6.10	5.00	6.41	0.18	0.08	2.22	0.07	8.70	m2	9.35
width not exceeding 300mm	2.14	10.00	2.35	0.11	0.02	1.16	0.04	3.55	m	3.82
Work to isolated beams										
width exceeding 300mm	6.10	5.00	6.41	0.23	0.08	2.68	0.07	9.16	m2	9.85
width not exceeding 300mm	2.14	10.00	2.35	0.14	0.02	1.43	0.04	3.83	m	4.11
Work to isolated columns										
width exceeding 300mm	6.10	5.00	6.41	0.23	0.08	2.68	0.07	9.16	m2	9.85
width not exceeding 300mm	2.14	10.00	2.35	0.14	0.02	1.43	0.04	3.83	m	4.11

SURFACE FINISHES

Labour hourly rates: (except Specialists) Craft Operatives £9.23 Labourer £7.02 Rates are national average prices. Refer to REGIONAL VARIATIONS for indicative levels of overall pricing in regions	MATERIALS			LABOUR				RATES		
	Del to Site £	Waste %	Material Cost £	Craft Optve Hrs	Lab Hrs	Labour Cost £	Sunds £	Nett Rate £	Unit	Gross Rate (+7.5%) £
M30: METAL MESH LATHING/ANCHORED REINFORCEMENT FOR PLASTERED COATINGS Cont.										
Extra cost of fixing lathing to brickwork, blockwork or concrete with cartridge fired nails in lieu of to timber with staples										
Work to walls										
width exceeding 300mm	-	-	-	0.10	-	0.92	0.09	1.01	m2	1.09
width not exceeding 300mm	-	-	-	0.06	-	0.55	0.03	0.58	m	0.63
Extra cost of fixing lathing to steel with tying wire in lieu of to timber with staples										
Work to walls										
width exceeding 300mm	-	-	-	0.15	-	1.38	0.10	1.48	m2	1.60
width not exceeding 300mm	-	-	-	0.09	-	0.83	0.03	0.86	m	0.93
Work to ceilings										
width exceeding 300mm	-	-	-	0.15	-	1.38	0.10	1.48	m2	1.60
width not exceeding 300mm	-	-	-	0.09	-	0.83	0.03	0.86	m	0.93
Work to isolated beams										
width exceeding 300mm	-	-	-	0.15	-	1.38	0.10	1.48	m2	1.60
width not exceeding 300mm	-	-	-	0.90	-	8.31	0.03	8.34	m	8.96
Work to isolated columns										
width exceeding 300mm	-	-	-	0.15	-	1.38	0.10	1.48	m2	1.60
width not exceeding 300mm	-	-	-	0.09	-	0.83	0.03	0.86	m	0.93
Expamet galvanised arch formers and fix to brickwork or blockwork with galvanised nails										
Arch corners										
380mm radius	12.87	5.00	13.51	0.33	0.17	4.24	0.13	17.88	nr	19.22
460mm radius	15.68	5.00	16.46	0.33	0.17	4.24	0.16	20.86	nr	22.43
610mm radius	19.61	5.00	20.59	0.33	0.17	4.24	0.20	25.03	nr	26.91
760mm radius	27.86	5.00	29.25	0.33	0.17	4.24	0.28	33.77	nr	36.31
Semi-circular arches										
380mm radius	25.35	5.00	26.62	0.67	0.34	8.57	0.25	35.44	nr	38.10
410mm radius	25.76	5.00	27.05	0.67	0.34	8.57	0.26	35.88	nr	38.57
420mm radius	26.61	5.00	27.94	0.67	0.34	8.57	0.27	36.78	nr	39.54
460mm radius	31.39	5.00	32.96	0.67	0.34	8.57	0.31	41.84	nr	44.98
610mm radius	38.78	5.00	40.72	0.67	0.34	8.57	0.39	49.68	nr	53.41
760mm radius	54.87	5.00	57.61	0.67	0.34	8.57	0.55	66.73	nr	71.74
Elliptical arches										
1220mm wide x 340mm rise	49.28	5.00	51.74	0.67	0.34	8.57	0.49	60.80	nr	65.37
1370mm wide x 360mm rise	51.67	5.00	54.25	0.67	0.34	8.57	0.52	63.34	nr	68.10
1520mm wide x 380mm rise	54.87	5.00	57.61	0.67	0.34	8.57	0.55	66.73	nr	71.74
1830mm wide x 410mm rise	59.29	5.00	62.25	1.00	0.50	12.74	0.59	75.58	nr	81.25
2130mm wide x 430mm rise	64.83	5.00	68.07	1.00	0.50	12.74	0.65	81.46	nr	87.57
2440mm wide x 440mm rise	68.04	5.00	71.44	1.00	0.50	12.74	0.68	84.86	nr	91.23
3050mm wide x 520mm rise	69.44	5.00	72.91	1.00	0.50	12.74	0.69	86.34	nr	92.82
Spandrel arches										
760mm wide x 180mm radius x 220mm rise	32.34	5.00	33.96	0.67	0.34	8.57	0.32	42.85	nr	46.06
910mm wide x 180mm radius x 240mm rise	32.89	5.00	34.53	0.67	0.34	8.57	0.34	43.45	nr	46.70
1220mm wide x 230mm radius x 290mm rise	43.55	5.00	45.73	0.67	0.34	8.57	0.44	54.74	nr	58.84
1520mm wide x 230mm radius x 330mm rise	48.45	5.00	50.87	0.67	0.34	8.57	0.48	59.92	nr	64.42
1830mm wide x 230mm radius x 360mm rise	53.35	5.00	56.02	0.67	0.34	8.57	0.53	65.12	nr	70.00
2130mm wide x 230mm radius x 370mm rise	58.12	5.00	61.03	0.67	0.34	8.57	0.58	70.18	nr	75.44
2440mm wide x 230mm radius x 390mm rise	61.60	5.00	64.68	1.33	0.67	16.98	0.62	82.28	nr	88.45
3050mm wide x 230mm radius x 440mm rise	64.83	5.00	68.07	1.33	0.67	16.98	0.65	85.70	nr	92.13
Bulls-eyes										
230mm radius	32.34	5.00	33.96	0.17	0.09	2.20	0.32	36.48	nr	39.21
Soffit strips										
155mm wide	3.22	5.00	3.38	0.09	0.05	1.18	0.03	4.59	m	4.94
Make-up pieces										
600mm long	7.84	5.00	8.23	0.17	0.09	2.20	0.08	10.51	nr	11.30
Expamet galvanised circular window formers and fix to brickwork or blockwork with galvanised nails										
Circular windows										
600mm diameter	29.15	5.00	30.61	1.33	0.67	16.98	0.29	47.88	nr	51.47
M30: METAL MESH LATHING/ANCHORED REINFORCEMENT FOR PLASTERED COATINGS; STAIRCASE AREAS										
Lathing; BB galvanised expanded metal lath, 9mm mesh x 0.500mm thick x 1.11 kg/m2; butt joints; fixing with galvanised staples										
Work to walls										
width exceeding 300mm	3.79	5.00	3.98	0.19	0.08	2.32	0.07	6.36	m2	6.84
width not exceeding 300mm	1.33	10.00	1.46	0.11	0.02	1.16	0.04	2.66	m	2.86
Work to ceilings										
width exceeding 300mm	3.79	5.00	3.98	0.23	0.08	2.68	0.08	6.74	m2	7.25
width not exceeding 300mm	1.33	10.00	1.46	0.14	0.02	1.43	0.04	2.94	m	3.16

Labour hourly rates: (except Specialists) Craft Operatives £9.23 Labourer £7.02 Rates are national average prices. Refer to REGIONAL VARIATIONS for indicative levels of overall pricing in regions	MATERIALS			LABOUR				RATES		
	Del to Site £	Waste %	Material Cost £	Craft Optve Hrs	Lab Hrs	Labour Cost £	Sunds £	Nett Rate £	Unit	Gross Rate (+7.5%) £
M30: METAL MESH LATHING/ANCHORED REINFORCEMENT FOR PLASTERED COATINGS; STAIRCASE AREAS Cont.										
Lathing; BB galvanised expanded metal lath 9mm mesh x 0.725mm thick x 1.61 kg/m2; butt joints; fixing with galvanised staples										
Work to walls										
width exceeding 300mm	4.43	5.00	4.65	0.19	0.08	2.32	0.07	7.04	m2	7.56
width not exceeding 300mm	1.55	10.00	1.71	0.11	0.02	1.16	0.04	2.90	m	3.12
Work to ceilings										
width exceeding 300mm	4.43	5.00	4.65	0.23	0.08	2.68	0.06	7.40	m2	7.95
width not exceeding 300mm	1.55	10.00	1.71	0.14	0.02	1.43	0.03	3.17	m	3.41
Lathing; galvanised expanded metal lath 9mm mesh x 0.950mm thick x 2.50 kg/m2; butt joints; fixing with galvanised staples										
Work to walls										
width exceeding 300mm	7.20	5.00	7.56	0.19	0.08	2.32	0.05	9.93	m2	10.67
width not exceeding 300mm	2.52	10.00	2.77	0.11	0.02	1.16	0.03	3.96	m	4.25
Work to ceilings										
width exceeding 300mm	7.20	5.00	7.56	0.23	0.08	2.68	0.06	10.30	m2	11.08
width not exceeding 300mm	2.52	10.00	2.77	0.14	0.02	1.43	0.03	4.23	m	4.55
Lathing; Expamet galvanised Rib-Lath, 0.400mm thick x 1.78 kg/m2; butt joints; fixing with galvanised staples										
Work to walls										
width exceeding 300mm	4.76	5.00	5.00	0.19	0.08	2.32	0.07	7.38	m2	7.94
width not exceeding 300mm	1.67	10.00	1.84	0.11	0.02	1.16	0.04	3.03	m	3.26
Work to ceilings										
width exceeding 300mm	4.76	5.00	5.00	0.23	0.08	2.68	0.08	7.76	m2	8.34
width not exceeding 300mm	1.67	10.00	1.84	0.14	0.02	1.43	0.04	3.31	m	3.56
Lathing; Expamet galvanised Rib-Lath, 0.500mm thick x 2.25 kg/m2; butt joints; fixing with galvanised staples										
Work to walls										
width exceeding 300mm	5.47	5.00	5.74	0.19	0.08	2.32	0.07	8.13	m2	8.74
width not exceeding 300mm	1.91	10.00	2.10	0.11	0.02	1.16	0.04	3.30	m	3.54
Work to ceilings										
width exceeding 300mm	5.47	5.00	5.74	0.23	0.08	2.68	0.08	8.51	m2	9.15
width not exceeding 300mm	1.91	10.00	2.10	0.14	0.02	1.43	0.04	3.57	m	3.84
Lathing; Expamet stainless steel Rib-Lath, 0.300mm thick x 1.52 kg/m2; butt joints; fixing with stainless steel staples										
Work to walls										
width exceeding 300mm	12.09	5.00	12.69	0.19	0.08	2.32	0.07	15.08	m2	16.21
width not exceeding 300mm	4.23	10.00	4.65	0.11	0.02	1.16	0.04	5.85	m	6.29
Work to ceilings										
width exceeding 300mm	12.09	5.00	12.69	0.23	0.08	2.68	0.08	15.46	m2	16.62
width not exceeding 300mm	4.23	10.00	4.65	0.14	0.02	1.43	0.04	6.13	m	6.59
Lathing; Expamet galvanised Spraylath, 0.500mm thick x 2.25 kg/m2; butt joints; fixing with galvanised staples										
Work to walls										
width exceeding 300mm	6.66	5.00	6.99	0.19	0.08	2.32	0.07	9.38	m2	10.08
width not exceeding 300mm	2.33	10.00	2.56	0.11	0.02	1.16	0.04	3.76	m	4.04
Work to ceilings										
width exceeding 300mm	6.66	5.00	6.99	0.23	0.08	2.68	0.08	9.76	m2	10.49
width not exceeding 300mm	2.33	10.00	2.56	0.14	0.02	1.43	0.04	4.04	m	4.34
Lathing; galvanised Red-rib lath, 0.400mm thick x 1.78 kg/m2, butt joints; fixing with galvanised staples										
Work to walls										
width exceeding 300mm	6.10	5.00	6.41	0.19	0.08	2.32	0.07	8.79	m2	9.45
width not exceeding 300mm	2.14	10.00	2.35	0.11	0.02	1.16	0.04	3.55	m	3.82
Work to ceilings										
width exceeding 300mm	6.10	5.00	6.41	0.23	0.08	2.68	0.08	9.17	m2	9.86
width not exceeding 300mm	2.14	10.00	2.35	0.11	0.02	1.43	0.04	3.83	m	4.11
M31: FIBROUS PLASTER										
PLASTERBOARD COVE - Coves										
Gyproc cove; fixing with adhesive										
100mm girth	0.81	10.00	0.89	0.15	0.08	1.95	0.09	2.93	m	3.15
extra; ends	-	-	-	0.05	-	0.46	-	0.46	nr	0.50
extra; internal angles	-	-	-	0.25	-	2.31	-	2.31	nr	2.48
extra; external angles	-	-	-	0.25	-	2.31	-	2.31	nr	2.48

SURFACE FINISHES

281

Labour hourly rates: (except Specialists) Craft Operatives £9.23 Labourer £7.02 Rates are national average prices. Refer to REGIONAL VARIATIONS for indicative levels of overall pricing in regions	MATERIALS			LABOUR				RATES		
	Del to Site £	Waste %	Material Cost £	Craft Optve Hrs	Lab Hrs	Labour Cost £	Sunds £	Nett Rate £	Unit	Gross Rate (+7.5%) £
M31: FIBROUS PLASTER Cont.										
PLASTERBOARD COVE - Coves Cont.										
Gyproc cove; fixing with adhesive Cont.										
127mm girth	0.81	10.00	0.89	0.15	0.08	1.95	0.15	2.99	m	3.21
extra; ends	-	-	-	0.05	-	0.46	-	0.46	nr	0.50
extra; internal angles	-	-	-	0.25	-	2.31	-	2.31	nr	2.48
extra; external angles	-	-	-	0.25	-	2.31	-	2.31	nr	2.48
Gyproc cove; fixing with nails to timber										
100mm girth	0.81	10.00	0.89	0.10	0.05	1.27	0.04	2.21	m	2.37
extra; ends	-	-	-	0.05	-	0.46	-	0.46	nr	0.50
extra; internal angles	-	-	-	0.25	-	2.31	-	2.31	nr	2.48
extra; external angles	-	-	-	0.25	-	2.31	-	2.31	nr	2.48
127mm girth	0.81	10.00	0.89	0.10	0.05	1.27	0.04	2.21	m	2.37
extra; ends	-	-	-	0.05	-	0.46	-	0.46	nr	0.50
extra; internal angles	-	-	-	0.25	-	2.31	-	2.31	nr	2.48
extra; external angles	-	-	-	0.25	-	2.31	-	2.31	nr	2.48
PLASTER VENTILATORS - Fibrous plaster ventilator; fixing in plastered wall										
Plain										
229 x 79mm	0.64	5.00	0.67	0.13	-	1.20	-	1.87	nr	2.01
229 x 152mm	0.80	5.00	0.84	0.15	-	1.38	-	2.22	nr	2.39
229 x 229mm	1.06	5.00	1.11	0.18	-	1.66	-	2.77	nr	2.98
Flyproof										
229 x 79mm	1.18	5.00	1.24	0.18	-	1.66	-	2.90	nr	3.12
229 x 152mm	1.43	5.00	1.50	0.20	-	1.85	-	3.35	nr	3.60
229 x 229mm	1.74	5.00	1.83	0.23	-	2.12	-	3.95	nr	4.25
M40: STONE/CONCRETE/QUARRY/CERAMIC TILING/MOSAIC										
CLAY TILE PAVING - Clay floor quarries, B.S.6431, terracotta; 3mm joints, symmetrical layout; bedding in 10mm cement mortar (1:3); pointing in cement mortar (1:3); on cement and sand base										
150 x 150 x 12.5mm units to floors level or to falls only not exceeding 15 degrees from horizontal										
plain	10.88	5.00	11.42	0.80	0.50	10.89	1.31	23.63	m2	25.40
150 x 150 x 20mm units to floors level or to falls only not exceeding 15 degrees from horizontal										
plain	15.39	5.00	16.16	1.00	0.60	13.44	1.31	30.91	m2	33.23
225 x 225 x 25mm units to floors level or to falls only not exceeding 15 degrees from horizontal										
plain	26.56	5.00	27.89	0.70	0.45	9.62	1.31	38.82	m2	41.73
150 x 150 x 12.5mm units to floors to falls and crossfalls and to slopes not exceeding 15 degrees from horizontal										
plain	10.88	5.00	11.42	0.88	0.50	11.63	1.31	24.37	m2	26.19
150 x 150 x 20mm units to floors to falls and crossfalls and to slopes not exceeding 15 degrees from horizontal										
plain	15.39	5.00	16.16	1.10	0.60	14.37	1.31	31.83	m2	34.22
225 x 225 x 25mm units to floors to falls and crossfalls and to slopes not exceeding 15 degrees from horizontal										
plain	26.56	5.00	27.89	0.75	0.45	10.08	1.31	39.28	m2	42.23
Extra for pointing with tinted mortar										
150 x 150 x 12.5mm tiles	1.32	10.00	1.45	0.10	-	0.92	-	2.38	m2	2.55
150 x 150 x 20mm tiles	1.32	10.00	1.45	0.10	-	0.92	-	2.38	m2	2.55
225 x 225 x 25mm tiles	1.32	10.00	1.45	0.07	-	0.65	-	2.10	m2	2.26
150 x 150 x 12.5mm units to skirtings on brickwork or blockwork base										
height 150mm; square top edge	1.77	5.00	1.86	0.30	0.08	3.33	0.29	5.48	m	5.89
height 150mm; rounded top edge	2.30	5.00	2.42	0.30	0.08	3.33	0.29	6.04	m	6.49
height 150mm; rounded top edge and cove at bottom	3.63	5.00	3.81	0.30	0.08	3.33	0.29	7.43	m	7.99
150 x 150 x 20mm units to skirtings on brickwork or blockwork base										
height 150mm; square top edge	2.51	5.00	2.64	0.38	0.09	4.14	0.29	7.06	m	7.59
height 150mm; rounded top edge	2.92	5.00	3.07	0.38	0.09	4.14	0.29	7.50	m	8.06
CLAY TILE PAVING - Clay floor quarries, B.S.6431, blended; 3mm joints symmetrical layout; bedding in 10mm cement mortar (1:3); pointing in cement mortar (1:3); on cement and sand base										
150 x 150 x 12.5mm units to floors level or to falls only not exceeding 15 degrees from horizontal										
plain	14.66	5.00	15.39	0.80	0.50	10.89	1.31	27.60	m2	29.67
150 x 150 x 20mm units to floors level or to falls only not exceeding 15 degrees from horizontal										
plain	16.51	5.00	17.34	1.00	0.60	13.44	1.31	32.09	m2	34.49

Labour hourly rates: (except Specialists) Craft Operatives £9.23 Labourer £7.02 Rates are national average prices. Refer to REGIONAL VARIATIONS for indicative levels of overall pricing in regions	MATERIALS			LABOUR				RATES		
	Del to Site £	Waste %	Material Cost £	Craft Optve Hrs	Lab Hrs	Labour Cost £	Sunds £	Nett Rate £	Unit	Gross Rate (+7.5%) £
M40: STONE/CONCRETE/QUARRY/CERAMIC TILING/MOSAIC Cont.										
CLAY TILE PAVING - Clay floor quarries, B.S.6431, blended; 3mm joints symmetrical layout; bedding in 10mm cement mortar (1:3); pointing in cement mortar (1:3);on cement and sand base Cont.										
225 x 225 x 25mm units to floors level or to falls only not exceeding 15 degrees from horizontal										
plain	29.88	5.00	31.37	0.70	0.45	9.62	1.31	42.30	m2	45.48
150 x 150 x 12.5mm units to floors to falls and crossfalls and to slopes not exceeding 15 degrees from horizontal										
plain	14.66	5.00	15.39	0.88	0.50	11.63	1.31	28.34	m2	30.46
150 x 150 x 20mm units to floors to falls and crossfalls and to slopes not exceeding 15 degrees from horizontal										
plain	16.51	5.00	17.34	1.10	0.60	14.37	1.31	33.01	m2	35.49
225 x 225 x 25mm units to floors to falls and crossfalls and to slopes not exceeding 15 degrees from horizontal										
plain	29.88	5.00	31.37	0.75	0.45	10.08	1.31	42.77	m2	45.97
Extra for pointing with tinted mortar										
150 x 150 x 12.5mm tiles	1.32	10.00	1.45	0.10	-	0.92	-	2.38	m2	2.55
150 x 150 x 20mm tiles	1.32	10.00	1.45	0.10	-	0.92	-	2.38	m2	2.55
225 x 225 x 25mm tiles	1.32	10.00	1.45	0.07	-	0.65	-	2.10	m2	2.26
150 x 150 x 12.5mm units to skirtings on brickwork or blockwork base										
height 150mm; square top edge	2.39	5.00	2.51	0.30	0.08	3.33	0.29	6.13	m	6.59
height 150mm; rounded top edge	2.73	5.00	2.87	0.30	0.08	3.33	0.29	6.49	m	6.97
height 150mm; rounded top edge and cove at bottom	3.83	5.00	4.02	0.30	0.08	3.33	0.29	7.64	m	8.22
150 x 150 x 20mm units to skirtings on brickwork or blockwork base										
height 150mm; square top edge	2.69	5.00	2.82	0.38	0.09	4.14	0.29	7.25	m	7.80
height 150mm; rounded top edge	2.92	5.00	3.07	0.38	0.09	4.14	0.29	7.50	m	8.06
CLAY TILE PAVING - Clay floor quarries, B.S.6431, dark; 3mm joints, symmetrical layout; bedding in 10mm cement mortar (1:3); pointing in cement mortar (1:3); on cement and sand base										
150 x 150 x 12.5mm units to floors level or to falls only not exceeding 15degrees from horizontal										
plain	19.01	5.00	19.96	0.80	0.50	10.89	1.31	32.16	m2	34.58
150 x 150 x 12.5mm units to floors to falls and crossfalls and to slopes not exceeding 15 degrees from horizontal										
plain	19.01	5.00	19.96	0.88	0.50	11.63	1.31	32.90	m2	35.37
150 x 150 x 12.5mm units to skirtings on brickwork or blockwork base										
height 150mm; square top edge	3.09	5.00	3.24	0.30	0.08	3.33	0.29	6.87	m	7.38
height 150mm; rounded top edge	3.61	5.00	3.79	0.30	0.08	3.33	0.29	7.41	m	7.97
height 150mm; rounded top edge and cove at bottom	4.99	5.00	5.24	0.30	0.08	3.33	0.29	8.86	m	9.52
CLAY TILE PAVING - Ceramic floor tiles, B.S.6431, fully vitrified, red; 3mm joints, symmetrical layout; bedding in 10mm cement mortar (1:3); pointing in cement mortar (1:3);on cement and sand base										
100 x 100 x 9mm units to floors level or to falls only not exceeding 15 degrees from horizontal										
plain	13.85	5.00	14.54	1.60	0.90	21.09	1.31	36.94	m2	39.71
152 x 152 x 12mm units to floors level or to falls only not exceeding 15 degrees from horizontal										
plain	15.20	5.00	15.96	0.90	0.55	12.17	1.31	29.44	m2	31.65
200 x 200 x 12mm units to floors level or to falls only not exceeding 15 degrees from horizontal										
plain	16.40	5.00	17.22	0.60	0.40	8.35	1.31	26.88	m2	28.89
100 x 100 x 9mm units to floors to falls and crossfalls and to slopes not exceeding 15 degrees from horizontal										
plain	13.85	5.00	14.54	1.76	0.90	22.56	1.31	38.42	m2	41.30
152 x 152 x 12mm units to floors to falls and crossfalls and to slopes not exceeding 15 degrees from horizontal										
plain	15.20	5.00	15.96	0.99	0.55	13.00	1.31	30.27	m2	32.54
200 x 200 x 12mm units to floors to falls and crossfalls and to slopes not exceeding 15 degrees from horizontal										
plain	16.40	5.00	17.22	0.66	0.40	8.90	1.31	27.43	m2	29.49

SURFACE FINISHES

Labour hourly rates: (except Specialists) Craft Operatives £9.23 Labourer £7.02 Rates are national average prices. Refer to REGIONAL VARIATIONS for indicative levels of overall pricing in regions	MATERIALS			LABOUR				RATES		
	Del to Site £	Waste %	Material Cost £	Craft Optve Hrs	Lab Hrs	Labour Cost £	Sunds £	Nett Rate £	Unit	Gross Rate (+7.5%) £
M40: STONE/CONCRETE/QUARRY/CERAMIC TILING/MOSAIC Cont.										
CLAY TILE PAVING - Ceramic floor tiles, B.S.6431, fully vitrified, red; 3mm joints, symmetrical layout; bedding in 10mm cement mortar (1:3); pointing in cement mortar (1:3);on cement and sand base Cont.										
Extra for pointing with tinted mortar										
100 x 100 x 9mm tiles	1.32	10.00	**1.45**	0.15	-	**1.38**	-	**2.84**	m2	**3.05**
152 x 152 x 12mm tiles	1.32	10.00	**1.45**	0.10	-	**0.92**	-	**2.38**	m2	**2.55**
200 x 200 x 12mm tiles	1.32	10.00	**1.45**	0.08	-	**0.74**	-	**2.19**	m2	**2.35**
100 x 100 x 9mm units to skirtings on brickwork or blockwork base										
height 100mm; square top edge	1.39	5.00	**1.46**	0.36	0.09	**3.95**	0.19	**5.60**	m	**6.02**
height 100mm; rounded top edge	1.85	5.00	**1.94**	0.36	0.09	**3.95**	0.19	**6.09**	m	**6.54**
152 x 152 x 12mm units to skirtings on brickwork or blockwork base										
height 152mm; square top edge	2.31	5.00	**2.43**	0.33	0.08	**3.61**	0.29	**6.32**	m	**6.80**
height 152mm; rounded top edge	3.03	5.00	**3.18**	0.33	0.08	**3.61**	0.29	**7.08**	m	**7.61**
height 152mm; rounded top edge and cove at bottom	6.04	5.00	**6.34**	0.33	0.08	**3.61**	0.29	**10.24**	m	**11.01**
200 x 200 x 12mm units to skirtings on brickwork or blockwork base										
height 200mm; square top edge	3.28	5.00	**3.44**	0.30	0.08	**3.33**	0.39	**7.16**	m	**7.70**
height 200mm; rounded top edge	5.14	5.00	**5.40**	0.30	0.08	**3.33**	0.39	**9.12**	m	**9.80**
CLAY TILE PAVING - Ceramic floor tiles, B.S.6431, fully vitrified, cream; 3mm joints, symmetrical layout; bedding in 10mm cement mortar (1:3); pointing in cement mortar (1:3); on cement and sand base										
100 x 100 x 9mm units to floors level or to falls only not exceeding 15 degrees from horizontal										
plain...	14.75	5.00	**15.49**	1.60	0.90	**21.09**	1.31	**37.88**	m2	**40.72**
152 x 152 x 12mm units to floors level or to falls only not exceeding 15 degrees from horizontal										
plain...	15.95	5.00	**16.75**	0.90	0.55	**12.17**	1.31	**30.23**	m2	**32.49**
200 x 200 x 12mm units to floors level or to falls only not exceeding 15 degrees from horizontal										
plain...	17.05	5.00	**17.90**	0.60	0.40	**8.35**	1.31	**27.56**	m2	**29.63**
100 x 100 x 9mm units to floors to falls and crossfalls and to slopes not exceeding 15 degrees from horizontal										
plain...	14.75	5.00	**15.49**	1.76	0.90	**22.56**	1.31	**39.36**	m2	**42.31**
152 x 152 x 12mm units to floors to falls and crossfalls and to slopes not exceeding 15 degrees from horizontal										
plain...	15.95	5.00	**16.75**	0.99	0.55	**13.00**	1.31	**31.06**	m2	**33.39**
200 x 200 x 12mm units to floors to falls and crossfalls and to slopes not exceeding 15 degrees from horizontal										
plain...	17.05	5.00	**17.90**	0.66	0.40	**8.90**	1.31	**28.11**	m2	**30.22**
Extra for pointing with tinted mortar										
100 x 100 x 9mm tiles	1.32	10.00	**1.45**	0.15	-	**1.38**	-	**2.84**	m2	**3.05**
152 x 152 x 12mm tiles	1.32	10.00	**1.45**	0.10	-	**0.92**	-	**2.38**	m2	**2.55**
200 x 200 x 12mm tiles	1.32	10.00	**1.45**	0.08	-	**0.74**	-	**2.19**	m2	**2.35**
100 x 100 x 9mm units to skirtings on brickwork or blockwork base										
height 100mm; square top edge	1.48	5.00	**1.55**	0.36	0.09	**3.95**	0.19	**5.70**	m	**6.13**
height 100mm; rounded top edge	1.94	5.00	**2.04**	0.36	0.09	**3.95**	0.19	**6.18**	m	**6.65**
152 x 152 x 12mm units to skirtings on brickwork or blockwork base										
height 152mm; square top edge	2.42	5.00	**2.54**	0.33	0.08	**3.61**	0.29	**6.44**	m	**6.92**
height 152mm; rounded top edge	3.09	5.00	**3.24**	0.33	0.08	**3.61**	0.29	**7.14**	m	**7.68**
height 152mm; rounded top edge and cove at bottom	6.04	5.00	**6.34**	0.33	0.08	**3.61**	0.29	**10.24**	m	**11.01**
200 x 200 x 12mm units to skirtings on brickwork or blockwork base										
height 200mm; square top edge	3.41	5.00	**3.58**	0.30	0.08	**3.33**	0.39	**7.30**	m	**7.85**
height 200mm; rounded top edge	5.14	5.00	**5.40**	0.30	0.08	**3.33**	0.39	**9.12**	m	**9.80**
CLAY TILE PAVING - Ceramic floor tiles, B.S.6431, fully vitrified, black; 3mm joints, symmetrical layout; bedding in 10mm cement mortar (1:3); pointing in cement mortar (1:3); on cement and sand base										
100 x 100 x 9mm units to floors level or to falls only not exceeding 15 degrees from horizontal										
plain...	18.10	5.00	**19.00**	1.60	0.90	**21.09**	1.31	**41.40**	m2	**44.51**
152 x 152 x 12mm units to floors level or to falls only not exceeding 15 degrees from horizontal										
plain...	19.00	5.00	**19.95**	0.90	0.55	**12.17**	1.31	**33.43**	m2	**35.94**

Labour hourly rates: (except Specialists) Craft Operatives £9.23 Labourer £7.02 Rates are national average prices. Refer to REGIONAL VARIATIONS for indicative levels of overall pricing in regions	MATERIALS			LABOUR				RATES		
	Del to Site £	Waste %	Material Cost £	Craft Optve Hrs	Lab Hrs	Labour Cost £	Sunds £	Nett Rate £	Unit	Gross Rate (+7.5%) £
M40: STONE/CONCRETE/QUARRY/CERAMIC TILING/MOSAIC Cont.										
CLAY TILE PAVING - Ceramic floor tiles, B.S.6431, fully vitrified, black; 3mm joints, symmetrical layout; bedding in 10mm cement mortar (1:3); pointing in cement mortar (1:3); on cement and sand base Cont.										
200 x 200 x 12mm units to floors level or to falls only not exceeding 15 degrees from horizontal										
plain	19.80	5.00	**20.79**	0.60	0.40	**8.35**	1.31	**30.45**	m2	**32.73**
100 x 100 x 9mm units to floors to falls and crossfalls and to slopes not exceeding 15 degrees from horizontal										
plain	18.10	5.00	**19.00**	1.76	0.90	**22.56**	1.31	**42.88**	m2	**46.09**
152 x 152 x 12mm units to floors to falls and crossfalls and to slopes not exceeding 15 degrees from horizontal										
plain	19.00	5.00	**19.95**	0.99	0.55	**13.00**	1.31	**34.26**	m2	**36.83**
200 x 200 x 12mm units to floors to falls and crossfalls and to slopes not exceeding 15 degrees from horizontal										
plain	19.80	5.00	**20.79**	0.66	0.40	**8.90**	1.31	**31.00**	m2	**33.32**
Extra for pointing with tinted mortar										
100 x 100 x 9mm tiles	1.32	10.00	**1.45**	0.15	-	**1.38**	-	**2.84**	m2	**3.05**
152 x 152 x 12mm tiles	1.32	10.00	**1.45**	0.10	-	**0.92**	-	**2.38**	m2	**2.55**
200 x 200 x 12mm tiles	1.32	10.00	**1.45**	0.08	-	**0.74**	-	**2.19**	m2	**2.35**
100 x 100 x 9mm units to skirtings on brickwork or blockwork base										
height 100mm; square top edge	1.81	5.00	**1.90**	0.36	0.09	**3.95**	0.19	**6.05**	m	**6.50**
height 100mm; rounded top edge	2.28	5.00	**2.39**	0.36	0.09	**3.95**	0.19	**6.54**	m	**7.03**
152 x 152 x 12mm units to skirtings on brickwork or blockwork base										
height 152mm; square top edge	2.89	5.00	**3.03**	0.33	0.08	**3.61**	0.29	**6.93**	m	**7.45**
height 152mm; rounded top edge	3.31	5.00	**3.48**	0.33	0.08	**3.61**	0.29	**7.37**	m	**7.93**
height 152mm; rounded top edge and cove at bottom	6.28	5.00	**6.59**	0.33	0.08	**3.61**	0.29	**10.49**	m	**11.28**
200 x 200 x 12mm units to skirtings on brickwork or blockwork base										
height 200mm; square top edge	3.96	5.00	**4.16**	0.30	0.08	**3.33**	0.39	**7.88**	m	**8.47**
height 200mm; rounded top edge	5.14	5.00	**5.40**	0.30	0.08	**3.33**	0.39	**9.12**	m	**9.80**
TILE WALL LININGS - Ceramic tiles, B.S.6431, glazed white; 2mm joints, symmetrical layout; bedding in 10mm cement mortar (1:3); pointing with neat white cement; on brickwork or blockwork base										
108 x 108 x 6.5mm units to walls										
plain, width exceeding 300mm	18.00	5.00	**18.90**	1.30	0.75	**17.26**	1.28	**37.44**	m2	**40.25**
plain, width not exceeding 300mm	5.94	10.00	**6.53**	0.78	0.47	**10.50**	0.38	**17.41**	m	**18.72**
152 x 152 x 5.5mm units to walls										
plain, width exceeding 300mm	13.10	5.00	**13.76**	1.00	0.60	**13.44**	1.14	**28.34**	m2	**30.46**
plain, width not exceeding 300mm	4.03	10.00	**4.43**	0.60	0.36	**8.07**	0.34	**12.84**	m	**13.80**
TILE WALL LININGS - Ceramic tiles, B.S.6431, glazed white; 2mm joints; symmetrical layout; fixing with thin bed adhesive; pointing with neat white cement; on plaster base										
108 x 108 x 6.5mm units to walls										
plain, width exceeding 300mm	18.00	5.00	**18.90**	1.30	0.75	**17.26**	1.35	**37.51**	m2	**40.33**
plain, width not exceeding 300mm	5.94	10.00	**6.53**	0.78	0.47	**10.50**	0.40	**17.43**	m	**18.74**
152 x 152 x 5.5mm units to walls										
plain, width exceeding 300mm	13.10	5.00	**13.76**	1.00	0.60	**13.44**	1.28	**28.48**	m2	**30.61**
plain, width not exceeding 300mm	4.03	10.00	**4.43**	0.60	0.36	**8.07**	0.38	**12.88**	m	**13.84**
TILE WALL LININGS - Ceramic tiles, B.S.6431, glazed white; 2mm joints, symmetrical layout; fixing with thick bed adhesive; pointing with neat white cement; on cement and sand base										
108 x 108 x 6.5mm units to walls										
plain, width exceeding 300mm	18.00	5.00	**18.90**	1.30	0.75	**17.26**	2.99	**39.15**	m2	**42.09**
plain, width not exceeding 300mm	5.94	10.00	**6.53**	0.78	0.47	**10.50**	0.75	**17.78**	m	**19.12**
152 x 152 x 5.5mm units to walls										
plain, width exceeding 300mm	13.10	5.00	**13.76**	1.00	0.60	**13.44**	2.43	**29.63**	m2	**31.85**
plain, width not exceeding 300mm	4.05	10.00	**4.46**	0.60	0.36	**8.07**	0.73	**13.25**	m	**14.24**
TILE WALL LININGS - Ceramic tiles, B.S.6431, glazed light colour; 2mm joints symmetrical layout; bedding in 10mm cement mortar (1:3); pointing with neat white cement; on cement and sand base										
108 x 108 x 6.5mm units to walls										
plain, width exceeding 300mm	18.00	5.00	**18.90**	1.30	0.75	**17.26**	1.28	**37.44**	m2	**40.25**

SURFACE FINISHES

Labour hourly rates: (except Specialists) Craft Operatives £9.23 Labourer £7.02 Rates are national average prices. Refer to REGIONAL VARIATIONS for indicative levels of overall pricing in regions	MATERIALS			LABOUR				RATES		
	Del to Site £	Waste %	Material Cost £	Craft Optve Hrs	Lab Hrs	Labour Cost £	Sunds £	Nett Rate £	Unit	Gross Rate (+7.5%) £
M40: STONE/CONCRETE/QUARRY/CERAMIC TILING/MOSAIC Cont.										
TILE WALL LININGS - Ceramic tiles, B.S.6431, glazed light colour; 2mm joints symmetrical layout; bedding in 10mm cement mortar (1:3); pointing with neat white cement; on cement and sand base Cont.										
108 x 108 x 6.5mm units to walls Cont.										
Plain, width not exceeding 300mm	5.94	10.00	6.53	0.78	0.47	10.50	0.38	17.41	m	18.72
152 x 152 x 5.5mm units to walls										
Plain, width exceeding 300mm	13.15	5.00	13.81	1.00	0.60	13.44	1.14	28.39	m2	30.52
Plain, width not exceeding 300mm	4.05	10.00	4.46	0.60	0.36	8.07	0.34	12.86	m	13.82
TILE WALL LININGS - Ceramic tiles, B.S.6431, glazed light colour; 2mm joints, symmetrical layout; fixing with thin bed adhesive; pointing with neat white cement; on plaster base										
108 x 108 x 6.5mm units to walls										
Plain, width exceeding 300mm	18.00	5.00	18.90	1.30	0.75	17.26	1.55	37.71	m2	40.54
Plain, width not exceeding 300mm	5.94	10.00	6.53	0.78	0.47	10.50	0.81	17.84	m	19.18
152 x 152 x 5.5mm units to walls										
Plain, width exceeding 300mm	13.15	5.00	13.81	1.00	0.60	13.44	2.56	29.81	m2	32.05
Plain, width not exceeding 300mm	4.05	10.00	4.46	0.60	0.36	8.07	0.77	13.29	m	14.29
TILE WALL LININGS - Ceramic tiles, B.S.6431, glazed light colour; 2mm joints, symmetrical layout; fixing with thick bed adhesive; pointing with neat white cement; on cement and sand base										
108 x 108 x 6.5mm units to walls										
Plain, width exceeding 300mm	22.20	5.00	23.31	1.30	0.75	17.26	1.28	41.85	m2	44.99
Plain, width not exceeding 300mm	7.33	10.00	8.06	0.78	0.47	10.50	0.38	18.94	m	20.36
152 x 152 x 5.5mm units to walls										
plain, width exceeding 300mm	16.25	5.00	17.06	1.00	0.60	13.44	1.14	31.64	m2	34.02
plain, width not exceeding 300mm	5.01	10.00	5.51	0.60	0.36	8.07	0.34	13.92	m	14.96
TILE WALL LININGS - Ceramic tiles, B.S.6431, glazed dark colour; 2mm joints, symmetrical layout; bedding in 10mm cement mortar (1:3); pointing with neat white cement; on brickwork or blockwork base										
108 x 108 x 6.5mm units to walls										
plain, width exceeding 300mm	22.20	5.00	23.31	1.30	0.75	17.26	1.55	42.12	m2	45.28
plain, width not exceeding 300mm	7.33	10.00	8.06	0.78	0.47	10.50	0.47	19.03	m	20.46
152 x 152 x 5.5mm units to walls										
plain, width exceeding 300mm	16.25	5.00	17.06	1.00	0.60	13.44	1.42	31.92	m2	34.32
plain, width not exceeding 300mm	5.01	10.00	5.51	0.60	0.36	8.07	0.42	14.00	m	15.05
TILE WALL LININGS - Ceramic tiles, B.S.6431, glazed dark colour; 2mm joints, symmetrical layout; fixing with thin bed adhesive; pointing with neat white cement; on plaster base										
108 x 108 x 6.5mm units to walls										
plain, width exceeding 300mm	22.20	5.00	23.31	1.30	0.75	17.26	1.55	42.12	m2	45.28
plain, width not exceeding 300mm	7.33	10.00	8.06	0.78	0.47	10.50	0.47	19.03	m	20.46
152 x 152 x 5.5mm units to walls										
plain, width exceeding 300mm	16.25	5.00	17.06	1.00	0.60	13.44	1.42	31.92	m2	34.32
plain, width not exceeding 300mm	5.01	10.00	5.51	0.60	0.36	8.07	0.42	14.00	m	15.05
TILE WALL LININGS - Ceramic tiles, B.S.6431, glazed dark colour; 2mm joints, symmetrical layout; fixing with thick bed adhesive; pointing with neat white cement; on cement and sand base										
108 x 108 x 6.5mm units to walls										
plain, width exceeding 300mm	22.20	5.00	23.31	1.30	0.75	17.26	2.70	43.27	m2	46.52
plain, width not exceeding 300mm	7.33	10.00	8.06	0.78	0.47	10.50	0.81	19.37	m	20.82
152 x 152 x 5.5mm units to walls										
plain, width exceeding 300mm	16.25	5.00	17.06	1.00	0.60	13.44	2.56	33.06	m2	35.54
plain, width not exceeding 300mm	5.01	10.00	5.51	0.60	0.36	8.07	0.77	14.35	m	15.42
TILE WALL LININGS - Extra over ceramic tiles, B.S.6431, glazed white; 2mm joints, symmetrical layout; bedding or fixing in any material to general surfaces on any base										
Special tiles										
rounded edge	0.03	5.00	0.03	-	-	-	-	0.03	m	0.03
external angle bead	2.28	5.00	2.39	0.10	-	0.92	-	3.32	m	3.57
internal angle	-	-	-	0.07	-	0.65	-	0.65	nr	0.69
external angle	-	-	-	0.07	-	0.65	-	0.65	nr	0.69
internal angle bead	2.28	5.00	2.39	0.10	-	0.92	-	3.32	m	3.57
internal angle	-	-	-	0.07	-	0.65	£	0.65	nr	0.69
external angle	-	-	-	0.07	-	0.65	-	0.65	nr	0.69

Labour hourly rates: (except Specialists) Craft Operatives £9.23 Labourer £7.02 Rates are national average prices. Refer to REGIONAL VARIATIONS for indicative levels of overall pricing in regions	MATERIALS			LABOUR				RATES		
	Del to Site £	Waste %	Material Cost £	Craft Optve Hrs	Lab Hrs	Labour Cost £	Sunds £	Nett Rate £	Unit	Gross Rate (+7.5%) £
M40: STONE/CONCRETE/QUARRY/CERAMIC TILING/MOSAIC Cont.										
TILE WALL LININGS - Extra over ceramic tiles, B.S.6431, glazed light colour; 2mm joints, symmetrical layout; bedding or fixing in any material to general surfaces on any base										
Special tiles										
rounded edge	0.21	5.00	0.22	-	-	-	-	0.22	m	0.24
external angle bead	2.28	5.00	2.39	0.10	-	0.92	-	3.32	m	3.57
internal angle	-	-	-	0.07	-	0.65	-	0.65	nr	0.69
external angle	-	-	-	0.07	-	0.65	-	0.65	nr	0.69
internal angle bead	2.28	5.00	2.39	0.10	-	0.92	-	3.32	m	3.57
internal angle	-	-	-	0.07	-	0.65	-	0.65	nr	0.69
external angle	-	-	-	0.07	-	0.65	-	0.65	nr	0.69
TILE WALL LININGS - Extra over ceramic tiles, B.S.6431, glazed dark colour; 2mm joints, symmetrical layout; bedding or fixing in any material to general surfaces on any base										
Special tiles										
rounded edge	-	5.00	-	-	-	-	-	-	m	-
external angle bead	2.28	5.00	2.39	0.10	-	0.92	-	3.32	m	3.57
internal angle	-	-	-	0.07	-	0.65	-	0.65	nr	0.69
external angle	-	-	-	0.07	-	0.65	-	0.65	nr	0.69
internal angle bead	2.28	5.00	2.39	0.10	-	0.92	-	3.32	m	3.57
internal angle	-	-	-	0.07	-	0.65	-	0.65	nr	0.69
external angle	-	-	-	0.07	-	0.65	-	0.65	nr	0.69
TILE CILLS - Ceramic tiles, B.S.6431, glazed white; 2mm joints, symmetrical layout; bedding in 10mm cement mortar (1:3); pointing with neat white cement; on brickwork or blockwork base - cills										
108 x 108 x 4mm units to cills on brickwork or blockwork base										
width 150mm; rounded angle	3.00	5.00	3.15	0.39	0.11	4.37	0.29	7.81	m	8.40
width 225mm; rounded angle	4.98	5.00	5.23	0.54	0.17	6.18	0.43	11.84	m	12.72
width 300mm; rounded angle	5.97	5.00	6.27	0.69	0.23	7.98	0.58	14.83	m	15.94
152 x 152 x 5.5mm units to cills on brickwork or blockwork base										
width 150mm; rounded angle	2.78	5.00	2.92	0.33	0.09	3.68	0.26	6.86	m	7.37
width 225mm; rounded angle	3.78	5.00	3.97	0.44	0.14	5.04	0.39	9.40	m	10.11
width 300mm; rounded angle	4.77	5.00	5.01	0.55	0.18	6.34	0.51	11.86	m	12.75
TILE CILLS - Clay floor quarries, B.S.6431, terracotta; 3mm joints, symmetrical layout; bedding in 10mm cement mortar (1:3); pointing in cement mortar (1:3); on cement and sand base - cills										
152 x 152 x 12.5mm units to cills on brickwork or blockwork base										
width 152mm; rounded angle	2.30	5.00	2.42	0.30	0.08	3.33	0.30	6.05	m	6.50
width 225mm; rounded angle	3.18	5.00	3.34	0.40	0.12	4.53	0.44	8.31	m	8.94
width 300mm; rounded angle	4.07	5.00	4.27	0.51	0.17	5.90	0.59	10.76	m	11.57
M40: STONE/CONCRETE/QUARRY/CERAMIC TILING/MOSAIC; PLANT ROOMS										
CLAY TILE PAVING - Clay floor quarries, B.S.6431, terracotta; 3mm joints, symmetrical layout; bedding in 10mm cement mortar (1:3); pointing in cement mortar (1:3); on cement and sand base										
150 x 150 x 12.5mm units to floors level or to falls only not exceeding 15 degrees from horizontal										
plain	10.88	5.00	11.42	0.88	0.50	11.63	1.31	24.37	m2	26.19
150 x 150 x 20mm units to floors level or to falls only not exceeding 15 degrees from horizontal										
plain	15.39	5.00	16.16	1.10	0.60	14.37	1.31	31.83	m2	34.22
225 x 225 x 25mm units to floors level or to falls only not exceeding 15 degrees from horizontal										
plain	26.56	5.00	27.89	0.75	0.45	10.08	1.31	39.28	m2	42.23
150 x 150 x 12.5mm units to floors to falls and crossfalls and to slopes not exceeding 15 degrees from horizontal										
plain	10.88	5.00	11.42	0.96	0.50	12.37	1.31	25.10	m2	26.99
150 x 150 x 20mm units to floors to falls and crossfalls and to slopes not exceeding 15 degrees from horizontal										
plain	15.39	5.00	16.16	1.20	0.60	15.29	1.31	32.76	m2	35.21
225 x 225 x 25mm units to floors to falls and crossfalls and to slopes not exceeding 15 degrees from horizontal										
plain	26.56	5.00	27.89	0.80	0.45	10.54	1.31	39.74	m2	42.72

SURFACE FINISHES

Labour hourly rates: (except Specialists) Craft Operatives £9.23 Labourer £7.02 Rates are national average prices. Refer to REGIONAL VARIATIONS for indicative levels of overall pricing in regions	MATERIALS			LABOUR				RATES		
	Del to Site £	Waste %	Material Cost £	Craft Optve Hrs	Lab Hrs	Labour Cost £	Sunds £	Nett Rate £	Unit	Gross Rate (+7.5%) £
M40: STONE/CONCRETE/QUARRY/CERAMIC TILING/MOSAIC; PLANT ROOMS Cont.										
CLAY TILE PAVING - Clay floor quarries, B.S.6431, blended; 3mm joints symmetrical layout; bedding in 10mm cement mortar (1:3); pointing in cement mortar (1:3);on cement and sand base										
150 x 150 x 12.5mm units to floors level or to falls only not exceeding 15 degrees from horizontal plain..	14.66	5.00	15.39	0.88	0.50	11.63	1.31	28.34	m2	30.46
150 x 150 x 20mm units to floors level or to falls only not exceeding 15 degrees from horizontal plain..	16.51	5.00	17.34	1.10	0.60	14.37	1.31	33.01	m2	35.49
225 x 225 x 25mm units to floors level or to falls only not exceeding 15 degrees from horizontal plain..	29.88	5.00	31.37	0.75	0.45	10.08	1.31	42.77	m2	45.97
150 x 150 x 12.5mm units to floors to falls and crossfalls and to slopes not exceeding 15 degrees from horizontal plain..	14.66	5.00	15.39	0.96	0.50	12.37	1.31	29.07	m2	31.25
150 x 150 x 20mm units to floors to falls and crossfalls and to slopes not exceeding 15 degrees from horizontal plain..	16.51	5.00	17.34	1.20	0.60	15.29	1.31	33.93	m2	36.48
225 x 225 x 25mm units to floors to falls and crossfalls and to slopes not exceeding 15 degrees from horizontal plain..	29.88	5.00	31.37	0.80	0.45	10.54	1.31	43.23	m2	46.47
CLAY TILE PAVING - Clay floor quarries, B.S.6431, dark; 3mm joints, symmetrical layout; bedding in 10mm cement mortar (1:3); pointing in cement mortar (1:3); on cement and sand base										
150 x 150 x 12.5mm units to floors level or to falls only not exceeding 15degrees from horizontal plain..	19.01	5.00	19.96	0.88	0.50	11.63	1.31	32.90	m2	35.37
150 x 150 x 12.5mm units to floors to falls and crossfalls and to slopes not exceeding 15 degrees from horizontal plain..	19.01	5.00	19.96	0.96	0.50	12.37	1.31	33.64	m2	36.16
M41: TERRAZZO TILING/IN SITU TERRAZZO										
Terrazzo tiles, B.S.4131, aggregate size random, ground, grouted and polished, standard colour range; 3mm joints, symmetrical layout; bedding in 40mm cement mortar (1:3); grouting with white cement; in-situ margins										
300 x 300 x 28mm units to floors on concrete base; level or to falls only not exceeding 15 degrees from horizontal plain..	27.70	5.00	29.09	1.10	0.65	14.72	3.49	47.29	m2	50.84
Terrazzo, white cement and white marble chippings (2:5); polished; on cement and sand base										
6mm work to walls										
width exceeding 300mm	8.45	5.00	8.87	3.60	1.80	45.86	-	54.74	m2	58.84
width not exceeding 300mm	2.65	10.00	2.92	2.16	0.54	23.73	-	26.64	m	28.64
16mm work to floors; one coat; floated level and to falls only not exceeding 15 degrees from horizontal; laid in bays, average size 610 x 610mm...	17.50	5.00	18.38	2.25	1.13	28.70	-	47.08	m2	50.61
6mm work to skirtings										
height 75mm......................................	1.50	5.00	1.58	1.25	0.14	12.52	-	14.10	m	15.15
height 75mm; curved to 3000mm radius	1.50	5.00	1.58	1.67	0.14	16.40	-	17.97	m	19.32
16mm work to skirtings on brickwork or blockwork base										
height 150mm.....................................	3.15	5.00	3.31	1.60	0.27	16.66	-	19.97	m	21.47
height 150mm; curved to 3000mm radius	3.15	5.00	3.31	2.13	0.27	21.56	-	24.86	m	26.73
Rounded angles and intersections										
10 - 100mm radius	-	-	-	0.18	-	1.66	-	1.66	m	1.79
Coves										
25mm girth......................................	-	-	-	0.40	-	3.69	-	3.69	m	3.97
40mm girth......................................	-	-	-	0.60	-	5.54	-	5.54	m	5.95
Accessories										
Plastic dividing strips; setting in bed and finishing 6 x 16mm; flat section	1.75	-	1.75	0.25	-	2.31	-	4.06	m	4.36

Labour hourly rates: (except Specialists) Craft Operatives £9.23 Labourer £7.02 Rates are national average prices. Refer to REGIONAL VARIATIONS for indicative levels of overall pricing in regions	MATERIALS			LABOUR				RATES		
	Del to Site £	Waste %	Material Cost £	Craft Optve Hrs	Lab Hrs	Labour Cost £	Sunds £	Nett Rate £	Unit	Gross Rate (+7.5%) £
M41: TERRAZZO TILING/IN SITU TERRAZZO; STAIRCASE AREAS										
Terrazzo tiles, B.S.4131, aggregate size random, ground, grouted and polished, standard colour range; 3mm joints, symmetrical layout; bedding in 40mm cement mortar (1:3); grouting with white cement; in-situ margins										
300 x 300 x 28mm units to floors on concrete base; level or to falls only not exceeding 15 degrees from horizontal										
plain....................	27.70	5.00	**29.09**	1.65	0.65	**19.79**	3.49	**52.37**	m2	**56.30**
305 x 305 x 28mm units to treads on concrete base										
width 292mm; rounded nosing	13.30	5.00	**13.97**	0.75	0.19	**8.26**	1.02	**23.24**	m	**24.98**
305 x 305 x 28mm units to plain risers on concrete base										
height 165mm	6.65	5.00	**6.98**	0.50	0.11	**5.39**	0.58	**12.95**	m	**13.92**
Terrazzo, white cement and white marble chippings (2:5); polished; on cement and sand base										
16mm work to treads; one coat										
width 279mm	4.55	5.00	**4.78**	3.00	0.32	**29.94**	-	**34.71**	m	**37.32**
extra; two line carborundum non-slip inlay	7.65	5.00	**8.03**	0.25	-	**2.31**	-	**10.34**	m	**11.12**
16mm work to undercut risers; one coat										
height 178mm	3.45	5.00	**3.62**	2.00	0.20	**19.86**	-	**23.49**	m	**25.25**
6mm work to strings; one coat										
height 150mm	1.95	5.00	**2.05**	1.80	0.27	**18.51**	-	**20.56**	m	**22.10**
height 200mm	2.15	5.00	**2.26**	2.50	0.36	**25.60**	-	**27.86**	m	**29.95**
M42: WOOD BLOCK/COMPOSITION BLOCK/ PARQUET FLOORING										
Maple wood blocks, tongued and grooved joints; symmetrical herringbone pattern layout, two block plain borders; fixing with adhesive; sanding, one coat sealer										
70 x 230 x 20mm units to floors on cement and sand base; level or to falls only not exceeding 15 degrees from horizontal										
plain.....................	28.75	5.00	**30.19**	1.50	1.25	**22.62**	3.95	**56.76**	m2	**61.01**
Merbau wood blocks, tongued and grooved joints; symmetrical herringbone pattern layout, two block plain borders; fixing with adhesive; sanding, one coat sealer										
70 x 230 x 20mm units to floors on cement and sand base; level or to falls only not exceeding 15 degrees from horizontal										
plain.....................	23.30	5.00	**24.47**	1.50	1.25	**22.62**	3.95	**51.03**	m2	**54.86**
Oak wood blocks, tongued and grooved joints; symmetrical herringbone pattern layout, two block plain borders; fixing with adhesive; sanding, one coat sealer										
70 x 230 x 20mm units to floors on cement and sand base; level or to falls only not exceeding 15 degrees from horizontal										
plain.....................	33.85	5.00	**35.54**	1.50	1.25	**22.62**	3.95	**62.11**	m2	**66.77**
M50: RUBBER/PLASTICS/CORK/LINO/CARPET TILING/SHEETING										
RUBBER FLOORING - Rubber floor tiles, B.S.1711; butt joints, symmetrical layout; fixing with adhesive; on cement and sand base										
610 x 610 x 2mm units to floors level or to falls only not exceeding 15 degrees from horizontal										
width exceeding 300mm	24.85	5.00	**26.09**	0.35	0.23	**4.85**	1.25	**32.19**	m2	**34.60**
width not exceeding 300mm	7.46	10.00	**8.21**	0.21	0.07	**2.43**	0.37	**11.01**	m	**11.83**
610 x 610 x 2.5mm units to floors level or to falls only not exceeding 15 degrees from horizontal										
width exceeding 300mm	29.20	5.00	**30.66**	0.35	0.23	**4.85**	1.25	**36.76**	m2	**39.51**
width not exceeding 300mm	8.76	10.00	**9.64**	0.21	0.07	**2.43**	0.37	**12.44**	m	**13.37**
RUBBER FLOORING - Smooth finish rubber matting, B.S.1711; butt joints; fixing with adhesive; on cement and sand base										
2mm work to floors level or to falls only not exceeding 15 degrees from horizontal										
width exceeding 300mm	24.10	5.00	**25.31**	0.15	0.13	**2.30**	1.25	**28.85**	m2	**31.02**
width not exceeding 300mm	7.23	10.00	**7.95**	0.09	0.04	**1.11**	0.39	**9.45**	m	**10.16**
2.5mm work to floors level or to falls only not exceeding 15 degrees from horizontal										
width exceeding 300mm	27.85	5.00	**29.24**	0.15	0.13	**2.30**	1.31	**32.85**	m2	**35.31**
width not exceeding 300mm	8.36	10.00	**9.20**	0.09	0.04	**1.11**	0.39	**10.70**	m	**11.50**

SURFACE FINISHES

Labour hourly rates: (except Specialists) Craft Operatives £9.23 Labourer £7.02 Rates are national average prices. Refer to REGIONAL VARIATIONS for indicative levels of overall pricing in regions	MATERIALS			LABOUR				RATES		
	Del to Site £	Waste %	Material Cost £	Craft Optve Hrs	Lab Hrs	Labour Cost £	Sunds £	Nett Rate £	Unit	Gross Rate (+7.5%) £
M50: RUBBER/PLASTICS/CORK/LINO/CARPET TILING/SHEETING Cont.										
RUBBER FLOORING - Smooth finish rubber matting, B.S.1711; butt joints; fixing with adhesive; on cement and sand base Cont.										
4mm work to floors level or to falls only not exceeding 15 degrees from horizontal										
width exceeding 300mm	31.85	5.00	33.44	0.15	0.13	2.30	1.31	37.05	m2	39.83
width not exceeding 300mm	9.56	10.00	10.52	0.09	0.04	1.11	0.39	12.02	m	12.92
PVC FLOORING - P.V.C. floor tiles, B.S.3260; butt joints, symmetrical layout; fixing with adhesive; two coats sealer; on cement and sand base										
300 x 300 x 2mm units to floors level or to falls only not exceeding 15 degrees from horizontal										
width exceeding 300mm	4.50	5.00	4.72	0.45	0.23	5.77	2.75	13.24	m2	14.24
width not exceeding 300mm	1.35	10.00	1.49	0.27	0.07	2.98	0.85	5.32	m	5.72
300 x 300 x 2.5mm units to floors level or to falls only not exceeding 15 degrees from horizontal										
width exceeding 300mm	5.45	5.00	5.72	0.45	0.23	5.77	2.75	14.24	m2	15.31
width not exceeding 300mm	1.64	10.00	1.80	0.27	0.07	2.98	0.85	5.64	m	6.06
PVC FLOORING - Fully flexible PVC heavy duty floor tiles, B.S.3261 type A; butt joints, symmetrical layout; fixing with adhesive; two coats sealer; on cement and sand base										
300 x 300 x 2mm units to floors level or to falls only not exceeding 15 degrees from horizontal										
width exceeding 300mm	8.15	5.00	8.56	0.45	0.23	5.77	2.75	17.08	m2	18.36
width not exceeding 300mm	2.45	10.00	2.69	0.27	0.07	2.98	0.85	6.53	m	7.02
300 x 300 x 2.5mm units to floors level or to falls only not exceeding 15 degrees from horizontal										
width exceeding 300mm	9.45	5.00	9.92	0.45	0.23	5.77	2.75	18.44	m2	19.82
width not exceeding 300mm	2.84	10.00	3.12	0.27	0.07	2.98	0.85	6.96	m	7.48
PVC FLOORING - Fully flexible P.V.C. heavy duty sheet, B.S.3261 Type A; butt joints; welded; fixing with adhesive; two coats sealer; on cement and sand base										
2mm work to floors level or to falls only not exceeding 15 degrees from horizontal										
width exceeding 300mm	8.15	5.00	8.56	0.30	0.15	3.82	2.85	15.23	m2	16.37
width not exceeding 300mm	2.45	10.00	2.69	0.18	0.05	2.01	0.85	5.56	m	5.97
2.5mm work to floors level or to falls only not exceeding 15 degrees from horizontal										
width exceeding 300mm	9.45	5.00	9.92	0.30	0.15	3.82	2.85	16.59	m2	17.84
width not exceeding 300mm	2.84	10.00	3.12	0.18	0.05	2.01	0.85	5.99	m	6.44
3mm work to floors level or to falls only not exceeding 15 degrees from horizontal										
width exceeding 300mm	10.40	5.00	10.92	0.30	0.15	3.82	2.85	17.59	m2	18.91
width not exceeding 300mm	3.12	10.00	3.43	0.18	0.05	2.01	0.85	6.29	m	6.77
PVC FLOORING - P.V.C. coved skirting; fixing with adhesive										
Skirtings on plaster base										
height 75mm	1.25	5.00	1.31	0.20	0.01	1.92	0.21	3.44	m	3.70
height 100mm	1.45	5.00	1.52	0.20	0.02	1.99	0.27	3.78	m	4.06
CORK FLOORING - Cork tiles, B.S.8203; butt joints; symmetrical layout; fixing with adhesive; two coats seal; on cement and sand base										
305 x 305 x 4.8mm units to floors level or to falls only not exceeding 15 degrees from horizontal										
width exceeding 300mm	10.15	5.00	10.66	0.75	0.23	8.54	1.20	20.39	m2	21.92
width not exceeding 300mm	3.10	10.00	3.41	0.45	0.07	4.64	0.36	8.41	m	9.05
305 x 305 x 6.4mm units to floors level or to falls only not exceeding 15 degrees from horizontal										
width exceeding 300mm	11.85	5.00	12.44	0.75	0.23	8.54	1.20	22.18	m2	23.84
width not exceeding 300mm	3.61	10.00	3.97	0.45	0.07	4.64	0.36	8.98	m	9.65
305 x 305 x 8.0mm units to floors level or to falls only not exceeding 15 degrees from horizontal										
width exceeding 300mm	15.25	5.00	16.01	0.75	0.23	8.54	1.12	25.67	m2	27.59
width not exceeding 300mm	4.05	10.00	4.46	0.45	0.07	4.64	0.36	9.46	m	10.17
LINOLEUM FLOORING - Linoleum tiles, B.S.6826; butt joints, symmetrical layout; fixing with adhesive; on cement and sand base										
500 x 500 x 2.5mm units to floors level or to falls only not exceeding 15 degrees from horizontal										
width exceeding 300mm	14.35	5.00	15.07	0.35	0.23	4.85	0.79	20.70	m2	22.26
width not exceeding 300mm	4.31	10.00	4.74	0.21	0.07	2.43	0.24	7.41	m	7.97

Labour hourly rates: (except Specialists) Craft Operatives £9.23 Labourer £7.02 Rates are national average prices. Refer to REGIONAL VARIATIONS for indicative levels of overall pricing in regions	MATERIALS			LABOUR				RATES		
	Del to Site	Waste	Material Cost	Craft Optve	Lab	Labour Cost	Sunds	Nett Rate	Unit	Gross Rate (+7.5%)
	£	.%	£	Hrs	Hrs	£	£	£		£
M50: RUBBER/PLASTICS/CORK/LINO/CARPET TILING/SHEETING **Cont.**										
LINOLEUM FLOORING - Linoleum sheet, B.S.6826; butt **joints; laying loose; on cement and sand base**										
2.5mm work to floors level or to falls only not exceeding 15 degrees from horizontal										
width exceeding 300mm	10.80	5.00	11.34	0.15	0.13	2.30	–	13.64	m2	14.66
width not exceeding 300mm	3.24	10.00	3.56	0.09	0.04	1.11	–	4.68	m	5.03
3.2mm work to floors level or to falls only not exceeding 15 degrees from horizontal										
width exceeding 300mm	13.55	5.00	14.23	0.15	0.13	2.30	–	16.52	m2	17.76
width not exceeding 300mm	4.07	10.00	4.48	0.09	0.04	1.11	–	5.59	m	6.01
4.5mm work to floors level or to falls only not exceeding 15 degrees from horizontal										
width exceeding 300mm	17.05	5.00	17.90	0.15	0.13	2.30	–	20.20	m2	21.71
width not exceeding 300mm	5.12	10.00	5.63	0.09	0.04	1.11	–	6.74	m	7.25
CARPET TILING - Carpet tiles; Marley Floors Ltd. **'Marleytex' nylon needleloom; butt joints,** **symmetrical layout; fixing with adhesive; on cement** **and sand base**										
500 x 500mm units to floors; level or to falls only not exceeding 15 degrees from horizontal										
heavy contract grade										
width exceeding 300mm........................	7.65	5.00	8.03	0.35	0.23	4.85	0.50	13.38	m2	14.38
textured heavy contract grade										
width exceeding 300mm.....................	8.20	5.00	8.61	0.35	0.23	4.85	0.50	13.96	m2	15.00
CARPET SHEETING - Fitted carpeting; Marley Floors **Ltd. 'Marleytex' nylon needleloom; fixing with** **adhesive; on cement and sand base**										
Work to floors; level or to falls only not exceeding 15 degrees from horizontal										
heavy contract grade										
width exceeding 300mm........................	7.10	5.00	7.46	0.20	0.13	2.76	0.50	10.71	m2	11.52
textured heavy contract grade										
width exceeding 300mm.....................	7.95	5.00	8.35	0.20	0.13	2.76	0.50	11.61	m2	12.48
M50: RUBBER/PLASTICS/CORK/LINO/CARPET **TILING/SHEETING; STAIRCASE AREAS**										
STAIR NOSINGS AND TREADS - Accessories										
Heavy duty aluminium alloy stair nosings with anti- slip inserts; 46mm wide with single line insert; fixing to timber with screws										
11.5mm drop	7.58	5.00	7.96	0.40	–	3.69	0.80	12.45	m	13.38
22mm drop	8.82	5.00	9.26	0.40	–	3.69	0.80	13.75	m	14.78
25mm drop	9.48	5.00	9.95	0.40	–	3.69	0.80	14.45	m	15.53
32mm drop	10.73	5.00	11.27	0.40	–	3.69	0.80	15.76	m	16.94
33mm drop	10.73	5.00	11.27	0.40	–	3.69	0.80	15.76	m	16.94
38mm drop	11.71	5.00	12.30	0.40	–	3.69	0.80	16.79	m	18.05
46mm drop	12.33	5.00	12.95	0.40	–	3.69	0.80	17.44	m	18.75
Heavy duty aluminium alloy stair nosings with anti- slip inserts; 80mm wide with two line inserts; fixing to timber with screws										
22mm drop	12.63	5.00	13.26	0.55	–	5.08	0.12	18.46	m	19.84
25mm drop	15.77	5.00	16.56	0.55	–	5.08	0.12	21.75	m	23.39
32mm drop	17.07	5.00	17.92	0.55	–	5.08	0.12	23.12	m	24.85
33mm drop	17.07	5.00	17.92	0.55	–	5.08	0.12	23.12	m	24.85
51mm drop	17.68	5.00	18.56	0.55	–	5.08	0.12	23.76	m	25.54
63mm drop	19.27	5.00	20.23	0.55	–	5.08	0.12	25.43	m	27.34
Heavy duty aluminium alloy stair nosings with anti- slip inserts; 46mm wide with single line insert; fixing to masonry with screws										
11.5mm drop	7.58	5.00	7.96	0.65	–	6.00	0.12	14.08	m	15.13
22mm drop	8.82	5.00	9.26	0.65	–	6.00	0.12	15.38	m	16.53
25mm drop	9.48	5.00	9.95	0.65	–	6.00	0.12	16.07	m	17.28
32mm drop	10.73	5.00	11.27	0.65	–	6.00	0.12	17.39	m	18.69
33mm drop	10.73	5.00	11.27	0.65	–	6.00	0.12	17.39	m	18.69
38mm drop	11.71	5.00	12.30	0.65	–	6.00	0.12	18.41	m	19.80
46mm drop	12.33	5.00	12.95	0.65	–	6.00	0.12	19.07	m	20.50
Heavy duty aluminium alloy stair nosings with anti- slip inserts; 80mm wide with two line inserts; fixing to masonry with screws										
22mm drop	12.63	5.00	13.26	0.80	–	7.38	0.18	20.83	m	22.39
25mm drop	15.77	5.00	16.56	0.80	–	7.38	0.18	24.12	m	25.93
32mm drop	17.07	5.00	17.92	0.80	–	7.38	0.18	25.49	m	27.40
33mm drop	17.07	5.00	17.92	0.80	–	7.38	0.18	25.49	m	27.40
51mm drop	17.68	5.00	18.56	0.80	–	7.38	0.18	26.13	m	28.09
63mm drop	19.27	5.00	20.23	0.80	–	7.38	0.18	27.80	m	29.88
Aluminium stair nosings with anti-slip inserts; 46mm wide with single line insert; fixing to timber with screws										
22mm drop	6.29	5.00	6.60	0.40	–	3.69	0.08	10.38	m	11.15
28mm drop	6.29	5.00	6.60	0.40	–	3.69	0.08	10.38	m	11.15

Labour hourly rates: (except Specialists) Craft Operatives £9.23 Labourer £7.02 Rates are national average prices. Refer to REGIONAL VARIATIONS for indicative levels of overall pricing in regions	MATERIALS			LABOUR				RATES		
	Del to Site £	Waste %	Material Cost £	Craft Optve Hrs	Lab Hrs	Labour Cost £	Sunds £	Nett Rate £	Unit	Gross Rate (+7.5%) £
M50: RUBBER/PLASTICS/CORK/LINO/CARPET TILING/SHEETING; STAIRCASE AREAS Cont.										
STAIR NOSINGS AND TREADS - Accessories Cont.										
Aluminium stair nosings with anti-slip inserts; 46mm wide with single line insert; fixing to timber with screws Cont.										
33mm drop	8.19	5.00	8.60	0.40	-	3.69	0.08	12.37	m	13.30
Aluminium stair nosings with anti-slip inserts; 66mm wide with single line insert; fixing to timber with screws										
22mm drop	8.82	5.00	9.26	0.45	-	4.15	0.12	13.53	m	14.55
25mm drop	9.48	5.00	9.95	0.45	-	4.15	0.12	14.23	m	15.29
32mm drop	10.73	5.00	11.27	0.45	-	4.15	0.12	15.54	m	16.71
33mm drop	11.33	5.00	11.90	0.45	-	4.15	0.12	16.17	m	17.38
Aluminium stair nosings with anti-slip inserts; 80mm wide with two line inserts; fixing to timber with screws										
22mm drop	10.12	5.00	10.63	0.55	-	5.08	0.12	15.82	m	17.01
28mm drop	11.33	5.00	11.90	0.55	-	5.08	0.12	17.09	m	18.37
33mm drop	12.02	5.00	12.62	0.55	-	5.08	0.12	17.82	m	19.15
46mm drop	12.62	5.00	13.25	0.55	-	5.08	0.12	18.45	m	19.83
51mm drop	13.27	5.00	13.93	0.55	-	5.08	0.12	19.13	m	20.56
Aluminium alloy stair nosings with anti-slip inserts; 46mm wide with single line insert; fixing to masonry with screws										
22mm drop	6.29	5.00	6.60	0.65	-	6.00	0.12	12.72	m	13.68
28mm drop	6.29	5.00	6.60	0.65	-	6.00	0.12	12.72	m	13.68
33mm drop	8.19	5.00	8.60	0.65	-	6.00	0.12	14.72	m	15.82
Aluminium alloy stair nosings with anti-slip inserts; 66mm wide with single line insert; fixing to masonry with screws										
22mm drop	8.82	5.00	9.26	0.70	-	6.46	0.18	15.90	m	17.09
25mm drop	9.48	5.00	9.95	0.70	-	6.46	0.18	16.59	m	17.84
32mm drop	10.73	5.00	11.27	0.70	-	6.46	0.18	17.91	m	19.25
33mm drop	11.33	5.00	11.90	0.70	-	6.46	0.18	18.54	m	19.93
Aluminium alloy stair nosings with anti-slip inserts; 80mm wide with two line inserts; fixing to masonry with screws										
22mm drop	10.12	5.00	10.63	0.80	-	7.38	0.18	18.19	m	19.55
28mm drop	11.33	5.00	11.90	0.80	-	7.38	0.18	19.46	m	20.92
33mm drop	12.02	5.00	12.62	0.80	-	7.38	0.18	20.18	m	21.70
46mm drop	12.63	5.00	13.26	0.80	-	7.38	0.18	20.83	m	22.39
51mm drop	13.26	5.00	13.92	0.80	-	7.38	0.18	21.49	m	23.10
M51: EDGE FIXED CARPETING										
Underlay to carpeting; Tredair; fixing with tacks; on timber base										
Work to floors; level or to falls only not exceeding 15 degrees from horizontal										
width exceeding 300mm	1.15	5.00	1.21	0.10	0.05	1.27	0.25	2.73	m2	2.94
Fitted carpeting; Tufted wool/nylon, p.c. £16.00 m2; fixing with tackless grippers; on timber base										
Work to floors; level or to falls only not exceeding 15 degrees from horizontal										
width exceeding 300mm	16.00	5.00	16.80	0.20	0.13	2.76	0.25	19.81	m2	21.29
Fitted carpeting; Axminster wool/nylon, P.C. £22.50 m2; fixing with tackless grippers; on timber base										
Work to floors; level or to falls only not exceeding 15 degrees from horizontal										
width exceeding 300mm	22.50	5.00	23.63	0.20	0.13	2.76	0.25	26.63	m2	28.63
Fitted carpeting; Wilton wool/nylon, p.c. £28.00 m2; fixing with tackless grippers; on timber base										
Work to floors; level or to falls only not exceeding 15 degrees from horizontal										
width exceeding 300mm	28.00	5.00	29.40	0.20	0.13	2.76	0.25	32.41	m2	34.84
M51: EDGE FIXED CARPETING; STAIRCASE AREAS										
STAIR NOSINGS AND TREADS - Accessories										
Aluminium alloy carpet nosings with anti-slip inserts; with single line insert; fixing to timber with screws										
65mm wide x 24mm drop	14.35	5.00	15.07	0.45	-	4.15	0.12	19.34	m	20.79
116mm wide overall with carpet gripper x 42mm drop	20.50	5.00	21.52	0.50	-	4.62	0.18	26.32	m	28.29
Aluminium alloy carpet nosings with anti-slip inserts; with two line inserts; fixing to timber with screws										
80mm wide x 31mm drop	16.40	5.00	17.22	0.55	-	5.08	0.12	22.42	m	24.10

Labour hourly rates: (except Specialists) Craft Operatives £9.23 Labourer £7.02 Rates are national average prices. Refer to REGIONAL VARIATIONS for indicative levels of overall pricing in regions	MATERIALS			LABOUR				RATES		
	Del to Site £	Waste %	Material Cost £	Craft Optve Hrs	Lab Hrs	Labour Cost £	Sunds £	Nett Rate £	Unit	Gross Rate (+7.5%) £
M51: EDGE FIXED CARPETING; STAIRCASE AREAS Cont.										
STAIR NOSINGS AND TREADS - Accessories Cont.										
Aluminium alloy carpet nosings with anti-slip inserts; with single line insert; fixing to masonry with screws										
65mm wide x 24mm drop	14.35	5.00	15.07	0.70	-	6.46	0.18	21.71	m	23.34
116mm wide overall with carpet gripper x 42mm drop	20.50	5.00	21.52	0.75	-	6.92	0.24	28.69	m	30.84
Aluminium alloy carpet nosings with anti-slip inserts; with two line inserts; fixing to masonry with screws										
80mm wide x 31mm drop	16.40	5.00	17.22	0.80	-	7.38	0.18	24.78	m	26.64
M52: DECORATIVE PAPERS/FABRICS										
Lining paper (prime cost sum for supply and allowance for waste included elsewhere); sizing; applying adhesive; hanging; butt joints										
Plaster walls and columns										
exceeding 0.50m2	-	-	-	0.12	-	1.11	0.07	1.18	m2	1.27
Plaster ceilings and beams										
exceeding 0.50m2	-	-	-	0.13	-	1.20	0.07	1.27	m2	1.37
Pulp paper (prime cost sum for supply and allowance for waste included elsewhere); sizing; applying adhesive; hanging; butt joints										
Plaster walls and columns										
exceeding 0.50m2	-	-	-	0.28	-	2.58	0.11	2.69	m2	2.90
Plaster ceilings and beams										
exceeding 0.50m2	-	-	-	0.31	-	2.86	0.11	2.97	m2	3.19
Washable paper (prime cost sum for supply and allowance for waste included elsewhere); sizing; applying adhesive; hanging; butt joints										
Plaster walls and columns										
exceeding 0.50m2	-	-	-	0.22	-	2.03	0.07	2.10	m2	2.26
Plaster ceilings and beams										
exceeding 0.50m2	-	-	-	0.24	-	2.22	0.07	2.29	m2	2.46
Vinyl coated paper (prime cost sum for supply and allowance for waste included elsewhere); sizing; applying adhesive; hanging; butt joints										
Plaster walls and columns										
exceeding 0.50m2	-	-	-	0.22	-	2.03	0.07	2.10	m2	2.26
Plaster ceilings and beams										
exceeding 0.50m2	-	%	-	0.24	-	2.22	0.07	2.29	m2	2.46
Embossed or textured paper (prime cost sum for supply and allowance for waste included elsewhere); sizing; applying adhesive; hanging; butt joints										
Plaster walls and columns										
exceeding 0.50m2	-	-	-	0.22	-	2.03	0.10	2.13	m2	2.29
Plaster ceilings and beams										
exceeding 0.50m2	-	-	-	0.24	-	2.22	0.10	2.32	m2	2.49
Woodchip paper (prime cost sum for supply and allowance for waste included elsewhere); sizing; applying adhesive; hanging; butt joints										
Plaster walls and columns										
exceeding 0.50m2	-	-	-	0.22	-	2.03	0.06	2.09	m2	2.25
Plaster ceilings and beams										
exceeding 0.50m2	-	-	-	0.24	-	2.22	0.07	2.29	m2	2.46
Paper border strips (prime cost sum for supply and allowance for waste included elsewhere); sizing; applying adhesive; hanging; butt joints										
Border strips										
25mm wide to papered walls and columns	-	-	-	0.08	-	0.74	0.02	0.76	m	0.82
75mm wide to papered walls and columns	-	-	-	0.09	-	0.83	0.02	0.85	m	0.91
Lining paper p.c. £1.00 roll; sizing; applying adhesive; hanging; butt joints										
Plaster walls and columns										
exceeding 0.50m2	0.21	12.00	0.24	0.12	-	1.11	0.07	1.41	m2	1.52
Plaster ceilings and beams										
exceeding 0.50m2	0.21	12.00	0.24	0.13	-	1.20	0.07	1.51	m2	1.62

Labour hourly rates: (except Specialists) Craft Operatives £9.23 Labourer £7.02 Rates are national average prices. Refer to REGIONAL VARIATIONS for indicative levels of overall pricing in regions	MATERIALS			LABOUR				RATES		
	Del to Site £	Waste %	Material Cost £	Craft Optve Hrs	Lab Hrs	Labour Cost £	Sunds £	Nett Rate £	Unit	Gross Rate (+7.5%) £
M52: DECORATIVE PAPERS/FABRICS Cont.										
Pulp paper p.c. £3.00 roll; sizing; applying adhesive; hanging; butt joints										
Plaster walls and columns exceeding 0.50m2	0.63	30.00	0.82	0.22	–	2.03	0.10	2.95	m2	3.17
Plaster ceilings and beams exceeding 0.50m2	0.63	25.00	0.79	0.24	–	2.22	0.10	3.10	m2	3.34
Pulp paper (24' drop pattern match) p.c. £4.00 roll; sizing; applying adhesive; hanging; butt joints										
Plaster walls and columns exceeding 0.50m2	0.84	40.00	1.18	0.28	–	2.58	0.11	3.87	m2	4.16
Plaster ceilings and beams exceeding 0.50m2	0.84	35.00	1.13	0.30	–	2.77	0.11	4.01	m2	4.31
Washable paper p.c. £4.00 roll; sizing; applying adhesive; hanging; butt joints										
Plaster walls and columns exceeding 0.50m2	0.84	25.00	1.05	0.22	–	2.03	0.07	3.15	m2	3.39
Plaster ceilings and beams exceeding 0.50m2	0.84	20.00	1.01	0.24	–	2.22	0.07	3.29	m2	3.54
Vinyl coated paper p.c. £6.50 roll; sizing; applying adhesive; hanging; butt joints										
Plaster walls and columns exceeding 0.50m2	1.37	25.00	1.71	0.22	–	2.03	0.07	3.81	m2	4.10
Plaster ceilings and beams exceeding 0.50m2	1.37	20.00	1.64	0.24	–	2.22	0.07	3.93	m2	4.22
Vinyl coated paper p.c. £7.00 roll; sizing; applying adhesive; hanging; butt joints										
Plaster walls and columns exceeding 0.50m2	1.47	25.00	1.84	0.22	–	2.03	0.07	3.94	m2	4.23
Plaster ceilings and beams exceeding 0.50m2	1.47	20.00	1.76	0.24	–	2.22	0.07	4.05	m2	4.35
Embossed paper p.c. £4.50 roll; sizing; applying adhesive; hanging; butt joints										
Plaster walls and columns exceeding 0.50m2	0.95	25.00	1.19	0.22	–	2.03	0.10	3.32	m2	3.57
Plaster ceilings and beams exceeding 0.50m2	0.95	20.00	1.14	0.24	–	2.22	0.10	3.46	m2	3.71
Textured paper p.c. £5.00 roll; sizing; applying adhesive; hanging; butt joints										
Plaster walls and columns exceeding 0.50m2	1.05	25.00	1.31	0.22	–	2.03	0.10	3.44	m2	3.70
Plaster ceilings and beams exceeding 0.50m2	1.05	20.00	1.26	0.24	–	2.22	0.10	3.58	m2	3.84
Woodchip paper p.c. £1.50 roll; sizing; applying adhesive; hanging; butt joints										
Plaster walls and columns exceeding 0.50m2	0.32	25.00	0.40	0.22	–	2.03	0.06	2.49	m2	2.68
Plaster ceilings and beams exceeding 0.50m2	0.32	20.00	0.38	0.24	–	2.22	0.07	2.67	m2	2.87
Paper border strips p.c. £0.25m; applying adhesive; hanging; butt joints										
Border strips 25mm wide to papered walls and columns	0.26	10.00	0.29	0.08	–	0.74	0.02	1.04	m	1.12
Paper border strips p.c. £0.50m; applying adhesive; hanging; butt joints										
Border strips 75mm wide to papered walls and columns	0.53	10.00	0.58	0.09	–	0.83	0.02	1.43	m	1.54
Hessian wall covering (prime cost sum for supply and allowance for waste included elsewhere); sizing; applying adhesive; hanging; butt joints										
Plaster walls and columns exceeding 0.50m2	–	–	–	0.40	–	3.69	0.18	3.87	m2	4.16

Labour hourly rates: (except Specialists) Craft Operatives £9.23 Labourer £7.02 Rates are national average prices. Refer to REGIONAL VARIATIONS for indicative levels of overall pricing in regions	MATERIALS			LABOUR				RATES		
	Del to Site	Waste	Material Cost	Craft Optve	Lab	Labour Cost	Sunds	Nett Rate	Unit	Gross Rate (+7.5%)
	£	%	£	Hrs	Hrs	£	£	£		£
M52: DECORATIVE PAPERS/FABRICS Cont.										
Textile hessian paper backed wall covering (prime cost sum for supply and allowance for waste included elsewhere); sizing; applying adhesive; hanging; butt joints										
Plaster walls and columns										
exceeding 0.50m2	-	-	-	0.33	-	3.05	0.17	3.22	m2	3.46
Hessian wall covering p.c. £4.00m2; sizing; applying adhesive; hanging; butt joints										
Plaster walls and columns										
exceeding 0.50m2	4.73	30.00	6.15	0.40	-	3.69	0.18	10.02	m2	10.77
Textile hessian paper backed wall covering p.c. £7.50m2; sizing; applying adhesive; hanging; butt joints										
Plaster walls and columns										
exceeding 0.50m2	8.40	25.00	10.50	0.33	-	3.05	0.17	13.72	m2	14.74
M52: DECORATIVE PAPERS/FABRICS - REDECORATIONS										
Stripping existing paper; lining paper (prime cost sum for supply and allowance for waste included elsewhere); sizing; applying adhesive hanging; butt joints										
Hard building board walls and columns										
exceeding 0.50m2	-	-	-	0.29	-	2.68	0.52	3.20	m2	3.44
Plaster walls and columns										
exceeding 0.50m2	-	-	-	0.25	-	2.31	0.44	2.75	m2	2.95
Hard building board ceilings and beams										
exceeding 0.50m2	-	-	-	0.32	-	2.95	0.53	3.48	m2	3.74
Plaster ceilings and beams										
exceeding 0.50m2	-	-	-	0.27	-	2.49	0.44	2.93	m2	3.15
Stripping existing paper; lining paper and cross lining (prime cost sum for supply and allowance for waste included elsewhere); sizing; applying adhesive; hanging; butt joints										
Hard building board walls and columns										
exceeding 0.50m2	-	-	-	0.41	-	3.78	0.59	4.37	m2	4.70
Plaster walls and columns										
exceeding 0.50m2	-	-	-	0.37	-	3.42	0.51	3.93	m2	4.22
Hard building board ceilings and beams										
exceeding 0.50m2	-	-	-	0.45	-	4.15	0.60	4.75	m2	5.11
Plaster ceilings and beams										
exceeding 0.50m2	-	-	-	0.40	-	3.69	0.51	4.20	m2	4.52
Stripping existing paper; pulp paper (prime cost sum for supply and allowance for waste included elsewhere); applying adhesive; hanging; butt joints										
Hard building board walls and columns										
exceeding 0.50m2	-	-	-	0.39	-	3.60	0.55	4.15	m2	4.46
Plaster walls and columns										
exceeding 0.50m2	-	-	-	0.35	-	3.23	0.47	3.70	m2	3.98
Hard building board ceilings and beams										
exceeding 0.50m2	-	-	-	0.43	-	3.97	0.56	4.53	m2	4.87
Plaster ceilings and beams										
exceeding 0.50m2	-	-	-	0.38	-	3.51	0.47	3.98	m2	4.28
Stripping existing washable or vinyl coated paper; washable paper (prime cost sum for supply and allowance for waste included elsewhere); applying adhesive; hanging; butt joints										
Hard building board walls and columns										
exceeding 0.50m2	-	-	-	0.40	-	3.69	0.53	4.22	m2	4.54
Plaster walls and columns										
exceeding 0.50m2	-	-	-	0.36	-	3.32	0.44	3.76	m2	4.05
Hard building board ceilings and beams										
exceeding 0.50m2	-	-	-	0.44	-	4.06	0.53	4.59	m2	4.94
Plaster ceilings and beams										
exceeding 0.50m2	-	-	-	0.39	-	3.60	0.45	4.05	m2	4.35

SURFACE FINISHES

Labour hourly rates: (except Specialists) Craft Operatives £9.23 Labourer £7.02 Rates are national average prices. Refer to REGIONAL VARIATIONS for indicative levels of overall pricing in regions	MATERIALS			LABOUR				RATES		
	Del to Site £	Waste %	Material Cost £	Craft Optve Hrs	Lab Hrs	Labour Cost £	Sunds £	Nett Rate £	Unit	Gross Rate (+7.5%) £
M52: DECORATIVE PAPERS/FABRICS - REDECORATIONS Cont.										
Stripping existing washable or vinyl coated paper; vinyl coated paper (prime cost sum for supply and allowance for waste included elsewhere); sizing; applying adhesive; hanging; butt joints										
Hard building board walls and columns exceeding 0.50m2	-	-	-	0.40	-	3.69	0.53	4.22	m2	4.54
Plaster walls and columns exceeding 0.50m2	-	-	-	0.36	-	3.32	0.44	3.76	m2	4.05
Hard building board ceilings and beams exceeding 0.50m2	-	-	-	0.44	-	4.06	0.53	4.59	m2	4.94
Plaster ceilings and beams exceeding 0.50m2	-	-	-	0.39	-	3.60	0.45	4.05	m2	4.35
Stripping existing paper; embossed or textured paper; (prime cost sum for supply and allowance for waste included elsewhere); sizing; applying adhesive; hanging; butt joints										
Hard building board walls and columns exceeding 0.50m2	-	-	-	0.40	-	3.69	0.55	4.24	m2	4.56
Plaster walls and columns exceeding 0.50m2	-	-	-	0.36	-	3.32	0.47	3.79	m2	4.08
Hard building board ceilings and beams exceeding 0.50m2	-	-	-	0.44	-	4.06	0.56	4.62	m2	4.97
Plaster ceilings and beams exceeding 0.50m2	-·	-	-	0.39	-	3.60	0.47	4.07	m2	4.37
Stripping existing paper; woodchip paper (prime cost sum for supply and allowance for waste included elsewhere); sizing; applying adhesive; hanging; butt joints										
Hard building board walls and columns exceeding 0.50m2	-	-	-	0.40	-	3.69	0.52	4.21	m2	4.53
Plaster walls and columns exceeding 0.50m2	-	-	-	0.36	-	3.32	0.49	3.81	m2	4.10
Hard building board ceilings and beams exceeding 0.50m2	-	-	-	0.44	-	4.06	0.58	4.64	m2	4.99
Plaster ceilings and beams exceeding 0.50m2	-	-	-	0.39	-	3.60	0.49	4.09	m2	4.40
Stripping existing varnished or painted paper; woodchip paper (prime cost sum for supply and allowance for waste included elsewhere); sizing; applying adhesive; hanging; butt joints										
Hard building board walls and columns exceeding 0.50m2	-	-	-	0.52	-	4.80	0.59	5.39	m2	5.79
Plaster walls and columns exceeding 0.50m2	-	-	-	0.48	-	4.43	0.46	4.89	m2	5.26
Hard building board ceilings and beams exceeding 0.50m2	-	-	-	0.57	-	5.26	0.55	5.81	m2	6.25
Plaster ceilings and beams exceeding 0.50m2	-	-	-	0.53	-	4.89	0.47	5.36	m2	5.76
Stripping existing paper; lining paper p.c. £1.00 roll; sizing; applying adhesive; hanging; butt joints										
Hard building board walls and columns exceeding 0.50m2	0.21	12.00	0.24	0.29	-	2.68	0.52	3.43	m2	3.69
Plaster walls and columns exceeding 0.50m2	0.21	12.00	0.24	0.25	-	2.31	0.44	2.98	m2	3.21
Hard building board ceilings and beams exceeding 0.50m2	0.21	12.00	0.24	0.32	-	2.95	0.53	3.72	m2	4.00
Plaster ceilings and beams exceeding 0.50m2	0.21	12.00	0.24	0.27	-	2.49	0.44	3.17	m2	3.40
Stripping existing paper; lining paper p.c. £1.00 roll and cross lining P.C. £1.00 roll; sizing; applying adhesive; hanging; butt joints										
Hard building board walls and columns exceeding 0.50m2	0.42	12.00	0.47	0.41	-	3.78	0.59	4.84	m2	5.21
Plaster walls and columns exceeding 0.50m2	0.42	12.00	0.47	0.37	-	3.42	0.51	4.40	m2	4.73

Labour hourly rates: (except Specialists) Craft Operatives £9.23 Labourer £7.02 Rates are national average prices. Refer to REGIONAL VARIATIONS for indicative levels of overall pricing in regions	MATERIALS			LABOUR				RATES		
	Del to Site £	Waste %	Material Cost £	Craft Optve Hrs	Lab Hrs	Labour Cost £	Sunds £	Nett Rate £	Unit	Gross Rate (+7.5%) £
M52: DECORATIVE PAPERS/FABRICS - REDECORATIONS Cont.										
Stripping existing paper; lining paper p.c. £1.00 roll and cross lining P.C. £1.00 roll; sizing; applying adhesive; hanging; butt joints Cont.										
Hard building board ceilings and beams exceeding 0.50m2	0.42	12.00	**0.47**	0.45	-	**4.15**	0.60	**5.22**	m2	**5.62**
Plaster ceilings and beams exceeding 0.50m2	0.42	12.00	**0.47**	0.40	-	**3.69**	0.51	**4.67**	m2	**5.02**
Stripping existing paper; pulp paper p.c. £3.00 roll; sizing; applying adhesive; hanging; butt joints										
Hard building board walls and columns exceeding 0.50m2	0.63	30.00	**0.82**	0.39	-	**3.60**	0.55	**4.97**	m2	**5.34**
Plaster walls and columns exceeding 0.50m2	0.63	30.00	**0.82**	0.35	-	**3.23**	0.47	**4.52**	m2	**4.86**
Hard building board ceilings and beams exceeding 0.50m2	0.63	25.00	**0.79**	0.43	-	**3.97**	0.56	**5.32**	m2	**5.72**
Plaster ceilings and beams exceeding 0.50m2	0.63	25.00	**0.79**	0.38	-	**3.51**	0.47	**4.76**	m2	**5.12**
Stripping existing paper; pulp paper with 24' drop pattern; p.c. £4.00 roll; sizing; applying adhesive; hanging; butt joints										
Hard building board walls and columns exceeding 0.50m2	0.84	40.00	**1.18**	0.45	-	**4.15**	0.56	**5.89**	m2	**6.33**
Plaster walls and columns exceeding 0.50m2	0.84	40.00	**1.18**	0.41	-	**3.78**	0.48	**5.44**	m2	**5.85**
Hard building board ceilings and beams exceeding 0.50m2	0.84	35.00	**1.13**	0.49	-	**4.52**	0.57	**6.23**	m2	**6.69**
Plaster ceilings and beams exceeding 0.50m2	0.84	35.00	**1.13**	0.44	-	**4.06**	0.48	**5.68**	m2	**6.10**
Stripping existing washable or vinyl coated paper; washable paper p.c. £3.75 roll; sizing; applying adhesive; hanging; butt joints										
Hard building board walls and columns exceeding 0.50m2	0.84	25.00	**1.05**	0.40	-	**3.69**	0.53	**5.27**	m2	**5.67**
Plaster walls and columns exceeding 0.50m2	0.84	25.00	**1.05**	0.36	-	**3.32**	0.44	**4.81**	m2	**5.17**
Hard building board ceilings and beams exceeding 0.50m2	0.84	20.00	**1.01**	0.44	-	**4.06**	0.53	**5.60**	m2	**6.02**
Plaster ceilings and beams exceeding 0.50m2	0.84	20.00	**1.01**	0.39	-	**3.60**	0.45	**5.06**	m2	**5.44**
Stripping existing washable or vinyl coated paper; vinyl coated paper P.C. £6.50 roll; sizing; applying adhesive; hanging; butt joints										
Hard building board walls and columns exceeding 0.50m2	1.37	25.00	**1.71**	0.40	-	**3.69**	0.53	**5.93**	m2	**6.38**
Plaster walls and columns exceeding 0.50m2	1.37	25.00	**1.71**	0.36	-	**3.32**	0.44	**5.48**	m2	**5.89**
Hard building board ceilings and beams exceeding 0.50m2	1.37	20.00	**1.64**	0.44	-	**4.06**	0.53	**6.24**	m2	**6.70**
Plaster ceilings and beams exceeding 0.50m2	1.37	20.00	**1.64**	0.39	-	**3.60**	0.45	**5.69**	m2	**6.12**
Stripping existing washable or vinyl coated paper; vinyl coated paper P.C. £8.00 roll; sizing; applying adhesive; hanging; butt joints										
Hard building board walls and columns exceeding 0.50m2	1.47	25.00	**1.84**	0.40	-	**3.69**	0.53	**6.06**	m2	**6.51**
Plaster walls and columns exceeding 0.50m2	1.47	25.00	**1.84**	0.36	-	**3.32**	0.44	**5.60**	m2	**6.02**
Hard building board ceilings and beams exceeding 0.50m2	1.47	20.00	**1.76**	0.44	-	**4.06**	0.53	**6.36**	m2	**6.83**
Plaster ceilings and beams exceeding 0.50m2	1.47	20.00	**1.76**	0.39	-	**3.60**	0.45	**5.81**	m2	**6.25**
Stripping existing paper; embossed paper p.c. £4.50 roll; sizing; applying adhesive; hanging; butt joints										
Hard building board walls and columns exceeding 0.50m2	0.95	25.00	**1.19**	0.39	-	**3.60**	0.55	**5.34**	m2	**5.74**

Labour hourly rates: (except Specialists) Craft Operatives £9.23 Labourer £7.02 Rates are national average prices. Refer to REGIONAL VARIATIONS for indicative levels of overall pricing in regions	MATERIALS			LABOUR				RATES		
	Del to Site £	Waste %	Material Cost £	Craft Optve Hrs	Lab Hrs	Labour Cost £	Sunds £	Nett Rate £	Unit	Gross Rate (+7.5%) £
M52: DECORATIVE PAPERS/FABRICS - REDECORATIONS Cont.										
Stripping existing paper; embossed paper p.c. £4.50 roll; sizing; applying adhesive; hanging; butt joints Cont.										
Plaster walls and columns exceeding 0.50m2	0.95	25.00	**1.19**	0.35	-	**3.23**	0.47	**4.89**	m2	**5.25**
Hard building board ceilings and beams exceeding 0.50m2	0.95	20.00	**1.14**	0.43	-	**3.97**	0.56	**5.67**	m2	**6.09**
Plaster ceilings and beams exceeding 0.50m2	0.95	20.00	**1.14**	0.38	-	**3.51**	0.47	**5.12**	m2	**5.50**
Stripping existing paper; textured paper p.c. £5.00 roll; sizing; applying adhesive; hanging; butt joints										
Hard building board walls and columns exceeding 0.50m2	1.05	25.00	**1.31**	0.39	-	**3.60**	0.55	**5.46**	m2	**5.87**
Plaster walls and columns exceeding 0.50m2	1.05	25.00	**1.31**	0.35	-	**3.23**	0.47	**5.01**	m2	**5.39**
Hard building board ceilings and beams exceeding 0.50m2	1.05	20.00	**1.26**	0.43	-	**3.97**	0.56	**5.79**	m2	**6.22**
Plaster ceilings and beams exceeding 0.50m2	1.05	20.00	**1.26**	0.38	-	**3.51**	0.47	**5.24**	m2	**5.63**
Stripping existing paper; woodchip paper p.c. £1.50 roll; sizing; applying adhesive; hanging; butt joints										
Hard building board walls and columns exceeding 0.50m2	0.32	25.00	**0.40**	0.39	-	**3.60**	0.52	**4.52**	m2	**4.86**
Plaster walls and columns exceeding 0.50m2	0.32	25.00	**0.40**	0.35	-	**3.23**	0.44	**4.07**	m2	**4.38**
Hard building board ceilings and beams exceeding 0.50m2	0.32	20.00	**0.38**	0.43	-	**3.97**	0.53	**4.88**	m2	**5.25**
Plaster ceilings and beams exceeding 0.50m2	0.32	20.00	**0.38**	0.38	-	**3.51**	0.44	**4.33**	m2	**4.66**
Washing down old distempered or painted surfaces; lining paper p.c. £1.00 roll; sizing; applying adhesive; hanging; butt joints										
Hard building board walls and columns exceeding 0.50m2	0.21	12.00	**0.24**	0.20	-	**1.85**	0.08	**2.16**	m2	**2.32**
Plaster walls and columns exceeding 0.50m2	0.21	12.00	**0.24**	0.20	-	**1.85**	0.08	**2.16**	m2	**2.32**
Hard building board ceilings and beams exceeding 0.50m2	0.21	12.00	**0.24**	0.21	-	**1.94**	0.08	**2.25**	m2	**2.42**
Plaster ceilings and beams exceeding 0.50m2	0.21	12.00	**0.24**	0.21	-	**1.94**	0.08	**2.25**	m2	**2.42**
Lining papered walls and columns exceeding 0.50m2	0.21	12.00	**0.24**	0.20	-	**1.85**	0.08	**2.16**	m2	**2.32**
Lining papered ceilings and beams exceeding 0.50m2	0.21	12.00	**0.24**	0.21	-	**1.94**	0.08	**2.25**	m2	**2.42**
Washing down old distempered or painted surfaces; lining paper p.c. £1.00 roll; and cross lining p.c. £1.00 roll; sizing; applying adhesive; hanging; butt joints										
Hard building board walls and columns exceeding 0.50m2	0.42	12.00	**0.47**	0.32	-	**2.95**	0.17	**3.59**	m2	**3.86**
Plaster walls and columns exceeding 0.50m2	0.42	12.00	**0.47**	0.32	-	**2.95**	0.17	**3.59**	m2	**3.86**
Hard building board ceilings and beams exceeding 0.50m2	0.42	12.00	**0.47**	0.34	-	**3.14**	0.17	**3.78**	m2	**4.06**
Plaster ceilings and beams exceeding 0.50m2	0.42	12.00	**0.47**	0.34	-	**3.14**	0.17	**3.78**	m2	**4.06**
Washing down old distempered or painted surfaces; pulp paper p.c. £2.75 roll; sizing; applying adhesive; hanging; butt joints										
Hard building board walls and columns exceeding 0.50m2	0.63	30.00	**0.82**	0.30	-	**2.77**	0.11	**3.70**	m2	**3.98**
Plaster walls and columns exceeding 0.50m2	0.63	30.00	**0.82**	0.30	-	**2.77**	0.11	**3.70**	m2	**3.98**
Hard building board ceilings and beams exceeding 0.50m2	0.63	25.00	**0.79**	0.32	-	**2.95**	0.11	**3.85**	m2	**4.14**

Labour hourly rates: (except Specialists) Craft Operatives £9.23 Labourer £7.02 Rates are national average prices. Refer to REGIONAL VARIATIONS for indicative levels of overall pricing in regions	MATERIALS			LABOUR				RATES		
	Del to Site £	Waste %	Material Cost £	Craft Optve Hrs	Lab Hrs	Labour Cost £	Sunds £	Nett Rate £	Unit	Gross Rate (+7.5%) £
M52: DECORATIVE PAPERS/FABRICS - REDECORATIONS Cont.										
Washing down old distempered or painted surfaces; pulp paper p.c. £2.75 roll; sizing; applying adhesive; hanging; butt joints Cont.										
Plaster ceilings and beams exceeding 0.50m2	0.63	25.00	0.79	0.32	-	2.95	0.11	3.85	m2	4.14
Washing down old distempered or painted surfaces; pulp paper with 24' drop pattern; p.c. £4.00 roll; sizing; applying adhesive; hanging; butt joints										
Hard building board walls and columns exceeding 0.50m2	0.84	40.00	1.18	0.36	-	3.32	0.12	4.62	m2	4.97
Plaster walls and columns exceeding 0.50m2	0.84	40.00	1.18	0.36	-	3.32	0.12	4.62	m2	4.97
Hard building board ceilings and beams exceeding 0.50m2	0.84	35.00	1.13	0.38	-	3.51	0.12	4.76	m2	5.12
Plaster ceilings and beams exceeding 0.50m2	0.84	35.00	1.13	0.38	-	3.51	0.12	4.76	m2	5.12
Washing down old distempered or painted surfaces; washable paper p.c. £4.00 roll; sizing; applying adhesive; hanging; butt joints										
Hard building board walls and columns exceeding 0.50m2	0.84	25.00	1.05	0.30	-	2.77	0.08	3.90	m2	4.19
Plaster walls and columns exceeding 0.50m2	0.84	25.00	1.05	0.30	-	2.77	0.08	3.90	m2	4.19
Hard building board ceilings and beams exceeding 0.50m2	0.84	20.00	1.01	0.32	-	2.95	0.09	4.05	m2	4.36
Plaster ceilings and beams exceeding 0.50m2	0.84	20.00	1.01	0.32	-	2.95	0.09	4.05	m2	4.36
Washing down old distempered or painted surfaces; vinyl coated paper p.c. £6.50 roll; sizing; applying adhesive; hanging; butt joints										
Hard building board walls and columns exceeding 0.50m2	1.37	25.00	1.71	0.30	-	2.77	0.08	4.56	m2	4.90
Plaster walls and columns exceeding 0.50m2	1.37	25.00	1.71	0.30	-	2.77	0.08	4.56	m2	4.90
Hard building board ceilings and beams exceeding 0.50m2	1.37	20.00	1.64	0.32	-	2.95	0.09	4.69	m2	5.04
Plaster ceilings and beams exceeding 0.50m2	1.37	20.00	1.64	0.32	-	2.95	0.09	4.69	m2	5.04
Washing down old distempered or painted surfaces; vinyl coated paper p.c. £7.00 roll; sizing; applying adhesive; hanging; butt joints										
Hard building board walls and columns exceeding 0.50m2	1.47	25.00	1.84	0.30	-	2.77	0.08	4.69	m2	5.04
Plaster walls and columns exceeding 0.50m2	1.47	25.00	1.84	0.30	-	2.77	0.08	4.69	m2	5.04
Hard building board ceilings and beams exceeding 0.50m2	1.47	20.00	1.76	0.32	-	2.95	0.09	4.81	m2	5.17
Plaster ceilings and beams exceeding 0.50m2	1.47	20.00	1.76	0.32	-	2.95	0.09	4.81	m2	5.17
Washing down old distempered or painted surfaces; embossed paper p.c. £4.50 roll; sizing; applying adhesive; hanging; butt joints										
Hard building board walls and columns exceeding 0.50m2	0.95	25.00	1.19	0.30	-	2.77	0.11	4.07	m2	4.37
Plaster walls and columns exceeding 0.50m2	0.95	25.00	1.19	0.30	-	2.77	0.11	4.07	m2	4.37
Hard building board ceilings and beams exceeding 0.50m2	0.95	20.00	1.14	0.32	-	2.95	0.11	4.20	m2	4.52
Plaster ceilings and beams exceeding 0.50m2	0.95	20.00	1.14	0.32	-	2.95	0.11	4.20	m2	4.52
Washing down old distempered or painted surfaces; textured paper p.c. £5.00 roll; sizing; applying adhesive; hanging; butt joints										
Hard building board walls and columns exceeding 0.50m2	1.05	25.00	1.31	0.30	-	2.77	0.11	4.19	m2	4.51

Labour hourly rates: (except Specialists) Craft Operatives £9.23 Labourer £7.02 Rates are national average prices. Refer to REGIONAL VARIATIONS for indicative levels of overall pricing in regions	MATERIALS			LABOUR				RATES		
	Del to Site £	Waste %	Material Cost £	Craft Optve Hrs	Lab Hrs	Labour Cost £	Sunds £	Nett Rate £	Unit	Gross Rate (+7.5%) £
M52: DECORATIVE PAPERS/FABRICS - REDECORATIONS Cont..										
Washing down old distempered or painted surfaces; textured paper p.c. £5.00 roll; sizing; applying adhesive; hanging; butt joints Cont.										
Plaster walls and columns exceeding 0.50m2	1.05	25.00	**1.31**	0.30	-	2.77	0.11	4.19	m2	4.51
Hard building board ceilings and beams exceeding 0.50m2	1.05	20.00	**1.26**	0.32	-	2.95	0.11	4.32	m2	4.65
Plaster ceilings and beams exceeding 0.50m2	1.05	20.00	**1.26**	0.32	-	2.95	0.11	4.32	m2	4.65
Washing down old distempered or painted surfaces; woodchip paper p.c. £1.50 roll; sizing; applying adhesive; hanging; butt joints										
Hard building board walls and columns exceeding 0.50m2	0.32	25.00	**0.40**	0.30	-	2.77	0.08	3.25	m2	3.49
Plaster walls and columns exceeding 0.50m2	0.32	25.00	**0.40**	0.30	-	2.77	0.08	3.25	m2	3.49
Hard building board ceilings and beams exceeding 0.50m2	0.32	20.00	**0.38**	0.32	-	2.95	0.08	3.42	m2	3.67
Plaster ceilings and beams exceeding 0.50m2	0.32	20.00	**0.38**	0.32	-	2.95	0.08	3.42	m2	3.67
Stripping existing paper; hessian wall covering (prime cost sum for supply and allowance for waste included elsewhere); sizing; applying adhesive; hanging; butt joints										
Hard building board walls and columns exceeding 0.50m2	-	-	-	0.57	-	5.26	0.64	5.90	m2	6.34
Plaster walls and columns exceeding 0.50m2	-	-	-	0.53	-	4.89	0.56	5.45	m2	5.86
Stripping existing paper; textile hessian paper backed wall covering (prime cost sum for supply and allowance for waste included elsewhere); sizing; applying adhesive; hanging; butt joints										
Hard building board walls and columns exceeding 0.50m2	-	-	-	0.50	-	4.62	0.63	5.25	m2	5.64
Plaster walls and columns exceeding 0.50m2	-	-	-	0.46	-	4.25	0.54	4.79	m2	5.14
Stripping existing hessian; hessian surfaced wall paper (prime cost sum for supply and allowance for waste included elsewhere); sizing; applying adhesive; hanging; butt joints										
Hard building board walls and columns exceeding 0.50m2	-	-	-	0.94	-	8.68	0.71	9.39	m2	10.09
Plaster walls and columns exceeding 0.50m2	-	-	-	0.90	-	8.31	0.62	8.93	m2	9.60
Stripping existing hessian; textile hessian paper backed wall covering (prime cost sum for supply and allowance for waste included elsewhere); sizing; applying adhesive; hanging; butt joints										
Hard building board walls and columns exceeding 0.50m2	-	-	-	0.87	-	8.03	0.69	8.72	m2	9.37
Plaster walls and columns exceeding 0.50m2	-	-	-	0.83	-	7.66	0.61	8.27	m2	8.89
Stripping existing hessian, paper backed; hessian surfaced wall paper (prime cost sum for supply and allowance for waste included elsewhere); sizing; applying adhesive; hanging; butt joints										
Hard building board walls and columns exceeding 0.50m2	-	-	-	0.57	-	5.26	0.64	5.90	m2	6.34
Plaster walls and columns exceeding 0.50m2	-	-	-	0.53	-	4.89	0.56	5.45	m2	5.86
Stripping existing hessian, paper backed; textile hessian paper backed wall covering (prime cost sum for supply and allowance for waste included elsewhere); sizing; applying adhesive; hanging; butt joints										
Hard building board walls and columns exceeding 0.50m2	-	-	-	0.50	-	4.62	0.63	5.25	m2	5.64

Labour hourly rates: (except Specialists) Craft Operatives £9.23 Labourer £7.02 Rates are national average prices. Refer to REGIONAL VARIATIONS for indicative levels of overall pricing in regions	MATERIALS			LABOUR				RATES		
	Del to Site £	Waste %	Material Cost £	Craft Optve Hrs	Lab Hrs	Labour Cost £	Sunds £	Nett Rate £	Unit	Gross Rate (+7.5%) £
M52: DECORATIVE PAPERS/FABRICS - REDECORATIONS Cont.										
Stripping existing hessian, paper backed; textile hessian paper backed wall covering (prime cost sum for supply and allowance for waste included elsewhere); sizing; applying adhesive; hanging; butt joints Cont.										
Plaster walls and columns exceeding 0.50m2	-	-	-	0.46	-	4.25	0.54	4.79	m2	5.14
Stripping existing paper; hessian surfaced wall paper p.c. £4.50m2; sizing; applying adhesive; hanging; butt joints										
Hard building board walls and columns exceeding 0.50m2	4.73	30.00	6.15	0.57	-	5.26	0.64	12.05	m2	12.95
Plaster walls and columns exceeding 0.50m2	4.73	30.00	6.15	0.53	-	4.89	0.56	11.60	m2	12.47
Stripping existing paper; textile hessian paper backed wall covering P.C. £8.00m2; sizing; applying adhesive; hanging; butt joints										
Hard building board walls and columns exceeding 0.50m2	8.40	25.00	10.50	0.50	-	4.62	0.63	15.74	m2	16.93
Plaster walls and columns exceeding 0.50m2	8.40	25.00	10.50	0.46	-	4.25	0.54	15.29	m2	16.43
Stripping existing hessian; hessian surfaced wall paper p.c. £4.50m2; sizing; applying adhesive; hanging ; butt joints										
Hard building board walls and columns exceeding 0.50m2	4.73	30.00	6.15	0.94	-	8.68	0.71	15.54	m2	16.70
Plaster walls and columns exceeding 0.50m2	4.73	30.00	6.15	0.90	-	8.31	0.54	15.00	m2	16.12
Stripping existing hessian; textile hessian paper backed wall covering P.C. £8.00m2; sizing; applying adhesive; hanging; butt joints										
Hard building board walls and columns exceeding 0.50m2	8.40	25.00	10.50	0.87	-	8.03	0.69	19.22	m2	20.66
Plaster walls and columns exceeding 0.50m2	8.40	25.00	10.50	0.83	-	7.66	0.61	18.77	m2	20.18
Stripping existing hessian, paper backed; hessian surfaced wall paper p.c. £4.50m2; sizing; applying adhesive; hanging; butt joints										
Hard building board walls and columns exceeding 0.50m2	4.73	30.00	6.15	0.57	-	5.26	0.64	12.05	m2	12.95
Plaster walls and columns exceeding 0.50m2	4.73	30.00	6.15	0.53	-	4.89	0.56	11.60	m2	12.47
Stripping existing hessian, paper backed; textile hessian paper backed wall covering p.c. £8.00m2; sizing; applying adhesive; hanging; butt joints										
Hard building board walls and columns exceeding 0.50m2	8.40	25.00	10.50	0.50	-	4.62	0.63	15.74	m2	16.93
Plaster walls and columns exceeding 0.50m2	8.40	25.00	10.50	0.46	-	4.25	0.54	15.29	m2	16.43
Expanded polystyrene sheet 2mm thick; sizing; applying adhesive; hanging; butt joints										
Plaster walls and columns exceeding 0.50m2	0.30	25.00	0.38	0.33	-	3.05	0.35	3.77	m2	4.05
Plaster ceilings and beams exceeding 0.50m2	0.31	25.00	0.39	0.39	-	3.60	0.35	4.34	m2	4.66
M60: PAINTING/CLEAR FINISHING - EMULSION PAINTING										
Mist coat, one full coat emulsion paint										
Concrete general surfaces girth exceeding 300mm	0.31	10.00	0.34	0.15	-	1.38	0.03	1.76	m2	1.89
Concrete general surfaces 3.50 - 5.00m above floor girth exceeding 300mm	0.31	10.00	0.34	0.17	-	1.57	0.03	1.94	m2	2.09
Plaster general surfaces girth exceeding 300mm	0.27	10.00	0.30	0.13	-	1.20	0.02	1.52	m2	1.63
isolated surfaces, girth not exceeding 300mm	0.08	10.00	0.09	0.07	-	0.65	0.01	0.74	m	0.80
Plaster general surfaces 3.50 - 5.00m above floor girth exceeding 300mm	0.27	10.00	0.30	0.15	-	1.38	0.03	1.71	m2	1.84

SURFACE FINISHES

SURFACE FINISHES – MAJOR WORKS

Labour hourly rates: (except Specialists) Craft Operatives £9.23 Labourer £7.02 Rates are national average prices. Refer to REGIONAL VARIATIONS for indicative levels of overall pricing in regions	MATERIALS			LABOUR				RATES		
	Del to Site £	Waste %	Material Cost £	Craft Optve Hrs	Lab Hrs	Labour Cost £	Sunds £	Nett Rate £	Unit	Gross Rate (+7.5%) £
M60: PAINTING/CLEAR FINISHING - EMULSION PAINTING Cont.										
Mist coat, one full coat emulsion paint Cont.										
Plasterboard general surfaces										
girth exceeding 300mm	0.27	10.00	0.30	0.13	-	1.20	0.02	1.52	m2	1.63
Plasterboard general surfaces 3.50 - 5.00m above floor										
girth exceeding 300mm	0.27	10.00	0.30	0.15	-	1.38	0.03	1.71	m2	1.84
Brickwork general surfaces										
girth exceeding 300mm	0.40	10.00	0.44	0.17	-	1.57	0.03	2.04	m2	2.19
Paper covered general surfaces										
girth exceeding 300mm	0.29	10.00	0.32	0.14	-	1.29	0.02	1.63	m2	1.75
Paper covered general surfaces 3.50 - 5.00m above floor										
girth exceeding 300mm	0.29	10.00	0.32	0.16	-	1.48	0.03	1.83	m2	1.96
Mist coat, two full coats emulsion paint										
Concrete general surfaces										
girth exceeding 300mm	0.53	10.00	0.58	0.21	-	1.94	0.04	2.56	m2	2.75
Concrete general surfaces 3.50 - 5.00m above floor										
girth exceeding 300mm	0.53	10.00	0.58	0.24	-	2.22	0.04	2.84	m2	3.05
Plaster general surfaces										
girth exceeding 300mm	0.44	10.00	0.48	0.19	-	1.75	0.03	2.27	m2	2.44
isolated surfaces, girth not exceeding 300mm	0.13	10.00	0.14	0.10	-	0.92	0.02	1.09	m	1.17
Plaster general surfaces 3.50 - 5.00m above floor										
girth exceeding 300mm	0.44	10.00	0.48	0.22	-	2.03	0.04	2.55	m2	2.75
Plasterboard general surfaces										
girth exceeding 300mm	0.44	10.00	0.48	0.19	-	1.75	0.03	2.27	m2	2.44
Plasterboard general surfaces 3.50 - 5.00m above floor										
girth exceeding 300mm	0.44	10.00	0.48	0.22	-	2.03	0.04	2.55	m2	2.75
Brickwork general surfaces										
girth exceeding 300mm	0.66	10.00	0.73	0.24	-	2.22	0.04	2.98	m2	3.20
Paper covered general surfaces										
girth exceeding 300mm	0.49	10.00	0.54	0.20	-	1.85	0.03	2.42	m2	2.60
Paper covered general surfaces 3.50 - 5.00m above floor										
girth exceeding 300mm	0.49	10.00	0.54	0.30	-	2.77	0.05	3.36	m2	3.61
M60: PAINTING/CLEAR FINISHING - CEMENT PAINTING										
One coat Snowcem; external work										
Cement rendered general surfaces										
girth exceeding 300mm	0.26	10.00	0.29	0.11	-	1.02	0.02	1.32	m2	1.42
isolated surfaces, girth not exceeding 300mm	0.08	10.00	0.09	0.04	-	0.37	0.01	0.47	m	0.50
Concrete general surfaces										
girth exceeding 300mm	0.35	10.00	0.39	0.11	-	1.02	0.02	1.42	m2	1.53
Brickwork general surfaces										
girth exceeding 300mm	0.35	10.00	0.39	0.13	-	1.20	0.02	1.60	m2	1.73
Rough cast general surfaces										
girth exceeding 300mm	0.40	10.00	0.44	0.17	-	1.57	0.03	2.04	m2	2.19
Two coats Snowcem; external work										
Cement rendered general surfaces										
girth exceeding 300mm	0.47	10.00	0.52	0.22	-	2.03	0.04	2.59	m2	2.78
isolated surfaces, girth not exceeding 300mm	0.14	10.00	0.15	0.08	-	0.74	0.01	0.90	m	0.97
Concrete general surfaces										
girth exceeding 300mm	0.63	10.00	0.69	0.22	-	2.03	0.04	2.76	m2	2.97
Brickwork general surfaces										
girth exceeding 300mm	0.63	10.00	0.69	0.24	-	2.22	0.04	2.95	m2	3.17
Rough cast general surfaces										
girth exceeding 300mm	0.71	10.00	0.78	0.33	-	3.05	0.06	3.89	m2	4.18
One coat sealer, one coat Snowcem; external work										
Cement rendered general surfaces										
girth exceeding 300mm	0.78	10.00	0.86	0.18	-	1.66	0.03	2.55	m2	2.74
isolated surfaces, girth not exceeding 300mm	0.23	10.00	0.25	0.06	-	0.55	0.01	0.82	m	0.88
Concrete general surfaces										
girth exceeding 300mm	0.87	10.00	0.96	0.18	-	1.66	0.03	2.65	m2	2.85

Labour hourly rates: (except Specialists) Craft Operatives £9.23 Labourer £7.02 Rates are national average prices. Refer to REGIONAL VARIATIONS for indicative levels of overall pricing in regions	MATERIALS			LABOUR				RATES		
	Del to Site £	Waste %	Material Cost £	Craft Optve Hrs	Lab Hrs	Labour Cost £	Sunds £	Nett Rate £	Unit	Gross Rate (+7.5%) £
M60: PAINTING/CLEAR FINISHING - CEMENT PAINTING Cont.										
One coat sealer, one coat Snowcem; external work Cont.										
Brickwork general surfaces										
girth exceeding 300mm	0.87	10.00	0.96	0.21	-	1.94	0.04	2.94	m2	3.16
Rough cast general surfaces										
girth exceeding 300mm	0.92	10.00	1.01	0.27	-	2.49	0.05	3.55	m2	3.82
One coat sealer, two coats Snowcem; external work										
Cement rendered general surfaces										
girth exceeding 300mm	0.99	10.00	1.09	0.34	-	3.14	0.06	4.29	m2	4.61
isolated surfaces, girth not exceeding 300mm	0.30	10.00	0.33	0.11	-	1.02	0.02	1.37	m	1.47
Concrete general surfaces										
girth exceeding 300mm	1.15	10.00	1.26	0.34	-	3.14	0.06	4.46	m2	4.80
Brickwork general surfaces										
girth exceeding 300mm	1.15	10.00	1.26	0.39	-	3.60	0.07	4.93	m2	5.30
Rough cast general surfaces										
girth exceeding 300mm	1.23	10.00	1.35	0.51	-	4.71	0.09	6.15	m2	6.61
One coat textured masonry paint; external work										
Cement rendered general surfaces										
girth exceeding 300mm	0.58	10.00	0.64	0.11	-	1.02	0.02	1.67	m2	1.80
isolated surfaces, girth not exceeding 300mm	0.17	10.00	0.19	0.04	-	0.37	0.01	0.57	m	0.61
Concrete general surfaces										
girth exceeding 300mm	0.65	10.00	0.71	0.11	-	1.02	0.02	1.75	m2	1.88
Brickwork general surfaces										
girth exceeding 300mm	0.72	10.00	0.79	0.13	-	1.20	0.02	2.01	m2	2.16
Rough cast general surfaces										
girth exceeding 300mm	1.16	10.00	1.28	0.17	-	1.57	0.03	2.88	m2	3.09
Two coats textured masonry paint; external work										
Cement rendered general surfaces										
girth exceeding 300mm	1.04	10.00	1.14	0.22	-	2.03	0.04	3.21	m2	3.46
isolated surfaces, girth not exceeding 300mm	0.31	10.00	0.34	0.08	-	0.74	0.01	1.09	m	1.17
Concrete general surfaces										
girth exceeding 300mm	1.16	10.00	1.28	0.22	-	2.03	0.04	3.35	m2	3.60
Brickwork general surfaces										
girth exceeding 300mm	1.30	10.00	1.43	0.24	-	2.22	0.04	3.69	m2	3.96
Rough cast general surfaces										
girth exceeding 300mm	2.08	10.00	2.29	0.33	-	3.05	0.06	5.39	m2	5.80
One coat stabilising solution, one coat textured masonry paint; external work										
Cement rendered general surfaces										
girth exceeding 300mm	1.09	10.00	1.20	0.18	-	1.66	0.03	2.89	m2	3.11
isolated surfaces, girth not exceeding 300mm	0.33	10.00	0.36	0.06	-	0.55	0.01	0.93	m	1.00
Concrete general surfaces										
girth exceeding 300mm	1.21	10.00	1.33	0.18	-	1.66	0.03	3.02	m2	3.25
Brickwork general surfaces										
girth exceeding 300mm	1.34	10.00	1.47	0.21	-	1.94	0.04	3.45	m2	3.71
Rough cast general surfaces										
girth exceeding 300mm	1.81	10.00	1.99	0.27	-	2.49	0.05	4.53	m2	4.87
One coat stabilising solution, two coats textured masonry paint; external work										
Cement rendered general surfaces										
girth exceeding 300mm	1.55	10.00	1.71	0.34	-	3.14	0.06	4.90	m2	5.27
isolated surfaces, girth not exceeding 300mm	0.46	10.00	0.51	0.11	-	1.02	0.02	1.54	m	1.66
Concrete general surfaces										
girth exceeding 300mm	1.72	10.00	1.89	0.34	-	3.14	0.06	5.09	m2	5.47
Brickwork general surfaces										
girth exceeding 300mm	1.92	10.00	2.11	0.39	-	3.60	0.07	5.78	m2	6.22
Rough cast general surfaces										
girth exceeding 300mm	2.74	10.00	3.01	0.51	-	4.71	0.09	7.81	m2	8.40
M60: PAINTING/CLEAR FINISHING - PRESERVATIVE TREATMENT										
One coat creosote B.S.144; external work (sawn timber)										
Wood general surfaces										
girth exceeding 300mm	0.07	10.00	0.08	0.10	-	0.92	0.02	1.02	m2	1.10

SURFACE FINISHES

Labour hourly rates: (except Specialists) Craft Operatives £9.23 Labourer £7.02 Rates are national average prices. Refer to REGIONAL VARIATIONS for indicative levels of overall pricing in regions	MATERIALS			LABOUR				RATES		
	Del to Site £	Waste %	Material Cost £	Craft Optve Hrs	Lab Hrs	Labour Cost £	Sunds £	Nett Rate £	Unit	Gross Rate (+7.5%) £
M60: PAINTING/CLEAR FINISHING - PRESERVATIVE TREATMENT Cont.										
One coat creosote B.S.144; external work (sawn timber) Cont.										
Wood general surfaces Cont.										
isolated surfaces, girth not exceeding 300mm	0.02	10.00	0.02	0.04	-	0.37	0.01	0.40	m	0.43
Two coats creosote B.S.144; external work (sawn timber)										
Wood general surfaces										
girth exceeding 300mm	0.15	10.00	0.17	0.20	-	1.85	0.03	2.04	m2	2.19
isolated surfaces, girth not exceeding 300mm	0.04	10.00	0.04	0.07	-	0.65	0.01	0.70	m	0.75
One coat wood preservative, internal work (sawn timber)										
Wood general surfaces										
girth exceeding 300mm	0.39	10.00	0.43	0.11	-	1.02	0.02	1.46	m2	1.57
isolated surfaces, girth not exceeding 300mm	0.12	10.00	0.13	0.04	-	0.37	0.01	0.51	m2	0.55
One coat wood preservative, internal work (wrought timber)										
Wood general surfaces										
girth exceeding 300mm	0.35	10.00	0.39	0.09	-	0.83	0.02	1.24	m2	1.33
isolated surfaces, girth not exceeding 300mm	0.10	10.00	0.11	0.03	-	0.28	0.01	0.40	m	0.43
One coat wood preservative, external work (sawn timber)										
Wood general surfaces										
girth exceeding 300mm	0.39	10.00	0.43	0.10	-	0.92	0.02	1.37	m2	1.47
isolated surfaces, girth not exceeding 300mm	0.12	10.00	0.13	0.04	-	0.37	0.01	0.51	m	0.55
One coat wood preservative, external work (wrought timber)										
Wood general surfaces										
girth exceeding 300mm	0.35	10.00	0.39	0.08	-	0.74	0.01	1.13	m2	1.22
isolated surfaces, girth not exceeding 300mm	0.10	10.00	0.11	0.02	-	0.18	-	0.29	m	0.32
Two coats wood preservative, internal work (sawn timber)										
Wood general surfaces										
girth exceeding 300mm	0.54	10.00	0.59	0.22	-	2.03	0.04	2.66	m2	2.86
isolated surfaces, girth not exceeding 300mm	0.16	10.00	0.18	0.08	-	0.74	0.01	0.92	m	0.99
Two coats wood preservative, internal work (wrought timber)										
Wood general surfaces										
girth exceeding 300mm	0.43	10.00	0.47	0.18	-	1.66	0.03	2.16	m2	2.33
isolated surfaces, girth not exceeding 300mm	0.13	10.00	0.14	0.05	-	0.46	0.01	0.61	m	0.66
Two coats wood preservative, external work (sawn timber)										
Wood general surfaces										
girth exceeding 300mm	0.54	10.00	0.59	0.20	-	1.85	0.03	2.47	m2	2.66
isolated surfaces, girth not exceeding 300mm	0.16	10.00	0.18	0.07	-	0.65	0.01	0.83	m	0.89
Two coats wood preservative, external work (wrought timber)										
Wood general surfaces										
girth exceeding 300mm	0.43	10.00	0.47	0.16	-	1.48	0.03	1.98	m2	2.13
isolated surfaces, girth not exceeding 300mm	0.13	10.00	0.14	0.05	-	0.46	0.01	0.61	m	0.66
Two coats Sadolins Classic; two coats Sadolins Holdex										
Wood general surfaces										
girth exceeding 300mm	2.61	10.00	2.87	0.54	-	4.98	0.08	7.94	m2	8.53
isolated surfaces, girth not exceeding 300mm	0.78	10.00	0.86	0.19	-	1.75	0.03	2.64	m	2.84
Wood glazed doors										
girth exceeding 300mm; panes, area not exceeding 0.10m2	1.31	10.00	1.44	0.89	-	8.21	0.13	9.79	m2	10.52
girth exceeding 300mm; panes, area 0.10 - 0.50m2 .	0.91	10.00	1.00	0.64	-	5.91	0.65	7.56	m2	8.13
girth exceeding 300mm; panes, area 0.50 - 1.00m2 .	0.65	10.00	0.71	0.54	-	4.98	0.08	5.78	m2	6.21
girth exceeding 300mm; panes, area exceeding 1.00m2	0.53	10.00	0.58	0.48	-	4.43	0.07	5.08	m2	5.46
Wood partially glazed doors										
girth exceeding 300mm; panes, area 0.50 - 1.00m2 .	1.57	10.00	1.73	0.54	-	4.98	0.08	6.79	m2	7.30
girth exceeding 300mm; panes, area exceeding 1.00m2	1.43	10.00	1.57	0.52	-	4.80	0.08	6.45	m2	6.94
Wood windows and screens										
girth exceeding 300mm; panes, area not exceeding 0.10m2	1.43	10.00	1.57	0.98	-	9.05	0.15	10.77	m2	11.58
girth exceeding 300mm; panes, area 0.10 - 0.50m2 .	1.03	10.00	1.13	0.71	-	6.55	0.11	7.80	m2	8.38
girth exceeding 300mm; panes, area 0.50 - 1.00m2 .	0.78	10.00	0.86	0.60	-	5.54	0.09	6.49	m2	6.97

Labour hourly rates: (except Specialists) Craft Operatives £9.23 Labourer £7.02 Rates are national average prices. Refer to REGIONAL VARIATIONS for indicative levels of overall pricing in regions	MATERIALS			LABOUR				RATES		
	Del to Site £	Waste %	Material Cost £	Craft Optve Hrs	Lab Hrs	Labour Cost £	Sunds £	Nett Rate £	Unit	Gross Rate (+7.5%) £
M60: PAINTING/CLEAR FINISHING - PRESERVATIVE TREATMENT Cont.										
Two coats Sadolins Classic; two coats Sadolins Holdex Cont.										
Wood windows and screens Cont.										
girth exceeding 300mm; panes, area exceeding 1.00m2	0.65	10.00	0.71	0.52	-	4.80	0.08	5.59	m2	6.01
One coat Sadolins Classic; two coats Sadolins Prestige; external work										
Wood general surfaces										
girth exceeding 300mm	1.80	10.00	1.98	0.50	-	4.62	0.08	6.67	m2	7.18
isolated surfaces, girth not exceeding 300mm	0.54	10.00	0.59	0.17	-	1.57	0.03	2.19	m	2.36
Wood glazed doors										
girth exceeding 300mm; panes, area not exceeding 0.10m2	0.90	10.00	0.99	0.78	-	7.20	0.12	8.31	m2	8.93
girth exceeding 300mm; panes, area 0.10 - 0.50m2 .	0.63	10.00	0.69	0.56	-	5.17	0.08	5.94	m2	6.39
girth exceeding 300mm; panes, area 0.50 - 1.00m2 .	0.46	10.00	0.51	0.47	-	4.34	0.08	4.92	m2	5.29
girth exceeding 300mm; panes, area exceeding 1.00m2	0.36	10.00	0.40	0.42	-	3.88	0.06	4.33	m2	4.66
Wood partially glazed doors										
girth exceeding 300mm; panes, area 0.50 - 1.00m2 .	1.08	10.00	1.19	0.47	-	4.34	0.07	5.60	m2	6.02
girth exceeding 300mm; panes, area exceeding 1.00m2	1.00	10.00	1.10	0.45	-	4.15	0.07	5.32	m2	5.72
Wood windows and screens										
girth exceeding 300mm; panes, area not exceeding 0.10m2	1.00	10.00	1.10	0.89	-	8.21	0.13	9.44	m2	10.15
girth exceeding 300mm; panes, area 0.10 - 0.50m2 .	0.72	10.00	0.79	0.65	-	6.00	0.13	6.92	m2	7.44
girth exceeding 300mm; panes, area 0.50 - 1.00m2 .	0.54	10.00	0.59	0.55	-	5.08	0.08	5.75	m2	6.18
girth exceeding 300mm; panes, area exceeding 1.00m2	0.46	10.00	0.51	0.49	-	4.52	0.07	5.10	m2	5.48
Wood railings fences and gates; open type										
girth exceeding 300mm	1.80	10.00	1.98	0.84	-	7.75	0.13	9.86	m2	10.60
isolated surfaces, girth not exceeding 300mm	0.54	10.00	0.59	0.25	-	2.31	0.04	2.94	m	3.16
Wood railings fences and gates; close type										
girth exceeding 300mm	1.80	10.00	1.98	0.84	-	7.75	0.13	9.86	m2	10.60
M60: PAINTING/CLEAR FINISHING - OIL PAINTING WALLS AND CEILINGS										
One coat primer, one undercoat, one coat full gloss finish										
Concrete general surfaces										
girth exceeding 300mm	0.85	10.00	0.94	0.30	-	2.77	0.05	3.75	m2	4.04
Concrete general surfaces 3.50 - 5.00m above floor										
girth exceeding 300mm	0.85	10.00	0.94	0.33	-	3.05	0.06	4.04	m2	4.34
Brickwork general surfaces										
girth exceeding 300mm	1.04	10.00	1.14	0.34	-	3.14	0.06	4.34	m2	4.67
Plasterboard general surfaces										
girth exceeding 300mm	0.66	10.00	0.73	0.27	-	2.49	0.05	3.27	m2	3.51
Plasterboard general surfaces 3.50 - 5.00m above floor										
girth exceeding 300mm	0.66	10.00	0.73	0.30	-	2.77	0.05	3.54	m2	3.81
Plaster general surfaces										
girth exceeding 300mm	0.76	10.00	0.84	0.27	-	2.49	0.05	3.38	m2	3.63
isolated surfaces, girth not exceeding 300mm	0.23	10.00	0.25	0.12	-	1.11	0.02	1.38	m	1.48
Plaster general surfaces 3.50 - 5.00m above floor										
girth exceeding 300mm	0.76	10.00	0.84	0.30	-	2.77	0.05	3.65	m2	3.93
One coat primer, two undercoats, one coat full gloss finish										
Concrete general surfaces										
girth exceeding 300mm	1.11	10.00	1.22	0.40	-	3.69	0.07	4.98	m2	5.36
Concrete general surfaces 3.50 - 5.00m above floor										
girth exceeding 300mm	1.11	10.00	1.22	0.44	-	4.06	0.08	5.36	m2	5.76
Brickwork general surfaces										
girth exceeding 300mm	1.35	10.00	1.49	0.45	-	4.15	0.08	5.72	m2	6.15
Plasterboard general surfaces										
girth exceeding 300mm	0.86	10.00	0.95	0.36	-	3.32	0.06	4.33	m2	4.65
Plasterboard general surfaces 3.50 - 5.00m above floor										
girth exceeding 300mm	0.86	10.00	0.95	0.40	-	3.69	0.07	4.71	m2	5.06
Plaster general surfaces										
girth exceeding 300mm	0.98	10.00	1.08	0.36	-	3.32	0.06	4.46	m2	4.80
isolated surfaces, girth not exceeding 300mm	0.30	10.00	0.33	0.16	-	1.48	0.03	1.84	m	1.97

Labour hourly rates: (except Specialists) Craft Operatives £9.23 Labourer £7.02. Rates are national average prices. Refer to REGIONAL VARIATIONS for indicative levels of overall pricing in regions	MATERIALS			LABOUR				RATES		
	Del to Site £	Waste %	Material Cost £	Craft Optve Hrs	Lab Hrs	Labour Cost £	Sunds £	Nett Rate £	Unit	Gross Rate (+7.5%) £

M60: PAINTING/CLEAR FINISHING - OIL PAINTING WALLS AND CEILINGS Cont.

One coat primer, two undercoats, one coat full gloss finish Cont.

	Del to Site £	Waste %	Material Cost £	Craft Optve Hrs	Lab Hrs	Labour Cost £	Sunds £	Nett Rate £	Unit	Gross Rate £
Plaster general surfaces 3.50 - 5.00m above floor girth exceeding 300mm	0.98	10.00	1.08	0.40	-	3.69	0.07	4.84	m2	5.20

One coat primer, one undercoat, one coat eggshell finish

Concrete general surfaces girth exceeding 300mm	0.90	10.00	0.99	0.30	-	2.77	0.05	3.81	m2	4.09
Concrete general surfaces 3.50 - 5.00m above floor girth exceeding 300mm	0.90	10.00	0.99	0.33	-	3.05	0.06	4.10	m2	4.40
Brickwork general surfaces girth exceeding 300mm	1.10	10.00	1.21	0.34	-	3.14	0.06	4.41	m2	4.74
Plasterboard general surfaces girth exceeding 300mm	0.70	10.00	0.77	0.27	-	2.49	0.05	3.31	m2	3.56
Plasterboard general surfaces 3.50 - 5.00m above floor girth exceeding 300mm	0.70	10.00	0.77	0.30	-	2.77	0.05	3.59	m2	3.86
Plaster general surfaces girth exceeding 300mm	0.80	10.00	0.88	0.27	-	2.49	0.04	3.41	m2	3.67
isolated surfaces, girth not exceeding 300mm	0.24	10.00	0.26	0.12	-	1.11	0.02	1.39	m	1.50
Plaster general surfaces 3.50 - 5.00m above floor girth exceeding 300mm	0.80	10.00	0.88	0.30	-	2.77	0.05	3.70	m2	3.98

One coat primer, two undercoats, one coat eggshell finish

Concrete general surfaces girth exceeding 300mm	1.16	10.00	1.28	0.40	-	3.69	0.07	5.04	m2	5.42
Concrete general surfaces 3.50 - 5.00m above floor girth exceeding 300mm	1.16	10.00	1.28	0.44	-	4.06	0.08	5.42	m2	5.82
Brickwork general surfaces girth exceeding 300mm	1.41	10.00	1.55	0.45	-	4.15	0.08	5.78	m2	6.22
Plasterboard general surfaces girth exceeding 300mm	0.90	10.00	0.99	0.36	-	3.32	0.06	4.37	m2	4.70
Plasterboard general surfaces 3.50 - 5.00m above floor girth exceeding 300mm	0.90	10.00	0.99	0.40	-	3.69	0.07	4.75	m2	5.11
Plaster general surfaces girth exceeding 300mm	1.03	10.00	1.13	0.36	-	3.32	0.06	4.52	m2	4.85
isolated surfaces, girth not exceeding 300mm	0.31	10.00	0.34	0.16	-	1.48	0.03	1.85	m	1.99
Plaster general surfaces 3.50 - 5.00m above floor girth exceeding 300mm	1.03	10.00	1.13	0.40	-	3.69	0.07	4.89	m2	5.26

M60: PAINTING/CLEAR FINISHING - SPRAY PAINTING

Spray one coat primer, one undercoat, one coat full gloss finish

Concrete general surfaces girth exceeding 300mm	0.95	15.00	1.09	0.26	-	2.40	0.05	3.54	m2	3.81
Brickwork general surfaces girth exceeding 300mm	1.14	15.00	1.31	0.30	-	2.77	0.05	4.13	m2	4.44
Plaster general surfaces girth exceeding 300mm	0.85	15.00	0.98	0.24	-	2.22	0.04	3.23	m2	3.48

Spray one coat primer, two undercoats, one coat full gloss finish

Concrete general surfaces girth exceeding 300mm	1.23	15.00	1.41	0.31	-	2.86	0.05	4.33	m2	4.65
Brickwork general surfaces girth exceeding 300mm	1.48	15.00	1.70	0.35	-	3.23	0.06	4.99	m2	5.37
Plaster general surfaces girth exceeding 300mm	1.11	15.00	1.28	0.28	-	2.58	0.04	3.90	m2	4.19

Spray one coat primer, one basecoat, one coat multicolour finish

Concrete general surfaces girth exceeding 300mm	1.17	15.00	1.35	0.39	-	3.60	0.07	5.02	m2	5.39
Brickwork general surfaces girth exceeding 300mm	1.40	15.00	1.61	0.41	-	3.78	0.07	5.46	m2	5.87

Labour hourly rates: (except Specialists) Craft Operatives £9.23 Labourer £7.02 Rates are national average prices. Refer to REGIONAL VARIATIONS for indicative levels of overall pricing in regions	MATERIALS			LABOUR				RATES		
	Del to Site £	Waste %	Material Cost £	Craft Optve Hrs	Lab Hrs	Labour Cost £	Sunds £	Nett Rate £	Unit	Gross Rate (+7.5%) £
M60: PAINTING/CLEAR FINISHING - SPRAY PAINTING Cont.										
Spray one coat primer, one basecoat, one coat multicolour finish Cont.										
Plaster general surfaces girth exceeding 300mm	1.05	15.00	1.21	0.37	-	3.42	0.06	4.68	m2	5.03
Spray one coat primer, one basecoat, one coat multicolour finish, one coat glaze										
Concrete general surfaces girth exceeding 300mm	1.44	15.00	1.66	0.43	-	3.97	0.08	5.70	m2	6.13
Brickwork general surfaces girth exceeding 300mm	1.73	15.00	1.99	0.45	-	4.15	0.08	6.22	m2	6.69
Plaster general surfaces girth exceeding 300mm	1.30	15.00	1.50	0.41	-	3.78	0.07	5.35	m2	5.75
M60: PAINTING/CLEAR FINISHING - CHLORINATED RUBBER PAINTING										
One coat primer, two coats chlorinated rubber paint										
Concrete general surfaces girth exceeding 300mm	3.99	10.00	4.39	0.30	-	2.77	0.05	7.21	m2	7.75
Brickwork general surfaces girth exceeding 300mm	4.82	10.00	5.30	0.32	-	2.95	0.06	8.32	m2	8.94
Plasterboard general surfaces girth exceeding 300mm	3.26	10.00	3.59	0.29	-	2.68	0.05	6.31	m2	6.79
Plaster general surfaces girth exceeding 300mm .... isolated surfaces, girth not exceeding 300mm	3.99 1.20	10.00 10.00	4.39 1.32	0.29 0.13	- -	2.68 1.20	0.05 0.02	7.12 2.54	m2 m	7.65 2.73
M60: PAINTING/CLEAR FINISHING - PLASTIC FINISH										
Textured plastic coating - Stippled finish										
Concrete general surfaces girth exceeding 300mm	0.48	10.00	0.53	0.25	-	2.31	0.04	2.88	m2	3.09
Concrete general surfaces 3.50 - 5.00m above floor girth exceeding 300mm	0.48	10.00	0.53	0.28	-	2.58	0.05	3.16	m2	3.40
Brickwork general surfaces girth exceeding 300mm	0.59	10.00	0.65	0.30	-	2.77	0.05	3.47	m2	3.73
Plasterboard general surfaces girth exceeding 300mm	0.40	10.00	0.44	0.20	-	1.85	0.03	2.32	m2	2.49
Plasterboard general surfaces 3.50 - 5.00m above floor girth exceeding 300mm	0.40	10.00	0.44	0.22	-	2.03	0.04	2.51	m2	2.70
Plaster general surfaces girth exceeding 300mm	0.40	10.00	0.44	0.20	-	1.85	0.03	2.32	m2	2.49
Plaster general surfaces 3.50 - 5.00m above floor girth exceeding 300mm	0.40	10.00	0.44	0.22	-	2.03	0.04	2.51	m2	2.70
Textured plastic coating - Combed Finish										
Concrete general surfaces girth exceeding 300mm	0.52	10.00	0.57	0.30	-	2.77	0.05	3.39	m2	3.65
Concrete general surfaces 3.50 - 5.00m above floor girth exceeding 300mm	0.52	10.00	0.57	0.33	-	3.05	0.06	3.68	m2	3.95
Brickwork general surfaces girth exceeding 300mm	0.62	10.00	0.68	0.35	-	3.23	0.06	3.97	m2	4.27
Plasterboard general surfaces girth exceeding 300mm	0.42	10.00	0.46	0.25	-	2.31	0.04	2.81	m2	3.02
Plasterboard general surfaces 3.50 - 5.00m above floor girth exceeding 300mm	0.42	10.00	0.46	0.27	-	2.49	0.05	3.00	m2	3.23
Plaster general surfaces girth exceeding 300mm	0.42	10.00	0.46	0.25	-	2.31	0.04	2.81	m2	3.02
Plaster general surfaces 3.50 - 5.00m above floor girth exceeding 300mm	0.42	10.00	0.46	0.27	-	2.49	0.05	3.00	m2	3.23
M60: PAINTING/CLEAR FINISHING - OIL PAINTING METALWORK										
One undercoat, one coat full gloss finish on ready primed metal surfaces										
Iron or steel structural work girth exceeding 300mm	0.45	10.00	0.50	0.28	-	2.58	0.05	3.13	m2	3.36

Labour hourly rates: (except Specialists) Craft Operatives £9.23 Labourer £7.02 Rates are national average prices. Refer to REGIONAL VARIATIONS for indicative levels of overall pricing in regions	MATERIALS			LABOUR				RATES		
	Del to Site £	Waste %	Material Cost £	Craft Optve Hrs	Lab Hrs	Labour Cost £	Sunds £	Nett Rate £	Unit	Gross Rate (+7.5%) £
M60: PAINTING/CLEAR FINISHING - OIL PAINTING METALWORK Cont.										
One undercoat, one coat full gloss finish on ready primed metal surfaces Cont.										
Iron or steel structural work Cont.										
isolated surfaces, girth not exceeding 300mm	0.14	10.00	0.15	0.12	-	1.11	0.02	1.28	m	1.38
Iron or steel structural members of roof trusses, lattice girders, purlins and the like										
girth exceeding 300mm	0.45	10.00	0.50	0.36	-	3.32	0.06	3.88	m2	4.17
isolated surfaces, girth not exceeding 300mm	0.14	10.00	0.15	0.12	-	1.11	0.02	1.28	m	1.38
Two undercoats, one coat full gloss finish on ready primed metal surfaces										
Iron or steel structural work										
girth exceeding 300mm	0.68	10.00	0.75	0.42	-	3.88	0.07	4.69	m2	5.05
isolated surfaces, girth not exceeding 300mm	0.20	10.00	0.22	0.14	-	1.29	0.02	1.53	m	1.65
Iron or steel structural members of roof trusses, lattice girders, purlins and the like										
girth exceeding 300mm	0.68	10.00	0.75	0.54	-	4.98	0.09	5.82	m2	6.26
isolated surfaces, girth not exceeding 300mm	0.20	10.00	0.22	0.18	-	1.66	0.03	1.91	m	2.05
One coat primer, one undercoat, one coat full gloss finish on metal surfaces										
Iron or steel general surfaces										
girth exceeding 300mm	0.73	10.00	0.80	0.39	-	3.60	0.07	4.47	m2	4.81
isolated surfaces, girth not exceeding 300mm	0.22	10.00	0.24	0.13	-	1.20	0.02	1.46	m	1.57
Galvanised glazed doors, windows or screens										
girth exceeding 300mm; panes, area not exceeding 0.10m2	0.20	10.00	0.22	0.72	-	6.65	0.13	7.00	m2	7.52
girth exceeding 300mm; panes, area 0.10 - 0.50m2 .	0.17	10.00	0.19	0.51	-	4.71	0.09	4.98	m2	5.36
girth exceeding 300mm; panes, area 0.50 - 1.00m2 .	0.15	10.00	0.17	0.43	-	3.97	0.08	4.21	m2	4.53
girth exceeding 300mm; panes, area exceeding 1.00m2	0.13	10.00	0.14	0.39	-	3.60	0.07	3.81	m2	4.10
Iron or steel structural work										
girth exceeding 300mm	0.73	10.00	0.80	0.42	-	3.88	0.07	4.75	m2	5.11
isolated surfaces, girth not exceeding 300mm	0.22	10.00	0.24	0.18	-	1.66	0.03	1.93	m	2.08
Iron or steel structural members of roof trusses, lattice girders, purlins and the like										
girth exceeding 300mm	0.73	10.00	0.80	0.54	-	4.98	0.09	5.88	m2	6.32
isolated surfaces, girth not exceeding 300mm	0.22	10.00	0.24	0.18	-	1.66	0.03	1.93	m	2.08
Iron or steel services										
girth exceeding 300mm	0.73	10.00	0.80	0.54	-	4.98	0.09	5.88	m2	6.32
isolated surfaces, girth not exceeding 300mm	0.22	10.00	0.24	0.18	-	1.66	0.03	1.93	m	2.08
isolated areas not exceeding 0.50m2 irrespective of girth	0.37	10.00	0.41	0.54	-	4.98		5.48	nr	5.89
Copper services										
girth exceeding 300mm	0.73	10.00	0.80	0.54	-	4.98	0.09	5.88	m2	6.32
isolated surfaces, girth not exceeding 300mm	0.22	10.00	0.24	0.18	-	1.66	0.03	1.93	m	2.08
Galvanised services										
girth exceeding 300mm	0.73	10.00	0.80	0.54	-	4.98	0.09	5.88	m2	6.32
isolated surfaces, girth not exceeding 300mm	0.22	10.00	0.24	0.18	-	1.66	0.03	1.93	m	2.08
One coat primer, one undercoat, one coat full gloss finish on metal surfaces ; external work										
Iron or steel general surfaces										
girth exceeding 300mm	0.73	10.00	0.80	0.42	-	3.88	0.07	4.75	m2	5.11
isolated surfaces, girth not exceeding 300mm	0.22	10.00	0.24	0.16	-	1.48	0.03	1.75	m	1.88
Galvanised glazed doors, windows or screens										
girth exceeding 300mm; panes, area not exceeding 0.10m2	0.20	10.00	0.22	0.75	-	6.92	0.13	7.27	m2	7.82
girth exceeding 300mm; panes, area 0.10 - 0.50m2 .	0.17	10.00	0.19	0.54	-	4.98	0.09	5.26	m2	5.66
girth exceeding 300mm; panes, area 0.50 - 1.00m2 .	0.15	10.00	0.17	0.46	-	4.25	0.08	4.49	m2	4.83
girth exceeding 300mm; panes, area exceeding 1.00m2	0.13	10.00	0.14	0.42	-	3.88	0.07	4.09	m2	4.40
Iron or steel structural work										
girth exceeding 300mm	0.73	10.00	0.80	0.42	-	3.88	0.07	4.75	m2	5.11
isolated surfaces, girth not exceeding 300mm	0.22	10.00	0.24	0.18	-	1.66	0.03	1.93	m	2.08
Iron or steel structural members of roof trusses, lattice girders, purlins and the like										
girth exceeding 300mm	0.73	10.00	0.80	0.54	-	4.98	0.09	5.88	m2	6.32
isolated surfaces, girth not exceeding 300mm	0.22	10.00	0.24	0.18	-	1.66	0.03	1.93	m	2.08
Iron or steel railings, fences and gates; plain open type										
girth exceeding 300mm	0.73	10.00	0.80	0.39	-	3.60	0.07	4.47	m2	4.81
isolated surfaces, girth not exceeding 300mm	0.22	10.00	0.24	0.13	-	1.20	0.02	1.46	m	1.57
Iron or steel railings, fences and gates; close type										
girth exceeding 300mm	0.73	10.00	0.80	0.33	-	3.05	0.06	3.91	m2	4.20

Labour hourly rates: (except Specialists) Craft Operatives £9.23 Labourer £7.02. Rates are national average prices. Refer to REGIONAL VARIATIONS for indicative levels of overall pricing in regions	MATERIALS			LABOUR				RATES		
	Del to Site £	Waste %	Material Cost £	Craft Optve Hrs	Lab Hrs	Labour Cost £	Sunds £	Nett Rate £	Unit	Gross Rate (+7.5%) £

M60: PAINTING/CLEAR FINISHING - OIL PAINTING METALWORK Cont.

One coat primer, one undercoat, one coat full gloss finish on metal surfaces ; external work Cont.

	Del to Site £	Waste %	Material Cost £	Craft Optve Hrs	Lab Hrs	Labour Cost £	Sunds £	Nett Rate £	Unit	Gross Rate £
Iron or steel railings, fences and gates; ornamental type										
girth exceeding 300mm	0.73	10.00	0.80	0.66	-	6.09	0.12	7.01	m2	7.54
Iron or steel eaves gutters										
girth exceeding 300mm	0.73	10.00	0.80	0.45	-	4.15	0.08	5.04	m2	5.41
isolated surfaces, girth not exceeding 300mm	0.22	10.00	0.24	0.15	-	1.38	0.03	1.66	m	1.78
Galvanised eaves gutters										
girth exceeding 300mm	0.73	10.00	0.80	0.45	-	4.15	0.08	5.04	m2	5.41
isolated surfaces, girth not exceeding 300mm	0.22	10.00	0.24	0.15	-	1.38	0.03	1.66	m	1.78
Iron or steel services										
girth exceeding 300mm	0.73	10.00	0.80	0.54	-	4.98	0.09	5.88	m2	6.32
isolated surfaces, girth not exceeding 300mm	0.22	10.00	0.24	0.18	-	1.66	0.03	1.93	m	2.08
isolated areas not exceeding 0.50m2 irrespective of girth	0.37	10.00	0.41	0.54	-	4.98	0.09	5.48	nr	5.89
Copper services										
girth exceeding 300mm	0.73	10.00	0.80	0.54	-	4.98	0.09	5.88	m2	6.32
isolated surfaces, girth not exceeding 300mm	0.22	10.00	0.24	0.18	-	1.66	0.03	1.93	m	2.08
Galvanised services										
girth exceeding 300mm	0.73	10.00	0.80	0.54	-	4.98	0.09	5.88	m2	6.32
isolated surfaces, girth not exceeding 300mm	0.22	10.00	0.24	0.18	-	1.66	0.03	1.93	m	2.08

One coat primer, two undercoats, one coat full gloss finish on metal surfaces

	Del to Site £	Waste %	Material Cost £	Craft Optve Hrs	Lab Hrs	Labour Cost £	Sunds £	Nett Rate £	Unit	Gross Rate £
Iron or steel general surfaces										
girth exceeding 300mm	0.96	10.00	1.06	0.52	-	4.80	0.09	5.95	m2	6.39
isolated surfaces, girth not exceeding 300mm	0.29	10.00	0.32	0.17	-	1.57	0.03	1.92	m	2.06
Galvanised glazed doors, windows or screens										
girth exceeding 300mm; panes, area not exceeding 0.10m2	0.26	10.00	0.29	0.96	-	8.86	0.17	9.32	m2	10.02
girth exceeding 300mm; panes, area 0.10 - 0.50m2	0.22	10.00	0.24	0.68	-	6.28	0.12	6.64	m2	7.14
girth exceeding 300mm; panes, area 0.50 - 1.00m2	0.19	10.00	0.21	0.57	-	5.26	0.10	5.57	m2	5.99
girth exceeding 300mm; panes, area exceeding 1.00m2	0.17	10.00	0.19	0.52	-	4.80	0.09	5.08	m2	5.46
Iron or steel structural work										
girth exceeding 300mm	0.96	10.00	1.06	0.56	-	5.17	0.10	6.32	m2	6.80
isolated surfaces, girth not exceeding 300mm	0.29	10.00	0.32	0.22	-	2.03	0.04	2.39	m	2.57
Iron or steel structural members of roof trusses, lattice girders, purlins and the like										
girth exceeding 300mm	0.96	10.00	1.06	0.72	-	6.65	0.13	7.83	m2	8.42
isolated surfaces, girth not exceeding 300mm	0.29	10.00	0.32	0.22	-	2.03	0.04	2.39	m	2.57
Iron or steel services										
girth exceeding 300mm	0.96	10.00	1.06	0.63	-	5.81	0.11	6.98	m2	7.50
isolated surfaces, girth not exceeding 300mm	0.29	10.00	0.32	0.21	-	1.94	0.04	2.30	m	2.47
isolated areas not exceeding 0.50m2 irrespective of girth	0.48	10.00	0.53	0.63	-	5.81	0.11	6.45	nr	6.94
Copper services										
girth exceeding 300mm	0.96	10.00	1.06	0.63	-	5.81	0.11	6.98	m2	7.50
isolated surfaces, girth not exceeding 300mm	0.29	10.00	0.32	0.21	-	1.94	0.04	2.30	m	2.47
Galvanised services										
girth exceeding 300mm	0.96	10.00	1.06	0.63	-	5.81	0.11	6.98	m2	7.50
isolated surfaces, girth not exceeding 300mm	0.29	10.00	0.32	0.21	-	1.94	0.04	2.30	m	2.47

One coat primer, two undercoats, one coat full gloss finish on metal surfaces; external work

	Del to Site £	Waste %	Material Cost £	Craft Optve Hrs	Lab Hrs	Labour Cost £	Sunds £	Nett Rate £	Unit	Gross Rate £
Iron or steel general surfaces										
girth exceeding 300mm	0.96	10.00	1.06	0.56	-	5.17	0.10	6.32	m2	6.80
isolated surfaces, girth not exceeding 300mm	0.29	10.00	0.32	0.19	-	1.75	0.03	2.10	m	2.26
Galvanised glazed doors, windows or screens										
girth exceeding 300mm; panes, area not exceeding 0.10m2	0.26	10.00	0.29	1.00	-	9.23	0.17	9.69	m2	10.41
girth exceeding 300mm; panes, area 0.10 - 0.50m2	0.22	10.00	0.24	0.72	-	6.65	0.13	7.02	m2	7.54
girth exceeding 300mm; panes, area 0.50 - 1.00m2	0.19	10.00	0.21	0.61	-	5.63	0.11	5.95	m2	6.40
girth exceeding 300mm; panes, area exceeding 1.00m2	0.17	10.00	0.19	0.56	-	5.17	0.10	5.46	m2	5.86
Iron or steel structural work										
girth exceeding 300mm	0.96	10.00	1.06	0.56	-	5.17	0.10	6.32	m2	6.80
isolated surfaces, girth not exceeding 300mm	0.29	10.00	0.32	0.22	-	2.03	0.04	2.39	m	2.57
Iron or steel structural members of roof trusses, lattice girders, purlins and the like										
girth exceeding 300mm	0.96	10.00	1.06	0.72	-	6.65	0.13	7.83	m2	8.42
isolated surfaces, girth not exceeding 300mm	0.29	10.00	0.32	0.22	-	2.03	0.04	2.39	m	2.57

SURFACE FINISHES

Labour hourly rates: (except Specialists) Craft Operatives £9.23 Labourer £7.02 Rates are national average prices. Refer to REGIONAL VARIATIONS for indicative levels of overall pricing in regions	MATERIALS			LABOUR				RATES		
	Del to Site £	Waste %	Material Cost £	Craft Optve Hrs	Lab Hrs	Labour Cost £	Sunds £	Nett Rate £	Unit	Gross Rate (+7.5%) £
M60: PAINTING/CLEAR FINISHING - OIL PAINTING METALWORK Cont.										
One coat primer, two undercoats, one coat full gloss finish on metal surfaces; external work Cont.										
Iron or steel railings, fences and gates; plain open type										
girth exceeding 300mm	0.96	10.00	1.06	0.52	–	4.80	0.09	5.95	m2	6.39
isolated surfaces, girth not exceeding 300mm	0.29	10.00	0.32	0.17	–	1.57	0.03	1.92	m	2.06
Iron or steel railings, fences and gates; close type										
girth exceeding 300mm	0.96	10.00	1.06	0.44	–	4.06	0.08	5.20	m2	5.59
Iron or steel railings, fences and gates; ornamental type										
girth exceeding 300mm	0.96	10.00	1.06	0.88	–	8.12	0.15	9.33	m2	10.03
Iron or steel eaves gutters										
girth exceeding 300mm	0.96	10.00	1.06	0.60	–	5.54	0.10	6.69	m2	7.20
isolated surfaces, girth not exceeding 300mm	0.29	10.00	0.32	0.20	–	1.85	0.03	2.19	m	2.36
Galvanised eaves gutters										
girth exceeding 300mm	0.96	10.00	1.06	0.60	–	5.54	0.10	6.69	m2	7.20
isolated surfaces, girth not exceeding 300mm	0.29	10.00	0.32	0.20	–	1.85	0.03	2.19	m	2.36
Iron or steel services										
girth exceeding 300mm	0.96	10.00	1.06	0.63	–	5.81	0.11	6.98	m2	7.50
isolated surfaces, girth not exceeding 300mm	0.29	10.00	0.32	0.21	–	1.94	0.04	2.30	m	2.47
isolated areas not exceeding 0.50m2 irrespective of girth	0.48	10.00	0.53	0.63	–	5.81	0.11	6.45	nr	6.94
Copper services										
girth exceeding 300mm	0.96	10.00	1.06	0.63	–	5.81	0.11	6.98	m2	7.50
isolated surfaces, girth not exceeding 300mm	0.29	10.00	0.32	0.21	–	1.94	0.04	2.30	m	2.47
Galvanised services										
girth exceeding 300mm	0.96	10.00	1.06	0.63	–	5.81	0.11	6.98	m2	7.50
isolated surfaces, girth not exceeding 300mm	0.29	10.00	0.32	0.21	–	1.94	0.03	2.29	m	2.46
One coat primer, one undercoat, two coats full gloss finish on metal surfaces										
Iron or steel general surfaces										
girth exceeding 300mm	0.96	10.00	1.06	0.52	–	4.80	0.09	5.95	m2	6.39
isolated surfaces, girth not exceeding 300mm	0.29	10.00	0.32	0.17	–	1.57	0.03	1.92	m	2.06
Galvanised glazed doors, windows or screens										
girth exceeding 300mm; panes, area not exceeding 0.10m2	0.26	10.00	0.29	0.96	–	8.86	0.17	9.32	m2	10.02
girth exceeding 300mm; panes, area 0.10 - 0.50m2 .	0.22	10.00	0.24	0.68	–	6.28	0.12	6.64	m2	7.14
girth exceeding 300mm; panes, area 0.50 - 1.00m2 .	0.19	10.00	0.21	0.57	–	5.26	0.10	5.57	m2	5.99
girth exceeding 300mm; panes, area exceeding 1.00m2	0.17	10.00	0.19	0.52	–	4.80	0.09	5.08	m2	5.46
Iron or steel structural work										
girth exceeding 300mm	0.96	10.00	1.06	0.56	–	5.17	0.10	6.32	m2	6.80
isolated surfaces, girth not exceeding 300mm	0.29	10.00	0.32	0.22	–	2.03	0.04	2.39	m	2.57
Iron or steel structural members of roof trusses, lattice girders, purlins and the like										
girth exceeding 300mm	0.96	10.00	1.06	0.72	–	6.65	0.13	7.83	m2	8.42
isolated surfaces, girth not exceeding 300mm	0.29	10.00	0.32	0.22	–	2.03	0.04	2.39	m	2.57
Iron or steel services										
girth exceeding 300mm	0.96	10.00	1.06	0.63	–	5.81	0.11	6.98	m2	7.50
isolated surfaces, girth not exceeding 300mm	0.29	10.00	0.32	0.21	–	1.94	0.04	2.30	m	2.47
isolated areas not exceeding 0.50m2 irrespective of girth	0.48	10.00	0.53	0.63	–	5.81	0.11	6.45	nr	6.94
Copper services										
girth exceeding 300mm	0.96	10.00	1.06	0.63	–	5.81	0.11	6.98	m2	7.50
isolated surfaces, girth not exceeding 300mm	0.29	10.00	0.32	0.21	–	1.94	0.04	2.30	m	2.47
Galvanised services										
girth exceeding 300mm	0.96	10.00	1.06	0.63	–	5.81	0.11	6.98	m2	7.50
isolated surfaces, girth not exceeding 300mm	0.29	10.00	0.32	0.21	–	1.94	0.04	2.30	m	2.47
One coat primer, one undercoat, two coats full gloss finish on metal surfaces; external work										
Iron or steel general surfaces										
girth exceeding 300mm	0.96	10.00	1.06	0.56	–	5.17	0.10	6.32	m2	6.80
isolated surfaces, girth not exceeding 300mm	0.29	10.00	0.32	0.19	–	1.75	0.03	2.10	m	2.26
Galvanised glazed doors, windows or screens										
girth exceeding 300mm; panes, area not exceeding 0.10m2	0.26	10.00	0.29	1.00	–	9.23	0.17	9.69	m2	10.41
girth exceeding 300mm; panes, area 0.10 - 0.50m2 .	0.22	10.00	0.24	0.72	–	6.65	0.13	7.02	m2	7.54
girth exceeding 300mm; panes, area 0.50 - 1.00m2 .	0.19	10.00	0.21	0.61	–	5.63	0.11	5.95	m2	6.40
girth exceeding 300mm; panes, area exceeding 1.00m2	0.17	10.00	0.19	0.56	–	5.17	0.10	5.46	m2	5.86
Iron or steel structural work										
girth exceeding 300mm	0.96	10.00	1.06	0.56	–	5.17	0.10	6.32	m2	6.80
isolated surfaces, girth not exceeding 300mm	0.29	10.00	0.32	0.22	–	2.03	0.04	2.39	m2	2.57

Labour hourly rates: (except Specialists) Craft Operatives £9.23 Labourer £7.02 Rates are national average prices. Refer to REGIONAL VARIATIONS for indicative levels of overall pricing in regions	MATERIALS			LABOUR				RATES		
	Del to Site £	Waste %	Material Cost £	Craft Optve Hrs	Lab Hrs	Labour Cost £	Sunds £	Nett Rate £	Unit	Gross Rate (+7.5%) £
M60: PAINTING/CLEAR FINISHING - OIL PAINTING METALWORK Cont.										
One coat primer, one undercoat, two coats full gloss finish on metal surfaces; external work Cont.										
Iron or steel structural members of roof trusses, lattice girders, purlins and the like										
girth exceeding 300mm	0.96	10.00	1.06	0.72	-	6.65	0.13	7.83	m2	8.42
isolated surfaces, girth not exceeding 300mm	0.29	10.00	0.32	0.22	-	2.03	0.04	2.39	m2	2.57
Iron or steel railings, fences and gates; plain open type										
girth exceeding 300mm	0.96	10.00	1.06	0.52	-	4.80	0.09	5.95	m2	6.39
isolated surfaces, girth not exceeding 300mm	0.29	10.00	0.32	0.17	-	1.57	0.03	1.92	m2	2.06
Iron or steel railings, fences and gates; close type										
girth exceeding 300mm	0.96	10.00	1.06	0.44	-	4.06	0.08	5.20	m2	5.59
Iron or steel railings, fences and gates; ornamental type										
girth exceeding 300mm	0.96	10.00	1.06	0.88	-	8.12	0.15	9.33	m2	10.03
Iron or steel eaves gutters										
girth exceeding 300mm	0.96	10.00	1.06	0.60	-	5.54	0.10	6.69	m2	7.20
isolated surfaces, girth not exceeding 300mm	0.29	10.00	0.32	0.20	-	1.85	0.03	2.19	m	2.36
Galvanised eaves gutters										
girth exceeding 300mm	0.96	10.00	1.06	0.60	-	5.54	0.10	6.69	m2	7.20
isolated surfaces, girth not exceeding 300mm	0.29	10.00	0.32	0.20	-	1.85	0.03	2.19	m	2.36
Iron or steel services										
girth exceeding 300mm	0.96	10.00	1.06	0.63	-	5.81	0.11	6.98	m2	7.50
isolated surfaces, girth not exceeding 300mm	0:29	10.00	0.32	0.21	-	1.94	0.04	2.30	m	2.47
isolated areas not exceeding 0.50m2 irrespective of girth	0.48	10.00	0.53	0.63	-	5.81	0.11	6.45	nr	6.94
Copper services										
girth exceeding 300mm	0.96	10.00	1.06	0.63	-	5.81	0.11	6.98	m2	7.50
isolated surfaces, girth not exceeding 300mm	0.29	10.00	0.32	0.21	-	1.94	0.04	2.30	m	2.47
Galvanised services										
girth exceeding 300mm	0.96	10.00	1.06	0.63	-	5.81	0.11	6.98	m2	7.50
isolated surfaces, girth not exceeding 300mm	0.29	10.00	0.32	0.21	-	1.94	0.04	2.30	m	2.47
M60: PAINTING/CLEAR FINISHING - METALLIC PAINTING										
One coat aluminium metallic paint										
Iron or steel radiators; panel type										
girth exceeding 300mm	0.40	10.00	0.44	0.14	-	1.29	0.02	1.75	m2	1.88
Iron or steel radiators; column type										
girth exceeding 300mm	0.40	10.00	0.44	0.20	-	1.85	0.03	2.32	m2	2.49
Iron or steel services										
isolated surfaces, girth not exceeding 300mm	0.12	10.00	0.13	0.05	-	0.46	0.01	0.60	m	0.65
Copper services										
isolated surfaces, girth not exceeding 300mm	0.12	10.00	0.13	0.05	-	0.46	0.01	0.60	m	0.65
Galvanised services										
isolated surfaces, girth not exceeding 300mm	0.12	10.00	0.13	0.05	-	0.46	0.01	0.60	m	0.65
Two coats aluminium metallic paint										
Iron or steel radiators; panel type										
girth exceeding 300mm	0.75	10.00	0.82	0.28	-	2.58	0.05	3.46	m2	3.72
Iron or steel radiators; column type										
girth exceeding 300mm	0.75	10.00	0.82	0.40	-	3.69	0.07	4.59	m2	4.93
Iron or steel services										
isolated surfaces, girth not exceeding 300mm	0.22	10.00	0.24	0.10	-	0.92	0.02	1.19	m	1.27
Copper services										
isolated surfaces, girth not exceeding 300mm	0.22	10.00	0.24	0.10	-	0.92	0.02	1.19	m	1.27
Galvanised services										
isolated surfaces, girth not exceeding 300mm	0.22	10.00	0.24	0.10	-	0.92	0.02	1.19	m	1.27
One coat gold or bronze metallic paint										
Iron or steel radiators; panel type										
girth exceeding 300mm	0.56	10.00	0.62	0.14	-	1.29	0.02	1.93	m2	2.07
Iron or steel radiators; column type										
girth exceeding 300mm	0.56	10.00	0.62	0.20	-	1.85	0.03	2.49	m2	2.68
Iron or steel services										
isolated surfaces, girth not exceeding 300mm	0.17	10.00	0.19	0.05	-	0.46	0.01	0.66	m	0.71
Copper services										
isolated surfaces, girth not exceeding 300mm	0.17	10.00	0.19	0.05	-	0.46	0.01	0.66	m	0.71

SURFACE FINISHES

SURFACE FINISHES – MAJOR WORKS

Labour hourly rates: (except Specialists) Craft Operatives £9.23 Labourer £7.02 Rates are national average prices. Refer to REGIONAL VARIATIONS for indicative levels of overall pricing in regions	MATERIALS			LABOUR				RATES		
	Del to Site	Waste	Material Cost	Craft Optve	Lab	Labour Cost	Sunds	Nett Rate	Unit	Gross Rate (+7.5%)
	£	%	£	Hrs	Hrs	£	£	£		£
M60: PAINTING/CLEAR FINISHING - METALLIC PAINTING Cont.										
One coat gold or bronze metallic paint Cont.										
Galvanised services										
isolated surfaces, girth not exceeding 300mm	0.17	10.00	0.19	0.05	-	0.46	0.01	0.66	m	0.71
Two coats gold or bronze metallic paint										
Iron or steel radiators; panel type										
girth exceeding 300mm	1.05	10.00	1.16	0.28	-	2.58	0.05	3.79	m2	4.07
Iron or steel radiators; column type										
girth exceeding 300mm	1.05	10.00	1.16	0.40	-	3.69	0.07	4.92	m2	5.29
Iron or steel services										
isolated surfaces, girth not exceeding 300mm	0.31	10.00	0.34	0.10	-	0.92	0.02	1.28	m	1.38
Copper services										
isolated surfaces, girth not exceeding 300mm	0.31	10.00	0.34	0.10	-	0.92	0.02	1.28	m	1.38
Galvanised services										
isolated surfaces, girth not exceeding 300mm	0.31	10.00	0.34	0.10	-	0.92	0.02	1.28	m	1.38
M60: PAINTING/CLEAR FINISHING - BITUMINOUS PAINT										
One coat black bitumen paint; external work										
Iron or steel general surfaces										
girth exceeding 300mm	0.21	10.00	0.23	0.13	-	1.20	0.02	1.45	m2	1.56
Iron or steel eaves gutters										
girth exceeding 300mm	0.21	10.00	0.23	0.14	-	1.29	0.02	1.54	m2	1.66
Iron or steel services										
girth exceeding 300mm	0.21	10.00	0.23	0.15	-	1.38	0.03	1.65	m2	1.77
isolated surfaces, girth not exceeding 300mm	0.06	10.00	0.07	0.05	-	0.46	0.01	0.54	m	0.58
Two coats black bitumen paint; external work										
Iron or steel general surfaces										
girth exceeding 300mm	0.42	10.00	0.46	0.26	-	2.40	0.05	2.91	m2	3.13
Iron or steel eaves gutters										
girth exceeding 300mm	0.42	10.00	0.46	0.28	-	2.58	0.05	3.10	m2	3.33
isolated surfaces, girth not exceeding 300mm	0.13	10.00	0.14	0.09	-	0.83	0.02	0.99	m	1.07
Iron or steel services										
girth exceeding 300mm	0.42	10.00	0.46	0.30	-	2.77	0.05	3.28	m2	3.53
isolated surfaces, girth not exceeding 300mm	0.13	10.00	0.14	0.09	-	0.83	0.02	0.99	m	1.07
M60: PAINTING/CLEAR FINISHING - OIL PAINTING WOODWORK										
One coat primer; carried out on site before fixing members										
Wood general surfaces										
girth exceeding 300mm	0.25	10.00	0.28	0.16	-	1.48	0.02	1.77	m2	1.90
isolated surfaces, girth not exceeding 300mm	0.08	10.00	0.09	0.06	-	0.55	0.01	0.65	m	0.70
One undercoat, one coat full gloss finish on ready primed wood surfaces										
Wood general surfaces										
girth exceeding 300mm	0.45	10.00	0.50	0.32	-	2.95	0.06	3.51	m2	3.77
isolated surfaces, girth not exceeding 300mm	0.14	10.00	0.15	0.11	-	1.02	0.02	1.19	m	1.28
Wood glazed doors										
girth exceeding 300mm; panes, area not exceeding 0.10m2	0.23	10.00	0.25	0.54	-	4.98	0.09	5.33	m2	5.73
girth exceeding 300mm; panes, area 0.10 - 0.50m2 .	0.16	10.00	0.18	0.38	-	3.51	0.07	3.75	m2	4.03
girth exceeding 300mm; panes, area 0.50 - 1.00m2 .	0.11	10.00	0.12	0.32	-	2.95	0.06	3.13	m2	3.37
girth exceeding 300mm; panes, area exceeding 1.00m2	0.09	10.00	0.10	0.29	-	2.68	0.05	2.83	m2	3.04
Wood partially glazed doors										
girth exceeding 300mm; panes, area not exceeding 0.10m2	0.34	10.00	0.37	0.43	-	3.97	0.08	4.42	m2	4.75
girth exceeding 300mm; panes, area 0.10 - 0.50m2 .	0.30	10.00	0.33	0.35	-	3.23	0.06	3.62	m2	3.89
girth exceeding 300mm; panes, area 0.50 - 1.00m2 .	0.27	10.00	0.30	0.32	-	2.95	0.06	3.31	m2	3.56
girth exceeding 300mm; panes, area exceeding 1.00m2	0.25	10.00	0.28	0.31	-	2.86	0.05	3.19	m2	3.43
Wood windows and screens										
girth exceeding 300mm; panes, area not exceeding 0.10m2	0.25	10.00	0.28	0.59	-	5.45	0.10	5.82	m2	6.26
girth exceeding 300mm; panes, area 0.10 - 0.50m2 .	0.18	10.00	0.20	0.42	-	3.88	0.07	4.14	m2	4.46
girth exceeding 300mm; panes, area 0.50 - 1.00m2 .	0.14	10.00	0.15	0.35	-	3.23	0.06	3.44	m2	3.70
girth exceeding 300mm; panes, area exceeding 1.00m2	0.11	10.00	0.12	0.32	-	2.95	0.06	3.13	m2	3.37

Labour hourly rates: (except Specialists) Craft Operatives £9.23 Labourer £7.02 Rates are national average prices. Refer to REGIONAL VARIATIONS for indicative levels of overall pricing in regions	MATERIALS			LABOUR				RATES		
	Del to Site £	Waste %	Material Cost £	Craft Optve Hrs	Lab Hrs	Labour Cost £	Sunds £	Nett Rate £	Unit	Gross Rate (+7.5%) £
M60: PAINTING/CLEAR FINISHING - OIL PAINTING WOODWORK Cont.										
One undercoat, one coat full gloss finish on ready primed wood surfaces; external work										
Wood general surfaces										
girth exceeding 300mm	0.45	10.00	**0.50**	0.34	-	**3.14**	0.06	**3.69**	m2	**3.97**
isolated surfaces, girth not exceeding 300mm	0.14	10.00	**0.15**	0.12	-	**1.11**	0.02	**1.28**	m	**1.38**
Wood glazed doors										
girth exceeding 300mm; panes, area not exceeding 0.10m2	0.23	10.00	**0.25**	0.54	-	**4.98**	0.09	**5.33**	m2	**5.73**
girth exceeding 300mm; panes, area 0.10 - 0.50m2 .	0.16	10.00	**0.18**	0.38	-	**3.51**	0.07	**3.75**	m2	**4.03**
girth exceeding 300mm; panes, area 0.50 - 1.00m2 .	0.11	10.00	**0.12**	0.32	-	**2.95**	0.06	**3.13**	m2	**3.37**
girth exceeding 300mm; panes, area exceeding 1.00m2	0.09	10.00	**0.10**	0.29	-	**2.68**	0.05	**2.83**	m2	**3.04**
Wood partially glazed doors										
girth exceeding 300mm; panes, area not exceeding 0.10m2	0.34	10.00	**0.37**	0.43	-	**3.97**	0.08	**4.42**	m2	**4.75**
girth exceeding 300mm; panes, area 0.10 - 0.50m2 .	0.30	10.00	**0.33**	0.35	-	**3.23**	0.06	**3.62**	m2	**3.89**
girth exceeding 300mm; panes, area 0.50 - 1.00m2 .	0.27	10.00	**0.30**	0.32	-	**2.95**	0.06	**3.31**	m2	**3.56**
girth exceeding 300mm; panes, area exceeding 1.00m2	0.25	10.00	**0.28**	0.31	-	**2.86**	0.05	**3.19**	m2	**3.43**
Wood windows and screens										
girth exceeding 300mm; panes, area not exceeding 0.10m2	0.25	10.00	**0.28**	0.61	-	**5.63**	0.11	**6.02**	m2	**6.47**
girth exceeding 300mm; panes, area 0.10 - 0.50m2 .	0.18	10.00	**0.20**	0.44	-	**4.06**	0.08	**4.34**	m2	**4.66**
girth exceeding 300mm; panes, area 0.50 - 1.00m2 .	0.14	10.00	**0.15**	0.37	-	**3.42**	0.06	**3.63**	m2	**3.90**
girth exceeding 300mm; panes, area exceeding 1.00m2	0.11	10.00	**0.12**	0.34	-	**3.14**	0.06	**3.32**	m2	**3.57**
Wood railings fences and gates; open type										
girth exceeding 300mm	0.45	10.00	**0.50**	0.25	-	**2.31**	0.04	**2.84**	m2	**3.06**
isolated surfaces, girth not exceeding 300mm	0.14	10.00	**0.15**	0.08	-	**0.74**	0.01	**0.90**	m	**0.97**
Wood railings fences and gates; close type										
girth exceeding 300mm	0.45	10.00	**0.50**	0.22	-	**2.03**	0.04	**2.57**	m2	**2.76**
Two undercoats, one coat full gloss finish on ready primed wood surfaces										
Wood general surfaces										
girth exceeding 300mm	0.68	10.00	**0.75**	0.46	-	**4.25**	0.08	**5.07**	m2	**5.45**
isolated surfaces, girth not exceeding 300mm	0.20	10.00	**0.22**	0.15	-	**1.38**	0.03	**1.63**	m	**1.76**
Wood glazed doors										
girth exceeding 300mm; panes, area not exceeding 0.10m2	0.34	10.00	**0.37**	0.76	-	**7.01**	0.13	**7.52**	m2	**8.08**
girth exceeding 300mm; panes, area 0.10 - 0.50m2 .	0.24	10.00	**0.26**	0.55	-	**5.08**	0.10	**5.44**	m2	**5.85**
girth exceeding 300mm; panes, area 0.50 - 1.00m2 .	0.17	10.00	**0.19**	0.46	-	**4.25**	0.08	**4.51**	m2	**4.85**
girth exceeding 300mm; panes, area exceeding 1.00m2	0.14	10.00	**0.15**	0.41	-	**3.78**	0.07	**4.01**	m2	**4.31**
Wood partially glazed doors										
girth exceeding 300mm; panes, area not exceeding 0.10m2	0.51	10.00	**0.56**	0.61	-	**5.63**	0.11	**6.30**	m2	**6.77**
girth exceeding 300mm; panes, area 0.10 - 0.50m2 .	0.44	10.00	**0.48**	0.51	-	**4.71**	0.09	**5.28**	m2	**5.68**
girth exceeding 300mm; panes, area 0.50 - 1.00m2 .	0.41	10.00	**0.45**	0.46	-	**4.25**	0.08	**4.78**	m2	**5.14**
girth exceeding 300mm; panes, area exceeding 1.00m2	0.37	10.00	**0.41**	0.44	-	**4.06**	0.08	**4.55**	m2	**4.89**
Wood windows and screens										
girth exceeding 300mm; panes, area not exceeding 0.10m2	0.37	10.00	**0.41**	0.84	-	**7.75**	0.15	**8.31**	m2	**8.93**
girth exceeding 300mm; panes, area 0.10 - 0.50m2 .	0.27	10.00	**0.30**	0.61	-	**5.63**	0.11	**6.04**	m2	**6.49**
girth exceeding 300mm; panes, area 0.50 - 1.00m2 .	0.20	10.00	**0.22**	0.51	-	**4.71**	0.09	**5.02**	m2	**5.39**
girth exceeding 300mm; panes, area exceeding 1.00m2	0.17	10.00	**0.19**	0.45	-	**4.15**	0.08	**4.42**	m2	**4.75**
Two undercoats, one coat full gloss finish on ready primed wood surfaces; external work										
Wood general surfaces										
girth exceeding 300mm	0.68	10.00	**0.75**	0.49	-	**4.52**	0.09	**5.36**	m2	**5.76**
isolated surfaces, girth not exceeding 300mm	0.20	10.00	**0.22**	0.17	-	**1.57**	0.03	**1.82**	m	**1.96**
Wood glazed doors										
girth exceeding 300mm; panes, area not exceeding 0.10m2	0.34	10.00	**0.37**	0.76	-	**7.01**	0.13	**7.52**	m2	**8.08**
girth exceeding 300mm; panes, area 0.10 - 0.50m2 .	0.24	10.00	**0.26**	0.55	-	**5.08**	0.10	**5.44**	m2	**5.85**
girth exceeding 300mm; panes, area 0.50 - 1.00m2 .	0.17	10.00	**0.19**	0.46	-	**4.25**	0.08	**4.51**	m2	**4.85**
girth exceeding 300mm; panes, area exceeding 1.00m2	0.14	10.00	**0.15**	0.41	-	**3.78**	0.07	**4.01**	m2	**4.31**
Wood partially glazed doors										
girth exceeding 300mm; panes, area not exceeding 0.10m2	0.51	10.00	**0.56**	0.61	-	**5.63**	0.11	**6.30**	m2	**6.77**
girth exceeding 300mm; panes, area 0.10 - 0.50m2 .	0.44	10.00	**0.48**	0.51	-	**4.71**	0.09	**5.28**	m2	**5.68**
girth exceeding 300mm; panes, area 0.50 - 1.00m2 .	0.41	10.00	**0.45**	0.46	-	**4.25**	0.08	**4.78**	m2	**5.14**
girth exceeding 300mm; panes, area exceeding 1.00m2	0.37	10.00	**0.41**	0.44	-	**4.06**	0.08	**4.55**	m2	**4.89**

Labour hourly rates: (except Specialists) Craft Operatives £9.23 Labourer £7.02 Rates are national average prices. Refer to REGIONAL VARIATIONS for indicative levels of overall pricing in regions	MATERIALS			LABOUR				RATES		
	Del to Site £	Waste %	Material Cost £	Craft Optve Hrs	Lab Hrs	Labour Cost £	Sunds £	Nett Rate £	Unit	Gross Rate (+7.5%) £
M60: PAINTING/CLEAR FINISHING - OIL PAINTING WOODWORK Cont.										
Two undercoats, one coat full gloss finish on ready primed wood surfaces; external work Cont.										
Wood windows and screens										
girth exceeding 300mm; panes, area not exceeding										
0.10m2	0.37	10.00	0.41	0.87	–	8.03	0.15	8.59	m2	9.23
girth exceeding 300mm; panes, area 0.10 - 0.50m2 .	0.27	10.00	0.30	0.64	–	5.91	0.11	6.31	m2	6.79
girth exceeding 300mm; panes, area 0.50 - 1.00m2 .	0.20	10.00	0.22	0.54	–	4.98	0.09	5.29	m2	5.69
girth exceeding 300mm; panes, area exceeding										
1.00m2	0.17	10.00	0.19	0.48	–	4.43	0.08	4.70	m2	5.05
Wood railings fences and gates; open type										
girth exceeding 300mm	0.68	10.00	0.75	0.38	–	3.51	0.07	4.33	m2	4.65
isolated surfaces, girth not exceeding 300mm	0.20	10.00	0.22	0.13	–	1.20	0.02	1.44	m	1.55
Wood railings fences and gates; close type										
girth exceeding 300mm	0.68	10.00	0.75	0.33	–	3.05	0.06	3.85	m2	4.14
One coat primer, one undercoat, one coat full gloss finish on wood surfaces										
Wood general surfaces										
girth exceeding 300mm	0.74	10.00	0.81	0.46	–	4.25	0.08	5.14	m2	5.53
isolated surfaces, girth not exceeding 300mm	0.22	10.00	0.24	0.15	–	1.38	0.03	1.66	m	1.78
Wood glazed doors										
girth exceeding 300mm; panes, area not exceeding										
0.10m2	0.37	10.00	0.41	0.76	–	7.01	0.13	7.55	m2	8.12
girth exceeding 300mm; panes, area 0.10 - 0.50m2 .	0.26	10.00	0.29	0.55	–	5.08	0.10	5.46	m2	5.87
girth exceeding 300mm; panes, area 0.50 - 1.00m2 .	0.19	10.00	0.21	0.46	–	4.25	0.08	4.53	m2	4.87
girth exceeding 300mm; panes, area exceeding										
1.00m2	0.15	10.00	0.17	0.41	–	3.78	0.07	4.02	m2	4.32
Wood partially glazed doors										
girth exceeding 300mm; panes, area not exceeding										
0.10m2	0.56	10.00	0.62	0.61	–	5.63	0.11	6.36	m2	6.83
girth exceeding 300mm; panes, area 0.10 - 0.50m2 .	0.48	10.00	0.53	0.51	–	4.71	0.09	5.33	m2	5.72
girth exceeding 300mm; panes, area 0.50 - 1.00m2 .	0.44	10.00	0.48	0.46	–	4.25	0.08	4.81	m2	5.17
girth exceeding 300mm; panes, area exceeding										
1.00m2	0.41	10.00	0.45	0.44	–	4.06	0.08	4.59	m2	4.94
Wood windows and screens										
girth exceeding 300mm; panes, area not exceeding										
0.10m2	0.41	10.00	0.45	0.84	–	7.75	0.15	8.35	m2	8.98
girth exceeding 300mm; panes, area 0.10 - 0.50m2 .	0.30	10.00	0.33	0.61	–	5.63	0.11	6.07	m2	6.53
girth exceeding 300mm; panes, area 0.50 - 1.00m2 .	0.22	10.00	0.24	0.51	–	4.71	0.09	5.04	m2	5.42
girth exceeding 300mm; panes, area exceeding										
1.00m2	0.19	10.00	0.21	0.45	–	4.15	0.08	4.44	m2	4.78
One coat primer, one undercoat, one coat full gloss finish on wood surfaces; external work										
Wood general surfaces										
girth exceeding 300mm	0.74	10.00	0.81	0.49	–	4.52	0.09	5.43	m2	5.83
isolated surfaces, girth not exceeding 300mm	0.22	10.00	0.24	0.17	–	1.57	0.03	1.84	m	1.98
Wood glazed doors										
girth exceeding 300mm; panes, area not exceeding										
0.10m2	0.37	10.00	0.41	0.76	–	7.01	0.13	7.55	m2	8.12
girth exceeding 300mm; panes, area 0.10 - 0.50m2 .	0.26	10.00	0.29	0.55	–	5.08	0.10	5.46	m2	5.87
girth exceeding 300mm; panes, area 0.50 - 1.00m2 .	0.19	10.00	0.21	0.46	–	4.25	0.08	4.53	m2	4.87
girth exceeding 300mm; panes, area exceeding										
1.00m2	0.15	10.00	0.17	0.41	–	3.78	0.07	4.02	m2	4.32
Wood partially glazed doors										
girth exceeding 300mm; panes, area not exceeding										
0.10m2	0.56	10.00	0.62	0.61	–	5.63	0.11	6.36	m2	6.83
girth exceeding 300mm; panes, area 0.10 - 0.50m2 .	0.48	10.00	0.53	0.51	–	4.71	0.09	5.33	m2	5.72
girth exceeding 300mm; panes, area 0.50 - 1.00m2 .	0.44	10.00	0.48	0.46	–	4.25	0.08	4.81	m2	5.17
girth exceeding 300mm; panes, area exceeding										
1.00m2	0.41	10.00	0.45	0.44	–	4.06	0.08	4.59	m2	4.94
Wood windows and screens										
girth exceeding 300mm; panes, area not exceeding										
0.10m2	0.41	10.00	0.45	0.87	–	8.03	0.15	8.63	m2	9.28
girth exceeding 300mm; panes, area 0.10 - 0.50m2 .	0.30	10.00	0.33	0.64	–	5.91	0.11	6.35	m2	6.82
girth exceeding 300mm; panes, area 0.50 - 1.00m2 .	0.22	10.00	0.24	0.54	–	4.98	0.09	5.32	m2	5.71
girth exceeding 300mm; panes, area exceeding										
1.00m2	0.19	10.00	0.21	0.48	–	4.43	0.08	4.72	m2	5.07
Wood railings fences and gates; open type										
girth exceeding 300mm	0.74	10.00	0.81	0.38	–	3.51	0.07	4.39	m2	4.72
isolated surfaces, girth not exceeding 300mm	0.22	10.00	0.24	0.13	–	1.20	0.02	1.46	m	1.57
Wood railings fences and gates; close type										
girth exceeding 300mm	0.74	10.00	0.81	0.33	–	3.05	0.06	3.92	m2	4.21
One coat primer, two undercoats, one coat full gloss finish on wood surfaces										
Wood general surfaces										
girth exceeding 300mm	0.97	10.00	1.07	0.60	–	5.54	0.10	6.71	m2	7.21
isolated surfaces, girth not exceeding 300mm	0.29	10.00	0.32	0.20	–	1.85	0.03	2.19	m	2.36

	MATERIALS			LABOUR				RATES		
Labour hourly rates: (except Specialists) Craft Operatives £9.23 Labourer £7.02 Rates are national average prices. Refer to REGIONAL VARIATIONS for indicative levels of overall pricing in regions	Del to Site £	Waste %	Material Cost £	Craft Optve Hrs	Lab Hrs	Labour Cost £	Sunds £	Nett Rate £	Unit	Gross Rate (+7.5%) £
M60: PAINTING/CLEAR FINISHING – OIL PAINTING WOODWORK Cont.										
One coat primer, two undercoats, one coat full gloss finish on wood surfaces Cont.										
Wood glazed doors										
girth exceeding 300mm; panes, area not exceeding 0.10m2	0.48	10.00	0.53	1.00	-	9.23	0.17	9.93	m2	10.67
girth exceeding 300mm; panes, area 0.10 - 0.50m2	0.34	10.00	0.37	0.72	-	6.65	0.13	7.15	m2	7.69
girth exceeding 300mm; panes, area 0.50 - 1.00m2 .	0.24	10.00	0.26	0.60	-	5.54	0.10	5.90	m2	6.34
girth exceeding 300mm; panes, area exceeding 1.00m2	0.19	10.00	0.21	0.54	-	4.98	0.09	5.28	m2	5.68
Wood partially glazed doors										
girth exceeding 300mm; panes, area not exceeding 0.10m2	0.73	10.00	0.80	0.80	-	7.38	0.14	8.33	m2	8.95
girth exceeding 300mm; panes, area 0.10 - 0.50m2	0.63	10.00	0.69	0.66	-	6.09	0.12	6.90	m2	7.42
girth exceeding 300mm; panes, area 0.50 - 1.00m2 .	0.58	10.00	0.64	0.60	-	5.54	0.10	6.28	m2	6.75
girth exceeding 300mm; panes, area exceeding 1.00m2	0.53	10.00	0.58	0.57	-	5.26	0.10	5.94	m2	6.39
Wood windows and screens										
girth exceeding 300mm; panes, area not exceeding 0.10m2	0.53	10.00	0.58	1.10	-	10.15	0.19	10.93	m2	11.75
girth exceeding 300mm; panes, area 0.10 - 0.50m2 .	0.39	10.00	0.43	0.79	-	7.29	0.14	7.86	m2	8.45
girth exceeding 300mm; panes, area 0.50 - 1.00m2 .	0.29	10.00	0.32	0.66	-	6.09	0.12	6.53	m2	7.02
girth exceeding 300mm; panes, area exceeding 1.00m2	0.24	10.00	0.26	0.59	-	5.45	0.10	5.81	m2	6.25
One coat primer, two undercoats, one coat full gloss finish on wood surfaces; external work										
Wood general surfaces										
girth exceeding 300mm	0.97	10.00	1.07	0.64	-	5.91	0.11	7.08	m2	7.62
isolated surfaces, girth not exceeding 300mm	0.29	10.00	0.32	0.22	-	2.03	0.04	2.39	m2	2.57
Wood glazed doors										
girth exceeding 300mm; panes, area not exceeding 0.10m2	0.48	10.00	0.53	0.10	-	0.92	0.17	1.62	m2	1.74
girth exceeding 300mm; panes, area 0.10 - 0.50m2	0.34	10.00	0.37	0.72	-	6.65	0.13	7.15	m2	7.69
girth exceeding 300mm; panes, area 0.50 - 1.00m2 .	0.24	10.00	0.26	0.60	-	5.54	0.10	5.90	m2	6.34
girth exceeding 300mm; panes, area exceeding 1.00m2	0.19	10.00	0.21	0.08	-	0.74	0.09	1.04	m2	1.12
Wood partially glazed doors										
girth exceeding 300mm; panes, area not exceeding 0.10m2	0.73	10.00	0.80	0.80	-	7.38	0.14	8.33	m2	8.95
girth exceeding 300mm; panes, area 0.10 - 0.50m2	0.63	10.00	0.69	0.66	-	6.09	0.12	6.90	m2	7.42
girth exceeding 300mm; panes, area 0.50 - 1.00m2 .	0.58	10.00	0.64	0.60	-	5.54	0.10	6.28	m2	6.75
girth exceeding 300mm; panes, area exceeding 1.00m2	0.53	10.00	0.58	0.57	-	5.26	0.10	5.94	m2	6.39
Wood windows and screens										
girth exceeding 300mm; panes, area not exceeding 0.10m2	0.53	10.00	0.58	1.14	-	10.52	0.20	11.31	m2	12.15
girth exceeding 300mm; panes, area 0.10 - 0.50m2 .	0.39	10.00	0.43	0.83	-	7.66	0.15	8.24	m2	8.86
girth exceeding 300mm; panes, area 0.50 - 1.00m2 .	0.29	10.00	0.32	0.70	-	6.46	0.12	6.90	m2	7.42
girth exceeding 300mm; panes, area exceeding 1.00m2	0.24	10.00	0.26	0.63	-	5.81	0.11	6.19	m2	6.65
Wood railings fences and gates; open type										
girth exceeding 300mm	0.97	10.00	1.07	0.50	-	4.62	0.09	5.77	m2	6.20
isolated surfaces, girth not exceeding 300mm	0.29	10.00	0.32	0.17	-	1.57	0.03	1.92	m	2.06
Wood railings fences and gates; close type										
girth exceeding 300mm	0.97	10.00	1.07	0.44	-	4.06	0.08	5.21	m2	5.60
One coat primer, one undercoat, two coats full gloss finish on wood surfaces										
Wood general surfaces										
girth exceeding 300mm	0.97	10.00	1.07	0.60	-	5.54	0.10	6.71	m2	7.21
isolated surfaces, girth not exceeding 300mm	0.29	10.00	0.32	0.20	-	1.85	0.03	2.19	m	2.36
Wood glazed doors										
girth exceeding 300mm; panes, area not exceeding 0.10m2	0.48	10.00	0.53	1.00	-	9.23	0.17	9.93	m2	10.67
girth exceeding 300mm; panes, area 0.10 - 0.50m2	0.34	10.00	0.37	0.72	-	6.65	0.13	7.15	m2	7.69
girth exceeding 300mm; panes, area 0.50 - 1.00m2 .	0.24	10.00	0.26	0.60	-	5.54	0.10	5.90	m2	6.34
girth exceeding 300mm; panes, area exceeding 1.00m2	0.19	10.00	0.21	0.54	-	4.98	0.09	5.28	m2	5.68
Wood partially glazed doors										
girth exceeding 300mm; panes, area not exceeding 0.10m2	0.73	10.00	0.80	0.80	-	7.38	0.14	8.33	m2	8.95
girth exceeding 300mm; panes, area 0.10 - 0.50m2	0.63	10.00	0.69	0.66	-	6.09	0.12	6.90	m2	7.42
girth exceeding 300mm; panes, area 0.50 - 1.00m2 .	0.58	10.00	0.64	0.60	-	5.54	0.10	6.28	m2	6.75
girth exceeding 300mm; panes, area exceeding 1.00m2	0.53	10.00	0.58	0.57	-	5.26	0.10	5.94	m2	6.39
Wood windows and screens										
girth exceeding 300mm; panes, area not exceeding 0.10m2	0.53	10.00	0.58	1.10	-	10.15	0.19	10.93	m2	11.75
girth exceeding 300mm; panes, area 0.10 - 0.50m2 .	0.39	10.00	0.43	0.79	-	7.29	0.14	7.86	m2	8.45
girth exceeding 300mm; panes, area 0.50 - 1.00m2 .	0.29	10.00	0.32	0.66	-	6.09	0.12	6.53	m2	7.02

SURFACE FINISHES

Labour hourly rates: (except Specialists) Craft Operatives £9.23 Labourer £7.02 Rates are national average prices. Refer to REGIONAL VARIATIONS for indicative levels of overall pricing in regions	MATERIALS			LABOUR				RATES		
	Del to Site £	Waste %	Material Cost £	Craft Optve Hrs	Lab Hrs	Labour Cost £	Sunds £	Nett Rate £	Unit	Gross Rate (+7.5%) £
M60: PAINTING/CLEAR FINISHING - OIL PAINTING WOODWORK Cont.										
One coat primer, one undercoat, two coats full gloss finish on wood surfaces Cont.										
Wood windows and screens Cont.										
girth exceeding 300mm; panes, area exceeding 1.00m2	0.24	10.00	0.26	0.59	-	5.45	0.10	5.81	m2	6.25
One coat primer, one undercoat, two coats full gloss finish wood surfaces; external work										
Wood general surfaces										
girth exceeding 300mm	0.97	10.00	1.07	0.64	-	5.91	0.11	7.08	m2	7.62
isolated surfaces, girth not exceeding 300mm	0.29	10.00	0.32	0.22	-	2.03	0.04	2.39	m	2.57
Wood glazed doors										
girth exceeding 300mm; panes, area not exceeding 0.10m2	0.48	10.00	0.53	1.00	-	9.23	0.17	9.93	m2	10.67
girth exceeding 300mm; panes, area 0.10 - 0.50m2 .	0.34	10.00	0.37	0.72	-	6.65	0.13	7.15	m2	7.69
girth exceeding 300mm; panes, area 0.50 - 1.00m2 .	0.24	10.00	0.26	0.60	-	5.54	0.10	5.90	m2	6.34
girth exceeding 300mm; panes, area exceeding 1.00m2	0.19	10.00	0.21	0.54	-	4.98	0.09	5.28	m2	5.68
Wood partially glazed doors										
girth exceeding 300mm; panes, area not exceeding 0.10m2	0.73	10.00	0.80	0.80	-	7.38	0.14	8.33	m2	8.95
girth exceeding 300mm; panes, area 0.10 - 0.50m2 .	0.63	10.00	0.69	0.66	-	6.09	0.12	6.90	m2	7.42
girth exceeding 300mm; panes, area 0.50 - 1.00m2 .	0.58	10.00	0.64	0.60	-	5.54	0.10	6.28	m2	6.75
girth exceeding 300mm; panes, area exceeding 1.00m2	0.53	10.00	0.58	0.57	-	5.26	0.10	5.94	m2	6.39
Wood windows and screens										
girth exceeding 300mm; panes, area not exceeding 0.10m2	0.53	10.00	0.58	1.14	-	10.52	0.20	11.31	m2	12.15
girth exceeding 300mm; panes, area 0.10 - 0.50m2 .	0.39	10.00	0.43	0.83	-	7.66	0.15	8.24	m2	8.86
girth exceeding 300mm; panes, area 0.50 - 1.00m2 .	0.29	10.00	0.32	0.70	-	6.46	0.12	6.90	m2	7.42
girth exceeding 300mm; panes, area exceeding 1.00m2	0.24	10.00	0.26	0.63	-	5.81	0.11	6.19	m2	6.65
Wood railings fences and gates; open type										
girth exceeding 300mm	0.97	10.00	1.07	0.50	-	4.62	0.09	5.77	m2	6.20
isolated surfaces, girth not exceeding 300mm	0.29	10.00	0.32	0.17	-	1.57	0.03	1.92	m	2.06
Wood railings fences and gates; close type										
girth exceeding 300mm	0.97	10.00	1.07	0.44	-	4.06	0.08	5.21	m2	5.60
M60: PAINTING/CLEAR FINISHING - POLYURETHANE LACQUER										
Two coats polyurethane lacquer										
Wood general surfaces										
girth exceeding 300mm	0.86	10.00	0.95	0.40	-	3.69	0.07	4.71	m2	5.06
isolated surfaces, girth not exceeding 300mm	0.26	10.00	0.29	0.13	-	1.20	0.02	1.51	m	1.62
Wood glazed doors										
girth exceeding 300mm; panes, area not exceeding 0.10m2	0.43	10.00	0.47	0.67	-	6.18	0.12	6.78	m2	7.29
girth exceeding 300mm; panes, area 0.10 - 0.50m2 .	0.30	10.00	0.33	0.48	-	4.43	0.08	4.84	m2	5.20
girth exceeding 300mm; panes, area 0.50 - 1.00m2 .	0.21	10.00	0.23	0.40	-	3.69	0.07	3.99	m2	4.29
girth exceeding 300mm; panes, area exceeding 1.00m2	0.17	10.00	0.19	0.36	-	3.32	0.06	3.57	m2	3.84
Wood partially glazed doors										
girth exceeding 300mm; panes, area not exceeding 0.10m2	0.64	10.00	0.70	0.54	-	4.98	0.09	5.78	m2	6.21
girth exceeding 300mm; panes, area 0.10 - 0.50m2 .	0.56	10.00	0.62	0.44	-	4.06	0.08	4.76	m2	5.11
girth exceeding 300mm; panes, area 0.50 - 1.00m2 .	0.52	10.00	0.57	0.40	-	3.69	0.07	4.33	m2	4.66
girth exceeding 300mm; panes, area exceeding 1.00m2	0.47	10.00	0.52	0.38	-	3.51	0.07	4.09	m2	4.40
Wood windows and screens										
girth exceeding 300mm; panes, area not exceeding 0.10m2	0.47	10.00	0.52	0.74	-	6.83	0.13	7.48	m2	8.04
girth exceeding 300mm; panes, area 0.10 - 0.50m2 .	0.34	10.00	0.37	0.53	-	4.89	0.09	5.36	m2	5.76
girth exceeding 300mm; panes, area 0.50 - 1.00m2 .	0.26	10.00	0.29	0.44	-	4.06	0.08	4.43	m2	4.76
girth exceeding 300mm; panes, area exceeding 1.00m2	0.21	10.00	0.23	0.40	-	3.69	0.07	3.99	m2	4.29
Two coats polyurethane lacquer; external work										
Wood general surfaces										
girth exceeding 300mm	0.86	10.00	0.95	0.42	-	3.88	0.07	4.89	m2	5.26
isolated surfaces, girth not exceeding 300mm	0.26	10.00	0.29	0.14	-	1.29	0.02	1.60	m	1.72
Wood glazed doors										
girth exceeding 300mm; panes, area not exceeding 0.10m2	0.43	10.00	0.47	0.67	-	6.18	0.12	6.78	m2	7.29
girth exceeding 300mm; panes, area 0.10 - 0.50m2 .	0.30	10.00	0.33	0.48	-	4.43	0.08	4.84	m2	5.20
girth exceeding 300mm; panes, area 0.50 - 1.00m2 .	0.21	10.00	0.23	0.40	-	3.69	0.07	3.99	m2	4.29
girth exceeding 300mm; panes, area exceeding 1.00m2	0.17	10.00	0.19	0.36	-	3.32	0.06	3.57	m2	3.84

Labour hourly rates: (except Specialists) Craft Operatives £9.23 Labourer £7.02 Rates are national average prices. Refer to REGIONAL VARIATIONS for indicative levels of overall pricing in regions	MATERIALS			LABOUR				RATES		
	Del to Site £	Waste %	Material Cost £	Craft Optve Hrs	Lab Hrs	Labour Cost £	Sunds £	Nett Rate £	Unit	Gross Rate (+7.5%) £
M60: PAINTING/CLEAR FINISHING - POLYURETHANE LACQUER Cont.										
Two coats polyurethane lacquer; external work Cont.										
Wood partially glazed doors										
girth exceeding 300mm; panes, area not exceeding 0.10m2	0.64	10.00	0.70	0.54	-	4.98	0.09	5.78	m2	6.21
girth exceeding 300mm; panes, area 0.10 - 0.50m2 .	0.56	10.00	0.62	0.44	-	4.06	0.08	4.76	m2	5.11
girth exceeding 300mm; panes, area 0.50 - 1.00m2 .	0.52	10.00	0.57	0.40	-	3.69	0.07	4.33	m2	4.66
girth exceeding 300mm; panes, area exceeding 1.00m2	0.47	10.00	0.52	0.38	-	3.51	0.07	4.09	m2	4.40
Wood windows and screens										
girth exceeding 300mm; panes, area not exceeding 0.10m2	0.47	10.00	0.52	0.76	-	7.01	0.13	7.66	m2	8.24
girth exceeding 300mm; panes, area 0.10 - 0.50m2 .	0.34	10.00	0.37	0.55	-	5.08	0.10	5.55	m2	5.97
girth exceeding 300mm; panes, area 0.50 - 1.00m2 .	0.26	10.00	0.29	0.46	-	4.25	0.08	4.61	m2	4.96
girth exceeding 300mm; panes, area exceeding 1.00m2	0.21	10.00	0.23	0.42	-	3.88	0.07	4.18	m2	4.49
Three coats polyurethane lacquer										
Wood general surfaces										
girth exceeding 300mm	1.29	10.00	1.42	0.47	-	4.34	0.08	5.84	m2	6.27
isolated surfaces, girth not exceeding 300mm	0.39	10.00	0.43	0.16	-	1.48	0.03	1.94	m	2.08
Wood glazed doors										
girth exceeding 300mm; panes, area not exceeding 0.10m2	0.64	10.00	0.70	0.78	-	7.20	0.14	8.04	m2	8.65
girth exceeding 300mm; panes, area 0.10 - 0.50m2 .	0.45	10.00	0.50	0.56	-	5.17	0.10	5.76	m2	6.20
girth exceeding 300mm; panes, area 0.50 - 1.00m2 .	0.32	10.00	0.35	0.47	-	4.34	0.08	4.77	m2	5.13
girth exceeding 300mm; panes, area exceeding 1.00m2	0.26	10.00	0.29	0.42	-	3.88	0.07	4.23	m2	4.55
Wood partially glazed doors .										
girth exceeding 300mm; panes, area not exceeding 0.10m2	0.97	10.00	1.07	0.63	-	5.81	0.11	6.99	m2	7.52
girth exceeding 300mm; panes, area 0.10 - 0.50m2 .	0.84	10.00	0.92	0.57	-	5.26	0.10	6.29	m2	6.76
girth exceeding 300mm; panes, area 0.50 - 1.00m2 .	0.77	10.00	0.85	0.47	-	4.34	0.08	5.27	m2	5.66
girth exceeding 300mm; panes, area exceeding 1.00m2	0.71	10.00	0.78	0.45	-	4.15	0.08	5.01	m2	5.39
Wood windows and screens										
girth exceeding 300mm; panes, area not exceeding 0.10m2	0.71	10.00	0.78	0.86	-	7.94	0.15	8.87	m2	9.53
girth exceeding 300mm; panes, area 0.10 - 0.50m2 .	0.52	10.00	0.57	0.62	-	5.72	0.11	6.40	m2	6.88
girth exceeding 300mm; panes, area 0.50 - 1.00m2 .	0.39	10.00	0.43	0.52	-	4.80	0.09	5.32	m2	5.72
girth exceeding 300mm; panes, area exceeding 1.00m2	0.32	10.00	0.35	0.46	-	4.25	0.08	4.68	m2	5.03
Three coats polyurethane lacquer; external work										
Wood general surfaces										
girth exceeding 300mm	1.29	10.00	1.42	0.50	-	4.62	0.09	6.12	m2	6.58
isolated surfaces, girth not exceeding 300mm	0.39	10.00	0.43	0.17	-	1.57	0.03	2.03	m	2.18
Wood glazed doors										
girth exceeding 300mm; panes, area not exceeding 0.10m2	0.64	10.00	0.70	0.78	-	7.20	0.14	8.04	m2	8.65
girth exceeding 300mm; panes, area 0.10 - 0.50m2 .	0.45	10.00	0.50	0.56	-	5.17	0.10	5.76	m2	6.20
girth exceeding 300mm; panes, area 0.50 - 1.00m2 .	0.32	10.00	0.35	0.47	-	4.34	0.08	4.77	m2	5.13
girth exceeding 300mm; panes, area exceeding 1.00m2	0.26	10.00	0.29	0.42	-	3.88	0.07	4.23	m2	4.55
Wood partially glazed doors										
girth exceeding 300mm; panes, area not exceeding 0.10m2	0.97	10.00	1.07	0.63	-	5.81	0.11	6.99	m2	7.52
girth exceeding 300mm; panes, area 0.10 - 0.50m2 .	0.84	10.00	0.92	0.57	-	5.26	0.10	6.29	m2	6.76
girth exceeding 300mm; panes, area 0.50 - 1.00m2 .	0.77	10.00	0.85	0.47	-	4.34	0.08	5.27	m2	5.66
girth exceeding 300mm; panes, area exceeding 1.00m2	0.71	10.00	0.78	0.45	-	4.15	0.08	5.01	m2	5.39
Wood windows and screens										
girth exceeding 300mm; panes, area not exceeding 0.10m2	0.71	10.00	0.78	0.89	-	8.21	0.16	9.16	m2	9.84
girth exceeding 300mm; panes, area 0.10 - 0.50m2 .	0.52	10.00	0.57	0.65	-	6.00	0.11	6.68	m2	7.18
girth exceeding 300mm; panes, area 0.50 - 1.00m2 .	0.39	10.00	0.43	0.55	-	5.08	0.10	5.61	m2	6.03
girth exceeding 300mm; panes, area exceeding 1.00m2	0.32	10.00	0.35	0.49	-	4.52	0.09	4.96	m2	5.34
M60: PAINTING/CLEAR FINISHING - FIRE RETARDANT PAINTS AND VARNISHES										
Two coats fire retardant paint										
Wood general surfaces										
girth exceeding 300mm	2.62	10.00	2.88	0.40	-	3.69	0.07	6.64	m2	7.14
isolated surfaces, girth not exceeding 300mm	0.79	10.00	0.87	0.13	-	1.20	0.02	2.09	m	2.25
Two coats fire retardant varnish, one overcoat varnish										
Wood general surfaces										
girth exceeding 300mm	2.35	10.00	2.59	0.54	-	4.98	0.09	7.66	m2	8.23
isolated surfaces, girth not exceeding 300mm	0.71	10.00	0.78	0.18	-	1.66	0.03	2.47	m	2.66

SURFACE FINISHES

Labour hourly rates: (except Specialists) Craft Operatives £9.23 Labourer £7.02 Rates are national average prices. Refer to REGIONAL VARIATIONS for indicative levels of overall pricing in regions	MATERIALS			LABOUR				RATES		
	Del to Site £	Waste %	Material Cost £	Craft Optve Hrs	Lab Hrs	Labour Cost £	Sunds £	Nett Rate £	Unit	Gross Rate (+7.5%) £
M60: PAINTING/CLEAR FINISHING - OILING HARDWOOD										
Two coats raw linseed oil										
Wood general surfaces										
girth exceeding 300mm	0.92	10.00	1.01	0.44	-	4.06	0.08	5.15	m2	5.54
isolated surfaces, girth not exceeding 300mm	0.28	10.00	0.31	0.15	-	1.38	0.03	1.72	m	1.85
M60: PAINTING/CLEAR FINISHING - FRENCH AND WAX POLISHING										
Stain; two coats wax polish										
Wood general surfaces										
girth exceeding 300mm	1.61	10.00	1.77	0.51	-	4.71	0.07	6.55	m2	7.04
isolated surfaces, girth not exceeding 300mm	0.48	10.00	0.53	0.17	-	1.57	0.02	2.12	m	2.28
Open grain French polish										
Wood general surfaces										
girth exceeding 300mm	1.99	10.00	2.19	2.40	-	22.15	0.34	24.68	m2	26.53
isolated surfaces, girth not exceeding 300mm	0.59	10.00	0.65	0.80	-	7.38	0.12	8.15	m	8.76
Stain; body in; fully French polish										
Wood general surfaces										
girth exceeding 300mm	2.84	10.00	3.12	3.60	-	33.23	0.50	36.85	m2	39.62
isolated surfaces, girth not exceeding 300mm	0.86	10.00	0.95	1.20	-	11.08	0.17	12.19	m	13.11
M60: PAINTING/CLEAR FINISHING - ROAD MARKINGS										
One coat spirit based road marking paint; external work										
Concrete general surfaces										
isolated surfaces, 50mm wide	0.06	5.00	0.06	0.10	-	0.92	0.01	1.00	m	1.07
isolated surfaces, 100mm wide	0.13	5.00	0.14	0.12	-	1.11	0.02	1.26	m	1.36
Two coats spirit based road marking paint; external work										
Concrete general surfaces										
isolated surfaces, 50mm wide	0.16	5.00	0.17	0.15	-	1.38	0.02	1.57	m	1.69
isolated surfaces, 100mm wide	0.28	5.00	0.29	0.18	-	1.66	0.02	1.98	m	2.12
Prime and apply non-reflective self adhesive road marking tape; external work										
Concrete general surfaces										
isolated surfaces, 50mm wide	1.16	5.00	1.22	0.10	-	0.92	0.01	2.15	m	2.31
isolated surfaces, 100mm wide	2.20	5.00	2.31	0.13	-	1.20	0.02	3.53	m	3.79
Prime and apply reflective self adhesive road marking tape; external work										
Concrete general surfaces										
isolated surfaces, 100mm wide	3.36	5.00	3.53	0.13	-	1.20	0.02	4.75	m	5.10
M60: PAINTING/CLEAR FINISHING (REDECORATIONS) EMULSION PAINTING										
Generally										

Note
  the following rates include for the cost of all
  preparatory work, e.g. washing down, etc.										

Two coats emulsion paint, existing emulsion painted surfaces										
Concrete general surfaces										
girth exceeding 300mm	0.44	10.00	0.48	0.29	-	2.68	0.05	3.21	m2	3.45
Plaster general surfaces										
girth exceeding 300mm	0.35	10.00	0.39	0.25	-	2.31	0.04	2.73	m2	2.94
Plasterboard general surfaces										
girth exceeding 300mm	0.35	10.00	0.39	0.26	-	2.40	0.05	2.83	m2	3.05
Brickwork general surfaces										
girth exceeding 300mm	0.53	10.00	0.58	0.29	-	2.68	0.05	3.31	m2	3.56
Paper covered general surfaces										
girth exceeding 300mm	0.40	10.00	0.44	0.26	-	2.40	0.05	2.89	m2	3.11
Two full coats emulsion paint, existing washable distempered surfaces										
Concrete general surfaces										
girth exceeding 300mm	0.44	10.00	0.48	0.30	-	2.77	0.05	3.30	m2	3.55
Plaster general surfaces										
girth exceeding 300mm	0.35	10.00	0.39	0.26	-	2.40	0.05	2.83	m2	3.05
Plasterboard general surfaces										
girth exceeding 300mm	0.35	10.00	0.39	0.27	-	2.49	0.05	2.93	m2	3.15

Labour hourly rates: (except Specialists) Craft Operatives £9.23 Labourer £7.02 Rates are national average prices. Refer to REGIONAL VARIATIONS for indicative levels of overall pricing in regions	MATERIALS			LABOUR				RATES		
	Del to Site £	Waste %	Material Cost £	Craft Optve Hrs	Lab Hrs	Labour Cost £	Sunds £	Nett Rate £	Unit	Gross Rate (+7.5%) £
M60: PAINTING/CLEAR FINISHING (REDECORATIONS) – EMULSION PAINTING Cont.										
Two full coats emulsion paint, existing washable distempered surfaces Cont.										
Brickwork general surfaces girth exceeding 300mm	0.53	10.00	**0.58**	0.30	–	**2.77**	0.05	**3.40**	m2	**3.66**
Paper covered general surfaces girth exceeding 300mm	0.40	10.00	**0.44**	0.27	–	**2.49**	0.05	**2.98**	m2	**3.21**
Two full coats emulsion paint, existing textured plastic coating surfaces										
Concrete general surfaces girth exceeding 300mm	0.44	10.00	**0.48**	0.32	–	**2.95**	0.06	**3.50**	m2	**3.76**
Plaster general surfaces girth exceeding 300mm	0.35	10.00	**0.39**	0.28	–	**2.58**	0.05	**3.02**	m2	**3.25**
Plasterboard general surfaces girth exceeding 300mm	0.35	10.00	**0.39**	0.29	–	**2.68**	0.05	**3.11**	m2	**3.35**
Brickwork general surfaces girth exceeding 300mm	0.53	10.00	**0.58**	0.32	–	**2.95**	0.06	**3.60**	m2	**3.87**
Paper covered general surfaces girth exceeding 300mm	0.40	10.00	**0.44**	0.29	–	**2.68**	0.05	**3.17**	m2	**3.40**
Mist coat, two full coats emulsion paint, existing non-washable distempered surfaces										
Concrete general surfaces girth exceeding 300mm	0.53	10.00	**0.58**	0.35	–	**3.23**	0.06	**3.87**	m2	**4.16**
Plaster general surfaces girth exceeding 300mm	0.44	10.00	**0.48**	0.31	–	**2.86**	0.05	**3.40**	m2	**3.65**
Plasterboard general surfaces girth exceeding 300mm	0.44	10.00	**0.48**	0.32	–	**2.95**	0.06	**3.50**	m2	**3.76**
Brickwork general surfaces girth exceeding 300mm	0.66	10.00	**0.73**	0.35	–	**3.23**	0.06	**4.02**	m2	**4.32**
Paper covered general surfaces girth exceeding 300mm	0.49	10.00	**0.54**	0.32	–	**2.95**	0.06	**3.55**	m2	**3.82**
Two full coats emulsion paint, existing gloss painted surfaces										
Concrete general surfaces girth exceeding 300mm	0.44	10.00	**0.48**	0.30	–	**2.77**	0.05	**3.30**	m2	**3.55**
Plaster general surfaces girth exceeding 300mm	0.35	10.00	**0.39**	0.26	–	**2.40**	0.05	**2.83**	m2	**3.05**
Plasterboard general surfaces girth exceeding 300mm	0.35	10.00	**0.39**	0.27	–	**2.49**	0.05	**2.93**	m2	**3.15**
Brickwork general surfaces girth exceeding 300mm	0.53	10.00	**0.58**	0.30	–	**2.77**	0.05	**3.40**	m2	**3.66**
Paper covered general surfaces girth exceeding 300mm	0.40	10.00	**0.44**	0.27	–	**2.49**	0.05	**2.98**	m2	**3.21**
M60: PAINTING/CLEAR FINISHING (REDECORATIONS) – CEMENT PAINTING										
Generally										
Note the following rates include for the cost of all preparatory work, e.g. washing down, etc. **One coat sealer, one coat Snowcem, existing undecorated surfaces; external work**										
Cement rendered general surfaces girth exceeding 300mm	0.78	10.00	**0.86**	0.32	–	**2.95**	0.06	**3.87**	m2	**4.16**
Concrete general surfaces girth exceeding 300mm	0.87	10.00	**0.96**	0.38	–	**3.51**	0.07	**4.53**	m2	**4.87**
Brickwork general surfaces girth exceeding 300mm	0.87	10.00	**0.96**	0.38	–	**3.51**	0.07	**4.53**	m2	**4.87**
Rough cast general surfaces girth exceeding 300mm	0.92	10.00	**1.01**	0.47	–	**4.34**	0.08	**5.43**	m2	**5.84**
One coat sealer, two coats Snowcem, existing undecorated surfaces; external work										
Cement rendered general surfaces girth exceeding 300mm	0.99	10.00	**1.09**	0.43	–	**3.97**	0.08	**5.14**	m2	**5.52**

SURFACE FINISHES

Labour hourly rates: (except Specialists) Craft Operatives £9.23 Labourer £7.02 Rates are national average prices. Refer to REGIONAL VARIATIONS for indicative levels of overall pricing in regions	MATERIALS			LABOUR				RATES		
	Del to Site £	Waste %	Material Cost £	Craft Optve Hrs	Lab Hrs	Labour Cost £	Sunds £	Nett Rate £	Unit	Gross Rate (+7.5%) £
M60: PAINTING/CLEAR FINISHING (REDECORATIONS) – CEMENT PAINTING Cont.										
One coat sealer, two coats Snowcem, existing undecorated surfaces; external work Cont.										
Concrete general surfaces girth exceeding 300mm	1.15	10.00	1.26	0.52	–	4.80	0.09	6.15	m2	6.62
Brickwork general surfaces girth exceeding 300mm	1.15	10.00	1.26	0.52	–	4.80	0.09	6.15	m2	6.62
Rough cast general surfaces girth exceeding 300mm	1.23	10.00	1.35	0.64	–	5.91	0.11	7.37	m2	7.92
One coat Snowcem, existing cement painted surfaces; external work										
Cement rendered general surfaces girth exceeding 300mm	0.26	10.00	0.29	0.24	–	2.22	0.04	2.54	m2	2.73
Concrete general surfaces girth exceeding 300mm	0.35	10.00	0.39	0.30	–	2.77	0.05	3.20	m2	3.44
Brickwork general surfaces girth exceeding 300mm	0.35	10.00	0.39	0.30	–	2.77	0.05	3.20	m2	3.44
Rough cast general surfaces girth exceeding 300mm	0.40	10.00	0.44	0.39	–	3.60	0.07	4.11	m2	4.42
Two coats Snowcem, existing cement painted surfaces; external work										
Cement rendered general surfaces girth exceeding 300mm	0.47	10.00	0.52	0.35	–	3.23	0.06	3.81	m2	4.09
Concrete general surfaces girth exceeding 300mm	0.63	10.00	0.69	0.44	–	4.06	0.08	4.83	m2	5.20
Brickwork general surfaces girth exceeding 300mm	0.63	10.00	0.69	0.44	–	4.06	0.08	4.83	m2	5.20
Rough cast general surfaces girth exceeding 300mm	0.71	10.00	0.78	0.56	–	5.17	0.10	6.05	m2	6.50
One coat stabilising solution, one coat textured masonry paint, existing undecorated surfaces; external work										
Cement rendered general surfaces girth exceeding 300mm	1.09	10.00	1.20	0.32	–	2.95	0.06	4.21	m2	4.53
Concrete general surfaces girth exceeding 300mm	1.21	10.00	1.33	0.38	–	3.51	0.07	4.91	m2	5.28
Brickwork general surfaces girth exceeding 300mm	1.34	10.00	1.47	0.38	–	3.51	0.07	5.05	m2	5.43
Rough cast general surfaces girth exceeding 300mm	1.81	10.00	1.99	0.47	–	4.34	0.08	6.41	m2	6.89
One coat stabilising solution, two coats textured masonry paint, existing undecorated surfaces; external work										
Cement rendered general surfaces girth exceeding 300mm	1.55	10.00	1.71	0.43	–	3.97	0.08	5.75	m2	6.19
Concrete general surfaces girth exceeding 300mm	1.72	10.00	1.89	0.52	–	4.80	0.09	6.78	m2	7.29
Brickwork general surfaces girth exceeding 300mm	1.92	10.00	2.11	0.52	–	4.80	0.09	7.00	m2	7.53
Rough cast general surfaces girth exceeding 300mm	2.74	10.00	3.01	0.64	–	5.91	0.11	9.03	m2	9.71
One coat textured masonry paint, existing cement painted surfaces; external work										
Cement rendered general surfaces girth exceeding 300mm	0.58	10.00	0.64	0.24	–	2.22	0.04	2.89	m2	3.11
Concrete general surfaces girth exceeding 300mm	0.65	10.00	0.71	0.30	–	2.77	0.05	3.53	m2	3.80
Brickwork general surfaces girth exceeding 300mm	0.72	10.00	0.79	0.30	–	2.77	0.05	3.61	m2	3.88
Rough cast general surfaces girth exceeding 300mm	1.16	10.00	1.28	0.39	–	3.60	0.07	4.95	m2	5.32

Labour hourly rates: (except Specialists) Craft Operatives £9.23 Labourer £7.02 Rates are national average prices. Refer to REGIONAL VARIATIONS for indicative levels of overall pricing in regions	MATERIALS			LABOUR				RATES		
	Del to Site	Waste	Material Cost	Craft Optve	Lab	Labour Cost	Sunds	Nett Rate	Unit	Gross Rate (+7.5%)
	£	%	£	Hrs	Hrs	£	£	£		£
M60: PAINTING/CLEAR FINISHING (REDECORATIONS) – CEMENT PAINTING Cont.										
Two coats textured masonry paint, existing cement painted surfaces; external work										
Cement rendered general surfaces girth exceeding 300mm	1.04	10.00	**1.14**	0.35	–	**3.23**	0.06	**4.43**	m2	4.77
Concrete general surfaces girth exceeding 300mm	1.16	10.00	**1.28**	0.44	–	**4.06**	0.08	**5.42**	m2	5.82
Brickwork general surfaces girth exceeding 300mm	1.30	10.00	**1.43**	0.44	–	**4.06**	0.08	**5.57**	m2	5.99
Rough cast general surfaces girth exceeding 300mm	2.08	10.00	**2.29**	0.56	–	**5.17**	0.10	**7.56**	m2	8.12
M60: PAINTING/CLEAR FINISHING (REDECORATIONS) – OIL PAINTING WALLS AND CEILINGS										
Generally										

Note
    the following rates include for the cost of all
    preparatory work, e.g. washing down, etc.
**One undercoat, one coat full gloss finish, existing
gloss painted surfaces**										

Concrete general surfaces girth exceeding 300mm	0.51	10.00	**0.56**	0.35	–	**3.23**	0.06	**3.85**	m2	4.14
Brickwork general surfaces girth exceeding 300mm	0.62	10.00	**0.68**	0.36	–	**3.32**	0.06	**4.06**	m2	4.37
Plasterboard general surfaces girth exceeding 300mm	0.40	10.00	**0.44**	0.32	–	**2.95**	0.06	**3.45**	m2	3.71
Plaster general surfaces girth exceeding 300mm	0.45	10.00	**0.50**	0.31	–	**2.86**	0.05	**3.41**	m2	3.66
One undercoat, one coat eggshell finish, existing gloss painted surfaces										
Concrete general surfaces girth exceeding 300mm	0.56	10.00	**0.62**	0.35	–	**3.23**	0.06	**3.91**	m2	4.20
Brickwork general surfaces girth exceeding 300mm	0.69	10.00	**0.76**	0.36	–	**3.32**	0.06	**4.14**	m2	4.45
Plasterboard general surfaces girth exceeding 300mm	0.44	10.00	**0.48**	0.32	–	**2.95**	0.06	**3.50**	m2	3.76
Plaster general surfaces girth exceeding 300mm	0.50	10.00	**0.55**	0.31	–	**2.86**	0.05	**3.46**	m2	3.72
Two undercoats, one coat full gloss finish, existing gloss painted surfaces										
Concrete general surfaces girth exceeding 300mm	0.77	10.00	**0.85**	0.44	–	**4.06**	0.08	**4.99**	m2	5.36
Brickwork general surfaces girth exceeding 300mm	0.94	10.00	**1.03**	0.44	–	**4.06**	0.08	**5.18**	m2	5.56
Plasterboard general surfaces girth exceeding 300mm	0.60	10.00	**0.66**	0.41	–	**3.78**	0.07	**4.51**	m2	4.85
Plaster general surfaces girth exceeding 300mm	0.68	10.00	**0.75**	0.40	–	**3.69**	0.07	**4.51**	m2	4.85
Two undercoats, one coat eggshell finish, existing gloss painted surfaces										
Concrete general surfaces girth exceeding 300mm	0.82	10.00	**0.90**	0.44	–	**4.06**	0.08	**5.04**	m2	5.42
Brickwork general surfaces girth exceeding 300mm	1.00	10.00	**1.10**	0.44	–	**4.06**	0.08	**5.24**	m2	5.63
Plasterboard general surfaces girth exceeding 300mm	0.64	10.00	**0.70**	0.41	–	**3.78**	0.07	**4.56**	m2	4.90
Plaster general surfaces girth exceeding 300mm	0.73	10.00	**0.80**	0.40	–	**3.69**	0.07	**4.57**	m2	4.91
One coat primer, one undercoat, one coat full gloss finish, existing washable distempered or emulsion painted surfaces										
Concrete general surfaces girth exceeding 300mm	0.85	10.00	**0.94**	0.44	–	**4.06**	0.08	**5.08**	m2	5.46
Brickwork general surfaces girth exceeding 300mm	1.04	10.00	**1.14**	0.44	–	**4.06**	0.08	**5.29**	m2	5.68

SURFACE FINISHES

Labour hourly rates: (except Specialists) Craft Operatives £9.23 Labourer £7.02 Rates are national average prices. Refer to REGIONAL VARIATIONS for indicative levels of overall pricing in regions	MATERIALS			LABOUR				RATES		
	Del to Site £	Waste %	Material Cost £	Craft Optve Hrs	Lab Hrs	Labour Cost £	Sunds £	Nett Rate £	Unit	Gross Rate (+7.5%) £
M60: PAINTING/CLEAR FINISHING (REDECORATIONS) - OIL PAINTING WALLS AND CEILINGS Cont.										
One coat primer, one undercoat, one coat full gloss finish, existing washable distempered or emulsion painted surfaces Cont.										
Plasterboard general surfaces										
girth exceeding 300mm	0.66	10.00	0.73	0.41	-	3.78	0.07	4.58	m2	4.92
Plaster general surfaces										
girth exceeding 300mm	0.76	10.00	0.84	0.40	-	3.69	0.07	4.60	m2	4.94
One coat primer, one undercoat, one coat eggshell finish, existing washable distempered or emulsion painted surfaces										
Concrete general surfaces										
girth exceeding 300mm	0.90	10.00	0.99	0.44	-	4.06	0.08	5.13	m2	5.52
Brickwork general surfaces										
girth exceeding 300mm	1.10	10.00	1.21	0.44	-	4.06	0.08	5.35	m2	5.75
Plasterboard general surfaces										
girth exceeding 300mm	0.70	10.00	0.77	0.41	-	3.78	0.07	4.62	m2	4.97
Plaster general surfaces										
girth exceeding 300mm	0.80	10.00	0.88	0.40	-	3.69	0.07	4.64	m2	4.99
One coat primer, two undercoats, one coat full gloss finish, existing washable distempered or emulsion painted surfaces										
Concrete general surfaces										
girth exceeding 300mm	1.11	10.00	1.22	0.53	-	4.89	0.09	6.20	m2	6.67
Brickwork general surfaces										
girth exceeding 300mm	1.35	10.00	1.49	0.54	-	4.98	0.09	6.56	m2	7.05
Plasterboard general surfaces										
girth exceeding 300mm	0.86	10.00	0.95	0.50	-	4.62	0.09	5.65	m2	6.07
Plaster general surfaces										
girth exceeding 300mm	0.98	10.00	1.08	0.49	-	4.52	0.09	5.69	m2	6.12
One coat primer, two undercoats, one coat eggshell finish, existing washable distempered or emulsion painted surfaces										
Concrete general surfaces										
girth exceeding 300mm	1.16	10.00	1.28	0.53	-	4.89	0.09	6.26	m2	6.73
Brickwork general surfaces										
girth exceeding 300mm	1.41	10.00	1.55	0.54	-	4.98	0.09	6.63	m2	7.12
Plasterboard general surfaces										
girth exceeding 300mm	0.90	10.00	0.99	0.50	-	4.62	0.09	5.70	m2	6.12
Plaster general surfaces										
girth exceeding 300mm	1.03	10.00	1.13	0.49	-	4.52	0.09	5.75	m2	6.18
One coat primer, one undercoat, one coat full gloss finish, existing non-washable distempered surfaces										
Concrete general surfaces										
girth exceeding 300mm	0.85	10.00	0.94	0.44	-	4.06	0.08	5.08	m2	5.46
Brickwork general surfaces										
girth exceeding 300mm	1.04	10.00	1.14	0.44	-	4.06	0.08	5.29	m2	5.68
Plasterboard general surfaces										
girth exceeding 300mm	0.66	10.00	0.73	0.41	-	3.78	0.07	4.58	m2	4.92
Plaster general surfaces										
girth exceeding 300mm	0.76	10.00	0.84	0.40	-	3.69	0.07	4.60	m2	4.94
One coat primer, one undercoat, one coat eggshell finish, existing non-washable distempered surfaces										
Concrete general surfaces										
girth exceeding 300mm	0.90	10.00	0.99	0.44	-	4.06	0.08	5.13	m2	5.52
Brickwork general surfaces										
girth exceeding 300mm	1.10	10.00	1.21	0.44	-	4.06	0.08	5.35	m2	5.75
Plasterboard general surfaces										
girth exceeding 300mm	0.70	10.00	0.77	0.41	-	3.78	0.07	4.62	m2	4.97
Plaster general surfaces										
girth exceeding 300mm	0.80	10.00	0.88	0.40	-	3.69	0.07	4.64	m2	4.99
One coat primer, two undercoats, one coat full gloss finish, existing non-washable distempered surfaces										
Concrete general surfaces										
girth exceeding 300mm	1.11	10.00	1.22	0.53	-	4.89	0.09	6.20	m2	6.67

Labour hourly rates: (except Specialists) Craft Operatives £9.23 Labourer £7.02 Rates are national average prices. Refer to REGIONAL VARIATIONS for indicative levels of overall pricing in regions	MATERIALS			LABOUR				RATES		
	Del to Site £	Waste %	Material Cost £	Craft Optve Hrs	Lab Hrs	Labour Cost £	Sunds £	Nett Rate £	Unit	Gross Rate (+7.5%) £
M60: PAINTING/CLEAR FINISHING (REDECORATIONS) - OIL PAINTING WALLS AND CEILINGS Cont.										
One coat primer, two undercoats, one coat full gloss finish, existing non-washable distempered surfaces Cont.										
Brickwork general surfaces girth exceeding 300mm	1.35	10.00	**1.49**	0.54	-	**4.98**	0.09	**6.56**	m2	7.05
Plasterboard general surfaces girth exceeding 300mm	0.86	10.00	**0.95**	0.50	-	**4.62**	0.09	**5.65**	m2	6.07
Plaster general surfaces girth exceeding 300mm	0.98	10.00	**1.08**	0.49	-	**4.52**	0.09	**5.69**	m2	6.12
One coat primer, two undercoats, one coat eggshell finish, existing non-washable distempered surfaces										
Concrete general surfaces girth exceeding 300mm	1.16	10.00	**1.28**	0.53	-	**4.89**	0.09	**6.26**	m2	6.73
Brickwork general surfaces girth exceeding 300mm	1.41	10.00	**1.55**	0.54	-	**4.98**	0.09	**6.63**	m2	7.12
Plasterboard general surfaces girth exceeding 300mm	0.90	10.00	**0.99**	0.50	-	**4.62**	0.09	**5.70**	m2	6.12
Plaster general surfaces girth exceeding 300mm	1.03	10.00	**1.13**	0.49	-	**4.52**	0.09	**5.75**	m2	6.18
M60: PAINTING/CLEAR FINISHING (REDECORATIONS) - PLASTIC FINISH										
Generally										

Note
  the following rates include for the cost of all
  preparatory work, e.g. washing down, etc.
Textured plastic coating - stippled finish, existing washable distempered or emulsion painted surfaces										

Concrete general surfaces girth exceeding 300mm	0.52	10.00	**0.57**	0.36	-	**3.32**	0.06	**3.95**	m2	4.25
Brickwork general surfaces girth exceeding 300mm	0.62	10.00	**0.68**	0.41	-	**3.78**	0.07	**4.54**	m2	4.88
Plasterboard general surfaces girth exceeding 300mm	0.42	10.00	**0.46**	0.31	-	**2.86**	0.05	**3.37**	m2	3.63
Plaster general surfaces girth exceeding 300mm	0.42	10.00	**0.46**	0.31	-	**2.86**	0.05	**3.37**	m2	3.63
Textured plastic coating - combed finish, existing washable distempered or emulsion painted surfaces										
Concrete general surfaces girth exceeding 300mm	0.52	10.00	**0.57**	0.41	-	**3.78**	0.07	**4.43**	m2	4.76
Brickwork general surfaces girth exceeding 300mm	0.62	10.00	**0.68**	0.46	-	**4.25**	0.08	**5.01**	m2	5.38
Plasterboard general surfaces girth exceeding 300mm	0.42	10.00	**0.46**	0.36	-	**3.32**	0.06	**3.84**	m2	4.13
Plaster general surfaces girth exceeding 300mm	0.42	10.00	**0.46**	0.36	-	**3.32**	0.06	**3.84**	m2	4.13
Textured plastic coating - stippled finish, existing non-washable distempered surfaces										
Concrete general surfaces girth exceeding 300mm	0.52	10.00	**0.57**	0.41	-	**3.78**	0.07	**4.43**	m2	4.76
Brickwork general surfaces girth exceeding 300mm	0.62	10.00	**0.68**	0.46	-	**4.25**	0.08	**5.01**	m2	5.38
Plasterboard general surfaces girth exceeding 300mm	0.42	10.00	**0.46**	0.36	-	**3.32**	0.06	**3.84**	m2	4.13
Plaster general surfaces girth exceeding 300mm	0.42	10.00	**0.46**	0.36	-	**3.32**	0.06	**3.84**	m2	4.13
Textured plastic coating - combed finish, existing non-washable distempered surfaces										
Concrete general surfaces girth exceeding 300mm	0.52	10.00	**0.57**	0.46	-	**4.25**	0.08	**4.90**	m2	5.27
Brickwork general surfaces girth exceeding 300mm	0.62	10.00	**0.68**	0.51	-	**4.71**	0.09	**5.48**	m2	5.89
Plasterboard general surfaces girth exceeding 300mm	0.42	10.00	**0.46**	0.41	-	**3.78**	0.07	**4.32**	m2	4.64

SURFACE FINISHES

Labour hourly rates: (except Specialists) Craft Operatives £9.23 Labourer £7.02 Rates are national average prices. Refer to REGIONAL VARIATIONS for indicative levels of overall pricing in regions	MATERIALS			LABOUR				RATES		
	Del to Site £	Waste %	Material Cost £	Craft Optve Hrs	Lab Hrs	Labour Cost £	Sunds £	Nett Rate £	Unit	Gross Rate (+7.5%) £
M60: PAINTING/CLEAR FINISHING (REDECORATIONS) – **PLASTIC FINISH Cont.**										
Textured plastic coating - combed finish, existing **non-washable distempered surfaces Cont.**										
Plaster general surfaces										
girth exceeding 300mm	0.42	10.00	0.46	0.41	–	3.78	0.07	4.32	m2	4.64
M60: PAINTING/CLEAR FINISHING (REDECORATIONS) – **CLEAN OUT GUTTERS**										
Generally										

Note
    the following rates include for the cost of all
    preparatory work, e.g. washing down, etc.
Clean out gutters prior to repainting, staunch joints
with red lead, bituminous compound or mastic										

Iron or steel eaves gutters										
generally..	–	–	–	0.05	–	0.46	–	0.46	m	0.50
M60: PAINTING/CLEAR FINISHING (REDECORATIONS) – **OIL** **PAINTING METALWORK**										
Generally										

Note
    the following rates include for the cost of all
    preparatory work, e.g. washing down, etc.
One undercoat, one coat full gloss finish, existing
gloss painted metal surfaces										

Iron or steel general surfaces										
girth exceeding 300mm	0.45	10.00	0.50	0.42	–	3.88	0.07	4.44	m2	4.77
isolated surfaces, girth not exceeding 300mm	0.14	10.00	0.15	0.14	–	1.29	0.02	1.47	m	1.58
Galvanised glazed doors, windows or screens										
girth exceeding 300mm; panes, area not exceeding 0.10m2	0.12	10.00	0.13	0.79	–	7.29	0.14	7.56	m2	8.13
girth exceeding 300mm; panes, area 0.10 - 0.50m2 .	0.10	10.00	0.11	0.58	–	5.35	0.10	5.56	m2	5.98
girth exceeding 300mm; panes, area 0.50 - 1.00m2 .	0.09	10.00	0.10	0.47	–	4.34	0.08	4.52	m2	4.86
girth exceeding 300mm; panes, area exceeding 1.00m2	0.08	10.00	0.09	0.42	–	3.88	0.07	4.03	m2	4.34
Iron or steel structural work										
girth exceeding 300mm	0.45	10.00	0.50	0.46	–	4.25	0.08	4.82	m2	5.18
isolated surfaces, girth not exceeding 300mm	0.14	10.00	0.15	0.19	–	1.75	0.03	1.94	m	2.08
Iron or steel structural members of roof trusses, lattice girders, purlins and the like										
girth exceeding 300mm	0.45	10.00	0.50	0.59	–	5.45	0.10	6.04	m2	6.49
isolated surfaces, girth not exceeding 300mm	0.14	10.00	0.15	0.20	–	1.85	0.03	2.03	m	2.18
Iron or steel services										
girth exceeding 300mm	0.45	10.00	0.50	0.59	–	5.45	0.10	6.04	m2	6.49
isolated surfaces, girth not exceeding 300mm	0.14	10.00	0.15	0.20	–	1.85	0.03	2.03	m	2.18
isolated areas not exceeding 0.50m2 irrespective of girth	0.45	10.00	0.50	0.59	–	5.45	0.10	6.04	nr	6.49
Copper services										
girth exceeding 300mm	0.45	10.00	0.50	0.59	–	5.45	0.10	6.04	m2	6.49
isolated surfaces, girth not exceeding 300mm	0.14	10.00	0.15	0.20	–	1.85	0.03	2.03	m	2.18
Galvanised services										
girth exceeding 300mm	0.45	10.00	0.50	0.59	–	5.45	0.10	6.04	m2	6.49
isolated surfaces, girth not exceeding 300mm	0.14	10.00	0.15	0.20	–	1.85	0.03	2.03	m	2.18
One undercoat, one coat full gloss finish, existing **gloss painted metal surfaces; external work**										
Iron or steel general surfaces										
girth exceeding 300mm	0.45	10.00	0.50	0.45	–	4.15	0.08	4.73	m2	5.08
isolated surfaces, girth not exceeding 300mm	0.14	10.00	0.15	0.15	–	1.38	0.03	1.57	m	1.69
Galvanised glazed doors, windows or screens										
girth exceeding 300mm; panes, area not exceeding 0.10m2	0.12	10.00	0.13	0.81	–	7.48	0.14	7.75	m2	8.33
girth exceeding 300mm; panes, area 0.10 - 0.50m2 .	0.10	10.00	0.11	0.60	–	5.54	0.10	5.75	m2	6.18
girth exceeding 300mm; panes, area 0.50 - 1.00m2 .	0.09	10.00	0.10	0.50	–	4.62	0.09	4.80	m2	5.16
girth exceeding 300mm; panes, area exceeding 1.00m2	0.08	10.00	0.09	0.44	–	4.06	0.08	4.23	m2	4.55
Iron or steel structural work										
girth exceeding 300mm	0.45	10.00	0.50	0.46	–	4.25	0.08	4.82	m2	5.18
isolated surfaces, girth not exceeding 300mm	0.14	10.00	0.15	0.19	–	1.75	0.03	1.94	m	2.08
Iron or steel structural members of roof trusses, lattice girders, purlins and the like										
girth exceeding 300mm	0.45	10.00	0.50	0.59	–	5.45	0.10	6.04	m2	6.49
isolated surfaces, girth not exceeding 300mm	0.14	10.00	0.15	0.20	–	1.85	0.03	2.03	m	2.18
Iron or steel railings, fences and gates; plain open type										
girth exceeding 300mm	0.45	10.00	0.50	0.44	–	4.06	0.08	4.64	m2	4.98

Labour hourly rates: (except Specialists) Craft Operatives £9.23 Labourer £7.02 Rates are national average prices. Refer to REGIONAL VARIATIONS for indicative levels of overall pricing in regions	MATERIALS			LABOUR				RATES		
	Del to Site £	Waste %	Material Cost £	Craft Optve Hrs	Lab Hrs	Labour Cost £	Sunds £	Nett Rate £	Unit	Gross Rate (+7.5%) £
M60: PAINTING/CLEAR FINISHING (REDECORATIONS) - OIL PAINTING METALWORK Cont.										
One undercoat, one coat full gloss finish, existing gloss painted metal surfaces; external work Cont.										
Iron or steel railings, fences and gates; plain open type Cont.										
isolated surfaces, girth not exceeding 300mm	0.14	10.00	0.15	0.15	–	1.38	0.03	1.57	m	1.69
Iron or steel railings, fences and gates; close type										
girth exceeding 300mm	0.45	10.00	0.50	0.40	–	3.69	0.07	4.26	m2	4.58
Iron or steel railings, fences and gates; ornamental type										
girth exceeding 300mm	0.45	10.00	0.50	0.80	–	7.38	0.14	8.02	m2	8.62
Iron or steel eaves gutters										
girth exceeding 300mm	0.45	10.00	0.50	0.48	–	4.43	0.08	5.01	m2	5.38
isolated surfaces, girth not exceeding 300mm	0.14	10.00	0.15	0.16	–	1.48	0.03	1.66	m	1.79
Galvanised eaves gutters										
girth exceeding 300mm	0.45	10.00	0.50	0.48	–	4.43	0.08	5.01	m2	5.38
isolated surfaces, girth not exceeding 300mm	0.14	10.00	0.15	0.16	–	1.48	0.03	1.66	m	1.79
Iron or steel services										
girth exceeding 300mm	0.45	10.00	0.50	0.59	–	5.45	0.10	6.04	m2	6.49
isolated surfaces, girth not exceeding 300mm	0.14	10.00	0.15	0.20	–	1.85	0.03	2.03	m	2.18
isolated areas not exceeding 0.50m2 irrespective of girth	0.23	10.00	0.25	0.59	–	5.45	0.10	5.80	nr	6.23
Copper services										
girth exceeding 300mm	0.45	10.00	0.50	0.59	–	5.45	0.10	6.04	m2	6.49
isolated surfaces, girth not exceeding 300mm	0.14	10.00	0.15	0.20	–	1.85	0.03	2.03	m	2.18
Galvanised services										
girth exceeding 300mm	0.45	10.00	0.50	0.59	–	5.45	0.10	6.04	m2	6.49
isolated surfaces, girth not exceeding 300mm	0.14	10.00	0.15	0.20	–	1.85	0.03	2.03	m	2.18
Two coats full gloss finish, existing gloss painted metal surfaces										
Iron or steel general surfaces										
girth exceeding 300mm	0.45	10.00	0.50	0.43	–	3.97	0.08	4.54	m2	4.88
isolated surfaces, girth not exceeding 300mm	0.14	10.00	0.15	0.14	–	1.29	0.02	1.47	m	1.58
Galvanised glazed doors, windows or screens										
girth exceeding 300mm; panes, area not exceeding 0.10m2	0.12	10.00	0.13	0.79	–	7.29	0.14	7.56	m2	8.13
girth exceeding 300mm; panes, area 0.10 - 0.50m2 .	0.10	10.00	0.11	0.58	–	5.35	0.10	5.56	m2	5.98
girth exceeding 300mm; panes, area 0.50 - 1.00m2 .	0.09	10.00	0.10	0.48	–	4.43	0.08	4.61	m2	4.96
girth exceeding 300mm; panes, area exceeding 1.00m2	0.08	10.00	0.09	0.42	–	3.88	0.07	4.03	m2	4.34
Iron or steel structural work										
girth exceeding 300mm	0.45	10.00	0.50	0.46	–	4.25	0.08	4.82	m2	5.18
isolated surfaces, girth not exceeding 300mm	0.14	10.00	0.15	0.19	–	1.75	0.03	1.94	m	2.08
Iron or steel structural members of roof trusses, lattice girders, purlins and the like										
girth exceeding 300mm	0.45	10.00	0.50	0.59	–	5.45	0.10	6.04	m2	6.49
isolated surfaces, girth not exceeding 300mm	0.14	10.00	0.15	0.20	–	1.85	0.03	2.03	m	2.18
Iron or steel services										
girth exceeding 300mm	0.45	10.00	0.50	0.59	–	5.45	0.10	6.04	m2	6.49
isolated surfaces, girth not exceeding 300mm	0.14	10.00	0.15	0.20	–	1.85	0.03	2.03	m	2.18
isolated areas not exceeding 0.50m2 irrespective of girth	0.23	10.00	0.25	0.59	–	5.45	0.10	5.80	nr	6.23
Copper services										
girth exceeding 300mm	0.45	10.00	0.50	0.59	–	5.45	0.10	6.04	m2	6.49
isolated surfaces, girth not exceeding 300mm	0.14	10.00	0.15	0.20	–	1.85	0.03	2.03	m	2.18
Galvanised services										
girth exceeding 300mm	0.45	10.00	0.50	0.59	–	5.45	0.10	6.04	m2	6.49
isolated surfaces, girth not exceeding 300mm	0.14	10.00	0.15	0.20	–	1.85	0.03	2.03	m	2.18
Two coats full gloss finish, existing gloss painted metal surfaces; external work										
Iron or steel general surfaces										
girth exceeding 300mm	0.45	10.00	0.50	0.45	–	4.15	0.08	4.73	m2	5.08
isolated surfaces, girth not exceeding 300mm	0.14	10.00	0.15	0.15	–	1.38	0.03	1.57	m	1.69
Galvanised glazed doors, windows or screens										
girth exceeding 300mm; panes, area not exceeding 0.10m2	0.12	10.00	0.13	0.81	–	7.48	0.14	7.75	m2	8.33
girth exceeding 300mm; panes, area 0.10 - 0.50m2 .	0.10	10.00	0.11	0.60	–	5.54	0.10	5.75	m2	6.18
girth exceeding 300mm; panes, area 0.50 - 1.00m2 .	0.09	10.00	0.10	0.50	–	4.62	0.09	4.80	m2	5.16
girth exceeding 300mm; panes, area exceeding 1.00m2	0.09	10.00	0.10	0.44	–	4.06	0.08	4.24	m2	4.56
Iron or steel structural work										
girth exceeding 300mm	0.45	10.00	0.50	0.46	–	4.25	0.08	4.82	m2	5.18
isolated surfaces, girth not exceeding 300mm	0.14	10.00	0.15	0.19	–	1.75	0.03	1.94	m	2.08

SURFACE FINISHES

Labour hourly rates: (except Specialists) Craft Operatives £9.23 Labourer £7.02 Rates are national average prices. Refer to REGIONAL VARIATIONS for indicative levels of overall pricing in regions	MATERIALS			LABOUR				RATES		
	Del to Site £	Waste %	Material Cost £	Craft Optve Hrs	Lab Hrs	Labour Cost £	Sunds £	Nett Rate £	Unit	Gross Rate (+7.5%) £
M60: PAINTING/CLEAR FINISHING (REDECORATIONS) - OIL PAINTING METALWORK Cont.										
Two coats full gloss finish, existing gloss painted metal surfaces; external work Cont.										
Iron or steel structural members of roof trusses, lattice girders, purlins and the like										
girth exceeding 300mm	0.45	10.00	0.50	0.59	-	5.45	0.10	6.04	m2	6.49
isolated surfaces, girth not exceeding 300mm	0.14	10.00	0.15	0.20	-	1.85	0.03	2.03	m	2.18
Iron or steel railings, fences and gates; plain open type										
girth exceeding 300mm	0.45	10.00	0.50	0.44	-	4.06	0.08	4.64	m2	4.98
isolated surfaces, girth not exceeding 300mm	0.14	10.00	0.15	0.15	-	1.38	0.03	1.57	m	1.69
Iron or steel railings, fences and gates; close type										
girth exceeding 300mm	0.45	10.00	0.50	0.40	-	3.69	0.07	4.26	m2	4.58
Iron or steel railings, fences and gates; ornamental type										
girth exceeding 300mm	0.45	10.00	0.50	0.67	-	6.18	0.12	6.80	m2	7.31
Iron or steel eaves gutters										
girth exceeding 300mm	0.45	10.00	0.50	0.48	-	4.43	0.08	5.01	m2	5.38
isolated surfaces, girth not exceeding 300mm	0.14	10.00	0.15	0.16	-	1.48	0.03	1.66	m	1.79
Galvanised eaves gutters										
girth exceeding 300mm	0.45	10.00	0.50	0.48	-	4.43	0.08	5.01	m2	5.38
isolated surfaces, girth not exceeding 300mm	0.14	10.00	0.15	0.16	-	1.48	0.03	1.66	m	1.79
Iron or steel services										
girth exceeding 300mm	0.45	10.00	0.50	0.59	-	5.45	0.10	6.04	m2	6.49
isolated surfaces, girth not exceeding 300mm	0.14	10.00	0.15	0.20	-	1.85	0.03	2.03	m	2.18
isolated areas not exceeding 0.50m2 irrespective of girth	0.23	10.00	0.25	0.59	-	5.45	0.10	5.80	nr	6.23
Copper services										
girth exceeding 300mm	0.45	10.00	0.50	0.59	-	5.45	0.10	6.04	m2	6.49
isolated surfaces, girth not exceeding 300mm	0.14	10.00	0.15	0.20	-	1.85	0.03	2.03	m	2.18
Galvanised services										
girth exceeding 300mm	0.45	10.00	0.50	0.59	-	5.45	0.10	6.04	m2	6.49
isolated surfaces, girth not exceeding 300mm	0.14	10.00	0.15	0.20	-	1.85	0.03	2.03	m	2.18
Two undercoats, one coat full gloss finish, existing gloss painted metal surfaces										
Iron or steel general surfaces										
girth exceeding 300mm	0.68	10.00	0.75	0.56	-	5.17	0.10	6.02	m2	6.47
isolated surfaces, girth not exceeding 300mm	0.20	10.00	0.22	0.18	-	1.66	0.03	1.91	m	2.05
Galvanised glazed doors, windows or screens										
girth exceeding 300mm; panes, area not exceeding 0.10m2	0.19	10.00	0.21	1.03	-	9.51	0.18	9.90	m2	10.64
girth exceeding 300mm; panes, area 0.10 - 0.50m2 .	0.15	10.00	0.17	0.75	-	6.92	0.13	7.22	m2	7.76
girth exceeding 300mm; panes, area 0.50 - 1.00m2 .	0.14	10.00	0.15	0.63	-	5.81	0.11	6.08	m2	6.53
girth exceeding 300mm; panes, area exceeding 1.00m2	0.12	10.00	0.13	0.55	-	5.08	0.10	5.31	m2	5.71
Iron or steel structural work										
girth exceeding 300mm	0.68	10.00	0.75	0.60	-	5.54	0.10	6.39	m2	6.86
isolated surfaces, girth not exceeding 300mm	0.20	10.00	0.22	0.25	-	2.31	0.04	2.57	m	2.76
Iron or steel structural members of roof trusses, lattice girders, purlins and the like										
girth exceeding 300mm	0.68	10.00	0.75	0.77	-	7.11	0.13	7.99	m2	8.58
isolated surfaces, girth not exceeding 300mm	0.20	10.00	0.22	0.26	-	2.40	0.05	2.67	m	2.87
Iron or steel services										
girth exceeding 300mm	0.68	10.00	0.75	0.77	-	7.11	0.13	7.99	m2	8.58
isolated surfaces, girth not exceeding 300mm	0.20	10.00	0.22	0.26	-	2.40	0.05	2.67	m	2.87
isolated areas not exceeding 0.50m2 irrespective of girth	0.34	10.00	0.37	0.77	-	7.11	0.13	7.61	nr	8.18
Copper services										
girth exceeding 300mm	0.68	10.00	0.75	0.77	-	7.11	0.13	7.99	m2	8.58
isolated surfaces, girth not exceeding 300mm	0.20	10.00	0.22	0.26	-	2.40	0.05	2.67	m	2.87
Galvanised services										
girth exceeding 300mm	0.68	10.00	0.75	0.77	-	7.11	0.13	7.99	m2	8.58
isolated surfaces, girth not exceeding 300mm	0.20	10.00	0.22	0.26	-	2.40	0.05	2.67	m	2.87
Two undercoats, one coat full gloss finish, existing gloss painted metal surfaces; external work										
Iron or steel general surfaces										
girth exceeding 300mm	0.68	10.00	0.75	0.59	-	5.45	0.10	6.29	m2	6.77
isolated surfaces, girth not exceeding 300mm	0.20	10.00	0.22	0.20	-	1.85	0.03	2.10	m	2.25
Galvanised glazed doors, windows or screens										
girth exceeding 300mm; panes, area not exceeding 0.10m2	0.19	10.00	0.21	1.06	-	9.78	0.19	10.18	m2	10.95
girth exceeding 300mm; panes, area 0.10 - 0.50m2 .	0.15	10.00	0.17	0.78	-	7.20	0.14	7.50	m2	8.07
girth exceeding 300mm; panes, area 0.50 - 1.00m2 .	0.14	10.00	0.15	0.66	-	6.09	0.12	6.37	m2	6.84

Labour hourly rates: (except Specialists) Craft Operatives £9.23 Labourer £7.02 Rates are national average prices. Refer to REGIONAL VARIATIONS for indicative levels of overall pricing in regions	MATERIALS			LABOUR				RATES		
	Del to Site	Waste	Material Cost	Craft Optve	Lab	Labour Cost	Sunds	Nett Rate	Unit	Gross Rate (+7.5%)
	£	%	£	Hrs	Hrs	£	£	£		£
M60: PAINTING/CLEAR FINISHING (REDECORATIONS) - OIL PAINTING METALWORK Cont.										
Two undercoats, one coat full gloss finish, existing gloss painted metal surfaces; external work Cont.										
Galvanised glazed doors, windows or screens Cont. girth exceeding 300mm; panes, area exceeding 1.00m2	0.12	10.00	**0.13**	0.58	-	**5.35**	0.10	**5.59**	m2	**6.00**
Iron or steel structural work girth exceeding 300mm	0.68	10.00	**0.75**	0.60	-	**5.54**	0.10	**6.39**	m2	**6.86**
isolated surfaces, girth not exceeding 300mm	0.20	10.00	**0.22**	0.25	-	**2.31**	0.04	**2.57**	m	**2.76**
Iron or steel structural members of roof trusses, lattice girders, purlins and the like girth exceeding 300mm	0.68	10.00	**0.75**	0.77	-	**7.11**	0.13	**7.99**	m2	**8.58**
isolated surfaces, girth not exceeding 300mm	0.20	10.00	**0.22**	0.26	-	**2.40**	0.05	**2.67**	m	**2.87**
Iron or steel railings, fences and gates; plain open type girth exceeding 300mm	0.68	10.00	**0.75**	0.57	-	**5.26**	0.10	**6.11**	m2	**6.57**
isolated surfaces, girth not exceeding 300mm	0.20	10.00	**0.22**	0.19	-	**1.75**	0.03	**2.00**	m	**2.15**
Iron or steel railings, fences and gates; close type girth exceeding 300mm	0.68	10.00	**0.75**	0.51	-	**4.71**	0.09	**5.55**	m2	**5.96**
Iron or steel railings, fences and gates; ornamental type girth exceeding 300mm	0.68	10.00	**0.75**	0.97	-	**8.95**	0.17	**9.87**	m2	**10.61**
Iron or steel eaves gutters girth exceeding 300mm	0.68	10.00	**0.75**	0.62	-	**5.72**	0.11	**6.58**	m2	**7.07**
isolated surfaces, girth not exceeding 300mm	0.20	10.00	**0.22**	0.21	-	**1.94**	0.04	**2.20**	m	**2.36**
Galvanised eaves gutters girth exceeding 300mm	0.68	10.00	**0.75**	0.62	-	**5.72**	0.11	**6.58**	m2	**7.07**
isolated surfaces, girth not exceeding 300mm	0.20	10.00	**0.22**	0.21	-	**1.94**	0.04	**2.20**	m	**2.36**
Iron or steel services girth exceeding 300mm	0.68	10.00	**0.75**	0.77	-	**7.11**	0.13	**7.99**	m2	**8.58**
isolated surfaces, girth not exceeding 300mm	0.20	10.00	**0.22**	0.26	-	**2.40**	0.05	**2.67**	m	**2.87**
isolated areas not exceeding 0.50m2 irrespective of girth	0.34	10.00	**0.37**	0.77	-	**7.11**	0.13	**7.61**	nr	**8.18**
Copper services girth exceeding 300mm	0.68	10.00	**0.75**	0.77	-	**7.11**	0.13	**7.99**	m2	**8.58**
isolated surfaces, girth not exceeding 300mm	0.20	10.00	**0.22**	0.26	-	**2.40**	0.05	**2.67**	m	**2.87**
Galvanised services girth exceeding 300mm	0.68	10.00	**0.75**	0.77	-	**7.11**	0.13	**7.99**	m2	**8.58**
isolated surfaces, girth not exceeding 300mm	0.20	10.00	**0.22**	0.26	-	**2.40**	0.05	**2.67**	m	**2.87**
One undercoat, two coats full gloss finish, existing gloss painted metal surfaces										
Iron or steel general surfaces girth exceeding 300mm	0.68	10.00	**0.75**	0.56	-	**5.17**	0.10	**6.02**	m2	**6.47**
isolated surfaces, girth not exceeding 300mm	0.20	10.00	**0.22**	0.18	-	**1.66**	0.03	**1.91**	m	**2.05**
Galvanised glazed doors, windows or screens girth exceeding 300mm; panes, area not exceeding 0.10m2	0.19	10.00	**0.21**	1.03	-	**9.51**	0.18	**9.90**	m2	**10.64**
girth exceeding 300mm; panes, area 0.10 - 0.50m2 .	0.15	10.00	**0.17**	0.75	-	**6.92**	0.13	**7.22**	m2	**7.76**
girth exceeding 300mm; panes, area 0.50 - 1.00m2 .	0.14	10.00	**0.15**	0.63	-	**5.81**	0.11	**6.08**	m2	**6.53**
girth exceeding 300mm; panes, area exceeding 1.00m2	0.12	10.00	**0.13**	0.55	-	**5.08**	0.10	**5.31**	m2	**5.71**
Iron or steel structural work girth exceeding 300mm	0.68	10.00	**0.75**	0.60	-	**5.54**	0.10	**6.39**	m2	**6.86**
isolated surfaces, girth not exceeding 300mm	0.20	10.00	**0.22**	0.25	-	**2.31**	0.04	**2.57**	m	**2.76**
Iron or steel structural members of roof trusses, lattice girders, purlins and the like girth exceeding 300mm	0.68	10.00	**0.75**	0.77	-	**7.11**	0.13	**7.99**	m2	**8.58**
isolated surfaces, girth not exceeding 300mm	0.20	10.00	**0.22**	0.26	-	**2.40**	0.05	**2.67**	m	**2.87**
Iron or steel services girth exceeding 300mm	0.68	10.00	**0.75**	0.77	-	**7.11**	0.13	**7.99**	m2	**8.58**
isolated surfaces, girth not exceeding 300mm	0.20	10.00	**0.22**	0.26	-	**2.40**	0.05	**2.67**	m	**2.87**
isolated areas not exceeding 0.50m2 irrespective of girth	0.34	10.00	**0.37**	0.77	-	**7.11**	0.13	**7.61**	nr	**8.18**
Copper services girth exceeding 300mm	0.68	10.00	**0.75**	0.77	-	**7.11**	0.13	**7.99**	m2	**8.58**
isolated surfaces, girth not exceeding 300mm	0.20	10.00	**0.22**	0.26	-	**2.40**	0.05	**2.67**	m	**2.87**
Galvanised services girth exceeding 300mm	0.68	10.00	**0.75**	0.77	-	**7.11**	0.13	**7.99**	m2	**8.58**
isolated surfaces, girth not exceeding 300mm	0.20	10.00	**0.22**	0.26	-	**2.40**	Sunds 0.05	**2.67**	m	**2.87**
One undercoat, two coats full gloss finish, existing gloss painted metal surfaces; external work										
Iron or steel general surfaces girth exceeding 300mm	0.68	10.00	**0.75**	0.59	-	**5.45**	0.10	**6.29**	m2	**6.77**

Labour hourly rates: (except Specialists) Craft Operatives £9.23 Labourer £7.02 Rates are national average prices. Refer to REGIONAL VARIATIONS for indicative levels of overall pricing in regions	MATERIALS			LABOUR				RATES		
	Del to Site £	Waste %	Material Cost £	Craft Optve Hrs	Lab Hrs	Labour Cost £	Sunds £	Nett Rate £	Unit	Gross Rate (+7.5%) £
M60: PAINTING/CLEAR FINISHING (REDECORATIONS) - OIL PAINTING METALWORK Cont.										
One undercoat, two coats full gloss finish, existing gloss painted metal surfaces; external work Cont...										
Iron or steel general surfaces Cont.										
isolated surfaces, girth not exceeding 300mm	0.20	10.00	0.22	0.20	-	1.85	0.03	2.10	m	2.25
Galvanised glazed doors, windows or screens										
girth exceeding 300mm; panes, area not exceeding										
0.10m2......................	0.19	10.00	0.21	1.06	-	9.78	0.19	10.18	m2	10.95
girth exceeding 300mm; panes, area 0.10 - 0.50m2 .	0.15	10.00	0.17	0.78	-	7.20	0.14	7.50	m2	8.07
girth exceeding 300mm; panes, area 0.50 - 1.00m2 .	0.14	10.00	0.15	0.66	-	6.09	0.12	6.37	m2	6.84
girth exceeding 300mm; panes, area exceeding										
1.00m2......................	0.12	10.00	0.13	0.58	-	5.35	0.10	5.59	m2	6.00
Iron or steel structural work										
girth exceeding 300mm	0.68	10.00	0.75	0.60	-	5.54	0.10	6.39	m2	6.86
isolated surfaces, girth not exceeding 300mm	0.20	10.00	0.22	0.25	-	2.31	0.04	2.57	m	2.76
Iron or steel structural members of roof trusses, lattice girders, purlins and the like										
girth exceeding 300mm	0.68	10.00	0.75	0.77	-	7.11	0.13	7.99	m2	8.58
isolated surfaces, girth not exceeding 300mm	0.20	10.00	0.22	0.26	-	2.40	0.05	2.67	m	2.87
Iron or steel railings, fences and gates; plain open type										
girth exceeding 300mm	0.68	10.00	0.75	0.57	-	5.26	0.10	6.11	m2	6.57
isolated surfaces, girth not exceeding 300mm	0.20	10.00	0.22	0.19	-	1.75	0.03	2.00	m	2.15
Iron or steel railings, fences and gates; close type										
girth exceeding 300mm	0.68	10.00	0.75	0.51	-	4.71	0.09	5.55	m2	5.96
Iron or steel railings, fences and gates; ornamental type										
girth exceeding 300mm	0.68	10.00	0.75	0.97	-	8.95	0.17	9.87	m2	10.61
Iron or steel eaves gutters										
girth exceeding 300mm	0.68	10.00	0.75	0.62	-	5.72	0.11	6.58	m2	7.07
isolated surfaces, girth not exceeding 300mm	0.20	10.00	0.22	0.21	-	1.94	0.04	2.20	m	2.36
Galvanised eaves gutters										
girth exceeding 300mm	0.68	10.00	0.75	0.62	-	5.72	0.11	6.58	m2	7.07
isolated surfaces, girth not exceeding 300mm	0.20	10.00	0.22	0.21	-	1.94	0.04	2.20	m	2.36
Iron or steel services										
girth exceeding 300mm	0.68	10.00	0.75	0.77	-	7.11	0.13	7.99	m2	8.58
isolated surfaces, girth not exceeding 300mm	0.20	10.00	0.22	0.26	-	2.40	0.05	2.67	m	2.87
isolated areas not exceeding 0.50m2 irrespective of girth......................	0.34	10.00	0.37	0.77	-	7.11	0.13	7.61	nr	8.18
Copper services										
girth exceeding 300mm	0.68	10.00	0.75	0.77	-	7.11	0.13	7.99	m2	8.58
isolated surfaces, girth not exceeding 300mm	0.20	10.00	0.22	0.26	-	2.40	0.05	2.67	m	2.87
Galvanised services										
girth exceeding 300mm	0.68	10.00	0.75	0.77	-	7.11	0.13	7.99	m2	8.58
isolated surfaces, girth not exceeding 300mm	0.20	10.00	0.22	0.26	-	2.40	0.05	2.67	m	2.87
M60: PAINTING/CLEAR FINISHING (REDECORATIONS) - OIL PAINTING WOODWORK										
Generally										
Note the following rates include for the cost of all preparatory work, e.g. washing down, etc.										
One undercoat, one coat full gloss finish, existing gloss painted wood surfaces										
Wood general surfaces										
girth exceeding 300mm	0.45	10.00	0.50	0.41	-	3.78	0.07	4.35	m2	4.68
isolated surfaces, girth not exceeding 300mm	0.14	10.00	0.15	0.14	-	1.29	0.02	1.47	m	1.58
Wood glazed doors										
girth exceeding 300mm; panes, area not exceeding										
0.10m2	0.23	10.00	0.25	0.72	-	6.65	0.13	7.03	m2	7.56
girth exceeding 300mm; panes, area 0.10 - 0.50m2 .	0.16	10.00	0.18	0.52	-	4.80	0.09	5.07	m2	5.45
girth exceeding 300mm; panes, area 0.50 - 1.00m2 .	0.11	10.00	0.12	0.43	-	3.97	0.08	4.17	m2	4.48
girth exceeding 300mm; panes, area exceeding										
1.00m2	0.09	10.00	0.10	0.37	-	3.42	0.06	3.57	m2	3.84
Wood partially glazed doors										
girth exceeding 300mm; panes, area not exceeding										
0.10m2	0.34	10.00	0.37	0.61	-	5.63	0.11	6.11	m2	6.57
girth exceeding 300mm; panes, area 0.10 - 0.50m2 .	0.30	10.00	0.33	0.49	-	4.52	0.09	4.94	m2	5.31
girth 0pceeding 300mm; panes, area 0.50 - 1.00m2 .	0.27	10.00	0.30	0.43	-	3.97	0.08	4.35	m2	4.67
girth exceeding 300mm; panes, area exceeding										
1.00m2	0.25	10.00	0.28	0.39	-	3.60	0.07	3.94	m2	4.24
Wood windows and screens										
girth excerding 300mm; panes, area not exceeding										
0.10m2	0.25	10.00	0.28	0.77	-	7.11	0.13	7.51	m2	8.08
girth exceeding 300mm; panes, area 0.10 - 0.50m2 .	0.18	10.00	0.20	0.56	-	5.17	0.10	5.47	m2	5.88
girth exceeding 300mm; panes, area 0.50 - 1.00m2 .	0.14	10.00	0.15	0.46	-	4.25	0.08	4.48	m2	4.82

Labour hourly rates: (except Specialists) Craft Operatives £9.23 Labourer £7.02 Rates are national average prices. Refer to REGIONAL VARIATIONS for indicative levels of overall pricing in regions	MATERIALS			LABOUR				RATES		
	Del to Site £	Waste %	Material Cost £	Craft Optve Hrs	Lab Hrs	Labour Cost £	Sunds £	Nett Rate £	Unit	Gross Rate (+7.5%) £
M60: PAINTING/CLEAR FINISHING (REDECORATIONS) - OIL PAINTING WOODWORK Cont.										
One undercoat, one coat full gloss finish, existing gloss painted wood surfaces Cont.										
Wood windows and screens Cont.										
girth exceeding 300mm; panes, area exceeding 1.00m2	0.11	10.00	0.12	0.40	-	3.69	0.07	3.88	m2	4.17
One undercoat, one coat full gloss finish, existing gloss painted wood surfaces; external work										
Wood general surfaces										
girth exceeding 300mm	0.45	10.00	0.50	0.44	-	4.06	0.08	4.64	m2	4.98
isolated surfaces, girth not exceeding 300mm	0.14	10.00	0.15	0.15	-	1.38	0.03	1.57	m	1.69
Wood glazed doors										
girth exceeding 300mm; panes, area not exceeding 0.10m2	0.23	10.00	0.25	0.73	-	6.74	0.13	7.12	m2	7.65
girth exceeding 300mm; panes, area 0.10 - 0.50m2	0.16	10.00	0.18	0.53	-	4.89	0.09	5.16	m2	5.54
girth exceeding 300mm; panes, area 0.50 - 1.00m2	0.11	10.00	0.12	0.44	-	4.06	0.08	4.26	m2	4.58
girth exceeding 300mm; panes, area exceeding 1.00m2	0.09	10.00	0.10	0.38	-	3.51	0.07	3.68	m2	3.95
Wood partially glazed doors										
girth exceeding 300mm; panes, area not exceeding 0.10m2	0.34	10.00	0.37	0.62	-	5.72	0.11	6.21	m2	6.67
girth exceeding 300mm; panes, area 0.10 - 0.50m2	0.30	10.00	0.33	0.50	-	4.62	0.09	5.04	m2	5.41
girth exceeding 300mm; panes, area 0.50 - 1.00m2	0.27	10.00	0.30	0.44	-	4.06	0.08	4.44	m2	4.77
girth exceeding 300mm; panes, area exceeding 1.00m2	0.25	10.00	0.28	0.40	-	3.69	0.07	4.04	m2	4.34
Wood windows and screens										
girth exceeding 300mm; panes, area not exceeding 0.10m2	0.25	10.00	0.28	0.80	-	7.38	0.14	7.80	m2	8.38
girth exceeding 300mm; panes, area 0.10 - 0.50m2	0.18	10.00	0.20	0.59	-	5.45	0.10	5.74	m2	6.17
girth exceeding 300mm; panes, area 0.50 - 1.00m2	0.14	10.00	0.15	0.49	-	4.52	0.09	4.77	m2	5.12
girth exceeding 300mm; panes, area exceeding 1.00m2	0.11	10.00	0.12	0.43	-	3.97	0.08	4.17	m2	4.48
Wood railings fences and gates; open type										
girth exceeding 300mm	0.45	10.00	0.50	0.37	-	3.42	0.06	3.97	m2	4.27
isolated surfaces, girth not exceeding 300mm	0.14	10.00	0.15	0.12	-	1.11	0.02	1.28	m	1.38
Wood railings fences and gates; close type										
girth exceeding 300mm	0.45	10.00	0.50	0.34	-	3.14	0.06	3.69	m2	3.97
Two coats full gloss finish, existing gloss painted wood surfaces										
Wood general surfaces										
girth exceeding 300mm	0.45	10.00	0.50	0.41	-	3.78	0.07	4.35	m2	4.68
isolated surfaces, girth not exceeding 300mm	0.14	10.00	0.15	0.14	-	1.29	0.02	1.47	m	1.58
Wood glazed doors										
girth exceeding 300mm; panes, area not exceeding 0.10m2	0.23	10.00	0.25	0.72	-	6.65	0.13	7.03	m2	7.56
girth exceeding 300mm; panes, area 0.10 - 0.50m2	0.16	10.00	0.18	0.52	-	4.80	0.09	5.07	m2	5.45
girth exceeding 300mm; panes, area 0.50 - 1.00m2	0.11	10.00	0.12	0.43	-	3.97	0.08	4.17	m2	4.48
girth exceeding 300mm; panes, area exceeding 1.00m2	0.09	10.00	0.10	0.37	-	3.42	0.06	3.57	m2	3.84
Wood partially glazed doors										
girth exceeding 300mm; panes, area not exceeding 0.10m2	0.34	10.00	0.37	0.61	-	5.63	0.11	6.11	m2	6.57
girth exceeding 300mm; panes, area 0.10 - 0.50m2	0.30	10.00	0.33	0.49	-	4.52	0.09	4.94	m2	5.31
girth exceeding 300mm; panes, area 0.50 - 1.00m2	0.27	10.00	0.30	0.43	-	3.97	0.08	4.35	m2	4.67
girth exceeding 300mm; panes, area exceeding 1.00m2	0.25	10.00	0.28	0.39	-	3.60	0.07	3.94	m2	4.24
Wood windows and screens										
girth exceeding 300mm; panes, area not exceeding 0.10m2	0.25	10.00	0.28	0.77	-	7.11	0.13	7.51	m2	8.08
girth exceeding 300mm; panes, area 0.10 - 0.50m2	0.18	10.00	0.20	0.56	-	5.17	0.10	5.47	m2	5.88
girth exceeding 300mm; panes, area 0.50 - 1.00m2	0.14	10.00	0.15	0.46	-	4.25	0.08	4.48	m2	4.82
girth exceeding 300mm; panes, area exceeding 1.00m2	0.11	10.00	0.12	0.40	-	3.69	0.07	3.88	m2	4.17
Two coats full gloss finish, existing gloss painted wood surfaces; external work										
Wood general surfaces										
girth exceeding 300mm	0.45	10.00	0.50	0.44	-	4.06	0.08	4.64	m2	4.98
isolated surfaces, girth not exceeding 300mm	0.14	10.00	0.15	0.15	-	1.38	0.03	1.57	m	1.69
Wood glazed doors										
girth exceeding 300mm; panes, area not exceeding 0.10m2	0.23	10.00	0.25	0.73	-	6.74	0.13	7.12	m2	7.65
girth exceeding 300mm; panes, area 0.10 - 0.50m2	0.16	10.00	0.18	0.53	-	4.89	0.09	5.16	m2	5.54
girth exceeding 300mm; panes, area 0.50 - 1.00m2	0.11	10.00	0.12	0.44	-	4.06	0.08	4.26	m2	4.58
girth exceeding 300mm; panes, area exceeding 1.00m2	0.09	10.00	0.10	0.38	-	3.51	0.07	3.68	m2	3.95

Labour hourly rates: (except Specialists) Craft Operatives £9.23 Labourer £7.02 Rates are national average prices. Refer to REGIONAL VARIATIONS for indicative levels of overall pricing in regions	MATERIALS			LABOUR				RATES		
	Del to Site	Waste	Material Cost	Craft Optve	Lab	Labour Cost	Sunds	Nett Rate	Unit	Gross Rate (+7.5%)
	£	%	£	Hrs	Hrs	£	£	£		£
M60: PAINTING/CLEAR FINISHING (REDECORATIONS) - OIL PAINTING WOODWORK Cont.										
Two coats full gloss finish, existing gloss painted wood surfaces; external work Cont.										
Wood partially glazed doors										
girth exceeding 300mm; panes, area not exceeding 0.10m2	0.34	10.00	0.37	0.62	-	5.72	0.11	6.21	m2	6.67
girth exceeding 300mm; panes, area 0.10 - 0.50m2 .	0.30	10.00	0.33	0.50	-	4.62	0.09	5.04	m2	5.41
girth exceeding 300mm; panes, area 0.50 - 1.00m2 .	0.27	10.00	0.30	0.44	-	4.06	0.08	4.44	m2	4.77
girth exceeding 300mm; panes, area exceeding 1.00m2	0.25	10.00	0.28	0.40	-	3.69	0.07	4.04	m2	4.34
Wood windows and screens										
girth exceeding 300mm; panes, area not exceeding 0.10m2	0.25	10.00	0.28	0.80	-	7.38	0.14	7.80	m2	8.38
girth exceeding 300mm; panes, area 0.10 - 0.50m2 .	0.18	10.00	0.20	0.59	-	5.45	0.10	5.74	m2	6.17
girth exceeding 300mm; panes, area 0.50 - 1.00m2 .	0.14	10.00	0.15	0.49	-	4.52	0.09	4.77	m2	5.12
girth exceeding 300mm; panes, area exceeding 1.00m2	0.11	10.00	0.12	0.43	-	3.97	0.08	4.17	m2	4.48
Wood railings fences and gates; open type										
girth exceeding 300mm	0.45	10.00	0.50	0.37	-	3.42	0.06	3.97	m2	4.27
isolated surfaces, girth not exceeding 300mm	0.14	10.00	0.15	0.12	-	1.11	0.02	1.28	m	1.38
Wood railings fences and gates; close type										
girth exceeding 300mm	0.45	10.00	0.50	0.34	-	3.14	0.06	3.69	m2	3.97
Two undercoats, one coat full gloss finish, existing gloss painted wood surfaces										
Wood general surfaces										
girth exceeding 300mm	0.68	10.00	0.75	0.55	-	5.08	0.10	5.92	m2	6.37
isolated surfaces, girth not exceeding 300mm	0.20	10.00	0.22	0.18	-	1.66	0.03	1.91	m	2.05
Wood glazed doors										
girth exceeding 300mm; panes, area not exceeding 0.10m2	0.34	10.00	0.37	0.94	-	8.68	0.16	9.21	m2	9.90
girth exceeding 300mm; panes, area 0.10 - 0.50m2 .	0.24	10.00	0.26	0.69	-	6.37	0.12	6.75	m2	7.26
girth exceeding 300mm; panes, area 0.50 - 1.00m2 .	0.17	10.00	0.19	0.57	-	5.26	0.10	5.55	m2	5.96
girth exceeding 300mm; panes, area exceeding 1.00m2	0.14	10.00	0.15	0.49	-	4.52	0.09	4.77	m2	5.12
Wood partially glazed doors										
girth exceeding 300mm; panes, area not exceeding 0.10m2	0.51	10.00	0.56	0.79	-	7.29	0.14	7.99	m2	8.59
girth exceeding 300mm; panes, area 0.10 - 0.50m2 .	0.44	10.00	0.48	0.65	-	6.00	0.11	6.59	m2	7.09
girth exceeding 300mm; panes, area 0.50 - 1.00m2 .	0.41	10.00	0.45	0.57	-	5.26	0.10	5.81	m2	6.25
girth exceeding 300mm; panes, area exceeding 1.00m2	0.37	10.00	0.41	0.52	-	4.80	0.09	5.30	m2	5.69
Wood windows and screens										
girth exceeding 300mm; panes, area not exceeding 0.10m2	0.37	10.00	0.41	1.02	-	9.41	0.18	10.00	m2	10.75
girth exceeding 300mm; panes, area 0.10 - 0.50m2 .	0.27	10.00	0.30	0.75	-	6.92	0.13	7.35	m2	7.90
girth exceeding 300mm; panes, area 0.50 - 1.00m2 .	0.20	10.00	0.22	0.62	-	5.72	0.11	6.05	m2	6.51
girth exceeding 300mm; panes, area exceeding 1.00m2	0.17	10.00	0.19	0.53	-	4.89	0.09	5.17	m2	5.56
Two undercoats, one coat full gloss finish, existing gloss painted wood surfaces; external work										
Wood general surfaces										
girth exceeding 300mm	0.68	10.00	0.75	0.59	-	5.45	0.10	6.29	m2	6.77
isolated surfaces, girth not exceeding 300mm	0.20	10.00	0.22	0.20	-	1.85	0.03	2.10	m	2.25
Wood glazed doors										
girth exceeding 300mm; panes, area not exceeding 0.10m2	0.34	10.00	0.37	0.95	-	8.77	0.17	9.31	m2	10.01
girth exceeding 300mm; panes, area 0.10 - 0.50m2 .	0.24	10.00	0.26	0.70	-	6.46	0.12	6.84	m2	7.36
girth exceeding 300mm; panes, area 0.50 - 1.00m2 .	0.17	10.00	0.19	0.58	-	5.35	0.10	5.64	m2	6.06
girth exceeding 300mm; panes, area exceeding 1.00m2	0.14	10.00	0.15	0.50	-	4.62	0.09	4.86	m2	5.22
Wood partially glazed doors										
girth exceeding 300mm; panes, area not exceeding 0.10m2	0.51	10.00	0.56	0.80	-	7.38	0.14	8.09	m2	8.69
girth exceeding 300mm; panes, area 0.10 - 0.50m2 .	0.44	10.00	0.48	0.66	-	6.09	0.12	6.70	m2	7.20
girth exceeding 300mm; panes, area 0.50 - 1.00m2 .	0.41	10.00	0.45	0.58	-	5.35	0.10	5.90	m2	6.35
girth exceeding 300mm; panes, area exceeding 1.00m2	0.37	10.00	0.41	0.53	-	4.89	0.09	5.39	m2	5.79
Wood windows and screens										
girth exceeding 300mm; panes, area not exceeding 0.10m2	0.37	10.00	0.41	1.05	-	9.69	0.18	10.28	m2	11.05
girth exceeding 300mm; panes, area 0.10 - 0.50m2 .	0.27	10.00	0.30	0.79	-	7.29	0.14	7.73	m2	8.31
girth exceeding 300mm; panes, area 0.50 - 1.00m2 .	0.20	10.00	0.22	0.68	-	6.28	0.12	6.62	m2	7.11
girth exceeding 300mm; panes, area exceeding 1.00m2	0.17	10.00	0.19	0.57	Lab	5.26	0.10	5.55	m2	5.96
Wood railings fences and gates; open type										
girth exceeding 300mm	0.68	10.00	0.75	0.50	-	4.62	0.09	5.45	m2	5.86
isolated surfaces, girth not exceeding 300mm	0.20	10.00	0.22	0.17	-	1.57	0.03	1.82	m	1.96

Labour hourly rates: (except Specialists) Craft Operatives £9.23 Labourer £7.02 Rates are national average prices. Refer to REGIONAL VARIATIONS for indicative levels of overall pricing in regions	MATERIALS			LABOUR				RATES		
	Del to Site £	Waste %	Material Cost £	Craft Optve Hrs	Lab Hrs	Labour Cost £	Sunds £	Nett Rate £	Unit	Gross Rate (+7.5%) £
M60: PAINTING/CLEAR FINISHING (REDECORATIONS) - OIL PAINTING WOODWORK Cont.										
Two undercoats, one coat full gloss finish, existing gloss painted wood surfaces; external work Cont.										
Wood railings fences and gates; close type										
girth exceeding 300mm	0.68	10.00	0.75	0.45	-	4.15	0.08	4.98	m2	5.36
One undercoat, two coats full gloss finish, existing gloss painted wood surfaces										
Wood general surfaces										
girth exceeding 300mm	0.68	10.00	0.75	0.57	-	5.26	0.10	6.11	m2	6.57
isolated surfaces, girth not exceeding 300mm	0.20	10.00	0.22	0.18	-	1.66	0.03	1.91	m	2.05
Wood glazed doors										
girth exceeding 300mm; panes, area not exceeding 0.10m2	0.34	10.00	0.37	0.94	-	8.68	0.16	9.21	m2	9.90
girth exceeding 300mm; panes, area 0.10 - 0.50m2 .	0.24	10.00	0.26	0.69	-	6.37	0.12	6.75	m2	7.26
girth exceeding 300mm; panes, area 0.50 - 1.00m2 .	0.17	10.00	0.19	0.57	-	5.26	0.10	5.55	m2	5.96
girth exceeding 300mm; panes, area exceeding 1.00m2	0.14	10.00	0.15	0.49	-	4.52	0.09	4.77	m2	5.12
Wood partially glazed doors										
girth exceeding 300mm; panes, area not exceeding 0.10m2	0.51	10.00	0.56	0.79	-	7.29	0.14	7.99	m2	8.59
girth exceeding 300mm; panes, area 0.10 - 0.50m2 .	0.44	10.00	0.48	0.65	-	6.00	0.11	6.59	m2	7.09
girth exceeding 300mm; panes, area 0.50 - 1.00m2 .	0.41	10.00	0.45	0.57	-	5.26	0.10	5.81	m2	6.25
girth exceeding 300mm; panes, area exceeding 1.00m2	0.37	10.00	0.41	0.52	-	4.80	0.09	5.30	m2	5.69
Wood windows and screens										
girth exceeding 300mm; panes, area not exceeding 0.10m2	0.37	10.00	0.41	1.02	-	9.41	0.18	10.00	m2	10.75
girth exceeding 300mm; panes, area 0.10 - 0.50m2 .	0.27	10.00	0.30	0.75	-	6.92	0.13	7.35	m2	7.90
girth exceeding 300mm; panes, area 0.50 - 1.00m2 .	0.20	10.00	0.22	0.62	-	5.72	0.11	6.05	m2	6.51
girth exceeding 300mm; panes, area exceeding 1.00m2	0.17	10.00	0.19	0.53	-	4.89	0.09	5.17	m2	5.56
One undercoat, two coats full gloss finish, existing gloss painted wood surfaces; external work										
Wood general surfaces										
girth exceeding 300mm	0.68	10.00	0.75	0.59	-	5.45	0.10	6.29	m2	6.77
isolated surfaces, girth not exceeding 300mm	0.20	10.00	0.22	0.20	-	1.85	0.03	2.10	m	2.25
Wood glazed doors										
girth exceeding 300mm; panes, area not exceeding 0.10m2	0.34	10.00	0.37	0.95	-	8.77	0.17	9.31	m2	10.01
girth exceeding 300mm; panes, area 0.10 - 0.50m2 .	0.24	10.00	0.26	0.70	-	6.46	0.12	6.84	m2	7.36
girth exceeding 300mm; panes, area 0.50 - 1.00m2 .	0.17	10.00	0.19	0.58	-	5.35	0.10	5.64	m2	6.06
girth exceeding 300mm; panes, area exceeding 1.00m2	0.14	10.00	0.15	0.50	-	4.62	0.09	4.86	m2	5.22
Wood partially glazed doors										
girth exceeding 300mm; panes, area not exceeding 0.10m2	0.51	10.00	0.56	0.80	-	7.38	0.14	8.09	m2	8.69
girth exceeding 300mm; panes, area 0.10 - 0.50m2 .	0.44	10.00	0.48	0.66	-	6.09	0.12	6.70	m2	7.20
girth exceeding 300mm; panes, area 0.50 - 1.00m2 .	0.41	10.00	0.45	0.58	-	5.35	0.10	5.90	m2	6.35
girth exceeding 300mm; panes, area exceeding 1.00m2	0.37	10.00	0.41	0.53	-	4.89	0.09	5.39	m2	5.79
Wood windows and screens										
girth exceeding 300mm; panes, area not exceeding 0.10m2	0.37	10.00	0.41	1.06	-	9.78	0.18	10.37	m2	11.15
girth exceeding 300mm; panes, area 0.10 - 0.50m2 .	0.27	10.00	0.30	0.79	-	7.29	0.14	7.73	m2	8.31
girth exceeding 300mm; panes, area 0.50 - 1.00m2 .	0.20	10.00	0.22	0.66	-	6.09	0.12	6.43	m2	6.91
girth exceeding 300mm; panes, area exceeding 1.00m2	0.17	10.00	0.19	0.57	-	5.26	0.10	5.55	m2	5.96
Wood railings fences and gates; open type										
girth exceeding 300mm	0.68	10.00	0.75	0.50	-	4.62	0.09	5.45	m2	5.86
isolated surfaces, girth not exceeding 300mm	0.20	10.00	0.22	0.17	-	1.57	0.03	1.82	m	1.96
Wood railings fences and gates; close type										
girth exceeding 300mm	0.68	10.00	0.75	0.45	-	4.15	0.08	4.98	m2	5.36
Burn off, one coat primer, one undercoat, one coat full gloss finish, existing painted wood surfaces										
Wood general surfaces										
girth exceeding 300mm	0.74	10.00	0.81	0.82	-	7.57	0.36	8.74	m2	9.40
isolated surfaces, girth not exceeding 300mm	0.22	10.00	0.24	0.24	-	2.22	0.10	2.56	m	2.75
Wood glazed doors										
girth exceeding 300mm; panes, area not exceeding 0.10m2	0.37	10.00	0.41	1.36	-	12.55	0.59	13.55	m2	14.57
girth exceeding 300mm; panes, area 0.10 - 0.50m2 .	0.26	10.00	0.29	0.98	-	9.05	0.43	9.76	m2	10.49
girth exceeding 300mm; panes, area 0.50 - 1.00m2 .	0.19	10.00	0.21	0.82	-	7.57	0.36	8.14	m2	8.75
girth exceeding 300mm; panes, area exceeding 1.00m2	0.15	10.00	0.17	0.73	-	6.74	0.32	7.22	m2	7.76
Wood partially glazed doors										
girth exceeding 300mm; panes, area not exceeding 0.10m2	0.56	10.00	0.62	1.09	-	10.06	0.48	11.16	m2	11.99

SURFACE FINISHES

Labour hourly rates: (except Specialists) Craft Operatives £9.23 Labourer £7.02 Rates are national average prices. Refer to REGIONAL VARIATIONS for indicative levels of overall pricing in regions	MATERIALS			LABOUR				RATES		
	Del to Site £	Waste %	Material Cost £	Craft Optve Hrs	Lab Hrs	Labour Cost £	Sunds £	Nett Rate £	Unit	Gross Rate (+7.5%) £
M60: PAINTING/CLEAR FINISHING (REDECORATIONS) - OIL PAINTING WOODWORK Cont.										
Burn off, one coat primer, one undercoat, one coat full gloss finish, existing painted wood surfaces Cont.	..									
Wood partially glazed doors Cont.										
girth exceeding 300mm; panes, area 0.10 - 0.50m2 .	0.48	10.00	0.53	0.90	-	8.31	0.39	9.22	m2	9.92
girth exceeding 300mm; panes, area 0.50 - 1.00m2 .	0.44	10.00	0.48	0.82	-	7.57	0.36	8.41	m2	9.04
girth exceeding 300mm; panes, area exceeding 1.00m2	0.41	10.00	0.45	0.78	-	7.20	0.34	7.99	m2	8.59
Wood windows and screens										
girth exceeding 300mm; panes, area not exceeding 0.10m2	0.41	10.00	0.45	1.50	-	13.85	0.66	14.96	m2	16.08
girth exceeding 300mm; panes, area 0.10 - 0.50m2 .	0.30	10.00	0.33	1.07	-	9.88	0.47	10.68	m2	11.48
girth exceeding 300mm; panes, area 0.50 - 1.00m2 .	0.22	10.00	0.24	0.90	-	8.31	0.39	8.94	m2	9.61
girth exceeding 300mm; panes, area exceeding 1.00m2	0.19	10.00	0.21	0.80	-	7.38	0.35	7.94	m2	8.54
Burn off, one coat primer, one undercoat, one coat full gloss finish, existing painted wood surfaces; external work										
Wood general surfaces										
girth exceeding 300mm	0.74	10.00	0.81	0.86	-	7.94	0.38	9.13	m2	9.82
isolated surfaces, girth not exceeding 300mm	0.22	10.00	0.24	0.26	-	2.40	0.11	2.75	m	2.96
Wood glazed doors										
girth exceeding 300mm; panes, area not exceeding 0.10m2	0.37	10.00	0.41	1.36	-	12.55	0.59	13.55	m2	14.57
girth exceeding 300mm; panes, area 0.10 - 0.50m2 .	0.26	10.00	0.29	0.98	-	9.05	0.43	9.76	m2	10.49
girth exceeding 300mm; panes, area 0.50 - 1.00m2 .	0.19	10.00	0.21	0.82	-	7.57	0.36	8.14	m2	8.75
girth exceeding 300mm; panes, area exceeding 1.00m2	0.15	10.00	0.17	0.73	-	6.74	0.32	7.22	m2	7.76
Wood partially glazed doors										
girth exceeding 300mm; panes, area not exceeding 0.10m2	0.56	10.00	0.62	1.09	-	10.06	0.48	11.16	m2	11.99
girth exceeding 300mm; panes, area 0.10 - 0.50m2 .	0.48	10.00	0.53	0.90	-	8.31	0.39	9.22	m2	9.92
girth exceeding 300mm; panes, area 0.50 - 1.00m2 .	0.44	10.00	0.48	0.82	-	7.57	0.36	8.41	m2	9.04
girth exceeding 300mm; panes, area exceeding 1.00m2	0.41	10.00	0.45	0.78	-	7.20	0.34	7.99	m2	8.59
Wood windows and screens										
girth exceeding 300mm; panes, area not exceeding 0.10m2	0.41	10.00	0.45	1.54	-	14.21	0.67	15.34	m2	16.49
girth exceeding 300mm; panes, area 0.10 - 0.50m2 .	0.30	10.00	0.33	1.11	-	10.25	0.49	11.07	m2	11.90
girth exceeding 300mm; panes, area 0.50 - 1.00m2 .	0.22	10.00	0.24	0.94	-	8.68	0.41	9.33	m2	10.03
girth exceeding 300mm; panes, area exceeding 1.00m2	0.19	10.00	0.21	0.84	-	7.75	0.37	8.33	m2	8.96
Wood railings fences and gates; open type										
girth exceeding 300mm	0.74	10.00	0.81	0.74	-	6.83	0.32	7.96	m2	8.56
isolated surfaces, girth not exceeding 300mm	0.22	10.00	0.24	0.22	-	2.03	0.10	2.37	m	2.55
Wood railings fences and gates; close type										
girth exceeding 300mm	0.74	10.00	0.81	0.69	-	6.37	0.30	7.48	m2	8.04
Burn off, one coat primer, two undercoats, one coat full gloss finish, existing painted wood surfaces										
Wood general surfaces										
girth exceeding 300mm	0.97	10.00	1.07	0.96	-	8.86	0.42	10.35	m2	11.12
isolated surfaces, girth not exceeding 300mm	0.29	10.00	0.32	0.29	-	2.68	0.13	3.13	m	3.36
Wood glazed doors										
girth exceeding 300mm; panes, area not exceeding 0.10m2	0.48	10.00	0.53	1.60	-	14.77	0.70	16.00	m2	17.20
girth exceeding 300mm; panes, area 0.10 - 0.50m2 .	0.34	10.00	0.37	1.15	-	10.61	0.50	11.49	m2	12.35
girth exceeding 300mm; panes, area 0.50 - 1.00m2 .	0.24	10.00	0.26	0.96	-	8.86	0.42	9.54	m2	10.26
girth exceeding 300mm; panes, area exceeding 1.00m2	0.19	10.00	0.21	0.86	-	7.94	0.38	8.53	m2	9.17
Wood partially glazed doors										
girth exceeding 300mm; panes, area not exceeding 0.10m2	0.73	10.00	0.80	1.28	-	11.81	0.56	13.18	m2	14.17
girth exceeding 300mm; panes, area 0.10 - 0.50m2 .	0.63	10.00	0.69	1.06	-	9.78	0.46	10.94	m2	11.76
girth exceeding 300mm; panes, area 0.50 - 1.00m2 .	0.58	10.00	0.64	0.96	-	8.86	0.42	9.92	m2	10.66
girth exceeding 300mm; panes, area exceeding 1.00m2	0.53	10.00	0.58	0.91	-	8.40	0.40	9.38	m2	10.09
Wood windows and screens										
girth exceeding 300mm; panes, area not exceeding 0.10m2	0.53	10.00	0.58	1.76	-	16.24	0.77	17.60	m2	18.92
girth exceeding 300mm; panes, area 0.10 - 0.50m2 .	0.39	10.00	0.43	1.27	-	11.72	0.55	12.70	m2	13.65
girth exceeding 300mm; panes, area 0.50 - 1.00m2 .	0.29	10.00	0.32	1.06	-	9.78	0.46	10.56	m2	11.36
girth exceeding 300mm; panes, area exceeding 1.00m2	0.24	10.00	0.26	0.94	-	8.68	Sunds	9.35	m2	10.05

Labour hourly rates: (except Specialists) Craft Operatives £9.23 Labourer £7.02 Rates are national average prices. Refer to REGIONAL VARIATIONS for indicative levels of overall pricing in regions	MATERIALS			LABOUR			RATES		
	Del to Site £	Waste %	Material Cost £	Craft Optve Hrs	Lab Hrs	Labour Cost £	Sunds £	Nett Rate £	Unit	Gross Rate (+7.5%) £

M60: PAINTING/CLEAR FINISHING (REDECORATIONS) - OIL PAINTING WOODWORK Cont.										
Burn off, one coat primer, two undercoats, one coat full gloss finish, existing painted wood surfaces; external work										
Wood general surfaces										
girth exceeding 300mm	0.97	10.00	1.07	1.01	-	9.32	0.44	10.83	m2	11.64
isolated surfaces, girth not exceeding 300mm	0.29	10.00	0.32	0.32	-	2.95	0.14	3.41	m	3.67
Wood glazed doors										
girth exceeding 300mm; panes, area not exceeding										
0.10m2	0.48	10.00	0.53	1.60	-	14.77	0.70	16.00	m2	17.20
girth exceeding 300mm; panes, area 0.10 - 0.50m2 .	0.34	10.00	0.37	1.15	-	10.61	0.50	11.49	m2	12.35
girth exceeding 300mm; panes, area 0.50 - 1.00m2 .	0.24	10.00	0.26	0.96	-	8.86	0.42	9.54	m2	10.26
girth exceeding 300mm; panes, area exceeding										
1.00m2	0.19	10.00	0.21	0.86	-	7.94	0.38	8.53	m2	9.17
Wood partially glazed doors										
girth exceeding 300mm; panes, area not exceeding										
0.10m2	0.73	10.00	0.80	1.28	-	11.81	0.56	13.18	m2	14.17
girth exceeding 300mm; panes, area 0.10 - 0.50m2 .	0.63	10.00	0.69	1.06	-	9.78	0.46	10.94	m2	11.76
girth exceeding 300mm; panes, area 0.50 - 1.00m2 .	0.58	10.00	0.64	0.96	-	8.86	0.42	9.92	m2	10.66
girth exceeding 300mm; panes, area exceeding										
1.00m2	0.53	10.00	0.58	0.91	-	8.40	0.40	9.38	m2	10.09
Wood windows and screens										
girth exceeding 300mm; panes, area not exceeding										
0.10m2	0.53	10.00	0.58	1.81	-	16.71	0.79	18.08	m2	19.44
girth exceeding 300mm; panes, area 0.10 - 0.50m2 .	0.39	10.00	0.43	1.32	-	12.18	0.58	13.19	m2	14.18
girth exceeding 300mm; panes, area 0.50 - 1.00m2 .	0.29	10.00	0.32	1.11	-	10.25	0.49	11.05	m2	11.88
girth exceeding 300mm; panes, area exceeding										
1.00m2	0.24	10.00	0.26	0.99	-	9.14	0.43	9.83	m2	10.57
Wood railings fences and gates; open type										
girth exceeding 300mm	0.97	10.00	1.07	0.86	-	7.94	0.38	9.38	m2	10.09
isolated surfaces, girth not exceeding 300mm	0.29	10.00	0.32	0.26	-	2.40	0.11	2.83	m	3.04
Wood railings fences and gates; close type										
girth exceeding 300mm	0.97	10.00	1.07	0.80	-	7.38	0.35	8.80	m2	9.46
Burn off, one coat primer, one undercoat, two coats full gloss finish, existing painted wood surfaces										
Wood general surfaces										
girth exceeding 300mm	0.97	10.00	1.07	0.96	-	8.86	0.42	10.35	m2	11.12
isolated surfaces, girth not exceeding 300mm	0.29	10.00	0.32	0.29	-	2.68	0.13	3.13	m	3.36
Wood glazed doors										
girth exceeding 300mm; panes, area not exceeding										
0.10m2	0.48	10.00	0.53	1.60	-	14.77	0.70	16.00	m2	17.20
girth exceeding 300mm; panes, area 0.10 - 0.50m2 .	0.34	10.00	0.37	1.15	-	10.61	0.50	11.49	m2	12.35
girth exceeding 300mm; panes, area 0.50 - 1.00m2 .	0.24	10.00	0.26	0.96	-	8.86	0.42	9.54	m2	10.26
girth exceeding 300mm; panes, area exceeding										
1.00m2	0.19	10.00	0.21	0.86	-	7.94	0.38	8.53	m2	9.17
Wood partially glazed doors										
girth exceeding 300mm; panes, area not exceeding										
0.10m2	0.73	10.00	0.80	1.28	-	11.81	0.56	13.18	m2	14.17
girth exceeding 300mm; panes, area 0.10 - 0.50m2 .	0.63	10.00	0.69	1.06	-	9.78	0.46	10.94	m2	11.76
girth exceeding 300mm; panes, area 0.50 - 1.00m2 .	0.58	10.00	0.64	0.96	-	8.86	0.42	9.92	m2	10.66
girth exceeding 300mm; panes, area exceeding										
1.00m2	0.53	10.00	0.58	0.91	-	8.40	0.40	9.38	m2	10.09
Wood windows and screens										
girth exceeding 300mm; panes, area not exceeding										
0.10m2	0.53	10.00	0.58	1.76	-	16.24	0.77	17.60	m2	18.92
girth exceeding 300mm; panes, area 0.10 - 0.50m2 .	0.39	10.00	0.43	1.27	-	11.72	0.55	12.70	m2	13.65
girth exceeding 300mm; panes, area 0.50 - 1.00m2 .	0.29	10.00	0.32	1.06	-	9.78	0.46	10.56	m2	11.36
girth exceeding 300mm; panes, area exceeding										
1.00m2	0.24	10.00	0.26	0.94	-	8.68	0.41	9.35	m2	10.05
Burn off, one coat primer, one undercoat, two coats full gloss finish, existing painted wood surfaces; external work										
Wood general surfaces										
girth exceeding 300mm	0.97	10.00	1.07	1.01	-	9.32	0.44	10.83	m2	11.64
isolated surfaces, girth not exceeding 300mm	0.29	10.00	0.32	0.32	-	2.95	0.14	3.41	m	3.67
Wood glazed doors										
girth exceeding 300mm; panes, area not exceeding										
0.10m2	0.48	10.00	0.53	1.60	-	14.77	0.70	16.00	m2	17.20
girth exceeding 300mm; panes, area 0.10 - 0.50m2 .	0.34	10.00	0.37	1.15	-	10.61	0.50	11.49	m2	12.35
girth exceeding 300mm; panes, area 0.50 - 1.00m2 .	0.24	10.00	0.26	0.96	-	8.86	0.42	9.54	m2	10.26
girth exceeding 300mm; panes, area exceeding										
1.00m2	0.19	10.00	0.21	0.86	-	7.94	0.38	8.53	m2	9.17
Wood partially glazed doors										
girth exceeding 300mm; panes, area not exceeding										
0.10m2	0.73	10.00	0.80	1.28	-	11.81	0.56	13.18	m2	14.17
girth exceeding 300mm; panes, area 0.10 - 0.50m2 .	0.63	10.00	0.69	1.06	-	9.78	0.46	10.94	m2	11.76
girth exceeding 300mm; panes, area 0.50 - 1.00m2 .	0.58	10.00	0.64	0.96	-	8.86	0.42	9.92	m2	10.66
girth exceeding 300mm; panes, area exceeding										
1.00m2	0.53	10.00	0.58	0.91	-	8.40	0.40	9.38	m2	10.09

Labour hourly rates: (except Specialists) Craft Operatives £9.23 Labourer £7.02 Rates are national average prices. Refer to REGIONAL VARIATIONS for indicative levels of overall pricing in regions	MATERIALS			LABOUR				RATES		
	Del to Site £	Waste %	Material Cost £	Craft Optve Hrs	Lab Hrs	Labour Cost £	Sunds £	Nett Rate £	Unit	Gross Rate (+7.5%) £
M60: PAINTING/CLEAR FINISHING (REDECORATIONS) - OIL PAINTING WOODWORK Cont.										
Burn off, one coat primer, one undercoat, two coats full gloss finish, existing painted wood surfaces; external work Cont.										
Wood windows and screens										
girth exceeding 300mm; panes, area not exceeding 0.10m2	0.53	10.00	0.58	1.81	-	16.71	0.79	18.08	m2	19.44
girth exceeding 300mm; panes, area 0.10 - 0.50m2 .	0.39	10.00	0.43	1.32	-	12.18	0.58	13.19	m2	14.18
girth exceeding 300mm; panes, area 0.50 - 1.00m2 .	0.29	10.00	0.32	1.11	-	10.25	0.49	11.05	m2	11.88
girth exceeding 300mm; panes, area exceeding 1.00m2	0.24	10.00	0.26	0.99	-	9.14	0.43	9.83	m2	10.57
Wood railings fences and gates; open type										
girth exceeding 300mm	0.97	10.00	1.07	0.86	-	7.94	0.38	9.38	m2	10.09
isolated surfaces, girth not exceeding 300mm	0.29	10.00	0.32	0.26	-	2.40	0.11	2.83	m	3.04
Wood railings fences and gates; close type										
girth exceeding 300mm	0.97	10.00	1.07	0.80	-	7.38	0.35	8.80	m2	9.46
M60: PAINTING/CLEAR FINISHING (REDECORATIONS) - POLYURETHANE LACQUER										
Generally										
Note the following rates include for the cost of all preparatory work, e.g. washing down, etc.										
Two coats polyurethane lacquer, existing lacquered surfaces										
Wood general surfaces										
girth exceeding 300mm	0.86	10.00	0.95	0.49	-	4.52	0.09	5.56	m2	5.98
isolated surfaces, girth not exceeding 300mm	0.26	10.00	0.29	0.16	-	1.48	0.03	1.79	m	1.93
Wood glazed doors										
girth exceeding 300mm; panes, area not exceeding 0.10m2	0.43	10.00	0.47	0.82	-	7.57	0.14	8.18	m2	8.80
girth exceeding 300mm; panes, area 0.10 - 0.50m2 .	0.30	10.00	0.33	0.60	-	5.54	0.10	5.97	m2	6.42
girth exceeding 300mm; panes, area 0.50 - 1.00m2 .	0.21	10.00	0.23	0.50	-	4.62	0.09	4.94	m2	5.31
girth exceeding 300mm; panes, area exceeding 1.00m2	0.17	10.00	0.19	0.43	-	3.97	0.08	4.24	m2	4.55
Wood partially glazed doors										
girth exceeding 300mm; panes, area not exceeding 0.10m2	0.64	10.00	0.70	0.69	-	6.37	0.12	7.19	m2	7.73
girth exceeding 300mm; panes, area 0.10 - 0.50m2 .	0.56	10.00	0.62	0.56	-	5.17	0.10	5.88	m2	6.33
girth exceeding 300mm; panes, area 0.50 - 1.00m2 .	0.52	10.00	0.57	0.50	-	4.62	0.09	5.28	m2	5.67
girth exceeding 300mm; panes, area exceeding 1.00m2	0.47	10.00	0.52	0.45	-	4.15	0.08	4.75	m2	5.11
Wood windows and screens										
girth exceeding 300mm; panes, area not exceeding 0.10m2	0.47	10.00	0.52	0.89	-	8.21	0.16	8.89	m2	9.56
girth exceeding 300mm; panes, area 0.10 - 0.50m2 .	0.34	10.00	0.37	0.65	-	6.00	0.11	6.48	m2	6.97
girth exceeding 300mm; panes, area 0.50 - 1.00m2 .	0.26	10.00	0.29	0.54	-	4.98	0.09	5.36	m2	5.76
girth exceeding 300mm; panes, area exceeding 1.00m2	0.21	10.00	0.23	0.47	-	4.34	0.08	4.65	m2	5.00
Two coats polyurethane lacquer, existing lacquered surfaces; external work										
Wood general surfaces										
girth exceeding 300mm	0.86	10.00	0.95	0.52	-	4.80	0.09	5.84	m2	6.27
isolated surfaces, girth not exceeding 300mm	0.26	10.00	0.29	0.18	-	1.66	0.03	1.98	m	2.13
Wood glazed doors										
girth exceeding 300mm; panes, area not exceeding 0.10m2	0.43	10.00	0.47	0.85	-	7.85	0.15	8.47	m2	9.10
girth exceeding 300mm; panes, area 0.10 - 0.50m2 .	0.30	10.00	0.33	0.62	-	5.72	0.11	6.16	m2	6.62
girth exceeding 300mm; panes, area 0.50 - 1.00m2 .	0.21	10.00	0.23	0.51	-	4.71	0.09	5.03	m2	5.41
girth exceeding 300mm; panes, area exceeding 1.00m2	0.17	10.00	0.19	0.44	-	4.06	0.08	4.33	m2	4.65
Wood partially glazed doors										
girth exceeding 300mm; panes, area not exceeding 0.10m2	0.64	10.00	0.70	0.72	-	6.65	0.13	7.48	m2	8.04
girth exceeding 300mm; panes, area 0.10 - 0.50m2 .	0.56	10.00	0.62	0.58	-	5.35	0.10	6.07	m2	6.52
girth exceeding 300mm; panes, area 0.50 - 1.00m2 .	0.52	10.00	0.57	0.51	-	4.71	0.09	5.37	m2	5.77
girth exceeding 300mm; panes, area exceeding 1.00m2	0.47	10.00	0.52	0.46	-	4.25	0.08	4.84	m2	5.21
Wood windows and screens										
girth exceeding 300mm; panes, area not exceeding 0.10m2	0.47	10.00	0.52	0.94	-	8.68	0.16	9.35	m2	10.05
girth exceeding 300mm; panes, area 0.10 - 0.50m2 .	0.34	10.00	0.37	0.69	-	6.37	0.12	6.86	m2	7.38
girth exceeding 300mm; panes, area 0.50 - 1.00m2 .	0.26	10.00	0.29	0.57	-	5.26	0.10	5.65	m2	6.07
girth exceeding 300mm; panes, area exceeding 1.00m2	0.21	10.00	0.23	0.50	-	4.62	0.09	4.94	m2	5.31

Labour hourly rates: (except Specialists) Craft Operatives £9.23 Labourer £7.02 Rates are national average prices. Refer to REGIONAL VARIATIONS for indicative levels of overall pricing in regions	MATERIALS			LABOUR				RATES		
	Del to Site £	Waste %	Material Cost £	Craft Optve Hrs	Lab Hrs	Labour Cost £	Sunds £	Nett Rate £	Unit	Gross Rate (+7.5%) £
M60: PAINTING/CLEAR FINISHING (REDECORATIONS) – POLYURETHANE LACQUER Cont.										
Three coats polyurethane lacquer, existing lacquered surfaces										
Wood general surfaces										
girth exceeding 300mm	1.29	10.00	**1.42**	0.56	–	**5.17**	0.10	**6.69**	m2	**7.19**
isolated surfaces, girth not exceeding 300mm	0.39	10.00	**0.43**	0.19	–	**1.75**	0.03	**2.21**	m	**2.38**
Wood glazed doors										
girth exceeding 300mm; panes, area not exceeding 0.10m2	0.64	10.00	**0.70**	0.93	–	**8.58**	0.16	**9.45**	m2	**10.16**
girth exceeding 300mm; panes, area 0.10 - 0.50m2 .	0.45	10.00	**0.50**	0.68	–	**6.28**	0.12	**6.89**	m2	**7.41**
girth exceeding 300mm; panes, area 0.50 - 1.00m2 .	0.32	10.00	**0.35**	0.57	–	**5.26**	0.10	**5.71**	m2	**6.14**
girth exceeding 300mm; panes, area exceeding 1.00m2	0.26	10.00	**0.29**	0.49	–	**4.52**	0.09	**4.90**	m2	**5.27**
Wood partially glazed doors										
girth exceeding 300mm; panes, area not exceeding 0.10m2	0.97	10.00	**1.07**	0.78	–	**7.20**	0.14	**8.41**	m2	**9.04**
girth exceeding 300mm; panes, area 0.10 - 0.50m2 .	0.84	10.00	**0.92**	0.69	–	**6.37**	0.12	**7.41**	m2	**7.97**
girth exceeding 300mm; panes, area 0.50 - 1.00m2 .	0.77	10.00	**0.85**	0.57	–	**5.26**	0.10	**6.21**	m2	**6.67**
girth exceeding 300mm; panes, area exceeding 1.00m2	0.71	10.00	**0.78**	0.51	–	**4.71**	0.09	**5.58**	m2	**6.00**
Wood windows and screens										
girth exceeding 300mm; panes, area not exceeding 0.10m2	0.71	10.00	**0.78**	1.01	–	**9.32**	0.18	**10.28**	m2	**11.05**
girth exceeding 300mm; panes, area 0.10 - 0.50m2 .	0.52	10.00	**0.57**	0.74	–	**6.83**	0.13	**7.53**	m2	**8.10**
girth exceeding 300mm; panes, area 0.50 - 1.00m2 .	0.39	10.00	**0.43**	0.62	–	**5.72**	0.11	**6.26**	m2	**6.73**
girth exceeding 300mm; panes, area exceeding 1.00m2	0.32	10.00	**0.35**	0.53	–	**4.89**	0.09	**5.33**	m2	**5.73**
Three coats polyurethane lacquer, existing lacquered surfaces; external work										
Wood general surfaces										
girth exceeding 300mm	1.29	10.00	**1.42**	0.60	–	**5.54**	0.10	**7.06**	m2	**7.59**
isolated surfaces, girth not exceeding 300mm	0.39	10.00	**0.43**	0.20	–	**1.85**	0.03	**2.31**	m	**2.48**
Wood glazed doors										
girth exceeding 300mm; panes, area not exceeding 0.10m2	0.64	10.00	**0.70**	0.96	–	**8.86**	0.17	**9.73**	m2	**10.46**
girth exceeding 300mm; panes, area 0.10 - 0.50m2 .	0.45	10.00	**0.50**	0.70	–	**6.46**	0.12	**7.08**	m2	**7.61**
girth exceeding 300mm; panes, area 0.50 - 1.00m2 .	0.32	10.00	**0.35**	0.58	–	**5.35**	0.10	**5.81**	m2	**6.24**
girth exceeding 300mm; panes, area exceeding 1.00m2	0.26	10.00	**0.29**	0.50	–	**4.62**	0.09	**4.99**	m2	**5.37**
Wood partially glazed doors										
girth exceeding 300mm; panes, area not exceeding 0.10m2	0.97	10.00	**1.07**	0.81	–	**7.48**	0.14	**8.68**	m2	**9.33**
girth exceeding 300mm; panes, area 0.10 - 0.50m2 .	0.84	10.00	**0.92**	0.71	–	**6.55**	0.12	**7.60**	m2	**8.17**
girth exceeding 300mm; panes, area 0.50 - 1.00m2 .	0.77	10.00	**0.85**	0.58	–	**5.35**	0.10	**6.30**	m2	**6.77**
girth exceeding 300mm; panes, area exceeding 1.00m2	0.71	10.00	**0.78**	0.53	–	**4.89**	0.09	**5.76**	m2	**6.20**
Wood windows and screens										
girth exceeding 300mm; panes, area not exceeding 0.10m2	0.71	10.00	**0.78**	1.07	–	**9.88**	0.19	**10.85**	m2	**11.66**
girth exceeding 300mm; panes, area 0.10 - 0.50m2 .	0.52	10.00	**0.57**	0.79	–	**7.29**	0.14	**8.00**	m2	**8.60**
girth exceeding 300mm; panes, area 0.50 - 1.00m2 .	0.39	10.00	**0.43**	0.66	–	**6.09**	0.12	**6.64**	m2	**7.14**
girth exceeding 300mm; panes, area exceeding 1.00m2	0.32	10.00	**0.35**	0.57	–	**5.26**	0.10	**5.71**	m2	**6.14**
M60: PAINTING/CLEAR FINISHING (REDECORATIONS) – FRENCH AND WAX POLISHING										
Generally										
Note the following rates include for the cost of all preparatory work, e.g. washing down, etc.										
Two coats wax polish, existing polished surfaces										
Wood general surfaces										
girth exceeding 300mm	0.48	10.00	**0.53**	0.40	–	**3.69**	0.07	**4.29**	m2	**4.61**
isolated surfaces, girth not exceeding 300mm	0.16	10.00	**0.18**	0.13	–	**1.20**	0.02	**1.40**	m	**1.50**
Strip old polish, oil, two coats wax polish, existing polished surfaces										
Wood general surfaces										
girth exceeding 300mm	1.39	10.00	**1.53**	1.12	–	**10.34**	0.20	**12.07**	m2	**12.97**
isolated surfaces, girth not exceeding 300mm	0.43	10.00	**0.47**	0.34	–	**3.14**	0.06	**3.67**	m	**3.95**
Strip old polish, oil, stain, two coats wax polish, existing polished surfaces										
Wood general surfaces										
girth exceeding 300mm	2.30	10.00	**2.53**	1.23	–	**11.35**	0.22	**14.10**	m2	**15.16**
isolated surfaces, girth not exceeding 300mm	0.71	10.00	**0.78**	0.38	–	**3.51**	0.07	**4.36**	m	**4.69**

SURFACE FINISHES

Labour hourly rates: (except Specialists) Craft Operatives £9.23 Labourer £7.02 Rates are national average prices. Refer to REGIONAL VARIATIONS for indicative levels of overall pricing in regions	MATERIALS			LABOUR				RATES		
	Del to Site £	Waste %	Material Cost £	Craft Optve Hrs	Lab Hrs	Labour Cost £	Sunds £	Nett Rate £	Unit	Gross Rate (+7.5%) £
M60: PAINTING/CLEAR FINISHING (REDECORATIONS) – FRENCH AND WAX POLISHING Cont.										
Open grain French polish existing polished surfaces										
Wood general surfaces										
girth exceeding 300mm	1.99	10.00	2.19	2.40	–	22.15	0.42	24.76	m2	26.62
isolated surfaces, girth not exceeding 300mm	0.58	10.00	0.64	0.80	–	7.38	0.14	8.16	m	8.77
Fully French polish existing polished surfaces										
Wood general surfaces										
girth exceeding 300mm	1.99	10.00	2.19	1.40	–	12.92	0.24	15.35	m2	16.50
isolated surfaces, girth not exceeding 300mm	0.58	10.00	0.64	0.47	–	4.34	0.08	5.06	m	5.44
Strip old polish, oil, open grain French polish, existing polished surfaces										
Wood general surfaces										
girth exceeding 300mm	2.73	10.00	3.00	3.12	–	28.80	0.55	32.35	m2	34.78
isolated surfaces, girth not exceeding 300mm	0.81	10.00	0.89	1.04	–	9.60	0.18	10.67	m	11.47
Strip old polish, oil, stain, open grain French polish, existing polished surfaces										
Wood general surfaces										
girth exceeding 300mm	3.65	10.00	4.01	3.23	–	29.81	0.56	34.39	m2	36.97
isolated surfaces, girth not exceeding 300mm	1.09	10.00	1.20	1.08	–	9.97	0.19	11.36	m	12.21
Strip old polish, oil, body in, fully French polish existing polished surfaces										
Wood general surfaces										
girth exceeding 300mm	2.62	10.00	2.88	4.21	–	38.86	0.74	42.48	m2	45.67
isolated surfaces, girth not exceeding 300mm	0.81	10.00	0.89	1.40	–	12.92	0.24	14.05	m	15.11
Strip old polish, oil, stain, body in, fully French polish, existing polished surfaces										
Wood general surfaces										
girth exceeding 300mm	3.65	10.00	4.01	4.32	–	39.87	0.76	44.65	m2	48.00
isolated surfaces, girth not exceeding 300mm	1.09	10.00	1.20	1.44	–	13.29	0.25	14.74	m	15.85
M60: PAINTING/CLEAR FINISHING (REDECORATIONS) – WATER REPELLENT										
One coat silicone based water repellent, existing surfaces										
Cement rendered general surfaces										
girth exceeding 300mm	0.79	10.00	0.87	0.11	–	1.02	0.02	1.90	m2	2.05
Stone general surfaces										
girth exceeding 300mm	1.84	10.00	2.02	0.12	–	1.11	0.02	3.15	m2	3.39
Brickwork general surfaces										
girth exceeding 300mm	1.84	10.00	2.02	0.12	–	1.11	0.02	3.15	m2	3.39
M60: PAINTING/CLEAR FINISHING (REDECORATIONS) – REMOVAL OF MOULD GROWTH										
Apply fungicide to existing decorated and infected surfaces										
Cement rendered general surfaces										
girth exceeding 300mm	0.10	10.00	0.11	0.20	–	1.85	0.03	1.99	m2	2.13
Concrete general surfaces										
girth exceeding 300mm	0.11	10.00	0.12	0.20	–	1.85	0.03	2.00	m2	2.15
Plaster general surfaces										
girth exceeding 300mm	0.10	10.00	0.11	0.20	–	1.85	0.03	1.99	m2	2.13
Brickwork general surfaces										
girth exceeding 300mm	0.14	10.00	0.15	0.22	–	2.03	0.04	2.22	m2	2.39
Apply fungicide to existing decorated and infected surfaces; external work										
Cement rendered general surfaces										
girth exceeding 300mm	0.10	10.00	0.11	0.22	–	2.03	0.04	2.18	m2	2.34
Concrete general surfaces										
girth exceeding 300mm	0.11	10.00	0.12	0.22	–	2.03	0.04	2.19	m2	2.36
Plaster general surfaces										
girth exceeding 300mm	0.10	10.00	0.11	0.22	–	2.03	0.04	2.18	m2	2.34
Brickwork general surfaces										
girth exceeding 300mm	0.14	10.00	0.15	0.24	–	2.22	0.04	2.41	m2	2.59

Labour hourly rates: (except Specialists) Craft Operatives £9.23 Labourer £7.02 Rates are national average prices. Refer to REGIONAL VARIATIONS for indicative levels of overall pricing in regions	MATERIALS			LABOUR				RATES		
	Del to Site £	Waste %	Material Cost £	Craft Optve Hrs	Lab Hrs	Labour Cost £	Sunds £	Nett Rate £	Unit	Gross Rate (+7.5%) £
M60: PAINTING/CLEAR FINISHING; STAIRCASE AREAS										
Mist coat, one full coat emulsion paint										
Concrete general surfaces girth exceeding 300mm	0.31	10.00	0.34	0.21	–	1.94	0.04	2.32	m2	2.49
Plaster general surfaces girth exceeding 300mm	0.27	10.00	0.30	0.19	–	1.75	0.03	2.08	m2	2.24
Plasterboard general surfaces girth exceeding 300mm	0.27	10.00	0.30	0.19	–	1.75	0.03	2.08	m2	2.24
Brickwork general surfaces girth exceeding 300mm	0.40	10.00	0.44	0.23	–	2.12	0.04	2.60	m2	2.80
Paper covered general surfaces girth exceeding 300mm	0.29	10.00	0.32	0.20	–	1.85	0.03	2.19	m2	2.36
Mist coat, two full coats emulsion paint										
Concrete general surfaces girth exceeding 300mm	0.53	10.00	0.58	0.30	–	2.77	0.05	3.40	m2	3.66
Plaster general surfaces girth exceeding 300mm	0.44	10.00	0.48	0.28	–	2.58	0.05	3.12	m2	3.35
Plasterboard general surfaces girth exceeding 300mm	0.44	10.00	0.48	0.28	–	2.58	0.05	3.12	m2	3.35
Brickwork general surfaces girth exceeding 300mm	0.66	10.00	0.73	0.33	–	3.05	0.06	3.83	m2	4.12
Paper covered general surfaces girth exceeding 300mm	0.49	10.00	0.54	0.29	–	2.68	0.05	3.27	m2	3.51
One coat textured masonry paint										
Cement rendered general surfaces girth exceeding 300mm	0.58	10.00	0.64	0.14	–	1.29	0.02	1.95	m2	2.10
Concrete general surfaces girth exceeding 300mm	0.65	10.00	0.71	0.14	–	1.29	0.02	2.03	m2	2.18
Brickwork general surfaces girth exceeding 300mm	0.72	10.00	0.79	0.16	–	1.48	0.03	2.30	m2	2.47
Rough cast general surfaces girth exceeding 300mm	1.16	10.00	1.28	0.20	–	1.85	0.03	3.15	m2	3.39
Two coats textured masonry paint										
Cement rendered general surfaces girth exceeding 300mm	1.04	10.00	1.14	0.28	–	2.58	0.05	3.78	m2	4.06
Concrete general surfaces girth exceeding 300mm	1.16	10.00	1.28	0.28	–	2.58	0.05	3.91	m2	4.20
Brickwork general surfaces girth exceeding 300mm	1.30	10.00	1.43	0.30	–	2.77	0.05	4.25	m2	4.57
Rough cast general surfaces girth exceeding 300mm	2.08	10.00	2.29	0.36	–	3.32	0.06	5.67	m2	6.10
One coat stabilising solution, one coat textured masonry paint										
Cement rendered general surfaces girth exceeding 300mm	1.09	10.00	1.20	0.24	–	2.22	0.04	3.45	m2	3.71
Concrete general surfaces girth exceeding 300mm	1.21	10.00	1.33	0.24	–	2.22	0.04	3.59	m2	3.86
Brickwork general surfaces girth exceeding 300mm	1.34	10.00	1.47	0.27	–	2.49	0.05	4.02	m2	4.32
Rough cast general surfaces girth exceeding 300mm	1.81	10.00	1.99	0.33	–	3.05	0.06	5.10	m2	5.48
One coat sealer, two coats Sandtex										
Cement rendered general surfaces girth exceeding 300mm	2.31	10.00	2.54	0.38	–	3.51	0.07	6.12	m2	6.58
Concrete general surfaces girth exceeding 300mm	2.49	10.00	2.74	0.38	–	3.51	0.07	6.32	m2	6.79
Brickwork general surfaces girth exceeding 300mm	2.88	10.00	3.17	0.41	–	3.78	0.07	7.02	m2	7.55
Rough cast general surfaces girth exceeding 300mm	2.71	10.00	2.98	0.49	–	4.52	0.09	7.59	m2	8.16

SURFACE FINISHES

SURFACE FINISHES – MAJOR WORKS

Labour hourly rates: (except Specialists) Craft Operatives £9.23 Labourer £7.02 Rates are national average prices. Refer to REGIONAL VARIATIONS for indicative levels of overall pricing in regions	MATERIALS			LABOUR				RATES		
	Del to Site £	Waste %	Material Cost £	Craft Optve Hrs	Lab Hrs	Labour Cost £	Sunds £	Nett Rate £	Unit	Gross Rate (+7.5%) £
M60: PAINTING/CLEAR FINISHING; STAIRCASE AREAS Cont.										
One coat primer, one undercoat, one coat full gloss finish										
Concrete general surfaces girth exceeding 300mm	0.85	10.00	0.94	0.39	–	3.60	0.07	4.60	m2	4.95
Brickwork general surfaces girth exceeding 300mm	1.04	10.00	1.14	0.43	–	3.97	0.08	5.19	m2	5.58
Plasterboard general surfaces girth exceeding 300mm	0.66	10.00	0.73	0.36	–	3.32	0.06	4.11	m2	4.42
Plaster general surfaces girth exceeding 300mm	0.76	10.00	0.84	0.36	–	3.32	0.06	4.22	m2	4.54
One coat primer, two undercoats, one coat full gloss finish										
Concrete general surfaces girth exceeding 300mm	1.11	10.00	1.22	0.52	–	4.80	0.09	6.11	m2	6.57
Brickwork general surfaces girth exceeding 300mm	1.35	10.00	1.49	0.57	–	5.26	0.10	6.85	m2	7.36
Plasterboard general surfaces girth exceeding 300mm	0.86	10.00	0.95	0.48	–	4.43	0.08	5.46	m2	5.87
Plaster general surfaces girth exceeding 300mm	0.98	10.00	1.08	0.48	–	4.43	0.08	5.59	m2	6.01
One coat primer, one undercoat, one coat eggshell finish										
Concrete general surfaces girth exceeding 300mm	0.90	10.00	0.99	0.39	–	3.60	0.07	4.66	m2	5.01
Brickwork general surfaces girth exceeding 300mm	1.10	10.00	1.21	0.43	–	3.97	0.08	5.26	m2	5.65
Plasterboard general surfaces girth exceeding 300mm	0.70	10.00	0.77	0.36	–	3.32	0.06	4.15	m2	4.46
Plaster general surfaces girth exceeding 300mm	0.80	10.00	0.88	0.36	–	3.32	0.06	4.26	m2	4.58
One coat primer, two undercoats, one coat eggshell finish										
Concrete general surfaces girth exceeding 300mm	1.16	10.00	1.28	0.52	–	4.80	0.09	6.17	m2	6.63
Brickwork general surfaces girth exceeding 300mm	1.41	10.00	1.55	0.57	–	5.26	0.10	6.91	m2	7.43
Plasterboard general surfaces girth exceeding 300mm	0.90	10.00	0.99	0.48	–	4.43	0.08	5.50	m2	5.91
Plaster general surfaces girth exceeding 300mm	1.03	10.00	1.13	0.48	–	4.43	0.08	5.64	m2	6.07
Textured plastic coating - Stippled finish										
Concrete general surfaces girth exceeding 300mm	0.52	10.00	0.57	0.34	–	3.14	0.06	3.77	m2	4.05
Brickwork general surfaces girth exceeding 300mm	0.62	10.00	0.68	0.44	–	4.06	0.08	4.82	m2	5.18
Plasterboard general surfaces girth exceeding 300mm	0.42	10.00	0.46	0.26	–	2.40	0.05	2.91	m2	3.13
Plaster general surfaces girth exceeding 300mm	0.42	10.00	0.46	0.26	–	2.40	0.05	2.91	m2	3.13
Textured plastic coating - Combed Finish										
Concrete general surfaces girth exceeding 300mm	0.52	10.00	0.57	0.39	–	3.60	0.07	4.24	m2	4.56
Brickwork general surfaces girth exceeding 300mm	0.62	10.00	0.68	0.54	–	4.98	0.09	5.76	m2	6.19
Plasterboard general surfaces girth exceeding 300mm	0.42	10.00	0.46	0.31	–	2.86	0.05	3.37	m2	3.63
Plaster general surfaces girth exceeding 300mm	0.42	10.00	0.46	0.31	–	2.86	0.05	3.37	m2	3.63

Labour hourly rates: (except Specialists) Craft Operatives £9.23 Labourer £7.02 Rates are national average prices. Refer to REGIONAL VARIATIONS for indicative levels of overall pricing in regions	MATERIALS			LABOUR			RATES		
	Del to Site £	Waste %	Material Cost £	Craft Optve Hrs	Lab Hrs	Labour Cost £	Sunds £	Nett Rate £	Unit	Gross Rate (+7.5%) £

M60: PAINTING/CLEAR FINISHING; STAIRCASE AREAS – REDECORATIONS

Generally

Note
the following rates include for the cost of all preparatory work, e.g. washing down, etc.

Two coats emulsion paint, existing emulsion painted surfaces

	Del to Site	Waste	Material Cost	Craft Optve	Lab	Labour Cost	Sunds	Nett Rate	Unit	Gross Rate
Concrete general surfaces girth exceeding 300mm	0.44	10.00	0.48	0.35	–	3.23	0.06	3.77	m2	4.06
Plaster general surfaces girth exceeding 300mm	0.35	10.00	0.39	0.31	–	2.86	0.05	3.30	m2	3.54
Plasterboard general surfaces girth exceeding 300mm	0.35	10.00	0.39	0.32	–	2.95	0.06	3.40	m2	3.65
Brickwork general surfaces girth exceeding 300mm	0.53	10.00	0.58	0.35	–	3.23	0.06	3.87	m2	4.16
Paper covered general surfaces girth exceeding 300mm	0.40	10.00	0.44	0.32	–	2.95	0.06	3.45	m2	3.71

Two full coats emulsion paint, existing washable distempered surfaces

Concrete general surfaces girth exceeding 300mm	0.44	10.00	0.48	0.36	–	3.32	0.06	3.87	m2	4.16
Plaster general surfaces girth exceeding 300mm	0.35	10.00	0.39	0.32	–	2.95	0.06	3.40	m2	3.65
Plasterboard general surfaces girth exceeding 300mm	0.35	10.00	0.39	0.33	–	3.05	0.06	3.49	m2	3.75
Brickwork general surfaces girth exceeding 300mm	0.53	10.00	0.58	0.36	–	3.32	0.06	3.97	m2	4.26
Paper covered general surfaces girth exceeding 300mm	0.40	10.00	0.44	0.33	–	3.05	0.06	3.55	m2	3.81

Two full coats emulsion paint, existing textured plastic coating surfaces

Concrete general surfaces girth exceeding 300mm	0.44	10.00	0.48	0.39	–	3.60	0.07	4.15	m2	4.47
Plaster general surfaces girth exceeding 300mm	0.35	10.00	0.39	0.34	–	3.14	0.06	3.58	m2	3.85
Plasterboard general surfaces girth exceeding 300mm	0.35	10.00	0.39	0.35	–	3.23	0.06	3.68	m2	3.95
Brickwork general surfaces girth exceeding 300mm	0.53	10.00	0.58	0.38	–	3.51	0.07	4.16	m2	4.47
Paper covered general surfaces girth exceeding 300mm	0.40	10.00	0.44	0.35	–	3.23	0.06	3.73	m2	4.01

Mist coat, two full coats emulsion paint, existing non-washable distempered surfaces

Concrete general surfaces girth exceeding 300mm	0.53	10.00	0.58	0.44	–	4.06	0.08	4.72	m2	5.08
Plaster general surfaces girth exceeding 300mm	0.44	10.00	0.48	0.40	–	3.69	0.07	4.25	m2	4.56
Plasterboard general surfaces girth exceeding 300mm	0.44	10.00	0.48	0.41	–	3.78	0.07	4.34	m2	4.66
Brickwork general surfaces girth exceeding 300mm	0.66	10.00	0.73	0.44	–	4.06	0.08	4.87	m2	5.23
Paper covered general surfaces girth exceeding 300mm	0.49	10.00	0.54	0.41	–	3.78	0.07	4.39	m2	4.72

Two full coats emulsion paint, existing gloss painted surfaces

Concrete general surfaces girth exceeding 300mm	0.44	10.00	0.48	0.36	–	3.32	0.06	3.87	m2	4.16
Plaster general surfaces girth exceeding 300mm	0.35	10.00	0.39	0.32	–	2.95	0.06	3.40	m2	3.65
Plasterboard general surfaces girth exceeding 300mm	0.35	10.00	0.39	0.33	–	3.05	0.06	3.49	m2	3.75
Brickwork general surfaces girth exceeding 300mm	0.53	10.00	0.58	0.36	–	3.32	0.06	3.97	m2	4.26

SURFACE FINISHES

Labour hourly rates: (except Specialists) Craft Operatives £9.23 Labourer £7.02 Rates are national average prices. Refer to REGIONAL VARIATIONS for indicative levels of overall pricing in regions	MATERIALS			LABOUR				RATES		
	Del to Site £	Waste %	Material Cost £	Craft Optve Hrs	Lab Hrs	Labour Cost £	Sunds £	Nett Rate £	Unit	Gross Rate (+7.5%) £
M60: PAINTING/CLEAR FINISHING; STAIRCASE AREAS – REDECORATIONS Cont.										
Two full coats emulsion paint, existing gloss painted surfaces Cont.										
Paper covered general surfaces girth exceeding 300mm	0.40	10.00	0.44	0.33	-	3.05	0.06	3.55	m2	3.81
Generally										
Note the following rates include for the cost of all preparatory work, e.g. washing down, etc.										
One undercoat, one coat full gloss finish, existing gloss painted surfaces										
Concrete general surfaces girth exceeding 300mm	0.51	10.00	0.56	0.41	-	3.78	0.07	4.42	m2	4.75
Brickwork general surfaces girth exceeding 300mm	0.62	10.00	0.68	0.42	-	3.88	0.07	4.63	m2	4.98
Plasterboard general surfaces girth exceeding 300mm	0.40	10.00	0.44	0.38	-	3.51	0.07	4.02	m2	4.32
Plaster general surfaces girth exceeding 300mm	0.45	10.00	0.50	0.37	-	3.42	0.06	3.97	m2	4.27
One undercoat, one coat eggshell finish, existing gloss painted surfaces										
Concrete general surfaces girth exceeding 300mm	0.56	10.00	0.62	0.41	-	3.78	0.07	4.47	m2	4.81
Brickwork general surfaces girth exceeding 300mm	0.69	10.00	0.76	0.42	-	3.88	0.07	4.71	m2	5.06
Plasterboard general surfaces girth exceeding 300mm	0.44	10.00	0.48	0.38	-	3.51	0.07	4.06	m2	4.37
Plaster general surfaces girth exceeding 300mm	0.50	10.00	0.55	0.32	-	2.95	0.06	3.56	m2	3.83
Two undercoats, one coat full gloss finish, existing gloss painted surfaces										
Concrete general surfaces girth exceeding 300mm	0.77	10.00	0.85	0.53	-	4.89	0.09	5.83	m2	6.27
Brickwork general surfaces girth exceeding 300mm	0.94	10.00	1.03	0.53	-	4.89	0.09	6.02	m2	6.47
Plasterboard general surfaces girth exceeding 300mm	0.60	10.00	0.66	0.50	-	4.62	0.09	5.37	m2	5.77
Plaster general surfaces girth exceeding 300mm	0.68	10.00	0.75	0.49	-	4.52	0.09	5.36	m2	5.76
Two undercoats, one coat eggshell finish, existing gloss painted surfaces										
Concrete general surfaces girth exceeding 300mm	0.82	10.00	0.90	0.53	-	4.89	0.09	5.88	m2	6.33
Brickwork general surfaces girth exceeding 300mm	1.00	10.00	1.10	0.53	-	4.89	0.09	6.08	m2	6.54
Plasterboard general surfaces girth exceeding 300mm	0.64	10.00	0.70	0.50	-	4.62	0.09	5.41	m2	5.81
Plaster general surfaces girth exceeding 300mm	0.73	10.00	0.80	0.49	-	4.52	0.09	5.42	m2	5.82
One coat primer, one undercoat, one coat full gloss finish, existing washable distempered or emulsion painted surfaces										
Concrete general surfaces girth exceeding 300mm	0.85	10.00	0.94	0.53	-	4.89	0.09	5.92	m2	6.36
Brickwork general surfaces girth exceeding 300mm	1.04	10.00	1.14	0.53	-	4.89	0.09	6.13	m2	6.59
Plasterboard general surfaces girth exceeding 300mm	0.66	10.00	0.73	0.50	-	4.62	0.09	5.43	m2	5.84
Plaster general surfaces girth exceeding 300mm	0.76	10.00	0.84	0.49	-	4.52	0.09	5.45	m2	5.86
One coat primer, one undercoat, one coat eggshell finish, existing washable distempered or emulsion painted surfaces										
Concrete general surfaces girth exceeding 300mm	0.90	10.00	0.99	0.53	-	4.89	0.09	5.97	m2	6.42

Labour hourly rates: (except Specialists) Craft Operatives £9.23 Labourer £7.02 Rates are national average prices. Refer to REGIONAL VARIATIONS for indicative levels of overall pricing in regions	MATERIALS			LABOUR				RATES		
	Del to Site £	Waste %	Material Cost £	Craft Optve Hrs	Lab Hrs	Labour Cost £	Sunds £	Nett Rate £	Unit	Gross Rate (+7.5%) £
M60: PAINTING/CLEAR FINISHING; STAIRCASE AREAS – REDECORATIONS Cont.										
One coat primer, one undercoat, one coat eggshell finish, existing washable distempered or emulsion painted surfaces Cont.										
Brickwork general surfaces girth exceeding 300mm	1.10	10.00	1.21	0.53	–	4.89	0.09	6.19	m2	6.66
Plasterboard general surfaces girth exceeding 300mm	0.70	10.00	0.77	0.50	–	4.62	0.09	5.47	m2	5.89
Plaster general surfaces girth exceeding 300mm	0.80	10.00	0.88	0.49	–	4.52	0.09	5.49	m2	5.90
One coat primer, two undercoats, one coat full gloss finish, existing washable distempered or emulsion painted surfaces										
Concrete general surfaces girth exceeding 300mm	1.11	10.00	1.22	0.65	–	6.00	0.11	7.33	m2	7.88
Brickwork general surfaces girth exceeding 300mm	1.35	10.00	1.49	0.66	–	6.09	0.12	7.70	m2	8.27
Plasterboard general surfaces girth exceeding 300mm	0.86	10.00	0.95	0.62	–	5.72	0.11	6.78	m2	7.29
Plaster general surfaces girth exceeding 300mm	0.98	10.00	1.08	0.61	–	5.63	0.11	6.82	m2	7.33
One coat primer, two undercoats, one coat eggshell finish, existing washable distempered or emulsion painted surfaces										
Concrete general surfaces girth exceeding 300mm	1.16	10.00	1.28	0.65	–	6.00	0.11	7.39	m2	7.94
Brickwork general surfaces girth exceeding 300mm	1.41	10.00	1.55	0.66	–	6.09	0.12	7.76	m2	8.35
Plasterboard general surfaces girth exceeding 300mm	0.90	10.00	0.99	0.62	–	5.72	0.11	6.82	m2	7.33
Plaster general surfaces girth exceeding 300mm	1.03	10.00	1.13	0.61	–	5.63	0.11	6.87	m2	7.39
One coat primer, one undercoat, one coat full gloss finish, existing non-washable distempered surfaces										
Concrete general surfaces girth exceeding 300mm	0.85	10.00	0.94	0.76	–	7.01	0.09	8.04	m2	8.64
Brickwork general surfaces girth exceeding 300mm	1.04	10.00	1.14	0.53	–	4.89	0.09	6.13	m2	6.59
Plasterboard general surfaces girth exceeding 300mm	0.66	10.00	0.73	0.50	–	4.62	0.09	5.43	m2	5.84
Plaster general surfaces girth exceeding 300mm	0.76	10.00	0.84	0.49	–	4.52	0.09	5.45	m2	5.86
One coat primer, one undercoat, one coat eggshell finish, existing non-washable distempered surfaces										
Concrete general surfaces girth exceeding 300mm	0.90	10.00	0.99	0.53	–	4.89	0.09	5.97	m2	6.42
Brickwork general surfaces girth exceeding 300mm	1.10	10.00	1.21	0.53	–	4.89	0.09	6.19	m2	6.66
Plasterboard general surfaces girth exceeding 300mm	0.70	10.00	0.77	0.50	–	4.62	0.09	5.47	m2	5.89
Plaster general surfaces girth exceeding 300mm	0.80	10.00	0.88	0.49	–	4.52	0.09	5.49	m2	5.90
One coat primer, two undercoats, one coat full gloss finish, existing non-washable distempered surfaces										
Concrete general surfaces girth exceeding 300mm	1.11	10.00	1.22	0.65	–	6.00	0.11	7.33	m2	7.88
Brickwork general surfaces girth exceeding 300mm	1.35	10.00	1.49	0.66	–	6.09	0.12	7.70	m2	8.27
Plasterboard general surfaces girth exceeding 300mm	0.86	10.00	0.95	0.62	–	5.72	0.11	6.78	m2	7.29
Plaster general surfaces girth exceeding 300mm	0.98	10.00	1.08	0.61	–	5.63	0.11	6.82	m2	7.33

SURFACE FINISHES

Labour hourly rates: (except Specialists) Craft Operatives £9.23 Labourer £7.02. Rates are national average prices. Refer to REGIONAL VARIATIONS for indicative levels of overall pricing in regions	MATERIALS			LABOUR				RATES		
	Del to Site £	Waste %	Material Cost £	Craft Optve Hrs	Lab Hrs	Labour Cost £	Sunds £	Nett Rate £	Unit	Gross Rate (+7.5%) £
M60: PAINTING/CLEAR FINISHING; STAIRCASE AREAS – REDECORATIONS Cont.										
One coat primer, two undercoats, one coat eggshell finish, existing non-washable distempered surfaces										
Concrete general surfaces girth exceeding 300mm	1.16	10.00	1.28	0.65	–	6.00	0.11	7.39	m2	7.94
Brickwork general surfaces girth exceeding 300mm	1.41	10.00	1.55	0.66	–	6.09	0.12	7.76	m2	8.35
Plasterboard general surfaces girth exceeding 300mm	0.90	10.00	0.99	0.62	–	5.72	0.11	6.82	m2	7.33
Plaster general surfaces girth exceeding 300mm	1.03	10.00	1.13	0.61	–	5.63	0.11	6.87	m2	7.39
Generally										
Note the following rates include for the cost of all preparatory work, e.g. washing down, etc. **Textured plastic coating - stippled finish, existing washable distempered or emulsion painted surfaces**										
Concrete general surfaces girth exceeding 300mm	0.52	10.00	0.57	0.40	–	3.69	0.07	4.33	m2	4.66
Brickwork general surfaces girth exceeding 300mm	0.62	10.00	0.68	0.40	–	3.69	0.07	4.44	m2	4.78
Plasterboard general surfaces girth exceeding 300mm	0.42	10.00	0.46	0.37	–	3.42	0.06	3.94	m2	4.23
Plaster general surfaces girth exceeding 300mm	0.42	10.00	0.46	0.36	–	3.32	0.06	3.84	m2	4.13
Textured plastic coating - combed finish, existing washable distempered or emulsion painted surfaces										
Concrete general surfaces girth exceeding 300mm	0.52	10.00	0.57	0.45	–	4.15	0.08	4.81	m2	5.17
Brickwork general surfaces girth exceeding 300mm	0.62	10.00	0.68	0.45	–	4.15	0.08	4.92	m2	5.28
Plasterboard general surfaces girth exceeding 300mm	0.42	10.00	0.46	0.42	–	3.88	0.07	4.41	m2	4.74
Plaster general surfaces girth exceeding 300mm	0.42	10.00	0.46	0.41	–	3.78	0.07	4.32	m2	4.64
Textured plastic coating - stippled finish, existing non-washable distempered surfaces										
Concrete general surfaces girth exceeding 300mm	0.52	10.00	0.57	0.51	–	4.71	0.09	5.37	m2	5.77
Brickwork general surfaces girth exceeding 300mm	0.62	10.00	0.68	0.51	–	4.71	0.09	5.48	m2	5.89
Plasterboard general surfaces girth exceeding 300mm	0.42	10.00	0.46	0.48	–	4.43	0.08	4.97	m2	5.35
Plaster general surfaces girth exceeding 300mm	0.42	10.00	0.46	0.47	–	4.34	0.08	4.88	m2	5.25
Textured plastic coating - combed finish, existing non-washable distempered surfaces										
Concrete general surfaces girth exceeding 300mm	0.52	10.00	0.57	0.56	–	5.17	0.10	5.84	m2	6.28
Brickwork general surfaces girth exceeding 300mm	0.62	10.00	0.68	0.56	–	5.17	0.10	5.95	m2	6.40
Plasterboard general surfaces girth exceeding 300mm	0.42	10.00	0.46	0.53	–	4.89	0.09	5.44	m2	5.85
Plaster general surfaces girth exceeding 300mm	0.42	10.00	0.46	0.52	–	4.80	0.09	5.35	m2	5.75

Labour hourly rates: (except Specialists) Craft Operatives £9.23 Labourer £7.02 Rates are national average prices. Refer to REGIONAL VARIATIONS for indicative levels of overall pricing in regions	MATERIALS			LABOUR				RATES		
	Del to Site £	Waste %	Material Cost £	Craft Optve Hrs	Lab Hrs	Labour Cost £	Sunds £	Nett Rate £	Unit	Gross Rate (+7.5%) £
N10: GENERAL FIXTURES/FURNISHING/EQUIPMENT										
MIRRORS - B.S.952, clear float, SG, silvered and protected with copper backing										
6mm thick; fixing to masonry with brass screws, chromium plated dome covers, rubber sleeves and washers										
holes 6mm diameter -4; edges polished										
254 x 400mm	29.12	5.00	30.58	0.50	-	4.62	0.38	35.57	nr	38.24
300 x 460mm	33.54	5.00	35.22	0.50	-	4.62	0.38	40.21	nr	43.23
360 x 500mm	37.84	5.00	39.73	0.50	-	4.62	0.38	44.73	nr	48.08
460 x 560mm	45.85	5.00	48.14	0.65	-	6.00	0.38	54.52	nr	58.61
460 x 600mm	47.84	5.00	50.23	0.65	-	6.00	0.38	56.61	nr	60.86
460 x 900mm	61.52	5.00	64.60	0.70	-	6.46	0.38	71.44	nr	76.79
500 x 680mm	53.87	5.00	56.56	0.65	-	6.00	0.38	62.94	nr	67.66
600 x 900mm	72.25	5.00	75.86	0.75	-	6.92	0.38	83.17	nr	89.40
holes 6mm diameter -4; edges bevelled										
254 x 400mm	35.08	5.00	36.83	0.50	-	4.62	0.38	41.83	nr	44.97
300 x 460mm	40.45	5.00	42.47	0.50	-	4.62	0.38	47.47	nr	51.03
360 x 500mm	45.67	5.00	47.95	0.50	-	4.62	0.38	52.95	nr	56.92
460 x 560mm	55.13	5.00	57.89	0.65	-	6.00	0.38	64.27	nr	69.09
460 x 600mm	57.51	5.00	60.39	0.65	-	6.00	0.38	66.77	nr	71.77
460 x 900mm	73.89	5.00	77.58	0.70	-	6.46	0.38	84.43	nr	90.76
500 x 680mm	64.61	5.00	67.84	0.65	-	6.00	0.38	74.22	nr	79.79
600 x 900mm	86.32	5.00	90.64	0.75	-	6.92	0.38	97.94	nr	105.28
CURTAIN TRACKS - Fixing only curtain track										
Metal or plastic track with fittings										
fixing with screws to softwood	-	-	-	0.40	-	3.69	-	3.69	m	3.97
fixing with screws to hardwood	-	-	-	0.60	-	5.54	-	5.54	m	5.95
BLINDS - Internal blinds										
Venetian blinds; stove enamelled aluminium alloy slats 25mm wide; plain colours										
1200mm drop; fixing to timber with screws										
1000mm wide	28.00	2.50	28.70	1.00	0.13	10.14	0.16	39.00	nr	41.93
2000mm wide	42.95	2.50	44.02	1.50	0.19	15.18	0.24	59.44	nr	63.90
3000mm wide	57.87	2.50	59.32	2.00	0.25	20.22	0.33	79.86	nr	85.85
Venetian blinds; stove enamelled aluminium alloy slats 50mm wide; plain colours										
1200mm drop; fixing to timber with screws										
1000mm wide	25.82	2.50	26.47	1.15	0.14	11.60	0.19	38.25	nr	41.12
2000mm wide	39.83	2.50	40.83	1.70	0.21	17.17	0.28	58.27	nr	62.64
3000mm wide	54.15	2.50	55.50	2.25	0.28	22.73	0.37	78.61	nr	84.50
Venetian blinds; stove enamelled aluminium alloy slats 25mm wide; plain colours with single cord control										
1200mm drop; fixing to timber with screws										
1000mm wide	39.83	2.50	40.83	1.00	0.13	10.14	0.16	51.13	nr	54.96
2000mm wide	62.22	2.50	63.78	1.50	0.19	15.18	0.24	79.19	nr	85.13
3000mm wide	84.32	2.50	86.43	2.00	0.25	20.22	0.33	106.97	nr	115.00
Venetian blinds; stove enamelled aluminium alloy slats 50mm wide; plain colours with single cord control										
1200mm drop; fixing to timber with screws										
1000mm wide	36.09	2.50	36.99	1.15	0.14	11.60	0.19	48.78	nr	52.44
2000mm wide	54.76	2.50	56.13	1.70	0.21	17.17	0.28	73.57	nr	79.09
3000mm wide	75.29	2.50	77.17	2.25	0.28	22.73	0.37	100.28	nr	107.80

FURNITURE/EQUIPMENT

Labour hourly rates: (except Specialists) Craft Operatives £9.23 Labourer £7.02 Rates are national average prices. Refer to REGIONAL VARIATIONS for indicative levels of overall pricing in regions	MATERIALS			LABOUR				RATES		
	Del to Site	Waste	Material Cost	Craft Optve	Lab	Labour Cost	Sunds	Nett Rate	Unit	Gross Rate (+7.5%)
	£	%	£	Hrs	Hrs	£	£	£		£
N10: GENERAL FIXTURES/FURNISHING/EQUIPMENT Cont.										
BLINDS - Internal blinds										
Venetian blinds; stove enamelled aluminium alloy slats 25mm wide; plain colours with channels for dimout										
1200mm drop; fixing to timber with screws										
1000mm wide.........................	67.83	2.50	69.53	2.00	0.25	20.22	0.33	90.07	nr	96.83
2000mm wide.........................	83.38	2.50	85.46	2.50	0.31	25.25	0.41	111.13	nr	119.46
3000mm wide.........................	98.31	2.50	100.77	3.00	0.38	30.36	0.49	131.62	nr	141.49
Venetian blinds; stove enamelled aluminium alloy slats 50mm wide; plain colours with channels for dimout										
1200mm drop; fixing to timber with screws										
1000mm wide.........................	65.33	2.50	66.96	2.25	0.28	22.73	0.37	90.07	nr	96.82
2000mm wide.........................	80.27	2.50	82.28	2.80	0.35	28.30	0.46	111.04	nr	119.37
3000mm wide.........................	94.58	2.50	96.94	3.35	0.42	33.87	0.55	131.36	nr	141.22
Roller blinds; automatic ratchet action; fire resistant material										
1200mm drop; fixing to timber with screws										
1000mm wide.........................	52.27	2.50	53.58	1.00	0.13	10.14	0.16	63.88	nr	68.67
2000mm wide.........................	64.10	2.50	65.70	1.50	0.19	15.18	0.24	81.12	nr	87.21
3000mm wide.........................	75.90	2.50	77.80	2.00	0.25	20.22	0.33	98.34	nr	105.72
Roller blinds; automatic ratchet action; holland type material										
1200mm drop; fixing to timber with screws										
1000mm wide.........................	50.41	2.50	51.67	1.00	0.13	10.14	0.16	61.97	nr	66.62
2000mm wide.........................	60.36	2.50	61.87	1.50	0.19	15.18	0.24	77.29	nr	83.08
3000mm wide.........................	70.28	2.50	72.04	2.00	0.25	20.22	0.33	92.58	nr	99.53
Roller blinds; self acting roller; blackout material										
1200mm drop; fixing to timber with screws										
1000mm wide.........................	46.67	2.50	47.84	1.15	0.14	11.60	0.19	59.62	nr	64.10
2000mm wide.........................	56.63	2.50	58.05	1.70	0.21	17.17	0.28	75.49	nr	81.15
3000mm wide.........................	66.58	2.50	68.24	2.25	0.28	22.73	0.37	91.35	nr	98.20
Roller blinds; 100% blackout; natural anodised box and channels										
1200mm drop; fixing to timber with screws										
1000mm wide.........................	154.32	2.50	158.18	2.00	0.25	20.22	0.33	178.72	nr	192.13
2000mm wide.........................	192.23	2.50	197.04	2.75	0.34	27.77	0.45	225.26	nr	242.15
3000mm wide.........................	235.20	2.50	241.08	3.50	0.44	35.39	0.57	277.04	nr	297.82
Vertical louvre blinds; 89mm wide louvres in standard material										
1200mm drop; fixing to timber with screws										
1000mm wide.........................	56.00	2.50	57.40	0.90	0.11	9.08	0.15	66.63	nr	71.63
2000mm wide.........................	98.31	2.50	100.77	1.35	0.17	13.65	0.22	114.64	nr	123.24
3000mm wide.........................	140.63	2.50	144.15	1.80	0.23	18.23	0.29	162.66	nr	174.86
Vertical louvre blinds; 127mm wide louvres in standard material										
1200mm drop; fixing to timber with screws										
1000mm wide.........................	50.41	2.50	51.67	0.95	0.12	9.61	0.15	61.43	nr	66.04
2000mm wide.........................	87.74	2.50	89.93	1.40	0.18	14.19	0.23	104.35	nr	112.18
3000mm wide.........................	125.69	2.50	128.83	1.85	0.23	18.69	0.30	147.82	nr	158.91
BLINDS - External blinds; manually operated										
Venetian blinds; stove enamelled aluminium slats 80mm wide; natural anodised side guides; excluding boxing										
1200mm drop; fixing to timber with screws										
1000mm wide.........................	-	-	Spclist	-	-	Spclist	-	242.69	nr	260.89
2000mm wide.........................	-	-	Spclist	-	-	Spclist	-	361.20	nr	388.29
3000mm wide.........................	-	-	Spclist	-	-	Spclist	-	486.50	nr	522.99
Queensland awnings; acrylic material; natural anodised arms and boxing										
1000mm projection										
1000mm long.........................	-	-	Spclist	-	-	Spclist	-	568.89	nr	611.56
2000mm long.........................	-	-	Spclist	-	-	Spclist	-	853.54	nr	917.56
3000mm long.........................	-	-	Spclist	-	-	Spclist	-	1137.78	nr	1223.11
Rollscreen vertical drop roller blinds; mesh material; natural anodised side guides and boxing										
1200mm drop; fixing to timber with screws										
1000mm wide.........................	-	-	Spclist	-	-	Spclist	-	237.01	nr	254.79
2000mm wide.........................	-	-	Spclist	-	-	Spclist	-	355.56	nr	382.23
3000mm wide.........................	-	-	Spclist	-	-	Spclist	-	473.87	nr	509.41
Quandrant canopy; acrylic material; natural anodised frames										
1000mm projection										
1000mm long.........................	-	-	Spclist	-	-	Spclist	-	313.79	nr	337.32
2000mm long.........................	-	-	Spclist	-	-	Spclist	-	479.72	nr	515.70
3000mm long.........................	-	-	Spclist	-	-	Spclist	-	652.42	nr	701.35
Foldaway awning; acrylic material; natural anodised arms and front rail										
2000mm projection										
1000mm long.........................	-	-	Spclist	-	-	Spclist	-	509.07	nr	547.25
2000mm long.........................	-	-	Spclist	-	-	Spclist	-	747.23	nr	803.27

Labour hourly rates: (except Specialists) Craft Operatives £9.23 Labourer £7.02 Rates are national average prices. Refer to REGIONAL VARIATIONS for indicative levels of overall pricing in regions	MATERIALS			LABOUR				RATES		
	Del to Site	Waste	Material Cost	Craft Optve	Lab	Labour Cost	Sunds	Nett Rate	Unit	Gross Rate (+7.5%)
	£	%	£	Hrs	Hrs	£	£	£		£
N10: GENERAL FIXTURES/FURNISHING/EQUIPMENT Cont.										
BLINDS - External blinds; manually operated Cont.										
Foldaway awning; acrylic material; natural anodised arms and front rail Cont. 2000mm projection Cont.										
3000mm long....................	-	-	Spclist	-	-	Spclist	-	971.86	nr	1044.75
BLINDS - External blinds; electrically operated										
Venetian blinds; stove enamelled aluminium slats 80mm wide; natural anodised side guides; excluding boxing 1200mm drop; fixing to timber with screws										
1000mm wide....................	-	-	Spclist	-	-	Spclist	-	557.60	nr	599.42
2000mm wide....................	-	-	Spclist	-	-	Spclist	-	681.77	nr	732.90
3000mm wide....................	-	-	Spclist	-	-	Spclist	-	799.16	nr	859.10
Queensland awnings; acrylic material; natural anodised arms and boxing 1000mm projection										
1000mm long....................	-	-	Spclist	-	-	Spclist	-	652.42	nr	701.35
2000mm long....................	-	-	Spclist	-	-	Spclist	-	931.22	nr	1001.06
3000mm long....................	-	-	Spclist	-	-	Spclist	-	1185.19	nr	1274.08
Rollscreen vertical drop roller blinds; mesh material; natural anodised side guides and boxing 1200mm drop; fixing to timber with screws										
1000mm wide....................	-	-	Spclist	-	-	Spclist	-	451.50	nr	485.36
2000mm wide....................	-	-	Spclist	-	-	Spclist	-	568.89	nr	611.56
3000mm wide....................	-	-	Spclist	-	-	Spclist	-	687.41	nr	738.97
Quandrant canopy; acrylic material; natural anodised frames 1000mm projection										
1000mm long....................	-	-	Spclist	-	-	Spclist	-	408.61	nr	439.26
2000mm long....................	-	-	Spclist	-	-	Spclist	-	568.89	nr	611.56
3000mm long....................	-	-	Spclist	-	-	Spclist	-	758.52	nr	815.41
Foldaway awning; acrylic material; natural anodised arms and front rail 2000mm projection										
1000mm long....................	-	-	Spclist	-	-	Spclist	-	746.11	nr	802.07
2000mm long....................	-	-	Spclist	-	-	Spclist	-	977.50	nr	1050.81
3000mm long....................	-	-	Spclist	-	-	Spclist	-	1241.63	nr	1334.75
STORAGE SYSTEMS - Vista two tone coloured steel boltless storage systems; Welconstruct Co. Ltd.; units 900mm wide x 1800mm high; assembling										
Open shelving with top, bottom, three shelves and braces; placing in position										
300mm deep code G106-001	113.18	2.50	116.01	1.00	0.50	12.74	-	128.75	nr	138.41
450mm deep code G106-002	135.32	2.50	138.70	1.20	0.60	15.29	-	153.99	nr	165.54
600mm deep code G106-003	151.96	2.50	155.76	1.40	0.70	17.84	-	173.60	nr	186.61
Closed shelving with top, bottom, and three shelves; placing in position										
300mm deep code G106-011	151.30	2.50	155.08	1.10	0.55	14.01	-	169.10	nr	181.78
extra for additional shelf code G106-301	8.10	2.50	8.30	0.20	0.10	2.55	-	10.85	nr	11.66
450mm deep code G106-012	177.66	2.50	182.10	1.30	0.65	16.56	-	198.66	nr	213.56
extra for additional shelf code G106-302	12.36	2.50	12.67	0.20	0.10	2.55	-	15.22	nr	16.36
600mm deep code G106-013	198.72	2.50	203.69	1.50	0.75	19.11	-	222.80	nr	239.51
extra for additional shelf code G106-303	15.44	2.50	15.83	0.20	0.10	2.55	-	18.37	nr	19.75
Unit comprising top, bottom, one shelf, six 300 x 300mm and four 450 x 450mm bins; placing in position										
300mm deep	184.55	2.50	189.16	1.10	0.55	14.01	-	203.18	nr	218.42
450mm deep	219.30	2.50	224.78	1.30	0.65	16.56	-	241.34	nr	259.45
600mm deep	248.80	2.50	255.02	1.50	0.75	19.11	-	274.13	nr	294.69
Unit comprising top, bottom, one shelf, eight 225 x 225mm, three 300 x 300mm and two 450 x 450mm bins and twelve 150x 150mm drawers; placing in position										
300mm deep	230.47	2.50	236.23	2.00	1.00	25.48	-	261.71	nr	281.34
450mm deep	279.67	2.50	286.66	2.20	1.10	28.03	-	314.69	nr	338.29
Unit comprising top, bottom, one shelf, three 300 x 225mm and fifteen 300 x 300mm bins; placing in position										
300mm deep	189.52	2.50	194.26	1.10	0.55	14.01	-	208.27	nr	223.89
450mm deep	233.50	2.50	239.34	1.30	0.65	16.56	-	255.90	nr	275.09
600mm deep	271.11	2.50	277.89	1.50	0.75	19.11	-	297.00	nr	319.27
Unit comprising top, bottom three 300 x 225mm, six 300 x 300mm and four 450 x 450mm bins; placing in position										
450mm deep	218.02	2.50	223.47	1.30	0.65	16.56	-	240.03	nr	258.03
600mm deep	248.97	2.50	255.19	1.50	0.75	19.11	-	274.30	nr	294.88
Extra over unit for pair of lockable doors code G106-401	178.78	2.50	183.25	0.20	0.10	2.55	-	185.80	nr	199.73

FURNITURE/EQUIPMENT

Labour hourly rates: (except Specialists) Craft Operatives £9.23 Labourer £7.02 Rates are national average prices. Refer to REGIONAL VARIATIONS for indicative levels of overall pricing in regions	MATERIALS			LABOUR				RATES		
	Del to Site £	Waste %	Material Cost £	Craft Optve Hrs	Lab Hrs	Labour Cost £	Sunds £	Nett Rate £	Unit	Gross Rate (+7.5%) £

N10: GENERAL FIXTURES/FURNISHING/EQUIPMENT Cont.

SHELVING SYSTEMS - Spanwel wide access boltless steel shelving systems; Welconstruct Co Ltd.; upright frames connected with shelf beams and with shelves; assembling

Description	Del to Site £	Waste %	Material Cost £	Craft Optve Hrs	Lab Hrs	Labour Cost £	Sunds £	Nett Rate £	Unit	Gross Rate £
Unit 1800mm high with 2 upright frames and three standard duty steel shelves including top and bottom; placing in position										
1200mm long x 650mm deep code G108-201	194.40	2.50	199.26	1.40	0.70	17.84	–	217.10	nr	233.38
extra; additional standard duty steel shelf and shelf beam code G108-225	43.36	2.50	44.44	0.20	0.10	2.55	–	46.99	nr	50.52
1200mm long x 800mm deep code G108-202	210.06	2.50	215.31	1.50	0.75	19.11	–	234.42	nr	252.00
extra; additional standard duty steel shelf and shelf beam G108-226	45.90	2.50	47.05	0.20	0.10	2.55	–	49.60	nr	53.32
1200mm long x 950mm deep code G108-203	220.32	2.50	225.83	1.60	0.80	20.38	–	246.21	nr	264.68
extra; additional standard duty steel shelf and shelf beam code G108-227	48.70	2.50	49.92	0.20	0.10	2.55	–	52.47	nr	56.40
2400mm long x 650mm deep code G108-204	284.04	2.50	291.14	1.50	0.75	19.11	–	310.25	nr	333.52
extra; additional standard duty steel shelf and shelf beam code G108-228	71.71	2.50	73.50	0.20	0.10	2.55	–	76.05	nr	81.75
2400mm long x 800mm deep code G108-205	302.40	2.50	309.96	1.60	0.80	20.38	–	330.34	nr	355.12
extra; additional standard duty steel shelf and shelf beam code G108-229	76.68	2.50	78.60	0.20	0.10	2.55	–	81.14	nr	87.23
2400mm long x 950mm deep code G108-206	320.76	2.50	328.78	1.70	0.85	21.66	–	350.44	nr	376.72
extra; additional standard duty steel shelf and shelf beam code G108-230	82.08	2.50	84.13	0.20	0.10	2.55	–	86.68	nr	93.18
Unit 2400mm high with 3 upright frames and five standard duty steel shelves including top and bottom; placing in position										
2400mm long x 650mm deep	527.05	2.50	540.23	2.80	1.40	35.67	–	575.90	nr	619.09
2400mm long x 800mm deep	521.70	2.50	534.74	3.00	1.50	38.22	–	572.96	nr	615.93
2400mm long x 950mm deep	549.50	2.50	563.24	3.20	1.60	40.77	–	604.01	nr	649.31

SHELVING SUPPORT SYSTEMS - Spur patent shelf supports in steel with white enamelled finish; Spur Shelving Ltd; Spur Steel-Lok components

Description	Del to Site £	Waste %	Material Cost £	Craft Optve Hrs	Lab Hrs	Labour Cost £	Sunds £	Nett Rate £	Unit	Gross Rate £
Wall uprights										
type 9011, 430mm long; fixing to masonry with screws	3.12	2.50	3.20	0.25	–	2.31	–	5.51	nr	5.92
type 9013, 1000mm long; fixing to masonry with screws	6.30	2.50	6.46	0.45	–	4.15	–	10.61	nr	11.41
type 9016, 1600mm long; fixing to masonry with screws	9.24	2.50	9.47	0.55	–	5.08	–	14.55	nr	15.64
type 9020, 2400mm long; fixing to masonry with screws	14.04	2.50	14.39	0.70	–	6.46	–	20.85	nr	22.42
Straight brackets										
type 9001, 120mm long	1.76	2.50	1.80	0.05	–	0.46	–	2.27	nr	2.44
type 9003, 270mm long	2.87	2.50	2.94	0.05	–	0.46	–	3.40	nr	3.66
type 9004, 360mm long	3.78	2.50	3.87	0.07	–	0.65	–	4.52	nr	4.86
type 9006, 470mm long	6.21	2.50	6.37	0.07	–	0.65	–	7.01	nr	7.54

MAT RIMS - Matwells in Aluminium

Description	Del to Site £	Waste %	Material Cost £	Craft Optve Hrs	Lab Hrs	Labour Cost £	Sunds £	Nett Rate £	Unit	Gross Rate £
Mat frames; welded fabrication 34 x 26 x 6mm angle section; angles mitred; plain lugs -4, welded on; mat space										
610 x 457mm	29.20	2.50	29.93	1.00	1.00	16.25	–	46.18	nr	49.64
762 x 457mm	33.45	2.50	34.29	1.25	1.25	20.31	–	54.60	nr	58.69
914 x 610mm	41.85	2.50	42.90	1.50	1.50	24.38	–	67.27	nr	72.32

MAT RIMS - Matwells in polished brass

Description	Del to Site £	Waste %	Material Cost £	Craft Optve Hrs	Lab Hrs	Labour Cost £	Sunds £	Nett Rate £	Unit	Gross Rate £
Mat frames; brazed fabrication 38 x 38 x 6mm angle section; angles mitred; plain lugs -4, welded on; mat space										
610 x 457mm	93.30	2.50	95.63	1.25	1.25	20.31	–	115.94	nr	124.64
762 x 457mm	107.00	2.50	109.68	1.50	1.50	24.38	–	134.05	nr	144.10
914 x 610mm	135.90	2.50	139.30	1.75	1.75	28.44	–	167.74	nr	180.32

CLOTHES LOCKERS - Individual compartment clothes lockers; Welconstruct Co. Ltd.; steel, stove enamelled finish; standard colour doors and with cam locks 442 series

Description	Del to Site £	Waste %	Material Cost £	Craft Optve Hrs	Lab Hrs	Labour Cost £	Sunds £	Nett Rate £	Unit	Gross Rate £
2 compartment; placing in position										
305 x 305 x 1830mm code GA442-021	67.50	2.50	69.19	–	0.50	3.51	–	72.70	nr	78.15
305 x 460 x 1830mm code G442-022	70.74	2.50	72.51	–	0.55	3.86	–	76.37	nr	82.10
380 x 380 x 1830mm code G442-024	70.74	2.50	72.51	–	0.60	4.21	–	76.72	nr	82.47
460 x 305 x 1830mm code G442-023	70.74	2.50	72.51	–	0.65	4.56	–	77.07	nr	82.85
460 X 460 X1830mm code G442-025	83.92	2.50	86.02	–	0.70	4.91	–	90.93	nr	97.75
3 compartment; placing in position										
305 x 305 x 1830mm code G442-026	79.38	2.50	81.36	–	0.50	3.51	–	84.87	nr	91.24
305 x 460 x 1830mm code G442-027	83.38	2.50	85.46	–	0.55	3.86	–	89.33	nr	96.02
380 x 380 x 1830mm code G442-029	83.38	2.50	85.46	–	0.60	4.21	–	89.68	nr	96.40
460 x 305 x 1830mm code G442-028	83.38	2.50	85.46	–	0.65	4.56	–	90.03	nr	96.78
460 x 460 x 1830mm code G442-030	82.48	2.50	84.54	–	0.70	4.91	–	89.46	nr	96.17

Labour hourly rates: (except Specialists) Craft Operatives £9.23 Labourer £7.02 Rates are national average prices. Refer to REGIONAL VARIATIONS for indicative levels of overall pricing in regions	MATERIALS			LABOUR				RATES		
	Del to Site £	Waste %	Material Cost £	Craft Optve Hrs	Lab Hrs	Labour Cost £	Sunds £	Nett Rate £	Unit	Gross Rate (+7.5%) £
N10: GENERAL FIXTURES/FURNISHING/EQUIPMENT Cont.										
CLOTHES LOCKERS - Individual compartment clothes lockers; Welconstruct Co. Ltd.; steel, stove enamelled finish; standard colour doors and with cam locks 442 series Cont.										
4 compartment; placing in position										
305 x 305 x 1830mm code G442-031	85.43	2.50	87.57	–	0.50	3.51	–	91.08	nr	97.91
305 x 460 x 1830mm code G442-032	91.26	2.50	93.54	–	0.50	3.51	–	97.05	nr	104.33
380 x 380 x 1830mm code G442-034	91.26	2.50	93.54	–	0.60	4.21	–	97.75	nr	105.09
460 x 305 x 1830mm code G442-033	91.26	2.50	93.54	–	0.65	4.56	–	98.10	nr	105.46
460 x 460 x 1830mm code G442-035	95.80	2.50	98.19	–	0.70	4.91	–	103.11	nr	110.84
6 compartment; placing in position										
305 x 305 x 1830mm code G442-036	97.96	2.50	100.41	–	0.50	3.51	–	103.92	nr	111.71
305 x 460 x 1830mm code G442-037	107.89	2.50	110.59	–	0.55	3.86	–	114.45	nr	123.03
380 x 380 x 1830mm code G442-039	107.89	2.50	110.59	–	0.60	4.21	–	114.80	nr	123.41
460 x 305 x 1830mm code G442-038	107.89	2.50	110.59	–	0.65	4.56	–	115.15	nr	123.79
460 x 460 x 1830mm code G442-040	112.21	2.50	115.02	–	0.70	4.91	–	119.93	nr	128.92
CLOTHES LOCKERS - Swimming bath lockers; Welconstruct Co Ltd.; galvanised steel, stove enamelled finish; standard colour doors with mastered series cam locks 452 series										
1 compartment; placing in position										
305 x 305 x 1830mm code G452-001	77.98	2.50	79.93	–	0.50	3.51	–	83.44	nr	89.70
2 compartment; placing in position										
305 x 305 x 1830mm code G452-002	101.09	2.50	103.62	–	0.50	3.51	–	107.13	nr	115.16
305 x 460 x 1830mm code G452-003	106.06	2.50	108.71	–	0.55	3.86	–	112.57	nr	121.02
3 compartment; placing in position										
305 x 305 x 1830mm code G452-004	119.02	2.50	122.00	–	0.50	3.51	–	125.51	nr	134.92
305 x 460 x 1830mm code G452-005	124.85	2.50	127.97	–	0.55	3.86	–	131.83	nr	141.72
4 compartment; placing in position										
305 x 305 x 1830mm code G452-006	127.98	2.50	131.18	–	0.50	3.51	–	134.69	nr	144.79
305 x 460 x 1830mm code G452-007	136.94	2.50	140.36	–	0.55	3.86	–	144.22	nr	155.04
CLOAK ROOM EQUIPMENT - Static square tube double sided coat racks; Welconstruct Co. Ltd.; hardwood seats; in assembled units										
Racks 1500mm long x 1675mm high x 610mm deep; placing in position; 40mm square tube										
10 hooks and angle framed steel mesh single shoe tray, reference G456-102 AND 109; placing in position	337.12	2.50	345.55	–	0.50	3.51	–	349.06	nr	375.24
10 hooks, reference G456-102; placing in position	295.92	2.50	303.32	–	0.50	3.51	–	306.83	nr	329.84
10 hooks and 5 baskets G456-102 and 106	336.58	2.50	344.99	–	0.50	3.51	–	348.50	nr	374.64
10 hooks and 10 baskets G456-102 and 107	367.52	2.50	376.71	–	0.50	3.51	–	380.22	nr	408.73
CLOAK ROOM EQUIPMENT - Static square tube single sided coat racks; Welconstruct Co. Ltd.; mobile; in assembled units										
Racks 1500mm long x 1825mm high x 600mm deep; placing in position; 40mm square tube										
15 hangers and top tray, reference G456-103	286.20	2.50	293.36	–	0.50	3.51	–	296.87	nr	319.13
CLOAK ROOM EQUIPMENT - Free-standing bench seats; Welconstruct Co. Ltd.; steel square tube framing; hardwood seats; placing in position										
Bench seat with shelf 1500mm long x 300mm deep and 450mm high; placing in position; 40mm square tube										
single sided 300mm deep, reference G456-115	144.29	2.50	147.90	–	0.50	3.51	–	151.41	nr	162.76
double sided 600mm deep G456-116	232.74	2.50	238.56	–	0.50	3.51	–	242.07	nr	260.22
CLOAK ROOM EQUIPMENT - Wall rack Welconstruct Co. Ltd; hardwood										
1500mm long; fixing to masonry										
5 hooks, reference G456-029	32.02	2.50	32.82	–	0.60	4.21	–	37.03	nr	39.81
HAT AND COAT RAILS - Rails in softwood, wrought										
25 x 75mm; chamfers -2	1.25	10.00	1.38	0.25	0.03	2.52	0.04	3.93	m	4.23
25 x 100mm; chamfers -2	1.68	10.00	1.85	0.25	0.03	2.52	0.04	4.41	m	4.74
25 x 75mm; chamfers -2; fixing to masonry with screws	1.25	10.00	1.38	0.50	0.06	5.04	0.08	6.49	m	6.98
25 x 100mm; chamfers -2; fixing to masonry with screws	1.68	10.00	1.85	0.50	0.06	5.04	0.08	6.96	m	7.49
HAT AND COAT RAILS - Rails in Afromosia, wrought, selected for transparent finish										
25 x 75mm; chamfers -2	3.86	10.00	4.25	0.50	0.06	5.04	0.08	9.36	m	10.06
25 x 100mm; chamfers -2	5.16	10.00	5.68	0.50	0.06	5.04	0.08	10.79	m	11.60
25 x 75mm; chamfers -2; fixing to masonry with screws	3.86	10.00	4.25	1.00	0.12	10.07	0.16	14.48	m	15.56
25 x 100mm; chamfers -2; fixing to masonry with screws	5.16	10.00	5.68	1.00	0.12	10.07	0.16	15.91	m	17.10

FURNITURE/EQUIPMENT

Labour hourly rates: (except Specialists) Craft Operatives £9.23 Labourer £7.02 Rates are national average prices. Refer to REGIONAL VARIATIONS for indicative levels of overall pricing in regions	MATERIALS			LABOUR				RATES		
	Del to Site	Waste	Material Cost	Craft Optve	Lab	Labour Cost	Sunds	Nett Rate	Unit	Gross Rate (+7.5%)
	£	%	£	Hrs	Hrs	£	£	£		£
N10: GENERAL FIXTURES/FURNISHING/EQUIPMENT Cont.										
HAT AND COAT RAILS - Hat and coat hooks										
B.M.A. finish	2.00	2.50	2.05	0.25	–	2.31	–	4.36	nr	4.68
chromium plated	2.00	2.50	2.05	0.25	–	2.31	–	4.36	nr	4.68
SAA finish	1.00	2.50	1.02	0.25	–	2.31	–	3.33	nr	3.58
N11: DOMESTIC KITCHEN FITTINGS										
Standard melamine finish on chipboard units, with backs										
Wall units										
fixing to masonry with screws										
400 x 300 x 600mm............................	56.27	2.50	57.68	1.75	0.22	17.70	–	75.37	nr	81.03
400 x 300 x 900mm............................	73.20	2.50	75.03	1.85	0.23	18.69	–	93.72	nr	100.75
500 x 300 x 600mm............................	60.43	2.50	61.94	1.90	0.25	19.29	–	81.23	nr	87.33
500 x 300 x 900mm............................	76.62	2.50	78.54	1.95	0.25	19.75	–	98.29	nr	105.66
600 x 300 x 600mm............................	66.02	2.50	67.67	1.95	0.25	19.75	–	87.42	nr	93.98
600 x 300 x 900mm............................	79.87	2.50	81.87	2.05	0.26	20.75	–	102.61	nr	110.31
1000 x 300 x 600mm...........................	106.07	2.50	108.72	2.35	0.30	23.80	–	132.52	nr	142.46
1000 x 300 x 900mm...........................	129.28	2.50	132.51	2.50	0.31	25.25	–	157.76	nr	169.60
1200 x 300 x 600mm...........................	115.20	2.50	118.08	2.50	0.31	25.25	–	143.33	nr	154.08
1200 x 300 x 900mm...........................	131.17	2.50	134.45	2.70	0.35	27.38	–	161.83	nr	173.96
Floor units on plinths and with plastic faced worktops; without drawers										
fixing to masonry with screws										
400 x 600 x 900mm............................	106.50	2.50	109.16	2.00	0.25	20.22	–	129.38	nr	139.08
500 x 600 x 900mm............................	116.99	2.50	119.91	2.20	0.28	22.27	–	142.19	nr	152.85
600 x 600 x 900mm............................	125.25	2.50	128.38	2.40	0.30	24.26	–	152.64	nr	164.09
1000 x 600 x 900mm...........................	209.27	2.50	214.50	3.20	0.40	32.34	–	246.85	nr	265.36
1200 x 600 x 900mm...........................	227.41	2.50	233.10	3.50	0.44	35.39	–	268.49	nr	288.63
Floor units on plinths and with plastic faced worktops; with one drawer										
fixing to masonry with screws										
500 x 600 x 900mm............................	146.01	2.50	149.66	2.20	0.28	22.27	–	171.93	nr	184.83
600 x 600 x 900mm............................	154.27	2.50	158.13	2.40	0.30	24.26	–	182.38	nr	196.06
1000 x 600 x 900mm...........................	251.37	2.50	257.65	3.20	0.40	32.34	–	290.00	nr	311.75
1200 x 600 x 900mm...........................	271.64	2.50	278.43	3.50	0.44	35.39	–	313.82	nr	337.36
Floor units on plinths and with plastic faced worktops; with four drawers										
fixing to masonry with screws										
500 x 600 x 900mm............................	203.66	2.50	208.75	2.20	0.28	22.27	–	231.02	nr	248.35
600 x 600 x 900mm............................	215.28	2.50	220.66	2.40	0.30	24.26	–	244.92	nr	263.29
Sink units on plinths										
1200 x 600 x 900mm, with one drawer; fixing to masonry with screws	192.90	2.50	197.72	3.50	0.44	35.39	–	233.12	nr	250.60
1200 x 600 x 900mm, without drawer; fixing to masonry with screws	147.37	2.50	151.05	3.50	0.44	35.39	–	186.45	nr	200.43
1500 x 600 x 900mm, without drawer; fixing to masonry with screws	230.78	2.50	236.55	3.75	0.47	37.91	–	274.46	nr	295.05
Store cupboards on plinths										
fixing to masonry with screws										
500 x 600 x 1950mm without shelves	169.92	2.50	174.17	3.60	0.45	36.39	–	210.56	nr	226.35
500 x 600 x 1950mm with shelves	181.76	2.50	186.30	3.60	0.45	36.39	–	222.69	nr	239.39
600 x 600 x 1950mm without shelves	184.97	2.50	189.59	4.00	0.50	40.43	–	230.02	nr	247.28
600 x 600 x 1950mm with shelves	199.48	2.50	204.47	4.00	0.50	40.43	–	244.90	nr	263.26
Laminated plastic faced units complete with ironmongery and with backs										
Wall units										
400 x 300 x 600mm, with one door; fixing to masonry with screws	112.49	2.50	115.30	1.75	0.22	17.70	–	133.00	nr	142.97
500 x 300 x 600mm, with one door; fixing to masonry with screws	120.84	2.50	123.86	1.85	0.23	18.69	–	142.55	nr	153.24
1000 x 300 x 600mm, with two doors; fixing to masonry with screws	212.13	2.50	217.43	2.35	0.30	23.80	–	241.23	nr	259.32
1200 x 300 x 600mm, with two doors; fixing to masonry with screws	230.39	2.50	236.15	2.50	0.31	25.25	–	261.40	nr	281.01
Floor units on plinths and with worktops										
500 x 600 x 900mm, with one drawer and one cupboard; fixing to masonry with screws	292.03	2.50	299.33	2.20	0.28	22.27	–	321.60	nr	345.72
500 x 600 x 900mm, with four drawers; fixing to masonry with screws	407.32	2.50	417.50	2.40	0.30	24.26	–	441.76	nr	474.89
1000 x 600 x 900mm, with two drawers and two cupboards; fixing to masonry with screws	502.73	2.50	515.30	3.20	0.40	32.34	–	547.64	nr	588.72
1500 x 600 x 900mm, with three drawers and three cupboards; fixing to masonry with screws	649.52	2.50	665.76	3.85	0.48	38.91	–	704.66	nr	757.51
Sink units on plinths										
1000 x 600 x 900mm, with one drawer and two cupboards, fixing to masonry with screws	385.63	2.50	395.27	3.65	0.47	36.99	–	432.26	nr	464.68
1500 x 600 x 900mm, with two drawers and three cupboards; fixing to masonry with screws	576.82	2.50	591.24	3.75	0.50	38.12	–	629.36	nr	676.57

Labour hourly rates: (except Specialists) Craft Operatives £9.23 Labourer £7.02 Rates are national average prices. Refer to REGIONAL VARIATIONS for indicative levels of overall pricing in regions	MATERIALS			LABOUR				RATES		
	Del to Site £	Waste %	Material Cost £	Craft Optve Hrs	Lab Hrs	Labour Cost £	Sunds £	Nett Rate £	Unit	Gross Rate (+7.5%) £
N11: DOMESTIC KITCHEN FITTINGS Cont.										
Laminated plastic faced units complete with ironmongery and with backs Cont.										
Tall units on plinths										
500 x 600 x 1950mm, broom cupboard; fixing to masonry with screws	354.69	2.50	363.56	3.75	0.50	38.12	–	401.68	nr	431.81
600 x 600 x 1950mm, larder unit; fixing to masonry with screws	370.09	2.50	379.34	3.75	0.50	38.12	–	417.46	nr	448.77
Hardwood veneered units complete with ironmongery but without backs										
Wardrobe units										
fixing to masonry with screws										
1000 x 600 x 2175mm	474.00	2.50	485.85	4.75	0.60	48.05	–	533.90	nr	573.95
1500 x 600 x 2175mm	595.50	2.50	610.39	4.75	0.60	48.05	–	658.44	nr	707.83
Pre-Assembled kitchen units, white melamine finish; Module 500 range; Magnet PLC										
Drawerline units; fixing to masonry with screws										
floor units										
300mm	69.62	2.50	71.36	1.25	0.31	13.71	–	85.07	nr	91.45
500mm	80.35	2.50	82.36	1.25	0.34	13.92	–	96.28	nr	103.50
600mm	84.18	2.50	86.28	1.25	0.34	13.92	–	100.21	nr	107.72
1000mm	130.14	2.50	133.39	1.75	0.45	19.31	–	152.71	nr	164.16
hob floor units										
1000mm	109.46	2.50	112.20	1.75	0.44	19.24	–	131.44	nr	141.30
sink units										
1000mm	98.73	2.50	101.20	1.75	0.44	19.24	–	120.44	nr	129.47
Tall units; fixing to masonry with screws										
floor units										
500mm	150.05	2.50	153.80	2.50	0.30	25.18	–	178.98	nr	192.41
corner floor units										
1000mm	103.33	2.50	105.91	1.75	0.38	18.82	–	124.73	nr	134.09
sink/floor units										
1000mm	98.73	2.50	101.20	1.75	0.38	18.82	–	120.02	nr	129.02
Drawer units; fixing to masonry with screws										
four drawer floor units										
500mm	145.55	2.50	149.19	1.25	0.46	14.77	–	163.96	nr	176.25
Wall units; fixing to masonry with screws										
standard units 600mm high										
300mm	37.46	2.50	38.40	2.00	0.32	20.71	–	59.10	nr	63.54
500mm	46.65	2.50	47.82	2.00	0.35	20.92	–	68.73	nr	73.89
600mm	72.69	2.50	74.51	2.00	0.37	21.06	–	95.56	nr	102.73
1000mm	80.35	2.50	82.36	2.60	0.47	27.30	–	109.66	nr	117.88
standard corner units 900mm high										
600 x 600mm	130.91	2.50	134.18	3.00	0.46	30.92	–	165.10	nr	177.48
Worktops; fixing with screws										
round front edge										
28mm	12.74	25.00	15.93	1.50	0.19	15.18	–	31.10	m	33.44
40mm	19.38	25.00	24.23	1.75	0.22	17.70	–	41.92	m	45.07
Plinth										
Minster 500	18.98	25.00	23.73	0.25	0.03	2.52	–	26.24	m	28.21
Cornice										
Minster 500	22.10	25.00	27.63	0.25	0.03	2.52	–	30.14	m	32.40
Pelmet										
Minster 500	17.84	25.00	22.30	0.25	0.03	2.52	–	24.82	m	26.68
Worktop trim										
coloured	5.28	25.00	6.60	0.25	0.03	2.52	–	9.12	m	9.80
N13: SANITARY APPLIANCES/FITTINGS										
Attendance on sanitary fittings										
Note: Sanitary fittings are usually included as p.c. items or provisional sums, the following are the allowances for attendance by the main contractor i.e. unloading, storing and distributing fittings for fixing by the plumber and returning empty cases and packings										
Sinks, fireclay										
610 x 457 x 254mm	–	–	–	–	1.00	7.02	0.54	7.56	nr	8.13
762 x 508 x 254mm	–	–	–	–	1.25	8.78	0.68	9.46	nr	10.16
add if including tubular stands	–	–	–	–	0.75	5.26	0.45	5.71	nr	6.14
Wash basins, 559 x 406mm; earthenware or fireclay										
single	–	–	–	–	0.80	5.62	0.45	6.07	nr	6.52
range of 4 with cover overlaps between basins	–	–	–	–	2.50	17.55	1.22	18.77	nr	20.18
Combined sinks and drainers; stainless steel										
1067 x 533mm	–	–	–	–	1.40	9.83	0.80	10.63	nr	11.43
1600 x 610mm	–	–	–	–	1.70	11.93	0.95	12.88	nr	13.85
add if including tubular stands	–	–	–	–	0.75	5.26	0.45	5.71	nr	6.14

FURNITURE/EQUIPMENT

FURNITURE/EQUIPMENT – MAJOR WORKS

Labour hourly rates: (except Specialists) Craft Operatives £9.23 Labourer £7.02 Rates are national average prices. Refer to REGIONAL VARIATIONS for indicative levels of overall pricing in regions	MATERIALS			LABOUR				RATES		
	Del to Site £	Waste %	Material Cost £	Craft Optve Hrs	Lab Hrs	Labour Cost £	Sunds £	Nett Rate £	Unit	Gross Rate (+7.5%) £
N13: SANITARY APPLIANCES/FITTINGS Cont.										
Attendance on sanitary fittings Cont.										
Combined sinks and drainers; porcelain enamel	..									
1067 x 533mm	-	-	-	-	1.40	9.83	0.80	10.63	nr	11.43
1600 x 610mm	-	-	-	-	1.70	11.93	0.95	12.88	nr	13.85
Baths excluding panels; pressed steel										
694 x 1688 x 570mm overall	-	-	-	-	2.50	17.55	1.18	18.73	nr	20.13
W.C. suites; china										
complete with WWP, flush pipe etc., ...	-	-	-	-	1.40	9.83	0.80	10.63	nr	11.43
Block pattern urinals; with glazed ends, back and channel in one piece with separate tread including flushing cistern etc.,										
single stall	-	-	-	-	2.75	19.31	1.35	20.66	nr	22.20
range of four with loose overlaps	-	-	-	-	6.00	42.12	3.15	45.27	nr	48.67
Wall urinals, bowl type, with flushing cistern, etc.,										
single...	-	-	-	-	1.40	9.83	0.80	10.63	nr	11.43
range of 3 with two divisions	-	-	-	-	3.30	23.17	1.62	24.79	nr	26.64
Slab urinals with one return end, flushing cistern, sparge pipe, etc.,										
1219mm long x 1067mm high	-	-	-	-	2.75	19.31	1.80	21.11	nr	22.69
1829mm long x 1067mm high	-	-	-	-	3.75	26.32	2.70	29.02	nr	31.20

This section continues
on the next page

Labour hourly rates: (except Specialists) Craft Operatives £12.44 Labourer £7.02 Rates are national average prices. Refer to REGIONAL VARIATIONS for indicative levels of overall pricing in regions	MATERIALS			LABOUR				RATES		
	Del to Site £	Waste %	Material Cost £	Craft Optve Hrs	Lab Hrs	Labour Cost £	Sunds £	Nett Rate £	Unit	Gross Rate (+7.5%) £
N13: SANITARY APPLIANCES/FITTINGS Cont.										
Fix only appliances (prime cost sum for supply included elsewhere)										
Sink units, stainless steel; combined overflow and waste outlet, plug and chain; pair pillar taps										
fixing on base unit with metal clips										
1000 x 600mm.....................	-	-	-	3.00	-	37.32	0.80	38.12	nr	40.98
1600 x 600mm.....................	-	-	-	3.50	-	43.54	0.80	44.34	nr	47.67
Sinks; white glazed fireclay; waste outlet, plug, chain and stay; cantilever brackets; pair bib taps										
fixing brackets to masonry with screws; sealing at back with white sealant										
610 x 457 x 254mm.................	-	-	-	4.00	-	49.76	1.30	51.06	nr	54.89
762 x 508 x 254mm.................	-	-	-	4.00	-	49.76	1.30	51.06	nr	54.89
Sinks; white glazed fireclay; waste outlet, plug, chain and stay; legs and screw-to-wall bearers; pair bib taps										
fixing legs to masonry with screws; sealing at back with mastic sealant										
610 x 457 x 254mm.................	-	-	-	4.50	-	55.98	1.30	57.28	nr	61.58
762 x 508 x 254mm.................	-	-	-	4.50	-	55.98	1.30	57.28	nr	61.58
Wash basins; vitreous china; waste outlet, plug, chain and stay; screw to wall brackets; pair pillar taps										
560 x 406mm; fixing brackets to masonry with screws; sealing at back with mastic sealant	-	-	-	3.00	-	37.32	1.30	38.62	nr	41.52
range of four, each 559 x 406mm with overlap strips; fixing brackets to masonry with screws; sealing overlap strips and at back with mastic sealant	-	-	-	12.75	-	158.61	4.95	163.56	nr	175.83
Wash basins; vitreous china; waste outlet, plug, chain and stay; legs and screw-to-wall bearers										
560 x 406mm; fixing legs to masonry with screws; sealing at back with mastic sealant	-	-	-	3.50	-	43.54	1.30	44.84	nr	48.20
range of 4, each 559 x 406mm with overlap strips; fixing legs to masonry with screws; sealing overlap strips and at back with mastic sealant ...	-	-	-	13.75	-	171.05	4.95	176.00	nr	189.20
Baths; enamelled pressed steel; combined overflow and waste outlet, plug and chain; metal cradles; pair pillar taps										
700 x 1700 x 570mm overall; sealing at walls with mastic sealant	-	-	-	5.50	-	68.42	2.90	71.32	nr	76.67
High level W.C. suites; vitreous china pan; plastics cistern with valveless fittings, ball valve, chain and pull handle; flush pipe; plastics seat and cover										
fixing pan and cistern brackets to masonry with screws; bedding pan in mastic; jointing pan to drain with cement mortar (1:2) and gaskin joint ..	-	-	-	4.50	-	55.98	1.75	57.73	nr	62.06
fixing pan and cistern brackets to masonry with screws; bedding pan in mastic; jointing pan to soil pipe with Multikwik connector	-	-	-	4.50	-	55.98	1.75	57.73	nr	62.06
Low level W.C. suites; vitreous china pan; plastics cistern with valveless fittings and ball valve; flush pipe; plastics seat and cover										
fixing pan and cistern brackets to masonry with screws; bedding pan in mastic; jointing pan to drain with cement mortar (1:2) and gaskin joint ..	-	-	-	4.50	-	55.98	1.75	57.73	nr	62.06
fixing pan and cistern brackets to masonry with screws; bedding pan in mastic; jointing pan to soil pipe with Multikwik connector	-	-	-	4.50	-	55.98	1.75	57.73	nr	62.06
Wall urinals; vitreous china bowl; automatic cistern; spreads and flush pipe; waste outlet										
single bowl; fixing bowl, cistern and pipe brackets to masonry with screws	-	-	-	4.50	-	55.98	1.35	57.33	nr	61.63
range of three bowls; fixing bowls, divisions, cistern and pipe brackets to masonry with screws .	-	-	-	11.50	-	143.06	2.75	145.81	nr	156.75
Single stall urinals; white glazed fireclay in one piece; white glazed fireclay tread; plastics automatic cistern; spreader and flush pipe; waste outlet and domed grating										
bedding stall and tread in cement mortar (1:4) and jointing tread with waterproof jointing compound; fixing cistern and pipe brackets to masonry with screws	-	-	-	7.00	-	87.08	3.30	90.38	nr	97.16
Slab urinals; range of four; white glazed fireclay back, ends, divisions, channel and tread; automatic cistern; flush and sparge pipes; waste outlet and domed grating										
bedding back, ends, channel and tread in cement mortar (1:4) and jointing with waterproof jointing compound; fixing divisions, cistern and pipe brackets to masonry with screws	-	-	-	19.00	-	236.36	10.50	246.86	nr	265.37

FURNITURE/EQUIPMENT

Labour hourly rates: (except Specialists) Craft Operatives £12.44 Labourer £7.02 Rates are national average prices. Refer to REGIONAL VARIATIONS for indicative levels of overall pricing in regions	MATERIALS			LABOUR				RATES		
	Del to Site	Waste	Material Cost	Craft Optve	Lab	Labour Cost	Sunds	Nett Rate	Unit	Gross Rate (+7.5%)
	£	%	£	Hrs	Hrs	£	£	£		£
N13: SANITARY APPLIANCES/FITTINGS Cont.										
Fix only appliances (prime cost sum for supply included elsewhere) Cont.										
Showers; white glazed fireclay tray 760 x 760 x 180mm; waste outlet; recessed valve and spray head										
bedding tray in cement mortar (1:4); fixing valve head to masonry with screws; sealing at walls with										
mastic sealant	-	-	-	4.75	-	59.09	4.25	63.34	nr	68.09
Shower curtains with rails, hooks, supports and fittings										
straight, 914mm long x 1829mm drop; fixing supports to masonry with screws	-	-	-	1.00	-	12.44	0.33	12.77	nr	13.73
angled, 1676mm girth x 1829mm drop; fixing supports to masonry with screws	-	-	-	1.50	-	18.66	0.45	19.11	nr	20.54
Supply and fix appliances										
Note the following prices include joints to copper services and wastes but do not include traps										
Sink units; single bowl with single drainer and back ledge; stainless steel; B.S.1244 Part 2 Type A; 38mm waste, plug and chain to B.S.3380 Part 1 with combined overflow; pair 13mm pillar taps to B.S.1010 Part 2										
fixing on base unit with metal clips										
1000 x 600mm.....................................	71.00	2.50	72.78	3.00	-	37.32	0.80	110.90	nr	119.21
1200 x 600mm.....................................	78.00	2.50	79.95	3.50	-	43.54	0.80	124.29	nr	133.61
Sink units; single bowl with double drainer and back ledge; stainless steel; B.S.1244 Part 2 Type B; 38mm waste, plug and chain to B.S.3380 Part 1 with combined overflow; pair 13mm pillar taps to B.S.1010 Part 2										
fixing on base unit with metal clips										
1500 x 600mm.....................................	88.00	2.50	90.20	3.00	-	37.32	1.05	128.57	nr	138.21
Sinks; white glazed fireclay; B.S.1206 with weir overflow; 38mm slotted waste, chain, stay and plug to B.S.3380 Part 1; painted cantilever brackets										
screwing chain stay; fixing brackets to masonry with screws; sealing at back with mastic sealant										
455 x 380 x 205mm............................	65.75	2.50	67.39	3.50	-	43.54	1.30	112.23	nr	120.65
610 x 455 x 255mm............................	92.75	2.50	95.07	3.50	-	43.54	1.30	139.91	nr	150.40
760 x 455 x 255mm............................	141.75	2.50	145.29	3.50	-	43.54	1.30	190.13	nr	204.39
915 x 610 x 305mm............................	407.75	2.50	417.94	4.50	-	55.98	1.30	475.22	nr	510.87
Sinks; white glazed fireclay; B.S.1206 with wier overflow; 38mm slotted waste, chain, stay and plug to B.S.3380 Part 1; plastic coated cantilever brackets										
screwing chain stay; fixing brackets to masonry with screws; sealing at back with mastic sealant										
455 x 380 x 205mm............................	72.00	2.50	73.80	3.50	-	43.54	1.30	118.64	nr	127.54
610 x 455 x 255mm............................	99.00	2.50	101.47	3.50	-	43.54	1.30	146.32	nr	157.29
760 x 455 x 255mm............................	148.00	2.50	151.70	3.50	-	43.54	1.30	196.54	nr	211.28
915 x 610 x 305mm............................	414.00	2.50	424.35	4.50	-	55.98	1.30	481.63	nr	517.75
Sinks; white glazed fireclay; B.S.1206 with wier overflow; 38mm slotted waste, chain, stay and plug to B.S.3380 Part 1; painted legs and screw-to-wall bearers										
screwing chain stay; fixing legs to masonry with screws; sealing at back with mastic sealant										
455 x 380 x 205mm............................	81.50	2.50	83.54	3.50	-	43.54	1.30	128.38	nr	138.01
610 x 455 x 255mm............................	108.50	2.50	111.21	3.50	-	43.54	1.30	156.05	nr	167.76
760 x 455 x 255mm............................	157.50	2.50	161.44	3.50	-	43.54	1.30	206.28	nr	221.75
915 x 610 x 305mm............................	423.50	2.50	434.09	4.50	-	55.98	1.30	491.37	nr	528.22
Wash basins; white vitreous china; B.S.5506 Part 3; 32mm slotted waste, chain, stay and plug to B.S.3380 Part 1; painted cantilever brackets; pair 13 pillar taps to B.S.1010 Part 2										
sealing at back with mastic sealant										
560 x 406mm.....................................	38.25	2.50	39.21	3.50	-	43.54	1.30	84.05	nr	90.35
Wash basins; white vitreous china; B.S.5506 Part 3; 32mm slotted waste, chain, stay and plug to B.S.3380 Part 1; plastic coated cantilever brackets; pair 13mm pillar taps to B.S.1010 Part 2										
sealing at back with mastic sealant										
560 x 406mm.....................................	43.50	2.50	44.59	3.50	-	43.54	1.30	89.43	nr	96.13

Labour hourly rates: (except Specialists) Craft Operatives £12.44 Labourer £7.02 Rates are national average prices. Refer to REGIONAL VARIATIONS for indicative levels of overall pricing in regions	MATERIALS			LABOUR				RATES		
	Del to Site £	Waste %	Material Cost £	Craft Optve Hrs	Lab Hrs	Labour Cost £	Sunds £	Nett Rate £	Unit	Gross Rate (+7.5%) £
N13: SANITARY APPLIANCES/FITTINGS Cont.										
Supply and fix appliances Cont.										
Wash basins; white vitreous china; B.S.5506 Part 3; 32mm slotted waste, chain, stay and plug to B.S.3380 Part 1; painted legs and screw-to-wall bearers; pair 13mm taps to B.S.1010 Part 2 fixing legs to masonry with screws; sealing at back with mastic sealant 560 x 406mm.....................	45.25	2.50	46.38	4.10	–	51.00	1.30	98.69	nr	106.09
Wash basins; white vitreous china; B.S.5506 Part 3; 32mm slotted waste, chain, stay and plug to B.S.3380 Part 1; plastic coated legs and screw-to-wall bearers; pair 13mm pillar taps to B.S.1010 Part 2 fixing legs to masonry with screws; sealing at back with mastic sealant 560 x 406mm.....................	54.25	2.50	55.61	4.10	–	51.00	1.30	107.91	nr	116.00
Wash basins; white vitreous china; B.S.5506 Part 3; 32mm slotted waste, chain, stay and plug to B.S.3380 Part 1; painted towel rail brackets; pair 13mm pillar taps to B.S.1010 Part 2 fixing brackets to masonry with screws; sealing at back with mastic sealant 560 x 406mm.....................	36.50	2.50	37.41	4.00	–	49.76	1.30	88.47	nr	95.11
Wash basins; white vitreous china; B.S.5506 Part 3; 32mm slotted waste, chain, stay and plug to B.S.3380 Part 1; plastic coated towel rail brackets; pair 13mm pillar taps to B.S.1010 Part 2 fixing brackets to masonry with screws; sealing at back with mastic sealant 560 x 406mm.....................	39.25	2.50	40.23	4.00	–	49.76	1.30	91.29	nr	98.14
Wash basins; coloured vitreous china; B.S.5506 Part 3; 32mm slotted waste, chain, stay and plug to B.S.3380 Part 1; painted build-in brackets; pair 13mm pillar taps to B.S.1010 Part 2 building in brackets to masonry; sealing at back with mastic sealant 560 x 406mm.....................	60.25	2.50	61.76	3.50	–	43.54	1.30	106.60	nr	114.59
Wash basins; coloured vitreous china; B.S.5506 Part 3; 32mm slotted waste, chain, stay and plug to B.S.3380 Part 1; plastic coated build-in brackets; pair 13mm pillar taps to B.S.1010 Part 2 building in brackets to masonry; sealing at back with mastic sealant 560 x 406mm.....................	65.50	2.50	67.14	3.50	–	43.54	1.30	111.98	nr	120.38
Wash basins; coloured vitreous china; B.S.5506 Part 3; 32mm slotted waste, chain, stay and plug to B.S.3380 Part 1; painted legs and screw-to-wall bearers; pair 13mm pillar taps to B.S.1010 Part 2 fixing legs to masonry with screws; sealing at back with mastic sealant 560 x 406mm.....................	67.25	2.50	68.93	4.10	–	51.00	1.30	121.24	nr	130.33
Wash basins; coloured vitreous china; B.S.5506 Part 3; 32mm slotted waste, chain, stay and plug to B.S.3380 Part 1; plastic coated legs and screw-to-wall bearers; pair 13mm pillar taps to B.S.1010 Part 2 fixing legs to masonry with screws; sealing at back with mastic sealant 560 x 406mm.....................	75.75	2.50	77.64	4.10	–	51.00	1.30	129.95	nr	139.69
Wash basins; coloured vitreous china; B.S.5506 Part 3; 32mm slotted waste, chain, stay and plug to B.S.3380 Part 1; painted towel rail brackets; pair 13mm pillar taps to B.S.1010 Part 2 fixing brackets to masonry with screws; sealing at back with mastic sealant 560 x 406mm.....................	58.50	2.50	59.96	4.00	–	49.76	1.30	111.02	nr	119.35
Wash basins; coloured vitreous china; B.S.5506 Part 3; 32mm slotted waste, chain, stay and plug to B.S.3380 Part 1; plastic coated towel rail brackets; pair 13mm pillar taps to B.S.1010 Part 2 fixing brackets to masonry with screws; sealing at back with mastic sealant 560 x 406mm.....................	61.25	2.50	62.78	4.00	–	49.76	1.30	113.84	nr	122.38
Wash basins, angle type; white vitreous china; B.S.5506 Part 3; 32mm slotted waste, chain, stay and plug to B.S.3380 Part 1; concealed brackets; pair 13mm pillar taps to B.S.1010 Part 2 fixing brackets to masonry with screws; sealing at back with mastic sealant 457 x 431mm.....................	64.00	2.50	65.60	4.00	–	49.76	1.30	116.66	nr	125.41

FURNITURE/EQUIPMENT

Labour hourly rates: (except Specialists) Craft Operatives £12.44 Labourer £7.02 Rates are national average prices. Refer to REGIONAL VARIATIONS for indicative levels of overall pricing in regions	MATERIALS			LABOUR				RATES		
	Del to Site £	Waste %	Material Cost £	Craft Optve Hrs	Lab Hrs	Labour Cost £	Sunds £	Nett Rate £	Unit	Gross Rate (+7.5%) £
N13: SANITARY APPLIANCES/FITTINGS Cont.										
Supply and fix appliances Cont.										
Wash basins, angle type; coloured vitreous china; B.S.5506 Part 3; 32mm slotted waste, chain, stay and plug to B.S.3380 Part 1; concealed brackets; pair 13mm pillar taps to B.S.1010 Part 2 fixing brackets to masonry with screws; sealing at back with mastic sealant	..									
457 x 431mm..................................	76.00	2.50	77.90	4.00	–	49.76	1.30	128.96	nr	138.63
Wash basins with pedestal; white vitreous china; B.S.5506 Part 3; 32mm slotted waste, chain, stay and plug to B.S.3380 Part 1; pair 13mm pillar taps to B.S.1010 Part 2 fixing basin to masonry with screws; sealing at back with mastic sealant										
560 x 406mm..................................	61.75	2.50	63.29	4.50	–	55.98	1.30	120.57	nr	129.62
Wash basins with pedestal; coloured vitreous china; B.S.5506 Part 3; 32mm slotted waste, chain, stay and plug to B.S.3380 Part 1; pair 13mm pillar taps to B.S.1010 Part 2 fixing basin to masonry with screws; sealing at back with mastic sealant										
560 x 406mm..................................	88.25	2.50	90.46	3.75	–	46.65	1.30	138.41	nr	148.79
Baths; white enamelled pressed steel; B.S.1390; 32mm overflow with front grid and 38mm waste, chain and plug to B.S.3380 Part 1 with combined overflow; metal cradles with adjustable feet; pair 19mm pillar taps to B.S.1010 Part 2 sealing at walls with mastic sealant										
700 x 1700 x 570mm overall..................	107.50	2.50	110.19	5.50	–	68.42	2.90	181.51	nr	195.12
724 x 1500 x 570mm overall..................	104.50	2.50	107.11	5.50	–	68.42	2.90	178.43	nr	191.81
Baths; coloured enamelled pressed steel; B.S.1390; 32mm overflow with front grid and 38mm waste, chain and plug to B.S.3380 Part 1 with combined overflow; metal cradles with adjustable feet; pair 19mm pillar taps to B.S.1010 Part 2 sealing at walls with mastic sealant										
700 x 1700 x 570mm overall..................	118.75	2.50	121.72	5.50	–	68.42	2.90	193.04	nr	207.52
Baths; white acrylic; B.S.4305; 32mm overflow with front grid and 38mm waste, chain and plug to B.S.3380 Part 1 with combined overflow; metal cradles with adjustable feet and wall fixing brackets; pair 19mm pillar taps to B.S.1010 Part 2 700 x 1700 x 570mm overall; fixing brackets to masonry with screws; sealing at walls with mastic sealant....................................	124.50	2.50	127.61	5.00	–	62.20	2.90	192.71	nr	207.17
Baths; coloured acrylic; B.S.4305; 32mm overflow with front grid and 38mm waste, chain and plug to B.S.3380 Part 1 with combined overflow; metal cradles with adjustable feet and wall fixing brackets; pair 19mm pillar taps to B.S.1010 Part 2 700 x 1700 x 570mm overall; fixing brackets to masonry with screws; sealing at walls with mastic sealant....................................	124.50	2.50	127.61	5.00	–	62.20	2.90	192.71	nr	207.17
White vitreous china pans to B.S.5503 pan with horizontal outlet; fixing pan to timber with screws; bedding pan in mastic; jointing pan to soil pipe with Multikwik connector	25.25	2.50	25.88	2.50	–	31.10	0.70	57.68	nr	62.01
pan with horizontal outlet; fixing pan to masonry with screws; bedding pan in mastic; jointing pan to soil pipe with Multikwik connector	25.25	2.50	25.88	3.00	–	37.32	1.00	64.20	nr	69.02
pan with S or P trap conversion bend to B.S.5627; fixing pan to timber with screws; bedding pan in mastic; jointing pan to drain with cement mortar (1:2) and gaskin joint	25.50	2.50	26.14	2.75	–	34.21	0.70	61.05	nr	65.63
pan with S or P trap conversion bend to B.S.5627; fixing pan to masonry with screws; bedding pan in mastic; jointing pan to drain with cement mortar (1:2) and gaskin joint	25.50	2.50	26.14	3.25	–	40.43	1.00	67.57	nr	72.64
pan with S or P trap conversion bend to B.S.5627; fixing pan to timber with screws; bedding pan in mastic; jointing pan to soil pipe with Multikwik connector	28.50	2.50	29.21	2.50	–	31.10	0.70	61.01	nr	65.59
pan with S or P trap conversion bend to B.S.5627; fixing pan to masonry with screws; bedding pan in mastic; jointing pan to soil pipe with Multikwik connector	28.50	2.50	29.21	3.00	–	37.32	1.00	67.53	nr	72.60
Coloured vitreous china pans to B.S.5503 pan with horizontal outlet; fixing pan to timber with screws; bedding pan in mastic; jointing pan to soil pipe with Multikwik connector	48.00	2.50	49.20	2.50	–	31.10	0.70	81.00	nr	87.08

Labour hourly rates: (except Specialists) Craft Operatives £12.44 Labourer £7.02 Rates are national average prices. Refer to REGIONAL VARIATIONS for indicative levels of overall pricing in regions	MATERIALS			LABOUR				RATES		
	Del to Site £	Waste %	Material Cost £	Craft Optve Hrs	Lab Hrs	Labour Cost £	Sunds £	Nett Rate £	Unit	Gross Rate (+7.5%) £
N13: SANITARY APPLIANCES/FITTINGS Cont.										
Supply and fix appliances Cont.										
Coloured vitreous china pans to B.S.5503 Cont. pan with horizontal outlet; fixing pan to masonry with screws; bedding pan in mastic; jointing pan to soil pipe with Multikwik connector	48.00	2.50	49.20	3.00	-	37.32	1.00	87.52	nr	94.08
Plastics W.C. seats; B.S.1254, type 1 black; fixing to pan	12.00	2.50	12.30	0.25	-	3.11	-	15.41	nr	16.57
Plastics W.C. seats and covers; B.S.1254, type 1 black; fixing to pan	14.50	2.50	14.86	0.25	-	3.11	-	17.97	nr	19.32
coloured; fixing to pan	24.00	2.50	24.60	0.25	-	3.11	-	27.71	nr	29.79
white; fixing to pan	27.00	2.50	27.68	0.25	-	3.11	-	30.79	nr	33.09
9 litre black plastics cistern with valveless fittings, chain and pull handle to B.S.1125; 13mm piston type high pressure ball valve to B.S.1212 Part 1 with 127mm plastics float to B.S.2456 fixing cistern brackets to masonry with screws ...	41.00	2.50	42.02	1.00	-	12.44	0.33	54.80	nr	58.90
Plastics flush pipe to B.S.1125; adjustable for back wall fixing	3.25	2.50	3.33	0.75	-	9.33	0.33	12.99	nr	13.97
for side wall fixing	4.50	2.50	4.61	0.75	-	9.33	0.33	14.27	nr	15.34
High level W.C. suites; white vitreous china pan to B.S.5503; 9 litre black plastics cistern with valveless fittings, chain and pull handle and plastics flush pipe to B.S.1125; 13mm piston type high pressure ball valve to B.S.1212 Part 1 with 127mm plastics float to B.S.2456; black plastics seat and cover to B.S.1254 type 1 pan with S or P trap conversion bend to B.S.5627; fixing pan to timber with screws; fixing cistern brackets to masonry with screws; bedding pan in mastic; jointing pan to soil pipe with Multikwik connector	84.50	2.50	86.61	4.00	-	49.76	1.75	138.12	nr	148.48
pan with S or P trap conversion bend to B.S.5627; fixing pan and cistern brackets to masonry with screws; bedding pan in mastic; jointing pan to soil pipe with Multikwik connector	84.50	2.50	86.61	4.50	-	55.98	1.75	144.34	nr	155.17
Low level W.C. suites; white vitreous china pan to B.S.5503; 9 litre white vitreous china cistern with valveless fittings and plastics flush bend to B.S.1125; 13mm piston type high pressure ball valve to B.S.1212 Part 1 with 127mm plastics float to B.S.2456; white plastics seat and cover to B.S.1254 type 1 pan with S or P trap conversion bend to B.S.5627; fixing pan to timber with screws; fixing cistern brackets to masonry with screws; bedding pan in mastic; jointing pan to soil pipe with Multikwik connector	87.25	2.50	89.43	4.00	-	49.76	1.75	140.94	nr	151.51
pan with S or P trap conversion bend to B.S.5627; fixing pan and cistern brackets to masonry with screws; bedding pan in mastic; jointing pan to soil pipe with Multikwik connector	87.25	2.50	89.43	4.50	-	55.98	1.75	147.16	nr	158.20
Low level W.C. suites; coloured vitreous china pan to B.S.5503; 9 litre coloured vitreous china cistern with valveless fittings and plastics flush bend to B.S.1125; 13mm piston type high pressure ball valve to B.S.1212 Part 1 with 127mm plastics float to B.S.2456; coloured plastics seat and cover to B.S.1254 type 1 pan with S or P trap conversion bend to B.S.5627; fixing pan to timber with screws; fixing cistern brackets to masonry with screws; bedding pan in mastic; jointing pan to soil pipe with Multikwik connector	127.50	2.50	130.69	4.00	-	49.76	1.75	182.20	nr	195.86
pan with S or P trap conversion bend to B.S.5627; fixing pan and cistern brackets to masonry with screws; bedding pan in mastic; jointing pan to soil pipe with Multikwik connector	127.50	2.50	130.69	4.50	-	55.98	1.75	188.42	nr	202.55
Wall urinals; white vitreous china bowl to B.S.5520; 4.5 litre white vitreous china automatic cistern to B.S.1876; spreader and stainless steel flush pipe; 32mm plastics waste outlet fixing bowl, cistern and pipe brackets to masonry with screws	184.25	2.50	188.86	4.50	-	55.98	1.35	246.19	nr	264.65

Labour hourly rates: (except Specialists) Craft Operatives £12.44 Labourer £7.02 Rates are national average prices. Refer to REGIONAL VARIATIONS for indicative levels of overall pricing in regions	MATERIALS			LABOUR				RATES		
	Del to Site £	Waste %	Material Cost £	Craft Optve Hrs	Lab Hrs	Labour Cost £	Sunds £	Nett Rate £	Unit	Gross Rate (+7.5%) £

N13: SANITARY APPLIANCES/FITTINGS Cont.

Supply and fix appliances Cont.

Wall urinals; range of 2; white vitreous china bowls to B.S.5520; white vitreous china divisions; 9 litre white plastics automatic cistern to B.S.1876; stainless steel flush pipes and spreaders; 32mm plastics waste outlets										
fixing bowls, divisions, cistern and pipe brackets to masonry with screws	362.75	2.50	371.82	9.00	-	111.96	2.00	485.78	nr	522.21
Wall urinals; range of 3; white vitreous china bowls to B.S.5520; white vitreous china divisions; 14 litre white plastics automatic cistern to B.S.1876; stainless steel flush pipes and spreaders; 32mm plastics waste outlets										
fixing bowls, divisions, cistern and pipe brackets to masonry with screws	537.50	2.50	550.94	13.50	-	167.94	2.65	721.53	nr	775.64
1200mm long stainless steel wall hung trough urinal with plastic; white; automatic cistern to B.S.1876; spreader and stainless steel flush pipe; waste outlet and domed grating										
fixing urinal cistern and pipe brackets to masonry with screws	370.00	2.50	379.25	5.00	-	62.20	3.30	444.75	nr	478.11
1800mm long stainless steel wall hung trough urinal with plastic; white; automatic cistern to B.S.1876; spreader and stainless steel flush pipe; waste outlet and domed grating										
fixing urinal cistern and pipe brackets to masonry with screws	417.50	2.50	427.94	7.00	-	87.08	5.70	520.72	nr	559.77
2400mm long stainless steel wall hung trough urinal with plastic; white; automatic cistern to B.S.1876; spreader and stainless steel flush pipe; waste outlet and domed grating										
fixing urinal cistern and pipe brackets to masonry with screws	520.00	2.50	533.00	10.00	-	124.40	8.00	665.40	nr	715.30
Showers; white glazed Armastone tray, 760 x 760 x 180mm; 38mm waste outlet; surface mechanical valve										
bedding tray in cement mortar (1:4); fixing valve and head to masonry with screws; sealing at walls with mastic sealant	176.00	2.50	180.40	4.75	-	59.09	4.25	243.74	nr	262.02
Showers; white glazed Armastone tray, 760 x 760 x 180mm; 38mm waste outlet; recessed thermostatic valve and swivel spray head										
bedding tray in cement mortar (1:4); fixing valve and head to masonry with screws; sealing at walls with mastic sealant	294.50	2.50	301.86	4.75	-	59.09	4.25	365.20	nr	392.59
Showers; coloured glazed acrylic tray, 760 x 760 x 180mm; 38mm waste outlet; recessed thermostatic valve and swivel spray head										
bedding tray in cement mortar (1:4); fixing valve and head to masonry with screws; sealing at walls with mastic sealant	331.00	2.50	339.27	4.75	-	59.09	4.25	402.62	nr	432.81
Showers; white moulded acrylic tray with removable front and side panels, 760 x 760 x 260mm with adjustable metal cradle; 38mm waste outlet; surface fixing mechanical valve, flexible tube hand spray and slide bar										
fixing cradle, valve and slide bar to masonry with screws; sealing at walls with mastic sealant	245.50	2.50	251.64	4.75	-	59.09	2.85	313.58	nr	337.10
Showers; white moulded acrylic tray with removable front and side panels, 760 x 760 x 260mm with adjustable metal cradle; 38mm waste outlet; surface fixing thermostatic valve, flexible tube hand spray and slide bar										
fixing cradle, valve and slide bar to masonry with screws; sealing at walls with mastic sealant	279.00	2.50	285.98	4.75	-	59.09	2.85	347.92	nr	374.01
Shower curtains; nylon; anodised aluminium rail, glider hooks, end and suspension fittings										
straight, 914mm long x 1829mm drop	30.00	2.50	30.75	1.00	-	12.44	0.33	43.52	nr	46.78
angled, 1676mm girth x 1829mm drop	42.25	2.50	43.31	1.50	-	18.66	0.50	62.47	nr	67.15
Shower curtains; heavy duty plastic; anodised aluminium rail, glider hooks, end and suspension fitting; fixing supports to masonry with screws										
straight, 914mm long x 1829mm drop	34.00	2.50	34.85	1.00	-	12.44	0.33	47.62	nr	51.19
angled, 1676mm girth x 1829mm drop	45.75	2.50	46.89	1.50	-	18.66	0.50	66.05	nr	71.01
3.2mm gloss finish enamelled hardboard bath panels; polished aluminium angle corner strips										
to front, 1830mm long x 610mm high; fixing to timber with detachable dome head screws	7.80	2.50	8.00	1.00	0.12	13.28	0.50	21.78	nr	23.41

Labour hourly rates: (except Specialists) Craft Operatives £12.44 Labourer £7.02 Rates are national average prices. Refer to REGIONAL VARIATIONS for indicative levels of overall pricing in regions	MATERIALS			LABOUR				RATES		
	Del to Site £	Waste %	Material Cost £	Craft Optve Hrs	Lab Hrs	Labour Cost £	Sunds £	Nett Rate £	Unit	Gross Rate (+7.5%) £
N13: SANITARY APPLIANCES/FITTINGS Cont.										
Supply and fix appliances Cont.										
3.2mm gloss finish enamelled hardboard bath panels; polished aluminium angle corner strips Cont.										
to front and one end, 2590mm girth x 610mm high; fixing panels to timber with detachable dome head screws; fixing cover strips to timber with screws	13.28	2.50	**13.61**	2.00	0.25	**26.64**	1.00	**41.25**	nr	**44.34**
to front and two ends, 3350mm girth x 610mm high; fixing panels to timber with detachable dome head screws; fixing coverstrips to timber with screws .	20.79	2.50	**21.31**	3.00	0.37	**39.92**	1.50	**62.73**	nr	**67.43**

**This section continues
on the next page**

FURNITURE/EQUIPMENT

Labour hourly rates: (except Specialists) Craft Operatives £9.23 Labourer £7.02 Rates are national average prices. Refer to REGIONAL VARIATIONS for indicative levels of overall pricing in regions	MATERIALS			LABOUR				RATES		
	Del to Site £	Waste %	Material Cost £	Craft Optve Hrs	Lab Hrs	Labour Cost £	Sunds £	Nett Rate £	Unit	Gross Rate (+7.5%) £
N15: SIGNS/NOTICES										
Signwriting in gloss paint; one coat										
Letters or numerals, Helvetica medium style, on painted or varnished surfaces										
50mm high	0.01	5.00	0.01	0.14	–	1.29	0.01	1.31	nr	1.41
extra; shading	0.01	5.00	0.01	0.08	–	0.74	0.01	0.76	nr	0.82
extra; outline	0.01	5.00	0.01	0.12	–	1.11	0.01	1.13	nr	1.21
100mm high	0.01	5.00	0.01	0.28	–	2.58	0.02	2.61	nr	2.81
extra; shading	0.01	5.00	0.01	0.16	–	1.48	0.01	1.50	nr	1.61
extra; outline	0.01	5.00	0.01	0.24	–	2.22	0.02	2.25	nr	2.41
150mm high	0.02	5.00	0.02	0.42	–	3.88	0.03	3.93	nr	4.22
extra; shading	0.01	5.00	0.01	0.24	–	2.22	0.02	2.25	nr	2.41
extra; outline	0.01	5.00	0.01	0.36	–	3.32	0.03	3.36	nr	3.62
200mm high	0.02	5.00	0.02	0.56	–	5.17	0.04	5.23	nr	5.62
extra; shading	0.01	5.00	0.01	0.32	–	2.95	0.02	2.98	nr	3.21
extra; outline	0.01	5.00	0.01	0.48	–	4.43	0.04	4.48	nr	4.82
300mm high	0.03	5.00	0.03	0.84	–	7.75	0.06	7.84	nr	8.43
extra; shading	0.02	5.00	0.02	0.48	–	4.43	0.04	4.49	nr	4.83
extra; outline	0.02	5.00	0.02	0.72	–	6.65	0.05	6.72	nr	7.22
stops	–	–	–	0.04	–	0.37	–	0.37	nr	0.40
Signwriting in gloss paint; two coats										
Letters or numerals, Helvetica medium style, on painted or varnished surfaces										
50mm high	0.01	5.00	0.01	0.25	–	2.31	0.02	2.34	nr	2.51
100mm high	0.02	5.00	0.02	0.50	–	4.62	0.04	4.68	nr	5.03
150mm high	0.02	5.00	0.02	0.75	–	6.92	0.05	6.99	nr	7.52
200mm high	0.03	5.00	0.03	1.00	–	9.23	0.07	9.33	nr	10.03
250mm high	0.04	5.00	0.04	1.25	–	11.54	0.09	11.67	nr	12.54
300mm high	0.04	5.00	0.04	1.50	–	13.85	0.11	14.00	nr	15.05
stops	–	–	–	0.07	–	0.65	0.01	0.66	nr	0.71

Building fabric sundries

Labour hourly rates: (except Specialists) Craft Operatives £9.23 Labourer £7.02 Rates are national average prices. Refer to REGIONAL VARIATIONS for indicative levels of overall pricing in regions	MATERIALS			LABOUR				RATES		
	Del to Site £	Waste %	Material Cost £	Craft Optve Hrs	Lab Hrs	Labour Cost £	Sunds £	Nett Rate £	Unit	Gross Rate (+7.5%) £
P10: SUNDRY INSULATION/PROOFING WORK/ FIRE STOPS										
Waterproof reinforced building paper B.S.1521; grade A1F; 150mm lapped joints										
Across members at 450mm centres vertical	0.59	5.00	**0.62**	0.17	0.02	**1.71**	0.02	**2.35**	m2	2.53
Waterproof reinforced building paper; reflection (thermal) grade, single sided; 150mm lapped joints										
Across members at 450mm centres vertical	0.88	5.00	**0.92**	0.17	0.02	**1.71**	0.02	**2.65**	m2	2.85
Waterproof reinforced building paper; reflection (thermal) grade, double sided; 150mm lapped joints										
Across members at 450mm centres vertical	1.31	5.00	**1.38**	0.17	0.02	**1.71**	0.02	**3.11**	m2	3.34
British Sisalkraft Ltd Insulex 714 vapour control layer; 150mm laps sealed with tape										

Across members at 450mm centres
   horizontal; fixing to timber with stainless steel
   staples ....
	1.32	10.00	**1.45**	0.09	0.01	**0.90**	0.05	**2.40**	m2	2.58

Plain areas horizontal; laid loose	1.32	10.00	**1.45**	0.05	0.01	**0.53**	-	**1.98**	m2	2.13
British Sisalkraft Ltd SK860 vapour control layer; 100mm laps sealed with tape										
Plain areas horizontal; laid loose	1.02	10.00	**1.12**	0.07	0.01	**0.72**	0.03	**1.87**	m2	2.01
Mineral wool insulation quilt; butt joints										
60mm thick; across members at 450mm centres horizontal; laid loose	0.85	15.00	**0.98**	0.09	0.01	**0.90**	-	**1.88**	m2	2.02
Paper faced mineral wool insulation quilt; butt joints										
60mm thick; between members at 450mm centres vertical	1.65	15.00	**1.90**	0.20	0.02	**1.99**	0.03	**3.91**	m2	4.21
Glass fibre insulation quilt; butt joints										
60mm thick; across members at 450mm centres horizontal; laid loose	0.88	15.00	**1.01**	0.09	0.01	**0.90**	-	**1.91**	m2	2.06
80mm thick; across members at 450mm centres horizontal; laid loose	1.24	15.00	**1.43**	0.09	0.01	**0.90**	-	**2.33**	m2	2.50
100mm thick; across members at 450mm centres horizontal; laid loose	1.54	15.00	**1.77**	0.09	0.01	**0.90**	-	**2.67**	m2	2.87
60mm thick; between members at 450mm centres horizontal; laid loose	0.88	10.00	**0.97**	0.15	0.02	**1.52**	-	**2.49**	m2	2.68
80mm thick; between members at 450mm centres horizontal; laid loose	1.24	10.00	**1.36**	0.15	0.02	**1.52**	-	**2.89**	m2	3.11
100mm thick; between members at 450mm centres horizontal; laid loose	1.54	10.00	**1.69**	0.15	0.02	**1.52**	-	**3.22**	m2	3.46
Paper faced glass fibre insulation quilt; butt joints										
60mm thick; across members at 450mm centres horizontal; laid loose	1.65	15.00	**1.90**	0.09	0.01	**0.90**	-	**2.80**	m2	3.01

BUILDING FABRIC SUNDRIES – MAJOR WORKS

Labour hourly rates: (except Specialists) Craft Operatives £9.23 Labourer £7.02 Rates are national average prices. Refer to REGIONAL VARIATIONS for indicative levels of overall pricing in regions	MATERIALS			LABOUR				RATES		
	Del to Site £	Waste %	Material Cost £	Craft Optve Hrs	Lab Hrs	Labour Cost £	Sunds £	Nett Rate £	Unit	Gross Rate (+7.5%) £
P10: SUNDRY INSULATION/PROOFING WORK/ FIRE STOPS Cont.										
Paper faced glass fibre insulation quilt; butt joints Cont.										
80mm thick; across members at 450mm centres										
horizontal; laid loose	2.00	15.00	**2.30**	0.09	0.01	**0.90**	-	**3.20**	m2	**3.44**
Owens Corning Building Products (UK) Ltd; Crown wool insulation quilt; butt joints										
100mm thick; between members at 450mm centres										
horizontal; laid loose	1.12	10.00	**1.23**	0.10	0.01	**0.99**	-	**2.23**	m2	**2.39**
150mm thick; between members at 450mm centres										
horizontal; laid loose	1.69	10.00	**1.86**	0.10	0.01	**0.99**	-	**2.85**	m2	**3.07**
200mm thick; between members at 450mm centres										
horizontal; laid loose	2.37	10.00	**2.61**	0.20	0.03	**2.06**	-	**4.66**	m2	**5.01**
100mm thick; between members at 450mm centres; 50mm thick across members at 450mm centres										
horizontal; laid loose	1.69	10.00	**1.86**	0.28	0.05	**2.94**	-	**4.79**	m2	**5.15**
100mm thick; between members at 450mm centres; 100mm thick across members at 450mm centres										
horizontal; laid loose	2.26	10.00	**2.49**	0.36	0.05	**3.67**	-	**6.16**	m2	**6.62**
Rockwool Rollbatts; insulation quilt; butt joints										
100mm thick; between members at 450mm centres										
horizontal; laid loose	1.16	10.00	**1.28**	0.10	0.01	**0.99**	-	**2.27**	m2	**2.44**
150mm thick; between members at 450mm centres										
horizontal; laid loose	1.79	10.00	**1.97**	0.15	0.02	**1.52**	-	**3.49**	m2	**3.76**
100mm thick; between members at 450mm centres; 100mm thick across members at 450mm centres										
horizontal; laid loose	2.32	10.00	**2.55**	0.20	0.03	**2.06**	-	**4.61**	m2	**4.95**
Glass fibre medium density insulation board; butt joints										
30mm thick; plain areas										
horizontal; bedding in bitumen	1.53	10.00	**1.68**	0.20	0.02	**1.99**	1.50	**5.17**	m2	**5.56**
50mm thick; plain areas										
horizontal; bedding in bitumen	2.73	10.00	**3.00**	0.22	0.03	**2.24**	1.50	**6.74**	m2	**7.25**
75mm thick; plain areas										
horizontal; bedding in bitumen	3.98	10.00	**4.38**	0.25	0.03	**2.52**	1.50	**8.40**	m2	**9.03**
30mm thick; across members at 450mm centres										
vertical	1.53	10.00	**1.68**	0.20	0.02	**1.99**	0.03	**3.70**	m2	**3.98**
50mm thick; across members at 450mm centres										
vertical	2.73	10.00	**3.00**	0.22	0.03	**2.24**	0.03	**5.27**	m2	**5.67**
75mm thick; across members at 450mm centres										
vertical	3.98	10.00	**4.38**	0.25	0.03	**2.52**	0.03	**6.93**	m2	**7.45**
Dow Construction Products; Styrofoam Floormate 200; butt joints										
25mm thick; plain areas										
horizontal; laid loose	4.33	10.00	**4.76**	0.10	0.01	**0.99**	-	**5.76**	m2	**6.19**
35mm thick; plain areas										
horizontal; laid loose	5.92	10.00	**6.51**	0.15	0.02	**1.52**	-	**8.04**	m2	**8.64**
50mm thick; plain areas										
horizontal; laid loose	8.30	10.00	**9.13**	0.20	0.03	**2.06**	-	**11.19**	m2	**12.03**
Sempatap latex foam sheeting; butt joints										
5mm thick; plain areas										
soffit; fixing with adhesive	7.28	15.00	**8.37**	1.25	1.25	**20.31**	2.40	**31.08**	m2	**33.42**
vertical; fixing with adhesive	7.28	15.00	**8.37**	1.00	1.00	**16.25**	2.40	**27.02**	m2	**29.05**
Sempafloor SBR latex foam sheeting with coated non woven polyester surface; butt joints										
4.5mm thick; plain areas										
horizontal; fixing with adhesive	8.13	15.00	**9.35**	0.75	0.75	**12.19**	2.05	**23.59**	m2	**25.36**
Expanded polystyrene sheeting; butt joints										
13mm thick; plain areas										
horizontal; laid loose	0.62	5.00	**0.65**	0.08	0.04	**1.02**	-	**1.67**	m2	**1.80**
vertical; fixing with adhesive	0.62	5.00	**0.65**	0.23	0.12	**2.97**	1.00	**4.62**	m2	**4.96**
19mm thick; plain areas										
horizontal; laid loose	0.96	5.00	**1.01**	0.09	0.05	**1.18**	-	**2.19**	m2	**2.35**
vertical; fixing with adhesive	0.96	5.00	**1.01**	0.24	0.12	**3.06**	1.00	**5.07**	m2	**5.45**

Labour hourly rates: (except Specialists) Craft Operatives £9.23 Labourer £7.02 Rates are national average prices. Refer to REGIONAL VARIATIONS for indicative levels of overall pricing in regions	MATERIALS			LABOUR				RATES		
	Del to Site £	Waste %	Material Cost £	Craft Optve Hrs	Lab Hrs	Labour Cost £	Sunds £	Nett Rate £	Unit	Gross Rate (+7.5%) £
P10: SUNDRY INSULATION/PROOFING WORK/ FIRE STOPS Cont.										
Expanded polystyrene sheeting; butt joints Cont.										
25mm thick; plain areas										
horizontal; laid loose	1.22	5.00	**1.28**	0.10	0.05	**1.27**	-	**2.56**	m2	2.75
vertical; fixing with adhesive	1.22	5.00	**1.28**	0.25	0.13	**3.22**	1.00	**5.50**	m2	5.91
50mm thick; plain areas										
horizontal; laid loose	2.41	5.00	**2.53**	0.12	0.06	**1.53**	-	**4.06**	m2	4.36
vertical; fixing with adhesive	2.41	5.00	**2.53**	0.27	0.14	**3.47**	1.00	**7.01**	m2	7.53
75mm thick; plain areas										
horizontal; laid loose	3.63	5.00	**3.81**	0.15	0.08	**1.95**	-	**5.76**	m2	6.19
vertical; fixing with adhesive	3.63	5.00	**3.81**	0.30	0.15	**3.82**	1.00	**8.63**	m2	9.28
Expanded polystyrene sheeting, non-flammable; butt joints										
13mm thick; plain areas										
horizontal; laid loose	0.97	5.00	**1.02**	0.08	0.04	**1.02**	-	**2.04**	m2	2.19
vertical; fixing with adhesive	0.97	5.00	**1.02**	0.23	0.12	**2.97**	1.00	**4.98**	m2	5.36
19mm thick; plain areas										
horizontal; laid loose	1.15	5.00	**1.21**	0.09	0.05	**1.18**	-	**2.39**	m2	2.57
vertical; fixing with adhesive	1.15	5.00	**1.21**	0.24	0.12	**3.06**	1.00	**5.27**	m2	5.66
25mm thick; plain areas										
horizontal; laid loose	1.39	5.00	**1.46**	0.10	0.05	**1.27**	-	**2.73**	m2	2.94
vertical; fixing with adhesive	1.39	5.00	**1.46**	0.25	0.13	**3.22**	1.00	**5.68**	m2	6.11
50mm thick; plain areas										
horizontal; laid loose	3.34	5.00	**3.51**	0.12	0.06	**1.53**	-	**5.04**	m2	5.41
vertical; fixing with adhesive	3.34	5.00	**3.51**	0.27	0.14	**3.47**	1.00	**7.98**	m2	8.58
75mm thick; plain areas										
horizontal; laid loose	5.02	5.00	**5.27**	0.15	0.08	**1.95**	-	**7.22**	m2	7.76
vertical; fixing with adhesive	5.02	5.00	**5.27**	0.30	0.15	**3.82**	1.00	**10.09**	m2	10.85
Perforated zinc sheeting; butt joints										
Nr 7 gauge; across members at 450mm centres										
soffit	17.91	20.00	**21.49**	1.25	0.16	**12.66**	0.20	**34.35**	m2	36.93
vertical	17.91	20.00	**21.49**	0.75	0.09	**7.55**	0.12	**29.17**	m2	31.35
Nr 8 gauge; across members at 450mm centres										
soffit	19.64	20.00	**23.57**	1.25	0.16	**12.66**	0.20	**36.43**	m2	39.16
vertical	19.64	20.00	**23.57**	0.75	0.09	**7.55**	0.12	**31.24**	m2	33.59
Nr 9 gauge; across members at 450mm centres										
soffit	21.95	20.00	**26.34**	1.25	0.16	**12.66**	0.20	**39.20**	m2	42.14
vertical	21.95	20.00	**26.34**	0.75	0.09	**7.55**	0.12	**34.01**	m2	36.57
Galvanised wire netting, B.S.1485; butt joints										
13mm mesh x Nr 22 gauge; across members at 450mm centres										
soffit	2.31	10.00	**2.54**	-	0.12	**0.84**	0.03	**3.41**	m2	3.67
vertical	2.31	10.00	**2.54**	-	0.08	**0.56**	0.03	**3.13**	m2	3.37
25mm mesh x Nr 19 gauge; across members at 450mm centres										
soffit	1.85	10.00	**2.04**	-	0.12	**0.84**	0.03	**2.91**	m2	3.13
vertical	1.85	10.00	**2.04**	-	0.08	**0.56**	0.03	**2.63**	m2	2.82
38mm mesh x Nr 19 gauge; across members at 450mm centres										
soffit	1.33	10.00	**1.46**	-	0.12	**0.84**	0.03	**2.34**	m2	2.51
vertical	1.33	10.00	**1.46**	-	0.08	**0.56**	0.03	**2.05**	m2	2.21
50mm mesh x Nr 19 gauge; across members at 450mm centres										
soffit	1.05	10.00	**1.16**	-	0.12	**0.84**	0.03	**2.03**	m2	2.18
vertical	1.05	10.00	**1.16**	-	0.08	**0.56**	0.03	**1.75**	m2	1.88
P11: FOAMED/FIBRE/BEAD CAVITY WALL INSULATION										
Expanded polystyrene pellet insulation										
305mm walls										
filling to 50mm wide cavity	3.40	10.00	**3.74**	0.20	-	**1.85**	0.35	**5.94**	m2	6.38
P20: UNFRAMED ISOLATED TRIMS/SKIRTINGS/SUNDRY ITEMS										
Softwood, wrought - skirtings, architraves, picture rails and cover fillets to B.S.1186 Part 3										
Skirtings, picture rails, architraves and the like; (finished sizes)										
13 x 45mm; reference 13CA45	0.62	10.00	**0.68**	0.22	0.22	**3.58**	0.03	**4.29**	m	4.61
20 x 70mm; reference 20CA70	0.97	10.00	**1.07**	0.24	0.03	**2.43**	0.03	**3.52**	m	3.79
13 x 45mm; reference 13CP45	0.62	10.00	**0.68**	0.23	0.03	**2.33**	0.03	**3.05**	m	3.27
13 x 70mm; reference 13CP70	0.80	10.00	**0.88**	0.23	0.03	**2.33**	0.03	**3.24**	m	3.49
13 x 70mm; reference 13CS70	0.80	10.00	**0.88**	0.25	0.03	**2.52**	0.03	**3.43**	m	3.69
13 x 95mm; reference 13CS95	1.03	10.00	**1.13**	0.23	0.03	**2.33**	0.03	**3.50**	m	3.76

BUILDING FABRIC SUNDRIES

Labour hourly rates: (except Specialists) Craft Operatives £9.23 Labourer £7.02 Rates are national average prices. Refer to REGIONAL VARIATIONS for indicative levels of overall pricing in regions	MATERIALS			LABOUR				RATES		
	Del to Site £	Waste %	Material Cost £	Craft Optve Hrs	Lab Hrs	Labour Cost £	Sunds £	Nett Rate £	Unit	Gross Rate (+7.5%) £
P20: UNFRAMED ISOLATED TRIMS/SKIRTINGS/SUNDRY ITEMS Cont.										
Softwood, wrought - skirtings, architraves, picture rails and cover fillets to B.S.1186 Part 3 Cont.										
Skirtings, picture rails, architraves and the like; (finished sizes) Cont.										
20 x 120mm; reference 20CS120	1.59	10.00	1.75	0.27	0.03	2.70	0.04	4.49	m	4.83
9 x 33mm; reference 9RA33	0.46	10.00	0.51	0.20	0.02	1.99	0.03	2.52	m	2.71
13 x 45mm; reference 13RA45	0.62	10.00	0.68	0.23	0.03	2.33	0.03	3.05	m	3.27
20 x 70mm; reference 20RA70	1.03	10.00	1.13	0.24	0.03	2.43	0.03	3.59	m	3.86
13 x 70mm; reference 13RS70	0.83	10.00	0.91	0.25	0.03	2.52	0.03	3.46	m	3.72
13 x 95mm; reference 13RS95	1.07	10.00	1.18	0.23	0.03	2.33	0.03	3.54	m	3.81
20 x 120mm; reference 20RS120	1.59	10.00	1.75	0.27	0.03	2.70	0.04	4.49	m	4.83
13 x 45mm; reference 13RP45	0.65	10.00	0.71	0.23	0.03	2.33	0.03	3.08	m	3.31
13 x 70mm; reference 13RP70	0.87	10.00	0.96	0.25	0.03	2.52	0.03	3.51	m	3.77
Cover fillets, stops, trims, beads, nosings and the like; (finished sizes)										
11 x 33mm; reference C33	0.46	10.00	0.51	0.19	0.02	1.89	0.02	2.42	m	2.60
11 x 45mm; reference C45	0.48	10.00	0.53	0.19	0.02	1.89	0.02	2.44	m	2.63
13 x 33mm; reference HR33	0.54	10.00	0.59	0.19	0.02	1.89	0.02	2.51	m	2.70
20 x 45mm; reference HR45	0.72	10.00	0.79	0.21	0.03	2.15	0.03	2.97	m	3.19
11 x 11mm; reference Q11	0.24	10.00	0.26	0.19	0.02	1.89	0.02	2.18	m	2.34
13 x 13mm; reference Q13	0.38	10.00	0.42	0.19	0.02	1.89	0.02	2.33	m	2.51
20 x 20mm; reference Q20	0.48	10.00	0.53	0.19	0.02	1.89	0.02	2.44	m	2.63
13 x 13mm; reference S13	0.38	10.00	0.42	0.19	0.02	1.89	0.02	2.33	m	2.51
20 x 20mm; reference S20	0.48	10.00	0.53	0.19	0.02	1.89	0.02	2.44	m	2.63
27 x 27mm; reference S27	0.62	10.00	0.68	0.21	0.03	2.15	0.03	2.86	m	3.08
12 x 95mm; reference 12HRCS95	1.14	10.00	1.25	0.25	0.03	2.52	0.04	3.81	m	4.10
15 x 95mm; reference 15HRCS95	1.21	10.00	1.33	0.27	0.03	2.70	0.04	4.07	m	4.38
Softwood, wrought - skirtings, architraves, picture rails and cover fillets										
Skirtings, picture rails, architraves and the like; (finished sizes)										
19 x 100mm	1.52	10.00	1.67	0.25	0.03	2.52	0.04	4.23	m	4.55
19 x 150mm	2.13	10.00	2.34	0.28	0.04	2.87	0.05	5.26	m	5.65
25 x 100mm	1.94	10.00	2.13	0.27	0.03	2.70	0.04	4.88	m	5.24
25 x 150mm	2.73	10.00	3.00	0.31	0.04	3.14	0.05	6.20	m	6.66
25 x 50mm; splays -1	1.10	10.00	1.21	0.22	0.03	2.24	0.04	3.49	m	3.75
25 x 63mm; splays -1	1.28	10.00	1.41	0.24	0.03	2.43	0.04	3.87	m	4.16
25 x 75mm; splays -1	1.64	10.00	1.80	0.25	0.03	2.52	0.04	4.36	m	4.69
32 x 100mm; splays -1	2.37	10.00	2.61	0.30	0.04	3.05	0.05	5.71	m	6.13
38 x 75mm; splays -1	2.18	10.00	2.40	0.28	0.04	2.87	0.05	5.31	m	5.71
15 x 50mm; mouldings -1	0.72	10.00	0.79	0.21	0.03	2.15	0.04	2.98	m	3.20
15 x 125mm; mouldings -1	1.34	10.00	1.47	0.25	0.03	2.52	0.04	4.03	m	4.33
19 x 50mm; mouldings -1	0.91	10.00	1.00	0.22	0.03	2.24	0.04	3.28	m	3.53
19 x 125mm; mouldings -1	1.70	10.00	1.87	0.27	0.03	2.70	0.04	4.61	m	4.96
25 x 50mm; mouldings -1	1.10	10.00	1.21	0.22	0.03	2.24	0.04	3.49	m	3.75
25 x 63mm; mouldings -1	1.28	10.00	1.41	0.24	0.03	2.43	0.04	3.87	m	4.16
25 x 75mm; mouldings -1	1.64	10.00	1.80	0.25	0.03	2.52	0.04	4.36	m	4.69
32 x 100mm; mouldings -1	2.37	10.00	2.61	0.30	0.04	3.05	0.04	5.71	m	6.13
38 x 75mm; mouldings -1	2.18	10.00	2.40	0.28	0.04	2.87	0.05	5.31	m	5.71
15 x 125mm; chamfers -1	1.34	10.00	1.47	0.25	0.03	2.52	0.04	4.03	m	4.33
19 x 125mm; chamfers -1	1.70	10.00	1.87	0.26	0.03	2.61	0.04	4.52	m	4.86
25 x 75mm; chamfers -2	1.70	10.00	1.87	0.25	0.03	2.52	0.04	4.43	m	4.76
25 x 100mm; chamfers -2	2.07	10.00	2.28	0.26	0.03	2.61	0.04	4.93	m	5.30
Cover fillets, stops, trims, beads, nosings and the like; (finished sizes)										
12 x 50mm; mouldings -1	0.62	10.00	0.68	0.21	0.03	2.15	0.04	2.87	m	3.09
19 x 19mm; mouldings -1	0.48	10.00	0.53	0.19	0.02	1.89	0.03	2.45	m	2.64
19 x 50mm; mouldings -1	0.91	10.00	1.00	0.22	0.03	2.24	0.04	3.28	m	3.53
25 x 25mm; mouldings -1	0.62	10.00	0.68	0.21	0.03	2.15	0.04	2.87	m	3.09
25 x 38mm; mouldings -1	0.86	10.00	0.95	0.22	0.03	2.24	0.04	3.23	m	3.47
19 x 19mm; chamfers -1	0.48	10.00	0.53	0.19	0.02	1.89	0.03	2.45	m	2.64
25 x 25mm; chamfers -1	0.62	10.00	0.68	0.21	0.03	2.15	0.04	2.87	m	3.09
25 x 38mm; chamfers -1	0.86	10.00	0.95	0.22	0.03	2.24	0.04	3.23	m	3.47
Afrormosia, wrought - skirtings, architraves, picture rails and cover fillets										
Skirtings, picture rails, architraves and the like; (finished sizes)										
19 x 100mm	6.03	10.00	6.63	0.42	0.05	4.23	0.10	10.96	m	11.78
19 x 150mm	7.81	10.00	8.59	0.48	0.06	4.85	0.10	13.54	m	14.56
25 x 100mm	7.44	10.00	8.18	0.46	0.06	4.67	0.10	12.95	m	13.92
25 x 150mm	10.06	10.00	11.07	0.48	0.06	4.85	0.10	16.02	m	17.22
extra; ends	1.54	-	1.54	0.05	0.01	0.53	0.10	2.17	nr	2.33
extra; angles	2.50	-	2.50	0.10	0.01	0.99	0.10	3.59	nr	3.86
extra; mitres	2.50	-	2.50	0.10	0.01	0.99	0.10	3.59	nr	3.86
extra; intersections	2.50	-	2.50	0.10	0.01	0.99	0.10	3.59	nr	3.86
25 x 50mm; splays -1	4.36	10.00	4.80	0.37	0.05	3.77	0.10	8.66	m	9.31
25 x 63mm; splays -1	5.13	10.00	5.64	0.41	0.05	4.14	0.10	9.88	m	10.62
25 x 75mm; splays -1	6.34	10.00	6.97	0.42	0.05	4.23	0.10	11.30	m	12.15
32 x 100mm; splays -1	8.71	10.00	9.58	0.48	0.06	4.85	0.10	14.53	m	15.62
38 x 75mm; splays -1	8.08	10.00	8.89	0.48	0.06	4.85	0.10	13.84	m	14.88
extra; ends	1.21	-	1.21	0.05	0.01	0.53	0.10	1.84	nr	1.98
extra; angles	2.05	-	2.05	0.10	0.01	0.99	0.10	3.14	nr	3.38
extra; mitres	2.05	-	2.05	0.10	0.01	0.99	0.10	3.14	nr	3.38
extra; intersections	2.05	-	2.05	0.10	0.01	0.99	0.10	3.14	nr	3.38
15 x 50mm; mouldings -1	2.83	10.00	3.11	0.36	0.05	3.67	0.10	6.89	m	7.40
15 x 125mm; mouldings -1	5.39	10.00	5.93	0.42	0.05	4.23	0.10	10.26	m	11.03

Labour hourly rates: (except Specialists) Craft Operatives £9.23 Labourer £7.02 Rates are national average prices. Refer to REGIONAL VARIATIONS for indicative levels of overall pricing in regions	MATERIALS			LABOUR				RATES		
	Del to Site £	Waste %	Material Cost £	Craft Optve Hrs	Lab Hrs	Labour Cost £	Sunds £	Nett Rate £	Unit	Gross Rate (+7.5%) £
P20: UNFRAMED ISOLATED TRIMS/SKIRTINGS/SUNDRY ITEMS Cont.										
Afrormosia, wrought - skirtings, architraves, picture rails and cover fillets Cont.										
Skirtings, picture rails, architraves and the like; (finished sizes) Cont.										
19 x 50mm; mouldings -1	3.53	10.00	3.88	0.37	0.05	3.77	0.10	7.75	m	8.33
19 x 125mm; mouldings -1	6.79	10.00	7.47	0.46	0.06	4.67	0.10	12.24	m	13.15
25 x 50mm; mouldings -1	4.36	10.00	4.80	0.37	0.05	3.77	0.10	8.66	m	9.31
25 x 63mm; mouldings -1	5.13	10.00	5.64	0.41	0.05	4.14	0.10	9.88	m	10.62
25 x 75mm; mouldings -1	6.34	10.00	6.97	0.42	0.05	4.23	0.10	11.30	m	12.15
32 x 100mm; mouldings -1	8.71	10.00	9.58	0.48	0.06	4.85	0.10	14.53	m	15.62
38 x 75mm; mouldings -1	8.08	10.00	8.89	0.48	0.06	4.85	0.10	13.84	m	14.88
extra; ends	1.21	-	1.21	0.05	0.01	0.53	0.10	1.84	nr	1.98
extra; angles	2.05	-	2.05	0.10	0.01	0.99	0.10	3.14	nr	3.38
extra; mitres	2.05	-	2.05	0.10	0.01	0.99	0.10	3.14	nr	3.38
extra; intersections	2.05	-	2.05	0.10	0.01	0.99	0.10	3.14	nr	3.38
15 x 125mm; chamfers -1	5.39	10.00	5.93	0.42	0.05	4.23	0.10	10.26	m	11.03
19 x 125mm; chamfers -1	6.79	10.00	7.47	0.46	0.06	4.67	0.10	12.24	m	13.15
25 x 75mm; chamfers -2	6.54	10.00	7.19	0.42	0.05	4.23	0.10	11.52	m	12.39
25 x 100mm; chamfers -2	7.81	10.00	8.59	0.44	0.06	4.48	0.10	13.17	m	14.16
Cover fillets, stops, trims, beads, nosings and the like; (finished sizes)										
12 x 50mm; mouldings -1	2.38	10.00	2.62	0.36	0.05	3.67	0.08	6.37	m	6.85
19 x 19mm; mouldings -1	1.60	10.00	1.76	0.32	0.04	3.23	0.08	5.07	m	5.45
19 x 50mm; mouldings -1	3.53	10.00	3.88	0.37	0.05	3.77	0.08	7.73	m	8.31
25 x 25mm; mouldings -1	2.44	10.00	2.68	0.36	0.05	3.67	0.08	6.44	m	6.92
25 x 38mm; mouldings -1	3.26	10.00	3.59	0.37	0.05	3.77	0.08	7.43	m	7.99
19 x 19mm; chamfers -1	1.60	10.00	1.76	0.32	0.04	3.23	0.08	5.07	m	5.45
25 x 25mm; chamfers -1	2.44	10.00	2.68	0.36	0.05	3.67	0.08	6.44	m	6.92
25 x 38mm; chamfers -1	3.26	10.00	3.59	0.37	0.05	3.77	0.08	7.43	m	7.99
Medium density fibreboard, B.S.1142, skirtings, architraves, picture rails and cover fillets to B.S.1186 Part 3										
Skirtings, picture rails, architraves and the like; (finished sizes)										
13 x 45mm; reference 13CA45	1.08	10.00	1.19	0.28	0.04	2.87	0.06	4.11	m	4.42
20 x 70mm; reference 20CA70	1.69	10.00	1.86	0.30	0.04	3.05	0.06	4.97	m	5.34
13 x 45mm; reference 13CP45	1.08	10.00	1.19	0.29	0.04	2.96	0.06	4.21	m	4.52
13 x 70mm; reference 13CP70	1.36	10.00	1.50	0.29	0.04	2.96	0.06	4.51	m	4.85
13 x 70mm; reference 13CS70	1.36	10.00	1.50	0.31	0.04	3.14	0.06	4.70	m	5.05
13 x 95mm; reference 13CS95	1.79	10.00	1.97	0.29	0.04	2.96	0.06	4.99	m	5.36
20 x 120mm; reference 20CS120	2.77	10.00	3.05	0.34	0.04	3.42	0.06	6.53	m	7.02
9 x 33mm; reference 9RA33	0.81	10.00	0.89	0.25	0.03	2.52	0.06	3.47	m	3.73
13 x 45mm; reference 13RA45	1.08	10.00	1.19	0.29	0.04	2.96	0.06	4.21	m	4.52
20 x 70mm; reference 20RA70	1.79	10.00	1.97	0.30	0.04	3.05	0.06	5.08	m	5.46
13 x 70mm; reference 13RS70	1.44	10.00	1.58	0.31	0.04	3.14	0.06	4.79	m	5.15
13 x 95mm; reference 13RS95	1.87	10.00	2.06	0.29	0.04	2.96	0.06	5.07	m	5.46
20 x 120mm; reference 20RS120	2.77	10.00	3.05	0.34	0.04	3.42	0.06	6.53	m	7.02
13 x 45mm; reference 13RP45	1.13	10.00	1.24	0.29	0.04	2.96	0.06	4.26	m	4.58
13 x 70mm; reference 13RP70	1.52	10.00	1.67	0.31	0.04	3.14	0.06	4.87	m	5.24
Cover fillets, stops, trims, beads, nosings and the like; (finished sizes)										
11 x 33mm; reference C33	0.81	10.00	0.89	0.24	0.02	2.36	0.05	3.30	m	3.54
11 x 45mm; reference C45	0.85	10.00	0.94	0.24	0.02	2.36	0.05	3.34	m	3.59
13 x 33mm; reference HR33	0.95	10.00	1.04	0.24	0.02	2.36	0.05	3.45	m	3.71
20 x 45mm; reference HR45	1.28	10.00	1.41	0.26	0.04	2.68	0.05	4.14	m	4.45
11 x 11mm; reference Q11	0.43	10.00	0.47	0.24	0.02	2.36	0.05	2.88	m	3.09
13 x 13mm; reference Q13	0.66	10.00	0.73	0.24	0.02	2.36	0.05	3.13	m	3.37
20 x 20mm; reference Q20	0.85	10.00	0.94	0.24	0.02	2.36	0.05	3.34	m	3.59
13 x 13mm; reference S13	0.66	10.00	0.73	0.24	0.02	2.36	0.05	3.13	m	3.37
20 x 20mm; reference S20	0.85	10.00	0.94	0.24	0.02	2.36	0.05	3.34	m	3.59
27 x 27mm; reference S27	1.08	10.00	1.19	0.26	0.04	2.68	0.05	3.92	m	4.21
12 x 95mm; reference 12HRCS95	2.00	10.00	2.20	0.31	0.04	3.14	0.05	5.39	m	5.80
15 x 95mm; reference 15HRCS95	2.10	10.00	2.31	0.34	0.04	3.42	0.05	5.78	m	6.21
Softwood, wrought - cappings										
Cover fillets, stops, trims, beads, nosings and the like; (finished sizes)										
25 x 50mm; level; rebates -1; mouldings -1	1.10	10.00	1.21	0.22	0.03	2.24	0.04	3.49	m	3.75
50 x 75mm; level; rebates -1; mouldings -1	3.16	10.00	3.48	0.31	0.04	3.14	0.05	6.67	m	7.17
50 x 100mm; level; rebates -1; mouldings -1	3.95	10.00	4.34	0.34	0.04	3.42	0.05	7.81	m	8.40
25 x 50mm; ramped; rebates -1; mouldings -1	1.10	10.00	1.21	0.33	0.04	3.33	0.05	4.59	m	4.93
50 x 75mm; ramped; rebates -1; mouldings -1	3.16	10.00	3.48	0.46	0.06	4.67	0.08	8.22	m	8.84
50 x 100mm; ramped; rebates -1; mouldings -1	3.95	10.00	4.34	0.51	0.06	5.13	0.08	9.55	m	10.27
Afrormosia, wrought - cappings										
Cover fillets, stops, trims, beads, nosings and the like; (finished sizes)										
25 x 50mm; level; rebates -1; mouldings -1	4.49	10.00	4.94	0.37	0.05	3.77	0.06	8.77	m	9.42
50 x 75mm; level; rebates -1; mouldings -1	11.54	10.00	12.69	0.53	0.07	5.38	0.09	18.17	m	19.53
extra; ends	1.73	-	1.73	0.08	0.01	0.81	0.01	2.55	nr	2.74
extra; angles	2.89	-	2.89	0.13	0.02	1.34	0.02	4.25	nr	4.57
extra; mitres	2.89	-	2.89	0.13	0.02	1.34	0.02	4.25	nr	4.57
extra; rounded corners not exceeding 300mm girth .	11.54	-	11.54	0.53	0.07	5.38	0.09	17.01	nr	18.29
50 x 100mm; level; rebates -1; mouldings -1	14.35	10.00	15.79	0.58	0.07	5.84	0.09	21.72	m	23.35
extra; ends	2.18	-	2.18	0.09	0.01	0.90	0.01	3.09	nr	3.32
extra; angles	3.59	-	3.59	0.14	0.02	1.43	0.02	5.04	nr	5.42

BUILDING FABRIC SUNDRIES

Labour hourly rates: (except Specialists) Craft Operatives £9.23 Labourer £7.02 Rates are national average prices. Refer to REGIONAL VARIATIONS for indicative levels of overall pricing in regions	MATERIALS			LABOUR				RATES		
	Del to Site £	Waste %	Material Cost £	Craft Optve Hrs	Lab Hrs	Labour Cost £	Sunds £	Nett Rate £	Unit	Gross Rate (+7.5%) £
P20: UNFRAMED ISOLATED TRIMS/SKIRTINGS/SUNDRY ITEMS Cont.										
Afrormosia, wrought - cappings Cont.										
Cover fillets, stops, trims, beads, nosings and the like; (finished sizes) Cont.										
extra; mitres	3.59	-	3.59	0.14	0.02	1.43	0.02	5.04	nr	5.42
extra; rounded corners not exceeding 300mm girth .	14.35	-	14.35	0.58	0.07	5.84	0.09	20.28	nr	21.81
25 x 50mm; ramped; rebates -1; mouldings -1	4.49	10.00	4.94	0.55	0.07	5.57	0.09	10.60	m	11.39
50 x 75mm; ramped; rebates -1; mouldings -1	11.54	10.00	12.69	0.80	0.10	8.09	0.13	20.91	m	22.48
extra; ends	1.73	-	1.73	0.12	0.02	1.25	0.02	3.00	nr	3.22
extra; angles	2.89	-	2.89	0.20	0.03	2.06	0.03	4.98	nr	5.35
extra; mitres	2.89	-	2.89	0.20	0.03	2.06	0.03	4.98	nr	5.35
extra; rounded corners not exceeding 300mm girth .	11.54	-	11.54	0.80	0.10	8.09	0.13	19.76	nr	21.24
50 x 100mm; ramped; rebates -1; mouldings -1	14.35	10.00	15.79	0.90	0.11	9.08	0.15	25.01	m	26.89
extra; ends	2.18	-	2.18	0.13	0.02	1.34	0.02	3.54	nr	3.81
extra; angles	3.59	-	3.59	0.22	0.03	2.24	0.04	5.87	nr	6.31
extra; mitres	3.59	-	3.59	0.22	0.03	2.24	0.04	5.87	nr	6.31
extra; rounded corners not exceeding 300mm girth .	14.35	-	14.35	0.90	0.11	9.08	0.15	23.58	nr	25.35
Sapele; wrought - cappings										
Cover fillets, stops, trims, beads, nosings and the like; (finished sizes)										
25 x 50mm; level; rebates -1; mouldings -1	3.26	10.00	3.59	0.33	0.04	3.33	0.05	6.96	m	7.48
50 x 75mm; level; rebates -1; mouldings -1	8.14	10.00	8.95	0.46	0.06	4.67	0.08	13.70	m	14.73
extra; ends	1.21	-	1.21	0.07	0.01	0.72	0.01	1.94	nr	2.08
extra; angles	2.05	-	2.05	0.12	0.02	1.25	0.02	3.32	nr	3.57
extra; mitres	2.05	-	2.05	0.12	0.02	1.25	0.02	3.32	nr	3.57
extra; rounded corners not exceeding 300mm girth .	8.14	-	8.14	0.46	0.06	4.67	0.08	12.89	nr	13.85
50 x 100mm; level; rebates -1; mouldings -1	9.98	10.00	10.98	0.50	0.06	5.04	0.08	16.09	m	17.30
extra; ends	1.52	-	1.52	0.08	0.01	0.81	0.01	2.34	nr	2.51
extra; angles	2.52	-	2.52	0.13	0.02	1.34	0.02	3.88	nr	4.17
extra; mitres	2.52	-	2.52	0.13	0.02	1.34	0.02	3.88	nr	4.17
extra; rounded corners not exceeding 300mm girth .	9.98	-	9.98	0.50	0.06	5.04	0.08	15.10	nr	16.23
25 x 50mm; ramped; rebates -1; mouldings -1	3.26	10.00	3.59	0.50	0.06	5.04	0.08	8.70	m	9.35
50 x 75mm; ramped; rebates -1; mouldings -1	8.14	10.00	8.95	0.70	0.09	7.09	0.11	16.16	m	17.37
extra; ends	1.21	-	1.21	0.10	0.01	0.99	0.02	2.22	nr	2.39
extra; angles	2.05	-	2.05	0.17	0.02	1.71	0.03	3.79	nr	4.07
extra; mitres	2.05	-	2.05	0.17	0.02	1.71	0.03	3.79	nr	4.07
extra; rounded corners not exceeding 300mm girth .	8.14	-	8.14	0.70	0.09	7.09	0.11	15.34	nr	16.49
50 x 100mm; ramped; rebates -1; mouldings -1	9.98	10.00	10.98	0.75	0.09	7.55	0.12	18.65	m	20.05
extra; ends	1.52	-	1.52	0.11	0.01	1.09	0.02	2.63	nr	2.82
extra; angles	2.52	-	2.52	0.19	0.02	1.89	0.03	4.44	nr	4.78
extra; mitres	2.52	-	2.52	0.19	0.02	1.89	0.03	4.44	nr	4.78
extra; rounded corners not exceeding 300mm girth .	9.98	-	9.98	0.75	0.09	7.55	0.12	17.65	nr	18.98
Softwood, wrought - window boards										
Cover fillets, stops, trims, beads, nosings and the like; (finished sizes)										
mouldings -1; tongued on										
25 x 50mm..........................	1.28	10.00	1.41	0.25	0.03	2.52	0.04	3.97	m	4.26
25 x 75mm..........................	2.07	10.00	2.28	0.26	0.03	2.61	0.04	4.93	m	5.30
32 x 50mm..........................	1.59	10.00	1.75	0.28	0.04	2.87	0.05	4.66	m	5.01
32 x 75mm..........................	2.44	10.00	2.68	0.29	0.04	2.96	0.05	5.69	m	6.12
rounded edges -1; tongued on										
25 x 50mm..........................	1.28	10.00	1.41	0.25	0.03	2.52	0.04	3.97	m	4.26
25 x 75mm..........................	2.07	10.00	2.28	0.26	0.03	2.61	0.04	4.93	m	5.30
32 x 50mm..........................	1.59	10.00	1.75	0.28	0.04	2.87	0.05	4.66	m	5.01
32 x 75mm..........................	2.44	10.00	2.68	0.29	0.04	2.96	0.05	5.69	m	6.12
Window boards										
25 x 150mm; mouldings -1	2.80	10.00	3.08	0.30	0.04	3.05	0.05	6.18	m	6.64
32 x 150mm; mouldings -1	3.29	10.00	3.62	0.33	0.04	3.33	0.05	7.00	m	7.52
25 x 150mm; rounded edges -1	2.80	10.00	3.08	0.30	0.04	3.05	0.05	6.18	m	6.64
32 x 150mm; rounded edges -1	3.29	10.00	3.62	0.33	0.04	3.33	0.05	7.00	m	7.52
Afrormosia, wrought - window boards										
Cover fillets, stops, trims, beads, nosings and the like; (finished sizes)										
mouldings -1; tongued on										
25 x 50mm..........................	4.16	10.00	4.58	0.38	0.05	3.86	0.06	8.49	m	9.13
25 x 75mm..........................	6.46	10.00	7.11	0.40	0.05	4.04	0.06	11.21	m	12.05
32 x 50mm..........................	5.01	10.00	5.51	0.42	0.05	4.23	0.07	9.81	m	10.54
32 x 75mm..........................	7.67	10.00	8.44	0.44	0.06	4.48	0.07	12.99	m	13.96
rounded edges -1; tongued on										
25 x 50mm..........................	4.16	10.00	4.58	0.38	0.05	3.86	0.06	8.49	m	9.13
25 x 75mm..........................	6.46	10.00	7.11	0.40	0.05	4.04	0.06	11.21	m	12.05
32 x 50mm..........................	5.01	10.00	5.51	0.42	0.05	4.23	0.07	9.81	m	10.54
32 x 75mm..........................	7.67	10.00	8.44	0.44	0.06	4.48	0.07	12.99	m	13.96
Window boards										
25 x 150mm; mouldings -1	10.27	10.00	11.30	0.52	0.07	5.29	0.09	16.68	m	17.93
32 x 150mm; mouldings -1	12.20	10.00	13.42	0.58	0.07	5.84	0.09	19.35	m	20.81
25 x 150mm; rounded edges -1	10.27	10.00	11.30	0.52	0.07	5.29	0.09	16.68	m	17.93
32 x 150mm; rounded edges -1	12.20	10.00	13.42	0.58	0.07	5.84	0.09	19.35	m	20.81
Sapele; wrought - window boards										
Cover fillets, stops, trims, beads, nosings and the like; (finished sizes)										
mouldings -1; tongued on										
25 x 50mm..........................	3.05	10.00	3.36	0.37	0.05	3.77	0.06	7.18	m	7.72

Labour hourly rates: (except Specialists) Craft Operatives £9.23 Labourer £7.02 Rates are national average prices. Refer to REGIONAL VARIATIONS for indicative levels of overall pricing in regions	MATERIALS			LABOUR				RATES		
	Del to Site	Waste	Material Cost	Craft Optve	Lab	Labour Cost	Sunds	Nett Rate	Unit	Gross Rate (+7.5%)
	£	%	£	Hrs	Hrs	£	£	£		£
P20: UNFRAMED ISOLATED TRIMS/SKIRTINGS/SUNDRY ITEMS Cont.										
Sapele; wrought - window boards Cont.										
Cover fillets, stops, trims, beads, nosings and the like; (finished sizes) Cont.										
mouldings -1; tongued on Cont.										
25 x 75mm	4.67	10.00	5.14	0.39	0.05	3.95	0.06	9.15	m	9.83
32 x 50mm	3.57	10.00	3.93	0.41	0.05	4.14	0.07	8.13	m	8.74
32 x 75mm	5.57	10.00	6.13	0.43	0.05	4.32	0.07	10.52	m	11.31
rounded edges -1; tongued on										
25 x 50mm	3.05	10.00	3.36	0.37	0.05	3.77	0.06	7.18	m	7.72
25 x 75mm	4.67	10.00	5.14	0.39	0.05	3.95	0.06	9.15	m	9.83
32 x 50mm	3.57	10.00	3.93	0.41	0.05	4.14	0.07	8.13	m	8.74
32 x 75mm	5.57	10.00	6.13	0.43	0.05	4.32	0.07	10.52	m	11.31
Window boards										
25 x 150mm; mouldings -1	7.40	10.00	8.14	0.45	0.06	4.57	0.07	12.78	m	13.74
32 x 150mm; mouldings -1	8.77	10.00	9.65	0.50	0.06	5.04	0.08	14.76	m	15.87
25 x 150mm; rounded edges -1	7.40	10.00	8.14	0.45	0.06	4.57	0.07	12.78	m	13.74
32 x 150mm; rounded edges -1	8.77	10.00	9.65	0.50	0.06	5.04	0.08	14.76	m	15.87
Moisture resistant medium density fibreboard, B.S.1142										
Window boards										
25 x 225mm; nosed and tongued	5.56	10.00	6.12	0.30	0.04	3.05	0.05	9.22	m	9.91
25 x 250mm; nosed and tongued	6.18	10.00	6.80	0.32	0.04	3.23	0.05	10.08	m	10.84
Softwood, wrought - isolated shelves and worktops										
Cover fillets, stops, trims, beads, nosings and the like; (finished sizes)										
19 x 12mm	0.28	10.00	0.31	0.18	0.35	4.12	0.30	4.73	m	5.08
19 x 16mm	0.34	10.00	0.37	0.19	0.02	1.89	0.03	2.30	m	2.47
19 x 18mm	0.39	10.00	0.43	0.19	0.02	1.89	0.03	2.35	m	2.53
19 x 22mm	0.43	10.00	0.47	0.20	0.03	2.06	0.03	2.56	m	2.75
19 x 25mm	0.48	10.00	0.53	0.20	0.03	2.06	0.03	2.61	m	2.81
Isolated shelves and worktops										
25 x 150mm	1.83	10.00	2.01	0.20	0.03	2.06	0.03	4.10	m	4.41
25 x 225mm	2.73	10.00	3.00	0.25	0.03	2.52	0.04	5.56	m	5.98
25 x 300mm	3.85	10.00	4.24	0.30	0.04	3.05	0.05	7.33	m	7.88
32 x 150mm	2.32	10.00	2.55	0.22	0.03	2.24	0.04	4.83	m	5.20
32 x 225mm	3.53	10.00	3.88	0.27	0.03	2.70	0.04	6.63	m	7.12
32 x 300mm	4.67	10.00	5.14	0.33	0.04	3.33	0.05	8.51	m	9.15
25 x 450mm; cross tongued	8.21	2.50	8.42	0.35	0.04	3.51	0.06	11.99	m	12.89
25 x 600mm; cross tongued	10.94	2.50	11.21	0.45	0.06	4.57	0.07	15.86	m	17.05
32 x 450mm; cross tongued	10.51	2.50	10.77	0.40	0.06	4.11	0.07	14.96	m	16.08
32 x 600mm; cross tongued	13.98	2.50	14.33	0.50	0.06	5.04	0.08	19.45	m	20.90
38 x 450mm; cross tongued	12.46	2.50	12.77	0.45	0.06	4.57	0.07	17.42	m	18.72
38 x 600mm; cross tongued	16.86	2.50	17.28	0.55	0.07	5.57	0.09	22.94	m	24.66
25 x 450mm overall; 25 x 50mm slats spaced 25mm apart	4.26	10.00	4.69	1.00	0.13	10.14	0.16	14.99	m	16.11
25 x 450mm overall; 25 x 50mm slats spaced 32mm apart	3.65	10.00	4.01	0.90	0.11	9.08	0.15	13.24	m	14.24
25 x 600mm overall; 25 x 50mm slats spaced 25mm apart	5.48	10.00	6.03	1.25	0.16	12.66	0.20	18.89	m	20.31
25 x 600mm overall; 25 x 50mm slats spaced 32mm apart	4.86	10.00	5.35	1.15	0.14	11.60	0.19	17.13	m	18.42
25 x 900mm overall; 25 x 50mm slats spaced 25mm apart	7.91	10.00	8.70	1.90	0.24	19.22	0.31	28.23	m	30.35
25 x 900mm overall; 25 x 50mm slats spaced 32mm apart	7.30	10.00	8.03	1.80	0.23	18.23	0.29	26.55	m	28.54
Afrormosia, wrought - isolated shelves and worktops										
Cover fillets, stops, trims, beads, nosings and the like; (finished sizes)										
19 x 12mm	1.00	10.00	1.10	0.31	0.04	3.14	0.05	4.29	m	4.61
19 x 16mm	1.18	10.00	1.30	0.33	0.04	3.33	0.05	4.67	m	5.03
19 x 18mm	1.36	10.00	1.50	0.33	0.04	3.33	0.05	4.87	m	5.24
19 x 22mm	1.47	10.00	1.62	0.35	0.04	3.51	0.06	5.19	m	5.58
19 x 25mm	1.71	10.00	1.88	0.35	0.04	3.51	0.06	5.45	m	5.86
Isolated shelves and worktops										
25 x 150mm	7.66	10.00	8.43	0.35	0.04	3.51	0.06	12.00	m	12.90
25 x 225mm	11.50	10.00	12.65	0.44	0.06	4.48	0.07	17.20	m	18.49
25 x 300mm	15.33	10.00	16.86	0.53	0.07	5.38	0.09	22.34	m	24.01
32 x 150mm	9.73	10.00	10.70	0.38	0.05	3.86	0.06	14.62	m	15.72
32 x 225mm	14.44	10.00	15.88	0.47	0.06	4.76	0.08	20.72	m	22.28
32 x 300mm	19.45	10.00	21.40	0.56	0.07	5.66	0.09	27.15	m	29.18
25 x 450mm; cross tongued	31.83	2.50	32.63	0.60	0.08	6.10	0.10	38.83	m	41.74
25 x 600mm; cross tongued	42.44	2.50	43.50	0.80	0.10	8.09	0.13	51.72	m	55.60
32 x 450mm; cross tongued	40.66	2.50	41.68	0.70	0.09	7.09	0.11	48.88	m	52.55
32 x 600mm; cross tongued	54.22	2.50	55.58	0.90	0.11	9.08	0.15	64.80	m	69.67
38 x 450mm; cross tongued	48.33	2.50	49.54	0.80	0.10	8.09	0.13	57.75	m	62.09
38 x 600mm; cross tongued	64.84	2.50	66.46	1.00	0.13	10.14	0.16	76.76	m	82.52
Plywood B.S.6566, II/III grade, INT bonded; butt joints - isolated shelves and worktops										
Cover fillets, stops, trims, beads, nosings and the like										
50 x 6.5mm	0.21	20.00	0.25	0.16	0.02	1.62	0.05	1.92	m	2.06

Labour hourly rates: (except Specialists) Craft Operatives £9.23 Labourer £7.02 Rates are national average prices. Refer to REGIONAL VARIATIONS for indicative levels of overall pricing in regions	MATERIALS			LABOUR				RATES		
	Del to Site	Waste	Material Cost	Craft Optve	Lab	Labour Cost	Sunds	Nett Rate	Unit	Gross Rate (+7.5%)
	£	%	£	Hrs	Hrs	£	£	£		£

P20: UNFRAMED ISOLATED TRIMS/SKIRTINGS/SUNDRY ITEMS Cont.

Plywood B.S.6566, II/III grade, INT bonded; butt joints - isolated shelves and worktops Cont.

Cover fillets, stops, trims, beads, nosings and the like Cont.

	Del to Site	Waste	Material Cost	Craft Optve	Lab	Labour Cost	Sunds	Nett Rate	Unit	Gross Rate
50 x 12mm	0.29	20.00	0.35	0.16	0.02	1.62	0.05	2.02	m	2.17
50 x 15mm	0.36	20.00	0.43	0.17	0.02	1.71	0.05	2.19	m	2.36
50 x 19mm	0.42	20.00	0.50	0.18	0.02	1.80	0.05	2.36	m	2.53
100 x 6.5mm	0.32	20.00	0.38	0.16	0.02	1.62	0.05	2.05	m	2.21
100 x 12mm	0.57	20.00	0.68	0.17	0.02	1.71	0.05	2.44	m	2.63
100 x 15mm	0.74	20.00	0.89	0.18	0.02	1.80	0.05	2.74	m	2.95
100 x 19mm	0.83	20.00	1.00	0.19	0.02	1.89	0.05	2.94	m	3.16

Isolated shelves and worktops

	Del to Site	Waste	Material Cost	Craft Optve	Lab	Labour Cost	Sunds	Nett Rate	Unit	Gross Rate
6.5 x 150mm	0.69	20.00	0.83	0.18	0.02	1.80	0.05	2.68	m	2.88
6.5 x 225mm	1.06	20.00	1.27	0.23	0.02	2.26	0.05	3.59	m	3.85
6.5 x 300mm	1.41	20.00	1.69	0.28	0.04	2.87	0.05	4.61	m	4.95
6.5 x 450mm	2.10	20.00	2.52	0.33	0.04	3.33	0.05	5.90	m	6.34
6.5 x 600mm	2.79	20.00	3.35	0.38	0.05	3.86	0.05	7.26	m	7.80
12 x 150mm	1.30	20.00	1.56	0.19	0.02	1.89	0.05	3.50	m	3.77
12 x 225mm	1.94	20.00	2.33	0.24	0.03	2.43	0.05	4.80	m	5.16
12 x 300mm	2.61	20.00	3.13	0.29	0.04	2.96	0.05	6.14	m	6.60
12 x 450mm	3.89	20.00	4.67	0.34	0.04	3.42	0.05	8.14	m	8.75
12 x 600mm	5.18	20.00	6.22	0.39	0.05	3.95	0.05	10.22	m	10.98
15 x 150mm	1.56	20.00	1.87	0.23	0.03	2.33	0.05	4.26	m	4.57
15 x 225mm	2.35	20.00	2.82	0.25	0.03	2.52	0.05	5.39	m	5.79
15 x 300mm	3.14	20.00	3.77	0.30	0.04	3.05	0.05	6.87	m	7.38
15 x 450mm	4.70	20.00	5.64	0.35	0.04	3.51	0.05	9.20	m	9.89
15 x 600mm	6.27	20.00	7.52	0.40	0.05	4.04	0.05	11.62	m	12.49
19 x 150mm	1.73	20.00	2.08	0.20	0.03	2.06	0.05	4.18	m	4.50
19 x 225mm	2.60	20.00	3.12	0.25	0.03	2.52	0.05	5.69	m	6.11
19 x 300mm	3.44	20.00	4.13	0.30	0.04	3.05	0.05	7.23	m	7.77
19 x 450mm	5.16	20.00	6.19	0.35	0.04	3.51	0.05	9.75	m	10.48
19 x 600mm	6.88	20.00	8.26	0.40	0.05	4.04	0.05	12.35	m	13.28

Blockboard, B.S.3444, 2/2 grade, INT bonded, butt joints - isolated shelves and worktops

Isolated shelves and worktops

	Del to Site	Waste	Material Cost	Craft Optve	Lab	Labour Cost	Sunds	Nett Rate	Unit	Gross Rate
16 x 150mm	2.00	20.00	2.40	0.20	0.03	2.06	0.07	4.53	m	4.87
16 x 225mm	3.00	20.00	3.60	0.25	0.03	2.52	0.07	6.19	m	6.65
16 x 300mm	4.00	20.00	4.80	0.30	0.04	3.05	0.07	7.92	m	8.51
16 x 450mm	6.00	20.00	7.20	0.35	0.04	3.51	0.07	10.78	m	11.59
16 x 600mm	7.97	20.00	9.56	0.40	0.05	4.04	0.07	13.68	m	14.70
18 x 150mm	2.06	20.00	2.47	0.20	0.03	2.06	0.07	4.60	m	4.94
18 x 225mm	3.09	20.00	3.71	0.25	0.03	2.52	0.07	6.30	m	6.77
18 x 300mm	4.11	20.00	4.93	0.30	0.04	3.05	0.07	8.05	m	8.66
18 x 450mm	6.16	20.00	7.39	0.35	0.04	3.51	0.07	10.97	m	11.80
18 x 600mm	8.23	20.00	9.88	0.40	0.05	4.04	0.07	13.99	m	15.04
22 x 150mm	2.78	20.00	3.34	0.20	0.03	2.06	0.07	5.46	m	5.87
22 x 225mm	4.17	20.00	5.00	0.25	0.03	2.52	0.07	7.59	m	8.16
22 x 300mm	5.37	20.00	6.44	0.30	0.04	3.05	0.07	9.56	m	10.28
22 x 450mm	8.34	20.00	10.01	0.35	0.04	3.51	0.07	13.59	m	14.61
22 x 600mm	11.12	20.00	13.34	0.40	0.05	4.04	0.07	17.46	m	18.77
25 x 150mm	3.01	20.00	3.61	0.20	0.03	2.06	0.07	5.74	m	6.17
25 x 225mm	4.49	20.00	5.39	0.25	0.03	2.52	0.07	7.98	m	8.57
25 x 300mm	6.00	20.00	7.20	0.30	0.04	3.05	0.07	10.32	m	11.09
25 x 450mm	9.00	20.00	10.80	0.35	0.04	3.51	0.07	14.38	m	15.46
25 x 600mm	12.00	20.00	14.40	0.40	0.05	4.04	0.07	18.51	m	19.90

Wood chipboard B.S.5669, type I standard; butt joints - isolated shelves and worktops

Isolated shelves and worktops

	Del to Site	Waste	Material Cost	Craft Optve	Lab	Labour Cost	Sunds	Nett Rate	Unit	Gross Rate
12 x 150mm	0.43	20.00	0.52	0.19	0.02	1.89	0.05	2.46	m	2.64
12 x 225mm	0.62	20.00	0.74	0.24	0.03	2.43	0.05	3.22	m	3.46
12 x 300mm	0.84	20.00	1.01	0.29	0.04	2.96	0.05	4.02	m	4.32
12 x 450mm	1.25	20.00	1.50	0.34	0.04	3.42	0.05	4.97	m	5.34
12 x 600mm	1.68	20.00	2.02	0.39	0.05	3.95	0.05	6.02	m	6.47
18 x 150mm	0.56	20.00	0.67	0.20	0.03	2.06	0.05	2.78	m	2.99
18 x 225mm	0.83	20.00	1.00	0.25	0.03	2.52	0.05	3.56	m	3.83
18 x 300mm	1.09	20.00	1.31	0.30	0.04	3.05	0.05	4.41	m	4.74
18 x 450mm	1.64	20.00	1.97	0.35	0.04	3.51	0.05	5.53	m	5.94
18 x 600mm	2.18	20.00	2.62	0.40	0.05	4.04	0.05	6.71	m	7.21
25 x 150mm	1.03	20.00	1.24	0.20	0.03	2.06	0.05	3.34	m	3.59
25 x 225mm	1.54	20.00	1.85	0.25	0.03	2.52	0.05	4.42	m	4.75
25 x 300mm	2.05	20.00	2.46	0.30	0.04	3.05	0.05	5.56	m	5.98
25 x 450mm	3.07	20.00	3.68	0.35	0.04	3.51	0.05	7.25	m	7.79
25 x 600mm	4.09	20.00	4.91	0.40	0.05	4.04	0.05	9.00	m	9.68

Blockboard, B.S.3444, 2/2 grade, INT bonded, faced with 1.5mm laminated plastic sheet, B.S.EN438, classified HGS, with 1.2mm laminated plastic sheet balance veneer; butt joints

Isolated shelves and worktops

	Del to Site	Waste	Material Cost	Craft Optve	Lab	Labour Cost	Sunds	Nett Rate	Unit	Gross Rate
16 x 150mm	22.00	2.50	22.55	0.40	0.05	4.04	0.09	26.68	m	28.68
16 x 225mm	29.35	2.50	30.08	0.50	0.06	5.04	0.09	35.21	m	37.85
16 x 300mm	36.70	2.50	37.62	0.60	0.08	6.10	0.09	43.81	m	47.09
16 x 450mm	49.90	2.50	51.15	0.70	0.09	7.09	0.09	58.33	m	62.71
16 x 600mm	58.70	2.50	60.17	0.80	0.10	8.09	0.09	68.34	m	73.47
18 x 150mm	22.00	2.50	22.55	0.40	0.05	4.04	0.09	26.68	m	28.68
18 x 225mm	29.35	2.50	30.08	0.50	0.06	5.04	0.09	35.21	m	37.85

Labour hourly rates: (except Specialists) Craft Operatives £9.23 Labourer £7.02 Rates are national average prices. Refer to REGIONAL VARIATIONS for indicative levels of overall pricing in regions	MATERIALS			LABOUR				RATES		
	Del to Site	Waste	Material Cost	Craft Optve	Lab	Labour Cost	Sunds	Nett Rate	Unit	Gross Rate (+7.5%)
	£	%	£	Hrs	Hrs	£	£	£		£
P20: UNFRAMED ISOLATED TRIMS/SKIRTINGS/SUNDRY ITEMS Cont.										
Blockboard, B.S.3444, 2/2 grade, INT bonded, faced with 1.5mm laminated plastic sheet, B.S.EN438, classified HGS, with 1.2mm laminated plastic sheet balance veneer; butt joints Cont.										
Isolated shelves and worktops Cont.										
18 x 300mm	36.70	2.50	37.62	0.60	0.08	6.10	0.09	43.81	m	47.09
18 x 450mm	49.90	2.50	51.15	0.70	0.09	7.09	0.09	58.33	m	62.71
18 x 600mm	58.20	2.50	59.66	0.80	0.10	8.09	0.09	67.83	m	72.92
22 x 150mm	23.48	2.50	24.07	0.40	0.05	4.04	0.09	28.20	m	30.32
22 x 225mm	32.00	2.50	32.80	0.50	0.06	5.04	0.09	37.93	m	40.77
22 x 300mm	38.15	2.50	39.10	0.60	0.08	6.10	0.09	45.29	m	48.69
22 x 450mm	51.37	2.50	52.65	0.70	0.09	7.09	0.09	59.84	m	64.32
22 x 600mm	60.09	2.50	61.59	0.80	0.10	8.09	0.09	69.77	m	75.00
25 x 150mm	25.00	2.50	25.63	0.40	0.05	4.04	0.09	29.76	m	31.99
25 x 225mm	32.29	2.50	33.10	0.50	0.06	5.04	0.09	38.22	m	41.09
25 x 300mm	39.69	2.50	40.68	0.60	0.08	6.10	0.09	46.87	m	50.39
25 x 450mm	52.86	2.50	54.18	0.70	0.09	7.09	0.09	61.36	m	65.97
25 x 600mm	61.64	2.50	63.18	0.80	0.10	8.09	0.09	71.36	m	76.71
Laminated plastic sheet, B.S.EN 438 classification HGS										
Cover fillets, stops, trims, beads, nosings and the like; fixing with adhesive										
1.5 x 16mm	0.50	20.00	0.60	0.20	0.03	2.06	0.06	2.72	m	2.92
1.5 x 18mm	0.56	20.00	0.67	0.21	0.03	2.15	0.07	2.89	m	3.11
1.5 x 22mm	0.61	20.00	0.73	0.23	0.03	2.33	0.08	3.15	m	3.38
1.5 x 25mm	0.67	20.00	0.80	0.25	0.03	2.52	0.09	3.41	m	3.67
1.5 x 75mm	1.28	20.00	1.54	0.30	0.04	3.05	0.28	4.87	m	5.23
Isolated shelves and worktops; fixing with adhesive										
1.5 x 150mm	2.54	20.00	3.05	0.35	0.04	3.51	0.56	7.12	m	7.65
1.5 x 225mm	3.81	20.00	4.57	0.40	0.05	4.04	0.84	9.46	m	10.16
1.5 x 300mm	5.16	20.00	6.19	0.45	0.06	4.57	1.12	11.89	m	12.78
1.5 x 450mm	7.70	20.00	9.24	0.53	0.07	5.38	1.69	16.31	m	17.54
1.5 x 600mm	10.24	20.00	12.29	0.60	0.08	6.10	2.25	20.64	m	22.19
P21: IRONMONGERY										
Fix only ironmongery (prime cost sum for supply included elsewhere)										
To softwood										
butt hinges; 50mm	-	-	-	0.15	-	1.38	-	1.38	nr	1.49
butt hinges; 100mm	-	-	-	0.15	-	1.38	-	1.38	nr	1.49
rising hinges; 100mm	-	-	-	0.25	-	2.31	-	2.31	nr	2.48
tee hinges; 150mm	-	-	-	0.40	-	3.69	-	3.69	nr	3.97
tee hinges; 300mm	-	-	-	0.50	-	4.62	-	4.62	nr	4.96
tee hinges; 450mm	-	-	-	0.60	-	5.54	-	5.54	nr	5.95
hook and band hinges; 300mm	-	-	-	0.80	-	7.38	-	7.38	nr	7.94
hook and band hinges; 450mm	-	-	-	1.00	-	9.23	-	9.23	nr	9.92
hook and band hinges; 900mm	-	-	-	1.50	-	13.85	-	13.85	nr	14.88
collinge hinges; 600mm	-	-	-	0.90	-	8.31	-	8.31	nr	8.93
collinge hinges; 750mm	-	-	-	1.15	-	10.61	-	10.61	nr	11.41
collinge hinges; 900mm	-	-	-	1.40	-	12.92	-	12.92	nr	13.89
single action floor springs and top centres	-	-	-	2.50	-	23.07	-	23.07	nr	24.81
double action floor springs and top centres	-	-	-	3.00	-	27.69	-	27.69	nr	29.77
coil springs	-	-	-	0.25	-	2.31	-	2.31	nr	2.48
overhead door closers	-	-	-	1.50	-	13.85	-	13.85	nr	14.88
concealed overhead door closers	-	-	-	2.50	-	23.07	-	23.07	nr	24.81
Perko door closers	-	-	-	1.50	-	13.85	-	13.85	nr	14.88
door selectors	-	-	-	2.00	0.25	20.22	-	20.22	nr	21.73
cabin hooks	-	-	-	0.20	-	1.85	-	1.85	nr	1.98
fanlight catches	-	-	-	0.35	-	3.23	-	3.23	nr	3.47
roller catches	-	-	-	0.35	-	3.23	-	3.23	nr	3.47
casement fasteners	-	-	-	0.35	-	3.23	-	3.23	nr	3.47
sash fasteners	-	-	-	1.00	-	9.23	-	9.23	nr	9.92
sash screws	-	-	-	0.50	-	4.62	-	4.62	nr	4.96
mortice latches	-	-	-	0.65	-	6.00	-	6.00	nr	6.45
night latches	-	-	-	1.00	-	9.23	-	9.23	nr	9.92
Norfolk latches	-	-	-	0.80	-	7.38	-	7.38	nr	7.94
rim latches	-	-	-	0.85	-	7.85	-	7.85	nr	8.43
Suffolk latches	-	-	-	0.80	-	7.38	-	7.38	nr	7.94
budget locks	-	-	-	0.70	-	6.46	-	6.46	nr	6.95
cylinder locks	-	-	-	1.00	-	9.23	-	9.23	nr	9.92
dead locks	-	-	-	0.80	-	7.38	-	7.38	nr	7.94
mortice locks	-	-	-	1.00	-	9.23	-	9.23	nr	9.92
rim locks	-	-	-	0.80	-	7.38	-	7.38	nr	7.94
automatic coin collecting locks	-	-	-	2.00	-	18.46	-	18.46	nr	19.84
casement stays	-	-	-	0.25	-	2.31	-	2.31	nr	2.48
quadrant stays	-	-	-	0.25	-	2.31	-	2.31	nr	2.48
barrel bolts; 150mm	-	-	-	0.30	-	2.77	-	2.77	nr	2.98
barrel bolts; 250mm	-	-	-	0.40	-	3.69	-	3.69	nr	3.97
door bolts; 150mm	-	-	-	0.30	-	2.77	-	2.77	nr	2.98
door bolts; 250mm	-	-	-	0.40	-	3.69	-	3.69	nr	3.97
monkey tail bolts										
300mm	-	-	-	0.35	-	3.23	-	3.23	nr	3.47
450mm	-	%	-	0.40	-	3.69	-	3.69	nr	3.97
600mm	-	-	-	0.50	-	4.62	-	4.62	nr	4.96
flush bolts; 200mm	-	-	-	0.70	-	6.46	-	6.46	nr	6.95
flush bolts; 450mm	-	-	-	1.00	-	9.23	-	9.23	nr	9.92
indicating bolts	-	-	-	1.00	-	9.23	-	9.23	nr	9.92

BUILDING FABRIC SUNDRIES

Labour hourly rates: (except Specialists) Craft Operatives £9.23 Labourer £7.02 Rates are national average prices. Refer to REGIONAL VARIATIONS for indicative levels of overall pricing in regions	MATERIALS			LABOUR				RATES		
	Del to Site	Waste	Material Cost	Craft Optve	Lab	Labour Cost	Sunds	Nett Rate	Unit	Gross Rate (+7.5%)
	£	%	£	Hrs	Hrs	£	£	£		£
P21: IRONMONGERY Cont.										
Fix only ironmongery (prime cost sum for supply included elsewhere) Cont.										
To softwood Cont.										
monkey tail bolts Cont.										
panic bolts; to single door	-	-	-	1.50	-	13.85	-	13.85	nr	14.88
panic bolts; to double doors	-	-	-	2.00	-	18.46	-	18.46	nr	19.84
knobs	-	-	-	0.20	-	1.85	-	1.85	nr	1.98
lever handles	-	-	-	0.20	-	1.85	-	1.85	nr	1.98
sash lifts	-	-	-	0.20	-	1.85	-	1.85	nr	1.98
pull handles; 150mm	-	-	-	0.15	-	1.38	-	1.38	nr	1.49
pull handles; 225mm	-	-	-	0.20	-	1.85	-	1.85	nr	1.98
pull handles; 300mm	-	-	-	0.25	-	2.31	-	2.31	nr	2.48
back plates	-	-	-	0.25	-	2.31	-	2.31	nr	2.48
escutcheon plates	-	-	-	0.20	-	1.85	-	1.85	nr	1.98
kicking plates	-	-	-	0.50	-	4.62	-	4.62	nr	4.96
letter plates	-	-	-	1.50	-	13.85	-	13.85	nr	14.88
push plates; 225mm	-	-	-	0.20	-	1.85	-	1.85	nr	1.98
push plates; 300mm	-	-	-	0.25	-	2.31	-	2.31	nr	2.48
shelf brackets	-	-	-	0.25	-	2.31	-	2.31	nr	2.48
sash cleats	-	-	-	0.20	-	1.85	-	1.85	nr	1.98
To softwood and brickwork										
cabin hooks	-	-	-	0.25	-	2.31	-	2.31	nr	2.48
Fix only ironmongery (prime cost sum for supply included elsewhere)										
To hardwood or the like										
butt hinges; 50mm	-	-	-	0.20	-	1.85	-	1.85	nr	1.98
butt hinges; 100mm	-	-	-	0.20	-	1.85	-	1.85	nr	1.98
rising hinges; 100mm	-	-	-	0.35	-	3.23	-	3.23	nr	3.47
tee hinges; 150mm	-	-	-	0.55	-	5.08	-	5.08	nr	5.46
tee hinges; 300mm	-	-	-	0.70	-	6.46	-	6.46	nr	6.95
tee hinges; 450mm	-	-	-	0.80	-	7.38	-	7.38	nr	7.94
hook and band hinges; 300mm	-	-	-	1.10	-	10.15	-	10.15	nr	10.91
hook and band hinges; 450mm	-	-	-	1.40	-	12.92	-	12.92	nr	13.89
hook and band hinges; 900mm	-	-	-	2.00	-	18.46	-	18.46	nr	19.84
collinge hinges; 600mm	-	-	-	1.30	-	12.00	-	12.00	nr	12.90
collinge hinges; 750mm	-	-	-	1.70	-	15.69	-	15.69	nr	16.87
collinge hinges; 900mm	-	-	-	2.05	-	18.92	-	18.92	nr	20.34
single action floor springs and top centres	-	-	-	3.75	-	34.61	-	34.61	nr	37.21
double action floor springs and top centres	-	-	-	4.50	-	41.53	-	41.53	nr	44.65
coil springs	-	-	-	0.40	-	3.69	-	3.69	nr	3.97
overhead door closers	-	-	-	2.25	-	20.77	-	20.77	nr	22.33
concealed overhead door closers	-	-	-	3.75	-	34.61	-	34.61	nr	37.21
Perko door closers	-	-	-	2.25	-	20.77	-	20.77	nr	22.33
door selectors	-	-	-	3.00	0.38	30.36	-	30.36	nr	32.63
cabin hooks	-	-	-	0.30	-	2.77	-	2.77	nr	2.98
fanlight catches	-	-	-	0.50	-	4.62	-	4.62	nr	4.96
roller catches	-	-	-	0.50	-	4.62	-	4.62	nr	4.96
casement fasteners	-	-	-	0.50	-	4.62	-	4.62	nr	4.96
sash fasteners	-	-	-	1.50	-	13.85	-	13.85	Unr	14.88
sash screws	-	-	-	0.75	-	6.92	-	6.92	nr	7.44
mortice latches	-	%	-	1.00	Hrs	9.23	-	9.23	nr	9.92
night latches	-	-	-	1.50	-	13.85	-	13.85	nr	14.88
Norfolk latches	-	-	-	1.20	-	11.08	-	11.08	nr	11.91
rim latches	-	-	-	0.75	-	6.92	-	6.92	nr	7.44
Suffolk latches	-	-	-	1.20	-	11.08	-	11.08	nr	11.91
budget locks	-	-	-	1.00	-	9.23	-	9.23	nr	9.92
cylinder locks	-	-	-	1.50	-	13.85	-	13.85	nr	14.88
dead locks	-	-	-	1.20	-	11.08	-	11.08	nr	11.91
mortice locks	-	-	-	1.50	-	13.85	-	13.85	nr	14.88
rim locks	-	-	-	1.20	-	11.08	-	11.08	nr	11.91
automatic coin collecting locks	-	-	-	3.00	-	27.69	-	27.69	nr	29.77
casement stays	-	-	-	0.35	-	3.23	-	3.23	nr	3.47
quadrant stays	-	-	-	0.35	-	3.23	-	3.23	nr	3.47
barrel bolts; 150mm	-	-	-	0.40	-	3.69	-	3.69	nr	3.97
barrel bolts; 250mm	-	-	-	0.55	-	5.08	-	5.08	nr	5.46
door bolts; 150mm	-	-	-	0.40	-	3.69	-	3.69	nr	3.97
door bolts; 250mm	-	-	-	0.55	-	5.08	-	5.08	nr	5.46
monkey tail bolts										
300mm	-	-	-	0.50	-	4.62	-	4.62	nr	4.96
450mm	-	-	-	0.60	-	5.54	-	5.54	nr	5.95
600mm	-	-	-	0.70	-	6.46	-	6.46	nr	6.95
flush bolts; 200mm	-	-	-	1.00	-	9.23	-	9.23	nr	9.92
flush bolts; 450mm	-	-	-	1.50	-	13.85	-	13.85	nr	14.88
indicating bolts	-	-	-	1.50	-	13.85	-	13.85	nr	14.88
panic bolts; to single door	-	-	-	2.25	-	20.77	-	20.77	nr	22.33
panic bolts; to double doors	-	-	-	3.00	-	27.69	-	27.69	nr	29.77
knobs	-	-	-	0.30	-	2.77	-	2.77	nr	2.98
lever handles	-	-	-	0.30	-	2.77	-	2.77	nr	2.98
sash lifts	-	-	-	0.30	-	2.77	-	2.77	nr	2.98
pull handles; 150mm	-	-	-	0.20	-	1.85	-	1.85	nr	1.98
pull handles; 225mm	-	-	-	0.30	-	2.77	-	2.77	nr	2.98
pull handles; 300mm	-	-	-	0.35	-	3.23	-	3.23	nr	3.47
back plates	-	-	-	0.35	-	3.23	-	3.23	nr	3.47
escutcheon plates	-	-	-	0.30	-	2.77	-	2.77	Unr	2.98
kicking plates	-	-	-	0.75	-	6.92	-	6.92	nr	7.44
letter plates	£	%	-	2.25	Hrs	20.77	-	20.77	nr	22.33
push plates; 225mm	-	-	-	0.30	-	2.77	-	2.77	nr	2.98
push plates; 300mm	-	-	-	0.35	-	3.23	-	3.23	nr	3.47
shelf brackets	-	-	-	0.40	-	3.69	-	3.69	nr	3.97
sash cleats	-	-	-	0.30	-	2.77	-	2.77	nr	2.98

Labour hourly rates: (except Specialists) Craft Operatives £9.23 Labourer £7.02 Rates are national average prices. Refer to REGIONAL VARIATIONS for indicative levels of overall pricing in regions	MATERIALS			LABOUR				RATES		
	Del to Site £	Waste %	Material Cost £	Craft Optve Hrs	Lab Hrs	Labour Cost £	Sunds £	Nett Rate £	Unit	Gross Rate (+7.5%) £
P21: IRONMONGERY Cont.										
Fix only ironmongery (prime cost sum for supply included elsewhere) Cont.										
To hardwood and brickwork										
cabin hooks	-	-	-	0.35	-	3.23	-	3.23	nr	3.47
Water bars; steel; galvanized										
Water bars; to concrete										
flat section; setting in groove in mastic										
25 x 3 x 900mm long...........................	2.10	5.00	2.21	0.60	0.08	6.10	0.40	8.70	nr	9.36
40 x 3 x 900mm long...........................	3.34	5.00	3.51	0.60	0.08	6.10	0.50	10.11	nr	10.86
40 x 6 x 900mm long...........................	4.80	5.00	5.04	0.60	0.08	6.10	0.50	11.64	nr	12.51
50 x 6 x 900mm long...........................	6.47	5.00	6.79	0.60	0.08	6.10	0.50	13.39	nr	14.40
Water bars; to hardwood										
flat section; setting in groove in mastic										
25 x 3 x 900mm long...........................	2.10	5.00	2.21	0.50	0.06	5.04	0.40	7.64	nr	8.21
40 x 3 x 900mm long...........................	3.34	5.00	3.51	0.50	0.06	5.04	0.50	9.04	nr	9.72
40 x 6 x 900mm long...........................	4.80	5.00	5.04	0.50	0.06	5.04	0.50	10.58	nr	11.37
50 x 6 x 900mm long...........................	6.47	5.00	6.79	0.50	0.06	5.04	0.50	12.33	nr	13.25
Sliding door gear, P.C Henderson Ltd.										
To softwood										
interior straight sliding door gear sets for commercial and domestic doors; Senator single door set comprising track, hangers, end stops and bottom guide; for doors 20 - 35mm thick, maximum weight 25kg maximum 900mm wide	17.71	2.50	18.15	1.50	-	13.85	-	32.00	nr	34.40
extra for pelmet 1855mm long	14.08	2.50	14.43	0.65	-	6.00	-	20.43	nr	21.96
extra for pelmet end cap	1.54	2.50	1.58	0.10	-	0.92	-	2.50	nr	2.69
extra for Doorseal deflector guide	6.59	2.50	6.75	0.35	-	3.23	-	9.99	nr	10.73
interior straight sliding door gear sets for commercial and domestic doors; Phantom single door set comprising top assembly, hangers, adjustable nylon guide and door stops; for doors 30 - 50mm thick, maximum weight 45kg, 610 - 915mm wide	28.57	2.50	29.28	1.75	-	16.15	-	45.44	nr	48.84
extra for soffit fixing bracket and bolt	1.16	2.50	1.19	0.20	-	1.85	-	3.04	nr	3.26
extra for pelmet 1550mm long	11.77	2.50	12.06	0.55	-	5.08	-	17.14	nr	18.43
extra for pelmet 1855mm long	14.08	2.50	14.43	0.65	-	6.00	-	20.43	nr	21.96
extra for pelmet end cap	1.54	2.50	1.58	0.10	-	0.92	-	2.50	nr	2.69
extra for Doorseal deflector guide	6.59	2.50	6.75	0.35	-	3.23	-	9.99	nr	10.73
interior straight sliding door gear sets for commercial and domestic doors; Marathon Junior Nr J2 single door set comprising top assembly, hangers, end stops, inverted guide channel and nylon guide; for doors 32 - 50mm thick, maximum weight, 55kg, 400 - 750mm wide; forming groove for guide channel	26.56	2.50	27.22	2.40	-	22.15	-	49.38	nr	53.08
interior straight sliding door gear sets for commercial and domestic doors; Marathon Junior Nr J3 single door set comprising top assembly, hangers, end stops, inverted guide channel and nylon guide; for doors 32 - 50mm thick, maximum weight, 55kg, 750 - 900mm wide; forming groove for guide channel	28.03	2.50	28.73	2.60	-	24.00	-	52.73	nr	56.68
interior straight sliding door gear sets for commercial and domestic doors; Marathon Junior Nr J4 single door set comprising top assembly, hangers, end stops, inverted guide channel and nylon guide; for doors 32 - 50mm thick, maximum weight, 55kg, 900 - 1050mm wide; forming groove for guide channel	30.34	2.50	31.10	2.80	-	25.84	-	56.94	nr	61.21
interior straight sliding door gear sets for commercial and domestic doors; Marathon Junior Nr J5 single door set comprising top assembly, hangers, end stops, inverted guide channel and nylon guide; for doors 32 - 50mm thick, maximum weight, 55kg, 1050 - 1200mm wide; forming groove for guide channel	34.29	2.50	35.15	3.00	-	27.69	-	62.84	nr	67.55
interior straight sliding door gear sets for commercial and domestic doors; Marathon Junior Nr J6 single door set comprising top assembly, hangers, end stops, inverted guide channel and nylon guide; for doors 32 - 50mm thick, maximum weight, 55kg, 1200 - 1500mm wide; forming groove for guide channel	56.91	2.50	58.33	3.20	-	29.54	-	87.87	nr	94.46
extra for soffit fixing bracket and bolt	1.16	2.50	1.19	0.20	-	1.85	-	3.04	nr	3.26
extra for pelmet 1550mm long	11.77	2.50	12.06	0.55	-	5.08	-	17.14	nr	18.43
extra for pelmet 1855mm long	14.08	2.50	14.43	0.65	-	6.00	-	20.43	nr	21.96
extra for pelmet end cap	1.54	2.50	1.58	0.10	-	0.92	-	2.50	nr	2.69
extra for Doorseal deflector guide	6.59	2.50	6.75	0.35	-	3.23	-	9.99	nr	10.73
interior straight sliding door gear sets for commercial and domestic doors; Marathon Senior Nr S3 single door set comprising top assembly, hangers, end stops, inverted guide channel and nylon guide; for doors 32 - 50mm thick maximum weight 90kg, 750 - 900mm wide; forming groove for guide channel	37.61	2.50	38.55	2.80	-	25.84	-	64.39	nr	69.22

Labour hourly rates: (except Specialists) Craft Operatives £9.23 Labourer £7.02 Rates are national average prices. Refer to REGIONAL VARIATIONS for indicative levels of overall pricing in regions	MATERIALS			LABOUR				RATES		
	Del to Site £	Waste %	Material Cost £	Craft Optve Hrs	Lab Hrs	Labour Cost £	Sunds £	Nett Rate £	Unit	Gross Rate (+7.5%) £

P21: IRONMONGERY Cont.

Sliding door gear, P.C Henderson Ltd. Cont.

To softwood Cont.

	Del to Site £	Waste %	Material Cost £	Craft Optve Hrs	Lab Hrs	Labour Cost £	Sunds £	Nett Rate £	Unit	Gross Rate (+7.5%) £
interior straight sliding door gear sets for commercial and domestic doors; Marathon Senior Nr S4 single door set comprising top assembly, hangers, end stops, inverted guide channel and nylon guide; for doors 32 - 50mm thick maximum weight 90kg, 900 - 1050mm wide; forming groove for guide channel	41.76	2.50	42.80	3.00	-	27.69	-	70.49	nr	75.78
interior straight sliding door gear sets for commercial and domestic doors; Marathon Senior Nr S5 single door set comprising top assembly, hangers, end stops, inverted guide channel and nylon guide; for doors 32 - 50mm thick maximum weight 90kg, 1050 - 1200mm wide; forming groove for guide channel	46.08	2.50	47.23	3.15	-	29.07	-	76.31	nr	82.03
interior straight sliding door gear sets for commercial and domestic doors; Marathon Senior Nr S6 single door set comprising top assembly, hangers, end stops, inverted guide channel and nylon guide; for doors 32 - 50mm thick maximum weight 90kg, 1200 - 1500mm wide; forming groove for guide channel	73.89	2.50	75.74	3.30	-	30.46	-	106.20	nr	114.16
extra for soffit fixing bracket and bolt	1.16	2.50	1.19	0.20	-	1.85	-	3.04	nr	3.26
extra for pelmet 1550mm long	11.77	2.50	12.06	0.55	-	5.08	-	17.14	nr	18.43
extra for pelmet 1855mm long	14.08	2.50	14.43	0.65	-	6.00	-	20.43	nr	21.96
extra for pelmet end cap	1.54	2.50	1.58	0.10	-	0.92	-	2.50	nr	2.69
extra for Doorseal deflector guide	6.59	2.50	6.75	0.30	-	2.77	-	9.52	nr	10.24
interior straight sliding door gear sets for wardrobe and cupboard doors; Single Top Nr ST12 single door set comprising track, hangers, guides and safety stop; for door 16 - 35mm thick, maximum weight 25kg, maximum 900mm wide; one door to 600mm wide opening	10.81	2.50	11.08	1.25	-	11.54	-	22.62	nr	24.31
interior straight sliding door gear sets for wardrobe and cupobard doors; Single Top Nr ST15 single door set comprising track, hangers, guides and safety stop; for door 16 - 35mm thick, maximum weight 25kg, maximum 900mm wide; one door to 750mm wide opening	12.18	2.50	12.48	1.35	-	12.46	-	24.95	nr	26.82
interior straight sliding door gear sets for wardrobe and cupboard doors; Single Top Nr ST18 single door set comprising track, hangers, guides and safety stop; for door 16 - 35mm thick, maximum weight 25kg, maximum 900mm wide; one door to 900mm wide opening	13.79	2.50	14.13	1.45	-	13.38	-	27.52	nr	29.58
interior straight sliding door gear sets for wardrobe and cupboard doors; Double Top Nr W12 bi-passing door set comprising double track section, hangers, guides and safety stop; for doors 16 - 35mm thick, maximum weight 25kg, maximum 900mm wide; two doors in 1200mm wide opening	18.54	2.50	19.00	2.00	-	18.46	-	37.46	nr	40.27
interior straight sliding door gear sets for wardrobe and cupboard doors; Double Top Nr W15 bi-passing door set comprising double track section, hangers, guides and safety stop; for doors 16 - 35mm thick, maximum weight 25kg, maximum 900mm wide; two doors in 1500mm wide opening	20.67	2.50	21.19	2.10	-	19.38	-	40.57	nr	43.61
interior straight sliding door gear sets for wardrobe and cupboard doors; Double Top Nr W18 bi-passing door set comprising double track section, hangers, guides and safety stop; for doors 16 - 35mm thick, maximum weight 25kg, maximum 900mm wide; two doors in 1800mm wide opening	22.93	2.50	23.50	2.20	-	20.31	-	43.81	nr	47.09
interior straight sliding door gear sets for wardrobe and cupboard doors; Double Top Nr W24 bi-passing door set comprising double track section, hangers, guides and safety stop; for doors 16 - 35mm thick, maximum weight 25kg, maximum 900mm wide; three doors in 2400mm wide opening	30.06	2.50	30.81	2.40	-	22.15	-	52.96	nr	56.94
interior straight sliding door gear sets for wardrobe and cupboard doors; Bi-Fold Nr B10-2 folding door set comprising top guide track, top and bottom pivots; top guide and hinges; for doors 20 - 35mm thick; maximum weight 14kg each leaf, maximum 530mm wide; two doors in 1065mm wide opening	17.05	2.50	17.48	3.00	-	27.69	-	45.17	nr	48.55
interior straight sliding door gear sets for wardrobes and cupboard doors; Bi-Fold Nr B15-4 folding door set comprising top guide track, top and bottom pivots, top guide and hinges; for doors 20 - 35mm thick, maximum weight 14kg each leaf, maximum 530mm wide; four doors in 1525mm wide opening with aligner	28.93	2.50	29.65	6.00	-	55.38	-	85.03	nr	91.41
interior straight sliding door gear sets for wardrobes and cupboard doors; Bi-Fold Nr B20-4 folding door set comprising top guide track, top and bottom pivots, top guide and hinges; for doors 20 - 35mm thick, maximum weight 14kg each leaf, maximum 530mm wide; four doors in 2135mm wide opening with aligner	31.67	2.50	32.46	6.50	-	59.99	-	92.46	nr	99.39

Labour hourly rates: (except Specialists) Craft Operatives £9.23 Labourer £7.02 Rates are national average prices. Refer to REGIONAL VARIATIONS for indicative levels of overall pricing in regions	MATERIALS			LABOUR				RATES		
	Del to Site £	Waste %	Material Cost £	Craft Optve Hrs	Lab Hrs	Labour Cost £	Sunds £	Nett Rate £	Unit	Gross Rate (+7.5%) £

P21: IRONMONGERY Cont.

Sliding door gear, P.C Henderson Ltd. Cont.

To softwood Cont.

	Del to Site	Waste	Material Cost	Craft Optve	Lab	Labour Cost	Sunds	Nett Rate	Unit	Gross Rate
interior straight sliding door gear sets for built in cupboard doors; Slipper Nr SS4 double passing door set comprising two top tracks, sliders, safety stop, flush pulls and guides; for doors 16 - 30mm thick, maximum weight 9kg, maximum 900mm wide; two doors in 1200mm wide opening	13.07	2.50	13.40	1.85	–	17.08	–	30.47	nr	32.76
interior straight sliding door gear sets for built in cupboard doors; Slipper Nr SS5 double passing door set comprising two top tracks, sliders, safety stop, flush pulls and guides; for doors 16 - 30mm thick, maximum weight 9kg, maximum 900mm wide; two doors in 1500mm wide opening	15.38	2.50	15.76	2.00	–	18.46	–	34.22	nr	36.79
interior straight sliding door gear sets for built in cupboard doors; Slipper Nr SS6 double passing door set comprising two top tracks, sliders, safety stop, flush pulls and guides; for doors 16 - 30mm thick, maximum weight 9kg, maximum 900mm wide; two doors in 1800mm wide opening	17.18	2.50	17.61	2.15	–	19.84	–	37.45	nr	40.26
interior straight sliding door gear sets for cupboards, book cases and cabinet work; Loretto Nr D4 bi-passing door set comprising two top guide channels, two bottom rails, nylon guides and bottom rollers; for doors 20 - 45mm thick; maximum weight 23kg, maximum 900mm wide; two doors in 1200mm wide opening; forming grooves for top guide channels and bottom rails; forming sinkings for bottom rollers	22.15	2.50	22.70	2.50	–	23.07	–	45.78	nr	49.21
interior straight sliding door gear sets for cupboards, book cases and cabinet work; Loretto Nr D5 bi-passing door set comprising two top guide channels, two bottom rails, nylon guides and bottom rollers; for doors 20 - 45mm thick; maximum weight 23kg, maximum 900mm wide; two doors in 1500mm wide opening; forming grooves for top guide channels and bottom rails; forming sinkings for bottom rollers	24.65	2.50	25.27	2.60	–	24.00	–	49.26	nr	52.96
interior straight sliding door gear sets for cupboards, book cases and cabinet work; Loretto Nr D6 bi-passing door set comprising two top guide channels, two bottom rails, nylon guides and bottom rollers; for doors 20 - 45mm thick; maximum weight 23kg, maximum 900mm wide; two doors in 1800mm wide opening; forming grooves for top guide channels and bottom rails; forming sinkings for bottom rollers	27.85	2.50	28.55	2.70	–	24.92	–	53.47	nr	57.48
extra for retractable roller guide bolt	5.23	2.50	5.36	0.35	–	3.23	–	8.59	nr	9.24
extra for flush pull	2.27	2.50	2.33	0.25	–	2.31	–	4.63	nr	4.98
extra for safety stop	0.71	2.50	0.73	0.10	–	0.92	–	1.65	nr	1.77
extra for cylinder lock	10.94	2.50	11.21	0.60	–	5.54	–	16.75	nr	18.01
interior straight sliding/folding room divider gear sets; Husky Folding Nr HF-2 folding door set comprising top track, brackets, top and bottom pivots, hangers and hinges, bottom channel and roller guides; for doors 20 - 40mm thick, maximum weight 25kg, maximum 600mm wide, maximum 2400mm high; two doors in 1200mm wide opening; forming groove for bottom channel	47.04	2.50	48.22	5.80	–	53.53	–	101.75	nr	109.38
interior straight sliding/folding room divider gear sets; Husky Folding Nr HF-4 folding door set comprising top track, brackets, top and bottom pivots, hangers and hinges, bottom channel and roller guides; for doors 20 - 40mm thick, maximum weight 25kg, maximum 600mm wide, maximum 2400mm high; four doors in 2400mm wide opening; forming groove for bottom channel	89.30	2.50	91.53	11.50	–	106.15	–	197.68	nr	212.50
interior straight sliding door gear sets for glass panels; Zenith 2 Nr 12 double passing door set comprising double top guide, bottom rail, glass rail with rubber glazing strip, end caps and bottom rollers; for panels 6mm thick, maximum weight 16kg or 1m2 per panel; two panels nr 1200mm wide opening	33.63	2.50	34.47	1.50	–	13.85	–	48.32	nr	51.94
extra for dust seal	3.16	2.50	3.24	0.30	–	2.77	–	6.01	m	6.46
interior straight sliding door gear sets for glass panels; Zenith 2 Nr 15 double passing door set comprising double top guide, bottom rail, glass rail with rubber glazing strip, end caps and bottom rollers; for panels 6mm thick, maximum weight 16kg or 1m2 per panel; two panels nr 1500mm wide opening	38.73	2.50	39.70	1.65	–	15.23	–	54.93	nr	59.05
extra for dust seal	3.16	2.50	3.24	0.30	–	2.77	–	6.01	m	6.46
interior straight sliding door gear sets for glass panels; Zenith 2 Nr 18 double passing door set comprising double top guide, bottom rail, glass rail with rubber glazing strip, end caps and bottom rollers; for panels 6mm thick, maximum weight 16kg or 1m2 per panel; two panels nr 1800mm wide opening	43.36	2.50	44.44	1.80	–	16.61	–	61.06	nr	65.64
extra for finger pull	2.27	2.50	2.33	0.40	–	3.69	–	6.02	nr	6.47
extra for cylinder lock	10.94	2.50	11.21	1.00	–	9.23	–	20.44	nr	21.98
extra for dust seal	3.16	2.50	3.24	0.30	–	2.77	–	6.01	m	6.46

BUILDING FABRIC SUNDRIES

Labour hourly rates: (except Specialists) Craft Operatives £9.23 Labourer £7.02 Rates are national average prices. Refer to REGIONAL VARIATIONS for indicative levels of overall pricing in regions	MATERIALS			LABOUR				RATES		
	Del to Site £	Waste %	Material Cost £	Craft Optve Hrs	Lab Hrs	Labour Cost £	Sunds £	Nett Rate £	Unit	Gross Rate (+7.5%) £
P21: IRONMONGERY Cont.										
Sliding door gear, P.C Henderson Ltd. Cont.										
To softwood and concrete										
exterior straight sliding bottom roller gear; Sterling 225 components comprising top guide reference 900, top guide brackets for face fixing for single run reference 1/900, top guide rollers for face fixing reference 54/900, bottom rollers reference 5 and bottom rail reference 299; for timber doors 44 - 50mm thick, maximum weight 225kg, maximum 3300mm high; single door in 1200mm wide opening; forming notches for bottom rollers; casting bottom rail into concrete	222.55	2.50	228.11	6.15	–	56.76	–	284.88	nr	306.24
exterior staight sliding bottom roller gear; Sterling 225 components comprising top guide reference 900, top guide brackets for soffit fixing for single run reference 3/900, top guide rollers for edge fixing reference 203/900, bottom rollers reference 5 and bottom rail reference 299; for timber doors 44 - 50mm thick, maximum weight 225kg, maximum 3300mm high; single door in 1200mm wide opening; forming notches for bottom rollers; casting bottom rail into concrete	240.73	2.50	246.75	6.35	–	58.61	–	305.36	nr	328.26
exterior straight sliding bottom roller gear; Sterling 225 components comprising top guide reference 900, top guide brackets for face fixing for single run reference 1/900, top guide rollers for face fixing reference 54/900, bottom rollers reference 5 and bottom rail reference 299; for timber doors 44 - 50mm thick, maximum weight 225kg, maximum 3300mm high; two doors in 2400mm wide opening; forming notches for bottom rollers; casting bottom rails into concrete	438.51	2.50	449.47	12.00	–	110.76	–	560.23	nr	602.25
exterior straight sliding bottom roller gear; Sterling 225 components comprising top guides reference 900, top guide brackets for face fixing for double run reference 5/900, top guide rollers for face fixing reference 54/900, bottom rollers reference 5 and bottom rails reference 299; for timber doors 44 - 50mm thick, maximum weight 225kg, maximum 3300 high; two doors in 2400mm wide opening; forming notches for bottom rollers; casting bottom rails into concrete	498.77	2.50	511.24	12.35	–	113.99	–	625.23	nr	672.12
exterior straight sliding bottom roller gear; Sterling 350 components comprising top guide reference 99, top guide brackets for face fixing for single run reference 31 and 31x, top guide rollers reference 53/99, bottom rollers reference 2 and bottom rail reference 299; for timber doors 44 - 50mm thick, maximum weight 350kg, maximum 4000mm high; single door in 2400mm wide opening; forming notches for bottom rollers; casting bottom rail into concrete	493.33	2.50	505.66	12.90	–	119.07	–	624.73	nr	671.59
exterior straight sliding bottom roller gear; Sterling 350 components comprising top guide reference 99, top guide brackets for face fixing for single run reference 31 and 31X, top guide rollers reference 53/99, bottom rollers reference 2 and bottom rail reference 299; for timber doors 44 - 50mm thick, maximum weight 350kg, maximum 4000mm high; two doors in 3600mm wide opening; forming notches for bottom rollers; casting bottom rail into concrete	828.18	2.50	848.88	16.33	–	150.73	–	999.61	nr	1074.58
exterior straight sliding bottom roller gear; Sterling 800 components comprising top guide reference 99, top guide brackets for face fixing for single run reference 31 and 31X, top guide rollers reference 53/99, bottom rollers reference 3 and bottom rail reference 298; for timber doors 54 - 63mm thick, maximum weight 800kg, maximum 5200mm high; single door in 3600mm wide opening; forming notches for bottom rollers; casting bottom rail into concrete	713.90	2.50	731.75	14.65	–	135.22	–	866.97	nr	931.99
exterior straight sliding bottom roller gear, Sterling 800 components comprising top guide, reference 99, top guide brackets for face fixing for single run reference 31 and 31X, top guide rollers reference 53/99, bottom rollers reference 3 and bottom rail reference 298; for timber doors 54 - 63mm thick, maximum weight 800kg, maximum 5200mm high; two doors in 5400mm wide opening; forming notches for bottom rollers; casting bottom rail into concrete	1175.40	2.50	1204.79	23.55	–	217.37	–	1422.15	nr	1528.81
Hinges; standard quality										
To softwood										
backflap hinges; steel										
25mm	0.23	2.50	0.24	0.15	–	1.38	–	1.62	nr	1.74
38mm	0.27	2.50	0.28	0.15	–	1.38	–	1.66	nr	1.79
50mm	0.35	2.50	0.36	0.15	–	1.38	–	1.74	nr	1.87
63mm	1.04	2.50	1.07	0.15	–	1.38	–	2.45	nr	2.63
75mm	1.65	2.50	1.69	0.15	–	1.38	–	3.08	nr	3.31

BUILDING FABRIC SUNDRIES – MAJOR WORKS

Labour hourly rates: (except Specialists) Craft Operatives £9.23 Labourer £7.02 Rates are national average prices. Refer to REGIONAL VARIATIONS for indicative levels of overall pricing in regions	MATERIALS			LABOUR				RATES		
	Del to Site £	Waste %	Material Cost £	Craft Optve Hrs	Lab Hrs	Labour Cost £	Sunds £	Nett Rate £	Unit	Gross Rate (+7.5%) £
P21: IRONMONGERY Cont.										
Hinges; standard quality Cont.										
To softwood Cont.										
butt hinges; steel; light medium pattern										
38mm	0.14	2.50	0.14	0.15	-	1.38	-	1.53	nr	1.64
50mm	0.14	2.50	0.14	0.15	-	1.38	-	1.53	nr	1.64
63mm	0.16	2.50	0.16	0.15	-	1.38	-	1.55	nr	1.66
75mm	0.16	2.50	0.16	0.15	-	1.38	-	1.55	nr	1.66
100mm	0.33	2.50	0.34	0.15	-	1.38	-	1.72	nr	1.85
butt hinges; steel; strong pattern										
75mm	0.64	2.50	0.66	0.15	-	1.38	-	2.04	nr	2.19
100mm	0.70	2.50	0.72	0.15	-	1.38	-	2.10	nr	2.26
butt hinges; cast iron; light										
50mm	0.84	2.50	0.86	0.15	-	1.38	-	2.25	nr	2.41
63mm	0.94	2.50	0.96	0.15	-	1.38	-	2.35	nr	2.52
75mm	1.04	2.50	1.07	0.15	-	1.38	-	2.45	nr	2.63
100mm	1.13	2.50	1.16	0.15	-	1.38	-	2.54	nr	2.73
butt hinges; brass, brass pin										
38mm	0.79	2.50	0.81	0.15	-	1.38	-	2.19	nr	2.36
50mm	0.91	2.50	0.93	0.15	-	1.38	-	2.32	nr	2.49
63mm	1.07	2.50	1.10	0.15	-	1.38	-	2.48	nr	2.67
75mm	1.16	2.50	1.19	0.15	-	1.38	-	2.57	nr	2.77
100mm	2.68	2.50	2.75	0.15	-	1.38	-	4.13	nr	4.44
butt hinges; brass, steel washers, steel pin										
75mm	1.85	2.50	1.90	0.15	-	1.38	-	3.28	nr	3.53
100mm	3.17	2.50	3.25	0.15	-	1.38	-	4.63	nr	4.98
butt hinges; brass, B.M.A., steel washers, steel pin										
75mm	3.50	2.50	3.59	0.15	-	1.38	-	4.97	nr	5.34
100mm	5.69	2.50	5.83	0.15	-	1.38	-	7.22	nr	7.76
butt hinges; brass, chromium plated, steel washers, steel pin										
75mm	3.50	2.50	3.59	0.15	-	1.38	-	4.97	nr	5.34
100mm	6.12	2.50	6.27	0.15	-	1.38	-	7.66	nr	8.23
butt hinges; aluminium, stainless steel pin and washers										
75mm	2.74	2.50	2.81	0.15	-	1.38	-	4.19	nr	4.51
100mm	3.39	2.50	3.47	0.15	-	1.38	-	4.86	nr	5.22
rising hinges; steel										
75mm	1.02	2.50	1.05	0.25	-	2.31	-	3.35	nr	3.60
100mm	1.37	2.50	1.40	0.25	-	2.31	-	3.71	nr	3.99
rising hinges; cast iron										
75mm	1.97	2.50	2.02	0.25	-	2.31	-	4.33	nr	4.65
100mm	2.89	2.50	2.96	0.25	-	2.31	-	5.27	nr	5.66
spring hinges; steel, japanned; single action										
100mm	12.78	2.50	13.10	0.50	-	4.62	-	17.71	nr	19.04
125mm	14.37	2.50	14.73	0.55	-	5.08	-	19.81	nr	21.29
150mm	15.85	2.50	16.25	0.60	-	5.54	-	21.78	nr	23.42
spring hinges; steel, japanned; double action										
75mm	13.82	2.50	14.17	0.60	-	5.54	-	19.70	nr	21.18
100mm	15.85	2.50	16.25	0.60	-	5.54	-	21.78	nr	23.42
125mm	18.58	2.50	19.04	0.65	-	6.00	-	25.04	nr	26.92
150mm	21.31	2.50	21.84	0.70	-	6.46	-	28.30	nr	30.43
tee hinges; japanned; light										
150mm	0.35	2.50	0.36	0.40	-	3.69	-	4.05	nr	4.35
230mm	0.52	2.50	0.53	0.40	-	3.69	-	4.22	nr	4.54
300mm	0.68	2.50	0.70	0.45	-	4.15	-	4.85	nr	5.21
375mm	0.97	2.50	0.99	0.50	-	4.62	-	5.61	nr	6.03
450mm	1.34	2.50	1.37	0.55	-	5.08	-	6.45	nr	6.93
tee hinges; self colour; heavy										
150mm	0.62	2.50	0.64	0.45	-	4.15	-	4.79	nr	5.15
230mm	0.84	2.50	0.86	0.45	-	4.15	-	5.01	nr	5.39
300mm	1.34	2.50	1.37	0.50	-	4.62	-	5.99	nr	6.44
375mm	2.03	2.50	2.08	0.55	-	5.08	-	7.16	nr	7.69
450mm	2.56	2.50	2.62	0.60	-	5.54	-	8.16	nr	8.77
600mm	4.81	2.50	4.93	0.70	-	6.46	-	11.39	nr	12.25
hook and band hinges; on plate; heavy										
300mm	2.46	2.50	2.52	0.80	-	7.38	-	9.91	nr	10.65
450mm	4.04	2.50	4.14	1.00	-	9.23	-	13.37	nr	14.37
610mm	6.40	2.50	6.56	1.20	-	11.08	-	17.64	nr	18.96
914mm	11.21	2.50	11.49	1.50	-	13.85	-	25.34	nr	27.24
collinge hinges; cup for wood; best quality										
610mm	30.37	2.50	31.13	0.90	-	8.31	-	39.44	nr	42.39
762mm	37.43	2.50	38.37	1.15	-	10.61	-	48.98	nr	52.65
914mm	47.80	2.50	48.99	1.40	-	12.92	-	61.92	nr	66.56
parliament hinges; steel; 100mm	3.44	2.50	3.53	0.25	-	2.31	-	5.83	nr	6.27
cellar flap hinges; wrought, welded, plain joint; 450mm	22.40	2.50	22.96	0.50	-	4.62	-	27.57	nr	29.64
To hardwood or the like										
backflap hinges; steel										
25mm	0.22	2.50	0.23	0.20	-	1.85	-	2.07	nr	2.23
38mm	0.27	2.50	0.28	0.20	-	1.85	-	2.12	nr	2.28
50mm	0.35	2.50	0.36	0.20	-	1.85	-	2.20	nr	2.37
63mm	1.04	2.50	1.07	0.20	-	1.85	-	2.91	nr	3.13
75mm	1.65	2.50	1.69	0.20	-	1.85	-	3.54	nr	3.80
butt hinges; steel; light medium pattern										
38mm	0.13	2.50	0.13	0.20	-	1.85	-	1.98	nr	2.13
50mm	0.13	2.50	0.13	0.20	-	1.85	-	1.98	nr	2.13
63mm	0.16	2.50	0.16	0.20	-	1.85	-	2.01	nr	2.16
75mm	0.16	2.50	0.16	0.20	-	1.85	-	2.01	nr	2.16
100mm	0.32	2.50	0.33	0.20	-	1.85	-	2.17	nr	2.34
butt hinges; steel; strong pattern										
75mm	0.64	2.50	0.66	0.20	-	1.85	-	2.50	nr	2.69

BUILDING FABRIC SUNDRIES

Labour hourly rates: (except Specialists) Craft Operatives £9.23 Labourer £7.02 Rates are national average prices. Refer to REGIONAL VARIATIONS for indicative levels of overall pricing in regions	MATERIALS			LABOUR				RATES		
	Del to Site £	Waste %	Material Cost £	Craft Optve Hrs	Lab Hrs	Labour Cost £	Sunds £	Nett Rate £	Unit	Gross Rate (+7.5%) £
P21: IRONMONGERY Cont.										
Hinges; standard quality Cont.										
To hardwood or the like Cont.										
butt hinges; steel; strong pattern Cont.										
100mm	0.70	2.50	0.72	0.20	-	1.85	-	2.56	nr	2.76
butt hinges; cast iron; light										
50mm	0.84	2.50	0.86	0.20	-	1.85	-	2.71	nr	2.91
63mm	0.94	2.50	0.96	0.20	-	1.85	-	2.81	nr	3.02
75mm	1.04	2.50	1.07	0.20	-	1.85	-	2.91	nr	3.13
100mm	1.13	2.50	1.16	0.20	-	1.85	-	3.00	nr	3.23
butt hinges; brass, brass pin										
38mm	0.79	2.50	0.81	0.20	-	1.85	-	2.66	nr	2.85
50mm	0.91	2.50	0.93	0.20	-	1.85	-	2.78	nr	2.99
63mm	1.07	2.50	1.10	0.20	-	1.85	-	2.94	nr	3.16
75mm	1.16	2.50	1.19	0.20	-	1.85	-	3.04	nr	3.26
100mm	2.68	2.50	2.75	0.20	-	1.85	-	4.59	nr	4.94
butt hinges; brass, steel washers, steel pin										
75mm	1.85	2.50	1.90	0.20	-	1.85	-	3.74	nr	4.02
100mm	3.17	2.50	3.25	0.20	-	1.85	-	5.10	nr	5.48
butt hinges; brass, B.M.A., steel washers, steel pin										
75mm	3.50	2.50	3.59	0.20	-	1.85	-	5.43	nr	5.84
100mm	5.69	2.50	5.83	0.20	-	1.85	-	7.68	nr	8.25
butt hinges; brass, chromium plated, steel washers, steel pin										
75mm	3.50	2.50	3.59	0.20	-	1.85	-	5.43	nr	5.84
100mm	6.12	2.50	6.27	0.20	-	1.85	-	8.12	nr	8.73
butt hinges; aluminium, stainless steel pin and washers										
75mm	2.74	2.50	2.81	0.20	-	1.85	-	4.65	nr	5.00
100mm	3.39	2.50	3.47	0.20	-	1.85	-	5.32	nr	5.72
rising hinges; steel										
75mm	1.02	2.50	1.05	0.35	-	3.23	-	4.28	nr	4.60
100mm	1.37	2.50	1.40	0.35	-	3.23	-	4.63	nr	4.98
rising hinges; cast iron										
75mm	1.97	2.50	2.02	0.35	-	3.23	-	5.25	nr	5.64
100mm	2.89	2.50	2.96	0.35	-	3.23	-	6.19	nr	6.66
spring hinges; steel, japanned; single action										
100mm	12.78	2.50	13.10	0.75	-	6.92	-	20.02	nr	21.52
125mm	14.37	2.50	14.73	0.80	-	7.38	-	22.11	nr	23.77
150mm	15.85	2.50	16.25	0.85	-	7.85	-	24.09	nr	25.90
spring hinges; steel, japanned; double action										
75mm	13.82	2.50	14.17	0.80	-	7.38	-	21.55	nr	23.17
100mm	15.85	2.50	16.25	0.85	-	7.85	-	24.09	nr	25.90
125mm	18.58	2.50	19.04	0.95	-	8.77	-	27.81	nr	29.90
150mm	21.31	2.50	21.84	1.00	-	9.23	-	31.07	nr	33.40
tee hinges; japanned; light										
150mm	0.35	2.50	0.36	0.55	-	5.08	-	5.44	nr	5.84
230mm	0.52	2.50	0.53	0.55	-	5.08	-	5.61	nr	6.03
300mm	0.68	2.50	0.70	0.65	-	6.00	-	6.70	nr	7.20
375mm	1.01	2.50	1.04	0.75	-	6.92	-	7.96	nr	8.55
450mm	1.34	2.50	1.37	0.80	-	7.38	-	8.76	nr	9.41
tee hinges; self colour; heavy										
150mm	0.62	2.50	0.64	0.60	-	5.54	-	6.17	nr	6.64
230mm	0.84	2.50	0.86	0.60	-	5.54	-	6.40	nr	6.88
300mm	1.34	2.50	1.37	0.70	-	6.46	-	7.83	nr	8.42
375mm	2.03	2.50	2.08	0.80	-	7.38	-	9.46	nr	10.17
450mm	2.56	2.50	2.62	0.85	-	7.85	-	10.47	nr	11.25
600mm	4.81	2.50	4.93	0.95	-	8.77	-	13.70	nr	14.73
hook and band hinges; on plate; heavy										
300mm	2.46	2.50	2.52	1.10	-	10.15	-	12.67	nr	13.63
450mm	4.04	2.50	4.14	1.40	-	12.92	-	17.06	nr	18.33
610mm	6.40	2.50	6.56	1.75	-	16.15	-	22.71	nr	24.42
914mm	11.21	2.50	11.49	2.00	-	18.46	-	29.95	nr	32.20
collinge hinges; cup for wood; best quality										
610mm	30.37	2.50	31.13	1.30	-	12.00	-	43.13	nr	46.36
762mm	37.43	2.50	38.37	1.70	-	15.69	-	54.06	nr	58.11
914mm	47.80	2.50	48.99	2.05	-	18.92	-	67.92	nr	73.01
parliament hinges; steel; 100mm	3.44	2.50	3.53	0.35	-	3.23	-	6.76	nr	7.26
cellar flap hinges; wrought, welded, plain joint; 450mm	22.40	2.50	22.96	0.75	-	6.92	-	29.88	nr	32.12
Floor springs; standard quality										
To softwood										
single action floor springs and top centres; B.M.A.	185.76	2.50	190.40	2.50	-	23.07	-	213.48	nr	229.49
single action floor springs and top centres; chromium plated	185.76	2.50	190.40	2.50	-	23.07	-	213.48	nr	229.49
double action floor springs and top centres; B.M.A.	224.00	2.50	229.60	3.00	-	27.69	-	257.29	nr	276.59
double action floor springs and top centres; chromium plated	224.00	2.50	229.60	3.00	-	27.69	-	257.29	nr	276.59
To hardwood or the like										
single action floor springs and top centres; B.M.A.	185.76	2.50	190.40	3.75	-	34.61	-	225.02	nr	241.89
single action floor springs and top centres; chromium plated	185.76	2.50	190.40	3.75	-	34.61	-	225.02	nr	241.89
double action floor springs and top centres; B.M.A.	224.00	2.50	229.60	4.50	-	41.53	-	271.13	nr	291.47
double action floor springs and top centres; chromium plated	224.00	2.50	229.60	4.50	-	41.53	-	271.13	nr	291.47

BUILDING FABRIC SUNDRIES – MAJOR WORKS

Labour hourly rates: (except Specialists) Craft Operatives £9.23 Labourer £7.02 Rates are national average prices. Refer to REGIONAL VARIATIONS for indicative levels of overall pricing in regions	MATERIALS			LABOUR				RATES		
	Del to Site £	Waste %	Material Cost £	Craft Optve Hrs	Lab Hrs	Labour Cost £	Sunds £	Nett Rate £	Unit	Gross Rate (+7.5%) £

P21: IRONMONGERY Cont.

Door closers; standard quality

To softwood

coil door springs; japanned	3.93	2.50	4.03	0.25	–	2.31	–	6.34	nr	6.81
overhead door closers; liquid check and spring; 'Briton' 2000 series; silver; light doors	63.89	2.50	65.49	1.50	–	13.85	–	79.33	nr	85.28
overhead door closers; liquid check and spring; 'Briton' 2000 series; silver; medium doors	74.30	2.50	76.16	1.50	–	13.85	–	90.00	nr	96.75
overhead door closers; liquid check and spring; 'Briton' 2000 series; silver; heavy doors	95.61	2.50	98.00	1.50	–	13.85	–	111.85	nr	120.23
concealed overhead door closers; liquid check and spring; medium doors	106.54	2.50	109.20	2.50	–	23.07	–	132.28	nr	142.20
Perko door closers; brass plate	6.66	2.50	6.83	1.50	–	13.85	–	20.67	nr	22.22

To hardwood or the like

coil door springs; japanned	3.93	2.50	4.03	0.40	–	3.69	–	7.72	nr	8.30
overhead door closers; liquid check and spring; 'Briton' 2000 series; silver; light doors	63.89	2.50	65.49	2.25	–	20.77	–	86.25	nr	92.72
overhead door closers; liquid check and spring; 'Briton' 2000 series; silver; medium doors	74.30	2.50	76.16	2.25	–	20.77	–	96.92	nr	104.19
overhead door closers; liquid check and spring; 'Briton' 2000 series; silver; heavy doors	95.61	2.50	98.00	2.25	–	20.77	–	118.77	nr	127.68
concealed overhead door closers; liquid check and spring; medium doors	106.54	2.50	109.20	3.75	–	34.61	–	143.82	nr	154.60
Perko door closers; brass plate	6.66	2.50	6.83	2.25	–	20.77	–	27.59	nr	29.66

Door selectors; standard quality

To softwood or the like

Union 8815 door selector; aluminium anodised silver finish	47.87	2.50	49.07	2.00	0.25	20.22	–	69.28	nr	74.48
Close Rite door selector; satin nickel plated	35.74	2.50	36.63	2.25	0.28	22.73	–	59.37	nr	63.82

To hardwood or the like

Union 8815 door selector; aluminium anodised silver finish	47.87	2.50	49.07	3.00	0.38	30.36	–	79.42	nr	85.38
Close Rite door selector; satin nickel plated	35.74	2.50	36.63	3.50	0.44	35.39	–	72.03	nr	77.43

Locks and latches; standard quality

To softwood

magnetic catches; 6lb pull	0.52	2.50	0.53	0.20	–	1.85	–	2.38	nr	2.56
magnetic catches; 9lb pull	0.64	2.50	0.66	0.20	–	1.85	–	2.50	nr	2.69
magnetic catches; 13lb pull	1.06	2.50	1.09	0.20	–	1.85	–	2.93	nr	3.15
roller catches; mortice; nylon; 18mm	1.06	2.50	1.09	0.35	–	3.23	–	4.32	nr	4.64
roller catches; mortice; nylon; 27mm	1.06	2.50	1.09	0.35	–	3.23	–	4.32	nr	4.64
roller catches; surface, adjustable; nylon; 16mm	1.85	2.50	1.90	0.25	–	2.31	–	4.20	nr	4.52
roller catches; mortice, adjustable; satin chrome plated; 25 x 22 x 10mm	4.53	2.50	4.64	0.40	–	3.69	–	8.34	nr	8.96
roller catches; double ball; brass; 43mm	0.69	2.50	0.71	0.30	–	2.77	–	3.48	nr	3.74
mortice latches; stamped steel case; 75mm	1.02	2.50	1.05	0.65	–	6.00	–	7.04	nr	7.57
mortice latches; locking; stamped steel case; 75mm	2.03	2.50	2.08	0.65	–	6.00	–	8.08	Unit	8.69
cylinder mortice latches; 'Union' key operated outside, knob inside, bolt held by slide; B.M.A.; for end set to suit 13mm rebate	39.94	2.50	40.94	0.75	–	6.92	–	47.86	nr	51.45
cylinder mortice latches; 'Union' key operated outside, knob inside, bolt held by slide; chromium plated; for end set to suit 13mm rebate	37.94	2.50	38.89	0.75	–	6.92	–	45.81	nr	49.25
cylinder rim night latches; 'Legge' 707; chromium plated	13.01	2.50	13.34	1.00	–	9.23	–	22.57	nr	24.26
cylinder rim night latches; 'Union' 1022; chromium plated	15.19	2.50	15.57	1.00	–	9.23	–	24.80	nr	26.66
cylinder rim night latches; 'Yale' 88; chromium plated	16.88	2.50	17.30	1.00	–	9.23	–	26.53	nr	28.52
Suffolk latches; japanned; medium; size Nr 2	2.03	2.50	2.08	0.80	–	7.38	–	9.46	nr	10.17
Suffolk latches; japanned; medium; size Nr 3	2.62	2.50	2.69	0.80	–	7.38	–	10.07	nr	10.82
Suffolk latches; japanned; heavy; size Nr 4	4.26	2.50	4.37	0.80	–	7.38	–	11.75	nr	12.63
Suffolk latches; galvanised; size Nr 4	4.53	2.50	4.64	0.80	–	7.38	–	12.03	nr	12.93
cupboard locks; 1 lever; japanned; 63mm	3.99	2.50	4.09	0.75	–	6.92	–	11.01	nr	11.84
cupboard locks; 2 lever; brass; 50mm	5.41	2.50	5.55	0.75	–	6.92	–	12.47	nr	13.40
cupboard locks; 2 lever; brass; 63mm	5.41	2.50	5.55	0.75	–	6.92	–	12.47	nr	13.40
cupboard locks; 4 lever; brass; 63mm	5.98	2.50	6.13	0.75	–	6.92	–	13.05	nr	14.03
rim dead lock; japanned case; 100mm	3.44	2.50	3.53	0.80	–	7.38	–	10.91	nr	11.73
rim dead lock; japanned case; 125mm	3.99	2.50	4.09	0.80	–	7.38	–	11.47	nr	12.33
rim dead lock; japanned case; 150mm	4.81	2.50	4.93	0.80	–	7.38	–	12.31	nr	13.24
mortice dead locks; japanned case; 75mm	5.03	2.50	5.16	0.80	–	7.38	–	12.54	nr	13.48
mortice locks; three levers	3.93	2.50	4.03	1.00	–	9.23	–	13.26	nr	14.25
mortice locks; five levers	10.38	2.50	10.64	1.00	–	9.23	–	19.87	nr	21.36
rim locks; japanned case; 150 x 100mm	4.81	2.50	4.93	0.80	–	7.38	–	12.31	nr	13.24
rim locks; japanned case; 150 x 75mm	5.36	2.50	5.49	0.80	–	7.38	–	12.88	nr	13.84
rim locks; japanned case; strong pattern; 150 x 100mm	7.21	2.50	7.39	0.80	–	7.38	–	14.77	nr	15.88
padlocks, 'Squire', anti-pilfer bolt, warded spring; zinc plated; 32mm	1.37	2.50	1.40	–	–	–	–	1.40	nr	1.51
padlocks, 'Squire', anti-pilfer bolt, warded spring; zinc plated; 38mm	2.13	2.50	2.18	–	–	–	–	2.18	nr	2.35
padlocks, 'Squire', anti-pilfer bolt, warded spring; zinc plated; 44mm	2.35	2.50	2.41	–	–	–	–	2.41	nr	2.59
padlocks; 'Squire', anti-pilfer bolt, 4 pin tumblers; zinc plated; 32mm	4.32	2.50	4.43	–	–	–	–	4.43	nr	4.76
padlocks; 'Squire', anti-pilfer bolt, 4 pin tumblers; zinc plated; 38mm	5.69	2.50	5.83	–	–	–	–	5.83	nr	6.27
padlocks; 'Squire', anti-pilfer bolt, 4 pin tumblers; zinc plated; 44mm	6.72	2.50	6.89	–	–	–	–	6.89	nr	7.40

BUILDING FABRIC SUNDRIES

BUILDING FABRIC SUNDRIES – MAJOR WORKS

	Del to Site £	Waste %	Material Cost £	Craft Optve Hrs	Lab Hrs	Labour Cost £	Sunds £	Nett Rate £	Unit	Gross Rate (+7.5%) £
P21: IRONMONGERY Cont.										
Locks and latches; standard quality Cont.										
To softwood Cont.										
padlocks; 'Squire', 4 levers; galvanised case	3.39	2.50	3.47	-	-	-	-	3.47	nr	3.74
padlocks, 'Squire', 4 levers; brass case	4.37	2.50	4.48	-	-	-	-	4.48	nr	4.82
padlocks; 'Squire', strong pattern; chromium plated	29.28	2.50	30.01	-	-	-	-	30.01	nr	32.26
hasps and staples; japanned wire; light; 75mm	0.56	2.50	0.57	0.30	-	2.77	-	3.34	nr	3.59
hasps and staples; japanned wire; light; 100mm ...	0.66	2.50	0.68	0.30	-	2.77	-	3.45	nr	3.70
hasps and staples; galvanised wire; heavy; 75mm .	1.22	2.50	1.25	0.30	-	2.77	-	4.02	nr	4.32
hasps and staples; galvanised wire; heavy; 100mm	1.32	2.50	1.35	0.30	-	2.77	-	4.12	nr	4.43
hasps and staples; galvanised wire; heavy; 150mm .	1.37	2.50	1.40	0.30	-	2.77	-	4.17	nr	4.49
locking bars; japanned; 200mm	5.14	2.50	5.27	0.45	-	4.15	-	9.42	nr	10.13
locking bars; japanned; 250mm	5.36	2.50	5.49	0.45	-	4.15	-	9.65	nr	10.37
locking bars; japanned; 300mm	6.78	2.50	6.95	0.45	-	4.15	-	11.10	nr	11.94
locking bars; japanned; 350mm	7.98	2.50	8.18	0.45	-	4.15	-	12.33	nr	13.26
To hardwood or the like										
magnetic catches; 6lb pull	0.52	2.50	0.53	0.30	-	2.77	-	3.30	nr	3.55
magnetic catches; 9lb pull	0.64	2.50	0.66	0.30	-	2.77	-	3.42	nr	3.68
magnetic catches; 13lb pull	1.06	2.50	1.09	0.30	-	2.77	-	3.86	nr	4.14
roller catches; mortice; nylon; 18mm	1.06	2.50	1.09	0.50	-	4.62	-	5.70	nr	6.13
roller catches; mortice; nylon; 27mm	1.06	2.50	1.09	0.50	-	4.62	-	5.70	nr	6.13
roller catches; surface, adjustable; nylon; 16mm .	1.85	2.50	1.90	0.40	-	3.69	-	5.59	nr	6.01
roller catches; mortice, adjustable; satin chrome plated; 25 x 22 x 10mm	4.53	2.50	4.64	0.60	-	5.54	-	10.18	nr	10.94
roller catches; double ball; brass; 43mm	0.69	2.50	0.71	0.45	-	4.15	-	4.86	nr	5.23
mortice latches; stamped steel case; 75mm	1.02	2.50	1.05	1.00	-	9.23	-	10.28	nr	11.05
mortice latches; locking; stamped steel case; 75mm	2.03	2.50	2.08	1.00	-	9.23	-	11.31	nr	12.16
cylinder mortice latches; 'Union' key operated outside, knob inside, bolt held by slide; B.M.A.; for end set to suit 13mm rebate	39.94	2.50	40.94	1.10	-	10.15	-	51.09	nr	54.92
cylinder mortice latches; 'Union' key operated outside, knob inside, bolt held by slide; chromium plated; for end set to suit 13mm rebate	39.94	2.50	40.94	1.10	-	10.15	-	51.09	nr	54.92
cylinder rim night latches; 'Legge' 707; chromium plated	13.01	2.50	13.34	1.50	-	13.85	-	27.18	nr	29.22
cylinder rim night latches; 'Union' 1022; chromium plated	15.19	2.50	15.57	1.50	-	13.85	-	29.41	nr	31.62
cylinder rim night latches; 'Yale' 88; chromium plated	16.88	2.50	17.30	1.50	-	13.85	-	31.15	nr	33.48
Suffolk latches; japanned; medium; size Nr 2	2.03	2.50	2.08	1.20	-	11.08	-	13.16	nr	14.14
Suffolk latches; japanned; medium; size Nr 3	2.62	2.50	2.69	1.20	-	11.08	-	13.76	nr	14.79
Suffolk latches; japanned; medium; size Nr 4	4.26	2.50	4.37	1.20	-	11.08	-	15.44	nr	16.60
Suffolk latches; japanned; heavy; size Nr 4	4.53	2.50	4.64	1.20	-	11.08	-	15.72	nr	16.90
Suffolk latches; galvanised; size Nr 4	3.39	2.50	3.47	1.20	-	11.08	-	14.55	nr	15.64
cupboard locks; 1 lever; japanned; 63mm	3.99	2.50	4.09	1.10	-	10.15	-	14.24	nr	15.31
cupboard locks; 2 lever; brass; 50mm	5.41	2.50	5.55	1.10	-	10.15	-	15.70	nr	16.88
cupboard locks; 2 lever; brass; 63mm	5.41	2.50	5.55	1.10	-	10.15	-	15.70	nr	16.88
cupboard locks; 4 lever; brass; 63mm	5.98	2.50	6.13	1.10	-	10.15	-	16.28	nr	17.50
rim dead lock; japanned case; 100mm	3.44	2.50	3.53	1.20	-	11.08	-	14.60	nr	15.70
rim dead lock; japanned case; 125mm	3.99	2.50	4.09	1.20	-	11.08	-	15.17	nr	16.30
rim dead lock; japanned case; 150mm	4.81	2.50	4.93	1.20	-	11.08	-	16.01	nr	17.21
mortice dead locks; japanned case; 75mm	5.03	2.50	5.16	1.20	-	11.08	-	16.23	nr	17.45
mortice locks; three levers	3.93	2.50	4.03	1.50	-	13.85	-	17.87	nr	19.21
mortice locks; five levers	10.38	2.50	10.64	1.50	-	13.85	-	24.48	nr	26.32
rim locks; japanned case; 150 x 100mm	4.81	2.50	4.93	1.20	-	11.08	-	16.01	nr	17.21
rim locks; japanned case; 150 x 75mm	5.36	2.50	5.49	1.20	-	11.08	-	16.57	nr	17.81
rim locks; japanned case; strong pattern; 150 x 100mm	7.21	2.50	7.39	1.20	-	11.08	-	18.47	nr	19.85
padlocks, 'Squire', anti-pilfer bolt, warded spring; zinc plated; 32mm	1.37	2.50	1.40	-	-	-	-	1.40	nr	1.51
padlocks, 'Squire', anti-pilfer bolt, warded spring; zinc plated; 38mm	2.13	2.50	2.18	-	-	-	-	2.18	nr	2.35
padlocks, 'Squire', anti-pilfer bolt, warded spring; zinc plated; 44mm	2.35	2.50	2.41	-	-	-	-	2.41	nr	2.59
padlocks; 'Squire', anti-pilfer bolt, 4 pin tumblers; zinc plated; 32mm	4.32	2.50	4.43	-	-	-	-	4.43	nr	4.76
padlocks; 'Squire', anti-pilfer bolt, 4 pin tumblers; zinc plated; 38mm	5.69	2.50	5.83	-	-	-	-	5.83	nr	6.27
padlocks; 'Squire', anti-pilfer bolt, 4 pin tumblers; zinc plated; 44mm	6.72	2.50	6.89	-	-	-	-	6.89	nr	7.40
padlocks; 'Squire', 4 levers; galvanised case	3.39	2.50	3.47	-	-	-	-	3.47	nr	3.74
padlocks; 'Squire', 4 levers; brass case	4.37	2.50	4.48	-	-	-	-	4.48	nr	4.82
padlocks; 'Squire', strong pattern; chromium plated	29.28	2.50	30.01	-	-	-	-	30.01	nr	32.26
hasps and staples; japanned wire; light; 75mm	0.56	2.50	0.57	0.50	-	4.62	-	5.19	nr	5.58
hasps and staples; japanned wire; light; 100mm ...	0.66	2.50	0.68	0.50	-	4.62	-	5.29	nr	5.69
hasps and staples; galvanised wire; heavy; 75mm ..	1.22	2.50	1.25	0.50	-	4.62	-	5.87	nr	6.31
hasps and staples; galvanised wire; heavy; 100mm .	1.32	2.50	1.35	0.50	-	4.62	-	5.97	nr	6.42
hasps and staples; galvanised wire; heavy; 150mm .	1.37	2.50	1.40	0.50	-	4.62	-	6.02	nr	6.47
locking bars; japanned; 200mm	5.14	2.50	5.27	0.60	-	5.54	-	10.81	nr	11.62
locking bars; japanned; 250mm	5.36	2.50	5.49	0.60	-	5.54	-	11.03	nr	11.86
locking bars; japanned; 300mm	6.78	2.50	6.95	0.60	-	5.54	-	12.49	nr	13.42
locking bars; japanned; 350mm	7.98	2.50	8.18	0.60	-	5.54	-	13.72	nr	14.75
Bolts; standard quality										
To softwood										
barrel bolts; japanned, steel barrel; medium										
100mm	1.04	2.50	1.07	0.25	-	2.31	-	3.37	nr	3.63
150mm	1.32	2.50	1.35	0.30	-	2.77	-	4.12	nr	4.43
200mm	1.70	2.50	1.74	0.35	-	3.23	-	4.97	nr	5.35

Labour hourly rates: (except Specialists) Craft Operatives £9.23 Labourer £7.02. Rates are national average prices. Refer to REGIONAL VARIATIONS for indicative levels of overall pricing in regions

Labour hourly rates: (except Specialists) Craft Operatives £9.23 Labourer £7.02 Rates are national average prices. Refer to REGIONAL VARIATIONS for indicative levels of overall pricing in regions	MATERIALS			LABOUR				RATES		
	Del to Site £	Waste %	Material Cost £	Craft Optve Hrs	Lab Hrs	Labour Cost £	Sunds £	Nett Rate £	Unit	Gross Rate (+7.5%) £

P21: IRONMONGERY Cont.

Bolts; standard quality Cont.

To softwood Cont.
barrel bolts; japanned, steel barrel; heavy

100mm	2.62	2.50	2.69	0.25	-	2.31	-	4.99	nr	5.37
150mm	3.17	2.50	3.25	0.30	-	2.77	-	6.02	nr	6.47
200mm	3.83	2.50	3.93	0.35	-	3.23	-	7.16	nr	7.69
250mm	4.53	2.50	4.64	0.40	-	3.69	-	8.34	nr	8.96
300mm	5.36	2.50	5.49	0.45	-	4.15	-	9.65	nr	10.37
barrel bolts; extruded brass, round brass shoot; 25mm wide										
75mm	1.04	2.50	1.07	0.25	-	2.31	-	3.37	nr	3.63
100mm	1.26	2.50	1.29	0.25	-	2.31	-	3.60	nr	3.87
150mm	1.58	2.50	1.62	0.30	-	2.77	-	4.39	nr	4.72
barrel bolts; extruded brass; B.M.A., round brass shoot; 25mm wide										
75mm	1.42	2.50	1.46	0.25	-	2.31	-	3.76	nr	4.05
100mm	1.70	2.50	1.74	0.25	-	2.31	-	4.05	nr	4.35
150mm	2.13	2.50	2.18	0.30	-	2.77	-	4.95	nr	5.32
barrel bolts; extruded brass, chromium plated, round brass shoot; 25mm wide										
75mm	1.42	2.50	1.46	0.25	-	2.31	-	3.76	nr	4.05
100mm	1.70	2.50	1.74	0.25	-	2.31	-	4.05	nr	4.35
150mm	2.13	2.50	2.18	0.30	-	2.77	-	4.95	nr	5.32
barrel bolts; extruded aluminium, S.A.A., round aluminium shoot; 25mm wide										
75mm	1.04	2.50	1.07	0.25	-	2.31	-	3.37	nr	3.63
100mm	1.14	2.50	1.17	0.25	-	2.31	-	3.48	nr	3.74
150mm	1.32	2.50	1.35	0.30	-	2.77	-	4.12	nr	4.43
monkey tail bolts										
japanned; 300mm	7.98	2.50	8.18	0.35	-	3.23	-	11.41	nr	12.27
japanned; 450mm	10.65	2.50	10.92	0.40	-	3.69	-	14.61	nr	15.70
japanned; 610mm	12.78	2.50	13.10	0.50	-	4.62	-	17.71	nr	19.04
flush bolts; B.M.A.										
100mm	6.88	2.50	7.05	0.50	-	4.62	-	11.67	nr	12.54
150mm	9.94	2.50	10.19	0.60	-	5.54	-	15.73	nr	16.91
200mm	17.05	2.50	17.48	0.70	-	6.46	-	23.94	nr	25.73
flush bolts; S.C.P.										
100mm	7.43	2.50	7.62	0.50	-	4.62	-	12.23	nr	13.15
150mm	10.11	2.50	10.36	0.60	-	5.54	-	15.90	nr	17.09
200mm	15.95	2.50	16.35	0.70	-	6.46	-	22.81	nr	24.52
lever action flush bolts; B.M.A.										
150mm	8.52	2.50	8.73	0.60	-	5.54	-	14.27	nr	15.34
200 x 25mm	9.07	2.50	9.30	0.70	-	6.46	-	15.76	nr	16.94
lever action flush bolts; S.A.A.										
150mm	6.88	2.50	7.05	0.60	-	5.54	-	12.59	nr	13.53
200 x 25mm	7.11	2.50	7.29	0.70	-	6.46	-	13.75	nr	14.78
necked bolts; extruded brass, round brass shoot; 25mm wide										
75mm	1.32	2.50	1.35	0.25	-	2.31	-	3.66	nr	3.94
100mm	1.43	2.50	1.47	0.25	-	2.31	-	3.77	nr	4.06
150mm	1.91	2.50	1.96	0.30	-	2.77	-	4.73	nr	5.08
necked bolts; extruded aluminium, S.A.A., round aluminium shoot; 25mm wide										
75mm	0.94	2.50	0.96	0.25	-	2.31	-	3.27	nr	3.52
100mm	1.04	2.50	1.07	0.25	-	2.31	-	3.37	nr	3.63
150mm	1.32	2.50	1.35	0.30	-	2.77	-	4.12	nr	4.43
indicating bolts; S.A.A.	5.69	2.50	5.83	1.00	-	9.23	-	15.06	nr	16.19
indicating bolts; B.M.A.	5.41	2.50	5.55	1.00	-	9.23	-	14.78	nr	15.88
panic bolts; to single door; iron, bronzed or silver	53.26	2.50	54.59	1.50	-	13.85	-	68.44	nr	73.57
panic bolts; to single door; aluminium box, steel shoots and cross rail, anodised silver ..	52.99	2.50	54.31	1.50	-	13.85	-	68.16	nr	73.27
panic bolts; to double doors; iron bronzed or silver	63.38	2.50	64.96	2.00	-	18.46	-	83.42	nr	89.68
panic bolts; to double doors; aluminium box, steel shoots and cross rail, anodised silver ..	63.38	2.50	64.96	2.00	-	18.46	-	83.42	nr	89.68
padlock bolts; galvanised; heavy										
150mm	3.12	2.50	3.20	0.35	-	3.23	-	6.43	nr	6.91
200mm	3.33	2.50	3.41	0.35	-	3.23	-	6.64	nr	7.14
250mm	3.99	2.50	4.09	0.50	-	4.62	-	8.70	nr	9.36
300mm	4.92	2.50	5.04	0.50	-	4.62	-	9.66	nr	10.38

To hardwood or the like
barrel bolts; japanned, steel barrel; medium

100mm	1.04	2.50	1.07	0.35	-	3.23	-	4.30	nr	4.62
150mm	1.32	2.50	1.35	0.40	-	3.69	-	5.04	nr	5.42
200mm	1.70	2.50	1.74	0.50	-	4.62	-	6.36	nr	6.83
barrel bolts; japanned, steel barrel; heavy										
100mm	2.62	2.50	2.69	0.35	-	3.23	-	5.92	nr	6.36
150mm	3.17	2.50	3.25	0.40	-	3.69	-	6.94	nr	7.46
200mm	3.83	2.50	3.93	0.50	-	4.62	-	8.54	nr	9.18
250mm	4.53	2.50	4.64	0.55	-	5.08	-	9.72	nr	10.45
300mm	5.36	2.50	5.49	0.65	-	6.00	-	11.49	nr	12.36
barrel bolts; extruded brass, round brass shoot; 25mm wide										
75mm	1.04	2.50	1.07	0.35	-	3.23	-	4.30	nr	4.62
100mm	1.26	2.50	1.29	0.35	Lab	3.23	-	4.52	nr	4.86
150mm	1.58	2.50	1.62	0.40	-	3.69	-	5.31	nr	5.71
barrel bolts; extruded brass; B.M.A., round brass shoot; 25mm wide										
75mm	1.42	2.50	1.46	0.35	-	3.23	-	4.69	nr	5.04
100mm	1.70	2.50	1.74	0.35	-	3.23	-	4.97	nr	5.35
150mm	2.13	2.50	2.18	0.40	-	3.69	-	5.88	nr	6.32

Labour hourly rates: (except Specialists) Craft Operatives £9.23 Labourer £7.02 Rates are national average prices. Refer to REGIONAL VARIATIONS for indicative levels of overall pricing in regions	MATERIALS			LABOUR				RATES		
	Del to Site	Waste	Material Cost	Craft Optve	Lab	Labour Cost	Sunds	Nett Rate	Unit	Gross Rate (+7.5%)
	£	%	£	Hrs	Hrs	£	£	£		£
P21: IRONMONGERY Cont.										
Bolts; standard quality Cont.										
To hardwood or the like Cont.										
barrel bolts; extruded brass, chromium plated, round brass shoot; 25mm wide										
75mm	1.42	2.50	1.46	0.35	–	3.23	–	4.69	nr	5.04
100mm	1.70	2.50	1.74	0.35	–	3.23	–	4.97	nr	5.35
150mm	2.13	2.50	2.18	0.40	–	3.69	–	5.88	nr	6.32
barrel bolts; extruded aluminium, S.A.A., round aluminium shoot; 25mm wide										
75mm	1.04	2.50	1.07	0.35	–	3.23	–	4.30	nr	4.62
100mm	1.14	2.50	1.17	0.35	–	3.23	–	4.40	nr	4.73
150mm	1.32	2.50	1.35	0.40	–	3.69	–	5.04	nr	5.42
monkey tail bolts										
japanned; 300mm	7.98	2.50	8.18	0.50	–	4.62	–	12.79	nr	13.75
japanned; 450mm	10.65	2.50	10.92	0.60	–	5.54	–	16.45	nr	17.69
japanned; 610mm	12.78	2.50	13.10	0.70	–	6.46	–	19.56	nr	21.03
flush bolts; B.M.A.										
100mm	6.88	2.50	7.05	0.75	–	6.92	–	13.97	nr	15.02
150mm	9.94	2.50	10.19	0.90	–	8.31	–	18.50	nr	19.88
200mm	17.05	2.50	17.48	1.00	–	9.23	–	26.71	nr	28.71
flush bolts; S.C.P.										
100mm	7.43	2.50	7.62	0.75	–	6.92	–	14.54	nr	15.63
150mm	10.11	2.50	10.36	0.90	–	8.31	–	18.67	nr	20.07
200mm	15.95	2.50	16.35	1.00	–	9.23	–	25.58	nr	27.50
lever action flush bolts; B.M.A.										
150mm	8.52	2.50	8.73	0.90	–	8.31	–	17.04	nr	18.32
200 x 25mm	9.07	2.50	9.30	1.00	–	9.23	–	18.53	nr	19.92
lever action flush bolts; S.A.A.										
150mm	6.88	2.50	7.05	0.90	–	8.31	–	15.36	nr	16.51
200 x 25mm	7.11	2.50	7.29	1.00	–	9.23	–	16.52	nr	17.76
necked bolts; extruded brass, round brass shoot; 25mm wide										
75mm	1.32	2.50	1.35	0.35	–	3.23	–	4.58	nr	4.93
100mm	1.43	2.50	1.47	0.35	–	3.23	–	4.70	nr	5.05
150mm	1.91	2.50	1.96	0.40	–	3.69	–	5.65	nr	6.07
necked bolts; extruded aluminium, S.A.A., round aluminium shoot; 25mm wide										
75mm	0.94	2.50	0.96	0.35	–	3.23	–	4.19	nr	4.51
100mm	1.04	2.50	1.07	0.35	–	3.23	–	4.30	nr	4.62
150mm	1.32	2.50	1.35	0.40	–	3.69	–	5.04	nr	5.42
indicating bolts; S.A.A.	5.69	2.50	5.83	1.50	–	13.85	–	19.68	nr	21.15
indicating bolts; B.M.A.	5.41	2.50	5.55	1.50	–	13.85	–	19.39	nr	20.84
panic bolts; to single door; iron, bronzed or silver	53.26	2.50	54.59	2.25	–	20.77	–	75.36	nr	81.01
panic bolts; to single door; aluminium box, steel shoots and cross rail, anodised silver	52.99	2.50	54.31	2.25	–	20.77	–	75.08	nr	80.71
panic bolts; to double doors; iron bronzed or silver	63.38	2.50	64.96	3.00	–	27.69		92.65	nr	99.60
panic bolts; to double doors; aluminium box, steel shoots and cross rail, anodised silver	63.38	2.50	64.96	3.00	–	27.69	–	92.65	nr	99.60
padlock bolts; galvanised; heavy										
150mm	3.12	2.50	3.20	0.50	–	4.62	–	7.81	nr	8.40
200mm	3.33	2.50	3.41	0.50	–	4.62	–	8.03	nr	8.63
250mm	3.99	2.50	4.09	0.75	–	6.92	–	11.01	nr	11.84
300mm	4.92	2.50	5.04	0.75	–	6.92	–	11.97	nr	12.86
Door handles; standard quality										
To softwood										
lever handles; spring action; B.M.A., best quality; 41 x 150mm	24.60	2.50	25.22	0.20	–	1.85	–	27.06	nr	29.09
lever handles; spring action; chromium plated, housing quality; 41 x 150mm	6.40	2.50	6.56	0.20	–	1.85	–	8.41	nr	9.04
lever handles; spring action; chromium plated, best quality; 41 x 150mm	22.40	2.50	22.96	0.20	–	1.85	–	24.81	nr	26.67
lever handles; spring action; S.A.A. housing quality; 41 x 150mm	3.88	2.50	3.98	0.20	–	1.85	–	5.82	nr	6.26
lever handles; spring action; S.A.A., best quality; 41 x 150mm	14.76	2.50	15.13	0.20	–	1.85	–	16.98	nr	18.25
lever handles; spring action; plastic/nylon, housing quality; 41 x 150mm	3.12	2.50	3.20	0.15	–	1.38	–	4.58	nr	4.93
pull handles; B.M.A.; 150mm	8.96	2.50	9.18	0.15	–	1.38	–	10.57	nr	11.36
pull handles; B.M.A.; 225mm	16.79	2.50	17.21	0.20	–	1.85	–	19.06	nr	20.48
pull handles; B.M.A.; 300mm	18.90	2.50	19.37	0.25	–	2.31	–	21.68	nr	23.31
pull handles; S.A.A.; 150mm	3.44	2.50	3.53	0.15	–	1.38	–	4.91	nr	5.28
pull handles; S.A.A.; 225mm	5.36	2.50	5.49	0.20	–	1.85	–	7.34	nr	7.89
pull handles; S.A.A.; 300mm	6.88	2.50	7.05	0.25	–	2.31	–	9.36	nr	10.06
pull handles on 50 x 250mm plate, lettered; B.M.A.; 225mm	38.80	2.50	39.77	0.40	–	3.69	–	43.46	nr	46.72
pull handles on 50 x 250mm plate, lettered; S.A.A.; 225mm	27.16	2.50	27.84	0.40	–	3.69	–	31.53	nr	33.90
To hardwood or the like										
lever handles; spring action; B.M.A., best quality; 41 x 150mm	24.60	2.50	25.22	0.30	–	2.77	–	27.98	nr	30.08
lever handles; spring action; chromium plated, housing quality; 41 x 150mm	6.40	2.50	6.56	0.30	–	2.77	–	9.33	nr	10.03
lever handles; spring action; chromium plated, best quality; 41 x 150mm	22.40	2.50	22.96	0.30	–	2.77	–	25.73	nr	27.66
lever handles; spring action; S.A.A. housing quality; 41 x 150mm	3.88	2.50	3.98	0.30	–	2.77	–	6.75	nr	7.25
lever handles; spring action; S.A.A., best quality; 41 x 150mm	14.76	2.50	15.13	0.30	–	2.77	–	17.90	nr	19.24

Labour hourly rates: (except Specialists) Craft Operatives £9.23 Labourer £7.02 Rates are national average prices. Refer to REGIONAL VARIATIONS for indicative levels of overall pricing in regions	MATERIALS			LABOUR				RATES		
	Del to Site £	Waste %	Material Cost £	Craft Optve Hrs	Lab Hrs	Labour Cost £	Sunds £	Nett Rate £	Unit	Gross Rate (+7.5%) £

P21: IRONMONGERY Cont.

Door handles; standard quality Cont.

To hardwood or the like Cont.

	Del to Site	Waste	Material Cost	Craft Optve	Lab	Labour Cost	Sunds	Nett Rate	Unit	Gross Rate
lever handles; spring action; plastic/nylon, housing quality; 41 x 150mm	3.12	2.50	3.20	0.20	-	1.85	-	5.04	nr	5.42
pull handles; B.M.A.; 150mm	8.96	2.50	9.18	0.20	-	1.85	-	11.03	nr	11.86
pull handles; B.M.A.; 225mm	16.79	2.50	17.21	0.30	-	2.77	-	19.98	nr	21.48
pull handles; B.M.A.; 300mm	18.90	2.50	19.37	0.35	-	3.23	-	22.60	nr	24.30
pull handles; S.A.A.; 150mm	3.44	2.50	3.53	0.20	-	1.85	-	5.37	nr	5.77
pull handles; S.A.A.; 225mm	5.36	2.50	5.49	0.30	-	2.77	-	8.26	nr	8.88
pull handles; S.A.A.; 300mm	6.88	2.50	7.05	0.35	-	3.23	-	10.28	nr	11.05
pull handles on 50 x 250mm plate, lettered; B.M.A.; 225mm	38.80	2.50	39.77	0.60	-	5.54	-	45.31	nr	48.71
pull handles on 50 x 250mm plate, lettered; S.A.A.; 225mm	27.16	2.50	27.84	0.60	-	5.54	-	33.38	nr	35.88

Door furniture; standard quality

To softwood

	Del to Site	Waste	Material Cost	Craft Optve	Lab	Labour Cost	Sunds	Nett Rate	Unit	Gross Rate
knobs; real B.M.A.; surface fixing	8.25	2.50	8.46	0.20	-	1.85	-	10.30	nr	11.07
knobs; S.A.A.; surface fixing	4.81	2.50	4.93	0.20	-	1.85	-	6.78	nr	7.28
knobs; real B.M.A.; secret fixing	13.34	2.50	13.67	0.30	-	2.77	-	16.44	nr	17.68
knobs; S.A.A.; secret fixing	6.94	2.50	7.11	0.30	-	2.77	-	9.88	nr	10.62

To hardwood or the like

	Del to Site	Waste	Material Cost	Craft Optve	Lab	Labour Cost	Sunds	Nett Rate	Unit	Gross Rate
knobs; real B.M.A.; surface fixing	8.25	2.50	8.46	0.30	-	2.77	-	11.23	nr	12.07
knobs; S.A.A.; surface fixing	4.81	2.50	4.93	0.30	-	2.77	-	7.70	nr	8.28
knobs; real B.M.A.; secret fixing	13.34	2.50	13.67	0.45	-	4.15	-	17.83	nr	19.16
knobs; S.A.A.; secret fixing	6.94	2.50	7.11	0.45	-	4.15	-	11.27	nr	12.11

Window furniture; standard quality

To softwood

	Del to Site	Waste	Material Cost	Craft Optve	Lab	Labour Cost	Sunds	Nett Rate	Unit	Gross Rate
fanlight catches; brass	4.92	2.50	5.04	0.35	-	3.23	-	8.27	nr	8.89
fanlight catches; B.M.A.	5.36	2.50	5.49	0.35	-	3.23	-	8.72	nr	9.38
fanlight catches; chromium plated	5.46	2.50	5.60	0.35	-	3.23	-	8.83	nr	9.49
casement fasteners; wedge plate; black malleable iron	1.04	2.50	1.07	0.35	-	3.23	-	4.30	nr	4.62
casement fasteners; wedge plate; B.M.A.	2.41	2.50	2.47	0.35	-	3.23	-	5.70	nr	6.13
casement fasteners; wedge plate; chromium plated	2.41	2.50	2.47	0.35	-	3.23	-	5.70	nr	6.13
casement fasteners; wedge plate; S.A.A.	1.91	2.50	1.96	0.35	-	3.23	-	5.19	nr	5.58
sash fasteners; brass; 70mm	5.52	2.50	5.66	1.00	-	9.23	-	14.89	nr	16.00
sash fasteners; B.M.A.; 70mm	5.90	2.50	6.05	1.00	-	9.23	-	15.28	nr	16.42
sash fasteners; chromium plated; 70mm	6.50	2.50	6.66	1.00	-	9.23	-	15.89	nr	17.08
casement stays; two pins; grey malleable iron										
200mm	1.37	2.50	1.40	0.35	-	3.23	-	4.63	nr	4.98
250mm	1.47	2.50	1.51	0.35	-	3.23	-	4.74	nr	5.09
300mm	1.58	2.50	1.62	0.35	-	3.23	-	4.85	nr	5.21
casement stays; two pins; B.M.A.										
200mm	3.17	2.50	3.25	0.35	-	3.23	-	6.48	nr	6.97
250mm	3.44	2.50	3.53	0.35	-	3.23	-	6.76	nr	7.26
300mm	3.72	2.50	3.81	0.35	-	3.23	-	7.04	nr	7.57
casement stays; two pins; chromium plated										
200mm	3.17	2.50	3.25	0.35	-	3.23	-	6.48	nr	6.97
250mm	3.44	2.50	3.53	0.35	-	3.23	-	6.76	nr	7.26
300mm	3.72	2.50	3.81	0.35	-	3.23	-	7.04	nr	7.57
casement stays; two pins; S.A.A.										
200mm	1.75	2.50	1.79	0.35	-	3.23	-	5.02	nr	5.40
250mm	1.91	2.50	1.96	0.35	-	3.23	-	5.19	nr	5.58
300mm	2.03	2.50	2.08	0.35	-	3.23	-	5.31	nr	5.71
sash lifts; polished brass; 50mm	1.09	2.50	1.12	0.20	-	1.85	-	2.96	nr	3.19
sash lifts; B.M.A.; 50mm	1.32	2.50	1.35	0.20	-	1.85	-	3.20	nr	3.44
sash lifts; chromium plated; 50mm	1.47	2.50	1.51	0.20	-	1.85	-	3.35	nr	3.60
flush lifts; brass; 75mm	1.42	2.50	1.46	0.50	-	4.62	-	6.07	nr	6.53
flush lifts; B.M.A.; 75mm	1.70	2.50	1.74	0.50	-	4.62	-	6.36	nr	6.83
flush lifts; chromium plated; 75mm	1.70	2.50	1.74	0.50	-	4.62	-	6.36	nr	6.83
sash cleats; polished brass; 76mm	1.70	2.50	1.74	0.20	-	1.85	-	3.59	nr	3.86
sash cleats; B.M.A.; 76mm	1.75	2.50	1.79	0.20	-	1.85	-	3.64	nr	3.91
sash cleats; chromium plated; 76mm	2.13	2.50	2.18	0.20	-	1.85	-	4.03	nr	4.33
sash pulleys; frame and wheel	2.18	2.50	2.23	0.50	-	4.62	-	6.85	nr	7.36

To hardwood or the like

	Del to Site	Waste	Material Cost	Craft Optve	Lab	Labour Cost	Sunds	Nett Rate	Unit	Gross Rate
fanlight catches; brass	4.92	2.50	5.04	0.50	-	4.62	-	9.66	nr	10.38
fanlight catches; B.M.A.	5.36	2.50	5.49	0.50	-	4.62	-	10.11	nr	10.87
fanlight catches; chromium plated	5.46	2.50	5.60	0.50	-	4.62	-	10.21	nr	10.98
casement fasteners; wedge plate; black malleable iron	1.04	2.50	1.07	0.50	-	4.62	-	5.68	nr	6.11
casement fasteners; wedge plate; B.M.A.	2.41	2.50	2.47	0.50	-	4.62	-	7.09	nr	7.62
casement fasteners; wedge plate; chromium plated	2.41	2.50	2.47	0.50	-	4.62	-	7.09	nr	7.62
casement fasteners; wedge plate; S.A.A.	1.91	2.50	1.96	0.50	-	4.62	-	6.57	nr	7.07
sash fasteners; brass; 70mm	5.52	2.50	5.66	1.50	-	13.85	-	19.50	nr	20.97
sash fasteners; B.M.A.; 70mm	5.90	2.50	6.05	1.50	-	13.85	-	19.89	nr	21.38
sash fasteners; chromium plated; 70mm	6.50	2.50	6.66	1.50	-	13.85	-	20.51	nr	22.05
casement stays; two pins; grey malleable iron										
200mm	1.37	2.50	1.40	0.50	-	4.62	-	6.02	nr	6.47
250mm	1.47	2.50	1.51	0.50	-	4.62	-	6.12	nr	6.58
300mm	1.58	2.50	1.62	0.50	-	4.62	-	6.23	nr	6.70
casement stays; two pins; B.M.A.										
200mm	3.17	2.50	3.25	0.50	-	4.62	-	7.86	nr	8.45
250mm	3.44	2.50	3.53	0.50	-	4.62	-	8.14	nr	8.75
300mm	3.72	2.50	3.81	0.50	-	4.62	-	8.43	nr	9.06
casement stays; two pins; chromium plated										
200mm	3.17	2.50	3.25	0.50	-	4.62	-	7.86	nr	8.45
250mm	3.44	2.50	3.53	0.50	-	4.62	-	8.14	nr	8.75

Labour hourly rates: (except Specialists) Craft Operatives £9.23 Labourer £7.02 Rates are national average prices. Refer to REGIONAL VARIATIONS for indicative levels of overall pricing in regions	MATERIALS			LABOUR				RATES		
	Del to Site	Waste	Material Cost	Craft Optve	Lab	Labour Cost	Sunds	Nett Rate	Unit	Gross Rate (+7.5%)
	£	%	£	Hrs	Hrs	£	£	£		£
P21: IRONMONGERY Cont.										
Window furniture; standard quality Cont.										
To hardwood or the like Cont.										
casement stays; two pins; chromium plated Cont.										
300mm	3.72	2.50	3.81	0.50	–	4.62	–	8.43	nr	9.06
casement stays; two pins; S.A.A.										
200mm	1.75	2.50	1.79	0.50	–	4.62	–	6.41	nr	6.89
250mm	1.91	2.50	1.96	0.50	–	4.62	–	6.57	nr	7.07
300mm	2.03	2.50	2.08	0.50	–	4.62	–	6.70	nr	7.20
sash lifts; polished brass; 50mm........	1.09	2.50	1.12	0.30	–	2.77	–	3.89	nr	4.18
sash lifts; B.M.A.; 50mm........	1.32	2.50	1.35	0.30	–	2.77	–	4.12	nr	4.43
sash lifts; chromium plated; 50mm........	1.47	2.50	1.51	0.30	–	2.77	–	4.28	nr	4.60
flush lifts; brass; 75mm........	1.42	2.50	1.46	0.75	–	6.92	–	8.38	nr	9.01
flush lifts; B.M.A.; 75mm........	1.70	2.50	1.74	0.75	–	6.92	–	8.66	nr	9.31
flush lifts; chromium plated; 75mm........	1.70	2.50	1.74	0.75	–	6.92	–	8.66	nr	9.31
sash cleats; polished brass; 76mm........	1.70	2.50	1.74	0.30	–	2.77	–	4.51	nr	4.85
sash cleats; B.M.A.; 76mm........	1.75	2.50	1.79	0.30	–	2.77	–	4.56	nr	4.90
sash cleats; chromium plated; 76mm........	2.13	2.50	2.18	0.30	–	2.77	–	4.95	nr	5.32
sash pulleys; frame and wheel	2.18	2.50	2.23	0.75	–	6.92	–	9.16	nr	9.84
Window furniture; Willan Building Services Ltd										
To softwood or the like										
Glidevale frame vent; aluminium with epoxy paint finish										
reference TV FV2; 198mm long; fixing with screws and clip on covers	6.44	2.50	6.60	0.20	0.03	2.06	0.15	8.81	nr	9.47
reference TV FV4; 358mm long; fixing with screws and clip on covers	7.76	2.50	7.95	0.30	0.04	3.05	0.18	11.18	nr	12.02
Glidevale canopy grille; aluminium with epoxy paint finish										
reference TV CG2; 155 x 15mm slot size; fixing with screws and clip on covers	3.22	2.50	3.30	0.20	0.03	2.06	0.07	5.43	nr	5.83
reference TV CG4; 308 x 15mm slot size; fixing with screws and clip on covers	4.55	2.50	4.66	0.30	0.03	2.98	0.11	7.75	nr	8.33
Glidevale condensation drainage channel; aluminium with powder coated finish										
reference CDN2500; to suit 4 or 6mm glazing; fixing with screws	20.27	2.50	20.78	0.40	0.05	4.04	0.48	25.30	m	27.20
To hardwood or the like										
Glidevale frame vent; aluminium with epoxy paint finish										
reference TV FV2; 198mm long; fixing with screws and clip on covers	6.44	2.50	6.60	0.30	0.04	3.05	0.15	9.80	nr	10.54
reference TV FV4; 358mm long; fixing with screws and clip on covers	7.76	2.50	7.95	0.45	0.06	4.57	0.18	12.71	nr	13.66
Glidevale canopy grille; aluminium with epoxy paint finish										
reference TV CG2; 155 x 15mm slot size; fixing with screws and clip on covers	3.22	2.50	3.30	0.30	0.04	3.05	0.07	6.42	nr	6.90
reference TV CG4; 308 x 15mm slot size; fixing with screws and clip on covers	4.55	2.50	4.66	0.45	0.06	4.57	0.11	9.35	nr	10.05
Glidevale condensation drainage channel; aluminium with powder coated finish										
reference CDN2500; to suit 4 or 6mm glazing; fixing with screws	20.27	2.50	20.78	0.60	0.08	6.10	0.48	27.36	m	29.41
Letter plates; standard quality										
To softwood										
letter plates; plain; real B.M.A.; 356mm wide	43.71	2.50	44.80	1.50	–	13.85	–	58.65	nr	63.05
letter plates; plain; anodised silver; 356mm wide	6.88	2.50	7.05	1.50	–	13.85	–	20.90	nr	22.46
letter plates; gravity flap; B.M.A.; 254mm wide ..	8.30	2.50	8.51	1.50	–	13.85	–	22.35	nr	24.03
letter plates; gravity flap; chromium plated; 254mm wide	8.30	2.50	8.51	1.50	–	13.85	–	22.35	nr	24.03
letter plates; gravity flap, two numerals; anodised silver; 254mm wide	7.21	2.50	7.39	1.60	–	14.77	–	22.16	nr	23.82
postal knockers; S.A.A.; frame 260 x 83mm; opening 203 x 44m	9.56	2.50	9.80	1.60	–	14.77	–	24.57	nr	26.41
To hardwood or the like										
letter plates; plain; real B.M.A.; 356mm wide	43.71	2.50	44.80	2.25	–	20.77	–	65.57	nr	70.49
letter plates; plain; anodised silver; 356mm wide	6.88	2.50	7.05	2.25	–	20.77	–	27.82	nr	29.91
letter plates; gravity flap; B.M.A.; 254mm wide ..	8.30	2.50	8.51	2.25	–	20.77	–	29.27	nr	31.47
letter plates; gravity flap; chromium plated; 254mm wide	8.30	2.50	8.51	2.25	–	20.77	–	29.27	nr	31.47
letter plates; gravity flap, two numerals; anodised silver; 254mm wide	7.21	2.50	7.39	2.35	–	21.69	–	29.08	nr	31.26
postal knockers; S.A.A.; frame 260 x 83mm; opening 203 x 44m	9.56	2.50	9.80	2.40	–	22.15	–	31.95	nr	34.35
Security locks; standard quality										
To hardwood or the like										
window catches; locking; polished aluminium	4.53	2.50	4.64	1.00	–	9.23	–	13.87	nr	14.91
dual screws; sanded brass	2.08	2.50	2.13	0.80	–	7.38	–	9.52	nr	10.23
window stops; locking; brass	4.81	2.50	4.93	0.80	–	7.38	–	12.31	nr	13.24

Labour hourly rates: (except Specialists) Craft Operatives £9.23 Labourer £7.02 Rates are national average prices. Refer to REGIONAL VARIATIONS for indicative levels of overall pricing in regions	MATERIALS			LABOUR				RATES		
	Del to Site £	Waste %	Material Cost £	Craft Optve Hrs	Lab Hrs	Labour Cost £	Sunds £	Nett Rate £	Unit	Gross Rate (+7.5%) £
P21: IRONMONGERY Cont.										
Security locks; standard quality Cont.										
To hardwood or the like Cont.										
mortice latches; locking	53.00	2.50	54.33	3.00	–	27.69	–	82.02	nr	88.17
double cylinder automatic deadlatches; B.M.A.	31.69	2.50	32.48	3.00	–	27.69	–	60.17	nr	64.69
double cylinder automatic deadlatches; chromium plated..	30.60	2.50	31.36	3.00	–	27.69	–	59.06	nr	63.48
mortice deadlocks; Chubb	24.60	2.50	25.22	3.00	–	27.69	–	52.91	nr	56.87
mortice locks; two bolt	30.60	2.50	31.36	3.00	–	27.69	–	59.06	nr	63.48
metal window locks; white	5.46	2.50	5.60	1.25	–	11.54	–	17.13	nr	18.42
security door chains; steel chain; brass	4.43	2.50	4.54	0.50	–	4.62	–	9.16	nr	9.84
security door chains; steel chain; chromium plated	4.92	2.50	5.04	0.50	–	4.62	–	9.66	nr	10.38
security mortice bolts; loose keys; chromium plated; 16mm diameter	4.15	2.50	4.25	1.00	–	9.23	–	13.48	nr	14.50
security mortice bolts; loose keys; chromium plated; 32mm diameter	5.52	2.50	5.66	1.10	–	10.15	–	15.81	nr	17.00
security hinge bolts; silver anodised	4.48	2.50	4.59	1.50	–	13.85	–	18.44	nr	19.82
door viewers; chromium plated	4.15	2.50	4.25	1.25	–	11.54	–	15.79	nr	16.98
Shelf brackets; standard quality										
To softwood										
shelf brackets; grey finished										
100 x 75mm	0.16	2.50	0.16	0.20	–	1.85	–	2.01	nr	2.16
150 x 125mm	0.22	2.50	0.23	0.20	–	1.85	–	2.07	nr	2.23
225 x 175mm	0.35	2.50	0.36	0.20	–	1.85	–	2.20	nr	2.37
300 x 250mm	0.52	2.50	0.53	0.20	–	1.85	–	2.38	nr	2.56
350 x 300mm............................	0.68	2.50	0.70	0.20	–	1.85	–	2.54	nr	2.73
Cabin hooks; standard quality										
To softwood										
cabin hooks and eyes; black japanned										
100mm	1.04	2.50	1.07	0.20	–	1.85	–	2.91	nr	3.13
150mm	1.32	2.50	1.35	0.20	–	1.85	–	3.20	nr	3.44
200mm	1.91	2.50	1.96	0.20	–	1.85	–	3.80	nr	4.09
250mm	2.13	2.50	2.18	0.20	–	1.85	–	4.03	nr	4.33
cabin hooks and eyes; polished brass										
100mm	4.37	2.50	4.48	0.20	–	1.85	–	6.33	nr	6.80
150mm	4.92	2.50	5.04	0.20	–	1.85	–	6.89	nr	7.41
To hardwood or the like										
cabin hooks and eyes; black japanned										
100mm	1.04	2.50	1.07	0.30	–	2.77	–	3.84	nr	4.12
150mm	1.32	2.50	1.35	0.30	–	2.77	–	4.12	nr	4.43
200mm	1.91	2.50	1.96	0.30	–	2.77	–	4.73	nr	5.08
250mm	2.13	2.50	2.18	0.30	–	2.77	–	4.95	nr	5.32
cabin hooks and eyes; polished brass										
100mm	4.37	2.50	4.48	0.30	–	2.77	–	7.25	nr	7.79
150mm	4.92	2.50	5.04	0.30	–	2.77	–	7.81	nr	8.40
To softwood and brickwork										
cabin hooks and eyes; black japanned										
100mm	1.04	2.50	1.07	0.25	–	2.31	–	3.37	nr	3.63
150mm	1.32	2.50	1.35	0.25	–	2.31	–	3.66	nr	3.94
200mm	1.91	2.50	1.96	0.25	–	2.31	–	4.27	nr	4.59
250mm	2.13	2.50	2.18	0.25	–	2.31	–	4.49	nr	4.83
cabin hooks and eyes; polished brass										
100mm	4.37	2.50	4.48	0.25	–	2.31	–	6.79	nr	7.30
150mm	4.92	2.50	5.04	0.25	–	2.31	–	7.35	nr	7.90
To hardwood and brickwork										
cabin hooks and eyes; black japanned										
100mm	1.04	2.50	1.07	0.35	–	3.23	–	4.30	nr	4.62
150mm	1.32	2.50	1.35	0.35	–	3.23	–	4.58	nr	4.93
200mm	1.91	2.50	1.96	0.35	–	3.23	–	5.19	nr	5.58
250mm	2.13	2.50	2.18	0.35	–	3.23	–	5.41	nr	5.82
cabin hooks and eyes; polished brass										
100mm	4.37	2.50	4.48	0.35	–	3.23	–	7.71	nr	8.29
150mm	4.92	2.50	5.04	0.35	–	3.23	–	8.27	nr	8.89
Draught seals and strips; standard quality										
To softwood										
draught excluders; plastic foam, self-adhesive										
900mm long..........................	0.16	10.00	0.18	0.15	–	1.38	–	1.56	nr	1.68
2000mm long.........................	0.33	10.00	0.36	0.30	–	2.77	–	3.13	nr	3.37
draught excluders; aluminium section, rubber tubing										
900mm long..........................	6.83	10.00	7.51	0.17	0.02	1.71	–	9.22	nr	9.91
2000mm long.........................	15.46	10.00	17.01	0.33	0.04	3.33	–	20.33	nr	21.86
draught excluders; aluminium section, vinyl seal										
900mm long..........................	6.83	10.00	7.51	0.17	0.02	1.71	–	9.22	nr	9.91
2000mm long.........................	15.46	10.00	17.01	0.33	0.04	3.33	–	20.33	nr	21.86
draught excluders; plastic moulding, nylon brush; 900mm long........................	5.85	10.00	6.43	0.17	0.02	1.71	–	8.14	nr	8.76
draught excluders; aluminium base, flexible arch; 900mm long.....................	15.57	10.00	17.13	0.25	0.03	2.52	–	19.65	nr	21.12
draught excluders; aluminium threshold, flexible arch; 900mm long	15.57	10.00	17.13	0.30	0.04	3.05	–	20.18	nr	21.69

BUILDING FABRIC SUNDRIES

Labour hourly rates: (except Specialists) Craft Operatives £9.23 Labourer £7.02 Rates are national average prices. Refer to REGIONAL VARIATIONS for indicative levels of overall pricing in regions	MATERIALS			LABOUR				RATES		
	Del to Site £	Waste %	Material Cost £	Craft Optve Hrs	Lab Hrs	Labour Cost £	Sunds £	Nett Rate £	Unit	Gross Rate (+7.5%) £
P21: IRONMONGERY Cont.										
Draught seals and strips; Sealmaster Ltd										
Threshold seals										
reference BDA; fixing to masonry with screws	8.72	2.50	**8.94**	0.40	0.05	**4.04**	0.20	**13.18**	m	**14.17**
reference BDB; fixing to masonry with screws	11.18	2.50	**11.46**	0.40	0.05	**4.04**	0.25	**15.75**	m	**16.93**
reference BDWB (weather board); fixing to timber										
with screws	13.58	2.50	**13.92**	0.60	0.08	**6.10**	0.30	**20.32**	m	**21.84**
reference TTM (stop seal); fixing to timber with										
screws	8.25	2.50	**8.46**	0.30	0.04	**3.05**	0.20	**11.71**	m	**12.58**
Drawer pulls; standard quality										
To softwood										
drawer pulls; brass; 100mm	1.42	2.50	**1.46**	0.15	-	**1.38**	-	**2.84**	nr	**3.05**
drawer pulls; B.M.A.; 100mm	1.85	2.50	**1.90**	0.15	-	**1.38**	-	**3.28**	nr	**3.53**
drawer pulls; chromium plated; 100mm	1.80	2.50	**1.85**	0.15	-	**1.38**	-	**3.23**	nr	**3.47**
drawer pulls; S.A.A.; 100mm	1.09	2.50	**1.12**	0.15	-	**1.38**	-	**2.50**	nr	**2.69**
To hardwood or the like										
drawer pulls; brass; 100mm	1.42	2.50	**1.46**	0.25	-	**2.31**	-	**3.76**	nr	**4.05**
drawer pulls; B.M.A.; 100mm	1.85	2.50	**1.90**	0.25	-	**2.31**	-	**4.20**	nr	**4.52**
drawer pulls; chromium plated; 100mm	1.80	2.50	**1.85**	0.25	-	**2.31**	-	**4.15**	nr	**4.46**
drawer pulls; S.A.A.; 100mm	1.09	2.50	**1.12**	0.25	-	**2.31**	-	**3.42**	nr	**3.68**
Hooks; standard quality										
To softwood										
cup hooks; polished brass; 25mm	0.16	2.50	**0.16**	0.05	-	**0.46**	-	**0.63**	nr	**0.67**
hat and coat hooks; B.M.A.	1.74	2.50	**1.78**	0.25	-	**2.31**	-	**4.09**	nr	**4.40**
hat and coat hooks; chromium plated	1.74	2.50	**1.78**	0.25	-	**2.31**	-	**4.09**	nr	**4.40**
hat and coat hooks; S.A.A.	0.76	2.50	**0.78**	0.25	-	**2.31**	-	**3.09**	nr	**3.32**
To hardwood or the like										
cup hooks; polished brass; 25mm	0.16	2.50	**0.16**	0.07	-	**0.65**	-	**0.81**	nr	**0.87**
hat and coat hooks; B.M.A.	1.74	2.50	**1.78**	0.40	-	**3.69**	-	**5.48**	nr	**5.89**
hat and coat hooks; chromium plated	1.74	2.50	**1.78**	0.40	-	**3.69**	-	**5.48**	nr	**5.89**
hat and coat hooks; S.A.A.	0.76	2.50	**0.78**	0.40	-	**3.69**	-	**4.47**	nr	**4.81**
P30: TRENCHES/PIPEWAYS/PITS FOR BURIED ENGINEERING **SERVICES**										
Site preparation										
Lifting turf for preservation										
stacking on site average 50 metres distance for										
immediate use; watering	-	-	**-**	-	0.50	**3.51**	-	**3.51**	m2	**3.77**
Re-laying turf										
taking from stack average 50 metres distance,										
laying on prepared bed, watering, maintaining	-	-	**-**	-	0.50	**3.51**	-	**3.51**	m2	**3.77**

This section continues
on the next page

Labour hourly rates: (except Specialists) Craft Operatives £9.23 Labourer £7.02. Rates are national average prices. Refer to REGIONAL VARIATIONS for indicative levels of overall pricing in regions	PLANT & TRANSPORT			LABOUR				RATES		
	Plant Cost £	Trans Cost £	P & T Cost £	Craft Optve Hrs	Lab Hrs	Labour Cost £	Sunds £	Nett Rate £	Unit	Gross Rate (+7.5%) £
P30: TRENCHES/PIPEWAYS/PITS FOR BURIED ENGINEERING SERVICES Cont.										
Excavating trenches to receive services not exceeding 200mm nominal size - by machine										
Excavations commencing from natural ground level; compacting; backfilling with excavated material										
not exceeding 1m deep; average 750mm deep	2.78	0.32	3.10	-	0.20	1.40	0.86	5.36	m	5.77
not exceeding 1m deep; average 750mm deep; next to roadways	3.72	0.32	4.04	-	0.20	1.40	0.86	6.30	m	6.78
Excavations commencing from existing ground level; levelling and grading backfilling to receive turf										
not exceeding 1m deep; average 750mm deep	3.93	0.32	4.25	-	0.20	1.40	0.86	6.51	m	7.00
Extra over excavating trenches irrespective of depth; breaking out existing materials										
hard rock	22.05	-	22.05	-	-	-	-	22.05	m3	23.70
concrete	18.31	-	18.31	-	-	-	-	18.31	m3	19.68
reinforced concrete	27.50	-	27.50	-	-	-	-	27.50	m3	29.56
brickwork, blockwork or stonework	12.22	-	12.22	-	-	-	-	12.22	m3	13.14
Extra over excavating trenches irrespective of depth; breaking out existing hard pavings										
concrete 150mm thick	2.89	-	2.89	-	-	-	-	2.89	m2	3.11
reinforced concrete 200mm thick	5.79	-	5.79	-	-	-	-	5.79	m2	6.22
concrete 150mm thick; reinstating	3.33	-	3.33	-	-	-	13.17	16.50	m2	17.74
macadam paving 75mm thick	0.50	-	0.50	-	-	-	-	0.50	m2	0.54
macadam paving 75mm thick; reinstating	0.50	-	0.50	-	-	-	16.70	17.20	m2	18.49
concrete flag paving 50mm thick	-	-	-	-	0.26	1.83	-	1.83	m2	1.96
concrete flag paving 50mm thick; re-instating	-	-	-	0.50	0.76	9.95	11.13	21.08	m2	22.66
Extra over excavating trenches irrespective of depth; excavating next existing services										
electricity services -1	-	-	-	-	5.00	35.10	-	35.10	m	37.73
gas services -1	-	-	-	-	5.00	35.10	-	35.10	m	37.73
water services -1	-	-	-	-	5.00	35.10	-	35.10	m	37.73
Extra over excavating trenches irrespective of depth; excavating around existing services crossing trench										
electricity services; cables crossing -1	-	-	-	-	7.50	52.65	-	52.65	nr	56.60
gas services services crossing -1	-	-	-	-	7.50	52.65	-	52.65	nr	56.60
water services services crossing -1	-	-	-	-	7.50	52.65	-	52.65	nr	56.60
Site preparation										
Lifting turf for preservation										
stacking on site average 50 metres distance for immediate use; watering	-	-	-	-	0.50	3.51	-	3.51	m2	3.77
Re-laying turf										
taking from stack average 50 metres distance, laying on prepared bed, watering, maintaining	-	-	-	-	0.50	3.51	-	3.51	m2	3.77
Excavating trenches to receive services not exceeding 200mm nominal size - by machine										
Excavations commencing from natural ground level; compacting; backfilling with excavated material										
not exceeding 1m deep; average 750mm deep	-	-	-	-	1.15	8.07	4.67	12.74	m	13.70
not exceeding 1m deep; average 750mm deep; next to roadways	-	-	-	-	1.15	8.07	5.59	13.66	m	14.69
Excavations commencing from existing ground level; levelling and grading backfilling to receive turf										
not exceeding 1m deep; average 750mm deep	-	-	-	-	1.15	8.07	5.57	13.64	m	14.67
Extra over excavating trenches irrespective of depth; breaking out existing materials										
hard rock	-	-	-	-	13.35	93.72	-	93.72	m3	100.75
concrete	-	-	-	-	10.00	70.20	-	70.20	m3	75.47
reinforced concrete	-	-	-	-	15.00	105.30	-	105.30	m3	113.20
brickwork, blockwork or stonework	-	-	-	-	7.00	49.14	-	49.14	m3	52.83
Extra over excavating trenches irrespective of depth; breaking out existing hard pavings										
concrete 150mm thick	-	-	-	-	1.50	10.53	-	10.53	m2	11.32
reinforced concrete 150mm thick	-	-	-	-	3.00	21.06	-	21.06	m2	22.64
concrete 150mm thick; reinstating	-	-	-	-	3.00	21.06	13.17	34.23	m2	36.80
macadam paving 75mm thick	-	-	-	-	0.25	1.75	-	1.75	m2	1.89
macadam paving 75mm thick; reinstating	-	-	-	-	0.25	1.75	16.70	18.45	m2	19.84
concrete flag paving 50mm thick	-	-	-	-	0.26	1.83	-	1.83	m2	1.96
concrete flag paving 50mm thick; re-instating	-	-	-	-	0.76	5.34	11.13	16.47	m2	17.71
Extra over excavating trenches irrespective of depth; excavating next existing services										
electricity services -1	-	-	-	-	5.00	35.10	-	35.10	m	37.73
gas services -1	-	-	-	-	5.00	35.10	-	35.10	m	37.73
water services -1	-	-	-	-	5.00	35.10	-	35.10	m	37.73
Extra over excavating trenches irrespective of depth; excavating around existing services crossing trench										
electricity services; cables crossing -1	-	-	-	-	7.50	52.65	-	52.65	nr	56.60
gas services services crossing -1	-	-	-	-	7.50	52.65	-	52.65	nr	56.60
water services services crossing -1	-	-	-	-	7.50	52.65	-	52.65	nr	56.60

BUILDING FABRIC SUNDRIES

Labour hourly rates: (except Specialists) Craft Operatives £9.23 Labourer £7.02 Rates are national average prices. Refer to REGIONAL VARIATIONS for indicative levels of overall pricing in regions	MATERIALS			LABOUR				RATES		
	Del to Site £	Waste %	Material Cost £	Craft Optve Hrs	Lab Hrs	Labour Cost £	Sunds £	Nett Rate £	Unit	Gross Rate (+7.5%) £
P30: TRENCHES/PIPEWAYS/PITS FOR BURIED ENGINEERING SERVICES Cont.										
Underground ducts; vitrified clayware, B.S.65 extra strength; flexible joints										
Straight										
100mm nominal size single way duct; laid in position in trench	2.71	5.00	2.85	0.16	0.02	1.62	-	4.46	m	4.80
100mm nominal size bonded to form 2 way duct; laid in position in trench	5.42	5.00	5.69	0.25	0.03	2.52	-	8.21	m	8.82
100mm nominal size bonded to form 4 way duct; laid in position in trench	10.85	5.00	11.39	0.35	0.04	3.51	-	14.90	m	16.02
100mm nominal size bonded to form 6 way duct; laid in position in trench	16.27	5.00	17.08	0.45	0.06	4.57	-	21.66	m	23.28
extra; providing and laying in position nylon draw wire	0.11	10.00	0.12	0.02	-	0.18	-	0.31	m	0.33
Stop cock pits, valve chambers and the like										
Stop cock pits; half brick thick sides of common bricks, B.S.3921, Category M, 215 x 102.5 x 65mm, compressive strength 20.5 N/mm2, in cement mortar (1:3); 100mm thick base and top of plain in-situ concrete, B.S.5328, ordinary prescribed mix ST4, 20mm agg										
600mm deep in clear; (surface boxes included elsewhere)										
internal size 225 x 225mm	18.35	5.00	19.27	1.50	1.50	24.38	2.28	45.92	nr	49.37
internal size 338 x 338mm	26.52	5.00	27.85	2.25	2.25	36.56	3.44	67.85	nr	72.94
Stop cock guards; vitrified clay; including all excavation backfilling, disposal of surplus excavated material, earthwork support and compaction of ground; surrounding in 150mm thick plain in-situ concrete, B.S.5328 ordinary prescribed mix ST4, 20mm aggre										
750mm deep in clear; (surface boxes included elsewhere)										
150mm diameter	21.14	5.00	22.20	1.15	1.15	18.69	1.83	42.71	nr	45.92
Cast iron surface boxes and covers, coated to B.S.4147; bedding in cement mortar (1:3)										
Surface boxes, B.S.5834 Part 2 marked S.V. light grade, hinged lid; 150 x 150mm overall top size, 76mm deep	11.45	-	11.45	0.75	0.38	9.59	0.20	21.24	nr	22.83
Surface boxes, B.S.750, marked 'FIRE HYDRANT' medium grade; minimum clear opening 230 x 380mm, minimum depth 100mm	34.96	-	34.96	1.25	0.63	15.96	0.61	51.53	nr	55.39
Surface boxes, B.S.5834 Part 2 marked 'W' or WATER heavy grade; double triangular cover; minimum clear opening 300 x 300mm, minimum depth 150mm	75.96	-	75.96	1.50	0.75	19.11	0.98	96.05	nr	103.25
Accessories/Sundry items										
Stop cock keys										
tee	15.15	-	15.15	-	-	-	-	15.15	nr	16.29
P31: HOLES/CHASES/COVERS/SUPPORTS FOR SERVICES; BUILDERS WORK IN CONNECTION WITH MECHANICAL SERVICES										
Cutting holes for services										
For ducts through concrete; making good										
rectangular ducts not exceeding 1.00m girth										
150mm thick	-	-	-	1.60	-	14.77	0.51	15.28	nr	16.42
200mm thick	-	-	-	1.90	-	17.54	0.61	18.15	nr	19.51
300mm thick	-	-	-	2.20	-	20.31	0.71	21.02	nr	22.59
rectangular ducts 1.00-2.00m girth										
150mm thick	-	-	-	3.20	-	29.54	1.13	30.67	nr	32.97
200mm thick	-	-	-	3.80	-	35.07	1.23	36.30	nr	39.03
300mm thick	-	-	-	4.40	-	40.61	1.43	42.04	nr	45.20
rectangular ducts 2.00-3.00m girth										
150mm thick	-	-	-	4.00	-	36.92	1.30	38.22	nr	41.09
200mm thick	-	-	-	4.75	-	43.84	1.54	45.38	nr	48.79
300mm thick	-	-	-	5.50	-	50.77	1.76	52.52	nr	56.46
For pipes through concrete; making good										
pipes not exceeding 55mm nominal size										
150mm thick	-	-	-	0.90	-	8.31	0.32	8.63	nr	9.27
200mm thick	-	-	-	1.15	-	10.61	0.41	11.02	nr	11.85
300mm thick	-	-	-	1.45	-	13.38	0.51	13.89	nr	14.94
pipes 55 - 110mm nominal size										
150mm thick	-	-	-	1.12	-	10.34	0.40	10.74	nr	11.54
200mm thick	-	-	-	1.36	-	12.55	0.47	13.02	nr	14.00
300mm thick	-	-	-	1.80	-	16.61	0.64	17.25	nr	18.55
pipes exceeding 110mm nominal size										
150mm thick	-	-	-	1.35	-	12.46	0.45	12.91	nr	13.88
200mm thick	-	-	-	1.72	-	15.88	0.57	16.45	nr	17.68
300mm thick	-	-	-	2.18	-	20.12	0.74	20.86	nr	22.43
For ducts through reinforced concrete; making good										
rectangular ducts not exceeding 1.00m girth										
150mm thick	-	-	-	2.40	-	22.15	0.74	22.89	nr	24.61

Labour hourly rates: (except Specialists) Craft Operatives £9.23 Labourer £7.02 Rates are national average prices. Refer to REGIONAL VARIATIONS for indicative levels of overall pricing in regions	MATERIALS			LABOUR				RATES		
	Del to Site	Waste	Material Cost	Craft Optve	Lab	Labour Cost	Sunds	Nett Rate	Unit	Gross Rate (+7.5%)
	£	%	£	Hrs	Hrs	£	£	£		£
P31: HOLES/CHASES/COVERS/SUPPORTS FOR SERVICES; BUILDERS WORK IN CONNECTION WITH MECHANICAL SERVICES Cont.										
Cutting holes for services Cont.										
For ducts through reinforced concrete; making good Cont.										
rectangular ducts not exceeding 1.00m girth Cont.										
200mm thick.....................	-	-	-	2.85	-	26.31	0.87	27.18	nr	29.21
300mm thick.....................	-	-	-	3.30	-	30.46	1.02	31.48	nr	33.84
rectangular ducts 1.00-2.00m girth										
150mm thick.....................	-	-	-	4.80	-	44.30	1.50	45.80	nr	49.24
200mm thick.....................	-	-	-	5.70	-	52.61	1.77	54.38	nr	58.46
300mm thick.....................	-	-	-	6.60	-	60.92	2.06	62.98	nr	67.70
rectangular ducts 2.00-3.00m girth										
150mm thick.....................	-	-	-	6.00	-	55.38	1.86	57.24	nr	61.53
200mm thick.....................	-	-	-	7.00	-	64.61	2.21	66.82	nr	71.83
300mm thick.....................	-	-	-	8.25	-	76.15	2.52	78.67	nr	84.57
For pipes through reinforced concrete; making good										
pipes not exceeding 55mm nominal size										
150mm thick.....................	-	-	-	1.35	-	12.46	0.46	12.92	nr	13.89
200mm thick.....................	-	-	-	1.75	-	16.15	0.60	16.75	nr	18.01
300mm thick.....................	-	-	-	2.20	-	20.31	0.75	21.06	nr	22.64
pipes 55 - 110mm nominal size										
150mm thick.....................	-	-	-	1.70	-	15.69	0.58	16.27	nr	17.49
200mm thick.....................	-	-	-	2.05	-	18.92	0.70	19.62	nr	21.09
300mm thick.....................	-	-	-	2.70	-	24.92	0.92	25.84	nr	27.78
pipes exceeding 110mm nominal size										
150mm thick.....................	-	-	-	2.00	-	18.46	0.68	19.14	nr	20.58
200mm thick.....................	-	-	-	2.60	-	24.00	0.88	24.88	nr	26.74
300mm thick.....................	-	-	-	3.20	-	29.54	1.09	30.63	nr	32.92
For ducts through brickwork; making good										
rectangular ducts not exceeding 1.00m girth										
102.5mm thick..................	-	-	-	1.20	-	11.08	0.35	11.43	nr	12.28
215mm thick....................	-	-	-	1.40	-	12.92	0.42	13.34	nr	14.34
327.5mm thick..................	-	-	-	1.65	-	15.23	0.50	15.73	nr	16.91
rectangular ducts 1.00-2.00m girth										
102.5mm thick..................	-	-	-	2.40	-	22.15	0.70	22.85	nr	24.57
215mm thick....................	-	-	-	2.80	-	25.84	0.84	26.68	nr	28.69
327.5mm thick..................	-	-	-	3.30	-	30.46	1.00	31.46	nr	33.82
rectangular ducts 2.00-3.00m girth										
102.5mm thick..................	-	-	-	3.00	-	27.69	0.90	28.59	nr	30.73
215mm thick....................	-	-	-	3.50	-	32.31	1.05	33.35	nr	35.86
327.5mm thick..................	-	-	-	4.10	-	37.84	1.25	39.09	nr	42.02
For pipes through brickwork; making good										
pipes not exceeding 55mm nominal size										
102.5mm thick..................	-	-	-	0.40	-	3.69	0.14	3.83	nr	4.12
215mm thick....................	-	-	-	0.66	-	6.09	0.22	6.31	nr	6.79
327.5mm thick..................	-	-	-	1.00	-	9.23	0.34	9.57	nr	10.29
pipes 55 - 110mm nominal size										
102.5mm thick..................	-	-	-	0.53	-	4.89	0.18	5.07	nr	5.45
215mm thick....................	-	-	-	0.88	-	8.12	0.30	8.42	nr	9.05
327.5mm thick..................	-	-	-	1.33	-	12.28	0.45	12.73	nr	13.68
pipes exceeding 110mm nominal size										
102.5mm thick..................	-	-	-	0.67	-	6.18	0.23	6.41	nr	6.90
215mm thick....................	-	-	-	1.10	-	10.15	0.37	10.52	nr	11.31
327.5mm thick..................	-	-	-	1.67	-	15.41	0.56	15.97	nr	17.17
For ducts through brickwork; making good fair face or facings one side										
rectangular ducts not exceeding 1.00m girth										
102.5mm thick..................	-	-	-	1.45	-	13.38	0.55	13.93	nr	14.98
215mm thick....................	-	-	-	1.65	-	15.23	0.65	15.88	nr	17.07
327.5mm thick..................	-	-	-	1.90	-	17.54	0.70	18.24	nr	19.60
rectangular ducts 1.00-2.00m girth										
102.5mm thick..................	-	-	-	2.70	-	24.92	1.10	26.02	nr	27.97
215mm thick....................	-	-	-	3.10	-	28.61	1.30	29.91	nr	32.16
327.5mm thick..................	-	-	-	3.60	-	33.23	1.40	34.63	nr	37.23
rectangular ducts 2.00-3.00m girth										
102.5mm thick..................	-	-	-	3.40	-	31.38	1.40	32.78	nr	35.24
215mm thick....................	-	-	-	3.90	-	36.00	1.65	37.65	nr	40.47
327.5mm thick..................	-	-	-	4.50	-	41.53	1.75	43.28	nr	46.53
For pipes through brickwork; making good fair face or facings one side										
pipes not exceeding 55mm nominal size										
102.5mm thick..................	-	-	-	0.50	-	4.62	0.18	4.79	nr	5.15
215mm thick....................	-	-	-	0.76	-	7.01	0.26	7.27	nr	7.82
327.5mm thick..................	-	-	-	1.10	-	10.15	0.37	10.52	nr	11.31
pipes 55 - 110mm nominal size										
102.5mm thick..................	-	-	-	0.68	-	6.28	0.22	6.50	nr	6.98
215mm thick....................	-	-	-	1.03	-	9.51	0.33	9.84	nr	10.57
327.5mm thick..................	-	-	-	1.48	-	13.66	0.49	14.15	nr	15.21
pipes exceeding 110mm nominal size										
102.5mm thick..................	-	-	-	0.87	-	8.03	0.27	8.30	nr	8.92
215mm thick....................	-	-	-	1.30	-	12.00	0.40	12.40	nr	13.33
327.5mm thick..................	-	-	-	1.87	-	17.26	0.59	17.85	nr	19.19
For ducts through blockwork; making good										
rectangular ducts not exceeding 1.00m girth										
100mm thick....................	-	-	-	0.90	-	8.31	0.35	8.66	nr	9.31
140mm thick....................	-	-	-	1.00	-	9.23	0.42	9.65	nr	10.37

Labour hourly rates: (except Specialists) Craft Operatives £9.23 Labourer £7.02 Rates are national average prices. Refer to REGIONAL VARIATIONS for indicative levels of overall pricing in regions	MATERIALS			LABOUR				RATES		
	Del to Site	Waste	Material Cost	Craft Optve	Lab	Labour Cost	Sunds	Nett Rate	Unit	Gross Rate (+7.5%)
	£	%	£	Hrs	Hrs	£	£	£		£

P31: HOLES/CHASES/COVERS/SUPPORTS FOR SERVICES; BUILDERS WORK IN CONNECTION WITH MECHANICAL SERVICES Cont.

Cutting holes for services Cont.

	Del to Site	Waste	Material Cost	Craft Optve	Lab	Labour Cost	Sunds	Nett Rate	Unit	Gross Rate
For ducts through blockwork; making good Cont.										
rectangular ducts not exceeding 1.00m girth Cont.										
190mm thick	-	-	-	1.10	-	10.15	0.50	10.65	nr	11.45
rectangular ducts 1.00-2.00m girth										
100mm thick	-	-	-	1.80	-	16.61	0.70	17.31	nr	18.61
140mm thick	-	-	-	1.95	-	18.00	0.84	18.84	nr	20.25
190mm thick	-	-	-	2.10	-	19.38	1.00	20.38	nr	21.91
rectangular ducts 2.00-3.00m girth										
100mm thick	-	-	-	2.25	-	20.77	0.90	21.67	nr	23.29
140mm thick	-	-	-	2.45	-	22.61	1.05	23.66	nr	25.44
190mm thick	-	-	-	2.65	-	24.46	1.25	25.71	nr	27.64
For pipes through blockwork; making good										
pipes not exceeding 55mm nominal size										
100mm thick	-	-	-	0.30	-	2.77	0.13	2.90	nr	3.12
140mm thick	-	-	-	0.40	-	3.69	0.21	3.90	nr	4.19
190mm thick	-	-	-	0.50	-	4.62	0.32	4.93	nr	5.31
pipes 55 - 110mm nominal size										
100mm thick	-	-	-	0.40	-	3.69	0.17	3.86	nr	4.15
140mm thick	-	-	-	0.53	-	4.89	0.28	5.17	nr	5.56
190mm thick	-	-	-	0.66	-	6.09	0.43	6.52	nr	7.01
pipes exceeding 110mm nominal size										
100mm thick	-	-	-	0.50	-	4.62	0.22	4.84	nr	5.20
140mm thick	-	-	-	0.68	-	6.28	0.35	6.63	nr	7.12
190mm thick	-	-	-	0.83	-	7.66	0.54	8.20	nr	8.82
For ducts through existing brickwork with plaster finish; making good each side										
rectangular ducts not exceeding 1.00m girth										
102.5mm thick	-	-	-	1.95	-	18.00	0.60	18.60	nr	19.99
215mm thick	-	-	-	2.25	-	20.77	0.75	21.52	nr	23.13
327.5mm thick	-	-	-	2.55	-	23.54	0.85	24.39	nr	26.22
rectangular ducts 1.00-2.00m girth										
102.5mm thick	-	-	-	3.60	-	33.23	1.20	34.43	nr	37.01
215mm thick	-	-	-	4.20	-	38.77	1.50	40.27	nr	43.29
327.5mm thick	-	-	-	4.80	-	44.30	1.70	46.00	nr	49.45
rectangular ducts 2.00-3.00m girth										
102.5mm thick	-	-	-	4.50	-	41.53	1.80	43.34	nr	46.59
215mm thick	-	-	-	5.25	-	48.46	2.25	50.71	nr	54.51
327.5mm thick	-	-	-	6.05	-	55.84	2.55	58.39	nr	62.77
For ducts through existing brickwork with plaster finish; making good one side										
rectangular ducts not exceeding 1.00m girth										
102.5mm thick	-	-	-	1.75	-	16.15	0.54	16.69	nr	17.94
215mm thick	-	-	-	2.05	-	18.92	0.68	19.60	nr	21.07
327.5mm thick	-	-	-	2.30	-	21.23	0.76	21.99	nr	23.64
rectangular ducts 1.00-2.00m girth										
102.5mm thick	-	-	-	3.25	-	30.00	1.08	31.08	nr	33.41
215mm thick	-	-	-	3.80	-	35.07	1.35	36.42	nr	39.16
327.5mm thick	-	-	-	4.35	-	40.15	1.55	41.70	nr	44.83
rectangular ducts 2.00-3.00m girth										
102.5mm thick	-	-	-	4.05	-	37.38	1.60	38.98	nr	41.91
215mm thick	-	-	-	4.75	-	43.84	2.00	45.84	nr	49.28
327.5mm thick	-	-	-	5.45	-	50.30	2.30	52.60	nr	56.55
For pipes through existing brickwork with plaster finish; making good each side										
pipes not exceeding 55mm nominal size										
102.5mm thick	-	-	-	0.65	-	6.00	0.22	6.22	nr	6.69
215mm thick	-	-	-	1.00	-	9.23	0.31	9.54	nr	10.26
327.5mm thick	-	-	-	1.45	-	13.38	0.44	13.82	nr	14.86
pipes 55 - 110mm nominal size										
102.5mm thick	-	-	-	0.95	-	8.77	0.26	9.03	nr	9.71
215mm thick	-	-	-	1.35	-	12.46	0.40	12.86	nr	13.83
327.5mm thick	-	-	-	1.95	-	18.00	0.59	18.59	nr	19.98
pipes exceeding 110mm nominal size										
102.5mm thick	-	-	-	1.20	-	11.08	0.32	11.40	nr	12.25
215mm thick	-	-	-	1.75	-	16.15	0.48	16.63	nr	17.88
327.5mm thick	-	-	-	2.50	-	23.07	0.71	23.79	nr	25.57
For pipes through existing brickwork with plaster finish; making good one side										
pipes not exceeding 55mm nominal size										
102.5mm thick	-	-	-	0.60	-	5.54	0.18	5.72	nr	6.15
215mm thick	-	-	-	0.90	-	8.31	0.26	8.57	nr	9.21
327.5mm thick	-	-	-	1.30	-	12.00	0.37	12.37	nr	13.30
pipes 55 - 110mm nominal size										
102.5mm thick	-	-	-	0.85	-	7.85	0.22	8.07	nr	8.67
215mm thick	-	-	-	1.20	-	11.08	0.33	11.41	nr	12.26
327.5mm thick	-	-	-	1.75	-	16.15	0.49	16.64	nr	17.89
pipes exceeding 110mm nominal size										
102.5mm thick	-	-	-	1.10	-	10.15	0.27	10.42	nr	11.20
215mm thick	-	-	-	1.60	-	14.77	0.40	15.17	nr	16.31
327.5mm thick	-	-	-	2.25	-	20.77	0.54	21.31	nr	22.91
For pipes through softwood										
pipes not exceeding 55mm nominal size										
19mm thick	-	-	-	0.10	-	0.92	-	0.92	nr	0.99
50mm thick	-	-	-	0.13	-	1.20	-	1.20	nr	1.29

Labour hourly rates: (except Specialists) Craft Operatives £9.23 Labourer £7.02 Rates are national average prices. Refer to REGIONAL VARIATIONS for indicative levels of overall pricing in regions	MATERIALS			LABOUR				RATES		
	Del to Site £	Waste %	Material Cost £	Craft Optve Hrs	Lab Hrs	Labour Cost £	Sunds £	Nett Rate £	Unit	Gross Rate (+7.5%) £

P31: HOLES/CHASES/COVERS/SUPPORTS FOR SERVICES; BUILDERS WORK IN CONNECTION WITH MECHANICAL SERVICES Cont.

Cutting holes for services Cont.

For pipes through softwood Cont.
pipes 55 - 110mm nominal size

19mm thick	-	-	-	0.15	-	1.38	-	1.38	nr	1.49
50mm thick	-	-	-	0.20	-	1.85	-	1.85	nr	1.98

For pipes through plywood
pipes not exceeding 55mm nominal size

13mm thick	-	-	-	0.10	-	0.92	-	0.92	nr	0.99
19mm thick	-	-	-	0.13	-	1.20	-	1.20	nr	1.29

pipes 55 - 110mm nominal size

13mm thick	-	-	-	0.15	-	1.38	-	1.38	nr	1.49
19mm thick	-	-	-	0.20	-	1.85	-	1.85	nr	1.98

For pipes through existing plasterboard with skim finish; making good one side
pipes not exceeding 55mm nominal size

13mm thick	-	-	-	0.20	-	1.85	-	1.85	nr	1.98

pipes 55 - 110mm nominal size

13mm thick	-	-	-	0.30	-	2.77	-	2.77	nr	2.98

Cutting chases for services

In concrete; making good

15mm nominal size pipes -1	-	-	-	0.30	-	2.77	-	2.77	m	2.98
22mm nominal size pipes -1	-	-	-	0.35	-	3.23	-	3.23	m	3.47

In brickwork

15mm nominal size pipes -1	-	-	-	0.25	-	2.31	-	2.31	m	2.48
22mm nominal size pipes -1	-	-	-	0.30	-	2.77	-	2.77	m	2.98

In brickwork; making good fair face or facings

15mm nominal size pipes -1	-	-	-	0.30	-	2.77	0.11	2.88	m	3.09
22mm nominal size pipes -1	-	-	-	0.35	-	3.23	0.17	3.40	m	3.66

In existing brickwork with plaster finish; making good

15mm nominal size pipes -1	-	-	-	0.40	-	3.69	0.22	3.91	m	4.21
22mm nominal size pipes -1	-	-	-	0.50	-	4.62	0.34	4.96	m	5.33

In blockwork; making good

15mm nominal size pipes -1	-	-	-	0.20	-	1.85	-	1.85	m	1.98
22mm nominal size pipes -1	-	-	-	0.25	-	2.31	-	2.31	m	2.48

Pipe and duct sleeves

Steel pipes B.S.1387 Table 2; casting into concrete; making good
100mm long; to the following nominal size steel pipes

15mm	0.55	5.00	0.58	0.10	0.05	1.27	-	1.85	nr	1.99
20mm	0.61	5.00	0.64	0.12	0.06	1.53	-	2.17	nr	2.33
25mm	0.75	5.00	0.79	0.13	0.06	1.62	-	2.41	nr	2.59
32mm	0.88	5.00	0.92	0.14	0.07	1.78	-	2.71	nr	2.91
40mm	1.01	5.00	1.06	0.16	0.08	2.04	-	3.10	nr	3.33
50mm	1.28	5.00	1.34	0.20	0.10	2.55	-	3.89	nr	4.18

150mm long; to the following nominal size steel pipes

15mm	0.68	5.00	0.71	0.12	0.06	1.53	-	2.24	nr	2.41
20mm	0.76	5.00	0.80	0.14	0.07	1.78	-	2.58	nr	2.78
25mm	0.95	5.00	1.00	0.15	0.08	1.95	-	2.94	nr	3.16
32mm	1.12	5.00	1.18	0.16	0.08	2.04	-	3.21	nr	3.46
40mm	1.28	5.00	1.34	0.17	0.09	2.20	-	3.54	nr	3.81
50mm	1.66	5.00	1.74	0.21	0.10	2.64	-	4.38	nr	4.71

200mm long; to the following nominal size steel pipes

15mm	0.80	5.00	0.84	0.13	0.06	1.62	-	2.46	nr	2.65
20mm	0.90	5.00	0.94	0.15	0.08	1.95	-	2.89	nr	3.11
25mm	1.14	5.00	1.20	0.16	0.08	2.04	-	3.24	nr	3.48
32mm	1.36	5.00	1.43	0.17	0.09	2.20	-	3.63	nr	3.90
40mm	1.56	5.00	1.64	0.18	0.09	2.29	-	3.93	nr	4.23
50mm	2.05	5.00	2.15	0.22	0.11	2.80	-	4.96	nr	5.33

250mm long; to the following nominal size steel pipes

15mm	0.93	5.00	0.98	0.15	0.08	1.95	-	2.92	nr	3.14
20mm	1.04	5.00	1.09	0.17	0.09	2.20	-	3.29	nr	3.54
25mm	1.34	5.00	1.41	0.18	0.09	2.29	-	3.70	nr	3.98
32mm	1.61	5.00	1.69	0.19	0.10	2.46	-	4.15	nr	4.46
40mm	1.84	5.00	1.93	0.20	0.11	2.62	-	4.55	nr	4.89
50mm	2.44	5.00	2.56	0.24	0.12	3.06	-	5.62	nr	6.04

100mm long; to the following nominal size copper pipes

15mm	0.61	5.00	0.64	0.10	0.05	1.27	-	1.91	nr	2.06
22mm	0.67	5.00	0.70	0.12	0.06	1.53	-	2.23	nr	2.40
28mm	0.83	5.00	0.87	0.13	0.06	1.62	-	2.49	nr	2.68
32mm	0.97	5.00	1.02	0.14	0.07	1.78	-	2.80	nr	3.01
45mm	1.11	5.00	1.17	0.16	0.08	2.04	-	3.20	nr	3.44
54mm	1.41	5.00	1.48	0.20	0.10	2.55	-	4.03	nr	4.33

150mm long; to the following nominal size copper pipes

15mm	0.75	5.00	0.79	0.12	0.06	1.53	-	2.32	nr	2.49

Labour hourly rates: (except Specialists) Craft Operatives £9.23 Labourer £7.02 Rates are national average prices. Refer to REGIONAL VARIATIONS for indicative levels of overall pricing in regions	MATERIALS			LABOUR				RATES		
	Del to Site £	Waste %	Material Cost £	Craft Optve Hrs	Lab Hrs	Labour Cost £	Sunds £	Nett Rate £	Unit	Gross Rate (+7.5%) £
P31: HOLES/CHASES/COVERS/SUPPORTS FOR SERVICES; BUILDERS WORK IN CONNECTION WITH MECHANICAL SERVICES Cont.										
Pipe and duct sleeves Cont.										
Steel pipes B.S.1387 Table 2; casting into concrete; making good Cont.										
150mm long; to the following nominal size copper pipes Cont.										
22mm	0.84	5.00	0.88	0.14	0.07	1.78	-	2.67	nr	2.87
28mm	1.05	5.00	1.10	0.15	0.08	1.95	-	3.05	nr	3.28
32mm	1.23	5.00	1.29	0.16	0.08	2.04	-	3.33	nr	3.58
45mm	1.41	5.00	1.48	0.17	0.09	2.20	-	3.68	nr	3.96
54mm	1.83	5.00	1.92	0.21	0.10	2.64	-	4.56	nr	4.90
200mm long; to the following nominal size copper pipes										
15mm	0.88	5.00	0.92	0.13	0.06	1.62	-	2.55	nr	2.74
22mm	0.99	5.00	1.04	0.15	0.08	1.95	-	2.99	nr	3.21
28mm	1.25	5.00	1.31	0.16	0.08	2.04	-	3.35	nr	3.60
32mm	1.50	5.00	1.58	0.17	0.09	2.20	-	3.78	nr	4.06
45mm	1.72	5.00	1.81	0.18	0.09	2.29	-	4.10	nr	4.41
54mm	2.26	5.00	2.37	0.22	0.11	2.80	-	5.18	nr	5.56
250mm long; to the following nominal size copper pipes										
15mm	1.03	5.00	1.08	0.15	0.08	1.95	-	3.03	nr	3.25
22mm	1.14	5.00	1.20	0.17	0.09	2.20	-	3.40	nr	3.65
28mm	1.47	5.00	1.54	0.18	0.09	2.29	-	3.84	nr	4.12
32mm	1.77	5.00	1.86	0.19	0.10	2.46	-	4.31	nr	4.64
45mm	2.01	5.00	2.11	0.20	0.10	2.55	-	4.66	nr	5.01
54mm	2.68	5.00	2.81	0.24	0.12	3.06	-	5.87	nr	6.31
Steel pipes B.S.1387 Table 2; building into blockwork; bedding and pointing in cement mortar (1:3); making good										
100mm long; to the following nominal size steel pipes										
15mm	0.55	5.00	0.58	0.15	0.08	1.95	-	2.52	nr	2.71
20mm	0.61	5.00	0.64	0.16	0.08	2.04	-	2.68	nr	2.88
25mm	0.75	5.00	0.79	0.17	0.09	2.20	-	2.99	nr	3.21
32mm	0.88	5.00	0.92	0.18	0.09	2.29	-	3.22	nr	3.46
40mm	1.01	5.00	1.06	0.19	0.10	2.46	-	3.52	nr	3.78
50mm	1.28	5.00	1.34	0.20	0.10	2.55	-	3.89	nr	4.18
140mm long; to the following nominal size steel pipes										
15mm	0.65	5.00	0.68	0.17	0.09	2.20	-	2.88	nr	3.10
20mm	0.73	5.00	0.77	0.18	0.09	2.29	-	3.06	nr	3.29
25mm	0.91	5.00	0.96	0.19	0.10	2.46	-	3.41	nr	3.67
32mm	1.07	5.00	1.12	0.20	0.10	2.55	-	3.67	nr	3.95
40mm	1.23	5.00	1.29	0.21	0.11	2.71	-	4.00	nr	4.30
50mm	1.59	5.00	1.67	0.22	0.11	2.80	-	4.47	nr	4.81
190mm long; to the following nominal size steel pipes										
15mm	0.78	5.00	0.82	0.19	0.10	2.46	-	3.27	nr	3.52
20mm	0.87	5.00	0.91	0.20	0.10	2.55	-	3.46	nr	3.72
25mm	1.11	5.00	1.17	0.21	0.11	2.71	-	3.88	nr	4.17
32mm	1.32	5.00	1.39	0.22	0.11	2.80	-	4.19	nr	4.50
40mm	1.50	5.00	1.58	0.23	0.12	2.97	-	4.54	nr	4.88
50mm	1.97	5.00	2.07	0.24	0.12	3.06	-	5.13	nr	5.51
215mm long; to the following nominal size steel pipes										
15mm	0.84	5.00	0.88	0.20	0.10	2.55	-	3.43	nr	3.69
20mm	0.94	5.00	0.99	0.21	0.11	2.71	-	3.70	nr	3.97
25mm	1.21	5.00	1.27	0.22	0.11	2.80	-	4.07	nr	4.38
32mm	1.44	5.00	1.51	0.23	0.12	2.97	-	4.48	nr	4.81
40mm	1.64	5.00	1.72	0.24	0.12	3.06	-	4.78	nr	5.14
50mm	2.17	5.00	2.28	0.25	0.13	3.22	-	5.50	nr	5.91
100mm long; to the following nominal size copper pipes										
15mm	0.61	5.00	0.64	0.15	0.08	1.95	-	2.59	nr	2.78
22mm	0.66	5.00	0.69	0.16	0.08	2.04	-	2.73	nr	2.94
28mm	0.83	5.00	0.87	0.17	0.09	2.20	-	3.07	nr	3.30
32mm	0.97	5.00	1.02	0.18	0.09	2.29	-	3.31	nr	3.56
45mm	1.01	5.00	1.06	0.19	0.10	2.46	-	3.52	nr	3.78
54mm	1.41	5.00	1.48	0.20	0.10	2.55	-	4.03	nr	4.33
140mm long; to the following nominal size copper pipes										
15mm	0.72	5.00	0.76	0.17	0.09	2.20	-	2.96	nr	3.18
22mm	0.80	5.00	0.84	0.18	0.09	2.29	-	3.13	nr	3.37
28mm	1.01	5.00	1.06	0.19	0.10	2.46	-	3.52	nr	3.78
32mm	1.18	5.00	1.24	0.20	0.10	2.55	-	3.79	nr	4.07
45mm	1.35	5.00	1.42	0.21	0.11	2.71	-	4.13	nr	4.44
54mm	1.75	5.00	1.84	0.22	0.11	2.80	-	4.64	nr	4.99
190mm long; to the following nominal size copper pipes										
15mm	0.86	5.00	0.90	0.19	0.10	2.46	-	3.36	nr	3.61
22mm	0.96	5.00	1.01	0.20	0.10	2.55	-	3.56	nr	3.82
28mm	1.22	5.00	1.28	0.21	0.11	2.71	-	3.99	nr	4.29
32mm	1.45	5.00	1.52	0.22	0.11	2.80	-	4.33	nr	4.65
45mm	1.65	5.00	1.73	0.23	0.12	2.97	-	4.70	nr	5.05
54mm	2.17	5.00	2.28	0.24	0.12	3.06	-	5.34	nr	5.74
215mm long; to the following nominal size copper pipes										
15mm	0.92	5.00	0.97	0.20	0.10	2.55	-	3.51	nr	3.78
22mm	1.03	5.00	1.08	0.21	0.11	2.71	-	3.79	nr	4.08
28mm	1.33	5.00	1.40	0.22	0.11	2.80	-	4.20	nr	4.51

Labour hourly rates: (except Specialists) Craft Operatives £9.23 Labourer £7.02 Rates are national average prices. Refer to REGIONAL VARIATIONS for indicative levels of overall pricing in regions	MATERIALS			LABOUR				RATES		
	Del to Site	Waste	Material Cost	Craft Optve	Lab	Labour Cost	Sunds	Nett Rate	Unit	Gross Rate (+7.5%)
	£	%	£	Hrs	Hrs	£	£	£		£
P31: HOLES/CHASES/COVERS/SUPPORTS FOR SERVICES; BUILDERS WORK IN CONNECTION WITH MECHANICAL SERVICES Cont.										
Pipe and duct sleeves Cont.										
Steel pipes B.S.1387 Table 2; building into blockwork; bedding and pointing in cement mortar (1:3); making good Cont.										
215mm long; to the following nominal size copper pipes Cont.										
32mm	1.58	5.00	1.66	0.23	0.12	2.97	-	4.62	nr	4.97
45mm	1.80	5.00	1.89	0.24	0.12	3.06	-	4.95	nr	5.32
54mm	2.39	5.00	2.51	0.25	0.13	3.22	-	5.73	nr	6.16
Ends of supports for equipment, fittings, appliances and ancillaries										
Fix only; casting into concrete; making good										
holderbat or bracket	-	-	-	0.25	-	2.31	0.28	2.59	nr	2.78
Fix only; building into brickwork										
holderbat or bracket	-	-	-	0.10	-	0.92	-	0.92	nr	0.99
Fix only; cutting and pinning to brickwork; making good										
holderbat or bracket	-	-	-	0.12	-	1.11	0.09	1.20	nr	1.29
Fix only; cutting and pinning to brickwork; making good fair face or facings one side										
holderbat or bracket	-	-	-	0.15	-	1.38	0.13	1.51	nr	1.63
Cavity fixings for ends of supports for pipes and ducts										
Rawlnut Multi-purpose fixings, plated pan head screws; The Rawlplug Co. Ltd; fixing to soft building board, drilling										
M4 with 30mm long screw; product code 09 - 130										
at 300mm centres	1.07	5.00	1.12	0.17	-	1.57	-	2.69	m	2.89
at 450mm centres	0.71	5.00	0.75	0.11	-	1.02	-	1.76	m	1.89
isolated	0.32	5.00	0.34	0.05	-	0.46	-	0.80	nr	0.86
M5 with 30mm long screw; product code 09 - 235										
at 300mm centres	1.30	5.00	1.37	0.20	-	1.85	-	3.21	m	3.45
at 450mm centres	0.87	5.00	0.91	0.13	-	1.20	-	2.11	m	2.27
isolated	0.39	5.00	0.41	0.06	-	0.55	-	0.96	nr	1.04
M5 with 50mm long screw; product code 09 - 317										
at 300mm centres	1.43	5.00	1.50	0.20	-	1.85	-	3.35	m	3.60
at 450mm centres	0.96	5.00	1.01	0.13	-	1.20	-	2.21	m	2.37
isolated	0.43	5.00	0.45	0.06	-	0.55	-	1.01	nr	1.08
Rawlnut Multi-purpose fixings, plated pan head screws; The Rawlplug Co. Ltd; fixing to sheet metal, drilling										
M4 with 30mm long screw; product code 09 - 130										
at 300mm centres	1.07	5.00	1.12	0.33	-	3.05	-	4.17	m	4.48
at 450mm centres	0.71	5.00	0.75	0.22	-	2.03	-	2.78	m	2.98
isolated	0.32	5.00	0.34	0.10	-	0.92	-	1.26	nr	1.35
M5 with 30mm long screw; product code 09 - 235										
at 300mm centres	1.30	5.00	1.37	0.40	-	3.69	-	5.06	m	5.44
at 450mm centres	0.87	5.00	0.91	0.27	-	2.49	-	3.41	m	3.66
isolated	0.39	5.00	0.41	0.12	-	1.11	-	1.52	nr	1.63
M5 with 50mm long screw; product code 09 - 317										
at 300mm centres	1.43	5.00	1.50	0.40	-	3.69	-	5.19	m	5.58
at 450mm centres	0.96	5.00	1.01	0.27	-	2.49	-	3.50	m	3.76
isolated	0.43	5.00	0.45	0.12	-	1.11	-	1.56	nr	1.68
Interset high performance cavity fixings, plated pan head screws; The Rawlplug Co. Ltd; fixing to soft building board, drilling										
M4 with 40mm long screw; product code 41 - 620										
at 300mm centres	1.30	5.00	1.37	0.17	-	1.57	-	2.93	m	3.15
at 450mm centres	0.87	5.00	0.91	0.11	-	1.02	-	1.93	m	2.07
isolated	0.39	5.00	0.41	0.05	-	0.46	-	0.87	nr	0.94
M5 with 40mm long screw; product code 41 - 636										
at 300mm centres	1.53	5.00	1.61	0.20	-	1.85	-	3.45	m	3.71
at 450mm centres	1.02	5.00	1.07	0.13	-	1.20	-	2.27	m	2.44
isolated	0.46	5.00	0.48	0.06	-	0.55	-	1.04	nr	1.11
M5 with 55mm long screw; product code 41 - 652										
at 300mm centres	1.60	5.00	1.68	0.20	-	1.85	-	3.53	m	3.79
at 450mm centres	1.07	5.00	1.12	0.13	-	1.20	-	2.32	m	2.50
isolated	0.48	5.00	0.50	0.06	-	0.55	-	1.06	nr	1.14
Spring toggles plated pan head screws; The Rawlplug Co. Ltd; fixing to soft building board, drilling										
M5 with 80mm long screw; product code 94 - 439										
at 300mm centres	1.07	5.00	1.12	0.20	-	1.85	-	2.97	m	3.19
at 450mm centres	0.71	5.00	0.75	0.13	-	1.20	-	1.95	m	2.09
isolated	0.32	5.00	0.34	0.06	-	0.55	-	0.89	nr	0.96
M6 with 60mm long screw; product code 94 - 442										
at 300mm centres	1.23	5.00	1.29	0.20	-	1.85	-	3.14	m	3.37
at 450mm centres	0.82	5.00	0.86	0.13	-	1.20	-	2.06	m	2.22
isolated	0.37	5.00	0.39	0.06	-	0.55	-	0.94	nr	1.01
M6 with 80mm long screw; product code 94 - 464										
at 300mm centres	1.53	5.00	1.61	0.20	-	1.85	-	3.45	m	3.71
at 450mm centres	1.02	5.00	1.07	0.13	-	1.20	-	2.27	m	2.44

BUILDING FABRIC SUNDRIES

Labour hourly rates: (except Specialists) Craft Operatives £9.23 Labourer £7.02 Rates are national average prices. Refer to REGIONAL VARIATIONS for indicative levels of overall pricing in regions	MATERIALS			LABOUR				RATES		
	Del to Site £	Waste %	Material Cost £	Craft Optve Hrs	Lab Hrs	Labour Cost £	Sunds £	Nett Rate £	Unit	Gross Rate (+7.5%) £
P31: HOLES/CHASES/COVERS/SUPPORTS FOR SERVICES; **BUILDERS WORK IN CONNECTION WITH MECHANICAL SERVICES** **Cont.**										
Cavity fixings for ends of supports for pipes and **ducts Cont.**										
Spring toggles plated pan head screws; The Rawlplug Co. Ltd; fixing to soft building board, drilling Cont.										
M6 with 80mm long screw; product code 94 - 464 Cont.										
isolated..	0.46	5.00	0.48	0.06	–	0.55	–	1.04	nr	1.11
Trench covers and frames										
Duct covers in cast iron, coated; continuous covers in 610mm lengths; steel bearers										
22mm deep, for pedestrian traffic; bedding frames to concrete in cement mortar (1:3); nominal width										
150mm	85.30	–	85.30	0.50	0.50	8.13	0.68	94.11	m	101.16
225mm	88.90	–	88.90	0.55	0.55	8.94	0.68	98.52	m	105.91
300mm	103.70	–	103.70	0.65	0.65	10.56	0.68	114.94	m	123.56
375mm	118.25	–	118.25	0.70	0.70	11.38	0.68	130.31	m	140.08
450mm	133.25	–	133.25	0.75	0.75	12.19	0.68	146.12	m	157.08
Pendock Profiles Ltd. floor ducting profiles; **galvanised mild steel tray section with 12mm plywood** **cover board; fixing tray section to masonry with** **nails; fixing cover board with screws, countersinking**										
For 50mm screeds										
reference FDT 100/50, 100mm wide	6.83	2.50	7.00	0.35	–	3.23	0.10	10.33	m	11.11
extra; stop end	1.99	2.50	2.04	0.10	–	0.92	–	2.96	nr	3.18
extra; corner	8.62	2.50	8.84	0.25	–	2.31	0.15	11.29	nr	12.14
extra; tee	8.62	2.50	8.84	0.30	–	2.77	0.15	11.75	nr	12.64
reference FDT 150/50, 150mm wide	9.18	2.50	9.41	0.40	–	3.69	0.10	13.20	m	14.19
extra; stop end	1.99	2.50	2.04	0.12	–	1.11	–	3.15	nr	3.38
extra; corner	9.60	2.50	9.84	0.30	–	2.77	0.15	12.76	nr	13.72
extra; tee	9.60	2.50	9.84	0.35	–	3.23	0.15	13.22	nr	14.21
reference FDT 200/50, 200mm wide	11.20	2.50	11.48	0.50	–	4.62	0.10	16.20	m	17.41
extra; stop end	1.99	2.50	2.04	0.15	–	1.38	–	3.42	nr	3.68
extra; corner	11.17	2.50	11.45	0.35	–	3.23	0.15	14.83	nr	15.94
extra; tee	11.17	2.50	11.45	0.40	–	3.69	0.15	15.29	nr	16.44
For 70mm screeds										
reference FDT 100/70, 100mm wide	7.53	2.50	7.72	0.35	–	3.23	0.10	11.05	m	11.88
extra; stop end	1.99	2.50	2.04	0.10	–	0.92	–	2.96	nr	3.18
extra; corner	8.62	2.50	8.84	0.25	–	2.31	0.15	11.29	nr	12.14
extra; tee	8.62	2.50	8.84	0.30	–	2.77	0.15	11.75	nr	12.64
reference FDT 150/70, 150mm wide	9.77	2.50	10.01	0.40	–	3.69	0.10	13.81	m	14.84
extra; stop end	1.99	2.50	2.04	0.12	–	1.11	–	3.15	nr	3.38
extra; corner	9.60	2.50	9.84	0.30	–	2.77	0.15	12.76	nr	13.72
extra; tee	9.60	2.50	9.84	0.35	–	3.23	0.15	13.22	nr	14.21
reference FDT 200/70, 200mm wide	11.69	2.50	11.98	0.50	–	4.62	0.10	16.70	m	17.95
extra; stop end	1.99	2.50	2.04	0.15	–	1.38	–	3.42	nr	3.68
extra; corner	11.17	2.50	11.45	0.35	–	3.23	0.15	14.83	nr	15.94
extra; tee	11.17	2.50	11.45	0.40	–	3.69	0.15	15.29	nr	16.44
Casings, bearers and supports for equipment										
Softwood, sawn										
25mm boarded platforms	4.63	10.00	5.09	1.10	0.14	11.14	0.15	16.38	m2	17.61
50 x 50mm bearers	0.46	10.00	0.51	0.16	0.02	1.62	0.02	2.14	m	2.30
50 x 75mm bearers	0.66	10.00	0.73	0.19	0.02	1.89	0.02	2.64	m	2.84
50 x 100mm bearers	0.92	10.00	1.01	0.23	0.03	2.33	0.03	3.38	m	3.63
38 x 50mm bearers; framed	0.37	10.00	0.41	0.25	0.03	2.52	0.03	2.96	m	3.18
Softwood, wrought										
25mm boarded platforms; tongued and grooved	9.26	10.00	10.19	1.20	0.15	12.13	0.17	22.48	m2	24.17
19mm boarded sides; tongued and grooved	6.95	10.00	7.64	1.10	0.14	11.14	0.15	18.93	m2	20.35
extra for holes for pipes not exceeding 55mm nominal size	–	–	–	0.25	0.35	4.76	–	4.76	nr	5.12
19mm boarded cover; tongued and grooved; ledged; sectional	13.31	10.00	14.64	1.20	0.15	12.13	0.17	26.94	m2	28.96
Fibreboard										
13mm sides	8.00	10.00	8.80	2.10	0.26	21.21	0.29	30.30	m2	32.57
extra for holes for pipes not exceeding 55mm nominal size	–	–	–	0.20	–	1.85	–	1.85	nr	1.98
Vermiculite insulation										
packing around tank	59.26	10.00	65.19	4.80	0.60	48.52	–	113.70	m3	122.23
Slag wool insulation										
packing around tank	29.49	10.00	32.44	6.00	0.75	60.65	–	93.08	m3	100.07
P31: HOLES/CHASES/COVERS/SUPPORTS FOR SERVICES; **BUILDERS WORK IN CONNECTION WITH ELECTRICAL SERVICES**										
Cutting or forming holes, mortices, sinkings and **chases - New buildings**										
Concealed steel conduits; making good										
luminaire points	–	–	–	0.70	0.35	8.92	0.35	9.27	nr	9.96
socket outlet points	–	–	–	0.60	0.30	7.64	0.30	7.94	nr	8.54

Labour hourly rates: (except Specialists) Craft Operatives £9.23 Labourer £7.02 Rates are national average prices. Refer to REGIONAL VARIATIONS for indicative levels of overall pricing in regions	MATERIALS			LABOUR				RATES		
	Del to Site £	Waste %	Material Cost £	Craft Optve Hrs	Lab Hrs	Labour Cost £	Sunds £	Nett Rate £	Unit	Gross Rate (+7.5%) £
P31: HOLES/CHASES/COVERS/SUPPORTS FOR SERVICES; BUILDERS WORK IN CONNECTION WITH ELECTRICAL SERVICES Cont.										
Cutting or forming holes, mortices, sinkings and chases - New buildings Cont.										
Concealed steel conduits; making good Cont.										
fitting outlet points	-	-	-	0.60	0.30	7.64	0.30	7.94	nr	8.54
equipment and control gear points	-	-	-	0.80	0.40	10.19	0.41	10.60	nr	11.40
Exposed p.v.c. conduits; making good										
luminaire points	-	-	-	0.25	0.13	3.22	0.12	3.34	nr	3.59
socket outlet points	-	-	-	0.20	0.10	2.55	0.10	2.65	nr	2.85
fitting outlet points	-	-	-	0.20	0.10	2.55	0.10	2.65	nr	2.85
equipment and control gear points	-	-	-	0.30	0.15	3.82	0.14	3.96	nr	4.26
Cutting mortices, sinking and the like for services - Existing buildings										
In existing concrete; making good										
75 x 75 x 35mm	-	-	-	0.45	0.13	5.07	0.50	5.57	nr	5.98
150 x 75 x 35mm	-	-	-	0.60	0.15	6.59	0.75	7.34	nr	7.89
In existing brickwork										
75 x 75 x 35mm	-	-	-	0.25	0.10	3.01	0.30	3.31	nr	3.56
150 x 75 x 35mm	-	-	-	0.40	0.13	4.60	0.45	5.05	nr	5.43
In existing brickwork with plaster finish; making good										
75 x 75 x 35mm	-	-	-	0.30	0.18	4.03	0.50	4.53	nr	4.87
150 x 75 x 35mm	-	-	-	0.45	0.20	5.56	0.70	6.26	nr	6.73
In existing blockwork; making good										
75 x 75 x 35mm	-	-	-	0.25	0.08	2.87	0.33	3.20	nr	3.44
150 x 75 x 35mm	-	-	-	0.35	0.10	3.93	0.36	4.29	nr	4.61
In existing blockwork with plaster finish; making good										
75 x 75 x 35mm	-	-	-	0.40	0.13	4.60	0.45	5.05	nr	5.43
150 x 75 x 35mm	-	-	-	0.46	0.15	5.30	0.40	5.70	nr	6.13
Cutting chases for services										
In existing concrete; making good										
20mm nominal size conduits -1	-	-	-	0.40	0.20	5.10	0.35	5.45	m	5.85
20mm nominal size conduits -3	-	-	-	0.50	0.25	6.37	0.41	6.78	m	7.29
20mm nominal size conduits -6	-	-	-	0.60	0.30	7.64	0.47	8.11	m	8.72
In existing brickwork										
20mm nominal size conduits -1	-	-	-	0.30	0.15	3.82	0.35	4.17	m	4.48
20mm nominal size conduits -3	-	-	-	0.40	0.20	5.10	0.41	5.51	m	5.92
20mm nominal size conduits -6	-	-	-	0.50	0.25	6.37	0.47	6.84	m	7.35
In existing brickwork with plaster finish; making good										
20mm nominal size conduits -1	-	-	-	0.45	0.23	5.77	0.46	6.23	m	6.70
20mm nominal size conduits -3	-	-	-	0.55	0.28	7.04	0.52	7.56	m	8.13
20mm nominal size conduits -6	-	-	-	0.65	0.33	8.32	0.58	8.90	m	9.56
In existing blockwork										
20mm nominal size conduits -1	-	-	-	0.23	0.12	2.97	0.35	3.32	m	3.56
20mm nominal size conduits -3	-	-	-	0.30	0.15	3.82	0.41	4.23	m	4.55
20mm nominal size conduits -6	-	-	-	0.38	0.19	4.84	0.47	5.31	m	5.71
In existing blockwork with plaster finish; making good										
20mm nominal size conduits -1	-	-	-	0.38	0.19	4.84	0.46	5.30	m	5.70
20mm nominal size conduits -3	-	-	-	0.45	0.23	5.77	0.52	6.29	m	6.76
20mm nominal size conduits -6	-	-	-	0.53	0.27	6.79	0.58	7.37	m	7.92
Lifting and replacing floorboards										
For cables or conduits										
in groups 1 - 3	-	-	-	0.16	0.08	2.04	0.10	2.14	m	2.30
in groups 3 - 6	-	-	-	0.20	0.10	2.55	0.11	2.66	m	2.86
in groups exceeding 6	-	-	-	0.25	0.13	3.22	0.12	3.34	m	3.59

BUILDING FABRIC SUNDRIES

Regional Factors

Laxton's Building Price Book is based upon national average prices (=1.00). Indicative levels of tender pricing in regions as at the second quarter 1999 are given on this map and the factors shown may be applied to adjust overall pricing.

MAJOR WORKS – TENDER VALUES	SMALL WORKS – TENDER VALUES
The measured rates are based on contracts valued in the range of £250,000 to £1,000.00. As a guide to pricing works of larger value and for cost planning or budgetary purposes, the following adjustments may be applied to overall contract values.	The measured rates are based on contracts valued in the range of £25,000 to £75,000. As a guide to pricing works of smaller or larger value and for cost planning or budgetary purposes, the following adjustments may be applied to overall contract values.
Contract Value £1,000,000 to £2,000,000.........deduct 2.5% £2,000,000 to £3,000,000.........deduct 5.0% £3,000,000 to £5,000,000.........deduct 7.5%	Contract Value £5,000 to £15,000..............................add 20.0% £15,000 to £25,000............................add 10.0% £25,000 to £75,000.......................rate as shown £75,000 to £100,000........................deduct 5.0% £100,000 to £150,000.......................deduct7.5% £150,000 to £250,000.......................deduct10.%

Paving/Planting/Fencing/Site furniture

Labour hourly rates: (except Specialists) Craft Operatives £9.23 Labourer £7.02 Rates are national average prices. Refer to REGIONAL VARIATIONS for indicative levels of overall pricing in regions	PLANT & TRANSPORT			LABOUR				RATES		
	Plant Cost	Trans Cost	P & T Cost	Craft Optve	Lab	Labour Cost	Sunds	Nett Rate	Unit	Gross Rate (+7.5%)
	£	£	£	Hrs	Hrs	£	£	£		£
Q10: KERBS/EDGINGS/CHANNELS/PAVING ACCESSORIES										
Excavating - by machine										
Trenches width not exceeding 0.30m										
maximum depth not exceeding 0.25m	4.90	-	4.90	-	1.71	12.00	-	16.90	m3	18.17
maximum depth not exceeding 1.00m	5.10	-	5.10	-	1.80	12.64	-	17.74	m3	19.07
Extra over any types of excavating irrespective of depth										
excavating below ground water level	4.32	-	4.32	-	-	-	-	4.32	m3	4.64
excavating in running silt, running sand or liquid mud	7.42	-	7.42	-	-	-	-	7.42	m3	7.98
Breaking out existing materials; extra over any types of excavating irrespective of depth										
hard rock	22.05	-	22.05	-	-	-	-	22.05	m3	23.70
concrete	18.31	-	18.31	-	-	-	-	18.31	m3	19.68
reinforced concrete	27.50	-	27.50	-	-	-	-	27.50	m3	29.56
brickwork, blockwork or stonework	12.22	-	12.22	-	-	-	-	12.22	m3	13.14
100mm diameter drain and concrete bed under	1.72	-	1.72	-	-	-	-	1.72	m	1.85
Breaking out existing hard pavings; extra over any types of excavating irrespective of depth										
concrete 150mm thick	2.89	-	2.89	-	-	-	-	2.89	m2	3.11
reinforced concrete 200mm thick	5.79	-	5.79	-	-	-	-	5.79	m2	6.22
brickwork, blockwork or stonework 100mm thick	1.36	-	1.36	-	-	-	-	1.36	m2	1.46
coated macadam or asphalt 75mm thick	0.49	-	0.49	-	-	-	-	0.49	m2	0.53
Surface treatments										
Compacting										
bottoms of excavations	0.15	-	0.15	-	-	-	-	0.15	m2	0.16
Excavating - by hand										
Trenches width not exceeding 0.30m										
maximum depth not exceeding 0.25m	-	-	-	-	4.25	29.84	-	29.84	m3	32.07
maximum depth not exceeding 1.00m	-	-	-	-	4.50	31.59	-	31.59	m3	33.96
Extra over any types of excavating irrespective of depth										
excavating below ground water level	-	-	-	-	1.50	10.53	-	10.53	m3	11.32
excavating in running silt, running sand or liquid mud	-	-	-	-	4.50	31.59	-	31.59	m3	33.96
Breaking out existing materials; extra over any types of excavating irrespective of depth										
hard rock	-	-	-	-	13.35	93.72	-	93.72	m3	100.75
concrete	-	-	-	-	10.00	70.20	-	70.20	m3	75.47
reinforced concrete	-	-	-	-	15.00	105.30	-	105.30	m3	113.20
brickwork, blockwork or stonework	-	-	-	-	7.00	49.14	-	49.14	m3	52.83
100mm diameter drain and concrete bed under	-	-	-	-	0.80	5.62	-	5.62	m	6.04
Breaking out existing hard pavings; extra over any types of excavating irrespective of depth										
concrete 150mm thick	-	-	-	-	1.50	10.53	-	10.53	m2	11.32
reinforced concrete 200mm thick	-	-	-	-	3.00	21.06	-	21.06	m2	22.64
brickwork, blockwork or stonework 100mm thick	-	-	-	-	0.70	4.91	-	4.91	m2	5.28
coated macadam or asphalt 75mm thick	-	-	-	-	0.25	1.75	-	1.75	m2	1.89

PAVING/FENCING/SITE FURNITURE

Labour hourly rates: (except Specialists) Craft Operatives £9.23 Labourer £7.02 Rates are national average prices. Refer to REGIONAL VARIATIONS for indicative levels of overall pricing in regions	MATERIALS			LABOUR				RATES		
	Del to Site £	Waste %	Material Cost £	Craft Optve Hrs	Lab Hrs	Labour Cost £	Sunds £	Nett Rate £	Unit	Gross Rate (+7.5%) £
Q10: KERBS/EDGINGS/CHANNELS/PAVING ACCESSORIES Cont.										
Earthwork support										

Earthwork support
  maximum depth not exceeding 1.00m; distance
  between opposing faces not exceeding 2.00m ....
	0.63	-	0.63	-	0.20	1.40	-	2.03	m2	2.19

Earthwork support; unstable ground
  maximum depth not exceeding 1.00m; distance
  between opposing faces not exceeding 2.00m ....
	1.13	-	1.13	-	0.36	2.53	-	3.66	m2	3.93

Earthwork support; next to roadways
  maximum depth not exceeding 1.00m; distance
  between opposing faces not exceeding 2.00m ....
	1.24	-	1.24	-	0.40	2.81	-	4.05	m2	4.35

Disposal										

Disposal of excavated material
  depositing on site in temporary spoil heaps where
  directed										

25m	-	-	-	-	1.25	8.78	-	8.78	m3	9.43
50m	-	-	-	-	1.50	10.53	-	10.53	m3	11.32
100m	-	-	-	-	2.00	14.04	-	14.04	m3	15.09
extra for removing beyond 100m and not exceeding 400m..............................	-	-	-	-	1.55	10.88	2.18	13.06	m3	14.04
extra for removing beyond 100m and not exceeding 800m..............................	-	-	-	-	1.55	10.88	2.84	13.72	m3	14.75
depositing on site in permanent spoil heaps average 25m distant........................	-	-	-	-	1.25	8.78	-	8.78	m3	9.43
removing from site to tip										
a) Inert....................................	-	-	-	-	1.35	9.48	13.00	22.48	m3	24.16
b) Active...................................	-	-	-	-	1.35	9.48	22.00	31.48	m3	33.84
c) Contaminated (guide price - always seek a quote for specialist disposal cost............	-	-	-	-	1.35	9.48	50.00	59.48	m3	63.94
Surface treatments										
Compacting bottoms of excavations	-	-	-	-	0.12	0.84	-	0.84	m2	0.91
Precast concrete; standard or stock pattern units; B.S.7263 Part 1; bedding, jointing and pointing in cement mortar (1:3); on plain in-situ concrete foundation; B.S.5328 ordinary prescribed mix ST4, 20mm aggregate										
Kerbs; Figs. 1, 2, 6 and 7; concrete foundation and haunching; formwork										
125 x 255mm kerb; 400 x 200mm foundation	9.71	5.00	10.20	-	1.09	7.65	-	17.85	m	19.19
125 x 255mm kerb; 400 x 200mm foundation; curved 10.00m radius	12.22	5.00	12.83	-	1.16	8.14	-	20.97	m	22.55
150 x 305mm kerb; 400 x 200mm foundation	11.67	5.00	12.25	-	1.16	8.14	-	20.40	m	21.93
150 x 305mm kerb; 400 x 200mm foundation; curved 10.00m radius	14.53	5.00	15.26	-	1.26	8.85	-	24.10	m	25.91
Edgings; Figs. 10, 11, 12 and 13; concrete foundation and haunching; formwork										
50 x 150mm edging; 300 x 150mm foundation	5.63	5.00	5.91	-	0.78	5.48	-	11.39	m	12.24
50 x 200mm edging; 300 x 150mm foundation	6.07	5.00	6.37	-	0.90	6.32	-	12.69	m	13.64
50 x 250mm edging; 300 x 150mm foundation	6.28	5.00	6.59	-	1.02	7.16	-	13.75	m	14.79
Channels; Fig. 8; concrete foundation and haunching; formwork										
255 x 125mm channel; 450 x 150mm foundation	8.64	5.00	9.07	-	0.86	6.04	-	15.11	m	16.24
255 x 125mm channel; 450 x 150mm foundation; curved 10.00 radius	12.24	5.00	12.85	-	0.93	6.53	-	19.38	m	20.83
Quadrants; Fig. 14; concrete foundation and haunching; formwork										
305 x 305 x 150mm quadrant; 500 x 500 x 150mm foundation	9.85	5.00	10.34	-	0.61	4.28	-	14.62	nr	15.72
305 x 305 x 255mm quadrant; 500 x 500 x 150mm foundation	9.93	5.00	10.43	-	0.69	4.84	-	15.27	nr	16.42
455 x 455 x 150mm quadrant; 650 x 650 x 150mm foundation	11.58	5.00	12.16	-	0.85	5.97	-	18.13	nr	19.49
455 x 455 x 255mm quadrant; 650 x 650 x 150mm foundation	11.74	5.00	12.33	-	0.94	6.60	-	18.93	nr	20.35
Granite; standard units; B.S.435; bedding jointing and pointing in cement mortar (1:3); on plain in-situ concrete foundation; B.S.5328 ordinary prescribed mix ST4, 20mm aggregate										

Excavating
  Disposal of excavated material; removing from site
  to tip										

a) Inert....................................	-	-	-	-	1.35	9.48	13.00	22.48	m3	24.16
b) Active...................................	-	-	-	-	1.35	9.48	22.00	31.48	m3	33.84
c) Contaminated (guide price - always seek a quote for specialist disposal cost............	-	-	-	-	1.35	9.48	50.00	59.48	m3	63.94
Edge kerb; concrete foundation and haunching; formwork										
150 x 300mm; 300 x 200mm foundation	18.06	2.50	18.51	0.50	0.65	9.18	-	27.69	m	29.77

Labour hourly rates: (except Specialists) Craft Operatives £9.23 Labourer £7.02 Rates are national average prices. Refer to REGIONAL VARIATIONS for indicative levels of overall pricing in regions	MATERIALS			LABOUR				RATES		
	Del to Site £	Waste %	Material Cost £	Craft Optve Hrs	Lab Hrs	Labour Cost £	Sunds £	Nett Rate £	Unit	Gross Rate (+7.5%) £
Q10: KERBS/EDGINGS/CHANNELS/PAVING ACCESSORIES Cont.										
Granite; standard units; B.S.435; bedding jointing and pointing in cement mortar (1:3); on plain in-situ concrete foundation; B.S.5328 ordinary prescribed mix ST4, 20mm aggregate Cont.										
Edge kerb; concrete foundation and haunching; formwork Cont.										
150 x 300mm; 300 x 200mm foundation; curved external radius exceeding 1000mm	22.74	2.50	**23.31**	0.80	0.65	**11.95**	-	**35.26**	m	37.90
200 x 300mm; 350 x 200mm foundation	23.45	2.50	**24.04**	0.60	0.73	**10.66**	-	**34.70**	m	37.30
200 x 300mm; 350 x 200mm foundation; curved external radius exceeding 1000mm	29.52	2.50	**30.26**	0.90	0.73	**13.43**	-	**43.69**	m	46.97
Flat kerb; concrete foundation and haunching; formwork										
300 x 150mm; 450 x 200mm foundation	25.20	2.50	**25.83**	0.50	0.73	**9.74**	-	**35.57**	m	38.24
300 x 150mm; 450 x 200mm foundation; curved external radius exceeding 1000mm	32.19	2.50	**32.99**	0.80	0.73	**12.51**	-	**45.50**	m	48.92
300 x 200mm; 450 x 200mm foundation	31.38	2.50	**32.16**	0.60	0.75	**10.80**	-	**42.97**	m	46.19
300 x 200mm; 450 x 200mm foundation; curved external radius exceeding 1000mm	40.19	2.50	**41.19**	0.90	0.75	**13.57**	-	**54.77**	m	58.87
Q20: GRANULAR SUB-BASES TO ROADS/PAVINGS										
Hard, dry, broken brick or stone to be obtained off site; wheeling not exceeding 25m - by machine										
Filling to make up levels; depositing in layers 150mm maximum thickness										
average thickness not exceeding 0.25m	12.40	25.00	**15.50**	-	-	**-**	2.90	**18.40**	m3	19.78
average thickness exceeding 0.25m	12.40	25.00	**15.50**	-	-	**-**	2.40	**17.90**	m3	19.24
Surface packing to filling to vertical or battered faces	-	-	**-**	-	-	**-**	0.50	**0.50**	m2	0.54
Compacting with 680 kg vibratory roller filling; blinding with sand, ashes or similar fine material	0.18	40.00	**0.25**	-	0.05	**0.35**	0.18	**0.78**	m2	0.84
Compacting with 6 - 8 tonnes smooth wheeled roller filling; blinding with sand, ashes or similar fine material	0.18	40.00	**0.25**	-	0.05	**0.35**	0.35	**0.95**	m2	1.02
Hard, dry, broken brick or stone to be obtained off site; wheeling not exceeding 50m - by machine										
Filling to make up levels; depositing in layers 150mm maximum thickness										
average thickness not exceeding 0.25m	12.40	25.00	**15.50**	-	-	**-**	3.70	**19.20**	m3	20.64
average thickness exceeding 0.25m	12.40	25.00	**15.50**	-	-	**-**	2.90	**18.40**	m3	19.78
Surface packing to filling to vertical or battered faces	-	-	**-**	-	-	**-**	0.50	**0.50**	m2	0.54
Compacting with 680 kg vibratory roller filling; blinding with sand, ashes or similar fine material	0.18	40.00	**0.25**	-	0.05	**0.35**	0.18	**0.78**	m2	0.84
Compacting with 6 - 8 tonnes smooth wheeled roller filling; blinding with sand, ashes or similar fine material	0.18	40.00	**0.25**	-	0.05	**0.35**	0.35	**0.95**	m2	1.02
MOT Type 1 to be obtained off site; wheeling not exceeding 25m - by machine										
Filling to make up levels; depositing in layers 150mm maximum thickness										
average thickness not exceeding 0.25m	14.50	25.00	**18.13**	-	-	**-**	2.90	**21.02**	m3	22.60
average thickness exceeding 0.25m	14.50	25.00	**18.13**	-	-	**-**	2.40	**20.52**	m3	22.06
Surface packing to filling to vertical or battered faces	-	-	**-**	-	-	**-**	0.50	**0.50**	m2	0.54
Compacting with 680 kg vibratory roller filling; blinding with sand, ashes or similar fine material	0.18	40.00	**0.25**	-	0.05	**0.35**	0.18	**0.78**	m2	0.84
Compacting with 6 - 8 tonnes smooth wheeled roller filling; blinding with sand, ashes or similar fine material	0.18	40.00	**0.25**	-	0.05	**0.35**	0.35	**0.95**	m2	1.02
MOT Type 1 to be obtained off site; wheeling not exceeding 50m - by machine										
Filling to make up levels; depositing in layers 150mm maximum thickness										
average thickness not exceeding 0.25m	14.50	25.00	**18.13**	-	-	**-**	3.70	**21.82**	m3	23.46
average thickness exceeding 0.25m	14.50	25.00	**18.13**	-	-	**-**	2.90	**21.02**	m3	22.60
Surface packing to filling to vertical or battered faces	-	-	**-**	-	-	**-**	0.50	**0.50**	m2	0.54

PAVING/FENCING/SITE FURNITURE

Labour hourly rates: (except Specialists) Craft Operatives £9.23 Labourer £7.02 Rates are national average prices. Refer to REGIONAL VARIATIONS for indicative levels of overall pricing in regions	MATERIALS			LABOUR				RATES		
	Del to Site £	Waste %	Material Cost £	Craft Optve Hrs	Lab Hrs	Labour Cost £	Sunds £	Nett Rate £	Unit	Gross Rate (+7.5%) £
Q20: GRANULAR SUB-BASES TO ROADS/PAVINGS Cont.										
MOT Type 1 to be obtained off site; wheeling not exceeding 50m - by machine Cont.										
Compacting with 680 kg vibratory roller filling; blinding with sand, ashes or similar fine material	0.18	40.00	0.25	-	0.05	0.35	0.18	0.78	m2	0.84
Compacting with 6 - 8 tonnes smooth wheeled roller filling; blinding with sand, ashes or similar fine material	0.18	40.00	0.25	-	0.05	0.35	0.35	0.95	m2	1.02
MOT Type 2 to be obtained off site; wheeling not exceeding 25m - by machine										
Filling to make up levels; depositing in layers 150mm maximum thickness										
average thickness not exceeding 0.25m	14.20	25.00	17.75	-	-	-	2.90	20.65	m3	22.20
average thickness exceeding 0.25m	14.20	25.00	17.75	-	-	-	2.40	20.15	m3	21.66
Surface packing to filling to vertical or battered faces	-	-	-	-	-	-	0.50	0.50	m2	0.54
Compacting with 680 kg vibratory roller filling; blinding with sand, ashes or similar fine material	0.18	40.00	0.25	-	0.05	0.35	0.18	0.78	m2	0.84
Compacting with 6 - 8 tonnes smooth wheeled roller filling; blinding with sand, ashes or similar fine material	0.18	40.00	0.25	-	0.05	0.35	0.35	0.95	m2	1.02
MOT Type 2 to be obtained off site; wheeling not exceeding 50m - by machine										
Filling to make up levels; depositing in layers 150mm maximum thickness										
average thickness not exceeding 0.25m	14.20	25.00	17.75	-	-	-	3.70	21.45	m3	23.06
average thickness exceeding 0.25m	14.20	25.00	17.75	-	-	-	2.90	20.65	m3	22.20
Surface packing to filling to vertical or battered faces	-	-	-	-	-	-	0.50	0.50	m2	0.54
Compacting with 680 kg vibratory roller filling; blinding with sand, ashes or similar fine material	0.18	40.00	0.25	-	0.05	0.35	0.18	0.78	m2	0.84
Compacting with 6 - 8 tonnes smooth wheeled roller filling; blinding with sand, ashes or similar fine material	0.18	40.00	0.25	-	0.05	0.35	0.35	0.95	m2	1.02
Ashes to be obtained off site; wheeling not exceeding 25m - by machine										
Filling to make up levels; depositing in layers 150mm maximum thickness										
average thickness not exceeding 0.25m	12.50	40.00	17.50	-	-	-	2.90	20.40	m3	21.93
Compacting with 680 kg vibratory roller filling	-	-	-	-	-	-	0.18	0.18	m2	0.19
Compacting with 6 - 8 tonnes smooth wheeled roller filling	-	-	-	-	-	-	0.35	0.35	m2	0.38
Ashes to be obtained off site; wheeling not exceeding 50m - by machine										
Filling to make up levels; depositing in layers 150mm maximum thickness										
average thickness not exceeding 0.25m	12.50	40.00	17.50	-	-	-	3.70	21.20	m3	22.79
Compacting with 680 kg vibratory roller filling	-	-	-	-	-	-	0.18	0.18	m2	0.19
Compacting with 6 - 8 tonnes smooth wheeled roller filling	-	-	-	-	-	-	0.35	0.35	m2	0.38
Hoggin to be obtained off site; wheeling not exceeding 25m - by machine										
Filling to make up levels; depositing in layers 150mm maximum thickness										
average thickness not exceeding 0.25m	8.00	33.00	10.64	-	-	-	2.90	13.54	m3	14.56
average thickness exceeding 0.25m	8.00	33.00	10.64	-	-	-	2.40	13.04	m3	14.02
Compacting with 680 kg vibratory roller filling	-	-	-	-	-	-	0.18	0.18	m2	0.19
Compacting with 6 - 8 tonnes smooth wheeled roller filling	-	-	-	-	-	-	0.35	0.35	m2	0.38

Labour hourly rates: (except Specialists) Craft Operatives £9.23 Labourer £7.02 Rates are national average prices. Refer to REGIONAL VARIATIONS for indicative levels of overall pricing in regions	MATERIALS			LABOUR				RATES		
	Del to Site £	Waste %	Material Cost £	Craft Optve Hrs	Lab Hrs	Labour Cost £	Sunds £	Nett Rate £	Unit	Gross Rate (+7.5%) £
Q20: GRANULAR SUB-BASES TO ROADS/PAVINGS Cont.										
Hoggin to be obtained off site; wheeling not exceeding 50m - by machine										
Filling to make up levels; depositing in layers 150mm maximum thickness										
average thickness not exceeding 0.25m	8.00	33.00	10.64	-	-	-	3.70	14.34	m3	15.42
average thickness exceeding 0.25m	8.00	33.00	10.64	-	-	-	2.90	13.54	m3	14.56
Compacting with 680 kg vibratory roller										
filling...	-	-	-	-	-	-	0.18	0.18	m2	0.19
Compacting with 6 - 8 tonnes smooth wheeled roller										
filling...	-	-	-	-	-	-	0.35	0.35	m2	0.38
Hard, dry, broken brick or stone to be obtained off site; wheeling not exceeding 25m - by hand										
Filling to make up levels; depositing in layers 150mm maximum thickness										
average thickness not exceeding 0.25m	12.40	25.00	15.50	-	1.50	10.53	-	26.03	m3	27.98
average thickness exceeding 0.25m	12.40	25.00	15.50	-	1.25	8.78	-	24.27	m3	26.10
Compacting with 680 kg vibratory roller										
filling; blinding with sand, ashes or similar fine material	0.18	40.00	0.25	-	0.05	0.35	0.18	0.78	m2	0.84
Compacting with 6 - 8 tonnes smooth wheeled roller										
filling; blinding with sand, ashes or similar fine material	0.18	40.00	0.25	-	0.05	0.35	0.35	0.95	m2	1.02
Hard, dry, broken brick or stone to be obtained off site; wheeling not exceeding 50m - by hand										
Filling to make up levels; depositing in layers 150mm maximum thickness										
average thickness not exceeding 0.25m	12.40	25.00	15.50	-	1.75	12.29	-	27.79	m3	29.87
average thickness exceeding 0.25m	12.40	25.00	15.50	-	1.50	10.53	-	26.03	m3	27.98
Compacting with 680 kg vibratory roller										
filling; blinding with sand, ashes or similar fine material	0.18	40.00	0.25	-	0.05	0.35	0.18	0.78	m2	0.84
Compacting with 6 - 8 tonnes smooth wheeled roller										
filling; blinding with sand, ashes or similar fine material	0.18	40.00	0.25	-	0.05	0.35	0.35	0.95	m2	1.02
MOT Type 1 to be obtained off site; wheeling not exceeding 25m - by hand										
Filling to make up levels; depositing in layers 150mm maximum thickness										
average thickness not exceeding 0.25m	14.50	25.00	18.13	-	1.50	10.53	-	28.66	m3	30.80
average thickness exceeding 0.25m	14.50	25.00	18.13	-	1.25	8.78	-	26.90	m3	28.92
Compacting with 680 kg vibratory roller										
filling; blinding with sand, ashes or similar fine material	0.18	40.00	0.25	-	0.05	0.35	0.18	0.78	m2	0.84
Compacting with 6 - 8 tonnes smooth wheeled roller										
filling; blinding with sand, ashes or similar fine material	0.18	40.00	0.25	-	0.05	0.35	0.35	0.95	m2	1.02
MOT Type 1 to be obtained off site; wheeling not exceeding 50m - by hand										
Filling to make up levels; depositing in layers 150mm maximum thickness										
average thickness not exceeding 0.25m	14.50	25.00	18.13	-	1.75	12.29	-	30.41	m3	32.69
average thickness exceeding 0.25m	14.50	25.00	18.13	-	1.50	10.53	-	28.66	m3	30.80
Compacting with 680 kg vibratory roller										
filling; blinding with sand, ashes or similar fine material	0.18	40.00	0.25	-	0.05	0.35	0.18	0.78	m2	0.84
Compacting with 6 - 8 tonnes smooth wheeled roller										
filling; blinding with sand, ashes or similar fine material	0.18	40.00	0.25	-	0.05	0.35	0.35	0.95	m2	1.02
MOT Type 2 to be obtained off site; wheeling not exceeding 25m - by hand										
Filling to make up levels; depositing in layers 150mm maximum thickness										
average thickness not exceeding 0.25m	14.20	25.00	17.75	-	1.50	10.53	-	28.28	m3	30.40
average thickness exceeding 0.25m	14.20	25.00	17.75	-	1.25	8.78	-	26.52	m3	28.51
Compacting with 680 kg vibratory roller										
filling; blinding with sand, ashes or similar fine material	0.18	40.00	0.25	-	0.05	0.35	0.18	0.78	m2	0.84
Compacting with 6 - 8 tonnes smooth wheeled roller										
filling; blinding with sand, ashes or similar fine material	0.18	40.00	0.25	-	0.05	0.35	0.35	0.95	m2	1.02

Labour hourly rates: (except Specialists) Craft Operatives £9.23 Labourer £7.02 Rates are national average prices. Refer to REGIONAL VARIATIONS for indicative levels of overall pricing in regions	MATERIALS			LABOUR				RATES		
	Del to Site £	Waste %	Material Cost £	Craft Optve Hrs	Lab Hrs	Labour Cost £	Sunds £	Nett Rate £	Unit	Gross Rate (+7.5%) £
Q20: GRANULAR SUB-BASES TO ROADS/PAVINGS Cont.										
MOT Type 2 to be obtained off site; wheeling not exceeding 50m - by hand										
Filling to make up levels; depositing in layers 150mm maximum thickness										
average thickness not exceeding 0.25m	13.52	25.00	**16.90**	-	1.75	**12.29**	-	**29.18**	m3	31.37
average thickness exceeding 0.25m	13.52	25.00	**16.90**	-	1.50	**10.53**	-	**27.43**	m3	29.49
Compacting with 680 kg vibratory roller										
filling; blinding with sand, ashes or similar fine material	0.18	40.00	**0.25**	-	0.05	**0.35**	0.18	**0.78**	m2	0.84
Compacting with 6 - 8 tonnes smooth wheeled roller										
filling; blinding with sand, ashes or similar fine material	0.18	40.00	**0.25**	-	0.05	**0.35**	0.35	**0.95**	m2	1.02
Ashes to be obtained off site; wheeling not exceeding 25m - by hand										
Filling to make up levels; depositing in layers 150mm maximum thickness										
average thickness not exceeding 0.25m	12.50	40.00	**17.50**	-	1.60	**11.23**	-	**28.73**	m3	30.89
Compacting with 680 kg vibratory roller										
filling	-	-	**-**	-	-	**-**	0.18	**0.18**	m2	0.19
Compacting with 6 - 8 tonnes smooth wheeled roller										
filling	-	-	**-**	-	-	**-**	0.35	**0.35**	m2	0.38
Ashes to be obtained off site; wheeling not exceeding 50m - by hand										
Filling to make up levels; depositing in layers 150mm maximum thickness										
average thickness not exceeding 0.25m	12.50	40.00	**17.50**	-	1.85	**12.99**	-	**30.49**	m3	32.77
Compacting with 680 kg vibratory roller										
filling	-	-	**-**	-	-	**-**	0.18	**0.18**	m2	0.19
Compacting with 6 - 8 tonnes smooth wheeled roller										
filling	-	-	**-**	-	-	**-**	0.35	**0.35**	m2	0.38
Hoggin to be obtained off site; wheeling not exceeding 25m - by hand										
Filling to make up levels; depositing in layers 150mm maximum thickness										
average thickness not exceeding 0.25m	8.00	33.00	**10.64**	-	1.50	**10.53**	-	**21.17**	m3	22.76
average thickness exceeding 0.25m	8.00	33.00	**10.64**	-	1.25	**8.78**	-	**19.41**	m3	20.87
Compacting with 680 kg vibratory roller										
filling	-	-	**-**	-	-	**-**	0.18	**0.18**	m2	0.19
Compacting with 6 - 8 tonnes smooth wheeled roller										
filling	-	-	**-**	-	-	**-**	0.35	**0.35**	m2	0.38
Hoggin to be obtained off site; wheeling not exceeding 50m - by hand										
Filling to make up levels; depositing in layers 150mm maximum thickness										
average thickness not exceeding 0.25m	8.00	33.00	**10.64**	-	1.75	**12.29**	-	**22.93**	m3	24.64
average thickness exceeding 0.25m	8.00	33.00	**10.64**	-	1.50	**10.53**	-	**21.17**	m3	22.76
Compacting with 680 kg vibratory roller										
filling	-	-	**-**	-	-	**-**	0.18	**0.18**	m2	0.19
Compacting with 6 - 8 tonnes smooth wheeled roller										
filling	-	-	**-**	-	-	**-**	0.35	**0.35**	m2	0.38
Q21: IN-SITU CONCRETE ROADS/PAVINGS										
Plain in-situ concrete; mix 1:8, all in aggregate										
Beds; poured on or against earth or unblinded hardcore										
thickness not exceeding 150mm	56.91	7.50	**61.18**	-	4.25	**29.84**	1.50	**92.51**	m3	99.45
Plain in-situ concrete; mix 1:6; all in aggregate										
Beds; poured on or against earth or unblinded hardcore										
thickness 150 - 450mm	63.35	7.50	**68.10**	-	4.00	**28.08**	1.50	**97.68**	m3	105.01
thickness not exceeding 150mm	63.35	7.50	**68.10**	-	4.25	**29.84**	1.50	**99.44**	m3	106.89
Plain in-situ concrete; B.S.5328, ordinary prescribed mix ST4, 20mm aggregate										
Beds										
thickness not exceeding 150mm	49.60	5.00	**52.08**	Hrs	4.25	**29.84**	1.50	**83.42**	m3	89.67
Beds; poured on or against earth or unblinded hardcore										
thickness not exceeding 150mm	49.60	7.50	**53.32**	-	4.25	**29.84**	1.50	**84.66**	m3	91.00

Labour hourly rates: (except Specialists) Craft Operatives £9.23 Labourer £7.02 Rates are national average prices. Refer to REGIONAL VARIATIONS for indicative levels of overall pricing in regions	MATERIALS			LABOUR				RATES		
	Del to Site £	Waste %	Material Cost £	Craft Optve Hrs	Lab Hrs	Labour Cost £	Sunds £	Nett Rate £	Unit	Gross Rate (+7.5%) £
Q21: IN-SITU CONCRETE ROADS/PAVINGS Cont.										
Reinforced in-situ concrete; B.S.5328, designed mix C20, 20mm aggregate, minimum cement content 240 kg/m3; vibrated										
Beds										
thickness 150 - 450mm	49.20	5.00	**51.66**	-	4.25	**29.84**	0.50	**82.00**	m3	88.14
thickness not exceeding 150mm	49.20	5.00	**51.66**	-	4.50	**31.59**	0.50	**83.75**	m3	90.03
Formwork and basic finish										
Edges of beds										
height not exceeding 250mm	0.89	12.50	**1.00**	0.53	0.11	**5.66**	0.22	**6.89**	m	7.40
height 250 - 500mm	1.57	12.50	**1.77**	0.96	0.19	**10.19**	0.41	**12.37**	m	13.30
Formwork steel forms										
Edges of beds										
height not exceeding 250mm	0.55	15.00	**0.63**	0.36	0.07	**3.81**	0.22	**4.67**	m	5.02
height 250 - 500mm	0.90	15.00	**1.03**	0.64	0.13	**6.82**	0.41	**8.26**	m	8.88
Reinforcement fabric; B.S.4483, hard drawn plain round steel; welded; including laps, tying wire, all cutting and bending, and spacers and chairs which are at the discretion of the Contractor										
Reference C283, 2.61 kg/m2; 100mm side laps; 400mm end laps										
generally	0.92	15.00	**1.06**	0.05	0.02	**0.60**	0.07	**1.73**	m2	1.86
Reference C385, 3.41 kg/m2; 100mm side laps; 400mm end laps										
generally	1.20	15.00	**1.38**	0.05	0.02	**0.60**	0.09	**2.07**	m2	2.23
Reference C503, 4.34 kg/m2; 100mm side laps; 400mm end laps										
generally	1.48	15.00	**1.70**	0.06	0.02	**0.69**	0.11	**2.51**	m2	2.69
Reference C636, 5.55 kg/m2; 100mm side laps; 400mm end laps										
generally	1.97	15.00	**2.27**	0.06	0.02	**0.69**	0.14	**3.10**	m2	3.33
Reference C785, 6.72 kg/m2; 100mm side laps; 400mm end laps										
generally	2.38	15.00	**2.74**	0.07	0.02	**0.79**	0.17	**3.69**	m2	3.97
Formed joints										
Sealant to joint; Grace Construction Products. ' Servimastic 96' including preparation, cleaners, primers and sealers										
10 x 25mm	2.44	5.00	**2.56**	-	0.10	**0.70**	-	**3.26**	m	3.51
12.5 x 25mm	3.04	5.00	**3.19**	-	0.11	**0.77**	-	**3.96**	m	4.26
Worked finish on in-situ concrete										
Tamping by mechanical means										
level surfaces	-	-	-	-	0.15	**1.05**	0.27	**1.32**	m2	1.42
sloping surfaces	-	-	-	-	0.30	**2.11**	0.27	**2.38**	m2	2.55
surfaces to falls	-	-	-	-	0.40	**2.81**	0.34	**3.15**	m2	3.38
Trowelling										
level surfaces	-	-	-	-	0.33	**2.32**	-	**2.32**	m2	2.49
sloping surfaces	-	-	-	-	0.40	**2.81**	-	**2.81**	m2	3.02
surfaces to falls	-	-	-	-	0.50	**3.51**	-	**3.51**	m2	3.77
Rolling with an indenting roller										
level surfaces	-	-	-	-	0.10	**0.70**	0.27	**0.97**	m2	1.04
sloping surfaces	-	-	-	-	0.15	**1.05**	0.27	**1.32**	m2	1.42
surfaces to falls	-	-	-	-	0.30	**2.11**	0.34	**2.45**	m2	2.63
Accessories cast into in-situ concrete										
Mild steel dowels; half coated with bitumen										
16mm diameter x 400mm long	0.22	10.00	**0.24**	-	0.06	**0.42**	0.18	**0.84**	nr	0.91
20mm diameter x 500mm long; with plastic compression cap	0.43	10.00	**0.47**	-	0.08	**0.56**	0.18	**1.21**	nr	1.31
Q22: COATED MACADAM/ASPHALT ROADS/PAVINGS										
Coated macadam, B.S.4987; base course 28mm nominal size aggregate, 50mm thick; wearing course 6mm nominal size aggregate, 20mm thick; rolled with 3 - 4 tonne roller; external										
70mm roads on concrete base; to falls and crossfalls, and slopes not exceeding 15 degrees from horizontal										
generally	-	-	**Spclist**	-	-	**Spclist**	-	**12.62**	m2	13.57
70mm pavings on concrete base; to falls and crossfalls, and slopes not exceeding 15 degrees from horizontal										
generally	-	-	**Spclist**	-	-	**Spclist**	-	**14.02**	m2	15.07

Labour hourly rates: (except Specialists) Craft Operatives £9.23 Labourer £7.02 Rates are national average prices. Refer to REGIONAL VARIATIONS for indicative levels of overall pricing in regions	MATERIALS			LABOUR				RATES		
	Del to Site £	Waste %	Material Cost £	Craft Optve Hrs	Lab Hrs	Labour Cost £	Sunds £	Nett Rate £	Unit	Gross Rate (+7.5%) £
Q22: COATED MACADAM/ASPHALT ROADS/PAVINGS Cont.										
Coated macadam, B.S.4987; base course 28mm nominal size aggregate, 50mm thick; wearing course 6mm nominal size aggregate, 20mm thick; rolled with 3 - 4 tonne roller; dressing surface with coated grit brushed on and lightly rolled; external										
70mm roads on concrete base; to falls and crossfalls, and slopes not exceeding 15 degrees from horizontal generally...	-	-	Spclist	-	-	Spclist	-	15.29	m2	16.44
70mm pavings on concrete base; to falls and crossfalls, and slopes not exceeding 15 degrees from horizontal generally...	-	-	Spclist	-	-	Spclist	-	16.67	m2	17.92
Coated macadam, B.S.4987; base course 40mm nominal size aggregate, 65mm thick; wearing course 10mm nominal size aggregate, 25mm thick; rolled with 6 - 8 tonne roller; external										
90mm roads on concrete base; to falls and crossfalls, and slopes not exceeding 15 degrees from horizontal generally...	-	-	Spclist	-	-	Spclist	-	15.29	m2	16.44
90mm pavings on concrete base; to falls and crossfalls, and slopes not exceeding 15 degrees from horizontal generally...	-	-	Spclist	-	-	Spclist	-	16.78	m2	18.04
Coated macadam, B.S.4987; base course 40mm nominal size aggregate, 65mm thick; wearing course 10mm nominal size aggregate, 25mm thick; rolled with 6 - 8 tonne roller; dressing surface with coated grit brushed on and lightly rolled; external										
90mm roads on concrete base; to falls and crossfalls, and slopes not exceeding 15 degrees from horizontal generally...	-	-	Spclist	-	-	Spclist	-	18.22	m2	19.59
90mm pavings on concrete base; to falls and crossfalls, and slopes not exceeding 15 degrees from horizontal generally...	-	-	Spclist	-	-	Spclist	-	18.99	m2	20.41
Fine cold asphalt, B.S.4987; single course 6mm nominal size aggregate; rolled with 3 - 4 tonne roller; external										
12mm roads on concrete base; to falls and crossfalls, and slopes not exceeding 15 degrees from horizontal generally...	-	-	Spclist	-	-	Spclist	-	5.75	m2	6.18
19mm roads on concrete base; to falls and crossfalls, and slopes not exceeding 15 degrees from horizontal generally...	-	-	Spclist	-	-	Spclist	-	6.71	m2	7.21
25mm roads on concrete base; to falls and crossfalls, and slopes not exceeding 15 degrees from horizontal generally...	-	-	Spclist	-	-	Spclist	-	7.90	m2	8.49
12mm pavings on concrete base; to falls and crossfalls, and slopes not exceeding 15 degrees from horizontal generally...	-	-	Spclist	-	-	Spclist	-	6.10	m2	6.56
19mm pavings on concrete base; to falls and crossfalls, and slopes not exceeding 15 degrees from horizontal generally...	-	-	Spclist	-	-	Spclist	-	7.66	m2	8.23
25mm pavings on concrete base; to falls and crossfalls, and slopes not exceeding 15 degrees from horizontal generally...	-	-	Spclist	-	-	Spclist	-	8.73	m2	9.38
GRAVEL/HOGGIN/WOODCHIP ROADS/PAVINGS										
Gravel paving; first layer clinker 75mm thick; intermediate layer coarse gravel 75mm thick; wearing layer blinding gravel 38mm thick; well water and roll										
188mm roads on blinded hardcore base; to falls and crossfalls and slopes not exceeding 15 degrees from horizontal generally...	4.60	10.00	5.06	-	0.44	3.09	-	8.15	m2	8.76
Gravel paving; single layer blinding gravel; well water and roll										
50mm pavings on blinded hardcore base; to falls and crossfalls and slopes not exceeding 15 degrees from horizontal generally...	1.63	10.00	1.79	-	0.13	0.91	-	2.71	m2	2.91

Labour hourly rates: (except Specialists) Craft Operatives £9.23 Labourer £7.02 Rates are national average prices. Refer to REGIONAL VARIATIONS for indicative levels of overall pricing in regions	MATERIALS			LABOUR				RATES		
	Del to Site £	Waste %	Material Cost £	Craft Optve Hrs	Lab Hrs	Labour Cost £	Sunds £	Nett Rate £	Unit	Gross Rate (+7.5%) £
GRAVEL/HOGGIN/WOODCHIP ROADS/PAVINGS Cont.										
Gravel paving; first layer clinker 50mm thick; **wearing layer blinding gravel 38mm thick; well water** **and roll**										
88mm pavings on blinded hardcore base; to falls and crossfalls and slopes not exceeding 15 degrees from horizontal										
generally..	1.95	10.00	2.15	–	0.24	1.68	–	3.83	m2	4.12
extra; 'Colas' and 10mm shingle dressing; one coat	1.12	5.00	1.18	–	0.15	1.05	–	2.23	m2	2.40
extra; 'Colas' and 10mm shingle dressing; two coats	1.92	5.00	2.02	–	0.25	1.75	–	3.77	m2	4.05
Q25: SLAB/BRICK/BLOCK/SETT/COBBLE PAVINGS										
York stone paving; 13mm thick bedding, jointing and **pointing in cement mortar (1:3)**										
Paving on blinded hardcore base										
50mm thick; to falls and crossfalls and to slopes not exceeding 15 degrees from horizontal	49.62	5.00	52.10	0.27	0.96	9.23	1.02	62.35	m2	67.03
75mm thick; to falls and crossfalls and to slopes not exceeding 15 degrees from horizontal	59.00	5.00	61.95	0.33	0.33	5.36	1.08	68.39	m2	73.52
extra; rubbed top surface	14.90	5.00	15.65	–	–	–	–	15.65	m2	16.82
Treads on concrete base										
50mm thick; 300mm wide	57.11	5.00	59.97	0.33	0.33	5.36	0.27	65.60	m	70.52
Precast concrete flags, B.S.7263 Part 1, natural **finish; 6mm joints, symmetrical layout; bedding in** **13mm cement mortar (1:3); jointing and pointing with** **lime and sand (1:2)**										
600 x 450 x 50mm units to pavings on sand, granular or blinded hardcore base 50mm thick; to falls and crossfalls and to slopes not exceeding 15 degrees from horizontal	6.65	5.00	6.98	0.27	0.27	4.39	1.08	12.45	m2	13.38
600 x 450 x 50mm units to pavings on sand, granular or blinded hardcore base Extra over for red or buff coloured	2.81	5.00	2.95	–	–	–	–	2.95	m2	3.17
600 x 600 x 50mm units to pavings on sand, granular or blinded hardcore base 50mm thick; to falls and crossfalls and to slopes not exceeding 15 degrees from horizontal	5.53	5.00	5.81	0.25	0.25	4.06	1.01	10.88	m2	11.69
600 x 600 x 50mm units to pavings on sand, granular or blinded hardcore base Extra over for red or buff coloured	2.53	5.00	2.66	–	–	–	–	2.66	m2	2.86
600 x 750 x 50mm units to pavings on sand, granular or blinded hardcore base 50mm thick; to falls and crossfalls and to slopes not exceeding 15 degrees from horizontal	5.28	5.00	5.54	0.23	0.23	3.74	1.03	10.31	m2	11.08
600 x 750 x 50mm units to pavings on sand, granular or blinded hardcore base Extra over for red or buff coloured	2.24	5.00	2.35	–	–	–	–	2.35	m2	2.53
600 x 900 x 50mm units to pavings on sand, granular or blinded hardcore base 50mm thick; to falls and crossfalls and to slopes not exceeding 15 degrees from horizontal	4.93	5.00	5.18	0.21	0.21	3.41	1.04	9.63	m2	10.35
600 x 900 x 50mm units to pavings on sand, granular or blinded hardcore base Extra over for red or buff coloured	1.87	5.00	1.96	–	–	–	–	1.96	m2	2.11
600 x 450 x 63mm units to pavings on sand, granular or blinded hardcore base 63mm thick; to falls and crossfalls and to slopes not exceeding 15 degrees from horizontal	7.60	5.00	7.98	0.27	0.27	4.39	1.06	13.43	m2	14.43
600 x 600 x 63mm units to pavings on sand, granular or blinded hardcore base 63mm thick; to falls and crossfalls and to slopes not exceeding 15 degrees from horizontal	6.42	5.00	6.74	0.25	0.25	4.06	1.00	11.80	m2	12.69
600 x 750 x 63mm units to pavings on sand , granular or blinded hardcore base 63mm thick; to falls and crossfalls and to slopes not exceeding 15 degrees from horizontal	5.90	5.00	6.20	0.23	0.23	3.74	1.01	10.94	m2	11.76
600 x 900 x 63mm units to pavings on sand, granular or blindcd hardcore base 63mm thick; to falls and crossfalls and to slopes not exceeding 15 degrees from horizontal	5.45	5.00	5.72	0.21	0.21	3.41	1.02	10.15	m2	10.92

PAVING/FENCING/SITE FURNITURE

Labour hourly rates: (except Specialists) Craft Operatives £9.23 Labourer £7.02 Rates are national average prices. Refer to REGIONAL VARIATIONS for indicative levels of overall pricing in regions	MATERIALS			LABOUR				RATES		
	Del to Site £	Waste %	Material Cost £	Craft Optve Hrs	Lab Hrs	Labour Cost £	Sunds £	Nett Rate £	Unit	Gross Rate (+7.5%) £
Q25: SLAB/BRICK/BLOCK/SETT/COBBLE PAVINGS Cont.										
Precast concrete flags, B.S.7263 Part 1, natural finish; 6mm joints, symmetrical layout; bedding in 13mm lime mortar (1:3); jointing and pointing with lime and sand (1:2)										
600 x 450 x 50mm units to pavings on sand, granular or blinded hardcore base										
50mm thick; to falls and crossfalls and to slopes not exceeding 15 degrees from horizontal	6.04	5.00	**6.34**	0.27	0.27	**4.39**	0.67	**11.40**	m2	**12.25**
601 x 450 x 50mm units to pavings on sand, granular or blinded hardcore base										
Extra over for red or buff coloured	2.81	5.00	**2.95**	-	-	**-**	-	**2.95**	m2	**3.17**
600 x 600 x 50mm units to pavings on sand, granular or blinded hardcore base										
50mm thick; to falls and crossfalls and to slopes not exceeding 15 degrees from horizontal	4.94	5.00	**5.19**	0.25	0.25	**4.06**	0.60	**9.85**	m2	**10.59**
600 x 600 x 50mm units to pavings on sand, granular or blinded hardcore base										
Extra over for red or buff coloured	2.53	5.00	**2.66**	-	-	**-**	-	**2.66**	m2	**2.86**
600 x 750 x 50mm units to pavings on sand, granular or blinded hardcore base										
50mm thick; to falls and crossfalls and to slopes not exceeding 15 degrees from horizontal	4.67	5.00	**4.90**	0.23	0.23	**3.74**	0.61	**9.25**	m2	**9.94**
600 x 750 x 50mm units to pavings on sand, granular or blinded hardcore base										
Extra over for red or buff coloured	2.24	5.00	**2.35**	-	-	**-**	-	**2.35**	m2	**2.53**
600 x 900 x 50mm units to pavings on sand, granular or blinded hardcore base										
50mm thick; to falls and crossfalls and to slopes not exceeding 15 degrees from horizontal	4.32	5.00	**4.54**	0.21	0.21	**3.41**	0.62	**8.57**	m2	**9.21**
600 x 900 x 50mm units to pavings on sand, granular or blinded hardcore base										
Extra over for red or buff coloured	1.87	5.00	**1.96**	-	-	**-**	-	**1.96**	m2	**2.11**
600 x 450 x 63mm units to pavings on sand, granular or blinded hardcore base										
63mm thick; to falls and crossfalls and to slopes not exceeding 15 degrees from horizontal	7.60	5.00	**7.98**	0.27	0.27	**4.39**	1.06	**13.43**	m2	**14.43**
600 x 600 x 63mm units to pavings on sand, granular or blinded hardcore base										
63mm thick; to falls and crossfalls and to slopes not exceeding 15 degrees from horizontal	6.42	5.00	**6.74**	0.25	0.25	**4.06**	1.00	**11.80**	m2	**12.69**
600 x 750 x 63mm units to pavings on sand, granular or blinded hardcore base										
63mm thick; to falls and crossfalls and to slopes not exceeding 15 degrees from horizontal	5.90	5.00	**6.20**	0.23	0.23	**3.74**	1.01	**10.94**	m2	**11.76**
600 x 900 x 63mm units to pavings on sand , granular or blinded hardcore base										
63mm thick; to falls and crossfalls and to slopes not exceeding 15 degrees from horizontal	5.45	5.00	**5.72**	0.21	0.21	**3.41**	1.02	**10.15**	m2	**10.92**
Precast concrete flags, B.S.7263 Part 1, natural finish; 6mm joints, symmetrical layout; bedding in 25mm sand; jointing and pointing with lime and sand (1:2)										
600 x 450 x 50mm units to pavings on sand, granular or blinded hardcore base										
50mm thick; to falls and crossfalls and to slopes not exceeding 15 degrees from horizontal	6.04	5.00	**6.34**	0.27	0.27	**4.39**	0.63	**11.36**	m2	**12.21**
600 x 450 x 50mm units to pavings on sand, granular or blinded hardcore base										
Extra over for red or buff coloured	2.81	5.00	**2.95**	-	-	**-**	-	**2.95**	m2	**3.17**
600 x 600 x 50mm units to pavings on sand, granular or blinded hardcore base										
50mm thick; to falls and crossfalls and to slopes not exceeding 15 degrees from horizontal	4.92	5.00	**5.17**	0.25	0.25	**4.06**	0.54	**9.77**	m2	**10.50**
600 x 600 x 50mm units to pavings on sand, granular or blinded hardcore base										
Extra over for red or buff coloured	2.53	5.00	**2.66**	-	-	**-**	-	**2.66**	m2	**2.86**
600 x 750 x 50mm units to pavings on sand, granular or blinded hardcore base										
50mm thick; to falls and crossfalls and to slopes not exceeding 15 degrees from horizontal	4.67	5.00	**4.90**	0.23	0.23	**3.74**	0.55	**9.19**	m2	**9.88**
600 x 750 x 50mm units to pavings on sand, granular or blinded hardcore base										
Extra over for red or buff coloured	2.24	5.00	**2.35**	-	-	**-**	-	**2.35**	m2	**2.53**

Labour hourly rates: (except Specialists) Craft Operatives £9.23 Labourer £7.02 Rates are national average prices. Refer to REGIONAL VARIATIONS for indicative levels of overall pricing in regions	MATERIALS			LABOUR				RATES		
	Del to Site £	Waste %	Material Cost £	Craft Optve Hrs	Lab Hrs	Labour Cost £	Sunds £	Nett Rate £	Unit	Gross Rate (+7.5%) £
Q25: SLAB/BRICK/BLOCK/SETT/COBBLE PAVINGS Cont.										
Precast concrete flags, B.S.7263 Part 1, natural finish; 6mm joints, symmetrical layout; bedding in 25mm sand; jointing and pointing with lime and sand (1:2) Cont.										
600 x 900 x 50mm units to pavings on sand, granular or blinded hardcore base										
50mm thick; to falls and crossfalls and to slopes not exceeding 15 degrees from horizontal	4.32	5.00	4.54	0.21	0.21	3.41	0.56	8.51	m2	9.15
600 x 900 x 50mm units to pavings on sand, granular or blinded hardcore base										
Extra over for red or buff coloured	1.87	5.00	1.96	-	-	-	-	1.96	m2	2.11
Keyblok precast concrete paving blocks; Marshalls Mono Ltd Driveline 50; in standard units; bedding on sand; covering with sand, compacting with plate vibrator, sweeping off surplus										
200 x 100mm units to pavings on 50mm sand base; natural colour										
symmetrical half bond layout; level and to falls only										
50mm thick...........................	7.50	10.00	8.25	-	1.05	7.37	1.40	17.02	m2	18.30
laid in straight herringbone pattern; level and to falls only										
50mm thick...................................	7.50	10.00	8.25	-	1.15	8.07	1.40	17.72	m2	19.05
Brick paving on concrete base; 13mm thick bedding, jointing and pointing in cement mortar (1:3)										
Paving to falls and crossfalls and to slopes not exceeding 15 degrees from horizontal										
25mm thick Brick paviors P.C. £300.00 per 1000 ...	18.00	5.00	18.90	2.50	1.40	32.90	1.01	52.81	m2	56.77
65mm thick Facing bricks P.C. £210.00 per 1000 ...	12.60	5.00	13.23	2.50	1.40	32.90	1.01	47.14	m2	50.68
50mm thick Staffordshire blue chequered paviors P.C. £300.00 per 1000	18.00	5.00	18.90	2.50	1.40	32.90	0.96	52.76	m2	56.72
50mm thick Accrington non slip paviors P.C. £300.00 per 1000	18.00	5.00	18.90	2.50	1.40	32.90	0.96	52.76	m2	56.72
Granite sett paving on concrete base; 13mm thick bedding and grouting in cement mortar (1:3)										
Paving to falls and crossfalls and to slopes not exceeding 15 degrees from horizontal										
125mm thick; new	28.09	2.50	28.79	1.65	2.00	29.27	2.23	60.29	m2	64.81
150mm thick; new	34.41	2.50	35.27	1.80	2.15	31.71	2.55	69.53	m2	74.74
100mm thick; reclaimed	31.13	2.50	31.91	1.50	1.80	26.48	1.91	60.30	m2	64.82
150mm thick; reclaimed	27.09	2.50	27.77	1.65	2.00	29.27	2.23	59.27	m2	63.71
200mm thick; reclaimed	45.14	2.50	46.27	1.80	2.15	31.71	2.55	80.53	m2	86.56
Cobble paving set in concrete bed (measured separately) tight butted, dry grouted with cement and sand (1:3) watered and brushed										
Paving level and to falls only										
plain...................................	12.00	2.50	12.30	1.65	1.65	26.81	2.55	41.66	m2	44.79
set to pattern	12.00	2.50	12.30	2.00	2.00	32.50	2.55	47.35	m2	50.90
Crazy paving on blinded hardcore base, bedding in 38mm thick sand; pointing in cement mortar (1:4)										
Paving to falls and crossfalls and to slopes not exceeding 15 degrees from horizontal										
50mm thick precast concrete flag	4.32	2.50	4.43	0.40	0.40	6.50	2.15	13.08	m2	14.06
38-50mm thick York stone	49.62	2.50	50.86	0.60	0.60	9.75	3.10	63.71	m2	68.49
38mm thick Westmorland green slate	96.50	2.50	98.91	0.60	0.60	9.75	2.15	110.81	m2	119.12
Grass concrete paving; voids filled with vegetable soil and sown with grass seed										
Precast concrete perforated slabs										
120mm thick units on 25mm thick sand bed; level and to falls only	18.20	2.50	18.66	0.28	0.28	4.55	1.22	24.43	m2	26.26
Q30: SEEDING/TURFING										
Cultivating										
Surfaces of natural ground										
digging over one spit deep; removing debris; weeding	-	-	-	-	0.25	1.75	-	1.75	m2	1.89
Surfaces of filling										
digging over one spit deep	-	-	-	-	0.30	2.11	-	2.11	m2	2.26
Surface applications										
Bone meal, 0.06 kg/m2; raking in										
general surfaces	0.21	5.00	0.22	-	0.05	0.35	-	0.57	m2	0.61
Selective weedkiller, 0.03 kg/m2; applying by spreader										
general surfaces	0.11	5.00	0.12	-	0.05	0.35	-	0.47	m2	0.50

PAVING/FENCING/SITE FURNITURE

Labour hourly rates: (except Specialists) Craft Operatives £9.23 Labourer £7.02 Rates are national average prices. Refer to REGIONAL VARIATIONS for indicative levels of overall pricing in regions	MATERIALS			LABOUR				RATES		
	Del to Site	Waste	Material Cost	Craft Optve	Lab	Labour Cost	Sunds	Nett Rate	Unit	Gross Rate (+7.5%)
	£	%	£	Hrs	Hrs	£	£	£		£
Q30: SEEDING/TURFING Cont.										
Surface applications Cont.										
Pre-seeding fertilizer at the rate of 0.05 kg/m2 general surfaces	0.07	5.00	0.07	-	0.05	0.35	-	0.42	m2	0.46
General grass fertilizer at the rate of 0.06 kg/m2 general surfaces	0.08	5.00	0.08	-	0.05	0.35	-	0.44	m2	0.47
Seeding										
Grass seed, 0.07 kg/m2, raking in; rolling; maintaining for 12 months after laying general surfaces	0.31	5.00	0.33	-	0.20	1.40	-	1.73	m2	1.86
Stone pick, roll and cut grass with a rotary cutter and remove arisings general surfaces	-	-	-	-	0.10	0.70	-	0.70	m2	0.75
Turfing										
Take turf from stack, wheel not exceeding 100m, lay, roll and water, maintaining for 12 months after laying general surfaces	-	-	-	-	0.50	3.51	-	3.51	m2	3.77
Imported turf; cultivated; lay, roll and water; maintaining for 12 months after laying general surfaces	2.45	5.00	2.57	-	0.35	2.46	-	5.03	m2	5.41
Q31: PLANTING										
Planting trees, shrubs and hedge plants										
Excavate or form pit, hole or trench, dig over ground in bottom, spread and pack around roots with finely broken soil, refill with top soil with one third by volume of farmyard manure incorporated, water in, remove surplus excavated material and provide labelling										
small tree	0.67	-	0.67	-	0.75	5.26	0.39	6.33	nr	6.80
medium tree	0.84	-	0.84	-	1.50	10.53	0.56	11.93	nr	12.82
large tree	1.06	-	1.06	-	3.00	21.06	0.72	22.84	nr	24.55
shrub	0.50	-	0.50	-	0.50	3.51	3.42	7.43	nr	7.99
hedge plant	0.50	-	0.50	-	0.25	1.75	1.71	3.96	nr	4.26
60mm diameter treated softwood tree stake, pointed and driven into ground and with two PVC tree ties secured around tree and nailed to stake										
2100mm long	2.84	-	2.84	-	0.33	2.32	1.45	6.61	nr	7.10
2400mm long	3.01	-	3.01	-	0.33	2.32	1.45	6.78	nr	7.28
2700mm long	3.33	-	3.33	-	0.33	2.32	1.45	7.10	nr	7.63
Planting herbaceous plants, bulbs, corms and tubers										
Provide planting bed of topsoil with one third by volume of farmyard manure incorporated										
150mm thick	4.12	5.00	4.33	-	0.75	5.26	-	9.59	m2	10.31
225mm thick	6.20	5.00	6.51	-	1.12	7.86	-	14.37	m2	15.45
Form hole and plant										
herbaceous plants	0.55	5.00	0.58	-	0.20	1.40	-	1.98	nr	2.13
bulbs, corms and tubers	0.12	5.00	0.13	-	0.10	0.70	-	0.83	nr	0.89
Mulching after planting										
25mm peat	0.88	-	0.88	-	0.10	0.70	-	1.58	m2	1.70
75mm farmyard manure	0.33	-	0.33	-	0.10	0.70	-	1.03	m2	1.11
Q40: FENCING										
POST AND WIRE FENCING										
Fencing; strained wire; B.S.1722 Part 3; concrete posts and struts; 4.00mm diameter galvanised mild steel line wires; galvanised steel fittings and accessories; backfilling around posts in concrete mix ST4										
900mm high fencing; type SC90; wires -3; posts with rounded tops										
posts at 2743mm centres; 600mm into ground	1.96	2.50	2.01	-	0.75	5.26	0.45	7.72	m	8.30
extra; end posts with struts -1	14.90	2.50	15.27	-	1.00	7.02	2.48	24.77	nr	26.63
extra; angle posts with struts -2	24.38	2.50	24.99	-	1.50	10.53	3.72	39.24	nr	42.18
1050mm high fencing; type SC105A; wires-5; posts rounded tops										
posts at 2743mm centres; 600mm into ground	2.39	2.50	2.45	-	0.80	5.62	0.45	8.52	m	9.15
extra; end posts with struts -1	19.32	2.50	19.80	-	1.10	7.72	2.48	30.00	nr	32.26
extra; angle posts with struts -2	29.66	2.50	30.40	-	1.65	11.58	3.72	45.70	nr	49.13

Labour hourly rates: (except Specialists) Craft Operatives £9.23 Labourer £7.02 Rates are national average prices. Refer to REGIONAL VARIATIONS for indicative levels of overall pricing in regions	MATERIALS			LABOUR				RATES		
	Del to Site £	Waste %	Material Cost £	Craft Optve Hrs	Lab Hrs	Labour Cost £	Sunds £	Nett Rate £	Unit	Gross Rate (+7.5%) £
CHAIN LINK FENCING										
Fencing; chain link; B.S.1722 Part 1; concrete posts; galvanised mesh, line and tying wires; galvanised steel fittings and accessories; backfilling around posts in concrete mix ST4										
900mm high fencing; type GLC 90; posts with rounded tops										
posts at 2743mm centres; 600mm into ground	3.65	5.00	3.83	-	1.00	7.02	0.45	11.30	m	12.15
extra; end posts with struts -1	14.90	5.00	15.65	-	1.20	8.42	2.48	26.55	nr	28.54
extra; angle posts with struts -2	24.38	5.00	25.60	-	1.60	11.23	3.72	40.55	nr	43.59
1200mm high fencing; type GLC 120; posts with rounded tops										
posts at 2743mm centres; 600mm into ground	4.43	5.00	4.65	-	1.10	7.72	0.45	12.82	m	13.79
extra; end posts with struts -1	20.48	5.00	21.50	-	1.40	9.83	2.48	33.81	nr	36.35
extra; angle posts with struts -2	32.77	5.00	34.41	-	1.80	12.64	3.72	50.76	nr	54.57
1400mm high fencing; type GLC 140A; posts with rounded tops										
posts at 2743mm centres; 600mm into ground	5.21	5.00	5.47	-	1.20	8.42	0.45	14.34	m	15.42
extra; end posts with struts -1	22.46	5.00	23.58	-	1.55	10.88	2.48	36.94	nr	39.71
extra; angle posts with struts -2	35.72	5.00	37.51	-	2.00	14.04	3.72	55.27	nr	59.41
1800mm high fencing; type GLC 180; posts with rounded tops										
posts at 2743mm centres; 600mm into ground	6.72	5.00	7.06	-	1.45	10.18	0.45	17.68	m	19.01
extra; end posts with struts -1	25.86	5.00	27.15	-	1.65	11.58	2.48	41.22	nr	44.31
extra; angle posts with struts -2	40.65	5.00	42.68	-	2.15	15.09	3.72	61.50	nr	66.11
Fencing; chain link; B.S.1722 Part 1; concrete posts; plastics coated Grade A mesh, line and tying wire; galvanised steel fittings and accessories; backfilling around posts in concrete mix ST4										
900mm high fencing; type PLC 90A; posts with rounded tops										
posts at 2743mm centres; 600mm into ground	4.00	5.00	4.20	-	1.00	7.02	0.45	11.67	m	12.55
extra; end posts with struts -1	14.90	5.00	15.65	-	1.20	8.42	2.48	26.55	nr	28.54
extra; angle posts with struts -2	24.38	5.00	25.60	-	1.60	11.23	3.72	40.55	nr	43.59
1200mm high fencing; type PLC 120; posts with rounded tops										
posts at 2743mm centres; 600mm into ground	5.04	5.00	5.29	-	1.10	7.72	0.45	13.46	m	14.47
extra; end posts with struts -1	20.48	5.00	21.50	-	1.40	9.83	2.48	33.81	nr	36.35
extra; angle posts with struts -2	32.77	5.00	34.41	-	1.80	12.64	3.72	50.76	nr	54.57
1400mm high fencing; type PLC 140A; posts with rounded tops										
posts at 2743mm centres; 600mm into ground	5.69	5.00	5.97	-	1.20	8.42	0.45	14.85	m	15.96
extra; end posts with struts -1	22.46	5.00	23.58	-	1.55	10.88	2.48	36.94	nr	39.71
extra; angle posts with struts -2	35.72	5.00	37.51	-	2.00	14.04	3.72	55.27	nr	59.41
1800mm high fencing; type PLC 180; posts with rounded tops										
posts at 2743mm centres; 600mm into ground	7.16	5.00	7.52	-	1.45	10.18	0.45	18.15	m	19.51
extra; end posts with struts -1	25.86	5.00	27.15	-	1.65	11.58	2.48	41.22	nr	44.31
extra; angle posts with struts -2	40.65	5.00	42.68	-	2.15	15.09	3.72	61.50	nr	66.11
Fencing; chain link; B.S.1722 Part 1; rolled steel angle posts; galvanised mesh, line and tying wires; galvanised steel fittings and accessories; backfilling around posts in concrete mix ST4										
900mm high fencing; type GLS 90; posts with rounded tops										
posts at 2743mm centres; 600mm into ground	3.43	2.50	3.52	-	0.65	4.56	0.54	8.62	m	9.27
extra; end posts with struts -1	17.84	2.50	18.29	-	1.00	7.02	2.98	28.29	nr	30.41
extra; angle posts with struts -2	25.14	2.50	25.77	-	1.60	11.23	4.46	41.46	nr	44.57
1200mm high fencing; type GLS 120; posts with rounded tops										
posts at 2743mm centres; 600mm into ground	4.26	2.50	4.37	-	0.75	5.26	0.54	10.17	m	10.93
extra; end posts with struts -1	21.90	2.50	22.45	-	1.10	7.72	2.98	33.15	nr	35.64
extra; angle posts with struts -2	31.88	2.50	32.68	-	1.75	12.29	4.46	49.42	nr	53.13
1400mm high fencing; type GLS 140A; posts with rounded tops										
posts at 2743mm centres; 600mm into ground	4.96	2.50	5.08	-	0.80	5.62	0.54	11.24	m	12.08
extra; end posts with struts -1	23.34	2.50	23.92	-	1.50	10.53	2.98	37.43	nr	40.24
extra; angle posts with struts -2	34.12	2.50	34.97	-	1.90	13.34	4.46	52.77	nr	56.73
1800mm high fencing; type GLS 180; posts with rounded tops										
posts at 2743mm centres; 600mm into ground	6.13	2.50	6.28	-	1.00	7.02	0.54	13.84	m	14.88
extra; end posts with struts -1	31.45	2.50	32.24	-	1.60	11.23	2.90	46.45	nr	49.93
extra; angle posts with struts -2	45.26	2.50	46.39	-	2.10	14.74	4.46	65.59	nr	70.51

PAVING/FENCING/SITE FURNITURE

Labour hourly rates: (except Specialists) Craft Operatives £9.23 Labourer £7.02 Rates are national average prices. Refer to REGIONAL VARIATIONS for indicative levels of overall pricing in regions	MATERIALS			LABOUR				RATES		
	Del to Site £	Waste %	Material Cost £	Craft Optve Hrs	Lab Hrs	Labour Cost £	Sunds £	Nett Rate £	Unit	Gross Rate (+7.5%) £

CHAIN LINK FENCING Cont.

Fencing; chain link; B.S.1722 Part 1; rolled steel angle posts; plastics coated Grade A mesh, line and tying wires; galvanised steel fittings and accessories; backfilling around posts in concrete mix ST4

900mm high fencing; type PLS 90A; posts with rounded tops

Description	Del to Site	Waste	Material Cost	Craft Optve	Lab	Labour Cost	Sunds	Nett Rate	Unit	Gross Rate
posts at 2743mm centres; 600mm into ground	3.93	2.50	4.03	–	0.65	4.56	0.54	9.13	m	9.82
extra; end posts with struts -1	17.08	2.50	17.51	–	1.00	7.02	2.98	27.51	nr	29.57
extra; angle posts with struts -2	25.14	2.50	25.77	–	1.60	11.23	4.46	41.46	nr	44.57
1200mm high fencing; type PLS 120; posts with rounded tops										
posts at 2743mm centres; 600mm into ground	5.06	2.50	5.19	–	0.75	5.26	0.54	10.99	m	11.82
extra; end posts with struts -1	21.90	2.50	22.45	–	1.10	7.72	2.98	33.15	nr	35.64
extra; angle posts with struts -2	31.88	2.50	32.68	–	1.75	12.29	4.46	49.42	nr	53.13
1400mm high fencing; type PLS 140A; posts with rounded tops										
posts at 2743mm centres; 600mm into ground	5.65	2.50	5.79	–	0.80	5.62	0.54	11.95	m	12.84
extra; end posts with struts -1	23.34	2.50	23.92	–	1.50	10.53	2.98	37.43	nr	40.24
extra; angle posts with struts -2	34.12	2.50	34.97	–	1.90	13.34	4.46	52.77	nr	56.73
1800mm high fencing; type PLS 180; posts with rounded tops										
posts at 2743mm centres; 600mm into ground	6.75	2.50	6.92	–	1.00	7.02	0.54	14.48	m	15.56
extra; end posts with struts -1	31.45	2.50	32.24	–	1.60	11.23	2.98	46.45	nr	49.93
extra; angle posts with struts -2	45.26	2.50	46.39	–	2.10	14.74	4.46	65.59	nr	70.51

CHESTNUT FENCING

Fencing; cleft chestnut pale; B.S.1722 Part 4; sweet chestnut posts and struts; galvanised accessories; backfilling around posts in concrete mix ST4

Description	Del to Site	Waste	Material Cost	Craft Optve	Lab	Labour Cost	Sunds	Nett Rate	Unit	Gross Rate
900mm high fencing; type CW90										
posts at 2000mm centres; 600mm into ground	3.01	5.00	3.16	–	0.25	1.75	0.45	5.37	m	5.77
extra; end posts with struts - 1	4.04	5.00	4.24	–	0.30	2.11	2.48	8.83	nr	9.49
extra; angle posts with struts - 2	5.18	5.00	5.44	–	0.40	2.81	3.72	11.97	nr	12.86
1200mm high fencing; type CW 120										
posts at 2000mm centres; 600mm into ground	3.55	5.00	3.73	–	0.17	1.19	0.45	5.37	m	5.77
extra; end posts with struts - 1	4.87	5.00	5.11	–	0.45	3.16	2.48	10.75	nr	11.56
extra; angle posts with struts - 2	6.20	5.00	6.51	–	0.60	4.21	3.72	14.44	nr	15.53
1500mm high fencing; type CW 150										
posts at 2000mm centres; 600mm into ground	5.80	5.00	6.09	–	0.22	1.54	0.45	8.08	m	8.69
extra; end posts with struts - 1	6.19	5.00	6.50	–	0.50	3.51	2.48	12.49	nr	13.43
extra; angle posts with struts - 2	7.91	5.00	8.31	–	0.65	4.56	3.72	16.59	nr	17.83

BOARDED FENCING

Fencing; close boarded; B.S.1722 Part 5; sawn softwood posts, rails, pales, gravel boards and centre stumps, pressure impregnated with preservative; backfilling around posts in concrete mix ST4

Description	Del to Site	Waste	Material Cost	Craft Optve	Lab	Labour Cost	Sunds	Nett Rate	Unit	Gross Rate
1200mm high fencing; type BW 120										
posts at 3000mm centres; 600mm into ground	11.52	5.00	12.10	1.25	–	11.54	1.52	25.15	m	27.04
1500mm high fencing; type BW 150										
posts at 3000mm centres; 750mm into ground	15.27	5.00	16.03	1.50	–	13.85	2.00	31.88	m	34.27
1800mm high fencing; type BW 180A										
posts at 3000mm centres; 750mm into ground	16.86	5.00	17.70	1.67	–	15.41	2.00	35.12	m	37.75

Fencing; wooden palisade; B.S.1722 Part 6; softwood posts, rails, pales and stumps; pressure impregnated with preservative; backfilling around posts in concrete mix ST4

Description	Del to Site	Waste	Material Cost	Craft Optve	Lab	Labour Cost	Sunds	Nett Rate	Unit	Gross Rate
1050mm high fencing; type WPW 105; 75 x 19mm rectangular pales with pointed tops										
posts at 3000mm centres; 600mm into ground	9.28	5.00	9.74	1.00	–	9.23	1.41	20.38	m	21.91
1200mm high fencing; type WPW 120; 75 x 19mm rectangular pales with pointed tops										
posts at 3000mm centres; 600mm into ground	10.58	5.00	11.11	1.25	–	11.54	1.41	24.06	m	25.86

STEEL RAILINGS

Ornamental steel railings; Ranalah Gates Ltd.; primed at works; fixing in brick openings

Description	Del to Site	Waste	Material Cost	Craft Optve	Lab	Labour Cost	Sunds	Nett Rate	Unit	Gross Rate
Railing										
reference 200, 356mm high	28.25	1.50	28.67	0.75	0.38	9.59	0.21	38.47	m	41.36
reference 220, 279mm high	51.90	1.50	52.68	0.75	0.38	9.59	0.21	62.48	m	67.16
reference 230, 432mm high	43.14	1.50	43.79	0.75	0.38	9.59	0.21	53.59	m	57.61
reference 240, 838mm high	59.35	1.50	60.24	1.00	0.50	12.74	0.26	73.24	m	78.73

Labour hourly rates: (except Specialists) Craft Operatives £9.23 Labourer £7.02. Rates are national average prices. Refer to REGIONAL VARIATIONS for indicative levels of overall pricing in regions	MATERIALS			LABOUR				RATES		
	Del to Site £	Waste %	Material Cost £	Craft Optve Hrs	Lab Hrs	Labour Cost £	Sunds £	Nett Rate £	Unit	Gross Rate (+7.5%) £

GUARD RAILS

Galvanised mild steel guard rail of tubing to B.S.1387 medium grade with Kee Klamp fittings

Rail or standard
40mm diameter ref. 7-2-G	5.73	5.00	6.02	0.25	–	2.31	0.30	8.62	m	9.27
extra; flanged end	1.15	5.00	1.21	–	–	–	–	1.21	nr	1.30
extra; bend No.15-7	4.42	5.00	4.64	0.15	–	1.38	–	6.03	nr	6.48
extra; three-way intersection No.20-7	6.68	5.00	7.01	0.15	–	1.38	–	8.40	nr	9.03
extra; three-way intersection No.25-7	6.42	5.00	6.74	0.15	–	1.38	–	8.13	nr	8.73
extra; four-way intersection No.21-7	5.22	5.00	5.48	0.15	–	1.38	–	6.87	nr	7.38
extra; four-way intersection No.26-7	4.95	5.00	5.20	0.15	–	1.38	–	6.58	nr	7.08
extra; five-way intersection No.35-7	7.43	5.00	7.80	0.15	–	1.38	–	9.19	nr	9.87
extra; five-way intersection No.40-7	11.08	5.00	11.63	0.15	–	1.38	–	13.02	nr	13.99
extra; floor plate No.61-7	4.03	5.00	4.23	0.15	–	1.38	–	5.62	nr	6.04

Infill panel fixed with clips
50 x 50mm welded mesh	19.60	5.00	20.58	0.80	–	7.38	0.20	28.16	m2	30.28

PRECAST CONCRETE POSTS

Precast concrete fence posts; B.S.1722 excavating holes, backfilling around posts in concrete mix C20, disposing of surplus materials, earthwork support

Fence posts for three wires, housing pattern
100 x 100mm tapering intermediate post 1570mm long; 600mm into ground	4.42	2.50	4.53	0.25	0.25	4.06	1.94	10.53	nr	11.32
100 x 100mm square strainer post (end or intermediate) 1570mm long; 600mm into ground	5.41	2.50	5.55	0.25	0.25	4.06	1.94	11.55	nr	12.41

Clothes line post
125 x 125mm tapering, 2670mm long	12.29	2.50	12.60	0.33	0.33	5.36	1.94	19.90	nr	21.39

Close boarded fence post
94 x 100mm, 2745mm long	7.32	2.50	7.50	0.40	0.40	6.50	1.94	15.94	nr	17.14

Chain link fence post for 1800mm high fencing
125 x 75mm tapering intermediate post, 2620mm long	7.98	2.50	8.18	0.33	0.33	5.36	1.94	15.48	nr	16.64
125 x 125mm end post, 1 strut, 2620mm long	17.94	2.50	18.39	0.66	0.66	10.73	3.89	33.00	nr	35.48
125 x 125mm angle post, 2 strut, 2620mm long	24.81	2.50	25.43	0.99	0.99	16.09	5.83	47.35	nr	50.90
125 x 125mm gate post, 2620mm long	13.38	2.50	13.71	0.33	0.33	5.36	1.94	21.02	nr	22.59
150 x 150mm gate post, 2620mm long	20.46	2.50	20.97	0.33	0.33	5.36	1.94	28.27	nr	30.39

TIMBER GATES

Gates; impregnated wrought softwood; featheredge pales; including ring latch and heavy hinges

Gates (posts included elsewhere)
900 x 1150mm	53.31	2.50	54.64	0.50	0.50	8.13	0.17	62.94	nr	67.66
900 x 1450mm	55.11	2.50	56.49	0.50	0.50	8.13	0.17	64.78	nr	69.64

STEEL GATES

Ornamental steel gates; Ranalah Gates Ltd.; primed at works; fixing in brick openings

Popular, 1118mm high overall
single gate reference 50, for 914mm wide opening	32.11	1.50	32.59	1.50	0.75	19.11	0.27	51.97	nr	55.87

Diana, 965mm high overall
single gate reference 800, for 914mm wide opening	37.96	1.50	38.53	1.50	0.75	19.11	0.27	57.91	nr	62.25
single gate reference 801, for 1016mm wide opening	40.34	1.50	40.95	1.50	0.75	19.11	0.87	60.93	nr	65.49
double gates reference 810, for 2438mm wide opening	84.88	1.50	86.15	2.25	1.13	28.70	0.49	115.34	nr	123.99

Popular arch gate 1568mm high overall
single gate reference 55, for 914mm wide opening	58.90	1.50	59.78	1.50	0.75	19.11	0.27	79.16	nr	85.10

Thrifty, 914mm high overall
single gate reference 20, for 914mm wide opening	26.36	1.50	26.76	1.50	0.75	19.11	0.27	46.14	nr	49.60

Elite, 914mm high overall
single gate reference 40, for 914mm wide opening	34.10	1.50	34.61	1.50	0.75	19.11	0.27	53.99	nr	58.04
double gates reference 41, for 2134mm wide opening	78.98	1.50	80.16	2.25	1.13	28.70	0.49	109.35	nr	117.56
double gates reference 42, for 2438mm wide opening	83.35	1.50	84.60	2.25	1.13	28.70	0.49	113.79	nr	122.32

Super, 965mm high overall
single gate reference 60, for 914mm wide opening	39.72	1.50	40.32	2.25	1.13	28.70	0.49	69.51	nr	74.72
double gates reference 62, for 2286mm wide opening	100.99	1.50	102.50	2.25	1.13	28.70	0.49	131.69	nr	141.57
double gates reference 63, for 2438mm wide opening	104.63	1.50	106.20	2.25	1.13	28.70	0.49	135.39	nr	145.54

Concord, 914mm high overall
single gate reference 160, for 914mm wide opening	44.94	1.50	45.61	2.25	1.13	28.70	0.49	74.80	nr	80.41
double gates reference 162, for 2286mm wide opening	103.27	1.50	104.82	2.25	1.13	28.70	0.49	134.01	nr	144.06
double gates reference 163, for 2438mm wide opening	109.56	1.50	111.20	2.25	1.13	28.70	0.49	140.39	nr	150.92

Regent, 914mm high overall
single gate reference 150, for 914mm wide opening	50.44	1.50	51.20	2.25	1.13	28.70	0.49	80.39	nr	86.42
double gates reference 151, for 2134mm wide opening	109.74	1.50	111.39	2.25	1.13	28.70	0.40	140.49	nr	151.02

PAVING/FENCING/SITE FURNITURE

Labour hourly rates: (except Specialists) Craft Operatives £9.23 Labourer £7.02 Rates are national average prices. Refer to REGIONAL VARIATIONS for indicative levels of overall pricing in regions	MATERIALS			LABOUR				RATES		
	Del to Site £	Waste %	Material Cost £	Craft Optve Hrs	Lab Hrs	Labour Cost £	Sunds £	Nett Rate £	Unit	Gross Rate (+7.5%) £
STEEL GATES Cont.										
Ornamental steel gates; Ranalah Gates Ltd.; primed at works; fixing in brick openings Cont.										
Regent, 914mm high overall Cont.										
double gates reference 152, for 2286mm wide opening	112.80	1.50	114.49	2.25	1.13	28.70	0.49	143.68	nr	154.46
double gates reference 153, for 2438mm wide opening	125.79	1.50	127.68	2.25	1.13	28.70	0.49	156.87	nr	168.63
Brighton, 991mm high overall										
single gate reference 600, for 914mm wide opening	57.08	1.50	57.94	1.50	0.75	19.11	0.27	77.32	nr	83.11
double gates reference 602, for 2286mm wide opening	145.37	1.50	147.55	2.25	1.13	28.70	0.49	176.74	nr	190.00
double gates reference 603, for 2438mm wide opening	154.79	1.50	157.11	2.25	1.13	28.70	0.49	186.30	nr	200.27
Preston, 1067mm high overall										
single gate reference 425, for 914mm wide opening	66.95	1.50	67.95	1.50	0.75	19.11	0.27	87.33	nr	93.88
single gate reference 426, for 1016mm wide opening	73.76	1.50	74.87	1.50	0.75	19.11	0.27	94.25	nr	101.31
double gates reference 430, for 2134mm wide opening	150.81	1.50	153.07	2.25	1.13	28.70	0.49	182.26	nr	195.93
double gates reference 431, for 2438mm wide opening	167.55	1.50	170.06	2.25	1.13	28.70	0.49	199.25	nr	214.20
Extra cost of fixing gates to and including pair of steel gate posts; backfilling around posts in concrete mix ST4										
for single or double gates; posts reference 901, size 40 x 40mm	34.94	1.50	35.46	0.33	0.17	4.24	2.08	41.78	nr	44.92
for single or double gates; posts reference 902, size 50 x 50mm	43.04	1.50	43.69	0.33	0.17	4.24	2.27	50.19	nr	53.96
for single or double gates; posts reference 903, size 70 x 70mm	56.72	1.50	57.57	0.66	0.33	8.41	2.63	68.61	nr	73.75
for single or double gates; posts reference 904, size 100 x 100mm	79.87	1.50	81.07	0.66	0.33	8.41	3.17	92.65	nr	99.59
Gates in chain link fencing										
Gate with 40 x 40 x 5mm painted angle iron framing, braces and rails, infilled with 50mm x 10 1/2 G mesh										
1200 x 1800mm; galvanised mesh	98.71	1.50	100.19	6.00	6.00	97.50	–	197.69	nr	212.52
1200 x 1800mm; plastic coated mesh	103.03	1.50	104.58	6.00	6.00	97.50	–	202.08	nr	217.23
Gate post of 80 x 80 x 6mm painted angle iron; backfilling around posts in concrete mix C30										
to suit 1800mm high gate	40.32	1.50	40.92	1.50	1.50	24.38	2.68	67.98	nr	73.08
Q50: SITE/STREET FURNITURE/EQUIPMENT										
Plain ordinary portland cement precast concrete bollards; setting in ground, excavating hole, removing surplus spoil, filling with concrete mix ST4; working around base										
230mm diameter										
455mm high above ground level; Woodhouse; Smooth Grey	33.70	–	33.70	0.50	0.50	8.13	2.08	43.91	nr	47.20
455mm high above ground level; Woodhouse; Smooth White	57.30	–	57.30	0.50	0.50	8.13	3.10	68.53	nr	73.66
150mm diameter										
915mm high above ground level; Bridgford	46.50	–	46.50	0.90	0.90	14.63	2.08	63.20	nr	67.95
915mm high above ground level; Bridgford; Smooth White	77.00	–	77.00	0.90	0.90	14.63	3.10	94.72	nr	101.83
Precast concrete plant containers; setting in position										
Plant container (soil filling included elsewhere)										
710mm diameter x 500mm high; Strada	91.70	–	91.70	0.80	0.80	13.00	–	104.70	nr	112.55
1066mm diameter x 760mm high; Shirley	386.60	–	386.60	1.00	1.00	16.25	–	402.85	nr	433.06
500 x 500 x 500mm high; Strada	73.80	–	73.80	0.60	0.60	9.75	–	83.55	nr	89.82
Precast concrete litter bin with galvanised wire plastic coated inner basket Plastic Coated; setting in position										
Plain concrete										
43 litres capacity; Elsworth	148.00	–	148.00	1.00	1.00	16.25	–	164.25	nr	176.57
65 litres capacity; Newstead	139.60	–	139.60	1.00	1.00	16.25	–	155.85	nr	167.54
190 litres capacity; Gransden	215.35	–	215.35	1.00	1.00	16.25	–	231.60	nr	248.97
433 litres capacity; Shirley	511.90	–	511.90	1.00	1.00	16.25	–	528.15	nr	567.76
Benches and seats with precast concrete supports and hardwood slats to seats and back rests; fixed or free-standing										
Bench; placing in position										
2000mm long; Kelvin; Smooth Grey	160.85	–	160.85	0.50	0.50	8.13	–	168.97	nr	181.65
2000mm long; Kelvin; Smooth White	174.95	–	174.95	0.60	0.60	9.75	–	184.70	nr	198.55

Labour hourly rates: (except Specialists) Craft Operatives £9.23 Labourer £7.02 Rates are national average prices. Refer to REGIONAL VARIATIONS for indicative levels of overall pricing in regions	MATERIALS			LABOUR				RATES		
	Del to Site	Waste	Material Cost	Craft Optve	Lab	Labour Cost	Sunds	Nett Rate	Unit	Gross Rate (+7.5%)
	£	%	£	Hrs	Hrs	£	£	£		£
Q50: SITE/STREET FURNITURE/EQUIPMENT Cont.										
Steel cycle stands, Mawrob Co. (Engineers) Ltd; galvanised steel construction; free standing										
Single sided										
reference TR5 for 5 cycles	175.00	-	175.00	1.00	0.25	10.98	-	185.99	nr	199.93
reference TR6 for 6 cycles	192.00	-	192.00	1.00	0.25	10.98	-	202.99	nr	218.21
reference TR8 for 8 cycles	230.50	-	230.50	1.25	0.30	13.64	-	244.14	nr	262.45
reference TR10 for 10 cycles	297.50	-	297.50	1.25	0.30	13.64	-	311.14	nr	334.48
Double sided										
reference DTR10 for 10 cycles	282.00	-	282.00	1.50	0.40	16.65	-	298.65	nr	321.05
reference DTR12 for 12 cycles	312.50	-	312.50	1.50	0.40	16.65	-	329.15	nr	353.84
reference DTR20 for 20 cycles	483.50	-	483.50	2.00	0.50	21.97	-	505.47	nr	543.38
Steel front wheel cycle supports; Mawrob Co. (Engineers) Ltd; galvanised steel construction, pillar mounted; bolting base plate to concrete										
Single sided										
reference MW/MC 6/S for 6 cycles	261.00	-	261.00	4.00	1.00	43.94	-	304.94	nr	327.81
reference MW/MC 9/S for 9 cycles	353.00	-	353.00	4.00	1.00	43.94	-	396.94	nr	426.71
reference MW/MC 12/S for 12 cycles	455.00	-	455.00	6.00	1.50	65.91	-	520.91	nr	559.98
Double sided										
reference MW/MC 6/D for 12 cycles	429.00	-	429.00	4.00	1.00	43.94	-	472.94	nr	508.41
reference MW/MC 9/D for 18 cycles	558.00	-	558.00	4.00	1.00	43.94	-	601.94	nr	647.09
reference MW/MC 12/D for 24 cycles	748.50	-	748.50	6.00	1.50	65.91	-	814.41	nr	875.49
Steel cycle shelters, Mawrob Co. (Engineers) Ltd; steel angle framing; with red oxide priming; corrugated galvanised steel roofing; assembling; bolting to concrete										
Horizontal loading, single sided										
reference MW/AV10 for 10 cycles	505.00	-	505.00	8.00	2.00	87.88	-	592.88	nr	637.35
reference MW/AK10 for 10 cycles	588.50	-	588.50	8.00	2.00	87.88	-	676.38	nr	727.11
reference MW/AW8 for 8 cycles	491.50	-	491.50	6.00	1.50	65.91	-	557.41	nr	599.22
reference MW/AW10 for 10 cycles	588.50	-	588.50	8.00	2.00	87.88	-	676.38	nr	727.11
reference MW/AW15 for 15 cycles	807.00	-	807.00	10.00	2.50	109.85	-	916.85	nr	985.61
reference MW/RS12 for 12 cycles	550.50	-	550.50	8.00	2.00	87.88	-	638.38	nr	686.26
reference MW/RS22 for 22 cycles	895.00	-	895.00	12.00	3.00	131.82	-	1026.82	nr	1103.83
Horizontal loading, double sided										
reference MW/AL12 for 12 cycles	847.00	-	847.00	10.00	2.50	109.85	-	956.85	nr	1028.61
reference MW/AL16 for 16 cycles	1058.50	-	1058.50	12.00	3.00	131.82	-	1190.32	nr	1279.59
reference MV/AL20 for 20 cycles	1217.50	-	1217.50	14.00	3.50	153.79	-	1371.29	nr	1474.14

PAVING/FENCING/SITE FURNITURE

Estimating and Tendering for Construction Work
Second edition

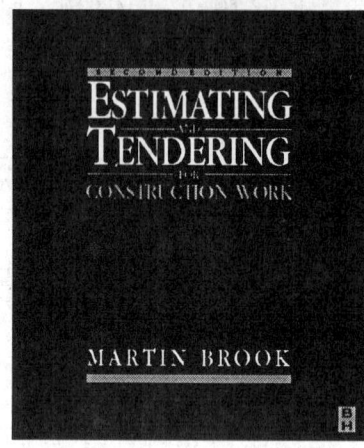

Estimating and Tendering for Construction Work takes a practical approach to estimating from a contractor's point of view. It explains the estimator's function within the construction team, the techniques and procedures used in building up an estimate, and how to convert an estimate into a tender.

This new edition has been written to reflect recent changes in procurement. These include the recommendations of Sir Michael Latham in his 1994 report 'Constructing the Team,' and new terminology introduced by the 6th edition of the CIOB Code of Estimating Practice 1997. The role of the estimator is covered in detail, from early cost studies through to the preparation of the estimate, and the handover of construction budgets for successful tenders.

The book includes copious examples of an estimator's data sheets and pricing notes and deals with co-ordinated project information. The chapter dealing with computer-aided estimating has been completely re-written to reflect the advantages of electronic exchange of information.

Paperback, 1998, £19.99, 288pp, isbn 0 7506 3404 7

Disposal systems

Labour hourly rates: (except Specialists) Craft Operatives £12.44 Labourer £7.02 Rates are national average prices. Refer to REGIONAL VARIATIONS for indicative levels of overall pricing in regions	MATERIALS			LABOUR				RATES		
	Del to Site £	Waste %	Material Cost £	Craft Optve Hrs	Lab Hrs	Labour Cost £	Sunds £	Nett Rate £	Unit	Gross Rate (+7.5%) £
R10: RAINWATER PIPEWORK/GUTTERS										
CAST IRON ROUND RAINWATER PIPES (B.S.460) - Cast iron pipes and fittings, B.S.460 Type A sockets; dry joints										
Pipes; straight										
75mm; in or on supports (included elsewhere)	12.65	5.00	13.28	0.67	–	8.33	–	21.62	m	23.24
extra; shoes	10.35	2.00	10.56	0.84	–	10.45	–	21.01	nr	22.58
extra; bends	8.88	2.00	9.06	0.67	–	8.33	–	17.39	nr	18.70
extra; offset bends 75mm projection	11.19	2.00	11.41	0.67	–	8.33	–	19.75	nr	21.23
extra; offset bends 150mm projection	11.19	2.00	11.41	0.67	–	8.33	–	19.75	nr	21.23
extra; offset bends 225mm projection	13.03	2.00	13.29	0.67	–	8.33	–	21.63	nr	23.25
extra; offset bends 300mm projection	15.95	2.00	16.27	0.67	–	8.33	–	24.60	nr	26.45
extra; branches	15.54	2.00	15.85	0.67	–	8.33	–	24.19	nr	26.00
100mm; in or on supports (included elsewhere)	17.27	5.00	18.13	0.80	–	9.95	–	28.09	m	30.19
extra; shoes	13.95	2.00	14.23	1.00	–	12.44	–	26.67	nr	28.67
extra; bends	12.53	2.00	12.78	0.80	–.	9.95	–	22.73	nr	24.44
extra; offset bends 75mm projection	21.10	2.00	21.52	0.80	–	9.95	–	31.47	nr	33.83
extra; offset bends 150mm projection	21.10	2.00	21.52	0.80	–	9.95	–	31.47	nr	33.83
extra; offset bends 225mm projection	25.55	2.00	26.06	0.80	–	9.95	–	36.01	nr	38.71
extra; offset bends 300mm projection	25.55	2.00	26.06	0.80	–	9.95	–	36.01	nr	38.71
extra; branches	18.45	2.00	18.82	0.80	–	9.95	–	28.77	nr	30.93
CAST IRON ROUND RAINWATER PIPES (B.S.460) - Cast iron pipes and fittings, B.S.460 Type A sockets; ears cast on; dry joints										
Pipes; straight										
63mm; ears fixing to masonry with galvanized pipe nails and distance pieces	13.62	5.00	14.30	0.57	–	7.09	–	21.39	m	23.00
extra; shoes	12.13	2.00	12.37	0.71	–	8.83	–	21.20	nr	22.80
extra; bends	7.31	2.00	7.46	0.57	–	7.09	–	14.55	nr	15.64
extra; offset bends 75mm projection	11.19	2.00	11.41	0.57	–	7.09	–	18.50	nr	19.89
extra; offset bends 150mm projection	11.19	2.00	11.41	0.57	–	7.09	–	18.50	nr	19.89
extra; offset bends 225mm projection	13.03	2.00	13.29	0.57	–	7.09	–	20.38	nr	21.91
extra; offset bends 300mm projection	15.24	2.00	15.54	0.57	–	7.09	–	22.64	nr	24.33
extra; branches	14.09	2.00	14.37	0.57	–	7.09	–	21.46	nr	23.07
75mm; ears fixing to masonry with galvanized pipe nails and distance pieces	13.62	5.00	14.30	0.67	–	8.33	–	22.64	m	24.33
extra; shoes	12.13	2.00	12.37	0.84	–	10.45	–	22.82	nr	24.53
extra; bends	8.88	2.00	9.06	0.67	–	8.33	–	17.39	nr	18.70
extra; offset bends 75mm projection	11.19	2.00	11.41	0.67	–	8.33	–	19.75	nr	21.23
extra; offset bends 150mm projection	11.19	2.00	11.41	0.67	–	8.33	–	19.75	nr	21.23
extra; offset bends 225mm projection	13.03	2.00	13.29	0.67	–	8.33	–	21.63	nr	23.25
extra; offset bends 300mm projection	15.95	2.00	16.27	0.67	–	8.33	–	24.60	nr	26.45
extra; branches	15.54	2.00	15.85	0.67	–	8.33	–	24.19	nr	26.00
100mm; ears fixing to masonry with galvanized pipe nails and distance pieces	18.25	5.00	19.16	0.80	–	9.95	–	29.11	m	31.30
extra; shoes	15.75	2.00	16.07	1.00	–	12.44	–	28.50	nr	30.64
extra; bends	12.53	2.00	12.78	0.80	–	9.95	–	22.73	nr	24.44
extra; offset bends 75mm projection	21.10	2.00	21.52	0.80	–	9.95	–	31.47	nr	33.83
extra; offset bends 150mm projection	21.10	2.00	21.52	0.80	–	9.95	–	31.47	nr	33.83
extra; offset bends 225mm projection	25.55	2.00	26.06	0.80	–	9.95	–	36.01	nr	38.71
extra; offset bends 300mm projection	25.55	2.00	26.06	0.80	–	9.95	–	36.01	nr	38.71
extra; branches	18.45	2.00	18.82	0.80	–	9.95	–	28.77	nr	30.93
CAST IRON ROUND RAINWATER PIPES (B.S.460) - Pipe supports										
Steel holderbats										
for 75mm pipes; for building in (fixing included elsewhere)	5.44	2.00	5.55	–	–	–	–	5.55	nr	5.96
for 100mm pipes; for building in (fixing included elsewhere)	5.60	2.00	5.71	–	–	–	–	5.71	nr	6.14
for 75mm pipes; fixing to masonry with galvanized screws	4.55	2.00	4.64	–	–	–	–	4.64	nr	4.99
for 100mm pipes; fixing to masonry with galvanized screws	4.71	2.00	4.80	0.17	–	2.11	–	6.92	nr	7.44

Labour hourly rates: (except Specialists) Craft Operatives £12.44 Labourer £7.02 Rates are national average prices. Refer to REGIONAL VARIATIONS for indicative levels of overall pricing in regions	MATERIALS			LABOUR				RATES		
	Del to Site £	Waste %	Material Cost £	Craft Optve Hrs	Lab Hrs	Labour Cost £	Sunds £	Nett Rate £	Unit	Gross Rate (+7.5%) £
R10: RAINWATER PIPEWORK/GUTTERS Cont.										
CAST IRON RECTANGULAR RAINWATER PIPES - Cast iron rectangular pipes and fittings; ears cast on; dry joints										
Pipes; straight										
100 x 75mm; ears fixing to masonry with galvanised pipe nails and distance pieces	52.13	5.00	54.74	1.00	–	12.44	–	67.18	m	72.21
extra; shoes (front)	44.91	2.00	45.81	1.00	–	12.44	–	58.25	nr	62.62
extra; bends (front)	35.80	2.00	36.52	0.75	–	9.33	–	45.85	nr	49.28
extra; offset bends (front) 150 projection	45.66	2.00	46.57	0.75	–	9.33	–	55.90	nr	60.10
extra; offset bends (front) 300 projection	64.07	2.00	65.35	0.75	–	9.33	–	74.68	nr	80.28
ALUMINIUM RAINWATER PIPES (B.S.2997) - Cast aluminium pipes and fittings B.S.2997 Part 2 Section B, ears cast on; dry joints										
Pipes; straight										
63mm; ears fixing to masonry with galvanized pipe nails	9.54	5.00	10.02	0.50	–	6.22	–	16.24	m	17.45
extra; shoes	6.25	2.00	6.38	0.63	–	7.84	–	14.21	nr	15.28
extra; bends	6.90	2.00	7.04	0.50	–	6.22	–	13.26	nr	14.25
extra; offset bends 75mm projection	13.28	2.00	13.55	0.50	–	6.22	–	19.77	nr	21.25
extra; offset bends 150mm projection	16.47	2.00	16.80	0.50	–	6.22	–	23.02	nr	24.75
extra; offset bends 225mm projection	18.68	2.00	19.05	0.50	–	6.22	–	25.27	nr	27.17
extra; offset bends 300mm projection	20.91	2.00	21.33	0.50	–	6.22	–	27.55	nr	29.61
75mm; ears fixing to masonry with galvanized pipe nails	11.09	5.00	11.64	0.67	–	8.33	–	19.98	m	21.48
extra; shoes	8.60	2.00	8.77	0.84	–	10.45	–	19.22	nr	20.66
extra; bends	9.05	2.00	9.23	0.67	–	8.33	–	17.57	nr	18.88
extra; offset bends 75mm projection	16.10	2.00	16.42	0.67	–	8.33	–	24.76	nr	26.61
extra; offset bends 150mm projection	18.51	2.00	18.88	0.67	–	8.33	–	27.22	nr	29.26
extra; offset bends 225mm projection	20.65	2.00	21.06	0.67	–	8.33	–	29.40	nr	31.60
extra; offset bends 300mm projection	22.92	2.00	23.38	0.67	–	8.33	–	31.71	nr	34.09
100mm; ears fixing to masonry with galvanized pipe nails	16.68	5.00	17.51	0.80	–	9.95	–	27.47	m	29.53
extra; shoes	10.22	2.00	10.42	1.00	–	12.44	–	22.86	nr	24.58
extra; bends	13.22	2.00	13.48	0.80	–	9.95	–	23.44	nr	25.19
extra; offset bends 75mm projection	18.22	2.00	18.58	0.80	–	9.95	–	28.54	nr	30.68
extra; offset bends 150mm projection	21.15	2.00	21.57	0.80	–	9.95	–	31.52	nr	33.89
extra; offset bends 225mm projection	23.94	2.00	24.42	0.80	–	9.95	–	34.37	nr	36.95
extra; offset bends 300mm projection	26.73	2.00	27.26	0.80	–	9.95	–	37.22	nr	40.01
UNPLASTICIZED PVC RAINWATER PIPES (B.S.4576) - PVC-U pipes and fittings, B.S.4576 Part 1; push fit joints; pipework and supports self coloured grey										
Pipes; straight										
68mm; in standard holderbats fixing to masonry with galvanized screws	2.35	5.00	2.47	0.40	–	4.98	–	7.44	m	8.00
extra; shoes	2.99	2.00	3.05	0.31	–	3.86	–	6.91	nr	7.42
extra; bends	2.34	2.00	2.39	0.25	–	3.11	–	5.50	nr	5.91
extra; offset bends (spigot)	1.30	2.00	1.33	0.25	–	3.11	–	4.44	nr	4.77
extra; offset bends (socket)	1.40	2.00	1.43	0.25	–	3.11	–	4.54	nr	4.88
extra; angled branches	4.67	2.00	4.76	0.25	–	3.11	–	7.87	nr	8.46
CAST IRON EAVES GUTTERS (B.S.460) - Cast iron half round gutters and fittings, B.S.460; bolted and mastic joints										
Gutters; straight										
100mm; in standard fascia brackets fixing to timber with galvanized screws	9.08	5.00	9.53	0.50	–	6.22	–	15.75	m	16.94
extra; stopped ends	1.76	2.00	1.80	0.50	–	6.22	–	8.02	nr	8.62
extra; running outlets	5.10	2.00	5.20	0.50	–	6.22	–	11.42	nr	12.28
extra; stopped ends with outlet	3.83	2.00	3.91	0.50	–	6.22	–	10.13	nr	10.89
extra; angles	5.24	2.00	5.34	0.50	–	6.22	–	11.56	nr	12.43
114mm; in standard fascia brackets fixing to timber with galvanized screws	9.44	5.00	9.91	0.50	–	6.22	–	16.13	m	17.34
extra; stopped ends	2.28	2.00	2.33	0.50	–	6.22	–	8.55	nr	9.19
extra; running outlets	5.56	2.00	5.67	0.50	–	6.22	–	11.89	nr	12.78
extra; stopped ends with outlet	4.21	2.00	4.29	0.50	–	6.22	–	10.51	nr	11.30
extra; angles	5.39	2.00	5.50	0.50	–	6.22	–	11.72	nr	12.60
125mm; in standard fascia brackets fixing to timber with galvanized screws	10.74	5.00	11.28	0.57	–	7.09	–	18.37	m	19.75
extra; stopped ends	2.28	2.00	2.33	0.57	–	7.09	–	9.42	nr	10.13
extra; running outlets	6.37	2.00	6.50	0.57	–	7.09	–	13.59	nr	14.61
extra; stopped ends with outlet	5.66	2.00	5.77	0.57	–	7.09	–	12.86	nr	13.83
extra; angles	6.37	2.00	6.50	0.57	–	7.09	–	13.59	nr	14.61
150mm; in standard fascia brackets fixing to timber with galvanized screws	17.25	5.00	18.11	0.75	–	9.33	–	27.44	m	29.50
extra; stopped ends	3.04	2.00	3.10	0.75	–	9.33	–	12.43	nr	13.36
extra; running outlets	11.01	2.00	11.23	0.75	–	9.33	–	20.56	nr	22.10
extra; stopped ends with outlet	10.82	2.00	11.04	0.75	–	9.33	–	20.37	nr	21.89
extra; angles	11.62	2.00	11.85	0.75	–	9.33	–	21.18	nr	22.77
CAST IRON EAVES GUTTERS (B.S.460) - Cast iron OG gutters and fittings, B.S.460; bolted and mastic joints										
Gutters; straight										
100mm; fixing to timber with galvanized screws ...	8.29	5.00	8.70	0.50	–	6.22	–	14.92	m	16.04
extra; stopped ends	1.59	2.00	1.62	0.50	–	6.22	–	7.84	nr	8.43
extra; running outlets	5.47	2.00	5.58	0.50	–	6.22	–	11.80	nr	12.68
extra; stopped ends with outlet	4.16	2.00	4.24	0.50	–	6.22	–	10.46	nr	11.25

Labour hourly rates: (except Specialists) Craft Operatives £12.44 Labourer £7.02 Rates are national average prices. Refer to REGIONAL VARIATIONS for indicative levels of overall pricing in regions	MATERIALS			LABOUR				RATES		
	Del to Site	Waste	Material Cost	Craft Optve	Lab	Labour Cost	Sunds	Nett Rate	Unit	Gross Rate (+7.5%)
	£	%	£	Hrs	Hrs	£	£	£		£

R10: RAINWATER PIPEWORK/GUTTERS Cont.

CAST IRON EAVES GUTTERS (B.S.460) - Cast iron OG gutters and fittings, B.S.460; bolted and mastic joints Cont.

Gutters; straight Cont.

	Del to Site	Waste	Material Cost	Craft Optve	Lab	Labour Cost	Sunds	Nett Rate	Unit	Gross Rate
extra; angles	5.47	2.00	5.58	0.50	–	6.22	–	11.80	nr	12.68
114mm; fixing to timber with galvanized screws	9.14	5.00	9.60	0.50	–	6.22	–	15.82	m	17.00
extra; stopped ends	2.11	2.00	2.15	0.50	–	6.22	–	8.37	nr	9.00
extra; running outlets	5.93	2.00	6.05	0.50	–	6.22	–	12.27	nr	13.19
extra; stopped ends with outlet	4.16	2.00	4.24	0.50	–	6.22	–	10.46	nr	11.25
extra; angles	5.93	2.00	6.05	0.50	–	6.22	–	12.27	nr	13.19
125mm; fixing to timber with galvanized screws	9.63	5.00	10.11	0.57	–	7.09	–	17.20	m	18.49
extra; stopped ends	2.11	2.00	2.15	0.57	–	7.09	–	9.24	nr	9.94
extra; running outlets	6.47	2.00	6.60	0.57	–	7.09	–	13.69	nr	14.72
extra; stopped ends with outlet	4.16	2.00	4.24	0.57	–	7.09	–	11.33	nr	12.18
extra; angles	6.47	2.00	6.60	0.57	–	7.09	–	13.69	nr	14.72

CAST IRON MOULDED SECTIONS (STOCK SECTIONS) - Cast iron moulded (stock sections) gutters and fittings; bolted and mastic joints

Gutters; straight

	Del to Site	Waste	Material Cost	Craft Optve	Lab	Labour Cost	Sunds	Nett Rate	Unit	Gross Rate
100 x 75mm; in standard fascia brackets fixing to timber with galvanized screws	16.33	5.00	17.15	0.75	–	9.33	–	26.48	m	28.46
extra; stopped ends	5.07	2.00	5.17	0.75	–	9.33	–	14.50	nr	15.59
extra; running outlets	13.31	2.00	13.58	0.75	–	9.33	–	22.91	nr	24.62
extra; angles	13.31	2.00	13.58	0.75	–	9.33	–	22.91	nr	24.62
extra; clips	5.56	2.00	5.67	0.75	–	9.33	–	15.00	nr	16.13
125 x 100mm; in standard fascia brackets fixing to timber with galvanized screws	22.80	5.00	23.94	1.00	–	12.44	–	36.38	m	39.11
extra; stopped ends	6.58	2.00	6.71	1.00	–	12.44	–	19.15	nr	20.59
extra; running outlets	19.12	2.00	19.50	1.00	–	12.44	–	31.94	nr	34.34
extra; angles	19.12	2.00	19.50	1.00	–	12.44	–	31.94	nr	34.34
extra; clips	6.58	2.00	6.71	1.00	–	12.44	–	19.15	nr	20.59

CAST IRON BOX GUTTERS - Cast iron box gutters and fittings; bolted and mastic joints

Gutters; straight

	Del to Site	Waste	Material Cost	Craft Optve	Lab	Labour Cost	Sunds	Nett Rate	Unit	Gross Rate
100 x 75mm; in or on supports (included elsewhere)	21.38	5.00	22.45	0.75	–	9.33	–	31.78	m	34.16
extra; stopped ends	2.92	2.00	2.98	0.75	–	9.33	–	12.31	nr	13.23
extra; running outlets	10.04	2.00	10.24	0.75	–	9.33	–	19.57	nr	21.04
extra; angles	10.04	2.00	10.24	0.75	–	9.33	–	19.57	nr	21.04
extra; clips	3.77	2.00	3.85	0.75	–	9.33	–	13.18	nr	14.16

ALUMINIUM GUTTERS (B.S.2997) - Cast aluminium half round gutters and fittings, B.S.2997 Part 2 Section A; bolted and mastic joints

Gutters; straight

	Del to Site	Waste	Material Cost	Craft Optve	Lab	Labour Cost	Sunds	Nett Rate	Unit	Gross Rate
102mm; in standard fascia brackets fixing to timber with galvanized screws	9.48	5.00	9.95	0.50	–	6.22	–	16.17	m	17.39
extra; stopped ends	2.22	2.00	2.26	0.50	–	6.22	–	8.48	nr	9.12
extra; running outlets	5.27	2.00	5.38	0.50	–	6.22	–	11.60	nr	12.47
extra; angles	4.58	2.00	4.67	0.50	–	6.22	–	10.89	nr	11.71
127mm; in standard fascia brackets fixing to timber with galvanized screws	12.66	5.00	13.29	0.57	–	7.09	–	20.38	m	21.91
extra; stopped ends	2.94	2.00	3.00	0.57	–	7.09	–	10.09	nr	10.85
extra; running outlets	6.02	2.00	6.14	0.57	–	7.09	–	13.23	nr	14.22
extra; angles	5.85	2.00	5.97	0.57	–	7.09	–	13.06	nr	14.04

UNPLASTICISED PVC GUTTERS (B.S.4576) - PVC-U half round gutters and fittings, B.S.4576, Part 1; push fit connector joints; gutterwork and supports self coloured grey

Gutters; straight

	Del to Site	Waste	Material Cost	Craft Optve	Lab	Labour Cost	Sunds	Nett Rate	Unit	Gross Rate
105mm; in standard fascia brackets fixing to timber with galvanized screws	2.94	5.00	3.09	0.33	–	4.11	–	7.19	m	7.73
extra; stopped ends	1.06	2.00	1.08	0.33	–	4.11	–	5.19	nr	5.58
extra; running outlets	2.22	2.00	2.26	0.33	–	4.11	–	6.37	nr	6.85
extra; angles	2.43	2.00	2.48	0.33	–	4.11	–	6.58	nr	7.08
114mm; in standard fascia brackets fixing to timber with galvanized screws	3.16	5.00	3.32	0.33	–	4.11	–	7.42	m	7.98
extra; stopped ends	1.27	2.00	1.30	0.33	–	4.11	–	5.40	nr	5.81
extra; running outlets	2.38	2.00	2.43	0.33	–	4.11	–	6.53	nr	7.02
extra; angles	2.59	2.00	2.64	0.33	–	4.11	–	6.75	nr	7.25

GRP GUTTERS - Corofil Ogee GRP gutters and fittings; Corofil GRP Division, Pre-Formed Components Ltd.; bolted and neomastic twinseal extrusion joints

Gutters; straight

	Del to Site	Waste	Material Cost	Craft Optve	Lab	Labour Cost	Sunds	Nett Rate	Unit	Gross Rate
102mm; fixing to timber with galvanized screws	16.11	–	16.11	0.40	–	4.98	–	21.09	m	22.67
extra; stopped ends	10.40	–	10.40	0.40	–	4.98	–	15.38	nr	16.53
extra; running outlets	23.40	–	23.40	0.40	–	4.98	–	28.38	nr	30.50
extra; angles	26.00	–	26.00	0.40	–	4.98	–	30.98	nr	33.30
140mm; fixing to timber with galvanized screws	22.61	–	22.61	0.50	–	6.22	–	28.83	m	30.99
extra; stopped ends	10.40	–	10.40	0.50	–	6.22	–	16.62	nr	17.87
extra; running outlets	23.40	–	23.40	0.50	–	6.22	–	29.62	nr	31.84
extra; angles	26.00	–	26.00	0.50	–	6.22	–	32.22	nr	34.64
205mm; fixing to timber with galvanized screws	25.42	–	25.42	0.50	–	6.22	–	31.64	m	34.01
extra; stopped ends	10.40	–	10.40	0.50	–	6.22	–	16.62	nr	17.87
extra; running outlets	23.40	–	23.40	0.50	–	6.22	–	29.62	nr	31.84

DISPOSAL SYSTEMS

Labour hourly rates: (except Specialists) Craft Operatives £12.44 Labourer £7.02 Rates are national average prices. Refer to REGIONAL VARIATIONS for indicative levels of overall pricing in regions	MATERIALS			LABOUR				RATES		
	Del to Site £	Waste %	Material Cost £	Craft Optve Hrs	Lab Hrs	Labour Cost £	Sunds £	Nett Rate £	Unit	Gross Rate (+7.5%) £
R10: RAINWATER PIPEWORK/GUTTERS Cont.										
GRP GUTTERS - Corofil Ogee GRP gutters and fittings; Corofil GRP Division, Pre-Formed Components Ltd.; bolted and neomastic twinseal extrusion joints Cont.										
Gutters; straight Cont.										
extra; angles	26.00	-	**26.00**	0.50	-	6.22	-	32.22	nr	34.64
Aluminium pipework ancillaries										
Rainwater heads; B.S.2997 square type										
258 x 190 x 178mm; 63mm outlet spigot; dry joint to pipe; fixing to masonry with galvanized screws	16.44	1.00	**16.60**	0.50	-	6.22	-	22.82	nr	24.54
258 x 190 x 178mm; 75mm outlet spigot; dry joint to pipe; fixing to masonry with galvanized screws	17.20	1.00	**17.37**	0.67	-	8.33	-	25.71	nr	27.63
258 x 190 x 178mm; 100mm outlet spigot; dry joint to pipe; fixing to masonry with galvanized screws	28.20	1.00	**28.48**	0.80	-	9.95	-	38.43	nr	41.32
RAINWATER HEADS - Cast iron pipework ancillaries										
Rainwater heads; B.S.460 hopper type (flat pattern)										
210 x 160 x 185mm; 63mm outlet spigot; dry joint to pipe; fixing to masonry with galvanized screws	9.57	1.00	**9.67**	0.50	-	6.22	-	15.89	nr	17.08
210 x 160 x 185mm; 75mm outlet spigot; dry joint to pipe; fixing to masonry with galvanized screws	9.57	1.00	**9.67**	0.67	-	8.33	-	18.00	nr	19.35
250 x 215 x 215mm; 100mm outlet spigot; dry joint to pipe; fixing to masonry with galvanized screws	23.71	1.00	**23.95**	0.80	-	9.95	-	33.90	nr	36.44
Rainwater heads; B.S.460 square type (flat pattern)										
250 x 180 x 175mm; 63mm outlet spigot; dry joint to pipe; fixing to masonry with galvanized screws	20.73	1.00	**20.94**	0.50	-	6.22	-	27.16	nr	29.19
250 x 180 x 175mm; 75mm outlet spigot; dry joint to pipe; fixing to masonry with galvanized screws	20.73	1.00	**20.94**	0.67	-	8.33	-	29.27	nr	31.47
300 x 250 x 200mm; 100mm outlet spigot; dry joint to pipe; fixing to masonry with galvanized screws	41.78	1.00	**42.20**	0.80	-	9.95	-	52.15	nr	56.06
RAINWATER HEADS - Plastics pipework ancillaries self coloured black										
Rainwater heads; B.S.4576 square type										
236 x 152 x 187mm; 68mm outlet spigot; push fit joint to plastics pipe; fixing to masonry with galvanized screws	6.14	1.00	**6.20**	0.50	-	6.22	-	12.42	nr	13.35
252 x 195 x 210mm; 110mm outlet spigot; push fit joint to plastics pipe; fixing to masonry with galvanized screws	12.31	1.00	**12.43**	0.80	-	9.95	-	22.39	nr	24.06
ROOF OUTLETS - Cast iron pipework ancillaries										
Roof outlets; luting flange for asphalt										
flat grating; 75mm outlet spigot; coupling joint to cast iron pipe	58.12	1.00	**58.70**	0.80	-	9.95	-	68.65	nr	73.80
flat grating; 100mm outlet spigot; coupling joint to cast iron pipe	69.16	1.00	**69.85**	1.00	-	12.44	-	82.29	nr	88.46
domical grating; 75mm outlet spigot; coupling joint to cast iron pipe	59.46	1.00	**60.05**	0.80	-	9.95	-	70.01	nr	75.26
domical grating; 100mm outlet spigot; coupling joint to cast iron pipe	70.47	1.00	**71.17**	1.00	-	12.44	-	83.61	nr	89.89
ROOF OUTLETS - Plastics pipework ancillaries, self coloured grey										
Roof outlets; B.S.4576; luting flange for asphalt										
domical grating; 75mm outlet spigot; push fit joint to plastics pipe	19.70	1.00	**19.90**	0.37	-	4.60	-	24.50	nr	26.34
domical grating; 100mm outlet spigot; push fit joint to plastics pipe	19.70	1.00	**19.90**	0.37	-	4.60	-	24.50	nr	26.34
R11: FOUL DRAINAGE ABOVE GROUND										
CAST IRON SOIL AND WASTE PIPES - Cast iron pipes and fittings, B.S.416; bolted and synthetic rubber gasket couplings										
Pipes; straight										
75mm; in or on supports (included elsewhere)	11.74	5.00	**12.33**	0.67	-	8.33	-	20.66	m	22.21
extra; pipes with inspection door bolted and sealed............................	24.39	2.00	**24.88**	0.67	-	8.33	-	33.21	nr	35.70
extra; short radius bends	13.09	2.00	**13.35**	0.67	-	8.33	-	21.69	nr	23.31
extra; short radius bends with inspection door bolted and sealed	24.87	2.00	**25.37**	0.67	-	8.33	-	33.70	nr	36.23
extra; offset bends 150mm projection	14.95	2.00	**15.25**	0.67	-	8.33	-	23.58	nr	25.35
extra; single angled branches	22.20	2.00	**22.64**	1.00	-	12.44	-	35.08	nr	37.72
extra; single angled branches with inspection door bolted and sealed	33.99	2.00	**34.67**	1.00	-	12.44	-	47.11	nr	50.64
extra; double angled branches	35.49	2.00	**36.20**	1.33	-	16.55	-	52.74	nr	56.70
extra; roof connectors	27.08	2.00	**27.62**	0.67	-	8.33	-	35.96	nr	38.65
boss pipes with one threaded boss	21.70	2.00	**22.13**	0.67	-	8.33	-	30.47	nr	32.75
boss pipes with two threaded bosses	27.54	2.00	**28.09**	0.67	-	8.33	-	36.43	nr	39.16
100mm; in or on supports (included elsewhere)	14.34	5.00	**15.06**	0.80	-	9.95	-	25.01	m	26.88
extra; pipes with inspection door bolted and sealed............................	26.91	2.00	**27.45**	0.80	-	9.95	-	37.40	nr	40.21
extra; short radius bends	17.70	2.00	**18.05**	0.80	-	9.95	-	28.01	nr	30.11
extra; short radius bends with inspection door bolted and sealed........................	30.11	2.00	**30.71**	0.80	-	9.95	-	40.66	nr	43.71

Labour hourly rates: (except Specialists) Craft Operatives £12.44 Labourer £7.02 Rates are national average prices. Refer to REGIONAL VARIATIONS for indicative levels of overall pricing in regions	Del to Site £	Waste %	Material Cost £	Craft Optve Hrs	Lab Hrs	Labour Cost £	Sunds £	Nett Rate £	Unit	Gross Rate (+7.5%) £

R11: FOUL DRAINAGE ABOVE GROUND Cont.

CAST IRON SOIL AND WASTE PIPES - Cast iron pipes and fittings, B.S.416; bolted and synthetic rubber gasket couplings Cont.

	Del to Site £	Waste %	Material Cost £	Craft Optve Hrs	Lab Hrs	Labour Cost £	Sunds £	Nett Rate £	Unit	Gross Rate £
Pipes; straight Cont.										
extra; offset bends 150mm projection	20.53	2.00	20.94	0.80	-	9.95	-	30.89	nr	33.21
extra; offset bends 300mm projection	24.60	2.00	25.09	0.80	-	9.95	-	35.04	nr	37.67
extra; single angled branches	30.35	2.00	30.96	1.20	-	14.93	-	45.88	nr	49.33
extra; single angled branches with inspection door bolted and sealed	42.77	2.00	43.63	1.20	-	14.93	-	58.55	nr	62.94
extra; double angled branches	40.99	2.00	41.81	1.60	-	19.90	-	61.71	nr	66.34
extra; double angled branches with inspection door bolted and sealed	53.41	2.00	54.48	1.60	-	19.90	-	74.38	nr	79.96
extra; single anti-syphon branches, angled branch	31.57	2.00	32.20	1.20	-	14.93	-	47.13	nr	50.66
extra; straight wc connectors	18.68	2.00	19.05	0.80	-	9.95	-	29.01	nr	31.18
extra; roof connectors	25.98	2.00	26.50	0.80	-	9.95	-	36.45	nr	39.19
boss pipes with one threaded boss	26.46	2.00	26.99	0.80	-	9.95	-	36.94	nr	39.71
boss pipes with two threaded bosses	32.29	2.00	32.94	0.80	-	9.95	-	42.89	nr	46.10
50mm; in standard wall fixing bracket (fixing included elsewhere)	14.75	5.00	15.49	0.55	-	6.84	-	22.33	m	24.00
extra; pipes with inspection door bolted and sealed	23.90	2.00	24.38	0.55	-	6.84	-	31.22	nr	33.56
extra; short radius bends	12.60	2.00	12.85	0.55	-	6.84	-	19.69	nr	21.17
extra; short radius bends with inspection door bolted and sealed	24.38	2.00	24.87	0.55	-	6.84	-	31.71	nr	34.09
extra; single angled branches	21.22	2.00	21.64	0.82	-	10.20	-	31.85	nr	34.23
extra; single angled branches with inspection door bolted and sealed	33.01	2.00	33.67	0.82	-	10.20	-	43.87	nr	47.16
extra; roof connectors	26.59	2.00	27.12	0.55	-	6.84	-	33.96	nr	36.51

CAST IRON SOIL AND WASTE PIPES - Cast iron Stanton SMU lightweight pipes and fittings, stainless steel couplings

	Del to Site £	Waste %	Material Cost £	Craft Optve Hrs	Lab Hrs	Labour Cost £	Sunds £	Nett Rate £	Unit	Gross Rate £
Pipes; straight										
50mm; in or on supports (included elsewhere)	8.64	5.00	9.07	0.55	-	6.84	-	15.91	m	17.11
extra; short pipe with access door	16.70	2.00	17.03	0.55	-	6.84	-	23.88	nr	25.67
extra; bends	8.17	2.00	8.33	0.55	-	6.84	-	15.18	nr	16.31
extra; offset bend 75mm projection	10.49	2.00	10.70	0.55	-	6.84	-	17.54	nr	18.86
extra; offset bend 150mm projection	12.67	2.00	12.92	0.55	-	6.84	-	19.77	nr	21.25
extra; single angled branch	14.32	2.00	14.61	0.82	-	10.20	-	24.81	nr	26.67
extra; plug	2.10	2.00	2.14	0.27	-	3.36	-	5.50	nr	5.91
extra; universal plug with one inlet	6.21	2.00	6.33	0.14	-	1.74	-	8.08	nr	8.68
extra; stepping ring for connection to other pipe materials	3.62	2.00	3.69	0.14	-	1.74	-	5.43	nr	5.84
extra; step coupling for connection to B.S.416 pipe	4.26	2.00	4.35	0.40	-	4.98	-	9.32	nr	10.02
75mm; in or on supports (included elsewhere)	9.93	5.00	10.43	0.67	-	8.33	-	18.76	m	20.17
extra; short pipe with access door	18.29	2.00	18.66	0.67	-	8.33	-	26.99	nr	29.01
extra; bends	9.13	2.00	9.31	0.67	-	8.33	-	17.65	nr	18.97
extra; offset bend 75mm projection	13.30	2.00	13.57	0.67	-	8.33	-	21.90	nr	23.54
extra; offset bend 150mm projection	16.61	2.00	16.94	0.67	-	8.33	-	25.28	nr	27.17
extra; single angled branch	15.40	2.00	15.71	1.00	-	12.44	-	28.15	nr	30.26
extra; diminishing piece 75/50	11.04	2.00	11.26	0.67	-	8.33	-	19.60	nr	21.07
extra; plug	2.22	2.00	2.26	0.33	-	4.11	-	6.37	nr	6.85
extra; universal plug with one inlet	8.85	2.00	9.03	0.17	-	2.11	-	11.14	nr	11.98
extra; push fit gasket to suit 32 DN pipe	1.46	2.00	1.49	-	-	-	-	1.49	nr	1.60
extra; push fit gasket to suit 40 DN pipe	1.46	2.00	1.49	-	-	-	-	1.49	nr	1.60
extra; stepping ring for connection to other pipe materials	4.21	2.00	4.29	0.17	-	2.11	-	6.41	nr	6.89
extra; step coupling for connection to B.S.416 pipe	4.76	2.00	4.86	-	-	-	-	4.86	nr	5.22
100mm; in or on supports (included elsewhere)	12.11	5.00	12.72	-	-	-	-	12.72	m	13.67
extra; short pipe with access door	20.74	2.00	21.15	0.80	-	9.95	-	31.11	nr	33.44
extra; bends	11.22	2.00	11.44	0.80	-	9.95	-	21.40	nr	23.00
extra; long radius bend	20.93	2.00	21.35	0.80	-	9.95	-	31.30	nr	33.65
extra; offset bend 75mm projection	17.19	2.00	17.53	0.80	-	9.95	-	27.49	nr	29.55
extra; offset bend 150mm projection	21.79	2.00	22.23	0.80	-	9.95	-	32.18	nr	34.59
extra; single angled branch	19.81	2.00	20.21	1.20	-	14.93	-	35.13	nr	37.77
extra; double angled branch	30.07	2.00	30.67	1.60	-	19.90	-	50.58	nr	54.37
extra; diminishing piece 100/50	13.65	2.00	13.92	0.80	-	9.95	-	23.88	nr	25.67
extra; diminishing piece 100/75	13.67	2.00	13.94	0.80	-	9.95	-	23.90	nr	25.69
extra; plug	2.59	2.00	2.64	0.40	-	4.98	-	7.62	nr	8.19
extra; universal plug with three inlets	11.14	2.00	11.36	0.40	-	4.98	-	16.34	nr	17.56
extra; bossed end cap with inlet	5.33	2.00	5.44	0.40	-	4.98	-	10.41	nr	11.19
extra; push fit gasket to suit 32 DN pipe	1.46	2.00	1.49	-	-	-	-	1.49	nr	1.60
extra; push fit gasket to suit 40 DN pipe	1.46	2.00	1.49	-	-	-	-	1.49	nr	1.60
extra; stepping ring for connection to other pipe materials	4.84	2.00	4.94	0.20	-	2.49	-	7.42	nr	7.98
extra; step coupling for connection to B.S.416 pipe	6.08	2.00	6.20	-	-	-	-	6.20	nr	6.67
extra; traditional joint connector	14.72	2.00	15.01	-	-	-	-	15.01	nr	16.14
extra; W.C. connector	8.32	2.00	8.49	0.80	-	9.95	-	18.44	nr	19.82
extra; roof connector (asphalt)	30.08	2.00	30.68	0.80	-	9.95	-	40.63	nr	43.68

CAST IRON SOIL AND WASTE PIPES - Pipe supports

	Del to Site £	Waste %	Material Cost £	Craft Optve Hrs	Lab Hrs	Labour Cost £	Sunds £	Nett Rate £	Unit	Gross Rate £
Galvanised steel vertical bracket										
for 50mm pipes; for fixing bolt(fixing included elsewhere)	4.09	2.00	4.17	-	-	-	-	4.17	nr	4.48
for 75mm pipes; for fixing bolt(fixing included elsewhere)	4.45	2.00	4.54	-	-	-	-	4.54	nr	4.88
for 100mm pipes; for fixing bolt(fixing included elsewhere)	4.86	2.00	4.96	-	-	-	-	4.96	nr	5.33

DISPOSAL SYSTEMS

Labour hourly rates: (except Specialists) Craft Operatives £12.44 Labourer £7.02 Rates are national average prices. Refer to REGIONAL VARIATIONS for indicative levels of overall pricing in regions	MATERIALS			LABOUR				RATES		
	Del to Site	Waste	Material Cost	Craft Optve	Lab	Labour Cost	Sunds	Nett Rate	Unit	Gross Rate (+7.5%)
	£	%	£	Hrs	Hrs	£	£	£		£
R11: FOUL DRAINAGE ABOVE GROUND Cont.										
CAST IRON SOIL AND WASTE PIPES - Pipe supports Cont.										
Aluminium wall hook - type 101										
for 50mm pipes; for M8 fixing bolt (fixing included elsewhere)	3.15	2.00	3.21	-	-	-	-	3.21	nr	3.45
for 75mm pipes; for M8 fixing bolt (fixing included elsewhere)	3.72	2.00	3.79	-	-	-	-	3.79	nr	4.08
for 100mm pipes; for M8 fixing bolt (fixing included elsewhere)	4.30	2.00	4.39	-	-	-	-	4.39	nr	4.71
UNPLASTICIZED MPVC WASTE PIPES (SOLVENT WELDED) - **MuPVC pipes and fittings, B.S.5255 Section Three;** **solvent welded joints**										
Pipes; straight										
36mm; in standard plastics pipe brackets fixing to timber with screws	1.54	10.00	1.69	-	-	-	-	1.69	m	1.82
extra; straight expansion couplings	1.04	2.00	1.06	-	-	-	-	1.06	nr	1.14
extra; connections to plastics pipe socket; socket adaptor; solvent welded joint	1.06	2.00	1.08	0.25	-	3.11	-	4.19	nr	4.51
extra; connections to plastics pipe boss; boss adaptor; solvent welded joint	1.06	2.00	1.08	0.25	-	3.11	-	4.19	nr	4.51
extra; fittings; one end	1.09	2.00	1.11	0.13	-	1.62	-	2.73	nr	2.93
extra; fittings; two ends	0.92	2.00	0.94	0.25	-	3.11	-	4.05	nr	4.35
extra; fittings; three ends	1.31	2.00	1.34	0.25	-	3.11	-	4.45	nr	4.78
42mm; in standard plastics pipe brackets fixing to timber with screws	1.65	10.00	1.81	0.33	-	4.11	-	5.92	m	6.36
extra; straight expansion couplings	1.25	2.00	1.27	0.25	-	3.11	-	4.38	nr	4.71
extra; connections to plastics pipe socket; socket adaptor; solvent welded joint	1.06	2.00	1.08	0.25	-	3.11	-	4.19	nr	4.51
extra; connections to plastics pipe boss; boss adaptor; solvent welded joint	1.06	2.00	1.08	0.25	-	3.11	-	4.19	nr	4.51
extra; fittings; one end	1.29	2.00	1.32	0.13	-	1.62	-	2.93	nr	3.15
extra; fittings; two ends	1.03	2.00	1.05	0.25	-	3.11	-	4.16	nr	4.47
extra; fittings; three ends	1.65	2.00	1.68	0.25	-	3.11	-	4.79	nr	5.15
55mm; in standard plastics pipe brackets fixing to timber with screws	2.98	10.00	3.28	0.40	-	4.98	-	8.25	m	8.87
extra; straight expansion couplings	1.70	2.00	1.73	0.33	-	4.11	-	5.84	nr	6.28
extra; connections to plastics pipe socket; socket adaptor; solvent welded joint	1.46	2.00	1.49	0.33	-	4.11	-	5.59	nr	6.01
extra; connections to plastics pipe boss; boss adaptor; solvent welded joint	1.46	2.00	1.49	0.33	-	4.11	-	5.59	nr	6.01
extra; fittings; one end	2.09	2.00	2.13	0.17	-	2.11	-	4.25	nr	4.57
extra; fittings; two ends	1.71	2.00	1.74	0.33	-	4.11	-	5.85	nr	6.29
extra; fittings; three ends	3.26	2.00	3.33	0.33	-	4.11	-	7.43	nr	7.99
36mm; in standard plastics pipe brackets fixing to masonry with screws	1.54	10.00	1.69	0.39	-	4.85	-	6.55	m	7.04
42mm; in standard plastics pipe brackets fixing to masonry with screws	1.65	10.00	1.81	0.39	-	4.85	-	6.67	m	7.17
55mm; in standard plastics pipe brackets fixing to masonry with screws	2.98	10.00	3.28	0.46	-	5.72	-	9.00	m	9.68
POLYPROPYLENE WASTE PIPES ('O' RING) - Polypropylene **pipes and fittings; butyl ring joints**										
Pipes; straight										
36mm; in standard plastics pipe brackets fixing to timber with screws	0.57	10.00	0.63	0.33	-	4.11	-	4.73	m	5.09
extra; connections to plastics pipe socket; socket adaptor; solvent welded joint	1.06	2.00	1.08	0.25	-	3.11	-	4.19	nr	4.51
extra; connections to plastics pipe boss; boss adaptor; solvent welded joint	1.06	2.00	1.08	0.25	-	3.11	-	4.19	nr	4.51
extra; fittings; one end	0.55	2.00	0.56	0.13	-	1.62	-	2.18	nr	2.34
extra; fittings; two ends	0.96	2.00	0.98	0.25	-	3.11	-	4.09	nr	4.40
extra; fittings; three ends	1.42	2.00	1.45	0.25	-	3.11	-	4.56	nr	4.90
42mm; in standard plastics pipe brackets fixing to timber with screws	0.69	10.00	0.76	0.33	-	4.11	-	4.86	m	5.23
extra; connections to plastics pipe socket; socket adaptor; solvent welded joint	1.06	2.00	1.08	0.33	-	4.11	-	5.19	nr	5.58
extra; connections to plastics pipe boss; boss adaptor; solvent welded joint	1.06	2.00	1.08	0.33	-	4.11	-	5.19	nr	5.58
extra; fittings; one end	0.68	2.00	0.69	0.17	-	2.11	-	2.81	nr	3.02
extra; fittings; two ends	0.10	2.00	0.10	0.33	-	4.11	-	4.21	nr	4.52
extra; fittings; three ends	1.47	2.00	1.50	0.33	-	4.11	-	5.60	nr	6.02
36mm; in standard plastics pipe brackets fixing to masonry with screws	0.57	10.00	0.63	0.39	-	4.85	-	5.48	m	5.89
42mm; in standard plastics pipe brackets fixing to masonry with screws	0.69	10.00	0.76	0.39	-	4.85	-	5.61	m	6.03
UNPLASTICIZED SOIL AND VENTILATING PIPES - PVC-U **pipes and fittings, B.S.4514; rubber ring joints;** **pipework self coloured grey**										
Pipes; straight										
82mm; in or on supports (included elsewhere)	3.96	5.00	4.16	0.44	-	5.47	-	9.63	m	10.35
extra; pipes with inspection door bolted and sealed	9.34	2.00	9.53	0.33	-	4.11	-	13.63	nr	14.65
extra; short radius bends	5.24	2.00	5.34	0.33	-	4.11	-	9.45	nr	10.16
extra; single angled branches	8.23	2.00	8.39	0.33	-	4.11	-	12.50	nr	13.44
extra; straight wc connectors	4.82	2.00	4.92	0.33	-	4.11	-	9.02	nr	9.70
pipes with one boss socket	5.98	2.00	6.10	0.33	-	4.11	-	10.20	nr	10.97
110mm; in or on supports (included elsewhere)	3.77	5.00	3.96	0.50	-	6.22	-	10.18	m	10.94

Labour hourly rates: (except Specialists) Craft Operatives £12.44 Labourer £7.02 Rates are national average prices. Refer to REGIONAL VARIATIONS for indicative levels of overall pricing in regions	MATERIALS			LABOUR				RATES		
	Del to Site	Waste	Material Cost	Craft Optve	Lab	Labour Cost	Sunds	Nett Rate	Unit	Gross Rate (+7.5%)
	£	%	£	Hrs	Hrs	£	£	£		£

R11: FOUL DRAINAGE ABOVE GROUND Cont.

UNPLASTICIZED SOIL AND VENTILATING PIPES - PVC-U
pipes and fittings, B.S.4514; rubber ring joints;
pipework self coloured grey Cont.

	Del to Site £	Waste %	Material Cost £	Craft Optve Hrs	Lab Hrs	Labour Cost £	Sunds £	Nett Rate £	Unit	Gross Rate £
Pipes; straight Cont.										
extra; connections to cast iron pipe socket caulking bush; caulked lead and hempen spun yarn joint	13.64	2.00	13.91	1.25	–	15.55	–	29.46	nr	31.67
extra; pipes with inspection door bolted and sealed	9.21	2.00	9.39	0.50	–	6.22	–	15.61	nr	16.79
extra; short radius bends	6.31	2.00	6.44	0.50	–	6.22	–	12.66	nr	13.61
extra; single angled branches	8.21	2.00	8.37	0.50	–	6.22	–	14.59	nr	15.69
extra; straight wc connectors	5.52	2.00	5.63	0.50	–	6.22	–	11.85	nr	12.74
pipes with one boss socket	4.62	2.00	4.71	0.50	–	6.22	–	10.93	nr	11.75
160mm; in or on supports (included elsewhere)	9.95	5.00	10.45	0.75	–	9.33	–	19.78	m	21.26
extra; pipes with inspection door bolted and sealed	20.35	2.00	20.76	0.75	–	9.33	–	30.09	nr	32.34
extra; short radius bends	16.59	2.00	16.92	0.75	–	9.33	–	26.25	nr	28.22
extra; single angled branches	27.32	2.00	27.87	0.75	–	9.33	–	37.20	nr	39.99
pipes with one boss socket	9.42	2.00	9.61	0.75	–	9.33	–	18.94	nr	20.36

UNPLASTICIZED SOIL AND VENTILATING PIPES - Plastics
pipework ancillaries, self coloured grey

	Del to Site £	Waste %	Material Cost £	Craft Optve Hrs	Lab Hrs	Labour Cost £	Sunds £	Nett Rate £	Unit	Gross Rate £
Weathering aprons										
fitted to 82mm pipes	1.42	2.00	1.45	0.33	–	4.11	–	5.55	nr	5.97
fitted to 110mm pipes	1.51	2.00	1.54	0.50	–	6.22	–	7.76	nr	8.34
fitted to 160mm pipes	4.14	2.00	4.22	0.75	–	9.33	–	13.55	nr	14.57

UNPLASTICIZED SOIL AND VENTILATING PIPES - Pipe
supports

	Del to Site £	Waste %	Material Cost £	Craft Optve Hrs	Lab Hrs	Labour Cost £	Sunds £	Nett Rate £	Unit	Gross Rate £
Plastics coated metal holderbats										
for 82mm pipes; for building in (fixing included elsewhere)	3.03	2.00	3.09	–	–	–	–	3.09	nr	3.32
for 110mm pipes; for building in (fixing included elsewhere)	3.00	2.00	3.06	–	–	–	–	3.06	nr	3.29
for 82mm pipes; fixing to timber with galvanized screws	2.43	2.00	2.48	0.17	–	2.11	–	4.59	nr	4.94
for 110mm pipes; fixing to timber with galvanized screws	2.40	2.00	2.45	–	–	–	–	2.45	nr	2.63
for 160mm pipes; fixing to timber with galvanized screws	3.71	2.00	3.78	–	–	–	–	3.78	nr	4.07

WIRE BALLOON GRATINGS - Copper pipework ancillaries

	Del to Site £	Waste %	Material Cost £	Craft Optve Hrs	Lab Hrs	Labour Cost £	Sunds £	Nett Rate £	Unit	Gross Rate £
Wire balloon gratings										
fitted to 50mm pipes	1.95	3.00	2.01	0.10	–	1.24	–	3.25	nr	3.50
fitted to 75mm pipes	2.30	3.00	2.37	0.10	–	1.24	–	3.61	nr	3.88
fitted to 100mm pipes	2.65	3.00	2.73	0.10	–	1.24	–	3.97	nr	4.27

WIRE BALLOON GRATINGS - Steel pipework ancillaries,
galvanized

	Del to Site £	Waste %	Material Cost £	Craft Optve Hrs	Lab Hrs	Labour Cost £	Sunds £	Nett Rate £	Unit	Gross Rate £
Wire balloon gratings										
fitted to 50mm pipes	1.55	2.00	1.58	0.10	–	1.24	–	2.83	nr	3.04
fitted to 75mm pipes	1.75	2.00	1.78	0.10	–	1.24	–	3.03	nr	3.26
fitted to 100mm pipes	2.00	2.00	2.04	0.10	–	1.24	–	3.28	nr	3.53

PLASTIC TRAPS - Plastics pipework ancillaries self
coloured white

	Del to Site £	Waste %	Material Cost £	Craft Optve Hrs	Lab Hrs	Labour Cost £	Sunds £	Nett Rate £	Unit	Gross Rate £
Traps; B.S.3943										
P, two piece, 75mm seal, inlet with coupling nut, outlet with seal ring socket										
36mm outlet	2.37	2.00	2.42	0.33	–	4.11	–	6.52	nr	7.01
42mm outlet	2.73	2.00	2.78	0.33	–	4.11	–	6.89	nr	7.41
S, two piece, 75mm seal, inlet with coupling nut, outlet with seal ring socket										
36mm outlet	3.00	2.00	3.06	0.33	–	4.11	–	7.17	nr	7.70
42mm outlet	3.52	2.00	3.59	0.33	–	4.11	–	7.70	nr	8.27
P, two piece, 75mm seal, with flexible polypropylene pipe for overflow connection, inlet with coupling nut, outlet with seal ring socket										
42mm outlet	5.08	2.00	5.18	0.50	–	6.22	–	11.40	nr	12.26

COPPER TRAPS - Copper pipework ancillaries

	Del to Site £	Waste %	Material Cost £	Craft Optve Hrs	Lab Hrs	Labour Cost £	Sunds £	Nett Rate £	Unit	Gross Rate £
Traps; B.S.1184 Section Two (Solid Drawn)										
P, two piece, 75mm seal, inlet with coupling nut, compression outlet										
35mm outlet	12.74	2.00	12.99	0.33	–	4.11	–	17.10	nr	18.38
42mm outlet	15.02	2.00	15.32	0.33	–	4.11	–	19.43	nr	20.88
54mm outlet	53.88	2.00	54.96	0.70	–	8.71	–	63.67	nr	68.44
S, two piece, 75mm seal, inlet with coupling nut, compression outlet										
35mm outlet	13.51	2.00	13.78	0.33	–	4.11	–	17.89	nr	19.23
42mm outlet	16.24	2.00	16.56	0.33	–	4.11	–	20.67	nr	22.22
54mm outlet	56.91	2.00	58.05	0.70	–	8.71	–	66.76	nr	71.76
bath, two piece, 75mm seal, overflow connection, cleaning eye and plug, inlet with coupling nut, compression outlet										
42mm outlet	22.17	2.00	22.61	0.50	–	6.22	–	28.83	nr	31.00

DISPOSAL SYSTEMS

Labour hourly rates: (except Specialists) Craft Operatives £9.23 Labourer £7.02 Rates are national average prices. Refer to REGIONAL VARIATIONS for indicative levels of overall pricing in regions	PLANT & TRANSPORT			LABOUR				RATES		
	Plant Cost £	Trans Cost £	P & T Cost £	Craft Optve Hrs	Lab Hrs	Labour Cost £	Sunds £	Nett Rate £	Unit	Gross Rate (+7.5%) £
R12: DRAINAGE BELOW GROUND										
EXCAVATION FOR DRAINS - BY MACHINE - Excavating trenches by machine to receive pipes not exceeding 200mm nominal size; disposing of surplus excavated material by removing from site										
Excavations commencing from natural ground level average depth										
500mm	1.33	2.34	3.67	-	0.23	1.61	0.63	5.91	m	6.36
750mm	2.14	2.34	4.48	-	0.34	2.39	0.95	7.82	m	8.40
1000mm	2.97	2.34	5.31	-	0.45	3.16	1.26	9.73	m	10.46
1250mm	3.83	2.34	6.17	-	0.61	4.28	1.90	12.35	m	13.28
1500mm	4.62	2.34	6.96	-	0.73	5.12	2.28	14.36	m	15.44
1750mm	5.48	2.34	7.82	-	0.86	6.04	2.66	16.52	m	17.76
2000mm	6.27	2.34	8.61	-	0.98	6.88	3.04	18.53	m	19.92
2250mm	8.03	2.34	10.37	-	1.69	11.86	4.05	26.28	m	28.26
2500mm	8.92	2.34	11.26	-	1.88	13.20	4.50	28.96	m	31.13
2750mm	9.88	2.34	12.22	-	2.06	14.46	4.95	31.63	m	34.00
3000mm	10.77	2.34	13.11	-	2.25	15.80	5.40	34.31	m	36.88
3250mm	11.73	2.34	14.07	-	2.44	17.13	5.85	37.05	m	39.83
3500mm	12.62	2.34	14.96	-	2.63	18.46	6.30	39.72	m	42.70
3750mm	13.58	2.34	15.92	-	2.82	19.80	6.75	42.47	m	45.65
4000mm	14.47	2.34	16.81	-	3.00	21.06	7.20	45.07	m	48.45
4250mm	15.43	2.34	17.77	-	4.25	29.84	8.42	56.02	m	60.23
4500mm	16.32	2.34	18.66	-	4.50	31.59	8.91	59.16	m	63.60
4750mm	17.28	2.34	19.62	-	4.75	33.34	9.41	62.38	m	67.05
5000mm	18.17	2.34	20.51	-	6.25	43.88	9.90	74.28	m	79.86
5250mm	19.13	2.34	21.47	-	6.56	46.05	10.40	77.92	m	83.77
5500mm	20.02	2.34	22.36	-	6.87	48.23	10.89	81.48	m	87.59
5750mm	20.98	2.34	23.32	-	7.18	50.40	11.39	85.11	m	91.50
6000mm	21.87	2.34	24.21	-	7.50	52.65	11.88	88.74	m	95.40
EXCAVATION FOR DRAINS - BY MACHINE - Excavating trenches by machine to receive pipes 225mm nominal size; disposing of surplus excavated material by removing from site										
Excavations commencing from natural ground level average depth										
500mm	1.42	2.47	3.89	-	0.24	1.68	0.63	6.20	m	6.67
750mm	2.28	2.47	4.75	-	0.36	2.53	0.95	8.23	m	8.84
1000mm	3.15	2.47	5.62	-	0.47	3.30	1.26	10.18	m	10.94
1250mm	4.00	2.47	6.47	-	0.64	4.49	1.90	12.86	m	13.83
1500mm	4.92	2.47	7.39	-	0.77	5.41	2.28	15.08	m	16.21
1750mm	5.79	2.47	8.26	-	0.90	6.32	2.66	17.24	m	18.53
2000mm	6.64	2.47	9.11	-	1.03	7.23	3.04	19.38	m	20.83
2250mm	8.45	2.47	10.92	-	1.77	12.43	4.05	27.40	m	29.45
2500mm	9.49	2.47	11.96	-	1.97	13.83	4.50	30.29	m	32.56
2750mm	10.45	2.47	12.92	-	2.16	15.16	4.95	33.03	m	35.51
3000mm	11.41	2.47	13.88	-	2.36	16.57	5.40	35.85	m	38.54
3250mm	12.37	2.47	14.84	-	2.56	17.97	5.85	38.66	m	41.56
3500mm	13.41	2.47	15.88	-	2.76	19.38	6.30	41.56	m	44.67
3750mm	14.37	2.47	16.84	-	2.96	20.78	6.75	44.37	m	47.70
4000mm	15.33	2.47	17.80	-	3.15	22.11	7.20	47.11	m	50.65
4250mm	16.29	2.47	18.76	-	4.46	31.31	8.42	58.49	m	62.88
4500mm	17.33	2.47	19.80	-	4.72	33.13	8.91	61.84	m	66.48
4750mm	18.29	2.47	20.76	-	4.98	34.96	9.41	65.13	m	70.01
5000mm	19.25	2.47	21.72	-	6.56	46.05	9.90	77.67	m	83.50
5250mm	20.22	2.47	22.69	-	6.89	48.37	10.40	81.46	m	87.57
5500mm	21.25	2.47	23.72	-	7.20	50.54	10.89	85.15	m	91.54
5750mm	22.21	2.47	24.68	-	7.54	52.93	11.39	89.00	m	95.68
6000mm	23.18	2.47	25.65	-	7.85	55.11	11.88	92.64	m	99.58
EXCAVATION FOR DRAINS - BY MACHINE - Excavating trenches by machine to receive pipes 300mm nominal size; disposing of surplus excavated material by removing from site										
Excavations commencing from natural ground level average depth										
500mm	1.58	2.73	4.31	-	0.25	1.75	0.63	6.70	m	7.20
750mm	2.57	2.73	5.30	-	0.44	3.09	0.95	9.34	m	10.04
1000mm	3.56	2.73	6.29	-	0.50	3.51	1.26	11.06	m	11.89
1250mm	4.55	2.73	7.28	-	0.67	4.70	1.90	13.88	m	14.92
1500mm	5.54	2.73	8.27	-	0.80	5.62	2.28	16.17	m	17.38
1750mm	6.53	2.73	9.26	-	0.95	6.67	2.66	18.59	m	19.98
2000mm	7.52	2.73	10.25	-	1.10	7.72	3.04	21.01	m	22.59
2250mm	9.59	2.73	12.32	-	1.86	13.06	4.05	29.43	m	31.63
2500mm	10.70	2.73	13.43	-	2.07	14.53	4.50	32.46	m	34.90
2750mm	11.91	2.73	14.64	-	2.25	15.80	4.95	35.38	m	38.04
3000mm	12.92	2.73	15.65	-	2.48	17.41	5.40	38.46	m	41.34
3250mm	14.03	2.73	16.76	-	2.68	18.81	5.85	41.42	m	44.53
3500mm	15.14	2.73	17.87	-	2.90	20.36	6.30	44.53	m	47.87
3750mm	16.25	2.73	18.98	-	3.10	21.76	6.75	47.49	m	51.05
4000mm	17.36	2.73	20.09	-	3.30	23.17	7.20	50.46	m	54.24
4250mm	18.47	2.73	21.20	-	4.68	32.85	7.65	61.70	m	66.33
4500mm	19.58	2.73	22.31	-	4.95	34.75	8.91	65.97	m	70.92
4750mm	20.69	2.73	23.42	-	5.23	36.71	9.41	69.54	m	74.76
5000mm	21.80	2.73	24.53	-	6.88	48.30	9.90	82.73	m	88.93
5250mm	22.91	2.73	25.64	-	7.22	50.68	10.40	86.72	m	93.23
5500mm	24.02	2.73	26.75	-	7.55	53.00	10.89	90.64	m	97.44
5750mm	25.13	2.73	27.86	-	7.90	55.46	11.39	94.71	m	101.81
6000mm	26.24	2.73	28.97	-	8.25	57.91	11.88	98.77	m	106.17

DISPOSAL SYSTEMS – MAJOR WORKS

Labour hourly rates: (except Specialists) Craft Operatives £9.23 Labourer £7.02 Rates are national average prices. Refer to REGIONAL VARIATIONS for indicative levels of overall pricing in regions	PLANT & TRANSPORT			LABOUR				RATES		
	Plant Cost	Trans Cost	P & T Cost	Craft Optve	Lab	Labour Cost	Sunds	Nett Rate	Unit	Gross Rate (+7.5%)
	£	£	£	Hrs	Hrs	£	£	£		£

R12: DRAINAGE BELOW GROUND Cont.

EXCAVATION FOR DRAINS - BY MACHINE - Excavating trenches by machine to receive pipes 400mm nominal size; disposing of surplus excavated material by removing from site

Excavations commencing from natural ground level
average depth

2250mm	11.19	3.25	14.44	–	2.03	14.25	4.05	32.74	m	35.20
2500mm	12.45	3.25	15.70	–	2.26	15.87	4.50	36.07	m	38.77
2750mm	13.78	3.25	17.03	–	2.47	17.34	4.95	39.32	m	42.27
3000mm	15.04	3.25	18.29	–	2.70	18.95	5.40	42.64	m	45.84
3250mm	16.37	3.25	19.62	–	2.92	20.50	5.85	45.97	m	49.42
3500mm	17.63	3.25	20.88	–	3.16	22.18	6.30	49.36	m	53.07
3750mm	18.96	3.25	22.21	–	3.38	23.73	6.75	52.69	m	56.64
4000mm	20.22	3.25	23.47	–	3.60	25.27	7.20	55.94	m	60.14
4250mm	21.55	3.25	24.80	–	5.10	35.80	8.42	69.02	m	74.20
4500mm	22.81	3.25	26.06	–	5.40	37.91	8.91	72.88	m	78.34
4750mm	24.41	3.25	27.66	–	5.70	40.01	9.41	77.08	m	82.87
5000mm	25.40	3.25	28.65	–	7.50	52.65	9.90	91.20	m	98.04
5250mm	26.73	3.25	29.98	–	7.85	55.11	10.40	95.49	m	102.65
5500mm	27.99	3.25	31.24	–	8.25	57.91	10.89	100.05	m	107.55
5750mm	29.32	3.25	32.57	–	8.60	60.37	11.39	104.33	m	112.16
6000mm	30.58	3.25	33.83	–	9.00	63.18	11.98	108.99	m	117.16

EXCAVATION FOR DRAINS - BY MACHINE - Excavating trenches by machine to receive pipes 450mm nominal size; disposing of surplus excavated material by removing from site

Excavations commencing from natural ground level
average depth

2250mm	11.98	3.38	15.36	–	2.11	14.81	4.05	34.22	m	36.79
2500mm	13.39	3.38	16.77	–	2.35	16.50	4.50	37.77	m	40.60
2750mm	14.72	3.38	18.10	–	2.58	18.11	4.95	41.16	m	44.25
3000mm	16.13	3.38	19.51	–	2.81	19.73	5.40	44.64	m	47.98
3250mm	17.53	3.38	20.91	–	3.05	21.41	5.85	48.17	m	51.78
3500mm	18.94	3.38	22.32	–	3.29	23.10	6.30	51.72	m	55.59
3750mm	20.27	3.38	23.65	–	3.52	24.71	6.75	55.11	m	59.24
4000mm	21.68	3.38	25.06	–	3.75	26.32	7.20	58.59	m	62.98
4250mm	23.08	3.38	26.46	–	5.30	37.21	8.42	72.09	m	77.49
4500mm	24.49	3.38	27.87	–	5.63	39.52	8.91	76.30	m	82.03
4750mm	25.82	3.38	29.20	–	5.95	41.77	9.41	80.38	m	86.41
5000mm	27.23	3.38	30.61	–	7.80	54.76	9.90	95.27	m	102.41
5250mm	28.63	3.38	32.01	–	8.20	57.56	10.40	99.97	m	107.47
5500mm	30.04	3.38	33.42	–	8.60	60.37	10.89	104.68	m	112.53
5750mm	31.37	3.38	34.75	–	8.95	62.83	11.39	108.97	m	117.14
6000mm	32.78	3.38	36.16	–	9.35	65.64	11.88	113.68	m	122.20

Excavating trenches by hand to receive pipes not exceeding 200mm nominal size; disposing of surplus excavated material on site

Excavations commencing from natural ground level
average depth

500mm	–	–	–	–	1.15	8.07	2.97	11.04	m	11.87
750mm	–	–	–	–	2.05	14.39	3.29	17.68	m	19.01
1000mm	–	–	–	–	2.95	20.71	3.60	24.31	m	26.13
1250mm	–	–	–	–	3.85	27.03	4.24	31.27	m	33.61
1500mm	–	–	–	–	4.75	33.34	4.62	37.97	m	40.81
1750mm	–	–	–	–	5.65	39.66	5.00	44.66	m	48.01
2000mm	–	–	–	–	6.50	45.63	5.38	51.01	m	54.84
2250mm	–	–	–	–	7.65	53.70	6.39	60.09	m	64.60
2500mm	–	–	–	–	8.80	61.78	6.84	68.62	m	73.76
2750mm	–	–	–	–	9.95	69.85	7.29	77.14	m	82.92
3000mm	–	–	–	–	11.10	77.92	7.74	85.66	m	92.09
3250mm	–	–	–	–	12.25	86.00	8.19	94.19	m	101.25
3500mm	–	–	–	–	13.45	94.42	8.64	103.06	m	110.79
3750mm	–	–	–	–	14.60	102.49	9.09	111.58	m	119.95
4000mm	–	–	–	–	15.80	110.92	9.54	120.46	m	129.49

EXCAVATION FOR DRAINS - BY HAND - Excavating trenches by hand to receive pipes 225mm nominal size; disposing of surplus excavated material on site

Excavations commencing from natural ground level
average depth

500mm	–	–	–	–	1.28	8.99	3.10	12.09	m	12.99
750mm	–	–	–	–	2.23	15.65	3.42	19.07	m	20.51
1000mm	–	–	–	–	3.18	22.32	3.73	26.05	m	28.01
1250mm	–	–	–	–	4.13	28.99	4.37	33.36	m	35.86
1500mm	–	–	–	–	5.08	35.66	4.75	40.41	m	43.44
1750mm	–	–	–	–	6.03	42.33	5.13	47.46	m	51.02
2000mm	–	–	–	–	6.95	48.79	5.51	54.30	m	58.37
2250mm	–	–	–	–	8.23	57.77	6.52	64.29	m	69.12
2500mm	–	–	–	–	9.45	66.34	6.97	73.31	m	78.81
2750mm	–	–	–	–	10.70	75.11	7.42	82.53	m	88.72
3000mm	–	–	–	–	11.95	83.89	7.87	91.76	m	98.64
3250mm	–	–	–	–	13.20	92.66	8.32	100.98	m	108.56
3500mm	–	–	–	–	14.45	101.44	8.77	110.21	m	118.47
3750mm	–	–	–	–	15.70	110.21	9.22	119.43	m	128.39
4000mm	–	–	–	–	17.00	119.34	9.67	129.01	m	138.69

DISPOSAL SYSTEMS

Labour hourly rates: (except Specialists) Craft Operatives £9.23 Labourer £7.02 Rates are national average prices. Refer to REGIONAL VARIATIONS for indicative levels of overall pricing in regions	PLANT & TRANSPORT			LABOUR				RATES		
	Plant Cost £	Trans Cost £	P & T Cost £	Craft Optve Hrs	Lab Hrs	Labour Cost £	Sunds £	Nett Rate £	Unit	Gross Rate (+7.5%) £
R12: DRAINAGE BELOW GROUND Cont.										
EXCAVATION FOR DRAINS - BY HAND - Excavating trenches **by hand to receive pipes 300mm nominal size;** **disposing of surplus excavated material on site**										
Excavations commencing from natural ground level average depth										
500mm	-	-	-	-	1.40	9.83	3.36	13.19	m	14.18
750mm	-	-	-	-	2.40	16.85	3.68	20.53	m	22.07
1000mm	-	-	-	-	3.40	23.87	3.99	27.86	m	29.95
1250mm	-	-	-	-	4.40	30.89	4.63	35.52	m	38.18
1500mm	-	-	-	-	5.40	37.91	5.01	42.92	m	46.14
1750mm	-	-	-	-	6.40	44.93	5.39	50.32	m	54.09
2000mm	-	-	-	-	7.40	51.95	5.77	57.72	m	62.05
2250mm	-	-	-	-	8.80	61.78	6.78	68.56	m	73.70
2500mm	-	-	-	-	10.10	70.90	7.23	78.13	m	83.99
2750mm	-	-	-	-	11.45	80.38	7.68	88.06	m	94.66
3000mm	-	-	-	-	12.75	89.50	8.13	97.64	m	104.96
3250mm	-	-	-	-	14.10	98.98	8.58	107.56	m	115.63
3500mm	-	-	-	-	15.45	108.46	9.03	117.49	m	126.30
3750mm	-	-	-	-	16.80	117.94	9.48	127.42	m	136.97
4000mm	-	-	-	-	18.15	127.41	9.93	137.34	m	147.64

This section continues
on the next page

420

Labour hourly rates: (except Specialists) Craft Operatives £9.23 Labourer £7.02 Rates are national average prices. Refer to REGIONAL VARIATIONS for indicative levels of overall pricing in regions	MATERIALS			LABOUR				RATES		
	Del to Site	Waste	Material Cost	Craft Optve	Lab	Labour Cost	Sunds	Nett Rate	Unit	Gross Rate (+7.5%)
	£	%	£	Hrs	Hrs	£	£	£		£

R12: DRAINAGE BELOW GROUND Cont.

BEDS, BENCHINGS AND COVERS - Granular material, 10mm nominal size pea shingle, to be obtained off site

Beds

	Del to Site	Waste	Material Cost	Craft Optve	Lab	Labour Cost	Sunds	Nett Rate	Unit	Gross Rate
400 x 150mm	1.58	10.00	1.74	–	0.15	1.05	–	2.79	m	3.00
450 x 150mm	1.85	10.00	2.04	–	0.16	1.12	–	3.16	m	3.40
525 x 150mm	2.11	10.00	2.32	–	0.17	1.19	–	3.51	m	3.78
600 x 150mm	2.38	10.00	2.62	–	0.22	1.54	–	4.16	m	4.47
700 x 150mm	2.90	10.00	3.19	–	0.24	1.68	–	4.87	m	5.24
750 x 150mm	2.98	10.00	3.28	–	0.27	1.90	–	5.17	m	5.56

Beds and surrounds

	Del to Site	Waste	Material Cost	Craft Optve	Lab	Labour Cost	Sunds	Nett Rate	Unit	Gross Rate
400 x 150mm bed; 150mm thick surround to 100mm internal diameter pipes -1	3.96	10.00	4.36	–	0.46	3.23	–	7.59	m	8.15
450 x 150mm bed; 150mm thick surround to 150mm internal diameter pipes -1	4.88	10.00	5.37	–	0.52	3.65	–	9.02	m	9.69
525 x 150mm bed; 150mm thick surround to 225mm internal diameter pipes -1	6.36	10.00	7.00	–	0.68	4.77	–	11.77	m	12.65
600 x 150mm bed; 150mm thick surround to 300mm internal diameter pipes -1	7.64	10.00	8.40	–	0.84	5.90	–	14.30	m	15.37
700 x 150mm bed; 150mm thick surround to 400mm internal diameter pipes -1	9.50	10.00	10.45	–	1.00	7.02	–	17.47	m	18.78
750 x 150mm bed; 150mm thick surround to 450mm internal diameter pipes -1	10.65	10.00	11.72	–	1.16	8.14	–	19.86	m	21.35

Beds and filling to half pipe depth

	Del to Site	Waste	Material Cost	Craft Optve	Lab	Labour Cost	Sunds	Nett Rate	Unit	Gross Rate
400 x 150mm overall; to 100mm internal diameter pipes - 1	1.58	10.00	1.74	–	0.12	0.84	–	2.58	m	2.77
450 x 225mm overall; to 150mm internal diameter pipes - 1	2.67	10.00	2.94	–	0.20	1.40	–	4.34	m	4.67
525 x 263mm overall; to 225mm internal diameter pipes - 1	3.64	10.00	4.00	–	0.28	1.97	–	5.97	m	6.42
600 x 300mm overall; to 300mm internal diameter pipes -1	4.75	10.00	5.22	–	0.36	2.53	–	7.75	m	8.33
700 x 350mm overall; to 400mm internal diameter pipes -1	6.47	10.00	7.12	–	0.49	3.44	–	10.56	m	11.35
750 x 375mm overall; to 450mm internal diameter pipes -1	7.42	10.00	8.16	–	0.56	3.93	–	12.09	m	13.00

BEDS, BENCHINGS AND COVERS - Sand; to be obtained off site

Beds

	Del to Site	Waste	Material Cost	Craft Optve	Lab	Labour Cost	Sunds	Nett Rate	Unit	Gross Rate
400 x 50mm	0.52	5.00	0.55	–	0.04	0.28	–	0.83	m	0.89
400 x 100mm	1.03	5.00	1.08	–	0.08	0.56	–	1.64	m	1.77
450 x 50mm	0.52	5.00	0.55	–	0.05	0.35	–	0.90	m	0.96
450 x 100mm	1.29	5.00	1.35	–	0.09	0.63	–	1.99	m	2.14
525 x 50mm	0.77	5.00	0.81	–	0.06	0.42	–	1.23	m	1.32
525 x 100mm	1.29	5.00	1.35	–	0.11	0.77	–	2.13	m	2.29
600 x 50mm	0.77	5.00	0.81	–	0.06	0.42	–	1.23	m	1.32
600 x 100mm	1.55	5.00	1.63	–	0.12	0.84	–	2.47	m	2.66
700 x 50mm	1.03	5.00	1.08	–	0.07	0.49	–	1.57	m	1.69
700 x 100mm	1.80	5.00	1.89	–	0.14	0.98	–	2.87	m	3.09
750 x 50mm	1.03	5.00	1.08	–	0.08	0.56	–	1.64	m	1.77
750 x 100mm	2.06	5.00	2.16	–	0.15	1.05	–	3.22	m	3.46

BEDS, BENCHINGS AND COVERS - Plain in-situ concrete; B.S.5328, ordinary prescribed mix, ST3, 20mm aggregate

Beds

	Del to Site	Waste	Material Cost	Craft Optve	Lab	Labour Cost	Sunds	Nett Rate	Unit	Gross Rate
400 x 100mm	1.93	5.00	2.03	–	0.18	1.26	1.92	5.21	m	5.60
400 x 150mm	2.90	5.00	3.04	–	0.27	1.90	2.88	7.82	m	8.41
450 x 100mm	2.42	5.00	2.54	–	0.20	1.40	1.92	5.87	m	6.30
450 x 150mm	3.38	5.00	3.55	–	0.30	2.11	2.88	8.54	m	9.18
525 x 100mm	2.42	5.00	2.54	–	0.24	1.68	1.92	6.15	m	6.61
525 x 150mm	3.86	5.00	4.05	–	0.35	2.46	2.88	9.39	m	10.09
600 x 100mm	2.90	5.00	3.04	–	0.27	1.90	1.92	6.86	m	7.37
600 x 150mm	4.35	5.00	4.57	–	0.38	2.67	2.88	10.12	m	10.87
700 x 100mm	3.38	5.00	3.55	–	0.32	2.25	1.92	7.72	m	8.29
700 x 150mm	5.31	5.00	5.58	–	0.47	3.30	2.88	11.75	m	12.64
750 x 100mm	3.86	5.00	4.05	–	0.34	2.39	1.92	8.36	m	8.99
750 x 150mm	5.31	5.00	5.58	–	0.51	3.58	2.88	12.04	m	12.94

Beds and surrounds

	Del to Site	Waste	Material Cost	Craft Optve	Lab	Labour Cost	Sunds	Nett Rate	Unit	Gross Rate
400 x 100mm bed; 150mm thick surround to 100mm internal diameter pipes -1	6.28	5.00	6.59	–	0.72	5.05	3.84	15.49	m	16.65
400 x 150mm bed; 150mm thick surround to 100mm internal diameter pipes -1	7.25	5.00	7.61	–	0.81	5.69	4.31	17.61	m	18.93
450 x 100mm bed; 150mm thick surround to 150mm internal diameter pipes -1	7.73	5.00	8.12	–	0.91	6.39	4.31	18.81	m	20.23
450 x 150mm bed; 150mm thick surround to 150mm internal diameter pipes -1	8.69	5.00	9.12	–	1.01	7.09	4.79	21.00	m	22.58
525 x 100mm bed; 150mm thick surround to 225mm internal diameter pipes -1	10.14	5.00	10.65	–	1.24	8.70	4.79	24.14	m	25.95
525 x 150mm bed; 150mm thick surround to 225mm internal diameter pipes -1	11.59	5.00	12.17	–	1.36	9.55	5.51	27.23	m	29.27
600 x 100mm bed; 150mm thick surround to 300mm internal diameter pipes -1	12.56	5.00	13.19	–	1.62	11.37	5.76	30.32	m	32.59
600 x 150mm bed; 150mm thick surround to 300mm internal diameter pipes -1	14.01	5.00	14.71	–	1.76	12.36	6.23	33.30	m	35.79
700 x 100mm bed; 150mm thick surround to 400mm internal diameter pipes -1	15.94	5.00	16.74	–	2.21	15.51	6.71	38.96	m	41.88

DISPOSAL SYSTEMS

Labour hourly rates: (except Specialists) Craft Operatives £9.23 Labourer £7.02 Rates are national average prices. Refer to REGIONAL VARIATIONS for indicative levels of overall pricing in regions	MATERIALS			LABOUR				RATES		
	Del to Site	Waste	Material Cost	Craft Optve	Lab	Labour Cost	Sunds	Nett Rate	Unit	Gross Rate (+7.5%)
	£	%	£	Hrs	Hrs	£	£	£		£
R12: DRAINAGE BELOW GROUND Cont.										
BEDS, BENCHINGS AND COVERS - Plain in-situ concrete; **B.S.5328, ordinary prescribed mix, ST3, 20mm** **aggregate Cont.**										
Beds and surrounds Cont.										
700 x 150mm bed; 150mm thick surround to 400mm										
internal diameter pipes -1	17.39	5.00	18.26	-	2.36	16.57	7.20	42.03	m	45.18
750 x 100mm bed; 150mm thick surround to 450mm										
internal diameter pipes -1	17.87	5.00	18.76	-	2.53	17.76	7.20	43.72	m	47.00
750 x 150mm bed; 150mm thick surround to 450mm										
internal diameter pipes -1	19.32	5.00	20.29	-	2.70	18.95	7.67	46.91	m	50.43
Beds and haunchings										
400 x 150mm; to 100mm internal diameter pipes -1 .	13.86	5.00	14.55	-	0.40	2.81	2.88	20.24	m	21.76
450 x 150mm; to 150mm internal diameter pipes -1 .	4.35	5.00	4.57	-	0.46	3.23	2.88	10.68	m	11.48
525 x 150mm; to 225mm internal diameter pipes -1 .	5.80	5.00	6.09	-	0.62	4.35	2.88	13.32	m	14.32
600 x 150mm; to 300mm internal diameter pipes -1 .	7.73	5.00	8.12	-	0.81	5.69	2.88	16.68	m	17.93
700 x 150mm; to 400mm internal diameter pipes -1 .	9.66	5.00	10.14	-	1.10	7.72	2.88	20.75	m	22.30
750 x 150mm; to 450mm internal diameter pipes -1 .	10.63	5.00	11.16	-	1.35	9.48	2.88	23.52	m	25.28
Vertical casings										
400 x 400mm to 100mm internal diameter pipes - 1 .	7.25	5.00	7.61	-	0.80	5.62	21.23	34.46	m	37.04
450 x 450mm to 150mm internal diameter pipes - 1 .	8.69	5.00	9.12	-	1.01	7.09	23.89	40.10	m	43.11
525 x 525mm to 225mm internal diameter pipes - 1 .	13.52	5.00	14.20	-	1.38	9.69	28.13	52.01	m	55.91
600 x 600mm to 300mm internal diameter pipes - 1 .	14.01	5.00	14.71	-	1.80	12.64	31.85	59.20	m	63.64
700 x 700mm to 400mm internal diameter pipes - 1 .	17.39	5.00	18.26	-	2.45	17.20	37.16	72.62	m	78.06
750 x 750mm to 450mm internal diameter pipes - 1 .	19.32	5.00	20.29	-	2.81	19.73	39.81	79.82	m	85.81
VITRIFIED CLAY FLEXIBLE JOINT DRAINS - Drains; **vitrified clay pipes and fittings, B.S.EN295, normal;** **flexible mechanical joints**										
Pipework in trenches										
100mm	6.71	5.00	7.05	0.16	0.06	1.90	0.19	9.13	m	9.82
extra; bends	9.70	5.00	10.19	0.08	-	0.74	-	10.92	nr	11.74
extra; branches	13.47	5.00	14.14	0.16	-	1.48	-	15.62	nr	16.79
150mm	8.71	5.00	9.15	0.18	0.08	2.22	0.21	11.58	m	12.45
extra; bends	16.00	5.00	16.80	0.09	-	0.83	-	17.63	nr	18.95
extra; branches	20.89	5.00	21.93	0.18	-	1.66	-	23.60	nr	25.37
225mm	16.85	5.00	17.69	0.22	0.11	2.80	0.28	20.78	m	22.33
extra; bends	32.64	5.00	34.27	0.11	-	1.02	-	35.29	nr	37.93
extra; branches	49.08	5.00	51.53	0.22	-	2.03	-	53.56	nr	57.58
300mm	26.35	5.00	27.67	0.33	0.14	4.03	0.47	32.17	m	34.58
extra; bends	64.40	5.00	67.62	0.17	-	1.57	-	69.19	nr	74.38
extra; branches	101.18	5.00	106.24	0.33	-	3.05	-	109.28	nr	117.48
400mm	54.08	5.00	56.78	0.55	0.18	6.34	0.71	63.83	m	68.62
extra; bends	203.21	5.00	213.37	0.39	-	3.60	-	216.97	nr	233.24
450mm	70.24	5.00	73.75	0.70	0.20	7.87	0.86	82.48	m	88.66
extra; bends	267.58	5.00	280.96	0.70	-	6.46	-	287.42	nr	308.98
Pipework in trenches; vertical										
100mm	6.71	5.00	7.05	0.18	0.06	2.08	0.19	9.32	m	10.02
150mm	8.71	5.00	9.15	0.20	0.08	2.41	0.21	11.76	m	12.65
225mm	16.85	5.00	17.69	0.24	0.11	2.99	0.28	20.96	m	22.53
300mm	26.35	5.00	27.67	0.36	0.14	4.31	0.47	32.44	m	34.88
400mm	54.08	5.00	56.78	0.60	0.18	6.80	0.71	64.30	m	69.12
450mm	70.24	5.00	73.75	0.77	0.20	8.51	0.86	83.12	m	89.36
VITRIFIED CLAY SLEEVE DRAINS - Drains; vitrified clay **pipes and fittings, B.S.EN295; sleeve joints, push-** **fit polypropylene standard ring flexible couplings**										
Pipework in trenches										
100mm	3.86	5.00	4.05	0.16	0.06	1.90	0.22	6.17	m	6.63
extra; bends	7.11	5.00	7.47	0.08	-	0.74	-	8.20	nr	8.82
extra; branches	13.33	5.00	14.00	0.16	-	1.48	-	15.47	nr	16.63
150mm	7.38	5.00	7.75	0.18	0.08	2.22	0.26	10.23	m	11.00
extra; bends	10.87	5.00	11.41	0.09	-	0.83	-	12.24	nr	13.16
extra; branches	14.39	5.00	15.11	0.18	-	1.66	-	16.77	nr	18.03
Pipework in trenches; vertical										
100mm	3.86	5.00	4.05	0.18	0.06	2.08	0.22	6.36	m	6.83
150mm	7.38	5.00	7.75	0.20	0.08	2.41	0.26	10.42	m	11.20
VITRIFIED CLAY SLEEVE DRAINS - Drains; vitrified clay **pipes and fittings, B.S.EN295; sleeve joints, push-** **fit polypropylene neoprene ring flexible couplings**										
Pipework in trenches										
100mm..	4.46	5.00	4.68	0.16	0.06	1.90	0.21	6.79	m	7.30
150mm..	8.50	5.00	8.93	0.18	0.08	2.22	0.25	11.40	m	12.25
Pipework in trenches; vertical										
100mm..	4.46	5.00	4.68	0.18	0.06	2.08	0.21	6.98	m	7.50
150mm..	8.50	5.00	8.93	0.20	0.08	2.41	0.25	11.58	m	12.45
VITRIFIED CLAY DRAINS - Drains; vitrified clay pipes **and fittings, B.S.EN295, normal; cement mortar (1:2)** **and tarred gasket joints**										
Pipework in trenches										
100mm..	4.69	5.00	4.92	0.33	0.03	3.26	0.28	8.46	m	9.10
extra; double collars	5.66	5.00	5.94	0.10	-	0.92	-	6.87	nr	7.38
extra; bends	4.69	5.00	4.92	0.20	-	1.85	-	6.77	nr	7.28
extra; rest bends	7.72	5.00	8.11	0.20	-	1.85	-	9.95	nr	10.70

Labour hourly rates: (except Specialists) Craft Operatives £9.23 Labourer £7.02 Rates are national average prices. Refer to REGIONAL VARIATIONS for indicative levels of overall pricing in regions	MATERIALS			LABOUR				RATES		
	Del to Site £	Waste %	Material Cost £	Craft Optve Hrs	Lab Hrs	Labour Cost £	Sunds £	Nett Rate £	Unit	Gross Rate (+7.5%) £
R12: DRAINAGE BELOW GROUND Cont.										
VITRIFIED CLAY DRAINS - Drains; vitrified clay pipes and fittings, B.S.EN295, normal; cement mortar (1:2) and tarred gasket joints Cont.										
Pipework in trenches Cont.										
extra; branches	8.62	5.00	9.05	0.22	-	2.03	-	11.08	nr	11.91
150mm	7.23	5.00	7.59	0.44	0.04	4.34	0.36	12.29	m	13.22
extra; double collars	9.43	5.00	9.90	0.15	-	1.38	-	11.29	nr	12.13
extra; bends	7.23	5.00	7.59	0.27	-	2.49	-	10.08	nr	10.84
extra; rest bends	13.05	5.00	13.70	0.27	-	2.49	-	16.19	nr	17.41
extra; branches	14.28	5.00	14.99	0.82	-	7.57	-	22.56	nr	24.25
225mm	14.32	5.00	15.04	0.55	0.07	5.57	0.45	21.05	m	22.63
extra; double collars	22.07	5.00	23.17	0.23	-	2.12	-	25.30	nr	27.19
extra; bends	22.63	5.00	23.76	0.34	-	3.14	-	26.90	nr	28.92
300mm	24.00	5.00	25.20	0.80	0.10	8.09	0.66	33.95	m	36.49
Pipework in trenches; vertical										
100mm	4.69	5.00	4.92	0.50	0.03	4.83	0.28	10.03	m	10.78
150mm	7.23	5.00	7.59	0.48	0.04	4.71	0.36	12.66	m	13.61
225mm	14.32	5.00	15.04	0.60	0.07	6.03	0.45	21.52	m	23.13
300mm	24.00	5.00	25.20	0.88	0.10	8.82	0.66	34.68	m	37.29
PVC-U DRAINS - Drains; Wavin; PVC-U pipes and fittings, B.S.4660; ring seal joints										
Pipework in trenches										
110mm, ref. 4D.076	3.27	5.00	3.43	0.14	0.02	1.43	-	4.87	m	5.23
extra; connections to cast iron and clay sockets, ref. 4D.107	2.95	5.00	3.10	0.08	-	0.74	-	3.84	nr	4.12
extra; 45 degree bends, ref. 4D.163	6.78	5.00	7.12	0.14	0.02	1.43	-	8.55	nr	9.19
extra; branches, ref. 4D.210	9.98	5.00	10.48	0.14	0.02	1.43	-	11.91	nr	12.80
extra; connections cast iron and clay spigot, ref. 4D.128	6.94	5.00	7.29	0.08	-	0.74	-	8.03	nr	8.63
160mm, ref. 6D.076	7.23	5.00	7.59	0.20	0.03	2.06	-	9.65	m	10.37
extra; connections to cast iron and clay sockets, ref. 6D.107	11.58	5.00	12.16	0.10	-	0.92	-	13.08	nr	14.06
extra; 45 degree bends, ref. 6D.163	16.11	5.00	16.92	0.20	0.03	2.06	-	18.97	nr	20.40
extra; branches, ref. 6D.210	28.76	5.00	30.20	0.20	0.03	2.06	-	32.25	nr	34.67
extra; connections to cast iron and clay spigot, ref. 6D.128	14.03	5.00	14.73	0.10	-	0.92	-	15.65	nr	16.83
Pipework in trenches; vertical										
110mm, ref. 4D.076	3.27	5.00	3.43	0.28	0.02	2.72	-	6.16	m	6.62
160mm, ref. 6D.076	7.23	5.00	7.59	0.24	0.03	2.43	-	10.02	m	10.77
PVC-U DRAINS - Drains; PVC-U solid wall concentric external rib-reinforced Wavin 'Ultra-Rib' pipes and fittings with sealing rings to joints .										
Pipework in trenches										
150mm 6UR 046	3.99	5.00	4.19	0.20	0.02	1.99	-	6.18	m	6.64
extra; connectors 6UR 205	5.27	5.00	5.53	0.05	-	0.46	-	6.00	nr	6.44
extra; bends 6UR 563	7.04	5.00	7.39	0.20	-	1.85	-	9.24	nr	9.93
extra; branches 6UR 213	17.13	5.00	17.99	0.20	-	1.85	-	19.83	nr	21.32
extra; connections to clayware pipe ends 6UR 129 .	12.90	5.00	13.55	0.10	-	0.92	-	14.47	nr	15.55
225mm 9UR 046 .	8.97	5.00	9.42	0.30	0.03	2.98	-	12.40	m	13.33
extra; connectors 9UR 205	11.01	5.00	11.56	0.08	-	0.74	-	12.30	nr	13.22
extra; bends 9UR 563	28.45	5.00	29.87	0.30	-	2.77	-	32.64	nr	35.09
extra; branches 9UR 213	50.97	5.00	53.52	0.30	-	2.77	-	56.29	nr	60.51
extra; connections to clayware pipe ends 9UR 109 .	28.51	5.00	29.94	0.12	-	1.11	-	31.04	nr	33.37
300mm 12UR 043	18.09	5.00	18.99	0.38	0.05	3.86	-	22.85	m	24.57
extra; connectors 12UR 205	22.20	5.00	23.31	0.10	-	0.92	-	24.23	nr	26.05
extra; bends 12UR 563	46.36	5.00	48.68	0.38	-	3.51	-	52.19	nr	56.10
extra; branches 12UR 213	108.63	5.00	114.06	0.38	-	3.51	-	117.57	nr	126.39
extra; connections to clayware pipe ends 12UR 112	74.98	5.00	78.73	0.15	-	1.38	-	80.11	nr	86.12
Pipework in trenches; vertical										
150mm	3.99	5.00	4.19	0.40	0.02	3.83	-	8.02	m	8.62
225mm	8.97	5.00	9.42	0.38	0.03	3.72	-	13.14	m	14.12
300mm	18.09	5.00	18.99	0.48	0.05	4.78	-	23.78	m	25.56
CONCRETE DRAINS - Drains; unreinforced concrete pipes and fittings B.S.5911, Class H, flexible joints										
Pipework in trenches										
300mm	12.26	5.00	12.87	0.59	0.24	7.13	0.20	20.20	m	21.72
extra; bends	73.53	5.00	77.21	0.30	0.12	3.61	0.10	80.92	nr	86.99
extra; 100mm branches	40.38	5.00	42.40	0.24	0.06	2.64	0.16	45.20	nr	48.59
extra; 150mm branches	42.75	5.00	44.89	0.40	0.11	4.46	0.26	49.61	nr	53.33
extra; 225mm branches	49.88	5.00	52.37	0.56	0.16	6.29	0.47	59.14	nr	63.57
extra; 300mm branches	54.15	5.00	56.86	0.72	0.22	8.19	0.66	65.71	nr	70.64
375mm	16.82	5.00	17.66	0.72	0.30	8.75	0.23	26.64	m	28.64
extra; bends	100.89	5.00	105.93	0.36	0.15	4.38	0.12	110.43	nr	118.71
extra; 100mm branches	40.38	5.00	42.40	0.24	0.06	2.64	0.16	45.20	nr	48.59
extra; 150mm branches	42.75	5.00	44.89	0.10	0.11	4.46	0.26	49.61	nr	53.33
extra; 225mm branches	49.88	5.00	52.37	0.56	0.16	6.29	0.47	59.14	nr	63.57
extra; 300mm branches	54.15	5.00	56.86	0.72	0.22	8.19	0.66	65.71	nr	70.64
extra; 375mm branches	73.15	5.00	76.81	0.88	0.27	10.02	0.84	87.67	nr	94.24
450mm	20.05	5.00	21.05	0.85	0.37	10.44	0.27	31.77	m	34.15
extra; bends	120.27	5.00	126.28	0.43	0.18	5.23	0.16	131.68	nr	141.55
extra; 100mm branches	40.38	5.00	42.40	0.24	0.06	2.64	0.16	45.20	nr	48.59
extra; 150mm branches	42.75	5.00	44.89	0.40	0.11	4.46	0.26	49.61	nr	53.33
extra; 225mm branches	49.88	5.00	52.37	0.56	0.16	6.29	0.47	59.14	nr	63.57
extra; 300mm branches	54.15	5.00	56.86	0.72	0.22	8.19	0.66	65.71	nr	70.64

Labour hourly rates: (except Specialists) Craft Operatives £9.23 Labourer £7.02 Rates are national average prices. Refer to REGIONAL VARIATIONS for indicative levels of overall pricing in regions	MATERIALS			LABOUR				RATES		
	Del to Site £	Waste %	Material Cost £	Craft Optve Hrs	Lab Hrs	Labour Cost £	Sunds £	Nett Rate £	Unit	Gross Rate (+7.5%) £
R12: DRAINAGE BELOW GROUND Cont.										
CONCRETE DRAINS - Drains; unreinforced concrete pipes and fittings B.S.5911, Class H, flexible joints Cont.										
Pipework in trenches Cont.										
extra; 375mm branches	73.15	5.00	76.81	0.88	0.27	10.02	0.84	87.67	nr	94.24
extra; 450mm branches	90.25	5.00	94.76	1.04	0.33	11.92	1.03	107.71	nr	115.79
525mm	23.09	5.00	24.24	0.98	0.43	12.06	0.30	36.61	m	39.35
extra; bends	138.54	5.00	145.47	0.49	0.22	6.07	0.17	151.70	nr	163.08
extra; 100mm branches	40.38	5.00	42.40	0.24	0.06	2.64	0.16	45.20	nr	48.59
extra; 150mm branches	42.75	5.00	44.89	0.40	0.11	4.46	0.26	49.61	nr	53.33
extra; 225mm branches	49.88	5.00	52.37	0.56	0.16	6.29	0.47	59.14	nr	63.57
extra; 300mm branches	54.15	5.00	56.86	0.72	0.22	8.19	0.66	65.71	nr	70.64
extra; 375mm branches	73.15	5.00	76.81	0.88	0.27	10.02	0.84	87.67	nr	94.24
extra; 450mm branches	90.25	5.00	94.76	1.04	0.33	11.92	1.03	107.71	nr	115.79
600mm	29.45	5.00	30.92	1.10	0.50	13.66	0.38	44.97	m	48.34
extra; bends	176.70	5.00	185.54	0.55	0.25	6.83	0.20	192.57	nr	207.01
extra; 100mm branches	40.38	5.00	42.40	0.24	0.06	2.64	0.16	45.20	nr	48.59
extra; 150mm branches	42.75	5.00	44.89	0.40	0.11	4.46	0.26	49.61	nr	53.33
extra; 225mm branches	49.88	5.00	52.37	0.56	0.16	6.29	0.47	59.14	nr	63.57
extra; 300mm branches	54.15	5.00	56.86	0.72	0.22	8.19	0.66	65.71	nr	70.64
extra; 375mm branches	73.15	5.00	76.81	0.88	0.27	10.02	0.84	87.67	nr	94.24
extra; 450mm branches	91.20	5.00	95.76	1.04	0.33	11.92	1.03	108.71	nr	116.86
CONCRETE DRAINS - Drains; reinforced concrete pipes and fittings, B.S.5911, Class H, flexible joints										
Pipework in trenches										
450mm	32.00	5.00	33.60	0.85	0.37	10.44	0.26	44.30	m	47.63
extra; bends	256.00	5.00	268.80	0.43	0.18	5.23	0.15	274.18	nr	294.75
extra; 100mm branches	35.00	5.00	36.75	0.24	0.06	2.64	0.15	39.54	nr	42.50
extra; 150mm branches	40.00	5.00	42.00	0.40	0.11	4.46	0.25	46.71	nr	50.22
extra; 225mm branches	45.00	5.00	47.25	0.56	0.16	6.29	0.45	53.99	nr	58.04
extra; 300mm branches	69.00	5.00	72.45	0.72	0.22	8.19	0.63	81.27	nr	87.37
extra; 375mm branches	74.00	5.00	77.70	0.88	0.27	10.02	0.80	88.52	nr	95.16
extra; 450mm branches	96.00	5.00	100.80	1.04	0.33	11.92	0.98	113.70	nr	122.22
525mm	36.50	5.00	38.33	0.98	0.43	12.06	0.29	50.68	m	54.48
extra; bends	292.00	5.00	306.60	0.49	0.22	6.07	0.16	312.83	nr	336.29
extra; 100mm branches	35.00	5.00	36.75	0.24	0.06	2.64	0.15	39.54	nr	42.50
extra; 150mm branches	40.00	5.00	42.00	0.40	0.11	4.46	0.25	46.71	nr	50.22
extra; 225mm branches	45.00	5.00	47.25	0.56	0.16	6.29	0.45	53.99	nr	58.04
extra; 300mm branches	69.00	5.00	72.45	0.72	0.22	8.19	0.63	81.27	nr	87.37
extra; 375mm branches	74.00	5.00	77.70	0.88	0.27	10.02	0.80	88.52	nr	95.16
extra; 450mm branches	96.00	5.00	100.80	1.04	0.33	11.92	0.98	113.70	nr	122.22
600mm	39.75	5.00	41.74	1.10	0.50	13.66	0.36	55.76	m	59.94
extra; bends	318.00	5.00	333.90	0.55	0.25	6.83	0.19	340.92	nr	366.49
extra; 100mm branches	33.00	5.00	34.65	0.24	0.06	2.64	0.15	37.44	nr	40.24
extra; 150mm branches	40.00	5.00	42.00	0.40	0.11	4.46	0.25	46.71	nr	50.22
extra; 225mm branches	65.00	5.00	68.25	0.56	0.16	6.29	0.63	75.17	nr	80.81
extra; 300mm branches	69.00	5.00	72.45	0.72	0.22	8.19	0.63	81.27	nr	87.37
extra; 375mm branches	74.00	5.00	77.70	0.88	0.27	10.02	0.80	88.52	nr	95.16
extra; 450mm branches	96.00	5.00	100.80	1.04	0.33	11.92	0.98	113.70	nr	122.22

This section continues
on the next page

Labour hourly rates: (except Specialists) Craft Operatives £12.44 Labourer £7.02 Rates are national average prices. Refer to REGIONAL VARIATIONS for indicative levels of overall pricing in regions	MATERIALS			LABOUR				RATES		
	Del to Site £	Waste %	Material Cost £	Craft Optve Hrs	Lab Hrs	Labour Cost £	Sunds £	Nett Rate £	Unit	Gross Rate (+7.5%) £

R12: DRAINAGE BELOW GROUND Cont.

CAST IRON DRAINS - Drains; cast iron pipes and cast iron fittings B.S.437, coated; bolted cast iron coupling and synthetic rubber gasket joints

Pipework in trenches

100mm	19.43	5.00	20.40	1.00	-	12.44	-	32.84	m	35.30
extra; connectors, large socket for clayware	27.01	2.00	27.55	1.00	-	12.44	-	39.99	nr	42.99
extra; bends	35.18	2.00	35.88	1.00	-	12.44	-	48.32	nr	51.95
extra; branches	45.90	2.00	46.82	1.50	-	18.66	-	65.48	nr	70.39
150mm	34.30	5.00	36.02	1.50	-	18.66	-	54.67	m	58.78
extra; connectors, large socket for clayware	39.60	2.00	40.39	1.50	-	18.66	-	59.05	nr	63.48
extra; diminishing pieces, reducing to 100mm	37.37	2.00	38.12	1.50	-	18.66	-	56.78	nr	61.04
extra; bends	51.06	2.00	52.08	1.50	-	18.66	-	70.74	nr	76.05
extra; branches	79.69	2.00	81.28	2.25	-	27.99	-	109.27	nr	117.47
in runs not exceeding 3m										
100mm	26.26	5.00	27.57	1.50	-	18.66	-	46.23	m	49.70
150mm	42.57	5.00	44.70	2.25	-	27.99	-	72.69	m	78.14

Pipework in trenches; vertical

100mm	19.43	5.00	20.40	1.20	-	14.93	-	35.33	m	37.98
150mm	34.30	5.00	36.02	1.80	-	22.39	-	58.41	m	62.79

STAINLESS STEEL DRAINS - Drains; stainless steel pipes and fittings, B.M. Stainless Steel Drains Ltd; ring seal push fit joints (AIS 1316)

Pipework in trenches

75mm	19.95	2.00	20.35	0.60	-	7.46	-	27.81	m	29.90
extra; bends	25.46	2.00	25.97	0.60	-	7.46	-	33.43	nr	35.94
extra; branches	27.58	2.00	28.13	0.90	-	11.20	-	39.33	nr	42.28
110mm	25.35	5.00	26.62	0.75	-	9.33	-	35.95	m	38.64
extra; diminishing pieces, reducing to 75mm	26.12	2.00	26.64	0.75	-	9.33	-	35.97	nr	38.67
extra; bends	32.52	2.00	33.17	0.75	-	9.33	-	42.50	nr	45.69
extra; branches	33.25	2.00	33.91	1.12	-	13.93	-	47.85	nr	51.44
in runs not exceeding 3m										
75mm	22.41	5.00	23.53	0.90	-	11.20	-	34.73	m	37.33
110mm	28.30	5.00	29.72	1.12	-	13.93	-	43.65	m	46.92

Pipework in trenches; vertical

75mm	19.95	5.00	20.95	0.75	-	9.33	-	30.28	m	32.55
110mm	25.35	5.00	26.62	0.94	-	11.69	-	38.31	m	41.18

This section continues
on the next page

DISPOSAL SYSTEMS

Labour hourly rates: (except Specialists) Craft Operatives £9.23 Labourer £7.02 Rates are national average prices. Refer to REGIONAL VARIATIONS for indicative levels of overall pricing in regions	MATERIALS			LABOUR				RATES		
	Del to Site £	Waste %	Material Cost £	Craft Optve Hrs	Lab Hrs	Labour Cost £	Sunds £	Nett Rate £	Unit	Gross Rate (+7.5%) £
R12: DRAINAGE BELOW GROUND Cont.										
VITRIFIED CLAY GULLIES - Vitrified clay accessories										
Gullies; Supersleve; joint to pipe; bedding and surrounding in concrete to B.S.5328, ordinary prescribed mix ST3, 20mm aggregate; Hepworth Building Products										
100mm outlet, reference RG1/1, trapped, round; 255mm diameter top	18.78	5.00	19.72	0.60	0.08	6.10	5.32	31.14	nr	33.47
100mm outlet, reference SG2/1, trapped, square; 150 x 150 top	25.09	5.00	26.34	0.60	0.08	6.10	4.41	36.85	nr	39.62
100mm outlet, reference SG1/1, trapped, reversible, round; 100mm trap; hopper reference SH1, 100mm outlet, 150 x 150mm rebated top	18.91	5.00	19.86	1.00	0.10	9.93	7.26	37.05	nr	39.83
100mm outlet, reference SG1/1, trapped, reversible, round; 100mm trap; hopper reference SH2, 100mm outlet, 100mm horizontal inlet, 150 x 150mm rebated top	25.51	5.00	26.79	1.15	0.10	11.32	7.26	45.36	nr	48.76
100mm outlet, reference SG1/1, trapped, reversible, round; 100mm trap; raising pieces - 1, reference RRP2/2, 100mm diameter, 225mm high; hopper reference SH1, 100mm outlet, 150 x 150mm rebated top	25.14	5.00	26.40	1.40	0.15	13.98	9.67	50.04	nr	53.80
100mm outlet, reference SG1/1, trapped, reversible, round; 100mm trap; raising pieces - 1, reference RRP2/2, 100mm diameter, 225mm high; hopper reference SH2, 100mm outlet, 100mm horizontal inlet, 150 x 150mm rebated top	31.74	5.00	33.33	1.45	0.16	14.51	9.67	57.50	nr	61.82
100mm outlet, reference RGP5, trapped, round; 225mm diameter x 600mm deep; perforated galvanised steel bucket reference IBP3	71.71	5.00	75.30	2.50	0.25	24.83	9.19	109.32	nr	117.51
100mm outlet, reference RGR1, trapped, round; 300mm diameter x 600mm deep	49.40	5.00	51.87	2.50	0.33	25.39	10.15	87.41	nr	93.97
150mm outlet, reference RGR2, trapped, round; 300mm diameter x 600mm deep	53.02	5.00	55.67	2.50	0.33	25.39	10.15	91.21	nr	98.05
150mm outlet, reference RGR3, trapped, round; 400mm diameter x 750mm deep	60.64	5.00	63.67	4.00	0.45	40.08	16.92	120.67	nr	129.72
150mm outlet, reference RGR4, trapped, round; 450mm diameter x 900mm deep	80.17	5.00	84.18	5.50	0.50	54.27	23.21	161.66	nr	173.79
100mm outlet, reference RGU2, trapped; internal size 450 x 300 x 525mm deep; perforated tray and galvanised cover and frame	351.01	5.00	368.56	8.00	1.00	80.86	8.73	458.15	nr	492.51
100mm outlet, reference RGU1, trapped; internal size 600 x 450 x 600mm deep; perforated tray and galvanised cover and frame	444.06	5.00	466.26	10.00	1.50	102.83	16.46	585.55	nr	629.47
Rainwater shoes; Supersleve; joint to pipe; bedding and surrounding in concrete to B.S.5328, ordinary prescribed mix ST3, 20mm aggregate										
100mm outlet, reference RRW/S3/1 trapless, round; 100mm vertical inlet, 250 x 150mm rectangular access opening	19.78	5.00	20.77	0.50	0.05	4.97	3.87	29.61	nr	31.83
CONCRETE GULLIES - Unreinforced concrete accessories										
Gullies; B.S.5911; joint to pipe; bedding and surrounding in concrete to B.S.5328 ordinary prescribed mix ST3, 20mm aggregate										
150mm outlet, trapped, round; 375mm diameter x 750mm deep internally, stopper	22.91	5.00	24.06	3.30	0.50	33.97	13.06	71.08	nr	76.42
150mm outlet, trapped, round; 375mm diameter x 900mm deep internally, stopper	24.43	5.00	25.65	3.50	0.50	35.81	15.95	77.42	nr	83.22
150mm outlet, trapped, round; 450mm diameter x 750mm deep internally, stopper	24.25	5.00	25.46	3.50	0.60	36.52	16.46	78.44	nr	84.32
150mm outlet, trapped, round; 450mm diameter x 900mm deep internally, stopper	24.74	5.00	25.98	3.70	0.60	38.36	19.82	84.16	nr	90.47
150mm outlet, trapped, round; 450mm diameter x 1050mm deep internally, stopper	26.19	5.00	27.50	4.00	0.60	41.13	23.21	91.84	nr	98.73
150mm outlet, trapped, round; 450mm diameter x 1200mm deep internally, stopper	34.06	5.00	35.76	4.50	0.60	45.75	26.11	107.62	nr	115.69
PVC-U GULLIES - PVC-U accessories										
Gullies; Osma; joint to pipe; bedding and surrounding in concrete to B.S.5328, ordinary prescribed mix ST3, 20mm aggregate										
110mm outlet trap, base reference 4D.500, outlet bend with access reference 4D.569, inlet raising piece 300mm long, plain hopper reference 4D.503	33.55	5.00	35.23	1.00	0.20	10.63	2.91	48.77	nr	52.43
110mm outlet bottle gulley, reference 4D.900 trapped, round; 200mm diameter top with grating reference 4D.919	22.84	5.00	23.98	1.50	0.25	15.60	4.84	44.42	nr	47.75
Sealed rodding eyes; Osma; joint to pipe; surrounding in concrete to B.S.5328; ordinary prescribed mix ST3, 20mm aggregate										
110mm outlet, reference 4D.316, with cover	23.99	5.00	25.19	1.80	0.45	19.77	2.91	47.87	nr	51.46
GULLY GRATING AND SEALING PLATES - Cast iron accessories, painted black										
Gratings; Hepworth Building Products										
reference IG6C; 140mm diameter	2.12	-	2.12	-	0.03	0.21	-	2.33	nr	2.51
reference IG7C; 197mm diameter	3.24	-	3.24	-	0.03	0.21	-	3.45	nr	3.71
reference IG8C; 284mm diameter	7.06	-	7.06	-	0.03	0.21	-	7.27	nr	7.82

Labour hourly rates: (except Specialists) Craft Operatives £9.23 Labourer £7.02 Rates are national average prices. Refer to REGIONAL VARIATIONS for indicative levels of overall pricing in regions	MATERIALS			LABOUR				RATES		
	Del to Site £	Waste %	Material Cost £	Craft Optve Hrs	Lab Hrs	Labour Cost £	Sunds £	Nett Rate £	Unit	Gross Rate (+7.5%) £

R12: DRAINAGE BELOW GROUND Cont.

GULLY GRATING AND SEALING PLATES - Cast iron accessories, painted black Cont.

Gratings; Hepworth Building Products Cont.

reference IG2C; 150 x 150mm	2.12	-	2.12	-	0.03	0.21	-	2.33	nr	2.51
reference IG3C; 225 x 225mm	6.31	-	6.31	-	0.03	0.21	-	6.52	nr	7.01
reference IG4C; 300 x 300mm	14.21	-	14.21	-	0.03	0.21	-	14.42	nr	15.50

Gratings and frames; bedding frames in cement mortar (1:3); Hepworth Building Products

reference IH5C; 100mm diameter	7.61	-	7.61	1.00	-	9.23	0.22	17.06	nr	18.34
reference IH6C; 150mm diameter	13.16	-	13.16	1.25	-	11.54	0.31	25.01	nr	26.88
reference IH7C; 225mm diameter	26.29	-	26.29	1.70	-	15.69	0.42	42.40	nr	45.58
reference IH2C; 150 x 150mm with lock and key	8.46	-	8.46	1.25	-	11.54	0.22	20.22	nr	21.73
reference IH3C; 230 x 230mm with lock and key	15.48	-	15.48	1.75	-	16.15	0.31	31.94	nr	34.34
reference IH4C; 316 x 316mm with lock and key	41.05	-	41.05	1.90	-	17.54	0.36	58.95	nr	63.37

Sealing plates and frames; bedding frames in cement mortar (1:3); Hepworth Building Products

reference IS5C; 140mm diameter	6.64	-	6.64	1.00	-	9.23	0.22	16.09	nr	17.30
reference IS6C; 197mm diameter	9.54	-	9.54	1.25	-	11.54	0.31	21.39	nr	22.99
reference IS7C; 273mm diameter	15.27	-	15.27	1.70	-	15.69	0.42	31.38	nr	33.73
reference IS2C; 150 x 150mm	8.19	-	8.19	1.25	-	11.54	0.22	19.95	nr	21.44
reference IS3C; 225 x 225mm	14.90	-	14.90	1.75	-	16.15	0.31	31.36	nr	33.71
reference IS4C; 318 x 318mm	37.43	-	37.43	1.90	-	17.54	0.36	55.33	nr	59.48

GULLY GRATING AND SEALING PLATES - Cast iron accessories, galvanised

Gratings; Hepworth Building Products

reference IG6G; 140mm diameter	3.89	-	3.89	-	0.03	0.21	-	4.10	nr	4.41
reference IG7G; 197mm diameter	4.95	-	4.95	-	0.03	0.21	-	5.16	nr	5.55
reference IG8G; 4mm diameter	10.59	-	10.59	-	0.03	0.21	-	10.80	nr	11.61
reference IG2G; 150 x 150mm	3.17	-	3.17	-	0.03	0.21	-	3.38	nr	3.63
reference IG3G; 225 x 225mm	9.47	-,	9.47	-	0.03	0.21	-	9.68	nr	10.41
reference IG4G; 300 x 300mm	21.33	-	21.33	-	0.03	0.21	-	21.54	nr	23.16

Gratings and frames; bedding frames in cement mortar (1:3); Hepworth Building Products

reference IH6G; 193mm diameter	19.76	-	19.76	1.25	-	11.54	0.31	31.61	nr	33.98
reference IH7G; 265mm diameter	39.22	-	39.22	1.70	-	15.69	0.42	55.33	nr	59.48
reference IH2G; 150 x 150mm with lock and key	12.70	-	12.70	1.25	-	11.54	0.22	24.46	nr	26.29
reference IH3G; 225 x 225mm with lock and key	23.20	-	23.20	1.75	-	16.15	0.31	39.66	nr	42.64

Sealing plates and frames; bedding frames in cement mortar (1:3); Hepworth Building Products

reference IS7G; 273mm diameter	22.94	-	22.94	1.70	-	15.69	0.42	39.05	nr	41.98
reference IS2G; 150 x 150mm	12.29	-	12.29	1.25	-	11.54	0.22	24.05	nr	25.85

GULLY GRATING AND SEALING PLATES - Cast iron accessories, painted black

Gratings and frames; Drainage Systems Ltd.; bedding frames in cement mortar (1:3)

reference B6064; 300 x 300mm	24.79	-	24.79	1.00	-	9.23	0.56	34.58	nr	37.17
reference B5883; 325 x 325mm, 69kg	75.39	-	75.39	2.50	-	23.07	0.72	99.19	nr	106.62
reference B5885; 400 x 350mm, 87kg	101.76	-	101.76	3.33	-	30.74	0.89	133.39	nr	143.39
reference B5887; 500 x 350mm, 140kg	139.64	-	139.64	4.25	-	39.23	1.00	179.87	nr	193.36
reference B5884; 340 x 340mm, 80kg	111.35	-	111.35	2.75	-	25.38	0.72	137.45	nr	147.76

SURFACE DRAINAGE SYSTEMS - Precast polyester concrete ACO Drain surface drainage systems; Technologies Plc; bedding and haunching in concrete to B.S.5328 ordinary prescribed mix ST5, 20mm aggregate

K100 Channel Class C system with galvanised steel slotted lockable gratings

constant depth or complete with 0.6% fall	40.00	5.00	42.00	1.50	1.95	27.53	5.83	75.36	m	81.02
extra; end cap	4.00	5.00	4.20	-	0.20	1.40	-	5.60	nr	6.02
extra; drain union	1.80	5.00	1.89	-	0.15	1.05	-	2.94	nr	3.16

Q100 Channel Class D monolithic system

constant depth or complete with 0.6% fall	50.00	5.00	52.50	1.70	2.10	30.43	12.95	95.88	m	103.07
extra; end cap	4.00	5.00	4.20	-	0.20	1.40	-	5.60	nr	6.02
extra; drain union	1.80	5.00	1.89	-	0.15	1.05	-	2.94	nr	3.16

Raindrain constant depth system with galvanised steel slotted lockable gratings

invert depth 115mm	12.00	5.00	12.60	1.40	1.65	24.50	4.82	41.92	m	45.07
extra; end cap	2.00	5.00	2.10	-	0.20	1.40	-	3.50	nr	3.77
extra; drain union	1.80	5.00	1.89	-	0.15	1.05	-	2.94	nr	3.16
extra; sump unit	40.00	5.00	42.00	1.00	1.00	16.25	6.47	64.72	nr	69.57

DISPOSAL SYSTEMS

DISPOSAL SYSTEMS — MAJOR WORKS

Labour hourly rates: (except Specialists) Craft Operatives £9.23 Labourer £7.02 Rates are national average prices. Refer to REGIONAL VARIATIONS for indicative levels of overall pricing in regions	PLANT & TRANSPORT			LABOUR				RATES		
	Plant Cost £	Trans Cost £	P & T Cost £	Craft Optve Hrs	Lab Hrs	Labour Cost £	Sunds £	Nett Rate £	Unit	Gross Rate (+7.5%) £
R12: DRAINAGE BELOW GROUND; MANHOLES AND SOAKAWAYS										
Excavating - by machine										
Excavating pits commencing from natural ground level; maximum depth not exceeding										
0.25m	4.30	-	4.30	-	0.84	5.90	-	10.20	m3	10.96
1.00m	5.30	-	5.30	-	0.88	6.18	-	11.48	m3	12.34
2.00m	5.30	-	5.30	-	1.00	7.02	-	12.32	m3	13.24
4.00m	6.00	-	6.00	-	1.27	8.92	-	14.92	m3	16.03
Working space allowance to excavation pits	5.00	-	5.00	-	0.60	4.21	-	9.21	m2	9.90
Compacting bottoms of excavations	0.15	-	0.15	-	-	-	-	0.15	m2	0.16
Excavated material depositing on site in temporary spoil heaps where directed										
25m	1.61	-	1.61	-	-	-	-	1.61	m3	1.73
50m	1.69	-	1.69	-	-	-	-	1.69	m3	1.82
100m	1.84	-	1.84	-	-	-	-	1.84	m3	1.98
Excavating - by hand										
Excavating pits commencing from natural ground level; maximum depth not exceeding										
0.25m	-	-	-	-	3.35	23.52	-	23.52	m3	25.28
1.00m	-	-	-	-	3.50	24.57	-	24.57	m3	26.41
2.00m	-	-	-	-	4.00	28.08	-	28.08	m3	30.19
4.00m	-	-	-	-	5.10	35.80	-	35.80	m3	38.49
Working space allowance to excavation pits	-	-	-	-	3.12	21.90	-	21.90	m2	23.55
Compacting bottoms of excavations	-	-	-	-	0.12	0.84	-	0.84	m2	0.91
Excavated material depositing on site in temporary spoil heaps where directed										
25m	-	-	-	-	1.25	8.78	-	8.78	m3	9.43
50m	-	-	-	-	1.50	10.53	-	10.53	m3	11.32
100m	-	-	-	-	2.00	14.04	-	14.04	m3	15.09

This section continues on the next page

Labour hourly rates: (except Specialists) Craft Operatives £9.23 Labourer £7.02 Rates are national average prices. Refer to REGIONAL VARIATIONS for indicative levels of overall pricing in regions	MATERIALS			LABOUR				RATES		
	Del to Site	Waste	Material Cost	Craft Optve	Lab	Labour Cost	Sunds	Nett Rate	Unit	Gross Rate (+7.5%)
	£	%	£	Hrs	Hrs	£	£	£		£
R12: DRAINAGE BELOW GROUND; MANHOLES AND SOAKAWAYS Cont.										
Earthwork support										

Earthwork support
  distance between opposing faces not exceeding
  2.00m; maximum depth not exceeding										

1.00m	0.63	–	0.63	–	0.20	1.40	0.02	2.05	m2	2.21
2.00m	0.76	–	0.76	–	0.25	1.75	0.03	2.54	m2	2.74
4.00m	0.90	–	0.90	–	0.30	2.11	0.03	3.04	m2	3.26
distance between opposing faces 2.00 – 4.00m; maximum not exceeding										
1.00m	0.68	–	0.68	–	0.21	1.47	0.02	2.17	m2	2.34
2.00m	0.84	–	0.84	–	0.26	1.83	0.03	2.70	m2	2.90
4.00m	0.96	–	0.96	–	0.32	2.25	0.03	3.24	m2	3.48

Earthwork support; in unstable ground
  distance between opposing faces not exceeding
  2.00m; maximum depth not exceeding										

1.00m	1.05	–	1.05	–	0.33	2.32	0.04	3.41	m2	3.66
2.00m	1.27	–	1.27	–	0.42	2.95	0.04	4.26	m2	4.58
4.00m	1.49	–	1.49	–	0.50	3.51	0.05	5.05	m2	5.43
distance between opposing faces 2.00 – 4.00m; maximum depth not exceeding										
1.00m	1.18	–	1.18	–	0.35	2.46	0.04	3.68	m2	3.95
2.00m	1.40	–	1.40	–	0.43	3.02	0.05	4.47	m2	4.80
4.00m	1.61	–	1.61	–	0.53	3.72	0.06	5.39	m2	5.79
FILLING – Excavated material arising from excavations										
Filling to excavation average thickness exceeding 0.25m	–	–	–	–	1.33	9.34	–	9.34	m3	10.04
Excavated material obtained from on site spoil heaps 25m distant										
Filling to excavation average thickness not exceeding 0.25m	–	–	–	–	1.75	12.29	–	12.29	m3	13.21
average thickness exceeding 0.25m	–	–	–	–	1.50	10.53	–	10.53	m3	11.32
Excavated material obtained from on site spoil heaps 50m distant										
Filling to excavation average thickness not exceeding 0.25m	–	–	–	–	2.00	14.04	–	14.04	m3	15.09
average thickness exceeding 0.25m	–	–	–	–	1.85	12.99	–	12.99	m3	13.96
Excavated material obtained from on site spoil heaps 100m distant										
Filling to excavation average thickness not exceeding 0.25m	–	–	–	–	2.50	17.55	–	17.55	m3	18.87
average thickness exceeding 0.25m	–	–	–	–	2.25	15.80	–	15.80	m3	16.98
Hard, dry broken brick or stone, 100 – 75mm gauge, to be obtained off site										
Filling to excavation average thickness exceeding 0.25m	12.40	25.00	15.50	–	1.25	8.78	–	24.27	m3	26.10
CONCRETE WORK – Plain in-situ concrete; B.S.5328, ordinary prescribed mix ST3, 20mm aggregate										
Beds; poured on or against earth or unblinded hardcore										
thickness 150 – 450mm	48.30	5.00	50.72	–	4.00	28.08	–	78.80	m3	84.70
thickness not exceeding 150mm	48.30	5.00	50.72	–	4.30	30.19	–	80.90	m3	86.97
Benching in bottoms; rendering with 13mm cement and sand (1:2) before final set										
600 x 450 x average 225mm; trowelling	2.66	5.00	2.79	–	0.54	3.79	3.18	9.76	nr	10.50
675 x 675 x average 225mm; trowelling	5.27	5.00	5.53	–	0.92	6.46	5.30	17.29	nr	18.59
800 x 675 x average 225mm; trowelling	5.78	5.00	6.07	–	1.10	7.72	6.15	19.94	nr	21.44
800 x 800 x average 300mm; trowelling	9.28	5.00	9.74	–	1.73	12.14	7.26	29.15	nr	31.33
1025 x 800 x average 300mm; trowelling	12.38	5.00	13.00	–	2.21	15.51	9.33	37.84	nr	40.68
Reinforced in-situ concrete; B.S.5328, designed mix ST4, 20mm aggregate, minimum cement content 240 kg/m3; vibrated										
Slabs										
thickness 150 – 450mm	49.60	5.00	52.08	–	3.50	24.57	–	76.65	m3	82.40
thickness not exceeding 150mm	49.60	5.00	52.08	–	4.00	28.08	–	80.16	m3	86.17
Formwork and basic finish										
Edges of suspended slabs; plain vertical height not exceeding 250mm	1.09	12.50	1.23	0.75	0.15	7.98	0.24	9.44	m	10.15

Soffits of slabs; horizontal
  slab thickness not exceeding 200mm; height to
  soffit not exceeding 1.50m
	8.69	10.00	9.56	2.20	0.42	23.25	0.25	33.06	m2	35.54

slab thickness not exceeding 200mm; height to soffit 1.50 – 3.00m	8.69	10.00	9.56	2.00	0.42	21.41	0.30	31.27	m2	33.61

DISPOSAL SYSTEMS

Labour hourly rates: (except Specialists) Craft Operatives £9.23 Labourer £7.02 Rates are national average prices. Refer to REGIONAL VARIATIONS for indicative levels of overall pricing in regions	MATERIALS			LABOUR				RATES		
	Del to Site £	Waste %	Material Cost £	Craft Optve Hrs	Lab Hrs	Labour Cost £	Sunds £	Nett Rate £	Unit	Gross Rate (+7.5%) £
R12: DRAINAGE BELOW GROUND; MANHOLES AND SOAKAWAYS Cont.										
Formwork and basic finish Cont.										
Holes; rectangular										
girth 1.00 - 2.00m; depth not exceeding 250mm	1.95	15.00	2.24	0.50	0.05	4.97	-	7.21	nr	7.75
girth 2.00 - 3.00m; depth not exceeding 250mm	2.54	15.00	2.92	0.80	0.10	8.09	-	11.01	nr	11.83
Reinforcement fabric; B.S.4483; hard drawn plain round steel, welded; including laps, tying wire, all cutting and bending and spacers and chairs which are at the discretion of the Contractor										
Reference A142, 2.22kg/m2; 200mm side laps; 200mm end laps										
generally	0.78	15.00	0.90	0.25	0.02	2.45	0.06	3.40	m2	3.66
BRICKWORK - Common bricks, B.S.3921, Category M, 215 x 102.5 x 65mm, compressive strength 20.5 N/mm2; in cement mortar (1:3)										
Walls; vertical										
102mm thick; stretcher bond	8.79	5.00	9.23	1.33	1.45	22.45	1.98	33.66	m2	36.19
215mm thick; English bond	17.73	5.00	18.62	2.30	2.54	39.06	3.97	61.65	m2	66.27
Milton Hall Second hard stock bricks, B.S.3921, Category M, 215 x 102.5 x 65mm; in cement mortar (1:3)										
Walls; vertical										
102mm thick; stretcher bond	17.11	5.00	17.97	1.33	1.45	22.45	1.98	42.40	m2	45.58
215mm thick; English bond	34.51	5.00	36.24	2.30	2.54	39.06	3.97	79.27	m2	85.21
Engineering bricks, Category F, B.S.3921, 215 x 102.5 x 65mm, class B; in cement mortar (1:3)										
Walls; vertical										
102mm thick; stretcher bond	17.53	5.00	18.41	1.33	1.45	22.45	1.98	42.84	m2	46.05
215mm thick; English bond	35.36	5.00	37.13	2.30	2.54	39.06	3.97	80.16	m2	86.17
Common bricks, B.S.3921, Category M, 215 x 102.5 x 65mm, compressive strength 20.5 N/mm2; in cement mortar (1:3); flush smooth pointing as work proceeds										
Walls; vertical										
102mm thick; stretcher bond; facework one side ...	8.79	5.00	9.23	1.53	1.65	25.70	1.98	36.91	m2	39.68
215mm thick; English bond; facework one side	17.73	5.00	18.62	2.50	2.74	42.31	3.97	64.90	m2	69.76
Milton Hall Second hard stock bricks, B.S.3921, Category M, 215 x 102.5 x 65mm; in cement mortar (1:3); flush smooth pointing as work proceeds										
Walls; vertical										
102mm thick; stretcher bond; facework one side ...	17.11	5.00	17.97	1.53	1.65	25.70	1.98	45.65	m2	49.07
215mm thick; English bond; facework one side	34.51	5.00	36.24	2.50	2.74	42.31	3.97	82.52	m2	88.70
Engineering bricks, B.S.3921, Category F, 215 x 102.5 x 65mm, class B; in cement mortar (1:3); flush smooth pointing as work proceeds										
Walls; vertical										
102mm thick; stretcher bond; facework one side ...	17.53	5.00	18.41	1.53	1.65	25.70	1.98	46.09	m2	49.55
215mm thick; English bond; facework one side	35.36	5.00	37.13	2.50	2.74	42.31	3.97	83.41	m2	89.66
Accessories/sundry items for brick/block/stone walling										
Building in ends of pipes; 100mm diameter making good fair face one side										
102mm brickwork...............................	-	-	-	0.08	0.08	1.30	0.06	1.36	nr	1.46
215mm brickwork...............................	-	-	-	0.13	0.13	2.11	0.09	2.20	nr	2.37
Building in ends of pipes; 150mm diameter making good fair face one side										
102mm brickwork...............................	-	-	-	0.13	0.13	2.11	0.07	2.18	nr	2.35
215mm brickwork...............................	-	-	-	0.20	0.20	3.25	0.12	3.37	nr	3.62
Building in ends of pipes; 225mm diameter making good fair face one side										
102mm brickwork...............................	-	-	-	0.20	0.20	3.25	0.09	3.34	nr	3.59
215mm brickwork...............................	-	-	-	0.25	0.25	4.06	0.17	4.23	nr	4.55
Building in ends of pipes; 300mm diameter making good fair face one side										
102mm brickwork...............................	-	-	-	0.25	0.25	4.06	0.20	4.26	nr	4.58
215mm brickwork...............................	-	-	-	0.33	0.33	5.36	0.28	5.64	nr	6.07
Mortar, cement and sand (1:3)										
13mm work to walls on brickwork base; one coat; trowelled										
width exceeding 300mm	1.13	-	1.13	0.40	0.40	6.50	0.31	7.94	m2	8.54

Labour hourly rates: (except Specialists) Craft Operatives £9.23 Labourer £7.02 Rates are national average prices. Refer to REGIONAL VARIATIONS for indicative levels of overall pricing in regions	MATERIALS			LABOUR				RATES		
	Del to Site £	Waste %	Material Cost £	Craft Optve Hrs	Lab Hrs	Labour Cost £	Sunds £	Nett Rate £	Unit	Gross Rate (+7.5%) £
R12: DRAINAGE BELOW GROUND; MANHOLES AND SOAKAWAYS **Cont.**										
Mortar, cement and sand (1:3) Cont.										

13mm work to floors on concrete base; one coat;
trowelled
    level and to falls only not exceeding 15 degrees
    form horizontal ....
	1.13	–	1.13	0.60	0.60	9.75	0.31	11.19	m2	12.03

CHANNELS, VITRIFIED CLAY - Channels in bottoms; **vitrified clay, B.S.65, normal, glazed; cement mortar** **(1:3) joints; bedding in cement mortar (1:3)**										
Half section straight; 600mm effective length										
100mm	1.69	5.00	1.77	0.30	0.03	2.98	0.16	4.91	nr	5.28
150mm	2.84	5.00	2.98	0.40	0.04	3.97	0.26	7.21	nr	7.76
straight; 1000mm effective length										
100mm	2.36	5.00	2.48	0.50	0.05	4.97	0.16	7.60	nr	8.17
150mm	4.46	5.00	4.68	0.67	0.07	6.68	0.26	11.62	nr	12.49
225mm	10.42	5.00	10.94	0.84	0.08	8.31	0.32	19.58	nr	21.04
300mm	21.94	5.00	23.04	1.25	0.15	12.59	0.39	36.02	nr	38.72
curved; 500mm effective length										
100mm	2.99	5.00	3.14	0.30	0.03	2.98	0.16	6.28	nr	6.75
150mm	4.94	5.00	5.19	0.40	0.25	5.45	0.26	10.89	nr	11.71
225mm	16.49	5.00	17.31	0.50	0.05	4.97	0.32	22.60	nr	24.30
300mm	33.63	5.00	35.31	0.75	0.09	7.55	0.39	43.26	nr	46.50
curved; 900mm effective length										
100mm	2.99	5.00	3.14	0.60	0.06	5.96	0.16	9.26	nr	9.95
150mm	4.94	5.00	5.19	0.80	0.08	7.95	0.26	13.39	nr	14.40
225mm	16.49	5.00	17.31	1.00	0.10	9.93	0.32	27.57	nr	29.63
300mm	33.63	5.00	35.31	1.50	0.18	15.11	0.39	50.81	nr	54.62
Three quarter section branch bends										
100mm	6.76	5.00	7.10	0.33	0.03	3.26	0.16	10.51	nr	11.30
150mm	11.74	5.00	12.33	0.50	0.05	4.97	0.26	17.55	nr	18.87
225mm	41.36	5.00	43.43	1.00	0.10	9.93	0.32	53.68	nr	57.71
Channels in bottoms; P.V.C.; 'O' ring joints; bedding **in cement mortar (1:3)**										
For 610mm clear opening straight										
110mm	15.21	5.00	15.97	0.40	0.03	3.90	2.23	22.10	nr	23.76
160mm	28.79	5.00	30.23	0.50	0.04	4.90	5.01	40.14	nr	43.15
short bend										
110mm	7.02	5.00	7.37	0.20	0.01	1.92	2.11	11.40	nr	12.25
160mm	16.67	5.00	17.50	0.30	0.02	2.91	4.74	25.15	nr	27.04
long radius bend										
110mm	24.95	5.00	26.20	0.33	0.02	3.19	2.11	31.49	nr	33.86
160mm	54.34	5.00	57.06	0.50	0.04	4.90	4.74	66.69	nr	71.69
PRECAST CONCRETE INSPECTION CHAMBERS - Precast **concrete; standard or stock pattern units; B.S.5911;** **jointing and pointing in cement mortar (1:3)**										
Chamber or shaft sections 900mm internal diameter										
250mm high.....................	19.74	5.00	20.73	0.58	1.16	13.50	0.08	34.30	nr	36.88
500mm high.....................	39.47	5.00	41.44	0.67	1.34	15.59	0.08	57.11	nr	61.40
750mm high.....................	29.60	5.00	31.08	0.87	1.74	20.24	0.08	51.40	nr	55.26
1000mm high....................	39.47	5.00	41.44	0.93	1.86	21.64	0.08	63.16	nr	67.90
1050mm internal diameter										
305mm high.....................	27.51	5.00	28.89	0.81	1.62	18.85	0.15	47.88	nr	51.48
458mm high.....................	41.31	5.00	43.38	0.87	1.74	20.24	0.15	63.77	nr	68.55
610mm high.....................	27.51	5.00	28.89	0.93	1.86	21.64	0.15	50.68	nr	54.48
762mm high.....................	34.37	5.00	36.09	0.99	1.98	23.04	0.15	59.28	nr	63.72
914mm high.....................	41.22	5.00	43.28	1.04	2.08	24.20	0.15	67.63	nr	72.70
1200mm internal diameter										
250mm high.....................	28.39	5.00	29.81	0.99	1.98	23.04	0.21	53.06	nr	57.04
500mm high.....................	44.67	5.00	46.90	1.04	2.08	24.20	0.21	71.31	nr	76.66
750mm high.....................	44.33	5.00	46.55	1.22	2.44	28.39	0.21	75.15	nr	80.78
1000mm high....................	53.17	5.00	55.83	1.28	2.56	29.79	0.21	85.82	nr	92.26

Cutting holes
  cutting holes for pipes; 100mm diameter; making
  good; pointing ....
	30.00	–	30.00	–	0.08	0.56	0.13	30.69	nr	32.99

cutting holes for pipes; 150mm diameter; making good; pointing	30.00	–	30.00	–	0.10	0.70	0.15	30.85	nr	33.17
cutting holes for pipes; 225mm diameter; making good; pointing	45.00	–	45.00	–	0.12	0.84	0.18	46.02	nr	49.47
cutting holes for pipes; 300mm diameter; making good; pointing	60.00	–	60.00	–	0.15	1.05	0.23	61.28	nr	65.88

Step irons
  galvanised malleable cast iron step irons
  B.S.1247, cast in ....
	3.35	–	3.35	–	0.15	1.05	–	4.40	nr	4.73

Cover slabs, heavy duty
  for chamber or shaft sections 900mm internal
  diameter, 125mm thick; access opening 600mm x
  600mm...............................
	39.87	5.00	41.86	1.50	2.25	29.64	1.23	72.73	nr	78.19

DISPOSAL SYSTEMS

Labour hourly rates: (except Specialists) Craft Operatives £9.23 Labourer £7.02 Rates are national average prices. Refer to REGIONAL VARIATIONS for indicative levels of overall pricing in regions	MATERIALS			LABOUR				RATES		
	Del to Site £	Waste %	Material Cost £	Craft Optve Hrs	Lab Hrs	Labour Cost £	Sunds £	Nett Rate £	Unit	Gross Rate (+7.5%) £
R12: DRAINAGE BELOW GROUND; MANHOLES AND SOAKAWAYS Cont.										
PRECAST CONCRETE INSPECTION CHAMBERS - Precast **concrete; standard or stock pattern units; B.S.5911;** **jointing and pointing in cement mortar (1:3) Cont.**										
Cover slabs, heavy duty Cont.										
for chamber or shaft sections 1050mm internal diameter, 125mm thick; access opening 600mm x 600mm	45.77	5.00	48.06	2.00	3.00	39.52	1.45	89.03	nr	95.71
for chamber or shaft sections 1200mm internal diameter, 125mm thick; access opening 600mm x 600mm	60.53	5.00	63.56	2.50	3.75	49.40	1.62	114.58	nr	123.17
for chamber or shaft sections 1350mm internal diameter, 125mm thick; access opening 600mm x 600mm	88.81	5.00	93.25	3.00	4.50	59.28	1.84	154.37	nr	165.95
for chamber or shaft sections 1500mm internal diameter, 150mm thick; access opening 600mm x 600mm	109.67	5.00	115.15	3.50	5.25	69.16	2.01	186.32	nr	200.30
for chamber or shaft sections 1800mm internal diameter, 150mm thick access opening 600mm x 600mm	126.95	5.00	133.30	4.00	6.00	79.04	-	212.34	nr	228.26
for chamber or shaft sections 2100mm internal diameter, 150mm thick access opening 600mm x 600mm	262.47	5.00	275.59	4.50	6.75	88.92	-	364.51	nr	391.85
for chamber or shft sections 2400mm internal diameter, 150mm thick access opening 600mm x 600mm	457.46	5.00	480.33	5.00	7.50	98.80	-	579.13	nr	622.57
for chamber or shaft sections 2700mm internal diameter, 150mm thick access opening 600mm x 600mm	592.22	5.00	621.83	5.50	8.25	108.68	-	730.51	nr	785.30
for chamber or shaft sections 3000mm internal diameter, 150mm thick access opening 600mm x 600mm	697.51	5.00	732.39	6.00	9.00	118.56	-	850.95	nr	914.77
Precast concrete inspection chambers; B.S.5911; **jointing and pointing in cement mortar (1:3)**										
Rectangular inspection chambers, internal size 457 x 610mm										
100mm high	6.45	5.00	6.77	0.50	0.35	7.07	0.08	13.92	nr	14.97
150mm high	7.10	5.00	7.46	0.65	0.45	9.16	0.08	16.69	nr	17.95
250mm high	9.60	5.00	10.08	0.75	0.53	10.64	0.08	20.80	nr	22.36
extra; galvanised malleable cast iron step irons B.S.1247, built in	2.90	5.00	3.04	-	-	-	-	3.04	nr	3.27
Rectangular inspection chambers, internal size 760 x 610mm										
100mm high	9.05	5.00	9.50	0.81	0.58	11.55	0.16	21.21	nr	22.80
150mm high	8.55	5.00	8.98	1.05	0.75	14.96	0.16	24.09	nr	25.90
225mm high	11.50	5.00	12.07	1.21	0.88	17.35	0.16	29.58	nr	31.80
extra; galvanised malleable cast iron step irons B.S.1247, built in	2.90	5.00	3.04	-	-	-	-	3.04	nr	3.27
Rectangular inspection chambers, internal size 990 x 610mm										
150mm high	10.30	5.00	10.82	1.05	0.75	14.96	0.16	25.93	nr	27.88
Rectangular inspection chambers, internal size 1200 x 750mm										
150mm high	21.05	5.00	22.10	1.20	0.85	17.04	0.16	39.31	nr	42.25
200mm high	25.95	5.00	27.25	1.38	1.00	19.76	0.16	47.16	nr	50.70
250mm high	29.05	5.00	30.50	1.57	1.14	22.49	0.16	53.16	nr	57.14
Base units										
450 x 610mm internal size; 360mm deep	26.05	5.00	27.35	1.05	0.75	14.96	0.08	42.39	nr	45.57
760 x 610mm internal size; 360mm deep	29.90	5.00	31.40	1.05	0.75	14.96	0.16	46.51	nr	50.00
990 x 610mm internal size; 360mm deep	42.10	5.00	44.20	1.05	0.75	14.96	0.16	59.32	nr	63.77
GRP INSPECTION CHAMBERS										
Inspection chambers; Terrain; GRP with preformed benching; 4 nr pvc adaptors; A15 light duty single seal cast iron cover to B.S.EN124 and plastic frame set in position 475mm internal diameter; depth to invert										
450mm	161.44	5.00	169.51	5.00	0.80	51.77	-	221.28	nr	237.87
585mm	191.59	5.00	201.17	5.00	0.80	51.77	-	252.94	nr	271.91
750mm	221.74	5.00	232.83	5.35	0.85	55.35	-	288.17	nr	309.79
930mm	251.89	5.00	264.48	5.35	0.85	55.35	-	319.83	nr	343.82
POLYPROPYLENE INSPECTION CHAMBERS										
Osma; Shallow inspection chamber Polypropylene with preformed benching. A15 light duty single sealed cover and frame set in position, 250mm diameter; for 100mm diameter pipes ref 4D960, depth to invert										
600mm	53.31	5.00	55.98	2.00	0.10	19.16	-	75.14	nr	80.77
Osma; universal inspection chamber; Polypropylene; 450mm diameter										
preformed chamber base benching, ref. 40.922; for100mm pipes	49.15	5.00	51.61	1.10	0.10	10.86	-	62.46	nr	67.15
chamber shaft 230mm effective length, ref. 40.925	12.52	5.00	13.15	-	0.50	3.51	-	16.66	nr	17.91
cast iron cover and frame; single seal A15 light duty to suit, ref. 40.344	28.66	5.00	30.09	-	0.50	3.51	-	33.60	nr	36.12

Labour hourly rates: (except Specialists) Craft Operatives £12.44 Labourer £7.02 Rates are national average prices. Refer to REGIONAL VARIATIONS for indicative levels of overall pricing in regions	MATERIALS			LABOUR				RATES		
	Del to Site £	Waste %	Material Cost £	Craft Optve Hrs	Lab Hrs	Labour Cost £	Sunds £	Nett Rate £	Unit	Gross Rate (+7.5%) £
R12: DRAINAGE BELOW GROUND; MANHOLES AND SOAKAWAYS **Cont.**										
CAST IRON INSPECTION CHAMBERS										
100mm diameter straight coated cast iron bolted inspection chambers; coupling joints; bedding in cement mortar (1:3)										
with one branch one side, fig 110	90.01	2.00	**91.81**	2.50	–	**31.10**	–	**122.91**	nr	**132.13**
with one branch each side, fig 111	120.29	2.00	**122.70**	3.25	–	**40.43**	–	**163.13**	nr	**175.36**
with two branches one side, fig 210	169.97	2.00	**173.37**	3.25	–	**40.43**	–	**213.80**	nr	**229.83**
with two branches each side, fig 212	227.54	2.00	**232.09**	4.75	–	**59.09**	–	**291.18**	nr	**313.02**
100-150mm diameter straight coated cast iron bolted inspection chambers; coupling joints; bedding in cement mortar (1:3)										
with one branch one side, fig 110	127.15	2.00	**129.69**	3.25	–	**40.43**	–	**170.12**	nr	**182.88**
150mm diameter straight coated cast iron bolted inspection chambers; coupling joints; bedding in cement mortar (1:3)										
with one branch one side, fig 110	158.34	2.00	**161.51**	3.50	–	**43.54**	–	**205.05**	nr	**220.43**
with one branch each side, fig 111	184.47	2.00	**188.16**	4.50	–	**55.98**	–	**244.14**	nr	**262.45**
with two branches one side, fig 210	289.40	2.00	**295.19**	4.50	–	**55.98**	–	**351.17**	nr	**377.51**
with two branches each side, fig 212	356.09	2.00	**363.21**	6.50	–	**80.86**	–	**444.07**	nr	**477.38**

This section continues
on the next page

DISPOSAL SYSTEMS

Labour hourly rates: (except Specialists) Craft Operatives £9.23 Labourer £7.02 Rates are national average prices. Refer to REGIONAL VARIATIONS for indicative levels of overall pricing in regions	MATERIALS			LABOUR				RATES		
	Del to Site £	Waste %	Material Cost £	Craft Optve Hrs	Lab Hrs	Labour Cost £	Sunds £	Nett Rate £	Unit	Gross Rate (+7.5%) £
R12: DRAINAGE BELOW GROUND; MANHOLES AND SOAKAWAYS **Cont.**										
MANHOLE ACCESSORIES - Manhole step irons										
Step irons; B.S.1247 malleable cast iron; galvanised; building in to joints										
figure 1 general purpose pattern; 110mm tail	2.01	2.50	2.06	0.25	-	2.31	1.42	5.79	nr	6.22
figure 1 general purpose pattern; 225mm tail	2.51	2.50	2.57	0.25	-	2.31	1.42	6.30	nr	6.77
Manhole covers - A15 light duty										
Access covers; B.S.EN124; coated; bedding frame in cement mortar (1:3); bedding cover in grease and sand										
reference MC1-45/45; clear opening 450 x 450mm ...	30.66	2.50	31.43	0.75	0.81	12.61	0.38	44.42	nr	47.75
reference MC1-60/45; clear opening 600 x 450mm ...	30.83	2.50	31.60	1.00	1.08	16.81	0.38	48.79	nr	52.45
reference MC1R-60/60; clear opening 600 x 600mm ..	86.83	2.50	89.00	2.00	2.20	33.90	0.38	123.28	nr	132.53
reference MC2-45/45; clear opening 450 x 450mm ...	47.18	2.50	48.36	1.50	1.60	25.08	0.49	73.93	nr	79.47
reference MC2-60/45; clear opening 600 x 450mm ...	61.79	2.50	63.33	1.75	1.88	29.35	0.49	93.17	nr	100.16
reference MC2-60/60; clear opening 600 x 600mm ...	87.14	2.50	89.32	2.00	2.20	33.90	0.49	123.71	nr	132.99
reference MC2R-45/45; clear opening 450 x 450mm ..	78.36	2.50	80.32	1.80	1.94	30.23	0.49	111.04	nr	119.37
reference MC2R-60/45; clear opening 600 x 450mm ..	72.79	2.50	74.61	2.20	2.40	37.15	0.49	112.25	nr	120.67
reference MC2R-60/60; clear opening 600 x 600mm ..	104.99	2.50	107.61	2.60	2.85	44.01	0.49	152.11	nr	163.52
lifting keys	3.50	2.50	3.59	-	-	-	-	3.59	nr	3.86
Manhole covers - B125 medium duty										
Access covers; B.S.EN124; coated; bedding frame in cement mortar (1:3); bedding cover in grease and sand										
reference MB1-60; clear opening 600mm diameter ...	66.69	2.50	68.36	3.50	3.85	59.33	0.77	128.46	nr	138.09
reference MB2-50; clear opening 500mm diameter ...	74.49	2.50	76.35	2.75	3.00	46.44	0.56	123.35	nr	132.61
reference MB2-55; clear opening 550mm diameter ...	69.36	2.50	71.09	2.25	2.45	37.97	0.56	109.62	nr	117.84
reference MB2-60; clear opening 600mm diameter ...	57.05	2.50	58.48	2.75	3.00	46.44	0.56	105.48	nr	113.39
reference MB2-60/45; clear opening 600 x 450mm ...	51.74	2.50	53.03	2.75	3.00	46.44	0.83	100.31	nr	107.83
reference MB2-60/60; clear opening 600 x 600mm ...	63.16	2.50	64.74	3.00	3.25	50.51	1.04	116.28	nr	125.01
reference MB2R-60/45; clear opening 600 x 450mm ..	92.78	2.50	95.10	2.75	3.00	46.44	0.83	142.37	nr	153.05
reference MB2R-60/60; clear opening 600 x 600mm ..	105.55	2.50	108.19	3.00	3.25	50.51	1.04	159.73	nr	171.71
lifting keys	4.00	2.50	4.10	-	-	-	-	4.10	nr	4.41
Manhole covers - D400 heavy duty										
Access covers; B.S.EN124; coated; bedding frame in cement mortar (1:3); bedding cover in grease and sand										
reference MD-60; clear opening 600mm diameter	103.06	2.50	105.64	1.80	1.80	29.25	1.04	135.93	nr	146.12
reference MA-60; clear opening 600 diameter	69.03	2.50	70.76	1.80	1.80	29.25	1.04	101.05	nr	108.62
reference MA-T; clear opening 550 x 495mm	106.52	2.50	109.18	2.00	2.00	32.50	1.04	142.72	nr	153.43
lifting keys	4.00	2.50	4.10	-	-	-	-	4.10	nr	4.41
INTERCEPTING TRAPS - Vitrified clay intercepting **traps**										
Intercepting traps; B.S.65; joint to pipe; building into side of manhole; bedding and surrounding in concrete to B.S.5328 ordinary prescribed mix C15P, 20mm aggregate cleaning arm and stopper										
100mm outlet, ref. R1 1/1....................	35.54	5.00	37.32	1.25	1.00	18.56	14.20	70.07	nr	75.33
150mm outlet, ref.R1 1/2.....................	51.25	5.00	53.81	1.75	1.50	26.68	18.02	98.52	nr	105.90
225mm outlet, ref. R1 1/3....................	159.97	5.00	167.97	2.25	2.00	34.81	25.66	228.44	nr	245.57
FRESH AIR INLETS - Aluminium; fresh air inlets										
Air inlet valve, mice flap; set in cement mortar (1:3)										
100mm diameter	19.20	5.00	20.16	1.15	-	10.61	-	30.77	nr	33.08
SETTLEMENT TANKS - Fibreglass settlement tanks; **including cover and frame**										
Settlement tank; fibreglass; set in position 1000mm deep to invert of inlet, standard grade										
2700 litres capacity.........................	486.00	-	486.00	0.90	1.80	20.94	-	506.94	nr	544.96
3750 litres capacity.........................	648.00	-	648.00	1.00	2.00	23.27	-	671.27	nr	721.62
4500 litres capacity.........................	767.00	-	767.00	1.10	2.20	25.60	-	792.60	nr	852.04
7500 litres capacity.........................	1435.00	-	1435.00	1.20	2.40	27.92	-	1462.92	nr	1572.64
9000 litres capacity.........................	1563.00	-	1563.00	1.30	2.60	30.25	-	1593.25	nr	1712.74
1500mm deep to invert of inlet, heavy grade										
2700 litres capacity.........................	629.00	-	629.00	0.90	1.80	20.94	-	649.94	nr	698.69
3750 litres capacity.........................	830.00	-	830.00	1.00	2.00	23.27	-	853.27	nr	917.27
4500 litres capacity.........................	984.00	-	984.00	1.10	2.20	25.60	-	1009.60	nr	1085.32
7500 litres capacity.........................	1831.00	-	1831.00	1.20	2.40	27.92	-	1858.92	nr	1998.34
9000 litres capacity.........................	2002.00	-	2002.00	1.30	2.60	30.25	-	2032.25	nr	2184.67

Labour hourly rates: (except Specialists) Craft Operatives £9.23 Labourer £7.02 Rates are national average prices. Refer to REGIONAL VARIATIONS for indicative levels of overall pricing in regions	MATERIALS			LABOUR				RATES		
	Del to Site	Waste	Material Cost	Craft Optve	Lab	Labour Cost	Sunds	Nett Rate	Unit	Gross Rate (+7.5%)
	£	%	£	Hrs	Hrs	£	£	£		£
R12: DRAINAGE BELOW GROUND; WORK TO EXISTING DRAINS										
Work to disused drains										
Fly ash filling to disused drain										
100mm diameter	0.50	–	0.50	0.40	–	3.69	–	4.19	m	4.51
150mm diameter	1.01	–	1.01	0.60	–	5.54	–	6.55	m	7.04
225mm diameter	1.96	–	1.96	0.90	–	8.31	–	10.27	m	11.04
Sealing end of disused drain with plain in-situ concrete										
100mm diameter	0.51	–	0.51	0.20	–	1.85	–	2.36	nr	2.53
150mm diameter	1.02	–	1.02	0.30	–	2.77	–	3.79	nr	4.07
225mm diameter	1.54	–	1.54	0.40	–	3.69	–	5.23	nr	5.62
R13: LAND DRAINAGE										
Drains; clayware pipes, B.S.1196; butt joints										
Pipework in trenches										
75mm	0.95	5.00	1.00	–	0.22	1.54	–	2.54	m	2.73
100mm	1.69	5.00	1.77	–	0.24	1.68	–	3.46	m	3.72
150mm	3.62	5.00	3.80	–	0.30	2.11	–	5.91	m	6.35
Drains; porous concrete pipes, B.S.5911; butt joints										
Pipework in trenches										
150mm	5.24	5.00	5.50	–	0.28	1.97	–	7.47	m	8.03
225mm	6.71	5.00	7.05	–	0.36	2.53	–	9.57	m	10.29
300mm	11.03	5.00	11.58	–	0.42	2.95	–	14.53	m	15.62
375mm	15.03	5.00	15.78	–	0.48	3.37	–	19.15	m	20.59
450mm	21.14	5.00	22.20	–	0.54	3.79	–	25.99	m	27.94
525mm	31.32	5.00	32.89	–	0.60	4.21	–	37.10	m	39.88
600mm	38.90	5.00	40.84	–	0.66	4.63	–	45.48	m	48.89
Drains; OGEE pipes, B.S.5911 Part 110; butt joints										
Pipework in trenches										
150mm	4.28	5.00	4.49	–	0.28	1.97	–	6.46	m	6.94
225mm	6.92	5.00	7.27	–	0.36	2.53	–	9.79	m	10.53
300mm	11.36	5.00	11.93	–	0.42	2.95	–	14.88	m	15.99
375mm	15.43	5.00	16.20	–	0.48	3.37	–	19.57	m	21.04
450mm	19.60	5.00	20.58	–	0.54	3.79	–	24.37	m	26.20
525mm	26.48	5.00	27.80	–	0.60	4.21	–	32.02	m	34.42
600mm	35.61	5.00	37.39	–	0.66	4.63	–	42.02	m	45.18
750mm	53.16	5.00	55.82	–	0.72	5.05	–	60.87	m	65.44
Drains; clayware perforated pipes, butt joints; 150mm hardcore to sides and top										
Pipework in trenches										
100mm	4.21	5.00	4.42	–	0.42	2.95	1.05	8.42	m	9.05
150mm	7.66	5.00	8.04	–	0.54	3.79	1.40	13.23	m	14.23
225mm	14.08	5.00	14.78	–	0.60	4.21	1.87	20.87	m	22.43

VISIT OUR BUILDING AND CONSTRUCTION WEBSITE TODAY!

www.bh.com/construction

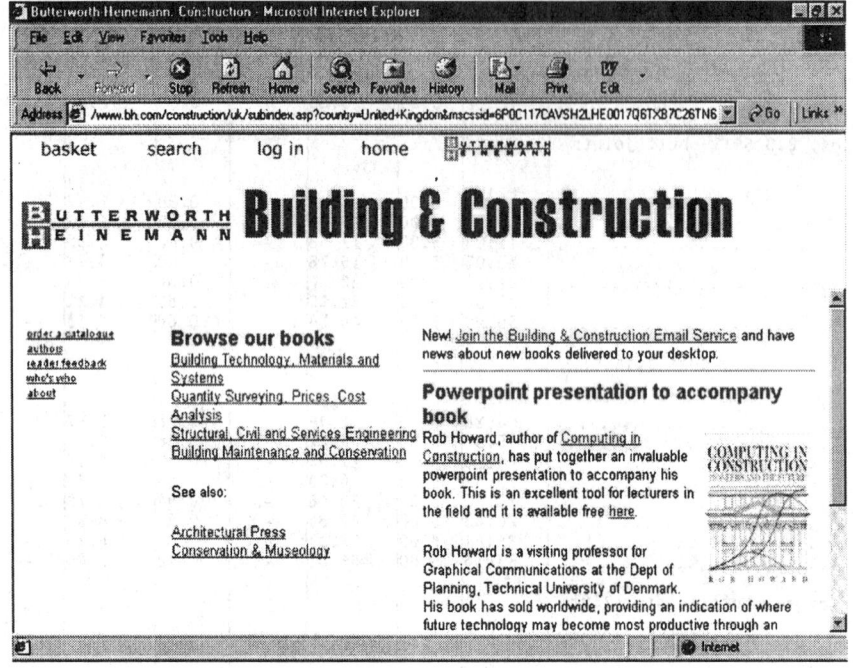

- FIND OUT MORE ABOUT OUR RANGE OF TITLES

- LATEST NEWS ON NEW TITLES AND PRICES

- HELPFUL SEARCH FACILITIES

- ORDER BOOKS ON-LINE

- BENEFIT FROM SPECIAL OFFERS

- REGULAR EMAIL UPDATES

- FIND OUT MORE ABOUT OUR AUTHORS

BOOKMARK THIS SITE TODAY!

Piped supply systems

Labour hourly rates: (except Specialists) Craft Operatives £12.44 Labourer £7.02 Rates are national average prices. Refer to REGIONAL VARIATIONS for indicative levels of overall pricing in regions	MATERIALS			LABOUR				RATES		
	Del to Site £	Waste %	Material Cost £	Craft Optve Hrs	Lab Hrs	Labour Cost £	Sunds £	Nett Rate £	Unit	Gross Rate (+7.5%) £
S115: HOT AND COLD WATER AND GAS										
BLACK AND GALVANISED STEEL TUBES AND FITTINGS - Steel pipes, B.S.1387 Table 2; steel fittings B.S.1740 Part 1; screwed and PTFE tape joints; pipework black										
Pipes, straight										
15mm; in malleable iron pipe brackets fixing to timber with screws	1.44	5.00	1.51	0.50	–	6.22	–	7.73	m	8.31
extra; made bends	–	–	–	0.17	–	2.11	–	2.11	nr	2.27
extra; connections to copper pipe ends;										
compression joint	0.49	2.00	0.50	0.10	–	1.24	–	1.74	nr	1.87
extra; fittings; one end	0.31	2.00	0.32	0.13	–	1.62	–	1.93	nr	2.08
extra; fittings; two ends	0.35	2.00	0.36	0.25	–	3.11	–	3.47	nr	3.73
extra; fittings; three ends	0.48	2.00	0.49	0.25	–	3.11	–	3.60	nr	3.87
20mm; in malleable iron pipe brackets fixing to timber with screws	1.67	5.00	1.75	0.57	–	7.09	–	8.84	m	9.51
extra; made bends	–	–	–	0.25	–	3.11	–	3.11	nr	3.34
extra; connections to copper pipe ends;										
compression joint	0.80	2.00	0.82	0.17	–	2.11	–	2.93	nr	3.15
extra; fittings; one end	0.35	2.00	0.36	0.17	–	2.11	–	2.47	nr	2.66
extra; fittings; two ends	0.48	2.00	0.49	0.33	–	4.11	–	4.59	nr	4.94
extra; fittings; three ends	0.70	2.00	0.71	0.33	–	4.11	–	4.82	nr	5.18
25mm; in malleable iron pipe brackets fixing to timber with screws	2.30	5.00	2.42	0.67	–	8.33	–	10.75	m	11.56
extra; made bends	–	–	–	0.33	–	4.11	–	4.11	nr	4.41
extra; connections to copper pipe ends;										
compression joint	1.56	2.00	1.59	0.25	–	3.11	–	4.70	nr	5.05
extra; fittings; one end	0.44	2.00	0.45	0.25	–	3.11	–	3.56	nr	3.83
extra; fittings; two ends	0.75	2.00	0.77	0.50	–	6.22	–	6.99	nr	7.51
extra; fittings; three ends	1.01	2.00	1.03	0.50	–	6.22	–	7.25	nr	7.79
32mm; in malleable iron pipe brackets fixing to timber with screws	2.88	5.00	3.02	0.80	–	9.95	–	12.98	m	13.95
extra; made bends	–	–	–	0.50	–	6.22	–	6.22	nr	6.69
extra; connections to copper pipe ends;										
compression joint	4.58	2.00	4.67	0.33	–	4.11	–	8.78	nr	9.44
extra; fittings; one end	0.64	2.00	0.65	0.33	–	4.11	–	4.76	nr	5.11
extra; fittings; two ends	1.23	2.00	1.25	0.67	–	8.33	–	9.59	nr	10.31
extra; fittings; three ends	1.67	2.00	1.70	0.67	–	8.33	–	10.04	nr	10.79
40mm; in malleable iron pipe brackets fixing to timber with screws	3.41	5.00	3.58	1.00	–	12.44	–	16.02	m	17.22
extra; made bends	–	–	–	0.67	–	8.33	–	8.33	nr	8.96
extra; connections to copper pipe ends;										
compression joint	7.36	2.00	7.51	0.50	–	6.22	–	13.73	nr	14.76
extra; fittings; one end	0.81	2.00	0.83	0.42	–	5.22	–	6.05	nr	6.50
extra; fittings; two ends	2.07	2.00	2.11	0.83	–	10.33	–	12.44	nr	13.37
extra; fittings; three ends	2.29	2.00	2.34	0.83	–	10.33	–	12.66	nr	13.61
50mm; in malleable iron pipe brackets fixing to timber with screws	4.74	5.00	4.98	1.33	–	16.55	–	21.52	m	23.14
extra; made bends	–	–	–	1.00	–	12.44	–	12.44	nr	13.37
extra; connections to copper pipe ends;										
compression joint	11.16	2.00	11.38	0.67	–	8.33	–	19.72	nr	21.20
extra; fittings; one end	1.58	2.00	1.61	0.50	–	6.22	–	7.83	nr	8.42
extra; fittings; two ends	2.42	2.00	2.47	1.00	–	12.44	–	14.91	nr	16.03
extra; fittings; three ends	3.30	2.00	3.37	1.00	–	12.44	–	15.81	nr	16.99
in malleable iron pipe brackets fixing to masonry with screws										
15mm	1.44	5.00	1.51	0.53	–	6.59	–	8.11	m	8.71
20mm	1.67	5.00	1.75	0.60	–	7.46	–	9.22	m	9.91
25mm	2.30	5.00	2.42	0.70	–	8.71	–	11.12	m	11.96
32mm	2.88	5.00	3.02	0.83	–	10.33	–	13.35	m	14.35
40mm	3.41	5.00	3.58	1.03	–	12.81	–	16.39	m	17.62
50mm	4.74	5.00	4.98	1.36	–	16.92	–	21.90	m	23.54
in malleable iron single rings, 165mm long screwed both ends steel tubes and malleable iron backplates fixing to timber with screws										
15mm	1.91	5.00	2.01	0.60	–	7.46	–	9.47	m	10.18
20mm	2.14	5.00	2.25	0.67	–	8.33	–	10.58	m	11.38
25mm	2.76	5.00	2.90	0.77	–	9.58	–	12.48	m	13.41
32mm	3.23	5.00	3.39	0.90	–	11.20	–	14.59	m	15.68
40mm	3.72	5.00	3.91	1.10	–	13.68	–	17.59	m	18.91

PIPED SUPPLY SYSTEMS – MAJOR WORKS

Labour hourly rates: (except Specialists) Craft Operatives £12.44 Labourer £7.02. Rates are national average prices. Refer to REGIONAL VARIATIONS for indicative levels of overall pricing in regions	Del to Site £	Waste %	Material Cost £	Craft Optve Hrs	Lab Hrs	Labour Cost £	Sunds £	Nett Rate £	Unit	Gross Rate (+7.5%) £
S115: HOT AND COLD WATER AND GAS Cont.										
BLACK AND GALVANISED STEEL TUBES AND FITTINGS - Steel pipes, B.S.1387 Table 2; steel fittings B.S.1740 Part 1; screwed and PTFE tape joints; pipework black Cont.										
Pipes, straight Cont.										
in malleable iron single rings, 165mm long screwed both ends steel tubes and malleable iron backplates fixing to timber with screws Cont.										
50mm	4.98	5.00	5.23	1.43	-	17.79	-	23.02	m	24.74
BLACK AND GALVANISED STEEL TUBES AND FITTINGS - Steel pipes, B.S.1387 Table 3; steel fittings B.S.1740 Part 1; screwed and PTFE tape joints; pipework black										
Pipes, straight										
15mm; in malleable iron pipe brackets fixing to timber with screws	1.66	5.00	1.74	0.50	-	6.22	-	7.96	m	8.56
extra; made bends	-	-	-	0.17	-	2.11	-	2.11	nr	2.27
extra; connections to copper pipe ends; compression joint	0.49	2.00	0.50	0.10	-	1.24	-	1.74	nr	1.87
extra; fittings; one end	0.31	2.00	0.32	0.13	-	1.62	-	1.93	nr	2.08
extra; fittings; two ends	0.35	2.00	0.36	0.25	-	3.11	-	3.47	nr	3.73
extra; fittings; three ends	0.48	2.00	0.49	0.25	-	3.11	-	3.60	nr	3.87
20mm; in malleable iron pipe brackets fixing to timber with screws	1.94	5.00	2.04	0.57	-	7.09	-	9.13	m	9.81
extra; made bends	-	-	-	0.39	-	4.85	-	4.85	nr	5.22
extra; connections to copper pipe ends; compression joint	0.80	2.00	0.82	0.17	-	2.11	-	2.93	nr	3.15
extra; fittings; one end	0.35	2.00	0.36	0.17	-	2.11	-	2.47	nr	2.66
extra; fittings; two ends	0.48	2.00	0.49	0.33	-	4.11	-	4.59	nr	4.94
extra; fittings; three ends	0.70	2.00	0.71	0.33	-	4.11	-	4.82	nr	5.18
25mm; in malleable iron pipe brackets fixing to timber with screws	2.73	5.00	2.87	0.67	-	8.33	-	11.20	m	12.04
extra; made bends	-	-	-	0.33	-	4.11	-	4.11	nr	4.41
extra; connections to copper pipe ends; compression joint	1.56	2.00	1.59	0.25	-	3.11	-	4.70	nr	5.05
extra; fittings; one end	0.44	2.00	0.45	0.25	-	3.11	-	3.56	nr	3.83
extra; fittings; two ends	0.75	2.00	0.77	0.50	-	6.22	-	6.99	nr	7.51
extra; fittings; three ends	1.01	2.00	1.03	0.50	-	6.22	-	7.25	nr	7.79
32mm; in malleable iron pipe brackets fixing to timber with screws	3.42	5.00	3.59	0.80	-	9.95	-	13.54	m	14.56
extra; made bends	-	-	-	0.50	-	6.22	-	6.22	nr	6.69
extra; connections to copper pipe ends; compression joint	4.58	2.00	4.67	0.33	-	4.11	-	8.78	nr	9.44
extra; fittings; one end	0.64	2.00	0.65	0.33	-	4.11	-	4.76	nr	5.11
extra; fittings; two ends	1.23	2.00	1.25	0.67	-	8.33	-	9.59	nr	10.31
extra; fittings; three ends	1.67	2.00	1.70	0.67	-	8.33	-	10.04	nr	10.79
40mm; in malleable iron pipe brackets fixing to timber with screws	4.04	5.00	4.24	1.00	-	12.44	-	16.68	m	17.93
extra; made bends	-	-	-	0.67	-	8.33	-	8.33	nr	8.96
extra; connections to copper pipe ends; compression joint	7.36	2.00	7.51	0.50	-	6.22	-	13.73	nr	14.76
extra; fittings; one end	0.81	2.00	0.83	0.42	-	5.22	-	6.05	nr	6.50
extra; fittings; two ends	2.07	2.00	2.11	0.83	-	10.33	-	12.44	nr	13.33
extra; fittings; three ends	2.29	2.00	2.34	0.83	-	10.33	-	12.66	nr	13.61
50mm; in malleable iron pipe brackets fixing to timber with screws	5.56	5.00	5.84	1.33	-	16.55	-	22.38	m	24.06
extra; made bends	-	-	-	1.00	-	12.44	-	12.44	nr	13.37
extra; connections to copper pipe ends; compression joint	11.16	2.00	11.38	0.67	-	8.33	-	19.72	nr	21.20
extra; fittings; one end	1.58	2.00	1.61	0.50	-	6.22	-	7.83	nr	8.42
extra; fittings; two ends	2.42	2.00	2.47	1.00	-	12.44	-	14.91	nr	16.03
extra; fittings; three ends	3.30	2.00	3.37	1.00	-	12.44	-	15.81	nr	16.99
in malleable iron pipe brackets fixing to masonry with screws										
15mm	1.66	5.00	1.74	0.53	-	6.59	-	8.34	m	8.96
20mm	1.94	5.00	2.04	0.60	-	7.46	-	9.50	m	10.21
25mm	2.73	5.00	2.87	0.70	-	8.71	-	11.57	m	12.44
32mm	3.42	5.00	3.59	0.83	-	10.33	-	13.92	m	14.96
40mm	4.04	5.00	4.24	1.03	-	12.81	-	17.06	m	18.33
50mm	5.80	5.00	6.09	1.36	-	16.92	-	23.01	m	24.73
in malleable iron single rings, 165mm long screwed both ends steel tubes and malleable iron backplates fixing to timber with screws										
15mm	2.13	5.00	2.24	0.60	-	7.46	-	9.70	m	10.43
20mm	2.41	5.00	2.53	0.67	-	8.33	-	10.87	m	11.68
25mm	3.19	5.00	3.35	0.77	-	9.58	-	12.93	m	13.90
32mm	3.77	5.00	3.96	0.90	-	11.20	-	15.15	m	16.29
40mm	4.35	5.00	4.57	1.10	-	13.68	-	18.25	m	19.62
50mm	5.80	5.00	6.09	1.43	-	17.79	-	23.88	m	25.67
BLACK AND GALVANISED STEEL TUBES AND FITTINGS - Steel pipes, B.S.1387 Table 2; steel fittings B.S.1740 Part 1; screwed and PTFE tape joints; pipework galvanised										
Pipes, straight										
15mm; in malleable iron pipe brackets fixing to timber with screws	2.13	5.00	2.24	0.50	-	6.22	-	8.46	m	9.09
extra; made bends	-	-	-	0.17	-	2.11	-	2.11	nr	2.27
extra; connections to copper pipe ends; compression joint	0.49	2.00	0.50	0.10	-	1.24	-	1.74	nr	1.87
extra; fittings; one end	0.43	2.00	0.44	0.13	-	1.62	-	2.06	nr	2.21
extra; fittings; two ends	0.50	2.00	0.51	0.25	-	3.11	-	3.62	nr	3.89
extra; fittings; three ends	0.68	2.00	0.69	0.25	-	3.11	-	3.80	nr	4.09

Labour hourly rates: (except Specialists) Craft Operatives £12.44 Labourer £7.02 Rates are national average prices. Refer to REGIONAL VARIATIONS for indicative levels of overall pricing in regions	MATERIALS			LABOUR				RATES		
	Del to Site £	Waste %	Material Cost £	Craft Optve Hrs	Lab Hrs	Labour Cost £	Sunds £	Nett Rate £	Unit	Gross Rate (+7.5%) £
S115: HOT AND COLD WATER AND GAS Cont.										
BLACK AND GALVANISED STEEL TUBES AND FITTINGS - Steel pipes, B.S.1387 Table 2; steel fittings B.S.1740 Part 1; screwed and PTFE tape joints; pipework galvanised Cont.										
Pipes, straight Cont.										
20mm; in malleable iron pipe brackets fixing to timber with screws	2.40	5.00	2.52	0.57	–	7.09	–	9.61	m	10.33
extra; made bends	–	–	–	0.25	–	3.11	–	3.11	nr	3.34
extra; connections to copper pipe ends;										
compression joint	0.80	2.00	0.82	0.17	–	2.11	–	2.93	nr	3.15
extra; fittings; one end	0.50	2.00	0.51	0.17	–	2.11	–	2.62	nr	2.82
extra; fittings; two ends	0.68	2.00	0.69	0.33	–	4.11	–	4.80	nr	5.16
extra; fittings; three ends	0.99	2.00	1.01	0.33	–	4.11	–	5.12	nr	5.50
25mm; in malleable iron pipe brackets fixing to timber with screws	3.24	5.00	3.40	0.67	–	8.33	–	11.74	m	12.62
extra; made bends	–	–	–	0.33	–	4.11	–	4.11	nr	4.41
extra; connections to copper pipe ends;										
compression joint	1.56	2.00	1.59	0.25	–	3.11	–	4.70	nr	5.05
extra; fittings; one end	0.62	2.00	0.63	0.25	–	3.11	–	3.74	nr	4.02
extra; fittings; two ends	1.06	2.00	1.08	0.50	–	6.22	–	7.30	nr	7.85
extra; fittings; three ends	1.43	2.00	1.46	0.50	–	6.22	–	7.68	nr	8.25
32mm; in malleable iron pipe brackets fixing to timber with screws	4.04	5.00	4.24	0.80	–	9.95	–	14.19	m	15.26
extra; made bends	–	–	–	0.50	–	6.22	–	6.22	nr	6.69
extra; connections to copper pipe ends;										
compression joint	4.58	2.00	4.67	0.33	–	4.11	–	8.78	nr	9.44
extra; fittings; one end	0.90	2.00	0.92	0.33	–	4.11	–	5.02	nr	5.40
extra; fittings; two ends	1.74	2.00	1.77	0.67	–	8.33	–	10.11	nr	10.87
extra; fittings; three ends	2.36	2.00	2.41	0.67	–	8.33	–	10.74	nr	11.55
40mm; in malleable iron pipe brackets fixing to timber with screws	4.81	5.00	5.05	1.00	–	12.44	–	17.49	m	18.80
extra; made bends	–	–	–	0.67	–	8.33	–	8.33	nr	8.96
extra; connections to copper pipe ends;										
compression joint	7.36	2.00	7.51	0.50	–	6.22	–	13.73	nr	14.76
extra; fittings; one end	1.15	2.00	1.17	0.42	–	5.22	–	6.40	nr	6.88
extra; fittings; two ends	2.92	2.00	2.98	0.83	–	10.33	–	13.30	nr	14.30
extra; fittings; three ends	3.23	2.00	3.29	0.83	–	10.33	–	13.62	nr	14.64
50mm; in malleable iron pipe brackets fixing to timber with screws	6.66	5.00	6.99	1.33	–	16.55	–	23.54	m	25.30
extra; made bends	–	–	–	1.00	–	12.44	–	12.44	nr	13.37
extra; connections to copper pipe ends;										
compression joint	11.16	2.00	11.38	0.67	–	8.33	–	19.72	nr	21.20
extra; fittings; one end	2.24	2.00	2.28	0.50	–	6.22	–	8.50	nr	9.14
extra; fittings; two ends	3.41	2.00	3.48	1.00	–	12.44	–	15.92	nr	17.11
extra; fittings; three ends	4.66	2.00	4.75	1.00	–	12.44	–	17.19	nr	18.48
in malleable iron pipe brackets fixing to masonry with screws										
15mm	2.13	5.00	2.24	0.53	–	6.59	–	8.83	m	9.49
20mm	2.40	5.00	2.52	0.60	–	7.46	–	9.98	m	10.73
25mm	3.24	5.00	3.40	0.70	–	8.71	–	12.11	m	13.02
32mm	4.04	5.00	4.24	0.83	–	10.33	–	14.57	m	15.66
40mm	4.81	5.00	5.05	1.03	–	12.81	–	17.86	m	19.20
50mm	6.66	5.00	6.99	1.36	–	16.92	–	23.91	m	25.70
in malleable iron single rings, 165mm long screwed both ends steel tubes and malleable iron backplates fixing to timber with screws										
15mm	2.69	5.00	2.82	0.60	–	7.46	–	10.29	m	11.06
20mm	2.96	5.00	3.11	0.67	–	8.33	–	11.44	m	12.30
25mm	3.78	5.00	3.97	0.77	–	9.58	–	13.55	m	14.56
32mm	4.44	5.00	4.66	0.90	–	11.20	–	15.86	m	17.05
40mm	5.14	5.00	5.40	1.10	–	13.68	–	19.08	m	20.51
50mm	6.88	5.00	7.22	1.43	–	17.79	–	25.01	m	26.89
BLACK AND GALVANISED STEEL TUBES AND FITTINGS - Steel pipes, B.S.1387 Table 3; steel fittings B.S.1740 Part 1; screwed and PTFE tape joints; pipework galvanised										
Pipes, straight										
15mm; in malleable iron pipe brackets fixing to timber with screws	2.45	5.00	2.57	0.50	–	6.22	–	8.79	m	9.45
extra; made bends	–	–	–	0.17	–	2.11	–	2.11	nr	2.27
extra; connections to copper pipe ends;										
compression joint	0.49	2.00	0.50	0.10	–	1.24	–	1.74	nr	1.87
extra; fittings; one end	0.43	2.00	0.44	0.13	–	1.62	–	2.06	nr	2.21
extra; fittings; two ends	0.50	2.00	0.51	0.25	–	3.11	–	3.62	nr	3.89
extra; fittings; three ends	0.68	2.00	0.69	0.25	–	3.11	–	3.80	nr	4.09
20mm; in malleable iron pipe brackets fixing to timber with screws	2.77	5.00	2.91	0.57	–	7.09	–	10.00	m	10.75
extra; made bends	–	–	–	0.25	–	3.11	–	3.11	nr	3.34
extra; connections to copper pipe ends;										
compression joint	0.80	2.00	0.82	0.17	–	2.11	–	2.93	nr	3.15
extra; fittings; one end	0.50	2.00	0.51	0.17	–	2.11	–	2.62	nr	2.82
extra; fittings; two ends	0.68	2.00	0.69	0.33	–	4.11	–	4.80	nr	5.16
extra; fittings; three ends	0.99	2.00	1.01	0.33	–	4.11	–	5.12	nr	5.50
25mm; in malleable iron pipe brackets fixing to timber with screws	3.82	5.00	4.01	0.67	–	8.33	–	12.35	m	13.27
extra; made bends	–	–	–	0.33	–	4.11	–	4.11	nr	4.41
extra; connections to copper pipe ends;										
compression joint	1.56	2.00	1.59	0.25	–	3.11	–	4.70	nr	5.05
extra; fittings; one end	0.62	2.00	0.63	0.25	–	3.11	–	3.74	nr	4.02
extra; fittings; two ends	1.06	2.00	1.08	0.50	–	6.22	–	7.30	nr	7.85
extra; fittings; three ends	1.43	2.00	1.46	0.50	–	6.22	–	7.68	nr	8.25

PIPED SUPPLY SYSTEMS

Labour hourly rates: (except Specialists) Craft Operatives £12.44 Labourer £7.02 Rates are national average prices. Refer to REGIONAL VARIATIONS for indicative levels of overall pricing in regions	MATERIALS			LABOUR				RATES		
	Del to Site £	Waste %	Material Cost £	Craft Optve Hrs	Lab Hrs	Labour Cost £	Sunds £	Nett Rate £	Unit	Gross Rate (+7.5%) £
S115: HOT AND COLD WATER AND GAS Cont.										
BLACK AND GALVANISED STEEL TUBES AND FITTINGS - Steel pipes, B.S.1387 Table 3; steel fittings B.S.1740 Part 1; screwed and PTFE tape joints; pipework galvanised Cont.										
Pipes, straight Cont.										
32mm; in malleable iron pipe brackets fixing to timber with screws	4.78	5.00	5.02	0.80	-	9.95	-	14.97	m	16.09
extra; made bends	-	-	-	0.50	-	6.22	-	6.22	nr	6.69
extra; connections to copper pipe ends; compression joint	4.58	2.00	4.67	0.33	-	4.11	-	8.78	nr	9.44
extra; fittings; one end	0.90	2.00	0.92	0.33	-	4.11	-	5.02	nr	5.40
extra; fittings; two ends	1.74	2.00	1.77	0.67	-	8.33	-	10.11	nr	10.87
extra; fittings; three ends	2.36	2.00	2.41	0.67	-	8.33	-	10.74	nr	11.55
40mm; in malleable iron pipe brackets fixing to timber with screws	5.68	5.00	5.96	1.00	-	12.44	-	18.40	m	19.78
extra; made bends	-	-	-	0.67	-	8.33	-	8.33	nr	8.96
extra; connections to copper pipe ends; compression joint	7.36	2.00	7.51	0.50	-	6.22	-	13.73	nr	14.76
extra; fittings; one end	1.15	2.00	1.17	0.42	-	5.22	-	6.40	nr	6.88
extra; fittings; two ends	2.92	2.00	2.98	0.83	-	10.33	-	13.30	nr	14.30
extra; fittings; three ends	3.23	2.00	3.29	0.83	-	10.33	-	13.62	nr	14.64
50mm; in malleable iron pipe brackets fixing to timber with screws	7.80	5.00	8.19	1.33	-	16.55	-	24.74	m	26.59
extra; made bends	-	-	-	1.00	-	12.44	-	12.44	nr	13.37
extra; connections to copper pipe ends; compression joint	11.16	2.00	11.38	0.67	-	8.33	-	19.72	nr	21.20
extra; fittings; one end	2.24	2.00	2.28	0.50	-	6.22	-	8.50	nr	9.14
extra; fittings; two ends	3.41	2.00	3.48	1.00	-	12.44	-	15.92	nr	17.11
extra; fittings; three ends	4.66	2.00	4.75	1.00	-	12.44	-	17.19	nr	18.48
in malleable iron pipe brackets fixing to masonry with screws										
15mm	2.45	5.00	2.57	0.53	-	6.59	-	9.17	m	9.85
20mm	2.77	5.00	2.91	0.60	-	7.46	-	10.37	m	11.15
25mm	3.82	5.00	4.01	0.70	-	8.71	-	12.72	m	13.67
32mm	4.78	5.00	5.02	0.83	-	10.33	-	15.34	m	16.50
40mm	5.68	5.00	5.96	1.03	-	12.81	-	18.78	m	20.19
50mm	7.80	5.00	8.19	1.36	-	16.92	-	25.11	m	26.99
in malleable iron single rings, 165mm long screwed both ends steel tubes and malleable iron backplates fixing to timber with screws										
15mm	3.01	5.00	3.16	0.60	-	7.46	-	10.62	m	11.42
20mm	3.33	5.00	3.50	0.67	-	8.33	-	11.83	m	12.72
25mm	4.36	5.00	4.58	0.77	-	9.58	-	14.16	m	15.22
32mm	5.18	5.00	5.44	0.90	-	11.20	-	16.64	m	17.88
40mm	6.01	5.00	6.31	1.10	-	13.68	-	19.99	m	21.49
50mm	8.02	5.00	8.42	1.43	-	17.79	-	26.21	m	28.18
Pipes; straight; in trenches										
15mm	2.05	5.00	2.15	0.22	-	2.74	-	4.89	m	5.26
20mm	2.32	5.00	2.44	0.25	-	3.11	-	5.55	m	5.96
25mm	3.31	5.00	3.48	0.29	-	3.61	-	7.08	m	7.61
32mm	4.11	5.00	4.32	0.33	-	4.11	-	8.42	m	9.05
40mm	4.79	5.00	5.03	0.40	-	4.98	-	10.01	m	10.76
50mm	6.64	5.00	6.97	0.50	-	6.22	-	13.19	m	14.18
LIGHT GAUGE STAINLESS STEEL TUBES AND FITTINGS - Stainless steel pipes, B.S.4127 Part 2; C.P. copper fittings, capillary, B.S.864 Part 2										
Pipes, straight										
15mm; in C.P. pipe brackets fixing to timber with screws	3.35	5.00	3.52	0.33	-	4.11	-	7.62	m	8.19
extra; fittings; one end	2.59	3.00	2.67	0.05	-	0.62	-	3.29	nr	3.54
extra; fittings; two ends	0.79	3.00	0.81	0.10	-	1.24	-	2.06	nr	2.21
extra; fittings; three ends	1.33	3.00	1.37	0.10	-	1.24	-	2.61	nr	2.81
22mm; in C.P. pipe brackets fixing to timber with screws	4.95	5.00	5.20	0.36	-	4.48	-	9.68	m	10.40
extra; fittings; one end	3.03	3.00	3.12	0.09	-	1.12	-	4.24	nr	4.56
extra; fittings; two ends	1.85	3.00	1.91	0.17	-	2.11	-	4.02	nr	4.32
extra; fittings; three ends	3.68	3.00	3.79	0.17	-	2.11	-	5.91	nr	6.35
28mm; in C.P. pipe brackets fixing to timber with screws	6.48	5.00	6.80	0.40	-	4.98	-	11.78	m	12.66
extra; fittings; one end	4.79	3.00	4.93	0.13	-	1.62	-	6.55	nr	7.04
extra; fittings; two ends	4.32	3.00	4.45	0.25	-	3.11	-	7.56	nr	8.13
extra; fittings; three ends	6.70	3.00	6.90	0.25	-	3.11	-	10.01	nr	10.76
35mm; in C.P. pipe brackets fixing to timber with screws	7.96	5.00	8.36	0.50	-	6.22	-	14.58	m	15.67
extra; fittings; one end	10.36	3.00	10.67	0.17	-	2.11	-	12.79	nr	13.74
extra; fittings; two ends	9.75	3.00	10.04	0.33	-	4.11	-	14.15	nr	15.21
extra; fittings; three ends	16.76	3.00	17.26	0.33	-	4.11	-	21.37	nr	22.97
42mm; in C.P. pipe brackets fixing to masonry with screws	11.34	5.00	11.91	0.67	-	8.33	-	20.24	m	21.76
extra; fittings; one end	15.91	3.00	16.39	0.25	-	3.11	-	19.50	nr	20.96
extra; fittings; two ends	17.37	3.00	17.89	0.50	-	6.22	-	24.11	nr	25.92
extra; fittings; three ends	26.25	3.00	27.04	0.50	-	6.22	-	33.26	nr	35.75
in C.P. pipe brackets fixing to masonry with screws										
15mm	3.35	5.00	3.52	0.36	-	4.48	-	8.00	m	8.60
22mm	4.95	5.00	5.20	0.39	-	4.85	-	10.05	m	10.80
28mm	6.48	5.00	6.80	0.43	-	5.35	-	12.15	m	13.06
35mm	7.96	5.00	8.36	0.53	-	6.59	-	14.95	m	16.07
42mm	11.34	5.00	11.91	0.70	-	8.71	-	20.61	m	22.16

Labour hourly rates: (except Specialists) Craft Operatives £12.44 Labourer £7.02 Rates are national average prices. Refer to REGIONAL VARIATIONS for indicative levels of overall pricing in regions	MATERIALS			LABOUR				RATES		
	Del to Site	Waste	Material Cost	Craft Optve	Lab	Labour Cost	Sunds	Nett Rate	Unit	Gross Rate (+7.5%)
	£	%	£	Hrs	Hrs	£	£	£		£

S115: HOT AND COLD WATER AND GAS Cont.

LIGHT GAUGE STAINLESS STEEL TUBES AND FITTINGS –
Stainless steel pipes, B.S.4127 Part 2; C.P. copper
alloy fittings, compression, B.S.864 Part 2, Type A

Pipes, straight

	Del to Site	Waste	Material Cost	Craft Optve	Lab	Labour Cost	Sunds	Nett Rate	Unit	Gross Rate
15mm; in C.P. pipe brackets fixing to timber with										
screws	3.40	5.00	3.57	0.33	-	4.11	-	7.68	m	8.25
extra; fittings; one end	1.13	3.00	1.16	0.05	-	0.62	-	1.79	nr	1.92
extra; fittings; two ends	0.86	3.00	0.89	0.10	-	1.24	-	2.13	nr	2.29
extra; fittings; three ends	1.27	3.00	1.31	0.10	-	1.24	-	2.55	nr	2.74
22mm; in C.P. pipe brackets fixing to timber with										
screws	4.96	5.00	5.21	0.36	-	4.48	-	9.69	m	10.41
extra; fittings; one end	1.35	3.00	1.39	0.09	-	1.12	-	2.51	nr	2.70
extra; fittings; two ends	1.47	3.00	1.51	0.17	-	2.11	-	3.63	nr	3.90
extra; fittings; three ends	2.07	3.00	2.13	0.17	-	2.11	-	4.25	nr	4.57
28mm; in C.P. pipe brackets fixing to timber with										
screws	6.73	5.00	7.07	0.40	-	4.98	-	12.04	m	12.95
extra; fittings; one end	3.86	3.00	3.98	0.13	-	1.62	-	5.59	nr	6.01
extra; fittings; two ends	4.64	3.00	4.78	0.25	-	3.11	-	7.89	nr	8.48
extra; fittings; three ends	8.22	3.00	8.47	0.25	-	3.11	-	11.58	nr	12.44
35mm; in C.P. pipe brackets fixing to timber with										
screws	8.55	5.00	8.98	0.50	-	6.22	-	15.20	m	16.34
extra; fittings; one end	6.66	3.00	6.86	0.17	-	2.11	-	8.97	nr	9.65
extra; fittings; two ends	11.15	3.00	11.48	0.33	-	4.11	-	15.59	nr	16.76
extra; fittings; three ends	14.72	3.00	15.16	0.33	-	4.11	-	19.27	nr	20.71
42mm; in C.P. pipe brackets fixing to masonry with										
screws	12.08	5.00	12.68	0.67	-	8.33	-	21.02	m	22.60
extra; fittings; one end	10.88	3.00	11.21	0.25	-	3.11	-	14.32	nr	15.39
extra; fittings; two ends	15.58	3.00	16.05	0.50	-	6.22	-	22.27	nr	23.94
extra; fittings; three ends	22.67	3.00	23.35	0.50	-	6.22	-	29.57	nr	31.79
in C.P. pipe brackets fixing to masonry with screws										
15mm	3.40	5.00	3.57	0.36	-	4.48	-	8.05	m	8.65
22mm	4.96	5.00	5.21	0.39	-	4.85	-	10.06	m	10.81
28mm	6.73	5.00	7.07	0.43	-	5.35	-	12.42	m	13.35
35mm	8.55	5.00	8.98	0.53	-	6.59	-	15.57	m	16.74
42mm	12.08	5.00	12.68	0.70	-	8.71	-	21.39	m	23.00

COPPER TUBING AND FITTINGS – Copper pipes, B.S.2871
Part 1 Table X; copper fittings, capillary, B.S.864
Part 2

Pipes, straight

	Del to Site	Waste	Material Cost	Craft Optve	Lab	Labour Cost	Sunds	Nett Rate	Unit	Gross Rate
15mm; in two piece copper spacing clips to timber										
with screws	0.72	5.00	0.76	0.33	-	4.11	-	4.86	m	5.23
extra; made bends	-	-	-	0.08	-	1.00	-	1.00	nr	1.07
extra; connections to iron pipe ends; screwed										
joint	1.25	3.00	1.29	0.10	-	1.24	-	2.53	nr	2.72
extra; fittings; one end	0.68	3.00	0.70	0.05	-	0.62	-	1.32	nr	1.42
extra; fittings; two ends	0.19	3.00	0.20	0.10	-	1.24	-	1.44	nr	1.55
extra; fittings; three ends	0.32	3.00	0.33	0.10	-	1.24	-	1.57	nr	1.69
22mm; in two piece copper spacing clips to timber										
with screws	1.35	5.00	1.42	0.36	-	4.48	-	5.90	m	6.34
extra; made bends	-	-	-	0.13	-	1.62	-	1.62	nr	1.74
extra; connections to iron pipe ends; screwed										
joint	2.21	3.00	2.28	0.17	-	2.11	-	4.39	nr	4.72
extra; fittings; one end	1.29	3.00	1.33	0.09	-	1.12	-	2.45	nr	2.63
extra; fittings; two ends	0.45	3.00	0.46	0.17	-	2.11	-	2.58	nr	2.77
extra; fittings; three ends	0.92	3.00	0.95	0.17	-	2.11	-	3.06	nr	3.29
28mm; in two piece copper spacing clips to timber										
with screws	1.84	5.00	1.93	0.40	-	4.98	-	6.91	m	7.43
extra; made bends	-	-	-	0.17	-	2.11	-	2.11	nr	2.27
extra; connections to iron pipe ends; screwed										
joint	3.50	3.00	3.61	0.25	-	3.11	-	6.71	nr	7.22
extra; fittings; one end	2.04	3.00	2.10	0.13	-	1.62	-	3.72	nr	4.00
extra; fittings; two ends	1.02	3.00	1.05	0.25	-	3.11	-	4.16	nr	4.47
extra; fittings; three ends	2.85	3.00	2.94	0.25	-	3.11	-	6.05	nr	6.50
35mm; in two piece copper spacing clips to timber										
with screws	4.54	5.00	4.77	0.50	-	6.22	-	10.99	m	11.81
extra; made bends	-	-	-	0.25	-	3.11	-	3.11	nr	3.34
extra; connections to iron pipe ends; screwed										
joint	6.17	3.00	6.36	0.33	-	4.11	-	10.46	nr	11.24
extra; fittings; one end	4.41	3.00	4.54	0.17	-	2.11	-	6.66	nr	7.16
extra; fittings; two ends	4.15	3.00	4.27	0.33	-	4.11	-	8.38	nr	9.01
extra; fittings; three ends	7.13	3.00	7.34	0.33	-	4.11	-	11.45	nr	12.31
42mm; in two piece copper spacing clips to timber										
with screws	5.71	5.00	6.00	0.67	-	8.33	-	14.33	m	15.41
extra; made bends	-	-	-	0.33	-	4.11	-	4.11	nr	4.41
extra; connections to iron pipe ends; screwed										
joint	7.95	3.00	8.19	0.50	-	6.22	-	14.41	nr	15.49
extra; fittings; one end	6.77	3.00	6.97	0.25	-	3.11	-	10.08	nr	10.84
extra; fittings; two ends	7.39	3.00	7.61	0.50	-	6.22	-	13.83	nr	14.87
extra; fittings; three ends	11.17	3.00	11.51	0.50	-	6.22	-	17.73	nr	19.05
54mm; in two piece copper spacing clips to timber										
with screws	7.77	5.00	8.16	1.00	-	12.44	-	20.60	m	22.14
extra; made bends	-	-	-	0.50	-	6.22	-	6.22	nr	6.69
extra; connections to iron pipe ends; screwed										
joint	12.07	3.00	12.43	0.67	-	8.33	-	20.77	nr	22.32
extra; fittings; one end	9.45	3.00	9.73	0.33	-	4.11	-	13.84	nr	14.88
extra; fittings; two ends	15.13	3.00	15.58	0.67	-	8.33	-	23.92	nr	25.71
extra; fittings; three ends	21.11	3.00	21.74	0.67	-	8.33	-	30.08	nr	32.33
in two piece copper spacing clips fixing to masonry with screws										
15mm	0.72	5.00	0.76	0.36	-	4.48	-	5.23	m	5.63

PIPED SUPPLY SYSTEMS

	MATERIALS			LABOUR				RATES		
Labour hourly rates: (except Specialists) Craft Operatives £12.44 Labourer £7.02. Rates are national average prices. Refer to REGIONAL VARIATIONS for indicative levels of overall pricing in regions	Del to Site £	Waste %	Material Cost £	Craft Optve Hrs	Lab Hrs	Labour Cost £	Sunds £	Nett Rate £	Unit	Gross Rate (+7.5%) £

S115: HOT AND COLD WATER AND GAS Cont.

COPPER TUBING AND FITTINGS - Copper pipes, B.S.2871 Part 1 Table X; copper fittings, capillary, B.S.864 Part 2 Cont.

Pipes, straight Cont.
in two piece copper spacing clips fixing to masonry with screws Cont.

	Del to Site	Waste	Material Cost	Craft Optve	Lab	Labour Cost	Sunds	Nett Rate	Unit	Gross Rate
22mm	1.35	5.00	1.42	0.39	-	4.85	-	6.27	m	6.74
28mm	1.84	5.00	1.93	0.43	-	5.35	-	7.28	m	7.83
35mm	4.54	5.00	4.77	0.53	-	6.59	-	11.36	m	12.21
42mm	5.71	5.00	6.00	0.70	-	8.71	-	14.70	m	15.81
54mm	8.23	5.00	8.64	1.03	-	12.81	-	21.45	m	23.06

in pressed brass pipe brackets fixing to timber with screws

15mm	0.99	5.00	1.04	0.33	-	4.11	-	5.14	m	5.53
22mm	1.67	5.00	1.75	0.36	-	4.48	-	6.23	m	6.70
28mm	2.21	5.00	2.32	0.40	-	4.98	-	7.30	m	7.84
35mm	4.92	5.00	5.17	0.50	-	6.22	-	11.39	m	12.24
42mm	6.07	5.00	6.37	0.67	-	8.33	-	14.71	m	15.81
54mm	8.23	5.00	8.64	1.00	-	12.44	-	21.08	m	22.66

in cast brass pipe brackets fixing to timber with screws

15mm	1.48	5.00	1.55	0.33	-	4.11	-	5.66	m	6.08
22mm	2.26	5.00	2.37	0.36	-	4.48	-	6.85	m	7.37
28mm	2.94	5.00	3.09	0.40	-	4.98	-	8.06	m	8.67
35mm	5.95	5.00	6.25	0.50	-	6.22	-	12.47	m	13.40
42mm	7.51	5.00	7.89	0.67	-	8.33	-	16.22	m	17.44
54mm	10.04	5.00	10.54	1.00	-	12.44	-	22.98	m	24.71

in cast brass single rings and back plates fixing to timber with screws

15mm	1.72	5.00	1.81	0.36	-	4.48	-	6.28	m	6.76
22mm	2.38	5.00	2.50	0.39	-	4.85	-	7.35	m	7.90
28mm	2.95	5.00	3.10	0.43	-	5.35	-	8.45	m	9.08
35mm	5.65	5.00	5.93	0.53	-	6.59	-	12.53	m	13.47
42mm	6.79	5.00	7.13	0.70	-	8.71	-	15.84	m	17.03
54mm	8.93	5.00	9.38	1.03	-	12.81	-	22.19	m	23.85
66.7mm	14.90	5.00	15.65	1.50	-	18.66	-	34.31	m	36.88
76.1mm	22.45	5.00	23.57	2.00	-	24.88	-	48.45	m	52.09
108mm	32.45	5.00	34.07	3.00	-	37.32	-	71.39	m	76.75

COPPER TUBING AND FITTINGS - Copper pipes, B.S.2871 Part 1 Table X; copper alloy fittings, compression, B.S.864 Part 2, Type A

Pipes, straight

15mm; in two piece copper spacing clips to timber with screws	0.79	5.00	0.83	0.33	-	4.11	-	4.93	m	5.30
extra; made bends	-	-	-	0.08	-	1.00	-	1.00	nr	1.07
extra; connections to iron pipe ends; screwed joint	0.49	3.00	0.50	0.10	-	1.24	-	1.75	nr	1.88
extra; fittings; one end	0.84	3.00	0.87	0.05	-	0.62	-	1.49	nr	1.60
extra; fittings; two ends	0.64	3.00	0.66	0.10	-	1.24	-	1.90	nr	2.05
extra; fittings; three ends	0.94	3.00	0.97	0.10	-	1.24	-	2.21	nr	2.38
22mm; in two piece copper spacing clips to timber with screws	1.45	5.00	1.52	0.36	-	4.48	-	6.00	m	6.45
extra; made bends	-	-	-	0.13	-	1.62	-	1.62	nr	1.74
extra; connections to iron pipe ends; screwed joint	0.80	3.00	0.82	0.17	-	2.11	-	2.94	nr	3.16
extra; fittings; one end	1.00	3.00	1.03	0.09	-	1.12	-	2.15	nr	2.31
extra; fittings; two ends	1.09	3.00	1.12	0.17	-	2.11	-	3.24	nr	3.48
extra; fittings; three ends	1.53	3.00	1.58	0.17	-	2.11	-	3.69	nr	3.97
28mm; in two piece copper spacing clips to timber with screws	2.19	5.00	2.30	0.40	-	4.98	-	7.28	m	7.82
extra; made bends	-	-	-	0.17	-	2.11	-	2.11	nr	2.27
extra; connections to iron pipe ends; screwed joint	1.56	3.00	1.61	0.25	-	3.11	-	4.72	nr	5.07
extra; fittings; one end	2.86	3.00	2.95	0.13	-	1.62	-	4.56	nr	4.91
extra; fittings; two ends	3.44	3.00	3.54	0.25	-	3.11	-	6.65	nr	7.15
extra; fittings; three ends	6.09	3.00	6.27	0.25	-	3.11	-	9.38	nr	10.09
35mm; in two piece copper spacing clips to timber with screws	5.24	5.00	5.50	0.50	-	6.22	-	11.72	m	12.60
extra; made bends	-	-	-	0.25	-	3.11	-	3.11	nr	3.34
extra; connections to iron pipe ends; screwed joint	4.58	3.00	4.72	0.33	-	4.11	-	8.82	nr	9.48
extra; fittings; one end	4.93	3.00	5.08	0.17	-	2.11	-	7.19	nr	7.73
extra; fittings; two ends	8.26	3.00	8.51	0.33	-	4.11	-	12.61	nr	13.56
extra; fittings; three ends	10.90	3.00	11.23	0.33	-	4.11	-	15.33	nr	16.48
42mm; in two piece copper spacing clips to timber with screws	6.64	5.00	6.97	0.67	-	8.33	-	15.31	m	16.45
extra; made bends	-	-	-	0.33	-	4.11	-	4.11	nr	4.41
extra; connections to iron pipe ends; screwed joint	7.36	3.00	7.58	0.50	-	6.22	-	13.80	nr	14.84
extra; fittings; one end	8.06	3.00	8.30	0.25	-	3.11	-	11.41	nr	12.27
extra; fittings; two ends	11.54	3.00	11.89	0.50	-	6.22	-	18.11	nr	19.46
extra; fittings; three ends	16.79	3.00	17.29	0.50	-	6.22	-	23.51	nr	25.28
54mm; in two piece copper spacing clips to timber with screws	8.77	5.00	9.21	1.00	-	12.44	-	21.65	m	23.27
extra; made bends	-	-	-	0.50	-	6.22	-	6.22	nr	6.69
extra; connections to iron pipe ends; screwed joint	11.16	3.00	11.49	0.67	-	8.33	-	19.83	nr	21.32
extra; fittings; one end	11.24	3.00	11.58	0.33	-	4.11	-	15.68	nr	16.86
extra; fittings; two ends	19.62	3.00	20.21	0.67	-	8.33	-	28.54	nr	30.68
extra; fittings; three ends	27.02	3.00	27.83	0.67	-	8.33	-	36.17	nr	38.88

Labour hourly rates: (except Specialists) Craft Operatives £12.44 Labourer £7.02 Rates are national average prices. Refer to REGIONAL VARIATIONS for indicative levels of overall pricing in regions	MATERIALS			LABOUR				RATES		
	Del to Site £	Waste %	Material Cost £	Craft Optve Hrs	Lab Hrs	Labour Cost £	Sunds £	Nett Rate £	Unit	Gross Rate (+7.5%) £

S115: HOT AND COLD WATER AND GAS Cont.

COPPER TUBING AND FITTINGS - Copper pipes, B.S.2871 Part 1 Table X; copper alloy fittings, compression, B.S.864 Part 2, Type A Cont.

Pipes, straight Cont.

in two piece copper spacing clips fixing to masonry with screws

	Del to Site £	Waste %	Material Cost £	Craft Optve Hrs	Lab Hrs	Labour Cost £	Sunds £	Nett Rate £	Unit	Gross Rate £
15mm	0.79	5.00	0.83	0.36	-	4.48	-	5.31	m	5.71
22mm	1.45	5.00	1.52	0.39	-	4.85	-	6.37	m	6.85
28mm	2.19	5.00	2.30	0.43	-	5.35	-	7.65	m	8.22
35mm	5.24	5.00	5.50	0.53	-	6.59	-	12.10	m	13.00
42mm	6.64	5.00	6.97	0.70	-	8.71	-	15.68	m	16.86
54mm	8.77	5.00	9.21	1.03	-	12.81	-	22.02	m	23.67

in pressed brass pipe brackets fixing to timber with screws

15mm	1.06	5.00	1.11	0.33	-	4.11	-	5.22	m	5.61
22mm	1.77	5.00	1.86	0.36	-	4.48	-	6.34	m	6.81
28mm	2.56	5.00	2.69	0.40	-	4.98	-	7.66	m	8.24
35mm	5.62	5.00	5.90	0.50	-	6.22	-	12.12	m	13.03
42mm	7.00	5.00	7.35	0.67	-	8.33	-	15.68	m	16.86
54mm	9.23	5.00	9.69	1.00	-	12.44	-	22.13	m	23.79

in cast brass pipe brackets fixing to timber with screws

15mm	1.55	5.00	1.63	0.33	-	4.11	-	5.73	m	6.16
22mm	2.36	5.00	2.48	0.36	-	4.48	-	6.96	m	7.48
28mm	3.29	5.00	3.45	0.40	-	4.98	-	8.43	m	9.06
35mm	6.65	5.00	6.98	0.50	-	6.22	-	13.20	m	14.19
42mm	8.44	5.00	8.86	0.67	-	8.33	-	17.20	m	18.49
54mm	11.04	5.00	11.59	1.00	-	12.44	-	24.03	m	25.83

in cast brass single rings and back plates fixing to timber with screws

15mm	1.79	5.00	1.88	0.33	-	4.11	-	5.98	m	6.43
22mm	2.48	5.00	2.60	0.36	-	4.48	-	7.08	m	7.61
28mm	3.30	5.00	3.46	0.40	-	4.98	-	8.44	m	9.07
35mm	6.35	5.00	6.67	0.50	-	6.22	-	12.89	m	13.85
42mm	7.72	5.00	8.11	0.67	-	8.33	-	16.44	m	17.67
54mm	9.93	5.00	10.43	1.00	-	12.44	-	22.87	m	24.58
66.7mm	20.84	5.00	21.88	1.50	-	18.66	-	40.54	m	43.58
76.1mm	29.66	5.00	31.14	2.00	-	24.88	-	56.02	m	60.22
108mm	45.55	5.00	47.83	3.00	-	37.32	-	85.15	m	91.53

COPPER TUBING AND FITTINGS - Copper pipes, B.S.2871 Part 1 Table X; copper alloy fittings, compression, B.S.864 Part 2 Type B

Pipes, straight

15mm; in two piece copper spacing clips to timber with screws	1.19	5.00	1.25	0.33	-	4.11	-	5.35	m	5.76
extra; made bends	-	-	-	0.08	-	1.00	-	1.00	nr	1.07
extra; connections to iron pipe ends; screwed joint	2.70	3.00	2.78	0.13	-	1.62	-	4.40	nr	4.73
extra; fittings; one end	2.95	3.00	3.04	0.07	-	0.87	-	3.91	nr	4.20
extra; fittings; two ends	3.50	3.00	3.61	0.13	-	1.62	-	5.22	nr	5.61
extra; fittings; three ends	4.90	3.00	5.05	0.13	-	1.62	-	6.66	nr	7.16
22mm; in two piece copper spacing clips to timber with screws	2.10	5.00	2.21	0.36	-	4.48	-	6.68	m	7.18
extra; made bends	-	-	-	0.13	-	1.62	-	1.62	nr	1.74
extra; connections to iron pipe ends; screwed joint	3.85	3.00	3.97	0.21	-	2.61	-	6.58	nr	7.07
extra; fittings; one end	3.60	3.00	3.71	0.11	-	1.37	-	5.08	nr	5.46
extra; fittings; two ends	5.70	3.00	5.87	0.21	-	2.61	-	8.48	nr	9.12
extra; fittings; three ends	8.25	3.00	8.50	0.21	-	2.61	-	11.11	nr	11.94
28mm; in two piece copper spacing clips to timber with screws	3.04	5.00	3.19	0.40	-	4.98	-	8.17	m	8.78
extra; made bends	-	-	-	0.17	-	2.11	-	2.11	nr	2.27
extra; connections to iron pipe ends; screwed joint	5.90	3.00	6.08	0.31	-	3.86	-	9.93	nr	10.68
extra; fittings; one end	6.75	3.00	6.95	0.16	-	1.99	-	8.94	nr	9.61
extra; fittings; two ends	9.65	3.00	9.94	0.31	-	3.86	-	13.80	nr	14.83
extra; fittings; three ends	13.90	3.00	14.32	0.31	-	3.86	-	18.17	nr	19.54
35mm; in two piece copper spacing clips to timber with screws	6.59	5.00	6.92	0.50	-	6.22	-	13.14	m	14.12
extra; made bends	-	-	-	0.25	-	3.11	-	3.11	nr	3.34
extra; connections to iron pipe ends; screwed joint	11.65	3.00	12.00	0.40	-	4.98	-	16.98	nr	18.25
extra; fittings; one end	17.90	3.00	18.44	0.20	-	2.49	-	20.93	nr	22.49
extra; fittings; two ends	18.55	3.00	19.11	0.40	-	4.98	-	24.08	nr	25.89
extra; fittings; three ends	25.20	3.00	25.96	0.40	-	4.98	-	30.93	nr	33.25
42mm; in two piece copper spacing clips to timber with screws	8.46	5.00	8.88	0.67	-	8.33	-	17.22	m	18.51
extra; made bends	-	-	-	0.33	-	4.11	-	4.11	nr	4.41
extra; connections to iron pipe ends; screwed joint	17.20	3.00	17.72	0.63	-	7.84	-	25.55	nr	27.47
extra; fittings; one end	25.05	3.00	25.80	0.32	-	3.98	-	29.78	nr	32.00
extra; fittings; two ends	27.80	3.00	28.63	0.63	-	7.84	-	36.47	nr	39.21
extra; fittings; three ends	41.00	3.00	42.23	0.63	-	7.84	-	50.07	nr	53.82
54mm; in two piece copper spacing clips to timber with screws	11.52	5.00	12.10	1.00	-	12.44	-	24.54	m	26.38
extra; made bends	-	-	-	0.50	-	6.22	-	6.22	nr	6.69
extra; connections to iron pipe ends; screwed joint	24.90	3.00	25.65	0.83	-	10.33	-	35.97	nr	38.67
extra; fittings; one end	38.45	3.00	39.60	0.42	-	5.22	-	44.83	nr	48.19
extra; fittings; two ends	44.25	3.00	45.58	0.83	-	10.33	-	55.90	nr	60.10

PIPED SUPPLY SYSTEMS

Labour hourly rates: (except Specialists) Craft Operatives £12.44 Labourer £7.02 Rates are national average prices. Refer to REGIONAL VARIATIONS for indicative levels of overall pricing in regions	MATERIALS			LABOUR				RATES		
	Del to Site	Waste	Material Cost	Craft Optve	Lab	Labour Cost	Sunds	Nett Rate	Unit	Gross Rate (+7.5%)
	£	%	£	Hrs	Hrs	£	£	£		£

S115: HOT AND COLD WATER AND GAS Cont.

COPPER TUBING AND FITTINGS - Copper pipes, B.S.2871 Part 1 Table X; copper alloy fittings, compression, B.S.864 Part 2 Type B Cont.

Pipes, straight Cont.

	Del to Site	Waste	Material Cost	Craft Optve	Lab	Labour Cost	Sunds	Nett Rate	Unit	Gross Rate
extra; fittings; three ends	63.75	3.00	65.66	0.83	–	10.33	–	75.99	nr	81.69
in two piece copper spacing clips fixing to masonry with screws										
15mm	1.19	5.00	1.25	0.36	–	4.48	–	5.73	m	6.16
22mm	2.10	5.00	2.21	0.39	–	4.85	–	7.06	m	7.59
28mm	3.04	5.00	3.19	0.43	–	5.35	–	8.54	m	9.18
35mm	6.59	5.00	6.92	0.53	–	6.59	–	13.51	m	14.53
42mm	8.46	5.00	8.88	0.70	–	8.71	–	17.59	m	18.91
54mm	11.52	5.00	12.10	1.03	–	12.81	–	24.91	m	26.78
in pressed brass pipe brackets fixing to timber with screws										
15mm	1.46	5.00	1.53	0.36	–	4.48	–	6.01	m	6.46
22mm	2.42	5.00	2.54	0.39	–	4.85	–	7.39	m	7.95
28mm	3.41	5.00	3.58	0.43	–	5.35	–	8.93	m	9.60
35mm	6.97	5.00	7.32	0.53	–	6.59	–	13.91	m	14.96
42mm	8.82	5.00	9.26	0.70	–	8.71	–	17.97	m	19.32
54mm	11.98	5.00	12.58	1.03	–	12.81	–	25.39	m	27.30
in cast brass pipe brackets fixing to timber with screws										
15mm	1.95	5.00	2.05	0.36	–	4.48	–	6.53	m	7.02
22mm	3.01	5.00	3.16	0.39	–	4.85	–	8.01	m	8.61
28mm	4.14	5.00	4.35	0.43	–	5.35	–	9.70	m	10.42
35mm	8.00	5.00	8.40	0.53	–	6.59	–	14.99	m	16.12
42mm	10.26	5.00	10.77	0.70	–	8.71	–	19.48	m	20.94
54mm	13.79	5.00	14.48	1.03	–	12.81	–	27.29	m	29.34
in cast brass single rings and back plates fixing to timber with screws										
15mm	2.19	5.00	2.30	0.36	–	4.48	–	6.78	m	7.29
22mm	3.13	5.00	3.29	0.39	–	4.85	–	8.14	m	8.75
28mm	4.15	5.00	4.36	0.43	–	5.35	–	9.71	m	10.43
35mm	7.70	5.00	8.09	0.53	–	6.59	–	14.68	m	15.78
42mm	9.54	5.00	10.02	0.70	–	8.71	–	18.73	m	20.13
54mm	12.68	5.00	13.31	1.03	–	12.81	–	26.13	m	28.09

COPPER TUBING AND FITTINGS - Copper pipes, B.S.2871 Part 1 Table Y; copper fittings, capillary, B.S.864 Part 2

Pipes; straight; in trenches

	Del to Site	Waste	Material Cost	Craft Optve	Lab	Labour Cost	Sunds	Nett Rate	Unit	Gross Rate
15mm	1.70	5.00	1.78	0.17	–	2.11	–	3.90	m	4.19
extra; made bends	–	–	–	1.20	–	14.93	–	14.93	nr	16.05
extra; connections to iron pipe ends; screwed joint	1.25	3.00	1.29	0.13	–	1.62	–	2.90	nr	3.12
extra; fittings; one end	0.68	3.00	0.70	0.07	–	0.87	–	1.57	nr	1.69
extra; fittings; two ends	0.19	3.00	0.20	0.13	–	1.62	–	1.81	nr	1.95
extra; fittings; three ends	0.32	3.00	0.33	0.13	–	1.62	–	1.95	nr	2.09
22mm	2.97	5.00	3.12	0.18	–	2.24	–	5.36	m	5.76
extra; made bends	–	–	–	0.13	–	1.62	–	1.62	nr	1.74
extra; connections to iron pipe ends; screwed joint	2.21	3.00	2.28	0.21	–	2.61	–	4.89	nr	5.26
extra; fittings; one end	1.29	3.00	1.33	0.11	–	1.37	–	2.70	nr	2.90
extra; fittings; two ends	0.45	3.00	0.46	0.21	–	2.61	–	3.08	nr	3.31
extra; fittings; three ends	0.92	3.00	0.95	0.21	–	2.61	–	3.56	nr	3.83
28mm	3.93	5.00	4.13	0.20	–	2.49	–	6.61	m	7.11
extra; made bends	–	–	–	0.17	–	2.11	–	2.11	nr	2.27
extra; connections to iron pipe ends; screwed joint	3.50	3.00	3.61	0.31	–	3.86	–	7.46	nr	8.02
extra; fittings; one end	2.04	3.00	2.10	0.16	–	1.99	–	4.09	nr	4.40
extra; fittings; two ends	1.02	3.00	1.05	0.31	–	3.86	–	4.91	nr	5.28
extra; fittings; three ends	2.85	3.00	2.94	0.31	–	3.86	–	6.79	nr	7.30
35mm	5.60	5.00	5.88	0.22	–	2.74	–	8.62	m	9.26
extra; made bends	–	–	–	0.25	–	3.11	–	3.11	nr	3.34
extra; connections to iron pipe ends; screwed joint	6.17	3.00	6.36	0.40	–	4.98	–	11.33	nr	12.18
extra; fittings; one end	4.41	3.00	4.54	0.20	–	2.49	–	7.03	nr	7.56
extra; fittings; two ends	4.15	3.00	4.27	0.40	–	4.98	–	9.25	nr	9.94
extra; fittings; three ends	7.13	3.00	7.34	0.40	–	4.98	–	12.32	nr	13.24
42mm	6.79	5.00	7.13	0.29	–	3.61	–	10.74	m	11.54
extra; made bends	–	–	–	0.33	–	4.11	–	4.11	nr	4.41
extra; connections to iron pipe ends; screwed joint	7.95	3.00	8.19	0.63	–	7.84	–	16.03	nr	17.23
extra; fittings; one end	6.77	3.00	6.97	0.32	–	3.98	–	10.95	nr	11.78
extra; fittings; two ends	7.39	3.00	7.61	0.63	–	7.84	–	15.45	nr	16.61
extra; fittings; three ends	11.17	3.00	11.51	0.63	–	7.84	–	19.34	nr	20.79
54mm	11.76	5.00	12.35	0.33	–	4.11	–	16.45	m	17.69
extra; made bends	–	–	–	0.50	–	6.22	–	6.22	nr	6.69
extra; connections to iron pipe ends; screwed joint	12.07	3.00	12.43	0.83	–	10.33	–	22.76	nr	24.46
extra; fittings; one end	9.45	3.00	9.73	0.42	–	5.22	–	14.96	nr	16.08
extra; fittings; two ends	15.13	3.00	15.58	0.83	–	10.33	–	25.91	nr	27.85
extra; fittings; three ends	21.11	3.00	21.74	0.83	–	10.33	–	32.07	nr	34.47

COPPER TUBING AND FITTINGS - Copper pipes, B.S.2871 Part 1 Table Y; non-dezincifiable fittings; compression, B.S.864 Part 2, Type B

Pipes; straight; in trenches

	Del to Site	Waste	Material Cost	Craft Optve	Lab	Labour Cost	Sunds	Nett Rate	Unit	Gross Rate
15mm	1.89	5.00	1.98	0.17	–	2.11	–	4.10	m	4.41
extra; made bends	–	–	–	0.08	–	1.00	–	1.00	nr	1.07

444

	MATERIALS			LABOUR				RATES		
Labour hourly rates: (except Specialists) Craft Operatives £12.44 Labourer £7.02 Rates are national average prices. Refer to REGIONAL VARIATIONS for indicative levels of overall pricing in regions	Del to Site £	Waste %	Material Cost £	Craft Optve Hrs	Lab Hrs	Labour Cost £	Sunds £	Nett Rate £	Unit	Gross Rate (+7.5%) £
S115: HOT AND COLD WATER AND GAS Cont.										
COPPER TUBING AND FITTINGS - Copper pipes, B.S.2871 Part 1 Table Y; non-dezincifiable fittings; compression, B.S.864 Part 2, Type B Cont.										
Pipes; straight; in trenches Cont.										
extra; connections to iron pipe ends; screwed joint	2.70	3.00	2.78	0.13	-	1.62	-	4.40	nr	4.73
extra; fittings; one end	2.95	3.00	3.04	0.07	-	0.87	-	3.91	nr	4.20
extra; fittings; two ends	3.50	3.00	3.61	0.13	-	1.62	-	5.22	nr	5.61
extra; fittings; three ends	4.90	3.00	5.05	0.13	-	1.62	-	6.66	nr	7.16
22mm	3.27	5.00	3.43	0.18	-	2.24	-	5.67	m	6.10
extra; made bends	-	-	-	0.13	-	1.62	-	1.62	nr	1.74
extra; connections to iron pipe ends; screwed joint	3.85	3.00	3.97	0.21	-	2.61	-	6.58	nr	7.07
extra; fittings; one end	3.60	3.00	3.71	0.11	-	1.37	-	5.08	nr	5.46
extra; fittings; two ends	5.70	3.00	5.87	0.21	-	2.61	-	8.48	nr	9.12
extra; fittings; three ends	8.25	3.00	8.50	0.21	-	2.61	-	11.11	nr	11.94
28mm	4.41	5.00	4.63	0.20	-	2.49	-	7.12	m	7.65
extra; made bends	-	-	-	0.17	-	2.11	-	2.11	nr	2.27
extra; connections to iron pipe ends; screwed joint	5.90	3.00	6.08	0.31	-	3.86	-	9.93	nr	10.68
extra; fittings; one end	6.75	3.00	6.95	0.16	-	1.99	-	8.94	nr	9.61
extra; fittings; two ends	9.65	3.00	9.94	0.31	-	3.86	-	13.80	nr	14.83
extra; fittings; three ends	13.90	3.00	14.32	0.31	-	3.86	-	18.17	nr	19.54
35mm	6.42	5.00	6.74	0.22	-	2.74	-	9.48	m	10.19
extra; made bends	-	-	-	0.25	-	3.11	-	3.11	nr	3.34
extra; connections to iron pipe ends; screwed joint	11.65	3.00	12.00	0.40	-	4.98	-	16.98	nr	18.25
extra; fittings; one end	17.90	3.00	18.44	0.20	-	2.49	-	20.93	nr	22.49
extra; fittings; two ends	18.55	3.00	19.11	0.40	-	4.98	-	24.08	nr	25.89
extra; fittings; three ends	25.20	3.00	25.96	0.40	-	4.98	-	30.93	nr	33.25
42mm	7.89	5.00	8.28	0.29	-	3.61	-	11.89	m	12.78
extra; made bends	-	-	-	0.33	-	4.11	-	4.11	nr	4.41
extra; connections to iron pipe ends; screwed joint	17.20	3.00	17.72	0.63	-	7.84	-	25.55	nr	27.47
extra; fittings; one end	25.05	3.00	25.80	0.32	-	3.98	-	29.78	nr	32.02
extra; fittings; two ends	27.80	3.00	28.63	0.63	-	7.84	-	36.47	nr	39.21
extra; fittings; three ends	41.00	3.00	42.23	0.63	-	7.84	-	50.07	nr	53.82
54mm	13.26	5.00	13.92	0.33	-	4.11	-	18.03	m	19.38
extra; made bends	-	-	-	0.50	-	6.22	-	6.22	nr	6.69
extra; connections to iron pipe ends; screwed joint	24.90	3.00	25.65	0.83	-	10.33	-	35.97	nr	38.67
extra; fittings; one end	38.45	3.00	39.60	0.42	-	5.22	-	44.83	nr	48.19
extra; fittings; two ends	44.25	3.00	45.58	0.83	-	10.33	-	55.90	nr	60.10
extra; fittings; three ends	65.75	3.00	67.72	0.83	-	10.33	-	78.05	nr	83.90
POLYBUTYLENE TUBES AND FITTINGS - Polybutylene pipes; Hepworth Hep20 flexible plumbing system B.S.7291 Class H; polybutylene slimline fittings										
Pipes, flexible; in ducts										
15mm; in pipe clips to timber	1.23	10.00	1.35	0.30	-	3.73	-	5.09	m	5.47
extra; made bend	-	-	-	0.01	-	0.12	-	0.12	nr	0.13
extra; connection to copper pipe ends	1.09	2.00	1.11	0.10	-	1.24	-	2.36	nr	2.53
extra; fittings; one end	1.33	2.00	1.36	0.05	-	0.62	-	1.98	nr	2.13
extra; fittings; two ends	1.44	2.00	1.47	0.10	-	1.24	-	2.71	nr	2.92
extra; fittings; three ends	2.06	2.00	2.10	0.10	-	1.24	-	3.35	nr	3.60
22mm; in pipe clips to timber	1.91	10.00	2.10	0.33	-	4.11	-	6.21	m	6.67
extra; made bend	-	-	-	0.01	-	0.12	-	0.12	nr	0.13
extra; connection to copper pipe ends	1.44	2.00	1.47	0.25	-	3.11	-	4.58	nr	4.92
extra; fittings; one end	1.68	2.00	1.71	0.13	-	1.62	-	3.33	nr	3.58
extra; fittings; two ends	2.03	2.00	2.07	0.25	-	3.11	-	5.18	nr	5.57
extra; fittings; three ends	2.60	2.00	2.65	0.25	-	3.11	-	5.76	nr	6.19
in pipe clips fixing to masonry with screws										
15mm	1.23	10.00	1.35	0.30	-	3.73	-	5.09	m	5.47
22mm	1.91	10.00	2.10	0.33	-	4.11	-	6.21	m	6.67
POLYBUTYLENE TUBES AND FITTINGS - Polybutylene pipes; Hepworth Hep20 flexible plumbing system B.S.7291 Class H; polybutylene demountable fittings										
Pipes, flexible										
15mm; in pipe clips to timber with screws	1.22	10.00	1.34	0.37	-	4.60	-	5.94	m	6.39
extra; connection to copper pipe ends	1.06	2.00	1.08	0.10	-	1.24	-	2.33	nr	2.50
extra; fittings; one end	1.30	2.00	1.33	0.05	-	0.62	-	1.95	nr	2.09
extra; fittings; two ends	1.42	2.00	1.45	0.10	-	1.24	-	2.69	nr	2.89
extra; fittings; three ends	2.03	2.00	2.07	0.10	-	1.24	-	3.31	nr	3.56
22mm; in pipe clips to timber with screws	1.90	10.00	2.09	0.43	-	5.35	-	7.44	m	8.00
extra; connection to copper pipe ends	1.40	2.00	1.43	0.17	-	2.11	-	3.54	nr	3.81
extra; fittings; one end	1.64	2.00	1.67	0.09	-	1.12	-	2.79	nr	3.00
extra; fittings; two ends	1.99	2.00	2.03	0.17	-	2.11	-	4.14	nr	4.46
extra; fittings; three ends	2.56	2.00	2.61	0.17	-	2.11	-	4.73	nr	5.08
in pipe clips fixing to masonry with screws										
15mm	1.22	10.00	1.34	0.33	-	4.11	-	5.45	m	5.86
22mm	1.90	10.00	2.09	0.37	-	4.60	-	6.69	m	7.19
Pipes, flexible										
28mm; in pipe clips to timber with screws	2.77	10.00	3.05	0.50	-	6.22	-	9.27	m	9.96
extra; fittings; two ends	4.32	2.00	4.41	0.33	-	4.11	-	8.51	nr	9.15
extra; fittings; three ends	6.02	2.00	6.14	0.33	-	4.11	-	10.25	nr	11.01
in pipe clips fixing to masonry with screws										
28mm	2.77	10.00	3.05	0.50	-	6.22	-	9.27	m	9.96

PIPED SUPPLY SYSTEMS

Labour hourly rates: (except Specialists) Craft Operatives £12.44 Labourer £7.02 Rates are national average prices. Refer to REGIONAL VARIATIONS for indicative levels of overall pricing in regions	MATERIALS			LABOUR				RATES		
	Del to Site £	Waste %	Material Cost £	Craft Optve Hrs	Lab Hrs	Labour Cost £	Sunds £	Nett Rate £	Unit	Gross Rate (+7.5%) £
S115: HOT AND COLD WATER AND GAS Cont.										
POLYETHYLENE TUBES AND FITTINGS - Polythene pipes, B.S.6572 Blue; copper alloy fittings, compression, B.S.864 Part 3, Type A										
Pipes; straight; in trenches										
20mm	0.32	10.00	0.35	0.14	-	1.74	-	2.09	m	2.25
extra; connections to copper pipe ends; compression joint	2.76	3.00	2.84	0.13	-	1.62	-	4.46	nr	4.79
extra; fittings; one end	2.69	3.00	2.77	0.07	-	0.87	-	3.64	nr	3.91
extra; fittings; two ends	3.77	3.00	3.88	0.13	-	1.62	-	5.50	nr	5.91
extra; fittings; three ends	5.15	3.00	5.30	0.13	-	1.62	-	6.92	nr	7.44
25mm	0.43	10.00	0.47	0.17	-	2.11	-	2.59	m	2.78
extra; connections to copper pipe ends; compression joint	4.19	3.00	4.32	0.21	-	2.61	-	6.93	nr	7.45
extra; fittings; one end	4.04	3.00	4.16	0.11	-	1.37	-	5.53	nr	5.94
extra; fittings; two ends	5.49	3.00	5.65	0.21	-	2.61	-	8.27	nr	8.89
extra; fittings; three ends	7.99	3.00	8.23	0.21	-	2.61	-	10.84	nr	11.66
32mm	0.72	10.00	0.79	0.20	-	2.49	-	3.28	m	3.53
extra; connections to copper pipe ends; compression joint (Type A)	7.18	3.00	7.40	0.31	-	3.86	-	11.25	nr	12.10
extra; fittings; one end	9.13	3.00	9.40	0.16	-	1.99	-	11.39	nr	12.25
extra; fittings; two ends	9.66	3.00	9.95	0.31	-	3.86	-	13.81	nr	14.84
extra; fittings; three ends	12.02	3.00	12.38	0.31	-	3.86	-	16.24	nr	17.45
50mm	1.68	10.00	1.85	0.33	-	4.11	-	5.95	m	6.40
extra; fittings; two ends	22.54	3.00	23.22	0.50	-	6.22	-	29.44	nr	31.64
extra; fittings; three ends	29.98	3.00	30.88	0.50	-	6.22	-	37.10	nr	39.88
63mm	2.47	10.00	2.72	0.45	-	5.60	-	8.31	m	8.94
extra; fittings; two ends	26.74	3.00	27.54	0.83	-	10.33	-	37.87	nr	40.71
extra; fittings; three ends	44.01	3.00	45.33	0.83	-	10.33	-	55.66	nr	59.83
UNPLASTICISED PVC TUBES AND FITTINGS - PVC-U pipes, B.S.3505 Class E; fittings, solvent welded joints, B.S.4346										
Pipes, straight										
3/8; in standard plastics pipe brackets fixing to timber with screws	1.16	5.00	1.22	0.33	-	4.11	-	5.32	m	5.72
extra; connections to iron pipe ends; screwed joint	0.65	2.00	0.66	0.10	-	1.24	-	1.91	nr	2.05
extra; fittings; one end	0.41	2.00	0.42	0.05	-	0.62	-	1.04	nr	1.12
extra; fittings; two ends	0.62	2.00	0.63	0.10	-	1.24	-	1.88	nr	2.02
extra; fittings; three ends	0.68	2.00	0.69	0.10	-	1.24	-	1.94	nr	2.08
1/2'; in standard plastics pipe brackets fixing to timber with screws	1.45	5.00	1.52	0.37	-	4.60	-	6.13	m	6.58
extra; connections to iron pipe ends; screwed joint	0.88	2.00	0.90	0.17	-	2.11	-	3.01	nr	3.24
extra; fittings; one end	0.49	2.00	0.50	0.09	-	1.12	-	1.62	nr	1.74
extra; fittings; two ends	0.70	2.00	0.71	0.17	-	2.11	-	2.83	nr	3.04
extra; fittings; three ends	0.81	2.00	0.83	0.17	-	2.11	-	2.94	nr	3.16
3/4'; in standard plastics pipe brackets fixing to timber with screws	1.79	5.00	1.88	0.43	-	5.35	-	7.23	m	7.77
extra; connections to iron pipe ends; screwed joint	0.95	2.00	0.97	0.25	-	3.11	-	4.08	nr	4.38
extra; fittings; one end	0.58	2.00	0.59	0.13	-	1.62	-	2.21	nr	2.37
extra; fittings; two ends	0.84	2.00	0.86	0.25	-	3.11	-	3.97	nr	4.26
extra; fittings; three ends	1.02	2.00	1.04	0.25	-	3.11	-	4.15	nr	4.46
1'; in standard plastics pipe brackets fixing to timber with screws	2.00	5.00	2.10	0.50	-	6.22	-	8.32	m	8.94
extra; connections to iron pipe ends; screwed joint	1.47	2.00	1.50	0.33	-	4.11	-	5.60	nr	6.02
extra; fittings; one end	0.65	2.00	0.66	0.17	-	2.11	-	2.78	nr	2.99
extra; fittings; two ends	1.16	2.00	1.18	0.33	-	4.11	-	5.29	nr	5.69
extra; fittings; three ends	1.54	2.00	1.57	0.33	-	4.11	-	5.68	nr	6.10
1 1/4'; in standard plastics pipe brackets fixing to timber with screws	2.85	5.00	2.99	0.60	-	7.46	-	10.46	m	11.24
extra; connections to iron pipe ends; screwed joint	2.07	2.00	2.11	0.50	-	6.22	-	8.33	nr	8.96
extra; fittings; one end	1.02	2.00	1.04	0.25	-	3.11	-	4.15	nr	4.46
extra; fittings; two ends	2.03	2.00	2.07	0.50	-	6.22	-	8.29	nr	8.91
extra; fittings; three ends	2.17	2.00	2.21	0.50	-	6.22	-	8.43	nr	9.07
1 1/2'; in standard plastics pipe brackets fixing to timber with screws	3.48	5.00	3.65	0.75	-	9.33	-	12.98	m	13.96
extra; connections to iron pipe ends; screwed joint	2.80	2.00	2.86	0.67	-	8.33	-	11.19	nr	12.03
extra; fittings; one end	1.72	2.00	1.75	0.33	-	4.11	-	5.86	nr	6.30
extra; fittings; two ends	2.63	2.00	2.68	0.67	-	8.33	-	11.02	nr	11.84
extra; fittings; three ends	3.15	2.00	3.21	0.67	-	8.33	-	11.55	nr	12.41
in standard plastics pipe brackets fixing to masonry with screws										
3/8'	1.16	5.00	1.22	0.39	-	4.85	-	6.07	m	6.52
1/2'	1.45	5.00	1.52	0.43	-	5.35	-	6.87	m	7.39
3/4'	1.79	5.00	1.88	0.49	-	6.10	-	7.98	m	8.57
1'	2.00	5.00	2.10	0.56	-	6.97	-	9.07	m	9.75
1 1/4'	2.85	5.00	2.99	0.66	-	8.21	-	11.20	m	12.04
1 1/2'	3.48	5.00	3.65	0.81	-	10.08	-	13.73	m	14.76
Pipes; straight; in trenches										
3/8'	0.63	5.00	0.66	0.17	-	2.11	-	2.78	m	2.98
1/2'	0.92	5.00	0.97	0.18	-	2.24	-	3.21	m	3.45
3/4'	1.29	5.00	1.35	0.20	-	2.49	-	3.84	m	4.13
1'	1.50	5.00	1.58	0.25	-	3.11	-	4.68	m	5.04
1 1/4'	2.25	5.00	2.36	0.29	-	3.61	-	5.97	m	6.42
1 1/2'	2.89	5.00	3.03	0.33	-	4.11	-	7.14	m	7.68

Labour hourly rates: (except Specialists) Craft Operatives £12.44 Labourer £7.02 Rates are national average prices. Refer to REGIONAL VARIATIONS for indicative levels of overall pricing in regions	MATERIALS			LABOUR				RATES		
	Del to Site £	Waste %	Material Cost £	Craft Optve Hrs	Lab Hrs	Labour Cost £	Sunds £	Nett Rate £	Unit	Gross Rate (+7.5%) £
S115: HOT AND COLD WATER AND GAS Cont.										
UNPLASTICISED OVERFLOW PIPES - PVC-U pipes and fittings; solvent welded joints; pipework self coloured white										
Pipes, straight										
19; in standard plastics pipe brackets fixing to timber with screws	0.85	5.00	0.89	0.25	-	3.11	-	4.00	m	4.30
extra; fittings; two ends	0.59	2.00	0.60	0.20	-	2.49	-	3.09	nr	3.32
extra; fittings; three ends	0.71	2.00	0.72	0.20	-	2.49	-	3.21	nr	3.45
extra; fittings; tank connector	0.87	2.00	0.89	0.20	-	2.49	-	3.38	nr	3.63
DUCTILE IRON PIPES AND FITTINGS - Ductile iron pipes and fittings, Tyton socketed flexible joints										
Pipes; straight; in trenches										
80mm	18.58	5.00	19.51	0.80	-	9.95	-	29.46	m	31.67
extra; bends, 90 degree	48.59	2.00	49.56	0.80	-	9.95	-	59.51	nr	63.98
extra; duckfoot bends, 90 degree	112.97	2.00	115.23	0.80	-	9.95	-	125.18	nr	134.57
extra; tees	66.57	2.00	67.90	1.20	-	14.93	-	82.83	nr	89.04
extra; hydrant tees	99.38	2.00	101.37	1.20	-	14.93	-	116.30	nr	125.02
extra; branches, 45 degree	190.74	2.00	194.55	1.20	-	14.93	-	209.48	nr	225.19
extra; flanged sockets	38.96	2.00	39.74	0.80	-	9.95	-	49.69	nr	53.42
extra; flanged spigots	36.91	2.00	37.65	0.80	-	9.95	-	47.60	nr	51.17
100mm	18.44	5.00	19.36	1.00	-	12.44	-	31.80	m	34.19
extra; bends, 90 degree	51.82	2.00	52.86	1.00	-	12.44	-	65.30	nr	70.19
extra; duckfoot bends, 90 degree	122.52	2.00	124.97	1.00	-	12.44	-	137.41	nr	147.72
extra; tees	69.74	2.00	71.13	1.50	-	18.66	-	89.79	nr	96.53
extra; hydrant tees	70.80	2.00	72.22	1.50	-	18.66	-	90.88	nr	97.69
extra; branches, 45 degree	271.98	2.00	277.42	1.50	-	18.66	-	296.08	nr	318.29
extra; flanged sockets	42.21	2.00	43.05	1.00	-	12.44	-	55.49	nr	59.66
extra; flanged spigots	39.00	2.00	39.78	1.00	-	12.44	-	52.22	nr	56.14
150mm	24.41	5.00	25.63	1.50	-	18.66	-	44.29	m	47.61
extra; bends, 90 degree	106.37	2.00	108.50	1.50	-	18.66	-	127.16	nr	136.69
extra; duckfoot bends, 90 degree	269.10	2.00	274.48	1.50	-	18.66	-	293.14	nr	315.13
extra; tees	106.01	2.00	108.13	2.25	-	27.99	-	136.12	nr	146.33
extra; hydrant tees	108.09	2.00	110.25	2.25	-	27.99	-	138.24	nr	148.61
extra; branches, 45 degree	344.25	2.00	351.13	2.25	-	27.99	-	379.13	nr	407.56
extra; flanged sockets	64.15	2.00	65.43	1.50	-	18.66	-	84.09	nr	90.40
extra; flanged spigots	45.21	2.00	46.11	1.50	-	18.66	-	64.77	nr	69.63
STOPCOCKS - Brass stopcocks										
Stopcocks; B.S.1010, crutch head; screwed and PTFE joints										
each end threaded internally										
13mm	3.49	2.00	3.56	0.25	-	3.11	-	6.67	nr	7.17
19mm	5.41	2.00	5.52	0.33	-	4.11	-	9.62	nr	10.35
25mm	12.33	2.00	12.58	0.50	-	6.22	-	18.80	nr	20.21
32mm	22.95	2.00	23.41	0.67	-	8.33	-	31.74	nr	34.12
40mm	25.57	2.00	26.08	0.75	-	9.33	-	35.41	nr	38.07
50mm	42.38	2.00	43.23	1.00	-	12.44	-	55.67	nr	59.84
Stopcocks; B.S.1010, crutch head; DZR; joints to polythene B.S.6572										
each end										
20mm	12.99	2.00	13.25	0.33	-	4.11	-	17.36	nr	18.66
25mm	20.65	2.00	21.06	0.50	-	6.22	-	27.28	nr	29.33
32mm	27.72	2.00	28.27	0.67	-	8.33	-	36.61	nr	39.35
50mm	69.31	2.00	70.70	0.75	-	9.33	-	80.03	nr	86.03
Stopcocks B.S.1010, crutch head; compression joints to copper (Type A)										
each end										
15mm	2.73	2.00	2.78	0.25	-	3.11	-	5.89	nr	6.34
22mm	4.75	2.00	4.84	0.33	-	4.11	-	8.95	nr	9.62
28mm	11.41	2.00	11.64	0.50	-	6.22	-	17.86	nr	19.20
35mm	34.28	2.00	34.97	0.67	-	8.33	-	43.30	nr	46.55
42mm	43.26	2.00	44.13	0.75	-	9.33	-	53.46	nr	57.46
54mm	65.30	2.00	66.61	1.00	-	12.44	-	79.05	nr	84.97
Stopcocks B.S.1010, crutch head; DZR; compression joints to copper (Type A)										
each end										
15mm	7.94	2.00	8.10	0.25	-	3.11	-	11.21	nr	12.05
22mm	13.03	2.00	13.29	0.33	-	4.11	-	17.40	nr	18.70
28mm	21.57	2.00	22.00	0.50	-	6.22	-	28.22	nr	30.34
35mm	39.69	2.00	40.48	0.67	-	8.33	-	48.82	nr	52.48
42mm	56.82	2.00	57.96	0.75	-	9.33	-	67.29	nr	72.33
54mm	77.39	2.00	78.94	1.00	-	12.44	-	91.38	nr	98.23
STOPCOCKS - Polybutylene stopcocks										
Stopcocks; fitted with Hep20 ends										
each end										
15mm	3.87	2.00	3.95	0.25	-	3.11	-	7.06	nr	7.59
22mm	4.62	2.00	4.71	0.33	-	4.11	-	8.82	nr	9.48
GATE VALVES - Brass gate valves										
Gate valves; B.S.5154 series B; screwed and PTFE joints										
each end threaded internally										
13mm	4.08	2.00	4.16	0.25	-	3.11	-	7.27	nr	7.82
19mm	4.97	2.00	5.07	0.33	-	4.11	-	9.17	nr	9.86

PIPED SUPPLY SYSTEMS

Labour hourly rates: (except Specialists) Craft Operatives £12.44 Labourer £7.02 Rates are national average prices. Refer to REGIONAL VARIATIONS for indicative levels of overall pricing in regions	MATERIALS			LABOUR				RATES		
	Del to Site £	Waste %	Material Cost £	Craft Optve Hrs	Lab Hrs	Labour Cost £	Sunds £	Nett Rate £	Unit	Gross Rate (+7.5%) £
S115: HOT AND COLD WATER AND GAS Cont.										
GATE VALVES - Brass gate valves Cont.										
Gate valves; B.S.5154 series B; screwed and PTFE joints Cont.										
each end threaded internally Cont.										
25mm	6.89	2.00	7.03	0.50	–	6.22	–	13.25	nr	14.24
32mm	10.07	2.00	10.27	0.67	–	8.33	–	18.61	nr	20.00
38mm	14.75	2.00	15.05	0.75	–	9.33	–	24.38	nr	26.20
51mm	21.15	2.00	21.57	1.00	–	12.44	–	34.01	nr	36.56
Gate valves; B.S.5154 series B; compression joints to copper (Type A)										
each end										
15mm	4.93	2.00	5.03	0.25	–	3.11	–	8.14	nr	8.75
22mm	5.98	2.00	6.10	0.33	–	4.11	–	10.20	nr	10.97
28mm	8.24	2.00	8.40	0.50	–	6.22	–	14.62	nr	15.72
35mm	15.51	2.00	15.82	0.67	–	8.33	–	24.16	nr	25.97
42mm	25.30	2.00	25.81	0.75	–	9.33	–	35.14	nr	37.77
54mm	39.18	2.00	39.96	1.00	–	12.44	–	52.40	nr	56.33
Gate valves; DZR; fitted with Hep20 ends										
each end										
15mm	3.62	2.00	3.69	0.25	–	3.11	–	6.80	nr	7.31
22mm	5.09	2.00	5.19	0.33	–	4.11	–	9.30	nr	9.99
BALL VALVES - Brass ball valves										
Float operated valves for low pressure; B.S.1212; copper float										
inlet threaded externally; fixing to steel										
13mm; part 1.....................	9.06	2.00	9.24	0.33	–	4.11	–	13.35	nr	14.35
13mm; part 2.....................	9.14	2.00	9.32	0.33	–	4.11	–	13.43	nr	14.44
19mm; part 1.....................	16.45	2.00	16.78	0.50	–	6.22	–	23.00	nr	24.72
25mm; part 1.....................	43.96	2.00	44.84	1.00	–	12.44	–	57.28	nr	61.58
Float operated valves for low pressure; B.S.1212; plastics float										
inlet threaded externally; fixing to steel										
13mm; part 1.....................	5.55	2.00	5.66	0.33	–	4.11	–	9.77	nr	10.50
13mm; part 2.....................	5.63	2.00	5.74	0.33	–	4.11	–	9.85	nr	10.59
19mm; part 1.....................	11.68	2.00	11.91	0.50	–	6.22	–	18.13	nr	19.49
25mm; part 1.....................	37.61	2.00	38.36	1.00	–	12.44	–	50.80	nr	54.61
Float operated valves for high pressure; B.S.1212; copper float										
inlet threaded externally; fixing to steel										
13mm; part 1.....................	7.78	2.00	7.94	0.33	–	4.11	–	12.04	nr	12.94
13mm; part 2.....................	9.01	2.00	9.19	0.33	–	4.11	–	13.30	nr	14.29
19mm; part 1.....................	16.29	2.00	16.62	0.50	–	6.22	–	22.84	nr	24.55
25mm; part 1.....................	43.96	2.00	44.84	1.00	–	12.44	–	57.28	nr	61.58
Float operated valves for high pressure; B.S.1212; plastics float										
inlet threaded externally; fixing to steel										
13mm; part 1.....................	4.27	2.00	4.36	0.33	–	4.11	–	8.46	nr	9.10
13mm; part 2.....................	5.50	2.00	5.61	0.33	–	4.11	–	9.72	nr	10.44
19mm; part 1.....................	11.50	2.00	11.73	0.50	–	6.22	–	17.95	nr	19.30
25mm; part 1.....................	37.61	2.00	38.36	1.00	–	12.44	–	50.80	nr	54.61
DRAINING TAPS - Brass draining taps										
Drain cocks; B.S.2879, square head Type2										
13mm...	1.62	2.00	1.65	0.10	–	1.24	–	2.90	nr	3.11
19mm...	11.70	2.00	11.93	0.13	–	1.62	–	13.55	nr	14.57
STORAGE CISTERNS - Galvanized water storage cisterns										
Cold water storage cisterns; galvanized steel; B.S.417 Part 2 Grade A with cover and byelaw 30 kit										
reference SCM45; 18 litres	84.95	–	84.95	1.50	–	18.66	–	103.61	nr	111.38
reference SCM70; 36 litres	104.00	–	104.00	1.50	–	18.66	–	122.66	nr	131.86
reference SCM90; 54 litres	115.25	–	115.25	1.50	–	18.66	–	133.91	nr	143.95
reference SCM110; 68 litres	121.95	–	121.95	1.75	–	21.77	–	143.72	nr	154.50
reference SCM135; 86 litres	129.50	–	129.50	1.75	–	21.77	–	151.27	nr	162.62
reference SCM180; 114 litres	144.10	–	144.10	1.75	–	21.77	–	165.87	nr	178.31
reference SCM230; 159 litres	180.60	–	180.60	2.00	–	24.88	–	205.48	nr	220.89
reference SCM270; 191 litres	192.80	–	192.80	2.00	–	24.88	–	217.68	nr	234.01
reference SCM320; 227 litres	210.25	–	210.25	2.50	–	31.10	–	241.35	nr	259.45
reference SCM450-1; 327 litres	247.40	–	247.40	3.00	–	37.32	–	284.72	nr	306.07
reference SCM450-2; 336 litres	240.35	–	240.35	3.00	–	37.32	–	277.67	nr	298.50
reference SCM570; 423 litres	323.85	–	323.85	4.00	–	49.76	–	373.61	nr	401.63
reference SCM680; 491 litres	365.75	–	365.75	4.00	–	49.76	–	415.51	nr	446.67
reference SCM910; 709 litres (excluding insulation)	358.60	–	358.60	4.00	–	49.76	–	408.36	nr	438.99
reference SCM1130; 841 litres (excluding insulation)	398.75	–	398.75	6.00	–	74.64	–	473.39	nr	508.89
reference SCM1600; 1227 litres (excluding insulation)	600.80	–	600.80	7.00	–	87.08	–	687.88	Unr	739.47
reference SCM2270; 1727 litres (excluding insulation)	707.45	%	707.45	8.00	–	99.52	£	806.97	nr	867.49
reference SCM2720; 2137 litres (excluding insulation)	761.60	–	761.60	9.00	–	111.96	–	873.56	nr	939.08
reference SCM4540; 3364 litres (excluding insulation)	1347.75	–	1347.75	11.00	–	136.84	–	1484.59	nr	1595.93

Labour hourly rates: (except Specialists) Craft Operatives £12.44 Labourer £7.02 Rates are national average prices. Refer to REGIONAL VARIATIONS for indicative levels of overall pricing in regions	MATERIALS			LABOUR				RATES		
	Del to Site	Waste	Material Cost	Craft Optve	Lab	Labour Cost	Sunds	Nett Rate	Unit	Gross Rate (+7.5%)
	£	%	£	Hrs	Hrs	£	£	£		£
S115: HOT AND COLD WATER AND GAS Cont.										
STORAGE CISTERNS - Galvanized water storage cisterns Cont.										
Cold water storage cisterns; galvanized steel; B.S.417 Part 2 Grade A with cover and byelaw 30 kit Cont.										
drilled holes for 13mm pipes	-	-	-	0.20	-	2.49	-	2.49	nr	2.67
drilled holes for 19mm pipes	-	-	-	0.20	-	2.49	-	2.49	nr	2.67
drilled holes for 25mm pipes	-	-	-	0.25	-	3.11	-	3.11	nr	3.34
drilled holed for 32mm pipes	-	-	-	0.25	-	3.11	-	3.11	nr	3.34
drilled holes for 38mm pipes	-	-	-	0.33	-	4.11	-	4.11	nr	4.41
drilled holes for 51mm pipes	-	-	-	0.33	-	4.11	-	4.11	nr	4.41
STORAGE CISTERNS - Plastics water storage cisterns										
Rectangular cold water storage cisterns; plastics; B.S.4213 with sealed lid and byelaw 30 kit										
18 litres	10.72	2.00	10.93	1.25	-	15.55	-	26.48	nr	28.47
114 litres	32.48	2.00	33.13	1.75	-	21.77	-	54.90	nr	59.02
182 litres	49.89	2.00	50.89	2.00	-	24.88	-	75.77	nr	81.45
227 litres	57.66	2.00	58.81	2.50	-	31.10	-	89.91	nr	96.66
drilled holes for 13mm pipes	-	-	-	0.13	-	1.62	-	1.62	nr	1.74
drilled holes for 19mm pipes	-	-	-	0.13	-	1.62	-	1.62	nr	1.74
drilled holes for 25mm pipes	-	-	-	0.17	-	2.11	-	2.11	nr	2.27
drilled holed for 32mm pipes	-	-	-	0.17	-	2.11	-	2.11	nr	2.27
drilled holes for 38mm pipes	-	-	-	0.22	-	2.74	-	2.74	nr	2.94
drilled holes for 51mm pipes	-	-	-	0.22	-	2.74	-	2.74	nr	2.94
STORAGE TANKS AND CYLINDERS - Galvanized hot water tanks										
Hot water storage tanks; galvanized steel; B.S.417 Part 2 Grade A										
reference TM114-1; 95 litres	126.50	-	126.50	3.00	-	37.32	-	163.82	nr	176.11
reference TM114-2; 95 litres	133.75	-	133.75	3.00	-	37.32	-	171.07	nr	183.90
reference TM136-1; 114 litres	136.00	-	136.00	3.00	-	37.32	-	173.32	nr	186.32
STORAGE TANKS AND CYLINDERS - Copper direct hot water cylinders; pre-insulated										
Direct hot water cylinders; copper cylinder, B.S.699 Grade 3										
reference 3; 116 litres; four bosses	90.75	2.00	92.56	2.00	-	24.88	-	117.44	nr	126.25
reference 7; 120 litres; four bosses	80.25	2.00	81.86	2.00	-	24.88	-	106.74	nr	114.74
reference 8; 144 litres; four bosses	84.00	2.00	85.68	2.00	-	24.88	-	110.56	nr	118.85
reference 9; 166 litres; four bosses	98.75	2.00	100.72	2.00	-	24.88	-	125.61	nr	135.03
STORAGE TANKS AND CYLINDERS - Copper indirect hot water cylinders; pre-insulated										
Double feed indirect hot water cylinders; copper cylinder, B.S.1566 Part 1 Grade 3										
reference 3; 114 litres; four bosses	83.50	2.00	85.17	2.00	-	24.88	-	110.05	nr	118.30
reference 7; 117 litres; four bosses	70.50	2.00	71.91	2.00	-	24.88	-	96.79	nr	104.05
reference 8; 140 litres; four bosses	90.50	2.00	92.31	2.00	-	24.88	-	117.19	nr	125.98
reference 9; 162 litres; four bosses	118.75	2.00	121.13	2.00	-	24.88	-	146.01	nr	156.96
Single feed indirect hot water cylinders; copper cylinder, B.S.1566 Part 2 Grade 3										
reference 3; 104 litres; four bosses	172.25	2.00	175.69	2.00	-	24.88	-	200.57	nr	215.62
reference 7; 108 litres; four bosses	172.25	2.00	175.69	2.00	-	24.88	-	200.57	nr	215.62
reference 8; 130 litres; four bosses	192.00	2.00	195.84	2.00	-	24.88	-	220.72	nr	237.27
reference 9; 152 litres; four bosses	234.75	2.00	239.44	2.00	-	24.88	-	264.32	nr	284.15
STORAGE TANKS AND CYLINDERS - Copper combination hot water storage units; pre-insulated										
Combination direct hot water storage units; copper unit, B.S.3198 with lid										
450mm diameter x 1200mm high; 115 litres hot water; 45 litres cold water	150.75	2.00	153.76	3.00	-	37.32	-	191.09	nr	205.42
500mm diameter x 1400mm high; 115 litres hot water; 115 litres cold water	183.00	2.00	186.66	3.00	-	37.32	-	223.98	nr	240.78
Combination double feed indirect hot water storage units; copper unit, B.S.3198 with lid										
450mm diameter x 1200mm high; 115 litres hot water; 45 litres cold water	194.00	2.00	197.88	3.00	-	37.32	-	235.20	nr	252.84
500mm diameter x 1400mm high; 115 litres hot water; 115 litres cold water	267.00	2.00	272.34	3.00	-	37.32	-	309.66	nr	332.88
Combination single feed indirect hot water storage units; copper unit, B.S.3198 with lid										
450mm diameter x 1200mm high; 115 litres hot water; 45 litres cold water	205.00	2.00	209.10	3.00	-	37.32	-	246.42	nr	264.90
500mm Diameter x 1400mm high; 115 litres hot water; 115 litres cold water	297.75	2.00	303.70	3.00	-	37.32	-	341.02	nr	366.60
PIPE INSULATION - Denso tape wrapping										
Insulation to pipework around one pipe										
15mm	0.53	5.00	0.56	0.17	-	2.11	-	2.67	m	2.87
22mm	0.63	5.00	0.66	0.18	-	2.24	-	2.90	m	3.12

PIPED SUPPLY SYSTEMS

Labour hourly rates: (except Specialists) Craft Operatives £12.44 Labourer £7.02 Rates are national average prices. Refer to REGIONAL VARIATIONS for indicative levels of overall pricing in regions	MATERIALS			LABOUR				RATES		
	Del to Site £	Waste %	Material Cost £	Craft Optve Hrs	Lab Hrs	Labour Cost £	Sunds £	Nett Rate £	Unit	Gross Rate (+7.5%) £
S115: HOT AND COLD WATER AND GAS Cont.										
PIPE INSULATION – Denso tape wrapping Cont.										
Insulation to pipework Cont.										
around one pipe Cont.										
28mm	0.73	5.00	0.77	0.20	-	2.49	-	3.25	m	3.50
35mm	1.06	5.00	1.11	0.22	-	2.74	-	3.85	m	4.14
42mm	1.45	5.00	1.52	0.25	-	3.11	-	4.63	m	4.98
54mm	1.88	5.00	1.97	0.33	-	4.11	-	6.08	m	6.54
PIPE INSULATION – Thermal insulation; glass fibre preformed lagging, butt joints in the running length; secured with metal bands										
19mm thick insulation to copper pipework										
around one pipe										
15mm	2.46	3.00	2.53	0.20	-	2.49	-	5.02	m	5.40
22mm	2.59	3.00	2.67	0.22	-	2.74	-	5.40	m	5.81
28mm	2.75	3.00	2.83	0.25	-	3.11	-	5.94	m	6.39
35mm	2.93	3.00	3.02	0.30	-	3.73	-	6.75	m	7.26
42mm	3.21	3.00	3.31	0.33	-	4.11	-	7.41	m	7.97
54mm	3.68	3.00	3.79	0.40	-	4.98	-	8.77	m	9.42
25mm thick insulation to copper pipework										
around one pipe										
15mm	2.75	3.00	2.83	0.20	-	2.49	-	5.32	m	5.72
22mm	2.82	3.00	2.90	0.22	-	2.74	-	5.64	m	6.06
28mm	3.02	3.00	3.11	0.25	-	3.11	-	6.22	m	6.69
35mm	3.35	3.00	3.45	0.30	-	3.73	-	7.18	m	7.72
42mm	3.63	3.00	3.74	0.33	-	4.11	-	7.84	m	8.43
54mm	4.26	3.00	4.39	0.40	-	4.98	-	9.36	m	10.07
PIPE INSULATION – Thermal insulation; foamed polyurethane preformed lagging, butt joints in the running length; secured with adhesive bands										
13mm thick insulation to copper pipework										
around one pipe										
15mm	1.59	3.00	1.64	0.17	-	2.11	-	3.75	m	4.03
22mm	1.90	3.00	1.96	0.18	-	2.24	-	4.20	m	4.51
28mm	2.18	3.00	2.25	0.20	-	2.49	-	4.73	m	5.09
35mm	2.38	3.00	2.45	0.22	-	2.74	-	5.19	m	5.58
42mm	2.74	3.00	2.82	0.25	-	3.11	-	5.93	m	6.38
54mm	3.60	3.00	3.71	0.33	-	4.11	-	7.81	m	8.40
INSULATION JACKETS AND LAGGING UNITS FOR CYLINDERS AND TANKS – Thermal insulation; glass fibre filled insulating jacket in strips, white pvc covering both sides; secured with metal straps and wire holder ring at top										
75mm thick insulation to equipment										
sides and tops of cylinders; overall size										
400mm diameter x 1050mm high...............	8.40	5.00	8.82	0.75	-	9.33	-	18.15	nr	19.51
450mm diameter x 900mm high................	7.55	5.00	7.93	0.75	-	9.33	-	17.26	nr	18.55
450mm diameter x 1050mm high...............	8.40	5.00	8.82	0.75	-	9.33	-	18.15	nr	19.51
450mm diameter x 1200mm high...............	10.00	5.00	10.50	0.75	-	9.33	-	19.83	nr	21.32
FUEL OIL STORAGE TANKS – Steel fuel oil storage tanks										
Oil storage tanks; carbon steel tank, B.S.799 Part 5 Type III, rectangular; 457mm diameter manhole cover, oil tight washer and screwed socket for fill, vent, sludge and draw off; painted one coat black bituminous paint										
14 gauge; 1520 x 610 x 1220mm; 1130 litre capacity	128.00	-	128.00	3.50	-	43.54	-	171.54	nr	184.41
14 gauge; 1830 x 610 x 1220mm; 1360 litre capacity	129.25	-	129.25	3.75	-	46.65	-	175.90	nr	189.09
12 gauge; 1830 x 1220 x 1220mm; 2730 litre capacity.................................	197.50	-	197.50	4.00	-	49.76	-	247.26	nr	265.80
1/8' plate; 2440 x 1520 x 1220mm; 4550 litre capacity.................................	500.50	-	500.50	4.50	-	55.98	-	556.48	nr	598.22

Mechanical heating/cooling/refrigeration systems

Labour hourly rates: (except Specialists) Craft Operatives £12.44 Labourer £7.02 Rates are national average prices. Refer to REGIONAL VARIATIONS for indicative levels of overall pricing in regions	MATERIALS			LABOUR				RATES		
	Del to Site £	Waste %	Material Cost £	Craft Optve Hrs	Lab Hrs	Labour Cost £	Sunds £	Nett Rate £	Unit	Gross Rate (+7.5%) £
T10: GAS/OIL FIRED BOILERS										
Gas flue pipes comprising galvanised steel outer skin and aluminium inner skin with air space between, B.S.715; socketed joints										
Pipes; straight										
100mm	11.79	2.00	12.03	0.67	–	8.33	–	20.36	m	21.89
extra; connections to appliance	2.93	2.00	2.99	0.50	–	6.22	–	9.21	nr	9.90
extra; adjustable pipes	6.90	2.00	7.04	0.50	–	6.22	–	13.26	nr	14.25
extra; terminals	8.99	2.00	9.17	0.50	–	6.22	–	15.39	nr	16.54
extra; bends, 90 degrees	7.88	2.00	8.04	0.50	–	6.22	–	14.26	nr	15.33
extra; bends, 45 degrees	7.88	2.00	8.04	0.50	–	6.22	–	14.26	nr	15.33
extra; tees	17.49	2.00	17.84	0.50	–	6.22	–	24.06	nr	25.86
125mm	14.13	2.00	14.41	0.80	–	9.95	–	24.36	m	26.19
extra; connections to appliance	3.27	2.00	3.34	0.67	–	8.33	–	11.67	nr	12.55
extra; adjustable pipes	7.75	2.00	7.91	0.67	–	8.33	–	16.24	nr	17.46
extra; terminals	9.87	2.00	10.07	0.67	–	8.33	–	18.40	nr	19.78
extra; bends, 90 degrees	9.31	2.00	9.50	0.67	–	8.33	–	17.83	nr	19.17
extra; bends, 45 degrees	9.31	2.00	9.50	0.67	–	8.33	–	17.83	nr	19.17
extra; tees	18.86	2.00	19.24	0.67	–	8.33	–	27.57	nr	29.64
150mm	15.77	2.00	16.09	1.00	–	12.44	–	28.53	m	30.66
extra; connections to appliance	3.73	2.00	3.80	1.00	–	12.44	–	16.24	nr	17.46
extra; adjustable pipes	9.78	2.00	9.98	1.00	–	12.44	–	22.42	nr	24.10
extra; terminals	12.67	2.00	12.92	1.00	–	12.44	–	25.36	nr	27.27
extra; bends, 90 degrees	11.66	2.00	11.89	1.00	–	12.44	–	24.33	nr	26.16
extra; bends, 45 degrees	11.66	2.00	11.89	1.00	–	12.44	–	24.33	nr	26.16
extra; tees	19.64	2.00	20.03	1.00	–	12.44	–	32.47	nr	34.91
Flue supports; galvanised steel wall bands fixing to masonry with screws										
for 100mm pipes	4.05	2.00	4.13	0.17	–	2.11	–	6.25	nr	6.71
for 125mm pipes	4.31	2.00	4.40	0.17	–	2.11	–	6.51	nr	7.00
for 150mm pipes	5.41	2.00	5.52	0.17	–	2.11	–	7.63	nr	8.21
Fire stop spacers										
for 100mm pipes	1.68	2.00	1.71	0.25	–	3.11	–	4.82	nr	5.19
for 125mm pipes	1.68	2.00	1.71	0.33	–	4.11	–	5.82	nr	6.26
for 150mm pipes	1.91	2.00	1.95	0.50	–	6.22	–	8.17	nr	8.78
Gas fired boilers										
Gas fired boilers for central heating and hot water; automatically controlled by thermostat with electrical control, gas governor and flame failure device										
free standing boiler; conventional flue; approximate output rating B Th U per hour										
40000	325.75	1.00	329.01	4.00	–	49.76	–	378.77	nr	407.18
50000	340.80	1.00	344.21	4.00	–	49.76	–	393.97	nr	423.52
60000	362.55	1.00	366.18	4.00	–	49.76	–	415.94	nr	447.13
80000	427.05	1.00	431.32	4.00	–	49.76	–	481.08	nr	517.16
100000	603.20	1.00	609.23	5.00	–	62.20	–	671.43	nr	721.79
free standing boiler; balanced flue; approximate output rating B Th U per hour										
40000	402.65	1.00	406.68	5.00	–	62.20	–	468.88	nr	504.04
50000	417.70	1.00	421.88	5.00	–	62.20	–	484.08	nr	520.38
60000	448.50	1.00	452.99	5.00	–	62.20	–	515.18	nr	553.82
80000	615.00	1.00	621.15	5.00	–	62.20	–	683.35	nr	734.60
100000	783.25	1.00	791.08	6.00	–	74.64	–	865.72	nr	930.65
wall mounted boiler; conventional flue; approximate output rating B Th U per hour										
40000	333.60	1.00	336.94	4.50	–	55.98	–	392.92	nr	422.38
50000	349.85	1.00	353.35	4.50	–	55.98	–	409.33	nr	440.03
60000	445.15	1.00	449.60	4.50	–	55.98	–	505.58	nr	543.50
wall mounted boiler; balanced flue; approximate output rating B Th U per hour										
40000	346.25	1.00	349.71	5.50	–	68.42	–	418.13	nr	449.49
50000	393.30	1.00	397.23	5.50	–	68.42	–	465.65	nr	500.58
60000	484.40	1.00	489.24	5.50	–	68.42	–	557.66	nr	599.49

Labour hourly rates: (except Specialists) Craft Operatives £12.44 Labourer £7.02 Rates are national average prices. Refer to REGIONAL VARIATIONS for indicative levels of overall pricing in regions	MATERIALS			LABOUR				RATES		
	Del to Site £	Waste %	Material Cost £	Craft Optve Hrs	Lab Hrs	Labour Cost £	Sunds £	Nett Rate £	Unit	Gross Rate (+7.5%) £
T10: GAS/OIL FIRED BOILERS Cont.										
Oil fired boilers										

Oil fired boilers for central heating and hot water;
automatically controlled by thermostat, with
electrical control box
  free standing boiler; conventional flue;
  approximate output rating BThU per hour										

40000 - 48000	667.70	1.00	674.38	4.00	-	49.76	-	724.14	nr	778.45
50000 - 65000	696.40	1.00	703.36	4.00	-	49.76	-	753.12	nr	809.61
70000 - 85000	794.35	1.00	802.29	4.00	-	49.76	-	852.05	nr	915.96
88000 - 110000	871.10	1.00	879.81	5.00	-	62.20	-	942.01	nr	1012.66
T11: COAL FIRED BOILERS										
Equipment										

Solid fuel boilers for central heating and hot water;
thermostatically controlled; with tools
  free standing boiler, gravity feed; approximate
  output rating B Th U per hour										

45000	1052.00	1.00	1062.52	4.00	-	49.76	-	1112.28	nr	1195.70
60000	1226.00	1.00	1238.26	4.00	-	49.76	-	1288.02	nr	1384.62
80000	1528.00	1.00	1543.28	5.00	-	62.20	-	1605.48	nr	1725.89
100000	1762.00	1.00	1779.62	6.00	-	74.64	-	1854.26	nr	1993.33
T31: LOW TEMPERATURE HOT WATER HEATING										
Accelerator pumps										
Variable head accelerator pumps for forced central heating; small bore indirect systems BSP connections; with valves										
20mm	37.70	2.00	38.45	0.75	-	9.33	-	47.78	nr	51.37
25mm	38.85	2.00	39.63	0.75	-	9.33	-	48.96	nr	52.63
Radiators										
Pressed steel radiators single panel with convector; air cock, plain plug 450mm high; fixing brackets to masonry with screws										
600mm long	15.05	1.00	15.20	3.00	-	37.32	-	52.52	nr	56.46
900mm long	22.45	1.00	22.67	3.00	-	37.32	-	59.99	nr	64.49
1200mm long	29.30	1.00	29.59	3.00	-	37.32	-	66.91	nr	71.93
1600mm long	37.95	1.00	38.33	3.00	-	37.32	-	75.65	nr	81.32
2400mm long	61.45	1.00	62.06	3.00	-	37.32	-	99.38	nr	106.84
600mm high; fixing brackets to masonry with screws										
600mm long	19.30	1.00	19.49	3.00	-	37.32	-	56.81	nr	61.07
900mm long	28.40	1.00	28.68	3.00	-	37.32	-	66.00	nr	70.95
1200mm long	37.20	1.00	37.57	3.00	-	37.32	-	74.89	nr	80.51
1600mm long	48.50	1.00	48.98	3.00	-	37.32	-	86.31	nr	92.78
2400mm long	79.00	1.00	79.79	3.00	-	37.32	-	117.11	nr	125.89
700mm high; fixing brackets to masonry with screws										
600mm long	23.00	1.00	23.23	3.00	-	37.32	-	60.55	nr	65.09
900mm long	33.40	1.00	33.73	3.00	-	37.32	-	71.05	nr	76.38
1200mm long	42.20	1.00	42.62	3.00	-	37.32	-	79.94	nr	85.94
1600mm long	62.00	1.00	62.62	3.00	-	37.32	-	99.94	nr	107.44
2400mm long	90.90	1.00	91.81	3.00	-	37.32	-	129.13	nr	138.81
Pressed steel radiators; double panel with single convector; air cock, plain plug 450mm high; fixing brackets to masonry with screws										
600mm long	25.35	1.00	25.60	3.00	-	37.32	-	62.92	nr	67.64
900mm long	37.80	1.00	38.18	3.00	-	37.32	-	75.50	nr	81.16
1200mm long	49.45	1.00	49.94	3.00	-	37.32	-	87.26	nr	93.81
1600mm long	64.85	1.00	65.50	3.00	-	37.32	-	102.82	nr	110.53
2400mm long	108.45	1.00	109.53	3.00	-	37.32	-	146.85	nr	157.87
600mm high; fixing brackets to masonry with screws										
600mm long	31.95	1.00	32.27	3.00	-	37.32	-	69.59	nr	74.81
900mm long	47.40	1.00	47.87	3.00	-	37.32	-	85.19	nr	91.58
1200mm long	62.00	1.00	62.62	3.00	-	37.32	-	99.94	nr	107.44
1600mm long	92.60	1.00	93.53	3.00	-	37.32	-	130.85	nr	140.66
2400mm long	136.20	1.00	137.56	3.00	-	37.32	-	174.88	nr	188.00
700mm high; fixing brackets to masonry with screws										
600mm long	36.25	1.00	36.61	3.00	-	37.32	-	73.93	nr	79.48
900mm long	53.00	1.00	53.53	3.00	-	37.32	-	90.85	nr	97.66
1200mm long	80.15	1.00	80.95	3.00	-	37.32	-	118.27	nr	127.14
1600mm long	105.05	1.00	106.10	3.00	-	37.32	-	143.42	nr	154.18
2400mm long	154.35	1.00	155.89	3.00	-	37.32	-	193.21	nr	207.70
Brass valves for radiators; self colour										
Valves for radiators, B.S.2767; wheel head inlet for copper; angle pattern										
15mm	4.62	2.00	4.71	0.25	-	3.11	-	7.82	nr	8.41
22mm	6.28	2.00	6.41	0.33	-	4.11	-	10.51	nr	11.30
inlet for copper; straight pattern										
15mm	5.18	2.00	5.29	0.25	-	3.11	-	8.40	nr	9.03
22mm	7.48	2.00	7.63	0.33	-	4.11	-	11.73	nr	12.61
Valves for radiators, B.S.2767; lock shield inlet for copper; angle pattern										
15mm	4.62	2.00	4.71	0.25	-	3.11	-	7.82	nr	8.41
22mm	6.28	2.00	6.41	0.33	-	4.11	-	10.51	nr	11.30

Labour hourly rates: (except Specialists) Craft Operatives £12.44 Labourer £7.02 Rates are national average prices. Refer to REGIONAL VARIATIONS for indicative levels of overall pricing in regions	MATERIALS			LABOUR				RATES		
	Del to Site £	Waste %	Material Cost £	Craft Optve Hrs	Lab Hrs	Labour Cost £	Sunds £	Nett Rate £	Unit	Gross Rate (+7.5%) £
T31: LOW TEMPERATURE HOT WATER HEATING Cont.										
Brass valves for radiators; self colour Cont.										
Valves for radiators, B.S.2767; lock shield Cont.										
inlet for copper; straight pattern										
15mm	5.18	2.00	5.28	0.25	–	3.11	–	8.39	nr	9.02
22mm	7.48	2.00	7.63	0.33	–	4.11	–	11.73	nr	12.61
Brass valves for radiators; chromium plated										
Valves for radiators, B.S.2767; wheel head										
inlet for copper; angle pattern										
15mm	6.28	2.00	6.41	0.25	–	3.11	–	9.52	nr	10.23
22mm	8.41	2.00	8.58	0.33	–	4.11	–	12.68	nr	13.63
inlet for copper; straight pattern										
15mm	7.48	2.00	7.63	0.25	–	3.11	–	10.74	nr	11.55
22mm	9.73	2.00	9.92	0.33	–	4.11	–	14.03	nr	15.08
Valves for radiators, B.S.2767; lock shield										
inlet for copper; angle pattern										
15mm	6.28	2.00	6.41	0.25	–	3.11	–	9.52	nr	10.23
22mm	8.41	2.00	8.58	0.33	–	4.11	–	12.68	nr	13.63
inlet for copper; straight pattern										
15mm	7.48	2.00	7.63	0.25	–	3.11	–	10.74	nr	11.55
22mm	9.73	2.00	9.92	0.33	–	4.11	–	14.03	nr	15.08
Brass thermostatic valves for radiators; chromium plated										
Thermostatic valves for radiators, one piece; ABS plastics head										
inlet for copper; angle pattern										
15mm	12.10	2.00	12.34	0.25	–	3.11	–	15.45	nr	16.61
inlet for copper; straight pattern										
15mm	12.10	2.00	12.34	0.25	–	3.11	–	15.45	nr	16.61
inlet for iron; angle pattern										
19mm	15.35	2.00	15.66	0.33	–	4.11	–	19.76	nr	21.24
Skirting heaters										
Pressed metal skirting heaters with fins on copper tube; straight										
900mm long	24.58	2.00	25.07	1.25	–	15.55	–	40.62	nr	43.67
1200mm long	32.55	2.00	33.20	1.67	–	20.77	–	53.98	nr	58.02
1500mm long	36.30	2.00	37.03	2.08	–	25.88	–	62.90	nr	67.62
1800mm long	40.05	2.00	40.85	2.50	–	31.10	–	71.95	nr	77.35
extra; internal corners	4.60	2.00	4.69	0.33	–	4.11	–	8.80	nr	9.46
extra; external corners	4.60	2.00	4.69	0.33	–	4.11	–	8.80	nr	9.46
extra; end stops	4.23	2.00	4.31	0.33	–	4.11	–	8.42	nr	9.05
extra; valve boxes	12.48	2.00	12.73	0.33	–	4.11	–	16.83	nr	18.10
T32: LOW TEMPERATURE HOT WATER HEATING (SMALL SCALE)										
Central heating installations - indicative prices										

Note
    the following are indicative prices for
    installation in two storey three bedroomed
    dwellings with a floor area of approximately 85m2
Installation; boiler, copper piping, pressed steel
radiators, heated towel rail; providing hot water to
sink , bath, lavatory basin; complete with all
necessary pumps, controls etc										

solid fuel fired	–	–	Spclist	–	–	Spclist	–	2950.00	it	3171.25
gas fired	–	–	Spclist	–	–	Spclist	–	2700.00	it	2902.50
oil fired, including oil storage tank	–	–	Spclist	–	–	Spclist	–	3350.00	it	3601.25

Faber and Kell's
Heating and Air Conditioning of Buildings
Eighth Edition

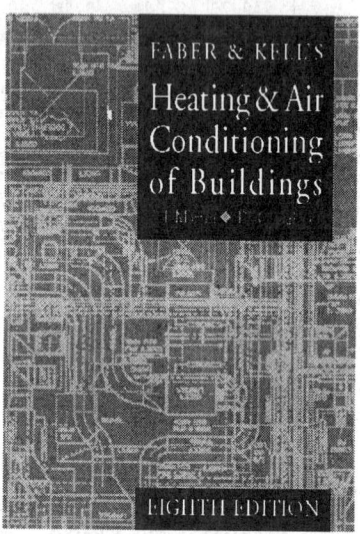

'Faber and Kells' has, for over fifty years, been accepted as the most practical and comprehensive book on heating and air conditioning design. It is regarded as the standard reference book for both students and practitioners.

In order to provide up-to-date information, this eight edition has been revised to include the latest changes to system design and covers many aspects in greater depth, whilst still retaining the character of previous editions.

Building services engineers, architects and others in the construction industry will find no better place for accessible and easily assimilated information on all aspects of the heating and air conditioning of buildings.

Paperback Edition, 1997, £35.00, isbn 0 7506 3778 1, 696pp

Ventilation/Air conditioning systems

Labour hourly rates: (except Specialists) Craft Operatives £14.38 Labourer £7.02 Rates are national average prices. Refer to REGIONAL VARIATIONS for indicative levels of overall pricing in regions	MATERIALS			LABOUR				RATES		
	Del to Site £	Waste %	Material Cost £	Craft Optve Hrs	Lab Hrs	Labour Cost £	Sunds £	Nett Rate £	Unit	Gross Rate (+7.5%) £
U10: GENERAL VENTILATION										
Ventilating fans										

Note
  the following prices exclude cutting holes in
  glass
Equipment										

Extract fans, window mounted; shutter and incorporated switch unit										
152mm diameter	193.71	2.50	**198.55**	1.25	–	**17.98**	–	**216.53**	nr	**232.77**
229mm diameter	301.55	2.50	**309.09**	1.50	–	**21.57**	–	**330.66**	nr	**355.46**
305mm diameter	385.42	2.50	**395.06**	1.50	–	**21.57**	–	**416.63**	nr	**447.87**

VISIT THE LAXTON'S WEBSITE TODAY!

www.laxtonsprices.co.uk

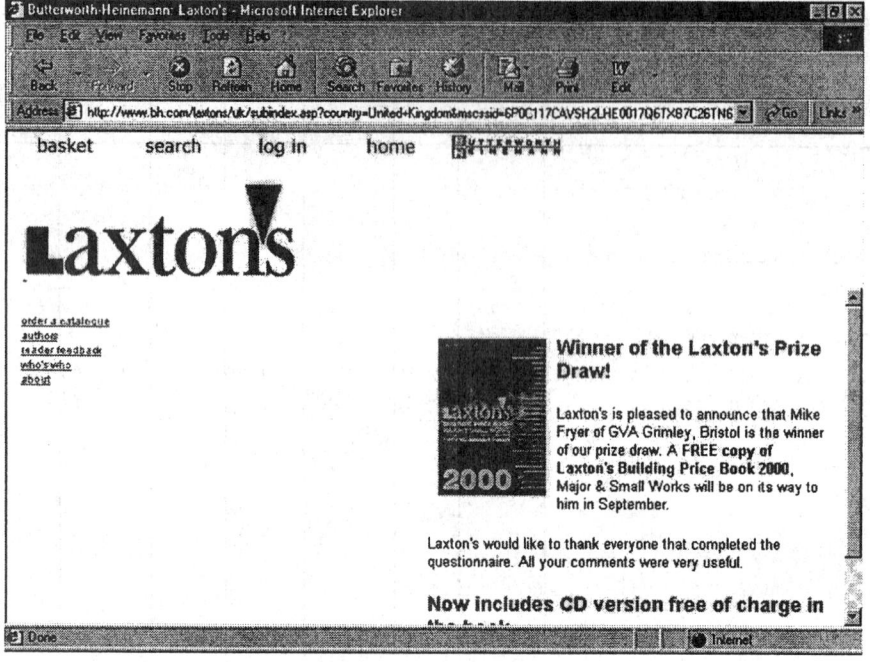

- FIND OUT MORE ABOUT BUTTERWORTH-HEINEMANN'S RANGE OF TITLES

- ORDER ON-LINE

- BENEFIT FROM SPECIAL OFFERS

- REGULAR EMAIL UPDATES

BOOKMARK THIS SITE TODAY!

Electrical supply/power/lighting systems

Labour hourly rates: (except Specialists) Craft Operatives £14.38 Labourer £7.02 Rates are national average prices. Refer to REGIONAL VARIATIONS for indicative levels of overall pricing in regions	MATERIALS			LABOUR				RATES		
	Del to Site £	Waste %	Material Cost £	Craft Optve Hrs	Lab Hrs	Labour Cost £	Sunds £	Nett Rate £	Unit	Gross Rate (+7.5%) £
V41: STREET/AREA/FLOOD LIGHTING										
Street lighting columns and set in ground; excavate hole, remove surplus spoil, filling with concrete mix ST4, working around base										
Aluminium; B.S.5649, nominal mounting height										
4.88m	234.82	-	234.82	4.75	9.50	135.00	21.29	391.11	nr	420.44
8.00m	578.41	-	578.41	5.50	11.00	156.31	33.12	767.84	nr	825.43
10.00m	734.98	-	734.98	6.25	12.50	177.63	72.16	984.76	nr	1058.62
12.00m	1030.33	-	1030.33	7.00	14.00	198.94	72.16	1301.43	nr	1399.04
Steel; B.S.5649, nominal mounting height										
4.00m	122.33	-	122.33	4.75	9.50	135.00	21.29	278.62	nr	299.51
5.00m	126.06	-	126.06	5.50	11.00	156.31	33.12	315.49	nr	339.15
6.00m	155.35	-	155.35	6.25	12.50	177.63	72.16	405.13	nr	435.52
8.00m	230.54	-	230.54	7.00	14.00	198.94	72.16	501.64	nr	539.26
GENERAL LIGHTING AND POWER - COMMERCIAL										
Steel trunking										
Straight lighting trunking; PVC lid										
50 x 50mm	5.06	5.00	5.31	0.40	-	5.75	0.50	11.57	m	12.43
extra; end cap	1.95	-	1.95	0.08	-	1.15	0.25	3.35	nr	3.60
extra; internal angle	9.58	-	9.58	0.25	-	3.60	0.25	13.43	nr	14.43
extra; external angle	9.58	-	9.58	0.25	-	3.60	0.25	13.43	nr	14.43
extra; tee	10.26	-	10.26	0.25	-	3.60	0.25	14.11	nr	15.16
extra; four way intersection	10.62	-	10.62	0.40	-	5.75	0.25	16.62	nr	17.87
Straight dado trunking, two compartment; steel lid										
150 x 30mm	18.12	5.00	19.03	0.45	-	6.47	0.50	26.00	m	27.95
extra; end cap	4.37	-	4.37	0.08	-	1.15	0.25	5.77	nr	6.20
extra; internal angle	20.24	-	20.24	0.35	-	5.03	0.25	25.52	nr	27.44
extra; external angle	20.24	-	20.24	0.35	-	5.03	0.25	25.52	nr	27.44
extra; single socket plate	7.96	-	7.96	0.16	-	2.30	0.25	10.51	nr	11.30
extra; twin socket plate	8.46	-	8.46	0.16	-	2.30	0.25	11.01	nr	11.84
Straight dado trunking, three compartment; steel lid										
200 x 38mm	23.94	5.00	25.14	0.50	-	7.19	0.75	33.08	m	35.56
extra; end cap	4.77	-	4.77	0.08	-	1.15	0.25	6.17	nr	6.63
extra; internal angle	24.97	-	24.97	0.40	-	5.75	0.25	30.97	nr	33.29
extra; external angle	24.97	-	24.97	0.40	-	5.75	0.25	30.97	nr	33.29
extra; single socket plate	8.22	-	8.22	0.16	-	2.30	0.25	10.77	nr	11.58
extra; twin socket plate	8.81	-	8.81	0.16	-	2.30	0.25	11.36	nr	12.21
Straight skirting trunking; two compartments										
200 x 38mm	23.90	5.00	25.09	0.50	-	7.19	0.75	33.03	m	35.51
extra; internal bend	23.61	-	23.61	0.40	-	5.75	0.25	29.61	nr	31.83
extra; external bend	23.61	-	23.61	0.40	-	5.75	0.25	29.61	nr	31.83
Straight underfloor trunking; two compartments										
150 x 25mm	16.26	5.00	17.07	0.70	-	10.07	-	27.14	m	29.17
extra; floor junction boxes; adjustable frame and trap	71.88	-	71.88	1.85	-	26.60	-	98.48	nr	105.87
extra; horizontal bend	15.88	-	15.88	0.50	-	7.19	-	23.07	nr	24.80
PVC trunking; white										
Straight mini-trunking; clip on lid										
16 x 16mm	1.78	5.00	1.87	0.20	-	2.88	0.50	5.25	m	5.64
extra; end cap	0.49	-	0.49	0.04	-	0.58	0.13	1.20	nr	1.28
extra; internal bend	0.50	-	0.50	0.08	-	1.15	0.13	1.78	nr	1.91
extra; external bend	0.50	-	0.50	0.08	-	1.15	0.13	1.78	nr	1.91
extra; flat bend	0.50	-	0.50	0.08	-	1.15	0.13	1.78	nr	1.91
extra; equal tee	0.83	-	0.83	0.17	-	2.44	0.13	3.40	nr	3.66
25 x 16mm	2.13	5.00	2.24	0.20	-	2.88	0.13	5.24	m	5.64
extra; end cap	0.50	-	0.50	0.04	-	0.58	0.13	1.21	nr	1.30
extra; internal bend	0.52	-	0.52	0.08	-	1.15	0.13	1.80	nr	1.94
extra; external bend	0.52	-	0.52	0.08	-	1.15	0.13	1.80	nr	1.94
extra; flat bend	0.52	-	0.52	0.08	-	1.15	0.13	1.80	nr	1.94

ELECTRICAL

Labour hourly rates: (except Specialists) Craft Operatives £14.38 Labourer £7.02 Rates are national average prices. Refer to REGIONAL VARIATIONS for indicative levels of overall pricing in regions	MATERIALS			LABOUR				RATES		
	Del to Site £	Waste %	Material Cost £	Craft Optve Hrs	Lab Hrs	Labour Cost £	Sunds £	Nett Rate £	Unit	Gross Rate (+7.5%) £
GENERAL LIGHTING AND POWER - COMMERCIAL Cont.										
PVC trunking; white Cont.										
Straight mini-trunking; clip on lid Cont.										
extra; equal tee	0.84	–	0.84	0.17	–	2.44	0.13	3.41	nr	3.67
40 x 16mm	2.64	5.00	2.77	0.25	–	3.60	0.13	6.50	m	6.98
extra; end cap	0.51	–	0.51	0.04	–	0.58	0.13	1.22	nr	1.31
extra; internal bend	0.61	–	0.61	0.08	–	1.15	0.13	1.89	nr	2.03
extra; external bend	0.61	–	0.61	0.08	–	1.15	0.13	1.89	nr	2.03
extra; flat bend	0.61	–	0.61	0.08	–	1.15	0.13	1.89	nr	2.03
extra; equal tee	0.94	–	0.94	0.17	–	2.44	0.13	3.51	nr	3.78
Straight dado trunking, three compartment; clip on lid										
145 x 40mm	20.84	5.00	21.88	0.40	–	5.75	0.50	28.13	m	30.24
extra; end caps	1.34	–	1.34	0.08	–	1.15	0.25	2.74	nr	2.95
extra; internal bend	4.76	–	4.76	0.30	–	4.31	0.25	9.32	nr	10.02
extra; external bend	4.76	–	4.76	0.30	–	4.31	0.25	9.32	nr	10.02
extra; tee	56.60	–	56.60	0.40	–	5.75	0.25	62.60	nr	67.30
extra; flat angle	41.69	–	41.69	0.30	–	4.31	0.25	46.25	nr	49.72
extra; mini trunking adaptor	1.30	–	1.30	0.20	–	2.88	0.25	4.43	nr	4.76
230 x 40mm	32.20	5.00	33.81	0.45	–	6.47	0.25	40.53	m	43.57
extra; end caps	4.27	–	4.27	0.08	–	1.15	0.25	5.67	nr	6.10
extra; internal bend	8.95	–	8.95	0.30	–	4.31	0.25	13.51	nr	14.53
extra; external bend	8.95	–	8.95	0.30	–	4.31	0.25	13.51	nr	14.53
extra; mini trunking adaptor	1.33	–	1.33	0.20	–	2.88	0.25	4.46	nr	4.79
Straight skirting trunking, three compartment; clip on lid										
100 x 40mm	9.68	5.00	10.16	0.35	–	5.03	0.50	15.70	m	16.87
extra; end cap	1.85	–	1.85	0.06	–	0.86	0.25	2.96	nr	3.19
extra; internal corner	2.38	–	2.38	0.25	–	3.60	0.25	6.22	nr	6.69
extra; external corner	2.38	–	2.38	0.25	–	3.60	0.25	6.22	nr	6.69
extra; mini trunking adaptor	1.67	–	1.67	0.16	–	2.30	0.25	4.22	nr	4.54
extra; socket adaptor, one gang	3.74	–	3.74	0.25	–	3.60	0.25	7.59	nr	8.15
extra; socket adaptor, two gang	5.37	–	5.37	0.25	–	3.60	0.25	9.21	nr	9.91
PVC heavy gauge conduit and fittings; push fit joints; spacer bar saddles at 600mm centres										
Conduits; straight										
20mm diameter; plugging to masonry; to surfaces ..	2.81	10.00	3.09	0.17	–	2.44	0.42	5.96	m	6.40
extra; small circular terminal boxes	4.32	2.50	4.43	0.10	–	1.44	0.27	6.14	nr	6.60
extra; small circular angle boxes	3.68	2.50	3.77	0.11	–	1.58	0.27	5.62	nr	6.05
extra; small circular three way boxes	4.26	2.50	4.37	0.13	–	1.87	0.27	6.51	nr	6.99
extra; small circular through way boxes	3.68	2.50	3.77	0.11	–	1.58	0.27	5.62	nr	6.05
25mm diameter; plugging to masonry; to surfaces ..	3.64	10.00	4.00	0.18	–	2.59	0.42	7.01	m	7.54
extra; small circular terminal boxes	5.38	2.50	5.51	0.10	–	1.44	0.27	7.22	nr	7.76
extra; small circular angle boxes	4.89	2.50	5.01	0.11	–	1.58	0.27	6.86	nr	7.38
extra; small circular three way boxes	5.69	2.50	5.83	0.13	–	1.87	0.27	7.97	nr	8.57
extra; small circular through way boxes	4.89	2.50	5.01	0.11	–	1.58	0.27	6.86	nr	7.38
20mm diameter; plugging to masonry; in chases	2.81	10.00	3.09	0.15	–	2.16	0.42	5.67	m	6.09
extra; small circular terminal boxes	4.32	2.50	4.43	0.10	–	1.44	0.27	6.14	nr	6.60
extra; small circular angle boxes	3.68	2.50	3.77	0.11	–	1.58	0.27	5.62	nr	6.05
extra; small circular three way boxes	4.26	2.50	4.37	0.13	–	1.87	0.27	6.51	nr	6.99
extra; small circular through way boxes	3.68	2.50	3.77	0.11	–	1.58	0.27	5.62	nr	6.05
25mm diameter; plugging to masonry; in chases	3.64	10.00	4.00	0.16	–	2.30	0.42	6.72	m	7.23
extra; small circular terminal boxes	5.38	2.50	5.51	0.10	–	1.44	0.27	7.22	nr	7.76
extra; small circular angle boxes	4.89	2.50	5.01	0.11	–	1.58	0.27	6.86	nr	7.38
extra; small circular three way boxes	5.69	2.50	5.83	0.13	–	1.87	0.27	7.97	nr	8.57
extra; small circular through way boxes	4.89	2.50	5.01	0.11	–	1.58	0.27	6.86	nr	7.38
Oval conduits; straight										
20mm (nominal size); plugging to masonry; in chases	2.10	10.00	2.31	0.15	–	2.16	0.42	4.89	m	5.25
25mm (nominal size); plugging to masonry in chases	2.73	10.00	3.00	0.16	–	2.30	0.42	5.72	m	6.15
20mm (nominal size); fixing to timber	2.10	10.00	2.31	0.13	–	1.87	0.21	4.39	m	4.72
25mm (nominal size); fixing to timber	2.73	10.00	3.00	0.15	–	2.16	0.21	5.37	m	5.77
Steel welded heavy gauge conduits and fittings, screwed joints; spacer bar saddles at 1000mm centres; enamelled black inside and outside by manufacturer										
Conduits; straight										
20mm diameter; plugging to masonry; to surfaces ..	4.25	10.00	4.67	0.33	–	4.75	0.25	9.67	m	10.40
extra; small circular terminal boxes	5.71	2.50	5.85	0.13	–	1.87	0.27	7.99	nr	8.59
extra; small circular angle boxes	6.90	2.50	7.07	0.16	–	2.30	0.27	9.64	nr	10.37
extra; small circular three way boxes	7.83	2.50	8.03	0.20	–	2.88	0.27	11.17	nr	12.01
extra; small circular through way boxes	6.90	2.50	7.07	0.16	–	2.30	0.27	9.64	nr	10.37
25mm diameter; plugging to masonry; to surfaces ..	5.67	10.00	6.24	0.42	–	6.04	0.25	12.53	m	13.47
extra; small circular terminal boxes	7.15	2.50	7.33	0.13	–	1.87	0.27	9.47	nr	10.18
extra; small circular angle boxes	8.68	2.50	8.90	0.16	–	2.30	0.27	11.47	nr	12.33
extra; small circular three way boxes	9.61	2.50	9.85	0.20	–	2.88	0.27	13.00	nr	13.97
extra; small circular through way boxes	8.68	2.50	8.90	0.16	–	2.30	0.27	11.47	nr	12.33
32mm diameter; plugging to masonry; to surfaces ..	7.33	10.00	8.06	0.50	–	7.19	0.25	15.50	m	16.67
Steel welded heavy gauge conduits and fittings, screwed joints; spacer bar saddles at 1000mm centres; galvanised inside and outside by manufacturer										
Conduits; straight										
20mm diameter; plugging to masonry; to surfaces ..	5.50	10.00	6.05	0.33	–	4.75	0.25	11.05	m	11.87
extra; small circular terminal boxes	6.85	2.50	7.02	0.13	–	1.87	0.27	9.16	nr	9.85
extra; small circular angle boxes	8.28	2.50	8.49	0.16	–	2.30	0.27	11.06	nr	11.89
extra; small circular three way boxes	9.40	2.50	9.63	0.20	–	2.88	0.27	12.78	nr	13.74

Labour hourly rates: (except Specialists) Craft Operatives £14.38 Labourer £7.02 Rates are national average prices. Refer to REGIONAL VARIATIONS for indicative levels of overall pricing in regions	MATERIALS			LABOUR				RATES		
	Del to Site	Waste	Material Cost	Craft Optve	Lab	Labour Cost	Sunds	Nett Rate	Unit	Gross Rate (+7.5%)
	£	%	£	Hrs	Hrs	£	£	£		£

GENERAL LIGHTING AND POWER – COMMERCIAL Cont.

Steel welded heavy gauge conduits and fittings, screwed joints; spacer bar saddles at 1000mm centres; galvanised inside and outside by manufacturer Cont.

Conduits; straight Cont.

extra; small circular through way boxes	8.28	2.50	8.49	0.16	–	2.30	0.27	11.06	nr	11.89
25mm diameter; plugging to masonry; to surfaces ..	7.27	10.00	8.00	0.42	–	6.04	0.25	14.29	m	15.36
extra; small circular terminal boxes	8.60	2.50	8.81	0.13	–	1.87	0.27	10.95	nr	11.78
extra; small circular angle boxes	10.42	2.50	10.68	0.16	–	2.30	0.27	13.25	nr	14.25
extra; small circular three way boxes	11.54	2.50	11.83	0.20	–	2.88	0.27	14.97	nr	16.10
extra; small circular through way boxes	10.42	2.50	10.68	0.16	–	2.30	0.27	13.25	nr	14.25
32mm diameter; plugging to masonry; to surfaces ..	9.93	10.00	10.92	0.50	–	7.19	0.25	18.36	m	19.74

Galvanised standard cable tray and fittings

Light duty; straight

50mm tray	2.46	5.00	2.58	0.25	–	3.60	0.09	6.27	m	6.74
extra; flat bend	6.28	–	6.28	0.20	–	2.88	0.25	9.41	nr	10.11
extra; tee	8.15	–	8.15	0.25	–	3.60	0.25	11.99	nr	12.89
extra; cross piece	13.75	–	13.75	0.40	–	5.75	0.50	20.00	nr	21.50
extra; riser bend	9.80	–	9.80	0.20	–	2.88	0.25	12.93	nr	13.90
100mm tray	3.68	5.00	3.86	0.33	–	4.75	0.09	8.70	m	9.35
extra; flat bend	6.78	–	6.78	0.23	–	3.31	0.25	10.34	nr	11.11
extra; tee	8.96	–	8.96	0.30	–	4.31	0.25	13.52	nr	14.54
extra; cross piece	14.65	–	14.65	0.41	–	5.90	0.50	21.05	nr	22.62
extra; riser bend	11.30	–	11.30	0.23	–	3.31	0.25	14.86	nr	15.97
extra; reducer 100 to 50mm	8.59	–	8.59	0.23	–	3.31	0.25	12.15	nr	13.06
150mm tray	4.81	5.00	5.05	0.42	–	6.04	0.09	11.18	m	12.02
extra; flat bend	7.92	–	7.92	0.25	–	3.60	0.25	11.77	nr	12.65
extra; tee	10.87	–	10.87	0.33	–	4.75	0.25	15.87	nr	17.06
extra; cross piece	17.20	–	17.20	0.42	–	6.04	0.50	23.74	nr	25.52
extra; riser bend	15.25	–	15.25	0.25	–	3.60	0.25	19.09	nr	20.53
extra; reducer 150 to 100mm	11.32	–	11.32	0.25	–	3.60	0.25	15.16	nr	16.30
300mm tray	11.46	5.00	12.03	0.50	–	7.19	0.09	19.31	m	20.76
extra; flat bend	14.60	–	14.60	0.33	–	4.75	0.25	19.60	nr	21.07
extra; tee	20.01	–	20.01	0.50	–	7.19	0.25	27.45	nr	29.51
extra; cross piece	30.75	–	30.75	0.58	–	8.34	0.50	39.59	nr	42.56
extra; riser bend	26.85	–	26.85	0.33	–	4.75	0.25	31.85	nr	34.23
extra; reducer 300 to 150mm	20.59	–	20.59	0.33	–	4.75	0.25	25.59	nr	27.50
450mm tray	23.15	5.00	24.31	0.58	–	8.34	0.09	32.74	m	35.19
extra; flat bend	23.37	–	23.37	0.50	–	7.19	0.25	30.81	nr	33.12
extra; tee	32.64	–	32.64	0.75	–	10.79	0.25	43.67	nr	46.95
extra; cross piece	50.75	–	50.75	1.00	–	14.38	0.50	65.63	nr	70.55
extra; riser bend	38.03	–	38.03	0.50	–	7.19	0.25	45.47	nr	48.88
extra; reducer 450 to 300mm	34.18	–	34.18	0.50	–	7.19	0.25	41.62	nr	44.74

Medium duty; straight

75mm tray	4.78	5.00	5.02	0.26	–	3.74	0.09	8.85	m	9.51
extra; flat bend	23.65	–	23.65	0.21	–	3.02	0.25	26.92	nr	28.94
extra; tee	31.15	–	31.15	0.26	–	3.74	0.25	35.14	nr	37.77
extra; cross piece	49.43	–	49.43	0.42	–	6.04	0.50	55.97	nr	60.17
extra; riser bend	12.85	–	12.85	0.21	–	3.02	0.25	16.12	nr	17.33
100mm tray	4.85	5.00	5.09	0.34	–	4.89	0.09	10.07	m	10.83
extra; flat bend	23.65	–	23.65	0.24	–	3.45	0.25	27.35	nr	29.40
extra; tee	31.15	–	31.15	0.31	–	4.46	0.25	35.86	nr	38.55
extra; cross piece	48.08	–	48.08	0.43	–	6.18	0.50	54.76	nr	58.87
extra; riser bend	11.76	–	11.76	0.24	–	3.45	0.25	15.46	nr	16.62
extra; reducer 100 to 75mm	10.91	–	10.91	0.24	–	3.45	0.25	14.61	nr	15.71
150mm tray	5.81	5.00	6.10	0.44	–	6.33	0.09	12.52	m	13.46
extra; flat bend	26.65	–	26.65	0.26	–	3.74	0.25	30.64	nr	32.94
extra; tee	35.43	–	35.43	0.34	–	4.89	0.25	40.57	nr	43.61
extra; cross piece	53.32	–	53.32	0.44	–	6.33	0.50	60.15	nr	64.66
extra; riser bend	15.42	–	15.42	0.26	–	3.74	0.25	19.41	nr	20.86
extra; reducer 150 to 100mm	9.47	–	9.47	0.26	–	3.74	0.25	13.46	nr	14.47
300mm tray	12.15	5.00	12.76	0.52	–	7.48	0.09	20.33	m	21.85
extra; flat bend	37.92	–	37.92	0.34	–	4.89	0.25	43.06	nr	46.29
extra; tee	47.02	–	47.02	0.52	–	7.48	0.25	54.75	nr	58.85
extra; cross piece	72.38	–	72.38	0.61	–	8.77	0.50	81.65	nr	87.78
extra; riser bend	25.42	–	25.42	0.34	–	4.89	0.25	30.56	nr	32.85
extra; reducer 300 to 150mm	20.80	–	20.80	0.34	–	4.89	0.25	25.94	nr	27.88
450mm tray	17.67	5.00	18.55	0.61	–	8.77	0.09	27.42	m	29.47
extra; flat bend	50.17	–	50.17	0.52	–	7.48	0.25	57.90	nr	62.24
extra; tee	65.73	–	65.73	0.78	–	11.22	0.25	77.20	nr	82.99
extra; cross piece	98.51	–	98.51	1.05	–	15.10	0.50	114.11	nr	122.67
extra; riser bend	38.35	–	38.35	0.52	–	7.48	0.25	46.08	nr	49.53
extra; reducer 450 to 300mm	34.31	–	34.31	0.52	–	7.48	0.25	42.04	nr	45.19

Heavy duty; straight

75mm tray	8.72	5.00	9.16	0.27	–	3.88	0.09	13.13	m	14.11
extra; flat bend	34.76	–	34.76	0.22	–	3.16	0.25	38.17	nr	41.04
extra; tee	42.73	–	42.73	0.27	–	3.88	0.25	46.86	nr	50.38
extra; cross piece	69.03	–	69.03	0.44	–	6.33	0.50	75.86	nr	81.55
extra; riser bend	23.89	–	23.89	0.22	–	3.16	0.25	27.30	nr	29.35
100mm tray	8.86	5.00	9.30	0.36	–	5.18	0.09	14.57	m	15.66
extra; flat bend	35.22	–	35.22	0.25	–	3.60	0.25	39.06	nr	41.99
extra; tee	43.10	–	43.10	0.33	–	4.75	0.25	48.10	nr	51.70
extra; cross piece	69.58	–	69.58	0.45	–	6.47	0.50	76.55	nr	82.29
extra; riser bend	24.48	–	24.48	0.25	–	3.60	0.25	28.32	nr	30.45
extra; reducer 100 to 75mm	15.93	–	15.93	0.25	–	3.60	0.25	19.77	nr	21.26
150mm tray	9.81	5.00	10.30	0.46	–	6.61	0.09	17.01	m	18.28
extra; flat bend	38.22	–	38.22	0.27	–	3.88	0.25	42.35	nr	45.53
extra; tee	47.38	–	47.38	0.36	–	5.18	0.25	52.81	nr	56.77
extra; cross piece	75.36	–	75.36	0.46	–	6.61	0.50	82.47	nr	88.66
extra; riser bend	26.59	–	26.59	0.27	–	3.88	0.25	30.72	nr	33.03

ELECTRICAL

Labour hourly rates: (except Specialists) Craft Operatives £14.38 Labourer £7.02 Rates are national average prices. Refer to REGIONAL VARIATIONS for indicative levels of overall pricing in regions	MATERIALS			LABOUR				RATES		
	Del to Site £	Waste %	Material Cost £	Craft Optve Hrs	Lab Hrs	Labour Cost £	Sunds £	Nett Rate £	Unit	Gross Rate (+7.5%) £
GENERAL LIGHTING AND POWER - COMMERCIAL Cont.										
Galvanised standard cable tray and fittings Cont.										
Heavy duty; straight Cont.										
extra; reducer 150 to 100mm	15.98	-	15.98	0.27	-	3.88	0.25	20.11	nr	21.62
300mm tray..	14.28	5.00	14.99	0.55	-	7.91	0.09	22.99	m	24.72
extra; flat bend	43.87	-	43.87	0.36	-	5.18	0.25	49.30	nr	52.99
extra; tee	57.25	-	57.25	0.55	-	7.91	0.25	65.41	nr	70.31
extra; cross piece	90.61	-	90.61	0.64	-	9.20	0.50	100.31	nr	107.84
extra; riser bend	29.21	-	29.21	0.36	-	5.18	0.25	34.64	nr	37.23
extra; reducer 300 to 150mm	26.21	-	26.21	0.36	-	5.18	0.25	31.64	nr	34.01
450mm tray..	24.69	5.00	25.92	0.64	-	9.20	0.09	35.22	m	37.86
extra; flat bend	74.36	-	74.36	0.55	-	7.91	0.25	82.52	nr	88.71
extra; tee	87.15	-	87.15	0.83	-	11.94	0.50	99.59	nr	107.05
extra; cross piece	139.90	-	139.90	1.10	-	15.82	0.25	155.97	nr	167.67
extra; riser bend	51.31	-	51.31	0.55	-	7.91	0.25	59.47	nr	63.93
extra; reducer 450 to 300mm	40.98	-	40.98	0.55	-	7.91	0.25	49.14	nr	52.82
Supports for cable tray										
Cantilever arms; mild steel; hot dip galvanised; fixing to masonry										
65mm wide	3.96	2.50	4.06	0.33	-	4.75	0.25	9.05	nr	9.73
90mm wide	3.96	2.50	4.06	0.33	-	4.75	0.25	9.05	nr	9.73
120mm wide	3.96	2.50	4.06	0.33	-	4.75	0.25	9.05	nr	9.73
Stand-off bracket; mild steel; hot dip galvanised; fixing to masonry										
75mm wide	3.96	2.50	4.06	0.33	-	4.75	0.25	9.05	nr	9.73
100mm wide	5.10	2.50	5.23	0.33	-	4.75	0.25	10.22	nr	10.99
150mm wide	5.10	2.50	5.23	0.33	-	4.75	0.25	10.22	nr	10.99
300mm wide	11.24	2.50	11.52	0.42	-	6.04	0.25	17.81	nr	19.15
450mm wide	15.39	2.50	15.77	0.42	-	6.04	0.25	22.06	nr	23.72
PVC insulated cables; single core; reference 6491X; stranded copper conductors; to B.S.6004										
Drawn into conduits or ducts or laid or drawn into trunking										
1.5mm2	0.48	15.00	0.55	0.03	-	0.43	-	0.98	m	1.06
2.5mm2	0.71	15.00	0.82	0.03	-	0.43	-	1.25	m	1.34
4.0mm2	1.16	15.00	1.33	0.04	-	0.58	-	1.91	m	2.05
6.0mm2	1.69	15.00	1.94	0.05	-	0.72	-	2.66	m	2.86
10.0mm2	3.16	15.00	3.63	0.05	-	0.72	-	4.35	m	4.68
PVC insulated and PVC sheathed cables; multicore; copper conductors and bare earth continuity conductor 6242Y; to B.S.6004										
Drawn into conduits or ducts or laid or drawn into trunking										
1.00mm2; twin with bare earth	0.81	15.00	0.93	0.05	-	0.72	-	1.65	m	1.77
1.50mm2; twin with bare earth	1.01	15.00	1.16	0.05	-	0.72	-	1.88	m	2.02
2.5mm2; twin with bare earth	1.38	15.00	1.59	0.05	-	0.72	-	2.31	m	2.48
4.00mm2; twin with bare earth	4.27	15.00	4.91	0.06	-	0.86	-	5.77	m	6.21
6.00mm2; twin with bare earth	5.07	15.00	5.83	0.07	-	1.01	-	6.84	m	7.35
Fixed to timber with clips										
1.00mm2; twin with bare earth	0.81	15.00	0.93	0.07	-	1.01	0.06	2.00	m	2.15
1.50mm2; twin with bare earth	1.01	15.00	1.16	0.07	-	1.01	0.06	2.23	m	2.40
2.5mm2; twin with bare earth	1.38	15.00	1.59	0.07	-	1.01	0.07	2.66	m	2.86
4.00mm2; twin with bare earth	4.27	15.00	4.91	0.08	-	1.15	0.08	6.14	m	6.60
6.00mm2; twin with bare earth	5.07	15.00	5.83	0.09	-	1.29	0.08	7.20	m	7.75
PVC insulated SWA armoured and PVC sheathed cables; multicore; copper conductors to B.S.6346										
Laid and laced on cable tray										
16mm2; three core	10.83	5.00	11.37	0.17	-	2.44	0.11	13.93	m	14.97
Fixed to masonry with screwed clips at average 300mm centres; plugging										
16mm2; three core	10.83	5.00	11.37	0.25	-	3.60	0.94	15.91	m	17.10
Mineral insulated copper sheathed cables; PVC outer sheath; copper conductors; 600V light duty to B.S.6207										
Fixed to masonry with screwed clips at average 300mm centres; plugging										
1.5mm2; two core	1.97	10.00	2.17	0.17	-	2.44	0.94	5.55	m	5.97
extra; termination including gland, seal and shroud..	2.52	2.50	2.58	0.42	-	6.04	-	8.62	nr	9.27
2.5mm2; two core	2.89	10.00	3.18	0.17	-	2.44	0.94	6.56	m	7.06
extra; termination including gland, seal and shroud..	2.52	2.50	2.58	0.42	-	6.04	-	8.62	nr	9.27
4.0mm2; two core	3.66	10.00	4.03	0.20	-	2.88	0.94	7.84	m	8.43
extra; termination including gland, seal and shroud..	2.52	2.50	2.58	0.50	-	7.19	-	9.77	nr	10.51

Labour hourly rates: (except Specialists) Craft Operatives £14.38 Labourer £7.02 Rates are national average prices. Refer to REGIONAL VARIATIONS for indicative levels of overall pricing in regions	MATERIALS			LABOUR				RATES		
	Del to Site	Waste	Material Cost	Craft Optve	Lab	Labour Cost	Sunds	Nett Rate	Unit	Gross Rate (+7.5%)
	£	%	£	Hrs	Hrs	£	£	£		£
GENERAL LIGHTING AND POWER - COMMERCIAL Cont.										
HV switchgear										
Switch fuse 500V, metal cased, fixed to masonry with screws; plugging										
32 amp; single pole and neutral; short circuit rating 6.4Ka	72.94	2.50	**74.76**	1.60	–	**23.01**	0.50	**98.27**	nr	105.64
Consumer control units, 250V, metal cased, fitted with 100 amp D.P. isolator and MCB's , fixed to masonry with screws; plugging										
eight way, single pole and neutral	83.74	2.50	**85.83**	1.60	–	**23.01**	0.50	**109.34**	nr	117.54
Distribution boards 500V, metal cased, MCB pattern, fitted with 100 amp D.P. isolator, fixed to masonry with screws; plugging										
nine way, single pole and neutral, 20 amp	104.08	2.50	**106.68**	1.80	–	**25.88**	0.50	**133.07**	nr	143.05
twelve way, triple pole and neutral, 32 amp ...	462.42	2.50	**473.98**	8.00	–	**115.04**	0.50	**589.52**	nr	633.73
Luminaires										
Note for lighting fittings and lamps see General Lighting and Power - Domestic Section										
Accessories; white plastics, with boxes, fixing to masonry with screws; plugging										
Flush plate switches										
5 amp; one gang; one way; single pole	3.21	2.50	**3.29**	0.42	–	**6.04**	0.25	**9.58**	nr	10.30
5 amp; two gang; one way; single pole	4.65	2.50	**4.77**	0.43	–	**6.18**	0.25	**11.20**	nr	12.04
Surface plate switches										
5 amp; one gang; one way; single pole	3.01	2.50	**3.09**	0.42	–	**6.04**	0.25	**9.37**	nr	10.08
5 amp; two gang; one way; single pole	4.44	2.50	**4.55**	0.43	–	**6.18**	0.25	**10.98**	nr	11.81
Ceiling switches										
5 amp; one gang; one way; single pole	3.57	2.50	**3.66**	0.43	–	**6.18**	0.25	**10.09**	nr	10.85
Flush switched socket outlets										
13 amp; one gang	5.77	2.50	**5.91**	0.43	–	**6.18**	0.25	**12.35**	nr	13.27
13 amp; two gang	10.79	2.50	**11.06**	0.43	–	**6.18**	0.25	**17.49**	nr	18.81
Surface switched socket outlets										
13 amp; one gang	5.49	2.50	**5.63**	0.43	–	**6.18**	0.25	**12.06**	nr	12.97
13 amp; two gang	10.49	2.50	**10.75**	0.43	–	**6.18**	0.25	**17.19**	nr	18.47
Flush switched socket outlets with RCCB (residual current device) protected at 30 mAmp										
13 amp; one gang	60.67	2.50	**62.19**	0.43	–	**6.18**	0.25	**68.62**	nr	73.77
Flush fused connection units										
13 amp; one gang; switched; flexible outlet	9.13	2.50	**9.36**	0.50	–	**7.19**	0.25	**16.80**	nr	18.06
13 amp; one gang; switched; flexible outlet; pilot lamp	12.31	2.50	**12.62**	0.50	–	**7.19**	0.25	**20.06**	nr	21.56
Accessories, metalclad, with boxes, fixing to masonry with screws; plugging										
Surface plate switches										
5 amp; one gang; two way; single pole	4.32	2.50	**4.43**	0.42	–	**6.04**	0.25	**10.72**	nr	11.52
5 amp; two gang; two way; single pole	5.25	2.50	**5.38**	0.43	–	**6.18**	0.25	**11.81**	nr	12.70
Surface switched socket outlets										
13 amp; one gang	8.43	2.50	**8.64**	0.43	–	**6.18**	0.25	**15.07**	nr	16.20
13 amp; two gang	16.66	2.50	**17.08**	0.43	–	**6.18**	0.25	**23.51**	nr	25.27
Weatherproof accessories, with boxes, fixing to masonry with screws; plugging										
Surface plate switches										
5 amp; one gang; two way; single pole	5.72	2.50	**5.86**	0.43	–	**6.18**	0.25	**12.30**	nr	13.22
GENERAL LIGHTING AND POWER - DOMESTIC										
Electric wiring										
Note the following approximate prices of various types of installations are dependent on the number and disposition of points; lamps and fittings together with cutting and making good are excluded										
Electric wiring in new building										
Installation with PVC insulated and sheathed cables										
lighting points	16.37	7.50	**17.60**	3.25	–	**46.73**	1.65	**65.98**	nr	70.93
socket outlets; 5A	17.65	7.50	**18.97**	4.40	–	**63.27**	1.27	**83.52**	nr	89.78
socket outlets; 13A ring main	21.18	7.50	**22.77**	3.70	–	**53.21**	1.15	**77.12**	nr	82.91
socket outlets; 13A radial circuit	24.04	7.50	**25.84**	3.85	–	**55.36**	1.27	**82.48**	nr	88.66
cooker points; 45A	98.64	7.50	**106.04**	6.30	–	**90.59**	1.96	**198.59**	nr	213.49
immersion heater points	29.80	7.50	**32.03**	4.60	–	**66.15**	1.40	**99.58**	nr	107.05
shaver sockets (transformer)	51.91	7.50	**55.80**	1.65	–	**23.73**	1.40	**80.93**	nr	87.00
Installation with mineral insulated copper sheathed cables										
lighting points	26.69	7.50	**28.69**	6.20	–	**89.16**	0.82	**118.67**	nr	127.57
socket outlets; 5A	37.80	7.50	**40.63**	4.85	–	**69.74**	0.75	**111.13**	nr	119.46

ELECTRICAL

Labour hourly rates: (except Specialists) Craft Operatives £14.38 Labourer £7.02 Rates are national average prices. Refer to REGIONAL VARIATIONS for indicative levels of overall pricing in regions	MATERIALS			LABOUR				RATES		
	Del to Site £	Waste %	Material Cost £	Craft Optve Hrs	Lab Hrs	Labour Cost £	Sunds £	Nett Rate £	Unit	Gross Rate (+7.5%) £
GENERAL LIGHTING AND POWER - DOMESTIC Cont.										
Electric wiring in new building Cont.										
Installation with mineral insulated copper sheathed cables Cont.										
socket outlets; 13A ring main	33.17	7.50	35.66	4.15	–	59.68	0.63	95.96	nr	103.16
socket outlets; 13A radial circuit	39.07	7.50	42.00	5.55	–	79.81	0.75	122.56	nr	131.75
cooker points; 45A	120.11	7.50	129.12	9.20	–	132.30	1.44	262.85	nr	282.57
immersion heater points	51.50	7.50	55.36	6.60	–	94.91	0.88	151.15	nr	162.49
shaver sockets (transformer)	73.40	7.50	78.91	3.00	–	43.14	0.88	122.93	nr	132.14
Installation with black enamel heavy gauge conduit with PVC cables										
lighting points	53.10	7.50	57.08	4.65	–	66.87	2.51	126.46	nr	135.94
socket outlets; 5A	65.02	7.50	69.90	4.70	–	67.59	2.76	140.24	nr	150.76
socket outlets; 13A ring main	56.31	7.50	60.53	4.25	–	61.12	2.26	123.91	nr	133.20
socket outlets; 13A radial circuit	66.59	7.50	71.58	6.00	–	86.28	2.76	160.62	nr	172.67
cooker points; 45A	153.87	7.50	165.41	9.65	–	138.77	3.26	307.44	nr	330.50
immersion heater points	74.76	7.50	80.37	7.15	–	102.82	2.76	185.94	nr	199.89
shaver sockets (transformer)	94.72	7.50	101.82	2.30	–	33.07	2.51	137.41	nr	147.71
Installation with black enamel heavy gauge conduit with coaxial cable										
T.V. sockets	66.04	7.50	70.99	5.75	–	82.69	2.76	156.44	nr	168.17
Installation with black enamel heavy gauge conduit with draw wire										
telephone points	53.47	7.50	57.48	5.45	–	78.37	2.76	138.61	nr	149.01
Electric wiring in extension to existing building										
Installation with PVC insulated and sheathed cables										
lighting points	16.37	7.50	17.60	3.25	–	46.73	1.65	65.98	nr	70.93
socket outlets; 5A	18.83	7.50	20.24	4.40	–	63.27	1.27	84.78	nr	91.14
socket outlets; 13A ring main	21.18	7.50	22.77	3.70	–	53.21	1.15	77.12	nr	82.91
socket outlets; 13A radial circuit	24.04	7.50	25.84	3.85	–	55.36	1.27	82.48	nr	88.66
cooker points; 45A	98.64	7.50	106.04	6.30	–	90.59	1.96	198.59	nr	213.49
immersion heater points	29.80	7.50	32.03	4.60	–	66.15	1.40	99.58	nr	107.05
shaver sockets (transformer)	51.91	7.50	55.80	1.65	–	23.73	1.40	80.93	nr	87.00
Installation with mineral insulated copper sheathed cables										
lighting points	26.69	7.50	28.69	6.20	–	89.16	0.82	118.67	nr	127.57
socket outlets; 5A	37.80	7.50	40.63	4.85	–	69.74	0.75	111.13	nr	119.46
socket outlets; 13A ring main	33.17	7.50	35.66	4.15	–	59.68	0.63	95.96	nr	103.16
socket outlets; 13A radial circuit	39.07	7.50	42.00	5.55	–	79.81	0.75	122.56	nr	131.75
cooker points; 45A	120.11	7.50	129.12	9.20	–	132.30	1.44	262.85	nr	282.57
immersion heater points	51.50	7.50	55.36	6.60	–	94.91	0.88	151.15	nr	162.49
shaver sockets (transformer)	73.40	7.50	78.91	3.00	–	43.14	0.88	122.93	nr	132.14
Installation with black enamel heavy gauge conduit with PVC cables										
lighting points	53.10	7.50	57.08	4.65	–	66.87	2.51	126.46	nr	135.94
socket outlets; 5A	65.02	7.50	69.90	4.70	–	67.59	2.76	140.24	nr	150.76
socket outlets; 13A ring main	56.21	7.50	60.43	4.25	–	61.12	2.26	123.80	nr	133.09
socket outlets; 13A radial circuit	66.59	7.50	71.58	6.00	–	86.28	2.76	160.62	nr	172.67
cooker points; 45A	153.87	7.50	165.41	9.65	–	138.77	3.26	307.44	nr	330.50
immersion heater points	74.76	7.50	80.37	7.15	–	102.82	2.76	185.94	nr	199.89
shaver sockets (transformer)	94.72	7.50	101.82	2.30	–	33.07	2.51	137.41	nr	147.71
Installation with black enamel heavy gauge conduit with coaxial cable										
T.V. sockets	66.04	7.50	70.99	5.75	–	82.69	2.76	156.44	nr	168.17
Installation with black enamel heavy gauge conduit with draw wire										
telephone points	53.47	7.50	57.48	5.45	–	78.37	2.76	138.61	nr	149.01
Electric wiring in extending existing installation in existing building										
Installation with PVC insulated and sheathed cables										
lighting points	16.37	7.50	17.60	3.75	–	53.92	1.65	73.17	nr	78.66
socket outlets; 5A	17.65	7.50	18.97	5.05	–	72.62	1.27	92.86	nr	99.83
socket outlets; 13A ring main	21.18	7.50	22.77	4.25	–	61.12	1.15	85.03	nr	91.41
socket outlets; 13A radial circuit	24.04	7.50	25.84	4.45	–	63.99	1.27	91.10	nr	97.94
cooker points; 45A	98.64	7.50	106.04	7.25	–	104.26	1.96	212.25	nr	228.17
immersion heater points	29.80	7.50	32.03	5.30	–	76.21	1.40	109.65	nr	117.87
shaver sockets (transformer)	51.91	7.50	55.80	1.90	–	27.32	1.40	84.53	nr	90.86
Installation with mineral insulated copper sheathed cables										
lighting points	26.69	7.50	28.69	7.15	–	102.82	0.82	132.33	nr	142.25
socket outlets; 5A	37.80	7.50	40.63	5.60	–	80.53	0.75	121.91	nr	131.06
socket outlets; 13A ring main	33.17	7.50	35.66	4.80	–	69.02	0.63	105.31	nr	113.21
socket outlets; 13A radial circuit	39.07	7.50	42.00	6.40	–	92.03	0.75	134.78	nr	144.89
cooker points; 45A	120.11	7.50	129.12	10.60	–	152.43	1.44	282.99	nr	304.21
immersion heater points	51.50	7.50	55.36	7.60	–	109.29	0.88	165.53	nr	177.95
shaver sockets (transformer)	73.40	7.50	78.91	3.45	–	49.61	0.88	129.40	Unr	139.10
Installation with black enamel heavy gauge conduit with PVC cables										
lighting points	53.10	7.50	57.08	5.35	–	76.93	2.51	136.53	nr	146.76
socket outlets; 5A	65.02	7.50	69.90	5.40	–	77.65	2.76	150.31	nr	161.58
socket outlets; 13A ring main	56.21	7.50	60.43	4.90	–	70.46	2.26	133.15	nr	143.13

Labour hourly rates: (except Specialists) Craft Operatives £14.38 Labourer £7.02 Rates are national average prices. Refer to REGIONAL VARIATIONS for indicative levels of overall pricing in regions	MATERIALS			LABOUR				RATES		
	Del to Site	Waste	Material Cost	Craft Optve	Lab	Labour Cost	Sunds	Nett Rate	Unit	Gross Rate (+7.5%)
	£	%	£	Hrs	Hrs	£	£	£		£
GENERAL LIGHTING AND POWER - DOMESTIC Cont.										
Electric wiring in extending existing installation in existing building Cont.										
Installation with black enamel heavy gauge conduit with PVC cables Cont.										
socket outlets; 13A radial circuit	66.59	7.50	71.58	6.90	-	99.22	2.76	173.57	nr	186.58
cooker points; 45A	153.87	7.50	165.41	11.10	-	159.62	3.26	328.29	nr	352.91
immersion heater points	74.76	7.50	80.37	8.25	-	118.64	2.76	201.76	nr	216.89
shaver sockets (transformer)	94.72	7.50	101.82	2.65	-	38.11	2.51	142.44	nr	153.12
Installation with black enamel heavy gauge conduit with coaxial cable										
T.V. sockets	66.67	7.50	71.67	6.60	-	94.91	2.76	169.34	nr	182.04
Installation with black enamel heavy gauge conduit with draw wire										
telephone points	53.47	7.50	57.48	6.25	-	89.88	2.76	150.12	nr	161.37
LAMP FITTINGS, FANS, HEATERS etc., ALL INSTALLATIONS										
Fluorescent lamp fittings, mains voltage operations; switch start; including lamps										
Batten type; single tube										
1200mm; 36W	18.59	2.50	19.05	0.75	-	10.79	0.50	30.34	nr	32.62
1500mm; 58W	21.03	2.50	21.56	1.00	-	14.38	0.50	36.44	nr	39.17
1800mm; 70W	25.41	2.50	26.05	1.00	-	14.38	0.50	40.93	nr	43.99
2400mm; 100W	34.76	2.50	35.63	1.25	-	17.98	0.50	54.10	nr	58.16
Batten type; twin tube										
1200mm; 36W	35.66	2.50	36.55	0.75	-	10.79	0.50	47.84	nr	51.42
1500mm; 58W	42.35	2.50	43.41	1.00	-	14.38	0.50	58.29	nr	62.66
1800mm; 70W	46.44	2.50	47.60	1.00	-	14.38	0.50	62.48	nr	67.17
2400mm; 100W	60.84	2.50	62.36	1.25	-	17.98	0.50	80.84	nr	86.90
Metal trough reflector; single tube										
1200mm; 36W	31.22	2.50	32.00	1.00	-	14.38	0.50	46.88	nr	50.40
1500mm; 58W	34.19	2.50	35.04	1.20	-	17.26	0.50	52.80	nr	56.76
1800mm; 70W	40.29	2.50	41.30	1.20	-	17.26	0.50	59.05	nr	63.48
2400mm; 100W	57.34	2.50	58.77	1.45	-	20.85	0.50	80.12	nr	86.13
Metal trough reflector; twin tube										
1200mm; 36W	48.28	2.50	49.49	1.00	-	14.38	0.50	64.37	nr	69.19
1500mm; 58W	55.51	2.50	56.90	1.20	-	17.26	0.50	74.65	nr	80.25
1800mm; 70W	61.32	2.50	62.85	1.20	-	17.26	0.50	80.61	nr	86.65
2400mm; 100W	83.42	2.50	85.51	1.45	-	20.85	0.50	106.86	nr	114.87
Plastics diffused type; single tube										
1200mm; 36W	32.21	2.50	33.02	1.00	-	14.38	0.50	47.90	nr	51.49
1500mm; 58W	36.08	2.50	36.98	1.20	-	17.26	0.50	54.74	nr	58.84
1800mm; 70W	44.15	2.50	45.25	1.20	-	17.26	0.50	63.01	nr	67.74
2400mm; 100W	59.02	2.50	60.50	1.45	-	20.85	0.50	81.85	nr	87.98
Plastics diffused type; twin tube										
1200mm; 36W	60.13	2.50	61.63	1.00	-	14.38	0.50	76.51	nr	82.25
1500mm; 58W	73.04	2.50	74.87	1.20	-	17.26	0.50	92.62	nr	99.57
1800mm; 70W	82.51	2.50	84.57	1.20	-	17.26	0.50	102.33	nr	110.00
2400mm; 100W	105.37	2.50	108.00	1.45	-	20.85	0.50	129.36	nr	139.06
Lighting fittings complete with tungsten lamps										
Bulkhead; alloy										
100W G.L.S. lamp	23.75	2.50	24.34	0.65	-	9.35	0.25	33.94	nr	36.49
Wall glass; alloy; corner bracket										
100W G.L.S. lamp	26.05	2.50	26.70	0.75	-	10.79	0.25	37.74	nr	40.57
Ceiling sphere										
152mm; 60W G.L.S. lamp	14.00	2.50	14.35	0.65	-	9.35	0.25	23.95	nr	25.74
203mm; 100W G.L.S. lamp	20.40	2.50	20.91	0.65	-	9.35	0.25	30.51	nr	32.80
Down light; semi recessed										
175mm; 150W PAR38 lamp	17.32	2.50	17.75	0.65	-	9.35	0.25	27.35	nr	29.40
Lighting track; single circuit; surface mounted										
1250mm; starter pack	17.60	2.50	18.04	1.20	-	17.26	0.50	35.80	nr	38.48
plug-in fitting; 100W G.L.S. lamp	13.50	2.50	13.84	0.20	-	2.88	-	16.71	nr	17.97
Fan heaters										
3kw unit fan heaters										
for commercial application	152.06	2.50	155.86	1.50	-	21.57	0.50	177.93	nr	191.28
Thermostatic switch units comprising ON/OFF switch and selector switch between winter and summer (fan only) operation ...	51.86	2.50	53.16	1.00	-	14.38	0.25	67.79	nr	72.87
Tubular heaters										
60W per 305mm loading										
600mm long	10.80	2.50	11.07	1.00	-	14.38	0.75	26.20	nr	28.16
910mm long	14.00	2.50	14.35	1.00	-	14.38	0.75	29.48	nr	31.69
1220mm long	13.66	2.50	14.00	1.25	-	17.98	1.00	32.98	nr	35.45
1520mm long	17.85	2.50	18.30	1.25	-	17.98	1.00	37.27	nr	40.07

ELECTRICAL

Labour hourly rates: (except Specialists) Craft Operatives £14.38 Labourer £7.02 Rates are national average prices. Refer to REGIONAL VARIATIONS for indicative levels of overall pricing in regions	MATERIALS			LABOUR				RATES		
	Del to Site £	Waste %	Material Cost £	Craft Optve Hrs	Lab Hrs	Labour Cost £	Sunds £	Nett Rate £	Unit	Gross Rate (+7.5%) £
LAMP FITTINGS, FANS, HEATERS etc., ALL INSTALLATIONS Cont.										
Tubular heaters Cont.										
60W per 305mm loading Cont.										
1830mm long	18.24	2.50	18.70	1.50	-	21.57	1.50	41.77	nr	44.90
Two way mounting brackets										
pair	5.19	2.50	5.32	0.40	-	5.75	0.50	11.57	nr	12.44
Night storage heaters; installed complete in new building including wiring										
Plastics insulated and sheathed cabled										
1.7KW	204.93	2.50	210.05	6.50	-	93.47	6.43	309.95	nr	333.20
2.55KW	277.57	2.50	284.51	7.00	-	100.66	6.43	391.60	nr	420.97
3.4KW	321.60	2.50	329.64	7.50	-	107.85	6.72	444.21	nr	477.53
Mineral insulated copper sheathed cables										
1.7KW	249.33	2.50	255.56	10.00	-	143.80	5.53	404.89	nr	435.26
2.55KW	292.23	2.50	299.54	10.50	-	150.99	5.53	456.06	nr	490.26
3.4KW	336.17	2.50	344.57	11.00	-	158.18	5.77	508.52	nr	546.66
Bell equipment										
Transformers										
3-5-8V	14.47	2.50	14.83	0.40	-	5.75	0.24	20.82	nr	22.39
12V	18.52	2.50	18.98	0.40	-	5.75	0.24	24.98	nr	26.85
Bells for transformer operation										
chime 2-note	10.40	2.50	10.66	0.50	-	7.19	0.24	18.09	nr	19.45
76mm domestic type	5.73	2.50	5.87	0.50	-	7.19	0.24	13.30	nr	14.30
150mm round bell type	20.84	2.50	21.36	0.40	-	5.75	0.24	27.35	nr	29.40
Bell pushes										
domestic	1.45	2.50	1.49	0.25	-	3.60	-	5.08	nr	5.46
industrial	5.72	2.50	5.86	0.30	-	4.31	-	10.18	nr	10.94
Immersion heaters and thermostats - Note - the following prices exclude builders work										
2 or 3KW immersion heaters; without flanges										
305mm; non-withdrawable elements	22.91	2.50	23.48	0.65	-	9.35	4.65	37.48	nr	40.29
457mm; non-withdrawable elements	23.14	2.50	23.72	0.80	-	11.50	4.65	39.87	nr	42.86
686mm; non-withdrawable elements	23.25	2.50	23.83	0.95	-	13.66	4.65	42.14	nr	45.30
Thermostats										
immersion heater	10.04	2.50	10.29	0.50	-	7.19	-	17.48	nr	18.79
15 amp a.c. for air heating; without switch	15.16	2.50	15.54	0.40	-	5.75	0.25	21.54	nr	23.16
Water heaters - Note - the following prices exclude builders work										
Storage type units; free outlet										
7 litre	104.57	-	104.57	1.50	-	21.57	5.15	131.29	nr	141.14
23 litre	461.65	-	461.65	2.25	-	32.35	5.65	499.65	nr	537.13
55 litre	673.33	-	673.33	4.50	-	64.71	7.98	746.02	nr	801.97
Storage type units; multi-point										
50 litre	634.77	-	634.77	4.50	-	64.71	7.98	707.46	nr	760.52
75 litre	754.78	-	754.78	5.25	-	75.50	7.98	838.26	nr	901.12
125 litre	863.83	-	863.83	6.00	-	86.28	10.30	960.41	nr	1032.44
Shower units - Note - the following prices exclude Plumbers work										
Instantaneous shower units										
complete with fittings	155.24	2.50	159.12	4.00	-	57.52	7.73	224.37	nr	241.20
TESTING										
Testing; existing installations										
Point										
from	-	-	-	2.00	-	28.76	-	28.76	nr	30.92
to	-	-	-	3.50	-	50.33	-	50.33	nr	54.10
Complete installation; three bedroom house										
from	-	-	-	10.00	-	143.80	-	143.80	nr	154.59
to	-	-	-	16.00	-	230.08	-	230.08	nr	247.34

Communications/Security/Control systems

Labour hourly rates: (except Specialists) Craft Operatives £9.23 Labourer £7.02 Rates are national average prices. Refer to REGIONAL VARIATIONS for indicative levels of overall pricing in regions	MATERIALS			LABOUR				RATES		
	Del to Site £	Waste %	Material Cost £	Craft Optve Hrs	Lab Hrs	Labour Cost £	Sunds £	Nett Rate £	Unit	Gross Rate (+7.5%) £
W52: LIGHTNING PROTECTION										
Lightning conductors										

Note
  the following are indicative average prices
Air termination rods
  610mm long; in clips, fixing to masonry ....
	-	-	Spclist	-	-	Spclist	-	39.50	nr	42.46

Air termination tape fixing to masonry	-	-	Spclist	-	-	Spclist	-	9.50	m	10.21
Copper tape down conductors 19 x 3mm; in clips, fixing to masonry	-	-	Spclist	-	-	Spclist	-	14.35	m	15.43
25 x 3mm; in clips, fixing to masonry	-	-	Spclist	-	-	Spclist	-	15.35	m	16.50
Test clamps with securing bolts	-	-	Spclist	-	-	Spclist	-	22.00	nr	23.65
Copper earth rods; tape connectors 2438 x 16mm; driving into ground	-	-	Spclist	-	-	Spclist	-	140.00	nr	150.50

COMMUNICATIONS/SECURITY/CONTROL

Transport systems

Labour hourly rates: (except Specialists) Craft Operatives £9.23 Labourer £7.02 Rates are national average prices. Refer to REGIONAL VARIATIONS for indicative levels of overall pricing in regions	MATERIALS			LABOUR				RATES		
	Del to Site £	Waste %	Material Cost £	Craft Optve Hrs	Lab Hrs	Labour Cost £	Sunds £	Nett Rate £	Unit	Gross Rate (+7.5%) £
X10: LIFTS										
Generally										
Note the following are average indicative prices **Light passenger lifts; standard range**										
Electro Hydraulic drive; 630 kg, 8 person, 0.63m/s, 3 stop										
basic; primed car and entrances	-	-	Spclist	-	-	Spclist	-	15500.00	nr	16662.50
median; laminate walls, standard carpet floor,										
painted entrances	-	-	Spclist	-	-	Spclist	-	18200.00	nr	19565.00
ACVF drive; 1000 kg, 13 person, 0.63m/s, 4 stop										
basic; primed car and entrances	-	-	Spclist	-	-	Spclist	-	34600.00	nr	37195.00
median; laminate walls, standard carpet floor,										
painted entrances	-	-	Spclist	-	-	Spclist	-	35900.00	nr	38592.50
Extras on the above lifts										
800/900mm landing doors in cellulose paint finish	-	-	Spclist	-	-	Spclist	-	150.00	nr	161.25
800/900mm landing doors in brushed stainless steel	-	-	Spclist	-	-	Spclist	-	420.00	nr	451.50
800/900mm landing doors in brushed brass	-	-	Spclist	-	-	Spclist	-	1210.00	nr	1300.75
landing position indicators	-	-	Spclist	-	-	Spclist	-	237.00	nr	254.78
landing direction arrows and gongs	-	-	Spclist	-	-	Spclist	-	140.00	nr	150.50
car preference key switch	-	-	Spclist	-	-	Spclist	-	120.00	nr	129.00
car entrance safety (Progard L)	-	-	Spclist	-	-	Spclist	-	400.00	nr	430.00
fluorescent emergency light	-	-	Spclist	-	-	Spclist	-	372.00	nr	399.90
half height mirror to rear wall	-	-	Spclist	-	-	Spclist	-	420.00	nr	451.50
full height mirror to rear wall	-	-	Spclist	-	-	Spclist	-	610.00	nr	655.75
vandal resistant buttons; per car/entrance	-	-	Spclist	-	-	Spclist	-	136.00	nr	146.20
car bumper rail	-	-	Spclist	-	-	Spclist	-	820.00	nr	881.50
trailing cable for Warden Call System	-	-	Spclist	-	-	Spclist	-	349.00	nr	375.18
car telephone	-	-	Spclist	-	-	Spclist	-	274.00	nr	294.55
General purpose passenger lifts										
ACVF drive; 630 kg, 8 person, 1.0m/s, 4 stop										
median; laminate walls, standard carpet floor,										
painted entrances	-	-	Spclist	-	-	Spclist	-	35600.00	nr	38270.00
Variable speed a/c drive; 1000 kg, 13 person, 1.6m/s, 5 stop										
median; laminate walls, standard carpet floor,										
painted entrances	-	-	Spclist	-	-	Spclist	-	38200.00	nr	41065.00
Variable speed a/c drive; 1600 kg, 21 person, 1.6m/s, 6 stop										
median; laminate walls, standard carpet floor,										
painted entrances	-	-	Spclist	-	-	Spclist	-	92000.00	nr	98900.00
If a high quality architectural finish is required, e.g. veneered panelling walls, marble floors, special ceiling, etc, add to the above prices										
from	-	-	Spclist	-	-	Spclist	-	15000.00	nr	16125.00
to	-	-	Spclist	-	-	Spclist	-	42000.00	nr	45150.00
Bed/Passenger lifts										
Electro Hydraulic drive; 1800 kg, 24 person, 0.63m/s, 3 stop										
standard specification	-	-	Spclist	-	-	Spclist	-	48200.00	nr	51815.00
Variable speed a/c drive; 2000 kg, 26 person, 1.6m/s, 4 stop										
standard specification	-	-	Spclist	-	-	Spclist	-	87500.00	nr	94062.50

Labour hourly rates: (except Specialists) Craft Operatives £9.23 Labourer £7.02 Rates are national average prices. Refer to REGIONAL VARIATIONS for indicative levels of overall pricing in regions	MATERIALS			LABOUR				RATES		
	Del to Site £	Waste %	Material Cost £	Craft Optve Hrs	Lab Hrs	Labour Cost £	Sunds £	Nett Rate £	Unit	Gross Rate (+7.5%) £
X10: LIFTS Cont.										
Goods lifts (direct coupled)										
Electro Hydraulic drive; 1500 kg, 0.4m/s, 3 stop stainless steel car lining with chequer plate floor and galvanised shutters	-	-	Spclist	-	-	Spclist	-	38200.00	nr	41065.00
Electro Hydraulic drive; 2000 kg, 0.4m/s, 3 stop stainless steel car lining with chequer plate floor and galvanised shutters	-	-	Spclist	-	-	Spclist	-	39400.00	nr	42355.00
Service hoists										
Single speed a/c drive; 50 kg, 0.63m/s, 2 stop standard specification	-	-	Spclist	-	-	Spclist	-	5400.00	nr	5805.00
Single speed a/c drive; 250 kg, 0.4m/s, 2 stop standard specification	-	-	Spclist	-	-	Spclist	-	6850.00	nr	7363.75
Extra on the above hoists two hour fire resisting shutters; per landing	-	-	Spclist	-	-	Spclist	-	227.00	nr	244.03
X11: ESCALATORS										
Escalators										
30 degree inclination; 3.00m vertical rise department store specification	-	-	Spclist	-	-	Spclist	-	58000.00	nr	62350.00

Basic Prices of Materials

This section gives basic prices of materials delivered to site. Prices are exclusive of Value Added Tax

EXCAVATION AND EARTHWORK

Filling
broken brick or stone	m3	12.40
clinker ashes	m3	12.50
MOT Type 1	m3	14.50
MOT Type 2	m3	14.20

Earthwork support
timber	m3	210.00

CONCRETE WORK

Ready mixed concrete
Prescribed mix
ST1 - 40mm aggregate	m3	45.45
ST2 - 40mm aggregate	m3	45.98
ST3 - 40mm aggregate	m3	48.30
ST3 - 20mm aggregate	m3	48.63
ST4 - 20mm aggregate	m3	49.60
ST5 - 20mm aggregate	m3	50.83
P390 - 20mm aggregate	m3	52.00

Design mix
C 7.5 - 7.5 N/mm2 - 40mm	m3	46.00
C 10 - 10 N/mm2 - 40mm	m3	48.00
C 15 - 15 N/mm2 - 40mm	m3	48.60
C 15 - 15 N/mm2 - 20mm	m3	48.60
C 20 - 20 N/mm2 - 20mm	m3	49.20
C 25 - 25 N/mm2 - 20mm	m3	49.90
C 30 - 30 N/mm2 - 20mm	m3	51.10

Specified mix
1:12 cement concrete 40mm.	m3	50.45
1:3:6 cement concrete 40mm	m3	49.45
1:2:4 cement concrete 20mm	m3	58.06
1:1.5:3 cement concrete 20mm	m3	66.65

Cement in bags 5t and over	tonne	79.00

Washed graded shingle
40-5mm	tonne	18.90
20-5mm	tonne	17.90

Washed sharp sand for
concreting	tonne	17.90

"All in" ballast
40mm and down	tonne	15.00
20mm and down	tonne	14.30

Rapid hardening cement	tonne	85.00
Sulphate resisting cement	tonne	85.00

Waterproofing powder	18 kg	22.90
Waterproofing liquid	litre	0.82

Lightweight aggregate medium grade
10-5mm, 919 kg/m3	tonne	40.00
fine grade 5mm to dust, 1200 kg/m3	tonne	50.00

Mild steel rods cut, bent, bundled and labelled
25mm diameter	tonne	290.00
20mm diameter	tonne	290.00
16mm diameter	tonne	290.00
12mm diameter	tonne	290.00
10mm diameter	tonne	290.00
8mm diameter	tonne	300.00
6mm diameter	tonne	335.00

High yield steel rods cut, bent, bundled and labelled
25mm diameter	tonne	300.00
20mm diameter	tonne	300.00
16mm diameter	tonne	300.00
12mm diameter	tonne	300.00
10mm diameter	tonne	300.00
8mm diameter	tonne	310.00
6mm diameter	tonne	345.00

Stainless steel rods cut, bent, bundled and labelled
25mm diameter	tonne	3350.00
20mm diameter	tonne	3350.00
16mm diameter	tonne	3350.00
12mm diameter	tonne	3350.00
10mm diameter	tonne	3350.00
8mm diameter	tonne	3350.00

Fabric reinforcement, B.S. Ref 4483
A393	m2	2.15
A252	m2	1.38
A193	m2	1.06
A142	m2	0.78
A98	m2	0.55
B1131	m2	3.88
B785	m2	2.87
B503	m2	2.09
B385	m2	1.61
B283	m2	1.32
B196	m2	1.09
C785	m2	2.38
C503	m2	1.48
C385	m2	1.20
C283	m2	0.92
D98	m2	0.50
D49	m2	0.66
C636	m2	1.97

Formwork
timber	m3	220.00
19mm plywood	m2	6.50

Claymaster expanded polystyrene permanent formwork
2400 x 1200 x 75mm thick	nr	21.12
2400 x 1200 x 100mm thick	nr	28.16
2400 x 1200 x 150mm thick	nr	42.24

Bitumen/latex solution	litre	2.27
Bituminous emulsion	litre	1.44
Asphaltic bitumen	litre	1.73

Polythene sheeting
125 mu	m2	0.30
250 mu	m2	0.36
300 mu	m2	0.49

Waterproof building paper to BS 1521
Grade B2	m2	0.38
Grade B1F	m2	0.53

BRICKWORK AND BLOCKWORK

Common bricks
plain	1000	149.00
keyed	1000	153.60

Selected regrades	1000	108.10
Second hard stocks	1000	473.00

Engineering bricks
Class A	1000	499.00
Class B	1000	297.00

Cement (7.5 - 10.0 tonnes)
portland	tonne	91.60
sulphate resisting	tonne	130.00

Cement including unloading (7.5 - 10 tonnes)
portland	tonne	95.29
sulphate resisting	tonne	123.69

Hydrated lime	tonne	140.00

Hydrated lime including
unloading	tonne	143.69

Sand	tonne	10.00

203 x 3mm Butterfly pattern wall ties to BS 1243 Table 1
galvanized steel	1000	78.50
stainless steel	1000	85.00

203 x 19 x 3mm Vertical twisted wall ties to BS 1243 Table 3
galvanized steel	1000	241.00
stainless steel	1000	388.00

BRICKWORK AND BLOCKWORK (Cont'd)

Stainless steel furfix wall extension profiles

100mm wide 2338mm high.	nr	24.60
150mm wide 2338mm high.	nr	30.95
215mm wide 2338mm high.	nr	44.50

Staffordshire blue bricks

wirecuts	1000	385.00
pressed	1000	370.00
splayed plinth or bullnose.	1000	1830.00

Facing bricks including crane offloading

Claydon Red Multi	1000	133.40
Tudors	1000	135.70
Milton Buff Ridgefaced	1000	171.20
Heathers	1000	135.70
Brecken Grey	1000	159.30
Leicester Red Stock	1000	222.30
West Hoathly Medium multi-stock	1000	299.30
Himley dark brown rustic	1000	298.30
Holbrook smooth red	1000	272.30
Tonbridge hand made multi-coloured	1000	526.30
Waingroves smooth red	1000	202.01
Old English Russet	1000	135.15
First hard stocks	1000	370.00
Second hard stocks	1000	290.00

Thermalite blocks
Shield 2000 blocks

75mm	m2	5.69
90mm	m2	6.82
100mm	m2	7.58
140mm	m2	10.61
150mm	m2	11.37
190mm	m2	14.40
200mm	m2	15.16

Smooth faced blocks

100mm	m2	13.00
140mm	m2	18.20
150mm	m2	19.50
190mm	m2	24.70
200mm	m2	26.00
215mm	m2	27.95

Turbo blocks

100mm	m2	7.86
115mm	m2	9.04
125mm	m2	9.83
130mm	m2	10.22
150mm	m2	11.79
190mm	m2	14.93
200mm	m2	15.72
215mm	m2	16.90

Party Wall 650 blocks

215mm	m2	16.56

Hi-Strength 7 blocks

100mm	m2	9.52
140mm	m2	13.33
150mm	m2	14.28
190mm	m2	18.09
200mm	m2	19.04
215mm	m2	20.47

Trench blocks

255mm	m2	19.64
275mm	m2	21.18
305mm	m2	23.49
355mm	m2	27.34

Lignacite blocks
solid Fair Face blocks
3.5 N/mm2

100mm	m2	9.65
140mm	m2	13.41
190mm	m2	17.15
215mm	m2	19.89

Toplite blocks
solid GT1 blocks 2.8 N/mm2

115mm	m2	7.85
125mm	m2	8.54
130mm	m2	8.87
140mm	m2	9.56
150mm	m2	10.24
200mm	m2	13.65
215mm	m2	14.68

solid Standard blocks 3.5 N/mm2

75mm	m2	5.41
90mm	m2	5.95
100mm	m2	6.62
140mm	m2	9.26
150mm	m2	9.92
190mm	m2	12.57
200mm	m2	13.23
215mm	m2	14.23

Topcrete blocks
solid Standard blocks 7.0 N/mm2

75mm	m2	5.41
90mm	m2	5.95
100mm	m2	6.62
140mm	m2	9.26
190mm	m2	12.57
215mm	m2	14.23

hollow Standard blocks 7.0 N/mm2

140mm	m2	8.03
190mm	m2	10.93
215mm	m2	10.55

Hemelite blocks
solid Standard blocks
3.5 N/mm2

75mm	m2	4.50
90mm	m2	5.92
100mm	m2	5.22
140mm	m2	7.63
190mm	m2	11.00
215mm	m2	11.92

7.0 N/mm2

90mm	m2	6.32
100mm	m2	5.62
140mm	m2	7.84
190mm	m2	11.39
215mm	m2	12.40

Damp proof courses

bitumen/latex solution	litre	2.27
bituminous emulsion	litre	1.44

bituminous hessian based weighing

3.90kg/m2	m2	6.08
pitch polymer	m2	6.23
slates size 350 x 225mm	1000	1090.00

Type G general purpose Cavitray without lead

900mm long	nr	6.36
220 x 332mm external angle	nr	6.36
230 x 117mm internal angle.	nr	6.36

Type G general purpose Cavitray with 150mm Code 4 lead attached

900mm long	nr	10.45
220 x 332mm external angle	nr	11.13
230 x 117mm internal angle	nr	11.13

Type X gable abutment standard stepped Cavitray without lead to suit 40 degree pitched roof

intermediate tray	nr	1.90

Type X gable abutment standard stepped Cavitray without lead to suit 40 degree pitched roof (Cont'd)

Ridge tray	nr	3.71
catchment tray	nr	1.90
corner catchment angle tray	nr	3.71

Type X gable abutment standard stepped Cavitray with Code 4 lead for plain tiles to suit 40 degree pitched roof

intermediate tray	nr	4.02
ridge tray	nr	8.48
catchment tray	nr	4.02
corner catchment angle tray	nr	8.87

Type W

Cavity weep ventilator	nr	0.41
Extension duct	nr	0.42

Stainless steel mesh reinforcement

65mm	m	0.54
115mm	m	0.92
175mm	m	1.49
225mm	m	2.02

150 x 25 x 3mm frame cramp

galvanized iron	nr	0.18
stainless steel	nr	0.39

Air bricks
terra cotta, red or buff

225 x 75mm	nr	1.52
225 x 150mm	nr	2.11
225 x 225mm	nr	5.81

iron, light

225 x 75mm	nr	1.79
225 x 150mm	nr	3.40
225 x 225mm	nr	5.09

iron, light, sliding

225 x 75mm	nr	4.73
225 x 150mm	nr	8.66
225 x 225mm	nr	11.48

iron, light, louvre

225 x 75mm	nr	2.45
225 x 150mm	nr	4.04
225 x 225mm	nr	6.88

galvanized iron, light

225 x 75mm	nr	3.25
225 x 150mm	nr	5.95
225 x 225mm	nr	8.67

galvanized iron, light sliding

225 x 75mm	nr	7.37
225 x 150mm	nr	14.12
225 x 225mm	nr	17.89

galvanized iron, light, louvre

225 x 75mm	nr	4.04
225 x 150mm	nr	7.22
225 x 225mm	nr	12.14

64mm Fire bricks	1000	1225.00

230x114x76mm MPK supra

flue bricks	1000	838.00
Fireclay	tonne	532.00
MPK dribrik mortar	50kg	23.10

Cavity wall insulation
glass fibre slabs

50mm	m2	3.10
75mm	m2	4.24
expanded polystyrene pellets	m3	81.00

ROOFING

Welsh Blue/grey slates, 6.5mm thick

610 x 305mm	1000	4180.00
510 x 255mm	1000	1890.00
405 x 205mm	1000	870.00
510 x 305mm	1000	2160.00
460 x 305mm	1000	1830.00
460 x 255mm	1000	1480.00

Fibre cement slates

600 x 300mm	1000	1379.00
500 x 250mm	1000	1037.00
400 x 200mm	1000	664.00

ROOFING   (Cont'd)

Plain tiles

machine made	1000	355.00
hand made	1000	588.00

Sand-faced pantiles

machine made	1000	438.30

Concrete tiles

plain	1000	265.50
interlocking 418 x 330mm	1000	591.30

Treated softwood battens

25 x 19mm	m	0.20
38 x 19mm	m	0.25.
50 x 19mm	m	0.35
50 x 25mm	m	0.43

Woodwool slabs
standard quality

25mm	m2	4.65
38mm	m2	4.92
50mm	m2	6.32
75mm	m2	8.39
100mm	m2	11.71

pre-screeded finish

25mm	m2	5.75
38mm	m2	6.02
50mm	m2	7.42
75mm	m2	9.49
100mm	m2	12.81

channel reinforced, standard quality

50mm	m2	16.08
75mm	m2	23.60
100mm	m2	27.65

channel reinforced, standard quality with pretextured soffit finish

50mm	m2	18.74
75mm	m2	26.26
100mm	m2	30.31

Felt to BS 747
reinforced bitumen underfelt type 1F

15kg/10m2	m2	1.12

with 50mm glass fibre insulation

bonded on		5.12

fibre based roofing felt type 1B,

25kg/10m2	m2	1.07

mineral surfaced fibre based roofing felt type 1E,

38kg/10m2	m2	1.29

glass fibre based roofing felt

36kg/10m2	m2	0.69

mineral surfaced glass fibre based roofing felt type 3E 28kg/

10m2	m2	1.52

Type 1F

Reinforced underslating felt.	m2	1.12

Reinforced underslating felt, aluminium

foil surfaced	m2	3.17

Sheet lead in 1 Tonne lots	tonne	686.00

Sheet zinc, commercial quality in 150-250kg lots

12 Gauge 0.65mm thick	100kg	216.50
14 Gauge 0.80mm thick	100kg	216.50

Sheet copper in 150-250kg lots

0.70mm thick	100kg	306.00
0.60mm thick	100kg	324.00

Sheet aluminium, commercial purity S1B0 in 10 roll lots
0.60mm thick

150mm wide	8m roll	11.69
225mm wide	8m roll	16.06
300mm wide	8m roll	21.85
450mm wide	8m roll	30.40
600mm wide	8m roll	40.75
900mm wide	8m roll	58.05

WOODWORK

Softwood for carcassing

General Structural (GS) Grade1 Class SC3	m3	180.00
Special Structural (SS) Grade1 Class SC4	m3	200.00

Machine General Structural (MGS) Grade	m3	180.00
Machine Special Structural (MSS) Grade	m3	200.00

Timber for First and Second Fixings and Composite items

softwood	m3	300.00
British Columbian Pine	m3	918.00
English Oak	m3	1300.00
Japenese Oak	m3	2810.00
African Mahogany	m3	750.00
South American Mahogany	m3	1155.00
Sapele	m3	777.00
Iroko	m3	770.00
Idigbo	m3	740.00
Utile	m3	920.00
Afrormosia	m3	1080.00
Beech	m3	810.00
Teak	m3	4250.00
Walnut	m3	1960.00

Plywood
Marine, WBP

4mm	m2	4.97
6 or 6.5mm	m2	5.00
9mm	m2	6.59
12mm	m2	8.20
15mm	m2	10.08
18mm	m2	12.09
24 or 25mm	m2	16.12

Far Eastern Redwood/Whitewood/BB Quality, WBP

4mm	m2	2.98
6 or 6.5mm	m2	3.81
9mm	m2	5.46
12mm	m2	7.07
15mm	m2	8.89
18mm	m2	10.55
24 or 25mm	m2	14.66

Far Eastern Redwood/Whitewood B/BB Quality, MR

4mm	m2	2.83
6 or 6.5mm	m2	3.62
9mm	m2	5.19
12mm	m2	6.72
15mm	m2	8.45
18/19mm	m2	10.02

Gaboon, B/BB Quality, WBP

4mm	m2	3.50
6 or 6.5mm	m2	4.40
9mm	m2	6.37
12mm	m2	8.02
15mm	m2	9.36
18mm	m2	11.23
24 or 25mm	m2	15.93

Decorative plywood with balancing veneer one side, 2438 x 1219mm figured oak

4mm	m2	6.56
6 or 6.5mm	m2	8.25
9mm	m2	9.37
12mm	m2	10.36

sapele

4mm	m2	4.64
6 or 6.5mm	m2	5.79
9mm	m2	6.94
12mm	m2	7.91

teak

4mm	m2	7.27
6 or 6.5mm	m2	9.45
9mm	m2	11.46
12mm	m2	13.97

Blockboard
Birch faced, Finnish manufacturer

16mm	m2	9.62
18mm	m2	9.62
25mm	m2	14.64

Oak faced

18mm	m2	13.93

Oak faced both sides

18mm	m2	15.88

Afrormosia faced

18mm	m2	17.51

Teak faced

18mm	m2	21.69

Chipboard, wood base

12mm	m2	2.15
15mm	m2	2.48
18mm	m2	2.88
22mm	m2	3.57
25mm	m2	4.04

STRUCTURAL STEELWORK

BASIS PRICES OF ROLLED STEEL

Steel joists and beams in small quantities required for quick delivery can be obtained from local merchants at a basis cost of about £440 per tonne subject to the usual extras for size. For larger quantities obtained direct from the mills see below.

The following prices are extracted, by permission, from the lists issued by British Steel from whom complete details can be obtained.

Prices at December 1997 (list 5)
BS EN 10025 Grade S275JR

Basis prices ex basing point 10 tonnes and over in standard lengths 6000mm to 18500mm of one quality, one serial size of section and one thickness for one delivery to one destination.

Joists

Size mm	Weight kg/m		
76 x 76	12.65	tonne	415.00
89 x 89	19.35	tonne	365.00
102 x 102	23.06	tonne	365.00
114 x 114	26.79	tonne	365.00
127 x 114	26.79	tonne	365.00
127 x 114	29.79	tonne	365.00
152 x 127	37.20	tonne	375.00
203 x 152	52.09	tonne	375.00
254 x 203	81.85	tonne	385.00

Quantity extras

10 tonnes and over		Basis

under 10 tonnes to 5 tonnes

add per tonne		10.00
under 5 tonnes	add per tonne	25.00

BASIC PRICES OF MATERIALS

STRUCTURAL STEELWORK (Cont'd)

BASIS PRICES OF ROLLED STEEL (Cont'd)

Universal Beams

Size mm	Weight kg/m		

Universal Beams

914 x 419	388] 343]	tonne	440.00
914 x 305	289] 253] 224] 201]	tonne	435.00
838 x 292	226] 194] 176]	tonne	430.00
762 x 267	197] 173] 147]	tonne	430.00
686 x 254	170] 152] 140] 125]	tonne	430.00
610 x 305	238] 179] 149]	tonne	420.00
610 x 229	140] 125] 113] 101]	tonne	420.00
533 x 210	122] 109] 101] 92] 82]	tonne	405.00
457 x 191	98] 89] 82] 74] 67]	tonne	395.00
457 x 152	82] 74] 67] 60] 52]	tonne	395.00
406 x 178	74] 67] 60] 54]	tonne	400.00
406 x 140	46] 39]	tonne	400.00
356 x 171	67] 57] 51] 45]	tonne	400.00
356 x 127	39] 33]	tonne	400.00
305 x 165	54] 46] 40]	tonne	395.00

305 x 127	48] 42] 37]	tonne	395.00
305 x 102	33] 28] 25]	tonne	395.00
254 x 146	43] 37] 31]	tonne	380.00
254 x 102	28] 25] 22]	tonne	380.00
203 x 133	30] 25]	tonne	340.00
203 x 102	23]	tonne	360.00
178 x 102	19]	tonne	330.00
152 x 89	16]	tonne	370.00
127 x 76	13]	tonne	390.00

Universal columns

356 x 406	634] 551] 467] 393] 340] 287] 235]	tonne	440.00
356 x 368	202] 177] 153] 129]	tonne	440.00
305 x 305	283] 240] 198] 158] 137] 118] 97]	tonne	420.00
254 x 254	167] 132] 107] 89] 73]	tonne	400.00
203 x 203	86] 71] 60] 52] 46]	tonne	390.00
152 x 152	37] 30] 23]	tonne	360.00

Quantity extras
10 tonnes and over...................Basis
under 10 tonnes to 5 tonnes
add per tonne 10.00
under 5 tonnes..... ..add per tonne 25.00

METALWORK

Mild steel sections (BS 4848)
equal angles

25 x 25 x 3mm............	m	0.60
25 x 25 x 4mm............	m	0.74
25 x 25 x 5mm............	m	0.86
30 x 30 x 3mm............	m	0.74
30 x 30 x 4mm............	m	0.91
30 x 30 x 5mm............	m	1.08
40 x 40 x 3mm............	m	1.00
40 x 40 x 4mm............	m	1.22
40 x 40 x 5mm............	m	1.35
40 x 40 x 6mm............	m	1.76
45 x 45 x 3mm............	m	1.13
45 x 45 x 4mm............	m	1.38
45 x 45 x 5mm............	m	1.56
45 x 45 x 6mm............	m	2.03
50 x 50 x 3mm............	m	1.25
50 x 50 x 4mm............	m	1.52
50 x 50 x 5mm............	m	1.80
50 x 50 x 6mm............	m	2.39
50 x 50 x 8mm............	m	2.96

unequal angles

40 x 25 x 4mm............	m	0.93
60 x 30 x 5mm............	m	1.56
60 x 30 x 6mm............	m	2.13
65 x 50 x 5mm............	m	2.01
65 x 50 x 6mm............	m	2.67
65 x 50 x 8mm............	m	3.27

circular hollow sections

21.3mm dia. x 3.2mm	m	2.36
26.9mm dia. x 3.2mm	m	2.95
33.7mm dia. x 2.6mm	m	3.15
33.7mm dia. x 3.2mm	m	3.77
33.7mm dia. x 4.0mm	m	4.59
42.4mm dia. x 2.6mm	m	4.00
42.4mm dia. x 3.2mm	m	4.84
42.4mm dia. x 4.0mm	m	5.98
48.3mm dia. x 3.2mm	m	5.61
48.3mm dia. x 4.0mm	m	6.88
48.3mm dia. x 5.0mm	m	8.16

rectangular hollow sections

50 x 30 x 2.6mm........	m	4.38
50 x 30 x 3.2mm........	m	5.27
60 x 40 x 3.2mm........	m	7.24
60 x 40 x 4.0mm........	m	8.88
80 x 40 x 3.2mm........	m	8.22
80 x 40 x 4.0mm........	m	10.09
90 x 50 x 3.6mm........	m	11.75
90 x 50 x 5.0mm........	m	15.22
100 x 50 x 3.2mm.......	m	11.13
100 x 50 x 4.0mm.......	m	13.73
100 x 50 x 6.0mm.......	m	16.26

METALWORK (Cont'd)

Mild steel sections (BS 4848) (Cont'd.)
square hollow sections

20 x 20 x 2.0mm	m	2.22
20 x 20 x 2.6mm	m	2.78
30 x 30 x 2.6mm	m	4.38
30 x 30 x 3.2mm	m	5.27
40 x 40 x 2.6mm	m	5.11
40 x 40 x 3.2mm	m	5.68
40 x 40 x 4.0mm	m	7.82
50 x 50 x 3.2mm	m	7.22
50 x 50 x 4.0mm	m	8.88
50 x 50 x 5.0mm	m	10.47

Mild steel bars (BS 6722)
round

10mm diameter	m	0.42
12mm diameter	m	0.64
16mm diameter	m	1.10
20mm diameter	m	1.72
25mm diameter	m	2.69
30mm diameter	m	3.88
32mm diameter	m	4.40
35mm diameter	m	5.28
40mm diameter	m	6.90
42mm diameter	m	7.59
45mm diameter	m	8.72
48mm diameter	m	9.93
50mm diameter	m	10.77

square

10 x 10mm	m	0.57
12 x 12mm	m	0.77
14 x 14mm	m	1.10
16 x 16mm	m	1.40
18 x 18mm	m	1.79
20 x 20mm	m	2.20
22 x 22mm	m	2.65
24 x 24mm	m	3.17
25 x 25mm	m	3.42
28 x 28mm	m	4.30
30 x 30mm	m	4.94
32 x 32mm	m	5.61
35 x 35mm	m	6.73
38 x 38mm	m	7.91
40 x 40mm	m	8.77
42 x 42mm	m	9.68
45 x 45mm	m	11.12
50 x 50mm	m	13.59

Mild steel flat sheets

26 gauge	m2	2.99
24 gauge	m2	3.66
22 gauge	m2	4.68
20 gauge	m2	6.00

Galvanized mild steel flat sheets

26 gauge	m2	6.25
24 gauge	m2	12.35
22 gauge	m2	9.23
20 gauge	m2	11.26

Galvanized mild steel strip

18 gauge x 25mm wide	kg	2.49
17 gauge x 29mm wide	kg	2.49
16 gauge x 32mm wide	kg	2.49
16 gauge x 38mm wide	kg	2.49

PLUMBING AND MECHANICAL ENGINEERING INSTALLATIONS

Cast iron eaves gutters (BS 460)
half round

102mm	m	6.88
angle	nr	5.24
nozzle piece	nr	5.10
stop end	nr	1.76
stop end outlet	nr	3.83
114mm	m	7.16
angle	nr	5.39
nozzle piece	nr	5.56
stop end	nr	2.28

Cast iron eaves gutters(BS 460)
half round (Cont'd)
114mm (Cont'd)

stop end outlet	nr	4.21
127mm	m	8.38
angle	nr	6.37
nozzle piece	nr	6.37
stop end	nr	2.28
stop end outlet	nr	5.66
152mm	m	14.32
angle	nr	11.62
nozzle piece	nr	11.01
stop end	nr	3.04
stop end outlet	nr	10.82

Cast iron eaves gutters (BS 460)
ogee

102mm	m	7.67
angle	nr	5.47
nozzle piece	nr	5.47
stop end	nr	1.59
stop end outlet	nr	4.16
114mm	m	8.44
angle	nr	5.93
nozzle piece	nr	5.93
stop end	nr	2.11
stop end outlet	nr	4.16
127mm	nr	8.85
angle	nr	6.47
nozzle piece	nr	6.47
stop end	nr	2.11
stop end outlet	nr	4.94

galvanized mild steel bracket
for half round gutter

102mm	nr	1.42
114mm	nr	1.42
127mm	nr	1.42
152mm	nr	1.80

Unplasticised PVC gutters (BS 4576)
half round

105mm	m	1.94
angle	nr	2.43
outlet	nr	2.22
stop end	nr	1.06
114mm	m	1.92
angle	nr	2.59
outlet	nr	2.38
stop end	nr	1.27

joint clips

100mm	nr	1.46
114mm	nr	2.12

brackets

100mm	nr	0.55
114mm	nr	0.53

Cast iron round rainwater pipes (BS 460)
pipe without ears

76mm	m	12.65
102mm	m	17.27

pipe with ears

64mm	m	13.52
76mm	m	13.52
102mm	m	18.15

bend

64mm	nr	7.31
76mm	nr	8.88
102mm	nr	12.53

shoe (eared)

64mm	nr	11.93
76mm	nr	11.93
102mm	nr	15.55

offset 76mm projection

64mm	nr	11.19
76mm	nr	11.19
102mm	nr	21.10

offset 152mm projection

64mm	nr	11.19
76mm	nr	11.19
102mm	nr	21.10

Cast iron round rainwater pipes (BS 460) (Cont'd)
offset 229mm projection

64mm	nr	13.03
76mm	nr	13.03
102mm	nr	25.55

offset 305mm projection

64mm	nr	15.24
76mm	nr	15.95
102mm	nr	25.55

single branch

64mm	nr	14.09
76mm	nr	15.54
102mm	nr	18.45

mild steel holderbat for
screwing on 76mm

76mm	nr	4.45
102mm	nr	4.61

mild steel holderbat for building in

76mm	nr	5.44
102mm	nr	5.60

Unplasticised PVC rainwater pipes (BS 4576)

68mm pipe with slip joints	m	1.81
bend	nr	2.34
shoe	nr	2.02

offset bend

socket	nr	1.40
spigot	nr	1.30
single branch	nr	4.67
pipe clip	nr	0.87

Black steel tubes and fittings (BS 1387)
medium weight tubing

15mm	m	1.14
20mm	m	1.34
25mm	m	1.92
32mm	m	2.38
40mm	m	2.77
50mm	m	3.90

heavy weight tubing

15mm	m	1.36
20mm	m	1.61
25mm	m	2.35
32mm	m	2.92
40mm	m	3.40
50mm	m	4.72

long screw and backnut

15mm	nr	1.79
20mm	nr	2.01
25mm	nr	2.97
32mm	nr	3.93
40mm	nr	4.76
50mm	nr	7.09

Galvanized steel tubes and fittings (BS 1387)
medium weight tubing

15mm	m	1.73
20mm	m	1.95
25mm	m	2.73
32mm	m	3.37
40mm	m	3.92
50mm	m	5.50

heavy weight tubing

15mm	m	2.05
20mm	m	2.32
25mm	m	3.31
32mm	m	4.11
40mm	m	4.79
50mm	m	6.64

long screw and backnut

15mm	nr	2.12
20mm	nr	2.49
25mm	nr	3.52
32mm	nr	4.35
40mm	nr	5.33
50mm	nr	8.01

PLUMBING AND MECHANICAL ENGINEERING INSTALLATIONS (Cont'd)

Black malleable cast iron pipe fittings

90 degree elbow
15mm	nr	0.35
20mm	nr	0.48
25mm	nr	0.75
32mm	nr	1.23
40mm	nr	2.07
50mm	nr	2.42

45 degree elbow
15mm	nr	0.75
20mm	nr	0.92
25mm	nr	1.36
32mm	nr	2.51
40mm	nr	3.08
50mm	nr	4.22

tee
15mm	nr	0.48
20mm	nr	0.70
25mm	nr	1.01
32mm	nr	1.67
40mm	nr	2.29
50mm	nr	3.30

pitcher tee
15mm	nr	1.32
20mm	nr	1.63
25mm	nr	2.44
32mm	nr	3.34
40mm	nr	5.16
50mm	nr	7.25

cross
15mm	nr	1.14
20mm	nr	1.71
25mm	nr	2.18
32mm	nr	2.86
40mm	nr	3.85
50mm	nr	5.95

socket, parallel thread
15mm	nr	0.33
20mm	nr	0.40
25mm	nr	0.53
32mm	nr	0.90
40mm	nr	1.23
50mm	nr	1.85

reducing socket
15mm	nr	0.46
20mm	nr	0.48
25mm	nr	0.64
32mm	nr	1.14
40mm	nr	1.76
50mm	nr	2.48

cap
15mm	nr	0.31
20mm	nr	0.35
25mm	nr	0.44
32mm	nr	0.64
40mm	nr	0.81
50mm	nr	1.58

plug
15mm	nr	0.26
20mm	nr	0.31
25mm	nr	0.40
32mm	nr	0.64
40mm	nr	0.79
50mm	nr	1.36

straight union, female, standard pattern
15mm	nr	1.41
20mm	nr	1.54
25mm	nr	1.80
32mm	nr	2.72
40mm	nr	3.08
50mm	nr	5.10

Galvanized malleable cast iron pipe fittings

90 degree elbow
15mm	nr	0.50
20mm	nr	0.68
25mm	nr	1.06
32mm	nr	1.74
40mm	nr	2.92
50mm	nr	3.41

45 degree elbow
15mm	nr	1.06
20mm	nr	1.30
25mm	nr	1.93
32mm	nr	3.54
40mm	nr	4.35
50mm	nr	5.96

tee
15mm	nr	0.68
20mm	nr	0.99
25mm	nr	1.43
32mm	nr	2.36
40mm	nr	3.23
50mm	nr	4.66

pitcher tee
15mm	nr	1.86
20mm	nr	2.30
25mm	nr	3.45
32mm	nr	4.72
40mm	nr	7.30
50mm	nr	10.24

cross
15mm	nr	1.61
20mm	nr	2.42
25mm	nr	3.07
32mm	nr	4.04
40mm	nr	5.43
50mm	nr	8.44

socket, parallel thread
15mm	nr	0.47
20mm	nr	0.56
25mm	nr	0.75
32mm	nr	1.27
40mm	nr	1.74
50mm	nr	2.61

reducing socket
15mm	nr	0.65
20mm	nr	0.68
25mm	nr	0.90
32mm	nr	1.61
40mm	nr	2.48
50mm	nr	3.51

cap
15mm	nr	0.43
20mm	nr	0.50
25mm	nr	0.62
32mm	nr	0.90
40mm	nr	1.15
50mm	nr	2.24

plug
15mm	nr	0.37
20mm	nr	0.43
25mm	nr	0.56
32mm	nr	0.90
40mm	nr	1.12
50mm	nr	1.93

straight union, female, standard pattern
15mm	nr	1.99
20mm	nr	2.17
25mm	nr	2.54
32mm	nr	3.85
40mm	nr	4.35
50mm	nr	7.20

Hepworth flexible polybutylene pipes and fittings

pipe
15mm	m	0.75
22mm	m	1.46
28mm	m	2.01

straight connector
15mm	nr	0.91
22mm	nr	1.22
28mm	nr	3.13

elbow 90 degrees
15mm	nr	1.12
22mm	nr	1.63
28mm	nr	3.76

tee (Equal)
15mm	nr	1.58
22mm	nr	2.02
28mm	nr	5.18

straight tap connector
15mm	nr	1.24
15mm x 3/4"	nr	1.47

pipe support sleeves
15mm	nr	0.15
22mm	nr	0.18
28mm	nr	0.28

Copper tubing (BS 2871)

table X
15mm	m	0.58
22mm	m	1.18
28mm	m	1.59
35mm	m	4.01
42mm	m	4.91
54mm	m	6.35
66.7mm	m	9.32
76.1mm	m	13.14
108mm	m	19.29

table Y
15mm	m	1.69
22mm	m	2.95
28mm	m	3.89
35mm	m	5.46
42mm	m	6.58
54mm	m	11.33

Compression pipe fittings for copper tubing (BS 864 Part 2)

Fittings type A - all copper

straight coupling
15mm	nr	0.53
22mm	nr	0.91
28mm	nr	2.74
35mm	nr	6.27
42mm	nr	8.71
54mm	nr	12.46

bent coupling
15mm	nr	0.64
22mm	nr	1.09
28mm	nr	3.44
35mm	nr	8.26
42mm	nr	11.54
54mm	nr	19.62

tee (Equal)
15mm	nr	0.94
22mm	nr	1.53
28mm	nr	6.09
35mm	nr	10.90
42mm	nr	16.79
54mm	nr	27.02

Fittings type A - copper to male iron

straight coupling
15mm	nr	0.49
22mm	nr	0.80
28mm	nr	1.56
35mm	nr	4.58
42mm	nr	7.36
54mm	nr	11.16

bent coupling
15mm	nr	0.94
22mm	nr	1.22
28mm	nr	2.62
35mm	nr	7.89
42mm	nr	11.37
54mm	nr	18.58

tee (female iron branch)
15mm	nr	2.81
22mm	nr	4.39
28mm	nr	6.19
35mm	nr	13.18

PLUMBING AND MECHANICAL ENGINEERING INSTALLATIONS (Cont'd)

Compression pipe fittings for copper tubing (BS 864 Part 2) (Cont'd)
Fittings type A - copper to male iron (Cont'd)
back plate wall elbow

15mm	nr	2.06

tank coupling

15mm	nr	2.66
22mm	nr	3.00
28mm	nr	4.72
35mm	nr	7.21
42mm	nr	10.27
54mm	nr	13.93

Fittings type B - all copper
straight coupling

15mm	nr	2.95
22mm	nr	4.80
28mm	nr	7.85
35mm	nr	14.40
42mm	nr	19.60
54mm	nr	29.00

bent coupling

15mm	nr	3.50
22mm	nr	5.70
28mm	nr	9.65
35mm	nr	18.55
42mm	nr	27.80
54mm	nr	44.25

tee (Equal)

15mm	nr	4.90
22mm	nr	8.25
28mm	nr	13.90
35mm	nr	25.20
42mm	nr	41.00
54mm	nr	65.75

Fittings type B - copper to male iron
straight coupling

15mm	nr	2.70
22mm	nr	3.85
28mm	nr	5.90
35mm	nr	11.65
42mm	nr	17.20
54mm	nr	24.90

bent coupling

15mm	nr	3.80
22mm	nr	5.25
28mm	nr	8.95
35mm	nr	20.00

Capillary fittings for copper tubing (BS 864 Part 2)
all copper
straight coupling

15mm	nr	0.10
22mm	nr	0.28
28mm	nr	0.63
35mm	nr	2.09
42mm	nr	3.12
54mm	nr	6.45

bent coupling

15mm	nr	0.19
22mm	nr	0.45
28mm	nr	1.02
35mm	nr	4.15
42mm	nr	7.39
54mm	nr	15.13

tee (Equal)

15mm	nr	0.32
22mm	nr	0.92
28mm	nr	2.85
35mm	nr	7.13
42mm	nr	11.17
54mm	nr	21.11

Capillary fittings for copper tubing (BS 864 Part 2) (Cont'd)
copper to male iron
straight connector

15mm	nr	1.25
22mm	nr	2.21
28mm	nr	3.50
35mm	nr	6.17
42mm	nr	7.95
54mm	nr	12.07

bent connector

15mm	nr	2.44
22mm	nr	3.47
28mm	nr	5.20
35mm	nr	7.36
42mm	nr	8.75
54mm	nr	15.47

tee (1/2 " female iron branch)

15mm	nr	1.61
22mm	nr	2.23
28mm	nr	6.72
35mm	nr	11.12
42mm	nr	13.27

back plate wall elbow

15mm	nr	2.61
22mm	nr	5.49

tank coupling

15mm	nr	2.97
22mm	nr	4.54
28mm	nr	5.96
35mm	nr	7.82
42mm	nr	10.25
54mm	nr	15.65

black malleable iron screw on brackets

15mm	nr	0.51
20mm	nr	0.56
25mm	nr	0.65
32mm	nr	0.89
40mm	nr	1.19
50mm	nr	1.57

galvanized malleable iron screw on brackets

15mm	nr	0.71
20mm	nr	0.79
25mm	nr	0.92
32mm	nr	1.25
40mm	nr	1.67
50mm	nr	2.21

Pipe clips for copper tube
two piece copper spacing clips

15mm	100	11.04
22mm	100	11.74
28mm	100	15.88
35mm	100	23.47
42mm	100	43.49
54mm	100	55.92

pressed brass screw on brackets

15mm	100	65.66
22mm	100	74.86
28mm	100	89.59
35mm	100	98.80
42mm	100	115.36
54mm	100	146.65

cast brass screw on brackets

15mm	100	164.43
22mm	100	193.90
28mm	100	236.24
35mm	100	304.96
42mm	100	403.74
54mm	100	510.52

Polythene tubing
BS 6572, blue

20mm	m	0.20
25mm	m	0.26
32mm	m	0.42
50mm	m	0.95
63mm	m	1.47

Brass compression fittings for polythene tube
straight coupling

20mm	nr	3.07
25mm	nr	4.35
32mm	nr	7.54
50mm	nr	18.25
63mm	nr	24.93

straight coupling with one end threaded

20mm	nr	2.71
25mm	nr	3.76
32mm	nr	5.57
50mm	nr	14.18
63mm	nr	18.75

tee

20mm	nr	5.15
25mm	nr	7.99
32mm	nr	12.02
50mm	nr	29.98
63mm	nr	44.01

elbow

20mm	nr	3.77
25mm	nr	5.49
32mm	nr	9.66
50mm	nr	22.54
63mm	nr	26.74

tap connector

20mm	nr	4.31
25mm	nr	7.38

Stop cocks (BS 1010)
brass
copper x copper

15mm	nr	2.73
22mm	nr	4.75
28mm	nr	11.41
35mm	nr	34.28
42mm	nr	43.26
54mm	nr	65.30

DZR
copper x copper

15mm	nr	7.94
22mm	nr	13.03
28mm	nr	21.57
35mm	nr	39.69
42mm	nr	56.82
54mm	nr	77.39

DZR
polythene x polythene

20mm	nr	12.99
25mm	nr	20.65
32mm	nr	27.72
50mm	nr	69.31
63mm	nr	97.29

Gate valves (BS 5154)
brass, fullway, Series B ends screwed
female BSP thread

13mm	nr	4.08
19mm	nr	4.97
25mm	nr	6.89
32mm	nr	10.07
38mm	nr	14.75
50mm	nr	21.15

copper x copper

15mm	nr	4.93
22mm	nr	5.98
28mm	nr	8.24
35mm	nr	15.51
42mm	nr	25.30
54mm	nr	39.18

bronze, fullway, Series B
ends screwed female BSP thread

13mm	nr	10.06
19mm	nr	14.27
25mm	nr	18.47
32mm	nr	26.33
38mm	nr	36.03
50mm	nr	51.58

PLUMBING AND MECHANICAL ENGINEERING INSTALLATIONS (Cont'd)

Ball valves (BS 1212)
piston type with copper ball
low pressure

13mm	nr	9.06
19mm	nr	16.45
25mm	nr	43.96

high pressure

13mm	nr	7.78
19mm	nr	16.29
25mm	nr	43.96
32mm	nr	85.42

BS 1212 Part 2 type (brass body) with copper ball
low pressure

13mm	nr	9.14

high pressure

13mm	nr	9.01

Cast iron soil and waste pipes (BS 416)
for coupling joints
spigoted pipe

50mm	m	10.44
75mm	m	10.06
100mm	m	12.15

couplings with gaskets

50mm	nr	4.56
75mm	nr	5.05
100mm	nr	6.57

bend

51mm	nr	8.04
76mm	nr	8.04
100mm	nr	11.13

single junction

50mm	nr	12.10
75mm	nr	12.10
100mm	nr	17.21

double junction

75mm	nr	20.34
100mm	nr	21.28

anti-syphon junction

100mm	nr	18.03

access pipe

50mm	nr	19.34
75mm	nr	19.34
100mm	nr	20.34

offset 150mm projection

75mm	nr	9.90
100mm	nr	13.96

offset 305mm projection

100mm	nr	18.03

roof connector

50mm	nr	22.03
75mm	nr	22.03
100mm	nr	19.41

bossed pipe 150mm long tapped 32mm

75mm	nr	16.65
100mm	nr	19.89

bossed pipe 150mm long tapped 32mm and 38mm

75mm	nr	22.49
100mm	nr	25.72

bossed pipe 240mm long tapped two 38mm

100mm	nr	25.98

WC connector 305mm long

100mm	nr	12.11

wall fixing bracket

75mm	nr	3.05
100mm	nr	3.43

Cast iron Stanton SMU lightweight soil and waste pipes for coupling joints
spigoted pipe

50mm	m	7.14
75mm	m	8.26
100mm	m	9.93

Cast iron Stanton SMU lightweight soil and waste pipes for coupling joints (Cont'd)
stainless steel couplings with gaskets

50mm	nr	3.01
75mm	nr	3.33
100mm	nr	4.35

short pipe with access door

50mm	nr	13.69
75mm	nr	14.96
100mm	nr	16.39

bends

50mm	nr	5.16
75mm	nr	5.80
100mm	nr	6.87

offset bend 75mm projection

50mm	nr	7.48
75mm	nr	9.97
100mm	nr	12.84

offset bend 150mm projection

50mm	nr	9.62
75mm	nr	13.28
100mm	nr	17.44

single angled branch

50mm	nr	8.30
75mm	nr	8.74
100mm	nr	11.11

Universal plug number of inlets 1

50mm	nr	6.21
75mm	nr	8.85

Universal plug no of inlets 3

100mm	nr	11.14

Traditional joint connector

100mm	nr	14.72

W.C. Connector

100mm	nr	8.32

Roof Connector

100mm	nr	30.08

Unplasticised MPVC waste pipes (solvent welded) (BS 5255)
pipe

32mm	m	1.29
40mm	m	1.36
50mm	m	2.34

bend

32mm	nr	0.92
40mm	nr	1.03
50mm	nr	1.71

tee

32mm	nr	1.31
40mm	nr	1.65
50mm	nr	3.26

coupling

32mm	nr	0.72
40mm	nr	0.87
50mm	nr	1.08

expansion coupling

32mm	nr	1.04
40mm	nr	1.25
50mm	nr	1.70

access plug

32mm	nr	1.09
40mm	nr	1.29
50mm	nr	2.09

plastic clips

32mm	nr	0.19
40mm	nr	0.23
55mm	nr	0.58
solvent cement	litre	13.08

Unplasticised PVC soil and ventilating pipes (BS 4514) with rubber ring "push fit" system
Pipe

82mm	m	3.96
110mm	m	3.77

bend

82mm	nr	5.24
110mm	nr	6.31

Unplasticised PVC soil and ventilating pipes (BS 4514) with rubber ring "push fit" system (Cont'd)
single junction

82mm	nr	8.23
110mm	nr	8.21

double junction

110mm	nr	15.13

access pipe

82mm	nr	9.34
110mm	nr	9.21

boss for copper

82mm	nr	4.91
110mm	nr	4.62

boss for PVC

82mm	nr	4.91
110mm	nr	4.62

W.C. connector

82mm	nr	4.82
110mm	nr	5.52

weathering apron

82mm	nr	1.91
110mm	nr	1.51

drain connector

110mm	nr	8.61

plastic coated metal holderbat for screwing on

82mm	nr	2.33
110mm	nr	2.30

plastic coated metal holderbat for building in

82mm	nr	3.03
110mm	nr	3.00

Plastic traps (BS 3943)
traps with 'O' ring joint outlet
P trap

36mm	nr	2.37
42mm	nr	2.73

S trap

36mm	nr	3.00
42mm	nr	3.52

bath trap with overflow connection

42mm	nr	5.08

Copper traps (BS 1184)
two piece trap with compression joint outlet
P trap, 76mm seal

35mm	nr	12.74
42mm	nr	15.02
54mm	nr	53.88

S trap, 76mm seal

35mm	nr	13.51
42mm	nr	16.24
54mm	nr	56.91

bath trap with male iron overflow connection, 76mm seal

42mm	nr	22.17

PLUMBING AND MECHANICAL ENGINEERING INSTALLATIONS (Cont'd)

Wire balloon gratings
galvanized iron

51mm	nr	1.55
76mm	nr	1.75
102mm	nr	2.00

copper

51mm	nr	1.95
76mm	nr	2.30
102mm	nr	2.65

plastic

51mm	nr	0.98
76mm	nr	1.42
102mm	nr	1.51

Galvanized water storage cisterns inc lid (BS 417) and byelaw 30 kit
Grade A open top cistern, BS type

SCM45, size 460 x 310 x 310mm, actual capacity 18 litres.. nr		84.95
SCM70, size 610 x 310 x 380mm, actual capacity 36 litres.. nr		104.00
SCM90, size 610 x 410 x 380mm, actual capacity 55 litres.. nr		115.25
SCM110, size 610 x 430 x 430mm, actual capacity 68 litres... nr		121.95
SCM135, size 610 x 460 x 480mm, actual capacity 86 litres... nr		129.50
SCM180, size 690 x 510 x 510mm, actual capacity 114 litres.. nr		144.10
SCM230, size 740 x 560 x 560mm, actual capacity 159 litres.. nr		180.60
SCM270, size 760 x 580 x 610mm, actual capacity 191 litres.. nr		192.80
SCM320, size 910 x 610 x 580mm, actual capacity 227 litres. nr		210.25
SCM450/1, size 1220 x 610 x 610mm, actual capacity 327 litres.. nr		247.40
SCM450/2, size 970 x 690 x 690mm, actual capacity 336 litres. nr		240.35
SCM570, size 970 x 760 x 790mm, actual capacity 423 litres.. nr		323.85
SCM680, size 1090 x 860 x 740mm, actual capacity 491 litres.. nr		365.75
SCM910, size 1170 x 890 x 890mm, actual capacity 709 litres.. nr		358.60*
SCM1130, size 1520 x 910 x 910mm, actual capacity 841 litres.. nr		398.75*
SCM1600, size 1520 x 1140 x 910mm, actual capacity 1250 litres..nr		600.80*
SCM2270, size 1830 x 1220 x 1020mm, actual capacity 1728 litres..nr		707.45*
SCM2720, size 1830 x 1220 x 1220mm, actual capacity 2137 litres..nr		761.60*
SCM4540, size 2440 x 1520 x 1220mm, actual capacity 3364 litres nr		1347.75*

* Price excludes insulation.

Rectangular plastic water storage cisterns (BS 4213)
open top cistern, BS type

C4, minimum capacity to water line 18 litres	nr	3.29
C25, minimum capacity to water line 114 litres	nr	17.04
C40, minimum capacity to water line 182 litres	nr	29.57
C50, minimum capacity to water line 227 litres	nr	31.84

sealed lid for cistern, byelaw 30 kit

18 litre	nr	7.43
114 litre	nr	15.44
182 litre	nr	20.32
227 litre	nr	25.82

Pipe insulation
13mm Armaflex/Class O for iron pipes

15mm	m	1.90

Pipe insulation (Cont'd)
13mm Armaflex/Class O for iron pipes

20mm	m	2.18
25mm	m	2.38
32mm	m	2.74
40mm	m	3.27
50mm	m	4.15

13mm Armaflex/Class O for copper pipes

15mm	m	1.59
22mm	m	1.90
28mm	m	2.18
35mm	m	2.38
42mm	m	2.74
54mm	m	3.60

19mm rigid section fibre glass for iron pipes

15mm	m	2.59
20mm	m	2.75
25mm	m	2.93
32mm	m	3.21
40mm	m	3.45
50mm	m	3.91

19mm rigid section fibre glass for copper pipes

15mm	m	2.46
22mm	m	2.59
28mm	m	2.75
35mm	m	2.93
42mm	m	3.21
54mm	m	3.68

25mm rigid section fibre glass for iron pipes

15mm	m	2.82
20mm	m	3.02
25mm	m	3.35
32mm	m	3.63
40mm	m	3.87
50mm	m	4.41

25mm rigid section fibre glass for copper pipes

15mm	m	2.75
22mm	m	2.82
28mm	m	3.02
35mm	m	3.35
42mm	m	3.63
54mm	m	4.26

ELECTRICAL INSTALLATIONS

Steel conduits
black enamelled, heavy gauge, welded, screwed

20mm	100m	240.72
25mm	100m	329.93
32mm	100m	457.11
1.5"	100m	585.11
2"	100m	945.64

galvanized, heavy gauge, welded, screwed

20mm	100m	363.47
25mm	100m	487.19
32mm	100m	616.75
1.5"	100m	786.26
2"	100m	1163.13

PVC Conduits
heavy gauge super high impact

16mm	100m	167.53
20mm	100m	167.53
25mm	100m	226.33
32mm	100m	363.90
38mm	100m	469.60
50mm	100m	780.94

oval section (nominal sizes)

16mm	100m	61.11
20mm	100m	96.85
25mm	100m	134.58

bending springs, heavy gauge

16mm	nr	10.31
20mm	nr	10.31
25mm	nr	15.30

PVC conduits, heavy gauge super high impact (Cont'd)
bending springs, heavy gauge (Cont'd)

32mm	nr	21.77
38mm	nr	34.68
50mm	nr	47.79

adhesive vinyl cement

1/4 litre	nr	7.39

Steel conduit fittings
bends, black enamel, screwed

20mm	10	22.65
25mm	10	36.06
32mm	10	63.32
1.5"	10	96.80
2"	10	192.23

inspection elbows, black enamel, screwed

20mm	10	43.84
25mm	10	54.40
32mm	10	114.99
1.5"	10	164.08
2"	10	318.49

inspection tees, black enamel, screwed

20mm	10	47.96
25mm	10	75.60
32mm	10	207.94

if galvanized, add 20% to above prices
black enamel standard circular boxes screwed to BS 31, 51mm centres, without covers, with 4mm tapped hole in base, for 20mm conduit

back outlet	10	71.84
terminal	10	35.09
terminal and back outlet	10	85.96
angle	10	41.75
angle tangent	10	88.31
through way	10	41.75
branch "U"	10	59.45
through way and back outlet	10	101.03
three way	10	45.88
three way tangent	10	100.71
branch three way	10	86.35
four way	10	58.24
twin through way	10	122.10
extra for light steel covers	10	5.70
extra for heavy steel covers	10	6.61
extra for rubber gaskets	10	11.20

black enamel standard circular boxes screwed to BS 31, 51mm centres, without covers, with 4mm tapped hole in base for 25mm conduit

back outlet	10	108.85
terminal	10	48.63
terminal and back outlet	10	123.55
angle	10	58.02
angle tangent	10	128.57
through way	10	58.02
branch "U"	10	128.41
through way and back outlet	10	138.79
three way	10	61.32
three way tangent	10	135.46
branch three way	10	143.19
four way	10	85.50
extra for light steel covers	10	5.70
extra for heavy steel covers	10	6.61
extra for rubber gaskets	10	11.20

32mm deep looping in boxes standard 4 hole pattern with 20mm knockouts and 51mm centre fixing lugs.

	10	18.30

saddles bar

20mm	10	5.17
25mm	10	6.47
32mm	10	18.47
1.5"	10	18.73
2"	10	22.74

BASIC PRICES OF MATERIALS - MAJOR WORKS

ELECTRICAL INSTALLATIONS (Cont'd)

Steel conduit fittings (Cont'd)

saddles, distance
20mm	10	16.63
25mm	10	21.58
32mm	10	30.37
1.5"	10	53.10
2"	10	75.34

brass bushes, hexagon smooth bore
20mm	10	7.06
25mm	10	12.66
32mm	10	25.54
1.5"	10	43.23
2"	10	71.79

brass bushes, ring
20mm	10	3.92
25mm	10	5.26
32mm	10	9.74
1.5"	10	16.91
2"	10	34.16

locknuts, circular pattern
20mm	10	2.35
25mm	10	4.19
32mm	10	9.90
1.5"	10	13.28
2"	10	27.14

couplers, solid
20mm	10	5.12
25mm	10	5.91
32mm	10	18.19
1.5"	10	24.31
2"	10	39.68
standard hook plates	10	24.77

standard pendant plates, internal thread
20mm	10	35.50
25mm	10	50.50

adaptable steel boxes, plain sides with flanged lid and screws
75 x 75 x 37mm deep	10	33.10
100 x 100 x 37mm deep	10	34.30
150 x 75 x 37mm deep	10	35.60
75 x 75 x 50mm deep	10	34.30
100 x 100 x 50mm deep	10	40.30
150 x 75 x 50mm deep	10	40.80
150 x 150 x 50mm deep	10	57.40
225 x 150 x 50mm deep	10	86.90
225 x 225 x 50mm deep	10	110.20
75 x 75 x 75mm deep	10	44.00
100 x 100 x 75mm deep	10	50.40
150 x 75 x 75mm deep	10	50.40
150 x 150 x 75mm deep	10	70.10
225 x 150 x 75mm deep	10	99.00
225 x 225 x 75mm deep	10	123.20
300 x 300 x 75mm deep	10	171.70
150 x 150 x 100mm deep	10	104.40
225 x 150 x 100mm deep	10	110.70
225 x 225 x 100mm deep	10	141.30
300 x 300 x 100mm deep	10	275.00

PVC conduit fittings

bends
20mm	10	23.51
25mm	10	31.66
32mm	10	53.54
38mm	10	93.20
50mm	10	163.52

inspection elbows
20mm	10	15.17

inspection tees
20mm	10	21.12
25mm	10	42.32

standard circular boxes, M4 threaded inserts, no earth terminal, for 20mm conduit
back outlet	10	22.46
terminal	10	16.29
terminal and back outlet	10	26.51
angle	10	18.51
through way	10	18.51
branch "U"	10	33.96

PVC conduit fittings (Cont'd)
standard circular boxes, M4 threaded inserts, no earth terminal, for 20mm conduit (Cont'd)
through way and back outlet	10	29.15
three way	10	20.26
branch three way	10	41.97
four way	10	23.78
twin through way	10	43.66
extra for standard cover	10	4.90
extra for overlapping cover	10	8.56
extra for rubber gasket	10	5.42
extra for brass earthing terminal	10	12.57

standard circular boxes, M4 threaded inserts, no earth terminal, for 25mm conduit
back outlet	10	26.89
terminal	10	25.48
terminal and back outlet	10	38.94
angle	10	27.66
through way	10	27.66
through way and back outlet	10	41.34
three way	10	30.43
four way	10	34.63
extra for standard cover	10	4.90
extra for overlapping cover	10	8.56
extra for rubber gasket	10	5.42
extra for brass earthing terminal	10	12.57

32mm deep looping in boxes with four 20mm knockouts, no earth terminal ... 10 ... 12.94

couplers, heavy gauge
20mm	10	3.98
25mm	10	5.42
32mm	10	12.31
38mm	10	23.95
50mm	10	41.99

couplers, expansion
20mm	10	9.51
25mm	10	12.31

saddles, spacer bar
20mm	10	5.95
25mm	10	7.60
32mm	10	15.37
38mm	10	27.25
50mm	10	46.25

adaptors, female thread to push-in with male bushes
20mm	10	6.13
25mm	10	8.76
32mm	10	17.13
38mm	10	35.84
50mm	10	62.40

adaptors, male thread to push-in with lock rings
20mm	10	6.45
25mm	10	8.76
32mm	10	17.13
38mm	10	35.84
50mm	10	62.40

Cables for use with conduit systems
single PVC insulated only 300/ 500V Class 6491X
1.0mm2 nominal	100m	32.30
1.5mm2 nominal	100m	52.50
2.5mm2 nominal	100m	76.00
4.0mm2 nominal	100m	125.10
6.0mm2 nominal	100m	181.70
10.0mm2 nominal	100m	339.80
16.0mm2 nominal	100m	524.80

Cables for use where conduit systems are not required
single PVC insulated and sheathed cable 300/500V Class 6181Y
1.0mm2 nominal	100m	52.60
1.5mm2 nominal	100m	69.90
2.5mm2 nominal	100m	116.50
4.0mm2 nominal	100m	189.90
6.0mm2 nominal	100m	252.60
10.0mm2 nominal	100m	401.70
16.0mm2 nominal	100m	522.50
25mm2 nominal	100m	1001.00

twin PVC insulated and sheathed cable 300/500V Class 6192Y
1.0mm2 nominal	100m	93.60
1.5mm2 nominal	100m	120.00
2.5mm2 nominal	100m	169.40
4.0mm2 nominal	100m	251.20
6.0mm2 nominal	100m	343.00
10.0mm2 nominal	100m	557.00
16.0mm2 nominal	100m	867.10

twin and earth PVC insulated and sheathed cable 300/500V Class 6242Y
1.0mm2 nominal	100m	80.40
1.5mm2 nominal	100m	100.40
2.5mm2 nominal	100m	137.30
4.0mm2 nominal	100m	427.00
6.0mm2 nominal	100m	507.00
10.0mm2 nominal	100m	818.50
16.0mm2 nominal	100m	1304.60

triple PVC insulated and sheathed cable 300/500V Class 6193Y
1.0mm2 nominal	100m	160.80
1.5mm2 nominal	100m	222.90
2.5mm2 nominal	100m	364.40
4.0mm2 nominal	100m	416.20
6.0mm2 nominal	100m	565.70
10.0mm2 nominal	100m	875.60
16.0mm2 nominal	100m	1380.40

triple and earth PVC insulated and sheathed cable 300/500V Class 6243Y
1.0mm2 nominal	100m	193.20
1.5mm2 nominal	100m	305.00

twin TRS flexible cords Class 3182
0.75mm2 nominal	100m	126.40
1.00mm2 nominal	100m	156.30
1.50mm2 nominal	100m	217.70
2.5mm2 nominal	100m	319.10

triple TRS flexible cords Class 3183
0.75mm2 nominal	100m	162.20
1.00mm2 nominal	100m	190.40
1.50mm2 nominal	100m	261.60
2.5mm2 nominal	100m	387.40

twin circular PVC flexible cords Class 3182Y
0.75mm2 nominal	100m	73.80
1.00mm2 nominal	100m	95.40
1.50mm2 nominal	100m	135.70
2.5mm2 nominal	100m	287.80

triple circular PVC flexible cords Class 3183Y
0.75mm2 nominal	100m	71.60
1.00mm2 nominal	100m	87.10
1.50mm2 nominal	100m	120.80
2.5mm2 nominal	100m	247.10
bell wire, PVC insulated only, twin (figure 8)	100m	35.40

MICS cable (Two complete terminations are required for each length of cable)
600V grade, 2 core
1.0mm2 conductor	100m	193.00
complete termination with earth tail	nr	4.84
1.5mm2 conductor	100m	227.50
complete termination with earth tail	nr	4.84
2.5mm2 conductor	100m	289.00
complete termination with earth tail	nr	4.84

ELECTRICAL INSTALLATIONS (Cont'd)

MICS cable (Two complete terminations are required for each length of cable)
600V grade, 2 core (Contd)

4.0mm2 conductor........ 100m		384.50
complete termination with earth tail.................... nr		4.84

600V grade, 3 core

1.0mm2 conductor........... 100m		203.00
complete termination with earth tail.................... nr		4.84
1.5mm2 conductor.......... 100m		257.52
complete termination with earth tail.................... nr		4.84
2.5mm2 conductor.......... 100m		404.00
complete termination with earth tail.................... nr		4.84

600V grade, 4 core

1.0mm2 conductor.......... 100m		243.00
complete termination with earth tail.................... nr		4.84
1.5mm2 conductor.......... 100m		311.00
complete termination with earth tail.................... nr		4.84
2.5mm2 conductor.......... 100m		490.50
complete termination with earth tail.................... nr		4.84

1000V grade, 2 core

1.5mm2 conductor.......... 100m		303.00
complete termination with earth tail.................... nr		4.84
2.5mm2 conductor......... 100m		372.00
complete termination with earth tail.................... nr		4.84
4.0mm2 conductor......... 100m		469.00
complete termination with earth tail.................... nr		10.27
6.0mm2 conductor.......... 100m		625.00
complete termination with earth tail.................... nr		10.27
10.0mm2 conductor......... 100m		809.00
complete termination with earth tail.................... nr		17.22
16.0mm2 conductor........ 100m		1165.00
complete termination with earth tail.................... nr		30.79
25.0mm2 conductor........ 100m		1635.00
complete termination with earth tail.................... nr		30.79

1000V grade, 3 core

1.5mm2 conductor........... 100m		335.00
complete termination with earth tail.......................nr		4.84
2.5mm2........................ 100m		472.00
complete termination with earth tail........................ nr		4.84
4.0mm2 conductor...........100m		535.00
complete termination with earth tail........................ nr		10.27
6.0mm2 conductor.......... 100m		690.00
complete termination with earth tail........................ nr		17.22
10.0mm2 conductor........ 100m		998.00
complete termination with earth tail........................ nr		30.79
16.0mm2 conductor....... 100m		1401.00
complete termination with earth tail........................ nr		26.62
25.0mm2 conductor........ 100m		2150.00
complete termination with earth tail........................ nr		30.79

1000V grade, 4 core

1.5mm2 conductor...........100m		417.00
complete termination with earth tail........................ nr		4.84
2.5mm2 conductor...........100m		524.00
complete termination with earth tail........................ nr		4.84
4.0mm2 conductor...........100m		655.00

MICS cable (Two complete terminations are required for each length of cable)
1000V grade, 4 core (Cont'd)
4.0mm2 conductor (Cont'd)

complete termination with earth tail........................ nr		10.27
6.0mm2 conductor...........100m		874.00
complete termination with earth tail........................ nr		17.22
10.0mm2 conductor.........100m		1240.00
complete termination with earth tail........................ nr		30.79
16.0mm2 conductor........100m		1809.00
complete termination with earth tail........................ nr		30.79
25.0mm2 conductor.........100m		2628.00
complete termination with earth tail........................ nr		30.79

Switch and fuse gear
metal cased switch fuses, H.R.C. pattern
SP and N, 500V

20A............................. nr		58.09
32A............................. nr		72.94
63A............................. nr		114.52
100A............................. nr		183.04

TP and N, 500V

20A............................. nr		76.02
32A............................. nr		100.97
63A............................. nr		169.65
100A............................. nr		293.81

metal cased consumers control units
250V SP and N, 60A D.P. main switch, MCB included

four way........................ nr		49.98
six way........................ nr		66.52
eight way......................nr		83.74

metal cased consumers control units
250V SP and N, fitted with 80A D.P.
R.C.C.B (Residual Current Device), 30m A trip, M.C.B. pattern

four way.......................nr		95.61
eight way......................nr		138.48
ten way.........................nr		156.69

Lighting switches
400W Tungsten load dimmer, flush, plastic with plaster depth box, white, one gang..................nr ... 22.50
5A surface metalclad, aluminium finish, two way switch

one gang........................ nr		4.32
two gang..........................nr		5.25
three gang........................nr		8.18
four gang........................nr		11.51
six gang..........................nr		13.92

5A surface plastic with moulded box, white

one gang........................ nr		3.01
two gang..........................nr		4.44
three gang....................... nr		5.98
four gang........................nr		10.65

5A flush plastic with plaster depth box, white

one gang........................ nr		3.21
two gang......................... nr		4.65
three gang.......................nr		6.18
four gang........................nr		10.95

5A flush brass with standard depth box, matt chrome or satin brass, two way switch

one gang........................ nr		9.23
two gang..........................nr		10.70
three gang...................... nr		15.62
intermediate.....................nr		9.69

surface plastic ceiling switch, ivory

5A...............................nr		3.57
15A............................. nr		11.47

Lighting Switches (Cont'd)
surfaceplasticceiling switch, ivory (Cont'd)

45A......................................nr		11.47

splashproof all insulated switch, 5A

one gang, two way..............nr		5.72
two gang, two way.............. nr		12.18

Socket outlets and plugs
three pin socket outlets with boxes or back plates as required
flush, insulated, white

5A plain................................nr		7.89
5A switched........................ nr		8.92
13A single plain...................nr		4.72
13A single switched............nr		5.77
13A twin switched...............nr		10.79

surface, insulated, white

5A plain............................. nr		7.67
5A switched........................ nr		8.67
13A single plain...................nr		4.58
13A single switched............nr		5.49
13A twin switched...............nr		10.49

metalclad, aluminium, surface

5A plain............................. nr		9.36
5A switched........................ nr		10.26
13A single plain...................nr		6.25
13A single switched............nr		8.43
13A twin switched...............nr		16.66

flush, metalclad, matt, chrome or satin brass

5A switched........................nr		14.37
13A single plain................. nr		16.33
13A single switched............nr		10.53
13A twin switched...............nr		17.97

plugs, white

5A................................. nr		2.73
13A................................. nr		2.24

three pin socket outlets R.C.C.B. (Residual Current Device)
protected at 30mA trip
flush insulated, white 1 gang

13A switched....................nr		60.67

surface, metalclad, 1 gang

13A switched....................nr		65.13

Power accessories
water heater switch complete with box, flush, white, 20A............ nr ... 12.34
cooker unit, white finish, metal box, with pilot lamps and auxiliary 13A socket, flush or surface......... nr ... 42.37
20A double, pole switches with boxes
flush, insulated, white............ nr ... 7.75
flush, insulated, white with pilot lamp and flex outlet........ nr ... 11.29
surface, metalclad with pilot lamp and flex outlet.............. nr ... 12.20
flush, metalclad with pilot lamp and flex outlet, BMA or matt chrome..............................nr ... 16.61
13A fused connection units (spurs) with switches and boxes
flush, insulated, white............ nr ... 9.13
flush, insulated, white with pilot lamp and flex outlet........ nr ... 12.31
surface, metalclad with pilot lamp and flex outlet.............. nr ... 11.85
flush, metalclad with pilot lamp and flex outlet, BMA or matt chrome.............................. nr ... 18.38

Wiring accessories
earthing and bonding clamps for pipes
13mm-32mm, conductor

1 x 10mm2...........................nr		0.66

32mm-50mm, conductor

2 x 16mm2............................nr		1.10

50mm-75mm, conductor

2 x 16mm2............................nr		1.66

ELECTRICAL INSTALLATIONS (Cont'd)

Wiring accessories (Cont'd)
junction boxes, bakelite, white or brown

small	nr	1.58
large	nr	1.84

connectors, block type
single pole, porcelain

5A	100	54.30
15A	100	85.10

earthing and bonding clamps for pipes
double pole, porcelain

5A	100	97.01
15A	100	164.77

cable clips, plastic, push fit, with pins

5mm	100	1.83
7mm	100	1.90
10mm	100	2.49

cable clips, for flat twin and earth cables, plastic, push fit, with pins

1.0 and 1.5mm2	1000	19.80
2.5mm2	1000	22.10
4.0mm2	1000	27.41
6.0mm2	1000	28.80
10.0mm2	1000	37.57
16.0mm2	1000	52.50

ceiling roses
surface, bakelite, white

2 plate and earth	nr	1.93
3 plate and earth	nr	3.49

surface, plug type

3 plate and earth	nr	4.26

Lamp holders
bakelite white

cord grip, standard pattern	nr	1.94
batten holder with HO shield	nr	4.75

Electric lamps
240V electric lamps, bayonet cap GLS

25W	nr	1.21
40W	nr	0.84
60W	nr	0.84
100W	nr	0.84
150W	nr	1.76
200W	nr	2.89
150W PAR 38 E.S. cap	nr	9.57

Fluorescent lamps
fluorescent lamps suitable for operation on either instant start or switch start circuits
450mm long, 15W

white	nr	7.19
warm white	nr	8.24

600mm long, 20W

white	nr	6.91
warm white	nr	7.45

600mm long, 40W

white	nr	7.62

900mm long, 30W

white	nr	7.32
warm white	nr	7.62

1200mm long, 40W

white	nr	7.04
warm white	nr	7.57

1500mm long, 65/80W

white	nr	7.77
warm white	nr	8.31

1800mm long, 75/85W

white	nr	11.20
warm white	nr	11.94

2400mm long, 125W

white	nr	13.02
warm white	nr	13.98

2D lamp, 16W 2 pin

warm polylux 2700	nr	6.58

2D lamp, 28W 2 pin

warm polylux 2700	nr	9.89

Fluorescent Lamps (Cont'd)
Powersaver fluorescent lamps suitable for operation on switch start circuits only
1200mm long, 36W

white	nr	5.12
warm white	nr	6.01

Underground Services Warning Tape 150mm wide x 100 micron thick in 365m rolls

for electric cables (yellow)	roll	54.75
for telephone cables (green)	roll	52.49

FLOOR WALL AND CEILING FINISHINGS

Portland cement, 7.5 tonnes to 10 tonnes	tonne	80.25
Sand	tonne	19.75
Granite chippings	tonne	36.90
Surface hardener	25 litre	37.10
Floor sealer	25 litre	137.00
Bonding fluid	5 litre	22.65

Gypsum plaster to BS 1191 Part 1 Class B

universal one coat plaster	tonne	163.00
Multi finish plaster	tonne	102.50
board finish plaster	tonne	102.50
hardwall plaster	tonne	139.50

Lightweight gypsum plaster to BS 1191 Part 2

browning plaster	tonne	141.50
bonding plaster	tonne	139.50
finish plaster	tonne	109.00

Renovating plaster

Thistle undercoat	tonne	146.00
Thistle finish	tonne	134.50

Tyrolean finish

Cullarend mixture	tonne	395.00
machine	hour	1.25

Plaster beads and stops
Expamet, galvanized
angle bead

reference 550	m	0.58

plaster stop bead for the following thickness plaster

10mm reference 562	m	0.68
13mm reference 563	m	0.68
16mm reference 565	m	0.87
19mm reference 566	m	0.87

depth gauge bead

reference 569	m	0.55

thin coat plaster angle bead for the following thickness plaster

3mm reference 553	m	0.70
6mm reference 554	m	0.91

thin coat plaster stop bead for the following thickness plaster

3mm reference 560	m	0.58
6mm reference 561	m	0.75
render stop reference 570	m	0.70

plasterboard edging bead for the following thickness plasterboard

9.5mm reference 567	m	0.99
12.7mm reference 568	m	0.99

Expamet, stainless steel

angle bead reference 545	m	1.69
render stop reference 547	m	1.49

Clay floor quarries to BS 6431
Terracotta

150 x 150 x 12.5mm	1000	253.00
150 x 150 x 20mm	1000	358.00
225 x 225 x 25mm	1000	1328.00

Blended

150 x 150 x 12.5mm	1000	341.00
150 x 150 x 20mm	1000	384.00
225 x 225 x 25mm	1000	1494.00

Dark

150 x 150 x 12.5mm	1000	442.00

With one rounded edge
Terracotta

150 x 150 x 12.5mm	1000	328.00
150 x 150 x 20mm	1000	417.00
225 x 225 x 25mm	1000	1490.00

Blended

150 x 150 x 12.5mm	1000	390.00
150 x 150 x 20mm	1000	417.00
225 x 225 x 25mm	1000	1667.00

Dark

150 x 150 x 12.5mm	1000	515.00

Rounded edge cove skirting tiles
Terracotta

150 x 100 x 12.5mm	1000	430.00
150 x 138 x 12.5mm	1000	518.00

Blended

150 x 100 x 12.5mm	1000	488.00
150 x 138 x 12.5mm	1000	547.00

Dark

150 x 138 x 12.5mm	1000	713.00

Ceramic floor tiles to BS 6431
plain
red

100 x 100 x 9mm	m2	13.85
152 x 152 x 12mm	m2	15.20
200 x 200 x 12mm	m2	16.40

Cream

100 x 100 x 9mm	m2	14.75
152 x 152 x 12mm	m2	15.95
200 x 200 x 12mm	m2	17.05

black

100 x 100 x 9mm	m2	18.10
152 x 152 x 12mm	m2	19.00
200 x 200 x 12mm	m2	19.80

with one rounded edge
red

100 x 100 x 9mm	m2	18.50
152 x 152 x 12mm	m2	19.95
200 x 200 x 12mm	m2	25.70

Cream

100 x 100 x 9mm	m2	19.40
152 x 152 x 12mm	m2	20.30
200 x 200 x 12mm	m2	25.70

black

100 x 100 x 9mm	m2	22.75
152 x 152 x 12mm	m2	21.75
200 x 200 x 12mm	m2	25.70

rounded edge cove skirting tiles
red

152 x 100 x 9mm	100	54.35
152 x 112 x 12mm	100	72.40
152 x 152 x 12mm	100	86.25

Cream

152 x 100 x 9mm	100	61.15
152 x 112 x 12mm	100	72.40
152 x 152 x 12mm	100	86.25

black

152 x 100 x 9mm	100	68.25
152 x 112 x 12mm	100	76.85
152 x 152 x 12mm	100	89.65

Glazed ceramic tiles to BS 6431
white

108 x 108 x 4mm	m2	18.00
152 x 152 x 5.5mm	m2	13.10

light colours

108 x 108 x 4mm	m2	18.00
152 x 152 x 5.5mm	m2	13.15

FLOOR WALL AND CEILING FINISHINGS (Cont'd)

Glazed ceramic tiles to BS 6431 (Cont'd)
dark colours

108 x 108 x 6.5mm.........	m2	22.20
152 x 152 x 5.5mm..........	m2	16.25

Gypsum plasterboard to BS 1230 Part 2
baseboard

9.5mm............................	m2	1.61
9.5mm insulating............	m2	2.19

lath

9.5mm............................	m2	1.87
12.7mm..........................	m2	2.22

wallboard

9.5mm............................	m2	1.61
9.5mm insulating..........	m2	2.19
12.5mm..........................	m2	1.87
12.5mm insulating..........	m2	2.45
15mm............................	m2	2.06
15mm insulating............	m2	2.69

plank with square edges

19mm............................	m2	3.06
joint Tape 50mm wide.....	150m	3.10

Plastic faced plasterboard

9.5mm with aluminium foil backing............................	m2	4.52
12.5mm with aluminium foil backing.....................	m2	4.87
PVC batten trim...............	m	0.77
plastic coated nail caps....	1000	8.85

Hardboard
standard

3.2mm............................	m2	0.90
4.8mm............................	m	1.50
6.0mm............................	m2	1.95

oil tempered

3.2mm............................	m2	1.50
4.8mm............................	m2	2.11
6.0mm............................	m2	2.77

flame retardent

3.2mm............................	m2	5.25
6.4mm............................	m2	6.85

Low density medium board
type LME

6mm............................	m2	6.44
9mm............................	m2	8.24
12mm..........................	m2	10.51

type LMN

6mm............................	m2	5.51
9mm............................	m2	6.85

Medium density fibre board
type MDF

6mm............................	m2	3.40
12mm..........................	m2	4.33
18mm..........................	m2	5.25
25mm..........................	m2	7.93

Insulating board 13mm thick

plain type SBN..............	m2	1.86
bitumen impregnated type SBI.............................	m2	2.20

Chipboard, wood base
flooring grade square edge

18mm..........................	m2	3.83
22mm..........................	m2	5.15

flooring grade t & g edge

18mm..........................	m2	4.50
22mm..........................	m2	5.80

Asbestos - free board
Masterboard

6mm............................	m2	6.85
9mm............................	m2	13.85
12mm..........................	m2	18.30

Supalux

6mm............................	m2	10.10
9mm............................	m2	15.00
12mm..........................	m2	21.80

Vermiculux

20mm..........................	m2	16.60
30mm..........................	m2	27.15
40mm..........................	m2	41.20
50mm..........................	m2	54.85
60mm..........................	m2	65.40

Expanded polystyrene
sheet polystyrene, density 16 kg/m3

13mm..........................	m2	0.62
19mm..........................	m2	0.96
25mm..........................	m2	1.22
50mm..........................	m2	2.41
75mm..........................	m2	3.63

sheet polystyrene, non-flammable

13mm..........................	m2	0.97
19mm..........................	m2	1.15
25mm..........................	m2	1.39
50mm..........................	m2	3.34
75mm..........................	m2	5.02

Cellular partitions

57mm panels.................	m2	6.64
63mm panels.................	m2	9.02

Laminated partitions
plasterboard components

50mm..........................	m2	6.94
65mm..........................	m2	9.36

Metal stud partitions
studs

48mm..........................	m	0.90

floor and ceiling channels

50mm..........................	m	0.82
12.7mm, tapered edge plaster- board with one face for direct decoration......................	m2	1.91

Lightweight aggregate, medium grade

10-5mm, 799kg/m3........	m3	37.00

Levelling screeding

compound......................	kg	0.62

Metal lathing
BB galvanized expanded metal lath, 9mm mesh

0.500mm thick x 1.11kg/ m2..................................	m2	3.79
0.725mm thick x 1.61kg/ m2..................................	m2	4.43

Expamet galvanized expanded metal lath, 9mm mesh

0.950mm thick x 2.50kg/ m2..................................	m2	7.20

Expamet galvanized Rib Lath

0.400mm thick x 1.35kg/ m2..................................	m2	4.76
0.500mm thick x 2.25kg/ m2..................................	m2	5.47

Expamet stainless steel Rib-Lath

0.300mm thick x 1.52kg/ m2..................................	m2	12.09

Expamet galvanized Spray-lath

0.500mm thick x 2.25kg/ m2..................................	m2	6.66

Metal lathing (Cont'd)
BB galvanized expanded metal lath, Expamet galvanized Red-rib lath

0.500mm thick x 2.25kg..	m2	6.10

Chicken wire
galvanized

13mm mesh x 22 gauge..	m2	2.58
25mm mesh x 22 gauge..	m2	1.29
38mm mesh x 19 gauge..	m2	1.44
50mm mesh x 19 gauge..	m2	1.08

GLAZING

Float glass
GG quality

3mm............................	m2	18.20
4mm............................	m2	18.20
5mm............................	m2	26.60
6mm............................	m2	26.60
10mm..........................	m2	53.88

obscured ground

3mm............................	m2	43.89
4mm............................	m2	48.16

Rough cast glass

6mm............................	m2	34.03

Georgian wired cast glass

7mm............................	m2	28.02

Georgian safety wired cast glass

7mm............................	m2	36.45

Patterned glass
white

4mm............................	m2	22.60
6mm............................	m2	34.03

tinted

4mm............................	m2	33.38
6mm............................	m2	37.77

Thick float glass
GG quality

12mm..........................	m2	62.06
15mm..........................	m2	103.26
19mm..........................	m2	146.27
25mm..........................	m2	232.19

Polished georgian wired glass

6mm............................	m2	60.56

Polished georgian safety wired glass

6mm............................	m2	71.31

Antisun float glass
grey

4mm............................	m2	34.71
6mm............................	m2	50.08
10mm..........................	m2	89.54
12mm..........................	m2	123.59

bronze

4mm............................	m2	34.71.
6mm............................	m2	50.08
10mm..........................	m2	89.54
12mm..........................	m2	123.59

green

6mm............................	m2	64.20

Toughened safety glass
clear float glasses

4mm............................	m2	26.85
5mm............................	m2	40.27
6mm............................	m2	40.27
10mm..........................	m2	63.89
12mm..........................	m2	117.00

solar control glasses, antisun
bronze

4mm............................	m2	49.22
6mm	m2	51.36
10mm..........................	m2	101.65
12mm..........................	m2	133.75

BASIC PRICES OF MATERIALS - MAJOR WORKS

GLAZING (Contd)

Toughened safety glass (Contd)
solar control glasses, antisun grey

4mm	m2	49.22
6mm	m2	51.36
10mm	m2	101.65
12mm	m2	133.75

white patterned glasses

4mm	m2	54.12
6mm	m2	56.50

Clear laminated glass
safety glass - float quality

4.4mm	m2	40.89
6.4mm	m2	34.28
8.8mm	m2	61.70
10.8mm	m2	72.17

anti-bandit glass - float quality

7.5mm	m2	69.42
9.5mm	m2	68.02
11.5mm	m2	73.83

Clear polycarbonate sheet
standard sheet with latex paper
masked both sides
Standard

2mm	m2	39.45
3mm	m2	58.93
4mm	m2	78.81
5mm	m2	98.60
6mm	m2	118.05
8mm	m2	157.40
9.5mm	m2	187.01
12mm	m2	236.06

Abrasion resistant

3mm	m2	91.39
4mm	m2	108.84
5mm	m2	130.50
6mm	m2	145.85
8mm	m2	176.53
9.5mm	m2	199.44

Sundries

linseed oil putty	25 kg	13.10
metal casement putty	25 kg	13.77
washleather strip	m	0.60
rubber glazing strip	m	0.56

PAINTING AND DECORATING

Creosote	200 ltr	134.00
Wood preservative	5 litre	20.65

Emulsion paint
matt

brilliant white	5 litre	8.20
colours	5 litre	9.05

silk

brilliant white	5 litre	10.65
colours	5 litre	12.40

Priming, paint

wood primer	5 litre	17.05
primer sealer	5 litre	18.00
aluminium sealer and wood primer	5 litre	21.15
alkali-resisting primer	5 litre	19.90
universal primer	5 litre	21.50
red oxide primer	5 litre	17.55
calcium plumbate primer	5 litre	21.50
zinc phosphate primer	5 litre	38.95
acrylic primer	5 litre	16.65

Undercoat paint

brilliant white	5 litre	13.30
colours	5 litre	13.75

Gloss paint

brilliant white	5 litre	13.30
colours	5 litre	13.75

Eggshell paint

brilliant white	5 litre	15.05
colours	5 litre	17.25

Plastic finish

coating compound	25 kg	14.95

Cement painting
cement paint

cream or white	40 kg	40.00
colours	40 kg	40.00
sealer	5 litre	16.50

textured masonry paint

white	10 litre	22.05
colours	10 litre	25.45
stabilising solution	20 litre	56.75

Mulitcolour painting

primer	5 litre	11.55
basecoat	5 litre	19.50
finish	5 litre	24.50

Chlorinated rubber

Primer	5 litre	43.75
finish	5 litre	44.00

Metallic paint

aluminium	5 litre	23.70
gold or bronze	5 litre	33.20
Bituminous paint	5 litre	10.05

Polyurethane lacquer

clear	5 litre	25.55

Linseed oil

raw	5 litre	12.95
boiled	5 litre	14.00

Fire retardant paints and varnishes

fire retardant paint	5 litre	40.30
fire retardant varnish	5 litre	29.20
overcoat varnish	5 litre	42.95

French polishing

clear polish	1 litre	7.00
button polish	1 litre	7.00
garnet polish	1 litre	7.00
methylated spirit	1 litre	1.75

DRAINAGE

Land drains
clay pipe in 300mm lengths

75mm	m	0.95
100mm	m	1.70
150mm	m	3.62

porous concrete pipe B.S.1194

150mm	m	4.15
225mm	m	5.35
300mm	m	8.75

perforated clay pipe

100mm	m	4.22
150mm	m	7.67
225mm	m	14.09

Vitrified clay flexible joint drains
(B.S.EN295)
pipe

100mm	m	6.69
150mm	m	8.72
225mm	m	16.85
300mm	m	26.35
400mm	m	54.08
450mm	m	70.25

Vitrified clay flexible joint drains
(B.S. EN295) (Cont'd)
90 degree bend

100mm	nr	9.71
150mm	nr	16.00
225mm	nr	32.64
300mm	nr	64.40
400mm	nr	145.13
450mm	nr	267.59

45 degree bend

100mm	nr	9.71
150mm	nr	16.00
225mm	nr	32.64
300mm	nr	64.40
400mm	nr	145.13
450mm	nr	191.12

30 degree bend

100mm	nr	9.08
150mm	nr	16.00
225mm	nr	32.64
300mm	nr	64.40

15 degree bend

100mm	nr	9.08
150mm	nr	16.00
225mm	nr	32.64
300mm	nr	64.40

rest bend

100mm	nr	11.53
150mm	nr	19.10
225mm	nr	45.47

90 degree curved square junction

100mm	nr	13.47
150mm	nr	20.90
225mm	nr	49.08
300mm	nr	101.18

45 degree oblique junction
with supersleve arm

150mm	nr	41.87
225mm	nr	89.72
300mm	nr	190.40

45 degree oblique junction

100mm	nr	13.47
150mm	nr	20.90
225mm	nr	49.08
300mm	nr	101.18

tumbling bay junction

100mm	nr	13.47
150mm	nr	20.90
225mm	nr	49.08
300mm	nr	121.44
400mm	nr	267.50

double collar

100mm	nr	8.63
150mm	nr	14.24
225mm	nr	31.13
300mm	nr	50.60
lubricant	2.5kg	7.05

Vitrified clay sleeve drains (B.S. EN295)
pipe(supersleve)

100mm	m	2.69
150mm	m	5.45

couplings with standard rings

100mm	nr	1.87
150mm	nr	3.38

couplings with neoprene rings

100mm	nr	2.83
150mm	nr	5.33

bend(plain ended)

100mm	nr	3.64
150mm	nr	7.49

rest bend

100mm(single socket)	nr	8.21

150mm(plain ended)
single junction(plain ended)

100 x 100mm	nr	7.87
150 x 100mm	nr	10.04
150 x 150mm	nr	11.82

DRAINAGE (Cont'd)

Vitrified clay drains (B.S. EN295)
British Standard(unjointed)
pipe

100mm	m	4.70
150mm	m	7.23
225mm	m	14.32
300mm	m	24.00

bend

100mm	nr	4.70
150mm	nr	7.23

rest bend

100mm	nr	7.72
150mm	nr	13.05

single junction

100mm	nr	8.63
150mm	nr	14.28

double junction

150mm	nr	17.05

double collar

100mm	nr	5.66
150mm	nr	9.43
225mm	nr	22.07

Drains PVC-U B.S.4660
pipe

110mm	m	2.81
160mm	m	6.29

coupling with double socket

110mm	nr	2.81
160mm	nr	5.68

45 degree bend

110mm	nr	6.78
160mm	nr	16.12

single junction

110mm	nr	9.57
160mm	nr	29.10

Drains: PVC-U solid wall
concentric external rib
reinforced Wavin "Ultra-Rib"
pipe

150mm 6UR 046	m	3.11
225mm 9UR 046	m	7.13
300mm 12UR 043	m	10.70

connectors

150mm 6UR 205	nr	5.27
225mm 9UR 205	nr	11.01
300mm 12UR 205	nr	22.20

bends

150mm 6UR 563	nr	7.04
225mm 9UR 563	nr	28.45
300mm 12UR 563	nr	46.37

branches

150mm 6UR 213	nr	17.13
225mm 9UR 213	nr	50.97
300mm 12UR 213	nr	108.63

connection to clayware pipe

150mm 6UR 129	nr	12.90
225mm 9UR 109	nr	28.51
300mm 12UR 112	nr	74.98

Concrete drains (BS 5911)
Class H, unreinforced, flexible
joints
pipe

300mm	m	11.99
375mm	m	16.20
450mm	m	18.82
525mm	m	22.68
600mm	m	27.57

bend

300mm	nr	85.92
375mm	nr	129.60
450mm	nr	150.56
525mm	nr	181.44
600mm	nr	220.56

Class H, reinforced, flexible joints
pipe

450mm	m	32.00
525mm	m	36.50
600mm	m	39.75

bend

450mm	nr	256.00
525mm	nr	292.00
600mm	nr	318.75

junction:extra cost on pipe up to 600mm diameter

100mm	nr	35.00
150mm	nr	40.00
225mm	nr	45.00
300mm	nr	69.00
375mm	nr	74.00
450mm	nr	96.00

Spun iron drains (BS 437) for
coupling joints
spigoted pipe

100mm	m	16.02
150mm	m	30.17

couplings with gaskets

100mm	nr	10.24
150mm	nr	12.40

medium radius bend

100mm	nr	19.16
150mm	nr	38.66

equal branch

100mm	nr	25.42
150mm	nr	54.89

connector with large socket for clayware

100mm	nr	16.77
150mm	nr	27.20

Vitrified clay manhole channels
British Standard Surface Water
Quality
half round, straight, 600mm long

100mm	nr	1.69
150mm	nr	2.84

half round, straight, 1000mm long

100mm	nr	2.36
150mm	nr	4.46
225mm	nr	10.43
300mm	nr	21.95

half round, curved, 500mm girth

100mm	nr	2.39
150mm	nr	4.14
225mm	nr	16.07
300mm	nr	32.75

half round, curved, 900mm girth

100mm	nr	2.39
150mm	nr	4.14
225mm	nr	16.07
300mm	nr	32.75

three-quarter section branch channel bend

100mm	nr	6.76
150mm	nr	11.75
225mm	nr	41.36

Precast concrete inspection chambers
Unreinforced
Chamber ring 900mm diameter

305mm high	nr	18.93
458mm high	nr	28.43
610mm high	nr	23.66
762mm high	nr	29.56
914mm high	nr	35.45

Chamber ring 1050mm diameter

305mm high	nr	21.63
458mm high	nr	32.49
610mm high	nr	27.04

Precast concrete inspection chambers
Unreinforced (Cont'd)

762mm high	nr	33.78
914mm high	nr	40.52

Chamber ring 1200mm diameter

305mm high	nr	28.39
458mm high	nr	42.63
610mm high	nr	35.48
762mm high	nr	44.43
914mm high	nr	53.17

Chamber ring 1350mm diameter

458mm high	nr	61.93
610mm high	nr	51.55
762mm high	nr	64.40
914mm high	nr	77.24

Chamber ring 1500mm diameter

500mm high	nr	77.93
750mm high	nr	77.93
1000mm high	nr	103.90

Chamber ring 1800mm diameter

500mm high	nr	90.40
750mm high	nr	90.40
1000mm high	nr	120.53

Cover slab, heavy duty

900mm diameter	nr	39.25
1050mm diameter	nr	45.05
1200mm diameter	nr	59.59
1350mm diameter	nr	87.29
1500mm diameter	nr	103.90

Cast iron inspection chambers for
coupling joints
straight
with one branch one side

100mm	nr	69.53
100-150mm	nr	104.51
150mm	nr	133.54

with one branch each side,

100mm	nr	89.57
150mm	nr	147.27

with two branches one side,

100mm	nr	139.25
150mm	nr	251.65

with two branches each side,

100mm	nr	176.34
150mm	nr	294.09

Manhole step irons (BS 1247)
galvanized
with 225mm tail, weighing

2.16kg	nr	2.75

Cast iron manhole covers and frames
(B.S. EN 124)
Light duty, single seal, solid top

MC1-45/45, 450 x 450mm	nr	30.66
MC1-60/45, 600 x 450mm	nr	30.83
MC1-60/60, 600 x 600mm	nr	65.53

Light duty, single seal, recessed top

MC1R-60/60, 600 x 600mm	nr	86.83

Light duty, double seal, solid top

MC2-45/45, 450 x 450mm	nr	47.18
MC2-60/45, 600 x 450mm	nr	61.79
MC2-60/60, 600 x 600mm	nr	87.14

Light duty, double seal, recessed top

MC2R-45/45, 450 x 450mm	nr	78.36
MC2R-60/45, 600 x 450mm	nr	72.79
MC2R-60/60, 600 x 600mm	nr	104.99

Medium duty,Class 2, single seal solid top

MB2-50, 500mm diameter	nr	74.79
MB2-55, 550mm diameter	nr	77.10
MB2-60, 600mm diameter	nr	57.05
MB2-60/45, 600 x 450mm	nr	51.75

DRAINAGE (Cont'd)

Cast iron manhole covers and frames
(B.S.EN124) (Cont'd)
Medium duty,Class 2, single seal solid top

MB2-60/60, 600 x 600mm... nr		63.16

Medium duty, Class 2, single seal recessed top

MB2R-60/45, 600 x 450mm.nr		92.78
MB2R-60/60, 600 x 600mm.nr		105.55
MB1-60, 600mm diameter... nr		66.69

Heavy duty

MA-50, 500mm diameter......nr		110.24
MA-60, 600mm diameter......nr		64.08
MA-T, 490 x 495mm.............nr		106.53
lifting keys............................nr		.90

FENCING

Post and wire fencing
galvanized line wire

3.15mm diameter................ m		0.05
4.00mm diameter................ m		0.08
5.00mm diameter................ m		0.13

painted angle iron posts
intermediate post

for 900mm high fencing....... nr		2.54
for 1200mm high fencing..... nr		3.05
for 1400mm high fencing...... nr		3.30

end straining post with one strut

for 900mm high fencing....... nr		15.07
for 1200mm high fencing...... nr		19.33
for 1400mm high fencing...... nr		20.60

intermediate straining post with two struts

for 900mm high fencing....... nr		23.36
for 1200mm high fencing...... nr		29.88
for 1400mm high fencing...... nr		31.86

corner straining post with two struts

for 900mm high fencing....... nr		22.19
for 1200mm high fencing...... nr		28.13
for 1400mm high fencing...... nr		30.11

reinforced concrete posts
intermediate post

for 900mm high fencing....... nr		4.42
for 1050mm high fencing...... nr		5.01

end straining post with one strut

for 900mm high fencing....... nr		9.71
for 1050mm high fencing...... nr		13.94

intermediate straining post with two struts

for 900mm high fencing....... nr		14.01
for 1050mm high fencing...... nr		18.90

corner staining post with two struts

for 900mm high fencing....... nr		14.01
for 1050mm high fencing....... nr		18.90

Chain link fencing
galvanized chain link mesh
50mm x 14 G (2.00mm)

900mm............................... m		0.68
1200mm............................. m		0.97

50mm x 12 1/2 G (2.50mm)

900mm............................... m		1.04
1200mm............................. m		1.40
1400mm............................. m		1.62
1800mm............................. m		2.04

50mm x 10 1/2 G (3.00mm)

900mm............................... m		1.45
1200mm............................. m		1.83
1400mm............................. m		2.30
1800mm............................. m		2.82

plastic coated chain link mesh
50mm x 12 1/2 G

900mm............................... m		1.77
1200mm............................. m		2.41
1400mm............................. m		2.78
1800mm............................. m		3.24

50mm x 10 1/2 G

900mm............................. m		2.52
1200mm........................... m		3.29

Chain link fencing (Cont'd)
plastic coated chain link mesh (Cont'd)
50mm x 10 1/2 G

1400mm............................. m		3.69
1800mm............................. m		4.10

painted angle iron posts
intermediate post

for 900mm high fencing... nr		2.54
for 1200mm high fencing.. nr		3.05
for 1400mm high fencing.. nr		3.30
for 1800mm high fencing.. nr		4.31

end straining post with one strut

for 900mm high fencing....... nr		15.07
for 1200mm high fencing...... nr		19.33
for 1400mm high fencing...... nr		20.60
for 1800mm high fencing...... nr		27.75

intermediate straining post with two struts

for 900mm high fencing....... nr		23.36
for 1200mm high fencing...... nr		29.88
for 1400mm high fencing...... nr		31.86
for 1800mm high fencing....... nr		41.69

corner straining post with two struts

for 900mm high fencing....... nr		22.19
for 1200mm high fencing...... nr		28.13
for 1400mm high fencing...... nr		30.11
for 1800mm high fencing....... nr		39.94

reinforced concrete posts
intermediate post

for 900mm high fencing....... nr		4.41
for 1200mm high fencing...... nr		5.23
for 1400mm high fencing...... nr		6.00
for 1800mm high fencing....... nr		7.98

end straining post with one strut

for 900mm high fencing....... nr		9.71
for 1200mm high fencing...... nr		13.16
for 1400mm high fencing...... nr		14.84
for 1800mm high fencing....... nr		17.94

intermediate straining post with two struts

for 900mm high fencing....... nr		14.01
for 1200mm high fencing...... nr		18.12
for 1400mm high fencing....... nr		20.49
for 1800mm high fencing....... nr		24.81

corner straining post with two struts

for 900mm high fencing........ nr		14.01
for 1200mm high fencing....... nr		18.12
for 1400mm high fencing....... nr		20.49
for 1800mm high fencing....... nr		24.81

line wire
galvanized

3.15mm diameter.................. m		0.05
4.00mm diameter.................. m		0.08

plastic coated

2.50mm/3.55mm................... m		0.06
3.00mm/4.00mm................... m		0.08
3.55mm/4.75mm................... m		0.19

Chestnut fencing
fencing with three rows of galvanized steel wire
900mm high, spacing between pales

75mm.................................. m		2.34

1200mm high, spacing between pales

75mm.................................. m		2.74

1500mm high, spacing between pales

75mm.................................. m		4.33

Softwood post 75-100mm girth with pointed end

1350mm long....................... nr		0.94
1650mm long....................... nr		1.16
2100mm long....................... nr		1.34

Boarded fencing
softwood post 125 x 100mm

for 1200mm high fence......... nr		7.65
for 1500mm high fence......... nr		8.87
for 1800mm high fence......... nr		10.03

two ex 75mm dia. arris rail.. m		0.78
25 x 100mm gravel board..... m		1.14

Boarded fencing (Cont'd)
softwood post 125 x 100 (Cont'd)
75mm dia. centre stump

600mm long........................ nr		0.68

ex 22 x 100mm feather edge pales

for 1200mm high fence......... nr		0.42
for 1500mm high fence......... nr		0.58
for 1800mm high fence......... nr		0.66

Panel fencing
preservative treated featherboard fencing panels

1800 x 1200mm.................. nr		12.26
1800 x 1500mm.................. nr		14.31
1800 x 1800mm.................. nr		16.39

preservative treated lap fencing panels

1800 x 1200mm.................. nr		11.04
1800 x 1500mm.................. nr		12.86
1800 x 1800mm.................. nr		14.66

75 x 75mm preservative treated fence post with cap

1800mm long...................... nr		5.46
2100mm long...................... nr		6.32
2400mm long...................... nr		7.16

EXTERNAL WORKS

Fine cold asphalt (BS 4987)
with bitumen coated granite,
limestone or blast furnace slag

aggregate..........................tonne		65.48

Bitumen macadam

10mm aggregate............. tonne		31.04
40mm aggregate............. tonne		25.22
bitumen coated grit..........tonne		29.10

Precast concrete kerbs and channels
(BS 7263 Part 1)
edging, figs 10 to 13 inclusive
straight

50 x 150mm..................... m		1.46
50 x 205mm..................... m		1.82
50 x 255mm..................... m		1.96

kerbs, figs 2, 5, 7 and 8
straight

125 x 255mm.................... m		2.50
150 x 305mm.................... m		4.46

curved

125 x 255mm.................... m		3.80
150 x 305mm.................... m		6.29

channel, fig 8
straight

225 x 125mm.................... m		3.24

curved

255 x 125mm.................... m		4.00

quadrant, fig 14

305 x 305 x 150mm.......... nr		3.46
305 x 305 x 225mm.......... nr		5.62
455 x 455 x 150mm......... nr		6.58
455 x 455 x 225mm........ nr		7.12

Granite kerbs (BS 435) or equivalent
Edge Kerb
straight

150 x 300mm.................... m		18.95
200 x 300mm.................... m		22.22

curved external radius not exceeding 1000mm

150 x 300mm.................... m		29.12
200 x 300mm.................... m		33.63

curved external radius exceeding 1000mm

150 x 300mm.................... m		24.76
200 x 300mm.................... m		25.60

Flat Kerb
Straight

300 x 150........................... m		18.40
300 x 200........................... m		22.17

Curved external radius not exceeding

EXTERNAL WORKS (Cont'd)

1000mm
300 x 150mm..................... m   26.43

Granite kerbs (BS 435)
or equivalent (Cont'd)
Curved external radius not exceeding
1000mm (Cont'd)
300 x 200mm..................... m   31.68
Curved external radius exceeding
100mm
300 x 150mm..................... m   24.82
300 x 200mm..................... m   29.78

Gravel paths
clinker............................. m3
12.00
blinding gravel
10 or 20mm..................... m3
17.00

Precast concrete flag paving (BS 7263)
Part 1
50mm plain flags
600 x 450mm.................... m2   4.78
600 x 600mm.................... m2   4.34
600 x 750mm.................... m2   3.85
600 x 900mm.................... m2   3.83

Granite sett paving
new granite setts to BS 435
100mm........................... m2   19.03
125mm........................... m2   23.78
150mm........................... m2   28.53
reclaimed granite setts
100mm........................... m2   15.46
125mm........................... m2   19.33
150mm........................... m2   23.19

Guard rails
40mm galvanized steel medium weight
tubing to BS 1387 ref
7-2-G............................. m   5.62
Kee Klamp fittings
bend No 15-7.................. nr   4.33
three-way intersection
No 20-7........................... nr   6.55
three-way intersection
No 25-7........................... nr   6.29
No 21-7........................... nr   5.12
four-way intersection
No 26-7........................... nr   4.88
five-way intersection
No 35-7........................... nr   7.28
five-way intersection
No 40-7........................... nr   10.86
floor plate No 61-7........... nr   3.95
51 x 51mm welded mesh
infill panel........................ m2   20.95

VISIT OUR BUILDING AND CONSTRUCTION WEBSITE TODAY!

www.bh.com/construction

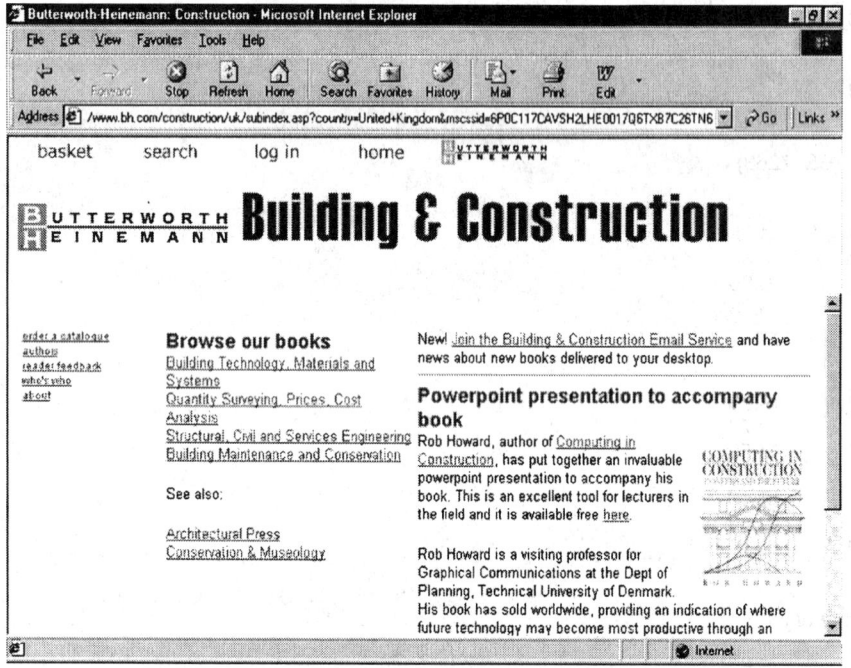

- FIND OUT MORE ABOUT OUR RANGE OF TITLES

- LATEST NEWS ON NEW TITLES AND PRICES

- HELPFUL SEARCH FACILITIES

- ORDER BOOKS ON-LINE

- BENEFIT FROM SPECIAL OFFERS

- REGULAR EMAIL UPDATES

- FIND OUT MORE ABOUT OUR AUTHORS

BOOKMARK THIS SITE TODAY!

Composite Prices for Approximate Estimating

The purpose of this section is to provide an easy reference to costs of "composite" items of work which would usually be measured for Approximate Estimates. Rates for items which are not included here may be found by reference to the preceding pages of this volume. Allowance has been made in the rates for the cost of works incidental to and forming part of the described item although not described therein, i.e. the cost of formwork to edges of slabs has been allowed in the rate for slabs; the cost of finishings to reveals has been allowed in the rate for finishings to walls.

The rates are based on rates contained in the preceding Major Works section of the book and include the cost of materials, Labour fixing or laying and allow 7.5% for overheads and profit: Preliminaries are not included within the rates. The rates are related to a housing or simple commercial building contract in the range of £250,000 to £1,000,000. For contracts of a higher or lower value, rates should be adjusted by the percentage adjustments given in Essential Information at the beginning of this Volume. Landfill tax is included for inert waste.

The following cost plans have been included in both the Major works and Small works sections, and priced accordingly as examples on how to construct a Cost Plan, for information and to demonstrate how to use the Approximate Estimating section.

COST PLAN

COST PLAN FOR AN OFFICE DEVELOPMENT

The Cost Plan that follows gives an example of an Approximate Estimate for an office building as the attached sketch drawings. The estimate can be amended by using the examples of Composite Prices following or the prices included within the preceding pages. Details of the specification used are indicated within the elements. The cost plan is based on a fully measured bill of quantities priced using the detailed rates within this current edition of Laxtons. Preliminaries are shown separately.

Floor area = 932m²

Description	Item	Quantity	Rate	Item Total	Cost per m²	Total
1.0 SUBSTRUCTURE						
Note: Rates allow for excavation by machine.						
1.1 Substructure (271m²)						
Clear site vegetation excavate 150 topsoil and remove spoil.	Site clearance	288m²	0.65	187.20	0.20	
300 x 600 reinforced concrete perimeter ground beams with attached 180 x 300 toes; damp proof membrane; 50 concrete blinding; Jablite; 102.5 thick common and facing bricks from toe to top of ground beam; 75 cavity filled with plain concrete; cavity weeps; damp proof courses (excavation included in ground slab excavation)	Ground beam	81m	140.74	11399.94	12.23	
600 x 450 reinforced concrete internal ground beams, mix C30; 50 thick concrete blinding; formwork; Jablite protection (excavation included in ground slab excavation)	Ground beam	34m	105.88	3599.92	3.86	
Excavate to reduce levels; level and compact bottoms; 100 hardcore filling, blinded; damp proof membrane; 50 Jablite insulation; 150 reinforced concrete ground slab 2 layers A252 fabric reinforcement	Ground floor slab	271m²	49.34	13371.14	14.35	

OFFICE DEVELOPMENT (Cont'd)

	Item	Quantity	Item Rate	Cost total	per m²	Total
1.0 SUBSTRUCTURE (Cont'd)						
1.1 Substructure (Cont'd)						
1900 x 1900 x 1150 lift pit: excavate; level and compact bottoms; working space; earthwork support; 50 concrete blinding; damp proof membrane; 200 reinforced concrete slab, 200 reinforced concrete walls, formwork; Bituthene tanking and protection board	Lift pit	1nr	1550.09	1550.09	1.66	
1500 x 1500 x 1000 stanchion bases: excavate; level and compact bottoms; earthwork support; reinforced concrete foundations; 102.5 bricks from top of foundation to ground slab level	Stanchion base	13nr	470.89	6121.57	6.57	36229.86
(Element unit rate = £133.69 /m²)				36229.86	38.87	
2.0 SUPERSTRUCTURE						
2.1 Frame (932m²)						
300 x 300 reinforced concrete columns; formwork; two coats RIW to exterior faces	Columns	111m	61.07	6778.77	7.27	
29.5m girth x 1.0m high steel frame to support roof over high level louvres	Frame	1nr	6000.00	6000.00	6.44	
300 x 550 reinforced concrete beams; formwork	Beams	61m	84.87	5177.07	5.55	17955.84
(Element unit rate = £19.26/m²)				17955.84	19.26	
2.2 Upper Floors (614m²)						
275 reinforced concrete floor slab; formwork; 300 x 400/675 x 175/300 x 150 reinforced concrete attached beams; formwork	Floor slab	614m²	76.07	46,706.98	50.11	
	Attached beams	273m	64.73	17671.29	18.96	64378.27
(Element unit rate = £104.85 /m²)				64378.27	69.07	
2.3 Roof						
2.3.1 Roof Structure (403m² on plan)						
Pitched roof trusses, wall plates; 160 thick insulation and Netlon support	Roof structure	403m²	31.03	12505.09	13.42	
100 thick blockwork walls in roof space	Blockwork	27m²	21.61	583.47	0.63	
Lift shaft capping internal size 1900 x 1900;200 thick blockwork walls and 150 reinforced concrete slab	Lift shaft	1nr	468.23	468.23	0.50	
2.3.2 Roof coverings						
Concrete interlocking roof tiles, underlay and battens	Roof finish	465m²	15.53	7221.45	7.75	
Ridge	Ridge	22m	14.30	314.60	0.34	
Hip	Hip	50m	14.77	738.50	0.79	
Valley	Valley	6m	28.91	173.46	0.19	
Abutments to cavity walls, lead flashings and cavity tray damp proof courses	Abutments	9m	50.26	452.34	0.49	

OFFICE DEVELOPMENT (Cont'd)	Item	Quantity	Item Rate	Cost total	per m²	Total

2.0 SUPERSTRUCTURE (Cont'd)

2.3 Roof (Cont'd)

2.3.2 Roof Coverings (Cont'd)

	Item	Quantity	Item Rate	Cost total	per m²	Total
Eaves framing, 25 x 225 softwood fascia 50 wide Resoplan soffit and ventilator, eaves tile ventilator and decoration; 225 x 160 pressed Aluminium gutters and fittings	Eaves	102m	133.31	19574.82	21.00	

2.33 Roof Drainage

100 x 100 Aluminium rainwater pipes and rainwater heads; GRP hopper heads	Rain water pipes	58m	55.94	3244.52	3.48	45276.48
(Element unit rate = £112.35/m²)				45276.48	48.59	

2.4 Stairs

Reinforced in-situ concrete staircase blockwork support walls screed and vinyl floor finishes; plaster paint to soffit's; to wall						
3nr 1200 wide flights, 1nr half landing and 1nr quarter landing; 3100 rise; handrail one side	Stairs	1nr	2970.83	2970.83	3.19	
4nr 1200 wide flights, 2nr half landings; 6200 rise; plastics coated steel balustrade one side and handrail one side	Stairs	1nr	11673.69	11673.69	12.53	
4nr 1200 wide flights, 2nr half landings; 6200 rise; plastics coated steel and glass in-fill panels balustrade one side and handrail one side	Stairs	1nr	12522.14	12522.14	13.44	27166.66
				27166.66	29.16	

2.5 External Walls (614m²)

Cavity walls, 102.5 thick facing bricks (PC £250 per 1000), 75 wide insulated cavity, 140 thick lightweight concrete blockwork	Facing brick wall	455m²	85.37	38843.35	41.68	
Cavity walls 102.5 thick facing bricks (PC £250 per 1000), built against concrete beams and columns	Facings	159m²	70.69	11239.71	12.06	
880 x 725 x 150 cast stonework pier cappings	Stonework	9nr	77.25	695.25	0.75	
780 x 1300 x 250 cast stonework spandrels	Stonework	9nr	154.5	1390.00	1.49	
6500 x 2500 attached main entrance canopy	Canopy	1nr	6704.76	6704.76	7.19	
Milled lead sheet on impregnated softwood boarding and framing, and building paper	Lead sheet	4m²	161.77	647.08	0.69	
Expansion joint in facing brickwork	Expansion joint	118m	12.79	1509.22	1.62	61029.87
(Element unit rate = £99.40/m²)				61029.87	65.48	

2.6 Windows and External Doors (214m²)

2.6.1 Windows (190m²)

Polyester powder coated Aluminium vertical pivot double glazed windows	Windows	190m²	417.70	79363.00	85.15	

OFFICE DEVELOPMENT (Cont'd)

	Item	Quantity	Item Rate	Cost total	per m²	Total
2.0 SUPERSTRUCTURE (Cont'd)						
2.6 Windows and External Doors (214m²) (Cont'd)						
2.6.1 Windows (190m²) (Cont'd)						
Polyester powder coated Aluminium fixed louvre panels	Louvres	35m²	173.07	6057.45	6.50	
Heads 150 x 150 x 10 stainless steel angle bolted to concrete beam, cavity damp proof course, weep vents, plaster and decoration	Heads	98m	60.90	5968.20	6.40	
Reveals closing cavity with facing bricks, damp proof course, plaster and decoration	Reveals	99m	15.31	1515.59	1.63	
Works to reveals of openings comprising closing cavity with facing bricks, damp proof course, 13 thick cement and sand render with angle beads, and ceramic tile finish	Reveals	27m	27.35	738.45	0.79	
Sills closing cavity with blockwork, damp proof course, and 19 x 150 Ash veneered MDF window boards	Sills	8m	27.35	216.24	0.23	
Sills 230 x 150 cast stonework sills and damp proof course	Sills	36m	38.55	1387.80	1.49	
Sills 230 x 150 cast stonework sills, damp proof course, and 19 x 150 Ash veneered MDF window boards	Sills	36m	201.66	7259.76	7.79	
Sills closing cavity with blockwork, damp proof course, and ceramic tile sill finish	Sills	9m	31.32	281.88	0.30	
Surrounds to 900 diameter circular openings stainless steel radius lintels, 102.5 x 102.5 facing brick on edge surround, 13 thick plaster, and eggshell paint decoration	Lintels	8m	117.43	939.44	1.01	
300 x 100 cast stonework surrounds to openings	Stonework	20m	48.91	978.20	1.05	
2.6.2 External Doors (24m²)						
Polyester powder coated Aluminium double glazed screens with integral double entrance doors	Entrance screens	24m²	389.33	9343.92	10.03	114050.03
(Element unit rate = £532.94/m²)				114049.93	122.37	
2.7 Internal Walls and Partitions (481m²)						
100 thick blockwork walls	Block walls	231m²	22.48	5192.88	5.57	
200 thick blockwork walls	Block walls	113m²	39.96	4515.48	4.84	
215 thick common brickwork walls	Brick walls	84m²	64.34	5404.56	5.80	
Demountable partitions	Partitions	42m²	25.86	1086.12	1.17	
Extra over demountable partitions for single doors	Partitions	2nr	24.87	49.74	0.05	
Extra over demountable partitions for double doors	Partitions	1nr	29.85	29.85	0.03	
WC cubicles	Cubicles	9nr	119.78	1078.02	1.16	

OFFICE DEVELOPMENT (Cont'd)	Item	Quantity	Item Rate	Cost total	per m²	Total
2.0 SUPERSTRUCTURE (Cont'd)						
2.7 Internal Walls and Partitions (481m²) (Cont'd)						
Screens in Softwood glazed with 6 thick GWPP glass	Screens	11m²	212.61	2338.71	2.51	
Duct panels in Formica Beautyboard and 50 x 75 softwood framework	Ducts	27m²	50.64	1367.28	1.47	
Duct panels in veneered MDF board and 50 x 75 softwood framework	Ducts	26m²	46.02	<u>1196.52</u>	<u>1.28</u>	22259.16
(Element unit rate = £46.28/m²)				<u>22259.16</u>	<u>23.88</u>	
2.8 Internal Doors (84m²)						
900 x 2100 half hour fire resistance single door sets comprising doors, Softwood frames, ironmongery and decoration	Doors	45m²	252.67	11370.15	12.20	
900 x 2100 one hour fire resistance single door sets comprising doors, Softwood frames, ironmongery and decoration	Doors	9m²	421.62	3794.58	4.07	
900 x 2100 half hour fire resistance single door sets comprising doors with 2nr vision panels, Softwood frames, ironmongery and decoration	Doors	11m²	293.93	3233.23	3.47	
150 x 2100 half hour fire resistance double door sets comprising Ash veneered doors with 2nr vision panels, Ash frames, ironmongery and decoration	Doors	19m²	285.53	5425.07	5.82	
100 wide lintels	Lintels	32m	8.64	276.48	0.30	
200 wide lintels	Lintels	7m	17.86	125.02	0.13	
215 wide lintels	Lintels	9m	21.77	<u>195.93</u>	<u>0.21</u>	24420.46
(Element unit rate = £290.72 /m²)				<u>24420.46</u>	<u>26.20</u>	
3.0 INTERNAL FINISHES						
3.1 Wall Finishes (1086m²)						
13 thick plaster to blockwork walls with eggshell paint finish	Plaster work	703m²	11.95	8400.85	9.01	
13 thick plaster to blockwork walls with lining paper and vinyl wallpaper finish	Plaster work	124m²	15.08	1869.92	2.01	
10 thick plaster to concrete walls with eggshell paint finish	Plaster work	75m²	16.17	1212.75	1.30	
13 thick cement and sand render to blockwork walls with 150 x 150 coloured ceramic tiles finish	Tiling	184m²	37.92	<u>6977.28</u>	<u>7.49</u>	18460.80
(Element unit rate = £17.00/m²)				<u>18460.80</u>	<u>19.81</u>	
3.2 Floor Finishes (779m²)						
75 thick sand and cement screed with 300 x 300 x 2.5 vinyl tile finish	Vinyl Tiling	19m²	30.61	581.59	0.62	
75 thick sand and cement screed with 2 thick sheet vinyl finish	Sheet vinyl	108m²	27.95	3018.60	3.24	
75 thick sand and cement screed with 300 x 300 x 6 carpet tile finish	Carpet Tiling	587m²	25.63	15044.81	16.14	

APPROXIMATE ESTIMATING

OFFICE DEVELOPMENT (Cont'd)

3.0 INTERNAL FINISHES (Cont'd)

3.2 Floor Finishes (779m²) (Cont'd)

	Item	Quantity	Item Rate	Cost total	per m²	Total
75 thick sand and cement screed with fitted carpet finish	Carpeting	17m²	39.49	671.33	0.72	
60 thick sand and cement screed with 150 x 150 x 13 ceramic tile finish	Ceramic Tiling	46m²	46.60	2143.60	2.30	
Entrance mats comprising brass Matt frame, 60 thick cement and sand screed and matting	Matting	2m²	190.45	380.90	0.41	
150 high x 2.5 thick vinyl tile skirting	Skirting	45m	4.06	182.70	0.20	
119 high x 2 thick sheet vinyl skirtings with 20 x 30 lacquered Hardwood cover fillet	Skirting	116m	14.79	1715.64	1.84	
25 x 50 Hardwood skirtings with lacquer finish	Skirting	214m	11.08	2371.12	2.54	
150 x 150 x 13 ceramic tile skirtings on 13 thick sand and cement render	Skirting	74m	12.38	<u>916.12</u>	<u>0.98</u>	27026.41
(Element unit rate = £34.69/m²)				<u>27026.41</u>	<u>28.99</u>	

3.3 Ceiling Finishes (809m²)

	Item	Quantity	Item Rate	Cost total	per m²	Total
Suspended ceiling systems to offices	Suspended	587m²	21.31	12508.97	13.42	
Suspended ceiling systems to toilets	Suspended	46m²	27.57	1268.22	1.36	
Suspended ceiling systems to reception area	Suspended	18m²	27.69	498.42	0.53	
Suspended ceiling plain edge trims and 25 x 38 painted softwood shadow battens	Edge Trim	300m	8.92	2676.00	2.87	
Gyproc MF suspended ceiling system with eggshell paint finish	Suspended	127m²	22.93	2912.11	3.12	
12.5 Gyproc Square edge board and skim on softwood supports with eggshell paint finish	Ceiling board	29m²	43.53	1262.37	1.35	
10 thick plaster to concrete with eggshell paint finish	Plastering	2m²	45.16	90.32	0.10	
Plaster coving	Cove	234m	4.66	<u>1090.44</u>	<u>1.17</u>	22306.85
(Element unit rate = £27.57/m²)				<u>22306.85</u>	<u>23.92</u>	

4.0 FITTINGS AND FURNISHINGS

4.1 Fittings

	Item	Quantity	Item Rate	Cost total	per m²	Total
2100 x 500 x 900 reception counter in Softwood	Counter	1nr	962.80	962.80	1.03	
2700 x 500 x 900 vanity units in Formica faced chipboard	Units	6nr	623.01	3738.06	4.01	
600 x 400 x 6 thick mirrors	Mirror	15nr	71.75	1076.25	1.15	
Vertical fabric blinds	Blinds	161m²	46.93	7555.73	8.11	
900 x 250 x 19 blockboard lipped shelves on Spur support system	Shelves	12nr	37.23	446.76	0.48	
1200 x 600 x 900 Sink base units in melamine finished particle board	Sink base unit	3nr	129.40	<u>388.20</u>	<u>0.42</u>	14167.80
				<u>14167.80</u>	<u>15.20</u>	

OFFICE DEVELOPMENT (Cont'd)

	Item	Quantity	Item Rate	Cost total	per m²	Total
5.0 SERVICES						
5.1 Sanitary Appliances						
1200 x 600 stainless steel sink units with pillar taps	Sink	3nr	119.22	357.66	0.38	
White glazed fireclay cleaner's sink units with pillar taps	Cleaners sink	3nr	221.75	665.25	0.71	
Vitreous china vanity basins with pillar taps	Vanity basins	12nr	206.00	2472.00	2.65	
Vitreous china wash hand basin and pedestal with pillar taps	Wash hand basins	1nr	148.79	148.79	0.16	
Vitreous china wash hand basin with pillar taps	Wash hand basins	1nr	90.35	90.35	0.10	
Vitreous china disabled wash hand basin with pillar taps	Wash hand basins	1nr	125.81	125.81	0.13	
Close coupled vitreous china WC suites	WC suite	11nr	175.64	1932.04	2.07	
Showers comprising shower tray, mixer valve and fittings, and corner shower cubicle	Shower	1nr	744.09	744.09	0.80	
Vitreous china urinals with automatic cistern	Urinals	3nr	264.65	793.95	0.85	
Disabled grab rails set	Grab rails	1nr	373.07	373.07	0.40	
Toilet roll holders	Accessories	11nr	14.42	158.62	0.17	
Paper towel dispensers	Accessories	8nr	23.69	189.52	0.20	
Paper towel waste bins	Accessories	8nr	36.05	288.40	0.31	
Sanitary towel disposal units	Accessories	3nr	679.80	2039.40	2.19	
Hand dryers	Accessories	8nr	236.94	1895.52	2.03	12274.47
				12274.47	13.15	
5.2 Services Equipment		1nr	2000.00	2000.00	2.15	2000.00
5.3 Disposal Installations						
Upvc waste pipes and fittings to	Internal Drainage	37nr	116.51	4310.87	4.63	4310.87
5.4 Water Installations						
Hot and cold water supplies		37nr	194.86	7209.82	7.74	7209.82
5.5 Heat Source						
Gas boiler for radiators and hot water from indirect storage tank		1nr	71229.03	71229.03	76.43	71229.03.
5.6 Space Heating and Air Treatment						
Low temperature hot water heating by radiators		1nr	2369.00	2369.00	2.54	2369.00
5.7 Ventilation Installations						
Extractor fan to toilet and kitchen areas		9nr	618.00	5562.00	5.97	5562.00

APPROXIMATE ESTIMATING

OFFICE DEVELOPMENT (Cont'd)

	Item	Quantity	Item Rate	Cost total	per m²	Total
5.0 SERVICES (Cont'd)						
5.8 Electrical Installations						
Lighting & Power		1nr	86492.40	86492.40	92.80	86492.40
5.9 Gas Installations						
To boiler room		1nr	1854.00	1854.00	1.99	1854.00
5.10 Lift and Conveyor Installations						
8 person 3 stop hydraulic lift		1nr	23690.00	23690.00	25.42	23690.00
5.11 Protective Installations						
Burglar Alarm System		1nr	5768.00	5768.00	6.19	5768.00
5.12 Communication Installations						
Cabling for telephones		1nr	7725.00	7725.00	8.29	7725.00
5.13 Special Installations						
Door entry to receptions on all floors		1nr	1300.00	1300.00	1.39	1300.00
5.14 Builders Work in Connection With Services						
Builders work in connection with plumbing, mechanical and electrical installations		1nr	5506.77	5506.77	5.91	5506.77
Builders Profit and Attendance on Services		1nr	1000.00	1000.00	1.07	1000.00
6.0 EXTERNAL WORKS						
6.1 Site Works						
6.1.1 Site Preparation						
Clearing site vegetation, filling root voids with excavated material	Site clearance	1716m²	0.59	1012.44	1.09	
Excavate topsoil 150 deep and preserve on site	Clear topsoil	214m³	4.53	969.42	1.04	
Dispose of surplus topsoil off site	Disposal	70m³	13.98	978.60	1.05	
6.1.2 Surface Treatments						
Excavate to reduce levels 400 deep; dispose of excavated material off site; level and compact bottoms; herbicide; 250 thick granular material filling, levelled, compacted and blinded; 70 thick two coat coated macadam pavings; white road lines and markings	Pavings	719m²	29.43	21160.17	22.70	
Excavate to reduce levels 400 deep; dispose of excavated material off site; level and compact bottoms; herbicide; 150 thick granular material filling, levelled, compacted and blinded; 200 x 100 x 65 thick clay brick paviors	Pavings	40m²	32.82	1312.80	1.41	

OFFICE DEVELOPMENT (Cont'd)	Item	Quantity	Item Rate	Cost total	per m²	Total
6.0 EXTERNAL WORKS (Cont'd)						
6.1.2 Surface Treatments (Cont'd)						
Excavate to reduce levels 400 deep; dispose of excavated material off site; level and compact bottoms; herbicide; 150 thick granular material filling, levelled, compacted and blinded; 900 x 600 x 50 thick concrete paving flags	Pavings	149m²	20.72	3087.28	3.31	
125 x 255 Precast concrete road kerbs and foundations	Kerbs	265m	19.19	5083.35	5.46	
50 x 150 Precast concrete edgings and foundations	Kerbs	48m	12.22	586.56	0.63	
Excavate to reduce levels 400 deep; dispose of excavated material off site; prepare sub-soil; herbicide; 150 thick preserved topsoil filling; rotavate; turf	Landscaping	55m²	16.37	900.35	0.97	
Excavate to reduce levels 400 deep; dispose of excavated material off site; prepare sub-soil; herbicide; 450 thick preserved topsoil filling; cultivate; plant 450 - 600 size herbaceous plants - 3/m2, and 3m high trees - 1nr per 22m2; fertiliser; 25 thick peat mulch	Planting	397m²	25.63	10175.11	10.92	
6.1.3 Site enclosure and division						
Palisade fencing	Fencing	62m	21.88	1356.56	1.46	
1800 high fencing	Fencing	96m	20.27	1945.92	2.09	
Extra for 1000 x 2000 gate	Gate	1nr	212.13	<u>212.13</u>	0.23	48782.69
				<u>48782.69</u>	<u>52.36</u>	
6.2 Drainage						
Excavate trench 250 - 500 deep; lay 100 diameter vitrified clay pipe and fittings bedded and surrounded in granular material	Pipes in trenches	44m	28.39	1249.16	1.34	
Excavate trench 500 - 750 deep; lay 100 diameter vitrified clay pipe and fittings bedded and surrounded in granular material	Pipes in trenches	59m	30.50	1799.50	1.93	
Excavate trench 750 - 1000 deep; lay 100 diameter vitrified clay pipe and fittings bedded and surrounded in granular material	Pipes in trenches	49m	32.33	1584.17	1.70	
Excavate trench 1000 -1250 deep; lay 100 diameter vitrified clay pipe and fittings bedded and surrounded in granular material	Pipes in trenches	4m	33.80	135.20	0.15	
Excavate trench 250 - 500 deep; lay 100 diameter vitrified clay pipe and fittings bedded and surrounded in in-situ concrete	Pipes in trenches	8m	37.85	302.80	0.32	
Excavate trench 1000 - 1250 deep; lay 150 diameter vitrified clay pipe and fittings bedded and surrounded in granular material	Pipes in trenches	23m	51.56	1185.88	1.27	
Excavate trench 1500 - 1750 deep; lay 150 diameter vitrified clay pipe and fittings bedded and surrounded in granular material	Pipes in trenches	10m	57.20	572.00	0.61	
100 diameter vertical vitrified clay pipe and fittings cased in in-situ concrete	Pipes vertical	3m	69.36	208.08	0.22	

APPROXIMATE ESTIMATING

OFFICE DEVELOPMENT (Cont'd)	Item	Quantity	Item Rate	Cost total	per m²	Total
6.0 EXTERNAL WORKS Cont'd						
6.2 Drainage (Cont'd)						
100 outlet vitrified clay yard gully with cast iron grating, surrounded in in-situ concrete	Gully	1nr	54.60	54.60	0.06	
100 outlet vitrified clay road gully with brick kerb and cast iron grating, surrounded in in-situ concrete	Gully	7nr	107.71	753.97	0.81	
Inspection chambers 600 x 900 x 600 deep internally comprising excavation works, in-situ concrete bases, suspended slabs and benchings, engineering brick walls and kerbs, clay channels, Galvanised iron step irons and grade A access covers and frames	Inspection chamber	4nr.	559.54	2237.76	2.40	
Inspection chambers 600 x 900 x 600 deep internally comprising excavation works, in-situ concrete bases, suspended slabs and benchings, engineering brick walls and kerbs, clay channels, Galvanised iron step irons and grade B access covers and frames	Inspection chamber	3nr	559.54	1678.62	1.80	11761.74
				11761.74	12.61	
6.3 External Services						
Excavate trench for 4nr ducts; lay 4nr 100 diameter uPVC ducts and fittings; lay warning tape; backfill trench		20m	37.12	742.40	0.80	
Sewer connection charges		1nr	5000.00	5000.00	5.36	
Gas main and meter charges		1nr	4000.00	4000.00	4.29	
Electricity main and meter charges		1nr	4000.00	4000.00	4.29	
Water main and meter charges		1nr	3000.00	3000.00	3.22	16742.40
				16742.40	17.96	
PRELIMINARIES						
Site staff, accommodation, lighting, water safety health and welfare, rubbish disposal, cleaning, drying, protection, security, plant transport, temporary works and scaffolding with a contract period of 38 weeks	Item			65300	70.09	65300.00

			TOTAL		£869629.38	
			Cost per m2 =		£939.65	

OFFICE DEVELOPMENT (Cont'd)

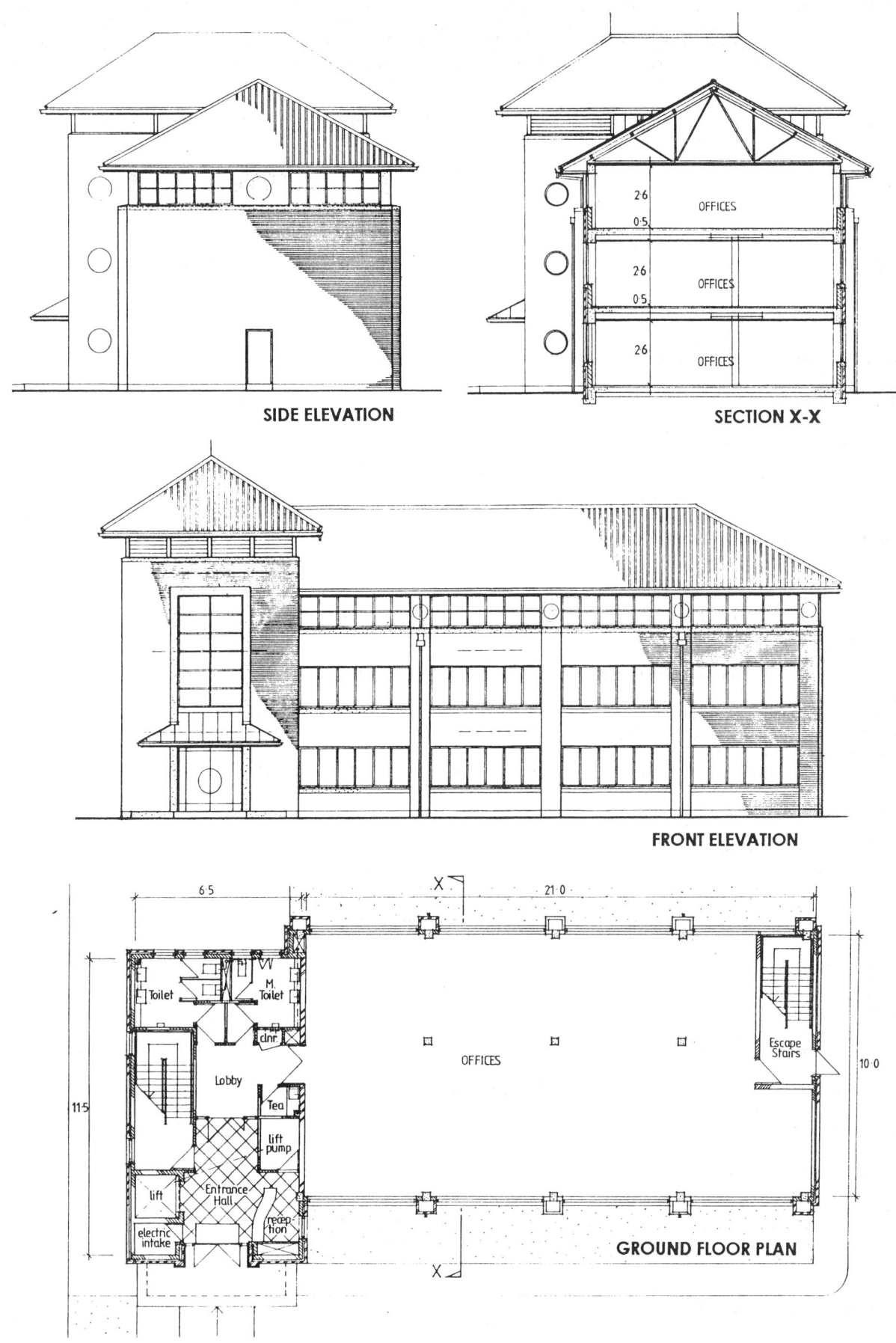

SIDE ELEVATION

SECTION X-X

FRONT ELEVATION

GROUND FLOOR PLAN

APPROXIMATE ESTIMATING

OFFICE DEVELOPMENT
© Cotterell Thomas & Thomas Architects

COST PLAN FOR A DETACHED HOUSE

The Cost Plan that follows gives an example of an Approximate Estimate for a detached house with garage and porch as the attached sketch drawings. The estimate can be amended by using the examples of Composite Prices following or the prices included within the preceding pages. Details of the specification used are indicated within the elements. The cost plan is based on a fully measured bill of quantities priced using the detailed rates within this current edition of Laxton's. Preliminaries are shown separately

Gross Internal Floor area =

House - 120m² (Porch - 5m²)
Garage - 22m²
Total   142m²

Description	Item	Quantity	Rate	Item Total	Cost per m²	Total
1.0 SUBSTRUCTURE						
Note: Rates allow for excavation by machine.						
1.1 Substructure (86m²)						
Clear site vegetation excavate 150 topsoil and remove spoil.	Site clearance	113m²	2.96	334.48	2.36	
Excavate trench commencing at reduced level, average 800 deep, concrete to within 150 of ground level, construct cavity brick wall 300 high and lay pitch polymer damp proof course.	Strip foundation	33m	100.75	3324.75	23.41	
Excavate trench commencing at reduced level, average 800 mm deep, concrete to within 150 of ground level, construct 103 brick wall 300 high and lay pitch polymer damp proof course.	Strip foundation	26m	76.68	1993.68	14.04	
Excavate trench commencing at reduced level, average 800 mm deep, concrete to within 150 of ground level, construct 215 brick wall 300 high and lay pitch polymer damp proof course.	Strip foundation	2m	170.69	341.38	2.40	
Excavate to reduce levels average 150 deep, lay 150 mm bed of hardcore blinded, DPM and 150 concrete ground slab reinforced with Ref. A 193 Fabric reinforcement.	Ground floor slab	85m²	43.35	3684.75	25.95	9679.04
(Element unit rate = £112.55/m²)				9679.04	68.16	
2.0 SUPERSTRUCTURE						
2.2 Upper Floors (55m²)						
175 deep floor joists with 22 chipboard finish.	Softwood floor	49m²	29.63	1451.87	10.22	
175 deep floor joists with 22 plywood finish.	Softwood floor	6m²	50.32	301.92	2.13	1753.79
(Element unit rate = £31.89/m²)				1753.79	12.53	
2.3 Roof(110m² on plan)						
2.3.1 Roof Structure 35 degree pitched roof with 50 x 100 rafters and ceiling joists, 100 glass fibre insulation.	Pitched roof	110m²	36.10	3971.00	27.96	
Gable end 275 cavity brick and block wall facings PC £250/1000.	Gable wall	29m²	79.51	2305.79	16.24	
Gable end 102 brick wall facings £250/1000	Gable wall	8m²	56.41	451.28	3.18	

DETACHED HOUSE (Cont'd)	Item	Quantity	Rate	Item total	Cost per m²	Total

2.0 SUPERSTRUCTURE(Cont'd)

2.3 Roof (Cont'd)

2.3.2 Roof Coverings

	Item	Quantity	Rate	Item total	Cost per m²	Total
Interlocking concrete tiles, felt underlay, battens.	Roof tiling	143m²	15.33	2220.79	15.64	
Verge, 150 under cloak.	Verge	38m	8.85	336.30	2.37	
Ridge, half round.	Ridge	14m	14.30	200.20	1.41	
Valley trough tiles.	Valley	2m	28.91	57.82	0.41	
Abutment, lead flashings cavity trays.	Abutment	11m	52.51	577.61	4.07	
Sheet lead cladding.	Lead roofing	2m²	129.85	259.70	1.83	
Eaves soffit, fascia and 100 PVC gutter.	Eaves	29m	36.30	1052.70	7.41	

2.3.3 Roof Drainage

	Item	Quantity	Rate	Item total	Cost per m²	Total
68 pvcu rainwater pipe.	Rainwater pipe	24m	15.98	383.52	2.70	11816.71
(Element unit rate = £107.42/m²)				11816.71	83.22	

2.4 Stairs (1 Nr)

	Item	Quantity	Rate	Item total	Cost per m²	Total
Softwood staircase 2670 rise, balustrade one side, 910 wide two flights with half landing.	Stairs	1nr	1148.80	1148.80	8.09	1122.20
				1148.80	8.09	

2.5 External Walls (152m2)

	Item	Quantity	Rate	Item total	Cost per m²	Total
Hollow wall, light weight blocks one skin facings PC £250/1000 one skin ties.	Hollow wall	127m²	78.28	9941.56	70.01	
Facing brick wall PC £250/1000 102 thick.	½ B Facing wall	25m²	53.81	1345.25	9.47	
Facing brick wall PC £250/1000 215 thick.	1 B Facing wall	6m²	97.22	291.66	2.05	11578.47
(Element unit rate = £76.17/m²)				11578.47	81.53	

2.6 Windows and External Doors (35m2)

2.6.1 Windows (16m2)

	Item	Quantity	Rate	Item total	Cost per m²	Total
Hardwood double glazed casement windows.	Hardwood windows	16m²	365.06	5840.96	41.13	
Galvanised steel lintel, DPC brick soldier course.	Lintels	16m	29.94	479.04	3.37	
Softwood window board, DPC close cavity.	Window board	15m	6.64	99.60	0.70	
Close cavity at jamb, DPC facings to reveal.	Jamb	32m	10.88	346.16	2.45	

2.6.2 External Doors (19m2)

	Item	Quantity	Rate	Item total	Cost per m²	Total
Hardwood glazed doors, frames, ironmongery PC £50.	Hardwood doors	6m²	572.07	3432.42	24.17	
Hardwood patio doors.	Patio doors	9m²	356.52	3208.68	22.60	
Galvanised steel garage door.	Garage door	4m²	121.71	486.84	3.43	
Galvanised steel lintel, DPC, brick soldier course.	Lintels	12m	28.96	347.52	2.45	

DETACHED HOUSE (Cont'd)	Item	Quantity	Rate	Item total	Cost per m²	Total
2.0 SUPERSTRUCTURE (Cont'd)						
2.6 Windows and External Doors (35m2) (Cont'd)						
Close cavity at jamb, DPC facings to reveal.	Jamb	4m	10.88	43.52	0.31	14286.74
(Element unit rate = £408.19/m²)				14286.74	100.61	
2.7 Internal Walls and Partitions (153m2)						
Light weight concrete block 100 thick.	Block walls	104m²	23.30	2423.20	17.06	
75 timber stud partition plasterboard and skim both sides.	Stud partition	53m²	45.82	2428.46	17.10	4851.66
(Element unit rate = £31.71/m²)				4851.66	34.16	
2.8 Internal Doors (40m2)						
12 hardboard embossed panel door, linings ironmongery PC £20.00.	Flush doors	18m²	166.62	2999.16	21.12	
3 softwood 2GG glazed door, linings, ironmongery PC £20.00	Glazed doors	5m²	196.50	982.50	6.92	
6 plywood faced door, linings, ironmongery PC £20.00	Flush doors	5m²	194.35	971.75	6.84	
1 plywood faced half hour fire check flush door, frame, ironmongery PC £60.00.	Fire doors	2m²	193.15	386.30	2.92	
Plywood ceiling hatch.	Hatch	1nr	58.04	58.04	0.41	
Precast concrete lintels.	Lintels	10m	14.07	140.70	0.99	5538.45
(Element unit rate = £138.46/m²)				5538.45	39.00	
3.0 INTERNAL FINISHES						
3.1 Wall Finishes (264m2)						
13 light weight plaster.	Plaster work	264m²	10.22	2698.08	19.00	
152 x 152 wall tiling coloured PC £20.00 supply.	Tiling	13m²	31.99	415.87	2.93	3113.95
(Element unit rate = £11.80/m²)				3113.95	21.93	
3.2 Floor Finishes (63m2)						
22 Chipboard, 50 insulation.	Board flooring	63m²	24.41	1537.83	10.83	
3 PVC sheet flooring.	Sheet vinyl	18m²	17.82	320.76	2.26	
25 x 125 softwood skirting.	Skirting	131m	7.59	994.29	7.00	2852.88
(Element unit rate = £45.28/m²)				2852.88	20.09	
3.3 Ceiling Finishes (115m2)						
9.5 Plasterboard and set.	Plasterboard	109m²	10.05	1095.45	7.71	
Plywood finish	Plywood	6m²	31.67	190.02	1.34	1285.47
(Element unit rate = £11.18/m²)				1285.47	9.05	

DETACHED HOUSE (Cont'd)	Item	Quantity	Rate	Item total	Cost per m²	Total

4.0 FITTINGS AND FURNISHINGS

4.1 Fittings

Kitchen units 600 base unit		2nr	164.00	328.00	2.31	
1000 base unit		6nr	265.24	1591.44	11.21	
500 broom cupboard		1nr	226.22	226.22	1.59	
600 wall unit		3nr	93.91	281.73	1.98	
1000 wall unit		2nr	142.37	284.74	2.01	
Work tops		10m	45.00	450.00	3.17	
Gas fire surround		1nr	525.00	525.00	3.70	3687.13
				3687.13	25.97	

5.0 SERVICES

5.1 Sanitary Appliances

Lavatory basin with pedestal.		3nr	135.68	407.04	2.87	
Bath with side panel.		1nr	229.65	229.65	1.62	
WC. suite.		3nr	172.80	518.40	3.65	
Stainless steel sink and drainer.		1nr	118.21	119.21	0.84	
Shower		1nr	498.13	498.13	3.51	

5.2 Services Equipment

Electric oven built in.		1nr	530.00	530.00	3.73	
Kitchen Extractor hood		1nr	270.00	270.00	1.90	
Gas hob built in.		1nr	330.00	330.00	2.32	
Gas fire with balanced flue.		1nr	350.00	350.00	2.46	

5.3 Disposal Installations

Upvc soil and vent pipe.		15m	49.61	744.15	5.24	
Automatic air admittance valve.		3nr	71.32	213.96	1.51	

5.4 Water Installations

Cold water to 9 No. draw off points.		item	804.12	804.12	5.66	
Hot water to 7 No. draw off points.		item	873.41	873.41	6.15	

5.5 Heat Source

Boiler pump and controls.		item	679.87	679.87	4.79	

5.6 Space Heating

Low temperature hot water radiator system.		item	3476.25	3476.25	24.48	

5.7 Ventilating Systems

Extract fan units.		3nr	273.67	821.01	5.78	

5.8 Electrical Installations

Electrical Installation comprising 16Nr lighting points, 30Nr outlet points, cooker points, immersion heater point and 9Nr fittings points to include smoke alarms, bell, telephone and aerial installation.		item	4218.16	4218.16	29.71	

5.9 Gas Installation

Installation for boiler, hob and fire.		item	500.00	500.00	3.52	

5.10 Builders Work

Building work for service installation.		item	1153.71	1153.71	8.12	16737.07
(Element unit rate = £117.86/m²)				16737.07	117.86	

APPROXIMATE ESTIMATING

DETACHED HOUSE (Cont'd)

	Item	Quantity	Rate	Item total	Cost per m²	Total
6.0 EXTERNAL WORKS						
6.1 Site Work						
Excavate and lay 50 Precast concrete flag paving on hardcore.	Paving	120m²	36.20	4344.00	30.59	
Clear site vegetation excavate 150 topsoil and remove spoil	Site clearance	479m²	2.96	1417.84	9.98	
Brick wall, excavation 600 deep, concrete foundation, 215 brickwork facing. PC £200/1000 damp proof course brick on edge coping 900 high.	Garden wall	13m	82.60	1073.80	7.56	
Plastic coated chain link fencing 900 high on angle iron posts	Fencing	92m	23.46	2158.32	15.20	
Turfed area's	Landscaping	103m²	24.29	2501.87	17.62	
6.2 DRAINAGE						
100 diameter clay pipes, excavation, 150 bed and surround, pea shingle, backfill 750 deep.		85m	38.82	3299.70	23.24	
Polypropylene inspection chambers; universal inspection chamber; polypropylene; 475 diameter; preformed chamber base benching; for 100mm pipe; chamber shaft, 175mm effective length; cast iron cover and frame; single seal A15 light duty to suit.		6nr	121.18	727.08	5.12	
6.3 EXTERNAL SERVICES						
100 diameter clay duct, excavation backfill, polythene warning tape 500 deep.		38m	16.59	630.42	4.44	
Provisional sums for External Services - Gas, electric, water, telephone.		item	4000.00	4000.00	28.17	20153.03
				20153.03	141.92	

PRELIMINARIES

Site staff, accommodation, lighting, water, safety and health and welfare, rubbish disposal cleaning, drying, protection, security, plant transport, temporary works and scaffolding with a contract period of 28 weeks				10000.00	70.42	10000.00

TOTAL	=	£118483.19
Cost per m² =		£834.39

DETACHED HOUSE (Cont'd)

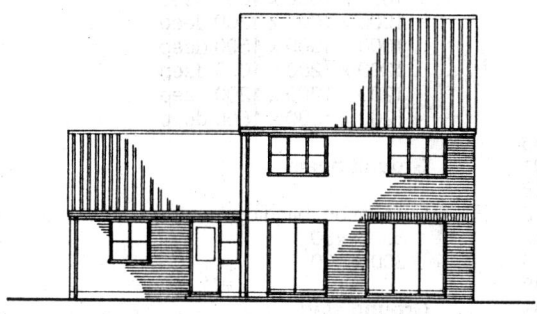

REAR ELEVATION

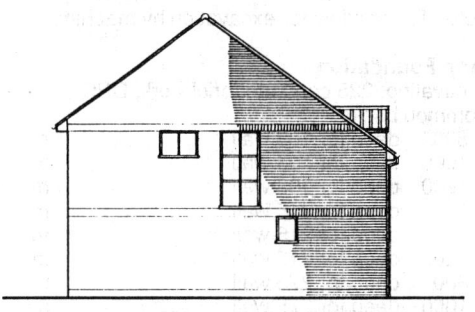

SIDE ELEVATION

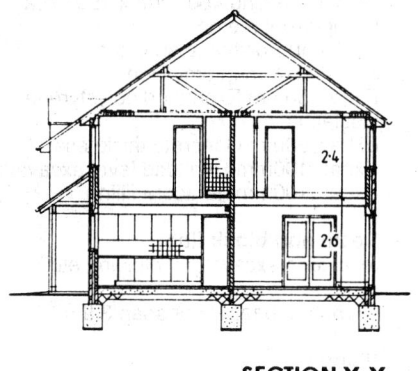

SECTION X-X

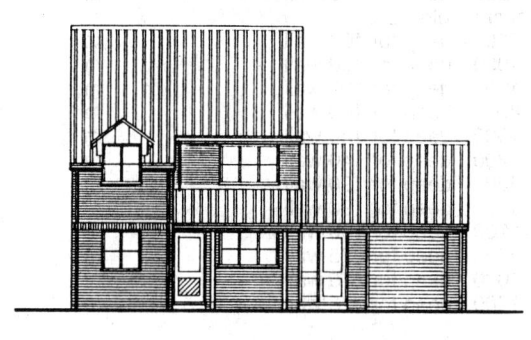

FRONT ELEVATION

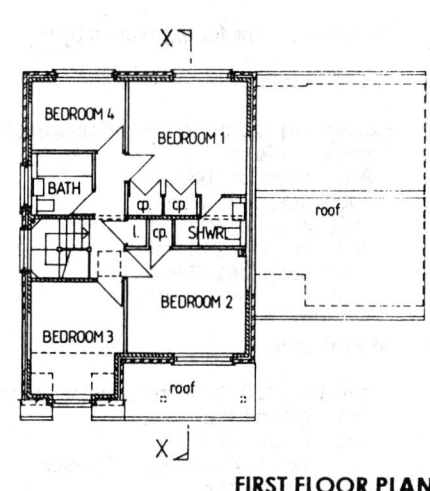

FIRST FLOOR PLAN

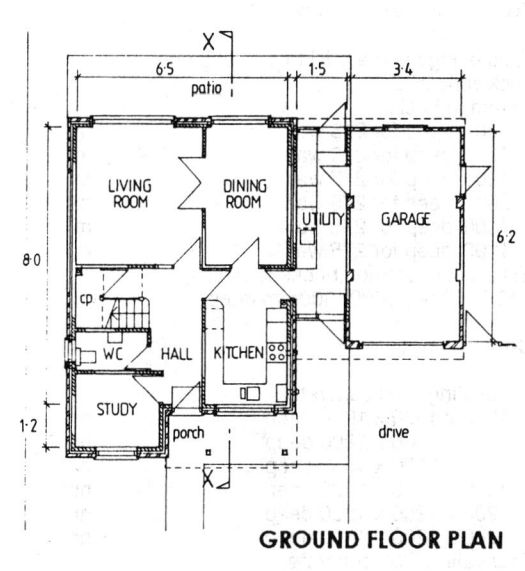

GROUND FLOOR PLAN

APPROXIMATE ESTIMATING

4 - BEDROOM DETACHED HOUSE
© Cotterell Thomas & Thomas Architects

COMPOSITE PRICES

The following composite prices are based on the detailed rates herein and may be used to amend the previous example cost plans or to create new approximate budget estimates.

SUBSTRUCTURE £

Note: Rates allow for excavation by machine.

Strip Foundation
Excavating, 225 concrete, brickwork, DPC
Common bricks

800	deep for 103 wall	m	42.88
1000	deep for 103 wall	m	51.91
1200	deep for 103 wall	m	67.48
800	deep for 215 wall	m	64.43
1000	deep for 215 wall	m	79.28
1200	deep for 215 wall	m	95.79
800	deep for 275 wall	m	76.48
1000	deep for 275 wall	m	94.89
1200	deep for 275 wall	m	115.10

Engineering bricks Class B

800	deep for 215 wall	m	81.73
1000	deep for 215 wall	m	101.36
1200	deep for 215 wall	m	122.64
800	deep for 275 wall	m	94.02
1000	deep for 275 wall	m	117.27
1200	deep for 275 wall	m	142.32

Concrete blocks

800	deep for 100 wall	m	33.96
1000	deep for 100 wall	m	40.52
1200	deep for 100 wall	m	48.88
800	deep for 140 wall	m	38.57
1000	deep for 140 wall	m	46.36
1200	deep for 140 wall	m	55.95
800	deep for 190 wall	m	43.40
1000	deep for 190 wall	m	52.41
1200	deep for 190 wall	m	63.22
800	deep for 275 wall *	m	55.54
1000	deep for 275 wall *	m	67.82
1200	deep for 275 wall *	m	81.89

* Thermalite Trenchblocks

Fabric reinforcement ref. A193

350 wide	m	0.92
600 wide	m	1.58
750 wide	m	1.97

Trench Fill Foundation

Excavating concrete fill to 150 of ground level brickwork DPC
Common bricks

800	deep for 215 wall	m	49.77
1000	deep for 215 wall	m	58.31
1200	deep for 215 wall	m	68.64
800	deep for 275 wall	m	46.69
1000	deep for 275 wall	m	55.23
1200	deep for 275 wall	m	65.01

Extra over common bricks for facings
(PC £250 per 1000) 300mm deep. | m | 3.36

Column base

Excavating, 450 concrete

1000 x 1000 x 1000 deep	nr	63.61
1000 x 1000 x 1200 deep	nr	71.88
1000 x 1000 x 1500 deep	nr	79.89
1200 x 1200 x 1000 deep	nr	89.11
1200 x 1200 x 1200 deep	nr	100.36
1200 x 1200 x 1500 deep	nr	111.11

Excavating, 750 concrete

1000 x 1000 x 1000 deep	nr	89.76
1000 x 1000 x 1200 deep	nr	98.32
1000 x 1000 x 1500 deep	nr	106.33
1200 x 1200 x 1000 deep	nr	127.18
1200 x 1200 x 1200 deep	nr	138.13
1200 x 1200 x 1500 deep	nr	149.18

Excavating, 450 reinforced concrete

1000 x 1000 x 1000 deep	nr	115.90
1000 x 1000 x 1200 deep	nr	125.12

SUBSTRUCTURE (Cont'd)
Column bases (Cont'd) £

1000 x 1000 x 1500 deep	nr	133.14
1200 x 1200 x 1000 deep	nr	165.78
1200 x 1200 x 1200 deep	nr	177.03
1200 x 1200 x 1500 deep	nr	187.79

Excavating, 750 reinforced concrete

1000 x 1000 x 1000 deep	nr	178.50
1000 x 1000 x 1200 deep	nr	187.06
1000 x 1000 x 1500 deep	nr	195.08
1200 x 1200 x 1000 deep	nr	254.98
1200 x 1200 x 1200 deep	nr	266.23
1200 x 1200 x 1500 deep	nr	276.98

Ground beam

Excavation, blinding, reinforced concrete

300 x 450	m	63.40
300 x 600	m	78.66

Ground slab

Excavation, 150 hardcore, blinding, concrete

150 ground slab	m²	19.42
200 ground slab	m²	23.44
225 ground slab	m²	25.45
Extra: 50 blinding	m²	3.93
Extra: Fabric reinforcement A142	m²	1.87
Extra: Fabric reinforcement A193	m²	2.24
Slab thickening 450 wide x 150 thick	m²	6.37
Damp proof membrane		
250mu Polythene sheeting	m²	1.07
2 coats Bitumen emulsion	m²	3.12
Extra: 50mm Expanded polystyrene insulation	m²	5.41
Extra: 50mm Concrete thickness	m²	3.97
Extra: 100mm Reduced level excavation	m²	1.51
Extra: 100mm hardcore filling	m²	2.06

Beam and block floor
(excludes excavation, blinding etc.)

155mm beam floor span 3.9m	m²	21.15

Piling

On/off site charge	nr	4085.00
450 bored cast in-site piles n/e 6m deep	m	23.93
610 bored cast in-site piles n/e 6m deep	m	36.18
Cutting off tops of piles	nr	8.19

INSIDE EXISTING BUILDING

Note: rates allow for excavation by hand.

Strip foundation

Excavating 225 concrete, brickwork, DPC
Common bricks

800	deep for 103 wall	m	51.98
1000	deep for 103 wall	m	63.22
1200	deep for 103 wall	m	86.08
800	deep for 215 wall	m	73.39
1000	deep for 215 wall	m	90.50
1200	deep for 215 wall	m	110.78

Ground slab

Excavation, 150 hardcore, blinding, concrete

150 ground slab	m²	26.90
200 ground slab	m²	31.87
Slab thickening 450 wide x 150 thick	m	8.33
Damp proof membrane		
250 mu polythene sheeting	m²	1.07
2 coats bitumen emulsion	m²	4.40

FRAME

£

Reinforced concrete

225 x 225 column	m	39.49
300 x 300 column	m	57.86
225 x 250 attached beam	m	44.33
300 x 450 attached beam	m	80.68
400 x 650 attached beam	m	129.36
Casing to 152 x 152 steel column	m	28.67
Casing to 203 x 203 steel column	m	37.08
Casing to 146 x 254 steel beam	m	36.83
Casing to 165 x 305 steel beam	m	43.33
Extra: wrought formwork	m²	2.63

Steel

Columns and beams	tonne	913.75
Roof trusses	tonne	1505.00
Decorating	m²	6.24
Purlins and cladding rails	tonne	935.25
Small sections	tonne	913.75
Decorating	m²	6.24

UPPER FLOORS

In-situ reinforced concrete

150 thick	m²	44.16
200 thick	m²	48.57
225 thick	m²	50.77

Hollow prestressed concrete

125 thick	m²	31.22
150 thick	m²	32.67
200 thick	m²	36.81

Beam and block

200 thick	m²	5.70
250 thick	m²	38.21

Softwood joist

175 thick	m²	13.87
200 thick	m²	15.49
225 thick	m²	16.85
250 thick	m²	18.65

Floor boarding

19 tongued and grooved softwood	m²	16.29
25 tongued and grooved softwood	m²	17.27
18 tongued and grooved chipboard	m²	9.20
22 tongued and grooved chipboard	m²	11.13
15 tongued and grooved plywood	m²	9.47
18 tongued and grooved plywood	m²	11.22

ROOF

Roof structure

Reinforced in-situ concrete

150 thick	m²	54.22
200 thick	m²	58.63
50 lightweight screed to falls	m²	14.25
50 cement sand screed to falls	m²	11.24
50 vermiculite screed to falls	m²	15.29
50 flat glass fibre insulation board	m²	7.25
Extra for vapour barrier	m²	1.07

Softwood roof joists

50 x 175 joists @ 400 centres	m²	15.24
50 x 225 joists @ 400 centres	m²	18.44
50 x 175 joists @ 600 centres	m²	10.25
50 x 225 joists @ 600 centres	m²	13.00
18 chipboard	m²	6.73
25 softwood tongued grooved boarding	m²	19.02
18 plywood decking	m²	15.80
24 plywood decking	m²	21.15
Pitched roof with 50 x 100 rafters and ceiling joist	m²	27.04
Softwood trussed rafters @ 600 centres	m	16.02

Roof coverings

Three layer fibre based felt	m²	14.88
Three layer high performance felt	m²	31.72
20 two coat mastic asphalt	m²	17.95
50 lightweight screed to falls	m²	14.25
50 cement sand screed to falls	m²	10.25
60 Vermiculite screed to falls	m²	15.29
50 flat glass fibre insulation board	m²	7.25

ROOF (Cont'd)

Roof Coverings (Cont'd)

£

Extra for vapour barrier	m²	1.07
381 x 227 concrete interlocking tiles (on slope)	m²	18.82
268 x 165 plain concrete tiles (on slope)	m²	37.98
268 x 165 machine made clay tiles (on slope)	m²	39.05
265 x 165 hand made clay tiles (on slope)	m²	53.76
265 x 165 hand made tiles	m²	51.49
600 x 300 fibre cement slates (on slope)	m²	31.30
610 x 305 natural slates (on slope)	m²	67.60
100 insulation quilt (on slope)	m²	2.87

Eaves soffit, fascia, gutter and decoration

150 x 225 with 100 pvc-u gutter	m	27.50
150 x 225 with 114 pvc-u gutter	m	27.74
150 x 225 with 100 half round cast iron gutter	m	38.83
150 x 225 with 114 half round cast iron gutter	m	36.89
150 x 225 with 125 half round cast iron gutter	m	39.20
150 x 225 with 150 half round cast iron gutter	m	48.72
150 x 225 with 100 og cast iron gutter	m	35.59
150 x 225 with 114 og cast iron gutter	m	35.59
150 x 225 with 125 og cast iron gutter	m	37.95
200 x 300 with 100 pvc-u gutter	m	30.47
200 x 300 with 114 pvc-u gutter	m	30.71
200 x 300 with 100 half round cast iron gutter	m	39.45
200 x 300 with 114 half round cast iron gutter	m	39.86
200 x 300 with 125 half round cast iron gutter	m	42.17
200 x 300 with 150 half round cast iron gutter	m	51.69
200 x 300 with 100 og cast iron gutter	m	38.56
200 x 300 with 114 og cast iron gutter	m	39.52
200 x 300 with 125 og cast iron gutter	m	40.92
Verge undercloak, large board and decoration 200 x 6	m	5.78

Hip rafter and capping

225 x 50 rafter concrete capping tiles	m	23.47
225 x 50 rafter clay capping tiles	m	27.67
225 x 50 rafter Clay machine made bonnet hip tiles	m	27.67
225 x 50 rafter clay hand made bonnet hip tile	m	27.67

Ridge board and capping

200 x 25 ridge board concrete capping tile	m	21.51
200 x 25 ridge board hand made clay tile	m	25.71

Rainwater pipe

68 pvc-u pipe	m	7.49
63 cast iron pipe and decoration	m	25.12
75 cast iron pipe and decoration	m	26.33
100 cast iron pipe and decoration	m	33.13

STAIRS

Reinforced concrete 2670 rise, mild steel balustrade one side, straight flight

1000 wide granolithic finish	nr	1733.52
1500 wide granolithic finish	nr	1961.87
1000 wide pvc sheet finish non slip nosings	nr	2125.19
1500 wide pvc sheet finish non slip nosings	nr	2481.97
1000 wide terrazzo finish	nr	2253.27
1500 wide terrazzo finish	nr	2611.58

Reinforced concrete 2670 rise, mild steel balustrade one side in two flights with half landing.

1000 wide granolithic finish	nr	2100.28
1500 wide granolithic finish	nr	2317.40
1000 wide pvc sheet finish non slip nosings	nr	2459.12
1500 wide pvc sheet finish non slip		

APPROXIMATE ESTIMATING

nosings	nr	2798.63
1000 wide terrazzo finish	nr	2615.24

Stairs (Cont'd) £

1500 wide terrazzo finish	nr	3307.30

Softwood staircase 2670 rise, balustrade one side.

910 wide straight flight	nr	834.21
910 wide one flight with three winders	nr	1017.73
910 wide two flight with half landing	nr	1017.73

Mild steel staircase 2670 rise, balustrade both sides.

920 wide straight flight	nr	1988.75
920 wide two flights with quarter landing	nr	3767.88
910 wide two flights with half landing	nr	4563.38

EXTERNAL WALLS

Common bricks

102 wall	m²	34.46
215 wall	m²	61.80
327 wall	m²	82.92

Hollow wall with Galvanised ties

275 common bricks both skins	m²	70.77
275 facings one side P. C. £150 per 1000	m²	75.31
275 facings one side P. C. £200 per 1000	m²	78.64
275 facings one side P. C. £250 per 1000	m²	81.97
275 facings one side P. C. £300 per 1000	m²	85.30

Facing bricks

102 wall P.C. £150 per 1000	m²	39.00
102 wall P.C. £200 per 1000	m²	42.33
102 wall P.C. £250 per 1000	m²	45.66
102 wall P.C. £300 per 1000	m²	48.99
215 wall P.C. £150 per 1000	m²	66.40
215 wall P.C. £200 per 1000	m²	73.12
215 wall P.C. £250 per 1000	m²	79.84
215 wall P.C. £300 per 1000	m²	86.55

Light weight concrete block 4.0n/mm²

100 wall	m²	22.15
140 wall	m²	27.70
190 wall	m²	33.25

Hollow wall in light weight blocks with Galvanised ties.

275 facings one side P. C. £150 per 1000	m²	63.00
275 facings one side P. C. £200 per 1000	m²	66.33
275 facings one side P. C. £250 per 1000	m²	69.66
275 facings one side P. C. £300 per 1000	m²	72.99

Extra for:

50 mm glass fibre slab insulation	m²	8.19
13 mm two coat plain render painted	m²	9.64
15 mm pebbled dash render	m²	11.92

WINDOWS AND EXTERNAL DOORS

Windows

Softwood glazed casement window

type 110v size 631 x 1050	nr	142.75
type 212c size 1220 x 1220	nr	242.03
type 312c size 1769 x 1200	nr	311.21
type 312ww size 1769 x 1200	nr	384.20
average cost	m²	168.87
purpose made in panes 0.10-0.50m² (double glazed)	m²	207.48

Hardwood double glazed standard casement windows

type X 107A size 630 x 750	nr	242.34
type X 112A size 630 x 1200	nr	304.85
type X 210A size 1200 x 1050	nr	435.17
type X 310AE size 1770 x 1050	nr	559.05
average cost	m²	354.59

Afrormosia double glazed casement windows purpose made in pane 0.l0-0.50m² — m² 392.08

European Oak double glazed casement windows purpose made in panes 0.10-0.50m² — m² 451.02

Softwood glazed double hung sash window

type GS3 size 825 x 1094	nr	398.43
type GS4 size 825 x 1394	nr	451.24
type GSW4 size 1051 x 1394	nr	500.84

average cost	m²	383.92

Galvanised steel windows

type NC05F size 508 x 1067	nr	123.30

WINDOWS & EXTERNAL DOORS (Cont'd)

Windows (Cont'd)

type ND2F size 997 x 1218	nr	257.48
type NC02F size 997 x 1067	nr	244.48
type NC04F size 1486 x 1067	nr	365.84
type ND4 size 1486 x 1218	nr	370.68

Upvc windows tilt / turn factory double glazed

size 600 x 1050 window	nr	406.24
size 900 x 1200 window	nr	471.80
size 900 x 1500 window	nr	525.44

Doors

44 External plywood flush door, 50 x 100 softwood frame, architrave one side, ironmongery P. C. £50.00 per leaf and decorate

762 x 1981 door	nr	237.04
838 x 1981 door	nr	232.68
1372 x 1981 pair doors	nr	383.80

44 Softwood external panel door glazed panels,50 x 100 softwood frame, architrave one side, ironmongery P.C. £50.00 decorated.

type 2XG size	762 x 1981	nr	267.77
type 2XG size	838 x 1981	nr	300.42
type 2XGG size	762 x 1981	nr	323.18
type 2XGG size	838 x 1981	nr	339.94
type KXT size	762 x 1981	nr	364.73
type KXT size	838 x 1981	nr	359.42

44 Hardwood Magnet Alicante door, bevelled glass 50 x 100 hardwood frame architrave one side, iron mongery P.C. £50.00 decorated.

Size 838 x 1981	nr	445.54

Galvanised steel garage door

2135 x 1980	nr	488.27

Opening in cavity wall

Galvanised steel lintel, DPC brick soldier course	m	48.84
Softwood window board DPC close cavity	m	12.77
Close cavity at jamb, DPC, facings to reveal	m	6.28

INTERNAL WALLS

Common bricks

102 wall	m²	34.46
215 wall	m²	61.80

Light weight concrete block 4.0 n/m²

75 wall	m²	16.76
90 wall	m²	21.29
100 wall	m²	22.15
140 wall	m²	27.70
190 wall	m²	33.25

Timber stud partition with plasterboard both sides

75 partition, direct decoration	m²	32.91
100 partition direct decoration	m²	35.78
75 partition with skim coat plaster and decorate	m²	40.85
100 partition with skim coat plaster and decorate	m²	43.64
75 partition with plastic faced plasterboard	m²	46.41
100 partition with plastic faced plasterboard	m²	49.20
Extra for 75 insulation	m²	3.11

Cellular core partition

57 partition direct decoration	m²	31.04
63 partition direct decoration	m²	35.45
57 partition with skim coat plaster and decorate	m²	38.90
63 partition with skim coat plaster and decorate	m²	43.31

INTERNAL DOORS

		£
34 flush door 38 x 150 lining, stop, architrave both sides, iron mongery PC £20.00 per leaf, decorated.		
762 x 1981 hardboard faced	nr	197.44
838 x 1981 hardboard faced	nr	204.02
1372 x 1981 pair hardboard faced	nr	281.20
762 x 1981 plywood faced	nr	206.16
838 x 1981 plywood faced	nr	213.04
1372 x 1981 pair plywood faced	nr	298.46
44 half hour fire check flush door 50 x 100 frame architrave both sides, iron mongery PC £60.00 per leaf, decorated.		
762 x 1981 plywood faced	nr	237.15
838 x 1981 plywood faced	nr	251.38
1372 x 1981 pair plywood faced	nr	397.55
Extra 6mm georgian polished wired glass		
508 x 508 panel	nr	35.29
584 x 584 panel	nr	41.50
Precast concrete lintels 100 x 145	m	8.48

WALL FINISHES

Hardwall plaster in two coats		
13 thick	m²	7.96
13 thick emulsion painted	m²	10.39
13 thick oil painted	m²	11.58
13 thick textured plastic coating stipple finish	m²	10.44
Lightweight plaster in two coats		
13 thick	m²	6.97
13 thick emulsion painted	m²	9.40
13 thick oil painted	m²	10.59
13 thick textured plastic coating stipple finish	m²	9.45
Plasterboard for direct decoration		
9.5 thick	m²	5.46
9.5 thick emulsion painted	m²	7.09
9.5 thick oil painted	m²	9.08
9.5 thick textured plastic coating stipple finish	m²	7.94
12.5 thick	m²	6.39
12.5 thick emulsion painted	m²	8.82
12.5 thick oil painted	m²	10.01
12.5 thick textured plastic coating stipple finish	m²	8.87
Plasterboard with 3 plaster skim coat finish		
9.5 thick plasterboard and skim	m²	9.39
9.5 thick plasterboard and skim emulsion painted	m²	11.82
9.5 thick plasterboard and skim oil painted	m²	13.01
9.5 thick textured plastic coating stipple finish	m²	7.94
12.5 thick plasterboard and skim	m²	10.32
12.5 thick plasterboard and skim emulsion painted	m²	12.75
12.5 thick plasterboard and skim oil painted	m²	13.94
12.5 thick textured plastic coating stipple finish	m²	8.87
Extra wall paper PC £4.00 per roll	m²	4.15
Extra vinyl coated wall paper PC £7.00 per roll	m²	4.23
Extra wall tiling 108 x 108 x 6.5 white glazed	m²	40.26
Extra wall tiling 152 x 152 x 5.5 white glazed	m²	30.56
Extra wall tiling 152 x 152 x 5.5 light colour	m²	31.99

FLOOR FINISHES

Cement sand bed		
25 thick	m²	6.43
32 thick	m²	7.56
38 thick	m²	8.43
50 thick	m²	10.25
25 thick with quarry tile paving	m²	36.05
32 thick with quarry tile paving	m²	37.18
38 thick with quarry tile paving	m²	38.05
50 thick with quarry tile paving	m²	39.87
25 thick with 3 pvc sheet flooring	m²	25.33
32 thick with 3 pvc sheet flooring	m²	26.46
38 thick with 3 pvc sheet flooring	m²	27.33
50 thick with 3 pvc sheet flooring	m²	29.15

FLOOR FINISHES (Cont'd)

		£
25 thick with 2.5 pvc tile flooring	m²	21.72
32 thick with 2.5 pvc tile flooring	m²	22.85
38 thick with 2.5 pvc tile flooring	m²	23.72
50 thick with 2.5 pvc tile flooring	m²	25.54
25 thick with 20 oak wood block flooring	m²	73.11
32 thick with 20 oak wood block flooring	m²	74.24
38 thick with 20 oak wood block flooring	m²	75.11
50 thick with 20 oak wood block flooring	m²	76.93
25 thick with 20 maple wood block flooring	m²	67.36
32 thick with 20 maple wood block flooring	m²	68.49
38 thick with 20 maple wood block flooring	m²	69.36
50 thick with 20 maple wood block flooring	m²	71.18
25 thick with 20 merbau wood block flooring	m²	61.20
32 thick with 20 merbau wood block flooring	m²	62.33
38 thick with 20 merbau wood block flooring	m²	63.20
50 thick with 20 merbau wood block flooring	m²	65.02
25 thick with carpet pc £25 / m² and underlay	m²	41.26
32 thick with carpet pc £25 / m² and underlay	m²	42.39
38 thick with carpet pc £25 / m² and underlay	m²	43.26
50 thick with carpet pc £25 / m² and underlay	m²	45.08
3 PVC sheet flooring	m²	18.90
Floor boarding and insulation to concrete		
22 tongued and grooved chipboard and 50 insulation board	m²	24.99
Softwood skirting decorated		
13 x 95 to BS 1186 reference 13RS95	m	5.08
20 x 70 to BS 1186 reference 20CA70	m	5.06
20 x 120 to BS 1186 reference 20CS120	m	6.10
19 x 100 square	m	5.82
19 x 150 square	m	6.92
25 x 100 square	m	6.51
25 x 150 square	m	7.93
15 x 125 moulded	m	5.60
19 x 125 moulded	m	6.23

CEILING FINISHES

Hardwall plaster in two coats		
13 thick	m²	8.85
13 thick emulsion painted	m²	11.89
13 thick oil painted	m²	14.60
13 thick textured plastic coatings stipple finish	m²	12.24
Lightweight plaster in two coats		
10 thick	m²	8.07
10 thick emulsion painted	m²	11.11
10 thick oil painted	m²	13.82
10 thick textured plastic coatings stipple finish	m²	11.46
Plasterboard for direct decoration		
9.5 thick	m²	7.70
9.5 thick emulsion painted	m²	10.74
9.5 thick oil painted	m²	13.45
9.5 thick textured plastic coatings stipple finish	m²	11.09
12.5 thick	m²	8.74
12.5 thick emulsion painted	m²	11.78
12.5 thick oil painted	m²	14.49
12.5 thick textured plastic coating stipple finish	m²	12.13
Plasterboard with 3 plaster skim coat finish		
9.5 thick plasterboard and skim	m²	12.43
9.5 thick plasterboard and skim emulsion painted	m²	15.47
9.5 thick plasterboard and skim oil painted	m²	18.75
9.5 thick textured plastic coating stipple finish	m²	15.82
12.5 thick plasterboard and skim	m²	13.47
12.5 thick plasterboard and skim emulsion painted	m²	16.51
12.5 thick plasterboard and skim oil painted	m²	19.22
12.5 thick textured plastic coating stipple finish	m²	16.86

APPROXIMATE ESTIMATING

Suspended ceiling

		£
300 x 300 x 15.8 plain mineral fibre tile	m²	25.22

CEILING FINISHES (Cont'd)

		£
300 x 300 x 15.8 textured mineral fibre tile	m²	27.70
300 x 300 patterned tile P.C. £15.00	m²	31.60
600 x 600 x 15.8 plain mineral fibre tile	m²	21.31
600 x 600 x 15.8 drilled mineral fibre tile	m²	23.79
600 x 600 patterned tile P.C. £15.00	m²	27.69

FITTINGS AND FURNISHINGS

Preassembled kitchen units

500 floor unit	nr	152.77
600 floor unit	nr	164.00
1000 floor unit	nr	265.24
500 floor drawer unit	nr	248.35
1000 sink unit	nr	250.47
500 wall unit	nr	87.26
600 wall unit	nr	93.91
1000 wall unit	nr	142.37
28 work top	m	33.38
40 work top	m	45.00
plinth	m	28.20
cornice	m	32.39
pelmet	m	26.67

SANITARY APPLIANCES

Sanitary appliance with allowance for waste fittings, traps, overflows, taps and builders work. Domestic building with plastic wastes

lavatory basin	nr	93.11
bath with panels to front and one end	nr	237.50
low level w.c. suite	nr	146.35
stainless steel sink and drainer	nr	123.19
shower	nr	426.68

Commercial building with copper wastes

lavatory basin	nr	103.79
low level w.c. suite	nr	146.35
bowl type urinal	nr	279.31
stainless steel sink	nr	135.80
drinking fountain	nr	377.99
shower	nr	447.14

DISPOSAL INSTALLATIONS

Soil and vent pipe for ground and first floor

plastic wastes and SVP	nr	166.23
copper wastes and cast iron SVP	nr	599.73

WATER INSTALLATIONS

Hot and cold water services domestic

Building	nr	889.59
gas instantaneous sink water heater	nr	621.03
electric instantaneous sink water heater	nr	561.55
undersink water heater	nr	610.22
electric shower unit	nr	241.20

HEATING INSTALLATIONS

Boiler, radiators, pipework, controls and pumps

100m² domestic building, solid fuel	m²	37.31
100m² domestic building, gas	m²	34.15
100m² domestic building, oil	m²	42.37
150m² domestic building, solid fuel	m²	33.30
150m² domestic building, gas	m²	29.95
150m² domestic building, oil	m²	37.05
500m² commercial building, gas	m²	49.75
500m² commercial building, oil	m²	57.90
1000m² commercial building, gas	m²	59.02
1000m² commercial building, oil	m²	69.67
1500m² commercial building, gas	m²	68.29
1500m² commercial building, oil	m²	80.03
2000m² commercial building, gas	m²	74.20
2000m² commercial building, oil	m²	87.56
Underfloor heating with polymer pipework and insulation - excludes screed.	m²	25.68

ELECTRICAL INSTALLATIONS

Concealed installation with lighting, power, and ancillary cooker, immersion heater etc.

100m² domestic building, PVC insulated and sheathed cables	m²	41.50
100m² domestic building, MICS cables	m²	57.05

ELECTRICAL INSTALLATIONS (Cont'd)

		£
100m² domestic building, steel conduit	m²	66.47
150m² domestic building, PVC insulated and sheathed cables	m²	37.24
150m² domestic building, MICS cables	m²	50.39
150m² domestic building, steel conduit	m²	58.16
500m² commercial building, light fittings, steel conduit	m²	83.09
1000m² commercial building, light fittings, steel conduit	m²	88.28
1500m² commercial building, light fittings, steel conduit	m²	93.47
2000m² commercial building, light fittings, steel conduit	m²	98.66

LIFT AND ESCALATOR INSTALLATIONS

Light passenger lift, laminate walls, stainless steel doors, landing indicators, phone, fluorescent emergency lighting.

3 stop, 8 person	nr	22753.47
4 stop, 13 person	nr	42487.25

General purpose passenger lift, laminate walls stainless steel doors; landing indicators phone; fluorescent emergency lighting.

4 stop, 8 person	nr	42164.75
5 stop, 13 person	nr	45666.03
6 stop, 21 person	nr	104207.31

Service hoist

2 stop, 50 kg	nr	5805.00

Escalator 30 degree inclination department store specification

3.00 m rise	nr	62350.00

SITE WORK

Pavings

250 excavation, 150 hardcore

150 concrete road tamped	m²	26.17
200 concrete road tamped	m²	30.10
150 concrete and 70 bitumen macadam	m²	40.72
200 concrete and 70 bitumen macadam	m²	44.65
Extra for Ref. A252 fabric reinforcement	m²	2.76
Extra for Ref. C636 fabric reinforcement	m²	3.63

Walls

Brick wall, excavation 600 deep concrete foundation, 215 wall with piers at 3m centres in facings PC £200 / 1000 damp proof course.

900 high brick on edge coping	m	115.82
1800 high brick on edge coping	m	188.67
900 high 300 x 100 splayed Precast concrete coping	m	116.20
1800 high 300 x 100 splayed Precast concrete coping	m	189.05

Fences

Plastic coated chain link fencing, posts at 2743 centres.

900 high 38 x 38 angle iron posts two line wires	m	9.80
1200 high 38 x 38 angle iron posts two line wires	m	11.79
1400 high 38 x 38 angle iron posts three line wires	m	12.82
1800 high 38 x 38 angle iron posts three line wires	m	15.53
900 high concrete posts two line wires	m	12.51
1200 high concrete posts two line wires	m	14.44
1400 high concrete posts three line wires	m	15.92
1800 high concrete posts three line wires	m	19.46

Close boarded fencing, 100 x 100 posts @ 2743 centres 25 x 100 pales, 25 x 150 gravel boards, creosote, post holes filled with concrete.

1200 high two arris rails	m	27.00

1500 high three arris rails	m	34.22
1800 high three arris rails	m	37.70

DRAINAGE

Drains

Vitrified clay flexible joint drain, excavation and backfill.

100 diameter 500 deep	m	13.73
100 diameter 750 deep	m	15.85
100 diameter 1000 deep	m	18.17
100 diameter 1500 deep	m	23.61
150 diameter 500 deep	m	18.28
150 diameter 750 deep	m	20.44
150 diameter 1000 deep	m	22.79
150 diameter 1500 deep	m	28.31

Vitrified clay flexible joint drain, excavation 150 bed and surround pea shingle and backfill.

100 diameter 500 deep	m	23.40
100 diameter 750 deep	m	25.65
100 diameter 1000 deep	m	27.97
100 diameter 1500 deep	m	33.41
150 diameter 500 deep	m	29.82
150 diameter 750 deep	m	32.13
150 diameter 1000 deep	m	34.48
150 diameter 1500 deep	m	40.01

Vitrified clay flexible joint drain, excavation 150 bed and haunch concrete and backfill.

100 diameter 500 deep	m	24.85
100 diameter 750 deep	m	27.03
100 diameter 1000 deep	m	29.35
100 diameter 1500 deep	m	35.02
150 diameter 500 deep	m	30.48
150 diameter 750 deep	m	32.69
150 diameter 1000 deep	m	35.04
150 diameter 1500 deep	m	40.56

Vitrified clay flexible joint drain, excavation 150 bed and surround concrete and back fill

100 diameter 500 deep	m	31.80
100 diameter 750 deep	m	34.14
100 diameter 1000 deep	m	36.46
100 diameter 1500 deep	m	41.90
150 diameter 500 deep	m	40.34
150 diameter 750 deep	m	42.65
150 diameter 1000 deep	m	45.00
150 diameter 1500 deep	m	50.53

Upvc drain, excavation 50 mm bed sand and back fill.

110 diameter 500 deep	m	13.44
110 diameter 750 deep	m	15.58
110 diameter 1000 deep	m	17.90
110 diameter 1500 deep	m	23.34
160 diameter 500 deep	m	18.88
160 diameter 750 deep	m	21.05
160 diameter 1000 deep	m	23.40
160 diameter 1500 deep	m	28.93

Upvc drain, excavation bed, and surround pea shingle and back fill.

110 diameter 500 deep	m	22.00
110 diameter 750 deep	m	24.25
110 diameter 1000 deep	m	26.57
110 diameter 1500 deep	m	32.01
160 diameter 500 deep	m	29.20
160 diameter 750 deep	m	31.51
160 diameter 1000 deep	m	33.86
160 diameter 1500 deep	m	39.38

Upvc drain, excavation bed and haunch concrete and back fill

110 diameter 500 deep	m	23.54
110 diameter 750 deep	m	25.73
110 diameter 1000 deep	m	28.04
110 diameter 1500 deep	m	33.49
160 diameter 500 deep	m	30.13
160 diameter 750 deep	m	32.37
160 diameter 1000 deep	m	34.72
160 diameter 1500 deep	m	40.24

Upvc drain, excavation bed and surround concrete and back fill

110 diameter 500 deep	m	30.49
110 diameter 750 deep	m	32.74
110 diameter 1000 deep	m	35.06

110 diameter 1500 deep	m	40.50
160 diameter 500 deep	m	39.72
160 diameter 750 deep	m	42.03
160 diameter 1000 deep	m	44.38
160 diameter 1500 deep	m	49.90

DRAINAGE

Drains (Cont'd)

Cast iron drain, excavation and back fill

100 diameter 500 deep	m	41.11
100 diameter 750 deep	m	41.25
100 diameter 1000 deep	m	45.55
100 diameter 1500 deep	m	50.99
150 diameter 500 deep	m	64.13
150 diameter 750 deep	m	66.29
150 diameter 1000 deep	m	68.64
150 diameter 1500 deep	m	74.16

Cast iron drain, excavation, bed and haunch concrete and back fill.

100 diameter 500 deep	m	52.30
100 diameter 750 deep	m	52.50
100 diameter 1000 deep	m	56.80
100 diameter 1500 deep	m	62.64
150 diameter 500 deep	m	75.56
150 diameter 750 deep	m	77.79
150 diameter 1000 deep	m	80.14
150 diameter 1500 deep	m	85.67

Cast iron drain, excavation, bed and surround concrete and back fill.

100 diameter 500 deep	m	59.27
100 diameter 750 deep	m	59.54
100 diameter 1000 deep	m	63.84
100 diameter 1500 deep	m	69.28
150 diameter 500 deep	m	86.20
150 diameter 750 deep	m	88.51
150 diameter 1000 deep	m	90.86
150 diameter 1500 deep	m	96.38

Brick manhole

Excavation, 150 concrete bed, 215 class B engineering brick walls, 100 clayware main channel, concrete benching, step irons and cast iron manhole cover.

600 x 450 with two branches 750 deep	nr	392.48
600 x 450 with two branches 1000 deep	nr	495.09
600 x 450 with two branches 1500 deep	nr	696.66
750 x 450 with three branches 750 deep	nr	527.16
750 x 450 with three branches 1000 deep	nr	623.09
750 x 450 with three branches 1500 deep	nr	834.56
900 x 600 with five branches 750 deep	nr	706.31
900 x 600 with five branches 1000 deep	nr	821.24
900 x 600 with five branches 1500 deep	nr	1082.31

APPROXIMATE ESTIMATING

Other available titles

Laxton's General Specification Volumes 1 & 2
(0 7506 3352 0)

Laxton's General Specification – electronic version
(0 7506 3693 9)

Laxton's Guide to Term Maintenance Contracts
(0 7506 2977 0)

Laxton's Measurement Rules for Contractors Quantities
(0 7506 2977 0)

Laxton's Trades Price Book: Small Works Repairs and Maintenance
(0 7506 2978 9)

Laxton's Guide to Budget Estimating
(0 7506 2967 3)

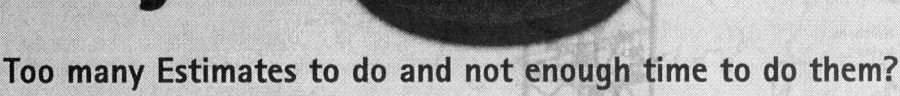
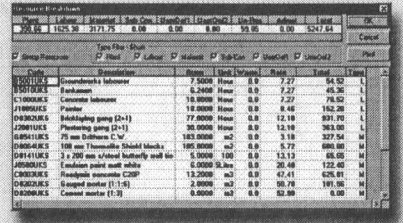

Regional Factors

Laxton's Building Price Book is based upon national average prices (=1.00). Indicative levels of tender pricing in regions as at the second quarter 1999 are given on this map and the factors shown may be applied to adjust overall pricing.

MAJOR WORKS – TENDER VALUES	SMALL WORKS – TENDER VALUES
The measured rates are based on contracts valued in the range of £250,000 to £1,000.00.	The measured rates are based on contracts valued in the range of £25,000 to £75,000. As a guide to pricing works of smaller or larger value and for cost planning or budgetary purposes, the following adjustments may be applied to overall contract values.
As a guide to pricing works of larger value and for cost planning or budgetary purposes, the following adjustments may be applied to overall contract values.	
	Contract Value
Contract Value	£5,000 to £15,000.............................add 20.0%
£1,000,000 to £2,000,000.........deduct 2.5%	£15,000 to £25,000............................add 10.0%
£2,000,000 to £3,000,000.........deduct 5.0%	£25,000 to £75,000.......................rate as shown
£3,000,000 to £5,000,000.........deduct 7.5%	£75,000 to £100,000.......................deduct 5.0%
	£100,000 to £150,000......................deduct7.5%
	£150,000 to £250,000......................deduct10.%

VISIT THE LAXTON'S WEBSITE TODAY!

www.laxtonsprices.co.uk

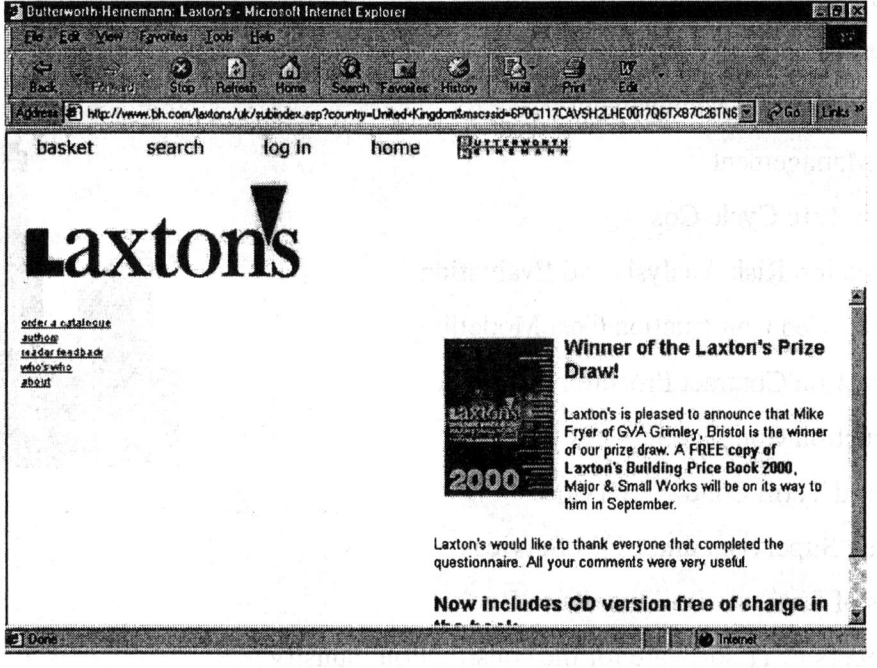

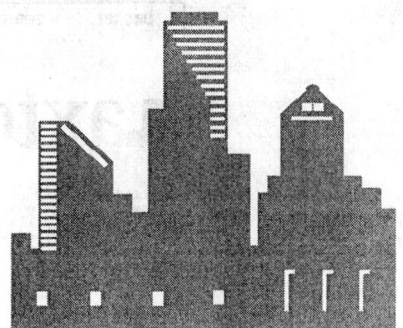

SMALL WORKS

The measured rates in this section are intended to apply to Contracts within the price range £25,000 to £75,000 based on National Average prices.

Adjustment to overall contract values for Regional Variations and contract values from £5,000 to £250,000 can be made as noted in the "Essential Information" section.

Regional Factors

Scotland
0.95

Northern Ireland
0.75

Northern
0.94

Yorks & Humber
0.94

North West
1.00

East Midlands
0.93

West Midlands
0.94

East Anglia
0.96

Wales
0.94

South East
1.06

G.L 1.18

South West
0.99

Laxton's Building Price Book is based upon national average prices (=1.00). Indicative levels of tender pricing in regions as at the second quarter 1999 are given on this map and the factors shown may be applied to adjust overall pricing.

MAJOR WORKS – TENDER VALUES	SMALL WORKS – TENDER VALUES
The measured rates are based on contracts valued in the range of £250,000 to £1,000.00. As a guide to pricing works of larger value and for cost planning or budgetary purposes, the following adjustments may be applied to overall contract values.	The measured rates are based on contracts valued in the range of £25,000 to £75,000. As a guide to pricing works of smaller or larger value and for cost planning or budgetary purposes, the following adjustments may be applied to overall contract values.
Contract Value £1,000,000 to £2,000,000.........deduct 2.5% £2,000,000 to £3,000,000.........deduct 5.0% £3,000,000 to £5,000,000.........deduct 7.5%	Contract Value £5,000 to £15,000.............................add 20.0% £15,000 to £25,000...........................add 10.0% £25,000 to £75,000......................rate as shown £75,000 to £100,000.......................deduct 5.0% £100,000 to £150,000.....................deduct7.5% £150,000 to £250,000.....................deduct10.%

Existing site/Buildings/Services

Labour hourly rates: (except Specialists) Craft Operatives £9.23 Labourer £7.02 Rates are national average prices. Refer to REGIONAL VARIATIONS for indicative levels of overall pricing in regions	MATERIALS			LABOUR				RATES		
	Del to Site £	Waste %	Material Cost £	Craft Optve Hrs	Lab Hrs	Labour Cost £	Sunds £	Nett Rate £	Unit	Gross Rate (+12.5%) £
C20: DEMOLITION										
Pulling down outbuildings										
Demolishing individual structures										
timber outbuilding 2.50 x 2.00 x 3.00m maximum high	-	-	-	-	11.00	77.22	59.50	136.72	nr	153.81
outbuilding with half brick walls, one brick piers and felt covered timber roof, 2.50 x 2.00 x 3.00m maximum high	-	-	-	-	22.00	154.44	119.00	273.44	nr	307.62
greenhouse with half brick dwarf walls and timber framing to upper walls and roof, 3.00 x 2.00 x 3.00m maximum high	-	-	-	-	18.00	126.36	102.00	228.36	nr	256.90
Demolishing individual structures; setting aside materials for re-use										
metal framed greenhouse, 3.00 x 2.00 x 3.00m maximum high	-	-	-	4.00	7.00	86.06	-	86.06	nr	96.82
prefabricated concrete garage, 5.40 x 2.60 x 2.40m maximum high	-	-	-	6.00	11.00	132.60	-	132.60	nr	149.18
Demolishing individual structures; making good structures										
timber framed lean-to outbuilding and remove flashings and make good facing brick wall at ridge and vertical abutments, 3.00 x 1.90 x 2.40m maximum high	-	-	-	1.50	12.00	98.08	63.00	161.09	nr	181.22
Unit rates for pricing the above and similar work										
pull down building 3.00 x 1.90 x 2.40m maximum high	-	-	-	-	10.00	70.20	59.50	129.70	nr	145.91
remove flashing	-	-	-	-	0.12	0.84	0.34	1.18	m	1.33
make good facing brick wall at ridge	-	-	-	0.20	0.20	3.25	0.31	3.56	m	4.00
make good facing brick wall at vertical abutment	-	-	-	0.17	0.17	2.76	0.31	3.07	m	3.46
make good rendered wall at ridge	-	-	-	0.20	0.20	3.25	0.72	3.97	m	4.47
make good rendered wall at vertical abutment	-	-	-	0.20	0.20	3.25	0.72	3.97	m	4.47
Demolishing individual structures; making good structures										
lean-to outbuilding with half brick walls and slate covered timber roof and remove flashings, hack off plaster to house wall and make good with rendering to match existing, 1.50 x 1.00 x 2.00m maximum high	-	-	-	4.00	12.50	124.67	40.04	164.71	nr	185.30
Unit rates for pricing the above and similar work										
pull down building with half brick walls, 1.50 x 1.00 x 2.00m maximum high	-	-	-	-	5.30	37.21	27.20	64.41	nr	72.46
pull down building with one brick walls, 1.50 x 1.00 x 2.00m maximum high	-	-	-	-	10.00	70.20	51.00	121.20	nr	136.35
hack off plaster or rendering and make good to match existing brick facings	-	-	-	1.00	2.00	23.27	1.02	24.29	m2	27.33
hack off plaster or rendering and make good to match existing rendering	-	-	-	1.30	2.30	28.15	4.11	32.26	m2	36.29
Removal of old work										
Demolishing parts of structures										
in-situ plain concrete bed										
75mm thick	-	-	-	-	0.90	6.32	2.55	8.87	m2	9.98
100mm thick	-	-	-	-	1.20	8.42	3.40	11.82	m2	13.30
150mm thick	-	-	-	-	1.80	12.64	5.10	17.74	m2	19.95
200mm thick	-	-	-	-	2.40	16.85	6.80	23.65	m2	26.60
in-situ reinforced concrete flat roof										
100mm thick	-	-	-	-	1.90	13.34	3.40	16.74	m2	18.83
150mm thick	-	-	-	-	2.85	20.01	5.10	25.11	m2	28.25
200mm thick	-	-	-	-	3.80	26.68	6.80	33.48	m2	37.66
225mm thick	-	-	-	-	4.28	30.05	7.65	37.70	m2	42.41
in-situ reinforced concrete upper floor										
100mm thick	-	-	-	-	1.90	13.34	3.40	16.74	m2	18.83
150mm thick	-	-	-	-	2.85	20.01	5.10	25.11	m2	28.25

Labour hourly rates: (except Specialists) Craft Operatives £9.23 Labourer £7.02 Rates are national average prices. Refer to REGIONAL VARIATIONS for indicative levels of overall pricing in regions	MATERIALS			LABOUR				RATES		
	Del to Site	Waste	Material Cost	Craft Optve	Lab	Labour Cost	Sunds	Nett Rate	Unit	Gross Rate (+12.5%)
	£	%	£	Hrs	Hrs	£	£	£		£
C20: DEMOLITION Cont.										
Removal of old work Cont.										
Demolishing parts of structures Cont.										
in-situ reinforced concrete upper floor Cont.										
200mm thick............................	-	-	-	-	3.80	26.68	6.80	33.48	m2	37.66
225mm thick............................	-	-	-	-	4.28	30.05	7.65	37.70	m2	42.41
in-situ reinforced concrete beam.............	-	-	-	-	19.00	133.38	34.00	167.38	m3	188.30
in-situ reinforced concrete column...........	-	-	-	-	18.00	126.36	34.00	160.36	m3	180.41
in-situ reinforced concrete wall										
100mm thick............................	-	-	-	-	1.80	12.64	3.40	16.04	m2	18.04
150mm thick............................	-	-	-	-	2.70	18.95	5.10	24.05	m2	27.06
200mm thick............................	-	-	-	-	3.60	25.27	6.80	32.07	m2	36.08
225mm thick............................	-	-	-	-	4.05	28.43	7.65	36.08	m2	40.59
in-situ reinforced concrete casing to beam....	-	-	-	-	16.00	112.32	34.00	146.32	m3	164.61
in-situ reinforced concrete casing to column ..	-	-	-	-	15.25	107.06	34.00	141.06	m3	158.69
brick internal walls in lime mortar										
102mm thick............................	-	-	-	-	0.63	4.42	3.47	7.89	m2	8.88
215mm thick............................	-	-	-	-	1.30	9.13	7.31	16.44	m2	18.49
327mm thick............................	-	-	-	-	2.00	14.04	11.12	25.16	m2	28.31
brick internal walls in cement mortar										
102mm thick............................	-	-	-	-	0.95	6.67	3.47	10.14	m2	11.41
215mm thick............................	-	-	-	-	1.94	13.62	7.31	20.93	m2	23.54
327mm thick............................	-	-	-	-	2.98	20.92	11.12	32.04	m2	36.04
reinforced brick internal walls in cement lime mortar										
102mm thick............................	-	-	-	-	0.95	6.67	3.47	10.14	m2	11.41
215mm thick............................	-	-	-	-	1.94	13.62	7.31	20.93	m2	23.54
hollow clay block internal walls in cement-lime mortar										
50mm thick.............................	-	-	-	-	0.30	2.11	1.70	3.81	m2	4.28
75mm thick.............................	-	-	-	-	0.40	2.81	2.55	5.36	m2	6.03
100mm thick............................	-	-	-	-	0.52	3.65	3.40	7.05	m2	7.93
hollow clay block internal walls in cement mortar										
50mm thick.............................	-	-	-	-	0.32	2.25	1.70	3.95	m2	4.44
75mm thick.............................	-	-	-	-	0.43	3.02	2.55	5.57	m2	6.26
100mm thick............................	-	-	-	-	0.57	4.00	3.40	7.40	m2	8.33
concrete block internal walls in cement lime mortar										
75mm thick.............................	-	-	-	-	0.35	2.46	2.55	5.01	m2	5.63
100mm thick............................	-	-	-	-	0.53	3.72	3.40	7.12	m2	8.01
190mm thick............................	-	-	-	-	0.70	4.91	6.46	11.37	m2	12.80
215mm thick............................	-	-	-	-	1.41	9.90	7.31	17.21	m2	19.36
concrete block internal walls in cement mortar										
75mm thick.............................	-	-	-	-	0.56	3.93	2.55	6.48	m2	7.29
100mm thick............................	-	-	-	-	0.73	5.12	3.40	8.52	m2	9.59
190mm thick............................	-	-	-	-	0.91	6.39	6.46	12.85	m2	14.45
215mm thick............................	-	-	-	-	1.67	11.72	7.31	19.03	m2	21.41
if internal walls plastered, add per side.....	-	-	-	-	0.12	0.84	0.68	1.52	m2	1.71
if internal walls rendered, add per side......	-	-	-	-	0.18	1.26	0.68	1.94	m2	2.19
brick external walls in lime mortar										
102mm thick............................	-	-	-	-	0.48	3.37	3.47	6.84	m2	7.69
215mm thick............................	-	-	-	-	0.98	6.88	7.31	14.19	m2	15.96
327mm thick............................	-	-	-	-	1.51	10.60	11.12	21.72	m2	24.44
brick external walls in cement mortar										
102mm thick............................	-	-	-	-	0.63	4.42	3.47	7.89	m2	8.88
215mm thick............................	-	-	-	-	1.30	9.13	7.31	16.44	m2	18.49
327mm thick............................	-	-	-	-	2.00	14.04	11.12	25.16	m2	28.31
reinforced brick external walls in cement lime mortar										
102mm thick............................	-	-	-	-	0.63	4.42	3.47	7.89	m2	8.88
215mm thick............................	-	-	-	-	1.30	9.13	7.31	16.44	m2	18.49
if external walls plastered, add per side.....	-	-	-	-	0.09	0.63	0.68	1.31	m2	1.48
if external walls rendered or rough cast, add per side.................................	-	-	-	-	0.12	0.84	0.68	1.52	m2	1.71
clean old bricks in lime mortar and stack for re-use, per thousand......................	-	-	-	-	15.00	105.30	-	105.30	nr	118.46
clean old bricks in cement mortar and stack for re-use, per thousand....................	-	-	-	-	25.00	175.50	-	175.50	nr	197.44
rough rubble walling 600mm thick, in lime mortar	-	-	-	-	3.75	26.32	20.40	46.73	m2	52.57
random rubble walling 350mm thick, in lime mortar	-	-	-	-	2.20	15.44	11.90	27.34	m2	30.76
random rubble walling 500mm thick, in lime mortar	-	-	-	-	3.10	21.76	17.00	38.76	m2	43.61
dressed stone walling 100mm thick, in gauged mortar	-	-	-	2.00	2.30	34.61	3.40	38.01	m2	42.76
dressed stone walling 200mm thick, in gauged mortar	-	-	-	3.90	4.50	67.59	6.80	74.39	m2	83.69
stone copings 300 x 50mm.................	-	-	-	0.20	0.26	3.67	0.51	4.18	m	4.70
stone copings 375 x 100mm................	-	-	-	0.40	0.52	7.34	1.28	8.62	m	9.70
stone staircases........................	-	-	-	-	15.00	105.30	34.00	139.30	m3	156.71
stone steps............................	-	-	-	-	14.75	103.55	34.00	137.54	m3	154.74
structural timbers										
50 x 75mm.............................	-	-	-	-	0.13	0.91	0.13	1.04	m	1.17
50 x 100mm............................	-	-	-	-	0.15	1.05	0.17	1.22	m	1.38
50 x 225mm............................	-	-	-	-	0.24	1.68	0.38	2.06	m	2.32
75 x 100mm............................	-	-	-	-	0.18	1.26	0.26	1.52	m	1.71
75 x 150mm............................	-	-	-	-	0.20	1.40	0.38	1.78	m	2.01
75 x 225mm............................	-	-	-	-	0.31	2.18	0.57	2.75	m	3.09
steelwork										
steel beams, joists and lintels not exceeding 10 kg/m...............................	-	-	-	50.00	50.00	812.50	34.00	846.50	t	952.31
steel beams, joists and lintels 10 - 20 kg/m..	-	-	-	46.00	46.00	747.50	34.00	781.50	t	879.19
steel beams, joists and lintels 20 - 50 kg/m..	-	-	-	40.00	40.00	650.00	34.00	684.00	t	769.50

Labour hourly rates: (except Specialists) Craft Operatives £9.23 Labourer £7.02 Rates are national average prices. Refer to REGIONAL VARIATIONS for indicative levels of overall pricing in regions	MATERIALS			LABOUR				RATES		
	Del to Site	Waste	Material Cost	Craft Optve	Lab	Labour Cost	Sunds	Nett Rate	Unit	Gross Rate (+12.5%)
	£	%	£	Hrs	Hrs	£	£	£		£
C20: DEMOLITION Cont.										
Removal of old work Cont.										
Demolishing parts of structures Cont.										
steelwork Cont.										
steel columns and stanchions not exceeding 10 kg/m	-	-	-	60.00	60.00	975.00	34.00	1009.00	t	1135.13
steel columns and stanchions 10 - 20 kg/m	-	-	-	56.00	56.00	910.00	34.00	944.00	t	1062.00
steel columns and stanchions 20 - 50 kg/m	-	-	-	50.00	50.00	812.50	34.00	846.50	t	952.31
steel purlins and rails not exceeding 10 kg/m	-	-	-	50.00	50.00	812.50	34.00	846.50	t	952.31
steel purlins and rails 10 - 20 kg/m	-	-	-	50.00	50.00	812.50	34.00	846.50	t	952.31
C30: SHORING/FACADE RETENTION										
Support of structures not to be demolished										
Provide and erect timber raking shore complete with sole piece, cleats, needles and 25mm brace boarding with										
two 150 x 150mm rakers (total length 8.00m) and 50 x 175mm wall piece	122.00	10.00	134.20	4.00	4.00	65.00	19.52	218.72	nr	246.06
weekly cost of maintaining last	-	-	-	1.00	1.00	16.25	0.98	17.23	nr	19.38
three 225 x 225mm rakers (total length 21.00m) and 50 x 250mm wall piece	461.00	10.00	507.10	23.20	23.20	377.00	84.87	968.97	nr	1090.09
weekly cost of maintaining last	-	-	-	1.50	1.50	24.38	4.19	28.57	nr	32.14
Provide and erect timber flying shore of 100 x 150mm main member, 50 x 175mm wall pieces and 50 x 100mm straining pieces and 75 x 100mm struts, distance between wall faces										
4.00m	124.00	10.00	136.40	15.00	15.00	243.75	4.19	384.34	nr	432.38
weekly cost of maintaining last	-	-	-	3.50	3.50	56.88	0.98	57.85	nr	65.09
5.00m	144.00	10.00	158.40	17.80	17.80	289.25	4.88	452.53	nr	509.10
weekly cost of maintaining last	-	-	-	3.00	3.00	48.75	1.12	49.87	nr	56.10
6.00m	157.00	10.00	172.70	5.10	20.60	191.69	5.41	369.80	nr	416.02
weekly cost of maintaining last	-	-	-	3.50	3.50	56.88	1.28	58.16	nr	65.42
C40: CLEANING MASONRY/CONCRETE										
Cleaning surfaces										
Thoroughly clean existing concrete surfaces prior to applying damp proof membrane										
floors	-	-	-	-	0.19	1.33	0.02	1.35	m2	1.52
walls	-	-	-	-	0.21	1.47	0.02	1.49	m2	1.68
Thoroughly clean existing concrete surfaces, fill in nail holes and small surface imperfections and leave smooth										
walls	-	-	-	0.35	0.20	4.63	0.04	4.67	m2	5.26
soffits	-	-	-	0.50	0.29	6.65	0.04	6.69	m2	7.53
Thoroughly clean existing brick or block surfaces prior to applying damp proof membrane										
walls	-	-	-	-	0.25	1.75	-	1.75	m2	1.97
Clean out air brick										
225 x 150mm	-	-	-	-	0.20	1.40	-	1.40	nr	1.58
C41: REPAIRING/RENOVATING/CONSERVING MASONRY										
Cut out decayed bricks and replace with new bricks in gauged mortar (1:2:9) and make good surrounding work										
Fair faced common bricks										
singly	0.20	10.00	0.22	0.17	0.17	2.76	0.14	3.12	nr	3.51
small patches	12.00	10.00	13.20	5.80	5.80	94.25	5.98	113.43	m2	127.61
Picked stock facing bricks										
singly	0.68	10.00	0.75	0.17	0.17	2.76	0.14	3.65	nr	4.11
small patches	40.80	10.00	44.88	5.80	5.80	94.25	5.98	145.11	m2	163.25
Cut out defective wall and re-build in gauged mortar (1:2:9) and tooth and bond to surrounding work (25% new bricks allowed)										
Half brick wall; stretcher bond										
common bricks	3.00	10.00	3.30	4.00	4.00	65.00	3.43	71.73	m2	80.70
One brick wall; English bond										
common bricks	6.00	10.00	6.60	7.90	7.90	128.38	7.37	142.35	m2	160.14
common bricks faced one side with picked stock facings	17.48	10.00	19.23	8.30	8.30	134.88	7.37	161.47	m2	181.66
picked stock facing bricks faced both sides	20.40	10.00	22.44	8.80	8.80	143.00	7.37	172.81	m2	194.41
One brick wall; Flemish bond										
common bricks	6.00	10.00	6.60	7.90	7.90	128.38	3.43	138.41	m2	155.71
common bricks faced one side with picked stock facings	17.48	10.00	19.23	8.30	8.30	134.88	7.37	161.47	m2	181.66
picked stock facing bricks faced both sides	20.40	10.00	22.44	8.80	8.80	143.00	7.37	172.81	m2	194.41
275mm hollow wall; skins in stretcher bond										
two half brick skins in common bricks	6.00	10.00	6.60	8.16	8.16	132.60	7.51	146.71	m2	165.05
one half brick skin in common bricks and one half brick skin in picked stock facings	14.55	10.00	16.00	8.60	8.60	139.75	7.51	163.26	m2	183.67

Labour hourly rates: (except Specialists) Craft Operatives £9.23 Labourer £7.02 Rates are national average prices. Refer to REGIONAL VARIATIONS for indicative levels of overall pricing in regions	MATERIALS			LABOUR				RATES		
	Del to Site £	Waste %	Material Cost £	Craft Optve Hrs	Lab Hrs	Labour Cost £	Sunds £	Nett Rate £	Unit	Gross Rate (+12.5%) £
C41: REPAIRING/RENOVATING/CONSERVING MASONRY Cont.										
Cut out defective wall and re-build in gauged mortar (1:2:9) and tooth and bond to surrounding work (25% new bricks allowed) Cont.										
275mm hollow wall; skins in stretcher bond Cont.										
one 100mm skin in concrete blocks and one half brick skin in picked stock facings	12.58	10.00	13.84	8.26	8.26	134.22	7.02	155.08	m2	174.47
Cut out crack in brickwork and stitch across with new bricks in gauged mortar(1:2:9); average 450mm wide										
Half brick wall										
common bricks	5.20	10.00	5.72	1.90	1.90	30.88	3.33	39.92	m	44.92
common bricks fair faced one side	5.20	10.00	5.72	2.30	2.30	37.38	3.33	46.42	m	52.23
picked stock facing bricks faced both sides	17.68	10.00	19.45	2.40	2.40	39.00	3.33	61.78	m	69.50
One brick wall										
common bricks	10.40	10.00	11.44	3.90	3.90	63.38	7.17	81.98	m	92.23
common bricks fair faced one side	10.40	10.00	11.44	4.40	4.40	71.50	7.17	90.11	m	101.37
common bricks faced one side with picked stock facings	29.12	10.00	32.03	4.60	4.60	74.75	7.17	113.95	m	128.20
picked stock facing bricks faced both sides	35.36	10.00	38.90	4.70	4.70	76.38	7.17	122.44	m	137.75
Cut out defective arch; re-build in picked stock facing bricks in gauged mortar (1:2:9) including centering										
Brick-on-edge flat arch										
102mm on soffit	4.76	10.00	5.24	1.00	1.00	16.25	0.84	22.33	m	25.12
215mm on soffit	9.06	10.00	9.97	0.95	0.95	15.44	1.47	26.87	m	30.23
Brick-on-end flat arch										
102mm on soffit	9.06	10.00	9.97	0.95	0.95	15.44	1.47	26.87	m	30.23
215mm on soffit	18.13	10.00	19.94	1.90	1.90	30.88	2.94	53.76	m	60.48
Segmental arch in two half brick rings										
102mm on soffit	9.06	10.00	9.97	2.90	2.90	47.13	1.47	58.56	m	65.88
215mm on soffit	18.13	10.00	19.94	2.90	2.90	47.13	2.94	70.01	m	78.76
Semi-circular arch in two half brick rings										
102mm on soffit	9.06	10.00	9.97	2.90	2.90	47.13	1.47	58.56	m	65.88
215mm on soffit	18.13	10.00	19.94	2.90	2.90	47.13	2.94	70.01	m	78.76
Cut out defective arch; replace with precast concrete, B.S.5328, designed mix C20, 20mm aggregate lintel; wedge and pin up to brickwork over with slates in cement mortar (1:3) and extend external rendering over										
Remove defective brick-on-end flat arch replace with 102 x 215mm precast concrete lintel reinforced with one 12mm mild steel bar	15.18	-	15.18	1.90	1.90	30.88	2.31	48.37	m	54.41
Cut out defective external sill and re-build										
Shaped brick-on-edge sill in picked stock facing bricks in gauged mortar (1:2:9) 225mm wide	9.06	10.00	9.97	1.00	1.00	16.25	2.05	28.27	m	31.80
Roofing tile sill set weathering and projecting in cement mortar (1:3) two courses	5.93	10.00	6.52	1.20	1.20	19.50	1.06	27.08	m	30.47
Take down defective brick-on-edge coping and re-build in picked stock facing bricks in gauged mortar (1:2:9)										
Coping to one brick wall	9.06	10.00	9.97	0.60	0.60	9.75	2.05	21.77	m	24.49
Coping with cement fillets both sides to one brick wall; with oversailing course	13.60	10.00	14.96	0.67	0.67	10.89	3.08	28.93	m	32.54
to one brick wall; with single course tile creasing	12.03	10.00	13.23	1.09	1.09	17.71	2.58	33.53	m	37.72
to one brick wall; with double course tile creasing	14.99	10.00	16.49	1.38	1.38	22.43	3.64	42.55	m	47.87
Cut out defective air brick and replace with new; bed and point in gauged mortar (1:2:9)										
Cast iron, light, square hole										
225 x 75mm	3.55	-	3.55	0.45	0.45	7.31	0.16	11.02	nr	12.40
225 x 150mm	6.51	-	6.51	0.45	0.45	7.31	0.26	14.08	nr	15.84
Terra cotta										
215 x 65mm	1.67	-	1.67	0.45	0.45	7.31	0.16	9.14	nr	10.29
215 x 140mm	2.31	-	2.31	0.45	0.45	7.31	0.26	9.88	nr	11.12
Rake out joints of old brickwork 20mm deep and re-point in cement mortar (1:3)										
For turned in edge of lead flashing flush pointing; horizontal	-	-	-	0.50	0.50	8.13	0.24	8.37	m	9.41
flush pointing; stepped	-	-	-	0.85	0.85	13.81	0.34	14.15	m	15.92

4

Labour hourly rates: (except Specialists) Craft Operatives £9.23 Labourer £7.02 Rates are national average prices. Refer to REGIONAL VARIATIONS for indicative levels of overall pricing in regions	MATERIALS			LABOUR				RATES		
	Del to Site £	Waste %	Material Cost £	Craft Optve Hrs	Lab Hrs	Labour Cost £	Sunds £	Nett Rate £	Unit	Gross Rate (+12.5%) £
C41: REPAIRING/RENOVATING/CONSERVING MASONRY Cont.										
Rake out joints of old brickwork 20mm deep and re-point in cement mortar (1:3) Cont.										
For turned in edge of lead flashing Cont.										
weathered pointing; horizontal	-	-	-	0.55	0.55	8.94	0.24	9.18	m	10.32
weathered pointing; stepped	-	-	-	0.90	0.90	14.63	0.34	14.97	m	16.84
ironed in pointing; horizontal	-	-	-	0.55	0.55	8.94	0.24	9.18	m	10.32
ironed in pointing; stepped	-	-	-	0.90	0.90	14.63	0.34	14.97	m	16.84
Rake out joints of old brickwork 20mm deep and re-point in cement-lime mortar(1:1:6)										
Generally; English bond										
flush pointing	-	-	-	1.00	0.66	13.86	0.69	14.55	m2	16.37
weathered pointing	-	-	-	1.00	0.66	13.86	0.69	14.55	m2	16.37
ironed in pointing	-	-	-	1.20	0.80	16.69	0.69	17.38	m2	19.55
Generally; Flemish bond										
flush pointing	-	-	-	1.00	0.66	13.86	0.69	14.55	m2	16.37
weathered pointing	-	-	-	1.00	0.66	13.86	0.69	14.55	m2	16.37
ironed in pointing	-	-	-	1.20	0.80	16.69	0.69	17.38	m2	19.55
Isolated areas not exceeding 1.00m2; English bond										
flush pointing	-	-	-	1.50	1.00	20.86	0.69	21.56	m2	24.25
weathered pointing	-	-	-	1.50	1.00	20.86	0.69	21.56	m2	24.25
ironed in pointing	-	-	-	1.80	1.20	25.04	0.69	25.73	m2	28.94
Isolated areas not exceeding 1.00m2; Flemish bond										
flush pointing	-	-	-	1.50	1.00	20.86	0.69	21.56	m2	24.25
weathered pointing	-	-	-	1.50	1.00	20.86	0.69	21.56	m2	24.25
ironed in pointing	-	-	-	1.80	1.20	25.04	0.69	25.73	m2	28.94
For using coloured mortar										
add	-	-	-	-	-	-	0.50	0.50	m2	0.56
Rake out joints of old brickwork 20mm deep and re-point in cement mortar (1:3)										
Generally; English bond										
flush pointing	-	-	-	1.20	0.80	16.69	0.69	17.38	m2	19.55
weathered pointing	-	-	-	1.20	0.80	16.69	0.69	17.38	m2	19.55
Generally; Flemish bond										
flush pointing	-	-	-	1.20	0.80	16.69	0.69	17.38	m2	19.55
weathered pointing	-	-	-	1.20	0.80	16.69	0.69	17.38	m2	19.55
Isolated areas not exceeding 1.00m2; English bond										
flush pointing	-	-	-	1.80	1.20	25.04	0.69	25.73	m2	28.94
weathered pointing	-	-	-	1.80	1.20	25.04	0.69	25.73	m2	28.94
Isolated areas not exceeding 1.00m2; Flemish bond										
flush pointing	-	-	-	1.80	1.20	25.04	0.69	25.73	m2	28.94
weathered pointing	-	-	-	1.80	1.20	25.04	0.69	25.73	m2	28.94
Repairing/Renovating stone										
Note restoration and repair of stonework is work for a specialist; prices being obtained for specific projects. Some firms that specialise in this class of work are included within the list at the end of section F-Masonry										
Repairs in Portland stonework and set in gauged mortar (1:2:9) and point to match existing										
Cut out decayed stones in facings of walls, piers or the like, prepare for and set new 75mm thick stone facings										
single stone	388.00	-	388.00	15.00	15.00	243.75	6.75	638.50	m2	718.31
areas of two or more adjacent stones	388.00	-	388.00	12.00	12.00	195.00	6.75	589.75	m2	663.47
Take out decayed stone sill, prepare for and set new sunk weathered and throated sill										
175 x 100mm	211.00	-	211.00	1.80	1.80	29.25	0.91	241.16	m	271.31
250 x 125mm	268.00	-	268.00	2.50	2.50	40.63	1.54	310.17	m	348.94
Take out sections of decayed stone coping, prepare for and set new weathered and twice throated coping										
300 x 75mm	274.00	-	274.00	1.40	1.40	22.75	1.14	297.89	m	335.13
300 x 100mm	332.00	-	332.00	1.70	1.70	27.63	1.51	361.13	m	406.28
Repairs in York stonework and set in gauged mortar (1:2:9) and point to match existing										
Cut out decayed stones in facings of walls, piers or the like, prepare for and set new 75mm thick stone facings										
single stone	266.00	-	266.00	17.00	17.00	276.25	6.75	549.00	m2	617.63
areas of two or more adjacent stones	266.00	-	266.00	13.75	13.75	223.44	6.75	496.19	m2	558.21
Take out decayed stone sill, prepare for and set new sunk weathered and throated sill										
175 x 100mm	120.00	-	120.00	1.80	1.80	29.25	1.14	150.39	m	169.19
250 x 125mm	161.00	-	161.00	2.50	2.50	40.63	1.51	203.13	m	228.53

Labour hourly rates: (except Specialists) Craft Operatives £9.23 Labourer £7.02 Rates are national average prices. Refer to REGIONAL VARIATIONS for indicative levels of overall pricing in regions	MATERIALS			LABOUR				RATES		
	Del to Site £	Waste %	Material Cost £	Craft Optve Hrs	Lab Hrs	Labour Cost £	Sunds £	Nett Rate £	Unit	Gross Rate (+12.5%) £
C41: REPAIRING/RENOVATING/CONSERVING MASONRY Cont.										
Repairs in York stonework and set in gauged mortar (1:2:9) and point to match existing Cont.										
Take out sections of decayed stone coping, prepare for and set new weathered and twice throated coping										
300 x 75mm	110.00	-	110.00	1.40	1.40	22.75	1.14	133.89	m	150.63
300 x 100mm	139.00	-	139.00	1.70	1.70	27.63	1.51	168.13	m	189.15
Rake out decayed mortar joints and re-point in gauged mortar (1:2:9)										
Re-point rubble walling										
coursed	-	-	-	1.25	1.25	20.31	1.12	21.43	m2	24.11
uncoursed	-	-	-	1.32	1.32	21.45	1.12	22.57	m2	25.39
squared	-	-	-	1.25	1.25	20.31	1.12	21.43	m2	24.11
Re-point stonework										
ashlar	-	-	-	1.20	1.20	19.50	1.12	20.62	m2	23.20
coursed block	-	-	-	1.20	1.20	19.50	1.12	20.62	m2	23.20
C42: REPAIRING/ RENOVATING/CONSERVING CONCRETE										
Re-bed loose copings										
Take off loose concrete coping and clean and remove old bedding mortar										
re-bed, joint and point in cement mortar (1:3)	-	-	-	1.00	1.00	16.25	1.00	17.25	m	19.41
Repairing cracks in concrete										
Repair cracks; clean out all dust and debris and fill with mortar mixed with a bonding agent										
up to 5mm wide	-	-	-	0.50	0.50	8.13	0.48	8.61	m	9.68
Repair cracks; cutting out to form groove, treat with bonding agent and fill with fine concrete mixed with a bonding agent										
25 x 25mm deep	-	-	-	0.50	0.50	8.13	0.87	8.99	m	10.12
40 x 40mm deep	-	-	-	0.60	0.60	9.75	2.06	11.81	m	13.29
C45: DAMP PROOF COURSE RENEWAL/INSERTION										
Insert damp proof course in old wall by cutting out one course of brickwork in alternate lengths not exceeding 1.00m and replace old bricks with 50mm bricks in cement mortar (1:3)										
B.S.6398 Class D, bitumen with hessian base laminated with lead in										
one brick wall	4.56	10.00	5.02	3.00	3.00	48.75	5.53	59.30	m	66.71
one and a half brick wall	6.83	10.00	7.51	4.00	4.00	65.00	8.31	80.82	m	90.93
254mm hollow wall	4.56	10.00	5.02	3.50	3.50	56.88	5.53	67.42	m	75.85
Hyload, pitch polymer in										
one brick wall	1.85	10.00	2.04	3.00	3.00	48.75	5.53	56.31	m	63.35
one and a half brick wall	2.77	10.00	3.05	4.00	4.00	65.00	8.31	76.36	m	85.90
254mm hollow wall	1.85	10.00	2.04	3.50	3.50	56.88	5.53	64.44	m	72.50
Two courses slates in cement mortar (1:3) in										
one brick wall	6.11	10.00	6.72	3.00	3.00	48.75	5.53	61.00	m	68.63
one and a half brick wall	9.17	10.00	10.09	4.00	4.00	65.00	8.31	83.40	m	93.82
254mm hollow wall	6.11	10.00	6.72	3.50	3.50	56.88	5.53	69.13	m	77.77
0.6mm bitumen coated copper in										
one brick wall	10.75	10.00	11.82	3.00	3.00	48.75	5.53	66.11	m	74.37
one and a half brick wall	16.13	10.00	17.74	4.00	4.00	65.00	8.31	91.05	m	102.43
254mm hollow wall	10.75	10.00	11.82	3.50	3.50	56.88	5.53	74.23	m	83.51
Insert damp proof course in old wall by hand sawing in 600mm lengths										
B.S.6398 Class D, bitumen with hessian base laminated with lead in										
one brick wall	4.56	10.00	5.02	2.50	2.50	40.63	8.27	53.91	m	60.65
one and a half brick wall	6.83	10.00	7.51	3.50	3.50	56.88	12.41	76.80	m	86.40
254mm hollow wall	4.56	10.00	5.02	3.00	3.00	48.75	8.27	62.04	m	69.79
Hyload, pitch polymer in										
one brick wall	1.85	10.00	2.04	2.50	2.50	40.63	8.27	50.93	m	57.30
one and a half brick wall	2.77	10.00	3.05	3.50	3.50	56.88	12.41	72.33	m	81.37
254mm hollow wall	1.85	10.00	2.04	3.00	3.00	48.75	8.27	59.06	m	66.44
Two courses slates in cement mortar (1:3) in										
one brick wall	6.11	10.00	6.72	2.50	2.50	40.63	8.27	55.62	m	62.57
one and a half brick wall	9.17	10.00	10.09	3.50	3.50	56.88	12.41	79.37	m	89.29
254mm hollow wall	6.11	10.00	6.72	3.00	3.00	48.75	8.27	63.74	m	71.71
0.6mm bitumen coated copper in										
one brick wall	10.75	10.00	11.82	2.50	2.50	40.63	8.27	60.72	m	68.31
one and a half brick wall	16.13	10.00	17.74	3.50	3.50	56.88	12.41	87.03	m	97.91
254mm hollow wall	10.75	10.00	11.82	3.00	3.00	48.75	8.27	68.84	m	77.45

Labour hourly rates: (except Specialists) Craft Operatives £9.23 Labourer £7.02 Rates are national average prices. Refer to REGIONAL VARIATIONS for indicative levels of overall pricing in regions	MATERIALS			LABOUR				RATES		
	Del to Site £	Waste %	Material Cost £	Craft Optve Hrs	Lab Hrs	Labour Cost £	Sunds £	Nett Rate £	Unit	Gross Rate (+12.5%) £
C45: DAMP PROOF COURSE RENEWAL/INSERTION Cont.										
Insert damp proof course in old wall by machine sawing in 600mm lengths										
B.S.6398 Class D, bitumen with hessian base laminated with lead in										
one brick wall	4.56	10.00	5.02	2.00	2.00	32.50	10.98	48.50	m	54.56
one and a half brick wall	6.83	10.00	7.51	3.00	3.00	48.75	16.46	72.72	m	81.81
254mm hollow wall	4.56	10.00	5.02	2.50	2.50	40.63	10.98	56.62	m	63.70
Hyload, pitch polymer in										
one brick wall	1.85	10.00	2.04	2.00	2.00	32.50	10.98	45.52	m	51.20
one and a half brick wall	2.77	10.00	3.05	3.00	3.00	48.75	16.46	68.26	m	76.79
254mm hollow wall	1.85	10.00	2.04	2.50	2.50	40.63	10.98	53.64	m	60.34
Two courses slates in cement mortar (1:3) in										
one brick wall	6.11	10.00	6.72	2.00	2.00	32.50	10.98	50.20	m	56.48
one and a half brick wall	9.17	10.00	10.09	3.00	3.00	48.75	16.46	75.30	m	84.71
254mm hollow wall	6.11	10.00	6.72	2.50	2.50	40.63	10.98	58.33	m	65.62
0.6mm bitumen coated copper in										
one brick wall	10.75	10.00	11.82	2.00	2.00	32.50	10.98	55.31	m	62.22
one and a half brick wall	16.13	10.00	17.74	3.00	3.00	48.75	16.46	82.95	m	93.32
254mm hollow wall	10.75	10.00	11.82	2.50	2.50	40.63	10.98	63.43	m	71.36
Insert cavity tray in old wall by cutting out by hand in short lengths; insert individual trays and make good with picked stock facing bricks in gauged mortar (1:2:9) including pinning up with slates										
Type E Cavitray by Cavity Trays Ltd in										
half brick skin of cavity wall	7.53	10.00	8.28	1.75	1.75	28.44	7.02	43.74	m	49.21
external angle	5.30	10.00	5.83	1.00	1.00	16.25	6.45	28.53	nr	32.10
internal angle	5.30	10.00	5.83	1.25	1.25	20.31	6.45	32.59	nr	36.67
Type X Cavity Trays Ltd to suit 40 degree pitched roof complete with attached code 4 lead flashing and dress over tiles in										
half brick skin of cavity wall	34.17	10.00	37.59	4.25	4.25	69.06	14.05	120.70	m	135.79
ridge tray	8.48	10.00	9.33	1.00	1.00	16.25	3.27	28.85	nr	32.45
catchment tray	4.02	10.00	4.42	0.50	0.50	8.13	1.72	14.27	nr	16.05
corner catchment tray	8.87	10.00	9.76	1.00	1.00	16.25	3.27	29.28	nr	32.94
Damp proofing old brick wall by silicone injection method (excluding removal and reinstatement of plaster, etc.)										
Damp proofing										
one brick wall	7.10	-	7.10	1.00	1.00	16.25	0.75	24.10	m	27.11
one and a half brick wall	10.60	-	10.60	1.50	1.50	24.38	1.12	36.09	m	40.61
254mm hollow wall	7.10	-	7.10	1.35	1.35	21.94	0.75	29.79	m	33.51
C50: REPAIRING/RENOVATING/CONSERVING METAL										
Replace defective arch bar										
Take out defective arch bar including cutting away brickwork as necessary, replace with new mild steel arch bar primed all round and make good all work disturbed										
32 x 6mm flat arch bar	1.47	-	1.47	1.03	1.03	16.74	0.71	18.92	m	21.28
50 x 6mm flat arch bar	2.28	-	2.28	1.12	1.12	18.20	1.04	21.52	m	24.21
50 x 50 x 6mm angle arch bar	4.26	-	4.26	1.27	1.27	20.64	1.85	26.75	m	30.09
75 x 50 x 6mm angle arch bar	5.33	-	5.33	1.44	1.44	23.40	2.33	31.06	m	34.94
Replace defective baluster										
Cut out defective baluster, clean core and bottom rails, provide new mild steel baluster and weld on										
13mm diameter x 686mm long	0.86	-	0.86	0.92	0.92	14.95	1.48	17.29	nr	19.45
13mm diameter x 762mm long	0.86	-	0.86	0.94	0.94	15.28	1.48	17.61	nr	19.82
13 x 13 x 686mm long	1.17	-	1.17	0.94	0.94	15.28	1.75	18.20	nr	20.47
13 x 13 x 762mm long	1.17	-	1.17	0.95	0.95	15.44	1.75	18.36	nr	20.65
Repair damaged weld										
Clean off damaged weld including removing remains of weld and cleaning off surrounding paint back to bright metal and re-weld connection between 10mm rail and										
13mm diameter rod	-	-	-	0.50	0.50	8.13	2.18	10.31	nr	11.59
13 x 13mm bar	-	-	-	0.53	0.53	8.61	2.59	11.20	nr	12.60
25 x 25mm bar	-	-	-	0.60	0.60	9.75	3.25	13.00	nr	14.63
C51: REPAIRING/RENOVATING/CONSERVING TIMBER										
Cut out rotten or infected structural timber members including cutting away wall, etc. and shoring up adjacent work and replace with new pressure impregnated timber										
Plates and bed in cement mortar (1:3)										
50 x 75mm	1.26	10.00	1.39	0.30	0.30	4.88	0.17	6.43	m	7.23
75 x 100mm	2.70	10.00	2.97	0.50	0.50	8.13	0.33	11.43	m	12.85

Labour hourly rates: (except Specialists) Craft Operatives £9.23 Labourer £7.02 Rates are national average prices. Refer to REGIONAL VARIATIONS for indicative levels of overall pricing in regions	MATERIALS			LABOUR				RATES		
	Del to Site £	Waste %	Material Cost £	Craft Optve Hrs	Lab Hrs	Labour Cost £	Sunds £	Nett Rate £	Unit	Gross Rate (+12.5%) £
C51: REPAIRING/RENOVATING/CONSERVING TIMBER Cont.										
Cut out rotten or infected structural timber members including cutting away wall, etc. and shoring up adjacent work and replace with new pressure impregnated timber Cont.										
Floor or roof joists										
50 x 175mm	2.93	10.00	3.22	0.57	0.57	9.26	0.38	12.87	m	14.47
50 x 225mm	3.77	10.00	4.15	0.67	0.67	10.89	0.49	15.52	m	17.47
75 x 175mm	4.74	10.00	5.21	0.79	0.79	12.84	0.58	18.63	m	20.96
Rafters										
38 x 100mm	1.64	10.00	1.80	0.35	0.35	5.69	0.17	7.66	m	8.62
50 x 100mm	1.67	10.00	1.84	0.38	0.38	6.17	0.22	8.23	m	9.26
Ceiling joists and collars										
50 x 100mm	1.67	10.00	1.84	0.38	0.38	6.17	0.22	8.23	m	9.26
50 x 150mm	2.51	10.00	2.76	0.52	0.52	8.45	0.33	11.54	m	12.98
Purlins, ceiling beams and struts										
75 x 100mm	2.70	10.00	2.97	0.55	0.55	8.94	0.33	12.24	m	13.77
100 x 225mm	8.52	10.00	9.37	1.23	1.23	19.99	0.99	30.35	m	34.14
Roof trusses										
75 x 100mm	2.70	10.00	2.97	0.92	0.92	14.95	0.33	18.25	m	20.53
75 x 150mm	4.05	10.00	4.46	1.37	1.37	22.26	0.49	27.21	m	30.61
Hangers in roof										
25 x 75mm	0.78	10.00	0.86	0.30	0.30	4.88	0.09	5.82	m	6.55
38 x 100mm	1.64	10.00	1.80	0.45	0.45	7.31	0.17	9.29	m	10.45
Cut out rotten or infected timber and prepare timber under to receive new										
Roof boarding; provide new 25mm pressure impregnated softwood boarding										
flat boarding; areas not exceeding 0.50m2	11.00	10.00	12.10	1.96	1.96	31.85	1.09	45.04	m2	50.67
flat boarding; areas 0.50 - 5.00m2	11.00	10.00	12.10	1.63	1.63	26.49	1.09	39.68	m2	44.64
sloping boarding; areas not exceeding 0.50m2	11.00	10.00	12.10	2.34	2.34	38.02	1.09	51.22	m2	57.62
sloping boarding; areas 0.50 - 5.00m2	11.00	10.00	12.10	1.89	1.89	30.71	1.09	43.90	m2	49.39
Gutter boarding and bearers; provide new 25mm pressure impregnated softwood boarding and 50 x 50mm bearers										
over 300mm wide	14.36	10.00	15.80	2.06	2.06	33.48	1.31	50.58	m2	56.90
200mm	3.88	15.00	4.46	0.55	0.55	8.94	0.44	13.84	m	15.57
225mm	4.38	15.00	5.04	0.62	0.62	10.07	0.48	15.59	m	17.54
Gutter sides; provide new 19mm pressure impregnated softwood boarding										
150mm	1.27	15.00	1.46	0.46	0.46	7.47	0.12	9.06	m	10.19
200mm	1.76	15.00	2.02	0.50	0.50	8.13	0.17	10.32	m	11.61
225mm	2.15	15.00	2.47	0.57	0.57	9.26	0.19	11.93	m	13.42
Eaves or verge soffit and bearers and provide new 25mm pressure impregnated softwood boarding and 25 x 50mm bearers										
over 300mm wide	14.31	10.00	15.74	2.10	2.10	34.13	1.31	51.18	m2	57.57
225mm	4.37	10.00	4.81	0.67	0.67	10.89	0.48	16.17	m	18.20
Fascia; provide new pressure impregnated softwood										
25 x 150mm	2.28	10.00	2.51	0.46	0.46	7.47	0.17	10.15	m	11.42
25 x 200mm	3.00	10.00	3.30	0.51	0.51	8.29	0.22	11.81	m	13.28
Barge board; provide new pressure impregnated softwood										
25 x 225mm	3.41	10.00	3.75	0.55	0.55	8.94	0.25	12.94	m	14.56
25 x 250mm	3.75	10.00	4.13	0.57	0.57	9.26	0.28	13.67	m	15.38
Take up floor boarding, remove all nails and clean joists under, re-lay boarding, clear away and replace damaged boards and sand surface (20% new softwood boards allowed)										
Square edged boarding										
25mm	2.70	15.00	3.11	1.73	1.73	28.11	0.44	31.66	m2	35.61
32mm	3.58	15.00	4.12	1.76	1.76	28.60	0.55	33.27	m2	37.43
Tongued and grooved boarding										
25mm	2.86	20.00	3.43	1.87	1.87	30.39	0.44	34.26	m2	38.54
32mm	3.79	20.00	4.55	1.89	1.89	30.71	0.55	35.81	m2	40.29
Take up floor boarding, remove all nails and clean joists under, re-lay boarding, destroy and replace infected boards with new pressure impregnated boards and sand surface (20% new softwood boards allowed)										
Square edged boarding										
25mm	2.97	15.00	3.42	1.81	1.81	29.41	0.44	33.27	m2	37.43
32mm	3.94	15.00	4.53	1.83	1.83	29.74	0.55	34.82	m2	39.17
Tongued and grooved boarding										
25mm	3.15	20.00	3.78	1.94	1.94	31.52	0.44	35.74	m2	40.21
32mm	4.17	20.00	5.00	1.95	1.95	31.69	0.55	37.24	m2	41.90

Labour hourly rates: (except Specialists) Craft Operatives £9.23 Labourer £7.02 Rates are national average prices. Refer to REGIONAL VARIATIONS for indicative levels of overall pricing in regions	MATERIALS			LABOUR				RATES		
	Del to Site £	Waste %	Material Cost £	Craft Optve Hrs	Lab Hrs	Labour Cost £	Sunds £	Nett Rate £	Unit	Gross Rate (+12.5%) £
C51: REPAIRING/RENOVATING/CONSERVING TIMBER Cont.										
Take up worn or damaged floor boarding, remove all nails and clean joists under and replace with new softwood boarding										
Square edged boarding										
25mm in areas not exceeding 0.50m2	13.50	15.00	15.53	1.90	1.90	30.88	1.09	47.49	m2	53.43
32mm in areas not exceeding 0.50m2	17.90	15.00	20.59	1.94	1.94	31.52	1.39	53.50	m2	60.19
25mm in areas 0.50 - 3.00m2	13.50	15.00	15.53	1.57	1.57	25.51	1.09	42.13	m2	47.39
32mm in areas 0.50 - 3.00m2	17.90	15.00	20.59	1.61	1.61	26.16	1.39	48.14	m2	54.15
Tongued and grooved boarding										
25mm in areas not exceeding 0.50m2	14.85	20.00	17.82	2.00	2.00	32.50	1.09	51.41	m2	57.84
32mm in areas not exceeding 0.50m2	19.70	20.00	23.64	2.06	2.06	33.48	1.39	58.51	m2	65.82
25mm in areas 0.50 - 3.00m2	14.85	20.00	17.82	1.82	1.82	29.57	1.09	48.48	m2	54.55
32mm in areas 0.50 - 3.00m2	19.70	20.00	23.64	1.87	1.87	30.39	1.39	55.42	m2	62.34
Punch down all nails, remove all tacks, etc. and fill nail holes										
Surface of existing flooring										
generally	-	-	-	0.50	0.10	5.32	0.03	5.35	m2	6.02
generally and machine sand in addition	-	-	-	1.00	0.30	11.34	1.18	12.52	m2	14.08
Note										
removal of old floor coverings included elsewhere										
Easing and overhauling doors										
Ease and adjust door and oil ironmongery										
762 x 1981mm	-	-	-	0.33	0.06	3.47	0.02	3.49	nr	3.92
914 x 2134mm	-	-	-	0.39	0.06	4.02	0.02	4.04	nr	4.55
Take off door, shave 12mm off bottom edge and re-hang										
762 x 1981mm	-	-	-	0.66	0.11	6.86	-	6.86	nr	7.72
914 x 2134mm	-	-	-	0.79	0.13	8.20	-	8.20	nr	9.23
Adapting doors										
Take off door and re-hang on opposite hand, piece out as necessary and oil ironmongery										
762 x 1981mm	-	-	-	2.50	0.25	24.83	0.10	24.93	nr	28.05
914 x 2134mm	-	-	-	2.65	0.44	27.55	0.10	27.65	nr	31.10
Refixing removed doors and frames										
Doors										
762 x 1981mm	-	-	-	2.00	0.34	20.85	0.14	20.99	nr	23.61
914 x 2134mm	-	-	-	2.25	0.38	23.44	0.14	23.58	nr	26.52
Doors and frames; oil existing ironmongery										
762 x 1981mm	-	-	-	4.00	0.66	41.55	0.35	41.90	nr	47.14
914 x 2134mm	-	-	-	4.50	0.75	46.80	0.35	47.15	nr	53.04
Repairing doors and frames										
Piece in door where ironmongery removed										
rim lock	-	-	-	1.00	0.15	10.28	0.16	10.44	nr	11.75
rim lock or latch and furniture	-	-	-	1.50	0.25	15.60	0.39	15.99	nr	17.99
mortice lock	-	-	-	1.50	0.25	15.60	0.39	15.99	nr	17.99
mortice lock or latch and furniture	-	-	-	2.00	0.32	20.71	0.46	21.17	nr	23.81
butt hinge	-	-	-	0.50	0.08	5.18	0.22	5.40	nr	6.07
Piece in frame where ironmongery removed										
lock or latch keep	-	-	-	0.50	0.08	5.18	0.22	5.40	nr	6.07
butt hinge	-	-	-	0.50	0.08	5.18	0.22	5.40	nr	6.07
Take off door, dismantle as necessary, cramp and re-wedge with new glued wedges, pin and re-hang										
762 x 1981mm, panelled	-	-	-	5.00	0.84	52.05	0.16	52.21	nr	58.73
914 x 2134mm, panelled	-	-	-	6.00	1.00	62.40	0.16	62.56	nr	70.38
762 x 1981mm, glazed panelled	-	-	-	5.50	0.90	57.08	0.16	57.24	nr	64.40
914 x 2134mm, glazed panelled	-	-	-	6.50	1.08	67.58	0.16	67.74	nr	76.20
Take off softwood panelled door 762 x 1981mm, dismantle as necessary and replace the following damaged members and re-hang										
38 x 125mm top rail	5.25	10.00	5.78	2.15	0.36	22.37	0.45	28.60	nr	32.17
38 x 125mm hanging stile	13.60	10.00	14.96	2.36	0.39	24.52	1.14	40.62	nr	45.70
38 x 125mm locking stile	13.60	10.00	14.96	2.36	0.39	24.52	1.14	40.62	nr	45.70
38 x 200mm middle rail	8.40	10.00	9.24	4.23	0.71	44.03	0.74	54.01	nr	60.76
38 x 200mm bottom rail	8.40	10.00	9.24	2.20	0.37	22.90	0.74	32.88	nr	36.99
Take off afromosia panelled door 762 x 1981mm, dismantle as necessary and replace the following damaged members and re-hang										
30 x 125mm top rail	24.00	10.00	26.40	5.20	0.87	54.10	0.45	80.95	nr	91.07
38 x 125mm hanging stile	61.00	10.00	67.10	5.36	0.89	55.72	1.14	123.96	nr	139.46
38 x 125mm locking stile	61.00	10.00	67.10	5.36	0.89	55.72	1.14	123.96	nr	139.46
38 x 200mm middle rail	38.00	10.00	41.80	10.40	1.73	108.14	0.74	150.68	nr	169.51
38 x 200mm bottom rail	38.00	10.00	41.80	5.30	0.88	55.10	0.74	97.64	nr	109.84
Take off damaged softwood stop and replace with new										
12 x 38mm	0.46	10.00	0.51	0.35	0.06	3.65	0.10	4.26	m	4.79
25 (fin) x 50mm glue and screw on	0.81	10.00	0.89	0.45	0.08	4.72	0.12	5.73	m	6.44

Labour hourly rates: (except Specialists) Craft Operatives £9.23 Labourer £7.02 Rates are national average prices. Refer to REGIONAL VARIATIONS for indicative levels of overall pricing in regions	MATERIALS			LABOUR				RATES		
	Del to Site	Waste	Material Cost	Craft Optve	Lab	Labour Cost	Sunds	Nett Rate	Unit	Gross Rate (+12.5%)
	£	%	£	Hrs	Hrs	£	£	£		£
C51: REPAIRING/RENOVATING/CONSERVING TIMBER Cont.										
Repairing doors and frames Cont.										
Take off damaged afromosia stop and replace with new										
12 x 38mm	1.60	10.00	1.76	0.45	0.08	4.72	0.10	6.58	m	7.40
25 (fin) x 50mm glue and screw on	3.20	10.00	3.52	0.69	0.12	7.21	0.12	10.85	m	12.21
Flush facing old doors										
Take off panelled door, remove ironmongery, cover both sides with 3.2mm hardboard; refix ironmongery, adjust stops and re-hang										
762 x 1981mm	5.40	10.00	5.94	3.10	0.50	32.12	0.25	38.31	nr	43.10
838 x 1981mm	5.40	10.00	5.94	3.15	0.51	32.65	0.25	38.84	nr	43.70
Take off panelled door, remove ironmongery, cover one side with 3.2mm hardboard and other side with 6mm fire resistant sheeting screwed on; refix ironmongery, adjust stops and re-hang										
762 x 1981mm	41.04	10.00	45.14	4.00	0.67	41.62	0.82	87.59	nr	98.54
838 x 1981mm	41.04	10.00	45.14	4.05	0.68	42.16	0.82	88.12	nr	99.13
Take off panelled door, remove ironmongery, cover one side with 3.2mm hardboard, fill panels on other side with 6mm fire resistant sheeting and cover with 6mm fire resistant sheeting screwed on; refix ironmongery, adjust stops and re-hang										
762 x 1981mm	60.21	10.00	66.23	5.00	0.83	51.98	1.05	119.26	nr	134.16
838 x 1981mm	60.21	10.00	66.23	5.05	0.84	52.51	1.05	119.79	nr	134.76
Take off flush door, remove ironmongery, cover one side with 6mm fire resistant sheeting screwed on; refix ironmongery and re-hang										
762 x 1981mm	38.34	10.00	42.17	2.50	0.42	26.02	0.53	68.73	nr	77.32
838 x 1981mm	38.34	10.00	42.17	2.50	0.43	26.09	0.53	68.80	nr	77.40
Easing and overhauling windows										
Ease and adjust and oil ironmongery										
casement opening light	-	-	-	0.41	0.07	4.28	0.02	4.30	nr	4.83
double hung sash and renew sash lines with best quality flax cord	1.06	10.00	1.17	1.64	0.27	17.03	0.05	18.25	nr	20.53
Refixing removed windows										
Windows										
casement opening light and oil ironmongery	-	-	-	1.03	0.17	10.70	0.04	10.74	nr	12.08
double hung sash and provide new best quality flax sash cord	0.99	10.00	1.09	1.70	0.28	17.66	0.05	18.80	nr	21.15
Single or multi-light timber casements and frames										
440 x 920mm	-	-	-	4.00	0.66	41.55	0.04	41.59	nr	46.79
1225 x 1070mm	-	-	-	6.00	1.00	62.40	0.05	62.45	nr	70.26
2395 x 1225mm	-	-	-	8.00	1.33	83.18	0.06	83.24	nr	93.64
Repairing windows										
Remove damaged members of cased frame and replace with new softwood										
parting bead	0.47	10.00	0.52	1.16	0.19	12.04	0.05	12.61	m	14.18
pulley stile	1.79	10.00	1.97	1.68	0.28	17.47	0.14	19.58	m	22.03
stop bead	0.68	10.00	0.75	0.62	0.10	6.42	0.05	7.22	m	8.13
Take out sash not exceeding 50mm thick, dismantle as necessary, replace the following damaged members in softwood and re-hang with new best quality flax sash cord										
glazing bar	2.38	10.00	2.62	3.14	0.52	32.63	0.05	35.30	m	39.71
stile	2.66	10.00	2.93	3.03	0.50	31.48	0.09	34.49	m	38.80
rail	4.17	10.00	4.59	2.80	0.45	29.00	0.26	33.85	m	38.08
Take out casement opening light not exceeding 50mm thick, dismantle as necessary, replace the following damaged members in softwood and re-hang										
glazing bar	1.32	10.00	1.45	1.97	0.32	20.43	0.05	21.93	m	24.67
hanging stile	1.60	10.00	1.76	1.85	0.30	19.18	0.10	21.04	m	23.67
shutting stile	1.60	10.00	1.76	1.85	0.30	19.18	0.10	21.04	m	23.67
rail	3.13	10.00	3.44	1.61	0.27	16.76	0.26	20.46	m	23.02
Take out sash, cramp and re-wedge with new glued wedges, pin and re-hang										
sliding sash	-	-	-	5.65	0.95	58.82	0.87	59.69	nr	67.15
opening casement light	-	-	-	5.00	0.85	52.12	0.87	52.99	nr	59.61
Repair corner of sash with 100 x 100mm angle repair plate										
let in flush	0.38	-	0.38	1.00	0.17	10.42	0.17	10.97	nr	12.35
Carefully rake out and dry crack in sill and fill with two part wood repair system	-	-	-	0.50	0.08	5.18	1.50	6.68	m	7.51

Labour hourly rates: (except Specialists) Craft Operatives £9.23 Labourer £7.02 Rates are national average prices. Refer to REGIONAL VARIATIONS for indicative levels of overall pricing in regions	MATERIALS			LABOUR				RATES		
	Del to Site	Waste	Material Cost	Craft Optve	Lab	Labour Cost	Sunds	Nett Rate	Unit	Gross Rate (+12.5%)
	£	%	£	Hrs	Hrs	£	£	£		£
C51: REPAIRING/RENOVATING/CONSERVING TIMBER Cont.										
Repairs to softwood skirtings, picture rails, architraves, etc.										
Take off and refix										
skirting	-	-	-	0.40	0.07	4.18	0.04	4.22	m	4.75
dado or picture rail	-	-	-	0.35	0.06	3.65	0.03	3.68	m	4.14
architrave	-	-	-	0.26	0.05	2.75	0.03	2.78	m	3.13
Cut out damaged section and piece in new										
19 x 100mm square skirting	1.04	10.00	1.14	0.69	0.18	7.63	0.20	8.98	m	10.10
25 x 150mm square skirting	2.03	10.00	2.23	0.71	0.12	7.40	0.24	9.87	m	11.10
19 x 125mm moulded skirting	3.45	10.00	3.79	0.70	0.11	7.23	0.23	11.26	m	12.67
19 x 50mm moulded dado or picture rail	1.85	10.00	2.04	0.74	0.12	7.67	0.10	9.81	m	11.03
25 x 50mm moulded or splayed architrave	2.50	10.00	2.75	0.51	0.09	5.34	0.12	8.21	m	9.24
38 x 75mm moulded or splayed architrave	3.59	10.00	3.95	0.52	0.09	5.43	0.23	9.61	m	10.81
Cut out rotten or infected skirting and grounds and replace with new treated softwood										
19 x 100mm square skirting	1.48	10.00	1.63	1.00	0.17	10.42	0.27	12.32	m	13.86
25 x 150mm square skirting	2.47	10.00	2.72	1.07	0.18	11.14	0.37	14.23	m	16.01
19 x 125mm moulded skirting	3.89	10.00	4.28	1.03	0.17	10.70	0.31	15.29	m	17.20
Repairs to staircase members in softwood										
Cut out damaged members of staircase and piece in new										
32 x 275mm tread	5.80	10.00	6.38	4.33	0.72	45.02	0.46	51.86	m	58.34
38 x 275mm tread	6.88	10.00	7.57	4.85	0.81	50.45	0.53	58.55	m	65.87
19 x 200mm riser	2.82	10.00	3.10	2.25	0.38	23.44	0.18	26.72	m	30.06
25 x 200mm riser	3.63	10.00	3.99	3.03	0.51	31.55	0.27	35.81	m	40.29
Cut back old tread for a width of 125mm and provide and fix new nosing with glued and dowelled joints to tread										
32mm with rounded nosing	3.98	10.00	4.38	2.16	0.36	22.46	0.76	27.60	m	31.05
38mm with rounded nosing	4.60	10.00	5.06	2.55	0.43	26.56	0.85	32.47	m	36.52
32mm with moulded nosing	4.73	10.00	5.20	2.54	0.42	26.39	0.76	32.36	m	36.40
38mm with moulded nosing	5.37	10.00	5.91	3.06	0.51	31.82	0.90	38.63	m	43.46
Replace missing square bar baluster 914mm high; clean out old mortices										
25 x 25mm	0.55	-	0.55	0.56	0.09	5.80	-	6.35	nr	7.14
38 x 38mm	1.06	-	1.06	0.75	0.13	7.84	-	8.90	nr	10.01
Cut out damaged square bar baluster 914mm high and piece in new										
25 x 25mm	0.55	-	0.55	0.75	0.03	7.13	0.03	7.71	nr	8.68
38 x 38mm	1.00	-	1.00	0.91	0.15	9.45	0.05	10.50	nr	11.82
Cut out damaged handrail and piece in new										
50mm mopstick	20.30	10.00	22.33	1.95	0.33	20.32	3.20	45.85	m	51.58
75 x 100mm moulded	33.20	10.00	36.52	3.97	0.66	41.28	3.47	81.27	m	91.42
Refixing removed handrail										
to new balustrade	-	-	-	0.69	0.12	7.21	-	7.21	m	8.11
Repairs to staircase members in oak										
Cut out damaged members of staircase and piece in new										
32 x 275mm tread	39.00	10.00	42.90	6.09	1.02	63.37	0.46	106.73	m	120.07
38 x 275mm tread	47.00	10.00	51.70	7.18	1.20	74.70	0.53	126.93	m	142.79
19 x 200mm riser	21.00	10.00	23.10	3.34	0.56	34.76	0.18	58.04	m	65.29
25 x 200mm riser	26.00	10.00	28.60	4.56	0.76	47.42	0.29	76.31	m	85.85
Cut back old tread for a width of 125mm and provide and fix new nosing with glued and dowelled joints to tread										
32mm with rounded nosing	23.00	10.00	25.30	2.23	0.37	23.18	0.76	49.24	m	55.40
38mm with rounded nosing	26.00	10.00	28.60	2.92	0.49	30.39	0.90	59.89	m	67.38
32mm with moulded nosing	24.00	10.00	26.40	3.20	0.53	33.26	0.76	60.42	m	67.97
38mm with moulded nosing	27.00	10.00	29.70	3.95	0.66	41.09	0.90	71.69	m	80.65
Replace missing square bar baluster 914mm high; clean out old mortices										
25 x 25mm	3.40	-	3.40	0.94	0.16	9.80	-	13.20	nr	14.85
38 x 38mm	6.90	-	6.90	0.99	0.17	10.33	-	17.23	nr	19.38
Cut out damaged square bar baluster 914mm high and piece in new										
25 x 25mm	3.40	-	3.40	1.21	0.20	12.57	0.03	16.00	nr	18.00
38 x 38mm	6.00	-	6.00	1.32	0.22	13.73	0.05	19.78	nr	22.25
Cut out damaged handrail and piece in new										
50mm mopstick	43.00	10.00	47.30	3.64	0.61	37.88	3.20	88.38	m	99.43
75 x 100mm moulded	68.00	10.00	74.80	5.50	0.92	57.22	3.47	135.49	m	152.43
Refixing removed handrail										
to new balustrade	-	-	-	1.35	0.23	14.08	-	14.08	m	15.83
Remove broken ironmongery and fix only new to softwood										
Door ironmongery										
75mm butt hinges	-	-	-	0.23	0.04	2.40	0.01	2.41	nr	2.72
100mm butt hinges	-	-	-	0.23	0.04	2.40	0.01	2.41	nr	2.72

Labour hourly rates: (except Specialists) Craft Operatives £9.23 Labourer £7.02 Rates are national average prices. Refer to REGIONAL VARIATIONS for indicative levels of overall pricing in regions	MATERIALS			LABOUR				RATES		
	Del to Site £	Waste %	Material Cost £	Craft Optve Hrs	Lab Hrs	Labour Cost £	Sunds £	Nett Rate £	Unit	Gross Rate (+12.5%) £
C51: REPAIRING/RENOVATING/CONSERVING TIMBER Cont.										
Remove broken ironmongery and fix only new to softwood Cont.										
Door ironmongery Cont.										
tee hinges up to 300mm	-	-	-	0.62	0.10	6.42	0.01	6.43	nr	7.24
single action floor spring hinges	-	-	-	3.00	0.50	31.20	1.30	32.50	nr	36.56
barrel bolt up to 300mm	-	-	-	0.41	0.07	4.28	0.01	4.29	nr	4.82
lock or latch furniture	-	-	-	0.44	0.07	4.55	0.01	4.56	nr	5.13
rim lock	-	-	-	0.80	0.13	8.30	0.01	8.31	nr	9.34
rim latch	-	-	-	0.80	0.13	8.30	0.01	8.31	nr	9.34
mortice lock	-	-	-	1.22	0.20	12.66	0.01	12.67	nr	14.26
mortice latch	-	-	-	1.08	0.18	11.23	0.01	11.24	nr	12.65
pull handle	-	-	-	0.40	0.07	4.18	0.03	4.21	nr	4.74
Window ironmongery										
sash fastener	-	-	-	1.09	0.18	11.32	0.01	11.33	nr	12.75
sash lift	-	-	-	0.23	0.04	2.40	0.01	2.41	nr	2.72
casement stay	-	-	-	0.40	0.07	4.18	0.01	4.19	nr	4.72
casement fastener	-	-	-	0.40	0.07	4.18	0.01	4.19	nr	4.72
Sundry ironmongery										
hat and coat hook	-	-	-	0.29	0.05	3.03	0.01	3.04	nr	3.42
toilet roll holder	-	-	-	0.29	0.05	3.03	0.01	3.04	nr	3.42
Remove broken ironmongery and fix only new to hardwood										
Door ironmongery										
75mm butt hinges	-	-	-	0.30	0.05	3.12	0.01	3.13	nr	3.52
100mm butt hinges	-	-	-	0.30	0.05	3.12	0.01	3.13	nr	3.52
tee hinges up to 300mm	-	-	-	0.86	0.14	8.92	0.01	8.93	nr	10.05
single action floor spring hinges	-	-	-	4.20	0.70	43.68	1.30	44.98	nr	50.60
barrel bolt up to 300mm	-	-	-	0.60	0.10	6.24	0.01	6.25	nr	7.03
lock or latch furniture	-	-	-	0.61	0.10	6.33	0.01	6.34	nr	7.14
rim lock	-	-	-	1.15	0.19	11.95	0.01	11.96	nr	13.45
rim latch	-	-	-	1.15	0.19	11.95	0.01	11.96	nr	13.45
mortice lock	-	-	-	1.77	0.30	18.44	0.01	18.45	nr	20.76
mortice latch	-	-	-	1.58	0.26	16.41	0.01	16.42	nr	18.47
pull handle	-	-	-	0.57	0.10	5.96	0.03	5.99	nr	6.74
Window ironmongery										
sash fastener	-	-	-	1.57	0.26	16.32	0.01	16.33	nr	18.37
sash lift	-	-	-	0.31	0.05	3.21	0.01	3.22	nr	3.63
casement stay	-	-	-	0.56	0.09	5.80	0.01	5.81	nr	6.54
casement fastener	-	-	-	0.56	0.09	5.80	0.01	5.81	nr	6.54
Sundry ironmongery										
hat and coat hook	-	-	-	0.42	0.07	4.37	0.01	4.38	nr	4.93
toilet roll holder	-	-	-	0.42	0.07	4.37	0.01	4.38	nr	4.93
C52: FUNGUS/BEETLE ERADICATION										
Protective treatment of existing timbers										
Treat with two coats of spray applied preservative										
boarding	0.56	10.00	0.62	-	0.15	1.05	0.04	1.71	m2	1.92
structural timbers	0.84	10.00	0.92	-	0.20	1.40	0.05	2.38	m2	2.68
Treat with two coats of brush applied preservative										
boarding	0.84	10.00	0.92	-	0.26	1.83	0.03	2.78	m2	3.13
structural timbers	1.26	10.00	1.39	-	0.80	5.62	0.05	7.05	m2	7.93
Insecticide treatment of existing timbers										
Treat worm infected timbers with spray applied proprietary insecticide										
boarding	0.84	10.00	0.92	-	0.24	1.68	0.05	2.66	m2	2.99
structural timbers	1.12	10.00	1.23	-	0.28	1.97	0.06	3.26	m2	3.66
joinery timbers	1.12	10.00	1.23	-	0.32	2.25	0.06	3.54	m2	3.98
Treat worm infected timbers with brush applied proprietary insecticide										
boarding	1.12	10.00	1.23	-	0.36	2.53	0.03	3.79	m2	4.26
structural timbers	1.31	10.00	1.44	-	0.42	2.95	0.05	4.44	m2	4.99
joinery timbers	1.31	10.00	1.44	-	0.48	3.37	0.05	4.86	m2	5.47
Treatment of wall surfaces										
Treat surfaces of concrete and brickwork adjoining areas where infected timbers removed										
with a blow lamp	-	-	-	-	0.25	1.75	0.03	1.78	m2	2.01
C90: ALTERATIONS - SPOT ITEMS										
Removal of fittings and fixtures										
Removing fittings and fixtures										
cupboard with doors, frames, architraves, etc	-	-	-	-	0.59	4.14	1.04	5.18	m2	5.83
worktop with legs and bearers	-	-	-	-	0.56	3.93	1.39	5.32	m2	5.99
shelving exceeding 300mm wide with bearers and brackets	-	-	-	-	0.45	3.16	0.88	4.04	m2	4.54
shelving not exceeding 300mm wide with bearers and brackets	-	-	-	-	0.20	1.40	0.28	1.68	m	1.89

Labour hourly rates: (except Specialists) Craft Operatives £9.23 Labourer £7.02 Rates are national average prices. Refer to REGIONAL VARIATIONS for indicative levels of overall pricing in regions	MATERIALS			LABOUR				RATES		
	Del to Site £	Waste %	Material Cost £	Craft Optve Hrs	Lab Hrs	Labour Cost £	Sunds £	Nett Rate £	Unit	Gross Rate (+12.5%) £
C90: ALTERATIONS - SPOT ITEMS Cont.										
Removal of fittings and fixtures Cont.										
Removing fittings and fixtures Cont.										
draining board and bearers										
500 x 600mm.........................	-	-	-	-	0.28	1.97	0.53	2.50	nr	2.81
600 x 900mm.........................	-	-	-	-	0.37	2.60	0.94	3.54	nr	3.98
wall cupboard unit										
510 x 305 x 305mm..................	-	-	-	-	0.47	3.30	0.84	4.14	nr	4.66
510 x 305 x 610mm..................	-	-	-	-	0.65	4.56	1.63	6.19	nr	6.97
1020 x 305 x 610mm.................	-	-	-	-	0.94	6.60	3.31	9.91	nr	11.15
1220 x 305 x 610mm.................	-	-	-	-	1.15	8.07	3.94	12.01	nr	13.51
floor cupboard unit										
510 x 510 x 915mm..................	-	-	-	-	0.26	1.83	4.14	5.97	nr	6.71
610 x 510 x 915mm..................	-	-	-	-	1.21	8.49	4.94	13.43	nr	15.11
1020 x 510 x 915mm.................	-	-	-	-	1.65	11.58	8.28	19.86	nr	22.35
1220 x 510 x 915mm.................	-	-	-	-	1.93	13.55	9.92	23.47	nr	26.40
sink base unit										
585 x 510 x 895mm..................	-	-	-	-	0.92	6.46	4.67	11.13	nr	12.52
1070 x 510 x 895mm.................	-	-	-	-	1.65	11.58	8.50	20.08	nr	22.59
tall cupboard unit										
510 x 510 x 1505mm.................	-	-	-	-	1.56	10.95	6.82	17.77	nr	19.99
610 x 510 x 1505mm.................	-	-	-	-	1.77	12.43	8.15	20.58	nr	23.15
bath panel and bearers............	-	-	-	-	0.47	3.30	0.88	4.18	nr	4.70
steel or concrete clothes line post with concrete base.....................	-	-	-	-	0.50	3.51	1.74	5.25	nr	5.91

**This section continues
on the next page**

Labour hourly rates: (except Specialists) Craft Operatives £12.44 Labourer £7.02 Rates are national average prices. Refer to REGIONAL VARIATIONS for indicative levels of overall pricing in regions	MATERIALS			LABOUR				RATES		
	Del to Site	Waste	Material Cost	Craft Optve	Lab	Labour Cost	Sunds	Nett Rate	Unit	Gross Rate (+12.5%)
	£	%	£	Hrs	Hrs	£	£	£		£
C90: ALTERATIONS - SPOT ITEMS Cont.										
Removal of plumbing and electrical installations										
Removing plumbing and electrical installations										
100mm cast iron rainwater gutters	-	-	-	0.20	-	2.49	0.17	2.66	m	2.99
75mm cast iron rainwater pipes	-	-	-	0.25	-	3.11	0.19	3.30	m	3.71
100mm cast iron rainwater pipes	-	-	-	0.25	-	3.11	0.32	3.43	m	3.86
cast iron rainwater head	-	-	-	0.30	-	3.73	0.53	4.26	nr	4.79
cast iron soil and vent pipe with caulked joints										
50mm	-	-	-	0.35	-	4.35	0.10	4.45	m	5.01
75mm	-	-	-	0.35	-	4.35	0.19	4.54	m	5.11
100mm	-	-	-	0.35	-	4.35	0.32	4.67	m	5.26
150mm	-	-	-	0.50	-	6.22	0.76	6.98	m	7.85
100mm asbestos cement soil and vent pipe	-	-	-	0.25	-	3.11	0.32	3.43	m	3.86
150mm asbestos cement soil and vent pipe	-	-	-	0.25	-	3.11	0.76	3.87	m	4.35
100mm p.v.c. soil and vent pipe	-	-	-	0.25	-	3.11	0.32	3.43	m	3.86
150mm p.v.c. soil and vent pipe	-	-	-	0.25	-	3.11	0.76	3.87	m	4.35
100mm lead soil and vent pipe	-	-	-	0.35	-	4.35	0.32	4.67	m	5.26
150mm lead soil and vent pipe	-	-	-	0.35	-	4.35	0.76	5.11	m	5.75
w.c. suites	-	-	-	1.55	-	19.28	3.26	22.54	nr	25.36
lavatory basins	-	-	-	1.35	-	16.79	1.96	18.75	nr	21.10
baths	-	-	-	1.55	-	19.28	6.53	25.81	nr	29.04
glazed ware sinks	-	-	-	1.35	-	16.79	3.26	20.05	nr	22.56
stainless steel sinks	-	-	-	1.35	-	16.79	1.96	18.75	nr	21.10
stainless steel sinks with single drainer	-	-	-	1.35	-	16.79	2.62	19.41	nr	21.84
stainless steel sinks with double drainer	-	-	-	1.35	-	16.79	3.26	20.05	nr	22.56
hot and cold water, waste pipes, etc. 8 - 25mm diameter										
copper	-	-	-	0.15	-	1.87	0.04	1.91	m	2.14
lead	-	-	-	0.15	-	1.87	0.04	1.91	m	2.14
polythene or p.v.c.	-	-	-	0.15	-	1.87	0.04	1.91	m	2.14
stainless steel	-	-	-	0.25	-	3.11	0.04	3.15	m	3.54
steel	-	-	-	0.25	-	3.11	0.04	3.15	m	3.54
hot and cold water, waste pipes, etc. 32 - 50mm diameter										
copper	-	-	-	0.15	-	1.87	0.10	1.97	m	2.21
lead	-	-	-	0.15	-	1.87	0.10	1.97	m	2.21
polythene or p.v.c.	-	-	-	0.15	-	1.87	0.10	1.97	m	2.21
stainless steel	-	-	-	0.25	-	3.11	0.10	3.21	m	3.61
steel	-	-	-	0.25	-	3.11	0.10	3.21	m	3.61
cold water cisterns up to 454 litres capacity .	-	-	-	2.00	-	24.88	14.82	39.70	nr	44.66
hot water tanks or cylinders up to 227 litre capacity	-	-	-	2.00	-	24.88	7.40	32.28	nr	36.31
wall type radiators up to 914mm long, 610 - 914mm high	-	-	-	1.35	-	16.79	1.31	18.10	nr	20.37
wall type radiators 914 - 1829mm long, 610 - 914mm high	-	-	-	1.95	-	24.26	2.62	26.88	nr	30.24
column type radiators up to 914mm long, 610 - 914mm high	-	-	-	1.35	-	16.79	6.85	23.64	nr	26.60
column type radiators 914 - 1829mm long, 610 - 914mm high	-	-	-	1.95	-	24.26	13.70	37.96	nr	42.70
pipe insulation from pipes 8 - 25mm diameter ..	-	-	-	0.10	-	1.24	0.10	1.34	m	1.51
pipe insulation from pipes 32 - 50mm diameter .	-	-	-	0.13	-	1.62	0.19	1.81	m	2.03
water tank or calorifier insulation from tanks up to 227 litres	-	-	-	0.50	-	6.22	7.40	13.62	nr	15.32
water tank or calorifier insulation from tanks 227 - 454 litres	-	-	-	0.75	-	9.33	14.82	24.15	nr	27.17
Removing plumbing and electrical installations; setting aside for re-use										
cast iron soil and vent pipe with caulked joints										
50mm	-	-	-	0.55	-	6.84	-	6.84	m	7.70
75mm	-	-	-	0.60	-	7.46	-	7.46	m	8.40
100mm	-	-	-	0.80	-	9.95	-	9.95	m	11.20
150mm	-	-	-	1.00	-	12.44	-	12.44	m	13.99

This section continues
on the next page

Labour hourly rates: (except Specialists) Craft Operatives £14.38 Labourer £7.02 Rates are national average prices. Refer to REGIONAL VARIATIONS for indicative levels of overall pricing in regions	MATERIALS			LABOUR				RATES		
	Del to Site £	Waste %	Material Cost £	Craft Optve Hrs	Lab Hrs	Labour Cost £	Sunds £	Nett Rate £	Unit	Gross Rate (+12.5%) £
C90: ALTERATIONS - SPOT ITEMS Cont.										
Removal of plumbing and electrical installations Cont.										
Removing plumbing and electrical installations										
wall mounted electric fire	-	-	-	2.00	-	28.76	1.74	30.50	nr	34.31
1.5 kW night storage heater	-	-	-	2.50	-	35.95	5.22	41.17	nr	46.32
2 KW night storage heater	-	-	-	2.50	-	35.95	6.96	42.91	nr	48.27
3 KW night storage heater	-	-	-	2.50	-	35.95	10.44	46.39	nr	52.19
Removing plumbing and electrical installations; extending and making good finishings										
lighting points	-	-	-	1.25	-	17.98	1.42	19.40	nr	21.82
flush type switches	-	-	-	1.25	-	17.98	1.42	19.40	nr	21.82
surface mounted type switches	-	-	-	1.00	-	14.38	1.14	15.52	nr	17.46
flush type socket outlets	-	-	-	1.25	-	17.98	1.42	19.40	nr	21.82
surface mounted type socket outlets	-	-	-	1.00	-	14.38	1.14	15.52	nr	17.46
flush type fitting points	-	-	-	1.25	-	17.98	1.42	19.40	nr	21.82
surface mounted type fitting points	-	-	-	1.00	-	14.38	1.14	15.52	nr	17.46
surface mounted p.v.c. insulated and sheathed cables	-	-	-	0.04	-	0.58	0.06	0.64	m	0.71
surface mounted mineral insulated copper sheathed cables	-	-	-	0.10	-	1.44	0.12	1.56	m	1.75
conduits up to 25mm diameter with junction boxes	-	-	-	0.15	-	2.16	0.17	2.33	m	2.62
surface mounted cable trunking up to 100 x 100mm	-	-	-	0.55	-	7.91	0.62	8.53	m	9.60

This section continues
on the next page

EXISTING SITE/BUILDINGS/SERVICES

Labour hourly rates: (except Specialists) Craft Operatives £9.23 Labourer £7.02 Rates are national average prices. Refer to REGIONAL VARIATIONS for indicative levels of overall pricing in regions	MATERIALS			LABOUR				RATES		
	Del to Site £	Waste %	Material Cost £	Craft Optve Hrs	Lab Hrs	Labour Cost £	Sunds £	Nett Rate £	Unit	Gross Rate (+12.5%) £
C90: ALTERATIONS - SPOT ITEMS Cont.										
Removal of floor, wall and ceiling finishings										
Removing finishings										
cement and sand to floors	-	-	-	-	0.86	6.04	1.74	7.78	m2	8.75
granolithic to floors	-	-	-	-	0.92	6.46	1.74	8.20	m2	9.22
granolithic to treads and risers	-	-	-	-	1.43	10.04	1.74	11.78	m2	13.25
plastic or similar tiles to floors	-	-	-	-	0.20	1.40	0.18	1.58	m2	1.78
plastic or similar tiles to floors; cleaning off for new	-	-	-	-	0.47	3.30	0.18	3.48	m2	3.91
plastic or similar tiles to floors and screed under	-	-	-	-	1.05	7.37	1.92	9.29	m2	10.45
linoleum and underlay to floors	-	-	-	-	0.07	0.49	1.74	2.23	m2	2.51
ceramic or quarry tiles to floors	-	-	-	-	1.21	8.49	1.04	9.53	m2	10.73
ceramic or quarry tiles to floors and screed under	-	-	-	-	2.39	16.78	2.78	19.56	m2	22.00
extra; cleaning and keying surface of concrete under	-	-	-	-	0.53	3.72	0.04	3.76	m2	4.23
asphalt to floors on loose underlay	-	-	-	-	0.35	2.46	0.70	3.16	m2	3.55
asphalt to floors keyed to concrete or screed	-	-	-	-	0.76	5.34	0.70	6.04	m2	6.79
granolithic skirtings	-	-	-	-	0.33	2.32	0.18	2.50	m	2.81
plastic or similar skirtings; cleaning off for new	-	-	-	-	0.13	0.91	0.04	0.95	m	1.07
timber skirtings	-	-	-	-	0.12	0.84	0.18	1.02	m	1.15
ceramic or quarry tile skirtings	-	-	-	-	0.43	3.02	0.20	3.22	m	3.62
plaster to walls	-	-	-	-	0.60	4.21	0.70	4.91	m2	5.53
rendering to walls	-	-	-	-	0.87	6.11	0.70	6.81	m2	7.66
rough cast to walls	-	-	-	-	0.90	6.32	0.88	7.20	m2	8.10
match boarding linings to walls with battens	-	-	-	-	0.38	2.67	1.04	3.71	m2	4.17
plywood or similar sheet linings to walls with battens	-	-	-	-	0.31	2.18	1.04	3.22	m2	3.62
insulating board linings to walls with battens ...	-	-	-	-	0.31	2.18	1.04	3.22	m2	3.62
plasterboard to walls	-	-	-	-	0.31	2.18	0.53	2.71	m2	3.04
plasterboard dry linings to walls	-	-	-	-	0.50	3.51	0.53	4.04	m2	4.54
plasterboard and skim to walls	-	-	-	-	0.60	4.21	0.70	4.91	m2	5.53
lath and plaster to walls	-	-	-	-	0.44	3.09	0.88	3.97	m2	4.46
metal lath and plaster to walls	-	-	-	-	0.33	2.32	0.88	3.20	m2	3.60
ceramic tiles to walls	-	-	-	-	0.75	5.26	0.53	5.79	m2	6.52
ceramic tiles to walls and backing under	-	-	-	-	1.11	7.79	0.88	8.67	m2	9.76
asphalt coverings to walls keyed to concrete or brickwork	-	-	-	-	0.66	4.63	0.70	5.33	m2	6.00
plaster to ceilings	-	-	-	-	0.74	5.19	0.70	5.89	m2	6.63
match boarding linings to ceilings with battens ..	-	-	-	-	0.54	3.79	1.04	4.83	m2	5.43
plywood or similar sheet linings to ceilings with battens	-	-	-	-	0.38	2.67	1.04	3.71	m2	4.17
insulating board linings to ceilings with battens	-	-	-	-	0.26	1.83	1.04	2.87	m2	3.22
plasterboard to ceilings	-	-	-	-	0.33	2.32	0.53	2.85	m2	3.20
plasterboard and skim to ceilings	-	-	-	-	0.40	2.81	0.70	3.51	m2	3.95
lath and plaster to ceilings	-	-	-	-	0.59	4.14	0.88	5.02	m2	5.65
metal lath and plaster to ceilings	-	-	-	-	0.47	3.30	0.88	4.18	m2	4.70
hacking surfaces of concrete as key for new finishings										
ceiling...........................	-	-	-	-	0.67	4.70	0.04	4.74	m2	5.34
floor.............................	-	-	-	-	0.53	3.72	0.04	3.76	m2	4.23
wall..............................	-	-	-	-	0.53	3.72	0.04	3.76	m2	4.23
hacking surfaces of brick wall and raking out joints as key for new finishings	-	-	-	-	0.60	4.21	0.04	4.25	m2	4.78
Removing finishings; extending and making good finishings										
plaster cornices up to 100mm girth on face	0.28	10.00	0.31	0.78	0.78	12.68	0.07	13.05	m	14.68
plaster cornices 100 - 200mm girth on face	0.46	10.00	0.51	0.88	0.88	14.30	0.18	14.99	m	16.86
plaster cornices 200 - 300mm girth on face	0.81	10.00	0.89	0.97	0.97	15.76	0.53	17.18	m	19.33
plaster ceiling roses up to 300mm diameter	0.14	10.00	0.15	0.41	0.41	6.66	0.14	6.96	nr	7.83
plaster ceiling roses 300 - 450mm diameter	0.28	10.00	0.31	0.64	0.64	10.40	0.28	10.99	nr	12.36
plaster ceiling roses 450 - 600mm diameter	0.55	10.00	0.60	0.90	0.90	14.63	0.55	15.78	nr	17.75
Removing finishings; carefully handling and disposing toxic or other special waste by approved method										
asbestos cement sheet linings to walls	-	-	-	2.00	2.00	32.50	2.78	35.28	m2	39.69
asbestos cement sheet linings to ceilings	-	-	-	3.00	3.00	48.75	2.78	51.53	m2	57.97
Removal of roof coverings										
Removing coverings										
felt to roofs	-	-	-	-	0.28	1.97	0.70	2.67	m2	3.00
felt skirtings to roofs	-	-	-	-	0.33	2.32	0.07	2.39	m	2.68
asphalt to roofs	-	-	-	-	0.35	2.46	0.88	3.34	m2	3.75
asphalt skirtings to roofs; on expanded metal reinforcement	-	-	-	-	0.20	1.40	0.14	1.54	m	1.74
asphalt skirtings to roofs; keyed to concrete, brickwork, etc......................	-	-	-	-	0.30	2.11	0.14	2.25	m	2.53
asphalt coverings to roofs; on expanded metal reinforcement; per 100mm of width	-	-	-	-	0.13	0.91	0.11	1.02	m	1.15
asphalt coverings to roofs; keyed to concrete, brickwork, etc.; per 100mm of width	-	-	-	-	0.20	1.40	0.11	1.51	m	1.70
slate to roofs	-	-	-	-	0.34	2.39	0.70	3.09	m2	3.47
tiles to roofs	-	-	-	-	0.31	2.18	1.04	3.22	m2	3.62
extra; removing battens	-	-	-	-	0.28	1.97	0.70	2.67	m2	3.00
extra; removing counter battens	-	-	-	-	0.10	0.70	0.14	0.84	m2	0.95
extra; removing underfelt	-	-	-	-	0.20	1.40	0.28	1.68	m2	1.89
lead to roofs	-	-	-	-	0.48	3.37	0.70	4.07	m2	4.58
lead flashings to roofs; per 25mm of girth	-	-	-	-	0.05	0.35	0.04	0.39	m	0.44
zinc to roofs	-	-	-	-	0.34	2.39	0.70	3.09	m2	3.47
zinc flashings to roofs; per 25mm of girth	-	-	-	-	0.03	0.21	0.04	0.25	m	0.28
copper to roofs	-	-	-	-	0.34	2.39	0.70	3.09	m2	3.47
copper flashings to roofs; per 25mm of girth	-	-	-	-	0.03	0.21	0.04	0.25	m	0.28

Labour hourly rates: (except Specialists) Craft Operatives £9.23 Labourer £7.02 Rates are national average prices. Refer to REGIONAL VARIATIONS for indicative levels of overall pricing in regions	MATERIALS			LABOUR				RATES		
	Del to Site	Waste	Material Cost	Craft Optve	Lab	Labour Cost	Sunds	Nett Rate	Unit	Gross Rate (+12.5%)
	£	%	£	Hrs	Hrs	£	£	£		£
C90: ALTERATIONS - SPOT ITEMS Cont.										
Removal of roof coverings Cont.										
Removing coverings Cont.										
corrugated metal sheeting to roofs	-	-	-	-	0.31	2.18	1.04	3.22	m2	3.62
corrugated translucent sheeting to roofs	-	-	-	-	0.31	2.18	1.04	3.22	m2	3.62
board roof decking	-	-	-	-	0.31	2.18	1.04	3.22	m2	3.62
woodwool roof decking	-	-	-	-	0.39	2.74	1.04	3.78	m2	4.25
Removing coverings; setting aside for re-use										
slate to roofs	-	-	-	-	0.60	4.21	-	4.21	m2	4.74
tiles to roofs	-	-	-	-	0.60	4.21	-	4.21	m2	4.74
clean and stack 405 x 205mm slates; per 100	-	-	-	-	1.50	10.53	-	10.53	nr	11.85
clean and stack 510 x 255mm slates; per 100	-	-	-	-	1.89	13.27	-	13.27	nr	14.93
clean and stack 610 x 305mm slates; per 100	-	-	-	-	2.24	15.72	-	15.72	nr	17.69
clean and stack concrete tiles; per 100	-	-	-	-	2.24	15.72	-	15.72	nr	17.69
clean and stack plain tiles; per 100	-	-	-	-	1.50	10.53	-	10.53	nr	11.85
Removing coverings; carefully handling and disposing toxic or other special waste by approved method										
asbestos cement sheeting to roofs	-	-	-	2.00	2.00	32.50	2.78	35.28	m2	39.69
asbestos cement roof decking	-	-	-	2.00	2.00	32.50	2.78	35.28	m2	39.69
Removal of woodwork										
Removing										
stud partitions plastered both sides	-	-	-	-	0.66	4.63	2.62	7.25	m2	8.16
roof boarding	-	-	-	-	0.19	1.33	0.88	2.21	m2	2.49
roof boarding; prepare joists for new	-	-	-	-	0.45	3.16	0.88	4.04	m2	4.54
gutter boarding and the like	-	-	-	-	0.19	1.33	0.88	2.21	m2	2.49
weather boarding and battens	-	-	-	-	0.24	1.68	1.04	2.72	m2	3.07
tilting fillets, angle fillets and the like	-	-	-	-	0.09	0.63	0.07	0.70	m	0.79
fascia boards 150 mm wide	-	-	-	-	0.10	0.70	0.14	0.84	m	0.95
barge boards 200mm wide	-	-	-	-	0.10	0.70	0.18	0.88	m	0.99
soffit boards 300mm wide	-	-	-	-	0.10	0.70	0.28	0.98	m	1.10
floor boarding	-	-	-	-	0.33	2.32	0.88	3.20	m2	3.60
handrails and brackets	-	-	-	-	0.10	0.70	0.55	1.25	m	1.41
balustrades complete down to and with cappings to aprons or strings	-	-	-	0.58	0.10	6.06	1.39	7.45	m	8.38
newel posts; cut off flush with landing or string	-	-	-	0.48	0.08	4.99	0.35	5.34	nr	6.01
ends of treads projecting beyond face of cut outer string including scotia under; make good treads where balusters removed	-	-	-	0.90	0.15	9.36	1.04	10.40	m	11.70
dado or picture rails with grounds	-	-	-	-	0.10	0.70	0.14	0.84	m	0.95
architrave	-	-	-	-	0.06	0.42	0.14	0.56	m	0.63
window boards and bearers	-	-	-	-	0.12	0.84	0.28	1.12	m	1.26
Removing; setting aside for reuse										
handrails and brackets	-	-	-	0.25	0.04	2.59	-	2.59	m	2.91
Removal of windows and doors										
Removing										
metal windows with internal and external sills; in conjunction with demolition										
997 x 923mm.....................	-	-	-	-	0.40	2.81	1.67	4.48	nr	5.04
1486 x 923mm....................	-	-	-	-	0.50	3.51	2.40	5.91	nr	6.65
1486 x 1513mm...................	-	-	-	-	0.60	4.21	3.94	8.15	nr	9.17
1994 x 1513mm...................	-	-	-	-	0.70	4.91	5.26	10.17	nr	11.45
metal windows with internal and external sills; preparatory to filling openings; cut out lugs										
997 x 923mm.....................	-	-	-	-	1.30	9.13	1.67	10.80	nr	12.15
1486 x 923mm....................	-	-	-	-	1.60	11.23	2.40	13.63	nr	15.34
1486 x 1513mm...................	-	-	-	-	2.00	14.04	3.94	17.98	nr	20.23
1994 x 1513mm...................	-	-	-	-	2.30	16.15	5.26	21.41	nr	24.08
wood single or multi-light casements and frames with internal and external sills; in conjunction with demolition										
440 x 920mm.....................	-	-	-	-	0.28	1.97	1.08	3.05	nr	3.43
1225 x 1070mm...................	-	-	-	-	0.29	2.04	3.41	5.45	nr	6.13
2395 x 1225mm...................	-	-	-	-	0.30	2.11	7.66	9.77	nr	10.99
wood single or multi-light casements and frames with internal and external sills; preparatory to filling openings; cut out fixing cramps										
440 x 920mm.....................	-	-	-	-	1.25	8.78	1.08	9.86	nr	11.09
1225 x 1070mm...................	-	-	-	-	1.70	11.93	3.41	15.34	nr	17.26
2395 x 1225mm...................	-	-	-	-	2.52	17.69	7.66	25.35	nr	28.52
wood cased frames and sashes with internal and external sills complete with accessories and weights; in conjunction with demolition										
610 x 1225mm....................	-	-	-	-	0.61	4.28	2.62	6.90	nr	7.76
915 x 1525mm....................	-	-	-	-	0.66	4.63	4.87	9.50	nr	10.69
1370 x 1525mm...................	-	-	-	-	0.75	5.26	7.27	12.54	nr	14.10
wood cased frames and sashes with internal and external sills complete with accessories and weights; preparatory to filling openings; cut out fixing cramps										
610 x 1225mm....................	-	-	-	-	1.79	12.57	2.62	15.19	nr	17.08
915 x 1525mm....................	-	-	-	-	2.56	17.97	4.87	22.84	nr	25.70
1370 x 1525mm...................	-	-	-	-	4.11	28.85	7.27	36.12	nr	40.64
single internal doors	-	-	-	-	0.23	1.61	2.09	3.70	nr	4.17
single internal doors and frames; in conjunction with demolition....................	-	-	-	-	0.18	1.26	4.70	5.96	nr	6.71
single internal doors and frames; preparatory to filling openings; cut out fixing cramps	-	-	-	-	1.00	7.02	4.70	11.72	nr	13.19

Labour hourly rates: (except Specialists) Craft Operatives £9.23 Labourer £7.02 Rates are national average prices. Refer to REGIONAL VARIATIONS for indicative levels of overall pricing in regions	MATERIALS			LABOUR				RATES		
	Del to Site £	Waste %	Material Cost £	Craft Optve Hrs	Lab Hrs	Labour Cost £	Sunds £	Nett Rate £	Unit	Gross Rate (+12.5%) £
C90: ALTERATIONS – SPOT ITEMS Cont.										
Removal of windows and doors Cont.										
Removing Cont.										
wood cased frames and sashes with internal and external sills complete with accessories and weights; preparatory to filling openings; cut out fixing cramps Cont.										
double internal doors..........................	-	-	-	-	0.47	3.30	4.18	7.48	nr	8.41
double internal doors and frames; in conjunction with demolition...................	-	-	-	-	0.19	1.33	7.14	8.47	nr	9.53
double internal doors and frames; preparatory to filling openings; cut out fixing cramps	-	-	-	-	1.20	8.42	7.14	15.56	nr	17.51
single external doors.........................	-	-	-	-	0.25	1.75	3.48	5.24	nr	5.89
single external doors and frames; in conjunction with demolition...................	-	-	-	-	0.18	1.26	6.26	7.52	nr	8.46
single external doors and frames; preparatory to filling openings; cut out fixing cramps	-	-	-	-	1.00	7.02	6.26	13.28	nr	14.94
double external doors.........................	-	-	-	-	0.47	3.30	4.18	7.48	nr	8.41
double external doors and frames; in conjunction with demolition...................	-	-	-	-	0.19	1.33	7.14	8.47	nr	9.53
double external doors and frames; preparatory to filling openings; cut out fixing cramps	-	-	-	-	1.20	8.42	7.14	15.56	nr	17.51
single door frames; in conjunction with demolition....................................	-	-	-	-	0.17	1.19	2.62	3.81	nr	4.29
single door frames; preparatory to filling openings; cut out fixing cramps	-	-	-	-	0.77	5.41	2.62	8.03	nr	9.03
double door frames; in conjunction with demolition....................................	-	-	-	-	0.18	1.26	2.96	4.22	nr	4.75
double door frames; preparatory to filling openings; cut out fixing cramps	-	-	-	-	0.82	5.76	2.96	8.72	nr	9.81
Removing; setting aside for reuse										
metal windows with internal and external sills; in conjunction with demolition										
997 x 923mm..............................	-	-	-	0.70	0.50	9.97	-	9.97	nr	11.22
1486 x 923mm..............................	-	-	-	0.85	0.65	12.41	-	12.41	nr	13.96
1486 x 1513mm.............................	-	-	-	0.90	0.75	13.57	-	13.57	nr	15.27
1994 x 1513mm.............................	-	-	-	1.00	0.85	15.20	-	15.20	nr	17.10
metal windows with internal and external sills; preparatory to filling openings; cut out lugs										
997 x 923mm..............................	-	-	-	0.70	1.40	16.29	-	16.29	nr	18.33
1486 x 923mm..............................	-	-	-	0.85	1.75	20.13	-	20.13	nr	22.65
1486 x 1513mm.............................	-	-	-	0.90	2.15	23.40	-	23.40	nr	26.32
1994 x 1513mm.............................	-	-	-	1.00	2.45	26.43	-	26.43	nr	29.73
wood single or multi-light casements and frames with internal and external sills; in conjunction with demolition										
440 x 920mm..............................	-	-	-	0.50	0.36	7.14	-	7.14	nr	8.03
1225 x 1070mm.............................	-	-	-	0.83	0.43	10.68	-	10.68	nr	12.01
2395 x 1225mm.............................	-	-	-	1.00	0.46	12.46	-	12.46	nr	14.02
wood single or multi-light casements and frames with internal and external sills; preparatory to filling openings; cut out fixing cramps										
440 x 920mm..............................	-	-	-	0.50	1.33	13.95	-	13.95	nr	15.70
1225 x 1070mm.............................	-	-	-	0.83	1.84	20.58	-	20.58	nr	23.15
2395 x 1225mm.............................	-	-	-	1.00	2.68	28.04	-	28.04	nr	31.55
wood cased frames and sashes with internal and external sills complete with accessories and weights; in conjunction with demolition										
610 x 1225mm.............................	-	-	-	0.97	0.78	14.43	-	14.43	nr	16.23
915 x 1525mm.............................	-	-	-	1.63	0.93	21.57	-	21.57	nr	24.27
1370 x 1525mm............................	-	-	-	1.94	1.07	25.42	-	25.42	nr	28.59
wood cased frames and sashes with internal and external sills complete with accessories and weights; preparatory to filling openings; cut out fixing cramps										
610 x 1225mm.............................	-	-	-	0.97	1.96	22.71	-	22.71	nr	25.55
915 x 1525mm.............................	-	-	-	1.63	2.83	34.91	-	34.91	nr	39.28
1370 x 1525mm............................	-	-	-	1.94	4.43	49.00	-	49.00	nr	55.13
note: the above rates assume that windows and sashes will be re-glazed when re-used										
single internal doors....................	-	-	-	0.50	0.31	6.79	-	6.79	nr	7.64
single internal doors and frames; in conjunction with demolition...................	-	-	-	1.00	0.34	11.62	-	11.62	nr	13.07
single internal doors and frames; preparatory to filling openings; cut out fixing cramps	-	-	-	1.00	1.16	17.37	-	17.37	nr	19.54
double internal doors....................	-	-	-	1.00	0.63	13.65	-	13.65	nr	15.36
double internal doors and frames; in conjunction with demolition...................	-	-	-	1.47	0.43	16.59	-	16.59	nr	18.66
double internal doors and frames; preparatory to filling openings; cut out fixing cramps	-	-	-	1.47	1.45	23.75	-	23.75	nr	26.72
single external doors....................	-	-	-	0.50	0.33	6.93	-	6.93	nr	7.80
single external doors and frames; in conjunction with demolition...................	-	-	-	1.00	0.34	11.62	-	11.62	nr	13.07
single external doors and frames; preparatory to filling openings; cut out fixing cramps	-	-	-	1.00	1.16	17.37	-	17.37	nr	19.54
double external doors....................	-	-	-	1.00	0.63	13.65	-	13.65	nr	15.36
double external doors and frames; in conjunction with demolition...................	-	-	-	1.47	0.44	16.66	-	16.66	nr	18.74
double external doors and frames; preparatory to filling openings; cut out fixing cramps	-	-	-	1.47	1.45	23.75	-	23.75	nr	26.72
single door frames; in conjunction with demolition....................................	-	-	-	0.50	0.25	6.37	-	6.37	nr	7.17

Labour hourly rates: (except Specialists) Craft Operatives £9.23 Labourer £7.02 Rates are national average prices. Refer to REGIONAL VARIATIONS for indicative levels of overall pricing in regions	MATERIALS			LABOUR				RATES		
	Del to Site	Waste	Material Cost	Craft Optve	Lab	Labour Cost	Sunds	Nett Rate	Unit	Gross Rate (+12.5%)
	£	%	£	Hrs	Hrs	£	£	£		£

C90: ALTERATIONS - SPOT ITEMS Cont.

Removal of windows and doors Cont.

Removing; setting aside for reuse Cont.
  note: the above rates assume that windows and sashes will be re-glazed when re-used Cont.

	Del to Site	Waste	Material Cost	Craft Optve	Lab	Labour Cost	Sunds	Nett Rate	Unit	Gross Rate
single door frames; preparatory to filling openings; cut out fixing cramps	-	-	-	0.50	0.85	10.58	-	10.58	nr	11.90
double door frames; in conjunction with demolition	-	-	-	0.50	0.26	6.44	-	6.44	nr	7.25
double door frames; preparatory to filling openings; cut out fixing cramps	-	-	-	0.50	0.90	10.93	-	10.93	nr	12.30

Removal of ironmongery

Removing door ironmongery (piecing in doors and frames included elsewhere)

	Del to Site	Waste	Material Cost	Craft Optve	Lab	Labour Cost	Sunds	Nett Rate	Unit	Gross Rate
butt hinges	-	-	-	-	0.06	0.42	0.01	0.43	nr	0.49
tee hinges up to 300mm	-	-	-	-	0.20	1.40	0.01	1.41	nr	1.59
floor spring hinges and top centres	-	-	-	-	0.92	6.46	0.20	6.66	nr	7.49
barrel bolts up to 200mm	-	-	-	-	0.13	0.91	0.01	0.92	nr	1.04
flush bolts up to 200mm	-	-	-	-	0.17	1.19	0.01	1.20	nr	1.35
indicating bolts	-	-	-	-	0.23	1.61	0.01	1.62	nr	1.83
double panic bolts	-	-	-	-	0.69	4.84	0.11	4.95	nr	5.57
single panic bolts	-	-	-	-	0.58	4.07	0.07	4.14	nr	4.66
Norfolk or Suffolk latches	-	-	-	-	0.23	1.61	0.01	1.62	nr	1.83
cylinder rim night latches	-	-	-	-	0.23	1.61	0.01	1.62	nr	1.83
rim locks or latches and furniture	-	-	-	-	0.23	1.61	0.01	1.62	nr	1.83
mortice dead locks	-	-	-	-	0.17	1.19	0.01	1.20	nr	1.35
mortice locks or latches and furniture	-	-	-	-	0.29	2.04	0.01	2.05	nr	2.30
Bales catches	-	-	-	-	0.12	0.84	0.01	0.85	nr	0.96
overhead door closers, surface fixed	-	-	-	-	0.29	2.04	0.11	2.15	nr	2.41
pull handles	-	-	-	-	0.12	0.84	0.01	0.85	nr	0.96
push plates	-	-	-	-	0.12	0.84	0.01	0.85	nr	0.96
kicking plates	-	-	-	-	0.17	1.19	0.04	1.23	nr	1.39
letter plates	-	-	-	-	0.12	0.84	0.01	0.85	nr	0.96

Removing window ironmongery

	Del to Site	Waste	Material Cost	Craft Optve	Lab	Labour Cost	Sunds	Nett Rate	Unit	Gross Rate
sash centres	-	-	-	-	0.12	0.84	0.01	0.85	nr	0.96
sash fasteners	-	-	-	-	0.17	1.19	0.01	1.20	nr	1.35
sash lifts	-	-	-	-	0.07	0.49	0.01	0.50	nr	0.56
sash screws	-	-	-	-	0.17	1.19	0.01	1.20	nr	1.35
casement fasteners	-	-	-	-	0.12	0.84	0.01	0.85	nr	0.96
casement stays	-	-	-	-	0.12	0.84	0.01	0.85	nr	0.96
quadrant stays	-	-	-	-	0.12	0.84	0.01	0.85	nr	0.96
fanlight catches	-	-	-	-	0.13	0.91	0.01	0.92	nr	1.04
curtain tracks	-	-	-	-	0.15	1.05	0.07	1.12	m	1.26

Removing sundry ironmongery

	Del to Site	Waste	Material Cost	Craft Optve	Lab	Labour Cost	Sunds	Nett Rate	Unit	Gross Rate
hat and coat hooks	-	-	-	-	0.07	0.49	0.01	0.50	nr	0.56
cabin hooks and eyes	-	-	-	-	0.12	0.84	0.01	0.85	nr	0.96
shelf brackets	-	-	-	-	0.07	0.49	0.04	0.53	nr	0.60
toilet roll holders	-	-	-	-	0.09	0.63	0.04	0.67	nr	0.76
towel rollers	-	-	-	-	0.12	0.84	0.07	0.91	nr	1.03

Removal of metalwork

Removing

	Del to Site	Waste	Material Cost	Craft Optve	Lab	Labour Cost	Sunds	Nett Rate	Unit	Gross Rate
balustrades 1067mm high	-	-	-	-	0.81	5.69	0.90	6.59	m	7.41
wire mesh screens with timber or metal beads screwed on										
generally	-	-	-	-	0.62	4.35	0.88	5.23	m2	5.89
225 x 225mm	-	-	-	-	0.20	1.40	0.07	1.47	nr	1.66
305 x 305mm	-	-	-	-	0.22	1.54	0.11	1.65	nr	1.86
guard bars of vertical bars at 100mm centres welded to horizontal fixing bars at 450mm centres, fixed with bolts	-	-	-	-	0.62	4.35	0.88	5.23	m2	5.89
guard bars of vertical bars at 100mm centres welded to horizontal fixing bars at 450mm centres, fixed with screws	-	-	-	-	0.39	2.74	0.88	3.62	m2	4.07
extra; piecing in softwood after removal of bolts	-	-	-	0.25	0.04	2.59	0.09	2.68	nr	3.01
small bracket	-	-	-	-	0.18	1.26	0.07	1.33	nr	1.50

Removing; making good finishings

	Del to Site	Waste	Material Cost	Craft Optve	Lab	Labour Cost	Sunds	Nett Rate	Unit	Gross Rate
balustrades 1067mm high; making good mortices in treads	-	-	-	0.97	0.97	15.76	1.14	16.90	m	19.02
guard bars of vertical bars at 100mm centres welded to horizontal fixing bars at 450mm centres fixed in mortices; making good concrete or brickwork and plaster	-	-	-	0.60	1.42	15.51	1.19	16.70	m2	18.78
small bracket built or cast in; making good concrete or brickwork and plaster	-	-	-	0.25	0.25	4.06	0.47	4.53	nr	5.10

Removal of manholes

Removing
  manholes overall size; remove cover and frame; break up brick sides and concrete bottom; fill in void with hardcore

	Del to Site	Waste	Material Cost	Craft Optve	Lab	Labour Cost	Sunds	Nett Rate	Unit	Gross Rate
overall size 914 x 1067mm and 600mm deep to invert	1.92	30.00	2.50	-	10.00	70.20	3.48	76.18	nr	85.70
overall size 914 x 1219mm and 900mm deep to invert	3.84	30.00	4.99	-	11.50	80.73	3.83	89.55	nr	100.75
manhole covers and frames; clean off brickwork	-	-	-	-	0.50	3.51	0.62	4.13	nr	4.65

Labour hourly rates: (except Specialists) Craft Operatives £9.23 Labourer £7.02 Rates are national average prices. Refer to REGIONAL VARIATIONS for indicative levels of overall pricing in regions	MATERIALS			LABOUR				RATES		
	Del to Site £	Waste %	Material Cost £	Craft Optve Hrs	Lab Hrs	Labour Cost £	Sunds £	Nett Rate £	Unit	Gross Rate (+12.5%) £
C90: ALTERATIONS - SPOT ITEMS Cont.										
Removal of manholes Cont.										
Removing Cont.										
manholes overall size; remove cover and frame; break up brick sides and concrete bottom; fill in void with hardcore Cont.										
fresh air inlets; clean out pipe socket	-	-	-	-	0.50	3.51	0.20	3.71	nr	4.17
Removal of fencing										
Removing										
chestnut pale fencing with posts										
610mm high	-	-	-	-	0.30	2.11	0.55	2.66	m	2.99
914mm high	-	-	-	-	0.34	2.39	0.84	3.23	m	3.63
1219mm high	-	-	-	-	0.38	2.67	1.12	3.79	m	4.26
close boarded fencing with posts										
1219mm high	-	-	-	-	0.68	4.77	1.28	6.05	m	6.81
1524mm high	-	-	-	-	0.81	5.69	1.60	7.29	m	8.20
1829mm high	-	-	-	-	0.91	6.39	1.92	8.31	m	9.35
chain link fencing with posts										
914mm high	-	-	-	-	0.29	2.04	0.62	2.66	m	2.99
1219mm high	-	-	-	-	0.34	2.39	2.58	4.97	m	5.59
1524mm high	-	-	-	-	0.38	2.67	1.04	3.71	m	4.17
1829mm high	-	-	-	-	0.42	2.95	1.28	4.23	m	4.76
timber gate and posts	-	-	-	-	0.98	6.88	2.62	9.50	nr	10.69
Strutting for forming openings										
Strutting generally										
the cost of strutting in connection with forming openings is to be added to the cost of forming openings; the following are examples of costs for strutting for various sizes and locations of openings. Costs are based on three uses of timber										
Strutting for forming small openings in internal load bearing walls										
opening size 900 x 2000mm	8.91	5.00	9.36	2.00	0.33	20.78	0.66	30.79	nr	34.64
opening size 2000 x 2300mm	19.80	5.00	20.79	5.00	0.83	51.98	1.43	74.20	nr	83.47
Strutting for forming openings in external one brick walls										
opening size 900 x 2000mm	13.37	5.00	14.04	3.00	0.50	31.20	1.14	46.38	nr	52.18
opening size 1500 x 1200mm	13.37	5.00	14.04	3.00	0.50	31.20	1.14	46.38	nr	52.18
Strutting for forming openings in external 280mm brick cavity walls										
opening size 900 x 2000mm	18.32	5.00	19.24	4.00	0.66	41.55	1.28	62.07	nr	69.83
opening size 1500 x 1200mm	18.32	5.00	19.24	4.00	0.66	41.55	1.28	62.07	nr	69.83
Strutting for forming large openings in internal load bearing walls on the ground floor of an average two storey building with timber floors and pitched roof; with load bearing surface 450mm below ground floor; including cutting holes and making good										
opening size 3700 x 2300mm	123.75	5.00	129.94	20.00	20.00	325.00	15.16	470.10	nr	528.86
Unit rates for pricing the above and similar work										
plates, struts, braces and wedges in supports to floor and roof	1.73	5.00	1.82	0.28	0.28	4.55	0.22	6.59	m	7.41
dead shore and needle, sole plates, braces and wedges including cutting holes and making good	62.37	5.00	65.49	10.00	10.00	162.50	7.55	235.54	nr	264.98
Strutting for forming large openings in external one brick walls on the ground floor of an average two storey building as described above										
opening size 6000 x 2500mm	250.47	5.00	262.99	40.00	40.00	650.00	30.07	943.06	nr	1060.95
Unit rates for pricing the above and similar work										
strutting to window opening over new opening	12.47	5.00	13.09	2.00	2.00	32.50	1.49	47.08	nr	52.97
plates, struts, braces and wedges in supports to floor and roof	1.68	5.00	1.76	0.28	0.28	4.55	0.22	6.53	m	7.35
dead shore and needle, sole plates, braces and wedges including cutting holes and making good	98.11	5.00	103.02	15.70	15.70	255.13	11.85	369.99	nr	416.24
set of two raking shores with 50mm wall piece, wedges and dogs	25.05	5.00	26.30	4.00	4.00	65.00	2.99	94.29	nr	106.08
Strutting for forming large openings in external 280mm brick cavity walls on the ground floor of an average two storey building as above described										
opening size 6000 x 2500mm	274.53	5.00	288.26	43.90	43.90	713.38	33.06	1034.69	nr	1164.03
Unit rates for pricing the above and similar work										
strutting to window opening over new opening	24.95	5.00	26.20	4.00	4.00	65.00	2.99	94.19	nr	105.96
plates, struts, braces and wedges in supports to floor and roof	1.68	5.00	1.76	0.28	0.28	4.55	0.22	6.53	m	7.35
dead shore and needle, sole plates, braces and wedges including cutting holes and making good	103.95	5.00	109.15	16.70	16.70	271.38	12.63	393.15	nr	442.30
set of two raking shores with 50mm wall piece, wedges and dogs	25.34	5.00	26.61	4.00	4.00	65.00	2.99	94.60	nr	106.42

Labour hourly rates: (except Specialists) Craft Operatives £9.23 Labourer £7.02 Rates are national average prices. Refer to REGIONAL VARIATIONS for indicative levels of overall pricing in regions	MATERIALS			LABOUR				RATES		
	Del to Site £	Waste %	Material Cost £	Craft Optve Hrs	Lab Hrs	Labour Cost £	Sunds £	Nett Rate £	Unit	Gross Rate (+12.5%) £

C90: ALTERATIONS - SPOT ITEMS Cont.

Spot items

Spot items generally
    the following items are usually specified as spot items where all work in all trades is included in one item; to assist in adapting these items for varying circumstances, unit rates have been given for the component operation where possible while these items are not in accordance with SMM7 either by description or measurement, it is felt that the pricing information given will be of value to the reader

Breaking up and reinstatement of concrete floors for excavations

Break up concrete bed and 150mm hardcore bed under for a width of 760mm for excavation of trench and reinstate with new hardcore and in-situ concrete, B.S.5328, DESIGNED mix C20, 20mm aggregate; make good up to new wall and existing concrete bed both side

	Del to Site	Waste	Material Cost	Craft Optve	Lab	Labour Cost	Sunds	Nett Rate	Unit	Gross Rate
with 100mm plain concrete bed	9.20	10.00	10.12	-	2.70	18.95	6.46	35.53	m	39.98
with 150mm plain concrete bed	13.11	10.00	14.42	-	5.00	35.10	7.82	57.34	m	64.51
with 100mm concrete bed reinforced with steel fabric to B.S.4483, Reference A193	11.00	10.00	12.10	-	4.05	28.43	6.46	46.99	m	52.86
with 150mm concrete bed reinforced with steel fabric to B.S.4483, Reference A193	14.91	10.00	16.40	-	7.33	51.46	7.82	75.68	m	85.14

Openings through concrete

Form opening for door frame through 150mm reinforced concrete wall plastered both sides and with skirting both sides; take off skirtings, cut opening through wall, square up reveals, make good existing plaster up to new frame both sides and form fitted ends on existing skirtings up to new frame; extend cement and sand floor screed and vinyl floor covering through opening and make good up to existing

	Del to Site	Waste	Material Cost	Craft Optve	Lab	Labour Cost	Sunds	Nett Rate	Unit	Gross Rate
838 x 2032mm	15.80	-	15.80	22.50	28.00	404.24	30.20	450.24	nr	506.51

Unit rates for pricing the above and similar work

	Del to Site	Waste	Material Cost	Craft Optve	Lab	Labour Cost	Sunds	Nett Rate	Unit	Gross Rate
cut opening through wall										
100mm thick....................	-	-	-	-	5.30	37.21	7.11	44.32	m2	49.86
150mm thick....................	-	-	-	-	7.90	55.46	10.63	66.09	m2	74.35
200mm thick....................	-	-	-	-	10.50	73.71	14.15	87.86	m2	98.84
make good fair face around opening	-	-	-	-	0.80	5.62	0.50	6.12	m	6.88
square up reveals to opening										
100mm wide.....................	-	-	-	0.75	0.75	12.19	1.24	13.43	m	15.11
150mm wide.....................	-	-	-	1.07	1.07	17.39	1.75	19.14	m	21.53
200mm wide.....................	-	-	-	1.38	1.38	22.43	2.78	25.20	m	28.36
make good existing plaster up to new frame	0.27	10.00	0.30	0.55	0.55	8.94	-	9.23	m	10.39
13mm two coat hardwall plaster to reveal not exceeding 300mm wide...................	0.52	10.00	0.57	0.64	0.64	10.40	-	10.97	m	12.34
take off old skirting....................	-	-	-	0.22	-	2.03	0.10	2.13	m	2.40
form fitted end on existing skirting up to new frame......................................	-	-	-	0.38	-	3.51	-	3.51	nr	3.95
short length of old skirting up to 100mm long with mitre with existing one end...........	-	-	-	0.58	-	5.35	0.04	5.39	nr	6.07
short length of old skirting up to 200mm long with mitres with existing both ends..........	-	-	-	1.15	-	10.61	0.06	10.67	nr	12.01
38mm cement and sand (1:3) screeded bed in opening...	3.91	10.00	4.30	1.54	1.54	25.02	-	29.33	m2	32.99
vinyl or similar floor covering to match existing, fixed with adhesive, not exceeding 300mm wide...................	5.40	10.00	5.94	0.15	0.15	2.44	0.19	8.57	m	9.64
iroko threshold; twice oiled; plugged and screwed										
15 x 100mm.....................	3.90	10.00	4.29	0.90	-	8.31	0.05	12.65	m	14.23
15 x 150mm.....................	5.70	10.00	6.27	1.20	-	11.08	0.06	17.41	m	19.58
15 x 200mm.....................	6.90	10.00	7.59	1.55	-	14.31	0.09	21.99	m	24.73

Form opening for staircase through 150mm reinforced concrete suspended floor plastered on soffit and with screed and vinyl floor covering on top; take up floor covering and screed, cut opening through floor, square up edges of slab, make good ceiling plaster up to lining and make good screed and floor covering to lining

	Del to Site	Waste	Material Cost	Craft Optve	Lab	Labour Cost	Sunds	Nett Rate	Unit	Gross Rate
900 x 2134mm	59.60	-	59.60	19.00	39.00	449.15	39.00	547.75	nr	616.22

Unit rates for pricing the above and similar work

	Del to Site	Waste	Material Cost	Craft Optve	Lab	Labour Cost	Sunds	Nett Rate	Unit	Gross Rate
take up vinyl or similar floor covering	-	-	-	-	0.55	3.86	0.14	4.00	m2	4.50
hack up screed	-	-	-	-	1.50	10.53	1.29	11.82	m2	13.30
cut opening through 150mm thick slab	-	-	-	-	7.90	55.46	10.63	66.09	m2	74.35
cut opening through 200mm thick slab	-	-	-	-	10.50	73.71	14.15	87.86	m2	98.84
make good fair face around opening	-	-	-	-	0.80	5.62	0.47	6.09	m	6.85
square up edges of 150mm thick slab	-	-	-	1.07	1.07	17.39	1.85	19.24	m	21.64
square up edges of 200mm thick slab	-	-	-	1.38	1.38	22.43	2.47	24.90	m	28.01
make good existing plaster up to new lining	0.27	10.00	0.30	0.55	0.55	8.94	-	9.23	m	10.39
make good existing 38mm floor screed up to new lining	0.62	10.00	0.68	0.45	0.45	7.31	-	7.99	m	8.99
make good existing vinyl or similar floor covering up to new lining	7.20	10.00	7.92	0.67	0.67	10.89	0.30	19.11	m	21.50

Labour hourly rates: (except Specialists) Craft Operatives £9.23 Labourer £7.02 Rates are national average prices. Refer to REGIONAL VARIATIONS for indicative levels of overall pricing in regions	MATERIALS			LABOUR				RATES		
	Del to Site £	Waste %	Material Cost £	Craft Optve Hrs	Lab Hrs	Labour Cost £	Sunds £	Nett Rate £	Unit	Gross Rate (+12.5%) £

C90: ALTERATIONS - SPOT ITEMS Cont.

Filling holes in concrete

Fill holes in concrete structure where pipes removed with concrete to B.S.5328, designed mix C20, 20mm aggregate; formwork

wall; thickness 100mm; pipe diameter

50mm	-	-	-	0.60	0.30	7.64	1.58	9.22	nr	10.38
100mm	-	-	-	0.78	0.39	9.94	2.02	11.96	nr	13.45
150mm	-	-	-	0.98	0.49	12.49	2.43	14.92	nr	16.78

wall; thickness 200mm; pipe diameter

50mm	-	-	-	0.78	0.39	9.94	2.02	11.96	nr	13.45
100mm	-	-	-	0.98	0.49	12.49	2.43	14.92	nr	16.78
150mm	-	-	-	1.00	0.77	14.64	3.12	17.76	nr	19.97

floor; thickness 200mm; pipe diameter

50mm	-	-	-	0.74	0.37	9.43	1.83	11.26	nr	12.66
100mm	-	-	-	0.93	0.46	11.81	2.34	14.15	nr	15.92
150mm	-	-	-	0.95	0.48	12.14	2.96	15.10	nr	16.99

Make good fair face to filling and surrounding work

to one side of wall; pipe diameter

50mm	-	-	-	0.10	0.10	1.63	0.21	1.84	nr	2.06
100mm	-	-	-	0.20	0.20	3.25	0.32	3.57	nr	4.02
150mm	-	-	-	0.25	0.25	4.06	0.39	4.45	nr	5.01

to soffit of floor; pipe diameter

50mm	-	-	-	0.10	0.10	1.63	0.21	1.84	nr	2.06
100mm	-	-	-	0.20	0.20	3.25	0.32	3.57	nr	4.02
150mm	-	-	-	0.25	0.25	4.06	0.39	4.45	nr	5.01

Make good plaster to filling and surrounding work

to one side of wall; pipe diameter

50mm	-	-	-	0.19	0.19	3.09	0.21	3.30	nr	3.71
100mm	-	-	-	0.36	0.36	5.85	0.32	6.17	nr	6.94
150mm	-	-	-	0.45	0.45	7.31	0.39	7.70	nr	8.67

to soffit of floor; pipe diameter

50mm	-	-	-	0.20	0.20	3.25	0.21	3.46	nr	3.89
100mm	-	-	-	0.40	0.40	6.50	0.32	6.82	nr	7.67
150mm	-	-	-	0.50	0.50	8.13	0.39	8.52	nr	9.58

Make good floor screed and surrounding work

pipe diameter 50mm	-	-	-	0.19	0.19	3.09	0.09	3.18	nr	3.57
pipe diameter 100mm	-	-	-	0.36	0.36	5.85	0.19	6.04	nr	6.79
pipe diameter 150mm	-	-	-	0.45	0.45	7.31	0.26	7.57	nr	8.52

Make good vinyl or similar floor covering and surrounding work

pipe diameter 50mm	-	-	-	0.50	0.50	8.13	1.56	9.69	nr	10.90
pipe diameter 100mm	-	-	-	1.00	1.00	16.25	2.09	18.34	nr	20.63
pipe diameter 150mm	-	-	-	1.20	1.20	19.50	2.93	22.43	nr	25.23

Fill holes in concrete structure where metal sections removed with concrete to B.S.5328, designed mix C20, 20mm aggregate; formwork

wall; thickness 100mm; section not exceeding 250mm deep	-	-	-	0.60	0.60	9.75	1.86	11.61	nr	13.06
wall; thickness 200mm; section not exceeding 250mm deep	-	-	-	0.75	0.75	12.19	2.71	14.90	nr	16.76
wall; thickness 100mm; section 250 - 500mm deep	-	-	-	0.75	0.75	12.19	2.71	14.90	nr	16.76
wall; thickness 200mm; section 250 - 500mm deep	-	-	-	0.90	0.90	14.63	3.27	17.90	nr	20.13

Make good fair face to filling and surrounding work

to one side of wall; section not exceeding 250mm deep	-	-	-	0.22	0.22	3.58	0.18	3.75	nr	4.22
to one side of wall; section 250 - 500mm deep	-	-	-	0.26	0.26	4.22	0.28	4.50	nr	5.07

Make good plaster to filling and surrounding work

to one side of wall; section not exceeding 250mm deep	-	-	-	0.35	0.35	5.69	0.49	6.18	nr	6.95
to one side of wall; section 250 - 500mm deep	-	-	-	0.43	0.43	6.99	0.74	7.73	nr	8.69

Filling openings in concrete

Fill opening where door removed in 150mm reinforced concrete wall with concrete reinforced with 12mm mild steel bars at 300mm centres both ways tied to existing structure; hack off plaster to reveals, hack up floor covering and screed in opening, prepare edges of opening to form joint with filling, drill edges of opening and tie in new reinforcement, plaster filling both sides and extended skirting both sides; making good junction of new and existing plaster and skirtings and make good floor screed and vinyl covering up to filling both sides

838 x 2032mm	100.10	-	100.10	22.50	34.50	449.87	9.63	559.60	nr	629.54

Unit rates for pricing the above and similar work

hack off plaster to reveal not exceeding 100mm wide	-	-	-	-	0.24	1.68	0.07	1.75	m	1.97
hack off plaster to reveal 100 - 200mm wide	-	-	-	-	0.33	2.32	0.14	2.46	m	2.76
take up vinyl or similar floor covering and screed	-	-	-	-	2.00	14.04	1.43	15.47	m2	17.40
prepare edge of opening 100mm wide to form joint with filling	-	-	-	0.25	0.25	4.06	-	4.06	m	4.57
prepare edge of opening 200mm wide to form joint with filling	-	-	-	0.35	0.35	5.69	-	5.69	m	6.40

Labour hourly rates: (except Specialists) Craft Operatives £9.23 Labourer £7.02. Rates are national average prices. Refer to REGIONAL VARIATIONS for indicative levels of overall pricing in regions	MATERIALS			LABOUR				RATES		
	Del to Site £	Waste %	Material Cost £	Craft Optve Hrs	Lab Hrs	Labour Cost £	Sunds £	Nett Rate £	Unit	Gross Rate (+12.5%) £
C90: ALTERATIONS - SPOT ITEMS Cont.										
Filling openings in concrete Cont.										
Unit rates for pricing the above and similar work Cont.										
concrete to B.S.5328, designed mix C20, 20mm aggregate in walls 150 - 450m thick	90.60	5.00	95.13	-	12.00	84.24	0.41	179.78	m3	202.25
concrete to B.S.5328, designed mix C20, 20mm aggregate in walls not exceeding 150mm thick	90.60	5.00	95.13	-	12.00	84.24	0.41	179.78	m3	202.25
12mm mild steel bar reinforcement; straight	752.00	10.00	827.20	85.00	-	784.55	4.51	1616.26	t	1818.29
drill edge of opening and tie in new reinforcement	-	-	-	-	1.00	7.02	0.41	7.43	m	8.36
formwork and basic finish to wall, per side	6.76	10.00	7.44	2.50	2.50	40.63	0.98	49.04	m2	55.17
formwork and fine formed finish to wall, per side	8.06	10.00	8.87	2.75	2.75	44.69	1.10	54.65	m2	61.49
junction between new and existing fair face	-	-	-	-	0.30	2.11	-	2.11	m	2.37
13mm two coat hardwall plaster over 300mm wide ...	1.76	10.00	1.94	1.30	1.30	21.13	-	23.06	m2	25.94
25 x 100mm softwood chamfered skirting, primed all round	1.46	10.00	1.61	0.56	0.09	5.80	0.11	7.52	m	8.46
junction with existing	-	-	-	0.50	0.08	5.18	-	5.18	nr	5.82
make good floor screed and vinyl or similar floor covering up to filling not exceeding 300mm wide ..	7.00	10.00	7.70	1.12	1.12	18.20	0.32	26.22	m	29.50
Fill opening where staircase removed in 150mm reinforced concrete floor with concrete reinforced with 12mm mild steel bars at 225mm centres both ways tied to existing structure; remove timber lining, prepare edges of opening to form joint with filling, drill edges of opening tie in new reinforcement, plaster filling on soffit and extend cement and sand screed and vinyl floor covering to match existing; make good junction of new and existing plaster and vinyl floor covering										
900 x 2134mm	114.00	-	114.00	14.75	22.30	292.69	8.00	414.69	nr	466.52
Unit rates for pricing the above and similar work										
take off timber lining	-	-	-	-	0.24	1.68	0.10	1.78	m	2.01
prepare edge of opening 150mm wide to form joint with filling	-	-	-	0.30	0.30	4.88	-	4.88	m	5.48
prepare edge of opening 200mm wide to form joint with filling	-	-	-	0.35	0.35	5.69	-	5.69	m	6.40
concrete to B.S.5328, designed mix C20, 20mm aggregate in slabs 150 - 450mm thick	90.60	5.00	95.13	-	12.00	84.24	0.41	179.78	m3	202.25
concrete to B.S.5328, designed mix C20, 20mm aggregate in slabs not exceeding 150mm thick	90.60	5.00	95.13	-	12.00	84.24	0.41	179.78	m3	202.25
12mm mild steel bar reinforcement; straight	752.00	10.00	827.20	85.00	-	784.55	4.51	1616.26	t	1818.29
drill edge of opening and tie in new reinforcement	-	-	-	-	1.00	7.02	0.41	7.43	m	8.36
formwork and basic finish to soffit, slab thickness not exceeding 200mm; height to soffit 1.50 - 3.00m	6.72	10.00	7.39	2.25	2.25	36.56	0.97	44.92	m2	50.54
formwork and basic finish to soffit, slab thickness 200 - 300mm; height to soffit 1.50 - 3.00m	8.06	10.00	8.87	2.50	2.50	40.63	1.10	50.59	m2	56.91
formwork and fine formed finish to soffit, slab thickness not exceeding 200mm; height to soffit 1.50 - 3.00m	6.72	10.00	7.39	2.50	2.50	40.63	1.12	49.14	m2	55.28
formwork and fine formed finish to soffit, slab thickness 200 - 300mm; height to soffit 1.50 - 3.00m	8.06	10.00	8.87	2.75	2.75	44.69	1.16	54.71	m2	61.55
junction between new and existing fair face	-	-	-	-	0.30	2.11	-	2.11	m	2.37
10mm two coat lightweight plaster over 300mm wide	1.73	10.00	1.90	1.30	1.30	21.13	-	23.03	m2	25.91
38mm cement and sand (1:3) trowelled bed	3.91	10.00	4.30	0.63	0.63	10.24	-	14.54	m2	16.36
vinyl or similar floor covering to match existing fixed with adhesive exceeding 300mm wide	18.00	10.00	19.80	1.25	0.50	15.05	0.56	35.41	m2	39.83
Openings through brickwork or blockwork										
Form openings for door frames through 100mm block partition plastered both sides and with skirting both sides; take off skirtings, cut opening through partition, insert 100 x 150mm precast concrete lintel and wedge and pin up over, quoin up jambs, extend plaster to faces of lintel and make good junction with existing plaster; make good existing plaster up to new frame both sides and form fitted ends on existing skirtings up to new frame; extend softwood board flooring through opening on and with bearers and make good up to existing										
838 x 2032mm	19.14	-	19.14	20.00	17.20	305.34	12.70	337.18	nr	379.33
Unit rates for pricing the above and similar work										
cut opening through 75mm block partition plastered both sides	-	-	-	1.10	1.30	19.28	3.91	23.19	m2	26.09
cut opening through 100mm block partition plastered both sides	-	-	-	1.30	1.56	22.95	4.59	27.54	m2	30.98
make good fair face around opening	-	-	-	0.25	0.25	4.06	0.10	4.16	m	4.68
precast concrete, B.S.5328, designed mix C20, 20mm aggregate lintel 75 x 150 x 1200mm, reinforced with 1.01 kg of 12mm mild steel bars	6.98	10.00	7.68	0.80	0.80	13.00	0.25	20.93	nr	23.54
precast concrete, B.S.5328, designed mix C20, 20mm aggregate lintel 100 x 150 x 1200mm, reinforced with 1.01 kg of 12mm mild steel bars	9.63	10.00	10.59	1.00	1.00	16.25	0.33	27.17	nr	30.57
precast concrete, B.S.5328 designed mix C20, 20mm aggregate lintel 75 x 150 x 1200mm, reinforced with 1.01 kg of 12mm mild steel bars, fair finish all faces	8.33	10.00	9.16	0.80	0.80	13.00	0.25	22.41	nr	25.21

Labour hourly rates: (except Specialists) Craft Operatives £9.23 Labourer £7.02 Rates are national average prices. Refer to REGIONAL VARIATIONS for indicative levels of overall pricing in regions	MATERIALS			LABOUR				RATES		
	Del to Site £	Waste %	Material Cost £	Craft Optve Hrs	Lab Hrs	Labour Cost £	Sunds £	Nett Rate £	Unit	Gross Rate (+12.5%) £

C90: ALTERATIONS - SPOT ITEMS Cont.

Openings through brickwork or blockwork Cont.

Unit rates for pricing the above and similar work Cont.

	Del to Site	Waste	Material Cost	Craft Optve	Lab	Labour Cost	Sunds	Nett Rate	Unit	Gross Rate
precast concrete, B.S.5328 designed mix C20, 20mm aggregate lintel 100 x 150 x 1200mm, reinforced with 1.01 kg of 12mm mild steel bars, fair finish all faces	12.05	10.00	13.26	1.00	1.00	16.25	0.33	29.84	nr	33.56
wedge and pin up over lintel 75mm wide	-	-	-	0.50	0.50	8.13	0.62	8.74	m	9.84
wedge and pin up over lintel 100mm wide	-	-	-	0.66	0.66	10.73	0.82	11.55	m	12.99
quoin up 75mm wide jambs	-	-	-	0.52	0.52	8.45	0.62	9.07	m	10.20
quoin up 100mm wide jambs	-	-	-	0.68	0.68	11.05	0.82	11.87	m	13.35
extend 13mm hardwall plaster to face of lintel	1.76	10.00	1.94	2.13	2.13	34.61	0.06	36.61	m2	41.18
make good existing plaster up to new frame	0.27	10.00	0.30	0.55	0.55	8.94	-	9.23	m	10.39
take off old skirting	-	-	-	0.22	-	2.03	0.10	2.13	m	2.40
form fitted end on existing skirting up to new frame	-	-	-	0.38	-	3.51	-	3.51	nr	3.95
38mm cement and sand (1:3) screeded bed in opening	3.91	10.00	4.30	1.54	1.54	25.02	-	29.33	m2	32.99
extend 25mm softwood board flooring through opening on and with bearers	28.40	10.00	31.24	16.20	2.70	168.48	0.54	200.26	m2	225.29
iroko threshold; twice oiled; plugged and screwed										
19 x 75mm	3.45	10.00	3.79	0.80	-	7.38	0.07	11.25	m	12.66
19 x 100mm	3.90	10.00	4.29	0.90	-	8.31	0.08	12.68	m	14.26

Form openings for door frames through one brick wall plastered one side and with external rendering other side; take off skirting, cut opening through wall, insert 215 x 150mm precast concrete lintel and wedge and pin up over; quoin up jambs; extend plaster to face of lintel and make good junction with existing plaster; make good existing plaster up to new frame one side; extend external rendering to face of lintel and to reveals and make good up to existing rendering and new frame the other side; form fitted ends on existing skirting up to new frame

	Del to Site	Waste	Material Cost	Craft Optve	Lab	Labour Cost	Sunds	Nett Rate	Unit	Gross Rate
914 x 2082mm	33.70	-	33.70	28.00	25.40	436.75	28.10	498.55	nr	560.87

Unit rates for pricing the above and similar work

	Del to Site	Waste	Material Cost	Craft Optve	Lab	Labour Cost	Sunds	Nett Rate	Unit	Gross Rate
cut opening through half brick wall	-	-	-	1.95	2.40	34.85	4.76	39.61	m2	44.56
cut opening through one brick wall	-	-	-	3.80	4.70	68.07	9.01	77.08	m2	86.71
make good fair face around opening	-	-	-	0.50	0.50	8.13	0.62	8.74	m	9.84
make good facings to match existing around opening	2.38	10.00	2.62	0.50	0.50	8.13	0.62	11.36	m	12.78
precast concrete, B.S.5328, designed mix C20, 20mm aggregate lintel 102 x 150 x 1200mm, reinforced with 1.07 kg of 12mm mild steel bars	9.63	10.00	10.59	1.00	1.00	16.25	0.36	27.20	nr	30.60
precast concrete, B.S.5328, designed mix C20, 20mm aggregate lintel 215 x 150 x 1200mm, reinforced with 2.13 kg of 12mm mild steel bars	21.34	10.00	23.47	1.50	1.50	24.38	0.74	48.59	nr	54.66
precast concrete, B.S.5328, designed mix C20, 20mm aggregate lintel 102 x 150 x 1200mm, reinforced with 1.07 kg of 12mm mild steel bars, fair finish all faces	12.05	10.00	13.26	1.00	1.00	16.25	0.36	29.86	nr	33.60
precast concrete, B.S.5328, designed mix C20, 20mm aggregate lintel 215 x 150 x 1200mm, reinforced with 2.13 kg of 12mm mild steel bars, fair finish all faces	23.09	10.00	25.40	1.50	1.50	24.38	0.72	50.49	nr	56.81
wedge and pin up over lintel 102mm wide	-	-	-	0.66	0.66	10.73	0.82	11.55	m	12.99
wedge and pin up over lintel 215mm wide	-	-	-	1.00	1.00	16.25	1.75	18.00	m	20.25
quoin up half brick jambs	-	-	-	0.70	0.70	11.38	0.82	12.20	m	13.72
quoin up one brick jambs	-	-	-	1.27	1.27	20.64	1.75	22.39	m	25.19
facings to match existing to margin	1.02	10.00	1.12	0.67	0.67	10.89	0.82	12.83	m	14.43
make good existing plaster up to new frame	0.27	10.00	0.30	0.55	0.55	8.94	-	9.23	m	10.39
13mm two coat hardwall plaster to reveal not exceeding 300mm wide	0.53	10.00	0.58	0.78	0.78	12.68	0.02	13.28	m	14.94
make good existing external rendering up to new frame	0.21	10.00	0.23	0.47	0.47	7.64	-	7.87	m	8.85
13mm two coat cement and sand (1:3) external rendering to reveal not exceeding 300mm wide	0.42	10.00	0.46	0.74	0.74	12.03	0.02	12.51	m	14.07
take off old skirting	-	-	-	0.23	-	2.12	0.09	2.21	m	2.49
short length of old skirting up to 100mm long with mitre with existing one end	-	-	-	0.58	-	5.35	0.04	5.39	nr	6.07
short length of old skirting up to 300mm long with mitres with existing both ends	-	-	-	1.15	-	10.61	0.05	10.66	nr	12.00
38mm cement and sand (1:3) screeded bed in opening	3.91	10.00	4.30	1.54	1.54	25.02	-	29.33	m2	32.99
extend 25mm softwood board flooring through opening on and with bearers	5.36	10.00	5.90	3.65	0.61	37.97	0.14	44.01	nr	49.51
iroko threshold, twice oiled, plugged and screwed										
15 x 113mm	4.00	10.00	4.40	0.90	-	8.31	0.05	12.76	m	14.35
15 x 225mm	7.40	10.00	8.14	1.65	-	15.23	0.09	23.46	m	26.39

Form opening for window through 275mm hollow wall with two half brick skins, plastered one side and faced with picked stock facings the other; cut opening through wall, insert 275 x 225mm precast concrete boot lintel and wedge and pin up over; insert damp proof course and cavity gutter; quoin up jambs, close cavity at jambs with brickwork bonded to existing and with vertical damp proof course and close cavity at sill with one course of slates; face margin externally to match existing and extend plaster to face of lintel and to reveals and make good up to existing plaster and new frame

	Del to Site	Waste	Material Cost	Craft Optve	Lab	Labour Cost	Sunds	Nett Rate	Unit	Gross Rate
900 x 900mm	51.30	-	51.30	15.90	16.80	264.69	14.60	330.59	nr	371.92

	MATERIALS			LABOUR				RATES		
Labour hourly rates: (except Specialists) Craft Operatives £9.23 Labourer £7.02 Rates are national average prices. Refer to REGIONAL VARIATIONS for indicative levels of overall pricing in regions	Del to Site £	Waste %	Material Cost £	Craft Optve Hrs	Lab Hrs	Labour Cost £	Sunds £	Nett Rate £	Unit	Gross Rate (+12.5%) £

C90: ALTERATIONS - SPOT ITEMS Cont.

Openings through brickwork or blockwork Cont.

	Del to Site £	Waste %	Material Cost £	Craft Optve Hrs	Lab Hrs	Labour Cost £	Sunds £	Nett Rate £	Unit	Gross Rate (+12.5%) £
Unit rates for pricing the above and similar work										
cut opening through wall with two half brick skins	-	-	-	3.90	4.80	69.69	8.16	77.85	m2	87.58
cut opening through wall with one half brick and one 100mm block skins	-	-	-	3.25	3.96	57.80	8.16	65.96	m2	74.20
precast concrete, B.S.5328, designed mix C20, 20mm aggregate boot lintel 275 x 225 x 1200mm long reinforced with 3.60 kg of 12mm mild steel bars and 0.67 kg of 6mm mild steel links	31.99	10.00	35.19	2.30	2.30	37.38	0.82	73.38	nr	82.56
close cavity at jambs with blockwork bonded to existing and with 112.5mm wide bituminous hessian based damp proof course, B.S.6398, Class A	2.34	-	2.34	0.50	0.50	8.13	0.21	10.68	m	12.01
close cavity at jambs with brickwork bonded to existing and with 112.5mm wide bituminous hessian based damp proof course, B.S.6398, Class A	2.09	-	2.09	0.70	0.70	11.38	0.82	14.29	m	16.07
close cavity at jambs with slates in cement mortar (1:3) set vertically	3.84	10.00	4.22	0.44	0.44	7.15	0.21	11.58	m	13.03
close cavity at sill with one course of slates in cement mortar (1:3)	3.84	10.00	4.22	0.44	0.44	7.15	0.21	11.58	m	13.03
lead cored bituminous hessian based damp proof course and cavity tray, B.S.6398, Class F, width exceeding 225mm	20.28	10.00	22.31	0.80	0.80	13.00	0.21	35.52	m2	39.96
precast concrete, B.S.5328, designed mix C20, 20mm aggregate lintel 160 x 225 x 1200mm long reinforced with 2.13 kg of 12mm mild steel bars	23.34	10.00	25.67	1.50	1.50	24.38	0.62	50.67	nr	57.00
76 x 76 x 6mm mild steel angle arch bar, primed	6.39	5.00	6.71	0.50	0.50	8.13	0.62	15.45	m	17.39
take out three courses of facing bricks and build brick-on-end flat arch in picked stock facing bricks	9.06	5.00	9.51	1.00	1.00	16.25	1.81	27.57	m	31.02
For unit rates for finishes, quoining up jambs, etc see previous items										
Form opening for window where door removed, with window head at same level as old door head, through one brick wall plastered one side and faced the other with picked stock facings; remove old lintel and flat arch, cut away for and insert 102 x 225mm precast concrete lintel and wedge and pin up over; build brick-on-end flat arch in facing bricks on 76 x 76 x 6mm mild steel angle arch bar and point to match existing; cut away jambs of old door opening to extend width; quoin up jambs and face margins externally to match existing; fill in lower part of old opening with brickwork in gauged mortar (1:2:9), bond to existing and face externally and point to match existing										
1200 x 900mm; old door opening 914 x 2082mm; extend plaster to face of lintel, reveals and filling and make good up to existing plaster and new frame; extend skirting one side and make good junction with existing	110.60	-	110.60	22.00	21.50	353.99	25.00	489.59	nr	550.79
Unit rates for pricing the above and similar work										
take out old lintel	-	-	-	1.00	1.30	18.36	3.40	21.76	m	24.48
cut away one brick wall to increase width of opening	-	-	-	3.80	4.70	68.07	8.50	76.57	m2	86.14
cut away 275mm hollow wall of two half brick skins to increase width of opening	-	-	-	3.90	4.80	69.69	8.50	78.19	m2	87.97
For other unit rates for pricing the above and similar work see previous items and items in 'Filling openings in Brickwork and Blockwork' section										
Form opening for door frame where old window removed, with door head at same level as old window head, through one brick wall plastered one side and faced the other with picked stock facings; remove old lintel and flat arch, cut away for and insert 102 x 225mm precast concrete lintel and wedge and pin up over; build brick-on-end flat arch in facing bricks on 76 x 76 x 6mm mild steel angle arch bar and point, fill old opening at sides of new opening to reduce width with brickwork in gauged mortar (1:2:9), bond to existing, wedge and pin up at soffit, face externally and to margins and point to match existing; take off skirtings, cut away wall below sill of old opening, quion up jambs and face margins externally										
914 x 2082mm, old window opening 1200 x 900mm; extend plaster to face of lintel and filling and make good up to existing plaster and new frame; form fitted ends on existing skirting up to new frame	66.80	-	66.80	24.00	23.80	388.60	28.50	483.90	nr	544.38

	MATERIALS			LABOUR				RATES		
Labour hourly rates: (except Specialists) Craft Operatives £9.23 Labourer £7.02 Rates are national average prices. Refer to REGIONAL VARIATIONS for indicative levels of overall pricing in regions	Del to Site	Waste	Material Cost	Craft Optve	Lab	Labour Cost	Sunds	Nett Rate	Unit	Gross Rate (+12.5%)
	£	%	£	Hrs	Hrs	£	£	£		£

C90: ALTERATIONS - SPOT ITEMS Cont.

Openings through brickwork or blockwork Cont.

For unit rates for pricing the above and similar work
   see previous items and items in 'Filling openings
   in Brickwork and Blockwork' section

Air bricks in old walls

Cut opening through old one brick wall, render all
round in cement mortar (1:3); build in clay air brick
externally and fibrous plaster ventilator internally
and make good facings and plaster

Description	Del to Site	Waste	Material Cost	Craft Optve	Lab	Labour Cost	Sunds	Nett Rate	Unit	Gross Rate
225 x 150mm	3.82	-	3.82	1.75	1.75	28.44	0.55	32.81	nr	36.91
225 x 225mm	8.21	-	8.21	2.33	2.33	37.86	0.69	46.76	nr	52.61

Cut opening through old 275mm hollow brick wall and
seal cavity with slates; build in clay air brick
externally and fibrous plaster ventilator internally
and make good facings and plaster

Description	Del to Site	Waste	Material Cost	Craft Optve	Lab	Labour Cost	Sunds	Nett Rate	Unit	Gross Rate
225 x 150mm	7.34	-	7.34	2.14	2.14	34.77	0.55	42.66	nr	48.00
225 x 225mm	13.13	-	13.13	2.81	2.81	45.66	0.69	59.48	nr	66.92

Openings through rubble walling

Form opening for door frame through 600mm rough
rubble wall faced both sides; cut opening through
wall, insert 600 x 150mm precast concrete lintel
finished fair on all exposed faces and wedge and pin
up over; quoin and face up jambs and extend softwood
board flooring through opening on and with bearers
and make good up to existing

Description	Del to Site	Waste	Material Cost	Craft Optve	Lab	Labour Cost	Sunds	Nett Rate	Unit	Gross Rate
838 x 2032mm	78.30	-	78.30	41.30	43.70	687.97	61.10	827.37	nr	930.79

Unit rates for pricing the above and similar work

Description	Del to Site	Waste	Material Cost	Craft Optve	Lab	Labour Cost	Sunds	Nett Rate	Unit	Gross Rate
cut opening through 350mm random rubble wall	-	-	-	6.00	7.00	104.52	11.90	116.42	m2	130.97
cut opening through 600mm rough rubble wall	-	-	-	10.00	12.00	176.54	20.40	196.94	m2	221.56
precast concrete, B.S.5328, designed mix C20, 20mm aggregate lintel 350 x 150 x 1140mm reinforced with 3.04 kg of 12mm mild steel bars; fair finish all faces	34.20	10.00	37.62	2.25	2.25	36.56	0.72	74.90	nr	84.27
precast concrete, B.S.5328, designed mix C20, 20mm aggregate lintel 600 x 150 x 1140mm reinforced with 5.00 kg of 12mm mild steel bars; fair finish all faces	57.80	10.00	63.58	3.65	3.65	59.31	1.13	124.02	nr	139.53
wedge and pin up over lintel 350mm wide	-	-	-	2.00	2.00	32.50	2.06	34.56	m	38.88
wedge and pin up over lintel 600mm wide	-	-	-	3.00	3.00	48.75	3.61	52.36	m	58.91
quoin up 350mm wide jambs	-	-	-	2.00	2.00	32.50	2.06	34.56	m	38.88
quoin up 600mm wide jambs	-	-	-	3.00	3.00	48.75	3.61	52.36	m	58.91
extend 25mm softwood board flooring through opening on and with bearers	13.40	10.00	14.74	9.70	1.60	100.76	0.39	115.89	nr	130.38

Openings through stone faced walls

Form opening for window through wall of half brick
backing, plastered, and 100mm stone facing; cut
opening through wall, insert 102 x 150mm precast
concrete lintel and wedge and pin up over and build
in 76 x 76 x 6mm mild steel angle arch bar to support
stone facing, clean soffit of stone and point; quoin
up jambs and face up and point externally; extend
plaster to face of lintel and to reveals and make
good up to existing plaster and new frame

Description	Del to Site	Waste	Material Cost	Craft Optve	Lab	Labour Cost	Sunds	Nett Rate	Unit	Gross Rate
900 x 900mm	24.40	-	24.40	19.20	19.80	316.21	18.90	359.51	nr	404.45

Unit rates for pricing the above and similar work

Description	Del to Site	Waste	Material Cost	Craft Optve	Lab	Labour Cost	Sunds	Nett Rate	Unit	Gross Rate
cut opening through wall	-	-	-	5.15	5.85	88.60	8.16	96.76	m2	108.86
precast concrete, B.S.5328, designed mix C20, 20mm aggregate lintel 102 x 150 x 1200mm reinforced with 1.07 kg of 12mm mild steel bars	9.63	10.00	10.59	1.05	1.05	17.06	0.31	27.97	nr	31.46
76 x 76 x 6mm mild steel angle arch bar, primed	6.39	5.00	6.71	0.50	0.50	8.13	0.62	15.45	m	17.39
quoin up jambs	-	-	-	0.70	0.70	11.38	0.93	12.31	m	13.84
face stone jambs and point	-	-	-	2.10	2.10	34.13	2.58	36.70	m	41.29
clean stone head and point	-	-	-	2.10	2.10	34.13	0.31	34.44	m	38.74

For unit rates for finishes, etc.
   see previous items

Filling openings in brickwork or blockwork

Fill opening where door removed in 100mm block wall
with concrete blocks in gauged mortar (1:2:9);
provide 50mm preservative treated softwood sole plate
on timber floor in opening, bond to blockwork at
jambs and wedge and pin up at head; plaster filling
both sides and extend skirting both sides; make good
junction of new and existing plaster and skirtings

Description	Del to Site	Waste	Material Cost	Craft Optve	Lab	Labour Cost	Sunds	Nett Rate	Unit	Gross Rate
838 x 2032mm	48.30	-	48.30	14.10	8.15	187.36	8.48	244.14	nr	274.65

Unit rates for pricing the above and similar work

Description	Del to Site	Waste	Material Cost	Craft Optve	Lab	Labour Cost	Sunds	Nett Rate	Unit	Gross Rate
75mm blockwork in filling	7.21	10.00	7.93	1.00	1.25	18.00	1.13	27.07	m2	30.45
100mm blockwork in filling	8.65	10.00	9.52	1.20	1.50	21.61	1.55	32.67	m2	36.75
75mm blockwork in filling and fair face and flush smooth pointing one side	7.21	15.00	8.29	1.25	1.50	22.07	1.34	31.70	m2	35.66
100mm blockwork in filling and fair face and flush smooth pointing one side	8.65	15.00	9.95	1.45	1.75	25.67	1.65	37.27	m2	41.92

Labour hourly rates: (except Specialists) Craft Operatives £9.23 Labourer £7.02 Rates are national average prices. Refer to REGIONAL VARIATIONS for indicative levels of overall pricing in regions	MATERIALS			LABOUR				RATES		
	Del to Site £	Waste %	Material Cost £	Craft Optve Hrs	Lab Hrs	Labour Cost £	Sunds £	Nett Rate £	Unit	Gross Rate (+12.5%) £
C90: ALTERATIONS - SPOT ITEMS Cont.										
Filling openings in brickwork or blockwork Cont.										
Unit rates for pricing the above and similar work Cont.										
preservative treated softwood sole plate										
50 x 75mm..........................	1.10	10.00	1.21	0.30	0.05	3.12	0.21	4.54	m	5.11
50 x 100mm.........................	1.46	10.00	1.61	0.33	0.06	3.47	0.28	5.35	m	6.02
lead cored bituminous hessian based damp proof course, B.S.6398, Class F, width not exceeding 225mm	20.28	10.00	22.31	0.80	0.80	13.00	0.41	35.72	m2	40.18
wedge and pin up at head 75mm wide	-	-	-	0.10	0.10	1.63	0.41	2.04	m	2.29
wedge and pin up at head 100mm wide...........	-	-	-	0.12	0.12	1.95	0.62	2.57	m	2.89
cut pockets and bond 75mm filling to existing blockwork at jambs.............	-	-	-	0.40	0.40	6.50	0.72	7.22	m	8.12
cut pockets and bond 100mm filling to existing blockwork at jambs.............	-	-	-	0.50	0.50	8.13	1.02	9.14	m	10.29
12mm (average) cement and sand dubbing over 300mm wide on filling....................	1.24	10.00	1.36	0.54	0.54	8.78	-	10.14	m2	11.41
13mm two coat lightweight plaster over 300mm wide on filling......................	1.80	10.00	1.98	1.27	1.27	20.64	0.02	22.64	m2	25.47
19 x 100mm softwood chamfered skirting, primed all round.........................	1.13	10.00	1.24	0.50	0.08	5.18	0.11	6.53	m	7.35
junction with existing.....................	-	-	-	0.50	0.08	5.18	-	5.18	nr	5.82
make good vinyl or similar floor covering up to filling.................................	6.00	10.00	6.60	0.50	0.50	8.13	0.14	14.87	m	16.72
Fill opening where door removed in half brick wall with brickwork in gauged mortar (1:2:9); provide lead cored damp proof course in opening, lapped with existing, bond to existing brickwork at jambs and wedge and pin up at head; plaster filling both sides and extend skirting both sides; make good junction of new and existing plaster and skirtings										
838 x 2032mm	47.30	-	47.30	15.70	12.75	234.42	10.40	292.12	nr	328.63
Unit rates for pricing the above and similar work										
half brick filling in common bricks	12.00	10.00	13.20	2.00	2.00	32.50	2.58	48.28	m2	54.31
half brick filling in common bricks fair faced and flush pointed one side	12.00	10.00	13.20	2.33	2.33	37.86	2.78	53.84	m2	60.57
50 x 102mm softwood sole plate	1.56	10.00	1.72	0.33	0.06	3.47	0.25	5.43	m	6.11
lead cored bituminous hessian based damp proof course, B.S.6398, Class F, width not exceeding 225mm	20.28	10.00	22.31	0.80	0.80	13.00	0.41	35.72	m2	40.18
wedge and pin up at head 102mm wide	-	-	-	0.12	0.12	1.95	0.62	2.57	m	2.89
cut pockets and bond half brick filling to existing brickwork at jambs	-	-	-	0.50	0.50	8.13	1.03	9.15	m	10.30
For unit rates for finishes, etc see previous items										
Fill opening where door removed in one brick external wall with brickwork in gauged mortar (1:2:9) faced externally with picked stock facings pointed to match existing; remove old lintel and arch, provide lead cored damp proof course in opening lapped with existing, bond to existing brickwork at jambs and wedge and pin up at head; plaster filling one side and extend skirting; make good junction of new and existing plaster and skirting										
914 x 2082mm	171.50	-	171.50	19.65	18.85	313.70	25.70	510.90	nr	574.76
Unit rates for pricing the above and similar work										
remove old lintel and arch	-	-	-	1.46	1.64	24.99	2.04	27.03	m	30.41
one brick filling in common bricks	24.00	10.00	26.40	4.00	4.00	65.00	6.18	97.58	m2	109.78
one brick filling in common bricks faced one side with picked stock facings and point to match existing	67.20	10.00	73.92	5.00	5.00	81.25	6.18	161.35	m2	181.52
one brick filling in common bricks fair faced and flush smooth pointing one side	24.00	10.00	26.40	4.33	4.33	70.36	6.18	102.94	m2	115.81
two course slate damp proof course, width not exceeding 225mm	25.60	10.00	28.16	1.02	1.02	16.57	0.72	45.45	m2	51.14
lead cored bituminous hessian based damp proof course, B.S.6398, Class F, width not exceeding 225mm	20.28	10.00	22.31	0.80	0.80	13.00	0.31	35.62	m2	40.07
wedge and pin up at head 215mm wide	-	-	-	0.20	0.20	3.25	1.44	4.69	m	5.28
cut pockets and bond one brick filling to existing brickwork at jambs	-	-	-	1.00	1.00	16.25	2.37	18.62	m	20.95
12mm cement and sand (1:3) two coat external rendering over 300mm wide on filling	1.24	10.00	1.36	1.25	1.25	20.31	0.04	21.72	m2	24.43
For unit rates for internal finishes, etc. see previous items										
Fill opening where door removed in 275mm external hollow wall with inner skin in common bricks and outer skin in picked stock facings pointed to match existing, in gauged mortar (1:2:9); remove old lintel and arch, provide lead cored combined damp proof course and cavity gutter lapped with existing; form cavity with ties; bond to existing brickwork at jambs and wedge and pin up at head; plaster filling one side and extend skirting; make good junction of new and existing plaster and skirting										
914 x 2082mm	121.80	-	121.80	19.90	19.10	317.76	26.25	465.81	nr	524.04

Labour hourly rates: (except Specialists) Craft Operatives £9.23 Labourer £7.02 Rates are national average prices. Refer to REGIONAL VARIATIONS for indicative levels of overall pricing in regions	MATERIALS			LABOUR				RATES		
	Del to Site £	Waste %	Material Cost £	Craft Optve Hrs	Lab Hrs	Labour Cost £	Sunds £	Nett Rate £	Unit	Gross Rate (+12.5%) £
C90: ALTERATIONS - SPOT ITEMS Cont.										
Filling openings in brickwork or blockwork Cont.										
Unit rates for pricing the above and similar work										
half brick filling in common bricks	12.00	10.00	**13.20**	2.00	2.00	**32.50**	2.58	**48.28**	m2	**54.31**
half brick filling in common bricks and fair face and flush smooth pointing one side	12.00	10.00	**13.20**	2.33	2.33	**37.86**	2.78	**53.84**	m2	**60.57**
half brick filling in picked stock facings and point to match existing	40.80	10.00	**44.88**	2.95	2.95	**47.94**	2.78	**95.60**	m2	**107.55**
form 50mm wide cavity with B.S.1243 Fig. 3 galvanised steel twisted wall ties built in	0.60	10.00	**0.66**	0.20	0.20	**3.25**	-	**3.91**	m2	**4.40**
lead cored bituminous hessian based damp proof course and cavity tray, B.S.6398, Class F, width exceeding 225mm	20.28	10.00	**22.31**	0.80	0.80	**13.00**	0.31	**35.62**	m2	**40.07**
wedge and pin up at head 102mm wide	-	-	-	0.12	0.12	**1.95**	0.62	**2.57**	m	**2.89**
cut pockets and bond half brick wall in common bricks to existing brickwork at jambs	-	-	-	0.50	0.50	**8.13**	1.03	**9.15**	m	**10.30**
cut pockets and bond half brick wall in facing bricks to existing brickwork at jambs and make good facings	-	-	-	0.60	0.60	**9.75**	2.16	**11.91**	m	**13.40**
For unit rates for internal finishes, etc. see previous items										
Fillings openings in rubble walling										
Fill opening where window removed in 600mm rough rubble wall with rubble walling in lime mortar (1:3) pointed to match existing; bond to existing walling at jambs and wedge and pin up at head; plaster filling one side and make good junction of new and existing plaster										
900 x 900mm	119.90	-	**119.90**	12.00	12.00	**195.00**	15.90	**330.80**	nr	**372.15**
Unit rates for pricing the above and similar work										
rough rubble walling in filling										
600mm thick.....................	124.00	10.00	**136.40**	6.00	6.00	**97.50**	12.36	**246.26**	m2	**277.04**
random rubble walling in filling										
300mm thick.....................	68.20	10.00	**75.02**	5.80	5.80	**94.25**	6.18	**175.45**	m2	**197.38**
500mm thick.....................	113.00	10.00	**124.30**	10.90	10.90	**177.13**	10.30	**311.73**	m2	**350.69**
wedge and pin up at head										
300mm wide.....................	-	-	-	0.28	0.28	**4.55**	2.06	**6.61**	m	**7.44**
500mm wide.....................	-	-	-	0.47	0.47	**7.64**	3.50	**11.14**	m	**12.53**
600mm wide.....................	-	-	-	0.56	0.56	**9.10**	4.12	**13.22**	m	**14.87**
cut pockets and bond walling to existing at jambs										
300mm	5.15	10.00	**5.67**	0.83	0.83	**13.49**	0.62	**19.77**	m	**22.24**
500mm	8.25	10.00	**9.07**	1.38	1.38	**22.43**	0.93	**32.43**	m	**36.48**
600mm	9.35	10.00	**10.29**	1.75	1.75	**28.44**	1.13	**39.85**	m	**44.83**
For unit rates for internal finishes, etc. see previous items										
Work to chimney stacks										
SEAL AND VENTILATE TOP OF FLUE - remove chimney pot and flaunching and provide 50mm precast concrete, B.S.5328, designed mix C20, 20mm aggregate, weathered and throated capping and bed in cement mortar (1:3) to seal top of stack; cut opening through brick side of stack and build in 225 x 225mm clay air brick to ventilate flue and make good facings										
stack size 600 x 600mm with one flue	17.11	10.00	**18.82**	4.90	4.90	**79.63**	4.32	**102.77**	nr	**115.61**
stack size 825 x 600mm with two flues	28.84	10.00	**31.72**	9.00	9.00	**146.25**	7.96	**185.93**	nr	**209.18**
stack size 1050 x 600mm with three flues	40.57	10.00	**44.63**	13.10	13.10	**212.88**	10.91	**268.41**	nr	**301.96**
RENEW DEFECTIVE CHIMNEY POT - remove chimney pot and provide new clay chimney pot set and flaunched in cement mortar (1:3)										
150mm diameter x 600mm high	28.97	-	**28.97**	2.50	2.50	**40.63**	3.91	**73.50**	nr	**82.69**
REBUILD DEFECTIVE CHIMNEY STACK - pull down defective stack to below roof level for a height of 2.00m; prepare for raising and rebuild in common brickwork faced with picked stock facings in gauged mortar (1:2:9) pointed to match existing; parge and core flues; provide No. 4 lead flashings and soakers; provide 150mm diameter clay chimney pots 600mm high set and flaunched in cement mortar (1:3); make good roof tiling and all other work disturbed										
600 x 600mm with one flue	370.00	-	**370.00**	35.30	37.55	**589.42**	59.00	**1018.42**	nr	**1145.72**
825 x 600mm with two flues	476.00	-	**476.00**	47.15	47.90	**771.45**	82.00	**1329.45**	nr	**1495.63**
REMOVE DEFECTIVE CHIMNEY STACK - pull down stack to below roof level; remove all flashings and soakers; piece in 50 x 100mm rafters; extend roof tiling with machine made plain tiles to 100mm gauge nailed every fourth course with galvanised nails to and with 25 x 19mm battens; bond new tiling to existing										
850 x 600mm for a height of 2.00m	34.90	-	**34.90**	26.00	22.30	**396.53**	36.00	**467.43**	nr	**525.85**

Labour hourly rates: (except Specialists) Craft Operatives £9.23 Labourer £7.02 Rates are national average prices. Refer to REGIONAL VARIATIONS for indicative levels of overall pricing in regions	MATERIALS			LABOUR				RATES		
	Del to Site	Waste	Material Cost	Craft Optve	Lab	Labour Cost	Sunds	Nett Rate	Unit	Gross Rate (+12.5%)
	£	%	£	Hrs	Hrs	£	£	£		£
C90: ALTERATIONS - SPOT ITEMS Cont.										
Work to chimney breasts and fireplaces										
TAKE OUT FIREPLACE AND FILL IN OPENING - remove fire surround, fire back and hearth; hack up screed to hearth and extend 25mm tongued and grooved softwood floor boarding over hearth on and with bearers; fill in opening where fireplace removed with 75mm concrete blocks in gauged mortar (1:2:9) bonded to existing at jambs and wedged and pinned up to soffits; form 225 x 225mm opening and provide and build in 225 x 225mm fibrous plaster louvered air vent; plaster filling with 12mm two coat lightweight plaster on 12mm (average) dubbing plaster; extend 19 x 100mm softwood chamfered skirting over filling and join with existing; make good all new finishings to existing										
fireplace opening 570 x 685mm, hearth 1030 x 405mm	35.00	-	35.00	8.00	7.45	126.14	10.30	171.44	nr	192.87
fireplace opening 800 x 760mm, hearth 1260 x 405mm	42.00	-	42.00	9.65	8.95	151.90	12.00	205.90	nr	231.64
REMOVE CHIMNEY BREAST - pull down chimney breast for full height from ground to roof level (two storeys) including removing two fire surrounds and hearths complete; make out brickwork where flues removed and make out brickwork where breasts removed; extend 50 x 175mm floor joists and 25mm tongued and grooved softwood floor boarding (ground and first floors); extend 50 x 100mm ceiling joists (first floor); plaster walls with 12mm two coat lightweight plaster on 12mm (average) plaster dubbing; extend ceiling plaster with expanded metal lathing and 19mm three coat lightweight plaster; run 19 x 100mm softwood chamfered skirting to walls to join with existing; make good all new finishings up to existing										
chimney breast 2.00m wide and 7.50m high including gathering in roof space	145.00	-	145.00	62.00	84.00	1161.94	144.00	1450.94	nr	1632.31
Form trap door in ceiling										
Form new trap door in ceiling with 100mm deep joists and lath and plaster finish; cut away and trim ceiling joists around opening and insert new trimming and trimmer joists; provide 32x 127mm softwood lining with 15 x 25mm stop and 20 x 70mm architrave; provide and place in position 18mm blockboard trap door, lipped all round; make good existing ceiling plaster around openings										
600 x 900mm	39.40	-	39.40	21.85	5.60	240.99	9.10	289.49	nr	325.67
Unit rates for pricing the above and similar work										
cut away plastered plasterboard ceiling	-	-	-	1.50	0.33	16.16	0.85	17.01	m2	19.14
cut away lath and plaster ceiling	-	-	-	2.00	0.41	21.34	1.02	22.36	m2	25.15
cut away and trim existing 100mm deep joists around opening										
664 x 964mm................................	-	-	-	0.95	0.16	9.89	0.71	10.60	nr	11.93
964 x 1444mm...............................	-	-	-	1.90	0.33	19.85	1.43	21.28	nr	23.94
cut away and trim existing 150mm deep joists around opening										
664 x 964mm................................	-	-	-	1.43	0.24	14.88	1.09	15.97	nr	17.97
964 x 1444mm...............................	-	-	-	2.86	0.48	29.77	2.14	31.91	nr	35.90
softwood trimming or trimmer joists										
75 x 100mm.................................	2.46	10.00	2.71	0.95	0.16	9.89	0.36	12.96	m	14.58
75 x 150mm.................................	3.68	10.00	4.05	1.00	0.17	10.42	0.39	14.86	m	16.72
softwood lining										
32 x 127mm.................................	2.10	10.00	2.31	1.50	0.25	15.60	0.11	18.02	m	20.27
32 x 150mm.................................	2.18	10.00	2.40	1.60	0.27	16.66	0.08	19.14	m	21.53
15 x 25mm softwood stop....................	0.45	10.00	0.50	0.33	0.05	3.40	0.05	3.94	m	4.43
19 x 50mm softwood twice rounded architrave to Patt. 18.................................	0.59	10.00	0.65	0.33	0.05	3.40	0.05	4.10	m	4.61
18mm blockboard trap door lipped all round										
600 x 900mm...............................	11.03	10.00	12.13	3.00	0.50	31.20	0.17	43.50	nr	48.94
900 x 1280mm..............................	25.24	10.00	27.76	6.00	1.00	62.40	0.30	90.46	nr	101.77
75mm steel butt hinge to softwood............	0.60	-	0.60	0.66	0.11	6.86	0.17	7.63	nr	8.59
make good plastered plasterboard ceiling around opening...............................	-	-	-	0.75	0.75	12.19	1.23	13.42	m	15.09
make good lath and plaster ceiling around opening...............................	-	-	-	0.83	0.83	13.49	1.51	15.00	m	16.87
Openings through stud partitions										
Form opening for door frame through lath and plaster finished 50 x 100mm stud partition; take off skirtings, cut away plaster, cut away and trim studding around opening and insert new studding at head and jambs of opening; make good existing lath and plaster up to new frame both sides and form fitted ends on existing skirtings up to new frame; extend softwood board flooring through opening on and with bearers and make good up to existing										
838 x 2032mm..............................	11.80	-	11.80	18.10	6.10	209.88	16.10	237.79	Unr	267.51
Unit rates for pricing the above and similar work										
cut away plastered plasterboard	-	-	-	1.00	0.25	10.98	0.85	11.84	m2	13.31
cut away lath and plaster	-	-	-	1.10	0.26	11.98	1.02	13.00	m2	14.62

Labour hourly rates: (except Specialists) Craft Operatives £9.23 Labourer £7.02 Rates are national average prices. Refer to REGIONAL VARIATIONS for indicative levels of overall pricing in regions	MATERIALS			LABOUR				RATES		
	Del to Site	Waste	Material Cost	Craft Optve	Lab	Labour Cost	Sunds	Nett Rate	Unit	Gross Rate (+12.5%)
	£	%	£	Hrs	Hrs	£	£	£		£
C90: ALTERATIONS - SPOT ITEMS Cont.										
Openings through stud partitions Cont.										
Unit rates for pricing the above and similar work Cont.										
cut away and trim existing 75mm studding around opening										
838 x 2032mm	-	-	-	2.05	0.36	21.45	2.18	23.63	nr	26.58
914 x 2032mm	-	-	-	2.05	0.36	21.45	2.18	23.63	nr	26.58
cut away and trim existing 100mm studding around opening										
838 x 2032mm	-	-	-	2.55	0.45	26.70	2.85	29.55	nr	33.24
914 x 2032mm	-	-	-	2.55	0.45	26.70	2.85	29.55	nr	33.24
50 x 75mm softwood studding	1.10	10.00	1.21	0.50	0.08	5.18	0.11	6.50	m	7.31
50 x 100mm softwood studding	1.46	10.00	1.61	0.60	0.10	6.24	0.12	7.97	m	8.96
make good plastered plasterboard around opening	-	-	-	0.75	0.75	12.19	1.23	13.42	m	15.09
make good lath and plaster around opening	-	-	-	0.83	0.83	13.49	1.51	15.00	m	16.87
take off skirting	-	-	-	0.21	-	1.94	0.18	2.12	m	2.38
form fitted end on existing skirting	-	-	-	0.57	-	5.26	-	5.26	nr	5.92
extend 25mm softwood tongued and grooved board flooring through opening on and with bearers	3.42	10.00	3.76	1.50	0.25	15.60	0.06	19.42	nr	21.85
Filling openings in stud partitions										
Fill opening where door removed in stud partition with 50 x 100mm studding and sole plate, covered on both sides with 9.5mm plasterboard baseboard with 5mm one coat board finish plaster; extend skirting on both sides; make good junction of new and existing plaster and skirtings										
838 x 2032mm	31.10	-	31.10	16.00	5.36	185.31	3.00	219.41	nr	246.83
Unit rates for pricing the above and similar work										
50 x 75mm softwood sole plate	1.10	10.00	1.21	0.50	0.08	5.18	0.11	6.50	m	7.31
50 x 100mm softwood sole plate	1.46	10.00	1.61	0.60	0.10	6.24	0.14	7.99	m	8.98
50 x 75mm softwood studding	1.10	10.00	1.21	0.50	0.08	5.18	0.11	6.50	m	7.31
50 x 100mm softwood studding	1.46	10.00	1.61	0.60	0.10	6.24	0.14	7.99	m	8.98
9.5mm plasterboard baseboard to filling	1.83	10.00	2.01	0.50	0.10	5.32	0.28	7.61	m2	8.56
9.5mm plasterboard baseboard to filling including packing out not exceeding 10mm	1.83	10.00	2.01	1.00	0.17	10.42	0.37	12.81	m2	14.41
5mm one coat board finish plaster on plasterboard to filling	0.96	10.00	1.06	0.91	0.91	14.79	0.03	15.87	m2	17.86
10mm two coat hardwall plaster on plasterboard to filling	1.46	10.00	1.61	1.27	1.27	20.64	0.03	22.27	m2	25.06
19 x 100mm softwood chamfered skirting, primed all round	1.24	10.00	1.36	0.54	0.09	5.62	0.18	7.16	m	8.05
junction with existing	-	-	-	0.57	-	5.26	-	5.26	nr	5.92
25 x 150mm softwood chamfered skirting, primed all round	2.29	10.00	2.52	0.55	0.09	5.71	0.24	8.47	m	9.53
junction with existing	-	-	-	0.57	-	5.26	-	5.26	nr	5.92
Work to existing drains										
BREAK UP CONCRETE PAVING FOR EXCAVATING DRAIN TRENCH AND REINSTATE - break up 100mm concrete paving and 150mm hardcore bed under for excavation of drain trench and reinstate with new hardcore and concrete, B.S.5328, designed mix C20, 20mm aggregate and make good up to existing paving both sides										
400mm wide	4.90	10.00	5.39	-	3.27	22.96	3.40	31.75	m	35.71
TAKE UP FLAG PAVING FOR EXCAVATING DRAIN TRENCH AND REINSTATE - carefully take up 50mm precast concrete flag paving and set aside for re-use and break up 150mm hardcore bed under for excavation of drain trench and reinstate with new hardcore and re-lay salvaged flag paving on and with 25mm sand bed and grout in lime sand (1:3) and make good up to existing paving on both sides										
400mm wide	1.95	10.00	2.15	1.47	1.47	23.89	3.13	29.16	m	32.81
INSERT NEW JUNCTION IN EXISTING DRAIN -excavate for and trace and expose existing drain, break into glazed vitrified clay drain and insert junction with 100mm branch, short length of new pipe and double collar and joint to existing drain; support earthwork, make good concrete bed and haunching and backfill										
existing 100mm diameter drain, invert depth										
600mm	21.52	10.00	23.67	3.00	12.00	111.93	8.65	144.25	nr	162.28
750mm	21.52	10.00	23.67	3.00	14.25	127.72	9.35	160.75	nr	180.84
900mm	21.52	10.00	23.67	3.00	16.50	143.52	10.25	177.44	nr	199.62
existing 150mm diameter drain invert depth										
750mm	35.05	10.00	38.56	4.50	15.75	152.10	10.80	201.46	nr	226.64
900mm	35.05	10.00	38.56	4.50	18.00	167.90	12.25	218.70	nr	246.04
1050mm	35.05	10.00	38.56	4.50	20.25	183.69	13.40	235.65	nr	265.10
1200mm	35.05	10.00	38.56	4.50	22.50	199.49	14.65	252.69	nr	284.28

Labour hourly rates: (except Specialists) Craft Operatives £9.23 Labourer £7.02 Rates are national average prices. Refer to REGIONAL VARIATIONS for indicative levels of overall pricing in regions	MATERIALS			LABOUR				RATES		
	Del to Site	Waste	Material Cost	Craft Optve	Lab	Labour Cost	Sunds	Nett Rate	Unit	Gross Rate (+12.5%)
	£	%	£	Hrs	Hrs	£	£	£		£
C90: ALTERATIONS - SPOT ITEMS Cont.										
Work to existing drains Cont.										
REPAIR DEFECTIVE DRAIN - excavate for and trace and expose existing drain, break out fractured glazed vitrified clay drain pipe, replace with new pipe and double collars and joint to existing drain; support earthwork, make good concrete bed and haunching and backfill										
existing 100mm diameter drain; single pipe length; invert depth										
450mm	21.03	10.00	23.13	3.00	11.50	108.42	6.55	138.10	nr	155.37
600mm	21.03	10.00	23.13	3.00	13.50	122.46	7.90	153.49	nr	172.68
750mm	21.03	10.00	23.13	3.00	16.00	140.01	9.65	172.79	nr	194.39
add for each additional pipe length; invert depth										
450mm	11.74	10.00	12.91	2.00	6.75	65.84	6.25	85.01	nr	95.64
600mm	11.74	10.00	12.91	2.00	7.60	71.81	7.55	92.28	nr	103.81
750mm	11.74	10.00	12.91	2.00	8.80	80.24	9.05	102.20	nr	114.97
Work to existing manholes										
RAISE TOP OF EXISTING MANHOLE - take off cover and frame and set aside for re-use, prepare level bed on existing one brick sides of manhole internal size 610 x 457mm and raise with Class B engineering brickwork in cement mortar (1:3) finished with a fair face and flush pointed; refix salvaged cover and frame, bed frame in cement mortar and cover in grease and sand										
raising 150mm high	16.74	10.00	18.41	5.30	5.30	86.13	3.17	107.71	nr	121.17
raising 225mm high	25.11	10.00	27.62	6.75	6.75	109.69	4.70	142.01	nr	159.76
raising 300mm high	33.48	10.00	36.83	8.50	8.50	138.13	6.02	180.97	nr	203.59
INSERT NEW BRANCH BEND IN BOTTOM OF EXISTING MANHOLE - break into bottom and one brick side of manhole, insert new glazed vitrified clay three quarter section branch bend to discharge over existing main channel, build in end of new drain and make good benching and side of manhole to match existing										
100mm diameter branch bend	7.66	10.00	8.43	5.75	5.75	93.44	2.27	104.13	nr	117.15
EXTEND EXISTING MANHOLE - take off cover and frame and set aside; break up one brick end wall of 457mm wide (internal) manhole and excavate for and extend manhole 225mm; support earthwork, level and compact bottom, part backfill and remove surplus spoil from site; extend bottom with 150mm concrete, BS 5328, ordinary prescribed mix C15P, 20mmaggregate bed; extend one brick sides and end with Class B engineering brickwork in cement mortar (1:3) fair faced and flush pointed and bond to existing; extend 100mm main channel and insert 100mm three quarter section branch channel bend; extend benching and build in end of new 100mm drain to one brick side, corbel over end of manhole										
manholes of the following invert depths; refix salvaged cover and frame, bed frame in cement mortar and cover in grease and sand										
450mm	42.26	10.00	46.49	17.60	17.90	288.11	8.76	343.35	nr	386.27
600mm	47.37	10.00	52.11	19.80	20.20	324.56	11.26	387.93	nr	436.42
750mm	52.49	10.00	57.74	22.00	22.50	361.01	13.76	432.51	nr	486.57
900mm	57.60	10.00	63.36	24.20	24.80	397.46	16.26	477.08	nr	536.72
Temporary weatherproof coverings										
Providing and erecting and clearing away on completion; for scaffold tube framework see preliminaries										
corrugated iron sheeting	7.40	10.00	8.14	0.15	0.50	4.89	0.79	13.82	m2	15.55
flexible reinforced plastic sheeting	2.73	10.00	3.00	0.10	0.25	2.68	0.64	6.32	m2	7.11
tarpaulins	4.10	10.00	4.51	0.10	0.25	2.68	0.64	7.83	m2	8.81
Temporary internal screens										
Providing and erecting and clearing away on completion; temporary screen of 50 x 50mm softwood framing and cover with										
reinforced building paper	2.54	10.00	2.79	0.43	0.30	6.07	0.22	9.09	m2	10.23
heavy duty polythene sheeting	1.78	10.00	1.96	0.43	0.30	6.07	0.20	8.23	m2	9.26
3mm hardboard	2.65	10.00	2.92	0.53	0.32	7.14	0.33	10.38	m2	11.68
Providing and erecting and clearing away on completion; temporary dustproof screen of 50 x 75mm softwood framing securely fixed to walls, floor and ceiling, the joints and edges of lining sealed with masking tape and cover with										
3mm hardboard lining one side	3.45	10.00	3.79	1.10	0.18	11.42	0.44	15.65	m2	17.61
3mm hardboard lining both sides	4.30	10.00	4.73	1.30	0.22	13.54	0.55	18.82	m2	21.18
6mm plywood lining one side	7.29	10.00	8.02	1.15	0.19	11.95	0.49	20.46	m2	23.01
6mm plywood lining both sides	12.38	10.00	13.62	1.40	0.23	14.54	0.77	28.92	m2	32.54
providing 35mm hardboard faced flush door size 838 x 1981mm in dustproof screen with 25 x 87mm softwood frame, 18 x 25mm softwood stop, pair of 100mm butts, pull handle and ball catch	135.00	10.00	148.50	1.50	0.25	15.60	2.78	166.88	nr	187.74

Labour hourly rates: (except Specialists) Craft Operatives £9.23 Labourer £7.02 Rates are national average prices. Refer to REGIONAL VARIATIONS for indicative levels of overall pricing in regions	MATERIALS			LABOUR				RATES		
	Del to Site	Waste	Material Cost	Craft Optve	Lab	Labour Cost	Sunds	Nett Rate	Unit	Gross Rate (+12.5%)
	£	%	£	Hrs	Hrs	£	£	£		£

C90: ALTERATIONS - SPOT ITEMS Cont.

Temporary dustproof corridors

Note
for prices of walls and doors, see dustproof
screens above
Providing and erecting and clearing away on
completion; ceiling of 50 x 75mm softwood joists at
450mm centres, the joints and edges sealed with
masking tape and cover with

3mm hardboard lining to soffit	3.69	10.00	4.06	1.10	0.18	11.42	0.44	15.92	m2	17.91
6mm plywood lining to soffit	7.53	10.00	8.28	1.15	0.19	11.95	0.55	20.78	m2	23.38

Temporary timber balustrades

Providing and erecting and clearing away on
completion; temporary softwood balustrade consisting
of 50 x 75mm plate, 50 x 50mm standards at 900mm
centres, four 25 x 150mm intermediate rails and 50 x
50mm handrail

1150mm high	6.83	10.00	7.51	1.00	0.26	11.06	1.26	19.83	m	22.31

Temporary steel balustrades

Providing and erecting and clearing away on
completion; temporary steel balustrade constructed of
50mm diameter galvanised scaffold tubing and fittings
with standards at 900mm centres intermediate rail,
handrail, plastic mesh infill and with plates on ends
of standards fixed to floor

1150mm high	6.50	-	6.50	0.18	0.90	7.98	0.30	14.78	m	16.63
weekly cost of hire and maintenance	0.65	-	0.65	0.02	0.10	0.89	0.04	1.58	m	1.77

Temporary fillings to openings in external walls

Providing and erecting and clearing away on
completion; temporary filling to window or door
opening of 50 x 100mm softwood framing and cover with

corrugated iron sheeting	16.05	15.00	18.46	0.45	0.45	7.31	1.31	27.08	m2	30.47
25mm softwood boarding	12.42	12.50	13.97	0.60	0.60	9.75	1.53	25.25	m2	28.41
12mm external quality plywood	12.40	10.00	13.64	0.50	0.50	8.13	0.99	22.75	m2	25.60

REPAIRING/RENOVATING ASPHALT WORK

Cut out crack in old covering and make good with new material to match existing

Floor or tanking

20mm thick in two coats, horizontal	-	-	-	0.66	0.66	10.73	0.50	11.23	m	12.63
30mm thick in three coats, horizontal	-	-	-	0.90	0.90	14.63	0.75	15.38	m	17.30
30mm thick in three coats, vertical	-	-	-	1.19	1.19	19.34	0.75	20.09	m	22.60

Cut out detached blister in old covering and make good with new material to match existing

Floor or tanking

20mm thick in two coats, horizontal	-	-	-	0.30	0.30	4.88	0.25	5.13	nr	5.77
30mm thick in three coats, horizontal	-	-	-	0.35	0.35	5.69	0.35	6.04	nr	6.79
30mm thick in three coats, vertical	-	-	-	0.48	0.48	7.80	0.35	8.15	nr	9.17

Cut out crack in old covering and make good with new material to match existing

Roof covering

20mm thick in two coats	-	-	-	0.66	0.66	10.73	0.50	11.23	m	12.63
Cut out detached blister in old covering and make good with new material to match existing										
20mm thick in two coats.....................	-	-	-	0.30	0.30	4.88	0.25	5.13	nr	5.77

REPAIRING/RENOVATING ROOF COVERINGS

Slate roofing repairs to roofs covered with 610 x 305mm slates

Remove damaged slates and replace with new; sloping
or vertical

one slate	6.32	5.00	6.64	0.32	0.32	5.20	0.09	11.93	nr	13.42
patch of 10 slates	63.20	5.00	66.36	1.10	1.10	17.88	0.70	84.94	nr	95.55
patch of slates 1.00 - 3.00m2	77.42	5.00	81.29	1.25	1.25	20.31	0.76	102.36	m2	115.16

Examine battens, remove defective and provide 20% new

19 x 38mm	0.21	10.00	0.23	0.09	0.09	1.46	0.05	1.74	m2	1.96
25 x 50mm	0.38	10.00	0.42	0.09	0.09	1.46	0.07	1.95	m2	2.19

Re-cover roof with slates previously removed and
stacked and fix with slate nails

with 75mm lap	-	-	-	0.77	0.77	12.51	0.09	12.60	m2	14.18
extra for providing 20% new slates	15.48	5.00	16.25	-	-	-	0.18	16.43	m2	18.49

Remove double course at eaves and

refix.....................................	-	-	-	0.24	0.24	3.90	0.31	4.21	m	4.74
extra for providing 20% new slates	5.25	5.00	5.51	-	-	-	0.09	5.60	m	6.30

Labour hourly rates: (except Specialists) Craft Operatives £9.23 Labourer £7.02 Rates are national average prices. Refer to REGIONAL VARIATIONS for indicative levels of overall pricing in regions	MATERIALS			LABOUR				RATES		
	Del to Site £	Waste %	Material Cost £	Craft Optve Hrs	Lab Hrs	Labour Cost £	Sunds £	Nett Rate £	Unit	Gross Rate (+12.5%) £
REPAIRING/RENOVATING ROOF COVERINGS Cont.										
Slate roofing repairs to roofs covered with 610 x 305mm slates Cont.										
Remove double course at verge and										
replace with new slates and bed and point in										
mortar..	23.64	5.00	24.82	0.29	0.29	4.71	0.58	30.11	m	33.88
Slate roofing repairs for roofs covered with 510 x 255mm slates										
Remove damaged slates and replace with new; sloping or vertical										
one slate...................................	3.52	5.00	3.70	0.32	0.32	5.20	0.09	8.99	nr	10.11
patch of 10 slates.......................	35.20	5.00	36.96	1.00	1.00	16.25	0.59	53.80	nr	60.52
patch of slates 1.00 - 3.00m2	63.36	5.00	66.53	1.49	1.49	24.21	0.75	91.49	m2	102.93
Examine battens, remove defective and provide 20% new										
19 x 38mm.................................	0.24	10.00	0.26	0.11	0.11	1.79	0.07	2.12	m2	2.39
25 x 50mm.................................	0.45	10.00	0.50	0.11	0.11	1.79	0.09	2.37	m2	2.67
Re-cover roof with slates previously removed and stacked and fix with slate nails										
with 75mm lap...........................	-	-	-	1.00	1.00	16.25	0.13	16.38	m2	18.43
extra for providing 20% new slates	12.67	5.00	13.30	-	-	-	0.18	13.48	m2	15.17
Remove double course at eaves and										
refix......................................	-	-	-	0.25	0.25	4.06	0.53	4.59	m	5.17
extra for providing 20% new slates	3.94	5.00	4.14	-	-	-	0.08	4.22	m	4.74
Remove double course at verge and										
replace with new slates and bed and point in										
mortar....................................	16.29	5.00	17.10	0.29	0.29	4.71	0.57	22.39	m	25.19
Slate roofing repairs for roofs covered with 405 x 205mm slates										
Remove damaged slates and replace with new; sloping or vertical										
one slate...................................	1.76	5.00	1.85	0.32	0.32	5.20	0.08	7.13	nr	8.02
patch of 10 slates.......................	17.60	5.00	18.48	0.90	0.90	14.63	0.51	33.62	nr	37.82
patch of slates 1.00 - 3.00m2	52.80	5.00	55.44	2.07	2.07	33.64	0.83	89.91	m2	101.15
Examine battens, remove defective and provide 20% new										
19 x 38mm.................................	0.32	10.00	0.35	0.13	0.13	2.11	0.08	2.54	m2	2.86
25 x 50mm.................................	0.60	10.00	0.66	0.13	0.13	2.11	0.09	2.86	m2	3.22
Re-cover roof with slates previously removed and stacked and fix with slate nails										
with 75mm lap...........................	-	-	-	1.14	1.14	18.52	0.19	18.72	m2	21.05
extra for providing 20% new slates	10.56	5.00	11.09	-	-	-	0.18	11.27	m2	12.68
Remove double course at eaves and										
refix......................................	-	-	-	0.26	0.26	4.22	0.52	4.75	m	5.34
extra for providing 20% new slates	2.39	5.00	2.51	-	-	-	0.08	2.59	m	2.91
Remove double course at verge and										
replace with new slates and bed and point in										
mortar....................................	10.26	5.00	10.77	0.30	0.30	4.88	0.63	16.28	m	18.31
Tile roofing repairs for roofs covered with machine made clay plain tiles laid to 100mm gauge										
Remove damaged tiles and replace with new, sloping or vertical										
one tile....................................	0.49	10.00	0.54	0.32	0.32	5.20	0.03	5.77	nr	6.49
patch of 10 tiles.........................	4.94	10.00	5.43	0.90	0.90	14.63	0.25	20.31	nr	22.85
patch of tiles 1.00 - 3.00m2	31.12	10.00	34.23	1.72	1.72	27.95	1.60	63.78	m2	71.75
Examine battens, remove defective and provide 20% new										
19 x 25mm.................................	0.38	10.00	0.42	0.13	0.13	2.11	0.08	2.61	m2	2.94
19 x 38mm.................................	0.51	10.00	0.56	0.14	0.14	2.27	0.09	2.93	m2	3.29
19 x 50mm.................................	0.83	10.00	0.91	0.14	0.14	2.27	0.11	3.30	m2	3.71
Re-cover roof with tiles previously removed and stacked										
nail every fourth course with galvanised nails ...	-	-	-	1.00	1.00	16.25	0.17	16.42	m2	18.47
extra for nailing every course	-	-	-	0.20	0.20	3.25	0.45	3.70	m2	4.16
extra for providing 20% new tiles	6.22	10.00	6.84	-	-	-	0.18	7.02	m2	7.90
Tile roofing repairs to roofs covered with machine made clay plain tiles laid to 90mm gauge										
Remove damaged tiles and replace with new, sloping or vertical										
one tile....................................	0.49	10.00	0.54	0.32	0.32	5.20	0.03	5.77	nr	6.49
patch of 10 tiles.........................	4.94	10.00	5.43	0.90	0.90	14.63	0.25	20.31	nr	22.85
patch of tiles 1.00 - 3.00m2	34.58	10.00	38.04	2.05	2.05	33.31	1.78	73.13	m2	82.27
Examine battens, remove defective and provide 20% new										
19 x 25mm.................................	0.44	10.00	0.48	0.14	0.14	2.27	0.09	2.85	m2	3.21
19 x 38mm.................................	0.54	10.00	0.59	0.15	0.15	2.44	0.11	3.14	m2	3.53
19 x 50mm.................................	0.93	10.00	1.02	0.15	0.15	2.44	0.12	3.58	m2	4.03

Labour hourly rates: (except Specialists) Craft Operatives £9.23 Labourer £7.02 Rates are national average prices. Refer to REGIONAL VARIATIONS for indicative levels of overall pricing in regions	MATERIALS			LABOUR				RATES		
	Del to Site	Waste	Material Cost	Craft Optve	Lab	Labour Cost	Sunds	Nett Rate	Unit	Gross Rate (+12.5%)
	£	%	£	Hrs	Hrs	£	£	£		£
REPAIRING/RENOVATING ROOF COVERINGS Cont.										
Tile roofing repairs to roofs covered with machine made clay plain tiles laid to 90mm gauge Cont.										
Re-cover roof with tiles previously removed and stacked										
nail every fourth course with galvanised nails ...	-	-	-	1.15	1.15	18.69	0.19	18.88	m2	21.24
extra for nailing every course	-	-	-	0.23	0.23	3.74	0.47	4.21	m2	4.73
extra for providing 20% new tiles	6.92	10.00	7.61	-	-	-	0.18	7.79	m2	8.77
Tile roofing repairs for roofs covered with 381 x 227mm concrete interlocking tiles laid to 75mm laps										
Remove damaged tiles and replace with new, sloping or vertical										
one tile....................................	0.51	10.00	0.56	0.32	0.32	5.20	0.07	5.83	nr	6.56
patch of 10 tiles	5.12	10.00	5.63	0.90	0.90	14.63	0.73	20.99	nr	23.61
patch of tiles 1.00 - 3.00m2	7.17	10.00	7.89	0.85	0.85	13.81	1.16	22.86	m2	25.72
Examine battens, remove defective and provide 20% new										
19 x 25mm	0.13	10.00	0.14	0.05	0.05	0.81	0.07	1.03	m2	1.15
19 x 38mm	0.18	10.00	0.20	0.06	0.06	0.97	0.07	1.24	m2	1.40
19 x 50mm	0.25	10.00	0.28	0.07	0.07	1.14	0.08	1.49	m2	1.68
Re-cover roof with tiles previously removed and stacked										
nail every fourth course with galvanised nails ...	-	-	-	0.50	0.50	8.13	0.02	8.14	m2	9.16
extra for nailing every course	-	-	-	0.05	0.05	0.81	0.06	0.87	m2	0.98
extra for providing 20% new tiles	1.44	10.00	1.58	-	-	-	0.22	1.80	m2	2.03
Tile roofing repairs generally										
Remove damaged clay tiles and provide new										
half round ridge or hip and bed in mortar	11.00	10.00	12.10	0.60	0.60	9.75	1.09	22.94	m	25.81
bonnet hip	15.50	10.00	17.05	0.75	0.75	12.19	0.30	29.54	m	33.23
trough valley	15.50	10.00	17.05	0.75	0.75	12.19	0.30	29.54	m	33.23
vertical angle	15.50	10.00	17.05	0.75	0.75	12.19	0.30	29.54	m	33.23
Remove double course and replace with new clay tiles and bed and point in mortar										
at eaves	5.93	10.00	6.52	0.23	0.23	3.74	0.45	10.71	m	12.05
at verge	12.35	10.00	13.59	0.38	0.38	6.17	1.39	21.15	m	23.79
Rake out defective pointing and re-point in cement mortar (1:3)										
ridge or hip tiles	-	-	-	0.34	0.34	5.53	0.17	5.70	m	6.41
Hack off defective cement mortar fillet and renew in cement mortar (1:3)	-	-	-	0.34	0.34	5.53	0.34	5.87	m	6.60
Remove defective hip hook and replace with new	1.14	-	1.14	0.34	0.34	5.53	0.07	6.74	nr	7.58
Corrugated fibre cement roofing repairs										
Remove damaged sheets and provide and fix new with screws and washers to timber purlins and ease and adjust edges of adjoining sheets as necessary										
one sheet; 1825mm long Profile 3 natural grey	14.07	10.00	15.48	1.23	1.23	19.99	1.99	37.45	nr	42.14
one sheet; 1825mm long Profile 3 standard colour .	16.88	10.00	18.57	1.23	1.23	19.99	1.99	40.55	nr	45.61
one sheet; 1825mm long Profile 6 natural grey	20.59	10.00	22.65	1.78	1.78	28.93	2.94	54.51	nr	61.33
one sheet; 1825mm long Profile 6 standard colour .	24.71	10.00	27.18	1.78	1.78	28.93	2.94	59.05	nr	66.43
over one sheet; Profile 3 natural grey	13.27	10.00	14.60	0.89	0.89	14.46	1.88	30.94	m2	34.81
over one sheet; Profile 3 standard colour	15.92	10.00	17.51	0.89	0.89	14.46	1.88	33.85	m2	38.09
over one sheet; Profile 6 natural grey	19.42	10.00	21.36	1.29	1.29	20.96	2.77	45.09	m2	50.73
over one sheet; Profile 6 standard colour	23.30	10.00	25.63	1.29	1.29	20.96	2.77	49.36	m2	55.53
Remove damaged sheets and provide and fix new with hook bolts and washers to steel purlins and ease and adjust edges of adjoining sheets as necessary										
one sheet; 1825mm long Profile 3 natural grey	14.07	10.00	15.48	1.37	1.37	22.26	1.99	39.73	nr	44.70
one sheet; 1825mm long Profile 3 standard colour .	16.88	10.00	18.57	1.37	1.37	22.26	1.99	42.82	nr	48.17
one sheet; 1825mm long Profile 6 natural grey	20.59	10.00	22.65	1.98	1.98	32.17	2.94	57.76	nr	64.98
one sheet; 1825mm long Profile 6 standard colour .	24.71	10.00	27.18	1.98	1.98	32.17	2.94	62.30	nr	70.08
over one sheet; Profile 3 natural grey	13.27	10.00	14.60	1.03	1.03	16.74	1.88	33.21	m2	37.37
over one sheet; Profile 3 standard colour	15.92	10.00	17.51	1.03	1.03	16.74	1.88	36.13	m2	40.65
over one sheet; Profile 6 natural grey	19.42	10.00	21.36	1.49	1.49	24.21	2.77	48.34	m2	54.39
over one sheet; Profile 6 standard colour	23.30	10.00	25.63	1.49	1.49	24.21	2.77	52.61	m2	59.19
Felt roofing repairs for roofs covered with bituminous felt roofing, B.S.747										
Sweep clean, cut out defective layer, re-bond adjacent felt and cover with one layer of felt bonded in hot bitumen; patches not exceeding 1.00m2										
fibre based roofing felt type 1B	1.42	15.00	1.63	0.88	0.88	14.30	2.38	18.31	m2	20.60
for each defective underlayer cut out and replaced, add	1.42	15.00	1.63	0.88	0.88	14.30	2.38	18.31	m2	20.60
fibre based mineral surfaced roofing felt type 1E	1.72	15.00	1.98	1.05	1.05	17.06	2.38	21.42	m2	24.10
high performance polyester based roofing felt type 5B	3.38	15.00	3.89	1.02	1.02	16.57	2.38	22.84	m2	25.70
for each defective underlayer cut out and replaced, add	3.38	15.00	3.89	1.02	1.02	16.57	2.38	22.84	m2	25.70
high performance polyester based mineral surfaced roofing felt type 5E	3.90	15.00	4.49	1.13	1.13	18.36	2.38	25.23	m2	28.38

Labour hourly rates: (except Specialists) Craft Operatives £9.23 Labourer £7.02 Rates are national average prices. Refer to REGIONAL VARIATIONS for indicative levels of overall pricing in regions	MATERIALS			LABOUR				RATES		
	Del to Site £	Waste %	Material Cost £	Craft Optve Hrs	Lab Hrs	Labour Cost £	Sunds £	Nett Rate £	Unit	Gross Rate (+12.5%) £

REPAIRING/RENOVATING ROOF COVERINGS Cont.

Felt roofing repairs for roofs covered with bituminous felt roofing, B.S.747 Cont.

Sweep clean, cut out defective layer, re-bond adjacent felt and cover with one layer of felt bonded in hot bitumen; patches 1.00 - 3.00m2

	Del to Site	Waste	Material Cost	Craft Optve	Lab	Labour Cost	Sunds	Nett Rate	Unit	Gross Rate
fibre based roofing felt type 1B	1.42	15.00	1.63	0.43	0.43	6.99	2.38	11.00	m2	12.38
for each defective underlayer cut out and replaced, add	1.42	15.00	1.63	0.43	0.43	6.99	2.38	11.00	m2	12.38
fibre based mineral surfaced roofing felt type 1E	1.72	15.00	1.98	0.56	0.56	9.10	2.38	13.46	m2	15.14
high performance polyester based roofing felt type 5B	3.38	15.00	3.89	0.53	0.53	8.61	2.38	14.88	m2	16.74
for each defective underlayer cut out and replaced, add	3.38	15.00	3.89	0.53	0.53	8.61	2.38	14.88	m2	16.74
high performance polyester based mineral surfaced roofing felt type 5E	3.90	15.00	4.49	0.65	0.65	10.56	2.38	17.43	m2	19.61

Sweep clean surface of existing roof, prime with hot bitumen and dress with

	Del to Site	Waste	Material Cost	Craft Optve	Lab	Labour Cost	Sunds	Nett Rate	Unit	Gross Rate
13mm layer of limestone or granite chippings	1.80	10.00	1.98	-	0.23	1.61	1.33	4.92	m2	5.54
13mm layer of pea shingle	1.25	10.00	1.38	-	0.36	2.53	1.33	5.23	m2	5.89
extra; removing existing chippings or shingle	-	-	-	-	0.24	1.68	0.40	2.08	m2	2.35

Liquid bitumen proofing (black) brush applied on old roof covering including cleaning old covering

Corrugated asbestos roofing

	Del to Site	Waste	Material Cost	Craft Optve	Lab	Labour Cost	Sunds	Nett Rate	Unit	Gross Rate
one coat	1.94	10.00	2.13	-	0.21	1.47	-	3.61	m2	4.06
two coats	3.63	10.00	3.99	-	0.25	1.75	-	5.75	m2	6.47

Felt roofing

	Del to Site	Waste	Material Cost	Craft Optve	Lab	Labour Cost	Sunds	Nett Rate	Unit	Gross Rate
one coat	1.68	10.00	1.85	-	0.17	1.19	-	3.04	m2	3.42
two coats	3.11	10.00	3.42	-	0.21	1.47	-	4.90	m2	5.51

For top coat in green in lieu black

	Del to Site	Waste	Material Cost	Craft Optve	Lab	Labour Cost	Sunds	Nett Rate	Unit	Gross Rate
add	0.52	10.00	0.57	-	-	-	-	0.57	m2	0.64

This section continues
on the next page

Labour hourly rates: (except Specialists) Craft Operatives £12.44 Labourer £7.02 Rates are national average prices. Refer to REGIONAL VARIATIONS for indicative levels of overall pricing in regions	MATERIALS			LABOUR				RATES		
	Del to Site £	Waste %	Material Cost £	Craft Optve Hrs	Lab Hrs	Labour Cost £	Sunds £	Nett Rate £	Unit	Gross Rate (+12.5%) £
REPAIRING/RENOVATING ROOF COVERINGS Cont.										
Repairs to lead roofing										
Repair crack										
clean out and fill with copper bit solder	-	-	-	0.80	-	9.95	0.30	10.25	m	11.53
Turn back flashing										
and re-dress	-	-	-	0.48	-	5.97	-	5.97	m	6.72
Repairs to lead flashings, etc.; replace with new lead										
Remove flashing and provide new code Nr4 lead; wedge into groove										
150mm girth	2.42	5.00	2.54	1.06	-	13.19	0.26	15.99	m	17.99
240mm girth	3.86	5.00	4.05	1.49	-	18.54	0.26	22.85	m	25.70
300mm girth	4.83	5.00	5.07	1.82	-	22.64	0.26	27.97	m	31.47
Remove stepped flashing and provide new code Nr 4 lead; wedge into groove										
180mm girth	2.90	5.00	3.04	1.54	-	19.16	0.26	22.46	m	25.27
240mm girth	3.86	5.00	4.05	1.94	-	24.13	0.26	28.45	m	32.00
300mm girth	4.83	5.00	5.07	2.34	-	29.11	0.26	34.44	m	38.75
Remove apron flashing and provide new code Nr 5 lead; wedge into groove										
150mm girth	3.01	5.00	3.16	1.06	-	13.19	0.26	16.61	m	18.68
300mm girth	6.01	5.00	6.31	1.82	-	22.64	0.26	29.21	m	32.86
450mm girth	9.02	5.00	9.47	2.58	-	32.10	0.40	41.97	m	47.21
Remove lining to valley gutter and provide new code Nr 5 lead										
240mm girth	4.81	5.00	5.05	1.49	-	18.54	-	23.59	m	26.53
300mm girth	6.01	5.00	6.31	1.82	-	22.64	-	28.95	m	32.57
450mm girth	9.02	5.00	9.47	2.58	-	32.10	-	41.57	m	46.76
Remove gutter lining and provide new code Nr 6 lead										
360mm girth	8.54	5.00	8.97	2.12	-	26.37	-	35.34	m	39.76
420mm girth	9.96	5.00	10.46	2.43	-	30.23	-	40.69	m	45.77
450mm girth	10.67	5.00	11.20	2.58	-	32.10	-	43.30	m	48.71
Repairs to zinc flashings, etc.; replace with 0.8mm zinc										
Remove flashing and provide new; wedge into groove										
150mm girth	2.31	5.00	2.43	1.12	-	13.93	0.18	16.54	m	18.61
240mm girth	3.70	5.00	3.88	1.60	-	19.90	0.18	23.97	m	26.97
300mm girth	4.62	5.00	4.85	1.94	-	24.13	0.26	29.24	m	32.90
Remove stepped flashing and provide new; wedge into groove										
180mm girth	2.77	5.00	2.91	1.65	-	20.53	0.18	23.61	m	26.57
240mm girth	3.70	5.00	3.88	2.08	-	25.88	0.18	29.94	m	33.68
300mm girth	4.62	5.00	4.85	2.52	-	31.35	0.26	36.46	m	41.02
Remove apron flashing and provide new; wedge into groove										
150mm girth	2.31	5.00	2.43	1.12	-	13.93	0.18	16.54	m	18.61
300mm girth	4.62	5.00	4.85	1.94	-	24.13	0.26	29.24	m	32.90
450mm girth	6.93	5.00	7.28	2.76	-	34.33	0.32	41.93	m	47.17
Remove lining to valley gutter and provide new										
240mm girth	3.70	5.00	3.88	1.60	-	19.90	-	23.79	m	26.76
300mm girth	4.62	5.00	4.85	1.94	-	24.13	-	28.98	m	32.61
450mm girth	6.93	5.00	7.28	2.76	-	34.33	-	41.61	m	46.81
Remove gutter lining and provide new										
360mm girth	5.54	5.00	5.82	2.27	-	28.24	-	34.06	m	38.31
420mm girth	6.47	5.00	6.79	2.61	-	32.47	-	39.26	m	44.17
450mm girth	6.93	5.00	7.28	2.76	-	34.33	-	41.61	m	46.81
Repairs to copper flashings, etc.; replace with 0.6mm copper										
Remove flashing and provide new; wedge into groove										
150mm girth	2.68	5.00	2.81	1.06	-	13.19	0.36	16.36	m	18.41
240mm girth	4.28	5.00	4.49	1.49	-	18.54	0.36	23.39	m	26.31
300mm girth	5.36	5.00	5.63	1.82	-	22.64	0.36	28.63	m	32.21
Remove stepped flashing and provide new; wedge into groove										
180mm girth	3.21	5.00	3.37	1.54	-	19.16	0.36	22.89	m	25.75
240mm girth	4.28	5.00	4.49	1.94	-	24.13	0.36	28.99	m	32.61
300mm girth	5.36	5.00	5.63	2.34	-	29.11	0.36	35.10	m	39.48
Remove apron flashing and provide new; wedge into groove										
150mm girth	2.68	5.00	2.81	1.06	-	13.19	0.36	16.36	m	18.41
300mm girth	5.36	5.00	5.63	1.82	-	22.64	Sunds	28.63	m	32.21
450mm girth	8.03	5.00	8.43	2.58	-	32.10	0.54	41.07	m	46.20
Remove lining to valley gutter and provide new										
240mm girth	4.28	5.00	4.49	1.49	-	18.54	-	23.03	m	25.91
300mm girth	5.36	5.00	5.63	1.82	-	22.64	-	28.27	m	31.80
450mm girth	8.03	5.00	8.43	2.58	-	32.10	-	40.53	m	45.59

Labour hourly rates: (except Specialists) Craft Operatives £12.44 Labourer £7.02 Rates are national average prices. Refer to REGIONAL VARIATIONS for indicative levels of overall pricing in regions	MATERIALS			LABOUR				RATES		
	Del to Site	Waste	Material Cost	Craft Optve	Lab	Labour Cost	Sunds	Nett Rate	Unit	Gross Rate (+12.5%)
	£	%	£	Hrs	Hrs	£	£	£		£
REPAIRING/RENOVATING ROOF COVERINGS Cont.										
Repairs to copper flashings, etc.; replace with 0.6mm copper Cont.										
Remove gutter lining and provide new										
360mm girth	6.43	5.00	6.75	2.12	-	26.37	-	33.12	m	37.26
420mm girth	7.50	5.00	7.88	2.43	-	30.23	-	38.10	m	42.87
450mm girth	8.03	5.00	8.43	2.58	-	32.10	-	40.53	m	45.59
Repairs to aluminium flashings etc; replace with 0.60mm aluminium										
Remove flashings and provide new; wedge into groove										
150mm girth	1.57	5.00	1.65	1.06	-	13.19	0.16	14.99	m	16.87
240mm girth	2.93	5.00	3.08	1.49	-	18.54	0.16	21.77	m	24.49
300mm girth	2.93	5.00	3.08	1.82	-	22.64	0.16	25.88	m	29.11
Remove stepped flashing and provide new; wedge into groove										
180mm girth	2.16	5.00	2.27	1.54	-	19.16	0.16	21.59	m	24.28
240mm girth	2.93	5.00	3.08	1.94	-	24.13	0.16	27.37	m	30.79
300mm girth	2.93	5.00	3.08	2.34	-	29.11	0.16	32.35	m	36.39
Remove apron flashing and provide new; wedge into groove										
150mm girth	1.57	5.00	1.65	1.06	-	13.19	0.16	14.99	m	16.87
300mm girth	2.93	5.00	3.08	1.82	-	22.64	0.16	25.88	m	29.11
450mm girth	4.09	5.00	4.29	2.56	-	31.85	0.24	36.38	m	40.93
Remove lining to valley gutter and provide new										
240mm girth	2.93	5.00	3.08	1.49	-	18.54	-	21.61	m	24.31
300mm girth	2.93	5.00	3.08	1.82	-	22.64	-	25.72	m	28.93
450mm girth	4.09	5.00	4.29	2.58	-	32.10	-	36.39	m	40.94
Remove gutter lining and provide new										
360mm girth	4.09	5.00	4.29	2.12	-	26.37	-	30.67	m	34.50
420mm girth	4.09	5.00	4.29	2.43	-	30.23	-	34.52	m	38.84
450mm girth	4.09	5.00	4.29	2.58	-	32.10	-	36.39	m	40.94
REPAIRING/RENOVATING PLUMBING										
Repairs to rainwater goods										
Clean out and make good loose fixings										
eaves gutter	-	-	-	0.50	-	6.22	-	6.22	m	7.00
rainwater pipe	-	-	-	0.50	-	6.22	-	6.22	m	7.00
rainwater head	-	-	-	0.80	-	9.95	-	9.95	nr	11.20
Clean out defective joint to existing eaves gutter and re-make joint										
with jointing compound and new bolt	0.17	-	0.17	0.80	-	9.95	-	10.12	nr	11.39
Remove remains of old balloon grating and provide new plastic grating in outlet or end of pipe										
50mm diameter	1.80	2.00	1.84	0.55	-	6.84	-	8.68	nr	9.76
75mm diameter	1.85	2.00	1.89	0.55	-	6.84	-	8.73	nr	9.82
100mm diameter	1.92	2.00	1.96	0.55	-	6.84	-	8.80	nr	9.90
Repair or replacement of stopcocks, valves, etc.										
Turn off water supply drain down as necessary and re-new washer to the following up to 19mm										
main stopcock	0.80	-	0.80	2.00	-	24.88	-	25.68	nr	28.89
service stopcock	0.80	-	0.80	2.00	-	24.88	-	25.68	nr	28.89
bib or pillar cock	0.80	-	0.80	2.00	-	24.88	-	25.68	nr	28.89
supatap	0.80	-	0.80	2.00	-	24.88	-	25.68	nr	28.89
draining tap	0.80	-	0.80	2.00	-	24.88	-	25.68	nr	28.89
Turn off water supply, drain down as necessary, take out old valves and re-new with joints to copper										
13mm brass stopcock (B.S.1010)	3.55	2.00	3.62	2.00	-	24.88	-	28.50	nr	32.06
13mm chromium plated sink pillar cock (B.S.1010)	13.25	2.00	13.52	2.00	-	24.88	-	38.40	nr	43.19
13mm brass bib cock (B.S.1010)	8.19	2.00	8.35	2.00	-	24.88	-	33.23	nr	37.39
13mm drain tap (B.S.2879) type 2	2.11	2.00	2.15	2.00	-	24.88	-	27.03	nr	30.41
19mm drain tap (B.S.2879) type 2	15.21	2.00	15.51	2.00	-	24.88	-	40.39	nr	45.44
13mm high pressure ball valve (B.S.1212) piston type with copper ball	10.11	2.00	10.31	2.00	-	24.88	-	35.19	nr	39.59
13mm high pressure ball valve (B.S.1212) piston type with plastic ball	5.55	2.00	5.66	2.00	-	24.88	-	30.54	nr	34.36
13mm low pressure ball valve (B.S.1212) piston type with copper ball	11.78	2.00	12.02	2.00	-	24.88	-	36.90	nr	41.51
13mm low pressure ball valve (B.S.1212) piston type with plastic ball	7.22	2.00	7.36	2.00	-	24.88	-	32.24	nr	36.27
19mm high pressure ball valve (B.S.1212) piston type with copper ball	21.18	2.00	21.60	2.50	-	31.10	-	52.70	nr	59.29
19mm high pressure ball valve (B.S.1212) piston type with plastic ball	14.97	2.00	15.27	2.50	-	31.10	-	46.37	nr	52.17
19mm low pressure ball valve (B.S.1212) piston type with copper ball	21.39	2.00	21.82	2.50	-	31.10	-	52.92	nr	59.53
19mm low pressure ball valve (B.S.1212) piston type with plastic ball	15.18	2.00	15.48	2.50	-	31.10	-	46.58	nr	52.41

Labour hourly rates: (except Specialists) Craft Operatives £12.44 Labourer £7.02 Rates are national average prices. Refer to REGIONAL VARIATIONS for indicative levels of overall pricing in regions	MATERIALS			LABOUR				RATES		
	Del to Site £	Waste %	Material Cost £	Craft Optve Hrs	Lab Hrs	Labour Cost £	Sunds £	Nett Rate £	Unit	Gross Rate (+12.5%) £
REPAIRING/RENOVATING PLUMBING Cont.										
Replacement of traps										
Take out old plastic trap and replace with new 76mm seal trap (B.S.3943) with 'O'' ring joint outlet										
36mm diameter P trap	3.08	2.00	3.14	1.25	-	15.55	-	18.69	nr	21.03
42mm diameter P trap	3.55	2.00	3.62	1.50	-	18.66	-	22.28	nr	25.07
36mm diameter S trap	3.90	2.00	3.98	1.25	-	15.55	-	19.53	nr	21.97
42mm diameter S trap	4.58	2.00	4.67	1.50	-	18.66	-	23.33	nr	26.25
42mm diameter bath trap with overflow connection .	6.60	2.00	6.73	1.75	-	21.77	-	28.50	nr	32.06
Take out old copper trap and replace with new 76mm seal two piece trap (B.S.1184) with compression joint										
35mm diameter P trap	15.34	2.00	15.65	1.25	-	15.55	-	31.20	nr	35.10
42mm diameter P trap	19.53	2.00	19.92	1.50	-	18.66	-	38.58	nr	43.40
35mm diameter S trap	17.58	2.00	17.93	1.25	-	15.55	-	33.48	nr	37.67
42mm diameter S trap	21.11	2.00	21.53	1.50	-	18.66	-	40.19	nr	45.22
42mm diameter bath trap with male iron overflow connection	28.82	2.00	29.40	1.75	-	21.77	-	51.17	nr	57.56
Preparation of pipe for insertion of new pipe or fitting										
Cut into cast iron pipe and take out a length up to 1.00m for insertion of new pipe or fitting										
50mm diameter	-	-	-	3.70	-	46.03	-	46.03	nr	51.78
64 and 76mm diameter	-	-	-	4.30	-	53.49	-	53.49	nr	60.18
89 and 102mm diameter	-	-	-	4.90	-	60.96	-	60.96	nr	68.58
Cut into cast iron pipe with caulked joint and take out a length up to 1.00m for insertion of new pipe or fitting										
50mm diameter	-	-	-	3.70	-	46.03	-	46.03	nr	51.78
64 and 76mm diameter	-	-	-	4.30	-	53.49	-	53.49	nr	60.18
89 and 102mm diameter	-	-	-	4.90	-	60.96	-	60.96	nr	68.58
Cut into asbestos cement pipe and take out a length up to 1.00m for insertion of new pipe or fitting										
50mm diameter	-	-	-	3.15	-	39.19	-	39.19	nr	44.08
64 and 76mm diameter	-	-	-	3.75	-	46.65	-	46.65	nr	52.48
89 and 102mm diameter	-	-	-	4.35	-	54.11	-	54.11	nr	60.88
Cut into polythene or p.v.c. pipe and take out a length up to 1.00m for insertion of new pipe or fitting										
up to 25mm diameter	-	-	-	2.50	-	31.10	-	31.10	nr	34.99
32 - 50mm diameter	-	-	-	3.15	-	39.19	-	39.19	nr	44.08
64 and 76mm diameter	-	-	-	3.75	-	46.65	-	46.65	nr	52.48
89 and 102mm diameter	-	-	-	4.40	-	54.74	-	54.74	nr	61.58
Cut into steel pipe and take out a length up to 1.00m for insertion of new pipe or fitting										
up to 25mm diameter	-	-	-	3.05	-	37.94	-	37.94	nr	42.68
32 - 50mm diameter	-	-	-	3.70	-	46.03	-	46.03	nr	51.78
64 and 76mm diameter	-	-	-	4.30	-	53.49	-	53.49	nr	60.18
89 and 102mm diameter	-	-	-	4.90	-	60.96	-	60.96	nr	68.58
Cut into stainless steel pipe and take out a length up to 1.00m for insertion of new pipe or fitting										
up to 25mm diameter	-	-	-	2.90	-	36.08	-	36.08	nr	40.59
32 - 50mm diameter	-	-	-	3.50	-	43.54	-	43.54	nr	48.98
64 and 76mm diameter	-	-	-	4.10	-	51.00	-	51.00	nr	57.38
89 and 102mm diameter	-	-	-	4.75	-	59.09	-	59.09	nr	66.48
Cut into copper pipe and take out a length up to 1.00m for insertion of new pipe or fitting										
up to 25mm diameter	-	-	-	2.90	-	36.08	-	36.08	nr	40.59
32 - 50mm diameter	-	-	-	3.50	-	43.54	-	43.54	nr	48.98
64 and 76mm diameter	-	-	-	4.10	-	51.00	-	51.00	nr	57.38
89 and 102mm diameter	-	-	-	4.75	-	59.09	-	59.09	nr	66.48
Cut into lead pipe and take out a length up to 1.00m for insertion of new pipe or fitting										
up to 25mm diameter	-	-	-	3.05	-	37.94	-	37.94	nr	42.68
32 - 50mm diameter	-	-	-	3.70	-	46.03	-	46.03	nr	51.78
64 and 76mm diameter	-	-	-	4.30	-	53.49	-	53.49	nr	60.18
89 and 102mm diameter	-	-	-	4.90	-	60.96	-	60.96	nr	68.58
Removal of traps, valves, etc., for re-use										
Take out the following, including unscrewing or uncoupling, for re-use and including draining down as necessary										
trap up to 25mm diameter	-	-	-	0.70	-	8.71	-	8.71	nr	9.80
trap 32 - 50mm diameter	-	-	-	0.85	-	10.57	-	10.57	nr	11.90
stop valve up to 25mm diameter	-	-	-	2.00	-	24.88	-	24.88	nr	27.99
stop valve 32 - 50mm diameter	-	-	-	2.25	-	27.99	-	27.99	nr	31.49
radiator valve up to 25mm diameter	-	-	-	2.00	-	24.88	-	24.88	nr	27.99
radiator valve 32 - 50mm diameter	-	-	-	2.25	-	27.99	-	27.99	nr	31.49
tap up to 25mm diameter	-	-	-	2.00	-	24.88	-	24.88	nr	27.99
tap 32 - 50mm diameter	-	-	-	2.25	-	27.99	-	27.99	nr	31.49
extra; burning out one soldered joint up to 25mm diameter	-	-	-	0.50	-	6.22	-	6.22	nr	7.00
extra; burning out one soldered joint 32 - 50mm diameter	-	-	-	0.80	-	9.95	-	9.95	nr	11.20

Labour hourly rates: (except Specialists) Craft Operatives £12.44 Labourer £7.02. Rates are national average prices. Refer to REGIONAL VARIATIONS for indicative levels of overall pricing in regions	MATERIALS			LABOUR				RATES		
	Del to Site £	Waste %	Material Cost £	Craft Optve Hrs	Lab Hrs	Labour Cost £	Sunds £	Nett Rate £	Unit	Gross Rate (+12.5%) £

REPAIRING/RENOVATING PLUMBING Cont.

Replacement of old sanitary fittings

	Del to Site £	Waste %	Material Cost £	Craft Optve Hrs	Lab Hrs	Labour Cost £	Sunds £	Nett Rate £	Unit	Gross Rate £
Disconnect trap, valves and services pipes, remove old fitting and provide and fix new including re-connecting trap, valves and services pipes										
560 x 405mm vitreous china lavatory basin with 32mm chromium plated waste outlet, plug, chain and stay and pair of cast iron brackets	24.25	5.00	25.46	6.00	–	74.64	–	100.10	nr	112.62
600 x 1000mm stainless steel sink with single drainer with 38mm chromium plated waste outlet, plug, chain and stay	55.50	5.00	58.27	6.00	–	74.64	–	132.91	nr	149.53
600 x 1500mm stainless steel sink with double drainer with 38mm chromium plated waste outlet, overflow plug, chain and stay	75.00	5.00	78.75	6.50	–	80.86	–	159.61	nr	179.56
1500mm pressed steel vitreous enamelled rectangular top bath with cradles and with 38mm chromium plated waste outlet, plug, chain and stay and 32mm chromium plated overflow and including removing old cast iron bath	94.75	5.00	99.49	8.00	–	99.52	–	199.01	nr	223.88
1700mm pressed steel vitreous enamelled rectangular top bath with cradles and with 38mm chromium plated waste outlet, plug, chain and stay and 32mm chromium plated overflow and including removing old cast iron bath	98.25	5.00	103.16	8.00	–	99.52	–	202.68	nr	228.02
Take off old W.C. seat and provide and fix new ring pattern black plastic seat and cover	16.75	5.00	17.59	1.50	–	18.66	–	36.25	nr	40.78
Disconnect supply pipe, overflow and flush pipe and take out old W.C. cistern and provide and fix new 9 litre black plastic W.C. cistern with cover and connect pipe and ball valve										
low level	56.75	5.00	59.59	5.00	–	62.20	–	121.79	nr	137.01
high level	48.25	5.00	50.66	5.00	–	62.20	–	112.86	nr	126.97
Disconnect flush pipe, take off seat and remove old W.C. pan, provide and fix new white china pan, connect flush pipe, form cement joint with drain and refix old seat										
S or P trap pan	36.75	5.00	38.59	5.00	–	62.20	–	100.79	nr	113.39
Disconnect and remove old high level W.C. suite complete and provide and fix new suite with P or S trap pan, plastic ring seat and cover, plastic flush pipe or bend, 9 litre plastic W.C. cistern with cover, chain and pull or lever handle and 13mm low pressure ball valve; adapt supply pipe and overflow as necessary and connect to new suite and cement joint to drain										
new high level suite	99.00	5.00	103.95	8.50	–	105.74	–	209.69	nr	235.90
new low level suite (vitreous china cistern)	100.50	5.00	105.53	9.00	–	111.96	–	217.49	nr	244.67
Disconnect and remove old low level W.C. suite complete and provide and fix new suite with P or S trap pan, plastic ring seat and cover, plastic flush bend, 9 litre vitreous china W.C. cistern with cover, lever handle and 13mm low pressure ball valve with plastic ball; re-connect supply pipe and overflow and cement joint to drain										
new low level suite	100.50	5.00	105.53	8.50	–	105.74	–	211.26	nr	237.67

Replacement of old water tanks

	Del to Site £	Waste %	Material Cost £	Craft Optve Hrs	Lab Hrs	Labour Cost £	Sunds £	Nett Rate £	Unit	Gross Rate £
Disconnect all pipework, set aside ball valve and take out cold water storage cistern in roof space; replace with new galvanised water storage cistern (B.S.417 Grade A); reconnect all pipework and ball valve; remove metal filings and clean inside of tank										
type SC40 size 690 x 510 x 510mm (114 litre)	84.50	2.00	86.19	9.00	–	111.96	0.30	198.45	nr	223.26
type SC70 size 910 x 610 x 580mm (227 litre)	147.25	2.00	150.19	9.00	–	111.96	0.50	262.65	nr	295.49
Disconnect all pipework and take out hot water tank from first floor level; replace with new galvanised hot water tank (B.S.417 Grade A); reconnect all pipework										
type T25/1 size 610 x 430 x 430mm (95 litre)	145.40	2.00	148.31	9.00	–	111.96	–	260.27	nr	292.80
type T30/1 size 410 x 460 x 480mm (114 litre)	156.40	2.00	159.53	9.00	–	111.96	–	271.49	nr	305.42

This section continues
on the next page

Labour hourly rates: (except Specialists) Craft Operatives £14.38 Labourer £7.02 Rates are national average prices. Refer to REGIONAL VARIATIONS for indicative levels of overall pricing in regions	MATERIALS			LABOUR				RATES		
	Del to Site £	Waste %	Material Cost £	Craft Optve Hrs	Lab Hrs	Labour Cost £	Sunds £	Nett Rate £	Unit	Gross Rate (+12.5%) £
REPAIRING/RENOVATING ELECTRICAL INSTALLATIONS										
Generally										

Note
   the following approximate estimates of costs are
   dependent on the number and disposition of the
   points. Lamps and fittings together with cutting
   and making good are excluded										

Strip out and re-wire complete										
Provide new lamp holders, flush switches, flush socket outlets, fitting outlets and consumer unit with circuit breakers; a three bedroom house with ten lighting points and associated switches, sixteen 13 amp socket outlets, three 13 amp fitting outlets, one 30 amp cooker control panel, three earth bonding points and one eight way consumer unit										
PVC cables in existing conduit	953.93	7.50	1025.47	52.85	–	759.98	30.00	1815.46	nr	2042.39
PVC cables and conduit	1057.73	7.50	1137.06	99.00	–	1423.62	30.00	2590.68	nr	2914.51
Strip out and re-wire with PVC insulated and sheathed cable in existing conduit										
Point to junction box, re-using existing lamp holder, socket outlet etc										
5 amp socket outlet	9.00	7.50	9.68	0.85	–	12.22	0.45	22.35	nr	25.14
13 amp socket or fitting outlet	12.00	7.50	12.90	0.85	–	12.22	0.45	25.57	nr	28.77
lighting point or switch	5.28	7.50	5.68	0.98	–	14.09	0.45	20.22	nr	22.75
Strip out and re-wire with PVC insulated and sheathed cable and new conduit										
Point to junction box, re-using existing lamp holder, socket outlet etc										
5 amp socket outlet	9.92	7.50	10.66	1.20	–	17.26	0.50	28.42	nr	31.97
13 amp socket outlet	12.92	7.50	13.89	1.20	–	17.26	0.50	31.65	nr	35.60
lighting point or switch	7.38	7.50	7.93	1.30	–	18.69	0.30	26.93	nr	30.29
Luminaires and accessories										
Take out damaged and renew										
lamp holder, batten type	5.09	7.50	5.47	0.56	–	8.05	0.25	13.77	nr	15.50
lamp holder, pendant type with rose	2.08	7.50	2.24	1.06	–	15.24	0.25	17.73	nr	19.94
switch, flush type	3.69	7.50	3.97	0.75	–	10.79	0.45	15.20	nr	17.10
switch, surface mounted type	3.46	7.50	3.72	0.75	–	10.79	0.45	14.95	nr	16.82
13 amp socket outlet, flush type	5.43	7.50	5.84	0.75	–	10.79	0.45	17.07	nr	19.21
13 amp socket outlet, surface mounted type	5.26	7.50	5.65	0.88	–	12.65	0.45	18.76	nr	21.10
5 amp socket outlet, flush type	9.07	7.50	9.75	0.70	–	10.07	0.45	20.27	nr	22.80
5 amp socket outlet, surface mounted type	8.82	7.50	9.48	0.88	–	12.65	0.45	22.59	nr	25.41

This section continues on the next page

Labour hourly rates: (except Specialists) Craft Operatives £9.23 Labourer £7.02 Rates are national average prices. Refer to REGIONAL VARIATIONS for indicative levels of overall pricing in regions	MATERIALS			LABOUR				RATES		
	Del to Site £	Waste %	Material Cost £	Craft Optve Hrs	Lab Hrs	Labour Cost £	Sunds £	Nett Rate £	Unit	Gross Rate (+12.5%) £

REPAIRING/RENOVATING FINISHINGS

Repairs to old wall and ceiling finishings

	Del to Site £	Waste %	Material Cost £	Craft Optve Hrs	Lab Hrs	Labour Cost £	Sunds £	Nett Rate £	Unit	Gross Rate £
Cut out damaged two coat plaster to wall in patches and make out in new plaster to match existing										
under 1.00m2	2.26	10.00	2.49	2.13	2.13	34.61	0.52	37.62	m2	42.32
1.00 - 2.00m2	2.26	10.00	2.49	1.76	1.76	28.60	0.52	31.61	m2	35.56
2.00 - 4.00m2	2.26	10.00	2.49	1.39	1.39	22.59	0.52	25.59	m2	28.79
Cut out damaged three coat plaster to wall in patches and make out in new plaster to match existing										
under 1.00m2	3.00	10.00	3.30	2.50	2.50	40.63	0.80	44.73	m2	50.32
1.00 - 2.00m2	3.00	10.00	3.30	2.06	2.06	33.48	0.80	37.58	m2	42.27
2.00 - 4.00m2	3.00	10.00	3.30	1.62	1.62	26.32	0.80	30.43	m2	34.23
Cut out damaged two coat rendering to wall in patches and make out in new rendering to match existing										
under 1.00m2	1.34	10.00	1.47	1.93	1.93	31.36	0.52	33.36	m2	37.53
1.00 - 2.00m2	1.34	10.00	1.47	1.60	1.60	26.00	0.52	27.99	m2	31.49
2.00 - 4.00m2	1.34	10.00	1.47	1.26	1.26	20.48	0.52	22.47	m2	25.28
Cut out damaged three coat rendering to wall in patches and make out in new rendering to match existing										
under 1.00m2	2.06	10.00	2.27	2.28	2.28	37.05	0.80	40.12	m2	45.13
1.00 - 2.00m2	2.06	10.00	2.27	1.87	1.87	30.39	0.80	33.45	m2	37.64
2.00 - 4.00m2	2.06	10.00	2.27	1.47	1.47	23.89	0.80	26.95	m2	30.32
Cut out damaged two coat plaster to ceiling in patches and make out in new plaster to match existing										
under 1.00m2	2.26	10.00	2.49	2.38	2.38	38.67	0.52	41.68	m2	46.89
1.00 - 2.00m2	2.26	10.00	2.49	1.96	1.96	31.85	0.52	34.86	m2	39.21
2.00 - 4.00m2	2.26	10.00	2.49	1.54	1.54	25.02	0.52	28.03	m2	31.53
Cut out damaged three coat plaster to ceiling in patches and make out in new plaster to match existing										
under 1.00m2	3.00	10.00	3.30	2.79	2.79	45.34	0.80	49.44	m2	55.62
1.00 - 2.00m2	3.00	10.00	3.30	2.30	2.30	37.38	0.80	41.48	m2	46.66
2.00 - 4.00m2	3.00	10.00	3.30	1.81	1.81	29.41	0.80	33.51	m2	37.70
Cut out damaged 9.5mm plasterboard and two coat plaster to wall in patches and make out in new plasterboard and plaster to match existing										
under 1.00m2	4.07	10.00	4.48	2.23	2.23	36.24	0.89	41.60	m2	46.81
1.00 - 2.00m2	4.07	10.00	4.48	1.78	1.78	28.93	0.89	34.29	m2	38.58
2.00 - 4.00m2	4.07	10.00	4.48	1.36	1.36	22.10	0.89	27.47	m2	30.90
Cut out damaged 12.5mm plasterboard and two coat plaster to wall in patches and make out in new plasterboard and plaster to match existing										
under 1.00m2	4.40	10.00	4.84	2.28	2.28	37.05	0.99	42.88	m2	48.24
1.00 - 2.00m2	4.40	10.00	4.84	1.83	1.83	29.74	0.99	35.57	m2	40.01
2.00 - 4.00m2	4.40	10.00	4.84	1.38	1.38	22.43	0.99	28.25	m2	31.79
Cut out damaged 9.5mm plasterboard and two coat plaster to ceiling in patches and make out in new plasterboard and plaster to match existing										
under 1.00m2	4.10	10.00	4.51	2.54	2.54	41.27	0.89	46.67	m2	52.51
1.00 - 2.00m2	4.10	10.00	4.51	2.03	2.03	32.99	0.89	38.39	m2	43.19
2.00 - 4.00m2	4.10	10.00	4.51	1.55	1.55	25.19	0.89	30.59	m2	34.41
Cut out damaged 12.5mm plasterboard and two coat plaster to ceiling in patches and make out in new plasterboard and plaster to match existing										
under 1.00m2	4.40	10.00	4.84	2.56	2.56	41.60	0.99	47.43	m2	53.36
1.00 - 2.00m2	4.40	10.00	4.84	2.06	2.06	33.48	0.99	39.31	m2	44.22
2.00 - 4.00m2	4.40	10.00	4.84	1.57	1.57	25.51	0.99	31.34	m2	35.26
Cut out damaged lath and plaster to wall in patches and make out with new metal lathing and three coat plaster to match existing										
under 1.00m2	8.67	10.00	9.54	2.53	2.53	41.11	0.99	51.64	m2	58.09
1.00 - 2.00m2	8.67	10.00	9.54	2.14	2.14	34.77	0.99	45.30	m2	50.96
2.00 - 4.00m2	8.67	10.00	9.54	1.73	1.73	28.11	0.99	38.64	m2	43.47
Cut out damaged lath and plaster to ceiling in patches and make out with new metal lathing and three coat plaster to match existing										
under 1.00m2	8.67	10.00	9.54	2.87	2.87	46.64	0.99	57.16	m2	64.31
1.00 - 2.00m2	8.67	10.00	9.54	2.43	2.43	39.49	0.99	50.01	m2	56.27
2.00 - 4.00m2	8.67	10.00	9.54	1.95	1.95	31.69	0.99	42.21	m2	47.49
Cut out crack in plaster, form dovetailed key and make good with plaster										
not exceeding 50mm wide	0.11	10.00	0.12	0.20	0.20	3.25	0.03	3.40	m	3.83
Make good plaster where the following removed										
small pipe	-	-	-	0.16	0.16	2.60	0.03	2.63	nr	2.96
small steel section	-	-	-	0.21	0.21	3.41	0.04	3.45	nr	3.88
Cut out damaged plaster moulding or cornice and make out with new to match existing										
up to 100mm girth on face	0.69	10.00	0.76	1.25	0.63	15.96	0.21	16.93	m	19.05
100 - 150mm girth on face	1.14	10.00	1.25	1.86	0.93	23.70	0.41	25.36	m	28.53
150 - 200mm girth on face	2.26	10.00	2.49	2.25	1.13	28.70	0.74	31.93	m	35.92

Labour hourly rates: (except Specialists) Craft Operatives £9.23 Labourer £7.02 Rates are national average prices. Refer to REGIONAL VARIATIONS for indicative levels of overall pricing in regions	MATERIALS			LABOUR				RATES		
	Del to Site	Waste	Material Cost	Craft Optve	Lab	Labour Cost	Sunds	Nett Rate	Unit	Gross Rate (+12.5%)
	£	%	£	Hrs	Hrs	£	£	£		£
REPAIRING/RENOVATING FINISHINGS Cont.										
Repairs to old wall and ceiling finishings Cont.										
Remove damaged fibrous plaster louvered ventilator, replace with new and make good										
225 x 75mm	1.26	10.00	1.39	0.50	0.50	8.13	0.10	9.61	nr	10.81
225 x 150mm	1.51	10.00	1.66	0.60	0.60	9.75	0.17	11.58	nr	13.03
Take out damaged 152 x 152 x 6mm white glazed wall tiles and renew to match existing										
isolated tile	0.45	5.00	0.47	0.32	0.32	5.20	0.06	5.73	nr	6.45
patch of 5 tiles	2.25	5.00	2.36	0.60	0.60	9.75	0.30	12.41	nr	13.96
patch 0.50 - 1.00m2	19.80	5.00	20.79	2.32	2.32	37.70	2.35	60.84	m2	68.44
patch 1.00 - 2.00m2	19.80	5.00	20.79	1.66	1.66	26.98	2.35	50.12	m2	56.38
Repairs to old floor finishings										
Take out damaged 150 x 150 x 12.5mm clay floor tiles and renew to match existing										
isolated tile	0.29	5.00	0.30	0.32	0.21	4.43	0.06	4.79	nr	5.39
patch of 5 tiles	1.45	5.00	1.52	0.65	0.45	9.16	0.30	10.98	nr	12.35
patch 0.50 - 1.00m2	13.05	5.00	13.70	2.40	1.61	33.45	2.35	49.51	m2	55.70
patch 1.00 - 2.00m2	13.05	5.00	13.70	1.90	1.27	26.45	2.35	42.50	m2	47.82
Cut out damaged 25mm granolithic paving in patches and make out in new granolithic to match existing										
under 1.00m2	6.26	5.00	6.57	1.50	1.50	24.38	1.10	32.05	m2	36.05
1.00 - 2.00m2	6.26	5.00	6.57	1.25	1.25	20.31	1.10	27.99	m2	31.48
2.00 - 4.00m2	6.26	5.00	6.57	1.07	1.07	17.39	1.10	25.06	m2	28.19
Cut out damaged 32mm granolithic paving in patches and make out in new granolithic to match existing										
under 1.00m2	9.02	5.00	9.47	1.68	1.68	27.30	1.29	38.06	m2	42.82
1.00 - 2.00m2	9.02	5.00	9.47	1.38	1.38	22.43	1.29	33.19	m2	37.33
2.00 - 4.00m2	9.02	5.00	9.47	1.15	1.15	18.69	1.29	29.45	m2	33.13
Cut out damaged 38mm granolithic paving in patches and make out in new granolithic to match existing										
under 1.00m2	9.50	5.00	9.97	1.80	1.80	29.25	1.52	40.74	m2	45.84
1.00 - 2.00m2	9.50	5.00	9.97	1.50	1.50	24.38	1.52	35.87	m2	40.35
2.00 - 4.00m2	9.50	5.00	9.97	1.25	1.25	20.31	1.52	31.81	m2	35.78
Cut out damaged 50mm granolithic paving in patches and make out in new granolithic to match existing										
under 1.00m2	12.51	5.00	13.14	2.12	2.12	34.45	1.85	49.44	m2	55.61
1.00 - 2.00m2	12.51	5.00	13.14	1.73	1.73	28.11	1.85	43.10	m2	48.49
2.00 - 4.00m2	12.51	5.00	13.14	1.45	1.45	23.56	1.85	38.55	m2	43.37
Cut out damaged granolithic paving to tread 275mm wide and make out in new granolithic to match existing										
25mm thick	1.74	5.00	1.83	0.44	0.44	7.15	0.31	9.29	m	10.45
32mm thick	2.19	5.00	2.30	0.48	0.48	7.80	0.35	10.45	m	11.76
38mm thick	2.63	5.00	2.76	0.50	0.50	8.13	0.41	11.30	m	12.71
50mm thick	3.50	5.00	3.67	0.59	0.59	9.59	0.51	13.77	m	15.49
Cut out damaged granolithic covering to plain riser 175mm wide and make out in new granolithic to match existing										
13mm thick	0.59	5.00	0.62	0.80	0.80	13.00	2.82	16.44	m	18.49
19mm thick	0.84	5.00	0.88	0.84	0.84	13.65	2.84	17.37	m	19.54
25mm thick	1.09	5.00	1.14	0.88	0.88	14.30	2.87	18.31	m	20.60
32mm thick	1.40	5.00	1.47	0.93	0.93	15.11	2.89	19.47	m	21.91
Cut out damaged granolithic covering to undercut riser 175mm wide and make out in new granolithic to match existing										
13mm thick	0.59	5.00	0.62	0.87	0.87	14.14	2.82	17.58	m	19.77
19mm thick	0.84	5.00	0.88	0.93	0.93	15.11	2.84	18.83	m	21.19
25mm thick	1.09	5.00	1.14	0.98	0.98	15.93	2.87	19.94	m	22.43
32mm thick	1.40	5.00	1.47	1.04	1.04	16.90	2.89	21.26	m	23.92
Cut out damaged granolithic skirting 150mm wide and make out in new granolithic to match existing										
13mm thick	0.50	5.00	0.53	0.72	0.72	11.70	2.41	14.64	m	16.46
19mm thick	0.72	5.00	0.76	0.76	0.76	12.35	2.44	15.55	m	17.49
25mm thick	0.95	5.00	1.00	0.80	0.80	13.00	2.46	16.46	m	18.51
32mm thick	1.19	5.00	1.25	0.82	0.82	13.32	2.49	17.06	m	19.20
Cut out crack in granolithic paving not exceeding 50mm wide, form dovetailed key and make good to match existing										
25mm thick	0.34	5.00	0.36	0.60	0.60	9.75	0.05	10.16	m	11.43
32mm thick	0.42	5.00	0.44	0.66	0.66	10.73	0.06	11.23	m	12.63
38mm thick	0.50	5.00	0.53	0.69	0.69	11.21	0.08	11.82	m	13.29
50mm thick	0.67	5.00	0.70	0.79	0.79	12.84	0.09	13.63	m	15.33
REPAIRING GLAZING										
Resecure glass										
Remove decayed putties, paint one coat on edge of rebate and re-putty										
wood window	0.13	10.00	0.14	0.69	-	6.37	0.02	6.53	m	7.35
metal window	0.13	10.00	0.14	0.76	-	7.01	0.02	7.18	m	8.08

	MATERIALS			LABOUR				RATES		
Labour hourly rates: (except Specialists) Craft Operatives £9.23 Labourer £7.02 Rates are national average prices. Refer to REGIONAL VARIATIONS for indicative levels of overall pricing in regions	Del to Site £	Waste %	Material Cost £	Craft Optve Hrs	Lab Hrs	Labour Cost £	Sunds £	Nett Rate £	Unit	Gross Rate (+12.5%) £

REPAIRING GLAZING Cont.

Resecure glass Cont.

Remove beads, remove decayed bedding materials, paint one coat on edge of rebate, re-bed glass and re-fix beads

	Del to Site	Waste	Material Cost	Craft Optve	Lab	Labour Cost	Sunds	Nett Rate	Unit	Gross Rate
wood window	0.13	10.00	0.14	1.22	–	11.26	0.02	11.42	m	12.85
metal window	0.13	10.00	0.14	1.32	–	12.18	0.02	12.35	m	13.89

REPAIRING/RENOVATING FENCING

Repairs to softwood fencing

Remove defective timber fence post and replace with new 127 x 102mm post; letting post 450mm into ground; backfilling around in concrete mix C20; securing ends of existing arris rails and gravel boards to new post

	Del to Site	Waste	Material Cost	Craft Optve	Lab	Labour Cost	Sunds	Nett Rate	Unit	Gross Rate
to suit fencing 914mm high	11.33	–	11.33	2.04	2.04	33.15	0.41	44.89	nr	50.50
to suit fencing 1219mm high	12.17	–	12.17	2.06	2.06	33.48	0.47	46.12	nr	51.88
to suit fencing 1524mm high	13.01	–	13.01	2.08	2.08	33.80	0.55	47.36	nr	53.28
to suit fencing 1829mm high	13.86	–	13.86	2.10	2.10	34.13	0.64	48.63	nr	54.70

Remove defective pale from palisade fence and replace with new 19 x 75mm pointed pale; fixing with galvanised nails

	Del to Site	Waste	Material Cost	Craft Optve	Lab	Labour Cost	Sunds	Nett Rate	Unit	Gross Rate
764mm long	0.72	–	0.72	0.55	0.09	5.71	0.18	6.61	nr	7.43
1069mm long	1.01	–	1.01	0.56	0.09	5.80	0.22	7.03	nr	7.91
1374mm long	1.30	–	1.30	0.58	0.10	6.06	0.25	7.61	nr	8.56

Remove defective pales from close boarded fence and replace with new 100mm wide feather edged pales; fixing with galvanised nails

	Del to Site	Waste	Material Cost	Craft Optve	Lab	Labour Cost	Sunds	Nett Rate	Unit	Gross Rate
singly	1.63	–	1.63	0.25	0.04	2.59	0.21	4.43	m	4.98
areas up to 1.00m2	9.94	20.00	11.93	1.32	0.22	13.73	1.75	27.41	m2	30.83
areas over 1.00m2	9.94	20.00	11.93	0.98	0.16	10.17	1.75	23.85	m2	26.83

Remove defective gravel board and centre stump and replace with new 25 x 150mm gravel board and 50 x 50mm centre stump; fixing with galvanised nails

	Del to Site	Waste	Material Cost	Craft Optve	Lab	Labour Cost	Sunds	Nett Rate	Unit	Gross Rate
securing to fence posts	2.44	5.00	2.56	0.53	0.09	5.52	0.21	8.30	m	9.33

Repairs to oak fencing

Remove defective timber fence post and replace with new 127 x 102mm post; letting post 450mm into ground; backfilling around in concrete mix C20; securing ends of existing arris rails and gravel boards to new post

	Del to Site	Waste	Material Cost	Craft Optve	Lab	Labour Cost	Sunds	Nett Rate	Unit	Gross Rate
to suit fencing 914mm high	20.70	–	20.70	2.17	2.17	35.26	0.41	56.37	nr	63.42
to suit fencing 1219mm high	24.16	–	24.16	2.21	2.21	35.91	0.47	60.54	nr	68.11
to suit fencing 1524mm high	27.62	–	27.62	2.25	2.25	36.56	0.55	64.73	nr	72.82
to suit fencing 1829mm high	31.08	–	31.08	2.29	2.29	37.21	0.64	68.93	nr	77.55

Remove defective pale from palisade fence and replace with new 19 x 75mm pointed pale; fixing with galvanised nails

	Del to Site	Waste	Material Cost	Craft Optve	Lab	Labour Cost	Sunds	Nett Rate	Unit	Gross Rate
764mm long	1.73	–	1.73	0.70	0.12	7.30	0.18	9.21	nr	10.37
1069mm long	2.40	–	2.40	0.76	0.13	7.93	0.22	10.55	nr	11.87
1374mm long	3.10	–	3.10	0.83	0.14	8.64	0.25	11.99	nr	13.49

Remove defective pales from close boarded fence and replace with new 100mm wide feather edged pales; fixing with galvanised nails

	Del to Site	Waste	Material Cost	Craft Optve	Lab	Labour Cost	Sunds	Nett Rate	Unit	Gross Rate
singly	3.88	–	3.88	0.27	0.05	2.84	0.21	6.93	m	7.80
areas up to 1.00m2	23.68	20.00	28.42	2.56	0.43	26.65	1.75	56.81	m2	63.92
areas over 1.00m2	23.68	20.00	28.42	2.07	0.35	21.56	1.75	51.73	m2	58.20

Remove defective gravel board and centre stump and replace with new 25 x 150mm gravel board and 50 x 50mm centre stump; fixing with galvanised nails

	Del to Site	Waste	Material Cost	Craft Optve	Lab	Labour Cost	Sunds	Nett Rate	Unit	Gross Rate
securing to fence posts	5.78	5.00	6.07	1.10	0.18	11.42	0.21	17.70	m	19.91

Repairs to posts
100 x 100mm precast concrete spur 1219mm long; setting into ground; backfilling around in concrete mix C20

	Del to Site	Waste	Material Cost	Craft Optve	Lab	Labour Cost	Sunds	Nett Rate	Unit	Gross Rate
bolting to existing timber post	12.44	–	12.44	0.76	0.76	12.35	1.66	26.45	nr	29.76

REPAIRING/RENOVATING EXTERNAL PAVINGS

Repairs to precast concrete flag paving

Take up uneven paving and set aside for re-use; level up and consolidate existing hardcore and relay salvaged paving

	Del to Site	Waste	Material Cost	Craft Optve	Lab	Labour Cost	Sunds	Nett Rate	Unit	Gross Rate
bedding on 25mm sand bed; grouting in lime mortar (1:3)	0.75	10.00	0.82	0.67	0.93	12.71	0.12	13.66	m2	15.36

Repairs to York stone paving

Take up old paving; re-square; relay random in random sizes

	Del to Site	Waste	Material Cost	Craft Optve	Lab	Labour Cost	Sunds	Nett Rate	Unit	Gross Rate
bedding, jointing and pointing in cement mortar (1:3)	3.71	5.00	3.90	1.10	1.10	17.88	0.90	22.67	m2	25.50

Labour hourly rates: (except Specialists) Craft Operatives £9.23 Labourer £7.02 Rates are national average prices. Refer to REGIONAL VARIATIONS for indicative levels of overall pricing in regions	MATERIALS			LABOUR				RATES		
	Del to Site	Waste	Material Cost	Craft Optve	Lab	Labour Cost	Sunds	Nett Rate	Unit	Gross Rate (+12.5%)
.	£	%	£	Hrs	Hrs	£	£	£		£
REPAIRING/RENOVATING EXTERNAL PAVINGS Cont.										
Repairs to granite sett paving										
Take up, clean and stack old paving	-	-	-	-	0.90	6.32	-	6.32	m2	7.11
Take up old paving, clean and re-lay bedding and grouting in cement mortar (1:3)	-	-	-	1.65	2.55	33.13	1.24	34.37	m2	38.67
Clean out joints of old paving and re-grout in cement mortar (1:3)	-	-	-	-	0.55	3.86	2.00	5.86	m2	6.59

Groundwork

GROUNDWORK

	PLANT & TRANSPORT			LABOUR				RATES		
Labour hourly rates: (except Specialists) Craft Operatives £9.23 Labourer £7.02 Rates are national average prices. Refer to REGIONAL VARIATIONS for indicative levels of overall pricing in regions	Plant Cost £	Trans Cost £	P & T Cost £	Craft Optve Hrs	Lab Hrs	Labour Cost £	Sunds £	Nett Rate £	Unit	Gross Rate (+12.5%) £
D20: EXCAVATING AND FILLING										
Site preparation										
Removing trees; filling voids left by removal of roots with selected material arising from excavation										
girth 0.60 - .50m	65.00	-	65.00	-	11.00	77.22	-	142.22	nr	160.00
girth 1.50 - 3.00m	120.00	-	120.00	-	23.00	161.46	-	281.46	nr	316.64
girth exceeding 3.00m	175.00	-	175.00	-	35.00	245.70	-	420.70	nr	473.29
Clearing site vegetation; filling voids left by removal of roots with selected material arising from excavation										
bushes, scrub, undergrowth, hedges, trees and tree stumps not exceeding 600mm girth.	0.61	-	0.61	-	-	-	-	0.61	m2	0.69
Lifting turf for preservation										
stacking on site average 100m distant for future use; watering	-	-	-	-	0.60	4.21	-	4.21	m2	4.74
Excavating - by machine										
Top soil for preservation										
average 150mm deep	0.28	-	0.28	-	-	-	-	0.28	m2	0.32
To reduce levels										
maximum depth not exceeding										
0.25m	2.17	-	2.17	-	-	-	-	2.17	m3	2.44
1.00m	1.14	-	1.14	-	-	-	-	1.14	m3	1.28
2.00m	1.31	-	1.31	-	-	-	-	1.31	m3	1.47
4.00m	1.48	-	1.48	-	-	-	-	1.48	m3	1.67
6.00m	1.82	-	1.82	-	-	-	-	1.82	m3	2.05
Basements and the like										
maximum depth not exceeding										
0.25m	2.17	-	2.17	-	0.45	3.16	-	5.33	m3	6.00
1.00m	1.37	-	1.37	-	0.50	3.51	-	4.88	m3	5.49
2.00m	1.60	-	1.60	-	0.55	3.86	-	5.46	m3	6.14
4.00m	2.28	-	2.28	-	0.70	4.91	-	7.19	m3	8.09
6.00m	3.19	-	3.19	-	0.85	5.97	-	9.16	m3	10.30
Trenches width not exceeding 0.30m										
maximum depth not exceeding										
0.25m	5.60	-	5.60	-	2.00	14.04	-	19.64	m3	22.09
1.00m	5.80	-	5.80	-	2.10	14.74	-	20.54	m3	23.11
Trenches width exceeding 0.30m										
maximum depth not exceeding										
0.25m	5.59	-	5.59	-	0.65	4.56	-	10.15	m3	11.42
1.00m	4.79	-	4.79	-	0.70	4.91	-	9.70	m3	10.92
2.00m	4.79	-	4.79	-	0.80	5.62	-	10.41	m3	11.71
4.00m	5.70	-	5.70	-	1.00	7.02	-	12.72	m3	14.31
6.00m	5.70	-	5.70	-	1.20	8.42	-	14.12	m3	15.89
commencing 1.50m below existing ground level, maximum depth not exceeding										
0.25m	4.80	-	4.80	-	1.00	7.02	-	11.82	m3	13.30
1.00m	4.80	-	4.80	-	1.05	7.37	-	12.17	m3	13.69
2.00m	4.80	-	4.80	-	1.30	9.13	-	13.93	m3	15.67
commencing 3.00m below existing ground level, maximum depth not exceeding										
0.25m	4.80	-	4.80	-	1.45	10.18	-	14.98	m3	16.85
1.00m	5.70	-	5.70	-	1.55	10.88	-	16.58	m3	18.65
2.00m	5.70	-	5.70	-	1.75	12.29	-	17.98	m3	20.23
For pile caps and ground beams between piles										
maximum depth not exceeding										
0.25m	5.81	-	5.81	-	0.95	6.67	-	12.48	m3	14.04
1.00m	5.70	-	5.70	-	1.00	7.02	-	12.72	m3	14.31
2.00m	5.70	-	5.70	-	1.15	8.07	-	13.77	m3	15.49

45

Labour hourly rates: (except Specialists) Craft Operatives £9.23 Labourer £7.02 Rates are national average prices. Refer to REGIONAL VARIATIONS for indicative levels of overall pricing in regions	PLANT & TRANSPORT			LABOUR				RATES		
	Plant Cost £	Trans Cost £	P & T Cost £	Craft Optve Hrs	Lab Hrs	Labour Cost £	Sunds £	Nett Rate £	Unit	Gross Rate (+12.5%) £
D20: EXCAVATING AND FILLING Cont.										
Excavating - by machine Cont.										
Pits										
maximum depth not exceeding										
0.25m	4.90	-	4.90	-	0.95	6.67	-	11.57	m3	13.02
1.00m	6.04	-	6.04	-	1.00	7.02	-	13.06	m3	14.69
2.00m.	6.04	-	6.04	-	1.15	8.07	-	14.11	m3	15.88
4.00m	6.84	-	6.84	-	1.45	10.18	-	17.02	m3	19.15
Extra over any types of excavating irrespective of depth										
excavating below ground water level	4.96	-	4.96	-	-	-	-	4.96	m3	5.58
excavating in running silt, running sand or liquid mud	8.61	-	8.61	-	-	-	-	8.61	m3	9.69
excavating below ground water level in running silt, running sand or liquid mud	11.84	-	11.84	-	-	-	-	11.84	m3	13.32
excavating in heavy soil or clay	1.10	-	1.10	-	-	-	-	1.10	m3	1.24
excavating in gravel	1.72	-	1.72	-	-	-	-	1.72	m3	1.94
excavating in brash, loose rock or chalk	2.06	-	2.06	-	-	-	-	2.06	m3	2.32
Breaking out existing materials; extra over any types of excavating irrespective of depth										
sandstone	17.00	-	17.00	-	-	-	-	17.00	m3	19.13
hard rock	25.58	-	25.58	-	-	-	-	25.58	m3	28.78
concrete	21.24	-	21.24	-	-	-	-	21.24	m3	23.90
reinforced concrete	31.90	-	31.90	-	-	-	-	31.90	m3	35.89
brickwork, blockwork or stonework	14.24	-	14.24	-	-	-	-	14.24	m3	16.02
drain and concrete bed under										
100mm diameter....................................	1.99	-	1.99	-	-	-	-	1.99	m	2.24
150mm diameter....................................	2.40	-	2.40	-	-	-	-	2.40	m	2.70
225mm diameter....................................	2.78	-	2.78	-	-	-	-	2.78	m	3.13
Breaking out existing hard pavings; extra over any types of excavating irrespective of depth										
concrete 150mm thick	3.35	-	3.35	-	-	-	-	3.35	m2	3.77
reinforced concrete 200mm thick	6.72	-	6.72	-	-	-	-	6.72	m2	7.56
coated macadam or asphalt 75mm thick	0.56	-	0.56	-	-	-	-	0.56	m2	0.63
Working space allowance to excavation; including additional earthwork support, disposal (excluding landfill tax) and backfilling										
reduce levels, basements and the like	5.02	-	5.02	-	0.30	2.11	-	7.13	m2	8.02
pits	5.70	-	5.70	-	0.70	4.91	-	10.61	m2	11.94
trenches	5.02	-	5.02	-	0.50	3.51	-	8.53	m2	9.60
Compacting										
bottoms of excavations	0.17	-	0.17	-	-	-	-	0.17	m2	0.19
Compacting with 680 kg vibratory roller										
ground	0.20	-	0.20	-	-	-	-	0.20	m2	0.23
Compacting with 6-8 tonnes smooth wheeled roller										
ground	0.40	-	0.40	-	-	-	-	0.40	m2	0.45
Trimming										
sides of cuttings; vertical or battered	0.49	-	0.49	-	-	-	-	0.49	m2	0.55
sides of embankments; vertical or battered	0.49	-	0.49	-	-	-	-	0.49	m2	0.55
Disposal of excavated material										
depositing on site in temporary spoil heaps where directed										
25m	-	1.86	1.86	-	0.12	0.84	-	2.70	m3	3.04
50m	-	1.94	1.94	-	0.12	0.84	-	2.78	m3	3.13
100m	-	2.12	2.12	-	0.12	0.84	-	2.96	m3	3.33
400m	-	2.51	2.51	-	0.12	0.84	-	3.35	m3	3.77
800m	-	3.26	3.26	-	0.12	0.84	-	4.10	m3	4.62
1200m	-	3.48	3.48	-	0.12	0.84	-	4.32	m3	4.86
1600m	-	3.69	3.69	-	0.12	0.84	-	4.53	m3	5.10
extra for each additional 1600m...............	-	0.57	0.57	-	-	-	-	0.57	m3	0.64
removing from site to tip a) Inert............	-	14.00	14.00	-	-	-	-	14.00	m3	15.75
removing from site to tip b) Active...........	-	23.00	23.00	-	-	-	-	23.00	m3	25.88
removing from site to tip c) Contaminated (Guide price - always seek a quotation for specialist disposal costs.)................	-	50.00	50.00	-	-	-	-	50.00	m3	56.25
Disposal of preserved top soil										
depositing on site in temporary spoil heaps where directed										
25m	-	1.86	1.86	-	0.12	0.84	-	2.70	m3	3.04
50m	-	1.94	1.94	-	0.12	0.84	-	2.78	m3	3.13
100m	-	2.12	2.12	-	0.12	0.84	-	2.96	m3	3.33
Excavating - by hand										
Top soil for preservation										
average 150mm deep	-	-	-	-	0.33	2.32	-	2.32	m2	2.61
To reduce levels										
maximum depth not exceeding										
0.25m	-	-	-	-	2.75	19.31	-	19.31	m3	21.72
1.00m	-	-	-	-	2.90	20.36	-	20.36	m3	22.90
2.00m	-	-	-	-	3.30	23.17	-	23.17	m3	26.06
4.00m	-	-	-	-	4.30	30.19	-	30.19	m3	33.96
6.00m	-	-	-	-	5.35	37.56	-	37.56	m3	42.25

Labour hourly rates: (except Specialists) Craft Operatives £9.23 Labourer £7.02 Rates are national average prices. Refer to REGIONAL VARIATIONS for indicative levels of overall pricing in regions	PLANT & TRANSPORT			LABOUR				RATES		
	Plant Cost	Trans Cost	P & T Cost	Craft Optve	Lab	Labour Cost	Sunds	Nett Rate	Unit	Gross Rate (+12.5%)
	£	£	£	Hrs	Hrs	£	£	£		£

D20: EXCAVATING AND FILLING Cont.

Excavating - by hand Cont.

	Plant	Trans	P & T	Craft	Lab	Labour Cost	Sunds	Nett Rate	Unit	Gross Rate
Basements and the like										
maximum depth not exceeding										
0.25m	-	-	-	-	2.85	20.01	-	20.01	m3	22.51
1.00m	-	-	-	-	3.05	21.41	-	21.41	m3	24.09
2.00m	-	-	-	-	3.40	23.87	-	23.87	m3	26.85
4.00m	-	-	-	-	4.40	30.89	-	30.89	m3	34.75
6.00m	-	-	-	-	5.35	37.56	-	37.56	m3	42.25
Trenches width not exceeding 0.30m										
maximum depth not exceeding										
0.25m	-	-	-	-	4.70	32.99	-	32.99	m3	37.12
1.00m	-	-	-	-	4.95	34.75	-	34.75	m3	39.09
Trenches width exceeding 0.30m										
maximum depth not exceeding										
0.25m	-	-	-	-	3.14	22.04	-	22.04	m3	24.80
1.00m	-	-	-	-	3.30	23.17	-	23.17	m3	26.06
2.00m	-	-	-	-	3.85	27.03	-	27.03	m3	30.41
4.00m	-	-	-	-	4.80	33.70	-	33.70	m3	37.91
6.00m	-	-	-	-	6.05	42.47	-	42.47	m3	47.78
commencing 1.50m below existing ground level,										
maximum depth not exceeding										
0.25m	-	-	-	-	4.70	32.99	-	32.99	m3	37.12
1.00m	-	-	-	-	4.95	34.75	-	34.75	m3	39.09
2.00m	-	-	-	-	6.20	43.52	-	43.52	m3	48.96
commencing 3.00m below existing ground level,										
maximum depth not exceeding										
0.25m	-	-	-	-	7.00	49.14	-	49.14	m3	55.28
1.00m	-	-	-	-	7.30	51.25	-	51.25	m3	57.65
2.00m	-	-	-	-	8.40	58.97	-	58.97	m3	66.34
For pile caps and ground beams between piles										
maximum depth not exceeding										
0.25m	-	-	-	-	3.70	25.97	-	25.97	m3	29.22
1.00m	-	-	-	-	3.85	27.03	-	27.03	m3	30.41
2.00m	-	-	-	-	4.40	30.89	-	30.89	m3	34.75
Pits										
maximum depth not exceeding										
0.25m	-	-	-	-	3.70	25.97	-	25.97	m3	29.22
1.00m	-	-	-	-	3.85	27.03	-	27.03	m3	30.41
2.00m	-	-	-	-	4.40	30.89	-	30.89	m3	34.75
4.00m	-	-	-	-	5.60	39.31	-	39.31	m3	44.23
Extra over any types of excavating irrespective of depth										
excavating below ground water level	-	-	-	-	1.65	11.58	-	11.58	m3	13.03
excavating in running silt, running sand or liquid mud	-	-	-	-	4.95	34.75	-	34.75	m3	39.09
excavating below ground water level in running silt, running sand or liquid mud	-	-	-	-	6.60	46.33	-	46.33	m3	52.12
excavating in heavy soil or clay	-	-	-	-	0.70	4.91	-	4.91	m3	5.53
excavating in gravel	-	-	-	-	1.80	12.64	-	12.64	m3	14.22
excavating in brash, loose rock or chalk	-	-	-	-	3.70	25.97	-	25.97	m3	29.22
Breaking out existing materials; extra over any types of excavating irrespective of depth										
sandstone	-	-	-	-	7.30	51.25	-	51.25	m3	57.65
hard rock	-	-	-	-	14.70	103.19	-	103.19	m3	116.09
concrete	-	-	-	-	11.00	77.22	-	77.22	m3	86.87
reinforced concrete	-	-	-	-	16.50	115.83	-	115.83	m3	130.31
brickwork, blockwork or stonework	-	-	-	-	7.70	54.05	-	54.05	m3	60.81
drain and concrete bed under										
100mm diameter	-	-	-	-	0.90	6.32	-	6.32	m	7.11
150mm diameter	-	-	-	-	1.10	7.72	-	7.72	m	8.69
225mm diameter	-	-	-	-	1.25	8.78	-	8.78	m	9.87
Breaking out existing hard pavings; extra over any types of excavating irrespective of depth										
concrete 150mm thick	-	-	-	-	1.65	11.58	-	11.58	m2	13.03
reinforced concrete 200mm thick	-	-	-	-	3.30	23.17	-	23.17	m2	26.06
coated macadam or asphalt 75mm thick	-	-	-	-	0.30	2.11	-	2.11	m2	2.37
Working space allowance to excavation; including additional earthwork support, disposal (excluding landfill tax) and backfilling										
reduce levels, basements and the like	-	-	-	-	2.85	20.01	-	20.01	m2	22.51
pits	-	-	-	-	3.45	24.22	-	24.22	m2	27.25
trenches	-	-	-	-	3.10	21.76	-	21.76	m2	24.48
Compacting										
bottoms of excavations	-	-	-	-	0.15	1.05	-	1.05	m2	1.18
Trimming										
sides of cuttings; vertical or battered	-	-	-	-	0.22	1.54	-	1.54	m2	1.74
sides of embankments; vertical or battered	-	-	-	-	0.22	1.54	-	1.54	m2	1.74
Disposal of excavated material										
depositing on site in temporary spoil heaps where directed										
25m	-	-	-	-	1.40	9.83	-	9.83	m3	11.06
50m	-	-	-	-	1.65	11.58	-	11.58	m3	13.03

GROUNDWORK

Labour hourly rates: (except Specialists) Craft Operatives £9.23 Labourer £7.02 Rates are national average prices. Refer to REGIONAL VARIATIONS for indicative levels of overall pricing in regions	PLANT & TRANSPORT			LABOUR				RATES		
	Plant Cost £	Trans Cost £	P & T Cost £	Craft Optve Hrs	Lab Hrs	Labour Cost £	Sunds £	Nett Rate £	Unit	Gross Rate (+12.5%) £
D20: EXCAVATING AND FILLING Cont.										
Excavating - by hand Cont.										
Disposal of excavated material Cont.										
depositing on site in temporary spoil heaps where directed Cont.										
100m	-	-	-	-	2.20	15.44	-	15.44	m3	17.37
400m	-	2.51	2.51	-	1.72	12.07	-	14.58	m3	16.41
800m	-	3.26	3.26	-	1.72	12.07	-	15.33	m3	17.25
1200m	-	3.48	3.48	-	1.72	12.07	-	15.55	m3	17.50
1600m	-	3.69	3.69	-	1.72	12.07	-	15.76	m3	17.73
extra for each additional 1600m										
25m	-	0.57	0.57	-	-	-	-	0.57	m3	0.64
removing from site to tip a) Inert	-	14.00	14.00	-	1.50	10.53	-	24.53	m3	27.60
removing from site to tip b) Active	-	23.00	23.00	-	1.50	10.53	-	33.53	m3	37.72
removing from site to tip c) Contaminated (Guide price - always seek a quotation for specialist disposal costs.)	-	50.00	50.00	-	1.50	10.53	-	60.53	m3	68.10
Disposal of preserved top soil										
depositing on site in temporary spoil heaps where directed										
25m	-	-	-	-	1.40	9.83	-	9.83	m3	11.06
50m	-	-	-	-	1.65	11.58	-	11.58	m3	13.03
100m	-	-	-	-	2.20	15.44	-	15.44	m3	17.37
Excavating inside existing building - by hand										
To reduce levels										
maximum depth not exceeding										
0.25m	-	-	-	-	3.85	27.03	-	27.03	m3	30.41
1.00m	-	-	-	-	4.10	28.78	-	28.78	m3	32.38
2.00m	-	-	-	-	4.65	32.64	-	32.64	m3	36.72
4.00m	-	-	-	-	6.00	42.12	-	42.12	m3	47.38
Trenches width not exceeding 0.30m										
maximum depth not exceeding										
0.25m	-	-	-	-	6.65	46.68	-	46.68	m3	52.52
1.00m	-	-	-	-	7.00	49.14	-	49.14	m3	55.28
Trenches width exceeding 0.30m										
maximum depth not exceeding										
0.25m	-	-	-	-	4.45	31.24	-	31.24	m3	35.14
1.00m	-	-	-	-	4.70	32.99	-	32.99	m3	37.12
2.00m	-	-	-	-	5.50	38.61	-	38.61	m3	43.44
4.00m	-	-	-	-	6.80	47.74	-	47.74	m3	53.70
Pits										
maximum depth not exceeding										
0.25m	-	-	-	-	5.20	36.50	-	36.50	m3	41.07
1.00m	-	-	-	-	5.40	37.91	-	37.91	m3	42.65
2.00m	-	-	-	-	6.20	43.52	-	43.52	m3	48.96
4.00m	-	-	-	-	7.85	55.11	-	55.11	m3	62.00
Extra over any types of excavating irrespective of depth										
excavating below ground water level	-	-	-	-	2.35	16.50	-	16.50	m3	18.56
Breaking out existing materials; extra over any types of excavating irrespective of depth										
concrete	-	-	-	-	12.00	84.24	-	84.24	m3	94.77
reinforced concrete	-	-	-	-	17.30	121.45	-	121.45	m3	136.63
brickwork, blockwork or stonework	-	-	-	-	7.80	54.76	-	54.76	m3	61.60
drain and concrete bed under										
100mm diameter	-	-	-	-	1.25	8.78	-	8.78	m	9.87
150mm diameter	-	-	-	-	1.50	10.53	-	10.53	m	11.85
Breaking out existing hard pavings; extra over any types of excavating irrespective of depth										
concrete 150mm thick	-	-	-	-	1.65	11.58	-	11.58	m2	13.03
reinforced concrete 200mm thick	-	-	-	-	3.30	23.17	-	23.17	m2	26.06
Working space allowance to excavation; including additional earthwork support, disposal (excluding landfill tax) and backfilling										
pits	-	-	-	-	4.40	30.89	-	30.89	m2	34.75
trenches	-	-	-	-	4.10	28.78	-	28.78	m2	32.38
Compacting										
bottoms of excavations	-	-	-	-	0.15	1.05	-	1.05	m2	1.18
Disposal of excavated material										
depositing on site in temporary spoil heaps where directed										
25m	-	-	-	-	1.95	13.69	-	13.69	m3	15.40
50m	-	-	-	-	2.15	15.09	-	15.09	m3	16.98
removing from site to tip a) Inert	-	14.00	14.00	-	2.50	17.55	-	31.55	m3	35.49
removing from site to tip b) Active	-	23.00	23.00	-	2.50	17.55	-	40.55	m3	45.62
removing from site to tip c) Contaminated (Guide price - always seek a quotation for specialist disposal costs.)	-	50.00	50.00	-	2.50	17.55	-	67.55	m3	75.99

	MATERIALS			LABOUR				RATES		
Labour hourly rates: (except Specialists) Craft Operatives £9.23 Labourer £7.02 Rates are national average prices. Refer to REGIONAL VARIATIONS for indicative levels of overall pricing in regions	Del to Site	Waste	Material Cost	Craft Optve	Lab	Labour Cost	Sunds	Nett Rate	Unit	Gross Rate (+12.5%)
	£	%	£	Hrs	Hrs	£	£	£		£

D20: EXCAVATING AND FILLING Cont.

Earthwork support

Earthwork support
  distance between opposing faces not exceeding
  2.00m; maximum depth not exceeding										

1.00m	0.94	-	0.94	-	0.21	1.47	0.03	2.44	m2	2.75
2.00m	1.18	-	1.18	-	0.26	1.83	0.04	3.05	m2	3.43
4.00m	1.34	-	1.34	-	0.32	2.25	0.05	3.64	m2	4.09
6.00m	1.52	-	1.52	-	0.37	2.60	0.05	4.17	m2	4.69
distance between opposing faces 2.00 - 4.00m; maximum depth not exceeding										
1.00m	1.02	-	1.02	-	0.22	1.54	0.04	2.60	m2	2.93
2.00m	1.25	-	1.25	-	0.27	1.90	0.04	3.19	m2	3.58
4.00m	1.48	-	1.48	-	0.34	2.39	0.05	3.92	m2	4.41
6.00m	1.67	-	1.67	-	0.39	2.74	0.06	4.47	m2	5.03
distance between opposing faces exceeding 4.00m; maximum depth not exceeding										
1.00m	1.16	-	1.16	-	0.23	1.61	0.04	2.81	m2	3.17
2.00m	1.35	-	1.35	-	0.29	2.04	0.05	3.44	m2	3.87
4.00m	1.60	-	1.60	-	0.35	2.46	0.06	4.12	m2	4.63
6.00m	1.80	-	1.80	-	0.41	2.88	0.06	4.74	m2	5.33

Earthwork support; unstable ground
  distance between opposing faces not exceeding
  2.00m; maximum depth not exceeding										

1.00m	1.60	-	1.60	-	0.35	2.46	0.06	4.12	m2	4.63
2.00m	1.92	-	1.92	-	0.44	3.09	0.07	5.08	m2	5.71
4.00m	2.22	-	2.22	-	0.53	3.72	0.08	6.02	m2	6.77
6.00m	2.53	-	2.53	-	0.62	4.35	0.09	6.97	m2	7.84
distance between opposing faces 2.00 - 4.00m; maximum depth not exceeding										
1.00m	1.76	-	1.76	-	0.37	2.60	0.06	4.42	m2	4.97
2.00m	2.10	-	2.10	-	0.45	3.16	0.07	5.33	m2	6.00
4.00m	2.45	-	2.45	-	0.56	3.93	0.08	6.46	m2	7.27
6.00m	2.81	-	2.81	-	0.65	4.56	0.09	7.46	m2	8.40
distance between opposing faces exceeding 4.00m; maximum depth not exceeding										
1.00m	1.92	-	1.92	-	0.37	2.60	0.07	4.59	m2	5.16
2.00m	2.26	-	2.26	-	0.49	3.44	0.08	5.78	m2	6.50
4.00m	2.69	-	2.69	-	0.58	4.07	0.09	6.85	m2	7.71
6.00m	3.05	-	3.05	-	0.68	4.77	0.10	7.92	m2	8.91

Earthwork support; next to roadways
  distance between opposing faces not exceeding
  2.00m; maximum depth not exceeding										

1.00m	1.86	-	1.86	-	0.42	2.95	0.06	4.87	m2	5.48
2.00m	2.26	-	2.26	-	0.53	3.72	0.08	6.06	m2	6.82
4.00m	2.69	-	2.69	-	0.63	4.42	0.09	7.20	m2	8.10
6.00m	3.05	-	3.05	-	0.74	5.19	0.10	8.34	m2	9.39
distance between opposing faces 2.00 - 4.00m; maximum depth not exceeding										
1.00m	2.10	-	2.10	-	0.44	3.09	0.07	5.26	m2	5.92
2.00m	2.51	-	2.51	-	0.55	3.86	0.09	6.46	m2	7.27
4.00m	2.96	-	2.96	-	0.67	4.70	0.10	7.76	m2	8.73
6.00m	3.34	-	3.34	-	0.78	5.48	0.12	8.94	m2	10.05
distance between opposing faces exceeding 4.00m; maximum depth not exceeding										
1.00m	2.26	-	2.26	-	0.46	3.23	0.08	5.57	m2	6.27
2.00m	2.74	-	2.74	-	0.59	4.14	0.09	6.97	m2	7.84
4.00m	3.20	-	3.20	-	0.69	4.84	0.11	8.15	m2	9.17
6.00m	3.66	-	3.66	-	0.82	5.76	0.13	9.55	m2	10.74

Earthwork support; left in
  distance between opposing faces not exceeding
  2.00m; maximum depth not exceeding										

1.00m	3.38	-	3.38	-	0.17	1.19	0.06	4.63	m2	5.21
2.00m	3.38	-	3.38	-	0.21	1.47	0.07	4.92	m2	5.54
4.00m	3.38	-	3.38	-	0.25	1.75	0.08	5.21	m2	5.87
6.00m	3.38	-	3.38	-	0.29	2.04	0.09	5.51	m2	6.19
distance between opposing faces 2.00 - 4.00m; maximum depth not exceeding										
1.00m	3.38	-	3.38	-	0.18	1.26	0.06	4.70	m2	5.29
2.00m	3.38	-	3.38	-	0.22	1.54	0.08	5.00	m2	5.63
4.00m	3.38	-	3.38	-	0.27	1.90	0.09	5.37	m2	6.04
6.00m	3.38	-	3.38	-	0.32	2.25	0.11	5.74	m2	6.45
distance between opposing faces exceeding 4.00m; maximum depth not exceeding										
1.00m	4.50	-	4.50	-	0.19	1.33	0.07	5.90	m2	6.64
2.00m	4.50	-	4.50	-	0.23	1.61	0.08	6.19	m2	6.97
4.00m	4.50	-	4.50	-	0.27	1.90	0.10	6.50	m2	7.31
6.00m	4.50	-	4.50	-	0.33	2.32	0.11	6.93	m2	7.79

Earthwork support; next to roadways; left in
  distance between opposing faces not exceeding
  2.00m; maximum depth not exceeding										

1.00m	11.25	-	11.25	-	0.34	2.39	0.11	13.75	m2	15.47
2.00m	11.25	-	11.25	-	0.42	2.95	0.14	14.34	m2	16.13
4.00m	11.25	-	11.25	-	0.51	3.58	0.16	14.99	m2	16.86
6.00m	11.25	-	11.25	-	0.59	4.14	0.18	15.57	m2	17.52
distance between opposing faces 2.00 - 4.00m; maximum depth not exceeding										
1.00m	13.50	-	13.50	-	0.36	2.53	0.13	16.16	m2	18.18
2.00m	13.50	-	13.50	-	0.44	3.09	0.15	16.74	m2	18.83
4.00m	13.50	-	13.50	-	0.54	3.79	0.18	17.47	m2	19.65

GROUNDWORK

Labour hourly rates: (except Specialists) Craft Operatives £9.23 Labourer £7.02 Rates are national average prices. Refer to REGIONAL VARIATIONS for indicative levels of overall pricing in regions	MATERIALS			LABOUR				RATES		
	Del to Site	Waste	Material Cost	Craft Optve	Lab	Labour Cost	Sunds	Nett Rate	Unit	Gross Rate (+12.5%)
	£	%	£	Hrs	Hrs	£	£	£		£
D20: EXCAVATING AND FILLING Cont.										
Earthwork support Cont.										
Earthwork support; next to roadways; left in Cont. distance between opposing faces 2.00 - 4.00m; maximum depth not exceeding Cont.										
6.00m	13.50	–	13.50	–	0.62	4.35	0.20	18.05	m2	20.31
distance between opposing faces exceeding 4.00m; maximum depth not exceeding										
1.00m	16.88	–	16.88	–	0.37	2.60	0.14	19.62	m2	22.07
2.00m	16.88	–	16.88	–	0.47	3.30	0.16	20.34	m2	22.88
4.00m	16.88	–	16.88	–	0.56	3.93	0.19	21.00	m2	23.63
6.00m	16.88	–	16.88	–	0.65	4.56	0.22	21.66	m2	24.37
Earthwork support; unstable ground; left in distance between opposing faces not exceeding 2.00m; maximum depth not exceeding										
1.00m	7.88	–	7.88	–	0.27	1.90	0.10	9.88	m2	11.11
2.00m	7.88	–	7.88	–	0.36	2.53	0.12	10.53	m2	11.84
4.00m	7.88	–	7.88	–	0.42	2.95	0.13	10.96	m2	12.33
6.00m	7.88	–	7.88	–	0.49	3.44	0.15	11.47	m2	12.90
distance between opposing faces 2.00 - 4.00m; maximum depth not exceeding										
1.00m	9.00	–	9.00	–	0.29	2.04	0.11	11.15	m2	12.54
2.00m	9.00	–	9.00	–	0.36	2.53	0.13	11.66	m2	13.11
4.00m	9.00	–	9.00	–	0.44	3.09	0.15	12.24	m2	13.77
6.00m	9.00	–	9.00	–	0.53	3.72	0.17	12.89	m2	14.50
distance between opposing faces exceeding 4.00m; maximum depth not exceeding										
1.00m	11.25	–	11.25	–	0.32	2.25	0.12	13.62	m2	15.32
2.00m	11.25	–	11.25	–	0.40	2.81	0.14	14.20	m2	15.97
4.00m	11.25	–	11.25	–	0.46	3.23	0.16	14.64	m2	16.47
6.00m	11.25	–	11.25	–	0.55	3.86	0.18	15.29	m2	17.20
Earthwork support; inside existing building distance between opposing faces not exceeding 2.00m; maximum depth not exceeding										
1.00m	0.98	–	0.98	–	0.23	1.61	0.03	2.62	m2	2.95
2.00m	1.22	–	1.22	–	0.29	2.04	0.04	3.30	m2	3.71
4.00m	1.50	–	1.50	–	0.32	2.25	0.05	3.80	m2	4.27
distance between opposing faces 2.00 - 4.00m; maximum depth not exceeding										
1.00m	1.08	–	1.08	–	0.24	1.68	0.04	2.80	m2	3.16
2.00m	1.34	–	1.34	–	0.30	2.11	0.05	3.50	m2	3.93
4.00m	1.52	–	1.52	–	0.37	2.60	0.05	4.17	m2	4.69
distance between opposing faces exceeding 4.00m; maximum depth not exceeding										
1.00m	1.21	–	1.21	–	0.25	1.75	0.04	3.00	m2	3.38
2.00m	1.45	–	1.45	–	0.33	2.32	0.05	3.82	m2	4.29
4.00m	1.67	–	1.67	–	0.38	2.67	0.06	4.40	m2	4.95
Earthwork support; unstable ground; inside existing building distance between opposing faces not exceeding 2.00m; maximum depth not exceeding										
1.00m	1.67	–	1.67	–	0.38	2.67	0.06	4.40	m2	4.95
2.00m	2.02	–	2.02	–	0.48	3.37	0.07	5.46	m2	6.14
4.00m	2.34	–	2.34	–	0.58	4.07	0.08	6.49	m2	7.30
distance between opposing faces 2.00 - 4.00m; maximum depth not exceeding										
1.00m	1.84	–	1.84	–	0.41	2.88	0.06	4.78	m2	5.38
2.00m	2.20	–	2.20	–	0.49	3.44	0.07	5.71	m2	6.42
4.00m	2.58	–	2.58	–	0.61	4.28	0.09	6.95	m2	7.82
distance between opposing faces exceeding 4.00m; maximum depth not exceeding										
1.00m	2.02	–	2.02	–	0.43	3.02	0.06	5.10	m2	5.74
2.00m	2.38	–	2.38	–	0.55	3.86	0.08	6.32	m2	7.11
4.00m	2.81	–	2.81	–	0.64	4.49	0.09	7.39	m2	8.32
Excavated material arising from excavation - by machine										
Filling to excavation average thickness exceeding 0.25m	–	–	–	–	–	–	2.20	2.20	m3	2.48
Compacting filling	–	–	–	–	–	–	0.22	0.22	m2	0.25
Selected excavated material obtained from on site spoil heaps 25m distant - by machine										
Filling to make up levels; depositing in layers 150mm maximum thickness										
average thickness not exceeding 0.25m	–	–	–	–	–	–	3.05	3.05	m3	3.43
average thickness exceeding 0.25m	–	–	–	–	–	–	2.50	2.50	m3	2.81
Compacting with 680 kg vibratory roller filling	–	–	–	–	–	–	0.21	0.21	m2	0.24
Compacting with 6 - 8 tonnes smooth wheeled roller filling	–	–	–	–	–	–	0.41	0.41	m2	0.46
Trimming sides of cuttings; vertical or battered	–	–	–	–	–	–	0.95	0.95	m2	1.07
sides of embankments; vertical or battered	–	–	–	–	–	–	0.95	0.95	m2	1.07

Labour hourly rates: (except Specialists) Craft Operatives £9.23 Labourer £7.02 Rates are national average prices. Refer to REGIONAL VARIATIONS for indicative levels of overall pricing in regions	MATERIALS			LABOUR			RATES			
	Del to Site £	Waste %	Material Cost £	Craft Optve Hrs	Lab Hrs	Labour Cost £	Sunds £	Nett Rate £	Unit	Gross Rate (+12.5%) £
D20: EXCAVATING AND FILLING Cont.										
Selected excavated material obtained from on site spoil heaps 50m distant - by machine										
Filling to make up levels; depositing in layers 150mm maximum thickness										
average thickness not exceeding 0.25m	-	-	-	-	-	-	3.90	3.90	m3	4.39
average thickness exceeding 0.25m	-	-	-	-	-	-	3.05	3.05	m3	3.43
Compacting with 680 kg vibratory roller										
filling	-	-	-	-	-	-	0.21	0.21	m2	0.24
Compacting with 6 - 8 tonnes smooth wheeled roller										
filling	-	-	-	-	-	-	0.41	0.41	m2	0.46
Trimming										
sides of cuttings; vertical or battered	-	-	-	-	-	-	0.95	0.95	m2	1.07
sides of embankments; vertical or battered	-	-	-	-	-	-	0.95	0.95	m2	1.07
Preserved topsoil obtained from on site spoil heaps not exceeding 100m distant - by machine										
Filling to make up levels										
average thickness not exceeding 0.25m	-	-	-	-	-	-	5.90	5.90	m3	6.64
Imported topsoil; wheeling not exceeding 50m - by machine										
Filling to make up levels										
average thickness not exceeding 0.25m	9.98	10.00	10.98	-	-	-	5.05	16.03	m3	18.03
Hard, dry, broken brick or stone to be obtained off site; wheeling not exceeding 25m - by machine										
Filling to make up levels; depositing in layers 150mm maximum thickness										
average thickness not exceeding 0.25m	13.02	25.00	16.27	-	-	-	3.05	19.32	m3	21.74
average thickness exceeding 0.25m	13.02	25.00	16.27	-	-	-	2.50	18.77	m3	21.12
Surface packing to filling										
to vertical or battered faces	-	-	-	-	-	-	0.57	0.57	m2	0.64
Compacting with 680 kg vibratory roller										
filling; blinding with sand, ashes or similar fine material	0.20	40.00	0.28	-	0.05	0.35	0.21	0.84	m2	0.95
Compacting with 6 - 8 tonnes smooth wheeled roller										
filling; blinding with sand, ashes or similar fine material	0.20	40.00	0.28	-	0.05	0.35	0.41	1.04	m2	1.17
Hard, dry, broken brick or stone to be obtained off site; wheeling not exceeding 50m - by machine										
Filling to make up levels; depositing in layers 150mm maximum thickness										
average thickness not exceeding 0.25m	13.02	25.00	16.27	-	-	-	3.90	20.18	m3	22.70
average thickness exceeding 0.25m	13.02	25.00	16.27	-	-	-	3.05	19.32	m3	21.74
Surface packing to filling										
to vertical or battered faces	-	-	-	-	-	-	0.57	0.57	m2	0.64
Compacting with 680 kg vibratory roller										
filling; blinding with sand, ashes or similar fine material	0.20	40.00	0.28	-	0.05	0.35	0.21	0.84	m2	0.95
Compacting with 6 - 8 tonnes smooth wheeled roller										
filling; blinding with sand, ashes or similar fine material	0.20	40.00	0.28	-	0.05	0.35	0.41	1.04	m2	1.17
MOT Type 1 to be obtained off site; wheeling not exceeding 25m - by machine										
Filling to make up levels; depositing in layers 150mm maximum thickness										
average thickness not exceeding 0.25m	15.23	25.00	19.04	-	-	-	3.05	22.09	m3	24.85
average thickness exceeding 0.25m	15.23	25.00	19.04	-	-	-	2.40	21.44	m3	24.12
Surface packing to filling										
to vertical or battered faces	-	-	-	-	-	-	0.57	0.57	m2	0.64
Compacting with 680 kg vibratory roller										
filling; blinding with sand, ashes or similar fine material	0.20	40.00	0.28	-	0.05	0.35	0.21	0.84	m2	0.95
Compacting with 6 - 8 tonnes smooth wheeled roller										
filling; blinding with sand, ashes or similar fine material	0.20	40.00	0.28	-	0.05	0.35	0.41	1.04	m2	1.17
MOT Type 1 to be obtained off site; wheeling not exceeding 50m - by machine										
Filling to make up levels; depositing in layers 150mm maximum thickness										
average thickness not exceeding 0.25m	15.23	25.00	19.04	-	-	-	3.90	22.94	m3	25.80
average thickness exceeding 0.25m	15.23	25.00	19.04	-	-	-	3.05	22.09	m3	24.85

GROUNDWORK

Labour hourly rates: (except Specialists) Craft Operatives £9.23 Labourer £7.02 Rates are national average prices. Refer to REGIONAL VARIATIONS for indicative levels of overall pricing in regions	MATERIALS			LABOUR				RATES		
	Del to Site £	Waste %	Material Cost £	Craft Optve Hrs	Lab Hrs	Labour Cost £	Sunds £	Nett Rate £	Unit	Gross Rate (+12.5%) £
D20: EXCAVATING AND FILLING Cont.										
MOT Type 1 to be obtained off site; wheeling not exceeding 50m - by machine Cont.										
Surface packing to filling										
to vertical or battered faces	-	-	-	-	-	-	0.57	0.57	m2	0.64
Compacting with 680 kg vibratory roller										
filling; blinding with sand, ashes or similar fine material	0.20	40.00	0.28	-	0.05	0.35	0.21	0.84	m2	0.95
Compacting with 6 - 8 tonnes smooth wheeled roller										
filling; blinding with sand, ashes or similar fine material	0.20	40.00	0.28	-	0.05	0.35	0.41	1.04	m2	1.17
MOT Type 2 to be obtained off site; wheeling not exceeding 25m - by machine										
Filling to make up levels; depositing in layers 150mm maximum thickness										
average thickness not exceeding 0.25m	14.90	25.00	18.63	-	-	-	3.05	21.68	m3	24.38
average thickness exceeding 0.25m	14.90	25.00	18.63	-	-	-	2.50	21.13	m3	23.77
Surface packing to filling										
to vertical or battered faces	-	-	-	-	-	-	0.57	0.57	m2	0.64
Compacting with 680 kg vibratory roller										
filling; blinding with sand, ashes or similar fine material	0.20	40.00	0.28	-	0.05	0.35	0.21	0.84	m2	0.95
Compacting with 6 - 8 tonnes smooth wheeled roller										
filling; blinding with sand, ashes or similar fine material	0.20	40.00	0.28	-	0.05	0.35	0.41	1.04	m2	1.17
MOT Type 2 to be obtained off site; wheeling not exceeding 50m - by machine										
Filling to make up levels; depositing in layers 150mm maximum thickness										
average thickness not exceeding 0.25m	14.90	25.00	18.63	-	-	-	3.90	22.52	m3	25.34
average thickness exceeding 0.25m	14.90	25.00	18.63	-	-	-	3.05	21.68	m3	24.38
Surface packing to filling										
to vertical or battered faces	-	-	-	-	-	-	0.57	0.57	m2	0.64
Compacting with 680 kg vibratory roller										
filling; blinding with sand, ashes or similar fine material	0.20	40.00	0.28	-	0.05	0.35	0.21	0.84	m2	0.95
Compacting with 6 - 8 tonnes smooth wheeled roller										
filling; blinding with sand, ashes or similar fine material	0.20	40.00	0.28	-	0.05	0.35	0.41	1.04	m2	1.17
Ashes to be obtained off site; wheeling not exceeding 25m - by machine										
Filling to make up levels; depositing in layers 150mm maximum thickness										
average thickness not exceeding 0.25m	13.00	40.00	18.20	-	-	-	3.05	21.25	m3	23.91
Compacting with 680 kg vibratory roller										
filling...................................	-	-	-	-	-	-	0.21	0.21	m2	0.24
Compacting with 6 - 8 tonnes smooth wheeled roller										
filling...................................	-	-	-	-	-	-	0.41	0.41	m2	0.46
Ashes to be obtained off site; wheeling not exceeding 50m - by machine										
Filling to make up levels; depositing in layers 150mm maximum thickness										
average thickness not exceeding 0.25m	13.00	40.00	18.20	-	-	-	3.90	22.10	m3	24.86
Compacting with 680 kg vibratory roller										
filling...................................	-	-	-	-	-	-	0.21	0.21	m2	0.24
Compacting with 6 - 8 tonnes smooth wheeled roller										
filling...................................	-	-	-	-	-	-	0.41	0.41	m2	0.46
Sand to be obtained off site; wheeling not exceeding 25m - by machine										
Filling to make up levels; depositing in layers 150mm maximum thickness										
average thickness not exceeding 0.25m	9.70	33.00	12.90	-	-	-	3.05	15.95	m3	17.94
average thickness exceeding 0.25m	9.70	33.00	12.90	-	-	-	2.50	15.40	m3	17.33
Compacting with 680 kg vibratory roller										
filling...................................	-	-	-	-	-	-	0.21	0.21	m2	0.24
Compacting with 6 - 8 tonnes smooth wheeled roller										
filling...................................	-	-	-	-	-	-	0.41	0.41	m2	0.46

Labour hourly rates: (except Specialists) Craft Operatives £9.23 Labourer £7.02 Rates are national average prices. Refer to REGIONAL VARIATIONS for indicative levels of overall pricing in regions	MATERIALS			LABOUR				RATES		
	Del to Site £	Waste %	Material Cost £	Craft Optve Hrs	Lab Hrs	Labour Cost £	Sunds £	Nett Rate £	Unit	Gross Rate (+12.5%) £
D20: EXCAVATING AND FILLING Cont.										
Sand to be obtained off site; wheeling not exceeding 50m - by machine										
Filling to make up levels; depositing in layers 150mm maximum thickness										
average thickness not exceeding 0.25m	9.70	33.00	12.90	-	-	-	3.90	16.80	m3	18.90
average thickness exceeding 0.25m	9.70	33.00	12.90	-	-	-	3.05	15.95	m3	17.94
Compacting with 680 kg vibratory roller										
filling	-	-	-	-	-	-	0.21	0.21	m2	0.24
Compacting with 6 - 8 tonnes smooth wheeled roller										
filling	-	-	-	-	-	-	0.41	0.41	m2	0.46
Hoggin to be obtained off site; wheeling not exceeding 25m - by machine										
Filling to make up levels; depositing in layers 150mm maximum thickness										
average thickness not exceeding 0.25m	8.40	33.00	11.17	-	-	-	3.05	14.22	m3	16.00
average thickness exceeding 0.25m	8.40	33.00	11.17	-	-	-	2.50	13.67	m3	15.38
Compacting with 680 kg vibratory roller										
filling	-	-	-	-	-	-	0.21	0.21	m2	0.24
Compacting with 6 - 8 tonnes smooth wheeled roller										
filling	-	-	-	-	-	-	0.41	0.41	m2	0.46
Hoggin to be obtained off site; wheeling not exceeding 50m - by machine										
Filling to make up levels; depositing in layers 150mm maximum thickness										
average thickness not exceeding 0.25m	8.40	33.00	11.17	-	-	-	3.90	15.07	m3	16.96
average thickness exceeding 0.25m	8.40	33.00	11.17	-	-	-	3.05	14.22	m3	16.00
Compacting with 680 kg vibratory roller										
filling	-	-	-	-	-	-	0.21	0.21	m2	0.24
Compacting with 6 - 8 tonnes smooth wheeled roller										
filling	-	-	-	-	-	-	0.41	0.41	m2	0.46
Excavated material arising from excavation - by hand										
Filling to excavation										
average thickness exceeding 0.25m	-	-	-	-	1.44	10.11	-	10.11	m3	11.37
Compacting										
filling	-	-	-	-	-	-	0.47	0.47	m2	0.53
Selected excavated material obtained from on site spoil heaps 25m distant - by hand										
Filling to make up levels; depositing in layers 150mm maximum thickness										
average thickness not exceeding 0.25m	-	-	-	-	1.85	12.99	-	12.99	m3	14.61
average thickness exceeding 0.25m	-	-	-	-	1.60	11.23	-	11.23	m3	12.64
Compacting with 680 kg vibratory roller										
filling	-	-	-	-	-	-	0.21	0.21	m2	0.24
Compacting with 6 - 8 tonnes smooth wheeled roller										
filling	-	-	-	-	-	-	0.41	0.41	m2	0.46
Trimming										
sides of cuttings; vertical or battered	-	-	-	-	0.22	1.54	-	1.54	m2	1.74
sides of embankments; vertical or battered	-	-	-	-	0.22	1.54	-	1.54	m2	1.74
Selected excavated material obtained from on site spoil heaps 50m distant - by hand										
Filling to make up levels; depositing in layers 150mm maximum thickness										
average thickness not exceeding 0.25m	-	-	-	-	2.10	14.74	-	14.74	m3	16.58
average thickness exceeding 0.25m	-	-	-	-	1.95	13.69	-	13.69	m3	15.40
Compacting with 680 kg vibratory roller										
filling	-	-	-	-	-	-	0.21	0.21	m2	0.24
Compacting with 6 - 8 tonnes smooth wheeled roller										
filling	-	-	-	-	-	-	0.41	0.41	m2	0.46
Trimming										
sides of cuttings; vertical or battered	-	-	-	-	0.22	1.54	-	1.54	m2	1.74
sides of embankments; vertical or battered	-	-	-	-	0.22	1.54	-	1.54	m2	1.74
Preserved topsoil obtained from on site spoil heaps not exceeding 100m distant - by hand										
Filling to make up levels										
average thickness not exceeding 0.25m	-	-	-	-	4.40	30.89	-	30.89	m3	34.75
Filling to external planters										
average thickness not exceeding 0.25m	-	-	-	-	5.50	38.61	-	38.61	m3	43.44

GROUNDWORK

Labour hourly rates: (except Specialists) Craft Operatives £9.23 Labourer £7.02 Rates are national average prices. Refer to REGIONAL VARIATIONS for indicative levels of overall pricing in regions	MATERIALS			LABOUR				RATES		
	Del to Site £	Waste %	Material Cost £	Craft Optve Hrs	Lab Hrs	Labour Cost £	Sunds £	Nett Rate £	Unit	Gross Rate (+12.5%) £
D20: EXCAVATING AND FILLING Cont.										
Imported topsoil; wheeling not exceeding 50m - by hand										
Filling to make up levels										
average thickness not exceeding 0.25m	10.00	10.00	11.00	-	3.85	27.03	-	38.03	m3	42.78
Filling to external planters										
average thickness exceeding 0.25m	10.00	10.00	11.00	-	4.95	34.75	-	45.75	m3	51.47
Hard, dry, broken brick or stone to be obtained off site; wheeling not exceeding 25m - by hand										
Filling to make up levels; depositing in layers 150mm maximum thickness										
average thickness not exceeding 0.25m	13.00	25.00	16.25	-	1.58	11.09	-	27.34	m3	30.76
average thickness exceeding 0.25m	13.00	25.00	16.25	-	1.31	9.20	-	25.45	m3	28.63
Surface packing to filling										
to vertical or battered faces	-	-	-	-	0.22	1.54	-	1.54	m2	1.74
Compacting with 680 kg vibratory roller filling; blinding with sand, ashes or similar fine material	0.20	40.00	0.28	-	0.05	0.35	0.21	0.84	m2	0.95
Compacting with 6 - 8 tonnes smooth wheeled roller filling; blinding with sand, ashes or similar fine material	0.20	40.00	0.28	-	0.05	0.35	0.41	1.04	m2	1.17
Hard, dry, broken brick or stone to be obtained off site; wheeling not exceeding 50m - by hand										
Filling to make up levels; depositing in layers 150mm maximum thickness										
average thickness not exceeding 0.25m	13.00	25.00	16.25	-	1.85	12.99	-	29.24	m3	32.89
average thickness exceeding 0.25m	13.00	25.00	16.25	-	1.58	11.09	-	27.34	m3	30.76
Surface packing to filling										
to vertical or battered faces	-	-	-	-	0.22	1.54	-	1.54	m2	1.74
Compacting with 680 kg vibratory roller filling; blinding with sand, ashes or similar fine material	0.20	40.00	0.28	-	0.05	0.35	0.21	0.84	m2	0.95
Compacting with 6 - 8 tonnes smooth wheeled roller filling; blinding with sand, ashes or similar fine material	0.20	40.00	0.28	-	0.05	0.35	0.41	1.04	m2	1.17
MOT Type 1 to be obtained off site; wheeling not exceeding 25m - by hand										
Filling to make up levels; depositing in layers 150mm maximum thickness										
average thickness not exceeding 0.25m	15.20	25.00	19.00	-	1.58	11.09	-	30.09	m3	33.85
average thickness exceeding 0.25m	15.20	25.00	19.00	-	1.31	9.20	-	28.20	m3	31.72
Surface packing to filling										
to vertical or battered faces	-	-	-	-	0.22	1.54	-	1.54	m2	1.74
Compacting with 680 kg vibratory roller filling; blinding with sand, ashes or similar fine material	0.20	40.00	0.28	-	0.05	0.35	0.21	0.84	m2	0.95
Compacting with 6 - 8 tonnes smooth wheeled roller filling; blinding with sand, ashes or similar fine material	0.20	40.00	0.28	-	0.05	0.35	0.41	1.04	m2	1.17
MOT Type 1 to be obtained off site; wheeling not exceeding 50m - by hand										
Filling to make up levels; depositing in layers 150mm maximum thickness										
average thickness not exceeding 0.25m	15.20	25.00	19.00	-	1.85	12.99	-	31.99	m3	35.99
average thickness exceeding 0.25m	15.20	25.00	19.00	-	1.58	11.09	-	30.09	m3	33.85
Surface packing to filling										
to vertical or battered faces	-	-	-	-	0.22	1.54	-	1.54	m2	1.74
Compacting with 680 kg vibratory roller filling; blinding with sand, ashes or similar fine material	0.20	40.00	0.28	-	0.05	0.35	0.21	0.84	m2	0.95
Compacting with 6 - 8 tonnes smooth wheeled roller filling; blinding with sand, ashes or similar fine material	0.20	40.00	0.28	-	0.05	0.35	0.41	1.04	m2	1.17
MOT Type 2 to be obtained off site; wheeling not exceeding 25m - by hand										
Filling to make up levels; depositing in layers 150mm maximum thickness										
average thickness not exceeding 0.25m	14.90	25.00	18.63	-	1.58	11.09	-	29.72	m3	33.43
average thickness exceeding 0.25m	14.90	25.00	18.63	-	1.31	9.20	-	27.82	m3	31.30

Labour hourly rates: (except Specialists) Craft Operatives £9.23 Labourer £7.02. Rates are national average prices. Refer to REGIONAL VARIATIONS for indicative levels of overall pricing in regions	MATERIALS			LABOUR				RATES		
	Del to Site £	Waste %	Material Cost £	Craft Optve Hrs	Lab Hrs	Labour Cost £	Sunds £	Nett Rate £	Unit	Gross Rate (+12.5%) £
D20: EXCAVATING AND FILLING Cont.										
MOT Type 2 to be obtained off site; wheeling not exceeding 25m - by hand Cont.										
Surface packing to filling to vertical or battered faces	-	-	-	-	0.22	1.54	-	1.54	m2	1.74
Compacting with 680 kg vibratory roller filling; blinding with sand, ashes or similar fine material	0.20	40.00	0.28	-	0.05	0.35	0.21	0.84	m2	0.95
Compacting with 6 - 8 tonnes smooth wheeled roller filling; blinding with sand, ashes or similar fine material	0.20	40.00	0.28	-	0.05	0.35	0.41	1.04	m2	1.17
MOT Type 2 to be obtained off site; wheeling not exceeding 50m - by hand										
Filling to make up levels; depositing in layers 150mm maximum thickness										
average thickness not exceeding 0.25m	14.90	25.00	18.63	-	1.85	12.99	-	31.61	m3	35.56
average thickness exceeding 0.25m	14.90	25.00	18.63	-	1.58	11.09	-	29.72	m3	33.43
Surface packing to filling to vertical or battered faces	-	-	-	-	0.22	1.54	-	1.54	m2	1.74
Compacting with 680 kg vibratory roller filling; blinding with sand, ashes or similar fine material	0.20	40.00	0.28	-	0.05	0.35	0.21	0.84	m2	0.95
Compacting with 6 - 8 tonnes smooth wheeled roller filling; blinding with sand, ashes or similar fine material	0.20	40.00	0.28	-	0.05	0.35	0.41	1.04	m2	1.17
Ashes to be obtained off site; wheeling not exceeding 25m - by hand										
Filling to make up levels; depositing in layers 150mm maximum thickness										
average thickness not exceeding 0.25m	13.00	40.00	18.20	-	1.70	11.93	-	30.13	m3	33.90
Compacting with 680 kg vibratory roller filling	-	-	-	-	-	-	0.21	0.21	m2	0.24
Compacting with 6 - 8 tonnes smooth wheeled roller filling	-	-	-	-	-	-	0.41	0.41	m2	0.46
Ashes to be obtained off site; wheeling not exceeding 50m - by hand										
Filling to make up levels; depositing in layers 150mm maximum thickness										
average thickness not exceeding 0.25m	13.00	40.00	18.20	-	1.94	13.62	-	31.82	m3	35.80
Compacting with 680 kg vibratory roller filling	-	-	-	-	-	-	0.21	0.21	m2	0.24
Compacting with 6 - 8 tonnes smooth wheeled roller filling	-	-	-	-	-	-	0.41	0.41	m2	0.46
Sand to be obtained off site; wheeling not exceeding 25m - by hand										
Filling to make up levels; depositing in layers 150mm maximum thickness										
average thickness not exceeding 0.25m	9.70	33.00	12.90	-	1.84	12.92	-	25.82	m3	29.05
average thickness exceeding 0.25m	9.70	33.00	12.90	-	1.58	11.09	-	23.99	m3	26.99
Compacting with 680 kg vibratory roller filling	-	-	-	-	-	-	0.21	0.21	m2	0.24
Compacting with 6 - 8 tonnes smooth wheeled roller filling	-	-	-	-	-	-	0.41	0.41	m2	0.46
Sand to be obtained off site; wheeling not exceeding 50m - by hand										
Filling to make up levels; depositing in layers 150mm maximum thickness										
average thickness not exceeding 0.25m	9.70	33.00	12.90	-	2.10	14.74	-	27.64	m3	31.10
average thickness exceeding 0.25m	9.70	33.00	12.90	-	1.84	12.92	-	25.82	m3	29.05
Compacting with 680 kg vibratory roller filling	-	-	-	-	-	-	0.21	0.21	m2	0.24
Compacting with 6 - 8 tonnes smooth wheeled roller filling	-	-	-	-	-	-	0.41	0.41	m2	0.46
Hoggin to be obtained off site; wheeling not exceeding 25m - by hand										
Filling to make up levels; depositing in layers 150mm maximum thickness										
average thickness not exceeding 0.25m	8.40	33.00	11.17	-	1.55	10.88	-	22.05	m3	24.81
average thickness exceeding 0.25m	8.40	33.00	11.17	-	1.30	9.13	-	20.30	m3	22.84

Labour hourly rates: (except Specialists) Craft Operatives £9.23 Labourer £7.02. Rates are national average prices. Refer to REGIONAL VARIATIONS for indicative levels of overall pricing in regions	MATERIALS			LABOUR				RATES		
	Del to Site £	Waste %	Material Cost £	Craft Optve Hrs	Lab Hrs	Labour Cost £	Sunds £	Nett Rate £	Unit	Gross Rate (+12.5%) £
D20: EXCAVATING AND FILLING Cont.										
Hoggin to be obtained off site; wheeling not exceeding 25m – by hand Cont.										
Compacting with 680 kg vibratory roller filling	-	-	-	-	-	-	0.21	0.21	m2	0.24
Compacting with 6 – 8 tonnes smooth wheeled roller filling	-	-	-	-	-	-	0.41	0.41	m2	0.46
Hoggin to be obtained off site; wheeling not exceeding 50m – by hand										
Filling to make up levels; depositing in layers 150mm maximum thickness										
average thickness not exceeding 0.25m	8.40	33.00	11.17	-	1.80	12.64	-	23.81	m3	26.78
average thickness exceeding 0.25m	8.40	33.00	11.17	-	1.55	10.88	-	22.05	m3	24.81
Compacting with 680 kg vibratory roller filling	-	-	-	-	-	-	0.21	0.21	m2	0.24
Compacting with 6 – 8 tonnes smooth wheeled roller filling	-	-	-	-	-	-	0.41	0.41	m2	0.46
Excavated material arising from excavation; work inside existing building – by hand										
Filling to excavation										
average thickness exceeding 0.25m	-	-	-	-	2.80	19.66	-	19.66	m3	22.11
Compacting filling	-	-	-	-	-	-	0.21	0.21	m2	0.24
Selected excavated material obtained from on site spoil heaps 25m distant; work inside existing building – by hand										
Filling to make up levels; depositing in layers 150mm maximum thickness										
average thickness not exceeding 0.25m	-	-	-	-	3.26	22.89	-	22.89	m3	25.75
average thickness exceeding 0.25m	-	-	-	-	3.08	21.62	-	21.62	m3	24.32
Compacting filling	-	-	-	-	-	-	0.21	0.21	m2	0.24
Selected excavated material obtained from on site spoil heaps 50m distant; work inside existing building – by hand										
Filling to make up levels; depositing in layers 150mm maximum thickness										
average thickness not exceeding 0.25m	-	-	-	-	3.52	24.71	-	24.71	m3	27.80
average thickness exceeding 0.25m	-	-	-	-	3.36	23.59	-	23.59	m3	26.54
Compacting filling	-	-	-	-	-	-	0.21	0.21	m2	0.24
Hard, dry, broken brick or stone to be obtained off site; wheeling not exceeding 25m; work inside existing building – by hand										
Filling to make up levels; depositing in layers 150mm maximum thickness										
average thickness not exceeding 0.25m	13.00	25.00	16.25	-	2.99	20.99	-	37.24	m3	41.89
average thickness exceeding 0.25m	13.00	25.00	16.25	-	2.81	19.73	-	35.98	m3	40.47
Surface packing to filling to vertical or battered faces	-	-	-	-	0.22	1.54	-	1.54	m2	1.74
Compacting filling; blinding with sand, ashes or similar fine material	0.20	40.00	0.28	-	0.05	0.35	0.21	0.84	m2	0.95
Hard, dry, broken brick or stone to be obtained off site; wheeling not exceeding 50m; work inside existing building – by hand										
Filling to make up levels; depositing in layers 150mm maximum thickness										
average thickness not exceeding 0.25m	13.00	25.00	16.25	-	3.26	22.89	-	39.14	m3	44.03
average thickness exceeding 0.25m	13.00	25.00	16.25	-	2.99	20.99	-	37.24	m3	41.89
Surface packing to filling to vertical or battered faces	-	-	-	-	0.22	1.54	-	1.54	m2	1.74
Compacting filling; blinding with sand, ashes or similar fine material	0.20	40.00	0.28	-	0.05	0.35	0.21	0.84	m2	0.95
Ashes to be obtained off site; wheeling not exceeding 25m; work inside existing building – by hand										
Filling to make up levels; depositing in layers 150mm maximum thickness										
average thickness not exceeding 0.25m	13.00	40.00	18.20	-	3.10	21.76	-	39.96	m3	44.96

Labour hourly rates: (except Specialists) Craft Operatives £9.23 Labourer £7.02 Rates are national average prices. Refer to REGIONAL VARIATIONS for indicative levels of overall pricing in regions	MATERIALS			LABOUR				RATES		
	Del to Site £	Waste %	Material Cost £	Craft Optve Hrs	Lab Hrs	Labour Cost £	Sunds £	Nett Rate £	Unit	Gross Rate (+12.5%) £
D20: EXCAVATING AND FILLING Cont.										
Ashes to be obtained off site; wheeling not exceeding 25m; work inside existing building - by hand Cont.										
Compacting										
filling	-	-	-	-	-	-	0.21	0.21	m2	0.24
Ashes to be obtained off site; wheeling not exceeding 50m; work inside existing building - by hand										
Filling to make up levels; depositing in layers 150mm maximum thickness										
average thickness not exceeding 0.25m	13.00	40.00	18.20	-	3.36	23.59	-	41.79	m3	47.01
Compacting										
filling	-	-	-	-	-	-	0.21	0.21	m2	0.24
Sand to be obtained off site; wheeling not exceeding 25m; work inside existing building - by hand										
Filling to make up levels; depositing in layers 150mm maximum thickness										
average thickness not exceeding 0.25m	9.70	33.00	12.90	-	3.26	22.89	-	35.79	m3	40.26
average thickness exceeding 0.25m	9.70	33.00	12.90	-	2.99	20.99	-	33.89	m3	38.13
Compacting										
filling	-	-	-	-	-	-	0.21	0.21	m2	0.24
Sand to be obtained off site; wheeling not exceeding 50m; work inside existing building - by hand										
Filling to make up levels; depositing in layers 150mm maximum thickness										
average thickness not exceeding 0.25m	9.70	33.00	12.90	-	3.52	24.71	-	37.61	m3	42.31
average thickness exceeding 0.25m	9.70	33.00	12.90	-	3.26	22.89	-	35.79	m3	40.26
Compacting										
filling	-	-	-	-	-	-	0.21	0.21	m2	0.24
D30: CAST IN PLACE PILING										
Prices include for a 21 N/mm2 concrete mix, nominal reinforcement and a minimum number of 50 piles on any one contract. The working loads sizes and lengths given below will depend on the nature of the soils in which the piles will be founded as well as structure to be supported (SIMPLEX PILING LTD.)										
On/Off site charge										
in addition to the following prices add approximately	-	-	Spclist	-	-	Spclist	-	4160.00	Sm	4680.00
Short auger piles										
up to 6m long; 450mm nominal diameter; 8 tonnes normal working load	-	-	Spclist	-	-	Spclist	-	24.36	m	27.41
up to 6m long; 610mm nominal diameter; 30 tonnes normal working load	-	-	Spclist	-	-	Spclist	-	36.85	m	41.46
Bored cast in-situ piles										
up to 15m long; 450mm nominal diameter; 40 tonnes normal working load	-	-	Spclist	-	-	Spclist	-	22.48	m	25.29
up to 15m long; 610mm nominal diameter; 120 tonnes normal working load	-	-	Spclist	-	-	Spclist	-	34.98	m	39.35
Auger piles										
up to 20m long; 450mm nominal diameter	-	-	Spclist	-	-	Spclist	-	22.80	m	25.65
up to 30m long; 1200mm nominal diameter	-	-	Spclist	-	-	Spclist	-	113.06	m	127.19
Large diameter auger piles (plain shaft)										
up to 20m long; 610mm nominal diameter; 150 tonnes normal working load	-	-	Spclist	-	-	Spclist	-	34.69	m	39.03
up to 30m long; 1525mm nominal diameter; 600 tonnes normal working load	-	-	Spclist	-	-	Spclist	-	175.89	m	197.88
Large diameter auger piles (belled base)										
up to 20m long; 610mm nominal diameter; 150 tonnes normal working load	-	-	Spclist	-	-	Spclist	-	47.48	m	53.41
up to 30m long; 1525mm nominal diameter; 1000 tonnes normal working load	-	-	Spclist	-	-	Spclist	-	226.19	m	254.46
Boring through obstructions, rock like formations, etc										
undertaken on a time basis at a rate per piling rig per hour										
Cutting off tops of piles; including preparation and integration of reinforcement into pile cap or ground beam and disposal										
450-610mm nominal diameter	-	-	Spclist	-	-	Spclist	-	8.34	m	9.38
610-1525mm nominal diameter	-	-	Spclist	-	-	Spclist	-	32.14	m	36.16

GROUNDWORK

Labour hourly rates: (except Specialists) Craft Operatives £9.23 Labourer £7.02 Rates are national average prices. Refer to REGIONAL VARIATIONS for indicative levels of overall pricing in regions	MATERIALS			LABOUR				RATES		
	Del to Site £	Waste %	Material Cost £	Craft Optve Hrs	Lab Hrs	Labour Cost £	Sunds £	Nett Rate £	Unit	Gross Rate (+12.5%) £
D32: STEEL PILING										
Mild steel Universal Bearing Piles 'H' section, to B.S.EN10025 Grade S275 and high yield steel Grade S275JR in lengths 9 - 15m supplied, handled, pitched and driven vertically with landbased plant (SIMPLEX PILING LTD.)										
Note the following prices are based on quantities of 25 - 150 tonnes										
On/Off site charge										
in addition to the following prices, add										
approximately	-	-	Spclist	-	-	Spclist	-	3460.00	Sm	3892.50
Mild steel piles										
203 x 203 x 45 kg/m, SWL 40 tonnes	-	-	Spclist	-	-	Spclist	-	33.46	m	37.64
203 x 203 x 54 kg/m, SWL 50 tonnes	-	-	Spclist	-	-	Spclist	-	37.74	m	42.46
254 x 254 x 63 kg/m, SWL 60 tonnes	-	-	Spclist	-	-	Spclist	-	42.66	m	47.99
254 x 254 x 71 kg/m, SWL 70 tonnes	-	-	Spclist	-	-	Spclist	-	45.94	m	51.68
305 x 305 x 79 kg/m, SWL 75 tonnes	-	-	Spclist	-	-	Spclist	-	54.10	m	60.86
254 x 254 x 85 kg/m, SWL 80 tonnes	-	-	Spclist	-	-	Spclist	-	55.68	m	62.64
305 x 305 x 88 kg/m, SWL 85 tonnes	-	-	Spclist	-	-	Spclist	-	58.66	m	65.99
305 x 305 x 95 kg/m, SWL 90 tonnes	-	-	Spclist	-	-	Spclist	-	62.37	m	70.17
356 x 368 x 109 kg/m, SWL 105 tonnes	-	-	Spclist	-	-	Spclist	-	74.73	m	84.07
305 x 305 x 110 kg/m, SWL 105 tonnes	-	-	Spclist	-	-	Spclist	-	70.30	m	79.09
305 x 305 x 126 kg/m, SWL 120 tonnes	-	-	Spclist	-	-	Spclist	-	80.20	m	90.22
305 x 305 x 149 kg/m, SWL 140 tonnes	-	-	Spclist	-	-	Spclist	-	88.51	m	99.57
356 x 368 x 133 kg/m, SWL 130 tonnes	-	-	Spclist	-	-	Spclist	-	92.64	m	104.22
356 x 368 x 152 kg/m, SWL 140 tonnes	-	-	Spclist	-	-	Spclist	-	98.32	m	110.61
356 x 368 x 174 kg/m, SWL 165 tonnes	-	-	Spclist	-	-	Spclist	-	113.07	m	127.20
305 x 305 x 186 kg/m, SWL 175 tonnes	-	-	Spclist	-	-	Spclist	-	122.94	m	138.31
305 x 305 x 223 kg/m, SWL 210 tonnes	-	-	Spclist	-	-	Spclist	-	134.55	m	151.37
Extra for high yield steel										
add......................................	-	-	Spclist	-	-	Spclist	-	36.83	T	41.43
For quiet piling, add										
to terminal charge, lump sum	-	-	Spclist	-	-	Spclist	-	734.73	Sm	826.57
plus for all piles	-	-	Spclist	-	-	Spclist	-	1.78	m	2.00
Mild steel sheet piling to B.S.EN10025 Grade S275 or high yield steel Grade S275JR in lengths 4.5-15m supplied, handled, pitched and driven by land based plant in one visit (SIMPLEX PILING LTD.)										
On/Off site charge										
in addition to the following prices, add										
approximately	-	-	Spclist	-	-	Spclist	-	3300.00	Sm	3712.50
Mild steel sheet piling										
Larssen Section 6W	-	-	Spclist	-	-	Spclist	-	72.53	m2	81.60
extra for high yield steel	-	-	Spclist	-	-	Spclist	-	3.97	m2	4.47
extra for one coat L.B.V.(Lowca Varnish to										
B.S.1070 Type 2) before driving	-	-	Spclist	-	-	Spclist	-	2.90	m2	3.26
extra for corners	-	-	Spclist	-	-	Spclist	-	28.99	m	32.61
extra for junctions	-	-	Spclist	-	-	Spclist	-	39.53	m	44.47
Frodingham Section 1N and Larssen Section 9W	-	-	Spclist	-	-	Spclist	-	80.11	m2	90.12
extra for high yield steel	-	-	Spclist	-	-	Spclist	-	3.30	m2	3.71
extra for one coat L.B.V.(Lowca Varnish to										
B.S.1070 Type 2) before driving	-	-	Spclist	-	-	Spclist	-	2.90	m2	3.26
extra for corners	-	-	Spclist	-	-	Spclist	-	28.99	m	32.61
extra for junctions	-	-	Spclist	-	-	Spclist	-	39.53	m	44.47
Frodingham Section 2N and Larssen Section 12W	-	-	Spclist	-	-	Spclist	-	89.17	m2	100.32
extra for high yield steel	-	-	Spclist	-	-	Spclist	-	4.22	m2	4.75
extra for one coat L.B.V.(Lowca Varnish to										
B.S.1070 Type 2) before driving	-	-	Spclist	-	-	Spclist	-	2.90	m2	3.26
extra for corners	-	-	Spclist	-	-	Spclist	-	28.99	m	32.61
extra for junctions	-	-	Spclist	-	-	Spclist	-	39.53	m	44.47
Frodingham Section 1BXN	-	-	Spclist	-	-	Spclist	-	98.83	m2	111.18
extra for high yield steel	-	-	Spclist	-	-	Spclist	-	6.54	m2	7.36
extra for one coat L.B.V.(Lowca Varnish to										
B.S.1070 Type 2) before driving	-	-	Spclist	-	-	Spclist	-	2.64	m2	2.97
extra for corners	-	-	Spclist	-	-	Spclist	-	31.63	m	35.58
extra for junctions	-	-	Spclist	-	-	Spclist	-	39.53	m	44.47
Frodingham Section 3N and Larssen Section 16W	-	-	Spclist	-	-	Spclist	-	99.71	m2	112.17
extra for high yield steel	-	-	Spclist	-	-	Spclist	-	5.41	m2	6.09
extra for one coat L.B.V.(Lowca Varnish to										
B.S.1070 Type 2) before driving	-	-	Spclist	-	-	Spclist	-	2.64	m2	2.97
extra for corners	-	-	Spclist	-	-	Spclist	-	31.63	m	35.58
extra for junctions	-	-	Spclist	-	-	Spclist	-	39.53	m	44.47
Larssen Section 20W	-	-	Spclist	-	-	Spclist	-	112.82	m2	126.92
extra for high yield steel	-	-	Spclist	-	-	Spclist	-	5.93	m2	6.67
extra for one coat L.B.V.(Lowca Varnish to										
B.S.1070 Type 2) before driving	-	-	Spclist	-	-	Spclist	-	2.97	m2	3.34
extra for corners	-	-	Spclist	-	-	Spclist	-	34.26	m	38.54
extra for junctions	-	-	Spclist	-	-	Spclist	-	46.11	m	51.87
Frodingham Section 4N and Larssen Section 25W	-	-	Spclist	-	-	Spclist	-	124.93	m2	140.55
extra for high yield steel	-	-	Spclist	-	-	Spclist	-	6.46	m2	7.27
extra for one coat L.B.V.(Lowca Varnish to										
B.S.1070 Type 2) before driving	-	-	Spclist	-	-	Spclist	-	3.30	m2	3.71
extra for corners	-	-	Spclist	-	-	Spclist	-	36.89	m	41.50
extra for junctions	-	-	Spclist	-	-	Spclist	-	46.11	m	51.87
Larssen Section 32W	-	-	Spclist	-	-	Spclist	-	147.50	m2	165.94
extra for high yield steel	-	-	Spclist	-	-	Spclist	-	7.18	m2	8.08
extra for one coat L.B.V.(Lowca Varnish to										
B.S.1070 Type 2) before driving	-	-	Spclist	-	-	Spclist	-	3.56	m2	4.00

Labour hourly rates: (except Specialists) Craft Operatives £9.23 Labourer £7.02 Rates are national average prices. Refer to REGIONAL VARIATIONS for indicative levels of overall pricing in regions	MATERIALS			LABOUR				RATES		
	Del to Site	Waste	Material Cost	Craft Optve	Lab	Labour Cost	Sunds	Nett Rate	Unit	Gross Rate (+12.5%)
	£	%	£	Hrs	Hrs	£	£	£		£
D32: STEEL PILING Cont.										
Mild steel sheet piling to B.S.EN10025 Grade S275 or high yield steel Grade S275JR in lengths 4.5-15m supplied, handled, pitched and driven by land based plant in one visit (SIMPLEX PILING LTD.) Cont.										
Mild steel sheet piling Cont.										
extra for corners	-	-	Spclist	-	-	Spclist	-	39.53	m	44.47
extra for junctions	-	-	Spclist	-	-	Spclist	-	48.80	m	54.90
Frodingham Section 5	-	-	Spclist	-	-	Spclist	-	166.45	m2	187.26
extra for high yield steel	-	-	Spclist	-	-	Spclist	-	9.49	m2	10.68
extra for one coat L.B.V.(Lowca Varnish to B.S.1070 Type 2) before driving	-	-	Spclist	-	-	Spclist	-	6.19	m2	6.96
extra for corners	-	-	Spclist	-	-	Spclist	-	39.53	m	44.47
extra for junctions	-	-	Spclist	-	-	Spclist	-	52.71	m	59.30
For quiet piling, add										
lump sum	-	-	Spclist	-	-	Spclist	-	670.00	Sm	753.75
plus for all sections, from	-	-	Spclist	-	-	Spclist	-	3.06	m2	3.44
to	-	-	Spclist	-	-	Spclist	-	5.11	m2	5.75
D40: EMBEDDED RETAINING WALLS										
Embedded retaining walls; contiguous panel construction; panel lengths not exceeding 5m. Note:- the following prices are indicative only: firm quotations should always be obtained (KVAERNER CEMENTATION FOUNDATIONS)										
On/Off site charge										
in addition to the following prices, add for bringing plant to site, erecting and dismantling, maintaining and removing from site, approximately.	-	-	Spclist	-	-	Spclist	-	67500.00	Sm	75937.50
Excavation and Bentonite slurry and disposal										
600mm thick wall; maximum depth										
5m	-	-	Spclist	-	-	Spclist	-	203.00	m3	228.38
10m	-	-	Spclist	-	-	Spclist	-	203.00	m3	228.38
15m	-	-	Spclist	-	-	Spclist	-	203.00	m3	228.38
20m	-	-	Spclist	-	-	Spclist	-	203.00	m3	228.38
800mm thick wall; maximum depth										
5m	-	-	Spclist	-	-	Spclist	-	169.00	m3	190.13
10m	-	-	Spclist	-	-	Spclist	-	169.00	m3	190.13
15m	-	-	Spclist	-	-	Spclist	-	169.00	m3	190.13
20m	-	-	Spclist	-	-	Spclist	-	169.00	m3	190.13
1000mm thick wall; maximum depth										
5m	-	-	Spclist	-	-	Spclist	-	135.00	m3	151.88
10m	-	-	Spclist	-	-	Spclist	-	135.00	m3	151.88
15m	-	-	Spclist	-	-	Spclist	-	135.00	m3	151.88
20m	-	-	Spclist	-	-	Spclist	-	135.00	m3	151.88
Excavating through obstructions, rock like formations, etc										
undertaken on a time basis at a rate per rig per hour	-	-	Spclist	-	-	Spclist	-	394.00	hr	443.25
Reinforced concrete; B.S.5328, designed mix C25, 20mm aggregate, minimum cement content 400 kg/m3										
600mm thick wall	-	-	Spclist	-	-	Spclist	-	84.00	m3	94.50
800mm thick wall	-	-	Spclist	-	-	Spclist	-	84.00	m3	94.50
1000mm thick wall	-	-	Spclist	-	-	Spclist	-	84.00	m3	94.50
Reinforced in-situ concrete; sulphate resisting; B.S.5328, designed mix C25, 20mm aggregate, minimum cement content 400 kg/m3										
600mm thick wall	-	-	Spclist	-	-	Spclist	-	96.00	m3	108.00
800mm thick wall	-	-	Spclist	-	-	Spclist	-	96.00	m3	108.00
1000mm thick wall	-	-	Spclist	-	-	Spclist	-	96.00	m3	108.00
Reinforcement bars; B.S.4449 hot rolled plain round mild steel; including hooks tying wire, and spacers and chairs which are at the discretion of the Contractor										
16mm; straight	-	-	Spclist	-	-	Spclist	-	602.00	t	677.25
20mm; straight	-	-	Spclist	-	-	Spclist	-	602.00	t	677.25
25mm; straight	-	-	Spclist	-	-	Spclist	-	602.00	t	677.25
32mm; straight	-	-	Spclist	-	-	Spclist	-	602.00	t	677.25
40mm; straight	-	-	Spclist	-	-	Spclist	-	602.00	t	677.25
16mm; bent	-	-	Spclist	-	-	Spclist	-	602.00	t	677.25
20mm; bent	-	-	Spclist	-	-	Spclist	-	602.00	t	677.25
25mm; bent	-	-	Spclist	-	-	Spclist	-	602.00	t	677.25
32mm; bent	-	-	Spclist	-	-	Spclist	-	602.00	t	677.25
40mm; bent	-	-	Spclist	-	-	Spclist	-	602.00	t	677.25
Reinforcement bars; B.S.4449 hot rolled deformed high yield steel; including hooks tying wire, and spacers and chairs which are at the discretion of the Contractor										
16mm; straight	-	-	Spclist	-	-	Spclist	-	602.00	t	677.25
20mm; straight	-	-	Spclist	-	-	Spclist	-	602.00	t	677.25
25mm; straight	-	-	Spclist	-	-	Spclist	-	602.00	t	677.25
32mm; straight	-	-	Spclist	-	-	Spclist	-	602.00	t	677.25
40mm; straight	-	-	Spclist	-	-	Spclist	-	602.00	t	677.25
16mm; bent	-	-	Spclist	-	-	Spclist	-	602.00	t	677.25
20mm; bent	-	-	Spclist	-	-	Spclist	-	602.00	t	677.25

GROUNDWORK

Labour hourly rates: (except Specialists) Craft Operatives £9.23 Labourer £7.02 Rates are national average prices. Refer to REGIONAL VARIATIONS for indicative levels of overall pricing in regions	MATERIALS			LABOUR				RATES		
	Del to Site £	Waste %	Material Cost £	Craft Optve Hrs	Lab Hrs	Labour Cost £	Sunds £	Nett Rate £	Unit	Gross Rate (+12.5%) £
D40: EMBEDDED RETAINING WALLS Cont.										
Embedded retaining walls; contiguous panel construction; panel lengths not exceeding 5m. Note:- the following prices are indicative only: firm quotations should always be obtained (KVAERNER CEMENTATION FOUNDATIONS) Cont.										
Reinforcement bars; B.S.4449 hot rolled deformed high yield steel; including hooks tying wire, and spacers and chairs which are at the discretion of the Contractor Cont.										
25mm; bent	-	-	Spclist	-	-	Spclist	-	602.00	t	677.25
32mm; bent	-	-	Spclist	-	-	Spclist	-	602.00	t	677.25
40mm; bent	-	-	Spclist	-	-	Spclist	-	602.00	t	677.25
Guide walls; excavation, disposal and support; reinforced in-situ concrete; B.S.5328, designed mix C25, 20mm aggregate, minimum cement content 290 kg/m3; reinforced with one layer fabric B.S.4483 reference A252, 3.95 kg/m2 including laps, tying wire, all cutting and bending, and spacers and chairs which are at the discretion of the Contractor; formwork both sides										
both sides; 1000mm high, 600mm apart; propped top and bottom at 2000mm centres	-	-	Spclist	-	-	Spclist	-	208.00	m	234.00
both sides; 1000mm high, 800mm apart; propped top and bottom at 2000mm centres			Spclist			Spclist		208.00	m	234.00
both sides; 1000mm high, 1000mm apart; propped top and bottom at 2000mm centres	-	-	Spclist	-	-	Spclist	-	208.00	m	234.00
both sides; 1500mm high, 600mm apart; propped top and bottom at 2000mm centres	-	-	Spclist	-	-	Spclist	-	320.00	m	360.00
both sides; 1500mm high, 800mm apart; propped top and bottom at 2000mm centres	-	-	Spclist	-	-	Spclist	-	320.00	m	360.00
both sides; 1500mm high, 1000mm apart; propped top and bottom at 2000mm centres	-	-	Spclist	-	-	Spclist	-	320.00	m	360.00
D41: CRIB WALLS/GABIONS/REINFORCED EARTH										
Retaining Walls										
Betoflor precast concrete landscape retaining walls including soil filling to pockets but excluding excavation, concrete foundations, stone backfill to rear of wall and planting which are all deemed measured separately										
Betoflor interlocking units 500mm long x 250mm wide x 200mm modular deep in wall 250mm wide	41.61	5.00	43.69	-	5.00	35.10	-	78.79	m2	88.64
Extra over for colours	1.50	5.00	1.58	-	0.50	3.51	-	5.09	m2	5.72
Betoatlas interlocking units 250mm long x 500mm wide x 200mm modular deep in wall 500mm wide	53.37	5.00	56.04	-	7.00	49.14	-	105.18	m2	118.33
Extra over for colours	3.00	5.00	3.15	-	0.10	0.70	-	3.85	m2	4.33
Betonap 150/50 woven mesh geotextile as reinforcement to backfill	3.26	5.00	3.42	-	0.30	2.11	-	5.53	m2	6.22
graded stone filling behind Betoflor or Betoatlas.	15.23	5.00	15.99	-	4.00	28.08	-	44.07	m3	49.58
concrete haunching to base of blocks	4.15	5.00	4.36	-	0.45	3.16	-	7.52	m	8.46
Betoflor range; Betojard precast concrete vertical acoustic wall including soil filling to pockets but excluding excavation, concrete foundations, stone backfill to rear of wall and planting which are all deemed measured separately										
Betojard standard 500mm long x 250mm wide x 200mm deep units in wall 250mm wide	49.57	5.00	52.05	-	5.75	40.37	-	92.41	m2	103.97
Extra over for colours	1.50	5.00	1.58	-	0.05	0.35	-	1.93	m2	2.17
Betojard pivot 290mm x 250mm wide x 200mm deep ...	49.57	5.00	52.05	-	7.00	49.14	-	101.19	m2	113.84
Extra over for colours	1.50	5.00	1.58	-	0.05	0.35	-	1.93	m2	2.17
Extra over fill core to pivot block with reinforced concrete with 4 No 10mm diameter vertical mild steel bars and 6mm links	4.20	5.00	4.41	-	2.00	14.04	-	18.45	m	20.76
Extra over dowelled connection between pivot and standard block with concrete infill to standard block and 1 No 10mm diameter x 200mm long galvanised dowel debonded one end at 300mm centres vertically...	5.31	5.00	5.58	-	1.00	7.02	-	12.60	m	14.17
concrete haunching to base of blocks	4.15	5.00	4.36	-	0.45	3.16	-	7.52	m	8.46
D50: UNDERPINNING										
Information										
The work of underpinning in this section comprises work to be carried out in short lengths; prices are exclusive of shoring and other temporary supports										
Excavating										
Preliminary trenches										
maximum depth not exceeding										
1.00m	-	-	-	-	5.00	35.10	-	35.10	m3	39.49
2.00m	-	-	-	-	6.00	42.12	-	42.12	m3	47.38
4.00m	-	-	-	-	7.00	49.14	-	49.14	m3	55.28
Underpinning pits										
maximum depth not exceeding										
1.00m	-	-	-	-	5.85	41.07	-	41.07	m3	46.20

Labour hourly rates: (except Specialists) Craft Operatives £9.23 Labourer £7.02 Rates are national average prices. Refer to REGIONAL VARIATIONS for indicative levels of overall pricing in regions	MATERIALS			LABOUR				RATES		
	Del to Site	Waste	Material Cost	Craft Optve	Lab	Labour Cost	Sunds	Nett Rate	Unit	Gross Rate (+12.5%)
	£	%	£	Hrs	Hrs	£	£	£		£
D50: UNDERPINNING Cont.										
Excavating Cont.										
Underpinning pits Cont.										
maximum depth not exceeding Cont.										
2.00m	-	-	-	-	7.30	51.25	-	51.25	m3	57.65
4.00m	-	-	-	-	8.80	61.78	-	61.78	m3	69.50
Earthwork support										
preliminary trenches; distance between opposing faces not exceeding 2.00m; maximum depth not exceeding										
1.00m	1.35	-	1.35	-	0.53	3.72	0.02	5.09	m2	5.73
2.00m	1.70	-	1.70	-	0.67	4.70	0.03	6.43	m2	7.24
4.00m	1.94	-	1.94	-	0.72	5.05	0.03	7.02	m2	7.90
preliminary trenches; distance between opposing faces 2.00 - 4.00m; maximum depth not exceeding										
1.00m	1.47	-	1.47	-	0.55	3.86	0.02	5.35	m2	6.02
2.00m	1.86	-	1.86	-	0.70	4.91	0.03	6.80	m2	7.65
4.00m	2.04	-	2.04	-	0.84	5.90	0.04	7.98	m2	8.97
underpinning pits; distance between opposing faces not exceeding 2.00m; maximum depth not exceeding										
1.00m	1.59	-	1.59	-	0.61	4.28	0.02	5.89	m2	6.63
2.00m	1.97	-	1.97	-	0.77	5.41	0.03	7.41	m2	8.33
4.00m	2.21	-	2.21	-	0.83	5.83	0.04	8.08	m2	9.09
underpinning pits; distance between opposing faces 2.00 - 4.00m; maximum depth not exceeding										
1.00m	1.68	-	1.68	-	0.64	4.49	0.02	6.19	m2	6.97
2.00m	2.13	-	2.13	-	0.80	5.62	0.03	7.78	m2	8.75
4.00m	2.45	-	2.45	-	0.95	6.67	0.04	9.16	m2	10.30
Cutting away existing projecting foundations										
masonry; maximum width 103mm; maximum depth 150mm	-	-	-	-	0.90	6.32	-	6.32	m	7.11
masonry; maximum width 154mm; maximum depth 225mm	-	-	-	-	1.07	7.51	-	7.51	m	8.45
concrete; maximum width 253mm; maximum depth 190mm	-	-	-	-	1.29	9.06	-	9.06	m	10.19
concrete; maximum width 304; maximum depth 300mm	-	-	-	-	1.52	10.67	-	10.67	m	12.00
Preparing the underside of the existing work to receive the pinning up of the new work										
350mm wide	-	-	-	-	0.45	3.16	-	3.16	m	3.55
500mm wide	-	-	-	-	0.56	3.93	-	3.93	m	4.42
1000mm wide	-	-	-	-	1.13	7.93	-	7.93	m	8.92
Compacting										
bottoms of excavations	-	-	-	-	0.45	3.16	-	3.16	m2	3.55
Disposal of excavated material										
removing from site to tip (including tipping charges but excluding landfill tax)	-	-	-	-	3.38	23.73	11.70	35.43	m3	39.86
Excavated material arising from excavations										
Filling to excavations										
average thickness exceeding 0.25m	-	-	-	-	2.25	15.80	-	15.80	m3	17.77
Compacting										
filling	-	-	-	-	0.23	1.61	-	1.61	m2	1.82
Plain in-situ concrete; B.S.5328, ordinary prescribed mix ST3, 20mm aggregate										
Foundations; poured on or against earth or unblinded hardcore										
generally	53.10	7.50	57.08	-	5.34	37.49	-	94.57	m3	106.39
Plain in-situ concrete; B.S.5328, ordinary prescribed mix ST4, 20mm aggregate										
Foundations; poured on or against earth or unblinded hardcore										
generally	54.60	7.50	58.70	-	5.34	37.49	-	96.18	m3	108.20
Formwork and basic finish										
Sides of foundations; plain vertical										
height exceeding 1.00m	6.70	10.00	7.37	3.94	0.79	41.91	1.54	50.82	m2	57.17
height not exceeding 250mm	2.00	10.00	2.20	1.19	0.25	12.74	0.46	15.40	m	17.32
height 250 - 500mm	3.67	10.00	4.04	2.16	0.43	22.96	0.85	27.84	m	31.32
height 0.50m - 1.00m	6.77	10.00	7.45	4.03	0.81	42.88	1.56	51.89	m	58.38
Common bricks, B.S.3921, Category M, 215 x 102.5 x 65mm, compressive strength not less than 5.2 N/mm2; in cement mortar (1:3)										
Walls; vertical										
215mm thick; English bond	19.50	5.00	20.48	4.70	5.00	78.48	5.26	104.22	m2	117.24
327mm thick; English bond	29.34	5.00	30.81	5.70	6.15	95.78	8.32	134.91	m2	151.77
440mm thick; English bond	39.01	5.00	40.96	6.30	6.90	106.59	10.93	158.48	m2	178.29
Bonding to existing including extra material										
thickness of new work 215mm	2.13	-	2.13	1.32	1.32	21.45	0.18	23.76	m	26.73
thickness of new work 327mm	3.20	-	3.20	1.90	1.90	30.88	0.23	34.31	m	38.59
thickness of new work 440mm	4.26	-	4.26	2.50	2.50	40.63	0.28	45.16	m	50.81

GROUNDWORK – SMALL WORKS

Labour hourly rates: (except Specialists) Craft Operatives £9.23 Labourer £7.02 Rates are national average prices. Refer to REGIONAL VARIATIONS for indicative levels of overall pricing in regions	MATERIALS			LABOUR				RATES		
	Del to Site £	Waste %	Material Cost £	Craft Optve Hrs	Lab Hrs	Labour Cost £	Sunds £	Nett Rate £	Unit	Gross Rate (+12.5%) £
D50: UNDERPINNING Cont.										
Milton Hall' Second Hard Stock bricks, B.S.3921, Category M, 215 x 102.5 x 65mm, in cement mortar (1:3)										
Walls; vertical										
215mm thick; English bond	37.96	5.00	39.86	4.70	5.00	78.48	5.26	123.60	m2	139.05
327mm thick; English bond	57.10	5.00	59.95	5.70	6.15	95.78	8.32	164.06	m2	184.57
440mm thick; English bond	75.92	5.00	79.72	6.30	6.90	106.59	10.93	197.23	m2	221.89
Bonding to existing including extra material										
thickness of new work 215mm	4.16	-	4.16	1.90	1.90	30.88	0.18	35.22	m	39.62
thickness of new work 327mm	6.22	-	6.22	1.90	1.90	30.88	0.23	37.33	m	41.99
thickness of new work 440mm	8.28	-	8.28	2.50	2.50	40.63	0.28	49.19	m	55.33
Engineering bricks, B.S.3921, Category F, 215 x 102.5 x 65mm, class A; in cement mortar (1:3)										
Walls; vertical										
215mm thick; English bond	65.34	5.00	68.61	5.20	5.50	86.61	5.26	160.47	m2	180.53
327mm thick; English bond	98.27	5.00	103.18	6.30	6.75	105.53	8.32	217.04	m2	244.17
440mm thick; English bond	130.68	5.00	137.21	6.95	7.55	117.15	10.93	265.29	m2	298.46
Bonding to existing including extra material										
thickness of new work 215mm	7.12	-	7.12	1.45	1.45	23.56	0.18	30.86	m	34.72
thickness of new work 327mm	10.69	-	10.69	2.10	2.10	34.13	0.23	45.05	m	50.68
thickness of new work 440mm	14.27	-	14.27	2.75	2.75	44.69	0.28	59.24	m	66.64
Engineering bricks, B.S.3921, Category F, 215 x 102.5 x 65mm, Class B; in cement mortar (1:3)										
Walls; vertical										
215mm thick; English bond	38.90	5.00	40.84	5.20	5.50	86.61	5.26	132.71	m2	149.30
327mm thick; English bond	58.51	5.00	61.44	6.30	6.75	105.53	8.32	175.29	m2	197.20
440mm thick; English bond	77.79	5.00	81.68	6.95	7.55	117.15	10.93	209.76	m2	235.98
Bonding to existing including extra material										
thickness of new work 215mm	4.25	-	4.25	1.45	1.45	23.56	0.18	27.99	m	31.49
thickness of new work 327mm	6.36	-	6.36	2.10	2.10	34.13	0.23	40.72	m	45.80
thickness of new work 440mm	8.50	-	8.50	2.75	2.75	44.69	0.28	53.47	m	60.15
Sundry items										
Wedging and pinning up to underside of existing construction with two courses slates in cement mortar (1:3)										
215mm walls	7.34	15.00	8.44	0.60	0.60	9.75	0.14	18.33	m	20.62
327mm walls	10.22	15.00	11.75	0.87	0.87	14.14	0.18	26.07	m	29.33
440mm walls	13.29	15.00	15.28	1.14	1.14	18.52	0.20	34.01	m	38.26

In situ concrete/Large precast concrete

Labour hourly rates: (except Specialists) Craft Operatives £9.23 Labourer £7.02 Rates are national average prices. Refer to REGIONAL VARIATIONS for indicative levels of overall pricing in regions	MATERIALS			LABOUR				RATES		
	Del to Site £	Waste %	Material Cost £	Craft Optve Hrs	Lab Hrs	Labour Cost £	Sunds £	Nett Rate £	Unit	Gross Rate (+12.5%) £
E10: MIXING/CASTING/CURING IN-SITU CONCRETE – (READY MIXED)										
Plain in-situ concrete; B.S.5328, ordinary prescribed mix ST3, 20mm aggregate										
Foundations poured on or against earth or unblinded hardcore										
generally..........................	51.92	7.50	55.81	–	2.30	16.15	–	71.96	m3	80.95
Isolated foundations; poured on or against earth or unblinded hardcore										
generally..........................	51.92	7.50	55.81	–	2.75	19.31	–	75.12	m3	84.51
Beds; poured on or against earth or unblinded hardcore										
thickness exceeding 450mm	51.92	7.50	55.81	–	2.20	15.44	–	71.26	m3	80.17
thickness 150 - 450mm	51.92	7.50	55.81	–	2.50	17.55	–	73.36	m3	82.53
thickness not exceeding 150mm	51.92	7.50	55.81	–	3.10	21.76	–	77.58	m3	87.27
Filling hollow walls										
thickness not exceeding 150mm	51.92	5.00	54.52	–	5.50	38.61	–	93.13	m3	104.77
Plain in-situ concrete; B.S.5328, ordinary prescribed mix ST4, 20mm aggregate										
Foundations poured on or against earth or unblinded hardcore										
generally..........................	53.32	7.50	57.32	–	2.30	16.15	–	73.47	m3	82.65
Isolated foundations; poured on or against earth or unblinded hardcore										
generally..........................	53.32	7.50	57.32	–	2.75	19.31	–	76.62	m3	86.20
Beds; poured on or against earth or unblinded hardcore										
thickness exceeding 450mm	53.32	7.50	57.32	–	2.20	15.44	–	72.76	m3	81.86
thickness 150 - 450mm	53.32	7.50	57.32	–	2.50	17.55	–	74.87	m3	84.23
thickness not exceeding 150mm	53.32	7.50	57.32	–	3.10	21.76	–	79.08	m3	88.97
Filling hollow walls										
thickness not exceeding 150mm	53.32	5.00	55.99	–	5.50	38.61	–	94.60	m3	106.42
Plain in-situ concrete; B.S.5328, ordinary prescribed mix ST5, 20mm aggregate										
Foundations poured on or against earth or unblinded hardcore										
generally.........................	54.14	7.50	58.20	–	2.30	16.15	–	74.35	m3	83.64
Isolated foundations; poured on or against earth or unblinded hardcore										
generally.........................	54.14	7.50	58.20	–	2.75	19.31	–	77.51	m3	87.19
Beds; poured on or against earth or unblinded hardcore										
thickness exceeding 450mm	54.14	7.50	58.20	–	2.20	15.44	–	73.64	m3	82.85
thickness 150 - 450mm	54.14	7.50	58.20	–	2.50	17.55	–	75.75	m3	85.22
thickness not exceeding 150mm	54.14	7.50	58.20	–	3.10	21.76	–	79.96	m3	89.96
Filling hollow walls										
thickness not exceeding 150mm	54.14	5.00	66.86	–	5.50	38.61	–	95.46	m3	107.39
Reinforced in-situ concrete; B.S.5328, designed mix C15, 20mm aggregate, minimum cement content 220 kg/m3; vibrated										
Foundations										
generally..........................	52.25	5.00	54.86	–	2.60	18.25	0.50	73.61	m3	82.82

Labour hourly rates: (except Specialists) Craft Operatives £9.23 Labourer £7.02 Rates are national average prices. Refer to REGIONAL VARIATIONS for indicative levels of overall pricing in regions	MATERIALS			LABOUR				RATES		
	Del to Site £	Waste %	Material Cost £	Craft Optve Hrs	Lab Hrs	Labour Cost £	Sunds £	Nett Rate £	Unit	Gross Rate (+12.5%) £
E10: MIXING/CASTING/CURING IN-SITU CONCRETE – (READY MIXED) Cont.										
Reinforced in-situ concrete; B.S.5328, designed mix C15, 20mm aggregate, minimum cement content 220 kg/m3; vibrated Cont.										
Ground beams										
generally	52.25	5.00	**54.86**	-	3.60	**25.27**	0.50	**80.63**	m3	90.71
Isolated foundations										
generally	52.25	5.00	**54.86**	-	3.00	**21.06**	0.50	**76.42**	m3	85.98
Beds										
thickness exceeding 450mm	52.25	5.00	**54.86**	-	2.50	**17.55**	0.50	**72.91**	m3	82.03
thickness 150 - 450mm	52.25	5.00	**54.86**	-	2.75	**19.31**	0.50	**74.67**	m3	84.00
thickness not exceeding 150mm	52.25	5.00	**54.86**	-	3.50	**24.57**	0.50	**79.93**	m3	89.92
Slabs										
thickness exceeding 450mm	52.25	2.50	**53.56**	-	3.20	**22.46**	0.50	**76.52**	m3	86.09
thickness 150 - 450mm	52.25	2.50	**53.56**	-	3.45	**24.22**	0.50	**78.28**	m3	88.06
thickness not exceeding 150mm	52.25	2.50	**53.56**	-	4.55	**31.94**	0.50	**86.00**	m3	96.75
Walls										
thickness exceeding 450mm	52.25	2.50	**53.56**	-	3.50	**24.57**	0.50	**78.63**	m3	88.45
thickness 150 - 450mm	52.25	2.50	**53.56**	-	3.90	**27.38**	0.50	**81.43**	m3	91.61
thickness not exceeding 150mm	52.25	2.50	**53.56**	-	5.00	**35.10**	0.50	**89.16**	m3	100.30
Beams										
isolated	52.25	2.50	**53.56**	-	4.70	**32.99**	0.50	**87.05**	m3	97.93
isolated deep	52.25	2.50	**53.56**	-	4.80	**33.70**	0.50	**87.75**	m3	98.72
attached deep	52.25	2.50	**53.56**	-	4.70	**32.99**	0.50	**87.05**	m3	97.93
Beam casings										
isolated	52.25	2.50	**53.56**	-	5.00	**35.10**	0.50	**89.16**	m3	100.30
isolated deep	52.25	2.50	**53.56**	-	5.10	**35.80**	0.50	**89.86**	m3	101.09
attached deep	52.25	2.50	**53.56**	-	5.00	**35.10**	0.50	**89.16**	m3	100.30
Columns										
generally	52.25	2.50	**53.56**	-	6.30	**44.23**	0.50	**98.28**	m3	110.57
Column casings										
generally	52.25	2.50	**53.56**	-	6.60	**46.33**	0.50	**100.39**	m3	112.94
Staircases										
generally	52.25	2.50	**53.56**	-	5.50	**38.61**	0.50	**92.67**	m3	104.25
Upstands										
generally	52.25	2.50	**53.56**	-	7.70	**54.05**	0.50	**108.11**	m3	121.62
Reinforced in-situ concrete; B.S.5328, designed mix C20, 20mm aggregate, minimum cement content 240 kg/m3; vibrated										
Foundations										
generally	52.89	5.00	**55.53**	-	2.60	**18.25**	0.50	**74.29**	m3	83.57
Ground beams										
generally	52.89	5.00	**55.53**	-	3.60	**25.27**	0.50	**81.31**	m3	91.47
Isolated foundations										
generally	52.89	5.00	**55.53**	-	3.00	**21.06**	0.50	**77.09**	m3	86.73
Beds										
thickness exceeding 450mm	52.89	5.00	**55.53**	-	2.50	**17.55**	0.50	**73.58**	m3	82.78
thickness 150 - 450mm	52.89	5.00	**55.53**	-	2.75	**19.31**	0.50	**75.34**	m3	84.76
thickness not exceeding 150mm	52.89	5.00	**55.53**	-	3.50	**24.57**	0.50	**80.60**	m3	90.68
Slabs										
thickness exceeding 450mm	52.89	2.50	**54.21**	-	3.20	**22.46**	0.50	**77.18**	m3	86.82
thickness 150 - 450mm	52.89	2.50	**54.21**	-	3.45	**24.22**	0.50	**78.93**	m3	88.80
thickness not exceeding 150mm	52.89	2.50	**54.21**	-	4.55	**31.94**	0.50	**86.65**	m3	97.48
Walls										
thickness exceeding 450mm	52.89	2.50	**54.21**	-	3.50	**24.57**	0.50	**79.28**	m3	89.19
thickness 150 - 450mm	52.89	2.50	**54.21**	-	3.90	**27.38**	0.50	**82.09**	m3	92.35
thickness not exceeding 150mm	52.89	2.50	**54.21**	-	5.00	**35.10**	0.50	**89.81**	m3	101.04
Beams										
isolated	52.89	2.50	**54.21**	-	4.70	**32.99**	0.50	**87.71**	m3	98.67
isolated deep	52.89	2.50	**54.21**	-	4.80	**33.70**	0.50	**88.41**	m3	99.46
attached deep	52.89	2.50	**54.21**	-	4.70	**32.99**	0.50	**87.71**	m3	98.67
Beam casings										
isolated	52.89	2.50	**54.21**	-	5.00	**35.10**	0.50	**89.81**	m3	101.04
isolated deep	52.89	2.50	**54.21**	-	5.10	**35.80**	0.50	**90.51**	m3	101.83
attached deep	52.89	2.50	**54.21**	-	5.00	**35.10**	0.50	**89.81**	m3	101.04
Columns										
generally	52.89	2.50	**54.21**	-	6.30	**44.23**	0.50	**98.94**	m3	111.31
Column casings										
generally	52.89	2.50	**54.21**	-	6.60	**46.33**	0.50	**101.04**	m3	113.67
Staircases										
generally	52.89	2.50	**54.21**	-	5.50	**38.61**	0.50	**93.32**	m3	104.99

Labour hourly rates: (except Specialists) Craft Operatives £9.23 Labourer £7.02 Rates are national average prices. Refer to REGIONAL VARIATIONS for indicative levels of overall pricing in regions	MATERIALS			LABOUR				RATES		
	Del to Site £	Waste %	Material Cost £	Craft Optve Hrs	Lab Hrs	Labour Cost £	Sunds £	Nett Rate £	Unit	Gross Rate (+12.5%) £
E10: MIXING/CASTING/CURING IN-SITU CONCRETE - (READY MIXED) Cont.										
Reinforced in-situ concrete; B.S.5328, designed mix C20, 20mm aggregate, minimum cement content 240 kg/m3; vibrated Cont.										
Upstands										
generally	52.89	2.50	54.21	-	7.70	54.05	0.50	108.77	m3	122.36
Reinforced in-situ concrete; B.S.5328, designed mix C25, 20mm aggregate, minimum cement content 290 kg/m3; vibrated										
Foundations										
generally	53.64	5.00	56.32	-	2.60	18.25	0.50	75.07	m3	84.46
Ground beams										
generally	53.64	5.00	56.32	-	3.60	25.27	0.50	82.09	m3	92.36
Isolated foundations										
generally	53.64	5.00	56.32	-	3.00	21.06	0.50	77.88	m3	87.62
Beds										
thickness exceeding 450mm	53.64	5.00	56.32	-	2.50	17.55	0.50	74.37	m3	83.67
thickness 150 - 450mm	53.64	5.00	56.32	-	2.75	19.31	0.50	76.13	m3	85.64
thickness not exceeding 150mm	53.64	5.00	56.32	-	3.50	24.57	0.50	81.39	m3	91.57
Slabs										
thickness exceeding 450mm	53.64	2.50	54.98	-	3.20	22.46	0.50	77.94	m3	87.69
thickness 150 - 450mm	53.64	2.50	54.98	-	3.45	24.22	0.50	79.70	m3	89.66
thickness not exceeding 150mm	53.64	2.50	54.98	-	4.55	31.94	0.50	87.42	m3	98.35
Walls										
thickness exceeding 450mm	53.64	2.50	54.98	-	3.50	24.57	0.50	80.05	m3	90.06
thickness 150 - 450mm	53.64	2.50	54.98	-	3.90	27.38	0.50	82.86	m3	93.22
thickness not exceeding 150mm	53.64	2.50	54.98	-	5.00	35.10	0.50	90.58	m3	101.90
Beams										
isolated	53.64	2.50	54.98	-	4.70	32.99	0.50	88.47	m3	99.53
isolated deep	53.64	2.50	54.98	-	4.80	33.70	0.50	89.18	m3	100.32
attached deep	53.64	2.50	54.98	-	4.70	32.99	0.50	88.47	m3	99.53
Beam casings										
isolated	53.64	2.50	54.98	-	5.00	35.10	0.50	90.58	m3	101.90
isolated deep	53.64	2.50	54.98	-	5.10	35.80	0.50	91.28	m3	102.69
attached deep	53.64	2.50	54.98	-	5.00	35.10	0.50	90.58	m3	101.90
Columns										
generally	53.64	2.50	54.98	-	6.30	44.23	0.50	99.71	m3	112.17
Column casings										
generally	53.64	2.50	54.98	-	6.60	46.33	0.50	101.81	m3	114.54
Staircases										
generally	53.64	2.50	54.98	-	5.50	38.61	0.50	94.09	m3	105.85
Upstands										
generally	53.64	2.50	54.98	-	7.70	54.05	0.50	109.54	m3	123.23
Reinforced in-situ concrete; B.S.5328, designed mix C30, 20mm aggregate, minimum cement content 290 kg/m3; vibrated										
Foundations										
generally	54.93	5.00	57.68	-	2.60	18.25	0.50	76.43	m3	85.98
Ground beams										
generally	54.93	5.00	57.68	-	3.60	25.27	0.50	83.45	m3	93.88
Isolated foundations										
generally	54.93	5.00	57.68	-	3.00	21.06	0.50	79.24	m3	89.14
Beds										
thickness exceeding 450mm	54.93	5.00	57.68	-	2.50	17.55	0.50	75.73	m3	85.19
thickness 150 - 450mm	54.93	5.00	57.68	-	2.75	19.31	0.50	77.48	m3	87.17
thickness not exceeding 150mm	54.93	5.00	57.68	-	3.50	24.57	0.50	82.75	m3	93.09
Slabs										
thickness exceeding 450mm	54.93	2.50	56.30	-	3.20	22.46	0.50	79.27	m3	89.18
thickness 150 - 450mm	54.93	2.50	56.30	-	3.45	24.22	0.50	81.02	m3	91.15
thickness not exceeding 150mm	54.93	2.50	56.30	-	4.55	31.94	0.50	88.74	m3	99.84
Walls										
thickness exceeding 450mm	54.93	2.50	56.30	-	3.50	24.57	0.50	81.37	m3	91.54
thickness 150 - 450mm	54.93	2.50	56.30	-	3.90	27.38	0.50	84.18	m3	94.70
thickness not exceeding 150mm	54.93	2.50	56.30	-	5.00	35.10	0.50	91.90	m3	103.39
Beams										
isolated	54.93	2.50	56.30	-	4.70	32.99	0.50	89.80	m3	101.02
isolated deep	54.93	2.50	56.30	-	4.80	33.70	0.50	90.50	m3	101.81
attached deep	54.93	2.50	56.30	Hrs	4.70	32.99	0.50	89.80	m3	101.02
Beam casings										
isolated	54.93	2.50	56.30	-	5.00	35.10	0.50	91.90	m3	103.39
isolated deep	54.93	2.50	56.30	-	5.10	35.80	0.50	92.61	m3	104.18

Labour hourly rates: (except Specialists) Craft Operatives £9.23 Labourer £7.02 Rates are national average prices. Refer to REGIONAL VARIATIONS for indicative levels of overall pricing in regions	MATERIALS			LABOUR				RATES		
	Del to Site £	Waste %	Material Cost £	Craft Optve Hrs	Lab Hrs	Labour Cost £	Sunds £	Nett Rate £	Unit	Gross Rate (+12.5%) £
E10: MIXING/CASTING/CURING IN-SITU CONCRETE - (READY MIXED) Cont.										
Reinforced in-situ concrete; B.S.5328, designed mix C30, 20mm aggregate, minimum cement content 290 kg/m3; vibrated Cont.										
Beam casings Cont.										
attached deep	54.93	2.50	**56.30**	-	5.00	**35.10**	0.50	**91.90**	m3	**103.39**
Columns										
generally	54.93	2.50	**56.30**	-	6.30	**44.23**	0.50	**101.03**	m3	**113.66**
Column casings										
generally	54.93	2.50	**56.30**	-	6.60	**46.33**	0.50	**103.14**	m3	**116.03**
Upstands										
generally	54.93	2.50	**56.30**	-	7.70	**54.05**	0.50	**110.86**	m3	**124.71**
E10: MIXING/CASTING/CURING IN-SITU CONCRETE - (SITE MIXED)										
Plain in-situ concrete; mix 1:6, all in aggregate										
Foundations poured on or against earth or unblinded hardcore										
generally	64.74	7.50	**69.60**	-	3.40	**23.87**	1.80	**95.26**	m3	**107.17**
Isolated foundations; poured on or against earth or unblinded hardcore										
generally	64.74	7.50	**69.60**	-	3.90	**27.38**	1.80	**98.77**	m3	**111.12**
Beds; poured on or against earth or unblinded hardcore										
thickness 150 - 450mm	64.74	7.50	**69.60**	-	3.60	**25.27**	1.80	**96.67**	m3	**108.75**
thickness not exceeding 150mm	64.74	7.50	**69.60**	-	4.25	**29.84**	1.80	**101.23**	m3	**113.88**
Beds; sloping not exceeding 15 degrees; poured on or against earth or unblinded hardcore										
thickness 150 - 450mm	64.74	7.50	**69.60**	-	3.80	**26.68**	1.80	**98.07**	m3	**110.33**
thickness not exceeding 150mm	64.74	7.50	**69.60**	-	4.45	**31.24**	1.80	**102.63**	m3	**115.46**
Beds; sloping exceeding 15 degrees; poured on or against earth or unblinded hardcore										
thickness 150 - 450mm	64.74	7.50	**69.60**	-	4.00	**28.08**	1.80	**99.48**	m3	**111.91**
thickness not exceeding 150mm	64.74	7.50	**69.60**	-	4.70	**32.99**	1.80	**104.39**	m3	**117.44**
Plain in-situ concrete; mix 1:8, all in aggregate										
Foundations poured on or against earth or unblinded hardcore										
generally	57.86	7.50	**62.20**	-	3.40	**23.87**	1.80	**87.87**	m3	**98.85**
Isolated foundations; poured on or against earth or unblinded hardcore										
generally	57.86	7.50	**62.20**	-	3.90	**27.38**	1.80	**91.38**	m3	**102.80**
Beds; poured on or against earth or unblinded hardcore										
thickness 150 - 450mm	57.86	7.50	**62.20**	-	3.60	**25.27**	1.80	**89.27**	m3	**100.43**
thickness not exceeding 150mm	57.86	7.50	**62.20**	-	4.25	**29.84**	1.80	**93.83**	m3	**105.56**
Beds; sloping not exceeding 15 degrees; poured on or against earth or unblinded hardcore										
thickness 150 - 450mm	57.86	7.50	**62.20**	-	3.80	**26.68**	1.80	**90.68**	m3	**102.01**
thickness not exceeding 150mm	57.86	7.50	**62.20**	-	4.45	**31.24**	1.80	**95.24**	m3	**107.14**
Beds; sloping exceeding 15 degrees; poured on or against earth or unblinded hardcore										
thickness 150 - 450mm	57.86	7.50	**62.20**	-	4.00	**28.08**	1.80	**92.08**	m3	**103.59**
thickness not exceeding 150mm	57.86	7.50	**62.20**	-	4.70	**32.99**	1.80	**96.99**	m3	**109.12**
Plain in-situ concrete; mix 1:12, all in aggregate										
Foundations poured on or against earth or unblinded hardcore										
generally	50.98	7.50	**54.80**	-	3.40	**23.87**	1.80	**80.47**	m3	**90.53**
Isolated foundations; poured on or against earth or unblinded hardcore										
generally	50.98	7.50	**54.80**	-	3.90	**27.38**	1.80	**83.98**	m3	**94.48**
Beds; poured on or against earth or unblinded hardcore										
thickness 150 - 450mm	50.98	7.50	**54.80**	-	3.60	**25.27**	1.80	**81.88**	m3	**92.11**
thickness not exceeding 150mm	50.98	7.50	**54.80**	-	4.25	**29.84**	1.80	**86.44**	m3	**97.24**
Beds; sloping not exceeding 15 degrees; poured on or against earth or unblinded hardcore										
thickness 150 - 450mm	50.98	7.50	**54.80**	-	3.80	**26.68**	1.80	**83.28**	m3	**93.69**
thickness not exceeding 150mm	50.98	7.50	**54.80**	-	4.45	**31.24**	1.80	**87.84**	m3	**98.82**
Beds; sloping exceeding 15 degrees; poured on or against earth or unblinded hardcore										
thickness 150 - 450mm	50.98	7.50	**54.80**	-	4.00	**28.08**	1.80	**84.68**	m3	**95.27**
thickness not exceeding 150mm	50.98	7.50	**54.80**	-	4.70	**32.99**	1.80	**89.60**	m3	**100.80**

Labour hourly rates: (except Specialists) Craft Operatives £9.23 Labourer £7.02 Rates are national average prices. Refer to REGIONAL VARIATIONS for indicative levels of overall pricing in regions	MATERIALS			LABOUR				RATES		
	Del to Site £	Waste %	Material Cost £	Craft Optve Hrs	Lab Hrs	Labour Cost £	Sunds £	Nett Rate £	Unit	Gross Rate (+12.5%) £
E10: MIXING/CASTING/CURING IN-SITU CONCRETE - (SITE MIXED) Cont.										
Plain in-situ concrete; B.S.5328, ordinary prescribed mix C15P, 20mm aggregate										
Foundations poured on or against earth or unblinded hardcore										
generally......	65.64	7.50	70.56	-	3.40	23.87	1.80	96.23	m3	108.26
Isolated foundations; poured on or against earth or unblinded hardcore										
generally......	65.64	7.50	70.56	-	3.90	27.38	1.80	99.74	m3	112.21
Beds; poured on or against earth or unblinded hardcore										
thickness exceeding 450mm	65.64	7.50	70.56	-	3.30	23.17	1.80	95.53	m3	107.47
thickness 150 - 450mm	65.64	7.50	70.56	-	3.60	25.27	1.80	97.64	m3	109.84
thickness not exceeding 150mm	65.64	7.50	70.56	-	4.25	29.84	1.80	102.20	m3	114.97
Filling hollow walls										
thickness not exceeding 150mm	65.64	5.00	68.92	-	6.60	46.33	1.80	117.05	m3	131.69
Plain in-situ concrete; B.S.5328, ordinary prescribed mix C20P, 20mm aggregate										
Foundations poured on or against earth or unblinded hardcore										
generally......	67.73	7.50	72.81	-	3.40	23.87	1.80	98.48	m3	110.79
Isolated foundations; poured on or against earth or unblinded hardcore										
generally......	67.73	7.50	72.81	-	3.90	27.38	1.80	101.99	m3	114.74
Beds; poured on or against earth or unblinded hardcore										
thickness exceeding 450mm	67.73	7.50	72.81	-	3.30	23.17	1.80	97.78	m3	110.00
thickness 150 - 450mm	67.73	7.50	72.81	-	3.60	25.27	1.80	99.88	m3	112.37
thickness not exceeding 150mm	67.73	7.50	72.81	-	4.25	29.84	1.80	104.44	m3	117.50
Filling hollow walls										
thickness not exceeding 150mm	67.73	5.00	71.12	-	6.60	46.33	1.80	119.25	m3	134.15
Plain in-situ concrete; B.S.5328, ordinary prescribed mix C25P, 20mm aggregate										
Foundations poured on or against earth or unblinded hardcore										
generally......	70.69	7.50	75.99	-	3.40	23.87	1.80	101.66	m3	114.37
Isolated foundations; poured on or against earth or unblinded hardcore										
generally......	70.69	7.50	75.99	-	3.90	27.38	1.80	105.17	m3	118.32
Beds; poured on or against earth or unblinded hardcore										
thickness exceeding 450mm	70.69	7.50	75.99	-	3.30	23.17	1.80	100.96	m3	113.58
thickness 150 - 450mm	70.69	7.50	75.99	-	3.60	25.27	1.80	103.06	m3	115.95
thickness not exceeding 150mm	70.69	7.50	75.99	-	4.25	29.84	1.80	107.63	m3	121.08
Filling hollow walls										
thickness not exceeding 150mm	70.69	5.00	74.22	-	6.60	46.33	1.80	122.36	m3	137.65
Reinforced in-situ concrete; B.S.5328, designed mix C15, 20mm aggregate, minimum cement content 220 kg/m3; vibrated										
Foundations										
generally......	61.16	5.00	64.22	-	3.70	25.97	2.30	92.49	m3	104.05
Ground beams										
generally......	61.16	5.00	64.22	-	4.70	32.99	2.30	99.51	m3	111.95
Isolated foundations										
generally......	61.16	5.00	64.22	-	4.10	28.78	2.30	95.30	m3	107.21
Beds										
thickness exceeding 450mm	61.16	5.00	64.22	-	3.60	25.27	2.30	91.79	m3	103.26
thickness 150 - 450mm	61.16	5.00	64.22	-	3.90	27.38	2.30	93.90	m3	105.63
thickness not exceeding 150mm	61.16	5.00	64.22	-	4.50	31.59	2.30	98.11	m3	110.37
Reinforced in-situ concrete; B.S.5328, designed mix C20, 20mm aggregate, minimum cement content 240 kg/m3; vibrated										
Foundations										
generally......	62.33	5.00	65.45	-	3.70	25.97	2.30	93.72	m3	105.44
Ground beams										
generally......	62.33	Waste	65.45	-	4.70	32.99	2.30	100.74	m3	113.33
Isolated foundations										
generally......	62.33	5.00	65.45	-	4.10	28.78	2.30	96.53	m3	108.59
Beds										
thickness exceeding 450mm	62.33	5.00	65.45	-	3.60	25.27	2.30	93.02	m3	104.65

IN SITU CONCRETE/LARGE PRECAST CONCRETE

Labour hourly rates: (except Specialists) Craft Operatives £9.23 Labourer £7.02 Rates are national average prices. Refer to REGIONAL VARIATIONS for indicative levels of overall pricing in regions	MATERIALS			LABOUR				RATES		
	Del to Site	Waste	Material Cost	Craft Optve	Lab	Labour Cost	Sunds	Nett Rate	Unit	Gross Rate (+12.5%)
	£	%	£	Hrs	Hrs	£	£	£		£
E10: MIXING/CASTING/CURING IN-SITU CONCRETE - (SITE MIXED) Cont.										
Reinforced in-situ concrete; B.S.5328, designed mix C20, 20mm aggregate, minimum cement content 240 kg/m3; vibrated Cont.										
Beds Cont.										
thickness 150 - 450mm	62.33	5.00	65.45	-	3.90	27.38	2.30	95.12	m3	107.02
thickness not exceeding 150mm	62.33	5.00	65.45	-	4.50	31.59	2.30	99.34	m3	111.75
Reinforced in-situ concrete; B.S.5328, designed mix C25, 20mm aggregate, minimum cement content 290 kg/m3; vibrated										
Foundations										
generally	66.21	5.00	69.52	-	3.70	25.97	2.30	97.79	m3	110.02
Ground beams										
generally	66.21	5.00	69.52	-	4.70	32.99	2.30	104.81	m3	117.92
Isolated foundations										
generally	66.21	5.00	69.52	-	4.10	28.78	2.30	100.60	m3	113.18
Beds										
thickness exceeding 450mm	66.21	5.00	69.52	-	3.60	25.27	2.30	97.09	m3	109.23
thickness 150 - 450mm	66.21	5.00	69.52	-	3.90	27.38	2.30	99.20	m3	111.60
thickness not exceeding 150mm	66.21	5.00	69.52	-	4.50	31.59	2.30	103.41	m3	116.34
Cement and sand (1:3); for grouting and the like										
Grouting.										
stanchion bases	2.05	10.00	2.25	-	0.45	3.16	1.05	6.46	nr	7.27
E10: MIXING/CASTING/CURING IN-SITU CONCRETE - (SUNDRIES)										
The foregoing concrete is based on the use of Portland cement. For other cements and waterproofers ADD as follows										
Concrete 1:6										
rapid hardening cement	3.47	-	3.47	-	-	-	-	3.47	m3	3.90
sulphate resisting cement	4.35	-	4.35	-	-	-	-	4.35	m3	4.89
waterproofing powder	10.03	-	10.03	-	-	-	-	10.03	m3	11.28
waterproofing liquid	5.11	-	5.11	-	-	-	-	5.11	m3	5.75
Concrete 1:8										
rapid hardening cement	2.61	-	2.61	-	-	-	-	2.61	m3	2.94
sulphate resisting cement	3.27	-	3.27	-	-	-	-	3.27	m3	3.68
waterproofing powder	7.53	-	7.53	-	-	-	-	7.53	m3	8.47
waterproofing liquid	5.11	-	5.11	-	-	-	-	5.11	m3	5.75
Concrete 1:12										
rapid hardening cement	1.74	-	1.74	-	-	-	-	1.74	m3	1.96
sulphate resisting cement	2.17	-	2.17	-	-	-	-	2.17	m3	2.44
waterproofing powder	5.01	-	5.01	-	-	-	-	5.01	m3	5.64
waterproofing liquid	5.11	-	5.11	-	-	-	-	5.11	m3	5.75
Concrete ST3 and C15										
rapid hardening cement	2.32	-	2.32	-	-	-	-	2.32	m3	2.61
sulphate resisting cement	2.89	-	2.89	-	-	-	-	2.89	m3	3.25
waterproofing powder	6.69	-	6.69	-	-	-	-	6.69	m3	7.53
waterproofing liquid	5.11	-	5.11	-	-	-	-	5.11	m3	5.75
Concrete ST4 and C20										
rapid hardening cement	3.43	-	3.43	-	-	-	-	3.43	m3	3.86
sulphate resisting cement	4.30	-	4.30	-	-	-	-	4.30	m3	4.84
waterproofing powder	9.89	-	9.89	-	-	-	-	9.89	m3	11.13
waterproofing liquid	5.11	-	5.11	-	-	-	-	5.11	m3	5.75
Concrete ST5 and C25										
rapid hardening cement	4.63	-	4.63	-	-	-	-	4.63	m3	5.21
sulphate resisting cement	5.79	-	5.79	-	-	-	-	5.79	m3	6.51
waterproofing powder	13.36	-	13.36	-	-	-	-	13.36	m3	15.03
waterproofing liquid	5.11	-	5.11	-	-	-	-	5.11	m3	5.75
Reinforced in-situ lightweight concrete; 20.5 N/mm2; vibrated										
Walls										
thickness exceeding 450mm	81.37	2.50	83.40	-	4.85	34.05	2.30	119.75	m3	134.72
thickness 150 - 450mm	81.37	2.50	83.40	-	5.10	35.80	2.30	121.51	m3	136.69
thickness not exceeding 150mm	81.37	2.50	83.40	-	5.60	39.31	2.30	125.02	m3	140.64
Beam casings										
isolated..........................	81.37	2.50	83.40	-	6.05	42.47	2.30	128.18	m3	144.20
Reinforced in-situ lightweight concrete; 26.0 N/mm2; vibrated										
Slabs										
thickness exceeding 450mm	85.28	2.50	87.41	-	4.40	30.89	2.30	120.60	m3	135.68
thickness 150 - 450mm	85.28	2.50	87.41	-	4.50	31.59	2.30	121.30	m3	136.46
thickness not exceeding 150mm	85.28	2.50	87.41	-	5.10	35.80	2.30	125.51	m3	141.20

Labour hourly rates: (except Specialists) Craft Operatives £9.23 Labourer £7.02 Rates are national average prices. Refer to REGIONAL VARIATIONS for indicative levels of overall pricing in regions	MATERIALS			LABOUR				RATES		
	Del to Site £	Waste %	Material Cost £	Craft Optve Hrs	Lab Hrs	Labour Cost £	Sunds £	Nett Rate £	Unit	Gross Rate (+12.5%) £
E10: MIXING/CASTING/CURING IN-SITU CONCRETE – (SUNDRIES) Cont.										
Reinforced in-situ lightweight concrete; 41.5 N/mm2; vibrated										
Beams										
isolated	93.11	2.50	95.44	-	5.50	38.61	2.30	136.35	m3	153.39
Polythene sheeting as temporary protection to surface of concrete (use and waste)										
Polythene sheeting and laying										
125mu	0.28	10.00	0.31	-	0.03	0.21	-	0.52	m2	0.58
Waterproof building paper, B.S.1521; laying on hardcore to receive concrete										
Waterproof building paper and laying										
Grade B1F	0.40	25.00	0.50	-	0.06	0.42	-	0.92	m2	1.04
Grade B2	0.30	25.00	0.38	-	0.06	0.42	-	0.80	m2	0.90
E20: FORMWORK FOR IN-SITU CONCRETE										
Formwork and basic finish										
Sides of foundations; plain vertical										
height exceeding 1.00m	3.02	12.50	3.40	1.93	0.39	20.55	0.78	24.73	m2	27.82
height not exceeding 250mm	0.94	12.50	1.06	0.58	0.12	6.20	0.24	7.49	m	8.43
height 250 - 500mm	1.65	12.50	1.86	1.06	0.21	11.26	0.45	13.56	m	15.26
height 0.50m - 1.00m	3.09	12.50	3.48	1.97	0.40	20.99	0.78	25.25	m	28.40
Sides of foundations; plain vertical; left in										
height exceeding 1.00m	13.62	-	13.62	1.10	0.22	11.70	0.47	25.79	m2	29.01
height not exceeding 250mm	4.06	-	4.06	0.33	0.07	3.54	0.17	7.77	m	8.74
height 250 - 500mm	7.50	-	7.50	0.61	0.12	6.47	0.25	14.22	m	16.00
height 0.50m - 1.00m	13.95	-	13.95	1.13	0.23	12.04	0.49	26.48	m	29.80
Sides of ground beams and edges of beds; plain vertical										
height exceeding 1.00m	3.02	12.50	3.40	1.93	0.39	20.55	0.78	24.73	m2	27.82
height not exceeding 250mm	0.94	12.50	1.06	0.58	0.12	6.20	0.24	7.49	m	8.43
height 250 - 500mm	1.65	12.50	1.86	1.06	0.21	11.26	0.45	13.56	m	15.26
height 0.50m - 1.00m	3.09	12.50	3.48	1.97	0.40	20.99	0.78	25.25	m	28.40
Sides of ground beams and edges of beds; plain vertical; left in										
height exceeding 1.00m	13.62	-	13.62	1.10	0.22	11.70	0.45	25.77	m2	28.99
height not exceeding 250mm	4.06	-	4.06	0.33	0.07	3.54	0.17	7.77	m	8.74
height 250 - 500mm	7.50	-	7.50	0.61	0.12	6.47	0.25	14.22	m	16.00
height 0.50m - 1.00m	13.95	-	13.95	1.13	0.23	12.04	0.49	26.48	m	29.80
Edges of suspended slabs; plain vertical										
height not exceeding 250mm	1.13	12.50	1.27	0.73	0.14	7.72	0.27	9.26	m	10.42
height 250 - 500mm	2.11	12.50	2.37	1.32	0.26	14.01	0.54	16.92	m	19.04
extra; recesses; plain rectangular										
12 x 25mm	0.01	-	0.01	0.05	-	0.46	-	0.47	m	0.53
25 x 50mm	0.03	-	0.03	0.07	-	0.65	-	0.68	m	0.76
50 x 75mm	0.15	-	0.15	0.09	-	0.83	-	0.98	m	1.10
50 x 100mm	0.24	-	0.24	0.12	-	1.11	-	1.35	m	1.52
extra; rebates; plain rectangular										
12 x 25mm	0.01	-	0.01	0.07	-	0.65	-	0.66	m	0.74
25 x 50mm	0.03	-	0.03	0.08	-	0.74	-	0.77	m	0.86
50 x 75mm	0.16	-	0.16	0.11	-	1.02	-	1.18	m	1.32
50 x 100mm	0.25	-	0.25	0.14	-	1.29	-	1.54	m	1.73
extra; chamfers										
60mm wide	0.06	-	0.06	0.05	-	0.46	-	0.52	m	0.59
75mm wide	0.07	-	0.07	0.07	-	0.65	-	0.72	m	0.81
100mm wide	0.18	-	0.18	0.12	-	1.11	-	1.29	m	1.45
Edges of suspended slabs; plain vertical; curved 10m radius										
height not exceeding 250mm	1.70	12.50	1.91	1.05	0.22	11.24	0.47	13.62	m	15.32
Edges of suspended slabs; plain vertical; curved 1m radius										
height not exceeding 250mm	2.87	12.50	3.23	1.73	0.36	18.50	0.75	22.47	m	25.28
Edges of suspended slabs; plain vertical; curved 20m radius										
height not exceeding 250mm	1.28	12.50	1.44	0.78	0.17	8.39	0.30	10.13	m	11.40
Sides of upstands; plain vertical										
height not exceeding 250mm	1.34	12.50	1.51	0.79	0.17	8.49	0.30	10.29	m	11.58
height 250 - 500mm	1.57	12.50	1.77	0.92	0.20	9.90	0.40	12.06	m	13.57
extra; recesses; plain rectangular										
12 x 25mm	0.01	-	0.01	0.05	-	0.46	-	0.47	m	0.53
25 x 50mm	0.03	-	0.03	0.08	-	0.74	-	0.77	m	0.86
50 x 75mm	0.15	-	0.15	0.10	-	0.92	-	1.07	m	1.21
50 x 100mm	0.21	-	0.21	0.12	-	1.11	-	1.32	m	1.48
extra; rebates; plain rectangular										
12 x 25mm	0.01	-	0.01	0.08	-	0.74	-	0.75	m	0.84
25 x 50mm	0.03	-	0.03	0.09	-	0.83	-	0.86	m	0.97
50 x 75mm	0.16	-	0.16	0.11	-	1.02	-	1.18	m	1.32
50 x 100mm	0.24	-	0.24	0.14	-	1.29	-	1.53	m	1.72

Labour hourly rates: (except Specialists) Craft Operatives £9.23 Labourer £7.02 Rates are national average prices. Refer to REGIONAL VARIATIONS for indicative levels of overall pricing in regions	MATERIALS			LABOUR				RATES		
	Del to Site £	Waste %	Material Cost £	Craft Optve Hrs	Lab Hrs	Labour Cost £	Sunds £	Nett Rate £	Unit	Gross Rate (+12.5%) £
E20: FORMWORK FOR IN-SITU CONCRETE Cont.										
Formwork and basic finish Cont.										
Sides of upstands; plain vertical Cont.										
extra; chamfers										
60mm wide..........................	0.06	-	0.06	0.06	-	0.55	-	0.61	m	0.69
75mm wide..........................	0.07	-	0.07	0.08	-	0.74	-	0.81	m	0.91
100mm wide.........................	0.18	-	0.18	0.12	-	1.11	-	1.29	m	1.45
Machine bases and plinths; plain vertical										
height exceeding 1.00m	3.33	12.50	3.75	2.12	0.42	22.52	0.87	27.13	m2	30.52
height not exceeding 250mm	0.98	12.50	1.10	0.62	0.12	6.57	0.25	7.92	m	8.91
height 250 - 500mm	1.82	12.50	2.05	1.17	0.22	12.34	0.47	14.86	m	16.72
height 0.50m - 1.00m	5.21	12.50	5.86	2.18	0.43	23.14	0.91	29.91	m	33.65
Soffits of slabs; horizontal										
slab thickness not exceeding 200mm; height to soffit										
1.50 - 3.00m......................	9.13	10.00	10.04	1.65	0.33	17.55	1.09	28.68	m2	32.26
3.00 - 4.50m......................	9.13	10.00	10.04	1.89	0.39	20.18	1.49	31.72	m2	35.68
4.50 - 6.00m......................	9.13	10.00	10.04	2.13	0.44	22.75	1.90	34.69	m2	39.03
slab thickness 200 - 300mm; height to soffit										
1.50 - 3.00m......................	10.05	10.00	11.06	1.82	0.36	19.33	1.23	31.61	m2	35.56
3.00 - 4.50m......................	10.05	10.00	11.06	2.06	0.42	21.96	1.61	34.63	m2	38.96
4.50 - 6.00m......................	10.05	10.00	11.06	2.30	0.47	24.53	2.01	37.59	m2	42.29
Soffits of slabs; horizontal; left in										
slab thickness not exceeding 200mm; height to soffit										
not exceeding 1.50m...............	24.49	-	24.49	1.10	0.22	11.70	0.70	36.89	m2	41.50
1.50 - 3.00m......................	24.49	-	24.49	1.10	0.22	11.70	0.78	36.97	m2	41.59
Soffits of slabs; horizontal; with frequent uses of prefabricated panels										
slab thickness not exceeding 200mm; height to soffit										
1.50 - 3.00m......................	2.71	15.00	3.12	1.65	0.33	17.55	1.09	21.75	m2	24.47
3.00 - 4.50m......................	2.71	15.00	3.12	1.89	0.39	20.18	1.52	24.82	m2	27.92
4.50 - 6.00m......................	2.71	15.00	3.12	2.13	0.44	22.75	1.90	27.77	m2	31.24
slab thickness 200 - 300mm; height to soffit										
1.50 - 3.00m......................	2.97	15.00	3.42	1.82	0.36	19.33	1.23	23.97	m2	26.97
3.00 - 4.50m......................	2.97	15.00	3.42	2.06	0.42	21.96	1.61	26.99	m2	30.36
4.50 - 6.00m......................	2.97	15.00	3.42	2.30	0.47	24.53	2.01	29.95	m2	33.70
Soffits of slabs; sloping not exceeding 15 degrees										
slab thickness not exceeding 200mm; height to soffit										
1.50 - 3.00m......................	10.06	10.00	11.07	1.82	0.36	19.33	1.22	31.61	m2	35.56
3.00 - 4.50m......................	10.06	10.00	11.07	2.06	0.42	21.96	1.59	34.62	m2	38.95
4.50 - 6.00m......................	10.06	10.00	11.07	2.36	0.47	25.08	1.98	38.13	m2	42.89
Soffits of slabs; sloping exceeding 15 degrees										
slab thickness not exceeding 200mm; height to soffit										
1.50 - 3.00m......................	10.52	10.00	11.57	1.90	0.39	20.27	1.40	33.25	m2	37.40
3.00 - 4.50m......................	10.52	10.00	11.57	2.15	0.44	22.93	1.79	36.30	m2	40.83
4.50 - 6.00m......................	10.52	10.00	11.57	2.39	0.50	25.57	2.17	39.31	m2	44.23
Soffits of landings; horizontal										
slab thickness not exceeding 200mm; height to soffit										
not exceeding 1.50m...............	9.13	10.00	10.04	1.98	0.40	21.08	1.09	32.22	m2	36.24
1.50 - 3.00m......................	9.13	10.00	10.04	2.22	0.45	23.65	1.49	35.18	m2	39.58
3.00 - 4.50m......................	9.13	10.00	10.04	2.46	0.51	26.29	1.90	38.23	m2	43.01
Top formwork										
sloping exceeding 15 degrees	3.54	12.50	3.98	1.65	0.33	17.55	0.62	22.15	m2	24.92
Walls; vertical										
plain........................	6.41	15.00	7.37	1.76	0.35	18.70	1.01	27.08	m2	30.47
plain; height exceeding 3.00m above floor level ..	5.91	15.00	6.80	1.76	0.35	18.70	1.14	26.64	m2	29.97
interrupted......................	5.91	15.00	6.80	1.93	0.39	20.55	1.12	28.47	m2	32.03
interrupted; height exceeding 3.00m above floor level.........................	6.31	15.00	7.26	1.93	0.39	20.55	1.24	29.05	m2	32.68
extra; recesses; plain rectangular										
12 x 25mm......................	0.01	-	0.01	0.06	-	0.55	-	0.56	m	0.63
25 x 50mm......................	0.03	-	0.03	0.08	-	0.74	-	0.77	m	0.86
50 x 75mm......................	0.15	-	0.15	0.10	-	0.92	-	1.07	m	1.21
50 x 100mm.....................	0.21	-	0.21	0.12	-	1.11	-	1.32	m	1.48
Walls; vertical; curved 10m radius										
plain........................	7.89	15.00	9.07	2.64	0.53	28.09	1.54	38.70	m2	43.54
Walls; vertical; curved 1m radius										
plain........................	13.11	15.00	15.08	4.40	0.88	46.79	2.52	64.39	m2	72.43
Walls; vertical; curved 20m radius										
plain........................	5.90	15.00	6.79	1.98	0.46	21.50	1.14	29.43	m2	33.11
Walls; battered										
plain........................	6.28	15.00	7.22	2.11	0.42	22.42	1.24	30.89	m2	34.75
Beams; attached to slabs										
regular shaped; rectangular; height to soffit										
1.50 - 3.00m......................	5.53	10.00	6.08	2.48	0.45	26.05	0.97	33.10	m2	37.24

Labour hourly rates: (except Specialists) Craft Operatives £9.23 Labourer £7.02 Rates are national average prices. Refer to REGIONAL VARIATIONS for indicative levels of overall pricing in regions	MATERIALS			LABOUR				RATES		
	Del to Site £	Waste %	Material Cost £	Craft Optve Hrs	Lab Hrs	Labour Cost £	Sunds £	Nett Rate £	Unit	Gross Rate (+12.5%) £
E20: FORMWORK FOR IN-SITU CONCRETE Cont.										
Formwork and basic finish Cont.										
Beams; attached to slabs Cont.										
regular shaped; rectangular; height to soffit Cont.										
3.00 - 4.50m..................	5.53	10.00	6.08	2.48	0.45	26.05	1.19	33.32	m2	37.49
regular shaped; rectangular; with 50mm wide chamfers; height to soffit										
1.50 - 3.00m................	5.63	10.00	6.19	2.54	0.45	26.60	0.97	33.77	m2	37.99
3.00 - 4.50m................	5.63	10.00	6.19	2.54	0.45	26.60	1.19	33.99	m2	38.23
Beams; isolated										
regular shaped; rectangular; height to soffit										
1.50 - 3.00m................	5.80	10.00	6.38	2.59	0.47	27.21	0.97	34.56	m2	38.87
3.00 - 4.50m................	5.80	10.00	6.38	2.59	0.47	27.21	1.19	34.78	m2	39.12
regular shaped; rectangular; with 50mm wide chamfers; height to soffit										
1.50 - 3.00m................	5.91	10.00	6.50	2.65	0.47	27.76	0.97	35.23	m2	39.63
3.00 - 4.50m................	5.91	10.00	6.50	2.65	0.47	27.76	1.19	35.45	m2	39.88
Columns; attached to walls										
regular shaped; rectangular	3.84	10.00	4.22	2.53	0.46	26.58	0.91	31.72	m2	35.68
regular shaped; rectangular, height exceeding 3.00m above floor level	3.84	10.00	4.22	2.53	0.46	26.58	0.91	31.72	m2	35.68
Columns; isolated										
regular shaped; rectangular	3.84	10.00	4.22	2.48	0.45	26.05	0.97	31.24	m2	35.15
regular shaped; rectangular, height exceeding 3.00m above floor level	4.11	10.00	4.52	2.48	0.45	26.05	0.97	31.54	m2	35.48
regular shaped; rectangular; with 50mm wide chamfers	3.84	10.00	4.22	2.63	0.45	27.43	0.97	32.63	m2	36.71
regular shaped; rectangular; with 50mm wide chamfers; height exceeding 3.00m above floor level	3.84	10.00	4.22	2.63	0.45	27.43	0.97	32.63	m2	36.71
Column casings; attached to walls										
regular shaped; rectangular	3.84	10.00	4.22	2.53	0.46	26.58	0.91	31.72	m2	35.68
regular shaped; rectangular, height exceeding 3.00m above floor level	4.11	10.00	4.52	2.53	0.46	26.58	0.91	32.01	m2	36.01
Column casings; isolated										
regular shaped; rectangular	4.11	10.00	4.52	2.48	0.45	26.05	0.97	31.54	m2	35.48
regular shaped; rectangular, height exceeding 3.00m above floor level	3.84	10.00	4.22	2.48	0.45	26.05	0.97	31.24	m2	35.15
regular shaped; rectangular; with 50mm wide chamfers	3.84	10.00	4.22	2.63	0.45	27.43	0.97	32.63	m2	36.71
regular shaped; rectangular; with 50mm wide chamfers; height exceeding 3.00m above floor level	4.11	10.00	4.52	2.63	0.45	27.43	0.97	32.92	m2	37.04
Extra over a basic finish for a fine formed finish										
slabs	-	-	-	-	0.39	2.74	-	2.74	m2	3.08
walls	-	-	-	-	0.39	2.74	-	2.74	m2	3.08
beams	-	-	-	-	0.39	2.74	-	2.74	m2	3.08
columns	-	-	-	-	0.39	2.74	-	2.74	m2	3.08
Wall kickers										
straight	0.88	15.00	1.01	0.29	0.77	8.08	0.18	9.27	m	10.43
curved 2m radius	2.16	15.00	2.48	0.73	0.16	7.86	0.41	10.76	m	12.10
curved 10m radius	1.31	15.00	1.51	0.44	0.09	4.69	0.25	6.45	m	7.26
curved 20m radius	0.96	15.00	1.10	0.33	0.08	3.61	0.21	4.92	m	5.54
Suspended wall kickers										
straight	0.95	15.00	1.09	0.32	0.07	3.44	0.20	4.74	m	5.33
curved 2m radius	2.39	15.00	2.75	0.80	0.18	8.65	0.46	11.86	m	13.34
curved 10m radius	1.41	15.00	1.62	0.44	0.10	4.76	0.27	6.65	m	7.49
curved 20m radius	1.09	15.00	1.25	0.36	0.09	3.95	0.23	5.44	m	6.12
Wall ends, soffits and steps in walls; plain width not exceeding 250mm	1.79	15.00	2.06	0.58	0.12	6.20	0.32	8.57	m	9.65
Openings in walls; plain width not exceeding 250mm	1.79	15.00	2.06	0.58	0.12	6.20	0.32	8.57	m	9.65
Stairflights; 1000mm wide; 155mm thick waist; 178mm risers; includes formwork to soffits, risers and strings										
strings 300mm wide	13.63	10.00	14.99	6.66	1.33	70.81	2.30	88.10	m	99.11
string 300mm wide; junction with wall	13.63	10.00	14.99	6.66	1.33	70.81	2.30	88.10	m	99.11
Stairflights; 1500mm wide; 180mm thick waist; 178mm risers; includes formwork to soffits, risers and strings										
strings 325mm wide	17.06	10.00	18.77	8.32	1.66	88.45	2.88	110.09	m	123.85
string 325mm wide; junction with wall	17.06	10.00	18.77	8.32	1.66	88.45	2.88	110.09	m	123.85
Stairflights; 1000mm wide; 155mm thick waist; 178mm undercut risers; includes formwork to soffits, risers and strings										
strings 300mm wide	13.63	10.00	14.99	6.66	1.33	70.81	2.30	88.10	m	99.11
string 300mm wide; junction with wall	13.63	10.00	14.99	6.66	1.33	70.81	2.30	88.10	m	99.11
Stairflights; 1500mm wide; 180mm thick waist; 178mm undercut risers; includes formwork to soffits, risers and strings										
strings 325mm wide	17.05	10.00	18.75	8.32	1.66	88.45	2.88	110.08	m	123.84

IN SITU CONCRETE/LARGE PRECAST CONCRETE

Labour hourly rates: (except Specialists) Craft Operatives £9.23 Labourer £7.02 Rates are national average prices. Refer to REGIONAL VARIATIONS for indicative levels of overall pricing in regions	MATERIALS			LABOUR				RATES		
	Del to Site £	Waste %	Material Cost £	Craft Optve Hrs	Lab Hrs	Labour Cost £	Sunds £	Nett Rate £	Unit	Gross Rate (+12.5%) £
E20: FORMWORK FOR IN-SITU CONCRETE Cont.										
Formwork and basic finish Cont.										
Stairflights; 1500mm wide; 180mm thick waist; 178mm undercut risers; includes formwork to soffits, risers and strings Cont.										
string 325mm wide; junction with wall	17.05	10.00	18.75	8.32	1.66	88.45	2.88	110.08	m	123.84
Mortices; rectangular										
girth not exceeding 500mm; depth not exceeding 250mm	0.81	10.00	0.89	0.53	0.11	5.66	0.22	6.78	nr	7.62
girth 0.50 - 1.00m; depth 250 - 500mm	2.63	10.00	2.89	1.69	0.34	17.99	0.68	21.56	nr	24.25
girth 0.50 - 1.00m; depth 0.50 - 1.00m	5.19	10.00	5.71	3.39	0.68	36.06	1.38	43.15	nr	48.55
Mortices; circular										
girth not exceeding 500mm; depth not exceeding 250mm	1.63	10.00	1.79	1.06	0.22	11.33	0.43	13.55	nr	15.25
Holes; rectangular										
girth not exceeding 500mm; depth not exceeding 250mm	0.81	10.00	0.89	0.53	0.11	5.66	0.22	6.78	nr	7.62
girth not exceeding 500mm; depth 250 - 500mm	1.50	10.00	1.65	0.97	0.20	10.36	0.41	12.42	nr	13.97
girth 0.50 - 1.00m; depth not exceeding 250mm	1.42	10.00	1.56	0.94	0.19	10.01	0.39	11.96	nr	13.46
girth 0.50 - 1.00m; depth 250 - 500mm	2.63	10.00	2.89	1.69	0.34	17.99	0.68	21.56	nr	24.25
girth 1.00m - 2.00m; depth not exceeding 250mm ...	2.04	10.00	2.24	1.33	0.26	14.10	0.54	16.89	nr	19.00
girth 2.00m - 3.00m; depth not exceeding 250mm ...	2.66	10.00	2.93	1.73	0.34	18.35	0.71	21.99	nr	24.74
girth 3.00m - 4.00m; depth not exceeding 250mm ...	3.34	10.00	3.67	2.12	0.43	22.59	0.85	27.11	nr	30.50
girth 1.00m - 2.00m; depth 250 - 500mm	3.73	10.00	4.10	2.42	0.48	25.71	0.97	30.78	nr	34.63
girth 2.00m - 3.00m; depth 250 - 500mm	4.85	10.00	5.34	3.14	0.63	33.40	1.27	40.01	nr	45.01
girth 3.00m - 4.00m; depth 250 - 500mm	6.02	10.00	6.62	3.85	0.77	40.94	1.54	49.10	nr	55.24
Holes; circular										
girth not exceeding 500mm; depth not exceeding 250mm	1.63	10.00	1.79	1.06	0.22	11.33	0.43	13.55	nr	15.25
girth not exceeding 500mm; depth 250 - 500mm	2.97	10.00	3.27	1.94	0.40	20.71	0.78	24.76	nr	27.86
girth 0.50 - 1.00m; depth not exceeding 250mm	2.88	10.00	3.17	1.87	0.37	19.86	0.75	23.78	nr	26.75
girth 0.50 - 1.00m; depth 250 - 500mm	5.19	10.00	5.71	3.39	0.68	36.06	1.38	43.15	nr	48.55
girth 1.00m - 2.00m; depth not exceeding 250mm ...	4.10	10.00	4.51	2.66	0.53	28.27	1.13	33.91	nr	38.15
girth 2.00m - 3.00m; depth not exceeding 250mm ...	5.32	10.00	5.85	3.45	0.68	36.62	1.40	43.87	nr	49.35
girth 3.00m - 4.00m; depth not exceeding 250mm ...	6.69	10.00	7.36	4.25	0.86	45.26	1.72	54.34	nr	61.14
girth 1.00m - 2.00m; depth 250 - 500mm	7.45	10.00	8.20	4.84	0.97	51.48	1.88	61.56	nr	69.25
girth 2.00m - 3.00m; depth 250 - 500mm	9.67	10.00	10.64	6.27	1.25	66.65	2.52	79.80	nr	89.78
girth 3.00m - 4.00m; depth 250 - 500mm	17.11	10.00	18.82	7.70	1.54	81.88	3.12	103.82	nr	116.80
Formwork and basic finish; coating with retarding agent										
Sides of foundations; plain vertical										
height exceeding 1.00m	3.65	12.50	4.11	1.93	0.65	22.38	0.78	27.26	m2	30.67
height not exceeding 250mm	1.00	12.50	1.13	0.58	0.15	6.41	0.24	7.77	m	8.74
height 250 - 500mm	1.89	12.50	2.13	1.06	0.31	11.96	0.44	14.53	m	16.34
height 0.50m - 1.00m	3.58	12.50	4.03	1.97	0.59	22.32	0.78	27.13	m	30.52
Sides of ground beams and edges of beds; plain vertical										
height exceeding 1.00m	3.67	12.50	4.13	1.93	0.65	22.38	0.78	27.29	m2	30.70
height not exceeding 250mm	1.00	12.50	1.13	0.58	0.15	6.41	0.24	7.77	m	8.74
height 250 - 500mm	1.89	12.50	2.13	1.06	0.31	11.96	0.44	14.53	m	16.34
height 0.50m - 1.00m	3.58	12.50	4.03	1.97	0.59	22.32	0.78	27.13	m	30.52
Edges of suspended slabs; plain vertical										
height not exceeding 250mm	1.19	12.50	1.34	0.73	0.18	8.00	0.27	9.61	m	10.81
height 250 - 500mm	2.31	12.50	2.60	1.32	0.39	14.92	0.54	18.06	m	20.32
Edges of suspended slabs; plain vertical; curved 10m radius										
height not exceeding 250mm	2.94	12.50	3.31	1.82	0.40	19.61	1.17	24.08	m	27.09
Edges of suspended slabs; plain vertical; curved 1m radius										
height not exceeding 250mm	1.78	12.50	2.00	1.10	0.25	11.91	0.47	14.38	m	16.18
Edges of suspended slabs; plain vertical; curved 20m radius										
height not exceeding 250mm	1.35	12.50	1.52	0.81	0.20	8.88	0.30	10.70	m	12.04
Sides of upstands; plain vertical										
height not exceeding 250mm	1.47	12.50	1.65	0.83	0.20	9.06	0.30	11.02	m	12.40
height 250 - 500mm	1.81	12.50	2.04	0.97	0.30	11.06	0.41	13.51	m	15.19
Machine bases and plinths; plain vertical										
height exceeding 1.00m	3.95	12.50	4.44	2.12	0.68	24.34	0.87	29.65	m2	33.36
height not exceeding 250mm	1.17	12.50	1.32	0.64	0.15	6.96	0.25	8.53	m	9.59
height 250 - 500mm	2.05	12.50	2.31	1.17	0.32	13.05	0.47	15.82	m	17.80
height 0.50m - 1.00m	3.85	12.50	4.33	2.18	0.63	24.54	0.91	29.79	m	33.51
Soffits of slabs; horizontal										
slab thickness not exceeding 200mm; height to soffit										
1.50 - 3.00m..............................	9.77	10.00	10.75	1.65	0.59	19.37	1.25	31.37	m2	35.29
3.00 - 4.50m..............................	9.77	10.00	10.75	1.89	0.65	22.01	1.49	34.24	m2	38.53
4.50 - 6.00m..............................	9.77	10.00	10.75	2.13	0.76	25.00	1.90	37.64	m2	42.35
slab thickness 200 - 300mm; height to soffit										
1.50 - 3.00m..............................	10.69	10.00	11.76	1.82	0.63	21.22	1.23	34.21	m2	38.49
3.00 - 4.50m..............................	10.69	10.00	11.76	2.06	0.68	23.79	1.61	37.16	m2	41.80

Labour hourly rates: (except Specialists) Craft Operatives £9.23 Labourer £7.02 Rates are national average prices. Refer to REGIONAL VARIATIONS for indicative levels of overall pricing in regions	MATERIALS			LABOUR				RATES		
	Del to Site £	Waste %	Material Cost £	Craft Optve Hrs	Lab Hrs	Labour Cost £	Sunds £	Nett Rate £	Unit	Gross Rate (+12.5%) £

E20: FORMWORK FOR IN-SITU CONCRETE Cont.

Formwork and basic finish; coating with retarding agent Cont.

	Del to Site £	Waste %	Material Cost £	Craft Optve Hrs	Lab Hrs	Labour Cost £	Sunds £	Nett Rate £	Unit	Gross Rate £
Soffits of slabs; horizontal Cont.										
slab thickness 200 - 300mm; height to soffit Cont.										
4.50 - 6.00m	10.69	10.00	11.76	2.30	0.74	26.42	2.01	40.19	m2	45.22
Soffits of slabs; horizontal; with frequent uses of prefabricated panels										
slab thickness not exceeding 200mm; height to soffit										
1.50 - 3.00m	3.35	15.00	3.85	1.65	0.59	19.37	1.25	24.47	m2	27.53
3.00 - 4.50m	3.35	15.00	3.85	1.89	0.65	22.01	1.49	27.35	m2	30.77
4.50 - 6.00m	3.35	15.00	3.85	2.13	0.70	24.57	1.90	30.33	m2	34.12
slab thickness 200 - 300mm; height to soffit										
1.50 - 3.00m	3.99	15.00	4.59	1.82	0.86	22.84	1.23	28.65	m2	32.24
3.00 - 4.50m	3.99	15.00	4.59	2.06	0.91	25.40	1.61	31.60	m2	35.55
4.50 - 6.00m	3.99	15.00	4.59	2.30	0.97	28.04	2.01	34.64	m2	38.97
Soffits of slabs; sloping not exceeding 15 degrees										
slab thickness not exceeding 200mm; height to soffit										
1.50 - 3.00m	10.72	10.00	11.79	1.82	0.63	21.22	1.22	34.23	m2	38.51
3.00 - 4.50m	10.72	10.00	11.79	2.06	0.68	23.79	1.59	37.17	m2	41.82
4.50 - 6.00m	10.72	10.00	11.79	2.30	0.74	26.42	1.98	40.20	m2	45.22
Soffits of slabs; sloping exceeding 15 degrees										
slab thickness not exceeding 200mm; height to soffit										
1.50 - 3.00m	11.17	10.00	12.29	1.90	0.65	22.10	1.40	35.79	m2	40.26
3.00 - 4.50m	11.17	10.00	12.29	2.15	0.70	24.76	1.79	38.84	m2	43.69
4.50 - 6.00m	11.17	10.00	12.29	2.33	0.76	26.84	2.17	41.30	m2	46.46
Soffits of landings; horizontal										
slab thickness not exceeding 200mm; height to soffit										
not exceeding 1.50m	9.77	10.00	10.75	1.98	0.66	22.91	1.25	34.91	m2	39.27
1.50 - 3.00m	9.77	10.00	10.75	2.22	0.72	25.55	1.49	37.78	m2	42.50
3.00 - 4.50m	9.77	10.00	10.75	2.46	0.77	28.11	1.90	40.76	m2	45.85
Top formwork										
sloping exceeding 15 degrees	4.06	12.50	4.57	1.65	0.59	19.37	0.64	24.58	m2	27.65
Walls; vertical										
plain	5.94	15.00	6.83	1.76	0.62	20.60	1.01	28.44	m2	31.99
plain; height exceeding 3.00m above floor level	6.55	15.00	7.53	1.76	0.62	20.60	1.14	29.27	m2	32.93
interrupted	6.41	15.00	7.37	1.93	0.65	22.38	1.12	30.87	m2	34.73
interrupted; height exceeding 3.00m above floor level	6.86	15.00	7.89	1.93	0.65	22.38	1.24	31.51	m2	35.44
Walls; vertical; curved 10m radius										
plain	8.54	15.00	9.82	2.64	0.79	29.91	1.54	41.27	m2	46.43
Walls; vertical; curved 1m radius										
plain	13.72	15.00	15.78	4.40	1.14	48.61	2.52	66.91	m2	75.28
Walls; vertical; curved 20m radius										
plain	6.54	15.00	7.52	1.98	0.66	22.91	1.03	31.46	m2	35.39
Walls; battered										
plain	6.92	15.00	7.96	2.11	0.68	24.25	1.24	33.45	m2	37.63
Beams; attached to slabs										
regular shaped; rectangular; height to soffit										
1.50 - 3.00m	6.15	10.00	6.76	2.48	0.76	28.23	0.97	35.96	m2	40.46
3.00 - 4.50m	6.15	10.00	6.76	2.48	0.76	28.23	1.19	36.18	m2	40.70
regular shaped; rectangular; with 50mm wide chamfers; height to soffit										
1.50 - 3.00m	6.25	10.00	6.88	2.54	0.76	28.78	0.97	36.62	m2	41.20
3.00 - 4.50m	6.25	10.00	6.88	2.54	0.76	28.78	1.19	36.84	m2	41.45
Beams; isolated										
regular shaped; rectangular; height to soffit										
1.50 - 3.00m	6.39	10.00	7.03	2.59	0.78	29.38	0.97	37.38	m2	42.05
3.00 - 4.50m	6.39	10.00	7.03	2.59	0.78	29.38	1.19	37.60	m2	42.30
regular shaped; rectangular; with 50mm wide chamfers; height to soffit										
1.50 - 3.00m	6.12	10.00	6.73	2.65	0.78	29.94	0.97	37.64	m2	42.34
3.00 - 4.50m	6.12	10.00	6.73	2.65	0.78	29.94	1.19	37.86	m2	42.59
Columns; attached to walls										
regular shaped; rectangular	4.49	10.00	4.94	2.53	0.77	28.76	0.91	34.61	m2	38.93
regular shaped; rectangular, height exceeding 3.00m above floor level	4.49	10.00	4.94	2.48	0.76	28.23	0.97	34.13	m2	38.40
Columns; isolated										
regular shaped; rectangular	4.49	10.00	4.94	2.48	0.76	28.23	0.97	34.13	m2	38.40
regular shaped; rectangular, height exceeding 3.00m above floor level	4.74	10.00	5.21	2.63	0.76	29.61	0.97	35.79	m2	40.27
regular shaped; rectangular; with 50mm wide chamfers	4.74	10.00	5.21	2.63	0.76	29.61	0.92	35.74	m2	40.21
regular shaped; rectangular; with 50mm wide chamfers; height exceeding 3.00m above floor level	4.74	10.00	5.21	2.63	0.76	29.61	0.92	35.74	m2	40.21

Labour hourly rates: (except Specialists) Craft Operatives £9.23 Labourer £7.02 Rates are national average prices. Refer to REGIONAL VARIATIONS for indicative levels of overall pricing in regions	MATERIALS			LABOUR				RATES		
	Del to Site £	Waste %	Material Cost £	Craft Optve Hrs	Lab Hrs	Labour Cost £	Sunds £	Nett Rate £	Unit	Gross Rate (+12.5%) £
E20: FORMWORK FOR IN-SITU CONCRETE Cont.										
Formwork and basic finish; coating with retarding agent Cont.										
Column casings; attached to walls										
regular shaped; rectangular	4.49	10.00	4.94	2.53	0.77	28.76	0.91	34.61	m2	38.93
regular shaped; rectangular, height exceeding 3.00m above floor level	4.49	10.00	4.94	2.53	0.77	28.76	0.91	34.61	m2	38.93
Column casings; isolated										
regular shaped; rectangular	4.49	10.00	4.94	2.48	0.76	28.23	0.97	34.13	m2	38.40
regular shaped; rectangular, height exceeding 3.00m above floor level	4.49	10.00	4.94	2.48	0.76	28.23	0.97	34.13	m2	38.40
regular shaped; rectangular; with 50mm wide chamfers	4.74	10.00	5.21	2.63	0.76	29.61	0.97	35.79	m2	40.27
regular shaped; rectangular; with 50mm wide chamfers; height exceeding 3.00m above floor level	4.74	10.00	5.21	2.48	0.76	28.23	0.97	34.41	m2	38.71
Extra over a basic finish for a fine formed finish										
slabs	-	-	-	-	0.39	2.74	-	2.74	m2	3.08
walls	-	-	-	-	0.39	2.74	-	2.74	m2	3.08
beams	-	-	-	-	0.39	2.74	-	2.74	m2	3.08
columns	-	-	-	-	0.39	2.74	-	2.74	m2	3.08
Wall kickers										
straight	0.96	15.00	1.10	0.29	0.10	3.38	0.18	4.66	m	5.25
curved 2m radius	2.26	15.00	2.60	0.73	0.19	8.07	0.42	11.09	m	12.48
curved 10m radius	1.40	15.00	1.61	0.44	0.13	4.97	0.25	6.83	m	7.69
curved 20m radius	1.09	15.00	1.25	0.33	0.11	3.82	0.21	5.28	m	5.94
Suspended wall kickers										
straight	1.08	15.00	1.24	0.32	0.11	3.73	0.20	5.17	m	5.81
curved 2m radius	2.48	15.00	2.85	0.80	0.21	8.86	0.46	12.17	m	13.69
curved 10m radius	1.55	15.00	1.78	0.44	0.14	5.04	0.27	7.10	m	7.98
curved 20m radius	1.18	15.00	1.36	0.36	0.12	4.17	0.23	5.75	m	6.47
Wall ends, soffits and steps in walls; plain										
width not exceeding 250mm	2.04	15.00	2.35	0.58	0.19	6.69	0.32	9.35	m	10.52
Openings in walls; plain										
width not exceeding 250mm	2.04	15.00	2.35	0.58	0.19	6.69	0.32	9.35	m	10.52
Stairflights; 1000mm wide; 155mm thick waist; 178mm risers; includes formwork to soffits, risers and strings										
strings 300mm wide	15.04	10.00	16.54	6.66	1.91	74.88	2.30	93.72	m	105.44
string 300mm wide; junction with wall	15.04	10.00	16.54	6.66	1.91	74.88	2.30	93.72	m	105.44
Stairflights; 1500mm wide; 180mm thick waist; 178mm risers; includes formwork to soffits, risers and strings										
strings 325mm wide	18.80	10.00	20.68	8.32	2.39	93.57	2.88	117.13	m	131.77
string 325mm wide; junction with wall	18.80	10.00	20.68	8.32	2.39	93.57	2.88	117.13	m	131.77
Stairflights; 1000mm wide; 155mm thick waist; 178mm undercut risers; includes formwork to soffits, risers and strings										
strings 300mm wide	15.04	10.00	16.54	6.66	1.91	74.88	2.30	93.72	m	105.44
string 300mm wide; junction with wall	15.04	10.00	16.54	6.66	1.91	74.88	2.30	93.72	m	105.44
Stairflights; 1500mm wide; 180mm thick waist; 178mm undercut risers; includes formwork to soffits, risers and strings										
strings 325mm wide	18.80	10.00	20.68	8.32	2.39	93.57	2.88	117.13	m	131.77
string 325mm wide; junction with wall	18.80	10.00	20.68	8.32	2.39	93.57	2.88	117.13	m	131.77
Claymaster low density expanded polystyrene permanent formwork; fixing with Clayfix hooks - 3/m2										
Sides of foundations; plain vertical 75mm thick										
height exceeding 1.00m	7.70	5.00	8.09	-	0.07	0.49	0.63	9.21	m2	10.36
height not exceeding 250mm	1.92	5.00	2.02	-	0.02	0.14	0.21	2.37	m	2.66
height 250-500mm	3.85	5.00	4.04	-	0.08	0.56	0.21	4.81	m	5.42
height 0.50m-1.00m	5.78	5.00	6.07	-	0.08	0.56	0.42	7.05	m	7.93
Sides of foundations; plain vertical 100mm thick										
height exceeding 1.00m	10.27	5.00	10.78	-	0.09	0.63	0.63	12.05	m2	13.55
height not exceeding 250mm	2.57	5.00	2.70	-	0.03	0.21	0.21	3.12	m	3.51
height 250-500mm	5.13	5.00	5.39	-	0.05	0.35	0.21	5.95	m	6.69
height 0.50m-1.00m	7.71	5.00	8.10	-	0.10	0.70	0.42	9.22	m	10.37
Sides of foundations; plain vertical 150mm thick										
height exceeding 1.00m	12.83	5.00	13.47	-	0.11	0.77	0.63	14.87	m2	16.73
height not exceeding 250mm	3.21	5.00	3.37	-	0.03	0.21	0.21	3.79	m	4.26
height 250-500mm	6.92	5.00	7.27	-	0.07	0.49	0.21	7.97	m	8.96
height 0.50m-1.00m	9.63	5.00	10.11	-	0.12	0.84	0.42	11.37	m	12.80
To underside of foundations; laid on earth or hardcore; 75mm thick										
horizontal	7.70	5.00	8.09	-	0.05	0.35	-	8.44	m2	9.49
To underside of foundations; laid on earth or hardcore; 100mm thick										
horizontal	10.27	5.00	10.78	-	0.07	0.49	-	11.27	m2	12.68

Labour hourly rates: (except Specialists) Craft Operatives £9.23 Labourer £7.02 Rates are national average prices. Refer to REGIONAL VARIATIONS for indicative levels of overall pricing in regions	MATERIALS			LABOUR				RATES		
	Del to Site £	Waste %	Material Cost £	Craft Optve Hrs	Lab Hrs	Labour Cost £	Sunds £	Nett Rate £	Unit	Gross Rate (+12.5%) £
E20: FORMWORK FOR IN-SITU CONCRETE Cont.										
Claymaster low density expanded polystyrene permanent formwork; fixing with Clayfix hooks - 3/m2 Cont.										
To underside of foundations; laid on earth or hardcore; 150mm thick										
horizontal	12.83	5.00	13.47	–	0.08	0.56	–	14.03	m2	15.79
Expamet Hy-rib permanent shuttering and reinforcement										
Reference 2611, 4.86 kg/m2 to soffits of slabs; horizontal; one rib side laps; 150mm end laps										
slab thickness 75mm; strutting and supports at 750mm centres; height to soffit										
1.50 - 3.00m.....................	11.51	5.00	12.09	1.31	0.33	14.41	1.12	27.61	m2	31.07
3.00 - 4.50m.....................	11.51	5.00	12.09	1.51	0.37	16.53	1.42	30.04	m2	33.80
slab thickness 100mm; strutting and supports at 650mm centres; height to soffit										
1.50 - 3.00m.....................	11.51	5.00	12.09	1.33	0.33	14.59	1.30	27.98	m2	31.48
3.00 - 4.50m.....................	11.51	5.00	12.09	1.53	0.39	16.86	1.76	30.71	m2	34.54
slab thickness 125mm; strutting and supports at 550mm centres; height to soffit										
1.50 - 3.00m.....................	11.51	5.00	12.09	1.35	0.34	14.85	1.48	28.41	m2	31.96
3.00 - 4.50m.....................	11.51	5.00	12.09	1.55	0.39	17.04	2.11	31.24	m2	35.14
slab thickness 150mm; strutting and supports at 450mm centres; height to soffit										
1.50 - 3.00m.....................	11.51	5.00	12.09	1.38	0.34	15.12	1.64	28.85	m2	32.46
3.00 - 4.50m.....................	11.51	5.00	12.09	1.58	0.40	17.39	2.46	31.94	m2	35.93
Reference 2411, 6.34 kg/m2 to soffits of slabs; horizontal; one rib side laps; 150mm end laps										
slab thickness 75mm; strutting and supports at 850mm centres; height to soffit										
1.50 - 3.00m.....................	14.70	5.00	15.44	1.34	0.34	14.76	1.12	31.31	m2	35.22
3.00 - 4.50m.....................	14.70	5.00	15.44	1.54	0.39	16.95	2.46	34.85	m2	39.20
slab thickness 100mm; strutting and supports at 750mm centres; height to soffit										
1.50 - 3.00m.....................	14.70	5.00	15.44	1.36	0.34	14.94	1.30	31.67	m2	35.63
3.00 - 4.50m.....................	14.70	5.00	15.44	1.57	0.40	17.30	1.76	34.49	m2	38.81
slab thickness 125mm; strutting and supports at 650mm centres; height to soffit										
1.50 - 3.00m.....................	14.70	5.00	15.44	1.38	0.35	15.19	1.48	32.11	m2	36.12
3.00 - 4.50m.....................	14.70	5.00	15.44	1.60	0.40	17.58	2.11	35.12	m2	39.51
slab thickness 150mm; strutting and supports at 550mm centres; height to soffit										
1.50 - 3.00m.....................	14.70	5.00	15.44	1.41	0.35	15.47	1.64	32.55	m2	36.61
3.00 - 4.50m.....................	14.70	5.00	15.44	1.62	0.41	17.83	2.46	35.73	m2	40.19
slab thickness 175mm; strutting and supports at 450mm centres; height to soffit										
1.50 - 3.00m.....................	14.70	5.00	15.44	1.44	0.36	15.82	1.89	33.14	m2	37.29
3.00 - 4.50m.....................	14.70	5.00	15.44	1.66	0.42	18.27	2.94	36.65	m2	41.23
slab thickness 200mm; strutting and supports at 350mm centres; height to soffit										
1.50 - 3.00m.....................	14.70	5.00	15.44	1.47	0.37	16.17	2.14	33.74	m2	37.96
3.00 - 4.50m.....................	14.70	5.00	15.44	1.69	0.43	18.62	3.42	37.47	m2	42.16
Reference 2611, 4.86 kg/m2 to soffits of arched slabs; one rib side laps										
900mm span; 75mm rise; height to soffit										
1.50 - 3.00m.....................	11.51	5.00	12.09	1.44	0.36	15.82	0.83	28.73	m2	32.33
1200mm span; 75mm rise; one row strutting and supports per span; height to soffit										
1.50 - 3.00m.....................	11.51	5.00	12.09	1.44	0.36	15.82	1.12	29.02	m2	32.65
3.00 - 4.50m.....................	11.51	5.00	12.09	1.64	0.41	18.02	1.42	31.52	m2	35.46
Reference 2411, 6.34 kg/m2 to soffits of arched slabs; one rib side laps										
1500mm span; 100mm rise; two rows strutting and supports per span; height to soffit										
1.50 - 3.00m.....................	14.70	5.00	15.44	1.78	0.44	19.52	2.46	37.41	m2	42.09
3.00 - 4.50m.....................	14.70	5.00	15.44	2.02	0.51	22.22	3.26	40.92	m2	46.03
1800mm span; 150mm rise; two rows strutting and supports per span; height to soffit										
1.50 - 3.00m.....................	14.70	5.00	15.44	1.78	0.44	19.52	2.46	37.41	m2	42.09
3.00 - 4.50m.....................	14.70	5.00	15.44	2.02	0.51	22.22	3.26	40.92	m2	46.03
E30: REINFORCEMENT FOR IN-SITU CONCRETE										
Reinforcement bars; B.S.4449, hot rolled plain round mild steel including hooks and tying wire, and spacers and chairs which are at the discretion of the Contractor										
Straight										
6mm.....................	365.00	5.00	383.25	51.00	5.00	505.83	20.16	909.24	t	1022.90
8mm.....................	330.00	5.00	346.50	44.00	5.00	441.22	17.85	805.57	t	906.27
10mm.....................	320.00	5.00	336.00	36.00	5.00	367.38	14.49	717.87	t	807.60
12mm.....................	320.00	5.00	336.00	31.00	5.00	321.23	12.39	669.62	t	753.32
16mm.....................	320.00	5.00	336.00	24.00	5.00	256.62	10.19	602.81	t	678.16
20mm.....................	320.00	5.00	336.00	20.00	5.00	219.70	5.57	561.27	t	631.43
25mm.....................	320.00	5.00	336.00	20.00	5.00	219.70	5.57	561.27	t	631.43
Bent										
6mm.....................	365.00	5.00	383.25	51.00	5.00	505.83	20.16	909.24	t	1022.90
8mm.....................	330.00	5.00	346.50	44.00	5.00	441.22	17.85	805.57	t	906.27

Labour hourly rates: (except Specialists) Craft Operatives £9.23 Labourer £7.02 Rates are national average prices. Refer to REGIONAL VARIATIONS for indicative levels of overall pricing in regions	MATERIALS			LABOUR				RATES		
	Del to Site £	Waste %	Material Cost £	Craft Optve Hrs	Lab Hrs	Labour Cost £	Sunds £	Nett Rate £	Unit	Gross Rate (+12.5%) £
E30: REINFORCEMENT FOR IN-SITU CONCRETE Cont.										
Reinforcement bars; B.S.4449, hot rolled plain round mild steel including hooks and tying wire, and spacers and chairs which are at the discretion of the Contractor Cont.										
Bent Cont.										
10mm	320.00	5.00	336.00	36.00	5.00	367.38	14.49	717.87	t	807.60
12mm	320.00	5.00	336.00	31.00	5.00	321.23	12.39	669.62	t	753.32
16mm	320.00	5.00	336.00	24.00	5.00	256.62	10.19	602.81	t	678.16
20mm	320.00	5.00	336.00	20.00	5.00	219.70	5.57	561.27	t	631.43
25mm	320.00	5.00	336.00	20.00	5.00	219.70	5.57	561.27	t	631.43
Links										
6mm	380.00	5.00	399.00	66.00	5.00	644.28	20.16	1063.44	t	1196.37
8mm	345.00	5.00	362.25	66.00	5.00	644.28	17.85	1024.38	t	1152.43
Reinforcement bars; B.S.4449, hot rolled deformed high yield steel including hooks and tying wire, and spacers and chairs which are at the discretion of the Contractor										
Straight										
6mm	375.00	5.00	393.75	51.00	5.00	505.83	20.16	919.74	t	1034.71
8mm	340.00	5.00	357.00	44.00	5.00	441.22	17.85	816.07	t	918.08
10mm	330.00	5.00	346.50	36.00	5.00	367.38	14.49	728.37	t	819.42
12mm	330.00	5.00	346.50	31.00	5.00	321.23	12.39	680.12	t	765.13
16mm	330.00	5.00	346.50	24.00	5.00	256.62	10.19	613.31	t	689.97
20mm	330.00	5.00	346.50	20.00	5.00	219.70	5.57	571.77	t	643.24
25mm	330.00	5.00	346.50	20.00	5.00	219.70	5.57	571.77	t	643.24
Bent										
6mm	375.00	5.00	393.75	51.00	5.00	505.83	20.16	919.74	t	1034.71
8mm	340.00	5.00	357.00	44.00	5.00	441.22	17.85	816.07	t	918.08
10mm	330.00	5.00	346.50	36.00	5.00	367.38	14.49	728.37	t	819.42
12mm	330.00	5.00	346.50	31.00	5.00	321.23	12.39	680.12	t	765.13
16mm	330.00	5.00	346.50	24.00	5.00	256.62	10.19	613.31	t	689.97
20mm	330.00	5.00	346.50	20.00	5.00	219.70	5.57	571.77	t	643.24
25mm	330.00	5.00	346.50	20.00	5.00	219.70	5.57	571.77	t	643.24
Links										
6mm	390.00	5.00	409.50	66.00	5.00	644.28	20.16	1073.94	t	1208.18
8mm	355.00	5.00	372.75	66.00	5.00	644.28	17.85	1034.88	t	1164.24
Take delivery, cut, bend and fix reinforcing rods; including hooks and tying wire, and spacers and chairs which are at the discretion of the Contractor										
Straight										
6mm	-	-	-	55.00	5.00	542.75	20.16	562.91	t	633.27
8mm	-	-	-	-	48.00	336.96	17.85	354.81	t	399.16
10mm	-	-	-	40.00	5.00	404.30	14.49	418.79	t	471.14
12mm	-	-	-	34.00	5.00	348.92	12.39	361.31	t	406.47
16mm	-	-	-	26.00	5.00	275.08	10.19	285.27	t	320.93
20mm	-	-	-	22.00	5.00	238.16	5.57	243.73	t	274.20
25mm	-	-	-	22.00	5.00	238.16	5.57	243.73	t	274.20
Bent										
6mm	-	-	-	75.00	5.00	727.35	20.16	747.51	t	840.95
8mm	-	-	-	66.00	5.00	644.28	17.85	662.13	t	744.90
10mm	-	-	-	55.00	5.00	542.75	14.49	557.24	t	626.89
12mm	-	-	-	47.00	5.00	468.91	12.39	481.30	t	541.46
16mm	-	-	-	37.00	5.00	376.61	10.19	386.80	t	435.15
20mm	-	-	-	31.00	5.00	321.23	5.57	326.80	t	367.65
25mm	-	-	-	29.00	5.00	302.77	5.57	308.34	t	346.88
Links										
6mm	-	-	-	99.00	5.00	948.87	20.16	969.03	t	1090.16
8mm	-	-	-	99.00	5.00	948.87	17.85	966.72	t	1087.56
Take delivery and fix reinforcing rods supplied cut to length and bent; including tying wire, and spacers and chairs which are at the discretion of the Contractor										
Straight										
6mm	-	-	-	51.00	5.00	505.83	20.16	525.99	t	591.74
8mm	-	-	-	44.00	5.00	441.22	17.85	459.07	t	516.45
10mm	-	-	-	36.00	5.00	367.38	14.49	381.87	t	429.60
12mm	-	-	-	31.00	5.00	321.23	12.39	333.62	t	375.32
16mm	-	-	-	24.00	5.00	256.62	10.19	266.81	t	300.16
20mm	-	-	-	20.00	5.00	219.70	5.57	225.27	t	253.43
25mm	-	-	-	20.00	5.00	219.70	5.57	225.27	t	253.43
Bent										
6mm	-	-	-	51.00	5.00	505.83	20.16	525.99	t	591.74
8mm	-	-	-	44.00	5.00	441.22	17.85	459.07	t	516.45
10mm	-	-	-	36.00	5.00	367.38	14.49	381.87	t	429.60
12mm	-	-	-	31.00	5.00	321.23	12.39	333.62	t	375.32
16mm	-	-	-	24.00	5.00	256.62	10.19	266.81	t	300.16
20mm	-	-	-	20.00	5.00	219.70	5.57	225.27	t	253.43
25mm	£	-	£	20.00	5.00	219.70	5.57	225.27	t	253.43
Links										
6mm	-	-	-	66.00	5.00	644.28	20.16	664.44	t	747.50

Labour hourly rates: (except Specialists) Craft Operatives £9.23 Labourer £7.02 Rates are national average prices. Refer to REGIONAL VARIATIONS for indicative levels of overall pricing in regions	MATERIALS			LABOUR				RATES		
	Del to Site £	Waste %	Material Cost £	Craft Optve Hrs	Lab Hrs	Labour Cost £	Sunds £	Nett Rate £	Unit	Gross Rate (+12.5%) £
E30: REINFORCEMENT FOR IN-SITU CONCRETE Cont.										
Take delivery and fix reinforcing rods supplied cut to length and bent; including tying wire, and spacers and chairs which are at the discretion of the Contractor Cont.										
Links Cont.										
8mm	-	-	-	66.00	5.00	644.28	17.85	662.13	t	744.90
Reinforcement bars; high yield stainless steel B.S.6744, Type 2 (minimum yield stress 460 N/mm2); including hooks and tying wire, and spacers and chairs which are at the discretion of the Contractor										
Straight										
8mm	2982.00	5.00	3131.10	83.00	10.00	836.29	69.30	4036.69	t	4541.28
10mm	2982.00	5.00	3131.10	69.00	10.00	707.07	50.00	3888.17	t	4374.19
12mm	2982.00	5.00	3131.10	54.00	10.00	568.62	47.30	3747.02	t	4215.40
16mm	2982.00	5.00	3131.10	50.00	10.00	531.70	40.70	3703.50	t	4166.44
20mm	3150.00	5.00	3307.50	41.00	10.00	448.63	30.25	3786.38	t	4259.68
25mm	3203.00	5.00	3363.15	33.00	10.00	374.79	22.00	3759.94	t	4229.93
Bent										
8mm	3518.00	5.00	3693.90	83.00	10.00	836.29	69.30	4599.49	t	5174.43
10mm	3518.00	5.00	3693.90	69.00	10.00	707.07	50.00	4450.97	t	5007.34
12mm	3518.00	5.00	3693.90	54.00	10.00	568.62	47.30	4309.82	t	4848.55
16mm	3518.00	5.00	3693.90	50.00	10.00	531.70	40.70	4266.30	t	4799.59
20mm	3518.00	5.00	3693.90	41.00	10.00	448.63	30.25	4172.78	t	4694.38
25mm	3518.00	5.00	3693.90	33.00	10.00	374.79	22.00	4090.69	t	4602.03
Links										
6mm	2982.00	5.00	3131.10	130.00	10.00	1270.10	69.30	4470.50	t	5029.31
8mm	2982.00	5.00	3131.10	115.00	10.00	1131.65	50.00	4312.75	t	4851.84
Reinforcement bars; Tor Bar grade 250 mild steel ribbed and twisted bars; including hooks and tying wire, and spacers and chairs which are at the discretion of the Contractor										
Straight										
6mm	375.00	5.00	393.75	51.00	5.00	505.83	20.16	919.74	t	1034.71
8mm	330.00	5.00	346.50	44.00	5.00	441.22	17.85	805.57	t	906.27
10mm	320.00	5.00	336.00	36.00	5.00	367.38	14.49	717.87	t	807.60
12mm	320.00	5.00	336.00	31.00	5.00	321.23	12.39	669.62	t	753.32
16mm	320.00	5.00	336.00	24.00	5.00	256.62	10.19	602.81	t	678.16
20mm	320.00	5.00	336.00	20.00	5.00	219.70	5.57	561.27	t	631.43
25mm	320.00	5.00	336.00	20.00	5.00	219.70	5.57	561.27	t	631.43
Bent										
6mm	375.00	5.00	393.75	51.00	5.00	505.83	20.16	919.74	t	1034.71
8mm	330.00	5.00	346.50	44.00	5.00	441.22	17.85	805.57	t	906.27
10mm	320.00	5.00	336.00	36.00	5.00	367.38	14.49	717.87	t	807.60
12mm	320.00	5.00	336.00	31.00	5.00	321.23	12.39	669.62	t	753.32
16mm	320.00	5.00	336.00	24.00	5.00	256.62	10.19	602.81	t	678.16
20mm	320.00	5.00	336.00	20.00	5.00	219.70	5.57	561.27	t	631.43
25mm	320.00	5.00	336.00	20.00	5.00	219.70	5.57	561.27	t	631.43
Links										
6mm	380.00	5.00	399.00	66.00	5.00	644.28	20.16	1063.44	t	1196.37
8mm	345.00	5.00	362.25	66.00	5.00	644.28	17.85	1024.38	t	1152.43
Reinforcement bars; Tor Bar grade 460 high yield steel ribbed and twisted bars; including hooks and tying wire, and spacers and chairs which are at the discretion of the Contractor										
Straight										
6mm	375.00	5.00	393.75	51.00	5.00	505.83	20.16	919.74	t	1034.71
8mm	340.00	5.00	357.00	44.00	5.00	441.22	17.85	816.07	t	918.08
10mm	330.00	5.00	346.50	36.00	5.00	367.38	14.49	728.37	t	819.42
12mm	330.00	5.00	346.50	31.00	5.00	321.23	12.39	680.12	t	765.13
16mm	330.00	5.00	346.50	24.00	5.00	256.62	10.19	613.31	t	689.97
20mm	330.00	5.00	346.50	20.00	5.00	219.70	5.57	571.77	t	643.24
25mm	330.00	5.00	346.50	20.00	5.00	219.70	5.57	571.77	t	643.24
Bent										
6mm	375.00	5.00	393.75	51.00	5.00	505.83	20.16	919.74	t	1034.71
8mm	340.00	5.00	357.00	44.00	5.00	441.22	17.85	816.07	t	918.08
10mm	330.00	5.00	346.50	36.00	5.00	367.38	14.49	728.37	t	819.42
12mm	330.00	5.00	346.50	31.00	5.00	321.23	12.39	680.12	t	765.13
16mm	330.00	5.00	346.50	24.00	5.00	256.62	10.19	613.31	t	689.97
20mm	330.00	5.00	346.50	20.00	5.00	219.70	5.57	571.77	t	643.24
25mm	330.00	5.00	346.50	20.00	5.00	219.70	5.57	571.77	t	643.24
Links										
6mm	390.00	5.00	409.50	66.00	5.00	644.28	20.16	1073.94	t	1208.18
8mm	355.00	5.00	372.75	66.00	5.00	644.28	17.85	1034.88	t	1164.24

IN-SITU CONCRETE/LARGE PRECAST CONCRETE – SMALL WORKS

Labour hourly rates: (except Specialists) Craft Operatives £9.23 Labourer £7.02 Rates are national average prices. Refer to REGIONAL VARIATIONS for indicative levels of overall pricing in regions	MATERIALS			LABOUR				RATES		
	Del to Site £	Waste %	Material Cost £	Craft Optve Hrs	Lab Hrs	Labour Cost £	Sunds £	Nett Rate £	Unit	Gross Rate (+12.5%) £
E30: REINFORCEMENT FOR IN-SITU CONCRETE Cont.										
Reinforcement fabric; B.S.4483, hard drawn plain round steel; welded; including laps, tying wire, all cutting and bending, and spacers and chairs which are at the discretion of the Contractor										
Reference A98, 1.54 kg/m2; 200mm side laps; 200mm end laps										
generally	0.55	15.00	0.63	0.07	0.01	0.72	0.04	1.39	m2	1.56
strips in one width										
750mm wide..	0.55	15.00	0.63	0.11	0.01	1.09	0.04	1.76	m2	1.98
900mm wide..	0.55	15.00	0.63	0.10	0.01	0.99	0.04	1.67	m2	1.87
1050mm wide.......................................	0.55	15.00	0.63	0.09	0.01	0.90	0.04	1.57	m2	1.77
1200mm wide.......................................	0.55	15.00	0.63	0.08	0.01	0.81	0.04	1.48	m2	1.67
Reference A142, 2.22kg/m2; 200mm side laps; 200mm end laps										
generally	0.78	15.00	0.90	0.08	0.02	0.88	0.06	1.84	m2	2.07
strips in one width										
750mm wide..	0.78	15.00	0.90	0.12	0.02	1.25	0.06	2.21	m2	2.48
900mm wide..	0.78	15.00	0.90	0.11	0.02	1.16	0.06	2.11	m2	2.38
1050mm wide.......................................	0.78	15.00	0.90	0.10	0.02	1.06	0.06	2.02	m2	2.27
1200mm wide.......................................	0.78	15.00	0.90	0.10	0.02	1.06	0.06	2.02	m2	2.27
Reference A193, 3.02kg/m2; 200mm side laps; 200mm end laps										
generally	1.06	15.00	1.22	0.08	0.02	0.88	0.08	2.18	m2	2.45
strips in one width										
750mm wide..	1.06	15.00	1.22	0.12	0.02	1.25	0.08	2.55	m2	2.87
900mm wide..	1.06	15.00	1.22	0.11	0.02	1.16	0.08	2.45	m2	2.76
1050mm wide.......................................	1.06	15.00	1.22	0.10	0.02	1.06	0.08	2.36	m2	2.66
1200mm wide.......................................	1.06	15.00	1.22	0.10	0.02	1.06	0.08	2.36	m2	2.66
Reference A252, 3.95kg/m2; 200mm side laps; 200mm end laps										
generally	1.38	15.00	1.59	0.09	0.02	0.97	0.10	2.66	m2	2.99
strips in one width										
750mm wide..	1.38	15.00	1.59	0.14	0.02	1.43	0.10	3.12	m2	3.51
900mm wide..	1.38	15.00	1.59	0.13	0.02	1.34	0.10	3.03	m2	3.41
1050mm wide.......................................	1.38	15.00	1.59	0.12	0.02	1.25	0.10	2.94	m2	3.30
1200mm wide.......................................	1.38	15.00	1.59	0.11	0.02	1.16	0.10	2.84	m2	3.20
Reference A393, 6.16 kg/m2; 200mm side laps; 200mm end laps										
generally	2.15	15.00	2.47	0.10	0.20	2.33	0.15	4.95	m2	5.57
strips in one width										
750mm wide..	2.15	15.00	2.47	0.15	0.02	1.52	0.15	4.15	m2	4.67
900mm wide..	2.15	15.00	2.47	0.14	0.02	1.43	0.15	4.06	m2	4.56
1050mm wide.......................................	2.15	15.00	2.47	0.13	0.02	1.34	0.15	3.96	m2	4.46
1200mm wide.......................................	2.15	15.00	2.47	0.12	0.02	1.25	0.15	3.87	m2	4.35
Reference B1131, 10.90 kg/m2; 100mm side laps; 200mm end laps										
generally	3.88	15.00	4.46	0.14	0.30	3.40	0.27	8.13	m2	9.15
strips in one width										
750mm wide..	3.88	15.00	4.46	0.21	0.03	2.15	0.27	6.88	m2	7.74
900mm wide..	3.88	15.00	4.46	0.20	0.03	2.06	0.27	6.79	m2	7.64
1050mm wide.......................................	3.88	15.00	4.46	0.18	0.03	1.87	0.27	6.60	m2	7.43
1200mm wide.......................................	3.88	15.00	4.46	0.17	0.03	1.78	0.27	6.51	m2	7.33
Reference B196, 3.05 kg/m2; 100mm side laps; 200mm end laps										
generally	1.09	15.00	1.25	0.08	0.02	0.88	0.15	2.28	m2	2.57
strips in one width										
750mm wide..	1.09	15.00	1.25	0.12	0.02	1.25	0.15	2.65	m2	2.98
900mm wide..	1.09	15.00	1.25	0.11	0.02	1.16	0.15	2.56	m2	2.88
1050mm wide.......................................	1.09	15.00	1.25	0.10	0.02	1.06	0.15	2.47	m2	2.78
1200mm wide.......................................	1.09	15.00	1.25	0.10	0.02	1.06	0.15	2.47	m2	2.78
Reference B283, 3.73 kg/m2; 100mm side laps; 200mm end laps										
generally	1.32	15.00	1.52	0.08	0.02	0.88	0.09	2.49	m2	2.80
strips in one width										
750mm wide..	1.32	15.00	1.52	0.12	0.02	1.25	0.09	2.86	m2	3.21
900mm wide..	1.32	15.00	1.52	0.11	0.02	1.16	0.09	2.76	m2	3.11
1050mm wide.......................................	1.32	15.00	1.52	0.10	0.02	1.06	0.09	2.67	m2	3.01
1200mm wide.......................................	1.32	15.00	1.52	0.10	0.02	1.06	0.09	2.67	m2	3.01
Reference B385, 4.53 kg/m2; 100mm side laps; 200mm end laps										
generally	1.61	15.00	1.85	0.09	0.02	0.97	0.11	2.93	m2	3.30
strips in one width										
750mm wide..	1.61	15.00	1.85	0.14	0.02	1.43	0.11	3.39	m2	3.82
900mm wide..	1.61	15.00	1.85	0.13	0.02	1.34	0.11	3.30	m2	3.71
1050mm wide.......................................	1.61	15.00	1.85	0.12	0.02	1.25	0.11	3.21	m2	3.61
1200mm wide.......................................	1.61	15.00	1.85	0.11	0.02	1.16	0.11	3.12	m2	3.51
Reference B503, 5.93 kg/m2; 100mm side laps; 200mm end laps										
generally	2.09	15.00	2.40	0.10	0.02	1.06	0.15	3.62	m2	4.07
strips in one width										
750mm wide..	2.09	15.00	2.40	0.15	0.02	1.52	0.15	4.08	m2	4.59
900mm wide..	2.09	15.00	2.40	0.14	0.02	1.43	0.15	3.99	m2	4.48
1050mm wide.......................................	2.09	15.00	2.40	0.13	0.02	1.34	0.15	3.89	m2	4.38
1200mm welded.....................................	2.09	15.00	2.40	0.12	0.02	1.25	0.15	3.80	m2	4.28

Labour hourly rates: (except Specialists) Craft Operatives £9.23 Labourer £7.02 Rates are national average prices. Refer to REGIONAL VARIATIONS for indicative levels of overall pricing in regions	MATERIALS			LABOUR				RATES		
	Del to Site £	Waste %	Material Cost £	Craft Optve Hrs	Lab Hrs	Labour Cost £	Sunds £	Nett Rate £	Unit	Gross Rate (+12.5%) £
E30: REINFORCEMENT FOR IN-SITU CONCRETE Cont.										
Reinforcement fabric; B.S.4483, hard drawn plain round steel; welded; including laps, tying wire, all cutting and bending, and spacers and chairs which are at the discretion of the Contractor Cont.										
Reference B785, 8.14 kg/m2; 100mm side laps; 200mm end laps										
generally............................	2.87	15.00	3.30	0.11	0.03	1.23	0.20	4.73	m2	5.32
strips in one width										
750mm wide......................	2.87	15.00	3.30	0.16	0.03	1.69	0.20	5.19	m2	5.84
900mm wide......................	2.87	15.00	3.30	0.15	0.03	1.60	0.20	5.10	m2	5.73
1050mm wide.....................	2.87	15.00	3.30	0.14	0.03	1.50	0.20	5.00	m2	5.63
1200mm wide.....................	2.87	15.00	3.30	0.13	0.03	1.41	0.20	4.91	m2	5.52
Reference C283, 2.61 kg/m2; 100mm side laps; 400mm end laps										
generally............................	0.92	15.00	1.06	0.08	0.02	0.88	0.07	2.01	m2	2.26
strips in one width										
750mm wide......................	0.92	15.00	1.06	0.12	0.02	1.25	0.07	2.38	m2	2.67
900mm wide......................	0.92	15.00	1.06	0.11	0.02	1.16	0.07	2.28	m2	2.57
1050mm wide.....................	0.92	15.00	1.06	0.10	0.02	1.06	0.07	2.19	m2	2.47
1200mm wide.....................	0.92	15.00	1.06	0.10	0.02	1.06	0.07	2.19	m2	2.47
Reference C385, 3.41 kg/m2; 100mm side laps; 400mm end laps										
generally............................	1.20	15.00	1.38	0.08	0.02	0.88	0.09	2.35	m2	2.64
strips in one width										
750mm wide......................	1.20	15.00	1.38	0.12	0.02	1.25	0.09	2.72	m2	3.06
900mm wide......................	1.20	15.00	1.38	0.11	0.02	1.16	0.09	2.63	m2	2.95
1050mm wide.....................	1.20	15.00	1.38	0.10	0.02	1.06	0.09	2.53	m2	2.85
1200mm wide.....................	1.20	15.00	1.38	0.10	0.02	1.06	0.09	2.53	m2	2.85
Reference C503, 4.34 kg/m2; 100mm side laps; 400mm end laps										
generally............................	1.48	15.00	1.70	0.09	0.02	0.97	0.11	2.78	m2	3.13
strips in one width										
750mm wide......................	1.48	15.00	1.70	0.14	0.02	1.43	0.11	3.24	m2	3.65
900mm wide......................	1.48	15.00	1.70	0.13	0.02	1.34	0.11	3.15	m2	3.55
1050mm wide.....................	1.48	15.00	1.70	0.12	0.02	1.25	0.11	3.06	m2	3.44
1200mm wide.....................	1.48	15.00	1.70	0.11	0.02	1.16	0.11	2.97	m2	3.34
Reference C636, 5.55 kg/m2; 100mm side laps; 400mm end laps										
generally............................	1.97	15.00	2.27	0.10	0.02	1.06	0.14	3.47	m2	3.90
strips in one width										
750mm wide......................	1.97	15.00	2.27	0.15	0.02	1.52	0.14	3.93	m2	4.42
900mm wide......................	1.97	15.00	2.27	0.14	0.02	1.43	0.14	3.84	m2	4.32
1050mm wide.....................	1.97	15.00	2.27	0.13	0.02	1.34	0.14	3.75	m2	4.21
1200mm wide.....................	1.97	15.00	2.27	0.12	0.02	1.25	0.14	3.65	m2	4.11
Reference C785, 6.72 kg/m2; 100mm side laps; 400mm end laps										
generally............................	2.38	15.00	2.74	0.10	0.02	1.06	0.17	3.97	m2	4.47
strips in one width										
750mm wide......................	2.38	15.00	2.74	0.15	0.02	1.52	0.17	4.43	m2	4.99
900mm wide......................	2.38	15.00	2.74	0.14	0.02	1.43	0.17	4.34	m2	4.88
1050mm wide.....................	2.38	15.00	2.74	0.13	0.02	1.34	0.17	4.25	m2	4.78
1200mm wide.....................	2.38	15.00	2.74	0.12	0.02	1.25	0.17	4.16	m2	4.67
Reference D49, 0.77 kg/m2; 100mm side laps; 100mm end laps										
bent.................................	0.56	15.00	0.64	0.04	0.01	0.39	0.06	1.10	m2	1.23
Reference D98, 1.54 kg/m2; 200mm side laps; 200mm end laps										
bent.................................	0.50	15.00	0.57	0.04	0.01	0.39	0.06	1.03	m2	1.16
Expamet 76mm mesh expanded steel reinforcement (uncoated); including laps, tying wire, all cutting and bending, and spacers and chairs which are at the discretion of the Contractor										
Reference 8, 3.80 kg/m2; 150mm side laps; 150mm end laps										
generally............................	5.81	15.00	6.68	0.08	0.02	0.88	0.42	7.98	m2	8.98
strips in one width										
750mm wide......................	5.81	15.00	6.68	0.12	0.02	1.25	0.42	8.35	m2	9.39
900mm wide......................	5.81	15.00	6.68	0.11	0.02	1.16	0.42	8.26	m2	9.29
1050mm wide.....................	5.81	15.00	6.68	0.10	0.02	1.06	0.42	8.16	m2	9.19
1200mm wide.....................	5.81	15.00	6.68	0.10	0.02	1.06	0.42	8.16	m2	9.19
Expamet 'Securilath' flattened security mesh expanded steel reinforcement (uncoated); including laps										
Reference 2073F; 150mm side laps; 150mm end laps										
generally............................	16.76	15.00	19.27	0.12	0.02	1.25	0.20	20.72	m2	23.31

IN SITU CONCRETE/LARGE PRECAST CONCRETE

Labour hourly rates: (except Specialists) Craft Operatives £9.23 Labourer £7.02 Rates are national average prices. Refer to REGIONAL VARIATIONS for indicative levels of overall pricing in regions	MATERIALS			LABOUR				RATES		
	Del to Site	Waste	Material Cost	Craft Optve	Lab	Labour Cost	Sunds	Nett Rate	Unit	Gross Rate (+12.5%)
	£	%	£	Hrs	Hrs	£	£	£		£

E40: DESIGNED JOINTS IN IN-SITU CONCRETE

Formed joints

Incorporating 10mm thick Korkpak, Servicised Ltd; formwork; reinforcement laid continuously across joint

in concrete, depth not exceeding 150mm; horizontal	3.60	5.00	3.78	-	0.08	0.56	4.69	9.03	m	10.16
in concrete, depth 150 - 300mm; horizontal	5.90	5.00	6.20	-	0.09	0.63	8.21	15.04	m	16.92
in concrete, depth 300 - 450mm; horizontal	10.32	5.00	10.84	-	0.10	0.70	11.94	23.48	m	26.41

Incorporating 13mm thick Korkpak, Servicised Ltd; formwork; reinforcement laid continuously across joint

in concrete, depth not exceeding 150mm; horizontal	4.21	5.00	4.42	-	0.08	0.56	4.69	9.67	m	10.88
in concrete, depth 150 - 300mm; horizontal	6.06	5.00	6.36	-	0.09	0.63	8.21	15.20	m	17.11
in concrete, depth 300 - 450mm; horizontal	10.52	5.00	11.05	-	0.10	0.70	11.94	23.69	m	26.65

Incorporating 19mm thick Korkpak, Servicised Ltd; formwork; reinforcement laid continuously across joint

in concrete, depth not exceeding 150mm; horizontal	4.48	5.00	4.70	-	0.08	0.56	4.69	9.96	m	11.20
in concrete, depth 150 - 300mm; horizontal	8.39	5.00	8.81	-	0.09	0.63	8.21	17.65	m	19.86
in concrete, depth 300 - 450mm; horizontal	14.86	5.00	15.60	-	0.10	0.70	11.94	28.25	m	31.78

Incorporating 195mm wide 'Serviseal STD 195', external face type p.v.c waterstop, Servicised Ltd; heat welded joints; formwork; reinforcement laid continuously across joint

in concrete, depth not exceeding 150mm; horizontal	4.50	10.00	4.95	-	0.44	3.09	4.69	12.73	m	14.32
in concrete, depth 150 - 300mm; horizontal	4.50	10.00	4.95	-	0.44	3.09	8.21	16.25	m	18.28
in concrete, depth 300 - 450mm; horizontal	4.50	10.00	4.95	-	0.44	3.09	11.94	19.98	m	22.48
vertical L piece	10.90	10.00	11.99	-	0.22	1.54	-	13.53	nr	15.23
flat L piece	8.65	10.00	9.52	-	0.22	1.54	-	11.06	nr	12.44
flat T piece	12.64	10.00	13.90	-	0.28	1.97	-	15.87	nr	17.85
flat X piece	14.90	10.00	16.39	-	0.33	2.32	-	18.71	nr	21.04
in concrete, width not exceeding 150mm; vertical .	4.09	10.00	4.50	-	0.55	3.86	4.69	13.05	m	14.68
in concrete, width 150 - 300mm; vertical	4.09	10.00	4.50	-	0.55	3.86	8.21	16.57	m	18.64
in concrete, width 300 - 450mm; vertical	4.09	10.00	4.50	-	0.55	3.86	11.94	20.30	m	22.84
flat L piece	8.65	10.00	9.52	-	0.22	1.54	-	11.06	nr	12.44
flat T piece	12.64	10.00	13.90	-	0.28	1.97	-	15.87	nr	17.85
flat X piece	14.90	10.00	16.39	-	0.33	2.32	-	18.71	nr	21.04

Incorporating 195mm wide 'Serviseal EXP 195', external face type p.v.c waterstop, Servicised Ltd; heat welded joints; formwork; reinforcement laid continuously across joint

in concrete, depth not exceeding 150mm; horizontal	4.74	10.00	5.21	-	0.44	3.09	4.69	12.99	m	14.62
in concrete, depth 150 - 300mm; horizontal	4.74	10.00	5.21	-	0.44	3.09	8.21	16.51	m	18.58
in concrete, depth 300 - 450mm; horizontal	4.74	10.00	5.21	-	0.44	3.09	11.94	20.24	m	22.77
vertical L piece	11.96	10.00	13.16	-	0.22	1.54	-	14.70	nr	16.54
flat L piece	8.82	10.00	9.70	-	0.22	1.54	-	11.25	nr	12.65
flat T piece	12.83	10.00	14.11	-	0.28	1.97	-	16.08	nr	18.09
flat X piece	14.54	10.00	15.99	-	0.33	2.32	-	18.31	nr	20.60
in concrete, width not exceeding 150mm; vertical .	4.74	10.00	5.21	-	0.55	3.86	4.69	13.77	m	15.49
in concrete, width 150 - 300mm; vertical	4.74	10.00	5.21	-	0.55	3.86	8.21	17.29	m	19.45
in concrete, width 300 - 450mm; vertical	4.74	10.00	5.21	-	0.55	3.86	11.94	21.02	m	23.64
flat L piece	8.82	10.00	9.70	-	0.22	1.54	-	11.25	nr	12.65
flat T piece	12.83	10.00	14.11	-	0.28	1.97	-	16.08	nr	18.09
flat X piece	14.54	10.00	15.99	-	0.33	2.32	-	18.31	nr	20.60

Incorporating 240mm wide 'Serviseal HD240', heavy duty section external face type p.v.c. waterstop, Servicised Ltd; heat welded joints; formwork; reinforcement laid continuously across joint

in concrete, depth not exceeding 150mm; horizontal	5.90	10.00	6.49	-	0.50	3.51	4.69	14.69	m	16.53
in concrete, depth 150 - 300mm; horizontal	5.90	10.00	6.49	-	0.50	3.51	8.21	18.21	m	20.49
in concrete, depth 300 - 450mm; horizontal	5.90	10.00	6.49	-	0.50	3.51	11.94	21.94	m	24.68
vertical L piece	11.18	10.00	12.30	-	0.22	1.54	-	13.84	nr	15.57
flat L piece	9.88	10.00	10.87	-	0.22	1.54	-	12.41	nr	13.96
flat T piece	14.41	10.00	15.85	-	0.28	1.97	-	17.82	nr	20.04
flat X piece	16.67	10.00	18.34	-	0.33	2.32	-	20.65	nr	23.24
in concrete, width not exceeding 150mm; vertical .	5.90	10.00	6.49	-	0.61	4.28	4.69	15.46	m	17.39
in concrete, width 150 - 300mm; vertical	5.90	10.00	6.49	-	0.61	4.28	8.21	18.98	m	21.35
in concrete, width 300 - 450mm; vertical	5.90	10.00	6.49	-	0.61	4.28	11.94	22.71	m	25.55
flat L piece	9.88	10.00	10.87	-	0.22	1.54	-	12.41	nr	13.96
flat T piece	14.41	10.00	15.85	-	0.28	1.97	-	17.82	nr	20.04
flat X piece	16.67	10.00	18.34	-	0.33	2.32	-	20.65	nr	23.24

Incorporating 240mm wide 'Serviseal EXP240', external face type p.v.c waterstop, Servicised Ltd; heat welded joints; formwork; reinforcement laid continuously across joint

in concrete, depth not exceeding 150mm; horizontal	6.15	10.00	6.76	-	0.50	3.51	4.69	14.97	m	16.84
in concrete, depth 150 - 300mm; horizontal	6.15	10.00	6.76	-	0.50	3.51	8.21	18.48	m	20.80
in concrete, depth 300 - 450mm; horizontal	6.15	10.00	6.76	-	0.50	3.51	11.94	22.21	m	24.99
vertical L piece	12.44	10.00	13.68	-	0.22	1.54	-	15.22	nr	17.12
flat L piece	10.33	10.00	11.33	-	0.22	1.54	-	12.87	nr	14.48
flat T piece	15.04	10.00	16.54	-	0.28	1.97	-	18.51	nr	20.82
flat X piece	17.50	10.00	19.25	-	0.33	2.32	-	21.57	nr	24.27
in concrete, width not exceeding 150mm; vertical .	6.15	10.00	6.76	-	0.61	4.28	4.69	15.74	m	17.70
in concrete, width 150 - 300mm; vertical	6.15	10.00	6.76	-	0.61	4.28	8.21	19.26	m	21.66
in concrete, width 300 - 450mm; vertical	6.15	10.00	6.76	-	0.61	4.28	11.94	22.99	m	25.86
flat L piece	10.30	10.00	11.33	-	0.22	1.54	-	12.87	nr	14.48
flat T piece	15.04	10.00	16.54	-	0.28	1.97	-	18.51	nr	20.82
flat X piece	17.50	10.00	19.25	-	0.33	2.32	-	21.57	nr	24.26

Labour hourly rates: (except Specialists) Craft Operatives £9.23 Labourer £7.02 Rates are national average prices. Refer to REGIONAL VARIATIONS for indicative levels of overall pricing in regions	MATERIALS			LABOUR				RATES		
	Del to Site	Waste	Material Cost	Craft Optve	Lab	Labour Cost	Sunds	Nett Rate	Unit	Gross Rate (+12.5%)
	£	%	£	Hrs	Hrs	£	£	£		£

E40: DESIGNED JOINTS IN IN-SITU CONCRETE Cont.

Formed joints Cont.

Incorporating 320mm wide 'Kicker 320', external face type p.v.c waterstop, Servicised Ltd; heat welded joints; formwork; reinforcement laid continuously across joint

in concrete, depth not exceeding 150mm; horizontal	8.25	10.00	9.07	-	0.66	4.63	4.69	18.40	m	20.70
in concrete, depth 150 - 300mm; horizontal	8.25	10.00	9.07	-	0.66	4.63	8.21	21.92	m	24.66
in concrete, depth 300 - 450mm; horizontal	8.25	10.00	9.07	-	0.66	4.63	11.94	25.65	m	28.85
vertical L piece	11.33	10.00	12.46	-	0.28	1.97	-	14.43	nr	16.23
flat L piece	17.93	10.00	19.72	-	0.28	1.97	-	21.69	nr	24.40
flat T piece	25.25	10.00	27.77	-	0.33	2.32	-	30.09	nr	33.85
flat X piece	33.87	10.00	37.26	-	0.39	2.74	-	39.99	nr	44.99

Incorporating 170mm wide 'FD170' flat dumbell internally placed p.v.c. waterstop, Servicised Ltd; heat welded joints; formwork; reinforcement laid continuously across joint

in concrete, depth not exceeding 150mm; horizontal	3.05	10.00	3.36	-	0.39	2.74	4.69	10.78	m	12.13
in concrete, depth 150 - 300mm; horizontal	3.05	10.00	3.36	-	0.39	2.74	8.21	14.30	m	16.09
in concrete, depth 300 - 450mm; horizontal	3.05	10.00	3.36	-	0.39	2.74	11.94	18.03	m	20.29
flat L piece	5.92	10.00	6.51	-	0.17	1.19	-	7.71	nr	8.67
flat T piece	8.28	10.00	9.11	-	0.22	1.54	-	10.65	nr	11.98
flat X piece	9.31	10.00	10.24	-	0.28	1.97	-	12.21	nr	13.73
in concrete, width not exceeding 150mm; vertical	3.05	10.00	3.36	-	0.48	3.37	4.69	11.41	m	12.84
in concrete, width 150 - 300mm; vertical	3.05	10.00	3.36	-	0.48	3.37	8.21	14.93	m	16.80
in concrete, width 300 - 450mm; vertical	3.05	10.00	3.36	-	0.48	3.37	11.94	18.66	m	21.00
vertical L piece	9.69	10.00	10.66	-	0.17	1.19	-	11.85	nr	13.33
flat L piece	5.92	10.00	6.51	-	0.17	1.19	-	7.71	nr	8.67
vertical T piece	12.84	10.00	14.12	-	0.22	1.54	-	15.67	nr	17.63
flat T piece	8.28	10.00	9.11	-	0.22	1.54	-	10.65	nr	11.98
flat X piece	9.31	10.00	10.24	-	0.28	1.97	-	12.21	nr	13.73

Incorporating 210mm wide 'CB210' flat dumbell internally placed p.v.c. waterstop, Servicised Ltd; heat welded joints; formwork; reinforcement laid continuously across joint

in concrete, depth not exceeding 150mm; horizontal	4.73	10.00	5.20	-	0.44	3.09	4.69	12.98	m	14.60
in concrete, depth 150 - 300mm; horizontal	4.73	10.00	5.20	-	0.44	3.09	8.21	16.50	m	18.56
in concrete, depth 300 - 450mm; horizontal	4.73	10.00	5.20	-	0.44	3.09	11.94	20.23	m	22.76
flat L piece	7.11	10.00	7.82	-	0.22	1.54	-	9.37	nr	10.54
flat T piece	10.27	10.00	11.30	-	0.28	1.97	-	13.26	nr	14.92
flat X piece	11.62	10.00	12.78	-	0.33	2.32	-	15.10	nr	16.99
in concrete, width not exceeding 150mm; vertical	4.73	10.00	5.20	-	0.55	3.86	4.69	13.75	m	15.47
in concrete, width 150 - 300mm; vertical	4.73	10.00	5.20	-	0.55	3.86	8.21	17.27	m	19.43
in concrete, width 300 - 450mm; vertical	4.73	10.00	5.20	-	0.55	3.86	11.94	21.00	m	23.63
vertical L piece	10.10	10.00	11.11	-	0.22	1.54	-	12.65	nr	14.24
flat L piece	7.11	10.00	7.82	-	0.22	1.54	-	9.37	nr	10.54
vertical T piece	9.48	10.00	10.43	-	0.28	1.97	-	12.39	nr	13.94
flat T piece	10.27	10.00	11.30	-	0.28	1.97	-	13.26	nr	14.92
flat X piece	11.62	10.00	12.78	-	0.33	2.32	-	15.10	nr	16.99

Incorporating 250mm wide 'FD250' flat dumbell internally placed p.v.c. waterstop, Servicised Ltd; heat welded joints; formwork; reinforcement laid continuously across joint

in concrete, depth not exceeding 150mm; horizontal	5.27	10.00	5.80	-	0.50	3.51	4.69	14.00	m	15.75
in concrete, depth 150 - 300mm; horizontal	5.27	10.00	5.80	-	0.50	3.51	8.21	17.52	m	19.71
in concrete, depth 300 - 450mm; horizontal	5.27	10.00	5.80	-	0.50	3.51	11.94	21.25	m	23.90
flat L piece	8.20	10.00	9.02	-	0.22	1.54	-	10.56	nr	11.88
flat T piece	11.96	10.00	13.16	-	1.14	8.00	-	21.16	nr	23.80
flat X piece	13.78	10.00	15.16	-	1.38	9.69	-	24.85	nr	27.95
in concrete, width not exceeding 150mm; vertical	5.27	10.00	5.80	-	0.61	4.28	4.69	14.77	m	16.62
in concrete, width 150 - 300mm; vertical	5.27	10.00	5.80	-	0.61	4.28	8.21	18.29	m	20.58
in concrete, width 300 - 450mm; vertical	5.27	10.00	5.80	-	0.61	4.28	11.94	22.02	m	24.77
vertical L piece	8.59	10.00	9.45	-	0.22	1.54	-	10.99	nr	12.37
flat L piece	8.20	10.00	9.02	-	0.22	1.54	-	10.56	nr	11.88
vertical T piece	7.99	10.00	8.79	-	0.28	1.97	-	10.75	nr	12.10
flat T piece	11.96	10.00	13.16	-	0.28	1.97	-	15.12	nr	17.01
flat X piece	13.78	10.00	15.16	-	0.33	2.32	-	17.47	nr	19.66

Incorporating 160mm wide 'CB160' centre bulb type internally placed p.v.c. waterstop, Servicised Ltd; heat welded joints; formwork; reinforcement laid continuously across joint

in concrete, depth not exceeding 150mm; horizontal	3.30	10.00	3.63	-	0.39	2.74	4.69	11.06	m	12.44
in concrete, depth 150 - 300mm; horizontal	3.30	10.00	3.63	-	0.39	2.74	8.21	14.58	m	16.40
in concrete, depth 300 - 450mm; horizontal	3.30	10.00	3.63	-	0.39	2.74	11.94	18.31	m	20.60
flat L piece	5.96	10.00	6.56	-	0.17	1.19	-	7.75	nr	8.72
flat T piece	8.34	10.00	9.17	-	0.22	1.54	-	10.72	nr	12.06
flat X piece	9.38	10.00	10.32	-	0.28	1.97	-	12.28	nr	13.82
in concrete, width not exceeding 150mm; vertical	3.30	10.00	3.63	-	0.48	3.37	4.69	11.69	m	13.15
in concrete, width 150 - 300mm; vertical	3.30	10.00	3.63	-	0.48	3.37	8.21	15.21	m	17.11
in concrete, width 300 - 450mm; vertical	3.30	10.00	3.63	-	0.48	3.37	11.94	18.94	m	21.31
vertical L piece	9.91	10.00	10.90	-	0.17	1.19	-	12.09	nr	13.61
flat L piece	5.96	10.00	6.56	-	0.17	1.19	-	7.75	nr	8.72
vertical T piece	9.38	10.00	10.32	-	0.22	1.54	-	11.86	nr	13.35
flat T piece	8.34	10.00	9.17	-	0.22	1.54	-	10.72	nr	12.06
flat X piece	9.38	10.00	10.32	-	0.28	1.97	-	12.28	nr	13.82

IN SITU CONCRETE/LARGE PRECAST CONCRETE

Labour hourly rates: (except Specialists) Craft Operatives £9.23 Labourer £7.02 Rates are national average prices. Refer to REGIONAL VARIATIONS for indicative levels of overall pricing in regions	MATERIALS			LABOUR				RATES		
	Del to Site £	Waste %	Material Cost £	Craft Optve Hrs	Lab Hrs	Labour Cost £	Sunds £	Nett Rate £	Unit	Gross Rate (+12.5%) £
E40: DESIGNED JOINTS IN IN-SITU CONCRETE Cont.										
Formed joints Cont.										
Incorporating 210mm wide 'FD210' centre bulb type internally placed p.v.c. waterstop, Servicised Ltd; heat welded joints; formwork; reinforcement laid continuously across joint										
in concrete, depth not exceeding 150mm; horizontal	3.98	10.00	4.38	-	0.44	3.09	4.69	12.16	m	13.68
in concrete, depth 150 - 300mm; horizontal	3.98	10.00	4.38	-	0.44	3.09	8.21	15.68	m	17.64
in concrete, depth 300 - 450mm; horizontal	3.98	10.00	4.38	-	0.44	3.09	11.94	19.41	m	21.83
flat L piece.................................	7.17	10.00	7.89	-	0.22	1.54	-	9.43	nr	10.61
flat T piece.................................	10.44	10.00	11.48	-	0.28	1.97	-	13.45	nr	15.13
flat X piece.................................	11.76	10.00	12.94	-	0.33	2.32	-	15.25	nr	17.16
in concrete, width not exceeding 150mm; vertical .	3.98	10.00	4.38	-	0.55	3.86	4.69	12.93	m	14.55
in concrete, width 150 - 300mm; vertical	3.98	10.00	4.38	-	0.55	3.86	8.21	16.45	m	18.51
in concrete, width 300 - 450mm; vertical	3.98	10.00	4.38	-	0.55	3.86	11.94	20.18	m	22.70
vertical L piece	8.27	10.00	9.10	-	0.22	1.54	-	10.64	nr	11.97
flat L piece.................................	7.17	10.00	7.89	-	0.22	1.54	-	9.43	nr	10.61
vertical T piece	7.76	10.00	8.54	-	0.28	1.97	-	10.50	nr	11.81
flat T piece.................................	10.44	10.00	11.48	-	0.28	1.97	-	13.45	nr	15.13
flat X piece.................................	11.76	10.00	12.94	-	0.33	2.32	-	15.25	nr	17.16
Incorporating 260mm wide 'CB260' centre bulb type internally placed p.v.c. waterstop, Servicised Ltd; heat welded joints; formwork; reinforcement laid continuously across joint										
in concrete, depth not exceeding 150mm; horizontal	5.52	10.00	6.07	-	0.50	3.51	4.69	14.27	m	16.06
in concrete, depth 150 - 300mm; horizontal	5.52	10.00	6.07	-	0.50	3.51	8.21	17.79	m	20.02
in concrete, depth 300 - 450mm; horizontal	5.52	10.00	6.07	-	0.50	3.51	11.94	21.52	m	24.21
flat L piece.................................	8.26	10.00	9.09	-	0.22	1.54	-	10.63	nr	11.96
flat T piece.................................	12.07	10.00	13.28	-	1.14	8.00	-	21.28	nr	23.94
flat X piece.................................	13.89	10.00	15.28	-	1.38	9.69	-	24.97	nr	28.09
in concrete, width not exceeding 150mm; vertical .	5.52	10.00	6.07	-	0.61	4.28	4.69	15.04	m	16.92
in concrete, width 150 - 300mm; vertical	5.52	10.00	6.07	-	0.61	4.28	8.21	18.56	m	20.88
in concrete, width 300 - 450mm; vertical	5.52	10.00	6.07	-	0.61	4.28	11.94	22.29	m	25.08
vertical L piece	8.55	10.00	9.40	-	0.22	1.54	-	10.95	nr	12.32
flat L piece.................................	8.26	10.00	9.09	-	0.22	1.54	-	10.63	nr	11.96
vertical T piece	8.03	10.00	8.83	-	0.28	1.97	-	10.80	nr	12.15
flat T piece.................................	12.07	10.00	13.28	-	0.28	1.97		15.24	nr	17.15
flat X piece.................................	13.89	10.00	15.28	-	0.33	2.32	-	17.60	nr	19.80
Incorporating 325mm wide 'CB325' centre bulb type internally placed p.v.c. waterstop, Servicised Ltd; heat welded joints; formwork; reinforcement laid continuously across joint										
in concrete, depth not exceeding 150mm; horizontal	12.27	10.00	13.50	-	0.55	3.86	4.69	22.05	m	24.80
in concrete, depth 150 - 300mm; horizontal	12.27	10.00	13.50	-	0.55	3.86	8.21	25.57	m	28.76
in concrete, depth 300 - 450mm; horizontal	12.27	10.00	13.50	-	0.55	3.86	11.94	29.30	m	32.96
flat L piece.................................	16.02	10.00	17.62	-	0.28	1.97	-	19.59	nr	22.04
flat T piece.................................	19.50	10.00	21.45	-	0.33	2.32	-	23.77	nr	26.74
flat X piece.................................	22.58	10.00	24.84	-	0.39	2.74	-	27.58	nr	31.02
in concrete, width not exceeding 150mm; vertical .	12.27	10.00	13.50	-	0.66	4.63	4.69	22.82	m	25.67
in concrete, width 150 - 300mm; vertical	12.27	10.00	13.50	-	0.66	4.63	8.21	26.34	m	29.63
in concrete, width 300 - 450mm; vertical	12.27	10.00	13.50	-	0.66	4.63	11.94	30.07	m	33.83
vertical L piece	9.09	10.00	10.00	-	0.28	1.97	-	11.96	nr	13.46
flat L piece.................................	16.02	10.00	17.62	-	0.28	1.97	-	19.59	nr	22.04
vertical T piece	8.48	10.00	9.33	-	0.33	2.32	-	11.64	nr	13.10
flat T piece.................................	19.50	10.00	21.45	-	0.33	2.32	-	23.77	nr	26.74
flat X piece.................................	22.58	10.00	24.84	-	0.39	2.74	-	27.58	nr	31.02
Incorporating 10 x 150mm 'Servi-tite CJ' flat dumbell internally placed p.v.c waterstop, Servicised Ltd; heat welded joints; formwork; reinforcement laid continuously across joint										
in concrete, depth not exceeding 150mm; horizontal	7.28	10.00	8.01	-	0.39	2.74	4.69	15.44	m	17.37
in concrete, depth 150 - 300mm; horizontal	7.28	10.00	8.01	-	0.39	2.74	8.21	18.96	m	21.33
in concrete, depth 300 - 450mm; horizontal	7.28	10.00	8.01	-	0.39	2.74	11.94	22.69	m	25.52
flat L piece.................................	8.90	10.00	9.79	-	0.17	1.19	-	10.98	nr	12.36
flat T piece.................................	15.30	10.00	16.83	-	0.22	1.54	-	18.37	nr	20.67
flat X piece.................................	17.16	10.00	18.88	-	0.28	1.97	-	20.84	nr	23.45
in concrete, width not exceeding 150mm; vertical .	7.28	10.00	8.01	-	0.48	3.37	4.69	16.07	m	18.08
in concrete, width 150 - 300mm; vertical	7.28	10.00	8.01	-	0.48	3.37	8.21	19.59	m	22.04
in concrete, width 300 - 450mm; vertical	7.28	10.00	8.01	-	0.48	3.37	11.94	23.32	m	26.23
vertical L piece	10.23	10.00	11.25	-	0.17	1.19	-	12.45	nr	14.00
flat L piece.................................	8.90	10.00	9.79	-	0.17	1.19	-	10.98	nr	12.36
vertical T piece	9.70	10.00	10.67	-	0.22	1.54	-	12.21	nr	13.74
flat T piece.................................	15.30	10.00	16.83	-	0.22	1.54	-	18.37	nr	20.67
flat X piece.................................	17.16	10.00	18.88	-	0.28	1.97	-	20.84	nr	23.45
Incorporating 10 x 230mm 'Servi-tite CJ' flat dumbell internally placed p.v.c waterstop, Servicised Ltd; heat welded joints; formwork; reinforcement laid continuously across joint										
in concrete, depth not exceeding 150mm; horizontal	10.19	10.00	11.21	-	0.50	3.51	4.69	19.41	m	21.84
in concrete, depth 150 - 300mm; horizontal	10.19	10.00	11.21	-	0.50	3.51	8.21	22.93	m	25.80
in concrete, depth 300 - 450mm; horizontal	10.19	10.00	11.21	-	0.50	3.51	11.94	26.66	m	29.99
flat L piece.................................	12.84	10.00	14.12	-	0.22	1.54	-	15.67	nr	17.63
flat T piece.................................	22.50	10.00	24.75	-	1.14	8.00	-	32.75	nr	36.85
flat X piece.................................	25.67	10.00	28.24	-	1.38	9.69	-	37.92	nr	42.67
in concrete, width not exceeding 150mm; vertical .	10.19	10.00	11.21	-	0.61	4.28	4.69	20.18	m	22.70
in concrete, width 150 - 300mm; vertical	10.19	10.00	11.21	-	0.61	4.28	8.21	23.70	m	26.66
in concrete, width 300 - 450mm; vertical	10.19	10.00	11.21	-	0.61	4.28	11.94	27.43	m	30.86
vertical L piece	8.05	10.00	8.86	-	0.22	1.54	-	10.40	nr	11.70
flat L piece.................................	12.84	10.00	14.12	-	0.22	1.54	-	15.67	nr	17.63
vertical T piece	10.05	10.00	11.06	-	0.28	1.97	-	13.02	nr	14.65

Labour hourly rates: (except Specialists) Craft Operatives £9.23 Labourer £7.02 Rates are national average prices. Refer to REGIONAL VARIATIONS for indicative levels of overall pricing in regions	MATERIALS			LABOUR				RATES		
	Del to Site	Waste	Material Cost	Craft Optve	Lab	Labour Cost	Sunds	Nett Rate	Unit	Gross Rate (+12.5%)
	£	%	£	Hrs	Hrs	£	£	£		£
E40: DESIGNED JOINTS IN IN-SITU CONCRETE Cont.										
Formed joints Cont.										
Incorporating 10 x 230mm 'Servi-tite CJ' flat dumbell internally placed p.v.c waterstop, Servicised Ltd; heat welded joints; formwork; reinforcement laid continuously across joint Cont.										
flat T piece	22.50	10.00	24.75	-	0.28	1.97	-	26.72	nr	30.06
flat X piece	25.67	10.00	28.24	-	0.33	2.32	-	30.55	nr	34.37
Incorporating 10 x 305mm 'Servi-tite CJ' flat dumbell internally placed p.v.c waterstop, Servicised Ltd; heat welded joints; formwork; reinforcement laid continuously across joint										
in concrete, depth not exceeding 150mm; horizontal	16.39	10.00	18.03	-	0.55	3.86	4.69	26.58	m	29.90
in concrete, depth 150 - 300mm; horizontal	16.39	10.00	18.03	-	0.55	3.86	8.21	30.10	m	33.86
in concrete, depth 300 - 450mm; horizontal	16.39	10.00	18.03	-	0.55	3.86	11.94	33.83	m	38.06
flat L piece	28.75	10.00	31.63	-	0.28	1.97	-	33.59	nr	37.79
flat T piece	37.96	10.00	41.76	-	0.33	2.32	-	44.07	nr	49.58
flat X piece	48.46	10.00	53.31	-	0.39	2.74	-	56.04	nr	63.05
in concrete, width not exceeding 150mm; vertical	16.39	10.00	18.03	-	0.66	4.63	4.69	27.35	m	30.77
in concrete, width 150 - 300mm; vertical	16.39	10.00	18.03	-	0.66	4.63	8.21	30.87	m	34.73
in concrete, width 300 - 450mm; vertical	16.39	10.00	18.03	-	0.66	4.63	11.94	34.60	m	38.93
vertical L piece	27.35	10.00	30.09	-	0.28	1.97	-	32.05	nr	36.06
flat L piece	28.75	10.00	31.63	-	0.28	1.97	-	33.59	nr	37.79
vertical T piece	35.19	10.00	38.71	-	0.33	2.32	-	41.03	nr	46.15
flat T piece	37.96	10.00	41.76	-	0.33	2.32	-	44.07	nr	49.58
flat X piece	48.46	10.00	53.31	-	0.39	2.74	-	56.04	nr	63.05
Incorporating 10 x 150mm 'Servi-tite XJ' centre bulb type internally placed p.v.c. waterstop, Servicised Ltd; heat welded joints; formwork; reinforcement laid continuously across joint										
in concrete, depth not exceeding 150mm; horizontal	7.55	10.00	8.30	-	0.39	2.74	4.69	15.73	m	17.70
in concrete, depth 150 - 300mm; horizontal	7.55	10.00	8.30	-	0.39	2.74	8.21	19.25	m	21.66
in concrete, depth 300 - 450mm; horizontal	7.55	10.00	8.30	-	0.39	2.74	11.94	22.98	m	25.86
flat L piece	13.56	10.00	14.92	-	0.17	1.19	-	16.11	nr	18.12
flat T piece	19.87	10.00	21.86	-	0.22	1.54	-	23.40	nr	26.33
flat X piece	22.25	10.00	24.48	-	0.28	1.97	-	26.44	nr	29.75
in concrete, width not exceeding 150mm; vertical	7.55	10.00	8.30	-	0.48	3.37	4.69	16.36	m	18.41
in concrete, width 150 - 300mm; vertical	7.55	10.00	8.30	-	0.48	3.37	8.21	19.88	m	22.37
in concrete, width 300 - 450mm; vertical	7.55	10.00	8.30	-	0.48	3.37	11.94	23.61	m	26.57
vertical L piece	10.41	10.00	11.45	-	0.17	1.19	-	12.64	nr	14.22
flat L piece	13.56	10.00	14.92	-	0.17	1.19	-	16.11	nr	18.12
vertical T piece	9.79	10.00	10.77	-	0.22	1.54	-	12.31	nr	13.85
flat T piece	19.87	10.00	21.86	-	0.22	1.54	-	23.40	nr	26.33
flat X piece	22.25	10.00	24.48	-	0.28	1.97	-	26.44	nr	29.75
Incorporating 10 x 230mm 'Servi-tite XJ' centre bulb type internally placed p.v.c. waterstop, Servicised Ltd; heat welded joints; formwork; reinforcement laid continuously across joint										
in concrete, depth not exceeding 150mm; horizontal	10.59	10.00	11.65	-	0.50	3.51	4.69	19.85	m	22.33
in concrete, depth 150 - 300mm; horizontal	10.59	10.00	11.65	-	0.50	3.51	8.21	23.37	m	26.29
in concrete, depth 300 - 450mm; horizontal	10.59	10.00	11.65	-	0.50	3.51	11.94	27.10	m	30.49
flat L piece	16.65	10.00	18.32	-	0.22	1.54	-	19.86	nr	22.34
flat T piece	29.83	10.00	32.81	-	1.14	8.00	-	40.82	nr	45.92
flat X piece	33.30	10.00	36.63	-	1.38	9.69	-	46.32	nr	52.11
in concrete, width not exceeding 150mm; vertical	10.59	10.00	11.65	-	0.61	4.28	4.69	20.62	m	23.20
in concrete, width 150 - 300mm; vertical	10.59	10.00	11.65	-	0.61	4.28	8.21	24.14	m	27.16
in concrete, width 300 - 450mm; vertical	10.59	10.00	11.65	-	0.61	4.28	11.94	27.87	m	31.36
vertical L piece	9.35	10.00	10.29	-	0.22	1.54	-	11.83	nr	13.31
flat L piece	16.65	10.00	18.32	-	0.22	1.54	-	19.86	nr	22.34
vertical T piece	8.75	10.00	9.63	-	0.28	1.97	-	11.59	nr	13.04
flat T piece	29.83	10.00	32.81	-	0.28	1.97	-	34.78	nr	39.13
flat X piece	33.30	10.00	36.63	-	0.33	2.32	-	38.95	nr	43.81
Incorporating 10 x 305mm 'Servi-tite XJ' centre bulb type internally placed p.v.c. waterstop, Servicised Ltd; heat welded joints; formwork; reinforcement laid continuously across joint										
in concrete, depth not exceeding 150mm; horizontal	18.22	10.00	20.04	-	0.55	3.86	4.69	28.59	m	32.17
in concrete, depth 150 - 300mm; horizontal	18.22	10.00	20.04	-	0.55	3.86	8.21	32.11	m	36.13
in concrete, depth 300 - 450mm; horizontal	18.22	10.00	20.04	-	0.55	3.86	11.94	35.84	m	40.32
flat L piece	30.20	10.00	33.22	-	0.28	1.97	-	35.19	nr	39.58
flat T piece	46.18	10.00	50.80	-	0.33	2.32	-	53.11	nr	59.75
flat X piece	60.85	10.00	66.94	-	0.39	2.74	-	69.67	nr	78.38
in concrete, width not exceeding 150mm; vertical	18.22	10.00	20.04	-	0.66	4.63	4.69	29.37	m	33.04
in concrete, width 150 - 300mm; vertical	18.22	10.00	20.04	-	0.66	4.63	8.21	32.89	m	37.00
in concrete, width 300 - 450mm; vertical	18.22	10.00	20.04	-	0.66	4.63	11.94	36.62	m	41.19
vertical L piece	27.83	10.00	30.61	-	0.28	1.97	-	32.58	nr	36.65
flat L piece	30.20	10.00	33.22	-	0.28	1.97	-	35.19	nr	39.58
vertical T piece	36.95	10.00	40.65	-	0.33	2.32	-	42.96	nr	48.33
flat T piece	46.18	10.00	50.80	-	0.33	2.32	-	53.11	nr	59.75
flat X piece	60.85	10.00	66.94	-	0.39	2.74	-	69.67	nr	78.38
Incorporating Reference 2411, 6.34 kg/m2 Expamet Hy-rib permanent shuttering and reinforcement; 150mm end laps; temporary supports; formwork laid continuously across joint										
in concrete, depth not exceeding 150mm; horizontal	2.21	5.00	2.32	0.25	0.07	2.80	5.05	10.17	m	11.44
in concrete, depth 150 - 300mm; horizontal	4.41	5.00	4.63	0.25	0.07	2.80	8.83	16.26	m	18.29
in concrete, width not exceeding 150mm; vertical	2.21	5.00	2.32	0.39	0.10	4.30	5.05	11.67	m	13.13
in concrete, width 150 - 300mm; vertical	4.41	5.00	4.63	0.39	0.10	4.30	8.83	17.76	m	19.98

IN SITU CONCRETE/LARGE PRECAST CONCRETE

Labour hourly rates: (except Specialists) Craft Operatives £9.23 Labourer £7.02 Rates are national average prices. Refer to REGIONAL VARIATIONS for indicative levels of overall pricing in regions	MATERIALS			LABOUR				RATES		
	Del to Site £	Waste %	Material Cost £	Craft Optve Hrs	Lab Hrs	Labour Cost £	Sunds £	Nett Rate £	Unit	Gross Rate (+12.5%) £
E40: DESIGNED JOINTS IN IN-SITU CONCRETE Cont.										
Formed joints Cont.										
Incorporating Reference 2611, 4.86 kg/m2 Expamet Hy-rib permanent shuttering and reinforcement; 150mm end laps; temporary supports; formwork laid continuously across joint										
in concrete, depth not exceeding 150mm; horizontal	1.72	5.00	1.81	0.25	0.07	2.80	5.05	9.65	m	10.86
in concrete, depth 150 - 300mm; horizontal	3.45	5.00	3.62	0.25	0.07	2.80	8.83	15.25	m	17.16
in concrete, width not exceeding 150mm; vertical .	1.72	5.00	1.81	0.39	0.10	4.30	5.05	11.16	m	12.55
in concrete, width 150 - 300mm; vertical	3.45	5.00	3.62	0.39	0.10	4.30	8.83	16.75	m	18.85
Incorporating Reference 2811, 3.39 kg/m2 Expamet Hy-rib permanent shuttering and reinforcement; 150mm end laps; temporary supports; formwork laid continuously across joint										
in concrete, depth not exceeding 150mm; horizontal	1.52	5.00	1.60	0.25	0.07	2.80	5.05	9.44	m	10.63
in concrete, depth 150 - 300mm; horizontal	3.05	5.00	3.20	0.25	0.07	2.80	8.83	14.83	m	16.69
in concrete, width not exceeding 150mm; vertical .	1.52	5.00	1.60	0.39	0.10	4.30	5.05	10.95	m	12.32
in concrete, width 150 - 300mm; vertical	3.05	5.00	3.20	0.39	0.10	4.30	8.83	16.33	m	18.38
Sealant to joint; Servicised Ltd. 'Servigard DW'; including preparation, cleaners, primers and sealers										
10 x 25mm; horizontal	0.22	5.00	0.23	-	0.15	1.05	-	1.28	m	1.44
13 x 25mm; horizontal	0.30	5.00	0.32	-	0.20	1.40	-	1.72	m	1.93
19 x 25mm; horizontal	0.42	5.00	0.44	-	0.25	1.75	-	2.20	m	2.47
Sealant to joint; Servicised Ltd. 'Servimastic 96'; including preparation, cleaners, primers and sealers										
10 x 25mm; horizontal	0.48	5.00	0.50	-	0.11	0.77	-	1.28	m	1.44
13 x 25mm; horizontal	0.63	5.00	0.66	-	0.12	0.84	-	1.50	m	1.69
19 x 25mm; horizontal	0.92	5.00	0.97	-	0.13	0.91	-	1.88	m	2.11
E41: WORKED FINISHES/ CUTTING ON IN-SITU CONCRETE										
Worked finishes on in-situ concrete										
Vacuum dewatering; power floating and power trowelling										
surfaces	-	-	-	-	0.22	1.54	0.22	1.76	m2	1.98
Tamping unset concrete										
surfaces	-	-	-	-	0.22	1.54	0.29	1.83	m2	2.06
surfaces to falls	-	-	-	-	0.45	3.16	0.38	3.54	m2	3.98
Trowelling										
surfaces	-	-	-	-	0.36	2.53	-	2.53	m2	2.84
surfaces to falls	-	-	-	-	0.39	2.74	-	2.74	m2	3.08
Hacking; by hand										
surfaces	-	-	-	-	0.55	3.86	-	3.86	m2	4.34
to soffits	-	-	-	-	0.72	5.05	-	5.05	m2	5.69
Hacking; by machine										
surfaces	-	-	-	-	0.22	1.54	0.23	1.77	m2	2.00
to soffits	-	-	-	-	0.28	1.97	0.23	2.20	m2	2.47
Bush hammering										
surfaces	-	-	-	1.43	-	13.20	0.65	13.85	m2	15.58
to soffits	-	-	-	2.20	-	20.31	0.86	21.17	m2	23.81
Cutting on in-situ concrete										
Cutting chases										
depth not exceeding 50mm	-	-	-	-	0.19	1.33	-	1.33	m	1.50
Cutting chases; making good										
depth not exceeding 50mm	-	-	-	-	0.22	1.54	0.25	1.79	m	2.02
Cutting mortices										
50 x 50mm; depth not exceeding 100mm	-	-	-	-	0.44	3.09	-	3.09	nr	3.47
75 x 75mm; depth not exceeding 100mm	-	-	-	-	0.55	3.86	-	3.86	nr	4.34
100 x 100mm; depth 100 - 200mm	-	-	-	-	0.66	4.63	-	4.63	nr	5.21
38mm diameter; depth not exceeding 100mm .	-	-	-	-	0.39	2.74	-	2.74	nr	3.08
Cutting holes										
225 x 150mm; depth not exceeding 100mm ...	-	-	-	-	1.25	8.78	0.34	9.12	nr	10.25
300 x 150mm; depth not exceeding 100mm ...	-	-	-	-	1.45	10.18	0.38	10.56	nr	11.88
300 x 300mm; depth not exceeding 100mm ...	-	-	-	-	1.60	11.23	0.38	11.61	nr	13.06
225 x 150mm; depth 100 - 200mm	-	-	-	-	1.89	13.27	0.48	13.75	nr	15.47
300 x 150mm; depth 100 - 200mm	-	-	-	-	2.11	14.81	0.54	15.35	nr	17.27
300 x 300mm; depth 100 - 200mm	-	-	-	-	2.32	16.29	0.54	16.83	nr	18.93
225 x 150mm; depth 200 - 300mm	-	-	-	-	2.20	15.44	0.56	16.00	nr	18.00
300 x 150mm; depth 200 - 300mm	-	-	-	-	2.43	17.06	0.61	17.67	nr	19.88
300 x 300mm; depth 200 - 300mm	-	-	-	-	2.65	18.60	0.61	19.21	nr	21.61
50mm diameter; depth not exceeding 100mm .	-	-	-	-	0.77	5.41	0.20	5.61	nr	6.31
100mm diameter; depth not exceeding 100mm	-	-	-	-	0.94	6.60	0.26	6.86	nr	7.72
150mm diameter; depth not exceeding 100mm	-	-	-	-	1.03	7.23	0.26	7.49	nr	8.43
50mm diameter; depth 100 - 200mm	-	-	-	-	1.30	9.13	0.33	9.46	nr	10.64
100mm diameter; depth 100 - 200mm	-	-	-	-	1.51	10.60	0.38	10.98	nr	12.35
150mm diameter; depth 100 - 200mm	-	-	-	-	1.62	11.37	0.38	11.75	nr	13.22
50mm diameter; depth 200 - 300mm	-	-	-	-	1.60	11.23	0.43	11.66	nr	13.12
100mm diameter; depth 200 - 300mm	-	-	-	-	1.79	12.57	0.46	13.03	nr	14.65
150mm diameter; depth 200 - 300mm	-	-	-	-	1.97	13.83	0.46	14.29	nr	16.08

Labour hourly rates: (except Specialists) Craft Operatives £9.23 Labourer £7.02 Rates are national average prices. Refer to REGIONAL VARIATIONS for indicative levels of overall pricing in regions	MATERIALS			LABOUR				RATES		
	Del to Site	Waste	Material Cost	Craft Optve	Lab	Labour Cost	Sunds	Nett Rate	Unit	Gross Rate (+12.5%)
	£	%	£	Hrs	Hrs	£	£	£		£
E41: WORKED FINISHES/ CUTTING ON IN-SITU CONCRETE Cont.										
Cutting on in-situ concrete Cont.										
Grouting into mortices with cement mortar (1:1); around steel										
50 x 50mm; depth not exceeding 100mm	-	-	-	-	0.11	0.77	0.51	1.28	nr	1.44
75 x 75mm; depth not exceeding 100mm	-	-	-	-	0.13	0.91	0.51	1.42	nr	1.60
100 x 100mm; depth 100 - 200mm	-	-	-	-	0.17	1.19	0.65	1.84	nr	2.07
38mm diameter; depth not exceeding 100mm	-	-	-	-	0.09	0.63	0.37	1.00	nr	1.13
Cutting on reinforced in-situ concrete										
Cutting holes										
225 x 150mm; depth not exceeding 100mm	-	-	-	-	1.56	10.95	0.43	11.38	nr	12.80
300 x 150mm; depth not exceeding 100mm	-	-	-	-	1.80	12.64	0.46	13.10	nr	14.73
300 x 300mm; depth not exceeding 100mm	-	-	-	-	1.99	13.97	0.46	14.43	nr	16.23
225 x 150mm; depth 100 - 200mm	-	-	-	-	2.35	16.50	0.63	17.13	nr	19.27
300 x 150mm; depth 100 - 200mm	-	-	-	-	2.63	18.46	0.66	19.12	nr	21.51
300 x 300mm; depth 100 - 200mm	-	-	-	-	2.89	20.29	0.66	20.95	nr	23.57
225 x 150mm; depth 200 - 300mm	-	-	-	-	2.78	19.52	0.71	20.23	nr	22.75
300 x 150mm; depth 200 - 300mm	-	-	-	-	3.06	21.48	0.78	22.26	nr	25.04
300 x 300mm; depth 200 - 300mm	-	-	-	-	3.33	23.38	0.78	24.16	nr	27.18
50mm diameter; depth not exceeding 100mm	-	-	-	-	0.95	6.67	0.27	6.94	nr	7.81
100mm diameter; depth not exceeding 100mm	-	-	-	-	1.16	8.14	0.31	8.45	nr	9.51
150mm diameter; depth not exceeding 100mm	-	-	-	-	1.30	9.13	0.31	9.44	nr	10.62
50mm diameter; depth 100 - 200mm	-	-	-	-	1.62	11.37	0.44	11.81	nr	13.29
100mm diameter; depth 100 - 200mm	-	-	-	-	1.89	13.27	0.47	13.74	nr	15.46
150mm diameter; depth 100 - 200mm	-	-	-	-	2.02	14.18	0.47	14.65	nr	16.48
50mm diameter; depth 200 - 300mm	-	-	-	-	2.00	14.04	0.52	14.56	nr	16.38
100mm diameter; depth 200 - 300mm	-	-	-	-	2.24	15.72	0.58	16.30	nr	18.34
150mm diameter; depth 200 - 300mm	-	-	-	-	2.46	17.27	0.58	17.85	nr	20.08
E42: ACCESSORIES CAST INTO IN-SITU CONCRETE										
Cast in accessories										
Hardwood fillets										
38 x 25mm; dovetail	0.75	10.00	0.82	-	0.11	0.77	-	1.60	m	1.80
38 x 25mm; 150mm long	0.24	10.00	0.26	-	0.04	0.28	-	0.54	nr	0.61
38 x 25mm; 225mm long	0.27	10.00	0.30	-	0.06	0.42	-	0.72	nr	0.81
Galvanised steel dowels										
12mm diameter x 150mm long	0.53	10.00	0.58	-	0.09	0.63	-	1.21	nr	1.37
12mm diameter x 300mm long	0.78	10.00	0.86	-	0.08	0.56	-	1.42	nr	1.60
12mm diameter x 600mm long	1.18	10.00	1.30	-	0.12	0.84	-	2.14	nr	2.41
Galvanised expanded metal tie										
300 x 50mm; bending and temporarily fixing to formwork	0.55	10.00	0.60	-	0.13	0.91	-	1.52	nr	1.71
Galvanised steel dovetailed masonry slots with 3mm thick twisted tie to suit 50mm cavity										
100mm long; temporarily fixing to formwork	0.56	10.00	0.62	0.11	-	1.02	-	1.63	nr	1.84
Stainless steel channel with anchors at 250mm centres										
28 x 15mm; temporarily fixing to formwork	8.36	5.00	8.78	0.28	-	2.58	-	11.36	m	12.78
28 x 15 x 150mm long; temporarily fixing to formwork	1.45	5.00	1.52	0.12	-	1.11	-	2.63	nr	2.96
M10 bolt 50mm long with `T' head nut and washers .	2.86	5.00	3.00	0.05	-	0.46	-	3.46	nr	3.90
fishtailed tie to suit 50mm cavity	1.32	5.00	1.39	0.06	-	0.55	-	1.94	nr	2.18
Stainless steel angle drilled at 450mm centres										
70 x 90 x 5mm	21.60	5.00	22.68	0.80	-	7.38	-	30.06	m	33.82
Copper dovetailed masonry slots with 3mm thick twisted tie to suit 50mm cavity										
100mm long; temporarily fixing to formwork	5.90	10.00	6.49	0.11	-	1.02	-	7.51	nr	8.44
Bolt boxes; The Expanded Metal Co. Ltd										
reference 220 for use with poured concrete; 75mm diameter										
150mm long.......................................	1.22	10.00	1.34	0.28	-	2.58	0.57	4.50	nr	5.06
225mm long.......................................	1.50	10.00	1.65	0.33	-	3.05	0.57	5.27	nr	5.92
300mm long.......................................	1.68	10.00	1.85	0.39	-	3.60	0.57	6.02	nr	6.77
reference 220 for use with poured concrete; 100mm diameter										
375mm long.......................................	2.72	10.00	2.99	0.44	-	4.06	0.57	7.62	nr	8.58
450mm long.......................................	3.06	10.00	3.37	0.50	-	4.62	0.57	8.55	nr	9.62
600mm long.......................................	3.39	10.00	3.73	0.55	-	5.08	0.57	9.38	nr	10.55
reference 220 for use with vibrated concrete; wrap with single layer of thin polythene sheet; 75mm diameter										
150mm long.......................................	1.33	10.00	1.46	0.28	-	2.58	0.69	4.74	nr	5.33
225mm long.......................................	1.60	10.00	1.76	0.33	-	3.05	0.69	5.50	nr	6.18
300mm long.......................................	1.85	10.00	2.04	0.39	-	3.60	0.69	6.32	nr	7.12
reference 220 for use with vibrated concrete; wrap with single layer of thin polythene sheet; 100mm diameter										
375mm long.......................................	2.88	10.00	3.17	0.44	-	4.06	0.69	7.92	nr	8.91
450mm long.......................................	3.25	10.00	3.58	0.50	-	4.62	0.69	8.88	nr	9.99
600mm long.......................................	3.58	10.00	3.94	0.55	-	5.08	0.69	9.70	nr	10.92
Steel rag bolts										
M 10 x 100mm long................................	1.01	5.00	1.06	0.10	-	0.92	-	1.98	nr	2.23

Labour hourly rates: (except Specialists) Craft Operatives £9.23 Labourer £7.02 Rates are national average prices. Refer to REGIONAL VARIATIONS for indicative levels of overall pricing in regions	MATERIALS			LABOUR				RATES		
	Del to Site £	Waste %	Material Cost £	Craft Optve Hrs	Lab Hrs	Labour Cost £	Sunds £	Nett Rate £	Unit	Gross Rate (+12.5%) £
E42: ACCESSORIES CAST INTO IN-SITU CONCRETE Cont.										
Cast in accessories Cont.										
Steel rag bolts Cont.										
M 10 x 160mm long	1.09	5.00	1.14	0.12	–	1.11	–	2.25	nr	2.53
M 12 x 100mm long	1.12	5.00	1.18	0.10	–	0.92	–	2.10	nr	2.36
M 12 x 160mm long	1.30	5.00	1.37	0.12	–	1.11	–	2.47	nr	2.78
M 12 x 200mm long	1.40	5.00	1.47	0.14	–	1.29	–	2.76	nr	3.11
M 16 x 120mm long	1.74	5.00	1.83	0.10	–	0.92	–	2.75	nr	3.09
M 16 x 160mm long	2.00	5.00	2.10	0.12	–	1.11	–	3.21	nr	3.61
M 16 x 200mm long	2.22	5.00	2.33	0.14	–	1.29	–	3.62	nr	4.08
E60: PRECAST/COMPOSITE CONCRETE DECKING										
Precast concrete floors and hoist, bed and grout (BIRCHWOOD CONCRETE LTD)										
150mm thick floors of solid lightweight concrete units 600mm wide										
span between supports										
3000mm	31.29	2.50	32.07	–	0.70	4.91	4.74	41.73	m2	46.94
3600mm	31.29	2.50	32.07	–	0.72	5.05	4.68	41.81	m2	47.03
4200mm	31.29	2.50	32.07	–	0.69	4.84	4.55	41.47	m2	46.65
200mm thick floors of solid lightweight concrete units 600mm wide										
span between supports										
3000mm	41.49	2.50	42.53	–	0.67	4.70	4.86	52.09	m2	58.60
3600mm	41.49	2.50	42.53	–	0.75	5.26	4.81	52.60	m2	59.18
4200mm	41.49	2.50	42.53	–	0.73	5.12	4.68	52.33	m2	58.87
125mm thick floors of hollow prestressed concrete units 750mm wide										
span between supports										
3000mm	21.20	2.50	21.73	–	0.70	4.91	4.68	31.32	m2	35.24
3600mm	21.20	2.50	21.73	–	0.69	4.84	4.55	31.12	m2	35.01
4200mm	21.20	2.50	21.73	–	0.67	4.70	4.42	30.85	m2	34.71
4800mm	21.20	2.50	21.73	–	0.66	4.63	4.28	30.64	m2	34.47
150mm thick floors of hollow prestressed concrete units 750mm wide										
span between supports										
3000mm	22.32	2.50	22.88	–	0.73	5.12	4.68	32.68	m2	36.77
3600mm	22.32	2.50	22.88	–	0.72	5.05	4.55	32.48	m2	36.54
4200mm	22.32	2.50	22.88	–	0.69	4.84	4.42	32.14	m2	36.16
4800mm	22.32	2.50	22.88	–	0.67	4.70	4.28	31.86	m2	35.84
200mm thick floors of hollow prestressed concrete units 750mm wide										
span between supports										
3000mm	25.68	2.50	26.32	–	0.76	5.34	4.86	36.52	m2	41.08
3600mm	25.68	2.50	26.32	–	0.75	5.26	4.81	36.40	m2	40.95
4200mm	25.68	2.50	26.32	–	0.73	5.12	4.74	36.19	m2	40.71
4800mm	25.68	2.50	26.32	–	0.72	5.05	4.68	36.06	m2	40.56
Prestressed concrete ground floors; prestressed concrete beams and 100mm building blocks; hoist bed and grout										
155mm thick floors, distributed load 1.5 kN/m2										
span not exceeding 3.9m	12.80	2.50	13.12	–	1.15	8.07	0.45	21.64	m2	24.35
span not exceeding 4.8m	12.88	2.50	13.20	–	1.05	7.37	0.45	21.02	m2	23.65
span not exceeding 5.3m	13.26	2.50	13.59	–	1.00	7.02	0.45	21.06	m2	23.69
155mm thick floors, distributed load 3.0 kN/m2										
span not exceeding 3.4m	13.12	2.50	13.45	–	1.15	8.07	0.45	21.97	m2	24.72
span not exceeding 4.2m	13.19	2.50	13.52	–	1.05	7.37	0.45	21.34	m2	24.01
span not exceeding 4.6m	13.26	2.50	13.59	–	1.00	7.02	0.45	21.06	m2	23.69
155mm thick floors, distributed load 4.5 kN/m2										
span not exceeding 2.8m	13.50	2.50	13.84	–	1.15	8.07	0.45	22.36	m2	25.16
span not exceeding 3.8m	13.55	2.50	13.89	–	1.05	7.37	0.45	21.71	m2	24.42
span not exceeding 4.1m	13.63	2.50	13.97	–	1.00	7.02	0.45	21.44	m2	24.12
Hollow block floor with precast reinforced concrete planks at 650mm centres, precast hollow infill blocks, in-situ concrete, designed mix C25, filling between blocks and temporary supports and bearers										
Floor as described; height to soffit 1.50-3.00m										
200mm thick	28.12	2.50	28.82	–	0.61	4.28	3.73	36.84	m2	41.44
250mm thick	29.52	2.50	30.26	–	0.66	4.63	4.40	39.29	m2	44.20
Floor as described; with 50mm in-situ concrete, designed mix C25, topping; height to soffit 1.50-3.00m										
200mm total thickness	27.78	2.50	28.47	–	0.55	3.86	6.21	38.55	m2	43.36
250mm total thickness	28.12	2.50	28.82	–	0.61	4.28	6.44	39.55	m2	44.49
300mm total thickness	29.52	2.50	30.26	–	0.66	4.63	7.22	42.11	m2	47.38

Masonry

Labour hourly rates: (except Specialists) Craft Operatives £9.23 Labourer £7.02 Rates are national average prices. Refer to REGIONAL VARIATIONS for indicative levels of overall pricing in regions	MATERIALS			LABOUR				RATES		
	Del to Site £	Waste %	Material Cost £	Craft Optve Hrs	Lab Hrs	Labour Cost £	Sunds £	Nett Rate £	Unit	Gross Rate (+12.5%) £
F10: BRICK/BLOCK WALLING										
Common bricks, B.S.3921, Category M, 215 x 102.5 x 65mm, compressive strength 20.5 N/mm2; in cement-lime mortar (1:2:9)										
Walls; vertical										
102mm thick; stretcher bond	9.67	5.00	10.15	1.45	1.60	24.62	1.98	36.75	m2	41.34
215mm thick; English bond	19.50	5.00	20.48	2.35	2.65	40.29	4.75	65.52	m2	73.71
327mm thick; English bond	29.34	5.00	30.81	2.85	3.30	49.47	7.51	87.79	m2	98.76
extra; grooved bricks; 50% wall surface keyed	0.15	5.00	0.16	-	-	-	-	0.16	m2	0.18
extra; grooved bricks; 100% wall surface keyed	0.30	5.00	0.32	-	-	-	-	0.32	m2	0.35
Walls; building against concrete (ties measured separately); vertical										
102mm thick; stretcher bond	9.67	5.00	10.15	1.60	1.75	27.05	1.98	39.19	m2	44.08
Walls; building against old brickwork; tie new to old with 3mm diameter galvanised twisted wire ties - 6/m2; vertical										
102mm thick; stretcher bond	9.67	5.00	10.15	1.60	1.75	27.05	2.55	39.76	m2	44.73
215mm thick; English bond	19.50	5.00	20.48	2.65	2.95	45.17	5.32	70.96	m2	79.83
Walls; bonding to stonework; including extra material; vertical										
102mm thick; stretcher bond	9.67	5.00	10.15	1.60	1.75	27.05	1.98	39.19	m2	44.08
215mm thick; English bond	19.50	5.00	20.48	2.65	2.95	45.17	4.75	70.39	m2	79.19
327mm thick; English bond	29.34	5.00	30.81	3.15	3.60	54.35	7.51	92.66	m2	104.25
440mm thick; English bond	39.01	5.00	40.96	3.70	4.30	64.34	10.30	115.60	m2	130.05
Walls; bonding to old brickwork; cutting pockets; including extra material; vertical										
102mm thick; stretcher bond	15.21	5.00	15.97	3.65	3.80	60.37	5.37	81.71	m2	91.92
215mm thick; English bond	25.05	5.00	26.30	4.60	4.90	76.86	9.78	112.94	m2	127.06
Walls; curved 2m radius; including extra material; vertical										
102mm thick; stretcher bond	9.67	5.00	10.15	2.85	3.00	47.37	1.98	59.50	m2	66.94
215mm thick; English bond	19.50	5.00	20.48	5.20	5.50	86.61	4.75	111.83	m2	125.81
327mm thick; English bond	29.34	5.00	30.81	7.10	6.81	113.34	7.51	151.66	m2	170.61
Walls; curved 6m radius; including extra material; vertical										
102mm thick; stretcher bond	9.67	5.00	10.15	2.15	2.30	35.99	1.98	48.12	m2	54.14
215mm thick; English bond	19.50	5.00	20.48	3.75	4.05	63.04	4.75	88.27	m2	99.30
327mm thick; English bond	29.34	5.00	30.81	4.95	5.40	83.60	7.51	121.91	m2	137.15
Isolated piers; vertical										
215mm thick; English bond	19.50	5.00	20.48	4.70	5.00	78.48	4.75	103.71	m2	116.67
327mm thick; English bond	29.34	5.00	30.81	5.70	6.15	95.78	7.51	134.10	m2	150.86
440mm thick; English bond	39.01	5.00	40.96	6.60	7.20	111.46	10.30	162.72	m2	183.06
Chimney stacks; vertical										
440mm thick; English bond	39.01	5.00	40.96	6.60	7.20	111.46	10.30	162.72	m2	183.06
890mm thick; English bond	78.01	5.00	81.91	10.40	11.60	177.42	20.55	279.88	m2	314.87
Projections; vertical										
215mm wide x 112mm projection	2.19	5.00	2.30	0.30	0.35	5.23	0.48	8.01	m	9.01
215mm wide x 215mm projection	4.37	5.00	4.59	0.50	0.60	8.83	0.94	14.36	m	16.15
327mm wide x 112mm projection	3.28	5.00	3.44	0.45	0.55	8.01	0.70	12.16	m	13.68
327mm wide x 215mm projection	6.56	5.00	6.89	0.75	0.95	13.59	1.40	21.88	m	24.61
Projections; horizontal										
215mm wide x 112mm projection	2.19	5.00	2.30	0.30	0.35	5.23	0.48	8.01	m	9.01
215mm wide x 215mm projection	4.37	5.00	4.59	0.50	0.60	8.83	0.94	14.36	m	16.15
327mm wide x 112mm projection	3.28	5.00	3.44	0.45	0.55	8.01	0.70	12.16	m	13.68
327mm wide x 215mm projection	6.56	5.00	6.89	0.75	0.95	13.59	1.40	21.88	m	24.61
Projections; bonding to old brickwork; cutting pockets; including extra material; vertical										
215mm wide x 112mm projection	3.64	5.00	3.82	0.93	0.98	15.46	0.98	20.27	m	22.80

Labour hourly rates: (except Specialists) Craft Operatives £9.23 Labourer £7.02 Rates are national average prices. Refer to REGIONAL VARIATIONS for indicative levels of overall pricing in regions	MATERIALS			LABOUR				RATES		
	Del to Site £	Waste %	Material Cost £	Craft Optve Hrs	Lab Hrs	Labour Cost £	Sunds £	Nett Rate £	Unit	Gross Rate (+12.5%) £
F10: BRICK/BLOCK WALLING Cont.										
Common bricks, B.S.3921, Category M, 215 x 102.5 x 65mm, compressive strength 20.5 N/mm2; in cement-lime mortar (1:2:9) Cont.										
Projections; bonding to old brickwork; cutting pockets; including extra material; vertical Cont.										
215mm wide x 215mm projection	5.82	5.00	6.11	1.13	1.23	19.06	1.43	26.61	m	29.93
327mm wide x 112mm projection	5.47	5.00	5.74	1.40	1.50	23.45	1.46	30.66	m	34.49
327mm wide x 215mm projection	8.75	5.00	9.19	1.70	1.90	29.03	2.13	40.35	m	45.39
Projections; bonding to old brickwork; cutting pockets; including extra material; horizontal										
215mm wide x 112mm projection	3.64	5.00	3.82	0.93	0.93	15.11	0.98	19.91	m	22.40
215mm wide x 215mm projection	5.82	5.00	6.11	1.13	1.23	19.06	1.43	26.61	m	29.93
327mm wide x 112mm projection	5.47	5.00	5.74	1.40	1.50	23.45	1.46	30.66	m	34.49
327mm wide x 215mm projection	8.75	5.00	9.19	1.70	1.90	29.03	2.13	40.35	m	45.39
Arches including centering; flat										
112mm high on face; 215mm thick; width of exposed soffit 215mm; bricks-on-edge	2.19	5.00	2.30	1.50	1.25	22.62	2.07	26.99	m	30.36
215mm high on face; 112mm thick; width of exposed soffit 112mm; bricks-on-end	2.19	5.00	2.30	1.45	1.25	22.16	1.21	25.67	m	28.88
Arches including centering; semi-circular										
112mm high on face; 215mm thick; width of exposed soffit 215mm; bricks-on-edge	2.19	5.00	2.30	2.15	1.60	31.08	3.71	37.09	m	41.72
215mm high on face; 215mm thick; width of exposed soffit 215mm; bricks-on-edge	4.37	5.00	4.59	2.95	2.45	44.43	3.71	52.73	m	59.32
Closing cavities; vertical										
50mm wide with brickwork 102mm thick	1.07	5.00	1.12	0.40	0.40	6.50	0.24	7.86	m	8.85
75mm wide with brickwork 102mm thick	1.07	5.00	1.12	0.40	0.40	6.50	0.24	7.86	m	8.85
Closing cavities; horizontal										
50mm wide with slates	3.56	5.00	3.74	0.66	0.66	10.73	0.20	14.66	m	16.50
75mm wide with slates	3.56	5.00	3.74	0.66	0.66	10.73	0.20	14.66	m	16.50
50mm wide with brickwork 102mm thick	1.07	5.00	1.12	0.40	0.40	6.50	0.24	7.86	m	8.85
75mm wide with brickwork 102mm thick	1.07	5.00	1.12	0.40	0.40	6.50	0.24	7.86	m	8.85
Bonding ends to existing common brickwork; cutting pockets; extra material walls; bonding every third course										
102mm thick	0.73	5.00	0.77	0.37	0.37	6.01	0.15	6.93	m	7.80
215mm	1.45	5.00	1.52	0.66	0.66	10.73	0.19	12.44	m	13.99
327mm thick	2.19	5.00	2.30	0.95	0.95	15.44	0.24	17.98	m	20.22
Second hard stock bricks, B.S.3921, Category M, 215 x 102.5 x 65mm; in cement-lime mortar (1:2:9)										
Walls; vertical										
102mm thick; stretcher bond	18.82	5.00	19.76	1.45	1.60	24.62	1.98	46.36	m2	52.15
215mm thick; English bond	37.96	5.00	39.86	2.35	2.65	40.29	4.75	84.90	m2	95.51
327mm thick; English bond	57.10	5.00	59.95	2.85	3.30	49.47	7.51	116.94	m2	131.55
Walls; building against concrete (ties measured separately); vertical										
102mm thick; stretcher bond	18.82	5.00	19.76	1.60	1.75	27.05	1.98	48.79	m2	54.89
Walls; building against old brickwork; tie new to old with 3mm diameter galvanised twisted wire ties - 6/m2; vertical										
102mm thick; stretcher bond	18.82	5.00	19.76	1.60	1.75	27.05	2.55	49.36	m2	55.53
215mm thick; English bond	37.96	5.00	39.86	2.65	2.95	45.17	5.32	90.35	m2	101.64
Walls; bonding to stonework; including extra material; vertical										
102mm thick; stretcher bond	18.82	5.00	19.76	1.60	1.75	27.05	1.98	48.79	m2	54.89
215mm thick; English bond	37.96	5.00	39.86	2.65	2.95	45.17	4.75	89.78	m2	101.00
327mm thick; English bond	57.10	5.00	59.95	3.15	3.60	54.35	7.51	121.81	m2	137.04
440mm thick; English bond	75.92	5.00	79.72	3.70	4.30	64.34	10.30	154.35	m2	173.65
Walls; bonding to old brickwork; cutting pockets; including extra material; vertical										
102mm thick; stretcher bond	27.84	5.00	29.23	3.65	3.80	60.37	5.37	94.97	m2	106.84
215mm thick; English bond	48.74	5.00	51.18	4.60	4.90	76.86	9.78	137.81	m2	155.04
Walls; curved 2m radius; including extra material; vertical										
102mm thick; stretcher bond	18.82	5.00	19.76	2.85	3.00	47.37	1.98	69.11	m2	77.74
215mm thick; English bond	37.96	5.00	39.86	5.20	5.50	86.61	4.75	131.21	m2	147.62
327mm thick; English bond	57.10	5.00	59.95	7.10	6.81	113.34	7.51	180.80	m2	203.40
Walls; curved 6m radius; including extra material; vertical										
102mm thick; stretcher bond	18.82	5.00	19.76	2.15	2.30	35.99	1.98	57.73	m2	64.95
215mm thick; English bond	37.96	5.00	39.86	3.75	4.05	63.04	4.75	107.65	m2	121.11
327mm thick; English bond	57.10	5.00	59.95	4.95	5.40	83.60	7.51	151.06	m2	169.94
Isolated piers; vertical										
215mm thick; English bond	37.96	5.00	39.86	4.70	5.00	78.48	4.75	123.09	m2	138.48
327mm thick; English bond	57.10	5.00	59.95	5.70	6.15	95.78	7.51	163.25	m2	183.66
440mm thick; English bond	75.92	5.00	79.72	6.60	7.20	111.46	10.30	201.48	m2	226.66

Labour hourly rates: (except Specialists) Craft Operatives £9.23 Labourer £7.02 Rates are national average prices. Refer to REGIONAL VARIATIONS for indicative levels of overall pricing in regions	MATERIALS			LABOUR				RATES		
	Del to Site £	Waste %	Material Cost £	Craft Optve Hrs	Lab Hrs	Labour Cost £	Sunds £	Nett Rate £	Unit	Gross Rate (+12.5%) £
F10: BRICK/BLOCK WALLING Cont.										
Second hard stock bricks, B.S.3921, Category M, 215 x 102.5 x 65mm; in cement-lime mortar (1:2:9) Cont.										
Chimney stacks; vertical										
440mm thick; English bond	75.92	5.00	79.72	6.60	7.20	111.46	10.30	201.48	m2	226.66
890mm thick; English bond	151.81	5.00	159.40	10.40	11.60	177.42	20.55	357.37	m2	402.05
Projections; vertical										
215mm wide x 112mm projection	4.26	5.00	4.47	0.30	0.35	5.23	0.48	10.18	m	11.45
215mm wide x 215mm projection	8.50	5.00	8.93	0.50	0.60	8.83	0.94	18.69	m	21.03
327mm wide x 112mm projection	6.38	5.00	6.70	0.45	0.55	8.01	0.70	15.41	m	17.34
327mm wide x 215mm projection	12.76	5.00	13.40	0.75	0.95	13.59	1.40	28.39	m	31.94
Projections; horizontal										
215mm wide x 112mm projection	4.26	5.00	4.47	0.30	0.35	5.23	0.48	10.18	m	11.45
215mm wide x 215mm projection	8.50	5.00	8.93	0.50	0.60	8.83	0.94	18.69	m	21.03
327mm wide x 112mm projection	6.38	5.00	6.70	0.45	0.55	8.01	0.70	15.41	m	17.34
327mm wide x 215mm projection	12.76	5.00	13.40	0.75	0.95	13.59	1.40	28.39	m	31.94
Projections; bonding to old brickwork; cutting pockets; including extra material; vertical										
215mm wide x 112mm projection	7.10	5.00	7.46	0.93	0.98	15.46	0.98	23.90	m	26.89
215mm wide x 215mm projection	11.34	5.00	11.91	1.13	1.23	19.06	1.43	32.40	m	36.45
327mm wide x 112mm projection	10.64	5.00	11.17	1.40	1.50	23.45	1.46	36.08	m	40.59
327mm wide x 215mm projection	17.02	5.00	17.87	1.70	1.90	29.03	2.13	49.03	m	55.16
Projections; bonding to old brickwork; cutting pockets; including extra material; horizontal										
215mm wide x 112mm projection	7.10	5.00	7.46	0.93	0.93	15.11	0.98	23.55	m	26.49
215mm wide x 215mm projection	11.34	5.00	11.91	1.13	1.23	19.06	1.43	32.40	m	36.45
327mm wide x 112mm projection	10.64	5.00	11.17	1.40	1.50	23.45	1.46	36.08	m	40.59
327mm wide x 215mm projection	17.02	5.00	17.87	1.70	1.90	29.03	2.13	49.03	m	55.16
Arches including centering; flat										
112mm high on face; 215mm thick; width of exposed soffit 215mm; bricks-on-edge	4.26	5.00	4.47	1.50	1.25	22.62	2.07	29.16	m	32.81
215mm high on face; 112mm thick; width of exposed soffit 112mm; bricks-on-end	4.26	5.00	4.47	1.45	1.25	22.16	1.21	27.84	m	31.32
Arches including centering; semi-circular										
112mm high on face; 215mm thick; width of exposed soffit 215mm; bricks-on-edge	4.26	5.00	4.47	2.15	1.60	31.08	3.71	39.26	m	44.17
215mm high on face; 215mm thick; width of exposed soffit 215mm; bricks-on-edge	8.50	5.00	8.93	2.95	2.45	44.43	3.71	57.06	m	64.20
Closing cavities; vertical										
50mm wide with brickwork 102mm thick	2.08	5.00	2.18	0.40	0.40	6.50	0.24	8.92	m	10.04
75mm wide with brickwork 102mm thick	2.08	5.00	2.18	0.40	0.40	6.50	0.24	8.92	m	10.04
Closing cavities; horizontal										
50mm wide with slates	3.73	5.00	3.92	0.66	0.66	10.73	0.20	14.84	m	16.70
75mm wide with slates	3.73	5.00	3.92	0.66	0.66	10.73	0.20	14.84	m	16.70
50mm wide with brickwork 102mm thick	2.08	5.00	2.18	0.40	0.40	6.50	0.24	8.92	m	10.04
75mm wide with brickwork 102mm thick	2.08	5.00	2.18	0.40	0.40	6.50	0.24	8.92	m	10.04
Bonding ends to existing common brickwork; cutting pockets; extra material walls; bonding every third course										
102mm thick	1.42	5.00	1.49	0.37	0.37	6.01	0.15	7.65	m	8.61
215mm	2.84	5.00	2.98	0.66	0.66	10.73	0.19	13.90	m	15.63
327mm thick	4.26	5.00	4.47	0.95	0.95	15.44	0.24	20.15	m	22.67
Engineering bricks, B.S.3921, Category F, 215 x 102.5 x 65, Class B; in cement-lime mortar (1:2:9)										
Walls; vertical										
102mm thick; stretcher bond	19.28	5.00	20.24	1.60	1.75	27.05	1.98	49.28	m2	55.44
215mm thick; English bond	38.90	5.00	40.84	2.60	2.90	44.36	4.75	89.95	m2	101.19
327mm thick; English bond	58.51	5.00	61.44	3.15	3.60	54.35	7.51	123.29	m2	138.70
Walls; building against concrete (ties measured separately); vertical										
102mm thick; stretcher bond	19.28	5.00	20.24	1.75	1.90	29.49	1.98	51.71	m2	58.18
Walls; building against old brickwork; tie new to old with 3mm diameter galvanised twisted wire ties - 6/m2; vertical										
102mm thick; stretcher bond	19.28	5.00	20.24	1.75	1.90	29.49	2.55	52.28	m2	58.82
215mm thick; English bond	38.90	5.00	40.84	2.90	3.20	49.23	5.32	95.40	m2	107.32
Walls; bonding to stonework; including extra material; vertical										
102mm thick; stretcher bond	19.28	5.00	20.24	1.75	1.90	29.49	1.98	51.71	m2	58.18
215mm thick; English bond	38.90	5.00	40.84	2.90	3.20	49.23	4.75	94.83	m2	106.68
327mm thick; English bond	58.51	5.00	61.44	3.45	3.90	59.22	7.51	128.17	m2	144.19
440mm thick; English bond	77.79	5.00	81.68	4.05	4.65	70.02	10.30	162.00	m2	182.25
Walls; bonding to old brickwork; cutting pockets; including extra material; vertical										
102mm thick; stretcher bond	30.28	5.00	31.79	4.00	4.15	66.05	5.37	103.22	m2	116.12
215mm thick; English bond	49.90	5.00	52.40	5.05	5.35	84.17	9.78	146.34	m2	164.64

MASONRY

Labour hourly rates: (except Specialists) Craft Operatives £9.23 Labourer £7.02 Rates are national average prices. Refer to REGIONAL VARIATIONS for indicative levels of overall pricing in regions	MATERIALS			LABOUR				RATES		
	Del to Site £	Waste %	Material Cost £	Craft Optve Hrs	Lab Hrs	Labour Cost £	Sunds £	Nett Rate £	Unit	Gross Rate (+12.5%) £
F10: BRICK/BLOCK WALLING Cont.										
Engineering bricks, B.S.3921, Category F, 215 x 102.5 x 65, Class B; in cement-lime mortar (1:2:9) Cont.										
Walls; curved 2m radius; including extra material; vertical										
102mm thick; stretcher bond	19.28	5.00	20.24	3.15	3.30	52.24	1.98	74.46	m2	83.77
215mm thick; English bond	38.90	5.00	40.84	5.70	6.00	94.73	4.75	140.33	m2	157.87
327mm thick; English bond	58.51	5.00	61.44	7.80	8.25	129.91	7.51	198.85	m2	223.71
Walls; curved 6m radius; including extra material; vertical										
102mm thick; stretcher bond	19.28	5.00	20.24	2.35	2.50	39.24	1.98	61.46	m2	69.15
215mm thick; English bond	38.90	5.00	40.84	4.10	4.40	68.73	4.75	114.33	m2	128.62
327mm thick; English bond	58.51	5.00	61.44	5.45	5.90	91.72	7.51	160.67	m2	180.75
Isolated piers; vertical										
215mm thick; English bond	38.90	5.00	40.84	5.15	5.45	85.79	4.75	131.39	m2	147.81
327mm thick; English bond	58.51	5.00	61.44	6.25	6.70	104.72	7.51	173.67	m2	195.38
440mm thick; English bond	77.79	5.00	81.68	7.25	7.85	122.02	10.30	214.00	m2	240.75
Chimney stacks; vertical										
440mm thick; English bond	77.79	5.00	81.68	7.25	7.85	122.02	10.30	214.00	m2	240.75
890mm thick; English bond	155.57	5.00	163.35	11.45	12.65	194.49	20.55	378.38	m2	425.68
Projections; vertical										
215mm wide x 112mm projection	4.36	5.00	4.58	0.33	0.38	5.71	0.48	10.77	m	12.12
215mm wide x 215mm projection	8.72	5.00	9.16	0.55	0.65	9.64	0.94	19.74	m	22.20
327mm wide x 112mm projection	6.53	5.00	6.86	0.50	0.60	8.83	0.70	16.38	m	18.43
327mm wide x 215mm projection	13.08	5.00	13.73	0.83	1.03	14.89	1.40	30.03	m	33.78
Projections; horizontal										
215mm wide x 112mm projection	4.36	5.00	4.58	0.33	0.38	5.71	0.48	10.77	m	12.12
215mm wide x 215mm projection	8.72	5.00	9.16	0.55	0.65	9.64	0.94	19.74	m	22.20
327mm wide x 112mm projection	6.53	5.00	6.86	0.50	0.60	8.83	0.70	16.38	m	18.43
327mm wide x 215mm projection	13.08	5.00	13.73	0.83	1.03	14.89	1.40	30.03	m	33.78
Projections; bonding to old brickwork; cutting pockets; including extra material; vertical										
215mm wide x 112mm projection	8.50	5.00	8.93	1.02	1.07	16.93	0.98	26.83	m	30.18
215mm wide x 215mm projection	13.08	5.00	13.73	1.24	1.34	20.85	1.43	36.02	m	40.52
327mm wide x 112mm projection	10.79	5.00	11.33	1.54	1.64	25.73	1.46	38.52	m	43.33
327mm wide x 215mm projection	17.33	5.00	18.20	1.87	2.07	31.79	2.13	52.12	m	58.63
Projections; bonding to old brickwork; cutting pockets; including extra material; horizontal										
215mm wide x 112mm projection	8.50	5.00	8.93	1.02	1.07	16.93	0.98	26.83	m	30.18
215mm wide x 215mm projection	13.08	5.00	13.73	1.24	1.34	20.85	1.43	36.02	m	40.52
327mm wide x 112mm projection	10.79	5.00	11.33	1.54	1.64	25.73	1.46	38.52	m	43.33
327mm wide x 215mm projection	17.33	5.00	18.20	1.87	2.07	31.79	2.13	52.12	m	58.63
Arches including centering; flat										
112mm high on face; 215mm thick; width of exposed soffit 215mm; bricks-on-edge	4.36	5.00	4.58	1.60	1.30	23.89	2.07	30.54	m	34.36
215mm high on face; 112mm thick; width of exposed soffit 112mm; bricks-on-end	4.36	5.00	4.58	1.55	1.35	23.78	1.21	29.57	m	33.27
Arches including centering; semi-circular										
112mm high on face; 215mm thick; width of exposed soffit 215mm; bricks-on-edge	4.36	5.00	4.58	2.25	1.65	32.35	3.71	40.64	m	45.72
215mm high on face; 215mm thick; width of exposed soffit 215mm; bricks-on-edge	8.72	5.00	9.16	3.10	2.55	46.51	3.71	59.38	m	66.80
Closing cavities; vertical										
50mm wide with brickwork 102mm thick	2.12	5.00	2.23	0.44	0.44	7.15	0.24	9.62	m	10.82
75mm wide with brickwork 102mm thick	2.12	5.00	2.23	0.44	0.44	7.15	0.24	9.62	m	10.82
Closing cavities; horizontal										
50mm wide with slates	3.73	5.00	3.92	0.66	0.66	10.73	0.20	14.84	m	16.70
75mm wide with slates	3.73	5.00	3.92	0.66	0.66	10.73	0.20	14.84	m	16.70
50mm wide with brickwork 102mm thick	3.33	5.00	3.50	0.44	0.44	7.15	0.24	10.89	m	12.25
75mm wide with brickwork 102mm thick	3.33	5.00	3.50	0.44	0.44	7.15	0.24	10.89	m	12.25
Bonding ends to existing common brickwork; cutting pockets; extra material										
walls; bonding every third course										
102mm thick.................................	1.31	5.00	1.38	0.41	0.41	6.66	0.15	8.19	m	9.21
215mm......................................	2.62	5.00	2.75	0.73	0.73	11.86	0.19	14.80	m	16.65
327mm thick.................................	3.93	5.00	4.13	0.95	0.95	15.44	0.24	19.80	m	22.28
Engineering bricks, B.S.3921, Category F, 215 x 102.5 x 65mm, Class A; in cement-lime mortar (1:2:9)										
Walls; vertical										
102mm thick; stretcher bond	32.40	5.00	34.02	1.60	1.75	27.05	1.98	63.05	m2	70.93
215mm thick; English bond	65.34	5.00	68.61	2.60	2.90	44.36	4.75	117.71	m2	132.43
327mm thick; English bond	98.27	5.00	103.18	3.15	3.60	54.35	7.51	165.04	m2	185.67
Walls; building against concrete (ties measured separately); vertical										
102mm thick; stretcher bond	32.40	5.00	34.02	1.75	1.90	29.49	1.98	65.49	m2	73.68

Labour hourly rates: (except Specialists) Craft Operatives £9.23 Labourer £7.02 Rates are national average prices. Refer to REGIONAL VARIATIONS for indicative levels of overall pricing in regions	MATERIALS			LABOUR				RATES		
	Del to Site £	Waste %	Material Cost £	Craft Optve Hrs	Lab Hrs	Labour Cost £	Sunds £	Nett Rate £	Unit	Gross Rate (+12.5%) £
F10: BRICK/BLOCK WALLING Cont.										
Engineering bricks, B.S.3921, Category F, 215 x 102.5 x 65mm, Class A; in cement-lime mortar (1:2:9) Cont.										
Walls; building against old brickwork; tie new to old with 3mm diameter galvanised twisted wire ties – 6/m2; vertical										
102mm thick; stretcher bond	32.40	5.00	34.02	1.75	1.90	29.49	2.55	66.06	m2	74.32
215mm thick; English bond	65.34	5.00	68.61	2.90	3.20	49.23	5.32	123.16	m2	138.55
Walls; bonding to stonework; including extra material; vertical										
102mm thick; stretcher bond	32.40	5.00	34.02	1.75	1.90	29.49	1.98	65.49	m2	73.68
215mm thick; English bond	65.34	5.00	68.61	2.90	3.20	49.23	4.75	122.59	m2	137.91
327mm thick; English bond	98.27	5.00	103.18	3.45	3.90	59.22	7.51	169.91	m2	191.15
440mm thick; English bond	130.67	5.00	137.20	4.05	4.65	70.02	10.30	217.53	m2	244.72
Walls; bonding to old brickwork; cutting pockets; including extra material; vertical										
102mm thick; stretcher bond	50.85	5.00	53.39	4.00	4.15	66.05	5.37	124.82	m2	140.42
215mm thick; English bond	87.08	5.00	91.43	5.05	5.35	84.17	9.78	185.38	m2	208.56
Walls; curved 2m radius; including extra material; vertical										
102mm thick; stretcher bond	32.40	5.00	34.02	3.15	3.30	52.24	1.98	88.24	m2	99.27
215mm thick; English bond	65.34	5.00	68.61	5.70	6.00	94.73	4.75	168.09	m2	189.10
327mm thick; English bond	98.27	5.00	103.18	7.80	8.25	129.91	7.51	240.60	m2	270.68
Walls; curved 6m radius; including extra material; vertical										
102mm thick; stretcher bond	32.40	5.00	34.02	2.35	2.50	39.24	1.98	75.24	m2	84.65
215mm thick; English bond	65.34	5.00	68.61	4.10	4.40	68.73	4.75	142.09	m2	159.85
327mm thick; English bond	98.27	5.00	103.18	5.45	5.90	91.72	7.51	202.41	m2	227.72
Isolated piers; vertical										
215mm thick; English bond	65.34	5.00	68.61	5.15	5.45	85.79	4.75	159.15	m2	179.04
327mm thick; English bond	98.27	5.00	103.18	6.25	6.70	104.72	7.51	215.41	m2	242.34
440mm thick; English bond	130.67	5.00	137.20	7.25	7.85	122.02	10.30	269.53	m2	303.22
Chimney stacks; vertical										
440mm thick; English bond	130.67	5.00	137.20	7.25	7.85	122.02	10.30	269.53	m2	303.22
890mm thick; English bond	261.34	5.00	274.41	11.45	12.65	194.49	20.55	489.44	m2	550.62
Projections; vertical										
215mm wide x 112mm projection	7.33	5.00	7.70	0.33	0.38	5.71	0.48	13.89	m	15.63
215mm wide x 215mm projection	14.64	5.00	15.37	0.55	0.65	9.64	0.94	25.95	m	29.20
327mm wide x 112mm projection	10.98	5.00	11.53	0.50	0.60	8.83	0.70	21.06	m	23.69
327mm wide x 215mm projection	21.97	5.00	23.07	0.83	1.03	14.89	1.40	39.36	m	44.28
Projections; horizontal										
215mm wide x 112mm projection	7.33	5.00	7.70	0.33	0.38	5.71	0.48	13.89	m	15.63
215mm wide x 215mm projection	14.64	5.00	15.37	0.55	0.65	9.64	0.94	25.95	m	29.20
327mm wide x 112mm projection	10.98	5.00	11.53	0.50	0.60	8.83	0.70	21.06	m	23.69
327mm wide x 215mm projection	21.97	5.00	23.07	0.83	1.03	14.89	1.40	39.36	m	44.28
Projections; bonding to old brickwork; cutting pockets; including extra material; vertical										
215mm wide x 112mm projection	14.27	5.00	14.98	1.02	1.07	16.93	0.98	32.89	m	37.00
215mm wide x 215mm projection	21.41	5.00	22.48	1.24	1.34	20.85	1.43	44.76	m	50.36
327mm wide x 112mm projection	18.12	5.00	19.03	1.54	1.64	25.73	1.46	46.21	m	51.99
327mm wide x 215mm projection	17.18	5.00	18.04	1.87	2.07	31.79	2.13	51.96	m	58.46
Projections; bonding to old brickwork; cutting pockets; including extra material; horizontal										
215mm wide x 112mm projection	14.27	5.00	14.98	1.02	1.07	16.93	0.98	32.89	m	37.00
215mm wide x 215mm projection	21.41	5.00	22.48	1.24	1.34	20.85	1.43	44.76	m	50.36
327mm wide x 112mm projection	18.12	5.00	19.03	1.54	1.64	25.73	1.46	46.21	m	51.99
327mm wide x 215mm projection	17.18	5.00	18.04	1.87	2.07	31.79	2.13	51.96	m	58.46
Arches including centering; flat										
112mm high on face; 215mm thick; width of exposed soffit 215mm; bricks-on-edge	6.94	5.00	7.29	1.60	1.30	23.89	2.07	33.25	m	37.41
215mm high on face; 112mm thick; width of exposed soffit 112mm; bricks-on-end	6.94	5.00	7.29	1.55	1.35	23.78	1.21	32.28	m	36.32
Arches including centering; semi-circular										
112mm high on face; 215mm thick; width of exposed soffit 215mm; bricks-on-edge	6.94	5.00	7.29	2.25	1.65	32.35	3.71	43.35	m	48.77
215mm high on face; 215mm thick; width of exposed soffit 215mm; bricks-on-edge	13.88	5.00	14.57	3.10	2.55	46.51	3.71	64.80	m	72.90
Closing cavities; vertical										
50mm wide with brickwork 102mm thick	3.39	5.00	3.56	0.44	0.44	7.15	0.24	10.95	m	12.32
75mm wide with brickwork 102mm thick	3.39	5.00	3.56	0.44	0.44	7.15	0.24	10.95	m	12.32
Closing cavities; horizontal										
50mm wide with slates	3.73	5.00	3.92	0.66	0.66	10.73	0.20	14.84	m	16.70
75mm wide with slates	3.73	5.00	3.92	0.66	0.66	10.73	0.20	14.84	m	16.70
50mm wide with brickwork 102mm thick	3.56	5.00	3.74	0.44	0.44	7.15	0.24	11.13	Unit	12.52
75mm wide with brickwork 102mm thick	3.56	5.00	3.74	0.44	0.44	7.15	0.24	11.13	m	12.52
Bonding ends to existing common brickwork; cutting pockets; extra material walls; bonding every third course										
102mm thick	2.08	5.00	2.18	0.41	0.41	6.66	0.15	9.00	m	10.12

MASONRY

Labour hourly rates: (except Specialists) Craft Operatives £9.23 Labourer £7.02 Rates are national average prices. Refer to REGIONAL VARIATIONS for indicative levels of overall pricing in regions	MATERIALS			LABOUR				RATES		
	Del to Site £	Waste %	Material Cost £	Craft Optve Hrs	Lab Hrs	Labour Cost £	Sunds £	Nett Rate £	Unit	Gross Rate (+12.5%) £
F10: BRICK/BLOCK WALLING Cont.										
Engineering bricks, B.S.3921, Category F, 215 x 102.5 x 65mm, Class A; in cement-lime mortar (1:2:9) Cont.										
Bonding ends to existing common brickwork; cutting pockets; extra material Cont.										
walls; bonding every third course Cont.										
215mm	4.17	5.00	4.38	0.73	0.73	11.86	0.19	16.43	m	18.48
327mm thick.................................	6.25	5.00	6.56	0.95	0.95	15.44	0.24	22.24	m	25.02
For other mortar mixes, ADD as follows										
For each half brick thickness										
1:1:6 cement-lime mortar	-	-	-	-	-	-	0.05	0.05	m2	0.06
1:4 cement mortar	-	-	-	-	-	-	0.01	0.01	m2	0.01
1:3 cement mortar	-	-	-	-	-	-	0.17	0.17	m2	0.19
For using Sulphate Resisting cement in lieu of Portland cement, ADD as follows										
For each half brick thickness										
1:2:9 cement-lime mortar	-	-	-	-	-	-	0.15	0.15	m2	0.17
1:1:6 cement-lime mortar	-	-	-	-	-	-	0.23	0.23	m2	0.26
1:4 cement mortar	-	-	-	-	-	-	0.36	0.36	m2	0.41
1:3 cement mortar	-	-	-	-	-	-	0.46	0.46	m2	0.52
Staffordshire wirecut bricks, B.S.3921,Category F, 215 x 102.5 x 65mm; in cement mortar (1:3)										
Walls; vertical										
215mm thick; English bond	50.40	5.00	52.92	2.60	2.90	44.36	5.26	102.54	m2	115.35
Staffordshire pressed bricks, B.S.3921,Category F, 215 x 102.5 x 65mm; in cement mortar (1:3)										
Walls; vertical										
215mm thick; English bond	48.43	5.00	50.85	2.60	2.90	44.36	5.26	100.47	m2	113.03
Composite walling of Staffordshire bricks, B.S.3921, Category F, 215 x 102.5 x 65mm; in cement mortar (1:3); wirecut bricks backing; pressed bricks facing; weather struck pointing as work proceeds										
Walls; vertical										
English bond; facework one side	48.51	5.00	50.94	2.90	3.20	49.23	5.26	105.43	m2	118.60
Flemish bond; facework one side	48.68	5.00	51.11	2.90	3.20	49.23	5.26	105.61	m2	118.81
Common bricks, B.S.3921, Category M, 215 x 102.5 x 65mm, compressive strength 20.5 N/mm2; in cement-lime mortar (1:2:9); flush smooth pointing as work proceeds										
Walls; vertical										
stretcher bond; facework one side										
102mm thick..............................	9.67	5.00	10.15	1.75	1.90	29.49	1.98	41.62	m2	46.83
stretcher bond; facework both sides										
102mm thick..............................	9.67	5.00	10.15	2.05	2.20	34.37	1.98	46.50	m2	52.31
English bond; facework one side										
215mm thick..............................	19.50	5.00	20.48	2.65	2.95	45.17	4.75	70.39	m2	79.19
English bond; facework both sides										
215mm thick..............................	19.50	5.00	20.48	2.95	3.25	50.04	4.75	75.27	m2	84.68
extra; special bricks; vertical angles; single bullnose; B.S.4729, type BN.1.1.............	4.29	5.00	4.50	0.25	0.25	4.06	-	8.57	m	9.64
extra; special bricks; vertical angles; squint; B.S.4729, type AN.1.1......................	5.39	5.00	5.66	0.25	0.25	4.06	-	9.72	m	10.94
extra; special bricks; intersections; birdsmouth; B.S.4729, type AN.4.1.............	19.53	5.00	20.51	0.25	0.25	4.06	-	24.57	m	27.64
Engineering bricks, B.S.3921, Category F, 215 x 102.5 x 65mm, Class B; in cement-lime mortar (1:2:9); flush smooth pointing as work proceeds										
Walls; vertical										
stretcher bond; facework one side										
102mm thick..............................	19.28	5.00	20.24	1.90	2.05	31.93	1.98	54.15	m2	60.92
stretcher bond; facework both sides										
102mm thick..............................	19.28	5.00	20.24	2.20	2.35	36.80	1.98	59.03	m2	66.41
English bond; facework one side										
215mm thick..............................	38.90	5.00	40.84	2.90	3.20	49.23	4.75	94.83	m2	106.68
English bond; facework both sides										
215mm thick..............................	38.90	5.00	40.84	3.20	3.50	54.11	4.75	99.70	m2	112.16
Engineering bricks, B.S.3921, Category F, 215 x 102.5 x 65mm, Class A; in cement-lime mortar (1:2:9); flush smooth pointing as work proceeds										
Walls; vertical										
stretcher bond; facework one side										
102mm thick..............................	32.40	5.00	34.02	1.90	2.05	31.93	1.98	67.93	m2	76.42
stretcher bond; facework both sides										
102mm thick..............................	32.40	5.00	34.02	2.20	2.35	36.80	1.98	72.80	m2	81.90
English bond; facework one side										
215mm thick..............................	65.34	5.00	68.61	2.90	3.20	49.23	4.75	122.59	m2	137.91
English bond; facework both sides										
215mm thick..............................	65.34	5.00	68.61	3.20	3.50	54.11	4.75	127.46	m2	143.40

Labour hourly rates: (except Specialists) Craft Operatives £9.23 Labourer £7.02 Rates are national average prices. Refer to REGIONAL VARIATIONS for indicative levels of overall pricing in regions	MATERIALS			LABOUR				RATES		
	Del to Site £	Waste %	Material Cost £	Craft Optve Hrs	Lab Hrs	Labour Cost £	Sunds £	Nett Rate £	Unit	Gross Rate (+12.5%) £
F10: BRICK/BLOCK WALLING Cont.										
Facing bricks, second hard stocks, B.S.3921, Category M, 215 x 102.5 x 65mm; in cement-lime mortar (1:2:9); flush smooth pointing as work proceeds										
Walls; vertical										
stretcher bond; facework one side										
102mm thick.........................	34.23	5.00	35.94	1.75	1.90	29.49	1.98	67.41	m2	75.84
stretcher bond; facework both sides										
102mm thick.........................	34.23	5.00	35.94	2.05	2.20	34.37	1.98	72.29	m2	81.32
English bond; facework one side										
215mm thick.........................	69.04	5.00	72.49	2.65	2.95	45.17	4.75	122.41	m2	137.71
English bond; facework both sides										
215mm thick.........................	69.04	5.00	72.49	2.95	3.25	50.04	4.75	127.29	m2	143.20
Facing bricks p.c. £150.00 per 1000, 215 x 102.5 x 65mm; in cement-lime mortar (1:2:9); flush smooth pointing as work proceeds										
Walls; vertical										
stretcher bond; facework one side										
102mm thick.........................	8.85	5.00	9.29	1.75	1.90	29.49	1.98	40.76	m2	45.86
stretcher bond; facework both sides										
102mm thick.........................	8.85	5.00	9.29	2.05	2.20	34.37	1.98	45.64	m2	51.34
English bond; facework both sides										
215mm thick.........................	17.85	5.00	18.74	2.65	2.95	45.17	4.75	68.66	m2	77.24
Flemish bond; facework both sides										
215mm thick.........................	17.85	5.00	18.74	2.95	3.25	50.04	4.75	73.54	m2	82.73
extra; special bricks; vertical angles; squint; B.S.4729, type AN.1.1....................	18.88	5.00	19.82	0.25	0.25	4.06	-	23.89	m	26.87
extra; special bricks; intersections; birdsmouth; B.S.4729, type AN.4.1.............	58.52	5.00	61.45	0.25	0.25	4.06	-	65.51	m	73.70
Walls; building overhand; vertical										
stretcher bond; facework one side										
102mm thick.........................	8.85	5.00	9.29	2.15	2.30	35.99	1.98	47.26	m2	53.17
stretcher bond; facework both sides										
102mm thick.........................	8.85	5.00	9.29	2.45	2.60	40.87	1.98	52.14	m2	58.66
English bond; facework both sides										
215mm thick.........................	17.85	5.00	18.74	3.05	3.35	51.67	4.75	75.16	m2	84.56
Flemish bond; facework both sides										
215mm thick.........................	17.85	5.00	18.74	3.35	3.65	56.54	4.75	80.04	m2	90.04
Arches including centering; flat										
112mm high on face; 215mm thick; width of exposed soffit 112mm; bricks-on-edge	2.00	5.00	2.10	1.05	1.05	17.06	1.98	21.14	m	23.79
112mm high on face; 215mm thick; width of exposed soffit 215mm; bricks-on-edge	2.00	5.00	2.10	1.50	1.25	22.62	1.98	26.70	m	30.04
215mm high on face; 112mm thick; width of exposed soffit 112mm; bricks-on-end	2.00	5.00	2.10	1.45	1.25	22.16	1.21	25.47	m	28.65
Arches including centering; segmental										
215mm high on face; 215mm thick; width of exposed soffit 112mm; bricks-on-edge	4.00	5.00	4.20	2.40	2.15	37.24	3.71	45.16	m	50.80
215mm high on face; 215mm thick; width of exposed soffit 215mm; bricks-on-edge	4.00	5.00	4.20	2.95	2.45	44.43	3.71	52.34	m	58.88
Arches including centering; semi-circular										
215mm high on face; 215mm thick; width of exposed soffit 112mm; bricks-on-edge	4.00	5.00	4.20	2.40	2.15	37.24	3.71	45.16	m	50.80
215mm high on face; 215mm thick; width of exposed soffit 215mm; bricks-on-edge	4.00	5.00	4.20	2.95	2.45	44.43	3.71	52.34	m	58.88
Facework copings; bricks-on-edge; pointing top and each side										
215 x 102.5mm; horizontal	2.00	5.00	2.10	0.70	0.75	11.73	0.52	14.35	m	16.14
215 x 102.5mm; with oversailing course and cement fillets both sides; horizontal	2.00	5.00	2.10	1.10	1.15	18.23	0.90	21.23	m	23.88
327 x 102.5mm; horizontal	3.00	5.00	3.15	1.05	1.10	17.41	0.76	21.32	m	23.99
327 x 102.5mm; with oversailing course and cement fillets both sides; horizontal	3.00	5.00	3.15	1.50	1.55	24.73	1.14	29.02	m	32.64
extra; galvanised iron coping cramp at angle or end...	1.85	5.00	1.94	0.04	0.04	0.65	-	2.59	nr	2.92
Facing bricks p.c. £200.00 per 1000, 215 x 102.5 x 65mm; in cement-lime mortar (1:2:9); flush smooth pointing as work proceeds										
Walls; vertical										
stretcher bond; facework one side										
102mm thick.........................	11.80	5.00	12.39	1.75	1.90	29.49	1.98	43.86	m2	49.34
stretcher bond; facework both sides										
102mm thick.........................	11.80	5.00	12.39	2.05	2.20	34.37	1.98	48.74	m2	54.83
English bond; facework both sides										
215mm thick.........................	23.80	5.00	24.99	2.65	2.95	45.17	4.75	74.91	m2	84.27
Flemish bond; facework both sides										
215mm thick.........................	23.80	5.00	24.99	2.95	3.25	50.04	4.75	79.78	m2	89.76
extra; special bricks; vertical angles; squint; B.S.4729, type AN.1.1.....................	17.94	5.00	18.84	0.25	0.25	4.06	-	22.90	m	25.76
extra; special bricks; intersections; birdsmouth; B.S.4729, type AN.4.1.............	57.64	5.00	60.52	0.25	0.25	4.06	-	64.58	m	72.66
Walls; building overhand; vertical										
stretcher bond; facework one side										
102mm thick.........................	11.80	5.00	12.39	2.15	2.30	35.99	1.98	50.36	m2	56.66

MASONRY

Labour hourly rates: (except Specialists) Craft Operatives £9.23 Labourer £7.02 Rates are national average prices. Refer to REGIONAL VARIATIONS for indicative levels of overall pricing in regions	MATERIALS			LABOUR				RATES		
	Del to Site £	Waste %	Material Cost £	Craft Optve Hrs	Lab Hrs	Labour Cost £	Sunds £	Nett Rate £	Unit	Gross Rate (+12.5%) £
F10: BRICK/BLOCK WALLING Cont.										
Facing bricks p.c. £200.00 per 1000, 215 x 102.5 x 65mm; in cement-lime mortar (1:2:9); flush smooth pointing as work proceeds Cont.										
Walls; building overhand; vertical Cont.										
stretcher bond; facework both sides										
102mm thick.................................	11.80	5.00	**12.39**	2.45	2.60	**40.87**	1.98	**55.24**	m2	62.14
English bond; facework both sides										
215mm thick.................................	23.80	5.00	**24.99**	3.05	3.35	**51.67**	4.75	**81.41**	m2	91.58
Flemish bond; facework both sides										
215mm thick.................................	23.80	5.00	**24.99**	3.35	3.65	**56.54**	4.75	**86.28**	m2	97.07
Arches including centering; flat										
112mm high on face; 215mm thick; width of exposed soffit 112mm; bricks-on-edge	2.67	5.00	**2.80**	1.05	1.05	**17.06**	1.98	**21.85**	m	24.58
112mm high on face; 215mm thick; width of exposed soffit 215mm; bricks-on-edge	2.67	5.00	**2.80**	1.50	1.25	**22.62**	1.98	**27.40**	m	30.83
215mm high on face; 112mm thick; width of exposed soffit 112mm; bricks-on-end	2.67	5.00	**2.80**	1.45	1.25	**22.16**	~1.21	**26.17**	m	29.44
Arches including centering; segmental										
215mm high on face; 215mm thick; width of exposed soffit 112mm; bricks-on-edge	5.33	5.00	**5.60**	2.40	2.15	**37.24**	3.71	**46.55**	m	52.37
215mm high on face; 215mm thick; width of exposed soffit 215mm; bricks-on-edge	5.33	5.00	**5.60**	2.95	2.45	**44.43**	3.71	**53.73**	m	60.45
Arches including centering; semi-circular										
215mm high on face; 215mm thick; width of exposed soffit 112mm; bricks-on-edge	5.33	5.00	**5.60**	2.40	2.15	**37.24**	3.71	**46.55**	m	52.37
215mm high on face; 215mm thick; width of exposed soffit 215mm; bricks-on-edge	5.33	5.00	**5.60**	2.95	2.45	**44.43**	3.71	**53.73**	m	60.45
Facework copings; bricks-on-edge; pointing top and each side										
215 x 102.5mm; horizontal	2.67	5.00	**2.80**	0.70	0.75	**11.73**	0.52	**15.05**	m	16.93
215 x 102.5mm; with oversailing course and cement fillets both sides; horizontal	2.67	5.00	**2.80**	1.10	1.15	**18.23**	0.90	**21.93**	m	24.67
327 x 102.5mm; horizontal	4.00	5.00	**4.20**	1.05	1.10	**17.41**	0.76	**22.37**	m	25.17
327 x 102.5mm; with oversailing course and cement fillets both sides; horizontal	4.00	5.00	**4.20**	1.50	1.55	**24.73**	1.14	**30.07**	m	33.82
extra; galvanised iron coping cramp at angle or end	1.85	5.00	**1.94**	0.04	0.04	**0.65**	-	**2.59**	nr	2.92
Facing bricks p.c. £250.00 per 1000, 215 x 102.5 x 65mm; in cement-lime mortar (1:2:9); flush smooth pointing as work proceeds										
Walls; vertical										
stretcher bond; facework one side										
102mm thick.................................	14.75	5.00	**15.49**	1.75	1.90	**29.49**	1.98	**46.96**	m2	52.83
stretcher bond; facework both sides										
102mm thick.................................	14.75	5.00	**15.49**	2.05	2.20	**34.37**	1.98	**51.83**	m2	58.31
English bond; facework both sides										
215mm thick.................................	29.75	5.00	**31.24**	2.65	2.95	**45.17**	4.75	**81.16**	m2	91.30
Flemish bond; facework both sides										
215mm thick.................................	29.75	5.00	**31.24**	2.95	3.25	**50.04**	4.75	**86.03**	m2	96.78
extra; special bricks; vertical angles; squint; B.S.4729, type AN.1.1.............	16.86	5.00	**17.70**	0.25	0.25	**4.06**	-	**21.77**	m	24.49
extra; special bricks; intersections; birdsmouth; B.S.4729, type AN.4.1.............	56.63	5.00	**59.46**	0.25	0.25	**4.06**	-	**63.52**	m	71.46
Walls; building overhand; vertical										
stretcher bond; facework one side										
102mm thick.................................	14.75	5.00	**15.49**	2.15	2.30	**35.99**	1.98	**53.46**	m2	60.14
stretcher bond; facework both sides										
102mm thick.................................	14.75	5.00	**15.49**	2.45	2.60	**40.87**	1.98	**58.33**	m2	65.62
English bond; facework both sides										
215mm thick.................................	29.75	5.00	**31.24**	3.05	3.35	**51.67**	4.75	**87.66**	m2	98.61
Flemish bond; facework both sides										
215mm thick.................................	29.75	5.00	**31.24**	3.35	3.65	**56.54**	4.75	**92.53**	m2	104.10
Arches including centering; flat										
112mm high on face; 215mm thick; width of exposed soffit 112mm; bricks-on-edge	3.33	5.00	**3.50**	1.05	1.05	**17.06**	1.98	**22.54**	m	25.36
112mm high on face; 215mm thick; width of exposed soffit 215mm; bricks-on-edge	3.33	5.00	**3.50**	1.50	1.25	**22.62**	1.98	**28.10**	m	31.61
215mm high on face; 112mm thick; width of exposed soffit 112mm; bricks-on-end	3.33	5.00	**3.50**	1.45	1.25	**22.16**	1.21	**26.86**	m	30.22
Arches including centering; segmental										
215mm high on face; 215mm thick; width of exposed soffit 112mm; bricks-on-edge	6.67	5.00	**7.00**	2.40	2.15	**37.24**	3.71	**47.96**	m	53.95
215mm high on face; 215mm thick; width of exposed soffit 215mm; bricks-on-edge	6.67	5.00	**7.00**	2.95	2.45	**44.43**	3.71	**55.14**	m	62.03
Arches including centering; semi-circular										
215mm high on face; 215mm thick; width of exposed soffit 112mm; bricks-on-edge	6.67	5.00	**7.00**	2.40	2.15	**37.24**	3.71	**47.96**	m	53.95
215mm high on face; 215mm thick; width of exposed soffit 215mm; bricks-on-edge	6.67	5.00	**7.00**	2.95	2.45	**44.43**	3.71	**55.14**	m	62.03
Facework copings; bricks-on-edge; pointing top and each side										
215 x 102.5mm; horizontal	3.33	5.00	**3.50**	0.70	0.75	**11.73**	0.52	**15.74**	m	17.71

Labour hourly rates: (except Specialists) Craft Operatives £9.23 Labourer £7.02 Rates are national average prices. Refer to REGIONAL VARIATIONS for indicative levels of overall pricing in regions	MATERIALS			LABOUR				RATES		
	Del to Site £	Waste %	Material Cost £	Craft Optve Hrs	Lab Hrs	Labour Cost £	Sunds £	Nett Rate £	Unit	Gross Rate (+12.5%) £
F10: BRICK/BLOCK WALLING Cont.										
Facing bricks p.c. £250.00 per 1000, 215 x 102.5 x 65mm; in cement-lime mortar (1:2:9); flush smooth pointing as work proceeds Cont.										
Facework copings; bricks-on-edge; pointing top and each side Cont.										
215 x 102.5mm; with oversailing course and cement fillets both sides; horizontal	3.33	5.00	3.50	1.10	1.15	18.23	0.90	22.62	m	25.45
327 x 102.5mm; horizontal	5.00	5.00	5.25	1.05	1.10	17.41	0.76	23.42	m	26.35
327 x 102.5mm; with oversailing course and cement fillets both sides; horizontal	5.00	5.00	5.25	1.50	1.55	24.73	1.14	31.12	m	35.01
extra; galvanised iron coping cramp at angle or end	1.85	5.00	1.94	0.04	0.04	0.65	-	2.59	nr	2.92
Facing bricks p.c. £300.00 per 1000, 215 x 102.5 x 65mm; in cement-lime mortar (1:2:9); flush smooth pointing as work proceeds										
Walls; vertical										
stretcher bond; facework one side										
102mm thick	17.70	5.00	18.59	1.75	1.90	29.49	1.98	50.06	m2	56.31
stretcher bond; facework both sides										
102mm thick	17.70	5.00	18.59	2.05	2.20	34.37	1.98	54.93	m2	61.80
English bond; facework both sides										
215mm thick	35.70	5.00	37.48	2.65	2.95	45.17	4.75	87.40	m2	98.33
Flemish bond; facework both sides										
215mm thick	35.70	5.00	37.48	2.95	3.25	50.04	4.75	92.28	m2	103.81
extra; special bricks; vertical angles; squint; B.S.4729, type AN.1.1	15.93	5.00	16.73	0.25	0.25	4.06	-	20.79	m	23.39
extra; special bricks; intersections; birdsmouth; B.S.4729, type AN.4.1	55.69	5.00	58.47	0.25	0.25	4.06	-	62.54	m	70.35
Walls; building overhand; vertical										
stretcher bond; facework one side										
102mm thick	17.70	5.00	18.59	2.15	2.30	35.99	1.98	56.56	m2	63.62
stretcher bond; facework both sides										
102mm thick	17.70	5.00	18.59	2.45	2.60	40.87	1.98	61.43	m2	69.11
English bond; facework both sides										
215mm thick	35.70	5.00	37.48	3.05	3.35	51.67	4.75	93.90	m2	105.64
Flemish bond; facework both sides										
215mm thick	35.70	5.00	37.48	3.35	3.65	56.54	4.75	98.78	m2	111.13
Arches including centering; flat										
112mm high on face; 215mm thick; width of exposed soffit 112mm; bricks-on-edge	4.00	5.00	4.20	1.05	1.05	17.06	1.98	23.24	m	26.15
112mm high on face; 215mm thick; width of exposed soffit 215mm; bricks-on-edge	4.00	5.00	4.20	1.50	1.25	22.62	1.98	28.80	m	32.40
215mm high on face; 112mm thick; width of exposed soffit 112mm; bricks-on-end	4.00	5.00	4.20	1.45	1.25	22.16	1.21	27.57	m	31.01
Arches including centering; segmental										
215mm high on face; 215mm thick; width of exposed soffit 112mm; bricks-on-edge	8.00	5.00	8.40	2.40	2.15	37.24	3.71	49.35	m	55.52
215mm high on face; 215mm thick; width of exposed soffit 215mm; bricks-on-edge	8.00	5.00	8.40	2.95	2.45	44.43	3.71	56.54	m	63.60
Arches including centering; semi-circular										
215mm high on face; 215mm thick; width of exposed soffit 112mm; bricks-on-edge	8.00	5.00	8.40	2.40	2.15	37.24	3.71	49.35	m	55.52
215mm high on face; 215mm thick; width of exposed soffit 215mm; bricks-on-edge	8.00	5.00	8.40	2.95	2.45	44.43	3.71	56.54	m	63.60
Facework copings; bricks-on-edge; pointing top and each side										
215 x 102.5mm; horizontal	4.00	5.00	4.20	0.70	0.75	11.73	0.52	16.45	m	18.50
215 x 102.5mm; with oversailing course and cement fillets both sides; horizontal	4.00	5.00	4.20	1.10	1.15	18.23	0.90	23.33	m	26.24
327 x 102.5mm; horizontal	6.00	5.00	6.30	1.05	1.10	17.41	0.76	24.47	m	27.53
327 x 102.5mm; with oversailing course and cement fillets both sides; horizontal	6.00	5.00	6.30	1.50	1.55	24.73	1.14	32.17	m	36.19
extra; galvanised iron coping cramp at angle or end	1.85	5.00	1.94	0.04	0.04	0.65	-	2.59	nr	2.92
Facing bricks p.c. £350.00 per 1000, 215 x 102.5 x 65mm; in cement-lime mortar (1:2:9); flush smooth pointing as work proceeds										
Walls; vertical										
stretcher bond; facework one side										
102mm thick	20.65	5.00	21.68	1.75	1.90	29.49	1.98	53.15	m2	59.80
stretcher bond; facework both sides										
102mm thick	20.65	5.00	21.68	2.05	2.20	34.37	1.98	58.03	m2	65.28
English bond; facework both sides										
215mm thick	41.65	5.00	43.73	2.65	2.95	45.17	4.75	93.65	m2	105.36
Flemish bond; facework both sides										
215mm thick	41.65	5.00	43.73	2.95	3.25	50.04	4.75	98.53	m2	110.84
extra; special bricks; vertical angles; squint; B.S.4729, type AN1.1	14.92	5.00	15.67	0.25	0.25	4.06	-	19.73	m	22.19
extra; special bricks; intersections; birdsmouth; B.S.4729, type AN.4.1	54.74	5.00	57.48	0.25	0.25	4.06	-	61.54	m	69.23
Walls; building overhand; vertical										
stretcher bond; facework one side										
102mm thick	20.65	5.00	21.68	2.15	2.30	35.99	1.98	59.65	m2	67.11

MASONRY

Labour hourly rates: (except Specialists) Craft Operatives £9.23 Labourer £7.02 Rates are national average prices. Refer to REGIONAL VARIATIONS for indicative levels of overall pricing in regions	MATERIALS			LABOUR				RATES		
	Del to Site £	Waste %	Material Cost £	Craft Optve Hrs	Lab Hrs	Labour Cost £	Sunds £	Nett Rate £	Unit	Gross Rate (+12.5%) £
F10: BRICK/BLOCK WALLING Cont.										
Facing bricks p.c. £350.00 per 1000, 215 x 102.5 x 65mm; in cement-lime mortar (1:2:9); flush smooth pointing as work proceeds Cont.										
Walls; building overhand; vertical Cont.										
stretcher bond; facework both sides										
102mm thick	20.65	5.00	21.68	2.45	2.60	40.87	1.98	64.53	m2	72.59
English bond; facework both sides										
215mm thick	41.65	5.00	43.73	3.05	3.35	51.67	4.75	100.15	m2	112.67
Flemish bond; facework both sides										
215mm thick	41.65	5.00	43.73	3.35	3.65	56.54	4.75	105.03	m2	118.15
Arches including centering; flat										
112mm high on face; 215mm thick; width of exposed soffit 112mm; bricks-on-edge	4.67	5.00	4.90	1.05	1.05	17.06	1.98	23.95	m	26.94
112mm high on face; 215mm thick; width of exposed soffit 215mm; bricks-on-edge	4.67	5.00	4.90	1.50	1.25	22.62	1.98	29.50	m	33.19
215mm high on face; 112mm thick; width of exposed soffit 112mm; bricks-on-end	4.67	5.00	4.90	1.45	1.25	22.16	1.21	28.27	m	31.81
Arches including centering; segmental										
215mm high on face; 215mm thick; width of exposed soffit 112mm; bricks-on-edge	9.33	5.00	9.80	2.40	2.15	37.24	3.71	50.75	m	57.10
215mm high on face; 215mm thick; width of exposed soffit 215mm; bricks-on-edge	9.33	5.00	9.80	2.95	2.45	44.43	3.71	57.93	m	65.18
Arches including centering; semi-circular										
215mm high on face; 215mm thick; width of exposed soffit 112mm; bricks-on-edge	9.33	5.00	9.80	2.40	2.15	37.24	3.71	50.75	m	57.10
215mm high on face; 215mm thick; width of exposed soffit 215mm; bricks-on-edge	9.33	5.00	9.80	2.95	2.45	44.43	3.71	57.93	m	65.18
Facework copings; bricks-on-edge; pointing top and each side										
215 x 102.5mm; horizontal	4.67	5.00	4.90	0.70	0.75	11.73	0.52	17.15	m	19.29
215 x 102.5mm; with oversailing course and cement fillets both sides; horizontal	4.67	5.00	4.90	1.10	1.15	18.23	0.90	24.03	m	27.03
327 x 102.5mm; horizontal	7.00	5.00	7.35	1.05	1.10	17.41	0.76	25.52	m	28.71
327 x 102.5mm; with oversailing course and cement fillets both sides; horizontal	7.00	5.00	7.35	1.50	1.55	24.73	1.14	33.22	m	37.37
extra; galvanised iron coping cramp at angle or end	1.85	5.00	1.94	0.04	0.04	0.65	-	2.59	nr	2.92
Composite walling of bricks 215 x 102.5x 65mm; in cement-lime mortar (1:2:9); common bricks B.S.3921 Category M backing, compressive strength 20.5 N/mm2; facing bricks p.c. £150.00 per 1000; flush smooth pointing as work proceeds										
Walls; vertical										
English bond; facework one side										
215mm thick	18.10	5.00	19.00	2.65	2.95	45.17	4.75	68.92	m2	77.54
327mm thick	27.77	5.00	29.16	3.15	3.60	54.35	7.51	91.02	m2	102.39
Flemish bond; facework one side										
215mm thick	18.24	5.00	19.15	2.65	2.95	45.17	4.75	69.07	m2	77.70
327mm thick	27.91	5.00	29.31	3.15	3.60	54.35	7.51	91.16	m2	102.56
Walls; building overhand; vertical										
English bond; facework one side										
215mm thick	18.10	5.00	19.00	3.05	3.35	51.67	4.75	75.42	m2	84.85
327mm thick	27.77	5.00	29.16	3.55	4.00	60.85	7.51	97.52	m2	109.70
Flemish bond; facework one side										
215mm thick	18.24	5.00	19.15	3.05	3.35	51.67	4.75	75.57	m2	85.02
327mm thick	27.91	5.00	29.31	3.55	4.00	60.85	7.51	97.66	m2	109.87
Composite walling of bricks 215 x 102.5x 65mm; in cement-lime mortar (1:2:9); common bricks B.S.3921 Category M backing, compressive strength 20.5 N/mm2; facing bricks p.c. £200.00 per 1000; flush smooth pointing as work proceeds										
Walls; vertical										
English bond; facework one side										
215mm thick	22.55	5.00	23.68	2.65	2.95	45.17	4.75	73.60	m2	82.80
327mm thick	32.22	5.00	33.83	3.15	3.60	54.35	7.51	95.69	m2	107.65
Flemish bond; facework one side										
215mm thick	22.19	5.00	23.30	2.65	2.95	45.17	4.75	73.22	m2	82.37
327mm thick	31.86	5.00	33.45	3.15	3.60	54.35	7.51	95.31	m2	107.22
Walls; building overhand; vertical										
English bond; facework one side										
215mm thick	22.55	5.00	23.68	3.05	3.35	51.67	4.75	80.10	m2	90.11
327mm thick	32.22	5.00	33.83	3.55	4.00	60.85	7.51	102.19	m2	114.96
Flemish bond; facework one side										
215mm thick	22.19	5.00	23.30	3.05	3.35	51.67	4.75	79.72	m2	89.68
327mm thick	31.86	5.00	33.45	3.55	4.00	60.85	7.51	101.81	m2	114.54

Labour hourly rates: (except Specialists) Craft Operatives £9.23 Labourer £7.02 Rates are national average prices. Refer to REGIONAL VARIATIONS for indicative levels of overall pricing in regions	MATERIALS			LABOUR				RATES		
	Del to Site £	Waste %	Material Cost £	Craft Optve Hrs	Lab Hrs	Labour Cost £	Sunds £	Nett Rate £	Unit	Gross Rate (+12.5%) £
F10: BRICK/BLOCK WALLING Cont.										
Composite walling of bricks 215 x 102.5 x 65mm; in cement-lime mortar (1:2:9); common bricks B.S.3921 Category M backing, compressive strength 20.5 N/mm2; facing bricks p.c. £250.00 per 1000; flush smooth pointing as work proceeds										
Walls; vertical										
English bond; facework one side										
215mm thick....................	27.00	5.00	28.35	2.65	2.95	45.17	4.75	78.27	m2	88.05
327mm thick....................	36.67	5.00	38.50	3.15	3.60	54.35	7.51	100.36	m2	112.91
Flemish bond; facework one side										
215mm thick....................	26.14	5.00	27.45	2.65	2.95	45.17	4.75	77.37	m2	87.04
327mm thick....................	35.81	5.00	37.60	3.15	3.60	54.35	7.51	99.46	m2	111.89
Walls; building overhand; vertical										
English bond; facework one side										
215mm thick....................	27.00	5.00	28.35	3.05	3.35	51.67	4.75	84.77	m2	95.36
327mm thick....................	36.67	5.00	38.50	3.55	4.00	60.85	7.51	106.86	m2	120.22
Flemish bond; facework one side										
215mm thick....................	26.14	5.00	27.45	3.05	3.35	51.67	4.75	83.87	m2	94.35
327mm thick....................	35.81	5.00	37.60	3.55	4.00	60.85	7.51	105.96	m2	119.20
Composite walling of bricks 215 x 102.5 x 65mm; in cement-lime mortar (1:2:9); common bricks B.S.3921 Category M backing, compressive strength 20.5 N/mm2; facing bricks p.c. £300.00 per 1000; flush smooth pointing as work proceeds										
Walls; vertical										
English bond; facework one side										
215mm thick....................	31.45	5.00	33.02	2.65	2.95	45.17	4.75	82.94	m2	93.31
327mm thick....................	41.12	5.00	43.18	3.15	3.60	54.35	7.51	105.03	m2	118.16
Flemish bond; facework one side										
215mm thick....................	30.09	5.00	31.59	2.65	2.95	45.17	4.75	81.51	m2	91.70
327mm thick....................	43.71	5.00	45.90	3.15	3.60	54.35	7.51	107.75	m2	121.22
Walls; building overhand; vertical										
English bond; facework one side										
215mm thick....................	31.45	5.00	33.02	3.05	3.35	51.67	4.75	89.44	m2	100.62
327mm thick....................	41.12	5.00	43.18	3.55	4.00	60.85	7.51	111.53	m2	125.47
Flemish bond; facework one side										
215mm thick....................	30.09	5.00	31.59	3.05	3.35	51.67	4.75	88.01	m2	99.01
327mm thick....................	43.79	5.00	45.98	3.55	4.00	60.85	7.51	114.34	m2	128.63
Composite walling of bricks 215 x 102.5 x 65mm; in cement-lime mortar (1:2:9); common bricks B.S.3921 Category M backing, compressive strength 20.5 N/mm2; facing bricks p.c. £350.00 per 1000; flush smooth pointing as work proceeds										
Walls; vertical										
English bond; facework one side										
215mm thick....................	35.90	5.00	37.70	2.65	2.95	45.17	4.75	87.61	m2	98.57
327mm thick....................	45.57	5.00	47.85	3.15	3.60	54.35	7.51	109.71	m2	123.42
Flemish bond; facework one side										
215mm thick....................	34.04	5.00	35.74	2.65	2.95	45.17	4.75	85.66	m2	96.37
327mm thick....................	43.71	5.00	45.90	3.15	3.60	54.35	7.51	107.75	m2	121.22
Walls; building overhand; vertical										
English bond; facework one side										
215mm thick....................	35.90	5.00	37.70	3.05	3.35	51.67	4.75	94.11	m2	105.88
327mm thick....................	45.57	5.00	47.85	3.55	4.00	60.85	7.51	116.21	m2	130.73
Flemish bond; facework one side										
215mm thick....................	34.04	5.00	35.74	3.05	3.35	51.67	4.75	92.16	m2	103.68
327mm thick....................	43.71	5.00	45.90	3.55	4.00	60.85	7.51	114.25	m2	128.53
For each £10.00 difference in cost of 1000 facing bricks, add or deduct the following										
Wall in facing bricks										
102mm thick; English bond; facework one side	0.89	5.00	0.93	-	-	-	-	0.93	m2	1.05
215mm thick; facework both sides	1.19	5.00	1.25	-	-	-	-	1.25	m2	1.41
Composite wall; common brick backing; facing bricks facework										
215mm thick; English bond; facework one side	0.89	5.00	0.93	-	-	-	-	0.93	m2	1.05
215mm thick; Flemish bond; facework one side	0.79	5.00	0.83	-	-	-	-	0.83	m2	0.93
Extra over flush smooth pointing as work proceeds for the following types of pointing										
Pointing as work proceeds										
ironing in joints	-	-	-	0.03	0.03	0.49	-	0.49	m2	0.55
weathered pointing	-	-	-	0.03	0.03	0.49	-	0.49	m2	0.55
recessed pointing	-	-	-	0.04	0.04	0.65	-	0.65	m2	0.73
Raking out joints and pointing on completion										
flush smooth pointing with cement-lime mortar (1:2:9)	-	-	-	0.75	0.75	12.19	0.16	12.35	m2	13.89
flush smooth pointing with coloured cement-lime mortar (1:2:9)	-	-	-	0.75	0.75	12.19	0.22	12.41	m2	13.96
weathered pointing with cement-lime mortar (1:2:9)	-	-	-	0.83	0.83	13.49	0.16	13.65	m2	15.35
weathered pointing with coloured cement-lime mortar (1:2:9)	-	-	-	0.83	0.83	13.49	0.22	13.71	m2	15.42

MASONRY

Labour hourly rates: (except Specialists) Craft Operatives £9.23 Labourer £7.02 Rates are national average prices. Refer to REGIONAL VARIATIONS for indicative levels of overall pricing in regions	MATERIALS			LABOUR				RATES		
	Del to Site £	Waste %	Material Cost £	Craft Optve Hrs	Lab Hrs	Labour Cost £	Sunds £	Nett Rate £	Unit	Gross Rate (+12.5%) £
F10: BRICK/BLOCK WALLING Cont.										
Extra over cement-lime mortar (1:2:9) for bedding and pointing for using coloured cement-lime mortar (1:2:9) for the following										
Wall										
102mm thick; pointing both sides	-	-	-	-	-	-	1.76	1.76	m2	1.98
215mm thick; pointing both sides	-	-	-	-	-	-	3.50	3.50	m2	3.94
Aerated concrete blocks, Thermalite, 440 x 215mm, Shield blocks (4.0 N/mm2);in cement-lime mortar (1:1:6)										
Walls; vertical										
75mm thick; stretcher bond	6.26	5.00	6.57	0.70	0.80	12.08	0.66	19.31	m2	21.72
90mm thick; stretcher bond	7.50	5.00	7.88	0.80	0.90	13.70	0.81	22.39	m2	25.19
100mm thick; stretcher bond	8.34	5.00	8.76	0.80	0.90	13.70	0.81	23.27	m2	26.18
140mm thick; stretcher bond	11.67	5.00	12.25	0.90	1.05	15.68	1.21	29.14	m2	32.78
190mm thick; stretcher bond	15.84	5.00	16.63	0.95	1.10	16.49	1.63	34.75	m2	39.10
Closing cavities; vertical										
50mm wide with blockwork 100mm thick	2.09	5.00	2.19	0.25	0.25	4.06	0.20	6.46	m	7.26
75mm wide with blockwork 100mm thick	2.09	5.00	2.19	0.25	0.25	4.06	0.20	6.46	m	7.26
Closing cavities; horizontal										
50mm wide with blockwork 100mm thick	2.09	5.00	2.19	0.25	0.25	4.06	0.20	6.46	m	7.26
75mm wide with blockwork 100mm thick	2.09	5.00	2.19	0.25	0.25	4.06	0.20	6.46	m	7.26
Bonding ends to common brickwork; forming pockets; extra material										
walls; bonding every third course										
75mm	0.63	-	0.63	0.20	0.20	3.25	-	3.88	m	4.37
90mm	0.75	-	0.75	0.25	0.25	4.06	-	4.81	m	5.41
100mm	0.83	-	0.83	0.25	0.25	4.06	-	4.89	m	5.50
140mm	1.17	-	1.17	0.35	0.35	5.69	-	6.86	m	7.71
190mm	1.58	-	1.58	0.40	0.40	6.50	-	8.08	m	9.09
Bonding ends to existing common brickwork; cutting pockets; extra material										
walls; bonding every third course										
75mm	1.26	-	1.26	0.40	0.40	6.50	-	7.76	m	8.73
90mm	1.50	-	1.50	0.50	0.50	8.13	-	9.63	m	10.83
100mm	1.66	-	1.66	0.50	0.50	8.13	-	9.79	m	11.01
140mm	2.34	-	2.34	0.70	0.70	11.38	-	13.72	m	15.43
190mm	3.16	-	3.16	0.80	0.80	13.00	-	16.16	m	18.18
Bonding ends to existing concrete blockwork; cutting pockets; extra material										
walls; bonding every third course										
75mm	1.26	-	1.26	0.30	0.30	4.88	-	6.13	m	6.90
90mm	1.50	-	1.50	0.35	0.35	5.69	-	7.19	m	8.09
100mm	1.66	-	1.66	0.40	0.40	6.50	-	8.16	m	9.18
140mm	2.34	-	2.34	0.55	0.55	8.94	-	11.28	m	12.69
190mm	3.16	-	3.16	0.65	0.65	10.56	-	13.72	m	15.44
Aerated concrete blocks, Thermalite, 440 x 215mm, Turbo blocks (2.8 N/mm2); in cement-lime mortar (1:1:6)										
Walls; vertical										
100mm thick; stretcher bond	8.65	5.00	9.08	0.80	0.80	13.00	0.81	22.89	m2	25.75
115mm thick; stretcher bond	9.94	5.00	10.44	0.80	0.90	13.70	0.81	24.95	m2	28.07
125mm thick; stretcher bond	10.81	5.00	11.35	0.85	1.00	14.87	1.21	27.43	m2	30.85
130mm thick; stretcher bond	11.24	5.00	11.80	0.85	1.00	14.87	1.21	27.88	m2	31.36
150mm thick; stretcher bond	12.97	5.00	13.62	0.90	1.05	15.68	1.21	30.51	m2	34.32
190mm thick; stretcher bond	16.42	5.00	17.24	0.95	1.10	16.49	1.63	35.36	m2	39.78
200mm thick; stretcher bond	17.29	5.00	18.15	1.00	1.20	17.65	1.63	37.44	m2	42.12
215mm thick; stretcher bond	18.59	5.00	19.52	1.00	1.20	17.65	2.04	39.21	m2	44.12
Closing cavities; vertical										
50mm wide with blockwork 100mm thick	2.16	5.00	2.27	0.25	0.25	4.06	0.20	6.53	m	7.35
75mm wide with blockwork 100mm thick	2.16	5.00	2.27	0.25	0.25	4.06	0.20	6.53	m	7.35
Closing cavities; horizontal										
50mm wide with blockwork 100mm thick	2.16	5.00	2.27	0.25	0.25	4.06	0.20	6.53	m	7.35
75mm wide with blockwork 100mm thick	2.16	5.00	2.27	0.25	0.25	4.06	0.20	6.53	m	7.35
Aerated concrete blocks, Thermalite, 440 x 215mm, Party wall blocks (4.0 N/mm2); in cement-lime mortar (1:1:6)										
Walls; vertical										
215mm thick; stretcher bond	18.22	5.00	19.13	1.00	1.20	17.65	2.04	38.83	m2	43.68
Aerated concrete blocks, Thermalite, 440 x 215mm, Hi-Strength 7 blocks (7.0 N/mm2); in cement-lime mortar (1:1:6)										
Walls; vertical										
100mm thick; stretcher bond	10.47	5.00	10.99	0.80	0.90	13.70	0.81	25.51	m2	28.69
140mm thick; stretcher bond	14.66	5.00	15.39	0.90	1.05	15.68	1.21	32.28	m2	36.32
150mm thick; stretcher bond	15.71	5.00	16.50	0.90	1.05	15.68	1.21	33.38	m2	37.56
190mm thick; stretcher bond	19.81	5.00	20.80	0.95	1.10	16.49	1.63	38.92	m2	43.79
200mm thick; stretcher bond	20.94	5.00	21.99	1.00	1.20	17.65	1.63	41.27	m2	46.43
215mm thick; stretcher bond	22.52	5.00	23.65	1.00	1.20	17.65	2.04	43.34	m2	48.76

Labour hourly rates: (except Specialists) Craft Operatives £9.23 Labourer £7.02 Rates are national average prices. Refer to REGIONAL VARIATIONS for indicative levels of overall pricing in regions	MATERIALS			LABOUR				RATES		
	Del to Site £	Waste %	Material Cost £	Craft Optve Hrs	Lab Hrs	Labour Cost £	Sunds £	Nett Rate £	Unit	Gross Rate (+12.5%) £
F10: BRICK/BLOCK WALLING Cont.										
Aerated concrete blocks, Thermalite, 440 x 215mm, Hi-Strength 7 blocks (7.0 N/mm2); in cement-lime mortar (1:1:6) Cont.										
Closing cavities; vertical										
50mm wide with blockwork 100mm thick	2.62	5.00	2.75	0.25	0.25	4.06	0.20	7.01	m	7.89
75mm wide with blockwork 100mm thick	2.62	5.00	2.75	0.25	0.25	4.06	0.20	7.01	m	7.89
Closing cavities; horizontal										
50mm wide with blockwork 100mm thick	2.62	5.00	2.75	0.25	0.25	4.06	0.20	7.01	m	7.89
75mm wide with blockwork 100mm thick	2.62	5.00	2.75	0.25	0.25	4.06	0.20	7.01	m	7.89
Aerated concrete blocks, Thermalite, 440 x 215mm, Trenchblocks (4.0 N/mm2); in cement mortar (1:4)										
Walls; vertical										
255mm thick; stretcher bond	21.60	5.00	22.68	1.30	1.50	22.53	2.04	47.25	m2	53.16
275mm thick; stretcher bond	23.30	5.00	24.47	1.45	1.65	24.97	2.23	51.66	m2	58.12
305mm thick; stretcher bond	25.84	5.00	27.13	1.65	1.85	28.22	2.48	57.83	m2	65.06
355mm thick; stretcher bond	30.07	5.00	31.57	1.95	2.15	33.09	2.90	67.56	m2	76.01
Aerated concrete blocks, Tarmac Toplite foundation blocks; 440 x 215mm, (3.5 N/mm2); in cement mortar (1:4)										
Walls; vertical										
260mm thick; stretcher bond	18.93	5.00	19.88	1.40	1.40	22.75	2.10	44.73	m2	50.32
275mm thick; stretcher bond	20.03	5.00	21.03	1.60	1.60	26.00	2.23	49.26	m2	55.42
300mm thick; stretcher bond	21.85	5.00	22.94	1.70	1.70	27.63	2.43	53.00	m2	59.62
Tarmac Topblock, medium density concrete block, Hemelite, 440 x 215mm, solid Standard blocks (3.5 N/mm2); in cement-lime mortar (1:1:6)										
Walls; vertical										
75mm thick; stretcher bond	4.95	5.00	5.20	0.75	0.90	13.24	0.66	19.10	m2	21.49
90mm thick; stretcher bond	6.51	5.00	6.84	0.90	1.05	15.68	0.81	23.32	m2	26.24
100mm thick; stretcher bond	5.74	5.00	6.03	0.90	1.05	15.68	0.81	22.52	m2	25.33
140mm thick; stretcher bond	8.39	5.00	8.81	1.00	1.20	17.65	1.21	27.67	m2	31.13
190mm thick; stretcher bond	12.10	5.00	12.71	1.05	1.25	18.47	1.63	32.80	m2	36.90
215mm thick; stretcher bond	13.11	5.00	13.77	1.10	1.30	19.28	2.04	35.08	m2	39.47
Closing cavities; vertical										
50mm wide with blockwork 100mm thick	1.44	5.00	1.51	0.25	0.25	4.06	0.20	5.77	m	6.50
75mm wide with blockwork 100mm thick	1.44	5.00	1.51	0.25	0.25	4.06	0.20	5.77	m	6.50
Closing cavities; horizontal										
50mm wide with blockwork 100mm thick	1.44	5.00	1.51	0.25	0.25	4.06	0.20	5.77	m	6.50
75mm wide with blockwork 100mm thick	1.44	5.00	1.51	0.25	0.25	4.06	0.20	5.77	m	6.50
Bonding ends to common brickwork; forming pockets; extra material										
walls; bonding every third course										
75mm	0.50	-	0.50	0.25	0.25	4.06	-	4.56	m	5.13
90mm	0.65	-	0.65	0.30	0.30	4.88	-	5.53	m	6.22
100mm	0.57	-	0.57	0.30	0.30	4.88	-	5.45	m	6.13
140mm	0.84	-	0.84	0.40	0.40	6.50	-	7.34	m	8.26
190mm	1.21	-	1.21	0.45	0.45	7.31	-	8.52	m	9.59
215mm	1.31	-	1.31	0.55	0.55	8.94	-	10.25	m	11.53
Bonding ends to existing common brickwork; cutting pockets; extra material										
walls; bonding every third course										
75mm	0.99	-	0.99	0.50	0.50	8.13	-	9.12	m	10.25
90mm	1.30	-	1.30	0.60	0.60	9.75	-	11.05	m	12.43
100mm	1.14	-	1.14	0.60	0.60	9.75	-	10.89	m	12.25
140mm	1.67	-	1.67	0.80	0.80	13.00	-	14.67	m	16.50
190mm	2.42	-	2.42	0.90	0.90	14.63	-	17.05	m	19.18
215mm	2.62	-	2.62	1.10	1.10	17.88	-	20.50	m	23.06
Bonding ends to existing concrete blockwork; cutting pockets; extra material										
walls; bonding every third course										
75mm	0.99	-	0.99	0.40	0.40	6.50	-	7.49	m	8.43
90mm	1.30	-	1.30	0.45	0.45	7.31	-	8.61	m	9.69
100mm	1.14	-	1.14	0.45	0.45	7.31	-	8.45	m	9.51
140mm	1.67	-	1.67	0.60	0.60	9.75	-	11.42	m	12.85
190mm	2.42	-	2.42	0.70	0.70	11.38	-	13.80	m	15.52
215mm	2.62	-	2.62	0.85	0.85	13.81	-	16.43	m	18.49
Tarmac Topblock, medium density concrete block, Hemelite, 440 x 215mm, solid Standard blocks (7.0 N/mm2); in cement-lime mortar (1:1:6)										
Walls; vertical										
90mm thick; stretcher bond	6.95	5.00	7.30	0.90	1.05	15.68	0.81	23.79	m2	26.76
100mm thick; stretcher bond	6.18	5.00	6.49	0.90	1.05	15.68	0.81	22.98	m2	25.85
140mm thick; stretcher bond	8.62	5.00	9.05	1.00	1.20	17.65	1.21	27.91	m2	31.40
190mm thick; stretcher bond	12.53	5.00	13.16	1.05	1.25	18.47	1.63	33.25	m2	37.41
215mm thick; stretcher bond	13.64	5.00	14.32	1.10	1.30	19.28	2.04	35.64	m2	40.10
Closing cavities; vertical										
50mm wide with blockwork 100mm thick	1.55	5.00	1.63	0.25	0.25	4.06	0.20	5.89	m	6.63
75mm wide with blockwork 100mm thick	1.55	5.00	1.63	0.25	0.25	4.06	0.20	5.89	m	6.63

MASONRY

Labour hourly rates: (except Specialists) Craft Operatives £9.23 Labourer £7.02 Rates are national average prices. Refer to REGIONAL VARIATIONS for indicative levels of overall pricing in regions	MATERIALS			LABOUR				RATES		
	Del to Site £	Waste %	Material Cost £	Craft Optve Hrs	Lab Hrs	Labour Cost £	Sunds £	Nett Rate £	Unit	Gross Rate (+12.5%) £
F10: BRICK/BLOCK WALLING Cont.										
Tarmac Topblock, medium density concrete block, Hemelite, 440 x 215mm, solid Standard blocks (7.0 N/mm2); in cement-lime mortar (1:1:6) Cont.										
Closing cavities; horizontal										
50mm wide with blockwork 100mm thick	1.55	5.00	1.63	0.25	0.25	4.06	0.20	5.89	m	6.63
75mm wide with blockwork 100mm thick	1.55	5.00	1.63	0.25	0.25	4.06	0.20	5.89	m	6.63
Tarmac Topblock, fair face concrete blocks, Lignacite, 440 x 215mm, solid Standard blocks (7.0 N/mm2); in cement-lime mortar (1:1:6)										
Walls; vertical										
100mm thick; stretcher bond	10.62	5.00	11.15	0.95	1.10	16.49	0.81	28.45	m2	32.01
140mm thick; stretcher bond	14.75	5.00	15.49	1.10	1.25	18.93	1.21	35.63	m2	40.08
190mm thick; stretcher bond	18.87	5.00	19.81	1.15	1.35	20.09	1.63	41.53	m2	46.73
Closing cavities; vertical										
50mm wide with blockwork 100mm thick	2.65	5.00	2.78	0.25	0.25	4.06	0.20	7.04	m	7.93
75mm wide with blockwork 100mm thick	2.65	5.00	2.78	0.25	0.25	4.06	0.20	7.04	m	7.93
Closing cavities; horizontal										
50mm wide with blockwork 100mm thick	2.65	5.00	2.78	0.25	0.25	4.06	0.20	7.04	m	7.93
75mm wide with blockwork 100mm thick	2.65	5.00	2.78	0.25	0.25	4.06	0.20	7.04	m	7.93
Tarmac Topblock, dense concrete blocks, Topcrete, 440 x 215mm, solid Standard blocks (7.0 N/mm2); in cement-lime mortar (1:1:6)										
Walls; vertical										
75mm thick; stretcher bond	5.95	5.00	6.25	1.00	1.20	17.65	0.66	24.56	m2	27.63
90mm thick; stretcher bond	6.55	5.00	6.88	1.15	1.35	20.09	0.81	27.78	m2	31.25
100mm thick; stretcher bond	7.28	5.00	7.64	1.20	1.40	20.90	0.81	29.36	m2	33.03
140mm thick; stretcher bond	10.19	5.00	10.70	1.30	1.50	22.53	1.21	34.44	m2	38.74
190mm thick; stretcher bond	13.83	5.00	14.52	1.40	1.60	24.15	1.63	40.31	m2	45.34
215mm thick; stretcher bond	15.65	5.00	16.43	1.50	1.70	25.78	2.02	44.23	m2	49.76
Closing cavities; vertical										
50mm wide with blockwork 100mm thick	1.83	5.00	1.92	0.25	0.25	4.06	0.20	6.18	m	6.96
75mm wide with blockwork 100mm thick	1.83	5.00	1.92	0.25	0.25	4.06	0.20	6.18	m	6.96
Closing cavities; horizontal										
50mm wide with blockwork 100mm thick	1.83	5.00	1.92	0.25	0.25	4.06	0.20	6.18	m	6.96
75mm wide with blockwork 100mm thick	1.83	5.00	1.92	0.25	0.25	4.06	0.20	6.18	m	6.96
Tarmac Topblock, dense concrete blocks, Topcrete, 440 x 215mm cellular standard blocks (7.0 N/mm2); in cement-lime mortar (1:1:6)										
Walls; vertical										
100mm thick; stretcher bond	5.95	5.00	6.25	1.10	1.13	18.09	0.87	25.20	m2	28.35
Closing cavities; vertical										
50mm wide with blockwork 100mm thick	1.49	5.00	1.56	0.25	0.25	4.06	0.20	5.83	m	6.56
75mm wide with blockwork 100mm thick	1.49	5.00	1.56	0.25	0.25	4.06	0.20	5.83	m	6.56
Closing cavities; horizontal										
50mm wide with blockwork 100mm thick	1.49	5.00	1.56	0.25	0.25	4.06	0.20	5.83	m	6.56
75mm wide with blockwork 100mm thick	1.49	5.00	1.56	0.25	0.25	4.06	0.20	5.83	m	6.56
Tarmac Topblock, dense concrete blocks, Topcrete, 440 x 215mm hollow standard blocks (7.0 N/mm2); in cement-lime mortar (1:1:6)										
Walls; vertical										
215mm thick; stretcher bond	11.61	5.00	12.19	1.50	1.56	24.80	2.02	39.01	m2	43.88
Aerated concrete blocks, Thermalite, 440 x 215mm, Smooth Face blocks (4.0 N/mm2); in cement-lime mortar (1:1:6); flush smooth pointing as work proceeds										
Walls; vertical										
stretcher bond; facework one side										
100mm thick......................................	14.30	5.00	15.02	1.50	1.56	24.80	0.81	40.62	m2	45.70
140mm thick......................................	20.02	5.00	21.02	1.10	1.25	18.93	1.21	41.16	m2	46.30
190mm thick......................................	27.17	5.00	28.53	1.15	1.35	20.09	1.63	50.25	m2	56.53
215mm thick......................................	30.75	5.00	32.29	1.20	1.40	20.90	2.02	55.21	m2	62.11
stretcher bond; facework both sides										
100mm thick......................................	14.30	5.00	15.02	1.20	1.35	20.55	0.81	36.38	m2	40.93
140mm thick......................................	20.02	5.00	21.02	1.30	1.45	22.18	1.21	44.41	m2	49.96
190mm thick......................................	27.17	5.00	28.53	1.35	1.55	23.34	1.63	53.50	m2	60.19
215mm thick......................................	30.75	5.00	32.29	1.40	1.60	24.15	1.93	58.37	m2	65.67
Tarmac Topblock, fair face concrete blocks, Lignacite, 440 x 215mm, solid Standard blocks (7.0 N/mm2); in cement-lime mortar (1:1:6); flush smooth pointing as work proceeds										
Walls; vertical										
stretcher bond; facework one side										
100mm thick......................................	10.62	5.00	11.15	1.15	1.35	20.09	0.81	32.05	m2	36.06
140mm thick......................................	14.75	5.00	15.49	1.30	1.50	22.53	1.21	39.23	m2	44.13
190mm thick......................................	18.87	5.00	19.81	1.35	1.55	23.34	1.63	44.78	m2	50.38

Labour hourly rates: (except Specialists) Craft Operatives £9.23 Labourer £7.02 Rates are national average prices. Refer to REGIONAL VARIATIONS for indicative levels of overall pricing in regions	MATERIALS			LABOUR				RATES		
	Del to Site £	Waste %	Material Cost £	Craft Optve Hrs	Lab Hrs	Labour Cost £	Sunds £	Nett Rate £	Unit	Gross Rate (+12.5%) £
F10: BRICK/BLOCK WALLING Cont.										
Tarmac Topblock, fair face concrete blocks, Lignacite, 440 x 215mm, solid Standard blocks (7.0 N/mm2); in cement-lime mortar (1:1:6); flush smooth pointing as work proceeds Cont.										
Walls; vertical Cont.										
stretcher bond; facework both sides										
100mm thick..................	10.62	5.00	11.15	1.35	1.55	23.34	0.81	35.30	m2	39.72
140mm thick..................	14.75	5.00	15.49	1.50	1.70	25.78	1.21	42.48	m2	47.79
190mm thick..................	18.87	5.00	19.81	1.55	1.75	26.59	1.63	48.03	m2	54.04
Reconstructed stone blocks, Marshalls Mono Ltd. Cromwell coursed random length Split Face buff walling blocks in cement-lime mortar (1:1:6); flat recessed pointing as work proceeds										
Walls; vertical										
100mm thick; alternate courses 102mm and 140mm high blocks; stretcher bond; facework one side ...	26.29	5.00	27.60	1.90	2.05	31.93	0.73	60.26	m2	67.80
100mm thick; one course 102mm high blocks and two courses 140mm high blocks; stretcher bond; facework one side	26.29	5.00	27.60	1.90	2.05	31.93	0.73	60.26	m2	67.80
Reconstructed stone blocks, Marshalls Mono Ltd. Cromwell coursed random length Pitched Face buff walling blocks in cement-lime mortar (1:1:6); flat recessed pointing as work proceeds										
Walls; vertical										
90mm thick; alternate courses 102mm and 140mm high blocks; stretcher bond; facework one side	28.77	5.00	30.21	1.90	2.05	31.93	0.73	62.87	m2	70.72
90mm thick; one course 102mm high blocks and two courses 140mm high blocks; stretcher bond; facework one side	28.77	5.00	30.21	1.90	2.05	31.93	0.73	62.87	m2	70.72
Firebricks, 215 x 102.5 x 65mm; in fire cement										
Flue linings; bonding to surrounding brickwork with headers -4/m2										
112mm thick; stretcher bond	93.04	5.00	97.69	1.90	2.05	31.93	5.97	135.59	m2	152.54
Flue linings; built clear of main brickwork but with one header in each course set projecting to contact main work										
112mm thick; stretcher bond	93.04	5.00	97.69	1.90	2.05	31.93	5.97	135.59	m2	152.54
112mm thick; stretcher bond; in segmental top to flue	93.04	5.00	97.69	2.50	2.65	41.68	5.97	145.34	m2	163.51
HR Supra Flue lining bricks, 230 x 114 x 76mm, Hepworth Refractories, Sheffield, MPK 21 mortar										
Flue linings; bonding to surrounding brickwork with headers -4/m2										
114mm thick; stretcher bond	37.61	5.00	39.49	1.90	2.05	31.93	3.62	75.04	m2	84.42
230mm thick; English bond	67.36	5.00	70.73	3.15	3.40	52.94	6.46	130.13	m2	146.40
F11: GLASS BLOCK WALLING										
SCREENS AND PANELS										
Hollow glass blocks, white, cross ribbed, in cement mortar (1:3); continuous joints; flat recessed pointing as work proceeds										
Screens or panels; 190 x 190mm blocks; vertical										
80mm thick; facework both sides	259.00	10.00	284.90	3.30	3.75	56.78	29.00	370.68	m2	417.02
Screens or panels; 240 x 240mm blocks; vertical										
80mm thick; facework both sides	237.00	10.00	260.70	2.75	3.25	48.20	21.00	329.90	m2	371.13
F20: NATURAL STONE RUBBLE WALLING										
Stone rubble work; random stones of Yorkshire limestone; bedding and jointing in lime mortar (1:3); uncoursed										
Walls; tapering both sides; including extra material										
500mm thick	55.00	15.00	63.25	3.75	4.30	64.80	13.00	141.05	m2	158.68
600mm thick	60.50	15.00	69.58	4.50	5.15	77.69	15.00	162.26	m2	182.55
Stone rubble work; squared rubble face stones of Yorkshire limestone; bedding and jointing in lime mortar (1:3); irregular coursed; courses average 150mm high										
Walls; vertical; face stones 100 - 150mm on bed; bonding to brickwork; including extra material; scappled or axed face; weather struck pointing as work proceeds										
150mm thick; faced one side	38.50	15.00	44.27	3.00	3.45	51.91	4.00	100.18	m2	112.71
Walls; vertical; face stones 100 - 150mm on bed; bonding to brickwork; including extra material; hammer dressed face; weather struck pointing as work proceeds										
150mm thick; faced one side	38.50	15.00	44.27	3.30	3.80	57.13	4.00	105.41	m2	118.59

MASONRY

Labour hourly rates: (except Specialists) Craft Operatives £9.23 Labourer £7.02 Rates are national average prices. Refer to REGIONAL VARIATIONS for indicative levels of overall pricing in regions	MATERIALS			LABOUR				RATES		
	Del to Site £	Waste %	Material Cost £	Craft Optve Hrs	Lab Hrs	Labour Cost £	Sunds £	Nett Rate £	Unit	Gross Rate (+12.5%) £
F20: NATURAL STONE RUBBLE WALLING Cont.										
Stone rubble work; squared rubble face stones of Yorkshire limestone; bedding and jointing in lime mortar (1:3); irregular coursed; courses average 150mm high Cont.										
Walls; vertical; face stones 100 – 150mm on bed; bonding to brickwork; including extra material; rock worked face; pointing with a parallel joint as work proceeds										
150mm thick; faced one side	40.70	15.00	46.81	3.60	4.15	62.36	4.00	113.17	m2	127.31
Stone rubble work; squared rubble face stones of Yorkshire limestone; bedding and jointing in lime mortar (1:3); regular coursed; courses average 150mm high										
Walls; vertical; face stones 100 – 150mm on bed; bonding to brickwork; including extra material; scappled or axed face; weather struck pointing as work proceeds										
150mm thick; faced one side	60.50	15.00	69.58	3.20	3.70	55.51	4.00	129.09	m2	145.22
Walls; vertical; face stones 100 – 150mm on bed; bonding to brickwork; including extra material; hammer dressed face; weather struck pointing as work proceeds										
150mm thick; faced one side	60.50	15.00	69.58	3.50	4.00	60.38	4.00	133.96	m2	150.71
Walls; vertical; face stones 100 – 150mm on bed; bonding to brickwork; including extra material; rock worked face; pointing with a parallel joint as work proceeds										
150mm thick; faced one side	62.70	15.00	72.11	3.80	4.40	65.96	4.00	142.07	m2	159.83
Stone rubble work; random rubble backing and squared rubble face stones of Yorkshire limestone; bedding and jointing in lime mortar (1:3); irregular coursed; courses average 150mm high										
Walls; vertical; face stones 100 – 150mm on bed; scappled or axed face; weather struck pointing as work proceeds										
350mm thick; faced one side	68.20	15.00	78.43	4.00	4.60	69.21	9.00	156.64	m2	176.22
500mm thick; faced one side	86.90	15.00	99.94	4.75	5.45	82.10	13.00	195.04	m2	219.42
Walls; vertical; face stones 100 – 150mm on bed; hammer dressed face; weather struck pointing as work proceeds										
350mm thick; faced one side	68.20	15.00	78.43	4.30	4.95	74.44	9.00	161.87	m2	182.10
500mm thick; faced one side	86.90	15.00	99.94	5.05	5.80	87.33	13.00	200.26	m2	225.30
Walls; vertical; face stones 100 – 150mm on bed; rock worked face; pointing with a parallel joint as work proceeds										
350mm thick; faced one side	68.20	15.00	78.43	4.60	5.30	79.66	9.00	167.09	m2	187.98
500mm thick; faced one side	86.90	15.00	99.94	5.35	6.15	92.55	13.00	205.49	m2	231.17
Quoin stones; scappled or axed face; attached faced two adjacent faces										
250 x 200 x 350mm............	23.65	5.00	24.83	0.60	0.30	7.64	0.50	32.98	nr	37.10
250 x 200 x 500mm............	33.85	5.00	35.54	0.90	0.45	11.47	0.65	47.66	nr	53.62
250 x 250 x 350mm............	29.15	5.00	30.61	0.75	0.40	9.73	0.55	40.89	nr	46.00
250 x 250 x 500mm............	41.80	5.00	43.89	1.00	0.50	12.74	0.85	57.48	nr	64.67
380 x 200 x 350mm............	36.00	5.00	37.80	0.90	0.45	11.47	0.80	50.07	nr	56.32
380 x 200 x 500mm............	51.70	5.00	54.28	1.15	0.55	14.48	0.90	69.66	nr	78.37
Quoin stones; rock worked face; attached faced two adjacent faces										
250 x 200 x 350mm............	23.65	5.00	24.83	0.65	0.30	8.11	0.50	33.44	nr	37.62
250 x 200 x 500mm............	33.85	5.00	35.54	0.95	0.45	11.93	0.65	48.12	nr	54.13
250 x 250 x 350mm............	29.15	5.00	30.61	0.80	0.40	10.19	0.55	41.35	nr	46.52
250 x 250 x 500mm............	41.80	5.00	43.89	1.10	0.55	14.01	0.85	58.75	nr	66.10
380 x 200 x 350mm............	36.00	5.00	37.80	0.95	0.50	12.28	0.80	50.88	nr	57.24
380 x 200 x 500mm............	51.70	5.00	54.28	1.25	0.60	15.75	0.90	70.93	nr	79.80
Arches; relieving										
225mm high on face, 180mm wide on soffit	93.50	5.00	98.17	1.75	0.85	22.12	1.10	121.39	m	136.57
225mm high on face, 250mm wide on soffit	129.25	5.00	135.71	2.30	1.15	29.30	1.55	166.56	m	187.39
Grooves										
12 x 25mm	-	-	-	0.30	-	2.77	-	2.77	m	3.12
12 x 38mm	-	-	-	0.35	-	3.23	-	3.23	m	3.63
Stone rubble work; random rubble backing and squared rubble face stones of Yorkshire limestone; bedding and jointing in lime mortar (1:3); regular coursed; courses average 150mm high										
Walls; vertical; face stones 100 – 150mm on bed; scappled or axed face; weather struck pointing as work proceeds										
350mm thick; faced one side	86.90	15.00	99.94	4.85	4.85	78.81	9.00	187.75	m2	211.22

Labour hourly rates: (except Specialists) Craft Operatives £9.23 Labourer £7.02 Rates are national average prices. Refer to REGIONAL VARIATIONS for indicative levels of overall pricing in regions	MATERIALS			LABOUR				RATES		
	Del to Site £	Waste %	Material Cost £	Craft Optve Hrs	Lab Hrs	Labour Cost £	Sunds £	Nett Rate £	Unit	Gross Rate (+12.5%) £

F20: NATURAL STONE RUBBLE WALLING Cont.

Stone rubble work; random rubble backing and squared rubble face stones of Yorkshire limestone; bedding and jointing in lime mortar (1:3); regular coursed; courses average 150mm high Cont.

Walls; vertical; face stones 125 - 200mm on bed; scappled or axed face; weather struck pointing as work proceeds

500mm thick; faced one side	106.70	15.00	122.71	5.70	5.70	92.63	13.00	228.33	m2	256.87

Walls; vertical; face stones 100 - 150mm on bed; hammer dressed face; weather struck pointing as work proceeds

350mm thick; faced one side	86.90	15.00	99.94	5.15	5.15	83.69	9.00	192.62	m2	216.70
500mm thick; faced one side	106.70	15.00	122.71	6.00	6.00	97.50	13.00	233.21	m2	262.36

Walls; vertical; face stones 100 - 150mm on bed; rock worked face; pointing with a parallel joint as work proceeds

350mm thick; faced one side	86.90	15.00	99.94	5.50	5.50	89.38	9.00	198.31	m2	223.10
500mm thick; faced one side	106.70	15.00	122.71	6.40	6.40	104.00	13.00	239.71	m2	269.67

Natural stonework

Note
> stonework is work for a specialist; prices being obtained for specific projects. Some firms that specialise in this class of work are included within the list at the end of this section

Natural stonework dressings; natural Dorset limestone; Portland Whitbed; bedding and jointing in mason's mortar (1:3:12); flush smooth pointing as work proceeds; slurrying with weak lime mortar and cleaning down on completion

Walls; vertical; building against brickwork; B.S.1243 Fig 1 specification 3.5 wall ties - 4/m2 built in

50mm thick; plain and rubbed one side	121.00	2.50	124.03	4.60	3.00	63.52	6.05	193.59	m2	217.79
75mm thick; plain and rubbed one side	154.00	2.50	157.85	5.20	3.60	73.27	6.60	237.72	m2	267.43
100mm thick; plain and rubbed one side	203.50	2.50	208.59	5.75	4.25	82.91	7.15	298.64	m2	335.98

Natural stonework; natural Dorset limestone, Portland Whitbed; bedding and jointing in cement-lime mortar (1:2:9); flush smooth pointing as work proceeds

Lintels
plain and rubbed faces -3; splayed and rubbed faces -1

200 x 100mm	63.80	2.50	65.39	1.45	1.00	20.40	1.40	87.20	m	98.10
225 x 125mm	82.50	2.50	84.56	1.70	1.25	24.47	1.80	110.83	m	124.68

Sills
plain and rubbed faces -3; sunk weathered and rubbed faces -1; grooves -1; throats -1

200 x 75mm	88.00	2.50	90.20	1.15	0.75	15.88	2.10	108.18	m	121.70
250 x 75mm	106.70	2.50	109.37	1.40	0.90	19.24	2.50	131.11	m	147.50
300 x 75mm	123.20	2.50	126.28	1.60	1.15	22.84	2.70	151.82	m	170.80

Jamb stones; attached
plain and rubbed faces -3; splayed and rubbed faces -1; rebates -1; grooves -1

175 x 75mm	68.20	2.50	69.91	1.15	0.75	15.88	1.85	87.63	m	98.59
200 x 100mm	73.70	2.50	75.54	1.45	1.00	20.40	2.40	98.35	m	110.64

Band courses; moulded; horizontal

225 x 125mm	145.20	2.50	148.83	1.85	1.30	26.20	3.30	178.33	m	200.62
250 x 150mm	180.40	2.50	184.91	2.05	1.55	29.80	3.85	218.56	m	245.88
300 x 150mm	195.80	2.50	200.69	2.30	1.80	33.87	4.50	239.06	m	268.94

Copings; horizontal
plain and rubbed faces -2; weathered and rubbed faces -1; throats -2; cramped joints with stainless steel cramps

300 x 50mm	107.00	2.50	109.68	0.85	0.60	12.06	4.65	126.38	m	142.18
300 x 75mm	124.00	2.50	127.10	1.00	0.80	14.85	5.20	147.15	m	165.54
375 x 100mm	172.00	2.50	176.30	1.55	1.15	22.38	6.25	204.93	m	230.55

Natural stonework, natural Yorkshire sandstone, Bolton Wood; bedding and jointing in cement-lime mortar (1:2:9); flush smooth pointing as work proceeds

Sills
plain and rubbed faces -3; sunk weathered and rubbed faces -1; throats -1

175 x 100mm	63.80	2.50	65.39	1.45	1.00	20.40	1.10	86.90	m	97.76
250 x 125mm	99.00	2.50	101.47	1.60	1.15	22.84	1.85	126.17	m	141.94

Copings; horizontal
plain and rubbed faces -2; weathered and rubbed faces -2; throats -2; cramped joints with stainless steel cramps

300 x 75mm	63.25	2.50	64.83	1.05	0.80	15.31	4.25	84.39	m	94.94
300 x 100mm	77.00	2.50	78.92	1.25	0.90	17.86	4.75	101.53	m	114.22

MASONRY

Labour hourly rates: (except Specialists) Craft Operatives £9.23 Labourer £7.02 Rates are national average prices. Refer to REGIONAL VARIATIONS for indicative levels of overall pricing in regions	MATERIALS			LABOUR				RATES		
	Del to Site	Waste	Material Cost	Craft Optve	Lab	Labour Cost	Sunds	Nett Rate	Unit	Gross Rate (+12.5%)
	£	%	£	Hrs	Hrs	£	£	£		£
F20: NATURAL STONE RUBBLE WALLING Cont.										
Natural stonework, natural Yorkshire sandstone, Bolton Wood; bedding and jointing in cement-lime mortar (1:2:9); flush smooth pointing as work proceeds Cont.										
Kerbs; horizontal										
plain and sawn faces - 4										
150 x 150mm..........................	38.00	2.50	38.95	1.00	0.75	14.49	1.25	54.70	m	61.53
225 x 150mm..........................	57.20	2.50	58.63	1.40	1.15	21.00	1.85	81.47	m	91.66
Cover stones										
75 x 300; rough edges -2; plain and sawn face -1 .	34.00	2.50	34.85	1.00	0.75	14.49	2.45	51.80	m	58.27
Templates										
150 x 300; rough edges -2; plain and sawn face -1	63.50	2.50	65.09	1.70	1.40	25.52	3.00	93.61	m	105.31
Steps; plain										
plain and rubbed top and front										
225 x 75mm...........................	28.50	2.50	29.21	0.85	0.65	12.41	2.45	44.07	m	49.58
225 x 150mm..........................	57.20	2.50	58.63	1.50	1.15	21.92	3.00	83.55	m	93.99
300 x 150mm..........................	76.15	2.50	78.05	1.85	1.50	27.61	3.65	109.31	m	122.97
Landings										
75 x 900 x 900mm; sawn edges -4; plain and rubbed face -1..	102.50	2.50	105.06	2.30	2.00	35.27	3.65	143.98	nr	161.98
F22: CAST STONE ASHLAR WALLING/DRESSINGS										
Cast stonework dressings; simulated Dorset limestone; Portland Whitbed; bedding and jointing in cement-lime mortar (1:2:9); flush smooth pointing as work proceeds										
Walls; vertical building against brickwork; B.S.1243 Fig 1 specification 3.5 wall ties -4/m2 built in										
100mm thick; plain and rubbed one side	79.75	5.00	83.74	4.60	2.90	62.82	6.10	152.65	m2	171.74
Fair raking cutting										
100 thick	16.50	-	16.50	1.15	0.60	14.83	-	31.33	m	35.24
Cast stonework; simulated Dorset limestone; Portland Whitbed; bedding and jointing in cement-lime mortar (1:2:9); flush smooth pointing as work proceeds										
Lintels										
plain and rubbed faces -3; splayed and rubbed faces -1										
200 x 100mm..........................	25.30	5.00	26.57	1.15	0.70	15.53	1.00	43.09	m	48.48
225 x 125mm..........................	37.40	5.00	39.27	1.45	0.85	19.35	1.25	59.87	m	67.35
Sills										
plain and rubbed faces -3; sunk weathered and rubbed faces -1; grooves -1; throats -1										
200 x 75mm...........................	23.10	5.00	24.25	1.05	0.60	13.90	1.25	39.41	m	44.33
extra; stoolings.....................	5.50	5.00	5.78	-	-	-	-	5.78	nr	6.50
250 x 75mm...........................	25.30	5.00	26.57	1.15	0.70	15.53	1.60	43.69	m	49.16
extra; stoolings.....................	5.50	5.00	5.78	-	-	-	-	5.78	nr	6.50
300 x 75mm...........................	29.70	5.00	31.18	1.45	0.85	19.35	1.90	52.44	m	58.99
extra; stoolings.....................	5.50	5.00	5.78	-	-	-	-	5.78	nr	6.50
Jamb stones; attached										
plain and rubbed faces -3; splayed and rubbed faces -1; rebates -1; grooves -1										
175 x 75mm...........................	24.75	5.00	25.99	1.05	0.60	13.90	0.90	40.79	m	45.89
200 x 100mm..........................	31.90	5.00	33.49	1.15	0.70	15.53	1.90	50.92	m	57.29
Band courses; plain; horizontal										
225 x 125mm	24.75	5.00	25.99	1.60	1.15	22.84	1.60	50.43	m	56.73
extra; external return	17.60	5.00	18.48	-	-	-	-	18.48	nr	20.79
250 x 150mm	30.25	5.00	31.76	1.75	1.25	24.93	1.90	58.59	m	65.91
extra; external return	17.60	5.00	18.48	-	-	-	-	18.48	nr	20.79
300 x 150mm	38.50	5.00	40.42	1.85	1.40	26.90	2.20	69.53	m	78.22
extra; external return	17.60	5.00	18.48	-	-	-	-	18.48	nr	20.79
Copings; horizontal										
plain and rubbed faces -2; weathered and rubbed faces -2; throats -2; cramped joints with stainless steel cramps										
300 x 50mm	19.25	5.00	20.21	0.60	0.35	8.00	3.75	31.96	m	35.95
extra; internal angles	17.60	5.00	18.48	-	-	-	-	18.48	nr	20.79
extra; external angles	17.60	5.00	18.48	-	-	-	-	18.48	nr	20.79
300 x 75mm	22.00	5.00	23.10	0.70	0.40	9.27	4.00	36.37	m	40.92
extra; internal angles	17.60	5.00	18.48	-	-	-	-	18.48	nr	20.79
extra; external angles	17.60	5.00	18.48	-	-	-	-	18.48	nr	20.79
375 x 100mm	30.80	5.00	32.34	0.85	0.60	12.06	4.75	49.15	m	55.29
extra; internal angles	17.60	5.00	18.48	-	-	-	-	18.48	nr	20.79
extra; external angles	17.60	5.00	18.48	-	-	-	-	18.48	Unr	20.79
Steps; plain										
300 x 150mm; plain and rubbed tread and riser	30.00	5.00	31.50	1.45	1.05	20.75	2.50	54.75	m	61.60

Labour hourly rates: (except Specialists) Craft Operatives £9.23 Labourer £7.02 Rates are national average prices. Refer to REGIONAL VARIATIONS for indicative levels of overall pricing in regions	MATERIALS			LABOUR				RATES		
	Del to Site £	Waste %	Material Cost £	Craft Optve Hrs	Lab Hrs	Labour Cost £	Sunds £	Nett Rate £	Unit	Gross Rate (+12.5%) £
F22: CAST STONE ASHLAR WALLING/DRESSINGS Cont.										
Cast stonework; simulated Dorset limestone; Portland Whitbed; bedding and jointing in cement-lime mortar (1:2:9); flush smooth pointing as work proceeds Cont.										
Steps; spandril 250mm wide tread; 180mm high riser; plain and rubbed tread and riser; carborundum finish to tread	49.50	5.00	51.98	2.30	1.45	31.41	1.90	85.28	m	95.94
Landings 150 x 900 x 900mm; plain and rubbed top surface	79.75	5.00	83.74	1.75	1.15	24.23	2.50	110.46	nr	124.27
F30: ACCESSORIES/SUNDRY ITEMS FOR BRICK/BLOCK/STONE WALLING										
Forming cavities in hollow walls										
Cavity with B.S.1243 Fig 1 galvanised wire butterfly wall ties, -4/m2 built in width of cavity 50mm	0.35	-	0.35	0.10	0.10	1.63	-	1.98	m2	2.22
Cavity with B.S.1243 Fig 1 stainless steel wire butterfly wall ties, -4/m2 built in width of cavity 50mm	0.37	-	0.37	0.10	0.10	1.63	-	2.00	m2	2.24
Cavity with B.S.1243 Fig 3 galvanised steel twisted wall ties, -4/m2 built in width of cavity 50mm	1.06	-	1.06	0.10	0.10	1.63	-	2.69	m2	3.02
Cavity with B.S.1243 Fig 3 stainless steel twisted wall ties, -4/m2 built in width of cavity 50mm	1.71	-	1.71	0.10	0.10	1.63	-	3.34	m2	3.75
Cavity with B.S.1243 Fig 3 galvanised steel twisted wall ties -4/m2 built in; 50 fibreglass resin bonded slab cavity insulation width of cavity 50mm	3.41	10.00	3.75	0.35	0.35	5.69	1.01	10.45	m2	11.75
Cavity with B.S.1243 Fig 3 galvanised steel twisted wall ties -4/m2 built in; 75 fibreglass resin bonded slab cavity insulation width of cavity 75mm	4.66	10.00	5.13	0.40	0.40	6.50	1.01	12.64	m2	14.22
Damp proof courses										
B.S.6398 Class A, bitumen with hessian base; 100mm laps; bedding in cement mortar (1:3); no allowance made for laps										
width not exceeding 225mm; vertical	6.05	5.00	6.35	0.50	-	4.62	-	10.97	m2	12.34
width exceeding 225mm; vertical	6.05	5.00	6.35	0.45	-	4.15	-	10.51	m2	11.82
width not exceeding 225mm; horizontal	6.05	5.00	6.35	0.35	-	3.23	-	9.58	m2	10.78
width exceeding 225mm; horizontal	6.05	5.00	6.35	0.30	-	2.77	-	9.12	m2	10.26
cavity trays; width not exceeding 225mm; horizontal	6.05	5.00	6.35	0.55	-	5.08	-	11.43	m2	12.86
cavity trays; width exceeding 225mm; horizontal	6.05	5.00	6.35	0.50	-	4.62	-	10.97	m2	12.34
B.S.6398 Class B, bitumen with fibre base; 100mm laps; bedding in cement mortar (1:3); no allowance made for laps										
width not exceeding 225mm; vertical	4.07	5.00	4.27	0.50	-	4.62	-	8.89	m2	10.00
width exceeding 225mm; vertical	4.07	5.00	4.27	0.45	-	4.15	-	8.43	m2	9.48
width not exceeding 225mm; horizontal	4.07	5.00	4.27	0.35	-	3.23	-	7.50	m2	8.44
width exceeding 225mm; horizontal	4.07	5.00	4.27	0.30	-	2.77	-	7.04	m2	7.92
B.S.6398 Class D, bitumen with hessian base laminated with lead; 100mm laps; bedding in cement mortar (1:3); no allowance made for laps										
width not exceeding 225mm; vertical	15.18	5.00	15.94	0.50	-	4.62	-	20.55	m2	23.12
width exceeding 225mm; vertical	15.18	5.00	15.94	0.45	-	4.15	-	20.09	m2	22.60
width not exceeding 225mm; horizontal	15.18	5.00	15.94	0.35	-	3.23	-	19.17	m2	21.57
width exceeding 225mm; horizontal	15.18	5.00	15.94	0.30	-	2.77	-	18.71	m2	21.05
B.S.6398 Class E, bitumen with fibre base laminated with lead; 100mm laps; bedding in cement mortar (1:3); no allowance made for laps										
width not exceeding 225mm; vertical	14.52	5.00	15.25	0.50	-	4.62	-	19.86	m2	22.34
width exceeding 225mm; vertical	14.52	5.00	15.25	0.45	-	4.15	-	19.40	m2	21.82
width not exceeding 225mm; horizontal	14.52	5.00	15.25	0.35	-	3.23	-	18.48	m2	20.79
width exceeding 225mm; horizontal	14.52	5.00	15.25	0.30	-	2.77	-	18.02	m2	20.27
Hyload, pitch polymer; 100mm laps; bedding in cement lime mortar (1:1:6); no allowance made for laps										
width not exceeding 225mm; vertical	7.43	5.00	7.80	0.50	-	4.62	-	12.42	m2	13.97
width exceeding 225mm; vertical	7.43	5.00	7.80	0.45	-	4.15	-	11.96	m2	13.45
width not exceeding 225mm; horizontal	7.43	5.00	7.80	0.35	-	3.23	-	11.03	m2	12.41
width exceeding 225mm; horizontal	7.43	5.00	7.80	0.30	-	2.77	-	10.57	m2	11.89
Synthaprufe bituminous latex emulsion; two coats brushed on; blinded with sand										
width not exceeding 225mm; vertical	3.28	5.00	3.44	-	0.60	4.21	0.15	7.81	m2	8.78
width exceeding 225mm; vertical	3.28	5.00	3.44	-	0.35	2.46	0.15	6.05	m2	6.81

MASONRY

Labour hourly rates: (except Specialists) Craft Operatives £9.23 Labourer £7.02 Rates are national average prices. Refer to REGIONAL VARIATIONS for indicative levels of overall pricing in regions	MATERIALS			LABOUR				RATES		
	Del to Site £	Waste %	Material Cost £	Craft Optve Hrs	Lab Hrs	Labour Cost £	Sunds £	Nett Rate £	Unit	Gross Rate (+12.5%) £
F30: ACCESSORIES/SUNDRY ITEMS FOR BRICK/BLOCK/STONE WALLING Cont.										
Damp proof courses Cont.										
Synthaprufe bituminous latex emulsion; three coats brushed on; blinded with sand										
width not exceeding 225mm; vertical	4.70	5.00	4.93	-	0.90	6.32	0.18	11.43	m2	12.86
width exceeding 225mm; vertical	4.70	5.00	4.93	-	0.50	3.51	0.18	8.63	m2	9.70
Bituminous emulsion; two coats brushed on										
width not exceeding 225mm; vertical	1.62	5.00	1.70	-	0.55	3.86	0.15	5.71	m2	6.43
width exceeding 225mm; vertical	1.62	5.00	1.70	-	0.36	2.53	0.15	4.38	m2	4.93
Bituminous emulsion; three coats brushed on										
width not exceeding 225mm; vertical	2.07	5.00	2.17	-	0.75	5.26	0.18	7.62	m2	8.57
width exceeding 225mm; vertical	2.07	5.00	2.17	-	0.40	2.81	0.18	5.16	m2	5.81
One course slates in cement mortar (1:3)										
width not exceeding 225mm; vertical	15.40	15.00	17.71	1.30	1.30	21.13	-	38.84	m2	43.69
width exceeding 225mm; vertical	15.40	15.00	17.71	1.20	1.20	19.50	-	37.21	m2	41.86
width not exceeding 225mm; horizontal	15.40	15.00	17.71	0.80	0.80	13.00	-	30.71	m2	34.55
width exceeding 225mm; horizontal	15.40	15.00	17.71	0.70	0.70	11.38	-	29.09	m2	32.72
Two courses slates in cement mortar (1:3)										
width not exceeding 225mm; vertical	30.80	15.00	35.42	2.20	2.20	35.75	-	71.17	m2	80.07
width exceeding 225mm; vertical	30.80	15.00	35.42	2.00	2.00	32.50	-	67.92	m2	76.41
width not exceeding 225mm; horizontal	30.80	15.00	35.42	1.30	1.30	21.13	-	56.55	m2	63.61
width exceeding 225mm; horizontal	30.80	15.00	35.42	1.20	1.20	19.50	-	54.92	m2	61.78
Cavity trays										
Type G Cavitray by Cavity Trays Ltd in										
half brick skin of cavity wall	7.78	5.00	8.17	0.20	-	1.85	-	10.02	m	11.27
external angle	7.00	5.00	7.35	0.10	-	0.92	-	8.27	nr	9.31
internal angle	7.00	5.00	7.35	0.10	-	0.92	-	8.27	nr	9.31
Type X Cavitray by Cavity Trays Ltd to suit 40 degree pitched roof complete with attached code 4 lead flashing and dress over tiles in										
half brick skin of cavity wall	17.63	5.00	18.51	0.20	-	1.85	-	20.36	m	22.90
ridge tray	4.08	5.00	4.28	0.10	-	0.92	-	5.21	nr	5.86
catchment tray	2.09	5.00	2.19	0.10	-	0.92	-	3.12	nr	3.51
corner catchment angle tray	4.08	5.00	4.28	0.10	-	0.92	-	5.21	nr	5.86
Type W Cavity weep/ventilator by Cavity Trays Ltd in										
half brick skin of cavity wall	0.45	5.00	0.47	0.10	-	0.92	-	1.40	nr	1.57
extension duct	0.57	5.00	0.60	0.10	-	0.92	-	1.52	nr	1.71
Joint reinforcement										
Expamet grade 304/S15 Exmet reinforcement, stainless steel; 150mm laps; no allowance made for laps										
65mm wide	0.59	5.00	0.62	0.04	0.04	0.65	0.04	1.31	m	1.47
115mm wide	1.01	5.00	1.06	0.05	0.05	0.81	0.04	1.91	m	2.15
175mm wide	1.64	5.00	1.72	0.06	0.06	0.97	0.05	2.75	m	3.09
225mm wide	2.22	5.00	2.33	0.07	0.07	1.14	0.06	3.53	m	3.97
Bed Joint Reinforcement (UK) Ltd grade 304/S15 Brickspan reinforcement, stainless steel; 150mm laps; no allowance made for laps										
50mm wide	1.33	5.00	1.40	0.04	0.04	0.65	0.04	2.09	m	2.35
60mm wide	1.33	5.00	1.40	0.04	0.04	0.65	0.04	2.09	m	2.35
80mm wide	1.42	5.00	1.49	0.05	0.05	0.81	0.04	2.34	m	2.64
100mm wide	1.42	5.00	1.49	0.05	0.05	0.81	0.04	2.34	m	2.64
150mm wide	1.66	5.00	1.74	0.05	0.05	0.81	0.04	2.60	m	2.92
160mm wide	1.66	5.00	1.74	0.06	0.06	0.97	0.05	2.77	m	3.11
Bed Joint Reinforcement (UK) Ltd grade 304/S15 Wallspan reinforcement, stainless steel; 150mm laps; no allowance made for laps										
210mm wide	5.90	5.00	6.20	0.06	0.06	0.97	0.05	7.22	m	8.12
222.5mm wide	5.90	5.00	6.20	0.06	0.06	0.97	0.05	7.22	m	8.12
235mm wide	5.90	5.00	6.20	0.07	0.07	1.14	0.06	7.39	m	8.32
250mm wide	5.90	5.00	6.20	0.07	0.07	1.14	0.06	7.39	m	8.32
275mm wide	5.90	5.00	6.20	0.07	0.07	1.14	0.06	7.39	m	8.32
Pointing in flashings										
Cement mortar (1:3)										
horizontal	-	-	-	0.45	-	4.15	0.20	4.35	m	4.90
horizontal; in old wall	-	-	-	0.50	-	4.62	0.20	4.82	m	5.42
stepped	-	-	-	0.80	-	7.38	0.32	7.70	m	8.67
stepped; in old wall	-	-	-	0.85	-	7.85	0.32	8.17	m	9.19
Wedging and pinning										
Two courses slates in cement mortar (1:3)										
width of wall 215mm	7.07	15.00	8.13	0.50	0.50	8.13	0.11	16.37	m	18.41

Labour hourly rates: (except Specialists) Craft Operatives £9.23 Labourer £7.02 Rates are national average prices. Refer to REGIONAL VARIATIONS for indicative levels of overall pricing in regions	MATERIALS			LABOUR				RATES		
	Del to Site £	Waste %	Material Cost £	Craft Optve Hrs	Lab Hrs	Labour Cost £	Sunds £	Nett Rate £	Unit	Gross Rate (+12.5%) £
F30: ACCESSORIES/SUNDRY ITEMS FOR BRICK/BLOCK/STONE WALLING Cont.										
Joints										
Expansion joints in facing brickwork 13mm wide; vertical; filling with Servicised Ltd, Fibrepack filler, Vertiseal compound pointing one side; including preparation, cleaners, primers and sealers										
102mm thick wall	2.73	7.50	2.93	0.10	0.20	2.33	–	5.26	m	5.92
215mm thick wall	3.66	7.50	3.93	0.11	0.22	2.56	–	6.49	m	7.31
Expansion joints in blockwork 13mm wide; vertical; filling with Servicised Ltd, Fibrepack filler, Vertiseal compound pointing one side; including preparation, cleaners, primers and sealers										
100mm thick wall	2.73	7.50	2.93	0.10	0.20	2.33	–	5.26	m	5.92
Expansion joints in glass blockwork 10mm wide; vertical; in filling with compressible material, polysulphide sealant both sides including preparation, cleaners, primers and sealers										
80mm thick wall	2.61	7.50	2.81	0.10	0.20	2.33	–	5.13	m	5.77
Slates and tiles for creasing										
Nibless creasing tiles, red, machine made, 265 x 165 x 10mm; in cement-lime mortar (1:1:6)										
one course 253mm wide	4.33	5.00	4.55	0.50	0.50	8.13	0.21	12.88	m	14.49
one course 365mm wide	7.03	5.00	7.38	0.70	0.70	11.38	0.29	19.05	m	21.43
two courses 253mm wide	8.67	5.00	9.10	0.85	0.85	13.81	0.42	23.34	m	26.25
two courses 365mm wide	14.06	5.00	14.76	1.15	1.15	18.69	0.56	34.01	m	38.26
Slate and tile sills										
Clay plain roofing tiles, red, B.S.402, machine made, 265 x 165 x 13mm; in cement-lime mortar										
one course 150mm wide; set weathering	2.67	5.00	2.80	0.45	0.45	7.31	0.23	10.35	m	11.64
two courses 150mm wide; set weathering	5.34	5.00	5.61	0.75	0.75	12.19	0.40	18.19	m	20.47
Fires and fire parts										
Solid one piece or two piece firebacks; B.S.1251; bed and joint in fire cement; concrete filling at back										
fire size 400mm	30.00	2.50	30.75	1.00	1.00	16.25	1.05	48.05	nr	54.06
fire size 450mm	36.10	2.50	37.00	1.10	1.10	17.88	1.17	56.05	nr	63.05
Solid one piece or two piece firebacks with cut out for boiler; B.S.1251; bed and joint in fire cement; concrete filling at back										
fire size 400mm	36.45	2.50	37.36	1.00	1.00	16.25	1.05	54.66	nr	61.49
fire size 450mm	42.46	2.50	43.52	1.10	1.10	17.88	1.17	62.57	nr	70.39
Frets and stools; black vitreous enamelled; place in position										
fire size 400mm	25.39	–	25.39	0.33	0.33	5.36	–	30.75	nr	34.60
fire size 450mm	39.68	–	39.68	0.36	0.36	5.85	–	45.53	nr	51.22
Frets and stools; lustre finish; place in position										
fire size 400mm	27.08	–	27.08	0.33	0.33	5.36	–	32.44	nr	36.50
fire size 450mm	41.33	–	41.33	0.36	0.36	5.85	–	47.18	nr	53.08
Continuous burning open fire; vitreous finish; self contained open fire; plugging and screwing; sealing to opening										
fire size 400mm	127.05	–	127.05	3.50	3.50	56.88	2.15	186.07	nr	209.33
fire size 450mm	145.20	–	145.20	3.75	3.75	60.94	4.17	210.31	nr	236.60
Gas ignited smokeless fuel fires; assembling; plugging and screwing; bedding and jointing in fire cement; sealing to opening										
grate only; black										
fire size 400mm	82.50	–	82.50	2.50	2.50	40.63	–	123.13	nr	138.52
fire size 450mm	83.60	–	83.60	2.50	2.50	40.63	–	124.22	nr	139.75
grate only; colours										
fire size 400mm	83.60	–	83.60	2.50	2.50	40.63	–	124.22	nr	139.75
fire size 450mm	84.70	–	84.70	2.50	2.50	40.63	–	125.33	nr	140.99
boiler and self contained flue										
fire size 400mm	191.40	–	191.40	4.50	4.50	73.13	2.15	266.68	nr	300.01
fire size 450mm	203.50	–	203.50	4.50	4.50	73.13	4.17	280.80	nr	315.89
threefold brick sets										
fire size 400mm	25.41	–	25.41	1.00	1.00	16.25	1.05	42.71	nr	48.05
fire size 450mm	25.41	–	25.41	1.00	1.00	16.25	1.17	42.83	nr	48.18
extra; chrome finish to fire front; fire size 400 or 450mm	25.41	–	25.41	–	–	–	–	25.41	nr	28.59
extra; back boiler unit with Bower Barffed rustless boiler; fire size 400 or 450mm	198.00	–	198.00	–	–	–	–	198.00	nr	222.75
extra; back boiler unit with copper boiler; 400 or 450mm	302.50	–	302.50	–	–	–	–	302.50	nr	340.31

MASONRY

Labour hourly rates: (except Specialists) Craft Operatives £9.23 Labourer £7.02 Rates are national average prices. Refer to REGIONAL VARIATIONS for indicative levels of overall pricing in regions	MATERIALS			LABOUR				RATES		
	Del to Site £	Waste %	Material Cost £	Craft Optve Hrs	Lab Hrs	Labour Cost £	Sunds £	Nett Rate £	Unit	Gross Rate (+12.5%) £
F30: ACCESSORIES/SUNDRY ITEMS FOR BRICK/BLOCK/STONE WALLING Cont.										
Flue linings										
Clay flue linings, B.S.1181; rebated joints; jointed in cement mortar (1:3)										
150mm diameter, Type 2	17.53	5.00	18.41	0.40	0.45	6.85	0.30	25.56	m	28.75
terminal Type 6F	16.78	5.00	17.62	0.45	0.40	6.96	0.15	24.73	nr	27.82
185mm diameter, Type 2	22.48	5.00	23.60	0.50	0.40	7.42	0.39	31.42	m	35.34
terminal Type 6F	18.24	5.00	19.15	0.50	0.40	7.42	0.18	26.75	m	30.10
225mm diameter, Type 2	37.48	5.00	39.35	0.60	0.80	11.15	0.42	50.93	m	57.29
terminal Type 6F	19.64	5.00	20.62	0.60	0.50	9.05	0.22	29.89	nr	33.63
185 x 185mm, Type 1	24.15	5.00	25.36	0.60	0.80	11.15	0.39	36.90	m	41.51
terminal Type 4D	27.31	5.00	28.68	0.45	0.40	6.96	0.18	35.82	nr	40.29
Chimney pots										
Clay chimney pots; set and flaunched in cement mortar (1:3)										
tapered roll top; 600mm high	31.87	5.00	33.46	0.80	0.95	14.05	0.39	47.91	nr	53.89
tapered roll top; 750mm high	43.37	5.00	45.54	0.95	1.15	16.84	0.50	62.88	nr	70.74
tapered roll top; 900mm high	58.08	5.00	60.98	1.10	1.40	19.98	0.58	81.55	nr	91.74
Air bricks										
Clay air bricks, B.S.493, square hole pattern; opening with slate lintel over										
common brick wall 102mm thick; opening size										
225 x 75mm..........................	1.67	-	1.67	0.20	0.20	3.25	0.58	5.50	nr	6.19
225 x 150mm.........................	2.31	-	2.31	0.25	0.25	4.06	0.62	6.99	nr	7.87
225 x 225mm.........................	6.35	-	6.35	0.30	0.30	4.88	0.66	11.89	nr	13.37
common brick wall 215mm thick; opening size										
225 x 75mm..........................	1.67	-	1.67	0.30	0.30	4.88	1.10	7.64	nr	8.60
225 x 150mm.........................	2.31	-	2.31	0.38	0.38	6.17	1.16	9.64	nr	10.85
225 x 225mm.........................	6.35	-	6.35	0.45	0.45	7.31	1.23	14.89	nr	16.75
common brick wall 215mm thick; facework one side; opening size										
225 x 75mm..........................	1.67	-	1.67	0.40	0.40	6.50	1.10	9.27	nr	10.43
225 x 150mm.........................	2.31	-	2.31	0.50	0.50	8.13	1.16	11.60	nr	13.04
225 x 225mm.........................	6.35	-	6.35	0.60	0.60	9.75	1.23	17.33	nr	19.50
common brick wall 327mm thick; facework one side; opening size										
225 x 75mm..........................	1.67	-	1.67	0.60	0.60	9.75	1.78	13.20	nr	14.85
225 x 150mm.........................	2.31	-	2.31	0.70	0.70	11.38	1.87	15.56	nr	17.50
225 x 225mm.........................	6.35	-	6.35	0.80	0.80	13.00	1.93	21.28	nr	23.94
common brick wall 440mm thick; facework one side; opening size										
225 x 75mm..........................	1.67	-	1.67	0.80	0.80	13.00	2.37	17.04	nr	19.17
225 x 150mm.........................	2.31	-	2.31	0.90	0.90	14.63	2.63	19.57	nr	22.01
225 x 225mm.........................	6.35	-	6.35	1.00	1.00	16.25	3.08	25.68	nr	28.89
cavity wall 252mm thick with 102mm facing brick outer skin, 100mm block inner skin and 50mm cavity; sealing cavity with slates in cement mortar (1:3); rendering all round with cement mortar (1:3); opening size										
225 x 75mm..........................	1.67	-	1.67	0.60	0.60	9.75	2.13	13.55	nr	15.24
225 x 150mm.........................	2.31	-	2.31	0.70	0.70	11.38	2.63	16.32	nr	18.35
225 x 225mm.........................	6.35	-	6.35	0.80	0.80	13.00	3.08	22.43	nr	25.23
Gas flue blocks										
Marflex HP system precast refractory concrete gas flue blocks; bedding and jointing in Flue joint refractory mortar										
recess block reference HP1, 405 x 140 x 222mm	4.83	5.00	5.07	0.42	-	3.88	0.58	9.53	nr	10.72
cover block reference HP2, 385 x 140 x 222mm	6.96	5.00	7.31	0.51	-	4.71	0.58	12.60	nr	14.17
Standard block reference HP3, 355 x 140 x 72mm ...	4.68	5.00	4.91	0.34	-	3.14	0.58	8.63	nr	9.71
Standard block reference HP3, 355 x 140 x 112mm ..	4.68	5.00	4.91	0.34	-	3.14	0.58	8.63	nr	9.71
Standard block reference HP3, 355 x 140 x 222mm ..	4.68	5.00	4.91	0.34	-	3.14	0.58	8.63	nr	9.71
Vent unit for bathroom/kitchen extract fans reference HP1 and HP2	11.79	5.00	12.38	0.34	-	3.14	0.58	16.10	nr	18.11
standard block reference HP4, 280 x 140 x 72mm ..	4.51	5.00	4.74	0.28	-	2.58	0.52	7.84	nr	8.82
standard block reference HP4, 280 x 140 x 112mm ..	4.51	5.00	4.74	0.28	-	2.58	0.52	7.84	nr	8.82
standard block reference HP4, 280 x 140 x 222mm ..	4.51	5.00	4.74	0.28	-	2.58	0.52	7.84	nr	8.82
120mm side offset block reference HP5, 400 x 140 x 222mm	5.69	5.00	5.97	0.42	-	3.88	0.75	10.60	nr	11.93
70mm back offset block reference HP6, 280 x 210 x 222mm	14.64	5.00	15.37	0.39	-	3.60	0.75	19.72	nr	22.19
vertical exit block reference HP7, 280 x 181 x 222mm	9.37	5.00	9.84	0.39	-	3.60	0.58	14.02	nr	15.77
angled entry/exit block reference HP8, 280 x 140 x 230mm	9.70	5.00	10.19	0.45	-	4.15	0.79	15.13	nr	17.02
double rebate block used when in conjunction with a boiler reference HP9, 280 x 140 x 222mm	7.17	5.00	7.53	0.45	-	4.15	0.52	12.20	nr	13.73
corbel block reference HP10	9.16	5.00	9.62	0.45	-	4.15	0.58	14.35	nr	16.15
conversion unit for back boiler recess units reference HP25, 280 x 262 x 222mm	11.20	5.00	11.76	0.45	-	4.15	0.52	16.43	nr	18.49
Arch bars										
Steel flat arch bar										
30 x 6mm............................	1.73	-	1.73	0.80	-	7.38	-	9.11	m	10.25
50 x 6mm............................	2.66	-	2.66	0.80	-	7.38	-	10.04	m	11.30

MASONRY – SMALL WORKS

Labour hourly rates: (except Specialists) Craft Operatives £9.23 Labourer £7.02 Rates are national average prices. Refer to REGIONAL VARIATIONS for indicative levels of overall pricing in regions	MATERIALS			LABOUR				RATES		
	Del to Site £	Waste %	Material Cost £	Craft Optve Hrs	Lab Hrs	Labour Cost £	Sunds £	Nett Rate £	Unit	Gross Rate (+12.5%) £
F30: ACCESSORIES/SUNDRY ITEMS FOR BRICK/BLOCK/STONE WALLING Cont.										
Arch bars Cont.										
Steel angle arch bar										
50 x 50 x 6mm	4.97	-	4.97	0.80	-	7.38	-	12.35	m	13.90
75 x 50 x 6mm	6.31	-	6.31	0.90	-	8.31	-	14.62	m	16.44
Building in										
Building in metal windows; building in lugs; bedding in cement mortar (1:3), pointing with Secomastic standard mastic one side										
200mm high; lugs to brick jambs; plugging and screwing to brick sill and concrete head										
500mm wide	-	-	-	0.56	0.35	7.63	0.96	8.59	nr	9.66
600mm wide	-	-	-	0.61	0.38	8.30	1.19	9.49	nr	10.67
900mm wide	-	-	-	0.78	0.46	10.43	1.63	12.06	nr	13.57
1200mm wide	-	-	-	0.95	0.54	12.56	2.09	14.65	nr	16.48
1500mm wide	-	-	-	1.11	0.61	14.53	2.53	17.06	nr	19.19
1800mm wide	-	-	-	1.28	0.69	16.66	2.97	19.63	nr	22.08
500mm high; lugs to brick jambs; plugging and screwing to brick sill and concrete head										
500mm wide	-	-	-	0.67	0.39	8.92	1.46	10.38	nr	11.68
600mm wide	-	-	-	0.72	0.43	9.66	1.62	11.28	nr	12.69
900mm wide	-	-	-	0.89	0.50	11.72	2.08	13.80	nr	15.53
1200mm wide	-	-	-	1.07	0.58	13.95	2.51	16.46	nr	18.51
1500mm wide	-	-	-	1.24	0.68	16.22	2.97	19.19	nr	21.59
1800mm wide	-	-	-	1.41	0.76	18.35	3.39	21.74	nr	24.46
700mm high; lugs to brick jambs; plugging and screwing to brick sill and concrete head										
500mm wide	-	-	-	0.79	0.44	10.38	1.76	12.14	nr	13.66
600mm wide	-	-	-	0.84	0.47	11.05	1.90	12.95	nr	14.57
900mm wide	-	-	-	1.02	0.55	13.28	2.37	15.65	nr	17.60
1200mm wide	-	-	-	1.19	0.65	15.55	2.81	18.36	nr	20.65
1500mm wide	-	-	-	1.37	0.73	17.77	3.25	21.02	nr	23.65
1800mm wide	-	-	-	1.53	0.82	19.88	3.67	23.55	nr	26.49
900mm high; lugs to brick jambs; plugging and screwing to brick sill and concrete head										
500mm wide	-	-	-	0.89	0.47	11.51	2.04	13.55	nr	15.25
600mm wide	-	-	-	0.96	0.50	12.37	2.20	14.57	nr	16.39
900mm wide	-	-	-	1.13	0.60	14.64	2.64	17.28	nr	19.44
1200mm wide	-	-	-	1.31	0.69	16.94	3.08	20.02	nr	22.52
1500mm wide	-	-	-	1.49	0.79	19.30	3.53	22.83	nr	25.68
1800mm wide	-	-	-	1.66	0.87	21.43	3.95	25.38	nr	28.55
1100mm high; lugs to brick jambs; plugging and screwing to brick sill and concrete head										
500mm wide	-	-	-	1.01	0.52	12.97	2.35	15.32	nr	17.24
600mm wide	-	-	-	1.07	0.55	13.74	2.46	16.20	nr	18.22
900mm wide	-	-	-	1.25	0.65	16.10	2.93	19.03	nr	21.41
1200mm wide	-	-	-	1.43	0.74	18.39	3.38	21.77	nr	24.50
1500mm wide	-	-	-	1.62	0.84	20.85	3.83	24.68	nr	27.76
1800mm wide	-	-	-	1.80	0.93	23.14	4.28	27.42	nr	30.85
1300mm high; lugs to brick jambs; plugging and screwing to brick sill and concrete head										
500mm wide	-	-	-	1.12	0.57	14.34	2.61	16.95	nr	19.07
600mm wide	-	-	-	1.18	0.60	15.10	2.75	17.85	nr	20.09
900mm wide	-	-	-	1.37	0.69	17.49	3.20	20.69	nr	23.28
1200mm wide	-	-	-	1.55	0.79	19.85	3.65	23.50	nr	26.44
1500mm wide	-	-	-	1.73	0.90	22.29	4.11	26.40	nr	29.70
1800mm wide	-	-	-	1.92	0.99	24.67	4.55	29.22	nr	32.87
1500mm high; lugs to brick jambs; plugging and screwing to brick sill and concrete head										
500mm wide	-	-	-	1.24	0.61	15.73	2.92	18.65	nr	20.98
600mm wide	-	-	-	1.29	0.65	16.47	3.04	19.51	nr	21.95
900mm wide	-	-	-	1.49	0.74	18.95	3.50	22.45	nr	25.25
1200mm wide	-	-	-	1.67	0.85	21.38	3.94	25.32	nr	28.49
1500mm wide	-	-	-	1.86	0.95	23.84	4.37	28.21	nr	31.73
1800mm wide	-	-	-	2.05	1.06	26.36	4.84	31.20	nr	35.10
2100mm high; lugs to brick jambs; plugging and screwing to brick sill and concrete head										
500mm wide	-	-	-	1.46	0.69	18.32	3.77	22.09	nr	24.85
600mm wide	-	-	-	1.52	0.73	19.15	3.92	23.07	nr	25.96
900mm wide	-	-	-	1.72	0.84	21.77	4.36	26.13	nr	29.40
1200mm wide	-	-	-	1.91	0.95	24.30	4.84	29.14	nr	32.78
1500mm wide	-	-	-	2.11	1.07	26.99	5.25	32.24	nr	36.27
1800mm wide	-	-	-	2.30	1.18	29.51	5.72	35.23	nr	39.64
Building in factory glazed metal windows; screwing with galvanised screws; bedding in cement mortar (1:3), pointing with Secomastic standard mastic one side										
300mm high; plugging and screwing lugs to brick jambs and sill and concrete head										
600mm wide	-	-	-	0.74	0.48	10.20	1.32	11.52	nr	12.96
900mm wide	-	-	-	0.88	0.55	11.98	1.78	13.76	nr	15.48
1200mm wide	-	-	-	1.03	0.60	13.72	2.22	15.94	nr	17.93
1500mm wide	-	-	-	1.18	0.65	15.45	2.67	18.12	nr	20.39
1800mm wide	-	-	-	1.32	0.71	17.17	3.10	20.27	nr	22.80
2400mm wide	-	-	-	1.62	0.84	20.85	4.00	24.85	nr	27.96
3000mm wide	-	-	-	1.91	0.95	24.30	4.88	29.18	nr	32.83
700mm high; plugging and screwing lugs to brick jambs and sill and concrete head										
600mm wide	-	-	-	1.01	0.63	13.74	1.90	15.64	nr	17.60
900mm wide	-	-	-	1.23	0.76	16.69	2.37	19.06	nr	21.44

MASONRY

Labour hourly rates: (except Specialists) Craft Operatives £9.23 Labourer £7.02 Rates are national average prices. Refer to REGIONAL VARIATIONS for indicative levels of overall pricing in regions	MATERIALS			LABOUR				RATES		
	Del to Site	Waste	Material Cost	Craft Optve	Lab	Labour Cost	Sunds	Nett Rate	Unit	Gross Rate (+12.5%)
	£	%	£	Hrs	Hrs	£	£	£		£
F30: ACCESSORIES/SUNDRY ITEMS FOR BRICK/BLOCK/STONE WALLING Cont.										
Building in Cont.										
Building in factory glazed metal windows; screwing with galvanised screws; bedding in cement mortar (1:3), pointing with Secomastic standard mastic one side Cont.										
700mm high; plugging and screwing lugs to brick jambs and sill and concrete head Cont.										
1200mm wide	-	-	-	1.44	0.90	19.61	2.81	22.42	nr	25.22
1500mm wide	-	-	-	1.66	1.01	22.41	3.25	25.66	nr	28.87
1800mm wide	-	-	-	1.88	1.15	25.43	3.67	29.10	nr	32.73
2400mm wide	-	-	-	2.32	1.42	31.38	4.59	35.97	nr	40.47
3000mm wide	-	-	-	2.75	1.67	37.11	5.50	42.61	nr	47.93
900mm high; plugging and screwing lugs to brick jambs and sill and concrete head										
600mm wide	-	-	-	1.14	0.71	15.51	2.20	17.71	nr	19.92
900mm wide	-	-	-	1.40	0.87	19.03	2.64	21.67	nr	24.38
1200mm wide	-	-	-	1.65	1.04	22.53	3.08	25.61	nr	28.81
1500mm wide	-	-	-	1.90	1.20	25.96	3.53	29.49	nr	33.18
1800mm wide	-	-	-	2.16	1.37	29.55	3.95	33.50	nr	37.69
2400mm wide	-	-	-	2.67	1.70	36.58	4.86	41.44	nr	46.62
3000mm wide	-	-	-	3.17	2.03	43.51	5.76	49.27	nr	55.43
1100mm high; plugging and screwing lugs to brick jambs and sill and concrete head										
600mm wide	-	-	-	1.28	0.79	17.36	2.48	19.84	nr	22.32
900mm wide	-	-	-	1.56	0.98	21.28	2.93	24.21	nr	27.23
1200mm wide	-	-	-	1.86	1.18	25.45	3.37	28.82	nr	32.42
1500mm wide	-	-	-	2.14	1.39	29.51	3.83	33.34	nr	37.51
1800mm wide	-	-	-	2.44	1.59	33.68	4.28	37.96	nr	42.71
2400mm wide	-	-	-	3.01	2.00	41.82	5.16	46.98	nr	52.86
3000mm wide	-	-	-	3.59	2.39	49.91	6.06	55.97	nr	62.97
1300mm high; plugging and screwing lugs to brick jambs and sill and concrete head										
600mm wide	-	-	-	1.41	0.85	18.98	2.93	21.91	nr	24.65
900mm wide	-	-	-	1.74	1.10	23.78	3.20	26.98	nr	30.35
1200mm wide	-	-	-	2.06	1.34	28.42	3.66	32.08	nr	36.09
1500mm wide	-	-	-	2.39	1.56	33.01	4.11	37.12	nr	41.76
1800mm wide	-	-	-	2.72	1.81	37.81	4.55	42.36	nr	47.66
2400mm wide	-	-	-	3.37	2.28	47.11	5.41	52.52	nr	59.09
3000mm wide	-	-	-	4.02	2.76	56.48	6.30	62.78	nr	70.63
1500mm high; plugging and screwing lugs to brick jambs and sill and concrete head										
600mm wide	-	-	-	1.54	0.93	20.74	3.04	23.78	nr	26.76
900mm wide	-	-	-	1.91	1.21	26.12	3.50	29.62	nr	33.33
1200mm wide	-	-	-	2.27	1.48	31.34	3.94	35.28	nr	39.69
1500mm wide	-	-	-	2.64	1.75	36.65	4.40	41.05	nr	46.18
1800mm wide	-	-	-	2.99	2.03	41.85	4.84	46.69	nr	52.52
2400mm wide	-	-	-	3.72	2.58	52.45	5.73	58.18	nr	65.45
3000mm wide	-	-	-	4.44	3.12	62.88	6.61	69.49	nr	78.18
2100mm high; plugging and screwing lugs to brick jambs and sill and concrete head										
600mm wide	-	-	-	1.95	1.17	26.21	3.92	30.13	nr	33.90
900mm wide	-	-	-	2.43	1.54	33.24	4.36	37.60	nr	42.30
1200mm wide	-	-	-	2.89	1.92	40.15	4.83	44.98	nr	50.61
1500mm wide	-	-	-	3.36	2.30	47.16	5.25	52.41	nr	58.96
1800mm wide	-	-	-	3.83	2.69	54.23	5.69	59.92	nr	67.42
2400mm wide	-	-	-	4.77	3.45	68.25	6.59	74.84	nr	84.19
3000mm wide	-	-	-	5.70	4.21	82.17	7.47	89.64	nr	100.84
Building in wood windows; building in lugs; bedding in cement mortar (1:3), pointing with Secomastic standard mastic one side										
768mm high; lugs to brick jambs										
438mm wide	-	-	-	1.52	0.25	15.78	1.78	17.56	nr	19.76
641mm wide	-	-	-	1.75	0.28	18.12	2.10	20.22	nr	22.75
1225mm wide	-	-	-	2.39	0.32	24.31	2.99	27.30	nr	30.71
1809mm wide	-	-	-	3.05	0.35	30.61	3.89	34.50	nr	38.81
2394mm wide	-	-	-	3.68	0.38	36.63	4.81	41.44	nr	46.62
920mm high; lugs to brick jambs										
438mm wide	-	-	-	1.72	0.28	17.84	1.98	19.82	nr	22.30
641mm wide	-	-	-	1.94	0.32	20.15	2.29	22.44	nr	25.25
1225mm wide	-	-	-	2.61	0.35	26.55	3.20	29.75	nr	33.47
1809mm wide	-	-	-	3.27	0.41	33.06	4.11	37.17	nr	41.82
2394mm wide	-	-	-	3.95	0.44	39.55	5.01	44.56	nr	50.13
1073mm high; lugs to brick jambs										
438mm wide	-	-	-	1.92	0.32	19.97	2.22	22.19	nr	24.96
641mm wide	-	-	-	2.14	0.35	22.21	2.53	24.74	nr	27.83
1225mm wide	-	-	-	2.84	0.38	28.88	3.44	32.32	nr	36.36
1809mm wide	-	-	-	3.52	0.44	35.58	4.32	39.90	nr	44.89
2394mm wide	-	-	-	4.21	0.50	42.37	5.24	47.61	nr	53.56
1225mm high; lugs to brick jambs										
438mm wide	-	-	-	2.10	0.32	21.63	2.45	24.08	nr	27.09
641mm wide	-	-	-	2.35	0.35	24.15	2.74	26.89	nr	30.25
1225mm wide	-	-	-	3.06	0.41	31.12	3.65	34.77	nr	39.12
1809mm wide	-	-	-	3.75	0.50	38.12	4.51	42.63	nr	47.96
2394mm wide	-	-	-	4.46	0.57	45.17	5.43	50.60	nr	56.92
1378mm high; lugs to brick jambs										
438mm wide	-	-	-	2.30	0.32	23.48	2.67	26.15	nr	29.41
641mm wide	-	-	-	2.54	0.35	25.90	2.97	28.87	nr	32.48
1225mm wide	-	-	-	3.27	0.44	33.27	3.88	37.15	nr	41.79
1809mm wide	-	-	-	3.99	0.54	40.62	4.77	45.39	nr	51.06
2394mm wide	-	-	-	4.72	0.63	47.99	5.68	53.67	nr	60.38

Labour hourly rates: (except Specialists) Craft Operatives £9.23 Labourer £7.02 Rates are national average prices. Refer to REGIONAL VARIATIONS for indicative levels of overall pricing in regions	MATERIALS			LABOUR				RATES		
	Del to Site	Waste	Material Cost	Craft Optve	Lab	Labour Cost	Sunds	Nett Rate	Unit	Gross Rate (+12.5%)
	£	%	£	Hrs	Hrs	£	£	£		£

F30: ACCESSORIES/SUNDRY ITEMS FOR BRICK/BLOCK/STONE WALLING Cont.

Building in Cont.

Building in fireplace interior with slabbed tile surround and loose hearth tiles for 400mm opening; bed and joint interior in fire cement and fill with concrete at back; plug and screw on surround lugs; bed hearth tiles in cement mortar (1:3) and point in

	Del to Site	Waste	Material Cost	Craft Optve	Lab	Labour Cost	Sunds	Nett Rate	Unit	Gross Rate
with firebrick back	-	-	-	5.00	5.00	81.25	4.51	85.76	nr	96.48
with back boiler unit and self contained flue	-	-	-	6.00	6.00	97.50	5.43	102.93	nr	115.80

Proprietary items

Precast prestressed concrete lintels; Stressline Ltd; Roughcast; bedding in cement-lime mortar (1:1:6)

100 x 70mm										
900mm long	3.30	2.50	3.38	0.28	0.14	3.57	0.23	7.18	nr	8.08
1050mm long	3.83	2.50	3.93	0.31	0.16	3.98	0.26	8.17	nr	9.19
1200mm long	4.39	2.50	4.50	0.36	0.18	4.59	0.32	9.41	nr	10.58
1500mm long	5.47	2.50	5.61	0.44	0.22	5.61	0.39	11.60	nr	13.05
1800mm long	6.57	2.50	6.73	0.51	0.26	6.53	0.42	13.69	nr	15.40
2100mm long	7.67	2.50	7.86	0.58	0.29	7.39	0.48	15.73	nr	17.70
2400mm long	8.77	2.50	8.99	0.63	0.32	8.06	0.55	17.60	nr	19.80
2700mm long	9.86	2.50	10.11	0.69	0.35	8.83	0.59	19.52	nr	21.96
3000mm long	10.95	2.50	11.22	0.76	0.38	9.68	0.66	21.57	nr	24.26
150 x 70mm										
900mm long	4.47	2.50	4.58	0.29	0.15	3.73	0.24	8.55	nr	9.62
1050mm long	5.21	2.50	5.34	0.32	0.16	4.08	0.29	9.71	nr	10.92
1200mm long	5.96	2.50	6.11	0.37	0.19	4.75	0.32	11.18	nr	12.58
1500mm long	7.46	2.50	7.65	0.45	0.23	5.77	0.39	13.80	nr	15.53
1800mm long	8.93	2.50	9.15	0.52	0.26	6.62	0.43	16.21	nr	18.23
2100mm long	10.42	2.50	10.68	0.59	0.30	7.55	0.52	18.75	nr	21.10
2400mm long	11.91	2.50	12.21	0.66	0.33	8.41	0.55	21.17	nr	23.81
2700mm long	13.40	2.50	13.73	0.71	0.36	9.08	0.58	23.40	nr	26.32
3000mm long	14.89	2.50	15.26	0.78	0.39	9.94	0.68	25.88	nr	29.11
225 x 70mm										
900mm long	6.89	2.50	7.06	0.30	0.15	3.82	0.24	11.12	nr	12.51
1050mm long	8.01	2.50	8.21	0.33	0.17	4.24	0.29	12.74	nr	14.33
1200mm long	9.19	2.50	9.42	0.38	0.19	4.84	0.32	14.58	nr	16.40
1500mm long	11.48	2.50	11.77	0.46	0.23	5.86	0.39	18.02	nr	20.27
1800mm long	13.78	2.50	14.12	0.53	0.27	6.79	0.43	21.34	nr	24.01
2100mm long	16.08	2.50	16.48	0.60	0.30	7.64	0.52	24.65	nr	27.73
2400mm long	18.36	2.50	18.82	0.67	0.34	8.57	0.58	27.97	nr	31.47
2700mm long	20.66	2.50	21.18	0.72	0.36	9.17	0.65	31.00	nr	34.87
3000mm long	22.95	2.50	23.52	0.79	0.40	10.10	0.68	34.30	nr	38.59
255 x 70mm										
900mm long	7.32	2.50	7.50	0.30	0.15	3.82	0.28	11.61	nr	13.06
1050mm long	8.55	2.50	8.76	0.35	0.18	4.49	0.30	13.56	nr	15.25
1200mm long	9.79	2.50	10.03	0.38	0.19	4.84	0.33	15.21	nr	17.11
1500mm long	12.22	2.50	12.53	0.46	0.23	5.86	0.39	18.78	nr	21.12
1800mm long	14.66	2.50	15.03	0.54	0.27	6.88	0.46	22.37	nr	25.16
2100mm long	17.10	2.50	17.53	0.61	0.32	7.88	0.54	25.94	nr	29.19
2400mm long	19.54	2.50	20.03	0.69	0.35	8.83	0.58	29.43	nr	33.11
2700mm long	21.99	2.50	22.54	0.74	0.37	9.43	0.62	32.59	nr	36.66
3000mm long	24.43	2.50	25.04	0.81	0.41	10.35	0.68	36.08	nr	40.58
100 x 145mm										
900mm long	6.66	2.50	6.83	0.30	0.15	3.82	0.24	10.89	nr	12.25
1050mm long	7.76	2.50	7.95	0.33	0.17	4.24	0.28	12.47	nr	14.03
1200mm long	8.87	2.50	9.09	0.38	0.19	4.84	0.33	14.26	nr	16.05
1500mm long	11.09	2.50	11.37	0.46	0.23	5.86	0.39	17.62	nr	19.82
1800mm long	13.29	2.50	13.62	0.53	0.27	6.79	0.46	20.87	nr	23.48
2100mm long	15.52	2.50	15.91	0.60	0.30	7.64	0.52	24.07	nr	27.08
2400mm long	17.73	2.50	18.17	0.67	0.34	8.57	0.58	27.32	nr	30.74
2700mm long	19.95	2.50	20.45	0.72	0.36	9.17	0.65	30.27	nr	34.06
3000mm long	22.17	2.50	22.72	0.79	0.40	10.10	0.68	33.50	nr	37.69

Galvanised steel lintels; SUPERGALV (BIRTLEY) lintels reference CB 50; bedding in cement-lime mortar (1:1:6)

	Del to Site	Waste	Material Cost	Craft Optve	Lab	Labour Cost	Sunds	Nett Rate	Unit	Gross Rate
140mm deep x 750mm long	14.76	2.50	15.13	0.32	0.32	5.20	-	20.33	nr	22.87
140mm deep x 1200mm long	23.57	2.50	24.16	0.44	0.44	7.15	-	31.31	nr	35.22
140mm deep x 1350mm long	27.47	2.50	28.16	0.48	0.48	7.80	-	35.96	nr	40.45
150mm deep x 1500mm long	30.65	2.50	31.42	0.51	0.51	8.29	-	39.70	nr	44.67
150mm deep x 1650mm long	34.60	2.50	35.47	0.54	0.54	8.78	-	44.24	nr	49.77
140mm deep x 1800mm long	37.74	2.50	38.68	0.58	0.58	9.43	-	48.11	nr	54.12
165mm deep x 1950mm long	41.07	2.50	42.10	0.61	0.61	9.91	-	52.01	nr	58.51
165mm deep x 2100mm long	43.36	2.50	44.44	0.65	0.65	10.56	-	55.01	nr	61.88
190mm deep x 2250mm long	48.80	2.50	50.02	0.69	0.69	11.21	-	61.23	nr	68.89
190mm deep x 2400mm long	52.06	2.50	53.36	0.73	0.73	11.86	-	65.22	nr	73.38
200mm deep x 2550mm long	59.70	2.50	61.19	0.77	0.77	12.51	-	73.70	nr	82.92
200mm deep x 2850mm long	79.28	2.50	81.26	0.82	0.82	13.32	-	94.59	nr	106.41
210mm deep x 3000mm long	85.15	2.50	87.28	0.87	0.87	14.14	-	101.42	nr	114.09
225mm deep x 3300mm long	93.74	2.50	96.08	0.94	0.94	15.28	-	111.36	nr	125.28
215mm deep x 3600mm long	106.46	2.50	109.12	1.02	1.02	16.57	-	125.70	nr	141.41
215mm deep x 3900mm long	126.52	2.50	129.68	1.09	1.09	17.71	-	147.40	nr	165.82

Galvanised steel lintels; SUPERGALV (BIRTLEY) lintels reference CB 50 H.D.; bedding in cement-lime mortar (1:1:6)

	Del to Site	Waste	Material Cost	Craft Optve	Lab	Labour Cost	Sunds	Nett Rate	Unit	Gross Rate
165mm deep x 750mm long	17.05	2.50	17.48	0.33	0.33	5.36	-	22.84	nr	25.69
165mm deep x 1050mm long	22.35	2.50	22.91	0.40	0.40	6.50	-	29.41	nr	33.08
165mm deep x 1200mm long	26.00	2.50	26.65	0.44	0.44	7.15	-	33.80	nr	38.02
165mm deep x 1350mm long	31.02	2.50	31.80	0.48	0.48	7.80	-	39.60	nr	44.54
165mm deep x 1500mm long	34.77	2.50	35.64	0.51	0.51	8.29	-	43.93	nr	49.42

MASONRY

Labour hourly rates: (except Specialists) Craft Operatives £9.23 Labourer £7.02 Rates are national average prices. Refer to REGIONAL VARIATIONS for indicative levels of overall pricing in regions	MATERIALS			LABOUR				RATES		
	Del to Site £	Waste %	Material Cost £	Craft Optve Hrs	Lab Hrs	Labour Cost £	Sunds £	Nett Rate £	Unit	Gross Rate (+12.5%) £
F30: ACCESSORIES/SUNDRY ITEMS FOR BRICK/BLOCK/STONE WALLING Cont.										
Proprietary items Cont.										
Galvanised steel lintels; SUPERGALV (BIRTLEY) lintels reference CB 50 H.D.; bedding in cement-lime mortar (1:1:6) Cont.										
165mm deep x 1650mm long	40.45	2.50	41.46	0.54	0.54	8.78	–	50.24	nr	56.52
200mm deep x 1800mm long	45.41	2.50	46.55	0.58	0.58	9.43	–	55.97	nr	62.97
200mm deep x 2100mm long	53.58	2.50	54.92	0.65	0.65	10.56	–	65.48	nr	73.67
200mm deep x 2250mm long	66.17	2.50	67.82	0.69	0.69	11.21	–	79.04	nr	88.92
225mm deep x 2400mm long	78.28	2.50	80.24	0.73	0.73	11.86	–	92.10	nr	103.61
215mm deep x 2550mm long	85.06	2.50	87.19	0.77	0.77	12.51	–	99.70	nr	112.16
Galvanised steel lintels; SUPERGALV (BIRTLEY) lintels reference CB 70; bedding in cement-lime mortar (1:1:6)										
140mm deep x 750mm long	14.91	2.50	15.28	0.34	0.34	5.53	–	20.81	nr	23.41
140mm deep x 1200mm long	23.85	2.50	24.45	0.46	0.46	7.47	–	31.92	nr	35.91
140mm deep x 1350mm long	27.44	2.50	28.13	0.50	0.50	8.13	–	36.25	nr	40.78
150mm deep x 1500mm long	30.75	2.50	31.52	0.53	0.53	8.61	–	40.13	nr	45.15
150mm deep x 1650mm long	34.84	2.50	35.71	0.57	0.57	9.26	–	44.97	nr	50.60
140mm deep x 1800mm long	38.01	2.50	38.96	0.61	0.61	9.91	–	48.87	nr	54.98
165mm deep x 1950mm long	41.88	2.50	42.93	0.64	0.64	10.40	–	53.33	nr	59.99
165mm deep x 2100mm long	44.25	2.50	45.36	0.68	0.68	11.05	–	56.41	nr	63.46
190mm deep x 2250mm long	49.09	2.50	50.32	0.72	0.72	11.70	–	62.02	nr	69.77
190mm deep x 2400mm long	52.34	2.50	53.65	0.76	0.76	12.35	–	66.00	nr	74.25
200mm deep x 2550mm long	59.85	2.50	61.35	0.81	0.81	13.16	–	74.51	nr	83.82
200mm deep x 2850mm long	79.50	2.50	81.49	0.86	0.86	13.98	–	95.46	nr	107.40
210mm deep x 3000mm long	85.41	2.50	87.55	0.91	0.91	14.79	–	102.33	nr	115.12
225mm deep x 3300mm long	95.97	2.50	98.37	0.98	0.98	15.93	–	114.29	nr	128.58
215mm deep x 3600mm long	105.50	2.50	108.14	1.06	1.06	17.23	–	125.36	nr	141.03
215mm deep x 3900mm long	137.56	2.50	141.00	1.14	1.14	18.52	–	159.52	nr	179.46
Galvanised steel lintels; SUPERGALV (BIRTLEY) lintels reference CB 70 H.D.; bedding in cement-lime mortar (1:1:6)										
165mm deep x 750mm long	17.05	2.50	17.48	0.33	0.33	5.36	–	22.84	nr	25.69
165mm deep x 1050mm long	22.74	2.50	23.31	0.41	0.41	6.66	–	29.97	nr	33.72
165mm deep x 1200mm long	25.94	2.50	26.59	0.46	0.46	7.47	–	34.06	nr	38.32
165mm deep x 1350mm long	31.41	2.50	32.20	0.49	0.49	7.96	–	40.16	nr	45.18
165mm deep x 1500mm long	34.93	2.50	35.80	0.53	0.53	8.61	–	44.42	nr	49.97
165mm deep x 1650mm long	44.83	2.50	45.95	0.56	0.56	9.10	–	55.05	nr	61.93
200mm deep x 1800mm long	48.80	2.50	50.02	0.61	0.61	9.91	–	59.93	nr	67.42
200mm deep x 2100mm long	56.93	2.50	58.35	0.68	0.68	11.05	–	69.40	nr	78.08
200mm deep x 2250mm long	67.68	2.50	69.37	0.72	0.72	11.70	–	81.07	nr	91.21
225mm deep x 2400mm long	78.67	2.50	80.64	0.76	0.76	12.35	–	92.99	nr	104.61
225mm deep x 2550mm long	85.49	2.50	87.63	0.81	0.81	13.16	–	100.79	nr	113.39
Galvanised steel lintels; SUPERGALV (BIRTLEY) lintels reference CB 50/130; bedding in cement-lime mortar (1:1:6)										
140mm deep x 750mm long	14.80	2.50	15.17	0.36	0.36	5.85	–	21.02	nr	23.65
140mm deep x 1200mm long	23.62	2.50	24.21	0.48	0.48	7.80	–	32.01	nr	36.01
140mm deep x 1350mm long	27.04	2.50	27.72	0.52	0.52	8.45	–	36.17	nr	40.69
150mm deep x 1500mm long	30.30	2.50	31.06	0.55	0.55	8.94	–	39.99	nr	44.99
150mm deep x 1650mm long	34.40	2.50	35.26	0.60	0.60	9.75	–	45.01	nr	50.64
140mm deep x 1800mm long	37.54	2.50	38.48	0.63	0.63	10.24	–	48.72	nr	54.81
165mm deep x 1950mm long	40.87	2.50	41.89	0.67	0.67	10.89	–	52.78	nr	59.38
165mm deep x 2100mm long	44.15	2.50	45.25	0.71	0.71	11.54	–	56.79	nr	63.89
190mm deep x 2250mm long	51.37	2.50	52.65	0.76	0.76	12.35	–	65.00	nr	73.13
190mm deep x 2400mm long	54.79	2.50	56.16	0.79	0.79	12.84	–	69.00	nr	77.62
200mm deep x 2550mm long	59.59	2.50	61.08	0.85	0.85	13.81	–	74.89	nr	84.25
200mm deep x 2850mm long	83.98	2.50	86.08	0.90	0.90	14.63	–	100.70	nr	113.29
210mm deep x 3000mm long	88.40	2.50	90.61	0.95	0.95	15.44	–	106.05	nr	119.30
225mm deep x 3300mm long	96.95	2.50	99.37	1.02	1.02	16.57	–	115.95	nr	130.44
215mm deep x 3600mm long	106.77	2.50	109.44	1.12	1.12	18.20	–	127.64	nr	143.59
215mm deep x 3900mm long	153.34	2.50	157.17	1.18	1.18	19.18	–	176.35	nr	198.39
Galvanised steel lintels; SUPERGALV (BIRTLEY) lintels reference CB 50/130 H.D.; bedding in cement-lime mortar (1:1:6)										
165mm deep x 750mm long	20.71	2.50	21.23	0.36	0.36	5.85	–	27.08	nr	30.46
165mm deep x 1200mm long	30.99	2.50	31.76	0.48	0.48	7.80	–	39.56	nr	44.51
165mm deep x 1350mm long	38.79	2.50	39.76	0.52	0.52	8.45	–	48.21	nr	54.24
165mm deep x 1500mm long	42.73	2.50	43.80	0.55	0.55	8.94	–	52.74	nr	59.33
165mm deep x 1650mm long	48.96	2.50	50.18	0.60	0.60	9.75	–	59.93	nr	67.43
200mm deep x 1800mm long	52.98	2.50	54.30	0.63	0.63	10.24	–	64.54	nr	72.61
200mm deep x 1950mm long	65.98	2.50	67.63	0.67	0.67	10.89	–	78.52	nr	88.33
200mm deep x 2250mm long	75.91	2.50	77.81	0.76	0.76	12.35	–	90.16	nr	101.43
225mm deep x 2400mm long	82.04	2.50	84.09	0.79	0.79	12.84	–	96.93	nr	109.04
225mm deep x 2550mm long	89.95	2.50	92.20	0.85	0.85	13.81	–	106.01	nr	119.26
225mm deep x 2700mm long	95.02	2.50	97.40	0.92	0.92	14.95	–	112.35	nr	126.39
Galvanised steel lintels; SUPERGALV (BIRTLEY) lintels reference CB 50/150; bedding in cement-lime mortar (1:1:6)										
140mm deep x 750mm long	18.84	2.50	19.31	0.36	0.36	5.85	–	25.16	nr	28.31
140mm deep x 1200mm long	30.44	2.50	31.20	0.48	0.48	7.80	–	39.00	nr	43.88
140mm deep x 1350mm long	34.63	2.50	35.50	0.52	0.52	8.45	–	43.95	nr	49.44
150mm deep x 1500mm long	38.25	2.50	39.21	0.55	0.55	8.94	–	48.14	nr	54.16
150mm deep x 1650mm long	46.06	2.50	47.21	0.60	0.60	9.75	–	56.96	nr	64.08
140mm deep x 1800mm long	48.11	2.50	49.31	0.63	0.63	10.24	–	59.55	nr	66.99
165mm deep x 1950mm long	51.77	2.50	53.06	0.67	0.67	10.89	–	63.95	nr	71.95
165mm deep x 2100mm long	56.06	2.50	57.46	0.71	0.71	11.54	–	69.00	nr	77.62

Labour hourly rates: (except Specialists) Craft Operatives £9.23 Labourer £7.02. Rates are national average prices. Refer to REGIONAL VARIATIONS for indicative levels of overall pricing in regions	MATERIALS			LABOUR				RATES		
	Del to Site £	Waste %	Material Cost £	Craft Optve Hrs	Lab Hrs	Labour Cost £	Sunds £	Nett Rate £	Unit	Gross Rate (+12.5%) £

F30: ACCESSORIES/SUNDRY ITEMS FOR BRICK/BLOCK/STONE WALLING Cont.

Proprietary items Cont.

Galvanised steel lintels; SUPERGALV (BIRTLEY) lintels reference CB 50/150; bedding in cement-lime mortar (1:1:6) Cont.

Item	Del	Waste	Mat	Craft	Lab	LabCost	Sunds	Nett	Unit	Gross
190mm deep x 2250mm long	62.97	2.50	64.54	0.76	0.76	12.35	-	76.89	nr	86.51
190mm deep x 2400mm long	72.05	2.50	73.85	0.79	0.79	12.84	-	86.69	nr	97.52
200mm deep x 2550mm long	78.73	2.50	80.70	0.85	0.85	13.81	-	94.51	nr	106.32
200mm deep x 2850mm long	99.57	2.50	102.06	0.90	0.90	14.63	-	116.68	nr	131.27
210mm deep x 3000mm long	105.13	2.50	107.76	0.95	0.95	15.44	-	123.20	nr	138.60
225mm deep x 3300mm long	116.72	2.50	119.64	1.02	1.02	16.57	-	136.21	nr	153.24

Galvanised steel lintels; SUPERGALV (BIRTLEY) lintels reference Box 50; bedding in cement-lime mortar (1:1:6)

Item	Del	Waste	Mat	Craft	Lab	LabCost	Sunds	Nett	Unit	Gross
215mm deep x 3900mm long	137.30	2.50	140.73	1.17	1.17	19.01	-	159.75	nr	179.71
215mm deep x 4200mm long	152.69	2.50	156.51	1.21	1.21	19.66	-	176.17	nr	198.19
215mm deep x 4500mm long	176.49	2.50	180.90	1.25	1.25	20.31	-	201.21	nr	226.37
215mm deep x 4800mm long	183.70	2.50	188.29	1.29	1.29	20.96	-	209.26	nr	235.41

Galvanised steel lintels; SUPERGALV (BIRTLEY) lintels reference Box 50 H.D.; bedding in cement-lime mortar (1:1:6)

Item	Del	Waste	Mat	Craft	Lab	LabCost	Sunds	Nett	Unit	Gross
215mm deep x 3300mm long	136.50	2.50	139.91	0.98	0.98	15.93	-	155.84	nr	175.32
215mm deep x 3600mm long	148.91	2.50	152.63	1.06	1.06	17.23	-	169.86	nr	191.09

Galvanised steel lintels; SUPERGALV (BIRTLEY) lintels reference SB100; bedding in cement-lime mortar (1:1:6)

Item	Del	Waste	Mat	Craft	Lab	LabCost	Sunds	Nett	Unit	Gross
75mm deep x 750mm long	9.98	2.50	10.23	0.52	0.52	8.45	-	18.68	nr	21.01
75mm deep x 900mm long	11.83	2.50	12.13	0.55	0.55	8.94	-	21.06	nr	23.70
75mm deep x 1050mm long	13.81	2.50	14.16	0.59	0.59	9.59	-	23.74	nr	26.71
75mm deep x 1200mm long	15.58	2.50	15.97	0.62	0.62	10.07	-	26.04	nr	29.30
75mm deep x 1500mm long	19.51	2.50	20.00	0.69	0.69	11.21	-	31.21	nr	35.11
150mm deep x 1800mm long	29.13	2.50	29.86	0.75	0.75	12.19	-	42.05	nr	47.30
150mm deep x 1950mm long	33.16	2.50	33.99	0.79	0.79	12.84	-	46.83	nr	52.68
150mm deep x 2100mm long	35.21	2.50	36.09	0.84	0.84	13.65	-	49.74	nr	55.96
150mm deep x 2250mm long	37.88	2.50	38.83	0.89	0.89	14.46	-	53.29	nr	59.95
150mm deep x 2400mm long	40.82	2.50	41.84	0.94	0.94	15.28	-	57.12	nr	64.25
150mm deep x 2550mm long	44.05	2.50	45.15	1.00	1.00	16.25	-	61.40	nr	69.08
150mm deep x 2700mm long	46.80	2.50	47.97	1.12	1.12	18.20	-	66.17	nr	74.44
225mm deep x 2850mm long	70.80	2.50	72.57	1.07	1.07	17.39	-	89.96	nr	101.20
225mm deep x 3000mm long	75.05	2.50	76.93	1.09	1.09	17.71	-	94.64	nr	106.47

Galvanised steel lintels; SUPERGALV (BIRTLEY) lintels reference SB140; bedding in cement-lime mortar (1:1:6)

Item	Del	Waste	Mat	Craft	Lab	LabCost	Sunds	Nett	Unit	Gross
75mm deep x 750mm long	12.44	2.50	12.75	0.52	0.52	8.45	-	21.20	nr	23.85
75mm deep x 900mm long	14.80	2.50	15.17	0.55	0.55	8.94	-	24.11	nr	27.12
75mm deep x 1050mm long	17.44	2.50	17.88	0.59	0.59	9.59	-	27.46	nr	30.90
75mm deep x 1200mm long	20.24	2.50	20.75	0.62	0.62	10.07	-	30.82	nr	34.67
75mm deep x 1350mm long	22.63	2.50	23.20	0.66	0.66	10.73	-	33.92	nr	38.16
75mm deep x 1500mm long	26.08	2.50	26.73	0.69	0.69	11.21	-	37.94	nr	42.69
150mm deep x 1650mm long	30.87	2.50	31.64	0.71	0.71	11.54	-	43.18	nr	48.58
150mm deep x 1800mm long	34.05	2.50	34.90	0.75	0.75	12.19	-	47.09	nr	52.97
150mm deep x 1950mm long	37.58	2.50	38.52	0.79	0.79	12.84	-	51.36	nr	57.78
150mm deep x 2100mm long	40.78	2.50	41.80	0.84	0.84	13.65	-	55.45	nr	62.38
150mm deep x 2250mm long	45.45	2.50	46.59	0.89	0.89	14.46	-	61.05	nr	68.68
150mm deep x 2400mm long	47.89	2.50	49.09	0.94	0.94	15.28	-	64.36	nr	72.41
150mm deep x 2550mm long	50.56	2.50	51.82	1.00	1.00	16.25	-	68.07	nr	76.58
150mm deep x 2700mm long	53.53	2.50	54.87	1.06	1.06	17.23	-	72.09	nr	81.10
225mm deep x 2850mm long	75.42	2.50	77.31	1.09	1.09	17.71	-	95.02	nr	106.90
225mm deep x 3000mm long	78.51	2.50	80.47	1.12	1.12	18.20	-	98.67	nr	111.01

Galvanised steel lintels; SUPERGALV (BIRTLEY) lintels reference SBL200; bedding in cement-lime mortar (1:1:6)

Item	Del	Waste	Mat	Craft	Lab	LabCost	Sunds	Nett	Unit	Gross
150mm deep x 750mm long	16.87	2.50	17.29	0.52	0.52	8.45	-	25.74	nr	28.96
150mm deep x 900mm long	20.24	2.50	20.75	0.55	0.55	8.94	-	29.68	nr	33.39
150mm deep x 1050mm long	23.51	2.50	24.10	0.59	0.59	9.59	-	33.69	nr	37.90
150mm deep x 1200mm long	26.78	2.50	27.45	0.62	0.62	10.07	-	37.52	nr	42.22
150mm deep x 1350mm long	30.04	2.50	30.79	0.66	0.66	10.73	-	41.52	nr	46.71
150mm deep x 1500mm long	34.52	2.50	35.38	0.69	0.69	11.21	-	46.60	nr	52.42
150mm deep x 1650mm long	38.79	2.50	39.76	0.77	0.77	12.51	-	52.27	nr	58.81
150mm deep x 1800mm long	42.32	2.50	43.38	0.83	0.83	13.49	-	56.87	nr	63.97
150mm deep x 1950mm long	45.84	2.50	46.99	0.87	0.87	14.14	-	61.12	nr	68.76
150mm deep x 2100mm long	49.56	2.50	50.80	0.92	0.92	14.95	-	65.75	nr	73.97
150mm deep x 2250mm long	56.82	2.50	58.24	0.89	0.89	14.46	-	72.70	nr	81.79
150mm deep x 2400mm long	60.26	2.50	61.77	0.94	0.94	15.28	-	77.04	nr	86.67
150mm deep x 2550mm long	64.00	2.50	65.60	1.00	1.00	16.25	-	81.85	nr	92.08
150mm deep x 2700mm long	66.93	2.50	68.60	1.06	1.06	17.23	-	85.83	nr	96.56

Galvanised steel lintels; SUPERGALV (BIRTLEY) internal door lintels reference INT100; bedding in cement-lime mortar (1:1:6)

Item	Del	Waste	Mat	Craft	Lab	LabCost	Sunds	Nett	Unit	Gross
100mm wide, 900mm long	3.54	2.50	3.63	0.38	0.38	6.17	-	9.80	nr	11.03
100mm wide, 1050mm long	4.05	2.50	4.15	0.41	0.41	6.66	-	10.81	nr	12.17
100mm wide, 1200mm long	4.49	2.50	4.60	0.44	0.44	7.15	-	11.75	nr	13.22

Stainless steel Furfix wall extension profiles; single flange; plugging and screwing to brickwork and building ties in to joints of new walls

Item	Del	Waste	Mat	Craft	Lab	LabCost	Sunds	Nett	Unit	Gross
100mm	12.05	5.00	12.65	0.30	-	2.77	-	15.42	m	17.35

MASONRY

Labour hourly rates: (except Specialists) Craft Operatives £9.23 Labourer £7.02 Rates are national average prices. Refer to REGIONAL VARIATIONS for indicative levels of overall pricing in regions	MATERIALS			LABOUR				RATES		
	Del to Site £	Waste %	Material Cost £	Craft Optve Hrs	Lab Hrs	Labour Cost £	Sunds £	Nett Rate £	Unit	Gross Rate (+12.5%) £
F30: ACCESSORIES/SUNDRY ITEMS FOR BRICK/BLOCK/STONE WALLING Cont.										
Proprietary items Cont.										
Stainless steel Furfix wall extension profiles; single flange; plugging and screwing to brickwork and building ties in to joints of new walls Cont.										
150mm	15.16	5.00	15.92	0.35	–	3.23	–	19.15	m	21.54
215mm	21.79	5.00	22.88	0.40	–	3.69	–	26.57	m	29.89
Harris and Edgar Ltd; Hemax restraint channel ref. 36/8; fixing to steelwork with Hemax M8 screws complete with plate washer at 450mm centres; incorporating ties at 450mm centres										
100mm long ties reference BP36 fishtail	8.95	5.00	9.40	0.60	0.60	9.75	0.69	19.84	m	22.32
125mm long ties reference BP36 fishtail	9.05	5.00	9.50	0.60	0.60	9.75	0.81	20.06	m	22.57
150mm long ties reference BP36 fishtail	9.13	5.00	9.59	0.60	0.60	9.75	0.92	20.26	m	22.79
Thermabate insulated cavity closers; fixing to timber with nails										
Thermabate 50; vertical	5.13	5.00	5.39	0.40	0.40	6.50	–	11.89	m	13.37
Thermabate 75; vertical	5.56	5.00	5.84	0.40	0.40	6.50	–	12.34	m	13.88
Thermabate 50; horizontal	5.13	5.00	5.39	0.40	0.40	6.50	–	11.89	m	13.37
Thermabate 75; horizontal	5.56	5.00	5.84	0.40	0.40	6.50	–	12.34	m	13.88
Thermabate insulated cavity closers; fixing to masonry with PVC-U ties at 225mm centres										
Thermabate 50; vertical	5.81	5.00	6.10	0.40	0.40	6.50	–	12.60	m	14.18
Thermabate 75; vertical	6.24	5.00	6.55	0.40	0.40	6.50	–	13.05	m	14.68
Thermabate 50; horizontal	5.81	5.00	6.10	0.40	0.40	6.50	–	12.60	m	14.18
Thermabate 75; horizontal	6.24	5.00	6.55	0.40	0.40	6.50	–	13.05	m	14.68
F31: PRECAST CONCRETE SILLS/LINTELS/COPINGS/FEATURES										
Precast concrete sills; B.S.5642 Part 1; bedding in cement lime mortar (1:1:6)										
Sills; figure 2 or figure 4										
50 x 150 x 300mm splayed and grooved	1.18	2.50	1.21	0.17	0.17	2.76	0.12	4.09	nr	4.60
50 x 150 x 400mm splayed and grooved	1.54	2.50	1.58	0.22	0.22	3.58	0.14	5.29	nr	5.96
50 x 150 x 700mm splayed and grooved	2.72	2.50	2.79	0.33	0.33	5.36	0.21	8.36	nr	9.41
50 x 150 x 1300mm splayed and grooved	4.98	2.50	5.10	0.55	0.55	8.94	0.32	14.36	nr	16.16
extra for stooled end	1.40	–	1.40	–	–	–	–	1.40	nr	1.58
75 x 150 x 300mm splayed and grooved	1.55	2.50	1.59	0.19	0.19	3.09	0.12	4.80	nr	5.40
75 x 150 x 400mm splayed and grooved	2.07	2.50	2.12	0.24	0.24	3.90	0.14	6.16	nr	6.93
75 x 150 x 700mm splayed and grooved	3.56	2.50	3.65	0.38	0.38	6.17	0.21	10.03	nr	11.29
75 x 150 x 1300mm splayed and grooved	6.72	2.50	6.89	0.60	0.60	9.75	0.32	16.96	nr	19.08
extra for stooled end	1.50	–	1.50	–	–	–	–	1.50	nr	1.69
100 x 150 x 300mm splayed and grooved	2.02	2.50	2.07	0.22	0.22	3.58	0.12	5.77	nr	6.49
100 x 150 x 400mm splayed and grooved	2.63	2.50	2.70	0.28	0.28	4.55	0.14	7.39	nr	8.31
100 x 150 x 700mm splayed and grooved	3.64	2.50	3.73	0.44	0.44	7.15	0.21	11.09	nr	12.48
100 x 150 x 1300mm splayed and grooved	8.49	2.50	8.70	0.66	0.66	10.73	0.32	19.75	nr	22.22
extra for stooled end	1.60	–	1.60	–	–	–	–	1.60	nr	1.80
Precast concrete copings B.S.5642 Part 2; bedding in cement-lime mortar (1:1:6)										
Copings; figure 1; horizontal										
75 x 200mm; splayed; rebated joints	4.71	2.50	4.83	0.28	0.14	3.57	0.18	8.57	m	9.65
extra; stopped ends	1.28	–	1.28	–	–	–	–	1.28	nr	1.44
extra; internal angles	2.73	–	2.73	–	–	–	–	2.73	nr	3.07
100 x 300mm; splayed; rebated joints	8.93	2.50	9.15	0.55	0.28	7.04	0.29	16.49	m	18.55
extra; stopped ends	1.71	–	1.71	–	–	–	–	1.71	nr	1.92
extra; internal angles	3.39	–	3.39	–	–	–	–	3.39	nr	3.81
75 x 200mm; saddleback; rebated joints	4.71	2.50	4.83	0.28	0.14	3.57	0.18	8.57	m	9.65
extra; hipped ends	1.28	–	1.28	–	–	–	–	1.28	nr	1.44
extra; internal angles	2.73	–	2.73	–	–	–	–	2.73	nr	3.07
100 x 300mm; saddleback; rebated joints	9.08	2.50	9.31	0.55	0.28	7.04	0.29	16.64	m	18.72
extra; hipped ends	1.76	–	1.76	–	–	–	–	1.76	nr	1.98
extra; internal angles	3.58	–	3.58	–	–	–	–	3.58	nr	4.03
Precast concrete lintels; B.S.5328, designed mix C25, 20mm aggregate minimum cement content 360 kg/m3; vibrated; reinforcement bars B.S.4449 hot rolled plain round mild steel; bedding in cement-lime mortar (1:1:6)										
Lintels										
75 x 150 x 1200mm; reinforced with 1.20 kg of 12mm bars	4.69	2.50	4.81	0.20	0.10	2.55	0.12	7.48	nr	8.41
75 x 150 x 1650mm; reinforced with 1.60 kg of 12mm bars	6.41	2.50	6.57	0.28	0.14	3.57	0.16	10.30	nr	11.58
102 x 150 x 1200mm; reinforced with 1.20 kg of 12mm bars	6.01	2.50	6.16	0.33	0.17	4.24	0.12	10.52	nr	11.83
102 x 150 x 1650mm; reinforced with 1.60 kg of 12mm bars	8.17	2.50	8.37	0.45	0.22	5.70	0.16	14.23	nr	16.01
215 x 150 x 1200mm; reinforced with 2.40 kg of 12mm bars	12.44	2.50	12.75	0.66	0.33	8.41	0.20	21.36	nr	24.03
215 x 150 x 1650mm; reinforced with 3.20 kg of 12mm bars	16.39	2.50	16.80	0.91	0.53	12.12	0.26	29.18	nr	32.83
215 x 225 x 1200mm; reinforced with 2.40 kg of 12mm bars	16.69	2.50	17.11	0.86	0.43	10.96	0.21	28.27	nr	31.81
215 x 225 x 1650mm; reinforced with 3.20 kg of 12mm bars	22.77	2.50	23.34	1.18	0.58	14.96	0.28	38.58	nr	43.41

MASONRY – SMALL WORKS

Labour hourly rates: (except Specialists) Craft Operatives £9.23 Labourer £7.02. Rates are national average prices. Refer to REGIONAL VARIATIONS for indicative levels of overall pricing in regions	MATERIALS			LABOUR				RATES		
	Del to Site £	Waste %	Material Cost £	Craft Optve Hrs	Lab Hrs	Labour Cost £	Sunds £	Nett Rate £	Unit	Gross Rate (+12.5%) £
F31: PRECAST CONCRETE SILLS/LINTELS/COPINGS/FEATURES Cont.										
Precast concrete lintels; B.S.5328, designed mix C25, 20mm aggregate minimum cement content 360 kg/m3; vibrated; reinforcement bars B.S.4449 hot rolled plain round mild steel; bedding in cement-lime mortar (1:1:6) Cont.										
Lintels Cont.										
327 x 150 x 1200mm; reinforced with 3.60 kg of 12mm bars	18.46	2.50	18.92	0.86	0.43	10.96	0.23	30.11	nr	33.87
327 x 150 x 1650mm; reinforced with 4.80 kg of 12mm bars	25.21	2.50	25.84	1.18	0.59	15.03	0.28	41.15	nr	46.30
327 x 225 x 1200mm; reinforced with 3.60 kg of 12mm bars	26.51	2.50	27.17	1.32	0.66	16.82	0.25	44.24	nr	49.77
327 x 225 x 1650mm; reinforced with 4.80 kg of 12mm bars	36.20	2.50	37.10	1.82	0.91	23.19	0.34	60.63	nr	68.21
75 x 150 x 1200mm; reinforced with 1.20 kg of 12mm bars; fair finish two faces	5.94	2.50	6.09	0.20	0.10	2.55	0.12	8.76	nr	9.85
75 x 150 x 1650mm; reinforced with 1.60 kg of 12mm bars; fair finish two faces	8.11	2.50	8.31	0.28	0.14	3.57	0.16	12.04	nr	13.54
102 x 150 x 1200mm; reinforced with 1.20 kg of 12mm bars; fair finish two faces	7.24	2.50	7.42	0.33	0.17	4.24	0.12	11.78	nr	13.25
102 x 150 x 1650mm; reinforced with 1.60 kg of 12mm bars; fair finish two faces	9.87	2.50	10.12	0.45	0.22	5.70	0.16	15.97	nr	17.97
215 x 150 x 1200mm; reinforced with 2.40 kg of 12mm bars; fair finish two faces	13.67	2.50	14.01	0.66	0.33	8.41	0.20	22.62	nr	25.45
215 x 150 x 1650mm; reinforced with 3.20 kg of 12mm bars; fair finish two faces	18.10	2.50	18.55	0.91	0.45	11.56	0.26	30.37	nr	34.17
215 x 225 x 1200mm; reinforced with 2.40 kg of 12mm bars; fair finish two faces	18.96	2.50	19.43	0.86	0.43	10.96	0.21	30.60	nr	34.43
215 x 225 x 1650mm; reinforced with 3.20 kg of 12mm bars; fair finish two faces	25.25	2.50	25.88	1.18	0.58	14.96	0.28	41.12	nr	46.26
327 x 150 x 1200mm; reinforced with 3.60 kg of 12mm bars; fair finish two faces	20.15	2.50	20.65	0.86	0.43	10.96	0.21	31.82	nr	35.80
327 x 150 x 1650mm; reinforced with 4.80 kg of 12mm bars; fair finish two faces	26.86	2.50	27.53	1.18	0.59	15.03	0.28	42.84	nr	48.20
327 x 225 x 1200mm; reinforced with 3.60 kg of 12mm bars; fair finish two faces	29.04	2.50	29.77	1.32	0.66	16.82	0.24	46.82	nr	52.68
327 x 225 x 1650mm; reinforced with 4.80 kg of 12mm bars; fair finish two faces	38.70	2.50	39.67	1.82	0.91	23.19	0.33	63.18	nr	71.08
Boot lintels										
215 x 150 x 1200mm; reinforced with 3.60 kg of 12mm bars and 0.51 kg of 6mm links	11.74	2.50	12.03	0.66	0.33	8.41	0.19	20.63	nr	23.21
215 x 150 x 1800mm; reinforced with 5.19 kg of 12mm bars and 0.77 kg of 6mm links	17.56	2.50	18.00	1.00	0.33	11.55	0.28	29.83	nr	33.55
215 x 225 x 1200mm; reinforced with 3.60 kg of 12mm bars and 0.62 kg of 6mm links	16.38	2.50	16.79	0.86	0.43	10.96	0.20	27.95	nr	31.44
215 x 225 x 1800mm; reinforced with 5.19 kg of 12mm bars and 0.93 kg of 6mm links	24.63	2.50	25.25	1.30	0.65	16.56	0.29	42.10	nr	47.36
252 x 150 x 1200mm; reinforced with 3.60 kg of 12mm bars and 0.56 kg of 6mm links	12.49	2.50	12.80	0.79	0.40	10.10	0.22	23.12	nr	26.01
252 x 150 x 1800mm; reinforced with 5.19 kg of 12mm bars and 0.83 kg of 6mm links	20.21	2.50	20.72	1.19	0.59	15.13	0.33	36.17	nr	40.69
252 x 225 x 1200mm; reinforced with 3.60 kg of 12mm bars and 0.67 kg of 6mm links	19.47	2.50	19.96	0.86	0.43	10.96	0.24	31.15	nr	35.05
252 x 225 x 1800mm; reinforced with 5.19 kg of 12mm bars and 1.00 kg of 6mm links	27.67	2.50	28.36	1.30	0.65	16.56	0.33	45.25	nr	50.91
Precast concrete padstones; B.S.5328, designed mix C25, 20mm aggregate, minimum cement content 360 kg/m3; vibrated; reinforced at contractor's discretion; bedding in cement-lime mortar (1:1:6)										
Padstones										
215 x 215 x 75mm	2.34	2.50	2.40	0.07	0.03	0.86	0.11	3.37	nr	3.79
215 x 215 x 150mm	4.55	2.50	4.66	0.09	0.04	1.11	0.11	5.89	nr	6.62
327 x 215 x 150mm	7.05	2.50	7.23	0.12	0.06	1.53	0.12	8.88	nr	9.98
327 x 327 x 150mm	8.49	2.50	8.70	0.19	0.09	2.39	0.13	11.22	nr	12.62
440 x 215 x 150mm	7.52	2.50	7.71	0.19	0.09	2.39	0.14	10.23	nr	11.51
440 x 327 x 150mm	11.43	2.50	11.72	0.22	0.11	2.80	0.15	14.67	nr	16.50
440 x 440 x 150mm	12.44	2.50	12.75	0.22	0.11	2.80	0.16	15.71	nr	17.68

MASONRY

Structural/Carcassing metal/timber

Labour hourly rates: (except Specialists) Craft Operatives £9.23 Labourer £7.02 Rates are national average prices. Refer to REGIONAL VARIATIONS for indicative levels of overall pricing in regions	MATERIALS			LABOUR				RATES		
	Del to Site	Waste	Material Cost	Craft Optve	Lab	Labour Cost	Sunds	Nett Rate	Unit	Gross Rate (+12.5%)
	£	%	£	Hrs	Hrs	£	£	£		£
G10: STRUCTURAL STEEL FRAMING										
Prices generally for fabricated steelwork										

Note
    prices can vary considerably dependent upon the
    character of the work; the following are average
    prices for work fabricated and delivered to site
Framing, fabrication; including shop and site black
bolts, nuts and washers for structural framing to
structural framing connections										

Weldable steel, B.S.EN 10025 : 1993 Grade S275JR (formerly B.S.4360 Grade 43B), hot rolled sections B.S.4 Part 1 (Euronorm 54); welded fabrication in accordance with B.S.5950 Part 2										
Columns										
weight less than 40 Kg/m......................	935.00	-	935.00	-	-	-	-	935.00	t	1051.88
weight 40-100 Kg/m..........................	935.00	-	935.00	-	-	-	-	935.00	t	1051.88
weight exceeding 100 Kg/m..................	935.00	-	935.00	-	-	-	-	935.00	t	1051.88
Beams										
weight less than 40 Kg/m..................	930.00	-	930.00	-	-	-	-	930.00	t	1046.25
weight 40-100 Kg/m..........................	930.00	-	930.00	-	-	-	-	930.00	t	1046.25
weight exceeding 100 Kg/m..................	930.00	-	930.00	-	-	-	-	930.00	t	1046.25
Beams, castellated										
weight less than 40 Kg/m..................	1150.00	-	1150.00	-	-	-	-	1150.00	t	1293.75
weight 40-100 Kg/m..........................	1150.00	-	1150.00	-	-	-	-	1150.00	t	1293.75
weight exceeding 100 Kg/m..................	1150.00	-	1150.00	-	-	-	-	1150.00	t	1293.75
Beams, curved										
weight less than 40 Kg/m..................	1425.00	-	1425.00	-	-	-	-	1425.00	t	1603.13
weight 40-100 Kg/m..........................	1425.00	-	1425.00	-	-	-	-	1425.00	t	1603.13
weight exceeding 100 Kg/m..................	1425.00	-	1425.00	-	-	-	-	1425.00	t	1603.13
Bracings, tubular										
weight less than 40 Kg/m..................	980.00	-	980.00	-	-	-	-	980.00	t	1102.50
weight 40-100 Kg/m..........................	980.00	-	980.00	-	-	-	-	980.00	t	1102.50
weight exceeding 100 Kg/m..................	980.00	-	980.00	-	-	-	-	980.00	t	1102.50
Purlins and cladding rails										
weight less than 40 Kg/m..................	960.00	-	960.00	-	-	-	-	960.00	t	1080.00
weight 40-100 Kg/m..........................	960.00	-	960.00	-	-	-	-	960.00	t	1080.00
weight exceeding 100 Kg/m..................	960.00	-	960.00	-	-	-	-	960.00	t	1080.00
portal frames.....................	875.00	-	875.00	-	-	-	-	875.00	t	984.38
trusses 12m to 18m span................	1450.00	-	1450.00	-	-	-	-	1450.00	t	1631.25
trusses 6m to 12m span.................	1500.00	-	1500.00	-	-	-	-	1500.00	t	1687.50
trusses up to 6m span.................	1600.00	-	1600.00	-	-	-	-	1600.00	t	1800.00
Weldable steel B.S.EN 10210 : 1994 Grade S275JOH (formerly B.S.4360 Grade 43C), hot rolled sections B.S.4848 Part2 (Euronorm 57); welded fabrication in accordance with B.S.5950 Part 2										
column, square hollow section										
weight less than 40 Kg/m..................	1080.00	-	1080.00	-	-	-	-	1080.00	t	1215.00
weight 40-100 Kg/m..........................	1080.00	-	1080.00	-	-	-	-	1080.00	t	1215.00
weight exceeding 100 Kg/m..................	1080.00	-	1080.00	-	-	-	-	1080.00	t	1215.00
column, rectangular hollow section										
weight less than 40 Kg/m..................	1100.00	-	1100.00	-	-	-	-	1100.00	t	1237.50
weight 40-100 Kg/m..........................	1100.00	-	1100.00	-	-	-	-	1100.00	t	1237.50
weight exceeding 100 Kg/m..................	1100.00	-	1100.00	-	-	-	-	1100.00	t	1237.50
Holding down bolts or assemblies; mild steel										
rag or indented bolts; M 10 with nuts and washers										
100mm long..........................	1.20	5.00	1.26	0.09	-	0.83	-	2.09	nr	2.35
160mm long..........................	1.24	5.00	1.30	0.13	-	1.20	-	2.50	nr	2.81
rag or indented bolts; M 12 with nuts and washers										
100mm long..........................	1.29	5.00	1.35	0.12	-	1.11	-	2.46	nr	2.77
160mm long..........................	1.50	5.00	1.58	0.15	-	1.38	-	2.96	nr	3.33
200mm long..........................	1.68	5.00	1.76	0.16	-	1.48	-	3.24	nr	3.65
rag or indented bolts; M 16 with nuts and washers										
120mm long..........................	1.94	5.00	2.04	0.13	-	1.20	-	3.24	nr	3.64
160mm long..........................	2.23	5.00	2.34	0.16	-	1.48	-	3.82	nr	4.30
200mm long..........................	2.60	5.00	2.73	0.18	-	1.66	-	4.39	nr	4.94

Labour hourly rates: (except Specialists) Craft Operatives £9.23 Labourer £7.02 Rates are national average prices. Refer to REGIONAL VARIATIONS for indicative levels of overall pricing in regions	MATERIALS			LABOUR				RATES		
	Del to Site £	Waste %	Material Cost £	Craft Optve Hrs	Lab Hrs	Labour Cost £	Sunds £	Nett Rate £	Unit	Gross Rate (+12.5%) £
G10: STRUCTURAL STEEL FRAMING Cont.										
Framing, fabrication; including shop and site black bolts, nuts and washers for structural framing to structural framing connections Cont.										
Holding down bolts or assemblies; mild steel Cont. holding down bolt assembly; M 20 bolt with 100 x 100 x 10mm plate washer tack welded to head; with nuts and washers										
300mm long.............................	4.49	5.00	4.71	0.28	-	2.58	-	7.30	nr	8.21
350mm long.............................	5.12	5.00	5.38	0.32	-	2.95	-	8.33	nr	9.37
400mm long.............................	5.81	5.00	6.10	0.35	-	3.23	-	9.33	nr	10.50
450mm long.............................	6.65	5.00	6.98	0.55	-	5.08	-	12.06	nr	13.57
High strength friction grip bolts; B.S.4395 Part 1 - general grade M 16; with nuts and washers										
50mm long..............................	0.72	5.00	0.76	0.09	-	0.83	-	1.59	nr	1.79
60mm long..............................	0.79	5.00	0.83	0.09	-	0.83	-	1.66	nr	1.87
70mm long..............................	0.85	5.00	0.89	0.10	-	0.92	-	1.82	nr	2.04
80mm long..............................	0.92	5.00	0.97	0.10	-	0.92	-	1.89	nr	2.13
90mm long..............................	0.98	5.00	1.03	0.12	-	1.11	-	2.14	nr	2.40
100mm long.............................	1.01	5.00	1.06	0.12	-	1.11	-	2.17	nr	2.44
120mm long.............................	1.12	5.00	1.18	0.13	-	1.20	-	2.38	nr	2.67
140mm long.............................	1.27	5.00	1.33	0.14	-	1.29	-	2.63	nr	2.95
150mm long.............................	1.28	5.00	1.34	0.14	-	1.29	-	2.64	nr	2.97
160mm long.............................	1.41	5.00	1.48	0.15	-	1.38	-	2.87	nr	3.22
180mm long.............................	1.60	5.00	1.68	0.16	-	1.48	-	3.16	nr	3.55
200mm long.............................	4.06	5.00	4.26	0.16	-	1.48	-	5.74	nr	6.46
220mm long.............................	4.31	5.00	4.53	0.16	-	1.48	-	6.00	nr	6.75
M 20; with nuts and washers										
60mm long..............................	1.38	5.00	1.45	0.09	-	0.83	-	2.28	nr	2.56
70mm long..............................	1.47	5.00	1.54	0.10	-	0.92	-	2.47	nr	2.77
80mm long..............................	1.53	5.00	1.61	0.10	-	0.92	-	2.53	nr	2.85
90mm long..............................	1.61	5.00	1.69	0.12	-	1.11	-	2.80	nr	3.15
100mm long.............................	1.67	5.00	1.75	0.12	-	1.11	-	2.86	nr	3.22
120mm long.............................	1.86	5.00	1.95	0.13	-	1.20	-	3.15	nr	3.55
140mm long.............................	2.09	5.00	2.19	0.15	-	1.38	-	3.58	nr	4.03
150mm long.............................	2.32	5.00	2.44	0.15	-	1.38	-	3.82	nr	4.30
160mm long.............................	2.38	5.00	2.50	0.16	-	1.48	-	3.98	nr	4.47
180mm long.............................	2.72	5.00	2.86	0.17	-	1.57	-	4.43	nr	4.98
200mm long.............................	4.72	5.00	4.96	0.18	-	1.66	-	6.62	nr	7.44
220mm long.............................	5.09	5.00	5.34	0.20	-	1.85	-	7.19	nr	8.09
240mm long.............................	5.50	5.00	5.78	0.21	-	1.94	-	7.71	nr	8.68
260mm long.............................	5.88	5.00	6.17	0.23	-	2.12	-	8.30	nr	9.33
280mm long.............................	6.28	5.00	6.59	0.24	-	2.22	-	8.81	nr	9.91
300mm long.............................	6.68	5.00	7.01	0.25	-	2.31	-	9.32	nr	10.49
M 24; with nuts and washers										
70mm long..............................	2.46	5.00	2.58	0.13	-	1.20	-	3.78	nr	4.26
80mm long..............................	2.78	5.00	2.92	0.13	-	1.20	-	4.12	nr	4.63
90mm long..............................	2.83	5.00	2.97	0.14	-	1.29	-	4.26	nr	4.80
100mm long.............................	2.98	5.00	3.13	0.14	-	1.29	-	4.42	nr	4.97
120mm long.............................	3.35	5.00	3.52	0.16	-	1.48	-	4.99	nr	5.62
140mm long.............................	3.80	5.00	3.99	0.17	-	1.57	-	5.56	nr	6.25
150mm long.............................	3.91	5.00	4.11	0.17	-	1.57	-	5.67	nr	6.38
160mm long.............................	4.20	5.00	4.41	0.18	-	1.66	-	6.07	nr	6.83
180mm long.............................	4.77	5.00	5.01	0.20	-	1.85	-	6.85	nr	7.71
200mm long.............................	6.84	5.00	7.18	0.21	-	1.94	-	9.12	nr	10.26
220mm long.............................	7.45	5.00	7.82	0.23	-	2.12	-	9.95	nr	11.19
240mm long.............................	8.02	5.00	8.42	0.24	-	2.22	-	10.64	nr	11.97
260mm long.............................	8.63	5.00	9.06	0.25	-	2.31	-	11.37	nr	12.79
280mm long.............................	9.20	5.00	9.66	0.28	-	2.58	-	12.24	nr	13.77
300mm long.............................	9.82	5.00	10.31	0.29	-	2.68	-	12.99	nr	14.61
Framing, erection; in accordance with B.S.5950 Part 2										
Note the following prices for erection include for the site to be reasonably clear, ease of access and the erection carried out during normal working hours Permanent erection of fabricated steelwork on site with bolted connections										
framing...............................	-	-	-	25.00	-	230.75	-	230.75	t	259.59
Cold rolled zed purlins and cladding rails; Metsec, galvanised steel										
Purlin sleeved system (section only); fixing to cleats on frame members at 6000mm centres; hoist and fix 3.00m above ground level; fixing with bolts										
reference 14214; 3.03 kg/m	7.31	2.50	7.49	0.23	0.03	2.33	-	9.83	m	11.05
reference 14216; 3.47 kg/m	8.27	2.50	8.48	0.23	0.03	2.33	-	10.81	m	12.16
reference 17214; 3.66 kg/m	8.67	2.50	8.89	0.29	0.03	2.89	-	11.77	m	13.25
reference 17216; 4.11 kg/m	9.84	2.50	10.09	0.29	0.03	2.89	-	12.97	m	14.59
reference 20216; 4.49 kg/m	10.77	2.50	11.04	0.35	0.05	3.58	-	14.62	m	16.45
reference 20218; 5.03 kg/m	12.12	2.50	12.42	0.35	0.05	3.58	-	16.00	m	18.01
reference 20220; 5.57 kg/m	13.13	2.50	13.46	0.35	0.05	3.58	-	17.04	m	19.17
reference 23218; 5.73 kg/m	13.78	2.50	14.12	0.38	0.05	3.86	-	17.98	m	20.23
reference 23220; 6.34 kg/m	15.26	2.50	15.64	0.38	0.05	3.86	-	19.50	m	21.94
Purlin sleeved system (section only); fixing to cleats on frame members at 6000mm centres; hoist and fix 6.00m above ground level; fixing with bolts										
reference 14214; 3.03 kg/m	7.31	2.50	7.49	0.25	0.03	2.52	-	10.01	m	11.26

Labour hourly rates: (except Specialists) Craft Operatives £9.23 Labourer £7.02 Rates are national average prices. Refer to REGIONAL VARIATIONS for indicative levels of overall pricing in regions	MATERIALS			LABOUR				RATES		
	Del to Site	Waste	Material Cost	Craft Optve	Lab	Labour Cost	Sunds	Nett Rate	Unit	Gross Rate (+12.5%)
	£	%	£	Hrs	Hrs	£	£	£		£

G10: STRUCTURAL STEEL FRAMING Cont.

Cold rolled zed purlins and cladding rails; Metsec, galvanised steel Cont.

Purlin sleeved system (section only); fixing to cleats on frame members at 6000mm centres; hoist and fix 6.00m above ground level; fixing with bolts Cont.

	Del to Site	Waste	Material Cost	Craft Optve	Lab	Labour Cost	Sunds	Nett Rate	Unit	Gross Rate
reference 14216; 3.47 kg/m	8.27	2.50	8.48	0.25	0.03	2.52	-	10.99	m	12.37
reference 17214; 3.66 kg/m	8.67	2.50	8.89	0.32	0.05	3.30	-	12.19	m	13.72
reference 17216; 4.11 kg/m	9.84	2.50	10.09	0.32	0.05	3.30	-	13.39	m	15.06
reference 20216; 4.49 kg/m	10.77	2.50	11.04	0.38	0.05	3.86	-	14.90	m	16.76
reference 20218; 5.03 kg/m	12.12	2.50	12.42	0.38	0.05	3.86	-	16.28	m	18.32
reference 20220; 5.57 kg/m	13.13	2.50	13.46	0.38	0.05	3.86	-	17.32	m	19.48
reference 23218; 5.73 kg/m	13.78	2.50	14.12	0.41	0.06	4.21	-	18.33	m	20.62
reference 23220; 6.34 kg/m	15.26	2.50	15.64	0.41	0.06	4.21	-	19.85	m	22.33

Purlin overlap system (cleat only); fixing to cleats on frame members at 6000mm centres; hoist and fix 3.00m above ground level; fixing with bolts galvanised

reference 142	3.48	2.50	3.57	0.23	0.03	2.33	-	5.90	m	6.64
reference 172	3.77	2.50	3.86	0.23	0.03	2.33	-	6.20	m	6.97
reference 202	5.25	2.50	5.38	0.35	0.04	3.51	-	8.89	m	10.00
reference 232	6.97	2.50	7.14	0.38	0.05	3.86	-	11.00	m	12.38

Purlin overlap system (cleat only); fixing to cleats on frame members at 6000mm centres; hoist and fix 6.00m above ground level; fixing with bolts galvanised

reference 142	3.48	2.50	3.57	0.25	0.03	2.52	-	6.09	m	6.85
reference 172	3.77	2.50	3.86	0.32	0.05	3.30	-	7.17	m	8.06
reference 202	5.25	2.50	5.38	0.38	0.05	3.86	-	9.24	m	10.39
reference 232	6.97	2.50	7.14	0.41	0.06	4.21	-	11.35	m	12.77

Purlin Non-Continuous system (section only) ; fixing to cleats on frame members at 6000mm centres; hoist and fix 3.00m above ground level; fixing with bolts

reference 17215; 3.85 kg/m	9.22	2.50	9.45	0.29	0.03	2.89	-	12.34	m	13.88
reference 17216; 4.11 kg/m	9.84	2.50	10.09	0.29	0.03	2.89	-	12.97	m	14.59
reference 20216; 4.49 kg/m	10.77	2.50	11.04	0.35	0.05	3.58	-	14.62	m	16.45
reference 20218; 5.03 kg/m	12.12	2.50	12.42	0.35	0.05	3.58	-	16.00	m	18.01
reference 20220; 5.57 kg/m	13.33	2.50	13.66	0.35	0.05	3.58	-	17.24	m	19.40
reference 23218; 5.73 kg/m	13.78	2.50	14.12	0.38	0.05	3.86	-	17.98	m	20.23
reference 23223; 7.26 kg/m	17.56	2.50	18.00	0.38	0.05	3.86	-	21.86	m	24.59
reference 26223; 7.92 kg/m	19.22	2.50	19.70	0.46	0.06	4.67	-	24.37	m	27.41
reference 26229; 9.88 kg/m	23.84	2.50	24.44	0.46	0.06	4.67	-	29.10	m	32.74

Purlin Non-Continuous system (section only) ; fixing to cleats on frame members at 6000mm centres; hoist and fix 6.00m above ground level; fixing with bolts

reference 17215; 3.85 kg/m	9.22	2.50	9.45	0.32	0.05	3.30	-	12.76	m	14.35
reference 17216; 4.11 kg/m	9.84	2.50	10.09	0.32	0.05	3.30	-	13.39	m	15.06
reference 20216; 4.49 kg/m	10.77	2.50	11.04	0.38	0.05	3.86	-	14.90	m	16.76
reference 20218; 5.03 kg/m	12.12	2.50	12.42	0.38	0.05	3.86	-	16.28	m	18.32
reference 20220; 5.57 kg/m	13.33	2.50	13.66	0.38	0.05	3.86	-	17.52	m	19.71
reference 23218; 5.73 kg/m	13.78	2.50	14.12	0.41	0.06	4.21	-	18.33	m	20.62
reference 23223; 7.26 kg/m	17.56	2.50	18.00	0.41	0.06	4.21	-	22.20	m	24.98
reference 26223; 7.92 kg/m	19.22	2.50	19.70	0.51	0.07	5.20	-	24.90	m	28.01
reference 26229; 9.88 kg/m	23.84	2.50	24.44	0.51	0.07	5.20	-	29.63	m	33.34

Galvanised purlin cleats; weld on

Reference 142 for 142mm deep purlins	4.99	2.50	5.11	-	-	-	-	5.11	nr	5.75
Reference 172 for 172mm deep purlins	5.95	2.50	6.10	-	-	-	-	6.10	nr	6.86
Reference 202 for 202mm deep purlins	7.78	2.50	7.97	-	-	-	-	7.97	nr	8.97
Reference 232 for 232mm deep purlins	10.19	2.50	10.44	-	-	-	-	10.44	nr	11.75
Reference 262 for 262mm deep purlins	10.88	2.50	11.15	-	-	-	-	11.15	nr	12.55

Round Lok anti-sag rods; push fit to purlins

purlins at 1150mm centres	2.58	2.50	2.64	0.29	0.03	2.89	-	5.53	nr	6.22
purlins at 1350mm centres	3.00	2.50	3.08	0.29	0.03	2.89	-	5.96	nr	6.71
purlins at 1550mm centres	3.39	2.50	3.47	0.38	0.05	3.86	-	7.33	nr	8.25
purlins at 1700mm centres	3.79	2.50	3.88	0.38	0.05	3.86	-	7.74	nr	8.71
purlins at 1950mm centres	4.18	2.50	4.28	0.38	0.05	3.86	-	8.14	nr	9.16

Side rail sleeved system (section only); fixing to cleats on stanchions at 6000mm centres; hoist and fix 3.00m above ground level; fixing with bolts

reference 14214; 3.03 kg/m	7.31	2.50	7.49	0.23	0.03	2.33	-	9.83	m	11.05
reference 14216; 3.47 kg/m	8.27	2.50	8.48	0.23	0.03	2.33	-	10.81	m	12.16
reference 17215; 3.85 kg/m	9.22	2.50	9.45	0.29	0.03	2.89	-	12.34	m	13.88
reference 17216; 4.11 kg/m	9.84	2.50	10.09	0.29	0.03	2.89	-	12.97	m	14.59
reference 20216; 4.49 kg/m	10.77	2.50	11.04	0.35	0.05	3.58	-	14.62	m	16.45
reference 20218; 5.03 kg/m	12.12	2.50	12.42	0.35	0.05	3.58	-	16.00	m	18.01
reference 20220; 5.57 kg/m	13.33	2.50	13.66	0.35	0.05	3.58	-	17.24	m	19.40
reference 23218; 5.73 kg/m	13.78	2.50	14.12	0.38	0.05	3.86	-	17.98	m	20.23
reference 23223; 7.26 kg/m	17.66	2.50	18.10	0.38	0.05	3.86	-	21.96	m	24.70

Side rail sleeved system (section only); fixing to cleats on stanchions at 6000mm centres; hoist and fix 6.00m above ground level; fixing with bolts

reference 14214; 3.03 kg/m	7.31	2.50	7.49	0.25	0.03	2.52	-	10.01	m	11.26
reference 14216; 3.47 kg/m	8.27	2.50	8.48	0.25	0.03	2.52	-	10.99	m	12.37
reference 17215; 3.85 kg/m	9.22	2.50	9.45	0.32	0.05	3.30	-	12.76	m	14.35
reference 17216; 4.11 kg/m	9.84	2.50	10.09	0.32	0.05	3.30	-	13.39	m	15.06
reference 20216; 4.49 kg/m	10.77	2.50	11.04	0.38	0.05	3.86	-	14.90	m	16.76

STRUCTURAL/CARCASSING METAL/TIMBER

Labour hourly rates: (except Specialists) Craft Operatives £9.23 Labourer £7.02 Rates are national average prices. Refer to REGIONAL VARIATIONS for indicative levels of overall pricing in regions	MATERIALS			LABOUR				RATES		
	Del to Site £	Waste %	Material Cost £	Craft Optve Hrs	Lab Hrs	Labour Cost £	Sunds £	Nett Rate £	Unit	Gross Rate (+12.5%) £

G10: STRUCTURAL STEEL FRAMING Cont.

Cold rolled zed purlins and cladding rails; Metsec, galvanised steel Cont.

Side rail sleeved system (section only); fixing to cleats on stanchions at 6000mm centres; hoist and fix 6.00m above ground level; fixing with bolts Cont.

reference 20218; 5.03 kg/m	12.12	2.50	12.42	0.38	0.05	3.86	-	16.28	m	18.32
reference 20220; 5.57 kg/m	13.13	2.50	13.46	0.38	0.05	3.86	-	17.32	m	19.48
reference 23218; 5.73 kg/m	13.78	2.50	14.12	0.41	0.06	4.21	-	18.33	m	20.62
reference 23223; 7.26 kg/m	17.66	2.50	18.10	0.41	0.06	4.21	-	22.31	m	25.10

Side rail single span system (section only) ; fixing to cleats on stanchions at 6000mm centres; hoist and fix 3.00m above ground level; fixing with bolts

reference 14214; 3.03 kg/m	7.31	2.50	7.49	0.23	0.03	2.33	-	9.83	m	11.05
reference 14216; 3.47 kg/m	8.27	2.50	8.48	0.23	0.03	2.33	-	10.81	m	12.16
reference 17215; 3.85 kg/m	9.22	2.50	9.45	0.29	0.03	2.89	-	12.34	m	13.88
reference 17216; 4.11 kg/m	9.84	2.50	10.09	0.29	0.03	2.89	-	12.97	m	14.59
reference 20216; 4.49 kg/m	10.77	2.50	11.04	0.35	0.05	3.58	-	14.62	m	16.45
reference 20218; 5.03 kg/m	12.12	2.50	12.42	0.35	0.05	3.58	-	16.00	m	18.01
reference 20220; 5.57 kg/m	13.13	2.50	13.46	0.35	0.05	3.58	-	17.04	m	19.17
reference 23218; 5.73 kg/m	13.78	2.50	14.12	0.38	0.05	3.86	-	17.98	m	20.23
reference 23223; 7.26 kg/m	17.66	2.50	18.10	0.38	0.05	3.86	-	21.96	m	24.70

Side rail single span system (section only) ; fixing to cleats on stanchions at 6000mm centres; hoist and fix 6.00m above ground level; fixing with bolts

reference 14214; 3.03 kg/m	7.31	2.50	7.49	0.25	0.03	2.52	-	10.01	m	11.26
reference 14216; 3.47 kg/m	8.27	2.50	8.48	0.25	0.03	2.52	-	10.99	m	12.37
reference 17215; 3.85 kg/m	9.22	2.50	9.45	0.32	0.05	3.30	-	12.76	m	14.35
reference 17216; 4.11 kg/m	9.84	2.50	10.09	0.32	0.05	3.30	-	13.39	m	15.06
reference 20216; 4.49 kg/m	10.77	2.50	11.04	0.38	0.05	3.86	-	14.90	m	16.76
reference 20218; 5.03 kg/m	12.12	2.50	12.42	0.38	0.05	3.86	-	16.28	m	18.32
reference 20220; 5.57 kg/m	13.13	2.50	13.46	0.38	0.05	3.86	-	17.32	m	19.48
reference 23218; 5.73 kg/m	13.78	2.50	14.12	0.41	0.06	4.21	-	18.33	m	20.62
reference 23223; 7.26 kg/m	17.66	2.50	18.10	0.41	0.06	4.21	-	22.31	m	25.10

Galvanised side rail cleats; weld on

Reference 142 for 142mm wide rails	4.99	2.50	5.11	-	-	-	-	5.11	nr	5.75
Reference 172 for 172mm wide rails	5.95	2.50	6.10	-	-	-	-	6.10	nr	6.86
Reference 202 for 202mm wide rails	7.78	2.50	7.97	-	-	-	-	7.97	nr	8.97
Reference 232 for 232mm wide rails	10.19	2.50	10.44	-	-	-	-	10.44	nr	11.75

Galvanised side rail supports; 122-262 series; weld on

rail 1000mm long	7.53	2.50	7.72	0.40	0.05	4.04	-	11.76	nr	13.23
rail 1400mm long	8.23	2.50	8.44	0.40	0.05	4.04	-	12.48	nr	14.04
rail 1600mm long	9.13	2.50	9.36	0.52	0.07	5.29	-	14.65	nr	16.48
rail 1800mm long	10.98	2.50	11.25	0.52	0.07	5.29	-	16.55	nr	18.61

Diagonal tie wire ropes (assembled with end brackets); bolt on

1700mm long	14.70	2.50	15.07	0.35	0.05	3.58	-	18.65	nr	20.98
2200mm long	16.44	2.50	16.85	0.35	0.05	3.58	-	20.43	nr	22.99
2600mm long	18.20	2.50	18.66	0.46	0.06	4.67	-	23.32	nr	26.24
3600mm long	22.46	2.50	23.02	0.46	0.06	4.67	-	27.69	nr	31.15

Surface preparation at works

Note
   notwithstanding the requirement of SMM7 to measure painting on structural steelwork in m2, painting off site has been given in tonnes of structural steelwork in accordance with normal steelwork contractors practice
Blast cleaning

surfaces of steelwork	-	-	Spclist	-	-	Spclist	-	77.00	t	86.63

Surface treatment at works

One coat micaceous oxide primer, 75 microns

surfaces of steelwork	-	-	Spclist	-	-	Spclist	-	86.00	t	96.75

Two coats micaceous oxide primer, 150 microns

surfaces of steelwork	-	-	Spclist	-	-	Spclist	-	160.00	t	180.00

G12: ISOLATED STRUCTURAL METAL MEMBERS

Isolated unfabricated structural members

Weldable steel, B.S.EN 10025 : 1993 Grade S275 (formerly B.S.4360 Grade 43A), hot rolled sections plain member; beam

weight less than 40 Kg/m	860.00	-	860.00	50.00	-	461.50	-	1321.50	t	1486.69
weight 40-100 Kg/m	860.00	-	860.00	47.05	-	434.27	-	1294.27	t	1456.06
weight exceeding 100 Kg/m	860.00	-	860.00	41.20	-	380.28	-	1240.28	t	1395.31

Steel short span lattice joists primed at works; treated timber inserts in top and bottom chords; full depth end seatings for bolting to supporting structure; hoist and fix 3.00m above ground level

200mm deep, 8.2 kg/m	28.90	2.50	29.62	0.76	0.09	7.65	-	37.27	m	41.93
250mm deep, 8.5 kg/m	30.15	2.50	30.90	0.78	0.10	7.90	-	38.81	m	43.66
300mm deep, 10.7 kg/m	38.06	2.50	39.01	0.86	0.10	8.64	-	47.65	m	53.61

Labour hourly rates: (except Specialists) Craft Operatives £9.23 Labourer £7.02 Rates are national average prices. Refer to REGIONAL VARIATIONS for indicative levels of overall pricing in regions	MATERIALS			LABOUR				RATES		
	Del to Site £	Waste %	Material Cost £	Craft Optve Hrs	Lab Hrs	Labour Cost £	Sunds £	Nett Rate £	Unit	Gross Rate (+12.5%) £

G12: ISOLATED STRUCTURAL METAL MEMBERS Cont.

Isolated unfabricated structural members Cont.

Steel short span lattice joists primed at works;
treated timber inserts in top and bottom chords; full
depth end seatings for bolting to supporting
structure; hoist and fix 3.00m above ground level
Cont.

	Del to Site £	Waste %	Material Cost £	Craft Optve Hrs	Lab Hrs	Labour Cost £	Sunds £	Nett Rate £	Unit	Gross Rate (+12.5%) £
350mm deep, 11.6 kg/m	40.92	2.50	41.94	0.93	0.12	9.43	–	51.37	m	57.79
350mm deep, 12.8 kg/m	44.72	2.50	45.84	1.04	0.13	10.51	–	56.35	m	63.39

Steel short span lattice joists primed at works;
treated timber inserts in top and bottom chords; full
depth end seatings for bolting to supporting
structure; hoist and fix 6.00m above ground level

	Del to Site £	Waste %	Material Cost £	Craft Optve Hrs	Lab Hrs	Labour Cost £	Sunds £	Nett Rate £	Unit	Gross Rate (+12.5%) £
200mm deep, 8.2 kg/m	28.90	2.50	29.62	0.83	0.10	8.36	–	37.99	m	42.73
250mm deep, 8.5 kg/m	30.15	2.50	30.90	0.86	0.10	8.64	–	39.54	m	44.49
300mm deep, 10.7 kg/m	38.06	2.50	39.01	0.94	0.12	9.52	–	48.53	m	54.60
350mm deep, 11.6 kg/m	40.92	2.50	41.94	1.02	0.13	10.33	–	52.27	m	58.80
350mm deep, 12.8 kg/m	44.92	2.50	46.04	1.13	0.14	11.41	–	57.46	m	64.64

Steel intermediate span lattice joists primed at
works; treated timber inserts in top and bottom
chords; full depth end seatings for bolting to
supporting structure; hoist and fix 3.00m above
ground level

	Del to Site £	Waste %	Material Cost £	Craft Optve Hrs	Lab Hrs	Labour Cost £	Sunds £	Nett Rate £	Unit	Gross Rate (+12.5%) £
450mm deep, 12.2 kg/m	39.16	2.50	40.14	0.98	0.13	9.96	–	50.10	m	56.36
500mm deep, 15.8 kg/m	48.40	2.50	49.61	1.18	0.15	11.94	–	61.55	m	69.25
550mm deep, 19.4 kg/m	57.20	2.50	58.63	1.45	0.18	14.65	–	73.28	m	82.44
600mm deep, 22.5 kg/m	65.45	2.50	67.09	1.55	0.20	15.71	–	82.80	m	93.15
650mm deep, 29.7 kg/m	86.24	2.50	88.40	2.05	0.25	20.68	–	109.07	m	122.71

Steel intermediate span lattice joists primed at
works; treated timber inserts in top and bottom
chords; full depth end seatings for bolting to
supporting structure; hoist and fix 6.00m above
ground level

	Del to Site £	Waste %	Material Cost £	Craft Optve Hrs	Lab Hrs	Labour Cost £	Sunds £	Nett Rate £	Unit	Gross Rate (+12.5%) £
450mm deep, 12.2 kg/m	39.16	2.50	40.14	1.08	0.14	10.95	–	51.09	m	57.48
500mm deep, 15.8 kg/m	48.40	2.50	49.61	1.30	0.16	13.12	–	62.73	m	70.57
550mm deep, 19.4 kg/m	57.20	2.50	58.63	1.60	0.20	16.17	–	74.80	m	84.15
600mm deep, 22.5 kg/m	65.45	2.50	67.09	1.71	0.22	17.33	–	84.41	m	94.97
650mm deep, 29.7 kg/m	86.24	2.50	88.40	2.25	0.29	22.80	–	111.20	m	125.10

Steel long span lattice joists primed at works;
treated timber inserts in top and bottom chords; full
depth end seatings for bolting to supporting
structure; hoist and fix 3.00m above ground level

	Del to Site £	Waste %	Material Cost £	Craft Optve Hrs	Lab Hrs	Labour Cost £	Sunds £	Nett Rate £	Unit	Gross Rate (+12.5%) £
700mm deep, 39.2 kg/m	111.65	2.50	114.44	2.25	0.29	22.80	–	137.24	m	154.40
800mm deep, 44.1 kg/m	119.46	2.50	122.45	2.54	0.32	25.69	–	148.14	m	166.65
900mm deep, 45.3 kg/m	120.12	2.50	123.12	2.61	0.32	26.34	–	149.46	m	168.14
1000mm deep, 46.1 kg/m	122.65	2.50	125.72	2.66	0.33	26.87	–	152.58	m	171.66
1500mm deep, 54.2 kg/m	145.20	2.50	148.83	2.76	0.35	27.93	–	176.76	m	198.86

Steel long span lattice joists primed at works;
treated timber inserts in top and bottom chords; full
depth end seatings for bolting to supporting
structure; hoist and fix 6.00m above ground level

	Del to Site £	Waste %	Material Cost £	Craft Optve Hrs	Lab Hrs	Labour Cost £	Sunds £	Nett Rate £	Unit	Gross Rate (+12.5%) £
700mm deep, 39.2 kg/m	111.65	2.50	114.44	2.48	0.31	25.07	–	139.51	m	156.95
800mm deep, 44.1 kg/m	119.46	2.50	122.45	2.79	0.35	28.21	–	150.66	m	169.49
900mm deep, 45.3 kg/m	120.12	2.50	123.12	2.88	0.36	29.11	–	152.23	m	171.26
1000mm deep, 46.1 kg/m	122.65	2.50	125.72	2.92	0.37	29.55	–	155.27	m	174.67
1500mm deep, 54.2 kg/m	145.20	2.50	148.83	3.04	0.38	30.73	–	179.56	m	202.00

Fixing bolts; steel

Black bolts, B.S.4190 grade 4.6
  M 8; with nuts and washers

	Del to Site £	Waste %	Material Cost £	Craft Optve Hrs	Lab Hrs	Labour Cost £	Sunds £	Nett Rate £	Unit	Gross Rate (+12.5%) £
30mm long	0.13	5.00	0.14	0.07	–	0.65	–	0.78	nr	0.88
40mm long	0.14	5.00	0.15	0.07	–	0.65	–	0.79	nr	0.89
50mm long	0.15	5.00	0.16	0.07	–	0.65	–	0.80	nr	0.90
60mm long	0.17	5.00	0.18	0.07	–	0.65	–	0.82	nr	0.93
70mm long	0.18	5.00	0.19	0.08	–	0.74	–	0.93	nr	1.04
80mm long	0.20	5.00	0.21	0.08	–	0.74	–	0.95	nr	1.07
90mm long	0.23	5.00	0.24	0.08	–	0.74	–	0.98	nr	1.10
100mm long	0.25	5.00	0.26	0.08	–	0.74	–	1.00	nr	1.13
120mm long	0.30	5.00	0.32	0.09	–	0.83	–	1.15	nr	1.29
M 10; with nuts and washers										
40mm long	0.20	5.00	0.21	0.07	–	0.65	–	0.86	nr	0.96
50mm long	0.21	5.00	0.22	0.07	–	0.65	–	0.87	nr	0.97
60mm long	0.24	5.00	0.25	0.07	–	0.65	–	0.90	nr	1.01
70mm long	0.25	5.00	0.26	0.08	–	0.74	–	1.00	nr	1.13
80mm long	0.31	5.00	0.33	0.08	–	0.74	–	1.06	nr	1.20
90mm long	0.33	5.00	0.35	0.08	–	0.74	–	1.08	nr	1.22
100mm long	0.38	5.00	0.40	0.08	–	0.74	–	1.14	nr	1.28
120mm long	0.41	5.00	0.43	0.09	–	0.83	–	1.26	nr	1.42
140mm long	0.53	5.00	0.56	0.10	–	0.92	–	1.48	nr	1.66
150mm long	0.60	5.00	0.63	0.12	–	1.11	–	1.74	nr	1.95
M 12; with nuts and washers										
40mm long	0.29	5.00	0.30	0.08	–	0.74	–	1.04	nr	1.17
50mm long	0.30	5.00	0.32	0.08	–	0.74	–	1.05	nr	1.19
60mm long	0.32	5.00	0.34	0.08	–	0.74	–	1.07	nr	1.21
70mm long	0.37	5.00	0.39	0.08	–	0.74	–	1.13	nr	1.27
80mm long	0.41	5.00	0.43	0.08	–	0.74	–	1.17	nr	1.32
90mm long	0.42	5.00	0.44	0.08	–	0.74	–	1.18	nr	1.33
100mm long	0.46	5.00	0.48	0.09	–	0.83	–	1.31	nr	1.48

Labour hourly rates: (except Specialists) Craft Operatives £9.23 Labourer £7.02 Rates are national average prices. Refer to REGIONAL VARIATIONS for indicative levels of overall pricing in regions	MATERIALS			LABOUR				RATES		
	Del to Site	Waste	Material Cost	Craft Optve	Lab	Labour Cost	Sunds	Nett Rate	Unit	Gross Rate (+12.5%)
	£	%	£	Hrs	Hrs	£	£	£		£
G12: ISOLATED STRUCTURAL METAL MEMBERS Cont.										
Fixing bolts; steel Cont.										
Black bolts, B.S.4190 grade 4.6 Cont.										
M 12; with nuts and washers Cont.										
120mm long	0.55	5.00	0.58	0.09	-	0.83	-	1.41	nr	1.58
140mm long	0.68	5.00	0.71	0.12	-	1.11	-	1.82	nr	2.05
150mm long	0.86	5.00	0.90	0.12	-	1.11	-	2.01	nr	2.26
160mm long	1.24	5.00	1.30	0.12	-	1.11	-	2.41	nr	2.71
180mm long	1.30	5.00	1.37	0.13	-	1.20	-	2.56	nr	2.89
200mm long	1.36	5.00	1.43	0.14	-	1.29	-	2.72	nr	3.06
220mm long	1.43	5.00	1.50	0.15	-	1.38	-	2.89	nr	3.25
240mm long	1.49	5.00	1.56	0.16	-	1.48	-	3.04	nr	3.42
260mm long	1.56	5.00	1.64	0.17	-	1.57	-	3.21	nr	3.61
280mm long	1.64	5.00	1.72	0.18	-	1.66	-	3.38	nr	3.81
300mm long	1.70	5.00	1.78	0.20	-	1.85	-	3.63	nr	4.08
M 16; with nuts and washers										
50mm long	0.53	5.00	0.56	0.08	-	0.74	-	1.29	nr	1.46
60mm long	0.59	5.00	0.62	0.08	-	0.74	-	1.36	nr	1.53
70mm long	0.65	5.00	0.68	0.08	-	0.74	-	1.42	nr	1.60
80mm long	0.71	5.00	0.75	0.09	-	0.83	-	1.58	nr	1.77
90mm long	0.76	5.00	0.80	0.09	-	0.83	-	1.63	nr	1.83
100mm long	0.89	5.00	0.93	0.10	-	0.92	-	1.86	nr	2.09
120mm long	1.03	5.00	1.08	0.10	-	0.92	-	2.00	nr	2.26
140mm long	1.15	5.00	1.21	0.12	-	1.11	-	2.32	nr	2.60
150mm long	1.21	5.00	1.27	0.13	-	1.20	-	2.47	nr	2.78
160mm long	1.68	5.00	1.76	0.14	-	1.29	-	3.06	nr	3.44
180mm long	1.75	5.00	1.84	0.14	-	1.29	-	3.13	nr	3.52
200mm long	1.86	5.00	1.95	0.15	-	1.38	-	3.34	nr	3.75
220mm long	1.97	5.00	2.07	0.15	-	1.38	-	3.45	nr	3.88
M 20; with nuts and washers										
60mm long	1.00	5.00	1.05	0.08	-	0.74	-	1.79	nr	2.01
70mm long	1.10	5.00	1.16	0.08	-	0.74	-	1.89	nr	2.13
80mm long	1.25	5.00	1.31	0.09	-	0.83	-	2.14	nr	2.41
90mm long	1.26	5.00	1.32	0.10	-	0.92	-	2.25	nr	2.53
100mm long	1.44	5.00	1.51	0.10	-	0.92	-	2.44	nr	2.74
120mm long	1.69	5.00	1.77	0.12	-	1.11	-	2.88	nr	3.24
140mm long	1.97	5.00	2.07	0.13	-	1.20	-	3.27	nr	3.68
150mm long	2.02	5.00	2.12	0.14	-	1.29	-	3.41	nr	3.84
160mm long	2.48	5.00	2.60	0.15	-	1.38	-	3.99	nr	4.49
180mm long	2.62	5.00	2.75	0.15	-	1.38	-	4.14	nr	4.65
200mm long	2.80	5.00	2.94	0.16	-	1.48	-	4.42	nr	4.97
220mm long	2.90	5.00	3.04	0.17	-	1.57	-	4.61	nr	5.19
240mm long	3.07	5.00	3.22	0.18	-	1.66	-	4.88	nr	5.50
260mm long	3.16	5.00	3.32	0.20	-	1.85	-	5.16	nr	5.81
280mm long	3.24	5.00	3.40	0.21	-	1.94	-	5.34	nr	6.01
300mm long	3.43	5.00	3.60	0.23	-	2.12	-	5.72	nr	6.44
M 24; with nuts and washers										
70mm long	2.51	5.00	2.64	0.10	-	0.92	-	3.56	nr	4.00
80mm long	2.62	5.00	2.75	0.10	-	0.92	-	3.67	nr	4.13
90mm long	2.70	5.00	2.84	0.12	-	1.11	-	3.94	nr	4.44
100mm long	2.80	5.00	2.94	0.12	-	1.11	-	4.05	nr	4.55
120mm long	2.99	5.00	3.14	0.12	-	1.11	-	4.25	nr	4.78
140mm long	3.36	5.00	3.53	0.13	-	1.20	-	4.73	nr	5.32
150mm long	3.56	5.00	3.74	0.15	-	1.38	-	5.12	nr	5.76
160mm long	3.68	5.00	3.86	0.13	-	1.20	-	5.06	nr	5.70
180mm long	3.78	5.00	3.97	0.15	-	1.38	-	5.35	nr	6.02
200mm long	3.96	5.00	4.16	0.16	-	1.48	-	5.63	nr	6.34
220mm long	4.37	5.00	4.59	0.17	-	1.57	-	6.16	nr	6.93
240mm long	4.49	5.00	4.71	0.18	-	1.66	-	6.38	nr	7.17
260mm long	4.79	5.00	5.03	0.21	-	1.94	-	6.97	nr	7.84
280mm long	4.94	5.00	5.19	0.23	-	2.12	-	7.31	nr	8.22
300mm long	5.09	5.00	5.34	0.25	-	2.31	-	7.65	nr	8.61
Black bolts, B.S.4190 grade 4.6; galvanised										
M 8; with nuts and washers										
30mm long	0.18	5.00	0.19	0.07	-	0.65	-	0.84	nr	0.94
40mm long	0.20	5.00	0.21	0.07	-	0.65	-	0.86	nr	0.96
50mm long	0.22	5.00	0.23	0.07	-	0.65	-	0.88	nr	0.99
60mm long	0.24	5.00	0.25	0.07	-	0.65	-	0.90	nr	1.01
70mm long	0.26	5.00	0.27	0.08	-	0.74	-	1.01	nr	1.14
80mm long	0.28	5.00	0.29	0.08	-	0.74	-	1.03	nr	1.16
90mm long	0.31	5.00	0.33	0.08	-	0.74	-	1.06	nr	1.20
100mm long	0.38	5.00	0.40	0.08	-	0.74	-	1.14	nr	1.28
120mm long	0.46	5.00	0.48	0.09	-	0.83	-	1.31	nr	1.48
M 10; with nuts and washers										
40mm long	0.28	5.00	0.29	0.07	-	0.65	-	0.94	nr	1.06
50mm long	0.31	5.00	0.33	0.07	-	0.65	-	0.97	nr	1.09
60mm long	0.35	5.00	0.37	0.07	-	0.65	-	1.01	nr	1.14
70mm long	0.38	5.00	0.40	0.08	-	0.74	-	1.14	nr	1.28
80mm long	0.47	5.00	0.49	0.08	-	0.74	-	1.23	nr	1.39
90mm long	0.52	5.00	0.55	0.08	-	0.74	-	1.28	nr	1.44
100mm long	0.56	5.00	0.59	0.08	-	0.74	-	1.33	nr	1.49
120mm long	0.60	5.00	0.63	0.09	-	0.83	-	1.46	nr	1.64
140mm long	0.78	5.00	0.82	0.10	-	0.92	-	1.74	nr	1.96
150mm long	0.90	5.00	0.94	0.12	-	1.11	-	2.05	nr	2.31
M 12; with nuts and washers										
40mm long	0.47	5.00	0.49	0.08	-	0.74	-	1.23	nr	1.39
50mm long	0.49	5.00	0.51	0.08	-	0.74	-	1.25	nr	1.41
60mm long	0.53	5.00	0.56	0.08	-	0.74	-	1.29	nr	1.46
70mm long	0.59	5.00	0.62	0.08	-	0.74	-	1.36	nr	1.53
80mm long	0.64	5.00	0.67	0.08	-	0.74	-	1.41	nr	1.59
90mm long	0.66	5.00	0.69	0.08	-	0.74	-	1.43	nr	1.61
100mm long	0.72	5.00	0.76	0.09	-	0.83	-	1.59	nr	1.79
120mm long	0.89	5.00	0.93	0.09	-	0.83	-	1.77	nr	1.99

Labour hourly rates: (except Specialists) Craft Operatives £9.23 Labourer £7.02 Rates are national average prices. Refer to REGIONAL VARIATIONS for indicative levels of overall pricing in regions	MATERIALS			LABOUR				RATES		
	Del to Site	Waste	Material Cost	Craft Optve	Lab	Labour Cost	Sunds	Nett Rate	Unit	Gross Rate (+12.5%)
	£	%	£	Hrs	Hrs	£	£	£		£

G12: ISOLATED STRUCTURAL METAL MEMBERS Cont.

Fixing bolts; steel Cont.

Black bolts, B.S.4190 grade 4.6; galvanised Cont.

M 12; with nuts and washers Cont.

140mm long....................	1.04	5.00	1.09	0.12	–	1.11	–	2.20	nr	2.47
150mm long....................	1.32	5.00	1.39	0.12	–	1.11	–	2.49	nr	2.81
160mm long....................	1.86	5.00	1.95	0.12	–	1.11	–	3.06	nr	3.44
180mm long....................	1.96	5.00	2.06	0.13	–	1.20	–	3.26	nr	3.67
200mm long....................	2.04	5.00	2.14	0.14	–	1.29	–	3.43	nr	3.86
220mm long....................	2.15	5.00	2.26	0.15	–	1.38	–	3.64	nr	4.10
240mm long....................	2.23	5.00	2.34	0.16	–	1.48	–	3.82	nr	4.30
260mm long....................	2.36	5.00	2.48	0.17	–	1.57	–	4.05	nr	4.55
280mm long....................	2.44	5.00	2.56	0.18	–	1.66	–	4.22	nr	4.75
300mm long....................	2.54	5.00	2.67	0.20	–	1.85	–	4.51	nr	5.08

M 16; with nuts and washers

50mm long....................	0.85	5.00	0.89	0.08	–	0.74	–	1.63	nr	1.83
60mm long....................	0.92	5.00	0.97	0.08	–	0.74	–	1.70	nr	1.92
70mm long....................	1.00	5.00	1.05	0.08	–	0.74	–	1.79	nr	2.01
80mm long....................	1.08	5.00	1.13	0.09	–	0.83	–	1.96	nr	2.21
90mm long....................	1.14	5.00	1.20	0.09	–	0.83	–	2.03	nr	2.28
100mm long....................	1.20	5.00	1.26	0.10	–	0.92	–	2.18	nr	2.46
120mm long....................	1.56	5.00	1.64	0.10	–	0.92	–	2.56	nr	2.88
140mm long....................	1.74	5.00	1.83	0.12	–	1.11	–	2.93	nr	3.30
150mm long....................	1.85	5.00	1.94	0.13	–	1.20	–	3.14	nr	3.54
160mm long....................	2.50	5.00	2.63	0.14	–	1.29	–	3.92	nr	4.41
180mm long....................	2.65	5.00	2.78	0.14	–	1.29	–	4.07	nr	4.58
200mm long....................	2.82	5.00	2.96	0.15	–	1.38	–	4.35	nr	4.89
220mm long....................	2.94	5.00	3.09	0.15	–	1.38	–	4.47	nr	5.03

M 20; with nuts and washers

60mm long....................	1.54	5.00	1.62	0.08	–	0.74	–	2.36	nr	2.65
70mm long....................	1.69	5.00	1.77	0.08	–	0.74	–	2.51	nr	2.83
80mm long....................	1.91	5.00	2.01	0.09	–	0.83	–	2.84	nr	3.19
90mm long....................	1.92	5.00	2.02	0.10	–	0.92	–	2.94	nr	3.31
100mm long....................	2.20	5.00	2.31	0.10	–	0.92	–	3.23	nr	3.64
120mm long....................	2.54	5.00	2.67	0.12	–	1.11	–	3.77	nr	4.25
140mm long....................	2.97	5.00	3.12	0.13	–	1.20	–	4.32	nr	4.86
150mm long....................	3.04	5.00	3.19	0.14	–	1.29	–	4.48	nr	5.04
160mm long....................	3.76	5.00	3.95	0.15	–	1.38	–	5.33	nr	6.00
180mm long....................	3.88	5.00	4.07	0.15	–	1.38	–	5.46	nr	6.14
200mm long....................	4.13	5.00	4.34	0.16	–	1.48	–	5.81	nr	6.54
220mm long....................	4.30	5.00	4.51	0.17	–	1.57	–	6.08	nr	6.84
240mm long....................	4.56	5.00	4.79	0.18	–	1.66	–	6.45	nr	7.26
260mm long....................	4.67	5.00	4.90	0.20	–	1.85	–	6.75	nr	7.59
280mm long....................	4.79	5.00	5.03	0.21	–	1.94	–	6.97	nr	7.84
300mm long....................	5.08	5.00	5.33	0.23	–	2.12	–	7.46	nr	8.39

M 24; with nuts and washers

70mm long....................	3.71	5.00	3.90	0.10	–	0.92	–	4.82	nr	5.42
80mm long....................	3.88	5.00	4.07	0.10	–	0.92	–	5.00	nr	5.62
90mm long....................	4.00	5.00	4.20	0.12	–	1.11	–	5.31	nr	5.97
100mm long....................	4.15	5.00	4.36	0.12	–	1.11	–	5.47	nr	6.15
120mm long....................	4.38	5.00	4.60	0.12	–	1.11	–	5.71	nr	6.42
140mm long....................	4.97	5.00	5.22	0.13	–	1.20	–	6.42	nr	7.22
150mm long....................	5.28	5.00	5.54	0.15	–	1.38	–	6.93	nr	7.79
160mm long....................	5.45	5.00	5.72	0.13	–	1.20	–	6.92	nr	7.79
180mm long....................	5.58	5.00	5.86	0.15	–	1.38	–	7.24	nr	8.15
200mm long....................	5.64	5.00	5.92	0.16	–	1.48	–	7.40	nr	8.32
220mm long....................	6.16	5.00	6.47	0.17	–	1.57	–	8.04	nr	9.04
240mm long....................	6.29	5.00	6.60	0.18	–	1.66	–	8.27	nr	9.30
260mm long....................	6.61	5.00	6.94	0.21	–	1.94	–	8.88	nr	9.99
280mm long....................	6.78	5.00	7.12	0.23	–	2.12	–	9.24	nr	10.40
300mm long....................	7.00	5.00	7.35	0.25	–	2.31	–	9.66	nr	10.86

G20: CARPENTRY/TIMBER FRAMING/FIRST FIXING

Sasco Trussed Rafters in softwood, sawn, B.S.4978, GS grade, impregnated; note:- prices are guide prices only, based on contracts within 30 mile radius of Widnes and are subject to adjustment for quantity: firm quotations should always be obtained

Trussed rafters; Hydro-Nail plated joints

22.5 degree Standard duo pitch; 450mm overhangs; fixing with clips (included elsewhere); span over wall plates

5000mm	27.07	2.50	27.75	1.30	0.16	13.12	6.60	47.47	nr	53.40
6000mm	31.15	2.50	31.93	1.40	0.17	14.12	7.20	53.24	nr	59.90
7000mm	34.86	2.50	35.73	1.50	0.19	15.18	7.80	58.71	nr	66.05
8000mm	41.27	2.50	42.30	1.60	0.20	16.17	8.40	66.87	nr	75.23
9000mm	47.77	2.50	48.96	1.70	0.21	17.17	9.00	75.13	nr	84.52
10000mm	57.84	2.50	59.29	1.80	0.22	18.16	9.60	87.04	nr	97.92

35 degree Standard duo pitch; 450mm overhangs; fixing with clips (included elsewhere); span over wall plates

5000mm	28.14	2.50	28.84	1.30	0.16	13.12	6.60	48.57	nr	54.64
6000mm	32.52	2.50	33.33	1.40	0.17	14.12	7.20	54.65	nr	61.48
7000mm	36.24	2.50	37.15	1.50	0.19	15.18	7.80	60.12	nr	67.64
8000mm	43.50	2.50	44.59	1.60	0.20	16.17	8.40	69.16	nr	77.80
9000mm	48.41	2.50	49.62	1.70	0.21	17.17	9.00	75.79	nr	85.26

45 degree Standard duo pitch; 450mm overhangs; fixing with clips (included elsewhere); span over wall plates

5000mm	44.18	2.50	45.28	1.30	0.16	13.12	6.60	65.01	nr	73.13
6000mm	52.86	2.50	54.18	1.40	0.17	14.12	7.20	75.50	nr	84.93
7000mm	61.66	2.50	63.20	1.50	0.19	15.18	7.80	86.18	nr	96.95

STRUCTURAL/CARCASSING METAL/TIMBER

Labour hourly rates: (except Specialists) Craft Operatives £9.23 Labourer £7.02 Rates are national average prices. Refer to REGIONAL VARIATIONS for indicative levels of overall pricing in regions	MATERIALS			LABOUR				RATES		
	Del to Site	Waste	Material Cost	Craft Optve	Lab	Labour Cost	Sunds	Nett Rate	Unit	Gross Rate (+12.5%)
	£	%	£	Hrs	Hrs	£	£	£		£

G20: CARPENTRY/TIMBER FRAMING/FIRST FIXING Cont.

Sasco Trussed Rafters in softwood, sawn, B.S.4978, GS
grade, impregnated; note:- prices are guide prices
only, based on contracts within 30 mile radius of
Widnes and are subject to adjustment for quantity:
firm quotations should always be obtained Cont.

Trussed rafters; Hydro-Nail plated joints Cont.

	Del to Site	Waste	Material Cost	Craft Optve	Lab	Labour Cost	Sunds	Nett Rate	Unit	Gross Rate
45 degree Standard duo pitch; 450mm overhangs; fixing with clips (included elsewhere); span over wall plates Cont.										
8000mm	70.50	2.50	72.26	1.60	0.20	16.17	8.40	96.83	nr	108.94
9000mm	79.43	2.50	81.42	1.70	0.21	17.17	9.00	107.58	nr	121.03
22.5 degree Bobtail duo pitch; 450mm overhangs; fixing with clips (included elsewhere); span over wall plates										
4000mm	29.77	2.50	30.51	1.20	0.15	12.13	6.00	48.64	nr	54.72
5000mm	37.09	2.50	38.02	1.30	0.16	13.12	6.60	57.74	nr	64.96
6000mm	44.56	2.50	45.67	1.40	0.17	14.12	7.20	66.99	nr	75.36
7000mm	52.05	2.50	53.35	1.50	0.19	15.18	7.80	76.33	nr	85.87
8000mm	59.52	2.50	61.01	1.60	0.20	16.17	8.45	85.63	nr	96.33
9000mm	67.06	2.50	68.74	1.70	0.21	17.17	9.00	94.90	nr	106.76
35 degree Bobtail duo pitch; 450mm overhangs; fixing with clips (included elsewhere); span over wall plates										
4000mm	31.04	2.50	31.82	1.20	0.15	12.13	6.00	49.95	nr	56.19
5000mm	38.69	2.50	39.66	1.30	0.16	13.12	6.60	59.38	nr	66.80
6000mm	46.55	2.50	47.71	1.40	0.17	14.12	7.20	69.03	nr	77.66
7000mm	54.10	2.50	55.45	1.50	0.19	15.18	7.80	78.43	nr	88.24
8000mm	61.87	2.50	63.42	1.60	0.20	16.17	8.45	88.04	nr	99.04
9000mm	69.53	2.50	71.27	1.70	0.21	17.17	9.00	97.43	nr	109.61
45 degree Bobtail duo pitch; 450mm overhangs; fixing with clips (included elsewhere); span over wall plates										
4000mm	50.21	2.50	51.47	1.20	0.15	12.13	6.00	69.59	nr	78.29
5000mm	60.14	2.50	61.64	1.30	0.16	13.12	6.60	81.37	nr	91.54
6000mm	72.26	2.50	74.07	1.40	0.17	14.12	7.20	95.38	nr	107.30
7000mm	84.31	2.50	86.42	1.50	0.19	15.18	7.80	109.40	nr	123.07
8000mm	104.00	2.50	106.60	1.60	0.20	16.17	8.45	131.22	nr	147.62
9000mm	108.44	2.50	111.15	1.70	0.21	17.17	9.00	137.32	nr	154.48
22.5 degree Monopitch; 450mm overhangs; fixing with clips (included elsewhere); span over wall plates										
2000mm	14.57	2.50	14.93	1.00	0.13	10.14	4.80	29.88	nr	33.61
3000mm	19.30	2.50	19.78	1.10	0.14	11.14	5.40	36.32	nr	40.86
4000mm	24.03	2.50	24.63	1.20	0.15	12.13	6.00	42.76	nr	48.10
5000mm	29.37	2.50	30.10	1.30	0.16	13.12	6.60	49.83	nr	56.05
6000mm	37.92	2.50	38.87	1.40	0.17	14.12	7.20	60.18	nr	67.71
35 degree Monopitch; 450mm overhangs; fixing with clips (included elsewhere); span over wall plates										
2000mm	15.57	2.50	15.96	1.00	0.13	10.14	4.80	30.90	nr	34.76
3000mm	20.52	2.50	21.03	1.10	0.14	11.14	5.40	37.57	nr	42.26
4000mm	25.56	2.50	26.20	1.20	0.15	12.13	6.00	44.33	nr	49.87
5000mm	31.52	2.50	32.31	1.30	0.16	13.12	6.60	52.03	nr	58.53
6000mm	40.60	2.50	41.62	1.40	0.17	14.12	7.20	62.93	nr	70.80
45 degree Monopitch; 450mm overhangs; fixing with clips (included elsewhere); span over wall plates										
2000mm	16.63	2.50	17.05	1.00	0.13	10.14	4.80	31.99	nr	35.99
3000mm	31.68	2.50	32.47	1.10	0.14	11.14	5.40	49.01	nr	55.13
4000mm	37.46	2.50	38.40	1.20	0.15	12.13	6.00	56.53	nr	63.59
22.5 degree duo pitch Girder Truss; fixing with clips (included elsewhere); span over wall plates										
5000mm	109.33	2.50	112.06	1.30	0.16	13.12	6.60	131.79	nr	148.26
6000mm	123.76	2.50	126.85	1.40	0.17	14.12	7.20	148.17	nr	166.69
7000mm	134.13	2.50	137.48	1.50	0.19	15.18	7.80	160.46	nr	180.52
8000mm	155.57	2.50	159.46	1.60	0.20	16.17	8.45	184.08	nr	207.09
9000mm	172.97	2.50	177.29	1.70	0.21	17.17	9.00	203.46	nr	228.89
10000mm	203.82	2.50	208.92	1.80	0.22	18.16	9.60	236.67	nr	266.26
35 degree duo pitch Girder Truss; fixing with clips (included elsewhere); span over wall plates										
5000mm	114.53	2.50	117.39	1.30	0.16	13.12	6.60	137.12	nr	154.25
6000mm	127.57	2.50	130.76	1.40	0.17	14.12	7.20	152.07	nr	171.08
7000mm	139.48	2.50	142.97	1.50	0.19	15.18	7.80	165.95	nr	186.69
8000mm	160.93	2.50	164.95	1.60	0.20	16.17	8.45	189.58	nr	213.27
9000mm	178.30	2.50	182.76	1.70	0.21	17.17	9.00	208.92	nr	235.04
10000mm	209.06	2.50	214.29	1.80	0.22	18.16	9.60	242.04	nr	272.30
45 degree duo pitch Girder Truss; fixing with clips (included elsewhere); span over wall plates										
5000mm	155.05	2.50	158.93	1.30	0.16	13.12	6.60	178.65	nr	200.98
6000mm	188.48	2.50	193.19	1.40	0.17	14.12	7.20	214.51	nr	241.32
7000mm	215.93	2.50	221.33	1.50	0.19	15.18	7.80	244.31	nr	274.85
8000mm	241.87	2.50	247.92	1.60	0.20	16.17	8.45	272.54	nr	306.61
9000mm	268.58	2.50	275.29	1.70	0.21	17.17	9.00	301.46	nr	339.14
22.5 degree pitch gable ladder; 450mm overhang; span over wall plate										
5000mm	14.42	2.50	14.78	1.30	0.16	13.12	6.60	34.50	nr	38.82
6000mm	16.48	2.50	16.89	1.40	0.17	14.12	7.20	38.21	nr	42.98
7000mm	18.47	2.50	18.93	1.50	0.19	15.18	7.80	41.91	nr	47.15
8000mm	22.13	2.50	22.68	1.60	0.20	16.17	8.45	47.31	nr	53.22
9000mm	25.35	2.50	25.98	1.70	0.21	17.17	9.00	52.15	nr	58.67
10000mm	30.76	2.50	31.53	1.80	0.22	18.16	9.60	59.29	nr	66.70
35 degree pitch gable ladder; 450mm overhang; span over wall plate										
5000mm	14.97	2.50	15.34	1.30	0.16	13.12	6.60	35.07	nr	39.45
6000mm	17.31	2.50	17.74	1.40	0.17	14.12	7.20	39.06	nr	43.94

Labour hourly rates: (except Specialists) Craft Operatives £9.23 Labourer £7.02 Rates are national average prices. Refer to REGIONAL VARIATIONS for indicative levels of overall pricing in regions	MATERIALS			LABOUR				RATES		
	Del to Site £	Waste %	Material Cost £	Craft Optve Hrs	Lab Hrs	Labour Cost £	Sunds £	Nett Rate £	Unit	Gross Rate (+12.5%) £
G20: CARPENTRY/TIMBER FRAMING/FIRST FIXING Cont.										
Sasco Trussed Rafters in softwood, sawn, B.S.4978, GS grade, impregnated; note:- prices are guide prices only, based on contracts within 30 mile radius of Widnes and are subject to adjustment for quantity: firm quotations should always be obtained Cont.										
Trussed rafters; Hydro-Nail plated joints Cont.										
35 degree pitch gable ladder; 450mm overhang; span over wall plate Cont.										
7000mm	18.98	2.50	19.45	1.50	0.19	15.18	7.80	42.43	nr	47.74
8000mm	23.04	2.50	23.62	1.60	0.20	16.17	8.45	48.24	nr	54.27
9000mm	26.32	2.50	26.98	1.70	0.21	17.17	9.00	53.14	nr	59.79
10000mm	31.90	2.50	32.70	1.80	0.22	18.16	9.60	60.46	nr	68.01
45 degree pitch gable ladder; 450mm overhang; span over wall plate										
5000mm	23.50	2.50	24.09	1.30	0.16	13.12	6.60	43.81	nr	49.29
6000mm	28.14	2.50	28.84	1.40	0.17	14.12	7.20	50.16	nr	56.43
7000mm	32.73	2.50	33.55	1.50	0.19	15.18	7.80	56.53	nr	63.59
8000mm	36.79	2.50	37.71	1.60	0.20	16.17	8.45	62.33	nr	70.12
9000mm	41.83	2.50	42.88	1.70	0.21	17.17	9.00	69.04	nr	77.67
10000mm	46.10	2.50	47.25	1.80	0.22	18.16	9.60	75.01	nr	84.39
Softwood, wrot, B.S.4978, GS grade - laminated beams										
Glued laminated beams; B.S.4169										
65 x 150 x 4000mm	49.40	2.50	50.63	0.60	0.07	6.03	3.80	60.46	nr	68.02
65 x 175 x 4000mm	57.54	2.50	58.98	0.80	0.10	8.09	4.25	71.31	nr	80.23
65 x 200 x 4000mm	65.66	2.50	67.30	0.90	0.11	9.08	4.75	81.13	nr	91.27
65 x 225 x 4000mm	74.17	2.50	76.02	1.10	0.14	11.14	5.90	93.06	nr	104.69
65 x 250 x 4000mm	82.69	2.50	84.76	1.20	0.15	12.13	6.40	103.29	nr	116.20
65 x 250 x 6000mm	123.97	2.50	127.07	1.30	0.16	13.12	6.95	147.14	nr	165.53
65 x 275 x 4000mm	90.76	2.50	93.03	1.30	0.16	13.12	6.95	113.10	nr	127.24
65 x 275 x 6000mm	137.90	2.50	141.35	1.40	0.18	14.19	7.45	162.98	nr	183.36
65 x 300 x 4000mm	97.86	2.50	100.31	1.40	0.18	14.19	7.45	121.94	nr	137.18
65 x 300 x 6000mm	147.64	2.50	151.33	1.50	0.19	15.18	8.05	174.56	nr	196.38
65 x 325 x 4000mm	108.70	2.50	111.42	1.60	0.20	16.17	8.55	136.14	nr	153.16
65 x 325 x 6000mm	159.83	2.50	163.83	1.70	0.21	17.17	9.10	190.09	nr	213.85
65 x 325 x 8000mm	213.80	2.50	219.15	1.80	0.22	18.16	9.60	246.90	nr	277.77
90 x 150 x 4000mm	62.64	2.50	64.21	0.90	0.11	9.08	4.75	78.04	nr	87.79
90 x 175 x 4000mm	73.82	2.50	75.67	1.10	0.14	11.14	5.90	92.70	nr	104.29
90 x 200 x 4000mm	83.98	2.50	86.08	1.20	0.15	12.13	6.40	104.61	nr	117.68
90 x 225 x 4000mm	93.79	2.50	96.13	1.30	0.16	13.12	6.95	116.21	nr	130.73
90 x 250 x 4000mm	105.91	2.50	108.56	1.40	0.18	14.19	7.45	130.19	nr	146.47
90 x 250 x 6000mm	156.46	2.50	160.37	1.50	0.19	15.18	8.05	183.60	nr	206.55
90 x 275 x 4000mm	115.81	2.50	118.71	1.50	0.19	15.18	8.05	141.93	nr	159.68
90 x 275 x 6000mm	174.40	2.50	178.76	1.60	0.20	16.17	8.55	203.48	nr	228.92
90 x 300 x 4000mm	125.29	2.50	128.42	1.70	0.21	17.17	9.10	154.69	nr	174.02
90 x 300 x 6000mm	190.98	2.50	195.75	1.80	0.22	18.16	9.60	223.51	nr	251.45
90 x 325 x 4000mm	138.49	2.50	141.95	1.80	0.22	18.16	9.60	169.71	nr	190.92
90 x 325 x 6000mm	205.88	2.50	211.03	1.90	0.24	19.22	10.20	240.45	nr	270.50
90 x 325 x 8000mm	272.59	2.50	279.40	2.00	0.25	20.22	10.65	310.27	nr	349.05
90 x 350 x 4000mm	146.29	2.50	149.95	1.90	0.24	19.22	10.20	179.37	nr	201.79
90 x 350 x 6000mm	222.14	2.50	227.69	2.00	0.25	20.22	10.65	258.56	nr	290.88
90 x 350 x 8000mm	296.64	2.50	304.06	2.10	0.26	21.21	11.20	336.46	nr	378.52
90 x 375 x 4000mm	156.46	2.50	160.37	2.00	0.25	20.22	10.65	191.24	nr	215.14
90 x 375 x 6000mm	235.68	2.50	241.57	2.10	0.26	21.21	11.20	273.98	nr	308.23
90 x 375 x 8000mm	315.94	2.50	323.84	2.20	0.28	22.27	11.70	357.81	nr	402.54
90 x 375 x 10000mm	394.85	2.50	404.72	2.30	0.29	23.26	12.25	440.24	nr	495.27
90 x 400 x 4000mm	169.99	2.50	174.24	2.20	0.28	22.27	11.70	208.21	nr	234.24
90 x 400 x 6000mm	251.94	2.50	258.24	2.30	0.29	23.26	12.25	293.75	nr	330.47
90 x 400 x 8000mm	336.60	2.50	345.01	2.40	0.30	24.26	12.80	382.07	nr	429.83
90 x 400 x 10000mm	420.23	2.50	430.74	2.50	0.31	25.25	13.35	469.34	nr	528.00
90 x 425 x 6000mm	281.74	2.50	288.78	2.40	0.30	24.26	12.80	325.84	nr	366.57
90 x 425 x 8000mm	354.88	2.50	363.75	2.50	0.31	25.25	13.35	402.35	nr	452.65
90 x 425 x 10000mm	442.45	2.50	453.51	2.60	0.33	26.31	13.85	493.68	nr	555.39
90 x 425 x 12000mm	530.96	2.50	544.23	2.70	0.34	27.31	14.40	585.94	nr	659.18
90 x 450 x 6000mm	284.45	2.50	291.56	2.50	0.31	25.25	13.35	330.16	nr	371.43
90 x 450 x 8000mm	378.59	2.50	388.05	2.60	0.33	26.31	13.85	428.22	nr	481.75
90 x 450 x 10000mm	472.04	2.50	483.84	2.70	0.34	27.31	14.40	525.55	nr	591.24
90 x 450 x 12000mm	567.54	2.50	581.73	2.80	0.35	28.30	14.95	624.98	nr	703.10
115 x 250 x 4000mm	133.76	2.50	137.10	1.70	0.21	17.17	9.10	163.37	nr	183.79
115 x 250 x 6000mm	197.83	2.50	202.78	1.80	0.22	18.16	9.60	230.53	nr	259.35
115 x 275 x 4000mm	145.27	2.50	148.90	1.80	0.22	18.16	9.60	176.66	nr	198.74
115 x 275 x 6000mm	222.14	2.50	227.69	1.90	0.24	19.22	10.20	257.12	nr	289.25
115 x 300 x 4000mm	160.85	2.50	164.87	1.90	0.24	19.22	10.20	194.29	nr	218.58
115 x 300 x 6000mm	241.44	2.50	247.48	2.00	0.25	20.22	10.65	278.34	nr	313.13
115 x 325 x 4000mm	174.40	2.50	178.76	2.00	0.25	20.22	10.65	209.63	nr	235.83
115 x 325 x 6000mm	262.10	2.50	268.65	2.10	0.26	21.21	11.20	301.06	nr	338.69
115 x 325 x 8000mm	350.12	2.50	358.87	2.20	0.28	22.27	11.70	392.84	nr	441.95
115 x 350 x 4000mm	190.64	2.50	195.41	2.20	0.28	22.27	11.70	229.38	nr	258.05
115 x 350 x 6000mm	284.45	2.50	291.56	2.30	0.29	23.26	12.25	327.08	nr	367.96
115 x 350 x 8000mm	378.59	2.50	388.05	2.40	0.30	24.26	12.80	425.11	nr	478.25
115 x 375 x 4000mm	204.19	2.50	209.29	2.30	0.29	23.26	12.25	244.81	nr	275.41
115 x 375 x 6000mm	304.09	2.50	311.69	2.40	0.30	24.26	12.80	348.75	nr	392.34
115 x 375 x 8000mm	405.31	2.50	415.44	2.50	0.31	25.26	13.35	454.04	nr	510.80
115 x 400 x 4000mm	216.05	2.50	221.45	2.40	0.30	24.26	12.80	258.51	nr	290.82
115 x 400 x 6000mm	325.08	2.50	333.21	2.50	0.31	25.25	13.35	371.81	nr	418.28
115 x 400 x 8000mm	433.44	2.50	444.28	2.60	0.33	26.31	Sunds	484.44	Unit	545.00
115 x 400 x 10000mm	542.48	2.50	556.04	2.70	0.34	27.31	14.40	597.75	nr	672.47
115 x 425 x 6000mm	336.60	2.50	345.01	2.60	0.33	26.31	13.85	385.18	nr	433.33
115 x 425 x 8000mm	450.02	2.50	461.27	2.70	0.34	27.31	14.40	502.98	nr	565.85
115 x 425 x 10000mm	560.09	2.50	574.09	2.80	0.35	28.30	14.95	617.34	nr	694.51
115 x 425 x 12000mm	670.48	2.50	687.24	2.90	0.36	29.29	15.50	732.04	nr	823.54
115 x 450 x 6000mm	356.23	2.50	365.14	2.80	0.35	28.30	14.95	408.39	nr	459.44

STRUCTURAL/CARCASSING METAL/TIMBER

Labour hourly rates: (except Specialists) Craft Operatives £9.23 Labourer £7.02 Rates are national average prices. Refer to REGIONAL VARIATIONS for indicative levels of overall pricing in regions	MATERIALS			LABOUR				RATES		
	Del to Site £	Waste %	Material Cost £	Craft Optve Hrs	Lab Hrs	Labour Cost £	Sunds £	Nett Rate £	Unit	Gross Rate (+12.5%) £
G20: CARPENTRY/TIMBER FRAMING/FIRST FIXING Cont.										
Softwood, wrot, B.S.4978, GS grade - laminated beams Cont.										
Glued laminated beams; B.S.4169 Cont.										
115 x 450 x 8000mm	475.43	2.50	487.32	2.90	0.36	29.29	15.50	532.11	nr	598.62
115 x 450 x 10000mm	594.78	2.50	609.65	3.00	0.38	30.36	16.00	656.01	nr	738.01
115 x 450 x 12000mm	715.18	2.50	733.06	3.10	0.39	31.35	16.55	780.96	nr	878.58
115 x 475 x 6000mm	378.59	2.50	388.05	2.90	0.36	29.29	15.50	432.85	nr	486.96
115 x 475 x 8000mm	505.22	2.50	517.85	3.00	0.38	30.36	16.00	564.21	nr	634.07
115 x 475 x 10000mm	632.00	2.50	647.80	3.10	0.39	31.35	16.55	695.70	nr	782.66
115 x 475 x 12000mm	755.98	2.50	774.88	3.20	0.40	32.34	17.10	824.32	nr	927.36
115 x 475 x 14000mm	880.43	2.50	902.44	3.30	0.41	33.34	17.60	953.38	nr	1072.55
115 x 500 x 6000mm	399.58	2.50	409.57	3.00	0.38	30.36	16.00	455.93	nr	512.92
115 x 500 x 8000mm	530.29	2.50	543.55	3.10	0.39	31.35	16.55	591.45	nr	665.38
115 x 500 x 10000mm	666.08	2.50	682.73	3.20	0.40	32.34	17.10	732.18	nr	823.70
115 x 500 x 12000mm	798.44	2.50	818.40	3.30	0.41	33.34	17.60	869.34	nr	978.00
115 x 500 x 14000mm	929.87	2.50	953.12	3.40	0.42	34.33	18.10	1005.55	nr	1131.24
140 x 350 x 4000mm	229.25	2.50	234.98	2.40	0.30	24.26	12.80	272.04	nr	306.04
140 x 350 x 6000mm	347.09	2.50	355.77	2.50	0.31	25.25	13.35	394.37	nr	443.66
140 x 350 x 8000mm	461.88	2.50	473.43	2.60	0.33	26.31	13.85	513.59	nr	577.79
140 x 375 x 4000mm	244.14	2.50	250.24	2.50	0.31	25.25	13.35	288.84	nr	324.95
140 x 375 x 6000mm	382.99	2.50	392.56	2.60	0.33	26.31	13.85	432.73	nr	486.82
140 x 375 x 8000mm	490.33	2.50	502.59	2.70	0.34	27.31	14.40	544.30	nr	612.33
140 x 375 x 10000mm	612.24	2.50	627.55	2.80	0.35	28.30	14.95	670.80	nr	754.65
140 x 400 x 6000mm	387.38	2.50	397.06	2.80	0.35	28.30	14.95	440.32	nr	495.35
140 x 400 x 8000mm	520.13	2.50	533.13	2.90	0.36	29.29	15.50	577.93	nr	650.17
140 x 400 x 10000mm	648.13	2.50	664.33	3.00	0.38	30.36	16.00	710.69	nr	799.53
140 x 425 x 6000mm	418.54	2.50	429.00	2.90	0.36	29.29	15.50	473.80	nr	533.02
140 x 425 x 8000mm	560.09	2.50	574.09	3.00	0.38	30.36	16.00	620.45	nr	698.01
140 x 425 x 10000mm	700.27	2.50	717.78	3.10	0.39	31.35	16.55	765.68	nr	861.39
140 x 425 x 12000mm	840.47	2.50	861.48	3.20	0.40	32.34	17.10	910.93	nr	1024.79
140 x 450 x 6000mm	440.89	2.50	451.91	3.00	0.38	30.36	16.00	498.27	nr	560.55
140 x 450 x 8000mm	588.54	2.50	603.25	3.10	0.39	31.35	16.55	651.15	nr	732.55
140 x 450 x 10000mm	734.14	2.50	752.49	3.20	0.40	32.34	17.10	801.94	nr	902.18
140 x 450 x 12000mm	880.43	2.50	902.44	3.30	0.41	33.34	17.60	953.38	nr	1072.55
140 x 475 x 6000mm	461.88	2.50	473.43	3.10	0.39	31.35	16.55	521.33	nr	586.49
140 x 475 x 8000mm	615.28	2.50	630.66	3.20	0.40	32.34	17.10	680.11	nr	765.12
140 x 475 x 10000mm	770.38	2.50	789.64	3.30	0.41	33.34	17.60	840.58	nr	945.65
140 x 475 x 12000mm	925.13	2.50	948.26	3.40	0.42	34.33	18.10	1000.69	nr	1125.77
140 x 475 x 14000mm	1074.12	2.50	1100.97	3.50	0.44	35.39	18.65	1155.02	nr	1299.39
165 x 425 x 6000mm	493.04	2.50	505.37	3.10	0.39	31.35	16.55	553.27	nr	622.43
165 x 425 x 8000mm	658.28	2.50	674.74	3.20	0.40	32.34	17.10	724.18	nr	814.70
165 x 425 x 10000mm	822.70	2.50	843.27	3.30	0.41	33.34	17.60	894.20	nr	1005.98
165 x 425 x 12000mm	985.56	2.50	1010.20	3.40	0.42	34.33	18.10	1062.63	nr	1195.46
165 x 450 x 6000mm	514.04	2.50	526.89	3.20	0.40	32.34	17.10	576.34	nr	648.38
165 x 450 x 8000mm	686.74	2.50	703.91	3.30	0.41	33.34	17.60	754.85	nr	849.20
165 x 450 x 10000mm	858.08	2.50	879.53	3.40	0.42	34.33	18.10	931.96	nr	1048.46
165 x 450 x 12000mm	1030.76	2.50	1056.53	3.50	0.44	35.39	18.65	1110.57	nr	1249.39
165 x 475 x 6000mm	545.20	2.50	558.83	3.40	0.42	34.33	18.10	611.26	nr	687.67
165 x 475 x 8000mm	728.72	2.50	746.94	3.50	0.44	35.39	18.65	800.98	nr	901.10
165 x 475 x 10000mm	908.87	2.50	931.59	3.60	0.45	36.39	19.20	987.18	nr	1110.58
165 x 475 x 12000mm	1093.42	2.50	1120.76	3.70	0.46	37.38	19.75	1177.89	nr	1325.12
165 x 475 x 14000mm	1273.91	2.50	1305.76	3.80	0.48	38.44	20.25	1364.45	nr	1535.01
190 x 475 x 6000mm	631.54	2.50	647.33	3.60	0.45	36.39	19.20	702.92	nr	790.78
190 x 475 x 8000mm	840.12	2.50	861.12	3.70	0.46	37.38	19.75	918.25	nr	1033.03
190 x 475 x 10000mm	1050.42	2.50	1076.68	3.80	0.48	38.44	20.25	1135.37	nr	1277.30
190 x 475 x 12000mm	1260.37	2.50	1291.88	3.90	0.49	39.44	20.80	1352.12	nr	1521.13
190 x 475 x 14000mm	1470.32	2.50	1507.08	4.00	0.50	40.43	21.35	1568.86	nr	1764.97
Softwood, sawn, B.S.4978, GS grade - carcassing										
Floor members										
38 x 75mm	0.63	10.00	0.69	0.14	0.02	1.43	0.02	2.15	m	2.41
38 x 100mm	0.84	10.00	0.92	0.17	0.02	1.71	0.03	2.66	m	3.00
50 x 75mm	0.75	10.00	0.82	0.17	0.02	1.71	0.03	2.56	m	2.89
50 x 100mm	0.99	10.00	1.09	0.19	0.02	1.89	0.03	3.01	m	3.39
50 x 175mm	1.74	10.00	1.91	0.29	0.04	2.96	0.05	4.92	m	5.54
50 x 200mm	1.98	10.00	2.18	0.32	0.04	3.23	0.05	5.46	m	6.15
50 x 225mm	2.23	10.00	2.45	0.35	0.04	3.51	0.06	6.02	m	6.78
50 x 250mm	2.48	10.00	2.73	0.38	0.05	3.86	0.06	6.65	m	7.48
75 x 150mm	2.35	10.00	2.59	0.35	0.04	3.51	0.06	6.16	m	6.93
75 x 175mm	2.74	10.00	3.01	0.40	0.05	4.04	0.06	7.12	m	8.01
75 x 200mm	3.14	10.00	3.45	0.42	0.05	4.23	0.07	7.75	m	8.72
75 x 225mm	3.53	10.00	3.88	0.48	0.06	4.85	0.08	8.81	m	9.92
75 x 250mm	3.92	10.00	4.31	0.54	0.07	5.48	0.09	9.88	m	11.11
Wall or partition members										
38 x 75mm	0.63	10.00	0.69	0.22	0.03	2.24	0.04	2.97	m	3.35
38 x 100mm	0.84	10.00	0.92	0.25	0.03	2.52	0.04	3.48	m	3.92
50 x 75mm	0.75	10.00	0.82	0.25	0.03	2.52	0.04	3.38	m	3.81
50 x 100mm	0.99	10.00	1.09	0.29	0.04	2.96	0.05	4.10	m	4.61
75 x 100mm	1.57	10.00	1.73	0.36	0.04	3.60	0.06	5.39	m	6.06
38 x 75mm; fixing to masonry	0.63	10.00	0.69	0.43	0.05	4.32	0.07	5.08	m	5.72
38 x 100mm; fixing to masonry	0.84	10.00	0.92	0.47	0.06	4.76	0.08	5.76	m	6.48
50 x 75mm; fixing to masonry	0.75	10.00	0.82	0.47	0.06	4.76	0.08	5.66	m	6.37
50 x 100mm; fixing to masonry	0.99	10.00	1.09	0.52	0.07	5.29	0.09	6.47	m	7.28
75 x 100mm; fixing to masonry	1.57	10.00	1.73	0.50	0.07	5.11	0.08	6.91	m	7.78
Plates										
38 x 75mm	0.63	10.00	0.69	0.22	0.03	2.24	0.04	2.97	m	3.35
38 x 100mm	0.84	10.00	0.92	0.25	0.03	2.52	0.04	3.48	m	3.92
50 x 75mm	0.75	10.00	0.82	0.25	0.03	2.52	0.04	3.38	m	3.81
50 x 100mm	0.99	10.00	1.09	0.29	0.04	2.96	0.05	4.10	m	4.61
75 x 150mm	2.35	10.00	2.59	0.48	0.06	4.85	0.08	7.52	m	8.46

Labour hourly rates: (except Specialists) Craft Operatives £9.23 Labourer £7.02 Rates are national average prices. Refer to REGIONAL VARIATIONS for indicative levels of overall pricing in regions	MATERIALS			LABOUR				RATES		
	Del to Site	Waste	Material Cost	Craft Optve	Lab	Labour Cost	Sunds	Nett Rate	Unit	Gross Rate (+12.5%)
	£	%	£	Hrs	Hrs	£	£	£		£

G20: CARPENTRY/TIMBER FRAMING/FIRST FIXING Cont.

Softwood, sawn, B.S.4978, GS grade - carcassing Cont.

Plates Cont.

	Del to Site	Waste	Material Cost	Craft Optve	Lab	Labour Cost	Sunds	Nett Rate	Unit	Gross Rate
38 x 75mm; fixing by bolting	0.63	10.00	0.69	0.26	0.03	2.61	-	3.30	m	3.72
38 x 100mm; fixing by bolting	0.84	10.00	0.92	0.30	0.04	3.05	-	3.97	m	4.47
50 x 75mm; fixing by bolting	0.75	10.00	0.82	0.30	0.04	3.05	-	3.87	m	4.36
50 x 100mm; fixing by bolting	0.99	10.00	1.09	0.35	0.04	3.51	-	4.60	m	5.18
75 x 150mm; fixing by bolting	2.35	10.00	2.59	0.58	0.07	5.84	-	8.43	m	9.48

Roof members; flat

50 x 175mm	1.83	10.00	2.01	0.29	0.04	2.96	0.05	5.02	m	5.65
50 x 200mm	2.09	10.00	2.30	0.32	0.04	3.23	0.05	5.58	m	6.28
50 x 225mm	2.35	10.00	2.59	0.35	0.04	3.51	0.06	6.16	m	6.93
50 x 250mm	2.62	10.00	2.88	0.38	0.05	3.86	0.06	6.80	m	7.65
75 x 175mm	2.74	10.00	3.01	0.40	0.05	4.04	0.06	7.12	m	8.01
75 x 200mm	3.14	10.00	3.45	0.42	0.05	4.23	0.07	7.75	m	8.72
75 x 225mm	3.53	10.00	3.88	0.48	0.06	4.85	0.08	8.81	m	9.92
75 x 250mm	3.92	10.00	4.31	0.54	0.07	5.48	0.09	9.88	m	11.11

Roof members; pitched

25 x 75mm	0.42	10.00	0.46	0.17	0.02	1.71	0.03	2.20	m	2.48
25 x 125mm	0.69	10.00	0.76	0.19	0.02	1.89	0.03	2.68	m	3.02
32 x 175mm	1.23	10.00	1.35	0.26	0.03	2.61	0.04	4.00	m	4.50
32 x 225mm	1.58	10.00	1.74	0.30	0.04	3.05	0.05	4.84	m	5.44
38 x 75mm	0.63	10.00	0.69	0.18	0.02	1.80	0.03	2.52	m	2.84
38 x 100mm	0.84	10.00	0.92	0.22	0.03	2.24	0.04	3.21	m	3.61
50 x 75mm	0.75	10.00	0.82	0.22	0.03	2.24	0.04	3.11	m	3.49
50 x 100mm	0.99	10.00	1.09	0.24	0.03	2.43	0.04	3.55	m	4.00
50 x 125mm	1.24	10.00	1.36	0.28	0.04	2.87	0.05	4.28	m	4.81
50 x 150mm	1.43	10.00	1.57	0.31	0.04	3.14	0.05	4.77	m	5.36
50 x 225mm	2.23	10.00	2.45	0.43	0.05	4.32	0.07	6.84	m	7.70
50 x 250mm	2.48	10.00	2.73	0.48	0.06	4.85	0.08	7.66	m	8.62
75 x 100mm	1.57	10.00	1.73	0.31	0.04	3.14	0.05	4.92	m	5.53
75 x 150mm	2.35	10.00	2.59	0.42	0.05	4.23	0.07	6.88	m	7.74
100 x 150mm	3.14	10.00	3.45	0.52	0.07	5.29	0.09	8.84	m	9.94
100 x 225mm	4.71	10.00	5.18	0.72	0.09	7.28	0.12	12.58	m	14.15

Joist strutting; herringbone; depth of joist 175mm										
50 x 50mm	1.09	10.00	1.20	0.78	0.10	7.90	0.13	9.23	m	10.38

Joist strutting; herringbone; depth of joist 200mm										
50 x 50mm	1.14	10.00	1.25	0.78	0.10	7.90	0.13	9.29	m	10.45

Joist strutting; herringbone; depth of joist 225mm										
50 x 50mm	1.19	10.00	1.31	0.78	0.10	7.90	0.13	9.34	m	10.51

Joist strutting; herringbone; depth of joist 250mm										
50 x 50mm	1.24	10.00	1.36	0.78	0.10	7.90	0.13	9.40	m	10.57

Joist strutting; block; depth of joist 150mm										
50 x 150mm	1.49	10.00	1.64	0.78	0.10	7.90	0.13	9.67	m	10.88

Joist strutting; block; depth of joist 175mm										
50 x 175mm	1.74	10.00	1.91	0.84	0.11	8.53	0.14	10.58	m	11.90

Joist strutting; block; depth of joist 200mm										
50 x 200mm	1.98	10.00	2.18	0.84	0.11	8.53	0.14	10.84	m	12.20

Joist strutting; block; depth of joist 225mm										
50 x 225mm	2.23	10.00	2.45	0.90	0.11	9.08	0.15	11.68	m	13.14

Joist strutting; block; depth of joist 250mm										
50 x 250mm	2.48	10.00	2.73	0.90	0.11	9.08	0.15	11.96	m	13.45

Noggings to joists

50 x 50mm	0.50	10.00	0.55	0.34	0.04	3.42	0.05	4.02	m	4.52
50 x 75mm	0.75	10.00	0.82	0.40	0.05	4.04	0.06	4.93	m	5.54

Wrot surfaces

plain; 50mm wide	-	-	-	0.18	-	1.66	-	1.66	m	1.87
plain; 75mm wide	-	-	-	0.24	-	2.22	-	2.22	m	2.49
plain; 100mm wide	-	-	-	0.30	-	2.77	-	2.77	m	3.12
plain; 150mm wide	-	-	-	0.36	-	3.32	-	3.32	m	3.74
plain; 200mm wide	-	-	-	0.42	-	3.88	-	3.88	m	4.36

Softwood, sawn, B.S.4978, GS grade, impregnated - carcassing

Floor members

38 x 75mm	0.70	10.00	0.77	0.14	0.02	1.43	0.02	2.22	m	2.50
38 x 100mm	0.95	10.00	1.04	0.17	0.02	1.71	0.03	2.78	m	3.13
50 x 75mm	0.85	10.00	0.94	0.17	0.02	1.71	0.03	2.67	m	3.01
50 x 100mm	1.13	10.00	1.24	0.19	0.02	1.89	0.03	3.17	m	3.56
50 x 175mm	1.97	10.00	2.17	0.29	0.04	2.96	0.05	5.17	m	5.82
50 x 200mm	2.26	10.00	2.49	0.32	0.04	3.23	0.05	5.77	m	6.49
50 x 225mm	2.54	10.00	2.79	0.35	0.04	3.51	0.06	6.37	m	7.16
50 x 250mm	2.82	10.00	3.10	0.38	0.05	3.86	0.06	7.02	m	7.90
75 x 150mm	2.66	10.00	2.93	0.35	0.04	3.51	0.06	6.50	m	7.31
75 x 175mm	3.10	10.00	3.41	0.40	0.05	4.04	0.06	7.51	m	8.45
75 x 200mm	3.55	10.00	3.90	0.42	0.05	4.23	0.07	8.20	m	9.23
75 x 225mm	3.99	10.00	4.39	0.48	0.06	4.85	0.08	9.32	m	10.49
75 x 250mm	4.43	10.00	4.87	0.54	0.07	5.48	0.09	10.44	m	11.74

Labour hourly rates: (except Specialists) Craft Operatives £9.23 Labourer £7.02 Rates are national average prices. Refer to REGIONAL VARIATIONS for indicative levels of overall pricing in regions	MATERIALS			LABOUR				RATES		
	Del to Site £	Waste %	Material Cost £	Craft Optve Hrs	Lab Hrs	Labour Cost £	Sunds £	Nett Rate £	Unit	Gross Rate (+12.5%) £
G20: CARPENTRY/TIMBER FRAMING/FIRST FIXING Cont.										
Softwood, sawn, B.S.4978, GS grade, impregnated - carcassing Cont.										
Wall or partition members										
38 x 75mm	0.70	10.00	0.77	0.22	0.03	2.24	0.04	3.05	m	3.43
38 x 100mm	0.95	10.00	1.04	0.25	0.03	2.52	0.04	3.60	m	4.05
50 x 75mm	0.85	10.00	0.94	0.25	0.03	2.52	0.04	3.49	m	3.93
50 x 100mm	1.13	10.00	1.24	0.29	0.04	2.96	0.05	4.25	m	4.78
75 x 100mm	1.77	10.00	1.95	0.36	0.04	3.60	0.06	5.61	m	6.31
38 x 75mm; fixing to masonry	0.70	10.00	0.77	0.43	0.05	4.32	0.07	5.16	m	5.80
38 x 100mm; fixing to masonry	0.95	10.00	1.04	0.47	0.06	4.76	0.08	5.88	m	6.62
50 x 75mm; fixing to masonry	0.85	10.00	0.94	0.47	0.06	4.76	0.08	5.77	m	6.50
50 x 100mm; fixing to masonry	1.13	10.00	1.24	0.52	0.07	5.29	0.09	6.62	m	7.45
75 x 100mm; fixing to masonry	1.77	10.00	1.95	0.50	0.07	5.11	0.08	7.13	m	8.03
Plates										
38 x 75mm	0.70	10.00	0.77	0.22	0.03	2.24	0.04	3.05	m	3.43
38 x 100mm	0.95	10.00	1.04	0.25	0.03	2.52	0.04	3.60	m	4.05
50 x 75mm	0.85	10.00	0.94	0.25	0.03	2.52	0.04	3.49	m	3.93
50 x 100mm	1.13	10.00	1.24	0.29	0.04	2.96	0.05	4.25	m	4.78
75 x 150mm	2.81	10.00	3.09	0.48	0.06	4.85	0.08	8.02	m	9.03
38 x 75mm; fixing by bolting	0.70	10.00	0.77	0.26	0.03	2.61	-	3.38	m	3.80
38 x 100mm; fixing by bolting	0.95	10.00	1.04	0.30	0.04	3.05	-	4.09	m	4.61
50 x 75mm; fixing by bolting	0.85	10.00	0.94	0.30	0.04	3.05	-	3.98	m	4.48
50 x 100mm; fixing by bolting	1.13	10.00	1.24	0.35	0.04	3.51	-	4.75	m	5.35
75 x 150mm; fixing by bolting	2.81	10.00	3.09	0.58	0.07	5.84	-	8.94	m	10.05
Roof members; flat										
50 x 175mm	1.97	10.00	2.17	0.29	0.04	2.96	0.05	5.17	m	5.82
50 x 200mm	2.26	10.00	2.49	0.32	0.04	3.23	0.05	5.77	m	6.49
50 x 225mm	2.54	10.00	2.79	0.35	0.04	3.51	0.06	6.37	m	7.16
50 x 250mm	2.82	10.00	3.10	0.38	0.05	3.86	0.06	7.02	m	7.90
75 x 175mm	3.10	10.00	3.41	0.40	0.05	4.04	0.06	7.51	m	8.45
75 x 200mm	3.55	10.00	3.90	0.42	0.05	4.23	0.07	8.20	m	9.23
75 x 225mm	3.99	10.00	4.39	0.48	0.06	4.85	0.08	9.32	m	10.49
75 x 250mm	4.43	10.00	4.87	0.54	0.07	5.48	0.09	10.44	m	11.74
Roof members; pitched										
25 x 75mm	0.46	10.00	0.51	0.17	0.02	1.71	0.03	2.25	m	2.53
25 x 125mm	0.77	10.00	0.85	0.19	0.02	1.89	0.03	2.77	m	3.12
32 x 175mm	1.39	10.00	1.53	0.26	0.03	2.61	0.04	4.18	m	4.70
32 x 225mm	1.78	10.00	1.96	0.30	0.04	3.05	0.05	5.06	m	5.69
38 x 75mm	0.70	10.00	0.77	0.18	0.02	1.80	0.03	2.60	m	2.93
38 x 100mm	0.95	10.00	1.04	0.22	0.03	2.24	0.04	3.33	m	3.74
50 x 75mm	0.85	10.00	0.94	0.22	0.03	2.24	0.04	3.22	m	3.62
50 x 100mm	1.13	10.00	1.24	0.24	0.03	2.43	0.04	3.71	m	4.17
50 x 125mm	1.41	10.00	1.55	0.28	0.04	2.87	0.05	4.47	m	5.02
50 x 150mm	1.69	10.00	1.86	0.31	0.04	3.14	0.05	5.05	m	5.68
50 x 225mm	2.54	10.00	2.79	0.43	0.05	4.32	0.07	7.18	m	8.08
50 x 250mm	2.82	10.00	3.10	0.48	0.06	4.85	0.08	8.03	m	9.04
75 x 100mm	1.77	10.00	1.95	0.31	0.04	3.14	0.05	5.14	m	5.78
75 x 150mm	2.66	10.00	2.93	0.42	0.05	4.23	0.07	7.22	m	8.13
100 x 150mm	3.55	10.00	3.90	0.52	0.07	5.29	0.09	9.29	m	10.45
100 x 225mm	5.07	10.00	5.58	0.72	0.09	7.28	0.12	12.97	m	14.60
Joist strutting; herringbone; depth of joist 175mm										
50 x 50mm	1.24	10.00	1.36	0.78	0.10	7.90	0.13	9.40	m	10.57
Joist strutting; herringbone; depth of joist 200mm										
50 x 50mm	1.30	10.00	1.43	0.78	0.10	7.90	0.13	9.46	m	10.64
Joist strutting; herringbone; depth of joist 225mm										
50 x 50mm	1.35	10.00	1.49	0.78	0.10	7.90	0.13	9.52	m	10.71
Joist strutting; herringbone; depth of joist 250mm										
50 x 50mm	1.41	10.00	1.55	0.78	0.10	7.90	0.13	9.58	m	10.78
Joist strutting; block; depth of joist 150mm										
50 x 150mm	1.69	10.00	1.86	0.78	0.10	7.90	0.13	9.89	m	11.13
Joist strutting; block; depth of joist 175mm										
50 x 175mm	1.97	10.00	2.17	0.84	0.11	8.53	0.14	10.83	m	12.19
Joist strutting; block; depth of joist 200mm										
50 x 200mm	2.26	10.00	2.49	0.84	0.11	8.53	0.14	11.15	m	12.55
Joist strutting; block; depth of joist 225mm										
50 x 225mm	2.54	10.00	2.79	0.90	0.11	9.08	0.15	12.02	m	13.53
Joist strutting; block; depth of joist 250mm										
50 x 250mm	2.82	10.00	3.10	0.90	0.11	9.08	0.15	12.33	m	13.87
Noggings to joists										
50 x 50mm	0.56	10.00	0.62	0.34	0.04	3.42	0.05	4.09	m	4.60
50 x 75mm	0.85	10.00	0.94	0.40	0.05	4.04	0.06	5.04	m	5.67
Wrot surfaces										
plain; 50mm wide	-	-	-	0.18	-	1.66	-	1.66	m	1.87
plain; 75mm wide	-	-	-	0.24	-	2.22	-	2.22	m	2.49
plain; 100mm wide	-	-	-	0.30	-	2.77	-	2.77	m	3.12
plain; 150mm wide	-	-	-	0.36	-	3.32	-	3.32	m	3.74
plain; 200mm wide	-	-	-	0.42	-	3.88	-	3.88	m	4.36

Labour hourly rates: (except Specialists) Craft Operatives £9.23 Labourer £7.02 Rates are national average prices. Refer to REGIONAL VARIATIONS for indicative levels of overall pricing in regions	MATERIALS			LABOUR				RATES		
	Del to Site £	Waste %	Material Cost £	Craft Optve Hrs	Lab Hrs	Labour Cost £	Sunds £	Nett Rate £	Unit	Gross Rate (+12.5%) £

G20: CARPENTRY/TIMBER FRAMING/FIRST FIXING Cont.

Oak, sawn - carcassing

Floor members

38 x 75mm	4.07	10.00	4.48	0.24	0.03	2.43	0.04	6.94	m	7.81
38 x 100mm	5.43	10.00	5.97	0.30	0.04	3.05	0.05	9.07	m	10.21
50 x 75mm	5.36	10.00	5.90	0.30	0.04	3.05	0.05	9.00	m	10.12
50 x 100mm	7.15	10.00	7.87	0.33	0.04	3.33	0.05	11.24	m	12.65
50 x 175mm	12.51	10.00	13.76	0.51	0.06	5.13	0.08	18.97	m	21.34
50 x 200mm	14.30	10.00	15.73	0.56	0.07	5.66	0.09	21.48	m	24.17
50 x 225mm	16.08	10.00	17.69	0.61	0.08	6.19	0.10	23.98	m	26.98
50 x 250mm	17.88	10.00	19.67	0.66	0.08	6.65	0.11	26.43	m	29.74
75 x 150mm	16.08	10.00	17.69	0.61	0.08	6.19	0.10	23.98	m	26.98
75 x 175mm	18.78	10.00	20.66	0.70	0.09	7.09	0.11	27.86	m	31.34
75 x 200mm	21.45	10.00	23.59	0.73	0.09	7.37	0.12	31.08	m	34.97
75 x 225mm	24.15	10.00	26.57	0.84	0.10	8.46	0.14	35.16	m	39.56
75 x 250mm	26.81	10.00	29.49	0.94	0.12	9.52	0.15	39.16	m	44.05

Wall or partition members

38 x 75mm	4.07	10.00	4.48	0.38	0.05	3.86	0.06	8.40	m	9.44
38 x 100mm	5.43	10.00	5.97	0.44	0.06	4.48	0.07	10.53	m	11.84
50 x 75mm	5.36	10.00	5.90	0.44	0.06	4.48	0.07	10.45	m	11.75
50 x 100mm	7.15	10.00	7.87	0.51	0.06	5.13	0.08	13.07	m	14.71
75 x 100mm	10.73	10.00	11.80	0.63	0.08	6.38	0.10	18.28	m	20.56
38 x 75mm; fixing to masonry	4.07	10.00	4.48	0.64	0.08	6.47	0.10	11.05	m	12.43
38 x 100mm; fixing to masonry	5.43	10.00	5.97	0.71	0.09	7.19	0.12	13.28	m	14.94
50 x 75mm; fixing to masonry	5.36	10.00	5.90	0.71	0.09	7.19	0.12	13.20	m	14.85
50 x 100mm; fixing to masonry	7.15	10.00	7.87	0.78	0.10	7.90	0.13	15.90	m	17.88
75 x 100mm; fixing to masonry	10.73	10.00	11.80	0.94	0.12	9.52	0.15	21.47	m	24.16

Plates

38 x 75mm	4.07	10.00	4.48	0.31	0.04	3.14	0.05	7.67	m	8.63
38 x 100mm	5.43	10.00	5.97	0.38	0.05	3.86	0.06	9.89	m	11.13
50 x 75mm	5.36	10.00	5.90	0.38	0.05	3.86	0.06	9.81	m	11.04
50 x 100mm	7.15	10.00	7.87	0.42	0.05	4.23	0.07	12.16	m	13.68
75 x 150mm	16.08	10.00	17.69	0.73	0.09	7.37	0.12	25.18	m	28.32
38 x 75mm; fixing by bolting	4.07	10.00	4.48	0.46	0.06	4.67	-	9.14	m	10.29
38 x 100mm; fixing by bolting	5.43	10.00	5.97	0.53	0.07	5.38	-	11.36	m	12.78
50 x 75mm; fixing by bolting	5.36	10.00	5.90	0.53	0.07	5.38	-	11.28	m	12.69
50 x 100mm; fixing by bolting	7.15	10.00	7.87	0.60	0.08	6.10	-	13.96	m	15.71
75 x 150mm; fixing by bolting	16.08	10.00	17.69	0.74	0.09	7.46	-	25.15	m	28.29

Roof members; flat

50 x 175mm	12.51	10.00	13.76	0.51	0.06	5.13	0.08	18.97	m	21.34
50 x 200mm	14.30	10.00	15.73	0.56	0.07	5.66	0.09	21.48	m	24.17
50 x 225mm	16.08	10.00	17.69	0.61	0.08	6.19	0.10	23.98	m	26.98
50 x 250mm	17.88	10.00	19.67	0.66	0.08	6.65	0.11	26.43	m	29.74
75 x 175mm	18.78	10.00	20.66	0.70	0.09	7.09	0.11	27.86	m	31.34
75 x 200mm	21.45	10.00	23.59	0.73	0.09	7.37	0.12	31.08	m	34.97
75 x 225mm	24.15	10.00	26.57	0.84	0.10	8.46	0.14	35.16	m	39.56
75 x 250mm	25.95	10.00	28.55	0.94	0.12	9.52	0.15	38.21	m	42.99

Roof members; pitched

25 x 75mm	2.70	10.00	2.97	0.30	0.04	3.05	0.05	6.07	m	6.83
25 x 125mm	4.48	10.00	4.93	0.33	0.04	3.33	0.05	8.30	m	9.34
32 x 175mm	8.01	10.00	8.81	0.45	0.06	4.57	0.07	13.46	m	15.14
32 x 225mm	10.30	10.00	11.33	0.52	0.07	5.29	0.09	16.71	m	18.80
38 x 75mm	4.07	10.00	4.48	0.31	0.04	3.14	0.05	7.67	m	8.63
38 x 100mm	5.43	10.00	5.97	0.38	0.05	3.86	0.06	9.89	m	11.13
50 x 75mm	5.36	10.00	5.90	0.38	0.05	3.86	0.06	9.81	m	11.04
50 x 100mm	7.15	10.00	7.87	0.42	0.05	4.23	0.07	12.16	m	13.68
50 x 125mm	8.93	10.00	9.82	0.49	0.06	4.94	0.08	14.85	m	16.70
50 x 150mm	10.73	10.00	11.80	0.54	0.07	5.48	0.09	17.37	m	19.54
50 x 225mm	16.08	10.00	17.69	0.75	0.09	7.55	0.12	25.36	m	28.53
50 x 250mm	17.88	10.00	19.67	0.84	0.10	8.46	0.14	28.26	m	31.80
75 x 100mm	10.73	10.00	11.80	0.54	0.07	5.48	0.09	17.37	m	19.54
75 x 150mm	16.08	10.00	17.69	0.73	0.09	7.37	0.12	25.18	m	28.32
100 x 150mm	21.45	10.00	23.59	0.94	0.12	9.52	0.15	33.26	m	37.42
100 x 225mm	32.18	10.00	35.40	1.26	0.16	12.75	0.21	48.36	m	54.41

Joist strutting; herringbone; depth of joist 175mm 50 x 50mm	7.87	10.00	8.66	1.36	0.17	13.75	0.22	22.62	m	25.45
Joist strutting; herringbone; depth of joist 200mm 50 x 50mm	8.22	10.00	9.04	1.36	0.17	13.75	0.22	23.01	m	25.88
Joist strutting; herringbone; depth of joist 225mm 50 x 50mm	8.58	10.00	9.44	1.36	0.17	13.75	0.22	23.40	m	26.33
Joist strutting; herringbone; depth of joist 250mm 50 x 50mm	8.93	10.00	9.82	1.36	0.17	13.75	0.22	23.79	m	26.76
Joist strutting; block; depth of joist 150mm 50 x 150mm	10.73	10.00	11.80	1.36	0.17	13.75	0.22	25.77	m	28.99
Joist strutting; block; depth of joist 175mm 50 x 175mm	12.51	10.00	13.76	1.47	0.18	14.83	0.24	28.83	m	32.44
Joist strutting; block; depth of joist 200mm 50 x 200mm	14.30	10.00	15.73	1.47	0.18	14.83	0.24	30.80	m	34.65
Joist strutting; block; depth of joist 225mm 50 x 225mm	16.08	10.00	17.69	1.57	0.20	15.90	0.26	33.84	m	38.07

Labour hourly rates: (except Specialists) Craft Operatives £9.23 Labourer £7.02 Rates are national average prices. Refer to REGIONAL VARIATIONS for indicative levels of overall pricing in regions	MATERIALS			LABOUR				RATES		
	Del to Site £	Waste %	Material Cost £	Craft Optve Hrs	Lab Hrs	Labour Cost £	Sunds £	Nett Rate £	Unit	Gross Rate (+12.5%) £
G20: CARPENTRY/TIMBER FRAMING/FIRST FIXING Cont.										
Oak, sawn - carcassing Cont.										
Joist strutting; block; depth of joist 250mm										
50 x 250mm	17.88	10.00	19.67	1.57	0.20	15.90	0.26	35.82	m	40.30
Noggings to joists										
50 x 50mm	3.58	10.00	3.94	0.60	0.08	6.10	0.10	10.14	m	11.40
50 x 75mm	5.36	10.00	5.90	0.70	0.09	7.09	0.11	13.10	m	14.74
Wrot surfaces										
plain; 50mm wide	-	-	-	0.36	-	3.32	-	3.32	m	3.74
plain; 75mm wide	-	-	-	0.48	-	4.43	-	4.43	m	4.98
plain; 100mm wide	-	-	-	0.60	-	5.54	-	5.54	m	6.23
plain; 150mm wide	-	-	-	0.72	-	6.65	-	6.65	m	7.48
plain; 200mm wide	-	-	-	0.84	-	7.75	-	7.75	m	8.72
Stress grading softwood										

General Structural (SS) grade
   included in rates
Special Structural (SS) grade
   add 10% to materials prices										

Flame proofing treatment to softwood										
For timbers treated with proofing process to Class 1 add 100% to materials prices										
Softwood, sawn - supports										
Butt jointed supports										
width exceeding 300mm; 19 x 38mm at 300mm centres; fixing to masonry	1.06	10.00	1.17	1.20	0.15	12.13	0.20	13.49	m2	15.18
width exceeding 300mm; 25 x 50mm at 300mm centres; fixing to masonry	1.85	10.00	2.04	1.30	0.16	13.12	0.21	15.37	m2	17.29
Framed supports										

width exceeding 300mm; 38 x 38mm members at 400mm
   centres one way; 38 x 38mm subsidiary members at
   400mm centres one way ....
	1.58	10.00	1.74	1.20	0.15	12.13	0.21	14.08	m2	15.84

width exceeding 300mm; 38 x 38mm members at 500mm
   centres one way; 38 x 38mm subsidiary members at
   400mm centres one way ....
	1.43	10.00	1.57	0.95	0.12	9.61	0.15	11.33	m2	12.75

width exceeding 300mm; 50 x 50mm members at 400mm
   centres one way; 50 x 50mm subsidiary members at
   400mm centres one way ....
	2.75	10.00	3.02	1.30	0.16	13.12	0.21	16.36	m2	18.40

width exceeding 300mm; 50 x 50mm members at 500mm
   centres one way; 50 x 50mm subsidiary members at
   400mm centres one way ....
	2.48	10.00	2.73	1.10	0.14	11.14	0.18	14.04	m2	15.80

width exceeding 300mm; 19 x 38mm members at 300mm
   centres one way; 19 x 38mm subsidiary members at
   300mm centres one way; fixing to masonry ....
	1.06	10.00	1.17	2.70	0.34	27.31	0.44	28.91	m2	32.53

width exceeding 300mm; 25 x 50mm members at 300mm
   centres one way; 25 x 50mm subsidiary members at
   300mm centres one way; fixing to masonry ....
	1.85	10.00	2.04	2.80	0.35	28.30	0.46	30.80	m2	34.65

width not exceeding 300mm; 38 x 38mm members
   longitudinally; 38 x 38mm subsidiary members at
   400mm centres laterally ....
	0.88	10.00	0.97	0.45	0.06	4.57	0.07	5.61	m	6.31

width not exceeding 300mm; 50 x 50mm members
   longitudinally; 50 x 50mm subsidiary members at
   400mm centres laterally ....
	1.54	10.00	1.69	0.50	0.06	5.04	0.08	6.81	m	7.66

Individual supports										
6 x 38mm	0.09	10.00	0.10	0.10	0.01	0.99	0.02	1.11	m	1.25
12 x 38mm	0.14	10.00	0.15	0.10	0.01	0.99	0.02	1.17	m	1.31
19 x 38mm	0.15	10.00	0.17	0.11	0.01	1.09	0.02	1.27	m	1.43
25 x 50mm	0.28	10.00	0.31	0.12	0.02	1.25	0.02	1.58	m	1.77
38 x 50mm	0.42	10.00	0.46	0.13	0.02	1.34	0.02	1.82	m	2.05
50 x 50mm	0.55	10.00	0.60	0.14	0.02	1.43	0.02	2.06	m	2.31
50 x 150mm	1.65	10.00	1.81	0.20	0.03	2.06	0.03	3.90	m	4.39
50 x 300mm	3.30	10.00	3.63	0.28	0.04	2.87	0.05	6.55	m	7.36
6 x 38mm; fixing to masonry	0.09	10.00	0.10	0.28	0.04	2.87	0.05	3.01	m	3.39
12 x 38mm; fixing to masonry	0.14	10.00	0.15	0.28	0.04	2.87	0.05	3.07	m	3.45
19 x 38mm; fixing to masonry	0.15	10.00	0.17	0.29	0.04	2.96	0.05	3.17	m	3.57
25 x 50mm; fixing to masonry	0.28	10.00	0.31	0.30	0.04	3.05	0.05	3.41	m	3.83
38 x 50mm; fixing to masonry	0.42	10.00	0.46	0.32	0.04	3.23	0.05	3.75	m	4.21
50 x 50mm; fixing to masonry	0.55	10.00	0.60	0.34	0.04	3.42	0.05	4.07	m	4.58
Softwood, sawn, impregnated - supports										
Butt jointed supports										
width exceeding 300mm; 50mm wide x 25mm average depth at 450mm centres	1.36	10.00	1.50	0.36	0.05	3.67	0.06	5.23	m2	5.88
width exceeding 300mm; 50mm wide x 25mm average depth at 600mm centres	1.05	10.00	1.16	0.28	0.04	2.87	0.05	4.07	m2	4.58
width exceeding 300mm; 50mm wide x 50mm average depth at 450mm centres	2.73	10.00	3.00	0.43	0.05	4.32	0.07	7.39	m2	8.32
width exceeding 300mm; 50mm wide x 50mm average depth at 600mm centres	1.98	10.00	2.18	0.32	0.04	3.23	0.05	5.46	m2	6.15
width exceeding 300mm; 50mm wide x 63mm average depth at 450mm centres	3.47	10.00	3.82	0.48	0.06	4.85	0.08	8.75	m2	9.84
width exceeding 300mm; 50mm wide x 63mm average depth at 600mm centres	2.73	10.00	3.00	0.36	0.05	3.67	0.06	6.74	m2	7.58
width exceeding 300mm; 50mm wide x 75mm average depth at 450mm centres	4.08	10.00	4.49	0.52	0.07	5.29	0.09	9.87	m2	11.10

Labour hourly rates: (except Specialists) Craft Operatives £9.23 Labourer £7.02 Rates are national average prices. Refer to REGIONAL VARIATIONS for indicative levels of overall pricing in regions	MATERIALS			LABOUR				RATES		
	Del to Site £	Waste %	Material Cost £	Craft Optve Hrs	Lab Hrs	Labour Cost £	Sunds £	Nett Rate £	Unit	Gross Rate (+12.5%) £
G20: CARPENTRY/TIMBER FRAMING/FIRST FIXING Cont.										
Softwood, sawn, impregnated - supports Cont.										
Butt jointed supports Cont.										
width exceeding 300mm; 50mm wide x 75mm average depth at 600mm centres	3.09	10.00	3.40	0.40	0.05	4.04	0.06	7.50	m2	8.44
Individual supports										
25 x 25mm; 1 labours	0.20	10.00	0.22	0.13	0.02	1.34	0.02	1.58	m	1.78
25 x 25mm; 2 labours	0.23	10.00	0.25	0.13	0.02	1.34	0.02	1.61	m	1.81
25 x 38mm; 1 labours	0.30	10.00	0.33	0.13	0.02	1.34	0.02	1.69	m	1.90
25 x 38mm; 2 labours	0.35	10.00	0.39	0.13	0.02	1.34	0.02	1.75	m	1.96
triangular; extreme dimensions										
25 x 25mm....................	0.20	10.00	0.22	0.13	0.02	1.34	0.02	1.58	m	1.78
38 x 75mm....................	0.88	10.00	0.97	0.16	0.02	1.62	0.03	2.62	m	2.94
50 x 75mm....................	1.16	10.00	1.28	0.18	0.02	1.80	0.03	3.11	m	3.50
50 x 100mm...................	1.55	10.00	1.71	0.20	0.03	2.06	0.03	3.79	m	4.27
75 x 100mm...................	2.32	10.00	2.55	0.23	0.03	2.33	0.04	4.93	m	5.54
rounded roll for lead										
50 x 50mm....................	1.86	10.00	2.05	0.26	0.03	2.61	0.04	4.70	m	5.28
50 x 75mm....................	2.78	10.00	3.06	0.30	0.04	3.05	0.05	6.16	m	6.93
rounded roll for lead; birdsmouthed on to ridge or hip										
50 x 50mm....................	1.86	10.00	2.05	0.46	0.06	4.67	0.08	6.79	m	7.64
50 x 75mm....................	2.78	10.00	3.06	0.49	0.06	4.94	0.08	8.08	m	9.09
roll for zinc; 32 x 44mm..............	1.05	10.00	1.16	0.19	0.02	1.89	0.03	3.08	m	3.46
Softwood, wrot, impregnated - gutter boarding										
Gutter boards including sides, tongued and grooved joints										
width exceeding 300mm; 19mm thick	7.26	10.00	7.99	2.40	0.30	24.26	0.39	32.63	m2	36.71
width exceeding 300mm; 25mm thick	9.57	10.00	10.53	2.70	0.34	27.31	0.44	38.27	m2	43.06
width not exceeding 300mm; 19mm thick										
150mm wide.................	1.09	10.00	1.20	0.36	0.05	3.67	0.06	4.93	m	5.55
225mm wide.................	1.65	10.00	1.81	0.48	0.06	4.85	0.08	6.75	m	7.59
width not exceeding 300mm; 25mm thick										
150mm wide.................	1.45	10.00	1.60	0.40	0.05	4.04	0.06	5.70	m	6.41
225mm wide.................	2.15	10.00	2.37	0.53	0.07	5.38	0.09	7.84	m	8.82
cesspool 225 x 225 x 150mm deep.............	1.85	10.00	2.04	1.80	0.23	18.23	0.29	20.55	nr	23.12
Chimney gutter boards, butt joints										
width not exceeding 300mm; 25mm thick x 100mm										
average wide.................	1.24	10.00	1.36	0.60	0.08	6.10	0.10	7.56	m	8.51
gusset end.................	0.33	10.00	0.36	0.42	0.05	4.23	0.07	4.66	nr	5.24
width not exceeding 300mm; 25mm thick x 175mm										
average wide.................	1.85	10.00	2.04	0.78	0.10	7.90	0.13	10.07	m	11.32
gusset end.................	0.46	10.00	0.51	0.48	0.06	4.85	0.08	5.44	nr	6.12
Lier boards; tongued and grooved joints										
width exceeding 300mm; 19mm thick	7.26	10.00	7.99	1.10	0.14	11.14	0.18	19.30	m2	21.71
width exceeding 300mm; 25mm thick	9.57	10.00	10.53	1.20	0.15	12.13	0.20	22.86	m2	25.71
Valley sole boards; butt joints										
width not exceeding 300mm; 25mm thick										
100mm wide.................	0.83	10.00	0.91	0.55	0.07	5.57	0.09	6.57	m	7.39
150mm wide.................	1.24	10.00	1.36	0.60	0.08	6.10	0.10	7.56	m	8.51
Plywood B.S.6566, II/III grade, WBP bonded, butt joints - gutter boarding										
Gutter boards including sides, butt joints										
width exceeding 300mm; 18mm thick	11.61	10.00	12.77	1.80	0.23	18.23	0.29	31.29	m2	35.20
width exceeding 300mm; 25mm thick	16.13	10.00	17.74	2.00	0.25	20.22	0.33	38.29	m2	43.07
width not exceeding 300mm; 18mm thick										
150mm wide.................	1.74	10.00	1.91	0.24	0.03	2.43	0.04	4.38	m	4.93
225mm wide.................	2.61	10.00	2.87	0.26	0.03	2.61	0.04	5.52	m	6.21
width not exceeding 300mm; 25mm thick										
150mm wide.................	2.42	10.00	2.66	0.28	0.04	2.87	0.05	5.58	m	6.27
225mm wide.................	3.63	10.00	3.99	0.30	0.04	3.05	0.05	7.09	m	7.98
cesspool 225 x 225 x 150mm deep.............	3.22	10.00	3.54	1.80	0.23	18.23	0.29	22.06	nr	24.82
Softwood, wrot, impregnated - eaves and verge boarding										
Fascia and barge boards										
width not exceeding 300mm; 25mm thick										
150mm wide.................	1.18	10.00	1.30	0.26	0.03	2.61	0.04	3.95	m	4.44
200mm wide.................	1.57	10.00	1.73	0.27	0.03	2.70	0.04	4.47	m	5.03
Fascia and barge boards, tongued, grooved and veed joints										
width not exceeding 300mm; 25mm thick										
225mm wide.................	2.04	10.00	2.24	0.36	0.05	3.67	0.06	5.98	m	6.73
250mm wide.................	2.26	10.00	2.49	0.38	0.05	3.86	0.06	6.40	m	7.20
Eaves or verge soffit boards										
width exceeding 300mm; 25mm thick	7.84	10.00	8.62	1.20	0.15	12.13	0.20	20.95	m2	23.57
width not exceeding 300mm; 25mm thick										
225mm wide.................	1.76	10.00	1.94	0.36	0.05	3.67	0.06	5.67	m	6.38

STRUCTURAL/CARCASSING METAL/TIMBER

Labour hourly rates: (except Specialists) Craft Operatives £9.23 Labourer £7.02 Rates are national average prices. Refer to REGIONAL VARIATIONS for indicative levels of overall pricing in regions	MATERIALS			LABOUR				RATES		
	Del to Site £	Waste %	Material Cost £	Craft Optve Hrs	Lab Hrs	Labour Cost £	Sunds £	Nett Rate £	Unit	Gross Rate (+12.5%) £
G20: CARPENTRY/TIMBER FRAMING/FIRST FIXING Cont.										
Plywood B.S.6566, II/III grade, WBP bonded, butt joints - eaves and verge boarding										
Fascia and barge boards width not exceeding 300mm; 18mm thick										
150mm wide	1.74	10.00	1.91	0.26	0.03	2.61	0.04	4.56	m	5.13
225mm wide	2.61	10.00	2.87	0.27	0.03	2.70	0.04	5.61	m	6.32
250mm wide	2.89	10.00	3.18	0.28	0.04	2.87	0.05	6.09	m	6.86
Eaves or verge soffit boards width exceeding 300mm; 18mm thick	11.61	10.00	12.77	0.90	0.11	9.08	0.15	22.00	m2	24.75
width not exceeding 300mm; 18mm thick										
225mm wide	2.61	10.00	2.87	0.30	0.04	3.05	0.05	5.97	m	6.72
Fascia and bargeboards; Caradon Celuform Ltd										
Fascia reference 4247; butt joint with joint trims reference 4560; fixing to timber with Polytop screws										
150mm wide	9.53	5.00	10.01	0.24	0.03	2.43	0.53	12.96	m	14.58
200mm wide	11.47	5.00	12.04	0.24	0.03	2.43	0.53	15.00	m	16.87
250mm wide	15.08	5.00	15.83	0.24	0.03	2.43	1.06	19.32	m	21.73
Elite solid fascia board reference 4245; butt joint with joint trims reference 4560; fixing to timber with Polytop screws										
150mm wide	9.59	5.00	10.07	0.24	0.03	2.43	0.53	13.03	m	14.65
200mm wide	11.80	5.00	12.39	0.24	0.03	2.43	0.53	15.35	m	17.26
250mm wide	15.92	5.00	16.72	0.24	0.03	2.43	1.06	20.20	m	22.73
Supaliner fascia board reference 4270; butt joint with joint trims reference 4586; fixing to timber with Polytop screws										
150mm wide	7.72	5.00	8.11	0.35	0.04	3.51	0.53	12.15	m	13.67
170mm wide	8.89	5.00	9.33	0.35	0.04	3.51	0.53	13.38	m	15.05
225mm wide	11.29	5.00	11.85	0.35	0.04	3.51	0.53	15.90	m	17.88
250mm wide	12.46	5.00	13.08	0.40	0.06	4.11	1.06	18.26	m	20.54
Celuvent multi purpose/soffite board reference 4258; butt joint with panel joint trim reference 4570; fixing to timber with stainless steel nails; fixing soffite board channel reference 4572 to masonry with Polytop screws										
100mm wide	9.36	5.00	9.83	0.45	0.07	4.64	0.53	15.00	m	16.88
225mm wide	15.70	5.00	16.48	0.45	0.07	4.64	1.06	22.19	m	24.96
325mm wide	20.70	5.00	21.73	0.70	0.09	7.09	1.58	30.41	m	34.21
Multi purpose/soffite board reference 4558; butt joint with panel joint trim reference 4570; fixing to timber with stainless steel nails; fixing soffite board channel reference 4572 to masonry with Polytop screws										
125mm wide	10.51	5.00	11.04	0.45	0.07	4.64	0.53	16.21	m	18.24
175mm wide	12.78	5.00	13.42	0.45	0.07	4.64	0.53	18.59	m	20.92
200mm wide	13.31	5.00	13.98	0.45	0.07	4.64	1.06	19.68	m	22.14
300mm wide	19.38	5.00	20.35	0.70	0.09	7.09	1.58	29.02	m	32.65
Steel - tie rods and straps										
Tie rods										
19mm diameter; threaded	5.71	2.50	5.85	0.30	0.04	3.05	-	8.90	m	10.02
extra; with nut and washer	0.28	2.50	0.29	0.25	0.03	2.52	-	2.81	nr	3.16
25mm diameter; threaded	7.73	2.50	7.92	0.35	0.04	3.51	-	11.43	m	12.86
extra; with nut and washer	0.50	2.50	0.51	0.35	0.04	3.51	-	4.02	nr	4.53
Straps 3 x 38; holes -4; fixing with bolts (bolts included elsewhere)										
500mm long	1.14	2.50	1.17	0.16	0.02	1.62	-	2.79	nr	3.13
750mm long	1.68	2.50	1.72	0.23	0.03	2.33	-	4.06	nr	4.56
1000mm long	2.29	2.50	2.35	0.30	0.04	3.05	-	5.40	nr	6.07
6 x 50; holes -4; fixing with bolts (bolts included elsewhere)										
500mm long	2.50	2.50	2.56	0.19	0.02	1.89	-	4.46	nr	5.01
750mm long	3.70	2.50	3.79	0.28	0.03	2.79	-	6.59	nr	7.41
1000mm long	4.97	2.50	5.09	0.36	0.04	3.60	-	8.70	nr	9.79
Steel; galvanised - straps										
Straps and clips; BAT Building ProductsLtd. 30 x 2.5mm standard strapping; fixing with nails										
600mm long	0.80	5.00	0.84	0.18	0.02	1.80	0.03	2.67	nr	3.01
800mm long	1.18	5.00	1.24	0.20	0.03	2.06	0.03	3.33	nr	3.74
1000mm long	1.51	5.00	1.59	0.22	0.03	2.24	0.04	3.87	nr	4.35
1200mm long	1.83	5.00	1.92	0.23	0.03	2.33	0.04	4.29	nr	4.83
1600mm long	2.38	5.00	2.50	0.24	0.03	2.43	0.04	4.96	nr	5.59
30 x 2.5mm standard strapping; twists -1; fixing with nails										
600mm long	0.89	5.00	0.93	0.18	0.02	1.80	0.03	2.77	nr	3.11
800mm long	1.26	5.00	1.32	0.20	0.03	2.06	0.03	3.41	nr	3.84
1000mm long	1.59	5.00	1.67	0.22	0.03	2.24	0.04	3.95	nr	4.44
1200mm long	1.90	5.00	2.00	0.23	0.03	2.33	0.04	4.37	nr	4.91
1600mm long	2.45	5.00	2.57	0.24	0.03	2.43	0.04	5.04	nr	5.67

Labour hourly rates: (except Specialists) Craft Operatives £9.23 Labourer £7.02 Rates are national average prices. Refer to REGIONAL VARIATIONS for indicative levels of overall pricing in regions	MATERIALS			LABOUR				RATES		
	Del to Site	Waste	Material Cost	Craft Optve	Lab	Labour Cost	Sunds	Nett Rate	Unit	Gross Rate (+12.5%)
	£	%	£	Hrs	Hrs	£	£	£		£
G20: CARPENTRY/TIMBER FRAMING/FIRST FIXING Cont.										
Steel; galvanised - straps Cont.										
Straps and clips; BAT Building ProductsLtd. Cont. 30 x 5mm M 305 strapping; bends -1; nailing one end (building in included elsewhere)										
600mm long	1.70	5.00	1.78	0.10	0.01	0.99	0.02	2.80	nr	3.15
800mm long	2.18	5.00	2.29	0.10	0.01	0.99	0.02	3.30	nr	3.71
1000mm long	2.76	5.00	2.90	0.11	0.01	1.09	0.02	4.00	nr	4.50
1200mm long	3.33	5.00	3.50	0.11	0.01	1.09	0.02	4.60	nr	5.18
1600mm long	4.39	5.00	4.61	0.12	0.02	1.25	0.02	5.88	nr	6.61
30 x 5mm M305 strapping; bends -1; twists -1; nailing one end (building in included elsewhere)										
600mm long	1.76	5.00	1.85	0.10	0.01	0.99	0.02	2.86	nr	3.22
800mm long	2.23	5.00	2.34	0.10	0.01	0.99	0.02	3.35	nr	3.77
1000mm long	3.41	5.00	3.58	0.11	0.01	1.09	0.02	4.69	nr	5.27
1200mm long	4.46	5.00	4.68	0.11	0.01	1.09	0.02	5.79	nr	6.51
1600mm long	4.46	5.00	4.68	0.12	0.02	1.25	0.02	5.95	nr	6.69
truss clips; fixing with nails										
for 38mm thick members	0.31	5.00	0.33	0.24	0.03	2.43	0.04	2.79	nr	3.14
for 50mm thick members	0.34	5.00	0.36	0.24	0.03	2.43	0.04	2.82	nr	3.18
Steel; galvanised - joist hangers										
Joist hangers; BAT Building Products Ltd; (building in where required included elsewhere)										
SPH type S, for 50 x 100mm joist	1.59	2.50	1.63	-	-	-	-	1.63	nr	1.83
SPH type S, for 50 x 125mm joist	1.62	2.50	1.66	-	-	-	-	1.66	nr	1.87
SPH type S, for 50 x 150mm joist	1.62	2.50	1.66	-	-	-	-	1.66	nr	1.87
SPH type S, for 50 x 175mm joist	1.69	2.50	1.73	-	-	-	-	1.73	nr	1.95
SPH type S, for 50 x 200mm joist	1.88	2.50	1.93	-	-	-	-	1.93	nr	2.17
SPH type S, for 50 x 225mm joist	2.01	2.50	2.06	-	-	-	-	2.06	nr	2.32
SPH type S, for 50 x 250mm joist	2.58	2.50	2.64	-	-	-	-	2.64	nr	2.98
SPH type S, for 63 x 100mm joist	2.02	2.50	2.07	-	-	-	-	2.07	nr	2.33
SPH type S, for 63 x 125mm joist	2.06	2.50	2.11	-	-	-	-	2.11	nr	2.38
SPH type S, for 63 x 150mm joist	2.35	2.50	2.41	-	-	-	-	2.41	nr	2.71
SPH type S, for 63 x 175mm joist	2.21	2.50	2.27	-	-	-	-	2.27	nr	2.55
SPH type S, for 63 x 200mm joist	2.35	2.50	2.41	-	-	-	-	2.41	nr	2.71
SPH type S, for 63 x 225mm joist	2.52	2.50	2.58	-	-	-	-	2.58	nr	2.91
SPH type S, for 63 x 250mm joist	2.69	2.50	2.76	-	-	-	-	2.76	nr	3.10
SPH type R, for 50 x 100mm joist	2.40	2.50	2.46	-	-	-	-	2.46	nr	2.77
SPH type R, for 50 x 125mm joist	2.44	2.50	2.50	-	-	-	-	2.50	nr	2.81
SPH type R, for 50 x 150mm joist	2.52	2.50	2.58	-	-	-	-	2.58	nr	2.91
SPH type R, for 50 x 175mm joist	2.61	2.50	2.68	-	-	-	-	2.68	nr	3.01
SPH type R, for 50 x 200mm joist	2.78	2.50	2.85	-	-	-	-	2.85	nr	3.21
SPH type R, for 50 x 225mm joist	3.05	2.50	3.13	-	-	-	-	3.13	nr	3.52
SPH type R, for 50 x 250mm joist	3.61	2.50	3.70	-	-	-	-	3.70	nr	4.16
SPH type R, for 63 x 100mm joist	2.93	2.50	3.00	-	-	-	-	3.00	nr	3.38
SPH type R, for 63 x 125mm joist	2.98	2.50	3.05	-	-	-	-	3.05	nr	3.44
SPH type R, for 63 x 150mm joist	3.31	2.50	3.39	-	-	-	-	3.39	nr	3.82
SPH type R, for 63 x 175mm joist	3.11	2.50	3.19	-	-	-	-	3.19	nr	3.59
SPH type R, for 63 x 200mm joist	3.31	2.50	3.39	-	-	-	-	3.39	nr	3.82
SPH type R, for 63 x 225mm joist	3.53	2.50	3.62	-	-	-	-	3.62	nr	4.07
SPH type R, for 63 x 250mm joist	3.73	2.50	3.82	-	-	-	-	3.82	nr	4.30
SPH type ST for 50 x 100mm joist	4.20	2.50	4.30	-	-	-	-	4.30	nr	4.84
SPH type ST for 50 x 125mm joist	4.32	2.50	4.43	-	-	-	-	4.43	nr	4.98
SPH type ST for 50 x 150mm joist	4.12	2.50	4.22	-	-	-	-	4.22	nr	4.75
SPH type ST for 50 x 175mm joist	4.69	2.50	4.81	-	-	-	-	4.81	nr	5.41
SPH type ST for 50 x 200mm joist	4.93	2.50	5.05	-	-	-	-	5.05	nr	5.68
SPH type ST for 50 x 225mm joist	5.39	2.50	5.52	-	-	-	-	5.52	nr	6.22
SPH type ST for 50 x 250mm joist	6.53	2.50	6.69	-	-	-	-	6.69	nr	7.53
Speedy Minor type for the following size joists; fixing with nails										
38 x 100mm	0.45	2.50	0.46	0.18	0.02	1.80	0.03	2.29	nr	2.58
50 x 100mm	0.45	2.50	0.46	0.18	0.02	1.80	0.03	2.29	nr	2.58
Speedy short leg type for the following size joists fixing with nails										
38 x 100mm	0.82	2.50	0.84	0.18	0.02	1.80	0.03	2.67	nr	3.01
50 x 175mm	0.82	2.50	0.84	0.18	0.02	1.80	0.03	2.67	nr	3.01
Speedy Standard Leg type for the following size joists; fixing with nails										
38 x 100mm	0.89	2.50	0.91	0.18	0.02	1.80	0.03	2.74	nr	3.09
50 x 175mm	0.89	2.50	0.91	0.18	0.02	1.80	0.03	2.74	nr	3.09
63 x 225mm	0.93	2.50	0.95	0.18	0.02	1.80	0.03	2.79	nr	3.13
75 x 225mm	0.93	2.50	0.95	0.20	0.03	2.06	0.03	3.04	nr	3.42
100 x 225mm	1.02	2.50	1.05	0.20	0.03	2.06	0.03	3.13	nr	3.52
Steel; galvanised - joist struts										
Herringbone joist struts; BAT Building Products Ltd. to suit joists at the following centres; fixing with nails										
400mm	0.32	5.00	0.34	0.12	0.02	1.25	0.02	1.60	m	1.80
450mm	0.35	5.00	0.37	0.12	0.02	1.25	0.02	1.64	m	1.84
600mm	0.40	5.00	0.42	0.12	0.02	1.25	0.02	1.69	m	1.90
Steel; galvanised - truss plates and framing anchors										
Truss plates; BAT Building Products Ltd. fixing with nails										
51 x 114mm	0.61	5.00	0.64	0.40	0.05	4.04	0.06	4.74	nr	5.34
76 x 254mm	0.49	5.00	0.51	1.20	0.15	12.13	0.20	12.84	nr	14.45
114 x 152mm	0.49	5.00	0.51	1.20	0.15	12.13	0.20	12.84	nr	14.45
114 x 254mm	0.85	5.00	0.89	1.80	0.23	18.23	0.29	19.41	nr	21.84
152 x 152mm	0.68	5.00	0.71	1.90	0.24	19.22	0.31	20.25	nr	22.78

Labour hourly rates: (except Specialists) Craft Operatives £9.23 Labourer £7.02 Rates are national average prices. Refer to REGIONAL VARIATIONS for indicative levels of overall pricing in regions	MATERIALS			LABOUR				RATES		
	Del to Site £	Waste %	Material Cost £	Craft Optve Hrs	Lab Hrs	Labour Cost £	Sunds £	Nett Rate £	Unit	Gross Rate (+12.5%) £
G20: CARPENTRY/TIMBER FRAMING/FIRST FIXING Cont.										
Steel; galvanised - truss plates and framing anchors Cont.										
Framing anchors; BAT Building Products Ltd.										
type A; fixing with nails	0.41	5.00	0.43	0.25	0.03	2.52	0.04	2.99	nr	3.36
type B; fixing with nails	0.41	5.00	0.43	0.25	0.03	2.52	0.04	2.99	nr	3.36
type C; fixing with nails	0.41	5.00	0.43	0.25	0.03	2.52	0.04	2.99	nr	3.36
Steel - timber connectors										
Connectors B.S.1579										
split ring connectors table 1; 64mm diameter	1.09	5.00	1.14	0.10	0.01	0.99	-	2.14	nr	2.40
split ring connectors table 1; 100 mm diameter	3.01	5.00	3.16	0.12	0.02	1.25	-	4.41	nr	4.96
shear plate connectors, table 2; 67mm diameter	1.80	5.00	1.89	0.10	0.01	0.99	-	2.88	nr	3.24
single sided round toothed-plate connectors, table 4										
38mm diameter	0.19	5.00	0.20	0.10	0.01	0.99	-	1.19	nr	1.34
51mm diameter	0.25	5.00	0.26	0.10	0.01	0.99	-	1.26	nr	1.41
64mm diameter	0.34	5.00	0.36	0.10	0.01	0.99	-	1.35	nr	1.52
76mm diameter	0.48	5.00	0.50	0.10	0.01	0.99	-	1.50	nr	1.68
double sided round toothed-plate connector, table 4										
38mm diameter	0.25	5.00	0.26	0.10	0.01	0.99	-	1.26	nr	1.41
51mm diameter	0.28	5.00	0.29	0.10	0.01	0.99	-	1.29	nr	1.45
64mm diameter	0.36	5.00	0.38	0.10	0.01	0.99	-	1.37	nr	1.54
76mm diameter	0.53	5.00	0.56	0.10	0.01	0.99	-	1.55	nr	1.74
Cast iron - connectors										
Connectors B.S.1579										
shear plate connectors, table 2; 102mm diameter	7.18	5.00	7.54	0.14	0.02	1.43	-	8.97	nr	10.09
Steel - bolts and nuts										
Black bolts, B.S.4190 grade 4.6, hexagon head										
M 12; with nuts and washers										
100mm long	0.49	5.00	0.51	0.11	0.01	1.09	-	1.60	nr	1.80
120mm long	0.60	5.00	0.63	0.13	0.02	1.34	-	1.97	nr	2.22
160mm long	1.29	5.00	1.35	0.13	0.02	1.34	-	2.69	nr	3.03
180mm long	1.33	5.00	1.40	0.14	0.02	1.43	-	2.83	nr	3.18
200mm long	1.39	5.00	1.46	0.17	0.02	1.71	-	3.17	nr	3.57
M 16; with nuts and washers										
100mm long	0.68	5.00	0.71	0.11	0.01	1.09	-	1.80	nr	2.02
120mm long	1.06	5.00	1.11	0.13	0.02	1.34	-	2.45	nr	2.76
160mm long	1.72	5.00	1.81	0.14	0.02	1.43	-	3.24	nr	3.64
180mm long	1.81	5.00	1.90	0.16	0.02	1.62	-	3.52	nr	3.96
200mm long	1.92	5.00	2.02	0.17	0.02	1.71	-	3.73	nr	4.19
M 20; with nuts and washers										
100mm long	1.49	5.00	1.56	0.12	0.02	1.25	-	2.81	nr	3.16
120mm long	2.03	5.00	2.13	0.11	0.01	1.09	-	3.22	nr	3.62
160mm long	2.56	5.00	2.69	0.17	0.02	1.71	-	4.40	nr	4.95
180mm long	2.58	5.00	2.71	0.17	0.02	1.71	-	4.42	nr	4.97
200mm long	2.85	5.00	2.99	0.17	0.02	1.71	-	4.70	nr	5.29
M 12; with nuts and 38 x 38 x 3mm square plate washers										
100mm long	0.50	5.00	0.53	0.11	0.01	1.09	-	1.61	nr	1.81
120mm long	0.62	5.00	0.65	0.13	0.02	1.34	-	1.99	nr	2.24
160mm long	1.28	5.00	1.34	0.13	0.02	1.34	-	2.68	nr	3.02
180mm long	1.38	5.00	1.45	0.14	0.02	1.43	-	2.88	nr	3.24
200mm long	1.42	5.00	1.49	0.17	0.02	1.71	-	3.20	nr	3.60
M 12; with nuts and 50 x 50 x 3mm square plate washers										
100mm long	0.54	5.00	0.57	0.11	0.01	1.09	-	1.65	nr	1.86
120mm long	0.64	5.00	0.67	0.13	0.02	1.34	-	2.01	nr	2.26
160mm long	1.32	5.00	1.39	0.13	0.02	1.34	-	2.73	nr	3.07
180mm long	1.38	5.00	1.45	0.14	0.02	1.43	-	2.88	nr	3.24
200mm long	1.45	5.00	1.52	0.17	0.02	1.71	-	3.23	nr	3.64
Black bolts, B.S.4933 Grade 4.6, cup head, square neck										
M 6; with nuts and washers										
25mm long	0.07	5.00	0.07	0.08	0.01	0.81	-	0.88	nr	0.99
50mm long	0.08	5.00	0.08	0.08	0.01	0.81	-	0.89	nr	1.00
75mm long	0.12	5.00	0.13	0.10	0.01	0.99	-	1.12	nr	1.26
100mm long	0.17	5.00	0.18	0.10	0.01	0.99	-	1.17	nr	1.32
150mm long	0.24	5.00	0.25	0.11	0.01	1.09	-	1.34	nr	1.50
M 8; with nuts and washers										
25mm long	0.12	5.00	0.13	0.08	0.01	0.81	-	0.93	nr	1.05
50mm long	0.13	5.00	0.14	0.08	0.01	0.81	-	0.95	nr	1.06
75mm long	0.17	5.00	0.18	0.10	0.01	0.99	-	1.17	nr	1.32
100mm long	0.25	5.00	0.26	0.10	0.01	0.99	-	1.26	nr	1.41
150mm long	0.36	5.00	0.38	0.11	0.01	1.09	-	1.46	nr	1.65
M 10; with nuts and washers										
50mm long	0.21	5.00	0.22	0.08	0.01	0.81	-	1.03	nr	1.16
75mm long	0.24	5.00	0.25	0.10	0.01	0.99	-	1.25	nr	1.40
100mm long	0.36	5.00	0.38	0.10	0.01	0.99	-	1.37	nr	1.54
150mm long	0.52	5.00	0.55	0.12	0.02	1.25	-	1.79	nr	2.02
M 12; with nuts and washers										
50mm long	0.31	5.00	0.33	0.10	0.01	0.99	-	1.32	nr	1.48
75mm long	0.41	5.00	0.43	0.10	0.01	0.99	-	1.42	nr	1.60
100mm long	0.50	5.00	0.53	0.11	0.01	1.09	-	1.61	nr	1.81
150mm long	0.70	5.00	0.73	0.13	0.02	1.34	-	2.08	nr	2.33

Labour hourly rates: (except Specialists) Craft Operatives £9.23 Labourer £7.02 Rates are national average prices. Refer to REGIONAL VARIATIONS for indicative levels of overall pricing in regions	MATERIALS			LABOUR				RATES		
	Del to Site £	Waste %	Material Cost £	Craft Optve Hrs	Lab Hrs	Labour Cost £	Sunds £	Nett Rate £	Unit	Gross Rate (+12.5%) £

G20: CARPENTRY/TIMBER FRAMING/FIRST FIXING Cont.

Steel; galvanised - bolts and nuts

Black bolts, B.S.4190 grade 4.6, hexagon head
M 12; with nuts and washers

100mm long..................	0.72	5.00	0.76	0.11	0.01	1.09	–	1.84	nr	2.07
120mm long..................	0.89	5.00	0.93	0.13	0.02	1.34	–	2.27	nr	2.56
160mm long..................	1.86	5.00	1.95	0.13	0.02	1.34	–	3.29	nr	3.70
180mm long..................	1.96	5.00	2.06	0.14	0.02	1.43	–	3.49	nr	3.93
200mm long..................	2.04	5.00	2.14	0.17	0.02	1.71	–	3.85	nr	4.33

M 16; with nuts and washers

100mm long..................	1.20	5.00	1.26	0.11	0.01	1.09	–	2.35	nr	2.64
120mm long..................	1.56	5.00	1.64	0.13	0.02	1.34	–	2.98	nr	3.35
160mm long..................	2.50	5.00	2.63	0.14	0.02	1.43	–	4.06	nr	4.56
180mm long..................	2.65	5.00	2.78	0.16	0.02	1.62	–	4.40	nr	4.95
200mm long..................	2.82	5.00	2.96	0.17	0.02	1.71	–	4.67	nr	5.25

M 20; with nuts and washers

100mm long..................	2.20	5.00	2.31	0.12	0.02	1.25	–	3.56	nr	4.00
120mm long..................	2.96	5.00	3.11	0.11	0.01	1.09	–	4.19	nr	4.72
160mm long..................	3.74	5.00	3.93	0.17	0.02	1.71	–	5.64	nr	6.34
180mm long..................	3.88	5.00	4.07	0.17	0.02	1.71	–	5.78	nr	6.51
200mm long..................	4.13	5.00	4.34	0.17	0.02	1.71	–	6.05	nr	6.80

M 12; with nuts and 38 x 38 x 3mm square plate washers

100mm long..................	0.79	5.00	0.83	0.11	0.01	1.09	–	1.92	nr	2.15
120mm long..................	0.95	5.00	1.00	0.13	0.02	1.34	–	2.34	nr	2.63
160mm long..................	1.93	5.00	2.03	0.13	0.02	1.34	–	3.37	nr	3.79
180mm long..................	2.03	5.00	2.13	0.14	0.02	1.43	–	3.56	nr	4.01
200mm long..................	2.12	5.00	2.23	0.17	0.02	1.71	–	3.94	nr	4.43

M 12; with nuts and 50 x 50 x 3mm square plate washers

100mm long..................	0.84	5.00	0.88	0.11	0.01	1.09	–	1.97	nr	2.21
120mm long..................	0.99	5.00	1.04	0.13	0.02	1.34	–	2.38	nr	2.68
160mm long..................	1.98	5.00	2.08	0.13	0.02	1.34	–	3.42	nr	3.85
180mm long..................	2.09	5.00	2.19	0.14	0.02	1.43	–	3.63	nr	4.08
200mm long..................	2.23	5.00	2.34	0.17	0.02	1.71	–	4.05	nr	4.56

Steel - expanding bolts

Expanding bolts; bolt projecting Rawlbolts; drilling masonry
with nuts and washers; reference

44 - 505....................	0.76	5.00	0.80	0.24	0.03	2.43	–	3.22	nr	3.63
44 - 510....................	0.85	5.00	0.89	0.24	0.03	2.43	–	3.32	nr	3.73
44 - 515....................	0.89	5.00	0.93	0.24	0.03	2.43	–	3.36	nr	3.78
44 - 555....................	0.96	5.00	1.01	0.26	0.03	2.61	–	3.62	nr	4.07
44 - 560....................	1.01	5.00	1.06	0.26	0.03	2.61	–	3.67	nr	4.13
44 - 565....................	1.08	5.00	1.13	0.26	0.03	2.61	–	3.74	nr	4.21
44 - 605....................	1.31	5.00	1.38	0.30	0.04	3.05	–	4.43	nr	4.98
44 - 610....................	1.38	5.00	1.45	0.30	0.04	3.05	–	4.50	nr	5.06
44 - 615....................	1.43	5.00	1.50	0.30	0.04	3.05	–	4.55	nr	5.12
44 - 655....................	2.09	5.00	2.19	0.32	0.04	3.23	–	5.43	nr	6.11
44 - 660....................	2.23	5.00	2.34	0.32	0.04	3.23	–	5.58	nr	6.27
44 - 665....................	2.79	5.00	2.93	0.32	0.04	3.23	–	6.16	nr	6.93
44 - 705....................	4.94	5.00	5.19	0.36	0.05	3.67	–	8.86	nr	9.97
44 - 710....................	5.34	5.00	5.61	0.36	0.05	3.67	–	9.28	nr	10.44
44 - 715....................	5.60	5.00	5.88	0.36	0.05	3.67	–	9.55	nr	10.75
44 - 755....................	7.44	5.00	7.81	0.40	0.05	4.04	–	11.86	nr	13.34
44 - 760....................	8.03	5.00	8.43	0.40	0.05	4.04	–	12.47	nr	14.03
44 - 765....................	9.13	5.00	9.59	0.40	0.05	4.04	–	13.63	nr	15.33

Expanding bolts; Sleeve Anchor (Rawlok); bolt projecting type; drilling masonry
with nuts and washers; reference

69 - 504....................	0.22	5.00	0.23	0.19	0.02	1.89	–	2.13	nr	2.39
69 - 506....................	0.26	5.00	0.27	0.19	0.02	1.89	–	2.17	nr	2.44
69 - 508....................	0.28	5.00	0.29	0.22	0.03	2.24	–	2.54	nr	2.85
69 - 510....................	0.35	5.00	0.37	0.22	0.03	2.24	–	2.61	nr	2.93
69 - 512....................	0.46	5.00	0.48	0.22	0.03	2.24	–	2.72	nr	3.06
69 - 514....................	0.40	5.00	0.42	0.24	0.03	2.43	–	2.85	nr	3.20
69 - 516....................	0.52	5.00	0.55	0.24	0.03	2.43	–	2.97	nr	3.34
69 - 518....................	0.66	5.00	0.69	0.24	0.03	2.43	–	3.12	nr	3.51
69 - 520....................	0.61	5.00	0.64	0.26	0.03	2.61	–	3.25	nr	3.66
69 - 522....................	0.68	5.00	0.71	0.26	0.03	2.61	–	3.32	nr	3.74
69 - 524....................	0.95	5.00	1.00	0.26	0.03	2.61	–	3.61	nr	4.06
69 - 525....................	1.17	5.00	1.23	0.26	0.03	2.61	–	3.84	nr	4.32
69 - 526....................	1.24	5.00	1.30	0.30	0.04	3.05	–	4.35	nr	4.90
69 - 528....................	1.45	5.00	1.52	0.30	0.04	3.05	–	4.57	nr	5.14
69 - 530....................	2.09	5.00	2.19	0.30	0.04	3.05	–	5.24	nr	5.90
69 - 533....................	2.10	5.00	2.21	0.32	0.04	3.23	–	5.44	nr	6.12
69 - 534....................	2.71	5.00	2.85	0.32	0.04	3.23	–	6.08	nr	6.84
69 - 536....................	3.27	5.00	3.43	0.32	0.04	3.23	–	6.67	nr	7.50

Expanding bolts; loose bolt Rawlbolts; drilling masonry
with washers; reference

44 - 015....................	0.76	5.00	0.80	0.24	0.03	2.43	–	3.22	nr	3.63
44 - 020....................	0.80	5.00	0.84	0.24	0.03	2.43	–	3.27	nr	3.67
44 - 025....................	0.81	5.00	0.85	0.24	0.03	2.43	–	3.28	nr	3.69
44 - 055....................	0.96	5.00	1.01	0.26	0.03	2.61	–	3.62	nr	4.07
44 - 060....................	0.99	5.00	1.04	0.26	0.03	2.61	–	3.65	nr	4.11
44 - 065....................	1.05	5.00	1.10	0.26	0.03	2.61	–	3.71	nr	4.18
44 - 105....................	1.27	5.00	1.33	0.30	0.04	3.05	–	4.38	nr	4.93
44 - 110....................	1.31	5.00	1.38	0.30	0.04	3.05	–	4.43	nr	4.98
44 - 115....................	1.38	5.00	1.45	0.30	0.04	3.05	–	4.50	nr	5.06

Labour hourly rates: (except Specialists) Craft Operatives £9.23 Labourer £7.02 Rates are national average prices. Refer to REGIONAL VARIATIONS for indicative levels of overall pricing in regions	MATERIALS			LABOUR				RATES		
	Del to Site £	Waste %	Material Cost £	Craft Optve Hrs	Lab Hrs	Labour Cost £	Sunds £	Nett Rate £	Unit	Gross Rate (+12.5%) £
G20: CARPENTRY/TIMBER FRAMING/FIRST FIXING Cont.										
Steel – expanding bolts Cont.										
Expanding bolts; loose bolt Rawlbolts; drilling masonry Cont.										
with washers; reference Cont.										
44 - 120	1.43	5.00	1.50	0.30	0.04	3.05	-	4.55	nr	5.12
44 - 155	1.89	5.00	1.98	0.32	0.04	3.23	-	5.22	nr	5.87
44 - 160	2.09	5.00	2.19	0.32	0.04	3.23	-	5.43	nr	6.11
44 - 165	2.18	5.00	2.29	0.32	0.04	3.23	-	5.52	nr	6.21
44 - 170	2.30	5.00	2.42	0.32	0.04	3.23	-	5.65	nr	6.36
44 - 205	4.39	5.00	4.61	0.36	0.05	3.67	-	8.28	nr	9.32
44 - 210	5.07	5.00	5.32	0.36	0.05	3.67	-	9.00	nr	10.12
44 - 215	5.49	5.00	5.76	0.36	0.05	3.67	-	9.44	nr	10.62
44 - 255	8.61	5.00	9.04	0.40	0.05	4.04	-	13.08	nr	14.72
44 - 260	8.94	5.00	9.39	0.40	0.05	4.04	-	13.43	nr	15.11
Expanding bolts; Sleeve Anchor (Rawlok); loose bolt type with countersunk bolt; drilling masonry										
reference										
69 - 572	0.41	5.00	0.43	0.19	0.02	1.89	-	2.32	nr	2.62
69 - 574	0.44	5.00	0.46	0.19	0.02	1.89	-	2.36	nr	2.65
69 - 576	0.52	5.00	0.55	0.19	0.02	1.89	-	2.44	nr	2.75
59 - 578	0.61	5.00	0.64	0.22	0.03	2.24	-	2.88	nr	3.24
69 - 580	0.72	5.00	0.76	0.22	0.03	2.24	-	3.00	nr	3.37
69 - 582	0.86	5.00	0.90	0.24	0.03	2.43	-	3.33	nr	3.74
69 - 584	0.97	5.00	1.02	0.24	0.03	2.43	-	3.44	nr	3.87
Expanding bolts; Sleeve Anchor (Rawlok); loose bolt type with round head bolt; drilling masonry										
reference										
69 - 604	0.44	5.00	0.46	0.19	0.02	1.89	-	2.36	nr	2.65
69 - 608	0.55	5.00	0.58	0.19	0.02	1.89	-	2.47	nr	2.78
69 - 610	0.62	5.00	0.65	0.22	0.03	2.24	-	2.89	nr	3.25
69 - 612	0.70	5.00	0.73	0.22	0.03	2.24	-	2.98	nr	3.35
69 - 614	0.87	5.00	0.91	0.24	0.03	2.43	-	3.34	nr	3.76
69 - 616	0.96	5.00	1.01	0.24	0.03	2.43	-	3.43	nr	3.86
Spit Mega high performance safety anchors; loose nut version; drilling masonry										
with nuts and washers, reference										
E10 - 15/20	1.63	5.00	1.71	0.30	0.03	2.98	-	4.69	nr	5.28
E10 - 15/45	1.79	5.00	1.88	0.30	0.03	2.98	-	4.86	nr	5.47
E10 - 15/65	1.91	5.00	2.01	0.30	0.03	2.98	-	4.99	nr	5.61
E12 - 18/25	2.30	5.00	2.42	0.32	0.03	3.16	-	5.58	nr	6.28
E12 - 18/45	2.39	5.00	2.51	0.32	0.03	3.16	-	5.67	nr	6.38
E12 - 18/65	2.55	5.00	2.68	0.32	0.03	3.16	-	5.84	nr	6.57
E16 - 24/25	6.45	5.00	6.77	0.36	0.03	3.53	-	10.31	nr	11.59
E16 - 24/45	6.84	5.00	7.18	0.36	0.04	3.60	-	10.79	nr	12.13
E16 - 24/95	8.42	5.00	8.84	0.36	0.04	3.60	-	12.44	nr	14.00
E20 - 28/25	9.43	5.00	9.90	0.40	0.04	3.97	-	13.87	nr	15.61
E20 - 28/45	11.29	5.00	11.85	0.40	0.04	3.97	-	15.83	nr	17.81
E20 - 28/95	11.35	5.00	11.92	0.40	0.04	3.97	-	15.89	nr	17.88
Spit Mega high performance safety anchors; bolt head version; drilling masonry										
with washers, reference										
V10 - 15/20	1.62	5.00	1.70	0.30	0.03	2.98	-	4.68	nr	5.27
V12 - 18/25	2.30	5.00	2.42	0.32	0.03	3.16	-	5.58	nr	6.28
V16 - 24/25	4.97	5.00	5.22	0.36	0.04	3.60	-	8.82	nr	9.92
V20 - 28/25	9.43	5.00	9.90	0.40	0.04	3.97	-	13.87	nr	15.61
Spit Fix high performance through bolt BZP anchors; drilling masonry										
with nuts and washers, reference										
6/10	0.17	5.00	0.18	0.22	0.03	2.24	-	2.42	nr	2.72
6/20	0.18	5.00	0.19	0.22	0.03	2.24	-	2.43	nr	2.73
6/50	0.19	5.00	0.20	0.22	0.03	2.24	-	2.44	nr	2.75
8/10	0.20	5.00	0.21	0.24	0.03	2.43	-	2.64	nr	2.97
8/50	0.26	5.00	0.27	0.24	0.03	2.43	-	2.70	nr	3.04
8/90	0.30	5.00	0.32	0.24	0.03	2.43	-	2.74	nr	3.08
10/10	0.31	5.00	0.33	0.26	0.03	2.61	-	2.94	nr	3.30
10/25	0.32	5.00	0.34	0.26	0.03	2.61	-	2.95	nr	3.31
10/45	0.34	5.00	0.36	0.26	0.03	2.61	-	2.97	nr	3.34
12/10	0.41	5.00	0.43	0.30	0.03	2.98	-	3.41	nr	3.84
12/40	0.56	5.00	0.59	0.30	0.03	2.98	-	3.57	nr	4.01
12/80	0.65	5.00	0.68	0.30	0.03	2.98	-	3.66	nr	4.12
12/120	0.83	5.00	0.87	0.30	0.03	2.98	-	3.85	nr	4.33
12/160	1.40	5.00	1.47	0.30	0.03	2.98	-	4.45	nr	5.01
16/10	1.10	5.00	1.16	0.32	0.03	3.16	-	4.32	nr	4.86
16/45	1.12	5.00	1.18	0.32	0.03	3.16	-	4.34	nr	4.88
16/95	1.14	5.00	1.20	0.32	0.03	3.16	-	4.36	nr	4.91
20/20	1.20	5.00	1.26	0.36	0.03	3.53	-	4.79	nr	5.39
20/60	1.31	5.00	1.38	0.36	0.04	3.60	-	4.98	nr	5.60
20/115	1.45	5.00	1.52	0.36	0.04	3.60	-	5.13	nr	5.77
Steel; plated - expanding bolts										
Expanding bolts; hook Rawlbolts; drilling masonry with nuts and washers; reference										
44 - 401	1.02	5.00	1.07	0.24	0.03	2.43	-	3.50	nr	3.93
44 - 406	1.24	5.00	1.30	0.26	0.03	2.61	-	3.91	nr	4.40
44 - 411	2.01	5.00	2.11	0.30	0.04	3.05	-	5.16	nr	5.81
44 - 416	3.18	5.00	3.34	0.32	0.04	3.23	-	6.57	nr	7.40

Labour hourly rates: (except Specialists) Craft Operatives £9.23 Labourer £7.02 Rates are national average prices. Refer to REGIONAL VARIATIONS for indicative levels of overall pricing in regions	MATERIALS			LABOUR				RATES		
	Del to Site £	Waste %	Material Cost £	Craft Optve Hrs	Lab Hrs	Labour Cost £	Sunds £	Nett Rate £	Unit	Gross Rate (+12.5%) £

G20: CARPENTRY/TIMBER FRAMING/FIRST FIXING Cont.

Steel; plated - expanding bolts Cont.

Expanding bolts; eye Rawlbolts; drilling masonry
with nuts and washers; reference

44 - 432	2.27	5.00	2.38	0.24	0.03	2.43	-	4.81	nr	5.41
44 - 437	2.57	5.00	2.70	0.26	0.03	2.61	-	5.31	nr	5.97
44 - 442	3.47	5.00	3.64	0.30	0.04	3.05	-	6.69	nr	7.53
44 - 447	5.16	5.00	5.42	0.32	0.04	3.23	-	8.65	nr	9.73

Expanding bolts; bolt projecting Through Bolts;
drilling masonry
with nuts and washers; reference

56 - 102	0.33	5.00	0.35	0.22	0.03	2.24	-	2.59	nr	2.91
56 - 104	0.34	5.00	0.36	0.22	0.03	2.24	-	2.60	nr	2.92
56 - 108	0.36	5.00	0.38	0.22	0.03	2.24	-	2.62	nr	2.95
56 - 112	0.40	5.00	0.42	0.22	0.03	2.24	-	2.66	nr	2.99
56 - 114	0.47	5.00	0.49	0.24	0.03	2.43	-	2.92	nr	3.28
56 - 116	0.50	5.00	0.53	0.24	0.03	2.43	-	2.95	nr	3.32
56 - 120	0.52	5.00	0.55	0.24	0.03	2.43	-	2.97	nr	3.34
56 - 124	0.55	5.00	0.58	0.24	0.03	2.43	-	3.00	nr	3.38
56 - 126	0.66	5.00	0.69	0.24	0.03	2.43	-	3.12	nr	3.51
56 - 128	0.64	5.00	0.67	0.26	0.03	2.61	-	3.28	nr	3.69
56 - 129	0.61	5.00	0.64	0.26	0.03	2.61	-	3.25	nr	3.66
56 - 132	0.67	5.00	0.70	0.26	0.03	2.61	-	3.31	nr	3.73
56 - 136	0.71	5.00	0.75	0.26	0.03	2.61	-	3.36	nr	3.78
56 - 138	0.84	5.00	0.88	0.26	0.03	2.61	-	3.49	nr	3.93
56 - 139	0.95	5.00	1.00	0.30	0.04	3.05	-	4.05	nr	4.55
56 - 140	0.96	5.00	1.01	0.30	0.04	3.05	-	4.06	nr	4.57
56 - 144	1.03	5.00	1.08	0.30	0.04	3.05	-	4.13	nr	4.65
56 - 148	1.11	5.00	1.17	0.30	0.04	3.05	-	4.22	nr	4.74
56 - 150	1.16	5.00	1.22	0.30	0.04	3.05	-	4.27	nr	4.80
56 - 152	1.74	5.00	1.83	0.32	0.04	3.23	-	5.06	nr	5.69
56 - 153	1.54	5.00	1.62	0.32	0.04	3.23	-	4.85	nr	5.46
56 - 156	2.10	5.00	2.21	0.32	0.04	3.23	-	5.44	nr	6.12
56 - 158	2.73	5.00	2.87	0.32	0.04	3.23	-	6.10	nr	6.86
56 - 159	2.73	5.00	2.87	0.36	0.05	3.67	-	6.54	nr	7.36
56 - 160	2.97	5.00	3.12	0.36	0.05	3.67	-	6.79	nr	7.64
56 - 164	3.80	5.00	3.99	0.36	0.05	3.67	-	7.66	nr	8.62
56 - 166	4.73	5.00	4.97	0.36	0.05	3.67	-	8.64	nr	9.72
56 - 168	5.42	5.00	5.69	0.40	0.05	4.04	-	9.73	nr	10.95
56 - 172	6.68	5.00	7.01	0.40	0.05	4.04	-	11.06	nr	12.44

Stainless steel grade 316 - expanding bolts

Expanding bolts; bolt projecting Through Bolts;
drilling masonry
with nuts and washers; reference

56 - 604	1.55	5.00	1.63	0.22	0.03	2.24	-	3.87	nr	4.35
56 - 616	1.89	5.00	1.98	0.24	0.03	2.43	-	4.41	nr	4.96
56 - 624	2.34	5.00	2.46	0.24	0.03	2.43	-	4.88	nr	5.49
56 - 628	2.45	5.00	2.57	0.26	0.03	2.61	-	5.18	nr	5.83
56 - 636	2.89	5.00	3.03	0.26	0.03	2.61	-	5.64	nr	6.35
56 - 638	3.22	5.00	3.38	0.26	0.03	2.61	-	5.99	nr	6.74
56 - 640	4.44	5.00	4.66	0.30	0.04	3.05	-	7.71	nr	8.68
56 - 648	5.35	5.00	5.62	0.30	0.04	3.05	-	8.67	nr	9.75
56 - 650	5.95	5.00	6.25	0.30	0.04	3.05	-	9.30	nr	10.46
56 - 652	7.65	5.00	8.03	0.32	0.04	3.23	-	11.27	nr	12.68
56 - 658	11.06	5.00	11.61	0.32	0.04	3.23	-	14.85	nr	16.70
56 - 660	14.97	5.00	15.72	0.36	0.05	3.67	-	19.39	nr	21.82
56 - 666	22.22	5.00	23.33	0.36	0.05	3.67	-	27.00	nr	30.38
56 - 672	30.65	5.00	32.18	0.40	0.05	4.04	-	36.23	nr	40.75

Chemical anchors

Chemical anchors; Kemfix capsules and standard studs;
drilling masonry

capsule reference 60-428; stud reference 60-708; with nuts and washers	1.44	5.00	1.51	0.29	0.04	2.96	-	4.47	nr	5.03
capsule reference 60-430; stud reference 60-710; with nuts and washers	1.62	5.00	1.70	0.32	0.04	3.23	-	4.94	nr	5.55
capsule reference 60-432; stud reference 60-712; with nuts and washers	1.97	5.00	2.07	0.36	0.05	3.67	-	5.74	nr	6.46
capsule reference 60-603; stud reference 60-014; with nuts and washers	3.43	5.00	3.60	0.40	0.05	4.04	-	7.64	nr	8.60

Chemical anchors; Kemfix capsules and stainless steel
studs; drilling masonry

capsule reference 60-428; stud reference 60-906; with nuts and washers	2.54	5.00	2.67	0.29	0.04	2.96	-	5.62	nr	6.33
capsule reference 60-430; stud reference 60-911; with nuts and washers	3.39	5.00	3.56	0.32	0.04	3.23	-	6.79	nr	7.64
capsule reference 60-432; stud reference 60-916; with nuts and washers	4.66	5.00	4.89	0.36	0.05	3.67	-	8.57	nr	9.64
capsule reference 60-436; stud reference 60-921; with nuts and washers	7.93	5.00	8.33	0.40	0.05	4.04	-	12.37	nr	13.92
capsule reference 60-440; stud reference 60-926; with nuts and washers	12.24	5.00	12.85	0.42	0.05	4.23	-	17.08	nr	19.21
capsule reference 60-444; stud reference 60-931; with nuts and washers	19.79	5.00	20.78	0.46	0.06	4.67	-	25.45	nr	28.63

Chemical anchors; Kemfix capsules and standard
internal threaded sockets; drilling masonry

capsule reference 60-428; socket reference 60-622	1.98	5.00	2.08	0.36	0.05	3.67	-	5.75	nr	6.47
capsule reference 60-430; socket reference 60-624	2.29	5.00	2.40	0.40	0.05	4.04	-	6.45	nr	7.25

STRUCTURAL/CARCASSING METAL/TIMBER

Labour hourly rates: (except Specialists) Craft Operatives £9.23 Labourer £7.02 Rates are national average prices. Refer to REGIONAL VARIATIONS for indicative levels of overall pricing in regions	MATERIALS			LABOUR				RATES		
	Del to Site £	Waste %	Material Cost £	Craft Optve Hrs	Lab Hrs	Labour Cost £	Sunds £	Nett Rate £	Unit	Gross Rate (+12.5%) £
G20: CARPENTRY/TIMBER FRAMING/FIRST FIXING Cont.										
Chemical anchors Cont.										
Chemical anchors; Kemfix capsules and standard internal threaded sockets; drilling masonry Cont.										
capsule reference 60-432; socket reference 60-628	3.09	5.00	3.24	0.42	0.05	4.23	-	7.47	nr	8.41
capsule reference 60-436; socket reference 60-630	5.71	5.00	6.00	0.46	0.06	4.67	-	10.66	nr	12.00
Chemcial anchors; Kemfix capsules and stainless steel internal threaded sockets; drilling masonry										
capsule reference 60-428; socket reference 60-987	3.40	5.00	3.57	0.36	0.05	3.67	-	7.24	nr	8.15
capsule reference 60-430; socket reference 60-989	4.41	5.00	4.63	0.40	0.05	4.04	-	8.67	nr	9.76
capsule reference 60-432; socket reference 60-993	5.68	5.00	5.96	0.42	0.05	4.23	-	10.19	nr	11.47
capsule reference 60-436; socket reference 60-995	11.18	5.00	11.74	0.46	0.06	4.67	-	16.41	nr	18.46
Chemical anchors in low density material; Kemfix capsules, perforated sleeves and standard studs; drilling masonry										
capsule reference 60-428; sleeve reference 60-105; stud reference 60-708; with nuts and washers	2.12	5.00	2.23	0.25	0.03	2.52	-	4.74	nr	5.34
capsule reference 60-430; sleeve reference 60-107; stud reference 60-710; with nuts and washers	2.33	5.00	2.45	0.32	0.04	3.23	-	5.68	nr	6.39
capsule reference 60-432; sleeve reference 60-113; stud reference 60-712; with nuts and washers	2.79	5.00	2.93	0.36	0.05	3.67	-	6.60	nr	7.43
capsule reference 60-436; sleeve reference 60-117; stud reference 60-014; with nuts and washers	4.35	5.00	4.57	0.42	0.05	4.23	-	8.80	nr	9.89
Chemical anchors in low density material; Kemfix capsules, perforated sleeves and stainless steel studs; drilling masonry										
capsule reference 60-428; sleeve reference 60-105; stud reference 60-906; with nuts and washers	2.66	5.00	2.79	0.29	0.04	2.96	-	5.75	nr	6.47
capsule reference 60-430; sleeve reference 60-107; stud reference 60-917; with nuts and washers	2.97	5.00	3.12	0.32	0.04	3.23	-	6.35	nr	7.15
capsule reference 60-432; sleeve reference 60-113; stud reference 60-916; with nuts and washers	3.92	5.00	4.12	0.36	0.05	3.67	-	7.79	nr	8.76
capsule reference 60-436; sleeve reference 60-117; stud reference 60-921; with nuts and washers	6.71	5.00	7.05	0.42	0.05	4.23	-	11.27	nr	12.68
Chemical anchors in low density material; Kemfix capsules, perforated sleeves and standard internal threaded sockets; drilling masonry										
capsule reference 60-428; sleeve reference 60-105; socket reference 60-622	2.66	5.00	2.79	0.29	0.04	2.96	-	5.75	nr	6.47
capsule reference 60-430; sleeve reference 60-107; socket reference 60-624	2.97	5.00	3.12	0.32	0.04	3.23	-	6.35	nr	7.15
capsule reference 60-432; sleeve reference 60-113; socket reference 60-628	3.92	5.00	4.12	0.36	0.05	3.67	-	7.79	nr	8.76
capsule reference 60-436; sleeve reference 60-117; socket reference 60-630	6.71	5.00	7.05	0.40	0.05	4.04	-	11.09	nr	12.47
Chemical anchors in low density material; Kemfix capsules, perforated sleeves and stainless steel internal threaded sockets; drilling masonry										
capsule reference 60-428; sleeve reference 60-105; socket reference 60-687	4.08	5.00	4.28	0.29	0.04	2.96	-	7.24	nr	8.15
capsule reference 60-430; sleeve reference 60-107; socket reference 60-689	5.13	5.00	5.39	0.32	0.04	3.23	-	8.62	nr	9.70
capsule reference 60-432; sleeve reference 60-113; socket reference 60-993	6.50	5.00	6.83	0.36	0.05	3.67	-	10.50	nr	11.81
capsule reference 60-436; sleeve reference 60-117; socket reference 60-995	12.19	5.00	12.80	0.40	0.05	4.04	-	16.84	nr	18.95
Spit Maxi high performance chemical anchors; capsules and zinc coated steelstuds; drilling masonry										
capsule reference M8 stud reference SM8; with nuts and washers	1.21	5.00	1.27	0.28	0.03	2.79	-	4.07	nr	4.57
capsule reference M10; stud reference SM10; with nuts and washers	1.36	5.00	1.43	0.31	0.04	3.14	-	4.57	nr	5.14
capsule reference M12; stud reference SM12; with nuts and washers	1.65	5.00	1.73	0.35	0.03	3.44	-	5.17	nr	5.82
capsule reference M16; stud reference SM16; with nuts and washers	2.29	5.00	2.40	0.38	0.03	3.72	-	6.12	nr	6.89
capsule reference M20; stud reference SM20; with nuts and washers	4.09	5.00	4.29	0.40	0.04	3.97	-	8.27	nr	9.30
capsule reference M24; stud reference SM24; with nuts and washers	5.91	5.00	6.21	0.44	0.04	4.34	-	10.55	nr	11.87
Spit Maxi high performance chemical anchors; capsules and stainless steel studs Grade 316 (A4); drilling masonry										
capsule reference M8 stud reference SM8i; with nuts and washers	1.87	5.00	1.96	0.28	0.03	2.79	-	4.76	nr	5.35
capsule reference M10; stud reference SM10i; with nuts and washers	2.56	5.00	2.69	0.31	0.03	3.07	-	5.76	nr	6.48
capsule reference M12; stud reference SM12i; with nuts and washers	3.71	5.00	3.90	0.35	0.03	3.44	-	7.34	nr	8.25
capsule reference M16; stud reference SM16i; with nuts and washers	6.07	5.00	6.37	0.38	0.03	3.72	-	10.09	nr	11.35
capsule reference M20; stud reference SM20i; with nuts and washers	11.07	5.00	11.62	0.40	0.04	3.97	-	15.60	nr	17.55
capsule reference M24; stud reference SM24i; with nuts and washers	16.31	5.00	17.13	0.44	0.04	4.34	-	21.47	nr	24.15

Labour hourly rates: (except Specialists) Craft Operatives £9.23 Labourer £7.02 Rates are national average prices. Refer to REGIONAL VARIATIONS for indicative levels of overall pricing in regions	MATERIALS			LABOUR				RATES		
	Del to Site	Waste	Material Cost	Craft Optve	Lab	Labour Cost	Sunds	Nett Rate	Unit	Gross Rate (+12.5%)
	£	%	£	Hrs	Hrs	£	£	£		£
G30: METAL PROFILED SHEET DECKING										
Galvanised steel troughed decking; 0.7mm thick metal, 35mm overall depth; natural soffit and top surface; 150mm end laps and one corrugation side laps										
Decking; fixed to steel rails at 1200mm centres with self tapping screws; drilling holes sloping; 10 degrees pitch	–	–	Spclist	–	–	Spclist	–	24.50	m2	27.56
Extra over decking for raking cutting	–	–	Spclist	–	–	Spclist	–	6.20	m	6.97
holes 50mm diameter; formed on site	–	–	Spclist	–	–	Spclist	–	7.00	nr	7.88
Galvanised steel troughed decking; 0.7mm thick metal, 48mm overall depth; natural soffit and top surface; 150mm end laps and one corrugation side laps										
Decking; fixed to steel rails at 1500mm centres with self tapping screws; drilling holes sloping; 10 degrees pitch	–	–	Spclist	–	–	Spclist	–	25.55	m2	28.74
Extra over decking for raking cutting	–	–	Spclist	–	–	Spclist	–	6.60	m	7.42
holes 50mm diameter; formed on site	–	–	Spclist	–	–	Spclist	–	7.00	nr	7.88
Galvanised steel troughed decking; 0.7mm thick metal, 63mm overall depth; natural soffit and top surface; 150mm end laps and one corrugation side laps										
Decking; fixed to steel rails at 2000mm centres with self tapping screws; drilling holes sloping; 10 degrees pitch	–	–	Spclist	–	–	Spclist	–	27.40	m2	30.82
Extra over decking for raking cutting	–	–	Spclist	–	–	Spclist	–	7.00	m	7.88
holes 50mm diameter; formed on site	–	–	Spclist	–	–	Spclist	–	7.00	nr	7.88
Galvanised steel troughed decking; 0.7mm thick metal, 100mm overall depth;natural soffit and top surface; 150mm end laps and one corrugation side laps										
Decking; fixed to steel rails at 3500mm centres with self tapping screws; drilling holes sloping; 10 degrees pitch	–	–	Spclist	–	–	Spclist	–	31.00	m2	34.88
Extra over decking for raking cutting	–	–	Spclist	–	–	Spclist	–	7.00	m	7.88
holes 50mm diameter; formed on site	–	–	Spclist	–	–	Spclist	–	7.00	nr	7.88
Galvanised steel troughed decking; 0.9mm thick metal, 35mm overall depth; natural soffit and top surface; 150mm end laps and one corrugation side laps										
Decking; fixed to steel rails at 1500mm centres with self tapping screws; drilling holes sloping; 10 degrees pitch	–	–	Spclist	–	–	Spclist	–	25.60	m2	28.80
Extra over decking for raking cutting	–	–	Spclist	–	–	Spclist	–	6.50	m	7.31
holes 50mm diameter; formed on site	–	–	Spclist	–	–	Spclist	–	7.00	nr	7.88
Galvanised steel troughed decking; 0.9mm thick metal, 48mm overall depth; natural soffit and top surface; 150mm end laps and one corrugation side laps										
Decking; fixing to steel rails at 2000mm centres with self tapping screws; drilling holes sloping; 10 degrees pitch	–	–	Spclist	–	–	Spclist	–	26.80	m2	30.15
Extra over decking for raking cutting	–	–	Spclist	–	–	Spclist	–	7.00	m	7.88
holes 50mm diameter; formed on site	–	–	Spclist	–	–	Spclist	–	7.00	nr	7.88
Galvanised steel troughed decking; 0.9mm thick metal, 63mm overall depth; natural soffit and top surface; 150mm end laps and one corrugation side laps										
Decking; fixing to steel rails at 2500mm centres with self tapping screws; drilling holes sloping; 10 degrees pitch	–	–	Spclist	–	–	Spclist	–	28.80	m2	32.40
Extra over decking for raking cutting	–	–	Spclist	–	–	Spclist	–	7.15	m	8.04
holes 50mm diameter; formed on site	–	–	Spclist	–	–	Spclist	–	7.00	nr	7.88
Galvanised steel troughed decking; 0.9mm thick metal, 100mm overall depth;natural soffit and top surface; 150mm end laps and one corrugation side laps										
Decking; fixing to steel rails at 4000mm centres with self tapping screws; drilling holes sloping; 10 degrees pitch	–	–	Spclist	–	–	Spclist	–	32.20	m2	36.23
Extra over decking for raking cutting	–	–	Spclist	–	–	Spclist	–	8.20	m	9.22

Labour hourly rates: (except Specialists) Craft Operatives £9.23 Labourer £7.02 Rates are national average prices. Refer to REGIONAL VARIATIONS for indicative levels of overall pricing in regions	MATERIALS			LABOUR				RATES		
	Del to Site £	Waste %	Material Cost £	Craft Optve Hrs	Lab Hrs	Labour Cost £	Sunds £	Nett Rate £	Unit	Gross Rate (+12.5%) £
G30: METAL PROFILED SHEET DECKING Cont.										
Galvanised steel troughed decking; 0.9mm thick metal, 100mm overall depth;natural soffit and top surface; 150mm end laps and one corrugation side laps Cont.										
Extra over decking for Cont.										
holes 50mm diameter; formed on site	-	-	Spclist	-	-	Spclist	-	7.00	nr	7.88
Galvanised steel troughed decking; 1.2mm thick metal, 35mm overall depth; natural soffit and top surface; 150mm end laps and one corrugation side laps										
Decking; fixing to steel rails at 2000mm centres with self tapping screws; drilling holes										
sloping; 10 degrees pitch	-	-	Spclist	-	-	Spclist	-	29.00	m2	32.63
Extra over decking for										
raking cutting	-	-	Spclist	-	-	Spclist	-	7.00	m	7.88
holes 50mm diameter; formed on site	-	-	Spclist	-	-	Spclist	-	7.00	nr	7.88
Galvanised steel troughed decking; 1.2mm thick metal, 48mm overall depth; natural soffit and top surface; 150mm end laps and one corrugation side laps										
Decking; fixing to steel rails at 2500mm centres with self tapping screws; drilling holes										
sloping; 10 degrees pitch	-	-	Spclist	-	-	Spclist	-	28.90	m2	32.51
Extra over decking for										
raking cutting	-	-	Spclist	-	-	Spclist	-	7.25	m	8.16
holes 50mm diameter; formed on site	-	-	Spclist	-	-	Spclist	-	7.00	nr	7.88
Galvanised steel troughed decking; 1.2mm thick metal, 63mm overall depth; natural soffit and top surface; 150mm end laps and one corrugation side laps										
Decking; fixing to steel rails at 3000mm centres with self tapping screws; drilling holes										
sloping; 10 degrees pitch	-	-	Spclist	-	-	Spclist	-	31.00	m2	34.88
Extra over decking for										
raking cutting	-	-	Spclist	-	-	Spclist	-	7.80	m	8.78
holes 50mm diameter; formed on site	-	-	Spclist	-	-	Spclist	-	7.00	nr	7.88
Galvanised steel troughed decking; 1.2mm thick metal, 100mm overall depth;natural soffit and top surface; 150mm end laps and one corrugation side laps										
Decking; fixing to steel rails at 4500mm centres with self tapping screws; drilling holes										
sloping; 10 degrees pitch	-	-	Spclist	-	-	Spclist	-	36.60	m2	41.17
Extra over decking for										
raking cutting	-	-	Spclist	-	-	Spclist	-	8.55	m	9.62
holes 50mm diameter; formed on site	-	-	Spclist	-	-	Spclist	-	7.00	nr	7.88
Aluminium troughed decking; 0.9mm thick metal 35mm overall depth; natural soffit and top surface; 150mm end laps and one corrugation side laps										
Decking; fixing to steel rails at 900mm centres with self tapping screws; drilling holes										
sloping; 10 degrees pitch	-	-	Spclist	-	-	Spclist	-	29.30	m2	32.96
Extra over decking for										
raking cutting	-	-	Spclist	-	-	Spclist	-	7.50	m	8.44
holes 50mm diameter; formed on site	-	-	Spclist	-	-	Spclist	-	7.00	nr	7.88
Aluminium troughed decking; 0.9mm thick metal, 48mm overall depth; natural soffit and top surface; 150mm end laps and one corrugation side laps										
Decking; fixing to steel rails at 1200mm centres with self tapping screws; drilling holes										
sloping; 10 degrees pitch	-	-	Spclist	-	-	Spclist	-	31.40	m2	35.33
Extra over decking for										
raking cutting	-	-	Spclist	-	-	Spclist	-	7.90	m	8.89
holes 50mm diameter; formed on site	-	-	Spclist	-	-	Spclist	-	7.00	nr	7.88
Aluminium troughed decking; 0.9mm thick metal, 63mm overall depth; natural soffit and top surface; 150mm end laps and one corrugation side laps										
Decking; fixing to steel rails at 1500mm centres with self tapping screws; drilling holes										
sloping; 10 degrees pitch	-	-	Spclist	-	-	Spclist	-	33.20	m2	37.35
Extra over decking for										
raking cutting	-	-	Spclist	-	-	Spclist	-	8.50	m	9.56
holes 50mm diameter; formed on site	-	-	Spclist	-	-	Spclist	-	7.00	nr	7.88

Labour hourly rates: (except Specialists) Craft Operatives £9.23 Labourer £7.02 Rates are national average prices. Refer to REGIONAL VARIATIONS for indicative levels of overall pricing in regions	MATERIALS			LABOUR				RATES		
	Del to Site £	Waste %	Material Cost £	Craft Optve Hrs	Lab Hrs	Labour Cost £	Sunds £	Nett Rate £	Unit	Gross Rate (+12.5%) £
G30: METAL PROFILED SHEET DECKING Cont.										
Aluminium troughed decking; 0.9mm thick metal, 100mm overall depth; natural soffit and top surface; 150mm end laps and one corrugation side laps										
Decking; fixing to steel rails at 2900mm centres with self tapping screws; drilling holes										
sloping; 10 degrees pitch	-	-	Spclist	-	-	Spclist	-	37.15	m2	41.79
Extra over decking for										
raking cutting	-	-	Spclist	-	-	Spclist	-	9.50	m	10.69
holes 50mm diameter; formed on site	-	-	Spclist	-	-	Spclist	-	7.00	nr	7.88
Aluminium troughed decking; 1.2mm thick metal, 35mm overall depth; natural soffit and top surface; 150mm end laps and one corrugation side laps										
Decking; fixing to steel rails at 1200mm centres with self tapping screws; drilling holes										
sloping; 10 degrees pitch	-	-	Spclist	-	-	Spclist	-	31.50	m2	35.44
Extra over decking for										
raking cutting	-	-	Spclist	-	-	Spclist	-	7.90	m	8.89
holes 50mm diameter; formed on site	-	-	Spclist	-	-	Spclist	-	7.00	nr	7.88
Aluminium troughed decking; 1.2mm thick metal, 48mm overall depth; natural soffit and top surface; 150mm end laps and one corrugation side laps										
Decking; fixing to steel rails at 1700mm centres with self tapping screws; drilling holes										
sloping; 10 degrees pitch	-	-	Spclist	-	-	Spclist	-	33.50	m2	37.69
Extra over decking for										
raking cutting	-	-	Spclist	-	-	Spclist	-	8.55	m	9.62
holes 50mm diameter; formed on site	-	-	Spclist	-	-	Spclist	-	7.00	nr	7.88
Aluminium troughed decking; 1.2mm thick metal, 63mm overall depth; natural soffit and top surface; 150mm end laps and one corrugation side laps										
Decking; fixing to steel rails at 2000mm centres with self tapping screws; drilling holes										
sloping; 10 degrees pitch	-	-	Spclist	-	-	Spclist	-	35.50	m2	39.94
Extra over decking for										
raking cutting	-	-	Spclist	-	-	Spclist	-	9.25	m	10.41
holes 50mm diameter; formed on site	-	-	Spclist	-	-	Spclist	-	7.00	nr	7.88
Aluminium troughed decking; 1.2mm thick metal, 100mm overall depth; natural soffit and top surface; 150mm end laps and one corrugation side laps										
Decking; fixing to steel rails at 3000mm centres with self tapping screws; drilling holes										
sloping; 10 degrees pitch	-	-	Spclist	-	-	Spclist	-	39.20	m2	44.10
Extra over decking for										
raking cutting	-	-	Spclist	-	-	Spclist	-	9.80	m	11.03
holes 50mm diameter; formed on site	-	-	Spclist	-	-	Spclist	-	7.00	nr	7.88
G32: EDGE SUPPORTED/ REINFORCED WOODWOOL SLAB DECKING										
Channel reinforced woodwool slabs, B.S.1105 Type SB; nominal thickness 50mm; natural both sides; butt joints										
Decking; fixing to timber joists at 2000mm general spacing with galvanised mild steel nails and gripper plates										
sloping; 10 degrees pitch	17.69	5.00	18.57	0.50	0.06	5.04	0.25	23.86	m2	26.84
Decking; fixing to steel rails at 2000mm centres with sherardised clips										
sloping; 10 degrees pitch	17.69	5.00	18.57	0.38	0.05	3.86	0.49	22.92	m2	25.79
Extra over decking for										
raking cutting	3.54	-	3.54	0.30	0.04	3.05	-	6.59	m	7.41
holes 100mm diameter; formed on site	-	-	-	0.20	-	1.85	-	1.85	nr	2.08
Channel reinforced woodwool slabs, B.S.1105 Type SB; nominal thickness 75mm; natural both sides; butt joints										
Decking; fixing to timber joists at 3000mm general spacing with galvanised mild steel nails and gripper plates										
sloping; 10 degrees pitch	25.96	5.00	27.26	0.65	0.08	6.56	0.37	34.19	m2	38.46
Decking; fixing to steel rails at 3000mm centres with sherardised clips										
sloping; 10 degrees pitch	25.96	5.00	27.26	0.48	0.06	4.85	0.78	32.89	m2	37.00
Extra over decking for										
raking cutting	5.19	-	5.19	0.40	0.05	4.04	-	9.23	m	10.39

STRUCTURAL/CARCASSING METAL/TIMBER

Labour hourly rates: (except Specialists) Craft Operatives £9.23 Labourer £7.02 Rates are national average prices. Refer to REGIONAL VARIATIONS for indicative levels of overall pricing in regions	MATERIALS			LABOUR				RATES		
	Del to Site £	Waste %	Material Cost £	Craft Optve Hrs	Lab Hrs	Labour Cost £	Sunds £	Nett Rate £	Unit	Gross Rate (+12.5%) £
G32: EDGE SUPPORTED/ REINFORCED WOODWOOL SLAB DECKING Cont.										
Channel reinforced woodwool slabs, B.S.1105 Type SB; nominal thickness 75mm; natural both sides; butt joints Cont.										
Extra over decking for Cont.										
holes 100mm diameter; formed on site	-	-	-	0.25	-	2.31	-	2.31	nr	2.60
Channel reinforced woodwool slabs, B.S.1105 Type SB; nominal thickness 100mm; natural both sides; butt joints										
Decking; fixing to timber joists at 4000mm general spacing with galvanised mild steel nails and gripper plates										
sloping; 10 degrees pitch	30.42	5.00	31.94	0.80	0.10	8.09	0.49	40.52	m2	45.58
Decking; fixing to steel rails at 4000mm centres with sherardised clips										
sloping; 10 degrees pitch	30.42	5.00	31.94	0.60	0.08	6.10	0.99	39.03	m2	43.91
Extra over decking for										
raking cutting	6.08	-	6.08	0.50	0.06	5.04	-	11.12	m	12.51
holes 100mm diameter; formed on site	-	-	-	0.30	-	2.77	-	2.77	nr	3.12
Channel reinforced woodwool slabs, B.S.1105 Type SB; nominal thickness 50mm; pre-textured soffit, natural top surface; butt joints										
Decking; fixing to timber joists at 2000mm general spacing with galvanised mild steel nails and gripper plates										
sloping; 10 degrees pitch	20.61	5.00	21.64	0.50	0.06	5.04	0.25	26.93	m2	30.29
Decking; fixing to steel rails at 2000mm centres with sherardised clips										
sloping; 10 degrees pitch	20.61	5.00	21.64	0.38	0.05	3.86	0.49	25.99	m2	29.24
Extra over decking for										
raking cutting	4.12	-	4.12	0.30	0.04	3.05	-	7.17	m	8.07
holes 100mm diameter; formed on site	-	-	-	0.20	-	1.85	-	1.85	nr	2.08
Channel reinforced woodwool slabs, B.S.1105 Type SB; nominal thickness 75mm; pre-textured soffit, natural top surface; butt joints										
Decking; fixing to timber joists at 3000mm general spacing with galvanised mild steel nails and gripper plates										
sloping; 10 degrees pitch	28.89	5.00	30.33	0.65	0.08	6.56	0.37	37.27	m2	41.92
Decking; fixing to steel rails at 3000mm centres with sherardised clips										
sloping; 10 degrees pitch	28.89	5.00	30.33	0.48	0.06	4.85	0.78	35.97	m2	40.46
Extra over decking for										
raking cutting	5.78	-	5.78	0.40	0.05	4.04	-	9.82	m	11.05
holes 100mm diameter; formed on site	-	-	-	0.25	-	2.31	-	2.31	nr	2.60
Channel reinforced woodwool slabs, B.S.1105 Type SB; nominal thickness 100mm; pre-textured soffit, natural top surface; butt joints										
Decking; fixing to timber joists at 4000mm general spacing with galvanised mild steel nails and gripper plates										
sloping; 10 degrees pitch	33.34	5.00	35.01	0.80	0.10	8.09	0.49	43.58	m2	49.03
Decking; fixing to steel rails at 4000mm centres with sherardised clips										
sloping; 10 degrees pitch	33.34	5.00	35.01	0.60	0.08	6.10	0.99	42.10	m2	47.36
Extra over decking for										
raking cutting	6.67	-	6.67	0.50	0.06	5.04	-	11.71	m	13.17
holes 100mm diameter; formed on site	-	-	-	0.30	-	2.77	-	2.77	nr	3.12

Cladding/Covering

Labour hourly rates: (except Specialists) Craft Operatives £9.23 Labourer £7.02 Rates are national average prices. Refer to REGIONAL VARIATIONS for indicative levels of overall pricing in regions	MATERIALS			LABOUR				RATES		
	Del to Site £	Waste %	Material Cost £	Craft Optve Hrs	Lab Hrs	Labour Cost £	Sunds £	Nett Rate £	Unit	Gross Rate (+12.5%) £
H10: PATENT GLAZING										
Patent glazing with aluminium alloy bars 2000mm long with aluminium wings and seatings for glass, spaced at approximately 600mm centres, glazed with B.S.952 Georgian wired cast, 7mm thick										
Roof areas										
single tier	89.45	5.00	93.92	1.68	0.84	21.40	2.58	117.91	m2	132.64
multi-tier	95.93	5.00	100.73	1.95	1.95	31.69	2.58	134.99	m2	151.87
Vertical surfaces										
single tier	99.57	5.00	104.55	1.70	1.70	27.63	2.58	134.75	m2	151.60
multi-tier	106.15	5.00	111.46	1.95	1.95	31.69	2.58	145.72	m2	163.94
Patent glazing with aluminium alloy bars 2000mm long with aluminium wings and seatings for glass, spaced at approximately 600mm centres, glazed with B.S.952 Georgian wired polished, 6mm thick										
Roof areas										
single tier	139.78	5.00	146.77	1.90	1.90	30.88	2.58	180.22	m2	202.75
multi-tier	150.25	5.00	157.76	2.20	2.20	35.75	2.58	196.09	m2	220.60
Vertical surfaces										
single tier	150.01	5.00	157.51	1.90	1.90	30.88	2.58	190.97	m2	214.84
multi-tier	159.55	5.00	167.53	2.20	2.20	35.75	2.58	205.86	m2	231.59
Patent glazing with aluminium alloy bars 3000mm long with aluminium wings and seatings for glass, spaced at approximately 600mm centres, glazed with B.S.952 Georgian wired cast, 7mm thick										
Roof areas										
single tier	106.29	5.00	111.60	1.70	1.70	27.63	2.58	141.81	m2	159.54
multi-tier	113.12	5.00	118.78	1.95	1.95	31.69	2.58	153.04	m2	172.17
Vertical surfaces										
single tier	115.53	5.00	121.31	1.70	1.70	27.63	2.58	151.51	m2	170.45
multi-tier	123.35	5.00	129.52	1.95	1.95	31.69	2.58	163.79	m2	184.26
Patent glazing with aluminium alloy bars 3000mm long with aluminium wings and seatings for glass, spaced at approximately 600mm centres, glazed with B.S.952 Georgian wired polished, 6mm thick										
Roof areas										
single tier	155.82	5.00	163.61	1.90	1.90	30.88	2.58	197.07	m2	221.70
multi-tier	167.47	5.00	175.84	2.20	2.20	35.75	2.58	214.17	m2	240.95
Vertical surfaces										
single tier	165.90	5.00	174.19	1.90	1.90	30.88	2.58	207.65	m2	233.61
multi-tier	177.68	5.00	186.56	2.20	2.20	35.75	2.58	224.89	m2	253.01
Extra over patent glazing in roof areas with aluminium alloy bars 2000mm long with aluminium wings and seatings for glass, spaced at approximately 600mm centres, glazed with B.S.952 Georgian wired polished, 6mm thick										
Opening lights including opening gear										
600 x 600mm	214.65	5.00	225.30	5.00	5.00	81.25	-	306.63	nr	344.96
600 x 900mm	260.73	5.00	273.77	6.30	6.30	102.38	-	376.14	nr	423.16

Labour hourly rates: (except Specialists) Craft Operatives £9.23 Labourer £7.02 Rates are national average prices. Refer to REGIONAL VARIATIONS for indicative levels of overall pricing in regions	MATERIALS			LABOUR				RATES		
	Del to Site £	Waste %	Material Cost £	Craft Optve Hrs	Lab Hrs	Labour Cost £	Sunds £	Nett Rate £	Unit	Gross Rate (+12.5%) £
H10: PATENT GLAZING Cont.										
Extra over patent glazing in roof areas with aluminium alloy bars 2000mm long with aluminium wings and seatings for glass, spaced at approximately 600mm centres, glazed with B.S.952 Georgian wired cast, 7mm thick										
Opening lights including opening gear										
600 x 600mm	214.65	5.00	225.38	4.60	4.60	74.75	-	300.13	nr	337.65
600 x 900mm	260.73	5.00	273.77	5.75	5.75	93.44	-	367.20	nr	413.10
Extra over patent glazing in vertical surfaces with aluminium alloy bars 3000mm long with aluminium wings and seatings for glass, spaced at approximately 600mm centres, glazed with B.S.952 Georgian wired cast, 7mm thick										
Opening lights including opening gear										
600 x 600mm	313.26	5.00	328.92	2.30	2.30	37.38	-	366.30	nr	412.09
600 x 900mm	380.90	5.00	399.94	3.50	3.50	56.88	-	456.82	nr	513.92
Extra over patent glazing to vertical surfaces with aluminium alloy bars 3000mm long with aluminium wings and seatings for glass, spaced at approximately 600mm centres, glazed with B.S.952 Georgian wired polished, 6mm thick										
Opening lights including opening gear										
600 x 600mm	313.26	5.00	328.92	2.55	2.55	41.44	-	370.36	nr	416.66
600 x 900mm	380.71	5.00	399.75	3.75	3.75	60.94	-	460.68	nr	518.27
H21: TIMBER WEATHERBOARDING										
Softwood, wrought, impregnated										
Boarding to walls, shiplapped joints; 19mm thick, 150mm wide boards										
width exceeding 300mm; fixing to timber	7.52	10.00	8.27	0.85	0.10	8.55	0.14	16.96	m2	19.08
Boarding to walls, shiplapped joints; 25mm thick, 150mm wide boards										
width exceeding 300mm; fixing to timber	9.90	10.00	10.89	0.90	0.11	9.08	0.15	20.12	m2	22.63
Western Red Cedar										
Boarding to walls, shiplapped joints; 19mm thick, 150mm wide boards										
width exceeding 300mm; fixing to timber	25.08	10.00	27.59	0.90	0.12	9.15	0.15	36.89	m2	41.50
Boarding to walls, shiplapped joints; 25mm thick, 150mm wide boards										
width exceeding 300mm; fixing to timber	33.00	10.00	36.30	0.95	0.12	9.61	0.15	46.06	m2	51.82
Abutments										
19mm thick softwood boarding										
raking cutting	0.75	-	0.75	0.07	-	0.65	-	1.40	m	1.57
curved cutting	1.13	-	1.13	0.10	-	0.92	-	2.05	m	2.31
25mm thick softwood boarding										
raking cutting	0.99	-	0.99	0.08	-	0.74	-	1.73	m	1.94
curved cutting	1.49	-	1.49	0.12	-	1.11	-	2.60	m	2.92
19mm thick Western Red Cedar boarding										
raking cutting	2.51	-	2.51	0.09	-	0.83	-	3.34	m	3.76
curved cutting	3.76	-	3.76	0.13	-	1.20	-	4.96	m	5.58
25mm thick Western Red Cedar boarding										
raking cutting	3.30	-	3.30	0.10	-	0.92	-	4.22	m	4.75
curved cutting	4.95	-	4.95	0.15	-	1.38	-	6.33	m	7.13
H30: FIBRE CEMENT PROFILED SHEET CLADDING/COVERING/SIDING										
Roof coverings; corrugated reinforced cement Profile 3 sheeting, standard grey colour; lapped one and a half corrugations at sides and 150mm at ends										
Coverings; fixing to timber joists at 900mm general spacing with galvanised mild steel drive screws and washers; drilling holes										
pitch 30 degrees from horizontal	10.22	15.00	11.75	0.28	0.28	4.55	0.45	16.75	m2	18.85
Coverings; fixing to steel purlins at 900mm general spacing with galvanised hook bolts and washers; drilling holes										
pitch 30 degrees from horizontal	10.22	15.00	11.75	0.38	0.38	6.17	0.90	18.83	m2	21.18
Eaves										
eaves filler pieces	7.54	10.00	8.29	0.12	0.12	1.95	-	10.24	m	11.52
Ridges										
two piece plain angular adjustable ridge tiles ...	10.91	10.00	12.00	0.23	0.23	3.74	-	15.74	m	17.71

Labour hourly rates: (except Specialists) Craft Operatives £9.23 Labourer £7.02 Rates are national average prices. Refer to REGIONAL VARIATIONS for indicative levels of overall pricing in regions	MATERIALS			LABOUR				RATES		
	Del to Site £	Waste %	Material Cost £	Craft Optve Hrs	Lab Hrs	Labour Cost £	Sunds £	Nett Rate £	Unit	Gross Rate (+12.5%) £
H30: FIBRE CEMENT PROFILED SHEET CLADDING/COVERING/SIDING Cont.										
Roof coverings; corrugated reinforced cement Profile 3 sheeting, standard grey colour; lapped one and a half corrugations at sides and 150mm at ends Cont.										
Barge boards standard barge boards	7.98	10.00	8.78	0.12	0.12	1.95	-	10.73	m	12.07
Aprons/sills apron flashings	7.58	10.00	8.34	0.15	0.15	2.44	-	10.78	m	12.12
Finials standard one piece ridge cap finials	10.51	10.00	11.56	0.17	0.17	2.76	-	14.32	nr	16.11
Cutting raking	2.05	-	2.05	0.28	0.28	4.55	-	6.60	m	7.42
Holes for pipes, standards or the like	-	-	-	0.58	-	5.35	-	5.35	nr	6.02
Roof coverings; corrugated reinforced cement Profile 3 sheeting, standard coloured; lapped one and a half corrugations at sides and 150mm at ends										
Coverings; fixing to timber joists at 900mm general spacing with galvanised mild steel drive screws and washers; drilling holes pitch 30 degrees from horizontal	12.26	15.00	14.10	0.28	0.28	4.55	0.59	19.24	m2	21.64
Coverings; fixing to steel purlins at 900mm general spacing with galvanised hook bolts and washers; drilling holes pitch 30 degrees from horizontal	12.26	15.00	14.10	0.38	0.38	6.17	1.17	21.44	m2	24.12
Eaves eaves filler pieces	9.43	10.00	10.37	0.12	0.12	1.95	-	12.32	m	13.86
Ridges two piece plain angular adjustable ridge tiles ...	13.15	10.00	14.47	0.23	0.23	3.74	-	18.20	m	20.48
Barge boards standard barge boards	9.95	10.00	10.95	0.12	0.12	1.95	-	12.90	m	14.51
Aprons/sills apron flashings	9.45	10.00	10.40	0.15	0.15	2.44	-	12.83	m	14.44
Finials standard one piece ridge cap finials	13.13	10.00	14.44	0.17	0.17	2.76	-	17.21	nr	19.36
Cutting raking	2.45	-	2.45	0.28	0.28	4.55	-	7.00	m	7.88
Holes for pipes, standards or the like	-	-	-	0.58	-	5.35	-	5.35	nr	6.02
Roof coverings; corrugated reinforced cement Profile 6 sheeting, standard grey colour; lapped half a corrugation at sides and 150mm at ends										
Coverings; fixing to timber joists at 900mm general spacing with galvanised mild steel drive screws and washers; drilling holes pitch 30 degrees from horizontal	8.93	15.00	10.27	0.28	0.28	4.55	0.47	15.29	m2	17.20
Coverings; fixing to steel purlins at 900mm general spacing with galvanised hook bolts and washers; drilling holes pitch 30 degrees from horizontal	8.93	15.00	10.27	0.38	0.38	6.17	0.95	17.39	m2	19.57
Eaves eaves filler pieces	5.94	10.00	6.53	0.12	0.12	1.95	-	8.48	m	9.54
Ridges two piece plain angular adjustable ridge tiles ...	10.92	10.00	12.01	0.23	0.23	3.74	-	15.75	m	17.72
Barge boards standard barge boards	7.34	10.00	8.07	0.12	0.12	1.95	-	10.02	m	11.28
Aprons/sills apron flashings	6.40	10.00	7.04	0.15	0.15	2.44	-	9.48	m	10.66
Finials standard one piece ridge cap finials	1.79	10.00	1.97	0.17	0.17	2.76	-	4.73	nr	5.32
Cutting raking	1.61	-	1.61	0.28	0.28	4.55	-	6.16	m	6.93
Holes for pipes, standards or the like	-	-	-	0.58	-	5.35	-	5.35	nr	6.02

CLADDING/COVERING

Labour hourly rates: (except Specialists) Craft Operatives £9.23 Labourer £7.02 Rates are national average prices. Refer to REGIONAL VARIATIONS for indicative levels of overall pricing in regions	MATERIALS			LABOUR				RATES		
	Del to Site £	Waste %	Material Cost £	Craft Optve Hrs	Lab Hrs	Labour Cost £	Sunds £	Nett Rate £	Unit	Gross Rate (+12.5%) £
H30: FIBRE CEMENT PROFILED SHEET CLADDING/COVERING/SIDING Cont.										
Roof coverings; corrugated reinforced cement Profile 6 sheeting, standard coloured; lapped half a corrugation at sides and 150mm at ends										
Coverings; fixing to timber joists at 900mm general spacing with galvanised mild steel drive screws and washers; drilling holes										
pitch 30 degrees from horizontal	10.72	15.00	**12.33**	0.28	0.28	**4.55**	0.58	**17.46**	m2	**19.64**
Coverings; fixing to steel purlins at 900mm general spacing with galvanised hook bolts and washers; drilling holes										
pitch 30 degrees from horizontal	10.72	15.00	**12.33**	0.38	0.38	**6.17**	1.17	**19.67**	m2	**22.13**
Eaves										
eaves filler pieces	7.43	10.00	**8.17**	0.12	0.12	**1.95**	–	**10.12**	m	**11.39**
Ridges										
two piece plain angular adjustable ridge tiles ...	13.66	10.00	**15.03**	0.23	0.23	**3.74**	–	**18.76**	m	**21.11**
Barge boards										
standard barge boards	9.17	10.00	**10.09**	0.12	0.12	**1.95**	–	**12.04**	m	**13.54**
Aprons/sills										
apron flashings	8.01	10.00	**8.81**	0.15	0.15	**2.44**	–	**11.25**	m	**12.65**
Finials										
standard one piece ridge cap finials	8.89	10.00	**9.78**	0.17	0.17	**2.76**	–	**12.54**	nr	**14.11**
Cutting										
raking...	2.15	–	**2.15**	0.28	0.28	**4.55**	–	**6.70**	m	**7.54**
Holes										
for pipes, standards or the like	–	–	–	0.58	–	**5.35**	–	**5.35**	nr	**6.02**
H31: METAL PROFILED/FLAT SHEET CLADDING/COVERING/SIDING										
Wall cladding; PVC colour coated both sides galvanised steel profiled sheeting 0.70mm thick and with sheets secured at seams and laps										
Coverings; fixing to steel rails at 900mm general spacing with galvanised hook bolts and washers; drilling holes										
pitch 90 degrees from horizontal	10.25	5.00	**10.76**	0.40	0.40	**6.50**	1.05	**18.31**	m2	**20.60**
Vertical angles										
vertical corner flashings 500mm girth	7.52	5.00	**7.90**	0.17	0.17	**2.76**	–	**10.66**	m	**11.99**
Filler blocks										
polyethylene...................................	1.58	5.00	**1.66**	0.12	0.12	**1.95**	–	**3.61**	m	**4.06**
PVC..	1.97	5.00	**2.07**	0.12	0.12	**1.95**	–	**4.02**	m	**4.52**
black synthetic rubber	2.19	5.00	**2.30**	0.12	0.12	**1.95**	–	**4.25**	m	**4.78**
Cutting										
raking...	2.05	–	**2.05**	0.28	0.28	**4.55**	–	**6.60**	m	**7.42**
Roof coverings; PVC colour coated both sides galvanised steel profiled sheeting 0.70mm thick and with sheets secured at seams and laps										
Coverings; fixing to steel purlins at 900mm general spacing with galvanised hook bolts and washers; drilling holes										
pitch 30 degrees from horizontal	10.25	5.00	**10.76**	0.35	0.35	**5.69**	1.05	**17.50**	m2	**19.69**
Ridges										
ridge cappings 500mm girth	10.10	5.00	**10.61**	0.17	0.17	**2.76**	–	**13.37**	m	**15.04**
Flashings										
gable flashings 500mm girth	10.10	5.00	**10.61**	0.17	0.17	**2.76**	–	**13.37**	m	**15.04**
eaves flashings 500mm girth	3.29	5.00	**3.45**	0.17	0.17	**2.76**	–	**6.22**	m	**6.99**
Filler blocks										
polyethylene...................................	1.58	5.00	**1.66**	0.12	0.12	**1.95**	–	**3.61**	m	**4.06**
PVC..	1.97	5.00	**2.07**	0.12	0.12	**1.95**	–	**4.02**	m	**4.52**
black synthetic rubber	2.19	5.00	**2.30**	0.12	0.12	**1.95**	–	**4.25**	m	**4.78**
Cutting										
raking...	2.05	–	**2.05**	0.28	0.28	**4.55**	–	**6.60**	m	**7.42**
H32: PLASTICS PROFILED SHEET CLADDING/ COVERING/SIDING										
Roof coverings; standard corrugated glass fibre 1.3mm thick reinforced translucent sheeting; lapped one and a half corrugations at sides and 150mm at ends										
Coverings; fixing to timber joists at 900mm general spacing with galvanised mild steel drive screws and washers; drilling holes										
pitch 30 degrees from horizontal	13.96	5.00	**14.66**	0.28	0.28	**4.55**	0.32	**19.53**	m2	**21.97**

Labour hourly rates: (except Specialists) Craft Operatives £9.23 Labourer £7.02 Rates are national average prices. Refer to REGIONAL VARIATIONS for indicative levels of overall pricing in regions	MATERIALS			LABOUR				RATES		
	Del to Site £	Waste %	Material Cost £	Craft Optve Hrs	Lab Hrs	Labour Cost £	Sunds £	Nett Rate £	Unit	Gross Rate (+12.5%) £
H32: PLASTICS PROFILED SHEET CLADDING/ COVERING/SIDING Cont.										
Roof coverings; standard corrugated glass fibre 1.3mm thick reinforced translucent sheeting; lapped one and a half corrugations at sides and 150mm at ends Cont.										
Coverings; fixing to steel purlins at 900mm general spacing with galvanised hook bolts and washers; drilling holes										
pitch 30 degrees from horizontal	13.96	5.00	14.66	0.35	0.35	5.69	0.39	20.74	m2	23.33
Cutting										
raking	2.79	-	2.79	0.28	0.28	4.55	-	7.34	m	8.26
Roof coverings; fire resisting corrugated glass fibre reinforced translucent sheeting; lapped one and a half corrugations at sides and 150mm at ends										
Coverings; fixing to timber joists at 900mm general spacing with galvanised mild steel drive screws and washers; drilling holes										
pitch 30 degrees from horizontal	19.55	5.00	20.53	0.28	0.28	4.55	0.32	25.40	m2	28.57
Coverings; fixing to steel purlins at 900mm general spacing with galvanised hook bolts and washers; drilling holes										
pitch 30 degrees from horizontal	19.55	5.00	20.53	0.35	0.35	5.69	0.50	26.72	m2	30.05
Cutting										
raking	3.90	-	3.90	0.28	0.28	4.55	-	8.45	m	9.51
Roof coverings; standard vinyl corrugated sheeting; lapped one and a half corrugations at sides and 150mm at ends; 1.3mm thick										
Coverings; fixing to timber joists at 900mm general spacing with galvanised mild steel drive screws and washers; drilling holes										
pitch 30 degrees from horizontal	17.37	5.00	18.24	0.28	0.28	4.55	0.51	23.30	m2	26.21
Coverings; fixing to steel purlins at 900mm general spacing with galvanised hook bolts and washers; drilling holes										
pitch 30 degrees from horizontal	17.37	5.00	18.24	0.35	0.35	5.69	0.51	24.44	m2	27.49
Cutting										
raking	3.38	-	3.38	0.28	0.28	4.55	-	7.93	m	8.92
Wall cladding; Swish Products high impact rigid uPVC profiled sections; colour white; secured with starter sections and clips										
Coverings with shiplap profile code C002 giving 150mm cover; fixing to timber										
vertical cladding										
sections applied horizontally	36.90	5.00	38.74	0.42	0.42	6.83	1.87	47.44	m2	53.37
sections applied vertically	36.90	5.00	38.74	0.42	0.42	6.83	1.87	47.44	m2	53.37
Coverings with open V profile code C003 giving 150mm cover; fixing to timber										
vertical cladding										
sections applied horizontally	32.67	5.00	34.30	0.45	0.45	7.31	2.67	44.29	m2	49.82
sections applied vertically	32.67	5.00	34.30	0.45	0.45	7.31	2.67	44.29	m2	49.82
Coverings with open V profile code C269 giving 100mm cover; fixing to timber										
vertical cladding										
sections applied horizontally	36.90	5.00	38.74	0.38	0.38	6.17	1.87	46.79	m2	52.64
sections applied vertically	36.90	5.00	38.74	0.38	0.38	6.17	1.87	46.79	m2	52.64
Vertical angles										
section C030 for vertically applied section	3.74	5.00	3.93	0.11	0.11	1.79	0.14	5.85	m	6.59
H51: NATURAL STONE SLAB CLADDING/FEATURES										
English blue/grey slate facings (Best Quality); natural riven finish; bedding, jointing and pointing in gauged mortar (1:2:9)										
450 x 600 x 20mm units to walls on brickwork or blockwork base										
plain, width exceeding 300mm	119.20	2.50	122.18	2.00	2.00	32.50	0.40	155.08	m2	174.47
450 x 600 x 30mm units to walls on brickwork or blockwork base										
plain, width exceeding 300mm	180.20	2.50	184.71	2.60	2.60	42.25	0.53	227.49	m2	255.92
450 x 600 x 40mm units to walls on brickwork or blockwork base										
plain, width exceeding 300mm	229.70	2.50	235.44	2.90	2.90	47.13	0.69	283.26	m2	318.66
450 x 600 x 50mm units to walls on brickwork or blockwork base										
plain, width exceeding 300mm	300.05	2.50	307.55	3.45	3.45	56.06	0.85	364.46	m2	410.02

CLADDING/COVERING

Labour hourly rates: (except Specialists) Craft Operatives £9.23 Labourer £7.02 Rates are national average prices. Refer to REGIONAL VARIATIONS for indicative levels of overall pricing in regions	MATERIALS			LABOUR				RATES		
	Del to Site	Waste	Material Cost	Craft Optve	Lab	Labour Cost	Sunds	Nett Rate	Unit	Gross Rate (+12.5%)
	£	%	£	Hrs	Hrs	£	£	£		£
H51: NATURAL STONE SLAB CLADDING/FEATURES Cont.										
English blue/grey slate facings (Best quality); fine rubbed finish, one face; bedding, jointing and pointing in gauged mortar (1:2:9)										
750 x 1200 x 20mm units to walls on brickwork or blockwork base										
plain, width exceeding 300mm	129.40	2.50	132.63	2.30	2.30	37.38	1.30	171.31	m2	192.72
extra; rubbed square edges	9.40	-	9.40	-	-	-	-	9.40	m	10.57
extra; half rounded edges	11.70	-	11.70	-	-	-	-	11.70	m	13.16
extra; full rounded edges	17.45	-	17.45	-	-	-	-	17.45	m	19.63
extra; rebated joints	17.45	-	17.45	-	-	-	-	17.45	m	19.63
750 x 1200 x 30mm units to walls on brickwork or blockwork base										
plain, width exceeding 300mm	193.50	2.50	198.34	2.60	2.60	42.25	1.39	241.98	m2	272.22
extra; rubbed square edges	10.45	-	10.45	-	-	-	-	10.45	m	11.76
extra; half rounded edges	12.55	-	12.55	-	-	-	-	12.55	m	14.12
extra; full rounded edges	18.80	-	18.80	-	-	-	-	18.80	m	21.15
extra; rebated joints	18.80	-	18.80	-	-	-	-	18.80	m	21.15
750 x 1200 x 40mm units to walls on brickwork or blockwork base										
plain, width exceeding 300mm	239.35	2.50	245.33	2.90	2.90	47.13	1.54	294.00	m2	330.75
extra; rubbed square edges	12.05	-	12.05	-	-	-	-	12.05	m	13.56
extra; half rounded edges	13.75	-	13.75	-	-	-	-	13.75	m	15.47
extra; full rounded edges	20.25	-	20.25	-	-	-	-	20.25	m	22.78
extra; rebated joints	20.25	-	20.25	-	-	-	-	20.25	m	22.78
750 x 1200 x 50mm units to walls on brickwork or blockwork base										
plain, width exceeding 300mm	322.35	2.50	330.41	3.45	3.45	56.06	1.72	388.19	m2	436.72
extra; rubbed square edges	12.95	-	12.95	-	-	-	-	12.95	m	14.57
extra; half rounded edges	14.30	-	14.30	-	-	-	-	14.30	m	16.09
extra; full rounded edges	20.80	-	20.80	-	-	-	-	20.80	m	23.40
extra; rebated joints	20.80	-	20.80	-	-	-	-	20.80	m	23.40
English blue/grey slate facings (Best quality); fine rubbed finish, both faces; bedding, jointing and pointing in gauged mortar (1:2:9)										
750 x 1200 x 20mm units to walls on brickwork or blockwork base										
plain, width exceeding 300mm	164.85	2.50	168.97	2.90	2.90	47.13	2.16	218.26	m2	245.54
750 x 1200 x 30mm units to walls on brickwork or blockwork base										
plain, width exceeding 300mm	228.95	2.50	234.67	3.15	2.75	48.38	2.27	285.32	m2	320.99
750 x 1200 x 40mm units to walls on brickwork or blockwork base										
plain, width exceeding 300mm	274.80	2.50	281.67	3.45	3.00	52.90	2.44	337.01	m2	379.14
750 x 1200 x 50mm units to walls on brickwork or blockwork base										
plain, width exceeding 300mm	357.70	2.50	366.64	4.00	3.50	61.49	2.57	430.70	m2	484.54
Kirkstone green slate; bedding, jointing and pointing in gauged mortar (1:2:9)										
30mm thick units to sills on brickwork or blockwork base										
width 125mm; weathered, throated and grooved	54.91	2.50	56.28	0.65	0.65	10.56	0.57	67.42	m	75.84
width 190mm; weathered, throated and grooved	69.66	2.50	71.40	0.70	0.70	11.38	0.74	83.52	m	93.96
38mm thick units to sills on brickwork or blockwork base										
width 125mm	63.45	2.50	65.04	0.70	0.70	11.38	0.57	76.98	m	86.60
extra; stooling for jambs	29.25	-	29.25	-	-	-	-	29.25	nr	32.91
extra; notching for jambs and mullions	17.55	-	17.55	-	-	-	-	17.55	nr	19.74
width 190mm	81.89	2.50	83.94	0.85	0.85	13.81	0.74	98.49	m	110.80
extra; stooling for jambs	29.25	-	29.25	-	-	-	-	29.25	nr	32.91
extra; notching for jambs and mullions	17.55	-	17.55	-	-	-	-	17.55	nr	19.74
50mm thick units to sills on brickwork or blockwork base										
width 125mm	76.44	2.50	78.35	0.85	0.85	13.81	0.57	92.73	m	104.33
extra; stooling for jambs	29.25	-	29.25	-	-	-	-	29.25	nr	32.91
extra; notching for jambs and mullions	17.55	-	17.55	-	-	-	-	17.55	nr	19.74
width 190mm	103.41	2.50	106.00	1.20	1.20	19.50	0.74	126.24	m	142.01
extra; stooling for jambs	29.25	-	29.25	-	-	-	-	29.25	nr	32.91
extra; notching for jambs and mullions	17.55	-	17.55	-	-	-	-	17.55	nr	19.74
30mm units to window boards on brickwork or blockwork base										
width 230mm	63.36	2.50	64.94	0.85	0.85	13.81	0.93	79.69	m	89.65
25mm units to combined sills and window boards on brickwork or blockwork base										
width 360mm	95.96	2.50	98.36	1.15	1.15	18.69	1.30	118.35	m	133.14

Labour hourly rates: (except Specialists) Craft Operatives £9.23 Labourer £7.02 Rates are national average prices. Refer to REGIONAL VARIATIONS for indicative levels of overall pricing in regions	MATERIALS			LABOUR				RATES		
	Del to Site £	Waste %	Material Cost £	Craft Optve Hrs	Lab Hrs	Labour Cost £	Sunds £	Nett Rate £	Unit	Gross Rate (+12.5%) £
H60: PLAIN ROOF TILING										
Quantities required for 1m2 of tiling										
Plain tiles to 65mm lap										
tiles, 60nr										
battens, 10m										
Plain tiles to 85mm lap										
tiles, 70nr										
battens, 11m										
Roof coverings; clayware machine made plain tiles B.S.402, red, 265 x 165mm; fixing every fourth course with two galvanised nails per tile to 65mm lap; 38 x 19mm pressure impregnated softwood battens fixed with galvanised nails; underlay B.S.747 type 1F reinforced bitumen felt; 150mm laps; fixing with galvanised steel clout nails										
Pitched 50 degrees from horizontal										
generally................................	23.96	5.00	25.16	0.72	0.36	9.17	5.73	40.06	m2	45.07
holes...................................	-	-	-	0.58	0.29	7.39	-	7.39	nr	8.31
abutments; square.......................	2.40	-	2.40	0.28	0.14	3.57	-	5.97	m	6.71
abutments; raking.......................	3.42	-	3.42	0.42	0.21	5.35	-	8.77	m	9.87
double course at eaves; purpose made eaves tile ..	4.50	5.00	4.72	0.35	0.18	4.49	0.53	9.75	m	10.97
verges; bed and point in coloured cement-lime mortar (1:1:6)...........................	2.25	5.00	2.36	0.40	0.20	5.10	0.38	7.84	m	8.82
verges; single extra undercloak course plain tiles; bed and point in coloured cement-lime mortar (1:1:6)...........................	3.00	5.00	3.15	0.58	0.29	7.39	0.38	10.92	m	12.28
ridge tiles, half round; in 300mm effective lengths; butt jointed; bedding and pointing in coloured cement-lime mortar (1:1:6).............	16.75	5.00	17.59	0.35	0.18	4.49	0.38	22.46	m	25.27
hip tiles, half round; in 300mm effective lengths; butt jointed; bedding and pointing in coloured cement-lime mortar (1:1:6).....................	16.75	5.00	17.59	0.35	0.18	4.49	0.38	22.46	m	25.27
hip tiles; angular......................	16.75	5.00	17.59	0.35	0.18	4.49	0.38	22.46	m	25.27
hip tiles; bonnet pattern; bedding and pointing in coloured cement-lime mortar (1:1:6).............	16.75	5.00	17.59	0.35	0.18	4.49	0.38	22.46	m	25.27
valley tiles; angular...................	16.75	5.00	17.59	0.50	0.25	6.37	0.38	24.34	m	27.38
Roof coverings; clayware machine made plain tiles B.S.402, red, 265 x 165mm; fixing every fourth course with two galvanised nails per tile to 65mm lap; 50 x 19mm pressure impregnated softwood battens fixed with galvanised nails; underlay B.S.747 type 1F reinforced bitumen felt; 150mm laps; fixing with galvanised steel clout nails										
Pitched 50 degrees from horizontal										
generally..............................	23.96	5.00	25.16	0.70	0.35	8.92	6.85	40.93	m2	46.04
Roof coverings; clayware machine made plain tiles B.S.402, red, 265 x 165mm; fixing every course with two galvanised nails per tile to 65mm lap; 38 x 19mm pressure impregnated softwood battens fixed with galvanised nails; underlay B.S.747 type 1F reinforced bitumen felt; 150mm laps; fixing with galvanised steel clout nails										
Pitched 50 degrees from horizontal										
generally..............................	23.96	5.00	25.16	0.85	0.43	10.86	7.02	43.04	m2	48.42
Roof coverings; clayware machine made plain tiles B.S.402, red, 265 x 165mm; fixing every fourth course with two copper nails per tile to 65mm lap; 38 x19mm pressure impregnated softwood battens fixed with galvanised nails; underlay B.S.747 type 1F reinforced bitumen felt; 150mm laps; fixing with galvanised steel clout nails										
Pitched 50 degrees from horizontal										
generally..............................	23.96	5.00	25.16	0.70	0.35	8.92	6.48	40.56	m2	45.63
Roof coverings; clayware machine made plain tiles B.S.402, red, 265 x 165mm; fixing every course with two copper nails per tile to 65mm lap; 38 x 19mm pressure impregnated softwood battens fixed with galvanised nails; underlay B.S.747 type 1F reinforced bitumen felt; 150mm laps; fixing with galvanised steel clout nails										
Pitched 50 degrees from horizontal										
generally..............................	23.96	5.00	25.16	0.85	0.43	10.86	9.27	45.29	m2	50.95
Roof coverings; clayware machine made plain tiles B.S.402, red, 265 x 165mm; fixing every fourth course with two galvanised nails per tile to 85mm lap; 38 x 19mm pressure impregnated softwood battens fixed with galvanised nails; underlay B.S.747 type 1F reinforced bitumen felt; 150mm laps; fixing with galvanised steel clout nails										
Pitched 40 degrees from horizontal										
generally..............................	27.96	5.00	29.36	0.80	0.40	10.19	6.01	45.56	m2	51.26

CLADDING/COVERING

Labour hourly rates: (except Specialists) Craft Operatives £9.23 Labourer £7.02 Rates are national average prices. Refer to REGIONAL VARIATIONS for indicative levels of overall pricing in regions	MATERIALS			LABOUR				RATES		
	Del to Site £	Waste %	Material Cost £	Craft Optve Hrs	Lab Hrs	Labour Cost £	Sunds £	Nett Rate £	Unit	Gross Rate (+12.5%) £
H60: PLAIN ROOF TILING Cont.										
Wall coverings; clayware machine made plain tiles B.S. 402, red, 265 x 165mm;fixing every course with two galvanised nails per tile to 38mm lap; 38 x 19mm pressure impregnated softwood battens fixed with galvanised nails; underlay B.S. 747 type 1F reinforced bitumen felt; 150mm laps; fixing with galvanised steel clout nails										
Vertical										
generally..	25.61	5.00	26.89	0.85	0.43	10.86	6.37	44.12	m2	49.64
Roof coverings; clayware hand made plain tiles B.S.402, red, 265 x 165mm; fixing every fourth course with two galvanised nails per tile to 65mm lap; 38 x 19mm pressure impregnated softwood battens fixed with galvanised nails; underlay B.S.747 type 1F reinforced bitumen felt; 150mm laps; fixing with galvanised steel clout nails										
Pitched 50 degrees from horizontal										
generally..	39.69	5.00	41.67	0.72	0.36	9.17	5.73	56.58	m2	63.65
holes	-	-	-	0.58	0.29	7.39	-	7.39	nr	8.31
abutments; square	3.97	-	3.97	0.28	0.14	3.57	-	7.54	m	8.48
abutments; raking	5.67	-	5.67	0.42	0.21	5.35	-	11.02	m	12.40
double course at eaves; purpose made eaves tile ..	8.72	5.00	9.16	0.35	0.18	4.49	0.53	14.18	m	15.95
verges; bed and point in coloured cement-lime mortar (1:1:6)	4.36	5.00	4.58	0.40	0.20	5.10	0.38	10.05	m	11.31
verges; single extra undercloak course plain tiles; bed and point in coloured cement-lime mortar (1:1:6)	5.82	5.00	6.11	0.58	0.29	7.39	0.38	13.88	m	15.62
ridge tiles, half round; in 300mm effective lengths; butt jointed; bedding and pointing in coloured cement-lime mortar (1:1:6)	16.75	5.00	17.59	0.35	0.18	4.49	0.38	22.46	m	25.27
hip tiles, half round; in 300mm effective lengths; butt jointed; bedding and pointing in coloured cement-lime mortar (1:1:6)	16.75	5.00	17.59	0.35	0.18	4.49	0.38	22.46	m	25.27
hip tiles; angular	16.75	5.00	17.59	0.35	0.18	4.49	0.38	22.46	m	25.27
hip tiles; bonnet pattern; bedding and pointing in coloured cement-lime mortar (1:1:6)	16.75	5.00	17.59	0.35	0.18	4.49	0.38	22.46	m	25.27
valley tiles; angular	16.75	5.00	17.59	0.50	0.25	6.37	0.38	24.34	m	27.38
Roof coverings; clayware hand made plain tiles B.S.402, red, 265 x 165mm; fixing every fourth course with two galvanised nails per tile to 85mm lap; 38 x 19mm pressure impregnated softwood battens fixed with galvanised nails; underlay B.S.747 type 1F reinforced bitumen felt; 150mm laps; fixing with galvanised steel clout nails										
Pitched 40 degrees from horizontal										
generally..	46.31	5.00	48.63	0.80	0.40	10.19	6.01	64.83	m2	72.93
Fittings										
Hip irons galvanised mild steel; 32 x 3 x 380mm girth; scrolled; fixing with galvanised steel screws to timber..	2.01	5.00	2.11	0.12	0.06	1.53	-	3.64	nr	4.09
Lead soakers fixing only	-	-	-	0.35	0.18	4.49	-	4.49	nr	5.06
Roof coverings; Redland Plain granular faced tiles, 268 x 165mm; fixing every fifth course with two aluminium nails to each tile to 65mm lap; 32 x 19mm pressure impregnated softwood battens fixed with galvanised nails; underlay B.S.747 type 1F bitumen reinforced felt; 150mm laps; fixing with galvanised steel clout nails										
Pitched 40 degrees from horizontal										
generally..	17.92	5.00	18.82	0.70	0.35	8.92	4.78	32.51	m2	36.58
double course at eaves and nailing each tile with two aluminium nails	1.79	5.00	1.88	0.35	0.18	4.49	0.47	6.84	m	7.70
verges; Redland Dry Verge system with plain tiles and tile-and-a-half tiles in alternate courses, clips and aluminium nails	12.11	5.00	12.72	0.28	0.14	3.57	1.01	17.29	m	19.45
verges; plain tile undercloak; tiles and undercloak bedded and pointed in tinted cement mortar (1:3) and with plain tiles and tile-and-a-half tiles in alternate courses and aluminium nails	1.79	5.00	1.88	0.28	0.14	3.57	0.42	5.87	m	6.60
double course at top edges with clips and nailing each tile with two aluminium nails	1.81	5.00	1.90	0.34	0.17	4.33	2.05	8.28	m	9.32
ridge or hip tiles, Redland Plain angle; butt jointed; bedding and pointing in tinted cement mortar (1:3)	18.54	5.00	19.47	0.35	0.18	4.49	0.38	24.34	m	27.38
ridge or hip tiles, Redland half round; butt jointed; bedding and pointing in tinted cement mortar (1:3)	12.57	5.00	13.20	0.35	0.18	4.49	0.38	18.07	m	20.33
valley tiles , Redland Plain valley tiles	13.07	5.00	13.72	0.52	0.26	6.62	0.38	20.73	m	23.32

Labour hourly rates: (except Specialists) Craft Operatives £9.23 Labourer £7.02 Rates are national average prices. Refer to REGIONAL VARIATIONS for indicative levels of overall pricing in regions	MATERIALS			LABOUR				RATES		
	Del to Site £	Waste %	Material Cost £	Craft Optve Hrs	Lab Hrs	Labour Cost £	Sunds £	Nett Rate £	Unit	Gross Rate (+12.5%) £
H60: PLAIN ROOF TILING Cont.										
Roof coverings; Redland Plain granular faced tiles, 268 x 165mm; fixing every fifth course with two aluminium nails to each tile to 65mm lap; 32 x 25mm pressure impregnated softwood battens fixed with galvanised nails; underlay; B.S.747 type 1F bitumen reinforced felt; 150mm laps; fixing with galvanised steel clout nails										
Pitched 40 degrees from horizontal										
generally	17.92	5.00	18.82	0.70	0.35	8.92	-	27.73	m2	31.20
abutments; square	1.79	-	1.79	0.28	0.15	3.64	-	5.43	m	6.11
abutments; raking	2.57	-	2.57	0.43	0.21	5.44	-	8.01	m	9.01
abutments; curved to 3000mm radius	3.59	-	3.59	0.56	0.28	7.13	-	10.72	m	12.06
Roof coverings; Redland Plain granular faced tiles, 268 x 165mm; fixing each tile with two aluminium nails to 65mm lap; 32 x 19mm pressure impregnated softwood battens fixed with galvanised nails; underlay; B.S.747 type 1F bitumen reinforced felt; 150mm laps; fixing with galvanised steel clout nails										
Pitched 40 degrees from horizontal										
generally	17.92	5.00	18.82	0.83	0.42	10.61	6.87	36.30	m2	40.83
Roof coverings; Plain through coloured tiles, 268 x 165mm; fixing every fifth course with two aluminium nails to each tile to 65mm lap; 32 x 19mm pressure impregnated softwood battens fixed with galvanised nails; underlay; B.S.747 type 1F bitumen reinforced felt; 150mm laps; fixing with galvanised steel clout nails										
Pitched 40 degrees from horizontal										
generally	17.92	5.00	18.82	0.70	0.35	8.92	4.78	32.51	m2	36.58
double course at eaves and nailing each tile with two aluminium nails	1.79	5.00	1.88	0.34	0.17	4.33	0.47	6.68	m	7.52
verges; Redland Dry Verge system with plain tiles and tile-and-a-half tiles in alternate courses, clips and aluminium nails	12.11	5.00	12.72	0.28	0.14	3.57	1.01	17.29	m	19.45
verges; plain tile undercloak; tiles and undercloak bedded and pointed in tinted cement mortar (1:3) and with plain tiles and tile-and-a-half tiles in alternate courses and aluminium nails	1.79	5.00	1.88	0.28	0.14	3.57	0.42	5.87	m	6.60
double course at top edges with clips and nailing each tile with two aluminium nails	1.81	5.00	1.90	0.34	0.17	4.33	2.05	8.28	m	9.32
ridge or hip tiles, Redland Plain angle; butt jointed; bedding and pointing in tinted cement mortar (1:3)	18.54	5.00	19.47	0.35	0.18	4.49	0.38	24.34	m	27.38
ridge or hip tiles, Redland half round; butt jointed; bedding and pointing in tinted cement mortar (1:3)	12.57	5.00	13.20	0.35	0.18	4.49	0.38	18.07	m	20.33
valley tiles, Redland Plain valley tiles	13.07	5.00	13.72	0.52	0.26	6.62	0.38	20.73	m	23.32
Roof coverings; Redland Downland plain granular faced tiles, 268 x 165mm; fixing every fifth course with two aluminium nails to each tile to 65mm lap; 32 x 19mm pressure impregnated softwood battens fixed with galvanised nails; underlay; B.S.747 type 1F bitumen reinforced felt; 150mm laps; fixing with galvanised steel clout nails										
Pitched 40 degrees from horizontal										
generally	21.51	5.00	22.59	0.70	0.35	8.92	4.78	36.28	m2	40.82
double course at eaves and nailing each tile with two aluminium nails	2.15	5.00	2.26	0.34	0.17	4.33	0.47	7.06	m	7.94
verges; Redland Dry Verge system with plain tiles and tile-and-a-half tiles in alternate courses, clips and aluminium nails	12.29	5.00	12.90	0.28	0.14	3.57	1.01	17.48	m	19.67
verges; plain tile undercloak; tiles and undercloak bedded and pointed in tinted cement mortar (1:3) and with plain tiles and tile-and-a-half tiles in alternate courses and aluminium nails	2.15	5.00	2.26	0.28	0.14	3.57	0.42	6.24	m	7.03
double course at top edges with clips and nailing each tile with two aluminium nails	2.17	5.00	2.28	0.34	0.17	4.33	2.05	8.66	m	9.74
ridge or hip tiles, Redland Plain angle; butt jointed; bedding and pointing in tinted cement mortar (1:3)	18.54	5.00	19.47	0.35	0.18	4.49	0.38	24.34	m	27.38
ridge or hip tiles, Redland half round; butt jointed; bedding and pointing in tinted cement mortar (1:3)	12.57	5.00	13.20	0.35	0.18	4.49	0.38	18.07	m	20.33
valley tiles, Redland Plain valley tiles	11.01	5.00	11.56	0.52	0.26	6.62	0.38	18.57	m	20.89
Roof coverings; Redland Downland plain through coloured tiles, 268 x 165mm; fixing every fifth course with two aluminium nails to each tile to 65mm lap; 32 x 19mm pressure impregnated softwood battens fixed with galvanised nails; underlay; B.S.747 type 1F bitumen reinforced felt; 150mm laps; fixing with galvanised steel clout nails										
Pitched 40 degrees from horizontal										
generally	21.51	5.00	22.59	0.70	0.35	8.92	4.78	36.28	m2	40.82

CLADDING/COVERING

Labour hourly rates: (except Specialists) Craft Operatives £9.23 Labourer £7.02 Rates are national average prices. Refer to REGIONAL VARIATIONS for indicative levels of overall pricing in regions	MATERIALS			LABOUR				RATES		
	Del to Site £	Waste %	Material Cost £	Craft Optve Hrs	Lab Hrs	Labour Cost £	Sunds £	Nett Rate £	Unit	Gross Rate (+12.5%) £
H60: PLAIN ROOF TILING Cont.										
Roof coverings; Redland Downland plain through coloured tiles, 268 x 165mm; fixing every fifth course with two aluminium nails to each tile to 65mm lap; 32 x 19mm pressure impregnated softwood battens fixed with galvanised nails; underlay; B.S.747 type 1F bitumen reinforced felt; 150mm laps; fixing with galvanised steel clout nails Cont.										
Pitched 40 degrees from horizontal Cont.										
double course at eaves and nailing each tile with two aluminium nails	2.15	5.00	2.26	0.34	0.17	4.33	0.47	7.06	m	7.94
verges; Redland Dry Verge system with plain tiles and tile-and-a-half tiles in alternate courses, clips and aluminium nails	12.29	5.00	12.90	0.28	0.14	3.57	1.01	17.48	m	19.67
verges; plain tile undercloak; tiles and undercloak bedded and pointed in tinted cement mortar (1:3) and with plain tiles and tile-and-a-half tiles in alternate courses and aluminium nails	2.15	5.00	2.26	0.28	0.14	3.57	0.42	6.24	m	7.03
double course at top edges with clips and nailing each tile with two aluminium nails	2.17	5.00	2.28	0.34	0.17	4.33	2.05	8.66	m	9.74
ridge or hip tiles, Redland Plain angle; butt jointed; bedding and pointing in tinted cement mortar (1:3)	18.54	5.00	19.47	0.35	0.18	4.49	0.38	24.34	m	27.38
ridge or hip tiles, Redland half round; butt jointed; bedding and pointing in tinted cement mortar (1:3)	12.57	5.00	13.20	0.35	0.18	4.49	0.38	18.07	m	20.33
valley tiles, Redland Plain valley tiles	11.01	5.00	11.56	0.52	0.26	6.62	0.38	18.57	m	20.89
Fittings										
Ventilator tiles										
Red Vent ridge ventilation terminal 450mm long for										
half round ridge	52.20	5.00	54.81	1.45	0.73	18.51	0.65	73.97	nr	83.21
universal Delta ridge	64.01	5.00	67.21	1.68	0.84	21.40	1.17	89.78	nr	101.01
universal angle ridge	46.46	5.00	48.78	1.45	0.73	18.51	1.17	68.46	nr	77.02
Redvent eaves ventilators fixing with aluminium nails	9.14	5.00	9.60	0.56	0.28	7.13	0.03	16.76	m	18.86
Red Vent Thruvent tiles for										
Stonewold slates	35.33	5.00	37.10	0.56	0.28	7.13	0.56	44.79	nr	50.39
Delta tiles	34.54	5.00	36.27	0.56	0.28	7.13	0.56	43.96	nr	49.46
Regent tiles	34.54	5.00	36.27	0.56	0.28	7.13	0.56	43.96	nr	49.46
Grovebury double pantiles	34.54	5.00	36.27	0.56	0.28	7.13	0.56	43.96	nr	49.46
Norfolk pantiles	34.54	5.00	36.27	0.56	0.28	7.13	0.56	43.96	nr	49.46
Redland 49 tiles	34.54	5.00	36.27	0.56	0.28	7.13	0.56	43.96	nr	49.46
Renown tiles	34.54	5.00	36.27	0.56	0.28	7.13	0.56	43.96	nr	49.46
Redland 50 double roman tiles	34.54	5.00	36.27	0.56	0.28	7.13	0.56	43.96	nr	49.46
Redland plain tiles	52.31	5.00	54.93	0.56	0.28	7.13	0.27	62.33	nr	70.12
Redland Downland plain tiles	47.59	5.00	49.97	0.56	0.28	7.13	0.27	57.37	nr	64.55
Red line ventilation tile complete with underlay seal and fixing clips for										
Stonewold slates	43.65	5.00	45.83	1.12	0.56	14.27	0.56	60.66	nr	68.24
Regent tiles	43.65	5.00	45.83	1.12	0.56	14.27	0.56	60.66	nr	68.24
Grovebury double pantiles	43.65	5.00	45.83	1.12	0.56	14.27	0.56	60.66	nr	68.24
Renown tiles	43.65	5.00	45.83	1.12	0.56	14.27	0.56	60.66	nr	68.24
Redland 50 double roman tiles	43.65	5.00	45.83	1.12	0.56	14.27	0.56	60.66	nr	68.24
Gas terminals										
Gas Flue ridge terminal Mark III, 450mm long with sealing gasket and fixing brackets										
half round ridge	67.95	5.00	71.35	1.68	0.84	21.40	0.65	93.40	nr	105.08
universal Delta ridge	78.64	5.00	82.57	1.93	0.95	24.48	3.03	110.08	nr	123.85
universal angle ridge	61.09	5.00	64.14	1.68	0.84	21.40	3.03	88.58	nr	99.65
Gas Flue ridge terminal Mark III, 450mm long with sealing gasket and fixing brackets with 150mm R type adaptor for										
half round ridge	67.95	5.00	71.35	1.68	0.84	21.40	0.65	93.40	nr	105.08
universal Delta ridge	79.76	5.00	83.75	1.93	0.95	24.48	3.03	111.26	nr	125.17
universal angle ridge	62.21	5.00	65.32	1.68	0.84	21.40	3.03	89.75	nr	100.97
extra for extension adaptor and gasket	30.49	5.00	32.01	0.35	0.17	4.42	0.65	37.09	nr	41.72
Hip irons										
galvanised mild steel; 32 x 3 x 380mm girth; scrolled; fixing with galvanised steel screws to timber	2.01	-	2.01	0.11	0.06	1.44	-	3.45	nr	3.88
Lead soakers										
fixing only	-	-	-	0.34	0.17	4.33	-	4.33	nr	4.87

Labour hourly rates: (except Specialists) Craft Operatives £9.23 Labourer £7.02 Rates are national average prices. Refer to REGIONAL VARIATIONS for indicative levels of overall pricing in regions	MATERIALS			LABOUR				RATES		
	Del to Site £	Waste %	Material Cost £	Craft Optve Hrs	Lab Hrs	Labour Cost £	Sunds £	Nett Rate £	Unit	Gross Rate (+12.5%) £

H61: FIBRE CEMENT SLATING

Quantities required for 1m2 of slating

Slates to 70mm lap
  400 x 240mm, 25.3nr
Slates to 102mm lap
  600 x 300mm, 13nr
  500 x 250mm, 19.5nr

Roof coverings; asbestos-free cement slates, 600 x 300mm; centre fixing with copper nails and copper disc rivets to 102mm lap; 38 x 19mm pressure impregnated softwood battens fixed with galvanised nails; underlay; B.S.747 type 1F bitumen reinforced felt; 150mm laps; fixing with galvanised steel clout nails

	Del to Site	Waste	Material Cost	Craft Optve	Lab	Labour Cost	Sunds	Nett Rate	Unit	Gross Rate
Pitched 30 degrees from horizontal										
generally	20.17	5.00	21.18	0.60	0.30	7.64	4.15	32.97	m2	37.09
holes	-	-	-	0.58	0.29	7.39	-	7.39	nr	8.31
abutments; square	2.28	5.00	2.39	0.23	0.12	2.97	-	5.36	m	6.03
abutments; raking	3.04	-	3.04	0.28	0.14	3.57	-	6.61	m	7.43
abutments; curved to 3000mm radius	4.04	-	4.04	0.35	0.18	4.49	-	8.53	m	9.60
double course at eaves	5.43	5.00	5.70	0.17	0.09	2.20	0.70	8.60	m	9.68
verges; slate undercloak and point in cement mortar (1:3)	3.11	5.00	3.27	0.09	0.05	1.18	0.56	5.01	m	5.63
ridges or hips; asbestos free cement; fixing with nails	20.10	5.00	21.11	0.23	0.12	2.97	0.38	24.45	m	27.51

Roof coverings; asbestos-free cement slates, 600 x 300mm; centre fixing with copper nails and copper disc rivets to 102mm lap; 50 x 25mm pressure impregnated softwood battens fixed with galvanised nails; underlay; B.S.747 type 1F bitumen reinforced felt; 150mm laps; fixing with galvanised steel clout nails

	Del to Site	Waste	Material Cost	Craft Optve	Lab	Labour Cost	Sunds	Nett Rate	Unit	Gross Rate
Pitched 30 degrees from horizontal										
generally	20.17	5.00	21.18	0.64	0.32	8.15	4.96	34.29	m2	38.58

Roof coverings; asbestos-free cement slates, 600 x 300mm; centre fixing with copper nails and copper disc rivets to 102mm lap (close boarding on rafters included elsewhere); underlay; B.S.747 type 1B underslating felt; 150mm laps; fixing with galvanised steel clout nails

	Del to Site	Waste	Material Cost	Craft Optve	Lab	Labour Cost	Sunds	Nett Rate	Unit	Gross Rate
Pitched 30 degrees from horizontal										
generally	20.17	5.00	21.18	0.48	0.24	6.12	2.62	29.91	m2	33.65

Roof coverings; asbestos-free cement slates, 500 x 250mm; centre fixing with copper nails and copper disc rivets to 102mm lap; 38 x 19mm pressure impregnated softwood battens fixed with galvanised nails; underlay; B.S.747 type 1F bitumen reinforced felt; 150mm laps; fixing with galvanised steel clout nails

	Del to Site	Waste	Material Cost	Craft Optve	Lab	Labour Cost	Sunds	Nett Rate	Unit	Gross Rate
Pitched 30 degrees from horizontal										
generally	22.75	5.00	23.89	0.67	0.34	8.57	5.02	37.48	m2	42.16
holes	-	-	-	0.58	0.29	7.39	-	7.39	nr	8.31
abutments; square	2.28	-	2.28	0.23	0.12	2.97	-	5.25	m	5.90
abutments; raking	3.43	-	3.43	0.35	0.18	4.49	-	7.92	m	8.91
abutments; curved to 3000mm radius	4.56	-	4.56	0.46	0.23	5.86	-	10.42	m	11.72
double course at eaves	4.67	5.00	4.90	0.21	0.11	2.71	0.74	8.35	m	9.40
verges; slate undercloak and point in cement mortar (1:3)	2.91	5.00	3.06	0.12	0.06	1.53	0.61	5.19	m	5.84
ridges or hips; asbestos free cement; fixing with nails	20.10	5.00	21.11	0.23	0.12	2.97	0.38	24.45	m	27.51

Roof coverings; asbestos-free cement slates, 500 x 250mm; centre fixing with copper nails and copper disc rivets to 102mm lap; 50 x 25mm pressure impregnated softwood battens fixed with galvanised nails; underlay; B.S.747 type 1F bitumen reinforced felt; 150mm laps; fixing with galvanised steel clout nails

	Del to Site	Waste	Material Cost	Craft Optve	Lab	Labour Cost	Sunds	Nett Rate	Unit	Gross Rate
Pitched 30 degrees from horizontal										
generally	22.75	5.00	23.89	0.72	0.36	9.17	6.03	39.09	m2	43.98

Roof coverings; asbestos-free cement slates, 400 x 200mm; centre fixing with copper nails and copper disc rivets to 102mm lap; 38 x 19mm pressure impregnated softwood battens fixed with galvanised nails; underlay; B.S.747 type 1F bitumen reinforced felt; 150mm laps; fixing with galvanised steel clout nails

	Del to Site	Waste	Material Cost	Craft Optve	Lab	Labour Cost	Sunds	Nett Rate	Unit	Gross Rate
Pitched 30 degrees from horizontal										
generally	23.91	5.00	25.11	0.71	0.36	9.08	6.42	40.61	m2	45.68
holes	-	-	-	0.58	0.29	7.39	-	7.39	nr	8.31
abutments; square	2.39	-	2.39	0.29	0.15	3.73	-	6.12	m	6.88
abutments; raking	3.58	-	3.58	0.44	0.22	5.61	-	9.19	m	10.33
abutments; curved to 3000mm radius	4.77	-	4.77	0.58	0.29	7.39	0.83	12.16	m	13.68
double course at eaves	3.74	5.00	3.93	0.23	0.97	8.93	0.83	13.69	m	15.40

CLADDING/COVERING

Labour hourly rates: (except Specialists) Craft Operatives £9.23 Labourer £7.02 Rates are national average prices. Refer to REGIONAL VARIATIONS for indicative levels of overall pricing in regions	MATERIALS			LABOUR				RATES		
	Del to Site	Waste	Material Cost	Craft Optve	Lab	Labour Cost	Sunds	Nett Rate	Unit	Gross Rate (+12.5%)
	£	%	£	Hrs	Hrs	£	£	£		£
H61: FIBRE CEMENT SLATING Cont.										
Roof coverings; asbestos-free cement slates, 400 x 200mm; centre fixing with copper nails and copper disc rivets to 102mm lap; 38 x 19mm pressure impregnated softwood battens fixed with galvanised nails; underlay; B.S.747 type 1F bitumen reinforced felt; 150mm laps; fixing with galvanised steel clout nails Cont.										
Pitched 30 degrees from horizontal Cont.										
verges; slate undercloak and point in cement mortar (1:3)	2.25	5.00	2.36	0.14	0.07	1.78	0.65	4.80	m	5.40
ridges or hips; asbestos free cement; fixing with nails	20.10	5.00	21.11	0.23	0.12	2.97	0.38	24.45	m	27.51
Roof coverings; asbestos-free cement slates, 400 x 200mm; centre fixing with copper nails and copper disc rivets to 102mm lap; 50 x 25mm pressure impregnated softwood battens fixed with galvanised nails; underlay; B.S.747 type 1F bitumen reinforced felt; 150mm laps; fixing with galvanised steel clout nails										
Pitched 30 degrees from horizontal										
generally	23.91	5.00	25.11	0.80	0.40	10.19	7.64	42.94	m2	48.30
Fittings										
Hip irons										
galvanised mild steel; 32 x 3 x 380mm girth; scrolled; fixing with galvanised steel screws to timber	2.00	5.00	2.10	0.11	0.06	1.44	-	3.54	nr	3.98
Lead soakers										
fixing only	-	-	-	0.34	0.17	4.33	-	4.33	nr	4.87
H62: NATURAL SLATING										
Quantities required for 1m2 of slating										
Slates to 75mm lap 610 x 305mm, 12.25nr 510 x 255mm, 18nr 405 x 205mm, 30nr Roof coverings; blue/grey slates, 610 x 305mm, 6.5mm thick; fixing with slate nails to 75mm lap; 38 x 19mm pressure impregnated softwood battens fixed with galvanised nails; underlay; B.S.747 type 1F bitumen reinforced felt; 150mm laps; fixing with galvanised steel clout nails										
Pitched 30 degrees from horizontal										
generally	57.61	5.00	60.49	0.58	0.29	7.39	3.12	71.00	m2	79.87
holes	-	-	-	0.58	0.29	7.39	-	7.39	nr	8.31
abutments; square	5.76	-	5.76	0.17	0.09	2.20	-	7.96	m	8.96
abutments; raking	8.64	-	8.64	0.26	0.13	3.31	-	11.95	m	13.45
double course at eaves	14.63	5.00	15.36	0.16	0.08	2.04	0.74	18.14	m	20.41
verges; slate undercloak and point in cement mortar (1:3)	9.41	5.00	9.88	0.09	0.05	1.18	0.43	11.49	m	12.93
Roof coverings; blue/grey slates, 610 x 305mm, 6.5mm thick; fixing with aluminium nails to 75mm lap; 38 x 19mm pressure impregnated softwood battens fixed with galvanised nails; underlay; B.S.747 type 1F bitumen reinforced felt; 150mm laps; fixing with galvanised steel clout nails										
Pitched 30 degrees from horizontal										
generally	57.61	5.00	60.49	0.58	0.29	7.39	3.67	71.55	m2	80.49
Roof coverings; blue/grey slates, 610 x 305mm, 6.5mm thick; fixing with copper nails to 75mm laps; 38 x 19mm pressure impregnated softwood battens fixed with galvanised nails; underlay; B.S.747 type 1F bitumen reinforced felt; 150mm laps; fixing with galvanised steel clout nails										
Pitched 30 degrees from horizontal										
generally	57.61	5.00	60.49	0.58	0.29	7.39	3.95	71.83	m2	80.81
Roof coverings; blue/grey slates, 610 x 305mm, 6.5mm thick; fixing with slate nails to 75mm lap; 50 x 25mm pressure impregnated softwood battens fixed with galvanised nails; underlay; B.S.747 type 1F bitumen reinforced felt; 150mm laps; fixing with galvanised steel clout nails										
Pitched 30 degrees from horizontal										
generally	57.61	5.00	60.49	0.62	0.31	7.90	3.93	72.32	m2	81.36

Labour hourly rates: (except Specialists) Craft Operatives £9.23 Labourer £7.02 Rates are national average prices. Refer to REGIONAL VARIATIONS for indicative levels of overall pricing in regions	MATERIALS			LABOUR				RATES		
	Del to Site £	Waste %	Material Cost £	Craft Optve Hrs	Lab Hrs	Labour Cost £	Sunds £	Nett Rate £	Unit	Gross Rate (+12.5%) £

H62: NATURAL SLATING Cont.

Roof coverings; blue/grey slates, 610 x 305mm, 6.5mm thick; fixing with slate nails to 75mm lap; 50 x 25mm pressure impregnated softwood counterbattens at 1067mm centres and 38 x 19mm pressure impregnated softwood battens fixed with galvanised nails; underlay; B.S.747 type 1B underslating felt; 150mm laps; fixing with galvanised steel clout nails

	Del to Site	Waste	Material Cost	Craft Optve	Lab	Labour Cost	Sunds	Nett Rate	Unit	Gross Rate
Pitched 30 degrees from horizontal generally	57.61	5.00	60.49	0.62	0.31	7.90	3.65	72.04	m2	81.04

Roof coverings; blue/grey slates, 510 x 255mm, 6.5mm thick; fixing with slate nails to 75mm lap; 38 x 19mm pressure impregnated softwood battens fixed with galvanised nails; underlay; B.S.747 type 1F bitumen reinforced felt; 150mm laps; fixing with galvanised steel clout nails

	Del to Site	Waste	Material Cost	Craft Optve	Lab	Labour Cost	Sunds	Nett Rate	Unit	Gross Rate
Pitched 30 degrees from horizontal										
generally	38.08	5.00	39.98	0.67	0.34	8.57	3.67	52.22	m2	58.75
holes	-	-	-	0.58	0.29	7.39	-	7.39	nr	8.31
abutments; square	3.81	-	3.81	0.23	0.12	2.97	-	6.78	m	7.62
abutments; raking	5.72	-	5.72	0.35	0.18	4.49	-	10.21	m	11.49
double course at eaves	8.37	5.00	8.79	0.20	0.10	2.55	0.75	12.09	m	13.60
verges; slate undercloak and point in cement mortar (1:3)	5.23	5.00	5.49	0.12	0.06	1.53	0.44	7.46	m	8.39

Roof coverings; blue/grey slates, 510 x 255mm, 6.5mm thick; fixing with aluminium nails to 75mm lap; 38 x 19mm pressure impregnated softwood battens fixed with galvanised nails; underlay; B.S.747 type 1F bitumen reinforced felt; 150mm laps; fixing with galvanised steel clout nails

	Del to Site	Waste	Material Cost	Craft Optve	Lab	Labour Cost	Sunds	Nett Rate	Unit	Gross Rate
Pitched 30 degrees from horizontal generally	38.08	5.00	39.98	-	0.34	2.39	4.49	46.86	m2	52.72

Roof coverings; blue/grey slates, 510 x 255mm, 6.5mm thick; fixing with copper nails to 75mm lap; 38 x 19mm pressure impregnated softwood battens fixed with galvanised nails; underlay; B.S.747 type 1F bitumen reinforced felt; 150mm laps; fixing with galvanised steel clout nails

	Del to Site	Waste	Material Cost	Craft Optve	Lab	Labour Cost	Sunds	Nett Rate	Unit	Gross Rate
Pitched 30 degrees from horizontal generally	38.08	5.00	39.98	-	0.34	2.39	4.89	47.26	m2	53.17

Roof coverings; blue/grey slates, 510 x 255mm, 6.5mm thick; fixing with slate nails to 75mm lap; 50 x 25mm pressure impregnated softwood battens fixed with galvanised nails; underlay; B.S.747 type 1F bitumen reinforced felt; 150mm laps; fixing with galvanised steel clout nails

	Del to Site	Waste	Material Cost	Craft Optve	Lab	Labour Cost	Sunds	Nett Rate	Unit	Gross Rate
Pitched 30 degrees from horizontal generally	38.08	5.00	39.98	0.72	0.36	9.17	4.68	53.84	m2	60.57

Roof coverings; blue/grey slates, 405 x 205mm, 6.5mm thick; fixing with slate nails to 75mm lap; 38 x 19mm pressure impregnated softwood battens fixed with galvanised nails; underlay; B.S.747 type 1F bitumen reinforced felt; 150mm laps; fixing with galvanised steel clout nails

	Del to Site	Waste	Material Cost	Craft Optve	Lab	Labour Cost	Sunds	Nett Rate	Unit	Gross Rate
Pitched 30 degrees from horizontal										
generally	29.36	5.00	30.83	0.80	0.40	10.19	4.49	45.51	m2	51.20
holes	-	-	-	0.58	0.29	7.39	-	7.39	nr	8.31
abutments; square	2.94	-	2.94	0.29	0.15	3.73	-	6.67	m	7.50
abutments; raking	4.41	-	4.41	0.44	0.22	5.61	-	10.02	m	11.27
double course at eaves	4.89	5.00	5.13	0.23	0.12	2.97	0.39	8.49	m	9.55
verges; slate undercloak and point in cement mortar (1:3)	2.94	5.00	3.09	0.14	0.07	1.78	0.45	5.32	m	5.99

Roof coverings; blue/grey slates, 405 x 205mm, 6.5mm thick; fixing with aluminium nails to 75mm lap; 38 x 19mm pressure impregnated softwood battens fixed with galvanised nails; underlay; B.S.747 type 1F bitumen reinforced felt; 150mm laps; fixing with galvanised steel clout nails

	Del to Site	Waste	Material Cost	Craft Optve	Lab	Labour Cost	Sunds	Nett Rate	Unit	Gross Rate
Pitched 30 degrees from horizontal generally	29.36	5.00	30.83	0.80	0.40	10.19	5.84	46.86	m2	52.72

Roof coverings; blue/grey slates, 405 x205mm, 6.5mm thick; fixing with copper nails to 75mmlap; 38 x 19mm pressure impregnated softwood battens fixed with galvanised nails; underlay; B.S.747 type 1F bitumen reinforced felt; 150mm laps; fixing with galvanised steel clout nails

	Del to Site	Waste	Material Cost	Craft Optve	Lab	Labour Cost	Sunds	Nett Rate	Unit	Gross Rate
Pitched 30 degrees from horizontal generally	29.36	5.00	30.83	0.80	0.40	10.19	6.51	47.53	m2	53.47

CLADDING/COVERING

Labour hourly rates: (except Specialists) Craft Operatives £9.23 Labourer £7.02 Rates are national average prices. Refer to REGIONAL VARIATIONS for indicative levels of overall pricing in regions	MATERIALS			LABOUR				RATES		
	Del to Site £	Waste %	Material Cost £	Craft Optve Hrs	Lab Hrs	Labour Cost £	Sunds £	Nett Rate £	Unit	Gross Rate (+12.5%) £
H62: NATURAL SLATING Cont.										
Roof coverings; blue/grey slates, 405 x 205mm, 6.5mm thick; fixing with slate nails to 75mm lap; 50 x 25mm pressure impregnated softwood battens fixed with galvanised nails; underlay; B.S.747 type 1F bitumen reinforced felt; 150mm laps; fixing with galvanised steel clout nails										
Pitched 30 degrees from horizontal										
generally	29.36	5.00	30.83	0.87	0.44	11.12	5.70	47.65	m2	53.60
Roof coverings; Burlington green slates, best random fixed in diminishing courses with alloy nails to 75mm lap; 38 x 19mm pressure impregnated softwood battens fixed with galvanised nails; underlay; B.S.747 type 1F bitumen reinforced felt; 150mm laps; fixing with galvanised steel clout nails										
Pitched 30 degrees from horizontal										
generally	115.46	5.00	121.23	0.86	0.43	10.96	5.13	137.32	m2	154.48
abutments; square	11.54	-	11.54	0.23	0.12	2.97	-	14.51	m	16.32
abutments; raking	17.31	-	17.31	0.35	0.18	4.49	0.62	22.42	m	25.23
double course at eaves	25.22	5.00	26.48	0.23	0.12	2.97	0.97	30.42	m	34.22
verges; slate undercloak and point in cement mortar (1:3)	16.82	5.00	17.66	0.23	0.12	2.97	0.59	21.22	m	23.87
hips, valleys and angles; close mitred	19.62	5.00	20.60	0.40	0.20	5.10	0.62	26.32	m	29.61
Fittings										
Hip irons galvanised mild steel; 32 x 3 x 380mm girth; scrolled; fixing with galvanised steel screws to timber	2.01	5.00	2.11	0.11	0.06	1.44	-	3.55	nr	3.99
Lead soakers fixing only	-	-	-	0.34	0.17	4.33	-	4.33	nr	4.87
H63: RECONSTRUCTED STONE SLATING/ TILING										
Roof coverings; Redland Cambrian interlocking riven textured slates, 300 x 336mm; fixing every slate with two stainless steel nails and one stainless steel clip to 50mm laps; 38 x 25mm pressure impregnated softwood battens fixed with galvanised nails; underlay; B.S.747 type 1F bitumen reinforced felt; 150mm laps; fixing with galvanised steel clout nails										
Pitched 40 degrees from horizontal										
generally	26.36	5.00	27.68	0.54	0.27	6.88	5.02	39.58	m2	44.52
abutments; square	2.63	-	2.63	0.11	0.06	1.44	-	4.07	m	4.57
abutments; raking	3.95	-	3.95	0.17	0.09	2.20	-	6.15	m	6.92
abutments; curved to 3000mm radius	5.28	-	5.28	0.23	0.12	2.97	-	8.25	m	9.28
supplementary fixing at eaves with stainless steel eaves clip to each slate	0.54	5.00	0.57	0.05	0.03	0.67	-	1.24	m	1.39
verges; 150 x 6mm fibre cement undercloak; slates and undercloak bedded and pointed in tinted cement mortar (1:3) with full slates and slate and a half slates in alternate courses and stainless steel verge clips	5.93	5.00	6.23	0.23	0.12	2.97	4.20	13.39	m	15.07
verges; 150 x 6mm fibre cement undercloak; slates and undercloak bedded and pointed in tinted cement mortar (1:3) with verge slates and slate and a half verge slates in alternate courses and stainless steel verge clips	5.93	5.00	6.23	0.23	0.12	2.97	4.20	13.39	m	15.07
ridge or hip tiles, Redland third round; butt jointed; bedding and pointing in tinted cement mortar (1:3)	8.58	5.00	9.01	0.23	0.12	2.97	0.38	12.35	m	13.90
ridge or hip tiles, Redland half round; butt jointed; bedding and pointing in tinted cement mortar (1:3)	8.58	5.00	9.01	0.23	0.12	2.97	0.38	12.35	m	13.90
ridge or hip tiles, Redland universal angle; butt jointed; bedding and pointing in tinted cement mortar (1:3)	9.16	5.00	9.62	0.23	0.12	2.97	0.38	12.96	m	14.58
ridge tiles, Redland Dry Ridge system with half round ridge tiles; fixing with stainless steel batten straps and ring shanked fixing nails with neoprene washers and sleeves; polypropylene ridge seals and uPVC profile filler units	27.03	5.00	28.38	0.45	0.23	5.77	-	34.15	m	38.42
ridge tiles, Redland Dry Ridge system with universal angle ridge tiles; fixing with stainless steel batten straps and ring shanked fixing nails with neoprene washers and sleeves; polypropylene ridge seals and uPVC profile filler units	27.61	5.00	28.99	0.45	0.23	5.77	-	34.76	m	39.10
ridge tiles, Redland Dry Vent Ridge system with half round ridge tiles; fixing with stainless steel batten straps and ring shanked nails with neoprene washers and sleeves; polypropylene ridge seals, PVC air flow control units and uPVC ventilated profile filler units	27.03	5.00	28.38	0.45	0.23	5.77	-	34.15	m	38.42
ridge tiles, Redland Dry Vent Ridge system with Universal angle ridge tiles; fixing with stainless steel batten straps and ring shanked nails with neoprene washers and sleeves; polypropylene ridge seals, PVC air flow control units and uPVC ventilated profile filler units	27.61	5.00	28.99	0.45	0.23	5.77	-	34.76	m	39.10

Labour hourly rates: (except Specialists) Craft Operatives £9.23 Labourer £7.02 Rates are national average prices. Refer to REGIONAL VARIATIONS for indicative levels of overall pricing in regions	MATERIALS			LABOUR				RATES		
	Del to Site £	Waste %	Material Cost £	Craft Optve Hrs	Lab Hrs	Labour Cost £	Sunds £	Nett Rate £	Unit	Gross Rate (+12.5%) £

H63: RECONSTRUCTED STONE SLATING/ TILING Cont.

Roof coverings; Redland Cambrian interlocking riven textured slates, 300 x 336mm; fixing every slate with two stainless steel nails and one stainless steel clip to 50mm laps; 38 x 25mm pressure impregnated softwood battens fixed with galvanised nails; underlay; B.S.747 type 1F bitumen reinforced felt; 150mm laps; fixing with galvanised steel clout nails Cont.

Pitched 40 degrees from horizontal Cont.

	Del to Site	Waste	Material Cost	Craft Optve	Lab	Labour Cost	Sunds	Nett Rate	Unit	Gross Rate
Monoridge filler units in conjunction with top tile, bed tile in nonsetting mastic sealant and screw on through filler unit with screws, washers and caps	38.57	5.00	40.50	0.12	0.12	1.95	–	42.45	m	47.75
valley tiles, Redland Universal valley troughs; laid with 100mm laps	24.37	5.00	25.59	0.28	0.28	4.55	0.38	30.52	m	34.33

Roof coverings; Redland Cambrian interlocking riven textured slates, 300 x 336mm; fixing every slate with two stainless steel nails and one stainless steel clip to 90mm lap; 38 x 25mm pressure impregnated softwood battens fixed with galvanised nails; underlay; B.S.747 type 1F bitumen reinforced felt; 150mm laps; fixing with galvanised steel clout nails

Pitched 40 degrees from horizontal

	Del to Site	Waste	Material Cost	Craft Optve	Lab	Labour Cost	Sunds	Nett Rate	Unit	Gross Rate
generally	31.44	5.00	33.01	0.62	0.31	7.90	6.02	46.93	m2	52.80
abutments; square	3.14	–	3.14	0.11	0.06	1.44	–	4.58	m	5.15
abutments; raking	4.71	–	4.71	0.17	0.09	2.20	–	6.91	m	7.77
abutments; curved to 3000mm radius	6.29	–	6.29	0.23	0.12	2.97	–	9.26	m	10.41

Fittings

Ventilator tiles

	Del to Site	Waste	Material Cost	Craft Optve	Lab	Labour Cost	Sunds	Nett Rate	Unit	Gross Rate
Red Vent ridge ventilation terminal 450mm long for half round ridge	52.20	5.00	54.81	1.45	0.73	18.51	0.65	73.97	nr	83.21
universal angle ridge	52.09	5.00	54.69	1.45	0.73	18.51	0.65	73.85	nr	83.08
Red Vent eaves ventilators; fixing with stainless steel nails	12.68	5.00	13.31	0.45	0.23	5.77	0.65	19.73	m	22.20
Redland Cambrian Thruvent interlocking slate complete with weather cap, underlay seal and fixing clips	42.19	5.00	44.30	0.56	0.28	7.13	0.65	52.08	nr	58.59

Gas terminals

	Del to Site	Waste	Material Cost	Craft Optve	Lab	Labour Cost	Sunds	Nett Rate	Unit	Gross Rate
Gas Flue ridge terminal Mark III, 450mm long with sealing gasket and fixing brackets half round ridge	67.95	5.00	71.35	1.50	0.75	19.11	0.65	91.11	nr	102.50
Gas Flue ridge terminal Mark III, 450mm long with sealing gasket and fixing brackets with 150mm R type adaptor for half round ridge	67.95	5.00	71.35	1.50	0.75	19.11	31.14	121.60	nr	136.80

Hip irons

	Del to Site	Waste	Material Cost	Craft Optve	Lab	Labour Cost	Sunds	Nett Rate	Unit	Gross Rate
galvanised mild steel; 32 x 3 x 380mm girth; scrolled; fixing with galvanised steel screws to timber	2.01	5.00	2.11	0.11	0.06	1.44	–	3.55	nr	3.99

Lead soakers

	Del to Site	Waste	Material Cost	Craft Optve	Lab	Labour Cost	Sunds	Nett Rate	Unit	Gross Rate
fixing only	–	–	–	0.34	0.17	4.33	–	4.33	nr	4.87

Fittings

	Del to Site	Waste	Material Cost	Craft Optve	Lab	Labour Cost	Sunds	Nett Rate	Unit	Gross Rate
Timloc uPVC Mark 3 eaves ventilators; reference 1126/40 fixing with nails to timber at 400mm centres 330mm wide	3.15	5.00	3.31	0.28	0.14	3.57	0.29	7.16	m	8.06
Timloc uPVC Mark 3 eaves ventilators; reference 1123 fixing with nails to timber at 600mm centres 330mm wide	2.34	5.00	2.46	0.22	0.11	2.80	0.29	5.55	m	6.24
Timlock uPVC soffit ventilators; reference 1137, 10mm airflow; fixing with screws to timber type C	1.87	5.00	1.96	0.17	0.06	1.99	0.11	4.06	m	4.57
Timlock uPVC Mark 2 eaves ventilators; reference 1122; fitting between trusses at 600mm centres 300mm girth	2.44	5.00	2.56	0.22	0.11	2.80	0.29	5.65	m	6.36
Timlock polypropylene over-fascia ventilators; reference 3011; fixing with screws to timber to top of fascia	3.34	5.00	3.51	0.22	0.11	2.80	0.18	6.49	m	7.30

H65: SINGLE LAP ROOF TILING

Roof coverings; Redland Stonewold through coloured slates, 430 x 380mm; fixing every fourth course with galvanised nails to 75mm lap; 38 x 22mm pressure impregnated softwood battens fixed with galvanised nails; underlay; B.S.747 type 1F reinforced bitumen felt; 150mm laps; fixing with galvanised steel clout nails

	Del to Site	Waste	Material Cost	Craft Optve	Lab	Labour Cost	Sunds	Nett Rate	Unit	Gross Rate
Pitched 40 degrees from horizontal generally	11.33	5.00	11.90	0.48	0.24	6.12	3.27	21.28	m2	23.94

Labour hourly rates: (except Specialists) Craft Operatives £9.23 Labourer £7.02 Rates are national average prices. Refer to REGIONAL VARIATIONS for indicative levels of overall pricing in regions	MATERIALS			LABOUR				RATES		
	Del to Site	Waste	Material Cost	Craft Optve	Lab	Labour Cost	Sunds	Nett Rate	Unit	Gross Rate (+12.5%)
	£	%	£	Hrs	Hrs	£	£	£		£

H65: SINGLE LAP ROOF TILING Cont.

Roof coverings; Redland Stonewold through coloured slates, 430 x 380mm; fixing every fourth course with galvanised nails to 75mm lap; 38 x 22mm pressure impregnated softwood battens fixed with galvanised nails; underlay; B.S.747 type 1F reinforced bitumen felt; 150mm laps; fixing with galvanised steel clout nails Cont.

Pitched 40 degrees from horizontal Cont.

Description	Del to Site £	Waste %	Material Cost £	Craft Optve Hrs	Lab Hrs	Labour Cost £	Sunds £	Nett Rate £	Unit	Gross Rate (+12.5%) £
abutments; square	1.14	-	1.14	0.12	0.06	1.53	-	2.67	m	3.00
abutments; raking	1.70	-	1.70	0.17	0.09	2.20	-	3.90	m	4.39
abutments; curved to 3000mm radius	2.26	-	2.26	0.23	0.12	2.97	-	5.23	m	5.88
supplementary fixing eaves course with one clip per slate	0.41	-	0.41	0.03	0.02	0.42	-	0.83	m	0.93
verges; Redland Dry Verge system with half slates, full slates and clips	15.36	5.00	16.13	0.28	0.14	3.57	2.03	21.73	m	24.44
verges; Redland Dry Verge system with half slates, verge slates and clips	15.36	5.00	16.13	0.28	0.14	3.57	2.03	21.73	m	24.44
verges; 150 x 6mm fibre cement undercloak, slates and undercloak bedded and pointed in tinted cement mortar (1:3) and with half slates, full slates and clips	4.15	5.00	4.36	0.23	0.12	2.97	4.20	11.52	m	12.96
verges; 150 x 6mm fibre cement undercloak, slates and undercloak bedded and pointed in tinted cement mortar (1:3) and with half slates, verge slates and clips	4.15	5.00	4.36	0.23	0.12	2.97	4.20	11.52	m	12.96
ridge or hip tiles, Redland third round; butt jointed; bedding and pointing in tinted cement mortar (1:3)	8.58	5.00	9.01	0.23	0.12	2.97	0.38	12.35	m	13.90
ridge or hip tiles, Redland half round; butt jointed; bedding and pointing in tinted cement mortar (1:3)	8.58	5.00	9.01	0.23	0.12	2.97	0.38	12.35	m	13.90
ridge tiles, Redland universal angle; butt jointed; bedding and pointing in tinted cement mortar (1:3)	9.16	5.00	9.62	0.23	0.12	2.97	0.38	12.96	m	14.58
ridge tiles, Redland universal Stonewold type Monopitch; butt jointed; fixing with aluminium nails and bedding and pointing in tinted cement mortar (1:3)	21.51	5.00	22.59	0.23	0.12	2.97	0.38	25.93	m	29.17
ridge tiles, Redland universal half round type Monopitch; butt jointed; fixing with aluminium nails and bedding and pointing in tinted cement mortar (1:3)	25.51	5.00	26.79	0.23	0.12	2.97	0.38	30.13	m	33.90
ridge tiles, Redland Dry Ridge system with half round ridge tiles; fixing with stainless steel batten straps and ring shanked fixing nails with neoprene washers and sleeves; polypropylene ridge seals and uPVC profile filler units	27.03	5.00	28.38	0.45	0.23	5.77	-	34.15	m	38.42
ridge tiles, Redland Dry Ridge system with universal angle ridge tiles; fixing with stainless steel batten straps and ring shanked fixing nails with neoprene washers and sleeves; polypropylene ridge seals and uPVC profile filler units	27.61	5.00	28.99	0.45	0.23	5.77	-	34.76	m	39.10
ridge tiles, Redland Dry Vent Ridge system with half round ridge tiles; fixing with stainless steel batten straps and ring shanked nails with neoprene washers and sleeves; polypropylene ridge seals, PVC air flow control units and uPVC ventilated profile filler units	27.03	5.00	28.38	0.45	0.23	5.77	-	34.15	m	38.42
ridge tiles, Redland Dry Vent Ridge system with Universal angle ridge tiles; fixing with stainless steel batten straps and ring shanked nails with neoprene washers and sleeves; polypropylene ridge seals, PVC air flow control units and uPVC ventilated profile filler units	27.61	5.00	28.99	0.45	0.23	5.77	-	34.76	m	39.10
Monoridge filler units in conjunction with top tile, bed tile in nonsetting mastic sealant and screw on through filler unit with screws, washers and caps	38.57	5.00	40.50	0.23	0.12	2.97	0.38	43.84	m	49.32
valley tiles, Redland Universal valley troughs; laid with 100mm laps	24.37	5.00	25.59	0.56	0.28	7.13	0.38	33.10	m	37.24

Roof coverings; Redland Stonewold through coloured slates, 430 x 380mm; fixing every fourth course with galvanised nails to 75mm lap; 38 x 25mm pressured impregnated softwood battens fixed with galvanised nails; underlay; B.S.747 type 1F reinforced bitumen felt; 150mm laps; fixing with galvanised steel clout nails

Pitched 40 degrees from horizontal

Description	Del to Site £	Waste %	Material Cost £	Craft Optve Hrs	Lab Hrs	Labour Cost £	Sunds £	Nett Rate £	Unit	Gross Rate (+12.5%) £
generally	11.33	5.00	11.90	0.47	0.24	6.02	2.28	20.20	m2	22.72

Roof coverings; Redland Stonewold through coloured slates, 430 x 380mm; fixing each course with galvanised nails to 75mm; 38 x 22mm pressure impregnated softwood battens fixed with galvanised nails; underlay; B.S 747 type 1F reinforced bitumen felt; 150mm laps; fixing with galvanised steel clout nails

Pitched 40 degrees from horizontal

Description	Del to Site £	Waste %	Material Cost £	Craft Optve Hrs	Lab Hrs	Labour Cost £	Sunds £	Nett Rate £	Unit	Gross Rate (+12.5%) £
generally	11.33	5.00	11.90	0.52	0.26	6.62	4.93	23.45	m2	26.38

Labour hourly rates: (except Specialists) Craft Operatives £9.23 Labourer £7.02 Rates are national average prices. Refer to REGIONAL VARIATIONS for indicative levels of overall pricing in regions	MATERIALS			LABOUR				RATES		
	Del to Site £	Waste %	Material Cost £	Craft Optve Hrs	Lab Hrs	Labour Cost £	Sunds £	Nett Rate £	Unit	Gross Rate (+12.5%) £
H65: SINGLE LAP ROOF TILING Cont.										
Roof coverings; Redland Stonewold through coloured slates, 430 x 380mm; fixing every fourth course with galvanised nails to 75mm lap; 38 x 22mm pressure impregnated softwood battens fixed with galvanised nails; underlay, B.S.747 type 1F aluminium foil surfaced reinforced bitumen felt; 150mm laps; fixing with galvanised steel clout nails										
Pitched 40 degrees from horizontal generally..	11.33	5.00	**11.90**	0.48	0.24	**6.12**	5.92	**23.93**	m2	**26.92**
Roof coverings; Redland Stonewold through coloured slates, 430 x 380mm; fixing every fourth course with galvanised nails to 75mm lap; 38 x 22mm pressure impregnated softwood battens fixed with galvanised nails; underlay, B.S.747 type 1F reinforced bitumen felt with 50mm glass fibre insulation bonded on; 150mm laps; fixing with galvanised steel clout nails										
Pitched 40 degrees from horizontal generally..	11.33	5.00	**11.90**	0.58	0.29	**7.39**	7.76	**27.05**	m2	**30.43**
Roof coverings; Redland Delta through coloured tiles, 430 x 380mm; fixing every fourth course with galvanised nails to 75mm lap; 38 x 22mm pressure impregnated softwood battens fixed with galvanised nails; underlay; B.S.747 type 1F reinforced bitumen felt; 150mm laps; fixing with galvanised steel clout nails										
Pitched 40 degrees from horizontal generally..	15.46	5.00	**16.23**	0.47	0.24	**6.02**	3.27	**25.53**	m2	**28.72**
Reform eaves filler unit and eaves clip to each tile..................................	0.81	5.00	**0.85**	0.07	0.04	**0.93**	0.27	**2.05**	m	**2.30**
verges; 150 x 6mm fibre cement undercloak; tiles and undercloak bedded and pointed in tinted cement mortar (1:3) and with standard tiles and clips ...	5.66	5.00	**5.94**	0.23	0.12	**2.97**	4.20	**13.11**	m	**14.75**
verges; 150 x 6mm fibre cement undercloak; tiles and undercloak bedded and pointed in tinted cement mortar (1:3) and with verge tiles and clips	5.66	5.00	**5.94**	0.23	0.12	**2.97**	4.20	**13.11**	m	**14.75**
ridge tiles, Redland Delta cut ridge tiles; butt jointed; bedding and pointing in tinted cement mortar (1:3)	12.48	5.00	**13.10**	0.23	0.12	**2.97**	0.38	**16.45**	m	**18.51**
ridge tiles, Redland Delta type Monopitch ridge tiles; butt jointed; fixing with aluminium nails and bedding and pointing in tinted cement mortar (1:3)	19.07	5.00	**20.02**	0.23	0.12	**2.97**	0.44	**23.43**	m	**26.36**
Roof coverings; Redland Stonewold through coloured slates and Delta through coloured tiles, 430 x 380mm; alternating one tile and one slate in each course; fixing every fourth course with galvanised nails to 75mm lap; 38 x 22mm pressure impregnated softwood battens fixed with galvanised nails; underlay; B.S.747 type 1F reinforced bitumen felt; 150mm laps; fixing with galvanised steel clout nails										
Pitched 40 degrees from horizontal generally..	13.40	5.00	**14.07**	0.47	0.24	**6.02**	3.27	**23.36**	m2	**26.28**
Reform eaves filler unit to Delta tiles and eaves clip to each slate or tile	0.81	5.00	**0.85**	0.07	0.04	**0.93**	0.27	**2.05**	m	**2.30**
Roof coverings; Redland Stonewold through coloured slates and Delta through coloured tiles, 430 x 380mm; alternating two tiles and one slate in each course; fixing every fourth course with galvanised nails to 75mm lap; 38 x 22mm pressure impregnated softwood battens fixed with galvanised nails; underlay; B.S.747 type 1F reinforced bitumen felt; 150mm laps; fixing with galvanised steel clout nails										
Pitched 40 degrees from horizontal generally..	14.09	5.00	**14.79**	0.47	0.24	**6.02**	3.27	**24.09**	m2	**27.10**
abutments; square	1.41	–	**1.41**	0.11	0.06	**1.44**	–	**2.85**	m	**3.20**
abutments; raking	2.12	–	**2.12**	0.17	0.09	**2.20**	–	**4.32**	m	**4.86**
abutments; curved to 3000mm radius	2.81	–	**2.81**	0.23	0.12	**2.97**	–	**5.78**	m	**6.50**
Roof coverings; Redland Stonewold through coloured slates and Delta through coloured tiles, 430 x 380mm; alternating three tiles and one slate in each course; fixing every fourth course with galvanised nails to 75mm lap; 38 x 22mm pressure impregnated softwood battens fixed with galvanised nails; underlay; B.S.747 type 1F bitumen reinforced felt; 150mm laps; fixing with galvanised steel clout nails										
Pitched 40 degrees from horizontal generally..	14.42	5.00	**15.14**	0.47	0.24	**6.02**	3.27	**24.43**	m2	**27.49**

Labour hourly rates: (except Specialists) Craft Operatives £9.23 Labourer £7.02 Rates are national average prices. Refer to REGIONAL VARIATIONS for indicative levels of overall pricing in regions	MATERIALS			LABOUR				RATES		
	Del to Site £	Waste %	Material Cost £	Craft Optve Hrs	Lab Hrs	Labour Cost £	Sunds £	Nett Rate £	Unit	Gross Rate (+12.5%) £
H65: SINGLE LAP ROOF TILING Cont.										
Roof coverings; Redland Stonewold through coloured slates and Delta through coloured tiles, 430 x 380mm; alternating one tile and two slates in each course; fixing every fourth course with galvanised nails to 75mm lap; 38 x 22mm pressure impregnated softwood battens fixed with galvanised nails; underlay; B.S.747 type 1F bitumen reinforced felt; 150mm laps; fixing with galvanised steel clout nails										
Pitched 40 degrees from horizontal										
generally	12.71	5.00	**13.35**	0.47	0.24	**6.02**	3.27	**22.64**	m2	25.47
Roof coverings; Redland Regent granular faced tiles, 418 x 332mm; fixing every fourth course with galvanised nails to 75mm lap; 38 x 22mm pressure impregnated softwood battens fixed with galvanised nails; underlay; B.S.747 type 1F bitumen reinforced felt; 150mm laps; fixing with galvanised steel clout nails										
Pitched 40 degrees from horizontal										
generally	6.82	5.00	**7.16**	0.47	0.24	**6.02**	3.09	**16.27**	m2	18.31
reform eaves filler unit and eaves clip to each eaves tile	0.81	5.00	**0.85**	0.07	0.04	**0.93**	0.27	**2.05**	m	2.30
verges; cloaked verge tiles and aluminium nails ..	9.96	5.00	**10.46**	0.17	0.09	**2.20**	0.07	**12.73**	m	14.32
verges; half tiles, cloaked verge tiles and aluminium nails	11.36	5.00	**11.93**	0.23	0.12	**2.97**	0.11	**15.00**	m	16.88
verges; 150 x 6mm fibre cement undercloak; tiles and undercloak bedded and pointed in tinted cement mortar (1:3) and with standard tiles and clips ...	2.10	5.00	**2.21**	0.23	0.12	**2.97**	4.20	**9.37**	m	10.54
verges; 150 x 6mm fibre cement undercloak; tiles and undercloak bedded and pointed in tinted cement mortar (1:3) and with verge tiles and clips	2.10	5.00	**2.21**	0.23	0.12	**2.97**	4.20	**9.37**	m	10.54
verges; 150 x 6mm fibre cement undercloak; tiles and undercloak bedded and pointed in tinted cement mortar (1:3) and with half tiles, standard tiles and clips	5.88	5.00	**6.17**	0.23	0.12	**2.97**	4.20	**13.34**	m	15.01
verges; 150 x 6mm fibre cement undercloak; tiles and undercloak bedded and pointed in tinted cement mortar (1:3) and with half tiles, verge tiles and clips	5.88	5.00	**6.17**	0.23	0.12	**2.97**	4.20	**13.34**	m	15.01
ridge or hip tiles, Redland third round; butt jointed; bedding and pointing in tinted cement mortar (1:3); dentil slips in pan of each tile set in bedding	8.58	5.00	**9.01**	0.34	0.17	**4.33**	1.67	**15.01**	m	16.89
ridge or hip tiles, Redland half round; butt jointed; bedding and pointing in tinted cement mortar (1:3); dentil slips in pan of each tile set in bedding	8.58	5.00	**9.01**	0.34	0.17	**4.33**	1.67	**15.01**	m	16.89
ridge tiles, Redland universal half round type Monopitch; butt jointed; fixing with aluminium nails and bedding and pointing in tinted cement mortar (1:3); dentil slips in pan of each tile set in bedding	9.16	5.00	**9.62**	0.34	0.17	**4.33**	1.67	**15.62**	m	17.57
valley tiles, Redland universal valley troughs; laid with 100mm laps	24.37	5.00	**25.59**	0.34	0.17	**4.33**	0.38	**30.30**	m	34.09
Roof coverings; Redland Regent granular faced tiles, 418 x 332mm; fixing every fourth course with aluminium nails to 75mm lap; 38 x 22mm pressure impregnated softwood battens fixed with galvanised nails; underlay; B.S.747 type 1F bitumen reinforced felt; 150mm laps; fixing with galvanised steel clout nails										
Pitched 40 degrees from horizontal										
generally	6.82	5.00	**7.16**	0.47	0.24	**6.02**	3.26	**16.44**	m2	18.50
Roof coverings; Redland Regent granular faced tiles, 418 x 332mm; fixing every fourth course with galvanised nails to 100mm lap; 38 x 22mm pressure impregnated softwood battens fixed with galvanised nails; underlay; B.S.747 type 1F bitumen reinforced felt; 150mm laps; fixing with galvanised steel clout nails										
Pitched 40 degrees from horizontal										
generally	7.38	5.00	**7.75**	0.47	0.24	**6.02**	2.85	**16.62**	m2	18.70
abutments; square	0.74	-	**0.74**	0.11	0.06	**1.44**	-	**2.18**	m	2.45
abutments; raking	1.10	-	**1.10**	0.17	0.09	**2.20**	-	**3.30**	m	3.71
abutments; curved to 3000mm radius	1.47	-	**1.47**	0.23	0.12	**2.97**	-	**4.44**	m	4.99
Roof coverings; Redland Regent through coloured tiles, 418 x 332mm; fixing every fourth course with galvanised nails to 75mm lap; 38 x 22mm pressure impregnated softwood battens fixed with galvanised nails; underlay; B.S.747 type 1F bitumen reinforced felt; 150mm laps; fixing with galvanised steel clout nails										
Pitched 40 degrees from horizontal										
generally	6.82	5.00	**7.16**	0.47	0.24	**6.02**	3.09	**16.27**	m2	18.31
reform eaves filler unit and eaves clip to each eaves tile	0.81	5.00	**0.85**	0.07	0.04	**0.93**	0.27	**2.05**	m	2.30

Labour hourly rates: (except Specialists) Craft Operatives £9.23 Labourer £7.02 Rates are national average prices. Refer to REGIONAL VARIATIONS for indicative levels of overall pricing in regions	MATERIALS			LABOUR				RATES		
	Del to Site £	Waste %	Material Cost £	Craft Optve Hrs	Lab Hrs	Labour Cost £	Sunds £	Nett Rate £	Unit	Gross Rate (+12.5%) £

H65: SINGLE LAP ROOF TILING Cont.

Roof coverings; Redland Regent through coloured tiles, 418 x 332mm; fixing every fourth course with galvanised nails to 75mm lap; 38 x 22mm pressure impregnated softwood battens fixed with galvanised nails; underlay; B.S.747 type 1F bitumen reinforced felt; 150mm laps; fixing with galvanised steel clout nails Cont.

Pitched 40 degrees from horizontal Cont.										
verges; cloaked verge tiles and aluminium nails ..	9.96	5.00	10.46	0.17	0.09	2.20	0.07	12.73	m	14.32
verges; 150 x 6mm fibre cement undercloak; tiles and undercloak bedded and pointed in tinted cement mortar (1:3) and with standard tiles and clips ...	2.10	5.00	2.21	0.23	0.12	2.97	4.20	9.37	m	10.54
verges; 150 x 6mm fibre cement undercloak; tiles and undercloak bedded and pointed in tinted cement mortar (1:3) and with verge tiles and clips	2.10	5.00	2.21	0.23	0.12	2.97	4.20	9.37	m	10.54
verges; 150 x 6mm fibre cement undercloak; tiles and undercloak bedded and pointed in tinted cement mortar (1:3) and with half tiles, standard tiles and clips	5.88	5.00	6.17	0.23	0.12	2.97	4.20	13.34	m	15.01
verges; 150 x 6mm fibre cement undercloak; tiles and undercloak bedded and pointed in tinted cement mortar (1:3) and with half tiles, verge tiles and clips	5.88	5.00	6.17	0.23	0.12	2.97	4.20	13.34	m	15.01
valley tiles, Redland universal valley troughs; laid with 100mm laps	24.37	5.00	25.59	0.34	0.17	4.33	0.38	30.30	m	34.09

Roof coverings; Redland Regent through coloured tiles, 418 x 332mm; fixing every fourth course with galvanised nails to 100mm lap; 38 x 22mm pressure impregnated softwood battens fixed with galvanised nails; underlay; B.S.747 type 1F bitumen reinforced felt; 150mm laps; fixing with galvanised steel clout nails

Pitched 40 degrees from horizontal										
generally	7.38	5.00	7.75	0.47	0.24	6.02	2.85	16.62	m2	18.70
abutments; square	0.74	-	0.74	0.11	0.06	1.44	-	2.18	m	2.45
abutments; raking	1.10	-	1.10	0.17	0.09	2.20	-	3.30	m	3.71
abutments; curved to 3000mm radius	1.47	-	1.47	0.23	0.12	2.97	-	4.44	m	4.99

Roof coverings; Redland Grovebury granular faced double pantiles, 418 x 332mm; fixing every fourth course with galvanised nails to 75mm lap; 38 x 22mm pressure impregnated softwood battens fixed with galvanised nails; underlay; B.S.747 type 1F bitumen reinforced felt; 150mm laps; fixing with galvanised steel clout nails

Pitched 40 degrees from horizontal										
generally	6.82	5.00	7.16	0.47	0.24	6.02	3.09	16.27	m2	18.31
reform eaves filler unit and eaves clip to each eaves tile......................................	0.81	5.00	0.85	0.07	0.04	0.93	0.27	2.05	m	2.30
verges; cloaked verge tiles and aluminium nails ..	9.96	5.00	10.46	0.17	0.09	2.20	0.07	12.73	m	14.32
verges; half tiles, cloaked verge tiles and aluminium nails	11.36	5.00	11.93	0.23	0.12	2.97	0.11	15.00	m	16.88
verges; 150 x 6mm fibre cement undercloak; tiles and undercloak bedded and pointed in tinted cement mortar (1:3) and with standard tiles and clips ...	2.10	5.00	2.21	0.23	0.12	2.97	4.20	9.37	m	10.54
verges; 150 x 6mm fibre cement undercloak; tiles and undercloak bedded and pointed in tinted cement mortar (1:3) and with treble roll verge tiles and clips	2.10	5.00	2.21	0.23	0.12	2.97	4.20	9.37	m	10.54
ridge or hip tiles, Redland third round; butt jointed; bedding and pointing in tinted cement mortar (1:3)	8.58	5.00	9.01	0.34	0.17	4.33	1.67	15.01	m	16.89
ridge or hip tiles, Redland half round; butt jointed; bedding and pointing in tinted cement mortar (1:3)	8.58	5.00	9.01	0.34	0.17	4.33	1.67	15.01	m	16.89
valley tiles, Redland Universal valley troughs; laid with 100mm laps	24.37	5.00	25.59	0.34	0.17	4.33	0.38	30.30	m	34.09

Roof coverings; Redland Grovebury through coloured double pantiles, 418 x 332mm; fixing every fourth course with galvanised nails to 75mm lap; 38 x 22mm pressure impregnated softwood battens fixed with galvanised nails; underlay; B.S.747 type 1F bitumen reinforced felt; 150mm laps; fixing with galvanised steel clout nails

Pitched 40 degrees from horizontal										
generally	6.82	5.00	7.16	0.47	0.24	6.02	3.09	16.27	m2	18.31
reform eaves filler unit and eaves clip to each eaves tile......................................	0.81	5.00	0.85	0.07	0.04	0.93	0.27	2.05	m	2.30
verges; cloaked verge tiles and aluminium nails ..	9.96	5.00	10.46	0.17	0.09	2.20	0.07	12.73	m	14.32
verges; half tiles, cloaked verge tiles and aluminium nails	11.36	5.00	11.93	0.23	0.12	2.97	0.11	15.00	m	16.88
verges; 150 x 6mm fibre cement undercloak; tiles and undercloak bedded and pointed in tinted cement mortar (1:3) and with standard tiles and clips ...	2.10	5.00	2.21	0.23	0.12	2.97	4.20	9.37	m	10.54
verges; 150 x 6mm fibre cement undercloak; tiles and undercloak bedded and pointed in tinted cement mortar (1:3) and with treble roll verge tiles and clips	2.10	5.00	2.21	0.23	0.12	2.97	4.20	9.37	m	10.54

CLADDING/COVERING

Labour hourly rates: (except Specialists) Craft Operatives £9.23 Labourer £7.02 Rates are national average prices. Refer to REGIONAL VARIATIONS for indicative levels of overall pricing in regions	MATERIALS			LABOUR				RATES		
	Del to Site £	Waste %	Material Cost £	Craft Optve Hrs	Lab Hrs	Labour Cost £	Sunds £	Nett Rate £	Unit	Gross Rate (+12.5%) £

H65: SINGLE LAP ROOF TILING Cont.

Roof coverings; Redland Grovebury through coloured double pantiles, 418 x 332mm; fixing every fourth course with galvanised nails to 75mm lap; 38 x 22mm pressure impregnated softwood battens fixed with galvanised nails; underlay; B.S.747 type 1F bitumen reinforced felt; 150mm laps; fixing with galvanised steel clout nails Cont.

Pitched 40 degrees from horizontal Cont.

	Del to Site	Waste	Material Cost	Craft Optve	Lab	Labour Cost	Sunds	Nett Rate	Unit	Gross Rate
ridge or hip tiles, Redland third round; butt jointed; bedding and pointing in tinted cement mortar (1:3)............	8.58	5.00	9.01	0.34	0.17	4.33	1.67	15.01	m	16.89
ridge or hip tiles, Redland half round; butt jointed ; bedding and pointing in tinted cement mortar (1:3)............	8.58	5.00	9.01	0.34	0.17	4.33	1.67	15.01	m	16.89
valley tiles, Redland Universal valley troughs; laid with 100mm laps	24.37	5.00	25.59	0.34	0.17	4.33	0.38	30.30	m	34.09

Roof coverings; Redland Norfolk through coloured pantiles, 381 x 227mm; fixing every fourth course with galvanised nails to 75mm lap; 38 x 22mm pressure impregnated softwood battens fixed with galvanised nails; underlay; B.S.747 type 1F bitumen reinforced felt; 150mm laps; fixing with galvanised steel clout nails

Pitched 40 degrees from horizontal

	Del to Site	Waste	Material Cost	Craft Optve	Lab	Labour Cost	Sunds	Nett Rate	Unit	Gross Rate
generally	8.03	5.00	8.43	0.61	0.31	7.81	3.68	19.92	m2	22.41
abutments; square	0.80	-	0.80	0.11	0.06	1.44	-	2.24	m	2.52
abutments; raking	1.20	-	1.20	0.17	0.09	2.20	-	3.40	m	3.83
abutments; curved to 3000mm radius ...	1.61	-	1.61	0.23	0.12	2.97	-	4.58	m	5.15
reform eaves filler unit and aluminium nails to each eaves tile	0.64	5.00	0.67	0.07	0.04	0.93	0.27	1.87	m	2.10
verges; plain tile undercloak; tiles and undercloak bedded and pointed in tinted cement mortar (1:3) and with standard tiles and aluminium nails	1.47	5.00	1.54	0.11	0.06	1.44	0.42	3.40	m	3.83
verges; plain tile undercloak; tiles and undercloak bedded and pointed in tinted cement mortar (1:3) and with standard tiles and clips ...	1.47	5.00	1.54	0.11	0.06	1.44	3.86	6.84	m	7.70
verges; 150 x 6mm fibre cement undercloak; tiles and undercloak bedded and pointed in tinted cement mortar (1:3) and with standard tiles and aluminium nails	1.47	5.00	1.54	0.17	0.09	2.20	2.31	6.05	m	6.81
verges; 150 x 6mm fibre cement undercloak; tiles and undercloak bedded and pointed in tinted cement mortar (1:3) and with standard tiles and clips ...	1.47	5.00	1.54	0.17	0.09	2.20	5.72	9.46	m	10.65
ridge or hip tiles, Redland third round; butt jointed; bedding and pointing in tinted cement mortar (1:3)..............	8.58	5.00	9.01	0.34	0.17	4.33	0.38	13.72	m	15.44
ridge or hip tiles, Redland half round; butt jointed; bedding and pointing in tinted cement mortar (1:3)..............	8.58	5.00	9.01	0.34	0.17	4.33	0.38	13.72	m	15.44
valley tiles, Redland Universal valley troughs; laid with 100mm laps	24.37	5.00	25.59	0.34	0.17	4.33	0.38	30.30	m	34.09

Roof coverings; Redland Norfolk through coloured pantiles, 381 x 227mm; fixing every fourth course with galvanised nails to 75mm lap; 38 x 25mm pressure impregnated softwood battens fixed with galvanised nails; underlay; B.S.747 type 1F bitumen reinforced felt; 150mm laps; fixing with galvanised steel clout nails

Pitched 40 degrees from horizontal

	Del to Site	Waste	Material Cost	Craft Optve	Lab	Labour Cost	Sunds	Nett Rate	Unit	Gross Rate
generally.........................	8.03	5.00	8.43	0.61	0.31	7.81	3.86	20.10	m2	22.61

Roof coverings; Redland Norfolk through coloured pantiles, 381 x 227mm; fixing every fourth course with galvanised nails to 100mm lap; 38 x 22mm pressure impregnated softwood battens fixed with galvanised nails; underlay; B.S.747 type 1F bitumen reinforced felt; 150mm laps; fixing with galvanised steel clout nails

Pitched 40 degrees from horizontal

	Del to Site	Waste	Material Cost	Craft Optve	Lab	Labour Cost	Sunds	Nett Rate	Unit	Gross Rate
generally.........................	8.78	5.00	9.22	0.66	0.33	8.41	3.74	21.37	m2	24.04

Roof coverings; Redland 49 granular faced tiles, 381 x 227mm; fixing every fourth course with galvanised nails to 75mm lap; 38 x 22mm pressure impregnated softwood battens fixed with galvanised nails; underlay B.S.747 type 1F bitumen reinforced felt; 150mm laps; fixing with galvanised steel clout nails

Pitched 40 degrees from horizontal

	Del to Site	Waste	Material Cost	Craft Optve	Lab	Labour Cost	Sunds	Nett Rate	Unit	Gross Rate
generally.........................	7.79	5.00	8.18	0.61	0.31	7.81	3.81	19.80	m2	22.27
supplementary fixing eaves course with one aluminium nail per tile	-	-	-	0.02	0.01	0.25	0.03	0.28	m	0.32
verges; Redland Dry Verge system with standard tiles and clips	12.63	5.00	13.26	0.28	0.14	3.57	2.03	18.86	m	21.22
verges; Redland Dry Verge system with verge tiles and clips................	12.63	5.00	13.26	0.28	0.14	3.57	2.03	18.86	m	21.22

Labour hourly rates: (except Specialists) Craft Operatives £9.23 Labourer £7.02 Rates are national average prices. Refer to REGIONAL VARIATIONS for indicative levels of overall pricing in regions	MATERIALS			LABOUR				RATES		
	Del to Site £	Waste %	Material Cost £	Craft Optve Hrs	Lab Hrs	Labour Cost £	Sunds £	Nett Rate £	Unit	Gross Rate (+12.5%) £

H65: SINGLE LAP ROOF TILING Cont.

Roof coverings; Redland 49 granular faced tiles, 381 x 227mm; fixing every fourth course with galvanised nails to 75mm lap; 38 x 22mm pressure impregnated softwood battens fixed with galvanised nails; underlay B.S.747 type 1F bitumen reinforced felt; 150mm laps; fixing with galvanised steel clout nails Cont.

Pitched 40 degrees from horizontal Cont.

	Del to Site £	Waste %	Material Cost £	Craft Optve Hrs	Lab Hrs	Labour Cost £	Sunds £	Nett Rate £	Unit	Gross Rate £
verges; plain tile undercloak; tiles and undercloak bedded and pointed in tinted cement mortar (1:3) and with standard tiles and aluminium nails	1.43	5.00	1.50	0.11	0.06	1.44	0.45	3.39	m	3.81
verges; plain tile undercloak; tiles and undercloak bedded and pointed in tinted cement mortar (1:3) and with verge tiles and aluminium nails	1.43	5.00	1.50	0.11	0.06	1.44	0.45	3.39	m	3.81
verges; 150 x 6mm fibre cement undercloak; tiles and undercloak bedded and pointed in tinted cement mortar (1:3) and with standard tiles and aluminium nails	1.43	5.00	1.50	0.17	0.09	2.20	2.31	6.01	m	6.76
verges; 150 x 6mm fibre cement undercloak; tiles and undercloak bedded and pointed in tinted cement mortar (1:3) and with verge tiles and aluminium nails	1.43	5.00	1.50	0.17	0.09	2.20	2.31	6.01	m	6.76
ridge or hip tiles, Redland third round; butt jointed; bedding and pointing in tinted cement mortar (1:3)	8.58	5.00	9.01	0.34	0.17	4.33	0.38	13.72	m	15.44
ridge or hip tiles, Redland half round; butt jointed; bedding and pointing in tinted cement mortar (1:3)	8.58	5.00	9.01	0.34	0.17	4.33	0.38	13.72	m	15.44
valley tiles, Redland Universal valley troughs; laid with 100mm laps	24.37	5.00	25.59	0.34	0.17	4.33	0.38	30.30	m	34.09

Roof coverings; Redland 49 granular faced tiles, 381 x 227mm; fixing every fourth course with galvanised nails to 100mm lap; 38 x 22mm pressure impregnated softwood battens fixed with galvanised nails; underlay; B.S.747 type 1F bitumen reinforced felt; 150mm laps; fixing with galvanised steel clout nails

Pitched 40 degrees from horizontal

	Del to Site £	Waste %	Material Cost £	Craft Optve Hrs	Lab Hrs	Labour Cost £	Sunds £	Nett Rate £	Unit	Gross Rate £
generally	8.51	5.00	8.94	0.66	0.33	8.41	3.92	21.26	m2	23.92
abutments; square	0.86	–	0.86	0.11	0.06	1.44	–	2.30	m	2.58
abutments; raking	1.27	–	1.27	0.17	0.09	2.20	–	3.47	m	3.90
abutments; curved to 3000mm radius	1.70	–	1.70	0.23	0.12	2.97	–	4.67	m	5.25

Roof coverings; Redland 49 through coloured tiles, 381 x 227mm; fixing every fourth course with galvanised nails to 75mm lap; 38 x 22mm pressure impregnated softwood battens fixed with galvanised nails; underlay; B.S.747 type 1F bitumen reinforced felt; 150mm laps; fixing with galvanised steel clout nails

Pitched 40 degrees from horizontal

	Del to Site £	Waste %	Material Cost £	Craft Optve Hrs	Lab Hrs	Labour Cost £	Sunds £	Nett Rate £	Unit	Gross Rate £
generally	7.79	5.00	8.18	0.61	0.31	7.81	3.81	19.80	m2	22.27
abutments; square	0.78	–	0.78	0.11	0.06	1.44	–	2.22	m	2.49
abutments; raking	1.17	–	1.17	0.17	0.09	2.20	–	3.37	m	3.79
abutments; curved to 3000mm radius	1.55	–	1.55	0.23	0.12	2.97	–	4.52	m	5.08
supplementary fixing eaves course with one aluminium nail per tile	–	–	–	0.02	0.01	0.25	0.03	0.28	m	0.32
verges; Redland Dry Verge system with standard tiles and clips	12.63	5.00	13.26	0.28	0.14	3.57	2.03	18.86	m	21.22
verges; Redland Dry Verge system with verge tiles and clips	12.63	5.00	13.26	0.28	0.14	3.57	2.03	18.86	m	21.22
verges; plain tile undercloak; tiles and undercloak bedded and pointed in tinted cement mortar (1:3) and with standard tiles and aluminium nails	1.43	5.00	1.50	0.11	0.06	1.44	0.45	3.39	m	3.81
verges; plain tile undercloak; tiles and undercloak bedded and pointed in tinted cement mortar (1:3) and with verge tiles and aluminium nails	1.43	5.00	1.50	0.11	0.06	1.44	0.45	3.39	m	3.81
verges; 150 x 6mm fibre cement undercloak; tiles and undercloak bedded and pointed in tinted cement mortar (1:3) and with standard tiles and aluminium nails	1.43	5.00	1.50	0.17	0.09	2.20	2.31	6.01	m	6.76
verges; 150 x 6mm fibre cement undercloak; tiles and undercloak bedded and pointed in tinted cement mortar (1:3) and with verge tiles and aluminium nails	1.43	5.00	1.50	0.17	0.09	2.20	2.31	6.01	m	6.76
ridge or hip tiles, Redland third round; butt jointed; bedding and pointing in tinted cement mortar (1:3)	8.58	5.00	9.01	0.34	0.17	4.33	0.38	13.72	m	15.44
ridge or hip tiles, Redland half round; butt jointed; bedding and pointing in tinted cement mortar (1:3)	8.58	5.00	9.01	0.34	0.17	4.33	0.38	13.72	m	15.44
valley tiles, Redland Universal valley troughs; laid with 100mm laps	24.37	5.00	25.59	0.34	0.17	4.33	0.38	30.30	m	34.09

Labour hourly rates: (except Specialists) Craft Operatives £9.23 Labourer £7.02 Rates are national average prices. Refer to REGIONAL VARIATIONS for indicative levels of overall pricing in regions	MATERIALS			LABOUR				RATES		
	Del to Site £	Waste %	Material Cost £	Craft Optve Hrs	Lab Hrs	Labour Cost £	Sunds £	Nett Rate £	Unit	Gross Rate (+12.5%) £

H65: SINGLE LAP ROOF TILING Cont.

Roof coverings; Redland 49 through coloured tiles, 381 x 227mm; fixing every fourth course with galvanised nails to 100mm lap; 38 x 22mm pressure impregnated softwood battens fixed with galvanised nails; underlay; B.S.747 type 1F bitumen reinforced felt; 150mm laps; fixing with galvanised steel clout nails

Pitched 40 degrees from horizontal

generally	8.51	5.00	8.94	0.66	0.33	8.41	3.92	21.26	m2	23.92

Roof coverings; Redland Renown granular faced tiles, 418 x 330mm; fixing every fourth course with galvanised nails to 75mm lap; 38 x 22mm pressure impregnated softwood battens fixed with galvanised nails; underlay; B.S.747 type 1F bitumen reinforced felt; 150mm laps; fixing with galvanised steel clout nails

Pitched 40 degrees from horizontal

generally	6.46	5.00	6.78	0.47	0.24	6.02	3.09	15.90	m2	17.88
abutments; square	0.64	-	0.64	0.11	0.06	1.44	-	2.08	m	2.34
abutments; raking	0.97	-	0.97	0.17	0.09	2.20	-	3.17	m	3.57
abutments; curved to 3000mm radius	1.29	-	1.29	0.23	0.12	2.97	-	4.26	m	4.79
reform eaves filler unit and aluminium nails to each eaves tile	0.64	5.00	0.67	0.07	0.04	0.93	0.30	1.90	m	2.14
verges; cloaked verge tiles and aluminium nails	9.96	5.00	10.46	0.17	0.09	2.20	0.07	12.73	m	14.32
verges; half tiles, cloaked verge tiles and aluminium nails	11.28	5.00	11.84	0.23	0.12	2.97	0.11	14.92	m	16.78
verges; plain tile undercloak; tiles and undercloak bedded and pointed in tinted cement mortar (1:3) and with standard tiles and aluminium nails	1.99	5.00	2.09	0.11	0.06	1.44	0.45	3.98	m	4.47
verges; plain tile undercloak; tiles and undercloak bedded and pointed in tinted cement mortar (1:3) and with verge tiles and aluminium nails	1.99	5.00	2.09	0.11	0.06	1.44	0.45	3.98	m	4.47
verges; 150 x 6mm fibre cement undercloak; tiles and undercloak bedded and pointed in tinted cement mortar (1:3) and with standard tiles and aluminium nails	1.99	5.00	2.09	0.17	0.09	2.20	2.31	6.60	m	7.43
verges; 150 x 6mm fibre cement undercloak; tiles and undercloak bedded and pointed in tinted cement mortar (1:3) and with verge tiles and aluminium nails	1.99	5.00	2.09	0.17	0.09	2.20	2.31	6.60	m	7.43
ridge or hip tiles, Redland third round; butt jointed; bedding and pointing in tinted cement mortar (1:3)	8.58	5.00	9.01	0.34	0.17	4.33	0.38	13.72	m	15.44
ridge or hip tiles, Redland half round; butt jointed; bedding and pointing in tinted cement mortar (1:3)	8.58	5.00	9.01	0.34	0.17	4.33	0.38	13.72	m	15.44
valley tiles, Redland Universal valley troughs; laid with 100mm laps	24.37	5.00	25.59	0.34	0.17	4.33	0.38	30.30	m	34.09

Roof coverings; Redland 50 granular faced double roman tiles, 418 x 330mm; fixing every fourth course with galvanised nails to 75mm lap; 38 x 22mm pressure impregnated softwood battens fixed with galvanised nails; underlay; B.S.747 type 1F bitumen reinforced felt; 150mm laps; fixing with galvanised steel clout nails

Pitched 40 degrees from horizontal

generally	6.46	5.00	6.78	0.47	0.24	6.02	3.09	15.90	m2	17.88
reform eaves filler unit and aluminium nails to each eaves tile	0.64	5.00	0.67	0.07	0.04	0.93	0.27	1.87	m	2.10
verges; cloaked verge tiles and aluminium nails	9.96	5.00	10.46	0.17	0.09	2.20	0.07	12.73	m	14.32
verges; half tiles, cloaked verge tiles and aluminium nails	9.96	5.00	10.46	0.23	0.12	2.97	0.07	13.49	m	15.18
verges; plain tile undercloak; tiles and undercloak bedded and pointed in tinted cement mortar (1:3) and with standard tiles and aluminium nails	1.99	5.00	2.09	0.11	0.06	1.44	0.45	3.98	m	4.47
verges; plain tile undercloak; tiles and undercloak bedded and pointed in tinted cement mortar (1:3) and with treble roll verge tiles and aluminium nails	2.99	5.00	3.14	0.11	0.06	1.44	0.45	5.03	m	5.65
verges; 150 x 6mm fibre cement undercloak; tiles and undercloak bedded and pointed in tinted cement mortar (1:3) and with standard tiles and aluminium nails	1.99	5.00	2.09	0.17	0.09	2.20	2.31	6.60	m	7.43
verges; 150 x 6mm fibre cement undercloak; tiles and undercloak bedded and pointed in tinted cement mortar (1:3) and with treble roll verge tiles and aluminium nails	2.99	5.00	3.14	0.17	0.09	2.20	2.31	7.65	m	8.61
ridge or hip tiles, Redland third round; butt jointed; bedding and pointing in tinted cement mortar (1:3)	8.58	5.00	9.01	0.34	0.17	4.33	0.38	13.72	m	15.44
ridge or hip tiles, Redland half round; butt jointed; bedding and pointing in tinted cement mortar (1:3)	8.58	5.00	9.01	0.34	0.17	4.33	0.38	13.72	m	15.44
valley tiles, Redland Universal valley troughs; laid with 100mm laps	24.37	5.00	25.59	0.34	0.17	4.33	0.38	30.30	m	34.09

Labour hourly rates: (except Specialists) Craft Operatives £9.23 Labourer £7.02 Rates are national average prices. Refer to REGIONAL VARIATIONS for indicative levels of overall pricing in regions	MATERIALS			LABOUR				RATES		
	Del to Site £	Waste %	Material Cost £	Craft Optve Hrs	Lab Hrs	Labour Cost £	Sunds £	Nett Rate £	Unit	Gross Rate (+12.5%) £
H65: SINGLE LAP ROOF TILING Cont.										
Roof coverings; Redland 50 through coloured double roman tiles, 418 x 330mm; fixing every fourth course with galvanised nails to 75mm lap; 38 x 22mm pressure impregnated softwood battens fixed with galvanised nails; underlay; B.S.747 type 1F bitumen reinforced felt; 150mm laps; fixing with galvanised steel clout nails										
Pitched 40 degrees from horizontal										
generally	6.46	5.00	6.78	0.47	0.24	6.02	3.09	15.90	m2	17.88
reform eaves filler unit and aluminium nails to each eaves tile	0.64	5.00	0.67	0.07	0.04	0.93	0.27	1.87	m	2.10
verges; cloaked verge tiles and aluminium nails ..	9.96	5.00	10.46	0.17	0.09	2.20	0.07	12.73	m	14.32
verges; half tiles, cloaked verge tiles and aluminium nails	9.96	5.00	10.46	0.23	0.12	2.97	0.45	13.87	m	15.61
verges; plain tile undercloak; tiles and undercloak bedded and pointed in tinted cement mortar (1:3) and with standard tiles and aluminium nails	1.99	5.00	2.09	0.11	0.06	1.44	0.07	3.60	m	4.05
verges; plain tile undercloak; tiles and undercloak bedded and pointed in tinted cement mortar (1:3) and with treble roll verge tiles and aluminium nails	2.99	5.00	3.14	0.11	0.06	1.44	0.45	5.03	m	5.65
verges; 150 x 6mm fibre cement undercloak; tiles and undercloak bedded and pointed in tinted cement mortar (1:3) and with standard tiles and aluminium nails	1.99	5.00	2.09	0.17	0.09	2.20	2.31	6.60	m	7.43
verges; 150 x 6mm fibre cement undercloak; tiles and undercloak bedded and pointed in tinted cement mortar (1:3) and with treble roll verge tiles and aluminium nails	2.99	5.00	3.14	0.17	0.09	2.20	0.45	5.79	m	6.51
ridge or hip tiles, Redland third round; butt jointed; bedding and pointing in tinted cement mortar (1:3)	8.58	5.00	9.01	0.34	0.17	4.33	0.38	13.72	m	15.44
ridge or hip tiles, Redland half round; butt jointed; bedding and pointing in tinted cement mortar (1:3)	8.58	5.00	9.01	0.34	0.17	4.33	0.38	13.72	m	15.44
valley tiles, Redland Universal valley troughs; laid with 100mm laps	24.37	5.00	25.59	0.34	0.17	4.33	0.38	30.30	m	34.09
Fittings										
Ventilator tiles										
Red Vent ridge ventilation terminal 450mm long for										
half round ridge	52.20	5.00	54.81	1.45	0.73	18.51	0.65	73.97	nr	83.21
universal Delta ridge	72.00	5.00	75.60	1.68	0.84	21.40	0.65	97.65	nr	109.86
universal angle ridge	52.09	5.00	54.69	1.45	0.73	18.51	0.65	73.85	nr	83.08
Redvent eaves ventilators fixing with aluminium nails	9.14	5.00	9.60	0.56	0.28	7.13	0.03	16.76	m	18.86
Red Vent Thruvent tiles for										
Stonewold slates....................	39.71	5.00	41.70	0.56	0.28	7.13	0.56	49.39	nr	55.56
Delta tiles.....................	39.71	5.00	41.70	0.56	0.28	7.13	0.56	49.39	nr	55.56
Regent tiles.....................	38.81	5.00	40.75	0.56	0.28	7.13	0.56	48.44	nr	54.50
Grovebury double pantiles....................	38.81	5.00	40.75	0.56	0.28	7.13	0.56	48.44	nr	54.50
Norfolk pantiles.....................	38.81	5.00	40.75	0.56	0.28	7.13	0.56	48.44	nr	54.50
Redland 49 tiles.....................	38.81	5.00	40.75	0.56	0.28	7.13	0.56	48.44	nr	54.50
Renown tiles.....................	38.81	5.00	40.75	0.56	0.28	7.13	0.56	48.44	nr	54.50
Redland 50 double roman tiles	-	-	-	0.56	0.28	7.13	38.81	45.94	nr	51.69
Red line ventilation tile complete with underlay seal and fixing clips for										
Stonewold slates.....................	48.15	5.00	50.56	1.12	0.56	14.27	0.27	65.10	nr	73.23
Regent tiles.....................	48.15	5.00	50.56	1.12	0.56	14.27	0.27	65.10	nr	73.23
Grovebury double pantiles	48.15	5.00	50.56	1.12	0.56	14.27	0.27	65.10	nr	73.23
Renown tiles.....................	48.15	5.00	50.56	1.12	0.56	14.27	0.27	65.10	nr	73.23
Redland 50 double roman tiles	48.15	5.00	50.56	1.12	0.56	14.27	0.27	65.10	nr	73.23
Gas terminals										
Gas Flue ridge terminal Mark III, 450mm long with sealing gasket and fixing brackets										
half round ridge.....................	-	-	-	1.69	0.84	21.50	0.38	21.88	nr	24.61
universal Delta ridge.....................	-	-	-	1.91	0.96	24.37	0.38	24.75	nr	27.84
extra for 150mm extension adaptor.............	-	-	-	0.34	0.17	4.33	0.27	4.60	nr	5.18
Hip irons										
galvanised mild steel; 32 x 3 x 380mm girth; scrolled; fixing with galvanised steel screws to timber.........................	2.01	5.00	2.11	0.11	0.06	1.44	-	3.55	nr	3.99
Lead soakers										
fixing only.........................	-	-	-	0.34	0.17	4.33	-	4.33	nr	4.87

CLADDING/COVERING

Labour hourly rates: (except Specialists) Craft Operatives £12.44 Labourer £7.02 Rates are national average prices. Refer to REGIONAL VARIATIONS for indicative levels of overall pricing in regions	MATERIALS			LABOUR				RATES		
	Del to Site	Waste	Material Cost	Craft Optve	Lab	Labour Cost	Sunds	Nett Rate	Unit	Gross Rate (+12.5%)
	£	%	£	Hrs	Hrs	£	£	£		£
H71: LEAD SHEET COVERINGS/FLASHINGS										
Sheet lead										
Technical data										
Code 3, 1.32mm thick, 14.97 kg/m2, colour code green										
Code 4, 1.80mm thick, 20.41 kg/m2, colour code blue										
Code 5, 2.24mm thick, 25.40 kg/m2, colour code red										
Code 6, 2.65mm thick, 30.05 kg/m2, colour code black										
Code 7, 3.15mm thick, 35.72 kg/m2, colour code white										
Code 8, 3.55mm thick, 40.26 kg/m2, colour code orange										
Milled lead sheet, B.S.1178										
Nr 3 roof coverings; fixing to timber with milled lead cleats and galvanised screws										
pitch 7.5 degrees from horizontal	11.81	5.00	12.40	4.40	–	54.74	0.25	67.39	m2	75.81
pitch 40 degrees from horizontal	11.81	5.00	12.40	5.05	–	62.82	0.25	75.47	m2	84.91
pitch 75 degrees from horizontal	11.81	5.00	12.40	5.70	–	70.91	0.25	83.56	m2	94.00
Nr 4 roof coverings; fixing to timber with milled lead cleats and galvanised screws										
pitch 7.5 degrees from horizontal	16.10	5.00	16.91	5.05	–	62.82	0.30	80.03	m2	90.03
pitch 40 degrees from horizontal	16.10	5.00	16.91	5.70	–	70.91	0.30	88.11	m2	99.13
pitch 75 degrees from horizontal	16.10	5.00	16.91	6.35	–	78.99	0.30	96.20	m2	108.22
Nr 5 roof coverings; fixing to timber with milled lead cleats and galvanised screws										
pitch 7.5 degrees from horizontal	20.04	5.00	21.04	5.70	–	70.91	0.36	92.31	m2	103.85
pitch 40 degrees from horizontal	20.04	5.00	21.04	6.35	–	78.99	0.36	100.40	m2	112.95
pitch 75 degrees from horizontal	20.04	5.00	21.04	6.95	–	86.46	0.36	107.86	m2	121.34
extra; oil patination to surfaces	1.40	10.00	1.54	0.23	–	2.86	–	4.40	m2	4.95
Nr 3 wall coverings; fixing to timber with milled lead cleats and galvanised screws										
vertical	11.81	5.00	12.40	8.00	–	99.52	0.25	112.17	m2	126.19
Nr 4 wall coverings; fixing to timber with milled lead cleats and galvanised screws										
vertical	16.10	5.00	16.91	8.65	–	107.61	0.30	124.81	m2	140.41
Nr 5 wall coverings; fixing to timber with milled lead cleats and galvanised screws										
vertical	20.04	5.00	21.04	9.25	–	115.07	0.36	136.47	m2	153.53
Flashings; horizontal										
Nr 3; 150mm lapped joints; fixing to masonry with milled lead clips and lead wedges										
150mm girth	1.77	5.00	1.86	0.46	–	5.72	0.25	7.83	m	8.81
240mm girth	2.83	5.00	2.97	0.74	–	9.21	0.25	12.43	m	13.98
300mm girth	3.54	5.00	3.72	0.92	–	11.44	0.25	15.41	m	17.34
Nr 4; 150mm lapped joints; fixing to masonry with milled lead clips and lead wedges										
150mm girth	2.42	5.00	2.54	0.50	–	6.22	0.30	9.06	m	10.19
240mm girth	3.86	5.00	4.05	0.82	–	10.20	0.30	14.55	m	16.37
300mm girth	4.88	5.00	5.12	1.02	–	12.69	0.30	18.11	m	20.38
Nr 5; 150mm lapped joints; fixing to masonry with milled lead clips and lead wedges										
150mm girth	3.01	5.00	3.16	0.58	–	7.22	0.36	10.74	m	12.08
240mm girth	4.81	5.00	5.05	0.92	–	11.44	0.36	16.86	m	18.96
300mm girth	6.01	5.00	6.31	1.15	–	14.31	0.36	20.98	m	23.60
Nr 3; 150mm lapped joints; fixing to timber with copper nails										
150mm girth	1.77	5.00	1.86	0.46	–	5.72	0.29	7.87	m	8.85
240mm girth	2.83	5.00	2.97	0.74	–	9.21	0.29	12.47	m	14.03
300mm girth	3.54	5.00	3.72	0.92	–	11.44	0.29	15.45	m	17.38
Nr 4; 150mm lapped joints; fixing to timber with copper nails										
150mm girth	2.42	5.00	2.54	0.50	–	6.22	0.29	9.05	m	10.18
240mm girth	3.86	5.00	4.05	0.82	–	10.20	0.29	14.54	m	16.36
300mm girth	4.83	5.00	5.07	1.02	–	12.69	0.29	18.05	m	20.31
Nr 5; 150mm lapped joints; fixing to timber with copper nails										
150mm girth	3.01	5.00	3.16	0.58	–	7.22	0.29	10.67	m	12.00
240mm girth	4.81	5.00	5.05	0.92	–	11.44	0.29	16.79	m	18.88
300mm girth	6.01	5.00	6.31	1.15	–	14.31	0.29	20.91	m	23.52
Flashings; stepped										
Nr 3; 150mm lapped joints; fixing to masonry with milled lead clips and lead wedges										
180mm girth	2.13	5.00	2.24	0.74	–	9.21	0.35	11.79	m	13.27
240mm girth	2.83	5.00	2.97	0.98	–	12.19	0.35	15.51	m	17.45
300mm girth	3.54	5.00	3.72	1.24	–	15.43	0.35	19.49	m	21.93

Labour hourly rates: (except Specialists) Craft Operatives £12.44 Labourer £7.02 Rates are national average prices. Refer to REGIONAL VARIATIONS for indicative levels of overall pricing in regions	MATERIALS			LABOUR				RATES		
	Del to Site £	Waste %	Material Cost £	Craft Optve Hrs	Lab Hrs	Labour Cost £	Sunds £	Nett Rate £	Unit	Gross Rate (+12.5%) £
H71: LEAD SHEET COVERINGS/FLASHINGS Cont.										
Flashings; stepped Cont.										
Nr 4; 150mm lapped joints; fixing to masonry with milled lead clips and lead wedges										
180mm girth	2.90	5.00	3.04	0.82	–	10.20	0.46	13.71	m	15.42
240mm girth	3.86	5.00	4.05	1.10	–	13.68	0.46	18.20	m	20.47
300mm girth	4.83	5.00	5.07	1.36	–	16.92	0.46	22.45	m	25.26
Nr 5; 150mm lapped joints; fixing to masonry with milled lead clips and lead wedges										
180mm girth	3.61	5.00	3.79	0.92	–	11.44	0.55	15.79	m	17.76
240mm girth	4.81	5.00	5.05	1.22	–	15.18	0.55	20.78	m	23.37
300mm girth	6.01	5.00	6.31	1.52	–	18.91	0.55	25.77	m	28.99
Aprons; horizontal										
Nr 3; 150mm lapped joints; fixing to masonry with milled lead clips and lead wedges										
150mm girth	1.77	5.00	1.86	0.46	–	5.72	0.25	7.83	m	8.81
240mm girth	2.83	5.00	2.97	0.74	–	9.21	0.25	12.43	m	13.98
300mm girth	3.54	5.00	3.72	0.92	–	11.44	0.25	15.41	m	17.34
450mm girth	5.31	5.00	5.58	1.38	–	17.17	0.35	23.09	m	25.98
Nr 4; 150mm lapped joints; fixing to masonry with milled lead clips and lead wedges										
150mm girth	2.42	5.00	2.54	0.50	–	6.22	0.30	9.06	m	10.19
240mm girth	3.86	5.00	4.05	0.82	–	10.20	0.30	14.55	m	16.37
300mm girth	4.83	5.00	5.07	1.02	–	12.69	0.30	18.06	m	20.32
450mm girth	7.25	5.00	7.61	1.52	–	18.91	0.46	26.98	m	30.35
Nr 5; 150mm lapped joints; fixing to masonry with milled lead clips and lead wedges										
150mm girth	3.01	5.00	3.16	0.58	–	7.22	0.36	10.74	m	12.08
240mm girth	4.81	5.00	5.05	0.92	–	11.44	0.36	16.86	m	18.96
300mm girth	6.01	5.00	6.31	1.15	–	14.31	0.36	20.98	m	23.60
450mm girth	9.02	5.00	9.47	1.72	–	21.40	0.55	31.42	m	35.35
Hips; sloping; dressing over slating and tiling										
Nr 3; 150mm lapped joints; fixing to timber with milled lead clips										
240mm girth	2.83	5.00	2.97	0.74	–	9.21	0.25	12.43	m	13.98
300mm girth	3.54	5.00	3.72	0.92	–	11.44	0.25	15.41	m	17.34
450mm girth	5.31	5.00	5.58	1.38	–	17.17	0.35	23.09	m	25.98
Nr 4; 150mm lapped joints; fixing to timber with milled lead clips										
240mm girth	3.86	5.00	4.05	0.82	–	10.20	0.30	14.55	m	16.37
300mm girth	4.83	5.00	5.07	1.02	–	12.69	0.30	18.06	m	20.32
450mm girth	7.25	5.00	7.61	1.52	–	18.91	0.46	26.98	m	30.35
Nr 5; 150mm lapped joints; fixing to timber with milled lead clips										
240mm girth	4.81	5.00	5.05	0.92	–	11.44	0.36	16.86	m	18.96
300mm girth	6.01	5.00	6.31	1.15	–	14.31	0.36	20.98	m	23.60
450mm girth	9.02	5.00	9.47	1.72	–	21.40	0.55	31.42	m	35.35
Kerbs; horizontal										
Nr 3; 150mm lapped joints; fixing to timber with copper nails										
240mm girth	2.83	5.00	2.97	0.74	–	9.21	0.25	12.43	m	13.98
300mm girth	3.54	5.00	3.72	0.92	–	11.44	0.25	15.41	m	17.34
450mm girth	5.31	5.00	5.58	1.38	–	17.17	0.35	23.09	m	25.98
Nr 4; 150mm lapped joints; fixing to timber with copper nails										
240mm girth	3.86	5.00	4.05	0.82	–	10.20	0.30	14.55	m	16.37
300mm girth	4.83	5.00	5.07	1.02	–	12.69	0.30	18.06	m	20.32
450mm girth	7.25	5.00	7.61	1.52	–	18.91	0.46	26.98	m	30.35
Nr 5; 150mm lapped joints; fixing to timber with copper nails										
240mm girth	4.81	5.00	5.05	0.92	–	11.44	0.36	16.86	m	18.96
300mm girth	6.02	5.00	6.32	1.15	–	14.31	0.36	20.99	m	23.61
450mm girth	9.02	5.00	9.47	1.72	–	21.40	0.55	31.42	m	35.35
Ridges; horizontal; dressing over slating and tiling										
Nr 3; 150mm lapped joints; fixing to timber with milled lead clips										
240mm girth	2.83	5.00	2.97	0.74	–	9.21	0.25	12.43	m	13.98
300mm girth	3.54	5.00	3.72	0.92	–	11.44	0.25	15.41	m	17.34
450mm girth	5.31	5.00	5.58	1.38	–	17.17	0.35	23.09	m	25.98
Nr 4; 150mm lapped joints; fixing to timber with milled lead clips										
240mm girth	3.86	5.00	4.05	0.82	–	10.20	0.30	14.55	m	16.37
300mm girth	4.83	5.00	5.07	1.02	–	12.69	0.30	18.06	m	20.32
450mm girth	7.25	5.00	7.61	1.52	–	18.91	0.46	26.98	m	30.35

CLADDING/COVERING

Labour hourly rates: (except Specialists) Craft Operatives £12.44 Labourer £7.02 Rates are national average prices. Refer to REGIONAL VARIATIONS for indicative levels of overall pricing in regions	MATERIALS			LABOUR				RATES		
	Del to Site	Waste	Material Cost	Craft Optve	Lab	Labour Cost	Sunds	Nett Rate	Unit	Gross Rate (+12.5%)
	£	%	£	Hrs	Hrs	£	£	£		£
H71: LEAD SHEET COVERINGS/FLASHINGS Cont.										
Ridges; horizontal; dressing over slating and tiling Cont.										
Nr 5; 150mm lapped joints; fixing to timber with milled lead clips										
240mm girth	4.81	5.00	5.05	0.92	-	11.44	0.36	16.86	m	18.96
300mm girth	6.02	5.00	6.32	1.15	-	14.31	0.36	20.99	m	23.61
450mm girth	9.02	5.00	9.47	1.72	-	21.40	0.55	31.42	m	35.35
Valleys; sloping										
Nr 3; dressing over tilting fillets -1; 150mm lapped joints; fixing to timber with copper nails										
240mm girth	2.83	5.00	2.97	0.74	-	9.21	0.25	12.43	m	13.98
300mm girth	3.54	5.00	3.72	0.92	-	11.44	0.25	15.41	m	17.34
450mm girth	5.31	5.00	5.58	1.38	-	17.17	0.35	23.09	m	25.98
Nr 4; dressing over tilting fillets -1; 150mm lapped joints; fixing to timber with copper nails										
240mm girth	3.86	5.00	4.05	0.82	-	10.20	0.30	14.55	m	16.37
300mm girth	4.83	5.00	5.07	1.02	-	12.69	0.30	18.06	m	20.32
450mm girth	7.55	5.00	7.93	1.52	-	18.91	0.46	27.30	m	30.71
Nr 5; dressing over tilting fillets -1; 150mm lapped joints; fixing to timber with copper nails										
240mm girth	4.81	5.00	5.05	0.92	-	11.44	0.36	16.86	m	18.96
300mm girth	6.01	5.00	6.31	1.15	-	14.31	0.36	20.98	m	23.60
450mm girth	9.02	5.00	9.47	1.72	-	21.40	0.55	31.42	m	35.35
Cavity gutters										
Nr 3; 150mm lapped joints; bedding in cement mortar (1:3)										
225mm girth	2.66	5.00	2.79	0.74	-	9.21	0.46	12.46	m	14.02
345mm girth	4.07	5.00	4.27	1.10	-	13.68	0.69	18.65	m	20.98
Nr 3 edges										
welted	-	-	-	0.20	-	2.49	-	2.49	m	2.80
beaded	-	-	-	0.20	-	2.49	-	2.49	m	2.80
Nr 4 edges										
welted	-	-	-	0.20	-	2.49	-	2.49	m	2.80
beaded	-	-	-	0.20	-	2.49	-	2.49	m	2.80
Nr 5 edges										
welted	-	-	-	0.20	-	2.49	-	2.49	m	2.80
beaded	-	-	-	0.20	-	2.49	-	2.49	m	2.80
Nr 3 seams										
leadburned	0.94	-	0.94	0.23	-	2.86	-	3.80	m	4.28
Nr 4 seams										
leadburned	1.17	-	1.17	0.23	-	2.86	-	4.03	m	4.54
Nr 5 seams										
leadburned	1.45	-	1.45	0.23	-	2.86	-	4.31	m	4.85
Nr 3 dressings										
corrugated roofing; fibre cement; down corrugations	-	-	-	0.29	-	3.61	-	3.61	m	4.06
corrugated roofing; fibre cement; across corrugations	-	-	-	0.29	-	3.61	-	3.61	m	4.06
glass and glazing bars; timber	-	-	-	0.29	-	3.61	-	3.61	m	4.06
Nr 4 dressings										
corrugated roofing; fibre cement; down corrugations	-	-	-	0.29	-	3.61	-	3.61	m	4.06
corrugated roofing; fibre cement; across corrugations	-	-	-	0.29	-	3.61	-	3.61	m	4.06
glass and glazing bars; timber	-	-	-	0.29	-	3.61	-	3.61	m	4.06
Nr 5 dressings										
corrugated roofing; fibre cement; down corrugations	-	-	-	0.29	-	3.61	-	3.61	m	4.06
corrugated roofing; fibre cement; across corrugations	-	-	-	0.29	-	3.61	-	3.61	m	4.06
glass and glazing bars; timber	-	-	-	0.29	-	3.61	-	3.61	m	4.06
Nr 3 soakers and slates handed to others for fixing										
180 x 180mm	0.38	5.00	0.40	0.23	-	2.86	-	3.26	nr	3.67
180 x 300mm	0.64	5.00	0.67	0.23	-	2.86	-	3.53	nr	3.97
450 x 450mm	2.39	5.00	2.51	0.23	-	2.86	-	5.37	nr	6.04
Nr 4 soakers and slates handed to others for fixing										
180 x 180mm	0.52	5.00	0.55	0.23	-	2.86	-	3.41	nr	3.83
180 x 300mm	0.87	5.00	0.91	0.23	-	2.86	-	3.77	nr	4.25
450 x 450mm	2.36	5.00	2.48	0.23	-	2.86	-	5.34	nr	6.01
Nr 5 soakers and slates handed to others for fixing										
180 x 180mm	0.65	5.00	0.68	0.23	-	2.86	-	3.54	nr	3.99
180 x 300mm	1.08	5.00	1.13	0.23	-	2.86	-	4.00	nr	4.49

Labour hourly rates: (except Specialists) Craft Operatives £12.44 Labourer £7.02 Rates are national average prices. Refer to REGIONAL VARIATIONS for indicative levels of overall pricing in regions	MATERIALS			LABOUR				RATES		
	Del to Site £	Waste %	Material Cost £	Craft Optve Hrs	Lab Hrs	Labour Cost £	Sunds £	Nett Rate £	Unit	Gross Rate (+12.5%) £
H71: LEAD SHEET COVERINGS/FLASHINGS Cont.										
Cavity gutters Cont.										
Nr 5 soakers and slates Cont.										
handed to others for fixing Cont.										
450 x 450mm....................................	4.06	5.00	4.26	0.23	–	2.86	–	7.12	nr	8.01
Nr 3 collars around pipes, standards and the like										
150mm long; soldered joints to metal covering										
50mm diameter.................................	2.86	5.00	3.00	1.00	–	12.44	–	15.44	nr	17.37
100mm diameter................................	3.21	5.00	3.37	1.15	–	14.31	–	17.68	nr	19.89
Nr 4 collars around pipes, standards and the like										
150mm long; soldered joints to metal covering										
50mm diameter.................................	3.69	5.00	3.87	1.15	–	14.31	–	18.18	nr	20.45
100mm diameter................................	4.12	5.00	4.33	1.15	–	14.31	–	18.63	nr	20.96
Nr 5 collars around pipes, standards and the like										
150mm long; soldered joints to metal covering										
50mm diameter.................................	4.61	5.00	4.84	1.15	–	14.31	–	19.15	nr	21.54
100mm diameter................................	5.14	5.00	5.40	1.15	–	14.31	–	19.70	nr	22.17
Nr 3 dots										
cast lead..	0.49	5.00	0.51	0.77	–	9.58	–	10.09	nr	11.35
soldered	5.40	5.00	5.67	0.86	–	10.70	–	16.37	nr	18.41
Nr 4 dots										
cast lead..	0.65	5.00	0.68	0.77	–	9.58	–	10.26	nr	11.54
soldered	5.40	5.00	5.67	0.86	–	10.70	–	16.37	nr	18.41
Nr 5 dots										
cast lead..	0.77	5.00	0.81	0.77	–	9.58	–	10.39	nr	11.69
soldered	5.40	5.00	5.67	0.86	–	10.70	–	16.37	nr	18.41
H72: ALUMINIUM SHEET STRIP COVERINGS/FLASHINGS										
Aluminium sheet, B.S.1470 grade S1BO, commercial purity										
0.60mm thick roof coverings; fixing to timber with aluminium cleats and aluminium alloy screws										
pitch 7.5 degrees from horizontal	8.66	5.00	9.09	5.05	–	62.82	0.14	72.06	m2	81.06
pitch 40 degrees from horizontal	8.66	5.00	9.09	5.70	–	70.91	0.14	80.14	m2	90.16
pitch 75 degrees from horizontal	8.66	5.00	9.09	6.35	–	78.99	0.14	88.23	m2	99.26
Flashings; horizontal										
0.60mm thick; 150mm lapped joints; fixing to masonry with aluminium clips and wedges										
150mm girth..................................	1.57	5.00	1.65	0.51	–	6.34	0.14	8.13	m	9.15
240mm girth..................................	2.93	5.00	3.08	0.82	–	10.20	0.14	13.42	m	15.09
300mm girth..................................	2.93	5.00	3.08	1.02	–	12.69	0.14	15.91	m	17.89
Flashings; stepped										
0.60mm thick; 150mm lapped joints; fixing to masonry with aluminium clips and wedges										
180mm girth..................................	2.01	5.00	2.11	0.82	–	10.20	0.21	12.52	m	14.09
240mm girth..................................	2.93	5.00	3.08	1.09	–	13.56	0.21	16.85	m	18.95
300mm girth..................................	2.93	5.00	3.08	1.37	–	17.04	0.21	20.33	m	22.87
Aprons; horizontal										
0.60mm thick; 150mm lapped joints; fixing to masonry with aluminium clips and wedges										
150mm girth..................................	1.57	5.00	1.65	0.51	–	6.34	0.14	8.13	m	9.15
180mm girth..................................	2.16	5.00	2.27	0.82	–	10.20	0.14	12.61	m	14.18
240mm girth..................................	2.93	5.00	3.08	1.02	–	12.69	0.14	15.91	m	17.89
450mm girth..................................	4.09	5.00	4.29	1.53	–	19.03	0.21	23.54	m	26.48
Hips; sloping										
0.60mm thick; 150mm lapped joints; fixing to timber with aluminium clips										
240mm girth..................................	2.93	5.00	3.08	0.82	–	10.20	0.14	13.42	m	15.09
300mm girth..................................	2.93	5.00	3.08	1.02	–	12.69	0.14	15.91	m	17.89
450mm girth..................................	4.09	5.00	4.29	1.53	–	19.03	0.21	23.54	m	26.48
Kerbs; horizontal										
0.60mm thick; 150mm lapped joints; fixing to timber with aluminium clips										
240mm girth..................................	2.93	5.00	3.08	0.82	–	10.20	0.14	13.42	m	15.09
300mm girth..................................	2.93	5.00	3.08	1.02	–	12.69	0.14	15.91	m	17.89
450mm girth..................................	4.09	5.00	4.29	1.53	–	19.03	0.21	23.54	m	26.48
Ridges; horizontal										
0.60mm thick; 150mm lapped joints; fixing to timber with aluminium clips										
240mm girth..................................	2.93	5.00	3.08	0.82	–	10.20	0.14	13.42	m	15.09
300mm girth..................................	2.93	5.00	3.08	1.02	–	12.69	0.14	15.91	m	17.89
450mm girth..................................	4.09	5.00	4.29	1.53	–	19.03	0.21	23.54	m	26.48
Valleys; sloping										
0.60mm thick; 150mm lapped joints; fixing to timber with aluminium clips										
240mm girth..................................	2.93	5.00	3.08	0.82	–	10.20	0.14	13.42	m	15.09
300mm girth..................................	2.93	5.00	3.08	1.02	–	12.69	0.14	15.91	m	17.89
450mm girth..................................	4.09	5.00	4.29	1.53	–	19.03	0.21	23.54	m	26.48

CLADDING/COVERING

Labour hourly rates: (except Specialists) Craft Operatives £12.44 Labourer £7.02 Rates are national average prices. Refer to REGIONAL VARIATIONS for indicative levels of overall pricing in regions	MATERIALS			LABOUR				RATES		
	Del to Site £	Waste %	Material Cost £	Craft Optve Hrs	Lab Hrs	Labour Cost £	Sunds £	Nett Rate £	Unit	Gross Rate (+12.5%) £
H72: ALUMINIUM SHEET STRIP COVERINGS/FLASHINGS Cont.										
Aluminium sheet, B.S.1470 grade S1BO, commercial purity Cont.										
0.60mm thick edges										
welted.............	-	-	-	0.20	-	2.49	-	2.49	m	2.80
beaded.............	-	-	-	0.20	-	2.49	-	2.49	m	2.80
0.60mm thick soakers and slates										
handed to others for fixing										
180 x 180mm.........	0.46	5.00	0.48	0.23	-	2.86	-	3.34	nr	3.76
180 x 300mm.........	0.76	5.00	0.80	0.23	-	2.86	-	3.66	nr	4.12
H73: COPPER STRIP SHEET COVERINGS/FLASHINGS										
Copper sheet; B.S.2870										
0.60mm thick roof coverings; fixing to timber with										
copper cleats and copper nails										
pitch 7.5 degrees from horizontal	17.85	5.00	18.74	5.05	-	62.82	0.41	81.97	m2	92.22
pitch 40 degrees from horizontal	17.85	5.00	18.74	5.70	-	70.91	0.41	90.06	m2	101.32
pitch 75 degrees from horizontal	17.85	5.00	18.74	6.35	-	78.99	0.41	98.15	m2	110.41
0.70mm thick roof coverings; fixing to timber with										
copper cleats and copper nails										
pitch 7.5 degrees from horizontal	22.40	5.00	23.52	5.70	-	70.91	0.52	94.95	m2	106.82
pitch 40 degrees from horizontal	22.40	5.00	23.52	6.35	-	78.99	0.52	103.03	m2	115.91
pitch 75 degrees from horizontal	22.40	5.00	23.52	6.95	-	86.46	0.52	110.50	m2	124.31
Flashings; horizontal										
0.60mm thick; 150mm lapped joints; fixing to										
masonry with copper clips and wedges										
150mm girth.....................	2.68	5.00	2.81	0.51	-	6.34	0.41	9.57	m	10.76
240mm girth.....................	4.28	5.00	4.49	0.82	-	10.20	0.41	15.10	m	16.99
300mm girth.....................	5.36	5.00	5.63	1.02	-	12.69	0.41	18.73	m	21.07
0.70mm thick; 150mm lapped joints; fixing to										
masonry with copper clips and wedges										
150mm girth.....................	3.36	5.00	3.53	0.58	-	7.22	0.52	11.26	m	12.67
240mm girth.....................	5.38	5.00	5.65	0.92	-	11.44	0.52	17.61	m	19.82
300mm girth.....................	6.72	5.00	7.06	1.15	-	14.31	0.52	21.88	m	24.62
Flashings; stepped										
0.60mm thick; 150mm lapped joints; fixing to										
masonry with copper clips and wedges										
180mm girth.....................	3.21	5.00	3.37	0.82	-	10.20	0.62	14.19	m	15.97
240mm girth.....................	4.28	5.00	4.49	1.09	-	13.56	0.62	18.67	m	21.01
300mm girth.....................	5.36	5.00	5.63	1.37	-	17.04	0.62	23.29	m	26.20
0.70mm thick; 150mm lapped joints; fixing to										
masonry with copper clips and wedges										
180mm girth.....................	4.03	5.00	4.23	0.92	-	11.44	0.78	16.46	m	18.51
240mm girth.....................	5.58	5.00	5.86	1.23	-	15.30	0.78	21.94	m	24.68
300mm girth.....................	6.72	5.00	7.06	1.53	-	19.03	0.78	26.87	m	30.23
Aprons; horizontal										
0.60mm thick; 150mm lapped joints; fixing to										
timber with copper clips										
150mm girth.....................	2.68	5.00	2.81	0.51	-	6.34	0.41	9.57	m	10.76
240mm girth.....................	4.28	5.00	4.49	0.82	-	10.20	0.41	15.10	m	16.99
300mm girth.....................	5.36	5.00	5.63	1.02	-	12.69	0.41	18.73	m	21.07
450mm girth.....................	8.03	5.00	8.43	1.53	-	19.03	0.62	28.08	m	31.60
0.70mm thick; 150mm lapped joints; fixing to										
timber with copper clips										
150mm girth.....................	3.36	5.00	3.53	0.58	-	7.22	0.52	11.26	m	12.67
240mm girth.....................	5.38	5.00	5.65	0.92	-	11.44	0.52	17.61	m	19.82
300mm girth.....................	6.72	5.00	7.06	1.23	-	15.30	0.52	22.88	m	25.74
450mm girth.....................	10.08	5.00	10.58	1.53	-	19.03	0.78	30.40	m	34.20
Hips; sloping										
0.60mm thick; 150mm lapped joints; fixing to										
timber with copper clips										
240mm girth.....................	4.28	5.00	4.49	0.71	-	8.83	0.36	13.69	m	15.40
300mm girth.....................	5.36	5.00	5.63	1.02	-	12.69	0.41	18.73	m	21.07
450mm girth.....................	8.03	5.00	8.43	1.53	-	19.03	0.62	28.08	m	31.60
0.70mm thick; 150mm lapped joints; fixing to										
timber with copper clips										
240mm girth.....................	5.38	5.00	5.65	0.92	-	11.44	0.52	17.61	m	19.82
300mm girth.....................	6.72	5.00	7.06	1.23	-	15.30	0.52	22.88	m	25.74
450mm girth.....................	10.08	5.00	10.58	1.53	-	19.03	0.78	30.40	m	34.20
Kerbs; horizontal										
0.60mm thick; 150mm lapped joints; fixing to										
timber with copper clips										
240mm girth.....................	4.28	5.00	4.49	0.82	-	10.20	0.41	15.10	m	16.99
300mm girth.....................	5.36	5.00	5.63	1.02	-	12.69	0.41	18.73	m	21.07
450mm girth.....................	8.03	5.00	8.43	1.53	-	19.03	0.62	28.08	m	31.60
0.70mm thick; 150mm lapped joints; fixing to										
timber with copper clips										
240mm girth.....................	5.38	5.00	5.65	0.92	-	11.44	0.52	17.61	m	19.82
300mm girth.....................	6.72	5.00	7.06	1.23	-	15.30	0.52	22.88	m	25.74
450mm girth.....................	10.08	5.00	10.58	1.53	-	19.03	0.78	30.40	m	34.20
Ridges; horizontal										
0.60mm thick; 150mm lapped joints; fixing to										
timber with copper clips										
240mm girth.....................	4.28	5.00	4.49	0.82	-	10.20	0.41	15.10	m	16.99

Labour hourly rates: (except Specialists) Craft Operatives £12.44 Labourer £7.02 Rates are national average prices. Refer to REGIONAL VARIATIONS for indicative levels of overall pricing in regions	MATERIALS			LABOUR				RATES		
	Del to Site £	Waste %	Material Cost £	Craft Optve Hrs	Lab Hrs	Labour Cost £	Sunds £	Nett Rate £	Unit	Gross Rate (+12.5%) £
H73: COPPER STRIP SHEET COVERINGS/FLASHINGS Cont.										
Copper sheet; B.S.2870 Cont.										
Ridges; horizontal Cont.										
0.60mm thick; 150mm lapped joints; fixing to										
timber with copper clips Cont.										
300mm girth......................	4.98	5.00	5.23	1.02	-	12.69	0.41	18.33	m	20.62
450mm girth......................	8.03	5.00	8.43	1.53	-	19.03	0.62	28.08	m	31.60
0.70mm thick; 150mm lapped joints; fixing to										
timber with copper clips										
240mm girth......................	5.38	5.00	5.65	0.92	-	11.44	0.52	17.61	m	19.82
300mm girth......................	6.72	5.00	7.06	1.23	-	15.30	0.52	22.88	m	25.74
450mm girth......................	10.08	5.00	10.58	1.53	-	19.03	0.78	30.40	m	34.20
Valleys; sloping										
0.60mm thick; 150mm lapped joints; fixing to										
timber with copper clips										
240mm girth......................	4.28	5.00	4.49	0.82	-	10.20	0.41	15.10	m	16.99
300mm girth......................	5.36	5.00	5.63	1.02	-	12.69	0.41	18.73	m	21.07
450mm girth......................	8.03	5.00	8.43	1.53	-	19.03	0.62	28.08	m	31.60
0.70mm thick; 150mm lapped joints; fixing to										
timber with copper clips										
240mm girth......................	5.38	5.00	5.65	0.80	-	9.95	0.45	16.05	m	18.06
300mm girth......................	6.72	5.00	7.06	1.07	-	13.31	0.45	20.82	m	23.42
450mm girth......................	10.08	5.00	10.58	1.33	-	16.55	0.68	27.81	m	31.29
0.60mm thick edges										
welted	-	-	-	0.17	-	2.11	-	2.11	m	2.38
beaded	-	-	-	0.17	-	2.11	-	2.11	m	2.38
0.70mm thick edges										
welted	-	-	-	0.17	-	2.11	-	2.11	m	2.38
beaded	-	-	-	0.17	-	2.11	-	2.11	m	2.38
H74: ZINC STRIP SHEET COVERINGS/FLASHINGS										
Zinc alloy sheet; B.S.6561										
0.65mm thick roof coverings; 25mm standing seam;										
fixing to timber with zinc clips										
pitch 7.5 degrees from horizontal	12.55	5.00	13.18	5.05	-	62.82	0.15	76.15	m2	85.67
pitch 40 degrees from horizontal	12.55	5.00	13.18	5.70	-	70.91	0.15	84.24	m2	94.76
pitch 75 degrees from horizontal	12.55	5.00	13.18	6.35	-	78.99	0.15	92.32	m2	103.86
0.8mm thick roof coverings; roll cap; fixing to										
timber with zinc clips										
pitch 7.5 degrees from horizontal	15.40	5.00	16.17	5.70	-	70.91	0.20	87.28	m2	98.19
pitch 40 degrees from horizontal	15.40	5.00	16.17	6.35	-	78.99	0.20	95.36	m2	107.28
pitch 75 degrees from horizontal	15.40	5.00	16.17	6.95	-	86.46	0.20	102.83	m2	115.68
Flashings; horizontal										
0.8mm thick; 150mm lapped joints; fixing to										
masonry with zinc clips and wedges										
150mm girth......................	2.31	5.00	2.43	0.51	-	6.34	0.15	8.92	m	10.03
240mm girth......................	3.70	5.00	3.88	0.82	-	10.20	0.15	14.24	m	16.02
300mm girth......................	4.62	5.00	4.85	1.02	-	12.69	0.15	17.69	m	19.90
Flashings; stepped										
0.8mm thick; 150mm lapped joints; fixing to										
masonry with zinc clips and wedges										
180mm girth......................	2.77	5.00	2.91	0.71	-	8.83	0.20	11.94	m	13.43
240mm girth......................	3.70	5.00	3.88	1.09	-	13.56	0.23	17.67	m	19.88
300mm girth......................	4.62	5.00	4.85	1.37	-	17.04	0.23	22.12	m	24.89
Aprons; horizontal										
0.8mm thick; 150mm lapped joints; fixing to										
masonry with zinc clips and wedges										
150mm girth......................	2.31	5.00	2.43	0.92	-	11.44	0.31	14.18	m	15.95
240mm girth......................	3.70	5.00	3.88	0.82	-	10.20	0.15	14.24	m	16.02
300mm girth......................	4.62	5.00	4.85	1.02	-	12.69	0.15	17.69	m	19.90
450mm girth......................	6.93	5.00	7.28	1.53	-	19.03	0.23	26.54	m	29.86
Hips; sloping										
0.8mm thick; 150mm lapped joints; fixing to timber										
with zinc clips										
240mm girth......................	3.70	5.00	3.88	0.82	-	10.20	0.15	14.24	m	16.02
300mm girth......................	4.62	5.00	4.85	1.02	-	12.69	0.15	17.69	m	19.90
450mm girth......................	6.93	5.00	7.28	1.53	-	19.03	0.23	26.54	m	29.86
Kerbs; horizontal										
0.8mm thick; 150mm lapped joints; fixing to timber										
with zinc clips										
240mm girth......................	3.70	5.00	3.88	0.82	-	10.20	0.15	14.24	m	16.02
300mm girth......................	4.62	5.00	4.85	1.02	-	12.69	0.15	17.69	m	19.90
450mm girth......................	6.93	5.00	7.28	1.53	-	19.03	0.23	26.54	m	29.86
Ridges; horizontal										
0.8mm thick; 150mm lapped joints; fixing to timber										
with zinc clips										
240mm girth......................	3.70	5.00	3.88	0.82	-	10.20	0.15	14.24	m	16.02
300mm girth......................	4.62	5.00	4.85	1.02	-	12.69	0.15	17.69	m	19.90
450mm girth......................	6.93	5.00	7.28	1.53	-	19.03	0.23	26.54	m	29.86

Labour hourly rates: (except Specialists) Craft Operatives £12.44 Labourer £7.02 Rates are national average prices. Refer to REGIONAL VARIATIONS for indicative levels of overall pricing in regions	MATERIALS			LABOUR				RATES		
	Del to Site	Waste	Material Cost	Craft Optve	Lab	Labour Cost	Sunds	Nett Rate	Unit	Gross Rate (+12.5%)
	£	%	£	Hrs	Hrs	£	£	£		£
H74: ZINC STRIP SHEET COVERINGS/FLASHINGS Cont.										
Zinc alloy sheet; B.S.6561 Cont.										

Valleys; sloping
  0.8mm thick; 150mm lapped joints; fixing to timber
  with zinc clips										

240mm girth....................................	3.70	5.00	3.88	0.82	-	10.20	0.15	14.24	m	16.02
300mm girth....................................	4.62	5.00	4.85	1.02	-	12.69	0.15	17.69	m	19.90
450mm girth....................................	6.93	5.00	7.28	1.53	-	19.03	0.23	26.54	m	29.86
0.65mm thick edges										
welted..	-	-	-	0.20	-	2.49	-	2.49	m	2.80
beaded..	-	-	-	0.20	-	2.49	-	2.49	m	2.80
0.8mm thick edges										
welted..	-	-	-	0.20	-	2.49	-	2.49	m	2.80
beaded..	-	-	-	0.20	-	2.49	-	2.49	m	2.80
0.65mm thick soakers and slates handed to others for fixing										
180 x 180mm..................................	0.41	5.00	0.43	0.23	-	2.86	-	3.29	nr	3.70
180 x 300mm..................................	0.68	5.00	0.71	0.23	-	2.86	-	3.58	nr	4.02
450 x 450mm..................................	2.53	5.00	2.66	0.23	-	2.86	-	5.52	nr	6.21
0.8mm thick soakers and slates - handed to others for fixing										
180 x 180mm..................................	0.50	5.00	0.53	0.23	-	2.86	-	3.39	nr	3.81
180 x 300mm..................................	0.83	5.00	0.87	0.23	-	2.86	-	3.73	nr	4.20
450 x 450mm..................................	3.12	5.00	3.28	0.23	-	2.86	-	6.14	nr	6.90
0.65mm thick collars around pipes, standards and the like										
150mm long; soldered joints to metal covering										
50mm diameter...............................	2.52	5.00	2.65	1.00	-	12.44	-	15.09	nr	16.97
100mm diameter..............................	2.83	5.00	2.97	1.00	-	12.44	-	15.41	nr	17.34
0.8mm thick collars around pipes, standards and the like										
150mm long; soldered joints to metal covering										
50mm diameter...............................	3.11	5.00	3.27	1.00	-	12.44	-	15.71	nr	17.67
10Qmm diameter..............................	3.47	5.00	3.64	1.00	-	12.44	-	16.08	nr	18.09

Waterproofing

Labour hourly rates: (except Specialists) Craft Operatives £9.23 Labourer £7.02 Rates are national average prices. Refer to REGIONAL VARIATIONS for indicative levels of overall pricing in regions	MATERIALS			LABOUR				RATES		
	Del to Site £	Waste %	Material Cost . £	Craft Optve Hrs	Lab Hrs	Labour Cost £	Sunds £	Nett Rate £	Unit	Gross Rate (+12.5%) £
J20: MASTIC ASPHALT TANKING/ DAMP PROOF MEMBRANES										
Tanking and damp proofing B.S.6925 (limestone aggregate)										
Tanking and damp proofing 13mm thick; coats of asphalt -1; to concrete base; horizontal work subsequently covered										
width not exceeding 150mm.....................	-	-	Spclist	-	-	Spclist	-	33.15	m2	37.29
width 150 - 225mm...........................	-	-	Spclist	-	-	Spclist	-	27.80	m2	31.27
width 225 - 300mm...........................	-	-	Spclist	-	-	Spclist	-	22.20	m2	24.98
width exceeding 300mm	-	-	Spclist	-	-	Spclist	-	16.65	m2	18.73
Tanking and damp proofing 20mm thick; coats of asphalt -2; to concrete base; horizontal work subsequently covered										
width not exceeding 150mm.....................	-	-	Spclist	-	-	Spclist	-	40.00	m2	45.00
width 150 - 225mm...........................	-	-	Spclist	-	-	Spclist	-	33.55	m2	37.74
width 225 - 300mm...........................	-	-	Spclist	-	-	Spclist	-	26.65	m2	29.98
width exceeding 300mm.....................	-	-	Spclist	-	-	Spclist	-	20.25	m2	22.78
Tanking and damp proofing 30mm thick; coats of asphalt -3; to concrete base; horizontal work subsequently covered										
width not exceeding 150mm.....................	-	-	Spclist	-	-	Spclist	-	58.15	m2	65.42
width 150 - 225mm...........................	-	-	Spclist	-	-	Spclist	-	48.70	m2	54.79
width 225 - 300mm...........................	-	-	Spclist	-	-	Spclist	-	38.80	m2	43.65
width exceeding 300mm	-	-	Spclist	-	-	Spclist	-	29.10	m2	32.74
Tanking and damp proofing 20mm thick; coats of asphalt -2; to concrete base; vertical work subsequently covered										
width not exceeding 150mm.....................	-	-	Spclist	-	-	Spclist	-	68.75	m2	77.34
width 150 - 225mm...........................	-	-	Spclist	-	-	Spclist	-	57.40	m2	64.58
width 225 - 300mm.........................	-	-	Spclist	-	-	Spclist	-	45.85	m2	51.58
width exceeding 300mm.....................	-	-	Spclist	-	-	Spclist	-	34.35	m2	38.64
Tanking and damp proofing 20mm thick; coats of asphalt -3; to concrete base; vertical work subsequently covered										
width not exceeding 150mm.....................	-	-	Spclist	-	-	Spclist	-	95.20	m2	107.10
width 150 - 225mm...........................	-	-	Spclist	-	-	Spclist	-	79.55	m2	89.49
width 225 - 300mm...........................	-	-	Spclist	-	-	Spclist	-	63.45	m2	71.38
width exceeding 300mm.....................	-	-	Spclist	-	-	Spclist	-	47.60	m2	53.55
Internal angle fillets to concrete base; priming base with bitumen two coats; work subsequently covered	-	-	Spclist	-	-	Spclist	-	4.95	m	5.57
J21: MASTIC ASPHALT ROOFING/INSULATION/FINISHES										
Roofing; B.S.6925 (limestone aggregate); sand rubbing; B.S.747 type 4A sheathing felt isolating membrane, butt joints										
Roofing 20mm thick; coats of asphalt -2; to concrete base; pitch not exceeding 6 degrees from horizontal										
width exceeding 300mm	-	-	Spclist	-	-	Spclist	-	18.00	m2	20.25
extra; solid water check roll	-	-	Spclist	-	-	Spclist	-	12.35	m	13.89
Roofing 20mm thick; coats of asphalt -2; to concrete base; pitch 30 degrees from horizontal										
width exceeding 300mm	-	-	Spclist	-	-	Spclist	-	32.50	m2	36.56
Roofing 20mm thick; coats of asphalt -2; to concrete base; pitch 45 degrees from horizontal										
width exceeding 300mm	-	-	Spclist	-	-	Spclist	-	32.50	m2	36.56
Paint two coats of Solar reflective roof paint on surfaces of asphalt										
width exceeding 300mm	-	-	Spclist	-	-	Spclist	-	3.75	m2	4.22

Labour hourly rates: (except Specialists) Craft Operatives £9.23 Labourer £7.02 Rates are national average prices. Refer to REGIONAL VARIATIONS for indicative levels of overall pricing in regions	MATERIALS			LABOUR				RATES		
	Del to Site £	Waste %	Material Cost £	Craft Optve Hrs	Lab Hrs	Labour Cost £	Sunds £	Nett Rate £	Unit	Gross Rate (+12.5%) £
J21: MASTIC ASPHALT ROOFING/INSULATION/FINISHES Cont.										
Roofing; B.S.6925 (limestone aggregate); sand rubbing; B.S.747 type 4A sheathing felt isolating membrane, butt joints Cont.										
Skirtings 13mm thick; coats of asphalt -2; to brickwork base										
girth not exceeding 150mm	-	-	Spclist	-	-	Spclist	-	12.25	m	13.78
Coverings to kerbs 20mm thick; coats of asphalt -2; to concrete base										
girth 600mm	-	-	Spclist	-	-	Spclist	-	33.30	m	37.46
Jointing new roofing to existing										
20mm thick	-	-	Spclist	-	-	Spclist	-	9.20	m	10.35
Edge trim; aluminium; silver anodised priming with bituminous primer; butt joints with internal jointing sleeves; fixing with aluminium alloy screws to timber; working two coat asphalt into grooves -1										
63.5mm wide x 44.4mm face depth	6.13	5.00	6.44	0.40	-	3.69	1.38	11.51	m	12.95
63.5mm wide x 76.2mm face depth	7.61	5.00	7.99	0.40	-	3.69	1.44	13.12	m	14.76
104.8mm wide x 44.4mm face depth	7.74	5.00	8.13	0.40	-	3.69	1.44	13.26	m	14.92
104.8mm wide x 76.2mm face depth	11.20	5.00	11.76	0.50	-	4.62	1.64	18.02	m	20.27
76.2mm fixing arm at 10 degrees x 38.1mm face depth	8.11	5.00	8.52	0.40	-	3.69	1.44	13.65	m	15.35
Edge trim; glass fibre reinforced butt joints; fixing with stainless steel screws to timber; working two coat asphalt into grooves -1 .	9.17	5.00	9.63	0.35	-	3.23	1.64	14.50	m	16.31
Roof Ventilators										
plastic; setting in position	6.68	2.50	6.85	0.85	-	7.85	0.54	15.23	nr	17.14
aluminium; setting in position	7.43	2.50	7.62	0.85	-	7.85	0.54	16.00	nr	18.00
Roofing; B.S.6925 (limestone aggregate); covering with 13mm white spar chippings in hot bitumen; B.S.747 type 4A sheathing felt isolating membrane; butt joints										
Roofing 20mm thick; coats of asphalt -2; to concrete base; pitch not exceeding 6 degrees from horizontal										
width exceeding 300mm	-	-	Spclist	-	-	Spclist	-	21.35	m2	24.02
Roofing 20mm thick; coats of asphalt -2; to concrete base; pitch 30 degrees from horizontal										
width exceeding 300mm	-	-	Spclist	-	-	Spclist	-	34.05	m2	38.31
J30: LIQUID APPLIED TANKING/DAMP PROOF MEMBRANES										
Synthaprufe bituminous emulsion; blinding with sand										
Coverings, coats -2; to concrete base; horizontal										
width not exceeding 150mm	3.36	10.00	3.70	-	0.46	3.23	0.14	7.07	m2	7.95
width 150 - 225mm	3.36	10.00	3.70	-	0.38	2.67	0.14	6.50	m2	7.32
width 225 - 300mm	3.36	10.00	3.70	-	0.30	2.11	0.14	5.94	m2	6.68
width exceeding 300mm	3.36	10.00	3.70	-	0.23	1.61	0.14	5.45	m2	6.13
Coverings, coats -3; to concrete base; horizontal										
width not exceeding 150mm	5.03	10.00	5.53	-	0.64	4.49	0.16	10.19	m2	11.46
width 150 - 225mm	5.03	10.00	5.53	-	0.53	3.72	0.16	9.41	m2	10.59
width 225 - 300mm	5.03	10.00	5.53	-	0.42	2.95	0.16	8.64	m2	9.72
width exceeding 300mm	5.03	10.00	5.53	-	0.32	2.25	0.16	7.94	m2	8.93
Coverings; coats -2; to concrete base; vertical										
width not exceeding 150mm	3.36	10.00	3.70	-	0.69	4.84	0.14	8.68	m2	9.76
width 150 - 225mm	3.36	10.00	3.70	-	0.57	4.00	0.14	7.84	m2	8.82
width 225 - 300mm	3.36	10.00	3.70	-	0.45	3.16	0.14	7.00	m2	7.87
width exceeding 300mm	3.36	10.00	3.70	-	0.35	2.46	0.14	6.29	m2	7.08
Coverings; coats -3; to concrete base; vertical										
width not exceeding 150mm	5.03	10.00	5.53	-	0.96	6.74	0.16	12.43	m2	13.99
width 150 - 225mm	5.03	10.00	5.53	-	0.80	5.62	0.16	11.31	m2	12.72
width 225 - 300mm	5.03	10.00	5.53	-	0.63	4.42	0.16	10.12	m2	11.38
width exceeding 300mm	5.03	10.00	5.53	-	0.48	3.37	0.16	9.06	m2	10.20
Bituminous emulsion										
Coverings, coats -2; to concrete base; horizontal										
width not exceeding 150mm	1.59	10.00	1.75	-	0.39	2.74	0.14	4.63	m2	5.21
width 150 - 225mm	1.59	10.00	1.75	-	0.32	2.25	0.14	4.14	m2	4.65
width 225 - 300mm	1.59	10.00	1.75	-	0.25	1.75	0.14	3.64	m2	4.10
width exceeding 300mm	1.59	10.00	1.75	-	0.20	1.40	0.14	3.29	m2	3.70
Coverings, coats -3; to concrete base; horizontal										
width not exceeding 150mm	2.18	10.00	2.40	-	0.55	3.86	0.16	6.42	m2	7.22
width 150 - 225mm	2.18	10.00	2.40	-	0.45	3.16	0.16	5.72	m2	6.43
width 225 - 300mm	2.18	10.00	2.40	-	0.36	2.53	0.16	5.09	m2	5.72
width exceeding 300mm	2.18	10.00	2.40	-	0.28	1.97	0.16	4.52	m2	5.09
Coverings; coats -2; to concrete base; vertical										
width not exceeding 150mm	1.59	10.00	1.75	-	0.58	4.07	0.14	5.96	m2	6.71
width 150 - 225mm	1.59	10.00	1.75	-	0.48	3.37	0.14	5.26	m2	5.92
width 225 - 300mm	1.59	10.00	1.75	-	0.38	2.67	0.14	4.56	m2	5.13

Labour hourly rates: (except Specialists) Craft Operatives £9.23 Labourer £7.02 Rates are national average prices. Refer to REGIONAL VARIATIONS for indicative levels of overall pricing in regions	MATERIALS			LABOUR				RATES		
	Del to Site	Waste	Material Cost	Craft Optve	Lab	Labour Cost	Sunds	Nett Rate	Unit	Gross Rate (+12.5%)
	£	%	£	Hrs	Hrs	£	£	£		£
J30: LIQUID APPLIED TANKING/DAMP PROOF MEMBRANES Cont.										
Bituminous emulsion Cont.										
Coverings; coats -2; to concrete base; vertical Cont.										
width exceeding 300mm	1.59	10.00	1.75	-	0.30	2.11	0.14	4.00	m2	4.49
Coverings; coats -3; to concrete base; vertical										
width not exceeding 150mm	2.18	10.00	2.40	-	0.82	5.76	0.16	8.31	m2	9.35
width 150 - 225mm	2.18	10.00	2.40	-	0.68	4.77	0.16	7.33	m2	8.25
width 225 - 300mm	2.18	10.00	2.40	-	0.54	3.79	0.16	6.35	m2	7.14
width exceeding 300mm	2.18	10.00	2.40	-	0.42	2.95	0.16	5.51	m2	6.19
R.I.W. liquid asphaltic composition										
Coverings, coats -2; to concrete base; horizontal										
width not exceeding 150mm	3.36	10.00	3.70	-	0.46	3.23	0.14	7.07	m2	7.95
width 150 - 225mm	3.36	10.00	3.70	-	0.38	2.67	0.14	6.50	m2	7.32
width 225 - 300mm	3.36	10.00	3.70	-	0.30	2.11	0.14	5.94	m2	6.68
width exceeding 300mm	3.36	10.00	3.70	-	0.23	1.61	0.14	5.45	m2	6.13
Coverings, coats -3; to concrete base; horizontal										
width not exceeding 150mm	5.03	10.00	5.53	-	0.64	4.49	0.16	10.19	m2	11.46
width 150 - 225mm	5.03	10.00	5.53	-	0.53	3.72	0.16	9.41	m2	10.59
width 225 - 300mm	5.03	10.00	5.53	-	0.42	2.95	0.16	8.64	m2	9.72
width exceeding 300mm	5.03	10.00	5.53	-	0.32	2.25	0.16	7.94	m2	8.93
Coverings; coats -2; to concrete base; vertical										
width not exceeding 150mm	3.36	10.00	3.70	-	0.69	4.84	0.14	8.68	m2	9.76
width 150 - 225mm	3.36	10.00	3.70	-	0.57	4.00	0.14	7.84	m2	8.82
width 225 - 300mm	3.36	10.00	3.70	-	0.45	3.16	0.14	7.00	m2	7.87
width exceeding 300mm	3.36	10.00	3.70	-	0.35	2.46	0.14	6.29	m2	7.08
Coverings; coats -3; to concrete base; vertical										
width not exceeding 150mm	5.03	10.00	5.53	-	0.96	6.74	0.16	12.43	m2	13.99
width 150 - 225mm	5.03	10.00	5.53	-	0.80	5.62	0.16	11.31	m2	12.72
width 225 - 300mm	5.03	10.00	5.53	-	0.63	4.42	0.16	10.12	m2	11.38
width exceeding 300mm	5.03	10.00	5.53	-	0.48	3.37	0.16	9.06	m2	10.20
J40: FLEXIBLE SHEET TANKING/DAMP PROOFING MEMBRANES										
Polythene sheeting; 100mm welted laps										
Tanking and damp proofing; 125mu										
horizontal; on concrete base	0.33	20.00	0.40	-	0.06	0.42	-	0.82	m2	0.92
vertical; on concrete base	0.33	20.00	0.40	-	0.09	0.63	-	1.03	m2	1.16
Tanking and damp proofing; 250mu										
horizontal; on concrete base	0.40	20.00	0.48	-	0.09	0.63	-	1.11	m2	1.25
vertical; on concrete base	0.40	20.00	0.48	-	0.14	0.98	-	1.46	m2	1.65
Tanking and damp proofing; 300mu										
horizontal; on concrete base	0.55	20.00	0.66	-	0.11	0.77	-	1.43	m2	1.61
vertical; on concrete base	0.55	20.00	0.66	-	0.17	1.19	-	1.85	m2	2.09
Bituthene 500X self adhesive damp proof membrane; 25mm lapped joints										
Tanking and damp proofing										
horizontal; on concrete base	3.89	20.00	4.67	-	0.18	1.26	-	5.93	m2	6.67
vertical; on concrete base	3.89	20.00	4.67	-	0.27	1.90	-	6.56	m2	7.38
Bituthene 500X self adhesive damp proof membrane; 75mm lapped joints										
Tanking and damp proofing										
horizontal; on concrete base	5.43	20.00	6.52	-	0.23	1.61	-	8.13	m2	9.15
vertical; on concrete base; priming with Servicised B primer	5.43	20.00	6.52	-	0.35	2.46	0.26	9.23	m2	10.39
extra; Bituthene internal angle fillet; 40 x 40mm	5.12	20.00	6.14	-	0.12	0.84	-	6.99	m	7.86
British Sisalkraft Ltd 728 damp proof membrane; 150mm laps sealed with tape										
Tanking and damp proofing										
horizontal; on concrete base	1.27	20.00	1.52	-	0.08	0.56	0.11	2.20	m2	2.47
Visqueen 1200 super damp proof membrane; 150mm laps sealed with tape and mastic										
Tanking and damp proofing										
horizontal; on concrete base	0.55	20.00	0.66	-	0.08	0.56	0.11	1.33	m2	1.50
J41: BUILT UP FELT ROOF COVERINGS										
Roofing; felt B.S.747, comprising 3G Rubervent underlay 2.6 kg/m2, 1 layer Ruberfort HP180 underlay 3.4 kg/m2, 1 layer Ruberfort HP180 mineral surface top sheet 3.6 kg/m2										
Roof coverings; layers of felt -3; bonding with hot bitumen compound to timber base										
pitch 7.5 degrees from horizontal	6.83	15.00	7.85	0.46	0.06	4.67	3.28	15.80	m2	17.78
pitch 40 degrees from horizontal	6.83	15.00	7.85	0.76	0.09	7.65	3.28	18.78	m2	21.13
pitch 75 degrees from horizontal	6.83	15.00	7.85	0.92	0.12	9.33	3.28	20.47	m2	23.03

Labour hourly rates: (except Specialists) Craft Operatives £9.23 Labourer £7.02 Rates are national average prices. Refer to REGIONAL VARIATIONS for indicative levels of overall pricing in regions	MATERIALS			LABOUR				RATES		
	Del to Site £	Waste %	Material Cost £	Craft Optve Hrs	Lab Hrs	Labour Cost £	Sunds £	Nett Rate £	Unit	Gross Rate (+12.5%) £
J41: BUILT UP FELT ROOF COVERINGS Cont.										
Roofing; felt B.S.747, comprising 3G Rubervent underlay 2.6 kg/m2, 1 layer Ruberfort HP180 underlay 3.4 kg/m2, 1 layer Ruberfort HP180 mineral surface top sheet 3.6 kg/m2 Cont.										
Roof coverings; layers of felt -3; bonding with hot bitumen compound to cement and sand or concrete base										
pitch 7.5 degrees from horizontal	6.83	15.00	7.85	0.46	0.06	4.67	3.28	15.80	m2	17.78
pitch 40 degrees from horizontal	6.83	15.00	7.85	0.76	0.09	7.65	3.28	18.78	m2	21.13
pitch 75 degrees from horizontal	6.83	15.00	7.85	0.92	0.12	9.33	3.28	20.47	m2	23.03
Skirtings; layers of felt - 3; bonding with hot bitumen compound to brickwork base										
girth 0 - 200mm	1.37	15.00	1.58	0.28	0.03	2.79	0.65	5.02	m	5.65
girth 200 - 400mm	2.73	15.00	3.14	0.46	0.06	4.67	1.30	9.11	m	10.24
Collars around pipes; standard and the like; layers of felt -1; bonding with hot bitumen compound to metal base										
50mm diameter x 150mm long; hole in 3 layer covering	0.25	15.00	0.29	0.17	0.02	1.71	0.17	2.17	nr	2.44
100mm diameter x 150mm long; hole in 3 layer covering	0.40	15.00	0.46	0.22	0.03	2.24	0.25	2.95	nr	3.32
Eaves trim; aluminium; silver anodised butt joints over matching sleeve pieces 200mm long; fixing to timber base with aluminium alloy screws; bedding in mastic										
63.5mm wide x 44.4mm face depth	6.59	5.00	6.92	0.40	-	3.69	1.50	12.11	m	13.63
63.5mm wide x 47.6mm face depth	6.63	5.00	6.96	0.40	-	3.69	1.50	12.15	m	13.67
63.5mm wide x 76.2mm face depth	8.11	5.00	8.52	0.40	-	3.69	1.58	13.79	m	15.51
104.8mm wide x 44.4mm face depth	8.34	5.00	8.76	0.40	-	3.69	1.58	14.03	m	15.78
104.8mm wide x 76.2mm face depth	12.05	5.00	12.65	0.50	-	4.62	1.77	19.04	m	21.42
Eaves trim; glass fibre reinforced butt joints; fixing to timber base with stainless steel screws; bedding in mastic	9.87	5.00	10.36	0.35	-	3.23	1.77	15.36	m	17.28
Roof ventilators										
plastic; setting in position	7.18	2.50	7.36	0.85	-	7.85	0.55	15.76	nr	17.72
aluminium; setting in position	8.01	2.50	8.21	0.85	-	7.85	0.55	16.61	nr	18.68
Roofing; B.S.747, all layers type 3B -1.8 kg/m2, 75mm laps; fully bonding layers with hot bitumen bonding compound										
Roof coverings; layers of felt -2; bonding with hot bitumen compound to timber base										
pitch 7.5 degrees from horizontal	1.52	15.00	1.75	0.38	0.05	3.86	2.29	7.90	m2	8.88
pitch 40 degrees from horizontal	1.52	15.00	1.75	0.63	0.08	6.38	2.29	10.41	m2	11.72
pitch 75 degrees from horizontal	1.52	15.00	1.75	0.76	0.09	7.65	2.29	11.68	m2	13.15
Roof coverings; layers of felt -2; bonding with hot bitumen compound to cement and sand or concrete base										
pitch 7.5 degrees from horizontal	1.52	15.00	1.75	0.38	0.05	3.86	2.29	7.90	m2	8.88
pitch 40 degrees from horizontal	1.52	15.00	1.75	0.63	0.08	6.38	2.29	10.41	m2	11.72
pitch 75 degrees from horizontal	1.52	15.00	1.75	0.76	0.09	7.65	2.29	11.68	m2	13.15
Roof coverings; layers of felt -3; bonding with hot bitumen compound to timber base										
pitch 7.5 degrees from horizontal	2.28	15.00	2.62	0.55	0.07	5.57	3.42	11.61	m2	13.06
pitch 40 degrees from horizontal	2.28	15.00	2.62	0.92	0.12	9.33	3.42	15.38	m2	17.30
pitch 75 degrees from horizontal	2.28	15.00	2.62	1.10	0.14	11.14	3.42	17.18	m2	19.33
Roof coverings; layers of felt -3; bonding with hot bitumen compound to cement and sand or concrete base										
pitch 7.5 degrees from horizontal	2.28	15.00	2.62	0.55	0.07	5.57	3.42	11.61	m2	13.06
pitch 40 degrees from horizontal	2.28	15.00	2.62	0.92	0.12	9.33	3.42	15.38	m2	17.30
pitch 75 degrees from horizontal	2.28	15.00	2.62	1.10	0.14	11.14	3.42	17.18	m2	19.33
Roofing; felt B.S.747, bottom and intermediate layers type 3B -1.8 kg/m2, top layer type 3E -2.8 kg/m2, 75mm laps; fully bonding layers with hot bitumen bonding compound										
Roof coverings; layers of felt -2; bonding with hot bitumen compound to timber base										
pitch 7.5 degrees from horizontal	2.43	15.00	2.79	0.38	0.05	3.86	2.29	8.94	m2	10.06
pitch 40 degrees from horizontal	2.43	15.00	2.79	0.63	0.08	6.38	2.29	11.46	m2	12.89
pitch 75 degrees from horizontal	2.43	15.00	2.79	0.76	0.09	7.65	2.29	12.73	m2	14.32
Roof coverings; layers of felt -2; bonding with hot bitumen compound to cement and sand or concrete base										
pitch 7.5 degrees from horizontal	2.43	15.00	2.79	0.38	0.05	3.86	2.29	8.94	m2	10.06
pitch 40 degrees from horizontal	2.43	15.00	2.79	0.63	0.08	6.38	2.29	11.46	m2	12.89
pitch 75 degrees from horizontal	2.43	15.00	2.79	0.76	0.09	7.65	2.29	12.73	m2	14.32
Roof coverings; layers of felt -3; bonding with hot bitumen compound to timber base										
pitch 7.5 degrees from horizontal	3.19	15.00	3.67	0.55	0.07	5.57	3.42	12.66	m2	14.24
pitch 40 degrees from horizontal	3.19	15.00	3.67	0.92	0.12	9.33	3.42	16.42	m2	18.48
pitch 75 degrees from horizontal	3.19	15.00	3.67	1.10	0.14	11.14	3.42	18.22	m2	20.50

Labour hourly rates: (except Specialists) Craft Operatives £9.23 Labourer £7.02 Rates are national average prices. Refer to REGIONAL VARIATIONS for indicative levels of overall pricing in regions	MATERIALS			LABOUR				RATES		
	Del to Site	Waste	Material Cost	Craft Optve	Lab	Labour Cost	Sunds	Nett Rate	Unit	Gross Rate (+12.5%)
	£	%	£	Hrs	Hrs	£	£	£		£
J41: BUILT UP FELT ROOF COVERINGS Cont.										
Roofing; felt B.S.747, bottom and intermediate layers type 3B -1.8 kg/m2, top layer type 3E -2.8 kg/m2, 75mm laps; fully bonding layers with hot bitumen bonding compound Cont.										
Roof coverings; layers of felt -3; bonding with hot bitumen compound to cement and sand or concrete base										
pitch 7.5 degrees from horizontal	3.19	15.00	3.67	0.55	0.07	5.57	3.42	12.66	m2	14.24
pitch 40 degrees from horizontal	3.19	15.00	3.67	0.92	0.12	9.33	3.42	16.42	m2	18.48
pitch 75 degrees from horizontal	3.19	15.00	3.67	1.10	0.14	11.14	3.42	18.22	m2	20.50
Skirtings; layers of felt - 3; bonding with hot bitumen compound to brickwork base										
girth 0 - 200mm	0.64	15.00	0.74	0.28	0.03	2.79	0.65	4.18	m	4.70
girth 200 - 400mm	1.28	15.00	1.47	0.46	0.06	4.67	1.29	7.43	m	8.36
Collars around pipes; standard and the like; layers of felt -1; bonding with hot bitumen compound to metal base										
50mm diameter x 150mm long; hole in 3 layer covering	0.25	15.00	0.29	0.17	0.02	1.71	0.15	2.15	nr	2.42
100mm diameter x 150mm long; hole in 3 layer covering	0.40	15.00	0.46	0.22	0.03	2.24	0.24	2.94	nr	3.31
Roofing; sheet, Ruberoid Building Products Superflex Ultrabond system, bottom layer Rubervent B.S. 747 type 3G; intermediate layer Superbase Ultrabond; top layer Superflex Ultrabond slate surfaced cap sheet, 50mm side laps, 75mm end laps; partially bonding bottom layer, fully bonding other layers in Ruberoid 95/25 hot bonding bitumen										
Roof coverings; layers of sheet -3; bonding with hot bitumen compound to timber base										
pitch not exceeding 5.0 degrees from horizontal ..	15.92	15.00	18.31	0.55	0.28	7.04	1.06	26.41	m2	29.71
Roof coverings; layers of sheet -3; bonding with hot bitumen compound to cement and sand or concrete base										
pitch not exceeding 5.0 degrees from horizontal ..	15.92	15.00	18.31	0.55	0.28	7.04	1.06	26.41	m2	29.71
Roof coverings; layers of sheet -3; bonding with hot bitumen compound to metal base										
pitch not exceeding 5.0 degrees from horizontal ..	15.92	15.00	18.31	0.55	0.28	7.04	1.06	26.41	m2	29.71
Skirtings; layers of sheet -2; bonding with hot bitumen compound to brickwork base										
girth 0 - 200mm	3.18	15.00	3.66	0.28	0.05	2.94	0.46	7.05	m	7.93
girth 200 - 400mm	6.36	15.00	7.31	0.46	0.06	4.67	0.91	12.89	m	14.50
Roofing; sheet, Ruberoid Building Products Superflex Ultrabond system, bottom layer Rubervent B.S. 747 type 3G; intermediate layer Superbase Ultrabond; top layer Superflex Ultrabond slate surfaced cap sheet, 50mm side laps, 75mm end laps; partially bonding bottom layer, fully bonding other layers in Ruberoid 95/25 hot bonding bitumen; 25mm Kingspan Thermaroof TR21 (PIR) Board										
Roof coverings; layers of sheet -3; bonding with hot bitumen compound to timber base										
pitch not exceeding 5.0 degrees from horizontal ..	21.82	15.00	25.09	1.10	0.36	12.68	1.06	38.83	m2	43.69
Roof coverings; layers of sheet -3; bonding with hot bitumen compound to cement and sand or concrete base										
pitch not exceeding 5.0 degrees from horizontal ..	21.82	15.00	25.09	1.16	0.36	13.23	1.06	39.39	m2	44.31
Roof coverings; layers of sheet -3; bonding with hot bitumen compound to metal base										
pitch not exceeding 5.0 degrees from horizontal ..	21.82	15.00	25.09	1.14	0.36	13.05	1.06	39.20	m2	44.10
Roofing; sheet, Ruberoid Building Products Superflex Ultrabond system, bottom layer Rubervent B.S. 747 type 3G; intermediate layer Superbase Ultrabond; top layer Superflex Ultrabond slate surfaced cap sheet, 50mm side laps, 75mm end laps; partially bonding bottom layer, fully bonding other layers in Ruberoid 95/25 hot bonding bitumen; 30mm Kingspan Thermaroof TR21 (PIR) Board										
Roof coverings; layers of sheet -3; bonding with hot bitumen compound to timber base										
pitch not exceeding 5.0 degrees from horizontal ..	21.82	15.00	25.09	1.12	0.36	12.86	1.06	39.02	m2	43.90
Roofing; sheet, Ruberoid Building Products Superflex Ultrabond system, bottom layer Rubervent B.S. 747 type 3G; intermediate layer Superbase Ultrabond; top layer Superflex Ultrabond slate surfaced cap sheet, 50mm side laps, 75mm end laps; partially bonding bottom layer, fully bonding other layers in Ruberoid 95/25 hot bonding bitumen; 35mm Kingspan Thermaroof TR21 (PIR) Board										
Roof coverings; layers of sheet -3; bonding with hot bitumen compound to timber base										
pitch not exceeding 5.0 degrees from horizontal ..	22.62	15.00	26.01	1.15	0.36	13.14	1.06	40.21	m2	45.24

WATERPROOFING

Labour hourly rates: (except Specialists) Craft Operatives £9.23 Labourer £7.02 Rates are national average prices. Refer to REGIONAL VARIATIONS for indicative levels of overall pricing in regions	MATERIALS			LABOUR				RATES		
	Del to Site £	Waste %	Material Cost £	Craft Optve Hrs	Lab Hrs	Labour Cost £	Sunds £	Nett Rate £	Unit	Gross Rate (+12.5%) £
J41: BUILT UP FELT ROOF COVERINGS Cont.										
Roofing; sheet, Ruberoid Building Products Superflex Ultrabond system, bottom layer Rubervent B.S. 747 type 3G; intermediate layer Superbase Ultrabond; top layer Superflex Ultrabond slate surfaced cap sheet, 50mm side laps, 75mm end laps; partially bonding bottom layer, fully bonding other layers in Ruberoid 95/25 hot bonding bitumen; 40mm Kingspan Thermaroof TR21 (PIR) Board										
Roof coverings; layers of sheet -3; bonding with hot bitumen compound to timber base										
pitch not exceeding 5.0 degrees from horizontal ..	23.06	15.00	26.52	1.18	0.36	13.42	1.06	41.00	m2	46.12
Roofing; sheet, Ruberoid Building Products Superflex Ultrabond system, bottom layer Rubervent B.S. 747 type 3G; intermediate layer Superbase Ultrabond; top layer Superflex Ultrabond slate surfaced cap sheet, 50mm side laps, 75mm end laps; partially bonding bottom layer, fully bonding other layers in Ruberoid 95/25 hot bonding bitumen; 50mm Kingspan Thermaboard TR21 (PIR) Board										
Roof coverings; layers of sheet -3; bonding with hot bitumen compound to timber base										
pitch not exceeding 5.0 degrees from horizontal ..	23.98	15.00	27.58	1.20	0.36	13.60	1.06	42.24	m2	47.52
Roofing; sheet, Ruberoid Building Products Superflex Ultratorch system, bottom layer Superbase Ultratorch; top layer Superflex Ultratorch slate surfaced cap sheet, 75mm side and end laps; fully bonding both layers (by torching); SuperRock Torch, mineral wool insulation with bitumen primed surface; Superbar vapour control layer										
Roof coverings; layers of sheet -3; bonding to timber base										
pitch not exceeding 5.0 degrees from horizontal ..	16.02	15.00	18.42	0.55	0.28	7.04	2.07	27.54	m2	30.98
Roof coverings; layers of sheet -3; bonding to cement and sand or concrete base										
pitch not exceeding 5.0 degrees from horizontal ..	16.02	15.00	18.42	0.55	0.28	7.04	2.07	27.54	m2	30.98
Roof coverings; layers of sheet -3; bonding to metal base										
pitch not exceeding 5.0 degrees from horizontal ..	16.02	15.00	18.42	0.59	0.28	7.41	2.07	27.90	m2	31.39
Skirtings; layers of sheet -2; bonding with hot bitumen compound to brickwork base										
girth 0 - 200mm	3.20	15.00	3.68	0.28	0.05	2.94	0.48	7.10	m	7.98
girth 200 - 400mm	6.40	15.00	7.36	0.46	0.06	4.67	0.96	12.99	m	14.61
Roofing; sheet, Ruberoid Building Products Superflex Ultratorch system, bottom layer Superbase Ultratorch; top layer Superflex Ultratorch slate surfaced cap sheet, 75mm side and end laps; fully bonding both layers (by torching); SuperRock Torch, 25mm Kingspan Thermaroof TR21 (PIR Board) insulation with bitumen primed surface; Superbar vapour control layer										
Roof coverings; layers of sheet -3; bonding to timber base										
pitch not exceeding 5.0 degrees from horizontal ..	21.92	15.00	25.21	1.10	0.36	12.68	2.07	39.96	m2	44.95
Roof coverings; layers of sheet -3; bonding to cement and sand or concrete base										
pitch not exceeding 5.0 degrees from horizontal ..	21.92	15.00	25.21	1.10	0.36	12.68	2.07	39.96	m2	44.95
Roof coverings; layers of sheet -3; bonding to metal base										
pitch not exceeding 5.0 degrees from horizontal ..	21.92	15.00	25.21	1.15	0.36	13.14	2.07	40.42	m2	45.47
Roofing; sheet, Ruberoid Building Products Superflex Ultratorch system, bottom layer Superbase Ultratorch; top layer Superflex Ultratorch slate surfaced cap sheet, 75mm side and end laps; fully bonding both layers (by torching); SuperRock Torch, 30mm Kingspan Thermaroof TR21 (PIR Board) insulation with bitumen primed surface; Superbar vapour control layer										
Roof coverings; layers of sheet -3; bonding to timber base										
pitch not exceeding 5.0 degrees from horizontal ..	21.92	15.00	25.21	1.12	0.36	12.86	2.07	40.14	m2	45.16
Roofing; sheet, Ruberoid Building Products Superflex Ultratorch system, bottom layer Superbase Ultratorch; top layer Superflex Ultratorch slate surfaced cap sheet, 75mm side and end laps; fully bonding both layers (by torching); SuperRock Torch, 35mm Kingspan Thermaroof TR21 (PIR Board) insulation with bitumen primed surface; Superbar vapour control layer										
Roof coverings; layers of sheet -3; bonding to timber base										
pitch not exceeding 5.0 degrees from horizontal ..	22.72	15.00	26.13	1.15	0.36	13.14	2.07	41.34	m2	46.51

Labour hourly rates: (except Specialists) Craft Operatives £9.23 Labourer £7.02 Rates are national average prices. Refer to REGIONAL VARIATIONS for indicative levels of overall pricing in regions	MATERIALS			LABOUR				RATES		
	Del to Site £	Waste %	Material Cost £	Craft Optve Hrs	Lab Hrs	Labour Cost £	Sunds £	Nett Rate £	Unit	Gross Rate (+12.5%) £
J41: BUILT UP FELT ROOF COVERINGS Cont.										
Roofing; sheet, Ruberoid Building Products Superflex Ultratorch system, bottom layer Superbase Ultratorch; top layer Superflex Ultratorch slate surfaced cap sheet, 75mm side and end laps; fully bonding both layers (by torching); SuperRock Torch, 40mm Kingspan Thermaroof TR21 (PIR Board) insulation with bitumen primed surface; Superbar vapour control layer										
Roof coverings; layers of sheet -3; bonding to timber base										
pitch not exceeding 5.0 degrees from horizontal ..	23.16	15.00	26.63	1.18	0.36	13.42	2.07	42.12	m2	47.39
Roofing; sheet, Ruberoid Building Products Superflex Ultratorch system, bottom layer Superbase Ultratorch; top layer Superflex Ultratorch slate surfaced cap sheet, 75mm side and end laps; fully bonding both layers (by torching); SuperRock Torch, 50mm Kingspan Thermaroof TR21 (PIR Board) insulation with bitumen primed surface; Superbar vapour control layer										
Roof coverings; layers of sheet -3; bonding to timber base										
pitch not exceeding 5.0 degrees from horizontal ..	24.08	15.00	27.69	1.20	0.36	13.60	2.07	43.37	m2	48.79

WATERPROOFING

179

Linings/Sheathing/Dry partitioning

Labour hourly rates: (except Specialists) Craft Operatives £9.23 Labourer £7.02 Rates are national average prices. Refer to REGIONAL VARIATIONS for indicative levels of overall pricing in regions	MATERIALS			LABOUR				RATES		
	Del to Site	Waste	Material Cost	Craft Optve	Lab	Labour Cost	Sunds	Nett Rate	Unit	Gross Rate (+12.5%)
	£	%	£	Hrs	Hrs	£	£	£		£
K10: PLASTERBOARD DRY LINING/PARTITIONS/CEILINGS										
PLASTERBOARD FOR DIRECT DECORATION - Linings; tapered edge sheeting of one layer 9.5mm thick Gyproc wall board for direct decoration, B.S.1230, butt joints; fixing with galvanised nails to timber base; butt joints filled with joint filler tape and joint finish, spot filling										
Walls										
height 2100 - 2400mm	4.10	5.00	4.30	0.96	0.48	12.23	1.39	17.93	m	20.17
height 2400 - 2700mm	4.62	5.00	4.85	1.08	0.54	13.76	1.57	20.18	m	22.70
height 2700 - 3000mm	5.13	5.00	5.39	1.20	0.60	15.29	1.74	22.41	m	25.22
Beams; 3nr faces										
total girth 600 - 1200mm	2.05	10.00	2.25	1.20	0.24	12.76	0.70	15.72	m	17.68
Columns; 4nr faces										
total girth 600 - 1200mm	2.05	10.00	2.25	1.20	0.24	12.76	0.70	15.72	m	17.68
total girth 1200 - 1800mm	3.08	10.00	3.39	1.80	0.36	19.14	1.05	23.58	m	26.53
Reveals and soffits of openings and recesses										
width not exceeding 300mm	0.51	10.00	0.56	0.24	0.06	2.64	0.17	3.37	m	3.79
Ceilings										
generally..	1.71	5.00	1.80	0.48	0.20	5.83	0.58	8.21	m2	9.24
PLASTERBOARD FOR DIRECT DECORATION - Linings; tapered edge sheeting of one layer 12.5mm thick Gyproc wallboard for direct decoration, B.S.1230 butt joints; fixing with galvanised nails to timber base; butt joints filled with joint filler tape and joint finish, spot filling										
Walls										
height 2100 - 2400mm	4.78	5.00	5.02	1.13	0.56	14.36	1.41	20.79	m	23.39
height 2400 - 2700mm	5.37	5.00	5.64	1.27	0.63	16.14	1.58	23.36	m	26.28
height 2700 - 3000mm	5.97	5.00	6.27	1.41	0.71	18.00	1.76	26.03	m	29.28
Beams; 3nr faces										
total girth 600 - 1200mm	2.39	10.00	2.63	1.42	0.28	15.07	0.70	18.40	m	20.70
Columns; 4nr faces										
total girth 600 - 1200mm	2.39	10.00	2.63	1.42	0.28	15.07	0.70	18.40	m	20.70
total girth 1200 - 1800mm	3.58	10.00	3.94	2.12	0.42	22.52	1.05	27.50	m	30.94
Reveals and soffits of openings and recesses										
width not exceeding 300mm	0.60	10.00	0.66	0.28	0.07	3.08	0.18	3.92	m	4.41
Ceilings										
generally..	1.99	5.00	2.09	0.56	0.24	6.85	0.59	9.53	m2	10.72
PLASTERBOARD FOR DIRECT DECORATION - Linings; tapered edge sheeting of one layer 15mm thick Gyproc wall board for direct decoration, B.S.1230, butt joints; fixing with galvanised nails to timber base; butt joints filled with joint filler tape and joint finish, spot filling										
Walls										
height 2100 - 2400mm	5.23	5.00	5.49	1.30	0.65	16.56	1.41	23.46	m	26.40
height 2400 - 2700mm	5.89	5.00	6.18	1.46	0.73	18.60	1.58	26.36	m	29.66
height 2700 - 3000mm	6.54	5.00	6.87	1.62	0.81	20.64	1.76	29.27	m	32.92
Beams; 3nr faces										
total girth 600 - 1200mm	2.62	10.00	2.88	1.64	0.33	17.45	0.70	21.04	m	23.67
Columns; 4nr faces										
total girth 600 - 1200mm	2.62	10.00	2.88	1.64	0.33	17.45	0.70	21.04	m	23.67
total girth 1200 - 1800mm	3.92	10.00	4.31	2.46	0.49	26.15	1.05	31.51	m	35.45

Labour hourly rates: (except Specialists) Craft Operatives £9.23 Labourer £7.02 Rates are national average prices. Refer to REGIONAL VARIATIONS for indicative levels of overall pricing in regions	MATERIALS			LABOUR				RATES		
	Del to Site £	Waste %	Material Cost £	Craft Optve Hrs	Lab Hrs	Labour Cost £	Sunds £	Nett Rate £	Unit	Gross Rate (+12.5%) £
K10: PLASTERBOARD DRY LINING/PARTITIONS/CEILINGS Cont.										
PLASTERBOARD FOR DIRECT DECORATION - Linings; tapered edge sheeting of one layer 15mm thick Gyproc wall board for direct decoration, B.S.1230, butt joints; fixing with galvanised nails to timber base; butt joints filled with joint filler tape and joint finish, spot filling Cont.										
Reveals and soffits of openings and recesses										
width not exceeding 300mm	0.65	10.00	0.71	0.32	0.08	3.52	0.18	4.41	m	4.96
Ceilings										
generally	2.18	5.00	2.29	0.64	0.28	7.87	0.59	10.75	m2	12.10
PLASTERBOARD FOR DIRECT DECORATION - Linings; tapered edge sheeting of one layer 9.5mm thick Gyproc Duplex wallboard for direct decoration, B.S.1230 butt joints; fixing with galvanised nails to timber base; butt joints filled with joint filler tape and joint finish, spot filling										
Walls										
height 2100 - 2400mm	5.59	5.00	5.87	0.96	0.48	12.23	1.39	19.49	m	21.93
height 2400 - 2700mm	6.29	5.00	6.60	1.08	0.54	13.76	1.57	21.93	m	24.68
height 2700 - 3000mm	6.99	5.00	7.34	1.20	0.60	15.29	1.74	24.37	m	27.41
Beams; 3nr faces										
total girth 600 - 1200mm	2.80	10.00	3.08	1.20	0.24	12.76	0.70	16.54	m	18.61
Columns; 4nr faces										
total girth 600 - 1200mm	2.80	10.00	3.08	1.20	0.24	12.76	0.70	16.54	m	18.61
total girth 1200 - 1800mm	4.19	10.00	4.61	1.80	0.36	19.14	1.05	24.80	m	27.90
Reveals and soffits of openings and recesses										
width not exceeding 300mm	0.70	10.00	0.77	0.24	0.06	2.64	0.17	3.58	m	4.02
Ceilings										
generally	2.33	5.00	2.45	0.48	0.20	5.83	0.58	8.86	m2	9.97
PLASTERBOARD FOR DIRECT DECORATION - Linings; tapered edge sheeting of one layer 12.5mm thick Gyproc Duplex wallboard for direct decoration, B.S.1230, butt joints; fixing with galvanised nails to timber base; butt joints filled with joint filler tape and joint finish, spot filling										
Walls										
height 2100 - 2400mm	6.26	5.00	6.57	1.13	0.56	14.36	1.41	22.34	m	25.14
height 2400 - 2700mm	7.05	5.00	7.40	1.27	0.63	16.14	1.58	25.13	m	28.27
height 2700 - 3000mm	7.83	5.00	8.22	1.41	0.71	18.00	1.76	27.98	m	31.48
Beams; 3nr faces										
total girth 600 - 1200mm	3.13	10.00	3.44	1.42	0.28	15.07	0.70	19.22	m	21.62
Columns; 4nr faces										
total girth 600 - 1200mm	3.13	10.00	3.44	1.42	0.28	15.07	0.70	19.22	m	21.62
total girth 1200 - 1800mm	4.70	10.00	5.17	2.12	0.42	22.52	1.05	28.74	m	32.33
Reveals and soffits of openings and recesses										
width not exceeding 300mm	0.78	10.00	0.86	0.28	0.07	3.08	0.18	4.11	m	4.63
Ceilings										
generally	2.61	5.00	2.74	0.56	0.24	6.85	0.59	10.18	m2	11.46
PLASTERBOARD FOR DIRECT DECORATION - Linings; tapered edge sheeting of one layer 15mm thick Gyproc Duplex wallboard for direct decoration, B.S.1230, butt joints; fixing with galvanised nails to timber base; butt joints filled with joint filler tape and joint finish, spot filling										
Walls										
height 2100 - 2400mm	6.86	5.00	7.20	1.30	0.65	16.56	1.41	25.18	m	28.32
height 2400 - 2700mm	7.72	5.00	8.11	1.46	0.73	18.60	1.58	28.29	m	31.82
height 2700 - 3000mm	8.58	5.00	9.01	1.62	0.81	20.64	1.76	31.41	m	35.33
Beams; 3nr faces										
total girth 600 - 1200mm	3.43	10.00	3.77	1.64	0.33	17.45	0.70	21.93	m	24.67
Columns; 4nr faces										
total girth 600 - 1200mm	3.43	10.00	3.77	1.64	0.33	17.45	0.70	21.93	m	24.67
total girth 1200 - 1800mm	5.15	10.00	5.67	2.46	0.49	26.15	1.05	32.86	m	36.97
Reveals and soffits of openings and recesses										
width not exceeding 300mm	0.86	10.00	0.95	0.32	0.08	3.52	0.18	4.64	m	5.22
Ceilings										
generally	2.86	5.00	3.00	0.64	0.28	7.87	0.59	11.47	m2	12.90

Labour hourly rates: (except Specialists) Craft Operatives £9.23 Labourer £7.02 Rates are national average prices. Refer to REGIONAL VARIATIONS for indicative levels of overall pricing in regions	MATERIALS			LABOUR				RATES		
	Del to Site	Waste	Material Cost	Craft Optve	Lab	Labour Cost	Sunds	Nett Rate	Unit	Gross Rate (+12.5%)
	£	%	£	Hrs	Hrs	£	£	£		£

K10: PLASTERBOARD DRY LINING/PARTITIONS/CEILINGS Cont.

PLASTERBOARD FOR DIRECT DECORATION - Linings; tapered edge sheeting of one layer 19mm thick Gyproc plank for direct decoration, B.S.1230, butt joints; fixing with galvanised nails to timber base; butt joints filled with joint filler tape and joint finish, spot filling

	Del to Site	Waste	Material Cost	Craft Optve	Lab	Labour Cost	Sunds	Nett Rate	Unit	Gross Rate
Walls										
height 2100 - 2400mm	7.75	5.00	8.14	1.44	0.72	18.35	1.30	27.78	m	31.26
height 2400 - 2700mm	8.72	5.00	9.16	1.62	0.81	20.64	1.47	31.26	m	35.17
height 2700 - 3000mm	9.69	5.00	10.17	1.80	0.90	22.93	1.63	34.74	m	39.08
Beams; 3nr faces										
total girth 600 - 1200mm	3.88	10.00	4.27	1.80	0.36	19.14	0.65	24.06	m	27.07
Columns; 4nr faces										
total girth 600 - 1200mm	3.88	10.00	4.27	1.80	0.36	19.14	0.65	24.06	m	27.07
total girth 1200 - 1800mm	5.81	10.00	6.39	2.70	0.54	28.71	0.98	36.08	m	40.59
Reveals and soffits of openings and recesses										
width not exceeding 300mm	0.97	10.00	1.07	0.36	0.09	3.95	0.16	5.18	m	5.83
Ceilings										
generally	3.23	5.00	3.39	0.72	0.30	8.75	0.54	12.68	m2	14.27

PLASTERBOARD FOR DIRECT DECORATION - Linings; two layers of Gypsum wallboard, first layer square edge sheeting 9.5mm thick, second layer tapered edge sheeting 9.5mm thick for direct decoration, B.S.1230, butt joints; fixing with galvanised nails to timber base; butt joints of second layer filled with joint filler tape and joint finish, spot filling

	Del to Site	Waste	Material Cost	Craft Optve	Lab	Labour Cost	Sunds	Nett Rate	Unit	Gross Rate
Walls										
height 2100 - 2400mm	8.21	5.00	8.62	1.68	0.84	21.40	1.75	31.77	m	35.75
height 2400 - 2700mm	9.23	5.00	9.69	1.89	0.95	24.11	1.97	35.78	m	40.25
height 2700 - 3000mm	10.26	5.00	10.77	2.10	1.05	26.75	2.19	39.72	m	44.68
Beams; 3nr faces										
total girth 600 - 1200mm	4.10	10.00	4.51	2.10	0.42	22.33	0.87	27.71	m	31.18
Columns; 4nr faces										
total girth 600 - 1200mm	4.10	10.00	4.51	2.10	0.42	22.33	0.87	27.71	m	31.18
total girth 1200 - 1800mm	6.16	10.00	6.78	3.15	0.63	33.50	1.31	41.58	m	46.78
Reveals and soffits of openings and recesses										
width not exceeding 300mm	1.03	10.00	1.13	0.42	0.11	4.65	0.22	6.00	m	6.75
Ceilings										
generally	3.42	5.00	3.59	0.84	0.35	10.21	0.73	14.53	m2	16.35

PLASTERBOARD FOR DIRECT DECORATION - Linings; two layers of Gypsum wallboard, first layer square edge sheeting 9.5mm thick, second layer tapered edge sheeting 12.5mm thick for direct decoration, B.S.1230, butt joints; fixing with galvanised nails to timber base; butt joints of second layer filled with joint filler tape and joint finish, spot filling

	Del to Site	Waste	Material Cost	Craft Optve	Lab	Labour Cost	Sunds	Nett Rate	Unit	Gross Rate
Walls										
height 2100 - 2400mm	8.88	5.00	9.32	1.92	0.96	24.46	1.75	35.53	m	39.98
height 2400 - 2700mm	9.99	5.00	10.49	2.16	1.08	27.52	1.97	39.98	m	44.98
height 2700 - 3000mm	11.10	5.00	11.65	2.40	1.20	30.58	2.19	44.42	m	49.97
Beams; 3nr faces										
total girth 600 - 1200mm	4.44	10.00	4.88	2.40	0.48	25.52	0.87	31.28	m	35.19
Columns; 4nr faces										
total girth 600 - 1200mm	4.44	10.00	4.88	2.40	0.48	25.52	0.87	31.28	m	35.19
total girth 1200 - 1800mm	6.66	10.00	7.33	3.60	0.72	38.28	1.31	46.92	m	52.78
Reveals and soffits of openings and recesses										
width not exceeding 300mm	1.11	10.00	1.22	0.48	0.12	5.27	0.22	6.71	m	7.55
Ceilings										
generally	3.70	5.00	3.88	0.96	0.40	11.67	0.73	16.28	m2	18.32

PLASTERBOARD FOR DIRECT DECORATION - Linings; two layers of Gypsum wallboard, first layer square edge sheeting 12.5mm thick, second layer tapered edge sheeting 12.5mm thick for direct decoration, B.S.1230, butt joints; fixing with galvanised nails to timber base; butt joints of second layer filled with joint filler tape and joint finish, spot filling

	Del to Site	Waste	Material Cost	Craft Optve	Lab	Labour Cost	Sunds	Nett Rate	Unit	Gross Rate
Walls										
height 2100 - 2400mm	9.55	5.00	10.03	2.04	1.02	25.99	1.76	37.78	m	42.50
height 2400 - 2700mm	10.75	5.00	11.29	2.30	1.15	29.30	1.98	42.57	m	47.89
height 2700 - 3000mm	11.94	5.00	12.54	2.55	1.28	32.52	2.20	47.26	m	53.17
Beams; 3nr faces										
total girth 600 - 1200mm	4.78	10.00	5.26	2.56	0.51	27.21	0.88	33.35	m	37.52

Labour hourly rates: (except Specialists) Craft Operatives £9.23 Labourer £7.02 Rates are national average prices. Refer to REGIONAL VARIATIONS for indicative levels of overall pricing in regions	MATERIALS			LABOUR				RATES		
	Del to Site £	Waste %	Material Cost £	Craft Optve Hrs	Lab Hrs	Labour Cost £	Sunds £	Nett Rate £	Unit	Gross Rate (+12.5%) £
K10: PLASTERBOARD DRY LINING/PARTITIONS/CEILINGS Cont.										
PLASTERBOARD FOR DIRECT DECORATION - Linings; two layers of Gypsum wallboard, first layer square edge sheeting 12.5mm thick, second layer tapered edge sheeting 12.5mm thick for direct decoration, B.S.1230, butt joints; fixing with galvanised nails to timber base; butt joints of second layer filled with joint filler tape and joint finish, spot filling Cont.										
Columns; 4nr faces										
total girth 600 - 1200mm	4.78	10.00	5.26	2.56	0.51	27.21	0.88	33.35	m	37.52
total girth 1200 - 1800mm	7.16	10.00	7.88	3.83	0.77	40.76	1.32	49.95	m	56.20
Reveals and soffits of openings and recesses										
width not exceeding 300mm	1.19	10.00	1.31	0.51	0.13	5.62	0.22	7.15	m	8.04
Ceilings										
generally	3.98	5.00	4.18	1.02	0.43	12.43	0.73	17.34	m2	19.51
PLASTERBOARD FOR DIRECT DECORATION - Linings; two layers of Gypsum wallboard, first layer square edge sheeting 15mm thick, second layer tapered edge sheeting 15mm thick for direct decoration, B.S.1230, butt joints; fixing with galvanised nails to timber base; butt joints of second layer filled with joint filler tape and joint finish, spot filling										
Walls										
height 2100 - 2400mm	10.46	5.00	10.98	2.16	1.08	27.52	1.77	40.27	m	45.31
height 2400 - 2700mm	11.77	5.00	12.36	2.43	1.22	30.99	2.00	45.35	m	51.02
height 2700 - 3000mm	13.08	5.00	13.73	2.70	1.35	34.40	2.22	50.35	m	56.65
Beams; 3nr faces										
total girth 600 - 1200mm	5.23	10.00	5.75	2.70	0.54	28.71	0.89	35.35	m	39.77
Columns; 4nr faces										
total girth 600 - 1200mm	5.23	10.00	5.75	2.70	0.54	28.71	0.89	35.35	m	39.77
total girth 1200 - 1800mm	7.85	10.00	8.63	4.05	0.81	43.07	1.33	53.03	m	59.66
Reveals and soffits of openings and recesses										
width not exceeding 300mm	1.31	10.00	1.44	0.54	0.14	5.97	0.22	7.63	m	8.58
Ceilings										
generally	4.36	5.00	4.58	1.08	0.46	13.20	0.74	18.52	m2	20.83
PLASTERBOARD FOR DIRECT DECORATION - Linings; two layers of Gypsum wallboard, first layer Duplex square edge sheeting 9.5mm thick, second layer tapered edge sheeting 9.5mm thick for direct decoration, B.S.1230, butt joints; fixing with galvanised nails to timber base; butt joints of second layer filled with joint filler tape and joint finish, spot filling										
Walls										
height 2100 - 2400mm	9.70	5.00	10.19	1.68	0.84	21.40	1.75	33.34	m	37.51
height 2400 - 2700mm	10.91	5.00	11.46	1.89	0.95	24.11	1.97	37.54	m	42.23
height 2700 - 3000mm	12.12	5.00	12.73	2.10	1.05	26.75	2.19	41.67	m	46.88
Beams; 3nr faces										
total girth 600 - 1200mm	4.85	10.00	5.34	2.10	0.42	22.33	0.87	28.54	m	32.10
Columns; 4nr faces										
total girth 600 - 1200mm	4.85	10.00	5.34	2.10	0.42	22.33	0.87	28.54	m	32.10
total girth 1200 - 1800mm	7.27	10.00	8.00	3.15	0.63	33.50	1.31	42.80	m	48.15
Reveals and soffits of openings and recesses										
width not exceeding 300mm	1.21	10.00	1.33	0.42	0.11	4.65	0.22	6.20	m	6.97
Ceilings										
generally	4.04	5.00	4.24	0.84	0.35	10.21	0.73	15.18	m2	17.08
PLASTERBOARD FOR DIRECT DECORATION - Linings; two layers of Gypsum wallboard, first layer Duplex square edge sheeting 9.5mm thick, second layer tapered edge sheeting 12.5mm thick for direct decoration, B.S.1230, butt joints; fixing with galvanised nails to timber base; butt joints of second layer filled with joint filler tape and joint finish, spot filling										
Walls										
height 2100 - 2400mm	10.37	5.00	10.89	1.92	0.96	24.46	1.75	37.10	m	41.74
height 2400 - 2700mm	11.66	5.00	12.24	2.16	1.08	27.52	1.97	41.73	m	46.95
height 2700 - 3000mm	12.96	5.00	13.61	2.40	1.20	30.58	2.19	46.37	m	52.17
Beams; 3nr faces										
total girth 600 - 1200mm	5.18	10.00	5.70	2.40	0.48	25.52	0.87	32.09	m	36.10
Columns; 4nr faces										
total girth 600 - 1200mm	5.18	10.00	5.70	2.40	0.48	25.52	0.87	32.09	m	36.10
total girth 1200 - 1800mm	7.78	10.00	8.56	3.60	0.72	38.28	1.31	48.15	m	54.17
Reveals and soffits of openings and recesses										
width not exceeding 300mm	1.30	10.00	1.43	0.48	0.12	5.27	0.22	6.92	m	7.79

Labour hourly rates: (except Specialists) Craft Operatives £9.23 Labourer £7.02 Rates are national average prices. Refer to REGIONAL VARIATIONS for indicative levels of overall pricing in regions	MATERIALS			LABOUR				RATES		
	Del to Site £	Waste %	Material Cost £	Craft Optve Hrs	Lab Hrs	Labour Cost £	Sunds £	Nett Rate £	Unit	Gross Rate (+12.5%) £

K10: PLASTERBOARD DRY LINING/PARTITIONS/CEILINGS Cont.

PLASTERBOARD FOR DIRECT DECORATION - Linings; two layers of Gypsum wallboard, first layer Duplex square edge sheeting 9.5mm thick, second layer tapered edge sheeting 12.5mm thick for direct decoration, B.S.1230, butt joints; fixing with galvanised nails to timber base; butt joints of second layer filled with joint filler tape and joint finish, spot filling Cont.

Ceilings generally	4.32	5.00	4.54	0.96	0.40	11.67	0.73	16.93	m2	19.05

PLASTERBOARD FOR DIRECT DECORATION - Linings; two layers of Gypsum wallboard, first layer Duplex square edge sheeting 12.5mm thick, second layer tapered edge sheeting 12.5mm thick for direct decoration, B.S.1230, butt joints; fixing with galvanised nails to timber base; butt joints of second layer filled with joint filler tape and joint finish, spot filling

Walls										
height 2100 - 2400mm	11.04	5.00	11.59	2.04	1.02	25.99	1.76	39.34	m	44.26
height 2400 - 2700mm	12.42	5.00	13.04	2.30	1.15	29.30	1.98	44.32	m	49.86
height 2700 - 3000mm	13.80	5.00	14.49	2.55	1.28	32.52	2.20	49.21	m	55.36
Beams; 3nr faces										
total girth 600 - 1200mm	7.73	10.00	8.39	2.56	0.51	27.21	-	35.60	m	40.05
Columns; 4nr faces										
total girth 600 - 1200mm	5.52	10.00	6.07	2.56	0.51	27.21	0.88	34.16	m	38.43
total girth 1200 - 1800mm	8.28	10.00	9.11	3.83	0.77	40.76	1.32	51.18	m	57.58
Reveals and soffits of openings and recesses										
width not exceeding 300mm	1.38	10.00	1.52	0.51	0.13	5.62	0.22	7.36	m	8.28
Ceilings generally	4.60	5.00	4.83	1.02	0.43	12.43	0.73	17.99	m2	20.24

PLASTERBOARD FOR DIRECT DECORATION - Linings; two layers of Gypsum wallboard, first layer Duplex square edge sheeting 15mm thick, second layer tapered edge sheeting 15mm thick for direct decoration, B.S.1230, butt joints; fixing with galvanised nails to timber base; butt joints of second layer filled with joint filler tape and joint finish, spot filling

Walls										
height 2100 - 2400mm	12.10	5.00	12.71	2.16	1.08	27.52	1.77	41.99	m	47.24
height 2400 - 2700mm	13.61	5.00	14.29	2.43	1.22	30.99	2.00	47.28	m	53.19
height 2700 - 3000mm	15.12	5.00	15.88	2.70	1.35	34.40	2.22	52.49	m	59.06
Beams; 3nr faces										
total girth 600 - 1200mm	6.05	10.00	6.66	2.70	0.54	28.71	0.89	36.26	m	40.79
Columns; 4nr faces										
total girth 600 - 1200mm	6.05	10.00	6.66	2.70	0.54	28.71	0.89	36.26	m	40.79
total girth 1200 - 1800mm	9.07	10.00	9.98	4.05	0.81	43.07	1.33	54.37	m	61.17
Reveals and soffits of openings and recesses										
width not exceeding 300mm	1.51	10.00	1.66	0.54	0.14	5.97	0.22	7.85	m	8.83
Ceilings generally	5.04	5.00	5.29	1.08	0.46	13.20	0.74	19.23	m2	21.63

PLASTIC FACED PLASTERBOARD - Linings; square edge sheeting of one layer 9.5mm thick Gyproc industrial grade plastic faced plasterboard, B.S.1230, butt joints; fixing with galvanised nails with plastic coated nail caps to timber base; butt joints covered with Gyproc batten section and trim

Walls										
height 2100 - 2400mm	11.95	5.00	12.55	0.72	0.36	9.17	9.89	31.61	m	35.56
height 2400 - 2700mm	13.45	5.00	14.12	0.81	0.41	10.35	11.12	35.60	m	40.05
height 2700 - 3000mm	14.94	5.00	15.69	0.90	0.45	11.47	12.36	39.51	m	44.45
Beams; 3nr faces										
total girth 600 - 1200mm	5.98	10.00	6.58	0.90	0.18	9.57	4.94	21.09	m	23.72
Columns; 4nr faces										
total girth 600 - 1200mm	5.98	10.00	6.58	0.90	0.18	9.57	4.94	21.09	m	23.72
total girth 1200 - 1800mm	8.96	10.00	9.86	1.35	0.27	14.36	7.41	31.62	m	35.57
Reveals and soffits of openings and recesses										
width not exceeding 300mm	1.49	10.00	1.64	0.18	0.05	2.01	1.24	4.89	m	5.50
Ceilings										
generally	4.98	5.00	5.23	0.36	0.15	4.38	4.12	13.72	m2	15.44
3.50 - 5.00m above floor	4.98	5.00	5.23	0.40	0.15	4.75	4.12	14.09	m2	15.86

LININGS/SHEATHING/DRY PARTITIONING

Labour hourly rates: (except Specialists) Craft Operatives £9.23 Labourer £7.02 Rates are national average prices. Refer to REGIONAL VARIATIONS for indicative levels of overall pricing in regions	MATERIALS			LABOUR				RATES		
	Del to Site	Waste	Material Cost	Craft Optve	Lab	Labour Cost	Sunds	Nett Rate	Unit	Gross Rate (+12.5%)
	£	%	£	Hrs	Hrs	£	£	£		£
K10: PLASTERBOARD DRY LINING/PARTITIONS/CEILINGS Cont.										
PLASTIC FACED PLASTERBOARD - Linings; square edge sheeting of one layer 12.5mm thick Gyproc industrial grade plastic faced plasterboard, B.S.1230, butt joints; fixing with galvanised nails with plastic coated nail caps to timber base; butt joints covered with Gyproc batten section and trim										
Walls										
height 2100 - 2400mm	12.86	5.00	13.50	0.84	0.42	10.70	9.90	34.10	m	38.37
height 2400 - 2700mm	14.47	5.00	15.19	0.95	0.47	12.07	11.13	38.39	m	43.19
height 2700 - 3000mm	16.08	5.00	16.88	1.05	0.53	13.41	12.37	42.67	m	48.00
Beams; 3nr faces										
total girth 600 - 1200mm	6.43	10.00	7.07	1.06	0.21	11.26	4.95	23.28	m	26.19
Columns; 4nr faces										
total girth 600 - 1200mm	6.43	10.00	7.07	1.06	0.21	11.26	4.95	23.28	m	26.19
total girth 1200 - 1800mm	9.65	10.00	10.62	1.58	0.32	16.83	7.42	34.86	m	39.22
Reveals and soffits of openings and recesses										
width not exceeding 300mm	1.61	10.00	1.77	0.21	0.05	2.29	1.24	5.30	m	5.96
Ceilings										
generally	5.36	5.00	5.63	0.42	0.18	5.14	4.12	14.89	m2	16.75
3.50 - 5.00m above floor	5.36	5.00	5.63	0.46	0.18	5.51	4.12	15.26	m2	17.16
THERMAL BOARD - Linings; tapered edged sheeting of one layer 22mm thick Gyproc Thermal board vapour check grade for direct decoration, butt joints; fixing with galvanised nails to timber base; butt joints filled with joint filler tape and joint finish, spot filling										
Walls										
height 2100 - 2400mm	11.35	5.00	11.92	1.72	0.86	21.91	1.42	35.25	m	39.66
height 2400 - 2700mm	12.77	5.00	13.41	1.94	0.97	24.72	1.60	39.72	m	44.69
height 2700 - 3000mm	14.19	5.00	14.90	2.15	1.08	27.43	1.77	44.10	m	49.61
Reveals and soffits of openings and recesses										
width not exceeding 300mm	1.42	10.00	1.56	0.43	0.11	4.74	0.18	6.48	m	7.29
Ceilings										
generally	4.73	5.00	4.97	0.86	0.38	10.61	0.59	16.16	m2	18.18
THERMAL BOARD - Linings; tapered edged sheeting of one layer 30mm thick Gyproc thermal board vapour check grade for direct decoration, butt joints; fixing with galvanised nails to timber base; butt joints filled with joint filler tape and joint finish, spot filling										
Walls										
height 2100 - 2400mm	12.96	5.00	13.61	2.02	1.01	25.73	1.53	40.87	m	45.98
height 2400 - 2700mm	14.58	5.00	15.31	2.27	1.14	28.95	1.72	45.98	m	51.73
height 2700 - 3000mm	16.20	5.00	17.01	2.52	1.26	32.10	1.91	51.02	m	57.40
Reveals and soffits of openings and recesses										
width not exceeding 300mm	1.62	10.00	1.78	0.51	0.13	5.62	0.19	7.59	m	8.54
Ceilings										
generally	5.40	5.00	5.67	1.01	0.43	12.34	0.64	18.65	m2	20.98
THERMAL BOARD - Linings; tapered edged sheeting of one layer 40mm thick Gyproc Thermal board vapour check grade for direct decoration, butt joints; fixing with galvanised nails to timber base; butt joints filled with joint filler tape and joint finish, spot filling										
Walls										
height 2100 - 2400mm	20.09	5.00	21.09	2.30	1.15	29.30	1.53	51.93	m	58.42
height 2400 - 2700mm	22.60	5.00	23.73	2.59	1.30	33.03	1.72	58.48	m	65.79
height 2700 - 3000mm	25.11	5.00	26.37	2.87	1.44	36.60	1.91	64.87	m	72.98
Reveals and soffits of openings and recesses										
width not exceeding 300mm	2.51	10.00	2.76	0.58	0.14	6.34	0.19	9.29	m	10.45
Ceilings										
generally	8.37	5.00	8.79	1.15	0.49	14.05	0.64	23.48	m2	26.42
THERMAL BOARD - Linings; tapered edged sheeting of one layer 50mm thick Gyproc Thermal board vapour check grade for direct decoration, butt joints; fixing with galvanised nails to timber base; butt joints filled with joint filler tape and joint finish, spot filling										
Walls										
height 2100 - 2400mm	21.70	5.00	22.79	2.59	1.30	33.03	1.60	57.42	m	64.59
height 2400 - 2700mm	24.41	5.00	25.63	2.91	1.46	37.11	1.80	64.54	m	72.61
height 2700 - 3000mm	27.12	5.00	28.48	3.24	1.62	41.28	2.00	71.75	m	80.72

	MATERIALS			LABOUR				RATES		
Labour hourly rates: (except Specialists) Craft Operatives £9.23 Labourer £7.02 Rates are national average prices. Refer to REGIONAL VARIATIONS for indicative levels of overall pricing in regions	Del to Site	Waste	Material Cost	Craft Optve	Lab	Labour Cost	Sunds	Nett Rate	Unit	Gross Rate (+12.5%)
	£	%	£	Hrs	Hrs	£	£	£		£
K10: PLASTERBOARD DRY LINING/PARTITIONS/CEILINGS Cont.										
THERMAL BOARD - Linings; tapered edged sheeting of one layer 50mm thick Gyproc Thermal board vapour check grade for direct decoration, butt joints; fixing with galvanised nails to timber base; butt joints filled with joint filler tape and joint finish, spot filling Cont.										
Reveals and soffits of openings and recesses width not exceeding 300mm	2.71	10.00	2.98	0.65	0.16	7.12	0.20	10.30	m	11.59
Ceilings generally...	9.04	5.00	9.49	1.30	0.56	15.93	0.67	26.09	m2	29.35
CELLULAR PARTITIONS - Proprietary partitions; Paramount single leaf, tapered edge panels for direct decoration, butt joints; fixing with galvanised nails to 19 x 57mm (f sizes) impregnated wrot softwood sole plate with 19 x 37 x 300mm (f sizes) locating blocks at 900mm centres and 19x 37mm (f sizes) impregnated wrought softwood wall and ceiling battens; applying continuous beads of Gyproc acoustical sealant to perimeters of construction both sides; butt joints filled with joint filler tape and joint finish, spot filling; fixing to masonry with cartridge fired nails										
Height 2100 - 2400mm, 57mm thick boarded both sides	17.14	5.00	18.00	2.85	1.43	36.34	9.57	63.91	m	71.90
Height 2400 - 2700mm, 57mm thick boarded both sides	19.28	5.00	20.24	3.95	1.98	50.36	10.35	80.95	m	91.07
Height 2100 - 2400mm, 63mm thick boarded both sides	23.26	5.00	24.42	3.55	1.78	45.26	9.62	79.31	m	89.22
Height 2400 - 2700mm, 63mm thick boarded both sides	26.16	5.00	27.47	3.95	1.98	50.36	10.40	88.23	m	99.25
CELLULAR PARTITIONS - Angles to partitions; 37 x 19mm softwood batten, 37 x 37mm softwood batten, fixing together through partitions; one panel rebated by cutting back one face by full panel thickness; joints filled with joint filler tape and joint finish										
Plain 57mm thick partitions	0.68	10.00	0.75	0.30	0.15	3.82	0.93	5.50	m	6.19
CELLULAR PARTITIONS - Tee junctions to partitions; 37 x 19mm softwood batten, 37 x 37mm softwood batten, fixing together through partitions; joints filled with joint filler tape and joint finish										
Plain 57mm thick partitions	0.68	10.00	0.75	0.20	0.10	2.55	0.93	4.23	m	4.75
CELLULAR PARTITIONS - Fair ends to partitions										
57 mm thick partitions facing with 9.5mm Gyproc wallboard; fixing to 37 x 37mm softwood batten with galvanised nails	0.45	10.00	0.50	0.15	0.08	1.95	1.05	3.49	m	3.93
CELLULAR PARTITIONS - Fixings for heavy fittings										
Sinks insert 37 x 37mm softwood fixing block into core prior to erection of panels; not exceeding 150mm long	0.07	10.00	0.08	0.13	-	1.20	-	1.28	nr	1.44
insert 37 x 37mm softwood fixing block into core prior to erection panels; 150 - 300mm long	0.14	10.00	0.15	0.13	-	1.20	-	1.35	nr	1.52
Radiators insert 37 x 37mm softwood fixing block into core prior to erection of panels; not exceeding 150mm long	0.07	10.00	0.08	0.13	-	1.20	-	1.28	nr	1.44
insert 37 x 37mm softwood fixing block into core prior to erection panels; 150 - 300mm long	0.14	10.00	0.15	0.13	-	1.20	-	1.35	nr	1.52
LAMINATED PARTITIONS - Proprietary partitions; Gyproc laminated partition; tapered edge sheeting of one layer 12.5mm thick Gyproc wall board each side for direct decoration, B.S.1230, butt joints; applying continuous beads of Gyproc acoustical sealant to perimeters of construction on both sides, fixing with nails to timber frame comprising 25 x 38mm (f sizes) perimeter members with one layer 19mm thick Gyproc plank infill; butt joints filled with joint filler tape and joint finish, spot filling; fixing to masonry with cartridge fired nails										
Height 2100 - 2400mm, 50mm thick boarded both sides	17.30	5.00	18.16	3.60	1.80	45.86	7.93	71.96	m	80.95
Height 2400 - 2700mm, 50mm thick boarded both sides	19.47	5.00	20.44	3.95	1.98	50.36	8.44	79.24	m	89.15

LININGS/SHEATHING/DRY PARTITIONING

Labour hourly rates: (except Specialists) Craft Operatives £9.23 Labourer £7.02 Rates are national average prices. Refer to REGIONAL VARIATIONS for indicative levels of overall pricing in regions	MATERIALS			LABOUR				RATES		
	Del to Site £	Waste %	Material Cost £	Craft Optve Hrs	Lab Hrs	Labour Cost £	Sunds £	Nett Rate £	Unit	Gross Rate (+12.5%) £
K10: PLASTERBOARD DRY LINING/PARTITIONS/CEILINGS Cont.										
LAMINATED PARTITIONS - Proprietary partitions; Gyproc laminated partition; tapered edge sheeting of one layer 19mm thick Gyproc plank for direct decoration, B.S 1230, butt joints; applying continuous beads of Gyproc acoustical sealant to perimeters of construction on both sides, fixing with nails to timber frame comprising 25 x 38mm (f sizes) perimeter members with one layer 19mm thick Gyproc plank infill; butt joints filled with joint filler tape and joint finish, spot filling; fixing to masonry with cartridge fired nails										
Height 2100 - 2400mm, 65mm thick										
boarded both sides	23.26	5.00	**24.42**	3.80	1.90	48.41	7.93	**80.77**	m	90.86
Height 2400 - 2700mm, 65mm thick										
boarded both sides	26.16	5.00	**27.47**	4.20	2.10	53.51	8.44	**89.42**	m	100.59
Height 2700 - 3000mm, 65mm thick										
boarded both sides	29.07	5.00	**30.52**	4.60	2.30	58.60	8.93	**98.06**	m	110.31
LAMINATED PARTITIONS - Angles to partitions; 38 x 25mm softwood batten, fixing together through partitions; each panel rebated by cutting back core and inner layers; joints filled with joint filler tape and joint finish										
Plain										
50mm thick partitions	0.31	10.00	**0.34**	0.30	-	2.77	0.93	**4.04**	m	4.54
65mm thick partitions	0.31	10.00	**0.34**	0.30	-	2.77	0.93	**4.04**	m	4.54
LAMINATED PARTITIONS - Tee junctions to partitions; forming groove by cutting back one face layer; bedding end in groove in bonding compound										
Plain										
50mm thick partitions	-	-	-	0.25	-	2.31	1.09	**3.40**	m	3.82
65mm thick partitions	-	-	-	0.25	-	2.31	1.09	**3.40**	m	3.82
METAL STUD PARTITIONS - Proprietary partitions; Gyproc metal stud; tapered edge sheeting of one layer 12.5mm thick Gyproc wallboard for direct decoration, B.S.1230, butt joints; applying continuous beads of Gyproc acoustical sealant to perimeters of construction both sides, fixing with Pozidriv head screws to steel frame comprising 50mm head and floor channels, 48mm vertical studs at 600mm centres; butt joints filled with joint filler tape and joint finish, spot filling; fixing to masonry with cartridge fired nails										
Height 2100 - 2400mm, 75mm thick										
boarded both sides	15.87	5.00	**16.66**	2.85	1.43	36.34	6.64	**59.65**	m	67.10
METAL STUD PARTITIONS - Proprietary partitions; Gyproc metal stud; tapered edge sheeting of two layers 12.5mm thick Gyproc wallboard for direct decoration, B.S.1230, butt joints; applying continuous beads of Gyproc acoustical sealant to perimeters of construction both sides, fixing with Pozidriv head screws to steel frame comprising 50mm head and floor channels, 48mm vertical studs at 600mm centres; butt joints filled with joint filler tape and joint finish, spot filling; fixing to masonry with cartridge fired nails										
Height 2100 - 2400mm, 100mm thick										
boarded both sides	25.61	5.00	**26.89**	3.25	1.63	41.44	6.64	**74.97**	m	84.34
Height 2400 - 2700mm, 100mm thick										
boarded both sides	28.59	5.00	**30.02**	3.65	1.83	46.54	7.09	**83.65**	m	94.10
Height 2700 - 3000mm, 100mm thick										
boarded both sides	31.58	5.00	**33.16**	4.05	2.03	51.63	7.54	**92.33**	m	103.87
METAL STUD PARTITIONS - Angles to partitions; fixing end studs together through partition; extending sheeting to one side across end of partition; joints filled with joint filler tape and joint finish										
Plain										
75mm thick partitions	0.18	10.00	**0.20**	0.13	-	1.20	0.93	**2.33**	m	2.62
100mm thick partitions	0.49	10.00	**0.54**	0.13	-	1.20	0.93	**2.67**	m	3.00

LININGS/SHEATHING/DRY PARTITIONING – SMALL WORKS

Labour hourly rates: (except Specialists) Craft Operatives £9.23 Labourer £7.02 Rates are national average prices. Refer to REGIONAL VARIATIONS for indicative levels of overall pricing in regions	MATERIALS			LABOUR				RATES		
	Del to Site £	Waste %	Material Cost £	Craft Optve Hrs	Lab Hrs	Labour Cost £	Sunds £	Nett Rate £	Unit	Gross Rate (+12.5%) £
K10: PLASTERBOARD DRY LINING/PARTITIONS/CEILINGS: STAIRCASE AREAS										
PLASTERBOARD FOR DIRECT DECORATION - Linings; tapered edge sheeting of one layer 9.5mm thick Gyproc wall board for direct decoration, B.S.1230, butt joints; fixing with galvanised nails to timber base; butt joints filled with joint filler tape and joint finish, spot filling										
Walls										
height 2100 - 2400mm	4.10	5.00	4.30	1.20	0.48	14.45	1.39	20.14	m	22.66
height 2400 - 2700mm	4.62	5.00	4.85	1.35	0.54	16.25	1.57	22.67	m	25.51
height 2700 - 3000mm	5.13	5.00	5.39	1.50	0.60	18.06	1.74	25.18	m	28.33
Reveals and soffits of openings and recesses width not exceeding 300mm	0.51	10.00	0.56	0.27	0.06	2.91	0.17	3.64	m	4.10
Ceilings generally	1.71	5.00	1.80	0.54	0.20	6.39	0.58	8.76	m2	9.86
PLASTERBOARD FOR DIRECT DECORATION - Linings; tapered edge sheeting of one layer 12.5mm thick Gyproc wallboard for direct decoration, B.S.1230 butt joints; fixing with galvanised nails to timber base; butt joints filled with joint filler tape and joint finish, spot filling										
Walls										
height 2100 - 2400mm	4.78	5.00	5.02	1.42	0.56	17.04	1.41	23.47	m	26.40
height 2400 - 2700mm	5.37	5.00	5.64	1.59	0.63	19.10	1.58	26.32	m	29.61
height 2700 - 3000mm	5.97	5.00	6.27	1.77	0.71	21.32	1.76	29.35	m	33.02
Reveals and soffits of openings and recesses width not exceeding 300mm	0.60	10.00	0.66	0.32	0.07	3.44	0.18	4.29	m	4.82
Ceilings generally	1.99	5.00	2.09	0.63	0.24	7.50	0.59	10.18	m2	11.45
PLASTERBOARD FOR DIRECT DECORATION - Linings; tapered edge sheeting of one layer 15mm thick Gyproc wallboard for direct decoration, B.S.1230 butt joints; fixing with galvanised nails to timber base; butt joints filled with joint filler tape and joint finish, spot filling										
Walls										
height 2100 - 2400mm	5.23	5.00	5.49	1.64	0.65	19.70	1.41	26.60	m	29.93
height 2400 - 2700mm	5.89	5.00	6.18	1.84	0.73	22.11	1.58	29.87	m	33.61
height 2700 - 3000mm	6.54	5.00	6.87	1.98	0.81	23.96	1.76	32.59	m	36.66
Reveals and soffits of openings and recesses width not exceeding 300mm	0.65	10.00	0.71	0.37	0.08	3.98	0.18	4.87	m	5.48
Ceilings generally	2.18	5.00	2.29	0.72	0.28	8.61	0.59	11.49	m2	12.93
PLASTERBOARD FOR DIRECT DECORATION - Linings; tapered edge sheeting of one layer 9.5mm thick Gyproc Duplex wallboard for direct decoration, B.S.1230 butt joints; fixing with galvanised nails to timber base; butt joints filled with joint filler tape and joint finish, spot filling										
Walls										
height 2100 - 2400mm	5.59	5.00	5.87	1.20	0.48	14.45	1.39	21.71	m	24.42
height 2400 - 2700mm	6.29	5.00	6.60	1.35	0.54	16.25	1.57	24.43	m	27.48
height 2700 - 3000mm	6.99	5.00	7.34	1.50	0.60	18.06	1.74	27.14	m	30.53
Reveals and soffits of openings and recesses width not exceeding 300mm	0.70	10.00	0.77	0.27	0.06	2.91	0.17	3.85	m	4.33
Ceilings generally	2.33	5.00	2.45	0.54	0.20	6.39	0.58	9.41	m2	10.59
PLASTERBOARD FOR DIRECT DECORATION - Linings; tapered edge sheeting of one layer 12.5mm thick Gyproc Duplex wallboard for direct decoration, B.S.1230, butt joints; fixing with galvanised nails to timber base; butt joints filled with joint filler tape and joint finish, spot filling										
Walls										
height 2100 - 2400mm	6.26	5.00	6.57	1.42	0.56	17.04	1.41	25.02	m	28.15
height 2400 - 2700mm	7.05	5.00	7.40	1.59	0.63	19.10	1.58	28.08	m	31.59
height 2700 - 3000mm	7.83	5.00	8.22	1.77	0.71	21.32	1.76	31.30	m	35.22
Reveals and soffits of openings and recesses width not exceeding 300mm	0.78	10.00	0.86	0.32	0.07	3.44	0.18	4.48	m	5.04
Ceilings generally	2.61	5.00	2.74	0.63	0.24	7.50	0.59	10.83	m2	12.18

LININGS/SHEATHING/DRY PARTITIONING

Labour hourly rates: (except Specialists) Craft Operatives £9.23 Labourer £7.02 Rates are national average prices. Refer to REGIONAL VARIATIONS for indicative levels of overall pricing in regions	MATERIALS			LABOUR				RATES		
	Del to Site £	Waste %	Material Cost £	Craft Optve Hrs	Lab Hrs	Labour Cost £	Sunds £	Nett Rate £	Unit	Gross Rate (+12.5%) £
K10: PLASTERBOARD DRY LINING/PARTITIONS/CEILINGS: STAIRCASE AREAS Cont.										
PLASTERBOARD FOR DIRECT DECORATION - Linings; tapered edge sheeting of one layer 15mm thick Gyproc Duplex wallboard for direct decoration, B.S.1230 butt joints; fixing with galvanised nails to timber base; butt joints filled with joint filler tape and joint finish, spot filling										
Walls										
height 2100 - 2400mm	6.86	5.00	7.20	1.64	0.65	19.70	1.41	28.31	m	31.85
height 2400 - 2700mm	7.72	5.00	8.11	1.84	0.73	22.11	1.58	31.79	m	35.77
height 2700 - 3000mm	8.58	5.00	9.01	1.98	0.81	23.96	1.76	34.73	m	39.07
Reveals and soffits of openings and recesses										
width not exceeding 300mm	0.86	10.00	0.95	0.37	0.08	3.98	0.18	5.10	m	5.74
Ceilings										
generally	2.86	5.00	3.00	0.72	0.28	8.61	0.59	12.20	m2	13.73
PLASTERBOARD FOR DIRECT DECORATION - Linings; tapered edge sheeting of one layer 19mm thick Gyproc plank for direct decoration, B.S.1230, butt joints; fixing with galvanised nails to timber base; butt joints filled with joint filler tape and joint finish, spot filling										
Walls										
height 2100 - 2400mm	7.75	5.00	8.14	1.80	0.72	21.67	1.30	31.11	m	34.99
height 2400 - 2700mm	8.72	5.00	9.16	2.03	0.81	24.42	1.47	35.05	m	39.43
height 2700 - 3000mm	9.69	5.00	10.17	2.25	0.90	27.09	1.63	38.89	m	43.75
Reveals and soffits of openings and recesses										
width not exceeding 300mm	0.97	10.00	1.07	0.41	0.09	4.42	0.16	5.64	m	6.35
Ceilings										
generally	3.23	5.00	3.39	0.81	0.30	9.58	0.54	13.51	m2	15.20
PLASTERBOARD FOR DIRECT DECORATION - Linings; two layers of Gypsum wallboard, first layer square edge sheeting 9.5mm thick, second layer tapered edge sheeting 9.5mm thick for direct decoration, B.S.1230, butt joints; fixing with galvanised nails to timber base; butt joints of second layer filled with joint filler tape and joint finish, spot filling										
Walls										
height 2100 - 2400mm	8.21	5.00	8.62	2.11	0.84	25.37	1.75	35.74	m	40.21
height 2400 - 2700mm	9.23	5.00	9.69	2.38	0.95	28.64	1.97	40.30	m	45.34
height 2700 - 3000mm	10.26	5.00	10.77	2.64	1.05	31.74	2.19	44.70	m	50.29
Reveals and soffits of openings and recesses										
width not exceeding 300mm	1.03	10.00	1.13	0.47	0.11	5.11	0.22	6.46	m	7.27
Ceilings										
generally	3.42	5.00	3.59	0.95	0.35	11.23	0.73	15.55	m2	17.49
PLASTERBOARD FOR DIRECT DECORATION - Linings; two layers of Gypsum wallboard, first layer square edge sheeting 9.5mm thick, second layer tapered edge sheeting 12.5mm thick for direct decoration, B.S.1230, butt joints; fixing with galvanised nails to timber base; butt joints of second layer filled with joint filler tape and joint finish, spot filling										
Walls										
height 2100 - 2400mm	8.88	5.00	9.32	2.40	0.96	28.89	1.75	39.97	m	44.96
height 2400 - 2700mm	9.99	5.00	10.49	2.70	1.08	32.50	1.97	44.96	m	50.58
height 2700 - 3000mm	11.10	5.00	11.65	3.00	1.20	36.11	2.19	49.96	m	56.20
Reveals and soffits of openings and recesses										
width not exceeding 300mm	1.11	10.00	1.22	0.48	0.12	5.27	0.22	6.71	m	7.55
Ceilings										
generally	3.70	5.00	3.88	1.08	0.40	12.78	0.73	17.39	m2	19.57
PLASTERBOARD FOR DIRECT DECORATION - Linings; two layers of Gypsum wallboard, first layer square edge sheeting 12.5mm thick, second layer tapered edge sheeting 12.5mm thick for direct decoration, B.S.1230, butt joints; fixing with galvanised nails to timber base; butt joints of second layer filled with joint filler tape and joint finish, spot filling										
Walls										
height 2100 - 2400mm	9.55	5.00	10.03	2.54	1.02	30.60	1.76	42.39	m	47.69
height 2400 - 2700mm	10.75	5.00	11.29	2.86	1.15	34.47	1.98	47.74	m	53.71
height 2700 - 3000mm	11.94	5.00	12.54	3.18	1.28	38.34	2.20	53.07	m	59.71
Reveals and soffits of openings and recesses										
width not exceeding 300mm	1.19	10.00	1.31	0.57	0.13	6.17	0.22	7.70	m	8.67
Ceilings										
generally	3.98	5.00	4.18	1.15	0.43	13.63	0.73	18.54	m2	20.86

Labour hourly rates: (except Specialists) Craft Operatives £9.23 Labourer £7.02 Rates are national average prices. Refer to REGIONAL VARIATIONS for indicative levels of overall pricing in regions	MATERIALS			LABOUR				RATES		
	Del to Site £	Waste %	Material Cost £	Craft Optve Hrs	Lab Hrs	Labour Cost £	Sunds £	Nett Rate £	Unit	Gross Rate (+12.5%) £
K10: PLASTERBOARD DRY LINING/PARTITIONS/CEILINGS: STAIRCASE AREAS Cont.										
PLASTERBOARD FOR DIRECT DECORATION - Linings; two layers of Gypsum wallboard, first layer square edge sheeting 15mm thick, second layer tapered edge sheeting 15mm thick for direct decoration, B.S.1230, butt joints; fixing with galvanised nails to timber base; butt joints of second layer filled with joint filler tape and joint finish, spot filling										
Walls										
height 2100 - 2400mm	10.46	5.00	10.98	2.68	1.08	32.32	1.77	45.07	m	50.70
height 2400 - 2700mm	11.77	5.00	12.36	3.01	1.22	36.35	2.00	50.71	m	57.04
height 2700 - 3000mm	13.08	5.00	13.73	3.35	1.35	40.40	2.22	56.35	m	63.40
Reveals and soffits of openings and recesses width not exceeding 300mm	1.31	10.00	1.44	0.66	0.14	7.07	0.22	8.74	m	9.83
Ceilings generally..	4.36	5.00	4.58	1.22	0.46	14.49	0.74	19.81	m2	22.28
PLASTERBOARD FOR DIRECT DECORATION - Linings; two layers of Gypsum wallboard, first layer square edge sheeting 9.5mm thick, second layer tapered edge sheeting 9.5mm thick for direct decoration, B.S.1230, butt joints; finish with galvanised nails to timber base; butt joint on second layer filled with joint filler tape and joint finish, spot filling										
Walls										
height 2100 - 2400mm	9.70	5.00	10.19	2.11	0.84	25.37	1.75	37.31	m	41.97
height 2400 - 2700mm	15.32	5.00	16.09	2.38	0.95	28.64	-	44.73	m	50.32
height 2700 - 3000mm	12.12	5.00	12.73	2.64	1.05	31.74	2.19	46.65	m	52.49
Reveals and soffits of openings and recesses width not exceeding 300mm	1.21	10.00	1.33	0.47	0.11	5.11	0.22	6.66	m	7.49
Ceilings generally..	4.04	5.00	4.24	0.95	0.35	11.23	0.73	16.20	m2	18.22
PLASTERBOARD FOR DIRECT DECORATION - Linings; two layers of Gypsum wallboard, first layer Duplex square edge sheeting 9.5mm thick, second layer tapered edge sheeting 12.5mm thick for direct decoration, B.S.1230, butt joints; fixing with galvanised nails to timber base; butt joints of second layer filled with joint filler tape and joint finish, spot filling										
Walls										
height 2100 - 2400mm	10.37	5.00	10.89	2.40	0.96	28.89	1.75	41.53	m	46.72
height 2400 - 2700mm	11.66	5.00	12.24	2.70	1.08	32.50	1.97	46.72	m	52.56
height 2700 - 3000mm	12.96	5.00	13.61	3.00	1.20	36.11	2.19	51.91	m	58.40
Reveals and soffits of openings and recesses width not exceeding 300mm	1.30	10.00	1.43	0.48	0.12	5.27	0.22	6.92	m	7.79
Ceilings generally..	4.32	5.00	4.54	1.08	0.40	12.78	0.73	18.04	m2	20.30
PLASTERBOARD FOR DIRECT DECORATION - Linings; two layers of Gypsum wallboard, first layer Duplex square edge sheeting 12.5mm thick, second layer tapered edge sheeting 12.5mm thick for direct decoration, B.S.1230, butt joints; fixing with galvanised nails to timber base; butt joints of second layer filled with joint filler tape and joint finish, spot filling										
Walls										
height 2100 - 2400mm	11.04	5.00	11.59	2.54	1.02	30.60	1.76	43.96	m	49.45
height 2400 - 2700mm	12.42	5.00	13.04	2.86	1.15	34.47	1.98	49.49	m	55.68
height 2700 - 3000mm	13.80	5.00	14.49	3.18	1.28	38.34	2.20	55.03	m	61.91
Reveals and soffits of openings and recesses width not exceeding 300mm	1.38	10.00	1.52	0.57	0.13	6.17	0.22	7.91	m	8.90
Ceilings generally..	4.60	5.00	4.83	1.15	0.43	13.63	0.73	19.19	m2	21.59
PLASTERBOARD FOR DIRECT DECORATION - Linings; two layers of Gypsum wallboard, first layer Duplex square edge sheeting 15mm thick, second layer tapered edge sheeting 15mm thick for direct decoration, B.S.1230, butt joints; fixing with galvanised nails to timber base; butt joints of second layer filled with joint filler tape and joint finish, spot filling										
Walls										
height 2100 - 2400mm	12.10	5.00	12.71	2.68	1.08	32.32	1.77	46.79	m	52.64
height 2400 - 2700mm	13.61	5.00	14.29	3.01	1.22	36.35	2.00	52.64	m	59.22
height 2700 - 3000mm	15.12	5.00	15.88	3.35	1.35	40.40	2.22	58.49	m	65.81
Reveals and soffits of openings and recesses width not exceeding 300mm	1.51	10.00	1.66	0.66	0.14	7.07	0.22	8.96	m	10.08

Labour hourly rates: (except Specialists) Craft Operatives £9.23 Labourer £7.02 Rates are national average prices. Refer to REGIONAL VARIATIONS for indicative levels of overall pricing in regions	MATERIALS			LABOUR				RATES		
	Del to Site £	Waste %	Material Cost £	Craft Optve Hrs	Lab Hrs	Labour Cost £	Sunds £	Nett Rate £	Unit	Gross Rate (+12.5%) £
K10: PLASTERBOARD DRY LINING/PARTITIONS/CEILINGS: STAIRCASE AREAS Cont.										
PLASTERBOARD FOR DIRECT DECORATION - Linings; two layers of Gypsum wallboard, first layer Duplex square edge sheeting 15mm thick, second layer tapered edge sheeting 15mm thick for direct decoration, B.S.1230, butt joints; fixing with galvanised nails to timber base; butt joints of second layer filled with joint filler tape and joint finish, spot filling Cont.										
Ceilings										
generally...	5.04	5.00	5.29	1.22	0.46	14.49	0.74	20.52	m2	23.09
PLASTIC FACED - Linings; square edge sheeting of one layer 9.5mm thick Gyproc industrial grade plastic faced plasterboard, B.S.1230, butt joints; fixing with galvanised nails with plastic coated nail caps to timber base; butt joints covered with Gyproc batten section and trim										
Walls										
height 2100 - 2400mm	11.95	5.00	12.55	0.91	0.36	10.93	9.89	33.36	m	37.53
height 2400 - 2700mm	13.45	5.00	14.12	1.03	0.41	12.39	11.11	37.62	m	42.32
height 2700 - 3000mm	14.94	5.00	15.69	1.14	0.45	13.68	12.36	41.73	m	46.94
Reveals and soffits of openings and recesses										
width not exceeding 300mm	1.49	10.00	1.64	0.20	0.05	2.20	1.24	5.08	m	5.71
Ceilings										
generally...	4.98	5.00	5.23	0.41	0.15	4.84	4.12	14.19	m2	15.96
PLASTIC FACED - Linings; square edge sheeting of one layer 12.5mm thick Gyproc industrial grade plastic faced plasterboard, B.S.1230, butt joints; fixing with galvanised nails with plastic coated nail caps to timber base; butt joints covered with Gyproc batten section and trim										
Walls										
height 2100 - 2400mm	12.86	5.00	13.50	1.06	0.42	12.73	9.90	36.14	m	40.65
height 2400 - 2700mm	14.47	5.00	15.19	1.19	0.47	14.28	11.13	40.61	m	45.68
height 2700 - 3000mm	16.08	5.00	16.88	1.32	0.53	15.90	12.37	45.16	m	50.80
Reveals and soffits of openings and recesses										
width not exceeding 300mm	1.61	10.00	1.77	0.24	0.05	2.57	1.24	5.58	m	6.27
Ceilings										
generally...	5.36	5.00	5.63	0.47	0.18	5.60	4.12	15.35	m2	17.27
K11: RIGID SHEET FLOORING/SHEATHING/LININGS/CASINGS										
Douglas Fir plywood, unsanded select sheathing quality, WBP bonded; tongued and grooved joints long edges										
15mm thick sheeting to floors										
width exceeding 300mm	6.34	7.50	6.82	0.35	0.04	3.51	0.11	10.44	m2	11.74
18mm thick sheeting to floors										
width exceeding 300mm	8.03	7.50	8.63	0.36	0.04	3.60	0.11	12.35	m2	13.89
Birch faced plywood, BB quality, WBP bonded; tongued and grooved joints all edges										
12mm thick sheeting to floors										
width exceeding 300mm	19.03	7.50	20.46	0.32	0.04	3.23	0.11	23.80	m2	26.78
15mm thick sheeting to floors										
width exceeding 300mm	22.85	7.50	24.56	0.35	0.04	3.51	0.11	28.19	m2	31.71
18mm thick sheeting to floors										
width exceeding 300mm	27.71	7.50	29.79	0.36	0.04	3.60	0.11	33.50	m2	37.69
Wood chipboard, B.S.5669, type II flooring, square edges; butt joints										
18mm thick sheeting to floors										
width exceeding 300mm	5.64	7.50	6.06	0.35	0.04	3.51	0.11	9.68	m2	10.89
22mm thick sheeting to floors										
width exceeding 300mm	7.65	7.50	8.22	0.36	0.04	3.60	0.11	11.94	m2	13.43
Wood chipboard, B.S.5669, type II flooring; tongued and grooved joints										
18mm thick sheeting to floors										
width exceeding 300mm	5.96	7.50	6.41	0.35	0.04	3.51	0.11	10.03	m2	11.28
22mm thick sheeting to floors										
width exceeding 300mm	7.93	7.50	8.52	0.36	0.04	3.60	0.11	12.24	m2	13.77
Abutments										
12mm thick plywood linings										
raking cutting......................	1.90	-	1.90	0.04	-	0.37	-	2.27	m	2.55

Labour hourly rates: (except Specialists) Craft Operatives £9.23 Labourer £7.02 Rates are national average prices. Refer to REGIONAL VARIATIONS for indicative levels of overall pricing in regions	MATERIALS			LABOUR				RATES		
	Del to Site £	Waste %	Material Cost £	Craft Optve Hrs	Lab Hrs	Labour Cost £	Sunds £	Nett Rate £	Unit	Gross Rate (+12.5%) £
K11: RIGID SHEET FLOORING/SHEATHING/LININGS/CASINGS Cont.										
Abutments Cont.										
12mm thick plywood linings Cont.										
curved cutting	2.86	-	2.86	0.06	-	0.55	-	3.41	m	3.84
15mm thick plywood linings										
raking cutting	2.29	-	2.29	0.05	-	0.46	-	2.75	m	3.10
curved cutting	3.43	-	3.43	0.07	-	0.65	-	4.08	m	4.59
18mm thick plywood linings										
raking cutting	2.77	-	2.77	0.05	-	0.46	-	3.23	m	3.64
curved cutting	4.16	-	4.16	0.07	-	0.65	-	4.81	m	5.41
18mm thick wood chipboard linings										
raking cutting	0.56	-	0.56	0.04	-	0.37	-	0.93	m	1.05
curved cutting	0.85	-	0.85	0.06	-	0.55	-	1.40	m	1.58
22mm thick wood chipboard linings										
raking cutting	0.77	-	0.77	0.05	-	0.46	-	1.23	m	1.39
curved cutting	1.14	-	1.14	0.07	-	0.65	-	1.79	m	2.01
SOUND INSULATED FLOORS - Foam backed softwood battens covered with tongued and grooved chipboard or plywood panels										
19mm chipboard panels to level sub-floors										
width exceeding 300mm; battens at 450mm centres; elevation										
50mm	17.78	10.00	19.56	0.35	0.04	3.51	0.29	23.36	m2	26.28
75mm	20.43	10.00	22.47	0.35	0.04	3.51	0.32	26.30	m2	29.59
width exceeding 300mm; battens at 600mm centres; elevation										
50mm	27.18	10.00	29.90	0.35	0.04	3.51	0.25	33.66	m2	37.87
75mm	29.83	10.00	32.81	0.35	0.04	3.51	0.26	36.58	m2	41.16
19mm chipboard panels with improved moisture resistance to level sub-floors										
width exceeding 300mm; battens at 450mm centres; elevation										
50mm	19.35	10.00	21.29	0.35	0.04	3.51	0.29	25.09	m2	28.22
75mm	22.00	10.00	24.20	0.35	0.04	3.51	0.32	28.03	m2	31.54
width exceeding 300mm; battens at 600mm centres; elevation										
50mm	17.78	10.00	19.56	0.35	0.04	3.51	0.25	23.32	m2	26.23
75mm	20.43	10.00	22.47	0.35	0.04	3.51	0.26	26.24	m2	29.52
19mm plywood panels to level sub-floors										
width exceeding 300mm; battens at 450mm centres; elevation										
50mm	27.18	10.00	29.90	0.35	0.04	3.51	0.29	33.70	m2	37.91
75mm	29.83	10.00	32.81	0.35	0.04	3.51	0.32	36.64	m2	41.22
FIBREBOARDS AND HARDBOARDS - Standard hardboard, B.S.1142, type SHA; butt joints										
3.2mm thick linings to walls										
width exceeding 300mm	1.03	5.00	1.08	0.38	0.05	3.86	0.09	5.03	m2	5.66
width not exceeding 300mm	0.31	10.00	0.34	0.23	0.02	2.26	0.05	2.65	m	2.99
4.8mm thick linings to walls										
width exceeding 300mm	1.70	5.00	1.78	0.40	0.05	4.04	0.10	5.93	m2	6.67
width not exceeding 300mm	0.51	10.00	0.56	0.24	0.02	2.36	0.06	2.98	m	3.35
6.4mm thick linings to walls										
width exceeding 300mm	2.21	5.00	2.32	0.41	0.05	4.14	0.10	6.56	m2	7.38
width not exceeding 300mm	0.66	10.00	0.73	0.25	0.02	2.45	0.06	3.23	m	3.64
3.2mm thick sheeting to floors										
width exceeding 300mm	1.03	5.00	1.08	0.19	0.02	1.89	0.04	3.02	m2	3.39
3.2mm thick linings to ceilings										
width exceeding 300mm	1.03	5.00	1.08	0.48	0.06	4.85	0.11	6.04	m2	6.80
width not exceeding 300mm	0.31	10.00	0.34	0.29	0.02	2.82	0.07	3.23	m	3.63
4.8mm thick linings to ceilings										
width exceeding 300mm	1.70	5.00	1.78	0.50	0.06	5.04	0.12	6.94	m2	7.81
width not exceeding 300mm	0.51	10.00	0.56	0.30	0.02	2.91	0.07	3.54	m	3.98
6.4mm thick linings to ceilings										
width exceeding 300mm	2.21	5.00	2.32	0.51	0.06	5.13	0.12	7.57	m2	8.52
width not exceeding 300mm	0.66	10.00	0.73	0.31	0.02	3.00	0.07	3.80	m	4.27
FIBREBOARDS AND HARDBOARDS - Tempered hardboard, B.S.1142, Type THE; butt joints										
3.2mm thick linings to walls										
width exceeding 300mm	1.70	5.00	1.78	0.38	0.05	3.86	0.09	5.73	m2	6.45
width not exceeding 300mm	0.51	10.00	0.56	0.23	0.02	2.26	0.05	2.87	m	3.23
4.8mm thick linings to walls										
width exceeding 300mm	2.40	5.00	2.52	0.40	0.05	4.04	0.10	6.66	m2	7.50
width not exceeding 300mm	0.72	10.00	0.79	0.24	0.02	2.36	0.06	3.21	m	3.61

Labour hourly rates: (except Specialists) Craft Operatives £9.23 Labourer £7.02 Rates are national average prices. Refer to REGIONAL VARIATIONS for indicative levels of overall pricing in regions	MATERIALS			LABOUR				RATES		
	Del to Site £	Waste %	Material Cost £	Craft Optve Hrs	Lab Hrs	Labour Cost £	Sunds £	Nett Rate £	Unit	Gross Rate (+12.5%) £
K11: RIGID SHEET FLOORING/SHEATHING/LININGS/CASINGS Cont.										
FIBREBOARDS AND HARDBOARDS - Tempered hardboard, B.S.1142, Type THE; butt joints Cont.										
6.4mm thick linings to walls										
width exceeding 300mm	3.15	5.00	3.31	0.41	0.05	4.14	0.10	7.54	m2	8.49
width not exceeding 300mm	0.95	10.00	1.04	0.25	0.02	2.45	0.06	3.55	m	4.00
3.2mm thick linings to ceilings										
width exceeding 300mm	1.70	5.00	1.78	0.48	0.06	4.85	0.11	6.75	m2	7.59
width not exceeding 300mm	0.51	10.00	0.56	0.29	0.02	2.82	0.07	3.45	m	3.88
4.8mm thick linings to ceilings										
width exceeding 300mm	2.40	5.00	2.52	0.50	0.06	5.04	0.12	7.68	m2	8.64
width not exceeding 300mm	0.72	10.00	0.79	0.30	0.02	2.91	0.07	3.77	m	4.24
6.4mm thick linings to ceilings										
width exceeding 300mm	3.15	5.00	3.31	0.51	0.06	5.13	0.12	8.56	m2	9.63
width not exceeding 300mm	0.95	10.00	1.04	0.31	0.02	3.00	0.07	4.12	m	4.63
FIBREBOARDS AND HARDBOARDS - Flame retardant hardboard, B.S.1142 tested to B.S.476 Part 7 Class 1; butt joints										
3.2mm thick linings to walls										
width exceeding 300mm	5.51	5.00	5.79	0.38	0.05	3.86	0.09	9.73	m2	10.95
width not exceeding 300mm	1.65	10.00	1.81	0.23	0.02	2.26	0.05	4.13	m	4.64
4.8mm thick linings to walls										
width exceeding 300mm	6.37	5.00	6.69	0.41	0.05	4.14	0.10	10.92	m2	12.29
width not exceeding 300mm	1.91	10.00	2.10	0.25	0.02	2.45	0.06	4.61	m	5.19
6.4mm thick linings to walls										
width exceeding 300mm	7.16	5.00	7.52	0.42	0.05	4.23	0.10	11.85	m2	13.33
width not exceeding 300mm	2.15	10.00	2.37	0.25	0.02	2.45	0.06	4.87	m	5.48
3.2mm thick linings to ceilings										
width exceeding 300mm	5.51	5.00	5.79	0.48	0.06	4.85	0.11	10.75	m2	12.09
width not exceeding 300mm	1.65	10.00	1.81	0.29	0.02	2.82	0.07	4.70	m	5.29
4.8mm thick linings to ceilings										
width exceeding 300mm	6.37	5.00	6.69	0.51	0.06	5.13	0.12	11.94	m2	13.43
width not exceeding 300mm	1.91	10.00	2.10	0.31	0.02	3.00	0.07	5.17	m	5.82
6.4mm thick linings to ceilings										
width exceeding 300mm	7.16	5.00	7.52	0.53	0.07	5.38	0.13	13.03	m2	14.66
width not exceeding 300mm	2.15	10.00	2.37	0.32	0.02	3.09	0.07	5.53	m	6.22
FIBREBOARDS AND HARDBOARDS - Low density medium board, B.S.1142, type LMN; butt joints										
6mm thick linings to walls										
width exceeding 300mm	6.64	5.00	6.97	0.41	0.05	4.14	0.10	11.21	m2	12.61
width not exceeding 300mm	1.99	10.00	2.19	0.25	0.02	2.45	0.06	4.70	m	5.28
9mm thick linings to walls										
width exceeding 300mm	8.24	5.00	8.65	0.42	0.05	4.23	0.10	12.98	m2	14.60
width not exceeding 300mm	2.47	10.00	2.72	0.25	0.02	2.45	0.06	5.22	m	5.88
12mm thick linings to walls										
width exceeding 300mm	10.92	5.00	11.47	0.44	0.06	4.48	0.11	16.06	m2	18.07
width not exceeding 300mm	3.28	10.00	3.61	0.26	0.02	2.54	0.06	6.21	m	6.98
6mm thick linings to ceilings										
width exceeding 300mm	6.64	5.00	6.97	0.51	0.06	5.13	0.12	12.22	m2	13.75
width not exceeding 300mm	1.99	10.00	2.19	0.31	0.02	3.00	0.07	5.26	m	5.92
9mm thick linings to ceilings										
width exceeding 300mm	8.24	5.00	8.65	0.53	0.07	5.38	0.13	14.17	m2	15.94
width not exceeding 300mm	2.47	10.00	2.72	0.32	0.02	3.09	0.07	5.88	m	6.62
12mm thick linings to ceilings										
width exceeding 300mm	10.92	5.00	11.47	0.55	0.07	5.57	0.13	17.16	m2	19.31
width not exceeding 300mm	3.28	10.00	3.61	0.33	0.02	3.19	0.08	6.87	m	7.73
FIBREBOARDS AND HARDBOARDS - Low density medium board, B.S.1142, type LMN; butt joints										
6mm thick linings to walls										
width exceeding 300mm	5.67	5.00	5.95	0.41	0.05	4.14	0.10	10.19	m2	11.46
width not exceeding 300mm	1.70	10.00	1.87	0.25	0.02	2.45	0.06	4.38	m	4.93
9mm thick linings to walls										
width exceeding 300mm	7.00	5.00	7.35	0.42	0.05	4.23	0.10	11.68	m2	13.14
width not exceeding 300mm	2.10	10.00	2.31	0.29	0.02	2.82	0.06	5.19	m	5.84
6mm thick linings to ceilings										
width exceeding 300mm	5.67	5.00	5.95	0.51	0.06	5.13	0.12	11.20	m2	12.60
width not exceeding 300mm	1.70	10.00	1.87	0.31	0.02	3.00	0.07	4.94	m	5.56
9mm thick linings to ceilings										
width exceeding 300mm	7.00	5.00	7.35	0.53	0.07	5.38	0.13	12.86	m2	14.47
width not exceeding 300mm	2.10	10.00	2.31	0.32	0.02	3.09	0.07	5.47	m	6.16

Labour hourly rates: (except Specialists) Craft Operatives £9.23 Labourer £7.02 Rates are national average prices. Refer to REGIONAL VARIATIONS for indicative levels of overall pricing in regions	MATERIALS			LABOUR				RATES		
	Del to Site £	Waste %	Material Cost £	Craft Optve Hrs	Lab Hrs	Labour Cost £	Sunds £	Nett Rate £	Unit	Gross Rate (+12.5%) £
K11: RIGID SHEET FLOORING/SHEATHING/LININGS/CASINGS Cont.										
FIBREBOARDS AND HARDBOARDS – Medium density fibreboard, B.S.1142, type MDF; butt joints										
9mm thick linings to walls										
width exceeding 300mm	3.86	5.00	4.05	0.41	0.05	4.14	0.10	8.29	m2	9.32
width not exceeding 300mm	1.16	10.00	1.28	0.25	0.02	2.45	0.06	3.78	m	4.26
12mm thick linings to walls										
width exceeding 300mm	4.89	5.00	5.13	0.44	0.06	4.48	0.11	9.73	m2	10.94
width not exceeding 300mm	1.47	10.00	1.62	0.26	0.02	2.54	0.06	4.22	m	4.74
18mm thick linings to walls										
width exceeding 300mm	5.97	5.00	6.27	0.47	0.06	4.76	0.11	11.14	m2	12.53
width not exceeding 300mm	1.79	10.00	1.97	0.28	0.02	2.72	0.06	4.75	m	5.35
25mm thick linings to walls										
width exceeding 300mm	9.01	5.00	9.46	0.50	0.06	5.04	0.12	14.62	m2	16.44
width not exceeding 300mm	2.70	10.00	2.97	0.30	0.02	2.91	0.07	5.95	m	6.69
9mm thick linings to ceilings										
width exceeding 300mm	3.86	5.00	4.05	0.51	0.06	5.13	0.12	9.30	m2	10.46
width not exceeding 300mm	1.16	10.00	1.28	0.31	0.02	3.00	0.07	4.35	m	4.89
12mm thick linings to ceilings										
width exceeding 300mm	4.89	5.00	5.13	0.55	0.07	5.57	0.13	10.83	m2	12.19
width not exceeding 300mm	1.47	10.00	1.62	0.33	0.02	3.19	0.08	4.88	m	5.49
18mm thick linings to ceilings										
width exceeding 300mm	5.97	5.00	6.27	0.59	0.07	5.94	0.14	12.35	m2	13.89
width not exceeding 300mm	1.79	10.00	1.97	0.35	0.02	3.37	0.08	5.42	m	6.10
25mm thick linings to ceilings										
width exceeding 300mm	9.01	5.00	9.46	0.63	0.08	6.38	0.15	15.99	m2	17.99
width not exceeding 300mm	2.71	10.00	2.98	0.38	0.02	3.65	0.09	6.72	m	7.56
FIBREBOARDS AND HARDBOARDS – Insulating softboard, B.S.1142; type SBN butt joints										
13mm thick linings to walls										
width exceeding 300mm	2.10	5.00	2.21	0.44	0.06	4.48	0.11	6.80	m2	7.65
width not exceeding 300mm	0.63	10.00	0.69	0.26	0.02	2.54	0.06	3.29	m	3.70
13mm thick linings to ceilings										
width exceeding 300mm	2.10	5.00	2.21	0.55	0.07	5.57	0.13	7.90	m2	8.89
width not exceeding 300mm	0.63	10.00	0.69	0.33	0.02	3.19	0.08	3.96	m	4.45
FIBREBOARDS AND HARDBOARDS – Bitumen impregnated insulating board, B.S.1142, type SBI; butt joints										
13mm thick linings to walls										
width exceeding 300mm	2.51	5.00	2.64	0.44	0.06	4.48	0.11	7.23	m2	8.13
width not exceeding 300mm	0.75	10.00	0.82	0.26	0.02	2.54	0.06	3.43	m	3.85
13mm thick linings to ceilings										
width exceeding 300mm	2.51	5.00	2.64	0.55	0.07	5.57	0.13	8.33	m2	9.38
width not exceeding 300mm	0.75	10.00	0.82	0.33	0.02	3.19	0.08	4.09	m	4.60
ASBESTOS FREE BOARD – Masterboard Class 0 fire resisting boards; butt joints										
6mm thick linings to walls										
width exceeding 300mm	7.70	5.00	8.09	0.72	0.09	7.28	0.17	15.53	m2	17.47
width not exceeding 300mm	2.31	10.00	2.54	0.43	0.03	4.18	0.10	6.82	m	7.67
9mm thick linings to walls										
width exceeding 300mm	15.75	5.00	16.54	0.80	0.10	8.09	0.19	24.81	m2	27.92
width not exceeding 300mm	4.73	10.00	5.20	0.48	0.03	4.64	0.11	9.95	m	11.20
12mm thick linings to walls										
width exceeding 300mm	20.75	5.00	21.79	0.92	0.12	9.33	0.22	31.34	m2	35.26
width not exceeding 300mm	6.23	10.00	6.85	0.55	0.04	5.36	0.13	12.34	m	13.88
6mm thick linings to ceilings										
width exceeding 300mm	7.70	5.00	8.09	0.77	0.10	7.81	0.19	16.08	m2	18.09
width not exceeding 300mm	2.31	10.00	2.54	0.46	0.03	4.46	0.11	7.11	m	8.00
9mm thick linings to ceilings										
width exceeding 300mm	15.75	5.00	16.54	0.90	0.11	9.08	0.22	25.84	m2	29.07
width not exceeding 300mm	4.73	10.00	5.20	0.54	0.03	5.19	0.12	10.52	m	11.83
12mm thick linings to ceilings										
width exceeding 300mm	20.75	5.00	21.79	0.99	0.12	9.98	0.24	32.01	m2	36.01
width not exceeding 300mm	6.23	10.00	6.85	0.99	0.04	9.42	0.14	16.41	m	18.46
ASBESTOS FREE BOARD – Supalux fire resisting boards, Sanded; butt joints; fixing to timber with screws, countersinking										
6mm thick linings to walls										
width exceeding 300mm	11.45	5.00	12.02	0.72	0.09	7.28	0.17	19.47	m2	21.90
width not exceeding 300mm	3.44	10.00	3.78	0.43	0.03	4.18	0.10	8.06	m	9.07

LININGS/SHEATHING/DRY PARTITIONING

Labour hourly rates: (except Specialists) Craft Operatives £9.23 Labourer £7.02 Rates are national average prices. Refer to REGIONAL VARIATIONS for indicative levels of overall pricing in regions	MATERIALS			LABOUR				RATES		
	Del to Site £	Waste %	Material Cost £	Craft Optve Hrs	Lab Hrs	Labour Cost £	Sunds £	Nett Rate £	Unit	Gross Rate (+12.5%) £
K11: RIGID SHEET FLOORING/SHEATHING/LININGS/CASINGS Cont.										
ASBESTOS FREE BOARD - Supalux fire resisting boards, Sanded; butt joints; fixing to timber with screws, countersinking Cont.										
9mm thick linings to walls										
width exceeding 300mm	17.05	5.00	17.90	0.80	0.10	8.09	0.19	26.18	m2	29.45
width not exceeding 300mm	5.12	10.00	5.63	0.48	0.03	4.64	0.11	10.38	m	11.68
12mm thick linings to walls										
width exceeding 300mm	22.50	5.00	23.63	0.92	0.12	9.33	0.22	33.18	m2	37.33
width not exceeding 300mm	6.75	10.00	7.42	0.55	0.04	5.36	0.13	12.91	m	14.53
6mm thick linings to ceilings										
width exceeding 300mm	11.45	5.00	12.02	0.77	0.10	7.81	0.19	20.02	m2	22.52
width not exceeding 300mm	3.44	10.00	3.78	0.46	0.03	4.46	0.11	8.35	m	9.39
9mm thick linings to ceilings										
width exceeding 300mm	17.05	5.00	17.90	0.90	0.11	9.08	0.22	27.20	m2	30.60
width not exceeding 300mm	5.12	10.00	5.63	0.54	0.03	5.19	0.12	10.95	m	12.32
12mm thick linings to ceilings										
width exceeding 300mm	22.50	5.00	23.63	0.99	0.12	9.98	0.24	33.85	m2	38.08
width not exceeding 300mm	6.75	10.00	7.42	0.59	0.04	5.73	0.14	13.29	m	14.95
6mm thick casings to isolated beams or the like										
total girth not exceeding 600mm	11.45	10.00	12.60	1.44	0.18	14.55	0.34	27.49	m2	30.93
total girth 600 - 1200mm	11.45	5.00	12.02	0.86	0.11	8.71	0.21	20.94	m2	23.56
9mm thick casings to isolated beams or the like										
total girth not exceeding 600mm	17.05	10.00	18.75	1.66	0.21	16.80	0.40	35.95	m2	40.44
total girth 600 - 1200mm	17.05	5.00	17.90	1.00	0.13	10.14	0.24	28.29	m2	31.82
12mm thick casings to isolated beams or the like										
total girth not exceeding 600mm	22.50	10.00	24.75	1.84	0.23	18.60	0.44	43.79	m2	49.26
total girth 600 - 1200mm	22.50	5.00	23.63	1.09	0.14	11.04	0.26	34.93	m2	39.29
6mm thick casings to isolated columns or the like										
total girth not exceeding 600mm	11.45	10.00	12.60	1.44	0.18	14.55	0.34	27.49	m2	30.93
total girth 600 - 1200mm	11.45	5.00	12.02	0.86	0.11	8.71	0.21	20.94	m2	23.56
9mm thick casings to isolated columns or the like										
total girth not exceeding 600mm	17.05	10.00	18.75	1.66	0.21	16.80	0.40	35.95	m2	40.44
total girth 600 - 1200mm	17.05	5.00	17.90	1.00	0.13	10.14	0.24	28.29	m2	31.82
12mm thick casings to isolated columns or the like										
total girth not exceeding 600mm	22.50	10.00	24.75	1.84	0.23	18.60	0.44	43.79	m2	49.26
total girth 600 - 1200mm	22.50	5.00	23.63	1.09	0.14	11.04	0.26	34.93	m2	39.29
ASBESTOS FREE BOARD - Vermiculux fire resisting boards; butt joints; fixing to steel with self tapping screws, countersinking										
20mm thick casings to isolated beams or the like										
total girth not exceeding 600mm	18.25	10.00	20.07	1.88	0.23	18.97	0.45	39.49	m2	44.43
total girth 600 - 1200mm	18.25	5.00	19.16	1.13	0.14	11.41	0.27	30.85	m2	34.70
30mm thick casings to isolated beams or the like										
total girth not exceeding 600mm	29.90	10.00	32.89	2.06	0.26	20.84	0.49	54.22	m2	61.00
total girth 600 - 1200mm	29.90	5.00	31.40	1.25	0.16	12.66	0.30	44.36	m2	49.90
40mm thick casings to isolated beams or the like										
total girth not exceeding 600mm	45.35	10.00	49.88	2.24	0.28	22.64	0.54	73.07	m2	82.20
total girth 600 - 1200mm	45.35	5.00	47.62	1.37	0.17	13.84	0.33	61.79	m2	69.51
50mm thick casings to isolated beams or the like										
total girth not exceeding 600mm	60.35	10.00	66.39	2.42	0.30	24.44	0.58	91.41	m2	102.83
total girth 600 - 1200mm	60.35	5.00	63.37	1.49	0.19	15.09	0.36	78.81	m2	88.67
60mm thick casings to isolated beams or the like										
total girth not exceeding 600mm	71.95	10.00	79.14	2.60	0.33	26.31	0.62	106.08	m2	119.34
total girth 600 - 1200mm	71.95	5.00	75.55	1.61	0.20	16.26	0.39	92.20	m2	103.73
20mm thick casings to isolated columns or the like										
total girth 600 - 1200mm	18.25	10.00	20.07	1.88	0.23	18.97	0.45	39.49	m2	44.43
total girth 1200 - 1800mm	18.25	5.00	19.16	1.13	0.14	11.41	0.27	30.85	m2	34.70
30mm thick casings to isolated columns or the like										
total girth 600 - 1200mm	29.90	10.00	32.89	2.06	0.26	20.84	0.49	54.22	m2	61.00
total girth 1200 - 1800mm	29.90	5.00	31.40	1.25	0.16	12.66	0.30	44.36	m2	49.90
40mm thick casings to isolated columns or the like										
total girth 600 - 1200mm	45.35	10.00	49.88	2.24	0.28	22.64	0.54	73.07	m2	82.20
total girth 1200 - 1800mm	45.35	5.00	47.62	1.37	0.17	13.84	0.33	61.79	m2	69.51
50mm thick casings to isolated columns or the like										
total girth 600 - 1200mm	60.35	10.00	66.39	2.42	0.30	24.44	0.58	91.41	m2	102.83
total girth 1200 - 1800mm	60.35	5.00	63.37	1.49	0.19	15.09	Sunds	78.81	m2	88.67
60mm thick casings to isolated columns or the like										
total girth 600 - 1200mm	71.95	10.00	79.14	2.60	0.33	26.31	0.62	106.08	m2	119.34
total girth 1200 - 1800mm	71.95	5.00	75.55	1.61	0.20	16.26	0.39	92.20	m2	103.73

Labour hourly rates: (except Specialists) Craft Operatives £9.23 Labourer £7.02 Rates are national average prices. Refer to REGIONAL VARIATIONS for indicative levels of overall pricing in regions	MATERIALS			LABOUR				RATES		
	Del to Site £	Waste %	Material Cost £	Craft Optve Hrs	Lab Hrs	Labour Cost £	Sunds £	Nett Rate £	Unit	Gross Rate (+12.5%) £
K11: RIGID SHEET FLOORING/SHEATHING/LININGS/CASINGS Cont.										
ASBESTOS FREE BOARD - Vermiculux fire resisting boards; butt joints; fixing to steel with self tapping screws, countersinking Cont.										
50 x 25mm x 20g mild steel fixing angle										
fixing to masonry	1.29	5.00	1.35	0.48	0.08	4.99	0.12	6.47	m	7.27
fixing to steel	1.29	5.00	1.35	0.48	0.08	4.99	0.12	6.47	m	7.27
ASBESTOS FREE BOARD - Tacboard Class 0 fire resisting boards; butt joints										
6mm thick linings to walls										
width exceeding 300mm	7.44	5.00	7.81	0.57	0.07	5.75	0.14	13.70	m2	15.42
width not exceeding 300mm	2.23	10.00	2.45	0.34	0.02	3.28	0.08	5.81	m	6.54
9mm thick linings to walls										
width exceeding 300mm	13.71	5.00	14.40	0.67	0.08	6.75	0.16	21.30	m2	23.96
width not exceeding 300mm	4.11	10.00	4.52	0.40	0.02	3.83	0.09	8.44	m	9.50
12mm thick linings to walls										
width exceeding 300mm	17.85	5.00	18.74	0.73	0.09	7.37	0.17	26.28	m2	29.57
width not exceeding 300mm	5.36	10.00	5.90	0.44	0.03	4.27	0.10	10.27	m	11.55
6mm thick linings to ceilings										
width exceeding 300mm	7.44	5.00	7.81	0.72	0.09	7.28	0.17	15.26	m2	17.17
width not exceeding 300mm	2.23	10.00	2.45	0.43	0.03	4.18	0.10	6.73	m	7.57
ASBESTOS FREE BOARD - New Tacfire Class 1 fire resisting boards; butt joints										
6mm thick linings to walls										
width exceeding 300mm	9.95	5.00	10.45	0.57	0.07	5.75	0.14	16.34	m2	18.38
width not exceeding 300mm	2.99	10.00	3.29	0.34	0.02	3.28	0.08	6.65	m	7.48
9mm thick linings to walls										
width exceeding 300mm	15.22	5.00	15.98	0.67	0.08	6.75	0.16	22.89	m2	25.75
width not exceeding 300mm	4.57	10.00	5.03	0.40	0.02	3.83	0.09	8.95	m	10.07
12mm thick linings to walls										
width exceeding 300mm	20.09	5.00	21.09	0.73	0.09	7.37	0.17	28.63	m2	32.21
width not exceeding 300mm	6.03	10.00	6.63	0.44	0.03	4.27	0.10	11.00	m	12.38
6mm thick linings to ceilings; 6mm cover fillets										
width exceeding 300mm	9.95	5.00	10.45	0.72	0.09	7.28	1.27	18.99	m2	21.37
width not exceeding 300mm	2.99	10.00	3.29	0.43	0.03	4.18	0.43	7.90	m	8.89
9mm thick linings to ceilings; 9mm cover fillets										
width exceeding 300mm	15.22	5.00	15.98	0.80	0.10	8.09	2.30	26.37	m2	29.66
width not exceeding 300mm	4.57	10.00	5.03	0.48	0.03	4.64	0.74	10.41	m	11.71
12mm thick linings to ceilings; 9mm cover fillets										
width exceeding 300mm	20.09	5.00	21.09	0.92	0.11	9.26	2.33	32.69	m2	36.77
width not exceeding 300mm	6.03	10.00	6.63	0.55	0.03	5.29	0.76	12.68	m	14.27
ASBESTOS FREE BOARD - Vicuclad 900 fire resisting boards; butt joints; joints filled with c/v cement; fixing with screws; countersinking										
18mm thick casings to isolated beams or the like; including noggins										
total girth not exceeding 600mm	14.87	10.00	16.36	1.78	0.22	17.97	2.72	37.05	m2	41.68
total girth 600 - 1200mm	14.87	5.00	15.61	1.07	0.07	10.37	3.60	29.58	m2	33.28
20mm thick casings to isolated beams or the like; including noggins										
total girth not exceeding 600mm	15.48	10.00	17.03	1.88	0.23	18.97	2.87	38.87	m2	43.72
total girth 600 - 1200mm	15.48	5.00	16.25	1.13	0.14	11.41	3.76	31.43	m2	35.36
25mm thick casings to isolated beams or the like; including noggins										
total girth not exceeding 600mm	16.85	10.00	18.54	1.98	0.25	20.03	3.25	41.82	m2	47.04
total girth 600 - 1200mm	16.85	5.00	17.69	1.19	0.15	12.04	4.12	33.85	m2	38.08
30mm thick casings to isolated beams or the like; including noggins										
total girth not exceeding 600mm	20.45	10.00	22.50	2.06	0.26	20.84	3.83	47.16	m2	53.06
total girth 600 - 1200mm	20.45	5.00	21.47	1.25	0.16	12.66	4.93	39.06	m2	43.95
18mm thick casings to isolated columns or the like										
total girth not exceeding 600mm	14.87	10.00	16.36	1.78	0.22	17.97	1.65	35.98	m2	40.48
total girth 600 - 1200mm	14.87	5.00	15.61	1.07	0.07	10.37	1.47	27.45	m2	30.88
20mm thick casings to isolated columns or the like										
total girth not exceeding 600mm	15.40	10.00	17.03	1.88	0.23	18.97	1.81	37.81	m2	42.63
total girth 600 - 1200mm	15.48	5.00	16.25	1.13	0.14	11.41	1.63	29.30	m2	32.96
25mm thick casings to isolated columns or the like										
total girth not exceeding 600mm	16.85	10.00	18.54	1.98	0.25	20.03	2.19	40.76	m2	45.85
total girth 600 - 1200mm	16.85	5.00	17.69	1.19	0.15	12.04	2.00	31.73	m2	35.70
30mm thick casings to isolated columns or the like										
total girth not exceeding 600mm	20.45	10.00	22.50	2.06	0.26	20.84	2.54	45.87	m2	51.61
total girth 600 - 1200mm	20.45	5.00	21.47	1.25	0.16	12.66	2.35	36.48	m2	41.04

Labour hourly rates: (except Specialists) Craft Operatives £9.23 Labourer £7.02 Rates are national average prices. Refer to REGIONAL VARIATIONS for indicative levels of overall pricing in regions	MATERIALS			LABOUR				RATES		
	Del to Site £	Waste %	Material Cost £	Craft Optve Hrs	Lab Hrs	Labour Cost £	Sunds £	Nett Rate £	Unit	Gross Rate (+12.5%) £
K11: RIGID SHEET FLOORING/SHEATHING/LININGS/CASINGS Cont.										
DUCT FRONTS - Plywood B.S.6566, II/III grade, MR bonded; butt joints; fixing to timber with screws, countersinking - duct fronts										
12mm thick linings to walls										
width exceeding 300mm	7.39	20.00	8.87	1.15	0.14	11.60	0.19	20.66	m2	23.24
width exceeding 300mm; fixing to timber with brass screws and cups	7.39	20.00	8.87	1.70	0.21	17.17	0.65	26.68	m2	30.02
DUCT FRONTS - Blockboard, B.S.3444, 2/2 grade, MR bonded; butt joints - duct fronts										
18mm thick linings to walls										
width exceeding 300mm	10.58	20.00	12.70	1.40	0.17	14.12	0.23	27.04	m2	30.42
width exceeding 300mm; fixing to timber with brass screws and cups	10.58	20.00	12.70	2.00	0.25	20.22	0.65	33.56	m2	37.76
DUCT FRONTS - Extra over plywood B.S.6566, II/III grade, MR bonded; butt joints; 12mm thick linings to walls; width exceeding 300mm										
Access panels										
300 x 600mm	-	-	-	0.60	0.08	6.10	0.10	6.20	nr	6.97
600 x 600mm	-	-	-	0.75	0.09	7.55	0.12	7.67	nr	8.63
600 x 900mm	-	-	-	0.90	0.11	9.08	0.15	9.23	nr	10.38
DUCT FRONTS - Extra over blockboard, B.S.3444, 2/2 grade, MR bonded, butt joints; 18mm thick linings to walls; width exceeding 300mm										
Access panels										
300 x 600mm	-	-	-	0.70	0.09	7.09	0.11	7.20	nr	8.10
600 x 600mm	-	-	-	0.90	0.11	9.08	0.15	9.23	nr	10.38
600 x 900mm	-	-	-	1.25	0.16	12.66	0.21	12.87	nr	14.48
PIPE CASINGS - Standard hardboard, B.S.1142, type SHA; butt joints - pipe casings										
3mm thick linings to pipe ducts and casings										
width exceeding 300mm	1.16	20.00	1.39	0.90	0.11	9.08	0.15	10.62	m2	11.95
width not exceeding 300mm	0.34	20.00	0.41	0.30	0.04	3.05	0.05	3.51	m	3.95
PIPE CASINGS - Plywood B.S.6566, II/III grade, MR bonded; butt joints; fixing to timber with screws, countersinking - pipe casings										
12mm thick linings to pipe ducts and casings										
width exceeding 300mm	7.39	20.00	8.87	1.70	0.21	17.17	0.28	26.31	m2	29.60
width not exceeding 300mm	2.22	20.00	2.66	0.60	0.08	6.10	0.10	8.86	m	9.97
PIPE CASINGS - Supalux Class 0 fire resisting boards, sanded; butt joints; fixing to timber with screws, countersinking - pipe casings										
9mm thick linings to pipe ducts and casings										
width exceeding 300mm	17.99	20.00	21.59	1.70	0.21	17.17	0.28	39.03	m2	43.91
width not exceeding 300mm	5.40	20.00	6.48	0.60	0.08	6.10	0.10	12.68	m	14.26
PIPE CASINGS - Pendock Profiles Ltd preformed plywood casings; white melamine finish; butt joints; fixing to timber with polytop white screws, countersinking										
5mm thick casings; horizontal										
reference TK110, 45 x 110mm	6.19	20.00	7.43	0.46	0.06	4.67	0.23	12.32	m	13.87
extra; internal corner	5.96	2.50	6.11	0.30	0.04	3.05	0.23	9.39	nr	10.56
extra; external corner	7.76	2.50	7.95	0.30	0.04	3.05	0.23	11.23	nr	12.64
extra; stop end	5.38	2.50	5.51	0.20	0.02	1.99	0.12	7.62	nr	8.57
reference TK150, 45 x 150mm	7.82	20.00	9.38	0.48	0.06	4.85	0.23	14.47	m	16.27
extra; internal corner	7.04	2.50	7.22	0.30	0.04	3.05	0.23	10.50	nr	11.81
extra; external corner	8.92	2.50	9.14	0.30	0.04	3.05	0.23	12.42	nr	13.98
extra; stop end	7.53	2.50	7.72	0.20	0.02	1.99	0.12	9.82	nr	11.05
reference TK190, 45 x 190mm	10.27	20.00	12.32	0.50	0.06	5.04	0.23	17.59	m	19.79
extra; internal corner	7.08	2.50	7.26	0.30	0.04	3.05	0.23	10.54	nr	11.85
extra; external corner	8.97	2.50	9.19	0.30	0.04	3.05	0.23	12.47	nr	14.03
extra; stop end	8.97	2.50	9.19	0.20	0.02	1.99	0.12	11.30	nr	12.71
reference TK125, 25 x 125mm	6.18	20.00	7.42	0.46	0.06	4.67	0.23	12.31	m	13.85
extra; internal corner	5.97	2.50	6.12	0.30	0.04	3.05	0.23	9.40	nr	10.57
extra; external corner	7.76	2.50	7.95	0.30	0.04	3.05	0.23	11.23	nr	12.64
extra; stop end	5.38	2.50	5.51	0.20	0.02	1.99	0.12	7.62	nr	8.57
5mm thick casings; vertical										
reference TK110, 45 x 110mm	5.64	20.00	6.77	0.46	0.06	4.67	0.23	11.66	m	13.12
reference TK135, 45 x 135mm	8.22	20.00	9.86	0.47	0.06	4.76	0.23	14.85	m	16.71
reference TK150, 45 x 150mm	7.21	20.00	8.65	0.48	0.06	4.85	0.23	13.73	m	15.45
reference TK190, 45 x 190mm	9.55	20.00	11.46	0.50	0.06	5.04	0.23	16.73	m	18.82
5mm thick casings; horizontal										
reference MX 100/100, 100 x 100mm	10.32	20.00	12.38	0.44	0.06	4.48	0.23	17.10	m	19.23
extra; stop end	8.05	2.50	8.25	0.20	0.02	1.99	0.12	10.36	nr	11.65
8mm thick casings; horizontal										
reference MX 150/150, 150 x 150mm	15.62	20.00	18.74	0.52	0.06	5.22	0.23	24.19	m	27.22
extra; internal corner	11.05	2.50	11.33	0.30	0.04	3.05	0.23	14.61	nr	16.43

Labour hourly rates: (except Specialists) Craft Operatives £9.23 Labourer £7.02 Rates are national average prices. Refer to REGIONAL VARIATIONS for indicative levels of overall pricing in regions	MATERIALS			LABOUR				RATES		
	Del to Site £	Waste %	Material Cost £	Craft Optve Hrs	Lab Hrs	Labour Cost £	Sunds £	Nett Rate £	Unit	Gross Rate (+12.5%) £
K11: RIGID SHEET FLOORING/SHEATHING/LININGS/CASINGS Cont.										
PIPE CASINGS - Pendock Profiles Ltd preformed plywood casings; white melamine finish; butt joints; fixing to timber with polytop white screws, countersinking Cont.										
8mm thick casings; horizontal Cont.	16.77	2.50	17.19	0.30	0.04	3.05	0.23	20.47	nr	23.03
extra; external corner	8.95	2.50	9.17	0.20	0.02	1.99	0.12	11.28	nr	12.69
reference MX 200/150, 200 x 150mm	17.30	20.00	20.76	0.53	0.07	5.38	0.23	26.37	m	29.67
extra; stop end	8.95	2.50	9.17	0.20	0.02	1.99	0.12	11.28	nr	12.69
reference MX 200/200, 200 x 200mm	18.87	20.00	22.64	0.54	0.07	5.48	0.23	28.35	m	31.89
extra; internal corner	11.21	2.50	11.49	0.30	0.04	3.05	0.23	14.77	nr	16.62
extra; external corner	16.95	2.50	17.37	0.30	0.04	3.05	0.23	20.65	nr	23.24
extra; stop end	8.95	2.50	9.17	0.20	0.02	1.99	0.12	11.28	nr	12.69
reference MX 200/300, 200 x 300mm	22.15	20.00	26.58	0.55	0.07	5.57	0.23	32.38	m	36.43
extra; stop end	9.82	2.50	10.07	0.20	0.02	1.99	0.12	12.17	nr	13.69
reference MX 300/150, 300 x 150mm	20.94	20.00	25.13	0.55	0.07	5.57	0.23	30.93	m	34.79
extra; internal corner	11.90	2.50	12.20	0.30	0.04	3.05	0.23	15.48	nr	17.41
extra; external corner	17.69	2.50	18.13	0.30	0.04	3.05	0.23	21.41	nr	24.09
extra; stop end	9.59	2.50	9.83	0.20	0.02	1.99	0.12	11.94	nr	13.43
reference MX 300/300, 300 x 300mm	25.25	20.00	30.30	0.57	0.07	5.75	0.23	36.28	m	40.82
extra; internal corner	12.70	2.50	13.02	0.30	0.04	3.05	0.23	16.30	nr	18.33
extra; external corner	18.49	2.50	18.95	0.30	0.04	3.05	0.23	22.23	nr	25.01
extra; stop end	12.20	2.50	12.51	0.20	0.02	1.99	0.12	14.61	nr	16.44
5mm thick casings; vertical										
reference MX 100/100, 100 x 100mm	8.63	20.00	10.36	0.48	0.06	4.85	0.23	15.44	m	17.37
8mm thick casings; vertical										
reference MX 150/150, 150 x 150mm	13.92	20.00	16.70	0.52	0.06	5.22	0.23	22.15	m	24.92
reference MX 200/150, 200 x 150mm	15.54	20.00	18.65	0.53	0.07	5.38	0.23	24.26	m	27.29
reference MX 200/200, 200 x 200mm	17.20	20.00	20.64	0.54	0.07	5.48	0.23	26.35	m	29.64
reference MX 200/300, 200 x 300mm	20.47	20.00	24.56	0.55	0.07	5.57	0.23	30.36	m	34.16
reference MX 300/150, 300 x 150mm	19.26	20.00	23.11	0.56	0.07	5.66	0.23	29.00	m	32.63
reference MX 300/300, 300 x 300mm	23.56	20.00	28.27	0.57	0.07	5.75	0.23	34.25	m	38.54
ROOF BOARDING - Plywood B.S.6566, II/III grade, WBP bonded; butt joints										
18mm thick sheeting to roofs; external										
width exceeding 300mm	11.61	10.00	12.77	0.40	0.05	4.04	0.06	16.87	m2	18.98
extra; cesspool 225 x 225 x 150mm	2.32	10.00	2.55	1.70	0.21	17.17	0.28	20.00	nr	22.50
extra; cesspool 300 x 300 x 150mm	3.19	10.00	3.51	1.80	0.22	18.16	0.29	21.96	nr	24.70
width exceeding 300mm; sloping	11.61	10.00	12.77	0.50	0.06	5.04	0.08	17.89	m2	20.12
24mm thick sheeting to roofs; external										
width exceeding 300mm	16.13	10.00	17.74	0.45	0.06	4.57	0.07	22.39	m2	25.19
extra; cesspool 225 x 225 x 150mm	3.22	10.00	3.54	1.80	0.22	18.16	0.29	21.99	nr	24.74
extra; cesspool 300 x 300 x 150mm	4.43	10.00	4.87	1.90	0.24	19.22	0.31	24.40	nr	27.46
width exceeding 300mm; sloping	16.13	10.00	17.74	0.55	0.07	5.57	0.09	23.40	m2	26.33
ROOF BOARDING - Wood chipboard, B.S.5669, type 1 standard; butt joints										
12mm thick sheeting to roofs; external										
width exceeding 300mm	2.37	10.00	2.61	0.30	0.04	3.05	0.05	5.71	m2	6.42
extra; cesspool 225 x 225 x 150mm	0.47	10.00	0.52	1.55	0.19	15.64	0.25	16.41	nr	18.46
extra; cesspool 300 x 300 x 150mm	0.65	10.00	0.71	1.65	0.21	16.70	0.27	17.69	nr	19.90
width exceeding 300mm; sloping	2.37	10.00	2.61	0.40	0.05	4.04	0.06	6.71	m2	7.55
18mm thick sheeting to roofs; external										
width exceeding 300mm	3.17	10.00	3.49	0.35	0.04	3.51	0.06	7.06	m2	7.94
extra; cesspool 225 x 225 x 150mm	0.64	10.00	0.70	1.65	0.21	16.70	0.27	17.68	nr	19.89
extra; cesspool 300 x 300 x 150mm	0.87	10.00	0.96	1.75	0.22	17.70	0.28	18.93	nr	21.30
width exceeding 300mm; sloping	3.17	10.00	3.49	0.45	0.06	4.57	0.07	8.13	m2	9.15
ROOF BOARDING - Abutments										
18mm thick plywood linings										
raking cutting	1.17	-	1.17	0.35	-	3.23	-	4.40	m	4.95
curved cutting	1.74	-	1.74	0.55	-	5.08	-	6.82	m	7.67
rebates	-	-	-	0.22	-	2.03	-	2.03	m	2.28
grooves	-	-	-	0.22	-	2.03	-	2.03	m	2.28
chamfers	-	-	-	0.16	-	1.48	-	1.48	m	1.66
24mm thick plywood linings										
raking cutting	1.62	-	1.62	0.50	-	4.62	-	6.24	m	7.01
curved cutting	2.42	-	2.42	0.70	-	6.46	-	8.88	m	9.99
rebates	-	-	-	0.22	-	2.03	-	2.03	m	2.28
grooves	-	-	-	0.22	-	2.03	-	2.03	m	2.28
chamfers	-	-	-	0.16	-	1.48	-	1.48	m	1.66
12mm thick wood chipboard linings										
raking cutting	0.23	-	0.23	0.30	-	2.77	-	3.00	m	3.37
curved cutting	0.35	-	0.35	0.40	-	3.69	-	4.04	m	4.55
rebates	-	-	-	0.22	-	2.03	-	2.03	m	2.28
grooves	-	-	-	0.22	-	2.03	-	2.03	m	2.28
chamfers	-	-	-	0.16	-	1.48	-	1.48	m	1.66
18mm thick wood chipboard linings										
raking cutting	0.32	-	0.32	0.35	-	3.23	-	3.55	m	3.99
curved cutting	0.47	-	0.47	0.55	-	5.08	-	5.55	m	6.24
rebates	-	-	-	0.22	-	2.03	-	2.03	m	2.28

LININGS/SHEATHING/DRY PARTITIONING

Labour hourly rates: (except Specialists) Craft Operatives £9.23 Labourer £7.02 Rates are national average prices. Refer to REGIONAL VARIATIONS for indicative levels of overall pricing in regions	MATERIALS			LABOUR				RATES		
	Del to Site £	Waste %	Material Cost £	Craft Optve Hrs	Lab Hrs	Labour Cost £	Sunds £	Nett Rate £	Unit	Gross Rate (+12.5%) £
K11: RIGID SHEET FLOORING/SHEATHING/LININGS/CASINGS Cont.										
ROOF BOARDING - Abutments Cont.										
18mm thick wood chipboard linings Cont.										
grooves	-	-	-	0.22	-	2.03	-	2.03	m	2.28
chamfers	-	-	-	0.16	-	1.48	-	1.48	m	1.66
WOOD WOOL BUILDING SLABS - Wood wool slabs, B.S.1105 Type A; natural both sides; butt joints										
25mm thick linings to walls										
width exceeding 300mm	5.27	5.00	5.53	0.36	0.04	3.60	0.06	9.20	m2	10.35
38mm thick linings to walls										
width exceeding 300mm	5.87	5.00	6.16	0.42	0.05	4.23	0.07	10.46	m2	11.77
50mm thick linings to walls										
width exceeding 300mm	5.92	5.00	6.22	0.50	0.06	5.04	0.08	11.33	m2	12.75
25mm thick linings to ceilings										
width exceeding 300mm	5.27	5.00	5.53	0.40	0.05	4.04	0.06	9.64	m2	10.84
38mm thick linings to ceilings										
width exceeding 300mm	5.87	5.00	6.16	0.46	0.06	4.67	0.07	10.90	m2	12.26
50mm thick linings to ceilings										
width exceeding 300mm	5.92	5.00	6.22	0.55	0.07	5.57	0.08	11.86	m2	13.35
WOOD WOOL BUILDING SLABS - Wood wool slabs, B.S.1105 Type B; natural both sides; butt joints										
50mm thick sheeting to roofs; external										
width exceeding 300mm; fixing to timber with galvanised nails and gripper plates	6.05	5.00	6.35	0.50	0.06	5.04	0.14	11.53	m2	12.97
width exceeding 300mm; fixing to steel rails with sherardised clips	6.05	5.00	6.35	0.38	0.05	3.86	0.28	10.49	m2	11.80
75mm thick sheeting to roofs; external										
width exceeding 300mm; fixing to timber with galvanised nails and gripper plates	9.72	5.00	10.21	0.62	0.08	6.28	0.17	16.66	m2	18.74
width exceeding 300mm; fixing to steel rails with sherardised clips	9.72	5.00	10.21	0.46	0.06	4.67	0.35	15.22	m2	17.13
100mm thick sheeting to roofs; external										
width exceeding 300mm; fixing to timber with galvanised nails and gripper plates	13.66	5.00	14.34	0.80	0.10	8.09	0.22	22.65	m2	25.48
width exceeding 300mm; fixing to steel rails with sherardised clips	13.66	5.00	14.34	0.60	0.08	6.10	0.44	20.88	m2	23.49
WOOD WOOL BUILDING SLABS - Wood wool slabs, B.S.1105 Type B; natural soffit, pre-felted top surface; butt joints										
50mm thick sheeting to roofs; external										
width exceeding 300mm; fixing to timber with galvanised nails and gripper plates	10.89	5.00	11.43	0.50	0.06	5.04	0.14	16.61	m2	18.69
width exceeding 300mm; fixing to steel rails with sherardised clips	10.89	5.00	11.43	0.38	0.05	3.86	0.28	15.57	m2	17.52
75mm thick sheeting to roofs; external										
width exceeding 300mm; fixing to timber with galvanised nails and gripper plates	14.55	5.00	15.28	0.62	0.08	6.28	0.17	21.73	m2	24.45
width exceeding 300mm; fixing to steel rails with sherardised clips	14.55	5.00	15.28	0.46	0.06	4.67	0.35	20.29	m2	22.83
100mm thick sheeting to roofs; external										
width exceeding 300mm; fixing to timber with galvanised nails and gripper plates	18.48	5.00	19.40	0.80	0.10	8.09	0.22	27.71	m2	31.17
width exceeding 300mm; fixing to steel rails with sherardised clips	18.48	5.00	19.40	0.60	0.08	6.10	0.44	25.94	m2	29.19
WOOD WOOL BUILDING SLABS - Wood wool slabs, B.S.1105 Type B; natural soffit, pre-screeded top surface; butt joints										
50mm thick sheeting to roofs; external										
width exceeding 300mm; fixing to timber with galvanised nails and gripper plates	9.21	5.00	9.67	0.50	0.06	5.04	0.14	14.85	m2	16.70
width exceeding 300mm; fixing to steel rails with sherardised clips	9.21	5.00	9.67	0.38	0.05	3.86	0.28	13.81	m2	15.54
75mm thick sheeting to roofs; external										
width exceeding 300mm; fixing to timber with galvanised nails and gripper plates	12.88	5.00	13.52	0.62	0.08	6.28	0.17	19.98	m2	22.48
width exceeding 300mm; fixing to steel rails with sherardised clips	12.88	5.00	13.52	0.46	0.06	4.67	0.35	18.54	m2	20.86
100mm thick sheeting to roofs; external										
width exceeding 300mm; fixing to timber with galvanised nails and gripper plates	16.83	5.00	17.67	0.80	0.10	8.09	0.22	25.98	m2	29.22
width exceeding 300mm; fixing to steel rails with sherardised clips	16.83	5.00	17.67	0.60	0.08	6.10	0.44	24.21	m2	27.24

Labour hourly rates: (except Specialists) Craft Operatives £9.23 Labourer £7.02 Rates are national average prices. Refer to REGIONAL VARIATIONS for indicative levels of overall pricing in regions	MATERIALS			LABOUR				RATES		
	Del to Site	Waste	Material Cost	Craft Optve	Lab	Labour Cost	Sunds	Nett Rate	Unit	Gross Rate (+12.5%)
	£	%	£	Hrs	Hrs	£	£	£		£

K11: RIGID SHEET FLOORING/SHEATHING/ LININGS/CASINGS; STAIRCASE AREAS

ASBESTOS FREE BOARD - Masterboard Class 0 fire resisting boards; butt joints

	Del to Site	Waste	Material Cost	Craft Optve	Lab	Labour Cost	Sunds	Nett Rate	Unit	Gross Rate
6mm thick linings to walls										
width exceeding 300mm	7.70	5.00	8.09	0.72	0.09	7.28	0.17	15.53	m2	17.47
width not exceeding 300mm	2.31	10.00	2.54	0.43	0.03	4.18	0.10	6.82	m	7.67
9mm thick linings to walls										
width exceeding 300mm	15.75	5.00	16.54	0.80	0.10	8.09	0.19	24.81	m2	27.92
width not exceeding 300mm	4.73	10.00	5.20	0.48	0.03	4.64	0.11	9.95	m	11.20
12mm thick linings to walls										
width exceeding 300mm	20.75	5.00	21.79	0.92	0.12	9.33	0.22	31.34	m2	35.26
width not exceeding 300mm	6.23	10.00	6.85	0.55	0.04	5.36	0.13	12.34	m	13.88
6mm thick linings to ceilings										
width exceeding 300mm	7.70	5.00	8.09	0.77	0.10	7.81	0.19	16.08	m2	18.09
width not exceeding 300mm	2.31	10.00	2.54	0.46	0.03	4.46	0.11	7.11	m	8.00
9mm thick linings to ceilings										
width exceeding 300mm	15.75	5.00	16.54	0.90	0.11	9.08	0.22	25.84	m2	29.07
width not exceeding 300mm	4.73	10.00	5.20	0.54	0.03	5.19	0.12	10.52	m	11.83
12mm thick linings to ceilings										
width exceeding 300mm	20.75	5.00	21.79	0.99	0.12	9.98	0.24	32.01	m2	36.01
width not exceeding 300mm	6.23	10.00	6.85	0.59	0.04	5.73	0.14	12.72	m	14.31

ASBESTOS FREE BOARD - Supalux fire resisting boards, sanded; butt joints; fixing to timber with screws, countersinking

	Del to Site	Waste	Material Cost	Craft Optve	Lab	Labour Cost	Sunds	Nett Rate	Unit	Gross Rate
6mm thick linings to walls										
width exceeding 300mm	11.45	5.00	12.02	0.72	0.09	7.28	0.17	19.47	m2	21.90
width not exceeding 300mm	3.44	10.00	3.78	0.43	0.03	4.18	0.10	8.06	m	9.07
9mm thick linings to walls										
width exceeding 300mm	17.05	5.00	17.90	0.80	0.10	8.09	0.19	26.18	m2	29.45
width not exceeding 300mm	5.12	10.00	5.63	0.48	0.03	4.64	0.11	10.38	m	11.68
12mm thick linings to walls										
width exceeding 300mm	22.50	5.00	23.63	0.92	0.12	9.33	0.22	33.18	m2	37.33
width not exceeding 300mm	6.75	10.00	7.42	0.55	0.04	5.36	0.13	12.91	m	14.53
6mm thick linings to ceilings										
width exceeding 300mm	11.45	5.00	12.02	0.77	0.10	7.81	0.19	20.02	m2	22.52
width not exceeding 300mm	3.44	10.00	3.78	0.46	0.03	4.46	0.11	8.35	m	9.39
9mm thick linings to ceilings										
width exceeding 300mm	17.05	5.00	17.90	0.90	0.11	9.08	0.22	27.20	m2	30.60
width not exceeding 300mm	5.12	10.00	5.63	0.54	0.03	5.19	0.12	10.95	m	12.32
12mm thick linings to ceilings										
width exceeding 300mm	22.50	5.00	23.63	0.99	0.12	9.98	0.24	33.85	m2	38.08
width not exceeding 300mm	6.75	10.00	7.42	0.59	0.04	5.73	0.14	13.29	m	14.95

K12: UNDER PURLIN/INSIDE RAIL PANEL LININGS

PLASTIC FACED PLASTERBOARD - Paraclip metal grid fixing system; square edge sheeting of one layer 9.5mm thick Gyproc industrial grade board, B.S.1230, butt joints; laying in position in zinc coated mild steel grid; fixing to metal with self-tapping screws

	Del to Site	Waste	Material Cost	Craft Optve	Lab	Labour Cost	Sunds	Nett Rate	Unit	Gross Rate
9.5mm thick linings to walls										
width exceeding 300mm	4.98	5.00	5.23	1.00	0.50	12.74	2.63	20.60	m2	23.17
width not exceeding 300mm	1.49	10.00	1.64	0.60	0.15	6.59	1.58	9.81	m	11.04
9.5mm thick linings to ceilings										
width exceeding 300mm	4.98	5.00	5.23	1.25	0.63	15.96	2.63	23.82	m2	26.80
width not exceeding 300mm	1.49	10.00	1.64	0.75	0.19	8.26	1.58	11.48	m	12.91

K13: RIGID SHEET FINE LININGS/ PANELLING

Softwood, wrot - wall panelling

	Del to Site	Waste	Material Cost	Craft Optve	Lab	Labour Cost	Sunds	Nett Rate	Unit	Gross Rate
19mm thick panelled linings to walls; square framed; including grounds										
width exceeding 300mm	79.95	2.50	81.95	1.90	0.24	19.22	0.32	101.49	m2	114.18
width exceeding 300mm; fixing to masonry with screws	79.95	2.50	81.95	2.90	0.36	29.29	0.53	111.77	m2	125.74
25mm thick panelled linings to walls; square framed; including grounds										
width exceeding 300mm	81.00	2.50	83.03	2.05	0.26	20.75	0.35	104.12	m2	117.14
width exceeding 300mm; fixing to masonry with screws	81.00	2.50	83.03	3.05	0.38	30.82	0.56	114.40	m2	128.70
19mm thick panelled linings to walls; square framed; obstructed by integral services; including grounds										
width exceeding 300mm	79.95	2.50	81.95	2.40	0.30	24.26	0.43	106.64	m2	119.97
width exceeding 300mm; fixing to masonry with screws	79.95	2.50	81.95	3.40	0.42	34.33	0.64	116.92	m2	131.53

LININGS/SHEATHING/DRY PARTITIONING

Labour hourly rates: (except Specialists) Craft Operatives £9.23 Labourer £7.02. Rates are national average prices. Refer to REGIONAL VARIATIONS for indicative levels of overall pricing in regions	MATERIALS			LABOUR				RATES		
	Del to Site £	Waste %	Material Cost £	Craft Optve Hrs	Lab Hrs	Labour Cost £	Sunds £	Nett Rate £	Unit	Gross Rate (+12.5%) £
K13: RIGID SHEET FINE LININGS/ PANELLING Cont.										
Softwood, wrot - wall panelling Cont.										
25mm thick panelled linings to walls; square framed; obstructed by integral services; including grounds										
width exceeding 300mm	81.00	2.50	83.03	2.55	0.32	25.78	0.46	109.27	m2	122.93
width exceeding 300mm; fixing to masonry with screws	81.00	2.50	83.03	3.55	0.44	35.86	0.66	119.54	m2	134.48
19mm thick panelled linings to walls; moulded; including grounds										
width exceeding 300mm	87.50	2.50	89.69	2.05	0.26	20.75	0.35	110.78	m2	124.63
width exceeding 300mm; fixing to masonry with screws	87.50	2.50	89.69	3.05	0.38	30.82	0.56	121.07	m2	136.20
25mm thick panelled linings to walls; moulded; including grounds										
width exceeding 300mm	90.75	2.50	93.02	2.15	0.27	21.74	0.46	115.22	m2	129.62
width exceeding 300mm; fixing to masonry with screws	90.75	2.50	93.02	3.15	0.39	31.81	0.58	125.41	m2	141.09
19mm thick panelled linings to walls; moulded; obstructed by integral services; including grounds										
width exceeding 300mm	87.50	2.50	89.69	2.55	0.32	25.78	0.46	115.93	m2	130.42
width exceeding 300mm; fixing to masonry with screws	87.50	2.50	89.69	3.55	0.44	35.86	0.66	126.20	m2	141.98
25mm thick panelled linings to walls; moulded; obstructed by integral services; including grounds										
width exceeding 300mm	90.75	2.50	93.02	2.65	0.33	26.78	0.47	120.26	m2	135.30
width exceeding 300mm; fixing to masonry with screws	90.75	2.50	93.02	3.65	0.46	36.92	0.68	130.62	m2	146.94
Afromosia, wrot, selected for transparent finish - wall panelling										
19mm thick panelled linings to walls; square framed; including grounds										
width exceeding 300mm	223.50	2.50	229.09	2.70	0.34	27.31	0.47	256.87	m2	288.97
width exceeding 300mm; fixing to masonry with screws	223.50	2.50	229.09	3.70	0.46	37.38	0.67	267.14	m2	300.53
25mm thick panelled linings to walls; square framed; including grounds										
width exceeding 300mm	229.75	2.50	235.49	2.80	0.35	28.30	0.48	264.27	m2	297.31
width exceeding 300mm; fixing to masonry with screws	229.75	2.50	235.49	3.80	0.47	38.37	0.70	274.57	m2	308.89
19mm thick panelled linings to walls; square framed; obstructed by integral services; including grounds										
width exceeding 300mm	223.50	2.50	229.09	3.20	0.40	32.34	0.56	261.99	m2	294.74
width exceeding 300mm; fixing to masonry with screws	223.50	2.50	229.09	4.20	0.52	42.42	0.77	272.27	m2	306.31
25mm thick panelled linings to walls; square framed; obstructed by integral services; including grounds										
width exceeding 300mm	229.75	2.50	235.49	3.30	0.41	33.34	0.55	269.38	m2	303.05
width exceeding 300mm; fixing to masonry with screws	229.75	2.50	235.49	4.30	0.54	43.48	0.76	279.73	m2	314.70
19mm thick panelled linings to walls; moulded; including grounds										
width exceeding 300mm	240.10	2.50	246.10	2.90	0.36	29.29	0.48	275.88	m2	310.36
width exceeding 300mm; fixing to masonry with screws	240.10	2.50	246.10	3.90	0.49	39.44	0.67	286.21	m2	321.99
25mm thick panelled linings to walls; moulded; including grounds										
width exceeding 300mm	255.65	2.50	262.04	3.00	0.37	30.29	0.50	292.83	m2	329.43
width exceeding 300mm; fixing to masonry with screws	255.65	2.50	262.04	4.00	0.50	40.43	0.69	303.16	m2	341.06
19mm thick panelled linings to walls; moulded; obstructed by integral services; including grounds										
width exceeding 300mm	240.10	2.50	246.10	3.40	0.42	34.33	0.56	280.99	m2	316.12
width exceeding 300mm; fixing to masonry with screws	240.10	2.50	246.10	4.40	0.55	44.47	0.76	291.34	m2	327.75
25mm thick panelled linings to walls; moulded; obstructed by integral services; including grounds										
width exceeding 300mm	255.65	2.50	262.04	3.50	0.44	35.39	0.59	298.03	m2	335.28
width exceeding 300mm; fixing to masonry with screws	255.65	2.50	262.04	4.50	0.56	45.47	0.78	308.29	m2	346.82
Sapele, wrot, selected for transparent finish - wall panelling										
19mm thick panelled linings to walls; square framed; including grounds										
width exceeding 300mm	150.00	2.50	153.75	2.45	0.31	24.79	0.41	178.95	m2	201.32
width exceeding 300mm; fixing to masonry with screws	150.00	2.50	153.75	3.45	0.43	34.86	0.61	189.22	m2	212.87

Labour hourly rates: (except Specialists) Craft Operatives £9.23 Labourer £7.02 Rates are national average prices. Refer to REGIONAL VARIATIONS for indicative levels of overall pricing in regions	MATERIALS			LABOUR				RATES		
	Del to Site £	Waste %	Material Cost £	Craft Optve Hrs	Lab Hrs	Labour Cost £	Sunds £	Nett Rate £	Unit	Gross Rate (+12.5%) £
K13: RIGID SHEET FINE LININGS/ PANELLING Cont.										
Sapele, wrot, selected for transparent finish - wall panelling Cont.										
25mm thick panelled linings to walls; square framed; including grounds										
width exceeding 300mm	157.32	2.50	161.25	2.60	0.33	26.31	0.43	188.00	m2	211.50
width exceeding 300mm; fixing to masonry with screws	157.32	2.50	161.25	3.60	0.45	36.39	0.62	198.26	m2	223.04
19mm thick panelled linings to walls; square framed; obstructed by integral services; including grounds										
width exceeding 300mm	150.00	2.50	153.75	2.95	0.37	29.83	0.51	184.09	m2	207.10
width exceeding 300mm; fixing to masonry with screws	150.00	2.50	153.75	3.95	0.49	39.90	0.70	194.35	m2	218.64
25mm thick panelled linings to walls; square framed; obstructed by integral services; including grounds										
width exceeding 300mm	157.32	2.50	161.25	3.10	0.39	31.35	0.52	193.12	m2	217.26
width exceeding 300mm; fixing to masonry with screws	157.32	2.50	161.25	4.10	0.51	41.42	0.71	203.39	m2	228.81
19mm thick panelled linings to walls; moulded; including grounds										
width exceeding 300mm	165.60	2.50	169.74	2.60	0.33	26.31	0.43	196.48	m2	221.05
width exceeding 300mm; fixing to masonry with screws	165.60	2.50	169.74	3.60	0.45	36.39	0.62	206.75	m2	232.59
25mm thick panelled linings to walls; moulded; including grounds										
width exceeding 300mm	171.80	2.50	176.10	2.75	0.34	27.77	0.46	204.32	m2	229.86
width exceeding 300mm; fixing to masonry with screws	171.80	2.50	176.10	3.75	0.47	37.91	0.66	214.67	m2	241.50
19mm thick panelled linings to walls; moulded; obstructed by integral services; including grounds										
width exceeding 300mm	165.60	2.50	169.74	3.10	0.39	31.35	0.52	201.61	m2	226.81
width exceeding 300mm; fixing to masonry with screws	165.60	2.50	169.74	4.10	0.51	41.42	0.71	211.87	m2	238.36
25mm thick panelled linings to walls; moulded; obstructed by integral services; including grounds										
width exceeding 300mm	171.80	2.50	176.10	3.25	0.41	32.88	0.55	209.52	m2	235.71
width exceeding 300mm; fixing to masonry with screws	171.80	2.50	176.10	4.25	0.53	42.95	0.75	219.79	m2	247.27
VENEERED PLYWOOD PANELLING - Plywood, pre-finished, decorative veneers; butt joints - wall panelling Note. prices are for panelling with Afromosia, Ash, Beech, Cherry, Elm, Oak, Knotted Pine, Sapele or Teak faced veneers										
4mm thick linings to walls										
width exceeding 300mm	11.00	15.00	12.65	0.95	0.12	9.61	0.15	22.41	m2	25.21
width exceeding 300 mm; fixing to timber with adhesive	11.00	15.00	12.65	1.10	0.14	11.14	2.20	25.99	m2	29.23
width exceeding 300mm; fixing to plaster with adhesive	11.00	15.00	12.65	1.05	0.13	10.60	2.20	25.45	m2	28.64
VENEERED PLYWOOD PANELLING - Plywood, flame retardant, pre-finished, decorative, veneers not matched; random V-grooves on face; butt joints - wall panelling. Note. prices are for panelling with Afromosia, Ash, Beech, Cherry, Elm, Oak, Knotted Pine, Sapele or Teak faced veneers										
4mm thick linings to walls										
width exceeding 300mm	29.50	15.00	33.92	0.95	0.12	9.61	0.15	43.69	m2	49.15
width exceeding 300 mm; fixing to timber with adhesive	29.50	15.00	33.92	1.10	0.14	11.14	2.20	47.26	m2	53.17
width exceeding 300mm; fixing to plaster with adhesive	29.50	15.00	33.92	1.05	0.13	10.60	2.20	46.73	m2	52.57
K20: TIMBER BOARD FLOORING/SHEATHING/ LININGS/CASINGS										
Softwood, sawn; fixing to timber - boarded flooring										
Boarding to floors, square edges; 19mm thick, 75mm wide boards										
width exceeding 300mm	3.74	7.50	4.02	0.70	0.09	7.09	0.11	11.22	m2	12.63
Boarding to floors, square edges; 25mm thick, 125mm wide boards										
width exceeding 300mm	4.68	7.50	5.03	0.65	0.08	6.56	0.11	11.70	m2	13.16
Boarding to floors, square edges; 32mm thick, 150mm wide boards										
width exceeding 300mm	5.95	7.50	6.40	0.65	0.08	6.56	0.11	13.07	m2	14.70
Softwood, wrought; fixing to timber - boarded flooring										
Boarding to floors, square edges; 19mm thick, 75mm wide boards										
width exceeding 300mm	6.24	7.50	6.71	0.85	0.10	8.55	0.14	15.40	m2	17.32

LININGS/SHEATHING/DRY PARTITIONING

Labour hourly rates: (except Specialists) Craft Operatives £9.23 Labourer £7.02 Rates are national average prices. Refer to REGIONAL VARIATIONS for indicative levels of overall pricing in regions	MATERIALS			LABOUR				RATES		
	Del to Site £	Waste %	Material Cost £	Craft Optve Hrs	Lab Hrs	Labour Cost £	Sunds £	Nett Rate £	Unit	Gross Rate (+12.5%) £
K20: TIMBER BOARD FLOORING/SHEATHING/ LININGS/CASINGS Cont.										
Softwood, wrought; fixing to timber - boarded flooring Cont.										
Boarding to floors, square edges; 25mm thick, 125mm wide boards										
width exceeding 300mm	7.80	7.50	8.38	0.80	0.10	8.09	0.13	16.60	m2	18.68
Boarding to floors, square edges; 32mm thick, 150mm wide boards										
width exceeding 300mm	9.92	7.50	10.66	0.80	0.10	8.09	0.13	18.88	m2	21.24
Boarding to floors, tongued and grooved joints; 19mm thick, 75mm wide boards										
width exceeding 300mm	7.13	7.50	7.66	0.95	0.12	9.61	0.15	17.43	m2	19.60
Boarding to floors, tongued and grooved joints; 25mm thick, 125mm wide boards										
width exceeding 300mm	8.54	7.50	9.18	0.90	0.11	9.08	0.15	18.41	m2	20.71
Boarding to floors, tongued and grooved joints; 32mm thick, 150mm wide boards										
width exceeding 300mm	10.69	7.50	11.49	0.90	0.11	9.08	0.15	20.72	m2	23.31
Abutments										
19mm thick softwood boarding										
raking cutting	0.63	-	0.63	0.07	-	0.65	-	1.28	m	1.44
curved cutting	0.94	-	0.94	0.10	-	0.92	-	1.86	m	2.10
25mm thick softwood boarding										
raking cutting	0.78	-	0.78	0.07	-	0.65	-	1.43	m	1.60
curved cutting	1.17	-	1.17	0.10	-	0.92	-	2.09	m	2.35
32mm thick softwood boarding										
raking cutting	1.00	-	1.00	0.07	-	0.65	-	1.65	m	1.85
curved cutting	1.49	-	1.49	0.10	-	0.92	-	2.41	m	2.71
SURFACE TREATMENT OF EXISTING WOOD FLOORING - Sanding and sealing existing wood flooring										
Machine sanding										
width exceeding 300mm	-	-	-	0.55	-	5.08	2.20	7.28	m2	8.19
Prepare, one priming coat and one finish coat of seal										
width exceeding 300mm	-	-	-	0.18	-	1.66	1.10	2.76	m2	3.11
WALL AND CEILING BOARDING - Softwood, wrought; fixing to timber - wall and ceiling boarding										
Boarding to walls, tongued, grooved and veed joints; 19mm thick, 100mm wide boards										
width exceeding 300mm	10.43	10.00	11.47	1.20	0.15	12.13	0.20	23.80	m2	26.78
Boarding to walls, tongued, grooved and veed joints; 19mm thick, 150mm wide boards										
width exceeding 300mm	9.52	10.00	10.47	0.85	0.10	8.55	0.14	19.16	m2	21.55
Boarding to walls, tongued, grooved and veed joints; 25mm thick, 100mm wide boards										
width exceeding 300mm	13.60	10.00	14.96	1.25	0.16	12.66	0.20	27.82	m2	31.30
Boarding to walls, tongued, grooved and veed joints; 25mm thick, 150mm wide boards										
width exceeding 300mm	12.69	10.00	13.96	0.90	0.11	9.08	0.15	23.19	m2	26.09
Boarding to ceilings, tongued, grooved and veed joints; 19mm thick, 100mm wide boards										
width exceeding 300mm	10.43	10.00	11.47	1.80	0.22	18.16	0.29	29.92	m2	33.66
Boarding to ceilings, tongued, grooved and veed joints; 19mm thick, 150mm wide boards										
width exceeding 300mm	9.52	10.00	10.47	1.20	0.15	12.13	0.20	22.80	m2	25.65
Boarding to ceilings, tongued, grooved and veed joints; 25mm thick, 100mm wide boards										
width exceeding 300mm	13.60	10.00	14.96	1.85	0.23	18.69	0.30	33.95	m2	38.19
Boarding to ceilings, tongued, grooved and veed joints; 25mm thick, 150mm wide boards										
width exceeding 300mm	12.69	10.00	13.96	1.25	0.16	12.66	0.20	26.82	m2	30.17
WALL AND CEILING BOARDING - Western Red Cedar, wrought; fixing to timber; wall and ceiling boarding										
Boarding to walls, tongued, grooved and veed joints; 19mm thick, 100mm wide boards										
width exceeding 300mm	42.68	10.00	46.95	1.25	0.16	12.66	0.20	59.81	m2	67.28
Boarding to walls, tongued, grooved and veed joints; 19mm thick, 150mm wide boards										
width exceeding 300mm	36.30	10.00	39.93	0.90	0.12	9.15	0.15	49.23	m2	55.38

Labour hourly rates: (except Specialists) Craft Operatives £9.23 Labourer £7.02 Rates are national average prices. Refer to REGIONAL VARIATIONS for indicative levels of overall pricing in regions	MATERIALS			LABOUR				RATES		
	Del to Site	Waste	Material Cost	Craft Optve	Lab	Labour Cost	Sunds	Nett Rate	Unit	Gross Rate (+12.5%)
	£	%	£	Hrs	Hrs	£	£	£		£
K20: TIMBER BOARD FLOORING/SHEATHING/ LININGS/CASINGS Cont.										
WALL AND CEILING BOARDING - Western Red Cedar, wrought; fixing to timber; wall and ceiling boarding Cont.										
Boarding to walls, tongued, grooved and veed joints; 25mm thick, 100mm wide boards width exceeding 300mm	53.63	10.00	58.99	1.30	0.17	13.19	0.21	72.40	m2	81.44
Boarding to walls, tongued, grooved and veed joints; 25mm thick, 150mm wide boards width exceeding 300mm	48.29	10.00	53.12	0.95	0.12	9.61	0.15	62.88	m2	70.74
WALL AND CEILING BOARDING - Knotty Pine, wrought, selected for transparent finish; fixing to timber - wall and ceiling boarding										
Boarding to walls, tongued, grooved and veed joints; 19mm thick, 100mm wide boards width exceeding 300mm	36.08	10.00	39.69	1.55	0.19	15.64	0.25	55.58	m2	62.53
Boarding to walls, tongued, grooved and veed joints; 19mm thick, 150mm wide boards width exceeding 300mm	30.80	10.00	33.88	1.05	0.13	10.60	0.17	44.65	m2	50.24
Boarding to walls, tongued, grooved and veed joints; 25mm thick, 100mm wide boards width exceeding 300mm	45.38	10.00	49.92	1.60	0.20	16.17	0.26	66.35	m2	74.64
Boarding to walls, tongued, grooved and veed joints; 25mm thick, 150mm wide boards width exceeding 300mm	40.87	10.00	44.96	1.10	0.14	11.14	0.18	56.27	m2	63.31
Boarding to ceilings, tongued, grooved and veed joints; 19mm thick, 100mm wide boards width exceeding 300mm	36.08	10.00	39.69	2.20	0.28	22.27	0.36	62.32	m2	70.11
Boarding to ceilings, tongued, grooved and veed joints; 19mm thick, 150mm wide boards width exceeding 300mm	30.80	10.00	33.88	1.60	0.20	16.17	0.26	50.31	m2	56.60
Boarding to ceilings, tongued, grooved and veed joints; 25mm thick, 100mm wide boards width exceeding 300mm	45.38	10.00	49.92	2.25	0.29	22.80	0.37	73.09	m2	82.23
Boarding to ceilings, tongued, grooved and veed joints; 25mm thick, 150mm wide boards width exceeding 300mm	40.87	10.00	44.96	1.65	0.21	16.70	0.27	61.93	m2	69.67
WALL AND CEILING BOARDING - Sapele, wrought, selected for transparent finish; fixing to timber - wall and ceiling boarding										
Boarding to walls, tongued, grooved and veed joints; 19mm thick, 100mm wide boards width exceeding 300mm	48.79	10.00	53.67	1.75	0.22	17.70	0.28	71.65	m2	80.60
Boarding to walls, tongued, grooved and veed joints; 19mm thick, 150mm wide boards width exceeding 300mm	42.46	10.00	46.71	1.20	0.15	12.13	0.20	59.03	m2	66.41
Boarding to walls, tongued, grooved and veed joints; 25mm thick, 100mm wide boards width exceeding 300mm	61.82	10.00	68.00	1.80	0.22	18.16	0.29	86.45	m2	97.26
Boarding to walls, tongued, grooved and veed joints; 25mm thick, 150mm wide boards width exceeding 300mm	55.55	10.00	61.10	1.25	0.16	12.66	0.20	73.97	m2	83.21
Boarding to ceilings, tongued, grooved and veed joints; 19mm thick, 100mm wide boards width exceeding 300mm	48.79	10.00	53.67	2.40	0.30	24.26	0.39	78.32	m2	88.11
Boarding to ceilings, tongued, grooved and veed joints; 19mm thick, 150mm wide boards width exceeding 300mm	42.46	10.00	46.71	1.80	0.22	18.16	0.29	65.15	m2	73.30
Boarding to ceilings, tongued, grooved and veed joints; 25mm thick, 100mm wide boards width exceeding 300mm	61.82	10.00	68.00	2.45	0.31	24.79	0.40	93.19	m2	104.84
Boarding to ceilings, tongued, grooved and veed joints; 25mm thick, 150mm wide boards width exceeding 300mm	55.55	10.00	61.10	1.85	0.23	18.69	0.30	80.10	m2	90.11
WALL AND CEILING BOARDING - Abutments										
19mm thick softwood boarding raking cutting	1.05	-	1.05	0.07	-	0.65	-	1.70	m	1.91
curved cutting	1.56	-	1.56	0.11	-	1.02	-	2.58	m	2.90
25mm thick softwood boarding raking cutting	1.36	-	1.36	0.08	-	0.74	-	2.10	m	2.36
curved cutting	2.04	-	2.04	0.12	-	1.11	-	3.15	m	3.54

Labour hourly rates: (except Specialists) Craft Operatives £9.23 Labourer £7.02 Rates are national average prices. Refer to REGIONAL VARIATIONS for indicative levels of overall pricing in regions	MATERIALS			LABOUR				RATES		
	Del to Site £	Waste %	Material Cost £	Craft Optve Hrs	Lab Hrs	Labour Cost £	Sunds £	Nett Rate £	Unit	Gross Rate (+12.5%) £

K20: TIMBER BOARD FLOORING/SHEATHING/ LININGS/CASINGS Cont.

WALL AND CEILING BOARDING - Abutments Cont.

19mm thick Western Red Cedar boarding										
raking cutting	4.27	-	4.27	0.08	-	0.74	-	5.01	m	5.63
curved cutting	6.40	-	6.40	0.12	-	1.11	-	7.51	m	8.45
25mm thick Western Red Cedar boarding										
raking cutting	5.37	-	5.37	0.10	-	0.92	-	6.29	m	7.08
curved cutting	8.04	-	8.04	0.13	-	1.20	-	9.24	m	10.39
19 mm thick hardwood boarding										
raking cutting	4.88	-	4.88	0.17	-	1.57	-	6.45	m	7.26
curved cutting	7.32	-	7.32	0.25	-	2.31	-	9.63	m	10.83
25 mm thick hardwood boarding										
raking cutting	6.18	-	6.18	0.19	-	1.75	-	7.93	m	8.93
curved cutting	9.27	-	9.27	0.29	-	2.68	-	11.95	m	13.44

WALL AND CEILING BOARDING - Finished angles

External; 19mm thick softwood boarding										
tongued and mitred	-	-	-	0.80	-	7.38	-	7.38	m	8.31
External; 25mm thick softwood boarding										
tongued and mitred	-	-	-	0.85	-	7.85	-	7.85	m	8.83
External; 19mm thick hardwood boarding										
tongued and mitred	-	-	-	1.15	-	10.61	-	10.61	m	11.94
External; 25mm thick hardwood boarding										
tongued and mitred	-	-	-	1.20	-	11.08	-	11.08	m	12.46

ROOF BOARDING - Softwood, sawn, impregnated; fixing to timber - roof boarding

Boarding to roofs, butt joints; 19mm thick, 75mm wide boards; external										
width exceeding 300mm	4.29	7.50	4.61	0.75	0.09	7.55	0.12	12.29	m2	13.82
width exceeding 300mm; sloping	4.29	7.50	4.61	0.85	0.11	8.62	0.14	13.37	m2	15.04
Boarding to roofs, butt joints; 25mm thick, 125mm wide boards; external										
width exceeding 300mm	5.64	7.50	6.06	0.70	0.09	7.09	0.11	13.27	m2	14.92
width exceeding 300mm; sloping	5.64	7.50	6.06	0.80	0.10	8.09	0.13	14.28	m2	16.06

ROOF BOARDING - Softwood, wrought, impregnated; fixing to timber - roof boarding

Boarding to roofs, tongued and grooved joints; 19mm thick, 75mm wide boards external										
width exceeding 300mm	8.58	7.50	9.22	0.95	0.12	9.61	0.15	18.98	m2	21.36
extra; cross rebated and rounded drip; 50mm wide .	0.89	10.00	0.98	0.24	0.03	2.43	0.04	3.44	m	3.88
extra; dovetailed cesspool 225 x 225 x 150mm	1.43	10.00	1.57	1.70	0.21	17.17	0.28	19.02	nr	21.40
extra; dovetailed cesspool 300 x 300 x 150mm	1.79	10.00	1.97	1.90	0.24	19.22	0.31	21.50	nr	24.19
width exceeding 300mm; sloping	8.58	7.50	9.22	1.05	0.13	10.60	0.17	20.00	m2	22.50
Boarding to roofs, tongued and grooved joints; 25mm thick, 125mm wide boards external										
width exceeding 300mm	10.27	7.50	11.04	0.90	0.11	9.08	0.15	20.27	m2	22.80
extra; cross rebated and rounded drip; 50mm wide .	0.89	10.00	0.98	0.24	0.03	2.43	0.04	3.44	m	3.88
extra; dovetailed cesspool 225 x 225 x 150mm	1.79	10.00	1.97	1.80	0.23	18.23	0.29	20.49	nr	23.05
extra; dovetailed cesspool 300 x 300 x 150mm	2.43	10.00	2.67	2.00	0.25	20.22	0.33	23.22	nr	26.12
width exceeding 300mm; sloping	10.27	7.50	11.04	1.00	0.12	10.07	0.16	21.27	m2	23.93

ROOF BOARDING - Abutments

19mm thick softwood boarding										
raking cutting	0.86	-	0.86	0.40	-	3.69	-	4.55	m	5.12
curved cutting	1.29	-	1.29	0.65	-	6.00	-	7.29	m	8.20
rebates	-	-	-	0.22	-	2.03	-	2.03	m	2.28
grooves	-	-	-	0.22	-	2.03	-	2.03	m	2.28
chamfers	-	-	-	0.16	-	1.48	-	1.48	m	1.66
25mm thick softwood boarding										
raking cutting	1.02	-	1.02	0.55	-	5.08	-	6.10	m	6.86
curved cutting	1.54	-	1.54	0.80	-	7.38	-	8.92	m	10.04
rebates	-	-	-	0.22	-	2.03	-	2.03	m	2.28
grooves	-	-	-	0.22	-	2.03	-	2.03	m	2.28
chamfers	-	-	-	0.16	-	1.48	-	1.48	m	1.66

K21: TIMBER STRIP/ BOARD FINE FLOORING/ LININGS

STRIP FLOORING - Hardwood, wrought, selected for transparent finish; sanded, two coats sealer finish

Boarding to floors, tongued and grooved joints; 22mm thick, 75mm wide boards; fixing on and with 25 x 50mm impregnated softwood battens to masonry										
width exceeding 300 mm; secret fixing to timber ..	39.20	7.50	42.14	1.50	0.19	15.18	0.24	57.56	m2	64.75
Boarding to floors, tongued and grooved joints; 12mm thick, 75mm wide boards; overlay										
width exceeding 300 mm; secret fixing to timber ..	32.60	7.50	35.05	1.20	0.15	12.13	0.20	47.37	m2	53.30

Labour hourly rates: (except Specialists) Craft Operatives £9.23 Labourer £7.02 Rates are national average prices. Refer to REGIONAL VARIATIONS for indicative levels of overall pricing in regions	MATERIALS			LABOUR				RATES		
	Del to Site	Waste	Material Cost	Craft Optve	Lab	Labour Cost	Sunds	Nett Rate	Unit	Gross Rate (+12.5%)
	£	%	£	Hrs	Hrs	£	£	£		£
K21: TIMBER STRIP/ BOARD FINE FLOORING/ LININGS Cont.										
SEMI-SPRUNG FLOORS - Foam backed softwood battens covered with hardwood tongued and grooved strip flooring										
22mm Standard Beech strip flooring to level sub-floors										
width exceeding 300mm; battens at approximately 400mm centres; elevation										
75mm	-	-	Spclist	-	-	Spclist	-	56.47	m2	63.53
22mm 'Sylva Squash' Beech strip flooring to level sub-floors										
width exceeding 300mm; battens at approximately 430mm centres; elevation										
75mm	-	-	Spclist	-	-	Spclist	-	60.38	m2	67.93
20mm Prime Maple strip flooring to level sub-floors										
width exceeding 300mm; battens at approximately 300mm centres; elevation										
75mm	-	-	Spclist	-	-	Spclist	-	45.94	m2	51.68
20mm First Grade Maple strip flooring to level sub-floors										
width exceeding 300mm; battens at approximately 300mm centres; elevation										
75mm	-	-	Spclist	-	-	Spclist	-	49.88	m2	56.12
K40: DEMOUNTABLE SUSPENDED CEILINGS										
Suspended ceilings; 300 x 300mm bevelled, grooved and rebated asbestos-free fire resisting tiles; laying in position in metal suspension system of main channel members on wire or rod hangers										
Depth of suspension 150 - 500mm										
9mm thick linings; fixing hangers to masonry	16.40	5.00	17.22	0.90	0.45	11.47	4.80	33.49	m2	37.67
9mm thick linings not exceeding 300mm wide; fixing hangers to masonry	4.95	10.00	5.45	0.54	0.14	5.97	2.90	14.31	m	16.10
Suspended ceilings; 600 x 600mm bevelled, grooved and rebated asbestos-free fire resisting tiles; laying in position in metal suspension system of main channel members on wire or rod hangers										
Depth of suspension 150 - 500mm										
9mm thick linings; fixing hangers to masonry	16.25	5.00	17.06	0.72	0.36	9.17	2.95	29.19	m2	32.83
9mm thick linings not exceeding 300mm wide; fixing hangers to masonry	4.90	10.00	5.39	0.36	0.11	4.09	1.75	11.23	m	12.64
Suspended ceilings; 300 x 300mm bevelled, grooved and rebated plain mineral fibre tiles; laying in position in metal suspension system of main channel members on wire or rod hangers										
Depth of suspension 150 - 500mm										
15.8mm thick linings; fixing hangers to masonry ..	9.95	5.00	10.45	0.90	0.45	11.47	4.85	26.76	m2	30.11
15.8mm thick linings not exceeding 300mm wide; fixing hangers to masonry	3.00	10.00	3.30	0.54	0.14	5.97	2.90	12.17	m	13.69
Suspended ceilings; 600 x 600mm bevelled, grooved and rebated plain mineral fibre tiles; laying in position in metal suspension system of main channel members on wire or rod hangers										
Depth of suspension 150 - 500mm										
15.8mm thick linings; fixing hangers to masonry ..	9.95	5.00	10.45	0.72	0.36	9.17	2.95	22.57	m2	25.39
15.8mm thick linings not exceeding 300mm wide; fixing hangers to masonry	3.00	10.00	3.30	0.43	0.11	4.74	1.75	9.79	m	11.01
Suspended ceilings; 300 x 300mm bevelled, grooved and rebated textured mineral fibre tiles; laying in position in metal suspension system of main channel members on wire or rod hangers										
Depth of suspension 150 - 500mm										
15.8mm thick linings; fixing hangers to masonry ..	12.45	5.00	13.07	0.90	0.45	11.47	4.85	29.39	m2	33.06
15.8mm thick linings not exceeding 300mm wide; fixing hangers to masonry	3.75	10.00	4.13	0.54	0.14	5.97	2.90	12.99	m	14.62
Suspended ceilings; 600 x 600mm bevelled, grooved and rebated regular drilled mineral fibre tiles; laying in position in metal suspension system of main channel members on wire or rod hangers										
Depth of suspension 150 - 500mm										
15.8mm thick linings; fixing hangers to masonry ..	12.45	5.00	13.07	0.72	0.36	9.17	4.10	26.35	m2	29.64
15.8mm thick linings not exceeding 300mm wide; fixing hangers to masonry	3.75	10.00	4.13	0.43	0.11	4.74	2.45	11.32	m	12.73

	MATERIALS			LABOUR				RATES		
Labour hourly rates: (except Specialists) Craft Operatives £9.23 Labourer £7.02. Rates are national average prices. Refer to REGIONAL VARIATIONS for indicative levels of overall pricing in regions	Del to Site £	Waste %	Material Cost £	Craft Optve Hrs	Lab Hrs	Labour Cost £	Sunds £	Nett Rate £	Unit	Gross Rate (+12.5%) £

K40: DEMOUNTABLE SUSPENDED CEILINGS Cont.

Suspended ceilings; 300 x 300mm bevelled, grooved and rebated patterned tiles p.c. £15.00/m2; laying in position in metal suspension system of main channel members on wire or rod hangers

Depth of suspension 150 - 500mm

15.8mm thick linings; fixing hangers to masonry ..	15.00	5.00	15.75	0.90	0.45	11.47	4.85	32.07	m2	36.07
15.8mm thick linings not exceeding 300mm wide; fixing hangers to masonry	4.50	10.00	4.95	0.54	0.14	5.97	2.90	13.82	m	15.54

Suspended ceilings; 600 x 600mm bevelled, grooved and rebated patterned tiles p.c. £15.00/m2; laying in position in metal suspension system of main channel members on wire or rod hangers

Depth of suspension 150 - 500mm

15.8mm thick linings; fixing hangers to masonry ..	15.00	5.00	15.75	0.72	0.36	9.17	4.10	29.02	m2	32.65
15.8mm thick linings not exceeding 300mm wide; fixing hangers to masonry	4.50	10.00	4.95	0.43	0.11	4.74	2.45	12.14	m	13.66

Suspended ceiling; British Gypsum Ltd, 1200 x 600mm Glasroc GRG tiles; clip fastening into Quicklock fire rated grid system with Q417 clips on suspended wire or rod hangers

Depth of suspension 150-500mm

10mm thick linings; fixing hangers to masonry	5.40	10.00	5.94	0.43	0.11	4.74	4.10	14.78	m2	16.63
10mm thick linings not exceeding 300mm wide; fixing hangers to masonry	1.60	10.00	1.76	0.43	0.11	4.74	2.45	8.95	m	10.07

Edge trims

Plain

angle section; white stove enamelled aluminium; fixing to masonry with screws at 450mm centres ...	1.20	5.00	1.26	0.50	0.25	6.37	-	7.63	m	8.58

K41: RAISED ACCESS FLOORS

Flooring; Microfloor 'Bonded 600' light grade full access system; 600 x 600mm high density particle board panels, B.S.5669 and DIN 68761; 100 x 100mm precast lightweight concrete pedestals at 600mm centres fixed to sub-floor with epoxy resin adhesive

Thickness of panel 30mm
finished floor height

50mm	-	-	Spclist	-	-	Spclist	-	19.24	m2	21.65
75mm	-	-	Spclist	-	-	Spclist	-	19.38	m2	21.80
100mm	-	-	Spclist	-	-	Spclist	-	19.62	m2	22.07
125mm	-	-	Spclist	-	-	Spclist	-	19.83	m2	22.31
150mm	-	-	Spclist	-	-	Spclist	-	20.23	m2	22.76
175mm	-	-	Spclist	-	-	Spclist	-	20.55	m2	23.12
200mm	-	-	Spclist	-	-	Spclist	-	20.73	m2	23.32

Flooring; Microfloor 'Bonded 600' medium grade full access system; 600 x 600mm high density particle board panels, B.S.5669 and DIN 68761; 100 x 100mm precast lightweight concrete pedestals at 600mm centres fixed to sub-floor with epoxy resin adhesive

Thickness of panel 38mm
finished floor height

50mm	-	-	Spclist	-	-	Spclist	-	21.88	m2	24.61
75mm	-	-	Spclist	-	-	Spclist	-	21.88	m2	24.61
100mm	-	-	Spclist	-	-	Spclist	-	22.30	m2	25.09
125mm	-	-	Spclist	-	-	Spclist	-	22.49	m2	25.30
150mm	-	-	Spclist	-	-	Spclist	-	23.22	m2	26.12
175mm	-	-	Spclist	-	-	Spclist	-	23.41	m2	26.34
200mm	-	-	Spclist	-	-	Spclist	-	23.59	m2	26.54

Flooring; Microfloor 'Bonded 600' office loadings grade full access system; 600 x 600mm high density particle board panels, B.S. 5669 and DIN 68761; 100 x 100mm precast lightweight concrete pedestals at 600mm centres fixed to sub-floor with epoxy resin adhesive

Thickness of panel 30mm
finished floor height

50mm	-	-	Spclist	-	-	Spclist	-	14.98	m2	16.85
75mm	-	-	Spclist	-	-	Spclist	-	15.07	m2	16.95
100mm	-	-	Spclist	-	-	Spclist	-	15.40	m2	17.32
125mm	-	-	Spclist	-	-	Spclist	-	15.59	m2	17.54
150mm	-	-	Spclist	-	-	Spclist	-	15.97	m2	17.97
175mm	-	-	Spclist	-	-	Spclist	-	16.33	m2	18.37
200mm	-	-	Spclist	-	-	Spclist	-	16.50	m2	18.56

Windows/Doors/Stairs

	MATERIALS			LABOUR				RATES		
Labour hourly rates: (except Specialists) Craft Operatives £9.23 Labourer £7.02 Rates are national average prices. Refer to REGIONAL VARIATIONS for indicative levels of overall pricing in regions	Del to Site £	Waste %	Material Cost £	Craft Optve Hrs	Lab Hrs	Labour Cost £	Sunds £	Nett Rate £	Unit	Gross Rate (+12.5%) £
L10: WINDOWS/ROOFLIGHTS/SCREENS/LOUVRES										
TIMBER WINDOWS - Casements in softwood, wrought; **knotting and priming by manufacturer**										
Magnet Trade; without bars; softwood sub-sills; hinges; fasteners; fixing to masonry with galvanised steel cramps -4 nr, 25 x 3 x 150mm girth, flat section, holes -2										
488 x 900mm overall; N09V	64.32	2.50	65.93	1.55	0.77	19.71	1.62	87.26	nr	98.17
631 x 750mm overall; 107V	68.92	2.50	70.64	1.55	0.77	19.71	1.62	91.97	nr	103.47
631 x 900mm overall; 109V	69.42	2.50	71.16	1.70	0.85	21.66	1.62	94.43	nr	106.24
631 x 1050mm overall; 110V	71.22	2.50	73.00	1.80	0.90	22.93	1.62	97.55	nr	109.75
631 x 1200mm overall; 112v	73.52	2.50	75.36	1.90	0.95	24.21	1.62	101.18	nr	113.83
915 x 900mm overall; 2N09W	85.00	2.50	87.13	1.90	0.95	24.21	1.62	112.95	nr	127.07
915 x 1050mm overall; 2N10W	86.10	2.50	88.25	2.05	1.02	26.08	1.62	115.95	nr	130.45
915 x 1200mm overall; 2N12W	88.40	2.50	90.61	2.15	1.07	27.36	1.62	119.59	nr	134.53
1200 x 900mm overall; 209W	106.78	2.50	109.45	2.15	1.07	27.36	1.62	138.43	nr	155.73
1200 x 1050mm overall; 210C	104.48	2.50	107.09	2.30	1.15	29.30	1.62	138.01	nr	155.27
1200 x 1050mm overall; 210W	108.48	2.50	111.19	2.30	1.15	29.30	1.62	142.11	nr	159.88
1200 x 1050mm overall; 210CV	132.66	2.50	135.98	2.30	1.15	29.30	1.62	166.90	nr	187.76
1200 x 1200mm overall; 212C	109.08	2.50	111.81	2.40	1.20	30.58	1.62	144.00	nr	162.00
1200 x 1200mm overall; 212W	110.78	2.50	113.55	2.40	1.20	30.58	1.62	145.75	nr	163.96
1200 x 1200mm overall; 212CV	137.85	2.50	141.30	2.40	1.20	30.58	1.62	173.49	nr	195.18
1524 x 1050mm overall; 3NN10WW	149.34	2.50	153.07	2.50	1.25	31.85	1.62	186.54	nr	209.86
1769 x 1050mm overall; 310C	125.16	2.50	128.29	2.90	1.45	36.95	1.62	166.85	nr	187.71
1769 x 1050mm overall; 310WW	199.29	2.50	204.27	2.90	1.45	36.95	1.62	242.84	nr	273.19
1769 x 1050mm overall; 310CVC	181.51	2.50	186.05	2.90	1.45	36.95	1.62	224.61	nr	252.69
1769 x 1200mm overall; 312C	129.76	2.50	133.00	3.00	1.50	38.22	1.62	172.84	nr	194.45
1769 x 1200mm overall; 312WW	203.29	2.50	208.37	3.00	1.50	38.22	1.62	248.21	nr	279.24
1769 x 1200mm overall; 312CVC	188.40	2.50	193.11	3.00	1.50	38.22	1.62	232.95	nr	262.07
TIMBER WINDOWS - Fully reversible windows in **softwood, wrought; base coat stain by manufacturer**										
Boulton and Paul Hi-Profile; horizontal tilt; safety and reverse locking catch; locking fastener with high security espagnolette bolt; fully weatherstripped; fixing to masonry with galvanised steel cramps -4 nr, 25 x 3 x 150mm girth, flat section, holes										
600 x 900mm overall; R0906	230.52	2.50	236.28	1.60	0.80	20.38	1.62	258.29	nr	290.57
900 x 900mm overall; R0909	240.17	2.50	246.17	1.90	0.95	24.21	1.62	272.00	nr	306.00
900 x 1050mm overall; R0910	244.82	2.50	250.94	2.00	1.00	25.48	1.62	278.04	nr	312.80
900 x 1200mm overall; R0912	251.70	2.50	257.99	2.10	1.05	26.75	1.62	286.37	nr	322.16
1200 x 900mm overall; R1209	260.81	2.50	267.33	2.10	1.05	26.75	1.62	295.70	nr	332.67
1200 x 1050mm overall; R1210	266.90	2.50	273.57	2.25	1.12	28.63	1.62	303.82	nr	341.80
1200 x 1200mm overall; R1212	274.14	2.50	280.99	2.40	1.20	30.58	1.62	313.19	nr	352.34
TIMBER WINDOWS - Windows in hardwood, wrought, **preservative treated; one coat Redwood stain by** **manufacturer**										
Crosby Sarek Ltd. Alpha Energy Saving windows; glazing beads; weatherstripping, hinges; fasteners; fixing to masonry with galvanised steel cramps -4 nr, 25 x 3 x 150mm girth; flat section, holes -2; side hung-non bar type										
630 x 750mm overall; X107C	148.47	2.50	152.18	1.90	0.95	24.21	1.62	178.01	nr	200.26
630 x 900mm overall; X109C	156.17	2.50	160.07	2.20	1.10	28.03	1.62	189.72	nr	213.44
630 x 1050mm overall; X110C	163.97	2.50	168.07	2.45	1.22	31.18	1.62	200.87	nr	225.98
630 x 1200mm overall; X112C	171.86	2.50	176.16	2.65	1.32	33.73	1.62	211.50	nr	237.94
630 x 1350mm overall; X113C	185.88	2.50	190.53	2.90	1.45	36.95	1.62	229.09	nr	257.73
1200 x 750mm overall; X207C	198.41	2.50	203.37	2.60	1.30	33.12	1.62	238.11	nr	267.88
1200 x 900mm overall; X209C	209.57	2.50	214.81	2.90	1.45	36.95	1.62	253.38	nr	285.05
1200 x 1050mm overall; X210C	217.32	2.50	222.75	3.05	1.52	38.82	1.62	263.19	nr	296.09
1200 x 1200mm overall; X212C	228.23	2.50	233.94	3.30	1.65	42.04	1.62	277.60	nr	312.30
1200 x 1350mm overall; X213C	245.22	2.50	251.35	3.55	1.77	45.19	1.62	298.16	nr	335.43
1770 x 750mm overall; X307C	252.17	2.50	258.47	3.30	1.65	42.04	1.62	302.14	nr	339.90
1770 x 750mm overall; X307CC	313.75	2.50	321.59	3.30	1.65	42.04	1.62	365.26	nr	410.91
1770 x 900mm overall; X309C	255.34	2.50	261.72	3.55	1.77	45.19	1.62	308.54	nr	347.10
1770 x 900mm overall; X309CC	330.01	2.50	338.26	3.55	1.77	45.19	1.62	385.07	nr	433.21

Labour hourly rates: (except Specialists) Craft Operatives £9.23 Labourer £7.02 Rates are national average prices. Refer to REGIONAL VARIATIONS for indicative levels of overall pricing in regions	MATERIALS			LABOUR				RATES		
	Del to Site	Waste	Material Cost	Craft Optve	Lab	Labour Cost	Sunds	Nett Rate	Unit	Gross Rate (+12.5%)
	£	%	£	Hrs	Hrs	£	£	£		£

L10: WINDOWS/ROOFLIGHTS/SCREENS/LOUVRES Cont.

TIMBER WINDOWS - Windows in hardwood, wrought, preservative treated; one coat Redwood stain by manufacturer Cont.

Crosby Sarek Ltd. Alpha Energy Saving windows; glazing beads; weatherstripping, hinges; fasteners; fixing to masonry with galvanised steel cramps -4 nr, 25 x 3 x 150mm girth; flat section, holes -2; side hung-non bar type Cont.

Description	Del to Site £	Waste %	Material Cost £	Craft Optve Hrs	Lab Hrs	Labour Cost £	Sunds £	Nett Rate £	Unit	Gross Rate £
1770 x 1050mm overall; X310C	264.02	2.50	270.62	3.70	1.85	47.14	1.62	319.38	nr	359.30
1770 x 1050mm overall; X310CC	344.21	2.50	352.82	3.70	1.85	47.14	1.62	401.57	nr	451.77
1770 x 1200mm overall; X312C	271.40	2.50	278.19	4.00	2.00	50.96	1.62	330.76	nr	372.11
1770 x 1200mm overall; X312CC	358.79	2.50	367.76	4.00	2.00	50.96	1.62	420.34	nr	472.88
1770 x 1350mm overall; X313C	294.90	2.50	302.27	4.30	2.15	54.78	1.62	358.67	nr	403.51
1770 x 1350mm overall; X313CC	386.01	2.50	395.66	4.30	2.15	54.78	1.62	452.06	nr	508.57
2339 x 900mm overall; X409CMC	413.92	2.50	424.27	4.30	2.15	54.78	1.62	480.67	nr	540.75
2339 x 1050mm overall; X410CMC	433.71	2.50	444.55	4.60	2.30	58.60	1.62	504.78	nr	567.87
2339 x 1200mm overall; X412CMC	454.17	2.50	465.52	4.85	2.42	61.75	1.62	528.90	nr	595.01
2339 x 1350mm overall; X413CMC	486.18	2.50	498.33	5.10	2.55	64.97	1.62	564.93	nr	635.54

Crosby Sarek Ltd; Alpha Energy Saving windows; glazing beads; weatherstripping; hinges; fasteners; fixing to masonry with galvanised steel cramps -4nr, 25 x 3 x 150mm girth, flat section, holes -2; non bar side hung/vent type

Description	Del to Site £	Waste %	Material Cost £	Craft Optve Hrs	Lab Hrs	Labour Cost £	Sunds £	Nett Rate £	Unit	Gross Rate £
630 x 1050mm overall; X110T	234.25	2.50	240.11	2.45	1.22	31.18	1.62	272.90	nr	307.02
630 x 1200mm overall; X112T	242.25	2.50	248.31	2.65	1.32	33.73	1.62	283.65	nr	319.11
1200 x 750mm overall; X207CV	259.44	2.50	265.93	2.60	1.30	33.12	1.62	300.67	nr	338.25
1200 x 900mm overall; X209CV	270.70	2.50	277.47	2.90	1.45	36.95	1.62	316.03	nr	355.54
1200 x 1050mm overall; X210CV	279.84	2.50	286.84	3.05	1.52	38.82	1.62	327.28	nr	368.19
1200 x 1050mm overall; X210T	281.44	2.50	288.48	3.05	1.52	38.82	1.62	328.92	nr	370.03
1200 x 1200mm overall; X212CV	290.50	2.50	297.76	3.30	1.65	42.04	1.62	341.42	nr	384.10
1200 x 1200mm overall; X212T	291.19	2.50	298.47	3.30	1.65	42.04	1.62	342.13	nr	384.90
1200 x 1350mm overall; X213CV	306.82	2.50	314.49	3.55	1.77	45.19	1.62	361.30	nr	406.47
1200 x 1350mm overall; X213T	301.96	2.50	309.51	3.55	1.77	45.19	1.62	356.32	nr	400.86
1200 x 1500mm overall; X215T	310.22	2.50	317.98	3.80	1.90	48.41	1.62	368.01	nr	414.01
1770 x 900mm overall; X309CVC	390.61	2.50	400.38	3.55	1.77	45.19	1.62	447.19	nr	503.09
1770 x 1050mm overall; X310CVC	405.67	2.50	415.81	3.70	1.85	47.14	1.62	464.57	nr	522.64
1770 x 1050mm overall; X310CW	365.87	2.50	375.02	3.70	1.85	47.14	1.62	423.77	nr	476.75
1770 x 1050mm overall; X310WW	395.76	2.50	405.65	3.70	1.85	47.14	1.62	454.41	nr	511.21
1770 x 1200mm overall; X312CVC	420.93	2.50	431.45	4.00	2.00	50.96	1.62	484.03	nr	544.54
1770 x 1200mm overall; X312CW	376.70	2.50	386.12	4.00	2.00	50.96	1.62	438.70	nr	493.53
1770 x 1200mm overall; X312WW	406.79	2.50	416.96	4.00	2.00	50.96	1.62	469.54	nr	528.23
1770 x 1350mm overall; X313CVC	448.34	2.50	459.55	4.30	2.15	54.78	1.62	515.95	nr	580.44
1770 x 1350mm overall; X313CW	393.41	2.50	403.25	4.30	2.15	54.78	1.62	459.65	nr	517.10
2339 x 1050mm overall; X410CWC	501.68	2.50	514.22	4.60	2.30	58.60	1.62	574.45	nr	646.25
2339 x 1050mm overall; X410TT	535.53	2.50	548.92	4.60	2.30	58.60	1.62	609.14	nr	685.29
2339 x 1200mm overall; X412CWC	519.35	2.50	532.33	4.85	2.42	61.75	1.62	595.71	nr	670.17
2339 x 1200mm overall; X412TT	550.30	2.50	564.06	4.85	2.42	61.75	1.62	627.43	nr	705.86
2339 x 1350mm overall; X413CWC	550.30	2.50	564.06	5.10	2.55	64.97	1.62	630.65	nr	709.48
2339 x 1350mm overall; X413TT	566.23	2.50	580.39	5.10	2.55	64.97	1.62	646.98	nr	727.85
2339 x 1500mm overall; X415TT	582.85	2.50	597.42	5.30	2.65	67.52	1.62	666.56	nr	749.88

Crosby Sarek Ltd; Alpha Energy Saving windows; glazing beads; weatherstripping; hinges; fasteners; fixing to masonry with galvanised steel cramps -4nr, 25 x 3 x 150mm girth, flat section, holes -2; top hung type

Description	Del to Site £	Waste %	Material Cost £	Craft Optve Hrs	Lab Hrs	Labour Cost £	Sunds £	Nett Rate £	Unit	Gross Rate £
630 x 600mm overall; X106A	153.87	2.50	157.72	1.80	0.90	22.93	1.62	182.27	nr	205.05
630 x 750mm overall; X107A	162.30	2.50	166.36	1.90	0.95	24.21	1.62	192.18	nr	216.21
630 x 900mm overall; X109A	168.80	2.50	173.02	2.20	1.10	28.03	1.62	202.67	nr	228.00
630 x 1050mm overall; X110A	175.04	2.50	179.42	2.45	1.22	31.18	1.62	212.21	nr	238.74
630 x 1200mm overall; X112A	185.20	2.50	189.83	2.65	1.32	33.73	1.62	225.18	nr	253.32
915 x 600mm overall; X2N06A	193.83	2.50	198.68	2.15	1.07	27.36	1.62	227.65	nr	256.11
915 x 750mm overall; X2N07A	202.55	2.50	207.61	2.35	1.17	29.90	1.62	239.14	nr	269.03
915 x 900mm overall; X2N09A	220.04	2.50	225.54	2.60	1.30	33.12	1.62	260.29	nr	292.82
915 x 1050mm overall; X2N10A	229.59	2.50	235.33	2.80	1.40	35.67	1.62	272.62	nr	306.70
915 x 1200mm overall; X2N12A	237.46	2.50	243.40	3.00	1.50	38.22	1.62	283.24	nr	318.64
1200 x 600mm overall; X206A	224.14	2.50	229.74	2.45	1.22	31.18	1.62	262.54	nr	295.36
1200 x 750mm overall; X207A	232.82	2.50	238.64	2.60	1.30	33.12	1.62	273.38	nr	307.56
1200 x 900mm overall; X209A	249.20	2.50	255.43	2.90	1.45	36.95	1.62	294.00	nr	330.75
1200 x 1050mm overall; X210A	258.32	2.50	264.78	3.05	1.52	38.82	1.62	305.22	nr	343.37
1200 x 1200mm overall; X212A	266.42	2.50	273.08	3.30	1.65	42.04	1.62	316.74	nr	356.34
1770 x 600mm overall; X306AE	269.53	2.50	276.27	3.20	1.60	40.77	1.62	318.66	nr	358.49
1770 x 750mm overall; X307AE	280.34	2.50	287.35	3.30	1.65	42.04	1.62	331.01	nr	372.39
1770 x 900mm overall; X309AE	299.43	2.50	306.92	3.55	1.77	45.19	1.62	353.73	nr	397.94
1770 x 1050mm overall; X310AE	311.52	2.50	319.31	3.70	1.85	47.14	1.62	368.07	nr	414.07
1770 x 1200mm overall; X312AE	322.00	2.50	330.05	4.00	2.00	50.96	1.62	382.63	nr	430.46

Crosby Sarek Ltd; Alpha Energy Saving windows; glazing beads; weatherstripping; hinges; fasteners; fixing to masonry with galvanised steel cramps -4nr, 25 x 3 x 150mm girth, flat section, holes -2; design guide - horizontal glazing bar and fanlight type

Description	Del to Site £	Waste %	Material Cost £	Craft Optve Hrs	Lab Hrs	Labour Cost £	Sunds £	Nett Rate £	Unit	Gross Rate £
488 x 900mm overall; XHN09D	200.27	2.50	205.28	1.90	0.95	24.21	1.62	231.10	nr	259.99
488 x 1050mm overall; XHN10D	209.37	2.50	214.60	2.15	1.07	27.36	1.62	243.58	nr	274.03
488 x 1200mm overall; XHN12D	217.05	2.50	222.48	2.40	1.20	30.58	1.62	254.67	nr	286.51
488 x 1350mm overall; XHN13D	231.47	2.50	237.26	2.50	1.15	31.15	1.62	270.02	Unit	303.78
630 x 900mm overall; XH109D	211.68	2.50	216.97	2.20	1.10	28.03	1.62	246.62	nr	277.45
630 x 1050mm overall; XH110D	218.93	2.50	224.40	2.45	1.22	31.18	1.62	257.20	nr	289.35
630 x 1200mm overall; XH112D	226.00	2.50	231.65	2.65	1.32	33.73	1.62	267.00	nr	300.37
630 x 1350mm overall; XH113D	232.40	2.50	238.21	2.90	1.45	36.95	1.62	276.78	nr	311.37
915 x 900mm overall; H2N09D	243.07	2.50	249.15	2.60	1.30	33.12	1.62	283.89	nr	319.38
915 x 1050mm overall; H2N10D	237.97	2.50	243.92	2.80	1.40	35.67	1.62	281.21	nr	316.36

Labour hourly rates: (except Specialists) Craft Operatives £9.23 Labourer £7.02 Rates are national average prices. Refer to REGIONAL VARIATIONS for indicative levels of overall pricing in regions	MATERIALS			LABOUR				RATES		
	Del to Site £	Waste %	Material Cost £	Craft Optve Hrs	Lab Hrs	Labour Cost £	Sunds £	Nett Rate £	Unit	Gross Rate (+12.5%) £
L10: WINDOWS/ROOFLIGHTS/SCREENS/LOUVRES Cont.										
TIMBER WINDOWS - Windows in hardwood, wrought, preservative treated; one coat Redwood stain by manufacturer Cont.										
Crosby Sarek Ltd; Alpha Energy Saving windows; glazing beads; weatherstripping; hinges; fasteners; fixing to masonry with galvanised steel cramps -4nr, 25 x 3 x 150mm girth, flat section, holes -2; design guide - horizontal glazing bar and fanlight type Cont.										
915 x 1200mm overall; H2N12D	244.67	2.50	250.79	3.00	1.50	38.22	1.62	290.63	nr	326.96
915 x 1350mm overall; XH2N13D	254.55	2.50	260.91	3.25	1.62	41.37	1.62	303.90	nr	341.89
1200 x 1050mm overall; XH210CD	387.38	2.50	397.06	3.05	1.52	38.82	1.62	437.51	nr	492.19
1200 x 1200mm overall; XH212CD	402.70	2.50	412.77	3.30	1.65	42.04	1.62	456.43	nr	513.48
1200 x 1350mm overall; XH213CD	418.71	2.50	429.18	3.60	1.80	45.86	1.62	476.66	nr	536.24
1770 x 1050mm overall; XH310CDC	545.35	2.50	558.98	3.70	1.85	47.14	1.62	607.74	nr	683.71
1770 x 1200mm overall; XH312CDC	568.22	2.50	582.43	4.10	2.05	52.23	1.62	636.28	nr	715.81
1770 x 1350mm overall; XH313CDC	595.08	2.50	609.96	4.30	2.15	54.78	1.62	666.36	nr	749.65
Crosby Sarek Ltd; Alpha Energy Saving windows; glazing beads; weatherstripping; hinges; fasteners; fixing to masonry with galvanised steel cramps -4nr, 25 x 3 x 150mm girth, flat section, holes -2; design guide - horizontal glazing bar type										
630 x 900mm overall; XH109C	166.10	2.50	170.25	2.20	1.10	28.03	1.62	199.90	nr	224.89
630 x 1050mm overall; XH110C	172.78	2.50	177.10	2.45	1.22	31.18	1.62	209.90	nr	236.13
630 x 1200mm overall; XH112C	179.94	2.50	184.44	2.65	1.32	33.73	1.62	219.78	nr	247.26
630 x 1200mm overall; XH113C	192.81	2.50	197.63	2.90	1.45	36.95	1.62	236.20	nr	265.72
1200 x 900mm overall; XH209C	233.57	2.50	239.41	2.90	1.45	36.95	1.62	277.98	nr	312.72
1200 x 1050mm overall; XH210C	242.81	2.50	248.88	3.05	1.52	38.82	1.62	289.32	nr	325.49
1200 x 1200mm overall; XH212C	253.35	2.50	259.68	3.30	1.65	42.04	1.62	303.35	nr	341.64
1200 x 1350mm overall; XH213C	267.80	2.50	274.50	3.60	1.80	45.86	1.62	321.98	nr	362.23
1770 x 1050mm overall; XH310CC	379.87	2.50	389.37	3.70	1.85	47.14	1.62	438.12	nr	492.89
1770 x 1200mm overall; XH312CC	394.00	2.50	403.85	4.10	2.05	52.23	1.62	457.70	nr	514.92
1770 x 1350mm overall; XH313CC	420.43	2.50	430.94	4.30	2.15	54.78	1.62	487.34	nr	548.26
2339 x 1050mm overall; XH410CMC	484.35	2.50	496.46	4.60	2.30	58.60	1.62	556.68	nr	626.27
2339 x 1200mm overall; XH412CMC	502.66	2.50	515.23	4.85	2.42	61.75	1.62	578.60	nr	650.93
2339 x 1350mm overall; XH413CMC	534.11	2.50	547.46	5.15	2.57	65.58	1.62	614.66	nr	691.49
Crosby Sarek Ltd; Alpha Energy Saving windows; glazing beads; weatherstripping; hinges; fasteners; fixing to masonry with galvanised steel cramps -4nr, 25 x 3 x 150mm girth, flat section, holes -2; curved head sash -side hung type										
630 x 750mm overall; XS107C	173.85	2.50	178.20	1.90	0.95	24.21	1.62	204.02	nr	229.53
630 x 900mm overall; XS109C	181.54	2.50	186.08	2.20	1.10	28.03	1.62	215.73	nr	242.69
630 x 1050mm overall; XS110C	189.34	2.50	194.07	2.45	1.22	31.18	1.62	226.87	nr	255.23
630 x 1200mm overall; XS112C	197.21	2.50	202.14	2.65	1.32	33.73	1.62	237.49	nr	267.17
630 x 1350mm overall; XS113C	211.24	2.50	216.52	2.90	1.45	36.95	1.62	255.09	nr	286.97
915 x 750mm overall; XS2N07C	246.95	2.50	253.12	2.35	1.17	29.90	1.62	284.65	nr	320.23
915 x 900mm overall; XS2N09C	256.14	2.50	262.54	2.60	1.30	33.12	1.62	297.29	nr	334.45
915 x 1050mm overall; XS2N10C	264.95	2.50	271.57	2.80	1.40	35.67	1.62	308.87	nr	347.47
915 x 1200mm overall; XS2N12C	273.49	2.50	280.33	3.00	1.50	38.22	1.62	320.17	nr	360.19
915 x 1350mm overall; XS2N13C	281.32	2.50	288.35	3.25	1.62	41.37	1.62	331.34	nr	372.76
1200 x 750mm overall; XS207C	258.20	2.50	264.65	2.60	1.30	33.12	1.62	299.40	nr	336.82
1200 x 750mm overall; XS207CC	338.07	2.50	346.52	2.60	1.30	33.12	1.62	381.27	nr	428.92
1200 x 900mm overall; XS209C	269.36	2.50	276.09	2.90	1.45	36.95	1.62	314.66	nr	353.99
1200 x 900mm overall; XS209CC	353.11	2.50	361.94	2.90	1.45	36.95	1.62	400.50	nr	450.57
1200 x 1050mm overall; XS210C	277.10	2.50	284.03	3.05	1.52	38.82	1.62	324.47	nr	365.03
1200 x 1050mm overall; XS210CC	369.89	2.50	379.14	3.05	1.52	38.82	1.62	419.58	nr	472.03
1200 x 1200mm overall; XS212C	288.02	2.50	295.22	3.30	1.65	42.04	1.62	338.88	nr	381.24
1200 x 1200mm overall; XS212CC	384.35	2.50	393.96	3.30	1.65	42.04	1.62	437.62	nr	492.32
1200 x 1350mm overall; XS213C	305.02	2.50	312.65	3.60	1.80	45.86	1.62	360.13	nr	405.15
1200 x 1350mm overall; XS213CC	414.15	2.50	424.50	3.60	1.80	45.86	1.62	471.99	nr	530.99
TIMBER WINDOWS - Purpose made windows in softwood, wrought										
38mm moulded casements or fanlights										
in one pane	81.65	2.50	83.69	1.75	0.22	17.70	0.28	101.67	m2	114.38
divided into panes 0.10 - 0.50m2	93.90	2.50	96.25	1.75	0.22	17.70	0.28	114.22	m2	128.50
divided into panes not exceeding 0.10m2	106.15	2.50	108.80	1.75	0.22	17.70	0.28	126.78	m2	142.63
50mm moulded casements or fanlights										
in one pane	93.90	2.50	96.25	1.85	0.23	18.69	0.30	115.24	m2	129.64
divided into panes 0.10 - 0.50m2	106.15	2.50	108.80	1.85	0.23	18.69	0.30	127.79	m2	143.77
divided into panes not exceeding 0.10m2	118.39	2.50	121.35	1.85	0.23	18.69	0.30	140.34	m2	157.88
38mm moulded casements with semi-circular heads (measured square)										
in one pane	163.30	2.50	167.38	1.90	0.24	19.22	0.31	186.91	m2	210.28
divided into panes 0.10 - 0.50m2	175.55	2.50	179.94	1.90	0.24	19.22	0.31	199.47	m2	224.40
divided into panes not exceeding 0.10m2	187.80	2.50	192.50	1.90	0.24	19.22	0.31	212.03	m2	238.53
50mm moulded casements with semi-circular heads (measured square)										
in one pane	187.80	2.50	192.50	2.00	0.25	20.22	0.33	213.04	m2	239.67
divided into panes 0.10 - 0.50m2	200.04	2.50	205.04	2.00	0.25	20.22	0.33	225.59	m2	253.78
divided into panes not exceeding 0.10m2	212.29	2.50	217.60	2.00	0.25	20.22	0.33	238.14	m2	267.91
38mm bullseye casements										
457mm diameter in one pane	122.48	2.50	125.54	1.15	0.14	11.60	0.19	137.33	nr	154.50
762mm diameter in one pane	163.30	2.50	167.38	1.75	0.22	17.70	0.28	185.36	nr	208.53

Labour hourly rates: (except Specialists) Craft Operatives £9.23 Labourer £7.02 Rates are national average prices. Refer to REGIONAL VARIATIONS for indicative levels of overall pricing in regions	MATERIALS			LABOUR				RATES		
	Del to Site £	Waste %	Material Cost £	Craft Optve Hrs	Lab Hrs	Labour Cost £	Sunds £	Nett Rate £	Unit	Gross Rate (+12.5%) £
L10: WINDOWS/ROOFLIGHTS/SCREENS/LOUVRES Cont.										
TIMBER WINDOWS - Purpose made windows in softwood, wrought Cont.										
50mm bullseye casements										
457mm diameter in one pane	130.64	2.50	133.91	1.25	0.16	12.66	0.21	146.78	nr	165.12
762mm diameter in one pane	175.55	2.50	179.94	1.90	0.24	19.22	0.31	199.47	nr	224.40
Labours										
check throated edge	0.41	2.50	0.42	-	-	-	-	0.42	m	0.47
rebated and splayed bottom rail	0.41	2.50	0.42	-	-	-	-	0.42	m	0.47
rebated and beaded meeting stile	0.58	2.50	0.59	-	-	-	-	0.59	m	0.67
fitting and hanging casement or fanlight on butts (included elsewhere)										
38mm	-	-	-	0.75	0.09	7.55	-	7.55	nr	8.50
50mm	-	-	-	0.80	0.10	8.09	-	8.09	nr	9.10
fitting and hanging casement or fanlight on sash centres (included elsewhere)										
38mm	-	-	-	1.65	0.21	16.70	-	16.70	nr	18.79
50mm	-	-	-	1.80	0.22	18.16	-	18.16	nr	20.43
TIMBER WINDOWS - Purpose made windows in Afrormosia, wrought										
38mm moulded casements or fanlights										
in one pane	204.70	2.50	209.82	2.60	0.32	26.24	0.42	236.48	m2	266.04
divided into panes 0.10 - 0.50m2	235.41	2.50	241.30	2.60	0.32	26.24	0.42	267.96	m2	301.45
divided into panes not exceeding 0.10m2	266.11	2.50	272.76	2.60	0.32	26.24	0.42	299.43	m2	336.86
50mm moulded casements or fanlights										
in one pane	235.41	2.50	241.30	2.75	0.34	27.77	0.45	269.51	m2	303.20
divided into panes 0.10 - 0.50m2	266.11	2.50	272.76	2.75	0.34	27.77	0.45	300.98	m2	338.60
divided into panes not exceeding 0.10m2	296.82	2.50	304.24	2.75	0.34	27.77	0.45	332.46	m2	374.02
38mm moulded casements with semi-circular heads (measured square)										
in one pane	409.40	2.50	419.63	2.80	0.35	28.30	0.46	448.40	m2	504.45
divided into panes 0.10 - 0.50m2	440.11	2.50	451.11	2.80	0.35	28.30	0.46	479.87	m2	539.86
divided into panes not exceeding 0.10m2	470.81	2.50	482.58	2.80	0.35	28.30	0.46	511.34	m2	575.26
50mm moulded casements with semi-circular heads (measured square)										
in one pane	470.81	2.50	482.58	3.00	0.37	30.29	0.49	513.36	m2	577.53
divided into panes 0.10 - 0.50m2	501.52	2.50	514.06	3.00	0.37	30.29	0.49	544.84	m2	612.94
divided into panes not exceeding 0.10m2	532.22	2.50	545.53	3.00	0.37	30.29	0.49	576.30	m2	648.34
38mm bullseye casements										
457mm diameter in one pane	307.05	2.50	314.73	1.75	0.22	17.70	0.28	332.70	nr	374.29
762mm diameter in one pane	409.40	2.50	419.63	2.60	0.32	26.24	0.42	446.30	nr	502.09
50mm bullseye casements										
457mm diameter in one pane	327.52	2.50	335.71	1.90	0.24	19.22	0.31	355.24	nr	399.64
762mm diameter in one pane	440.11	2.50	451.11	2.80	0.35	28.30	0.46	479.87	nr	539.86
Labours										
check throated edge	1.02	2.50	1.05	-	-	-	-	1.05	m	1.18
rebated and splayed bottom rail	1.02	2.50	1.05	-	-	-	-	1.05	m	1.18
rebated and beaded meeting stile	1.44	2.50	1.48	-	-	-	-	1.48	m	1.66
fitting and hanging casement or fanlight on butts (included elsewhere)										
38mm	-	-	-	1.30	0.16	13.12	-	13.12	nr	14.76
50mm	-	-	-	1.40	0.17	14.12	-	14.12	nr	15.88
fitting and hanging casement or fanlight on sash centres (included elsewhere)										
38mm	-	-	-	2.85	0.36	28.83	-	28.83	nr	32.44
50mm	-	-	-	3.20	0.40	32.34	-	32.34	nr	36.39
TIMBER WINDOWS - Purpose made windows in European Oak, wrought										
38mm moulded casements or fanlights										
in one pane	244.95	2.50	251.07	3.45	0.43	34.86	0.56	286.50	m2	322.31
divided into panes 0.10 - 0.50m2	281.69	2.50	288.73	3.45	0.43	34.86	0.56	324.15	m2	364.67
divided into panes not exceeding 0.10m2	318.44	2.50	326.40	3.45	0.43	34.86	0.56	361.82	m2	407.05
50mm moulded casements or fanlights										
in one pane	281.69	2.50	288.73	3.70	0.46	37.38	0.60	326.71	m2	367.55
divided into panes 0.10 - 0.50m2	318.44	2.50	326.40	3.70	0.46	37.38	0.60	364.38	m2	409.93
divided into panes not exceeding 0.10m2	355.18	2.50	364.06	3.70	0.46	37.38	0.60	402.04	m2	452.29
38mm moulded casements with semi-circular heads (measured square)										
in one pane	489.90	2.50	502.15	3.80	0.48	38.44	0.62	541.21	m2	608.86
divided into panes 0.10 - 0.50m2	526.24	2.50	539.40	3.80	0.48	38.44	0.62	578.46	m2	650.77
divided into panes not exceeding 0.10m2	563.39	2.50	577.47	3.80	0.48	38.44	0.62	616.54	m2	693.61
50mm moulded casements with semi-circular heads (measured square)										
in one pane	563.39	2.50	577.47	4.00	0.50	40.43	0.65	618.55	m2	695.87
divided into panes 0.10 - 0.50m2	600.13	2.50	615.13	4.00	0.50	40.43	0.65	656.21	m2	738.24
divided into panes not exceeding 0.10m2	636.87	2.50	652.79	4.00	0.50	40.43	0.65	693.87	m2	780.61
38mm bullseye casements										
457mm diameter in one pane	367.43	2.50	376.62	2.30	0.29	23.26	0.37	400.25	nr	450.28
762mm diameter in one pane	489.90	2.50	502.15	3.45	0.43	34.86	0.56	537.57	nr	604.77

Labour hourly rates: (except Specialists) Craft Operatives £9.23 Labourer £7.02 Rates are national average prices. Refer to REGIONAL VARIATIONS for indicative levels of overall pricing in regions	MATERIALS			LABOUR				RATES		
	Del to Site	Waste	Material Cost	Craft Optve	Lab	Labour Cost	Sunds	Nett Rate	Unit	Gross Rate (+12.5%)
	£	%	£	Hrs	Hrs	£	£	£		£
L10: WINDOWS/ROOFLIGHTS/SCREENS/LOUVRES Cont.										
TIMBER WINDOWS - Purpose made windows in European Oak, wrought Cont.										
50mm bullseye casements										
457mm diameter in one pane	391.92	2.50	401.72	2.50	0.31	25.25	0.41	427.38	nr	480.80
762mm diameter in one pane	526.64	2.50	539.81	3.80	0.47	38.37	0.62	578.80	nr	651.15
Labours										
check throated edge	1.23	2.50	1.26	-	-	-	-	1.26	m	1.42
rebated and splayed bottom rail	1.23	2.50	1.26	-	-	-	-	1.26	m	1.42
rebated and beaded meeting stile	1.71	2.50	1.75	-	-	-	-	1.75	m	1.97
fitting and hanging casement or fanlight on butts (included elsewhere)										
38mm	-	-	-	1.45	0.18	14.65	-	14.65	nr	16.48
50mm	-	-	-	1.60	0.20	16.17	-	16.17	nr	18.19
fitting and hanging casement or fanlight on sash centres (included elsewhere)										
38mm	-	-	-	3.30	0.41	33.34	-	33.34	nr	37.50
50mm	-	-	-	3.60	0.45	36.39	-	36.39	nr	40.94
TIMBER WINDOWS - Sash windows in softwood, wrought, preservative treated; priming or staining by manufacturer										
Magnet Trade; weather stripping; spiral balances; brass plated fittings; fixing to masonry with galvanised steel cramps -4 nr, 25 x 3 x 150mm girth, flat section holes -2										
825 x 1094mm overall; SS3	246.34	2.50	252.50	3.80	1.90	48.41	1.55	302.46	nr	340.27
825 x 1394mm overall; SS4	261.73	2.50	268.27	4.15	2.07	52.84	1.55	322.66	nr	362.99
920 x 1225mm overall; SS2	275.11	2.50	281.99	4.15	2.07	52.84	1.55	336.37	nr	378.42
1051 x 1094mm overall; SSW3	261.93	2.50	268.48	4.15	2.07	52.84	1.55	322.86	nr	363.22
1051 x 1394mm overall; SSW4	280.31	2.50	287.32	4.50	2.25	57.33	1.55	346.20	nr	389.47
1051 x 1694mm overall; SSW5	280.31	2.50	287.32	4.80	2.40	61.15	1.55	350.02	nr	393.77
Magnet Trade; Georgian, open rebated for glass; weatherstripping; spiral balances; brass plated fittings; fixing to masonry with galvanised steel cramps -4 nr, 25 x 3 x 150mm girth, flat section, holes -2										
825 x 1094mm overall; GS3	273.91	2.50	280.76	2.60	1.30	33.12	1.55	315.43	nr	354.86
825 x 1394mm overall; GS4	297.49	2.50	304.93	3.00	1.50	38.22	1.55	344.70	nr	387.78
1051 x 1394mm overall; GSW4	328.56	2.50	336.77	2.60	1.30	33.12	1.40	371.30	nr	417.71
TIMBER WINDOWS - Sash windows mainly in softwood, wrought										
Cased frames; 25mm inside and outside linings; 32mm pulley stiles and head; 10mm beads, back linings, etc; 76mm oak sunk weathered and throated sills; 38mm moulded sashes										
in one pane	154.00	2.50	157.85	2.50	0.31	25.25	0.41	183.51	m2	206.45
divided into panes 0.50 - 1.00m2	161.70	2.50	165.74	2.50	0.31	25.25	0.41	191.40	m2	215.33
divided into panes 0.10 - 0.50m2	177.10	2.50	181.53	2.50	0.31	25.25	0.41	207.19	m2	233.09
divided into panes not exceeding 0.10m2	200.20	2.50	205.21	2.50	0.31	25.25	0.41	230.87	m2	259.72
Cased frames; 25mm inside and outside linings; 32mm pulley stiles and head; 10mm beads, back linings, etc; 76mm oak sunk weathered and throated sills; 50mm moulded sashes										
in one pane	177.10	2.50	181.53	2.75	0.34	27.77	0.45	209.75	m2	235.97
divided into panes 0.50 - 1.00m2	184.80	2.50	189.42	2.75	0.34	27.77	0.45	217.64	m2	244.84
divided into panes 0.10 - 0.50m2	200.20	2.50	205.21	2.75	0.34	27.77	0.45	233.42	m2	262.60
divided into panes not exceeding 0.10m2	223.30	2.50	228.88	2.75	0.34	27.77	0.45	257.10	m2	289.24
extra; windows in three lights with boxed mullions	30.80	2.50	31.57	0.75	0.09	7.55	0.12	39.24	nr	44.15
extra; windows with moulded horns	38.50	2.50	39.46	-	-	-	-	39.46	nr	44.40
extra; deep bottom rails and draught beads	3.85	2.50	3.95	-	-	-	-	3.95	m	4.44
Labours										
throats	0.39	2.50	0.40	-	-	-	-	0.40	m	0.45
grooves for jamb linings	0.46	2.50	0.47	-	-	-	-	0.47	m	0.53
fitting and hanging sashes in double hung windows; providing brass faced iron pulleys, best flax cords and iron weights										
38mm thick weighing 14 lbs per sash	38.50	2.50	39.46	1.55	0.19	15.64	-	55.10	nr	61.99
50mm thick weighing 20 lbs per sash	44.00	2.50	45.10	1.55	0.19	15.64	-	60.74	nr	68.33
Note for small frames and sashes, i.e. 1.25m2 and under, add 20% to the foregoing prices										
TIMBER WINDOWS - Sash windows in Afrormosia, wrought										
Cased frames; 25mm inside and outside linings; 32mm pulley stiles and head; 10mm beads, back linings, etc; 76mm oak sunk weathered and throated sills; 38mm moulded sashes										
in one pane	385.00	2.50	394.63	4.40	0.55	44.47	0.72	439.82	m2	494.80
divided into panes 0.50 - 1.00m2	404.25	2.50	414.36	4.40	0.55	44.47	0.72	459.55	Um2	516.99
divided into panes 0.10 - 0.50m2	442.75	2.50	453.82	4.40	0.55	44.47	0.72	499.01	m2	561.39
divided into panes not exceeding 0.10m2	500.50	2.50	513.01	4.40	0.55	44.47	0.72	558.21	m2	627.98

WINDOWS/DOORS/STAIRS

Labour hourly rates: (except Specialists) Craft Operatives £9.23 Labourer £7.02 Rates are national average prices. Refer to REGIONAL VARIATIONS for indicative levels of overall pricing in regions	MATERIALS			LABOUR				RATES		
	Del to Site £	Waste %	Material Cost £	Craft Optve Hrs	Lab Hrs	Labour Cost £	Sunds £	Nett Rate £	Unit	Gross Rate (+12.5%) £
L10: WINDOWS/ROOFLIGHTS/SCREENS/LOUVRES Cont.										
TIMBER WINDOWS - Sash windows in Afrormosia, wrought Cont.										
Cased frames; 25mm inside and outside linings; 32mm pulley stiles and head; 10mm beads, back linings, etc; 76mm oak sunk weathered and throated sills; 50mm moulded sashes										
in one pane	442.75	2.50	453.82	4.80	0.60	48.52	0.78	503.11	m2	566.00
divided into panes 0.50 - 1.00m2	462.00	2.50	473.55	4.80	0.60	48.52	0.78	522.85	m2	588.20
divided into panes 0.10 - 0.50m2	500.50	2.50	513.01	4.80	0.60	48.52	0.78	562.31	m2	632.60
divided into panes not exceeding 0.10m2	558.25	2.50	572.21	4.80	0.60	48.52	0.78	621.50	m2	699.19
extra; windows in three lights with boxed mullions	77.00	2.50	78.92	1.30	0.16	13.12	0.21	92.26	nr	103.79
extra; windows with moulded horns	96.25	2.50	98.66	-	-	-	-	98.66	nr	110.99
extra; deep bottom rails and draught beads	9.63	2.50	9.87	-	-	-	-	9.87	m	11.10
Labours										
throats	0.97	2.50	0.99	-	-	-	-	0.99	m	1.12
grooves for jamb linings	1.16	2.50	1.19	-	-	-	-	1.19	m	1.34
fitting and hanging sashes in double hung windows; providing brass faced iron pulleys, best flax cords and iron weights										
38mm thick weighing 14 lbs per sash...........	38.50	2.50	39.46	2.75	0.34	27.77	-	67.23	nr	75.64
50mm thick weighing 20 lbs per sash...........	44.00	2.50	45.10	2.75	0.34	27.77	-	72.87	nr	81.98
Note for small frames and sashes, i.e. 1.25m2 and under, add 20% to the foregoing prices										
TIMBER WINDOWS - Sash windows in European Oak, wrought										
Cased frames; 25mm inside and outside linings; 32mm pulley stiles and head; 10mm beads, back linings, etc; 76mm oak sunk weathered and throated sills; 38mm moulded sashes										
in one pane	462.00	2.50	473.55	5.00	0.62	50.50	0.81	524.86	m2	590.47
divided into panes 0.50 - 1.00m2	485.10	2.50	497.23	5.00	0.62	50.50	0.81	548.54	m2	617.11
divided into panes 0.10 - 0.50m2	531.30	2.50	544.58	5.00	0.62	50.50	0.81	595.89	m2	670.38
divided into panes not exceeding 0.10m2	600.60	2.50	615.62	5.00	0.62	50.50	0.81	666.93	m2	750.29
Cased frames; 25mm inside and outside linings; 32mm pulley stiles and head; 10mm beads, back linings, etc; 76mm oak sunk weathered and throated sills; 50mm moulded sashes										
in one pane	531.30	2.50	544.58	5.50	0.69	55.61	0.90	601.09	m2	676.23
divided into panes 0.50 - 1.00m2	554.40	2.50	568.26	5.50	0.69	55.61	0.90	624.77	m2	702.86
divided into panes 0.10 - 0.50m2	600.60	2.50	615.62	5.50	0.69	55.61	0.90	672.12	m2	756.14
divided into panes not exceeding 0.10m2	669.90	2.50	686.65	5.50	0.69	55.61	0.90	743.16	m2	836.05
extra; windows in three lights with boxed mullions	92.40	2.50	94.71	1.55	0.19	15.64	0.25	110.60	nr	124.43
extra; windows with moulded horns	115.50	2.50	118.39	-	-	-	-	118.39	nr	133.19
extra; deep bottom rails and draught beads	11.55	2.50	11.84	-	-	-	-	11.84	m	13.32
Labours										
throats	1.16	2.50	1.19	-	-	-	-	1.19	m	1.34
grooves for jamb linings	1.39	2.50	1.42	-	-	-	-	1.42	m	1.60
fitting and hanging sashes in double hung windows; providing brass faced iron pulleys, best flax cords and iron weights										
38mm thick weighing 14 lbs per sash...........	38.50	2.50	39.46	3.10	0.39	31.35	-	70.81	nr	79.66
50mm thick weighing 20 lbs per sash...........	44.00	2.50	45.10	3.10	0.39	31.35	-	76.45	nr	86.01
Note for small frames and sashes, i.e. 1.25m2 and under, add 20% to the foregoing prices										
TIMBER ROOFLIGHTS - Skylights in softwood, wrought										
Chamfered, straight bar										
38mm..	74.20	2.50	76.06	1.50	0.19	15.18	0.24	91.47	m2	102.91
50mm..	85.91	2.50	88.06	1.75	0.22	17.70	0.28	106.03	m2	119.29
63mm..	97.63	2.50	100.07	2.00	0.25	20.22	0.33	120.62	m2	135.69
Moulded, straight bar										
38mm..	78.10	2.50	80.05	1.50	0.19	15.18	0.24	95.47	m2	107.41
50mm..	89.82	2.50	92.07	1.75	0.22	17.70	0.28	110.04	m2	123.80
63mm..	101.53	2.50	104.07	2.00	0.25	20.22	0.33	124.61	m2	140.19
TIMBER ROOFLIGHTS - Skylights in Oak, wrought										
Chamfered, straight bar										
38mm..	222.59	2.50	228.15	2.30	0.29	23.26	0.37	251.79	m2	283.26
50mm..	257.73	2.50	264.17	2.70	0.34	27.31	0.44	291.92	m2	328.41
63mm..	292.88	2.50	300.20	3.10	0.39	31.35	0.50	332.05	m2	373.56
Moulded, straight bar										
38mm..	234.30	2.50	240.16	2.30	0.29	23.26	0.37	263.79	m2	296.77
50mm..	269.45	2.50	276.19	2.70	0.34	27.31	0.44	303.93	m2	341.93
63mm..	304.59	2.50	312.20	3.10	0.39	31.35	0.50	344.06	m2	387.06
TIMBER ROOFLIGHTS - Skylight kerbs in softwood, wrought										
Kerbs; dovetailed at angles										
38 x 225mm	10.12	2.50	10.37	0.31	0.04	3.14	0.05	13.57	m	15.26
50 x 225mm	12.13	2.50	12.43	0.35	0.04	3.51	0.05	15.99	m	17.99
38 x 225mm; chamfers -1 nr	10.62	2.50	10.89	0.31	0.04	3.14	0.05	14.08	m	15.84

Labour hourly rates: (except Specialists) Craft Operatives £9.23 Labourer £7.02 Rates are national average prices. Refer to REGIONAL VARIATIONS for indicative levels of overall pricing in regions	MATERIALS			LABOUR				RATES		
	Del to Site £	Waste %	Material Cost £	Craft Optve Hrs	Lab Hrs	Labour Cost £	Sunds £	Nett Rate £	Unit	Gross Rate (+12.5%) £
L10: WINDOWS/ROOFLIGHTS/SCREENS/LOUVRES Cont.										
TIMBER ROOFLIGHTS - Skylight kerbs in softwood, wrought Cont.										
Kerbs; dovetailed at angles Cont.										
50 x 225mm; chamfers -1 nr	12.64	2.50	12.96	0.35	0.04	3.51	0.05	16.52	m	18.58
Kerbs; in two thicknesses to circular skylights										
38 x 225mm	30.32	2.50	31.08	0.47	0.06	4.76	0.08	35.92	m	40.41
50 x 225mm	36.34	2.50	37.25	0.53	0.07	5.38	0.08	42.71	m	48.05
38 x 225mm; chamfers -1 nr	31.34	2.50	32.12	0.47	0.06	4.76	0.08	36.96	m	41.58
50 x 225mm; chamfers -1 nr	37.40	2.50	38.34	0.53	0.07	5.38	0.08	43.80	m	49.27
TIMBER ROOFLIGHTS - Skylight kerbs in Oak, wrought										
Kerbs; dovetailed at angles										
38 x 225mm	47.30	2.50	48.48	0.62	0.08	6.28	0.10	54.87	m	61.73
50 x 225mm	56.76	2.50	58.18	0.69	0.08	6.93	0.11	65.22	m	73.37
38 x 225mm; chamfers -1 nr	49.67	2.50	50.91	0.62	0.08	6.28	0.10	57.30	m	64.46
50 x 225mm; chamfers -1 nr	59.13	2.50	60.61	0.69	0.08	6.93	0.11	67.65	m	76.10
Kerbs; in two thicknesses to circular skylights										
38 x 225mm	141.90	2.50	145.45	0.94	0.12	9.52	0.15	155.12	m	174.51
50 x 225mm	170.28	2.50	174.54	1.06	0.13	10.70	0.17	185.40	m	208.58
38 x 225mm; chamfers -1 nr	146.63	2.50	150.30	0.94	0.12	9.52	0.15	159.96	m	179.96
50 x 225mm; chamfers -1 nr	175.01	2.50	179.39	1.06	0.13	10.70	0.17	190.25	m	214.03
TIMBER ROOFLIGHTS - Roof windows in Nordic red pine, wrought, treated										
Velux roof windows, The Velux Company Ltd; aluminium clad externally; hinges; fittings; factory glazed clear float double glazed sealed unit; type EDZ 0000 flashings; fixing to timber with screws; dressing flashings										
550 x 980mm overall, GGL104	186.83	2.50	191.50	5.50	2.75	70.07	-	261.57	nr	294.27
660 x 1180mm overall, GGL206	218.61	2.50	224.08	6.00	3.00	76.44	-	300.52	nr	338.08
660 x 1180mm overall, GHL206	284.51	2.50	291.62	6.00	3.00	76.44	-	368.06	nr	414.07
780 x 980mm overall, GGL304	216.86	2.50	222.28	6.00	3.00	76.44	-	298.72	nr	336.06
780 x 980mm overall, GHL304	278.48	2.50	285.44	6.00	3.00	76.44	-	361.88	nr	407.12
780 x 1400mm overall, GGL308	255.11	2.50	261.49	7.00	3.50	89.18	-	350.67	nr	394.50
780 x 1400mm overall, GHL308	346.65	2.50	355.32	7.00	3.50	89.18	-	444.50	nr	500.06
940 x 1600mm overall, GGL410	306.48	2.50	314.14	8.00	4.00	101.92	-	416.06	nr	468.07
1140 x 1180mm overall, GGL606	287.70	2.50	294.89	7.50	3.75	95.55	-	390.44	nr	439.25
1140 x 1180mm overall, GHL606	353.71	2.50	362.55	7.50	3.75	95.55	-	458.10	nr	515.37
1340 x 980mm overall, GGL804	292.23	2.50	299.54	7.50	3.75	95.55	-	395.09	nr	444.47
1340 x 980mm overall, GHL804	358.47	2.50	367.43	7.50	3.75	95.55	-	462.98	nr	520.85
1340 x 1400mm overall, GGL808	346.21	2.50	354.87	8.00	4.00	101.92	-	456.79	nr	513.88
1340 x 1400mm overall, GHL808	426.85	2.50	437.52	8.00	4.00	101.92	-	539.44	nr	606.87
Velux roof windows, The Velux Company Ltd; aluminium clad externally; hinges; fittings; factory glazed clear float double glazed sealed unit; type EDL 0000 flashings; fixing to timber with screws; dressing flashings										
550 x 980mm overall, GGL104	182.50	2.50	187.06	5.50	2.75	70.07	-	257.13	nr	289.27
660 x 1180mm overall, GGL206	214.04	2.50	219.39	6.00	3.00	76.44	-	295.83	nr	332.81
660 x 1180mm overall, GHL206	279.94	2.50	286.94	6.00	3.00	76.44	-	363.38	nr	408.80
780 x 980mm overall, GGL304	211.72	2.50	217.01	6.00	3.00	76.44	-	293.45	nr	330.13
780 x 980mm overall, GHL304	273.35	2.50	280.18	6.00	3.00	76.44	-	356.62	nr	401.20
780 x 1400mm overall, GGL308	249.09	2.50	255.32	7.00	3.50	89.18	-	344.50	nr	387.56
780 x 1400mm overall, GHL308	340.63	2.50	349.15	7.00	3.50	89.18	-	438.33	nr	493.12
940 x 1600mm overall, GGL410	300.79	2.50	308.31	8.00	4.00	101.92	-	410.23	nr	461.51
1140 x 1180mm overall, GGL606	281.20	2.50	288.23	7.50	3.75	95.55	-	383.78	nr	431.75
1140 x 1180mm overall, GHL606	347.24	2.50	355.92	7.50	3.75	95.55	-	451.47	nr	507.90
1340 x 980mm overall, GGL804	284.17	2.50	291.27	7.50	3.75	95.55	-	386.82	nr	435.18
1340 x 980mm overall, GHL804	350.36	2.50	359.12	7.50	3.75	95.55	-	454.67	nr	511.50
1340 x 1400mm overall, GGL808	337.85	2.50	346.30	8.00	4.00	101.92	-	448.22	nr	504.24
1340 x 1400mm overall, GHL808	418.50	2.50	428.96	8.00	4.00	101.92	-	530.88	nr	597.24
Velux roof windows, The Velux Company Ltd; aluminium clad externally; hinges; fittings; factory glazed clear float/heat absorbing glass double glazed sealed unit; type EDZ 0000 flashings; fixing to timber with screws; dressing flashings										
550 x 980mm overall, GGL104	222.48	2.50	228.04	5.50	2.75	70.07	-	298.11	nr	335.38
660 x 1180mm overall, GGL206	257.16	2.50	263.59	6.00	3.00	76.44	-	340.03	nr	382.53
660 x 1180mm overall, GHL206	322.07	2.50	330.12	6.00	3.00	76.44	-	406.56	nr	457.38
780 x 980mm overall, GGL304	254.23	2.50	260.59	6.00	3.00	76.44	-	337.03	nr	379.15
780 x 980mm overall, GHL304	314.91	2.50	322.78	6.00	3.00	76.44	-	399.22	nr	449.13
780 x 1400mm overall, GGL308	301.83	2.50	309.38	7.00	3.50	89.18	-	398.56	nr	448.38
780 x 1400mm overall, GHL308	391.88	2.50	401.68	7.00	3.50	89.18	-	490.86	nr	552.21
940 x 1600mm overall, GGL410	356.11	2.50	365.01	8.00	4.00	101.92	-	466.93	nr	525.30
1140 x 1180mm overall, GGL606	339.72	2.50	348.21	7.50	3.75	95.55	-	443.76	nr	499.23
1140 x 1180mm overall, GHL606	404.65	2.50	414.77	7.50	3.75	95.55	-	510.32	nr	574.11
1340 x 980mm overall, GGL804	343.91	2.50	352.51	7.50	3.75	95.55	-	448.06	nr	504.06
1340 x 980mm overall, GHL804	408.82	2.50	419.04	7.50	3.75	95.55	-	514.59	nr	578.91
1340 x 1400mm overall, GGL808	402.57	2.50	412.63	8.00	4.00	101.92	-	514.55	nr	578.87
1340 x 1400mm overall, GHL808	481.74	2.50	493.78	8.00	4.00	101.92	-	595.70	nr	670.17

Labour hourly rates: (except Specialists) Craft Operatives £9.23 Labourer £7.02 Rates are national average prices. Refer to REGIONAL VARIATIONS for indicative levels of overall pricing in regions	MATERIALS			LABOUR				RATES		
	Del to Site £	Waste %	Material Cost £	Craft Optve Hrs	Lab Hrs	Labour Cost £	Sunds £	Nett Rate £	Unit	Gross Rate (+12.5%) £

L10: WINDOWS/ROOFLIGHTS/SCREENS/LOUVRES Cont.

TIMBER ROOFLIGHTS - Roof windows in Nordic red pine, wrought, treated Cont.

Velux roof windows, The Velux Company Ltd; aluminium clad externally; hinges; fittings; factory glazed clear float/heat absorbing glass double glazed sealed unit; type EDL 0000 flashings; fixing to timber with screws; dressing flashings

550 x 980mm overall, GGL104	218.14	2.50	223.59	5.50	2.75	70.07	-	293.66	nr	330.37
660 x 1180mm overall, GGL206	252.59	2.50	258.90	6.00	3.00	76.44	-	335.34	nr	377.26
660 x 1180mm overall, GHL206	317.50	2.50	325.44	6.00	3.00	76.44	-	401.88	nr	452.11
780 x 980mm overall, GGL304	249.10	2.50	255.33	6.00	3.00	76.44	-	331.77	nr	373.24
780 x 980mm overall, GHL304	309.78	2.50	317.52	6.00	3.00	76.44	-	393.96	nr	443.21
780 x 1400mm overall, GGL308	295.82	2.50	303.22	7.00	3.50	89.18	-	392.40	nr	441.44
780 x 1400mm overall, GHL308	385.86	2.50	395.51	7.00	3.50	89.18	-	484.69	nr	545.27
940 x 1600mm overall, GGL410	350.41	2.50	359.17	8.00	4.00	101.92	-	461.09	nr	518.73
1140 x 1180mm overall, GGL606	333.46	2.50	341.80	7.50	3.75	95.55	-	437.35	nr	492.01
1140 x 1180mm overall, GHL606	398.17	2.50	408.12	7.50	3.75	95.55	-	503.67	nr	566.63
1340 x 980mm overall, GGL804	335.79	2.50	344.18	7.50	3.75	95.55	-	439.73	nr	494.70
1340 x 980mm overall, GHL804	400.72	2.50	410.74	7.50	3.75	95.55	-	506.29	nr	569.57
1340 x 1400mm overall, GGL808	394.20	2.50	404.06	8.00	4.00	101.92	-	505.98	nr	569.22
1340 x 1400mm overall, GHL808	473.38	2.50	485.21	8.00	4.00	101.92	-	587.13	nr	660.53

Velux roof windows, The Velux Company Ltd; aluminium clad externally; hinges; fittings; factory glazed clear float/clear laminated glass double glazed sealed unit; type EDZ 0000 flashings; fixing to timber with screws; dressing flashings

550 x 980mm overall, GGL104	230.93	2.50	236.70	5.50	2.75	70.07	-	306.77	nr	345.12
660 x 1180mm overall, GGL206	274.96	2.50	281.83	6.00	3.00	76.44	-	358.27	nr	403.06
660 x 1180mm overall, GHL206	339.80	2.50	348.30	6.00	3.00	76.44	-	424.74	nr	477.83
780 x 980mm overall, GGL304	270.85	2.50	277.62	6.00	3.00	76.44	-	354.06	nr	398.32
780 x 980mm overall, GHL304	331.13	2.50	339.41	6.00	3.00	76.44	-	415.85	nr	467.83
780 x 1400mm overall, GGL308	329.02	2.50	337.25	7.00	3.50	89.18	-	426.43	nr	479.73
780 x 1400mm overall, GHL308	418.61	2.50	429.08	7.00	3.50	89.18	-	518.26	nr	583.04
940 x 1600mm overall, GGL410	399.03	2.50	409.01	8.00	4.00	101.92	-	510.93	nr	574.79
1140 x 1180mm overall, GGL606	377.34	2.50	386.77	7.50	3.75	95.55	-	482.32	nr	542.61
1140 x 1180mm overall, GHL606	440.81	2.50	451.83	7.50	3.75	95.55	-	547.38	nr	615.80
1340 x 980mm overall, GGL804	379.83	2.50	389.33	7.50	3.75	95.55	-	484.88	nr	545.49
1340 x 980mm overall, GHL804	443.84	2.50	454.94	7.50	3.75	95.55	-	550.49	nr	619.30
1340 x 1400mm overall, GGL808	459.23	2.50	470.71	8.00	4.00	101.92	-	572.63	nr	644.21
1340 x 1400mm overall, GHL808	536.79	2.50	550.21	8.00	4.00	101.92	-	652.13	nr	733.65

Velux roof windows, The Velux Company Ltd; aluminium clad externally; hinges; fittings; factory glazed clear float/clear laminated glass double glazed sealed unit; type EDL 0000 flashings; fixing to timber with screws; dressing flashings

550 x 980mm overall, GGL104	226.60	2.50	232.26	5.50	2.75	70.07	-	302.33	nr	340.13
660 x 1180mm overall, GGL206	270.40	2.50	277.16	6.00	3.00	76.44	-	353.60	nr	397.80
660 x 1180mm overall, GHL206	334.87	2.50	343.24	6.00	3.00	76.44	-	419.68	nr	472.14
780 x 980mm overall, GGL304	265.71	2.50	272.35	6.00	3.00	76.44	-	348.79	nr	392.39
780 x 980mm overall, GHL304	326.00	2.50	334.15	6.00	3.00	76.44	-	410.59	nr	461.91
780 x 1400mm overall, GGL308	323.52	2.50	331.61	7.00	3.50	89.18	-	420.79	nr	473.39
780 x 1400mm overall, GHL308	412.61	2.50	422.93	7.00	3.50	89.18	-	512.11	nr	576.12
940 x 1600mm overall, GGL410	393.34	2.50	403.17	8.00	4.00	101.92	-	505.09	nr	568.23
1140 x 1180mm overall, GGL606	370.85	2.50	380.12	7.50	3.75	95.55	-	475.67	nr	535.13
1140 x 1180mm overall, GHL606	434.31	2.50	445.17	7.50	3.75	95.55	-	540.72	nr	608.31
1340 x 980mm overall, GGL804	371.72	2.50	381.01	7.50	3.75	95.55	-	476.56	nr	536.13
1340 x 980mm overall, GHL804	435.72	2.50	446.61	7.50	3.75	95.55	-	542.16	nr	609.93
1340 x 1400mm overall, GGL808	459.23	2.50	470.71	8.00	4.00	101.92	-	572.63	nr	644.21
1340 x 1400mm overall, GHL808	528.60	2.50	541.82	8.00	4.00	101.92	-	643.74	nr	724.20

TIMBER ROOFLIGHTS - Bedding and pointing frames

Bedding in cement mortar (1:3); pointing with Secomastic standard mastic one side

wood frames	1.10	10.00	1.21	0.20	-	1.85	-	3.06	m	3.44

Bedding in cement mortar (1:3); pointing with coloured two part polysulphide mastic one side

wood frames	2.20	10.00	2.42	0.20	-	1.85	-	4.27	m	4.80

METAL WINDOWS - Windows in galvanised steel

Crittall Windows Ltd. Duralife Homelight; weatherstripping; fixing to masonry with lugs
fixed lights

508 x 292mm, NG5	15.07	2.50	15.45	0.61	0.25	7.39	-	22.83	nr	25.69
508 x 923mm, NC5	20.65	2.50	21.17	0.98	0.35	11.50	-	32.67	nr	36.75
508 x 1218mm, ND5	23.61	2.50	24.20	1.23	0.41	14.23	-	38.43	nr	43.24
997 x 628mm, NE13	27.28	2.50	27.96	1.12	0.40	13.15	-	41.11	nr	46.25
997 x 923mm, NC13	31.96	2.50	32.76	1.24	0.44	14.53	-	47.29	nr	53.20
997 x 1218mm, ND13	37.08	2.50	38.01	1.50	0.51	17.43	-	55.43	nr	62.36
1486 x 628mm, NE14	32.70	2.50	33.52	1.50	0.53	17.57	-	51.08	nr	57.47
1486 x 923mm, NC14	36.72	2.50	37.64	1.63	0.58	19.12	-	56.75	nr	63.85
1486 x 1218mm, ND14	41.58	2.50	42.62	1.90	0.66	22.17	-	64.79	nr	72.89
top hung lights										
508 x 292mm, NG1	50.68	2.50	51.95	0.61	0.25	7.39	-	59.33	nr	66.75
508 x 457mm, NH1	58.81	2.50	60.28	0.74	0.25	8.59	-	68.87	nr	77.47
508 x 628mm, NE1	62.38	2.50	63.94	0.86	0.29	9.97	-	73.91	nr	83.15
997 x 628mm, NE13E	83.31	2.50	85.39	1.12	0.32	12.58	-	97.98	nr	110.22
997 x 923mm, NC13C	89.66	2.50	91.90	1.24	0.40	14.25	-	106.15	nr	119.42

Labour hourly rates: (except Specialists) Craft Operatives £9.23 Labourer £7.02 Rates are national average prices. Refer to REGIONAL VARIATIONS for indicative levels of overall pricing in regions	MATERIALS			LABOUR				RATES		
	Del to Site	Waste	Material Cost	Craft Optve	Lab	Labour Cost	Sunds	Nett Rate	Unit	Gross Rate (+12.5%)
	£	%	£	Hrs	Hrs	£	£	£		£
L10: WINDOWS/ROOFLIGHTS/SCREENS/LOUVRES Cont.										
METAL WINDOWS - Windows in galvanised steel Cont.										
Crittall Windows Ltd. Duralife Homelight; weatherstripping; fixing to masonry with lugs Cont.										
bottom hung lights										
508 x 628mm, NL1............................	69.30	2.50	71.03	0.86	0.32	10.18	-	81.22	nr	91.37
side hung lights										
508 x 628mm, NES1...........................	64.64	2.50	66.26	0.86	0.32	10.18	-	76.44	nr	86.00
508 x 923mm, NC1............................	69.45	2.50	71.19	0.98	0.35	11.50	-	82.69	nr	93.02
508 x 1067mm, NCO1.........................	73.63	2.50	75.47	1.10	0.38	12.82	-	88.29	nr	99.33
508 x 1218mm, ND1..........................	78.37	2.50	80.33	1.23	0.41	14.23	-	94.56	nr	106.38
mixed lights										
279 x 923mm, NC6F..........................	55.99	2.50	57.39	0.98	0.35	11.50	-	68.89	nr	77.50
508 x 923mm, NC5F..........................	61.38	2.50	62.91	0.98	0.35	11.50	-	74.42	nr	83.72
508 x 1067mm, NCO5F.......................	62.83	2.50	64.40	1.10	0.38	12.82	-	77.22	nr	86.87
997 x 292mm, NG2...........................	69.06	2.50	70.79	0.85	0.33	10.16	-	80.95	nr	91.07
997 x 457mm, NH2...........................	72.98	2.50	74.80	0.98	0.37	11.64	-	86.45	nr	97.25
997 x 628mm, NE2...........................	76.90	2.50	78.82	1.12	0.40	13.15	-	91.97	nr	103.46
997 x 628mm, NES2..........................	87.56	2.50	89.75	1.12	0.40	13.15	-	102.89	nr	115.76
997 x 923mm, NC2...........................	95.76	2.50	98.15	1.24	0.44	14.53	-	112.69	nr	126.77
997 x 923mm, NC2F..........................	127.12	2.50	130.30	1.24	0.44	14.53	-	144.83	nr	162.94
997 x 1067mm, NCO2........................	104.31	2.50	106.92	1.37	0.47	15.94	-	122.86	nr	138.22
997 x 1067mm, NCO2F.......................	138.68	2.50	142.15	1.37	0.47	15.94	-	158.09	nr	177.85
997 x 1218mm, ND2.........................	115.68	2.50	118.57	1.50	0.51	17.43	-	136.00	nr	153.00
997 x 1218mm, ND2F........................	140.15	2.50	143.65	1.50	0.51	17.43	-	161.08	nr	181.21
997 x 1513mm, NDV2FSB.....................	161.54	2.50	165.58	1.63	0.58	19.12	-	184.69	nr	207.78
1486 x 628mm, NE3.........................	104.27	2.50	106.88	1.50	0.51	17.43	-	124.30	nr	139.84
1486 x 923mm, NC4.........................	158.38	2.50	162.34	1.63	0.58	19.12	-	181.46	nr	204.14
1486 x 923mm, NC4F........................	193.85	2.50	198.70	1.63	0.58	19.12	-	217.81	nr	245.04
1486 x 1067mm, NCO4.......................	188.81	2.50	193.53	1.63	0.58	19.12	-	212.65	nr	239.23
1486 x 1067mm, NCO4F......................	213.36	2.50	218.69	1.77	0.61	20.62	-	239.31	nr	269.23
1486 x 1218mm, ND4........................	203.34	2.50	208.42	1.90	0.66	22.17	-	230.59	nr	259.42
1486 x 1218mm, ND4F.......................	216.43	2.50	221.84	1.90	0.66	22.17	-	244.01	nr	274.51
1486 x 1513mm, NDV4FSB....................	238.91	2.50	244.88	2.04	0.69	23.67	-	268.56	nr	302.13
1994 x 923mm, NC11F.......................	213.21	2.50	218.54	2.01	0.69	23.40	-	241.94	nr	272.18
1994 x 1218mm, ND11F......................	241.75	2.50	247.79	2.30	0.81	26.92	-	274.71	nr	309.05
1994 x 1513mm, NDV11FSB...................	265.55	2.50	272.19	2.47	0.86	28.84	-	301.02	nr	338.65
Extra for										
mullions										
292mm high..............................	8.65	2.50	8.87	0.23	0.12	2.97	-	11.83	nr	13.31
457mm high..............................	11.20	2.50	11.48	0.35	0.17	4.42	-	15.90	nr	17.89
628mm high..............................	12.09	2.50	12.39	0.46	0.23	5.86	-	18.25	nr	20.53
923mm high..............................	18.19	2.50	18.64	0.58	0.29	7.39	-	26.03	nr	29.29
1067mm high.............................	20.01	2.50	20.51	0.69	0.35	8.83	-	29.34	nr	33.00
1218mm high.............................	21.20	2.50	21.73	0.92	0.46	11.72	-	33.45	nr	37.63
1513mm high.............................	25.99	2.50	26.64	1.15	0.58	14.69	-	41.33	nr	46.49
2056mm high.............................	31.09	2.50	31.87	1.38	0.69	17.58	-	49.45	nr	55.63
transoms										
279mm wide..............................	8.65	2.50	8.87	0.23	0.12	2.97	-	11.83	nr	13.31
508mm wide..............................	12.10	2.50	12.40	0.35	0.17	4.42	-	16.83	nr	18.93
997mm wide..............................	20.43	2.50	20.94	0.69	0.35	8.83	-	29.77	nr	33.49
1486mm wide.............................	28.48	2.50	29.19	1.15	0.58	14.69	-	43.88	nr	49.36
Controlair Ventilators, permanent, mill finish										
508mm wide..............................	23.81	2.50	24.41	-	-	-	-	24.41	nr	27.46
997mm wide..............................	36.68	2.50	37.60	-	-	-	-	37.60	nr	42.30
1486mm wide.............................	42.80	2.50	43.87	-	-	-	-	43.87	nr	49.35
Controlair Ventilators, permanent, mill finish, flyscreen										
508mm wide..............................	28.93	2.50	29.65	-	-	-	-	29.65	nr	33.36
997mm wide..............................	46.46	2.50	47.62	-	-	-	-	47.62	nr	53.57
1486mm wide.............................	54.78	2.50	56.15	-	-	-	-	56.15	nr	63.17
Controlair Ventilators, adjustable										
508mm wide 1-LT.........................	31.33	2.50	32.11	-	-	-	-	32.11	nr	36.13
997mm wide 2-LT.........................	45.72	2.50	46.86	-	-	-	-	46.86	nr	52.72
Controlair Ventilators, adjustable, flyscreen										
508mm wide 1-LT.........................	35.98	2.50	36.88	-	-	-	-	36.88	nr	41.49
997mm wide 2-LT.........................	55.67	2.50	57.06	-	-	-	-	57.06	nr	64.19
locks with Parkes locking handles for										
side hung lights.......................	26.59	2.50	27.25	-	-	-	-	27.25	nr	30.66
horizontally pivoted lights............	26.59	2.50	27.25	-	-	-	-	27.25	nr	30.66
METAL WINDOWS - Windows in galvanised steel; polyester powder coated; white matt										
Crittall Windows Ltd. Duralife Homelight; weatherstripping; fixing to masonry with lugs										
fixed lights										
508 x 292mm, NG5..........................	18.71	2.50	19.18	0.61	0.25	7.39	-	26.56	nr	29.88
508 x 923mm, NC5..........................	25.62	2.50	26.26	0.98	0.35	11.50	-	37.76	nr	42.48
508 x 1218mm, ND5........................	29.30	2.50	30.03	1.23	0.41	14.23	-	44.26	nr	49.80
997 x 628mm, NE13........................	33.86	2.50	34.71	1.12	0.40	13.15	-	47.85	nr	53.83
997 x 923mm, NC13........................	39.65	2.50	40.64	1.24	0.44	14.53	-	55.18	nr	62.07
997 x 1218mm, ND13.......................	45.68	2.50	46.82	1.50	0.51	17.43	-	64.25	nr	72.28
1486 x 628mm, NE14.......................	40.66	2.50	41.68	1.50	0.53	17.57	-	59.24	nr	66.65
1486 x 923mm, NC14.......................	45.55	2.50	46.69	1.63	0.58	19.12	-	65.81	nr	74.03
1486 x 1218mm, ND14......................	51.56	2.50	52.85	1.90	0.66	22.17	-	75.02	nr	84.40
top hung lights										
508 x 292mm, NG1..........................	59.57	2.50	61.06	0.61	0.25	7.39	-	68.44	nr	77.00
508 x 457mm, NH1..........................	69.31	2.50	71.04	0.74	0.25	8.59	-	79.63	nr	89.58
508 x 628mm, NE1..........................	71.27	2.50	73.05	0.86	0.29	9.97	-	83.03	nr	93.40
997 x 628mm, NE13E........................	98.96	2.50	101.43	1.12	0.32	12.58	-	114.02	nr	128.27
997 x 923mm, NC13C........................	109.56	2.50	112.30	1.24	0.40	14.25	-	126.55	nr	142.37

WINDOWS/DOORS/STAIRS

Labour hourly rates: (except Specialists) Craft Operatives £9.23 Labourer £7.02 Rates are national average prices. Refer to REGIONAL VARIATIONS for indicative levels of overall pricing in regions	MATERIALS			LABOUR				RATES		
	Del to Site	Waste	Material Cost	Craft Optve	Lab	Labour Cost	Sunds	Nett Rate	Unit	Gross Rate (+12.5%)
	£	%	£	Hrs	Hrs	£	£	£		£
L10: WINDOWS/ROOFLIGHTS/SCREENS/LOUVRES Cont.										
METAL WINDOWS - Windows in galvanised steel; **polyester powder coated; white matt Cont.**										
Crittall Windows Ltd. Duralife Homelight; weatherstripping; fixing to masonry with lugs Cont.										
bottom hung lights										
508 x 628mm, NL1..........................	85.99	2.50	88.14	0.86	0.32	10.18	-	98.32	nr	110.61
side hung lights										
508 x 628mm, NES1	76.86	2.50	78.78	0.86	0.32	10.18	-	88.97	nr	100.09
508 x 923mm, NC1	82.66	2.50	84.73	0.98	0.35	11.50	-	96.23	nr	108.26
508 x 1067mm, NCO1	87.61	2.50	89.80	1.10	0.38	12.82	-	102.62	nr	115.45
508 x 1218mm, ND1	93.41	2.50	95.75	1.23	0.41	14.23	-	109.98	nr	123.72
mixed lights										
279 x 923mm, NC6F	64.97	2.50	66.59	0.98	0.35	11.50	-	78.10	nr	87.86
508 x 923mm, NC5F	72.89	2.50	74.71	0.98	0.35	11.50	-	86.21	nr	96.99
508 x 1067mm, NCO5F	74.80	2.50	76.67	1.10	0.38	12.82	-	89.49	nr	100.68
997 x 292mm, NG2	82.15	2.50	84.20	0.85	0.33	10.16	-	94.37	nr	106.16
997 x 457mm, NH2	86.45	2.50	88.61	0.98	0.37	11.64	-	100.25	nr	112.79
997 x 628mm, NE2	91.13	2.50	93.41	1.12	0.40	13.15	-	106.55	nr	119.87
997 x 628mm, NES2	104.77	2.50	107.39	1.12	0.40	13.15	-	120.53	nr	135.60
997 x 923mm, NC2	114.23	2.50	117.09	1.24	0.44	14.53	-	131.62	nr	148.02
997 x 923mm, NC2F	150.17	2.50	153.92	1.24	0.44	14.53	-	168.46	nr	189.52
997 x 1067mm, NCO2	123.98	2.50	127.08	1.37	0.47	15.94	-	143.02	nr	160.90
997 x 1067mm, NCO2F	163.89	2.50	167.99	1.37	0.47	15.94	-	183.93	nr	206.92
997 x 1218mm, ND2	138.04	2.50	141.49	1.50	0.51	17.43	-	158.92	nr	178.78
997 x 1218mm, ND2F	165.71	2.50	169.85	1.50	0.51	17.43	-	187.28	nr	210.69
997 x 1513mm, NDV2FSB	192.19	2.50	196.99	1.63	0.58	19.12	-	216.11	nr	243.13
1486 x 628mm, NE3	123.90	2.50	127.00	1.50	0.51	17.43	-	144.42	nr	162.48
1486 x 923mm, NC4	187.05	2.50	191.73	1.63	0.58	19.12	-	210.84	nr	237.20
1486 x 923mm, NC4F	228.71	2.50	234.43	1.63	0.58	19.12	-	253.54	nr	285.24
1486 x 1067mm, NCO4	223.80	2.50	229.40	1.63	0.58	19.12	-	248.51	nr	279.58
1486 x 1067mm, NCO4F	251.55	2.50	257.84	1.77	0.61	20.62	-	278.46	nr	313.27
1486 x 1218mm, ND4	241.32	2.50	247.35	1.90	0.66	22.17	-	269.52	nr	303.21
1486 x 1218mm, ND4F	255.28	2.50	261.66	1.90	0.66	22.17	-	283.83	nr	319.31
1486 x 1513mm, NDV4FSB	295.93	2.50	303.33	2.04	0.69	23.67	-	327.00	nr	367.88
1994 x 923mm, NC11F	252.05	2.50	258.35	2.01	0.69	23.40	-	281.75	nr	316.97
1994 x 1218mm, ND11F	285.89	2.50	293.04	2.30	0.81	26.92	-	319.95	nr	359.95
1994 x 1513mm, NDV11FSB	401.23	2.50	411.26	2.47	0.86	28.84	-	440.10	nr	495.11
Extra for										
mullions										
292mm high.....................	10.79	2.50	11.06	0.23	0.12	2.97	-	14.03	nr	15.78
457mm high	14.06	2.50	14.41	0.35	0.17	4.42	-	18.84	nr	21.19
628mm high	15.20	2.50	15.58	0.46	0.23	5.86	-	21.44	nr	24.12
923mm high	23.06	2.50	23.64	0.58	0.29	7.39	-	31.03	nr	34.90
1067mm high	25.15	2.50	25.78	0.69	0.35	8.83	-	34.60	nr	38.93
1218mm high	27.68	2.50	28.37	0.92	0.46	11.72	-	40.09	nr	45.10
1513mm high	26.48	2.50	27.14	1.15	0.58	14.69	-	41.83	nr	47.06
2056mm high	38.85	2.50	39.82	1.38	0.69	17.58	-	57.40	nr	64.58
transoms										
279mm wide	11.94	2.50	12.24	0.23	0.12	2.97	-	15.20	nr	17.10
508mm wide	15.19	2.50	15.57	0.35	0.17	4.42	-	19.99	nr	22.49
997mm wide	25.56	2.50	26.20	0.69	0.35	8.83	-	35.02	nr	39.40
1486mm wide	35.55	2.50	36.44	1.15	0.58	14.69	-	51.12	nr	57.52
Controlair Ventilators, permanent, mill finish										
508mm wide	29.85	2.50	30.60	-	-	-	-	30.60	nr	34.42
997mm wide	45.81	2.50	46.96	-	-	-	-	46.96	nr	52.82
1486mm wide	53.63	2.50	54.97	-	-	-	-	54.97	nr	61.84
Controlair Ventilators, permanent, mill finish, flyscreen										
508mm wide	34.91	2.50	35.78	-	-	-	-	35.78	nr	40.26
997mm wide	55.61	2.50	57.00	-	-	-	-	57.00	nr	64.13
1486mm wide	65.48	2.50	67.12	-	-	-	-	67.12	nr	75.51
Controlair Ventilators, adjustable										
508mm wide 1-LT..........................	38.86	2.50	39.83	-	-	-	-	39.83	nr	44.81
997mm wide 2-LT	57.20	2.50	58.63	-	-	-	-	58.63	nr	65.96
Controlair Ventilators, adjustable, flyscreen										
508mm wide 1-LT..........................	43.91	2.50	45.01	-	-	-	-	45.01	nr	50.63
997mm wide 2-LT..........................	66.94	2.50	68.61	-	-	-	-	68.61	nr	77.19
locks with Parkes locking handles for										
side hung lights..........................	26.38	2.50	27.04	-	-	-	-	27.04	nr	30.42
horizontally pivoted lights	26.38	2.50	27.04	-	-	-	-	27.04	nr	30.42
METAL WINDOWS - Windows in aluminium; polyester **powder finish; white matt**										
Kawneer Products 102 Casement range; factory glazed, 4mm, one pane clear glass; fixing to masonry with lugs										
fixed lights										
600 x 400mm...............................	53.00	2.50	54.33	0.81	0.35	9.93	-	64.26	nr	72.29
600 x 800mm	65.00	2.50	66.63	1.25	0.46	14.77	-	81.39	nr	91.57
600 x 1000mm	72.00	2.50	73.80	1.40	0.58	16.99	-	90.79	nr	102.14
600 x 1200mm	78.00	2.50	79.95	1.69	0.68	20.37	-	100.32	nr	112.86
600 x 1600mm	93.00	2.50	95.33	1.69	0.68	20.37	-	115.70	nr	130.16
800 x 400mm	57.00	2.50	58.42	0.97	0.39	11.69	-	70.12	nr	78.88
800 x 800mm	72.00	2.50	73.80	1.53	0.63	18.54	-	92.34	nr	103.89
800 x 1000mm	80.00	2.50	82.00	1.71	0.71	20.77	-	102.77	nr	115.61
800 x 1200mm	87.00	2.50	89.17	2.09	0.89	25.54	-	114.71	nr	129.05
800 x 1600mm.....'........................	104.00	2.50	106.60	2.09	0.89	25.54	-	132.14	nr	148.66
1200 x 400mm	69.00	2.50	70.72	1.13	0.44	13.52	-	84.24	nr	94.77
1200 x 800mm	87.00	2.50	89.17	1.81	0.76	22.04	-	111.22	nr	125.12
1200 x 1000mm	98.00	2.50	100.45	2.04	0.86	24.87	-	125.32	nr	140.98
1200 x 1200mm	107.00	2.50	109.68	2.48	1.91	36.30	-	145.97	nr	164.22

Labour hourly rates: (except Specialists) Craft Operatives £9.23 Labourer £7.02. Rates are national average prices. Refer to REGIONAL VARIATIONS for indicative levels of overall pricing in regions	MATERIALS			LABOUR				RATES		
	Del to Site £	Waste %	Material Cost £	Craft Optve Hrs	Lab Hrs	Labour Cost £	Sunds £	Nett Rate £	Unit	Gross Rate (+12.5%) £
L10: WINDOWS/ROOFLIGHTS/SCREENS/LOUVRES Cont.										
METAL WINDOWS - Windows in aluminium; polyester powder finish; white matt Cont.										
Kawneer Products 102 Casement range; factory glazed, 4mm, one pane clear glass; fixing to masonry with lugs Cont.										
fixed lights Cont.										
1200 x 1600mm	127.00	2.50	130.18	2.48	1.91	36.30	-	166.47	nr	187.28
1400 x 800mm	95.00	2.50	97.38	2.08	1.71	31.20	-	128.58	nr	144.65
1400 x 1000mm	106.00	2.50	108.65	2.35	1.85	34.68	-	143.33	nr	161.24
1400 x 1400mm	127.00	2.50	130.18	2.89	2.12	41.56	-	171.73	nr	193.20
1400 x 1600mm	151.00	2.50	154.78	2.89	2.12	41.56	-	196.33	nr	220.87
top hung casement										
600 x 800mm	139.00	2.50	142.47	1.25	0.52	15.19	-	157.66	nr	177.37
600 x 1000mm	150.00	2.50	153.75	1.40	0.58	16.99	-	170.74	nr	192.09
600 x 1200mm	160.00	2.50	164.00	1.69	0.68	20.37	-	184.37	nr	207.42
800 x 800mm	150.00	2.50	153.75	1.53	0.63	18.54	-	172.29	nr	193.83
800 x 1000mm	162.00	2.50	166.05	1.71	0.71	20.77	-	186.82	nr	210.17
800 x 1200mm	173.00	2.50	177.32	2.09	0.89	25.54	-	202.86	nr	228.22
top hung fanlights										
600 x400mm	138.00	2.50	141.45	0.81	0.35	9.93	-	151.38	nr	170.31
800 x 600mm	154.00	2.50	157.85	0.97	0.39	11.69	-	169.54	nr	190.73
1200 x 600mm	167.00	2.50	171.18	1.13	0.44	13.52	-	184.69	nr	207.78
Kawneer Products 102 Casement range; factory glazed, 4mm, one pane obscure glass; fixing to masonry with lugs										
fixed lights										
600 x 400mm	53.00	2.50	54.33	0.81	0.35	9.93	-	64.26	nr	72.29
600 x 800mm	66.00	2.50	67.65	1.25	0.46	14.77	-	82.42	nr	92.72
600 x 1000mm	73.00	2.50	74.83	1.40	0.58	16.99	-	91.82	nr	103.30
600 x 1200mm	80.00	2.50	82.00	1.69	0.68	20.37	-	102.37	nr	115.17
600 x 1600mm	95.00	2.50	97.38	1.69	0.68	20.37	-	117.75	nr	132.47
800 x 400mm	57.00	2.50	58.42	0.97	0.39	11.69	-	70.12	nr	78.88
800 x 800mm	73.00	2.50	74.83	1.53	0.63	18.54	-	93.37	nr	105.04
800 x 1000mm	82.00	2.50	84.05	1.71	0.71	20.77	-	104.82	nr	117.92
1200 x 400mm	70.00	2.50	71.75	1.13	0.44	13.52	-	85.27	nr	95.93
1200 x 800mm	89.00	2.50	91.22	1.81	0.76	22.04	-	113.27	nr	127.42
top hung casement										
600 x 800mm	140.00	2.50	143.50	1.25	0.52	15.19	-	158.69	nr	178.52
600 x 1000mm	151.00	2.50	154.78	1.40	0.58	16.99	-	171.77	nr	193.24
600 x 1200mm	161.00	2.50	165.03	1.69	0.68	20.37	-	185.40	nr	208.57
800 x 800mm	151.00	2.50	154.78	1.53	0.63	18.54	-	173.32	nr	194.98
800 x 1000mm	163.00	2.50	167.07	1.71	0.71	20.77	-	187.84	nr	211.32
800 x 1200mm	175.00	2.50	179.38	2.09	0.89	25.54	-	204.91	nr	230.53
top hung fanlights										
600 x400mm	140.00	2.50	143.50	0.81	0.35	9.93	-	153.43	nr	172.61
800 x 600mm	156.00	2.50	159.90	0.97	0.39	11.69	-	171.59	nr	193.04
1200 x 600mm	171.00	2.50	175.28	1.13	0.44	13.52	-	188.79	nr	212.39
METAL WINDOWS - Sliding windows in aluminium; polyester powder finish; white matt										
Kawneer Products Kingsley Equal series; factory double glazed, 18mm - 4/12/4 units with clear glass; fixing to masonry with lugs										
vertical sliders										
600 x 800mm	249.00	2.50	255.22	1.25	0.52	15.19	-	270.41	nr	304.21
600 x 1000mm	292.00	2.50	299.30	1.40	0.58	16.99	-	316.29	nr	355.83
600 x 1200mm	342.00	2.50	350.55	1.54	0.62	18.57	-	369.12	nr	415.26
800 x 800mm	446.00	2.50	457.15	1.53	0.63	18.54	-	475.69	nr	535.16
800 x 1000mm	494.00	2.50	506.35	1.71	0.71	20.77	-	527.12	nr	593.01
800 x 1200mm	552.00	2.50	565.80	1.91	0.81	23.32	-	589.12	nr	662.75
1200 x 800mm	885.00	2.50	907.13	2.04	0.86	24.87	-	931.99	nr	1048.49
1200 x 1000mm	942.00	2.50	965.55	2.25	0.98	27.65	-	993.20	nr	1117.35
1200 x 1200mm	1022.00	2.50	1047.55	2.48	1.08	30.47	-	1078.02	nr	1212.77
1400 x 800mm	1119.00	2.50	1146.97	2.62	1.14	32.19	-	1179.16	nr	1326.56
1400 x 1000mm	1193.00	2.50	1222.83	2.89	1.28	35.66	-	1258.49	nr	1415.80
1400 x 1200mm	1248.00	2.50	1279.20	2.89	1.11	34.47	-	1313.67	nr	1477.88
horizontal sliders										
1200 x 800mm	233.00	2.50	238.82	1.81	0.76	22.04	-	260.87	nr	293.47
1200 x 1000mm	246.00	2.50	252.15	2.04	0.86	24.87	-	277.02	nr	311.64
1200 x 1400mm	272.00	2.50	278.80	2.25	0.98	27.65	-	306.45	nr	344.75
1400 x 800mm	245.00	2.50	251.13	2.08	0.87	25.31	-	276.43	nr	310.98
1400 x 1000mm	261.00	2.50	267.52	2.35	1.01	28.78	-	296.31	nr	333.34
1400 x 1400mm	288.00	2.50	295.20	2.62	1.14	32.19	-	327.39	nr	368.31
1400 x 1600mm	304.00	2.50	311.60	2.89	1.28	35.66	-	347.26	nr	390.67
1600 x 800mm	260.00	2.50	266.50	2.37	1.00	28.90	-	295.40	nr	332.32
1600 x 1000mm	276.00	2.50	282.90	2.67	1.16	32.79	-	315.69	nr	355.15
1600 x 1600mm	322.00	2.50	330.05	3.28	1.48	40.66	-	370.71	nr	417.05
Window sills in galvanised pressed steel B.S.6510; fixing to metal										
for opening										
508mm wide; AWA	11.52	2.50	11.81	0.58	0.29	7.39	-	19.20	nr	21.60
508mm wide; AWB	22.49	2.50	23.05	0.58	0.29	7.39	-	30.44	nr	34.25
508mm wide; PS	13.13	2.50	13.46	0.58	0.29	7.39	-	20.85	nr	23.45
508mm wide; RPS	13.13	2.50	13.46	0.58	0.29	7.39	-	20.85	nr	23.45
628mm wide; AWA	12.84	2.50	13.16	0.69	0.35	8.83	-	21.99	nr	24.74
628mm wide; AWB	14.00	2.50	14.35	0.69	0.35	8.83	-	23.18	nr	26.07
628mm wide; PS	15.77	2.50	16.16	0.69	0.35	8.83	-	24.99	nr	28.11
628mm wide; RPS	15.77	2.50	16.16	0.69	0.35	8.83	-	24.99	nr	28.11
997mm wide; AWA	18.03	2.50	18.48	0.81	0.41	10.35	-	28.84	nr	32.44
997mm wide; AWB	20.29	2.50	20.80	0.81	0.41	10.35	-	31.15	nr	35.05

Labour hourly rates: (except Specialists) Craft Operatives £9.23 Labourer £7.02 Rates are national average prices. Refer to REGIONAL VARIATIONS for indicative levels of overall pricing in regions	MATERIALS			LABOUR				RATES		
	Del to Site £	Waste %	Material Cost £	Craft Optve Hrs	Lab Hrs	Labour Cost £	Sunds £	Nett Rate £	Unit	Gross Rate (+12.5%) £
L10: WINDOWS/ROOFLIGHTS/SCREENS/LOUVRES Cont.										
METAL WINDOWS - Sliding windows in aluminium; polyester powder finish; white matt Cont.										
Window sills in galvanised pressed steel B.S.6510; fixing to metal Cont.										
for opening Cont.										
997mm wide; PS	21.17	2.50	21.70	0.81	0.41	10.35	-	32.05	nr	36.06
997mm wide; RPS	21.17	2.50	21.70	0.81	0.41	10.35	-	32.05	nr	36.06
1237mm wide; AWA	20.29	2.50	20.80	0.92	0.46	11.72	-	32.52	nr	36.58
1237mm wide; AWB	21.88	2.50	22.43	0.92	0.46	11.72	-	34.15	nr	38.42
1237mm wide; PS	26.36	2.50	27.02	0.92	0.46	11.72	-	38.74	nr	43.58
1237mm wide; RPS	26.36	2.50	27.02	0.92	0.46	11.72	-	38.74	nr	43.58
1486mm wide; AWA	23.57	2.50	24.16	1.04	0.52	13.25	-	37.41	nr	42.08
1486mm wide; AWB	26.28	2.50	26.94	1.04	0.52	13.25	-	40.19	nr	45.21
1486mm wide; PS	28.42	2.50	29.13	1.04	0.52	13.25	-	42.38	nr	47.68
1486mm wide; RPS	28.42	2.50	29.13	1.04	0.52	13.25	-	42.38	nr	47.68
1846mm wide; AWA	26.73	2.50	27.40	1.15	0.58	14.69	-	42.08	nr	47.34
1846mm wide; AWB	30.87	2.50	31.64	1.15	0.58	14.69	-	46.33	nr	52.12
1846mm wide; PS	34.34	2.50	35.20	1.15	0.58	14.69	-	49.88	nr	56.12
1846mm wide; RPS	34.34	2.50	35.20	1.15	0.58	14.69	-	49.88	nr	56.12
METAL ROOFLIGHTS - Rooflights, mainly in mill finish aluminium										
Rooflights; factory glazed with one layer 6mm thick Georgian wired cast glass										
non-ventilating base frame; fixing to masonry with screws										
600 x 600mm	92.00	2.50	94.30	2.00	1.00	25.48	-	119.78	nr	134.75
600 x 900mm	112.00	2.50	114.80	2.30	1.15	29.30	-	144.10	nr	162.11
600 x 1200mm	118.50	2.50	121.46	2.30	1.15	29.30	-	150.76	nr	169.61
900 x 900mm	138.00	2.50	141.45	2.30	1.15	29.30	-	170.75	nr	192.10
900 x 1200mm	164.00	2.50	168.10	2.50	1.25	31.85	-	199.95	nr	224.94
1200 x 1200mm	198.00	2.50	202.95	2.80	1.40	35.67	-	238.62	nr	268.45
non-ventilating base frame and kerb; fixing to masonry with screws										
600 x 600mm	176.00	2.50	180.40	2.00	1.00	25.48	-	205.88	nr	231.62
600 x 900mm	242.00	2.50	248.05	2.30	1.15	29.30	-	277.35	nr	312.02
600 x 1200mm	256.00	2.50	262.40	2.30	1.15	29.30	-	291.70	nr	328.16
900 x 900mm	264.00	2.50	270.60	2.30	1.15	29.30	-	299.90	nr	337.39
900 x 1200mm	308.00	2.50	315.70	2.50	1.25	31.85	-	347.55	nr	390.99
1200 x 1200mm	367.00	2.50	376.18	2.80	1.40	35.67	-	411.85	nr	463.33
ventilating base frame; fixing to masonry with screws										
600 x 600mm	198.00	2.50	202.95	2.00	1.00	25.48	-	228.43	nr	256.98
600 x 900mm	223.00	2.50	228.57	2.30	1.15	29.30	-	257.88	nr	290.11
600 x 1200mm	250.00	2.50	256.25	2.30	1.15	29.30	-	285.55	nr	321.25
900 x 900mm	256.00	2.50	262.40	2.30	1.15	29.30	-	291.70	nr	328.16
900 x 1200mm	287.00	2.50	294.18	2.50	1.25	31.85	-	326.02	nr	366.78
1200 x 1200mm	333.00	2.50	341.32	2.80	1.40	35.67	-	377.00	nr	424.12
ventilating base frame and kerb; fixing to masonry with screws										
600 x 600mm	281.00	2.50	288.02	2.00	1.00	25.48	-	313.51	nr	352.69
600 x 900mm	328.00	2.50	336.20	2.30	1.15	29.30	-	365.50	nr	411.19
600 x 1200mm	373.00	2.50	382.32	2.30	1.15	29.30	--	411.63	nr	463.08
900 x 900mm	382.00	2.50	391.55	2.30	1.15	29.30	-	420.85	nr	473.46
900 x 1200mm	432.00	2.50	442.80	2.50	1.25	31.85	-	474.65	nr	533.98
1200 x 1200mm	505.00	2.50	517.63	2.80	1.40	35.67	-	553.30	nr	622.46
Rooflights; factory glazed with one layer 6mm thick rough cast glass, one layer 6mm thick Georgian wired cast glass, 6mm air space										
non-ventilating base frame; fixing to masonry with screws										
600 x 600mm	119.00	2.50	121.97	2.50	1.25	31.85	-	153.82	nr	173.05
600 x 900mm	152.00	2.50	155.80	2.80	1.40	35.67	-	191.47	nr	215.41
600 x 1200mm	184.00	2.50	188.60	3.20	1.60	40.77	-	229.37	nr	258.04
900 x 900mm	198.00	2.50	202.95	3.20	1.60	40.77	-.	243.72	nr	274.18
900 x 1200mm	243.00	2.50	249.07	3.70	1.85	47.14	-	296.21	nr	333.24
1200 x 1200mm	301.00	2.50	308.52	4.10	2.05	52.23	-	360.76	nr	405.85
non-ventilating base frame and kerb; fixing to masonry with screws										
600 x 600mm	204.00	2.50	209.10	2.50	1.25	31.85	-	240.95	nr	271.07
600 x 900mm	256.00	2.50	262.40	2.80	1.40	35.67	-	298.07	nr	335.33
600 x 1200mm	307.00	2.50	314.68	3.20	1.60	40.77	-	355.44	nr	399.87
900 x 900mm	322.00	2.50	330.05	3.20	1.60	40.77	-	370.82	nr	417.17
900 x 1200mm	387.00	2.50	396.68	3.70	1.85	47.14	-	443.81	nr	499.29
1200 x 1200mm	471.00	2.50	482.77	4.10	2.05	52.23	-	535.01	nr	601.89
ventilating base frame; fixing to masonry with screws										
600 x 600mm	217.00	2.50	222.43	2.50	1.25	31.85	-	254.28	nr	286.06
600 x 900mm	256.00	2.50	262.40	2.80	1.40	35.67	-	298.07	nr	335.33
600 x 1200mm	295.00	2.50	302.38	3.20	1.60	40.77	-	343.14	nr	386.04
900 x 900mm	322.00	2.50	330.05	3.20	1.60	40.77	-	370.82	nr	417.17
900 x 1200mm	353.00	2.50	361.82	3.70	1.85	47.14	-	408.96	nr	460.08
1200 x 1200mm	420.00	2.50	430.50	4.10	2.05	52.23	-	482.73	nr	543.08
ventilating base frame and kerb; fixing to masonry with screws										
600 x 600mm	301.00	2.50	308.52	2.50	1.25	31.85	-	340.38	nr	382.92
600 x 900mm	362.00	2.50	371.05	2.80	1.40	35.67	-	406.72	nr	457.56
600 x 1200mm	420.00	2.50	430.50	3.20	1.60	40.77	-	471.27	nr	530.18
900 x 900mm	447.00	2.50	458.18	3.20	1.60	40.77	-	498.94	nr	561.31
900 x 1200mm	498.00	2.50	510.45	3.70	1.85	47.14	-	557.59	nr	627.29
1200 x 1200mm	590.00	2.50	604.75	4.10	2.05	52.23	-	656.98	nr	739.11

Labour hourly rates: (except Specialists) Craft Operatives £9.23 Labourer £7.02 Rates are national average prices. Refer to REGIONAL VARIATIONS for indicative levels of overall pricing in regions	MATERIALS			LABOUR				RATES		
	Del to Site £	Waste %	Material Cost £	Craft Optve Hrs	Lab Hrs	Labour Cost £	Sunds £	Nett Rate £	Unit	Gross Rate (+12.5%) £
L10: WINDOWS/ROOFLIGHTS/SCREENS/LOUVRES Cont.										
METAL SCREENS, BORROWED LIGHTS, FRAMES AND GRILLES - Screens in mild steel										
Screens; 6 x 50 x 50mm angle framing to perimeter; 75 x 75 x 3mm mesh infill; welded connections										
2000 x 2000mm overall; fixing to masonry with screws	276.00	2.50	282.90	6.50	3.25	82.81	-	365.71	nr	411.42
3000 x 2000mm overall; fixing to masonry with screws	686.00	2.50	703.15	9.00	4.50	114.66	-	817.81	nr	920.04
Screens; 6 x 50 x 50mm angle framing to perimeter; 75 x 75 x 5mm mesh infill; welded connections										
2000 x 2000mm overall; fixing to masonry with screws	283.00	2.50	290.07	6.50	3.25	82.81	-	372.88	nr	419.50
Screens; 6 x 50 x 75mm angle framing to perimeter; 75 x 75 x 3mm mesh infill; welded connections										
3000 x 2000mm overall; fixing to masonry with screws	417.00	2.50	427.43	9.00	4.50	114.66	-	542.09	nr	609.85
Screens; 6 x 50 x 50mm angle framing to perimeter; 6 x 51 x 102mm tee mullion; 75 x 75 x 3mm mesh infill; welded connections										
2000 x 4000mm overall; mullions -1 nr; fixing to masonry with screws	714.00	2.50	731.85	9.00	4.50	114.66	-	846.51	nr	952.32
Screens; 38.1 x 38.1 x 3.2mm hollow section framing to perimeter; 75 x 75 x 3mm mesh infill; welded connections										
2000 x 2000mm overall; fixing to masonry with screws	371.00	2.50	380.27	6.50	3.25	82.81	-	463.08	nr	520.97
Screens; 38.1 x 38.1 x 3.2mm hollow section framing to perimeter; 75 x 75 x 5mm mesh infill; welded connections										
2000 x 2000mm overall; fixing to masonry with screws	391.00	2.50	400.77	6.50	3.25	82.81	-	483.58	nr	544.03
Screens; 76.2 x 38.1 x 4mm hollow section framing to perimeter; 75 x 75 x 3mm mesh infill; welded connections										
2000 x 2000mm overall; fixing to masonry with screws	405.00	2.50	415.13	6.50	3.25	82.81	-	497.94	nr	560.18
METAL SCREENS, BORROWED LIGHTS, FRAMES AND GRILLES - Grilles in mild steel										
Grilles; 13 x 51mm flat bar framing to perimeter; 13mm diameter vertical infill bars at 150mm centres; welded connections										
1000 x 1000mm overall; fixing to masonry with screws	276.00	2.50	282.90	4.50	2.25	57.33	-	340.23	nr	382.76
2000 x 1000mm overall; fixing to masonry with screws	371.00	2.50	380.27	6.50	3.25	82.81	-	463.08	nr	520.97
Grilles; 13 x 51mm flat bar framing to perimeter; 13mm diameter vertical infill bars at 300mm centres; welded connections										
1000 x 1000mm overall; fixing to masonry with screws	263.00	2.50	269.57	4.50	2.25	57.33	-	326.90	nr	367.77
Grilles; 13 x 51mm flat bar framing to perimeter; 18mm diameter vertical infill bars at 150mm centres; welded connections										
1000 x 1000mm overall; fixing to masonry with screws	270.00	2.50	276.75	4.50	2.25	57.33	-	334.08	nr	375.84
METAL SCREENS, BORROWED LIGHTS, FRAMES AND GRILLES - Glazing frames in mild steel										
Glazing frames; 13 x 38 x 3mm angle framing to perimeter; welded connections; 13 x 13mm glazing beads, fixed with screws										
500 x 500mm overall; fixing to timber with screws	162.00	2.50	166.05	3.50	1.75	44.59	-	210.64	nr	236.97
Glazing frames; 15 x 21 x 3mm angle framing to perimeter; welded connections; 13 x 13mm glazing beads, fixed with screws										
500 x 1000mm overall; fixing to timber with screws	263.00	2.50	269.57	3.50	1.75	44.59	-	314.17	nr	353.44
Glazing frames; 18 x 25 x 3mm angle framing to perimeter; welded connections; 9 x 25mm glazing beads, fixed with screws										
1000 x 1000mm overall; fixing to timber with screws	317.00	2.50	324.93	4.50	2.25	57.33	-	382.26	nr	430.04
Glazing frames; 18 x 25 x 3mm angle framing to perimeter; welded connections; 13 x 13mm glazing beads, fixed with screws										
1000 x 1000mm overall; fixing to timber with screws	338.00	2.50	346.45	4.50	2.25	57.33	-	403.78	nr	454.25

WINDOWS/DOORS/STAIRS

Labour hourly rates: (except Specialists) Craft Operatives £9.23 Labourer £7.02 Rates are national average prices. Refer to REGIONAL VARIATIONS for indicative levels of overall pricing in regions	MATERIALS			LABOUR				RATES		
	Del to Site £	Waste %	Material Cost £	Craft Optve Hrs	Lab Hrs	Labour Cost £	Sunds £	Nett Rate £	Unit	Gross Rate (+12.5%) £

L10: WINDOWS/ROOFLIGHTS/SCREENS/LOUVRES Cont.

METAL SCREENS, BORROWED LIGHTS, FRAMES AND GRILLES -
Glazing frames in mild steel Cont.

Glazing frames; 18 x 38 x 3mm angle framing to perimeter; welded connections; 9 x 25mm glazing beads, fixed with screws										
1000 x 1000mm overall; fixing to timber with screws	418.00	2.50	428.45	3.50	1.75	44.59	-	473.04	nr	532.17
Glazing frames; 18 x 25 x 3mm angle framing to perimeter; welded connections; 15 x 15 x 3mm channel glazing beads; fixed with screws										
1000 x 1000mm overall; fixing to timber with screws	418.00	2.50	428.45	3.50	1.75	44.59	-	473.04	nr	532.17
Glazing frames; 18 x 38 x 3mm angle framing to perimeter; welded connections; 15 x 15 x 3mm channel glazing beads; fixed with screws										
1000 x 1000mm overall; fixing to timber with screws	418.00	2.50	428.45	3.50	1.75	44.59	-	473.04	nr	532.17

METAL SCREENS, BORROWED LIGHTS, FRAMES AND GRILLES -
Glazing frames in aluminium

Glazing frames; 13 x 38 x 3mm angle framing to perimeter; welded connections; 13 x 13mm glazing beads, fixed with screws										
500 x 500mm overall; fixing to timber with screws	245.00	2.50	251.13	3.50	1.75	44.59	-	295.71	nr	332.68
Glazing frames; 15 x 21 x 3mm angle framing to perimeter; welded connections; 13 x 13mm glazing beads, fixed with screws										
500 x 1000mm overall; fixing to timber with screws	270.00	2.50	276.75	3.50	1.75	44.59	-	321.34	nr	361.51
Glazing frames; 18 x 25 x 3mm angle framing to perimeter; welded connections; 9 x 25mm glazing beads, fixed with screws										
1000 x 1000mm overall; fixing to timber with screws	284.00	2.50	291.10	4.50	2.25	57.33	-	348.43	nr	391.98
Glazing frames; 18 x 25 x 3mm angle framing to perimeter; welded connections; 13 x 13mm glazing beads, fixed with screws										
1000 x 1000mm overall; fixing to timber with screws	292.00	2.50	299.30	4.50	2.25	57.33	-	356.63	nr	401.21
Glazing frames; 18 x 38 x 3mm angle framing to perimeter; welded connections; 9 x 25mm glazing beads, fixed with screws										
1000 x 1000mm overall; fixing to timber with screws	338.00	2.50	346.45	4.50	2.25	57.33	-	403.78	nr	454.25
Glazing frames; 18 x 25 x 3mm angle framing to perimeter; welded connections; 15 x 15 x 3mm channel glazing beads; fixed with screws										
1000 x 1000mm overall; fixing to timber with screws	338.00	2.50	346.45	4.50	2.25	57.33	-	403.78	nr	454.25
Glazing frames; 18 x 38 x 3mm angle framing to perimeter; welded connections; 15 x 15 x 3mm channel glazing beads; fixed with screws										
1000 x 1000mm overall; fixing to timber with screws	376.00	2.50	385.40	4.50	2.25	57.33	-	442.73	nr	498.07

METAL SCREENS, BORROWED LIGHTS, FRAMES AND GRILLES -
Bedding and pointing frames

Pointing with Secomastic										
standard metal frames one side	0.83	10.00	0.91	0.14	-	1.29	-	2.21	m	2.48
Bedding in cement mortar (1:3); pointing with Secomastic standard mastic one side										
metal frames	0.92	10.00	1.01	0.20	-	1.85	-	2.86	m	3.22

PLASTICS WINDOWS - Windows in uPVC; white

Windows; factory glazed 20mm double glazed units; hinges; fastenings										
fixed light; fixing to masonry with cleats and screws; overall size										
600 x 600mm...........................	104.87	2.50	107.49	2.60	1.30	33.12	1.55	142.17	nr	159.94
600 x 900mm...........................	112.38	2.50	115.19	3.25	1.62	41.37	1.55	158.11	nr	177.87
600 x 1050mm...........................	119.87	2.50	122.87	3.55	1.77	45.19	1.55	169.61	nr	190.81
600 x 1200mm...........................	129.32	2.50	132.55	3.90	1.95	49.69	1.55	183.79	nr	206.76
600 x 1500mm...........................	151.78	2.50	155.57	4.55	2.27	57.93	1.55	215.06	nr	241.94
750 x 600mm...........................	108.61	2.50	111.33	2.95	1.47	37.55	1.55	150.42	nr	169.23
750 x 900mm...........................	127.37	2.50	130.55	3.60	1.80	45.86	1.55	177.97	nr	200.21
750 x 1050mm...........................	134.85	2.50	138.22	3.90	1.95	49.69	1.55	189.46	nr	213.14
750 x 1200mm...........................	142.34	2.50	145.90	4.20	2.10	53.51	1.55	200.96	nr	226.08
750 x 1500mm.........../.............	166.68	2.50	170.85	4.85	2.42	61.75	1.55	234.15	nr	263.42
900 x 600mm...........................	118.00	2.50	120.95	3.25	1.62	41.37	1.55	163.87	nr	184.35
900 x 900mm...........................	155.44	2.50	159.33	3.90	1.95	49.69	1.55	210.56	nr	236.88
900 x 1050mm...........................	149.82	2.50	153.57	4.10	2.05	52.23	1.55	207.35	nr	233.27
900 x 1200mm...........................	161.07	2.50	165.10	4.55	2.27	57.93	1.55	224.58	nr	252.65

Labour hourly rates: (except Specialists) Craft Operatives £9.23 Labourer £7.02 Rates are national average prices. Refer to REGIONAL VARIATIONS for indicative levels of overall pricing in regions	MATERIALS			LABOUR				RATES		
	Del to Site £	Waste %	Material Cost £	Craft Optve Hrs	Lab Hrs	Labour Cost £	Sunds £	Nett Rate £	Unit	Gross Rate (+12.5%) £

L10: WINDOWS/ROOFLIGHTS/SCREENS/LOUVRES Cont.

PLASTICS WINDOWS - Windows in uPVC; white Cont.

Windows; factory glazed 20mm double glazed units;
hinges; fastenings Cont.
  fixed light; fixing to masonry with cleats and
  screws; overall size Cont.

900 x 1500mm	189.25	2.50	193.98	5.15	2.57	65.58	1.55	261.11	nr	293.75
1200 x 600mm	129.32	2.50	132.55	3.65	1.82	46.47	1.55	180.57	nr	203.14
1200 x 900mm	155.53	2.50	159.42	4.30	2.15	54.78	1.55	215.75	nr	242.72
1200 x 1050mm	170.50	2.50	174.76	4.60	2.30	58.60	1.55	234.92	nr	264.28
1200 x 1200mm	185.49	2.50	190.13	5.00	2.50	63.70	1.55	255.38	nr	287.30
1200 x 1500mm	221.00	2.50	226.53	5.70	2.85	72.62	1.55	300.69	nr	338.28

fixed light with fixed light; fixing to masonry
with cleats and screws; overall size

600 x 1350mm	187.30	2.50	191.98	4.20	2.10	53.51	1.55	247.04	nr	277.92
600 x 1500mm	202.27	2.50	207.33	4.55	2.27	57.93	1.55	266.81	nr	300.16
600 x 1800mm	222.95	2.50	228.52	4.85	2.42	61.75	1.55	291.83	nr	328.31
600 x 2100mm	243.46	2.50	249.55	5.15	2.57	65.58	1.55	316.67	nr	356.26
750 x 1350mm	202.27	2.50	207.33	4.55	2.27	57.93	1.55	266.81	nr	300.16
750 x 1500mm	215.45	2.50	220.84	4.85	2.42	61.75	1.55	284.14	nr	319.66
750 x 1800mm	243.46	2.50	249.55	5.15	2.57	65.58	1.55	316.67	nr	356.26
750 x 2100mm	273.43	2.50	280.27	5.50	2.75	70.07	1.55	351.89	nr	395.87
900 x 1350mm	219.21	2.50	224.69	4.85	2.42	61.75	1.55	287.99	nr	323.99
900 x 1500mm	232.24	2.50	238.05	5.15	2.57	65.58	1.55	305.17	nr	343.32
900 x 1800mm	260.41	2.50	266.92	5.40	2.70	68.80	1.55	337.27	nr	379.42
900 x 2100mm	288.27	2.50	295.48	5.80	2.90	73.89	1.55	370.92	nr	417.28
1200 x 1350mm	247.21	2.50	253.39	5.40	2.70	68.80	1.55	323.74	nr	364.20
1200 x 1500mm	264.16	2.50	270.76	5.70	2.85	72.62	1.55	344.93	nr	388.05
1200 x 1800mm	297.86	2.50	305.31	6.05	3.02	77.04	1.55	383.90	nr	431.89
1200 x 2100mm	329.62	2.50	337.86	6.35	3.17	80.86	1.55	420.27	nr	472.81

overall, tilt/turn; fixing to masonry with cleats
and screws; overall size

600 x 600mm	305.35	2.50	312.98	2.60	1.30	33.12	1.55	347.66	nr	391.11
600 x 900mm	337.12	2.50	345.55	3.25	1.62	41.37	1.55	388.47	nr	437.03
600 x 1050mm	363.17	2.50	372.25	3.60	1.80	45.86	1.55	419.66	nr	472.12
600 x 1200mm	374.57	2.50	383.93	3.90	1.95	49.69	1.55	435.17	nr	489.57
600 x 1350mm	400.63	2.50	410.65	4.20	2.10	53.51	1.55	465.70	nr	523.92
600 x 1500mm	421.47	2.50	432.01	4.55	2.27	57.93	1.55	491.49	nr	552.92
750 x 600mm	317.39	2.50	325.32	2.95	1.47	37.55	1.55	364.42	nr	409.98
750 x 900mm	354.05	2.50	362.90	3.60	1.80	45.86	1.55	410.32	nr	461.60
750 x 1050mm	374.91	2.50	384.28	3.90	1.95	49.69	1.55	435.52	nr	489.96
750 x 1200mm	395.24	2.50	405.12	4.20	2.10	53.51	1.55	460.18	nr	517.70
750 x 1350mm	415.78	2.50	426.17	4.55	2.27	57.93	1.55	485.66	nr	546.36
750 x 1500mm	434.50	2.50	445.36	4.85	2.42	61.75	1.55	508.67	nr	572.25
900 x 900mm	374.57	2.50	383.93	3.90	1.95	49.69	1.55	435.17	nr	489.57
900 x 1050mm	397.05	2.50	406.98	4.10	2.05	52.23	1.55	460.76	nr	518.36
900 x 1200mm	417.73	2.50	428.17	4.55	2.27	57.93	1.55	487.66	nr	548.61
900 x 1350mm	438.24	2.50	449.20	4.85	2.42	61.75	1.55	512.50	nr	576.56
900 x 1500mm	464.46	2.50	476.07	5.15	2.57	65.58	1.55	543.20	nr	611.10
1200 x 1050mm	445.74	2.50	456.88	4.60	2.30	58.60	1.55	517.04	nr	581.67
1200 x 1200mm	471.95	2.50	483.75	5.00	2.50	63.70	1.55	549.00	nr	617.62
1200 x 1350mm	492.64	2.50	504.96	5.40	2.70	68.80	1.55	575.30	nr	647.21
1200 x 1500mm	518.86	2.50	531.83	5.70	2.85	72.62	1.55	606.00	nr	681.75

tilt/turn with fixed lights; fixing to masonry
with cleats and screws; overall size

600 x 1350mm	391.50	2.50	401.29	4.20	2.10	53.51	1.55	456.35	nr	513.39
600 x 1500mm	412.03	2.50	422.33	4.55	2.27	57.93	1.55	481.81	nr	542.04
600 x 2100mm	475.70	2.50	487.59	5.15	2.57	65.58	1.55	554.72	nr	624.06
750 x 1350mm	417.73	2.50	428.17	4.55	2.27	57.93	1.55	487.66	nr	548.61
750 x 1500mm	436.45	2.50	447.36	4.85	2.42	61.75	1.55	510.67	nr	574.50
750 x 2100mm	498.34	2.50	510.80	5.50	2.75	70.07	1.55	582.42	nr	655.22
900 x 1350mm	447.87	2.50	459.07	4.85	2.42	61.75	1.55	522.37	nr	587.67
900 x 1500mm	464.46	2.50	476.07	5.15	2.57	65.58	1.55	543.20	nr	611.10
900 x 2100mm	531.89	2.50	545.19	5.80	2.90	73.89	1.55	620.63	nr	698.21
1200 x 1350mm	505.67	2.50	518.31	5.40	2.70	68.80	1.55	588.66	nr	662.24
1200 x 1500mm	526.34	2.50	539.50	5.70	2.85	72.62	1.55	613.67	nr	690.37
1200 x 2100mm	582.36	2.50	596.92	6.35	3.17	80.86	1.55	679.33	nr	764.25

tilt/turn sash with fixed side light; fixing to
masonry with cleats and screws; overall size

1800 x 900mm	498.35	2.50	510.81	5.30	2.65	67.52	1.55	579.88	nr	652.37
1800 x 1050mm	531.89	2.50	545.19	5.65	2.82	71.95	1.55	618.68	nr	696.02
1800 x 1200mm	565.60	2.50	579.74	6.05	3.02	77.04	1.55	658.33	nr	740.62
1800 x 1350mm	599.31	2.50	614.29	6.50	3.25	82.81	1.55	698.65	nr	785.98
1800 x 1500mm	633.01	2.50	648.84	6.80	3.40	86.63	1.55	737.02	nr	829.14

tilt/turn sash with centre fixed light; fixing to
masonry with cleats and screws; overall size

2400 x 900mm	786.60	2.50	806.26	5.30	2.65	67.52	1.55	875.34	nr	984.75
2400 x 1050mm	843.12	2.50	864.20	6.90	3.45	87.91	1.55	953.65	nr	1072.86
2400 x 1200mm	898.96	2.50	921.43	7.30	3.65	93.00	1.55	1015.99	nr	1142.98
2400 x 1350mm	945.86	2.50	969.51	7.70	3.85	98.10	1.55	1069.15	nr	1202.80
2400 x 1500mm	983.33	2.50	1007.91	8.00	4.00	101.92	1.55	1111.38	nr	1250.31

**PLASTICS ROOFLIGHTS - Roof domelights in glass
reinforced plastics**

Domelights, translucent
  plain; fixing to masonry with screws

610 x 610mm	90.42	2.50	92.68	1.15	1.15	18.69	0.31	111.68	nr	125.64
914 x 914mm	116.17	2.50	119.07	2.30	2.30	37.38	0.63	157.08	nr	176.71
1219 x 1219mm	154.49	2.50	158.35	3.50	3.50	56.88	0.96	216.19	nr	243.21

  fixed ventilator; fixing to masonry with screws

610 x 610mm	160.76	2.50	164.78	1.15	1.15	18.69	0.31	183.78	nr	206.75
914 x 914mm	193.41	2.50	198.25	2.30	2.30	37.38	0.63	236.25	nr	265.78
1219 x 1219mm	238.64	2.50	244.61	3.50	3.50	56.88	0.96	302.44	nr	340.25

WINDOWS/DOORS/STAIRS

Labour hourly rates: (except Specialists) Craft Operatives £9.23 Labourer £7.02 Rates are national average prices. Refer to REGIONAL VARIATIONS for indicative levels of overall pricing in regions	MATERIALS			LABOUR				RATES		
	Del to Site	Waste	Material Cost	Craft Optve	Lab	Labour Cost	Sunds	Nett Rate	Unit	Gross Rate (+12.5%)
	£	%	£	Hrs	Hrs	£	£	£		£
L10: WINDOWS/ROOFLIGHTS/SCREENS/LOUVRES Cont.										
PLASTICS ROOFLIGHTS - Roof domelights in glass reinforced plastics Cont.										

Domelights, translucent Cont.
   controlled ventilator; fixing to masonry with
   screws										

610 x 610mm	180.24	2.50	184.75	1.15	1.15	18.69	0.31	203.74	nr	229.21
914 x 914mm	244.91	2.50	251.03	2.30	2.30	37.38	0.63	289.04	nr	325.17
1219 x 1219mm	328.50	2.50	336.71	3.50	3.50	56.88	0.96	394.55	nr	443.87
PLASTICS ROOFLIGHTS - Bedding and pointing frames										
Pointing with Secomastic standard mastic plastics frames one side	1.37	10.00	1.51	0.14	-	1.29	-	2.80	m	3.15
Bedding in cement mortar (1:3); pointing with Secomastic standard mastic one side plastics frames	1.79	10.00	1.97	0.20	-	1.85	-	3.81	m	4.29
L20: DOORS/SHUTTERS/HATCHES										
TIMBER DOORS - Doors in Scandinavian softwood, wrought										
Internal panelled doors; SA; Magnet Trade 762 x 1981 x 34mm (f sizes)	65.32	2.50	66.95	1.40	0.18	14.19	-	81.14	nr	91.28
Internal panelled doors; SA; glazed with bevelled glass; Magnet Trade 762 x 1981 x 34mm (f sizes)	182.98	2.50	187.55	4.85	0.61	49.05	-	236.60	nr	266.18
Internal panelled doors; Blenheim with timber; Magnet Trade										
686 x 1981 x 34mm (f sizes)	58.03	2.50	59.48	1.15	0.14	11.60	-	71.08	nr	79.96
762 x 1981 x 34mm (f sizes)	60.03	2.50	61.53	1.40	0.18	14.19	-	75.72	nr	85.18
Internal panelled doors; Blenheim with MDF panels; Magnet Trade										
686 x 1981 x 34mm (f sizes)	74.01	2.50	75.86	1.15	0.14	11.60	-	87.46	nr	98.39
762 x 1981 x 34mm (f sizes)	78.01	2.50	79.96	1.15	0.14	11.60	-	91.56	nr	103.00
Internal panelled doors; Victorian softwood panels; Magnet Trade										
686 x 1981 x 34mm (f sizes)	54.03	2.50	55.38	1.15	0.14	11.60	-	66.98	nr	75.35
762 x 1981 x 34mm (f sizes)	57.03	2.50	58.46	1.40	0.18	14.19	-	72.64	nr	81.72
TIMBER DOORS - Doors in Scandinavian softwood, wrought, preservative treated										
External ledged and braced doors; L & B Magnet Trade										
686 x 1981 x 44mm (f sizes)	76.31	2.50	78.22	1.25	0.16	12.66	-	90.88	nr	102.24
762 x 1981 x 44mm (f sizes)	83.00	2.50	85.08	1.50	0.16	14.97	-	100.04	nr	112.55
838 x 1981 x 44mm (f sizes)	83.00	2.50	85.08	1.75	0.22	17.70	-	102.77	nr	115.62
External framed, ledged and braced doors; YX; Magnet Trade										
686 x 1981 x 44mm (f sizes)	113.78	2.50	116.62	1.25	0.16	12.66	-	129.29	nr	145.45
762 x 1981 x 44mm (f sizes)	144.97	2.50	148.59	1.50	0.19	15.18	-	163.77	nr	184.24
813 x 1981 x 44mm (f sizes)	144.97	2.50	148.59	1.75	0.22	17.70	-	166.29	nr	187.08
838 x 1981 x 44mm (f sizes)	114.97	2.50	117.84	1.75	0.22	17.70	-	135.54	nr	152.48
External panelled doors; Cavendish; Magnet Trade										
762 x 1981 x 44mm (f sizes)	160.23	2.50	164.24	1.50	0.19	15.18	-	179.41	nr	201.84
838 x 1981 x 44mm (f sizes)	160.23	2.50	164.24	1.75	0.22	17.70	-	181.93	nr	204.67
External panelled doors; KXT; Magnet Trade										
762 x 1981 x 44mm (f sizes)	89.80	2.50	92.05	1.50	0.19	15.18	-	107.22	nr	120.63
External panelled doors; Kentucky; Magnet Trade										
838 x 1981 x 44mm (f sizes)	158.23	2.50	162.19	1.75	0.22	17.70	-	179.88	nr	202.37
External panelled doors; Pembroke; Magnet Trade										
838 x 1981 x 44mm (f sizes)	173.62	2.50	177.96	1.75	0.22	17.70	-	195.66	nr	220.11
External panelled doors; Stable; softwood; Magnet Trade										
762 x 1981 x 44mm (f sizes)	137.45	2.50	140.89	2.75	0.34	27.77	-	168.66	nr	189.74
838 x 1981 x 44mm (f sizes)	140.05	2.50	143.55	3.00	0.38	30.36	-	173.91	nr	195.65
External panelled doors; Stable upper door glazed with Geneva glazing panel, Magnet Trade										
762 x 1981 x 44mm (f sizes)	187.99	2.50	192.69	3.10	0.39	31.35	-	224.04	nr	252.05
838 x 1981 x 44mm (f sizes)	190.59	2.50	195.35	3.35	0.42	33.87	-	229.22	nr	257.88
External panelled doors; Stable upper door glazed with Salisbury glazing panel, Magnet Trade										
762 x 1981 x 44mm (f sizes)	197.68	2.50	202.62	3.10	0.39	31.35	-	233.97	nr	263.22
838 x 1981 x 44mm (f sizes)	200.28	2.50	205.29	3.35	0.42	33.87	Sunds	239.16	Unit	269.05
External panelled doors; 2XG; Magnet Trade										
762 x 1981 x 44mm (f sizes)	58.63	2.50	60.10	1.50	0.19	15.18	-	75.27	nr	84.68
813 x 2032 x 44mm (f sizes)	60.33	2.50	61.84	1.75	0.22	17.70	-	79.54	nr	89.48
838 x 1981 x 44mm (f sizes)	60.33	2.50	61.84	1.75	0.22	17.70	-	79.54	nr	89.48

	MATERIALS			LABOUR				RATES		
Labour hourly rates: (except Specialists) Craft Operatives £9.23 Labourer £7.02 Rates are national average prices. Refer to REGIONAL VARIATIONS for indicative levels of overall pricing in regions	Del to Site	Waste	Material Cost	Craft Optve	Lab	Labour Cost	Sunds	Nett Rate	Unit	Gross Rate (+12.5%)
	£	%	£	Hrs	Hrs	£	£	£		£

L20: DOORS/SHUTTERS/HATCHES Cont.

TIMBER DOORS - Doors in Scandinavian softwood, wrought, preservative treated Cont.

	Del to Site £	Waste %	Material Cost £	Craft Optve Hrs	Lab Hrs	Labour Cost £	Sunds £	Nett Rate £	Unit	Gross Rate £
External panelled doors; 2XGG; Magnet Trade										
762 x 1981 x 44mm (f sizes)	54.13	2.50	55.48	1.50	0.19	15.18	-	70.66	nr	79.49
838 x 1981 x 44mm (f sizes)	55.73	2.50	57.12	1.75	0.22	17.70	-	74.82	nr	84.17
External panelled doors; 2XG, glazed with 4mm clear tempered safety glass; Magnet Trade										
762 x 1981 x 44mm (f sizes)	77.00	2.50	78.92	1.95	0.24	19.68	-	98.61	nr	110.93
813 x 2032 x 44mm (f sizes)	83.79	2.50	85.88	2.20	0.28	22.27	-	108.16	nr	121.68
838 x 1981 x 44mm (f sizes)	82.70	2.50	84.77	2.20	0.28	22.27	-	107.04	nr	120.42
External panelled doors; 2XG, glazed with 4mm obscure tempered safety glass; Magnet Trade										
762 x 1981 x 44mm (f sizes)	81.00	2.50	83.03	1.95	0.24	19.68	-	102.71	nr	115.55
813 x 2032 x 44mm (f sizes)	85.00	2.50	87.13	2.20	0.28	22.27	-	109.40	nr	123.07
838 x 1981 x 44mm (f sizes)	85.00	2.50	87.13	2.20	0.28	22.27	-	109.40	nr	123.07
External panelled doors; 2XGG, glazed with 4mm obscure tempered safety glass; Magnet Trade										
762 x 1981 x 44mm (f sizes)	90.28	2.50	92.54	2.30	0.29	23.26	-	115.80	nr	130.28
838 x 1981 x 44mm (f sizes)	95.27	2.50	97.65	2.55	0.32	25.78	-	123.43	nr	138.86
Garage doors; MFL, pair, side hung; Magnet Trade										
1981 x 2134 x 44mm (f sizes) overall	306.87	2.50	314.54	4.00	0.50	40.43	-	354.97	nr	399.34
2134 x 2134 x 44mm (f sizes) overall	314.86	2.50	322.73	4.20	0.53	42.49	-	365.22	nr	410.87
Garage doors; 301, pair, side hung; Magnet Trade										
1981 x 2134 x 44mm (f sizes) overall	347.83	2.50	356.53	4.00	0.50	40.43	-	396.96	nr	446.58
2134 x 2134 x 44mm (f sizes) overall	257.02	2.50	263.45	4.20	0.53	42.49	-	305.93	nr	344.17

TIMBER DOORS - Doors in Hemlock, wrought

	Del to Site £	Waste %	Material Cost £	Craft Optve Hrs	Lab Hrs	Labour Cost £	Sunds £	Nett Rate £	Unit	Gross Rate £
Internal panelled doors; 2G; Magnet Trade										
762 x 1981 x 34mm (f sizes)	79.41	2.50	81.40	1.40	0.18	14.19	-	95.58	nr	107.53
Internal panelled doors; 2GG; Magnet Trade										
762 x 1981 x 34mm (f sizes)	75.71	2.50	77.60	1.40	0.18	14.19	-	91.79	nr	103.26
Internal panelled doors; SA; Magnet Trade										
686 x 1981 x 34mm (f sizes)	102.79	2.50	105.36	1.15	0.14	11.60	-	116.96	nr	131.58
762 x 1981 x 34mm (f sizes)	103.79	2.50	106.38	1.40	0.18	14.19	-	120.57	nr	135.64
813 x 2032 x 34mm (f sizes)	112.28	2.50	115.09	1.60	0.20	16.17	-	131.26	nr	147.67
838 x 1981 x 34mm (f sizes)	112.28	2.50	115.09	1.60	0.20	16.17	-	131.26	nr	147.67
Internal panelled doors; 10; Magnet Trade										
762 x 1981 x 34mm (f sizes)	60.53	2.50	62.04	1.40	0.18	14.19	-	76.23	nr	85.76
Internal panelled doors; 2G, glazed with 4mm clear tempered safety glass; Magnet Trade										
762 x 1981 x 34mm (f sizes)	101.23	2.50	103.76	1.85	0.23	18.69	-	122.45	nr	137.76
Internal panelled doors; 2GG, glazed with 4mm clear tempered safety glass; Magnet Trade										
762 x 1981 x 34mm (f sizes)	113.55	2.50	116.39	2.20	0.28	22.27	-	138.66	nr	155.99
Internal panelled doors; 10, glazed with 4mm clear tempered safety glass; Magnet Trade										
762 x 1981 x 34mm (f sizes)	99.53	2.50	102.02	1.95	0.24	19.68	-	121.70	nr	136.91
Internal panelled doors; 2G, glazed with 4mm obscure tempered safety glass; Magnet Trade										
762 x 1981 x 34mm (f sizes)	105.23	2.50	107.86	1.85	0.23	18.69	-	126.55	nr	142.37
Internal panelled doors; 2GG, glazed with 4mm obscure tempered safety glass; Magnet Trade										
762 x 1981 x 34mm (f sizes)	115.30	2.50	118.18	2.20	0.28	22.27	-	140.45	nr	158.01
Internal panelled doors; 10, glazed with 4mm obscure tempered safety glass; Magnet Trade										
762 x 1981 x 34mm (f sizes)	101.83	2.50	104.38	1.95	0.24	19.68	-	124.06	nr	139.57
Internal panelled doors; SA, glazed with bevelled glass; Magnet Trade										
762 x 1981 x 34mm (f sizes)	221.45	2.50	226.99	4.85	0.61	49.05	-	276.03	nr	310.54
813 x 2032 x 34mm (f sizes)	237.43	2.50	243.37	5.05	0.63	51.03	-	294.40	nr	331.20
838 x 1981 x 34mm (f sizes)	236.33	2.50	242.24	5.05	0.63	51.03	-	293.27	nr	329.93
External panelled doors; 2XG; Magnet Trade										
762 x 1981 x 44mm (f sizes)	87.50	2.50	89.69	1.50	0.19	15.18	-	104.87	nr	117.97
813 x 2032 x 44mm (f sizes)	91.60	2.50	93.89	1.75	0.22	17.70	-	111.59	nr	125.54
838 x 1981 x 44mm (f sizes)	91.60	2.50	93.89	1.75	0.22	17.70	-	111.59	nr	125.54
External panelled doors; 2XGG; Magnet Trade										
762 x 1981 x 44mm (f sizes)	83.31	2.50	85.39	1.50	0.19	15.18	-	100.57	nr	113.14
813 x 2032 x 44mm (f sizes)	87.40	2.50	89.58	1.75	0.22	17.70	-	107.28	nr	120.69
838 x 1981 x 44mm (f sizes)	87.40	2.50	89.58	1.75	0.22	17.70	-	107.28	nr	120.69
External panelled doors; KXT; Magnet Trade										
762 x 1981 x 44mm (f sizes)	147.04	2.50	150.72	1.50	0.19	15.18	-	165.89	nr	186.63
838 x 1981 x 44mm (f sizes)	151.04	2.50	154.82	1.75	0.22	17.70	-	172.51	nr	194.08

	MATERIALS			LABOUR				RATES		
Labour hourly rates: (except Specialists) Craft Operatives £9.23 Labourer £7.02 Rates are national average prices. Refer to REGIONAL VARIATIONS for indicative levels of overall pricing in regions	Del to Site	Waste	Material Cost	Craft Optve	Lab	Labour Cost	Sunds	Nett Rate	Unit	Gross Rate (+12.5%)
	£	%	£	Hrs	Hrs	£	£	£		£
L20: DOORS/SHUTTERS/HATCHES Cont.										
TIMBER DOORS - Doors in Hemlock, wrought Cont.										
External panelled doors; SA; Magnet Trade										
762 x 1981 x 44mm (f sizes)	111.98	2.50	114.78	1.50	0.19	15.18	-	129.96	nr	146.20
838 x 1981 x 44mm (f sizes)	115.87	2.50	118.77	1.75	0.22	17.70	-	136.46	nr	153.52
External panelled doors; 10; Magnet Trade										
762 x 1981 x 44mm (f sizes)	82.21	2.50	84.27	1.50	0.19	15.18	-	99.44	nr	111.87
External panelled doors; 2XG, glazed with 4mm clear tempered safety glass; Magnet Trade										
762 x 1981 x 44mm (f sizes)	105.87	2.50	108.52	1.95	0.24	19.68	-	128.20	nr	144.23
813 x 2032 x 44mm (f sizes)	115.06	2.50	117.94	2.20	0.28	22.27	-	140.21	nr	157.73
838 x 1981 x 44mm (f sizes)	116.26	2.50	119.17	2.20	0.28	22.27	-	141.44	nr	159.12
External panelled doors; 2XGG, glazed with 4mm clear tempered safety glass; Magnet Trade										
762 x 1981 x 44mm (f sizes)	114.25	2.50	117.11	2.30	0.29	23.26	-	140.37	nr	157.92
813 x 2032 x 44mm (f sizes)	124.64	2.50	127.76	2.55	0.32	25.78	-	153.54	nr	172.73
838 x 1981 x 44mm (f sizes)	123.54	2.50	126.63	2.55	0.32	25.78	-	152.41	nr	171.46
External panelled doors; 10, glazed with 4mm clear tempered safety glass; Magnet Trade										
762 x 1981 x 44mm (f sizes)	115.46	2.50	118.35	2.10	0.26	21.21	-	139.55	nr	157.00
External panelled doors; 2XG, glazed with 4mm obscure tempered safety glass; Magnet Trade										
762 x 1981 x 44mm (f sizes)	109.87	2.50	112.62	1.95	0.24	19.68	-	132.30	nr	148.84
813 x 2032 x 44mm (f sizes)	116.26	2.50	119.17	2.20	0.28	22.27	-	141.44	nr	159.12
838 x 1981 x 44mm (f sizes)	116.26	2.50	119.17	2.20	0.28	22.27	-	141.44	nr	159.12
External panelled doors; 2XGG, glazed with 4mm obscure tempered safety glass; Magnet Trade										
762 x 1981 x 44mm (f sizes)	119.45	2.50	122.44	2.30	0.29	23.26	-	145.70	nr	163.91
813 x 2032 x 44mm (f sizes)	126.94	2.50	130.11	2.55	0.32	25.78	-	155.90	nr	175.38
838 x 1981 x 44mm (f sizes)	126.94	2.50	130.11	2.55	0.32	25.78	-	155.90	nr	175.38
External panelled doors; 10, glazed with 4mm obscure tempered safety glass; Magnet Trade										
762 x 1981 x 44mm (f sizes)	117.76	2.50	120.70	2.10	0.26	21.21	-	141.91	nr	159.65
External panelled doors; SA, glazed with bevelled glass; Magnet Trade										
762 x 1981 x 44mm (f sizes)	233.03	2.50	238.86	4.95	0.62	50.04	-	288.90	nr	325.01
838 x 1981 x 44mm (f sizes)	243.32	2.50	249.40	5.20	0.65	52.56	-	301.96	nr	339.71
TIMBER DOORS - Doors in Hardwood, wrought										
External panelled doors; Alicante; Magnet Trade										
813 x 2032 x 44mm (f sizes)	229.06	2.50	234.79	2.65	0.33	26.78	-	261.56	nr	294.26
838 x 1981 x 44mm (f sizes)	229.06	2.50	234.79	2.65	0.33	26.78	-	261.56	nr	294.26
External panelled doors; Carolina, glazed in clear glass; Magnet Trade										
813 x 2032 x 44mm (f sizes)	249.93	2.50	256.18	2.65	0.33	26.78	-	282.95	nr	318.32
838 x 1981 x 44mm (f sizes)	249.93	2.50	256.18	2.65	0.33	26.78	-	282.95	nr	318.32
External panelled doors; Alicante, glazed with bevelled glass; Magnet Trade										
813 x 2032 x 44mm (f sizes)	286.99	2.50	294.16	4.70	0.59	47.52	-	341.69	nr	384.40
838 x 1981 x 44mm (f sizes)	286.99	2.50	294.16	4.70	0.59	47.52	-	341.69	nr	384.40
TIMBER DOORS - Doors in Hardwood (solid, laminated or veneered), wrought										
External panelled doors, fire resisting; FD20 2XGG; Magnet Trade										
838 x 1981 x 44mm (f sizes)	176.32	2.50	180.73	2.55	0.32	25.78	-	206.51	nr	232.32
External panelled doors; Airedale; Magnet Trade										
838 x 1981 x 44mm (f sizes)	415.07	2.50	425.45	2.65	0.33	26.78	-	452.22	nr	508.75
External panelled doors; Belvoir; Magnet Trade										
762 x 1981 x 44mm (f sizes)	162.53	2.50	166.59	2.40	0.30	24.26	-	190.85	nr	214.71
838 x 1981 x 44mm (f sizes)	162.53	2.50	166.59	2.65	0.33	26.78	-	193.37	nr	217.54
External panelled doors; Cadiz; Magnet Trade										
838 x 1981 x 44mm (f sizes)	322.77	2.50	330.84	2.65	0.33	26.78	-	357.62	nr	402.32
External panelled doors; Conway; Magnet Trade										
762 x 1981 x 44mm (f sizes)	221.07	2.50	226.60	2.40	0.30	24.26	-	250.85	nr	282.21
813 x 2032 x 44mm (f sizes)	221.07	2.50	226.60	2.65	0.33	26.78	-	253.37	nr	285.04
838 x 1981 x 44mm (f sizes)	221.07	2.50	226.60	2.65	0.33	26.78	-	253.37	nr	285.04
External panelled doors; Elizabethan; Magnet Trade										
813 x 2032 x 44mm (f sizes)	247.24	2.50	253.42	2.65	0.33	26.78	-	280.20	nr	315.22
838 x 1981 x 44mm (f sizes)	247.24	2.50	253.42	2.65	0.33	26.78	-	280.20	nr	315.22
External panelled doors; Manilla; Magnet Trade										
762 x 1981 x 44mm (f sizes)	239.65	2.50	245.64	2.40	0.30	24.26	£	269.90	nr	303.64
813 x 2032 x 44mm (f sizes)	239.65	2.50	245.64	2.65	0.33	26.78	-	272.42	nr	306.47
838 x 1981 x 44mm (f sizes)	239.65	2.50	245.64	2.65	0.33	26.78	-	272.42	nr	306.47

Labour hourly rates: (except Specialists) Craft Operatives £9.23 Labourer £7.02 Rates are national average prices. Refer to REGIONAL VARIATIONS for indicative levels of overall pricing in regions	MATERIALS			LABOUR				RATES		
	Del to Site £	Waste %	Material Cost £	Craft Optve Hrs	Lab Hrs	Labour Cost £	Sunds £	Nett Rate £	Unit	Gross Rate (+12.5%) £
L20: DOORS/SHUTTERS/HATCHES Cont.										
TIMBER DOORS - Doors in Hardwood (solid, laminated or veneered), wrought Cont.										
External panelled doors; Richmond; Magnet Trade										
838 x 1981 x 44mm (f sizes)	314.97	2.50	322.84	2.65	0.33	26.78	-	349.62	nr	393.32
External panelled doors; Rutland; Magnet Trade										
762 x 1981 x 44mm (f sizes)	160.84	2.50	164.86	2.40	0.30	24.26	-	189.12	nr	212.76
838 x 1981 x 44mm (f sizes)	160.84	2.50	164.86	2.65	0.33	26.78	-	191.64	nr	215.59
External panelled doors; Stable door; Magnet Trade										
762 x 1981 x 44mm (f sizes)	330.26	2.50	338.52	4.15	0.52	41.95	-	380.47	nr	428.03
838 x 1981 x 44mm (f sizes)	356.13	2.50	365.03	4.45	0.56	45.00	-	410.04	nr	461.29
External panelled doors; Stourbridge; Magnet Trade										
838 x 1981 x 44mm (f sizes)	319.37	2.50	327.35	2.65	0.33	26.78	-	354.13	nr	398.40
External panelled doors; Stuart; Magnet Trade										
762 x 1981 x 44mm (f sizes)	189.70	2.50	194.44	2.40	0.30	24.26	-	218.70	nr	246.04
838 x 1981 x 44mm (f sizes)	197.39	2.50	202.32	2.65	0.33	26.78	-	229.10	nr	257.74
External panelled doors; Belvoir, glazed with clear glass; Magnet Trade										
762 x 1981 x 44mm (f sizes)	180.90	2.50	185.42	2.90	0.36	29.29	-	214.72	nr	241.56
838 x 1981 x 44mm (f sizes)	184.89	2.50	189.51	3.10	0.39	31.35	-	220.86	nr	248.47
External panelled doors; Belvoir, glazed with obscure glass; Magnet Trade										
762 x 1981 x 44mm (f sizes)	184.89	2.50	189.51	2.90	0.36	29.29	-	218.81	nr	246.16
838 x 1981 x 44mm (f sizes)	187.19	2.50	191.87	3.10	0.39	31.35	-	223.22	nr	251.12
External panelled doors; Rutland, glazed with clear glass; Magnet Trade										
762 x 1981 x 44mm (f sizes)	191.79	2.50	196.58	3.20	0.40	32.34	-	228.93	nr	257.54
838 x 1981 x 44mm (f sizes)	196.98	2.50	201.90	3.45	0.43	34.86	-	236.77	nr	266.36
External panelled doors; Rutland, glazed with obscure glass; Magnet Trade										
762 x 1981 x 44mm (f sizes)	196.98	2.50	201.90	3.20	0.40	32.34	-	234.25	nr	263.53
838 x 1981 x 44mm (f sizes)	200.38	2.50	205.39	3.45	0.43	34.86	-	240.25	nr	270.28
External panelled doors; Stourbridge, glazed with clear glass; Magnet Trade										
838 x 1981 x 44mm (f sizes)	339.35	2.50	347.83	6.10	0.76	61.64	-	409.47	nr	460.66
External panelled doors; Stourbridge, glazed with obscure glass; Magnet Trade										
838 x 1981 x 44mm (f sizes)	349.34	2.50	358.07	6.10	0.76	61.64	-	419.71	nr	472.18
External panelled doors; Airedale, glazed with bevelled glass; Magnet Trade										
838 x 1981 x 44mm (f sizes)	470.02	2.50	481.77	4.70	0.59	47.52	-	529.29	nr	595.45
External panelled doors; Stuart, glazed with bevelled glass; Magnet Trade										
762 x 1981 x 44mm (f sizes)	278.10	2.50	285.05	5.85	0.73	59.12	-	344.17	nr	387.19
838 x 1981 x 44mm (f sizes)	285.79	2.50	292.93	6.10	0.76	61.64	-	354.57	nr	398.89
External panelled doors; Stable door, upper door glazed with bevelled glass; Magnet Trade										
762 x 1981 x 44mm (f sizes)	294.08	2.50	301.43	5.75	0.72	58.13	-	359.56	nr	404.50
838 x 1981 x 44mm (f sizes)	294.08	2.50	301.43	6.05	0.76	61.18	-	362.61	nr	407.93
External panelled doors; Cadiz, glazed with Geneva glazing panel; Magnet Trade										
838 x 1981 x 44mm (f sizes)	406.67	2.50	416.84	3.00	0.38	30.36	-	447.19	nr	503.09
External panelled doors; Cadiz, glazed with Salisbury glazing panel; Magnet Trade										
838 x 1981 x 44mm (f sizes)	416.36	2.50	426.77	3.00	0.38	30.36	-	457.13	nr	514.27
External panelled doors; Richmond, glazed with Malton glazing panel; Magnet Trade										
838 x 1981 x 44mm (f sizes)	467.71	2.50	479.40	3.35	0.42	33.87	-	513.27	nr	577.43
External panelled doors; Belvoir, glazed with clear glass double glazing unit; Magnet Trade										
762 x 1981 x 44mm (f sizes)	227.95	2.50	233.65	2.90	0.36	29.29	-	262.94	nr	295.81
838 x 1981 x 44mm (f sizes)	233.35	2.50	239.18	3.10	0.39	31.35	-	270.53	nr	304.35
External panelled doors; Belvoir, glazed with leaded glass double glazing unit; Magnet Trade										
762 x 1981 x 44mm (f sizes)	303.97	2.50	311.57	2.90	0.36	29.29	-	340.86	nr	383.47
838 x 1981 x 44mm (f sizes)	303.97	2.50	311.57	3.10	0.39	31.35	-	342.92	nr	385.79
External panelled doors; Belvoir, glazed with obscure glass double glazing unit; Magnet Trade										
762 x 1981 x 44mm (f sizes)	229.15	2.50	234.88	2.90	0.36	29.29	-	264.17	nr	297.19
838 x 1981 x 44mm (f sizes)	236.04	2.50	241.94	3.10	0.39	31.35	-	273.29	nr	307.45
External panelled doors; Rutland, glazed with clear glass double glazing unit; Magnet Trade										
762 x 1981 x 44mm (f sizes)	260.72	2.50	267.24	3.20	0.40	32.34	-	299.58	nr	337.03
838 x 1981 x 44mm (f sizes)	268.42	2.50	275.13	3.45	0.45	35.00	-	310.13	nr	348.90

Labour hourly rates: (except Specialists) Craft Operatives £9.23 Labourer £7.02 Rates are national average prices. Refer to REGIONAL VARIATIONS for indicative levels of overall pricing in regions	MATERIALS			LABOUR				RATES		
	Del to Site £	Waste %	Material Cost £	Craft Optve Hrs	Lab Hrs	Labour Cost £	Sunds £	Nett Rate £	Unit	Gross Rate (+12.5%) £
L20: DOORS/SHUTTERS/HATCHES Cont.										
TIMBER DOORS - Doors in Hardwood (solid, laminated or veneered), wrought Cont.										
External panelled doors; Rutland, glazed with leaded glass double glazing unit; Magnet Trade										
762 x 1981 x 44mm (f sizes)	360.59	2.50	369.60	3.20	0.40	32.34	-	401.95	nr	452.19
838 x 1981 x 44mm (f sizes)	378.40	2.50	387.86	3.45	0.45	35.00	-	422.86	nr	475.72
External panelled doors; Rutland, glazed with obscure glass double glazing unit; Magnet Trade										
762 x 1981 x 44mm (f sizes)	273.40	2.50	280.24	3.20	0.40	32.34	-	312.58	nr	351.65
838 x 1981 x 44mm (f sizes)	273.40	2.50	280.24	3.45	0.45	35.00	-	315.24	nr	354.64
Garage doors; Chevron; Magnet Trade										
2134 x 2134 x 44mm (f sizes)	856.92	2.50	878.34	5.45	0.68	55.08	-	933.42	nr	1050.10
TIMBER DOORS - Doors in softwood and fibreboard core, Mahogany or Sapele facings										
Internal panelled doors; Granada; Magnet Trade										
686 x 1981 x 34mm (f sizes)	288.30	2.50	295.51	2.20	0.28	22.27	-	317.78	nr	357.50
762 x 1981 x 34mm (f sizes)	293.30	2.50	300.63	2.40	0.30	24.26	-	324.89	nr	365.50
Internal panelled doors; Malaga; Magnet Trade										
762 x 1981 x 34mm (f sizes)	243.55	2.50	249.64	2.40	0.30	24.26	-	273.90	nr	308.13
Internal panelled doors; Palma; Magnet Trade										
762 x 1981 x 34mm (f sizes)	221.78	2.50	227.32	2.40	0.30	24.26	-	251.58	nr	283.03
TIMBER DOORS - Doors in Hardwood (solid, laminated or veneered), wrought										
Internal panelled doors; SA; Magnet Trade										
686 x 1981 x 34mm (f sizes)	171.62	2.50	175.91	2.20	0.28	22.27	-	198.18	nr	222.95
762 x 1981 x 34mm (f sizes)	172.32	2.50	176.63	2.40	0.30	24.26	-	200.89	nr	226.00
Internal panelled doors; Victorian; Magnet Trade										
686 x 1981 x 34mm (f sizes)	203.29	2.50	208.37	2.20	0.28	22.27	-	230.64	nr	259.47
762 x 1981 x 34mm (f sizes)	206.78	2.50	211.95	2.40	0.30	24.26	-	236.21	nr	265.73
Internal panelled doors; Windsor; Magnet Trade										
686 x 1981 x 34mm (f sizes)	203.29	2.50	208.37	2.20	0.28	22.27	-	230.64	nr	259.47
762 x 1981 x 34mm (f sizes)	206.78	2.50	211.95	2.40	0.30	24.26	-	236.21	nr	265.73
Internal panelled doors; 10; Magnet Trade										
762 x 1981 x 34mm (f sizes)	80.16	2.50	82.16	2.40	0.30	24.26	-	106.42	nr	119.72
Internal panelled doors; 10, glazed with 4mm clear tempered safety glass; Magnet Trade										
762 x 1981 x 34mm (f sizes)	117.86	2.50	120.81	3.00	0.38	30.36	-	151.16	nr	170.06
Internal panelled doors; 10, glazed with 4mm obscure tempered safety glass; Magnet Trade										
762 x 1981 x 34mm (f sizes)	118.16	2.50	121.11	3.00	0.38	30.36	-	151.47	nr	170.41
Internal panelled doors; SA, glazed with bevelled glass; Magnet Trade										
762 x 1981 x 34mm (f sizes)	264.71	2.50	271.33	5.85	0.73	59.12	-	330.45	nr	371.75
TIMBER DOORS - Doors in solid or laminated construction, Oak facings										
Internal panelled doors; Louis; Magnet Trade										
762 x 1981 x 34mm (f sizes)	250.04	2.50	256.29	2.40	0.30	24.26	-	280.55	nr	315.62
TIMBER DOORS - Doors in hardboard, embossed										
Internal panelled doors; Colinist; Magnet Trade										
610 x 1981 x 34mm (f sizes)	60.93	2.50	62.45	1.85	0.23	18.69	-	81.14	nr	91.29
686 x 1981 x 34mm (f sizes)	61.23	2.50	62.76	1.95	0.24	19.68	-	82.44	nr	92.75
762 x 1981 x 34mm (f sizes)	61.43	2.50	62.97	2.10	0.26	21.21	-	84.17	nr	94.70
Internal panelled doors; Sentinel; Magnet Trade										
686 x 1981 x 34mm (f sizes)	64.62	2.50	66.24	1.95	0.24	19.68	-	85.92	nr	96.66
762 x 1981 x 34mm (f sizes)	64.92	2.50	66.54	2.10	0.26	21.21	-	87.75	nr	98.72
TIMBER DOORS - Flush doors										
Internal; hardboard facings; Magnet Trade										
305 x 1981 x 34mm	23.96	2.50	24.56	1.15	0.14	11.60	-	36.16	nr	40.68
381 x 1981 x 34mm	24.86	2.50	25.48	1.25	0.16	12.66	-	38.14	nr	42.91
457 x 1981 x 34mm	24.86	2.50	25.48	1.40	0.18	14.19	-	39.67	nr	44.63
533 x 1981 x 34mm	24.86	2.50	25.48	1.50	0.19	15.18	-	40.66	nr	45.74
610 x 1981 x 34mm	24.86	2.50	25.48	1.60	0.20	16.17	-	41.65	nr	46.86
686 x 1981 x 34mm	25.36	2.50	25.99	1.75	0.22	17.70	-	43.69	nr	49.15
711 x 1981 x 34mm	25.76	2.50	26.40	1.85	0.23	18.69	-	45.09	nr	50.73
762 x 1981 x 34mm	25.76	2.50	26.40	1.95	0.24	19.68	-	46.09	nr	51.85
813 x 2032 x 34mm	27.06	2.50	27.74	2.20	0.28	22.27	-	50.01	Unt	56.26
838 x 1981 x 34mm	27.06	2.50	27.74	2.20	0.28	22.27	-	50.01	nr	56.26
Internal; plywood facings; Magnet Trade										
381 x 1981 x 34mm	33.66	2.50	34.50	1.25	0.16	12.66	-	47.16	nr	53.06
457 x 1981 x 34mm	33.66	2.50	34.50	1.40	0.18	14.19	-	48.69	nr	54.77
533 x 1981 x 34mm	33.66	2.50	34.50	1.50	0.19	15.18	-	49.68	nr	55.89

Labour hourly rates: (except Specialists) Craft Operatives £9.23 Labourer £7.02 Rates are national average prices. Refer to REGIONAL VARIATIONS for indicative levels of overall pricing in regions	MATERIALS			LABOUR				RATES		
	Del to Site £	Waste %	Material Cost £	Craft Optve Hrs	Lab Hrs	Labour Cost £	Sunds £	Nett Rate £	Unit	Gross Rate (+12.5%) £
L20: DOORS/SHUTTERS/HATCHES Cont.										
TIMBER DOORS - Flush doors Cont.										
Internal; plywood facings; Magnet Trade Cont.										
610 x 1981 x 34mm	33.66	2.50	34.50	1.60	0.20	16.17	-	50.67	nr	57.01
686 x 1981 x 34mm	34.05	2.50	34.90	1.75	0.22	17.70	-	52.60	nr	59.17
711 x 1981 x 34mm	34.55	2.50	35.41	1.85	0.23	18.69	-	54.10	nr	60.87
762 x 1981 x 34mm	34.55	2.50	35.41	1.95	0.24	19.68	-	55.10	nr	61.98
813 x 2032 x 34mm	36.15	2.50	37.05	2.20	0.28	22.27	-	59.33	nr	66.74
838 x 1981 x 34mm	36.15	2.50	37.05	2.20	0.28	22.27	-	59.33	nr	66.74
Internal; Sapele Showpiece; Magnet Trade										
381 x 1981 x 34mm	43.15	2.50	44.23	1.50	0.19	15.18	-	59.41	nr	66.83
457 x 1981 x 34mm	43.15	2.50	44.23	1.60	0.20	16.17	-	60.40	nr	67.95
533 x 1981 x 34mm	43.15	2.50	44.23	1.75	0.22	17.70	-	61.93	nr	69.67
610 x 1981 x 34mm	43.15	2.50	44.23	1.85	0.23	18.69	-	62.92	nr	70.78
686 x 1981 x 34mm	43.15	2.50	44.23	1.95	0.24	19.68	-	63.91	nr	71.90
762 x 1981 x 34mm	43.15	2.50	44.23	2.40	0.30	24.26	-	68.49	nr	77.05
838 x 1981 x 34mm	44.94	2.50	46.06	2.40	0.30	24.26	-	70.32	nr	79.11
External; MF1X, plywood facings; Magnet Trade										
762 x 1981 x 44mm	59.83	2.50	61.33	2.10	0.26	21.21	-	82.53	nr	92.85
838 x 1981 x 44mm	61.93	2.50	63.48	2.30	0.29	23.26	-	86.74	nr	97.59
External; MF2X, plywood facings; Magnet Trade										
762 x 1981 x 44mm; glazing aperture 457 x 457mm ..	75.21	2.50	77.09	2.10	0.26	21.21	-	98.30	nr	110.59
838 x 1981 x 44mm; glazing aperture 457 x 457mm ..	77.21	2.50	79.14	2.30	0.29	23.26	-	102.41	nr	115.21
External; MF4X, plywood facings; Magnet Trade										
762 x 1981 x 44mm; glazing aperture 559 x 864mm ..	75.21	2.50	77.09	2.10	0.26	21.21	-	98.30	nr	110.59
838 x 1981 x 44mm; glazing aperture 635 x 864mm ..	77.21	2.50	79.14	2.30	0.29	23.26	-	102.41	nr	115.21
Internal; fire resisting; Magnaseal; FD30; Magnet Trade										
610 x 1981 x 44mm	58.13	2.50	59.58	1.95	0.24	19.68	-	79.27	nr	89.17
686 x 1981 x 44mm	59.53	2.50	61.02	2.20	0.28	22.27	-	83.29	nr	93.70
762 x 1981 x 44mm	59.53	2.50	61.02	2.20	0.28	22.27	-	83.29	nr	93.70
813 x 2032 x 44mm	62.23	2.50	63.79	2.55	0.32	25.78	-	89.57	nr	100.76
838 x 1981 x 44mm	62.23	2.50	63.79	2.55	0.32	25.78	-	89.57	nr	100.76
Internal; fire resisting; plywood facings; FDG30; Magnet Trade										
762 x 1981 x 44mm; glazing aperture 508 x 508mm ..	74.91	2.50	76.78	2.20	0.28	22.27	-	99.05	nr	111.44
813 x 2032 x 44mm; glazing aperture 559 x 559mm ..	77.31	2.50	79.24	2.55	0.32	25.78	-	105.03	nr	118.15
838 x 1981 x 44mm; glazing aperture 584 x 584mm ..	77.31	2.50	79.24	2.55	0.32	25.78	-	105.03	nr	118.15
External; fire resisting; plywood facings; FD30; Magnet Trade										
762 x 1981 x 44mm	77.11	2.50	79.04	2.20	0.28	22.27	-	101.31	nr	113.97
838 x 1981 x 44mm	80.31	2.50	82.32	2.65	0.33	26.78	-	109.09	nr	122.73
Internal; fire resisting; Sapele facings; FD30; Magnet Trade										
762 x 1981 x 44mm	82.11	2.50	84.16	2.55	0.32	25.78	-	109.95	nr	123.69
838 x 1981 x 44mm	85.70	2.50	87.84	2.75	0.34	27.77	-	115.61	nr	130.06
TIMBER DOORS - Trap doors in softwood, wrought										
19mm matchboarding on 25 x 75mm ledges										
457 x 610mm	25.10	2.50	25.73	0.21	0.03	2.15	-	27.88	nr	31.36
610 x 610mm	31.04	2.50	31.82	0.22	0.03	2.24	-	34.06	nr	38.31
762 x 610mm	37.63	2.50	38.57	0.23	0.03	2.33	-	40.90	nr	46.02
TIMBER DOORS - Trap doors in B.C. Pine, wrought										
19mm matchboarding on 25 x 75mm ledges										
457 x 610mm	31.08	2.50	31.86	0.21	0.03	2.15	-	34.01	nr	38.26
610 x 610mm	43.94	2.50	45.04	0.22	0.03	2.24	-	47.28	nr	53.19
762 x 610mm	56.18	2.50	57.58	0.23	0.03	2.33	-	59.92	nr	67.41
TIMBER DOORS - Panelled doors										
Note the following prices for panelled doors are for doors to detail in moderate numbers: doors in large numbers to one pattern would cost considerably less										
TIMBER DOORS - Panelled doors in softwood, wrought										
38mm square framed (or chamfered or moulded one or both sides)										
two panel	70.35	2.50	72.11	1.00	0.13	10.14	-	82.25	m2	92.53
four panel	78.23	2.50	80.19	1.00	0.13	10.14	-	90.33	m2	101.62
six panel	86.41	2.50	88.57	1.00	0.13	10.14	-	98.71	m2	111.05
add if upper panels open moulded in small squares for glass	16.98	2.50	17.40	-	-	-	-	17.40	m2	19.58
50mm square framed (or chamfered or moulded one or both sides)										
two panel	78.23	2.50	80.19	1.15	0.14	11.60	-	91.78	m2	103.26
four panel	86.41	2.50	88.57	1.15	0.14	11.60	-	100.17	m2	112.69
six panel	94.60	2.50	96.97	1.15	0.14	11.60	-	108.56	m2	122.13
add if upper panels open moulded in small squares for glass	16.98	2.50	17.40	-	-	-	-	17.40	m2	19.58

WINDOWS/DOORS/STAIRS

Labour hourly rates: (except Specialists) Craft Operatives £9.23 Labourer £7.02 Rates are national average prices. Refer to REGIONAL VARIATIONS for indicative levels of overall pricing in regions	MATERIALS			LABOUR				RATES		
	Del to Site £	Waste %	Material Cost £	Craft Optve Hrs	Lab Hrs	Labour Cost £	Sunds £	Nett Rate £	Unit	Gross Rate (+12.5%) £
L20: DOORS/SHUTTERS/HATCHES Cont.										
TIMBER DOORS - Panelled doors in Afromosia, wrought										
38mm square framed (or chamfered or moulded one or both sides)										
two panel	208.01	2.50	213.21	1.75	0.22	17.70	-	230.91	m2	259.77
four panel	221.89	2.50	227.44	1.75	0.22	17.70	-	245.13	m2	275.78
six panel	235.76	2.50	241.65	1.75	0.22	17.70	-	259.35	m2	291.77
add if upper panels open moulded in small squares for glass	39.52	2.50	40.51	-	-	-	-	40.51	m2	45.57
50mm square framed (or chamfered or moulded one or both sides)										
two panel	228.83	2.50	234.55	2.00	0.25	20.22	-	254.77	m2	286.61
four panel	242.69	2.50	248.76	2.00	0.25	20.22	-	268.97	m2	302.59
six panel	256.56	2.50	262.97	2.00	0.25	20.22	-	283.19	m2	318.59
add if upper panels open moulded in small squares for glass	39.52	2.50	40.51	-	-	-	-	40.51	m2	45.57
TIMBER DOORS - Panelled doors in Sapele, wrought										
38mm square framed (or chamfered or moulded one or both sides)										
two panel	151.54	2.50	155.33	1.45	0.18	14.65	-	169.98	m2	191.22
four panel	163.37	2.50	167.45	1.45	0.18	14.65	-	182.10	m2	204.86
six panel	175.22	2.50	179.60	1.45	0.18	14.65	-	194.25	m2	218.53
add if upper panels open moulded in small squares for glass	28.41	2.50	29.12	-	-	-	-	29.12	m2	32.76
50mm square framed (or chamfered or moulded one or both sides)										
two panel	163.37	2.50	167.45	1.75	0.22	17.70	-	185.15	m2	208.30
four panel	175.22	2.50	179.60	1.75	0.22	17.70	-	197.30	m2	221.96
six panel	187.06	2.50	191.74	1.75	0.22	17.70	-	209.43	m2	235.61
add if upper panels open moulded in small squares for glass	28.41	2.50	29.12	-	-	-	-	29.12	m2	32.76
TIMBER DOORS - Garage doors in softwood, wrought										
Side hung; framed, tongued and grooved boarded										
2134 x 1981 x 44mm overall, pair	310.41	2.50	318.17	4.50	0.56	45.47	-	363.64	nr	409.09
2134 x 2134 x 44mm overall, pair	322.83	2.50	330.90	4.70	0.59	47.52	-	378.42	nr	425.73
Up and over; framed, tongued and grooved boarded; up and over door gear										
2134 x 1981 x 44mm	434.57	2.50	445.43	9.00	1.10	90.79	-	536.23	nr	603.25
TIMBER DOORS - Garage doors in Western Red Cedar, wrought										
Side hung; framed, tongued and grooved boarded										
2134 x 1981 x 44mm overall, pair	448.72	2.50	459.94	5.00	0.62	50.50	-	510.44	nr	574.25
2134 x 2134 x 44mm overall, pair	460.85	2.50	472.37	5.25	0.66	53.09	-	525.46	nr	591.14
Up and over; framed, tongued and grooved boarded; up and over door gear										
2134 x 1981 x 44mm	570.00	2.50	584.25	9.25	1.15	93.45	-	677.70	nr	762.41
TIMBER DOORS - Garage doors in Oak, wrought										
Side hung; framed, tongued and grooved boarded										
2134 x 1981 x 44mm overall, pair	1199.94	2.50	1229.94	7.80	0.98	78.87	-	1308.81	nr	1472.41
2134 x 2134 x 44mm overall, pair	1266.61	2.50	1298.28	8.10	1.01	81.85	-	1380.13	nr	1552.64
Up and over; framed, tongued and grooved boarded; up and over door gear										
2134 x 1981 x 44mm	1399.94	2.50	1434.94	13.00	1.62	131.36	-	1566.30	nr	1762.09
TIMBER DOORS - Patio doors in hard wood, wrought										
Double glazed clear toughened glass, Magnet Trade, one opening leaf, doors and frames treated with base coat stains, fixing frame to masonry with screws										
2073 x 1805mm overall, HP 6	1272.67	2.50	1304.49	12.00	1.50	121.29	-	1425.78	nr	1604.00
2073 X 2387mm overall, HP 8	1272.67	2.50	1304.49	12.00	1.50	121.29	-	1425.78	nr	1604.00
extra; ventilator head, for HP 6	1.00	2.50	1.02	-	-	-	-	1.02	nr	1.15
extra; ventilator head, for HP 8	1.00	2.50	1.02	-	-	-	-	1.02	nr	1.15
DOORSETS - Doorsets mainly in softwood, wrought										
Doorsets; 28mm thick jambs, head and transome with 12mm thick stop to suit 100mm thick wall; 14mm thick hardwood threshold; honeycomb core plywood faced flush door lipped two long edges; 6mm plywood transome panel fixed with pinned beads; 65mm snap in hinges; 57mm backset mortice latch; fixing frame to masonry with screws										
526 x 2040 x 40mm flush door -1 nr; basic dimensions										
600 x 2100mm	132.77	2.50	136.09	1.25	0.16	12.66	-	148.75	nr	167.34
600 x 2400mm	162.34	2.50	166.40	1.35	0.17	13.65	-	180.05	nr	202.56
626 x 2040 x 40mm flush door -1 nr; basic dimensions										
700 x 2100mm	132.77	2.50	136.09	1.25	0.16	12.66	-	148.75	nr	167.34
700 x 2400mm	162.34	2.50	166.40	1.35	0.17	13.65	-	180.05	nr	202.56

Labour hourly rates: (except Specialists) Craft Operatives £9.23 Labourer £7.02 Rates are national average prices. Refer to REGIONAL VARIATIONS for indicative levels of overall pricing in regions	MATERIALS			LABOUR				RATES		
	Del to Site £	Waste %	Material Cost £	Craft Optve Hrs	Lab Hrs	Labour Cost £	Sunds £	Nett Rate £	Unit	Gross Rate (+12.5%) £

L20: DOORS/SHUTTERS/HATCHES Cont.

DOORSETS - Doorsets mainly in softwood, wrought Cont.

Doorsets; 28mm thick jambs, head and transome with 12mm thick stop to suit 100mm thick wall; 14mm thick hardwood threshold; honeycomb core plywood faced flush door lipped two long edges; 6mm plywood transome panel fixed with pinned beads; 65mm snap in hinges; 57mm backset mortice latch; fixing frame to masonry with screws Cont.

	Del to Site £	Waste %	Material Cost £	Craft Optve Hrs	Lab Hrs	Labour Cost £	Sunds £	Nett Rate £	Unit	Gross Rate £
726 x 2040 x 40mm flush door -1 nr; basic dimensions										
800 x 2100mm....................	132.77	2.50	136.09	1.25	0.16	12.66	-	148.75	nr	167.34
800 x 2400mm....................	162.34	2.50	166.40	1.35	0.17	13.65	-	180.05	nr	202.56
826 x 2040 x 40mm flush door -1 nr; basic dimensions										
900 x 2100mm....................	138.52	2.50	141.98	1.25	0.16	12.66	-	154.64	nr	173.97
900 x 2400mm....................	165.80	2.50	169.94	1.35	0.17	13.65	-	183.60	nr	206.55

Note
actual frame size is 14mm narrower and 14mm shorter than basic size dimensions

ROLLER SHUTTERS - Roller shutters in wood

Note
roller shutters are always purpose made to order and the following prices are indicative only. Firm quotations should always be obtained

	Del to Site £	Waste %	Material Cost £	Craft Optve Hrs	Lab Hrs	Labour Cost £	Sunds £	Nett Rate £	Unit	Gross Rate £
Pole and hook operation										
2134 x 2134mm	875.00	-	875.00	12.00	12.00	195.00	25.00	1095.00	nr	1231.88

LININGS - Door frames and door lining sets in softwood, wrought

	Del to Site £	Waste %	Material Cost £	Craft Optve Hrs	Lab Hrs	Labour Cost £	Sunds £	Nett Rate £	Unit	Gross Rate £
Sets										
linings; fixing to masonry with screws										
32 x 113mm...................	5.32	2.50	5.45	0.45	0.06	4.57	0.09	10.12	m	11.38
32 x 150mm...................	6.39	2.50	6.55	0.45	0.06	4.57	0.09	11.21	m	12.62
32 x 225mm...................	9.56	2.50	9.80	0.70	0.08	7.02	0.13	16.95	m	19.07
32 x 330mm...................	15.02	2.50	15.40	0.70	0.08	7.02	0.13	22.55	m	25.37
38 x 113mm...................	6.17	2.50	6.32	0.45	0.06	4.57	0.09	10.99	m	12.36
38 x 150mm...................	7.47	2.50	7.66	0.45	0.06	4.57	0.09	12.32	m	13.86
38 x 225mm...................	10.99	2.50	11.26	0.70	0.08	7.02	0.13	18.42	m	20.72
38 x 330mm...................	17.75	2.50	18.19	0.70	0.08	7.02	0.13	25.35	m	28.51
linings										
32 x 113mm; labours -1.....	5.47	2.50	5.61	0.23	0.03	2.33	0.04	7.98	m	8.98
32 x 113mm; labours -2.....	5.54	2.50	5.68	0.23	0.03	2.33	0.04	8.05	m	9.06
32 x 113mm; labours -3.....	5.60	2.50	5.74	0.23	0.03	2.33	0.04	8.11	m	9.13
32 x 113mm; labours -4.....	5.68	2.50	5.82	0.23	0.03	2.33	0.04	8.20	m	9.22
38 x 113mm; labours -1.....	6.33	2.50	6.49	0.23	0.03	2.33	0.04	8.86	m	9.97
38 x 113mm; labours -2.....	6.39	2.50	6.55	0.23	0.03	2.33	0.04	8.92	m	10.04
38 x 113mm; labours -3.....	6.47	2.50	6.63	0.23	0.03	2.33	0.04	9.01	m	10.13
38 x 113mm; labours -4.....	6.53	2.50	6.69	0.23	0.03	2.33	0.04	9.07	m	10.20

Internal door lining sets; supplied unassembled; fixing to masonry with screws

	Del to Site £	Waste %	Material Cost £	Craft Optve Hrs	Lab Hrs	Labour Cost £	Sunds £	Nett Rate £	Unit	Gross Rate £
32 x 115mm rebated linings; assembling										
for 610 x 1981mm doors.....................	23.73	2.50	24.32	2.30	0.29	23.26	0.37	47.96	nr	53.95
for 686 x 1981mm doors.....................	23.73	2.50	24.32	2.30	0.29	23.26	0.37	47.96	nr	53.95
for 762 x 1981mm doors.....................	23.73	2.50	24.32	2.30	0.29	23.26	0.37	47.96	nr	53.95
32 x 140mm rebated linings; assembling										
for 610 x 1981mm doors.....................	25.88	2.50	26.53	2.30	0.29	23.26	0.37	50.16	nr	56.43
for 686 x 1981mm doors.....................	25.88	2.50	26.53	2.30	0.29	23.26	0.37	50.16	nr	56.43
for 762 x 1981mm doors.....................	25.88	2.50	26.53	2.30	0.29	23.26	0.37	50.16	nr	56.43

LININGS - Door frames and door lining sets in Afromosia, wrought, selected for transparent finish

	Del to Site £	Waste %	Material Cost £	Craft Optve Hrs	Lab Hrs	Labour Cost £	Sunds £	Nett Rate £	Unit	Gross Rate £
Sets										
linings; fixing to masonry with screws										
32 x 113mm...................	13.23	2.50	13.56	0.70	0.09	7.09	0.13	20.78	m	23.38
32 x 150mm...................	16.02	2.50	16.42	0.70	0.09	7.09	0.13	23.64	m	26.60
32 x 225mm...................	23.73	2.50	24.32	1.00	0.12	10.07	0.18	34.58	m	38.90
32 x 330mm...................	36.95	2.50	37.87	1.00	0.12	10.07	0.18	48.13	m	54.14
38 x 113mm...................	15.43	2.50	15.82	0.70	0.09	7.09	0.13	23.04	m	25.92
38 x 150mm...................	18.48	2.50	18.94	0.70	0.09	7.09	0.13	26.16	m	29.44
38 x 225mm...................	27.09	2.50	27.77	1.00	0.12	10.07	0.18	38.02	m	42.77
38 x 330mm...................	43.11	2.50	44.19	1.00	0.12	10.07	0.18	54.44	m	61.25
linings										
32 x 113mm; labours -1.....	13.42	2.50	13.76	0.35	0.04	3.51	0.07	17.34	m	19.50
32 x 113mm; labours -2.....	13.55	2.50	13.89	0.35	0.04	3.51	0.07	17.47	m	19.65
32 x 113mm; labours -3.....	13.67	2.50	14.01	0.35	0.04	3.51	0.07	17.59	m	19.79
32 x 113mm; labours -4.....	13.82	2.50	14.17	0.35	0.04	3.51	0.07	17.75	m	19.97
38 x 113mm; labours -1.....	15.62	2.50	16.01	0.35	0.04	3.51	0.07	19.59	m	22.04
38 x 113mm; labours -2.....	15.75	2.50	16.14	0.35	0.04	3.51	0.07	19.73	m	22.19
38 x 113mm; labours -3.....	15.88	2.50	16.28	0.35	0.04	3.51	0.07	19.86	m	22.34
38 x 113mm; labours -4.....	16.02	2.50	16.42	0.35	0.04	3.51	0.07	20.00	m	22.50

LININGS - Door frames and door lining sets in Sapele, wrought, selected for transparent finish

	Del to Site £	Waste %	Material Cost £	Craft Optve Hrs	Lab Hrs	Labour Cost £	Sunds £	Nett Rate £	Unit	Gross Rate £
Sets										
linings; fixing to masonry with screws										
32 x 113mm...................	10.11	2.50	10.36	0.65	0.08	6.56	0.12	17.04	m	19.17
32 x 150mm...................	12.13	2.50	12.43	0.65	0.08	6.56	0.12	19.11	m	21.50

Labour hourly rates: (except Specialists) Craft Operatives £9.23 Labourer £7.02 Rates are national average prices. Refer to REGIONAL VARIATIONS for indicative levels of overall pricing in regions	MATERIALS			LABOUR				RATES		
	Del to Site	Waste	Material Cost	Craft Optve	Lab	Labour Cost	Sunds	Nett Rate	Unit	Gross Rate (+12.5%)
	£	%	£	Hrs	Hrs	£	£	£		£

L20: DOORS/SHUTTERS/HATCHES Cont.

LININGS - Door frames and door lining sets in Sapele, wrought, selected for transparent finish Cont.

Sets Cont.

linings; fixing to masonry with screws Cont.

	Del to Site	Waste	Material Cost	Craft Optve	Lab	Labour Cost	Sunds	Nett Rate	Unit	Gross Rate
32 x 225mm	18.19	2.50	18.64	0.92	0.11	9.26	0.17	28.08	m	31.59
32 x 330mm	28.30	2.50	29.01	0.92	0.11	9.26	0.17	38.44	m	43.25
38 x 113mm	11.84	2.50	12.14	0.65	0.08	6.56	0.12	18.82	m	21.17
38 x 150mm	14.15	2.50	14.50	0.65	0.08	6.56	0.12	21.18	m	23.83
38 x 225mm	20.79	2.50	21.31	0.92	0.11	9.26	0.17	30.74	m	34.59
38 x 330mm	33.50	2.50	34.34	0.92	0.11	9.26	0.17	43.77	m	49.24

linings

	Del to Site	Waste	Material Cost	Craft Optve	Lab	Labour Cost	Sunds	Nett Rate	Unit	Gross Rate
32 x 113mm; labours -1	10.29	2.50	10.55	0.31	0.04	3.14	0.07	13.76	m	15.48
32 x 113mm; labours -2	10.40	2.50	10.66	0.31	0.04	3.14	0.07	13.87	m	15.61
32 x 113mm; labours -3	10.52	2.50	10.78	0.31	0.04	3.14	0.07	14.00	m	15.74
32 x 113mm; labours -4	10.63	2.50	10.90	0.31	0.04	3.14	0.07	14.11	m	15.87
38 x 113mm; labours -1	12.01	2.50	12.31	0.31	0.04	3.14	0.07	15.52	m	17.46
38 x 113mm; labours -2	12.13	2.50	12.43	0.31	0.04	3.14	0.07	15.65	m	17.60
38 x 113mm; labours -3	12.24	2.50	12.55	0.31	0.04	3.14	0.07	15.76	m	17.73
38 x 113mm; labours -4	12.36	2.50	12.67	0.31	0.04	3.14	0.07	15.88	m	17.87

FRAMES - Door frames and door lining sets in softwood, wrought - frames

Sets

	Del to Site	Waste	Material Cost	Craft Optve	Lab	Labour Cost	Sunds	Nett Rate	Unit	Gross Rate
38 x 75mm jambs	4.98	2.50	5.10	0.21	0.03	2.15	0.04	7.29	m	8.21
38 x 100mm jambs	5.76	2.50	5.90	0.21	0.03	2.15	0.04	8.09	m	9.10
50 x 75mm jambs	5.76	2.50	5.90	0.23	0.03	2.33	0.04	8.28	m	9.31
50 x 100mm jambs	6.60	2.50	6.76	0.23	0.03	2.33	0.04	9.14	m	10.28
50 x 125mm jambs	7.47	2.50	7.66	0.23	0.03	2.33	0.04	10.03	m	11.28
63 x 100mm jambs	7.47	2.50	7.66	0.23	0.03	2.33	0.04	10.03	m	11.28
75 x 100mm jambs	9.56	2.50	9.80	0.25	0.03	2.52	0.04	12.36	m	13.90
75 x 113mm jambs	10.92	2.50	11.19	0.25	0.03	2.52	0.04	13.75	m	15.47
100 x 100mm jambs	12.07	2.50	12.37	0.29	0.04	2.96	0.05	15.38	m	17.30
100 x 113mm jambs	13.87	2.50	14.22	0.31	0.04	3.14	0.05	17.41	m	19.58
100 x 125mm jambs	15.68	2.50	16.07	0.31	0.04	3.14	0.05	19.26	m	21.67
113 x 113mm jambs	15.68	2.50	16.07	0.31	0.04	3.14	0.05	19.26	m	21.67
113 x 125mm jambs	17.38	2.50	17.81	0.35	0.04	3.51	0.06	21.39	m	24.06
113 x 150mm jambs; labours -1	19.13	2.50	19.61	0.35	0.04	3.51	0.06	23.18	m	26.08
38 x 75mm jambs; labours -1	5.10	2.50	5.23	0.21	0.03	2.15	0.04	7.42	m	8.34
38 x 75mm jambs; labours -2	5.16	2.50	5.29	0.21	0.03	2.15	0.04	7.48	m	8.41
38 x 75mm jambs; labours -3	5.26	2.50	5.39	0.21	0.03	2.15	0.04	7.58	m	8.53
38 x 75mm jambs; labours -4	5.32	2.50	5.45	0.21	0.03	2.15	0.04	7.64	m	8.60
38 x 75mm heads	4.98	2.50	5.10	0.21	0.03	2.15	0.04	7.29	m	8.21
38 x 100mm heads	5.76	2.50	5.90	0.21	0.03	2.15	0.04	8.09	m	9.10
50 x 75mm heads	5.76	2.50	5.90	0.23	0.03	2.33	0.04	8.28	m	9.31
50 x 100mm heads	6.69	2.50	6.86	0.23	0.03	2.33	0.04	9.23	m	10.38
50 x 125mm heads	7.47	2.50	7.66	0.23	0.03	2.33	0.04	10.03	m	11.28
63 x 100mm heads	7.47	2.50	7.66	0.23	0.03	2.33	0.04	10.03	m	11.28
75 x 100mm heads	9.56	2.50	9.80	0.25	0.03	2.52	0.04	12.36	m	13.90
75 x 113mm heads	10.93	2.50	11.20	0.25	0.03	2.52	0.04	13.76	m	15.48
100 x 100mm heads	12.07	2.50	12.37	0.29	0.04	2.96	0.05	15.38	m	17.30
100 x 113mm heads	13.87	2.50	14.22	0.31	0.04	3.14	0.05	17.41	m	19.58
100 x 125mm heads	15.69	2.50	16.08	0.31	0.04	3.14	0.05	19.27	m	21.68
113 x 113mm heads	15.69	2.50	16.08	0.31	0.04	3.14	0.05	19.27	m	21.68
113 x 125mm heads	17.38	2.50	17.81	0.35	0.04	3.51	0.05	21.38	m	24.05
113 x 150mm heads	19.13	2.50	19.61	0.35	0.04	3.51	0.05	23.17	m	26.07
38 x 75mm heads; labours -1	5.10	2.50	5.23	0.21	0.03	2.15	0.03	7.41	m	8.33
38 x 75mm heads; labours -2	5.16	2.50	5.29	0.21	0.03	2.15	0.03	7.47	m	8.40
38 x 75mm heads; labours -3	5.26	2.50	5.39	0.21	0.03	2.15	0.03	7.57	m	8.52
38 x 75mm heads; labours -4	5.32	2.50	5.45	0.21	0.03	2.15	0.03	7.63	m	8.59
63 x 125mm sills	11.35	2.50	11.63	0.26	0.03	2.61	0.04	14.28	m	16.07
75 x 125mm sills	12.94	2.50	13.26	0.28	0.04	2.87	0.05	16.18	m	18.20
75 x 150mm sills	14.52	2.50	14.88	0.29	0.04	2.96	0.05	17.89	m	20.13
63 x 125mm sills; labours -1	11.50	2.50	11.79	0.26	0.03	2.61	0.04	14.44	m	16.24
63 x 125mm sills; labours -2	11.55	2.50	11.84	0.26	0.03	2.61	0.04	14.49	m	16.30
63 x 125mm sills; labours -3	11.65	2.50	11.94	0.26	0.03	2.61	0.04	14.59	m	16.42
63 x 125mm sills; labours -4	11.70	2.50	11.99	0.26	0.03	2.61	0.04	14.64	m	16.47
63 x 100mm mullions	7.47	2.50	7.66	0.06	0.01	0.62	-	8.28	m	9.32
75 x 100mm mullions	9.56	2.50	9.80	0.06	0.01	0.62	-	10.42	m	11.73
75 x 113mm mullions	10.92	2.50	11.19	0.06	0.01	0.62	-	11.82	m	13.29
100 x 100mm mullions	12.07	2.50	12.37	0.06	0.01	0.62	-	13.00	m	14.62
100 x 125mm mullions	15.69	2.50	16.08	0.06	0.01	0.62	-	16.71	m	18.79
63 x 100mm mullions; labours -1	7.62	2.50	7.81	0.06	0.01	0.62	-	8.43	m	9.49
63 x 100mm mullions; labours -2	7.69	2.50	7.88	0.06	0.01	0.62	-	8.51	m	9.57
63 x 100mm mullions; labours -3	7.78	2.50	7.97	0.06	0.01	0.62	-	8.60	m	9.67
63 x 100mm mullions; labours -4	7.83	2.50	8.03	0.06	0.01	0.62	-	8.65	m	9.73
63 x 100mm transoms	7.47	2.50	7.66	0.06	0.01	0.62	-	8.28	m	9.32
75 x 100mm transoms	9.56	2.50	9.80	0.06	0.01	0.62	-	10.42	m	11.73
75 x 113mm transoms	10.92	2.50	11.19	0.06	0.01	0.62	-	11.82	m	13.29
100 x 100mm transoms	12.07	2.50	12.37	0.06	0.01	0.62	-	13.00	m	14.62
100 x 125mm transoms	15.69	2.50	16.08	0.06	0.01	0.62	-	16.71	m	18.79
63 x 100mm transoms; labours -1	7.62	2.50	7.81	0.06	0.01	0.62	-	8.43	m	9.49
63 x 100mm transoms; labours -2	7.69	2.50	7.88	0.06	0.01	0.62	-	8.51	m	9.57
63 x 100mm transoms; labours -3	7.78	2.50	7.97	0.06	0.01	0.62	-	8.60	m	9.67
63 x 100mm transoms; labours -4	7.83	2.50	8.03	0.06	0.01	0.62	-	8.65	m	9.73

Internal door frame sets; supplied unassembled 38 x 63mm jambs and head; stops (supplied loose); assembling

	Del to Site	Waste	Material Cost	Craft Optve	Lab	Labour Cost	Sunds	Nett Rate	Unit	Gross Rate
for 686 x 1981mm doors	21.57	2.50	22.11	1.15	0.14	11.60	0.19	33.90	nr	38.13
for 762 x 1981mm doors	21.57	2.50	22.11	1.15	0.14	11.60	0.19	33.90	nr	38.13

Labour hourly rates: (except Specialists) Craft Operatives £9.23 Labourer £7.02 Rates are national average prices. Refer to REGIONAL VARIATIONS for indicative levels of overall pricing in regions	MATERIALS			LABOUR				RATES		
	Del to Site £	Waste %	Material Cost £	Craft Optve Hrs	Lab Hrs	Labour Cost £	Sunds £	Nett Rate £	Unit	Gross Rate (+12.5%) £
L20: DOORS/SHUTTERS/HATCHES Cont.										
FRAMES - Door frames and door lining sets in softwood, wrought - frames Cont.										
Internal door frame sets; supplied unassembled Cont.										
38 x 75mm jambs and head; stops (supplied loose); assembling										
for 686 x 1981mm doors	23.00	2.50	23.57	1.15	0.14	11.60	0.19	35.36	nr	39.78
for 762 x 1981mm doors	23.00	2.50	23.57	1.15	0.14	11.60	0.19	35.36	nr	39.78
50 x 100mm jambs and head; stops (supplied loose); assembling										
for 686 x 1981mm doors	35.94	2.50	36.84	1.15	0.14	11.60	0.19	48.63	nr	54.70
for 762 x 1981mm doors	35.94	2.50	36.84	1.15	0.14	11.60	0.19	48.63	nr	54.70
50 x 113mm jambs and head; stops (supplied loose); assembling										
for 686 x 1981mm doors	44.57	2.50	45.68	1.15	0.14	11.60	0.19	57.47	nr	64.66
for 762 x 1981mm doors	44.57	2.50	45.68	1.15	0.14	11.60	0.19	57.47	nr	64.66
External door frame sets; one coat external primer before delivery to site										
63 x 75mm jambs and head; rebates -1										
for 762 x 1981 x 50mm doors	38.81	2.50	39.78	1.15	0.14	11.60	0.19	51.57	nr	58.01
for 838 x 1981 x 50mm doors	40.25	2.50	41.26	1.15	0.14	11.60	0.19	53.04	nr	59.67
63 x 88mm jambs and head; rebates -1										
for 762 x 1981 x 50mm doors	40.25	2.50	41.26	1.15	0.14	11.60	0.19	53.04	nr	59.67
for 838 x 1981 x 50mm doors	41.70	2.50	42.74	1.15	0.14	11.60	0.19	54.53	nr	61.35
63 x 75mm jambs and head, rebates -1 nr; hardwood sill										
for 762 x 1981 x 50mm doors	86.25	2.50	88.41	1.45	0.18	14.65	0.24	103.29	nr	116.21
for 838 x 1981 x 50mm doors	87.69	2.50	89.88	1.45	0.18	14.65	0.24	104.77	nr	117.87
63 x 88mm jambs and head, rebates -1 nr; hardwood sill										
for 762 x 1981 x 50mm doors	89.10	2.50	91.33	1.45	0.18	14.65	0.24	106.21	nr	119.49
for 838 x 1981 x 50mm doors	90.56	2.50	92.82	1.45	0.18	14.65	0.24	107.71	nr	121.17
for 1168 x 1981 x 50mm doors	91.98	2.50	94.28	1.75	0.22	17.70	0.28	112.26	nr	126.29
External garage door frame sets; one coat external primer before delivery to site; supplied unassembled										
75 x 100mm jambs and head; assembling										
for 2134 x 1981mm side hung doors	63.45	2.50	65.04	2.50	0.31	25.25	0.41	90.70	nr	102.03
for 2134 x 2134mm side hung doors	67.66	2.50	69.35	2.50	0.31	25.25	0.41	95.01	nr	106.89
75 x 75mm jambs and head; assembling										
for 2134 x 1981mm up and over doors	58.93	2.50	60.40	2.50	0.31	25.25	0.41	86.06	nr	96.82
FRAMES - Door frames and door lining sets in Afromosia, wrought, selected for transparent finish - frames										
Sets										
38 x 75mm jambs	12.32	2.50	12.63	0.36	0.05	3.67	0.06	16.36	m	18.41
38 x 100mm jambs	14.19	2.50	14.54	0.36	0.05	3.67	0.06	18.28	m	20.56
50 x 75mm jambs	14.19	2.50	14.54	0.40	0.05	4.04	0.06	18.65	m	20.98
50 x 100mm jambs	16.67	2.50	17.09	0.40	0.05	4.04	0.06	21.19	m	23.84
50 x 125mm jambs	18.48	2.50	18.94	0.40	0.05	4.04	0.06	23.05	m	25.93
63 x 100mm jambs	18.48	2.50	18.94	0.40	0.05	4.04	0.06	23.05	m	25.93
75 x 100mm jambs	23.73	2.50	24.32	0.44	0.06	4.48	0.07	28.88	m	32.49
75 x 113mm jambs	27.09	2.50	27.77	0.44	0.06	4.48	0.07	32.32	m	36.36
100 x 100mm jambs	30.21	2.50	30.97	0.51	0.06	5.13	0.08	36.17	m	40.70
100 x 113mm jambs	34.50	2.50	35.36	0.54	0.07	5.48	0.09	40.93	m	46.04
100 x 125mm jambs	38.84	2.50	39.81	0.60	0.08	6.10	0.10	46.01	m	51.76
113 x 113mm jambs	38.84	2.50	39.81	0.54	0.07	5.48	0.09	45.38	m	51.05
113 x 125mm jambs	43.11	2.50	44.19	0.60	0.08	6.10	0.10	50.39	m	56.69
113 x 150mm jambs	47.47	2.50	48.66	0.60	0.08	6.10	0.10	54.86	m	61.71
38 x 75mm jambs; labours -1	12.51	2.50	12.82	0.36	0.05	3.67	0.06	16.56	m	18.63
38 x 75mm jambs; labours -2	12.64	2.50	12.96	0.36	0.05	3.67	0.06	16.69	m	18.78
38 x 75mm jambs; labours -3	12.84	2.50	13.16	0.36	0.05	3.67	0.06	16.89	m	19.01
38 x 75mm jambs; labours -4	13.04	2.50	13.37	0.36	0.05	3.67	0.06	17.10	m	19.24
38 x 75mm heads	12.32	2.50	12.63	0.36	0.05	3.67	0.06	16.36	m	18.41
38 x 100mm heads	14.19	2.50	14.54	0.36	0.05	3.67	0.06	18.28	m	20.56
50 x 75mm heads	14.19	2.50	14.54	0.40	0.05	4.04	0.06	18.65	m	20.98
50 x 100mm heads	16.67	2.50	17.09	0.40	0.05	4.04	0.06	21.19	m	23.84
50 x 125mm heads	18.48	2.50	18.94	0.40	0.05	4.04	0.06	23.05	m	25.93
63 x 100mm heads	18.48	2.50	18.94	0.40	0.05	4.04	0.06	23.05	m	25.93
75 x 100mm heads	23.73	2.50	24.32	0.44	0.06	4.48	0.07	28.88	m	32.49
75 x 113mm heads	27.09	2.50	27.77	0.44	0.06	4.48	0.07	32.32	m	36.36
100 x 100mm heads	30.21	2.50	30.97	0.51	0.06	5.13	0.08	36.17	m	40.70
100 x 113mm heads	34.50	2.50	35.36	0.54	0.07	5.48	0.09	40.93	m	46.04
100 x 125mm heads	38.84	2.50	39.81	0.60	0.08	6.10	0.10	46.01	m	51.76
113 x 113mm heads	38.84	2.50	39.81	0.54	0.07	5.48	0.09	45.38	m	51.05
113 x 125mm heads	43.11	2.50	44.19	0.60	0.08	6.10	0.10	50.39	m	56.69
113 x 150mm heads	47.47	2.50	48.66	0.60	0.08	6.10	0.10	54.86	m	61.71
38 x 75mm heads; labours -1	12.51	2.50	12.82	0.36	0.05	3.67	0.06	16.56	m	18.63
38 x 75mm heads; labours -2	12.64	2.50	12.96	0.36	0.05	3.67	0.06	16.69	m	18.78
38 x 75mm heads; labours -3	12.84	2.50	13.16	0.36	0.05	3.67	0.06	16.89	m	19.01
38 x 75mm heads; labours -4	13.04	2.50	13.37	0.36	0.05	3.67	0.06	17.10	m	19.24
63 x 125mm sills	28.01	2.50	28.71	0.46	0.06	4.67	0.00	33.46	m	37.64
75 x 125mm sills	32.03	2.50	32.83	0.51	0.06	5.13	0.08	38.04	m	42.79
75 x 150mm sills	36.06	2.50	36.96	0.52	0.07	5.29	0.09	42.34	m	47.64
63 x 125mm sills; labours -1	28.34	2.50	29.05	0.46	0.06	4.67	0.08	33.80	m	38.02
63 x 125mm sills; labours -2	28.47	2.50	29.18	0.46	0.06	4.67	0.08	33.93	m	38.17
63 x 125mm sills; labours -3	28.59	2.50	29.30	0.46	0.06	4.67	0.08	34.05	m	38.31
63 x 125mm sills; labours -4	28.72	2.50	29.44	0.46	0.06	4.67	0.08	34.19	m	38.46
63 x 100mm mullions	18.48	2.50	18.94	0.06	0.01	0.62	-	19.57	m	22.01
75 x 100mm mullions	23.73	2.50	24.32	0.06	0.01	0.62	-	24.95	m	28.07
75 x 113mm mullions	27.09	2.50	27.77	0.06	0.01	0.62	-	28.39	m	31.94

Labour hourly rates: (except Specialists) Craft Operatives £9.23 Labourer £7.02 Rates are national average prices. Refer to REGIONAL VARIATIONS for indicative levels of overall pricing in regions	MATERIALS			LABOUR				RATES		
	Del to Site	Waste	Material Cost	Craft Optve	Lab	Labour Cost	Sunds	Nett Rate	Unit	Gross Rate (+12.5%)
	£	%	£	Hrs	Hrs	£	£	£		£

L20: DOORS/SHUTTERS/HATCHES Cont.

FRAMES - Door frames and door lining sets in Afromosia, wrought, selected for transparent finish - frames Cont.

Sets Cont.

100 x 100mm mullions	30.21	2.50	30.97	0.06	0.01	0.62	-	31.59	m	35.54
100 x 125mm mullions	38.84	2.50	39.81	0.06	0.01	0.62	-	40.44	m	45.49
63 x 100mm mullions; labours -1	18.67	2.50	19.14	0.06	0.01	0.62	-	19.76	m	22.23
63 x 100mm mullions; labours -2	18.80	2.50	19.27	0.06	0.01	0.62	-	19.89	m	22.38
63 x 100mm mullions; labours -3	18.93	2.50	19.40	0.06	0.01	0.62	-	20.03	m	22.53
63 x 100mm mullions; labours -4	19.06	2.50	19.54	0.06	0.01	0.62	-	20.16	m	22.68
63 x 100mm transoms	18.48	2.50	18.94	0.06	0.01	0.62	-	19.57	m	22.01
75 x 100mm transoms	23.73	2.50	24.32	0.06	0.01	0.62	-	24.95	m	28.07
75 x 113mm transoms	27.09	2.50	27.77	0.06	0.01	0.62	-	28.39	m	31.94
100 x 100mm transoms	30.21	2.50	30.97	0.06	0.01	0.62	-	31.59	m	35.54
100 x 125mm transoms	38.84	2.50	39.81	0.06	0.01	0.62	-	40.44	m	45.49
63 x 100mm transoms; labours -1	18.67	2.50	19.14	0.06	0.01	0.62	-	19.76	m	22.23
63 x 100mm transoms; labours -2	18.80	2.50	19.27	0.06	0.01	0.62	-	19.89	m	22.38
63 x 100mm transoms; labours -3	18.93	2.50	19.40	0.06	0.01	0.62	-	20.03	m	22.53
63 x 100mm transoms; labours -4	19.06	2.50	19.54	0.06	0.01	0.62	-	20.16	m	22.68

FRAMES - Door frames and door lining sets in Sapele, wrought, selected for transparent finish - frames

Sets

38 x 75mm jambs	9.47	2.50	9.71	0.31	0.04	3.14	0.05	12.90	m	14.51
38 x 100mm jambs	10.91	2.50	11.18	0.31	0.04	3.14	0.05	14.37	m	16.17
50 x 75mm jambs	10.91	2.50	11.18	0.35	0.04	3.51	0.06	14.75	m	16.60
50 x 100mm jambs	12.71	2.50	13.03	0.35	0.04	3.51	0.06	16.60	m	18.67
50 x 125mm jambs	14.15	2.50	14.50	0.35	0.04	3.51	0.06	18.08	m	20.33
63 x 100mm jambs	14.15	2.50	14.50	0.35	0.04	3.51	0.06	18.08	m	20.33
75 x 100mm jambs	18.19	2.50	18.64	0.38	0.05	3.86	0.06	22.56	m	25.38
75 x 113mm jambs	20.79	2.50	21.31	0.38	0.05	3.86	0.06	25.23	m	28.38
100 x 100mm jambs	22.99	2.50	23.56	0.43	0.05	4.32	0.07	27.95	m	31.45
100 x 113mm jambs	26.33	2.50	26.99	0.46	0.06	4.67	0.08	31.74	m	35.70
100 x 125mm jambs	29.74	2.50	30.48	0.46	0.06	4.67	0.08	35.23	m	39.63
113 x 113mm jambs	29.74	2.50	30.48	0.46	0.06	4.67	0.08	35.23	m	39.63
113 x 125mm jambs	33.15	2.50	33.98	0.52	0.07	5.29	0.09	39.36	m	44.28
113 x 150mm jambs	36.39	2.50	37.30	0.52	0.07	5.29	0.09	42.68	m	48.02
38 x 75mm jambs; labours -1	9.65	2.50	9.89	0.31	0.04	3.14	0.05	13.08	m	14.72
38 x 75mm jambs; labours -2	9.76	2.50	10.00	0.31	0.04	3.14	0.05	13.20	m	14.85
38 x 75mm jambs; labours -3	9.88	2.50	10.13	0.31	0.04	3.14	0.05	13.32	m	14.98
38 x 75mm jambs; labours -4	9.99	2.50	10.24	0.31	0.04	3.14	0.05	13.43	m	15.11
38 x 75mm heads	9.47	2.50	9.71	0.31	0.04	3.14	0.05	12.90	m	14.51
38 x 100mm heads	10.91	2.50	11.18	0.31	0.04	3.14	0.05	14.37	m	16.17
50 x 75mm heads	10.91	2.50	11.18	0.35	0.04	3.51	0.06	14.75	m	16.60
50 x 100mm heads	12.71	2.50	13.03	0.35	0.04	3.51	0.06	16.60	m	18.67
50 x 125mm heads	14.15	2.50	14.50	0.35	0.04	3.51	0.06	18.08	m	20.33
63 x 100mm heads	14.15	2.50	14.50	0.35	0.04	3.51	0.06	18.08	m	20.33
75 x 100mm heads	14.73	2.50	15.10	0.38	0.05	3.86	0.06	19.02	m	21.39
75 x 113mm heads	20.79	2.50	21.31	0.38	0.05	3.86	0.06	25.23	m	28.38
100 x 100mm heads	22.99	2.50	23.56	0.43	0.05	4.32	0.07	27.95	m	31.45
100 x 113mm heads	26.33	2.50	26.99	0.46	0.06	4.67	0.08	31.74	m	35.70
100 x 125mm heads	29.74	2.50	30.48	0.46	0.06	4.67	0.08	35.23	m	39.63
113 x 113mm heads	29.74	2.50	30.48	0.46	0.06	4.67	0.08	35.23	m	39.63
113 x 125mm heads	33.15	2.50	33.98	0.52	0.07	5.29	0.09	39.36	m	44.28
113 x 150mm heads	37.05	2.50	37.98	0.52	0.07	5.29	0.09	43.36	m	48.78
38 x 75mm heads; labours -1	9.65	2.50	9.89	0.31	0.04	3.14	0.05	13.08	m	14.72
38 x 75mm heads; labours -2	9.76	2.50	10.00	0.31	0.04	3.14	0.05	13.20	m	14.85
38 x 75mm heads; labours -3	9.88	2.50	10.13	0.31	0.04	3.14	0.05	13.32	m	14.98
38 x 75mm heads; labours -4	9.99	2.50	10.24	0.31	0.04	3.14	0.05	13.43	m	15.11
63 x 125mm sills	21.60	2.50	22.14	0.40	0.05	4.04	0.06	26.24	m	29.52
75 x 125mm sills	24.54	2.50	25.15	0.42	0.05	4.23	0.07	29.45	m	33.13
75 x 150mm sills	27.72	2.50	28.41	0.43	0.05	4.32	0.07	32.80	m	36.90
63 x 125mm sills; labours -1	21.77	2.50	22.31	0.40	0.05	4.04	0.06	26.42	m	29.72
63 x 125mm sills; labours -2	21.89	2.50	22.44	0.40	0.05	4.04	0.06	26.54	m	29.86
63 x 125mm sills; labours -3	22.00	2.50	22.55	0.40	0.05	4.04	0.06	26.65	m	29.98
63 x 125mm sills; labours -4	22.12	2.50	22.67	0.40	0.05	4.04	0.06	26.78	m	30.12
63 x 100mm mullions	14.15	2.50	14.50	0.06	0.01	0.62	-	15.13	m	17.02
75 x 100mm mullions	14.73	2.50	15.10	0.06	0.01	0.62	-	15.72	m	17.69
75 x 113mm mullions	20.79	2.50	21.31	0.06	0.01	0.62	-	21.93	m	24.68
100 x 100mm mullions	22.99	2.50	23.56	0.06	0.01	0.62	-	24.19	m	27.21
100 x 125mm mullions	29.74	2.50	30.48	0.06	0.01	0.62	-	31.11	m	35.00
63 x 100mm mullions; labours -1	14.32	2.50	14.68	0.06	0.01	0.62	-	15.30	m	17.21
63 x 100mm mullions; labours -2	14.44	2.50	14.80	0.06	0.01	0.62	-	15.43	m	17.35
63 x 100mm mullions; labours -3	14.55	2.50	14.91	0.06	0.01	0.62	-	15.54	m	17.48
63 x 100mm mullions; labours -4	14.67	2.50	15.04	0.06	0.01	0.62	-	15.66	m	17.62
63 x 100mm transoms	14.15	2.50	14.50	0.06	0.01	0.62	-	15.13	m	17.02
75 x 100mm transoms	14.73	2.50	15.10	0.06	0.01	0.62	-	15.72	m	17.69
75 x 113mm transoms	20.79	2.50	21.31	0.06	0.01	0.62	-	21.93	m	24.68
100 x 100mm transoms	22.99	2.50	23.56	0.06	0.01	0.62	-	24.19	m	27.21
100 x 125mm transoms	29.74	2.50	30.48	0.06	0.01	0.62	-	31.11	m	35.00
63 x 100mm transoms; labours -1	14.32	2.50	14.68	0.06	0.01	0.62	-	15.30	m	17.21
63 x 100mm transoms; labours -2	14.44	2.50	14.80	0.06	0.01	0.62	-	15.43	m	17.35
63 x 100mm transoms; labours -3	14.55	2.50	14.91	0.06	0.01	0.62	-	15.54	m	17.48
63 x 100mm transoms; labours -4	14.67	2.50	15.04	0.06	0.01	0.62	-	15.66	m	17.62

FRAMES - Intumescent strips and smoke seals

Albi-Flex, white PVC sleeved self adhesive; Rentokil Ltd.

12 x 4mm intumescent strip, half hour application; setting into groove in timber frame or door	3.25	10.00	3.58	0.17	-	1.57	-	5.14	m	5.79

Labour hourly rates: (except Specialists) Craft Operatives £9.23 Labourer £7.02 Rates are national average prices. Refer to REGIONAL VARIATIONS for indicative levels of overall pricing in regions	MATERIALS			LABOUR				RATES		
	Del to Site £	Waste %	Material Cost £	Craft Optve Hrs	Lab Hrs	Labour Cost £	Sunds £	Nett Rate £	Unit	Gross Rate (+12.5%) £
L20: DOORS/SHUTTERS/HATCHES Cont.										
FRAMES – Intumescent strips and smoke seals Cont.										
Albi-Flex, white PVC sleeved self adhesive; Rentokil Ltd. Cont.										
18 x 4mm intumescent strip, one hour application; setting into groove in timber frame or door	4.55	10.00	5.00	0.17	–	1.57	–	6.57	m	7.40
22 x 4mm intumescent strip with integral cold smoke seal, half hour application; setting into groove in timber frame or door	6.89	10.00	7.58	0.17	–	1.57	–	9.15	m	10.29
28 x 4mm intumescent strip with integral cold smoke seal, one hour application; setting into groove in timber frame or door	8.00	10.00	8.80	0.17	–	1.57	–	10.37	m	11.67
12 x 4mm intumescent strip, half hour application; fixing to both sides of glass behind glazing beads	6.50	10.00	7.15	0.35	–	3.23	–	10.38	m	11.68
25 x 4mm intumescent strip, one hour application; fixing to both sides of glass behind glazing beads	11.69	10.00	12.86	0.35	–	3.23	–	16.09	m	18.10
10 x 4mm cold smoke seal; setting into groove in timber frame or door	3.56	10.00	3.92	0.17	–	1.57	–	5.49	m	6.17
Lorient Polyproducts Ltd; System 36										
15 x 12mm glazing channel reference LG1512; fitting over the edge of the pane	1.52	10.00	1.67	0.17	–	1.57	–	3.24	m	3.65
Lorient Polyproducts Ltd; System 90										
intumescent lining reference LX4402 to suit 44mm thick doors; fixing to timber with adhesive	3.30	10.00	3.63	0.35	–	3.23	–	6.86	m	7.72
intumescent lining reference LX5402 to suit 54mm thick doors; fixing to timber with adhesive	3.70	10.00	4.07	0.35	–	3.23	–	7.30	m	8.21
Mann McGowan Fabrications Ltd										
Pyroglaze 30 half hour application; fixing to both sides of glass behind glazing beads	2.67	10.00	2.94	0.40	–	3.69	–	6.63	m	7.46
Pyroglaze 60 one hour application; fixing to both sides of glass behind glazing beads	4.18	10.00	4.60	0.40	–	3.69	–	8.29	m	9.33
FRAMES – Bedding and pointing frames										
Bedding in cement mortar (1:3); pointing one side wood frames	0.22	10.00	0.24	0.14	–	1.29	–	1.53	m	1.73
Bedding in cement mortar (1:3); pointing each side wood frames	0.22	10.00	0.24	0.20	–	1.85	–	2.09	m	2.35
Bedding in cement mortar (1:3); pointing with Secomastic standard mastic one side wood frames one side	1.20	10.00	1.32	0.20	–	1.85	–	3.17	m	3.56
METAL DOORS – Doors and sidelights in galvanised steel										
Crittall Windows Ltd. Duralife Homelight; weatherstripping; fixing to masonry with lugs										
761 x 2056mm doors NA15	413.07	2.50	423.40	3.20	1.60	40.77	–	464.16	nr	522.19
997 x 2056mm doors NA2	622.02	2.50	637.57	3.45	1.70	43.78	–	681.35	nr	766.52
1143 x 2056mm doors NA25	634.37	2.50	650.23	3.65	1.85	46.68	–	696.91	nr	784.02
279 x 2056mm sidelights, NA6	69.62	2.50	71.36	1.40	0.70	17.84	–	89.20	nr	100.35
508 x 2056mm sidelights, NA5	84.61	2.50	86.73	1.70	0.85	21.66	–	108.38	nr	121.93
997 x 2056mm sidelights, NA13F	160.99	2.50	165.01	2.00	1.05	25.83	–	190.85	nr	214.70
METAL DOORS – Doors and sidelights in galvanised steel; polyester powder coated; white matt										
Crittall Windows Ltd. Duralife Homelight; weatherstripping; fixing to masonry with lugs										
761 x 2056mm doors NA15	520.72	2.50	533.74	3.20	1.60	40.77	–	574.51	nr	646.32
997 x 2056mm doors NA2	783.58	2.50	803.17	3.45	1.70	43.78	–	846.95	nr	952.82
1143 x 2056mm doors NA25	798.68	2.50	818.65	3.65	1.85	46.68	–	865.32	nr	973.49
279 x 2056mm sidelights, NA6	86.29	2.50	88.45	1.40	0.70	17.84	–	106.28	nr	119.57
508 x 2056mm sidelights, NA5	104.93	2.50	107.55	1.70	0.85	21.66	–	129.21	nr	145.36
997 x 2056mm sidelights, NA13F	195.72	2.50	200.61	2.00	1.05	25.83	–	226.44	nr	254.75
METAL DOORS – Garage doors; Birtley Durham Type in galvanised steel; one coat primer by manufacturer										
Overhead garage doors; tensioning device fixing to timber with screws; for opening size										
2135 x 1980mm	242.00	2.50	248.05	6.00	6.00	97.50	–	345.55	nr	388.74
2135 x 2135mm	254.00	2.50	260.35	6.00	6.00	97.50	–	357.85	nr	402.58
Overhead garage doors; counterbalanced by springs										
4270 x 1980mm	760.00	2.50	779.00	8.00	8.00	130.00	–	909.00	nr	1022.63
4270 x 2135mm	760.00	2.50	779.00	8.00	8.00	130.00	–	909.00	nr	1022.63
METAL DOORS – Patio doors in aluminium, with wrought hardwood subframes										
Double glazed clear toughened glass, Magnet Trade Magnastar, one opening leaf, fixing subframe to masonry with screws										
2085 x 1501mm overall, Bp5 white	416.58	2.50	427.00	8.00	8.00	130.00	–	837.25	nr	941.91
2085 x 1805mm overall, Bp6 white	533.28	2.50	546.61	8.00	8.00	130.00	–	837.25	nr	941.91
2085 x 2086mm overall, Bp7 white	593.08	2.50	607.91	8.00	8.00	130.00	–	837.25	nr	941.91
2085 x 2387mm overall, Bp8 white	702.38	2.50	719.94	8.00	8.00	130.00	–	837.25	nr	941.91

Labour hourly rates: (except Specialists) Craft Operatives £9.23 Labourer £7.02 Rates are national average prices. Refer to REGIONAL VARIATIONS for indicative levels of overall pricing in regions	MATERIALS			LABOUR				RATES		
	Del to Site £	Waste %	Material Cost £	Craft Optve Hrs	Lab Hrs	Labour Cost £	Sunds £	Nett Rate £	Unit	Gross Rate (+12.5%) £
L20: DOORS/SHUTTERS/HATCHES Cont.										
METAL DOORS - Patio doors and frames in aluminium										
Double glazed clear toughened glass, Magnet Trade Magnaplus, one opening leaf, fixing subframe to masonry with screws										
2090 x 1501mm overall, THP5 white	500.69	2.50	513.21	8.00	8.00	130.00	-	643.21	nr	723.61
2090 x 1805mm overall, THP6 white	572.33	2.50	586.64	8.00	8.00	130.00	-	716.64	nr	806.22
2090 x 2086mm overall, THP7 white	619.31	2.50	634.79	8.00	8.00	130.00	-	764.69	nr	860.39
2090 X 2387 overall, THP8 white	681.95	2.50	707.25	8.00	8.00	130.00	-	829.00	nr	932.63
extra; former for THP5	80.25	2.50	82.26	-	-	82.26	-	1.02	nr	92.54
extra; former for THP6	82.20	2.50	84.26	-	-	84.26	-	1.02	nr	94.79
extra; former for THP7	85.10	2.50	87.23	-	-	87.23	-	1.02	nr	98.13
extra; former for THP8	91.48	2.50	93.77	-	-	93.77	-	1.02	nr	105.49
ROLLER SHUTTERS - Roller shutters in steel, galvanised										
Note roller shutters are always purpose made to order and the following prices are indicative only. Firm quotations should always be obtained Crank handle operation										
4572 x 3658mm overall; fixing to masonry with screws	1260.00	2.50	1291.50	24.00	24.00	390.00	-	1681.50	nr	1891.69
Endless hand chain operation										
5486 x 4267mm overall; fixing to masonry with screws	1475.00	2.50	1511.88	24.00	24.00	390.00	-	1901.88	nr	2139.61
Electric motor operation										
6096 x 6096mm overall; fixing to masonry with screws	3025.00	2.50	3100.62	30.00	30.00	487.50	-	3588.12	nr	4036.64
ROLLER SHUTTERS - Roller shutters in aluminium										
Endless hand chain operation										
3553 x 3048mm overall; fixing to masonry with screws	1190.00	2.50	1219.75	24.00	24.00	390.00	-	1609.75	nr	1810.97
ROLLER SHUTTERS - Bedding and pointing frames										
Pointing with Secomastic standard mastic										
metal frames one side	0.88	5.00	0.92	0.12	0.06	1.53	-	2.45	m	2.76
Bedding in cement mortar (1:3); pointing with Secomastic standard mastic one side										
metal frames one side	1.01	5.00	1.06	0.20	0.10	2.55	-	3.61	m	4.06
PLASTICS/RUBBER DOORS - Flexible doors mainly in plastics and rubber										
Mancuna standard doors; Mandor Engineering Ltd; 43mm diameter steel tube frame; top and bottom plates with pivoting arrangements and spring unit; 8mm rubber panel doors with triangular perspex windows in each leaf										
fixing to masonry; for opening size										
1800 x 2100mm..................................	878.00	2.50	899.95	4.40	4.40	71.50	-	971.45	nr	1092.88
2440 x 2400mm..................................	1021.00	2.50	1046.53	5.50	5.50	89.38	-	1135.90	nr	1277.89
Mancuna standard doors; Mandor Engineering Ltd; 43mm diameter steel tube frame; top and bottom plates with pivoting arrangements and spring unit; 12mm rubber panel doors with triangular perspex windows in each leaf										
fixing to masonry; for opening size										
1800 x 2100mm..................................	1021.00	2.50	1046.53	4.95	4.95	80.44	-	1126.96	nr	1267.83
2440 x 2400mm..................................	1266.00	2.50	1297.65	6.05	6.05	98.31	-	1395.96	nr	1570.46
Mancuna standard doors; Mandor Engineering Ltd; 43mm diameter steel tube frame; top and bottom plates with pivoting arrangements and spring unit; 6mm clear flexible panel doors										
fixing to masonry; for opening size										
1800 x 2100mm..................................	666.00	2.50	682.65	3.85	3.85	62.56	-	745.21	nr	838.36
2440 x 2400mm..................................	732.00	2.50	750.30	4.40	4.40	71.50	-	821.80	nr	924.52
Mancuna standard doors; Mandor Engineering Ltd; 43mm diameter steel tube frame; top and bottom plates with pivoting arrangements and spring unit; 9mm clear flexible panel doors										
fixing to masonry; for opening size										
1800 x 2100mm..................................	764.00	2.50	783.10	4.13	4.13	67.11	-	850.21	nr	956.49
2440 x 2400mm..................................	863.00	2.50	884.58	4.68	4.68	76.05	-	960.63	nr	1080.70
Mandor heavy duty doors; Mandor Engineering Ltd; 63mm diameter steel tube frame; top and bottom plates with pivoting arrangements and spring unit; 12mm rubber panel doors with oval panel windows in each leaf										
fixing to masonry; for opening size										
2440 x 3050mm..................................	1605.00	2.50	1645.13	8.80	8.80	143.00	-	1788.13	nr	2011.64
2740 x 3350mm..................................	1943.00	2.50	1991.58	9.90	9.90	160.88	-	2152.45	nr	2421.51
3050 x 3660mm..................................	2058.00	2.50	2109.45	11.00	11.00	178.75	-	2288.20	nr	2574.22

Labour hourly rates: (except Specialists) Craft Operatives £9.23 Labourer £7.02 Rates are national average prices. Refer to REGIONAL VARIATIONS for indicative levels of overall pricing in regions	MATERIALS			LABOUR				RATES		
	Del to Site £	Waste %	Material Cost £	Craft Optve Hrs	Lab Hrs	Labour Cost £	Sunds £	Nett Rate £	Unit	Gross Rate (+12.5%) £
L20: DOORS/SHUTTERS/HATCHES Cont.										
PLASTICS/RUBBER DOORS - Flexible doors mainly in plastics and rubber Cont.										
Mandor heavy duty doors; Mandor Engineering Ltd; 63mm diameter steel tube frame; top and bottom plates with pivoting arrangements and spring unit; 9mm clear flexible panel doors										
fixing to masonry; for opening size										
2440 x 3050mm	1075.00	2.50	1101.88	6.60	6.60	107.25	-	1209.13	nr	1360.27
2740 x 3350mm	1229.00	2.50	1259.72	7.70	7.70	125.13	-	1384.85	nr	1557.96
3050 x 3660mm	1414.00	2.50	1449.35	8.80	8.80	143.00	-	1592.35	nr	1791.39
Manby aluminium doors; Mandor Engineering Ltd; extruded aluminium frame; pivot arrangement and spring unit; clear pvc panel doors										
fixing to masonry; for opening size										
1800 x 2100mm	727.00	2.50	745.17	4.13	4.13	67.11	-	812.29	nr	913.82
2440 x 2400mm	797.00	2.50	816.92	5.23	5.23	84.99	-	901.91	nr	1014.65
Manby aluminium doors; Mandor Engineering Ltd; extruded aluminium frame; pivot arrangement and spring unit; half coloured half clear pvc panel doors										
fixing to masonry; for opening size										
1800 x 2100mm	797.00	2.50	816.92	4.13	4.12	67.04	-	883.97	nr	994.46
2440 x 2400mm	874.00	2.50	895.85	5.23	5.23	84.99	-	980.84	nr	1103.44
PLASTICS/RUBBER DOORS - Flexible strip curtains mainly in pvc										
Mandor strip curtains; Mandor Engineering Ltd; 200mm wide x 2.0mm thick clear pvc strip curtains, suspended from 80 x 80 x 6mm steel angle; fixing angle to masonry with bolts										
double overlap; for opening size										
1800 x 2100mm	108.00	2.50	110.70	2.20	2.20	35.75	-	146.45	nr	164.76
2100 x 2440mm	148.00	2.50	151.70	2.53	2.53	41.11	-	192.81	nr	216.91
single overlap; for opening size										
1800 x 2100mm	87.00	2.50	89.17	1.98	1.98	32.17	-	121.35	nr	136.52
2100 x 2440mm	118.00	2.50	120.95	2.31	2.31	37.54	-	158.49	nr	178.30
Mandor strip curtains; Mandor Engineering Ltd; 300mm wide x 2.5mm thick clear pvc strip curtains, suspended from 80 x 80 x 6mm steel angle; fixing angle to masonry with bolts										
double overlap; for opening size										
2440 x 2740mm	210.00	2.50	215.25	2.90	2.90	47.13	-	262.38	nr	295.17
2740 x 2900mm	250.00	2.50	256.25	3.23	3.23	52.49	-	308.74	nr	347.33
single overlap; for opening size										
2440 x 2740mm	162.00	2.50	166.05	2.68	2.68	43.55	-	209.60	nr	235.80
2740 x 2900mm	193.00	2.50	197.82	3.01	3.01	48.91	-	246.74	nr	277.58
Wavespan 20T strip curtains; Mandor Engineering Ltd.; 305mm wide overlapping nylon reinforced clear pvc strips, stapled to 50 x 50mm softwood batten; fixing batten to masonry with screws; for opening size										
1000 x 2440mm	70.00	2.50	71.75	1.38	1.38	22.43	-	94.17	nr	105.95
1000 x 2740mm	80.00	2.50	82.00	1.65	1.65	26.81	-	108.81	nr	122.41
1000 x 3050mm	88.00	2.50	90.20	1.93	1.93	31.36	-	121.56	nr	136.76
2000 x 2440mm	140.00	2.50	143.50	2.75	2.75	44.69	-	188.19	nr	211.71
2000 x 2740mm	159.00	2.50	162.97	3.30	3.30	53.63	-	216.60	nr	243.68
2000 x 3050mm	172.00	2.50	176.30	3.90	3.90	63.38	-	239.68	nr	269.63
3000 x 2440mm	212.00	2.50	217.30	4.10	4.10	66.63	-	283.93	nr	319.42
3000 x 2740mm	238.00	2.50	243.95	5.00	5.00	81.25	-	325.20	nr	365.85
3000 x 3050mm	260.00	2.50	266.50	6.00	6.00	97.50	-	364.00	nr	409.50
L30: STAIRS/WALKWAYS/BALUSTRADES										
TIMBER STAIRS - Stairs in softwood, wrought										
Straight flight staircase and balustrade; 25mm treads; 19mm risers, glued; wedged and blocked; 25mm wall string; 38mm outer string; 75 x 75mm newels; 25 x 25mm balusters; 38 x 75mm hardwood handrail										
864mm wide x 2600mm rise overall; balustrade to one side; fixing to masonry with screws	666.75	2.50	683.42	15.00	1.90	151.79	2.40	837.61	nr	942.31
864mm wide x 2600mm rise overall with 3 nr winders at bottom; balustrade to one side; fixing to masonry with screws	800.10	2.50	820.10	20.00	2.50	202.15	3.25	1025.50	nr	1153.69
864mm wide x 2600mm rise overall with 3 nr winders at top; balustrade to one side; fixing to masonry with screws	800.10	2.50	820.10	20.00	2.50	202.15	3.25	1025.50	nr	1153.69
TIMBER STAIRS - Balustrades in softwood, wrought										
Isolated balustrades; 25 x 25mm balusters at 150mm centres, housed construction (handrail included elsewhere)										
914mm high; fixing to timber with screws	35.02	2.50	35.90	0.80	0.10	8.09	0.13	44.11	m	49.63
Isolated balustrades; 38 x 38mm balusters at 150mm centres, housed construction (handrail included elsewhere)										
914mm high; fixing to timber with screws	51.32	2.50	52.60	1.15	0.14	11.60	0.19	64.39	m	72.44

WINDOWS/DOORS/STAIRS

Labour hourly rates: (except Specialists) Craft Operatives £9.23 Labourer £7.02 Rates are national average prices. Refer to REGIONAL VARIATIONS for indicative levels of overall pricing in regions	MATERIALS			LABOUR				RATES		
	Del to Site £	Waste %	Material Cost £	Craft Optve Hrs	Lab Hrs	Labour Cost £	Sunds £	Nett Rate £	Unit	Gross Rate (+12.5%) £
L30: STAIRS/WALKWAYS/BALUSTRADES Cont.										
TIMBER STAIRS - Balustrades in softwood, wrought Cont.										
Isolated balustrades; 50 x 50mm balusters at 150mm centres, housed construction (handrail included elsewhere)										
914mm high; fixing to timber with screws	67.43	2.50	69.12	1.30	0.16	13.12	0.21	82.45	m	92.75
Isolated balustrades; 25 x 25mm balusters at 150mm centres, 50mm mopstick handrail, housed construction										
914mm high; fixing to timber with screws	58.70	2.50	60.17	1.20	0.15	12.13	0.20	72.50	m	81.56
extra; ramps	36.03	2.50	36.93	1.10	0.14	11.14	0.18	48.25	nr	54.28
extra; wreaths	72.04	2.50	73.84	2.15	0.27	21.74	0.35	95.93	nr	107.92
extra; bends	36.03	2.50	36.93	0.70	0.09	7.09	0.11	44.13	nr	49.65
Isolated balustrades; 38 x 38mm balusters at 150mm centres, 50mm mopstick handrail, housed construction										
914mm high; fixing to timber with screws	74.71	2.50	76.58	1.55	0.19	15.64	0.25	92.47	m	104.03
extra; ramps	36.03	2.50	36.93	1.10	0.14	11.14	0.18	48.25	nr	54.28
extra; wreaths	72.04	2.50	73.84	2.15	0.27	21.74	0.35	95.93	nr	107.92
extra; bends	36.03	2.50	36.93	0.70	0.09	7.09	0.11	44.13	nr	49.65
Isolated balustrades; 50 x 50mm balusters at 150mm centres, 50mm mopstick handrail, housed construction										
914mm high; fixing to timber with screws	90.71	2.50	92.98	1.75	0.22	17.70	0.28	110.95	m	124.82
extra; ramps	36.03	2.50	36.93	1.10	0.14	11.14	0.18	48.25	nr	54.28
extra; wreaths	72.04	2.50	73.84	2.15	0.27	21.74	0.35	95.93	nr	107.92
extra; bends	36.03	2.50	36.93	0.70	0.09	7.09	0.11	44.13	nr	49.65
Isolated balustrades; 25 x 25mm balusters at 150mm centres, 75 x 100mm moulded handrail; housed construction										
914mm high; fixing to timber with screws	72.04	2.50	73.84	1.25	0.16	12.66	0.20	86.70	m	97.54
extra; ramps	53.36	2.50	54.69	2.40	0.30	24.26	0.39	79.34	nr	89.26
extra; wreaths	106.72	2.50	109.39	4.80	0.60	48.52	0.78	158.68	nr	178.52
extra; bends	53.36	2.50	54.69	1.45	0.18	14.65	0.24	69.58	nr	78.28
Isolated balustrades; 38 x 38mm balusters at 150mm centres, 75 x 100mm moulded handrail; housed construction										
914mm high; fixing to timber with screws	88.04	2.50	90.24	1.60	0.20	16.17	0.26	106.67	m	120.01
extra; ramps	53.36	2.50	54.69	2.40	0.30	24.26	0.39	79.34	nr	89.26
extra; wreaths	106.72	2.50	109.39	4.80	0.60	48.52	0.78	158.68	nr	178.52
extra; bends	53.36	2.50	54.69	1.45	0.18	14.65	0.24	69.58	nr	78.28
Isolated balustrades; 50 x 50mm balusters at 150mm centres, 75 x 100mm moulded handrail; housed construction										
914mm high; fixing to timber with screws	104.06	2.50	106.66	1.80	0.23	18.23	0.29	125.18	m	140.83
extra; ramps	53.36	2.50	54.69	2.40	0.30	24.26	0.39	79.34	nr	89.26
extra; wreaths	106.72	2.50	109.39	4.80	0.60	48.52	0.78	158.68	nr	178.52
extra; bends	53.36	2.50	54.69	1.45	0.18	14.65	0.24	69.58	nr	78.28
TIMBER STAIRS - Balustrades in European Oak, wrought, selected for transparent finish										
Isolated balustrades; 25 x 25mm balusters at 150mm centres, 50mm mopstick handrail, housed construction										
914mm high; fixing to timber with screws	117.39	2.50	120.32	2.05	0.26	20.75	0.33	141.40	m	159.08
extra; ramps	72.05	2.50	73.85	1.80	0.23	18.23	0.29	92.37	nr	103.92
extra; wreaths	144.07	2.50	147.67	3.60	0.45	36.39	0.59	184.65	nr	207.73
extra; bends	72.05	2.50	73.85	1.30	0.16	13.12	0.21	87.18	nr	98.08
Isolated balustrades; 38 x 38mm balusters at 150mm centres, 50mm mopstick handrail, housed construction										
914mm high; fixing to timber with screws	149.42	2.50	153.16	2.60	0.26	25.82	0.41	179.39	m	201.81
extra; ramps	72.05	2.50	73.85	1.80	0.23	18.23	0.29	92.37	nr	103.92
extra; wreaths	144.07	2.50	147.67	3.60	0.45	36.39	0.59	184.65	nr	207.73
extra; bends	72.05	2.50	73.85	1.30	0.16	13.12	0.21	87.18	nr	98.08
Isolated balustrades; 50 x 50mm balusters at 150mm centres, 50mm mopstick handrail, housed construction										
914mm high; fixing to timber with screws	181.43	2.50	185.97	3.00	0.38	30.36	0.49	216.81	m	243.92
extra; ramps	72.05	2.50	73.85	1.80	0.23	18.23	0.29	92.37	nr	103.92
extra; wreaths	144.07	2.50	147.67	3.60	0.45	36.39	0.59	184.65	nr	207.73
extra; bends	72.05	2.50	73.85	1.30	0.16	13.12	0.21	87.18	nr	98.08
Isolated balustrades; 25 x 25mm balusters at 150mm centres, 75 x 100mm moulded handrail; housed construction										
914mm high; fixing to timber with screws	144.04	2.50	147.64	2.15	0.27	21.74	0.35	169.73	m	190.95
extra; ramps	106.72	2.50	109.39	3.70	0.46	37.38	0.60	147.37	nr	165.79
extra; wreaths	213.44	2.50	218.78	7.45	0.93	75.29	1.21	295.28	nr	332.19
extra; bends	106.72	2.50	109.39	2.60	0.33	26.31	0.42	136.12	nr	153.14
Isolated balustrades; 38 x 38mm balusters at 150mm centres, 75 x 100mm moulded handrail; housed construction										
914mm high; fixing to timber with screws	176.10	2.50	180.50	2.70	0.34	27.31	0.44	208.25	m	234.28
extra; ramps	106.72	2.50	109.39	3.70	0.46	37.38	0.60	147.37	nr	165.79
extra; wreaths	213.44	2.50	218.78	7.45	0.93	75.29	1.21	295.28	nr	332.19
extra; bends	106.72	2.50	109.39	2.60	0.33	26.31	0.42	136.12	nr	153.14

Labour hourly rates: (except Specialists) Craft Operatives £9.23 Labourer £7.02 Rates are national average prices. Refer to REGIONAL VARIATIONS for indicative levels of overall pricing in regions	MATERIALS			LABOUR				RATES		
	Del to Site £	Waste %	Material Cost £	Craft Optve Hrs	Lab Hrs	Labour Cost £	Sunds £	Nett Rate £	Unit	Gross Rate (+12.5%) £

L30: STAIRS/WALKWAYS/BALUSTRADES Cont.

TIMBER STAIRS - Balustrades in European Oak, wrought, selected for transparent finish Cont.

Isolated balustrades; 50 x 50mm balusters at 150mm centres, 75 x 100mm moulded handrail; housed construction

914mm high; fixing to timber with screws	208.11	2.50	213.31	3.10	0.39	31.35	0.50	245.16	m	275.81
extra; ramps	106.72	2.50	109.39	3.70	0.46	37.38	0.60	147.37	nr	165.79
extra; wreaths	213.44	2.50	218.78	7.45	0.93	75.29	1.21	295.28	nr	332.19
extra; bends	106.72	2.50	109.39	2.60	0.33	26.31	0.42	136.12	nr	153.14

TIMBER STAIRS - Handrails in softwood, wrought

Associated handrails

50mm mopstick; fixing through metal backgrounds with screws	20.01	2.50	20.51	0.40	0.05	4.04	0.06	24.61	m	27.69
extra; ramps	36.03	2.50	36.93	1.20	0.15	12.13	0.20	49.26	nr	55.42
extra; wreaths	72.04	2.50	73.84	2.40	0.30	24.26	0.39	98.49	nr	110.80
extra; bends	36.03	2.50	36.93	1.20	0.15	12.13	0.20	49.26	nr	55.42
75 x 100mm; moulded; fixing through metal backgrounds with screws	33.36	2.50	34.19	0.50	0.06	5.04	0.08	39.31	m	44.22
extra; ramps	53.36	2.50	54.69	1.45	0.18	14.65	0.24	69.58	nr	78.28
extra; wreaths	106.72	2.50	109.39	2.90	0.36	29.29	0.47	139.15	nr	156.55
extra; bends	53.36	2.50	54.69	1.45	0.18	14.65	0.24	69.58	nr	78.28

TIMBER STAIRS - Handrails in African Mahogany, wrought, selected for transparent finish

Associated handrails

50mm mopstick; fixing through metal backgrounds with screws	25.02	2.50	25.65	0.60	0.08	6.10	0.10	31.85	m	35.83
extra; ramps	45.03	2.50	46.16	1.80	0.23	18.23	0.29	64.67	nr	72.76
extra; wreaths	90.05	2.50	92.30	3.60	0.45	36.39	0.59	129.28	nr	145.44
extra; bends	45.03	2.50	46.16	1.80	0.23	18.23	0.29	64.67	nr	72.76
75 x 100mm; moulded; fixing through metal backgrounds with screws	41.70	2.50	42.74	0.70	0.09	7.09	0.11	49.95	m	56.19
extra; ramps	66.71	2.50	68.38	2.15	0.27	21.74	0.35	90.47	nr	101.78
extra; wreaths	133.40	2.50	136.74	4.30	0.54	43.48	0.70	180.91	nr	203.53
extra; bends	66.71	2.50	68.38	2.15	0.27	21.74	0.35	90.47	nr	101.78

TIMBER STAIRS - Handrails in European Oak, wrought, selected for transparent finish

Associated handrails

50mm mopstick; fixing through metal backgrounds with screws	40.04	2.50	41.04	0.60	0.08	6.10	0.10	47.24	m	53.15
extra; ramps	72.05	2.50	73.85	1.80	0.23	18.23	0.29	92.37	nr	103.92
extra; wreaths	144.07	2.50	147.67	3.60	0.45	36.39	0.59	184.65	nr	207.73
extra; bends	72.05	2.50	73.85	1.80	0.23	18.23	0.29	92.37	nr	103.92
75 x 100mm; moulded; fixing through metal backgrounds with screws	66.72	2.50	68.39	0.70	0.09	7.09	0.11	75.59	m	85.04
extra; ramps	106.72	2.50	109.39	2.15	0.27	21.74	0.35	131.48	nr	147.91
extra; wreaths	213.44	2.50	218.78	4.30	0.54	43.48	0.70	262.96	nr	295.83
extra; bends	106.72	2.50	109.39	2.15	0.26	21.67	0.35	131.41	nr	147.83

METAL STAIRS AND BALUSTRADES - Staircases in steel

Straight flight staircases; 180 x 10mm flat stringers, shaped ends to top and bottom; 6 x 250mm on plain raised pattern plate treads, with 50 x 50 x 6mm shelf angles and 40 x 40 x 6mm stiffening bars, bolted to stringers; welded, cleated and bolted connections

770mm wide x 3000mm going x 2600mm rise overall; fixing to masonry with 4 nr Rawlbolts	-	-	Spclist	-	-	Spclist	-	1605.00	nr	1805.63
920mm wide x 3000mm going x 2600mm rise overall; fixing to masonry with 4 nr Rawlbolts	-	-	Spclist	-	-	Spclist	-	1840.00	nr	2070.00

Straight flight staircases, 180 x 10mm flat stringers, shaped ends to top and bottom; 6 x 250mm tray treads, with three 6mm diameter reinforcing bars welded to inside, and 50 x 50 x 6mm shelf angles, bolted to stringers; welded, cleated and bolted connections

770mm wide x 3000mm going x 2600mm rise overall; fixing to masonry with 4 nr Rawlbolts	-	-	Spclist	-	-	Spclist	-	1520.00	nr	1710.00
920mm wide x 3000mm going x 2600mm rise overall; fixing to masonry with 4 nr Rawlbolts	-	-	Spclist	-	-	Spclist	-	1570.00	nr	1766.25

Straight flight staircases; 178 x 76mm channel stringers, shaped ends to top and bottom; 6 x 250mm on plain raised pattern plate treads, with 50 x 50 x 6mm shelf angles and 40 x 40 x 6mm stiffening bars, bolted to stringers; welded, cleated and bolted connections

770mm wide x 3000mm going x 2600mm rise overall; fixing to masonry with 4 nr Rawlbolts	-	-	Spclist	-	-	Spclist	-	1745.00	nr	1963.13
920mm wide x 3000mm going x 2600mm rise overall; fixing to masonry with 4 nr Rawlbolts	-	-	Spclist	-	-	Spclist	-	1990.00	nr	2238.75

WINDOWS/DOORS/STAIRS

Labour hourly rates: (except Specialists) Craft Operatives £9.23 Labourer £7.02 Rates are national average prices. Refer to REGIONAL VARIATIONS for indicative levels of overall pricing in regions	MATERIALS			LABOUR				RATES		
	Del to Site £	Waste %	Material Cost £	Craft Optve Hrs	Lab Hrs	Labour Cost £	Sunds £	Nett Rate £	Unit	Gross Rate (+12.5%) £
L30: STAIRS/WALKWAYS/BALUSTRADES Cont.										
METAL STAIRS AND BALUSTRADES - Staircases in steel Cont.										

Straight flight staircases and balustrades; 180 x 10mm flat stringers, shaped ends to top and bottom; 6 x 250mm on plain raised pattern plate treads with 50 x 50 x 6mm shelf angles and 40 x 40 x 6mm stiffening bars, bolted to stringers; 915mm high balustrade to both sides consisting of 25mm diameter solid bar handrail and 32mm diameter solid bar standards at 250mm centres with base plate welded on and bolted to face of stringer, and ball type joints at intersections; welded, cleated and bolted connections
    770mm wide x 3000mm going x 2600mm rise overall;
    fixing to masonry with 4 nr Rawlbolts ....
	-	-	Spclist	-	-	Spclist	-	2405.00	nr	2705.63

Straight flight staircases and balustrades; 180 x 10mm flat stringers, shaped ends to top and bottom; 6 x 250mm on plain raised pattern plate treads with 50 x 50 x 6mm shelf angles and 40 x 40 x 6mm stiffening bars, bolted to stringers; 915mm high balustrade to both sides consisting of 25mm diameter solid bar handrail and 32mm diameter solid bar standards at 250mm centres with base plate welded on and bolted to top of stringer, and ball type joints at intersections; welded, cleated and bolted connections
    770mm wide x 3000mm going x 2600mm rise overall;
    fixing to masonry with 4 nr Rawlbolts ....
	-	-	Spclist	-	-	Spclist	-	2450.00	nr	2756.25

Straight flight staircases and balustrades; 180 x 10mm flat stringers, shaped ends to top and bottom; 6 x 250mm on plain raised pattern plate treads with 50 x 50 x 6mm shelf angles and 40 x 40 x 6mm stiffening bars, bolted to stringers; 915mm high balustrade to one side consisting of 25mm diameter solid bar handrail and 32mm diameter solid bar standards at 250mm centres with base plate welded on and bolted to face of stringer, and ball type joints at intersections; welded, cleated and bolted connections
    770mm wide x 3000mm going x 2600mm rise overall;
    fixing to masonry with 4 nr Rawlbolts ....
	-	-	Spclist	-	-	Spclist	-	2190.00	nr	2463.75

Straight flight staircases and balustrades; 180 x 10mm flat stringers, shaped ends to top and bottom; 6 x 250mm on plain raised pattern plate treads with 50 x 50 x 6mm shelf angles and 40 x 40 x 6mm stiffening bars, bolted to stringers; 915mm high balustrade to one side consisting of 25mm diameter solid bar handrail and 32mm diameter solid bar standards at 250mm centres with base plate welded on and bolted to top of stringer, and ball type joints at intersections; welded, cleated and bolted connections
    770mm wide x 3000mm going x 2600mm rise overall;
    fixing to masonry with 4 nr Rawlbolts ....
	-	-	Spclist	-	-	Spclist	-	2190.00	nr	2463.75

Straight flight staircases and balustrades; 180 x 10mm flat stringers, shaped ends to top and bottom; 6 x 250mm on plain raised pattern plate treads with 50 x 50 x 6mm shelf angles and 40 x 40 x 6mm stiffening bars, bolted to stringers; 1070mm high balustrade to both sides consisting of 25mm diameter solid bar handrail and 32mm diameter solid bar standards at 250mm centres with base plate welded on and bolted to top of stringer, and ball type joints at intersections; welded, cleated and bolted connections
    770mm wide x 3000mm going x 2600mm rise overall;
    fixing to masonry with 4 nr Rawlbolts ....
	-	-	Spclist	-	-	Spclist	-	2405.00	nr	2705.63

Straight flight staircases and balustrades; 180 x 10mm flat stringers, shaped ends to top and bottom; 6 x 250mm on plain raised pattern plate treads with 50 x 50 x 6mm shelf angles and 40 x 40 x 6mm stiffening bars, bolted to stringers; 1070mm high balustrade to one side consisting of 25mm diameter solid bar handrail and 32mm diameter solid bar standards at 250mm centres with base plate welded on and bolted to top of stringer, and ball type joints at intersections; welded, cleated and bolted connections
    770mm wide x 3000mm going x 2600mm rise overall;
    fixing to masonry with 4 nr Rawlbolts ....
	-	-	Spclist	-	-	Spclist	-	2245.00	nr	2525.63

Straight flight staircases and balustrades; 180 x 10mm flat stringers, shaped ends to top and bottom; 6 x 250mm on plain raised pattern plate treads with 50 x 50 x 6mm shelf angles and 40 x 40 x 6mm stiffening bars, bolted to stringers; 915mm high balustrade to one side consisting of 25mm diameter solid bar handrail and intermediate rail, and 32mm diameter solid bar standards at 250mm centres with base plate welded on and bolted to top of stringer, and ball type joints at intersections
    770mm wide x 3000mm going x 2600mm rise overall;
    fixing to masonry with 4 nr Rawlbolts ....
	-	-	Spclist	-	-	Spclist	-	2070.00	nr	2328.75

Labour hourly rates: (except Specialists) Craft Operatives £9.23 Labourer £7.02 Rates are national average prices. Refer to REGIONAL VARIATIONS for indicative levels of overall pricing in regions	MATERIALS			LABOUR				RATES		
	Del to Site £	Waste %	Material Cost £	Craft Optve Hrs	Lab Hrs	Labour Cost £	Sunds £	Nett Rate £	Unit	Gross Rate (+12.5%) £

L30:   STAIRS/WALKWAYS/BALUSTRADES Cont.

METAL STAIRS AND BALUSTRADES - Staircases in steel Cont.

Straight flight staircases and balustrades; 180 x 10mm flat stringers, shaped ends to top and bottom; 6 x 250mm on plain raised pattern plate treads with 50 x 50 x 6mm shelf angles and 40 x 40 x 6mm stiffening bars, bolted to stringers; 1070mm high balustrade to one side consisting of 25mm diameter solid bar handrail and intermediate rail, and 32mm diameter solid bar standards at 250mm centres with base plate welded on and bolted to top of stringer, and ball type joints at intersections										
770mm wide x 3000mm going x 2600mm rise overall; fixing to masonry with 4 nr Rawlbolts	-	-	Spclist	-	-	Spclist	-	2250.00	nr	2531.25
Quarter landing staircases, in two flights; 180 x 10mm flat stringers, shaped ends to top and bottom; 6 x 250mm on plain raised pattern plate treads, with 50 x 50 x 6mm shelf angles and 40 x 40 x 6mm stiffening bars, bolted to stringers; 6mm on plain raised pattern plate landing, welded on; 100 x 8mm flat kicking plates welded on; welded, cleated and bolted connections										
770mm wide x 2000mm going first flight excluding landing x 1000mm going second flight excluding landing x 2600mm rise overall; 770 x 770mm landing overall; fixing to masonry with 4 nr Rawlbolts ...	-	-	Spclist	-	-	Spclist	-	3585.00	nr	4033.13
920mm wide x 2000mm going first flight excluding landing x 1000mm going second flight excluding landing x 2600mm rise overall; 920 x 920mm landing overall; fixing to masonry with 4 nr Rawlbolts ...	-	-	Spclist	-	-	Spclist	-	3770.00	nr	4241.25
Half landing staircases, in two flights; 180 x 10mm flat stringers, shaped ends to top and bottom; 6 x 250mm on plain raised pattern plate treads, with 50 x 50 x 6mm shelf angles and 40 x 40 x 6mm stiffening bars, bolted to stringers; 6mm on plain raised pattern plate landing, welded on; 100 x 8mm flat kicking plates welded on; welded, cleated and bolted connections										
770mm wide x 2000mm going first flight excluding landing x 1000mm going second flight excluding landing x 2600mm rise overall; 770 x 1640mm landing overall; fixing to masonry with 4 nr Rawlbolts	-	-	Spclist	-	-	Spclist	-	4030.00	nr	4533.75
920mm wide x 2000mm going first flight excluding landing x 1000mm going second flight excluding landing x 2600mm rise overall; 920 x 1940mm landing overall; fixing to masonry with 4 nr Rawlbolts	-	-	Spclist	-	-	Spclist	-	4565.00	nr	5135.63

METAL STAIRS AND BALUSTRADES - Balustrades in steel

Isolated balustrades; 6 x 38mm flat core rail; 13mm diameter balusters at 250mm centres; welded fabrication ground to smooth finish; casting into mortices in concrete; wedging in position; temporary wedges										
838mm high; level	111.43	2.50	114.22	1.10	0.55	14.01	-	128.23	m	144.26
extra; ramps	19.00	2.50	19.48	1.10	0.55	14.01	-	33.49	nr	37.68
extra; wreaths	31.75	2.50	32.54	2.20	1.10	28.03	-	60.57	nr	68.14
extra; bends	17.26	2.50	17.69	1.10	0.55	14.01	-	31.71	nr	35.67
838mm high; raking	114.98	2.50	117.85	1.40	0.70	17.84	-	135.69	m	152.65
914mm high; level	122.49	2.50	125.55	1.10	0.55	14.01	-	139.57	m	157.01
extra; ramps	19.00	2.50	19.48	1.10	0.55	14.01	-	33.49	nr	37.68
extra; wreaths	31.75	2.50	32.54	2.20	1.10	28.03	-	60.57	nr	68.14
extra; bends	17.26	2.50	17.69	1.10	0.55	14.01	-	31.71	nr	35.67
914mm high; raking	126.18	2.50	129.33	1.40	0.70	17.84	-	147.17	m	165.57
Isolated balustrades; 6 x 38mm flat core rail; 13 x 13mm balusters at 250mm centres; welded fabrication ground to smooth finish; casting into mortices in concrete; wedging in position; temporary wedges										
838mm high; level	126.56	2.50	129.72	1.10	0.55	14.01	-	143.74	m	161.71
838mm high; raking	120.24	2.50	123.25	1.10	0.55	14.01	-	137.26	m	154.42
914mm high; level	126.18	2.50	129.33	1.10	0.55	14.01	-	143.35	m	161.27
914mm high; raking	134.04	2.50	137.39	1.10	0.55	14.01	-	151.41	m	170.33
Isolated balustrades; 13 x 51mm rounded handrail; 13mm diameter balusters at 250mm centres; welded fabrication ground to smooth finish; casting into mortices in concrete; wedging in position; temporary wedges										
838mm high; level	133.52	2.50	136.86	1.10	0.55	14.01	-	150.87	m	169.73
extra; ramps	20.72	2.50	21.24	1.10	0.55	14.01	-	35.25	nr	39.66
extra; wreaths	35.06	2.50	35.94	2.20	1.10	28.03	-	63.96	nr	71.96
extra; bends	19.00	2.50	19.48	1.10	0.55	14.01	-	33.49	nr	37.68
838mm high; raking	137.27	2.50	140.70	1.10	0.55	14.01	-	154.72	m	174.06
914mm high; level	142.93	2.50	146.50	1.10	0.55	14.01	-	160.52	m	180.58
extra; ramps	20.72	2.50	21.24	1.10	0.55	14.01	-	35.25	nr	39.66
extra; wreaths	35.07	2.50	35.95	2.20	1.10	28.03	-	63.97	nr	71.97
extra; bends	19.00	2.50	19.48	1.10	0.55	14.01	-	33.49	nr	37.68
914mm high; raking	149.05	2.50	152.78	1.10	0.55	14.01	-	166.79	m	187.64

Labour hourly rates: (except Specialists)
Craft Operatives £9.23 Labourer £7.02
Rates are national average prices.
Refer to REGIONAL VARIATIONS for indicative levels
of overall pricing in regions

	MATERIALS			LABOUR				RATES		
	Del to Site £	Waste %	Material Cost £	Craft Optve Hrs	Lab Hrs	Labour Cost £	Sunds £	Nett Rate £	Unit	Gross Rate (+12.5%) £

L30: STAIRS/WALKWAYS/BALUSTRADES Cont.

METAL STAIRS AND BALUSTRADES - Balustrades in steel Cont.

Isolated balustrades; 6 x 38mm flat core rail; 10 x 51mm flat bottom rail; 13mm diameter infill balusters at 250mm centres; 25 x 25mm standards at 3000mm centres; welded fabrication ground to smooth finish; casting into mortices in concrete; wedging in position; temporary wedges

	Del to Site £	Waste %	Material Cost £	Craft Optve Hrs	Lab Hrs	Labour Cost £	Sunds £	Nett Rate £	Unit	Gross Rate (+12.5%) £
838mm high; level	147.43	2.50	151.12	1.10	0.55	14.01	-	165.13	m	185.77
extra; ramps	35.81	2.50	36.71	1.10	0.55	14.01	-	50.72	nr	57.06
extra; wreaths	54.54	2.50	55.90	2.20	1.10	28.03	-	83.93	nr	94.42
extra; bends	31.68	2.50	32.47	1.10	0.55	14.01	-	46.49	nr	52.30
838mm high; raking	149.05	2.50	152.78	1.10	0.55	14.01	-	166.79	m	187.64
914mm high; level	162.05	2.50	166.10	1.10	0.55	14.01	-	180.12	m	202.63
extra; ramps	35.86	2.50	36.76	1.10	0.55	14.01	-	50.77	nr	57.12
extra; wreaths	44.64	2.50	45.76	2.20	1.10	28.03	-	73.78	nr	83.01
extra; bends	31.70	2.50	32.49	1.10	0.55	14.01	-	46.51	nr	52.32
914mm high; raking	167.37	2.50	171.55	1.10	0.55	14.01	-	185.57	m	208.76

METAL STAIRS AND BALUSTRADES - Balustrades in steel tubing, B.S.1387, medium grade; galvanised after fabrication

Isolated balustrades; 40mm diameter handrails; 40mm diameter standards at 1500mm centres; welded fabrication ground to smooth finish; casting into mortices in concrete; wedging in position; temporary wedges

	Del to Site £	Waste %	Material Cost £	Craft Optve Hrs	Lab Hrs	Labour Cost £	Sunds £	Nett Rate £	Unit	Gross Rate (+12.5%) £
838mm high; level	64.04	2.50	65.64	0.85	0.40	10.65	-	76.29	m	85.83
838mm high; raking	66.88	2.50	68.55	0.85	0.40	10.65	-	79.21	m	89.11
914mm high; level	78.05	2.50	80.00	0.85	0.40	10.65	-	90.65	m	101.99
914mm high; raking	77.44	2.50	79.38	0.85	0.40	10.65	-	90.03	m	101.28

Isolated balustrades; 40mm diameter handrails; 40mm diameter standards at 1500mm centres, with 6 x 75mm diameter fixing plates welded on to end, holes -3 nr; welded fabrication ground to smooth finish; fixing to masonry with Rawlbolts

	Del to Site £	Waste %	Material Cost £	Craft Optve Hrs	Lab Hrs	Labour Cost £	Sunds £	Nett Rate £	Unit	Gross Rate (+12.5%) £
838mm high; level	78.36	2.50	80.32	1.10	0.55	14.01	-	94.33	m	106.12
838mm high; raking	82.36	2.50	84.42	1.10	0.55	14.01	-	98.43	m	110.74
914mm high; level	92.98	2.50	95.30	1.10	0.55	14.01	-	109.32	m	122.98
914mm high; raking	96.33	2.50	98.74	1.10	0.55	14.01	-	112.75	m	126.85

METAL STAIRS AND BALUSTRADES - Spiral staircases in steel; one coat red oxide primer before delivery to site

Crescent 'H' range domestic spiral staircases; model 15 'H'; Crescent of Cambridge Ltd.; comprising tread modules complete with centre column, tread support, wooden tread, tread baluster, handrail section, PVC handrail cover, baluster wicket infill panel and tread riser bar, and with attachment brackets for fixing to floor and newel extending from centre column; assembling and bolting together

	Del to Site £	Waste %	Material Cost £	Craft Optve Hrs	Lab Hrs	Labour Cost £	Sunds £	Nett Rate £	Unit	Gross Rate (+12.5%) £
1568mm diameter; for floor to floor height units between 2160mm and 2640mm; plywood treads with hardwood edgings; fixing base and ground plates to timber	1030.00	2.50	1055.75	31.50	16.00	403.07	-	1458.82	nr	1641.17
extra; additional tread modules, 180 - 220mm per rise	76.00	2.50	77.90	1.65	0.85	21.20	-	99.10	nr	111.48
extra; mini landing to match treads, at upper floor level	189.00	2.50	193.72	2.10	1.05	26.75	-	220.48	nr	248.04
1568mm diameter; for floor to floor height units between 2160mm and 2640mm; plywood treads with carpet recess; fixing base and ground plates to timber	1175.00	2.50	1204.38	31.50	16.00	403.07	-	1607.44	nr	1808.37
extra; additional tread modules, 180 - 220mm per rise	89.00	2.50	91.22	2.10	1.05	26.75	-	117.98	nr	132.73
1568mm diameter; for floor to floor height units between 2160mm and 2640mm; hardwood treads, plain or with hardwood surround providing carpet recess; fixing base and ground plates to timber	1335.00	2.50	1368.38	31.50	16.00	403.07	-	1771.44	nr	1992.87
extra; additional tread modules, 180 - 220mm per rise	100.80	2.50	103.32	2.10	1.05	26.75	-	130.07	nr	146.33
extra; mini landing to match treads, at upper floor level	228.00	2.50	233.70	2.10	1.05	26.75	-	260.45	nr	293.01
stairwell balustrading with straight bar infill and PVC handrail cover	83.60	2.50	85.69	31.50	16.00	403.07	-	488.76	m	549.85
stairwell balustrading to match staircase including PVC handrail cover	138.20	2.50	141.66	1.95	1.00	25.02	-	166.67	m	187.51

Labour hourly rates: (except Specialists) Craft Operatives £9.23 Labourer £7.02 Rates are national average prices. Refer to REGIONAL VARIATIONS for indicative levels of overall pricing in regions	MATERIALS			LABOUR				RATES		
	Del to Site £	Waste %	Material Cost £	Craft Optve Hrs	Lab Hrs	Labour Cost £	Sunds £	Nett Rate £	Unit	Gross Rate (+12.5%) £
L30: STAIRS/WALKWAYS/BALUSTRADES Cont.										
METAL STAIRS AND BALUSTRADES - Spiral staircases in steel; one coat red oxide primer before delivery to site Cont.										
Crescent 'H' range domestic spiral staircases; model 18 'H'; Crescent of Cambridge Ltd.; comprising tread modules complete with centre column, tread support, wooden tread, tread baluster, handrail section, PVC handrail cover, baluster wicket infill panel and tread riser bar, and with attachment brackets for fixing to floor and newel extending from centre column; assembling and bolting together										
1956 diameter; for floor to floor height units between 2160mm and 2640mm; plywood treads with hardwood edgings; fixing base and ground plates to timber	1066.00	2.50	1092.65	31.50	16.00	403.07	-	1495.71	nr	1682.68
extra; additional tread modules. 180-220mm per rise	79.15	2.50	81.13	1.65	0.85	21.20	-	102.33	nr	115.12
extra; mini landing to match treads, at upper floor level	196.70	2.50	201.62	2.10	1.05	26.75	-	228.37	nr	256.92
1956 diameter; for floor to floor height units between 2160mm and 2640mm; plywood treads with carpet recess fixing base and ground plates to timber	1214.00	2.50	1244.35	34.00	17.00	433.16	-	1677.51	nr	1887.20
extra; additional tread modules, 180-220mm per rise	91.00	2.50	93.28	1.65	0.85	21.20	-	114.47	nr	128.78
1956 diameter; for floor to floor height units between 2160mm and 2640mm; hardwood treads, plain or with hardwood surround providing carpet recess; fixing base and ground plates to timber	1361.00	2.50	1395.03	31.50	16.50	406.57	-	1801.60	nr	2026.80
extra; additional tread modules, 180-220mm per rise	104.00	2.50	106.60	1.65	0.85	21.20	-	127.80	nr	143.77
extra; mini landing to match treads, at upper floor level	234.00	2.50	239.85	2.10	1.05	26.75	-	266.60	nr	299.93
stairwell balustrading with straight bar infill and PVC handrail cover	83.60	2.50	85.69	1.95	1.00	25.02	-	110.71	m	124.55
stairwell balustrading to match staircase including PVC handrail cover	138.20	2.50	141.66	1.95	1.00	25.02	-	166.67	nr	187.51
Crescent 'H' range domestic spiral staircases; model 20 'H'; Crescent of Cambridge Ltd.; comprising tread modules complete with centre column, tread support, wooden tread, tread baluster, handrail section, PVC handrail cover, baluster wicket infill panel and tread riser bar, and with attachment brackets for fixing to floor and newel extending from centre column; assembling and bolting together										
2092mm diameter; for floor to floor height units between 2160mm and 2640mm; plywood treads with hardwood edgings; fixing base and ground plates to timber	1107.00	2.50	1134.68	31.50	16.00	403.07	-	1537.74	nr	1729.96
extra; additional tread modules, 180 - 220mm per rise	81.50	2.50	83.54	1.65	0.85	21.20	-	104.73	nr	117.83
extra; mini landing to match treads, at upper floor level	203.00	2.50	208.07	2.10	1.05	26.75	-	234.83	nr	264.18
2092mm diameter; for floor to floor height units between 2160mm and 2640mm; plywood treads with carpet recess; fixing base and ground plates to timber	1247.00	2.50	1278.18	31.50	16.00	403.07	-	1681.24	nr	1891.40
extra; additional tread modules, 180 - 220mm per rise	92.40	2.50	94.71	1.65	0.85	21.20	-	115.91	nr	130.39
2092mm diameter; for floor to floor height units between 2160mm and 2640mm; hardwood treads, plain or with hardwood surround providing carpet recess; fixing base and ground plates to timber	1384.00	2.50	1418.60	31.50	16.00	403.07	-	1821.67	nr	2049.37
extra; additional tread modules, 180 - 220mm per rise	108.00	2.50	110.70	1.65	0.85	21.20	-	131.90	nr	148.38
extra; mini landing to match treads, at upper floor level	240.00	2.50	246.00	2.10	1.05	26.75	-	272.75	nr	306.85
stairwell balustrading with straight bar infill and PVC handrail cover	83.60	2.50	85.69	1.95	1.00	25.02	-	110.71	m	124.55
stairwell balustrading to match staircase including PVC handrail cover	138.20	2.50	141.66	1.95	1.00	25.02	-	166.67	m	187.51
Crescent 'S' range shop, office, factory and domestic spiral staircases; model 15 'S'; Crescent of Cambridge Ltd.; comprising tread modules complete with tread/collar section, baluster bar and section of handrail, and with upper floor attachment details with flanged column tube section, newel and top handrail baluster; assembling and bolting together										
1548mm diameter; for floor to floor height units between 2520mm and 3080mm; plain or ribbed treads; fixing base and ground plates and upper floor attachment to masonry with expanding bolts	1136.00	2.50	1164.40	37.00	18.50	471.38		1635.78	nr	1840.25
extra; additional tread modules, 180 - 220mm per rise	76.00	2.50	77.90	1.65	0.85	21.20	-	99.10	nr	111.48
1548mm diameter; for floor to floor height units between 2520mm and 3080mm; recessed treads; fixing base and ground plates and upper floor attachment to masonry with expanding bolts	1214.00	2.50	1244.35	39.00	19.50	496.86	-	1741.21	nr	1958.86
extra; additional tread modules, 180 - 220mm per rise	80.20	2.50	82.20	1.65	0.85	21.20	-	103.40	nr	116.33

Labour hourly rates: (except Specialists) Craft Operatives £9.23 Labourer £7.02. Rates are national average prices. Refer to REGIONAL VARIATIONS for indicative levels of overall pricing in regions	MATERIALS			LABOUR				RATES		
	Del to Site £	Waste %	Material Cost £	Craft Optve Hrs	Lab Hrs	Labour Cost £	Sunds £	Nett Rate £	Unit	Gross Rate (+12.5%) £

L30: STAIRS/WALKWAYS/BALUSTRADES Cont.

METAL STAIRS AND BALUSTRADES - Spiral staircases in steel; one coat red oxide primer before delivery to site Cont.

Crescent 'S' range shop, office, factory and domestic spiral staircases; model 15 'S'; Crescent of Cambridge Ltd.; comprising tread modules complete with tread/collar section, baluster bar and section of handrail, and with upper floor attachment details with flanged column tube section, newel and top handrail baluster; assembling and bolting together Cont.

	Del to Site £	Waste %	Material Cost £	Craft Optve Hrs	Lab Hrs	Labour Cost £	Sunds £	Nett Rate £	Unit	Gross Rate £
1548mm diameter; for floor to floor height units between 2520mm and 3080mm; recessed treads with studded rubber tread profile covers; fixing base and ground plates and upper floor attachment to masonry with expanding bolts	1547.00	2.50	1585.68	39.00	19.50	496.86	-	2082.53	nr	2342.85
extra; additional tread modules, 180 - 220mm per rise	102.60	2.50	105.17	1.65	0.85	21.20	-	126.36	nr	142.16
extra; 825 x 825mm plain or Durbar plate landing at upper floor level, complete with column attachment section, column newel, top handrail, baluster and handrail section, and balustrading complete to one side	377.00	2.50	386.43	3.20	1.60	40.77	-	427.19	nr	480.59
extra; 825 x 825mm recessed landing at upper floor level complete with column attachment section, column newel, top handrail, baluster and handrail section, and balustrading complete to one side	377.00	2.50	386.43	3.20	1.60	40.77	-	427.19	nr	480.59
extra; stairwell balustrade including handrail balusters and ground rail	72.24	2.50	74.05	1.95	1.00	25.02	-	99.06	m	111.45
extra; additional tread balusters	7.80	2.50	8.00	0.85	0.45	11.00	-	19.00	nr	21.37
extra; riser bars	6.40	2.50	6.56	0.86	0.45	11.10	-	17.66	nr	19.86
extra; PVC handrail cover	3.90	2.50	4.00	0.32	0.15	4.01	-	8.00	m	9.00

Crescent 'S' range shop, office, factory and domestic spiral staircases; model 20 'S'; Crescent of Cambridge Ltd.; comprising tread modules complete with tread/collar section, baluster bar and section of handrail, and with upper floor attachment details with flanged column tube section, newel and top handrail baluster; assembling and bolting together

	Del to Site £	Waste %	Material Cost £	Craft Optve Hrs	Lab Hrs	Labour Cost £	Sunds £	Nett Rate £	Unit	Gross Rate £
2072mm diameter; for floor to floor height units between 2520mm and 3080mm; plain or ribbed treads; fixing base and ground plates and upper floor attachment to masonry with expanding bolts	1264.00	2.50	1295.60	39.00	19.50	496.86	-	1792.46	nr	2016.52
extra; additional tread modules, 180 - 220mm per rise	83.60	2.50	85.69	1.65	0.85	21.20	-	106.89	nr	120.25
2072mm diameter; for floor to floor height units between 2520mm and 3080mm; recessed treads; fixing base and ground plates and upper floor attachment to masonry with expanding bolts	1348.00	2.50	1381.70	39.00	19.50	496.86	-	1878.56	nr	2113.38
extra; additional tread modules, 180 - 220mm per rise	89.00	2.50	91.22	1.65	0.85	21.20	-	112.42	nr	126.47
2072mm diameter; for floor to floor height units between 2520mm and 3080mm; recessed treads with studded rubber tread profile covers; fixing base and ground plates and upper floor attachment to masonry with expanding bolts	1772.00	2.50	1816.30	39.00	19.50	496.86	-	2313.16	nr	2602.30
extra; additional tread modules, 180 - 220mm per rise	119.50	2.50	122.49	1.65	0.85	21.20	-	143.68	nr	161.64
extra; 1100 x 1100mm plain or Durbar plate landing at upper floor level, complete with column attachment section, column newel, top handrail, baluster and handrail section, and balustrading complete to one side	412.00	2.50	422.30	3.20	1.60	40.77	-	463.07	nr	520.95
extra; 1100 x 1100mm recessed landing at upper floor level complete with column attachment section, column newel, top handrail, baluster and handrail section, and balustrading complete to one side	447.74	2.50	458.93	3.70	1.85	47.14	-	506.07	nr	569.33
extra; stairwell balustrade including handrail balusters and ground rail	72.24	2.50	74.05	1.95	1.00	25.02	-	99.06	m	111.45
extra; additional tread balusters	7.80	2.50	8.00	0.85	0.45	11.00	-	19.00	nr	21.37
extra; riser bars	6.40	2.50	6.56	0.85	0.45	11.00	-	17.56	nr	19.76
extra; PVC handrail cover	3.90	2.50	4.00	0.32	0.16	4.08	-	8.07	m	9.08

METAL STAIRS AND BALUSTRADES - Spiral staircases in steel; hot dip galvanised finish

Crescent 'S' range shop, office, factory and domestic spiral staircases; model 15 'S'; Crescent of Cambridge Ltd.; comprising tread modules complete with tread/collar section, baluster bar and section of handrail, and with upper floor attachment details with flanged column tube section, newel and top handrail baluster; assembling and bolting together

	Del to Site £	Waste %	Material Cost £	Craft Optve Hrs	Lab Hrs	Labour Cost £	Sunds £	Nett Rate £	Unit	Gross Rate £
1548mm diameter; for floor to floor height units between 2520mm and 3080mm; plain or ribbed treads; fixing base and ground plates and upper floor attachment to masonry with expanding bolts	1214.00	2.50	1244.35	39.00	19.50	496.86	-	1741.21	nr	1958.86
extra; additional tread modules, 180 - 220mm per rise	81.50	2.50	83.54	1.65	0.85	21.20	-	104.73	nr	117.83

Labour hourly rates: (except Specialists) Craft Operatives £9.23 Labourer £7.02 Rates are national average prices. Refer to REGIONAL VARIATIONS for indicative levels of overall pricing in regions	MATERIALS			LABOUR				RATES		
	Del to Site £	Waste %	Material Cost £	Craft Optve Hrs	Lab Hrs	Labour Cost £	Sunds £	Nett Rate £	Unit	Gross Rate (+12.5%) £

L30: STAIRS/WALKWAYS/BALUSTRADES Cont.

METAL STAIRS AND BALUSTRADES - Spiral staircases in steel; hot dip galvanised finish Cont.

	Del to Site £	Waste %	Material Cost £	Craft Optve Hrs	Lab Hrs	Labour Cost £	Sunds £	Nett Rate £	Unit	Gross Rate £
Crescent 'S' range shop, office, factory and domestic spiral staircases; model 15 'S'; Crescent of Cambridge Ltd.; comprising tread modules complete with tread/collar section, baluster bar and section of handrail, and with upper floor attachment details with flanged column tube section, newel and top handrail baluster; assembling and bolting together Cont.										
1548mm diameter; for floor to floor height units between 2520mm and 3080mm; recessed treads; fixing base and ground plates and upper floor attachment to masonry with expanding bolts	1292.00	2.50	1324.30	39.00	19.50	496.86	-	1821.16	nr	2048.80
extra; additional tread modules, 180 - 220mm per rise	87.00	2.50	89.17	1.65	0.85	21.20	-	110.37	nr	124.17
extra; 825 x 825mm plain or Durbar plate landing at upper floor level, complete with column attachment section, column newel, top handrail, baluster and handrail section, and balustrading complete to one side	385.00	2.50	394.63	3.15	1.60	40.31	-	434.93	nr	489.30
extra; 825 x 825mm recessed landing at upper floor level complete with column attachment section, column newel, top handrail, baluster and handrail section, and balustrading complete to one side	414.00	2.50	424.35	3.70	1.85	47.14	-	471.49	nr	530.42
extra; stairwell balustrade including handrail balusters and ground rail	78.20	2.50	80.16	1.95	1.00	25.02	-	105.17	m	118.32
extra; additional tread balusters	8.90	2.50	9.12	0.85	0.45	11.00	-	20.13	nr	22.64
extra; riser bars	7.80	2.50	8.00	0.85	0.45	11.00	-	19.00	nr	21.37
Crescent 'S' range shop, office, factory and domestic spiral staircases; model 20 'S'; Crescent of Cambridge Ltd.; comprising tread modules complete with tread/collar section, baluster bar and section of handrail, and with upper floor attachment details with flanged column tube section, newel and top handrail baluster; assembling and bolting together										
2072mm diameter; for floor to floor height units between 2520mm and 3080mm; plain or ribbed treads; fixing base and ground plates and upper floor attachment to masonry with expanding bolts	1337.00	2.50	1370.43	39.00	19.50	496.86	-	1867.29	nr	2100.70
extra; additional tread modules, 180 - 220mm per rise	92.30	2.50	94.61	1.65	0.85	21.20	-	115.80	nr	130.28
2072mm diameter; for floor to floor height units between 2520mm and 3080mm; recessed treads; fixing base and ground plates and upper floor attachment to masonry with expanding bolts	1398.00	2.50	1432.95	39.00	19.50	496.86	-	1929.81	nr	2171.04
extra; additional tread modules, 180 - 220mm per rise	95.60	2.50	97.99	1.65	0.85	21.20	-	119.19	nr	134.08
extra; 1100 x 1100mm plain or Durbar plate landing at upper floor level, complete with column attachment section, column newel, top handrail, baluster and handrail section, and balustrading complete to one side	468.00	2.50	479.70	3.20	1.60	40.77	-	520.47	nr	585.53
extra; 1100 x 1100mm recessed landing at upper floor level complete with column attachment section, column newel, top handrail, baluster and handrail section, and balustrading complete to one side	509.00	2.50	521.73	3.70	1.85	47.14	-	568.86	nr	639.97
extra; stairwell balustrade including handrail balusters and ground rail	78.20	2.50	80.16	1.95	1.00	25.02	-	105.17	m	118.32
extra; additional tread balusters	8.90	2.50	9.12	0.85	0.45	11.00	-	20.13	nr	22.64
extra; riser bars	7.80	2.50	8.00	0.85	0.45	11.00	-	19.00	nr	21.37
Crescent modular spiral fire escape staircases; model 15; Crescent of Cambridge Ltd.; comprising tread modules with plain or ribbed treads, and complete with centre core, spacers, baluster bar and section of handrail, tread riser bars and standard landing with balustrade one side, bars at 115mm centres; assembling and bolting together										
1548mm diameter; ground to first floor assembly, with base and ground plates; for floor to floor heights between 2520mm and 3080mm; fixing base plates and landing to masonry with expanding bolts	1982.00	2.50	2031.55	42.00	21.00	535.08	-	2566.63	nr	2887.46
extra; base extension	128.00	2.50	131.20	2.10	1.05	26.75	-	157.95	nr	177.70
1548mm diameter; intermediate floor assembly; for floor to floor heights between 2520mm and 3080mm; fixing landing to masonry with expanding bolts	1904.00	2.50	1951.60	42.00	21.00	535.08	-	2486.68	nr	2797.52
1548mm diameter; top floor assembly with two additional lengths of balustrade to landing; for floor to floor heights between 2520mm and 3080mm; fixing landing to masonry with expanding bolts	2210.00	2.50	2265.25	31.50	16.00	403.07	-	2668.32	nr	3001.85
extra; additional tread modules 180 - 220mm per rise	103.00	2.50	105.58	1.65	0.85	21.20	-	126.77	nr	142.62
extra; shield type infill panel in lieu of additional balusters to tread module	152.00	2.50	155.80	1.05	0.55	13.55	-	169.35	nr	190.52
extra; 'wicket bar' type infill panel in lieu of additional balusters to tread module	731.00	2.50	749.27	1.05	0.55	13.55	-	762.83	nr	858.18
extra; PVC handrail cover	3.91	2.50	4.01	0.32	0.16	4.08	-	8.08	m	9.10

WINDOWS/DOORS/STAIRS

Labour hourly rates: (except Specialists) Craft Operatives £9.23 Labourer £7.02 Rates are national average prices. Refer to REGIONAL VARIATIONS for indicative levels of overall pricing in regions	MATERIALS			LABOUR				RATES		
	Del to Site £	Waste %	Material Cost £	Craft Optve Hrs	Lab Hrs	Labour Cost £	Sunds £	Nett Rate £	Unit	Gross Rate (+12.5%) £
L30: STAIRS/WALKWAYS/BALUSTRADES Cont.										
METAL STAIRS AND BALUSTRADES - Spiral staircases in steel; hot dip galvanised finish Cont.										
Crescent modular spiral fire escape staircases; model 20; Crescent of Cambridge Ltd.; comprising tread modules with plain or ribbed treads, and complete with centre core, spacers, baluster bar and section of handrail, tread riser bars and standard landing with balustrade one side, bars at 115mm centres; assembling and bolting together										
2072mm diameter; ground to first floor assembly, with base and ground plates; for floor to floor heights between 2520mm and 3080mm; fixing base plates and landing to masonry with expanding bolts	2355.00	2.50	2413.88	42.00	21.00	535.08	-	2948.95	nr	3317.57
extra; base extension	141.30	2.50	144.83	2.10	1.05	26.75	-	171.59	nr	193.03
2072mm diameter; intermediate floor assembly; for floor to floor heights between 2520mm and 3080mm; fixing landing to masonry with expanding bolts ...	2277.00	2.50	2333.93	31.50	16.00	403.07	-	2736.99	nr	3079.11
2072mm diameter; top floor assembly with two additional lengths of balustrade to landing; for floor to floor heights between 2520mm and 3080mm; fixing landing to masonry with expanding bolts ...	2585.00	2.50	2649.62	31.50	16.00	403.07	-	3052.69	nr	3434.28
extra; additional tread modules 180 - 220mm per rise	102.00	2.50	104.55	1.65	0.85	21.20	-	125.75	nr	141.46
extra; 'wicket bar' type infill panel in lieu of additional balusters to tread module	128.20	2.50	131.41	1.05	0.55	13.55	-	144.96	nr	163.08
extra; PVC handrail cover	3.91	2.50	4.01	0.32	0.16	4.08	-	8.08	m	9.10
METAL STAIRS AND BALUSTRADES - Ladders in steel										
Vertical ladders; 65 x 10mm flat stringers; 65 x 10 x 250mm girth stringer brackets, bent once, bolted to stringers; 20mm diameter solid bar rungs welded to stringers; fixing to masonry with expanding bolts										
395mm wide x 3000mm long overall; brackets -6 nr; rungs -7 nr	415.00	2.50	425.38	11.00	11.00	178.75	-	604.13	nr	679.64
395mm wide x 5000mm long overall; brackets -8 nr; rungs -14 nr	415.00	2.50	425.38	13.50	13.50	219.38	-	644.75	nr	725.34
470mm wide x 3000mm long overall; brackets -6 nr; rungs -7 nr	350.00	2.50	358.75	11.00	11.00	178.75	-	537.50	nr	604.69
470mm wide x 5000mm long overall; brackets -8 nr; rungs -14 nr	620.00	2.50	635.50	13.50	13.50	219.38	-	854.88	nr	961.73
395mm wide x 3000mm long overall; stringers rising and returning 900mm above top fixing points; brackets -6 nr; rungs -7 nr	350.00	2.50	358.75	11.00	11.00	178.75	-	537.50	nr	604.69
395mm wide x 5000mm long overall, stringers rising and returning 900mm above top fixing points; brackets -8 nr; rungs -14 nr	620.00	2.50	635.50	13.50	13.50	219.38	-	854.88	nr	961.73
470mm wide x 3000mm long overall, stringers rising and returning 900mm above top fixing points; brackets -6 nr; rungs -7 nr	620.00	2.50	635.50	11.00	11.00	178.75	-	814.25	nr	916.03
470mm wide x 5000mm long overall, stringers rising and returning 900mm above top fixing points; brackets -8 nr; rungs -14 nr	750.00	2.50	768.75	13.50	13.52	219.52	-	988.27	nr	1111.80
Vertical ladders; 65 x 10mm flat stringers; 65 x 10 x 250mm girth stringer brackets, bent once, bolted to stringers; 20mm diameter solid bar rungs welded to stringers; 65 x 10mm flat back hoops 900mm diameter, welded to stringers										
395mm wide x 3000mm long overall; brackets -6 nr; rungs -7 nr; hoops -7 nr	750.00	2.50	768.75	16.50	16.50	268.13	-	1036.88	nr	1166.48
395mm wide x 5000mm long overall; brackets -8 nr; rungs -14 nr; hoops -14 nr	915.00	2.50	937.88	21.00	21.00	341.25	-	1279.13	nr	1439.02
470mm wide x 3000mm long overall; brackets -6 nr; rungs -7 nr; hoops -7 nr	855.00	2.50	876.38	16.50	16.50	268.13	-	1144.50	nr	1287.56
470mm wide x 5000mm long overall; brackets -8 nr; rungs -14 nr; hoops -14 nr	1010.00	2.50	1035.25	21.00	21.00	341.25	-	1376.50	nr	1548.56
395mm wide x 3000mm long overall, stringers rising and returning 900mm above top fixing points; brackets -6 nr; rungs -7 nr; hoops -7 nr	725.00	2.50	743.13	16.50	16.50	268.13	-	1011.25	nr	1137.66
395mm wide x 5000mm long overall, stringers rising and returning 900mm above top fixing points; brackets -8 nr; rungs -14 nr; hoops -14 nr	900.00	2.50	922.50	21.00	21.00	341.25	-	1263.75	nr	1421.72
470mm wide x 3000mm long overall, stringers rising and returning 900mm above top fixing points; brackets -6 nr; rungs -7 nr; hoops -7 nr	820.00	2.50	840.50	16.50	16.50	268.13	-	1108.63	nr	1247.20
470mm wide x 5000mm long overall, stringers rising and returning 900mm above top fixing points; brackets -8 nr; rungs -14 nr; hoops -14 nr	1070.00	2.50	1096.75	21.00	21.00	341.25	-	1438.00	nr	1617.75

Labour hourly rates: (except Specialists) Craft Operatives £9.23 Labourer £7.02 Rates are national average prices. Refer to REGIONAL VARIATIONS for indicative levels of overall pricing in regions	MATERIALS			LABOUR				RATES		
	Del to Site £	Waste %	Material Cost £	Craft Optve Hrs	Lab Hrs	Labour Cost £	Sunds £	Nett Rate £	Unit	Gross Rate (+12.5%) £

L30: STAIRS/WALKWAYS/BALUSTRADES Cont.

METAL STAIRS AND BALUSTRADES - Ladders in steel Cont.

Ship type ladders; 75 x 10mm flat stringers, shaped and cleated ends to tops and bottoms; 65 x 10 x 350mm girth stringer brackets, bent once, bolted to stringers; 8 x 75mm on plain raised pattern plate treads with 50 x 50 x 6mm shelf angles, bolted to stringers; 25mm diameter solid bar handrails to both sides, with flattened ends bolted to stringers; 32mm diameter x 175mm long solid bar standards with three way ball type joint at intersection with handrail, and flattened ends bolted to stringers; fixing to masonry with expanding bolts

420mm wide x 3000mm long overall, handrails rising and returning 900mm above top of stringers; brackets -4 nr; treads -12 nr; standards -6 nr ...	565.00	2.50	579.13	11.00	11.00	178.75	–	757.88	nr	852.61
420mm wide x 5000mm long overall, handrails rising and returning 900mm above top of stringers; brackets -4 nr; treads -21 nr; standards -10 nr ..	600.00	2.50	615.00	13.00	13.00	211.25	–	826.25	nr	929.53
470mm wide x 3000mm long overall, handrails rising and returning 900mm above top of stringers; brackets -4 nr; treads -12 nr; standards -6 nr ...	660.00	2.50	676.50	11.00	11.00	178.75	–	855.25	nr	962.16
470mm wide x 5000mm long overall, handrails rising and returning 900mm above top of stringers; brackets -4 nr; treads -21 nr; standards -10 nr ..	900.00	2.50	922.50	13.00	13.00	211.25	–	1133.75	nr	1275.47

Ship type ladders; 75 x 10mm flat stringers, shaped and cleated ends to tops and bottoms; 65 x 10 x 350mm girth stringer brackets, bent once, bolted to stringers; 8 x 75mm on plain raised pattern plate treads with 50 x 50 x 6mm shelf angles, bolted to stringers; 25mm diameter solid bar handrails to both sides, with flattened ends bolted to stringers; 32mm diameter x 330mm long solid bar standards with three way ball type joint at intersection with handrail, and flattened ends bolted to stringers; fixing to masonry with expanding bolts

420mm wide x 3000mm long overall, handrails rising and returning 900mm above top of stringers; brackets -4 nr; treads -12 nr; standards -6 nr ...	570.00	2.50	584.25	11.00	11.00	178.75	–	763.00	nr	858.38
420mm wide x 5000mm long overall, handrails rising and returning 900mm above top of stringers; brackets -4 nr; treads -21 nr; standards -10 nr ..	820.00	2.50	840.50	13.00	13.00	211.25	–	1051.75	nr	1183.22
470mm wide x 3000mm long overall, handrails rising and returning 900mm above top of stringers; brackets -4 nr; treads -12 nr; standards -6 nr ...	710.00	2.50	727.75	11.00	11.00	178.75	–	906.50	nr	1019.81
470mm wide x 5000mm long overall, handrails rising and returning 900mm above top of stringers; brackets -4 nr; treads -21 nr; standards -10 nr ..	915.00	2.50	937.88	13.00	13.00	211.25	–	1149.13	nr	1292.77

L40: GENERAL GLAZING

Glass; B.S.952, clear float

3mm thick to wood rebates with B.S.544 putty										
not exceeding 0.15m2	19.56	5.00	20.54	0.96	0.10	9.56	0.32	30.42	m2	34.22
0.15 - 4.00m2	19.56	5.00	20.54	0.48	0.05	4.78	0.16	25.48	m2	28.66
4mm thick to wood rebates with B.S.544 putty										
not exceeding 0.15m2	19.56	5.00	20.54	0.96	0.10	9.56	0.32	30.42	m2	34.22
0.15 - 4.00m2	19.56	5.00	20.54	0.48	0.05	4.78	0.16	25.48	m2	28.66
5mm thick to wood rebates with B.S.544 putty										
not exceeding 0.15m2	28.60	5.00	30.03	1.08	0.10	10.67	0.32	41.02	m2	46.15
0.15 - 4.00m2	28.60	5.00	30.03	0.54	0.05	5.34	0.16	35.53	m2	39.97
6mm thick to wood rebates with B.S.544 putty										
not exceeding 0.15m2	28.60	5.00	30.03	1.08	0.10	10.67	0.32	41.02	m2	46.15
0.15 - 4.00m2	28.60	5.00	30.03	0.54	0.05	5.34	0.16	35.53	m2	39.97
3mm thick to wood rebates with bradded wood beads (included elsewhere) and B.S.544 putty										
not exceeding 0.15m2	19.56	5.00	20.54	1.20	0.10	11.78	0.32	32.64	m2	36.72
0.15 - 4.00m2	19.56	5.00	20.54	0.60	0.05	5.89	0.16	26.59	m2	29.91
4mm thick to wood rebates with bradded wood beads (included elsewhere) and B.S.544 putty										
not exceeding 0.15m2	19.56	5.00	20.54	1.20	0.10	11.78	0.32	32.64	m2	36.72
0.15 - 4.00m2	19.56	5.00	20.54	0.60	0.05	5.89	0.16	26.59	m2	29.91
5mm thick to wood rebates with bradded wood beads (included elsewhere) and B.S.544 putty										
not exceeding 0.15m2	28.60	5.00	30.03	1.32	0.10	12.89	0.32	43.24	m2	48.64
0.15 - 4.00m2	28.60	5.00	30.03	0.66	0.05	6.44	0.16	36.63	m2	41.21
6mm thick to wood rebates with bradded wood beads (included elsewhere) and B.S.544 putty										
not exceeding 0.15m2	28.60	5.00	30.03	1.32	0.10	12.89	0.32	43.24	m2	48.64
0.15 - 4.00m2	28.60	5.00	30.03	0.66	0.05	6.44	0.16	36.63	m2	41.21
3mm thick to wood rebates with screwed wood beads and glazing strip (included elsewhere)										
not exceeding 0.15m2	19.56	5.00	20.54	1.68	0.10	16.21	0.32	37.07	m2	41.70
0.15 - 4.00m2	19.56	5.00	20.54	0.84	0.05	8.10	0.16	28.80	m2	32.40

Labour hourly rates: (except Specialists) Craft Operatives £9.23 Labourer £7.02 Rates are national average prices. Refer to REGIONAL VARIATIONS for indicative levels of overall pricing in regions	MATERIALS			LABOUR				RATES		
	Del to Site £	Waste %	Material Cost £	Craft Optve Hrs	Lab Hrs	Labour Cost £	Sunds £	Nett Rate £	Unit	Gross Rate (+12.5%) £
L40: GENERAL GLAZING Cont.										
Glass; B.S.952, clear float Cont.										
4mm thick to wood rebates with screwed wood beads and glazing strip (included elsewhere)										
not exceeding 0.15m2	19.56	5.00	20.54	1.68	0.10	16.21	0.32	37.07	m2	41.70
0.15 - 4.00m2	19.56	5.00	20.54	0.84	0.05	8.10	0.16	28.80	m2	32.40
5mm thick to wood rebates with screwed wood beads and glazing strip (included elsewhere)										
not exceeding 0.15m2	28.60	5.00	30.03	1.80	0.10	17.32	0.32	47.67	m2	53.62
0.15 - 4.00m2	28.60	5.00	30.03	0.90	0.05	8.66	0.16	38.85	m2	43.70
6mm thick to wood rebates with screwed wood beads and glazing strip (included elsewhere)										
not exceeding 0.15m2	28.60	5.00	30.03	1.80	0.10	17.32	0.32	47.67	m2	53.62
0.15 - 4.00m2	28.60	5.00	30.03	0.90	0.05	8.66	0.16	38.85	m2	43.70
3mm thick to metal rebates with metal casement glazing compound										
not exceeding 0.15m2	19.56	5.00	20.54	1.08	0.10	10.67	0.37	31.58	m2	35.53
0.15 - 4.00m2	19.56	5.00	20.54	0.54	0.05	5.34	0.16	26.03	m2	29.29
4mm thick to metal rebates with metal casement glazing compound										
not exceeding 0.15m2	19.56	5.00	20.54	1.08	0.10	10.67	0.37	31.58	m2	35.53
0.15 - 4.00m2	19.56	5.00	20.54	0.54	0.05	5.34	0.16	26.03	m2	29.29
5mm thick to metal rebates with metal casement glazing compound										
not exceeding 0.15m2	28.60	5.00	30.03	1.20	0.10	11.78	0.37	42.18	m2	47.45
0.15 - 4.00m2	28.60	5.00	30.03	0.60	0.05	5.89	0.16	36.08	m2	40.59
6mm thick to metal rebates with metal casement glazing compound										
not exceeding 0.15m2	28.60	5.00	30.03	1.20	0.10	11.78	0.37	42.18	m2	47.45
0.15 - 4.00m2	28.60	5.00	30.03	0.60	0.05	5.89	0.16	36.08	m2	40.59
3mm thick to metal rebates with clipped metal beads and gaskets (included elsewhere)										
not exceeding 0.15m2	19.56	5.00	20.54	1.44	0.10	13.99	0.32	34.85	m2	39.21
0.15 - 4.00m2	19.56	5.00	20.54	0.72	0.05	7.00	0.16	27.69	m2	31.16
4mm thick to metal rebates with clipped metal beads and gaskets (included elsewhere)										
not exceeding 0.15m2	19.56	5.00	20.54	1.44	0.10	13.99	0.32	34.85	m2	39.21
0.15 - 4.00m2	19.56	5.00	20.54	0.72	0.05	7.00	0.16	27.69	m2	31.16
5mm thick to metal rebates with clipped metal beads and gaskets (included elsewhere)										
not exceeding 0.15m2	28.60	5.00	30.03	1.56	0.10	15.10	0.32	45.45	m2	51.13
0.15 - 4.00m2	28.60	5.00	30.03	0.78	0.05	7.55	0.16	37.74	m2	42.46
6mm thick to metal rebates with clipped metal beads and gaskets (included elsewhere)										
not exceeding 0.15m2	28.60	5.00	30.03	1.56	0.10	15.10	0.32	45.45	m2	51.13
0.15 - 4.00m2	28.60	5.00	30.03	0.78	0.05	7.55	0.16	37.74	m2	42.46
Glass; B.S.952, rough cast										
6mm thick to wood rebates with B.S.544 putty										
not exceeding 0.15m2	36.58	5.00	38.41	1.08	0.10	10.67	0.32	49.40	m2	55.57
0.15 - 4.00m2	36.58	5.00	38.41	0.54	0.05	5.34	0.16	43.90	m2	49.39
6mm thick to wood rebates with bradded wood beads (included elsewhere) and B.S.544 putty										
not exceeding 0.15m2	36.58	5.00	38.41	1.32	0.10	12.89	0.32	51.61	m2	58.07
0.15 - 4.00m2	36.58	5.00	38.41	0.66	0.05	6.44	0.16	45.01	m2	50.64
6mm thick to wood rebates with screwed wood beads and glazing strip (included elsewhere)										
not exceeding 0.15m2	36.58	5.00	38.41	1.80	0.10	17.32	0.32	56.05	m2	63.05
0.15 - 4.00m2	36.58	5.00	38.41	0.90	0.05	8.66	0.16	47.23	m2	53.13
6mm thick to metal rebates with metal casement glazing compound										
not exceeding 0.15m2	36.58	5.00	38.41	1.20	0.10	11.78	0.37	50.56	m2	56.88
0.15 - 4.00m2	36.58	5.00	38.41	0.60	0.05	5.89	0.18	44.48	m2	50.04
6mm thick to metal rebates with clipped metal beads and gaskets (included elsewhere)										
not exceeding 0.15m2	36.58	5.00	38.41	1.56	0.10	15.10	0.37	53.88	m2	60.61
0.15 - 4.00m2	36.58	5.00	38.41	0.78	0.05	7.55	0.18	46.14	m2	51.91
Glass; B.S.952, Pyroshield clear										
6mm thick to wood rebates with screwed beads and intumescent glazing strip (included elsewhere)										
not exceeding 0.15m2	65.10	5.00	68.36	2.16	0.10	20.64	-	88.99	m2	100.12
not exceeding 0.15m2; aligning panes with adjacent panes	65.10	8.50	70.63	2.40	0.10	22.85	-	93.49	m2	105.17
0.15 - 4.00m2	65.10	5.00	68.36	1.08	0.05	10.32	-	78.67	m2	88.51
0.15 - 4.00m2; aligning panes with adjacent panes	65.10	8.50	70.63	1.20	0.05	11.43	-	82.06	m2	92.32

Labour hourly rates: (except Specialists) Craft Operatives £9.23 Labourer £7.02 Rates are national average prices. Refer to REGIONAL VARIATIONS for indicative levels of overall pricing in regions	MATERIALS			LABOUR				RATES		
	Del to Site £	Waste %	Material Cost £	Craft Optve Hrs	Lab Hrs	Labour Cost £	Sunds £	Nett Rate £	Unit	Gross Rate (+12.5%) £

L40: GENERAL GLAZING Cont.

Glass; B.S.952, Pyroshield clear Cont.

	Del to Site £	Waste %	Material Cost £	Craft Optve Hrs	Lab Hrs	Labour Cost £	Sunds £	Nett Rate £	Unit	Gross Rate £
6mm thick to wood rebates with screwed beads and Pyroglazing strip (included elsewhere)										
not exceeding 0.15m2	65.10	5.00	68.36	2.16	0.10	20.64	-	88.99	m2	100.12
not exceeding 0.15m2; aligning panes with adjacent panes	65.10	8.50	70.63	2.40	0.10	22.85	-	93.49	m2	105.17
0.15 - 4.00m2	65.10	5.00	68.36	1.08	0.05	10.32	-	78.67	m2	88.51
0.15 - 4.00m2; aligning panes with adjacent panes	65.10	8.50	70.63	1.20	0.05	11.43	-	82.06	m2	92.32
6mm thick to wood rebates with screwed beads and Lorient System 90 (included elsewhere)										
not exceeding 0.15m2	65.10	5.00	68.36	2.40	0.10	22.85	-	91.21	m2	102.61
not exceeding 0.15m2; aligning panes with adjacent panes	65.10	8.50	70.63	2.64	0.10	25.07	-	95.70	m2	107.67
0.15 - 4.00m2	65.10	5.00	68.36	1.20	0.05	11.43	-	79.78	m2	89.75
0.15 - 4.00m2; aligning panes with adjacent panes	65.10	8.50	70.63	1.32	0.05	12.53	-	83.17	m2	93.56
6mm thick to metal rebates with screwed metal beads and mild steel flat strips (included elsewhere)										
not exceeding 0.15m2	65.10	5.00	68.36	2.16	0.10	20.64	-	88.99	m2	100.12
not exceeding 0.15m2; aligning panes with adjacent panes	65.10	8.50	70.63	2.40	0.10	22.85	-	93.49	m2	105.17
0.15 - 4.00m2	65.10	5.00	68.36	1.08	0.05	10.32	-	78.67	m2	88.51
0.15 - 4.00m2; aligning panes with adjacent panes	65.10	8.50	70.63	1.20	0.05	11.43	-	82.06	m2	92.32

Glass; B.S.952, Pyroshield safety clear

	Del to Site £	Waste %	Material Cost £	Craft Optve Hrs	Lab Hrs	Labour Cost £	Sunds £	Nett Rate £	Unit	Gross Rate £
6mm thick to wood rebates with screwed beads and intumescent glazing strip (included elsewhere)										
not exceeding 0.15m2	76.66	5.00	80.49	2.16	0.10	20.64	-	101.13	m2	113.77
not exceeding 0.15m2; aligning panes with adjacent panes	76.66	8.50	83.18	2.40	0.10	22.85	-	106.03	m2	119.28
0.15 - 4.00m2	76.66	5.00	80.49	1.08	0.05	10.32	-	90.81	m2	102.16
0.15 - 4.00m2; aligning panes with adjacent panes	76.66	8.50	83.18	1.20	0.05	11.43	-	94.60	m2	106.43
6mm thick to wood rebates with screwed beads and Pyroglazing strip (included elsewhere)										
not exceeding 0.15m2	76.66	5.00	80.49	2.16	0.10	20.64	-	101.13	m2	113.77
not exceeding 0.15m2; aligning panes with adjacent panes	76.66	8.50	83.18	2.40	0.10	22.85	-	106.03	m2	119.28
0.15 - 4.00m2	76.66	5.00	80.49	1.08	0.05	10.32	-	90.81	m2	102.16
0.15 - 4.00m2; aligning panes with adjacent panes	76.66	8.50	83.18	1.20	0.05	11.43	-	94.60	m2	106.43
6mm thick to wood rebates with screwed beads and Lorient System 90 (included elsewhere)										
not exceeding 0.15m2	76.66	5.00	80.49	2.40	0.10	22.85	-	103.35	m2	116.27
not exceeding 0.15m2; aligning panes with adjacent panes	76.66	8.50	83.18	2.64	0.10	25.07	-	108.25	m2	121.78
0.15 - 4.00m2	76.66	5.00	80.49	1.20	0.05	11.43	-	91.92	m2	103.41
0.15 - 4.00m2; aligning panes with adjacent panes	76.66	8.50	83.18	1.32	0.05	12.53	-	95.71	m2	107.67
6mm thick to metal rebates with screwed metal beads and mild steel flat strips (included elsewhere)										
not exceeding 0.15m2	76.66	5.00	80.49	2.16	0.10	20.64	-	101.13	m2	113.77
not exceeding 0.15m2; aligning panes with adjacent panes	76.66	8.50	83.18	2.40	0.10	22.85	-	106.03	m2	119.28
0.15 - 4.00m2	76.66	5.00	80.49	1.08	0.05	10.32	-	90.81	m2	102.16
0.15 - 4.00m2; aligning panes with adjacent panes	76.66	8.50	83.18	1.20	0.05	11.43	-	94.60	m2	106.43

Glass; B.S.952, Pyroshield texture

	Del to Site £	Waste %	Material Cost £	Craft Optve Hrs	Lab Hrs	Labour Cost £	Sunds £	Nett Rate £	Unit	Gross Rate £
7mm thick to wood rebates with screwed beads and intumescent glazing strip (included elsewhere)										
not exceeding 0.15m2	30.12	5.00	31.63	2.16	0.10	20.64	-	52.26	m2	58.80
not exceeding 0.15m2; aligning panes with adjacent panes	30.12	8.50	32.68	2.40	0.10	22.85	-	55.53	m2	62.48
0.15 - 4.00m2	30.12	5.00	31.63	1.08	0.05	10.32	-	41.95	m2	47.19
0.15 - 4.00m2; aligning panes with adjacent panes	30.12	8.50	32.68	1.20	0.05	11.43	-	44.11	m2	49.62
7mm thick to wood rebates with screwed beads and Pyroglazing strip (included elsewhere)										
not exceeding 0.15m2	30.12	5.00	31.63	2.16	0.10	20.64	-	52.26	m2	58.80
not exceeding 0.15m2; aligning panes with adjacent panes	30.12	8.50	32.68	2.40	0.10	22.85	-	55.53	m2	62.48
0.15 - 4.00m2	30.12	5.00	31.63	1.08	0.05	10.32	-	41.95	m2	47.19
0.15 - 4.00m2; aligning panes with adjacent panes	30.12	8.50	32.68	1.20	0.05	11.43	-	44.11	m2	49.62
7mm thick to wood rebates with screwed beads and Lorient System 90 (included elsewhere)										
not exceeding 0.15m2	30.12	5.00	31.63	2.40	0.10	22.85	-	54.48	m2	61.29
not exceeding 0.15m2; aligning panes with adjacent panes	30.12	8.50	32.68	2.64	0.10	25.07	-	57.75	m2	64.97
0.15 - 4.00m2	30.12	5.00	31.63	1.20	0.05	11.43	-	43.05	m2	48.43
0.15 - 4.00m2; aligning panes with adjacent panes	30.12	8.50	32.68	1.32	0.05	12.53	-	45.21	m2	50.87
7mm thick to metal rebates with screwed metal beads and mild steel flat strips (included elsewhere)										
not exceeding 0.15m2	30.12	5.00	31.63	2.16	0.10	20.64	-	52.26	m2	58.80
not exceeding 0.15m2; aligning panes with adjacent panes	30.12	8.50	32.68	2.40	0.10	22.85	-	55.53	m2	62.48
0.15 - 4.00m2	30.12	5.00	31.63	1.08	0.05	10.32	-	41.95	m2	47.19
0.15 - 4.00m2; aligning panes with adjacent panes	30.12	8.50	32.68	1.20	0.05	11.43	-	44.11	m2	49.62

WINDOWS/DOORS/STAIRS

Labour hourly rates: (except Specialists) Craft Operatives £9.23 Labourer £7.02 Rates are national average prices. Refer to REGIONAL VARIATIONS for indicative levels of overall pricing in regions	MATERIALS			LABOUR				RATES		
	Del to Site £	Waste %	Material Cost £	Craft Optve Hrs	Lab Hrs	Labour Cost £	Sunds £	Nett Rate £	Unit	Gross Rate (+12.5%) £
L40: GENERAL GLAZING Cont.										
Glass; B.S.952, Pyroshield safety texture										
7mm thick to wood rebates with screwed beads and intumescent glazing strip (included elsewhere)										
not exceeding 0.15m2	39.18	5.00	41.14	2.16	0.10	20.64	-	61.78	m2	69.50
not exceeding 0.15m2; aligning panes with adjacent panes	39.18	8.50	42.51	2.40	0.10	22.85	-	65.36	m2	73.53
0.15 - 4.00m2	39.18	5.00	41.14	1.08	0.05	10.32	-	51.46	m2	57.89
0.15 - 4.00m2; aligning panes with adjacent panes	39.18	8.50	42.51	1.20	0.05	11.43	-	53.94	m2	60.68
7mm thick to wood rebates with screwed beads and Pyroglazing strip (included elsewhere)										
not exceeding 0.15m2	39.18	5.00	41.14	2.16	0.10	20.64	-	61.78	m2	69.50
not exceeding 0.15m2; aligning panes with adjacent panes	39.18	8.50	42.51	2.40	0.10	22.85	-	65.36	m2	73.53
0.15 - 4.00m2	39.18	5.00	41.14	1.08	0.05	10.32	-	51.46	m2	57.89
0.15 - 4.00m2; aligning panes with adjacent panes	39.18	8.50	42.51	1.20	0.05	11.43	-	53.94	m2	60.68
7mm thick to wood rebates with screwed beads and Lorient System 90 (included elsewhere)										
not exceeding 0.15m2	39.18	5.00	41.14	2.40	0.10	22.85	-	63.99	m2	71.99
not exceeding 0.15m2; aligning panes with adjacent panes	39.18	8.50	42.51	2.64	0.10	25.07	-	67.58	m2	76.03
0.15 - 4.00m2	39.18	5.00	41.14	1.20	0.05	11.43	-	52.57	m2	59.14
0.15 - 4.00m2; aligning panes with adjacent panes	39.18	8.50	42.51	1.32	0.05	12.53	-	55.04	m2	61.93
7mm thick to metal rebates with screwed metal beads and mild steel flat strips (included elsewhere)										
not exceeding 0.15m2	39.18	5.00	41.14	2.16	0.10	20.64	-	61.78	m2	69.50
not exceeding 0.15m2; aligning panes with adjacent panes	39.18	8.50	42.51	2.40	0.10	22.85	-	65.36	m2	73.53
0.15 - 4.00m2	39.18	5.00	41.14	1.08	0.05	10.32	-	51.46	m2	57.89
0.15 - 4.00m2; aligning panes with adjacent panes	39.18	8.50	42.51	1.20	0.05	11.43	-	53.94	m2	60.68
Glass; B.S.952, white patterned										
4mm thick to wood rebates with B.S.544 putty										
not exceeding 0.15m2	24.30	5.00	25.52	0.96	0.10	9.56	0.32	35.40	m2	39.82
0.15 - 4.00m2	24.30	5.00	25.52	0.48	0.05	4.78	0.16	30.46	m2	34.26
6mm thick to wood rebates with B.S.544 putty										
not exceeding 0.15m2	36.58	5.00	38.41	1.08	0.10	10.67	0.32	49.40	m2	55.57
0.15 - 4.00m2	36.58	5.00	38.41	0.54	0.05	5.34	0.16	43.90	m2	49.39
4mm thick to wood rebates with bradded wood beads (included elsewhere) and B.S.544 putty										
not exceeding 0.15m2	24.30	5.00	25.52	1.20	0.10	11.78	0.32	37.61	m2	42.31
0.15 - 4.00m2	24.30	5.00	25.52	0.60	0.05	5.89	0.16	31.56	m2	35.51
6mm thick to wood rebates with bradded wood beads (included elsewhere) and B.S.544 putty										
not exceeding 0.15m2	36.58	5.00	38.41	1.32	0.10	12.89	0.32	51.61	m2	58.07
0.15 - 4.00m2	36.58	5.00	38.41	0.66	0.05	6.44	0.16	45.01	m2	50.64
4mm thick to wood rebates with screwed wood beads and glazing strip (included elsewhere)										
not exceeding 0.15m2	24.30	5.00	25.52	1.68	0.10	16.21	0.32	42.04	m2	47.30
0.15 - 4.00m2	24.30	5.00	25.52	0.84	0.05	8.10	0.16	33.78	m2	38.00
6mm thick to wood rebates with screwed wood beads and glazing strip (included elsewhere)										
not exceeding 0.15m2	36.58	5.00	38.41	1.80	0.10	17.32	0.32	56.05	m2	63.05
0.15 - 4.00m2	36.58	5.00	38.41	0.90	0.05	8.66	0.16	47.23	m2	53.13
4mm thick to metal rebates with metal casement glazing compound										
not exceeding 0.15m2	24.30	5.00	25.52	1.08	0.10	10.67	0.37	36.56	m2	41.12
0.15 - 4.00m2	24.30	5.00	25.52	0.54	0.05	5.34	0.18	31.03	m2	34.91
6mm thick to metal rebates with metal casement glazing compound										
not exceeding 0.15m2	36.58	5.00	38.41	1.20	0.10	11.78	0.37	50.56	m2	56.88
0.15 - 4.00m2	36.58	5.00	38.41	0.60	0.05	5.89	0.18	44.48	m2	50.04
4mm thick to metal rebates with clipped metal beads and gaskets (included elsewhere)										
not exceeding 0.15m2	24.30	5.00	25.52	1.44	0.10	13.99	0.32	39.83	m2	44.81
0.15 - 4.00m2	24.30	5.00	25.52	0.72	0.05	7.00	0.16	32.67	m2	36.76
6mm thick to metal rebates with clipped metal beads and gaskets (included elsewhere)										
not exceeding 0.15m2	36.58	5.00	38.41	1.56	0.10	15.10	0.32	53.83	m2	60.56
0.15 - 4.00m2	36.58	5.00	38.41	0.78	0.05	7.55	0.16	46.12	m2	51.88
Glass; B.S.952, tinted patterned										
4mm thick to wood rebates with B.S.544 putty										
not exceeding 0.15m2	35.88	5.00	37.67	0.96	0.10	9.56	0.32	47.56	m2	53.50
0.15 - 4.00m2	35.88	5.00	37.67	0.48	0.05	4.78	0.16	42.62	m2	47.94
6mm thick to wood rebates with B.S.544 putty										
not exceeding 0.15m2	40.60	5.00	42.63	1.08	0.10	10.67	0.32	53.62	m2	60.32
0.15 - 4.00m2	40.60	5.00	42.63	0.54	0.05	5.34	0.16	48.13	m2	54.14

Labour hourly rates: (except Specialists) Craft Operatives £9.23 Labourer £7.02 Rates are national average prices. Refer to REGIONAL VARIATIONS for indicative levels of overall pricing in regions	MATERIALS			LABOUR				RATES		
	Del to Site £	Waste %	Material Cost £	Craft Optve Hrs	Lab Hrs	Labour Cost £	Sunds £	Nett Rate £	Unit	Gross Rate (+12.5%) £
L40: GENERAL GLAZING Cont.										
Glass; B.S.952, tinted patterned Cont.										
4mm thick to wood rebates with bradded wood beads (included elsewhere) and B.S.544 putty										
not exceeding 0.15m2	35.88	5.00	37.67	1.20	0.10	11.78	0.32	49.77	m2	55.99
0.15 – 4.00m2	35.88	5.00	37.67	0.60	0.05	5.89	0.16	43.72	m2	49.19
6mm thick to wood rebates with bradded wood beads (included elsewhere) and B.S.544 putty										
not exceeding 0.15m2	40.60	5.00	42.63	1.32	0.10	12.89	0.32	55.84	m2	62.82
0.15 – 4.00m2	40.60	5.00	42.63	0.66	0.05	6.44	0.16	49.23	m2	55.39
4mm thick to wood rebates with screwed wood beads and glazing strip (included elsewhere)										
not exceeding 0.15m2	35.88	5.00	37.67	1.68	0.10	16.21	0.32	54.20	m2	60.98
0.15 – 4.00m2	35.88	5.00	37.67	0.84	0.05	8.10	0.16	45.94	m2	51.68
6mm thick to wood rebates with screwed wood beads and glazing strip (included elsewhere)										
not exceeding 0.15m2	40.60	5.00	42.63	1.80	0.10	17.32	0.32	60.27	m2	67.80
0.15 – 4.00m2	40.60	5.00	42.63	0.90	0.05	8.66	0.16	51.45	m2	57.88
4mm thick to metal rebates with metal casement glazing compound										
not exceeding 0.15m2	35.88	5.00	37.67	1.08	0.10	10.67	0.32	48.66	m2	54.75
0.15 – 4.00m2	35.88	5.00	37.67	0.54	0.05	5.34	0.18	43.19	m2	48.59
6mm thick to metal rebates with metal casement glazing compound										
not exceeding 0.15m2	40.60	5.00	42.63	1.20	0.10	11.78	0.32	54.73	m2	61.57
0.15 – 4.00m2	40.60	5.00	42.63	0.60	0.05	5.89	0.18	48.70	m2	54.79
4mm thick to metal rebates with clipped metal beads and gaskets (included elsewhere)										
not exceeding 0.15m2	35.88	5.00	37.67	1.44	0.10	13.99	0.32	51.99	m2	58.49
0.15 – 4.00m2	35.88	5.00	37.67	0.72	0.05	7.00	0.16	44.83	m2	50.43
6mm thick to metal rebates with clipped metal beads and gaskets (included elsewhere)										
not exceeding 0.15m2	40.60	5.00	42.63	1.56	0.10	15.10	0.32	58.05	m2	65.31
0.15 – 4.00m2	40.60	5.00	42.63	0.78	0.05	7.55	0.16	50.34	m2	56.63
Patterns available										

4mm Bronze tint
   Autumn; Cotswold; Everglade; Sycamore
4mm white
   Arctic
6mm white
   Deep Flemish
4 and 6mm white
   Autumn; Cotswold; Driftwood; Everglade; Flemish;
   Linkon; Mayflower; Reeded; Stippolyte										

Glass; B.S.952, antisun float; grey										
4mm thick to wood rebates with B.S.544 putty										
not exceeding 2400 x 1200mm	37.31	5.00	39.18	0.72	0.12	7.49	0.23	46.89	m2	52.76
6mm thick to wood rebates with B.S.544 putty										
not exceeding 5950 x 3150mm	35.84	5.00	37.63	0.78	0.12	8.04	0.23	45.90	m2	51.64
10mm thick to wood rebates with B.S.544 putty										
not exceeding 5950 x 3150mm	96.26	5.00	101.07	1.80	0.12	17.46	0.29	118.82	m2	133.67
12mm thick to wood rebates with B.S.544 putty										
not exceeding 5950 x 3150mm	132.86	5.00	139.50	2.40	0.12	22.99	0.36	162.86	m2	183.21
4mm thick to wood rebates with bradded wood beads (included elsewhere) and B.S.544 putty										
not exceeding 2400 x 1200mm	37.31	5.00	39.18	0.84	0.12	8.60	0.23	48.00	m2	54.00
6mm thick to wood rebates with bradded wood beads (included elsewhere) and B.S.544 putty										
not exceeding 5950 x 3150mm	53.84	5.00	56.53	0.90	0.12	9.15	0.23	65.91	m2	74.15
10mm thick to wood rebates with bradded wood beads (included elsewhere) and B.S.544 putty										
not exceeding 5950 x 3150mm	96.26	5.00	101.07	1.92	0.12	18.56	0.29	119.93	m2	134.92
12mm thick to wood rebates with bradded wood beads (included elsewhere) and B.S.544 putty										
not exceeding 5950 x 3150mm	132.86	5.00	139.50	2.52	0.12	24.10	0.36	163.97	m2	184.46
4mm thick to wood rebates with screwed wood beads and glazing strip (included elsewhere) not exceeding 2400 x 1200mm	37.31	5.00	39.18	1.08	0.12	10.81	–	49.99	m2	56.23
6mm thick to wood rebates with screwed wood beads and glazing strip (included elsewhere) not exceeding 5950 x 3150mm	53.84	5.00	56.53	1.14	0.12	11.36	–	67.90	m2	76.38
10mm thick to wood rebates with screwed wood beads and glazing strip (included elsewhere) not exceeding 5950 x 3150mm	96.26	5.00	101.07	2.04	0.12	19.67	–	120.74	m2	135.84

WINDOWS/DOORS/STAIRS

Labour hourly rates: (except Specialists) Craft Operatives £9.23 Labourer £7.02 Rates are national average prices. Refer to REGIONAL VARIATIONS for indicative levels of overall pricing in regions	MATERIALS			LABOUR				RATES		
	Del to Site	Waste	Material Cost	Craft Optve	Lab	Labour Cost	Sunds	Nett Rate	Unit	Gross Rate (+12.5%)
	£	%	£	Hrs	Hrs	£	£	£		£
L40: GENERAL GLAZING Cont.										
Glass; B.S.952, antisun float; grey Cont.										
12mm thick to wood rebates with screwed wood beads and glazing strip (included elsewhere)										
not exceeding 5950 x 3150mm	132.86	5.00	139.50	2.64	0.12	25.21	-	164.71	m2	185.30
Glass; B.S.952, antisun float; bronze										
4mm thick to wood rebates with B.S.544 putty										
not exceeding 2400 x 1200mm	37.81	5.00	39.70	0.72	0.12	7.49	0.23	47.42	m2	53.35
6mm thick to wood rebates with B.S.544 putty										
not exceeding 5950 x 3150mm	53.84	5.00	56.53	0.78	0.12	8.04	0.23	64.80	m2	72.90
10mm thick to wood rebates with B.S.544 putty										
not exceeding 5950 x 3150mm	96.26	5.00	101.07	1.80	0.12	17.46	0.29	118.82	m2	133.67
12mm thick to wood rebates with B.S.544 putty										
not exceeding 5950 x 3150mm	132.86	5.00	139.50	2.40	0.12	22.99	0.36	162.86	m2	183.21
4mm thick to wood rebates with bradded wood beads (included elsewhere) and B.S.544 putty										
not exceeding 2400 x 1200mm	37.31	5.00	39.18	0.84	0.12	8.60	0.23	48.00	m2	54.00
6mm thick to wood rebates with bradded wood beads (included elsewhere) and B.S.544 putty										
not exceeding 5950 x 3150mm	53.84	5.00	56.53	0.90	0.12	9.15	0.23	65.91	m2	74.15
10mm thick to wood rebates with bradded wood beads (included elsewhere) and B.S.544 putty										
not exceeding 5950 x 3150mm	96.26	5.00	101.07	1.92	0.12	18.56	0.29	119.93	m2	134.92
12mm thick to wood rebates with bradded wood beads (included elsewhere) and B.S.544 putty										
not exceeding 5950 x 3150mm	132.86	5.00	139.50	2.52	0.12	24.10	0.36	163.97	m2	184.46
4mm thick to wood rebates with screwed wood beads and glazing strip (included elsewhere)										
not exceeding 2400 x 1200mm	37.31	5.00	39.18	1.08	0.12	10.81	-	49.99	m2	56.23
6mm thick to wood rebates with screwed wood beads and glazing strip (included elsewhere)										
not exceeding 5950 x 3150mm	53.84	5.00	56.53	1.14	0.12	11.36	-	67.90	m2	76.38
10mm thick to wood rebates with screwed wood beads and glazing strip (included elsewhere)										
not exceeding 5950 x 3150mm	96.26	5.00	101.07	2.04	0.12	19.67	-	120.74	m2	135.84
12mm thick to wood rebates with screwed wood beads and glazing strip (included elsewhere)										
not exceeding 5950 x 3150mm	132.86	5.00	139.50	2.64	0.12	25.21	-	164.71	m2	185.30
Glass; B.S.952, antisun float; green										
6mm thick to wood rebates with B.S.544 putty										
not exceeding 3150 x 2050mm	69.02	5.00	72.47	0.78	0.12	8.04	0.23	80.74	m2	90.84
6mm thick to wood rebates with bradded wood beads (included elsewhere) and B.S.544 putty										
not exceeding 3150 x 2050mm	69.02	5.00	72.47	0.90	0.12	9.15	0.23	81.85	m2	92.08
6mm thick to wood rebates with screwed wood beads and glazing strip (included elsewhere)										
not exceeding 3150 x 2050mm	69.02	5.00	72.47	1.14	0.12	11.36	-	83.84	m2	94.32
Glass; B.S.952, clear float										
10mm thick to wood rebates with screwed wood beads and glazing strip (included elsewhere)										
not exceeding 5950 x 3150mm	57.92	5.00	60.82	2.70	0.12	25.76	-	86.58	m2	97.40
12mm thick to wood rebates with screwed wood beads and glazing strip (included elsewhere)										
not exceeding 5950 x 3150mm	66.71	5.00	70.05	3.30	0.12	31.30	-	101.35	m2	114.02
15mm thick to wood rebates with screwed wood beads and glazing strip (included elsewhere)										
not exceeding 2950 x 2000mm	111.00	5.00	116.55	4.20	0.12	39.61	-	156.16	m2	175.68
19mm thick to wood rebates with screwed wood beads and glazing strip (included elsewhere)										
not exceeding 2950 x 2000mm	157.24	5.00	165.10	4.80	0.12	45.15	-	210.25	m2	236.53
25mm thick to wood rebates with screwed wood beads and glazing strip (included elsewhere)										
not exceeding 2950 x 2000mm	249.60	5.00	262.08	5.40	0.12	50.68	-	312.76	m2	351.86
Glass; B.S.952, toughened clear float										
4mm thick to metal rebates with screwed metal beads and gaskets (included elsewhere)										
not exceeding 2400 x 1300mm	28.86	5.00	30.30	0.72	0.12	7.49	-	37.79	m2	42.51

Labour hourly rates: (except Specialists) Craft Operatives £9.23 Labourer £7.02 Rates are national average prices. Refer to REGIONAL VARIATIONS for indicative levels of overall pricing in regions	MATERIALS			LABOUR				RATES		
	Del to Site £	Waste %	Material Cost £	Craft Optve Hrs	Lab Hrs	Labour Cost £	Sunds £	Nett Rate £	Unit	Gross Rate (+12.5%) £
L40: GENERAL GLAZING Cont.										
Glass; B.S.952, toughened clear float Cont.										
5mm thick to metal rebates with screwed metal beads and gaskets (included elsewhere) not exceeding 2500 x 1520mm	43.29	5.00	45.45	0.72	0.12	7.49	–	52.94	m2	59.56
6mm thick to metal rebates with screwed metal beads and gaskets (included elsewhere) not exceeding 2500 x 1520mm	43.29	5.00	45.45	0.78	0.12	8.04	–	53.50	m2	60.18
10mm thick to metal rebates with screwed metal beads and gaskets (included elsewhere) not exceeding 2500 x 1520mm	68.68	5.00	72.11	1.80	0.12	17.46	–	89.57	m2	100.77
12mm thick to metal rebates with screwed metal beads and gaskets (included elsewhere) not exceeding 2500 x 1520mm	134.58	5.00	141.31	2.40	0.12	22.99	–	164.30	m2	184.84
Glass; B.S.952, toughened white patterned										
4mm thick to metal rebates with screwed metal beads and gaskets (included elsewhere) not exceeding 2100 x 1300mm	58.18	5.00	61.09	0.72	0.12	7.49	–	68.58	m2	77.15
6mm thick to metal rebates with screwed metal beads and gaskets (included elsewhere) not exceeding 2100 x 1300mm	60.74	5.00	63.78	0.78	0.12	8.04	–	71.82	m2	80.80
Glass; B.S.952, toughened tinted patterned										
4mm thick to metal rebates with screwed metal beads and gaskets (included elsewhere) not exceeding 2100 x 1300mm	58.18	5.00	61.09	0.72	0.12	7.49	–	68.58	m2	77.15
6mm thick to metal rebates with screwed metal beads and gaskets (included elsewhere) not exceeding 2100 x 1300mm	60.74	5.00	63.78	0.78	0.12	8.04	–	71.82	m2	80.80
Patterns available										
4mm tinted Everglade										
4mm white Reeded										
4mm tinted Autumn; Cotswold										
4 and 6mm white Autumn; Cotswold; Driftwood; Everglade; Flemish; Linkon; Mayflower; Stippolyte; Sycamore										
Glass; B.S.952, toughened antisun float; grey										
4mm thick to metal rebates with screwed metal beads and gaskets (included elsewhere) not exceeding 2100 x 1250mm	52.91	5.00	55.56	0.72	0.12	7.49	–	63.04	m2	70.92
6mm thick to metal rebates with screwed metal beads and gaskets (included elsewhere) not exceeding 2500 x 1520mm	55.21	5.00	57.97	0.78	0.12	8.04	–	66.01	m2	74.26
10mm thick to metal rebates with screwed metal beads and gaskets (included elsewhere) not exceeding 2500 x 1520mm	109.23	5.00	114.69	1.80	0.12	17.46	–	132.15	m2	148.67
12mm thick to metal rebates with screwed metal beads and gaskets (included elsewhere) not exceeding 2500 x 1520mm	143.78	5.00	150.97	2.40	0.12	22.99	–	173.96	m2	195.71
Glass; B.S.952, toughened antisun float; bronze										
4mm thick to metal rebates with screwed metal beads and gaskets (included elsewhere) not exceeding 2100 x 1250mm	52.91	5.00	55.56	0.72	0.12	7.49	–	63.04	m2	70.92
6mm thick to metal rebates with screwed metal beads and gaskets (included elsewhere) not exceeding 2500 x 1520mm	55.21	5.00	57.97	0.78	0.12	8.04	–	66.01	m2	74.26
10mm thick to metal rebates with screwed metal beads and gaskets (included elsewhere) not exceeding 2500 x 1520mm	109.23	5.00	114.69	1.80	0.12	17.46	–	132.15	m2	148.67
12mm thick to metal rebates with screwed metal beads and gaskets (included elsewhere) not exceeding 2500 x 1520mm	143.78	5.00	150.97	2.40	0.12	22.99	–	173.96	m2	195.71
Glass; B.S.952, laminated safety, clear float										
4.4mm thick to metal rebates with screwed metal beads and gaskets (included elsewhere) not exceeding 2100 x 1200mm	43.96	5.00	46.16	1.92	0.12	18.56	–	64.72	m2	72.81
6.4mm thick to metal rebates with screwed metal beads and gaskets (included elsewhere) not exceeding 3210 x 2000mm	36.85	5.00	38.69	2.40	0.12	22.99	–	61.69	m2	69.40

WINDOWS/DOORS/STAIRS

Labour hourly rates: (except Specialists) Craft Operatives £9.23 Labourer £7.02 Rates are national average prices. Refer to REGIONAL VARIATIONS for indicative levels of overall pricing in regions	MATERIALS			LABOUR				RATES		
	Del to Site £	Waste %	Material Cost £	Craft Optve Hrs	Lab Hrs	Labour Cost £	Sunds £	Nett Rate £	Unit	Gross Rate (+12.5%) £
L40: GENERAL GLAZING Cont.										
Glass; B.S.952, laminated anti-bandit, clear float										
7.5mm thick to metal rebates with screwed metal beads and gaskets (included elsewhere)										
not exceeding 3210 x 2000mm	74.63	5.00	78.36	3.00	0.12	28.53	-	106.89	m2	120.26
9.5mm thick to metal rebates with screwed metal beads and gaskets (included elsewhere)										
not exceeding 3600 x 2500mm	73.12	5.00	76.78	3.30	0.12	31.30	-	108.08	m2	121.59
11.5mm thick to metal rebates with screwed metal beads and gaskets (included elsewhere)										
not exceeding 3180 x 2000mm	79.37	5.00	83.34	3.90	0.12	36.84	-	120.18	m2	135.20
not exceeding 4500 x 2500mm	96.54	5.00	101.37	3.90	0.12	36.84	-	138.21	m2	155.48
Materials resembling glass; UVA stabilised polycarbonate sheet, latex paper masked both sides										
3mm thick to metal rebates with screwed metal beads and gaskets (included elsewhere)										
standard grade	63.35	5.00	66.52	1.20	0.12	11.92	-	78.44	m2	88.24
4mm thick to metal rebates with screwed metal beads and gaskets (included elsewhere)										
standard grade	84.72	5.00	88.96	1.20	0.12	11.92	-	100.87	m2	113.48
5mm thick to metal rebates with screwed metal beads and gaskets (included elsewhere)										
standard grade	105.99	5.00	111.29	1.20	0.12	11.92	-	123.21	m2	138.61
6mm thick to metal rebates with screwed metal beads and gaskets (included elsewhere)										
standard grade	126.90	5.00	133.25	1.38	0.12	13.58	-	146.82	m2	165.18
8mm thick to metal rebates with screwed metal beads and gaskets (included elsewhere)										
standard grade	169.21	5.00	177.67	1.62	0.12	15.80	-	193.47	m2	217.65
9.5mm thick to metal rebates with screwed metal beads and gaskets (included elsewhere)										
standard grade	201.04	5.00	211.09	1.80	0.12	17.46	-	228.55	m2	257.12
3mm thick to metal rebates with screwed metal beads and gaskets (included elsewhere)										
abrasion resistant hard coated grade	98.24	5.00	103.15	1.20	0.12	11.92	-	115.07	m2	129.45
4mm thick to metal rebates with screwed metal beads and gaskets (included elsewhere)										
abrasion resistant hard coated grade	117.00	5.00	122.85	1.20	0.12	11.92	-	134.77	m2	151.61
5mm thick to metal rebates with screwed metal beads and gaskets (included elsewhere)										
abrasion resistant hard coated grade	140.29	5.00	147.30	1.20	0.12	11.92	-	159.22	m2	179.13
6mm thick to metal rebates with screwed metal beads and gaskets (included elsewhere)										
abrasion resistant hard coated grade	15.79	5.00	16.58	1.38	0.12	13.58	-	30.16	m2	33.93
8mm thick to metal rebates with screwed metal beads and gaskets (included elsewhere)										
abrasion resistant hard coated grade	189.77	5.00	199.26	1.62	0.12	15.80	-	215.05	m2	241.94
9.5mm thick to metal rebates with screwed metal beads and gaskets (included elsewhere)										
abrasion resistant hard coated grade	214.40	5.00	225.12	1.80	0.12	17.46	-	242.58	m2	272.90
Materials resembling glass; Meshlite vandal resistant glazing; Georgian, clear, smooth both faces or crinkle one face										
3mm thick to wood rebates with screwed wood beads (included elsewhere) and B.S.544 putty										
general purpose grade	30.66	5.00	32.19	1.15	0.12	11.46	0.15	43.80	m2	49.27
fire retardant grade, class 2	36.98	5.00	38.83	1.15	0.12	11.46	0.15	50.44	m2	56.74
fire retardant grade, class 0	43.43	5.00	45.60	1.15	0.12	11.46	0.15	57.21	m2	64.36
4mm thick to wood rebates with screwed wood beads (included elsewhere) and B.S.544 putty										
general purpose grade	39.61	5.00	41.59	1.15	0.12	11.46	0.15	53.20	m2	59.85
fire retardant grade, class 2	47.83	5.00	50.22	1.15	0.12	11.46	0.15	61.83	m2	69.56
fire retardant grade, class 0	56.31	5.00	59.13	1.15	0.12	11.46	0.15	70.73	m2	79.57
6mm thick to wood rebates with screwed wood beads (included elsewhere) and B.S.544 putty										
general purpose grade	52.95	5.00	55.60	1.30	0.12	12.84	0.15	68.59	m2	77.16
fire retardant grade, class 2	67.10	5.00	70.45	1.30	0.12	12.84	0.15	83.45	m2	93.88
fire retardant grade, class 0	81.49	5.00	85.56	1.30	0.12	12.84	0.15	98.56	m2	110.88
3mm thick to metal rebates with screwed metal beads and gaskets (included elsewhere)										
general purpose grade	30.66	5.00	32.19	1.15	0.12	11.46	-	43.65	m2	49.11
fire retardant grade, class 2	36.98	5.00	38.83	1.15	0.12	11.46	-	50.29	m2	56.57
fire retardant grade, class 0	43.43	5.00	45.60	1.15	0.12	11.46	-	57.06	m2	64.19

Labour hourly rates: (except Specialists) Craft Operatives £9.23 Labourer £7.02 Rates are national average prices. Refer to REGIONAL VARIATIONS for indicative levels of overall pricing in regions	MATERIALS			LABOUR				RATES		
	Del to Site £	Waste %	Material Cost £	Craft Optve Hrs	Lab Hrs	Labour Cost £	Sunds £	Nett Rate £	Unit	Gross Rate (+12.5%) £
L40: GENERAL GLAZING Cont.										
Materials resembling glass; Meshlite vandal resistant glazing; Georgian, clear, smooth both faces or crinkle one face Cont.										
4mm thick to metal rebates with screwed metal beads and gaskets (included elsewhere)										
general purpose grade	39.61	5.00	41.59	1.15	0.12	11.46	-	53.05	m2	59.68
fire retardant grade, class 2	47.83	5.00	50.22	1.15	0.12	11.46	-	61.68	m2	69.39
fire retardant grade, class 0	56.31	5.00	59.13	1.15	0.12	11.46	-	70.58	m2	79.41
6mm thick to metal rebates with screwed metal beads and gaskets (included elsewhere)										
general purpose grade	52.95	5.00	55.60	1.30	0.12	12.84	-	68.44	m2	76.99
fire retardant grade, class 2	67.10	5.00	70.45	1.30	0.12	12.84	-	83.30	m2	93.71
fire retardant grade, class 0	81.49	5.00	85.56	1.30	0.12	12.84	-	98.41	m2	110.71
Materials resembling glass; Meshlite Vandal resistant glazing; plain, opaque colours or clear, smooth both faces or crinkle one face										
2mm thick to wood rebates with screwed wood beads (included elsewhere) and B.S.544 putty										
general purpose grade	23.22	5.00	24.38	1.15	0.12	11.46	0.15	35.99	m2	40.49
3mm thick to wood rebates with screwed wood beads (included elsewhere) and B.S.544 putty										
general purpose grade	27.79	5.00	29.18	1.15	0.12	11.46	0.15	40.79	m2	45.88
fire retardant grade, class 2	34.95	5.00	36.70	1.15	0.12	11.46	0.15	48.30	m2	54.34
fire retardant grade, class 0	42.10	5.00	44.20	1.15	0.12	11.46	0.15	55.81	m2	62.79
4mm thick to wood rebates with screwed wood beads (included elsewhere) and B.S.544 putty										
general purpose grade	36.98	5.00	38.83	1.15	0.12	11.46	0.15	50.44	m2	56.74
fire retardant grade, class 2	45.42	5.00	47.69	1.15	0.12	11.46	0.15	59.30	m2	66.71
fire retardant grade, class 0	54.93	5.00	57.68	1.15	0.12	11.46	0.15	69.28	m2	77.94
6mm thick to wood rebates with screwed wood beads (included elsewhere) and B.S.544 putty										
general purpose grade	48.83	5.00	51.27	1.30	0.12	12.84	0.15	64.26	m2	72.30
fire retardant grade, class 2	64.38	5.00	67.60	1.30	0.12	12.84	0.15	80.59	m2	90.66
fire retardant grade, class 0	79.94	5.00	83.94	1.30	0.12	12.84	0.15	96.93	m2	109.04
2mm thick to metal rebates with screwed metal beads and gaskets (included elsewhere)										
general purpose grade; plain	23.22	5.00	24.38	1.15	0.12	11.46	-	35.84	m2	40.32
3mm thick to metal rebates with screwed metal beads and gaskets (included elsewhere)										
general purpose grade	27.79	5.00	29.18	1.15	0.12	11.46	-	40.64	m2	45.72
fire retardant grade, class 2	34.95	5.00	36.70	1.15	0.12	11.46	-	48.15	m2	54.17
fire retardant grade, class 0	42.10	5.00	44.20	1.15	0.12	11.46	-	55.66	m2	62.62
4mm thick to metal rebates with screwed metal beads and gaskets (included elsewhere)										
general purpose grade	36.98	5.00	38.83	1.15	0.12	11.46	-	50.29	m2	56.57
fire retardant grade, class 2	45.42	5.00	47.69	1.15	0.12	11.46	-	59.15	m2	66.54
fire retardant grade, class 0	54.93	5.00	57.68	1.15	0.12	11.46	-	69.13	m2	77.78
6mm thick to metal rebates with screwed metal beads and gaskets (included elsewhere)										
general purpose grade	48.83	5.00	51.27	1.30	0.12	12.84	-	64.11	m2	72.13
fire retardant grade, class 2	64.38	5.00	67.60	1.30	0.12	12.84	-	80.44	m2	90.50
fire retardant grade, class 0	79.94	5.00	83.94	1.30	0.12	12.84	-	96.78	m2	108.88
Materials resembling glass; Meshlite Vandal resistant glazing; diamond, clear, smooth both faces or crinkle one face										
3mm thick to wood rebates with screwed wood beads (included elsewhere) and B.S.544 putty										
general purpose grade	31.24	5.00	32.80	1.15	0.12	11.46	0.15	44.41	m2	49.96
fire retardant grade, class 2	36.95	5.00	38.80	1.15	0.12	11.46	0.15	50.40	m2	56.70
fire retardant grade, class 0	42.79	5.00	44.93	1.15	0.12	11.46	0.15	56.54	m2	63.60
4mm thick to wood rebates with screwed wood beads (included elsewhere) and B.S.544 putty										
general purpose grade	40.89	5.00	42.93	1.15	0.12	11.46	0.15	54.54	m2	61.36
fire retardant grade, class 2	48.39	5.00	50.81	1.15	0.12	11.46	0.15	62.42	m2	70.22
fire retardant grade, class 0	55.88	5.00	58.67	1.15	0.12	11.46	0.15	70.28	m2	79.07
6mm thick to wood rebates with screwed wood beads (included elsewhere) and B.S.544 putty										
general purpose grade	53.77	5.00	56.46	1.30	0.12	12.84	0.15	69.45	m2	78.13
fire retardant grade, class 2	67.50	5.00	70.88	1.30	0.12	12.84	0.15	83.87	m2	94.35
fire retardant grade, class 0	80.93	5.00	84.98	1.30	0.12	12.84	0.15	97.97	m2	110.21
3mm thick to metal rebates with screwed metal beads and gaskets (included elsewhere)										
general purpose grade	31.24	5.00	32.80	1.15	0.12	11.46	-	44.26	m2	49.79
fire retardant grade, class 2	36.95	5.00	38.80	1.15	0.12	11.46	-	50.25	m2	56.54
fire retardant grade, class 0	42.79	5.00	44.93	1.15	0.12	11.46	-	56.39	m2	63.43

WINDOWS/DOORS/STAIRS

Labour hourly rates: (except Specialists) Craft Operatives £9.23 Labourer £7.02 Rates are national average prices. Refer to REGIONAL VARIATIONS for indicative levels of overall pricing in regions	MATERIALS			LABOUR				RATES		
	Del to Site £	Waste %	Material Cost £	Craft Optve Hrs	Lab Hrs	Labour Cost £	Sunds £	Nett Rate £	Unit	Gross Rate (+12.5%) £
L40: GENERAL GLAZING Cont.										
Materials resembling glass; Meshlite Vandal resistant glazing; diamond, clear, smooth both faces or crinkle one face Cont.										
4mm thick to metal rebates with screwed metal beads and gaskets (included elsewhere)										
general purpose grade	40.89	5.00	42.93	1.15	0.12	11.46	-	54.39	m2	61.19
fire retardant grade, class 2	48.39	5.00	50.81	1.15	0.12	11.46	-	62.27	m2	70.05
fire retardant grade, class 0	55.88	5.00	58.67	1.15	0.12	11.46	-	70.13	m2	78.90
6mm thick to metal rebates with screwed metal beads and gaskets (included elsewhere)										
general purpose grade	53.77	5.00	56.46	1.30	0.12	12.84	-	69.30	m2	77.96
fire retardant grade, class 2	67.50	5.00	70.88	1.30	0.12	12.84	-	83.72	m2	94.18
fire retardant grade, class 0	80.93	5.00	84.98	1.30	0.12	12.84	-	97.82	m2	110.05
Factory made double glazed hermetically sealed units										
Two panes B.S.952, clear float 3 or 4mm thick; to metal rebates with screwed metal beads and gaskets (included elsewhere)										
521mm wide x 421mm high	11.00	5.00	11.55	1.15	-	10.61	-	22.16	nr	24.94
521mm wide x 621mm high	16.00	5.00	16.80	1.45	-	13.38	-	30.18	nr	33.96
740mm wide x 740mm high	27.50	5.00	28.88	2.00	-	18.46	-	47.34	nr	53.25
848mm wide x 848mm high	36.00	5.00	37.80	2.30	-	21.23	-	59.03	nr	66.41
1048mm wide x 1048mm high	55.00	5.00	57.75	2.90	-	26.77	-	84.52	nr	95.08
1148mm wide x 1248mm high	71.50	5.00	75.08	3.45	-	31.84	-	106.92	nr	120.28
Two panes B.S.952, clear float 5 or 6mm thick; to metal rebates with screwed metal beads and gaskets (included elsewhere)										
521mm wide x 421mm high	16.50	5.00	17.32	1.25	-	11.54	-	28.86	nr	32.47
521mm wide x 621mm high	24.00	5.00	25.20	1.60	-	14.77	-	39.97	nr	44.96
740mm wide x 740mm high	41.25	5.00	43.31	2.20	-	20.31	-	63.62	nr	71.57
848mm wide x 848mm high	54.00	5.00	56.70	2.50	-	23.07	-	79.78	nr	89.75
1048mm wide x 1048mm high	82.50	5.00	86.63	3.20	-	29.54	-	116.16	nr	130.68
1148mm wide x 1248mm high	107.25	5.00	112.61	3.80	-	35.07	-	147.69	nr	166.15
Inner pane B.S.952, clear float 4mm thick; outer pane B.S.952, white patterned 4mm thick; to metal rebates with screwed metal beads and gaskets (included elsewhere)										
521mm wide x 421mm high	13.20	5.00	13.86	1.15	-	10.61	-	24.47	nr	27.53
521mm wide x 621mm high	19.20	5.00	20.16	1.45	-	13.38	-	33.54	nr	37.74
740mm wide x 740mm high	33.00	5.00	34.65	2.00	-	18.46	-	53.11	nr	59.75
848mm wide x 848mm high	43.20	5.00	45.36	2.30	-	21.23	-	66.59	nr	74.91
1048mm wide x 1048mm high	66.00	5.00	69.30	2.90	-	26.77	-	96.07	nr	108.08
1148mm wide x 1248mm high	85.80	5.00	90.09	3.45	-	31.84	-	121.93	nr	137.18
Inner pane B.S.952, clear float 4mm thick; outer pane B.S.952, white patterned 6mm thick; to metal rebates with screwed metal beads and gaskets (included elsewhere)										
521mm wide x 421mm high	19.80	5.00	20.79	1.25	-	11.54	-	32.33	nr	36.37
521mm wide x 621mm high	28.80	5.00	30.24	1.60	-	14.77	-	45.01	nr	50.63
740mm wide x 740mm high	49.50	5.00	51.98	2.20	-	20.31	-	72.28	nr	81.32
848mm wide x 848mm high	64.80	5.00	68.04	2.50	-	23.07	-	91.11	nr	102.50
1048mm wide x 1048mm high	99.00	5.00	103.95	3.20	-	29.54	-	133.49	nr	150.17
1148mm wide x 1248mm high	128.70	5.00	135.13	3.80	-	35.07	-	170.21	nr	191.49
Polyester window films; Durable Berkeley Company Ltd.,										
3M Scotchshield Safety films; applying to glass										
type SH4CLL, optically clear	-	-	Spclist	-	-	Spclist	-	28.92	m2	32.53
type SH4S1L, combination solar/safety	-	-	Spclist	-	-	Spclist	-	34.70	m2	39.04
3M Scotchtint Solar Control films; applying to glass										
type P18, silver	-	-	Spclist	-	-	Spclist	-	31.81	m2	35.79
type RE15S1X, external	-	-	Spclist	-	-	Spclist	-	40.48	m2	45.54
type RE35NEARL, neutral	-	-	Spclist	-	-	Spclist	-	31.81	m2	35.79
3M Scotchtint Plus All Seasons insulating films; applying to glass										
type LE20 S1AR (silver)	-	-	Spclist	-	-	Spclist	-	40.48	m2	45.54
type LE35 AMARL (bronze)	-	-	Spclist	-	-	Spclist	-	40.48	m2	45.54
type LE50 AMARL (bronze)	-	-	Spclist	-	-	Spclist	-	40.48	m2	45.54
Glass doors; glass, B.S.952, toughened clear, polished plate; fittings finish BMA, satin chrome or polished chrome; prices include the provision of floor springs but exclude handles										
12mm thick										
750mm wide x 2150mm high	1077.00	-	1077.00	9.00	9.00	146.25	-	1223.25	nr	1376.16
762mm wide x 2134mm high	1017.00	-	1017.00	9.00	9.00	146.25	-	1163.25	nr	1308.66
800mm wide x 2150mm high	1114.00	-	1114.00	9.00	9.00	146.25	-	1260.25	nr	1417.78
838mm wide x 2134mm high	1069.00	-	1069.00	9.00	9.00	146.25	-	1215.25	nr	1367.16
850mm wide x 2150mm high	1146.00	-	1146.00	9.00	9.00	146.25	-	1292.25	nr	1453.78
900mm wide x 2150mm high	1185.00	-	1185.00	9.00	9.00	146.25	-	1331.25	nr	1497.66
914mm wide x 2134mm high	1113.00	-	1113.00	9.00	9.00	146.25	-	1259.25	nr	1416.66
950mm wide x 2150mm high	1258.00	-	1258.00	10.00	10.00	162.50	-	1420.50	nr	1598.06
1000mm wide x 2150mm high	1250.00	-	1250.00	10.00	10.00	162.50	-	1412.50	nr	1589.06
1100mm wide x 2150mm high	1489.00	-	1489.00	11.00	11.00	178.75	-	1667.75	nr	1876.22

Labour hourly rates: (except Specialists) Craft Operatives £9.23 Labourer £7.02 Rates are national average prices. Refer to REGIONAL VARIATIONS for indicative levels of overall pricing in regions	MATERIALS			LABOUR				RATES		
	Del to Site £	Waste %	Material Cost £	Craft Optve Hrs	Lab Hrs	Labour Cost £	Sunds £	Nett Rate £	Unit	Gross Rate (+12.5%) £
L40: GENERAL GLAZING Cont.										
Glass doors; glass, B.S.952, toughened clear, polished plate; fittings finish BMA, satin chrome or polished chrome; prices include the provision of floor springs but exclude handles Cont.										
12mm thick Cont.										
1200mm wide x 2150mm high	1559.00	-	1559.00	11.00	11.00	178.75	-	1737.75	nr	1954.97
Drilling										
Hole through sheet or float glass; not exceeding 6mm thick										
6 - 15mm diameter	3.49	-	3.49	-	-	-	-	3.49	nr	3.93
16 - 38mm diameter	4.93	-	4.93	-	-	-	-	4.93	nr	5.55
exceeding 38mm diameter	9.95	-	9.95	-	-	-	-	9.95	nr	11.19
Hole through sheet or float glass; not exceeding 10mm thick										
6 - 15mm diameter	4.52	-	4.52	-	-	-	-	4.52	nr	5.09
16 - 38mm diameter	6.97	-	6.97	-	-	-	-	6.97	nr	7.84
exceeding 38mm diameter	12.08	-	12.08	-	-	-	-	12.08	nr	13.59
Hole through sheet or float glass; not exceeding 12mm thick										
6 - 15mm diameter	5.60	-	5.60	-	-	-	-	5.60	nr	6.30
16 - 38mm diameter	7.92	-	7.92	-	-	-	-	7.92	nr	8.91
exceeding 38mm diameter	14.33	-	14.33	-	-	-	-	14.33	nr	16.12
Hole through sheet or float glass; not exceeding 19mm thick										
6 - 15mm diameter	7.00	-	7.00	-	-	-	-	7.00	nr	7.88
16 - 38mm diameter	9.95	-	9.95	-	-	-	-	9.95	nr	11.19
exceeding 38mm diameter	17.69	-	17.69	-	-	-	-	17.69	nr	19.90
Hole through sheet or float glass; not exceeding 25mm thick										
6 - 15mm diameter	8.80	-	8.80	-	-	-	-	8.80	nr	9.90
16 - 38mm diameter	12.47	-	12.47	-	-	-	-	12.47	nr	14.03
exceeding 38mm diameter	21.98	-	21.98	-	-	-	-	21.98	nr	24.73
For wired and laminated glass add 50%										
For countersunk holes add 33 1/3%										
Bedding edges of panes										
Wash leather strips										
to edges of 3mm thick glass or the like	0.61	5.00	0.64	0.06	-	0.55	-	1.19	m	1.34
to edges of 6mm thick glass or the like	0.74	5.00	0.78	0.07	-	0.65	-	1.42	m	1.60
Rubber glazing strips										
to edges of 3mm thick glass or the like	0.57	5.00	0.60	0.06	-	0.55	-	1.15	m	1.30
to edges of 6mm thick glass or the like	0.68	5.00	0.71	0.07	-	0.65	-	1.36	m	1.53
Hacking out existing glass; preparing for re-glazing										
Float glass										
wood rebates	-	-	-	0.44	0.44	7.15	0.17	7.32	m	8.23
wood rebates and screwed wood beads; storing beads for re-use	-	-	-	0.52	0.52	8.45	0.17	8.62	m	9.70
metal rebates	-	-	-	0.46	0.46	7.47	0.17	7.64	m	8.60
metal rebates and screwed metal beads; storing beads for re-use	-	-	-	0.55	0.55	8.94	0.17	9.11	m	10.25
Float glass behind guard bars in position										
wood rebates and screwed wood beads; storing beads for re-use	-	-	-	0.85	0.85	13.81	0.17	13.98	m	15.73
metal rebates and screwed metal beads; storing beads for re-use	-	-	-	0.90	0.90	14.63	0.17	14.80	m	16.64

WINDOWS/DOORS/STAIRS

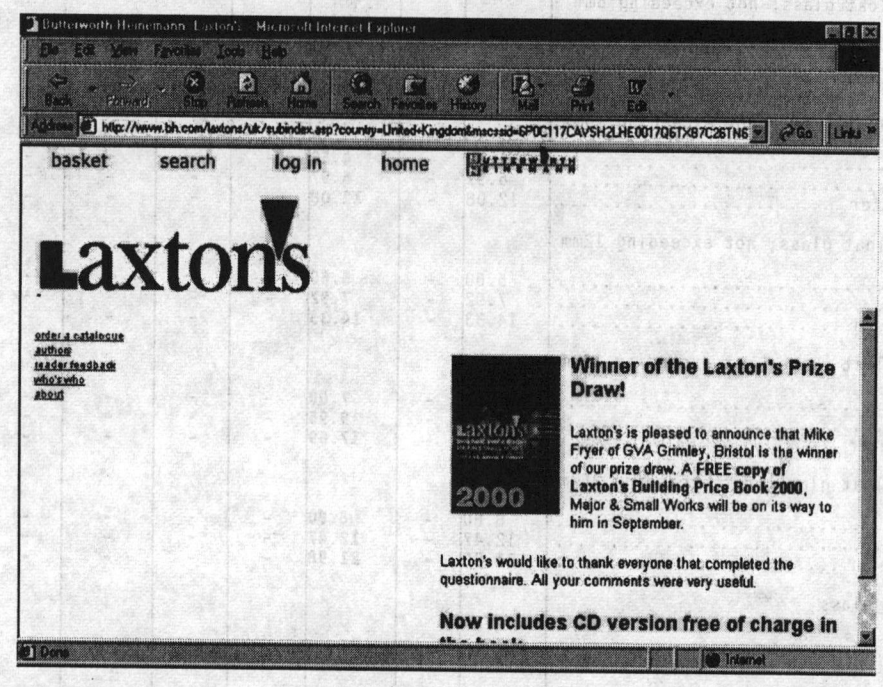

Surface finishes

Labour hourly rates: (except Specialists) Craft Operatives £9.23 Labourer £7.02 Rates are national average prices. Refer to REGIONAL VARIATIONS for indicative levels of overall pricing in regions	MATERIALS			LABOUR				RATES		
	Del to Site £	Waste %	Material Cost £	Craft Optve Hrs	Lab Hrs	Labour Cost £	Sunds £	Nett Rate £	Unit	Gross Rate (+12.5%) £
M10: CEMENT: SAND/CONCRETE SCREEDS/TOPPINGS										
Mortar, cement and sand (1:3) - screeds										
13mm work to walls on brickwork or blockwork base; one coat; screeded										
width exceeding 300mm	1.26	5.00	1.32	0.46	0.23	5.86	-	7.18	m2	8.08
width not exceeding 300mm	0.38	10.00	0.42	0.28	0.07	3.08	-	3.49	m	3.93
13mm work to walls on brickwork or blockwork base; one coat; trowelled										
width exceeding 300mm	1.26	5.00	1.32	0.61	0.23	7.24	-	8.57	m2	9.64
width not exceeding 300mm	0.38	10.00	0.42	0.37	0.07	3.91	-	4.32	m	4.87
13mm work to walls on brickwork or blockwork base; one coat; floated										
width exceeding 300mm	1.26	5.00	1.32	0.58	0.23	6.97	-	8.29	m2	9.33
width not exceeding 300mm	0.38	10.00	0.42	0.35	0.07	3.72	-	4.14	m	4.66
19mm work to floors on concrete base; one coat; screeded										
level and to falls only not exceeding 15 degrees from horizontal	1.85	5.00	1.94	0.26	0.17	3.59	0.20	5.74	m2	6.45
to falls and crossfalls and to slopes not exceeding 15 degrees from horizontal	1.85	5.00	1.94	0.38	0.17	4.70	0.20	6.84	m2	7.70
25mm work to floors on concrete base; one coat; screeded										
level and to falls only not exceeding 15 degrees from horizontal	2.43	5.00	2.55	0.29	0.18	3.94	0.20	6.69	m2	7.53
to falls and crossfalls and to slopes not exceeding 15 degrees from horizontal	2.43	5.00	2.55	0.41	0.18	5.05	0.20	7.80	m2	8.77
32mm work to floors on concrete base; one coat; screeded										
level and to falls only not exceeding 15 degrees from horizontal	3.11	5.00	3.27	0.32	0.29	4.99	0.20	8.45	m2	9.51
to falls and crossfalls and to slopes not exceeding 15 degrees from horizontal	3.11	5.00	3.27	0.44	0.20	5.47	0.20	8.93	m2	10.05
38mm work to floors on concrete base; one coat; screeded										
level and to falls only not exceeding 15 degrees from horizontal	3.70	5.00	3.88	0.34	0.21	4.61	0.20	8.70	m2	9.78
to falls and crossfalls and to slopes not exceeding 15 degrees from horizontal	3.70	5.00	3.88	0.46	0.21	5.72	0.20	9.80	m2	11.03
50mm work to floors on concrete base; one coat; screeded										
level and to falls only not exceeding 15 degrees from horizontal	4.86	5.00	5.10	0.40	0.24	5.38	0.20	10.68	m2	12.01
to falls and crossfalls and to slopes not exceeding 15 degrees from horizontal	4.86	5.00	5.10	0.52	0.24	6.48	0.20	11.79	m2	13.26
19mm work to floors on concrete base; one coat; trowelled										
level and to falls only not exceeding 15 degrees from horizontal	1.85	5.00	1.94	0.41	0.17	4.98	0.20	7.12	m2	8.01
to falls and crossfalls and to slopes not exceeding 15 degrees from horizontal	1.85	5.00	1.94	0.53	0.17	6.09	0.20	8.23	m2	9.26
25mm work to floors on concrete base; one coat; trowelled										
level and to falls only not exceeding 15 degrees from horizontal	2.43	5.00	2.55	0.44	0.18	5.32	0.20	8.08	m2	9.09
to falls and crossfalls and to slopes not exceeding 15 degrees from horizontal	2.43	5.00	2.55	0.56	0.18	6.43	0.20	9.18	m2	10.33

Labour hourly rates: (except Specialists) Craft Operatives £9.23 Labourer £7.02 Rates are national average prices. Refer to REGIONAL VARIATIONS for indicative levels of overall pricing in regions	MATERIALS			LABOUR				RATES		
	Del to Site	Waste	Material Cost	Craft Optve	Lab	Labour Cost	Sunds	Nett Rate	Unit	Gross Rate (+12.5%)
	£	%	£	Hrs	Hrs	£	£	£		£
M10: CEMENT: SAND/CONCRETE SCREEDS/TOPPINGS Cont.										
Mortar, cement and sand (1:3) - screeds Cont.										
32mm work to floors on concrete base; one coat; trowelled										
level and to falls only not exceeding 15 degrees from horizontal	3.11	5.00	3.27	0.47	0.20	5.74	0.20	9.21	m2	10.36
to falls and crossfalls and to slopes not exceeding 15 degrees from horizontal	3.11	5.00	3.27	0.59	0.20	6.85	0.20	10.32	m2	11.60
38mm work to floors on concrete base; one coat; trowelled										
level and to falls only not exceeding 15 degrees from horizontal	3.70	5.00	3.88	0.49	0.21	6.00	0.20	10.08	m2	11.34
to falls and crossfalls and to slopes not exceeding 15 degrees from horizontal	3.70	5.00	3.88	0.61	0.21	7.10	0.20	11.19	m2	12.59
50mm work to floors on concrete base; one coat; trowelled										
level and to falls only not exceeding 15 degrees from horizontal	4.86	5.00	5.10	0.55	0.24	6.76	0.20	12.06	m2	13.57
to falls and crossfalls and to slopes not exceeding 15 degrees from horizontal	4.86	5.00	5.10	0.67	0.24	7.87	0.20	13.17	m2	14.82
19mm work to floors on concrete base; one coat; floated										
level and to falls only not exceeding 15 degrees from horizontal	1.85	5.00	1.94	0.38	0.17	4.70	0.20	6.84	m2	7.70
to falls and crossfalls and to slopes not exceeding 15 degrees from horizontal	1.85	5.00	1.94	0.50	0.17	5.81	0.20	7.95	m2	8.94
25mm work to floors on concrete base; one coat; floated										
level and to falls only not exceeding 15 degrees from horizontal	2.43	5.00	2.55	0.41	0.18	5.05	0.20	7.80	m2	8.77
to falls and crossfalls and to slopes not exceeding 15 degrees from horizontal	2.43	5.00	2.55	0.53	0.18	6.16	0.20	8.91	m2	10.02
32mm work to floors on concrete base; one coat; floated										
level and to falls only not exceeding 15 degrees from horizontal	3.11	5.00	3.27	0.44	0.20	5.47	0.20	8.93	m2	10.05
to falls and crossfalls and to slopes not exceeding 15 degrees from horizontal	3.11	5.00	3.27	0.56	0.20	6.57	0.20	10.04	m2	11.29
38mm work to floors on concrete base; one coat; floated										
level and to falls only not exceeding 15 degrees from horizontal	3.70	5.00	3.88	0.46	0.21	5.72	0.20	9.80	m2	11.03
to falls and crossfalls and to slopes not exceeding 15 degrees from horizontal	3.70	5.00	3.88	0.58	0.21	6.83	0.20	10.91	m2	12.28
50mm work to floors on concrete base; one coat; floated										
level and to falls only not exceeding 15 degrees from horizontal	4.86	5.00	5.10	0.52	0.24	6.48	0.20	11.79	m2	13.26
to falls and crossfalls and to slopes not exceeding 15 degrees from horizontal	4.86	5.00	5.10	0.64	0.24	7.59	0.20	12.90	m2	14.51
Mortar, cement and sand (1:3) - paving										
25mm work to floors on concrete base; one coat; trowelled										
level and to falls only not exceeding 15 degrees from horizontal	2.43	5.00	2.55	0.44	0.18	5.32	0.20	8.08	m2	9.09
to falls and crossfalls and to slopes not exceeding 15 degrees from horizontal	2.43	5.00	2.55	0.56	0.18	6.43	0.20	9.18	m2	10.33
to slopes exceeding 15 degrees from horizontal	2.43	5.00	2.55	0.62	0.18	6.99	0.20	9.74	m2	10.95
32mm work to floors on concrete base; one coat; trowelled										
level and to falls only not exceeding 15 degrees from horizontal	3.11	5.00	3.27	0.47	0.20	5.74	0.20	9.21	m2	10.36
to falls and crossfalls and to slopes not exceeding 15 degrees from horizontal	3.11	5.00	3.27	0.59	0.20	6.85	0.20	10.32	m2	11.60
to slopes exceeding 15 degrees from horizontal	3.11	5.00	3.27	0.65	0.20	7.40	0.20	10.87	m2	12.23
38mm work to floors on concrete base; one coat; trowelled										
level and to falls only not exceeding 15 degrees from horizontal	3.70	5.00	3.88	0.49	0.21	6.00	0.20	10.08	m2	11.34
to falls and crossfalls and to slopes not exceeding 15 degrees from horizontal	3.70	5.00	3.88	0.61	0.21	7.10	0.20	11.19	m2	12.59
to slopes exceeding 15 degrees from horizontal	3.70	5.00	3.88	0.67	0.21	7.66	0.20	11.74	m2	13.21
50mm work to floors on concrete base; one coat; trowelled										
level and to falls only not exceeding 15 degrees from horizontal	4.86	5.00	5.10	0.55	0.24	6.76	0.20	12.06	m2	13.57
to falls and crossfalls and to slopes not exceeding 15 degrees from horizontal	4.86	5.00	5.10	0.67	0.24	7.87	0.20	13.17	m2	14.82
to slopes exceeding 15 degrees from horizontal	4.86	5.00	5.10	0.73	0.24	8.42	0.20	13.73	m2	15.44
If paving oil proofed, add										
25mm work	1.51	10.00	1.66	-	0.03	0.21	-	1.87	m2	2.11

Labour hourly rates: (except Specialists) Craft Operatives £9.23 Labourer £7.02 Rates are national average prices. Refer to REGIONAL VARIATIONS for indicative levels of overall pricing in regions	MATERIALS			LABOUR				RATES		
	Del to Site £	Waste %	Material Cost £	Craft Optve Hrs	Lab Hrs	Labour Cost £	Sunds £	Nett Rate £	Unit	Gross Rate (+12.5%) £
M10: CEMENT: SAND/CONCRETE SCREEDS/TOPPINGS Cont.										
Mortar, cement and sand (1:3) - paving Cont.										
If paving oil proofed, add Cont.										
32mm work	1.94	10.00	2.13	-	0.03	0.21	-	2.34	m2	2.64
38mm work	2.30	10.00	2.53	-	0.03	0.21	-	2.74	m2	3.08
50mm work	3.02	10.00	3.32	-	0.03	0.21	-	3.53	m2	3.97
19mm work to skirtings on brickwork or blockwork base										
height 100mm	0.28	10.00	0.31	0.80	0.05	7.74	-	8.04	m	9.05
height 150mm	0.42	10.00	0.46	0.90	0.05	8.66	-	9.12	m	10.26
25mm work to skirtings on brickwork or blockwork base										
height 100mm	0.36	10.00	0.40	0.85	0.05	8.20	-	8.59	m	9.67
height 150mm	0.55	10.00	0.60	0.95	0.05	9.12	-	9.72	m	10.94
Granolithic; cement and granite chippings (2:5); steel trowelled										
25mm work to floors on concrete base; one coat										
level and to falls only not exceeding 15 degrees from horizontal	4.61	5.00	4.84	0.48	0.21	5.90	0.41	11.16	m2	12.55
to falls and crossfalls and to slopes not exceeding 15 degrees from horizontal	4.61	5.00	4.84	0.60	0.21	7.01	0.41	12.26	m2	13.80
to slopes exceeding 15 degrees from horizontal	4.61	5.00	4.84	0.66	0.21	7.57	0.41	12.82	m2	14.42
32mm work to floors on concrete base; one coat										
level and to falls only not exceeding 15 degrees from horizontal	5.90	5.00	6.20	0.51	0.22	6.25	0.41	12.86	m2	14.46
to falls and crossfalls and to slopes not exceeding 15 degrees from horizontal	5.90	5.00	6.20	0.63	0.22	7.36	0.41	13.96	m2	15.71
to slopes exceeding 15 degrees from horizontal	5.90	5.00	6.20	0.69	0.22	7.91	0.41	14.52	m2	16.33
38mm work to floors on concrete base; one coat										
level and to falls only not exceeding 15 degrees from horizontal	7.00	5.00	7.35	0.55	0.24	6.76	0.41	14.52	m2	16.34
to falls and crossfalls and to slopes not exceeding 15 degrees from horizontal	7.00	5.00	7.35	0.67	0.24	7.87	0.41	15.63	m2	17.58
to slopes exceeding 15 degrees from horizontal	7.00	5.00	7.35	0.73	0.24	8.42	0.41	16.18	m2	18.21
50mm work to floors on concrete base; one coat										
level and to falls only not exceeding 15 degrees from horizontal	9.22	5.00	9.68	0.61	0.27	7.53	0.41	17.62	m2	19.82
to falls and crossfalls and to slopes not exceeding 15 degrees from horizontal	9.22	5.00	9.68	0.73	0.27	8.63	0.41	18.72	m2	21.06
to slopes exceeding 15 degrees from horizontal	9.22	5.00	9.68	0.79	0.27	9.19	0.41	19.28	m2	21.69
If paving tinted, add according to tint										
25mm work	4.01	10.00	4.41	-	0.03	0.21	-	4.62	m2	5.20
32mm work	5.13	10.00	5.64	-	0.03	0.21	-	5.85	m2	6.59
38mm work	6.10	10.00	6.71	-	0.04	0.28	-	6.99	m2	7.86
50mm work	8.02	10.00	8.82	-	0.05	0.35	-	9.17	m2	10.32
If carborundum trowelled into surface										
add	2.75	10.00	3.02	0.10	0.05	1.27	-	4.30	m2	4.84
25mm linings to channels on concrete base										
150mm girth on face; to falls	1.04	10.00	1.14	0.25	0.05	2.66	0.06	3.86	m	4.35
225mm girth on face; to falls	1.56	10.00	1.72	0.30	0.05	3.12	0.09	4.93	m	5.54
32mm linings to channels on concrete base										
150mm girth on face; to falls	1.33	10.00	1.46	0.28	0.05	2.94	0.06	4.46	m	5.02
225mm girth on face; to falls	1.99	10.00	2.19	0.33	0.05	3.40	0.09	5.68	m	6.39
38mm linings to channels on concrete base										
150mm girth on face; to falls	1.58	10.00	1.74	0.30	0.05	3.12	0.06	4.92	m	5.53
225mm girth on face; to falls	2.36	10.00	2.60	0.35	0.05	3.58	0.09	6.27	m	7.05
50mm linings to channels on concrete base										
150mm girth on face; to falls	2.07	10.00	2.28	0.35	0.05	3.58	0.06	5.92	m	6.66
225mm girth on face; to falls	3.11	10.00	3.42	0.40	0.05	4.04	0.09	7.55	m	8.50
13mm work to skirtings on concrete base										
height 150mm	0.54	10.00	0.59	0.85	0.05	8.20	-	8.79	m	9.89
19mm work to skirtings on concrete base										
height 150mm	0.79	10.00	0.87	0.90	0.05	8.66	-	9.53	m	10.72
25mm work to skirtings on concrete base										
height 150mm	1.04	10.00	1.14	0.95	0.05	9.12	-	10.26	m	11.55
32mm work to skirtings on concrete base										
height 150mm	1.33	10.00	1.46	1.00	0.05	9.58	-	11.04	m	12.42
Rounded angles and intersections										
10 - 100mm radius	-	-	-	-	0.20	1.40	-	1.40	m	1.58
Surface hardeners on paving										
Proprietary surface hardener (Sodium Silicate Base)										
two coats	0.84	5.00	0.88	-	0.23	1.61	-	2.50	m2	2.81
three coats	1.26	5.00	1.32	-	0.29	2.04	-	3.36	m2	3.78
Polyurethane floor sealer										
two coats	1.69	5.00	1.77	-	0.23	1.61	-	3.39	m2	3.81

SURFACE FINISHES

Labour hourly rates: (except Specialists) Craft Operatives £9.23 Labourer £7.02 Rates are national average prices. Refer to REGIONAL VARIATIONS for indicative levels of overall pricing in regions	MATERIALS			LABOUR				RATES		
	Del to Site £	Waste %	Material Cost £	Craft Optve Hrs	Lab Hrs	Labour Cost £	Sunds £	Nett Rate £	Unit	Gross Rate (+12.5%) £
M10: CEMENT: SAND/CONCRETE SCREEDS/TOPPINGS Cont.										
Surface hardeners on paving Cont.										
Polyurethane floor sealer Cont.										
three coats	2.46	5.00	2.58	-	0.29	2.04	-	4.62	m2	5.20
Nitoflor Lithurin concrete surface dressing										
one coat	0.39	10.00	0.43	-	0.21	1.47	-	1.90	m2	2.14
two coats	0.53	10.00	0.58	-	0.30	2.11	-	2.69	m2	3.03
If on old floors										
add for cleaning and degreasing	0.68	5.00	0.71	-	0.40	2.81	-	3.52	m2	3.96
Vermiculite screed consisting of cement and Vermiculite aggregate (1:6) finished with 20mm cement and sand (1:4) screeded bed										
45mm work to roofs on concrete base; two coats to falls and crossfalls and to slopes not exceeding 15 degrees from horizontal	4.62	5.00	4.85	0.55	0.60	9.29	0.20	14.34	m2	16.13
60mm work to roofs on concrete base; two coats to falls and crossfalls and to slopes not exceeding 15 degrees from horizontal	6.34	5.00	6.66	0.60	0.65	10.10	0.20	16.96	m2	19.08
70mm work to roofs on concrete base; two coats to falls and crossfalls and to slopes not exceeding 15 degrees from horizontal	7.49	5.00	7.86	0.65	0.70	10.91	0.20	18.98	m2	21.35
80mm work to roofs on concrete base; two coats to falls and crossfalls and to slopes not exceeding 15 degrees from horizontal	8.64	5.00	9.07	0.70	0.75	11.73	0.20	21.00	m2	23.62
Lightweight concrete screed consisting of cement and lightweight aggregate, medium grade, 10 - 5 gauge, 799 kg/m3 (1:10) finished with 15mm cement and sand (1:4) trowelled bed										
50mm work to roofs on concrete base; two coats to falls and crossfalls and to slopes not exceeding 15 degrees from horizontal	4.05	5.00	4.25	0.73	0.62	11.09	0.20	15.54	m2	17.49
75mm work to roofs on concrete base; two coats to falls and crossfalls and to slopes not exceeding 15 degrees from horizontal	6.00	5.00	6.30	0.78	0.67	11.90	0.20	18.40	m2	20.70
100mm work to roofs on concrete base; two coats to falls and crossfalls and to slopes not exceeding 15 degrees from horizontal	8.50	5.00	8.93	0.83	0.72	12.72	0.20	21.84	m2	24.57
Screed reinforcement										
Reinforcement; galvanised wire netting, B.S.1485, 13mm mesh, 22 gauge wire; 150mm laps; placing in position										
floors	2.68	5.00	2.81	0.05	-	0.46	-	3.28	m2	3.68
Reinforcement; galvanised wire netting, B.S.1485, 25mm mesh, 22 gauge wire; 150mm laps; placing in position										
floors	1.34	5.00	1.41	0.05	-	0.46	-	1.87	m2	2.10
Reinforcement; galvanised wire netting, B.S.1485, 38mm mesh, 19 gauge wire; 150mm laps; placing in position										
floors	1.49	5.00	1.56	0.05	-	0.46	-	2.03	m2	2.28
Reinforcement; galvanised wire netting, B.S.1485, 50mm mesh, 19 gauge wire; 150mm laps; placing in position										
floors	1.13	5.00	1.19	0.05	-	0.46	-	1.65	m2	1.85
Division strips										
Aluminium dividing strips; setting in bed										
5 x 25mm; flat section	4.60	5.00	4.83	0.21	-	1.94	-	6.77	m	7.61
Brass dividing strips; setting in bed										
5 x 25mm; flat section	5.70	5.00	5.99	0.21	-	1.94	-	7.92	m	8.91
M10: CEMENT: SAND/CONCRETE SCREEDS/TOPPINGS; STAIRCASE AREAS										
Mortar, cement and sand (1:3) - paving										
13mm work to walls on brickwork or blockwork base; one coat; screeded										
width exceeding 300mm	1.26	5.00	1.32	0.69	0.23	7.98	-	9.31	m2	10.47
width not exceeding 300mm	0.38	10.00	0.42	0.41	0.07	4.28	-	4.69	m	5.28
13mm work to walls on brickwork or blockwork base; one coat; trowelled										
width exceeding 300mm	1.26	5.00	1.32	0.92	0.23	10.11	-	11.43	m2	12.86
width not exceeding 300mm	0.38	10.00	0.42	0.55	0.07	5.57	-	5.99	m	6.73

Labour hourly rates: (except Specialists) Craft Operatives £9.23 Labourer £7.02 Rates are national average prices. Refer to REGIONAL VARIATIONS for indicative levels of overall pricing in regions	MATERIALS			LABOUR				RATES		
	Del to Site £	Waste %	Material Cost £	Craft Optve Hrs	Lab Hrs	Labour Cost £	Sunds £	Nett Rate £	Unit	Gross Rate (+12.5%) £
M10: CEMENT: SAND/CONCRETE SCREEDS/TOPPINGS; STAIRCASE AREAS Cont.										
Mortar, cement and sand (1:3) – paving Cont.										
13mm work to walls on brickwork or blockwork base; one coat; floated										
width exceeding 300mm	1.26	5.00	1.32	0.87	0.23	9.64	–	10.97	m2	12.34
width not exceeding 300mm	0.38	10.00	0.42	0.52	0.07	5.29	–	5.71	m	6.42
19mm work to floors on concrete base; one coat; screeded										
level and to falls only not exceeding 15 degrees from horizontal	1.85	5.00	1.94	0.39	0.17	4.79	0.20	6.94	m2	7.80
25mm work to floors on concrete base; one coat; screeded										
level and to falls only not exceeding 15 degrees from horizontal	2.43	5.00	2.55	0.44	0.18	5.32	0.20	8.08	m2	9.09
32mm work to floors on concrete base; one coat; screeded										
level and to falls only not exceeding 15 degrees from horizontal	3.11	5.00	3.27	0.48	0.29	6.47	0.20	9.93	m2	11.17
38mm work to floors on concrete base; one coat; screeded										
level and to falls only not exceeding 15 degrees from horizontal	3.70	5.00	3.88	0.51	0.21	6.18	0.20	10.27	m2	11.55
50mm work to floors on concrete base; one coat; screeded										
level and to falls only not exceeding 15 degrees from horizontal	4.86	5.00	5.10	0.60	0.24	7.22	0.20	12.53	m2	14.09
19mm work to floors on concrete base; one coat; trowelled										
level and to falls only not exceeding 15 degrees from horizontal	1.85	5.00	1.94	0.62	0.17	6.92	0.20	9.06	m2	10.19
25mm work to floors on concrete base; one coat; trowelled										
level and to falls only not exceeding 15 degrees from horizontal	2.43	5.00	2.55	0.66	0.18	7.36	0.20	10.11	m2	11.37
32mm work to floors on concrete base; one coat; trowelled										
level and to falls only not exceeding 15 degrees from horizontal	3.11	5.00	3.27	0.71	0.20	7.96	0.20	11.42	m2	12.85
38mm work to floors on concrete base; one coat; trowelled										
level and to falls only not exceeding 15 degrees from horizontal	3.70	5.00	3.88	0.74	0.21	8.30	0.20	12.39	m2	13.94
50mm work to floors on concrete base; one coat; trowelled										
level and to falls only not exceeding 15 degrees from horizontal	4.86	5.00	5.10	0.83	0.24	9.35	0.20	14.65	m2	16.48
19mm work to floors on concrete base; one coat; floated										
level and to falls only not exceeding 15 degrees from horizontal	1.85	5.00	1.94	0.57	0.17	6.45	0.20	8.60	m2	9.67
25mm work to floors on concrete base; one coat; floated										
level and to falls only not exceeding 15 degrees from horizontal	2.43	5.00	2.55	0.62	0.18	6.99	0.20	9.74	m2	10.95
32mm work to floors on concrete base; one coat; floated										
level and to falls only not exceeding 15 degrees from horizontal	3.11	5.00	3.27	0.66	0.20	7.50	0.20	10.96	m2	12.33
38mm work to floors on concrete base; one coat; floated										
level and to falls only not exceeding 15 degrees from horizontal	3.70	5.00	3.88	0.69	0.21	7.84	0.20	11.93	m2	13.42
50mm work to floors on concrete base; one coat; floated										
level and to falls only not exceeding 15 degrees from horizontal	4.86	5.00	5.10	0.78	0.24	8.88	0.20	14.19	m2	15.96
19mm work to treads on concrete base; screeded										
width 275mm	0.76	10.00	0.84	0.20	0.04	2.13	0.06	3.02	m	3.40
25mm work to treads on concrete base; screeded										
width 275mm	1.00	10.00	1.10	0.21	0.04	2.22	0.06	3.38	m	3.80
32mm work to treads on concrete base; screeded										
width 275mm	1.28	10.00	1.41	0.23	0.05	2.47	0.06	3.94	m	4.43
38mm work to treads on concrete base; screeded										
width 275mm	1.52	10.00	1.67	0.25	0.05	2.66	0.06	4.39	m	4.94

SURFACE FINISHES

Labour hourly rates: (except Specialists) Craft Operatives £9.23 Labourer £7.02 Rates are national average prices. Refer to REGIONAL VARIATIONS for indicative levels of overall pricing in regions	MATERIALS			LABOUR				RATES		
	Del to Site	Waste	Material Cost	Craft Optve	Lab	Labour Cost	Sunds	Nett Rate	Unit	Gross Rate (+12.5%)
	£	%	£	Hrs	Hrs	£	£	£		£
M10: CEMENT: SAND/CONCRETE SCREEDS/TOPPINGS; STAIRCASE AREAS Cont.										
Mortar, cement and sand (1:3) - paving Cont.										
19mm work to treads on concrete base; floated width 275mm	0.76	10.00	0.84	0.25	0.04	2.59	0.06	3.48	m	3.92
25mm work to treads on concrete base; floated width 275mm	1.00	10.00	1.10	0.26	0.04	2.68	0.06	3.84	m	4.32
32mm work to treads on concrete base; floated width 275mm	1.28	10.00	1.41	0.28	0.05	2.94	0.06	4.40	m	4.95
38mm work to treads on concrete base; floated width 275mm	1.52	10.00	1.67	0.30	0.05	3.12	0.06	4.85	m	5.46
19mm work to treads on concrete base; trowelled width 275mm	0.76	10.00	0.84	0.30	0.04	3.05	0.06	3.95	m	4.44
25mm work to treads on concrete base; trowelled width 275mm	1.00	10.00	1.10	0.31	0.04	3.14	0.06	4.30	m	4.84
32mm work to treads on concrete base; trowelled width 275mm	1.28	10.00	1.41	0.33	0.05	3.40	0.06	4.86	m	5.47
38mm work to treads on concrete base; trowelled width 275mm	1.52	10.00	1.67	0.35	0.05	3.58	0.06	5.31	m	5.98
19mm work to plain risers on concrete base; keyed height 175mm	0.49	10.00	0.54	0.35	0.03	3.44	-	3.98	m	4.48
25mm work to plain risers on concrete base; keyed height 175mm	0.64	10.00	0.70	0.40	0.03	3.90	-	4.61	m	5.18
32mm work to plain risers on concrete base; keyed height 175mm	0.82	10.00	0.90	0.45	0.03	4.36	-	5.27	m	5.92
19mm work to plain risers on concrete base; trowelled height 175mm	0.49	10.00	0.54	0.55	0.03	5.29	-	5.83	m	6.55
25mm work to plain risers on concrete base; trowelled height 175mm	0.64	10.00	0.70	0.60	0.03	5.75	-	6.45	m	7.26
32mm work to plain risers on concrete base; trowelled height 175mm	0.82	10.00	0.90	0.65	0.03	6.21	-	7.11	m	8.00
19mm work to undercut risers on concrete base; keyed height 175mm	0.49	10.00	0.54	0.40	0.03	3.90	-	4.44	m	5.00
25mm work to undercut risers on concrete base; keyed height 175mm	0.64	10.00	0.70	0.45	0.03	4.36	-	5.07	m	5.70
32mm work to undercut risers on concrete base; keyed height 175mm	0.82	10.00	0.90	0.50	0.03	4.83	-	5.73	m	6.44
19mm work to undercut risers on concrete base; trowelled height 175mm	0.49	10.00	0.54	0.60	0.03	5.75	-	6.29	m	7.07
25mm work to undercut risers on concrete base; trowelled height 175mm	0.64	10.00	0.70	0.65	0.03	6.21	-	6.91	m	7.78
32mm work to undercut risers on concrete base; trowelled height 175mm	0.82	10.00	0.90	0.70	0.03	6.67	-	7.57	m	8.52
Granolithic; cement and granite chippings (2:5); steel trowelled - paving										
25mm work to floors on concrete base; one coat level and to falls only not exceeding 15 degrees from horizontal	4.61	5.00	4.84	0.72	0.21	8.12	0.41	13.37	m2	15.04
32mm work to floors on concrete base; one coat level and to falls only not exceeding 15 degrees from horizontal	5.90	5.00	6.20	0.77	0.22	8.65	0.41	15.26	m2	17.16
38mm work to floors on concrete base; one coat level and to falls only not exceeding 15 degrees from horizontal	7.00	5.00	7.35	0.83	0.24	9.35	0.41	17.11	m2	19.24
50mm work to floors on concrete base; one coat level and to falls only not exceeding 15 degrees from horizontal	9.22	5.00	9.68	0.92	0.27	10.39	0.41	20.48	m2	23.04
25mm work to treads on concrete base width 275mm	1.90	10.00	2.09	0.30	0.06	3.19	0.11	5.39	m	6.06
32mm work to treads on concrete base width 275mm	2.43	10.00	2.67	0.35	0.06	3.65	0.11	6.43	m	7.24
38mm work to treads on concrete base width 275mm	2.89	10.00	3.18	0.40	0.07	4.18	0.11	7.47	m	8.41

Labour hourly rates: (except Specialists) Craft Operatives £9.23 Labourer £7.02 Rates are national average prices. Refer to REGIONAL VARIATIONS for indicative levels of overall pricing in regions	MATERIALS			LABOUR				RATES		
	Del to Site £	Waste %	Material Cost £	Craft Optve Hrs	Lab Hrs	Labour Cost £	Sunds £	Nett Rate £	Unit	Gross Rate (+12.5%) £
M10: CEMENT: SAND/CONCRETE SCREEDS/TOPPINGS; STAIRCASE AREAS Cont.										
Granolithic; cement and granite chippings (2:5); steel trowelled - paving Cont.										
50mm work to treads on concrete base width 275mm	3.80	10.00	4.18	0.45	0.07	4.64	0.11	8.93	m	10.05
13mm work to plain risers on concrete base height 175mm	0.63	10.00	0.69	0.55	0.05	5.43	-	6.12	m	6.89
19mm work to plain risers on concrete base height 175mm	0.92	10.00	1.01	0.60	0.05	5.89	-	6.90	m	7.76
25mm work to plain risers on concrete base height 175mm	1.21	10.00	1.33	0.65	0.05	6.35	-	7.68	m	8.64
32mm work to plain risers on concrete base height 175mm	1.55	10.00	1.71	0.70	0.05	6.81	-	8.52	m	9.58
13mm work to undercut risers on concrete base height 175mm	0.63	10.00	0.69	0.60	0.05	5.89	-	6.58	m	7.40
19mm work to undercut risers on concrete base height 175mm	0.92	10.00	1.01	0.65	0.05	6.35	-	7.36	m	8.28
25mm work to undercut risers on concrete base height 175mm	1.21	10.00	1.33	0.70	0.05	6.81	-	8.14	m	9.16
32mm work to undercut risers on concrete base height 175mm	1.55	10.00	1.71	0.75	0.05	7.27	-	8.98	m	10.10
13mm work to strings on concrete base height 300mm	1.08	10.00	1.19	0.75	0.05	7.27	-	8.46	m	9.52
19mm work to strings on concrete base height 300mm	1.58	10.00	1.74	0.80	0.05	7.74	-	9.47	m	10.66
25mm work to strings on concrete base height 300mm	2.07	10.00	2.28	0.85	0.06	8.27	-	10.54	m	11.86
32mm work to strings on concrete base height 300mm	2.65	10.00	2.92	0.90	0.07	8.80	-	11.71	m	13.18
13mm work to strings on brickwork or blockwork base height 275mm	0.99	10.00	1.09	0.70	0.05	6.81	-	7.90	m	8.89
19mm work to strings on brickwork or blockwork base height 275mm	1.44	10.00	1.58	0.75	0.05	7.27	-	8.86	m	9.96
25mm work to strings on brickwork or blockwork base height 275mm	1.90	10.00	2.09	0.80	0.06	7.81	-	9.90	m	11.13
32mm work to strings on brickwork or blockwork base height 275mm	2.43	10.00	2.67	0.85	0.07	8.34	-	11.01	m	12.39
13mm work to aprons on concrete base height 150mm	0.54	10.00	0.59	0.55	0.05	5.43	-	6.02	m	6.77
19mm work to aprons on concrete base height 150mm	0.79	10.00	0.87	0.60	0.05	5.89	-	6.76	m	7.60
25mm work to aprons on concrete base height 150mm	1.04	10.00	1.14	0.65	0.05	6.35	-	7.49	m	8.43
32mm work to aprons on concrete base height 150mm	1.33	10.00	1.46	0.70	0.05	6.81	-	8.28	m	9.31
M11: MASTIC ASPHALT FLOORING/FLOOR UNDERLAYS										
Paving; B.S.6925 (limestone aggregate) Grade III, brown, smooth floated finish										
Flooring and underlay 15mm thick; coats of asphalt - 1; to concrete base; flat										
width not exceeding 150mm	-	-	Spclist	-	-	Spclist	-	26.25	m2	29.53
width 150 - 225mm	-	-	Spclist	-	-	Spclist	-	22.00	m2	24.75
width 225 - 300mm	-	-	Spclist	-	-	Spclist	-	17.40	m2	19.57
width exceeding 300mm	-	-	Spclist	-	-	Spclist	-	13.15	m2	14.79
Skirtings 15mm thick; coats of asphalt - 1; no underlay; to brickwork base										
girth not exceeding 150mm	-	-	Spclist	-	-	Spclist	-	8.45	m	9.51
Paving; B.S.6925 (limestone aggregate) Grade III, red, smooth floated finish										
Flooring and underlay 15mm thick; coats of asphalt - 1; to concrete base; flat										
width not exceeding 150mm	-	-	Spclist	-	-	Spclist	-	25.25	m2	28.41
width 150 - 225mm	-	-	Spclist	-	-	Spclist	-	23.70	m2	26.66
width 225 - 300mm	-	-	Spclist	-	-	Spclist	-	18.95	m2	21.32
width exceeding 300mm	-	-	Spclist	-	-	Spclist	-	14.05	m2	15.81

SURFACE FINISHES

Labour hourly rates: (except Specialists) Craft Operatives £9.23 Labourer £7.02 Rates are national average prices. Refer to REGIONAL VARIATIONS for indicative levels of overall pricing in regions	MATERIALS			LABOUR				RATES		
	Del to Site £	Waste %	Material Cost £	Craft Optve Hrs	Lab Hrs	Labour Cost £	Sunds £	Nett Rate £	Unit	Gross Rate (+12.5%) £
M11: MASTIC ASPHALT FLOORING/FLOOR UNDERLAYS Cont.										
Paving; B.S.6925 (limestone aggregate) Grade III, red, smooth floated finish Cont.										
Skirtings 15mm thick; coats of asphalt - 1; no underlay; to brickwork base										
girth not exceeding 150mm	-	-	Spclist	-	-	Spclist	-	9.40	m	10.57
M12: TROWELLED BITUMEN/RESIN/RUBBER LATEX FLOORING										
One coat levelling screeds										
3mm work to floors on concrete base; one coat										
level and to falls only not exceeding 15 degrees										
from horizontal	2.76	5.00	2.90	0.46	0.05	4.60	-	7.49	m2	8.43
3mm work to floors on existing timber boarded base; one coat										
level and to falls only not exceeding 15 degrees										
from horizontal	2.76	5.00	2.90	0.58	0.06	5.77	-	8.67	m2	9.76
M20: PLASTERED/RENDERED/ROUGHCAST COATINGS										
PLASTERBOARD - Baseboarding; gypsum baseboard, Thistle; 5mm joints, filled with plaster and scrimmed; fixing with galvanised nails; internal										
9.5mm work to walls on timber base										
width exceeding 300mm	1.71	5.00	1.80	0.29	0.15	3.73	0.19	5.72	m2	6.43
width not exceeding 300mm	0.51	10.00	0.56	0.17	0.05	1.92	0.11	2.59	m	2.91
9.5mm work to ceilings on timber base										
width exceeding 300mm	1.71	5.00	1.80	0.35	0.15	4.28	0.19	6.27	m2	7.05
width not exceeding 300mm	0.51	10.00	0.56	0.21	0.05	2.29	0.11	2.96	m	3.33
9.5mm work to isolated beams on timber base										
width exceeding 300mm	1.71	5.00	1.80	0.44	0.15	5.11	0.19	7.10	m2	7.99
width not exceeding 300mm	0.51	10.00	0.56	0.26	0.05	2.75	0.11	3.42	m	3.85
9.5mm work to isolated columns on timber base										
width exceeding 300mm	1.71	5.00	1.80	0.44	0.15	5.11	0.19	7.10	m2	7.99
width not exceeding 300mm	0.51	10.00	0.56	0.26	0.05	2.75	0.11	3.42	m	3.85
PLASTERBOARD - Baseboarding; gypsum baseboard, Thistle Duplex; 5mm joints filled with plaster and scrimmed; fixing with galvanised nails; internal										
9.5mm work to walls on timber base										
width exceeding 300mm	2.33	5.00	2.45	0.29	0.15	3.73	0.19	6.37	m2	7.16
width not exceeding 300mm	0.70	10.00	0.77	0.17	0.05	1.92	0.11	2.80	m	3.15
9.5mm work to ceilings on timber base										
width exceeding 300mm	2.33	5.00	2.45	0.35	0.15	4.28	0.19	6.92	m2	7.79
width not exceeding 300mm	0.70	10.00	0.77	0.21	0.05	2.29	0.11	3.17	m	3.57
9.5mm work to isolated beams on timber base										
width exceeding 300mm	2.33	5.00	2.45	0.44	0.15	5.11	0.19	7.75	m2	8.72
width not exceeding 300mm	0.70	10.00	0.77	0.26	0.05	2.75	0.11	3.63	m	4.08
9.5mm work to isolated columns on timber base										
width exceeding 300mm	2.33	5.00	2.45	0.44	0.15	5.11	0.19	7.75	m2	8.72
width not exceeding 300mm	0.70	10.00	0.77	0.26	0.05	2.75	0.11	3.63	m	4.08
PLASTERBOARD - Baseboarding; Gyproc lath; 3mm joints, filled with plaster; fixing with galvanised nails; internal										
9.5mm work to walls on timber base										
width exceeding 300mm	2.01	5.00	2.11	0.29	0.15	3.73	0.15	5.99	m2	6.74
width not exceeding 300mm	0.60	10.00	0.66	0.17	0.05	1.92	0.09	2.67	m	3.00
12.5mm work to walls on timber base										
width exceeding 300mm	2.36	5.00	2.48	0.35	0.18	4.49	0.15	7.12	m2	8.01
width not exceeding 300mm	0.71	10.00	0.78	0.21	0.05	2.29	0.09	3.16	m	3.56
9.5mm work to ceilings on timber base										
width exceeding 300mm	2.01	5.00	2.11	0.35	0.15	4.28	0.15	6.54	m2	7.36
width not exceeding 300mm	0.60	10.00	0.66	0.21	0.05	2.29	0.09	3.04	m	3.42
12.5mm work to ceilings on timber base										
width exceeding 300mm	2.36	5.00	2.48	0.42	0.18	5.14	0.15	7.77	m2	8.74
width not exceeding 300mm	0.71	10.00	0.78	0.25	0.05	2.66	0.09	3.53	m	3.97
9.5mm work to isolated beams on timber base										
width exceeding 300mm	2.01	5.00	2.11	0.44	0.15	5.11	0.15	7.37	m2	8.30
width not exceeding 300mm	0.60	10.00	0.66	0.26	0.05	2.75	0.09	3.50	m	3.94
12.5mm work to isolated beams on timber base										
width exceeding 300mm	2.36	5.00	2.48	0.53	0.18	6.16	0.15	8.78	m2	9.88
width not exceeding 300mm	0.71	10.00	0.78	0.32	0.05	3.30	0.09	4.18	m	4.70
9.5mm work to isolated columns on timber base										
width exceeding 300mm	2.01	5.00	2.11	0.44	0.15	5.11	0.15	7.37	m2	8.30
width not exceeding 300mm	0.60	10.00	0.66	0.26	0.05	2.75	0.09	3.50	m	3.94

Labour hourly rates: (except Specialists) Craft Operatives £9.23 Labourer £7.02 Rates are national average prices. Refer to REGIONAL VARIATIONS for indicative levels of overall pricing in regions	MATERIALS			LABOUR				RATES		
	Del to Site	Waste	Material Cost	Craft Optve	Lab	Labour Cost	Sunds	Nett Rate	Unit	Gross Rate (+12.5%)
	£	%	£	Hrs	Hrs	£	£	£		£
M20: PLASTERED/RENDERED/ROUGHCAST COATINGS Cont.										
PLASTERBOARD - Baseboarding; Gyproc lath; 3mm joints, filled with plaster; fixing with galvanised nails; internal Cont.										
12.5mm work to isolated columns on timber base										
width exceeding 300mm	2.36	5.00	2.48	0.53	0.18	6.16	0.15	8.78	m2	9.88
width not exceeding 300mm	0.71	10.00	0.78	0.32	0.05	3.30	0.09	4.18	m	4.70
PLASTERBOARD - Baseboarding; Gyproc plank, square edge; 5mm joints, filled with plaster and scrimmed; internal										
19mm work to walls on timber base										
width exceeding 300mm	3.23	5.00	3.39	0.47	0.24	6.02	0.18	9.59	m2	10.79
width not exceeding 300mm	0.97	10.00	1.07	0.28	0.07	3.08	0.11	4.25	m	4.78
19mm work to ceilings on timber base										
width exceeding 300mm	3.23	5.00	3.39	0.56	0.24	6.85	0.18	10.43	m2	11.73
width not exceeding 300mm	0.97	10.00	1.07	0.34	0.07	3.63	0.11	4.81	m	5.41
19mm work to isolated beams on timber base										
width exceeding 300mm	3.23	5.00	3.39	0.71	0.24	8.24	0.18	11.81	m2	13.29
width not exceeding 300mm	0.97	10.00	1.07	0.43	0.07	4.46	0.11	5.64	m	6.34
19mm work to isolated columns on timber base										
width exceeding 300mm	3.23	5.00	3.39	0.71	0.24	8.24	0.18	11.81	m2	13.29
width not exceeding 300mm	0.97	10.00	1.07	0.43	0.07	4.46	0.11	5.64	m	6.34
PLASTERBOARD - Baseboarding; Gyproc square edge wallboard; 3mm joints filled with plaster and scrimmed; fixing with galvanised nails; internal										
9.5mm work to walls on timber base										
width exceeding 300mm	1.71	5.00	1.80	0.29	0.15	3.73	0.19	5.72	m2	6.43
width not exceeding 300mm	0.51	10.00	0.56	0.17	0.05	1.92	0.11	2.59	m	2.91
12.5mm work to walls on timber base										
width exceeding 300mm	1.99	5.00	2.09	0.35	0.18	4.49	0.19	6.77	m2	7.62
width not exceeding 300mm	0.60	10.00	0.66	0.21	0.05	2.29	0.11	3.06	m	3.44
15mm work to walls on timber base										
width exceeding 300mm	2.18	5.00	2.29	0.41	0.21	5.26	0.19	7.74	m2	8.70
width not exceeding 300mm	0.65	10.00	0.71	0.25	0.06	2.73	0.11	3.55	m	4.00
9.5mm work to ceilings on timber base										
width exceeding 300mm	1.71	5.00	1.80	0.35	0.15	4.28	0.19	6.27	m2	7.05
width not exceeding 300mm	0.51	10.00	0.56	0.21	0.05	2.29	0.11	2.96	m	3.33
12.5mm work to ceilings on timber base										
width exceeding 300mm	1.99	5.00	2.09	0.42	0.18	5.14	0.19	7.42	m2	8.35
width not exceeding 300mm	0.60	10.00	0.66	0.25	0.05	2.66	0.11	3.43	m	3.86
15mm work to ceilings on timber base										
width exceeding 300mm	2.18	5.00	2.29	0.49	0.25	6.28	0.19	8.76	m2	9.85
width not exceeding 300mm	0.65	10.00	0.71	0.29	0.08	3.24	0.11	4.06	m	4.57
9.5mm work to isolated beams on timber base										
width exceeding 300mm	1.71	5.00	1.80	0.44	0.15	5.11	0.19	7.10	m2	7.99
width not exceeding 300mm	0.51	10.00	0.56	0.26	0.05	2.75	0.11	3.42	m	3.85
12.5mm work to isolated beams on timber base										
width exceeding 300mm	1.99	5.00	2.09	0.53	0.18	6.16	0.19	8.44	m2	9.49
width not exceeding 300mm	0.60	10.00	0.66	0.32	0.05	3.30	0.11	4.07	m	4.58
15mm work to isolated beams on timber base										
width exceeding 300mm	2.18	5.00	2.29	0.62	0.21	7.20	0.19	9.68	m2	10.89
width not exceeding 300mm	0.65	10.00	0.71	0.37	0.06	3.84	0.11	4.66	m	5.24
9.5mm work to isolated columns on timber base										
width exceeding 300mm	1.71	5.00	1.80	0.44	0.15	5.11	0.19	7.10	m2	7.99
width not exceeding 300mm	0.51	10.00	0.56	0.26	0.05	2.75	0.11	3.42	m	3.85
12.5mm work to isolated columns on timber base										
width exceeding 300mm	1.99	5.00	2.09	0.53	0.18	6.16	0.19	8.44	m2	9.49
width not exceeding 300mm	0.60	10.00	0.66	0.32	0.05	3.30	0.11	4.07	m	4.58
15mm work to isolated columns on timber base										
width exceeding 300mm	2.18	5.00	2.29	0.62	0.21	7.20	0.19	9.68	m2	10.89
width not exceeding 300mm	0.65	10.00	0.71	0.37	0.06	3.84	0.11	4.66	m	5.24
PLASTERBOARD - Baseboarding; Gyproc square edge Duplex wallboard; 3mm joints filled with plaster and scrimmed; fixing with galvanised nails; internal										
9.5mm work to walls on timber base										
width exceeding 300mm	2.33	5.00	2.45	0.29	0.15	3.73	0.19	6.37	m2	7.16
width not exceeding 300mm	0.70	10.00	0.77	0.17	0.05	1.92	0.11	2.80	m	3.15
12.5mm work to walls on timber base										
width exceeding 300mm	2.61	5.00	2.74	0.35	0.18	4.49	0.19	7.42	m2	8.35
width not exceeding 300mm	0.78	10.00	0.86	0.21	0.05	2.29	0.11	3.26	m	3.66
15mm work to walls on timber base										
width exceeding 300mm	2.86	5.00	3.00	0.41	0.21	5.26	0.19	8.45	m2	9.51

Labour hourly rates: (except Specialists) Craft Operatives £9.23 Labourer £7.02 Rates are national average prices. Refer to REGIONAL VARIATIONS for indicative levels of overall pricing in regions	MATERIALS			LABOUR				RATES		
	Del to Site £	Waste %	Material Cost £	Craft Optve Hrs	Lab Hrs	Labour Cost £	Sunds £	Nett Rate £	Unit	Gross Rate (+12.5%) £
M20: PLASTERED/RENDERED/ROUGHCAST COATINGS Cont.										
PLASTERBOARD - Baseboarding; Gyproc square edge Duplex wallboard; 3mm joints filled with plaster and scrimmed; fixing with galvanised nails; internal Cont.										
15mm work to walls on timber base Cont.										
width not exceeding 300mm	0.86	10.00	0.95	0.25	0.06	2.73	0.11	3.78	m	4.26
9.5mm work to ceilings on timber base										
width exceeding 300mm	2.33	5.00	2.45	0.35	0.15	4.28	0.19	6.92	m2	7.79
width not exceeding 300mm	0.70	10.00	0.77	0.21	0.05	2.29	0.11	3.17	m	3.57
12.5mm work to ceilings on timber base										
width exceeding 300mm	2.61	5.00	2.74	0.42	0.18	5.14	0.19	8.07	m2	9.08
width not exceeding 300mm	0.78	10.00	0.86	0.25	0.05	2.66	0.11	3.63	m	4.08
15mm work to ceilings on timber base										
width exceeding 300mm	2.86	5.00	3.00	0.49	0.25	6.28	0.19	9.47	m2	10.65
width not exceeding 300mm	0.86	10.00	0.95	0.29	0.08	3.24	0.11	4.29	m	4.83
9.5mm work to isolated beams on timber base										
width exceeding 300mm	2.33	5.00	2.45	0.44	0.15	5.11	0.19	7.75	m2	8.72
width not exceeding 300mm	0.70	10.00	0.77	0.26	0.05	2.75	0.11	3.63	m	4.08
12.5mm work to isolated beams on timber base										
width exceeding 300mm	2.61	5.00	2.74	0.53	0.18	6.16	0.19	9.09	m2	10.22
width not exceeding 300mm	0.78	10.00	0.86	0.32	0.05	3.30	0.11	4.27	m	4.81
15mm work to isolated beams on timber base										
width exceeding 300mm	2.86	5.00	3.00	0.62	0.21	7.20	0.19	10.39	m2	11.69
width not exceeding 300mm	0.86	10.00	0.95	0.37	0.06	3.84	0.11	4.89	m	5.50
9.5mm work to isolated columns on timber base										
width exceeding 300mm	2.33	5.00	2.45	0.44	0.15	5.11	0.19	7.75	m2	8.72
width not exceeding 300mm	0.70	10.00	0.77	0.26	0.05	2.75	0.11	3.63	m	4.08
12.5mm work to isolated columns on timber base										
width exceeding 300mm	2.61	5.00	2.74	0.53	0.18	6.16	0.19	9.09	m2	10.22
width not exceeding 300mm	0.78	10.00	0.86	0.32	0.05	3.30	0.11	4.27	m	4.81
15mm work to isolated columns on timber base										
width exceeding 300mm	2.86	5.00	3.00	0.62	0.21	7.20	0.19	10.39	m2	11.69
width not exceeding 300mm	0.86	10.00	0.95	0.37	0.06	3.84	0.11	4.89	m	5.50
PLASTERBOARD - Baseboarding; two layers 9.5mm Gyproc square edge wallboard; second layer with 3mm joints filled with plaster and scrimmed; fixing with galvanised nails; internal										
19mm work to walls on timber base										
width exceeding 300mm	3.42	5.00	3.59	0.60	0.30	7.64	0.28	11.52	m2	12.95
width not exceeding 300mm	1.03	10.00	1.13	0.36	0.09	3.95	0.17	5.26	m	5.91
19mm work to ceilings on timber base										
width exceeding 300mm	3.42	5.00	3.59	0.72	0.30	8.75	0.28	12.62	m2	14.20
width not exceeding 300mm	1.03	10.00	1.13	0.43	0.09	4.60	0.17	5.90	m	6.64
19mm work to isolated beams on timber base										
width exceeding 300mm	3.42	5.00	3.59	0.80	0.30	9.49	0.28	13.36	m2	15.03
width not exceeding 300mm	1.03	10.00	1.13	0.48	0.09	5.06	0.17	6.37	m	7.16
19mm work to isolated columns on timber base										
width exceeding 300mm	3.42	5.00	3.59	0.80	0.30	9.49	0.28	13.36	m2	15.03
width not exceeding 300mm	1.03	10.00	1.13	0.48	0.09	5.06	0.17	6.37	m	7.16
PLASTERBOARD - Baseboarding; one layer 9.5mm and one layer 12.5mm Gyproc square edge wallboard; second layer with 3mm joints filled with plaster and scrimmed; fixing with galvanised nails, internal										
22mm work to walls on timber base										
width exceeding 300mm	3.70	5.00	3.88	0.65	0.33	8.32	0.28	12.48	m2	14.04
width not exceeding 300mm	1.11	10.00	1.22	0.39	0.10	4.30	0.17	5.69	m	6.40
22mm work to ceilings on timber base										
width exceeding 300mm	3.70	5.00	3.88	0.78	0.33	9.52	0.28	13.68	m2	15.39
width not exceeding 300mm	1.11	10.00	1.22	0.47	0.10	5.04	0.17	6.43	m	7.23
22mm work to isolated beams on timber base										
width exceeding 300mm	3.70	5.00	3.88	0.98	0.33	11.36	0.28	15.53	m2	17.47
width not exceeding 300mm	1.11	10.00	1.22	0.59	0.10	6.15	0.17	7.54	m	8.48
22mm work to isolated columns on timber base										
width exceeding 300mm	3.70	5.00	3.88	0.98	0.33	11.36	0.28	15.53	m2	17.47
width not exceeding 300mm	1.11	10.00	1.22	0.59	0.10	6.15	0.17	7.54	m	8.48
PLASTERBOARD - Baseboarding; two layers 12.5mm Gyproc square edge wallboard; second layer with 3mm joints filled with plaster and scrimmed; fixing with galvanised nails; internal										
25mm work to walls on timber base										
width exceeding 300mm	3.98	5.00	4.18	0.70	0.35	8.92	0.28	13.38	m2	15.05
width not exceeding 300mm	1.19	10.00	1.31	0.42	0.11	4.65	0.17	6.13	m	6.89

Labour hourly rates: (except Specialists) Craft Operatives £9.23 Labourer £7.02 Rates are national average prices. Refer to REGIONAL VARIATIONS for indicative levels of overall pricing in regions	MATERIALS			LABOUR				RATES		
	Del to Site £	Waste %	Material Cost £	Craft Optve Hrs	Lab Hrs	Labour Cost £	Sunds £	Nett Rate £	Unit	Gross Rate (+12.5%) £

M20: PLASTERED/RENDERED/ROUGHCAST COATINGS Cont.

PLASTERBOARD - Baseboarding; two layers 12.5mm Gyproc square edge wallboard; second layer with 3mm joints filled with plaster and scrimmed; fixing with galvanised nails; internal Cont.

25mm work to ceilings on timber base										
width exceeding 300mm	3.98	5.00	4.18	0.84	0.35	10.21	0.28	14.67	m2	16.50
width not exceeding 300mm	1.19	10.00	1.31	0.50	0.11	5.39	0.17	6.87	m	7.72
25mm work to isolated beams on timber base										
width exceeding 300mm	3.98	5.00	4.18	1.05	0.35	12.15	0.28	16.61	m2	18.68
width not exceeding 300mm	1.19	10.00	1.31	0.64	0.11	6.68	0.17	8.16	m	9.18
25mm work to isolated columns on timber base										
width exceeding 300mm	3.98	5.00	4.18	1.05	0.35	12.15	0.28	16.61	m2	18.68
width not exceeding 300mm	1.19	10.00	1.31	0.64	0.11	6.68	0.17	8.16	m	9.18

PLASTERBOARD - Baseboarding; two layers 15mm Gyproc square edge wallboard; second layer with 3mm joints filled with plaster and scrimmed; fixing with galvanised nails; internal

30mm work to walls on timber base										
width exceeding 300mm	4.36	5.00	4.58	0.75	0.38	9.59	0.28	14.45	m2	16.25
width not exceeding 300mm	1.31	10.00	1.44	0.45	0.11	4.93	0.17	6.54	m	7.35
30mm work to ceilings on timber base										
width exceeding 300mm	4.36	5.00	4.58	0.90	0.38	10.97	0.28	15.83	m2	17.81
width not exceeding 300mm	1.31	10.00	1.44	0.54	0.11	5.76	0.17	7.37	m	8.29
30mm work to isolated beams on timber base										
width exceeding 300mm	4.36	5.00	4.58	1.13	0.38	13.10	0.28	17.96	m2	20.20
width not exceeding 300mm	1.31	10.00	1.44	0.68	0.11	7.05	0.17	8.66	m	9.74
30mm work to isolated columns on timber base										
width exceeding 300mm	4.36	5.00	4.58	1.13	0.38	13.10	0.28	17.96	m2	20.20
width not exceeding 300mm	1.31	10.00	1.44	0.68	0.11	7.05	0.17	8.66	m	9.74

PLASTERBOARD - Baseboarding; two layers; first layer 9.5mm Gyproc Duplex wallboard, second layer 9.5mm Gyproc wallboard both layers with square edge boards; second layer with 3mm joints filled with plaster and scrimmed; fixing with galvanised nails; internal

19mm work to walls on timber base										
width exceeding 300mm	4.04	5.00	4.24	0.60	0.30	7.64	0.28	12.17	m2	13.69
width not exceeding 300mm	1.21	10.00	1.33	0.36	0.09	3.95	0.17	5.46	m	6.14
19mm work to ceilings on timber base										
width exceeding 300mm	4.04	5.00	4.24	0.72	0.30	8.75	0.28	13.27	m2	14.93
width not exceeding 300mm	1.21	10.00	1.33	0.43	0.09	4.60	0.17	6.10	m	6.86
19mm work to isolated beams on timber base										
width exceeding 300mm	4.04	5.00	4.24	0.80	0.30	9.49	0.28	14.01	m2	15.76
width not exceeding 300mm	1.21	10.00	1.33	0.48	0.09	5.06	0.17	6.56	m	7.38
19mm work to isolated columns on timber base										
width exceeding 300mm	4.04	5.00	4.24	0.80	0.30	9.49	0.28	14.01	m2	15.76
width not exceeding 300mm	1.21	10.00	1.33	0.48	0.09	5.06	0.17	6.56	m	7.38

PLASTERBOARD - Baseboarding; two layers; first layer 12.5mm Gyproc Duplex wallboard, second layer 9.5mm Gyproc wallboard, both layers with square edge boards; second layer with 3mm joints filled with plaster and scrimmed; fixing with galvanised nails; internal

22mm work to walls on timber base										
width exceeding 300mm	4.32	5.00	4.54	0.65	0.33	8.32	0.28	13.13	m2	14.77
width not exceeding 300mm	1.30	10.00	1.43	0.39	0.10	4.30	0.17	5.90	m	6.64
22mm work to ceilings on timber base										
width exceeding 300mm	4.32	5.00	4.54	0.78	0.33	9.52	0.28	14.33	m2	16.12
width not exceeding 300mm	1.30	10.00	1.43	0.47	0.10	5.04	0.17	6.64	m	7.47
22mm work to isolated beams on timber base										
width exceeding 300mm	4.32	5.00	4.54	0.98	0.33	11.36	0.28	16.18	m2	18.20
width not exceeding 300mm	1.30	10.00	1.43	0.59	0.10	6.15	0.17	7.75	m	8.72
22mm work to isolated columns on timber base										
width exceeding 300mm	4.32	5.00	4.54	0.98	0.33	11.36	0.28	16.18	m2	18.20
width not exceeding 300mm	1.30	10.00	1.43	0.59	0.10	6.15	0.17	7.75	m	8.72

PLASTERBOARD - Baseboarding; two layers; first layer 12.5mm Gyproc Duplex wallboard, second layer 12.5mm Gyproc wallboard, both layers with square edge boards; second layer with 3mm joints filled with plaster and scrimmed; fixing with galvanised nails; internal

25mm work to walls on timber base										
width exceeding 300mm	4.60	5.00	4.83	0.70	0.35	8.92	0.28	14.03	m2	15.78
width not exceeding 300mm	1.38	10.00	1.52	0.42	0.11	4.65	0.17	6.34	m	7.13

SURFACE FINISHES

Labour hourly rates: (except Specialists) Craft Operatives £9.23 Labourer £7.02 Rates are national average prices. Refer to REGIONAL VARIATIONS for indicative levels of overall pricing in regions	MATERIALS			LABOUR				RATES		
	Del to Site	Waste	Material Cost	Craft Optve	Lab	Labour Cost	Sunds	Nett Rate	Unit	Gross Rate (+12.5%)
	£	%	£	Hrs	Hrs	£	£	£		£
M20: PLASTERED/RENDERED/ROUGHCAST COATINGS Cont.										
PLASTERBOARD - Baseboarding; two layers; first layer 12.5mm Gyproc Duplex wallboard, second layer 12.5mm Gyproc wallboard, both layers with square edge boards; second layer with 3mm joints filled with plaster and scrimmed; fixing with galvanised nails; internal Cont.										
25mm work to ceilings on timber base										
width exceeding 300mm	4.60	5.00	4.83	0.84	0.35	10.21	0.28	15.32	m2	17.24
width not exceeding 300mm	1.38	10.00	1.52	0.50	0.11	5.39	0.17	7.08	m	7.96
25mm work to isolated beams on timber base										
width exceeding 300mm	4.60	5.00	4.83	1.05	0.35	12.15	0.28	17.26	m2	19.42
width not exceeding 300mm	1.38	10.00	1.52	0.64	0.11	6.68	0.17	8.37	m	9.41
25mm work to isolated columns on timber base										
width exceeding 300mm	4.60	5.00	4.83	1.05	0.35	12.15	0.28	17.26	m2	19.42
width not exceeding 300mm	1.38	10.00	1.52	0.64	0.11	6.68	0.17	8.37	m	9.41
PLASTERBOARD - Baseboarding; two layers; first layer 15mm Gyproc Duplex wallboard, second layer 15mm Gyproc wallboard, both layers with square edge boards; second layer with 3mm joints filled with plaster and scrimmed; fixing with galvanised nails; internal										
30mm work to walls on timber base										
width exceeding 300mm	5.04	5.00	5.29	0.75	0.38	9.59	0.28	15.16	m2	17.06
width not exceeding 300mm	1.51	10.00	1.66	0.45	0.11	4.93	0.17	6.76	m	7.60
30mm work to ceilings on timber base										
width exceeding 300mm	5.04	5.00	5.29	0.90	0.38	10.97	0.28	16.55	m2	18.61
width not exceeding 300mm	1.51	10.00	1.66	0.54	0.11	5.76	0.17	7.59	m	8.54
30mm work to isolated beams on timber base										
width exceeding 300mm	5.04	5.00	5.29	1.13	0.38	13.10	0.28	18.67	m2	21.00
width not exceeding 300mm	1.51	10.00	1.66	0.68	0.11	7.05	0.17	8.88	m	9.99
30mm work to isolated columns on timber base										
width exceeding 300mm	5.04	5.00	5.29	1.13	0.38	13.10	0.28	18.67	m2	21.00
width not exceeding 300mm	1.51	10.00	1.66	0.68	0.11	7.05	0.17	8.88	m	9.99
DAMP WALL TREATMENT - Newlath; fixing with masonry nails; internal										
Work to walls on brickwork or blockwork base										
width exceeding 300mm	6.40	5.00	6.72	0.40	0.20	5.10	1.60	13.42	m2	15.09
width not exceeding 300mm	1.92	10.00	2.11	0.24	0.06	2.64	0.48	5.23	m	5.88
PORTLAND CEMENT WORK - Plain face, first and finishing coats cement and sand (1:3), total 13mm thick; wood floated; external										
13mm work to walls on brickwork or blockwork base										
width exceeding 300mm	1.26	5.00	1.32	0.60	0.30	7.64	-	8.97	m2	10.09
width exceeding 300mm; dubbing average 6mm thick	1.85	5.00	1.94	0.84	0.44	10.84	-	12.78	m2	14.38
width exceeding 300mm; dubbing average 12mm thick	2.43	5.00	2.55	0.88	0.58	12.19	-	14.75	m2	16.59
width exceeding 300mm; dubbing average 19mm thick	3.11	5.00	3.27	0.92	0.74	13.69	-	16.95	m2	19.07
width exceeding 300mm; curved to 3000mm radius	1.26	5.00	1.32	0.80	0.30	9.49	-	10.81	m2	12.16
width not exceeding 300mm	0.38	10.00	0.42	0.36	0.09	3.95	-	4.37	m	4.92
width not exceeding 300mm; dubbing average 6mm thick	0.55	10.00	0.60	0.53	0.17	6.09	-	6.69	m	7.53
width not exceeding 300mm; dubbing average 12mm thick	0.73	10.00	0.80	0.53	0.17	6.09	-	6.89	m	7.75
width not exceeding 300mm; dubbing average 19mm thick	0.93	10.00	1.02	0.55	0.22	6.62	-	7.64	m	8.60
width not exceeding 300mm; curved to 3000mm radius	0.38	10.00	0.42	0.48	0.09	5.06	-	5.48	m	6.17
If plain face waterproofed										
add	0.55	10.00	0.60	-	0.02	0.14	-	0.75	m2	0.84
Rounded angles										
radius 10 - 100mm	-	-	-	0.30	-	2.77	-	2.77	m	3.12
Work forming flush skirting; 13mm thick										
height 150mm	0.28	50.00	0.42	0.60	0.05	5.89	-	6.31	m	7.10
height 225mm	0.43	50.00	0.65	0.65	0.07	6.49	-	7.14	m	8.03
height 150mm; curved to 3000mm radius	0.28	50.00	0.42	0.80	0.05	7.74	-	8.15	m	9.17
height 225mm; curved to 3000mm radius	0.43	50.00	0.65	0.86	0.07	8.43	-	9.07	m	10.21
Work forming projecting skirting; 13mm projection										
height 150mm	0.28	50.00	0.42	0.90	0.05	8.66	-	9.08	m	10.21
height 225mm	0.43	50.00	0.65	0.98	0.07	9.54	-	10.18	m	11.45
height 150mm; curved to 3000mm radius	0.28	50.00	0.42	1.20	0.05	11.43	-	11.85	m	13.33
height 225mm; curved to 3000mm radius	0.43	50.00	0.65	1.29	0.07	12.40	-	13.04	m	14.67
ROUGH CAST - Render and dry dash; first and finishing coats cement and sand (1:3), total 15mm thick; wood floated; dry dash of pea shingle; external										
15mm work to walls on brickwork or blockwork base										
width exceeding 300mm	1.64	5.00	1.72	0.85	0.43	10.86	-	12.59	m2	14.16
extra; spatterdash coat	0.12	5.00	0.13	0.23	0.12	2.97	-	3.09	m2	3.48
width exceeding 300mm; curved to 3000mm radius	1.64	5.00	1.72	1.15	0.43	13.63	-	15.36	m2	17.27

SURFACE FINISHES – SMALL WORKS

	MATERIALS			LABOUR				RATES		
Labour hourly rates: (except Specialists) Craft Operatives £9.23 Labourer £7.02 Rates are national average prices. Refer to REGIONAL VARIATIONS for indicative levels of overall pricing in regions	Del to Site £	Waste %	Material Cost £	Craft Optve Hrs	Lab Hrs	Labour Cost £	Sunds £	Nett Rate £	Unit	Gross Rate (+12.5%) £

M20: PLASTERED/RENDERED/ROUGHCAST COATINGS Cont.

ROUGH CAST - Render and dry dash; first and finishing coats cement and sand (1:3), total 15mm thick; wood floated; dry dash of pea shingle; external Cont.

15mm work to walls on brickwork or blockwork base Cont.

	Del to Site	Waste	Material Cost	Craft Optve	Lab	Labour Cost	Sunds	Nett Rate	Unit	Gross Rate
width not exceeding 300mm	0.49	10.00	0.54	0.51	0.13	5.62	-	6.16	m	6.93
extra; spatterdash coat	0.04	10.00	0.04	0.14	0.04	1.57	-	1.62	m	1.82
width not exceeding 300mm; curved to 3000mm radius	0.49	10.00	0.54	0.69	0.13	7.28	-	7.82	m	8.80

ROUGH CAST - Render and wet dash; first and finishing coats cement and sand (1:3), total 15mm thick; wood floated; wet dash of crushed stone or shingle and cement slurry

15mm work to walls on brickwork or blockwork base

	Del to Site	Waste	Material Cost	Craft Optve	Lab	Labour Cost	Sunds	Nett Rate	Unit	Gross Rate
width exceeding 300mm	2.24	5.00	2.35	0.95	0.48	12.14	-	14.49	m2	16.30
extra; spatterdash coat	0.52	5.00	0.55	0.28	0.14	3.57	-	4.11	m2	4.63
width exceeding 300mm; curved to 3000mm radius	2.24	5.00	2.35	1.24	0.48	14.81	-	17.17	m2	19.31
width not exceeding 300mm	0.67	10.00	0.74	0.57	0.14	6.24	-	6.98	m	7.85
extra; spatterdash coat	0.16	10.00	0.18	0.17	0.04	1.85	-	2.03	m	2.28
width not exceeding 300mm; curved to 3000mm radius	0.67	10.00	0.74	0.74	0.14	7.81	-	8.55	m	9.62

TYROLEAN FINISH - Render and Tyrolean finish; first and finishing coats cement and sand (1:3), total 15mm thick; wood floated; Tyrolean finish of 'Cullamix' mixture applied by machine

15mm work to walls on brickwork or blockwork base

	Del to Site	Waste	Material Cost	Craft Optve	Lab	Labour Cost	Sunds	Nett Rate	Unit	Gross Rate
width exceeding 300mm	3.50	5.00	3.67	1.05	0.30	11.80	0.55	16.02	m2	18.03
width not exceeding 300mm	1.05	10.00	1.16	0.63	0.09	6.45	0.33	7.93	m	8.92

HARDWALL PLASTERING - Plaster, B.S.1191, Part 1, Class B; finishing coat of board finish, 3mm thick; steel trowelled; internal

3mm work to walls on concrete or plasterboard base

	Del to Site	Waste	Material Cost	Craft Optve	Lab	Labour Cost	Sunds	Nett Rate	Unit	Gross Rate
width exceeding 300mm	0.45	5.00	0.47	0.33	0.17	4.24	0.01	4.72	m2	5.31
width exceeding 300mm; curved to 3000mm radius	0.45	5.00	0.47	0.44	0.17	5.25	0.01	5.74	m2	6.45
width not exceeding 300mm	0.13	10.00	0.14	0.20	0.05	2.20	-	2.34	m	2.63
width not exceeding 300mm; curved to 3000mm radius	0.13	10.00	0.14	0.27	0.05	2.84	-	2.99	m	3.36

3mm work to ceilings on concrete or plasterboard base

	Del to Site	Waste	Material Cost	Craft Optve	Lab	Labour Cost	Sunds	Nett Rate	Unit	Gross Rate
width exceeding 300mm	0.45	5.00	0.47	0.40	0.17	4.89	0.01	5.37	m2	6.04
width exceeding 300mm; 3.50 - 5.00m above floor	0.45	5.00	0.47	0.43	0.17	5.16	0.01	5.64	m2	6.35
width not exceeding 300mm	0.13	10.00	0.14	0.24	0.05	2.57	-	2.71	m	3.05
width not exceeding 300mm; 3.50 - 5.00m above floor	0.13	10.00	0.14	0.26	0.05	2.75	-	2.89	m	3.26

3mm work to isolated beams on concrete or plasterboard base

	Del to Site	Waste	Material Cost	Craft Optve	Lab	Labour Cost	Sunds	Nett Rate	Unit	Gross Rate
width exceeding 300mm	0.45	5.00	0.47	0.50	0.17	5.81	0.01	6.29	m2	7.08
width not exceeding 300mm	0.13	10.00	0.14	0.30	0.05	3.12	-	3.26	m	3.67

3mm work to isolated columns on concrete or plasterboard base

	Del to Site	Waste	Material Cost	Craft Optve	Lab	Labour Cost	Sunds	Nett Rate	Unit	Gross Rate
width exceeding 300mm	0.45	5.00	0.47	0.50	0.17	5.81	0.01	6.29	m2	7.08
width not exceeding 300mm	0.13	10.00	0.14	0.30	0.05	3.12	-	3.26	m	3.67

HARDWALL PLASTERING - Plaster; first coat of hardwall, 11mm thick; finishing coat of multi-finish 2mm thick; steel trowelled; internal

13mm work to walls on brickwork or blockwork base

	Del to Site	Waste	Material Cost	Craft Optve	Lab	Labour Cost	Sunds	Nett Rate	Unit	Gross Rate
width exceeding 300mm	1.64	5.00	1.72	0.60	0.30	7.64	0.04	9.41	m2	10.58
width exceeding 300mm; curved to 3000mm radius	1.64	5.00	1.72	0.80	0.30	9.49	0.04	11.25	m2	12.66
width not exceeding 300mm	0.49	10.00	0.54	0.36	0.09	3.95	0.01	4.50	m	5.07
width not exceeding 300mm; curved to 3000mm radius	0.49	10.00	0.54	0.48	0.09	5.06	0.01	5.61	m	6.31

13mm work to ceilings on concrete base

	Del to Site	Waste	Material Cost	Craft Optve	Lab	Labour Cost	Sunds	Nett Rate	Unit	Gross Rate
width exceeding 300mm	1.64	5.00	1.72	0.72	0.30	8.75	0.04	10.51	m2	11.83
width exceeding 300mm; 3.50 - 5.00m above floor	1.64	5.00	1.72	0.78	0.30	9.31	0.04	11.07	m2	12.45
width not exceeding 300mm	0.49	10.00	0.54	0.43	0.09	4.60	0.01	5.15	m	5.79
width not exceeding 300mm; 3.50 - 5.00m above floor	0.49	10.00	0.54	0.47	0.09	4.97	0.01	5.52	m	6.21

13mm work to isolated beams on concrete base

	Del to Site	Waste	Material Cost	Craft Optve	Lab	Labour Cost	Sunds	Nett Rate	Unit	Gross Rate
width exceeding 300mm	1.64	5.00	1.72	0.90	0.30	10.41	0.04	12.18	m2	13.70
width not exceeding 300mm	0.49	10.00	0.54	0.54	0.09	5.62	0.01	6.17	m	6.94

13mm work to isolated columns on concrete base

	Del to Site	Waste	Material Cost	Craft Optve	Lab	Labour Cost	Sunds	Nett Rate	Unit	Gross Rate
width exceeding 300mm	1.64	5.00	1.72	0.90	0.30	10.41	0.04	12.18	m2	13.70
width not exceeding 300mm	0.49	10.00	0.54	0.54	0.09	5.62	0.01	6.17	m	6.94

Rounded angles

	Del to Site	Waste	Material Cost	Craft Optve	Lab	Labour Cost	Sunds	Nett Rate	Unit	Gross Rate
radius 10 - 100mm	-	-	-	0.30	-	2.77	-	2.77	m	3.12

SURFACE FINISHES

Labour hourly rates: (except Specialists) Craft Operatives £9.23 Labourer £7.02 Rates are national average prices. Refer to REGIONAL VARIATIONS for indicative levels of overall pricing in regions	MATERIALS			LABOUR				RATES		
	Del to Site	Waste	Material Cost	Craft Optve	Lab	Labour Cost	Sunds	Nett Rate	Unit	Gross Rate (+12.5%)
	£	%	£	Hrs	Hrs	£	£	£		£
M20: PLASTERED/RENDERED/ROUGHCAST COATINGS Cont.										
HARDWALL PLASTERING - Plaster; Thistle Universal one coat plaster, 13mm thick; steel trowelled; internal										
Note										
the thickness is from the face of the metal lathing										
13mm work to walls on metal lathing base										
width exceeding 300mm	2.12	5.00	2.23	0.50	0.25	6.37	0.05	8.65	m2	9.73
width exceeding 300mm; curved to 3000mm radius	2.12	5.00	2.23	0.67	0.25	7.94	0.05	10.22	m2	11.49
width not exceeding 300mm	0.64	10.00	0.70	0.30	0.08	3.33	0.02	4.05	m	4.56
width not exceeding 300mm; curved to 3000mm radius	0.64	10.00	0.70	0.40	0.08	4.25	0.02	4.98	m	5.60
13mm work to ceilings on metal lathing base										
width exceeding 300mm	2.12	5.00	2.23	0.60	0.25	7.29	0.05	9.57	m2	10.77
width exceeding 300mm; 3.50 - 5.00m above floor	2.12	5.00	2.23	0.65	0.25	7.75	0.05	10.03	m2	11.28
width not exceeding 300mm	0.64	10.00	0.70	0.36	0.08	3.88	0.02	4.61	m	5.18
width not exceeding 300mm; 3.50 - 5.00m above floor	0.64	10.00	0.70	0.39	0.08	4.16	0.02	4.89	m	5.50
13mm work to isolated beams on metal lathing base										
width exceeding 300mm	2.12	5.00	2.23	0.75	0.25	8.68	0.05	10.95	m2	12.32
width not exceeding 300mm	0.64	10.00	0.70	0.45	0.08	4.72	0.02	5.44	m	6.12
13mm work to isolated columns on metal lathing base										
width exceeding 300mm	2.12	5.00	2.23	0.75	0.25	8.68	0.05	10.95	m2	12.32
width not exceeding 300mm	0.64	10.00	0.70	0.45	0.08	4.72	0.02	5.44	m	6.12
Rounded angles										
radius 10 - 100mm	-	-	-	0.22	-	2.03	-	2.03	m	2.28
LIGHTWEIGHT PLASTERING - Plaster, Carlite; pre-mixed; floating coat of browning, 11mm thick; finishing coat of finish, 2mm thick; steel trowelled; internal										
13mm work to walls on brickwork or blockwork base										
width exceeding 300mm	1.39	5.00	1.46	0.50	0.25	6.37	0.03	7.86	m2	8.84
width exceeding 300mm; curved to 3000mm radius	1.39	5.00	1.46	0.67	0.25	7.94	0.03	9.43	m2	10.61
width exceeding 300mm; dubbing average 6mm thick	2.00	5.00	2.10	0.78	0.48	10.57	0.05	12.72	m2	14.31
width exceeding 300mm; dubbing average 6mm thick; curved to 3000mm radius	2.00	5.00	2.10	0.94	0.48	12.05	0.05	14.20	m2	15.97
width exceeding 300mm; dubbing average 12mm thick	2.61	5.00	2.74	0.78	0.48	10.57	0.07	13.38	m2	15.05
width exceeding 300mm; dubbing average 12mm thick; curved to 3000mm radius	2.61	5.00	2.74	0.94	0.48	12.05	0.07	14.86	m2	16.71
width exceeding 300mm; dubbing average 19mm thick	3.33	5.00	3.50	0.80	0.62	11.74	0.08	15.31	m2	17.23
width exceeding 300mm; dubbing average 19mm thick; curved to 3000mm radius	3.33	5.00	3.50	0.97	0.62	13.31	0.08	16.88	m2	18.99
width not exceeding 300mm	0.42	10.00	0.46	0.30	0.08	3.33	0.01	3.80	m	4.28
width not exceeding 300mm; curved to 3000mm radius	0.42	10.00	0.46	0.40	0.08	4.25	0.01	4.73	m	5.32
width not exceeding 300mm; dubbing average 6mm thick	0.60	10.00	0.66	0.47	0.15	5.39	0.02	6.07	m	6.83
width not exceeding 300mm; dubbing average 6mm thick; curved to 3000mm radius	0.60	10.00	0.66	0.65	0.15	7.05	0.02	7.73	m	8.70
width not exceeding 300mm; dubbing average 12mm thick	0.78	10.00	0.86	0.47	0.15	5.39	0.02	6.27	m	7.05
width not exceeding 300mm; dubbing average 12mm thick; curved to 3000mm radius	0.78	10.00	0.86	0.65	0.15	7.05	0.02	7.93	m	8.92
width not exceeding 300mm; dubbing average 19mm thick	1.00	10.00	1.10	0.48	0.19	5.76	0.02	6.88	m	7.74
width not exceeding 300mm; dubbing average 19mm thick; curved to 3000mm radius	1.00	10.00	1.10	0.64	0.19	7.24	0.02	8.36	m	9.41
Rounded angles										
radius 10 - 100mm	-	-	-	0.30	-	2.77	-	2.77	m	3.12
LIGHTWEIGHT PLASTERING - Plaster, Carlite; pre-mixed; floating coat of bonding 8mm thick; finishing coat of finish, 2mm thick; steel trowelled; internal										
10mm work to walls on concrete or plasterboard base										
width exceeding 300mm	1.30	5.00	1.37	0.50	0.25	6.37	0.03	7.76	m2	8.74
width exceeding 300mm; curved to 3000mm radius	1.30	5.00	1.37	0.67	0.25	7.94	0.03	9.33	m2	10.50
width not exceeding 300mm	0.39	10.00	0.43	0.30	0.08	3.33	0.01	3.77	m	4.24
width not exceeding 300mm; curved to 3000mm radius	0.39	10.00	0.43	0.40	0.08	4.25	0.01	4.69	m	5.28
10mm work to ceilings on concrete or plasterboard base										
width exceeding 300mm	1.30	5.00	1.37	0.60	0.25	7.29	0.03	8.69	m2	9.77
width exceeding 300mm; 3.50 - 5.00m above floor	1.30	5.00	1.37	0.65	0.25	7.75	0.03	9.15	m2	10.29
width not exceeding 300mm	0.39	10.00	0.43	0.36	0.08	3.88	0.01	4.32	m	4.86
width not exceeding 300mm; 3.50 - 5.00m above floor	0.39	10.00	0.43	0.39	0.08	4.16	0.01	4.60	m	5.18
10mm work to isolated beams on concrete or plasterboard base										
width exceeding 300mm	1.30	5.00	1.37	0.75	0.25	8.68	0.03	10.07	m2	11.33
width not exceeding 300mm	0.39	10.00	0.43	0.45	0.08	4.72	0.01	5.15	m	5.80
10mm work to isolated columns on concrete or plasterboard base										
width exceeding 300mm	1.30	5.00	1.37	0.75	0.25	8.68	0.03	10.07	m2	11.33
width not exceeding 300mm	0.39	10.00	0.43	0.45	0.08	4.72	0.01	5.15	m	5.80
Rounded angles										
radius 10 - 100mm	-	-	-	0.30	-	2.77	-	2.77	m	3.12

Labour hourly rates: (except Specialists) Craft Operatives £9.23 Labourer £7.02 Rates are national average prices. Refer to REGIONAL VARIATIONS for indicative levels of overall pricing in regions	MATERIALS			LABOUR				RATES		
	Del to Site £	Waste %	Material Cost £	Craft Optve Hrs	Lab Hrs	Labour Cost £	Sunds £	Nett Rate £	Unit	Gross Rate (+12.5%) £

M20: PLASTERED/RENDERED/ROUGHCAST COATINGS Cont.

LIGHTWEIGHT PLASTERING - Plaster, Carlite; pre-mixed; floating coat of bonding 11mm thick; finishing coat of finish, 2mm thick; steel trowelled; internal

13mm work to ceilings on precast concrete beam and infill blocks base

width exceeding 300mm	1.99	5.00	2.09	0.60	0.25	7.29	0.05	9.43	m2	10.61
width not exceeding 300mm	0.60	10.00	0.66	0.36	0.08	3.88	0.01	4.55	m	5.12

LIGHTWEIGHT PLASTERING - Plaster, Carlite; pre-mixed; pricking up and floating coats metal lathing, 11mm thick; finishing coat of finish, 2mm thick; steel trowelled; internal

Note
the thickness is from the face of the metal lathing

13mm work to walls on metal lathing base

width exceeding 300mm	2.66	5.00	2.79	0.70	0.35	8.92	0.07	11.78	m2	13.25
width exceeding 300mm; curved to 3000mm radius	2.66	5.00	2.79	0.93	0.35	11.04	0.07	13.90	m2	15.64
width not exceeding 300mm	0.80	10.00	0.88	0.42	0.11	4.65	0.02	5.55	m	6.24
width not exceeding 300mm; curved to 3000mm radius	0.80	10.00	0.88	0.56	0.11	5.94	0.02	6.84	m	7.70

13mm work to ceilings on metal lathing base

width exceeding 300mm	2.66	5.00	2.79	0.84	0.35	10.21	0.07	13.07	m2	14.71
width exceeding 300mm; 3.50 - 5.00m above floor	2.66	5.00	2.79	0.91	0.35	10.86	0.07	13.72	m2	15.43
width not exceeding 300mm	0.80	10.00	0.88	0.50	0.11	5.39	0.02	6.29	m	7.07
width not exceeding 300mm; 3.50 - 5.00m above floor	0.80	10.00	0.88	0.55	0.11	5.85	0.02	6.75	m	7.59

13mm work to isolated beams on metal lathing base

width exceeding 300mm	2.66	5.00	2.79	1.05	0.35	12.15	0.07	15.01	m2	16.89
width not exceeding 300mm	0.80	10.00	0.88	0.63	0.11	6.59	0.02	7.49	m	8.42

13mm work to isolated columns on metal lathing base

width exceeding 300mm	2.66	5.00	2.79	1.05	0.35	12.15	0.07	15.01	m2	16.89
width not exceeding 300mm	0.80	10.00	0.88	0.63	0.11	6.59	0.02	7.49	m	8.42

Rounded angles

radius 10 - 100mm	-	-	-	0.40	-	3.69	-	3.69	m	4.15

RENOVATING PLASTER - Renovating plaster, Thistle; undercoat 11mm thick; finishing coat 2mm thick; steel trowelled; internal

13mm work to walls on existing brickwork or blockwork base

width exceeding 300mm	1.72	5.00	1.81	0.65	0.33	8.32	0.04	10.16	m2	11.43
width exceeding 300mm; curved to 3000mm radius	1.72	5.00	1.81	0.87	0.33	10.35	0.04	12.19	m2	13.72
width exceeding 300mm; dubbing average 6mm thick	2.46	5.00	2.58	0.90	0.51	11.89	0.06	14.53	m2	16.35
width exceeding 300mm; dubbing average 6mm thick; curved to 3000mm radius	2.46	5.00	2.58	1.20	0.51	14.66	0.06	17.30	m2	19.46
width exceeding 300mm; dubbing average 12mm thick	3.19	5.00	3.35	0.93	0.69	13.43	0.08	16.86	m2	18.96
width exceeding 300mm; dubbing average 12mm thick; curved to 3000mm radius	3.19	5.00	3.35	1.24	0.69	16.29	0.08	19.72	m2	22.18
width not exceeding 300mm	0.52	10.00	0.57	0.39	0.10	4.30	0.01	4.88	m	5.49
width not exceeding 300mm; curved to 3000mm radius	0.52	10.00	0.57	0.52	0.10	5.50	0.01	6.08	m	6.84
width not exceeding 300mm; dubbing average 6mm thick	0.74	10.00	0.81	0.54	0.15	6.04	0.02	6.87	m	7.73
width not exceeding 300mm; dubbing average 6mm thick; curved to 3000mm radius	0.74	10.00	0.81	0.72	0.15	7.70	0.02	8.53	m	9.60
width not exceeding 300mm; dubbing average 12mm thick	0.96	10.00	1.06	0.56	0.21	6.64	0.02	7.72	m	8.68
width not exceeding 300mm; dubbing average 12mm thick; curved to 3000mm radius	0.96	10.00	1.06	0.75	0.21	8.40	0.02	9.47	m	10.66

Rounded angles

radius 10 - 100mm	-	-	-	0.30	-	2.77	-	2.77	m	3.12

PLASTER BEADS AND STOPS - Accessories

Galvanised steel angle beads
Expamet fixing to brickwork or blockwork with masonry nails

Reference 550	0.61	10.00	0.67	0.10	-	0.92	-	1.59	m	1.79
Reference 553	0.73	10.00	0.80	0.10	-	0.92	-	1.73	m	1.94
Reference 554	0.96	10.00	1.06	0.10	-	0.92	-	1.98	m	2.23

Galvanised steel stop beads
Expamet fixing to brickwork or blockwork with masonry nails

Reference 560	0.61	10.00	0.67	0.10	-	0.92	-	1.59	m	1.79
Reference 561	0.79	10.00	0.87	0.10	-	0.92	-	1.79	m	2.02
Reference 562	0.71	10.00	0.78	0.10	-	0.92	-	1.70	m	1.92
Reference 563	0.71	10.00	0.78	0.10	-	0.92	-	1.70	m	1.92
Reference 565	0.91	10.00	1.00	0.10	-	0.92	-	1.92	m	2.16
Reference 566	0.91	10.00	1.00	0.10	-	0.92	-	1.92	m	2.16

Galvanised steel plasterboard edging beads
Expamet fixing to timber with galvanised nails

Reference 567	1.04	10.00	1.14	0.10	-	0.92	-	2.07	m	2.33
Reference 568	1.04	10.00	1.14	0.10	-	0.92	-	2.07	m	2.33

SURFACE FINISHES

SURFACE FINISHES – SMALL WORKS

Labour hourly rates: (except Specialists) Craft Operatives £9.23 Labourer £7.02 Rates are national average prices. Refer to REGIONAL VARIATIONS for indicative levels of overall pricing in regions	MATERIALS			LABOUR				RATES		
	Del to Site £	Waste %	Material Cost £	Craft Optve Hrs	Lab Hrs	Labour Cost £	Sunds £	Nett Rate £	Unit	Gross Rate (+12.5%) £
M20: PLASTERED/RENDERED/ROUGHCAST COATINGS Cont.										
PLASTER BEADS AND STOPS - Accessories Cont.										
Galvanised steel depth gauge beads										
Expamet fixing to brickwork or blockwork with masonry nails										
Reference 569	0.57	10.00	0.63	0.10	-	0.92	-	1.55	m	1.74
PLASTER BEADS AND STOPS - Accessories; external										
Galvanised steel stop beads										
Expamet fixing to brickwork or blockwork with masonry nails										
Reference 570	0.74	10.00	0.81	0.10	-	0.92	-	1.74	m	1.95
Stainless steel angle beads										
Expamet fixing to brickwork or blockwork with masonry nails										
Reference 545	1.77	10.00	1.95	0.10	-	0.92	-	2.87	m	3.23
Stainless steel stop beads										
Expamet fixing to brickwork or blockwork with masonry nails										
Reference 547	1.58	10.00	1.74	0.10	-	0.92	-	2.66	m	2.99
ANTI-CRACK STRIPS - Accessories - anti-crack strips										
Lathing; BB galvanised expanded metal lath, 9mm mesh X 0.500mm thick x 1.11 kg/m2; butt joints fixing with nails										
100mm wide to walls; one edge to timber; one edge to brickwork or blockwork	0.44	10.00	0.48	0.10	-	0.92	0.02	1.43	m	1.61
100mm wide to walls; one edge to timber; one edge to concrete	0.44	10.00	0.48	0.12	-	1.11	0.02	1.61	m	1.81
100mm wide to walls; one edge to brickwork or blockwork; one edge to concrete	0.44	10.00	0.48	0.14	-	1.29	0.02	1.80	m	2.02
Lathing; BB galvanised expanded metal lath, 9mm mesh x 0.725mm thick x 1.61 kg/m2; butt joints; fixing with nails										
100mm wide to walls; one edge to timber; one edge to brickwork or blockwork	0.52	10.00	0.57	0.10	-	0.92	0.02	1.51	m	1.70
100mm wide to walls; one edge to timber; one edge to concrete	0.52	10.00	0.57	0.12	-	1.11	0.02	1.70	m	1.91
100mm wide to walls; one edge to brickwork or blockwork; one edge to concrete	0.52	10.00	0.57	0.14	-	1.29	0.02	1.88	m	2.12
BONDING FLUID - Prepare and apply bonding fluid to receive plaster or cement rendering										
Work to walls on existing cement and sand base										
width exceeding 300mm	0.66	10.00	0.73	0.25	-	2.31	-	3.03	m2	3.41
Work to walls on existing glazed tiling base										
width exceeding 300mm	0.44	10.00	0.48	0.20	-	1.85	-	2.33	m2	2.62
Work to walls on existing painted surface										
width exceeding 300mm	0.53	10.00	0.58	0.23	-	2.12	-	2.71	m2	3.04
Work to walls on existing concrete base										
width exceeding 300mm	0.58	10.00	0.64	0.25	-	2.31	-	2.95	m2	3.31
Work to ceilings on existing cement and sand base										
width exceeding 300mm	0.66	10.00	0.73	0.31	-	2.86	-	3.59	m2	4.04
Work to ceilings on existing painted surface										
width exceeding 300mm	0.53	10.00	0.58	0.28	-	2.58	-	3.17	m2	3.56
Work to ceilings on existing concrete base										
width exceeding 300mm	0.58	10.00	0.64	0.31	-	2.86	-	3.50	m2	3.94
M20: PLASTERED/RENDERED/ROUGHCAST COATINGS; STAIRCASE AREAS										
PLASTERBOARD - Baseboarding; gypsum baseboard, Thistle; 5mm joints, filled with plaster and scrimmed; fixing with galvanised nails; internal										
9.5mm work to walls on timber base										
width exceeding 300mm	1.71	5.00	1.80	0.36	0.15	4.38	0.19	6.36	m2	7.16
width not exceeding 300mm	0.51	10.00	0.56	0.22	0.05	2.38	0.11	3.05	m	3.43
9.5mm work to ceilings on timber base										
width exceeding 300mm	1.71	5.00	1.80	0.44	0.15	5.11	0.19	7.10	m2	7.99
width not exceeding 300mm	0.51	10.00	0.56	0.26	0.05	2.75	0.11	3.42	m	3.85
PLASTERBOARD - Baseboarding; gypsum baseboard, Thistle Duplex; 5mm joints filled with plaster and scrimmed; fixing with galvanised nails; internal										
9.5mm work to walls on timber base										
width exceeding 300mm	2.33	5.00	2.45	0.36	0.15	4.38	0.19	7.01	m2	7.89
width not exceeding 300mm	0.70	10.00	0.77	0.22	0.05	2.38	0.11	3.26	m	3.67

SURFACE FINISHES – SMALL WORKS

Labour hourly rates: (except Specialists) Craft Operatives £9.23 Labourer £7.02 Rates are national average prices. Refer to REGIONAL VARIATIONS for indicative levels of overall pricing in regions	MATERIALS			LABOUR				RATES		
	Del to Site £	Waste %	Material Cost £	Craft Optve Hrs	Lab Hrs	Labour Cost £	Sunds £	Nett Rate £	Unit	Gross Rate (+12.5%) £

M20: PLASTERED/RENDERED/ROUGHCAST COATINGS; STAIRCASE AREAS Cont.

PLASTERBOARD - Baseboarding; gypsum baseboard, Thistle Duplex; 5mm joints filled with plaster and scrimmed; fixing with galvanised nails; internal Cont.

	Del to Site £	Waste %	Material Cost £	Craft Optve Hrs	Lab Hrs	Labour Cost £	Sunds £	Nett Rate £	Unit	Gross Rate £
9.5mm work to ceilings on timber base										
width exceeding 300mm	2.33	5.00	2.45	0.39	0.15	4.65	0.19	7.29	m2	8.20
width not exceeding 300mm	0.70	10.00	0.77	0.23	0.05	2.47	0.11	3.35	m	3.77
PLASTERBOARD - Baseboarding; Gyproc lath; 3mm joints, filled with plaster; fixing with galvanised nails; internal										
9.5mm work to walls on timber base										
width exceeding 300mm	2.01	5.00	2.11	0.36	0.15	4.38	0.15	6.64	m2	7.47
width not exceeding 300mm	0.60	10.00	0.66	0.22	0.05	2.38	0.09	3.13	m	3.52
12.5mm work to walls on timber base										
width exceeding 300mm	2.36	5.00	2.48	0.44	0.18	5.32	0.15	7.95	m2	8.95
width not exceeding 300mm	0.71	10.00	0.78	0.26	0.05	2.75	0.09	3.62	m	4.07
9.5mm work to ceilings on timber base										
width exceeding 300mm	2.01	5.00	2.11	0.39	0.15	4.65	0.15	6.91	m2	7.78
width not exceeding 300mm	0.60	10.00	0.66	0.23	0.05	2.47	0.09	3.22	m	3.63
12.5mm work to ceilings on timber base										
width exceeding 300mm	2.36	5.00	2.48	0.47	0.18	5.60	0.15	8.23	m2	9.26
width not exceeding 300mm	0.71	10.00	0.78	0.28	0.05	2.94	0.09	3.81	m	4.28
PLASTERBOARD - Baseboarding; Gyproc plank, square edge; 5mm joints, filled with plaster and scrimmed; internal										
19mm work to walls on timber base										
width exceeding 300mm	3.23	5.00	3.39	0.59	0.24	7.13	0.18	10.70	m2	12.04
width not exceeding 300mm	0.97	10.00	1.07	0.35	0.07	3.72	0.11	4.90	m	5.51
19mm work to ceilings on timber base										
width exceeding 300mm	3.23	5.00	3.39	0.63	0.24	7.50	0.18	11.07	m2	12.46
width not exceeding 300mm	0.97	10.00	1.07	0.38	0.07	4.00	0.11	5.18	m	5.82
PLASTERBOARD - Baseboarding; Gyproc square edge wallboard; 3mm joints filled with plaster and scrimmed; fixing with galvanised nails; internal										
9.5mm work to walls on timber base										
width exceeding 300mm	1.71	5.00	1.80	0.36	0.15	4.38	0.19	6.36	m2	7.16
width not exceeding 300mm	0.51	10.00	0.56	0.22	0.05	2.38	0.11	3.05	m	3.43
12.5mm work to walls on timber base										
width exceeding 300mm	1.99	5.00	2.09	0.44	0.18	5.32	0.19	7.60	m2	8.55
width not exceeding 300mm	0.60	10.00	0.66	0.26	0.05	2.75	0.11	3.52	m	3.96
15mm work to walls on timber base										
width exceeding 300mm	2.18	5.00	2.29	0.52	0.21	6.27	0.19	8.75	m2	9.85
width not exceeding 300mm	0.65	10.00	0.71	0.31	0.06	3.28	0.11	4.11	m	4.62
9.5mm work to ceilings on timber base										
width exceeding 300mm	1.71	5.00	1.80	0.39	0.15	4.65	0.19	6.64	m2	7.47
width not exceeding 300mm	0.51	10.00	0.56	0.23	0.05	2.47	0.11	3.14	m	3.54
12.5mm work to ceilings on timber base										
width exceeding 300mm	1.99	5.00	2.09	0.47	0.18	5.60	0.19	7.88	m2	8.87
width not exceeding 300mm	0.60	10.00	0.66	0.28	0.05	2.94	0.11	3.71	m	4.17
15mm work to ceilings on timber base										
width exceeding 300mm	2.18	5.00	2.29	0.55	0.25	6.83	0.19	9.31	m2	10.47
width not exceeding 300mm	0.65	10.00	0.71	0.33	0.08	3.61	0.11	4.43	m	4.99
PLASTERBOARD - Baseboarding; Gyproc square edge Duplex wallboard; 3mm joints filled with plaster and scrimmed; fixing with galvanised nails; internal										
9.5mm work to walls on timber base										
width exceeding 300mm	2.33	5.00	2.45	0.36	0.15	4.38	0.19	7.01	m2	7.89
width not exceeding 300mm	0.70	10.00	0.77	0.22	0.05	2.38	0.11	3.26	m	3.67
12.5mm work to walls on timber base										
width exceeding 300mm	2.61	5.00	2.74	0.44	0.18	5.32	0.19	8.26	m2	9.29
width not exceeding 300mm	0.78	10.00	0.86	0.26	0.05	2.75	0.11	3.72	m	4.18
15mm work to walls on timber base										
width exceeding 300mm	2.86	5.00	3.00	0.52	0.21	6.27	0.19	9.47	m2	10.65
width not exceeding 300mm	0.86	10.00	0.95	0.31	0.06	3.28	0.11	4.34	m	4.88
9.5mm work to ceilings on timber base										
width exceeding 300mm	2.33	5.00	2.45	0.39	0.15	4.65	0.19	7.29	m2	8.20
width not exceeding 300mm	0.70	10.00	0.77	0.23	0.05	2.47	0.11	3.35	m	3.77
12.5mm work to ceilings on timber base										
width exceeding 300mm	2.61	5.00	2.74	0.47	0.18	5.60	0.19	8.53	m2	9.60
width not exceeding 300mm	0.78	10.00	0.86	0.28	0.05	2.94	0.11	3.90	m	4.39

SURFACE FINISHES

Labour hourly rates: (except Specialists) Craft Operatives £9.23 Labourer £7.02 Rates are national average prices. Refer to REGIONAL VARIATIONS for indicative levels of overall pricing in regions	MATERIALS			LABOUR				RATES		
	Del to Site £	Waste %	Material Cost £	Craft Optve Hrs	Lab Hrs	Labour Cost £	Sunds £	Nett Rate £	Unit	Gross Rate (+12.5%) £
M20: PLASTERED/RENDERED/ROUGHCAST COATINGS; STAIRCASE AREAS Cont.										
PLASTERBOARD - Baseboarding; Gyproc square edge Duplex wallboard; 3mm joints filled with plaster and scrimmed; fixing with galvanised nails; internal Cont.										
15mm work to ceilings on timber base										
width exceeding 300mm	2.86	5.00	3.00	0.55	0.25	6.83	0.19	10.02	m2	11.28
width not exceeding 300mm	0.86	10.00	0.95	0.33	0.08	3.61	0.11	4.66	m	5.25
PLASTERBOARD - Baseboarding; two layers 9.5mm Gyproc square edge wallboard; second layer with 3mm joints filled with plaster and scrimmed; fixing with galvanised nails; internal										
19mm work to walls on timber base										
width exceeding 300mm	3.42	5.00	3.59	0.75	0.30	9.03	0.28	12.90	m2	14.51
width not exceeding 300mm	1.03	10.00	1.13	0.45	0.09	4.79	0.17	6.09	m	6.85
19mm work to ceilings on timber base										
width exceeding 300mm	3.42	5.00	3.59	0.81	0.30	9.58	0.28	13.45	m2	15.13
width not exceeding 300mm	1.03	10.00	1.13	0.49	0.09	5.15	0.17	6.46	m	7.26
PLASTERBOARD - Baseboarding; one layer 9.5mm and one layer 12.5mm Gyproc square edge wallboard; second layer with 3mm joints filled with plaster and scrimmed; fixing with galvanised nails, internal										
22mm work to walls on timber base										
width exceeding 300mm	3.70	5.00	3.88	0.81	0.33	9.79	0.28	13.96	m2	15.70
width not exceeding 300mm	1.11	10.00	1.22	0.49	0.10	5.22	0.17	6.62	m	7.44
22mm work to ceilings on timber base										
width exceeding 300mm	3.70	5.00	3.88	0.88	0.33	10.44	0.28	14.60	m2	16.43
width not exceeding 300mm	1.11	10.00	1.22	0.53	0.10	5.59	0.17	6.98	m	7.86
PLASTERBOARD - Baseboarding; two layers 12.5mm Gyproc square edge wallboard; second layer with 3mm joints filled with plaster and scrimmed; fixing with galvanised nails; internal										
25mm work to walls on timber base										
width exceeding 300mm	3.98	5.00	4.18	0.88	0.35	10.58	0.28	15.04	m2	16.92
width not exceeding 300mm	1.19	10.00	1.31	0.53	0.11	5.66	0.17	7.14	m	8.04
25mm work to ceilings on timber base										
width exceeding 300mm	3.98	5.00	4.18	0.95	0.35	11.23	0.28	15.68	m2	17.65
width not exceeding 300mm	1.19	10.00	1.31	0.57	0.11	6.03	0.17	7.51	m	8.45
PLASTERBOARD - Baseboarding; two layers 15mm Gyproc square edge wallboard; second layer with 3mm joints filled with plaster and scrimmed; fixing with galvanised nails; internal										
30mm work to walls on timber base										
width exceeding 300mm	4.36	5.00	4.58	0.95	0.38	11.44	0.28	16.29	m2	18.33
width not exceeding 300mm	1.31	10.00	1.44	0.57	0.11	6.03	0.17	7.64	m	8.60
30mm work to ceilings on timber base										
width exceeding 300mm	4.36	5.00	4.58	1.02	0.38	12.08	0.28	16.94	m2	19.06
width not exceeding 300mm	1.31	10.00	1.44	0.61	0.11	6.40	0.17	8.01	m	9.02
PLASTERBOARD - Baseboarding; two layers; first layer 9.5mm Gyproc Duplex wallboard, second layer 9.5mm Gyproc wallboard both layers with square edge boards; second layer with 3mm joints filled with plaster and scrimmed; fixing with galvanised nails; internal										
19mm work to walls on timber base										
width exceeding 300mm	4.04	5.00	4.24	0.75	0.30	9.03	0.28	13.55	m2	15.24
width not exceeding 300mm	1.21	10.00	1.33	0.45	0.09	4.79	0.17	6.29	m	7.07
19mm work to ceilings on timber base										
width exceeding 300mm	4.04	5.00	4.24	0.81	0.30	9.58	0.28	14.10	m2	15.87
width not exceeding 300mm	1.21	10.00	1.33	0.49	0.09	5.15	0.17	6.66	m	7.49
PLASTERBOARD - Baseboarding; two layers; first layer 12.5mm Gyproc Duplex wallboard, second layer 9.5mm Gyproc wallboard, both layers with square edge boards; second layer with 3mm joints filled with plaster and scrimmed; fixing with galvanised nails; internal										
22mm work to walls on timber base										
width exceeding 300mm	4.32	5.00	4.54	0.81	0.33	9.79	0.28	14.61	m2	16.44
width not exceeding 300mm	1.30	10.00	1.43	0.49	0.10	5.22	0.17	6.82	m	7.68
22mm work to ceilings on timber base										
width exceeding 300mm	4.32	5.00	4.54	0.88	0.33	10.44	0.28	15.26	m2	17.16
width not exceeding 300mm	1.30	10.00	1.43	0.53	0.10	5.59	0.17	7.19	m	8.09

Labour hourly rates: (except Specialists) Craft Operatives £9.23 Labourer £7.02 Rates are national average prices. Refer to REGIONAL VARIATIONS for indicative levels of overall pricing in regions	MATERIALS			LABOUR				RATES		
	Del to Site £	Waste %	Material Cost £	Craft Optve Hrs	Lab Hrs	Labour Cost £	Sunds £	Nett Rate £	Unit	Gross Rate (+12.5%) £
M20: PLASTERED/RENDERED/ROUGHCAST COATINGS; STAIRCASE AREAS Cont.										
PLASTERBOARD - Baseboarding; two layers; first layer 12.5mm Gyproc Duplex wallboard, second layer 12.5mm Gyproc wallboard, both layers with square edge boards; second layer with 3mm joints filled with plaster and scrimmed; fixing with galvanised nails; internal										
25mm work to walls on timber base										
width exceeding 300mm	4.60	5.00	4.83	0.88	0.35	10.58	0.28	15.69	m2	17.65
width not exceeding 300mm	1.38	10.00	1.52	0.53	0.11	5.66	0.17	7.35	m	8.27
25mm work to ceilings on timber base										
width exceeding 300mm	4.60	5.00	4.83	0.95	0.35	11.23	0.28	16.34	m2	18.38
width not exceeding 300mm	1.38	10.00	1.52	0.57	0.11	6.03	0.17	7.72	m	8.69
PLASTERBOARD - Baseboarding; two layers; first layer 15mm Gyproc Duplex wallboard, second layer 15mm Gyproc wallboard, both layers with square edge boards; second layer with 3mm joints filled with plaster and scrimmed; fixing with galvanised nails; internal										
30mm work to walls on timber base										
width exceeding 300mm	5.04	5.00	5.29	0.95	0.38	11.44	0.28	17.01	m2	19.13
width not exceeding 300mm	1.51	10.00	1.66	0.57	0.11	6.03	0.17	7.86	m	8.85
30mm work to ceilings on timber base										
width exceeding 300mm	5.04	5.00	5.29	1.02	0.38	12.08	0.28	17.65	m2	19.86
width not exceeding 300mm	1.51	10.00	1.66	0.61	0.11	6.40	0.17	8.23	m	9.26
HARDWALL PLASTERING - Plaster, B.S.1191, Part 1, Class B; finishing coat of board finish, 3mm thick; steel trowelled; internal										
3mm work to walls on concrete or plasterboard base										
width exceeding 300mm	0.45	5.00	0.47	0.41	0.17	4.98	0.01	5.46	m2	6.14
width not exceeding 300mm	0.13	10.00	0.14	0.25	0.05	2.66	-	2.80	m	3.15
3mm work to ceilings on concrete or plasterboard base										
width exceeding 300mm	0.45	5.00	0.47	0.45	0.17	5.35	0.01	5.83	m2	6.56
width not exceeding 300mm	0.13	10.00	0.14	0.27	0.05	2.84	-	2.99	m	3.36
HARDWALL PLASTERING - Plaster; first coat of hardwall, 11mm thick; finishing coat of multi-finish 2mm thick; steel trowelled; internal										
13mm work to walls on brickwork or blockwork base										
width exceeding 300mm	1.64	5.00	1.72	0.75	0.30	9.03	0.04	10.79	m2	12.14
width not exceeding 300mm	0.49	10.00	0.54	0.45	0.09	4.79	0.01	5.33	m	6.00
13mm work to ceilings on concrete base										
width exceeding 300mm	1.64	5.00	1.72	0.81	0.30	9.58	0.04	11.34	m2	12.76
width not exceeding 300mm	0.49	10.00	0.54	0.49	0.09	5.15	0.01	5.70	m	6.42
HARDWALL PLASTERING - Plaster; Thistle Universal one coat plaster, 13mm thick; steel trowelled; internal										
Note the thickness is from the face of the metal lathing										
13mm work to walls on metal lathing base										
width exceeding 300mm	2.12	5.00	2.23	0.63	0.25	7.57	0.05	9.85	m2	11.08
width not exceeding 300mm	0.64	10.00	0.70	0.38	0.08	4.07	0.02	4.79	m	5.39
13mm work to ceilings on metal lathing base										
width exceeding 300mm	2.12	5.00	2.23	0.68	0.25	8.03	0.05	10.31	m2	11.60
width not exceeding 300mm	0.64	10.00	0.70	0.41	0.08	4.35	0.02	5.07	m	5.70
LIGHTWEIGHT PLASTERING										
LIGHTWEIGHT PLASTERING - Plaster, Carlite; pre-mixed; floating coat of browning, 11mm thick; finishing coat of finish, 2mm thick; steel trowelled; internal										
13mm work to walls on brickwork or blockwork base										
width exceeding 300mm	1.39	5.00	1.46	0.63	0.25	7.57	0.03	9.06	m2	10.19
width not exceeding 300mm	0.42	10.00	0.46	0.38	0.08	4.07	0.01	4.54	m	5.11
LIGHTWEIGHT PLASTERING - Plaster, Carlite; pre-mixed; floating coat of bonding 8mm thick; finishing coat of finish, 2mm thick; steel trowelled; internal										
10mm work to walls on concrete or plasterboard base										
width exceeding 300mm	1.30	5.00	1.37	0.63	0.25	7.57	0.03	8.96	m2	10.09
width not exceeding 300mm	0.39	10.00	0.43	0.38	0.08	4.07	0.01	4.51	m	5.07
10mm work to ceilings on concrete or plasterboard base										
width exceeding 300mm	1.30	5.00	1.37	0.68	0.25	8.03	0.03	9.43	m2	10.60
width not exceeding 300mm	0.39	10.00	0.43	0.41	0.08	4.35	0.01	4.78	m	5.38

SURFACE FINISHES

Labour hourly rates: (except Specialists) Craft Operatives £9.23 Labourer £7.02 Rates are national average prices. Refer to REGIONAL VARIATIONS for indicative levels of overall pricing in regions	MATERIALS			LABOUR				RATES		
	Del to Site £	Waste %	Material Cost £	Craft Optve Hrs	Lab Hrs	Labour Cost £	Sunds £	Nett Rate £	Unit	Gross Rate (+12.5%) £
LIGHTWEIGHT PLASTERING Cont.										
LIGHTWEIGHT PLASTERING – Plaster, Carlite; pre-mixed; pricking up and floating coats metal lathing, 11mm thick; finishing coat of finish, 2mm thick; steel trowelled; internal										
Note the thickness is from the face of the metal lathing										
13mm work to walls on metal lathing base										
width exceeding 300mm	2.66	5.00	2.79	0.88	0.35	10.58	0.07	13.44	m2	15.12
width not exceeding 300mm	0.80	10.00	0.88	0.53	0.11	5.66	0.02	6.56	m	7.38
13mm work to ceilings on metal lathing base										
width exceeding 300mm	2.66	5.00	2.79	0.95	0.35	11.23	0.07	14.09	m2	15.85
width not exceeding 300mm	0.80	10.00	0.88	0.57	0.11	6.03	0.02	6.93	m	7.80
M30: METAL MESH LATHING/ANCHORED REINFORCEMENT FOR PLASTERED COATINGS										
Lathing; BB galvanised expanded metal lath, 9mm mesh x 0.500mm thick x 1.11 kg/m2; butt joints; fixing with galvanised staples										
Work to walls										
width exceeding 300mm	4.21	5.00	4.42	0.18	0.09	2.29	0.05	6.76	m2	7.61
width not exceeding 300mm	1.47	10.00	1.62	0.11	0.03	1.23	0.03	2.87	m	3.23
Work to ceilings										
width exceeding 300mm	4.21	5.00	4.42	0.23	0.09	2.75	0.07	7.25	m2	8.15
width not exceeding 300mm	1.47	10.00	1.62	0.14	0.03	1.50	0.04	3.16	m	3.55
Work to isolated beams										
width exceeding 300mm	4.21	5.00	4.42	0.27	0.09	3.12	0.07	7.61	m2	8.57
width not exceeding 300mm	1.47	10.00	1.62	0.16	0.03	1.69	0.04	3.34	m	3.76
Work to isolated columns										
width exceeding 300mm	4.21	5.00	4.42	0.27	0.09	3.12	0.07	7.61	m2	8.57
width not exceeding 300mm	1.47	10.00	1.62	0.16	0.03	1.69	0.04	3.34	m	3.76
Lathing; BB galvanised expanded metal lath 9mm mesh x 0.725mm thick x 1.61 kg/m2; butt joints; fixing with galvanised staples										
Work to walls										
width exceeding 300mm	4.93	5.00	5.18	0.18	0.09	2.29	0.07	7.54	m2	8.48
width not exceeding 300mm	1.73	10.00	1.90	0.11	0.03	1.23	0.04	3.17	m	3.57
Work to ceilings										
width exceeding 300mm	4.93	5.00	5.18	0.23	0.09	2.75	0.07	8.00	m2	9.00
width not exceeding 300mm	1.73	10.00	1.90	0.14	0.03	1.50	0.04	3.45	m	3.88
Work to isolated beams										
width exceeding 300mm	4.93	5.00	5.18	0.27	0.09	3.12	0.07	8.37	m2	9.42
width not exceeding 300mm	1.73	10.00	1.90	0.16	0.03	1.69	0.04	3.63	m	4.08
Work to isolated columns										
width exceeding 300mm	4.93	5.00	5.18	0.27	0.09	3.12	0.07	8.37	m2	9.42
width not exceeding 300mm	1.73	10.00	1.90	0.16	0.03	1.69	0.04	3.63	m	4.08
Lathing; galvanised expanded metal lath 9mm mesh x 0.950mm thick x 2.50 kg/m2; butt joints; fixing with galvanised staples										
Work to walls										
width exceeding 300mm	7.56	5.00	7.94	0.18	0.09	2.29	0.05	10.28	m2	11.57
width not exceeding 300mm	2.65	10.00	2.92	0.11	0.03	1.23	0.03	4.17	m	4.69
Work to ceilings										
width exceeding 300mm	7.56	5.00	7.94	0.23	0.09	2.75	0.07	10.76	m2	12.11
width not exceeding 300mm	2.65	10.00	2.92	0.14	0.03	1.50	0.04	4.46	m	5.02
Work to isolated beams										
width exceeding 300mm	7.56	5.00	7.94	0.27	0.09	3.12	0.07	11.13	m2	12.52
width not exceeding 300mm	2.65	10.00	2.92	0.16	0.03	1.69	0.04	4.64	m	5.22
Work to isolated columns										
width exceeding 300mm	7.56	5.00	7.94	0.27	0.09	3.12	0.07	11.13	m2	12.52
width not exceeding 300mm	2.65	10.00	2.92	0.16	0.03	1.69	0.04	4.64	m	5.22
Lathing; Expamet galvanised Rib-Lath, 0.400mm thick x 1.78 kg/m2; butt joints; fixing with galvanised staples										
Work to walls										
width exceeding 300mm	5.00	5.00	5.25	0.18	0.09	2.29	0.05	7.59	m2	8.54
width not exceeding 300mm	1.75	10.00	1.93	0.11	0.03	1.23	0.03	3.18	m	3.58
Work to ceilings										
width exceeding 300mm	5.00	5.00	5.25	0.23	0.09	2.75	0.07	8.07	m2	9.08
width not exceeding 300mm	1.75	10.00	1.93	0.14	0.03	1.50	0.04	3.47	m	3.90
Work to isolated beams										
width exceeding 300mm	5.00	5.00	5.25	0.27	0.09	3.12	0.07	8.44	m2	9.50
width not exceeding 300mm	1.75	10.00	1.93	0.16	0.03	1.69	0.04	3.65	m	4.11

Labour hourly rates: (except Specialists) Craft Operatives £9.23 Labourer £7.02 Rates are national average prices. Refer to REGIONAL VARIATIONS for indicative levels of overall pricing in regions	MATERIALS			LABOUR				RATES		
	Del to Site £	Waste %	Material Cost £	Craft Optve Hrs	Lab Hrs	Labour Cost £	Sunds £	Nett Rate £	Unit	Gross Rate (+12.5%) £
M30: METAL MESH LATHING/ANCHORED REINFORCEMENT FOR PLASTERED COATINGS Cont.										
Lathing; Expamet galvanised Rib-Lath, 0.400mm thick x 1.78 kg/m2; butt joints; fixing with galvanised staples Cont.										
Work to isolated columns										
width exceeding 300mm	5.00	5.00	5.25	0.27	0.09	3.12	0.07	8.44	m2	9.50
width not exceeding 300mm	1.75	10.00	1.93	0.16	0.03	1.69	0.04	3.65	m	4.11
Lathing; Expamet galvanised Rib-Lath, 0.500mm thick x 2.25 kg/m2; butt joints; fixing with galvanised staples										
Work to walls										
width exceeding 300mm	5.75	5.00	6.04	0.18	0.09	2.29	0.05	8.38	m2	9.43
width not exceeding 300mm	2.01	10.00	2.21	0.11	0.03	1.23	0.03	3.47	m	3.90
Work to ceilings										
width exceeding 300mm	5.75	5.00	6.04	0.23	0.09	2.75	0.07	8.86	m2	9.97
width not exceeding 300mm	2.01	10.00	2.21	0.14	0.03	1.50	0.04	3.75	m	4.22
Work to isolated beams										
width exceeding 300mm	5.75	5.00	6.04	0.27	0.09	3.12	0.07	9.23	m2	10.39
width not exceeding 300mm	2.01	10.00	2.21	0.16	0.03	1.69	0.04	3.94	m	4.43
Work to isolated columns										
width exceeding 300mm	5.75	5.00	6.04	0.27	0.09	3.12	0.07	9.23	m2	10.39
width not exceeding 300mm	2.01	10.00	2.21	0.16	0.03	1.69	0.04	3.94	m	4.43
Lathing; Expamet stainless steel Rib-Lath, 0.300mm thick x 1.52 kg/m2; butt joints; fixing with stainless steel staples										
Work to walls										
width exceeding 300mm	13.43	5.00	14.10	0.18	0.09	2.29	0.05	16.44	m2	18.50
width not exceeding 300mm	4.70	10.00	5.17	0.11	0.03	1.23	0.03	6.43	m	7.23
Work to ceilings										
width exceeding 300mm	13.43	5.00	14.10	0.23	0.09	2.75	0.07	16.93	m2	19.04
width not exceeding 300mm	4.70	10.00	5.17	0.14	0.03	1.50	0.04	6.71	m	7.55
Work to isolated beams										
width exceeding 300mm	13.43	5.00	14.10	0.27	0.09	3.12	0.07	17.30	m2	19.46
width not exceeding 300mm	4.70	10.00	5.17	0.16	0.03	1.69	0.04	6.90	m	7.76
Work to isolated columns										
width exceeding 300mm	13.43	5.00	14.10	0.27	0.09	3.12	0.07	17.30	m2	19.46
width not exceeding 300mm	4.70	10.00	5.17	0.16	0.03	1.69	0.04	6.90	m	7.76
Lathing; Expamet galvanised Spraylath, 0.500mm thick x 2.25 kg/m2; butt joints; fixing with galvanised staples										
Work to walls										
width exceeding 300mm	7.07	5.00	7.42	0.18	0.09	2.29	0.05	9.77	m2	10.99
width not exceeding 300mm	2.47	10.00	2.72	0.11	0.03	1.23	0.03	3.97	m	4.47
Work to ceilings										
width exceeding 300mm	7.07	5.00	7.42	0.23	0.09	2.75	0.07	10.25	m2	11.53
width not exceeding 300mm	2.47	10.00	2.72	0.14	0.03	1.50	0.04	4.26	m	4.79
Work to isolated beams										
width exceeding 300mm	7.07	5.00	7.42	0.27	0.09	3.12	0.07	10.62	m2	11.94
width not exceeding 300mm	2.47	10.00	2.72	0.16	0.03	1.69	0.04	4.44	m	5.00
Work to isolated columns										
width exceeding 300mm	7.07	5.00	7.42	0.27	0.09	3.12	0.07	10.62	m2	11.94
width not exceeding 300mm	2.47	10.00	2.72	0.16	0.03	1.69	0.04	4.44	m	5.00
Lathing; galvanised Red-rib lath, 0.400mm thick x 1.78 kg/m2, butt joints; fixing with galvanised staples										
Work to walls										
width exceeding 300mm	6.40	5.00	6.72	0.18	0.09	2.29	0.05	9.06	m2	10.20
width not exceeding 300mm	2.24	10.00	2.46	0.11	0.03	1.23	0.03	3.72	m	4.18
Work to ceilings										
width exceeding 300mm	6.40	5.00	6.72	0.23	0.09	2.75	0.07	9.54	m2	10.74
width not exceeding 300mm	2.24	10.00	2.46	0.14	0.03	1.50	0.04	4.01	m	4.51
Work to isolated beams										
width exceeding 300mm	6.40	5.00	6.72	0.27	0.09	3.12	0.07	9.91	m2	11.15
width not exceeding 300mm	2.24	10.00	2.46	0.16	0.03	1.69	0.04	4.19	m	4.72
Work to isolated columns										
width exceeding 300mm	6.40	5.00	6.72	0.27	0.09	3.12	0.07	9.91	m2	11.15
width not exceeding 300mm	2.24	10.00	2.46	0.16	0.03	1.69	0.04	4.19	m	4.72

SURFACE FINISHES

Labour hourly rates: (except Specialists) Craft Operatives £9.23 Labourer £7.02 Rates are national average prices. Refer to REGIONAL VARIATIONS for indicative levels of overall pricing in regions	MATERIALS			LABOUR				RATES		
	Del to Site £	Waste %	Material Cost £	Craft Optve Hrs	Lab Hrs	Labour Cost £	Sunds £	Nett Rate £	Unit	Gross Rate (+12.5%) £
M30: METAL MESH LATHING/ANCHORED REINFORCEMENT FOR PLASTERED COATINGS Cont.										
Extra cost of fixing lathing to brickwork, blockwork or concrete with cartridge fired nails in lieu of to timber with staples										
Work to walls										
width exceeding 300mm	-	-	-	0.12	-	1.11	0.09	1.20	m2	1.35
width not exceeding 300mm	-	-	-	0.07	-	0.65	0.03	0.68	m	0.76
Extra cost of fixing lathing to steel with tying wire in lieu of to timber with staples										
Work to walls										
Width exceeding 300mm	-	-	-	0.18	-	1.66	0.10	1.76	m2	1.98
width not exceeding 300mm	-	-	-	0.11	-	1.02	0.03	1.05	m	1.18
Work to ceilings										
width exceeding 300mm	-	-	-	0.18	-	1.66	0.10	1.76	m2	1.98
width not exceeding 300mm	-	-	-	0.11	-	1.02	0.03	1.05	m	1.18
Work to isolated beams										
width exceeding 300mm	-	-	-	0.18	-	1.66	0.10	1.76	m2	1.98
width not exceeding 300mm	-	-	-	0.11	-	1.02	0.03	1.05	m	1.18
Work to isolated columns										
width exceeding 300mm	-	-	-	0.18	-	1.66	0.10	1.76	m2	1.98
width not exceeding 300mm	-	-	-	0.11	-	1.02	0.03	1.05	m	1.18
Expamet galvanised arch formers and fix to brickwork or blockwork with galvanised nails										
Arch corners										
380mm radius	14.31	5.00	15.03	0.40	0.20	5.10	0.14	20.26	nr	22.79
460mm radius	17.43	5.00	18.30	0.40	0.20	5.10	0.17	23.57	nr	26.51
610mm radius	21.78	5.00	22.87	0.40	0.20	5.10	0.22	28.18	nr	31.71
760mm radius	30.96	5.00	32.51	0.40	0.20	5.10	0.31	37.91	nr	42.65
Semi-circular arches										
380mm radius	28.17	5.00	29.58	0.80	0.40	10.19	0.28	40.05	nr	45.06
410mm radius	28.62	5.00	30.05	0.80	0.40	10.19	0.29	40.53	nr	45.60
420mm radius	29.56	5.00	31.04	0.80	0.40	10.19	0.30	41.53	nr	46.72
460mm radius	34.86	5.00	36.60	0.80	0.40	10.19	0.35	47.15	nr	53.04
610mm radius	43.08	5.00	45.23	0.80	0.40	10.19	0.43	55.86	nr	62.84
760mm radius	60.97	5.00	64.02	0.80	0.40	10.19	0.61	74.82	nr	84.17
Elliptical arches										
1220mm wide x 340mm rise	54.76	5.00	57.50	0.80	0.40	10.19	0.55	68.24	nr	76.77
1370mm wide x 360mm rise	57.40	5.00	60.27	0.80	0.40	10.19	0.57	71.03	nr	79.91
1520mm wide x 380mm rise	60.97	5.00	64.02	0.80	0.40	10.19	0.61	74.82	nr	84.17
1830mm wide x 410mm rise	65.65	5.00	68.93	1.20	0.60	15.29	0.66	84.88	nr	95.49
2130mm wide x 430mm rise	72.04	5.00	75.64	1.20	0.60	15.29	0.72	91.65	nr	103.11
2440mm wide x 440mm rise	75.60	5.00	79.38	1.20	0.60	15.29	0.76	95.43	nr	107.36
3050mm wide x 520mm rise	77.16	5.00	81.02	1.20	0.60	15.29	0.77	97.08	nr	109.21
Spandrel arches										
760mm wide x 180mm radius x 220mm rise	35.58	5.00	37.36	0.80	0.40	10.19	0.36	47.91	nr	53.90
910mm wide x 180mm radius x 240mm rise	37.65	5.00	39.53	0.80	0.40	10.19	0.38	50.10	nr	56.37
1220mm wide x 230mm radius x 290mm rise	48.39	5.00	50.81	0.80	0.40	10.19	0.48	61.48	nr	69.17
1520mm wide x 230mm radius x 330mm rise	53.83	5.00	56.52	0.80	0.40	10.19	0.54	67.25	nr	75.66
1830mm wide x 230mm radius x 360mm rise	59.28	5.00	62.24	0.80	0.40	10.19	0.59	73.03	nr	82.15
2130mm wide x 230mm radius x 370mm rise	64.58	5.00	67.81	0.80	0.40	10.19	0.65	78.65	nr	88.48
2440mm wide x 230mm radius x 390mm rise	68.45	5.00	71.87	1.60	0.80	20.38	0.68	92.94	nr	104.55
3050mm wide x 230mm radius x 440mm rise	72.04	5.00	75.64	1.60	0.80	20.38	0.72	96.75	nr	108.84
Bulls-eyes										
230mm radius	35.58	5.00	37.36	0.20	0.10	2.55	0.36	40.27	nr	45.30
Soffit strips										
155mm wide	3.58	5.00	3.76	0.10	0.05	1.27	0.04	5.07	m	5.71
Make-up pieces										
600mm long	8.71	5.00	9.15	0.20	0.10	2.55	0.09	11.78	nr	13.26
Expamet galvanised circular window formers and fix to brickwork or blockwork with galvanised nails										
Circular windows										
600mm diameter	32.39	5.00	34.01	1.60	0.80	20.38	0.32	54.71	nr	61.55
M30: METAL MESH LATHING/ANCHORED REINFORCEMENT FOR PLASTERED COATINGS; STAIRCASE AREAS										
Lathing; BB galvanised expanded metal lath, 9mm mesh x 0.500mm thick x 1.11 kg/m2; butt joints; fixing with galvanised staples										
Work to walls										
width exceeding 300mm	4.21	5.00	4.42	0.23	0.09	2.75	0.07	7.25	Um2	8.15
width not exceeding 300mm	1.47	10.00	1.62	0.14	0.03	1.50	0.04	3.16	m	3.55
Work to ceilings										
width exceeding 300mm	4.21	5.00	4.42	0.29	0.09	3.31	0.08	7.81	m2	8.79
width not exceeding 300mm	1.47	10.00	1.62	0.17	0.03	1.78	0.04	3.44	m	3.87

	MATERIALS			LABOUR				RATES		
Labour hourly rates: (except Specialists) Craft Operatives £9.23 Labourer £7.02 Rates are national average prices. Refer to REGIONAL VARIATIONS for indicative levels of overall pricing in regions	Del to Site £	Waste %	Material Cost £	Craft Optve Hrs	Lab Hrs	Labour Cost £	Sunds £	Nett Rate £	Unit	Gross Rate (+12.5%) £
M30: METAL MESH LATHING/ANCHORED REINFORCEMENT FOR PLASTERED COATINGS; STAIRCASE AREAS Cont.										
Lathing; BB galvanised expanded metal lath 9mm mesh x 0.725mm thick x 1.61 kg/m2; butt joints; fixing with galvanised staples										
Work to walls										
width exceeding 300mm	4.93	5.00	5.18	0.23	0.09	2.75	0.07	8.00	m2	9.00
width not exceeding 300mm	1.73	10.00	1.90	0.14	0.03	1.50	0.04	3.45	m	3.88
Work to ceilings										
width exceeding 300mm	4.93	5.00	5.18	0.29	0.09	3.31	0.08	8.56	m2	9.64
width not exceeding 300mm	1.73	10.00	1.90	0.17	0.03	1.78	0.04	3.72	m	4.19
Lathing; galvanised expanded metal lath 9mm mesh x 0.950mm thick x 2.50 kg/m2; butt joints; fixing with galvanised staples										
Work to walls										
width exceeding 300mm	7.56	5.00	7.94	0.23	0.09	2.75	0.07	10.76	m2	12.11
width not exceeding 300mm	2.65	10.00	2.92	0.14	0.03	1.50	0.04	4.46	m	5.02
Work to ceilings										
width exceeding 300mm	7.56	5.00	7.94	0.29	0.09	3.31	0.08	11.33	m2	12.74
width not exceeding 300mm	2.65	10.00	2.92	0.17	0.03	1.78	0.04	4.73	m	5.33
Lathing; Expamet galvanised Rib-Lath, 0.400mm thick x 1.78 kg/m2; butt joints; fixing with galvanised staples										
Work to walls										
width exceeding 300mm	5.00	5.00	5.25	0.23	0.09	2.75	0.07	8.07	m2	9.08
width not exceeding 300mm	1.75	10.00	1.93	0.14	0.03	1.50	0.04	3.47	m	3.90
Work to ceilings										
width exceeding 300mm	5.00	5.00	5.25	0.29	0.09	3.31	0.08	8.64	m2	9.72
width not exceeding 300mm	1.75	10.00	1.93	0.17	0.03	1.78	0.04	3.74	m	4.21
Lathing; Expamet galvanised Rib-Lath, 0.500mm thick x 2.25 kg/m2; butt joints; fixing with galvanised staples										
Work to walls										
width exceeding 300mm	5.75	5.00	6.04	0.23	0.09	2.75	0.07	8.86	m2	9.97
width not exceeding 300mm	2.01	10.00	2.21	0.14	0.03	1.50	0.04	3.75	m	4.22
Work to ceilings										
width exceeding 300mm	5.75	5.00	6.04	0.29	0.09	3.31	0.08	9.43	m2	10.60
width not exceeding 300mm	2.01	10.00	2.21	0.17	0.03	1.78	0.04	4.03	m	4.53
Lathing; Expamet stainless steel Rib-Lath, 0.300mm thick x 1.52 kg/m2; butt joints; fixing with stainless steel staples										
Work to walls										
width exceeding 300mm	13.43	5.00	14.10	0.23	0.09	2.75	0.07	16.93	m2	19.04
width not exceeding 300mm	4.70	10.00	5.17	0.14	0.03	1.50	0.04	6.71	m	7.55
Work to ceilings										
width exceeding 300mm	13.43	5.00	14.10	0.29	0.09	3.31	0.08	17.49	m2	19.68
width not exceeding 300mm	4.70	10.00	5.17	0.17	0.03	1.78	0.04	6.99	m	7.86
Lathing; Expamet galvanised Spraylath, 0.500mm thick x 2.25 kg/m2; butt joints; fixing with galvanised staples										
Work to walls										
width exceeding 300mm	7.07	5.00	7.42	0.23	0.09	2.75	0.07	10.25	m2	11.53
width not exceeding 300mm	2.47	10.00	2.72	0.14	0.03	1.50	0.04	4.26	m	4.79
Work to ceilings										
width exceeding 300mm	7.07	5.00	7.42	0.29	0.09	3.31	0.08	10.81	m2	12.16
width not exceeding 300mm	2.47	10.00	2.72	0.17	0.03	1.78	0.04	4.54	m	5.10
Lathing; galvanised Red-rib lath, 0.400mm thick x 1.78 kg/m2, butt joints; fixing with galvanised staples										
Work to walls										
width exceeding 300mm	6.40	5.00	6.72	0.23	0.09	2.75	0.07	9.54	m2	10.74
width not exceeding 300mm	2.24	10.00	2.46	0.14	0.03	1.50	0.04	4.01	m	4.51
Work to ceilings										
width exceeding 300mm	6.40	5.00	6.72	0.29	0.09	3.31	0.08	10.11	m2	11.37
width not exceeding 300mm	2.24	10.00	2.46	0.17	0.03	1.78	0.04	4.28	m	4.82
M31: FIBROUS PLASTER										
PLASTERBOARD COVE - Coves										
Gyproc cove; fixing with adhesive										
100mm girth	0.90	10.00	0.99	0.18	0.09	2.29	0.11	3.39	m	3.82
extra; ends	-	-	-	0.06	-	0.55	-	0.55	nr	0.62
extra; internal angles	-	-	-	0.30	-	2.77	-	2.77	nr	3.12
extra; external angles	-	-	-	0.30	-	2.77	-	2.77	nr	3.12

Labour hourly rates: (except Specialists) Craft Operatives £9.23 Labourer £7.02 Rates are national average prices. Refer to REGIONAL VARIATIONS for indicative levels of overall pricing in regions	MATERIALS			LABOUR				RATES		
	Del to Site	Waste	Material Cost	Craft Optve	Lab	Labour Cost	Sunds	Nett Rate	Unit	Gross Rate (+12.5%)
	£	%	£	Hrs	Hrs	£	£	£		£

M31: FIBROUS PLASTER Cont.

PLASTERBOARD COVE - Coves Cont.

Gyproc cove; fixing with adhesive Cont.

127mm girth	0.90	10.00	0.99	0.18	0.09	2.29	0.18	3.46	m	3.90
extra; ends	-	-	-	0.06	-	0.55	-	0.55	nr	0.62
extra; internal angles	-	-	-	0.30	-	2.77	-	2.77	nr	3.12
extra; external angles	-	-	-	0.30	-	2.77	-	2.77	nr	3.12

Gyproc cove; fixing with nails to timber

100mm girth	0.90	10.00	0.99	0.12	0.06	1.53	0.04	2.56	m	2.88
extra; ends	-	-	-	0.06	-	0.55	-	0.55	nr	0.62
extra; internal angles	-	-	-	0.30	-	2.77	-	2.77	nr	3.12
extra; external angles	-	-	-	0.30	-	2.77	-	2.77	nr	3.12
127mm girth	0.90	10.00	0.99	0.12	0.06	1.53	0.04	2.56	m	2.88
extra; ends	-	-	-	0.06	-	0.55	-	0.55	nr	0.62
extra; internal angles	-	-	-	0.30	-	2.77	-	2.77	nr	3.12
extra; external angles	-	-	-	0.30	-	2.77	-	2.77	nr	3.12

PLASTER VENTILATORS - Fibrous plaster ventilator; fixing in plastered wall

Plain

229 x 79mm	0.70	5.00	0.73	0.15	-	1.38	-	2.12	nr	2.38
229 x 152mm	0.88	5.00	0.92	0.18	-	1.66	-	2.59	nr	2.91
229 x 229mm	1.14	5.00	1.20	0.20	-	1.85	-	3.04	nr	3.42

Flyproof

229 x 79mm	1.29	5.00	1.35	0.20	-	1.85	-	3.20	nr	3.60
229 x 152mm	1.54	5.00	1.62	0.23	-	2.12	-	3.74	nr	4.21
229 x 229mm	1.90	5.00	2.00	0.25	-	2.31	-	4.30	nr	4.84

M40: STONE/CONCRETE/QUARRY/CERAMIC TILING/MOSAIC

CLAY TILE PAVING - Clay floor quarries, B.S.6431, terracotta; 3mm joints, symmetrical layout; bedding in 10mm cement mortar (1:3); pointing in cement mortar (1:3); on cement and sand base

150 x 150 x 12.5mm units to floors level or to falls only not exceeding 15 degrees from horizontal

plain	12.73	5.00	13.37	0.95	0.60	12.98	1.46	27.81	m2	31.28

150 x 150 x 20mm units to floors level or to falls only not exceeding 15 degrees from horizontal

plain	17.97	5.00	18.87	1.20	0.73	16.20	1.46	36.53	m2	41.10

225 x 225 x 25mm units to floors level or to falls only not exceeding 15 degrees from horizontal

plain	31.00	5.00	32.55	0.85	0.55	11.71	1.46	45.72	m2	51.43

150 x 150 x 12.5mm units to floors to falls and crossfalls and to slopes not exceeding 15 degrees from horizontal

plain	12.73	5.00	13.37	1.05	0.60	13.90	1.46	28.73	m2	32.32

150 x 150 x 20mm units to floors to falls and crossfalls and to slopes not exceeding 15 degrees from horizontal

plain	17.97	5.00	18.87	1.30	0.73	17.12	1.46	37.45	m2	42.13

225 x 225 x 25mm units to floors to falls and crossfalls and to slopes not exceeding 15 degrees from horizontal

plain	31.00	5.00	32.55	0.65	0.43	9.02	1.46	43.03	m2	48.41

Extra for pointing with tinted mortar

150 x 150 x 12.5mm tiles	1.32	10.00	1.45	0.11	-	1.02	-	2.47	m2	2.78
150 x 150 x 20mm tiles	1.32	10.00	1.45	0.11	-	1.02	-	2.47	m2	2.78
225 x 225 x 25mm tiles	1.32	10.00	1.45	0.08	-	0.74	-	2.19	m2	2.46

150 x 150 x 12.5mm units to skirtings on brickwork or blockwork base

height 150mm; square top edge	2.07	5.00	2.17	0.36	0.09	3.95	0.32	6.45	m	7.25
height 150mm; rounded top edge	2.68	5.00	2.81	0.36	0.09	3.95	0.32	7.09	m	7.97
height 150mm; rounded top edge and cove at bottom	4.23	5.00	4.44	0.36	0.09	3.95	0.32	8.72	m	9.81

150 x 150 x 20mm units to skirtings on brickwork or blockwork base

height 150mm; square top edge	2.93	5.00	3.08	0.46	0.11	5.02	0.32	8.41	m	9.47
height 150mm; rounded top edge	3.40	5.00	3.57	0.46	0.11	5.02	0.32	8.91	m	10.02

CLAY TILE PAVING - Clay floor quarries, B.S.6431, blended; 3mm joints symmetrical layout; bedding in 10mm cement mortar (1:3); pointing in cement mortar (1:3);on cement and sand base

150 x 150 x 12.5mm units to floors level or to falls only not exceeding 15 degrees from horizontal

plain	17.11	5.00	17.97	0.95	0.60	12.98	1.46	32.41	m2	36.46

150 x 150 x 20mm units to floors level or to falls only not exceeding 15 degrees from horizontal

plain	19.31	5.00	20.28	1.20	0.73	16.20	1.46	37.94	m2	42.68

Labour hourly rates: (except Specialists) Craft Operatives £9.23 Labourer £7.02 Rates are national average prices. Refer to REGIONAL VARIATIONS for indicative levels of overall pricing in regions	MATERIALS			LABOUR				RATES		
	Del to Site £	Waste %	Material Cost £	Craft Optve Hrs	Lab Hrs	Labour Cost £	Sunds £	Nett Rate £	Unit	Gross Rate (+12.5%) £
M40: STONE/CONCRETE/QUARRY/CERAMIC TILING/MOSAIC Cont.										
CLAY TILE PAVING - Clay floor quarries, B.S.6431, blended; 3mm joints symmetrical layout; bedding in 10mm cement mortar (1:3); pointing in cement mortar (1:3);on cement and sand base Cont.										
225 x 225 x 25mm units to floors level or to falls only not exceeding 15 degrees from horizontal plain..................................	34.86	5.00	36.60	0.85	0.55	11.71	1.46	49.77	m2	55.99
150 x 150 x 12.5mm units to floors to falls and crossfalls and to slopes not exceeding 15 degrees from horizontal plain..................................	17.11	5.00	17.97	1.05	0.60	13.90	1.48	33.35	m2	37.52
150 x 150 x 20mm units to floors to falls and crossfalls and to slopes not exceeding 15 degrees from horizontal plain..................................	19.31	5.00	20.28	1.30	0.73	17.12	1.46	38.86	m2	43.72
225 x 225 x 25mm units to floors to falls and crossfalls and to slopes not exceeding 15 degrees from horizontal plain..................................	34.86	5.00	36.60	0.65	0.43	9.02	1.46	47.08	m2	52.97
Extra for pointing with tinted mortar										
150 x 150 x 12.5mm tiles	1.32	10.00	1.45	0.11	-	1.02	-	2.47	m2	2.78
150 x 150 x 20mm tiles	1.32	10.00	1.45	0.11	-	1.02	-	2.47	m2	2.78
225 x 225 x 25mm tiles	1.32	10.00	1.45	0.08	-	0.74	-	2.19	m2	2.46
150 x 150 x 12.5mm units to skirtings on brickwork or blockwork base										
height 150mm; square top edge	2.79	5.00	2.93	0.36	0.09	3.95	0.32	7.20	m	8.10
height 150mm; rounded top edge	3.19	5.00	3.35	0.36	0.09	3.95	0.32	7.62	m	8.58
height 150mm; rounded top edge and cove at bottom	4.47	5.00	4.69	0.36	0.09	3.95	0.32	8.97	m	10.09
150 x 150 x 20mm units to skirtings on brickwork or blockwork base										
height 150mm; square top edge	3.14	5.00	3.30	0.46	0.11	5.02	0.32	8.63	m	9.71
height 150mm; rounded top edge	3.40	5.00	3.57	0.46	0.11	5.02	0.32	8.91	m	10.02
CLAY TILE PAVING - Clay floor quarries, B.S.6431, dark; 3mm joints, symmetrical layout; bedding in 10mm cement mortar (1:3); pointing in cement mortar (1:3); on cement and sand base										
150 x 150 x 12.5mm units to floors level or to falls only not exceeding 15degrees from horizontal plain..................................	22.15	5.00	23.26	0.95	0.60	12.98	1.46	37.70	m2	42.41
150 x 150 x 12.5mm units to floors to falls and crossfalls and to slopes not exceeding 15 degrees from horizontal plain..................................	22.15	5.00	23.26	1.05	0.60	13.90	1.46	38.62	m2	43.45
150 x 150 x 12.5mm units to skirtings on brickwork or blockwork base										
height 150mm; square top edge	3.61	5.00	3.79	0.36	0.09	3.95	0.32	8.07	m	9.07
height 150mm; rounded top edge	4.20	5.00	4.41	0.36	0.09	3.95	0.32	8.68	m	9.77
height 150mm; rounded top edge and cove at bottom	5.82	5.00	6.11	0.36	0.09	3.95	0.32	10.39	m	11.68
CLAY TILE PAVING - Ceramic floor tiles, B.S.6431, fully vitrified, red; 3mm joints, symmetrical layout; bedding in 10mm cement mortar (1:3); pointing in cement mortar (1:3);on cement and sand base										
100 x 100 x 9mm units to floors level or to falls only not exceeding 15 degrees from horizontal plain..................................	16.65	5.00	17.48	1.95	1.10	25.72	1.46	44.66	m2	50.25
152 x 152 x 12mm units to floors level or to falls only not exceeding 15 degrees from horizontal plain..................................	18.25	5.00	19.16	1.10	0.68	14.93	1.46	35.55	m2	39.99
200 x 200 x 12mm units to floors level or to falls only not exceeding 15 degrees from horizontal plain..................................	19.70	5.00	20.68	0.75	0.50	10.43	1.46	32.58	m2	36.65
100 x 100 x 9mm units to floors to falls and crossfalls and to slopes not exceeding 15 degrees from horizontal plain..................................	16.65	5.00	17.48	2.15	1.10	27.57	1.46	46.51	m2	52.32
152 x 152 x 12mm units to floors to falls and crossfalls and to slopes not exceeding 15 degrees from horizontal plain..................................	18.25	5.00	19.16	1.21	0.68	15.94	1.46	36.56	m2	41.13
200 x 200 x 12mm units to floors to falls and crossfalls and to slopes not exceeding 15 degrees from horizontal plain..................................	19.70	5.00	20.68	0.83	0.50	11.17	1.46	33.32	m2	37.48

SURFACE FINISHES

Labour hourly rates: (except Specialists) Craft Operatives £9.23 Labourer £7.02 Rates are national average prices. Refer to REGIONAL VARIATIONS for indicative levels of overall pricing in regions	MATERIALS			LABOUR				RATES		
	Del to Site £	Waste %	Material Cost £	Craft Optve Hrs	Lab Hrs	Labour Cost £	Sunds £	Nett Rate £	Unit	Gross Rate (+12.5%) £
M40: STONE/CONCRETE/QUARRY/CERAMIC TILING/MOSAIC Cont.										
CLAY TILE PAVING - Ceramic floor tiles, B.S.6431, fully vitrified, red; 3mm joints, symmetrical layout; bedding in 10mm cement mortar (1:3); pointing in cement mortar (1:3);on cement and sand base Cont.										
Extra for pointing with tinted mortar										
100 x 100 x 9mm tiles	1.32	10.00	1.45	0.18	-	1.66	-	3.11	m2	3.50
152 x 152 x 12mm tiles	1.32	10.00	1.45	0.12	-	1.11	-	2.56	m2	2.88
200 x 200 x 12mm tiles	1.32	10.00	1.45	0.10	-	0.92	-	2.38	m2	2.67
100 x 100 x 9mm units to skirtings on brickwork or blockwork base										
height 100mm; square top edge	1.67	5.00	1.75	0.40	0.11	4.46	0.22	6.44	m	7.24
height 100mm; rounded top edge	2.23	5.00	2.34	0.40	0.11	4.46	0.22	7.03	m	7.90
152 x 152 x 12mm units to skirtings on brickwork or blockwork base										
height 152mm; square top edge	2.77	5.00	2.91	0.36	0.10	4.02	0.32	7.25	m	8.16
height 152mm; rounded top edge	3.63	5.00	3.81	0.36	0.10	4.02	0.32	8.16	m	9.18
height 152mm; rounded top edge and cove at bottom	7.04	5.00	7.39	0.36	0.10	4.02	0.32	11.74	m	13.20
200 x 200 x 12mm units to skirtings on brickwork or blockwork base										
height 200mm; square top edge	3.94	5.00	4.14	0.33	0.10	3.75	0.43	8.31	m	9.35
height 200mm; rounded top edge	6.17	5.00	6.48	0.33	0.10	3.75	0.43	10.66	m	11.99
CLAY TILE PAVING - Ceramic floor tiles, B.S.6431, fully vitrified, cream; 3mm joints, symmetrical layout; bedding in 10mm cement mortar (1:3); pointing in cement mortar (1:3); on cement and sand base										
100 x 100 x 9mm units to floors level or to falls only not exceeding 15 degrees from horizontal										
plain......................................	17.70	5.00	18.59	1.95	1.10	25.72	1.46	45.77	m2	51.49
152 x 152 x 12mm units to floors level or to falls only not exceeding 15 degrees from horizontal										
plain......................................	19.15	5.00	20.11	1.10	0.68	14.93	1.46	36.49	m2	41.06
200 x 200 x 12mm units to floors level or to falls only not exceeding 15 degrees from horizontal										
plain......................................	20.45	5.00	21.47	0.75	0.50	10.43	1.46	33.37	m2	37.54
100 x 100 x 9mm units to floors to falls and crossfalls and to slopes not exceeding 15 degrees from horizontal										
plain......................................	17.70	5.00	18.59	2.15	1.10	27.57	1.46	47.61	m2	53.56
152 x 152 x 12mm units to floors to falls and crossfalls and to slopes not exceeding 15 degrees from horizontal										
plain......................................	19.15	5.00	20.11	1.21	0.68	15.94	1.46	37.51	m2	42.20
200 x 200 x 12mm units to floors to falls and crossfalls and to slopes not exceeding 15 degrees from horizontal										
plain......................................	20.45	5.00	21.47	0.83	0.50	11.17	1.46	34.10	m2	38.37
Extra for pointing with tinted mortar										
100 x 100 x 9mm tiles	1.32	10.00	1.45	0.18	-	1.66	-	3.11	m2	3.50
152 x 152 x 12mm tiles	1.32	10.00	1.45	0.12	-	1.11	-	2.56	m2	2.88
200 x 200 x 12mm tiles	1.32	10.00	1.45	0.10	-	0.92	-	2.38	m2	2.67
100 x 100 x 9mm units to skirtings on brickwork or blockwork base										
height 100mm; square top edge	1.77	5.00	1.86	0.40	0.11	4.46	0.22	6.54	m	7.36
height 100mm; rounded top edge	2.33	5.00	2.45	0.40	0.11	4.46	0.22	7.13	m	8.02
152 x 152 x 12mm units to skirtings on brickwork or blockwork base										
height 152mm; square top edge	2.91	5.00	3.06	0.36	0.10	4.02	0.32	7.40	m	8.33
height 152mm; rounded top edge	3.70	5.00	3.88	0.36	0.10	4.02	0.32	8.23	m	9.26
height 152mm; rounded top edge and cove at bottom	7.05	5.00	7.40	0.36	0.10	4.02	0.32	11.75	m	13.22
200 x 200 x 12mm units to skirtings on brickwork or blockwork base										
height 200mm; square top edge	4.09	5.00	4.29	0.33	0.10	3.75	0.43	8.47	m	9.53
height 200mm; rounded top edge	6.17	5.00	6.48	0.33	0.10	3.75	0.43	10.66	m	11.99
CLAY TILE PAVING - Ceramic floor tiles, B.S.6431, fully vitrified, black; 3mm joints, symmetrical layout; bedding in 10mm cement mortar (1:3); pointing in cement mortar (1:3); on cement and sand base										
100 x 100 x 9mm units to floors level or to falls only not exceeding 15 degrees from horizontal										
plain......................................	21.75	5.00	22.84	1.95	1.10	25.72	1.46	50.02	m2	56.27
152 x 152 x 12mm units to floors level or to falls only not exceeding 15 degrees from horizontal										
plain......................................	22.80	5.00	23.94	1.10	0.68	14.93	1.46	40.33	m2	45.37

Labour hourly rates: (except Specialists) Craft Operatives £9.23 Labourer £7.02. Rates are national average prices. Refer to REGIONAL VARIATIONS for indicative levels of overall pricing in regions	MATERIALS			LABOUR				RATES		
	Del to Site £	Waste %	Material Cost £	Craft Optve Hrs	Lab Hrs	Labour Cost £	Sunds £	Nett Rate £	Unit	Gross Rate (+12.5%) £
M40: STONE/CONCRETE/QUARRY/CERAMIC TILING/MOSAIC Cont.										
CLAY TILE PAVING - Ceramic floor tiles, B.S.6431, fully vitrified, black; 3mm joints, symmetrical layout; bedding in 10mm cement mortar (1:3); pointing in cement mortar (1:3); on cement and sand base Cont.										
200 x 200 x 12mm units to floors level or to falls only not exceeding 15 degrees from horizontal										
plain	23.70	5.00	24.89	0.75	0.50	10.43	1.46	36.78	m2	41.37
100 x 100 x 9mm units to floors to falls and crossfalls and to slopes not exceeding 15 degrees from horizontal										
plain	21.75	5.00	22.84	2.15	1.10	27.57	1.46	51.86	m2	58.35
152 x 152 x 12mm units to floors to falls and crossfalls and to slopes not exceeding 15 degrees from horizontal										
plain	22.80	5.00	23.94	1.21	0.68	15.94	1.46	41.34	m2	46.51
200 x 200 x 12mm units to floors to falls and crossfalls and to slopes not exceeding 15 degrees from horizontal										
plain	23.70	5.00	24.89	0.83	0.50	11.17	1.46	37.52	m2	42.21
Extra for pointing with tinted mortar										
100 x 100 x 9mm tiles	1.32	10.00	1.45	0.18	-	1.66	-	3.11	m2	3.50
152 x 152 x 12mm tiles	1.32	10.00	1.45	0.12	-	1.11	-	2.56	m2	2.88
200 x 200 x 12mm tiles	1.32	10.00	1.45	0.10	-	0.92	-	2.38	m2	2.67
100 x 100 x 9mm units to skirtings on brickwork or blockwork base										
height 100mm; square top edge	2.18	5.00	2.29	0.40	0.11	4.46	0.22	6.97	m	7.84
height 100mm; rounded top edge	2.73	5.00	2.87	0.40	0.11	4.46	0.22	7.55	m	8.49
152 x 152 x 12mm units to skirtings on brickwork or blockwork base										
height 152mm; square top edge	3.47	5.00	3.64	0.36	0.10	4.02	0.32	7.99	m	8.99
height 152mm; rounded top edge	3.96	5.00	4.16	0.36	0.10	4.02	0.32	8.50	m	9.57
height 152mm; rounded top edge and cove at bottom	7.32	5.00	7.69	0.36	0.10	4.02	0.32	12.03	m	13.53
200 x 200 x 12mm units to skirtings on brickwork or blockwork base										
height 200mm; square top edge	4.74	5.00	4.98	0.33	0.10	3.75	0.43	9.15	m	10.30
height 200mm; rounded top edge	6.17	5.00	6.48	0.33	0.10	3.75	0.43	10.66	m	11.99
TILE WALL LININGS - Ceramic tiles, B.S.6431, glazed white; 2mm joints, symmetrical layout; bedding in 10mm cement mortar (1:3); pointing with neat white cement; on brickwork or blockwork base										
108 x 108 x 6.5mm units to walls										
plain, width exceeding 300mm	20.40	5.00	21.42	1.60	0.90	21.09	1.46	43.97	m2	49.46
plain, width not exceeding 300mm	6.73	10.00	7.40	0.95	0.57	12.77	0.44	20.61	m	23.19
152 x 152 x 5.5mm units to walls										
plain, width exceeding 300mm	14.85	5.00	15.59	1.25	0.73	16.66	1.30	33.55	m2	37.75
plain, width not exceeding 300mm	4.57	10.00	5.03	0.75	0.45	10.08	0.39	15.50	m	17.44
TILE WALL LININGS - Ceramic tiles, B.S.6431, glazed white; 2mm joints; symmetrical layout; fixing with thin bed adhesive; pointing with neat white cement; on plaster base										
108 x 108 x 6.5mm units to walls										
plain, width exceeding 300mm	20.40	5.00	21.42	1.60	0.90	21.09	1.50	44.01	m2	49.51
plain, width not exceeding 300mm	6.73	10.00	7.40	0.95	0.57	12.77	0.45	20.62	m	23.20
152 x 152 x 5.5mm units to walls										
plain, width exceeding 300mm	14.85	5.00	15.59	1.25	0.73	16.66	1.42	33.67	m2	37.88
plain, width not exceeding 300mm	4.57	10.00	5.03	0.75	0.45	10.08	0.43	15.54	m	17.48
TILE WALL LININGS - Ceramic tiles, B.S.6431, glazed white; 2mm joints, symmetrical layout; fixing with thick bed adhesive; pointing with neat white cement; on cement and sand base										
108 x 108 x 6.5mm units to walls										
plain, width exceeding 300mm	20.40	5.00	21.42	1.60	0.90	21.09	2.77	45.28	m2	50.94
plain, width not exceeding 300mm	6.73	10.00	7.40	0.95	0.57	12.77	0.83	21.00	m	23.63
152 x 152 x 5.5mm units to walls										
plain, width exceeding 300mm	14.85	5.00	15.59	1.25	0.73	16.66	2.68	34.93	m2	39.30
plain, width not exceeding 300mm	4.57	10.00	5.03	0.75	0.45	10.08	0.81	15.92	m	17.91
TILE WALL LININGS - Ceramic tiles, B.S.6431, glazed light colour; 2mm joints symmetrical layout; bedding in 10mm cement mortar (1:3); pointing with neat white cement; on cement and sand base										
108 x 108 x 6.5mm units to walls										
plain, width exceeding 300mm	20.40	5.00	21.42	1.60	0.90	21.09	1.46	43.97	m2	49.46
plain, width not exceeding 300mm	6.73	10.00	7.40	0.95	0.57	12.77	0.44	20.61	m	23.19

SURFACE FINISHES

Labour hourly rates: (except Specialists) Craft Operatives £9.23 Labourer £7.02. Rates are national average prices. Refer to REGIONAL VARIATIONS for indicative levels of overall pricing in regions	MATERIALS			LABOUR				RATES		
	Del to Site £	Waste %	Material Cost £	Craft Optve Hrs	Lab Hrs	Labour Cost £	Sunds £	Nett Rate £	Unit	Gross Rate (+12.5%) £
M40: STONE/CONCRETE/QUARRY/CERAMIC TILING/MOSAIC Cont.										
TILE WALL LININGS - Ceramic tiles, B.S.6431, glazed light colour; 2mm joints symmetrical layout; bedding in 10mm cement mortar (1:3); pointing with neat white cement; on cement and sand base Cont.										
152 x 152 x 5.5mm units to walls										
plain, width exceeding 300mm	14.95	5.00	15.70	1.25	0.73	16.66	1.30	33.66	m2	37.87
plain, width not exceeding 300mm	4.60	10.00	5.06	0.75	0.45	10.08	0.39	15.53	m	17.47
TILE WALL LININGS - Ceramic tiles, B.S.6431, glazed light colour; 2mm joints, symmetrical layout; fixing with thin bed adhesive; pointing with neat white cement; on plaster base										
108 x 108 x 6.5mm units to walls										
plain, width exceeding 300mm	20.40	5.00	21.42	1.60	0.90	21.09	1.50	44.01	m2	49.51
plain, width not exceeding 300mm	6.73	10.00	7.40	0.95	0.57	12.77	0.45	20.62	m	23.20
152 x 152 x 5.5mm units to walls										
plain, width exceeding 300mm	14.95	5.00	15.70	1.25	0.73	16.66	1.42	33.78	m2	38.00
plain, width not exceeding 300mm	4.60	10.00	5.06	0.75	0.45	10.08	0.43	15.57	m	17.52
TILE WALL LININGS - Ceramic tiles, B.S.6431, glazed light colour; 2mm joints, symmetrical layout; fixing with thick bed adhesive; pointing with neat white cement; on cement and sand base										
108 x 108 x 6.5mm units to walls										
plain, width exceeding 300mm	20.40	5.00	21.42	1.60	0.90	21.09	3.01	45.52	m2	51.21
plain, width not exceeding 300mm	6.73	10.00	7.40	0.95	0.57	12.77	0.90	21.07	m	23.71
152 x 152 x 5.5mm units to walls										
plain, width exceeding 300mm	14.95	5.00	15.70	1.25	0.73	16.66	2.85	35.21	m2	39.61
plain, width not exceeding 300mm	4.60	10.00	5.06	0.75	0.45	10.08	0.85	15.99	m	17.99
TILE WALL LININGS - Ceramic tiles, B.S.6431, glazed dark colour; 2mm joints, symmetrical layout; bedding in 10mm cement mortar (1:3); pointing with neat white cement; on brickwork or blockwork base										
108 x 108 x 6.5mm units to walls										
plain, width exceeding 300mm	25.15	5.00	26.41	1.60	0.90	21.09	1.46	48.95	m2	55.07
plain, width not exceeding 300mm	8.30	10.00	9.13	0.95	0.57	12.77	0.44	22.34	m	25.13
152 x 152 x 5.5mm units to walls										
plain, width exceeding 300mm	18.45	5.00	19.37	1.25	0.73	16.66	1.30	37.33	m2	42.00
plain, width not exceeding 300mm	5.68	10.00	6.25	0.75	0.45	10.08	0.39	16.72	m	18.81
TILE WALL LININGS - Ceramic tiles, B.S.6431, glazed dark colour; 2mm joints, symmetrical layout; fixing with thin bed adhesive; pointing with neat white cement; on plaster base										
108 x 108 x 6.5mm units to walls										
plain, width exceeding 300mm	25.15	5.00	26.41	1.60	0.90	21.09	1.50	48.99	m2	55.12
plain, width not exceeding 300mm	8.30	10.00	9.13	0.95	0.57	12.77	0.45	22.35	m	25.14
152 x 152 x 5.5mm units to walls										
plain, width exceeding 300mm	18.45	5.00	19.37	1.25	0.73	16.66	1.42	37.45	m2	42.14
plain, width not exceeding 300mm	5.68	10.00	6.25	0.75	0.45	10.08	0.43	16.76	m	18.85
TILE WALL LININGS - Ceramic tiles, B.S.6431, glazed dark colour; 2mm joints, symmetrical layout; fixing with thick bed adhesive; pointing with neat white cement; on cement and sand base										
108 x 108 x 6.5mm units to walls										
plain, width exceeding 300mm	25.15	5.00	26.41	1.60	0.90	21.09	3.01	50.50	m2	56.82
plain, width not exceeding 300mm	8.30	10.00	9.13	0.95	0.57	12.77	0.90	22.80	m	25.65
152 x 152 x 5.5mm units to walls										
plain, width exceeding 300mm	18.45	5.00	19.37	1.25	0.73	16.66	2.75	38.78	m2	43.63
plain, width not exceeding 300mm	5.68	10.00	6.25	0.75	0.45	10.08	0.82	17.15	m	19.29
TILE WALL LININGS - Extra over ceramic tiles, B.S.6431, glazed white; 2mm joints, symmetrical layout; bedding or fixing in any material to general surfaces on any base										
Special tiles										
rounded edge	0.16	5.00	0.17	-	-	-	-	0.17	m	0.19
external angle bead	2.44	5.00	2.56	0.14	-	1.29	-	3.85	m	4.34
internal angle	-	-	-	0.10	-	0.92	-	0.92	nr	1.04
external angle	-	-	-	0.10	-	0.92	-	0.92	nr	1.04
internal angle bead	2.44	5.00	2.56	0.14	-	1.29	-	3.85	m	4.34
internal angle	-	-	-	0.10	-	0.92	-	0.92	nr	1.04
external angle	-	-	-	0.10	-	0.92	-	0.92	nr	1.04

Labour hourly rates: (except Specialists) Craft Operatives £9.23 Labourer £7.02 Rates are national average prices. Refer to REGIONAL VARIATIONS for indicative levels of overall pricing in regions	MATERIALS			LABOUR				RATES		
	Del to Site £	Waste %	Material Cost £	Craft Optve Hrs	Lab Hrs	Labour Cost £	Sunds £	Nett Rate £	Unit	Gross Rate (+12.5%) £
M40: STONE/CONCRETE/QUARRY/CERAMIC TILING/MOSAIC Cont.										
TILE WALL LININGS - Extra over ceramic tiles, B.S.6431, glazed light colour; 2mm joints, symmetrical layout; bedding or fixing in any material to general surfaces on any base										
Special tiles										
rounded edge	0.06	5.00	0.06	-	-	-	-	0.06	m	0.07
external angle bead	2.44	5.00	2.56	0.14	-	1.29	-	3.85	m	4.34
internal angle	-	-	-	0.10	-	0.92	-	0.92	nr	1.04
external angle	-	-	-	0.10	-	0.92	-	0.92	nr	1.04
internal angle bead	2.44	5.00	2.56	0.14	-	1.29	-	3.85	m	4.34
internal angle	-	-	-	0.10	-	0.92	-	0.92	nr	1.04
external angle	-	-	-	0.10	-	0.92	-	0.92	nr	1.04
TILE WALL LININGS - Extra over ceramic tiles, B.S.6431, glazed dark colour; 2mm joints, symmetrical layout; bedding or fixing in any material to general surfaces on any base										
Special tiles										
rounded edge	0.16	5.00	0.17	-	-	-	-	0.17	m	0.19
external angle bead	2.44	5.00	2.56	0.14	-	1.29	-	3.85	m	4.34
internal angle	-	-	-	0.10	-	0.92	-	0.92	nr	1.04
external angle	-	-	-	0.10	-	0.92	-	0.92	nr	1.04
internal angle bead	2.44	5.00	2.56	0.14	-	1.29	-	3.85	m	4.34
internal angle	-	-	-	0.10	-	0.92	-	0.92	nr	1.04
external angle	-	-	-	0.10	-	0.92	-	0.92	nr	1.04
TILE CILLS - Ceramic tiles, B.S.6431, glazed white; 2mm joints, symmetrical layout; bedding in 10mm cement mortar (1:3); pointing with neat white cement; on brickwork or blockwork base - cills										
108 x 108 x 4mm units to cills on brickwork or blockwork base										
width 150mm; rounded angle	3.24	5.00	3.40	0.44	0.14	5.04	0.33	8.78	m	9.87
width 225mm; rounded angle	5.48	5.00	5.75	0.60	0.20	6.94	0.49	13.19	m	14.83
width 300mm; rounded angle	6.01	5.00	6.31	0.75	0.27	8.82	0.66	15.79	m	17.76
152 x 152 x 5.5mm units to cills on brickwork or blockwork base										
width 150mm; rounded angle	2.92	5.00	3.07	0.36	0.11	4.09	0.29	7.45	m	8.38
width 225mm; rounded angle	4.05	5.00	4.25	0.50	0.17	5.81	0.44	10.50	m	11.81
width 300mm; rounded angle	5.18	5.00	5.44	0.60	0.22	7.08	0.58	13.10	m	14.74
TILE CILLS - Clay floor quarries, B.S.6431, terracotta; 3mm joints, symmetrical layout; bedding in 10mm cement mortar (1:3); pointing in cement mortar (1:3); on cement and sand base - cills										
152 x 152 x 12.5mm units to cills on brickwork or blockwork base										
width 152mm; rounded angle	2.68	5.00	2.81	0.33	0.10	3.75	0.33	6.89	m	7.75
width 225mm; rounded angle	3.72	5.00	3.91	0.44	0.15	5.11	0.49	9.51	m	10.70
width 300mm; rounded angle	4.75	5.00	4.99	0.56	0.20	6.57	0.66	12.22	m	13.75
M40: STONE/CONCRETE/QUARRY/CERAMIC TILING/MOSAIC; PLANT ROOMS										
CLAY TILE PAVING - Clay floor quarries, B.S.6431, terracotta; 3mm joints, symmetrical layout; bedding in 10mm cement mortar (1:3); pointing in cement mortar (1:3); on cement and sand base										
150 x 150 x 12.5mm units to floors level or to falls only not exceeding 15 degrees from horizontal										
plain	12.73	5.00	13.37	1.05	0.60	13.90	1.46	28.73	m2	32.32
150 x 150 x 20mm units to floors level or to falls only not exceeding 15 degrees from horizontal										
plain	17.97	5.00	18.87	1.32	0.73	17.31	1.46	37.64	m2	42.34
225 x 225 x 25mm units to floors level or to falls only not exceeding 15 degrees from horizontal										
plain	31.00	5.00	32.55	0.66	0.55	9.95	1.46	43.96	m2	49.46
150 x 150 x 12.5mm units to floors to falls and crossfalls and to slopes not exceeding 15 degrees from horizontal										
plain	12.73	5.00	13.37	1.14	0.60	14.73	1.46	29.56	m2	33.26
150 x 150 x 20mm units to floors to falls and crossfalls and to slopes not exceeding 15 degrees from horizontal										
plain	17.97	5.00	18.87	1.44	0.73	18.42	1.46	38.74	m2	43.59
225 x 225 x 25mm units to floors to falls and crossfalls and to slopes not exceeding 15 degrees from horizontal										
plain	31.00	5.00	32.55	0.72	0.43	9.66	1.46	43.67	m2	49.13

SURFACE FINISHES

Labour hourly rates: (except Specialists) Craft Operatives £9.23 Labourer £7.02 Rates are national average prices. Refer to REGIONAL VARIATIONS for indicative levels of overall pricing in regions	MATERIALS			LABOUR				RATES		
	Del to Site £	Waste %	Material Cost £	Craft Optve Hrs	Lab Hrs	Labour Cost £	Sunds £	Nett Rate £	Unit	Gross Rate (+12.5%) £
M40: STONE/CONCRETE/QUARRY/CERAMIC TILING/MOSAIC; PLANT ROOMS Cont.										
CLAY TILE PAVING - Clay floor quarries, B.S.6431, blended; 3mm joints symmetrical layout; bedding in 10mm cement mortar (1:3); pointing in cement mortar (1:3);on cement and sand base										
150 x 150 x 12.5mm units to floors level or to falls only not exceeding 15 degrees from horizontal plain...	17.11	5.00	17.97	1.05	0.60	13.90	1.46	33.33	m2	37.50
150 x 150 x 20mm units to floors level or to falls only not exceeding 15 degrees from horizontal plain...	19.31	5.00	20.28	1.32	0.73	17.31	1.46	39.04	m2	43.92
225 x 225 x 25mm units to floors level or to falls only not exceeding 15 degrees from horizontal plain...	34.86	5.00	36.60	0.66	0.55	9.95	1.46	48.02	m2	54.02
150 x 150 x 12.5mm units to floors to falls and crossfalls and to slopes not exceeding 15 degrees from horizontal plain...	17.11	5.00	17.97	1.14	0.60	14.73	1.46	34.16	m2	38.43
150 x 150 x 20mm units to floors to falls and crossfalls and to slopes not exceeding 15 degrees from horizontal plain...	19.31	5.00	20.28	1.44	0.73	18.42	1.46	40.15	m2	45.17
225 x 225 x 25mm units to floors to falls and crossfalls and to slopes not exceeding 15 degrees from horizontal plain...	34.86	5.00	36.60	0.72	0.43	9.66	1.46	47.73	m2	53.69
CLAY TILE PAVING - Clay floor quarries, B.S.6431, dark; 3mm joints, symmetrical layout; bedding in 10mm cement mortar (1:3); pointing in cement mortar (1:3); on cement and sand base										
150 x 150 x 12.5mm units to floors level or to falls only not exceeding 15degrees from horizontal plain...	22.15	5.00	23.26	1.05	0.60	13.90	1.46	38.62	m2	43.45
150 x 150 x 12.5mm units to floors to falls and crossfalls and to slopes not exceeding 15 degrees from horizontal plain...	32.15	5.00	33.76	1.14	0.60	14.73	1.46	49.95	m2	56.20
M41: TERRAZZO TILING/IN SITU TERRAZZO										
Terrazzo tiles, B.S.4131, aggregate size random, ground, grouted and polished, standard colour range; 3mm joints, symmetrical layout; bedding in 40mm cement mortar (1:3); grouting with white cement; in-situ margins										
300 x 300 x 28mm units to floors on concrete base; level or to falls only not exceeding 15 degrees from horizontal plain...	27.70	5.00	29.09	1.10	0.65	14.72	3.89	47.69	m2	53.65
Terrazzo, white cement and white marble chippings (2:5); polished; on cement and sand base										
6mm work to walls width exceeding 300mm	8.95	5.00	9.40	4.30	2.15	54.78	-	64.18	m2	72.20
width not exceeding 300mm	2.70	10.00	2.97	2.58	0.65	28.38	-	31.35	m	35.26
16mm work to floors; one coat; floated level and to falls only not exceeding 15 degrees from horizontal; laid in bays, average size 610 x 610mm	17.95	5.00	18.85	2.70	1.35	34.40	-	53.25	m2	59.90
6mm work to skirtings height 75mm	1.55	5.00	1.63	1.50	0.16	14.97	-	16.60	m	18.67
height 75mm; curved to 3000mm radius	1.55	5.00	1.63	2.00	0.16	19.58	-	21.21	m	23.86
16mm work to skirtings on brickwork or blockwork base height 150mm.......................	3.35	5.00	3.52	1.90	0.32	19.78	-	23.30	m	26.21
height 150mm; curved to 3000mm radius	3.35	5.00	3.52	2.53	0.32	25.60	-	29.12	m	32.76
Rounded angles and intersections 10 - 100mm radius	-	-	-	0.21	-	1.94	-	1.94	m	2.18
Coves 25mm girth	-	-	-	0.48	-	4.43	-	4.43	m	4.98
40mm girth	-	-	-	0.72	-	6.65	-	6.65	m	7.48
Accessories										
Plastic dividing strips; setting in bed and finishing 6 x 16mm; flat section	1.80	-	1.80	0.30	-	2.77	-	4.57	m	5.14

Labour hourly rates: (except Specialists) Craft Operatives £9.23 Labourer £7.02 Rates are national average prices. Refer to REGIONAL VARIATIONS for indicative levels of overall pricing in regions	MATERIALS			LABOUR				RATES		
	Del to Site £	Waste %	Material Cost £	Craft Optve Hrs	Lab Hrs	Labour Cost £	Sunds £	Nett Rate £	Unit	Gross Rate (+12.5%) £
M41: TERRAZZO TILING/IN SITU TERRAZZO; STAIRCASE AREAS										
Terrazzo tiles, B.S.4131, aggregate size random, ground, grouted and polished, standard colour range; 3mm joints, symmetrical layout; bedding in 40mm cement mortar (1:3); grouting with white cement; in-situ margins										
300 x 300 x 28mm units to floors on concrete base; level or to falls only not exceeding 15 degrees from horizontal										
plain..........	27.70	5.00	29.09	2.05	0.65	23.48	3.89	56.46	m2	63.52
305 x 305 x 28mm units to treads on concrete base										
width 292mm; rounded nosing	13.30	5.00	13.97	0.95	0.19	10.10	1.14	25.21	m	28.36
305 x 305 x 28mm units to plain risers on concrete base										
height 165mm..........	6.65	5.00	6.98	0.65	0.11	6.77	0.64	14.39	m	16.19
Terrazzo, white cement and white marble chippings (2:5); polished; on cement and sand base										
16mm work to treads; one coat										
width 279mm..........	4.80	5.00	5.04	3.60	0.38	35.90	-	40.94	m	46.05
extra; two line carborundum non-slip inlay	8.10	5.00	8.51	0.30	-	2.77	-	11.27	m	12.68
16mm work to undercut risers; one coat										
height 178mm..........	3.60	5.00	3.78	2.40	0.24	23.84	-	27.62	m	31.07
6mm work to strings; one coat										
height 150mm	2.10	5.00	2.21	2.15	0.32	22.09	-	24.30	m	27.33
height 200mm	2.25	5.00	2.36	3.00	0.43	30.71	-	33.07	m	37.20
M42: WOOD BLOCK/COMPOSITION BLOCK/ PARQUET FLOORING										
Maple wood blocks, tongued and grooved joints; symmetrical herringbone pattern layout, two block plain borders; fixing with adhesive; sanding, one coat sealer										
70 x 230 x 20mm units to floors on cement and sand base; level or to falls only not exceeding 15 degrees from horizontal										
plain..........	30.25	5.00	31.76	1.80	1.45	26.79	4.53	63.09	m2	70.97
Merbau wood blocks, tongued and grooved joints; symmetrical herringbone pattern layout, two block plain borders; fixing with adhesive; sanding, one coat sealer										
70 x 230 x 20mm units to floors on cement and sand base; level or to falls only not exceeding 15 degrees from horizontal										
plain..........	34.85	5.00	36.59	1.80	1.45	26.79	4.53	67.92	m2	76.40
Oak wood blocks, tongued and grooved joints; symmetrical herringbone pattern layout, two block plain borders; fixing with adhesive; sanding, one coat sealer										
70 x 230 x 20mm units to floors on cement and sand base; level or to falls only not exceeding 15 degrees from horizontal										
plain..........	35.35	5.00	37.12	1.80	1.45	26.79	4.53	68.44	m2	77.00
M50: RUBBER/PLASTICS/CORK/LINO/CARPET TILING/SHEETING										
RUBBER FLOORING - Rubber floor tiles, B.S.1711; butt joints, symmetrical layout; fixing with adhesive; on cement and sand base										
610 x 610 x 2mm units to floors level or to falls only not exceeding 15 degrees from horizontal										
width exceeding 300mm	26.25	5.00	27.56	0.40	0.25	5.45	1.37	34.38	m2	38.68
width not exceeding 300mm	7.88	10.00	8.67	0.24	0.08	2.78	0.41	11.85	m	13.34
610 x 610 x 2.5mm units to floors level or to falls only not exceeding 15 degrees from horizontal										
width exceeding 300mm	30.80	5.00	32.34	0.40	0.25	5.45	1.37	39.16	m2	44.05
width not exceeding 300mm	9.24	10.00	10.16	0.24	0.08	2.78	0.41	13.35	m	15.02
RUBBER FLOORING - Smooth finish rubber matting, B.S.1711; butt joints; fixing with adhesive; on cement and sand base										
2mm work to floors level or to falls only not exceeding 15 degrees from horizontal										
width exceeding 300mm	25.20	5.00	26.46	0.20	0.15	2.90	1.37	30.73	m2	34.57
width not exceeding 300mm	7.56	10.00	8.32	0.12	0.05	1.46	0.41	10.18	m	11.46
2.5mm work to floors level or to falls only not exceeding 15 degrees from horizontal										
width exceeding 300mm	29.20	5.00	30.66	0.20	0.15	2.90	1.37	34.93	m2	39.30
width not exceeding 300mm	8.76	10.00	9.64	0.12	0.05	1.46	0.41	11.50	m	12.94

SURFACE FINISHES

Labour hourly rates: (except Specialists) Craft Operatives £9.23 Labourer £7.02 Rates are national average prices. Refer to REGIONAL VARIATIONS for indicative levels of overall pricing in regions	MATERIALS			LABOUR				RATES		
	Del to Site £	Waste %	Material Cost £	Craft Optve Hrs	Lab Hrs	Labour Cost £	Sunds £	Nett Rate £	Unit	Gross Rate (+12.5%) £
M50: RUBBER/PLASTICS/CORK/LINO/CARPET TILING/SHEETING Cont.										
RUBBER FLOORING - Smooth finish rubber matting, B.S.1711; butt joints; fixing with adhesive; on cement and sand base Cont.										
4mm work to floors level or to falls only not exceeding 15 degrees from horizontal										
width exceeding 300mm	33.50	5.00	35.17	0.20	0.15	2.90	1.37	39.44	m2	44.37
width not exceeding 300mm	10.05	10.00	11.06	0.12	0.05	1.46	0.41	12.92	m	14.54
PVC FLOORING - P.V.C. floor tiles, B.S.3260; butt joints, symmetrical layout; fixing with adhesive; two coats sealer; on cement and sand base										
300 x 300 x 2mm units to floors level or to falls only not exceeding 15 degrees from horizontal										
width exceeding 300mm	4.70	5.00	4.93	0.60	0.30	7.64	2.75	15.33	m2	17.25
width not exceeding 300mm	1.41	10.00	1.55	0.36	0.09	3.95	0.85	6.36	m	7.15
300 x 300 x 2.5mm units to floors level or to falls only not exceeding 15 degrees from horizontal										
width exceeding 300mm	5.70	5.00	5.99	0.60	0.30	7.64	2.75	16.38	m2	18.43
width not exceeding 300mm	1.71	10.00	1.88	0.36	0.09	3.95	0.85	6.69	m	7.52
PVC FLOORING - Fully flexible PVC heavy duty floor tiles, B.S.3261 type A; butt joints, symmetrical layout; fixing with adhesive; two coats sealer; on cement and sand base										
300 x 300 x 2mm units to floors level or to falls only not exceeding 15 degrees from horizontal										
width exceeding 300mm	8.60	5.00	9.03	0.60	0.30	7.64	2.75	19.42	m2	21.85
width not exceeding 300mm	2.58	10.00	2.84	0.36	0.09	3.95	0.85	7.64	m	8.60
300 x 300 x 2.5mm units to floors level or to falls only not exceeding 15 degrees from horizontal										
width exceeding 300mm	9.90	5.00	10.40	0.60	0.30	7.64	2.75	20.79	m2	23.39
width not exceeding 300mm	2.97	10.00	3.27	0.36	0.09	3.95	0.85	8.07	m	9.08
PVC FLOORING - Fully flexible P.V.C. heavy duty sheet,B.S.3261 Type A; butt joints; welded; fixing with adhesive; two coats sealer; on cement and sand base										
2mm work to floors level or to falls only not exceeding 15 degrees from horizontal										
width exceeding 300mm	8.60	5.00	9.03	0.40	0.20	5.10	2.85	16.98	m2	19.10
width not exceeding 300mm	2.58	10.00	2.84	0.24	0.06	2.64	0.86	6.33	m	7.13
2.5mm work to floors level or to falls only not exceeding 15 degrees from horizontal										
width exceeding 300mm	9.90	5.00	10.40	0.40	0.20	5.10	2.85	18.34	m2	20.63
width not exceeding 300mm	2.97	10.00	3.27	0.24	0.06	2.64	0.86	6.76	m	7.61
3mm work to floors level or to falls only not exceeding 15 degrees from horizontal										
width exceeding 300mm	10.90	5.00	11.45	0.40	0.20	5.10	2.85	19.39	m2	21.81
width not exceeding 300mm	3.27	10.00	3.60	0.24	0.06	2.64	0.86	7.09	m	7.98
PVC FLOORING - P.V.C. coved skirting; fixing with adhesive										
Skirtings on plaster base										
height 75mm	1.30	5.00	1.37	0.25	0.02	2.45	0.21	4.02	m	4.53
height 100mm	1.50	5.00	1.58	0.25	0.02	2.45	0.28	4.30	m	4.84
CORK FLOORING - Cork tiles, B.S.8203; butt joints; symmetrical layout; fixing with adhesive; two coats seal; on cement and sand base										
305 x 305 x 4.8mm units to floors level or to falls only not exceeding 15 degrees from horizontal										
width exceeding 300mm	11.55	5.00	12.13	0.90	0.28	10.27	1.50	23.90	m2	26.89
width not exceeding 300mm	3.52	10.00	3.87	0.54	0.08	5.55	0.45	9.87	m	11.10
305 x 305 x 6.4mm units to floors level or to falls only not exceeding 15 degrees from horizontal										
width exceeding 300mm	13.50	5.00	14.18	0.90	0.28	10.27	1.50	25.95	m2	29.19
width not exceeding 300mm	4.12	10.00	4.53	0.54	0.08	5.55	0.45	10.53	m	11.84
305 x 305 x 8.0mm units to floors level or to falls only not exceeding 15 degrees from horizontal										
width exceeding 300mm	17.30	5.00	18.16	0.90	0.28	10.27	1.50	29.94	m2	33.68
width not exceeding 300mm	5.28	10.00	5.81	0.54	0.08	5.55	0.45	11.80	m	13.28
LINOLEUM FLOORING - Linoleum tiles, B.S.6826; butt joints, symmetrical layout; fixing with adhesive; on cement and sand base										
500 x 500 x 2.5mm units to floors level or to falls only not exceeding 15 degrees from horizontal										
width exceeding 300mm	14.90	5.00	15.65	0.45	0.28	6.12	0.79	22.55	m2	25.37
width not exceeding 300mm	4.47	10.00	4.92	0.27	0.08	3.05	0.24	8.21	m	9.24

Labour hourly rates: (except Specialists) Craft Operatives £9.23 Labourer £7.02 Rates are national average prices. Refer to REGIONAL VARIATIONS for indicative levels of overall pricing in regions	MATERIALS			LABOUR				RATES		
	Del to Site £	Waste %	Material Cost £	Craft Optve Hrs	Lab Hrs	Labour Cost £	Sunds £	Nett Rate £	Unit	Gross Rate (+12.5%) £
M50: RUBBER/PLASTICS/CORK/LINO/CARPET TILING/SHEETING Cont.										
LINOLEUM FLOORING – Linoleum sheet, B.S.6826; butt joints; laying loose; on cement and sand base										
2.5mm work to floors level or to falls only not exceeding 15 degrees from horizontal										
width exceeding 300mm	11.35	5.00	11.92	0.20	0.15	2.90	–	14.82	m2	16.67
width not exceeding 300mm	3.41	10.00	3.75	0.12	0.05	1.46	–	5.21	m	5.86
3.2mm work to floors level or to falls only not exceeding 15 degrees from horizontal										
width exceeding 300mm	14.35	5.00	15.07	0.20	0.15	2.90	–	17.97	m2	20.21
width not exceeding 300mm	4.31	10.00	4.74	0.12	0.05	1.46	–	6.20	m	6.97
4.5mm work to floors level or to falls only not exceeding 15 degrees from horizontal										
width exceeding 300mm	17.85	5.00	18.74	0.20	0.15	2.90	–	21.64	m2	24.35
width not exceeding 300mm	5.36	10.00	5.90	0.12	0.05	1.46	–	7.35	m	8.27
CARPET TILING – Carpet tiles; Marley Floors Ltd. 'Marleytex' nylon needleloom; butt joints, symmetrical layout; fixing with adhesive; on cement and sand base										
500 x 500mm units to floors; level or to falls only not exceeding 15 degrees from horizontal										
heavy contract grade width exceeding 300mm	8.45	5.00	8.87	0.60	0.30	7.64	0.60	17.12	m2	19.26
textured heavy contract grade width exceeding 300mm	8.95	5.00	9.40	0.60	0.30	7.64	0.60	17.64	m2	19.85
CARPET SHEETING – Fitted carpeting; Marley Floors Ltd. 'Marleytex' nylon needleloom; fixing with adhesive; on cement and sand base										
Work to floors; level or to falls only not exceeding 15 degrees from horizontal										
heavy contract grade width exceeding 300mm	8.20	5.00	8.61	0.40	0.20	5.10	0.60	14.31	m2	16.09
textured heavy contract grade width exceeding 300mm	8.70	5.00	9.13	0.40	0.20	5.10	0.60	14.83	m2	16.68
M50: RUBBER/PLASTICS/CORK/LINO/CARPET TILING/SHEETING; STAIRCASE AREAS										
STAIR NOSINGS AND TREADS – Accessories										
Heavy duty aluminium alloy stair nosings with anti-slip inserts; 46mm wide with single line insert; fixing to timber with screws										
11.5mm drop	8.34	5.00	8.76	0.46	–	4.25	0.10	13.10	m	14.74
22mm drop	9.70	5.00	10.19	0.46	–	4.25	0.10	14.53	m	16.35
25mm drop	10.43	5.00	10.95	0.46	–	4.25	0.10	15.30	m	17.21
32mm drop	11.80	5.00	12.39	0.46	–	4.25	0.10	16.74	m	18.83
33mm drop	11.80	5.00	12.39	0.46	–	4.25	0.10	16.74	m	18.83
38mm drop	12.88	5.00	13.52	0.46	–	4.25	0.10	17.87	m	20.10
46mm drop	13.56	5.00	14.24	0.46	–	4.25	0.10	18.58	m	20.91
Heavy duty aluminium alloy stair nosings with anti-slip inserts; 80mm wide with two line inserts; fixing to timber with screws										
22mm drop	13.89	5.00	14.58	0.62	–	5.72	0.15	20.46	m	23.01
25mm drop	17.35	5.00	18.22	0.62	–	5.72	0.15	24.09	m	27.10
32mm drop	18.78	5.00	19.72	0.62	–	5.72	0.15	25.59	m	28.79
33mm drop	18.78	5.00	19.72	0.62	–	5.72	0.15	25.59	m	28.79
51mm drop	19.45	5.00	20.42	0.62	–	5.72	0.15	26.30	m	29.58
63mm drop	21.20	5.00	22.26	0.62	–	5.72	0.15	28.13	m	31.65
Heavy duty aluminium alloy stair nosings with anti-slip inserts; 46mm wide with single line insert; fixing to masonry with screws										
11.5mm drop	8.34	5.00	8.76	0.75	–	6.92	0.15	15.83	m	17.81
22mm drop	9.70	5.00	10.19	0.75	–	6.92	0.15	17.26	m	19.41
25mm drop	10.43	5.00	10.95	0.75	–	6.92	0.15	18.02	m	20.28
32mm drop	11.80	5.00	12.39	0.75	–	6.92	0.15	19.46	m	21.90
33mm drop	11.80	5.00	12.39	0.75	–	6.92	0.15	19.46	m	21.90
38mm drop	12.88	5.00	13.52	0.75	–	6.92	0.15	20.60	m	23.17
46mm drop	13.56	5.00	14.24	0.75	–	6.92	0.15	21.31	m	23.97
Heavy duty aluminium alloy stair nosings with anti-slip inserts; 80mm wide with two line inserts; fixing to masonry with screws										
22mm drop	13.89	5.00	14.58	0.90	–	8.31	0.21	23.10	m	25.99
25mm drop	17.35	5.00	18.22	0.90	–	8.31	0.21	26.73	m	30.08
32mm drop	18.78	5.00	19.72	0.90	–	8.31	0.21	28.24	m	31.77
33mm drop	18.78	5.00	19.72	0.90	–	8.31	0.21	28.24	m	31.77
51mm drop	19.45	5.00	20.42	0.90	–	8.31	0.21	28.94	m	32.56
63mm drop	21.12	5.00	22.18	0.90	Lab	8.31	0.21	30.69	m	34.53
Aluminium stair nosings with anti-slip inserts; 46mm wide with single line insert; fixing to timber with screws										
22mm drop	6.92	5.00	7.27	0.46	–	4.25	0.10	11.61	m	13.06
28mm drop	6.92	5.00	7.27	0.46	–	4.25	0.10	11.61	m	13.06

SURFACE FINISHES

Labour hourly rates: (except Specialists) Craft Operatives £9.23 Labourer £7.02 Rates are national average prices. Refer to REGIONAL VARIATIONS for indicative levels of overall pricing in regions	MATERIALS			LABOUR				RATES		
	Del to Site £	Waste %	Material Cost £	Craft Optve Hrs	Lab Hrs	Labour Cost £	Sunds £	Nett Rate £	Unit	Gross Rate (+12.5%) £
M50: RUBBER/PLASTICS/CORK/LINO/CARPET TILING/SHEETING; STAIRCASE AREAS Cont.										
STAIR NOSINGS AND TREADS - Accessories Cont.										
Aluminium stair nosings with anti-slip inserts; 46mm wide with single line insert; fixing to timber with screws Cont.										
33mm drop	9.01	5.00	9.46	0.46	–	4.25	0.10	13.81	m	15.53
Aluminium stair nosings with anti-slip inserts; 66mm wide with single line insert; fixing to timber with screws										
22mm drop	9.70	5.00	10.19	0.52	–	4.80	0.15	15.13	m	17.03
25mm drop	10.43	5.00	10.95	0.52	–	4.80	0.15	15.90	m	17.89
32mm drop	11.80	5.00	12.39	0.52	–	4.80	0.15	17.34	m	19.51
33mm drop	12.46	5.00	13.08	0.52	–	4.80	0.15	18.03	m	20.29
Aluminium stair nosings with anti-slip inserts; 80mm wide with two line inserts; fixing to timber with screws										
22mm drop	11.13	5.00	11.69	0.62	–	5.72	0.15	17.56	m	19.75
28mm drop	12.46	5.00	13.08	0.62	–	5.72	0.15	18.96	m	21.33
33mm drop	13.22	5.00	13.88	0.62	–	5.72	0.15	19.75	m	22.22
46mm drop	13.88	5.00	14.57	0.62	–	5.72	0.15	20.45	m	23.00
51mm drop	14.60	5.00	15.33	0.62	–	5.72	0.15	21.20	m	23.85
Aluminium alloy stair nosings with anti-slip inserts; 46mm wide with single line insert; fixing to masonry with screws										
22mm drop	6.92	5.00	7.27	0.75	–	6.92	0.15	14.34	m	16.13
28mm drop	6.92	5.00	7.27	0.75	–	6.92	0.15	14.34	m	16.13
33mm drop	9.01	5.00	9.46	0.75	–	6.92	0.15	16.53	m	18.60
Aluminium alloy stair nosings with anti-slip inserts; 66mm wide with single line insert; fixing to masonry with screws										
22mm drop	9.70	5.00	10.19	0.80	–	7.38	0.21	17.78	m	20.00
25mm drop	10.43	5.00	10.95	0.80	–	7.38	0.21	18.55	m	20.86
32mm drop	11.80	5.00	12.39	0.80	–	7.38	0.21	19.98	m	22.48
33mm drop	12.46	5.00	13.08	0.80	–	7.38	0.21	20.68	m	23.26
Aluminium alloy stair nosings with anti-slip inserts; 80mm wide with two line inserts; fixing to masonry with screws										
22mm drop	11.13	5.00	11.69	0.90	–	8.31	0.21	20.20	m	22.73
28mm drop	12.46	5.00	13.08	0.90	–	8.31	0.21	21.60	m	24.30
33mm drop	13.22	5.00	13.88	0.90	–	8.31	0.21	22.40	m	25.20
46mm drop	13.89	5.00	14.58	0.90	–	8.31	0.21	23.10	m	25.99
51mm drop	14.59	5.00	15.32	0.90	–	8.31	0.21	23.84	m	26.82
M51: EDGE FIXED CARPETING										
Underlay to carpeting; Tredair; fixing with tacks; on timber base										
Work to floors; level or to falls only not exceeding 15 degrees from horizontal										
width exceeding 300mm	1.70	5.00	1.78	0.15	0.08	1.95	0.30	4.03	m2	4.53
Fitted carpeting; Tufted wool/nylon, p.c. £16.00 m2; fixing with tackless grippers; on timber base										
Work to floors; level or to falls only not exceeding 15 degrees from horizontal										
width exceeding 300mm	15.40	5.00	16.17	0.40	0.20	5.10	0.30	21.57	m2	24.26
Fitted carpeting; Axminster wool/nylon, P.C. £22.50 m2; fixing with tackless grippers; on timber base										
Work to floors; level or to falls only not exceeding 15 degrees from horizontal										
width exceeding 300mm	20.50	5.00	21.52	0.40	0.20	5.10	0.30	26.92	m2	30.29
Fitted carpeting; Wilton wool/nylon, p.c. £28.00 m2; fixing with tackless grippers; on timber base										
Work to floors; level or to falls only not exceeding 15 degrees from horizontal										
width exceeding 300mm	25.75	5.00	27.04	0.40	0.20	5.10	0.30	32.43	m2	36.49
M51: EDGE FIXED CARPETING; STAIRCASE AREAS										
STAIR NOSINGS AND TREADS - Accessories										
Aluminium alloy carpet nosings with anti-slip inserts; with single line insert; fixing to timber with screws										
65mm wide x 24mm drop	15.75	5.00	16.54	0.52	–	4.80	0.15	21.49	m	24.17
116mm wide overall with carpet gripper x 42mm drop	22.50	5.00	23.63	0.58	–	5.35	0.21	29.19	m	32.84
Aluminium alloy carpet nosings with anti-slip inserts; with two line inserts; fixing to timber with screws										
80mm wide x 31mm drop	18.00	5.00	18.90	0.62	–	5.72	0.15	24.77	m	27.87

Labour hourly rates: (except Specialists) Craft Operatives £9.23 Labourer £7.02 Rates are national average prices. Refer to REGIONAL VARIATIONS for indicative levels of overall pricing in regions	MATERIALS			LABOUR				RATES		
	Del to Site	Waste	Material Cost	Craft Optve	Lab	Labour Cost	Sunds	Nett Rate	Unit	Gross Rate (+12.5%)
	£	%	£	Hrs	Hrs	£	£	£		£
M51: EDGE FIXED CARPETING; STAIRCASE AREAS Cont.										
STAIR NOSINGS AND TREADS - Accessories Cont.										
Aluminium alloy carpet nosings with anti-slip inserts; with single line insert; fixing to masonry with screws										
65mm wide x 24mm drop	15.75	5.00	16.54	0.80	-	7.38	0.21	24.13	m	27.15
116mm wide overall with carpet gripper x 42mm drop	22.50	5.00	23.63	0.85	-	7.85	0.28	31.75	m	35.72
Aluminium alloy carpet nosings with anti-slip inserts; with two line inserts; fixing to masonry with screws										
80mm wide x 31mm drop	18.00	5.00	18.90	0.90	-	8.31	0.21	27.42	m	30.84
M52: DECORATIVE PAPERS/FABRICS										
Lining paper (prime cost sum for supply and allowance for waste included elsewhere); sizing; applying adhesive; hanging; butt joints										
Plaster walls and columns										
exceeding 0.50m2	-	-	-	0.13	-	1.20	0.10	1.30	m2	1.46
Plaster ceilings and beams										
exceeding 0.50m2	-	-	-	0.14	-	1.29	0.10	1.39	m2	1.57
Pulp paper (prime cost sum for supply and allowance for waste included elsewhere); sizing; applying adhesive; hanging; butt joints										
Plaster walls and columns										
exceeding 0.50m2	-	-	-	0.31	-	2.86	0.14	3.00	m2	3.38
Plaster ceilings and beams										
exceeding 0.50m2	-	-	-	0.34	-	3.14	0.15	3.29	m2	3.70
Washable paper (prime cost sum for supply and allowance for waste included elsewhere); sizing; applying adhesive; hanging; butt joints										
Plaster walls and columns										
exceeding 0.50m2	-	-	-	0.24	-	2.22	0.09	2.31	m2	2.59
Plaster ceilings and beams										
exceeding 0.50m2	-	-	-	0.26	-	2.40	0.09	2.49	m2	2.80
Vinyl coated paper (prime cost sum for supply and allowance for waste included elsewhere); sizing; applying adhesive; hanging; butt joints										
Plaster walls and columns										
exceeding 0.50m2	-	-	-	0.24	-	2.22	0.09	2.31	m2	2.59
Plaster ceilings and beams										
exceeding 0.50m2	-	-	-	0.26	-	2.40	0.09	2.49	m2	2.80
Embossed or textured paper (prime cost sum for supply and allowance for waste included elsewhere); sizing; applying adhesive; hanging; butt joints										
Plaster walls and columns										
exceeding 0.50m2	-	-	-	0.24	-	2.22	0.13	2.35	m2	2.64
Plaster ceilings and beams										
exceeding 0.50m2	-	-	-	0.26	-	2.40	0.13	2.53	m2	2.85
Woodchip paper (prime cost sum for supply and allowance for waste included elsewhere); sizing; applying adhesive; hanging; butt joints										
Plaster walls and columns										
exceeding 0.50m2	-	-	-	0.24	-	2.22	0.08	2.30	m2	2.58
Plaster ceilings and beams										
exceeding 0.50m2	-	-	-	0.26	-	2.40	0.08	2.48	m2	2.79
Paper border strips (prime cost sum for supply and allowance for waste included elsewhere); sizing; applying adhesive; hanging; butt joints										
Border strips										
25mm wide to papered walls and columns	-	-	-	0.09	-	0.83	0.02	0.85	m	0.96
75mm wide to papered walls and columns	-	-	-	0.10	-	0.92	0.03	0.95	m	1.07
Lining paper p.c. £1.00 roll; sizing; applying adhesive; hanging; butt joints										
Plaster walls and columns										
exceeding 0.50m2	0.21	12.00	0.24	0.13	-	1.20	0.10	1.54	m2	1.73
Plaster ceilings and beams										
exceeding 0.50m2	0.21	12.00	0.24	0.14	-	1.29	0.10	1.63	m2	1.83

SURFACE FINISHES

Labour hourly rates: (except Specialists) Craft Operatives £9.23 Labourer £7.02. Rates are national average prices. Refer to REGIONAL VARIATIONS for indicative levels of overall pricing in regions	MATERIALS			LABOUR				RATES		
	Del to Site £	Waste %	Material Cost £	Craft Optve Hrs	Lab Hrs	Labour Cost £	Sunds £	Nett Rate £	Unit	Gross Rate (+12.5%) £
M52: DECORATIVE PAPERS/FABRICS Cont.										
Pulp paper p.c. £3.00 roll; sizing; applying adhesive; hanging; butt joints										
Plaster walls and columns exceeding 0.50m2	0.63	30.00	0.82	0.24	-	2.22	0.13	3.16	m2	3.56
Plaster ceilings and beams exceeding 0.50m2	0.63	25.00	0.79	0.26	-	2.40	0.13	3.32	m2	3.73
Pulp paper (24' drop pattern match) p.c. £4.00 roll; sizing; applying adhesive; hanging; butt joints										
Plaster walls and columns exceeding 0.50m2	0.84	40.00	1.18	0.31	-	2.86	0.14	4.18	m2	4.70
Plaster ceilings and beams exceeding 0.50m2	0.84	35.00	1.13	0.33	-	3.05	0.14	4.32	m2	4.86
Washable paper p.c. £4.00 roll; sizing; applying adhesive; hanging; butt joints										
Plaster walls and columns exceeding 0.50m2	0.84	25.00	1.05	0.24	-	2.22	0.09	3.36	m2	3.77
Plaster ceilings and beams exceeding 0.50m2	0.84	20.00	1.01	0.26	-	2.40	0.09	3.50	m2	3.94
Vinyl coated paper p.c. £6.50 roll; sizing; applying adhesive; hanging; butt joints										
Plaster walls and columns exceeding 0.50m2	1.37	25.00	1.71	0.24	-	2.22	0.09	4.02	m2	4.52
Plaster ceilings and beams exceeding 0.50m2	1.37	20.00	1.64	0.26	-	2.40	0.09	4.13	m2	4.65
Vinyl coated paper p.c. £7.00 roll; sizing; applying adhesive; hanging; butt joints										
Plaster walls and columns exceeding 0.50m2	1.47	25.00	1.84	0.24	-	2.22	0.09	4.14	m2	4.66
Plaster ceilings and beams exceeding 0.50m2	1.47	20.00	1.76	0.26	-	2.40	0.09	4.25	m2	4.79
Embossed paper p.c. £4.50 roll; sizing; applying adhesive; hanging; butt joints										
Plaster walls and columns exceeding 0.50m2	0.95	25.00	1.19	0.24	-	2.22	0.13	3.53	m2	3.97
Plaster ceilings and beams exceeding 0.50m2	0.95	20.00	1.14	0.26	-	2.40	0.13	3.67	m2	4.13
Textured paper p.c. £5.00 roll; sizing; applying adhesive; hanging; butt joints										
Plaster walls and columns exceeding 0.50m2	1.05	25.00	1.31	0.24	-	2.22	0.13	3.66	m2	4.11
Plaster ceilings and beams exceeding 0.50m2	1.05	20.00	1.26	0.26	-	2.40	0.13	3.79	m2	4.26
Woodchip paper p.c. £1.50 roll; sizing; applying adhesive; hanging; butt joints										
Plaster walls and columns exceeding 0.50m2	0.32	25.00	0.40	0.24	-	2.22	0.08	2.70	m2	3.03
Plaster ceilings and beams exceeding 0.50m2	0.32	20.00	0.38	0.26	-	2.40	0.08	2.86	m2	3.22
Paper border strips p.c. £0.25m; applying adhesive; hanging; butt joints										
Border strips 25mm wide to papered walls and columns	0.26	10.00	0.29	0.09	-	0.83	0.02	1.14	m	1.28
Paper border strips p.c. £0.50m; applying adhesive; hanging; butt joints										
Border strips 75mm wide to papered walls and columns	0.53	10.00	0.58	0.10	-	0.92	0.03	1.54	m	1.73
Hessian wall covering (prime cost sum for supply and allowance for waste included elsewhere); sizing; applying adhesive; hanging; butt joints										
Plaster walls and columns exceeding 0.50m2	-	-	-	0.44	-	4.06	0.25	4.31	m2	4.85

Labour hourly rates: (except Specialists) Craft Operatives £9.23 Labourer £7.02 Rates are national average prices. Refer to REGIONAL VARIATIONS for indicative levels of overall pricing in regions	MATERIALS			LABOUR				RATES		
	Del to Site £	Waste %	Material Cost £	Craft Optve Hrs	Lab Hrs	Labour Cost £	Sunds £	Nett Rate £	Unit	Gross Rate (+12.5%) £
M52: DECORATIVE PAPERS/FABRICS Cont.										
Textile hessian paper backed wall covering (prime cost sum for supply and allowance for waste included elsewhere); sizing; applying adhesive; hanging; butt joints										
Plaster walls and columns exceeding 0.50m2	-	-	-	0.36	-	3.32	0.24	3.56	m2	4.01
Hessian wall covering p.c. £4.00m2; sizing; applying adhesive; hanging; butt joints										
Plaster walls and columns exceeding 0.50m2	4.73	30.00	6.15	0.24	-	2.22	0.25	8.61	m2	9.69
Textile hessian paper backed wall covering p.c. £7.50m2; sizing; applying adhesive; hanging; butt joints										
Plaster walls and columns exceeding 0.50m2	8.40	25.00	10.50	0.36	-	3.32	0.24	14.06	m2	15.82
M52: DECORATIVE PAPERS/FABRICS - REDECORATIONS										
Stripping existing paper; lining paper (prime cost sum for supply and allowance for waste included elsewhere); sizing; applying adhesive hanging; butt joints										
Hard building board walls and columns exceeding 0.50m2	-	-	-	0.32	-	2.95	0.60	3.55	m2	4.00
Plaster walls and columns exceeding 0.50m2	-	-	-	0.28	-	2.58	0.50	3.08	m2	3.47
Hard building board ceilings and beams exceeding 0.50m2	-	-	-	0.35	-	3.23	0.60	3.83	m2	4.31
Plaster ceilings and beams exceeding 0.50m2	-	-	-	0.30	-	2.77	0.51	3.28	m2	3.69
Stripping existing paper; lining paper and cross lining (prime cost sum for supply and allowance for waste included elsewhere); sizing; applying adhesive; hanging; butt joints										
Hard building board walls and columns exceeding 0.50m2	-	-	-	0.45	-	4.15	0.69	4.84	m2	5.45
Plaster walls and columns exceeding 0.50m2	-	-	-	0.41	-	3.78	0.60	4.38	m2	4.93
Hard building board ceilings and beams exceeding 0.50m2	-	-	-	0.50	-	4.62	0.70	5.32	m2	5.98
Plaster ceilings and beams exceeding 0.50m2	-	-	-	0.44	-	4.06	0.61	4.67	m2	5.26
Stripping existing paper; pulp paper (prime cost sum for supply and allowance for waste included elsewhere); applying adhesive; hanging; butt joints										
Hard building board walls and columns exceeding 0.50m2	-	-	-	0.43	-	3.97	0.63	4.60	m2	5.17
Plaster walls and columns exceeding 0.50m2	-	-	-	0.39	-	3.60	0.54	4.14	m2	4.66
Hard building board ceilings and beams exceeding 0.50m2	-	-	-	0.47	-	4.34	0.64	4.98	m2	5.60
Plaster ceilings and beams exceeding 0.50m2	-	-	-	0.42	-	3.88	0.54	4.42	m2	4.97
Stripping existing washable or vinyl coated paper; washable paper (prime cost sum for supply and allowance for waste included elsewhere); applying adhesive; hanging; butt joints										
Hard building board walls and columns exceeding 0.50m2	-	-	-	0.44	-	4.06	0.59	4.65	m2	5.23
Plaster walls and columns exceeding 0.50m2	-	-	-	0.40	-	3.69	0.50	4.19	m2	4.72
Hard building board ceilings and beams exceeding 0.50m2	-	-	-	0.48	-	4.43	0.60	5.03	m2	5.66
Plaster ceilings and beams exceeding 0.50m2	-	-	-	0.43	-	3.97	0.50	4.47	m2	5.03

SURFACE FINISHES

Labour hourly rates: (except Specialists) Craft Operatives £9.23 Labourer £7.02 Rates are national average prices. Refer to REGIONAL VARIATIONS for indicative levels of overall pricing in regions	MATERIALS			LABOUR				RATES		
	Del to Site £	Waste %	Material Cost £	Craft Optve Hrs	Lab Hrs	Labour Cost £	Sunds £	Nett Rate £	Unit	Gross Rate (+12.5%) £
M52: DECORATIVE PAPERS/FABRICS - REDECORATIONS Cont.										
Stripping existing washable or vinyl coated paper; vinyl coated paper (prime cost sum for supply and allowance for waste included elsewhere); sizing; applying adhesive; hanging; butt joints										
Hard building board walls and columns										
exceeding 0.50m2	-	-	-	0.44	-	4.06	0.59	4.65	m2	5.23
Plaster walls and columns										
exceeding 0.50m2	-	-	-	0.40	-	3.69	0.50	4.19	m2	4.72
Hard building board ceilings and beams										
exceeding 0.50m2	-	-	-	0.48	-	4.43	0.60	5.03	m2	5.66
Plaster ceilings and beams										
exceeding 0.50m2	-	-	-	0.43	-	3.97	0.50	4.47	m2	5.03
Stripping existing paper; embossed or textured paper; (prime cost sum for supply and allowance for waste included elsewhere); sizing; applying adhesive; hanging; butt joints										
Hard building board walls and columns										
exceeding 0.50m2	-	-	-	0.43	-	3.97	0.63	4.60	m2	5.17
Plaster walls and columns										
exceeding 0.50m2	-	-	-	0.39	-	3.60	0.54	4.14	m2	4.66
Hard building board ceilings and beams										
exceeding 0.50m2	-	-	-	0.47	-	4.34	0.64	4.98	m2	5.60
Plaster ceilings and beams										
exceeding 0.50m2	-	-	-	0.42	-	3.88	0.55	4.43	m2	4.98
Stripping existing paper; woodchip paper (prime cost sum for supply included elsewhere); sizing; applying adhesive; hanging; butt joints										
Hard building board walls and columns										
exceeding 0.50m2	-	-	-	0.43	-	3.97	0.58	4.55	m2	5.12
Plaster walls and columns										
exceeding 0.50m2	-	-	-	0.39	-	3.60	0.57	4.17	m2	4.69
Hard building board ceilings and beams										
exceeding 0.50m2	-	-	-	0.47	-	4.34	0.67	5.01	m2	5.63
Plaster ceilings and beams										
exceeding 0.50m2	-	-	-	0.42	-	3.88	0.57	4.45	m2	5.00
Stripping existing varnished or painted paper; woodchip paper (prime cost sum for supply and allowance for waste included elsewhere); sizing; applying adhesive; hanging; butt joints										
Hard building board walls and columns										
exceeding 0.50m2	-	-	-	0.57	-	5.26	0.68	5.94	m2	6.68
Plaster walls and columns										
exceeding 0.50m2	-	-	-	0.53	-	4.89	0.51	5.40	m2	6.08
Hard building board ceilings and beams										
exceeding 0.50m2	-	-	-	0.63	-	5.81	0.61	6.42	m2	7.23
Plaster ceilings and beams										
exceeding 0.50m2	-	-	-	0.58	-	5.35	0.52	5.87	m2	6.61
Stripping existing paper; lining paper p.c. £1.00 roll; sizing; applying adhesive; hanging; butt joints										
Hard building board walls and columns										
exceeding 0.50m2	0.21	12.00	0.24	0.59	-	5.45	0.60	6.28	m2	7.07
Plaster walls and columns										
exceeding 0.50m2	0.21	12.00	0.24	0.28	-	2.58	0.50	3.32	m2	3.73
Hard building board ceilings and beams										
exceeding 0.50m2	0.21	12.00	0.24	0.35	-	3.23	0.60	4.07	m2	4.57
Plaster ceilings and beams										
exceeding 0.50m2	0.21	12.00	0.24	0.30	-	2.77	0.51	3.51	m2	3.95
Stripping existing paper; lining paper p.c. £1.00 roll and cross lining P.C. £1.00 roll; sizing; applying adhesive; hanging; butt joints										
Hard building board walls and columns										
exceeding 0.50m2	0.42	12.00	0.47	0.45	-	4.15	0.69	5.31	m2	5.98
Plaster walls and columns										
exceeding 0.50m2	0.42	12.00	0.47	0.41	-	3.78	0.60	4.85	m2	5.46

Labour hourly rates: (except Specialists) Craft Operatives £9.23 Labourer £7.02 Rates are national average prices. Refer to REGIONAL VARIATIONS for indicative levels of overall pricing in regions	MATERIALS			LABOUR				RATES		
	Del to Site £	Waste %	Material Cost £	Craft Optve Hrs	Lab Hrs	Labour Cost £	Sunds £	Nett Rate £	Unit	Gross Rate (+12.5%) £
M52: DECORATIVE PAPERS/FABRICS - REDECORATIONS Cont.										
Stripping existing paper; lining paper p.c. £1.00 roll and cross lining P.C. £1.00 roll; sizing; applying adhesive; hanging; butt joints Cont.										
Hard building board ceilings and beams exceeding 0.50m2	0.42	12.00	0.47	0.50	–	4.62	0.70	5.79	m2	6.51
Plaster ceilings and beams exceeding 0.50m2	0.42	12.00	0.47	0.44	–	4.06	0.61	5.14	m2	5.78
Stripping existing paper; pulp paper p.c. £3.00 roll; sizing; applying adhesive; hanging; butt joints										
Hard building board walls and columns exceeding 0.50m2	0.63	30.00	0.82	0.43	–	3.97	0.63	5.42	m2	6.10
Plaster walls and columns exceeding 0.50m2	0.63	30.00	0.82	0.39	–	3.60	0.54	4.96	m2	5.58
Hard building board ceilings and beams exceeding 0.50m2	0.63	25.00	0.79	0.47	–	4.34	0.64	5.77	m2	6.49
Plaster ceilings and beams exceeding 0.50m2	0.63	25.00	0.79	0.42	–	3.88	0.54	5.20	m2	5.85
Stripping existing paper; pulp paper with 24' drop pattern; p.c. £4.00 roll; sizing; applying adhesive; hanging; butt joints										
Hard building board walls and columns exceeding 0.50m2	0.84	40.00	1.18	0.50	–	4.62	0.64	6.43	m2	7.23
Plaster walls and columns exceeding 0.50m2	0.84	40.00	1.18	0.45	–	4.15	0.55	5.88	m2	6.61
Hard building board ceilings and beams exceeding 0.50m2	0.84	35.00	1.13	0.54	–	4.98	0.65	6.77	m2	7.61
Plaster ceilings and beams exceeding 0.50m2	0.84	35.00	1.13	0.48	–	4.43	0.55	6.11	m2	6.88
Stripping existing washable or vinyl coated paper; washable paper p.c. £3.75 roll; sizing; applying adhesive; hanging; butt joints										
Hard building board walls and columns exceeding 0.50m2	0.84	25.00	1.05	0.44	–	4.06	0.59	5.70	m2	6.41
Plaster walls and columns exceeding 0.50m2	0.84	25.00	1.05	0.40	–	3.69	0.50	5.24	m2	5.90
Hard building board ceilings and beams exceeding 0.50m2	0.84	20.00	1.01	0.48	–	4.43	0.60	6.04	m2	6.79
Plaster ceilings and beams exceeding 0.50m2	0.84	20.00	1.01	0.43	–	3.97	0.50	5.48	m2	6.16
Stripping existing washable or vinyl coated paper; vinyl coated paper P.C. £6.50 roll; sizing; applying adhesive; hanging; butt joints										
Hard building board walls and columns exceeding 0.50m2	1.37	25.00	1.71	0.44	–	4.06	0.59	6.36	m2	7.16
Plaster walls and columns exceeding 0.50m2	1.37	25.00	1.71	0.40	–	3.69	0.50	5.90	m2	6.64
Hard building board ceilings and beams exceeding 0.50m2	1.37	20.00	1.64	0.48	–	4.43	0.60	6.67	m2	7.51
Plaster ceilings and beams exceeding 0.50m2	1.37	20.00	1.64	0.43	–	3.97	0.50	6.11	m2	6.88
Stripping existing washable or vinyl coated paper; vinyl coated paper P.C. £8.00 roll; sizing; applying adhesive; hanging; butt joints										
Hard building board walls and columns exceeding 0.50m2	1.47	25.00	1.84	0.44	–	4.06	0.59	6.49	m2	7.30
Plaster walls and columns exceeding 0.50m2	1.47	25.00	1.84	0.40	–	3.69	0.50	6.03	m2	6.78
Hard building board ceilings and beams exceeding 0.50m2	1.47	20.00	1.76	0.48	–	4.43	0.60	6.79	m2	7.64
Plaster ceilings and beams exceeding 0.50m2	1.47	20.00	1.76	0.43	–	3.97	0.50	6.23	m2	7.01
Stripping existing paper; embossed paper p.c. £4.50 roll; sizing; applying adhesive; hanging; butt joints										
Hard building board walls and columns exceeding 0.50m2	0.95	25.00	1.19	0.43	–	3.97	0.63	5.79	m2	6.51

SURFACE FINISHES

Labour hourly rates: (except Specialists) Craft Operatives £9.23 Labourer £7.02 Rates are national average prices. Refer to REGIONAL VARIATIONS for indicative levels of overall pricing in regions	MATERIALS			LABOUR				RATES		
	Del to Site £	Waste %	Material Cost £	Craft Optve Hrs	Lab Hrs	Labour Cost £	Sunds £	Nett Rate £	Unit	Gross Rate (+12.5%) £
M52: DECORATIVE PAPERS/FABRICS - REDECORATIONS Cont.										
Stripping existing paper; embossed paper p.c. £4.50 roll; sizing; applying adhesive; hanging; butt joints Cont.										
Plaster walls and columns exceeding 0.50m2	0.95	25.00	1.19	0.39	-	3.60	0.54	5.33	m2	5.99
Hard building board ceilings and beams exceeding 0.50m2	0.95	20.00	1.14	0.47	-	4.34	0.64	6.12	m2	6.88
Plaster ceilings and beams exceeding 0.50m2	0.95	20.00	1.14	0.42	-	3.88	0.54	5.56	m2	6.25
Stripping existing paper; textured paper p.c. £5.00 roll; sizing; applying adhesive; hanging; butt joints										
Hard building board walls and columns exceeding 0.50m2	1.05	25.00	1.31	0.43	-	3.97	0.63	5.91	m2	6.65
Plaster walls and columns exceeding 0.50m2	1.05	25.00	1.31	0.39	-	3.60	0.54	5.45	m2	6.13
Hard building board ceilings and beams exceeding 0.50m2	1.05	20.00	1.26	0.47	-	4.34	0.64	6.24	m2	7.02
Plaster ceilings and beams exceeding 0.50m2	1.05	20.00	1.26	0.42	-	3.88	0.54	5.68	m2	6.39
Stripping existing paper; woodchip paper p.c. £1.50 roll; sizing; applying adhesive; hanging; butt joints										
Hard building board walls and columns exceeding 0.50m2	0.32	25.00	0.40	0.43	-	3.97	0.58	4.95	m2	5.57
Plaster walls and columns exceeding 0.50m2	0.32	25.00	0.40	0.39	-	3.60	0.49	4.49	m2	5.05
Hard building board ceilings and beams exceeding 0.50m2	0.32	20.00	0.38	0.47	-	4.34	0.59	5.31	m2	5.98
Plaster ceilings and beams exceeding 0.50m2	0.32	20.00	0.38	0.42	-	3.88	0.49	4.75	m2	5.34
Washing down old distempered or painted surfaces; lining paper p.c. £1.00 roll; sizing; applying adhesive; hanging; butt joints										
Hard building board walls and columns exceeding 0.50m2	0.21	12.00	0.24	0.22	-	2.03	0.11	2.38	m2	2.67
Plaster walls and columns exceeding 0.50m2	0.21	12.00	0.24	0.22	-	2.03	0.11	2.38	m2	2.67
Hard building board ceilings and beams exceeding 0.50m2	0.21	12.00	0.24	0.23	-	2.12	0.11	2.47	m2	2.78
Plaster ceilings and beams exceeding 0.50m2	0.21	12.00	0.24	0.23	-	2.12	0.11	2.47	m2	2.78
Lining papered walls and columns exceeding 0.50m2	0.21	12.00	0.24	0.22	-	2.03	0.11	2.38	m2	2.67
Lining papered ceilings and beams exceeding 0.50m2	0.21	12.00	0.24	0.23	-	2.12	0.11	2.47	m2	2.78
Washing down old distempered or painted surfaces; lining paper p.c. £1.00 roll; and cross lining p.c. £1.00 roll; sizing; applying adhesive; hanging; butt joints										
Hard building board walls and columns exceeding 0.50m2	0.42	12.00	0.47	0.35	-	3.23	0.24	3.94	m2	4.43
Plaster walls and columns exceeding 0.50m2	0.42	12.00	0.47	0.35	-	3.23	0.24	3.94	m2	4.43
Hard building board ceilings and beams exceeding 0.50m2	0.42	12.00	0.47	0.37	-	3.42	0.24	4.13	m2	4.64
Plaster ceilings and beams exceeding 0.50m2	0.42	12.00	0.47	0.37	-	3.42	0.24	4.13	m2	4.64
Washing down old distempered or painted surfaces; pulp paper p.c. £2.75 roll; sizing; applying adhesive; hanging; butt joints										
Hard building board walls and columns exceeding 0.50m2	0.63	30.00	0.82	0.33	-	3.05	0.14	4.00	m2	4.51
Plaster walls and columns exceeding 0.50m2	0.63	30.00	0.82	0.33	-	3.05	0.14	4.00	m2	4.51
Hard building board ceilings and beams exceeding 0.50m2	0.63	25.00	0.79	0.35	-	3.23	0.15	4.17	m2	4.69

Labour hourly rates: (except Specialists) Craft Operatives £9.23 Labourer £7.02 Rates are national average prices. Refer to REGIONAL VARIATIONS for indicative levels of overall pricing in regions	MATERIALS			LABOUR				RATES		
	Del to Site	Waste	Material Cost	Craft Optve	Lab	Labour Cost	Sunds	Nett Rate	Unit	Gross Rate (+12.5%)
	£	%	£	Hrs	Hrs	£	£	£		£
M52: DECORATIVE PAPERS/FABRICS - REDECORATIONS Cont.										
Washing down old distempered or painted surfaces; pulp paper p.c. £2.75 roll; sizing; applying adhesive; hanging; butt joints Cont.										
Plaster ceilings and beams exceeding 0.50m2	0.63	25.00	0.79	0.35	-	3.23	0.15	4.17	m2	4.69
Washing down old distempered or painted surfaces; pulp paper with 24' drop pattern; p.c. £4.00 roll; sizing; applying adhesive; hanging; butt joints										
Hard building board walls and columns exceeding 0.50m2	0.84	40.00	1.18	0.40	-	3.69	0.15	5.02	m2	5.65
Plaster walls and columns exceeding 0.50m2	0.84	40.00	1.18	0.40	-	3.69	0.15	5.02	m2	5.65
Hard building board ceilings and beams exceeding 0.50m2	0.84	35.00	1.13	0.42	-	3.88	0.16	5.17	m2	5.82
Plaster ceilings and beams exceeding 0.50m2	0.84	35.00	1.13	0.42	-	3.88	0.16	5.17	m2	5.82
Washing down old distempered or painted surfaces; washable paper p.c. £4.00 roll; sizing; applying adhesive; hanging; butt joints										
Hard building board walls and columns exceeding 0.50m2	0.84	25.00	1.05	0.33	-	3.05	0.10	4.20	m2	4.72
Plaster walls and columns exceeding 0.50m2	0.84	25.00	1.05	0.33	-	3.05	0.10	4.20	m2	4.72
Hard building board ceilings and beams exceeding 0.50m2	0.84	20.00	1.01	0.35	-	3.23	0.10	4.34	m2	4.88
Plaster ceilings and beams exceeding 0.50m2	0.84	20.00	1.01	0.35	-	3.23	0.10	4.34	m2	4.88
Washing down old distempered or painted surfaces; vinyl coated paper p.c. £6.50 roll; sizing; applying adhesive; hanging; butt joints										
Hard building board walls and columns exceeding 0.50m2	1.37	25.00	1.71	0.33	-	3.05	0.10	4.86	m2	5.47
Plaster walls and columns exceeding 0.50m2	1.37	25.00	1.71	0.33	-	3.05	0.10	4.86	m2	5.47
Hard building board ceilings and beams exceeding 0.50m2	1.37	20.00	1.64	0.35	-	3.23	0.10	4.97	m2	5.60
Plaster ceilings and beams exceeding 0.50m2	1.37	20.00	1.64	0.35	-	3.23	0.10	4.97	m2	5.60
Washing down old distempered or painted surfaces; vinyl coated paper p.c. £7.00 roll; sizing; applying adhesive; hanging; butt joints										
Hard building board walls and columns exceeding 0.50m2	1.47	25.00	1.84	0.33	-	3.05	0.10	4.98	m2	5.61
Plaster walls and columns exceeding 0.50m2	1.47	25.00	1.84	0.33	-	3.05	0.10	4.98	m2	5.61
Hard building board ceilings and beams exceeding 0.50m2	1.47	20.00	1.76	0.35	-	3.23	0.10	5.09	m2	5.73
Plaster ceilings and beams exceeding 0.50m2	1.47	20.00	1.76	0.35	-	3.23	0.10	5.09	m2	5.73
Washing down old distempered or painted surfaces; embossed paper p.c. £4.50 roll; sizing; applying adhesive; hanging; butt joints										
Hard building board walls and columns exceeding 0.50m2	0.95	25.00	1.19	0.33	-	3.05	0.14	4.37	m2	4.92
Plaster walls and columns exceeding 0.50m2	0.95	25.00	1.19	0.33	-	3.05	0.14	4.37	m2	4.92
Hard building board ceilings and beams exceeding 0.50m2	0.95	20.00	1.14	0.35	-	3.23	0.15	4.52	m2	5.09
Plaster ceilings and beams exceeding 0.50m2	0.95	20.00	1.14	0.35	-	3.23	0.15	4.52	m2	5.09
Washing down old distempered or painted surfaces; textured paper p.c. £5.00 roll; sizing; applying adhesive; hanging; butt joints										
Hard building board walls and columns exceeding 0.50m2	1.05	25.00	1.31	0.33	-	3.05	0.14	4.50	m2	5.06

Labour hourly rates: (except Specialists) Craft Operatives £9.23 Labourer £7.02 Rates are national average prices. Refer to REGIONAL VARIATIONS for indicative levels of overall pricing in regions	MATERIALS			LABOUR				RATES		
	Del to Site £	Waste %	Material Cost £	Craft Optve Hrs	Lab Hrs	Labour Cost £	Sunds £	Nett Rate £	Unit	Gross Rate (+12.5%) £
M52: DECORATIVE PAPERS/FABRICS - REDECORATIONS Cont.										
Washing down old distempered or painted surfaces; textured paper p.c. £5.00 roll; sizing; applying adhesive; hanging; butt joints Cont.										
Plaster walls and columns										
exceeding 0.50m2	1.05	25.00	1.31	0.33	–	3.05	0.14	4.50	m2	5.06
Hard building board ceilings and beams										
exceeding 0.50m2	1.05	20.00	1.26	0.35	–	3.23	0.15	4.64	m2	5.22
Plaster ceilings and beams										
exceeding 0.50m2	1.05	20.00	1.26	0.35	–	3.23	0.15	4.64	m2	5.22
Washing down old distempered or painted surfaces; woodchip paper p.c. £1.50 roll; sizing; applying adhesive; hanging; butt joints										
Hard building board walls and columns										
exceeding 0.50m2	0.32	25.00	0.40	0.33	–	3.05	0.09	3.54	m2	3.98
Plaster walls and columns										
exceeding 0.50m2	0.32	25.00	0.40	0.33	–	3.05	0.09	3.54	m2	3.98
Hard building board ceilings and beams										
exceeding 0.50m2	0.32	20.00	0.38	0.35	–	3.23	0.10	3.71	m2	4.18
Plaster ceilings and beams										
exceeding 0.50m2	0.32	20.00	0.38	0.35	–	3.23	0.10	3.71	m2	4.18
Stripping existing paper; hessian wall covering (prime cost sum for supply and allowance for waste included elsewhere); sizing; applying adhesive; hanging; butt joints										
Hard building board walls and columns										
exceeding 0.50m2	–	–	–	0.63	–	5.81	0.75	6.56	m2	7.39
Plaster walls and columns										
exceeding 0.50m2	–	–	–	0.58	–	5.35	0.66	6.01	m2	6.77
Stripping existing paper; textile hessian paper backed wall covering (prime cost sum for supply and allowance for waste included elsewhere); sizing; applying adhesive; hanging; butt joints										
Hard building board walls and columns										
exceeding 0.50m2	–	–	–	0.55	–	5.08	0.74	5.82	m2	6.54
Plaster walls and columns										
exceeding 0.50m2	–	–	–	0.51	–	4.71	0.65	5.36	m2	6.03
Stripping existing hessian; hessian surfaced wall paper (prime cost sum for supply and allowance for waste included elsewhere); sizing; applying adhesive; hanging; butt joints										
Hard building board walls and columns										
exceeding 0.50m2	–	–	–	1.03	–	9.51	0.82	10.33	m2	11.62
Plaster walls and columns										
exceeding 0.50m2	–	–	–	0.99	–	9.14	0.73	9.87	m2	11.10
Stripping existing hessian; textile hessian paper backed wall covering (prime cost sum for supply and allowance for waste included elsewhere); sizing; applying adhesive; hanging; butt joints										
Hard building board walls and columns										
exceeding 0.50m2	–	–	–	0.96	–	8.86	0.81	9.67	m2	10.88
Plaster walls and columns										
exceeding 0.50m2	–	–	–	0.91	–	8.40	0.71	9.11	m2	10.25
Stripping existing hessian, paper backed; hessian surfaced wall paper (prime cost sum for supply and allowance for waste included elsewhere); sizing; applying adhesive; hanging; butt joints										
Hard building board walls and columns										
exceeding 0.50m2	–	–	–	0.63	–	5.81	0.75	6.56	m2	7.39
Plaster walls and columns										
exceeding 0.50m2	–	–	–	0.58	–	5.35	0.66	6.01	m2	6.77
Stripping existing hessian, paper backed; textile hessian paper backed wall covering (prime cost sum for supply and allowance for waste included elsewhere); sizing; applying adhesive; hanging; butt joints										
Hard building board walls and columns										
exceeding 0.50m2	–	–	–	0.55	–	5.08	0.74	5.82	m2	6.54

Labour hourly rates: (except Specialists) Craft Operatives £9.23 Labourer £7.02. Rates are national average prices. Refer to REGIONAL VARIATIONS for indicative levels of overall pricing in regions	MATERIALS			LABOUR				RATES		
	Del to Site £	Waste %	Material Cost £	Craft Optve Hrs	Lab Hrs	Labour Cost £	Sunds £	Nett Rate £	Unit	Gross Rate (+12.5%) £
M52: DECORATIVE PAPERS/FABRICS – REDECORATIONS Cont.										
Stripping existing hessian, paper backed; textile hessian paper backed wall covering (prime cost sum for supply and allowance for waste included elsewhere); sizing; applying adhesive; hanging; butt joints Cont.										
Plaster walls and columns exceeding 0.50m2	-	-	-	0.51	-	4.71	0.65	5.36	m2	6.03
Stripping existing paper; hessian surfaced wall paper p.c. £4.50m2; sizing; applying adhesive; hanging; butt joints										
Hard building board walls and columns exceeding 0.50m2	4.73	30.00	6.15	0.63	-	5.81	0.75	12.71	m2	14.30
Plaster walls and columns exceeding 0.50m2	4.73	30.00	6.15	0.58	-	5.35	0.66	12.16	m2	13.68
Stripping existing paper; textile hessian paper backed wall covering P.C. £8.00m2; sizing; applying adhesive; hanging; butt joints										
Hard building board walls and columns exceeding 0.50m2	8.40	25.00	10.50	0.55	-	5.08	0.74	16.32	m2	18.36
Plaster walls and columns exceeding 0.50m2	8.40	25.00	10.50	0.51	-	4.71	0.65	15.86	m2	17.84
Stripping existing hessian; hessian surfaced wall paper p.c. £4.50m2; sizing; applying adhesive; hanging ; butt joints										
Hard building board walls and columns exceeding 0.50m2	4.73	30.00	6.15	1.03	-	9.51	0.82	16.48	m2	18.54
Plaster walls and columns exceeding 0.50m2	4.73	30.00	6.15	0.99	-	9.14	0.65	15.94	m2	17.93
Stripping existing hessian; textile hessian paper backed wall covering P.C. £8.00m2; sizing; applying adhesive; hanging; butt joints										
Hard building board walls and columns exceeding 0.50m2	8.40	25.00	10.50	0.96	-	8.86	0.81	20.17	m2	22.69
Plaster walls and columns exceeding 0.50m2	8.40	25.00	10.50	0.91	-	8.40	0.71	19.61	m2	22.06
Stripping existing hessian, paper backed; hessian surfaced wall paper p.c. £4.50m2; sizing; applying adhesive; hanging; butt joints										
Hard building board walls and columns exceeding 0.50m2	4.73	30.00	6.15	0.63	-	5.81	0.75	12.71	m2	14.30
Plaster walls and columns exceeding 0.50m2	4.73	30.00	6.15	0.58	-	5.35	0.66	12.16	m2	13.68
Stripping existing hessian, paper backed; textile hessian paper backed wall covering p.c. £8.00m2; sizing; applying adhesive; hanging; butt joints										
Hard building board walls and columns exceeding 0.50m2	8.40	25.00	10.50	0.55	-	5.08	0.74	16.32	m2	18.36
Plaster walls and columns exceeding 0.50m2	8.40	25.00	10.50	0.51	-	4.71	0.65	15.86	m2	17.84
Expanded polystyrene sheet 2mm thick; sizing; applying adhesive; hanging; butt joints										
Plaster walls and columns exceeding 0.50m2	0.38	25.00	0.47	0.36	-	3.32	0.39	4.19	m2	4.71
Plaster ceilings and beams exceeding 0.50m2	0.38	25.00	0.47	0.43	-	3.97	0.39	4.83	m2	5.44
M60: PAINTING/CLEAR FINISHING – EMULSION PAINTING										
Mist coat, one full coat emulsion paint										
Concrete general surfaces girth exceeding 300mm	0.31	10.00	0.34	0.17	-	1.57	0.03	1.94	m2	2.18
Concrete general surfaces 3.50 - 5.00m above floor girth exceeding 300mm	0.31	10.00	0.34	0.19	-	1.75	0.03	2.12	m2	2.39
Plaster general surfaces girth exceeding 300mm	0.27	10.00	0.30	0.14	-	1.29	0.02	1.61	m2	1.81
isolated surfaces, girth not rexceeding 300mm	0.08	10.00	0.09	0.08	-	0.74	0.01	0.84	m	0.94
Plaster general surfaces 3.50 - 5.00m above floor girth exceeding 300mm	0.27	10.00	0.30	0.17	-	1.57	0.03	1.90	m2	2.13

SURFACE FINISHES

301

Labour hourly rates: (except Specialists) Craft Operatives £9.23 Labourer £7.02 Rates are national average prices. Refer to REGIONAL VARIATIONS for indicative levels of overall pricing in regions	MATERIALS			LABOUR				RATES		
	Del to Site £	Waste %	Material Cost £	Craft Optve Hrs	Lab Hrs	Labour Cost £	Sunds £	Nett Rate £	Unit	Gross Rate (+12.5%) £
M60: PAINTING/CLEAR FINISHING - EMULSION PAINTING Cont.										
Mist coat, one full coat emulsion paint Cont.										
Plasterboard general surfaces girth exceeding 300mm	0.27	10.00	0.30	0.14	-	1.29	0.02	1.61	m2	1.81
Plasterboard general surfaces 3.50 - 5.00m above floor girth exceeding 300mm	0.27	10.00	0.30	0.17	-	1.57	0.03	1.90	m2	2.13
Brickwork general surfaces girth exceeding 300mm	0.40	10.00	0.44	0.19	-	1.75	0.03	2.22	m2	2.50
Paper covered general surfaces girth exceeding 300mm	0.29	10.00	0.32	0.15	-	1.38	0.03	1.73	m2	1.95
Paper covered general surfaces 3.50 - 5.00m above floor girth exceeding 300mm	0.29	10.00	0.32	0.18	-	1.66	0.03	2.01	m2	2.26
Mist coat, two full coats emulsion paint										
Concrete general surfaces girth exceeding 300mm	0.54	10.00	0.59	0.23	-	2.12	0.04	2.76	m2	3.10
Concrete general surfaces 3.50 - 5.00m above floor girth exceeding 300mm	0.54	10.00	0.59	0.26	-	2.40	0.05	3.04	m2	3.42
Plaster general surfaces girth exceeding 300mm .... isolated surfaces, girth not exceeding 300mm	0.45 0.13	10.00 10.00	0.50 0.14	0.21 0.11	- -	1.94 1.02	0.04 0.02	2.47 1.18	m2 m	2.78 1.33
Plaster general surfaces 3.50 - 5.00m above floor girth exceeding 300mm	0.45	10.00	0.50	0.24	-	2.22	0.04	2.75	m2	3.09
Plasterboard general surfaces girth exceeding 300mm	0.45	10.00	0.50	0.21	-	1.94	0.04	2.47	m2	2.78
Plasterboard general surfaces 3.50 - 5.00m above floor girth exceeding 300mm	0.45	10.00	0.50	0.24	-	2.22	0.04	2.75	m2	3.09
Brickwork general surfaces girth exceeding 300mm	0.67	10.00	0.74	0.26	-	2.40	0.05	3.19	m2	3.59
Paper covered general surfaces girth exceeding 300mm	0.49	10.00	0.54	0.22	-	2.03	0.04	2.61	m2	2.94
Paper covered general surfaces 3.50 - 5.00m above floor girth exceeding 300mm	0.49	10.00	0.54	0.33	-	3.05	0.06	3.64	m2	4.10
M60: PAINTING/CLEAR FINISHING - CEMENT PAINTING										
One coat Snowcem; external work										
Cement rendered general surfaces girth exceeding 300mm .... isolated surfaces, girth not exceeding 300mm	0.29 0.09	10.00 10.00	0.32 0.10	0.12 0.04	- -	1.11 0.37	0.02 0.01	1.45 0.48	m2 m	1.63 0.54
Concrete general surfaces girth exceeding 300mm	0.39	10.00	0.43	0.12	-	1.11	0.02	1.56	m2	1.75
Brickwork general surfaces girth exceeding 300mm	0.39	10.00	0.43	0.14	-	1.29	0.02	1.74	m2	1.96
Rough cast general surfaces girth exceeding 300mm	0.44	10.00	0.48	0.19	-	1.75	0.03	2.27	m2	2.55
Two coats Snowcem; external work										
Cement rendered general surfaces girth exceeding 300mm .... isolated surfaces, girth not exceeding 300mm	0.53 0.16	10.00 10.00	0.58 0.18	0.24 0.09	- -	2.22 0.83	0.04 0.02	2.84 1.03	m2 m	3.19 1.16
Concrete general surfaces girth exceeding 300mm	0.70	10.00	0.77	0.24	-	2.22	0.04	3.03	m2	3.40
Brickwork general surfaces girth exceeding 300mm	0.70	10.00	0.77	0.26	-	2.40	0.05	3.22	m2	3.62
Rough cast general surfaces girth exceeding 300mm	0.79	10.00	0.87	0.36	-	3.32	0.06	4.25	m2	4.78
One coat sealer, one coat Snowcem; external work										
Cement rendered general surfaces girth exceeding 300mm .... isolated surfaces, girth not exceeding 300mm	0.87 0.26	10.00 10.00	0.96 0.29	0.20 0.07	- -	1.85 0.65	0.03 0.01	2.83 0.94	m2 m	3.19 1.06
Concrete general surfaces girth exceeding 300mm	0.96	10.00	1.06	0.20	-	1.85	0.03	2.93	m2	3.30

Labour hourly rates: (except Specialists) Craft Operatives £9.23 Labourer £7.02 Rates are national average prices. Refer to REGIONAL VARIATIONS for indicative levels of overall pricing in regions	MATERIALS			LABOUR				RATES		
	Del to Site £	Waste %	Material Cost £	Craft Optve Hrs	Lab Hrs	Labour Cost £	Sunds £	Nett Rate £	Unit	Gross Rate (+12.5%) £
M60: PAINTING/CLEAR FINISHING – CEMENT PAINTING Cont.										
One coat sealer, one coat Snowcem; external work Cont.										
Brickwork general surfaces girth exceeding 300mm	0.96	10.00	1.06	0.23	–	2.12	0.04	3.22	m2	3.62
Rough cast general surfaces girth exceeding 300mm	1.02	10.00	1.12	0.30	–	2.77	0.05	3.94	m2	4.43
One coat sealer, two coats Snowcem; external work										

Cement rendered general surfaces
    girth exceeding 300mm ....
    isolated surfaces, girth not exceeding 300mm .....
	1.10
	0.33
		10.00
		10.00
			1.21
			0.36
				0.37
				0.12
					–
					–
						3.42
						1.11
							0.06
							0.02
								4.69
								1.49
									m2
									m
										5.27
										1.68

Concrete general surfaces girth exceeding 300mm	1.28	10.00	1.41	0.37	–	3.42	0.06	4.88	m2	5.49
Brickwork general surfaces girth exceeding 300mm	1.28	10.00	1.41	0.43	–	3.97	0.08	5.46	m2	6.14
Rough cast general surfaces girth exceeding 300mm	1.37	10.00	1.51	0.56	–	5.17	0.10	6.78	m2	7.62
One coat textured masonry paint; external work										

Cement rendered general surfaces
    girth exceeding 300mm ....
    isolated surfaces, girth not exceeding 300mm .....
	0.61
	0.18
		10.00
		10.00
			0.67
			0.20
				0.12
				0.04
					–
					–
						1.11
						0.37
							0.02
							0.01
								1.80
								0.58
									m2
									m
										2.02
										0.65

Concrete general surfaces girth exceeding 300mm	0.69	10.00	0.76	0.12	–	1.11	0.02	1.89	m2	2.12
Brickwork general surfaces girth exceeding 300mm	0.76	10.00	0.84	0.14	–	1.29	0.02	2.15	m2	2.42
Rough cast general surfaces girth exceeding 300mm	1.23	10.00	1.35	0.19	–	1.75	0.03	3.14	m2	3.53
Two coats textured masonry paint; external work										

Cement rendered general surfaces
    girth exceeding 300mm ....
    isolated surfaces, girth not exceeding 300mm .....
	1.11
	0.33
		10.00
		10.00
			1.22
			0.36
				0.24
				0.09
					–
					–
						2.22
						0.83
							0.04
							0.02
								3.48
								1.21
									m2
									m
										3.91
										1.37

Concrete general surfaces girth exceeding 300mm	1.23	10.00	1.35	0.24	–	2.22	0.04	3.61	m2	4.06
Brickwork general surfaces girth exceeding 300mm	1.38	10.00	1.52	0.26	–	2.40	0.05	3.97	m2	4.46
Rough cast general surfaces girth exceeding 300mm	2.21	10.00	2.43	0.36	–	3.32	0.06	5.81	m2	6.54
One coat stabilising solution, one coat textured masonry paint; external work										

Cement rendered general surfaces
    girth exceeding 300mm ....
    isolated surfaces, girth not exceeding 300mm .....
	1.14
	0.34
		10.00
		10.00
			1.25
			0.37
				0.20
				0.07
					–
					–
						1.85
						0.65
							0.03
							0.01
								3.13
								1.03
									m2
									m
										3.52
										1.16

Concrete general surfaces girth exceeding 300mm	1.28	10.00	1.41	0.20	–	1.85	0.03	3.28	m2	3.69
Brickwork general surfaces girth exceeding 300mm	1.41	10.00	1.55	0.23	–	2.12	0.04	3.71	m2	4.18
Rough cast general surfaces girth exceeding 300mm	1.91	10.00	2.10	0.30	–	2.77	0.05	4.92	m2	5.54
One coat stabilising solution, two coats textured masonry paint; external work										

Cement rendered general surfaces
    girth exceeding 300mm ....
    isolated surfaces, girth not exceeding 300mm .....
	1.63
	0.49
		10.00
		10.00
			1.79
			0.54
				0.37
				0.12
					–
					–
						3.42
						1.11
							0.06
							0.02
								5.27
								1.67
									m2
									m
										5.93
										1.87

Concrete general surfaces girth exceeding 300mm	1.82	10.00	2.00	0.37	–	3.42	0.06	5.48	m2	6.16
Brickwork general surfaces girth exceeding 300mm	2.03	10.00	2.23	0.43	–	3.97	0.08	6.28	m2	7.07
Rough cast general surfaces girth exceeding 300mm	2.90	10.00	3.19	0.56	–	5.17	0.10	8.46	m2	9.52
M60: PAINTING/CLEAR FINISHING – PRESERVATIVE TREATMENT										
One coat creosote B.S.144; external work (sawn timber)										
Wood general surfaces girth exceeding 300mm	0.08	10.00	0.09	0.11	–	1.02	0.02	1.12	m2	1.26

SURFACE FINISHES

Labour hourly rates: (except Specialists) Craft Operatives £9.23 Labourer £7.02 Rates are national average prices. Refer to REGIONAL VARIATIONS for indicative levels of overall pricing in regions	MATERIALS			LABOUR				RATES		
	Del to Site	Waste	Material Cost	Craft Optve	Lab	Labour Cost	Sunds	Nett Rate	Unit	Gross Rate (+12.5%)
	£	%	£	Hrs	Hrs	£	£	£		£
M60: PAINTING/CLEAR FINISHING - PRESERVATIVE TREATMENT Cont.										
One coat creosote B.S.144; external work (sawn timber) Cont.										
Wood general surfaces Cont.										
isolated surfaces, girth not exceeding 300mm	0.02	10.00	0.02	0.04	-	0.37	0.01	0.40	m	0.45
Two coats creosote B.S.144; external work (sawn timber)										
Wood general surfaces										
girth exceeding 300mm	0.16	10.00	0.18	0.22	-	2.03	0.04	2.25	m2	2.53
isolated surfaces, girth not exceeding 300mm	0.05	10.00	0.06	0.08	-	0.74	0.01	0.80	m	0.90
One coat wood preservative, internal work (sawn timber)										
Wood general surfaces										
girth exceeding 300mm	0.43	10.00	0.47	0.12	-	1.11	0.02	1.60	m2	1.80
isolated surfaces, girth not exceeding 300mm	0.13	10.00	0.14	0.04	-	0.37	0.01	0.52	m	0.59
One coat wood preservative, internal work (wrought timber)										
Wood general surfaces										
girth exceeding 300mm	0.39	10.00	0.43	0.10	-	0.92	0.02	1.37	m2	1.54
isolated surfaces, girth not exceeding 300mm	0.12	10.00	0.13	0.03	-	0.28	0.01	0.42	m	0.47
One coat wood preservative, external work (sawn timber)										
Wood general surfaces										
girth exceeding 300mm	0.92	10.00	1.01	0.11	-	1.02	0.02	2.05	m2	2.30
isolated surfaces, girth not exceeding 300mm	0.27	10.00	0.30	0.01	-	0.09	0.01	0.40	m	0.45
One coat wood preservative, external work (wrought timber)										
Wood general surfaces										
girth exceeding 300mm	0.72	10.00	0.79	0.09	-	0.83	0.02	1.64	m2	1.85
isolated surfaces, girth not exceeding 300mm	0.22	10.00	0.24	0.02	-	0.18	-	0.43	m	0.48
Two coats wood preservative, internal work (sawn timber)										
Wood general surfaces										
girth exceeding 300mm	0.60	10.00	0.66	0.24	-	2.22	0.04	2.92	m2	3.28
isolated surfaces, girth not exceeding 300mm	0.18	10.00	0.20	0.09	-	0.83	0.02	1.05	m	1.18
Two coats wood preservative, internal work (wrought timber)										
Wood general surfaces										
girth exceeding 300mm	0.48	10.00	0.53	0.22	-	2.03	0.04	2.60	m2	2.92
isolated surfaces, girth not exceeding 300mm	0.14	10.00	0.15	0.06	-	0.55	0.01	0.72	m	0.81
Two coats wood preservative, external work (sawn timber)										
Wood general surfaces										
girth exceeding 300mm	1.20	10.00	1.32	0.22	-	2.03	0.04	3.39	m2	3.81
isolated surfaces, girth not exceeding 300mm	0.36	10.00	0.40	0.08	-	0.74	0.01	1.14	m	1.29
Two coats wood preservative, external work (wrought timber)										
Wood general surfaces										
girth exceeding 300mm	0.96	10.00	1.06	0.18	-	1.66	0.03	2.75	m2	3.09
isolated surfaces, girth not exceeding 300mm	0.29	10.00	0.32	0.05	-	0.46	0.01	0.79	m	0.89
Two coats Sadolins Classic; two coats Sadolins Holdex										
Wood general surfaces										
girth exceeding 300mm	2.90	10.00	3.19	0.60	-	5.54	0.09	8.82	m2	9.92
isolated surfaces, girth not exceeding 300mm	0.87	10.00	0.96	0.22	-	2.03	0.03	3.02	m	3.39
Wood glazed doors										
girth exceeding 300mm; panes, area not exceeding 0.10m2	1.46	10.00	1.61	0.98	-	9.05	0.15	10.80	m2	12.15
girth exceeding 300mm; panes, area 0.10 - 0.50m2 .	1.02	10.00	1.12	0.71	-	6.55	0.11	7.79	m2	8.76
girth exceeding 300mm; panes, area 0.50 - 1.00m2 .	0.72	10.00	0.79	0.60	-	5.54	0.09	6.42	m2	7.22
girth exceeding 300mm; panes, area exceeding 1.00m2	0.59	10.00	0.65	0.52	-	4.80	0.08	5.53	m2	6.22
Wood partially glazed doors										
girth exceeding 300mm; panes, area 0.50 - 1.00m2 .	1.75	10.00	1.93	0.60	-	5.54	0.09	7.55	m2	8.50
girth exceeding 300mm; panes, area exceeding 1.00m2	1.59	10.00	1.75	0.58	-	5.35	0.09	7.19	m2	8.09
Wood windows and screens										
girth exceeding 300mm; panes, area not exceeding 0.10m2	1.59	10.00	1.75	1.09	-	10.06	0.16	11.97	m2	13.47
girth exceeding 300mm; panes, area 0.10 - 0.50m2 .	1.16	10.00	1.28	0.78	-	7.20	0.12	8.60	m2	9.67
girth exceeding 300mm; panes, area 0.50 - 1.00m2 .	0.87	10.00	0.96	0.66	-	6.09	0.10	7.15	m2	8.04

Labour hourly rates: (except Specialists) Craft Operatives £9.23 Labourer £7.02 Rates are national average prices. Refer to REGIONAL VARIATIONS for indicative levels of overall pricing in regions	MATERIALS			LABOUR				RATES		
	Del to Site £	Waste %	Material Cost £	Craft Optve Hrs	Lab Hrs	Labour Cost £	Sunds £	Nett Rate £	Unit	Gross Rate (+12.5%) £

M60: PAINTING/CLEAR FINISHING - PRESERVATIVE TREATMENT Cont.

Two coats Sadolins Classic; two coats Sadolins Holdex Cont.

Wood windows and screens Cont.
    girth exceeding 300mm; panes, area exceeding 1.00m2 ......

	Del to Site	Waste	Material Cost	Craft Optve	Lab	Labour Cost	Sunds	Nett Rate	Unit	Gross Rate
Wood windows and screens Cont. girth exceeding 300mm; panes, area exceeding 1.00m2	0.72	10.00	0.79	0.58	-	5.35	0.09	6.24	m2	7.01

One coat Sadolins Classic; two coats Sadolins Prestige; external work

	Del to Site	Waste	Material Cost	Craft Optve	Lab	Labour Cost	Sunds	Nett Rate	Unit	Gross Rate
Wood general surfaces										
girth exceeding 300mm	2.01	10.00	2.21	0.55	-	5.08	0.08	7.37	m2	8.29
isolated surfaces, girth not exceeding 300mm	0.60	10.00	0.66	0.19	-	1.75	0.03	2.44	m	2.75
Wood glazed doors										
girth exceeding 300mm; panes, area not exceeding 0.10m2	1.00	10.00	1.10	0.86	-	7.94	0.13	9.17	m2	10.31
girth exceeding 300mm; panes, area 0.10 - 0.50m2	0.70	10.00	0.77	0.62	-	5.72	0.09	6.58	m2	7.41
girth exceeding 300mm; panes, area 0.50 - 1.00m2	0.50	10.00	0.55	0.52	-	4.80	0.08	5.43	m2	6.11
girth exceeding 300mm; panes, area exceeding 1.00m2	0.40	10.00	0.44	0.46	-	4.25	0.07	4.76	m2	5.35
Wood partially glazed doors										
girth exceeding 300mm; panes, area 0.50 - 1.00m2	1.21	10.00	1.33	0.52	-	4.80	0.08	6.21	m2	6.99
girth exceeding 300mm; panes, area exceeding 1.00m2	1.11	10.00	1.22	0.50	-	4.62	0.08	5.92	m2	6.66
Wood windows and screens										
girth exceeding 300mm; panes, area not exceeding 0.10m2	1.11	10.00	1.22	0.98	-	9.05	0.15	10.42	m2	11.72
girth exceeding 300mm; panes, area 0.10 - 0.50m2	0.81	10.00	0.89	0.72	-	6.65	0.11	7.65	m2	8.60
girth exceeding 300mm; panes, area 0.50 - 1.00m2	0.60	10.00	0.66	0.61	-	5.63	0.09	6.38	m2	7.18
girth exceeding 300mm; panes, area exceeding 1.00m2	0.50	10.00	0.55	0.54	-	4.98	0.08	5.61	m2	6.32
Wood railings fences and gates; open type										
girth exceeding 300mm	2.01	10.00	2.21	0.55	-	5.08	0.08	7.37	m2	8.29
isolated surfaces, girth not exceeding 300mm	0.60	10.00	0.66	0.19	-	1.75	0.03	2.44	m	2.75
Wood railings fences and gates; close type										
girth exceeding 300mm	2.01	10.00	2.21	0.55	-	5.08	0.08	7.37	m2	8.29

M60: PAINTING/CLEAR FINISHING - OIL PAINTING WALLS AND CEILINGS

One coat primer, one undercoat, one coat full gloss finish

	Del to Site	Waste	Material Cost	Craft Optve	Lab	Labour Cost	Sunds	Nett Rate	Unit	Gross Rate
Concrete general surfaces										
girth exceeding 300mm	0.95	10.00	1.04	0.33	-	3.05	0.06	4.15	m2	4.67
Concrete general surfaces 3.50 - 5.00m above floor										
girth exceeding 300mm	0.95	10.00	1.04	0.36	-	3.32	0.06	4.43	m2	4.98
Brickwork general surfaces										
girth exceeding 300mm	1.17	10.00	1.29	0.37	-	3.42	0.06	4.76	m2	5.36
Plasterboard general surfaces										
girth exceeding 300mm	0.74	10.00	0.81	0.30	-	2.77	0.05	3.63	m2	4.09
Plasterboard general surfaces 3.50 - 5.00m above floor										
girth exceeding 300mm	0.74	10.00	0.81	0.33	-	3.05	0.06	3.92	m2	4.41
Plaster general surfaces										
girth exceeding 300mm	0.85	10.00	0.94	0.30	-	2.77	0.05	3.75	m2	4.22
isolated surfaces, girth not exceeding 300mm	0.25	10.00	0.28	0.13	-	1.20	0.02	1.49	m	1.68
Plaster general surfaces 3.50 - 5.00m above floor										
girth exceeding 300mm	0.85	10.00	0.94	0.33	-	3.05	0.06	4.04	m2	4.55

One coat primer, two undercoats, one coat full gloss finish

	Del to Site	Waste	Material Cost	Craft Optve	Lab	Labour Cost	Sunds	Nett Rate	Unit	Gross Rate
Concrete general surfaces										
girth exceeding 300mm	1.25	10.00	1.38	0.44	-	4.06	0.08	5.52	m2	6.21
Concrete general surfaces 3.50 - 5.00m above floor										
girth exceeding 300mm	1.25	10.00	1.38	0.48	-	4.43	0.08	5.89	m2	6.62
Brickwork general surfaces										
girth exceeding 300mm	1.53	10.00	1.68	0.50	-	4.62	0.09	6.39	m2	7.19
Plasterboard general surfaces										
girth exceeding 300mm	0.97	10.00	1.07	0.40	-	3.69	0.07	4.83	m2	5.43
Plasterboard general surfaces 3.50 - 5.00m above floor										
girth exceeding 300mm	0.97	10.00	1.07	0.44	-	4.06	0.08	5.21	m2	5.86
Plaster general surfaces										
girth exceeding 300mm	1.11	10.00	1.22	0.40	-	3.69	0.07	4.98	m2	5.61
isolated surfaces, girth not exceeding 300mm	0.33	10.00	0.36	0.18	-	1.66	0.03	2.05	m	2.31

SURFACE FINISHES

Labour hourly rates: (except Specialists) Craft Operatives £9.23 Labourer £7.02 Rates are national average prices. Refer to REGIONAL VARIATIONS for indicative levels of overall pricing in regions	MATERIALS			LABOUR				RATES		
	Del to Site	Waste	Material Cost	Craft Optve	Lab	Labour Cost	Sunds	Nett Rate	Unit	Gross Rate (+12.5%)
	£	%	£	Hrs	Hrs	£	£	£		£
M60: PAINTING/CLEAR FINISHING - OIL PAINTING WALLS AND CEILINGS Cont.										
One coat primer, two undercoats, one coat full gloss finish Cont.										
Plaster general surfaces 3.50 - 5.00m above floor										
girth exceeding 300mm	1.11	10.00	1.22	0.44	-	4.06	0.08	5.36	m2	6.03
One coat primer, one undercoat, one coat eggshell finish										
Concrete general surfaces										
girth exceeding 300mm	0.99	10.00	1.09	0.33	-	3.05	0.06	4.19	m2	4.72
Concrete general surfaces 3.50 - 5.00m above floor										
girth exceeding 300mm	0.99	10.00	1.09	0.36	-	3.32	0.06	4.47	m2	5.03
Brickwork general surfaces										
girth exceeding 300mm	1.21	10.00	1.33	0.37	-	3.42	0.06	4.81	m2	5.41
Plasterboard general surfaces										
girth exceeding 300mm	0.77	10.00	0.85	0.30	-	2.77	0.05	3.67	m2	4.12
Plasterboard general surfaces 3.50 - 5.00m above floor										
girth exceeding 300mm	0.77	10.00	0.85	0.33	-	3.05	0.06	3.95	m2	4.45
Plaster general surfaces										
girth exceeding 300mm	0.88	10.00	0.97	0.30	-	2.77	0.05	3.79	m2	4.26
isolated surfaces, girth not exceeding 300mm	0.26	10.00	0.29	0.13	-	1.20	0.02	1.51	m	1.69
Plaster general surfaces 3.50 - 5.00m above floor										
girth exceeding 300mm	0.88	10.00	0.97	0.33	-	3.05	0.06	4.07	m2	4.58
One coat primer, two undercoats, one coat eggshell finish										
Concrete general surfaces										
girth exceeding 300mm	1.28	10.00	1.41	0.44	-	4.06	0.08	5.55	m2	6.24
Concrete general surfaces 3.50 - 5.00m above floor										
girth exceeding 300mm	1.28	10.00	1.41	0.48	-	4.43	0.08	5.92	m2	6.66
Brickwork general surfaces										
girth exceeding 300mm	1.57	10.00	1.73	0.50	-	4.62	0.09	6.43	m2	7.24
Plasterboard general surfaces										
girth exceeding 300mm	1.00	10.00	1.10	0.40	-	3.69	0.07	4.86	m2	5.47
Plasterboard general surfaces 3.50 - 5.00m above floor										
girth exceeding 300mm	1.00	10.00	1.10	0.44	-	4.06	0.08	5.24	m2	5.90
Plaster general surfaces										
girth exceeding 300mm	1.14	10.00	1.25	0.40	-	3.69	0.07	5.02	m2	5.64
isolated surfaces, girth not exceeding 300mm	0.34	10.00	0.37	0.18	-	1.66	0.03	2.07	m	2.32
Plaster general surfaces 3.50 - 5.00m above floor										
girth exceeding 300mm	1.14	10.00	1.25	0.44	-	4.06	0.08	5.40	m2	6.07
M60: PAINTING/CLEAR FINISHING - SPRAY PAINTING										
Spray one coat primer, one undercoat, one coat full gloss finish										
Concrete general surfaces										
girth exceeding 300mm	1.10	15.00	1.26	0.29	-	2.68	0.05	3.99	m2	4.49
Brickwork general surfaces										
girth exceeding 300mm	1.32	15.00	1.52	0.33	-	3.05	0.06	4.62	m2	5.20
Plaster general surfaces										
girth exceeding 300mm	0.99	15.00	1.14	0.26	-	2.40	0.05	3.59	m2	4.04
Spray one coat primer, two undercoats, one coat full gloss finish										
Concrete general surfaces										
girth exceeding 300mm	1.42	15.00	1.63	0.34	-	3.14	0.06	4.83	m2	5.44
Brickwork general surfaces										
girth exceeding 300mm	1.71	15.00	1.97	0.39	-	3.60	0.07	5.64	m2	6.34
Plaster general surfaces										
girth exceeding 300mm	1.28	15.00	1.47	0.31	-	2.86	0.05	4.38	m2	4.93
Spray one coat primer, one basecoat, one coat multicolour finish										
Concrete general surfaces										
girth exceeding 300mm	1.25	15.00	1.44	0.43	-	3.97	0.08	5.49	m2	6.17
Brickwork general surfaces										
girth exceeding 300mm	1.51	15.00	1.74	0.45	-	4.15	0.08	5.97	m2	6.72

Labour hourly rates: (except Specialists) Craft Operatives £9.23 Labourer £7.02 Rates are national average prices. Refer to REGIONAL VARIATIONS for indicative levels of overall pricing in regions	MATERIALS			LABOUR				RATES		
	Del to Site £	Waste %	Material Cost £	Craft Optve Hrs	Lab Hrs	Labour Cost £	Sunds £	Nett Rate £	Unit	Gross Rate (+12.5%)
M60: PAINTING/CLEAR FINISHING - SPRAY PAINTING Cont.										
Spray one coat primer, one basecoat, one coat multicolour finish Cont.										
Plaster general surfaces girth exceeding 300mm	1.13	15.00	1.30	0.41	-	3.78	0.07	5.15	m2	5.80
Spray one coat primer, one basecoat, one coat multicolour finish, one coat glaze										
Concrete general surfaces girth exceeding 300mm	1.54	15.00	1.77	0.47	-	4.34	0.08	6.19	m2	6.96
Brickwork general surfaces girth exceeding 300mm	1.85	15.00	2.13	0.50	-	4.62	0.09	6.83	m2	7.69
Plaster general surfaces girth exceeding 300mm	1.39	15.00	1.60	0.45	-	4.15	0.08	5.83	m2	6.56
M60: PAINTING/CLEAR FINISHING - CHLORINATED RUBBER PAINTING										
One coat primer, two coats chlorinated rubber paint										
Concrete general surfaces girth exceeding 300mm	4.39	10.00	4.83	0.33	-	3.05	0.06	7.93	m2	8.93
Brickwork general surfaces girth exceeding 300mm	5.31	10.00	5.84	0.35	-	3.23	0.06	9.13	m2	10.27
Plasterboard general surfaces girth exceeding 300mm	3.59	10.00	3.95	0.32	-	2.95	0.06	6.96	m2	7.83
Plaster general surfaces girth exceeding 300mm	4.39	10.00	4.83	0.32	-	2.95	0.06	7.84	m2	8.82
isolated surfaces, girth not exceeding 300mm	1.32	10.00	1.45	0.14	-	1.29	0.02	2.76	m	3.11
M60: PAINTING/CLEAR FINISHING - PLASTIC FINISH										
Textured plastic coating - Stippled finish										
Concrete general surfaces girth exceeding 300mm	0.53	10.00	0.58	0.28	-	2.58	0.05	3.22	m2	3.62
Concrete general surfaces 3.50 - 5.00m above floor girth exceeding 300mm	0.53	10.00	0.58	0.31	-	2.86	0.05	3.49	m2	3.93
Brickwork general surfaces girth exceeding 300mm	0.64	10.00	0.70	0.33	-	3.05	0.06	3.81	m2	4.29
Plasterboard general surfaces girth exceeding 300mm	0.44	10.00	0.48	0.22	-	2.03	0.04	2.55	m2	2.87
Plasterboard general surfaces 3.50 - 5.00m above floor girth exceeding 300mm	0.44	10.00	0.48	0.24	-	2.22	0.04	2.74	m2	3.08
Plaster general surfaces girth exceeding 300mm	0.44	10.00	0.48	0.22	-	2.03	0.04	2.55	m2	2.87
Plaster general surfaces 3.50 - 5.00m above floor girth exceeding 300mm	0.44	10.00	0.48	0.24	-	2.22	0.04	2.74	m2	3.08
Textured plastic coating - Combed Finish										
Concrete general surfaces girth exceeding 300mm	0.57	10.00	0.63	0.33	-	3.05	0.06	3.73	m2	4.20
Concrete general surfaces 3.50 - 5.00m above floor girth exceeding 300mm	0.57	10.00	0.63	0.36	-	3.32	0.06	4.01	m2	4.51
Brickwork general surfaces girth exceeding 300mm	0.69	10.00	0.76	0.39	-	3.60	0.07	4.43	m2	4.98
Plasterboard general surfaces girth exceeding 300mm	0.46	10.00	0.51	0.28	-	2.58	0.05	3.14	m2	3.53
Plasterboard general surfaces 3.50 - 5.00m above floor girth exceeding 300mm	0.46	10.00	0.51	0.30	-	2.77	0.05	3.33	m2	3.74
Plaster general surfaces girth exceeding 300mm	0.46	10.00	0.51	0.28	-	2.58	0.05	3.14	m2	3.53
Plaster general surfaces 3.50 - 5.00m above floor girth exceeding 300mm	0.46	10.00	0.51	0.30	-	2.77	0.05	3.33	m2	3.74
M60: PAINTING/CLEAR FINISHING - OIL PAINTING METALWORK										
One undercoat, one coat full gloss finish on ready primed metal surfaces										
Iron or steel structural work girth exceeding 300mm	0.52	10.00	0.57	0.31	-	2.86	0.05	3.48	m2	3.92

SURFACE FINISHES

Labour hourly rates: (except Specialists) Craft Operatives £9.23 Labourer £7.02 Rates are national average prices. Refer to REGIONAL VARIATIONS for indicative levels of overall pricing in regions	MATERIALS			LABOUR				RATES		
	Del to Site £	Waste %	Material Cost £	Craft Optve Hrs	Lab Hrs	Labour Cost £	Sunds £	Nett Rate £	Unit	Gross Rate (+12.5%) £
M60: PAINTING/CLEAR FINISHING - OIL PAINTING - METALWORK Cont.										
One undercoat, one coat full gloss finish on ready primed metal surfaces Cont.										
Iron or steel structural work Cont.										
isolated surfaces, girth not exceeding 300mm	0.16	10.00	0.18	0.13	-	1.20	0.02	1.40	m	1.57
Iron or steel structural members of roof trusses, lattice girders, purlins and the like										
girth exceeding 300mm	0.52	10.00	0.57	0.40	-	3.69	0.07	4.33	m2	4.88
isolated surfaces, girth not exceeding 300mm	0.16	10.00	0.18	0.13	-	1.20	0.02	1.40	m	1.57
Two undercoats, one coat full gloss finish on ready primed metal surfaces										
Iron or steel structural work										
girth exceeding 300mm	0.78	10.00	0.86	0.46	-	4.25	0.08	5.18	m2	5.83
isolated surfaces, girth not exceeding 300mm	0.24	10.00	0.26	0.15	-	1.38	0.03	1.68	m	1.89
Iron or steel structural members of roof trusses, lattice girders, purlins and the like										
girth exceeding 300mm	0.78	10.00	0.86	0.59	-	5.45	0.10	6.40	m2	7.20
isolated surfaces, girth not exceeding 300mm	0.24	10.00	0.26	0.20	-	1.85	0.03	2.14	m	2.41
One coat primer, one undercoat, one coat full gloss finish on metal surfaces										
Iron or steel general surfaces										
girth exceeding 300mm	0.83	10.00	0.91	0.43	-	3.97	0.08	4.96	m2	5.58
isolated surfaces, girth not exceeding 300mm	0.25	10.00	0.28	0.14	-	1.29	0.02	1.59	m	1.79
Galvanised glazed doors, windows or screens										
girth exceeding 300mm; panes, area not exceeding 0.10m2	0.23	10.00	0.25	0.79	-	7.29	0.14	7.68	m2	8.65
girth exceeding 300mm; panes, area 0.10 - 0.50m2	0.19	10.00	0.21	0.56	-	5.17	0.10	5.48	m2	6.16
girth exceeding 300mm; panes, area 0.50 - 1.00m2 .	0.17	10.00	0.19	0.47	-	4.34	0.08	4.61	m2	5.18
girth exceeding 300mm; panes, area exceeding 1.00m2	0.15	10.00	0.17	0.43	-	3.97	0.08	4.21	m2	4.74
Iron or steel structural work										
girth exceeding 300mm	0.83	10.00	0.91	0.46	-	4.25	0.08	5.24	m2	5.89
isolated surfaces, girth not exceeding 300mm	0.25	10.00	0.28	0.20	-	1.85	0.03	2.15	m	2.42
Iron or steel structural members of roof trusses, lattice girders, purlins and the like										
girth exceeding 300mm	0.83	10.00	0.91	0.59	-	5.45	0.10	6.46	m2	7.27
isolated surfaces, girth not exceeding 300mm	0.25	10.00	0.28	0.20	-	1.85	0.03	2.15	m	2.42
Iron or steel services										
girth exceeding 300mm	0.83	10.00	0.91	0.59	-	5.45	0.10	6.46	m2	7.27
isolated surfaces, girth not exceeding 300mm	0.25	10.00	0.28	0.20	-	1.85	0.03	2.15	m	2.42
isolated areas not exceeding 0.50m2 irrespective of girth	0.42	10.00	0.46	0.59	-	5.45	0.10	6.01	nr	6.76
Copper services										
girth exceeding 300mm	0.83	10.00	0.91	0.59	-	5.45	0.10	6.46	m2	7.27
isolated surfaces, girth not exceeding 300mm	0.25	10.00	0.28	0.20	-	1.85	0.03	2.15	m	2.42
Galvanised services										
girth exceeding 300mm	0.83	10.00	0.91	0.59	-	5.45	0.10	6.46	m2	7.27
isolated surfaces, girth not exceeding 300mm	0.25	10.00	0.28	0.20	-	1.85	0.03	2.15	m	2.42
One coat primer, one undercoat, one coat full gloss finish on metal surfaces ; external work										
Iron or steel general surfaces										
girth exceeding 300mm	0.83	10.00	0.91	0.46	-	4.25	0.08	5.24	m2	5.89
isolated surfaces, girth not exceeding 300mm	0.25	10.00	0.28	0.18	-	1.66	0.03	1.97	m	2.21
Galvanised glazed doors, windows or screens										
girth exceeding 300mm; panes, area not exceeding 0.10m2	0.23	10.00	0.25	0.83	-	7.66	0.15	8.06	m2	9.07
girth exceeding 300mm; panes, area 0.10 - 0.50m2 .	0.19	10.00	0.21	0.59	-	5.45	0.10	5.75	m2	6.47
girth exceeding 300mm; panes, area 0.50 - 1.00m2 .	0.17	10.00	0.19	0.51	-	4.71	0.09	4.98	m2	5.61
girth exceeding 300mm; panes, area exceeding 1.00m2	0.15	10.00	0.17	0.46	-	4.25	0.08	4.49	m2	5.05
Iron or steel structural work										
girth exceeding 300mm	0.83	10.00	0.91	0.46	-	4.25	0.08	5.24	m2	5.89
isolated surfaces, girth not exceeding 300mm	0.25	10.00	0.28	0.20	-	1.85	0.03	2.15	m	2.42
Iron or steel structural members of roof trusses, lattice girders, purlins and the like										
girth exceeding 300mm	0.83	10.00	0.91	0.59	-	5.45	0.10	6.46	m2	7.27
isolated surfaces, girth not exceeding 300mm	0.25	10.00	0.28	0.20	-	1.85	0.03	2.15	m	2.42
Iron or steel railings, fences and gates; plain open type										
girth exceeding 300mm	0.83	10.00	0.91	0.43	-	3.97	0.08	4.96	m2	5.58
isolated surfaces, girth not exceeding 300mm	0.25	10.00	0.28	0.14	-	1.29	0.02	1.59	m	1.79
Iron or steel railings, fences and gates; close type										
girth exceeding 300mm	0.83	10.00	0.91	0.36	-	3.32	0.06	4.30	m2	4.83

Labour hourly rates: (except Specialists) Craft Operatives £9.23 Labourer £7.02 Rates are national average prices. Refer to REGIONAL VARIATIONS for indicative levels of overall pricing in regions	MATERIALS			LABOUR				RATES		
	Del to Site	Waste	Material Cost	Craft Optve	Lab	Labour Cost	Sunds	Nett Rate	Unit	Gross Rate (+12.5%)
	£	%	£	Hrs	Hrs	£	£	£		£
M60: PAINTING/CLEAR FINISHING - OIL PAINTING **METALWORK Cont.**										
One coat primer, one undercoat, one coat full gloss **finish on metal surfaces ; external work Cont.**										
Iron or steel railings, fences and gates; ornamental type										
girth exceeding 300mm	0.83	10.00	0.91	0.73	–	6.74	0.13	7.78	m2	8.75
Iron or steel eaves gutters										
girth exceeding 300mm	0.83	10.00	0.91	0.50	–	4.62	0.09	5.62	m2	6.32
isolated surfaces, girth not exceeding 300mm	0.25	10.00	0.28	0.17	–	1.57	0.03	1.87	m	2.11
Galvanised eaves gutters										
girth exceeding 300mm	0.83	10.00	0.91	0.50	–	4.62	0.09	5.62	m2	6.32
isolated surfaces, girth not exceeding 300mm	0.25	10.00	0.28	0.17	–	1.57	0.03	1.87	m	2.11
Iron or steel services										
girth exceeding 300mm	0.83	10.00	0.91	0.59	–	5.45	0.10	6.46	m2	7.27
isolated surfaces, girth not exceeding 300mm	0.25	10.00	0.28	0.20	–	1.85	0.03	2.15	m	2.42
isolated areas not exceeding 0.50m2 irrespective of girth	0.42	10.00	0.46	0.59	–	5.45	0.10	6.01	nr	6.76
Copper services										
girth exceeding 300mm	0.83	10.00	0.91	0.59	–	5.45	0.10	6.46	m2	7.27
isolated surfaces, girth not exceeding 300mm	0.25	10.00	0.28	0.20	–	1.85	0.03	2.15	m	2.42
Galvanised services										
girth exceeding 300mm	0.83	10.00	0.91	0.59	–	5.45	0.10	6.46	m2	7.27
isolated surfaces, girth not exceeding 300mm	0.25	10.00	0.28	0.20	–	1.85	0.03	2.15	m	2.42
One coat primer, two undercoats, one coat full gloss **finish on metal surfaces**										
Iron or steel general surfaces										
girth exceeding 300mm	1.10	10.00	1.21	0.57	–	5.26	0.10	6.57	m2	7.39
isolated surfaces, girth not exceeding 300mm	0.33	10.00	0.36	0.19	–	1.75	0.03	2.15	m	2.42
Galvanised glazed doors, windows or screens										
girth exceeding 300mm; panes, area not exceeding 0.10m2	0.30	10.00	0.33	1.06	–	9.78	0.19	10.30	m2	11.59
girth exceeding 300mm; panes, area 0.10 - 0.50m2 .	0.25	10.00	0.28	0.75	–	6.92	0.13	7.33	m2	8.24
girth exceeding 300mm; panes, area 0.50 - 1.00m2 .	0.22	10.00	0.24	0.63	–	5.81	0.11	6.17	m2	6.94
girth exceeding 300mm; panes, area exceeding 1.00m2	0.20	10.00	0.22	0.57	–	5.26	0.10	5.58	m2	6.28
Iron or steel structural work										
girth exceeding 300mm	1.10	10.00	1.21	0.62	–	5.72	0.11	7.04	m2	7.92
isolated surfaces, girth not exceeding 300mm	0.33	10.00	0.36	0.24	–	2.22	0.04	2.62	m	2.95
Iron or steel structural members of roof trusses, lattice girders, purlins and the like										
girth exceeding 300mm	1.10	10.00	1.21	0.79	–	7.29	0.14	8.64	m2	9.72
isolated surfaces, girth not exceeding 300mm	0.33	10.00	0.36	0.24	–	2.22	0.04	2.62	m	2.95
Iron or steel services										
girth exceeding 300mm	1.10	10.00	1.21	0.69	–	6.37	0.12	7.70	m2	8.66
isolated surfaces, girth not exceeding 300mm	0.33	10.00	0.36	0.23	–	2.12	0.04	2.53	m	2.84
isolated areas not exceeding 0.50m2 irrespective of girth	0.55	10.00	0.60	0.69	–	6.37	0.12	7.09	nr	7.98
Copper services										
girth exceeding 300mm	1.10	10.00	1.21	0.69	–	6.37	0.12	7.70	m2	8.66
isolated surfaces, girth not exceeding 300mm	0.33	10.00	0.36	0.23	–	2.12	0.04	2.53	m	2.84
Galvanised services										
girth exceeding 300mm	1.10	10.00	1.21	0.69	–	6.37	0.12	7.70	m2	8.66
isolated surfaces, girth not exceeding 300mm	0.33	10.00	0.36	0.23	–	2.12	0.04	2.53	m	2.84
One coat primer, two undercoats, one coat full gloss **finish on metal surfaces; external work**										
Iron or steel general surfaces										
girth exceeding 300mm	1.10	10.00	1.21	0.62	–	5.72	0.11	7.04	m2	7.92
isolated surfaces, girth not exceeding 300mm	0.33	10.00	0.36	0.21	–	1.94	0.04	2.34	m	2.63
Galvanised glazed doors, windows or screens										
girth exceeding 300mm; panes, area not exceeding 0.10m2	0.30	10.00	0.33	1.10	–	10.15	0.19	10.67	m2	12.01
girth exceeding 300mm; panes, area 0.10 - 0.50m2 .	0.25	10.00	0.28	0.79	–	7.29	0.14	7.71	m2	8.67
girth exceeding 300mm; panes, area 0.50 - 1.00m2 .	0.22	10.00	0.24	0.67	–	6.18	0.12	6.55	m2	7.36
girth exceeding 300mm; panes, area exceeding 1.00m2	0.20	10.00	0.22	0.62	–	5.72	0.11	6.05	m2	6.81
Iron or steel structural work										
girth exceeding 300mm	1.10	10.00	1.21	0.62	–	5.72	0.11	7.04	m2	7.92
isolated surfaces, girth not exceeding 300mm	0.33	10.00	0.36	0.24	–	2.22	0.04	2.62	m	2.95
Iron or steel structural members of roof trusses, lattice girders, purlins and the like										
girth exceeding 300mm .!......................	1.10	10.00	1.21	0.79	Hrs	7.29	0.14	8.64	m2	9.72
isolated surfaces, girth not exceeding 300mm	0.33	10.00	0.36	0.24	–	2.22	0.04	2.62	m	2.95

SURFACE FINISHES

SURFACE FINISHES – SMALL WORKS

Labour hourly rates: (except Specialists) Craft Operatives £9.23 Labourer £7.02. Rates are national average prices. Refer to REGIONAL VARIATIONS for indicative levels of overall pricing in regions	MATERIALS			LABOUR				RATES		
	Del to Site £	Waste %	Material Cost £	Craft Optve Hrs	Lab Hrs	Labour Cost £	Sunds £	Nett Rate £	Unit	Gross Rate (+12.5%) £
M60: PAINTING/CLEAR FINISHING - OIL PAINTING METALWORK Cont.										
One coat primer, two undercoats, one coat full gloss finish on metal surfaces; external work Cont.										
Iron or steel railings, fences and gates; plain open type										
girth exceeding 300mm	1.10	10.00	1.21	0.57	-	5.26	0.10	6.57	m2	7.39
isolated surfaces, girth not exceeding 300mm	0.33	10.00	0.36	0.19	-	1.75	0.03	2.15	m	2.42
Iron or steel railings, fences and gates; close type										
girth exceeding 300mm	1.10	10.00	1.21	0.48	-	4.43	0.08	5.72	m2	6.44
Iron or steel railings, fences and gates; ornamental type										
girth exceeding 300mm	1.10	10.00	1.21	0.97	-	8.95	0.17	10.33	m2	11.62
Iron or steel eaves gutters										
girth exceeding 300mm	1.10	10.00	1.21	0.66	-	6.09	0.12	7.42	m2	8.35
isolated surfaces, girth not exceeding 300mm	0.33	10.00	0.36	0.22	-	2.03	0.04	2.43	m	2.74
Galvanised eaves gutters										
girth exceeding 300mm	1.10	10.00	1.21	0.66	-	6.09	0.12	7.42	m2	8.35
isolated surfaces, girth not exceeding 300mm	0.33	10.00	0.36	0.22	-	2.03	0.04	2.43	m	2.74
Iron or steel services										
girth exceeding 300mm	1.10	10.00	1.21	0.69	-	6.37	0.12	7.70	m2	8.66
isolated surfaces, girth not exceeding 300mm	0.33	10.00	0.36	0.23	-	2.12	0.04	2.53	m	2.84
isolated areas not exceeding 0.50m2 irrespective of girth	0.55	10.00	0.60	0.69	-	6.37	0.12	7.09	nr	7.98
Copper services										
girth exceeding 300mm	1.10	10.00	1.21	0.69	-	6.37	0.12	7.70	m2	8.66
isolated surfaces, girth not exceeding 300mm	0.33	10.00	0.36	0.23	-	2.12	0.04	2.53	m	2.84
Galvanised services										
girth exceeding 300mm	1.10	10.00	1.21	0.69	-	6.37	0.12	7.70	m2	8.66
isolated surfaces, girth not exceeding 300mm	0.33	10.00	0.36	0.23	-	2.12	0.04	2.53	m	2.84
One coat primer, one undercoat, two coats full gloss finish on metal surfaces										
Iron or steel general surfaces										
girth exceeding 300mm	1.10	10.00	1.21	0.57	-	5.26	0.10	6.57	m2	7.39
isolated surfaces, girth not exceeding 300mm	0.33	10.00	0.36	0.19	-	1.75	0.03	2.15	m	2.42
Galvanised glazed doors, windows or screens										
girth exceeding 300mm; panes, area not exceeding 0.10m2	0.30	10.00	0.33	1.06	-	9.78	0.19	10.30	m2	11.59
girth exceeding 300mm; panes, area 0.10 - 0.50m2	0.25	10.00	0.28	0.75	-	6.92	0.13	7.33	m2	8.24
girth exceeding 300mm; panes, area 0.50 - 1.00m2	0.22	10.00	0.24	0.63	-	5.81	0.11	6.17	m2	6.94
girth exceeding 300mm; panes, area exceeding 1.00m2	0.20	10.00	0.22	0.57	-	5.26	0.10	5.58	m2	6.28
Iron or steel structural work										
girth exceeding 300mm	1.10	10.00	1.21	0.62	-	5.72	0.11	7.04	m2	7.92
isolated surfaces, girth not exceeding 300mm	0.33	10.00	0.36	0.24	-	2.22	0.04	2.62	m	2.95
Iron or steel structural members of roof trusses, lattice girders, purlins and the like										
girth exceeding 300mm	1.10	10.00	1.21	0.79	-	7.29	0.14	8.64	m2	9.72
isolated surfaces, girth not exceeding 300mm	0.33	10.00	0.36	0.24	-	2.22	0.04	2.62	m	2.95
Iron or steel services										
girth exceeding 300mm	1.10	10.00	1.21	0.69	-	6.37	0.12	7.70	m2	8.66
isolated surfaces, girth not exceeding 300mm	0.33	10.00	0.36	0.23	-	2.12	0.04	2.53	m	2.84
isolated areas not exceeding 0.50m2 irrespective of girth	0.55	10.00	0.60	0.69	-	6.37	0.12	7.09	nr	7.98
Copper services										
girth exceeding 300mm	1.10	10.00	1.21	0.69	-	6.37	0.12	7.70	m2	8.66
isolated surfaces, girth not exceeding 300mm	0.33	10.00	0.36	0.23	-	2.12	0.04	2.53	m	2.84
Galvanised services										
girth exceeding 300mm	1.10	10.00	1.21	0.69	-	6.37	0.12	7.70	m2	8.66
isolated surfaces, girth not exceeding 300mm	0.33	10.00	0.36	0.23	-	2.12	0.04	2.53	m	2.84
One coat primer, one undercoat, two coats full gloss finish on metal surfaces; external work										
Iron or steel general surfaces										
girth exceeding 300mm	1.10	10.00	1.21	0.62	-	5.72	0.11	7.04	m2	7.92
isolated surfaces, girth not exceeding 300mm	0.33	10.00	0.36	0.21	-	1.94	0.04	2.34	m	2.63
Galvanised glazed doors, windows or screens										
girth exceeding 300mm; panes, area not exceeding 0.10m2	0.30	10.00	0.33	1.10	-	10.15	0.19	10.67	m2	12.01
girth exceeding 300mm; panes, area 0.10 - 0.50m2	0.25	10.00	0.28	0.79	-	7.29	0.14	7.71	m2	8.67
girth exceeding 300mm; panes, area 0.50 - 1.00m2	0.22	10.00	0.24	0.67	-	6.18	0.12	6.55	m2	7.36
girth exceeding 300mm; panes, area exceeding 1.00m2	0.20	10.00	0.22	0.62	-	5.72	0.11	6.05	m2	6.81
Iron or steel structural work										
girth exceeding 300mm	1.10	10.00	1.21	0.62	-	5.72	0.11	7.04	m2	7.92
isolated surfaces, girth not exceeding 300mm	1.10	10.00	1.21	0.62	-	5.72	0.11	7.04	m2	7.92

Labour hourly rates: (except Specialists) Craft Operatives £9.23 Labourer £7.02 Rates are national average prices. Refer to REGIONAL VARIATIONS for indicative levels of overall pricing in regions	MATERIALS			LABOUR				RATES		
	Del to Site £	Waste %	Material Cost £	Craft Optve Hrs	Lab Hrs	Labour Cost £	Sunds £	Nett Rate £	Unit	Gross Rate (+12.5%) £
M60: PAINTING/CLEAR FINISHING - OIL PAINTING METALWORK Cont.										
One coat primer, one undercoat, two coats full gloss finish on metal surfaces; external work Cont.										
Iron or steel structural members of roof trusses, lattice girders, purlins and the like										
girth exceeding 300mm	1.10	10.00	1.21	0.62	-	5.72	0.11	7.04	m2	7.92
isolated surfaces, girth not exceeding 300mm	0.33	10.00	0.36	0.24	-	2.22	0.04	2.62	m2	2.95
Iron or steel railings, fences and gates; plain open type										
girth exceeding 300mm	1.10	10.00	1.21	0.57	-	5.26	0.10	6.57	m2	7.39
isolated surfaces, girth not exceeding 300mm	0.33	10.00	0.36	0.19	-	1.75	0.03	2.15	m2	2.42
Iron or steel railings, fences and gates; close type										
girth exceeding 300mm	1.10	10.00	1.21	0.48	-	4.43	0.08	5.72	m2	6.44
Iron or steel railings, fences and gates; ornamental type										
girth exceeding 300mm	1.10	10.00	1.21	0.97	-	8.95	0.17	10.33	m2	11.62
Iron or steel eaves gutters										
girth exceeding 300mm	1.10	10.00	1.21	0.66	-	6.09	0.12	7.42	m2	8.35
isolated surfaces, girth not exceeding 300mm	0.33	10.00	0.36	0.22	-	2.03	0.04	2.43	m	2.74
Galvanised eaves gutters										
girth exceeding 300mm	1.10	10.00	1.21	0.66	-	6.09	0.12	7.42	m2	8.35
isolated surfaces, girth not exceeding 300mm	0.33	10.00	0.36	0.22	-	2.03	0.04	2.43	m	2.74
Iron or steel services										
girth exceeding 300mm	1.10	10.00	1.21	0.69	-	6.37	0.12	7.70	m2	8.66
isolated surfaces, girth not exceeding 300mm	0.33	10.00	0.36	0.23	-	2.12	0.04	2.53	m	2.84
isolated areas not exceeding 0.50m2 irrespective of girth	0.55	10.00	0.60	0.69	-	6.37	0.12	7.09	nr	7.98
Copper services										
girth exceeding 300mm	1.10	10.00	1.21	0.69	-	6.37	0.12	7.70	m2	8.66
isolated surfaces, girth not exceeding 300mm	0.33	10.00	0.36	0.23	-	2.12	0.04	2.53	m	2.84
Galvanised services										
girth exceeding 300mm	1.10	10.00	1.21	0.69	-	6.37	0.12	7.70	m2	8.66
isolated surfaces, girth not exceeding 300mm	0.33	10.00	0.36	0.23	-	2.12	0.04	2.53	m	2.84
M60: PAINTING/CLEAR FINISHING - METALLIC PAINTING										
One coat aluminium metallic paint										
Iron or steel radiators; panel type										
girth exceeding 300mm	0.44	10.00	0.48	0.15	-	1.38	0.03	1.90	m2	2.14
Iron or steel radiators; column type										
girth exceeding 300mm	0.44	10.00	0.48	0.22	-	2.03	0.04	2.55	m2	2.87
Iron or steel services										
isolated surfaces, girth not exceeding 300mm	0.13	10.00	0.14	0.06	-	0.55	0.01	0.71	m	0.80
Copper services										
isolated surfaces, girth not exceeding 300mm	0.13	10.00	0.14	0.06	-	0.55	0.01	0.71	m	0.80
Galvanised services										
isolated surfaces, girth not exceeding 300mm	0.13	10.00	0.14	0.06	-	0.55	0.01	0.71	m	0.80
Two coats aluminium metallic paint										
Iron or steel radiators; panel type										
girth exceeding 300mm	0.83	10.00	0.91	0.31	-	2.86	0.05	3.82	m2	4.30
Iron or steel radiators; column type										
girth exceeding 300mm	0.83	10.00	0.91	0.44	-	4.06	0.08	5.05	m2	5.69
Iron or steel services										
isolated surfaces, girth not exceeding 300mm	0.25	10.00	0.28	0.11	-	1.02	0.02	1.31	m	1.47
Copper services										
isolated surfaces, girth not exceeding 300mm	0.25	10.00	0.28	0.11	-	1.02	0.02	1.31	m	1.47
Galvanised services										
isolated surfaces, girth not exceeding 300mm	0.25	10.00	0.28	0.11	-	1.02	0.02	1.31	m	1.47
One coat gold or bronze metallic paint										
Iron or steel radiators; panel type										
girth exceeding 300mm	0.59	10.00	0.65	0.15	-	1.38	0.03	2.06	m2	2.32
Iron or steel radiators; column type										
girth exceeding 300mm	0.59	10.00	0.65	0.22	-	2.03	0.04	2.72	m2	3.06
Iron or steel services										
isolated surfaces, girth not exceeding 300mm	0.18	10.00	0.20	0.06	-	0.55	0.01	0.76	m	0.86
Copper services										
isolated surfaces, girth not exceeding 300mm	0.18	10.00	0.20	0.06	-	0.55	0.01	0.76	m	0.86

SURFACE FINISHES

Labour hourly rates: (except Specialists) Craft Operatives £9.23 Labourer £7.02 Rates are national average prices. Refer to REGIONAL VARIATIONS for indicative levels of overall pricing in regions	MATERIALS			LABOUR				RATES		
	Del to Site £	Waste %	Material Cost £	Craft Optve Hrs	Lab Hrs	Labour Cost £	Sunds £	Nett Rate £	Unit	Gross Rate (+12.5%) £
M60: PAINTING/CLEAR FINISHING - METALLIC PAINTING Cont.										
One coat gold or bronze metallic paint Cont.										
Galvanised services										
isolated surfaces, girth not exceeding 300mm	0.18	10.00	0.20	0.06	-	0.55	0.01	0.76	m	0.86
Two coats gold or bronze metallic paint										
Iron or steel radiators; panel type										
girth exceeding 300mm	1.10	10.00	1.21	0.31	-	2.86	0.05	4.12	m2	4.64
Iron or steel radiators; column type										
girth exceeding 300mm	1.10	10.00	1.21	0.44	-	4.06	0.08	5.35	m2	6.02
Iron or steel services										
isolated surfaces, girth not exceeding 300mm	0.33	10.00	0.36	0.11	-	1.02	0.02	1.40	m	1.57
Copper services										
isolated surfaces, girth not exceeding 300mm	0.33	10.00	0.36	0.11	-	1.02	0.02	1.40	m	1.57
Galvanised services										
isolated surfaces, girth not exceeding 300mm	0.33	10.00	0.36	0.11	-	1.02	0.02	1.40	m	1.57
M60: PAINTING/CLEAR FINISHING - BITUMINOUS PAINT										
One coat black bitumen paint; external work										
Iron or steel general surfaces										
girth exceeding 300mm	0.26	10.00	0.29	0.14	-	1.29	0.02	1.60	m2	1.80
Iron or steel eaves gutters										
girth exceeding 300mm	0.26	10.00	0.29	0.15	-	1.38	0.03	1.70	m2	1.91
Iron or steel services										
girth exceeding 300mm	0.26	10.00	0.29	0.17	-	1.57	0.03	1.89	m2	2.12
isolated surfaces, girth not exceeding 300mm	0.08	10.00	0.09	0.06	-	0.55	0.01	0.65	m	0.73
Two coats black bitumen paint; external work										
Iron or steel general surfaces										
girth exceeding 300mm	0.53	10.00	0.58	0.29	-	2.68	0.05	3.31	m2	3.72
Iron or steel eaves gutters										
girth exceeding 300mm	0.53	10.00	0.58	0.31	-	2.86	0.05	3.49	m2	3.93
isolated surfaces, girth not exceeding 300mm	0.16	10.00	0.18	0.10	-	0.92	0.02	1.12	m	1.26
Iron or steel services										
girth exceeding 300mm	0.53	10.00	0.58	0.33	-	3.05	0.05	3.68	m2	4.14
isolated surfaces, girth not exceeding 300mm	0.16	10.00	0.18	0.10	-	0.92	0.02	1.12	m	1.26
M60: PAINTING/CLEAR FINISHING - OIL PAINTING WOODWORK										
One coat primer; carried out on site before fixing members										
Wood general surfaces										
girth exceeding 300mm	0.43	10.00	0.47	0.39	-	3.60	0.03	4.10	m2	4.62
isolated surfaces, girth not exceeding 300mm	0.28	10.00	0.31	0.77	-	7.11	0.05	7.47	m	8.40
One undercoat, one coat full gloss finish on ready primed wood surfaces										
Wood general surfaces										
girth exceeding 300mm	0.52	10.00	0.57	0.35	-	3.23	0.06	3.86	m2	4.35
isolated surfaces, girth not exceeding 300mm	0.16	10.00	0.18	0.12	-	1.11	0.02	1.30	m	1.47
Wood glazed doors										
girth exceeding 300mm; panes, area not exceeding 0.10m2	0.26	10.00	0.29	0.59	-	5.45	0.10	5.83	m2	6.56
girth exceeding 300mm; panes, area 0.10 - 0.50m2 .	0.18	10.00	0.20	0.42	-	3.88	0.07	4.14	m2	4.66
girth exceeding 300mm; panes, area 0.50 - 1.00m2 .	0.13	10.00	0.14	0.35	-	3.23	0.06	3.43	m2	3.86
girth exceeding 300mm; panes, area exceeding 1.00m2	0.10	10.00	0.11	0.32	-	2.95	0.06	3.12	m2	3.51
Wood partially glazed doors										
girth exceeding 300mm; panes, area not exceeding 0.10m2	0.39	10.00	0.43	0.47	-	4.34	0.08	4.85	m2	5.45
girth exceeding 300mm; panes, area 0.10 - 0.50m2 .	0.34	10.00	0.37	0.39	-	3.60	0.07	4.04	m2	4.55
girth exceeding 300mm; panes, area 0.50 - 1.00m2 .	0.31	10.00	0.34	0.35	-	3.23	0.06	3.63	m2	4.09
girth exceeding 300mm; panes, area exceeding 1.00m2	0.29	10.00	0.32	0.34	-	3.14	0.05	3.51	m2	3.95
Wood windows and screens										
girth exceeding 300mm; panes, area not exceeding 0.10m2	0.29	10.00	0.32	0.65	-	6.00	0.11	6.43	m2	7.23
girth exceeding 300mm; panes, area 0.10 - 0.50m2 .	0.21	10.00	0.23	0.46	-	4.25	0.08	4.56	m2	5.13
girth exceeding 300mm; panes, area 0.50 - 1.00m2 .	0.16	10.00	0.18	0.39	-	3.60	0.07	3.85	m2	4.33
girth exceeding 300mm; panes, area exceeding 1.00m2	0.13	10.00	0.14	0.35	-	3.23	0.06	3.43	m2	3.86

SURFACE FINISHES – SMALL WORKS

Labour hourly rates: (except Specialists) Craft Operatives £9.23 Labourer £7.02 Rates are national average prices. Refer to REGIONAL VARIATIONS for indicative levels of overall pricing in regions	MATERIALS			LABOUR				RATES		
	Del to Site £	Waste %	Material Cost £	Craft Optve Hrs	Lab Hrs	Labour Cost £	Sunds £	Nett Rate £	Unit	Gross Rate (+12.5%) £

M60: PAINTING/CLEAR FINISHING - OIL PAINTING WOODWORK Cont.

One undercoat, one coat full gloss finish on ready primed wood surfaces; external work

Wood general surfaces

girth exceeding 300mm	0.52	10.00	0.57	0.37	-	3.42	0.06	4.05	m2	4.55
isolated surfaces, girth not exceeding 300mm	0.16	10.00	0.18	0.13	-	1.20	0.02	1.40	m	1.57

Wood glazed doors

girth exceeding 300mm; panes, area not exceeding 0.10m2	0.26	10.00	0.29	0.59	-	5.45	0.10	5.83	m2	6.56
girth exceeding 300mm; panes, area 0.10 - 0.50m2 .	0.18	10.00	0.20	0.42	-	3.88	0.07	4.14	m2	4.66
girth exceeding 300mm; panes, area 0.50 - 1.00m2 .	0.13	10.00	0.14	0.35	-	3.23	0.06	3.43	m2	3.86
girth exceeding 300mm; panes, area exceeding 1.00m2	0.10	10.00	0.11	0.32	-	2.95	0.06	3.12	m2	3.51

Wood partially glazed doors

girth exceeding 300mm; panes, area not exceeding 0.10m2	0.39	10.00	0.43	0.47	-	4.34	0.08	4.85	m2	5.45
girth exceeding 300mm; panes, area 0.10 - 0.50m2 .	0.34	10.00	0.37	0.39	-	3.60	0.07	4.04	m2	4.55
girth exceeding 300mm; panes, area 0.50 - 1.00m2 .	0.31	10.00	0.34	0.35	-	3.23	0.06	3.63	m2	4.09
girth exceeding 300mm; panes, area exceeding 1.00m2	0.29	10.00	0.32	0.34	-	3.14	0.06	3.52	m2	3.96

Wood windows and screens

girth exceeding 300mm; panes, area not exceeding 0.10m2	0.29	10.00	0.32	0.67	-	6.18	0.12	6.62	m2	7.45
girth exceeding 300mm; panes, area 0.10 - 0.50m2 .	0.21	10.00	0.23	0.48	-	4.43	0.08	4.74	m2	5.33
girth exceeding 300mm; panes, area 0.50 - 1.00m2 .	0.16	10.00	0.18	0.41	-	3.78	0.07	4.03	m2	4.53
girth exceeding 300mm; panes, area exceeding 1.00m2	0.13	10.00	0.14	0.37	-	3.42	0.06	3.62	m2	4.07

Wood railings fences and gates; open type

girth exceeding 300mm	0.52	10.00	0.57	0.28	-	2.58	0.05	3.21	m2	3.61
isolated surfaces, girth not exceeding 300mm	0.16	10.00	0.18	0.09	-	0.83	0.02	1.03	m	1.16

Wood railings fences and gates; close type

girth exceeding 300mm	0.52	10.00	0.57	0.24	-	2.22	0.04	2.83	m2	3.18

Two undercoats, one coat full gloss finish on ready primed wood surfaces

Wood general surfaces

girth exceeding 300mm	0.78	10.00	0.86	0.51	-	4.71	0.09	5.66	m2	6.36
isolated surfaces, girth not exceeding 300mm	0.24	10.00	0.26	0.17	-	1.57	0.03	1.86	m	2.10

Wood glazed doors

girth exceeding 300mm; panes, area not exceeding 0.10m2	0.39	10.00	0.43	0.84	-	7.75	0.15	8.33	m2	9.37
girth exceeding 300mm; panes, area 0.10 - 0.50m2 .	0.27	10.00	0.30	0.61	-	5.63	0.11	6.04	m2	6.79
girth exceeding 300mm; panes, area 0.50 - 1.00m2 .	0.20	10.00	0.22	0.51	-	4.71	0.09	5.02	m2	5.64
girth exceeding 300mm; panes, area exceeding 1.00m2	0.16	10.00	0.18	0.45	-	4.15	0.08	4.41	m2	4.96

Wood partially glazed doors

girth exceeding 300mm; panes, area not exceeding 0.10m2	0.59	10.00	0.65	0.67	-	6.18	0.12	6.95	m2	7.82
girth exceeding 300mm; panes, area 0.10 - 0.50m2 .	0.51	10.00	0.56	0.56	-	5.17	0.10	5.83	m2	6.56
girth exceeding 300mm; panes, area 0.50 - 1.00m2 .	0.47	10.00	0.52	0.51	-	4.71	0.09	5.31	m2	5.98
girth exceeding 300mm; panes, area exceeding 1.00m2	0.43	10.00	0.47	0.48	-	4.43	0.08	4.98	m2	5.61

Wood windows and screens

girth exceeding 300mm; panes, area not exceeding 0.10m2	0.43	10.00	0.47	0.92	-	8.49	0.16	9.12	m2	10.27
girth exceeding 300mm; panes, area 0.10 - 0.50m2 .	0.31	10.00	0.34	0.67	-	6.18	0.12	6.65	m2	7.48
girth exceeding 300mm; panes, area 0.50 - 1.00m2 .	0.24	10.00	0.26	0.56	-	5.17	0.10	5.53	m2	6.22
girth exceeding 300mm; panes, area exceeding 1.00m2	0.20	10.00	0.22	0.50	-	4.62	0.09	4.92	m2	5.54

Two undercoats, one coat full gloss finish on ready primed wood surfaces; external work

Wood general surfaces

girth exceeding 300mm	0.78	10.00	0.86	0.54	-	4.98	0.09	5.93	m2	6.67
isolated surfaces, girth not exceeding 300mm	0.24	10.00	0.26	0.19	-	1.75	0.03	2.05	m	2.30

Wood glazed doors

girth exceeding 300mm; panes, area not exceeding 0.10m2	0.39	10.00	0.43	0.84	-	7.75	0.15	8.33	m2	9.37
girth exceeding 300mm; panes, area 0.10 - 0.50m2 .	0.27	10.00	0.30	0.61	-	5.63	0.11	6.04	m2	6.79
girth exceeding 300mm; panes, area 0.50 - 1.00m2 .	0.20	10.00	0.22	0.51	-	4.71	0.09	5.02	m2	5.64
girth exceeding 300mm; panes, area exceeding 1.00m2	0.16	10.00	0.18	0.45	-	4.15	0.00	4.41	m2	4.96

Wood partially glazed doors

girth exceeding 300mm; panes, area not exceeding 0.10m2	0.59	10.00	0.65	0.67	-	6.18	0.12	6.95	m2	7.82
girth exceeding 300mm; panes, area 0.10 - 0.50m2 .	0.51	10.00	0.56	0.56	-	5.17	0.10	5.83	m2	6.56
girth exceeding 300mm; panes, area 0.50 - 1.00m2 .	0.47	10.00	0.52	0.51	-	4.71	0.09	5.31	m2	5.98
girth exceeding 300mm; panes, area exceeding 1.00m2	0.43	10.00	0.47	0.48	-	4.43	0.08	4.98	m2	5.61

SURFACE FINISHES

Labour hourly rates: (except Specialists) Craft Operatives £9.23 Labourer £7.02 Rates are national average prices. Refer to REGIONAL VARIATIONS for indicative levels of overall pricing in regions	MATERIALS			LABOUR				RATES		
	Del to Site £	Waste %	Material Cost £	Craft Optve Hrs	Lab Hrs	Labour Cost £	Sunds £	Nett Rate £	Unit	Gross Rate (+12.5%) £

M60: PAINTING/CLEAR FINISHING - OIL PAINTING WOODWORK Cont.

Two undercoats, one coat full gloss finish on ready primed wood surfaces; external work Cont.

Wood windows and screens
  girth exceeding 300mm; panes, area not exceeding

0.10m2	0.43	10.00	0.47	0.96	-	8.86	0.17	9.50	m2	10.69
girth exceeding 300mm; panes, area 0.10 - 0.50m2 .	0.31	10.00	0.34	0.70	-	6.46	0.12	6.92	m2	7.79
girth exceeding 300mm; panes, area 0.50 - 1.00m2 .	0.24	10.00	0.26	0.59	-	5.45	0.10	5.81	m2	6.54
girth exceeding 300mm; panes, area exceeding 1.00m2	0.20	10.00	0.22	0.53	-	4.89	0.09	5.20	m2	5.85

Wood railings fences and gates; open type

girth exceeding 300mm	0.78	10.00	0.86	0.42	-	3.88	0.07	4.80	m2	5.41
isolated surfaces, girth not exceeding 300mm	0.24	10.00	0.26	0.14	-	1.29	0.02	1.58	m	1.77

Wood railings fences and gates; close type

girth exceeding 300mm	0.78	10.00	0.86	0.36	-	3.32	0.06	4.24	m2	4.77

One coat primer, one undercoat, one coat full gloss finish on wood surfaces

Wood general surfaces

girth exceeding 300mm	0.83	10.00	0.91	0.51	-	4.71	0.09	5.71	m2	6.42
isolated surfaces, girth not exceeding 300mm	0.25	10.00	0.28	0.17	-	1.57	0.03	1.87	m	2.11

Wood glazed doors
  girth exceeding 300mm; panes, area not exceeding

0.10m2	0.42	10.00	0.46	0.84	-	7.75	0.15	8.37	m2	9.41
girth exceeding 300mm; panes, area 0.10 - 0.50m2 .	0.29	10.00	0.32	0.61	-	5.63	0.11	6.06	m2	6.82
girth exceeding 300mm; panes, area 0.50 - 1.00m2 .	0.21	10.00	0.23	0.51	-	4.71	0.09	5.03	m2	5.66
girth exceeding 300mm; panes, area exceeding 1.00m2	0.17	10.00	0.19	0.45	-	4.15	0.08	4.42	m2	4.97

Wood partially glazed doors
  girth exceeding 300mm; panes, area not exceeding

0.10m2	0.63	10.00	0.69	0.67	-	6.18	0.12	7.00	m2	7.87
girth exceeding 300mm; panes, area 0.10 - 0.50m2 .	0.54	10.00	0.59	0.56	-	5.17	0.10	5.86	m2	6.60
girth exceeding 300mm; panes, area 0.50 - 1.00m2 .	0.50	10.00	0.55	0.51	-	4.71	0.09	5.35	m2	6.02
girth exceeding 300mm; panes, area exceeding 1.00m2	0.46	10.00	0.51	0.48	-	4.43	0.08	5.02	m2	5.64

Wood windows and screens
  girth exceeding 300mm; panes, area not exceeding

0.10m2	0.46	10.00	0.51	0.92	-	8.49	0.16	9.16	m2	10.30
girth exceeding 300mm; panes, area 0.10 - 0.50m2 .	0.33	10.00	0.36	0.67	-	6.18	0.12	6.67	m2	7.50
girth exceeding 300mm; panes, area 0.50 - 1.00m2 .	0.25	10.00	0.28	0.56	-	5.17	0.10	5.54	m2	6.24
girth exceeding 300mm; panes, area exceeding 1.00m2	0.21	10.00	0.23	0.50	-	4.62	0.09	4.94	m2	5.55

One coat primer, one undercoat, one coat full gloss finish on wood surfaces; external work

Wood general surfaces

girth exceeding 300mm	0.83	10.00	0.91	0.54	-	4.98	0.08	5.98	m2	6.72
isolated surfaces, girth not exceeding 300mm	0.25	10.00	0.28	0.19	-	1.75	0.03	2.06	m	2.32

Wood glazed doors
  girth exceeding 300mm; panes, area not exceeding

0.10m2	0.42	10.00	0.46	0.84	-	7.75	0.15	8.37	m2	9.41
girth exceeding 300mm; panes, area 0.10 - 0.50m2 .	0.29	10.00	0.32	0.61	-	5.63	0.11	6.06	m2	6.82
girth exceeding 300mm; panes, area 0.50 - 1.00m2 .	0.21	10.00	0.23	0.51	-	4.71	0.09	5.03	m2	5.66
girth exceeding 300mm; panes, area exceeding 1.00m2	0.17	10.00	0.19	0.45	-	4.15	0.08	4.42	m2	4.97

Wood partially glazed doors
  girth exceeding 300mm; panes, area not exceeding

0.10m2	0.63	10.00	0.69	0.67	-	6.18	0.12	7.00	m2	7.87
girth exceeding 300mm; panes, area 0.10 - 0.50m2 .	0.54	10.00	0.59	0.56	-	5.17	0.10	5.86	m2	6.60
girth exceeding 300mm; panes, area 0.50 - 1.00m2 .	0.50	10.00	0.55	0.51	-	4.71	0.09	5.35	m2	6.02
girth exceeding 300mm; panes, area exceeding 1.00m2	0.46	10.00	0.51	0.48	-	4.43	0.08	5.02	m2	5.64

Wood windows and screens
  girth exceeding 300mm; panes, area not exceeding

0.10m2	0.46	10.00	0.51	0.96	-	8.86	0.17	9.54	m2	10.73
girth exceeding 300mm; panes, area 0.10 - 0.50m2 .	0.33	10.00	0.36	0.70	-	6.46	0.12	6.94	m2	7.81
girth exceeding 300mm; panes, area 0.50 - 1.00m2 .	0.25	10.00	0.28	0.59	-	5.45	0.10	5.82	m2	6.55
girth exceeding 300mm; panes, area exceeding 1.00m2	0.21	10.00	0.23	0.53	-	4.89	0.09	5.21	m2	5.86

Wood railings fences and gates; open type

girth exceeding 300mm	0.83	10.00	0.91	0.42	-	3.88	0.07	4.86	m2	5.47
isolated surfaces, girth not exceeding 300mm	0.25	10.00	0.28	0.14	-	1.29	0.02	1.59	m	1.79

Wood railings fences and gates; close type

girth exceeding 300mm	0.83	10.00	0.91	0.36	-	3.32	0.06	4.30	m2	4.83

One coat primer, two undercoats, one coat full gloss finish on wood surfaces

Wood general surfaces

girth exceeding 300mm	1.10	10.00	1.21	0.66	-	6.09	0.12	7.42	m2	8.35
isolated surfaces, girth not exceeding 300mm	0.33	10.00	0.36	0.22	-	2.03	0.04	2.43	m	2.74

Labour hourly rates: (except Specialists) Craft Operatives £9.23 Labourer £7.02 Rates are national average prices. Refer to REGIONAL VARIATIONS for indicative levels of overall pricing in regions	MATERIALS			LABOUR				RATES		
	Del to Site £	Waste %	Material Cost £	Craft Optve Hrs	Lab Hrs	Labour Cost £	Sunds £	Nett Rate £	Unit	Gross Rate (+12.5%) £
M60: PAINTING/CLEAR FINISHING - OIL PAINTING WOODWORK Cont.										
One coat primer, two undercoats, one coat full gloss finish on wood surfaces Cont.										
Wood glazed doors										
girth exceeding 300mm; panes, area not exceeding 0.10m2	0.55	10.00	0.60	1.10	-	10.15	0.19	10.95	m2	12.32
girth exceeding 300mm; panes, area 0.10 - 0.50m2	0.38	10.00	0.42	0.79	-	7.29	0.14	7.85	m2	8.83
girth exceeding 300mm; panes, area 0.50 - 1.00m2	0.27	10.00	0.30	0.66	-	6.09	0.12	6.51	m2	7.32
girth exceeding 300mm; panes, area exceeding 1.00m2	0.22	10.00	0.24	0.59	-	5.45	0.10	5.79	m2	6.51
Wood partially glazed doors										
girth exceeding 300mm; panes, area not exceeding 0.10m2	0.82	10.00	0.90	0.88	-	8.12	0.15	9.17	m2	10.32
girth exceeding 300mm; panes, area 0.10 - 0.50m2	0.71	10.00	0.78	0.73	-	6.74	0.13	7.65	m2	8.61
girth exceeding 300mm; panes, area 0.50 - 1.00m2	0.66	10.00	0.73	0.66	-	6.09	0.12	6.94	m2	7.81
girth exceeding 300mm; panes, area exceeding 1.00m2	0.60	10.00	0.66	0.63	-	5.81	0.11	6.58	m2	7.41
Wood windows and screens										
girth exceeding 300mm; panes, area not exceeding 0.10m2	0.60	10.00	0.66	1.21	-	11.17	0.21	12.04	m2	13.54
girth exceeding 300mm; panes, area 0.10 - 0.50m2	0.44	10.00	0.48	0.87	-	8.03	0.15	8.66	m2	9.75
girth exceeding 300mm; panes, area 0.50 - 1.00m2	0.33	10.00	0.36	0.73	-	6.74	0.13	7.23	m2	8.13
girth exceeding 300mm; panes, area exceeding 1.00m2	0.27	10.00	0.30	0.65	-	6.00	0.11	6.41	m2	7.21
One coat primer, two undercoats, one coat full gloss finish on wood surfaces; external work										
Wood general surfaces										
girth exceeding 300mm	1.10	10.00	1.21	0.70	-	6.46	0.12	7.79	m2	8.76
isolated surfaces, girth not exceeding 300mm	0.33	10.00	0.36	0.24	-	2.22	0.04	2.62	m2	2.95
Wood glazed doors										
girth exceeding 300mm; panes, area not exceeding 0.10m2	0.55	10.00	0.60	1.10	-	10.15	0.19	10.95	m2	12.32
girth exceeding 300mm; panes, area 0.10 - 0.50m2	0.38	10.00	0.42	0.79	-	7.29	0.14	7.85	m2	8.83
girth exceeding 300mm; panes, area 0.50 - 1.00m2	0.27	10.00	0.30	0.66	-	6.09	0.12	6.51	m2	7.32
girth exceeding 300mm; panes, area exceeding 1.00m2	0.22	10.00	0.24	0.59	-	5.45	0.10	5.79	m2	6.51
Wood partially glazed doors										
girth exceeding 300mm; panes, area not exceeding 0.10m2	0.82	10.00	0.90	0.88	-	8.12	0.15	9.17	m2	10.32
girth exceeding 300mm; panes, area 0.10 - 0.50m2	0.71	10.00	0.78	0.73	-	6.74	0.13	7.65	m2	8.61
girth exceeding 300mm; panes, area 0.50 - 1.00m2	0.66	10.00	0.73	0.66	-	6.09	0.12	6.94	m2	7.81
girth exceeding 300mm; panes, area exceeding 1.00m2	0.60	10.00	0.66	0.63	-	5.81	0.11	6.58	m2	7.41
Wood windows and screens										
girth exceeding 300mm; panes, area not exceeding 0.10m2	0.60	10.00	0.66	1.25	-	11.54	0.22	12.42	m2	13.97
girth exceeding 300mm; panes, area 0.10 - 0.50m2	0.44	10.00	0.48	0.91	-	8.40	0.16	9.04	m2	10.17
girth exceeding 300mm; panes, area 0.50 - 1.00m2	0.33	10.00	0.36	0.77	-	7.11	0.13	7.60	m2	8.55
girth exceeding 300mm; panes, area exceeding 1.00m2	0.27	10.00	0.30	0.69	-	6.37	0.12	6.79	m2	7.63
Wood railings fences and gates; open type										
girth exceeding 300mm	1.10	10.00	1.21	0.55	-	5.08	0.10	6.39	m2	7.18
isolated surfaces, girth not exceeding 300mm	0.33	10.00	0.36	0.19	-	1.75	0.03	2.15	m	2.42
Wood railings fences and gates; close type										
girth exceeding 300mm	1.10	10.00	1.21	0.48	-	4.43	0.08	5.72	m2	6.44
One coat primer, one undercoat, two coats full gloss finish on wood surfaces										
Wood general surfaces										
girth exceeding 300mm	1.10	10.00	1.21	0.66	-	6.09	0.12	7.42	m2	8.35
isolated surfaces, girth not exceeding 300mm	0.33	10.00	0.36	0.22	-	2.03	0.04	2.43	m	2.74
Wood glazed doors										
girth exceeding 300mm; panes, area not exceeding 0.10m2	0.55	10.00	0.60	1.10	-	10.15	0.19	10.95	m2	12.32
girth exceeding 300mm; panes, area 0.10 - 0.50m2	0.38	10.00	0.42	0.79	-	7.29	0.14	7.85	m2	8.83
girth exceeding 300mm; panes, area 0.50 - 1.00m2	0.27	10.00	0.30	0.66	-	6.09	0.12	6.51	m2	7.32
girth exceeding 300mm; panes, area exceeding 1.00m2	0.22	10.00	0.24	0.59	-	5.45	0.10	5.79	m2	6.51
Wood partially glazed doors										
girth exceeding 300mm; panes, area not exceeding 0.10m2	0.82	10.00	0.90	0.88	-	8.12	0.15	9.17	m2	10.32
girth exceeding 300mm; panes, area 0.10 - 0.50m2	0.71	10.00	0.78	0.73	-	6.74	0.13	7.65	m2	8.61
girth exceeding 300mm; panes, area 0.50 - 1.00m2	0.66	10.00	0.73	0.66	-	6.09	0.12	6.94	m2	7.81
girth exceeding 300mm; panes, area exceeding 1.00m2	0.60	10.00	0.66	0.63	Lab	5.81	Sunds 0.11	6.58	Um2	7.41
Wood windows and screens										
girth exceeding 300mm; panes, area not exceeding 0.10m2	0.60	10.00	0.66	1.21	-	11.17	0.21	12.04	m2	13.54
girth exceeding 300mm; panes, area 0.10 - 0.50m2	0.44	10.00	0.48	0.87	-	8.03	0.15	8.66	m2	9.75
girth exceeding 300mm; panes, area 0.50 - 1.00m2	0.33	10.00	0.36	0.73	-	6.74	0.13	7.23	m2	8.13

Labour hourly rates: (except Specialists) Craft Operatives £9.23 Labourer £7.02 Rates are national average prices. Refer to REGIONAL VARIATIONS for indicative levels of overall pricing in regions	MATERIALS			LABOUR				RATES		
	Del to Site £	Waste %	Material Cost £	Craft Optve Hrs	Lab Hrs	Labour Cost £	Sunds £	Nett Rate £	Unit	Gross Rate (+12.5%) £
M60: PAINTING/CLEAR FINISHING - OIL PAINTING WOODWORK Cont.										
One coat primer, one undercoat, two coats full gloss finish on wood surfaces Cont.										
Wood windows and screens Cont.										
girth exceeding 300mm; panes, area exceeding										
1.00m2	0.27	10.00	0.30	0.65	-	6.00	0.11	6.41	m2	7.21
One coat primer, one undercoat, two coats full gloss finish wood surfaces; external work										
Wood general surfaces										
girth exceeding 300mm	1.10	10.00	1.21	0.70	-	6.46	0.12	7.79	m2	8.76
isolated surfaces, girth not exceeding 300mm	0.33	10.00	0.36	0.24	-	2.22	0.04	2.62	m	2.95
Wood glazed doors										
girth exceeding 300mm; panes, area not exceeding										
0.10m2	0.55	10.00	0.60	1.10	-	10.15	0.19	10.95	m2	12.32
girth exceeding 300mm; panes, area 0.10 - 0.50m2 .	0.38	10.00	0.42	0.79	-	7.29	0.14	7.85	m2	8.83
girth exceeding 300mm; panes, area 0.50 - 1.00m2 .	0.27	10.00	0.30	0.66	-	6.09	0.12	6.51	m2	7.32
girth exceeding 300mm; panes, area exceeding										
1.00m2	0.22	10.00	0.24	0.59	-	5.45	0.10	5.79	m2	6.51
Wood partially glazed doors										
girth exceeding 300mm; panes, area not exceeding										
0.10m2	0.82	10.00	0.90	0.88	-	8.12	0.15	9.17	m2	10.32
girth exceeding 300mm; panes, area 0.10 - 0.50m2 .	0.71	10.00	0.78	0.73	-	6.74	0.13	7.65	m2	8.61
girth exceeding 300mm; panes, area 0.50 - 1.00m2 .	0.66	10.00	0.73	0.66	-	6.09	0.12	6.94	m2	7.81
girth exceeding 300mm; panes, area exceeding										
1.00m2	0.60	10.00	0.66	0.63	-	5.81	0.11	6.58	m2	7.41
Wood windows and screens										
girth exceeding 300mm; panes, area not exceeding										
0.10m2	0.60	10.00	0.66	1.25	-	11.54	0.22	12.42	m2	13.97
girth exceeding 300mm; panes, area 0.10 - 0.50m2 .	0.44	10.00	0.48	0.91	-	8.40	0.16	9.04	m2	10.17
girth exceeding 300mm; panes, area 0.50 - 1.00m2 .	0.33	10.00	0.36	0.77	-	7.11	0.13	7.60	m2	8.55
girth exceeding 300mm; panes, area exceeding										
1.00m2	0.27	10.00	0.30	0.69	-	6.37	0.12	6.79	m2	7.63
Wood railings fences and gates; open type										
girth exceeding 300mm	1.10	10.00	1.21	0.55	-	5.08	0.10	6.39	m2	7.18
isolated surfaces, girth not exceeding 300mm	0.33	10.00	0.36	0.19	-	1.75	0.03	2.15	m	2.42
Wood railings fences and gates; close type										
girth exceeding 300mm	1.10	10.00	1.21	0.48	-	4.43	0.08	5.72	m2	6.44
M60: PAINTING/CLEAR FINISHING - POLYURETHANE LACQUER										
Two coats polyurethane lacquer										
Wood general surfaces										
girth exceeding 300mm	0.93	10.00	1.02	0.44	-	4.06	0.08	5.16	m2	5.81
isolated surfaces, girth not exceeding 300mm	0.28	10.00	0.31	0.14	-	1.29	0.02	1.62	m	1.82
Wood glazed doors										
girth exceeding 300mm; panes, area not exceeding										
0.10m2	0.47	10.00	0.52	0.74	-	6.83	0.13	7.48	m2	8.41
girth exceeding 300mm; panes, area 0.10 - 0.50m2 .	0.33	10.00	0.36	0.53	-	4.89	0.09	5.34	m2	6.01
girth exceeding 300mm; panes, area 0.50 - 1.00m2 .	0.23	10.00	0.25	0.44	-	4.06	0.08	4.39	m2	4.94
girth exceeding 300mm; panes, area exceeding										
1.00m2	0.19	10.00	0.21	0.40	-	3.69	0.07	3.97	m2	4.47
Wood partially glazed doors										
girth exceeding 300mm; panes, area not exceeding										
0.10m2	0.70	10.00	0.77	0.59	-	5.45	0.10	6.32	m2	7.11
girth exceeding 300mm; panes, area 0.10 - 0.50m2 .	0.61	10.00	0.67	0.48	-	4.43	0.08	5.18	m2	5.83
girth exceeding 300mm; panes, area 0.50 - 1.00m2 .	0.56	10.00	0.62	0.44	-	4.06	0.08	4.76	m2	5.35
girth exceeding 300mm; panes, area exceeding										
1.00m2	0.51	10.00	0.56	0.42	-	3.88	0.07	4.51	m2	5.07
Wood windows and screens										
girth exceeding 300mm; panes, area not exceeding										
0.10m2	0.51	10.00	0.56	0.81	-	7.48	0.14	8.18	m2	9.20
girth exceeding 300mm; panes, area 0.10 - 0.50m2 .	0.37	10.00	0.41	0.58	-	5.35	0.10	5.86	m2	6.59
girth exceeding 300mm; panes, area 0.50 - 1.00m2 .	0.28	10.00	0.31	0.48	-	4.43	0.08	4.82	m2	5.42
girth exceeding 300mm; panes, area exceeding										
1.00m2	0.23	10.00	0.25	0.44	-	4.06	0.08	4.39	m2	4.94
Two coats polyurethane lacquer; external work										
Wood general surfaces										
girth exceeding 300mm	0.93	10.00	1.02	0.46	-	4.25	0.08	5.35	m2	6.02
isolated surfaces, girth not exceeding 300mm	0.28	10.00	0.31	0.15	-	1.38	0.03	1.72	m	1.94
Wood glazed doors										
girth exceeding 300mm; panes, area not exceeding										
0.10m2	0.47	10.00	0.52	0.74	-	6.83	0.13	7.48	m2	8.41
girth exceeding 300mm; panes, area 0.10 - 0.50m2 .	0.33	10.00	0.36	0.53	-	4.89	0.09	5.34	Um2	6.01
girth exceeding 300mm; panes, area 0.50 - 1.00m2 .	0.23	10.00	0.25	0.44	-	4.06	0.08	4.39	m2	4.94
girth exceeding 300mm; panes, area exceeding										
1.00m2	0.19	10.00	0.21	0.40	-	3.69	0.07	3.97	m2	4.47

Labour hourly rates: (except Specialists) Craft Operatives £9.23 Labourer £7.02 Rates are national average prices. Refer to REGIONAL VARIATIONS for indicative levels of overall pricing in regions	MATERIALS			LABOUR				RATES		
	Del to Site	Waste	Material Cost	Craft Optve	Lab	Labour Cost	Sunds	Nett Rate	Unit	Gross Rate (+12.5%)
	£	%	£	Hrs	Hrs	£	£	£		£
M60: PAINTING/CLEAR FINISHING - POLYURETHANE LACQUER Cont.										
Two coats polyurethane lacquer; external work Cont.										
Wood partially glazed doors										
girth exceeding 300mm; panes, area not exceeding 0.10m2	0.70	10.00	0.77	0.59	-	5.45	0.10	6.32	m2	7.11
girth exceeding 300mm; panes, area 0.10 - 0.50m2	0.61	10.00	0.67	0.48	-	4.43	0.08	5.18	m2	5.83
girth exceeding 300mm; panes, area 0.50 - 1.00m2	0.56	10.00	0.62	0.44	-	4.06	0.08	4.76	m2	5.35
girth exceeding 300mm; panes, area exceeding 1.00m2	0.51	10.00	0.56	0.42	-	3.88	0.07	4.51	m2	5.07
Wood windows and screens										
girth exceeding 300mm; panes, area not exceeding 0.10m2	0.51	10.00	0.56	0.84	-	7.75	0.15	8.46	m2	9.52
girth exceeding 300mm; panes, area 0.10 - 0.50m2	0.37	10.00	0.41	0.61	-	5.63	0.11	6.15	m2	6.92
girth exceeding 300mm; panes, area 0.50 - 1.00m2	0.28	10.00	0.31	0.51	-	4.71	0.09	5.11	m2	5.74
girth exceeding 300mm; panes, area exceeding 1.00m2	0.23	10.00	0.25	0.46	-	4.25	0.08	4.58	m2	5.15
Three coats polyurethane lacquer										
Wood general surfaces										
girth exceeding 300mm	1.40	10.00	1.54	0.52	-	4.80	0.09	6.43	m2	7.23
isolated surfaces, girth not exceeding 300mm	0.42	10.00	0.46	0.18	-	1.66	0.03	2.15	m	2.42
Wood glazed doors										
girth exceeding 300mm; panes, area not exceeding 0.10m2	0.70	10.00	0.77	0.86	-	7.94	0.15	8.86	m2	9.97
girth exceeding 300mm; panes, area 0.10 - 0.50m2	0.49	10.00	0.54	0.62	-	5.72	0.11	6.37	m2	7.17
girth exceeding 300mm; panes, area 0.50 - 1.00m2	0.35	10.00	0.39	0.52	-	4.80	0.09	5.27	m2	5.93
girth exceeding 300mm; panes, area exceeding 1.00m2	0.28	10.00	0.31	0.46	-	4.25	0.08	4.63	m2	5.21
Wood partially glazed doors										
girth exceeding 300mm; panes, area not exceeding 0.10m2	1.05	10.00	1.16	0.69	-	6.37	0.12	7.64	m2	8.60
girth exceeding 300mm; panes, area 0.10 - 0.50m2	0.91	10.00	1.00	0.63	-	5.81	0.11	6.93	m2	7.79
girth exceeding 300mm; panes, area 0.50 - 1.00m2	0.84	10.00	0.92	0.52	-	4.80	0.09	5.81	m2	6.54
girth exceeding 300mm; panes, area exceeding 1.00m2	0.77	10.00	0.85	0.50	-	4.62	0.09	5.55	m2	6.25
Wood windows and screens										
girth exceeding 300mm; panes, area not exceeding 0.10m2	0.77	10.00	0.85	0.95	-	8.77	0.17	9.79	m2	11.01
girth exceeding 300mm; panes, area 0.10 - 0.50m2	0.56	10.00	0.62	0.68	-	6.28	0.12	7.01	m2	7.89
girth exceeding 300mm; panes, area 0.50 - 1.00m2	0.42	10.00	0.46	0.57	-	5.26	0.10	5.82	m2	6.55
girth exceeding 300mm; panes, area exceeding 1.00m2	0.35	10.00	0.39	0.51	-	4.71	0.09	5.18	m2	5.83
Three coats polyurethane lacquer; external work										
Wood general surfaces										
girth exceeding 300mm	1.40	10.00	1.54	0.55	-	5.08	0.10	6.72	m2	7.56
isolated surfaces, girth not exceeding 300mm	0.42	10.00	0.46	0.19	-	1.75	0.03	2.25	m	2.53
Wood glazed doors										
girth exceeding 300mm; panes, area not exceeding 0.10m2	0.70	10.00	0.77	0.86	-	7.94	0.15	8.86	m2	9.97
girth exceeding 300mm; panes, area 0.10 - 0.50m2	0.49	10.00	0.54	0.62	-	5.72	0.11	6.37	m2	7.17
girth exceeding 300mm; panes, area 0.50 - 1.00m2	0.35	10.00	0.39	0.52	-	4.80	0.09	5.27	m2	5.93
girth exceeding 300mm; panes, area exceeding 1.00m2	0.28	10.00	0.31	0.46	-	4.25	0.08	4.63	m2	5.21
Wood partially glazed doors										
girth exceeding 300mm; panes, area not exceeding 0.10m2	1.05	10.00	1.16	0.69	-	6.37	0.12	7.64	m2	8.60
girth exceeding 300mm; panes, area 0.10 - 0.50m2	0.91	10.00	1.00	0.63	-	5.81	0.11	6.93	m2	7.79
girth exceeding 300mm; panes, area 0.50 - 1.00m2	0.84	10.00	0.92	0.52	-	4.80	0.09	5.81	m2	6.54
girth exceeding 300mm; panes, area exceeding 1.00m2	0.77	10.00	0.85	0.50	-	4.62	0.09	5.55	m2	6.25
Wood windows and screens										
girth exceeding 300mm; panes, area not exceeding 0.10m2	0.77	10.00	0.85	0.98	-	9.05	0.17	10.06	m2	11.32
girth exceeding 300mm; panes, area 0.10 - 0.50m2	0.56	10.00	0.62	0.72	-	6.65	0.13	7.39	m2	8.32
girth exceeding 300mm; panes, area 0.50 - 1.00m2	0.42	10.00	0.46	0.61	-	5.63	0.11	6.20	m2	6.98
girth exceeding 300mm; panes, area exceeding 1.00m2	0.35	10.00	0.39	0.54	-	4.98	0.09	5.46	m2	6.14
M60: PAINTING/CLEAR FINISHING - FIRE RETARDANT PAINTS AND VARNISHES										
Two coats fire retardant paint										
Wood general surfaces										
girth exceeding 300mm	2.76	10.00	3.04	0.44	-	4.06	0.08	7.18	m2	8.07
isolated surfaces, girth not exceeding 300mm	0.83	10.00	0.91	0.14	-	1.29	0.02	2.23	m	2.50
Two coats fire retardant varnish, one overcoat varnish										
Wood general surfaces										
girth exceeding 300mm	2.49	10.00	2.74	0.59	-	5.45	0.10	8.28	m2	9.32
isolated surfaces, girth not exceeding 300mm	0.75	10.00	0.82	0.20	-	1.85	0.03	2.70	m	3.04

SURFACE FINISHES

SURFACE FINISHES – SMALL WORKS

Labour hourly rates: (except Specialists) Craft Operatives £9.23 Labourer £7.02 Rates are national average prices. Refer to REGIONAL VARIATIONS for indicative levels of overall pricing in regions	MATERIALS			LABOUR				RATES		
	Del to Site £	Waste %	Material Cost £	Craft Optve Hrs	Lab Hrs	Labour Cost £	Sunds £	Nett Rate £	Unit	Gross Rate (+12.5%) £
M60: PAINTING/CLEAR FINISHING - OILING HARDWOOD										
Two coats raw linseed oil										
Wood general surfaces										
girth exceeding 300mm	1.09	10.00	1.20	0.48	-	4.43	0.08	5.71	m2	6.42
isolated surfaces, girth not exceeding 300mm	0.33	10.00	0.36	0.17	-	1.57	0.03	1.96	m	2.21
M60: PAINTING/CLEAR FINISHING - FRENCH AND WAX POLISHING										
Stain; two coats wax polish										
Wood general surfaces										
girth exceeding 300mm	1.66	10.00	1.83	0.56	-	5.17	0.07	7.06	m2	7.95
isolated surfaces, girth not exceeding 300mm	0.49	10.00	0.54	0.19	-	1.75	0.02	2.31	m	2.60
Open grain French polish										
Wood general surfaces										
girth exceeding 300mm	2.05	10.00	2.25	2.64	-	24.37	0.36	26.98	m2	30.35
isolated surfaces, girth not exceeding 300mm	0.61	10.00	0.67	0.88	-	8.12	0.12	8.91	m	10.03
Stain; body in; fully French polish										
Wood general surfaces										
girth exceeding 300mm	2.93	10.00	3.22	3.96	-	36.55	0.54	40.31	m2	45.75
isolated surfaces, girth not exceeding 300mm	0.88	10.00	0.97	1.32	-	12.18	0.17	13.32	m	14.99
M60: PAINTING/CLEAR FINISHING - ROAD MARKINGS										
One coat spirit based road marking paint; external work										
Concrete general surfaces										
isolated surfaces, 50mm wide	0.07	5.00	0.07	0.12	-	1.11	0.02	1.20	m	1.35
isolated surfaces, 100mm wide	0.14	5.00	0.15	0.14	-	1.29	0.02	1.46	m	1.64
Two coats spirit based road marking paint; external work										
Concrete general surfaces										
isolated surfaces, 50mm wide	0.15	5.00	0.16	0.18	-	1.66	0.02	1.84	m	2.07
isolated surfaces, 100mm wide	0.22	5.00	0.23	0.21	-	1.94	0.03	2.20	m	2.47
Prime and apply non-reflective self adhesive road marking tape; external work										
Concrete general surfaces										
isolated surfaces, 50mm wide	1.14	5.00	1.20	0.12	-	1.11	0.02	2.32	m	2.62
isolated surfaces, 100mm wide..................	2.18	5.00	2.29	0.15	-	1.38	0.02	3.69	m	4.16
Prime and apply reflective self adhesive road marking tape; external work										
Concrete general surfaces										
isolated surfaces, 100mm wide	3.55	5.00	3.73	0.15	-	1.38	0.02	5.13	m	5.77
M60: PAINTING/CLEAR FINISHING (REDECORATIONS) - EMULSION PAINTING										
Generally										
Note the following rates include for the cost of all preparatory work, e.g. washing down, etc.										
Two coats emulsion paint, existing emulsion painted surfaces										
Concrete general surfaces										
girth exceeding 300mm	0.45	10.00	0.50	0.32	-	2.95	0.06	3.51	m2	3.95
Plaster general surfaces										
girth exceeding 300mm	0.36	10.00	0.40	0.28	-	2.58	0.05	3.03	m2	3.41
Plasterboard general surfaces										
girth exceeding 300mm	0.36	10.00	0.40	0.29	-	2.68	0.05	3.12	m2	3.51
Brickwork general surfaces										
girth exceeding 300mm	0.54	10.00	0.59	0.32	-	2.95	0.06	3.61	m2	4.06
Paper covered general surfaces										
girth exceeding 300mm	0.40	10.00	0.44	0.30	-	2.77	0.05	3.26	m2	3.67
Two full coats emulsion paint, existing washable distempered surfaces										
Concrete general surfaces										
girth exceeding 300mm	0.45	10.00	0.50	0.33	-	3.05	0.06	3.60	m2	4.05
Plaster general surfaces										
girth exceeding 300mm	0.36	10.00	0.40	0.29	-	2.68	0.05	3.12	m2	3.51
Plasterboard general surfaces										
girth exceeding 300mm	0.36	10.00	0.40	0.30	-	2.77	0.05	3.21	m2	3.62

Labour hourly rates: (except Specialists) Craft Operatives £9.23 Labourer £7.02 Rates are national average prices. Refer to REGIONAL VARIATIONS for indicative levels of overall pricing in regions	MATERIALS			LABOUR				RATES		
	Del to Site £	Waste %	Material Cost £	Craft Optve Hrs	Lab Hrs	Labour Cost £	Sunds £	Nett Rate £	Unit	Gross Rate (+12.5%) £
M60: PAINTING/CLEAR FINISHING (REDECORATIONS) – EMULSION PAINTING Cont.										
Two full coats emulsion paint, existing washable distempered surfaces Cont.										
Brickwork general surfaces girth exceeding 300mm	0.54	10.00	0.59	0.33	–	3.05	0.06	3.70	m2	4.16
Paper covered general surfaces girth exceeding 300mm	0.40	10.00	0.44	0.30	–	2.77	0.05	3.26	m2	3.67
Two full coats emulsion paint, existing textured plastic coating surfaces										
Concrete general surfaces girth exceeding 300mm	0.45	10.00	0.50	0.36	–	3.32	0.06	3.88	m2	4.36
Plaster general surfaces girth exceeding 300mm	0.36	10.00	0.40	0.32	–	2.95	0.06	3.41	m2	3.84
Plasterboard general surfaces girth exceeding 300mm	0.36	10.00	0.40	0.33	–	3.05	0.06	3.50	m2	3.94
Brickwork general surfaces girth exceeding 300mm	0.54	10.00	0.59	0.36	–	3.32	0.06	3.98	m2	4.47
Paper covered general surfaces girth exceeding 300mm	0.40	10.00	0.44	0.33	–	3.05	0.06	3.55	m2	3.99
Mist coat, two full coats emulsion paint, existing non-washable distempered surfaces										
Concrete general surfaces girth exceeding 300mm	0.54	10.00	0.59	0.38	–	3.51	0.07	4.17	m2	4.69
Plaster general surfaces girth exceeding 300mm	0.45	10.00	0.50	0.34	–	3.14	0.06	3.69	m2	4.15
Plasterboard general surfaces girth exceeding 300mm	0.45	10.00	0.50	0.35	–	3.23	0.06	3.79	m2	4.26
Brickwork general surfaces girth exceeding 300mm	0.67	10.00	0.74	0.38	–	3.51	0.07	4.31	m2	4.85
Paper covered general surfaces girth exceeding 300mm	0.49	10.00	0.54	0.35	–	3.23	0.06	3.83	m2	4.31
Two full coats emulsion paint, existing gloss painted surfaces										
Concrete general surfaces girth exceeding 300mm	0.45	10.00	0.50	0.33	–	3.05	0.06	3.60	m2	4.05
Plaster general surfaces girth exceeding 300mm	0.36	10.00	0.40	0.29	–	2.68	0.05	3.12	m2	3.51
Plasterboard general surfaces girth exceeding 300mm	0.36	10.00	0.40	0.30	–	2.77	0.05	3.21	m2	3.62
Brickwork general surfaces girth exceeding 300mm	0.54	10.00	0.59	0.33	–	3.05	0.06	3.70	m2	4.16
Paper covered general surfaces girth exceeding 300mm	0.40	10.00	0.44	0.30	–	2.77	0.05	3.26	m2	3.67
M60: PAINTING/CLEAR FINISHING (REDECORATIONS) – CEMENT PAINTING										
Generally										
Note the following rates include for the cost of all preparatory work, e.g. washing down, etc. **One coat sealer, one coat Snowcem, existing** **undecorated surfaces; external work**										
Cement rendered general surfaces girth exceeding 300mm	0.87	10.00	0.96	0.36	–	3.32	0.06	4.34	m2	4.88
Concrete general surfaces girth exceeding 300mm	0.96	10.00	1.06	0.43	–	3.97	0.08	5.10	m2	5.74
Brickwork general surfaces girth exceeding 300mm	0.96	10.00	1.06	0.43	–	3.97	0.08	5.10	m2	5.74
Rough cast general surfaces girth exceeding 300mm	1.02	10.00	1.12	0.52	–	4.80	0.09	6.01	m2	6.76
One coat sealer, two coats Snowcem, existing **undecorated surfaces; external work**										
Cement rendered general surfaces girth exceeding 300mm	1.10	10.00	1.21	0.47	–	4.34	0.08	5.63	m2	6.33

SURFACE FINISHES

Labour hourly rates: (except Specialists) Craft Operatives £9.23 Labourer £7.02 Rates are national average prices. Refer to REGIONAL VARIATIONS for indicative levels of overall pricing in regions	MATERIALS			LABOUR				RATES		
	Del to Site	Waste	Material Cost	Craft Optve	Lab	Labour Cost	Sunds	Nett Rate	Unit	Gross Rate (+12.5%)
	£	%	£	Hrs	Hrs	£	£	£		£

M60: PAINTING/CLEAR FINISHING (REDECORATIONS) – CEMENT PAINTING Cont.

One coat sealer, two coats Snowcem, existing undecorated surfaces; external work Cont.

	Del to Site £	Waste %	Material Cost £	Craft Optve Hrs	Lab Hrs	Labour Cost £	Sunds £	Nett Rate £	Unit	Gross Rate £
Concrete general surfaces girth exceeding 300mm	1.28	10.00	1.41	0.58	–	5.35	0.10	6.86	m2	7.72
Brickwork general surfaces girth exceeding 300mm	1.28	10.00	1.41	0.58	–	5.35	0.10	6.86	m2	7.72
Rough cast general surfaces girth exceeding 300mm	1.37	10.00	1.51	0.70	–	6.46	0.12	8.09	m2	9.10
One coat Snowcem, existing cement painted surfaces; external work										
Cement rendered general surfaces girth exceeding 300mm	0.29	10.00	0.32	0.27	–	2.49	0.05	2.86	m2	3.22
Concrete general surfaces girth exceeding 300mm	0.39	10.00	0.43	0.34	–	3.14	0.06	3.63	m2	4.08
Brickwork general surfaces girth exceeding 300mm	0.39	10.00	0.43	0.34	–	3.14	0.06	3.63	m2	4.08
Rough cast general surfaces girth exceeding 300mm	0.44	10.00	0.48	0.44	–	4.06	0.08	4.63	m2	5.20
Two coats Snowcem, existing cement painted surfaces; external work										
Cement rendered general surfaces girth exceeding 300mm	0.53	10.00	0.58	0.38	–	3.51	0.07	4.16	m2	4.68
Concrete general surfaces girth exceeding 300mm	0.70	10.00	0.77	0.49	–	4.52	0.09	5.38	m2	6.06
Brickwork general surfaces girth exceeding 300mm	0.70	10.00	0.77	0.49	–	4.52	0.09	5.38	m2	6.06
Rough cast general surfaces girth exceeding 300mm	0.79	10.00	0.87	0.60	–	5.54	0.10	6.51	m2	7.32
One coat stabilising solution, one coat textured masonry paint, existing undecorated surfaces; external work										
Cement rendered general surfaces girth exceeding 300mm	1.14	10.00	1.25	0.35	–	3.23	0.06	4.54	m2	5.11
Concrete general surfaces girth exceeding 300mm	1.28	10.00	1.41	0.42	–	3.88	0.07	5.35	m2	6.02
Brickwork general surfaces girth exceeding 300mm	1.41	10.00	1.55	0.42	–	3.88	0.07	5.50	m2	6.18
Rough cast general surfaces girth exceeding 300mm	1.91	10.00	2.10	0.52	–	4.80	0.09	6.99	m2	7.86
One coat stabilising solution, two coats textured masonry paint, existing undecorated surfaces; external work										
Cement rendered general surfaces girth exceeding 300mm	1.63	10.00	1.79	0.47	–	4.34	0.08	6.21	m2	6.99
Concrete general surfaces girth exceeding 300mm	1.82	10.00	2.00	0.57	–	5.26	0.10	7.36	m2	8.28
Brickwork general surfaces girth exceeding 300mm	2.03	10.00	2.23	0.57	–	5.26	0.10	7.59	m2	8.54
Rough cast general surfaces girth exceeding 300mm	2.90	10.00	3.19	0.70	–	6.46	0.12	9.77	m2	10.99
One coat textured masonry paint, existing cement painted surfaces; external work										
Cement rendered general surfaces girth exceeding 300mm	0.61	10.00	0.67	0.26	–	2.40	0.05	3.12	m2	3.51
Concrete general surfaces girth exceeding 300mm	0.69	10.00	0.76	0.33	–	3.05	0.06	3.86	m2	4.35
Brickwork general surfaces girth exceeding 300mm	0.76	10.00	0.84	0.33	–	3.05	0.06	3.94	m2	4.43
Rough cast general surfaces girth exceeding 300mm	1.23	10.00	1.35	0.43	–	3.97	0.08	5.40	m2	6.08

Labour hourly rates: (except Specialists) Craft Operatives £9.23 Labourer £7.02 Rates are national average prices. Refer to REGIONAL VARIATIONS for indicative levels of overall pricing in regions	MATERIALS			LABOUR				RATES		
	Del to Site £	Waste %	Material Cost £	Craft Optve Hrs	Lab Hrs	Labour Cost £	Sunds £	Nett Rate £	Unit	Gross Rate (+12.5%) £
M60: PAINTING/CLEAR FINISHING (REDECORATIONS) – CEMENT PAINTING Cont.										
Two coats textured masonry paint, existing cement painted surfaces; external work										
Cement rendered general surfaces girth exceeding 300mm	1.11	10.00	1.22	0.39	–	3.60	0.07	4.89	m2	5.50
Concrete general surfaces girth exceeding 300mm	1.23	10.00	1.35	0.48	–	4.43	0.08	5.86	m2	6.60
Brickwork general surfaces girth exceeding 300mm	1.38	10.00	1.52	0.48	–	4.43	0.08	6.03	m2	6.78
Rough cast general surfaces girth exceeding 300mm	2.21	10.00	2.43	0.62	–	5.72	0.11	8.26	m2	9.30
M60: PAINTING/CLEAR FINISHING (REDECORATIONS) – OIL PAINTING WALLS AND CEILINGS										
Generally										
Note the following rates include for the cost of all preparatory work, e.g. washing down, etc.										
One undercoat, one coat full gloss finish, existing gloss painted surfaces										
Concrete general surfaces girth exceeding 300mm	0.59	10.00	0.65	0.38	–	3.51	0.07	4.23	m2	4.75
Brickwork general surfaces girth exceeding 300mm	0.72	10.00	0.79	0.39	–	3.60	0.07	4.46	m2	5.02
Plasterboard general surfaces girth exceeding 300mm	0.46	10.00	0.51	0.35	–	3.23	0.06	3.80	m2	4.27
Plaster general surfaces girth exceeding 300mm	0.52	10.00	0.57	0.34	–	3.14	0.06	3.77	m2	4.24
One undercoat, one coat eggshell finish, existing gloss painted surfaces										
Concrete general surfaces girth exceeding 300mm	0.62	10.00	0.68	0.38	–	3.51	0.07	4.26	m2	4.79
Brickwork general surfaces girth exceeding 300mm	0.76	10.00	0.84	0.39	–	3.60	0.07	4.51	m2	5.07
Plasterboard general surfaces girth exceeding 300mm	0.48	10.00	0.53	0.35	–	3.23	0.06	3.82	m2	4.30
Plaster general surfaces girth exceeding 300mm	0.55	10.00	0.60	0.34	–	3.14	0.06	3.80	m2	4.28
Two undercoats, one coat full gloss finish, existing gloss painted surfaces										
Concrete general surfaces girth exceeding 300mm	0.88	10.00	0.97	0.48	–	4.43	0.08	5.48	m2	6.16
Brickwork general surfaces girth exceeding 300mm	1.08	10.00	1.19	0.48	–	4.43	0.08	5.70	m2	6.41
Plasterboard general surfaces girth exceeding 300mm	0.69	10.00	0.76	0.45	–	4.15	0.08	4.99	m2	5.62
Plaster general surfaces girth exceeding 300mm	0.78	10.00	0.86	0.44	–	4.06	0.08	5.00	m2	5.62
Two undercoats, one coat eggshell finish, existing gloss painted surfaces										
Concrete general surfaces girth exceeding 300mm	0.92	10.00	1.01	0.48	–	4.43	0.08	5.52	m2	6.21
Brickwork general surfaces girth exceeding 300mm	1.12	10.00	1.23	0.48	–	4.43	0.08	5.74	m2	6.46
Plasterboard general surfaces girth exceeding 300mm	0.71	10.00	0.78	0.45	–	4.15	0.08	5.01	m2	5.64
Plaster general surfaces girth exceeding 300mm	0.82	10.00	0.90	0.44	–	4.06	0.08	5.04	m2	5.67
One coat primer, one undercoat, one coat full gloss finish, existing washable distempered or emulsion painted surfaces										
Concrete general surfaces girth exceeding 300mm	0.95	10.00	1.04	0.48	–	4.43	0.08	5.56	m2	6.25
Brickwork general surfaces girth exceeding 300mm	1.17	10.00	1.29	0.48	–	4.43	0.08	5.80	m2	6.52

SURFACE FINISHES

Labour hourly rates: (except Specialists) Craft Operatives £9.23 Labourer £7.02 Rates are national average prices. Refer to REGIONAL VARIATIONS for indicative levels of overall pricing in regions	MATERIALS			LABOUR				RATES		
	Del to Site £	Waste %	Material Cost £	Craft Optve Hrs	Lab Hrs	Labour Cost £	Sunds £	Nett Rate £	Unit	Gross Rate (+12.5%) £
M60: PAINTING/CLEAR FINISHING (REDECORATIONS) – OIL PAINTING WALLS AND CEILINGS Cont.										
One coat primer, one undercoat, one coat full gloss finish, existing washable distempered or emulsion painted surfaces Cont.										
Plasterboard general surfaces girth exceeding 300mm	0.74	10.00	0.81	0.45	–	4.15	0.08	5.05	m2	5.68
Plaster general surfaces girth exceeding 300mm	0.85	10.00	0.94	0.44	–	4.06	0.08	5.08	m2	5.71
One coat primer, one undercoat, one coat eggshell finish, existing washable distempered or emulsion painted surfaces										
Concrete general surfaces girth exceeding 300mm	0.99	10.00	1.09	0.48	–	4.43	0.08	5.60	m2	6.30
Brickwork general surfaces girth exceeding 300mm	1.21	10.00	1.33	0.48	–	4.43	0.08	5.84	m2	6.57
Plasterboard general surfaces girth exceeding 300mm	0.77	10.00	0.85	0.45	–	4.15	0.08	5.08	m2	5.72
Plaster general surfaces girth exceeding 300mm	0.88	10.00	0.97	0.44	–	4.06	0.08	5.11	m2	5.75
One coat primer, two undercoats, one coat full gloss finish, existing washable distempered or emulsion painted surfaces										
Concrete general surfaces girth exceeding 300mm	1.25	10.00	1.38	0.58	–	5.35	0.10	6.83	m2	7.68
Brickwork general surfaces girth exceeding 300mm	1.53	10.00	1.68	0.59	–	5.45	0.10	7.23	m2	8.13
Plasterboard general surfaces girth exceeding 300mm	0.97	10.00	1.07	0.55	–	5.08	0.10	6.24	m2	7.02
Plaster general surfaces girth exceeding 300mm	1.11	10.00	1.22	0.54	–	4.98	0.09	6.30	m2	7.08
One coat primer, two undercoats, one coat eggshell finish, existing washable distempered or emulsion painted surfaces										
Concrete general surfaces girth exceeding 300mm	1.28	10.00	1.41	0.58	–	5.35	0.10	6.86	m2	7.72
Brickwork general surfaces girth exceeding 300mm	1.57	10.00	1.73	0.59	–	5.45	0.10	7.27	m2	8.18
Plasterboard general surfaces girth exceeding 300mm	1.00	10.00	1.10	0.55	–	5.08	0.10	6.28	m2	7.06
Plaster general surfaces girth exceeding 300mm	1.14	10.00	1.25	0.54	–	4.98	0.09	6.33	m2	7.12
One coat primer, one undercoat, one coat full gloss finish, existing non-washable distempered surfaces										
Concrete general surfaces girth exceeding 300mm	0.95	10.00	1.04	0.48	–	4.43	0.08	5.56	m2	6.25
Brickwork general surfaces girth exceeding 300mm	1.17	10.00	1.29	0.48	–	4.43	0.08	5.80	m2	6.52
Plasterboard general surfaces girth exceeding 300mm	0.74	10.00	0.81	0.45	–	4.15	0.08	5.05	m2	5.68
Plaster general surfaces girth exceeding 300mm	0.85	10.00	0.94	0.44	–	4.06	0.08	5.08	m2	5.71
One coat primer, one undercoat, one coat eggshell finish, existing non-washable distempered surfaces										
Concrete general surfaces girth exceeding 300mm	0.99	10.00	1.09	0.06	–	0.55	0.08	1.72	m2	1.94
Brickwork general surfaces girth exceeding 300mm	1.21	10.00	1.33	0.48	–	4.43	0.08	5.84	m2	6.57
Plasterboard general surfaces girth exceeding 300mm	0.77	10.00	0.85	0.45	–	4.15	0.08	5.08	m2	5.72
Plaster general surfaces girth exceeding 300mm	0.88	10.00	0.97	0.44	–	4.06	0.08	5.11	m2	5.75
One coat primer, two undercoats, one coat full gloss finish, existing non-washable distempered surfaces										
Concrete general surfaces girth exceeding 300mm	1.25	10.00	1.38	0.58	–	5.35	0.10	6.83	m2	7.68

Labour hourly rates: (except Specialists) Craft Operatives £9.23 Labourer £7.02 Rates are national average prices. Refer to REGIONAL VARIATIONS for indicative levels of overall pricing in regions	MATERIALS			LABOUR				RATES		
	Del to Site £	Waste %	Material Cost £	Craft Optve Hrs	Lab Hrs	Labour Cost £	Sunds £	Nett Rate £	Unit	Gross Rate (+12.5%) £
M60: PAINTING/CLEAR FINISHING (REDECORATIONS) - OIL PAINTING WALLS AND CEILINGS Cont.										
One coat primer, two undercoats, one coat full gloss finish, existing non-washable distempered surfaces Cont.										
Brickwork general surfaces girth exceeding 300mm	1.53	10.00	1.68	0.59	-	5.45	0.10	7.23	m2	8.13
Plasterboard general surfaces girth exceeding 300mm	0.97	10.00	1.07	0.55	-	5.08	0.10	6.24	m2	7.02
Plaster general surfaces girth exceeding 300mm	1.11	10.00	1.22	0.54	-	4.98	0.09	6.30	m2	7.08
One coat primer, two undercoats, one coat eggshell finish, existing non-washable distempered surfaces										
Concrete general surfaces girth exceeding 300mm	1.28	10.00	1.41	0.58	-	5.35	0.10	6.86	m2	7.72
Brickwork general surfaces girth exceeding 300mm	1.57	10.00	1.73	0.59	-	5.45	0.10	7.27	m2	8.18
Plasterboard general surfaces girth exceeding 300mm	1.00	10.00	1.10	0.55	-	5.08	0.10	6.28	m2	7.06
Plaster general surfaces girth exceeding 300mm	1.14	10.00	1.25	0.54	-	4.98	0.09	6.33	m2	7.12
M60: PAINTING/CLEAR FINISHING (REDECORATIONS) PLASTIC FINISH										
Generally										
Note the following rates include for the cost of all preparatory work, e.g. washing down, etc.										
Textured plastic coating - stippled finish, existing washable distempered or emulsion painted surfaces										
Concrete general surfaces girth exceeding 300mm	0.57	10.00	0.63	0.40	-	3.69	0.07	4.39	m2	4.94
Brickwork general surfaces girth exceeding 300mm	0.69	10.00	0.76	0.45	-	4.15	0.08	4.99	m2	5.62
Plasterboard general surfaces girth exceeding 300mm	0.46	10.00	0.51	0.34	-	3.14	0.06	3.70	m2	4.17
Plaster general surfaces girth exceeding 300mm	0.46	10.00	0.51	0.34	-	3.14	0.06	3.70	m2	4.17
Textured plastic coating - combed finish, existing washable distempered or emulsion painted surfaces										
Concrete general surfaces girth exceeding 300mm	0.57	10.00	0.63	0.45	-	4.15	0.08	4.86	m2	5.47
Brickwork general surfaces girth exceeding 300mm	0.69	10.00	0.76	0.50	-	4.62	0.09	5.46	m2	6.15
Plasterboard general surfaces girth exceeding 300mm	0.46	10.00	0.51	0.40	-	3.69	0.07	4.27	m2	4.80
Plaster general surfaces girth exceeding 300mm	0.46	10.00	0.51	0.40	-	3.69	0.07	4.27	m2	4.80
Textured plastic coating - stippled finish, existing non-washable distempered surfaces										
Concrete general surfaces girth exceeding 300mm	0.57	10.00	0.63	0.45	-	4.15	0.08	4.86	m2	5.47
Brickwork general surfaces girth exceeding 300mm	0.69	10.00	0.76	0.50	-	4.62	0.09	5.46	m2	6.15
Plasterboard general surfaces girth exceeding 300mm	0.46	10.00	0.51	0.40	-	3.69	0.07	4.27	m2	4.80
Plaster general surfaces girth exceeding 300mm	0.46	10.00	0.51	0.40	-	3.69	0.07	4.27	m2	4.80
Textured plastic coating - combed finish, existing non-washable distempered surfaces										
Concrete general surfaces girth exceeding 300mm	0.57	10.00	0.63	0.50	-	4.62	0.09	5.33	m2	6.00
Brickwork general surfaces girth exceeding 300mm	0.69	10.00	0.76	0.56	-	5.17	0.10	6.03	m2	6.78
Plasterboard general surfaces girth exceeding 300mm	0.46	10.00	0.51	0.45	-	4.15	0.08	4.74	m2	5.33

SURFACE FINISHES

Labour hourly rates: (except Specialists) Craft Operatives £9.23 Labourer £7.02 Rates are national average prices. Refer to REGIONAL VARIATIONS for indicative levels of overall pricing in regions	MATERIALS			LABOUR				RATES		
	Del to Site £	Waste %	Material Cost £	Craft Optve Hrs	Lab Hrs	Labour Cost £	Sunds £	Nett Rate £	Unit	Gross Rate (+12.5%) £
M60: PAINTING/CLEAR FINISHING (REDECORATIONS) - PLASTIC FINISH Cont.										
Textured plastic coating - combed finish, existing non-washable distempered surfaces Cont.										
Plaster general surfaces										
girth exceeding 300mm	0.46	10.00	0.51	0.45	-	4.15	0.08	4.74	m2	5.33
M60: PAINTING/CLEAR FINISHING (REDECORATIONS) - CLEAN OUT GUTTERS										
Generally										
Note the following rates include for the cost of all preparatory work, e.g. washing down, etc. Clean out gutters prior to repainting, staunch joints with red lead, bituminous compound or mastic										
Iron or steel eaves gutters										
generally	-	-	-	0.06	-	0.55	-	0.55	m	0.62
M60: PAINTING/CLEAR FINISHING (REDECORATIONS) - OIL PAINTING METALWORK										
Generally										
Note the following rates include for the cost of all preparatory work, e.g. washing down, etc. One undercoat, one coat full gloss finish, existing gloss painted metal surfaces										
Iron or steel general surfaces										
girth exceeding 300mm	0.52	10.00	0.57	0.47	-	4.34	0.08	4.99	m2	5.61
isolated surfaces, girth not exceeding 300mm	0.16	10.00	0.18	0.16	-	1.48	0.03	1.68	m	1.89
Galvanised glazed doors, windows or screens										
girth exceeding 300mm; panes, area not exceeding 0.10m2	0.14	10.00	0.15	0.88	-	8.12	0.15	8.43	m2	9.48
girth exceeding 300mm; panes, area 0.10 - 0.50m2 .	0.12	10.00	0.13	0.64	-	5.91	0.11	6.15	m2	6.92
girth exceeding 300mm; panes, area 0.50 - 1.00m2 .	0.10	10.00	0.11	0.53	-	4.89	0.09	5.09	m2	5.73
girth exceeding 300mm; panes, area exceeding 1.00m2	0.09	10.00	0.10	0.46	-	4.25	0.08	4.42	m2	4.98
Iron or steel structural work										
girth exceeding 300mm	0.52	10.00	0.57	0.52	-	4.80	0.09	5.46	m2	6.14
isolated surfaces, girth not exceeding 300mm	0.16	10.00	0.18	0.21	-	1.94	0.04	2.15	m	2.42
Iron or steel structural members of roof trusses, lattice girders, purlins and the like										
girth exceeding 300mm	0.52	10.00	0.57	0.65	-	6.00	0.11	6.68	m2	7.52
isolated surfaces, girth not exceeding 300mm	0.16	10.00	0.18	0.22	-	2.03	0.04	2.25	m	2.53
Iron or steel services										
girth exceeding 300mm	0.52	10.00	0.57	0.65	-	6.00	0.11	6.68	m2	7.52
isolated surfaces, girth not exceeding 300mm	0.16	10.00	0.18	0.22	-	2.03	0.04	2.25	m	2.53
isolated areas not exceeding 0.50m2 irrespective of girth	0.26	10.00	0.29	0.65	-	6.00	0.11	6.40	nr	7.19
Copper services										
girth exceeding 300mm	0.52	10.00	0.57	0.65	-	6.00	0.11	6.68	m2	7.52
isolated surfaces, girth not exceeding 300mm	0.16	10.00	0.18	0.22	-	2.03	0.04	2.25	m	2.53
Galvanised services										
girth exceeding 300mm	0.52	10.00	0.57	0.65	-	6.00	0.11	6.68	m2	7.52
isolated surfaces, girth not exceeding 300mm	0.16	10.00	0.18	0.22	-	2.03	0.04	2.25	m	2.53
One undercoat, one coat full gloss finish, existing gloss painted metal surfaces; external work										
Iron or steel general surfaces										
girth exceeding 300mm	0.52	10.00	0.57	0.50	-	4.62	0.09	5.28	m2	5.94
isolated surfaces, girth not exceeding 300mm	0.16	10.00	0.18	0.17	-	1.57	0.03	1.78	m	2.00
Galvanised glazed doors, windows or screens										
girth exceeding 300mm; panes, area not exceeding 0.10m2	0.14	10.00	0.15	0.90	-	8.31	0.16	8.62	m2	9.70
girth exceeding 300mm; panes, area 0.10 - 0.50m2 .	0.12	10.00	0.13	0.67	-	6.18	0.12	6.44	m2	7.24
girth exceeding 300mm; panes, area 0.50 - 1.00m2 .	0.10	10.00	0.11	0.55	-	5.08	0.10	5.29	m2	5.95
girth exceeding 300mm; panes, area exceeding 1.00m2	0.09	10.00	0.10	0.49	-	4.52	0.09	4.71	m2	5.30
Iron or steel structural work										
girth exceeding 300mm	0.52	10.00	0.57	0.52	-	4.80	0.09	5.46	m2	6.14
isolated surfaces, girth not exceeding 300mm	0.16	10.00	0.18	0.21	-	1.94	0.04	2.15	m	2.42
Iron or steel structural members of roof trusses, lattice girders, purlins and the like										
girth exceeding 300mm	0.52	10.00	0.57	0.65	-	6.00	0.11	6.68	m2	7.52
isolated surfaces, girth not exceeding 300mm	0.16	10.00	0.18	0.22	-	2.03	0.04	2.25	m	2.53
Iron or steel railings, fences and gates; plain open type										
girth exceeding 300mm	0.52	10.00	0.57	0.49	-	4.52	0.09	5.18	m2	5.83

Labour hourly rates: (except Specialists) Craft Operatives £9.23 Labourer £7.02 Rates are national average prices. Refer to REGIONAL VARIATIONS for indicative levels of overall pricing in regions	MATERIALS			LABOUR				RATES		
	Del to Site £	Waste %	Material Cost £	Craft Optve Hrs	Lab Hrs	Labour Cost £	Sunds £	Nett Rate £	Unit	Gross Rate (+12.5%) £
M60: PAINTING/CLEAR FINISHING (REDECORATIONS) - OIL PAINTING METALWORK Cont.										
One undercoat, one coat full gloss finish, existing gloss painted metal surfaces; external work Cont.										
Iron or steel railings, fences and gates; plain open type Cont.										
isolated surfaces, girth not exceeding 300mm	0.16	10.00	0.18	0.16	-	1.48	0.03	1.68	m	1.89
Iron or steel railings, fences and gates; close type										
girth exceeding 300mm	0.52	10.00	0.57	0.45	-	4.15	0.08	4.81	m2	5.41
Iron or steel railings, fences and gates; ornamental type										
girth exceeding 300mm	0.52	10.00	0.57	0.90	-	8.31	0.16	9.04	m2	10.17
Iron or steel eaves gutters										
girth exceeding 300mm	0.52	10.00	0.57	0.53	-	4.89	0.09	5.55	m2	6.25
isolated surfaces, girth not exceeding 300mm	0.16	10.00	0.18	0.17	-	1.57	0.03	1.78	m	2.00
Galvanised eaves gutters										
girth exceeding 300mm	0.52	10.00	0.57	0.53	-	4.89	0.09	5.55	m2	6.25
isolated surfaces, girth not exceeding 300mm	0.16	10.00	0.18	0.17	-	1.57	0.03	1.78	m	2.00
Iron or steel services										
girth exceeding 300mm	0.52	10.00	0.57	0.65	-	6.00	0.11	6.68	m2	7.52
isolated surfaces, girth not exceeding 300mm	0.16	10.00	0.18	0.22	-	2.03	0.04	2.25	m	2.53
isolated areas not exceeding 0.50m2 irrespective of girth..........	0.26	10.00	0.29	0.65	-	6.00	0.11	6.40	nr	7.19
Copper services										
girth exceeding 300mm	0.52	10.00	0.57	0.65	-	6.00	0.11	6.68	m2	7.52
isolated surfaces, girth not exceeding 300mm	0.16	10.00	0.18	0.22	-	2.03	0.04	2.25	m	2.53
Galvanised services										
girth exceeding 300mm	0.52	10.00	0.57	0.65	-	6.00	0.11	6.68	m2	7.52
isolated surfaces, girth not exceeding 300mm	0.16	10.00	0.18	0.22	-	2.03	0.04	2.25	m	2.53
Two coats full gloss finish, existing gloss painted metal surfaces										
Iron or steel general surfaces										
girth exceeding 300mm	0.52	10.00	0.57	0.47	-	4.34	0.08	4.99	m2	5.61
isolated surfaces, girth not exceeding 300mm	0.16	10.00	0.18	0.16	-	1.48	0.03	1.68	m	1.89
Galvanised glazed doors, windows or screens										
girth exceeding 300mm; panes, area not exceeding 0.10m2................	0.14	10.00	0.15	0.88	-	8.12	0.15	8.43	m2	9.48
girth exceeding 300mm; panes, area 0.10 - 0.50m2..	0.12	10.00	0.13	0.64	-	5.91	0.11	6.15	m2	6.92
girth exceeding 300mm; panes, area 0.50 - 1.00m2 .	0.10	10.00	0.11	0.53	-	4.89	0.09	5.09	m2	5.73
girth exceeding 300mm; panes, area exceeding 1.00m2................	0.09	10.00	0.10	0.46	-	4.25	0.08	4.42	m2	4.98
Iron or steel structural work										
girth exceeding 300mm	0.52	10.00	0.57	0.52	-	4.80	0.09	5.46	m2	6.14
isolated surfaces, girth not exceeding 300mm	0.16	10.00	0.18	0.21	-	1.94	0.04	2.15	m	2.42
Iron or steel structural members of roof trusses, lattice girders, purlins and the like										
girth exceeding 300mm	0.52	10.00	0.57	0.65	-	6.00	0.11	6.68	m2	7.52
isolated surfaces, girth not exceeding 300mm	0.16	10.00	0.18	0.22	-	2.03	0.04	2.25	m	2.53
Iron or steel services										
girth exceeding 300mm	0.52	10.00	0.57	0.65	-	6.00	0.11	6.68	m2	7.52
isolated surfaces, girth not exceeding 300mm	0.16	10.00	0.18	0.22	-	2.03	0.04	2.25	m	2.53
isolated areas not exceeding 0.50m2 irrespective of girth..........	0.26	10.00	0.29	0.65	-	6.00	0.11	6.40	nr	7.19
Copper services										
girth exceeding 300mm	0.52	10.00	0.57	0.65	-	6.00	0.11	6.68	m2	7.52
isolated surfaces, girth not exceeding 300mm	0.16	10.00	0.18	0.22	-	2.03	0.04	2.25	m	2.53
Galvanised services										
girth exceeding 300mm	0.52	10.00	0.57	0.65	-	6.00	0.11	6.68	m2	7.52
isolated surfaces, girth not exceeding 300mm	0.16	10.00	0.18	0.22	-	2.03	0.04	2.25	m	2.53
Two coats full gloss finish, existing gloss painted metal surfaces; external work										
Iron or steel general surfaces										
girth exceeding 300mm	0.52	10.00	0.57	0.50	-	4.62	0.09	5.28	m2	5.94
isolated surfaces, girth not exceeding 300mm	0.16	10.00	0.18	0.17	-	1.57	0.03	1.78	m	2.00
Galvanised glazed doors, windows or screens										
girth exceeding 300mm; panes, area not exceeding 0.10m2................	0.14	10.00	0.15	0.90	-	8.31	0.16	8.62	m2	9.70
girth exceeding 300mm; panes, area 0.10 - 0.50m2..	0.12	10.00	0.13	0.67	-	6.18	0.12	6.44	m2	7.24
girth exceeding 300mm; panes, area 0.50 - 1.00m2 .	0.10	10.00	0.11	0.55	-	5.08	0.10	5.29	m2	5.95
girth exceeding 300mm; panes, area exceeding 1.00m2................	0.09	10.00	0.10	0.49	-	4.52	0.09	4.71	m2	5.30
Iron or steel structural work										
girth exceeding 300mm	0.52	10.00	0.57	0.52	-	4.80	0.09	5.46	m2	6.14
isolated surfaces, girth not exceeding 300mm	0.16	10.00	0.18	0.21	-	1.94	0.04	2.15	m	2.42

SURFACE FINISHES

	MATERIALS			LABOUR				RATES		
Labour hourly rates: (except Specialists) Craft Operatives £9.23 Labourer £7.02 Rates are national average prices. Refer to REGIONAL VARIATIONS for indicative levels of overall pricing in regions	Del to Site	Waste	Material Cost	Craft Optve	Lab	Labour Cost	Sunds	Nett Rate	Unit	Gross Rate (+12.5%)
	£	%	£	Hrs	Hrs	£	£	£		£

M60: PAINTING/CLEAR FINISHING (REDECORATIONS) - OIL PAINTING METALWORK Cont.

Two coats full gloss finish, existing gloss painted metal surfaces; external work Cont.

	Del to Site	Waste	Material Cost	Craft Optve	Lab	Labour Cost	Sunds	Nett Rate	Unit	Gross Rate
Iron or steel structural members of roof trusses, lattice girders, purlins and the like										
girth exceeding 300mm	0.52	10.00	0.57	0.65	-	6.00	0.11	6.68	m2	7.52
isolated surfaces, girth not exceeding 300mm	0.16	10.00	0.18	0.22	-	2.03	0.04	2.25	m	2.53
Iron or steel railings, fences and gates; plain open type										
girth exceeding 300mm	0.52	10.00	0.57	0.49	-	4.52	0.09	5.18	m2	5.83
isolated surfaces, girth not exceeding 300mm	0.16	10.00	0.18	0.16	-	1.48	0.03	1.68	m	1.89
Iron or steel railings, fences and gates; close type										
girth exceeding 300mm	0.52	10.00	0.57	0.45	-	4.15	0.08	4.81	m2	5.41
Iron or steel railings, fences and gates; ornamental type										
girth exceeding 300mm	0.52	10.00	0.57	0.82	-	7.57	0.16	8.30	m2	9.34
Iron or steel eaves gutters										
girth exceeding 300mm	0.52	10.00	0.57	0.53	-	4.89	0.09	5.55	m2	6.25
isolated surfaces, girth not exceeding 300mm	0.16	10.00	0.18	0.17	-	1.57	0.03	1.78	m	2.00
Galvanised eaves gutters										
girth exceeding 300mm	0.52	10.00	0.57	0.53	-	4.89	0.09	5.55	m2	6.25
isolated surfaces, girth not exceeding 300mm	0.16	10.00	0.18	0.17	-	1.57	0.03	1.78	m	2.00
Iron or steel services										
girth exceeding 300mm	0.52	10.00	0.57	0.65	-	6.00	0.11	6.68	m2	7.52
isolated surfaces, girth not exceeding 300mm	0.16	10.00	0.18	0.22	-	2.03	0.04	2.25	m	2.53
isolated areas not exceeding 0.50m2 irrespective of girth	0.26	10.00	0.29	0.65	-	6.00	0.11	6.40	nr	7.19
Copper services										
girth exceeding 300mm	0.52	10.00	0.57	0.65	-	6.00	0.11	6.68	m2	7.52
isolated surfaces, girth not exceeding 300mm	0.16	10.00	0.18	0.22	-	2.03	0.04	2.25	m	2.53
Galvanised services										
girth exceeding 300mm	0.52	10.00	0.57	0.65	-	6.00	0.11	6.68	m2	7.52
isolated surfaces, girth not exceeding 300mm	0.16	10.00	0.18	0.22	-	2.03	0.04	2.25	m	2.53

Two undercoats, one coat full gloss finish, existing gloss painted metal surfaces

	Del to Site	Waste	Material Cost	Craft Optve	Lab	Labour Cost	Sunds	Nett Rate	Unit	Gross Rate
Iron or steel general surfaces										
girth exceeding 300mm	0.78	10.00	0.86	0.62	-	5.72	0.11	6.69	m2	7.53
isolated surfaces, girth not exceeding 300mm	0.24	10.00	0.26	0.21	-	1.94	0.04	2.24	m	2.52
Galvanised glazed doors, windows or screens										
girth exceeding 300mm; panes, area not exceeding 0.10m2	0.22	10.00	0.24	1.15	-	10.61	0.20	11.06	m2	12.44
girth exceeding 300mm; panes, area 0.10 - 0.50m2	0.18	10.00	0.20	0.83	-	7.66	0.15	8.01	m2	9.01
girth exceeding 300mm; panes, area 0.50 - 1.00m2	0.16	10.00	0.18	0.69	-	6.37	0.12	6.66	m2	7.50
girth exceeding 300mm; panes, area exceeding 1.00m2	0.14	10.00	0.15	0.61	-	5.63	0.11	5.89	m2	6.63
Iron or steel structural work										
girth exceeding 300mm	0.78	10.00	0.86	0.67	-	6.18	0.12	7.16	m2	8.06
isolated surfaces, girth not exceeding 300mm	0.24	10.00	0.26	0.27	-	2.49	0.05	2.81	m	3.16
Iron or steel structural members of roof trusses, lattice girders, purlins and the like										
girth exceeding 300mm	0.78	10.00	0.86	0.84	-	7.75	0.15	8.76	m2	9.86
isolated surfaces, girth not exceeding 300mm	0.24	10.00	0.26	0.28	-	2.58	0.05	2.90	m	3.26
Iron or steel services										
girth exceeding 300mm	0.78	10.00	0.86	0.84	-	7.75	0.15	8.76	m2	9.86
isolated surfaces, girth not exceeding 300mm	0.24	10.00	0.26	0.28	-	2.58	0.05	2.90	m	3.26
isolated areas not exceeding 0.50m2 irrespective of girth	0.39	10.00	0.43	0.84	-	7.75	0.15	8.33	nr	9.37
Copper services										
girth exceeding 300mm	0.78	10.00	0.86	0.84	-	7.75	0.15	8.76	m2	9.86
isolated surfaces, girth not exceeding 300mm	0.24	10.00	0.26	0.28	-	2.58	0.05	2.90	m	3.26
Galvanised services										
girth exceeding 300mm	0.78	10.00	0.86	0.84	-	7.75	0.15	8.76	m2	9.86
isolated surfaces, girth not exceeding 300mm	0.24	10.00	0.26	0.28	-	2.58	0.05	2.90	m	3.26

Two undercoats, one coat full gloss finish, existing gloss painted metal surfaces; external work

	Del to Site	Waste	Material Cost	Craft Optve	Lab	Labour Cost	Sunds	Nett Rate	Unit	Gross Rate
Iron or steel general surfaces										
girth exceeding 300mm	0.78	10.00	0.86	0.65	-	6.00	0.11	6.97	m2	7.84
isolated surfaces, girth not exceeding 300mm	0.24	10.00	0.26	0.22	-	2.03	0.04	2.33	m	2.63
Galvanised glazed doors, windows or screens										
girth exceeding 300mm; panes, area not exceeding 0.10m2	0.22	10.00	0.24	1.18	-	10.89	0.21	11.34	m2	12.76
girth exceeding 300mm; panes, area 0.10 - 0.50m2	0.18	10.00	0.20	0.86	-	7.94	0.15	8.29	m2	9.32
girth exceeding 300mm; panes, area 0.50 - 1.00m2	0.16	10.00	0.18	0.72	-	6.65	0.13	6.95	m2	7.82

Labour hourly rates: (except Specialists) Craft Operatives £9.23 Labourer £7.02 Rates are national average prices. Refer to REGIONAL VARIATIONS for indicative levels of overall pricing in regions	MATERIALS			LABOUR				RATES		
	Del to Site £	Waste %	Material Cost £	Craft Optve Hrs	Lab Hrs	Labour Cost £	Sunds £	Nett Rate £	Unit	Gross Rate (+12.5%) £
M60: PAINTING/CLEAR FINISHING (REDECORATIONS) - OIL PAINTING METALWORK Cont.										
Two undercoats, one coat full gloss finish, existing gloss painted metal surfaces; external work Cont.										
Galvanised glazed doors, windows or screens Cont.										
girth exceeding 300mm; panes, area exceeding 1.00m2	0.14	10.00	0.15	0.64	-	5.91	0.11	6.17	m2	6.94
Iron or steel structural work										
girth exceeding 300mm	0.78	10.00	0.86	0.67	-	6.18	0.12	7.16	m2	8.06
isolated surfaces, girth not exceeding 300mm	0.24	10.00	0.26	0.27	-	2.49	0.05	2.81	m	3.16
Iron or steel structural members of roof trusses, lattice girders, purlins and the like										
girth exceeding 300mm	0.78	10.00	0.86	0.84	-	7.75	0.15	8.76	m2	9.86
isolated surfaces, girth not exceeding 300mm	0.24	10.00	0.26	0.28	-	2.58	0.05	2.90	m	3.26
Iron or steel railings, fences and gates; plain open type										
girth exceeding 300mm	0.78	10.00	0.86	0.63	-	5.81	0.11	6.78	m2	7.63
isolated surfaces, girth not exceeding 300mm	0.24	10.00	0.26	0.21	-	1.94	0.04	2.24	m	2.52
Iron or steel railings, fences and gates; close type										
girth exceeding 300mm	0.78	10.00	0.86	0.57	-	5.26	0.10	6.22	m2	7.00
Iron or steel railings, fences and gates; ornamental type										
girth exceeding 300mm	0.78	10.00	0.86	1.06	-	9.78	0.19	10.83	m2	12.19
Iron or steel eaves gutters										
girth exceeding 300mm	0.78	10.00	0.86	0.68	-	6.28	0.12	7.25	m2	8.16
isolated surfaces, girth not exceeding 300mm	0.24	10.00	0.26	0.23	-	2.12	0.04	2.43	m	2.73
Galvanised eaves gutters										
girth exceeding 300mm	0.78	10.00	0.86	0.68	-	6.28	0.12	7.25	m2	8.16
isolated surfaces, girth not exceeding 300mm	0.24	10.00	0.26	0.23	-	2.12	0.04	2.43	m	2.73
Iron or steel services										
girth exceeding 300mm	0.78	10.00	0.86	0.84	-	7.75	0.15	8.76	m2	9.86
isolated surfaces, girth not exceeding 300mm	0.24	10.00	0.26	0.28	-	2.58	0.05	2.90	m	3.26
isolated areas not exceeding 0.50m2 irrespective of girth	0.39	10.00	0.43	0.84	-	7.75	0.15	8.33	nr	9.37
Copper services										
girth exceeding 300mm	0.78	10.00	0.86	0.84	-	7.75	0.15	8.76	m2	9.86
isolated surfaces, girth not exceeding 300mm	0.24	10.00	0.26	0.28	-	2.58	0.05	2.90	m	3.26
Galvanised services										
girth exceeding 300mm	0.78	10.00	0.86	0.84	-	7.75	0.15	8.76	m2	9.86
isolated surfaces, girth not exceeding 300mm	0.24	10.00	0.26	0.28	-	2.58	0.05	2.90	m	3.26
One undercoat, two coats full gloss finish, existing gloss painted metal surfaces										
Iron or steel general surfaces										
girth exceeding 300mm	0.78	10.00	0.86	0.62	-	5.72	0.11	6.69	m2	7.53
isolated surfaces, girth not exceeding 300mm	0.24	10.00	0.26	0.21	-	1.94	0.04	2.24	m	2.52
Galvanised glazed doors, windows or screens										
girth exceeding 300mm; panes, area not exceeding 0.10m2	0.22	10.00	0.24	1.16	-	10.71	0.20	11.15	m2	12.54
girth exceeding 300mm; panes, area 0.10 - 0.50m2	0.18	10.00	0.20	0.83	-	7.66	0.15	8.01	m2	9.01
girth exceeding 300mm; panes, area 0.50 - 1.00m2	0.16	10.00	0.18	0.69	-	6.37	0.12	6.66	m2	7.50
girth exceeding 300mm; panes, area exceeding 1.00m2	0.14	10.00	0.15	0.61	-	5.63	0.11	5.89	m2	6.63
Iron or steel structural work										
girth exceeding 300mm	0.78	10.00	0.86	0.67	-	6.18	0.12	7.16	m2	8.06
isolated surfaces, girth not exceeding 300mm	0.24	10.00	0.26	0.27	-	2.49	0.05	2.81	m	3.16
Iron or steel structural members of roof trusses, lattice girders, purlins and the like										
girth exceeding 300mm	0.78	10.00	0.86	0.84	-	7.75	0.15	8.76	m2	9.86
isolated surfaces, girth not exceeding 300mm	0.24	10.00	0.26	0.28	-	2.58	0.05	2.90	m	3.26
Iron or steel services										
girth exceeding 300mm	0.78	10.00	0.86	0.84	-	7.75	0.15	8.76	m2	9.86
isolated surfaces, girth not exceeding 300mm	0.24	10.00	0.26	0.28	-	2.58	0.05	2.90	m	3.26
isolated areas not exceeding 0.50m2 irrespective of girth	0.39	10.00	0.43	0.84	-	7.75	0.15	8.33	nr	9.37
Copper services										
girth exceeding 300mm	0.78	10.00	0.86	0.84	-	7.75	0.15	8.76	m2	9.86
isolated surfaces, girth not exceeding 300mm	0.24	10.00	0.26	0.28	-	2.58	0.05	2.90	m	3.26
Galvanised services										
girth exceeding 300mm	0.78	10.00	0.86	0.84	-	7.75	0.15	8.76	m2	9.86
isolated surfaces, girth not exceeding 300mm	0.24	10.00	0.26	0.28	-	2.58	0.05	2.90	m	3.26
One undercoat, two coats full gloss finish, existing gloss painted metal surfaces; external work										
Iron or steel general surfaces										
girth exceeding 300mm	0.78	10.00	0.86	0.65	-	6.00	0.11	6.97	m2	7.84

SURFACE FINISHES

Labour hourly rates: (except Specialists) Craft Operatives £9.23 Labourer £7.02. Rates are national average prices. Refer to REGIONAL VARIATIONS for indicative levels of overall pricing in regions	MATERIALS			LABOUR				RATES		
	Del to Site £	Waste %	Material Cost £	Craft Optve Hrs	Lab Hrs	Labour Cost £	Sunds £	Nett Rate £	Unit	Gross Rate (+12.5%) £
M60: PAINTING/CLEAR FINISHING (REDECORATIONS) - OIL PAINTING METALWORK Cont.										
One undercoat, two coats full gloss finish, existing gloss painted metal surfaces; external work Cont.										
Iron or steel general surfaces Cont.										
isolated surfaces, girth not exceeding 300mm	0.24	10.00	0.26	0.22	-	2.03	0.04	2.33	m	2.63
Galvanised glazed doors, windows or screens										
girth exceeding 300mm; panes, area not exceeding 0.10m2	0.22	10.00	0.24	1.18	-	10.89	0.21	11.34	m2	12.76
girth exceeding 300mm; panes, area 0.10 - 0.50m2 .	0.18	10.00	0.20	0.86	-	7.94	0.15	8.29	m2	9.32
girth exceeding 300mm; panes, area 0.50 - 1.00m2 .	0.16	10.00	0.18	0.72	-	6.65	0.13	6.95	m2	7.82
girth exceeding 300mm; panes, area exceeding 1.00m2	0.14	10.00	0.15	0.64	-	5.91	0.11	6.17	m2	6.94
Iron or steel structural work										
girth exceeding 300mm	0.78	10.00	0.86	0.67	-	6.18	0.12	7.16	m2	8.06
isolated surfaces, girth not exceeding 300mm	0.24	10.00	0.26	0.27	-	2.49	0.05	2.81	m	3.16
Iron or steel structural members of roof trusses, lattice girders, purlins and the like										
girth exceeding 300mm	0.78	10.00	0.86	0.84	-	7.75	0.15	8.76	m2	9.86
isolated surfaces, girth not exceeding 300mm	0.24	10.00	0.26	0.28	-	2.58	0.05	2.90	m	3.26
Iron or steel railings, fences and gates; plain open type										
girth exceeding 300mm	0.78	10.00	0.86	0.64	-	5.91	0.11	6.88	m2	7.73
isolated surfaces, girth not exceeding 300mm	0.24	10.00	0.26	0.21	-	1.94	0.04	2.24	m	2.52
Iron or steel railings, fences and gates; close type										
girth exceeding 300mm	0.78	10.00	0.86	0.57	-	5.26	0.10	6.22	m2	7.00
Iron or steel railings, fences and gates; ornamental type										
girth exceeding 300mm	0.78	10.00	0.86	1.06	-	9.78	0.19	10.83	m2	12.19
Iron or steel eaves gutters										
girth exceeding 300mm	0.78	10.00	0.86	0.68	-	6.28	0.12	7.25	m2	8.16
isolated surfaces, girth not exceeding 300mm	0.24	10.00	0.26	0.23	-	2.12	0.04	2.43	m	2.73
Galvanised eaves gutters										
girth exceeding 300mm	0.78	10.00	0.86	0.68	-	6.28	0.12	7.25	m2	8.16
isolated surfaces, girth not exceeding 300mm	0.24	10.00	0.26	0.23	-	2.12	0.04	2.43	m	2.73
Iron or steel services										
girth exceeding 300mm	0.78	10.00	0.86	0.84	-	7.75	0.15	8.76	m2	9.86
isolated surfaces, girth not exceeding 300mm	0.24	10.00	0.26	0.28	-	2.58	0.05	2.90	m	3.26
isolated areas not exceeding 0.50m2 irrespective of girth	0.39	10.00	0.43	0.84	-	7.75	0.15	8.33	nr	9.37
Copper services										
girth exceeding 300mm	0.78	10.00	0.86	0.84	-	7.75	0.15	8.76	m2	9.86
isolated surfaces, girth not exceeding 300mm	0.24	10.00	0.26	0.28	-	2.58	0.05	2.90	m	3.26
Galvanised services										
girth exceeding 300mm	0.78	10.00	0.86	0.84	-	7.75	0.15	8.76	m2	9.86
isolated surfaces, girth not exceeding 300mm	0.24	10.00	0.26	0.28	-	2.58	0.05	2.90	m	3.26
M60: PAINTING/CLEAR FINISHING (REDECORATIONS) - OIL PAINTING WOODWORK										
Generally										
Note										
the following rates include for the cost of all preparatory work, e.g. washing down, etc.										
One undercoat, one coat full gloss finish, existing gloss painted wood surfaces										
Wood general surfaces										
girth exceeding 300mm	0.52	10.00	0.57	0.45	-	4.15	0.08	4.81	m2	5.41
isolated surfaces, girth not exceeding 300mm	0.16	10.00	0.18	0.15	-	1.38	0.03	1.59	m	1.79
Wood glazed doors										
girth exceeding 300mm; panes, area not exceeding 0.10m2	0.26	10.00	0.29	0.79	-	7.29	0.14	7.72	m2	8.68
girth exceeding 300mm; panes, area 0.10 - 0.50m2 .	0.18	10.00	0.20	0.57	-	5.26	0.10	5.56	m2	6.25
girth exceeding 300mm; panes, area 0.50 - 1.00m2 .	0.13	10.00	0.14	0.48	-	4.43	0.09	4.66	m2	5.25
girth exceeding 300mm; panes, area exceeding 1.00m2	0.10	10.00	0.11	0.41	-	3.78	0.07	3.96	m2	4.46
Wood partially glazed doors										
girth exceeding 300mm; panes, area not exceeding 0.10m2	0.39	10.00	0.43	0.67	-	6.18	0.12	6.73	m2	7.57
girth exceeding 300mm; panes, area 0.10 - 0.50m2 .	0.34	10.00	0.37	0.54	-	4.98	0.10	5.46	m2	6.14
girth exceeding 300mm; panes, area 0.50 - 1.00m2 .	0.31	10.00	0.34	0.48	-	4.43	0.09	4.86	m2	5.47
girth exceeding 300mm; panes, area exceeding 1.00m2	0.29	10.00	0.32	0.43	-	3.97	0.08	4.37	m2	4.91
Wood windows and screens										
girth exceeding 300mm; panes, area not exceeding 0.10m2	0.29	10.00	0.32	0.85	-	7.85	0.14	8.30	m2	9.34
girth exceeding 300mm; panes, area 0.10 - 0.50m2 .	0.21	10.00	0.23	0.61	-	5.63	0.11	5.97	m2	6.72
girth exceeding 300mm; panes, area 0.50 - 1.00m2 .	0.16	10.00	0.18	0.52	-	4.80	0.09	5.07	m2	5.70

Labour hourly rates: (except Specialists) Craft Operatives £9.23 Labourer £7.02 Rates are national average prices. Refer to REGIONAL VARIATIONS for indicative levels of overall pricing in regions	MATERIALS			LABOUR				RATES		
	Del to Site £	Waste %	Material Cost £	Craft Optve Hrs	Lab Hrs	Labour Cost £	Sunds £	Nett Rate £	Unit	Gross Rate (+12.5%) £

M60: PAINTING/CLEAR FINISHING (REDECORATIONS) - OIL PAINTING WOODWORK Cont.

One undercoat, one coat full gloss finish, existing gloss painted wood surfaces Cont.

Wood windows and screens Cont.

girth exceeding 300mm; panes, area exceeding 1.00m2	0.13	10.00	0.14	0.44	-	4.06	0.08	4.28	m2	4.82

One undercoat, one coat full gloss finish, existing gloss painted wood surfaces; external work

Wood general surfaces

girth exceeding 300mm	0.52	10.00	0.57	0.48	-	4.43	0.08	5.08	m2	5.72
isolated surfaces, girth not exceeding 300mm	0.16	10.00	0.18	0.16	-	1.48	0.03	1.68	m	1.89

Wood glazed doors

girth exceeding 300mm; panes, area not exceeding 0.10m2	0.26	10.00	0.29	0.81	-	7.48	0.14	7.90	m2	8.89
girth exceeding 300mm; panes, area 0.10 - 0.50m2	0.18	10.00	0.20	0.59	-	5.45	0.10	5.74	m2	6.46
girth exceeding 300mm; panes, area 0.50 - 1.00m2	0.13	10.00	0.14	0.49	-	4.52	0.09	4.76	m2	5.35
girth exceeding 300mm; panes, area exceeding 1.00m2	0.10	10.00	0.11	0.42	-	3.88	0.07	4.06	m2	4.56

Wood partially glazed doors

girth exceeding 300mm; panes, area not exceeding 0.10m2	0.39	10.00	0.43	0.69	-	6.37	0.12	6.92	m2	7.78
girth exceeding 300mm; panes, area 0.10 - 0.50m2	0.34	10.00	0.37	0.56	-	5.17	0.10	5.64	m2	6.35
girth exceeding 300mm; panes, area 0.50 - 1.00m2	0.31	10.00	0.34	0.49	-	4.52	0.09	4.95	m2	5.57
girth exceeding 300mm; panes, area exceeding 1.00m2	0.29	10.00	0.32	0.44	-	4.06	0.08	4.46	m2	5.02

Wood windows and screens

girth exceeding 300mm; panes, area not exceeding 0.10m2	0.29	10.00	0.32	0.89	-	8.21	0.16	8.69	m2	9.78
girth exceeding 300mm; panes, area 0.10 - 0.50m2	0.21	10.00	0.23	0.65	-	6.00	0.11	6.34	m2	7.13
girth exceeding 300mm; panes, area 0.50 - 1.00m2	0.16	10.00	0.18	0.55	-	5.08	0.10	5.35	m2	6.02
girth exceeding 300mm; panes, area exceeding 1.00m2	0.13	10.00	0.14	0.47	-	4.34	0.08	4.56	m2	5.13

Wood railings fences and gates; open type

girth exceeding 300mm	0.52	10.00	0.57	0.41	-	3.78	0.07	4.43	m2	4.98
isolated surfaces, girth not exceeding 300mm	0.16	10.00	0.18	0.13	-	1.20	0.02	1.40	m	1.57

Wood railings fences and gates; close type

girth exceeding 300mm	0.52	10.00	0.57	0.37	-	3.42	0.06	4.05	m2	4.55

Two coats full gloss finish, existing gloss painted wood surfaces

Wood general surfaces

girth exceeding 300mm	0.52	10.00	0.57	0.45	-	4.15	0.08	4.81	m2	5.41
isolated surfaces, girth not exceeding 300mm	0.16	10.00	0.18	0.15	-	1.38	0.03	1.59	m	1.79

Wood glazed doors

girth exceeding 300mm; panes, area not exceeding 0.10m2	0.26	10.00	0.29	0.79	-	7.29	0.14	7.72	m2	8.68
girth exceeding 300mm; panes, area 0.10 - 0.50m2	0.18	10.00	0.20	0.57	-	5.26	0.10	5.56	m2	6.25
girth exceeding 300mm; panes, area 0.50 - 1.00m2	0.13	10.00	0.14	0.48	-	4.43	0.08	4.65	m2	5.24
girth exceeding 300mm; panes, area exceeding 1.00m2	0.10	10.00	0.11	0.41	-	3.78	0.07	3.96	m2	4.46

Wood partially glazed doors

girth exceeding 300mm; panes, area not exceeding 0.10m2	0.39	10.00	0.43	0.67	-	6.18	0.12	6.73	m2	7.57
girth exceeding 300mm; panes, area 0.10 - 0.50m2	0.34	10.00	0.37	0.54	-	4.98	0.09	5.45	m2	6.13
girth exceeding 300mm; panes, area 0.50 - 1.00m2	0.31	10.00	0.34	0.48	-	4.43	0.08	4.85	m2	5.46
girth exceeding 300mm; panes, area exceeding 1.00m2	0.29	10.00	0.32	0.43	-	3.97	0.08	4.37	m2	4.91

Wood windows and screens

girth exceeding 300mm; panes, area not exceeding 0.10m2	0.29	10.00	0.32	0.85	-	7.85	0.15	8.31	m2	9.35
girth exceeding 300mm; panes, area 0.10 - 0.50m2	0.21	10.00	0.23	0.61	-	5.63	0.11	5.97	m2	6.72
girth exceeding 300mm; panes, area 0.50 - 1.00m2	0.16	10.00	0.18	0.52	-	4.80	0.09	5.07	m2	5.70
girth exceeding 300mm; panes, area exceeding 1.00m2	0.13	10.00	0.14	0.44	-	4.06	0.08	4.28	m2	4.82

Two coats full gloss finish, existing gloss painted wood surfaces; external work

Wood general surfaces

girth exceeding 300mm	0.52	10.00	0.57	0.48	-	4.43	0.08	5.08	m2	5.72
isolated surfaces, girth not exceeding 300mm	0.16	10.00	0.18	0.16	-	1.48	0.03	1.68	m	1.89

Wood glazed doors

girth exceeding 300mm; panes, area not exceeding 0.10m2	0.26	10.00	0.29	0.81	-	7.48	0.14	7.90	m2	8.89
girth exceeding 300mm; panes, area 0.10 - 0.50m2	0.18	10.00	0.20	0.59	-	5.45	0.10	5.74	m2	6.46
girth exceeding 300mm; panes, area 0.50 - 1.00m2	0.13	10.00	0.14	0.49	-	4.52	0.09	4.76	m2	5.35
girth exceeding 300mm; panes, area exceeding 1.00m2	0.10	10.00	0.11	0.42	-	3.88	0.07	4.06	m2	4.56

SURFACE FINISHES

SURFACE FINISHES – SMALL WORKS

Labour hourly rates: (except Specialists)
Craft Operatives £9.23 Labourer £7.02
Rates are national average prices.
Refer to REGIONAL VARIATIONS for indicative levels
of overall pricing in regions

	MATERIALS			LABOUR				RATES		
	Del to Site £	Waste %	Material Cost £	Craft Optve Hrs	Lab Hrs	Labour Cost £	Sunds £	Nett Rate £	Unit	Gross Rate (+12.5%) £
M60: PAINTING/CLEAR FINISHING (REDECORATIONS) - OIL PAINTING WOODWORK Cont.										
Two coats full gloss finish, existing gloss painted wood surfaces; external work Cont.										
Wood partially glazed doors										
girth exceeding 300mm; panes, area not exceeding 0.10m2	0.39	10.00	0.43	0.69	-	6.37	0.12	6.92	m2	7.78
girth exceeding 300mm; panes, area 0.10 - 0.50m2	0.34	10.00	0.37	0.56	-	5.17	0.10	5.64	m2	6.35
girth exceeding 300mm; panes, area 0.50 - 1.00m2	0.31	10.00	0.34	0.49	-	4.52	0.09	4.95	m2	5.57
girth exceeding 300mm; panes, area exceeding 1.00m2	0.29	10.00	0.32	0.44	-	4.06	0.08	4.46	m2	5.02
Wood windows and screens										
girth exceeding 300mm; panes, area not exceeding 0.10m2	0.29	10.00	0.32	0.89	-	8.21	0.16	8.69	m2	9.78
girth exceeding 300mm; panes, area 0.10 - 0.50m2	0.21	10.00	0.23	0.65	-	6.00	0.11	6.34	m2	7.13
girth exceeding 300mm; panes, area 0.50 - 1.00m2	0.16	10.00	0.18	0.55	-	5.08	0.10	5.35	m2	6.02
girth exceeding 300mm; panes, area exceeding 1.00m2	0.13	10.00	0.14	0.47	-	4.34	0.08	4.56	m2	5.13
Wood railings fences and gates; open type										
girth exceeding 300mm	0.52	10.00	0.57	0.41	-	3.78	0.07	4.43	m2	4.98
isolated surfaces, girth not exceeding 300mm	0.16	10.00	0.18	0.13	-	1.20	0.02	1.40	m	1.57
Wood railings fences and gates; close type										
girth exceeding 300mm	0.52	10.00	0.57	0.37	-	3.42	0.06	4.05	m2	4.55
Two undercoats, one coat full gloss finish, existing gloss painted wood surfaces										
Wood general surfaces										
girth exceeding 300mm	0.78	10.00	0.86	0.61	-	5.63	0.11	6.60	m2	7.42
isolated surfaces, girth not exceeding 300mm	0.24	10.00	0.26	0.20	-	1.85	0.03	2.14	m	2.41
Wood glazed doors										
girth exceeding 300mm; panes, area not exceeding 0.10m2	0.39	10.00	0.43	1.04	-	9.60	0.18	10.21	m2	11.48
girth exceeding 300mm; panes, area 0.10 - 0.50m2	0.27	10.00	0.30	0.76	-	7.01	0.13	7.44	m2	8.37
girth exceeding 300mm; panes, area 0.50 - 1.00m2	0.20	10.00	0.22	0.64	-	5.91	0.11	6.24	m2	7.02
girth exceeding 300mm; panes, area exceeding 1.00m2	0.16	10.00	0.18	0.54	-	4.98	0.09	5.25	m2	5.91
Wood partially glazed doors										
girth exceeding 300mm; panes, area not exceeding 0.10m2	0.59	10.00	0.65	0.87	-	8.03	0.15	8.83	m2	9.93
girth exceeding 300mm; panes, area 0.10 - 0.50m2	0.51	10.00	0.56	0.71	-	6.55	0.12	7.23	m2	8.14
girth exceeding 300mm; panes, area 0.50 - 1.00m2	0.47	10.00	0.52	0.64	-	5.91	0.11	6.53	m2	7.35
girth exceeding 300mm; panes, area exceeding 1.00m2	0.43	10.00	0.47	0.57	-	5.26	0.10	5.83	m2	6.56
Wood windows and screens										
girth exceeding 300mm; panes, area not exceeding 0.10m2	0.43	10.00	0.47	1.12	-	10.34	0.20	11.01	m2	12.39
girth exceeding 300mm; panes, area 0.10 - 0.50m2	0.31	10.00	0.34	0.82	-	7.57	0.14	8.05	m2	9.06
girth exceeding 300mm; panes, area 0.50 - 1.00m2	0.24	10.00	0.26	0.69	-	6.37	0.12	6.75	m2	7.60
girth exceeding 300mm; panes, area exceeding 1.00m2	0.20	10.00	0.22	0.59	-	5.45	0.10	5.77	m2	6.49
Two undercoats, one coat full gloss finish, existing gloss painted wood surfaces; external work										
Wood general surfaces										
girth exceeding 300mm	0.78	10.00	0.86	0.65	-	6.00	0.11	6.97	m2	7.84
isolated surfaces, girth not exceeding 300mm	0.24	10.00	0.26	0.22	-	2.03	0.04	2.33	m	2.63
Wood glazed doors										
girth exceeding 300mm; panes, area not exceeding 0.10m2	0.39	10.00	0.43	1.26	-	11.63	0.22	12.28	m2	13.81
girth exceeding 300mm; panes, area 0.10 - 0.50m2	0.27	10.00	0.30	0.78	-	7.20	0.14	7.64	m2	8.59
girth exceeding 300mm; panes, area 0.50 - 1.00m2	0.20	10.00	0.22	0.65	-	6.00	0.11	6.33	m2	7.12
girth exceeding 300mm; panes, area exceeding 1.00m2	0.16	10.00	0.18	0.55	-	5.08	0.10	5.35	m2	6.02
Wood partially glazed doors										
girth exceeding 300mm; panes, area not exceeding 0.10m2	0.59	10.00	0.65	0.89	-	8.21	0.16	9.02	m2	10.15
girth exceeding 300mm; panes, area 0.10 - 0.50m2	0.51	10.00	0.56	0.73	-	6.74	0.13	7.43	m2	8.36
girth exceeding 300mm; panes, area 0.50 - 1.00m2	0.47	10.00	0.52	0.65	-	6.00	0.11	6.63	m2	7.45
girth exceeding 300mm; panes, area exceeding 1.00m2	0.43	10.00	0.47	0.58	-	5.35	0.10	5.93	m2	6.67
Wood windows and screens										
girth exceeding 300mm; panes, area not exceeding 0.10m2	0.43	10.00	0.47	1.18	-	10.89	0.21	11.57	m2	13.02
girth exceeding 300mm; panes, area 0.10 - 0.50m2	0.31	10.00	0.34	0.87	-	8.03	0.15	8.52	m2	9.59
girth exceeding 300mm; panes, area 0.50 - 1.00m2	0.24	10.00	0.26	0.73	-	6.74	0.13	7.13	m2	8.02
girth exceeding 300mm; panes, area exceeding 1.00m2	0.20	10.00	0.22	0.63	-	5.81	0.11	6.14	m2	6.91
Wood railings fences and gates; open type										
girth exceeding 300mm	0.78	10.00	0.86	0.55	-	5.08	0.10	6.03	m2	6.79
isolated surfaces, girth not exceeding 300mm	0.24	10.00	0.26	0.18	-	1.66	0.03	1.96	m	2.20

Labour hourly rates: (except Specialists) Craft Operatives £9.23 Labourer £7.02 Rates are national average prices. Refer to REGIONAL VARIATIONS for indicative levels of overall pricing in regions	MATERIALS			LABOUR				RATES		
	Del to Site £	Waste %	Material Cost £	Craft Optve Hrs	Lab Hrs	Labour Cost £	Sunds £	Nett Rate £	Unit	Gross Rate (+12.5%) £
M60: PAINTING/CLEAR FINISHING (REDECORATIONS) - OIL PAINTING WOODWORK Cont.										
Two undercoats, one coat full gloss finish, existing gloss painted wood surfaces; external work Cont.										
Wood railings fences and gates; close type										
girth exceeding 300mm	0.78	10.00	0.86	0.49	-	4.52	0.09	5.47	m2	6.15
One undercoat, two coats full gloss finish, existing gloss painted wood surfaces										
Wood general surfaces										
girth exceeding 300mm	0.78	10.00	0.86	0.61	-	5.63	0.11	6.60	m2	7.42
isolated surfaces, girth not exceeding 300mm	0.24	10.00	0.26	0.20	-	1.85	0.03	2.14	m	2.41
Wood glazed doors										
girth exceeding 300mm; panes, area not exceeding 0.10m2	0.39	10.00	0.43	1.04	-	9.60	0.18	10.21	m2	11.48
girth exceeding 300mm; panes, area 0.10 - 0.50m2 .	0.27	10.00	0.30	0.76	-	7.01	0.13	7.44	m2	8.37
girth exceeding 300mm; panes, area 0.50 - 1.00m2 .	0.20	10.00	0.22	0.64	-	5.91	0.11	6.24	m2	7.02
girth exceeding 300mm; panes, area exceeding 1.00m2	0.16	10.00	0.18	0.54	-	4.98	0.09	5.25	m2	5.91
Wood partially glazed doors										
girth exceeding 300mm; panes, area not exceeding 0.10m2	0.59	10.00	0.65	0.87	-	8.03	0.15	8.83	m2	9.93
girth exceeding 300mm; panes, area 0.10 - 0.50m2 .	0.51	10.00	0.56	0.71	-	6.55	0.12	7.23	m2	8.14
girth exceeding 300mm; panes, area 0.50 - 1.00m2 .	0.47	10.00	0.52	0.64	-	5.91	0.11	6.53	m2	7.35
girth exceeding 300mm; panes, area exceeding 1.00m2	0.43	10.00	0.47	0.57	-	5.26	0.10	5.83	m2	6.56
Wood windows and screens										
girth exceeding 300mm; panes, area not exceeding 0.10m2	0.43	10.00	0.47	1.12	-	10.34	0.20	11.01	m2	12.39
girth exceeding 300mm; panes, area 0.10 - 0.50m2 .	0.31	10.00	0.34	0.82	-	7.57	0.14	8.05	m2	9.06
girth exceeding 300mm; panes, area 0.50 - 1.00m2 .	0.24	10.00	0.26	0.69	-	6.37	0.12	6.75	m2	7.60
girth exceeding 300mm; panes, area exceeding 1.00m2	0.20	10.00	0.22	0.59	-	5.45	0.10	5.77	m2	6.49
One undercoat, two coats full gloss finish, existing gloss painted wood surfaces; external work										
Wood general surfaces										
girth exceeding 300mm	0.78	10.00	0.86	0.65	-	6.00	0.11	6.97	m2	7.84
isolated surfaces, girth not exceeding 300mm	0.24	10.00	0.26	0.22	-	2.03	0.04	2.33	m	2.63
Wood glazed doors										
girth exceeding 300mm; panes, area not exceeding 0.10m2	0.39	10.00	0.43	1.06	-	9.78	0.22	10.43	m2	11.74
girth exceeding 300mm; panes, area 0.10 - 0.50m2 .	0.27	10.00	0.30	0.78	-	7.20	0.14	7.64	m2	8.59
girth exceeding 300mm; panes, area 0.50 - 1.00m2 .	0.20	10.00	0.22	0.65	-	6.00	0.11	6.33	m2	7.12
girth exceeding 300mm; panes, area exceeding 1.00m2	0.16	10.00	0.18	0.55	-	5.08	0.10	5.35	m2	6.02
Wood partially glazed doors										
girth exceeding 300mm; panes, area not exceeding 0.10m2	0.59	10.00	0.65	0.89	-	8.21	0.16	9.02	m2	10.15
girth exceeding 300mm; panes, area 0.10 - 0.50m2 .	0.51	10.00	0.56	0.73	-	6.74	0.13	7.43	m2	8.36
girth exceeding 300mm; panes, area 0.50 - 1.00m2 .	0.47	10.00	0.52	0.65	-	6.00	0.11	6.63	m2	7.45
girth exceeding 300mm; panes, area exceeding 1.00m2	0.43	10.00	0.47	0.58	-	5.35	0.10	5.93	m2	6.67
Wood windows and screens										
girth exceeding 300mm; panes, area not exceeding 0.10m2	0.43	10.00	0.47	1.18	-	10.89	0.20	11.56	m2	13.01
girth exceeding 300mm; panes, area 0.10 - 0.50m2 .	0.31	10.00	0.34	0.87	-	8.03	0.15	8.52	m2	9.59
girth exceeding 300mm; panes, area 0.50 - 1.00m2 .	0.24	10.00	0.26	0.73	-	6.74	0.13	7.13	m2	8.02
girth exceeding 300mm; panes, area exceeding 1.00m2	0.20	10.00	0.22	0.63	-	5.81	0.11	6.14	m2	6.91
Wood railings fences and gates; open type										
girth exceeding 300mm	0.78	10.00	0.86	0.55	-	5.08	0.10	6.03	m2	6.79
isolated surfaces, girth not exceeding 300mm	0.24	10.00	0.26	0.18	-	1.66	0.03	1.96	m	2.20
Wood railings fences and gates; close type										
girth exceeding 300mm	0.78	10.00	0.86	0.49	-	4.52	0.09	5.47	m2	6.15
Burn off, one coat primer, one undercoat, one coat full gloss finish, existing painted wood surfaces										
Wood general surfaces										
girth exceeding 300mm	0.83	10.00	0.91	0.90	-	8.31	0.39	9.61	m2	10.81
isolated surfaces, girth not exceeding 300mm	0.25	10.00	0.28	0.26	-	2.40	0.11	2.78	m	3.13
Wood glazed doors										
girth exceeding 300mm; panes, area not exceeding 0.10m2	0.42	10.00	0.46	1.50	-	13.85	0.66	14.97	m2	16.84
girth exceeding 300mm; panes, area 0.10 - 0.50m2 .	0.29	10.00	0.32	1.08	-	9.97	0.47	10.76	m2	12.10
girth exceeding 300mm; panes, area 0.50 - 1.00m2 .	0.21	10.00	0.23	0.90	-	8.31	0.39	8.93	m2	10.04
girth exceeding 300mm; panes, area exceeding 1.00m2	0.17	10.00	0.19	0.80	-	7.38	0.35	7.92	m2	8.91
Wood partially glazed doors										
girth exceeding 300mm; panes, area not exceeding 0.10m2	0.63	10.00	0.69	1.20	-	11.08	0.52	12.29	m2	13.83

SURFACE FINISHES

	MATERIALS			LABOUR				RATES		
Labour hourly rates: (except Specialists) Craft Operatives £9.23 Labourer £7.02 Rates are national average prices. Refer to REGIONAL VARIATIONS for indicative levels of overall pricing in regions	Del to Site £	Waste %	Material Cost £	Craft Optve Hrs	Lab Hrs	Labour Cost £	Sunds £	Nett Rate £	Unit	Gross Rate (+12.5%) £

M60: PAINTING/CLEAR FINISHING (REDECORATIONS) – OIL PAINTING WOODWORK Cont.

Burn off, one coat primer, one undercoat, one coat full gloss finish, existing painted wood surfaces Cont.

	Del to Site	Waste	Material Cost	Craft Optve	Lab	Labour Cost	Sunds	Nett Rate	Unit	Gross Rate
Wood partially glazed doors Cont.										
girth exceeding 300mm; panes, area 0.10 - 0.50m2 .	0.54	10.00	0.59	0.99	–	9.14	0.43	10.16	m2	11.43
girth exceeding 300mm; panes, area 0.50 - 1.00m2 .	0.50	10.00	0.55	0.90	–	8.31	0.39	9.25	m2	10.40
girth exceeding 300mm; panes, area exceeding 1.00m2	0.46	10.00	0.51	0.86	–	7.94	0.38	8.82	m2	9.93
Wood windows and screens										
girth exceeding 300mm; panes, area not exceeding 0.10m2	0.46	10.00	0.51	1.65	–	15.23	0.72	16.46	m2	18.51
girth exceeding 300mm; panes, area 0.10 - 0.50m2 .	0.33	10.00	0.36	1.18	–	10.89	0.52	11.77	m2	13.25
girth exceeding 300mm; panes, area 0.50 - 1.00m2 .	0.25	10.00	0.28	0.99	–	9.14	0.43	9.84	m2	11.07
girth exceeding 300mm; panes, area exceeding 1.00m2	0.21	10.00	0.23	0.88	–	8.12	0.38	8.73	m2	9.83

Burn off, one coat primer, one undercoat, one coat full gloss finish, existing painted wood surfaces; external work

	Del to Site	Waste	Material Cost	Craft Optve	Lab	Labour Cost	Sunds	Nett Rate	Unit	Gross Rate
Wood general surfaces										
girth exceeding 300mm	0.83	10.00	0.91	0.95	–	8.77	0.42	10.10	m2	11.36
isolated surfaces, girth not exceeding 300mm	0.25	10.00	0.28	0.29	–	2.68	0.13	3.08	m	3.47
Wood glazed doors										
girth exceeding 300mm; panes, area not exceeding 0.10m2	0.42	10.00	0.46	1.50	–	13.85	0.66	14.97	m2	16.84
girth exceeding 300mm; panes, area 0.10 - 0.50m2 .	0.29	10.00	0.32	1.08	–	9.97	0.47	10.76	m2	12.10
girth exceeding 300mm; panes, area 0.50 - 1.00m2 .	0.21	10.00	0.23	0.90	–	8.31	0.39	8.93	m2	10.04
girth exceeding 300mm; panes, area exceeding 1.00m2	0.17	10.00	0.19	0.80	–	7.38	0.35	7.92	m2	8.91
Wood partially glazed doors										
girth exceeding 300mm; panes, area not exceeding 0.10m2	0.63	10.00	0.69	1.20	–	11.08	0.52	12.29	m2	13.83
girth exceeding 300mm; panes, area 0.10 - 0.50m2 .	0.54	10.00	0.59	0.99	–	9.14	0.43	10.16	m2	11.43
girth exceeding 300mm; panes, area 0.50 - 1.00m2 .	0.50	10.00	0.55	0.90	–	8.31	0.39	9.25	m2	10.40
girth exceeding 300mm; panes, area exceeding 1.00m2	0.46	10.00	0.51	0.86	–	7.94	0.38	8.82	m2	9.93
Wood windows and screens										
girth exceeding 300mm; panes, area not exceeding 0.10m2	0.46	10.00	0.51	1.69	–	15.60	0.74	16.84	m2	18.95
girth exceeding 300mm; panes, area 0.10 - 0.50m2 .	0.33	10.00	0.36	1.22	–	11.26	0.53	12.15	m2	13.67
girth exceeding 300mm; panes, area 0.50 - 1.00m2 .	0.25	10.00	0.28	1.03	–	9.51	0.45	10.23	m2	11.51
girth exceeding 300mm; panes, area exceeding 1.00m2	0.21	10.00	0.23	0.92	–	8.49	0.40	9.12	m2	10.26
Wood railings fences and gates; open type										
girth exceeding 300mm	0.83	10.00	0.91	0.81	–	7.48	0.35	8.74	m2	9.83
isolated surfaces, girth not exceeding 300mm	0.25	10.00	0.28	0.24	–	2.22	0.10	2.59	m	2.91
Wood railings fences and gates; close type										
girth exceeding 300mm	0.83	10.00	0.91	0.76	–	7.01	0.33	8.26	m2	9.29

Burn off, one coat primer, two undercoats, one coat full gloss finish, existing painted wood surfaces

	Del to Site	Waste	Material Cost	Craft Optve	Lab	Labour Cost	Sunds	Nett Rate	Unit	Gross Rate
Wood general surfaces										
girth exceeding 300mm	1.10	10.00	1.21	1.06	–	9.78	0.46	11.45	m2	12.89
isolated surfaces, girth not exceeding 300mm	0.33	10.00	0.36	0.32	–	2.95	0.14	3.46	m	3.89
Wood glazed doors										
girth exceeding 300mm; panes, area not exceeding 0.10m2	0.55	10.00	0.60	1.76	–	16.24	0.77	17.62	m2	19.82
girth exceeding 300mm; panes, area 0.10 - 0.50m2 .	0.38	10.00	0.42	1.27	–	11.72	0.55	12.69	m2	14.28
girth exceeding 300mm; panes, area 0.50 - 1.00m2 .	0.27	10.00	0.30	1.06	–	9.78	0.46	10.54	m2	11.86
girth exceeding 300mm; panes, area exceeding 1.00m2	0.22	10.00	0.24	0.95	–	8.77	0.42	9.43	m2	10.61
Wood partially glazed doors										
girth exceeding 300mm; panes, area not exceeding 0.10m2	0.82	10.00	0.90	1.41	–	13.01	0.62	14.54	m2	16.35
girth exceeding 300mm; panes, area 0.10 - 0.50m2 .	0.71	10.00	0.78	1.17	–	10.80	0.51	12.09	m2	13.60
girth exceeding 300mm; panes, area 0.50 - 1.00m2 .	0.66	10.00	0.73	1.06	–	9.78	0.46	10.97	m2	12.34
girth exceeding 300mm; panes, area exceeding 1.00m2	0.60	10.00	0.66	1.00	–	9.23	0.44	10.33	m2	11.62
Wood windows and screens										
girth exceeding 300mm; panes, area not exceeding 0.10m2	0.60	10.00	0.66	1.94	–	17.91	0.85	19.42	m2	21.84
girth exceeding 300mm; panes, area 0.10 - 0.50m2 .	0.44	10.00	0.48	1.40	–	12.92	0.61	14.02	m2	15.77
girth exceeding 300mm; panes, area 0.50 - 1.00m2 .	0.33	10.00	0.36	1.17	–	10.80	0.51	11.67	m2	13.13
girth exceeding 300mm; panes, area exceeding 1.00m2	0.27	10.00	0.30	1.03	–	9.51	0.45	10.25	m2	11.54

Labour hourly rates: (except Specialists) Craft Operatives £9.23 Labourer £7.02 Rates are national average prices. Refer to REGIONAL VARIATIONS for indicative levels of overall pricing in regions	MATERIALS			LABOUR				RATES		
	Del to Site £	Waste %	Material Cost £	Craft Optve Hrs	Lab Hrs	Labour Cost £	Sunds £	Nett Rate £	Unit	Gross Rate (+12.5%) £
M60: PAINTING/CLEAR FINISHING (REDECORATIONS) - OIL PAINTING WOODWORK Cont.										
Burn off, one coat primer, two undercoats, one coat full gloss finish, existing painted wood surfaces; external work										
Wood general surfaces										
girth exceeding 300mm	1.10	10.00	1.21	1.11	-	10.25	0.49	11.95	m2	13.44
isolated surfaces, girth not exceeding 300mm	0.33	10.00	0.36	0.35	-	3.23	0.15	3.74	m	4.21
Wood glazed doors										
girth exceeding 300mm; panes, area not exceeding 0.10m2	0.55	10.00	0.60	1.76	-	16.24	0.77	17.62	m2	19.82
girth exceeding 300mm; panes, area 0.10 - 0.50m2 .	0.38	10.00	0.42	1.27	-	11.72	0.55	12.69	m2	14.28
girth exceeding 300mm; panes, area 0.50 - 1.00m2 .	0.27	10.00	0.30	1.06	-	9.78	0.46	10.54	m2	11.86
girth exceeding 300mm; panes, area exceeding 1.00m2	0.22	10.00	0.24	0.95	-	8.77	0.42	9.43	m2	10.61
Wood partially glazed doors										
girth exceeding 300mm; panes, area not exceeding 0.10m2	0.82	10.00	0.90	1.41	-	13.01	0.62	14.54	m2	16.35
girth exceeding 300mm; panes, area 0.10 - 0.50m2 .	0.71	10.00	0.78	1.17	-	10.80	0.51	12.09	m2	13.60
girth exceeding 300mm; panes, area 0.50 - 1.00m2 .	0.66	10.00	0.73	1.06	-	9.78	0.46	10.97	m2	12.34
girth exceeding 300mm; panes, area exceeding 1.00m2	0.60	10.00	0.66	1.00	-	9.23	0.44	10.33	m2	11.62
Wood windows and screens										
girth exceeding 300mm; panes, area not exceeding 0.10m2	0.60	10.00	0.66	1.99	-	18.37	0.87	19.90	m2	22.38
girth exceeding 300mm; panes, area 0.10 - 0.50m2 .	0.44	10.00	0.48	1.45	-	13.38	0.63	14.50	m2	16.31
girth exceeding 300mm; panes, area 0.50 - 1.00m2 .	0.33	10.00	0.36	1.22	-	11.26	0.53	12.15	m2	13.67
girth exceeding 300mm; panes, area exceeding 1.00m2	0.27	10.00	0.30	1.09	-	10.06	0.48	10.84	m2	12.19
Wood railings fences and gates; open type										
girth exceeding 300mm	1.10	10.00	1.21	0.95	-	8.77	0.42	10.40	m2	11.70
isolated surfaces, girth not exceeding 300mm	0.33	10.00	0.36	0.29	-	2.68	0.13	3.17	m	3.57
Wood railings fences and gates; close type										
girth exceeding 300mm	1.10	10.00	1.21	0.88	-	8.12	0.38	9.71	m2	10.93
Burn off, one coat primer, one undercoat, two coats full gloss finish, existing painted wood surfaces										
Wood general surfaces										
girth exceeding 300mm	1.10	10.00	1.21	1.06	-	9.78	0.46	11.45	m2	12.89
isolated surfaces, girth not exceeding 300mm	0.33	10.00	0.36	0.32	-	2.95	0.14	3.46	m	3.89
Wood glazed doors										
girth exceeding 300mm; panes, area not exceeding 0.10m2	0.55	10.00	0.60	1.76	-	16.24	0.77	17.62	m2	19.82
girth exceeding 300mm; panes, area 0.10 - 0.50m2 .	0.38	10.00	0.42	1.27	-	11.72	0.55	12.69	m2	14.28
girth exceeding 300mm; panes, area 0.50 - 1.00m2 .	0.27	10.00	0.30	1.06	-	9.78	0.46	10.54	m2	11.86
girth exceeding 300mm; panes, area exceeding 1.00m2	0.22	10.00	0.24	0.95	-	8.77	0.42	9.43	m2	10.61
Wood partially glazed doors										
girth exceeding 300mm; panes, area not exceeding 0.10m2	0.82	10.00	0.90	1.41	-	13.01	0.62	14.54	m2	16.35
girth exceeding 300mm; panes, area 0.10 - 0.50m2 .	0.71	10.00	0.78	1.17	-	10.80	0.51	12.09	m2	13.60
girth exceeding 300mm; panes, area 0.50 - 1.00m2 .	0.66	10.00	0.73	1.06	-	9.78	0.46	10.97	m2	12.34
girth exceeding 300mm; panes, area exceeding 1.00m2	0.60	10.00	0.66	1.00	-	9.23	0.44	10.33	m2	11.62
Wood windows and screens										
girth exceeding 300mm; panes, area not exceeding 0.10m2	0.60	10.00	0.66	1.94	-	17.91	0.85	19.42	m2	21.84
girth exceeding 300mm; panes, area 0.10 - 0.50m2 .	0.44	10.00	0.48	1.40	-	12.92	0.61	14.02	m2	15.77
girth exceeding 300mm; panes, area 0.50 - 1.00m2 .	0.33	10.00	0.36	1.17	-	10.80	0.51	11.67	m2	13.13
girth exceeding 300mm; panes, area exceeding 1.00m2	0.27	10.00	0.30	1.03	-	9.51	0.45	10.25	m2	11.54
Burn off, one coat primer, one undercoat, two coats full gloss finish, existing painted wood surfaces; external work										
Wood general surfaces										
girth exceeding 300mm	1.10	10.00	1.21	1.11	-	10.25	0.49	11.95	m2	13.44
isolated surfaces, girth not exceeding 300mm	0.33	10.00	0.36	0.35	-	3.23	0.15	3.74	m	4.21
Wood glazed doors										
girth exceeding 300mm; panes, area not exceeding 0.10m2	0.55	10.00	0.60	1.76	-	16.24	0.77	17.62	m2	19.82
girth exceeding 300mm; panes, area 0.10 - 0.50m2 .	0.38	10.00	0.42	1.27	-	11.72	0.55	12.69	m2	14.28
girth exceeding 300mm; panes, area 0.50 - 1.00m2 .	0.27	10.00	0.30	1.06	-	9.78	0.46	10.54	m2	11.86
girth exceeding 300mm; panes, area exceeding 1.00m2	0.22	10.00	0.24	0.95	-	8.77	0.42	9.43	m2	10.61
Wood partially glazed doors										
girth exceeding 300mm; panes, area not exceeding 0.10m2	0.82	10.00	0.90	1.41	-	13.01	0.62	14.54	m2	16.35
girth exceeding 300mm; panes, area 0.10 - 0.50m2 .	0.71	10.00	0.78	1.17	-	10.80	0.51	12.09	m2	13.60
girth exceeding 300mm; panes, area 0.50 - 1.00m2 .	0.66	10.00	0.73	1.06	-	9.78	0.46	10.97	m2	12.34
girth exceeding 300mm; panes, area exceeding 1.00m2	0.60	10.00	0.66	1.00	-	9.23	0.44	10.33	m2	11.62

SURFACE FINISHES

Labour hourly rates: (except Specialists) Craft Operatives £9.23 Labourer £7.02 Rates are national average prices. Refer to REGIONAL VARIATIONS for indicative levels of overall pricing in regions	MATERIALS			LABOUR				RATES		
	Del to Site £	Waste %	Material Cost £	Craft Optve Hrs	Lab Hrs	Labour Cost £	Sunds £	Nett Rate £	Unit	Gross Rate (+12.5%) £

M60: PAINTING/CLEAR FINISHING (REDECORATIONS) - OIL PAINTING WOODWORK Cont.

Burn off, one coat primer, one undercoat, two coats full gloss finish, existing painted wood surfaces; external work Cont.

	Del to Site	Waste	Material Cost	Craft Optve	Lab	Labour Cost	Sunds	Nett Rate	Unit	Gross Rate
Wood windows and screens										
girth exceeding 300mm; panes, area not exceeding 0.10m2	0.60	10.00	0.66	1.99	-	18.37	0.87	19.90	m2	22.38
girth exceeding 300mm; panes, area 0.10 - 0.50m2	0.44	10.00	0.48	1.45	-	13.38	0.63	14.50	m2	16.31
girth exceeding 300mm; panes, area 0.50 - 1.00m2	0.33	10.00	0.36	1.22	-	11.26	0.53	12.15	m2	13.67
girth exceeding 300mm; panes, area exceeding 1.00m2	0.27	10.00	0.30	1.09	-	10.06	0.48	10.84	m2	12.19
Wood railings fences and gates; open type										
girth exceeding 300mm	1.10	10.00	1.21	0.95	-	8.77	0.42	10.40	m2	11.70
isolated surfaces, girth not exceeding 300mm	0.33	10.00	0.36	0.29	-	2.68	0.13	3.17	m	3.57
Wood railings fences and gates; close type										
girth exceeding 300mm	1.10	10.00	1.21	0.88	-	8.12	0.38	9.71	m2	10.93

M60: PAINTING/CLEAR FINISHING (REDECORATIONS) - POLYURETHANE LACQUER

Generally

Note
the following rates include for the cost of all preparatory work, e.g. washing down, etc.

Two coats polyurethane lacquer, existing lacquered surfaces

	Del to Site	Waste	Material Cost	Craft Optve	Lab	Labour Cost	Sunds	Nett Rate	Unit	Gross Rate
Wood general surfaces										
girth exceeding 300mm	0.93	10.00	1.02	0.54	-	4.98	0.09	6.10	m2	6.86
isolated surfaces, girth not exceeding 300mm	0.28	10.00	0.31	0.18	-	1.66	0.03	2.00	m	2.25
Wood glazed doors										
girth exceeding 300mm; panes, area not exceeding 0.10m2	0.47	10.00	0.52	0.92	-	8.49	0.16	9.17	m2	10.31
girth exceeding 300mm; panes, area 0.10 - 0.50m2	0.33	10.00	0.36	0.67	-	6.18	0.12	6.67	m2	7.50
girth exceeding 300mm; panes, area 0.50 - 1.00m2	0.23	10.00	0.25	0.55	-	5.08	0.10	5.43	m2	6.11
girth exceeding 300mm; panes, area exceeding 1.00m2	0.19	10.00	0.21	0.48	-	4.43	0.08	4.72	m2	5.31
Wood partially glazed doors										
girth exceeding 300mm; panes, area not exceeding 0.10m2	0.70	10.00	0.77	0.77	-	7.11	0.13	8.01	m2	9.01
girth exceeding 300mm; panes, area 0.10 - 0.50m2	0.61	10.00	0.67	0.62	-	5.72	0.11	6.50	m2	7.32
girth exceeding 300mm; panes, area 0.50 - 1.00m2	0.56	10.00	0.62	0.55	-	5.08	0.10	5.79	m2	6.52
girth exceeding 300mm; panes, area exceeding 1.00m2	0.51	10.00	0.56	0.50	-	4.62	0.09	5.27	m2	5.92
Wood windows and screens										
girth exceeding 300mm; panes, area not exceeding 0.10m2	0.51	10.00	0.56	0.99	-	9.14	0.17	9.87	m2	11.10
girth exceeding 300mm; panes, area 0.10 - 0.50m2	0.37	10.00	0.41	0.72	-	6.65	0.13	7.18	m2	8.08
girth exceeding 300mm; panes, area 0.50 - 1.00m2	0.28	10.00	0.31	0.59	-	5.45	0.10	5.85	m2	6.59
girth exceeding 300mm; panes, area exceeding 1.00m2	0.23	10.00	0.25	0.52	-	4.80	0.09	5.14	m2	5.79

Two coats polyurethane lacquer, existing lacquered surfaces; external work

	Del to Site	Waste	Material Cost	Craft Optve	Lab	Labour Cost	Sunds	Nett Rate	Unit	Gross Rate
Wood general surfaces										
girth exceeding 300mm	0.93	10.00	1.02	0.57	-	5.26	0.10	6.38	m2	7.18
isolated surfaces, girth not exceeding 300mm	0.28	10.00	0.31	0.19	-	1.75	0.03	2.09	m	2.35
Wood glazed doors										
girth exceeding 300mm; panes, area not exceeding 0.10m2	0.47	10.00	0.52	0.94	-	8.68	0.16	9.35	m2	10.52
girth exceeding 300mm; panes, area 0.10 - 0.50m2	0.33	10.00	0.36	0.68	-	6.28	0.12	6.76	m2	7.60
girth exceeding 300mm; panes, area 0.50 - 1.00m2	0.23	10.00	0.25	0.56	-	5.17	0.10	5.52	m2	6.21
girth exceeding 300mm; panes, area exceeding 1.00m2	0.19	10.00	0.21	0.49	-	4.52	0.09	4.82	m2	5.42
Wood partially glazed doors										
girth exceeding 300mm; panes, area not exceeding 0.10m2	0.70	10.00	0.77	0.79	-	7.29	0.14	8.20	m2	9.23
girth exceeding 300mm; panes, area 0.10 - 0.50m2	0.61	10.00	0.67	0.63	-	5.81	0.11	6.60	m2	7.42
girth exceeding 300mm; panes, area 0.50 - 1.00m2	0.56	10.00	0.62	0.56	-	5.17	0.10	5.88	m2	6.62
girth exceeding 300mm; panes, area exceeding 1.00m2	0.51	10.00	0.56	0.51	-	4.71	0.09	5.36	m2	6.03
Wood windows and screens										
girth exceeding 300mm; panes, area not exceeding 0.10m2	0.51	10.00	0.56	1.04	-	9.60	0.18	10.34	m2	11.63
girth exceeding 300mm; panes, area 0.10 - 0.50m2	0.37	10.00	0.41	0.76	-	7.01	0.13	7.55	m2	8.50
girth exceeding 300mm; panes, area 0.50 - 1.00m2	0.28	10.00	0.31	0.63	-	5.81	0.11	6.23	m2	7.01
girth exceeding 300mm; panes, area exceeding 1.00m2	0.23	10.00	0.25	0.55	-	5.08	0.10	5.43	m2	6.11

Labour hourly rates: (except Specialists) Craft Operatives £9.23 Labourer £7.02 Rates are national average prices. Refer to REGIONAL VARIATIONS for indicative levels of overall pricing in regions	MATERIALS			LABOUR				RATES		
	Del to Site £	Waste %	Material Cost £	Craft Optve Hrs	Lab Hrs	Labour Cost £	Sunds £	Nett Rate £	Unit	Gross Rate (+12.5%) £
M60: PAINTING/CLEAR FINISHING (REDECORATIONS) – POLYURETHANE LACQUER Cont.										
Three coats polyurethane lacquer, existing lacquered surfaces										
Wood general surfaces										
girth exceeding 300mm	1.40	10.00	1.54	0.62	–	5.72	0.11	7.37	m2	8.29
isolated surfaces, girth not exceeding 300mm	0.42	10.00	0.46	0.21	–	1.94	0.04	2.44	m	2.75
Wood glazed doors										
girth exceeding 300mm; panes, area not exceeding 0.10m2	0.70	10.00	0.77	1.04	–	9.60	0.18	10.55	m2	11.87
girth exceeding 300mm; panes, area 0.10 - 0.50m2 .	0.49	10.00	0.54	0.76	–	7.01	0.13	7.68	m2	8.64
girth exceeding 300mm; panes, area 0.50 - 1.00m2 .	0.35	10.00	0.39	0.63	–	5.81	0.11	6.31	m2	7.10
girth exceeding 300mm; panes, area exceeding 1.00m2	0.28	10.00	0.31	0.54	–	4.98	0.09	5.38	m2	6.05
Wood partially glazed doors										
girth exceeding 300mm; panes, area not exceeding 0.10m2	1.05	10.00	1.16	0.87	–	8.03	0.15	9.34	m2	10.50
girth exceeding 300mm; panes, area 0.10 - 0.50m2 .	0.91	10.00	1.00	0.77	–	7.11	0.13	8.24	m2	9.27
girth exceeding 300mm; panes, area 0.50 - 1.00m2 .	0.84	10.00	0.92	0.63	–	5.81	0.11	6.85	m2	7.71
girth exceeding 300mm; panes, area exceeding 1.00m2	0.77	10.00	0.85	0.58	–	5.35	0.10	6.30	m2	7.09
Wood windows and screens										
girth exceeding 300mm; panes, area not exceeding 0.10m2	0.77	10.00	0.85	1.13	–	10.43	0.20	11.48	m2	12.91
girth exceeding 300mm; panes, area 0.10 - 0.50m2 .	0.56	10.00	0.62	0.82	–	7.57	0.14	8.32	m2	9.37
girth exceeding 300mm; panes, area 0.50 - 1.00m2 .	0.42	10.00	0.46	0.68	–	6.28	0.12	6.86	m2	7.72
girth exceeding 300mm; panes, area exceeding 1.00m2	0.35	10.00	0.39	0.59	–	5.45	0.10	5.93	m2	6.67
Three coats polyurethane lacquer, existing lacquered surfaces; external work										
Wood general surfaces										
girth exceeding 300mm	1.40	10.00	1.54	0.66	–	6.09	0.12	7.75	m2	8.72
isolated surfaces, girth not exceeding 300mm	0.42	10.00	0.46	0.22	–	2.03	0.04	2.53	m	2.85
Wood glazed doors										
girth exceeding 300mm; panes, area not exceeding 0.10m2	0.70	10.00	0.77	1.06	–	9.78	0.19	10.74	m2	12.09
girth exceeding 300mm; panes, area 0.10 - 0.50m2 .	0.49	10.00	0.54	0.77	–	7.11	0.12	7.77	m2	8.74
girth exceeding 300mm; panes, area 0.50 - 1.00m2 .	0.35	10.00	0.39	0.64	–	5.91	0.11	6.40	m2	7.20
girth exceeding 300mm; panes, area exceeding 1.00m2	0.28	10.00	0.31	0.54	–	4.98	0.09	5.38	m2	6.05
Wood partially glazed doors										
girth exceeding 300mm; panes, area not exceeding 0.10m2	1.05	10.00	1.16	0.89	–	8.21	0.16	9.53	m2	10.72
girth exceeding 300mm; panes, area 0.10 - 0.50m2 .	0.91	10.00	1.00	0.78	–	7.20	0.14	8.34	m2	9.38
girth exceeding 300mm; panes, area 0.50 - 1.00m2 .	0.84	10.00	0.92	0.64	–	5.91	0.11	6.94	m2	7.81
girth exceeding 300mm; panes, area exceeding 1.00m2	0.77	10.00	0.85	0.59	–	5.45	0.10	6.39	m2	7.19
Wood windows and screens										
girth exceeding 300mm; panes, area not exceeding 0.10m2	0.77	10.00	0.85	1.18	–	10.89	0.21	11.95	m2	13.44
girth exceeding 300mm; panes, area 0.10 - 0.50m2 .	0.56	10.00	0.62	0.87	–	8.03	0.15	8.80	m2	9.90
girth exceeding 300mm; panes, area 0.50 - 1.00m2 .	0.42	10.00	0.46	0.73	–	6.74	0.13	7.33	m2	8.25
girth exceeding 300mm; panes, area exceeding 1.00m2	0.35	10.00	0.39	0.63	–	5.81	0.11	6.31	m2	7.10
M60: PAINTING/CLEAR FINISHING (REDECORATIONS) – FRENCH AND WAX POLISHING										
Generally										
Note the following rates include for the cost of all preparatory work, e.g. washing down, etc.										
Two coats wax polish, existing polished surfaces										
Wood general surfaces										
girth exceeding 300mm	0.48	10.00	0.53	0.40	–	3.69	0.07	4.29	m2	4.83
isolated surfaces, girth not exceeding 300mm	0.16	10.00	0.18	0.14	–	1.29	0.02	1.49	m	1.67
Strip old polish, oil, two coats wax polish, existing polished surfaces										
Wood general surfaces										
girth exceeding 300mm	1.40	10.00	1.54	1.23	–	11.35	0.22	13.11	m2	14.75
isolated surfaces, girth not exceeding 300mm	0.43	10.00	0.47	0.37	–	3.42	0.06	3.95	m	4.44
Strip old polish, oil, stain, two coats wax polish, existing polished surfaces										
Wood general surfaces										
girth exceeding 300mm	2.33	10.00	2.56	1.35	–	12.46	0.24	15.26	m2	17.17
isolated surfaces, girth not exceeding 300mm	0.71	10.00	0.78	0.42	–	3.88	0.07	4.73	m	5.32

SURFACE FINISHES

Labour hourly rates: (except Specialists) Craft Operatives £9.23 Labourer £7.02 Rates are national average prices. Refer to REGIONAL VARIATIONS for indicative levels of overall pricing in regions	MATERIALS			LABOUR				RATES		
	Del to Site £	Waste %	Material Cost £	Craft Optve Hrs	Lab Hrs	Labour Cost £	Sunds £	Nett Rate £	Unit	Gross Rate (+12.5%) £
M60: PAINTING/CLEAR FINISHING (REDECORATIONS) – FRENCH AND WAX POLISHING Cont.										
Open grain French polish existing polished surfaces										
Wood general surfaces										
girth exceeding 300mm	2.01	10.00	2.21	2.64	–	24.37	0.46	27.04	m2	30.42
isolated surfaces, girth not exceeding 300mm	0.59	10.00	0.65	0.88	–	8.12	0.15	8.92	m	10.04
Fully French polish existing polished surfaces										
Wood general surfaces										
girth exceeding 300mm	2.01	10.00	2.21	1.54	–	14.21	0.27	16.70	m2	18.78
isolated surfaces, girth not exceeding 300mm	0.59	10.00	0.65	0.52	–	4.80	0.09	5.54	m	6.23
Strip old polish, oil, open grain French polish, existing polished surfaces										
Wood general surfaces										
girth exceeding 300mm	2.76	10.00	3.04	3.43	–	31.66	0.60	35.29	m2	39.71
isolated surfaces, girth not exceeding 300mm	0.81	10.00	0.89	1.14	–	10.52	0.20	11.61	m	13.06
Strip old polish, oil, stain, open grain French polish, existing polished surfaces										
Wood general surfaces										
girth exceeding 300mm	3.68	10.00	4.05	3.55	–	32.77	0.62	37.43	m2	42.11
isolated surfaces, girth not exceeding 300mm	1.09	10.00	1.20	1.19	–	10.98	0.21	12.39	m	13.94
Strip old polish, oil, body in, fully French polish existing polished surfaces										
Wood general surfaces										
girth exceeding 300mm	2.65	10.00	2.92	4.63	–	42.73	0.81	46.46	m2	52.27
isolated surfaces, girth not exceeding 300mm	0.81	10.00	0.89	1.54	–	14.21	0.27	15.38	m	17.30
Strip old polish, oil, stain, body in, fully French polish, existing polished surfaces										
Wood general surfaces										
girth exceeding 300mm	3.68	10.00	4.05	4.74	–	43.75	0.83	48.63	m2	54.71
isolated surfaces, girth not exceeding 300mm	1.09	10.00	1.20	1.58	–	14.58	0.28	16.06	m	18.07
M60: PAINTING/CLEAR FINISHING (REDECORATIONS) – WATER REPELLENT										
One coat silicone based water repellent, existing surfaces										
Cement rendered general surfaces										
girth exceeding 300mm	0.85	10.00	0.94	0.12	–	1.11	0.02	2.06	m2	2.32
Stone general surfaces										
girth exceeding 300mm	2.00	10.00	2.20	0.13	–	1.20	0.02	3.42	m2	3.85
Brickwork general surfaces										
girth exceeding 300mm	2.00	10.00	2.20	0.13	–	1.20	0.02	3.42	m2	3.85
M60: PAINTING/CLEAR FINISHING (REDECORATIONS) – REMOVAL OF MOULD GROWTH										
Apply fungicide to existing decorated and infected surfaces										
Cement rendered general surfaces										
girth exceeding 300mm	0.10	10.00	0.11	0.22	–	2.03	0.04	2.18	m2	2.45
Concrete general surfaces										
girth exceeding 300mm	0.11	10.00	0.12	0.22	–	2.03	0.04	2.19	m2	2.47
Plaster general surfaces										
girth exceeding 300mm	0.10	10.00	0.11	0.22	–	2.03	0.04	2.18	m2	2.45
Brickwork general surfaces										
girth exceeding 300mm	0.14	10.00	0.15	0.24	–	2.22	0.04	2.41	m2	2.71
Apply fungicide to existing decorated and infected surfaces; external work										
Cement rendered general surfaces										
girth exceeding 300mm	0.10	10.00	0.11	0.24	–	2.22	0.04	2.37	m2	2.66
Concrete general surfaces										
girth exceeding 300mm	0.11	10.00	0.12	0.24	–	2.22	0.04	2.38	m2	2.67
Plaster general surfaces										
girth exceeding 300mm	0.10	10.00	0.11	0.24	–	2.22	0.04	2.37	m2	2.66
Brickwork general surfaces										
girth exceeding 300mm	0.14	10.00	0.15	0.26	–	2.40	0.05	2.60	m2	2.93

Labour hourly rates: (except Specialists) Craft Operatives £9.23 Labourer £7.02 Rates are national average prices. Refer to REGIONAL VARIATIONS for indicative levels of overall pricing in regions	MATERIALS			LABOUR				RATES		
	Del to Site £	Waste %	Material Cost £	Craft Optve Hrs	Lab Hrs	Labour Cost £	Sunds £	Nett Rate £	Unit	Gross Rate (+12.5%) £
M60: PAINTING/CLEAR FINISHING; STAIRCASE AREAS										
Mist coat, one full coat emulsion paint										
Concrete general surfaces girth exceeding 300mm	0.31	10.00	0.34	0.23	–	2.12	0.04	2.50	m2	2.82
Plaster general surfaces girth exceeding 300mm	0.27	10.00	0.30	0.21	–	1.94	0.04	2.28	m2	2.56
Plasterboard general surfaces girth exceeding 300mm	0.27	10.00	0.30	0.21	–	1.94	0.04	2.28	m2	2.56
Brickwork general surfaces girth exceeding 300mm	0.40	10.00	0.44	0.25	–	2.31	0.04	2.79	m2	3.14
Paper covered general surfaces girth exceeding 300mm	0.29	10.00	0.32	0.22	–	2.03	0.04	2.39	m2	2.69
Mist coat, two full coats emulsion paint										
Concrete general surfaces girth exceeding 300mm	0.54	10.00	0.59	0.33	–	3.05	0.06	3.70	m2	4.16
Plaster general surfaces girth exceeding 300mm	0.45	10.00	0.50	0.31	–	2.86	0.05	3.41	m2	3.83
Plasterboard general surfaces girth exceeding 300mm	0.45	10.00	0.50	0.31	–	2.86	0.05	3.41	m2	3.83
Brickwork general surfaces girth exceeding 300mm	0.67	10.00	0.74	0.36	–	3.32	0.06	4.12	m2	4.63
Paper covered general surfaces girth exceeding 300mm	0.49	10.00	0.54	0.32	–	2.95	0.06	3.55	m2	4.00
One coat textured masonry paint										
Cement rendered general surfaces girth exceeding 300mm	0.61	10.00	0.67	0.15	–	1.38	0.03	2.09	m2	2.35
Concrete general surfaces girth exceeding 300mm	0.69	10.00	0.76	0.15	–	1.38	0.03	2.17	m2	2.45
Brickwork general surfaces girth exceeding 300mm	0.76	10.00	0.84	0.18	–	1.66	0.03	2.53	m2	2.84
Rough cast general surfaces girth exceeding 300mm	1.23	10.00	1.35	0.22	–	2.03	0.04	3.42	m2	3.85
Two coats textured masonry paint										
Cement rendered general surfaces girth exceeding 300mm	1.11	10.00	1.22	0.31	–	2.86	0.05	4.13	m2	4.65
Concrete general surfaces girth exceeding 300mm	1.23	10.00	1.35	0.31	–	2.86	0.05	4.26	m2	4.80
Brickwork general surfaces girth exceeding 300mm	1.38	10.00	1.52	0.33	–	3.05	0.06	4.62	m2	5.20
Rough cast general surfaces girth exceeding 300mm	2.21	10.00	2.43	0.40	–	3.69	0.07	6.19	m2	6.97
One coat stabilising solution, one coat textured masonry paint										
Cement rendered general surfaces girth exceeding 300mm	1.14	10.00	1.25	0.26	–	2.40	0.05	3.70	m2	4.17
Concrete general surfaces girth exceeding 300mm	1.28	10.00	1.41	0.26	–	2.40	0.05	3.86	m2	4.34
Brickwork general surfaces girth exceeding 300mm	1.41	10.00	1.55	0.30	–	2.77	0.05	4.37	m2	4.92
Rough cast general surfaces girth exceeding 300mm	1.91	10.00	2.10	0.36	–	3.32	0.06	5.48	m2	6.17
One coat sealer, two coats Sandtex										
Cement rendered general surfaces girth exceeding 300mm	2.41	10.00	2.65	0.42	–	3.88	0.07	6.60	m2	7.42
Concrete general surfaces girth exceeding 300mm	2.59	10.00	2.85	0.42	–	3.88	0.07	6.80	m2	7.65
Brickwork general surfaces girth exceeding 300mm	2.99	10.00	3.29	0.45	–	4.15	0.08	7.52	m2	8.46
Rough cast general surfaces , girth exceeding 300mm	2.83	10.00	3.11	0.54	–	4.98	0.09	8.19	m2	9.21

SURFACE FINISHES

SURFACE FINISHES – SMALL WORKS

Labour hourly rates: (except Specialists) Craft Operatives £9.23 Labourer £7.02 Rates are national average prices. Refer to REGIONAL VARIATIONS for indicative levels of overall pricing in regions	MATERIALS			LABOUR				RATES		
	Del to Site £	Waste %	Material Cost £	Craft Optve Hrs	Lab Hrs	Labour Cost £	Sunds £	Nett Rate £	Unit	Gross Rate (+12.5%) £
M60: PAINTING/CLEAR FINISHING; STAIRCASE AREAS Cont.										
One coat primer, one undercoat, one coat full gloss finish										
Concrete general surfaces girth exceeding 300mm	0.95	10.00	1.04	0.43	–	3.97	0.08	5.09	m2	5.73
Brickwork general surfaces girth exceeding 300mm	1.17	10.00	1.29	0.47	–	4.34	0.08	5.71	m2	6.42
Plasterboard general surfaces girth exceeding 300mm	0.74	10.00	0.81	0.40	–	3.69	0.07	4.58	m2	5.15
Plaster general surfaces girth exceeding 300mm	0.85	10.00	0.94	0.40	–	3.69	0.07	4.70	m2	5.28
One coat primer, two undercoats, one coat full gloss finish										
Concrete general surfaces girth exceeding 300mm	1.25	10.00	1.38	0.57	–	5.26	0.10	6.74	m2	7.58
Brickwork general surfaces girth exceeding 300mm	1.53	10.00	1.68	0.63	–	5.81	0.11	7.61	m2	8.56
Plasterboard general surfaces girth exceeding 300mm	0.97	10.00	1.07	0.53	–	4.89	0.09	6.05	m2	6.81
Plaster general surfaces girth exceeding 300mm	1.11	10.00	1.22	0.53	–	4.89	0.09	6.20	m2	6.98
One coat primer, one undercoat, one coat eggshell finish										
Concrete general surfaces girth exceeding 300mm	0.99	10.00	1.09	0.43	–	3.97	0.08	5.14	m2	5.78
Brickwork general surfaces girth exceeding 300mm	1.21	10.00	1.33	0.47	–	4.34	0.08	5.75	m2	6.47
Plasterboard general surfaces girth exceeding 300mm	0.77	10.00	0.85	0.40	–	3.69	0.07	4.61	m2	5.19
Plaster general surfaces girth exceeding 300mm	0.88	10.00	0.97	0.40	–	3.69	0.07	4.73	m2	5.32
One coat primer, two undercoats, one coat eggshell finish										
Concrete general surfaces girth exceeding 300mm	1.28	10.00	1.41	0.57	–	5.26	0.10	6.77	m2	7.62
Brickwork general surfaces girth exceeding 300mm	1.57	10.00	1.73	0.63	–	5.81	0.11	7.65	m2	8.61
Plasterboard general surfaces girth exceeding 300mm	1.00	10.00	1.10	0.53	–	4.89	0.09	6.08	m2	6.84
Plaster general surfaces girth exceeding 300mm	1.14	10.00	1.25	0.53	–	4.89	0.09	6.24	m2	7.02
Textured plastic coating - Stippled finish										
Concrete general surfaces girth exceeding 300mm	0.57	10.00	0.63	0.37	–	3.42	0.06	4.10	m2	4.61
Brickwork general surfaces girth exceeding 300mm	0.69	10.00	0.76	0.48	–	4.43	0.08	5.27	m2	5.93
Plasterboard general surfaces girth exceeding 300mm	0.46	10.00	0.51	0.29	–	2.68	0.05	3.23	m2	3.64
Plaster general surfaces girth exceeding 300mm	0.46	10.00	0.51	0.29	–	2.68	0.05	3.23	m2	3.64
Textured plastic coating - Combed Finish										
Concrete general surfaces girth exceeding 300mm	0.57	10.00	0.63	0.43	–	3.97	0.08	4.68	m2	5.26
Brickwork general surfaces girth exceeding 300mm	0.69	10.00	0.76	0.59	–	5.45	0.10	6.30	m2	7.09
Plasterboard general surfaces girth exceeding 300mm	0.46	10.00	0.51	0.34	–	3.14	0.06	3.70	m2	4.17
Plaster general surfaces girth exceeding 300mm	0.42	10.00	0.46	0.34	–	3.14	0.06	3.66	m2	4.12

Labour hourly rates: (except Specialists) Craft Operatives £9.23 Labourer £7.02 Rates are national average prices. Refer to REGIONAL VARIATIONS for indicative levels of overall pricing in regions	MATERIALS			LABOUR				RATES		
	Del to Site £	Waste %	Material Cost £	Craft Optve Hrs	Lab Hrs	Labour Cost £	Sunds £	Nett Rate £	Unit	Gross Rate (+12.5%) £
M60: PAINTING/CLEAR FINISHING; STAIRCASE AREAS – REDECORATIONS										
Generally										

Note
   the following rates include for the cost of all
   preparatory work, e.g. washing down, etc.										

Two coats emulsion paint, existing emulsion painted surfaces										
Concrete general surfaces girth exceeding 300mm	0.45	10.00	0.50	0.39	-	3.60	0.07	4.16	m2	4.69
Plaster general surfaces girth exceeding 300mm	0.36	10.00	0.40	0.35	-	3.23	0.06	3.69	m2	4.15
Plasterboard general surfaces girth exceeding 300mm	0.36	10.00	0.40	0.36	-	3.32	0.06	3.78	m2	4.25
Brickwork general surfaces girth exceeding 300mm	0.54	10.00	0.59	0.39	-	3.60	0.07	4.26	m2	4.80
Paper covered general surfaces girth exceeding 300mm	0.40	10.00	0.44	0.36	-	3.32	0.06	3.82	m2	4.30
Two full coats emulsion paint, existing washable distempered surfaces										
Concrete general surfaces girth exceeding 300mm	0.45	10.00	0.50	0.40	-	3.69	0.07	4.26	m2	4.79
Plaster general surfaces girth exceeding 300mm	0.36	10.00	0.40	0.36	-	3.32	0.06	3.78	m2	4.25
Plasterboard general surfaces girth exceeding 300mm	0.36	10.00	0.40	0.37	-	3.42	0.06	3.87	m2	4.35
Brickwork general surfaces girth exceeding 300mm	0.54	10.00	0.59	0.40	-	3.69	0.07	4.36	m2	4.90
Paper covered general surfaces girth exceeding 300mm	0.40	10.00	0.44	0.37	-	3.42	0.06	3.92	m2	4.40
Two full coats emulsion paint, existing textured plastic coating surfaces										
Concrete general surfaces girth exceeding 300mm	0.45	10.00	0.50	0.44	-	4.06	0.08	4.64	m2	5.22
Plaster general surfaces girth exceeding 300mm	0.36	10.00	0.40	0.39	-	3.60	0.07	4.07	m2	4.57
Plasterboard general surfaces girth exceeding 300mm	0.36	10.00	0.40	0.40	-	3.69	0.07	4.16	m2	4.68
Brickwork general surfaces girth exceeding 300mm	0.54	10.00	0.59	0.43	-	3.97	0.08	4.64	m2	5.22
Paper covered general surfaces girth exceeding 300mm	0.40	10.00	0.44	0.40	-	3.69	0.07	4.20	m2	4.73
Mist coat, two full coats emulsion paint, existing non-washable distempered surfaces										
Concrete general surfaces girth exceeding 300mm	0.54	10.00	0.59	0.49	-	4.52	0.09	5.21	m2	5.86
Plaster general surfaces girth exceeding 300mm	0.45	10.00	0.50	0.45	-	4.15	0.08	4.73	m2	5.32
Plasterboard general surfaces girth exceeding 300mm	0.45	10.00	0.50	0.46	-	4.25	0.08	4.82	m2	5.42
Brickwork general surfaces girth exceeding 300mm	0.67	10.00	0.74	0.49	-	4.52	0.09	5.35	m2	6.02
Paper covered general surfaces girth exceeding 300mm	0.49	10.00	0.54	0.46	-	4.25	0.08	4.86	m2	5.47
Two full coats emulsion paint, existing gloss painted surfaces										
Concrete general surfaces girth exceeding 300mm	0.45	10.00	0.50	0.48	-	4.43	0.08	5.01	m2	5.63
Plaster general surfaces girth exceeding 300mm	0.36	10.00	0.40	0.34	-	3.14	0.06	3.59	m2	4.04
Plasterboard general surfaces girth exceeding 300mm	0.36	10.00	0.40	0.35	-	3.23	0.06	3.69	m2	4.15
Brickwork general surfaces girth exceeding 300mm	0.54	10.00	0.59	0.38	-	3.51	0.07	4.17	m2	4.69

SURFACE FINISHES

Labour hourly rates: (except Specialists) Craft Operatives £9.23 Labourer £7.02 Rates are national average prices. Refer to REGIONAL VARIATIONS for indicative levels of overall pricing in regions	MATERIALS			LABOUR				RATES		
	Del to Site £	Waste %	Material Cost £	Craft Optve Hrs	Lab Hrs	Labour Cost £	Sunds £	Nett Rate £	Unit	Gross Rate (+12.5%) £
M60: PAINTING/CLEAR FINISHING; STAIRCASE AREAS – REDECORATIONS Cont.										
Two full coats emulsion paint, existing gloss painted surfaces Cont.										
Paper covered general surfaces girth exceeding 300mm	0.40	10.00	0.44	0.35	–	3.23	0.06	3.73	m2	4.20
Generally										

Note
    the following rates include for the cost of all
    preparatory work, e.g. washing down, etc.										

One undercoat, one coat full gloss finish, existing gloss painted surfaces										
Concrete general surfaces girth exceeding 300mm	0.59	10.00	0.65	0.44	–	4.06	0.08	4.79	m2	5.39
Brickwork general surfaces girth exceeding 300mm	0.72	10.00	0.79	0.45	–	4.15	0.07	5.02	m2	5.64
Plasterboard general surfaces girth exceeding 300mm	0.46	10.00	0.51	0.41	–	3.78	0.07	4.36	m2	4.91
Plaster general surfaces girth exceeding 300mm	0.52	10.00	0.57	0.40	–	3.69	0.07	4.33	m2	4.88
One undercoat, one coat eggshell finish, existing gloss painted surfaces										
Concrete general surfaces girth exceeding 300mm	0.62	10.00	0.68	0.44	–	4.06	0.08	4.82	m2	5.43
Brickwork general surfaces girth exceeding 300mm	0.76	10.00	0.84	0.45	–	4.15	0.08	5.07	m2	5.70
Plasterboard general surfaces girth exceeding 300mm	0.48	10.00	0.53	0.41	–	3.78	0.07	4.38	m2	4.93
Plaster general surfaces girth exceeding 300mm	0.55	10.00	0.60	0.40	–	3.69	0.07	4.37	m2	4.91
Two undercoats, one coat full gloss finish, existing gloss painted surfaces										
Concrete general surfaces girth exceeding 300mm	0.88	10.00	0.97	0.57	–	5.26	0.10	6.33	m2	7.12
Brickwork general surfaces girth exceeding 300mm	1.08	10.00	1.19	0.57	–	5.26	0.10	6.55	m2	7.37
Plasterboard general surfaces girth exceeding 300mm	0.69	10.00	0.76	0.54	–	4.98	0.09	5.83	m2	6.56
Plaster general surfaces girth exceeding 300mm	0.78	10.00	0.86	0.53	–	4.89	0.09	5.84	m2	6.57
Two undercoats, one coat eggshell finish, existing gloss painted surfaces										
Concrete general surfaces girth exceeding 300mm	0.92	10.00	1.01	0.57	–	5.26	0.10	6.37	m2	7.17
Brickwork general surfaces girth exceeding 300mm	1.12	10.00	1.23	0.57	–	5.26	0.10	6.59	m2	7.42
Plasterboard general surfaces girth exceeding 300mm	0.71	10.00	0.78	0.54	–	4.98	0.09	5.86	m2	6.59
Plaster general surfaces girth exceeding 300mm	0.82	10.00	0.90	0.53	–	4.89	0.09	5.88	m2	6.62
One coat primer, one undercoat, one coat full gloss finish, existing washable distempered or emulsion painted surfaces										
Concrete general surfaces girth exceeding 300mm	0.95	10.00	1.04	0.57	–	5.26	0.10	6.41	m2	7.21
Brickwork general surfaces girth exceeding 300mm	1.17	10.00	1.29	0.57	–	5.26	0.10	6.65	m2	7.48
Plasterboard general surfaces girth exceeding 300mm	0.74	10.00	0.81	0.54	–	4.98	0.09	5.89	m2	6.62
Plaster general surfaces girth exceeding 300mm	0.85	10.00	0.94	0.53	–	4.89	0.09	5.92	m2	6.66
One coat primer, one undercoat, one coat eggshell finish, existing washable distempered or emulsion painted surfaces										
Concrete general surfaces girth exceeding 300mm	0.99	10.00	1.09	0.57	–	5.26	0.10	6.45	m2	7.26

Labour hourly rates: (except Specialists) Craft Operatives £9.23 Labourer £7.02 Rates are national average prices. Refer to REGIONAL VARIATIONS for indicative levels of overall pricing in regions	MATERIALS			LABOUR				RATES		
	Del to Site £	Waste %	Material Cost £	Craft Optve Hrs	Lab Hrs	Labour Cost £	Sunds £	Nett Rate £	Unit	Gross Rate (+12.5%) £
M60: PAINTING/CLEAR FINISHING; STAIRCASE AREAS – REDECORATIONS Cont.										
One coat primer, one undercoat, one coat eggshell finish, existing washable distempered or emulsion painted surfaces Cont.										
Brickwork general surfaces girth exceeding 300mm	1.21	10.00	1.33	0.57	–	5.26	0.10	6.69	m2	7.53
Plasterboard general surfaces girth exceeding 300mm	0.77	10.00	0.85	0.54	–	4.98	0.09	5.92	m2	6.66
Plaster general surfaces girth exceeding 300mm	0.88	10.00	0.97	0.53	–	4.89	0.09	5.95	m2	6.69
One coat primer, two undercoats, one coat full gloss finish, existing washable distempered or emulsion painted surfaces										
Concrete general surfaces girth exceeding 300mm	1.25	10.00	1.38	0.70	–	6.46	0.12	7.96	m2	8.95
Brickwork general surfaces girth exceeding 300mm	1.53	10.00	1.68	0.71	–	6.55	0.12	8.36	m2	9.40
Plasterboard general surfaces girth exceeding 300mm	0.97	10.00	1.07	0.67	–	6.18	0.12	7.37	m2	8.29
Plaster general surfaces girth exceeding 300mm	0.98	10.00	1.08	0.66	–	6.09	0.12	7.29	m2	8.20
One coat primer, two undercoats, one coat eggshell finish, existing washable distempered or emulsion painted surfaces										
Concrete general surfaces girth exceeding 300mm	1.16	10.00	1.28	0.70	–	6.46	0.12	7.86	m2	8.84
Brickwork general surfaces girth exceeding 300mm	1.41	10.00	1.55	0.71	–	6.55	0.12	8.22	m2	9.25
Plasterboard general surfaces girth exceeding 300mm	0.90	10.00	0.99	0.67	–	6.18	0.12	7.29	m2	8.21
Plaster general surfaces girth exceeding 300mm	1.03	10.00	1.13	0.66	–	6.09	0.12	7.34	m2	8.26
One coat primer, one undercoat, one coat full gloss finish, existing non-washable distempered surfaces										
Concrete general surfaces girth exceeding 300mm	0.95	10.00	1.04	0.57	–	5.26	0.10	6.41	m2	7.21
Brickwork general surfaces girth exceeding 300mm	1.17	10.00	1.29	0.57	–	5.26	0.10	6.65	m2	7.48
Plasterboard general surfaces girth exceeding 300mm	0.74	10.00	0.81	0.54	–	4.98	0.09	5.89	m2	6.62
Plaster general surfaces girth exceeding 300mm	0.85	10.00	0.94	0.53	–	4.89	0.09	5.92	m2	6.66
One coat primer, one undercoat, one coat eggshell finish, existing non-washable distempered surfaces										
Concrete general surfaces girth exceeding 300mm	0.99	10.00	1.09	0.57	–	5.26	0.10	6.45	m2	7.26
Brickwork general surfaces girth exceeding 300mm	1.21	10.00	1.33	0.57	–	5.26	0.10	6.69	m2	7.53
Plasterboard general surfaces girth exceeding 300mm	0.77	10.00	0.85	0.54	–	4.98	0.09	5.92	m2	6.66
Plaster general surfaces girth exceeding 300mm	0.88	10.00	0.97	0.53	–	4.89	0.09	5.95	m2	6.69
One coat primer, two undercoats, one coat full gloss finish, existing non-washable distempered surfaces										
Concrete general surfaces girth exceeding 300mm	1.25	10.00	1.38	0.70	–	6.46	0.12	7.96	m2	8.95
Brickwork general surfaces girth exceeding 300mm	1.53	10.00	1.68	0.71	–	6.55	0.12	8.36	m2	9.40
Plasterboard general surfaces girth exceeding 300mm	0.97	10.00	1.07	0.67	–	6.18	0.12	7.37	m2	8.29
Plaster general surfaces girth exceeding 300mm	0.98	10.00	1.08	0.66	–	6.09	0.12	7.29	m2	8.20

Labour hourly rates: (except Specialists) Craft Operatives £9.23 Labourer £7.02 Rates are national average prices. Refer to REGIONAL VARIATIONS for indicative levels of overall pricing in regions	MATERIALS			LABOUR				RATES		
	Del to Site	Waste	Material Cost	Craft Optve	Lab	Labour Cost	Sunds	Nett Rate	Unit	Gross Rate (+12.5%)
	£	%	£	Hrs	Hrs	£	£	£		£
M60: PAINTING/CLEAR FINISHING; STAIRCASE AREAS – REDECORATIONS Cont.										
One coat primer, two undercoats, one coat eggshell finish, existing non-washable distempered surfaces										
Concrete general surfaces										
girth exceeding 300mm	1.16	10.00	**1.28**	0.70	-	**6.46**	0.12	**7.86**	m2	**8.84**
Brickwork general surfaces										
girth exceeding 300mm	1.41	10.00	**1.55**	0.71	-	**6.55**	0.12	**8.22**	m2	**9.25**
Plasterboard general surfaces										
girth exceeding 300mm	0.90	10.00	**0.99**	0.67	-	**6.18**	0.12	**7.29**	m2	**8.21**
Plaster general surfaces										
girth exceeding 300mm	1.03	10.00	**1.13**	0.66	-	**6.09**	0.12	**7.34**	m2	**8.26**
Generally										
Note the following rates include for the cost of all preparatory work, e.g. washing down, etc.										
Textured plastic coating - stippled finish, existing washable distempered or emulsion painted surfaces										
Concrete general surfaces										
girth exceeding 300mm	0.57	10.00	**0.63**	0.50	-	**4.62**	0.09	**5.33**	m2	**6.00**
Brickwork general surfaces										
girth exceeding 300mm	0.69	10.00	**0.76**	0.55	-	**5.08**	0.10	**5.94**	m2	**6.68**
Plasterboard general surfaces										
girth exceeding 300mm	0.46	10.00	**0.51**	0.43	-	**3.97**	0.08	**4.55**	m2	**5.12**
Plaster general surfaces										
girth exceeding 300mm	0.46	10.00	**0.51**	0.42	-	**3.88**	0.07	**4.45**	m2	**5.01**
Textured plastic coating - combed finish, existing washable distempered or emulsion painted surfaces										
Concrete general surfaces										
girth exceeding 300mm	0.57	10.00	**0.63**	0.55	-	**5.08**	0.10	**5.80**	m2	**6.53**
Brickwork general surfaces										
girth exceeding 300mm	0.69	10.00	**0.76**	0.60	-	**5.54**	0.10	**6.40**	m2	**7.20**
Plasterboard general surfaces										
girth exceeding 300mm	0.46	10.00	**0.51**	0.48	-	**4.43**	0.08	**5.02**	m2	**5.64**
Plaster general surfaces										
girth exceeding 300mm	0.46	10.00	**0.51**	0.47	-	**4.34**	0.08	**4.92**	m2	**5.54**
Textured plastic coating - stippled finish, existing non-washable distempered surfaces										
Concrete general surfaces										
girth exceeding 300mm	0.57	10.00	**0.63**	0.55	-	**5.08**	0.10	**5.80**	m2	**6.53**
Brickwork general surfaces										
girth exceeding 300mm	0.69	10.00	**0.76**	0.60	-	**5.54**	0.10	**6.40**	m2	**7.20**
Plasterboard general surfaces										
girth exceeding 300mm	0.46	10.00	**0.51**	0.48	-	**4.43**	0.08	**5.02**	m2	**5.64**
Plaster general surfaces										
girth exceeding 300mm	0.46	10.00	**0.51**	0.47	-	**4.34**	0.08	**4.92**	m2	**5.54**
Textured plastic coating - combed finish, existing non-washable distempered surfaces										
Concrete general surfaces										
girth exceeding 300mm	0.57	10.00	**0.63**	0.60	-	**5.54**	0.10	**6.26**	m2	**7.05**
Brickwork general surfaces										
girth exceeding 300mm	0.69	10.00	**0.76**	0.65	-	**6.00**	0.11	**6.87**	m2	**7.73**
Plasterboard general surfaces										
girth exceeding 300mm	0.46	10.00	**0.51**	0.53	-	**4.89**	0.09	**5.49**	m2	**6.17**
Plaster general surfaces										
girth exceeding 300mm	0.46	10.00	**0.51**	0.52	-	**4.80**	0.09	**5.40**	m2	**6.07**

Furniture/Equipment

table

Labour hourly rates: (except Specialists) Craft Operatives £9.23 Labourer £7.02 Rates are national average prices. Refer to REGIONAL VARIATIONS for indicative levels of overall pricing in regions	MATERIALS			LABOUR				RATES		
	Del to Site £	Waste %	Material Cost £	Craft Optve Hrs	Lab Hrs	Labour Cost £	Sunds £	Nett Rate £	Unit	Gross Rate (+12.5%) £
N10: GENERAL FIXTURES/FURNISHING/EQUIPMENT										
MIRRORS - B.S.952, clear float, SG, silvered and protected with copper backing										
6mm thick; fixing to masonry with brass screws, chromium plated dome covers, rubber sleeves and washers										
holes 6mm diameter -4; edges polished										
254 x 400mm	31.30	5.00	32.87	0.50	-	4.62	0.41	37.89	nr	42.63
300 x 460mm	36.06	5.00	37.86	0.50	-	4.62	0.41	42.89	nr	48.25
360 x 500mm	40.68	5.00	42.71	0.50	-	4.62	0.41	47.74	nr	53.71
460 x 560mm	49.29	5.00	51.75	0.65	-	6.00	0.41	58.16	nr	65.43
460 x 600mm	51.43	5.00	54.00	0.65	-	6.00	0.41	60.41	nr	67.96
460 x 900mm	66.13	5.00	69.44	0.70	-	6.46	0.41	76.31	nr	85.85
500 x 680mm	57.91	5.00	60.81	0.65	-	6.00	0.41	67.22	nr	75.62
600 x 900mm	77.67	5.00	81.55	0.75	-	6.92	0.41	88.89	nr	100.00
holes 6mm diameter -4; edges bevelled										
254 x 400mm	37.71	5.00	39.60	0.50	-	4.62	0.41	44.62	nr	50.20
300 x 460mm	43.48	5.00	45.65	0.50	-	4.62	0.41	50.68	nr	57.01
360 x 500mm	49.10	5.00	51.56	0.50	-	4.62	0.41	56.58	nr	63.65
460 x 560mm	59.27	5.00	62.23	0.65	-	6.00	0.41	68.64	nr	77.22
460 x 600mm	61.82	5.00	64.91	0.65	-	6.00	0.41	71.32	nr	80.24
460 x 900mm	79.43	5.00	83.40	0.70	-	6.46	0.41	90.27	nr	101.56
500 x 680mm	69.46	5.00	72.93	0.65	-	6.00	0.41	79.34	nr	89.26
600 x 900mm	92.79	5.00	97.43	0.75	-	6.92	0.41	104.76	nr	117.86
CURTAIN TRACKS - Fixing only curtain track										
Metal or plastic track with fittings										
fixing with screws to softwood	-	-	-	0.50	-	4.62	-	4.62	m	5.19
fixing with screws to hardwood	-	-	-	0.70	-	6.46	-	6.46	m	7.27
BLINDS - Internal blinds										
Venetian blinds; stove enamelled aluminium alloy slats 25mm wide; plain colours										
1200mm drop; fixing to timber with screws										
1000mm wide	30.80	2.50	31.57	1.10	0.14	11.14	0.18	42.89	nr	48.25
2000mm wide	47.25	2.50	48.43	1.65	0.21	16.70	0.27	65.40	nr	73.58
3000mm wide	63.66	2.50	65.25	2.20	0.28	22.27	0.36	87.88	nr	98.87
Venetian blinds; stove enamelled aluminium alloy slats 50mm wide; plain colours										
1200mm drop; fixing to timber with screws										
1000mm wide	28.40	2.50	29.11	1.27	0.16	12.85	0.21	42.17	nr	47.44
2000mm wide	43.81	2.50	44.91	1.90	0.24	19.22	0.31	64.44	nr	72.49
3000mm wide	59.57	2.50	61.06	2.50	0.31	25.25	0.41	86.72	nr	97.56
Venetian blinds; stove enamelled aluminium alloy slats 25mm wide; plain colours with single cord control										
1200mm drop; fixing to timber with screws										
1000mm wide	43.81	2.50	44.91	1.10	0.14	11.14	0.18	56.22	nr	63.25
2000mm wide	68.44	2.50	70.15	1.65	0.21	16.70	0.27	87.12	nr	98.02
3000mm wide	92.75	2.50	95.07	2.20	0.28	22.27	0.36	117.70	nr	132.41
Venetian blinds; stove enamelled aluminium alloy slats 50mm wide; plain colours with single cord control										
1200mm drop; fixing to timber with screws										
1000mm wide	39.70	2.50	40.69	1.27	0.16	12.85	0.21	53.75	nr	60.47
2000mm wide	60.24	2.50	61.75	1.90	0.24	19.22	0.31	81.28	nr	91.44
3000mm wide	82.82	2.50	84.89	2.50	0.31	25.25	0.41	110.55	nr	124.37
Venetian blinds; stove enamelled aluminium alloy slats 25mm wide; plain colours with channels for dimout										
1200mm drop; fixing to timber with screws										
1000mm wide	74.61	2.50	76.48	2.20	0.28	22.27	0.36	99.11	nr	111.50
2000mm wide	91.72	2.50	94.01	2.75	0.34	27.77	0.45	122.23	nr	137.51

FURNITURE/EQUIPMENT

Labour hourly rates: (except Specialists) Craft Operatives £9.23 Labourer £7.02 Rates are national average prices. Refer to REGIONAL VARIATIONS for indicative levels of overall pricing in regions	MATERIALS			LABOUR				RATES		
	Del to Site £	Waste %	Material Cost £	Craft Optve Hrs	Lab Hrs	Labour Cost £	Sunds £	Nett Rate £	Unit	Gross Rate (+12.5%) £
N10: GENERAL FIXTURES/FURNISHING/EQUIPMENT Cont.										
BLINDS - Internal blinds Cont.										
Venetian blinds; stove enamelled aluminium alloy slats 25mm wide; plain colours with channels for dimout Cont.										
1200mm drop; fixing to timber with screws Cont.										
3000mm wide......................	108.14	2.50	110.84	3.30	0.41	33.34	0.54	144.72	nr	162.81
Venetian blinds; stove enamelled aluminium alloy slats 50mm wide; plain colours with channels for dimout										
1200mm drop; fixing to timber with screws										
1000mm wide......................	71.86	2.50	73.66	2.50	0.31	25.25	0.41	99.32	nr	111.73
2000mm wide......................	88.30	2.50	90.51	3.10	0.39	31.35	0.51	122.37	nr	137.66
3000mm wide......................	104.04	2.50	106.64	3.70	0.46	37.38	0.60	144.62	nr	162.70
Roller blinds; automatic ratchet action; fire resistant material										
1200mm drop; fixing to timber with screws										
1000mm wide......................	57.50	2.50	58.94	1.10	0.14	11.14	0.18	70.25	nr	79.03
2000mm wide......................	70.51	2.50	72.27	1.65	0.21	16.70	0.27	89.25	nr	100.40
3000mm wide......................	83.49	2.50	85.58	2.20	0.28	22.27	0.36	108.21	nr	121.73
Roller blinds; automatic ratchet action; holland type material										
1200mm drop; fixing to timber with screws										
1000mm wide......................	55.45	2.50	56.84	1.10	0.14	11.14	0.18	68.15	nr	76.67
2000mm wide......................	66.40	2.50	68.06	1.65	0.21	16.70	0.27	85.03	nr	95.66
3000mm wide......................	77.31	2.50	79.24	2.20	0.28	22.27	0.36	101.87	nr	114.61
Roller blinds; self acting roller; blackout material										
1200mm drop; fixing to timber with screws										
1000mm wide......................	51.34	2.50	52.62	1.27	0.16	12.85	0.21	65.68	nr	73.89
2000mm wide......................	62.29	2.50	63.85	1.90	0.24	19.22	0.31	83.38	nr	93.80
3000mm wide......................	73.24	2.50	75.07	2.50	0.31	25.25	0.41	100.73	nr	113.32
Roller blinds; 100% blackout; natural anodised box and channels										
1200mm drop; fixing to timber with screws										
1000mm wide......................	169.73	2.50	173.97	2.20	0.28	22.27	0.36	196.60	nr	221.18
2000mm wide......................	212.22	2.50	217.53	3.05	0.31	30.33	0.49	248.34	nr	279.39
3000mm wide......................	258.72	2.50	265.19	3.85	0.48	38.91	0.63	304.72	nr	342.81
Vertical louvre blinds; 89mm wide louvres in standard material										
1200mm drop; fixing to timber with screws										
1000mm wide......................	61.60	2.50	63.14	1.00	0.13	10.14	0.16	73.44	nr	82.62
2000mm wide......................	108.14	2.50	110.84	1.50	0.19	15.18	0.24	126.26	nr	142.05
3000mm wide......................	154.69	2.50	158.56	2.00	0.25	20.22	0.33	179.10	nr	201.49
Vertical louvre blinds; 127mm wide louvres in standard material										
1200mm drop; fixing to timber with screws										
1000mm wide......................	55.45	2.50	56.84	1.05	0.13	10.60	0.17	67.61	nr	76.06
2000mm wide......................	96.51	2.50	98.92	1.55	0.19	15.64	0.25	114.81	nr	129.16
3000mm wide......................	138.26	2.50	141.72	2.03	0.25	20.49	0.33	162.54	nr	182.86
BLINDS - External blinds; manually operated										
Venetian blinds; stove enamelled aluminium slats 80mm wide; natural anodised side guides; excluding boxing										
1200mm drop; fixing to timber with screws										
1000mm wide......................	-	-	Spclist	-	-	Spclist	-	266.96	nr	300.33
2000mm wide......................	-	-	Spclist	-	-	Spclist	-	397.32	nr	446.99
3000mm wide......................	-	-	Spclist	-	-	Spclist	-	535.15	nr	602.04
Queensland awnings; acrylic material; natural anodised arms and boxing										
1000mm projection										
1000mm long......................	-	-	Spclist	-	-	Spclist	-	625.78	nr	704.00
2000mm long......................	-	-	Spclist	-	-	Spclist	-	938.09	nr	1055.35
3000mm long......................	-	-	Spclist	-	-	Spclist	-	1251.56	nr	1408.01
Rollscreen vertical drop roller blinds; mesh material; natural anodised side guides and boxing										
1200mm drop; fixing to timber with screws										
1000mm wide......................	-	-	Spclist	-	-	Spclist	-	260.71	nr	293.30
2000mm wide......................	-	-	Spclist	-	-	Spclist	-	391.12	nr	440.01
3000mm wide......................	-	-	Spclist	-	-	Spclist	-	521.26	nr	586.42
Quandrant canopy; acrylic material; natural anodised frames										
1000mm projection										
1000mm long......................	-	-	Spclist	-	-	Spclist	-	345.17	nr	388.32
2000mm long......................	-	-	Spclist	-	-	Spclist	-	527.69	nr	593.65
3000mm long......................	-	-	Spclist	-	-	Spclist	-	717.66	nr	807.37
Foldaway awning; acrylic material; natural anodised arms and front rail										
2000mm projection										
1000mm long......................	-	-	Spclist	-	-	Spclist	-	559.98	nr	629.98
2000mm long......................	-	-	Spclist	-	-	Spclist	-	821.95	nr	924.69
3000mm long......................	-	-	Spclist	-	-	Spclist	-	1069.05	nr	1202.68

Labour hourly rates: (except Specialists) Craft Operatives £9.23 Labourer £7.02 Rates are national average prices. Refer to REGIONAL VARIATIONS for indicative levels of overall pricing in regions	MATERIALS			LABOUR				RATES		
	Del to Site £	Waste %	Material Cost £	Craft Optve Hrs	Lab Hrs	Labour Cost £	Sunds £	Nett Rate £	Unit	Gross Rate (+12.5%) £
N10: GENERAL FIXTURES/FURNISHING/EQUIPMENT Cont.										
BLINDS - External blinds; electrically operated										
Venetian blinds; stove enamelled aluminium slats 80mm wide; natural anodised side guides; excluding boxing 1200mm drop; fixing to timber with screws										
1000mm wide	-	-	Spclist	-	-	Spclist	-	613.36	nr	690.03
2000mm wide	-	-	Spclist	-	-	Spclist	-	749.95	nr	843.69
3000mm wide	-	-	Spclist	-	-	Spclist	-	879.08	nr	988.97
Queensland awnings; acrylic material; natural anodised arms and boxing 1000mm projection										
1000mm long	-	-	Spclist	-	-	Spclist	-	717.66	nr	807.37
2000mm long	-	-	Spclist	-	-	Spclist	-	1024.34	nr	1152.38
3000mm long	-	-	Spclist	-	-	Spclist	-	1303.71	nr	1466.67
Rollscreen vertical drop roller blinds; mesh material; natural anodised side guides and boxing 1200mm drop; fixing to timber with screws										
1000mm wide	-	-	Spclist	-	-	Spclist	-	496.65	nr	558.73
2000mm wide	-	-	Spclist	-	-	Spclist	-	625.78	nr	704.00
3000mm wide	-	-	Spclist	-	-	Spclist	-	756.15	nr	850.67
Quandrant canopy; acrylic material; natural anodised frames 1000mm projection										
1000mm long	-	-	Spclist	-	-	Spclist	-	449.47	nr	505.65
2000mm long	-	-	Spclist	-	-	Spclist	-	625.78	nr	704.00
3000mm long	-	-	Spclist	-	-	Spclist	-	834.37	nr	938.67
Foldaway awning; acrylic material; natural anodised arms and front rail 2000mm projection										
1000mm long	-	-	Spclist	-	-	Spclist	-	820.72	nr	923.31
2000mm long	-	-	Spclist	-	-	Spclist	-	1075.25	nr	1209.66
3000mm long	-	-	Spclist	-	-	Spclist	-	1365.79	nr	1536.51
STORAGE SYSTEMS - Vista two tone coloured steel boltless storage systems; Welconstruct Co. Ltd.; units 900mm wide x 1800mm high; assembling										
Open shelving with top, bottom, three shelves and braces; placing in position										
300mm deep code G106-001	113.18	2.50	116.01	1.20	0.60	15.29	-	131.30	nr	147.71
450mm deep code G106-002	135.32	2.50	138.70	1.40	0.70	17.84	-	156.54	nr	176.11
600mm deep code G106-003	151.96	2.50	155.76	1.60	0.80	20.38	-	176.14	nr	198.16
Closed shelving with top, bottom, and three shelves; placing in position										
300mm deep code G106-011	151.30	2.50	155.08	1.30	0.65	16.56	-	171.64	nr	193.10
extra for additional shelf code G106-301	8.10	2.50	8.30	0.22	0.11	2.80	-	11.11	nr	12.49
450mm deep code G106-012	177.66	2.50	182.10	1.50	0.75	19.11	-	201.21	nr	226.36
extra for additional shelf code G106-302	12.36	2.50	12.67	0.22	0.11	2.80	-	15.47	nr	17.41
600mm deep code G106-013	198.72	2.50	203.69	1.70	0.85	21.66	-	225.35	nr	253.51
extra for additional shelf code G106-303	15.44	2.50	15.83	0.22	0.11	2.80	-	18.63	nr	20.96
Unit comprising top, bottom, one shelf, six 300 x 300mm and four 450 x 450mm bins; placing in position										
300mm deep	184.55	2.50	189.16	1.30	0.65	16.56	-	205.73	nr	231.44
450mm deep	210.30	2.50	215.56	1.50	0.75	19.11	-	234.67	nr	264.00
600mm deep	248.80	2.50	255.02	1.70	0.85	21.66	-	276.68	nr	311.26
Unit comprising top, bottom, one shelf, eight 225 x 225mm, three 300 x 300mm and two 450 x 450mm bins and twelve 150x 150mm drawers; placing in position										
300mm deep	230.47	2.50	236.23	2.20	1.10	28.03	-	264.26	nr	297.29
450mm deep	279.67	2.50	286.66	2.40	1.20	30.58	-	317.24	nr	356.89
Unit comprising top, bottom, one shelf, three 300 x 225mm and fifteen 300 x 300mm bins; placing in position										
300mm deep	189.52	2.50	194.26	1.30	0.65	16.56	-	210.82	nr	237.17
450mm deep	233.50	2.50	239.34	1.50	0.75	19.11	-	258.45	nr	290.75
600mm deep	271.11	2.50	277.89	1.70	0.85	21.66	-	299.55	nr	336.99
Unit comprising top, bottom three 300 x 225mm, six 300 x 300mm and four 450 x 450mm bins; placing in position										
450mm deep	218.02	2.50	223.47	1.50	0.75	19.11	-	242.58	nr	272.90
600mm deep	248.97	2.50	255.19	1.70	0.85	21.66	-	276.85	nr	311.46
Extra over unit for pair of lockable doors code G106-401	178.78	2.50	183.25	0.22	0.11	2.80	-	186.05	nr	209.31
SHELVING SYSTEMS - Spanwel wide access boltless steel shelving systems; Welconstruct Co Ltd.; upright frames connected with shelf beams and with shelves; assembling										
Unit 1800mm high with 2 upright frames and three standard duty steel shelves including top and bottom; placing in position										
1200mm long x 650mm deep code G108-201	194.40	2.50	199.26	1.60	0.80	20.38	-	219.64	nr	247.10

FURNITURE/EQUIPMENT

	MATERIALS			LABOUR				RATES		
Labour hourly rates: (except Specialists) Craft Operatives £9.23 Labourer £7.02 Rates are national average prices. Refer to REGIONAL VARIATIONS for indicative levels of overall pricing in regions	Del to Site £	Waste %	Material Cost £	Craft Optve Hrs	Lab Hrs	Labour Cost £	Sunds £	Nett Rate £	Unit	Gross Rate (+12.5%) £

N10: GENERAL FIXTURES/FURNISHING/EQUIPMENT Cont.

SHELVING SYSTEMS - Spanwel wide access boltless steel shelving systems; Welconstruct Co Ltd.; upright frames connected with shelf beams and with shelves; assembling Cont.

Unit 1800mm high with 2 upright frames and three standard duty steel shelves including top and bottom; placing in position Cont.

	Del to Site £	Waste %	Material Cost £	Craft Optve Hrs	Lab Hrs	Labour Cost £	Sunds £	Nett Rate £	Unit	Gross Rate £
extra; additional standard duty steel shelf and shelf beam code G108-225	43.36	2.50	44.44	0.24	0.12	3.06	-	47.50	nr	53.44
1200mm long x 800mm deep code G108-202	210.06	2.50	215.31	1.70	0.85	21.66	-	236.97	nr	266.59
extra; additional standard duty steel shelf and shelf beam G108-226	45.90	2.50	47.05	0.24	0.12	3.06	-	50.11	nr	56.37
1200mm long x 950mm deep code G108-203	220.32	2.50	225.83	1.80	0.90	22.93	-	248.76	nr	279.86
extra; additional standard duty steel shelf and shelf beam code G108-227	48.70	2.50	49.92	0.24	0.12	3.06	-	52.98	nr	59.60
2400mm long x 650mm deep code G108-204	284.04	2.50	291.14	1.70	0.85	21.66	-	312.80	nr	351.90
extra; additional standard duty steel shelf and shelf beam code G108-228	71.11	2.50	72.89	0.24	0.12	3.06	-	75.95	nr	85.44
2400mm long x 800mm deep code G108-205	302.40	2.50	309.96	1.80	0.90	22.93	-	332.89	nr	374.50
extra; additional standard duty steel shelf and shelf beam code G108-229	76.68	2.50	78.60	0.24	0.12	3.06	-	81.65	nr	91.86
2400mm long x 950mm deep code G108-206	320.76	2.50	328.78	1.90	0.95	24.21	-	352.99	nr	397.11
extra; additional standard duty steel shelf and shelf beam code G108-230	82.08	2.50	84.13	0.24	0.12	3.06	-	87.19	nr	98.09

Unit 2400mm high with 3 upright frames and five standard duty steel shelves including top and bottom; placing in position

	Del to Site £	Waste %	Material Cost £	Craft Optve Hrs	Lab Hrs	Labour Cost £	Sunds £	Nett Rate £	Unit	Gross Rate £
2400mm long x 650mm deep	527.05	2.50	540.23	3.20	1.60	40.77	-	580.99	nr	653.62
2400mm long x 800mm deep	521.70	2.50	534.74	3.45	1.75	44.13	-	578.87	nr	651.23
2400mm long x 950mm deep	549.50	2.50	563.24	3.70	1.85	47.14	-	610.38	nr	686.67

SHELVING SYSTEMS - SHELVING SUPPORT SYSTEMS

Spur patent shelf supports in steel with white enamelled finish; Spur Shelving Ltd; Spur Steel-Lok components

Wall uprights

	Del to Site £	Waste %	Material Cost £	Craft Optve Hrs	Lab Hrs	Labour Cost £	Sunds £	Nett Rate £	Unit	Gross Rate £
type 9011, 430mm long; fixing to masonry with screws	3.28	2.50	3.36	0.25	-	2.31	-	5.67	nr	6.38
type 9013, 1000mm long; fixing to masonry with screws	6.62	2.50	6.79	0.45	-	4.15	-	10.94	nr	12.31
type 9016, 1600mm long; fixing to masonry with screws	9.70	2.50	9.94	0.55	-	5.08	-	15.02	nr	16.90
type 9020, 2400mm long; fixing to masonry with screws	14.74	2.50	15.11	0.70	-	6.46	-	21.57	nr	24.27

Straight brackets

	Del to Site £	Waste %	Material Cost £	Craft Optve Hrs	Lab Hrs	Labour Cost £	Sunds £	Nett Rate £	Unit	Gross Rate £
type 9001, 120mm long	1.85	2.50	1.90	0.05	-	0.46	-	2.36	nr	2.65
type 9003, 270mm long	3.02	2.50	3.10	0.05	-	0.46	-	3.56	nr	4.00
type 9004, 360mm long	3.97	2.50	4.07	0.07	-	0.65	-	4.72	nr	5.30
type 9006, 470mm long	6.52	2.50	6.68	0.07	-	0.65	-	7.33	nr	8.25

MAT RIMS - Matwells in Aluminium

Mat frames; welded fabrication
34 x 26 x 6mm angle section; angles mitred; plain lugs -4, welded on; mat space

	Del to Site £	Waste %	Material Cost £	Craft Optve Hrs	Lab Hrs	Labour Cost £	Sunds £	Nett Rate £	Unit	Gross Rate £
610 x 457mm	29.20	2.50	29.93	1.10	1.10	17.88	-	47.81	nr	53.78
762 x 457mm	33.45	2.50	34.29	1.40	1.40	22.75	-	57.04	nr	64.17
914 x 610mm	41.85	2.50	42.90	1.70	1.70	27.63	-	70.52	nr	79.34

MAT RIMS - Matwells in polished brass

Mat frames; brazed fabrication
38 x 38 x 6mm angle section; angles mitred; plain lugs -4, welded on; mat space

	Del to Site £	Waste %	Material Cost £	Craft Optve Hrs	Lab Hrs	Labour Cost £	Sunds £	Nett Rate £	Unit	Gross Rate £
610 x 457mm	93.30	2.50	95.63	1.40	1.40	22.75	-	118.38	nr	133.18
762 x 457mm	107.00	2.50	109.68	1.70	1.70	27.63	-	137.30	nr	154.46
914 x 610mm	135.90	2.50	139.30	1.90	1.90	30.88	-	170.17	nr	191.44

CLOTHES LOCKERS - Individual compartment clothes lockers; Welconstruct Co. Ltd.; steel, stove enamelled finish; standard colour doors and with cam locks 442 series

2 compartment; placing in position

	Del to Site £	Waste %	Material Cost £	Craft Optve Hrs	Lab Hrs	Labour Cost £	Sunds £	Nett Rate £	Unit	Gross Rate £
305 x 305 x 1830mm code GA442-021	65.60	2.50	67.24	-	0.60	4.21	-	71.45	nr	80.38
305 x 460 x 1830mm code G442-022	68.69	2.50	70.41	-	0.65	4.56	-	74.97	nr	84.34
380 x 380 x 1830mm code G442-024	68.69	2.50	70.41	-	0.70	4.91	-	75.32	nr	84.74
460 x 305 x 1830mm code G442-023	68.69	2.50	70.41	-	0.75	5.26	-	75.67	nr	85.13
460 X 460 X1830mm code G442-025	81.43	2.50	83.47	-	0.80	5.62	-	89.08	nr	100.22

3 compartment; placing in position

	Del to Site £	Waste %	Material Cost £	Craft Optve Hrs	Lab Hrs	Labour Cost £	Sunds £	Nett Rate £	Unit	Gross Rate £
305 x 305 x 1830mm code G442-026	77.00	2.50	78.92	-	0.60	4.21	-	83.14	nr	93.53
305 x 460 x 1830mm code G442-027	80.89	2.50	82.91	-	0.65	4.56	-	87.48	nr	98.41
380 x 380 x 1830mm code G442-029	80.89	2.50	82.91	-	0.70	4.91	-	87.83	nr	98.80
460 x 305 x 1830mm code G442-028	80.89	2.50	82.91	-	0.75	5.26	-	88.18	nr	99.20
460 x 460 x 1830mm code G442-030	85.43	2.50	87.57	-	0.80	5.62	-	93.18	nr	104.83

4 compartment; placing in position

	Del to Site £	Waste %	Material Cost £	Craft Optve Hrs	Lab Hrs	Labour Cost £	Sunds £	Nett Rate £	Unit	Gross Rate £
305 x 305 x 1830mm code G442-031	82.94	2.50	85.01	-	0.60	4.21	-	89.23	nr	100.38

Labour hourly rates: (except Specialists) Craft Operatives £9.23 Labourer £7.02 Rates are national average prices. Refer to REGIONAL VARIATIONS for indicative levels of overall pricing in regions	MATERIALS			LABOUR				RATES		
	Del to Site	Waste	Material Cost	Craft Optve	Lab	Labour Cost	Sunds	Nett Rate	Unit	Gross Rate (+12.5%)
	£	%	£	Hrs	Hrs	£	£	£		£
SHELVING SYSTEMS - SHELVING SUPPORT SYSTEMS Cont.										
CLOTHES LOCKERS - Individual compartment clothes lockers; Welconstruct Co. Ltd.; steel, stove enamelled finish; standard colour doors and with cam locks 442 series Cont.										
4 compartment; placing in position Cont.										
305 x 460 x 1830mm code G442-032	88.56	2.50	90.77	-	0.65	4.56	-	95.34	nr	107.25
380 x 380 x 1830mm code G442-034	88.56	2.50	90.77	-	0.70	4.91	-	95.69	nr	107.65
460 x 305 x 1830mm code G442-033	88.56	2.50	90.77	-	0.75	5.26	-	96.04	nr	108.04
460 x 460 x 1830mm code G442-035	92.99	2.50	95.31	-	0.80	5.62	-	100.93	nr	113.55
6 compartment; placing in position										
305 x 305 x 1830mm code G442-036	95.15	2.50	97.53	-	0.60	4.21	-	101.74	nr	114.46
305 x 460 x 1830mm code G442-037	104.65	2.50	107.27	-	0.65	4.56	-	111.83	nr	125.81
380 x 380 x 1830mm code G442-039	104.65	2.50	107.27	-	0.70	4.91	-	112.18	nr	126.20
460 x 305 x 1830mm code G442-038	104.65	2.50	107.27	-	0.75	5.26	-	112.53	nr	126.60
460 x 460 x 1830mm code G442-040	108.86	2.50	111.58	-	0.80	5.62	-	117.20	nr	131.85
CLOTHES LOCKERS - Swimming bath lockers; Welconstruct Co Ltd.; galvanised steel, stove enamelled finish; standard colour doors with mastered series cam locks 452 series										
1 compartment; placing in position										
305 x 305 x 1830mm code G452-001	75.71	2.50	77.60	-	0.60	4.21	-	81.81	nr	92.04
2 compartment; placing in position										
305 x 305 x 1830mm code G452-002	98.17	2.50	100.62	-	0.60	4.21	-	104.84	nr	117.94
305 x 460 x 1830mm code G452-003	103.03	2.50	105.61	-	0.65	4.56	-	110.17	nr	123.94
3 compartment; placing in position										
305 x 305 x 1830mm code G452-004	115.56	2.50	118.45	-	0.50	3.51	-	121.96	nr	137.20
305 x 460 x 1830mm code G452-005	121.28	2.50	124.31	-	0.55	3.86	-	128.17	nr	144.19
4 compartment; placing in position										
305 x 305 x 1830mm code G452-006	124.31	2.50	127.42	-	0.60	4.21	-	131.63	nr	148.08
305 x 460 x 1830mm code G452-007	132.95	2.50	136.27	-	0.65	4.56	-	140.84	nr	158.44
CLOAK ROOM EQUIPMENT - Static square tube double sided coat racks; Welconstruct Co. Ltd.; hardwood seats; in assembled units										
Racks 1500mm long x 1675mm high x 610mm deep; placing in position; 40mm square tube										
10 hooks and angle framed steel mesh single shoe tray, reference G456-102 AND 109; placing in position	344.90	2.50	353.52	-	0.60	4.21	-	357.73	nr	402.45
10 hooks, reference G456-102; placing in position	309.96	2.50	317.71	-	0.60	4.21	-	321.92	nr	362.16
10 hooks and 5 baskets G456-102 and 106	352.62	2.50	361.44	-	0.60	4.21	-	365.65	nr	411.35
10 hooks and 10 baskets G456-102 and 107	385.02	2.50	394.65	-	0.60	4.21	-	398.86	nr	448.71
CLOAK ROOM EQUIPMENT - Static square tube single sided coat racks; Welconstruct Co. Ltd.; mobile; in assembled units										
Racks 1500mm long x 1825mm high x 600mm deep; placing in position; 40mm square tube										
15 hangers and top tray, reference G456-103	300.24	2.50	307.75	-	0.60	4.21	-	311.96	nr	350.95
CLOAK ROOM EQUIPMENT - Free-standing bench seats; Welconstruct Co. Ltd.; steel square tube framing; hardwood seats; placing in position										
Bench seat with shelf 1500mm long x 300mm deep and 450mm high; placing in position; 40mm square tube										
single sided 300mm deep, reference G456-115	143.86	2.50	147.46	-	0.70	4.91	-	152.37	nr	171.42
double sided 600mm deep G456-116	232.20	2.50	238.01	-	0.70	4.91	-	242.92	nr	273.28
CLOAK ROOM EQUIPMENT - Wall rack Welconstruct Co. Ltd; hardwood										
1500mm long; fixing to masonry										
5 hooks, reference G456-029	32.02	2.50	32.82	-	0.70	4.91	-	37.73	nr	42.45
HAT AND COAT RAILS - Hat and coat rails										
Rails in softwood, wrought										
25 x 75mm; chamfers -2	1.38	10.00	1.52	0.29	0.04	2.96	0.05	4.53	m	5.09
25 x 100mm; chamfers -2	1.85	10.00	2.04	0.29	0.04	2.96	0.05	5.04	m	5.67
25 x 75mm; chamfers -2; fixing to masonry with screws	1.38	10.00	1.52	0.57	0.07	5.75	0.09	7.36	m	8.28
25 x 100mm; chamfers -2; fixing to masonry with screws	1.85	10.00	2.04	0.57	0.07	5.75	0.09	7.88	m	8.86
Rails in Afromosia, wrought, selected for transparent finish										
25 x 75mm; chamfers -2	4.25	10.00	4.67	0.57	0.07	5.75	0.09	10.52	m	11.83
25 x 100mm; chamfers -2	5.68	10.00	6.25	0.57	0.07	5.75	0.09	12.09	m	13.60
25 x 75mm; chamfers -2; fixing to masonry with screws	4.25	10.00	4.67	1.15	0.14	11.60	0.19	16.46	m	18.52
25 x 100mm; chamfers -2; fixing to masonry with screws	5.68	10.00	6.25	1.15	0.14	11.60	0.19	18.04	m	20.29

FURNITURE/EQUIPMENT

Labour hourly rates: (except Specialists) Craft Operatives £9.23 Labourer £7.02. Rates are national average prices. Refer to REGIONAL VARIATIONS for indicative levels of overall pricing in regions	MATERIALS			LABOUR				RATES		
	Del to Site £	Waste %	Material Cost £	Craft Optve Hrs	Lab Hrs	Labour Cost £	Sunds £	Nett Rate £	Unit	Gross Rate (+12.5%) £
SHELVING SYSTEMS - SHELVING SUPPORT SYSTEMS Cont.										
HAT AND COAT RAILS - Hat and coat rails Cont.										
Hat and coat hooks										
B.M.A. finish	2.20	2.50	2.25	0.40	-	3.69	-	5.95	nr	6.69
chromium plated	2.20	2.50	2.25	0.40	-	3.69	-	5.95	nr	6.69
SAA finish	1.10	2.50	1.13	0.40	-	3.69	-	4.82	nr	5.42
N11: DOMESTIC KITCHEN FITTINGS										
Standard melamine finish on chipboard units, with backs										
Wall units										
fixing to masonry with screws										
400 x 300 x 600mm.....................	61.90	2.50	63.45	2.00	0.25	20.22	-	83.66	nr	94.12
400 x 300 x 900mm.....................	80.52	2.50	82.53	2.10	0.26	21.21	-	103.74	nr	116.71
500 x 300 x 600mm.....................	66.47	2.50	68.13	2.20	0.28	22.27	-	90.40	nr	101.70
500 x 300 x 900mm.....................	84.28	2.50	86.39	2.25	0.28	22.73	-	109.12	nr	122.76
600 x 300 x 600mm.....................	72.62	2.50	74.44	2.25	0.28	22.73	-	97.17	nr	109.31
600 x 300 x 900mm.....................	87.86	2.50	90.06	2.35	0.29	23.73	-	113.78	nr	128.01
1000 x 300 x 600mm....................	116.68	2.50	119.60	2.70	0.34	27.31	-	146.90	nr	165.27
1000 x 300 x 900mm....................	142.21	2.50	145.77	2.90	0.36	29.29	-	175.06	nr	196.94
1200 x 300 x 600mm....................	126.72	2.50	129.89	2.90	0.36	29.29	-	159.18	nr	179.08
1200 x 300 x 900mm....................	144.29	2.50	147.90	3.10	0.39	31.35	-	179.25	nr	201.65
Floor units on plinths and with plastic faced worktops; without drawers										
fixing to masonry with screws										
400 x 600 x 900mm....................	117.15	2.50	120.08	2.25	0.28	22.73	-	142.81	nr	160.66
500 x 600 x 900mm....................	128.69	2.50	131.91	2.55	0.32	25.78	-	157.69	nr	177.40
600 x 600 x 900mm....................	137.79	2.50	141.23	2.75	0.34	27.77	-	169.00	nr	190.13
1000 x 600 x 900mm...................	230.20	2.50	235.96	3.70	0.46	37.38	-	273.34	nr	307.50
1200 x 600 x 900mm...................	250.15	2.50	256.40	4.00	0.50	40.43	-	296.83	nr	333.94
Floor units on plinths and with plastic faced worktops; with one drawer										
fixing to masonry with screws										
500 x 600 x 900mm....................	160.61	2.50	164.63	2.55	0.32	25.78	-	190.41	nr	214.21
600 x 600 x 900mm....................	169.70	2.50	173.94	2.75	0.34	27.77	-	201.71	nr	226.93
1000 x 600 x 900mm...................	276.51	2.50	283.42	3.70	0.46	37.38	-	320.80	nr	360.90
1200 x 600 x 900mm...................	298.80	2.50	306.27	4.00	0.50	40.43	-	346.70	nr	390.04
Floor units on plinths and with plastic faced worktops; with four drawers										
fixing to masonry with screws										
500 x 600 x 900mm....................	224.03	2.50	229.63	2.00	0.25	20.22	-	249.85	nr	281.08
600 x 600 x 900mm....................	236.81	2.50	242.73	2.75	0.34	27.77	-	270.50	nr	304.31
Sink units on plinths										
1200 x 600 x 900mm, with one drawer; fixing to masonry with screws	212.19	2.50	217.49	4.00	0.50	40.43	-	257.92	nr	290.17
1200 x 600 x 900mm, without drawer; fixing to masonry with screws	162.11	2.50	166.16	4.00	0.50	40.43	-	206.59	nr	232.42
1500 x 600 x 900mm, without drawer; fixing to masonry with screws	253.86	2.50	260.21	4.30	0.55	43.55	-	303.76	nr	341.73
Store cupboards on plinths										
fixing to masonry with screws										
500 x 600 x 1950mm without shelves	186.91	2.50	191.58	4.00	0.50	40.43	-	232.01	nr	261.01
500 x 600 x 1950mm with shelves	199.94	2.50	204.94	4.00	0.50	40.43	-	245.37	nr	276.04
600 x 600 x 1950mm without shelves	203.47	2.50	208.56	4.50	0.56	45.47	-	254.02	nr	285.78
600 x 600 x 1950mm with shelves	219.43	2.50	224.92	4.50	0.56	45.47	-	270.38	nr	304.18
Laminated plastic faced units complete with ironmongery and with backs										
Wall units										
400 x 300 x 600mm, with one door; fixing to masonry with screws	123.74	2.50	126.83	2.00	0.25	20.22	-	147.05	nr	165.43
500 x 300 x 600mm, with one door; fixing to masonry with screws	132.92	2.50	136.24	2.10	0.26	21.21	-	157.45	nr	177.13
1000 x 300 x 600mm, with two doors; fixing to masonry with screws	233.34	2.50	239.17	2.70	0.34	27.31	-	266.48	nr	299.79
1200 x 300 x 600mm, with two doors; fixing to masonry with screws	253.43	2.50	259.77	2.85	0.36	28.83	-	288.60	nr	324.67
Floor units on plinths and with worktops										
500 x 600 x 900mm, with one drawer and one cupboard; fixing to masonry with screws	321.23	2.50	329.26	2.20	0.28	22.27	-	351.53	nr	395.47
500 x 600 x 900mm, with four drawers; fixing to masonry with screws	448.05	2.50	459.25	2.50	0.31	25.25	-	484.50	nr	545.07
1000 x 600 x 900mm, with two drawers and two cupboards; fixing to masonry with screws	553.00	2.50	566.83	3.50	0.45	35.46	-	602.29	nr	677.58
1500 x 600 x 900mm, with three drawers and three cupboards; fixing to masonry with screws	714.47	2.50	732.33	4.20	0.53	42.49	-	774.82	nr	871.67
Sink units on plinths										
1000 x 600 x 900mm, with one drawer and two cupboards, fixing to masonry with screws	424.19	2.50	434.79	4.20	0.52	42.42	-	477.21	nr	536.86
1500 x 600 x 900mm', with two drawers and three cupboards; fixing to masonry with screws	634.50	2.50	650.36	4.30	0.54	43.48	-	693.84	nr	780.57

Labour hourly rates: (except Specialists) Craft Operatives £9.23 Labourer £7.02 Rates are national average prices. Refer to REGIONAL VARIATIONS for indicative levels of overall pricing in regions	MATERIALS			LABOUR				RATES		
	Del to Site £	Waste %	Material Cost £	Craft Optve Hrs	Lab Hrs	Labour Cost £	Sunds £	Nett Rate £	Unit	Gross Rate (+12.5%) £
N11: DOMESTIC KITCHEN FITTINGS Cont.										
Laminated plastic faced units complete with ironmongery and with backs Cont.										
Tall units on plinths										
500 x 600 x 1950mm, broom cupboard; fixing to masonry with screws	390.16	2.50	**399.91**	4.30	0.54	**43.48**	-	**443.39**	nr	**498.82**
600 x 600 x 1950mm, larder unit; fixing to masonry with screws	407.00	2.50	**417.18**	4.30	0.54	**43.48**	-	**460.65**	nr	**518.24**
Hardwood veneered units complete with ironmongery but without backs										
Wardrobe units										
fixing to masonry with screws										
1000 x 600 x 2175mm	521.40	2.50	**534.43**	4.75	0.60	**48.05**	-	**582.49**	nr	**655.30**
1500 x 600 x 2175mm	655.00	2.50	**671.38**	5.50	0.69	**55.61**	-	**726.98**	nr	**817.86**
Pre-Assembled kitchen units, white melamine finish; Module 500 range; Magnet PLC										
Drawerline units; fixing to masonry with screws										
floor units										
300mm	76.58	2.50	**78.49**	1.50	0.34	**16.23**	-	**94.73**	nr	**106.57**
500mm	88.39	2.50	**90.60**	1.50	0.38	**16.51**	-	**107.11**	nr	**120.50**
600mm	92.60	2.50	**94.92**	1.50	0.38	**16.51**	-	**111.43**	nr	**125.36**
1000mm	143.15	2.50	**146.73**	2.00	0.50	**21.97**	-	**168.70**	nr	**189.79**
hob floor units										
1000mm	120.41	2.50	**123.42**	2.00	0.50	**21.97**	-	**145.39**	nr	**163.56**
sink units										
1000mm	108.60	2.50	**111.32**	2.00	0.50	**21.97**	-	**133.29**	nr	**149.95**
Tall units; fixing to masonry with screws										
floor units										
500mm	165.06	2.50	**169.19**	3.00	0.31	**29.87**	-	**199.05**	nr	**223.93**
corner floor units										
1000mm	113.66	2.50	**116.50**	2.00	0.41	**21.34**	-	**137.84**	nr	**155.07**
sink/floor units										
1000mm	108.60	2.50	**111.32**	2.00	0.41	**21.34**	-	**132.65**	nr	**149.23**
Drawer units; fixing to masonry with screws										
four drawer floor units										
500mm	160.11	2.50	**164.11**	1.50	0.51	**17.43**	-	**181.54**	nr	**204.23**
Wall units; fixing to masonry with screws										
standard units 600mm high										
300mm	51.32	2.50	**52.60**	2.50	0.35	**25.53**	-	**78.14**	nr	**87.90**
500mm	51.32	2.50	**52.60**	2.50	0.40	**25.88**	-	**78.49**	nr	**88.30**
600mm	79.96	2.50	**81.96**	2.50	0.41	**25.95**	-	**107.91**	nr	**121.40**
1000mm	88.39	2.50	**90.60**	3.00	0.52	**31.34**	-	**121.94**	nr	**137.18**
standard corner units 900mm high										
600 x 600mm	144.00	2.50	**147.60**	3.50	0.51	**35.89**	-	**183.49**	nr	**206.42**
Worktops; fixing with screws										
round front edge										
28mm	14.01	25.00	**17.51**	1.50	0.19	**15.18**	-	**32.69**	m	**36.78**
40mm	21.32	25.00	**26.65**	1.75	0.22	**17.70**	-	**44.35**	m	**49.89**
Plinth										
Minster 500	20.87	25.00	**26.09**	0.25	0.03	**2.52**	-	**28.61**	m	**32.18**
Cornice										
Minster 500	24.31	25.00	**30.39**	0.25	0.03	**2.52**	-	**32.91**	m	**37.02**
Pelmet										
Minster 500	19.62	25.00	**24.52**	0.25	0.03	**2.52**	-	**27.04**	m	**30.42**
Worktop trim										
coloured	5.81	25.00	**7.26**	0.25	0.03	**2.52**	-	**9.78**	m	**11.00**
N13: SANITARY APPLIANCES/FITTINGS										
Attendance on sanitary fittings										
Note: Sanitary fittings are usually included as p.c. items or provisional sums, the following are the allowances for attendance by the main contractor i.e. unloading, storing and distributing fittings for fixing by the plumber and returning empty cases and packings										
Sinks, fireclay										
610 x 457 x 254mm	-	-	-	-	1.15	**8.07**	0.65	**8.72**	nr	**9.81**
762 x 508 x 254mm	-	-	-	-	1.40	**9.83**	0.82	**10.65**	nr	**11.98**
add if including tubular stands	-	-	-	-	0.85	**5.97**	0.54	**6.51**	nr	**7.32**
Wash basins, 559 x 406mm; earthenware or fireclay										
single	-	-	-	-	0.90	**6.32**	0.54	**6.86**	nr	**7.72**
range of 4 with cover overlaps between basins	-	-	-	-	2.75	**19.31**	1.46	**20.77**	nr	**23.36**
Combined sinks and drainers; stainless steel										
1067 x 533mm	-	-	-	-	1.60	**11.23**	0.96	**12.19**	nr	**13.72**
1600 x 610mm	-	-	-	-	1.90	**13.34**	1.14	**14.48**	nr	**16.29**
add if including tubular stands	-	-	-	-	0.85	**5.97**	0.54	**6.51**	nr	**7.32**

FURNITURE/EQUIPMENT

Labour hourly rates: (except Specialists) Craft Operatives £9.23 Labourer £7.02 Rates are national average prices. Refer to REGIONAL VARIATIONS for indicative levels of overall pricing in regions	MATERIALS			LABOUR				RATES		
	Del to Site	Waste	Material Cost	Craft Optve	Lab	Labour Cost	Sunds	Nett Rate	Unit	Gross Rate (+12.5%)
	£	%	£	Hrs	Hrs	£	£	£		£
N13: SANITARY APPLIANCES/FITTINGS Cont.										
Attendance on sanitary fittings Cont.										
Combined sinks and drainers; porcelain enamel										
1067 x 533mm	-	-	-	-	1.60	11.23	0.96	12.19	nr	13.72
1600 x 610mm	-	-	-	-	1.90	13.34	1.14	14.48	nr	16.29
Baths excluding panels; pressed steel										
694 x 1688 x 570mm overall	-	-	-	-	2.75	19.31	1.42	20.73	nr	23.32
W.C. suites; china										
complete with WWP, flush pipe etc., ...	-	-	-	-	1.60	11.23	0.96	12.19	nr	13.72
Block pattern urinals; with glazed ends, back and channel in one piece with separate tread including flushing cistern etc.,										
single stall	-	-	-	-	3.00	21.06	1.62	22.68	nr	25.52
range of four with loose overlaps	-	-	-	-	7.00	49.14	3.78	52.92	nr	59.53
Wall urinals, bowl type, with flushing cistern, etc.,										
single	-	-	-	-	1.60	11.23	0.96	12.19	nr	13.72
range of 3 with two divisions	-	-	-	-	3.75	26.32	1.94	28.27	nr	31.80
Slab urinals with one return end, flushing cistern, sparge pipe, etc.,										
1219mm long x 1067mm high	-	-	-	-	3.15	22.11	2.16	24.27	nr	27.31
1829mm long x 1067mm high	-	-	-	-	4.25	29.84	3.24	33.08	nr	37.21

This section continues on the next page

Labour hourly rates: (except Specialists) Craft Operatives £12.44 Labourer £7.02 Rates are national average prices. Refer to REGIONAL VARIATIONS for indicative levels of overall pricing in regions	MATERIALS			LABOUR				RATES		
	Del to Site £	Waste %	Material Cost £	Craft Optve Hrs	Lab Hrs	Labour Cost £	Sunds £	Nett Rate £	Unit	Gross Rate (+12.5%) £
N13: SANITARY APPLIANCES/FITTINGS Cont.										
Fix only appliances (prime cost sum for supply included elsewhere)										
Sink units, stainless steel; combined overflow and waste outlet, plug and chain; pair pillar taps fixing on base unit with metal clips										
1000 x 600mm...............................	-	-	-	3.60	-	44.78	1.00	45.78	nr	51.51
1600 x 600mm...............................	-	-	-	4.20	-	52.25	1.00	53.25	nr	59.90
Sinks; white glazed fireclay; waste outlet, plug, chain and stay; cantilever brackets; pair bib taps fixing brackets to masonry with screws; sealing at back with white sealant										
610 x 457 x 254mm.........................	-	-	-	4.80	-	59.71	1.60	61.31	nr	68.98
762 x 508 x 254mm.........................	-	-	-	4.80	-	59.71	1.60	61.31	nr	68.98
Sinks; white glazed fireclay; waste outlet, plug, chain and stay; legs and screw-to-wall bearers; pair bib taps fixing legs to masonry with screws; sealing at back with mastic sealant										
610 x 457 x 254mm.........................	-	-	-	5.40	-	67.18	1.60	68.78	nr	77.37
762 x 508 x 254mm.........................	-	-	-	5.40	-	67.18	1.60	68.78	nr	77.37
Wash basins; vitreous china; waste outlet, plug, chain and stay; screw to wall brackets; pair pillar taps										
560 x 406mm; fixing brackets to masonry with screws; sealing at back with mastic sealant	-	-	-	3.60	-	44.78	1.60	46.38	nr	52.18
range of four, each 559 x 406mm with overlap strips; fixing brackets to masonry with screws; sealing overlap strips and at back with mastic sealant...........................	-	-	-	15.30	-	190.33	6.00	196.33	nr	220.87
Wash basins; vitreous china; waste outlet, plug, chain and stay; legs and screw-to-wall bearers										
560 x 406mm; fixing legs to masonry with screws; sealing at back with mastic sealant..............	-	-	-	4.20	-	52.25	1.60	53.85	nr	60.58
range of 4, each 559 x 406mm with overlap strips; fixing legs to masonry with screws; sealing overlap strips and at back with mastic sealant ...	-	-	-	16.50	-	205.26	6.00	211.26	nr	237.67
Baths; enamelled pressed steel; combined overflow and waste outlet, plug and chain; metal cradles; pair pillar taps										
700 x 1700 x 570mm overall; sealing at walls with mastic sealant	-	-	-	6.60	-	82.10	3.50	85.60	nr	96.30
High level W.C. suites; vitreous china pan; plastics cistern with valveless fittings, ball valve, chain and pull handle; flush pipe; plastics seat and cover fixing pan and cistern brackets to masonry with screws; bedding pan in mastic; jointing pan to drain with cement mortar (1:2) and gaskin joint ..	-	-	-	5.40	-	67.18	2.10	69.28	nr	77.94
fixing pan and cistern brackets to masonry with screws; bedding pan in mastic; jointing pan to soil pipe with Multikwik connector	-	-	-	5.40	-	67.18	2.10	69.28	nr	77.94
Low level W.C. suites; vitreous china pan; plastics cistern with valveless fittings and ball valve; flush pipe; plastics seat and cover fixing pan and cistern brackets to masonry with screws; bedding pan in mastic; jointing pan to drain with cement mortar (1:2) and gaskin joint ..	-	-	-	5.40	-	67.18	2.10	69.28	nr	77.94
fixing pan and cistern brackets to masonry with screws; bedding pan in mastic; jointing pan to soil pipe with Multikwik connector	-	-	-	5.40	-	67.18	2.10	69.28	nr	77.94
Wall urinals; vitreous china bowl; automatic cistern; spreads and flush pipe; waste outlet										
single bowl; fixing bowl, cistern and pipe brackets to masonry with screws	-	-	-	5.40	-	67.18	1.65	68.83	nr	77.43
range of three bowls; fixing bowls, divisions, cistern and pipe brackets to masonry with screws .	-	-	-	16.20	-	201.53	3.30	204.83	nr	230.43
Single stall urinals; white glazed fireclay in one piece; white glazed fireclay tread; plastics automatic cistern; spreader and flush pipe; waste outlet and domed grating bedding stall and tread in cement mortar (1:4) and jointing tread with waterproof jointing compound; fixing cistern and pipe brackets to masonry with screws	-	-	-	8.40	-	104.50	4.20	108.70	nr	122.28
Slab urinals; range of four; white glazed fireclay back, ends, divisions, channel and tread; automatic cistern; flush and sparge pipes; waste outlet and domed grating bedding back, ends, channel and tread in cement mortar (1:4) and jointing with waterproof jointing compound; fixing divisions, cistern and pipe brackets to masonry with screws	-	-	-	22.80	-	283.63	13.00	296.63	nr	333.71

FURNITURE/EQUIPMENT

Labour hourly rates: (except Specialists) Craft Operatives £12.44 Labourer £7.02 Rates are national average prices. Refer to REGIONAL VARIATIONS for indicative levels of overall pricing in regions	MATERIALS			LABOUR				RATES		
	Del to Site £	Waste %	Material Cost £	Craft Optve Hrs	Lab Hrs	Labour Cost £	Sunds £	Nett Rate £	Unit	Gross Rate (+12.5%) £
N13: SANITARY APPLIANCES/FITTINGS Cont.										
Fix only appliances (prime cost sum for supply included elsewhere) Cont.										
Showers; white glazed fireclay tray 760 x 760 x 180mm; waste outlet; recessed valve and spray head bedding tray in cement mortar (1:4); fixing valve head to masonry with screws; sealing at walls with mastic sealant	-	-	-	5.70	-	70.91	5.00	75.91	nr	85.40
Shower curtains with rails, hooks, supports and fittings										
straight, 914mm long x 1829mm drop; fixing supports to masonry with screws	-	-	-	1.20	-	14.93	0.50	15.43	nr	17.36
angled, 1676mm girth x 1829mm drop; fixing supports to masonry with screws	-	-	-	1.80	-	22.39	0.65	23.04	nr	25.92
Supply and fix appliances										
Note the following prices include joints to copper services and wastes but do not include traps										
Sink units; single bowl with single drainer and back ledge; stainless steel; B.S.1244 Part 2 Type A; 38mm waste, plug and chain to B.S.3380 Part 1 with combined overflow; pair 13mm pillar taps to B.S.1010 Part 2										
fixing on base unit with metal clips										
1000 x 600mm..........................	81.75	2.50	83.79	3.00	-	37.32	0.80	121.91	nr	137.15
1200 x 600mm..........................	89.75	2.50	91.99	3.50	-	43.54	0.80	136.33	nr	153.38
Sink units; single bowl with double drainer and back ledge; stainless steel; B.S.1244 Part 2 Type B; 38mm waste, plug and chain to B.S.3380 Part 1 with combined overflow; pair 13mm pillar taps to B.S.1010 Part 2										
fixing on base unit with metal clips										
1500 x 600mm..........................	101.25	2.50	103.78	3.00	-	37.32	1.05	142.15	nr	159.92
Sinks; white glazed fireclay; B.S.1206 with weir overflow; 38mm slotted waste, chain, stay and plug to B.S.3380 Part 1; painted cantilever brackets screwing chain stay; fixing brackets to masonry with screws; sealing at back with mastic sealant										
455 x 380 x 205mm..............................	75.75	2.50	77.64	3.50	-	43.54	1.30	122.48	nr	137.79
610 x 455 x 255mm..............................	106.75	2.50	109.42	3.50	-	43.54	1.30	154.26	nr	173.54
760 x 455 x 255mm..............................	163.00	2.50	167.07	3.50	-	43.54	1.30	211.91	nr	238.40
915 x 610 x 305mm..............................	469.00	2.50	480.73	4.50	-	55.98	1.30	538.01	nr	605.26
Sinks; white glazed fireclay; B.S.1206 with wier overflow; 38mm slotted waste, chain, stay and plug to B.S.3380 Part 1; plastic coated cantilever brackets screwing chain stay; fixing brackets to masonry with screws; sealing at back with mastic sealant										
455 x 380 x 205mm..............................	83.00	2.50	85.08	3.50	-	43.54	1.30	129.91	nr	146.15
610 x 455 x 255mm..............................	114.00	2.50	116.85	3.50	-	43.54	1.30	161.69	nr	181.90
760 x 455 x 255mm..............................	170.25	2.50	174.51	3.50	-	43.54	1.30	219.35	nr	246.76
915 x 610 x 305mm..............................	476.25	2.50	488.16	4.50	-	55.98	1.30	545.44	nr	613.62
Sinks; white glazed fireclay; B.S.1206 with wier overflow; 38mm slotted waste, chain, stay and plug to B.S.3380 Part 1; painted legs and screw-to-wall bearers										
screwing chain stay; fixing legs to masonry with screws; sealing at back with mastic sealant										
455 x 380 x 205mm..............................	94.00	2.50	96.35	3.50	-	43.54	1.30	141.19	nr	158.84
610 x 455 x 255mm..............................	125.00	2.50	128.13	3.50	-	43.54	1.30	172.97	nr	194.59
760 x 455 x 255mm..............................	181.25	2.50	185.78	3.50	-	43.54	1.30	230.62	nr	259.45
915 x 610 x 305mm..............................	487.25	2.50	499.43	4.50	-	55.98	487.25	1042.66	nr	1172.99
Wash basins; white vitreous china; B.S.5506 Part 3; 32mm slotted waste, chain, stay and plug to B.S.3380 Part 1; painted cantilever brackets; pair 13 pillar taps to B.S.1010 Part 2										
sealing at back with mastic sealant										
560 x 406mm..........................	44.00	2.50	45.10	3.50	-	43.54	1.30	89.94	nr	101.18
Wash basins; white vitreous china; B.S.5506 Part 3; 32mm slotted waste, chain, stay and plug to B.S.3380 Part 1; plastic coated cantilever brackets; pair 13mm pillar taps to B.S.1010 Part 2										
sealing at back with mastic sealant										
560 x 406mm..........................	50.00	2.50	51.25	4.20	-	52.25	1.60	105.10	nr	118.24
Wash basins; white vitreous china; B.S.5506 Part 3; 32mm slotted waste, chain, stay and plug to B.S.3380 Part 1; painted legs and screw-to-wall bearers; pair 13mm taps to B.S.1010 Part 2										
fixing legs to masonry with screws; sealing at back with mastic sealant										
560 x 406mm............	52.00	2.50	53.30	4.90	-	60.96	1.60	115.86	nr	130.34

Labour hourly rates: (except Specialists) Craft Operatives £12.44 Labourer £7.02 Rates are national average prices. Refer to REGIONAL VARIATIONS for indicative levels of overall pricing in regions	MATERIALS			LABOUR				RATES		
	Del to Site £	Waste %	Material Cost £	Craft Optve Hrs	Lab Hrs	Labour Cost £	Sunds £	Nett Rate £	Unit	Gross Rate (+12.5%) £

N13: SANITARY APPLIANCES/FITTINGS Cont.

Supply and fix appliances Cont.

Wash basins; white vitreous china; B.S.5506 Part 3; 32mm slotted waste, chain, stay and plug to B.S.3380 Part 1; plastic coated legs and screw-to-wall bearers; pair 13mm pillar taps to B.S.1010 Part 2
  fixing legs to masonry with screws; sealing at
  back with mastic sealant
    560 x 406mm........................
	62.50	2.50	64.06	4.90	–	60.96	1.60	126.62	nr	142.45

Wash basins; white vitreous china; B.S.5506 Part 3; 32mm slotted waste, chain, stay and plug to B.S.3380 Part 1; painted towel rail brackets; pair 13mm pillar taps to B.S.1010 Part 2
  fixing brackets to masonry with screws; sealing at
  back with mastic sealant
    560 x 406mm........................
	42.00	2.50	43.05	4.80	–	59.71	1.60	104.36	nr	117.41

Wash basins; white vitreous china; B.S.5506 Part 3; 32mm slotted waste, chain, stay and plug to B.S.3380 Part 1; plastic coated towel rail brackets; pair 13mm pillar taps to B.S.1010 Part 2
  fixing brackets to masonry with screws; sealing at
  back with mastic sealant
    560 x 406mm........................
	45.25	2.50	46.38	4.80	–	59.71	1.60	107.69	nr	121.15

Wash basins; coloured vitreous china; B.S.5506 Part 3; 32mm slotted waste, chain, stay and plug to B.S.3380 Part 1; painted build-in brackets; pair 13mm pillar taps to B.S.1010 Part 2
  building in brackets to masonry; sealing at back
  with mastic sealant
    560 x 406mm........................
	69.25	2.50	70.98	4.20	–	52.25	1.60	124.83	nr	140.43

Wash basins; coloured vitreous china; B.S.5506 Part 3; 32mm slotted waste, chain, stay and plug to B.S.3380 Part 1; plastic coated build-in brackets; pair 13mm pillar taps to B.S.1010 Part 2
  building in brackets to masonry; sealing at back
  with mastic sealant
    560 x 406mm........................
	75.25	2.50	77.13	4.20	–	52.25	1.60	130.98	nr	147.35

Wash basins; coloured vitreous china; B.S.5506 Part 3; 32mm slotted waste, chain, stay and plug to B.S.3380 Part 1; painted legs and screw-to-wall bearers; pair 13mm pillar taps to B.S.1010 Part 2
  fixing legs to masonry with screws; sealing at
  back with mastic sealant
    560 x 406mm........................
	77.25	2.50	79.18	4.90	–	60.96	1.60	141.74	nr	159.45

Wash basins; coloured vitreous china; B.S.5506 Part 3; 32mm slotted waste, chain, stay and plug to B.S.3380 Part 1; plastic coated legs and screw-to-wall bearers; pair 13mm pillar taps to B.S.1010 Part 2
  fixing legs to masonry with screws; sealing at
  back with mastic sealant
    560 x 406mm........................
	87.00	2.50	89.17	4.90	–	60.96	1.60	151.73	nr	170.70

Wash basins; coloured vitreous china; B.S.5506 Part 3; 32mm slotted waste, chain, stay and plug to B.S.3380 Part 1; painted towel rail brackets; pair 13mm pillar taps to B.S.1010 Part 2
  fixing brackets to masonry with screws; sealing at
  back with mastic sealant
    560 x 406mm........................
	67.25	2.50	68.93	4.80	–	59.71	1.60	130.24	nr	146.52

Wash basins; coloured vitreous china; B.S.5506 Part 3; 32mm slotted waste, chain, stay and plug to B.S.3380 Part 1; plastic coated towel rail brackets; pair 13mm pillar taps to B.S.1010 Part 2
  fixing brackets to masonry with screws; sealing at
  back with mastic sealant
    560 x 406mm........................
	70.50	2.50	72.26	4.80	–	59.71	1.60	133.57	nr	150.27

Wash basins, angle type; white vitreous china; B.S.5506 Part 3; 32mm slotted waste, chain, stay and plug to B.S.3380 Part 1; concealed brackets; pair 13mm pillar taps to B.S.1010 Part 2
  fixing brackets to masonry with screws; sealing at
  back with mastic sealant
    457 x 431mm........................
	73.50	2.50	75.34	4.80	–	59.71	1.60	136.65	nr	153.73

Wash basins, angle type; coloured vitreous china; B.S.5506 Part 3; 32mm slotted waste, chain, stay and plug to B.S.3380 Part 1; concealed brackets; pair 13mm pillar taps to B.S.1010 Part 2
  fixing brackets to masonry with screws; sealing at
  back with mastic sealant
    457 x 431mm........................
	87.50	2.50	89.69	4.80	–	59.71	1.60	151.00	nr	169.87

FURNITURE/EQUIPMENT

Labour hourly rates: (except Specialists) Craft Operatives £12.44 Labourer £7.02 Rates are national average prices. Refer to REGIONAL VARIATIONS for indicative levels of overall pricing in regions	MATERIALS			LABOUR				RATES		
	Del to Site £	Waste %	Material Cost £	Craft Optve Hrs	Lab Hrs	Labour Cost £	Sunds £	Nett Rate £	Unit	Gross Rate (+12.5%) £
N13: SANITARY APPLIANCES/FITTINGS Cont.										
Supply and fix appliances Cont.										
Wash basins with pedestal; white vitreous china; B.S.5506 Part 3; 32mm slotted waste, chain, stay and plug to B.S.3380 Part 1; pair 13mm pillar taps to B.S.1010 Part 2										
fixing basin to masonry with screws; sealing at back with mastic sealant										
560 x 406mm....................................	71.00	2.50	72.78	4.50	-	55.98	1.60	130.35	nr	146.65
Wash basins with pedestal; coloured vitreous china; B.S.5506 Part 3; 32mm slotted waste, chain, stay and plug to B.S.3380 Part 1; pair 13mm pillar taps to B.S.1010 Part 2										
fixing basin to masonry with screws; sealing at back with mastic sealant										
560 x 406mm....................................	101.50	2.50	104.04	4.50	-	55.98	1.60	161.62	nr	181.82
Baths; white enamelled pressed steel; B.S.1390; 32mm overflow with front grid and 38mm waste, chain and plug to B.S.3380 Part 1 with combined overflow; metal cradles with adjustable feet; pair 19mm pillar taps to B.S.1010 Part 2										
sealing at walls with mastic sealant										
700 x 1700 x 570mm overall....................	123.75	2.50	126.84	6.60	-	82.10	3.50	212.45	nr	239.00
724 x 1500 x 570mm overall....................	120.25	2.50	123.26	6.60	-	82.10	3.50	208.86	nr	234.97
Baths; coloured enamelled pressed steel; B.S.1390; 32mm overflow with front grid and 38mm waste, chain and plug to B.S.3380 Part 1 with combined overflow; metal cradles with adjustable feet; pair 19mm pillar taps to B.S.1010 Part 2										
sealing at walls with mastic sealant										
700 x 1700 x 570mm overall....................	136.50	2.50	139.91	6.60	-	82.10	3.50	225.52	nr	253.71
Baths; white acrylic; B.S.4305; 32mm overflow with front grid and 38mm waste, chain and plug to B.S.3380 Part 1 with combined overflow; metal cradles with adjustable feet and wall fixing brackets; pair 19mm pillar taps to B.S.1010 Part 2										
700 x 1700 x 570mm overall; fixing brackets to masonry with screws; sealing at walls with mastic sealant...	143.25	2.50	146.83	6.00	-	74.64	3.50	224.97	nr	253.09
Baths; coloured acrylic; B.S.4305; 32mm overflow with front grid and 38mm waste, chain and plug to B.S.3380 Part 1 with combined overflow; metal cradles with adjustable feet and wall fixing brackets; pair 19mm pillar taps to B.S.1010 Part 2										
700 x 1700 x 570mm overall; fixing brackets to masonry with screws; sealing at walls with mastic sealant...	143.25	2.50	146.83	6.00	-	74.64	3.50	224.97	nr	253.09
White vitreous china pans to B.S.5503										
pan with horizontal outlet; fixing pan to timber with screws; bedding pan in mastic; jointing pan to soil pipe with Multikwik connector	29.00	2.50	29.73	3.00	-	37.32	0.90	67.94	nr	76.44
pan with horizontal outlet; fixing pan to masonry with screws; bedding pan in mastic; jointing pan to soil pipe with Multikwik connector	29.00	2.50	29.73	3.60	-	44.78	1.20	75.71	nr	85.17
pan with S or P trap conversion bend to B.S.5627; fixing pan to timber with screws; bedding pan in mastic; jointing pan to drain with cement mortar (1:2) and gaskin joint	29.50	2.50	30.24	3.30	-	41.05	0.90	72.19	nr	81.21
pan with S or P trap conversion bend to B.S.5627; fixing pan to masonry with screws; bedding pan in mastic; jointing pan to drain with cement mortar (1:2) and gaskin joint	29.50	2.50	30.24	3.90	-	48.52	1.20	79.95	nr	89.95
pan with S or P trap conversion bend to B.S.5627; fixing pan to timber with screws; bedding pan in mastic; jointing pan to soil pipe with Multikwik connector..	32.75	2.50	33.57	3.00	-	37.32	0.90	71.79	nr	80.76
pan with S or P trap conversion bend to B.S.5627; fixing pan to masonry with screws; bedding pan in mastic; jointing pan to soil pipe with Multikwik connector..	32.75	2.50	33.57	3.60	-	44.78	1.20	79.55	nr	89.50
Coloured vitreous china pans to B.S.5503										
pan with horizontal outlet; fixing pan to timber with screws; bedding pan in mastic; jointing pan to soil pipe with Multikwik connector	55.25	2.50	56.63	3.00	-	37.32	0.90	94.85	nr	106.71
pan with horizontal outlet; fixing pan to masonry with screws; bedding pan in mastic; jointing pan to soil pipe with Multikwik connector	55.25	2.50	56.63	3.60	-	44.78	1.20	102.62	nr	115.44
Plastics W.C. seats; B.S.1254, type 1										
black; fixing to pan	12.75	2.50	13.07	0.30	-	3.73	-	16.80	nr	18.90
Plastics W.C. seats and covers; B.S.1254, type 1										
black; fixing to pan	16.75	2.50	17.17	0.30	-	3.73	-	20.90	nr	23.51
coloured; fixing to pan	27.50	2.50	28.19	0.30	-	3.73	-	31.92	nr	35.91
white; fixing to pan	31.00	2.50	31.77	0.30	-	3.73	-	35.51	nr	39.95

Labour hourly rates: (except Specialists) Craft Operatives £12.44 Labourer £7.02 Rates are national average prices. Refer to REGIONAL VARIATIONS for indicative levels of overall pricing in regions	MATERIALS			LABOUR				RATES		
	Del to Site £	Waste %	Material Cost £	Craft Optve Hrs	Lab Hrs	Labour Cost £	Sunds £	Nett Rate £	Unit	Gross Rate (+12.5%) £
N13: SANITARY APPLIANCES/FITTINGS Cont.										
Supply and fix appliances Cont.										
9 litre black plastics cistern with valveless fittings, chain and pull handle to B.S.1125; 13mm piston type high pressure ball valve to B.S.1212 Part 1 with 127mm plastics float to B.S.2456										
fixing cistern brackets to masonry with screws ...	47.25	2.50	48.43	1.20	-	14.93	0.45	63.81	nr	71.79
Plastics flush pipe to B.S.1125; adjustable										
for back wall fixing	3.75	2.50	3.84	0.90	-	11.20	0.45	15.49	nr	17.43
for side wall fixing	5.25	2.50	5.38	0.90	-	11.20	0.45	17.03	nr	19.16
High level W.C. suites; white vitreous china pan to B.S.5503; 9 litre black plastics cistern with valveless fittings, chain and pull handle and plastics flush pipe to B.S.1125; 13mm piston type high pressure ball valve to B.S.1212 Part 1 with 127mm plastics float to B.S.2456; black plastics seat and cover to B.S.1254 type 1										
pan with S or P trap conversion bend to B.S.5627; fixing pan to timber with screws; fixing cistern brackets to masonry with screws; bedding pan in mastic; jointing pan to soil pipe with Multikwik connector	97.25	2.50	99.68	4.80	-	59.71	2.10	161.49	nr	181.68
pan with S or P trap conversion bend to B.S.5627; fixing pan and cistern brackets to masonry with screws; bedding pan in mastic; jointing pan to soil pipe with Multikwik connector	97.25	2.50	99.68	5.40	-	67.18	2.10	168.96	nr	190.08
Low level W.C. suites; white vitreous china pan to B.S.5503; 9 litre white vitreous china cistern with valveless fittings and plastics flush bend to B.S.1125; 13mm piston type high pressure ball valve to B.S.1212 Part 1 with 127mm plastics float to B.S.2456; white plastics seat and cover to B.S.1254 type 1										
pan with S or P trap conversion bend to B.S.5627; fixing pan to timber with screws; fixing cistern brackets to masonry with screws; bedding pan in mastic; jointing pan to soil pipe with Multikwik connector	100.50	2.50	103.01	4.80	-	59.71	2.10	164.82	nr	185.43
pan with S or P trap conversion bend to B.S.5627; fixing pan and cistern brackets to masonry with screws; bedding pan in mastic; jointing pan to soil pipe with Multikwik connector	100.50	2.50	103.01	5.40	-	67.18	2.10	172.29	nr	193.82
Low level W.C. suites; coloured vitreous china pan to B.S.5503; 9 litre coloured vitreous china cistern with valveless fittings and plastics flush bend to B.S.1125; 13mm piston type high pressure ball valve to B.S.1212 Part 1 with 127mm plastics float to B.S.2456; coloured plastics seat and cover to B.S.1254 type 1										
pan with S or P trap conversion bend to B.S.5627; fixing pan to timber with screws; fixing cistern brackets to masonry with screws; bedding pan in mastic; jointing pan to soil pipe with Multikwik connector	146.75	2.50	150.42	4.80	-	59.71	2.10	212.23	nr	238.76
pan with S or P trap conversion bend to B.S.5627; fixing pan and cistern brackets to masonry with screws; bedding pan in mastic; jointing pan to soil pipe with Multikwik connector	146.75	2.50	150.42	5.40	-	67.18	2.10	219.69	nr	247.16
Wall urinals; white vitreous china bowl to B.S.5520; 4.5 litre white vitreous china automatic cistern to B.S.1876; spreader and stainless steel flush pipe; 32mm plastics waste outlet										
fixing bowl, cistern and pipe brackets to masonry with screws	212.00	2.50	217.30	5.40	-	67.18	1.65	286.13	nr	321.89
Wall urinals; range of 2; white vitreous china bowls to B.S.5520; white vitreous china divisions; 9 litre white plastics automatic cistern to B.S.1876; stainless steel flush pipes and spreaders; 32mm plastics waste outlets										
fixing bowls, divisions, cistern and pipe brackets to masonry with screws	417.25	2.50	427.68	10.80	-	134.35	2.45	564.48	nr	635.04
Wall urinals; range of 3; white vitreous china bowls to B.S.5520; white vitreous china divisions; 14 litre white plastics automatic cistern to B.S.1876; stainless steel flush pipes and spreaders; 32mm plastics waste outlets										
fixing bowls, divisions, cistern and pipe brackets to masonry with screws	618.25	2.50	633.71	16.20	-	201.53	3.25	838.48	nr	943.29
1200mm long stainless steel wall hung trough urinal with plastic; white; automatic cistern to B.S.1876; spreader and stainless steel flush pipe; waste outlet and domed grating										
fixing urinal cistern and pipe brackets to masonry with screws	425.50	2.50	436.14	7.00	-	87.08	4.15	527.37	nr	593.29

FURNITURE/EQUIPMENT

Labour hourly rates: (except Specialists) Craft Operatives £12.44 Labourer £7.02 Rates are national average prices. Refer to REGIONAL VARIATIONS for indicative levels of overall pricing in regions	MATERIALS			LABOUR				RATES		
	Del to Site £	Waste %	Material Cost £	Craft Optve Hrs	Lab Hrs	Labour Cost £	Sunds £	Nett Rate £	Unit	Gross Rate (+12.5%) £
N13: SANITARY APPLIANCES/FITTINGS Cont.										
Supply and fix appliances Cont.										
1800mm long stainless steel wall hung trough urinal with plastic; white; automatic cistern to B.S.1876; spreader and stainless steel flush pipe; waste outlet and domed grating										
fixing urinal cistern and pipe brackets to masonry with screws	480.25	2.50	492.26	9.50	-	118.18	7.00	617.44	nr	694.62
2400mm long stainless steel wall hung trough urinal with plastic; white; automatic cistern to B.S.1876; spreader and stainless steel flush pipe; waste outlet and domed grating										
fixing urinal cistern and pipe brackets to masonry with screws	598.00	2.50	612.95	12.00	-	149.28	10.00	772.23	nr	868.76
Showers; white glazed Armastone tray, 760 x 760 x 180mm; 38mm waste outlet; surface mechanical valve										
bedding tray in cement mortar (1:4); fixing valve and head to masonry with screws; sealing at walls with mastic sealant	202.50	2.50	207.56	5.70	-	70.91	5.00	283.47	nr	318.90
Showers; white glazed Armastone tray, 760 x 760 x 180mm; 38mm waste outlet; recessed thermostatic valve and swivel spray head										
bedding tray in cement mortar (1:4); fixing valve and head to masonry with screws; sealing at walls with mastic sealant	338.75	2.50	347.22	5.70	-	70.91	5.00	423.13	nr	476.02
Showers; coloured glazed acrylic tray, 760 x 760 x 180mm; 38mm waste outlet; recessed thermostatic valve and swivel spray head										
bedding tray in cement mortar (1:4); fixing valve and head to masonry with screws; sealing at walls with mastic sealant	380.75	2.50	390.27	5.70	-	70.91	5.00	466.18	nr	524.45
Showers; white moulded acrylic tray with removable front and side panels, 760 x 760 x 260mm with adjustable metal cradle; 38mm waste outlet; surface fixing mechanical valve, flexible tube hand spray and slide bar										
fixing cradle, valve and slide bar to masonry with screws; sealing at walls with mastic sealant	282.25	2.50	289.31	5.70	-	70.91	3.50	363.71	nr	409.18
Showers; white moulded acrylic tray with removable front and side panels, 760 x 760 x 260mm with adjustable metal cradle; 38mm waste outlet; surface fixing thermostatic valve, flexible tube hand spray and slide bar										
fixing cradle, valve and slide bar to masonry with screws; sealing at walls with mastic sealant	320.75	2.50	328.77	5.70	-	70.91	3.50	403.18	nr	453.57
Shower curtains; nylon; anodised aluminium rail, glider hooks, end and suspension fittings										
straight, 914mm long x 1829mm drop	34.50	2.50	35.36	1.20	-	14.93	0.50	50.79	nr	57.14
angled, 1676mm girth x 1829mm drop	48.50	2.50	49.71	1.80	-	22.39	0.65	72.75	nr	81.85
Shower curtains; heavy duty plastic; anodised aluminium rail, glider hooks, end and suspension fitting; fixing supports to masonry with screws										
straight, 914mm long x 1829mm drop	39.00	2.50	39.98	1.20	-	14.93	0.50	55.40	nr	62.33
angled, 1676mm girth x 1829mm drop	52.50	2.50	53.81	1.80	-	22.39	0.65	76.85	nr	86.46
3.2mm gloss finish enamelled hardboard bath panels; polished aluminium angle corner strips										
to front, 1830mm long x 610mm high; fixing to timber with detachable dome head screws	8.58	2.50	8.79	1.10	0.13	14.60	0.55	23.94	nr	26.93
to front and one end, 2590mm girth x 610mm high; fixing panels to timber with detachable dome head screws; fixing cover strips to timber with screws	14.61	2.50	14.98	2.20	0.27	29.26	1.10	45.34	nr	51.01
to front and two ends, 3350mm girth x 610mm high; fixing panels to timber with detachable dome head screws; fixing coverstrips to timber with screws .	22.87	2.50	23.44	3.30	0.40	43.86	1.60	68.90	nr	77.51

This section continues on the next page

Labour hourly rates: (except Specialists) Craft Operatives £9.23 Labourer £7.02 Rates are national average prices. Refer to REGIONAL VARIATIONS for indicative levels of overall pricing in regions	MATERIALS			LABOUR				RATES		
	Del to Site £	Waste %	Material Cost £	Craft Optve Hrs	Lab Hrs	Labour Cost £	Sunds £	Nett Rate £	Unit	Gross Rate (+12.5%) £
N15: SIGNS/NOTICES										
Signwriting in gloss paint; one coat										
Letters or numerals, Helvetica medium style, on painted or varnished surfaces										
50mm high	0.01	5.00	0.01	0.15	-	1.38	0.01	1.41	nr	1.58
extra; shading	0.01	5.00	0.01	0.09	-	0.83	0.01	0.85	nr	0.96
extra; outline	0.01	5.00	0.01	0.13	-	1.20	0.01	1.22	nr	1.37
100mm high	0.01	5.00	0.01	0.31	-	2.86	0.02	2.89	nr	3.25
extra; shading	0.01	5.00	0.01	0.17	-	1.57	0.01	1.59	nr	1.79
extra; outline	0.01	5.00	0.01	0.26	-	2.40	0.02	2.43	nr	2.73
150mm high	0.02	5.00	0.02	0.45	-	4.15	0.03	4.20	nr	4.73
extra; shading	0.01	5.00	0.01	0.26	-	2.40	0.02	2.43	nr	2.73
extra; outline	0.01	5.00	0.01	0.39	-	3.60	0.03	3.64	nr	4.10
200mm high	0.02	5.00	0.02	0.60	-	5.54	0.04	5.60	nr	6.30
extra; shading	0.01	5.00	0.01	0.34	-	3.14	0.02	3.17	nr	3.56
extra; outline	0.01	5.00	0.01	0.52	-	4.80	0.04	4.85	nr	5.46
300mm high	0.03	5.00	0.03	0.90	-	8.31	0.06	8.40	nr	9.45
extra; shading	0.02	5.00	0.02	0.52	-	4.80	0.04	4.86	nr	5.47
extra; outline	0.02	5.00	0.02	0.77	-	7.11	0.06	7.19	nr	8.09
stops	-	-	-	0.04	-	0.37	-	0.37	nr	0.42
Signwriting in gloss paint; two coats										
Letters or numerals, Helvetica medium style, on painted or varnished surfaces										
50mm high	0.01	5.00	0.01	0.27	-	2.49	0.02	2.52	nr	2.84
100mm high	0.02	5.00	0.02	0.54	-	4.98	0.04	5.05	nr	5.68
150mm high	0.02	5.00	0.02	0.81	-	7.48	0.06	7.56	nr	8.50
200mm high	0.03	5.00	0.03	1.08	-	9.97	0.08	10.08	nr	11.34
250mm high	0.04	5.00	0.04	1.34	-	12.37	0.10	12.51	nr	14.07
300mm high	0.04	5.00	0.04	1.61	-	14.86	0.12	15.02	nr	16.90
stops	-	-	-	0.08	-	0.74	0.01	0.75	nr	0.84

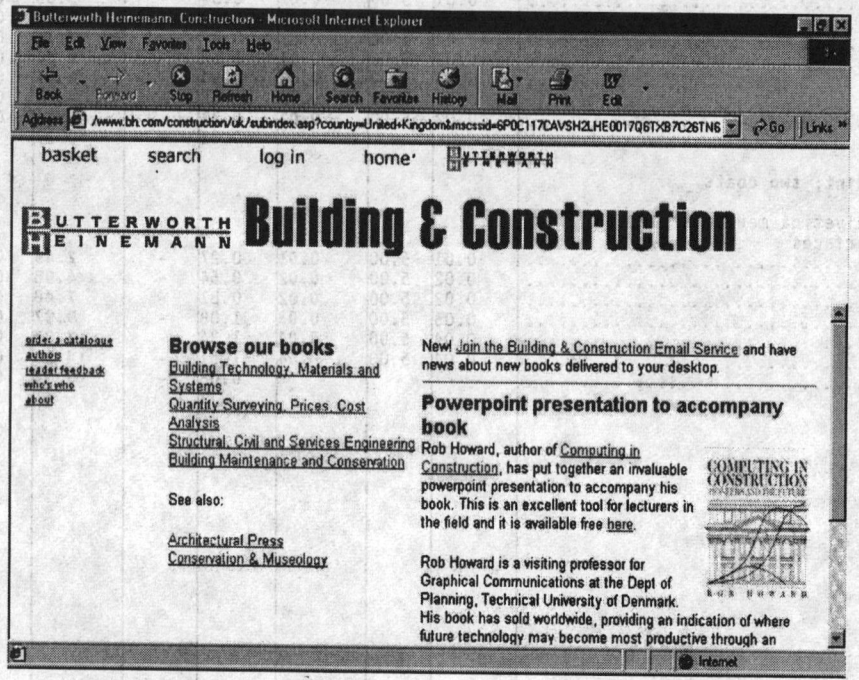

Building fabric sundries

Labour hourly rates: (except Specialists) Craft Operatives £9.23 Labourer £7.02 Rates are national average prices. Refer to REGIONAL VARIATIONS for indicative levels of overall pricing in regions	MATERIALS			LABOUR				RATES		
	Del to Site £	Waste %	Material Cost £	Craft Optve £	Lab £	Labour Cost £	Sunds £	Nett Rate £	Unit	Gross Rate (+12.5%) £
P10: SUNDRY INSULATION/PROOFING WORK/ FIRE STOPS										
Waterproof reinforced building paper B.S.1521; grade A1F; 150mm lapped joints										
Across members at 450mm centres vertical....................................	0.69	5.00	0.72	0.20	0.03	2.06	0.03	2.81	m2	3.16
Waterproof reinforced building paper; reflection (thermal) grade, single sided; 150mm lapped joints										
Across members at 450mm centres vertical....................................	1.01	5.00	1.06	0.20	0.03	2.06	0.03	3.15	m2	3.54
Waterproof reinforced building paper; reflection (thermal) grade, double sided; 150mm lapped joints										
Across members at 450mm centres vertical....................................	1.51	5.00	1.59	0.20	0.03	2.06	0.03	3.67	m2	4.13
British Sisalkraft Ltd Insulex 714 vapour control layer; 150mm laps sealed with tape										

Across members at 450mm centres
    horizontal; fixing to timber with stainless steel
    staples ....
	1.45	10.00	1.60	0.10	0.01	0.99	0.06	2.65	m2	2.98

Plain areas horizontal; laid loose	1.45	10.00	1.60	0.06	0.01	0.62	-	2.22	m2	2.50
British Sisalkraft Ltd SK860 vapour control layer; 100mm laps sealed with tape										
Plain areas horizontal; laid loose	1.03	10.00	1.13	0.08	0.01	0.81	0.04	1.98	m2	2.23
Mineral wool insulation quilt; butt joints										
60mm thick; across members at 450mm centres horizontal; laid loose	0.98	15.00	1.13	0.12	0.02	1.25	-	2.38	m2	2.67
Paper faced mineral wool insulation quilt; butt joints										
60mm thick; between members at 450mm centres vertical.....................................	1.90	15.00	2.19	0.25	0.03	2.52	0.03	4.73	m2	5.32
Glass fibre insulation quilt; butt joints										
60mm thick; across members at 450mm centres horizontal; laid loose	1.01	15.00	1.16	0.12	0.02	1.25	-	2.41	m2	2.71
80mm thick; across members at 450mm centres horizontal; laid loose	1.43	15.00	1.64	0.12	0.02	1.25	-	2.89	m2	3.25
100mm thick; across members at 450mm centres horizontal; laid loose	1.77	15.00	2.04	0.12	0.02	1.25	-	3.28	m2	3.69
60mm thick; between members at 450mm centres horizontal; laid loose	1.01	10.00	1.11	0.18	0.02	1.80	-	2.91	m2	3.28
80mm thick; between members at 450mm centres horizontal; laid loose	1.43	10.00	1.57	0.18	0.02	1.80	-	3.37	m2	3.80
100mm thick; between members at 450mm centres horizontal; laid loose	1.77	10.00	1.95	0.18	0.02	1.80	-	3.75	m2	4.22
Paper faced glass fibre insulation quilt; butt joints										
60mm thick; across members at 450mm centres horizontal; laid loose	1.90	15.00	2.19	0.12	0.02	1.25	-	3.43	m2	3.86

BUILDING FABRIC SUNDRIES – SMALL WORKS

Labour hourly rates: (except Specialists) Craft Operatives £9.23 Labourer £7.02 Rates are national average prices. Refer to REGIONAL VARIATIONS for indicative levels of overall pricing in regions	MATERIALS			LABOUR				RATES		
	Del to Site £	Waste %	Material Cost £	Craft Optve £	Lab £	Labour Cost £	Sunds £	Nett Rate £	Unit	Gross Rate (+12.5%) £
P10: SUNDRY INSULATION/PROOFING WORK/ FIRE STOPS Cont.										
Paper faced glass fibre insulation quilt; butt joints Cont.										
80mm thick; across members at 450mm centres horizontal; laid loose	2.30	15.00	2.65	0.12	0.02	1.25	-	3.89	m2	4.38
Owens Corning Building Products (UK) Ltd; Crown wool insulation quilt; butt joints										
100mm thick; between members at 450mm centres horizontal; laid loose	1.23	10.00	1.35	0.11	0.02	1.16	-	2.51	m2	2.82
150mm thick; between members at 450mm centres horizontal; laid loose	1.86	10.00	2.05	0.11	0.02	1.16	-	3.20	m2	3.60
200mm thick; between members at 450mm centres horizontal; laid loose	2.61	10.00	2.87	0.22	0.04	2.31	-	5.18	m2	5.83
100mm thick; between members at 450mm centres; 50mm thick across members at 450mm centres horizontal; laid loose	1.86	10.00	2.05	0.30	0.06	3.19	-	5.24	m2	5.89
100mm thick; between members at 450mm centres; 100mm thick across members at 450mm centres horizontal; laid loose	2.49	10.00	2.74	0.40	0.06	4.11	-	6.85	m2	7.71
Rockwool Rollbatts; insulation quilt; butt joints										
100mm thick; between members at 450mm centres horizontal; laid loose	1.28	10.00	1.41	0.11	0.02	1.16	-	2.56	m2	2.88
150mm thick; between members at 450mm centres horizontal; laid loose	1.97	10.00	2.17	0.17	0.03	1.78	-	3.95	m2	4.44
100mm thick; between members at 450mm centres; 100mm thick across members at 450mm centres horizontal; laid loose	2.55	10.00	2.81	0.22	0.04	2.31	-	5.12	m2	5.76
Glass fibre medium density insulation board; butt joints										
30mm thick; plain areas horizontal; bedding in bitumen	1.76	10.00	1.94	0.23	0.03	2.33	1.65	5.92	m2	6.66
50mm thick; plain areas horizontal; bedding in bitumen	3.14	10.00	3.45	0.25	0.03	2.52	1.65	7.62	m2	8.57
75mm thick; plain areas horizontal; bedding in bitumen	4.58	10.00	5.04	0.28	0.03	2.79	1.65	9.48	m2	10.67
30mm thick; across members at 450mm centres vertical	1.76	10.00	1.94	0.23	0.03	2.33	0.03	4.30	m2	4.84
50mm thick; across members at 450mm centres vertical	3.14	10.00	3.45	0.25	0.03	2.52	0.03	6.00	m2	6.75
75mm thick; across members at 450mm centres vertical	4.58	10.00	5.04	0.28	0.03	2.79	0.03	7.86	m2	8.85
Dow Construction Products; Styrofoam Floormate 200; butt joints										
25mm thick; plain areas horizontal; laid loose	4.76	10.00	5.24	0.12	0.02	1.25	-	6.48	m2	7.29
35mm thick; plain areas horizontal; laid loose	6.51	10.00	7.16	0.17	0.03	1.78	-	8.94	m2	10.06
50mm thick; plain areas horizontal; laid loose	9.13	10.00	10.04	0.22	0.04	2.31	-	12.35	m2	13.90
Sempatap latex foam sheeting; butt joints										

5mm thick; plain areas
    soffit; fixing with adhesive ....
    vertical; fixing with adhesive ....
	8.37
	8.37
		15.00
		15.00
			9.63
			9.63
				1.50
				1.20
					1.50
					1.20
						24.38
						19.50
							3.00
							3.00
								37.00
								32.13
									m2
									m2
										41.63
										36.14

Sempafloor SBR latex foam sheeting with coated non woven polyester surface; butt joints										
4.5mm thick; plain areas horizontal; fixing with adhesive	9.35	15.00	10.75	0.90	0.90	14.63	2.55	27.93	m2	31.42
Expanded polystyrene sheeting; butt joints										

13mm thick; plain areas
    horizontal; laid loose ....
    vertical; fixing with adhesive ....
	0.68
	0.68
		5.00
		5.00
			0.71
			0.71
				0.09
				0.27
					0.05
					0.14
						1.18
						3.47
							-
							1.10
								1.90
								5.29
									m2
									m2
										2.13
										5.95

19mm thick; plain areas
    horizontal; laid loose ....
    vertical; fixing with adhesive ....
	1.06
	0.14
		5.00
		5.00
			1.11
			0.15
				0.10
				-
					0.05
					0.28
						1.27
						1.97
							-
							1.06
								2.39
								3.17
									m2
									m2
										2.69
										3.57

Labour hourly rates: (except Specialists) Craft Operatives £9.23 Labourer £7.02 Rates are national average prices. Refer to REGIONAL VARIATIONS for indicative levels of overall pricing in regions	MATERIALS			LABOUR				RATES		
	Del to Site £	Waste %	Material Cost £	Craft Optve £	Lab £	Labour Cost £	Sunds £	Nett Rate £	Unit	Gross Rate (+12.5%) £
P10: SUNDRY INSULATION/PROOFING WORK/ FIRE STOPS Cont.										
Expanded polystyrene sheeting; butt joints Cont.										
25mm thick; plain areas										
horizontal; laid loose	1.34	5.00	1.41	0.12	0.06	1.53	-	2.94	m2	3.30
vertical; fixing with adhesive	1.34	5.00	1.41	0.30	0.15	3.82	1.10	6.33	m2	7.12
50mm thick; plain areas										
horizontal; laid loose	2.65	5.00	2.78	0.14	0.07	1.78	-	4.57	m2	5.14
vertical; fixing with adhesive	2.65	5.00	2.78	0.32	0.16	4.08	1.10	7.96	m2	8.95
75mm thick; plain areas										
horizontal; laid loose	3.99	5.00	4.19	0.18	0.09	2.29	-	6.48	m2	7.29
vertical; fixing with adhesive	3.99	5.00	4.19	0.36	0.18	4.59	1.10	9.88	m2	11.11
Expanded polystyrene sheeting, non-flammable; butt joints										
13mm thick; plain areas										
horizontal; laid loose	1.07	5.00	1.12	0.09	0.05	1.18	-	2.31	m2	2.59
vertical; fixing with adhesive	1.07	5.00	1.12	0.27	0.14	3.47	1.10	5.70	m2	6.41
19mm thick; plain areas										
horizontal; laid loose	1.27	5.00	1.33	0.10	0.05	1.27	-	2.61	m2	2.93
vertical; fixing with adhesive	1.27	5.00	1.33	0.28	0.14	3.57	1.10	6.00	m2	6.75
25mm thick; plain areas										
horizontal; laid loose	1.53	5.00	1.61	0.12	0.06	1.53	-	3.14	m2	3.53
vertical; fixing with adhesive	1.53	5.00	1.61	0.30	0.15	3.82	1.10	6.53	m2	7.34
50mm thick; plain areas										
horizontal; laid loose	3.67	5.00	3.85	0.14	0.07	1.78	-	5.64	m2	6.34
vertical; fixing with adhesive	3.67	5.00	3.85	0.32	0.16	4.08	1.10	9.03	m2	10.16
75mm thick; plain areas										
horizontal; laid loose	5.52	5.00	5.80	0.18	0.09	2.29	-	8.09	m2	9.10
vertical; fixing with adhesive	5.52	5.00	5.80	0.36	0.18	4.59	1.10	11.48	m2	12.92
Perforated zinc sheeting; butt joints										
Nr 7 gauge; across members at 450mm centres										
soffit	18.81	20.00	22.57	1.50	0.19	15.18	0.24	37.99	m2	42.74
vertical	18.81	20.00	22.57	1.00	0.13	10.14	0.16	32.87	m2	36.98
Nr 8 gauge; across members at 450mm centres										
soffit	20.62	20.00	24.74	1.50	0.19	15.18	0.24	40.16	m2	45.18
vertical	20.62	20.00	24.74	1.00	0.13	10.14	0.16	35.05	m2	39.43
Nr 9 gauge; across members at 450mm centres										
soffit	23.05	20.00	27.66	1.50	0.19	15.18	0.24	43.08	m2	48.46
vertical	23.05	20.00	27.66	1.00	0.13	10.14	0.16	37.96	m2	42.71
Galvanised wire netting, B.S.1485; butt joints										
13mm mesh x Nr 22 gauge; across members at 450mm centres										
soffit	2.54	10.00	2.79	-	0.10	0.70	0.03	3.53	m2	3.97
vertical	2.54	10.00	2.79	-	0.10	0.70	0.03	3.53	m2	3.97
25mm mesh x Nr 19 gauge; across members at 450mm centres										
soffit	2.04	10.00	2.24	-	0.10	0.70	0.03	2.98	m2	3.35
vertical	2.04	10.00	2.24	-	0.10	0.70	0.03	2.98	m2	3.35
38mm mesh x Nr 19 gauge; across members at 450mm centres										
soffit	1.46	10.00	1.61	-	0.10	0.70	0.03	2.34	m2	2.63
vertical	1.46	10.00	1.61	-	0.10	0.70	0.03	2.34	m2	2.63
50mm mesh x Nr 19 gauge; across members at 450mm centres										
soffit	1.16	10.00	1.28	-	0.10	0.70	0.03	2.01	m2	2.26
vertical	1.16	10.00	1.28	-	0.10	0.70	0.03	2.01	m2	2.26
P11: FOAMED/FIBRE/BEAD CAVITY WALL INSULATION										
Expanded polystyrene pellet insulation										
305mm walls										
filling to 50mm wide cavity	4.06	10.00	4.47	0.25	-	2.31	0.40	7.17	m2	8.07
P20: UNFRAMED ISOLATED TRIMS/SKIRTINGS/SUNDRY ITEMS										
Softwood, wrought - skirtings, architraves, picture rails and cover fillets to B.S.1186 Part 3										
Skirtings, picture rails, architraves and the like; (finished sizes)										
13 x 45mm; reference 13CA45	0.65	10.00	0.71	0.25	0.03	2.52	0.03	3.26	m	3.67
20 x 70mm; reference 20CA70	1.01	10.00	1.11	0.27	0.03	2.70	0.03	3.84	m	4.32
13 x 45mm; reference 13CP45	0.65	10.00	0.71	0.26	0.03	2.61	0.03	3.36	m	3.77
13 x 70mm; reference 13CP70	0.84	10.00	0.92	0.26	0.03	2.61	0.03	3.56	m	4.01
13 x 70mm; reference 13CS70	0.84	10.00	0.92	0.28	0.03	2.79	0.03	3.75	m	4.22
13 x 95mm; reference 13CS95	1.08	10.00	1.19	0.26	0.03	2.61	0.03	3.83	m	4.31

	MATERIALS			LABOUR				RATES		
Labour hourly rates: (except Specialists) Craft Operatives £9.23 Labourer £7.02 Rates are national average prices. Refer to REGIONAL VARIATIONS for indicative levels of overall pricing in regions	Del to Site £	Waste %	Material Cost £	Craft Optve £	Lab £	Labour Cost £	Sunds £	Nett Rate £	Unit	Gross Rate (+12.5%) £
P20: UNFRAMED ISOLATED TRIMS/SKIRTINGS/SUNDRY ITEMS Cont.										
Softwood, wrought - skirtings, architraves, picture rails and cover fillets to B.S.1186 Part 3 Cont.										
Skirtings, picture rails, architraves and the like; (finished sizes) Cont.										
20 x 120mm; reference 20CS120	1.66	10.00	1.83	0.31	0.03	3.07	0.04	4.94	m	5.56
9 x 33mm; reference 9RA33	0.48	10.00	0.53	0.23	0.02	2.26	0.03	2.82	m	3.17
13 x 45mm; reference 13RA45	0.65	10.00	0.71	0.26	0.03	2.61	0.03	3.36	m	3.77
20 x 70mm; reference 20RA70	1.08	10.00	1.19	0.27	0.03	2.70	0.03	3.92	m	4.41
13 x 70mm; reference 13RS70	0.87	10.00	0.96	0.28	0.03	2.79	0.03	3.78	m	4.25
13 x 95mm; reference 13RS95	1.12	10.00	1.23	0.26	0.03	2.61	0.03	3.87	m	4.36
20 x 120mm; reference 20RS120	1.66	10.00	1.83	0.31	0.03	3.07	0.04	4.94	m	5.56
13 x 45mm; reference 13RP45	0.68	10.00	0.75	0.26	0.03	2.61	0.03	3.39	m	3.81
13 x 70mm; reference 13RP70	0.91	10.00	1.00	0.28	0.03	2.79	0.03	3.83	m	4.30
Cover fillets, stops, trims, beads, nosings and the like; (finished sizes)										
11 x 33mm; reference C33	0.48	10.00	0.53	0.22	0.02	2.17	0.02	2.72	m	3.06
11 x 45mm; reference C45	0.51	10.00	0.56	0.22	0.02	2.17	0.02	2.75	m	3.10
13 x 33mm; reference HR33	0.56	10.00	0.62	0.22	0.02	2.17	0.02	2.81	m	3.16
20 x 45mm; reference HR45	0.76	10.00	0.84	0.24	0.03	2.43	0.03	3.29	m	3.70
11 x 11mm; reference Q11	0.25	10.00	0.28	0.22	0.02	2.17	0.02	2.47	m	2.77
13 x 13mm; reference Q13	0.40	10.00	0.44	0.22	0.02	2.17	0.02	2.63	m	2.96
20 x 20mm; reference Q20	0.51	10.00	0.56	0.22	0.02	2.17	0.02	2.75	m	3.10
13 x 13mm; reference S13	0.40	10.00	0.44	0.22	0.02	2.17	0.02	2.63	m	2.96
20 x 20mm; reference S20	0.51	10.00	0.56	0.22	0.02	2.17	0.02	2.75	m	3.10
27 x 27mm; reference S27	0.65	10.00	0.71	0.24	0.03	2.43	0.03	3.17	m	3.57
12 x 95mm; reference 12HRCS95	1.20	10.00	1.32	0.28	0.03	2.79	0.04	4.16	m	4.67
15 x 95mm; reference 15HRCS95	1.27	10.00	1.40	0.31	0.03	3.07	0.04	4.51	m	5.07
Softwood, wrought - skirtings, architraves, picture rails and cover fillets										
Skirtings, picture rails, architraves and the like; (finished sizes)										
19 x 100mm	1.60	10.00	1.76	0.29	0.04	2.96	0.05	4.77	m	5.36
19 x 150mm	2.23	10.00	2.45	0.32	0.04	3.23	0.05	5.74	m	6.45
25 x 100mm	2.04	10.00	2.24	0.31	0.04	3.14	0.05	5.44	m	6.12
25 x 150mm	2.86	10.00	3.15	0.36	0.05	3.67	0.06	6.88	m	7.74
25 x 50mm; splays -1	1.16	10.00	1.28	0.25	0.03	2.52	0.04	3.83	m	4.31
25 x 63mm; splays -1	1.34	10.00	1.47	0.28	0.04	2.87	0.05	4.39	m	4.94
25 x 75mm; splays -1	1.72	10.00	1.89	0.29	0.04	2.96	0.05	4.90	m	5.51
32 x 100mm; splays -1	2.49	10.00	2.74	0.34	0.04	3.42	0.05	6.21	m	6.98
38 x 75mm; splays -1	2.29	10.00	2.52	0.32	0.04	3.23	0.06	5.81	m	6.54
15 x 50mm; mouldings -1	0.76	10.00	0.84	0.24	0.03	2.43	0.04	3.30	m	3.71
15 x 125mm; mouldings -1	1.41	10.00	1.55	0.29	0.04	2.96	0.05	4.56	m	5.13
19 x 50mm; mouldings -1	0.96	10.00	1.06	0.25	0.03	2.52	0.04	3.61	m	4.07
19 x 125mm; mouldings -1	1.78	10.00	1.96	0.31	0.04	3.14	0.05	5.15	m	5.79
25 x 50mm; mouldings -1	1.16	10.00	1.28	0.25	0.03	2.52	0.04	3.83	m	4.31
25 x 63mm; mouldings -1	1.34	10.00	1.47	0.28	0.04	2.87	0.05	4.39	m	4.94
25 x 75mm; mouldings -1	1.72	10.00	1.89	0.29	0.04	2.96	0.05	4.90	m	5.51
32 x 100mm; mouldings -1	2.49	10.00	2.74	0.34	0.04	3.42	0.05	6.21	m	6.98
38 x 75mm; mouldings -1	2.29	10.00	2.52	0.32	0.04	3.23	0.06	5.81	m	6.54
15 x 125mm; chamfers -1	1.41	10.00	1.55	0.29	0.04	2.96	0.05	4.56	m	5.13
19 x 125mm; chamfers -1	1.78	10.00	1.96	0.30	0.04	3.05	0.05	5.06	m	5.69
25 x 75mm; chamfers -2	1.78	10.00	1.96	0.29	0.04	2.96	0.05	4.97	m	5.59
25 x 100mm; chamfers -2	2.17	10.00	2.39	0.30	0.04	3.05	0.05	5.49	m	6.17
Cover fillets, stops, trims, beads, nosings and the like; (finished sizes)										
12 x 50mm; mouldings -1	0.65	10.00	0.71	0.24	0.03	2.43	0.04	3.18	m	3.58
19 x 19mm; mouldings -1	0.51	10.00	0.56	0.22	0.03	2.24	0.04	2.84	m	3.20
19 x 50mm; mouldings -1	0.96	10.00	1.06	0.25	0.03	2.52	0.04	3.61	m	4.07
25 x 25mm; mouldings -1	0.65	10.00	0.71	0.24	0.03	2.43	0.04	3.18	m	3.58
25 x 38mm; mouldings -1	0.90	10.00	0.99	0.25	0.03	2.52	0.04	3.55	m	3.99
19 x 19mm; chamfers -1	0.51	10.00	0.56	0.22	0.03	2.24	0.04	2.84	m	3.20
25 x 25mm; chamfers -1	0.65	10.00	0.71	0.24	0.03	2.43	0.04	3.18	m	3.58
25 x 38mm; chamfers -1	0.90	10.00	0.99	0.25	0.03	2.52	0.04	3.55	m	3.99
Afrormosia, wrought - skirtings, architraves, picture rails and cover fillets										
Skirtings, picture rails, architraves and the like; (finished sizes)										
19 x 100mm	6.63	10.00	7.29	0.48	0.06	4.85	0.08	12.22	m	13.75
19 x 150mm	8.59	10.00	9.45	0.55	0.07	5.57	0.09	15.11	m	17.00
25 x 100mm	8.18	10.00	9.00	0.53	0.07	5.38	0.09	14.47	m	16.28
25 x 150mm	11.07	10.00	12.18	0.55	0.07	5.57	0.09	17.83	m	20.06
extra; ends	1.69	-	1.69	0.06	0.01	0.62	0.01	2.32	nr	2.61
extra; angles	2.75	-	2.75	0.11	0.02	1.16	0.02	3.93	nr	4.42
extra; mitres	2.75	-	2.75	0.11	0.02	1.16	0.02	3.93	nr	4.42
extra; intersections	2.75	-	2.75	0.11	0.02	1.16	0.02	3.93	nr	4.42
25 x 50mm; splays -1	4.80	10.00	5.28	0.43	0.05	4.32	0.07	9.67	m	10.88
25 x 63mm; splays -1	5.64	10.00	6.20	0.47	0.06	4.76	0.08	11.04	m	12.42
25 x 75mm; splays -1	6.97	10.00	7.67	0.48	0.06	4.85	0.08	12.60	m	14.17
32 x 100mm; splays -1	9.58	10.00	10.54	0.55	0.07	5.57	0.09	16.20	m	18.22
38 x 75mm; splays -1	8.89	10.00	9.78	0.55	0.07	5.57	0.09	15.44	m	17.37
extra; ends	1.33	-	1.33	0.06	0.01	0.62	0.01	1.96	nr	2.21
extra; angles	2.26	-	2.26	0.11	0.02	1.16	0.02	3.44	nr	3.87
extra; mitres	2.26	-	2.26	0.11	0.02	1.16	0.02	3.44	nr	3.87
extra; intersections	2.26	-	2.26	0.11	0.02	1.16	0.02	3.44	nr	3.87
15 x 50mm; mouldings -1	3.11	10.00	3.42	0.41	0.05	4.14	0.07	7.63	m	8.58
15 x 125mm; mouldings -1	5.93	10.00	6.52	0.48	0.06	4.85	0.08	11.45	m	12.89

Labour hourly rates: (except Specialists) Craft Operatives £9.23 Labourer £7.02 Rates are national average prices. Refer to REGIONAL VARIATIONS for indicative levels of overall pricing in regions	MATERIALS			LABOUR				RATES		
	Del to Site £	Waste %	Material Cost £	Craft Optve £	Lab £	Labour Cost £	Sunds £	Nett Rate £	Unit	Gross Rate (+12.5%) £
P20: UNFRAMED ISOLATED TRIMS/SKIRTINGS/SUNDRY ITEMS Cont.										
Afrormosia, wrought - skirtings, architraves, picture rails and cover fillets Cont.										
Skirtings, picture rails, architraves and the like; (finished sizes) Cont.										
19 x 50mm; mouldings -1	3.88	10.00	4.27	0.43	0.05	4.32	0.07	8.66	m	9.74
19 x 125mm; mouldings -1	7.47	10.00	8.22	0.53	0.07	5.38	0.09	13.69	m	15.40
25 x 50mm; mouldings -1	4.80	10.00	5.28	0.43	0.05	4.32	0.07	9.67	m	10.88
25 x 63mm; mouldings -1	5.64	10.00	6.20	0.47	0.06	4.76	0.08	11.04	m	12.42
25 x 75mm; mouldings -1	6.97	10.00	7.67	0.48	0.06	4.85	0.08	12.60	m	14.17
32 x 100mm; mouldings -1	9.58	10.00	10.54	0.55	0.07	5.57	0.09	16.20	m	18.22
38 x 75mm; mouldings -1	8.89	10.00	9.78	0.55	0.07	5.57	0.09	15.44	m	17.37
extra; ends	1.33	-	1.33	0.06	0.01	0.62	0.01	1.96	nr	2.21
extra; angles	2.26	-	2.26	0.11	0.02	1.16	0.02	3.44	nr	3.87
extra; mitres	2.26	-	2.26	0.11	0.02	1.16	0.02	3.44	nr	3.87
extra; intersections	2.26	-	2.26	0.11	0.02	1.16	0.02	3.44	nr	3.87
15 x 125mm; chamfers -1	5.93	10.00	6.52	0.48	0.06	4.85	0.08	11.45	m	12.89
19 x 125mm; chamfers -1	7.47	10.00	8.22	0.53	0.07	5.38	0.09	13.69	m	15.40
25 x 75mm; chamfers -2	7.19	10.00	7.91	0.48	0.06	4.85	0.08	12.84	m	14.45
25 x 100mm; chamfers -2	8.59	10.00	9.45	0.51	0.06	5.13	0.08	14.66	m	16.49
Cover fillets, stops, trims, beads, nosings and the like; (finished sizes)										
12 x 50mm; mouldings -1	2.62	10.00	2.88	0.41	0.05	4.14	0.07	7.09	m	7.97
19 x 19mm; mouldings -1	1.76	10.00	1.94	0.37	0.05	3.77	0.06	5.76	m	6.48
19 x 50mm; mouldings -1	3.88	10.00	4.27	0.43	0.05	4.32	0.07	8.66	m	9.74
25 x 25mm; mouldings -1	2.68	10.00	2.95	0.41	0.05	4.14	0.07	7.15	m	8.05
25 x 38mm; mouldings -1	3.59	10.00	3.95	0.43	0.05	4.32	0.07	8.34	m	9.38
19 x 19mm; chamfers -1	1.76	10.00	1.94	0.37	0.05	3.77	0.06	5.76	m	6.48
25 x 25mm; chamfers -1	2.68	10.00	2.95	0.41	0.05	4.14	0.07	7.15	m	8.05
25 x 38mm; chamfers -1	3.59	10.00	3.95	0.43	0.05	4.32	0.07	8.34	m	9.38
Medium density fibreboard, B.S.1142, skirtings, architraves, picture rails and cover fillets to B.S.1186 Part 3										
Skirtings, picture rails, architraves and the like; (finished sizes)										
13 x 45mm; reference 13CA45	1.19	10.00	1.31	0.32	0.05	3.30	0.07	4.68	m	5.27
20 x 70mm; reference 20CA70	1.86	10.00	2.05	0.34	0.05	3.49	0.07	5.61	m	6.31
13 x 45mm; reference 13CP45	1.19	10.00	1.31	0.33	0.05	3.40	0.07	4.78	m	5.37
13 x 70mm; reference 13CP70	1.50	10.00	1.65	0.33	0.05	3.40	0.07	5.12	m	5.76
13 x 70mm; reference 13CS70	1.50	10.00	1.65	0.35	0.05	3.58	0.07	5.30	m	5.96
13 x 95mm; reference 13CS95	1.97	10.00	2.17	0.33	0.05	3.40	0.07	5.63	m	6.34
20 x 120mm; reference 20CS120	3.05	10.00	3.36	0.38	0.05	3.86	0.07	7.28	m	8.19
9 x 33mm; reference 9RA33	0.89	10.00	0.98	0.28	0.03	2.79	0.07	3.84	m	4.32
13 x 45mm; reference 13RA45	1.19	10.00	1.31	0.33	0.05	3.40	0.07	4.78	m	5.37
20 x 70mm; reference 20RA70	1.97	10.00	2.17	0.34	0.05	3.49	0.07	5.73	m	6.44
13 x 70mm; reference 13RS70	1.58	10.00	1.74	0.35	0.05	3.58	0.07	5.39	m	6.06
13 x 95mm; reference 13RS95	2.06	10.00	2.27	0.33	0.05	3.40	0.07	5.73	m	6.45
20 x 120mm; reference 20RS120	3.05	10.00	3.36	0.38	0.05	3.86	0.07	7.28	m	8.19
13 x 45mm; reference 13RP45	1.24	10.00	1.36	0.33	0.05	3.40	0.07	4.83	m	5.43
13 x 70mm; reference 13RP70	1.67	10.00	1.84	0.35	0.05	3.58	0.07	5.49	m	6.17
Cover fillets, stops, trims, beads, nosings and the like; (finished sizes)										
11 x 33mm; reference C33	0.89	10.00	0.98	0.27	0.02	2.63	0.06	3.67	m	4.13
11 x 45mm; reference C45	0.94	10.00	1.03	0.27	0.02	2.63	0.06	3.73	m	4.19
13 x 33mm; reference HR33	1.05	10.00	1.16	0.27	0.02	2.63	0.06	3.85	m	4.33
20 x 45mm; reference HR45	1.41	10.00	1.55	0.29	0.04	2.96	0.06	4.57	m	5.14
11 x 11mm; reference Q11	0.47	10.00	0.52	0.27	0.02	2.63	0.06	3.21	m	3.61
13 x 13mm; reference Q13	0.73	10.00	0.80	0.27	-	2.49	0.06	3.36	m	3.77
20 x 20mm; reference Q20	0.96	10.00	1.06	0.27	0.02	2.63	0.06	3.75	m	4.22
13 x 13mm; reference S13	0.73	10.00	0.80	0.27	0.02	2.63	0.06	3.50	m	3.93
20 x 20mm; reference S20	0.94	10.00	1.03	0.27	0.02	2.63	0.06	3.73	m	4.19
27 x 27mm; reference S27	1.19	10.00	1.31	0.29	0.04	2.96	0.06	4.33	m	4.87
12 x 95mm; reference 12HRCS95	2.20	10.00	2.42	0.35	0.05	3.58	0.06	6.06	m	6.82
15 x 95mm; reference 15HRCS95	2.31	10.00	2.54	0.38	0.05	3.86	0.06	6.46	m	7.27
Softwood, wrought - cappings										
Cover fillets, stops, trims, beads, nosings and the like; (finished sizes)										
25 x 50mm; level; rebates -1; mouldings -1	1.16	10.00	1.28	0.25	0.03	2.52	0.04	3.83	m	4.31
50 x 75mm; level; rebates -1; mouldings -1	3.31	10.00	3.64	0.36	0.05	3.67	0.06	7.37	m	8.30
50 x 100mm; level; rebates -1; mouldings -1	4.14	10.00	4.55	0.39	0.05	3.95	0.06	8.56	m	9.64
25 x 50mm; ramped; rebates -1; mouldings -1	1.16	10.00	1.28	0.38	0.05	3.86	0.06	5.19	m	5.84
50 x 75mm; ramped; rebates -1; mouldings -1	3.31	10.00	3.64	0.53	0.07	5.38	0.09	9.11	m	10.25
50 x 100mm; ramped; rebates -1; mouldings -1	4.14	10.00	4.55	0.59	0.07	5.94	0.10	10.59	m	11.91
Afrormosia, wrought - cappings										
Cover fillets, stops, trims, beads, nosings and the like; (finished sizes)										
25 x 50mm; level; rebates -1; mouldings -1	4.94	10.00	5.43	0.43	0.05	4.32	0.07	9.82	m	11.05
50 x 75mm; level; rebates -1; mouldings -1	12.69	10.00	13.96	0.61	0.08	6.19	0.10	20.25	m	22.78
extra; ends	1.90	-	1.90	0.09	0.01	0.90	0.01	2.81	nr	3.16
extra; angles	3.18	-	3.18	0.15	0.02	1.52	0.02	4.72	nr	5.32
extra; mitres	3.18	-	3.18	0.15	0.02	1.52	0.02	4.72	nr	5.32
extra; rounded corners not exceeding 300mm girth	12.69	-	12.69	0.61	0.08	6.19	0.10	18.98	nr	21.35
50 x 100mm; level; rebates -1; mouldings -1	15.79	10.00	17.37	0.67	0.08	6.75	0.11	24.22	nr	27.25
extra; ends	2.40	-	2.40	0.10	0.01	0.99	0.02	3.41	nr	3.84
extra; angles	3.95	-	3.95	0.16	0.02	1.62	0.03	5.60	nr	6.30

BUILDING FABRIC SUNDRIES

Labour hourly rates: (except Specialists) Craft Operatives £9.23 Labourer £7.02 Rates are national average prices. Refer to REGIONAL VARIATIONS for indicative levels of overall pricing in regions	MATERIALS			LABOUR				RATES		
	Del to Site £	Waste %	Material Cost £	Craft Optve £	Lab £	Labour Cost £	Sunds £	Nett Rate £	Unit	Gross Rate (+12.5%) £

P20: UNFRAMED ISOLATED TRIMS/SKIRTINGS/SUNDRY ITEMS Cont.

Afrormosia, wrought - cappings Cont.

Cover fillets, stops, trims, beads, nosings and the like; (finished sizes) Cont.

extra; mitres	3.95	-	3.95	0.16	0.02	1.62	0.03	5.60	nr	6.30
extra; rounded corners not exceeding 300mm girth	15.79	-	15.79	0.67	0.08	6.75	0.11	22.65	nr	25.48
25 x 50mm; ramped; rebates -1; mouldings -1	4.94	10.00	5.43	0.63	0.08	6.38	0.10	11.91	m	13.40
50 x 75mm; ramped; rebates -1; mouldings -1	12.69	10.00	13.96	0.92	0.12	9.33	0.15	23.44	m	26.37
extra; ends	1.90	-	1.90	0.14	0.02	1.43	0.02	3.35	nr	3.77
extra; angles	3.18	-	3.18	0.23	0.03	2.33	0.04	5.55	nr	6.25
extra; mitres	3.18	-	3.18	0.23	0.03	2.33	0.04	5.55	nr	6.25
extra; rounded corners not exceeding 300mm girth	12.69	-	12.69	0.92	0.12	9.33	0.15	22.17	nr	24.95
50 x 100mm; ramped; rebates -1; mouldings -1	15.79	10.00	17.37	1.05	0.13	10.60	0.17	28.14	m	31.66
extra; ends	2.40	-	2.40	0.15	0.02	1.52	0.02	3.94	nr	4.44
extra; angles	3.95	-	3.95	0.25	0.03	2.52	0.04	6.51	nr	7.32
extra; mitres	3.95	-	3.95	0.25	0.03	2.52	0.04	6.51	nr	7.32
extra; rounded corners not exceeding 300mm girth	15.79	-	15.79	1.05	0.13	10.60	0.17	26.56	nr	29.88

Sapele; wrought - cappings

Cover fillets, stops, trims, beads, nosings and the like; (finished sizes)

25 x 50mm; level; rebates -1; mouldings -1	3.59	10.00	3.95	0.38	0.05	3.86	0.06	7.87	m	8.85
50 x 75mm; level; rebates -1; mouldings -1	8.95	10.00	9.85	0.53	0.07	5.38	0.08	15.31	m	17.22
extra; ends	1.33	-	1.33	0.08	0.01	0.81	0.01	2.15	nr	2.42
extra; angles	2.26	-	2.26	0.14	0.02	1.43	0.02	3.71	nr	4.18
extra; mitres	2.26	-	2.26	0.14	0.02	1.43	0.02	3.71	nr	4.18
extra; rounded corners not exceeding 300mm girth	8.95	-	8.95	0.53	0.07	5.38	0.09	14.42	nr	16.23
50 x 100mm; level; rebates -1; mouldings -1	10.98	10.00	12.08	0.57	0.07	5.75	0.09	17.92	m	20.16
extra; ends	1.67	-	1.67	0.09	0.01	0.90	0.01	2.58	nr	2.90
extra; angles	2.77	-	2.77	0.15	0.02	1.52	0.02	4.31	nr	4.85
extra; mitres	2.77	-	2.77	0.15	0.02	1.52	0.02	4.31	nr	4.85
extra; rounded corners not exceeding 300mm girth	10.98	-	10.98	0.57	0.07	5.75	0.09	16.82	nr	18.93
25 x 50mm; ramped; rebates -1; mouldings -1	3.59	10.00	3.95	0.57	0.07	5.75	0.09	9.79	m	11.02
50 x 75mm; ramped; rebates -1; mouldings -1	8.95	10.00	9.85	0.80	0.10	8.09	0.13	18.06	m	20.32
extra; ends	1.33	-	1.33	0.11	0.01	1.09	0.02	2.44	nr	2.74
extra; angles	2.26	-	2.26	0.20	0.03	2.06	0.03	4.35	nr	4.89
extra; mitres	2.26	-	2.26	0.20	0.03	2.06	0.03	4.35	nr	4.89
extra; rounded corners not exceeding 300mm girth	8.95	-	8.95	0.80	0.10	8.09	0.13	17.17	nr	19.31
50 x 100mm; ramped; rebates -1; mouldings -1	10.98	10.00	12.08	0.85	0.11	8.62	0.14	20.84	m	23.44
extra; ends	1.67	-	1.67	0.13	0.02	1.34	0.02	3.03	nr	3.41
extra; angles	2.77	-	2.77	0.22	0.03	2.24	0.04	5.05	nr	5.68
extra; mitres	2.77	-	2.77	0.22	0.03	2.24	0.04	5.05	nr	5.68
extra; rounded corners not exceeding 300mm girth	10.98	-	10.98	0.85	0.11	8.62	0.14	19.74	nr	22.20

Softwood, wrought - window boards

Cover fillets, stops, trims, beads, nosings and the like; (finished sizes)
mouldings -1; tongued on

25 x 50mm	1.34	10.00	1.47	0.29	0.04	2.96	0.05	4.48	m	5.04
25 x 75mm	2.17	10.00	2.39	0.30	0.04	3.05	0.05	5.49	m	6.17
32 x 50mm	1.66	10.00	1.83	0.32	0.04	3.23	0.05	5.11	m	5.75
32 x 75mm	2.55	10.00	2.81	0.33	0.04	3.33	0.05	6.18	m	6.95

rounded edges -1; tongued on

25 x 50mm	1.34	10.00	1.47	0.29	0.04	2.96	0.05	4.48	m	5.04
25 x 75mm	2.17	10.00	2.39	0.30	0.04	3.05	0.05	5.49	m	6.17
32 x 50mm	1.66	10.00	1.83	0.32	0.04	3.23	0.05	5.11	m	5.75
32 x 75mm	2.55	10.00	2.81	0.33	0.04	3.33	0.05	6.18	m	6.95

Window boards

25 x 150mm; mouldings -1	2.94	10.00	3.23	0.34	0.04	3.42	0.05	6.70	m	7.54
32 x 150mm; mouldings -1	3.44	10.00	3.78	0.38	0.05	3.86	0.06	7.70	m	8.67
25 x 150mm; rounded edges -1	2.94	10.00	3.23	0.34	0.04	3.42	0.05	6.70	m	7.54
32 x 150mm; rounded edges -1	3.44	10.00	3.78	0.38	0.05	3.86	0.06	7.70	m	8.67

Afrormosia, wrought - window boards

Cover fillets, stops, trims, beads, nosings and the like; (finished sizes)
mouldings -1; tongued on

25 x 50mm	4.58	10.00	5.04	0.44	0.06	4.48	0.07	9.59	m	10.79
25 x 75mm	7.11	10.00	7.82	0.46	0.06	4.67	0.08	12.57	m	14.14
32 x 50mm	5.51	10.00	6.06	0.48	0.06	4.85	0.08	10.99	m	12.37
32 x 75mm	8.44	10.00	9.28	0.50	0.06	5.04	0.08	14.40	m	16.20

rounded edges -1; tongued on

25 x 50mm	4.58	10.00	5.04	0.44	0.06	4.48	0.07	9.59	m	10.79
25 x 75mm	7.11	10.00	7.82	0.46	0.06	4.67	0.08	12.57	m	14.14
32 x 50mm	5.51	10.00	6.06	0.48	0.06	4.85	0.08	10.99	m	12.37
32 x 75mm	8.44	10.00	9.28	0.50	0.06	5.04	0.08	14.40	m	16.20

Window boards

25 x 150mm; mouldings -1	11.30	10.00	12.43	0.60	0.08	6.10	0.10	18.63	m	20.96
32 x 150mm; mouldings -1	13.42	10.00	14.76	0.67	0.08	6.75	0.11	21.62	m	24.32
25 x 150mm; rounded edges -1	11.30	10.00	12.43	0.60	0.08	6.10	0.10	18.63	m	20.96
32 x 150mm; rounded edges -1	13.42	10.00	14.76	0.67	0.08	6.75	0.11	21.62	m	24.32

Sapele; wrought - window boards

Cover fillets, stops, trims, beads, nosings and the like; (finished sizes)
mouldings -1; tongued on

25 x 50mm	3.36	10.00	3.70	0.43	0.05	4.32	0.07	8.09	m	9.10

	MATERIALS			LABOUR				RATES		
Labour hourly rates: (except Specialists) Craft Operatives £9.23 Labourer £7.02 Rates are national average prices. Refer to REGIONAL VARIATIONS for indicative levels of overall pricing in regions	Del to Site	Waste	Material Cost	Craft Optve	Lab	Labour Cost	Sunds	Nett Rate	Unit	Gross Rate (+12.5%)
	£	%	£	£	£	£	£	£		£
P20: UNFRAMED ISOLATED TRIMS/SKIRTINGS/SUNDRY ITEMS Cont.										
Sapele; wrought - window boards Cont.										
Cover fillets, stops, trims, beads, nosings and the like; (finished sizes) Cont.										
mouldings -1; tongued on Cont.										
25 x 75mm	5.14	10.00	5.65	0.45	0.06	4.57	0.07	10.30	m	11.59
32 x 50mm	3.93	10.00	4.32	0.47	0.06	4.76	0.08	9.16	m	10.31
32 x 75mm	6.13	10.00	6.74	0.49	0.06	4.94	0.08	11.77	m	13.24
rounded edges -1; tongued on										
25 x 50mm	3.36	10.00	3.70	0.43	0.05	4.32	0.07	8.09	m	9.10
25 x 75mm	5.14	10.00	5.65	0.45	0.06	4.57	0.07	10.30	m	11.59
32 x 50mm	3.93	10.00	4.32	0.47	0.06	4.76	0.08	9.16	m	10.31
32 x 75mm	6.13	10.00	6.74	0.49	0.06	4.94	0.08	11.77	m	13.24
Window boards										
25 x 150mm; mouldings -1	8.14	10.00	8.95	0.52	0.07	5.29	0.09	14.34	m	16.13
32 x 150mm; mouldings -1	9.65	10.00	10.62	0.57	0.07	5.75	0.09	16.46	m	18.51
25 x 150mm; rounded edges -1	8.14	10.00	8.95	0.52	0.07	5.29	0.09	14.34	m	16.13
32 x 150mm; rounded edges -1	9.65	10.00	10.62	0.57	0.07	5.75	0.09	16.46	m	18.51
Moisture resistant medium density fibreboard, B.S.1142										
Window boards										
25 x 225mm; nosed and tongued	6.12	10.00	6.73	0.34	0.05	3.49	0.06	10.28	m	11.57
25 x 250mm; nosed and tongued	6.80	10.00	7.48	0.36	0.05	3.67	0.06	11.21	m	12.62
Softwood, wrought - isolated shelves and worktops										
Cover fillets, stops, trims, beads, nosings and the like; (finished sizes)										
19 x 12mm	0.30	10.00	0.33	0.21	0.03	2.15	0.04	2.52	m	2.83
19 x 16mm	0.35	10.00	0.39	0.22	0.03	2.24	0.04	2.67	m	3.00
19 x 18mm	0.41	10.00	0.45	0.22	0.03	2.24	0.04	2.73	m	3.07
19 x 22mm	0.45	10.00	0.50	0.23	0.03	2.33	0.04	2.87	m	3.23
19 x 25mm	0.51	10.00	0.56	0.23	0.03	2.33	0.04	2.93	m	3.30
Isolated shelves and worktops										
25 x 150mm	1.91	10.00	2.10	0.23	0.03	2.33	0.04	4.47	m	5.03
25 x 225mm	2.86	10.00	3.15	0.29	0.04	2.96	0.05	6.15	m	6.92
25 x 300mm	3.83	10.00	4.21	0.34	0.04	3.42	0.05	7.68	m	8.64
32 x 150mm	2.43	10.00	2.67	0.25	0.03	2.52	0.04	5.23	m	5.88
32 x 225mm	3.70	10.00	4.07	0.31	0.04	3.14	0.05	7.26	m	8.17
32 x 300mm	4.90	10.00	5.39	0.38	0.05	3.86	0.06	9.31	m	10.47
25 x 450mm; cross tongued	8.60	2.50	8.81	0.40	0.05	4.04	0.06	12.92	m	14.53
25 x 600mm; cross tongued	11.46	2.50	11.75	0.52	0.07	5.29	0.09	17.13	m	19.27
32 x 450mm; cross tongued	11.01	2.50	11.29	0.46	0.06	4.67	0.08	16.03	m	18.04
32 x 600mm; cross tongued	14.64	2.50	15.01	0.57	0.07	5.75	0.09	20.85	m	23.45
38 x 450mm; cross tongued	13.06	2.50	13.39	0.52	0.07	5.29	0.09	18.77	m	21.11
38 x 600mm; cross tongued	17.46	2.50	17.90	0.63	0.08	6.38	0.10	24.37	m	27.42
25 x 450mm overall; 25 x 50mm slats spaced 25mm apart	4.47	10.00	4.92	1.15	0.14	11.60	0.19	16.70	m	18.79
25 x 450mm overall; 25 x 50mm slats spaced 32mm apart	3.83	10.00	4.21	1.05	0.13	10.60	0.17	14.99	m	16.86
25 x 600mm overall; 25 x 50mm slats spaced 25mm apart	5.74	10.00	6.31	1.45	0.18	14.65	0.24	21.20	m	23.85
25 x 600mm overall; 25 x 50mm slats spaced 32mm apart	5.09	10.00	5.60	1.35	0.17	13.65	0.22	19.47	m	21.91
25 x 900mm overall; 25 x 50mm slats spaced 25mm apart	8.28	10.00	9.11	2.20	0.27	22.20	0.36	31.67	m	35.63
25 x 900mm overall; 25 x 50mm slats spaced 32mm apart	7.65	10.00	8.41	2.05	0.26	20.75	0.33	29.49	m	33.18
Afrormosia, wrought - isolated shelves and worktops										
Cover fillets, stops, trims, beads, nosings and the like; (finished sizes)										
19 x 12mm	1.10	10.00	1.21	0.35	0.04	3.51	0.06	4.78	m	5.38
19 x 16mm	1.30	10.00	1.43	0.38	0.05	3.86	0.06	5.35	m	6.02
19 x 18mm	1.50	10.00	1.65	0.38	0.05	3.86	0.06	5.57	m	6.26
19 x 22mm	1.62	10.00	1.78	0.40	0.05	4.04	0.06	5.88	m	6.62
19 x 25mm	1.88	10.00	2.07	0.40	0.05	4.04	0.06	6.17	m	6.94
Isolated shelves and worktops										
25 x 150mm	8.43	10.00	9.27	0.40	0.05	4.04	0.06	13.38	m	15.05
25 x 225mm	12.65	10.00	13.91	0.51	0.06	5.13	0.08	19.12	m	21.51
25 x 300mm	16.86	10.00	18.55	0.61	0.08	6.19	0.10	24.84	m	27.94
32 x 150mm	10.70	10.00	11.77	0.44	0.06	4.48	0.07	16.32	m	18.36
32 x 225mm	15.88	10.00	17.47	0.54	0.07	5.48	0.09	23.03	m	25.91
32 x 300mm	21.40	10.00	23.54	0.64	0.08	6.47	0.10	30.11	m	33.87
25 x 450mm; cross tongued	35.01	2.50	35.89	0.70	0.09	7.09	0.11	43.09	m	48.47
25 x 600mm; cross tongued	46.68	2.50	47.85	0.90	0.11	9.08	0.15	57.08	m	64.21
32 x 450mm; cross tongued	44.73	2.50	45.85	0.80	0.10	8.09	0.13	54.06	m	60.82
32 x 600mm; cross tongued	59.64	2.50	61.13	1.00	0.13	10.14	0.16	71.43	m	80.36
38 x 450mm; cross tongued	53.16	2.50	54.49	0.90	0.11	9.08	0.15	63.72	m	71.68
38 x 600mm; cross tongued	71.32	2.50	73.10	1.10	0.14	11.14	0.18	84.42	m	94.97
Plywood B.S.6566, II/III grade, INT bonded; butt joints - isolated shelves and worktops										
Cover fillets, stops, trims, beads, nosings and the like										
50 x 6.5mm	0.21	20.00	0.25	0.17	0.02	1.71	0.06	2.02	m	2.27

BUILDING FABRIC SUNDRIES

Labour hourly rates: (except Specialists) Craft Operatives £9.23 Labourer £7.02 Rates are national average prices. Refer to REGIONAL VARIATIONS for indicative levels of overall pricing in regions	MATERIALS			LABOUR				RATES		
	Del to Site £	Waste %	Material Cost £	Craft Optve £	Lab £	Labour Cost £	Sunds £	Nett Rate £	Unit	Gross Rate (+12.5%) £
P20: UNFRAMED ISOLATED TRIMS/SKIRTINGS/SUNDRY ITEMS Cont.										
Plywood B.S.6566, II/III grade, INT bonded; butt joints - isolated shelves and worktops Cont.										
Cover fillets, stops, trims, beads, nosings and the like Cont.										
50 x 12mm	0.29	20.00	0.35	0.18	0.02	1.80	0.06	2.21	m	2.49
50 x 15mm	0.36	20.00	0.43	0.20	0.03	2.06	0.06	2.55	m	2.87
50 x 19mm	0.41	20.00	0.49	0.21	0.03	2.15	0.06	2.70	m	3.04
100 x 6.5mm	0.32	20.00	0.38	0.18	0.02	1.80	0.06	2.25	m	2.53
100 x 12mm	0.59	20.00	0.71	0.20	0.03	2.06	0.06	2.82	m	3.18
100 x 15mm	0.74	20.00	0.89	0.21	0.03	2.15	0.06	3.10	m	3.48
100 x 19mm	0.83	20.00	1.00	0.22	0.03	2.24	0.06	3.30	m	3.71
Isolated shelves and worktops										
6.5 x 150mm	0.69	20.00	0.83	0.21	0.03	2.15	0.06	3.04	m	3.42
6.5 x 225mm	1.06	20.00	1.27	0.26	0.03	2.61	0.06	3.94	m	4.44
6.5 x 300mm	1.41	20.00	1.69	0.32	0.04	3.23	0.06	4.99	m	5.61
6.5 x 450mm	2.10	20.00	2.52	0.38	0.05	3.86	0.06	6.44	m	7.24
6.5 x 600mm	2.79	20.00	3.35	0.44	0.06	4.48	0.06	7.89	m	8.88
12 x 150mm	1.30	20.00	1.56	0.22	0.03	2.24	0.06	3.86	m	4.34
12 x 225mm	1.94	20.00	2.33	0.28	0.04	2.87	0.06	5.25	m	5.91
12 x 300mm	2.61	20.00	3.13	0.33	0.04	3.33	0.06	6.52	m	7.33
12 x 450mm	3.89	20.00	4.67	0.39	0.05	3.95	0.06	8.68	m	9.76
12 x 600mm	5.18	20.00	6.22	0.45	0.06	4.57	0.06	10.85	m	12.21
15 x 150mm	1.56	20.00	1.87	0.23	0.03	2.33	0.06	4.27	m	4.80
15 x 225mm	2.35	20.00	2.82	0.29	0.04	2.96	0.06	5.84	m	6.57
15 x 300mm	3.14	20.00	3.77	0.34	0.04	3.42	0.06	7.25	m	8.15
15 x 450mm	4.70	20.00	5.64	0.40	0.05	4.04	0.06	9.74	m	10.96
15 x 600mm	6.27	20.00	7.52	0.46	0.06	4.67	0.06	12.25	m	13.78
19 x 150mm	1.73	20.00	2.08	0.23	0.03	2.33	0.06	4.47	m	5.03
19 x 225mm	2.60	20.00	3.12	0.29	0.04	2.96	0.06	6.14	m	6.90
19 x 300mm	3.44	20.00	4.13	0.34	0.04	3.42	0.06	7.61	m	8.56
19 x 450mm	5.16	20.00	6.19	0.40	0.05	4.04	0.06	10.30	m	11.58
19 x 600mm	6.88	20.00	8.26	0.46	0.06	4.67	0.06	12.98	m	14.61
Blockboard, B.S.3444, 2/2 grade, INT bonded, butt joints - isolated shelves and worktops										
Isolated shelves and worktops										
16 x 150mm	2.09	20.00	2.51	0.23	0.03	2.33	0.09	4.93	m	5.55
16 x 225mm	3.14	20.00	3.77	0.29	0.04	2.96	0.09	6.82	m	7.67
16 x 300mm	4.18	20.00	5.02	0.34	0.04	3.42	0.09	8.53	m	9.59
16 x 450mm	6.27	20.00	7.52	0.40	0.05	4.04	0.09	11.66	m	13.11
16 x 600mm	8.35	20.00	10.02	0.46	0.06	4.67	0.09	14.78	m	16.62
18 x 150mm	2.16	20.00	2.59	0.23	0.03	2.33	0.09	5.02	m	5.64
18 x 225mm	3.23	20.00	3.88	0.29	0.04	2.96	0.09	6.92	m	7.79
18 x 300mm	4.30	20.00	5.16	0.34	0.04	3.42	0.09	8.67	m	9.75
18 x 450mm	6.46	20.00	7.75	0.40	0.05	4.04	0.09	11.89	m	13.37
18 x 600mm	8.62	20.00	10.34	0.46	0.06	4.67	0.09	15.10	m	16.99
22 x 150mm	2.92	20.00	3.50	0.23	0.03	2.33	0.09	5.93	m	6.67
22 x 225mm	4.37	20.00	5.24	0.29	0.04	2.96	0.09	8.29	m	9.33
22 x 300mm	5.83	20.00	7.00	0.34	0.04	3.42	0.09	10.51	m	11.82
22 x 450mm	8.73	20.00	10.48	0.40	0.05	4.04	0.09	14.61	m	16.44
22 x 600mm	11.65	20.00	13.98	0.46	0.06	4.67	0.09	18.74	m	21.08
25 x 150mm	3.16	20.00	3.79	0.23	0.03	2.33	0.09	6.22	m	6.99
25 x 225mm	4.71	20.00	5.65	0.29	0.04	2.96	0.09	8.70	m	9.79
25 x 300mm	6.28	20.00	7.54	0.34	0.04	3.42	0.09	11.05	m	12.43
25 x 450mm	9.42	20.00	11.30	0.40	0.05	4.04	0.09	15.44	m	17.37
25 x 600mm	12.58	20.00	15.10	0.46	0.06	4.67	0.09	19.85	m	22.33
Wood chipboard B.S.5669, type I standard; butt joints - isolated shelves and worktops										
Isolated shelves and worktops										
12 x 150mm	0.47	20.00	0.56	0.22	0.03	2.24	0.06	2.87	m	3.22
12 x 225mm	0.68	20.00	0.82	0.28	0.04	2.87	0.06	3.74	m	4.21
12 x 300mm	0.92	20.00	1.10	0.33	0.04	3.33	0.06	4.49	m	5.05
12 x 450mm	1.38	20.00	1.66	0.39	0.05	3.95	0.06	5.67	m	6.38
12 x 600mm	1.85	20.00	2.22	0.45	0.06	4.57	0.06	6.85	m	7.71
18 x 150mm	0.62	20.00	0.74	0.23	0.03	2.33	0.06	3.14	m	3.53
18 x 225mm	0.91	20.00	1.09	0.29	0.04	2.96	0.06	4.11	m	4.62
18 x 300mm	1.20	20.00	1.44	0.34	0.04	3.42	0.06	4.92	m	5.53
18 x 450mm	1.80	20.00	2.16	0.40	0.05	4.04	0.06	6.26	m	7.05
18 x 600mm	2.40	20.00	2.88	0.46	0.06	4.67	0.06	7.61	m	8.56
25 x 150mm	1.13	20.00	1.36	0.23	0.03	2.33	0.06	3.75	m	4.22
25 x 225mm	1.69	20.00	2.03	0.29	0.04	2.96	0.06	5.05	m	5.68
25 x 300mm	2.26	20.00	2.71	0.34	0.04	3.42	0.06	6.19	m	6.96
25 x 450mm	3.38	20.00	4.06	0.40	0.05	4.04	0.06	8.16	m	9.18
25 x 600mm	4.50	20.00	5.40	0.46	0.06	4.67	0.06	10.13	m	11.39
Blockboard, B.S.3444, 2/2 grade, INT bonded, faced with 1.5mm laminated plastic sheet, B.S.EN438, classified HGS, with 1.2mm laminated plastic sheet balance veneer; butt joints										
Isolated shelves and worktops										
16 x 150mm	23.07	2.50	23.65	0.46	0.06	4.67	0.12	28.43	m	31.99
16 x 225mm	30.75	2.50	31.52	0.57	0.07	5.75	0.12	37.39	m	42.07
16 x 300mm	38.45	2.50	39.41	0.69	0.09	7.00	0.12	46.53	m	52.35
16 x 450mm	52.27	2.50	53.58	0.80	0.10	8.09	0.12	61.78	m	69.51
16 x 600mm	61.49	2.50	63.03	0.92	0.12	9.33	0.12	72.48	m	81.54
18 x 150mm	23.07	2.50	23.65	0.46	0.06	4.67	0.12	28.43	m	31.99
18 x 225mm	30.75	2.50	31.52	0.57	0.07	5.75	0.12	37.39	m	42.07

	MATERIALS			LABOUR				RATES		
Labour hourly rates: (except Specialists) Craft Operatives £9.23 Labourer £7.02 Rates are national average prices. Refer to REGIONAL VARIATIONS for indicative levels of overall pricing in regions	Del to Site £	Waste %	Material Cost £	Craft Optve £	Lab £	Labour Cost £	Sunds £	Nett Rate £	Unit	Gross Rate (+12.5%) £

P20: UNFRAMED ISOLATED TRIMS/SKIRTINGS/SUNDRY ITEMS Cont.

Blockboard, B.S.3444, 2/2 grade, INT bonded, faced with 1.5mm laminated plastic sheet, B.S.EN438, classified HGS, with 1.2mm laminated plastic sheet balance veneer; butt joints Cont.

Isolated shelves and worktops Cont.

	Del to Site £	Waste %	Material Cost £	Craft Optve £	Lab £	Labour Cost £	Sunds £	Nett Rate £	Unit	Gross Rate £
18 x 300mm	38.45	2.50	39.41	0.69	0.09	7.00	0.12	46.53	m	52.35
18 x 450mm	52.27	2.50	53.58	0.80	0.10	8.09	0.12	61.78	m	69.51
18 x 600mm	61.49	2.50	63.03	0.92	0.12	9.33	0.12	72.48	m	81.54
22 x 150mm	24.60	2.50	25.22	0.46	0.06	4.67	0.12	30.00	m	33.75
22 x 225mm	33.53	2.50	34.37	0.57	0.07	5.75	0.12	40.24	m	45.27
22 x 300mm	39.96	2.50	40.96	0.69	0.09	7.00	0.12	48.08	m	54.09
22 x 450mm	53.81	2.50	55.16	0.80	0.10	8.09	0.12	63.36	m	71.28
22 x 600mm	62.95	2.50	64.52	0.92	0.11	9.26	0.12	73.91	m	83.15
25 x 150mm	26.21	2.50	26.87	0.46	0.06	4.67	0.12	31.65	m	35.61
25 x 225mm	33.83	2.50	34.68	0.57	0.07	5.75	0.12	40.55	m	45.62
25 x 300mm	41.58	2.50	42.62	0.69	0.09	7.00	0.12	49.74	m	55.96
25 x 450mm	55.34	2.50	56.72	0.80	0.10	8.09	0.12	64.93	m	73.05
25 x 600mm	64.57	2.50	66.18	0.92	0.11	9.26	0.12	75.57	m	85.01

Laminated plastic sheet, B.S.EN 438 classification HGS

Cover fillets, stops, trims, beads, nosings and the like; fixing with adhesive

	Del to Site £	Waste %	Material Cost £	Craft Optve £	Lab £	Labour Cost £	Sunds £	Nett Rate £	Unit	Gross Rate £
1.5 x 16mm	0.55	20.00	0.66	0.23	0.03	2.33	0.07	3.06	m	3.45
1.5 x 18mm	0.62	20.00	0.74	0.24	0.03	2.43	0.08	3.25	m	3.66
1.5 x 22mm	0.67	20.00	0.80	0.26	0.03	2.61	0.09	3.50	m	3.94
1.5 x 25mm	0.74	20.00	0.89	0.29	0.04	2.96	0.10	3.95	m	4.44
1.5 x 75mm	1.41	20.00	1.69	0.34	0.04	3.42	0.30	5.41	m	6.09

Isolated shelves and worktops; fixing with adhesive

	Del to Site £	Waste %	Material Cost £	Craft Optve £	Lab £	Labour Cost £	Sunds £	Nett Rate £	Unit	Gross Rate £
1.5 x 150mm	2.79	20.00	3.35	0.40	0.05	4.04	0.60	7.99	m	8.99
1.5 x 225mm	4.19	20.00	5.03	0.46	0.06	4.67	0.90	10.60	m	11.92
1.5 x 300mm	5.68	20.00	6.82	0.52	0.07	5.29	1.20	13.31	m	14.97
1.5 x 450mm	8.47	20.00	10.16	0.61	0.08	6.19	1.80	18.16	m	20.43
1.5 x 600mm	11.26	20.00	13.51	0.69	0.09	7.00	2.40	22.91	m	25.78

P21: IRONMONGERY

Fix only ironmongery (prime cost sum for supply included elsewhere)

To softwood

	Del to Site	Waste	Material Cost	Craft Optve £	Lab £	Labour Cost £	Sunds	Nett Rate £	Unit	Gross Rate £
butt hinges; 50mm	-	-	-	0.17	-	1.57	-	1.57	nr	1.77
butt hinges; 100mm	-	-	-	0.17	-	1.57	-	1.57	nr	1.77
rising hinges; 100mm	-	-	-	0.30	-	2.77	-	2.77	nr	3.12
tee hinges; 150mm	-	-	-	0.45	-	4.15	-	4.15	nr	4.67
tee hinges; 300mm	-	-	-	0.55	-	5.08	-	5.08	nr	5.71
tee hinges; 450mm	-	-	-	0.65	-	6.00	-	6.00	nr	6.75
hook and band hinges; 300mm	-	-	-	0.90	-	8.31	-	8.31	nr	9.35
hook and band hinges; 450mm	-	-	-	1.10	-	10.15	-	10.15	nr	11.42
hook and band hinges; 900mm	-	-	-	1.65	-	15.23	-	15.23	nr	17.13
collinge hinges; 600mm	-	-	-	1.00	-	9.23	-	9.23	nr	10.38
collinge hinges; 750mm	-	-	-	1.25	-	11.54	-	11.54	nr	12.98
collinge hinges; 900mm	-	-	-	1.55	-	14.31	-	14.31	nr	16.09
single action floor springs and top centres	-	-	-	2.75	-	25.38	-	25.38	nr	28.56
double action floor springs and top centres	-	-	-	3.30	-	30.46	-	30.46	nr	34.27
coil springs	-	-	-	0.30	-	2.77	-	2.77	nr	3.12
overhead door closers	-	-	-	1.65	-	15.23	-	15.23	nr	17.13
concealed overhead door closers	-	-	-	2.75	-	25.38	-	25.38	nr	28.56
Perko door closers	-	-	-	1.65	-	15.23	-	15.23	nr	17.13
door selectors	-	-	-	2.20	0.28	22.27	-	22.27	nr	25.06
cabin hooks	-	-	-	0.25	-	2.31	-	2.31	nr	2.60
fanlight catches	-	-	-	0.40	-	3.69	-	3.69	nr	4.15
roller catches	-	-	-	0.40	-	3.69	-	3.69	nr	4.15
casement fasteners	-	-	-	0.40	-	3.69	-	3.69	nr	4.15
sash fasteners	-	-	-	1.10	-	10.15	-	10.15	nr	11.42
sash screws	-	-	-	0.55	-	5.08	-	5.08	nr	5.71
mortice latches	-	-	-	0.70	-	6.46	-	6.46	nr	7.27
night latches	-	-	-	1.10	-	10.15	-	10.15	nr	11.42
Norfolk latches	-	-	-	0.90	-	8.31	-	8.31	nr	9.35
rim latches	-	-	-	0.95	-	8.77	-	8.77	nr	9.86
Suffolk latches	-	-	-	0.90	-	8.31	-	8.31	nr	9.35
budget locks	-	-	-	0.75	-	6.92	-	6.92	nr	7.79
cylinder locks	-	-	-	1.10	-	10.15	-	10.15	nr	11.42
dead locks	-	-	-	0.90	-	8.31	-	8.31	nr	9.35
mortice locks	-	-	-	1.10	-	10.15	-	10.15	nr	11.42
rim locks	-	-	-	0.90	-	8.31	-	8.31	nr	9.35
automatic coin collecting locks	-	-	-	2.20	-	20.31	-	20.31	nr	22.84
casement stays	-	-	-	0.30	-	2.77	-	2.77	nr	3.12
quadrant stays	-	-	-	0.30	-	2.77	-	2.77	nr	3.12
barrel bolts; 150mm	-	-	-	0.35	-	3.23	-	3.23	nr	3.63
barrel bolts; 250mm	-	-	-	0.45	-	4.15	-	4.15	nr	4.67
door bolts; 150mm	-	-	-	0.35	-	3.23	-	3.23	nr	3.63
door bolts; 250mm	-	-	-	0.45	-	4.15	-	4.15	nr	4.67
monkey tail bolts										
300mm	-	-	-	0.40	-	3.69	-	3.69	nr	4.15
450mm	-	-	-	0.45	-	4.15	-	4.15	nr	4.67
600mm	-	-	-	0.55	-	5.08	-	5.08	nr	5.71
flush bolts; 200mm	-	-	-	0.75	-	6.92	-	6.92	nr	7.79
flush bolts; 450mm	-	-	-	1.10	-	10.15	-	10.15	nr	11.42
indicating bolts	-	-	-	1.10	-	10.15	-	10.15	nr	11.42

BUILDING FABRIC SUNDRIES

BUILDING FABRIC SUNDRIES — SMALL WORKS

Labour hourly rates: (except Specialists) Craft Operatives £9.23 Labourer £7.02 Rates are national average prices. Refer to REGIONAL VARIATIONS for indicative levels of overall pricing in regions	MATERIALS			LABOUR				RATES		
	Del to Site £	Waste %	Material Cost £	Craft Optve £	Lab £	Labour Cost £	Sunds £	Nett Rate £	Unit	Gross Rate (+12.5%) £
P21: IRONMONGERY Cont.										
Fix only ironmongery (prime cost sum for supply included elsewhere) Cont.										
To softwood Cont.										
monkey tail bolts Cont.										
panic bolts; to single door	-	-	-	1.65	-	15.23	-	15.23	nr	17.13
panic bolts; to double doors	-	-	-	2.20	-	20.31	-	20.31	nr	22.84
knobs	-	-	-	0.25	-	2.31	-	2.31	nr	2.60
lever handles	-	-	-	0.25	-	2.31	-	2.31	nr	2.60
sash lifts	-	-	-	0.20	-	1.85	-	1.85	nr	2.08
pull handles; 150mm	-	-	-	0.20	-	1.85	-	1.85	nr	2.08
pull handles; 225mm	-	-	-	0.25	-	2.31	-	2.31	nr	2.60
pull handles; 300mm	-	-	-	0.30	-	2.77	-	2.77	nr	3.12
back plates	-	-	-	0.30	-	2.77	-	2.77	nr	3.12
escutcheon plates	-	-	-	0.25	-	2.31	-	2.31	nr	2.60
kicking plates	-	-	-	0.55	-	5.08	-	5.08	nr	5.71
letter plates	-	-	-	1.65	-	15.23	-	15.23	nr	17.13
push plates; 225mm	-	-	-	0.25	-	2.31	-	2.31	nr	2.60
push plates; 300mm	-	-	-	0.30	-	2.77	-	2.77	nr	3.12
shelf brackets	-	-	-	0.30	-	2.77	-	2.77	nr	3.12
sash cleats	-	-	-	0.25	-	2.31	-	2.31	nr	2.60
To softwood and brickwork										
cabin hooks	-	-	-	0.30	-	2.77	-	2.77	nr	3.12
Fix only ironmongery (prime cost sum for supply included elsewhere)										
To hardwood or the like										
butt hinges; 50mm	-	-	-	0.25	-	2.31	-	2.31	nr	2.60
butt hinges; 100mm	-	-	-	0.25	-	2.31	-	2.31	nr	2.60
rising hinges; 100mm	-	-	-	0.40	-	3.69	-	3.69	nr	4.15
tee hinges; 150mm	-	-	-	0.60	-	5.54	-	5.54	nr	6.23
tee hinges; 300mm	-	-	-	0.75	-	6.92	-	6.92	nr	7.79
tee hinges; 450mm	-	-	-	0.90	-	8.31	-	8.31	nr	9.35
hook and band hinges; 300mm	-	-	-	1.20	-	11.08	-	11.08	nr	12.46
hook and band hinges; 450mm	-	-	-	1.55	-	14.31	-	14.31	nr	16.09
hook and band hinges; 900mm	-	-	-	2.20	-	20.31	-	20.31	nr	22.84
collinge hinges; 600mm	-	-	-	1.45	-	13.38	-	13.38	nr	15.06
collinge hinges; 750mm	-	-	-	1.85	-	17.08	-	17.08	nr	19.21
collinge hinges; 900mm	-	-	-	2.25	-	20.77	-	20.77	nr	23.36
single action floor springs and top centres	-	-	-	4.10	-	37.84	-	37.84	nr	42.57
double action floor springs and top centres	-	-	-	4.90	-	45.23	-	45.23	nr	50.88
coil springs	-	-	-	0.45	-	4.15	-	4.15	nr	4.67
overhead door closers	-	-	-	2.45	-	22.61	-	22.61	nr	25.44
concealed overhead door closers	-	-	-	4.10	-	37.84	-	37.84	nr	42.57
Perko door closers	-	-	-	2.45	-	22.61	-	22.61	nr	25.44
door selectors	-	-	-	3.30	0.40	33.27	-	33.27	nr	37.43
cabin hooks	-	-	-	0.35	-	3.23	-	3.23	nr	3.63
fanlight catches	-	-	-	0.55	-	5.08	-	5.08	nr	5.71
roller catches	-	-	-	0.55	-	5.08	-	5.08	nr	5.71
casement fasteners	-	-	-	0.55	-	5.08	-	5.08	nr	5.71
sash fasteners	-	-	-	1.65	-	15.23	-	15.23	nr	17.13
sash screws	-	-	-	0.80	-	7.38	-	7.38	nr	8.31
mortice latches	-	-	-	1.10	-	10.15	-	10.15	nr	11.42
night latches	-	-	-	1.65	-	15.23	-	15.23	nr	17.13
Norfolk latches	-	-	-	1.30	-	12.00	-	12.00	nr	13.50
rim latches	-	-	-	0.80	-	7.38	-	7.38	nr	8.31
Suffolk latches	-	-	-	1.30	-	12.00	-	12.00	nr	13.50
budget locks	-	-	-	1.10	-	10.15	-	10.15	nr	11.42
cylinder locks	-	-	-	1.65	-	15.23	-	15.23	nr	17.13
dead locks	-	-	-	1.30	-	12.00	-	12.00	nr	13.50
mortice locks	-	-	-	1.65	-	15.23	-	15.23	nr	17.13
rim locks	-	-	-	1.30	-	12.00	-	12.00	nr	13.50
automatic coin collecting locks	-	-	-	3.30	-	30.46	-	30.46	nr	34.27
casement stays	-	-	-	0.40	-	3.69	-	3.69	nr	4.15
quadrant stays	-	-	-	0.40	-	3.69	-	3.69	nr	4.15
barrel bolts; 150mm	-	-	-	0.45	-	4.15	-	4.15	nr	4.67
barrel bolts; 250mm	-	-	-	0.60	-	5.54	-	5.54	nr	6.23
door bolts; 150mm	-	-	-	0.45	-	4.15	-	4.15	nr	4.67
door bolts; 250mm	-	-	-	0.60	-	5.54	-	5.54	nr	6.23
monkey tail bolts										
300mm	-	-	-	0.55	-	5.08	-	5.08	nr	5.71
450mm	-	-	-	0.65	-	6.00	-	6.00	nr	6.75
600mm	-	-	-	0.75	-	6.92	-	6.92	nr	7.79
flush bolts; 200mm	-	-	-	1.10	-	10.15	-	10.15	nr	11.42
flush bolts; 450mm	-	-	-	1.65	-	15.23	-	15.23	nr	17.13
indicating bolts	-	-	-	1.65	-	15.23	-	15.23	nr	17.13
panic bolts; to single door	-	-	-	2.45	-	22.61	-	22.61	nr	25.44
panic bolts; to double doors	-	-	-	3.30	-	30.46	-	30.46	nr	34.27
knobs	-	-	-	0.35	-	3.23	-	3.23	nr	3.63
lever handles	-	-	-	0.35	-	3.23	-	3.23	nr	3.63
sash lifts	-	-	-	0.35	-	3.23	-	3.23	nr	3.63
pull handles; 150mm	-	-	-	0.30	-	2.77	-	2.77	nr	3.12
pull handles; 225mm	-	-	-	0.35	-	3.23	-	3.23	nr	3.63
pull handles; 300mm	-	-	-	0.40	-	3.69	-	3.69	nr	4.15
back plates	-	-	-	0.40	-	3.69	-	3.69	nr	4.15
escutcheon plates	-	-	-	0.35	-	3.23	-	3.23	nr	3.63
kicking plates	-	-	-	0.80	-	7.38	-	7.38	nr	8.31
letter plates	-	-	-	2.45	-	22.61	-	22.61	nr	25.44
push plates; 225mm	-	-	-	0.35	-	3.23	-	3.23	nr	3.63
push plates; 300mm	-	-	-	0.40	-	3.69	-	3.69	nr	4.15
shelf brackets	-	-	-	0.45	-	4.15	-	4.15	nr	4.67
sash cleats	-	-	-	0.35	-	3.23	-	3.23	nr	3.63

Labour hourly rates: (except Specialists) Craft Operatives £9.23 Labourer £7.02 Rates are national average prices. Refer to REGIONAL VARIATIONS for indicative levels of overall pricing in regions	MATERIALS			LABOUR				RATES		
	Del to Site	Waste	Material Cost	Craft Optve	Lab	Labour Cost	Sunds	Nett Rate	Unit	Gross Rate (+12.5%)
	£	%	£	£	£	£	£	£		£
P21: IRONMONGERY Cont.										
Fix only ironmongery (prime cost sum for supply included elsewhere) Cont.										
To hardwood and brickwork										
cabin hooks	-	-	-	0.40	-	3.69	-	3.69	nr	4.15
Water bars; steel; galvanized										
Water bars; to concrete										
flat section; setting in groove in mastic										
25 x 3 x 900mm long.........................	2.20	5.00	2.31	0.75	0.09	7.55	0.45	10.31	nr	11.60
40 x 3 x 900mm long.........................	3.50	5.00	3.67	0.75	0.09	7.55	0.55	11.78	nr	13.25
40 x 6 x 900mm long.........................	5.03	5.00	5.28	0.75	0.09	7.55	0.55	13.39	nr	15.06
50 x 6 x 900mm long.........................	6.78	5.00	7.12	0.75	0.09	7.55	0.55	15.22	nr	17.13
Water bars; to hardwood										
flat section; setting in groove in mastic										
25 x 3 x 900mm long.........................	2.20	5.00	2.31	0.65	0.08	6.56	0.45	9.32	nr	10.49
40 x 3 x 900mm long.........................	3.50	5.00	3.67	0.65	0.08	6.56	0.55	10.79	nr	12.13
40 x 6 x 900mm long.........................	5.03	5.00	5.28	0.65	0.08	6.56	0.55	12.39	nr	13.94
50 x 6 x 900mm long.........................	6.78	5.00	7.12	0.65	0.08	6.56	0.55	14.23	nr	16.01
Sliding door gear, P.C Henderson Ltd.										
To softwood										
interior straight sliding door gear sets for commercial and domestic doors; Senator single door set comprising track, hangers, end stops and bottom guide; for doors 20 - 35mm thick, maximum weight 25kg maximum 900mm wide	19.48	2.50	19.97	1.65	-	15.23	-	35.20	nr	39.60
extra for pelmet 1855mm long	15.49	2.50	15.88	0.70	-	6.46	-	22.34	nr	25.13
extra for pelmet end cap	1.69	2.50	1.73	0.11	-	1.02	-	2.75	nr	3.09
extra for Doorseal deflector guide	7.25	2.50	7.43	0.40	-	3.69	-	11.12	nr	12.51
interior straight sliding door gear sets for commercial and domestic doors; Phantom single door set comprising top assembly, hangers, adjustable nylon guide and door stops; for doors 30 - 50mm thick, maximum weight 45kg, 610 - 915mm wide	31.47	2.50	32.26	1.95	-	18.00	-	50.26	nr	56.54
extra for soffit fixing bracket and bolt	1.28	2.50	1.31	0.22	-	2.03	-	3.34	nr	3.76
extra for pelmet 1550mm long	12.95	2.50	13.27	0.60	-	5.54	-	18.81	nr	21.16
extra for pelmet 1855mm long	15.49	2.50	15.88	0.70	-	6.46	-	22.34	nr	25.13
extra for pelmet end cap	1.69	2.50	1.73	0.11	-	1.02	-	2.75	nr	3.09
extra for Doorseal deflector guide	7.25	2.50	7.43	0.40	-	3.69	-	11.12	nr	12.51
interior straight sliding door gear sets for commercial and domestic doors; Marathon Junior Nr J2 single door set comprising top assembly, hangers, end stops, inverted guide channel and nylon guide; for doors 32 - 50mm thick, maximum weight, 55kg, 400 - 750mm wide; forming groove for guide channel	29.22	2.50	29.95	2.65	-	24.46	-	54.41	nr	61.21
interior straight sliding door gear sets for commercial and domestic doors; Marathon Junior Nr J3 single door set comprising top assembly, hangers, end stops, inverted guide channel and nylon guide; for doors 32 - 50mm thick, maximum weight, 55kg, 750 - 900mm wide; forming groove for guide channel	30.83	2.50	31.60	2.85	-	26.31	-	57.91	nr	65.14
interior straight sliding door gear sets for commercial and domestic doors; Marathon Junior Nr J4 single door set comprising top assembly, hangers, end stops, inverted guide channel and nylon guide; for doors 32 - 50mm thick, maximum weight, 55kg, 900 - 1050mm wide; forming groove for guide channel	33.37	2.50	34.20	3.10	-	28.61	-	62.82	nr	70.67
interior straight sliding door gear sets for commercial and domestic doors; Marathon Junior Nr J5 single door set comprising top assembly, hangers, end stops, inverted guide channel and nylon guide; for doors 32 - 50mm thick, maximum weight, 55kg, 1050 - 1200mm wide; forming groove for guide channel	37.72	2.50	38.66	3.30	-	30.46	-	69.12	nr	77.76
interior straight sliding door gear sets for commercial and domestic doors; Marathon Junior Nr J6 single door set comprising top assembly, hangers, end stops, inverted guide channel and nylon guide; for doors 32 - 50mm thick, maximum weight, 55kg, 1200 - 1500mm wide; forming groove for guide channel	62.60	2.50	64.17	3.50	-	32.31	-	96.47	nr	108.53
extra for soffit fixing bracket and bolt	1.28	2.50	1.31	0.22	-	2.03	-	3.34	nr	3.76
extra for pelmet 1550mm long	12.95	2.50	13.27	0.60	-	5.54	-	18.81	nr	21.16
extra for pelmet 1855mm long	15.49	2.50	15.88	0.70	-	6.46	-	22.34	nr	25.13
extra for pelmet end cap	1.69	2.50	1.73	0.11	-	1.02	-	2.75	nr	3.09
extra for Doorseal deflector guide	7.25	2.50	7.43	0.40	-	3.69	-	11.12	nr	12.51
interior straight sliding door gear sets for commercial and domestic doors; Marathon Senior Nr S3 single door set comprising top assembly, hangers, end stops, inverted guide channel and nylon guide; for doors 32 - 50mm thick maximum weight 90kg, 750 - 900mm wide; forming groove for guide channel	41.37	2.50	42.40	3.10	-	28.61	-	71.02	nr	79.89

BUILDING FABRIC SUNDRIES

BUILDING FABRIC SUNDRIES – SMALL WORKS

Labour hourly rates: (except Specialists) Craft Operatives £9.23 Labourer £7.02 Rates are national average prices. Refer to REGIONAL VARIATIONS for indicative levels of overall pricing in regions	MATERIALS			LABOUR				RATES		
	Del to Site £	Waste %	Material Cost £	Craft Optve £	Lab £	Labour Cost £	Sunds £	Nett Rate £	Unit	Gross Rate (+12.5%) £
P21: IRONMONGERY Cont.										
Sliding door gear, P.C Henderson Ltd. Cont.										
To softwood Cont.										
interior straight sliding door gear sets for commercial and domestic doors; Marathon Senior Nr S4 single door set comprising top assembly, hangers, end stops, inverted guide channel and nylon guide; for doors 32 - 50mm thick maximum weight 90kg, 900 - 1050mm wide; forming groove for guide channel	45.94	2.50	47.09	3.30	-	30.46	-	77.55	nr	87.24
interior straight sliding door gear sets for commercial and domestic doors; Marathon Senior Nr S5 single door set comprising top assembly, hangers, end stops, inverted guide channel and nylon guide; for doors 32 - 50mm thick maximum weight 90kg, 1050 - 1200mm wide; forming groove for guide channel	50.69	2.50	51.96	3.50	-	32.31	-	84.26	nr	94.80
interior straight sliding door gear sets for commercial and domestic doors; Marathon Senior Nr S6 single door set comprising top assembly, hangers, end stops, inverted guide channel and nylon guide; for doors 32 - 50mm thick maximum weight 90kg, 1200 - 1500mm wide; forming groove for guide channel	81.28	2.50	83.31	3.65	-	33.69	-	117.00	nr	131.63
extra for soffit fixing bracket and bolt	1.28	2.50	1.31	0.22	-	2.03	-	3.34	nr	3.76
extra for pelmet 1550mm long	12.95	2.50	13.27	0.60	-	5.54	-	18.81	nr	21.16
extra for pelmet 1855mm long	15.49	2.50	15.88	0.70	-	6.46	-	22.34	nr	25.13
extra for pelmet end cap	1.69	2.50	1.73	0.11	-	1.02	-	2.75	nr	3.09
extra for Doorseal deflector guide	7.25	2.50	7.43	0.35	-	3.23	-	10.66	nr	11.99
interior straight sliding door gear sets for wardrobe and cupboard doors; Single Top Nr ST12 single door set comprising track, hangers, guides and safety stop; for door 16 - 35mm thick, maximum weight 25kg, maximum 900mm wide; one door to 600mm wide opening	11.89	2.50	12.19	1.40	-	12.92	-	25.11	nr	28.25
interior straight sliding door gear sets for wardrobe and cupoard doors; Single Top Nr ST15 single door set comprising track, hangers, guides and safety stop; for door 16 - 35mm thick, maximum weight 25kg, maximum 900mm wide; one door to 750mm wide opening	13.40	2.50	13.73	1.50	-	13.85	-	27.58	nr	31.03
interior straight sliding door gear sets for wardrobe and cupboard doors; Single Top Nr ST18 single door set comprising track, hangers, guides and safety stop; for door 16 - 35mm thick, maximum weight 25kg, maximum 900mm wide; one door to 900mm wide opening	15.17	2.50	15.55	1.60	-	14.77	-	30.32	nr	34.11
interior straight sliding door gear sets for wardrobe and cupboard doors; Double Top Nr W12 bi-passing door set comprising double track section, hangers, guides and safety stop; for doors 16 - 35mm thick, maximum weight 25kg, maximum 900mm wide; two doors in 1200mm wide opening	20.39	2.50	20.90	2.20	-	20.31	-	41.21	nr	46.36
interior straight sliding door gear sets for wardrobe and cupboard doors; Double Top Nr W15 bi-passing door set comprising double track section, hangers, guides and safety stop; for doors 16 - 35mm thick, maximum weight 25kg, maximum 900mm wide; two doors in 1500mm wide opening	22.74	2.50	23.31	2.30	-	21.23	-	44.54	nr	50.10
interior straight sliding door gear sets for wardrobe and cupboard doors; Double Top Nr W18 bi-passing door set comprising double track section, hangers, guides and safety stop; for doors 16 - 35mm thick, maximum weight 25kg, maximum 900mm wide; two doors in 1800mm wide opening	25.22	2.50	25.85	2.40	-	22.15	-	48.00	nr	54.00
interior straight sliding door gear sets for wardrobe and cupboard doors; Double Top Nr W24 bi-passing door set comprising double track section, hangers, guides and safety stop; for doors 16 - 35mm thick, maximum weight 25kg, maximum 900mm wide; three doors in 2400mm wide opening	33.07	2.50	33.90	2.65	-	24.46	-	58.36	nr	65.65
interior straight sliding door gear sets for wardrobe and cupboard doors; Bi-Fold Nr B10-2 folding door set comprising top guide track, top and bottom pivots; top guide and hinges; for doors 20 - 35mm thick; maximum weight 14kg each leaf, maximum 530mm wide; two doors in 1065mm wide opening	18.76	2.50	19.23	3.30	-	30.46	-	49.69	nr	55.90
interior straight sliding door gear sets for wardrobes and cupboard doors; Bi-Fold Nr B15-4 folding door set comprising top guide track, top and bottom pivots, top guide and hinges; for doors 20 - 35mm thick, maximum weight 14kg each leaf, maximum 530mm wide; four doors in 1525mm wide opening with aligner	31.82	2.50	32.62	6.60	-	60.92	-	93.53	nr	105.23
interior straight sliding door gear sets for wardrobes and cupboard doors; Bi-Fold Nr B20-4 folding door set comprising top guide track, top and bottom pivots, top guide and hinges; for doors 20 - 35mm thick, maximum weight 14kg each leaf, maximum 530mm wide; four doors in 2135mm wide opening with aligner	34.84	2.50	35.71	7.15	-	65.99	-	101.71	nr	114.42

Labour hourly rates: (except Specialists) Craft Operatives £9.23 Labourer £7.02 Rates are national average prices. Refer to REGIONAL VARIATIONS for indicative levels of overall pricing in regions	MATERIALS			LABOUR				RATES		
	Del to Site £	Waste %	Material Cost £	Craft Optve £	Lab £	Labour Cost £	Sunds £	Nett Rate £	Unit	Gross Rate (+12.5%) £

P21: IRONMONGERY Cont.

Sliding door gear, P.C Henderson Ltd. Cont.

To softwood Cont.

interior straight sliding door gear sets for built in cupboard doors; Slipper Nr SS4 double passing door set comprising two top tracks, sliders, safety stop, flush pulls and guides; for doors 16 - 30mm thick, maximum weight 9kg, maximum 900mm wide; two doors in 1200mm wide opening	14.38	2.50	14.74	2.05	-	18.92	-	33.66	nr	37.87
interior straight sliding door gear sets for built in cupboard doors; Slipper Nr SS5 double passing door set comprising two top tracks, sliders, safety stop, flush pulls and guides; for doors 16 - 30mm thick, maximum weight 9kg, maximum 900mm wide; two doors in 1500mm wide opening	16.92	2.50	17.34	2.20	-	20.31	-	37.65	nr	42.36
interior straight sliding door gear sets for built in cupboard doors; Slipper Nr SS6 double passing door set comprising two top tracks, sliders, safety stop, flush pulls and guides; for doors 16 - 30mm thick, maximum weight 9kg, maximum 900mm wide; two doors in 1800mm wide opening	18.90	2.50	19.37	2.35	-	21.69	-	41.06	nr	46.20
interior straight sliding door gear sets for cupboards, book cases and cabinet work; Loretto Nr D4 bi-passing door set comprising two top guide channels, two bottom rails, nylon guides and bottom rollers; for doors 20 - 45mm thick; maximum weight 23kg, maximum 900mm wide; two doors in 1200mm wide opening; forming grooves for top guide channels and bottom rails; forming sinkings for bottom rollers	24.37	2.50	24.98	2.75	-	25.38	-	50.36	nr	56.66
interior straight sliding door gear sets for cupboards, book cases and cabinet work; Loretto Nr D5 bi-passing door set comprising two top guide channels, two bottom rails, nylon guides and bottom rollers; for doors 20 - 45mm thick; maximum weight 23kg, maximum 900mm wide; two doors in 1500mm wide opening; forming grooves for top guide channels and bottom rails; forming sinkings for bottom rollers	27.12	2.50	27.80	2.85	-	26.31	-	54.10	nr	60.87
interior straight sliding door gear sets for cupboards, book cases and cabinet work; Loretto Nr D6 bi-passing door set comprising two top guide channels, two bottom rails, nylon guides and bottom rollers; for doors 20 - 45mm thick; maximum weight 23kg, maximum 900mm wide; two doors in 1800mm wide opening; forming grooves for top guide channels and bottom rails; forming sinkings for bottom rollers	30.64	2.50	31.41	3.00	-	27.69	-	59.10	nr	66.48
extra for retractable roller guide bolt	5.75	2.50	5.89	0.40	-	3.69	-	9.59	nr	10.78
extra for flush pull	2.50	2.50	2.56	0.30	-	2.77	-	5.33	nr	6.00
extra for safety stop	0.78	2.50	0.80	0.11	-	1.02	-	1.81	nr	2.04
extra for cylinder lock	12.03	2.50	12.33	0.65	-	6.00	-	18.33	nr	20.62
interior straight sliding/folding room divider gear sets; Husky Folding Nr HF-2 folding door set comprising top track, brackets, top and bottom pivots, hangers and hinges, bottom channel and roller guides; for doors 20 - 40mm thick, maximum weight 25kg, maximum 600mm wide, maximum 2400mm high; two doors in 1200mm wide opening; forming groove for bottom channel	51.74	2.50	53.03	6.40	-	59.07	-	112.11	nr	126.12
interior straight sliding/folding room divider gear sets; Husky Folding Nr HF-4 folding door set comprising top track, brackets, top and bottom pivots, hangers and hinges, bottom channel and roller guides; for doors 20 - 40mm thick, maximum weight 25kg, maximum 600mm wide, maximum 2400mm high; four doors in 2400mm wide opening; forming groove for bottom channel	97.93	2.50	100.38	12.65	-	116.76	-	217.14	nr	244.28
interior straight sliding door gear sets for glass panels; Zenith 2 Nr 12 double passing door set comprising double top guide, bottom rail, glass rail with rubber glazing strip, end caps and bottom rollers; for panels 6mm thick, maximum weight 16kg or 1m2 per panel; two panels nr 1200mm wide opening	36.99	2.50	37.91	1.65	-	15.23	-	53.14	nr	59.79
extra for dust seal	3.48	2.50	3.57	0.35	-	3.23	-	6.80	m	7.65
interior straight sliding door gear sets for glass panels; Zenith 2 Nr 15 double passing door set comprising double top guide, bottom rail, glass rail with rubber glazing strip, end caps and bottom rollers; for panels 6mm thick, maximum weight 16kg or 1m2 per panel; two panels nr 1500mm wide opening	42.60	2.50	43.66	1.80	-	16.61	-	60.28	nr	67.81
extra for dust seal	3.48	2.50	3.57	0.35	-	3.23	-	6.80	m	7.65
interior straight sliding door gear sets for glass panels; Zenith 2 Nr 18 double passing door set comprising double top guide, bottom rail, glass rail with rubber glazing strip, end caps and bottom rollers; for panels 6mm thick, maximum weight 16kg or 1m2 per panel; two panels nr 1800mm wide opening	47.70	2.50	48.89	2.00	-	18.46	-	67.35	nr	75.77
extra for finger pull	2.50	2.50	2.56	0.45	-	4.15	-	6.72	nr	7.56
extra for cylinder lock	12.03	2.50	12.33	1.10	-	10.15	-	22.48	nr	25.29
extra for dust seal	3.48	2.50	3.57	0.35	-	3.23	-	6.80	m	7.65

BUILDING FABRIC SUNDRIES

Labour hourly rates: (except Specialists) Craft Operatives £9.23 Labourer £7.02 Rates are national average prices. Refer to REGIONAL VARIATIONS for indicative levels of overall pricing in regions	MATERIALS			LABOUR				RATES		
	Del to Site £	Waste %	Material Cost £	Craft Optve £	Lab £	Labour Cost £	Sunds £	Nett Rate £	Unit	Gross Rate (+12.5%) £

P21: IRONMONGERY Cont.

Sliding door gear, P.C Henderson Ltd. Cont.

To softwood and concrete

exterior straight sliding bottom roller gear; Sterling 225 components comprising top guide reference 900, top guide brackets for face fixing for single run reference 1/900, top guide rollers for face fixing reference 54/900, bottom rollers reference 5 and bottom rail reference 299; for timber doors 44 - 50mm thick, maximum weight 225kg, maximum 3300mm high; single door in 1200mm wide opening; forming notches for bottom rollers; casting bottom rail into concrete	244.81	2.50	250.93	6.75	-	62.30	-	313.23	nr	352.39
exterior staight sliding bottom roller gear; Sterling 225 components comprising top guide reference 900, top guide brackets for soffit fixing for single run reference 3/900, top guide rollers for edge fixing reference 203/900, bottom rollers reference 5 and bottom rail reference 299; for timber doors 44 - 50mm thick, maximum weight 225kg, maximum 3300mm high; single door in 1200mm wide opening; forming notches for bottom rollers; casting bottom rail into concrete	264.80	2.50	271.42	7.00	-	64.61	-	336.03	nr	378.03
exterior straight sliding bottom roller gear; Sterling 225 components comprising top guide reference 900, top guide brackets for face fixing for single run reference 1/900, top guide rollers for face fixing reference 54/900, bottom rollers reference 5 and bottom rail reference 299; for timber doors 44 - 50mm thick, maximum weight 225kg, maximum 3300mm high; two doors in 2400mm wide opening; forming notches for bottom rollers; casting bottom rails into concrete	482.36	2.50	494.42	13.20	-	121.84	-	616.26	nr	693.29
exterior straight sliding bottom roller gear; Sterling 225 components comprising top guides reference 900, top guide brackets for face fixing for double run reference 5/900, top guide rollers for face fixing reference 54/900, bottom rollers reference 5 and bottom rails reference 299; for timber doors 44 - 50mm thick, maximum weight 225kg, maximum 3300 high; two doors in 2400mm wide opening; forming notches for bottom rollers; casting bottom rails into concrete	548.65	2.50	562.37	13.60	-	125.53	-	687.89	nr	773.88
exterior straight sliding bottom roller gear; Sterling 350 components comprising top guide reference 99, top guide brackets for face fixing for single run reference 31 and 31x, top guide rollers reference 53/99, bottom rollers reference 2 and bottom rail reference 299; for timber doors 44 - 50mm thick, maximum weight 350kg, maximum 4000mm high; single door in 2400mm wide opening; forming notches for bottom rollers; casting bottom rail into concrete	542.66	2.50	556.23	14.20	-	131.07	-	687.29	nr	773.20
exterior straight sliding bottom roller gear; Sterling 350 components comprising top guide reference 99, top guide brackets for face fixing for single run reference 31 and 31X, top guide rollers reference 53/99, bottom rollers reference 2 and bottom rail reference 299; for timber doors 44 - 50mm thick, maximum weight 350kg, maximum 4000mm high; two doors in 3600mm wide opening; forming notches for bottom rollers; casting bottom rail into concrete	911.00	2.50	933.77	18.00	-	166.14	-	1099.92	nr	1237.40
exterior straight sliding bottom roller gear; Sterling 800 components comprising top guide reference 99, top guide brackets for face fixing for single run reference 31 and 31X, top guide rollers reference 53/99, bottom rollers reference 3 and bottom rail reference 298; for timber doors 54 - 63mm thick, maximum weight 800kg, maximum 5200mm high; single door in 3600mm wide opening; forming notches for bottom rollers; casting bottom rail into concrete	785.29	2.50	804.92	16.10	-	148.60	-	953.53	nr	1072.72
exterior straight sliding bottom roller gear, Sterling 800 components comprising top guide, reference 99, top guide brackets for face fixing for single run reference 31 and 31X, top guide rollers reference 53/99, bottom rollers reference 3 and bottom rail reference 298; for timber doors 54 - 63mm thick, maximum weight 800kg, maximum 5200mm high; two doors in 5400mm wide opening; forming notches for bottom rollers; casting bottom rail into concrete	1292.94	2.50	1325.26	25.90	-	239.06	-	1564.32	nr	1759.86

Hinges; standard quality

To softwood
backflap hinges; steel

25mm	0.25	2.50	0.26	0.15	-	1.38	-	1.64	nr	1.85
38mm	0.30	2.50	0.31	0.17	-	1.57	-	1.88	nr	2.11
50mm	0.39	2.50	0.40	0.17	-	1.57	-	1.97	nr	2.21
63mm	1.14	2.50	1.17	0.17	-	1.57	-	2.74	nr	3.08
75mm	1.82	2.50	1.87	0.17	-	1.57	-	3.43	nr	3.86

BUILDING FABRIC SUNDRIES — SMALL WORKS

Labour hourly rates: (except Specialists) Craft Operatives £9.23 Labourer £7.02 Rates are national average prices. Refer to REGIONAL VARIATIONS for indicative levels of overall pricing in regions	MATERIALS			LABOUR				RATES		
	Del to Site	Waste	Material Cost	Craft Optve	Lab	Labour Cost	Sunds	Nett Rate	Unit	Gross Rate (+12.5%)
	£	%	£	£	£	£	£	£		£
P21: IRONMONGERY Cont.										
Hinges; standard quality Cont.										
To softwood Cont.										
butt hinges; steel; light medium pattern										
38mm	0.15	2.50	0.15	0.17	-	1.57	-	1.72	nr	1.94
50mm	0.15	2.50	0.15	0.17	-	1.57	-	1.72	nr	1.94
63mm	0.18	2.50	0.18	0.17	-	1.57	-	1.75	nr	1.97
75mm	0.18	2.50	0.18	0.17	-	1.57	-	1.75	nr	1.97
100mm	0.37	2.50	0.38	0.17	-	1.57	-	1.95	nr	2.19
butt hinges; steel; strong pattern										
75mm	0.70	2.50	0.72	0.17	-	1.57	-	2.29	nr	2.57
100mm	0.77	2.50	0.79	0.17	-	1.57	-	2.36	nr	2.65
butt hinges; cast iron; light										
50mm	0.92	2.50	0.94	0.17	-	1.57	-	2.51	nr	2.83
63mm	1.03	2.50	1.06	0.17	-	1.57	-	2.62	nr	2.95
75mm	1.14	2.50	1.17	0.17	-	1.57	-	2.74	nr	3.08
100mm	1.24	2.50	1.27	0.17	-	1.57	-	2.84	nr	3.20
butt hinges; brass, brass pin										
38mm	0.87	2.50	0.89	0.17	-	1.57	-	2.46	nr	2.77
50mm	1.00	2.50	1.02	0.17	-	1.57	-	2.59	nr	2.92
63mm	1.18	2.50	1.21	0.17	-	1.57	-	2.78	nr	3.13
75mm	1.28	2.50	1.31	0.17	-	1.57	-	2.88	nr	3.24
100mm	2.95	2.50	3.02	0.17	-	1.57	-	4.59	nr	5.17
butt hinges; brass, steel washers, steel pin										
75mm	2.04	2.50	2.09	0.17	-	1.57	-	3.66	nr	4.12
100mm	3.49	2.50	3.58	0.17	-	1.57	-	5.15	nr	5.79
butt hinges; brass, B.M.A., steel washers, steel pin										
75mm	3.85	2.50	3.95	0.17	-	1.57	-	5.52	nr	6.20
100mm	6.26	2.50	6.42	0.17	-	1.57	-	7.99	nr	8.98
butt hinges; brass, chromium plated, steel washers, steel pin										
75mm	3.85	2.50	3.95	0.17	-	1.57	-	5.52	nr	6.20
100mm	6.73	2.50	6.90	0.17	-	1.57	-	8.47	nr	9.53
butt hinges; aluminium, stainless steel pin and washers										
75mm	3.01	2.50	3.09	0.17	-	1.57	-	4.65	nr	5.24
100mm	3.73	2.50	3.82	0.17	-	1.57	-	5.39	nr	6.07
rising hinges; steel										
75mm	1.12	2.50	1.15	0.30	-	2.77	-	3.92	nr	4.41
100mm	1.51	2.50	1.55	0.30	-	2.77	-	4.32	nr	4.86
rising hinges; cast iron										
75mm	2.17	2.50	2.22	0.30	-	2.77	-	4.99	nr	5.62
100mm	3.18	2.50	3.26	0.30	-	2.77	-	6.03	nr	6.78
spring hinges; steel, japanned; single action										
100mm	14.06	2.50	14.41	0.57	-	5.26	-	19.67	nr	22.13
125mm	15.81	2.50	16.21	0.62	-	5.72	-	21.93	nr	24.67
150mm	17.44	2.50	17.88	0.67	-	6.18	-	24.06	nr	27.07
spring hinges; steel, japanned; double action										
75mm	15.20	2.50	15.58	0.67	-	6.18	-	21.76	nr	24.48
100mm	17.44	2.50	17.88	0.67	-	6.18	-	24.06	nr	27.07
125mm	20.44	2.50	20.95	0.77	-	7.11	-	28.06	nr	31.57
150mm	23.44	2.50	24.03	0.77	-	7.11	-	31.13	nr	35.02
tee hinges; japanned; light										
150mm	0.39	2.50	0.40	0.45	-	4.15	-	4.55	nr	5.12
230mm	0.57	2.50	0.58	0.50	-	4.62	-	5.20	nr	5.85
300mm	0.75	2.50	0.77	0.55	-	5.08	-	5.85	nr	6.58
375mm	1.07	2.50	1.10	0.60	-	5.54	-	6.63	nr	7.46
450mm	1.47	2.50	1.51	0.65	-	6.00	-	7.51	nr	8.44
tee hinges; self colour; heavy										
150mm	0.68	2.50	0.70	0.50	-	4.62	-	5.31	nr	5.98
230mm	0.92	2.50	0.94	0.55	-	5.08	-	6.02	nr	6.77
300mm	1.47	2.50	1.51	0.60	-	5.54	-	7.04	nr	7.93
375mm	2.23	2.50	2.29	0.65	-	6.00	-	8.29	nr	9.32
450mm	2.82	2.50	2.89	0.70	-	6.46	-	9.35	nr	10.52
600mm	5.29	2.50	5.42	0.75	-	6.92	-	12.34	nr	13.89
hook and band hinges; on plate; heavy										
300mm	2.71	2.50	2.78	0.90	-	8.31	-	11.08	nr	12.47
450mm	4.44	2.50	4.55	1.10	-	10.15	-	14.70	nr	16.54
610mm	7.04	2.50	7.22	1.30	-	12.00	-	19.22	nr	21.62
914mm	12.33	2.50	12.64	1.65	-	15.23	-	27.87	nr	31.35
collinge hinges; cup for wood; best quality										
610mm	33.41	2.50	34.25	1.00	-	9.23	-	43.48	nr	48.91
762mm	41.17	2.50	42.20	1.25	-	11.54	-	53.74	nr	60.45
914mm	52.58	2.50	53.89	1.55	-	14.31	-	68.20	nr	76.73
parliament hinges; steel; 100mm	3.78	2.50	3.87	0.30	-	2.77	-	6.64	nr	7.47
cellar flap hinges; wrought, welded, plain joint; 450mm	24.64	2.50	25.26	0.55	-	5.08	-	30.33	nr	34.12
To hardwood or the like										
backflap hinges; steel										
25mm	0.24	2.50	0.25	0.25	-	2.31	-	2.55	nr	2.87
38mm	0.30	2.50	0.31	0.25	-	2.31	-	2.62	nr	2.94
50mm	0.39	2.50	0.40	0.25	-	2.31	-	2.71	nr	3.05
63mm	1.14	2.50	1.17	0.25	-	2.31	-	3.48	nr	3.91
75mm	1.82	2.50	1.87	0.25	-	2.31	-	4.17	nr	4.69
butt hinges; steel; light medium pattern										
38mm	0.15	2.50	0.15	0.25	-	2.31	-	2.46	nr	2.77
50mm	0.15	2.50	0.15	0.25	-	2.31	-	2.46	nr	2.77
63mm	0.16	2.50	0.16	0.25	-	2.31	-	2.47	nr	2.78
75mm	0.16	2.50	0.16	0.25	-	2.31	-	2.47	nr	2.78
100mm	0.35	2.50	0.36	0.25	-	2.31	-	2.67	nr	3.00
butt hinges; steel; strong pattern										
75mm	0.70	2.50	0.72	0.25	-	2.31	-	3.02	nr	3.40

BUILDING FABRIC SUNDRIES

Labour hourly rates: (except Specialists) Craft Operatives £9.23 Labourer £7.02 Rates are national average prices. Refer to REGIONAL VARIATIONS for indicative levels of overall pricing in regions	MATERIALS			LABOUR				RATES		
	Del to Site £	Waste %	Material Cost £	Craft Optve £	Lab £	Labour Cost £	Sunds £	Nett Rate £	Unit	Gross Rate (+12.5%) £
P21: IRONMONGERY Cont.										
Hinges; standard quality Cont.										
To hardwood or the like Cont.										
butt hinges; steel; strong pattern Cont.										
100mm	0.77	2.50	0.79	0.25	-	2.31	-	3.10	nr	3.48
butt hinges; cast iron; light										
50mm	0.92	2.50	0.94	0.25	-	2.31	-	3.25	nr	3.66
63mm	1.03	2.50	1.06	0.25	-	2.31	-	3.36	nr	3.78
75mm	1.14	2.50	1.17	0.25	-	2.31	-	3.48	nr	3.91
100mm	1.24	2.50	1.27	0.25	-	2.31	-	3.58	nr	4.03
butt hinges; brass, brass pin										
38mm	0.87	2.50	0.89	0.25	-	2.31	-	3.20	nr	3.60
50mm	1.00	2.50	1.02	0.25	-	2.31	-	3.33	nr	3.75
63mm	1.18	2.50	1.21	0.25	-	2.31	-	3.52	nr	3.96
75mm	1.28	2.50	1.31	0.25	-	2.31	-	3.62	nr	4.07
100mm	2.95	2.50	3.02	0.25	-	2.31	-	5.33	nr	6.00
butt hinges; brass, steel washers, steel pin										
75mm	2.04	2.50	2.09	0.25	-	2.31	-	4.40	nr	4.95
100mm	3.49	2.50	3.58	0.25	-	2.31	-	5.88	nr	6.62
butt hinges; brass, B.M.A., steel washers, steel pin										
75mm	3.85	2.50	3.95	0.25	-	2.31	-	6.25	nr	7.04
100mm	6.73	2.50	6.90	0.25	-	2.31	-	9.21	nr	10.36
butt hinges; brass, chromium plated, steel washers, steel pin										
75mm	3.85	2.50	3.95	0.25	-	2.31	-	6.25	nr	7.04
100mm	6.73	2.50	6.90	0.25	-	2.31	-	9.21	nr	10.36
butt hinges; aluminium, stainless steel pin and washers										
75mm	3.01	2.50	3.09	0.25	-	2.31	-	5.39	nr	6.07
100mm	3.73	2.50	3.82	0.25	-	2.31	-	6.13	nr	6.90
rising hinges; steel										
75mm	1.12	2.50	1.15	0.40	-	3.69	-	4.84	nr	5.45
100mm	1.51	2.50	1.55	0.40	-	3.69	-	5.24	nr	5.89
rising hinges; cast iron										
75mm	2.17	2.50	2.22	0.40	-	3.69	-	5.92	nr	6.66
100mm	3.18	2.50	3.26	0.40	-	3.69	-	6.95	nr	7.82
spring hinges; steel, japanned; single action										
100mm	14.06	2.50	14.41	0.85	-	7.85	-	22.26	nr	25.04
125mm	15.81	2.50	16.21	0.90	-	8.31	-	24.51	nr	27.58
150mm	17.44	2.50	17.88	0.95	-	8.77	-	26.64	nr	29.98
spring hinges; steel, japanned; double action										
75mm	15.20	2.50	15.58	0.90	-	8.31	-	23.89	nr	26.87
100mm	17.44	2.50	17.88	0.95	-	8.77	-	26.64	nr	29.98
125mm	20.44	2.50	20.95	1.05	-	9.69	-	30.64	nr	34.47
150mm	23.44	2.50	24.03	1.10	-	10.15	-	34.18	nr	38.45
tee hinges; japanned; light										
150mm	0.39	2.50	0.40	0.60	-	5.54	-	5.94	nr	6.68
230mm	0.57	2.50	0.58	0.65	-	6.00	-	6.58	nr	7.41
300mm	0.75	2.50	0.77	0.75	-	6.92	-	7.69	nr	8.65
375mm	1.11	2.50	1.14	0.80	-	7.38	-	8.52	nr	9.59
450mm	1.47	2.50	1.51	0.90	-	8.31	-	9.81	nr	11.04
tee hinges; self colour; heavy										
150mm	0.68	2.50	0.70	0.65	-	6.00	-	6.70	nr	7.53
230mm	0.92	2.50	0.94	0.70	-	6.46	-	7.40	nr	8.33
300mm	1.47	2.50	1.51	0.80	-	7.38	-	8.89	nr	10.00
375mm	2.23	2.50	2.29	0.85	-	7.85	-	10.13	nr	11.40
450mm	2.82	2.50	2.89	0.90	-	8.31	-	11.20	nr	12.60
600mm	5.28	2.50	5.41	1.00	-	9.23	-	14.64	nr	16.47
hook and band hinges; on plate; heavy										
300mm	2.71	2.50	2.78	1.20	-	11.08	-	13.85	nr	15.59
450mm	4.44	2.50	4.55	1.55	-	14.31	-	18.86	nr	21.21
610mm	7.04	2.50	7.22	1.90	-	17.54	-	24.75	nr	27.85
914mm	12.33	2.50	12.64	2.20	-	20.31	-	32.94	nr	37.06
collinge hinges; cup for wood; best quality										
610mm	33.41	2.50	34.25	1.45	-	13.38	-	47.63	nr	53.58
762mm	41.17	2.50	42.20	1.85	-	17.08	-	59.27	nr	66.68
914mm	52.58	2.50	53.89	2.25	-	20.77	-	74.66	nr	83.99
parliament hinges; steel; 100mm	3.78	2.50	3.87	0.40	-	3.69	-	7.57	nr	8.51
cellar flap hinges; wrought, welded, plain joint; 450mm	24.64	2.50	25.26	0.85	-	7.85	-	33.10	nr	37.24
Floor springs; standard quality										
To softwood										
single action floor springs and top centres; B.M.A.	204.34	2.50	209.45	2.75	-	25.38	-	234.83	nr	264.18
single action floor springs and top centres; chromium plated	204.34	2.50	209.45	2.75	-	25.38	-	234.83	nr	264.18
double action floor springs and top centres; B.M.A.	246.40	2.50	252.56	3.30	-	30.46	-	283.02	nr	318.40
double action floor springs and top centres; chromium plated	246.40	2.50	252.56	3.30	-	30.46	-	283.02	nr	318.40
To hardwood or the like										
single action floor springs and top centres; B.M.A.	204.34	2.50	209.45	4.10	-	37.84	-	247.29	nr	278.20
single action floor springs and top centres; chromium plated	204.34	2.50	209.45	4.10	-	37.84	-	247.29	nr	278.20
double action floor springs and top centres; B.M.A.	246.40	2.50	252.56	4.90	-	45.23	-	297.79	nr	335.01
double action floor springs and top centres; chromium plated	246.40	2.50	252.56	4.90	-	45.23	-	297.79	nr	335.01

BUILDING FABRIC SUNDRIES – SMALL WORKS

Labour hourly rates: (except Specialists) Craft Operatives £9.23 Labourer £7.02 Rates are national average prices. Refer to REGIONAL VARIATIONS for indicative levels of overall pricing in regions	MATERIALS			LABOUR				RATES		
	Del to Site	Waste	Material Cost	Craft Optve	Lab	Labour Cost	Sunds	Nett Rate	Unit	Gross Rate (+12.5%)
	£	%	£	£	£	£	£	£		£
P21: IRONMONGERY Cont.										
Door closers; standard quality										
To softwood										
coil door springs; japanned	4.32	2.50	4.43	0.30	–	2.77	–	7.20	nr	8.10
overhead door closers; liquid check and spring; 'Briton' 2000 series; silver; light doors	70.28	2.50	72.04	1.65	–	15.23	–	87.27	nr	98.17
overhead door closers; liquid check and spring; 'Briton' 2000 series; silver; medium doors	81.73	2.50	83.77	1.65	–	15.23	–	99.00	nr	111.38
overhead door closers; liquid check and spring; 'Briton' 2000 series; silver; heavy doors	105.17	2.50	107.80	1.65	–	15.23	–	123.03	nr	138.41
concealed overhead door closers; liquid check and spring; medium doors	117.19	2.50	120.12	2.75	–	25.38	–	145.50	nr	163.69
Perko door closers; brass plate	7.33	2.50	7.51	1.65	–	15.23	–	22.74	nr	25.59
To hardwood or the like										
coil door springs; japanned	4.32	2.50	4.43	0.45	–	4.15	–	8.58	nr	9.65
overhead door closers; liquid check and spring; 'Briton' 2000 series; silver; light doors	70.28	2.50	72.04	2.45	–	22.61	–	94.65	nr	106.48
overhead door closers; liquid check and spring; 'Briton' 2000 series; silver; medium doors	81.73	2.50	83.77	2.45	–	22.61	–	106.39	nr	119.69
overhead door closers; liquid check and spring; 'Briton' 2000 series; silver; heavy doors	105.17	2.50	107.80	2.45	–	22.61	–	130.41	nr	146.71
concealed overhead door closers; liquid check and spring; medium doors	117.19	2.50	120.12	4.10	–	37.84	–	157.96	nr	177.71
Perko door closers; brass plate	7.33	2.50	7.51	2.45	–	22.61	–	30.13	nr	33.89
Door selectors; standard quality										
To softwood or the like										
Union 8815 door selector; aluminium anodised silver finish	52.66	2.50	53.98	2.20	0.28	22.27	–	76.25	nr	85.78
Close Rite door selector; satin nickel plated	39.31	2.50	40.29	2.50	0.31	25.25	–	65.54	nr	73.74
To hardwood or the like										
Union 8815 door selector; aluminium anodised silver finish	52.66	2.50	53.98	3.30	0.41	33.34	–	87.31	nr	98.23
Close Rite door selector; satin nickel plated	39.31	2.50	40.29	3.90	0.49	39.44	–	79.73	nr	89.70
Locks and latches; standard quality										
To softwood										
magnetic catches; 6lb pull	0.57	2.50	0.58	0.25	–	2.31	–	2.89	nr	3.25
magnetic catches; 9lb pull	0.70	2.50	0.72	0.25	–	2.31	–	3.02	nr	3.40
magnetic catches; 13lb pull	1.17	2.50	1.20	0.25	–	2.31	–	3.51	nr	3.95
roller catches; mortice; nylon; 18mm	1.17	2.50	1.20	0.40	–	3.69	–	4.89	nr	5.50
roller catches; mortice; nylon; 27mm	1.17	2.50	1.20	0.40	–	3.69	–	4.89	nr	5.50
roller catches; surface, adjustable; nylon; 16mm .	2.04	2.50	2.09	0.30	–	2.77	–	4.86	nr	5.47
roller catches; mortice, adjustable; satin chrome plated; 25 x 22 x 10mm	4.98	2.50	5.10	0.45	–	4.15	–	9.26	nr	10.42
roller catches; double ball; brass; 43mm	0.76	2.50	0.78	0.35	–	3.23	–	4.01	nr	4.51
mortice latches; stamped steel case; 75mm	1.12	2.50	1.15	0.70	–	6.46	–	7.61	nr	8.56
mortice latches; locking; stamped steel case; 75mm	2.23	2.50	2.29	0.70	–	6.46	–	8.75	nr	9.84
cylinder mortice latches; 'Union' key operated outside, knob inside, bolt held by slide; B.M.A.; for end set to suit 13mm rebate	43.93	2.50	45.03	0.80	–	7.38	–	52.41	nr	58.96
cylinder mortice latches; 'Union' key operated outside, knob inside, bolt held by slide; chromium plated; for end set to suit 13mm rebate	43.93	2.50	45.03	0.80	–	7.38	–	52.41	nr	58.96
cylinder rim night latches; 'Legge' 707; chromium plated	14.31	2.50	14.67	1.10	–	10.15	–	24.82	nr	27.92
cylinder rim night latches; 'Union' 1022; chromium plated	16.71	2.50	17.13	1.10	–	10.15	–	27.28	nr	30.69
cylinder rim night latches; 'Yale' 88; chromium plated	18.57	2.50	19.03	1.10	–	10.15	–	29.19	nr	32.84
Suffolk latches; japanned; medium; size Nr 2	2.23	2.50	2.29	0.90	–	8.31	–	10.59	nr	11.92
Suffolk latches; japanned; medium; size Nr 3	2.88	2.50	2.95	0.90	–	8.31	–	11.26	nr	12.67
Suffolk latches; japanned; heavy; size Nr 4	4.69	2.50	4.81	0.90	–	8.31	–	13.11	nr	14.75
Suffolk latches; galvanised; size Nr 4	4.98	2.50	5.10	0.90	–	8.31	–	13.41	nr	15.09
cupboard locks; 1 lever; japanned; 63mm	4.39	2.50	4.50	0.80	–	7.38	–	11.88	nr	13.37
cupboard locks; 2 lever; brass; 50mm	5.95	2.50	6.10	0.80	–	7.38	–	13.48	nr	15.17
cupboard locks; 2 lever; brass; 63mm	5.95	2.50	6.10	0.80	–	7.38	–	13.48	nr	15.17
cupboard locks; 4 lever; brass; 63mm	5.78	2.50	5.92	0.80	–	7.38	–	13.31	nr	14.97
rim dead lock; japanned case; 100mm	3.78	2.50	3.87	0.90	–	8.31	–	12.18	nr	13.70
rim dead lock; japanned case; 125mm	4.39	2.50	4.50	0.90	–	8.31	–	12.81	nr	14.41
rim dead lock; japanned case; 150mm	5.29	2.50	5.42	0.90	–	8.31	–	13.73	nr	15.45
mortice dead locks; japanned case; 75mm	5.53	2.50	5.67	0.90	–	8.31	–	13.98	nr	15.72
mortice locks; three levers	4.32	2.50	4.43	1.10	–	10.15	–	14.58	nr	16.40
mortice locks; five levers	11.42	2.50	11.71	1.10	–	10.15	–	21.86	nr	24.59
rim locks; japanned case; 150 x 100mm	5.29	2.50	5.42	0.90	–	8.31	–	13.73	nr	15.45
rim locks; japanned case; 150 x 75mm	5.90	2.50	6.05	0.90	–	8.31	–	14.35	nr	16.15
rim locks; japanned case; strong pattern; 150 x 100mm	7.93	2.50	8.13	0.90	–	8.31	–	16.44	nr	18.49
padlocks, 'Squire', anti-pilfer bolt, warded spring; zinc plated; 32mm	1.51	2.50	1.55	–	–	–	–	1.55	nr	1.74
padlocks, 'Squire', anti-pilfer bolt, warded spring; zinc plated; 38mm	2.34	2.50	2.40	–	–	–	–	2.40	nr	2.70
padlocks, 'Squire', anti-pilfer bolt, warded spring; zinc plated; 44mm	2.59	2.50	2.65	–	–	–	–	2.65	Unr	2.99
padlocks, 'Squire', anti-pilfer bolt, 4 pin tumblers; zinc plated; 32mm	4.75	2.50	4.87	–	–	–	–	4.87	nr	5.48
padlocks, 'Squire', anti-pilfer bolt, 4 pin tumblers; zinc plated; 38mm	6.26	2.50	6.42	–	–	–	–	6.42	nr	7.22
padlocks, 'Squire', anti-pilfer bolt, 4 pin tumblers; zinc plated; 44mm	7.39	2.50	7.57	–	–	–	–	7.57	nr	8.52

BUILDING FABRIC SUNDRIES

Labour hourly rates: (except Specialists) Craft Operatives £9.23 Labourer £7.02 Rates are national average prices. Refer to REGIONAL VARIATIONS for indicative levels of overall pricing in regions	MATERIALS			LABOUR				RATES		
	Del to Site	Waste	Material Cost	Craft Optve	Lab	Labour Cost	Sunds	Nett Rate	Unit	Gross Rate (+12.5%)
	£	%	£	£	£	£	£	£		£
P21: IRONMONGERY Cont.										
Locks and latches; standard quality Cont.										
To softwood Cont.										
padlocks; 'Squire', 4 levers; galvanised case	3.73	2.50	3.82	-	-	-	-	3.82	nr	4.30
padlocks, 'Squire', 4 levers; brass case	4.81	2.50	4.93	-	-	-	-	4.93	nr	5.55
padlocks; 'Squire', strong pattern; chromium plated..	32.21	2.50	33.02	-	-	-	-	33.02	nr	37.14
hasps and staples; japanned wire; light; 75mm	0.62	2.50	0.64	0.35	-	3.23	-	3.87	nr	4.35
hasps and staples; japanned wire; light; 100mm ...	0.73	2.50	0.75	0.35	-	3.23	-	3.98	nr	4.48
hasps and staples; galvanised wire; heavy; 75mm .	1.34	2.50	1.37	0.35	-	3.23	-	4.60	nr	5.18
hasps and staples; galvanised wire; heavy; 100mm	1.45	2.50	1.49	0.35	-	3.23	-	4.72	nr	5.31
hasps and staples; galvanised wire; heavy; 150mm .	1.51	2.50	1.55	0.35	-	3.23	-	4.78	nr	5.38
locking bars; japanned; 200mm	5.65	2.50	5.79	0.50	-	4.62	-	10.41	nr	11.71
locking bars; japanned; 250mm	5.90	2.50	6.05	0.50	-	4.62	-	10.66	nr	12.00
locking bars; japanned; 300mm	7.46	2.50	7.65	0.50	-	4.62	-	12.26	nr	13.79
locking bars; japanned; 350mm	8.78	2.50	9.00	0.50	-	4.62	-	13.61	nr	15.32
To hardwood or the like										
magnetic catches; 6lb pull	0.57	2.50	0.58	0.35	-	3.23	-	3.81	nr	4.29
magnetic catches; 9lb pull	0.70	2.50	0.72	0.35	-	3.23	-	3.95	nr	4.44
magnetic catches; 13lb pull	1.17	2.50	1.20	0.35	-	3.23	-	4.43	nr	4.98
roller catches; mortice; nylon; 18mm	1.17	2.50	1.20	0.55	-	5.08	-	6.28	nr	7.06
roller catches; mortice; nylon; 27mm	1.17	2.50	1.20	0.55	-	5.08	-	6.28	nr	7.06
roller catches; surface, adjustable; nylon; 16mm .	2.04	2.50	2.09	0.45	-	4.15	-	6.24	nr	7.03
roller catches; mortice, adjustable; satin chrome plated; 25 x 22 x 10mm	4.98	2.50	5.10	0.65	-	6.00	-	11.10	nr	12.49
roller catches; double ball; brass; 43mm	0.76	2.50	0.78	0.50	-	4.62	-	5.39	nr	6.07
mortice latches; stamped steel case; 75mm	1.12	2.50	1.15	1.10	-	10.15	-	11.30	nr	12.71
mortice latches; locking; stamped steel case; 75mm	2.23	2.50	2.29	1.10	-	10.15	-	12.44	nr	13.99
cylinder mortice latches; 'Union' key operated outside, knob inside, bolt held by slide; B.M.A.; for end set to suit 13mm rebate	43.93	2.50	45.03	1.20	-	11.08	-	56.10	nr	63.12
cylinder mortice latches; 'Union' key operated outside, knob inside, bolt held by slide; chromium plated; for end set to suit 13mm rebate	43.93	2.50	45.03	1.20	-	11.08	-	56.10	nr	63.12
cylinder rim night latches; 'Legge' 707; chromium plated..	14.31	2.50	14.67	1.65	-	15.23	-	29.90	nr	33.63
cylinder rim night latches; 'Union' 1022; chromium plated..	16.71	2.50	17.13	1.65	-	15.23	-	32.36	nr	36.40
cylinder rim night latches; 'Yale' 88; chromium plated..	18.57	2.50	19.03	1.65	-	15.23	-	34.26	nr	38.55
Suffolk latches; japanned; medium; size Nr 2	2.23	2.50	2.29	1.30	-	12.00	-	14.28	nr	16.07
Suffolk latches; japanned; medium; size Nr 3	2.88	2.50	2.95	1.30	-	12.00	-	14.95	nr	16.82
Suffolk latches; japanned; medium; size Nr 4	4.69	2.50	4.81	1.30	-	12.00	-	16.81	nr	18.91
Suffolk latches; japanned; heavy; size Nr 4	4.98	2.50	5.10	1.30	-	12.00	-	17.10	nr	19.24
Suffolk latches; galvanised; size Nr 4	3.73	2.50	3.82	1.30	-	12.00	-	15.82	nr	17.80
cupboard locks; 1 lever; japanned; 63mm	4.39	2.50	4.50	1.20	-	11.08	-	15.58	nr	17.52
cupboard locks; 2 lever; brass; 50mm	5.95	2.50	6.10	1.20	-	11.08	-	17.17	nr	19.32
cupboard locks; 2 lever; brass; 63mm	5.95	2.50	6.10	1.20	-	11.08	-	17.17	nr	19.32
cupboard locks; 4 lever; brass; 63mm	6.58	2.50	6.74	1.20	-	11.08	-	17.82	nr	20.05
rim dead lock; japanned case; 100mm	3.98	2.50	4.08	1.30	-	12.00	-	16.08	nr	18.09
rim dead lock; japanned case; 125mm	4.39	2.50	4.50	1.30	-	12.00	-	16.50	nr	18.56
rim dead lock; japanned case; 150mm	5.29	2.50	5.42	1.30	-	12.00	-	17.42	nr	19.60
mortice dead locks; japanned case; 75mm	5.53	2.50	5.67	1.30	-	12.00	-	17.67	nr	19.88
mortice locks; three levers	4.32	2.50	4.43	1.65	-	15.23	-	19.66	nr	22.11
mortice locks; five levers	11.42	2.50	11.71	1.65	-	15.23	-	26.93	nr	30.30
rim locks; japanned case; 150 x 100mm	5.29	2.50	5.42	1.30	-	12.00	-	17.42	nr	19.60
rim locks; japanned case; 150 x 75mm	5.90	2.50	6.05	1.30	-	12.00	-	18.05	nr	20.30
rim locks; japanned case; strong pattern; 150 x 100mm..	7.93	2.50	8.13	1.30	-	12.00	-	20.13	nr	22.64
padlocks, 'Squire', anti-pilfer bolt, warded spring; zinc plated; 32mm	1.51	2.50	1.55	-	-	-	-	1.55	nr	1.74
padlocks, 'Squire', anti-pilfer bolt, warded spring; zinc plated; 38mm	2.34	2.50	2.40	-	-	-	-	2.40	nr	2.70
padlocks, 'Squire', anti-pilfer bolt, warded spring; zinc plated; 44mm	2.59	2.50	2.65	-	-	-	-	2.65	nr	2.99
padlocks, 'Squire', anti-pilfer bolt, 4 pin tumblers; zinc plated; 32mm	4.75	2.50	4.87	-	-	-	-	4.87	nr	5.48
padlocks, 'Squire', anti-pilfer bolt, 4 pin tumblers; zinc plated; 38mm	6.26	2.50	6.42	-	-	-	-	6.42	nr	7.22
padlocks, 'Squire', anti-pilfer bolt, 4 pin tumblers; zinc plated; 44mm	7.39	2.50	7.57	-	-	-	-	7.57	nr	8.52
padlocks, 'Squire', 4 levers; galvanised case	3.73	2.50	3.82	-	-	-	-	3.82	nr	4.30
padlocks, 'Squire', 4 levers; brass case	4.81	2.50	4.93	-	-	-	-	4.93	nr	5.55
padlocks, 'Squire', strong pattern; chromium plated..	32.21	2.50	33.02	-	-	-	-	33.02	nr	37.14
hasps and staples; japanned wire; light; 75mm	0.62	2.50	0.64	0.50	-	4.62	-	5.25	nr	5.91
hasps and staples; japanned wire; light; 100mm ...	0.73	2.50	0.75	0.55	-	5.08	-	5.82	nr	6.55
hasps and staples; galvanised wire; heavy; 75mm ..	1.34	2.50	1.37	0.55	-	5.08	-	6.45	nr	7.26
hasps and staples; galvanised wire; heavy; 100mm .	1.45	2.50	1.49	0.55	-	5.08	-	6.56	nr	7.38
hasps and staples; galvanised wire; heavy; 150mm .	1.51	2.50	1.55	0.55	-	5.08	-	6.62	nr	7.45
locking bars; japanned; 200mm	5.65	2.50	5.79	0.65	-	6.00	-	11.79	nr	13.26
locking bars; japanned; 250mm	5.90	2.50	6.05	0.65	-	6.00	-	12.05	nr	13.55
locking bars; japanned; 300mm	7.46	2.50	7.65	0.65	-	6.00	-	13.65	nr	15.35
locking bars; japanned; 350mm	8.78	2.50	9.00	0.65	-	6.00	-	15.00	nr	16.87
Bolts; standard quality										
To softwood										
barrel bolts; japanned, steel barrel; medium										
100mm	1.14	2.50	1.17	0.30	-	2.77	-	3.94	nr	4.43
150mm	1.45	2.50	1.49	0.35	-	3.23	-	4.72	nr	5.31
200mm	1.87	2.50	1.92	0.40	-	3.69	-	5.61	nr	6.31

Labour hourly rates: (except Specialists) Craft Operatives £9.23 Labourer £7.02 Rates are national average prices. Refer to REGIONAL VARIATIONS for indicative levels of overall pricing in regions	MATERIALS			LABOUR				RATES		
	Del to Site £	Waste %	Material Cost £	Craft Optve £	Lab £	Labour Cost £	Sunds £	Nett Rate £	Unit	Gross Rate (+12.5%) £
P21: IRONMONGERY Cont.										
Bolts; standard quality Cont.										
To softwood Cont.										
barrel bolts; japanned, steel barrel; heavy										
100mm	2.88	2.50	2.95	0.30	-	2.77	-	5.72	nr	6.44
150mm	3.49	2.50	3.58	0.35	-	3.23	-	6.81	nr	7.66
200mm	4.21	2.50	4.32	0.40	-	3.69	-	8.01	nr	9.01
250mm	4.98	2.50	5.10	0.45	-	4.15	-	9.26	nr	10.42
300mm	5.90	2.50	6.05	0.50	-	4.62	-	10.66	nr	12.00
barrel bolts; extruded brass, round brass shoot; 25mm wide										
75mm	1.14	2.50	1.17	0.30	-	2.77	-	3.94	nr	4.43
100mm	1.45	2.50	1.49	0.30	-	2.77	-	4.26	nr	4.79
150mm	1.74	2.50	1.78	0.35	-	3.23	-	5.01	nr	5.64
barrel bolts; extruded brass; B.M.A., round brass shoot; 25mm wide										
75mm	1.56	2.50	1.60	0.30	-	2.77	-	4.37	nr	4.91
100mm	1.87	2.50	1.92	0.30	-	2.77	-	4.69	nr	5.27
150mm	2.34	2.50	2.40	0.35	-	3.23	-	5.63	nr	6.33
barrel bolts; extruded brass, chromium plated, round brass shoot; 25mm wide										
75mm	1.56	2.50	1.60	0.30	-	2.77	-	4.37	nr	4.91
100mm	1.87	2.50	1.92	0.30	-	2.77	-	4.69	nr	5.27
150mm	2.34	2.50	2.40	0.35	-	3.23	-	5.63	nr	6.33
barrel bolts; extruded aluminium, S.A.A., round aluminium shoot; 25mm wide										
75mm	1.14	2.50	1.17	0.30	-	2.77	-	3.94	nr	4.43
100mm	1.25	2.50	1.28	0.30	-	2.77	-	4.05	nr	4.56
150mm	1.45	2.50	1.49	0.35	-	3.23	-	4.72	nr	5.31
monkey tail bolts										
japanned; 300mm	8.78	2.50	9.00	0.40	-	3.69	-	12.69	nr	14.28
japanned; 450mm	11.72	2.50	12.01	0.45	-	4.15	-	16.17	nr	18.19
japanned; 610mm	14.06	2.50	14.41	0.55	-	5.08	-	19.49	nr	21.92
flush bolts; B.M.A.										
100mm	7.57	2.50	7.76	0.55	-	5.08	-	12.84	nr	14.44
150mm	10.93	2.50	11.20	0.65	-	6.00	-	17.20	nr	19.35
200mm	18.76	2.50	19.23	0.75	-	6.92	-	26.15	nr	29.42
flush bolts; S.C.P.										
100mm	8.17	2.50	8.37	0.55	-	5.08	-	13.45	nr	15.13
150mm	11.12	2.50	11.40	0.65	-	6.00	-	17.40	nr	19.57
200mm	17.55	2.50	17.99	0.75	-	6.92	-	24.91	nr	28.03
lever action flush bolts; B.M.A.										
150mm	9.37	2.50	9.60	0.65	-	6.00	-	15.60	nr	17.55
200 x 25mm	9.98	2.50	10.23	0.75	-	6.92	-	17.15	nr	19.30
lever action flush bolts; S.A.A.										
150mm	7.24	2.50	7.42	0.65	-	6.00	-	13.42	nr	15.10
200 x 25mm	7.82	2.50	8.02	0.75	-	6.92	-	14.94	nr	16.81
necked bolts; extruded brass, round brass shoot; 25mm wide										
75mm	1.45	2.50	1.49	0.30	-	2.77	-	4.26	nr	4.79
100mm	1.57	2.50	1.61	0.30	-	2.77	-	4.38	nr	4.93
150mm	2.10	2.50	2.15	0.35	-	3.23	-	5.38	nr	6.06
necked bolts; extruded aluminium, S.A.A., round aluminium shoot; 25mm wide										
75mm	1.03	2.50	1.06	0.30	-	2.77	-	3.82	nr	4.30
100mm	1.14	2.50	1.17	0.30	-	2.77	-	3.94	nr	4.43
150mm	1.45	2.50	1.49	0.35	-	3.23	-	4.72	nr	5.31
indicating bolts; S.A.A.	6.26	2.50	6.42	1.10	-	10.15	-	16.57	nr	18.64
indicating bolts; B.M.A.	6.03	2.50	6.18	1.10	-	10.15	-	16.33	nr	18.38
panic bolts; to single door; iron, bronzed or silver	58.59	2.50	60.05	1.65	-	15.23	-	75.28	nr	84.69
panic bolts; to single door; aluminium box, steel shoots and cross rail, anodised silver	58.29	2.50	59.75	1.65	-	15.23	-	74.98	nr	84.35
panic bolts; to double doors; iron bronzed or silver	67.72	2.50	69.41	2.20	-	20.31	-	89.72	nr	100.93
panic bolts; to double doors; aluminium box, steel shoots and cross rail, anodised silver	67.72	2.50	69.41	2.20	-	20.31	-	89.72	nr	100.93
padlock bolts; galvanised; heavy										
150mm	3.43	2.50	3.52	0.35	-	3.23	-	6.75	nr	7.59
200mm	3.66	2.50	3.75	0.35	-	3.23	-	6.98	nr	7.85
250mm	4.39	2.50	4.50	0.50	-	4.62	-	9.11	nr	10.25
300mm	0.50	2.50	0.51	-	-	-	-	0.51	nr	0.58
To hardwood or the like										
barrel bolts; japanned, steel barrel; medium										
100mm	1.14	2.50	1.17	0.40	-	3.69	-	4.86	nr	5.47
150mm	1.45	2.50	1.49	0.45	-	4.15	-	5.64	nr	6.34
200mm	1.87	2.50	1.92	0.55	-	5.08	-	6.99	nr	7.87
barrel bolts; japanned, steel barrel; heavy										
100mm	2.88	2.50	2.95	0.40	-	3.69	-	6.64	nr	7.47
150mm	3.49	2.50	3.58	0.45	-	4.15	-	7.73	nr	8.70
200mm	4.21	2.50	4.32	0.55	-	5.08	-	9.39	nr	10.57
250mm	4.98	2.50	5.10	0.60	-	5.54	-	10.64	nr	11.97
300mm	5.90	2.50	6.05	0.70	-	6.46	-	12.51	nr	14.07
barrel bolts; extruded brass, round brass shoot; 25mm wide										
75mm	1.14	2.50	1.17	0.40	-	3.69	-	4.86	nr	5.47
100mm	1.45	2.50	1.49	0.40	-	3.69	-	5.18	nr	5.83
150mm	1.74	2.50	1.78	0.45	-	4.15	-	5.94	nr	6.68
barrel bolts; extruded brass; B.M.A., round brass shoot; 25mm wide										
75mm	1.56	2.50	1.60	0.40	-	3.69	-	5.29	nr	5.95
100mm	1.87	2.50	1.92	0.40	-	3.69	-	5.61	nr	6.31
150mm	2.34	2.50	2.40	0.45	-	4.15	-	6.55	nr	7.37

BUILDING FABRIC SUNDRIES

Labour hourly rates: (except Specialists) Craft Operatives £9.23 Labourer £7.02 Rates are national average prices. Refer to REGIONAL VARIATIONS for indicative levels of overall pricing in regions	MATERIALS			LABOUR				RATES		
	Del to Site £	Waste %	Material Cost £	Craft Optve £	Lab £	Labour Cost £	Sunds £	Nett Rate £	Unit	Gross Rate (+12.5%) £
P21: IRONMONGERY Cont.										
Bolts; standard quality Cont.										
To hardwood or the like Cont.										
barrel bolts; extruded brass, chromium plated, round brass shoot; 25mm wide										
75mm	1.56	2.50	1.60	0.40	-	3.69	-	5.29	nr	5.95
100mm	1.87	2.50	1.92	0.40	-	3.69	-	5.61	nr	6.31
150mm	2.34	2.50	2.40	0.45	-	4.15	-	6.55	nr	7.37
barrel bolts; extruded aluminium, S.A.A., round aluminium shoot; 25mm wide										
75mm	1.14	2.50	1.17	0.40	-	3.69	-	4.86	nr	5.47
100mm	1.25	2.50	1.28	0.40	-	3.69	-	4.97	nr	5.59
150mm	1.45	2.50	1.49	0.45	-	4.15	-	5.64	nr	6.34
monkey tail bolts										
japanned; 300mm	8.78	2.50	9.00	0.55	-	5.08	-	14.08	nr	15.84
japanned; 450mm	11.72	2.50	12.01	0.65	-	6.00	-	18.01	nr	20.26
japanned; 610mm	14.06	2.50	14.41	0.75	-	6.92	-	21.33	nr	24.00
flush bolts; B.M.A.										
100mm	7.57	2.50	7.76	0.90	-	8.31	-	16.07	nr	18.07
150mm	10.93	2.50	11.20	1.00	-	9.23	-	20.43	nr	22.99
200mm	18.76	2.50	19.23	1.10	-	10.15	-	29.38	nr	33.05
flush bolts; S.C.P.										
100mm	8.17	2.50	8.37	0.90	-	8.31	-	16.68	nr	18.77
150mm	11.12	2.50	11.40	1.00	-	9.23	-	20.63	nr	23.21
200mm	17.55	2.50	17.99	1.10	-	10.15	-	28.14	nr	31.66
lever action flush bolts; B.M.A.										
150mm	9.37	2.50	9.60	1.00	-	9.23	-	18.83	nr	21.19
200 x 25mm	9.98	2.50	10.23	1.10	-	10.15	-	20.38	nr	22.93
lever action flush bolts; S.A.A.										
150mm	7.24	2.50	7.42	1.00	-	9.23	-	16.65	nr	18.73
200 x 25mm	7.82	2.50	8.02	1.10	-	10.15	-	18.17	nr	20.44
necked bolts; extruded brass, round brass shoot; 25mm wide										
75mm	1.45	2.50	1.49	0.40	-	3.69	-	5.18	nr	5.83
100mm	1.57	2.50	1.61	0.40	-	3.69	-	5.30	nr	5.96
150mm	2.10	2.50	2.15	0.45	-	4.15	-	6.31	nr	7.09
necked bolts; extruded aluminium, S.A.A., round aluminium shoot; 25mm wide										
75mm	1.03	2.50	1.06	0.40	-	3.69	-	4.75	nr	5.34
100mm	1.14	2.50	1.17	0.40	-	3.69	-	4.86	nr	5.47
150mm	1.45	2.50	1.49	0.45	-	4.15	-	5.64	nr	6.34
indicating bolts; S.A.A.	6.26	2.50	6.42	1.65	-	15.23	-	21.65	nr	24.35
indicating bolts; B.M.A.	6.03	2.50	6.18	1.65	-	15.23	-	21.41	nr	24.09
panic bolts; to single door; iron, bronzed or silver	58.59	2.50	60.05	2.45	-	22.61	-	82.67	nr	93.00
panic bolts; to single door; aluminium box, steel shoots and cross rail, anodised silver	58.29	2.50	59.75	2.45	-	22.61	-	82.36	nr	92.66
panic bolts; to double doors; iron bronzed or silver	67.72	2.50	69.41	3.30	-	30.46	-	99.87	nr	112.36
panic bolts; to double doors; aluminium box, steel shoots and cross rail, anodised silver	67.72	2.50	69.41	3.30	-	30.46	-	99.87	nr	112.36
padlock bolts; galvanised; heavy										
150mm	3.43	2.50	3.52	0.50	-	4.62	-	8.13	nr	9.15
200mm	3.66	2.50	3.75	0.50	-	4.62	-	8.37	nr	9.41
250mm	4.37	2.50	4.48	0.75	-	6.92	-	11.40	nr	12.83
300mm	5.41	2.50	5.55	0.75	-	6.92	-	12.47	nr	14.03
Door handles; standard quality										
To softwood										
lever handles; spring action; B.M.A., best quality; 41 x 150mm	27.06	2.50	27.74	0.25	-	2.31	-	30.04	nr	33.80
lever handles; spring action; chromium plated, housing quality; 41 x 150mm	7.04	2.50	7.22	0.25	-	2.31	-	9.52	nr	10.71
lever handles; spring action; chromium plated, best quality; 41 x 150mm	24.64	2.50	25.26	0.25	-	2.31	-	27.56	nr	31.01
lever handles; spring action; S.A.A. housing quality; 41 x 150mm	4.27	2.50	4.38	0.25	-	2.31	-	6.68	nr	7.52
lever handles; spring action; S.A.A., best quality; 41 x 150mm	16.27	2.50	16.68	0.25	-	2.31	-	18.98	nr	21.36
lever handles; spring action; plastic/nylon, housing quality; 41 x 150mm	3.43	2.50	3.52	0.25	-	2.31	-	5.82	nr	6.55
pull handles; B.M.A.; 150mm	9.86	2.50	10.11	0.20	-	1.85	-	11.95	nr	13.45
pull handles; B.M.A.; 225mm	18.47	2.50	18.93	0.25	-	2.31	-	21.24	nr	23.89
pull handles; B.M.A.; 300mm	20.79	2.50	21.31	0.30	-	2.77	-	24.08	nr	27.09
pull handles; S.A.A.; 150mm	3.78	2.50	3.87	0.20	-	1.85	-	5.72	nr	6.44
pull handles; S.A.A.; 225mm	3.90	2.50	4.00	0.25	-	2.31	-	6.30	nr	7.09
pull handles; S.A.A.; 300mm	7.57	2.50	7.76	0.30	-	2.77	-	10.53	nr	11.84
pull handles on 50 x 250mm plate, lettered; B.M.A.; 225mm	42.68	2.50	43.75	0.45	-	4.15	-	47.90	nr	53.89
pull handles on 50 x 250mm plate, lettered; S.A.A.; 225mm	29.88	2.50	30.63	0.45	-	4.15	-	34.78	nr	39.13
To hardwood or the like										
lever handles; spring action; B.M.A., best quality; 41 x 150mm	27.06	2.50	27.74	0.35	-	3.23	-	30.97	nr	34.84
lever handles; spring action; chromium plated, housing quality; 41 x 150mm	7.04	2.50	7.22	0.35	-	3.23	-	10.45	Unit	11.75
lever handles; spring action; chromium plated, best quality; 41 x 150mm	24.64	2.50	25.26	0.35	-	3.23	-	28.49	nr	32.05
lever handles; spring action; S.A.A. housing quality; 41 x 150mm	4.27	2.50	4.38	0.35	-	3.23	-	7.61	nr	8.56
lever handles; spring action; S.A.A., best quality; 41 x 150mm	16.27	2.50	16.68	0.35	-	3.23	-	19.91	nr	22.40

Labour hourly rates: (except Specialists) Craft Operatives £9.23 Labourer £7.02 Rates are national average prices. Refer to REGIONAL VARIATIONS for indicative levels of overall pricing in regions	MATERIALS			LABOUR				RATES		
	Del to Site £	Waste %	Material Cost £	Craft Optve £	Lab £	Labour Cost £	Sunds £	Nett Rate £	Unit	Gross Rate (+12.5%) £
P21: IRONMONGERY Cont.										
Door handles; standard quality Cont.										
To hardwood or the like Cont.										
lever handles; spring action; plastic/nylon,										
housing quality; 41 x 150mm	3.43	2.50	3.52	0.35	–	3.23	–	6.75	nr	7.59
pull handles; B.M.A.; 150mm	9.86	2.50	10.11	0.30	–	2.77	–	12.88	nr	14.48
pull handles; B.M.A.; 225mm	18.47	2.50	18.93	0.35	–	3.23	–	22.16	nr	24.93
pull handles; B.M.A.; 300mm	20.79	2.50	21.31	0.40	–	3.69	–	25.00	nr	28.13
pull handles; S.A.A.; 150mm	3.78	2.50	3.87	0.30	–	2.77	–	6.64	nr	7.47
pull handles; S.A.A.; 225mm	5.90	2.50	6.05	0.35	–	3.23	–	9.28	nr	10.44
pull handles; S.A.A.; 300mm	7.57	2.50	7.76	0.40	–	3.69	–	11.45	nr	12.88
pull handles on 50 x 250mm plate, lettered;										
B.M.A.; 225mm	42.68	2.50	43.75	0.65	–	6.00	–	49.75	nr	55.96
pull handles on 50 x 250mm plate, lettered;										
S.A.A.; 225mm	29.88	2.50	30.63	0.65	–	6.00	–	36.63	nr	41.20
Door furniture; standard quality										
To softwood										
knobs; real B.M.A.; surface fixing	9.08	2.50	9.31	0.25	–	2.31	–	11.61	nr	13.07
knobs; S.A.A.; surface fixing	5.29	2.50	5.42	0.25	–	2.31	–	7.73	nr	8.70
knobs; real B.M.A.; secret fixing	14.67	2.50	15.04	0.35	–	3.23	–	18.27	nr	20.55
knobs; S.A.A.; secret fixing	7.63	2.50	7.82	0.35	–	3.23	–	11.05	nr	12.43
To hardwood or the like										
knobs; real B.M.A.; surface fixing	9.08	2.50	9.31	0.35	–	3.23	–	12.54	nr	14.10
knobs; S.A.A.; surface fixing	5.29	2.50	5.42	0.35	–	3.23	–	8.65	nr	9.73
knobs; real B.M.A.; secret fixing	14.67	2.50	15.04	0.50	–	4.62	–	19.65	nr	22.11
knobs; S.A.A.; secret fixing	7.63	2.50	7.82	0.50	–	4.62	–	12.44	nr	13.99
Window furniture; standard quality										
To softwood										
fanlight catches; brass	5.41	2.50	5.55	0.40	–	3.69	–	9.24	nr	10.39
fanlight catches; B.M.A.	5.90	2.50	6.05	0.40	–	3.69	–	9.74	nr	10.96
fanlight catches; chromium plated	6.01	2.50	6.16	0.40	–	3.69	–	9.85	nr	11.08
casement fasteners; wedge plate; black malleable										
iron	1.14	2.50	1.17	0.40	–	3.69	–	4.86	nr	5.47
casement fasteners; wedge plate; B.M.A.	2.65	2.50	2.72	0.40	–	3.69	–	6.41	nr	7.21
casement fasteners; wedge plate; chromium plated .	2.65	2.50	2.72	0.40	–	3.69	–	6.41	nr	7.21
casement fasteners; wedge plate; S.A.A.	2.10	2.50	2.15	0.40	–	3.69	–	5.84	nr	6.58
sash fasteners; brass; 70mm	6.07	2.50	6.22	1.10	–	10.15	–	16.37	nr	18.42
sash fasteners; B.M.A.; 70mm	6.49	2.50	6.65	1.10	–	10.15	–	16.81	nr	18.91
sash fasteners; chromium plated; 70mm	7.15	2.50	7.33	1.10	–	10.15	–	17.48	nr	19.67
casement stays; two pins; grey malleable iron										
200mm	1.51	2.50	1.55	0.40	–	3.69	–	5.24	nr	5.89
250mm	1.62	2.50	1.66	0.40	–	3.69	–	5.35	nr	6.02
300mm	1.74	2.50	1.78	0.40	–	3.69	–	5.48	nr	6.16
casement stays; two pins; B.M.A.										
200mm	3.49	2.50	3.58	0.40	–	3.69	–	7.27	nr	8.18
250mm	3.78	2.50	3.87	0.40	–	3.69	–	7.57	nr	8.51
300mm	4.09	2.50	4.19	0.40	–	3.69	–	7.88	nr	8.87
casement stays; two pins; chromium plated										
200mm	3.49	2.50	3.58	0.40	–	3.69	–	7.27	nr	8.18
250mm	3.78	2.50	3.87	0.40	–	3.69	–	7.57	nr	8.51
300mm	4.09	2.50	4.19	0.40	–	3.69	–	7.88	nr	8.87
casement stays; two pins; S.A.A.										
200mm	1.93	2.50	1.98	0.40	–	3.69	–	5.67	nr	6.38
250mm	2.14	2.50	2.19	0.40	–	3.69	–	5.89	nr	6.62
300mm	2.23	2.50	2.29	0.40	–	3.69	–	5.98	nr	6.72
sash lifts; polished brass; 50mm.............	1.20	2.50	1.23	0.25	–	2.31	–	3.54	nr	3.98
sash lifts; B.M.A.; 50mm...................	1.45	2.50	1.49	0.25	–	2.31	–	3.79	nr	4.27
sash lifts; chromium plated; 50mm...........	1.62	2.50	1.66	0.25	–	2.31	–	3.97	nr	4.46
flush lifts; brass; 75mm....................	1.56	2.50	1.60	0.55	–	5.08	–	6.68	nr	7.51
flush lifts; B.M.A.; 75mm..................	1.87	2.50	1.92	0.55	–	5.08	–	6.99	nr	7.87
flush lifts; chromium plated; 75mm..........	1.87	2.50	1.92	0.55	–	5.08	–	6.99	nr	7.87
sash cleats; polished brass; 76mm...........	1.87	2.50	1.92	0.25	–	2.31	–	4.22	nr	4.75
sash cleats; B.M.A.; 76mm.................	1.93	2.50	1.98	0.25	–	2.31	–	4.29	nr	4.82
sash cleats; chromium plated; 76mm.........	2.34	2.50	2.40	0.25	–	2.31	–	4.71	nr	5.29
sash pulleys; frame and wheel...............	2.40	2.50	2.46	0.55	–	5.08	–	7.54	nr	8.48
To hardwood or the like										
fanlight catches; brass	5.41	2.50	5.55	0.55	–	5.08	–	10.62	nr	11.95
fanlight catches; B.M.A.	5.90	2.50	6.05	0.55	–	5.08	–	11.12	nr	12.51
fanlight catches; chromium plated	6.01	2.50	6.16	0.55	–	5.08	–	11.24	nr	12.64
casement fasteners; wedge plate; black malleable										
iron	1.14	2.50	1.17	0.55	–	5.08	–	6.25	nr	7.03
casement fasteners; wedge plate; B.M.A.	2.65	2.50	2.72	0.55	–	5.08	–	7.79	nr	8.77
casement fasteners; wedge plate; chromium plated .	2.65	2.50	2.72	0.55	–	5.08	–	7.79	nr	8.77
casement fasteners; wedge plate; S.A.A.	2.10	2.50	2.15	0.55	–	5.08	–	7.23	nr	8.13
sash fasteners; brass; 70mm	6.07	2.50	6.22	1.65	–	15.23	–	21.45	nr	24.13
sash fasteners; B.M.A.; 70mm	6.49	2.50	6.65	1.65	–	15.23	–	21.88	nr	24.62
sash fasteners; chromium plated; 70mm	7.15	2.50	7.33	1.65	–	15.23	–	22.56	nr	25.38
casement stays; two pins; grey malleable iron										
200mm	1.51	2.50	1.55	0.55	–	5.08	–	6.62	nr	7.45
250mm	1.62	2.50	1.66	0.55	–	5.08	–	6.74	nr	7.58
300mm	1.74	2.50	1.78	0.55	–	5.08	–	6.86	nr	7.72
casement stays; two pins; B.M.A.										
200mm	3.49	2.50	3.58	0.55	–	5.08	–	8.65	nr	9.74
250mm	3.78	2.50	3.87	0.55	–	5.08	–	8.95	nr	10.07
300mm	4.09	2.50	4.19	0.55	–	5.08	–	9.27	nr	10.43
casement stays; two pins; chromium plated										
200mm	3.49	2.50	3.58	0.55	–	5.08	–	8.65	nr	9.74
250mm	3.78	2.50	3.87	0.55	–	5.08	–	8.95	nr	10.07

Labour hourly rates: (except Specialists) Craft Operatives £9.23 Labourer £7.02 Rates are national average prices. Refer to REGIONAL VARIATIONS for indicative levels of overall pricing in regions	MATERIALS			LABOUR				RATES		
	Del to Site £	Waste %	Material Cost £	Craft Optve £	Lab £	Labour Cost £	Sunds £	Nett Rate £	Unit	Gross Rate (+12.5%) £
P21: IRONMONGERY Cont.										
Window furniture; standard quality Cont.										
To hardwood or the like Cont.										
casement stays; two pins; chromium plated Cont.										
300mm	4.09	2.50	4.19	0.55	–	5.08	–	9.27	nr	10.43
casement stays; two pins; S.A.A.										
200mm	1.93	2.50	1.98	0.55	–	5.08	–	7.05	nr	7.94
250mm	2.14	2.50	2.19	0.55	–	5.08	–	7.27	nr	8.18
300mm	2.23	2.50	2.29	0.55	–	5.08	–	7.36	nr	8.28
sash lifts; polished brass; 50mm..............	1.20	2.50	1.23	0.35	–	3.23	–	4.46	nr	5.02
sash lifts; B.M.A.; 50mm..................	1.45	2.50	1.49	0.35	–	3.23	–	4.72	nr	5.31
sash lifts; chromium plated; 50mm..........	1.62	2.50	1.66	0.35	–	3.23	–	4.89	nr	5.50
flush lifts; brass; 75mm.................	1.56	2.50	1.60	0.80	–	7.38	–	8.98	nr	10.11
flush lifts; B.M.A.; 75mm...............	1.87	2.50	1.92	0.80	–	7.38	–	9.30	nr	10.46
flush lifts; chromium plated; 75mm......	1.87	2.50	1.92	0.80	–	7.38	–	9.30	nr	10.46
sash cleats; polished brass; 76mm........	1.87	2.50	1.92	0.35	–	3.23	–	5.15	nr	5.79
sash cleats; B.M.A.; 76mm...............	1.93	2.50	1.98	0.35	–	3.23	–	5.21	nr	5.86
sash cleats; chromium plated; 76mm......	2.34	2.50	2.40	0.35	–	3.23	–	5.63	nr	6.33
sash pulleys; frame and wheel..............	2.40	2.50	2.46	0.80	–	7.38	–	9.84	nr	11.07
Window furniture; Willan Building Services Ltd										
To softwood or the like										
Glidevale frame vent; aluminium with epoxy paint finish										
reference TV FV2; 198mm long; fixing with screws and clip on covers	7.08	2.50	7.26	0.22	0.03	2.24	0.17	9.67	nr	10.88
reference TV FV4; 358mm long; fixing with screws and clip on covers	8.54	2.50	8.75	0.33	0.04	3.33	0.20	12.28	nr	13.82
Glidevale canopy grille; aluminium with epoxy paint finish										
reference TV CG2; 155 x 15mm slot size; fixing with screws and clip on covers	3.54	2.50	3.63	0.22	0.03	2.24	0.08	5.95	nr	6.69
reference TV CG4; 308 x 15mm slot size; fixing with screws and clip on covers	5.01	2.50	5.14	0.33	0.03	3.26	0.12	8.51	nr	9.58
Glidevale condensation drainage channel; aluminium with powder coated finish										
reference CDN2500; to suit 4 or 6mm glazing; fixing with screws	22.30	2.50	22.86	0.45	0.06	4.57	0.53	27.96	m	31.46
To hardwood or the like										
Glidevale frame vent; aluminium with epoxy paint finish										
reference TV FV2; 198mm long; fixing with screws and clip on covers	7.08	2.50	7.26	0.33	0.05	3.40	0.17	10.82	nr	12.18
reference TV FV4; 358mm long; fixing with screws and clip on covers	8.54	2.50	8.75	0.50	0.07	5.11	0.20	14.06	nr	15.82
Glidevale canopy grille; aluminium with epoxy paint finish										
reference TV CG2; 155 x 15mm slot size; fixing with screws and clip on covers	3.54	2.50	3.63	0.33	0.05	3.40	0.08	7.11	nr	7.99
reference TV CG4; 308 x 15mm slot size; fixing with screws and clip on covers	5.01	2.50	5.14	0.50	0.07	5.11	0.12	10.36	nr	11.66
Glidevale condensation drainage channel; aluminium with powder coated finish										
reference CDN2500; to suit 4 or 6mm glazing; fixing with screws	22.30	2.50	22.86	0.70	0.10	7.16	0.55	30.57	m	34.39
Letter plates; standard quality										
To softwood										
letter plates; plain; real B.M.A.; 356mm wide	48.08	2.50	49.28	1.65	–	15.23	–	64.51	nr	72.58
letter plates; plain; anodised silver; 356mm wide	7.57	2.50	7.76	1.65	–	15.23	–	22.99	nr	25.86
letter plates; gravity flap; B.M.A.; 254mm wide ..	9.13	2.50	9.36	1.65	–	15.23	–	24.59	nr	27.66
letter plates; gravity flap; chromium plated; 254mm wide	9.13	2.50	9.36	1.65	–	15.23	–	24.59	nr	27.66
letter plates; gravity flap, two numerals; anodised silver; 254mm wide	7.93	2.50	8.13	1.75	–	16.15	–	24.28	nr	27.32
postal knockers; S.A.A.; frame 260 x 83mm; opening 203 x 44m.......................................	10.52	2.50	10.78	1.75	–	16.15	–	26.94	nr	30.30
To hardwood or the like										
letter plates; plain; real B.M.A.; 356mm wide	48.08	2.50	49.28	2.50	–	23.07	–	72.36	nr	81.40
letter plates; plain; anodised silver; 356mm wide	7.57	2.50	7.76	2.50	–	23.07	–	30.83	nr	34.69
letter plates; gravity flap; B.M.A.; 254mm wide ..	9.13	2.50	9.36	2.50	–	23.07	–	32.43	nr	36.49
letter plates; gravity flap; chromium plated; 254mm wide	9.13	2.50	9.36	2.50	–	23.07	–	32.43	nr	36.49
letter plates; gravity flap, two numerals; anodised silver; 254mm wide	7.93	2.50	8.13	2.65	–	24.46	–	32.59	nr	36.66
postal knockers; S.A.A.; frame 260 x 83mm; opening 203 x 44m.......................................	10.52	2.50	10.78	2.65	–	24.46	–	35.24	nr	39.65
Security locks; standard quality										
To hardwood or the like										
window catches; locking; polished aluminium	4.98	2.50	5.10	1.20	–	11.08	–	16.18	nr	18.20
dual screws; sanded brass	2.29	2.50	2.35	1.00	–	9.23	–	11.58	nr	13.02
window stops; locking; brass	5.29	2.50	5.42	1.00	–	9.23	–	14.65	nr	16.48

Labour hourly rates: (except Specialists) Craft Operatives £9.23 Labourer £7.02 Rates are national average prices. Refer to REGIONAL VARIATIONS for indicative levels of overall pricing in regions	MATERIALS			LABOUR				RATES		
	Del to Site £	Waste %	Material Cost £	Craft Optve £	Lab £	Labour Cost £	Sunds £	Nett Rate £	Unit	Gross Rate (+12.5%) £
P21: IRONMONGERY Cont.										
Security locks; standard quality Cont.										
To hardwood or the like Cont.										
mortice latches; locking	58.30	2.50	59.76	3.50	-	32.31	-	92.06	nr	103.57
double cylinder automatic deadlatches; B.M.A.	34.86	2.50	35.73	3.50	-	32.31	-	68.04	nr	76.54
double cylinder automatic deadlatches; chromium plated...............	33.66	2.50	34.50	3.50	-	32.31	-	66.81	nr	75.16
mortice deadlocks; Chubb	27.06	2.50	27.74	3.50	-	32.31	-	60.04	nr	67.55
mortice locks; two bolt	33.66	2.50	34.50	3.50	-	32.31	-	66.81	nr	75.16
metal window locks; white	6.01	2.50	6.16	1.50	-	13.85	-	20.01	nr	22.51
security door chains; steel chain; brass	4.87	2.50	4.99	0.60	-	5.54	-	10.53	nr	11.85
security door chains; steel chain; chromium plated	5.41	2.50	5.55	0.60	-	5.54	-	11.08	nr	12.47
security mortice bolts; loose keys; chromium plated; 16mm diameter	4.57	2.50	4.68	1.20	-	11.08	-	15.76	nr	17.73
security mortice bolts; loose keys; chromium plated; 32mm diameter	6.07	2.50	6.22	1.30	-	12.00	-	18.22	nr	20.50
security hinge bolts; silver anodised	4.93	2.50	5.05	1.80	-	16.61	-	21.67	nr	24.38
door viewers; chromium plated	4.57	2.50	4.68	1.50	-	13.85	-	18.53	nr	20.85
Shelf brackets; standard quality										
To softwood										
shelf brackets; grey finished										
100 x 75mm...............	0.18	2.50	0.18	0.25	-	2.31	-	2.49	nr	2.80
150 x 125mm...............	0.24	2.50	0.25	0.25	-	2.31	-	2.55	nr	2.87
225 x 175mm...............	0.39	2.50	0.40	0.25	-	2.31	-	2.71	nr	3.05
300 x 250mm...............	0.57	2.50	0.58	0.25	-	2.31	-	2.89	nr	3.25
350 x 300mm...............	0.75	2.50	0.77	0.25	-	2.31	-	3.08	nr	3.46
Cabin hooks; standard quality										
To softwood										
cabin hooks and eyes; black japanned										
100mm	1.14	2.50	1.17	0.25	-	2.31	-	3.48	nr	3.91
150mm	1.45	2.50	1.49	0.25	-	2.31	-	3.79	nr	4.27
200mm	2.10	2.50	2.15	0.25	-	2.31	-	4.46	nr	5.02
250mm	2.34	2.50	2.40	0.25	-	2.31	-	4.71	nr	5.29
cabin hooks and eyes; polished brass										
100mm	4.81	2.50	4.93	0.25	-	2.31	-	7.24	nr	8.14
150mm	5.41	2.50	5.55	0.25	-	2.31	-	7.85	nr	8.83
To hardwood or the like										
cabin hooks and eyes; black japanned										
100mm	1.14	2.50	1.17	0.35	-	3.23	-	4.40	nr	4.95
150mm	1.45	2.50	1.49	0.35	-	3.23	-	4.72	nr	5.31
200mm	2.10	2.50	2.15	0.35	-	3.23	-	5.38	nr	6.06
250mm	2.34	2.50	2.40	0.35	-	3.23	-	5.63	nr	6.33
cabin hooks and eyes; polished brass										
100mm	4.81	2.50	4.93	0.35	-	3.23	-	8.16	nr	9.18
150mm	5.41	2.50	5.55	0.35	-	3.23	-	8.78	nr	9.87
To softwood and brickwork										
cabin hooks and eyes; black japanned										
100mm	1.14	2.50	1.17	0.30	-	2.77	-	3.94	nr	4.43
150mm	1.45	2.50	1.49	0.30	-	2.77	-	4.26	nr	4.79
200mm	2.10	2.50	2.15	0.30	-	2.77	-	4.92	nr	5.54
250mm	2.34	2.50	2.40	0.30	-	2.77	-	5.17	nr	5.81
cabin hooks and eyes; polished brass										
100mm	4.81	2.50	4.93	0.30	-	2.77	-	7.70	nr	8.66
150mm	5.41	2.50	5.55	0.30	-	2.77	-	8.31	nr	9.35
To hardwood and brickwork										
cabin hooks and eyes; black japanned										
100mm	1.14	2.50	1.17	0.40	-	3.69	-	4.86	nr	5.47
150mm	1.45	2.50	1.49	0.40	-	3.69	-	5.18	nr	5.83
200mm	2.10	2.50	2.15	0.40	-	3.69	-	5.84	nr	6.58
250mm	2.34	2.50	2.40	0.40	-	3.69	-	6.09	nr	6.85
cabin hooks and eyes; polished brass										
100mm	4.81	2.50	4.93	0.40	-	3.69	-	8.62	nr	9.70
150mm	5.41	2.50	5.55	0.40	-	3.69	-	9.24	nr	10.39
Draught seals and strips; standard quality										
To softwood										
draught excluders; plastic foam, self-adhesive										
900mm long...............	0.18	10.00	0.20	0.15	-	1.38	-	1.58	nr	1.78
2000mm long...............	0.36	10.00	0.40	0.30	-	2.77	-	3.17	nr	3.56
draught excluders; aluminium section, rubber tubing										
900mm long...............	7.51	10.00	8.26	0.17	0.02	1.71	-	9.97	nr	11.22
2000mm long...............	17.02	10.00	18.72	0.35	0.04	3.51	-	22.23	nr	25.01
draught excluders; aluminium section, vinyl seal										
900mm long...............	7.51	10.00	8.26	0.17	0.02	1.71	-	9.97	nr	11.22
2000mm long...............	17.02	10.00	18.72	0.35	0.04	3.51	-	22.23	nr	25.01
draught excluders; plastic moulding, nylon brush; 900mm long	6.44	10.00	7.08	0.17	0.02	1.71	-	8.79	nr	9.89
draught excluders; aluminium base, flexible arch; 900mm long...............	17.13	10.00	18.84	0.25	0.03	2.52	-	21.36	nr	24.03
draught excluders; aluminium threshold, flexible arch; 900mm long...................	17.13	10.00	18.84	0.30	0.04	3.05	-	21.89	nr	24.63

BUILDING FABRIC SUNDRIES

BUILDING FABRIC SUNDRIES - SMALL WORKS

Labour hourly rates: (except Specialists) Craft Operatives £9.23 Labourer £7.02 Rates are national average prices. Refer to REGIONAL VARIATIONS for indicative levels of overall pricing in regions	MATERIALS			LABOUR				RATES		
	Del to Site £	Waste %	Material Cost £	Craft Optve £	Lab £	Labour Cost £	Sunds £	Nett Rate £	Unit	Gross Rate (+12.5%) £
P21: IRONMONGERY Cont.										
Draught seals and strips; Sealmaster Ltd										
Threshold seals										
reference BDA; fixing to masonry with screws	9.59	2.50	9.83	0.45	0.06	4.57	0.22	14.62	m	16.45
reference BDB; fixing to masonry with screws	12.30	2.50	12.61	0.45	0.06	4.57	0.30	17.48	m	19.67
reference BDWB (weather board); fixing to timber with screws	14.94	2.50	15.31	0.70	0.10	7.16	0.35	22.83	m	25.68
reference TTM (stop seal); fixing to timber with screws	9.08	2.50	9.31	0.33	0.05	3.40	0.22	12.92	m	14.54
Drawer pulls; standard quality										
To softwood										
drawer pulls; brass; 100mm	1.56	2.50	1.60	0.20	-	1.85	-	3.44	nr	3.88
drawer pulls; B.M.A.; 100mm	2.04	2.50	2.09	0.20	-	1.85	-	3.94	nr	4.43
drawer pulls; chromium plated; 100mm	1.98	2.50	2.03	0.20	-	1.85	-	3.88	nr	4.36
drawer pulls; S.A.A.; 100mm	1.20	2.50	1.23	0.20	-	1.85	-	3.08	nr	3.46
To hardwood or the like										
drawer pulls; brass; 100mm	1.56	2.50	1.60	0.30	-	2.77	-	4.37	nr	4.91
drawer pulls; B.M.A.; 100mm	2.04	2.50	2.09	0.30	-	2.77	-	4.86	nr	5.47
drawer pulls; chromium plated; 100mm	1.98	2.50	2.03	0.30	-	2.77	-	4.80	nr	5.40
drawer pulls; S.A.A.; 100mm	1.20	2.50	1.23	0.30	-	2.77	-	4.00	nr	4.50
Hooks; standard quality										
To softwood										
cup hooks; polished brass; 25mm	0.18	2.50	0.18	0.07	-	0.65	-	0.83	nr	0.93
hat and coat hooks; B.M.A.	1.91	2.50	1.96	0.30	-	2.77	-	4.73	nr	5.32
hat and coat hooks; chromium plated	1.91	2.50	1.96	0.30	-	2.77	-	4.73	nr	5.32
hat and coat hooks; S.A.A.	0.84	2.50	0.86	0.30	-	2.77	-	3.63	nr	4.08
To hardwood or the like										
cup hooks; polished brass; 25mm	0.18	2.50	0.18	0.10	-	0.92	-	1.11	nr	1.25
hat and coat hooks; B.M.A.	1.91	2.50	1.96	0.45	-	4.15	-	6.11	nr	6.88
hat and coat hooks; chromium plated	1.91	2.50	1.96	0.45	-	4.15	-	6.11	nr	6.88
hat and coat hooks; S.A.A.	0.84	2.50	0.86	0.45	-	4.15	-	5.01	nr	5.64
P30: TRENCHES/PIPEWAYS/PITS FOR BURIED ENGINEERING SERVICES										
Site preparation										
Lifting turf for preservation										
stacking on site average 50 metres distance for immediate use; watering	-	-	-	-	0.60	4.21	-	4.21	m2	4.74
Re-laying turf										
taking from stack average 50 metres distance, laying on prepared bed, watering, maintaining	-	-	-	-	0.60	4.21	-	4.21	m2	4.74

This section continues
on the next page

Labour hourly rates: (except Specialists) Craft Operatives £9.23 Labourer £7.02 Rates are national average prices. Refer to REGIONAL VARIATIONS for indicative levels of overall pricing in regions	PLANT & TRANSPORT			LABOUR				RATES		
	Plant Cost £	Trans Cost £	P & T Cost £	Craft Optve £	Lab £	Labour Cost £	Sunds £	Nett Rate £	Unit	Gross Rate (+12.5%) £
P30: TRENCHES/PIPEWAYS/PITS FOR BURIED ENGINEERING SERVICES Cont.										
Excavating trenches to receive services not exceeding 200mm nominal size - by machine										
Excavations commencing from natural ground level; compacting; backfilling with excavated material										
not exceeding 1m deep; average 750mm deep	0.69	0.38	1.07	-	0.25	1.75	2.59	5.42	m	6.09
not exceeding 1m deep; average 750mm deep; next to roadways	0.69	0.38	1.07	-	0.25	1.75	2.59	5.42	m	6.09
Excavations commencing from existing ground level; levelling and grading backfilling to receive turf										
not exceeding 1m deep; average 750mm deep	0.69	0.38	1.07	-	0.25	1.75	2.48	5.30	m	5.97
Extra over excavating trenches irrespective of depth; breaking out existing materials										
hard rock..	25.58	-	25.58	-	-	-	-	25.58	m3	28.78
concrete...	21.24	-	21.24	-	-	-	-	21.24	m3	23.90
reinforced concrete	31.90	-	31.90	-	-	-	-	31.90	m3	35.89
brickwork, blockwork or stonework	14.24	-	14.24	-	-	-	-	14.24	m3	16.02
Extra over excavating trenches irrespective of depth; breaking out existing hard pavings										
concrete 150mm thick...............................	3.83	-	3.83	-	-	-	-	3.83	m2	4.31
reinforced concrete 200mm thick	5.69	-	5.69	-	-	-	-	5.69	m2	6.40
concrete 150mm thick; reinstating	3.83	-	3.83	-	-	-	13.06	16.89	m2	19.00
macadam paving 75mm thick	0.67	-	0.67	-	-	-	-	0.67	m2	0.75
macadam paving 75mm thick; reinstating	0.67	-	0.67	-	-	-	16.98	17.65	m2	19.86
concrete flag paving 50mm thick	-	-	-	-	0.30	2.11	-	2.11	m2	2.37
concrete flag paving 50mm thick; re-instating	-	-	-	0.60	0.90	11.86	10.97	22.83	m2	25.68
Extra over excavating trenches irrespective of depth; excavating next existing services										
electricity services -1	-	-	-	-	5.75	40.37	-	40.37	m	45.41
gas services -1	-	-	-	-	5.75	40.37	-	40.37	m	45.41
water services -1	-	-	-	-	5.75	40.37	-	40.37	m	45.41
Extra over excavating trenches irrespective of depth; excavating around existing services crossing trench										
electricity services; cables crossing -1	-	-	-	-	8.50	59.67	-	59.67	nr	67.13
gas services services crossing -1	-	-	-	-	8.50	59.67	-	59.67	nr	67.13
water services services crossing -1	-	-	-	-	8.50	59.67	-	59.67	nr	67.13
Site preparation										
Lifting turf for preservation										
stacking on site average 50 metres distance for immediate use; watering	-	-	-	-	0.60	4.21	-	4.21	m2	4.74
Re-laying turf										
taking from stack average 50 metres distance, laying on prepared bed, watering, maintaining	-	-	-	-	0.60	4.21	-	4.21	m2	4.74
Excavating trenches to receive services not exceeding 200mm nominal size - by machine										
Excavations commencing from natural ground level; compacting; backfilling with excavated material										
not exceeding 1m deep; average 750mm deep	-	-	-	-	1.20	8.42	2.59	11.01	m	12.39
not exceeding 1m deep; average 750mm deep; next to roadways	-	-	-	-	1.20	8.42	2.59	11.01	m	12.39
Excavations commencing from existing ground level; levelling and grading backfilling to receive turf										
not exceeding 1m deep; average 750mm deep	-	-	-	-	1.20	8.42	2.59	11.01	m	12.39
Extra over excavating trenches irrespective of depth; breaking out existing materials										
hard rock..	-	-	-	-	14.70	103.19	-	103.19	m3	116.09
concrete...	-	-	-	-	11.00	77.22	-	77.22	m3	86.87
reinforced concrete	-	-	-	-	16.50	115.83	-	115.83	m3	130.31
brickwork, blockwork or stonework	-	-	-	-	7.70	54.05	-	54.05	m3	60.81
Extra over excavating trenches irrespective of depth; breaking out existing hard pavings										
concrete 150mm thick...............................	-	-	-	-	1.65	11.58	-	11.58	m2	13.03
reinforced concrete 150mm thick	-	-	-	-	3.30	23.17	-	23.17	m2	26.06
concrete 150mm thick; reinstating	-	-	-	-	3.30	23.17	13.06	36.23	m2	40.75
macadam paving 75mm thick	-	-	-	-	0.30	2.11	-	2.11	m2	2.37
macadam paving 75mm thick; reinstating	-	-	-	-	0.30	2.11	16.98	19.09	m2	21.47
concrete flag paving 50mm thick	-	-	-	-	0.32	2.25	-	2.25	m2	2.53
concrete flag paving 50mm thick; re-instating	-	-	-	-	0.88	6.18	11.52	17.70	m2	19.91
Extra over excavating trenches irrespective of depth; excavating next existing services										
electricity services -1	-	-	-	-	5.75	40.37	-	40.37	m	45.41
gas services -1	-	-	-	-	5.75	40.37	-	40.37	m	45.41
water services -1	-	-	-	-	5.75	40.37	-	40.37	m	45.41
Extra over excavating trenches irrespective of depth; excavating around existing services crossing trench										
electricity services; cables crossing -1	-	-	-	-	8.50	59.67	-	59.67	nr	67.13
gas services services crossing -1	-	-	-	-	8.50	59.67	-	59.67	nr	67.13
water services services crossing -1	-	-	-	-	8.50	59.67	-	59.67	nr	67.13

BUILDING FABRIC SUNDRIES

BUILDING FABRIC SUNDRIES — SMALL WORKS

Labour hourly rates: (except Specialists) Craft Operatives £9.23 Labourer £7.02 Rates are national average prices. Refer to REGIONAL VARIATIONS for indicative levels of overall pricing in regions	MATERIALS			LABOUR				RATES		
	Del to Site £	Waste %	Material Cost £	Craft Optve £	Lab £	Labour Cost £	Sunds £	Nett Rate £	Unit	Gross Rate (+12.5%) £
P30: TRENCHES/PIPEWAYS/PITS FOR BURIED ENGINEERING SERVICES Cont.										
Underground ducts; vitrified clayware, B.S.65 extra strength; flexible joints										
Straight										
100mm nominal size single way duct; laid in position in trench	4.37	5.00	4.59	0.20	0.03	2.06	-	6.65	m	7.48
100mm nominal size bonded to form 2 way duct; laid in position in trench	8.74	5.00	9.18	0.30	0.04	3.05	-	12.23	m	13.76
100mm nominal size bonded to form 4 way duct; laid in position in trench	17.48	5.00	18.35	0.45	0.06	4.57	-	22.93	m	25.79
100mm nominal size bonded to form 6 way duct; laid in position in trench	26.22	5.00	27.53	0.55	0.07	5.57	-	33.10	m	37.24
extra; providing and laying in position nylon draw wire	0.13	10.00	0.14	0.02	-	0.18	-	0.33	m	0.37
Stop cock pits, valve chambers and the like										
Stop cock pits; half brick thick sides of common bricks, B.S.3921, Category M, 215 x 102.5 x 65mm, compressive strength 20.5 N/mm2, in cement mortar (1:3); 100mm thick base and top of plain in-situ concrete, B.S.5328, ordinary prescribed mix ST4, 20mm agg										
600mm deep in clear; (surface boxes included elsewhere)										
internal size 225 x 225mm..................	17.52	5.00	18.40	1.75	1.75	28.44	3.01	49.84	nr	56.07
internal size 338 x 338mm..................	24.35	5.00	25.57	2.55	2.55	41.44	4.55	71.56	nr	80.50
Stop cock guards; vitrified clay; including all excavation backfilling, disposal of surplus excavated material, earthwork support and compaction of ground; surrounding in 150mm thick plain in-situ concrete, B.S.5328 ordinary prescribed mix ST4, 20mm aggre										
750mm deep in clear; (surface boxes included elsewhere)										
150mm diameter...........................	14.32	5.00	15.04	1.30	1.30	21.13	1.90	38.06	nr	42.82
Cast iron surface boxes and covers, coated to B.S.4147; bedding in cement mortar (1:3)										
Surface boxes, B.S.5834 Part 2 marked S.V. light grade, hinged lid; 150 x 150mm overall top size, 76mm deep	10.24	-	10.24	0.85	0.43	10.86	0.20	21.30	nr	23.97
Surface boxes, B.S.750, marked 'FIRE HYDRANT' medium grade; minimum clear opening 230 x 380mm, minimum depth 100mm	35.93	-	35.93	1.40	0.70	17.84	0.60	54.37	nr	61.16
Surface boxes, B.S.5834 Part 2 marked 'W' or WATER heavy grade; double triangular cover; minimum clear opening 300 x 300mm, minimum depth 150mm ...	67.96	-	67.96	1.70	0.85	21.66	0.99	90.61	nr	101.93
Accessories/Sundry items										
Stop cock keys										
tee..	15.13	-	15.13	-	-	-	-	15.13	nr	17.02
P31: HOLES/CHASES/COVERS/SUPPORTS FOR SERVICES; BUILDERS WORK IN CONNECTION WITH MECHANICAL SERVICES										
Cutting holes for services										
For ducts through concrete; making good										
rectangular ducts not exceeding 1.00m girth										
150mm thick...........................	-	-	-	1.84	-	16.98	0.54	17.52	nr	19.71
200mm thick...........................	-	-	-	2.18	-	20.12	0.65	20.77	nr	23.37
300mm thick...........................	-	-	-	2.52	-	23.26	0.76	24.02	nr	27.02
rectangular ducts 1.00-2.00m girth										
150mm thick...........................	-	-	-	3.68	-	33.97	1.10	35.07	nr	39.45
200mm thick...........................	-	-	-	4.35	-	40.15	1.28	41.43	nr	46.61
300mm thick...........................	-	-	-	5.06	-	46.70	1.49	48.19	nr	54.22
rectangular ducts 2.00-3.00m girth										
150mm thick...........................	-	-	-	4.60	-	42.46	1.37	43.83	nr	49.31
200mm thick...........................	-	-	-	5.46	-	50.40	1.61	52.01	nr	58.51
300mm thick...........................	-	-	-	6.33	-	58.43	1.85	60.28	nr	67.81
For pipes through concrete; making good										
pipes not exceeding 55mm nominal size										
150mm thick...........................	-	-	-	1.04	-	9.60	0.34	9.94	nr	11.18
200mm thick...........................	-	-	-	1.32	-	12.18	0.43	12.61	nr	14.19
300mm thick...........................	-	-	-	1.67	-	15.41	0.55	15.96	nr	17.96
pipes 55 - 110mm nominal size										
150mm thick...........................	-	-	-	1.28	-	11.81	0.42	12.23	nr	13.76
200mm thick...........................	-	-	-	1.56	-	14.40	0.49	14.89	nr	16.75
300mm thick...........................	-	-	-	2.08	-	19.20	0.67	19.87	nr	22.35
pipes exceeding 110mm nominal size										
150mm thick...........................	-	-	-	1.55	-	14.31	0.49	14.80	nr	16.65
200mm thick...........................	-	-	-	1.98	-	18.28	0.64	18.92	nr	21.28
300mm thick...........................	-	-	-	2.50	-	23.07	0.81	23.89	nr	26.87
For ducts through reinforced concrete; making good										
rectangular ducts not exceeding 1.00m girth										
150mm thick...........................	-	-	-	2.76	-	25.47	0.81	26.28	nr	29.57

Labour hourly rates: (except Specialists) Craft Operatives £9.23 Labourer £7.02 Rates are national average prices. Refer to REGIONAL VARIATIONS for indicative levels of overall pricing in regions	MATERIALS			LABOUR				RATES		
	Del to Site £	Waste %	Material Cost £	Craft Optve £	Lab £	Labour Cost £	Sunds £	Nett Rate £	Unit	Gross Rate (+12.5%) £
P31: HOLES/CHASES/COVERS/SUPPORTS FOR SERVICES; BUILDERS WORK IN CONNECTION WITH MECHANICAL SERVICES Cont.										
Cutting holes for services Cont.										
For ducts through reinforced concrete; making good Cont.										
rectangular ducts not exceeding 1.00m girth Cont.										
200mm thick	-	-	-	3.28	-	30.27	0.99	31.26	nr	35.17
300mm thick	-	-	-	3.80	-	35.07	1.16	36.23	nr	40.76
rectangular ducts 1.00-2.00m girth										
150mm thick	-	-	-	5.52	-	50.95	1.62	52.57	nr	59.14
200mm thick	-	-	-	6.56	-	60.55	1.96	62.51	nr	70.32
300mm thick	-	-	-	7.60	-	70.15	2.26	72.41	nr	81.46
rectangular ducts 2.00-3.00m girth										
150mm thick	-	-	-	6.90	-	63.69	1.95	65.64	nr	73.84
200mm thick	-	-	-	8.05	-	74.30	2.30	76.60	nr	86.18
300mm thick	-	-	-	9.48	-	87.50	2.64	90.14	nr	101.41
For pipes through reinforced concrete; making good										
pipes not exceeding 55mm nominal size										
150mm thick	-	-	-	1.55	-	14.31	0.48	14.79	nr	16.63
200mm thick	-	-	-	2.00	-	18.46	0.62	19.08	nr	21.47
300mm thick	-	-	-	2.50	-	23.07	0.78	23.86	nr	26.84
pipes 55 - 110mm nominal size										
150mm thick	-	-	-	1.96	-	18.09	0.59	18.68	nr	21.02
200mm thick	-	-	-	2.36	-	21.78	0.72	22.50	nr	25.32
300mm thick	-	-	-	3.10	-	28.61	0.96	29.57	nr	33.27
pipes exceeding 110mm nominal size										
150mm thick	-	-	-	2.30	-	21.23	0.72	21.95	nr	24.69
200mm thick	-	-	-	3.00	-	27.69	0.91	28.60	nr	32.17
300mm thick	-	-	-	3.70	-	34.15	1.16	35.31	nr	39.72
For ducts through brickwork; making good										
rectangular ducts not exceeding 1.00m girth										
102.5mm thick	-	-	-	1.38	-	12.74	0.38	13.12	nr	14.76
215mm thick	-	-	-	1.60	-	14.77	0.46	15.23	nr	17.13
327.5mm thick	-	-	-	1.90	-	17.54	0.55	18.09	nr	20.35
rectangular ducts 1.00-2.00m girth										
102.5mm thick	-	-	-	2.76	-	25.47	0.77	26.24	nr	29.53
215mm thick	-	-	-	3.22	-	29.72	0.92	30.64	nr	34.47
327.5mm thick	-	-	-	3.80	-	35.07	1.10	36.17	nr	40.70
rectangular ducts 2.00-3.00m girth										
102.5mm thick	-	-	-	3.45	-	31.84	1.00	32.84	nr	36.95
215mm thick	-	-	-	4.02	-	37.10	1.15	38.25	nr	43.04
327.5mm thick	-	-	-	4.72	-	43.57	1.38	44.95	nr	50.56
For pipes through brickwork; making good										
pipes not exceeding 55mm nominal size										
102.5mm thick	-	-	-	0.46	-	4.25	0.15	4.40	nr	4.95
215mm thick	-	-	-	0.76	-	7.01	0.24	7.25	nr	8.16
327.5mm thick	-	-	-	1.15	-	10.61	0.37	10.98	nr	12.36
pipes 55 - 110mm nominal size										
102.5mm thick	-	-	-	0.60	-	5.54	0.20	5.74	nr	6.46
215mm thick	-	-	-	1.02	-	9.41	0.33	9.74	nr	10.96
327.5mm thick	-	-	-	1.53	-	14.12	0.49	14.61	nr	16.44
pipes exceeding 110mm nominal size										
102.5mm thick	-	-	-	0.77	-	7.11	0.25	7.36	nr	8.28
215mm thick	-	-	-	1.26	-	11.63	0.41	12.04	nr	13.54
327.5mm thick	-	-	-	1.92	-	17.72	0.62	18.34	nr	20.63
For ducts through brickwork; making good fair face or facings one side										
rectangular ducts not exceeding 1.00m girth										
102.5mm thick	-	-	-	1.67	-	15.41	0.61	16.02	nr	18.03
215mm thick	-	-	-	1.95	-	18.00	0.72	18.72	nr	21.06
327.5mm thick	-	-	-	2.18	-	20.12	0.77	20.89	nr	23.50
rectangular ducts 1.00-2.00m girth										
102.5mm thick	-	-	-	3.10	-	28.61	1.21	29.82	nr	33.55
215mm thick	-	-	-	3.55	-	32.77	1.43	34.20	nr	38.47
327.5mm thick	-	-	-	4.14	-	38.21	1.54	39.75	nr	44.72
rectangular ducts 2.00-3.00m girth										
102.5mm thick	-	-	-	3.90	-	36.00	1.54	37.54	nr	42.23
215mm thick	-	-	-	4.48	-	41.35	1.82	43.17	nr	48.57
327.5mm thick	-	-	-	5.18	-	47.81	1.93	49.74	nr	55.96
For pipes through brickwork; making good fair face or facings one side										
pipes not exceeding 55mm nominal size										
102.5mm thick	-	-	-	0.58	-	5.35	0.20	5.55	nr	6.25
215mm thick	-	-	-	0.87	-	8.03	0.29	8.32	nr	9.36
327.5mm thick	-	-	-	1.27	-	11.72	0.41	12.13	nr	13.65
pipes 55 - 110mm nominal size										
102.5mm thick	-	-	-	0.78	-	7.20	0.24	7.44	nr	8.37
215mm thick	-	-	-	1.18	-	10.89	0.36	11.25	nr	12.66
327.5mm thick	-	-	-	1.70	-	15.69	0.54	16.23	nr	18.26
pipes exceeding 110mm nominal size										
102.5mm thick	-	-	-	1.00	-	9.23	0.30	9.53	nr	10.72
215mm thick	-	-	-	1.50	-	13.85	0.44	14.29	nr	16.07
327.5mm thick	-	-	-	2.15	-	19.84	0.65	20.49	nr	23.06
For ducts through blockwork; making good										
rectangular ducts not exceeding 1.00m girth										
100mm thick	-	-	-	1.04	-	9.60	0.39	9.99	nr	11.24
140mm thick	-	-	-	1.15	-	10.61	0.46	11.07	nr	12.46

BUILDING FABRIC SUNDRIES – SMALL WORKS

Labour hourly rates: (except Specialists) Craft Operatives £9.23 Labourer £7.02 Rates are national average prices. Refer to REGIONAL VARIATIONS for indicative levels of overall pricing in regions	MATERIALS			LABOUR				RATES		
	Del to Site £	Waste %	Material Cost £	Craft Optve £	Lab £	Labour Cost £	Sunds £	Nett Rate £	Unit	Gross Rate (+12.5%) £
P31: HOLES/CHASES/COVERS/SUPPORTS FOR SERVICES; BUILDERS WORK IN CONNECTION WITH MECHANICAL SERVICES Cont.										
Cutting holes for services Cont.										
For ducts through blockwork; making good Cont.										
rectangular ducts not exceeding 1.00m girth Cont.										
190mm thick	-	-	-	1.26	-	11.63	0.55	12.18	nr	13.70
rectangular ducts 1.00-2.00m girth										
100mm thick	-	-	-	2.08	-	19.20	0.77	19.97	nr	22.46
140mm thick	-	-	-	2.24	-	20.68	0.92	21.60	nr	24.29
190mm thick	-	-	-	2.42	-	22.34	1.10	23.44	nr	26.37
rectangular ducts 2.00-3.00m girth										
100mm thick	-	-	-	2.58	-	23.81	0.99	24.80	nr	27.90
140mm thick	-	-	-	2.82	-	26.03	1.16	27.19	nr	30.59
190mm thick	-	-	-	3.05	-	28.15	1.38	29.53	nr	33.22
For pipes through blockwork; making good										
pipes not exceeding 55mm nominal size										
100mm thick	-	-	-	0.35	-	3.23	0.14	3.37	nr	3.79
140mm thick	-	-	-	0.46	-	4.25	0.23	4.48	nr	5.04
190mm thick	-	-	-	0.58	-	5.35	0.35	5.70	nr	6.42
pipes 55 - 110mm nominal size										
100mm thick	-	-	-	0.46	-	4.25	0.19	4.44	nr	4.99
140mm thick	-	-	-	0.61	-	5.63	0.31	5.94	nr	6.68
190mm thick	-	-	-	0.76	-	7.01	0.47	7.48	nr	8.42
pipes exceeding 110mm nominal size										
100mm thick	-	-	-	0.58	-	5.35	0.24	5.59	nr	6.29
140mm thick	-	-	-	0.78	-	7.20	0.39	7.59	nr	8.54
190mm thick	-	-	-	0.95	-	8.77	0.59	9.36	nr	10.53
For ducts through existing brickwork with plaster finish; making good each side										
rectangular ducts not exceeding 1.00m girth										
102.5mm thick	-	-	-	2.24	-	20.68	0.65	21.33	nr	23.99
215mm thick	-	-	-	2.58	-	23.81	0.82	24.63	nr	27.71
327.5mm thick	-	-	-	2.93	-	27.04	0.94	27.98	nr	31.48
rectangular ducts 1.00-2.00m girth										
102.5mm thick	-	-	-	4.14	-	38.21	1.32	39.53	nr	44.47
215mm thick	-	-	-	4.84	-	44.67	1.76	46.43	nr	52.24
327.5mm thick	-	-	-	5.52	-	50.95	1.88	52.83	nr	59.43
rectangular ducts 2.00-3.00m girth										
102.5mm thick	-	-	-	5.18	-	47.81	1.98	49.79	nr	56.02
215mm thick	-	-	-	6.04	-	55.75	2.48	58.23	nr	65.51
327.5mm thick	-	-	-	6.95	-	64.15	2.81	66.96	nr	75.33
For ducts through existing brickwork with plaster finish; making good one side										
rectangular ducts not exceeding 1.00m girth										
102.5mm thick	-	-	-	2.00	-	18.46	0.60	19.06	nr	21.44
215mm thick	-	-	-	2.35	-	21.69	0.75	22.44	nr	25.25
327.5mm thick	-	-	-	2.65	-	24.46	0.85	25.31	nr	28.47
rectangular ducts 1.00-2.00m girth										
102.5mm thick	-	-	-	3.74	-	34.52	1.18	35.70	nr	40.16
215mm thick	-	-	-	4.38	-	40.43	1.49	41.92	nr	47.16
327.5mm thick	-	-	-	5.00	-	46.15	1.70	47.85	nr	53.83
rectangular ducts 2.00-3.00m girth										
102.5mm thick	-	-	-	4.65	-	42.92	1.76	44.68	nr	50.26
215mm thick	-	-	-	5.46	-	50.40	2.20	52.60	nr	59.17
327.5mm thick	-	-	-	6.26	-	57.78	2.53	60.31	nr	67.85
For pipes through existing brickwork with plaster finish; making good each side										
pipes not exceeding 55mm nominal size										
102.5mm thick	-	-	-	0.75	-	6.92	0.24	7.16	nr	8.06
215mm thick	-	-	-	1.15	-	10.61	0.34	10.95	nr	12.32
327.5mm thick	-	-	-	1.66	-	15.32	0.48	15.80	nr	17.78
pipes 55 - 110mm nominal size										
102.5mm thick	-	-	-	1.10	-	10.15	0.29	10.44	nr	11.75
215mm thick	-	-	-	1.55	-	14.31	0.44	14.75	nr	16.59
327.5mm thick	-	-	-	2.24	-	20.68	0.65	21.33	nr	23.99
pipes exceeding 110mm nominal size										
102.5mm thick	-	-	-	1.38	-	12.74	0.35	13.09	nr	14.72
215mm thick	-	-	-	2.00	-	18.46	0.53	18.99	nr	21.36
327.5mm thick	-	-	-	2.88	-	26.58	0.78	27.36	nr	30.78
For pipes through existing brickwork with plaster finish; making good one side										
pipes not exceeding 55mm nominal size										
102.5mm thick	-	-	-	0.70	-	6.46	0.20	6.66	nr	7.49
215mm thick	-	-	-	1.04	-	9.60	0.29	9.89	nr	11.13
327.5mm thick	-	-	-	1.50	-	13.85	0.41	14.26	nr	16.04
pipes 55 - 110mm nominal size										
102.5mm thick	-	-	-	0.98	-	9.05	0.24	9.29	nr	10.45
215mm thick	-	-	-	1.38	-	12.74	0.36	13.10	nr	14.73
327.5mm thick	-	-	-	2.00	-	18.46	0.54	19.00	nr	21.38
pipes exceeding 110mm nominal size										
102.5mm thick	-	-	-	1.26	-	11.63	0.30	11.93	nr	13.42
215mm thick	-	-	-	1.84	-	16.98	0.44	17.42	nr	19.60
327.5mm thick	-	-	-	2.58	-	23.81	0.65	24.46	nr	27.52
For pipes through softwood										
pipes not exceeding 55mm nominal size										
19mm thick	-	-	-	0.12	-	1.11	-	1.11	nr	1.25
50mm thick	-	-	-	0.15	-	1.38	-	1.38	nr	1.56

Labour hourly rates: (except Specialists) Craft Operatives £9.23 Labourer £7.02 Rates are national average prices. Refer to REGIONAL VARIATIONS for indicative levels of overall pricing in regions	MATERIALS			LABOUR				RATES		
	Del to Site	Waste	Material Cost	Craft Optve	Lab	Labour Cost	Sunds	Nett Rate	Unit	Gross Rate (+12.5%)
	£	%	£	£	£	£	£	£		£
P31: HOLES/CHASES/COVERS/SUPPORTS FOR SERVICES; BUILDERS WORK IN CONNECTION WITH MECHANICAL SERVICES Cont.										
Cutting holes for services Cont.										
For pipes through softwood Cont. pipes 55 - 110mm nominal size										
19mm thick..........................	-	-	-	0.17	-	1.57	-	1.57	nr	1.77
50mm thick..........................	-	-	-	0.23	-	2.12	-	2.12	nr	2.39
For pipes through plywood pipes not exceeding 55mm nominal size										
13mm thick..........................	-	-	-	0.12	-	1.11	-	1.11	nr	1.25
19mm thick..........................	-	-	-	0.15	-	1.38	-	1.38	nr	1.56
pipes 55 - 110mm nominal size										
13mm thick..........................	-	-	-	0.17	-	1.57	-	1.57	nr	1.77
19mm thick..........................	-	-	-	0.23	-	2.12	-	2.12	nr	2.39
For pipes through existing plasterboard with skim finish; making good one side pipes not exceeding 55mm nominal size										
13mm thick..........................	-	-	-	0.23	-	2.12	-	2.12	nr	2.39
pipes 55 - 110mm nominal size										
13mm thick..........................	-	-	-	0.35	-	3.23	-	3.23	nr	3.63
Cutting chases for services										
In concrete; making good										
15mm nominal size pipes -1	-	-	-	0.35	-	3.23	-	3.23	m	3.63
22mm nominal size pipes -1	-	-	-	0.40	-	3.69	-	3.69	m	4.15
In brickwork										
15mm nominal size pipes -1	-	-	-	0.30	-	2.77	-	2.77	m	3.12
22mm nominal size pipes -1	-	-	-	0.35	-	3.23	-	3.23	m	3.63
In brickwork; making good fair face or facings										
15mm nominal size pipes -1	-	-	-	0.35	-	3.23	0.10	3.33	m	3.75
22mm nominal size pipes -1	-	-	-	0.40	-	3.69	0.15	3.84	m	4.32
In existing brickwork with plaster finish; making good										
15mm nominal size pipes -1	-	-	-	0.45	-	4.15	0.23	4.38	m	4.93
22mm nominal size pipes -1	-	-	-	0.55	-	5.08	0.37	5.45	m	6.13
In blockwork; making good										
15mm nominal size pipes -1	-	-	-	0.23	-	2.12	-	2.12	m	2.39
22mm nominal size pipes -1	-	-	-	0.29	-	2.68	-	2.68	m	3.01
Pipe and duct sleeves										

Steel pipes B.S.1387 Table 2; casting into concrete;
making good
  100mm long; to the following nominal size steel
  pipes										

15mm	0.59	5.00	0.62	0.12	0.06	1.53	-	2.15	nr	2.42
20mm	0.66	5.00	0.69	0.14	0.07	1.78	-	2.48	nr	2.79
25mm	0.81	5.00	0.85	0.15	0.08	1.95	-	2.80	nr	3.15
32mm	0.95	5.00	1.00	0.16	0.08	2.04	-	3.04	nr	3.42
40mm	1.09	5.00	1.14	0.18	0.09	2.29	-	3.44	nr	3.87
50mm	1.38	5.00	1.45	0.22	0.11	2.80	-	4.25	nr	4.78
150mm long; to the following nominal size steel pipes										
15mm	0.73	5.00	0.77	0.14	0.07	1.78	-	2.55	nr	2.87
20mm	0.82	5.00	0.86	0.16	0.08	2.04	-	2.90	nr	3.26
25mm	1.02	5.00	1.07	0.17	0.09	2.20	-	3.27	nr	3.68
32mm	1.20	5.00	1.26	0.18	0.09	2.29	-	3.55	nr	4.00
40mm	1.38	5.00	1.45	0.19	0.10	2.46	-	3.90	nr	4.39
50mm	1.78	5.00	1.87	0.23	0.12	2.97	-	4.83	nr	5.44
200mm long; to the following nominal size steel pipes										
15mm	0.86	5.00	0.90	0.15	0.08	1.95	-	2.85	nr	3.21
20mm	0.97	5.00	1.02	0.17	0.09	2.20	-	3.22	nr	3.62
25mm	1.23	5.00	1.29	0.18	0.09	2.29	-	3.58	nr	4.03
32mm	1.46	5.00	1.53	0.19	0.10	2.46	-	3.99	nr	4.49
40mm	1.68	5.00	1.76	0.20	0.10	2.55	-	4.31	nr	4.85
50mm	2.20	5.00	2.31	0.24	0.12	3.06	-	5.37	nr	6.04
250mm long; to the following nominal size steel pipes										
15mm	1.00	5.00	1.05	0.15	0.08	1.95	-	3.00	nr	3.37
20mm	1.12	5.00	1.18	0.19	0.10	2.46	-	3.63	nr	4.09
25mm	1.44	5.00	1.51	0.20	0.10	2.55	-	4.06	nr	4.57
32mm	1.73	5.00	1.82	0.21	0.11	2.71	-	4.53	nr	5.09
40mm	1.98	5.00	2.08	0.22	0.11	2.80	-	4.88	nr	5.49
50mm	2.62	5.00	2.75	0.26	0.13	3.31	-	6.06	nr	6.82
100mm long; to the following nominal size copper pipes										
15mm	0.61	5.00	0.64	0.12	0.06	1.53	-	2.17	nr	2.44
22mm	0.67	5.00	0.70	0.14	0.07	1.78	-	2.49	nr	2.80
28mm	0.83	5.00	0.87	0.15	0.08	1.95	-	2.82	nr	3.17
32mm	0.97	5.00	1.02	0.16	0.08	2.04	-	3.06	nr	3.44
45mm	1.11	5.00	1.17	0.18	0.09	2.29	-	3.46	nr	3.89
54mm	1.41	5.00	1.48	0.22	0.11	2.80	-	4.28	nr	4.82
150mm long; to the following nominal size copper pipes										
15mm	0.75	5.00	0.79	0.14	0.07	1.78	-	2.57	nr	2.89

BUILDING FABRIC SUNDRIES

Labour hourly rates: (except Specialists) Craft Operatives £9.23 Labourer £7.02 Rates are national average prices. Refer to REGIONAL VARIATIONS for indicative levels of overall pricing in regions	MATERIALS			LABOUR				RATES		
	Del to Site	Waste	Material Cost	Craft Optve	Lab	Labour Cost	Sunds	Nett Rate	Unit	Gross Rate (+12.5%)
	£	%	£	£	£	£	£	£		£
P31: HOLES/CHASES/COVERS/SUPPORTS FOR SERVICES; **BUILDERS WORK IN CONNECTION WITH MECHANICAL SERVICES** **Cont.**										
Pipe and duct sleeves Cont.										
Steel pipes B.S.1387 Table 2; casting into concrete; making good Cont.										
150mm long; to the following nominal size copper pipes Cont.										
22mm	0.84	5.00	0.88	0.16	0.08	2.04	–	2.92	nr	3.29
28mm	1.05	5.00	1.10	0.17	0.09	2.20	–	3.30	nr	3.72
32mm	1.23	5.00	1.29	0.18	0.09	2.29	–	3.58	nr	4.03
45mm	1.41	5.00	1.48	0.19	0.10	2.46	–	3.94	nr	4.43
54mm	1.83	5.00	1.92	0.23	0.12	2.97	–	4.89	nr	5.50
200mm long; to the following nominal size copper pipes										
15mm	0.88	5.00	0.92	0.15	0.08	1.95	–	2.87	nr	3.23
22mm	0.99	5.00	1.04	0.17	0.09	2.20	–	3.24	nr	3.65
28mm	1.25	5.00	1.31	0.18	0.09	2.29	–	3.61	nr	4.06
32mm	1.50	5.00	1.58	0.19	0.10	2.46	–	4.03	nr	4.53
45mm	1.72	5.00	1.81	0.20	0.10	2.55	–	4.35	nr	4.90
54mm	2.26	5.00	2.37	0.24	0.12	3.06	–	5.43	nr	6.11
250mm long; to the following nominal size copper pipes										
15mm	1.03	5.00	1.08	0.17	0.09	2.20	–	3.28	nr	3.69
22mm	1.14	5.00	1.20	0.19	0.10	2.46	–	3.65	nr	4.11
28mm	1.47	5.00	1.54	0.20	0.10	2.55	–	4.09	nr	4.60
32mm	1.77	5.00	1.86	0.21	0.11	2.71	–	4.57	nr	5.14
45mm	2.61	5.00	2.74	0.22	0.11	2.80	–	5.54	nr	6.24
54mm	2.68	5.00	2.81	0.26	0.13	3.31	–	6.13	nr	6.89
Steel pipes B.S.1387 Table 2; building into blockwork; bedding and pointing in cement mortar (1:3); making good										
100mm long; to the following nominal size steel pipes										
15mm	0.59	5.00	0.62	0.17	0.09	2.20	–	2.82	nr	3.17
20mm	0.66	5.00	0.69	0.18	0.09	2.29	–	2.99	nr	3.36
25mm	0.81	5.00	0.85	0.19	0.10	2.46	–	3.31	nr	3.72
32mm	0.95	5.00	1.00	0.20	0.10	2.55	–	3.55	nr	3.99
40mm	1.09	5.00	1.14	0.21	0.11	2.71	–	3.86	nr	4.34
50mm	1.38	5.00	1.45	0.22	0.11	2.80	–	4.25	nr	4.78
140mm long; to the following nominal size steel pipes										
15mm	0.70	5.00	0.73	0.19	0.10	2.46	–	3.19	nr	3.59
20mm	0.78	5.00	0.82	0.20	0.10	2.55	–	3.37	nr	3.79
25mm	0.98	5.00	1.03	0.21	0.11	2.71	–	3.74	nr	4.21
32mm	1.18	5.00	1.24	0.22	0.11	2.80	–	4.04	nr	4.55
40mm	1.32	5.00	1.39	0.23	0.12	2.97	–	4.35	nr	4.90
50mm	1.71	5.00	1.80	0.24	0.12	3.06	–	4.85	nr	5.46
190mm long; to the following nominal size steel pipes										
15mm	0.84	5.00	0.88	0.21	0.11	2.71	–	3.59	nr	4.04
20mm	0.94	5.00	0.99	0.22	0.11	2.80	–	3.79	nr	4.26
25mm	1.19	5.00	1.25	0.23	0.12	2.97	–	4.21	nr	4.74
32mm	1.42	5.00	1.49	0.24	0.12	3.06	–	4.55	nr	5.12
40mm	1.61	5.00	1.69	0.25	0.13	3.22	–	4.91	nr	5.52
50mm	2.12	5.00	2.23	0.26	0.13	3.31	–	5.54	nr	6.23
215mm long; to the following nominal size steel pipes										
15mm	0.90	5.00	0.94	0.22	0.11	2.80	–	3.75	nr	4.22
20mm	1.01	5.00	1.06	0.23	0.12	2.97	–	4.03	nr	4.53
25mm	1.30	5.00	1.37	0.24	0.12	3.06	–	4.42	nr	4.98
32mm	1.55	5.00	1.63	0.25	0.13	3.22	–	4.85	nr	5.45
40mm	1.76	5.00	1.85	0.26	0.13	3.31	–	5.16	nr	5.81
50mm	2.33	5.00	2.45	0.27	0.14	3.47	–	5.92	nr	6.66
100mm long; to the following nominal size copper pipes										
15mm	0.61	5.00	0.64	0.17	0.09	2.20	–	2.84	nr	3.20
22mm	0.66	5.00	0.69	0.18	0.09	2.29	–	2.99	nr	3.36
28mm	0.83	5.00	0.87	0.19	0.10	2.46	–	3.33	nr	3.74
32mm	0.97	5.00	1.02	0.20	0.10	2.55	–	3.57	nr	4.01
45mm	1.01	5.00	1.06	0.21	0.11	2.71	–	3.77	nr	4.24
54mm	1.41	5.00	1.48	0.22	0.11	2.80	–	4.28	nr	4.82
140mm long; to the following nominal size copper pipes										
15mm	0.72	5.00	0.76	0.19	0.10	2.46	–	3.21	nr	3.61
22mm	0.80	5.00	0.84	0.20	0.10	2.55	–	3.39	nr	3.81
28mm	1.01	5.00	1.06	0.21	0.11	2.71	–	3.77	nr	4.24
32mm	1.18	5.00	1.24	0.22	0.11	2.80	–	4.04	nr	4.55
45mm	1.35	5.00	1.42	0.23	0.12	2.97	–	4.38	nr	4.93
54mm	1.75	5.00	1.84	0.24	0.12	3.06	–	4.90	nr	5.51
190mm long; to the following nominal size copper pipes										
15mm	0.86	5.00	0.90	0.21	0.11	2.71	–	3.61	nr	4.07
22mm	0.96	5.00	1.01	0.22	0.11	2.80	–	3.81	nr	4.29
28mm	1.22	5.00	1.28	0.23	0.12	2.97	–	4.25	nr	4.78
32mm	1.45	5.00	1.52	0.24	0.12	3.06	–	4.58	nr	5.15
45mm	1.65	5.00	1.73	0.25	0.13	3.22	–	4.95	Unr	5.57
54mm	2.17	5.00	2.28	0.26	0.13	3.31	–	5.59	nr	6.29
215mm long; to the following nominal size copper pipes										
15mm	0.92	5.00	0.97	0.23	0.12	2.97	–	3.93	nr	4.42
22mm	1.03	5.00	1.08	0.24	0.12	3.06	–	4.14	nr	4.66
28mm	1.33	5.00	1.40	0.25	0.13	3.22	–	4.62	nr	5.19

Labour hourly rates: (except Specialists) Craft Operatives £9.23 Labourer £7.02 Rates are national average prices. Refer to REGIONAL VARIATIONS for indicative levels of overall pricing in regions	MATERIALS			LABOUR			RATES		
	Del to Site £	Waste %	Material Cost £	Craft Optve £	Lab £	Labour Cost £	Sunds £	Nett Rate £	Unit	Gross Rate (+12.5%) £

P31: HOLES/CHASES/COVERS/SUPPORTS FOR SERVICES; BUILDERS WORK IN CONNECTION WITH MECHANICAL SERVICES Cont.

Pipe and duct sleeves Cont.

Steel pipes B.S.1387 Table 2; building into blockwork; bedding and pointing in cement mortar (1:3); making good Cont.
215mm long; to the following nominal size copper pipes Cont.

	Del to Site £	Waste %	Material Cost £	Craft Optve £	Lab £	Labour Cost £	Sunds £	Nett Rate £	Unit	Gross Rate £
32mm	1.58	5.00	1.66	0.26	0.13	3.31	-	4.97	nr	5.59
45mm	1.80	5.00	1.89	0.27	0.14	3.47	-	5.36	nr	6.04
54mm	2.39	5.00	2.51	0.28	0.14	3.57	-	6.08	nr	6.84

Ends of supports for equipment, fittings, appliances and ancillaries

Fix only; casting into concrete; making good

holderbat or bracket	-	-	-	0.29	-	2.68	0.29	2.97	nr	3.34

Fix only; building into brickwork

holderbat or bracket	-	-	-	0.12	-	1.11	-	1.11	nr	1.25

Fix only; cutting and pinning to brickwork; making good

holderbat or bracket	-	-	-	0.14	-	1.29	0.09	1.38	nr	1.55

Fix only; cutting and pinning to brickwork; making good fair face or facings one side

holderbat or bracket	-	-	-	0.17	-	1.57	0.14	1.71	nr	1.92

Cavity fixings for ends of supports for pipes and ducts

Rawlnut Multi-purpose fixings, plated pan head screws; The Rawlplug Co. Ltd; fixing to soft building board, drilling

	Del to Site £	Waste %	Material Cost £	Craft Optve £	Lab £	Labour Cost £	Sunds £	Nett Rate £	Unit	Gross Rate £
M4 with 30mm long screw; product code 09 - 130										
at 300mm centres...........................	1.18	5.00	1.24	0.19	-	1.75	-	2.99	m	3.37
at 450mm centres...........................	0.78	5.00	0.82	0.12	-	1.11	-	1.93	m	2.17
isolated..................................	0.35	5.00	0.37	0.06	-	0.55	-	0.92	nr	1.04
M5 with 30mm long screw; product code 09 - 235										
at 300mm centres...........................	1.43	5.00	1.50	0.22	-	2.03	-	3.53	m	3.97
at 450mm centres...........................	0.96	5.00	1.01	0.14	-	1.29	-	2.30	m	2.59
isolated..................................	0.43	5.00	0.45	0.07	-	0.65	-	1.10	nr	1.23
M5 with 50mm long screw; product code 09 - 317										
at 300mm centres...........................	1.57	5.00	1.65	0.22	-	2.03	-	3.68	m	4.14
at 450mm centres...........................	1.06	5.00	1.11	0.14	-	1.29	-	2.41	m	2.71
isolated..................................	0.47	5.00	0.49	0.07	-	0.65	-	1.14	nr	1.28

Rawlnut Multi-purpose fixings, plated pan head screws; The Rawlplug Co. Ltd; fixing to sheet metal, drilling

	Del to Site £	Waste %	Material Cost £	Craft Optve £	Lab £	Labour Cost £	Sunds £	Nett Rate £	Unit	Gross Rate £
M4 with 30mm long screw; product code 09 - 130										
at 300mm centres...........................	1.18	5.00	1.24	0.36	-	3.32	-	4.56	m	5.13
at 450mm centres...........................	0.78	5.00	0.82	0.24	-	2.22	-	3.03	m	3.41
isolated..................................	0.35	5.00	0.37	0.11	-	1.02	-	1.38	nr	1.56
M5 with 30mm long screw; product code 09 - 235										
at 300mm centres...........................	1.43	5.00	1.50	0.44	-	4.06	-	5.56	m	6.26
at 450mm centres...........................	0.96	5.00	1.01	0.30	-	2.77	-	3.78	m	4.25
isolated..................................	0.43	5.00	0.45	0.13	-	1.20	-	1.65	nr	1.86
M5 with 50mm long screw; product code 09 - 317										
at 300mm centres...........................	1.57	5.00	1.65	0.44	-	4.06	-	5.71	m	6.42
at 450mm centres...........................	1.06	5.00	1.11	0.30	-	2.77	-	3.88	m	4.37
isolated..................................	0.47	5.00	0.49	0.13	-	1.20	-	1.69	nr	1.91

Interset high performance cavity fixings, plated pan head screws; The Rawlplug Co. Ltd; fixing to soft building board, drilling

	Del to Site £	Waste %	Material Cost £	Craft Optve £	Lab £	Labour Cost £	Sunds £	Nett Rate £	Unit	Gross Rate £
M4 with 40mm long screw; product code 41 - 620										
at 300mm centres...........................	1.57	5.00	1.65	0.19	-	1.75	-	3.40	m	3.83
at 450mm centres...........................	1.06	5.00	1.11	0.12	-	1.11	-	2.22	m	2.50
isolated..................................	0.47	5.00	0.49	0.06	-	0.55	-	1.05	nr	1.18
M5 with 40mm long screw; product code 41 - 636										
at 300mm centres...........................	1.68	5.00	1.76	0.22	-	2.03	-	3.79	m	4.27
at 450mm centres...........................	1.12	5.00	1.18	0.14	-	1.29	-	2.47	m	2.78
isolated..................................	0.51	5.00	0.54	0.07	-	0.65	-	1.18	nr	1.33
M5 with 55mm long screw; product code 41 - 652										
at 300mm centres...........................	1.76	5.00	1.85	0.22	-	2.03	-	3.88	m	4.36
at 450mm centres...........................	1.18	5.00	1.24	0.14	-	1.29	-	2.53	m	2.85
isolated..................................	0.53	5.00	0.56	0.07	-	0.65	-	1.20	nr	1.35

Spring toggles plated pan head screws; The Rawlplug Co. Ltd; fixing to soft building board, drilling

	Del to Site £	Waste %	Material Cost £	Craft Optve £	Lab £	Labour Cost £	Sunds £	Nett Rate £	Unit	Gross Rate £
M5 with 80mm long screw; product code 94 - 439										
at 300mm centres...........................	1.18	5.00	1.24	0.22	-	2.03	-	3.27	m	3.68
at 450mm centres...........................	0.78	5.00	0.82	0.14	-	1.29	-	2.11	m	2.38
isolated..................................	0.35	5.00	0.37	0.07	-	0.65	-	1.01	nr	1.14
M6 with 60mm long screw; product code 94 - 442										
at 300mm centres...........................	1.35	5.00	1.42	0.50	-	4.62	-	6.03	m	6.79
at 450mm centres...........................	0.90	5.00	0.94	0.50	-	4.62	-	5.56	m	6.25
isolated..................................	0.41	5.00	0.43	0.50	-	4.62	-	5.05	nr	5.68
M6 with 80mm long screw; product code 94 - 464										
at 300mm centres...........................	1.68	5.00	1.76	0.50	-	4.62	-	6.38	m	7.18
at 450mm centres...........................	1.12	5.00	1.18	0.50	-	4.62	-	5.79	m	6.51

BUILDING FABRIC SUNDRIES

Labour hourly rates: (except Specialists) Craft Operatives £9.23 Labourer £7.02 Rates are national average prices. Refer to REGIONAL VARIATIONS for indicative levels of overall pricing in regions	MATERIALS			LABOUR				RATES		
	Del to Site £	Waste %	Material Cost £	Craft Optve £	Lab £	Labour Cost £	Sunds £	Nett Rate £	Unit	Gross Rate (+12.5%) £
P31: HOLES/CHASES/COVERS/SUPPORTS FOR SERVICES; BUILDERS WORK IN CONNECTION WITH MECHANICAL SERVICES Cont.										
Cavity fixings for ends of supports for pipes and ducts Cont.										
Spring toggles plated pan head screws; The Rawlplug Co. Ltd; fixing to soft building board, drilling Cont.										
M6 with 80mm long screw; product code 94 - 464 Cont.										
isolated..........................	0.51	5.00	0.54	0.50	-	4.62	-	5.15	nr	5.79
Trench covers and frames										
Duct covers in cast iron, coated; continuous covers in 610mm lengths; steel bearers										
22mm deep, for pedestrian traffic; bedding frames to concrete in cement mortar (1:3); nominal width										
150mm	89.57	-	89.57	0.50	0.50	8.13	0.75	98.44	m	110.75
225mm	93.35	-	93.35	0.55	0.55	8.94	0.75	103.04	m	115.92
300mm	108.90	-	108.90	0.65	0.65	10.56	0.75	120.21	m	135.24
375mm	124.16	-	124.16	0.70	0.70	11.38	0.75	136.29	m	153.32
450mm	139.90	-	139.90	0.75	0.75	12.19	0.75	152.84	m	171.94
Pendock Profiles Ltd. floor ducting profiles; galvanised mild steel tray section with 12mm plywood cover board; fixing tray section to masonry with nails; fixing cover board with screws, countersinking										
For 50mm screeds										
reference FDT 100/50, 100mm wide	7.17	2.50	7.35	0.40	-	3.69	0.12	11.16	m	12.56
extra; stop end	2.08	2.50	2.13	0.12	-	1.11	-	3.24	nr	3.64
extra; corner	9.05	2.50	9.28	0.29	-	2.68	0.17	12.12	nr	13.64
extra; tee	9.05	2.50	9.28	0.35	-	3.23	0.17	12.68	nr	14.26
reference FDT 150/50, 150mm wide	9.63	2.50	9.87	0.46	-	4.25	0.12	14.24	m	16.02
extra; stop end	2.08	2.50	2.13	0.14	-	1.29	-	3.42	nr	3.85
extra; corner	10.08	2.50	10.33	0.35	-	3.23	0.17	13.73	nr	15.45
extra; tee	10.08	2.50	10.33	0.40	-	3.69	0.17	14.19	nr	15.97
reference FDT 200/50, 200mm wide	11.76	2.50	12.05	0.58	-	5.35	0.12	17.53	m	19.72
extra; stop end	2.08	2.50	2.13	0.17	-	1.57	-	3.70	nr	4.16
extra; corner	11.72	2.50	12.01	0.40	-	3.69	0.17	15.88	nr	17.86
extra; tee	11.72	2.50	12.01	0.46	-	4.25	0.17	16.43	nr	18.48
For 70mm screeds										
reference FDT 100/70, 100mm wide	7.90	2.50	8.10	0.40	-	3.69	0.12	11.91	m	13.40
extra; stop end	2.08	2.50	2.13	0.12	-	1.11	-	3.24	nr	3.64
extra; corner	9.05	2.50	9.28	0.29	-	2.68	0.17	12.12	nr	13.64
extra; tee	9.05	2.50	9.28	0.35	-	3.23	0.17	12.68	nr	14.26
reference FDT 150/70, 150mm wide	10.26	2.50	10.52	0.46	-	4.25	0.12	14.88	m	16.74
extra; stop end	2.08	2.50	2.13	0.14	-	1.29	-	3.42	nr	3.85
extra; corner	10.08	2.50	10.33	0.35	-	3.23	0.17	13.73	nr	15.45
extra; tee	10.08	2.50	10.33	0.40	-	3.69	0.17	14.19	nr	15.97
reference FDT 200/70, 200mm wide	12.27	2.50	12.58	0.58	-	5.35	0.12	18.05	m	20.31
extra; stop end	2.08	2.50	2.13	0.17	-	1.57	-	3.70	nr	4.16
extra; corner	11.72	2.50	12.01	0.40	-	3.69	0.17	15.88	nr	17.86
extra; tee	11.72	2.50	12.01	0.46	-	4.25	0.17	16.43	nr	18.48
Casings, bearers and supports for equipment										
Softwood, sawn										
25mm boarded platforms	5.09	10.00	5.60	1.25	0.16	12.66	0.17	18.43	m2	20.73
50 x 50mm bearers	0.51	10.00	0.56	0.18	0.02	1.80	0.02	2.38	m	2.68
50 x 75mm bearers	0.73	10.00	0.80	0.22	0.03	2.24	0.03	3.07	m	3.46
50 x 100mm bearers	1.01	10.00	1.11	0.26	0.03	2.61	0.03	3.75	m	4.22
38 x 50mm bearers; framed	0.41	10.00	0.45	0.29	0.04	2.96	0.04	3.45	m	3.88
Softwood, wrought										
25mm boarded platforms; tongued and grooved	10.19	10.00	11.21	1.40	0.17	14.12	0.20	25.52	m2	28.71
19mm boarded sides; tongued and grooved	7.65	10.00	8.41	1.25	0.16	12.66	0.17	21.25	m2	23.90
extra for holes for pipes not exceeding 55mm nominal size	-	-	-	0.30	0.04	3.05	-	3.05	nr	3.43
19mm boarded cover; tongued and grooved; ledged; sectional	14.64	10.00	16.10	1.40	0.17	14.12	0.20	30.42	m2	34.22
Fibreboard										
13mm sides	8.80	10.00	9.68	2.40	0.30	24.26	0.34	34.28	m2	38.56
extra for holes for pipes not exceeding 55mm nominal size	-	-	-	0.25	-	2.31	-	2.31	nr	2.60
Vermiculite insulation										
packing around tank	65.19	10.00	71.71	5.50	0.70	55.68	-	127.39	m3	143.31
Slag wool insulation										
packing around tank	32.44	10.00	35.68	7.00	0.85	70.58	-	106.26	m3	119.54
P31: HOLES/CHASES/COVERS/SUPPORTS FOR SERVICES; BUILDERS WORK IN CONNECTION WITH ELECTRICAL SERVICES										
Cutting or forming holes, mortices, sinkings and chases - New buildings										
Concealed steel conduits; making good										
luminaire points	-	-	-	0.80	0.40	10.19	0.38	10.57	nr	11.89
socket outlet points	-	-	-	0.70	0.35	8.92	0.33	9.25	nr	10.40

Labour hourly rates: (except Specialists) Craft Operatives £9.23 Labourer £7.02 Rates are national average prices. Refer to REGIONAL VARIATIONS for indicative levels of overall pricing in regions	MATERIALS			LABOUR				RATES		
	Del to Site £	Waste %	Material Cost £	Craft Optve £	Lab £	Labour Cost £	Sunds £	Nett Rate £	Unit	Gross Rate (+12.5%) £
P31: HOLES/CHASES/COVERS/SUPPORTS FOR SERVICES; BUILDERS WORK IN CONNECTION WITH ELECTRICAL SERVICES Cont.										
Cutting or forming holes, mortices, sinkings and chases - New buildings Cont.										
Concealed steel conduits; making good Cont.										
fitting outlet points	-	-	-	0.70	0.35	8.92	0.33	9.25	nr	10.40
equipment and control gear points	-	-	-	0.90	0.45	11.47	0.45	11.92	nr	13.41
Exposed p.v.c. conduits; making good										
luminaire points	-	-	-	0.28	0.14	3.57	0.13	3.70	nr	4.16
socket outlet points	-	-	-	0.23	0.12	2.97	0.11	3.08	nr	3.46
fitting outlet points	-	-	-	0.23	0.12	2.97	0.11	3.08	nr	3.46
equipment and control gear points	-	-	-	0.35	0.18	4.49	0.15	4.64	nr	5.22
Cutting mortices, sinking and the like for services - Existing buildings										
In existing concrete; making good										
75 x 75 x 35mm	-	-	-	0.35	0.18	4.49	0.15	4.64	nr	5.22
150 x 75 x 35mm	-	-	-	0.66	0.18	7.36	0.83	8.19	nr	9.21
In existing brickwork										
75 x 75 x 35mm	-	-	-	0.50	0.15	5.67	0.55	6.22	nr	7.00
150 x 75 x 35mm	-	-	-	0.44	0.15	5.11	0.50	5.61	nr	6.32
In existing brickwork with plaster finish; making good										
75 x 75 x 35mm	-	-	-	0.28	0.13	3.50	0.33	3.83	nr	4.31
150 x 75 x 35mm	-	-	-	0.50	0.23	6.23	0.77	7.00	nr	7.87
In existing blockwork; making good										
75 x 75 x 35mm	-	-	-	0.33	0.20	4.45	0.55	5.00	nr	5.62
150 x 75 x 35mm	-	-	-	0.39	0.12	4.44	0.26	4.70	nr	5.29
In existing blockwork with plaster finish; making good										
75 x 75 x 35mm	-	-	-	0.28	0.09	3.22	0.24	3.46	nr	3.89
150 x 75 x 35mm	-	-	-	0.35	0.18	4.49	0.44	4.93	nr	5.55
Cutting chases for services										
In existing concrete; making good										
20mm nominal size conduits -1	-	-	-	0.46	0.23	5.86	0.38	6.24	m	7.02
20mm nominal size conduits -3	-	-	-	0.58	0.29	7.39	0.45	7.84	m	8.82
20mm nominal size conduits -6	-	-	-	0.70	0.35	8.92	0.52	9.44	m	10.62
In existing brickwork										
20mm nominal size conduits -1	-	-	-	0.35	0.18	4.49	0.38	4.87	m	5.48
20mm nominal size conduits -3	-	-	-	0.46	0.23	5.86	0.45	6.31	m	7.10
20mm nominal size conduits -6	-	-	-	0.58	0.29	7.39	0.52	7.91	m	8.90
In existing brickwork with plaster finish; making good										
20mm nominal size conduits -1	-	-	-	0.52	0.26	6.62	0.50	7.12	m	8.02
20mm nominal size conduits -3	-	-	-	0.62	0.31	7.90	0.56	8.46	m	9.52
20mm nominal size conduits -6	-	-	-	0.75	0.38	9.59	0.62	10.21	m	11.49
In existing blockwork										
20mm nominal size conduits -1	-	-	-	0.26	0.13	3.31	0.38	3.69	m	4.15
20mm nominal size conduits -3	-	-	-	0.35	0.18	4.49	0.45	4.94	m	5.56
20mm nominal size conduits -6	-	-	-	0.44	0.22	5.61	0.52	6.13	m	6.89
In existing blockwork with plaster finish; making good										
20mm nominal size conduits -1	-	-	-	0.44	0.22	5.61	0.50	6.11	m	6.87
20mm nominal size conduits -3	-	-	-	0.52	0.26	6.62	0.56	7.18	m	8.08
20mm nominal size conduits -6	-	-	-	0.60	0.30	7.64	0.62	8.26	m	9.30
Lifting and replacing floorboards										
For cables or conduits										
in groups 1 - 3	-	-	-	0.18	0.09	2.29	0.11	2.40	m	2.70
in groups 3 - 6	-	-	-	0.22	0.11	2.80	0.12	2.92	m	3.29
in groups exceeding 6	-	-	-	0.28	0.14	3.57	0.13	3.70	m	4.16

BUILDING FABRIC SUNDRIES

Paving/Planting/Fencing/Site furniture

Labour hourly rates: (except Specialists) Craft Operatives £9.23 Labourer £7.02 Rates are national average prices. Refer to REGIONAL VARIATIONS for indicative levels of overall pricing in regions	PLANT & TRANSPORT			LABOUR				RATES		
	Plant Cost £	Trans Cost £	P & T Cost £	Craft Optve Hrs	Lab Hrs	Labour Cost £	Sunds £	Nett Rate £	Unit	Gross Rate (+12.5%) £
Q10: KERBS/EDGINGS/CHANNELS/PAVING ACCESSORIES										
Excavating - by machine										
Trenches width not exceeding 0.30m										
maximum depth not exceeding 0.25m	5.60	-	5.60	-	2.00	14.04	-	19.64	m3	22.09
maximum depth not exceeding 1.00m	5.80	-	5.80	-	2.10	14.74	-	20.54	m3	23.11
Extra over any types of excavating irrespective of depth										
excavating below ground water level	4.96	-	4.96	-	-	-	-	4.96	m3	5.58
excavating in running silt, running sand or liquid mud	8.61	-	8.61	-	-	-	-	8.61	m3	9.69
Breaking out existing materials; extra over any types of excavating irrespective of depth										
hard rock..................................	25.58	-	25.58	-	-	-	-	25.58	m3	28.78
concrete..................................	21.24	-	21.24	-	-	-	-	21.24	m3	23.90
reinforced concrete	31.90	-	31.90	-	-	-	-	31.90	m3	35.89
brickwork, blockwork or stonework	14.24	-	14.24	-	-	-	-	14.24	m3	16.02
100mm diameter drain and concrete bed under	1.99	-	1.99	-	-	-	-	1.99	m	2.24
Breaking out existing hard pavings; extra over any types of excavating irrespective of depth										
concrete 150mm thick	3.35	-	3.35	-	-	-	-	3.35	m2	3.77
reinforced concrete 200mm thick	6.72	-	6.72	-	-	-	-	6.72	m2	7.56
brickwork, blockwork or stonework 100mm thick	1.54	-	1.54	-	-	-	-	1.54	m2	1.73
coated macadam or asphalt 75mm thick	0.56	-	0.56	-	-	-	-	0.56	m2	0.63
Surface treatments										
Compacting										
bottoms of excavations	0.17	-	0.17	-	-	-	-	0.17	m2	0.19
Excavating - by hand										
Trenches width not exceeding 0.30m										
maximum depth not exceeding 0.25m	-	-	-	-	4.70	32.99	-	32.99	m3	37.12
maximum depth not exceeding 1.00m	-	-	-	-	4.95	34.75	-	34.75	m3	39.09
Extra over any types of excavating irrespective of depth										
excavating below ground water level	-	-	-	-	1.65	11.58	-	11.58	m3	13.03
excavating in running silt, running sand or liquid mud	-	-	-	-	6.60	46.33	-	46.33	m3	52.12
Breaking out existing materials; extra over any types of excavating irrespective of depth										
hard rock..................................	-	-	-	-	14.70	103.19	-	103.19	m3	116.09
concrete..................................	-	-	-	-	11.00	77.22	-	77.22	m3	86.87
reinforced concrete	-	-	-	-	16.50	115.83	-	115.83	m3	130.31
brickwork, blockwork or stonework	-	-	-	-	7.70	54.05	-	54.05	m3	60.81
100mm diameter drain and concrete bed under	-	-	-	-	0.90	6.32	-	6.32	m	7.11
Breaking out existing hard pavings; extra over any types of excavating irrespective of depth										
concrete 150mm thick	-	-	-	-	1.65	11.58	-	11.58	m2	13.03
reinforced concrete 200mm thick	-	-	-	-	3.30	23.17	-	23.17	m2	26.06
brickwork, blockwork or stonework 100mm thick	-	-	-	-	1.00	7.02	-	7.02	m2	7.90
coated macadam or asphalt 75mm thick	-	-	-	-	0.30	2.11	-	2.11	m2	2.37

Labour hourly rates: (except Specialists) Craft Operatives £9.23 Labourer £7.02 Rates are national average prices. Refer to REGIONAL VARIATIONS for indicative levels of overall pricing in regions	MATERIALS			LABOUR				RATES		
	Del to Site £	Waste %	Material Cost £	Craft Optve Hrs	Lab Hrs	Labour Cost £	Sunds £	Nett Rate £	Unit	Gross Rate (+12.5%) £
Q10: KERBS/EDGINGS/CHANNELS/PAVING ACCESSORIES Cont.										
Earthwork support										

Earthwork support
  maximum depth not exceeding 1.00m; distance
  between opposing faces not exceeding 2.00m ....
	0.94	-	0.94	-	0.21	1.47	-	2.41	m2	2.72

Earthwork support; unstable ground
  maximum depth not exceeding 1.00m; distance
  between opposing faces not exceeding 2.00m ....
	1.60	-	1.60	-	0.35	2.46	-	4.06	m2	4.56

Earthwork support; next to roadways
  maximum depth not exceeding 1.00m; distance
  between opposing faces not exceeding 2.00m ....
	1.86	-	1.86	-	0.44	3.09	-	4.95	m2	5.57

Disposal										

Disposal of excavated material
  depositing on site in temporary spoil heaps where
  directed										

25m	-	-	-	-	1.40	9.83	-	9.83	m3	11.06
50m	-	-	-	-	1.65	11.58	-	11.58	m3	13.03
100m	-	-	-	-	2.20	15.44	-	15.44	m3	17.37
extra for removing beyond 100m and not exceeding 400m...............................	-	-	-	-	1.72	12.07	2.64	14.71	m3	16.55
extra for removing beyond 100m and not exceeding 800m...............................	-	-	-	-	1.72	12.07	3.42	15.49	m3	17.43
depositing on site in permanent spoil heaps average 25m distant..........................	-	-	-	-	1.40	9.83	-	9.83	m3	11.06
removing from site to tip										
a) Inert....................	-	-	-	-	1.50	10.53	14.00	24.53	m3	27.60
b) Active...................	-	-	-	-	1.50	10.53	23.00	33.53	m3	37.72
c) Contaminated (guide price - always seek a quote for specialist disposal cost............	-	-	-	-	1.50	10.53	50.00	60.53	m3	68.10
Surface treatments										
Compacting bottoms of excavations	-	-	-	-	0.16	1.12	-	1.12	m2	1.26
Precast concrete; standard or stock pattern units; B.S.7263 Part 1; bedding, jointing and pointing in cement mortar (1:3); on plain in-situ concrete foundation; B.S.5328 ordinary prescribed mix ST4, 20mm aggregate										
Kerbs; Figs. 1, 2, 6 and 7; concrete foundation and haunching; formwork										
125 x 255mm kerb; 400 x 200mm foundation	9.68	5.00	10.16	-	1.20	8.42	-	18.59	m	20.91
125 x 255mm kerb; 400 x 200mm foundation; curved 10.00m radius	12.51	5.00	13.14	-	1.28	8.99	-	22.12	m	24.89
150 x 305mm kerb; 400 x 200mm foundation	11.64	5.00	12.22	-	1.28	8.99	-	21.21	m	23.86
150 x 305mm kerb; 400 x 200mm foundation; curved 10.00m radius	14.85	5.00	15.59	-	1.36	9.55	-	25.14	m	28.28
Edgings; Figs. 10, 11, 12 and 13; concrete foundation and haunching; formwork										
50 x 150mm edging; 300 x 150mm foundation	5.76	5.00	6.05	-	0.86	6.04	-	12.09	m	13.60
50 x 200mm edging; 300 x 150mm foundation	6.10	5.00	6.41	-	0.99	6.95	-	13.35	m	15.02
50 x 250mm edging; 300 x 150mm foundation	6.31	5.00	6.63	-	1.12	7.86	-	14.49	m	16.30
Channels; Fig. 8; concrete foundation and haunching; formwork										
255 x 125mm channel; 450 x 150mm foundation	9.64	5.00	10.12	-	0.95	6.67	-	16.79	m	18.89
255 x 125mm channel; 450 x 150mm foundation; curved 10.00 radius	12.57	5.00	13.20	-	1.03	7.23	-	20.43	m	22.98
Quadrants; Fig. 14; concrete foundation and haunching; formwork										
305 x 305 x 150mm quadrant; 500 x 500 x 150mm foundation	9.89	5.00	10.38	-	0.67	4.70	-	15.09	nr	16.97
305 x 305 x 255mm quadrant; 500 x 500 x 150mm foundation	9.97	5.00	10.47	-	0.76	5.34	-	15.80	nr	17.78
455 x 455 x 150mm quadrant; 650 x 650 x 150mm foundation	11.59	5.00	12.17	-	0.94	6.60	-	18.77	nr	21.11
455 x 455 x 255mm quadrant; 650 x 650 x 150mm foundation	11.76	5.00	12.35	-	1.03	7.23	-	19.58	nr	22.03
Granite; standard units; B.S.435; bedding jointing and pointing in cement mortar (1:3); on plain in-situ concrete foundation; B.S.5328 ordinary prescribed mix ST4, 20mm aggregate										

Excavating
  Disposal of excavated material; removing from site
  to tip										

a) Inert....................	-	-	-	-	1.50	10.53	14.00	24.53	m3	27.60
b) Active...................	-	-	-	-	1.50	10.53	23.00	33.53	m3	37.72
c) Contaminated (guide price - always seek a quote for specialist disposal cost............	-	-	-	-	1.50	10.53	50.00	60.53	m3	68.10
Edge kerb; concrete foundation and haunching; formwork										
150 x 300mm; 300 x 200mm foundation	19.42	2.50	19.91	0.55	0.72	10.13	-	30.04	m	33.79

Labour hourly rates: (except Specialists) Craft Operatives £9.23 Labourer £7.02 Rates are national average prices. Refer to REGIONAL VARIATIONS for indicative levels of overall pricing in regions	MATERIALS			LABOUR				RATES		
	Del to Site £	Waste %	Material Cost £	Craft Optve Hrs	Lab Hrs	Labour Cost £	Sunds £	Nett Rate £	Unit	Gross Rate (+12.5%) £
Q10: KERBS/EDGINGS/CHANNELS/PAVING ACCESSORIES Cont.										
Granite; standard units; B.S.435; bedding jointing and pointing in cement mortar (1:3); on plain in-situ concrete foundation; B.S.5328 ordinary prescribed mix ST4, 20mm aggregate Cont.										
Edge kerb; concrete foundation and haunching; formwork Cont.										
150 x 300mm; 300 x 200mm foundation; curved external radius exceeding 1000mm	24.80	2.50	25.42	0.88	0.72	13.18	-	38.60	m	43.42
200 x 300mm; 350 x 200mm foundation	25.32	2.50	25.95	0.66	0.80	11.71	-	37.66	m	42.37
200 x 300mm; 350 x 200mm foundation; curved external radius exceeding 1000mm	32.25	2.50	33.06	0.99	0.80	14.75	-	47.81	m	53.79
Flat kerb; concrete foundation and haunching; formwork										
300 x 150mm; 450 x 200mm foundation	27.12	2.50	27.80	0.55	0.80	10.69	-	38.49	m	43.30
300 x 150mm; 450 x 200mm foundation; curved external radius exceeding 1000mm	35.07	2.50	35.95	0.88	0.80	13.74	-	49.69	m	55.90
300 x 200mm; 450 x 200mm foundation	33.98	2.50	34.83	0.66	0.83	11.92	-	46.75	m	52.59
300 x 200mm; 450 x 200mm foundation; curved external radius exceeding 1000mm	42.44	2.50	43.50	0.99	0.83	14.96	-	58.47	m	65.77
Q20: GRANULAR SUB-BASES TO ROADS/PAVINGS										
Hard, dry, broken brick or stone to be obtained off site; wheeling not exceeding 25m - by machine										
Filling to make up levels; depositing in layers 150mm maximum thickness										
average thickness not exceeding 0.25m	13.02	25.00	16.27	-	-	-	3.05	19.32	m3	21.74
average thickness exceeding 0.25m	13.02	25.00	16.27	-	-	-	2.50	18.77	m3	21.12
Surface packing to filling to vertical or battered faces	-	-	-	-	-	-	0.57	0.57	m2	0.64
Compacting with 680 kg vibratory roller filling; blinding with sand, ashes or similar fine material	0.20	40.00	0.28	-	0.05	0.35	0.21	0.84	m2	0.95
Compacting with 6 - 8 tonnes smooth wheeled roller filling; blinding with sand, ashes or similar fine material	0.20	40.00	0.28	-	0.05	0.35	0.41	1.04	m2	1.17
Hard, dry, broken brick or stone to be obtained off site; wheeling not exceeding 50m - by machine										
Filling to make up levels; depositing in layers 150mm maximum thickness										
average thickness not exceeding 0.25m	13.02	25.00	16.27	-	-	-	3.90	20.18	m3	22.70
average thickness exceeding 0.25m	13.02	25.00	16.27	-	-	-	3.05	19.32	m3	21.74
Surface packing to filling to vertical or battered faces	-	-	-	-	-	-	0.57	0.57	m2	0.64
Compacting with 680 kg vibratory roller filling; blinding with sand, ashes or similar fine material	0.20	40.00	0.28	-	0.05	0.35	0.21	0.84	m2	0.95
Compacting with 6 - 8 tonnes smooth wheeled roller filling; blinding with sand, ashes or similar fine material	0.20	40.00	0.28	-	0.05	0.35	0.41	1.04	m2	1.17
MOT Type 1 to be obtained off site; wheeling not exceeding 25m - by machine										
Filling to make up levels; depositing in layers 150mm maximum thickness										
average thickness not exceeding 0.25m	15.23	25.00	19.04	-	-	-	3.05	22.09	m3	24.85
average thickness exceeding 0.25m	15.23	25.00	19.04	-	-	-	2.40	21.44	m3	24.12
Surface packing to filling to vertical or battered faces	-	-	-	-	-	-	0.57	0.57	m2	0.64
Compacting with 680 kg vibratory roller filling; blinding with sand, ashes or similar fine material	0.20	40.00	0.28	-	0.05	0.35	0.21	0.84	m2	0.95
Compacting with 6 - 8 tonnes smooth wheeled roller filling; blinding with sand, ashes or similar fine material	0.20	40.00	0.28	-	0.05	0.35	0.41	1.04	m2	1.17
MOT Type 1 to be obtained off site; wheeling not exceeding 50m - by machine										
Filling to make up levels; depositing in layers 150mm maximum thickness										
average thickness not exceeding 0.25m	14.90	25.00	18.63	-	-	-	3.05	21.68	m3	24.38
average thickness exceeding 0.25m	14.90	25.00	18.63	-	-	-	2.50	21.13	m3	23.77
Surface packing to filling to vertical or battered faces	-	-	-	-	-	-	0.57	0.57	m2	0.64

PAVING/FENCING/SITE FURNITURE

Labour hourly rates: (except Specialists) Craft Operatives £9.23 Labourer £7.02 Rates are national average prices. Refer to REGIONAL VARIATIONS for indicative levels of overall pricing in regions	MATERIALS			LABOUR				RATES		
	Del to Site	Waste	Material Cost	Craft Optve	Lab	Labour Cost	Sunds	Nett Rate	Unit	Gross Rate (+12.5%)
	£	%	£	Hrs	Hrs	£	£	£		£
Q20: GRANULAR SUB-BASES TO ROADS/PAVINGS Cont.										
MOT Type 1 to be obtained off site; wheeling not exceeding 50m - by machine Cont.										
Compacting with 680 kg vibratory roller filling; blinding with sand, ashes or similar fine material	0.20	40.00	0.28	-	0.05	0.35	0.21	0.84	m2	0.95
Compacting with 6 - 8 tonnes smooth wheeled roller filling; blinding with sand, ashes or similar fine material	0.20	40.00	0.28	-	0.05	0.35	0.41	1.04	m2	1.17
MOT Type 2 to be obtained off site; wheeling not exceeding 25m - by machine										
Filling to make up levels; depositing in layers 150mm maximum thickness										
average thickness not exceeding 0.25m	14.90	25.00	18.63	-	-	-	13.90	32.52	m3	36.59
average thickness exceeding 0.25m	14.90	25.00	18.63	-	-	-	3.05	21.68	m3	24.38
Surface packing to filling to vertical or battered faces	-	-	-	-	-	-	0.57	0.57	m2	0.64
Compacting with 680 kg vibratory roller filling; blinding with sand, ashes or similar fine material	0.20	40.00	0.28	-	0.05	0.35	0.21	0.84	m2	0.95
Compacting with 6 - 8 tonnes smooth wheeled roller filling; blinding with sand, ashes or similar fine material	0.20	40.00	0.28	-	0.05	0.35	0.41	1.04	m2	1.17
MOT Type 2 to be obtained off site; wheeling not exceeding 50m - by machine										
Filling to make up levels; depositing in layers 150mm maximum thickness										
average thickness not exceeding 0.25m	13.00	25.00	16.25	-	-	-	3.05	19.30	m3	21.71
average thickness exceeding 0.25m	14.90	25.00	18.63	-	-	-	0.21	18.84	m3	21.19
Surface packing to filling to vertical or battered faces	-	-	-	-	-	-	0.41	0.41	m2	0.46
Compacting with 680 kg vibratory roller filling; blinding with sand, ashes or similar fine material	0.20	40.00	0.28	-	0.05	0.35	0.21	0.84	m2	0.95
Compacting with 6 - 8 tonnes smooth wheeled roller filling; blinding with sand, ashes or similar fine material	0.20	40.00	0.28	-	0.05	0.35	0.41	1.04	m2	1.17
Ashes to be obtained off site; wheeling not exceeding 25m - by machine										
Filling to make up levels; depositing in layers 150mm maximum thickness										
average thickness not exceeding 0.25m	13.00	40.00	18.20	-	-	-	3.05	21.25	m3	23.91
Compacting with 680 kg vibratory roller filling	-	-	-	-	-	-	0.21	0.21	m2	0.24
Compacting with 6 - 8 tonnes smooth wheeled roller filling	-	-	-	-	-	-	0.41	0.41	m2	0.46
Ashes to be obtained off site; wheeling not exceeding 50m - by machine										
Filling to make up levels; depositing in layers 150mm maximum thickness										
average thickness not exceeding 0.25m	13.00	40.00	18.20	-	-	-	3.90	22.10	m3	24.86
Compacting with 680 kg vibratory roller filling	-	-	-	-	-	-	0.21	0.21	m2	0.24
Compacting with 6 - 8 tonnes smooth wheeled roller filling	-	-	-	-	-	-	0.41	0.41	m2	0.46
Hoggin to be obtained off site; wheeling not exceeding 25m - by machine										
Filling to make up levels; depositing in layers 150mm maximum thickness										
average thickness not exceeding 0.25m	8.40	33.00	11.17	-	-	-	3.05	14.22	m3	16.00
average thickness exceeding 0.25m	8.40	33.00	11.17	-	-	-	2.50	13.67	m3	15.38
Compacting with 680 kg vibratory roller filling	-	-	-	-	-	-	0.21	0.21	m2	0.24
Compacting with 6 - 8 tonnes smooth wheeled roller filling	-	-	-	-	-	-	0.41	0.41	m2	0.46

Labour hourly rates: (except Specialists) Craft Operatives £9.23 Labourer £7.02 Rates are national average prices. Refer to REGIONAL VARIATIONS for indicative levels of overall pricing in regions	MATERIALS			LABOUR				RATES		
	Del to Site £	Waste %	Material Cost £	Craft Optve Hrs	Lab Hrs	Labour Cost £	Sunds £	Nett Rate £	Unit	Gross Rate (+12.5%) £
Q20: GRANULAR SUB-BASES TO ROADS/PAVINGS Cont.										
Hoggin to be obtained off site; wheeling not exceeding 50m - by machine										
Filling to make up levels; depositing in layers 150mm maximum thickness										
average thickness not exceeding 0.25m	8.40	33.00	11.17	-	-	-	3.90	15.07	m3	16.96
average thickness exceeding 0.25m	8.40	33.00	11.17	-	-	-	3.05	14.22	m3	16.00
Compacting with 680 kg vibratory roller										
filling	-	-	-	-	-	-	0.21	0.21	m2	0.24
Compacting with 6 - 8 tonnes smooth wheeled roller										
filling	-	-	-	-	-	-	0.41	0.41	m2	0.46
Hard, dry, broken brick or stone to be obtained off site; wheeling not exceeding 25m - by hand										
Filling to make up levels; depositing in layers 150mm maximum thickness										
average thickness not exceeding 0.25m	13.00	25.00	16.25	-	1.58	11.09	-	27.34	m3	30.76
average thickness exceeding 0.25m	13.00	25.00	16.25	-	1.31	9.20	-	25.45	m3	28.63
Compacting with 680 kg vibratory roller filling; blinding with sand, ashes or similar fine material	0.20	40.00	0.28	-	0.05	0.35	0.21	0.84	m2	0.95
Compacting with 6 - 8 tonnes smooth wheeled roller filling; blinding with sand, ashes or similar fine material	0.20	40.00	0.28	-	0.05	0.35	0.41	1.04	m2	1.17
Hard, dry, broken brick or stone to be obtained off site; wheeling not exceeding 50m - by hand										
Filling to make up levels; depositing in layers 150mm maximum thickness										
average thickness not exceeding 0.25m	13.00	25.00	16.25	-	1.85	12.99	-	29.24	m3	32.89
average thickness exceeding 0.25m	13.00	25.00	16.25	-	1.58	11.09	-	27.34	m3	30.76
Compacting with 680 kg vibratory roller filling; blinding with sand, ashes or similar fine material	0.20	40.00	0.28	-	0.05	0.35	0.21	0.84	m2	0.95
Compacting with 6 - 8 tonnes smooth wheeled roller filling; blinding with sand, ashes or similar fine material	0.20	40.00	0.28	-	0.05	0.35	0.41	1.04	m2	1.17
MOT Type 1 to be obtained off site; wheeling not exceeding 25m - by hand										
Filling to make up levels; depositing in layers 150mm maximum thickness										
average thickness not exceeding 0.25m	15.20	25.00	19.00	-	1.58	11.09	-	30.09	m3	33.85
average thickness exceeding 0.25m	15.20	25.00	19.00	-	1.31	9.20	-	28.20	m3	31.72
Compacting with 680 kg vibratory roller filling; blinding with sand, ashes or similar fine material	0.20	40.00	0.28	-	0.05	0.35	0.21	0.84	m2	0.95
Compacting with 6 - 8 tonnes smooth wheeled roller filling; blinding with sand, ashes or similar fine material	0.20	40.00	0.28	-	0.05	0.35	0.41	1.04	m2	1.17
MOT Type 1 to be obtained off site; wheeling not exceeding 50m - by hand										
Filling to make up levels; depositing in layers 150mm maximum thickness										
average thickness not exceeding 0.25m	15.20	25.00	19.00	-	1.85	12.99	-	31.99	m3	35.99
average thickness exceeding 0.25m	15.20	25.00	19.00	-	1.58	11.09	-	30.09	m3	33.85
Compacting with 680 kg vibratory roller filling; blinding with sand, ashes or similar fine material	0.20	40.00	0.28	-	0.05	0.35	0.21	0.84	m2	0.95
Compacting with 6 - 8 tonnes smooth wheeled roller filling; blinding with sand, ashes or similar fine material	0.20	40.00	0.28	-	0.05	0.35	0.41	1.04	m2	1.17
MOT Type 2 to be obtained off site; wheeling not exceeding 25m - by hand										
Filling to make up levels; depositing in layers 150mm maximum thickness										
average thickness not exceeding 0.25m	14.90	25.00	18.63	-	1.58	11.09	-	29.72	m3	33.43
average thickness exceeding 0.25m	14.90	25.00	18.63	-	1.31	9.20	-	27.82	m3	31.30
Compacting with 680 kg vibratory roller filling; blinding with sand, ashes or similar fine material	0.20	40.00	0.28	-	0.05	0.35	0.21	0.84	m2	0.95
Compacting with 6 - 8 tonnes smooth wheeled roller filling; blinding with sand, ashes or similar fine material	0.20	40.00	0.28	-	0.05	0.35	0.41	1.04	m2	1.17

PAVING/FENCING/SITE FURNITURE

Labour hourly rates: (except Specialists) Craft Operatives £9.23 Labourer £7.02 Rates are national average prices. Refer to REGIONAL VARIATIONS for indicative levels of overall pricing in regions	MATERIALS			LABOUR				RATES		
	Del to Site £	Waste %	Material Cost £	Craft Optve Hrs	Lab Hrs	Labour Cost £	Sunds £	Nett Rate £	Unit	Gross Rate (+12.5%) £
Q20: GRANULAR SUB-BASES TO ROADS/PAVINGS Cont.										
MOT Type 2 to be obtained off site; wheeling not exceeding 50m – by hand										
Filling to make up levels; depositing in layers 150mm maximum thickness										
average thickness not exceeding 0.25m	14.90	25.00	18.63	-	1.85	12.99	-	31.61	m3	35.56
average thickness exceeding 0.25m	14.90	25.00	18.63	-	1.58	11.09	-	29.72	m3	33.43
Compacting with 680 kg vibratory roller										
filling; blinding with sand, ashes or similar fine material	0.20	40.00	0.28	-	0.05	0.35	0.21	0.84	m2	0.95
Compacting with 6 - 8 tonnes smooth wheeled roller										
filling; blinding with sand, ashes or similar fine material	0.20	40.00	0.28	-	0.05	0.35	0.41	1.04	m2	1.17
Ashes to be obtained off site; wheeling not exceeding 25m - by hand										
Filling to make up levels; depositing in layers 150mm maximum thickness										
average thickness not exceeding 0.25m	13.00	40.00	18.20	-	1.70	11.93	-	30.13	m3	33.90
Compacting with 680 kg vibratory roller										
filling	-	-	-	-	-	-	0.21	0.21	m2	0.24
Compacting with 6 - 8 tonnes smooth wheeled roller										
filling	-	-	-	-	-	-	0.41	0.41	m2	0.46
Ashes to be obtained off site; wheeling not exceeding 50m - by hand										
Filling to make up levels; depositing in layers 150mm maximum thickness										
average thickness not exceeding 0.25m	13.00	40.00	18.20	-	1.94	13.62	-	31.82	m3	35.80
Compacting with 680 kg vibratory roller										
filling	-	-	-	-	-	-	0.21	0.21	m2	0.24
Compacting with 6 - 8 tonnes smooth wheeled roller										
filling	-	-	-	-	-	-	0.41	0.41	m2	0.46
Hoggin to be obtained off site; wheeling not exceeding 25m - by hand										
Filling to make up levels; depositing in layers 150mm maximum thickness										
average thickness not exceeding 0.25m	8.40	33.00	11.17	-	1.55	10.88	-	22.05	m3	24.81
average thickness exceeding 0.25m	8.40	33.00	11.17	-	1.30	9.13	-	20.30	m3	22.84
Compacting with 680 kg vibratory roller										
filling	-	-	-	-	-	-	0.21	0.21	m2	0.24
Compacting with 6 - 8 tonnes smooth wheeled roller										
filling	-	-	-	-	-	-	0.41	0.41	m2	0.46
Hoggin to be obtained off site; wheeling not exceeding 50m - by hand										
Filling to make up levels; depositing in layers 150mm maximum thickness										
average thickness not exceeding 0.25m	8.40	33.00	11.17	-	1.80	12.64	-	23.81	m3	26.78
average thickness exceeding 0.25m	8.40	33.00	11.17	-	1.55	10.88	-	22.05	m3	24.81
Compacting with 680 kg vibratory roller										
filling	-	-	-	-	-	-	0.21	0.21	m2	0.24
Compacting with 6 - 8 tonnes smooth wheeled roller										
filling	-	-	-	-	-	-	0.41	0.41	m2	0.46
Q21: IN-SITU CONCRETE ROADS/PAVINGS										
Plain in-situ concrete; mix 1:8, all in aggregate										
Beds; poured on or against earth or unblinded hardcore										
thickness not exceeding 150mm	57.86	7.50	62.20	-	4.68	32.85	1.80	96.85	m3	108.96
Plain in-situ concrete; mix 1:6; all in aggregate										
Beds; poured on or against earth or unblinded hardcore										
thickness 150 - 450mm	64.74	7.50	69.60	-	4.40	30.89	1.80	102.28	m3	115.07
thickness not exceeding 150mm	64.74	7.50	69.60	-	4.68	32.85	64.74	167.19	m3	188.09
Plain in-situ concrete; B.S.5328, ordinary prescribed mix ST4, 20mm aggregate										
Beds										
thickness not exceeding 150mm	53.32	5.00	55.99	-	4.68	32.85	1.80	90.64	m3	101.97
Beds; poured on or against earth or unblinded hardcore										
thickness not exceeding 150mm	53.32	7.50	57.32	-	4.68	32.85	1.80	91.97	m3	103.47

Labour hourly rates: (except Specialists) Craft Operatives £9.23 Labourer £7.02 Rates are national average prices. Refer to REGIONAL VARIATIONS for indicative levels of overall pricing in regions	MATERIALS			LABOUR				RATES		
	Del to Site £	Waste %	Material Cost £	Craft Optve Hrs	Lab Hrs	Labour Cost £	Sunds £	Nett Rate £	Unit	Gross Rate (+12.5%) £
Q21: IN-SITU CONCRETE ROADS/PAVINGS Cont.										
Reinforced in-situ concrete; B.S.5328, designed mix C20, 20mm aggregate, minimum cement content 240 kg/m3; vibrated										
Beds										
thickness 150 - 450mm	52.89	5.00	55.53	–	4.68	32.85	0.50	88.89	m3	100.00
thickness not exceeding 150mm	52.89	5.00	55.53	–	4.95	34.75	0.50	90.78	m3	102.13
Formwork and basic finish										
Edges of beds										
height not exceeding 250mm	0.94	12.50	1.06	0.58	0.12	6.20	0.24	7.49	m	8.43
height 250 - 500mm	1.65	12.50	1.86	1.06	0.21	11.26	0.45	13.56	m	15.26
Formwork steel forms										
Edges of beds										
height not exceeding 250mm	0.66	15.00	0.76	0.40	0.08	4.25	0.24	5.25	m	5.91
height 250 - 500mm	1.16	15.00	1.33	0.70	0.14	7.44	0.45	9.23	m	10.38
Reinforcement fabric; B.S.4483, hard drawn plain round steel; welded; including laps, tying wire, all cutting and bending, and spacers and chairs which are at the discretion of the Contractor										
Reference C283, 2.61 kg/m2; 100mm side laps; 400mm end laps										
generally	0.92	15.00	1.06	0.06	0.02	0.69	0.07	1.82	m2	2.05
Reference C385, 3.41 kg/m2; 100mm side laps; 400mm end laps										
generally	1.20	15.00	1.38	0.06	0.02	0.69	0.09	2.16	m2	2.43
Reference C503, 4.34 kg/m2; 100mm side laps; 400mm end laps										
generally	1.48	15.00	1.70	0.07	0.02	0.79	0.11	2.60	m2	2.92
Reference C636, 5.55 kg/m2; 100mm side laps; 400mm end laps										
generally	1.97	15.00	2.27	0.07	0.02	0.79	0.14	3.19	m2	3.59
Reference C785, 6.72 kg/m2; 100mm side laps; 400mm end laps										
generally	2.38	15.00	2.74	0.08	0.02	0.88	0.17	3.79	m2	4.26
Formed joints										
Sealant to joint; Grace Construction Products. ' Servimastic 96' including preparation, cleaners, primers and sealers										
10 x 25mm	1.93	5.00	2.03	–	0.11	0.77	–	2.80	m	3.15
12.5 x 25mm	2.50	5.00	2.63	–	0.12	0.84	–	3.47	m	3.90
Worked finish on in-situ concrete										
Tamping by mechanical means										
level surfaces	–	–	–		0.16	1.12	0.29	1.41	m2	1.59
sloping surfaces	–	–	–		0.33	2.32	0.38	2.70	m2	3.03
surfaces to falls	–	–	–		0.44	3.09	0.38	3.47	m2	3.90
Trowelling										
level surfaces	–	–	–		0.36	2.53	–	2.53	m2	2.84
sloping surfaces	–	–	–		0.44	3.09	–	3.09	m2	3.47
surfaces to falls	–	–	–		0.55	3.86	–	3.86	m2	4.34
Rolling with an indenting roller										
level surfaces	–	–	–		0.11	0.77	0.28	1.05	m2	1.18
sloping surfaces	–	–	–		0.17	1.19	0.28	1.47	m2	1.66
surfaces to falls	–	–	–		0.33	2.32	0.36	2.68	m2	3.01
Accessories cast into in-situ concrete										
Mild steel dowels; half coated with bitumen										
16mm diameter x 400mm long	0.24	10.00	0.26	–	0.06	0.42	0.18	0.87	nr	0.97
20mm diameter x 500mm long; with plastic compression cap	0.46	10.00	0.51	–	0.08	0.56	0.18	1.25	nr	1.40
Q22: COATED MACADAM/ASPHALT ROADS/PAVINGS										
Coated macadam, B.S.4987; base course 28mm nominal size aggregate, 50mm thick; wearing course 6mm nominal size aggregate, 20mm thick; rolled with 3 - 4 tonne roller; external										
70mm roads on concrete base; to falls and crossfalls, and slopes not exceeding 15 degrees from horizontal										
generally	–	–	Spclist	–	–	Spclist	–	13.82	m2	15.55
70mm pavings on concrete base; to falls and crossfalls, and slopes not exceeding 15 degrees from horizontal										
generally	–	–	Spclist	–	–	Spclist	–	15.37	m2	17.29

PAVING/FENCING/SITE FURNITURE

Labour hourly rates: (except Specialists) Craft Operatives £9.23 Labourer £7.02 Rates are national average prices. Refer to REGIONAL VARIATIONS for indicative levels of overall pricing in regions	MATERIALS			LABOUR				RATES		
	Del to Site £	Waste %	Material Cost £	Craft Optve Hrs	Lab Hrs	Labour Cost £	Sunds £	Nett Rate £	Unit	Gross Rate (+12.5%) £
Q22: COATED MACADAM/ASPHALT ROADS/PAVINGS Cont.										
Coated macadam, B.S.4987; base course 28mm nominal size aggregate, 50mm thick; wearing course 6mm nominal size aggregate, 20mm thick; rolled with 3 - 4 tonne roller; dressing surface with coated grit brushed on and lightly rolled; external										
70mm roads on concrete base; to falls and crossfalls, and slopes not exceeding 15 degrees from horizontal										
generally	-	-	Spclist	-	-	Spclist	-	16.74	m2	18.83
70mm pavings on concrete base; to falls and crossfalls, and slopes not exceeding 15 degrees from horizontal										
generally	-	-	Spclist	-	-	Spclist	-	18.25	m2	20.53
Coated macadam, B.S.4987; base course 40mm nominal size aggregate, 65mm thick; wearing course 10mm nominal size aggregate, 25mm thick; rolled with 6 - 8 tonne roller; external										
90mm roads on concrete base; to falls and crossfalls, and slopes not exceeding 15 degrees from horizontal										
generally	-	-	Spclist	-	-	Spclist	-	16.74	m2	18.83
90mm pavings on concrete base; to falls and crossfalls, and slopes not exceeding 15 degrees from horizontal										
generally	-	-	Spclist	-	-	Spclist	-	18.36	m2	20.66
Coated macadam, B.S.4987; base course 40mm nominal size aggregate, 65mm thick; wearing course 10mm nominal size aggregate, 25mm thick; rolled with 6 - 8 tonne roller; dressing surface with coated grit brushed on and lightly rolled; external										
90mm roads on concrete base; to falls and crossfalls, and slopes not exceeding 15 degrees from horizontal										
generally	-	-	Spclist	-	-	Spclist	-	19.86	m2	22.34
90mm pavings on concrete base; to falls and crossfalls, and slopes not exceeding 15 degrees from horizontal										
generally	-	-	Spclist	-	-	Spclist	-	20.55	m2	23.12
Fine cold asphalt, B.S.4987; single course 6mm nominal size aggregate; rolled with 3 - 4 tonne roller; external										
12mm roads on concrete base; to falls and crossfalls, and slopes not exceeding 15 degrees from horizontal										
generally	-	-	Spclist	-	-	Spclist	-	6.30	m2	7.09
19mm roads on concrete base; to falls and crossfalls, and slopes not exceeding 15 degrees from horizontal										
generally	-	-	Spclist	-	-	Spclist	-	7.30	m2	8.21
25mm roads on concrete base; to falls and crossfalls, and slopes not exceeding 15 degrees from horizontal										
generally	-	-	Spclist	-	-	Spclist	-	8.67	m2	9.75
12mm pavings on concrete base; to falls and crossfalls, and slopes not exceeding 15 degrees from horizontal										
generally	-	-	Spclist	-	-	Spclist	-	6.68	m2	7.51
19mm pavings on concrete base; to falls and crossfalls, and slopes not exceeding 15 degrees from horizontal										
generally	-	-	Spclist	-	-	Spclist	-	8.38	m2	9.43
25mm pavings on concrete base; to falls and crossfalls, and slopes not exceeding 15 degrees from horizontal										
generally	-	-	Spclist	-	-	Spclist	-	9.59	m2	10.79
GRAVEL/HOGGIN/WOODCHIP ROADS/PAVINGS										
Gravel paving; first layer clinker 75mm thick; intermediate layer coarse gravel 75mm thick; wearing layer blinding gravel 38mm thick; well water and roll										
188mm roads on blinded hardcore base; to falls and crossfalls and slopes not exceeding 15 degrees from horizontal										
generally	4.54	10.00	4.99	-	0.44	3.09	-	8.08	m2	9.09
Gravel paving; single layer blinding gravel; well water and roll										
50mm pavings on blinded hardcore base; to falls and crossfalls and slopes not exceeding 15 degrees from horizontal										
generally	1.63	10.00	1.79	-	0.15	1.05	-	2.85	m2	3.20

Labour hourly rates: (except Specialists) Craft Operatives £9.23 Labourer £7.02 Rates are national average prices. Refer to REGIONAL VARIATIONS for indicative levels of overall pricing in regions	MATERIALS			LABOUR				RATES		
	Del to Site £	Waste %	Material Cost £	Craft Optve Hrs	Lab Hrs	Labour Cost £	Sunds £	Nett Rate £	Unit	Gross Rate (+12.5%) £
GRAVEL/HOGGIN/WOODCHIP ROADS/PAVINGS Cont.										
Gravel paving; first layer clinker 50mm thick; **wearing layer blinding gravel 38mm thick; well water** **and roll**										
88mm pavings on blinded hardcore base; to falls and crossfalls and slopes not exceeding 15 degrees from horizontal										
generally	1.96	10.00	2.16	-	0.26	1.83	-	3.98	m2	4.48
extra; 'Colas' and 10mm shingle dressing; one coat	1.10	5.00	1.16	-	0.17	1.19	-	2.35	m2	2.64
extra; 'Colas' and 10mm shingle dressing; two coats	1.87	5.00	1.96	-	0.27	1.90	-	3.86	m2	4.34
Q25: SLAB/BRICK/BLOCK/SETT/COBBLE PAVINGS										
York stone paving; 13mm thick bedding, jointing and **pointing in cement mortar (1:3)**										
Paving on blinded hardcore base										
50mm thick; to falls and crossfalls and to slopes not exceeding 15 degrees from horizontal	54.78	5.00	57.52	0.30	0.30	4.88	1.04	63.43	m2	71.36
75mm thick; to falls and crossfalls and to slopes not exceeding 15 degrees from horizontal	61.88	5.00	64.97	0.36	0.36	5.85	1.10	71.92	m2	80.91
extra; rubbed top surface	16.39	5.00	17.21	-	-	-	-	17.21	m2	19.36
Treads on concrete base										
50mm thick; 300mm wide	58.87	5.00	61.81	0.36	0.36	5.85	0.30	67.96	m	76.46
Precast concrete flags, B.S.7263 Part 1, natural **finish; 6mm joints, symmetrical layout; bedding in** **13mm cement mortar (1:3); jointing and pointing with** **lime and sand (1:2)**										
600 x 450 x 50mm units to pavings on sand, granular or blinded hardcore base										
50mm thick; to falls and crossfalls and to slopes not exceeding 15 degrees from horizontal	7.30	5.00	7.67	0.30	0.30	4.88	1.06	13.60	m2	15.30
600 x 450 x 50mm units to pavings on sand, granular or blinded hardcore base										
Extra over for red or buff coloured	3.11	5.00	3.27	-	-	-	-	3.27	m2	3.67
600 x 600 x 50mm units to pavings on sand, granular or blinded hardcore base										
50mm thick; to falls and crossfalls and to slopes not exceeding 15 degrees from horizontal	6.08	5.00	6.38	0.28	0.28	4.55	1.00	11.93	m2	13.43
600 x 600 x 50mm units to pavings on sand, granular or blinded hardcore base										
Extra over for red or buff coloured	2.81	5.00	2.95	-	-	-	-	2.95	m2	3.32
600 x 750 x 50mm units to pavings on sand, granular or blinded hardcore base										
50mm thick; to falls and crossfalls and to slopes not exceeding 15 degrees from horizontal	5.80	5.00	6.09	0.25	0.25	4.06	1.07	11.22	m2	12.63
600 x 750 x 50mm units to pavings on sand, granular or blinded hardcore base										
Extra over for red or buff coloured	2.44	5.00	2.56	-	-	-	-	2.56	m2	2.88
600 x 900 x 50mm units to pavings on sand, granular or blinded hardcore base										
50mm thick; to falls and crossfalls and to slopes not exceeding 15 degrees from horizontal	5.42	5.00	5.69	0.23	0.23	3.74	1.08	10.51	m2	11.82
600 x 900 x 50mm units to pavings on sand, granular or blinded hardcore base										
Extra over for red or buff coloured	2.07	5.00	2.17	-	-	-	-	2.17	m2	2.45
600 x 450 x 63mm units to pavings on sand, granular or blinded hardcore base										
63mm thick; to falls and crossfalls and to slopes not exceeding 15 degrees from horizontal	8.36	5.00	8.78	0.30	0.30	4.88	1.06	14.71	m2	16.55
600 x 600 x 63mm units to pavings on sand, granular or blinded hardcore base										
63mm thick; to falls and crossfalls and to slopes not exceeding 15 degrees from horizontal	7.05	5.00	7.40	0.28	0.28	4.55	1.00	12.95	m2	14.57
600 x 750 x 63mm units to pavings on sand , granular or blinded hardcore base										
63mm thick; to falls and crossfalls and to slopes not exceeding 15 degrees from horizontal	6.47	5.00	6.79	0.25	0.25	4.06	1.01	11.87	m2	13.35
600 x 900 x 63mm units to pavings on sand, granular or blinded hardcore base										
63mm thick; to falls and crossfalls and to slopes not exceeding 15 degrees from horizontal	5.97	5.00	6.27	0.23	0.23	3.74	1.02	11.03	m2	12.40

Labour hourly rates: (except Specialists) Craft Operatives £9.23 Labourer £7.02 Rates are national average prices. Refer to REGIONAL VARIATIONS for indicative levels of overall pricing in regions	MATERIALS			LABOUR				RATES		
	Del to Site £	Waste %	Material Cost £	Craft Optve Hrs	Lab Hrs	Labour Cost £	Sunds £	Nett Rate £	Unit	Gross Rate (+12.5%) £

Q25: SLAB/BRICK/BLOCK/SETT/COBBLE PAVINGS Cont.

Precast concrete flags, B.S.7263 Part 1, natural finish; 6mm joints, symmetrical layout; bedding in 13mm lime mortar (1:3); jointing and pointing with lime and sand (1:2)

	Del to Site £	Waste %	Material Cost £	Craft Optve Hrs	Lab Hrs	Labour Cost £	Sunds £	Nett Rate £	Unit	Gross Rate £
600 x 450 x 50mm units to pavings on sand, granular or blinded hardcore base										
50mm thick; to falls and crossfalls and to slopes not exceeding 15 degrees from horizontal	7.71	5.00	8.10	0.30	0.30	4.88	1.06	14.03	m2	15.78
601 x 450 x 50mm units to pavings on sand, granular or blinded hardcore base										
Extra over for red or buff coloured	3.11	5.00	3.27	-	-	-	-	3.27	m2	3.67
600 x 600 x 50mm units to pavings on sand, granular or blinded hardcore base										
50mm thick; to falls and crossfalls and to slopes not exceeding 15 degrees from horizontal	6.47	5.00	6.79	0.28	0.28	4.55	1.00	12.34	m2	13.89
600 x 600 x 50mm units to pavings on sand, granular or blinded hardcore base										
Extra over for red or buff coloured	2.81	5.00	2.95	-	-	-	-	2.95	m2	3.32
600 x 750 x 50mm units to pavings on sand, granular or blinded hardcore base										
50mm thick; to falls and crossfalls and to slopes not exceeding 15 degrees from horizontal	6.19	5.00	6.50	0.25	0.25	4.06	1.01	11.57	m2	13.02
600 x 750 x 50mm units to pavings on sand, granular or blinded hardcore base										
Extra over for red or buff coloured	2.44	5.00	2.56	-	-	-	-	2.56	m2	2.88
600 x 900 x 50mm units to pavings on sand, granular or blinded hardcore base										
50mm thick; to falls and crossfalls and to slopes not exceeding 15 degrees from horizontal	5.80	5.00	6.09	0.23	0.23	3.74	1.06	10.89	m2	12.25
600 x 900 x 50mm units to pavings on sand, granular or blinded hardcore base										
Extra over for red or buff coloured	2.07	5.00	2.17	-	-	-	-	2.17	m2	2.45
600 x 450 x 63mm units to pavings on sand, granular or blinded hardcore base										
63mm thick; to falls and crossfalls and to slopes not exceeding 15 degrees from horizontal	8.75	5.00	9.19	0.30	0.30	4.88	1.06	15.12	m2	17.01
600 x 600 x 63mm units to pavings on sand, granular or blinded hardcore base										
63mm thick; to falls and crossfalls and to slopes not exceeding 15 degrees from horizontal	7.44	5.00	7.81	0.28	0.28	4.55	1.00	13.36	m2	15.03
600 x 750 x 63mm units to pavings on sand, granular or blinded hardcore base										
63mm thick; to falls and crossfalls and to slopes not exceeding 15 degrees from horizontal	6.86	5.00	7.20	0.25	0.25	4.06	1.01	12.28	m2	13.81
600 x 900 x 63mm units to pavings on sand , granular or blinded hardcore base										
63mm thick; to falls and crossfalls and to slopes not exceeding 15 degrees from horizontal	6.36	5.00	6.68	0.23	0.23	3.74	1.02	11.44	m2	12.86

Precast concrete flags, B.S.7263 Part 1, natural finish; 6mm joints, symmetrical layout; bedding in 25mm sand; jointing and pointing with lime and sand (1:2)

	Del to Site £	Waste %	Material Cost £	Craft Optve Hrs	Lab Hrs	Labour Cost £	Sunds £	Nett Rate £	Unit	Gross Rate £
600 x 450 x 50mm units to pavings on sand, granular or blinded hardcore base										
50mm thick; to falls and crossfalls and to slopes not exceeding 15 degrees from horizontal	6.68	5.00	7.01	0.30	0.30	4.88	1.06	12.95	m2	14.57
600 x 450 x 50mm units to pavings on sand, granular or blinded hardcore base										
Extra over for red or buff coloured	3.11	5.00	3.27	-	-	-	-	3.27	m2	3.67
600 x 600 x 50mm units to pavings on sand, granular or blinded hardcore base										
50mm thick; to falls and crossfalls and to slopes not exceeding 15 degrees from horizontal	5.44	5.00	5.71	0.28	0.28	4.55	1.00	11.26	m2	12.67
600 x 600 x 50mm units to pavings on sand, granular or blinded hardcore base										
Extra over for red or buff coloured	2.81	5.00	2.95	-	-	-	-	2.95	m2	3.32
600 x 750 x 50mm units to pavings on sand, granular or blinded hardcore base										
50mm thick; to falls and crossfalls and to slopes not exceeding 15 degrees from horizontal	5.16	5.00	5.42	0.25	0.25	4.06	1.01	10.49	m2	11.80
600 x 750 x 50mm units to pavings on sand, granular or blinded hardcore base										
Extra over for red or buff coloured	2.44	5.00	2.56	-	-	-	-	2.56	m2	2.88

Labour hourly rates: (except Specialists) Craft Operatives £9.23 Labourer £7.02 Rates are national average prices. Refer to REGIONAL VARIATIONS for indicative levels of overall pricing in regions	MATERIALS			LABOUR				RATES		
	Del to Site	Waste	Material Cost	Craft Optve	Lab	Labour Cost	Sunds	Nett Rate	Unit	Gross Rate (+12.5%)
	£	%	£	Hrs	Hrs	£	£	£		£
Q25: SLAB/BRICK/BLOCK/SETT/COBBLE PAVINGS Cont.										
Precast concrete flags, B.S.7263 Part 1, natural finish; 6mm joints, symmetrical layout; bedding in 25mm sand; jointing and pointing with lime and sand (1:2) Cont.										
600 x 900 x 50mm units to pavings on sand, granular or blinded hardcore base										
50mm thick; to falls and crossfalls and to slopes not exceeding 15 degrees from horizontal	4.77	5.00	5.01	0.23	0.23	3.74	1.02	9.77	m2	10.99
600 x 900 x 50mm units to pavings on sand, granular or blinded hardcore base										
Extra over for red or buff coloured	2.07	5.00	2.17	-	-	-	-	2.17	m2	2.45
Keyblok precast concrete paving blocks; Marshalls Mono Ltd Driveline 50; in standard units; bedding on sand; covering with sand, compacting with plate vibrator, sweeping off surplus										
200 x 100mm units to pavings on 50mm sand base; natural colour										
symmetrical half bond layout; level and to falls only										
50mm thick..................................	7.50	10.00	8.25	-	1.35	9.48	1.42	19.15	m2	21.54
laid in straight herringbone pattern; level and to falls only										
50mm thick..................................	7.50	10.00	8.25	-	1.50	10.53	1.42	20.20	m2	22.73
Brick paving on concrete base; 13mm thick bedding, jointing and pointing in cement mortar (1:3)										
Paving to falls and crossfalls and to slopes not exceeding 15 degrees from horizontal										
25mm thick Brick paviors P.C. £300.00 per 1000 ...	18.00	5.00	18.90	2.75	1.55	36.26	1.04	56.20	m2	63.23
65mm thick Facing bricks P.C. £210.00 per 1000 ...	12.60	5.00	13.23	2.75	1.55	36.26	1.04	50.53	m2	56.85
50mm thick Staffordshire blue chequered paviors P.C. £300.00 per 1000	18.00	5.00	18.90	2.75	1.55	36.26	1.04	56.20	m2	63.23
50mm thick Accrington non slip paviors P.C. £300.00 per 1000	18.00	5.00	18.90	2.75	1.55	36.26	1.04	56.20	m2	63.23
Granite sett paving on concrete base; 13mm thick bedding and grouting in cement mortar (1:3)										
Paving to falls and crossfalls and to slopes not exceeding 15 degrees from horizontal										
125mm thick; new	32.93	2.50	33.75	1.80	2.20	32.06	2.31	68.12	m2	76.64
150mm thick; new	38.02	2.50	38.97	2.00	2.35	34.96	2.04	75.97	m2	85.46
100mm thick; reclaimed	35.90	2.50	36.80	1.65	2.00	29.27	1.98	68.05	m2	76.55
150mm thick; reclaimed	32.08	2.50	32.88	1.80	2.20	32.06	2.31	67.25	m2	75.66
200mm thick; reclaimed	52.80	2.50	54.12	2.00	2.35	34.96	2.64	91.72	m2	103.18
Cobble paving set in concrete bed (measured separately) tight butted, dry grouted with cement and sand (1:3) watered and brushed										
Paving level and to falls only										
plain	13.20	2.50	13.53	1.85	1.85	30.06	2.70	46.29	m2	52.08
set to pattern	13.20	2.50	13.53	2.20	2.20	35.75	2.70	51.98	m2	58.48
Crazy paving on blinded hardcore base, bedding in 38mm thick sand; pointing in cement mortar (1:4)										
Paving to falls and crossfalls and to slopes not exceeding 15 degrees from horizontal										
50mm thick precast concrete flag	4.80	2.50	4.92	0.50	0.50	8.13	2.35	15.40	m2	17.32
38-50mm thick York stone	53.00	2.50	54.33	0.70	0.70	11.38	3.40	69.10	m2	77.74
38mm thick Westmorland green slate	100.00	2.50	102.50	0.70	0.70	11.38	2.35	116.22	m2	130.75
Grass concrete paving; voids filled with vegetable soil and sown with grass seed										
Precast concrete perforated slabs										
120mm thick units on 25mm thick sand bed; level and to falls only	19.30	2.50	19.78	0.32	0.32	5.20	1.16	26.14	m2	29.41
Q30: SEEDING/TURFING										
Cultivating										
Surfaces of natural ground										
digging over one spit deep; removing debris; weeding..................................	-	-	-	-	0.28	1.97	-	1.97	m2	2.21
Surfaces of filling										
digging over one spit deep	-	-	-	-	0.33	2.32	-	2.32	m2	2.61
Surface applications										
Bone meal, 0.06 kg/m2; raking in										
general surfaces	0.21	5.00	0.22	-	0.06	0.42	-	0.64	m2	0.72
Selective weedkiller, 0.03 kg/m2; applying by spreader										
general surfaces	0.11	5.00	0.12	-	0.06	0.42	-	0.54	m2	0.60

PAVING/FENCING/SITE FURNITURE

Labour hourly rates: (except Specialists) Craft Operatives £9.23 Labourer £7.02 Rates are national average prices. Refer to REGIONAL VARIATIONS for indicative levels of overall pricing in regions	MATERIALS			LABOUR				RATES		
	Del to Site	Waste	Material Cost	Craft Optve	Lab	Labour Cost	Sunds	Nett Rate	Unit	Gross Rate (+12.5%)
	£	%	£	Hrs	Hrs	£	£	£		£
Q30: SEEDING/TURFING Cont.										
Surface applications Cont.										
Pre-seeding fertilizer at the rate of 0.05 kg/m2										
general surfaces	0.07	5.00	0.07	-	0.06	0.42	-	0.49	m2	0.56
General grass fertilizer at the rate of 0.06 kg/m2										
general surfaces	0.08	5.00	0.08	-	0.06	0.42	-	0.51	m2	0.57
Seeding										
Grass seed, 0.07 kg/m2, raking in; rolling; maintaining for 12 months after laying										
general surfaces	0.31	5.00	0.33	-	0.30	2.11	-	2.43	m2	2.74
Stone pick, roll and cut grass with a rotary cutter and remove arisings										
general surfaces	-	-	-	-	0.11	0.77	-	0.77	m2	0.87
Turfing										
Take turf from stack, wheel not exceeding 100m, lay, roll and water, maintaining for 12 months after laying										
general surfaces	-	-	-	-	0.70	4.91	-	4.91	m2	5.53
Imported turf; cultivated; lay, roll and water; maintaining for 12 months after laying										
general surfaces	2.65	5.00	2.78	-	0.50	3.51	-	6.29	m2	7.08
Q31: PLANTING										
Planting trees, shrubs and hedge plants										
Excavate or form pit, hole or trench, dig over ground in bottom, spread and pack around roots with finely broken soil, refill with top soil with one third by volume of farmyard manure incorporated, water in, remove surplus excavated material and provide labelling										
small tree	0.63	-	0.63	-	0.85	5.97	0.37	6.97	nr	7.84
medium tree	0.79	-	0.79	-	1.65	11.58	0.53	12.90	nr	14.52
large tree	1.00	-	1.00	-	3.30	23.17	0.68	24.85	nr	27.95
shrub	0.47	-	0.47	-	0.55	3.86	0.16	4.49	nr	5.05
hedge plant	0.47	-	0.47	-	0.28	1.97	0.16	2.60	nr	2.92
60mm diameter treated softwood tree stake, pointed and driven into ground and with two PVC tree ties secured around tree and nailed to stake										
2100mm long	1.90	-	1.90	-	0.35	2.46	1.38	5.74	nr	6.45
2400mm long	2.09	-	2.09	-	0.35	2.46	1.38	5.93	nr	6.67
2700mm long	2.45	-	2.45	-	0.35	2.46	1.38	6.29	nr	7.07
Planting herbaceous plants, bulbs, corms and tubers										
Provide planting bed of topsoil with one third by volume of farmyard manure incorporated										
150mm thick	3.94	5.00	4.14	-	0.80	5.62	-	9.75	m2	10.97
225mm thick	5.93	5.00	6.23	-	1.25	8.78	-	15.00	m2	16.88
Form hole and plant										
herbaceous plants	0.53	5.00	0.56	-	0.22	1.54	-	2.10	nr	2.36
bulbs, corms and tubers	0.11	5.00	0.12	-	0.11	0.77	-	0.89	nr	1.00
Mulching after planting										
25mm peat	0.84	-	0.84	-	0.11	0.77	-	1.61	m2	1.81
75mm farmyard manure	0.32	-	0.32	-	0.11	0.77	-	1.09	m2	1.23
Q40: FENCING										
POST AND WIRE FENCING										
Fencing; strained wire; B.S.1722 Part 3; concrete posts and struts; 4.00mm diameter galvanised mild steel line wires; galvanised steel fittings and accessories; backfilling around posts in concrete mix ST4										
900mm high fencing; type SC90; wires -3; posts with rounded tops										
posts at 2743mm centres; 600mm into ground	2.07	2.50	2.12	-	0.85	5.97	0.49	8.58	m	9.65
extra; end posts with struts -1	16.89	2.50	17.31	-	1.10	7.72	2.67	27.70	nr	31.17
extra; angle posts with struts -2	28.07	2.50	28.77	-	1.65	11.58	4.00	44.35	nr	49.90
1050mm high fencing; type SC105A; wires -5; posts rounded tops										
posts at 2743mm centres; 600mm into ground	2.32	2.50	2.38	-	0.90	6.32	0.49	9.19	m	10.33
extra; end posts with struts -1	22.45	2.50	23.01	-	1.20	8.42	2.67	34.11	nr	38.37
extra; angle posts with struts -2	35.42	2.50	36.31	-	1.80	12.64	4.00	52.94	nr	59.56

Labour hourly rates: (except Specialists) Craft Operatives £9.23 Labourer £7.02 Rates are national average prices. Refer to REGIONAL VARIATIONS for indicative levels of overall pricing in regions	MATERIALS			LABOUR				RATES		
	Del to Site £	Waste %	Material Cost £	Craft Optve Hrs	Lab Hrs	Labour Cost £	Sunds £	Nett Rate £	Unit	Gross Rate (+12.5%) £
CHAIN LINK FENCING										
Fencing; chain link; B.S.1722 Part 1; concrete posts; galvanised mesh, line and tying wires; galvanised steel fittings and accessories; backfilling around posts in concrete mix ST4										
900mm high fencing; type GLC 90; posts with rounded tops										
posts at 2743mm centres; 600mm into ground	3.86	5.00	4.05	-	1.10	7.72	0.49	12.27	m	13.80
extra; end posts with struts -1	15.72	5.00	16.51	-	1.30	9.13	2.67	28.30	nr	31.84
extra; angle posts with struts -2	25.74	5.00	27.03	-	1.85	12.99	4.00	44.01	nr	49.52
1200mm high fencing; type GLC 120; posts with rounded tops										
posts at 2743mm centres; 600mm into ground	4.51	5.00	4.74	-	1.20	8.42	0.49	13.65	m	15.36
extra; end posts with struts -1	21.62	5.00	22.70	-	1.60	11.23	2.67	36.60	nr	41.18
extra; angle posts with struts -2	34.59	5.00	36.32	-	2.05	14.39	4.00	54.71	nr	61.55
1400mm high fencing; type GLC 140A; posts with rounded tops										
posts at 2743mm centres; 600mm into ground	5.51	5.00	5.79	-	1.30	9.13	0.49	15.40	m	17.33
extra; end posts with struts -1	23.70	5.00	24.89	-	1.80	12.64	2.67	40.19	nr	45.21
extra; angle posts with struts -2	37.71	5.00	39.60	-	2.30	16.15	4.00	59.74	nr	67.21
1800mm high fencing; type GLC 180; posts with rounded tops										
posts at 2743mm centres; 600mm into ground	6.86	5.00	7.20	-	1.60	11.23	0.49	18.93	m	21.29
extra; end posts with struts -1	27.58	5.00	28.96	-	1.90	13.34	2.67	44.97	nr	50.59
extra; angle posts with struts -2	43.48	5.00	45.65	-	2.45	17.20	4.00	66.85	nr	75.21
Fencing; chain link; B.S.1722 Part 1; concrete posts; plastics coated Grade A mesh, line and tying wire; galvanised steel fittings and accessories; backfilling around posts in concrete mix ST4										
900mm high fencing; type PLC 90A; posts with rounded tops										
posts at 2743mm centres; 600mm into ground	4.22	5.00	4.43	-	1.10	7.72	0.49	12.64	m	14.22
extra; end posts with struts -1	15.72	5.00	16.51	-	1.40	9.83	2.67	29.00	nr	32.63
extra; angle posts with struts -2	25.74	5.00	27.03	-	1.85	12.99	4.00	44.01	nr	49.52
1200mm high fencing; type PLC 120; posts with rounded tops										
posts at 2743mm centres; 600mm into ground	5.15	5.00	5.41	-	1.20	8.42	0.49	14.32	m	16.11
extra; end posts with struts -1	21.62	5.00	22.70	-	1.60	11.23	2.67	36.60	nr	41.18
extra; angle posts with struts -2	34.59	5.00	36.32	-	2.05	14.39	4.00	54.71	nr	61.55
1400mm high fencing; type PLC 140A; posts with rounded tops										
posts at 2743mm centres; 600mm into ground	6.04	5.00	6.34	-	1.30	9.13	0.49	15.96	m	17.95
extra; end posts with struts -1	23.70	5.00	24.89	-	1.80	12.64	2.67	40.19	nr	45.21
extra; angle posts with struts -2	37.71	5.00	39.60	-	2.30	16.15	4.00	59.74	nr	67.21
1800mm high fencing; type PLC 180; posts with rounded tops										
posts at 2743mm centres; 600mm into ground	7.32	5.00	7.69	-	1.60	11.23	0.49	19.41	m	21.83
extra; end posts with struts -1	25.53	5.00	26.81	-	1.90	13.34	2.67	42.81	nr	48.17
extra; angle posts with struts -2	43.48	5.00	45.65	-	2.45	17.20	4.00	66.85	nr	75.21
Fencing; chain link; B.S.1722 Part 1; rolled steel angle posts; galvanised mesh, line and tying wires; galvanised steel fittings and accessories; backfilling around posts in concrete mix ST4										
900mm high fencing; type GLS 90; posts with rounded tops										
posts at 2743mm centres; 600mm into ground	3.79	2.50	3.88	-	0.70	4.91	0.58	9.38	m	10.55
extra; end posts with struts -1	18.89	2.50	19.36	-	1.15	8.07	3.20	30.64	nr	34.46
extra; angle posts with struts -2	26.62	2.50	27.29	-	1.85	12.99	4.80	45.07	nr	50.71
1200mm high fencing; type GLS 120; posts with rounded tops										
posts at 2743mm centres; 600mm into ground	3.95	2.50	4.05	-	0.80	5.62	0.58	10.24	m	11.53
extra; end posts with struts -1	23.19	2.50	23.77	-	1.25	8.78	3.20	35.74	nr	40.21
extra; angle posts with struts -2	33.75	2.50	34.59	-	2.00	14.04	4.80	53.43	nr	60.11
1400mm high fencing; type GLS 140A; posts with rounded tops										
posts at 2743mm centres; 600mm into ground	5.10	2.50	5.23	-	0.90	6.32	0.58	12.13	m	13.64
extra; end posts with struts -1	24.71	2.50	25.33	-	1.70	11.93	3.20	40.46	nr	45.52
extra; angle posts with struts -2	36.13	2.50	37.03	-	2.20	15.44	4.80	57.28	nr	64.44
1800mm high fencing; type GLS 180; posts with rounded tops										
posts at 2743mm centres; 600mm into ground	6.31	2.50	6.47	-	1.10	7.72	0.58	14.77	m	16.62
extra; end posts with struts -1	33.30	2.50	34.13	-	1.85	12.99	3.20	50.32	nr	56.61
extra; angle posts with struts -2	47.93	2.50	49.13	-	2.40	16.85	4.80	70.78	nr	79.62

	MATERIALS			LABOUR				RATES		
Labour hourly rates: (except Specialists) Craft Operatives £9.23 Labourer £7.02 Rates are national average prices. Refer to REGIONAL VARIATIONS for indicative levels of overall pricing in regions	Del to Site £	Waste %	Material Cost £	Craft Optve Hrs	Lab Hrs	Labour Cost £	Sunds £	Nett Rate £	Unit	Gross Rate (+12.5%) £

CHAIN LINK FENCING Cont.

Fencing; chain link; B.S.1722 Part 1; rolled steel angle posts; plastics coated Grade A mesh, line and tying wires; galvanised steel fittings and accessories; backfilling around posts in concrete mix ST4

900mm high fencing; type PLS 90A; posts with rounded tops

	Del to Site £	Waste %	Material Cost £	Craft Optve Hrs	Lab Hrs	Labour Cost £	Sunds £	Nett Rate £	Unit	Gross Rate £
posts at 2743mm centres; 600mm into ground	3.94	2.50	4.04	-	0.70	4.91	0.58	9.53	m	10.72
extra; end posts with struts -1	18.89	2.50	19.36	-	1.15	8.07	3.20	30.64	nr	34.46
extra; angle posts with struts -2	26.62	2.50	27.29	-	1.85	12.99	4.80	45.07	nr	50.71

1200mm high fencing; type PLS 120; posts with rounded tops

posts at 2743mm centres; 600mm into ground	5.07	2.50	5.20	-	0.85	5.97	0.58	11.74	m	13.21
extra; end posts with struts -1	23.19	2.50	23.77	-	1.25	8.78	2.20	34.74	nr	39.09
extra; angle posts with struts -2	33.75	2.50	34.59	-	2.00	14.04	4.80	53.43	nr	60.11

1400mm high fencing; type PLS 140A; posts with rounded tops

posts at 2743mm centres; 600mm into ground	5.67	2.50	5.81	-	0.90	6.32	0.58	12.71	m	14.30
extra; end posts with struts -1	24.71	2.50	25.33	-	1.70	11.93	3.20	40.46	nr	45.52
extra; angle posts with struts -2	36.13	2.50	37.03	-	2.20	15.44	4.80	57.28	nr	64.44

1800mm high fencing; type PLS 180; posts with rounded tops

posts at 2743mm centres; 600mm into ground	6.80	2.50	6.97	-	1.10	7.72	0.58	15.27	m	17.18
extra; end posts with struts -1	33.30	2.50	34.13	-	1.85	12.99	3.20	50.32	nr	56.61
extra; angle posts with struts -2	47.93	2.50	49.13	-	2.40	16.85	4.80	70.78	nr	79.62

CHESTNUT FENCING

Fencing; cleft chestnut pale; B.S.1722 Part 4; sweet chestnut posts and struts; galvanised accessories; backfilling around posts in concrete mix ST4

900mm high fencing; type CW90

posts at 2000mm centres; 600mm into ground	3.19	5.00	3.35	-	0.28	1.97	0.67	5.99	m	6.73
extra; end posts with struts - 1	4.28	5.00	4.49	-	0.35	2.46	2.67	9.62	nr	10.82
extra; angle posts with struts - 2	5.49	5.00	5.76	-	0.46	3.23	4.00	12.99	nr	14.62

1200mm high fencing; type CW 120

posts at 2000mm centres; 600mm into ground	3.76	5.00	3.95	-	0.20	1.40	0.67	6.02	m	6.77
extra; end posts with struts - 1	5.15	5.00	5.41	-	0.52	3.65	2.67	11.73	nr	13.19
extra; angle posts with struts - 2	6.57	5.00	6.90	-	0.70	4.91	4.00	15.81	nr	17.79

1500mm high fencing; type CW 150

posts at 2000mm centres; 600mm into ground	6.15	5.00	6.46	-	0.25	1.75	0.67	8.88	m	9.99
extra; end posts with struts - 1	6.56	5.00	6.89	-	0.58	4.07	2.67	13.63	nr	15.33
extra; angle posts with struts - 2	8.38	5.00	8.80	-	0.75	5.26	4.00	18.06	nr	20.32

BOARDED FENCING

Fencing; close boarded; B.S.1722 Part 5; sawn softwood posts, rails, pales, gravel boards and centre stumps, pressure impregnated with preservative; backfilling around posts in concrete mix ST4

1200mm high fencing; type BW 120

posts at 3000mm centres; 600mm into ground	12.19	5.00	12.80	1.44	-	13.29	1.64	27.73	m	31.20

1500mm high fencing; type BW 150

posts at 3000mm centres; 750mm into ground	15.91	5.00	16.71	1.73	-	15.97	2.15	34.82	m	39.18

1800mm high fencing; type BW 180A

posts at 3000mm centres; 750mm into ground	17.84	5.00	18.73	1.92	-	17.72	2.15	38.60	m	43.43

Fencing; wooden palisade; B.S.1722 Part 6; softwood posts, rails, pales and stumps; pressure impregnated with preservative; backfilling around posts in concrete mix ST4

1050mm high fencing; type WPW 105; 75 x 19mm rectangular pales with pointed tops

posts at 3000mm centres; 600mm into ground	9.54	5.00	10.02	1.15	-	10.61	1.51	22.14	m	24.91

1200mm high fencing; type WPW 120; 75 x 19mm rectangular pales with pointed tops

posts at 3000mm centres; 600mm into ground	10.84	5.00	11.38	1.44	-	13.29	1.51	26.18	m	29.46

STEEL RAILINGS

Ornamental steel railings; Ranalah Gates Ltd.; primed at works; fixing in brick openings

Railing

reference 200, 356mm high	28.25	1.50	28.67	0.75	0.38	9.59	0.21	38.47	m	43.28
reference 220, 279mm high	51.90	1.50	52.68	0.75	0.38	9.59	0.21	62.48	m	70.29
reference 230, 432mm high	43.14	1.50	43.79	0.75	0.38	9.59	0.21	53.59	m	60.29
reference 240, 838mm high	59.35	1.50	60.24	1.00	0.50	12.74	0.26	73.24	m	82.40

Labour hourly rates: (except Specialists) Craft Operatives £9.23 Labourer £7.02 Rates are national average prices. Refer to REGIONAL VARIATIONS for indicative levels of overall pricing in regions	MATERIALS			LABOUR				RATES		
	Del to Site	Waste	Material Cost	Craft Optve	Lab	Labour Cost	Sunds	Nett Rate	Unit	Gross Rate (+12.5%)
	£	%	£	Hrs	Hrs	£	£	£		£
GUARD RAILS										
Galvanised mild steel guard rail of tubing to B.S.1387 medium grade with Kee Klamp fittings										
Rail or standard										
40mm diameter ref. 7-2-G	5.73	5.00	6.02	0.30	–	2.77	0.30	9.09	m	10.22
extra; flanged end	1.15	5.00	1.21	–	–	–	–	1.21	nr	1.36
extra; bend No.15-7	4.42	5.00	4.64	0.18	–	1.66	–	6.30	nr	7.09
extra; three-way intersection No.20-7	6.68	5.00	7.01	0.18	–	1.66	–	8.68	nr	9.76
extra; three-way intersection No.25-7	6.42	5.00	6.74	0.18	–	1.66	–	8.40	nr	9.45
extra; four-way intersection No.21-7	5.22	5.00	5.48	0.18	–	1.66	–	7.14	nr	8.04
extra; four-way intersection No.26-7	4.95	5.00	5.20	0.18	–	1.66	–	6.86	nr	7.72
extra; five-way intersection No.35-7	7.43	5.00	7.80	0.18	–	1.66	–	9.46	nr	10.65
extra; five-way intersection No.40-7	11.08	5.00	11.63	0.18	–	1.66	–	13.30	nr	14.96
extra; floor plate No.61-7	4.03	5.00	4.23	0.18	–	1.66	–	5.89	nr	6.63
Infill panel fixed with clips										
50 x 50mm welded mesh	19.60	5.00	20.58	0.90	–	8.31	0.21	29.10	m2	32.73
PRECAST CONCRETE POSTS										
Precast concrete fence posts; B.S.1722 excavating holes, backfilling around posts in concrete mix C20, disposing of surplus materials, earthwork support										
Fence posts for three wires, housing pattern										
100 x 100mm tapering intermediate post 1570mm long; 600mm into ground	4.66	2.50	4.78	0.29	0.29	4.71	2.25	11.74	nr	13.21
100 x 100mm square strainer post (end or intermediate) 1570mm long; 600mm into ground	5.71	2.50	5.85	0.29	0.29	4.71	2.25	12.82	nr	14.42
Clothes line post										
125 x 125mm tapering, 2670mm long	12.97	2.50	13.29	0.38	0.38	6.17	2.25	21.72	nr	24.43
Close boarded fence post										
94 x 100mm, 2745mm long	7.72	2.50	7.91	0.46	0.46	7.47	2.25	17.64	nr	19.84
Chain link fence post for 1800mm high fencing										
125 x 75mm tapering intermediate post, 2620mm long	8.43	2.50	8.64	0.38	0.38	6.17	2.25	17.07	nr	19.20
125 x 125mm end post, 1 strut, 2620mm long	19.22	2.50	19.70	0.38	0.38	6.17	4.49	30.37	nr	34.16
125 x 125mm angle post, 2 strut, 2620mm long	26.76	2.50	27.43	1.14	1.14	18.52	6.74	52.69	nr	59.28
125 x 125mm gate post, 2620mm long	14.13	2.50	14.48	0.38	0.38	6.17	2.25	22.91	nr	25.77
150 x 150mm gate post, 2620mm long	21.59	2.50	22.13	0.38	0.38	6.17	2.25	30.55	nr	34.37
TIMBER GATES										
Gates; impregnated wrought softwood; featheredge pales; including ring latch and heavy hinges										
Gates (posts included elsewhere)										
900 x 1150mm	56.27	2.50	57.68	0.60	0.60	9.75	0.16	67.59	nr	76.04
900 x 1450mm	58.17	2.50	59.62	0.60	0.60	9.75	0.16	69.53	nr	78.23
STEEL GATES										
Ornamental steel gates; Ranalah Gates Ltd.; primed at works; fixing in brick openings										
Popular, 1118mm high overall										
single gate reference 50, for 914mm wide opening	32.11	1.50	32.59	1.65	0.85	21.20	0.26	54.05	nr	60.80
Diana, 965mm high overall										
single gate reference 800, for 914mm wide opening	37.96	1.50	38.53	1.65	0.85	21.20	0.26	59.99	nr	67.48
single gate reference 801, for 1016mm wide opening	40.34	1.50	40.95	1.65	0.85	21.20	0.26	62.40	nr	70.20
double gates reference 810, for 2438mm wide opening	84.88	1.50	86.15	2.50	1.25	31.85	0.47	118.47	nr	133.28
Popular arch gate 1568mm high overall										
single gate reference 55, for 914mm wide opening	58.90	1.50	59.78	1.65	0.85	21.20	0.26	81.24	nr	91.39
Thrifty, 914mm high overall										
single gate reference 20, for 914mm wide opening	26.36	1.50	26.76	1.65	0.85	21.20	0.26	48.21	nr	54.24
Elite, 914mm high overall										
single gate reference 40, for 914mm wide opening	34.10	1.50	34.61	1.65	0.85	21.20	0.26	56.07	nr	63.08
double gates reference 41, for 2134mm wide opening	78.98	1.50	80.16	2.50	1.25	31.85	0.47	112.48	nr	126.55
double gates reference 42, for 2438mm wide opening	83.35	1.50	84.60	2.50	1.25	31.85	0.47	116.92	nr	131.54
Super, 965mm high overall										
single gate reference 60, for 914mm wide opening	39.72	1.50	40.32	1.65	0.85	21.20	0.26	61.77	nr	69.49
double gates reference 62, for 2286mm wide opening	100.94	1.50	102.45	2.50	1.25	31.85	0.47	134.77	nr	151.62
double gates reference 63, for 2438mm wide opening	104.63	1.50	106.20	2.50	1.25	31.85	0.47	138.52	nr	155.83
Concord, 914mm high overall										
single gate reference 160, for 914mm wide opening	44.94	1.50	45.61	1.65	0.85	21.20	0.26	67.07	nr	75.45
double gates reference 162, for 2286mm wide opening	103.27	1.50	104.82	2.50	1.25	31.85	0.47	137.14	nr	154.28
double gates reference 163, for 2438mm wide opening	109.56	1.50	111.20	2.50	1.25	31.85	0.47	143.52	Unit	161.46
Regent, 914mm high overall										
single gate reference 150, for 914mm wide opening	50.44	1.50	51.20	1.65	0.85	21.20	0.26	72.65	nr	81.73
double gates reference 151, for 2134mm wide opening	109.74	1.50	111.39	2.50	1.25	31.85	0.47	143.71	nr	161.67

PAVING/FENCING/SITE FURNITURE

PAVING/PLANTING/FENCING/SITE FURNITURE – SMALL WORKS

Labour hourly rates: (except Specialists) Craft Operatives £9.23 Labourer £7.02 Rates are national average prices. Refer to REGIONAL VARIATIONS for indicative levels of overall pricing in regions	MATERIALS			LABOUR				RATES		
	Del to Site £	Waste %	Material Cost £	Craft Optve Hrs	Lab Hrs	Labour Cost £	Sunds £	Nett Rate £	Unit	Gross Rate (+12.5%) £
STEEL GATES Cont.										
Ornamental steel gates; Ranalah Gates Ltd.; primed at works; fixing in brick openings Cont.										
Regent, 914mm high overall Cont.										
double gates reference 152, for 2286mm wide opening	112.80	1.50	114.49	2.50	1.25	31.85	0.47	146.81	nr	165.16
double gates reference 153, for 2438mm wide opening	125.79	1.50	127.68	2.50	1.25	31.85	0.47	160.00	nr	180.00
Brighton, 991mm high overall										
single gate reference 600, for 914mm wide opening	57.08	1.50	57.94	1.65	0.85	21.20	0.26	79.39	nr	89.32
double gates reference 602, for 2286mm wide opening	145.37	1.50	147.55	2.50	1.25	31.85	0.47	179.87	nr	202.35
double gates reference 603, for 2438mm wide opening	154.79	1.50	157.11	2.50	1.25	31.85	0.47	189.43	nr	213.11
Preston, 1067mm high overall										
single gate reference 425, for 914mm wide opening	66.95	1.50	67.95	1.65	0.85	21.20	0.26	89.41	nr	100.59
single gate reference 426, for 1016mm wide opening	73.76	1.50	74.87	1.65	0.85	21.20	0.26	96.32	nr	108.36
double gates reference 430, for 2134mm wide opening	150.81	1.50	153.07	2.50	1.25	31.85	0.47	185.39	nr	208.57
double gates reference 431, for 2438mm wide opening	167.55	1.50	170.06	2.50	1.25	31.85	0.47	202.38	nr	227.68
Extra cost of fixing gates to and including pair of steel gate posts; backfilling around posts in concrete mix ST4										
for single or double gates; posts reference 901, size 40 x 40mm	35.09	1.50	35.62	0.35	0.20	4.63	2.28	42.53	nr	47.85
for single or double gates; posts reference 902, size 50 x 50mm	43.19	1.50	43.84	0.35	0.20	4.63	2.48	50.95	nr	57.32
for single or double gates; posts reference 903, size 70 x 70mm	56.87	1.50	57.72	0.75	0.35	9.38	2.86	69.96	nr	78.71
for single or double gates; posts reference 904, size 100 x 100mm	79.26	1.50	80.45	0.75	0.35	9.38	3.47	93.30	nr	104.96
Gates in chain link fencing										
Gate with 40 x 40 x 5mm painted angle iron framing, braces and rails, infilled with 50mm x 10 1/2 G mesh										
1200 x 1800mm; galvanised mesh	106.30	1.50	107.89	6.00	6.00	97.50	-	205.39	nr	231.07
1200 x 1800mm; plastic coated mesh	108.75	1.50	110.38	6.00	6.00	97.50	-	207.88	nr	233.87
Gate post of 80 x 80 x 6mm painted angle iron; backfilling around posts in concrete mix C30										
to suit 1800mm high gate	40.67	1.50	41.28	1.50	1.50	24.38	2.88	68.54	nr	77.10
Q50: SITE/STREET FURNITURE/EQUIPMENT										
Plain ordinary portland cement precast concrete bollards; setting in ground, excavating hole, removing surplus spoil, filling with concrete mix ST4; working around base										
230mm diameter										
455mm high above ground level; Woodhouse; Smooth Grey	33.70	-	33.70	0.65	0.65	10.56	3.10	47.36	nr	53.28
455mm high above ground level; Woodhouse; Smooth White	57.30	-	57.30	0.65	0.65	10.56	3.10	70.96	nr	79.83
150mm diameter										
915mm high above ground level; Bridgford	46.50	-	46.50	1.05	1.05	17.06	3.10	66.66	nr	75.00
915mm high above ground level; Bridgford; Smooth White	77.00	-	77.00	1.05	1.05	17.06	3.10	97.16	nr	109.31
Precast concrete plant containers; setting in position										
Plant container (soil filling included elsewhere)										
710mm diameter x 500mm high; Strada	91.70	-	91.70	0.95	0.95	15.44	-	107.14	nr	120.53
1066mm diameter x 760mm high; Shirley	386.60	-	386.60	1.15	1.15	18.69	-	405.29	nr	455.95
500 x 500 x 500mm high; Strada	73.80	-	73.80	0.75	0.75	12.19	-	85.99	nr	96.74
Precast concrete litter bin with galvanised wire plastic coated inner basket Plastic Coated; setting in position										
Plain concrete										
43 litres capacity; Elsworth	148.00	-	148.00	1.15	1.15	18.69	-	166.69	nr	187.52
65 litres capacity; Newstead	139.60	-	139.60	1.15	1.15	18.69	-	158.29	nr	178.07
190 litres capacity; Gransden	215.35	-	215.35	1.25	1.25	20.31	-	235.66	nr	265.12
433 litres capacity; Shirley	511.90	-	511.90	1.25	1.25	20.31	-	532.21	nr	598.74
Benches and seats with precast concrete supports and hardwood slats to seats and back rests; fixed or free-standing										
Bench; placing in position										
2000mm long; Kelvin; Smooth Grey	160.25	-	160.25	0.60	0.60	9.75	-	170.00	nr	191.25
2000mm long; Kelvin; Smooth White	174.95	-	174.95	0.60	0.60	9.75	-	184.70	nr	207.79

Labour hourly rates: (except Specialists) Craft Operatives £9.23 Labourer £7.02 Rates are national average prices. Refer to REGIONAL VARIATIONS for indicative levels of overall pricing in regions	MATERIALS			LABOUR				RATES		
	Del to Site £	Waste %	Material Cost £	Craft Optve Hrs	Lab Hrs	Labour Cost £	Sunds £	Nett Rate £	Unit	Gross Rate (+12.5%) £
Q50: SITE/STREET FURNITURE/EQUIPMENT Cont.										
Steel cycle stands, Mawrob Co. (Engineers) Ltd; **galvanised steel construction; free standing**										
Single sided										
reference TR5 for 5 cycles	175.00	-	175.00	1.20	0.30	13.18	-	188.18	nr	211.70
reference TR6 for 6 cycles	192.00	-	192.00	1.20	0.30	13.18	-	205.18	nr	230.83
reference TR8 for 8 cycles	230.50	-	230.50	1.50	0.35	16.30	-	246.80	nr	277.65
reference TR10 for 10 cycles	297.50	-	297.50	1.50	0.35	16.30	-	313.80	nr	353.03
Double sided										
reference DTR10 for 10 cycles	282.00	-	282.00	1.80	0.45	19.77	-	301.77	nr	339.49
reference DTR12 for 12 cycles	312.50	-	312.50	1.80	0.45	19.77	-	332.27	nr	373.81
reference DTR20 for 20 cycles	483.50	-	483.50	2.40	0.60	26.36	-	509.86	nr	573.60
Steel front wheel cycle supports; Mawrob Co. **(Engineers) Ltd; galvanised steel construction,** **pillar mounted; bolting base plate to concrete**										
Single sided										
reference MW/MC 6/S for 6 cycles	261.00	-	261.00	4.80	1.20	52.73	-	313.73	nr	352.94
reference MW/MC 9/S for 9 cycles	353.00	-	353.00	4.80	1.20	52.73	-	405.73	nr	456.44
reference MW/MC 12/S for 12 cycles	455.00	-	455.00	7.20	1.80	79.09	-	534.09	nr	600.85
Double sided										
reference MW/MC 6/D for 12 cycles	429.50	-	429.50	4.80	1.20	52.73	-	482.23	nr	542.51
reference MW/MC 9/D for 18 cycles	558.00	-	558.00	4.80	1.20	52.73	-	610.73	nr	687.07
reference MW/MC 12/D for 24 cycles	748.50	-	748.50	7.20	1.80	79.09	-	827.59	nr	931.04
Steel cycle shelters, Mawrob Co. (Engineers) Ltd; **steel angle framing; with red oxide priming;** **corrugated galvanised steel roofing; assembling;** **bolting to concrete**										
Horizontal loading, single sided										
reference MW/AV10 for 10 cycles	505.00	-	505.00	9.60	2.40	105.46	-	610.46	nr	686.76
reference MW/AK10 for 10 cycles	588.50	-	588.50	9.60	2.40	105.46	-	693.96	nr	780.70
reference MW/AW8 for 8 cycles	491.50	-	491.50	7.20	1.80	79.09	-	570.59	nr	641.92
reference MW/AW10 for 10 cycles	588.50	-	588.50	9.60	2.40	105.46	-	693.96	nr	780.70
reference MW/AW15 for 15 cycles	807.00	-	807.00	12.00	3.00	131.82	-	938.82	nr	1056.17
reference MW/RS12 for 12 cycles	550.50	-	550.50	9.60	2.40	105.46	-	655.96	nr	737.95
reference MW/RS22 for 22 cycles	895.00	-	895.00	14.40	3.60	158.18	-	1053.18	nr	1184.83
Horizontal loading, double sided										
reference MW/AL12 for 12 cycles	847.00	-	847.00	12.00	3.00	131.82	-	978.82	nr	1101.17
reference MW/AL16 for 16 cycles	1058.50	-	1058.50	14.40	3.60	158.18	-	1216.68	nr	1368.77
reference MV/AL20 for 20 cycles	1217.50	-	1217.50	16.80	4.20	184.55	-	1402.05	nr	1577.30

PAVING/FENCING/SITE FURNITURE

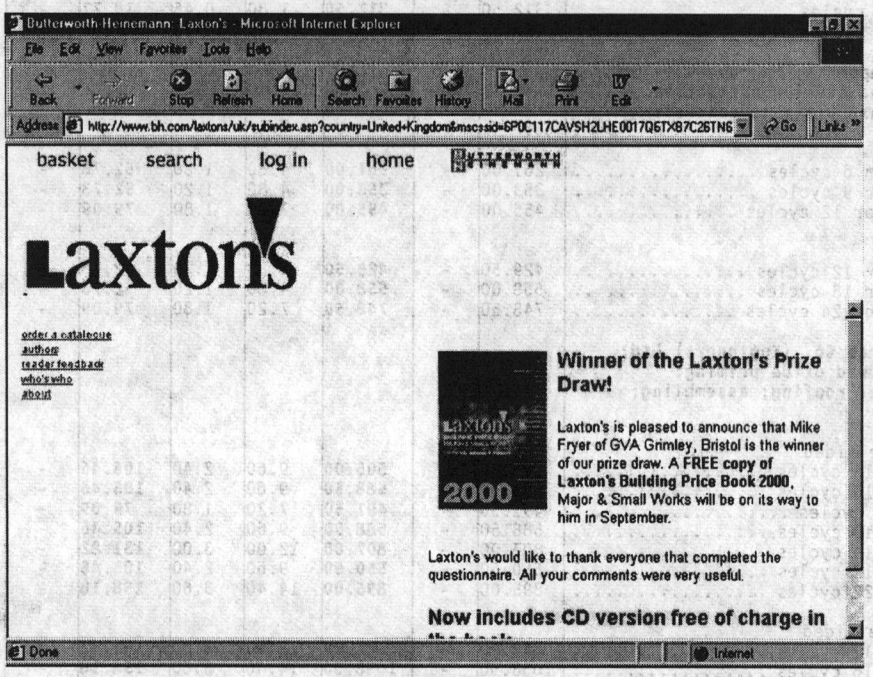

Disposal systems

Labour hourly rates: (except Specialists) Craft Operatives £12.44 Labourer £7.02 Rates are national average prices. Refer to REGIONAL VARIATIONS for indicative levels of overall pricing in regions	MATERIALS			LABOUR				RATES		
	Del to Site £	Waste %	Material Cost £	Craft Optve Hrs	Lab Hrs	Labour Cost £	Sunds £	Nett Rate £	Unit	Gross Rate (+12.5%) £
R10: RAINWATER PIPEWORK/GUTTERS										
CAST IRON ROUND RAINWATER PIPES (B.S.460) - Cast iron pipes and fittings, B.S.460 Type A sockets; dry joints										
Pipes; straight										
75mm; in or on supports (included elsewhere)	14.55	5.00	15.28	0.80	-	9.95	-	25.23	m	28.38
extra; shoes	11.90	2.00	12.14	1.00	-	12.44	-	24.58	nr	27.65
extra; bends	10.21	2.00	10.41	0.80	-	9.95	-	20.37	nr	22.91
extra; offset bends 75mm projection	12.87	2.00	13.13	0.80	-	9.95	-	23.08	nr	25.96
extra; offset bends 150mm projection	12.87	2.00	13.13	0.80	-	9.95	-	23.08	nr	25.96
extra; offset bends 225mm projection	14.98	2.00	15.28	0.80	-	9.95	-	25.23	nr	28.39
extra; offset bends 300mm projection	18.34	2.00	18.71	0.80	-	9.95	-	28.66	nr	32.24
extra; branches	17.87	2.00	18.23	0.80	-	9.95	-	28.18	nr	31.70
100mm; in or on supports (included elsewhere)	19.86	5.00	20.85	0.96	-	11.94	-	32.80	m	36.89
extra; shoes	16.04	2.00	16.36	1.20	-	14.93	-	31.29	nr	35.20
extra; bends	14.41	2.00	14.70	0.96	-	11.94	-	26.64	nr	29.97
extra; offset bends 75mm projection	24.27	2.00	24.76	0.96	-	11.94	-	36.70	nr	41.29
extra; offset bends 150mm projection	24.27	2.00	24.76	0.96	-	11.94	-	36.70	nr	41.29
extra; offset bends 225mm projection	29.38	2.00	29.97	0.96	-	11.94	-	41.91	nr	47.15
extra; offset bends 300mm projection	29.38	2.00	29.97	0.96	-	11.94	-	41.91	nr	47.15
extra; branches	21.22	2.00	21.64	0.96	-	11.94	-	33.59	nr	37.79
CAST IRON ROUND RAINWATER PIPES (B.S.460) - Cast iron pipes and fittings, B.S.460 Type A sockets; ears cast on; dry joints										
Pipes; straight										
63mm; ears fixing to masonry with galvanized pipe nails and distance pieces	15.66	5.00	16.44	0.75	-	9.33	-	25.77	m	28.99
extra; shoes	13.95	2.00	14.23	0.85	-	10.57	-	24.80	nr	27.90
extra; bends	8.41	2.00	8.58	0.68	-	8.46	-	17.04	nr	19.17
extra; offset bends 75mm projection	12.87	2.00	13.13	0.68	-	8.46	-	21.59	nr	24.28
extra; offset bends 150mm projection	12.87	2.00	13.13	0.68	-	8.46	-	21.59	nr	24.28
extra; offset bends 225mm projection	14.98	2.00	15.28	0.68	-	8.46	-	23.74	nr	26.71
extra; offset bends 300mm projection	17.53	2.00	17.88	0.68	-	8.46	-	26.34	nr	29.63
extra; branches	16.20	2.00	16.52	0.68	-	8.46	-	24.98	nr	28.11
75mm; ears fixing to masonry with galvanized pipe nails and distance pieces	15.66	5.00	16.44	0.80	-	9.95	-	26.40	m	29.69
extra; shoes	13.95	2.00	14.23	1.00	-	12.44	-	26.67	nr	30.00
extra; bends	10.21	2.00	10.41	0.80	-	9.95	-	20.37	nr	22.91
extra; offset bends 75mm projection	12.87	2.00	13.13	0.80	-	9.95	-	23.08	nr	25.96
extra; offset bends 150mm projection	12.87	2.00	13.13	0.80	-	9.95	-	23.08	nr	25.96
extra; offset bends 225mm projection	14.98	2.00	15.28	0.80	-	9.95	-	25.23	nr	28.39
extra; offset bends 300mm projection	18.34	2.00	18.71	0.80	-	9.95	-	28.66	nr	32.24
extra; branches	17.87	2.00	18.23	0.80	-	9.95	-	28.18	nr	31.70
100mm; ears fixing to masonry with galvanized pipe nails and distance pieces	20.99	5.00	22.04	0.96	-	11.94	-	33.98	m	38.23
extra; shoes	18.11	2.00	18.47	1.20	-	14.93	-	33.40	nr	37.58
extra; bends	14.41	2.00	14.70	0.96	-	11.94	-	26.64	nr	29.97
extra; offset bends 75mm projection	24.27	2.00	24.76	0.96	-	11.94	-	36.70	nr	41.29
extra; offset bends 150mm projection	24.27	2.00	24.76	0.96	-	11.94	-	36.70	nr	41.29
extra; offset bends 225mm projection	29.38	2.00	29.97	0.96	-	11.94	-	41.91	nr	47.15
extra; offset bends 300mm projection	29.38	2.00	29.97	0.96	-	11.94	-	41.91	nr	47.15
extra; branches	21.22	2.00	21.64	0.96	-	11.94	-	33.59	nr	37.79
CAST IRON ROUND RAINWATER PIPES (B.S.460) - Pipe supports										
Steel holderbats										
for 75mm pipes; for building in (fixing included elsewhere)	6.26	2.00	6.39	-	-	-	-	6.39	nr	7.18
for 100mm pipes; for building in (fixing included elsewhere)	6.44	2.00	6.57	-	-	-	-	6.57	nr	7.39
for 75mm pipes; fixing to masonry with galvanized screws	5.23	2.00	5.33	-	-	-	-	5.33	nr	6.00
for 100mm pipes; fixing to masonry with galvanized screws	5.42	2.00	5.53	0.20	-	2.49	-	8.02	nr	9.02

DISPOSAL SYSTEMS

Labour hourly rates: (except Specialists) Craft Operatives £12.44 Labourer £7.02 Rates are national average prices. Refer to REGIONAL VARIATIONS for indicative levels of overall pricing in regions	MATERIALS			LABOUR				RATES		
	Del to Site £	Waste %	Material Cost £	Craft Optve Hrs	Lab Hrs	Labour Cost £	Sunds £	Nett Rate £	Unit	Gross Rate (+12.5%) £

R10: RAINWATER PIPEWORK/GUTTERS Cont.

CAST IRON RECTANGULAR RAINWATER PIPES – Cast iron rectangular pipes and fittings; ears cast on; dry joints

Pipes; straight
100 x 75mm; ears fixing to masonry with galvanised

pipe nails and distance pieces	59.95	5.00	62.95	1.20	–	14.93	–	77.88	m	87.61
extra; shoes (front)	51.65	2.00	52.68	1.20	–	14.93	–	67.61	nr	76.06
extra; bends (front)	41.17	2.00	41.99	0.90	–	11.20	–	53.19	nr	59.84
extra; offset bends (front) 150 projection	52.51	2.00	53.56	0.90	–	11.20	–	64.76	nr	72.85
extra; offset bends (front) 300 projection	73.68	2.00	75.15	0.90	–	11.20	–	86.35	nr	97.14

ALUMINIUM RAINWATER PIPES (B.S.2997) – Cast aluminium pipes and fittings B.S.2997 Part 2 Section B, ears cast on; dry joints

Pipes; straight
63mm; ears fixing to masonry with galvanized pipe

nails	10.97	5.00	11.52	0.60	–	7.46	–	18.98	m	21.36
extra; shoes	7.16	2.00	7.30	0.75	–	9.33	–	16.63	nr	18.71
extra; bends	7.94	2.00	8.10	0.60	–	7.46	–	15.56	nr	17.51
extra; offset bends 75mm projection	15.27	2.00	15.58	0.60	–	7.46	–	23.04	nr	25.92
extra; offset bends 150mm projection	18.94	2.00	19.32	0.60	–	7.46	–	26.78	nr	30.13
extra; offset bends 225mm projection	21.48	2.00	21.91	0.60	–	7.46	–	29.37	nr	33.05
extra; offset bends 300mm projection	24.04	2.00	24.52	0.60	–	7.46	–	31.98	nr	35.98

75mm; ears fixing to masonry with galvanized pipe

nails	12.75	5.00	13.39	0.80	–	9.95	–	23.34	m	26.26
extra; shoes	9.89	2.00	10.09	1.00	–	12.44	–	22.53	nr	25.34
extra; bends	10.41	2.00	10.62	0.80	–	9.95	–	20.57	nr	23.14
extra; offset bends 75mm projection	18.52	2.00	18.89	0.80	–	9.95	–	28.84	nr	32.45
extra; offset bends 150mm projection	21.29	2.00	21.72	0.80	–	9.95	–	31.67	nr	35.63
extra; offset bends 225mm projection	23.75	2.00	24.23	0.80	–	9.95	–	34.18	nr	38.45
extra; offset bends 300mm projection	26.36	2.00	26.89	0.80	–	9.95	–	36.84	nr	41.44

100mm; ears fixing to masonry with galvanized pipe

nails	19.18	5.00	20.14	0.96	–	11.94	–	32.08	m	36.09
extra; shoes	11.75	2.00	11.98	1.20	–	14.93	–	26.91	nr	30.28
extra; bends	15.20	2.00	15.50	0.96	–	11.94	–	27.45	nr	30.88
extra; offset bends 75mm projection	20.95	2.00	21.37	0.96	–	11.94	–	33.31	nr	37.48
extra; offset bends 150mm projection	24.32	2.00	24.81	0.96	–	11.94	–	36.75	nr	41.34
extra; offset bends 225mm projection	27.53	2.00	28.08	0.96	–	11.94	–	40.02	nr	45.03
extra; offset bends 300mm projection	30.74	2.00	31.35	0.96	–	11.94	–	43.30	nr	48.71

UNPLASTICIZED PVC RAINWATER PIPES (B.S.4576) – PVC-U pipes and fittings, B.S.4576 Part 1; push fit joints; pipework and supports self coloured grey

Pipes; straight
68mm; in standard holderbats fixing to masonry

with galvanized screws	3.06	5.00	3.21	0.48	–	5.97	–	9.18	m	10.33
extra; shoes	3.76	2.00	3.84	0.37	–	4.60	–	8.44	nr	9.49
extra; bends	3.04	2.00	3.10	0.30	–	3.73	–	6.83	nr	7.69
extra; offset bends (spigot)	1.69	2.00	1.72	0.30	–	3.73	–	5.46	nr	6.14
extra; offset bends (socket)	1.82	2.00	1.86	0.30	–	3.73	–	5.59	nr	6.29
extra; angled branches	6.07	2.00	6.19	0.30	–	3.73	–	9.92	nr	11.16

CAST IRON EAVES GUTTERS (B.S.460) – Cast iron half round gutters and fittings, B.S.460; bolted and mastic joints

Gutters; straight
100mm; in standard fascia brackets fixing to

timber with galvanized screws	10.44	5.00	10.96	0.60	–	7.46	–	18.43	m	20.73
extra; stopped ends	2.02	2.00	2.06	0.60	–	7.46	–	9.52	nr	10.71
extra; running outlets	5.87	2.00	5.99	0.60	–	7.46	–	13.45	nr	15.13
extra; stopped ends with outlet	4.40	2.00	4.49	0.60	–	7.46	–	11.95	nr	13.45
extra; angles	6.03	2.00	6.15	0.60	–	7.46	–	13.61	nr	15.32

114mm; in standard fascia brackets fixing to

timber with galvanized screws	10.86	5.00	11.40	0.60	–	7.46	–	18.87	m	21.23
extra; stopped ends	2.62	2.00	2.67	0.60	–	7.46	–	10.14	nr	11.40
extra; running outlets	6.39	2.00	6.52	0.60	–	7.46	–	13.98	nr	15.73
extra; stopped ends with outlet	4.84	2.00	4.94	0.60	–	7.46	–	12.40	nr	13.95
extra; angles	6.20	2.00	6.32	0.60	–	7.46	–	13.79	nr	15.51

125mm; in standard fascia brackets fixing to

timber with galvanized screws	12.35	5.00	12.97	0.68	–	8.46	–	21.43	m	24.11
extra; stopped ends	2.62	2.00	2.67	0.68	–	8.46	–	11.13	nr	12.52
extra; running outlets	7.33	2.00	7.48	0.68	–	8.46	–	15.94	nr	17.93
extra; stopped ends with outlet	6.51	2.00	6.64	0.68	–	8.46	–	15.10	nr	16.99
extra; angles	7.33	2.00	7.48	0.68	–	8.46	–	15.94	nr	17.93

150mm; in standard fascia brackets fixing to

timber with galvanized screws	19.84	5.00	20.83	0.90	–	11.20	–	32.03	m	36.03
extra; stopped ends	3.50	2.00	3.57	0.90	–	11.20	–	14.77	nr	16.61
extra; running outlets	12.70	2.00	12.95	0.90	–	11.20	–	24.15	nr	27.17
extra; stopped ends with outlet	12.44	2.00	12.69	0.90	–	11.20	–	23.88	nr	26.87
extra; angles	13.36	2.00	13.63	0.90	–	11.20	–	24.82	nr	27.93

CAST IRON EAVES GUTTERS (B.S.460) – Cast iron OG gutters and fittings, B.S.460; bolted and mastic joints

Gutters; straight

100mm; fixing to timber with galvanized screws	9.53	5.00	10.01	0.60	–	7.46	–	17.47	m	19.65
extra; stopped ends	1.83	2.00	1.87	0.60	–	7.46	–	9.33	nr	10.50
extra; running outlets	6.29	2.00	6.42	0.60	–	7.46	–	13.88	nr	15.61
extra; stopped ends with outlet	4.78	2.00	4.88	0.60	–	7.46	–	12.34	nr	13.88

Labour hourly rates: (except Specialists) Craft Operatives £12.44 Labourer £7.02 Rates are national average prices. Refer to REGIONAL VARIATIONS for indicative levels of overall pricing in regions	MATERIALS			LABOUR				RATES		
	Del to Site	Waste	Material Cost	Craft Optve	Lab	Labour Cost	Sunds	Nett Rate	Unit	Gross Rate (+12.5%)
	£	%	£	Hrs	Hrs	£	£	£		£
R10: RAINWATER PIPEWORK/GUTTERS Cont.										
CAST IRON EAVES GUTTERS (B.S.460) - Cast iron OG gutters and fittings, B.S.460; bolted and mastic joints Cont.										
Gutters; straight Cont.										
extra; angles	6.29	2.00	6.42	0.60	-	7.46	-	13.88	nr	15.61
114mm; fixing to timber with galvanized screws ...	10.51	5.00	11.04	0.60	-	7.46	-	18.50	m	20.81
extra; stopped ends	2.43	2.00	2.48	0.60	-	7.46	-	9.94	nr	11.19
extra; running outlets	6.82	2.00	6.96	0.60	-	7.46	-	14.42	nr	16.22
extra; stopped ends with outlet	4.78	2.00	4.88	0.60	-	7.46	-	12.34	nr	13.88
extra; angles	6.82	2.00	6.96	0.60	-	7.46	-	14.42	nr	16.22
125mm; fixing to timber with galvanized screws ...	11.07	5.00	11.62	0.68	-	8.46	-	20.08	m	22.59
extra; stopped ends	2.43	2.00	2.48	0.68	-	8.46	-	10.94	nr	12.31
extra; running outlets	7.44	2.00	7.59	0.68	-	8.46	-	16.05	nr	18.05
extra; stopped ends with outlet	4.78	2.00	4.88	0.68	-	8.46	-	13.33	nr	15.00
extra; angles	7.44	2.00	7.59	0.68	-	8.46	-	16.05	nr	18.05
CAST IRON MOULDED SECTIONS (STOCK SECTIONS) - Cast iron moulded (stock sections) gutters and fittings; bolted and mastic joints										
Gutters; straight										
100 x 75mm; in standard fascia brackets fixing to timber with galvanized screws	18.78	5.00	19.72	0.90	-	11.20	-	30.91	m	34.78
extra; stopped ends	5.83	2.00	5.95	0.90	-	11.20	-	17.14	nr	19.29
extra; running outlets	15.31	2.00	15.62	0.90	-	11.20	-	26.81	nr	30.16
extra; angles	15.31	2.00	15.62	0.90	-	11.20	-	26.81	nr	30.16
extra; clips	6.39	2.00	6.52	0.90	-	11.20	-	17.71	nr	19.93
125 x 100mm; in standard fascia brackets fixing to timber with galvanized screws	26.22	5.00	27.53	1.20	-	14.93	-	42.46	m	47.77
extra; stopped ends	7.57	2.00	7.72	1.20	-	14.93	-	22.65	nr	25.48
extra; running outlets	21.99	2.00	22.43	1.20	-	14.93	-	37.36	nr	42.03
extra; angles	21.99	2.00	22.43	1.20	-	14.93	-	37.36	nr	42.03
extra; clips	7.57	2.00	7.72	1.20	-	14.93	-	22.65	nr	25.48
CAST IRON BOX GUTTERS - Cast iron box gutters and fittings; bolted and mastic joints										
Gutters; straight										
100 x 75mm; in or on supports (included elsewhere)	24.59	5.00	25.82	0.90	-	11.20	-	37.02	m	41.64
extra; stopped ends	3.36	2.00	3.43	0.90	-	11.20	-	14.62	nr	16.45
extra; running outlets	11.55	2.00	11.78	0.90	-	11.20	-	22.98	nr	25.85
extra; angles	11.55	2.00	11.78	0.90	-	11.20	-	22.98	nr	25.85
extra; clips	4.34	2.00	4.43	0.90	-	11.20	-	15.62	nr	17.58
ALUMINIUM GUTTERS (B.S.2997) - Cast aluminium half round gutters and fittings, B.S.2997 Part 2 Section A; bolted and mastic joints										
Gutters; straight										
102mm; in standard fascia brackets fixing to timber with galvanized screws	10.90	5.00	11.45	0.60	-	7.46	-	18.91	m	21.27
extra; stopped ends	2.55	2.00	2.60	0.60	-	7.46	-	10.07	nr	11.32
extra; running outlets	6.06	2.00	6.18	0.60	-	7.46	-	13.65	nr	15.35
extra; angles	5.27	2.00	5.38	0.60	-	7.46	-	12.84	nr	14.44
127mm; in standard fascia brackets fixing to timber with galvanized screws	14.56	5.00	15.29	0.68	-	8.46	-	23.75	m	26.72
extra; stopped ends	3.38	2.00	3.45	0.68	-	8.46	-	11.91	nr	13.40
extra; running outlets	6.92	2.00	7.06	0.68	-	8.46	-	15.52	nr	17.46
extra; angles	6.73	2.00	6.86	0.40	-	4.98	-	11.84	nr	13.32
UNPLASTICISED PVC GUTTERS (B.S.4576) - PVC-U half round gutters and fittings, B.S.4576, Part 1; push fit connector joints; gutterwork and supports self coloured grey										
Gutters; straight										
105mm; in standard fascia brackets fixing to timber with galvanized screws	3.82	5.00	4.01	0.40	-	4.98	-	8.99	m	10.11
extra; stopped ends	1.38	2.00	1.41	0.40	-	4.98	-	6.38	nr	7.18
extra; running outlets	2.89	2.00	2.95	0.40	-	4.98	-	7.92	nr	8.91
extra; angles	3.16	2.00	3.22	0.40	-	4.98	-	8.20	nr	9.22
114mm; in standard fascia brackets fixing to timber with galvanized screws	4.11	5.00	4.32	0.40	-	4.98	-	9.29	m	10.45
extra; stopped ends	1.65	2.00	1.68	0.40	-	4.98	-	6.66	nr	7.49
extra; running outlets	3.09	2.00	3.15	0.40	-	4.98	-	8.13	nr	9.14
extra; angles	3.37	2.00	3.44	0.40	-	4.98	-	8.41	nr	9.47
GRP GUTTERS - Corofil Ogee GRP gutters and fittings; Corofil GRP Division, Pre-Formed Components Ltd.; bolted and neomastic twinseal extrusion joints										
Gutters; straight										
102mm; fixing to timber with galvanized screws ...	16.11	-	16.11	0.50	-	6.22	-	22.33	m	25.12
extra; stopped ends	10.40	-	10.40	0.50	-	6.22	-	16.62	nr	18.70
extra; running outlets	23.40	-	23.40	0.50	-	6.22	-	29.62	nr	33.32
extra; angles	26.00	-	26.00	0.50	-	6.22	-	32.22	nr	36.25
140mm; fixing to timber with galvanized screws ...	22.61	-	22.61	0.50	-	6.22	-	28.83	m	32.43
extra; stopped ends	10.40	-	10.40	0.50	-	6.22	-	16.62	nr	18.70
extra; running outlets	23.40	-	23.40	0.50	-	6.22	-	29.62	nr	33.32
extra; angles	26.00	-	26.00	0.50	-	6.22	-	32.22	nr	36.25
205mm; fixing to timber with galvanized screws ...	26.42	-	26.42	0.60	-	7.46	-	33.88	m	38.12
extra; stopped ends	10.40	-	10.40	0.60	-	7.46	-	17.86	nr	20.10
extra; running outlets	23.40	-	23.40	0.60	-	7.46	-	30.86	nr	34.72

DISPOSAL SYSTEMS

Labour hourly rates: (except Specialists) Craft Operatives £12.44 Labourer £7.02 Rates are national average prices. Refer to REGIONAL VARIATIONS for indicative levels of overall pricing in regions	MATERIALS			LABOUR				RATES		
	Del to Site	Waste	Material Cost	Craft Optve	Lab	Labour Cost	Sunds	Nett Rate	Unit	Gross Rate (+12.5%)
	£	%	£	Hrs	Hrs	£	£	£		£
R10: RAINWATER PIPEWORK/GUTTERS Cont.										
GRP GUTTERS - Corofil Ogee GRP gutters and fittings; Corofil GRP Division, Pre-Formed Components Ltd.; bolted and neomastic twinseal extrusion joints Cont.										
Gutters; straight Cont.										
extra; angles	26.00	-	26.00	0.60	-	7.46	-	33.46	nr	37.65
Aluminium pipework ancillaries										
Rainwater heads; B.S.2997 square type										
258 x 190 x 178mm; 63mm outlet spigot; dry joint to pipe; fixing to masonry with galvanized screws	18.91	1.00	19.10	0.60	-	7.46	-	26.56	nr	29.88
258 x 190 x 178mm; 75mm outlet spigot; dry joint to pipe; fixing to masonry with galvanized screws	19.78	1.00	19.98	0.80	-	9.95	-	29.93	nr	33.67
258 x 190 x 178mm; 100mm outlet spigot; dry joint to pipe; fixing to masonry with galvanized screws	32.43	1.00	32.75	0.96	-	11.94	-	44.70	nr	50.28
RAINWATER HEADS - Cast iron pipework ancillaries										
Rainwater heads; B.S.460 hopper type (flat pattern)										
210 x 160 x 185mm; 63mm outlet spigot; dry joint to pipe; fixing to masonry with galvanized screws	11.01	1.00	11.12	0.80	-	9.95	-	21.07	nr	23.71
210 x 160 x 185mm; 75mm outlet spigot; dry joint to pipe; fixing to masonry with galvanized screws	11.01	1.00	11.12	0.80	-	9.95	-	21.07	nr	23.71
250 x 215 x 215mm; 100mm outlet spigot; dry joint to pipe; fixing to masonry with galvanized screws	27.27	1.00	27.54	0.96	-	11.94	-	39.49	nr	44.42
Rainwater heads; B.S.460 square type (flat pattern)										
250 x 180 x 175mm; 63mm outlet spigot; dry joint to pipe; fixing to masonry with galvanized screws	23.84	1.00	24.08	0.60	-	7.46	-	31.54	nr	35.49
250 x 180 x 175mm; 75mm outlet spigot; dry joint to pipe; fixing to masonry with galvanized screws	23.84	1.00	24.08	0.80	-	9.95	-	34.03	nr	38.28
300 x 250 x 200mm; 100mm outlet spigot; dry joint to pipe; fixing to masonry with galvanized screws	48.05	1.00	48.53	0.96	-	11.94	-	60.47	nr	68.03
RAINWATER HEADS - Plastics pipework ancillaries self coloured black										
Rainwater heads; B.S.4576 square type										
236 x 152 x 187mm; 68mm outlet spigot; push fit joint to plastics pipe; fixing to masonry with galvanized screws	7.98	1.00	8.06	0.60	-	7.46	-	15.52	nr	17.57
252 x 195 x 210mm; 110mm outlet spigot; push fit joint to plastics pipe; fixing to masonry with galvanized screws	16.00	1.00	16.16	0.96	-	11.94	-	28.10	nr	31.62
ROOF OUTLETS - Cast iron pipework ancillaries										
Roof outlets; luting flange for asphalt										
flat grating; 75mm outlet spigot; coupling joint to cast iron pipe	66.84	1.00	67.51	0.96	-	11.94	-	79.45	nr	89.38
flat grating; 100mm outlet spigot; coupling joint to cast iron pipe	79.53	1.00	80.33	1.20	-	14.93	-	95.25	nr	107.16
domical grating; 75mm outlet spigot; coupling joint to cast iron pipe	68.38	1.00	69.06	0.96	-	11.94	-	81.01	nr	91.13
domical grating; 100mm outlet spigot; coupling joint to cast iron pipe	81.04	1.00	81.85	1.20	-	14.93	-	96.78	nr	108.88
ROOF OUTLETS - Plastics pipework ancillaries, self coloured grey										
Roof outlets; B.S.4576; luting flange for asphalt										
domical grating; 75mm outlet spigot; push fit joint to plastics pipe	25.61	1.00	25.87	0.45	-	5.60	-	31.46	nr	35.40
domical grating; 100mm outlet spigot; push fit joint to plastics pipe	25.61	1.00	25.87	0.45	-	5.60	-	31.46	nr	35.40
R11: FOUL DRAINAGE ABOVE GROUND										
CAST IRON SOIL AND WASTE PIPES - Cast iron pipes and fittings, B.S.416; bolted and synthetic rubber gasket couplings										
Pipes; straight										
75mm; in or on supports (included elsewhere)	13.50	5.00	14.18	1.20	-	14.93	-	29.10	m	32.74
extra; pipes with inspection door bolted and sealed..	28.05	2.00	28.61	0.80	-	9.95	-	38.56	nr	43.38
extra; short radius bends	15.05	2.00	15.35	0.80	-	9.95	-	25.30	nr	28.47
extra; short radius bends with inspection door bolted and sealed	28.60	2.00	29.17	0.80	-	9.95	-	39.12	nr	44.01
extra; offset bends 150mm projection	17.19	2.00	17.53	0.80	-	9.95	-	27.49	nr	30.92
extra; single angled branches	25.55	2.00	26.06	1.20	-	14.93	-	40.99	nr	46.11
extra; single angled branches with inspection door bolted and sealed	39.09	2.00	39.87	1.20	-	14.93	-	54.80	nr	61.65
extra; double angled branches	40.81	2.00	41.63	1.50	-	18.66	-	60.29	nr	67.82
extra; roof connectors	31.14	2.00	31.76	1.50	-	18.66	-	50.42	nr	56.73
boss pipes with one threaded boss	24.96	2.00	25.46	0.80	-	9.95	-	35.41	nr	39.84
boss pipes with two threaded bosses	31.67	2.00	32.30	0.80	-	9.95	-	42.26	nr	47.54
100mm; in or on supports (included elsewhere)	16.49	5.00	17.31	0.96	-	11.94	-	29.26	m	32.91
extra; pipes with inspection door bolted and sealed..	30.95	2.00	31.57	0.96	-	11.94	-	43.51	nr	48.95
extra; short radius bends	20.36	2.00	20.77	0.96	-	11.94	-	32.71	nr	36.80
extra; short radius bends with inspection door bolted and sealed	34.63	2.00	35.32	0.96	-	11.94	-	47.27	nr	53.17

Labour hourly rates: (except Specialists) Craft Operatives £12.44 Labourer £7.02 Rates are national average prices. Refer to REGIONAL VARIATIONS for indicative levels of overall pricing in regions	MATERIALS			LABOUR				RATES		
	Del to Site £	Waste %	Material Cost £	Craft Optve Hrs	Lab Hrs	Labour Cost £	Sunds £	Nett Rate £	Unit	Gross Rate (+12.5%) £

R11: FOUL DRAINAGE ABOVE GROUND Cont.

CAST IRON SOIL AND WASTE PIPES - Cast iron pipes and fittings, B.S.416; bolted and synthetic rubber gasket couplings Cont.

Pipes; straight Cont.

	Del to Site £	Waste %	Material Cost £	Craft Optve Hrs	Lab Hrs	Labour Cost £	Sunds £	Nett Rate £	Unit	Gross Rate £
extra; offset bends 150mm projection	23.61	2.00	24.08	0.96	-	11.94	-	36.02	nr	40.53
extra; offset bends 300mm projection	28.29	2.00	28.86	0.96	-	11.94	-	40.80	nr	45.90
extra; single angled branches	34.89	2.00	35.59	1.44	-	17.91	-	53.50	nr	60.19
extra; single angled branches with inspection door bolted and sealed	49.12	2.00	50.10	1.44	-	17.91	-	68.02	nr	76.52
extra; double angled branches	47.14	2.00	48.08	1.92	-	23.88	-	71.97	nr	80.96
extra; double angled branches with inspection door bolted and sealed	61.42	2.00	62.65	1.92	-	23.88	-	86.53	nr	97.35
extra; single anti-syphon branches, angled branch	36.31	2.00	37.04	1.44	-	17.91	-	54.95	nr	61.82
extra; straight wc connectors	21.48	2.00	21.91	0.96	-	11.94	-	33.85	nr	38.08
extra; roof connectors	29.88	2.00	30.48	0.96	-	11.94	-	42.42	nr	47.72
boss pipes with one threaded boss	30.49	2.00	31.10	0.96	-	11.94	-	43.04	nr	48.42
boss pipes with two threaded bosses	37.13	2.00	37.87	0.96	-	11.94	-	49.81	nr	56.04
50mm; in standard wall fixing bracket (fixing included elsewhere)	16.96	5.00	17.81	0.66	-	8.21	-	26.02	m	29.27
extra; pipes with inspection door bolted and sealed	27.49	2.00	28.04	0.66	-	8.21	-	36.25	nr	40.78
extra; short radius bends	14.49	2.00	14.78	0.66	-	8.21	-	22.99	nr	25.86
extra; short radius bends with inspection door bolted and sealed	28.04	2.00	28.60	0.66	-	8.21	-	36.81	nr	41.41
extra; single angled branches	24.40	2.00	24.89	1.00	-	12.44	-	37.33	nr	41.99
extra; single angled branches with inspection door bolted and sealed	37.96	2.00	38.72	1.00	-	12.44	-	51.16	nr	57.55
extra; roof connectors	30.58	2.00	31.19	1.30	-	16.17	-	47.36	nr	53.28

CAST IRON SOIL AND WASTE PIPES - Cast iron Stanton SMU lightweight pipes and fittings, stainless steel couplings

Pipes; straight

	Del to Site £	Waste %	Material Cost £	Craft Optve Hrs	Lab Hrs	Labour Cost £	Sunds £	Nett Rate £	Unit	Gross Rate £
50mm; in or on supports (included elsewhere)	9.94	5.00	10.44	0.66	-	8.21	-	18.65	m	20.98
extra; short pipe with access door	19.21	2.00	19.59	0.66	-	8.21	-	27.80	nr	31.28
extra; bends	9.40	2.00	9.59	0.66	-	8.21	-	17.80	nr	20.02
extra; offset bend 75mm projection	12.06	2.00	12.30	0.66	-	8.21	-	20.51	nr	23.08
extra; offset bend 150mm projection	14.57	2.00	14.86	0.66	-	8.21	-	23.07	nr	25.96
extra; single angled branch	16.47	2.00	16.80	1.00	-	12.44	-	29.24	nr	32.89
extra; plug	2.42	2.00	2.47	0.33	-	4.11	-	6.57	nr	7.40
extra; universal plug with one inlet	7.14	2.00	7.28	0.17	-	2.11	-	9.40	nr	10.57
extra; stepping ring for connection to other pipe materials	4.16	2.00	4.24	0.17	-	2.11	-	6.36	nr	7.15
extra; step coupling for connection to B.S.416 pipe	4.90	2.00	5.00	0.50	-	6.22	-	11.22	nr	12.62
75mm; in or on supports (included elsewhere)	11.42	5.00	11.99	0.80	-	9.95	-	21.94	m	24.69
extra; short pipe with access door	21.03	2.00	21.45	0.80	-	9.95	-	31.40	nr	35.33
extra; bends	10.50	2.00	10.71	0.80	-	9.95	-	20.66	nr	23.24
extra; offset bend 75mm projection	15.30	2.00	15.61	0.80	-	9.95	-	25.56	nr	28.75
extra; offset bend 150mm projection	19.10	2.00	19.48	0.80	-	9.95	-	29.43	nr	33.11
extra; single angled branch	17.71	2.00	18.06	1.20	-	14.93	-	32.99	nr	37.12
extra; diminishing piece 75/50	12.70	2.00	12.95	0.80	-	9.95	-	22.91	nr	25.77
extra; plug	2.55	2.00	2.60	0.40	-	4.98	-	7.58	nr	8.52
extra; universal plug with one inlet	10.18	2.00	10.38	0.20	-	2.49	-	12.87	nr	14.48
extra; push fit gasket to suit 32 DN pipe	1.68	2.00	1.71	-	-	-	-	1.71	nr	1.93
extra; push fit gasket to suit 40 DN pipe	1.68	2.00	1.71	-	-	-	-	1.71	nr	1.93
extra; stepping ring for connection to other pipe materials	4.84	2.00	4.94	0.20	-	2.49	-	7.42	nr	8.35
extra; step coupling for connection to B.S.416 pipe	5.47	2.00	5.58	-	-	-	-	5.58	nr	6.28
100mm; in or on supports (included elsewhere)	13.93	5.00	14.63	-	-	-	-	14.63	m	16.45
extra; short pipe with access door	23.85	2.00	24.33	1.00	-	12.44	-	36.77	nr	41.36
extra; bends	12.90	2.00	13.16	1.00	-	12.44	-	25.60	nr	28.80
extra; long radius bend	24.07	2.00	24.55	1.00	-	12.44	-	36.99	nr	41.62
extra; offset bend 75mm projection	17.77	2.00	18.13	1.00	-	12.44	-	30.57	nr	34.39
extra; offset bend 150mm projection	25.06	2.00	25.56	1.00	-	12.44	-	38.00	nr	42.75
extra; single angled branch	22.76	2.00	23.22	1.50	-	18.66	-	41.88	nr	47.11
extra; double angled branch	34.58	2.00	35.27	2.00	-	24.88	-	60.15	nr	67.67
extra; diminishing piece 100/50	15.70	2.00	16.01	1.00	-	12.44	-	28.45	nr	32.01
extra; diminishing piece 100/75	15.72	2.00	16.03	1.00	-	12.44	-	28.47	nr	32.03
extra; plug	2.98	2.00	3.04	0.50	-	6.22	-	9.26	nr	10.42
extra; universal plug with three inlets	12.81	2.00	13.07	0.50	-	6.22	-	19.29	nr	21.70
extra; bossed end cap with inlet	6.13	2.00	6.25	0.50	-	6.22	-	12.47	nr	14.03
extra; push fit gasket to suit 32 DN pipe	1.68	2.00	1.71	-	-	-	-	1.71	nr	1.93
extra; push fit gasket to suit 40 DN pipe	1.68	2.00	1.71	-	-	-	-	1.71	nr	1.93
extra; stepping ring for connection to other pipe materials	5.57	2.00	5.68	0.25	-	3.11	-	8.79	nr	9.89
extra; step coupling for connection to B.S.416 pipe	6.99	2.00	7.13	-	-	-	-	7.13	nr	8.02
extra; traditional joint connector	16.93	2.00	17.27	-	-	-	-	17.27	nr	19.43
extra; W.C. connector	9.57	2.00	9.76	1.00	-	12.44	-	22.20	nr	24.98
extra; roof connector (asphalt)	34.59	2.00	35.28	1.00	-	12.44	-	47.72	nr	53.69

CAST IRON SOIL AND WASTE PIPES - Pipe supports

Galvanised steel vertical bracket

	Del to Site £	Waste %	Material Cost £	Craft Optve Hrs	Lab Hrs	Labour Cost £	Sunds £	Nett Rate £	Unit	Gross Rate £
for 50mm pipes; for fixing bolt(fixing included elsewhere)	4.70	2.00	4.79	-	-	-	-	4.79	nr	5.39
for 75mm pipes; for fixing bolt(fixing included elsewhere)	5.12	2.00	5.22	-	-	-	-	5.22	nr	5.88
for 100mm pipes; for fixing bolt(fixing included elsewhere)	5.59	2.00	5.70	-	-	-	-	5.70	nr	6.41

DISPOSAL SYSTEMS

Labour hourly rates: (except Specialists) Craft Operatives £12.44 Labourer £7.02 Rates are national average prices. Refer to REGIONAL VARIATIONS for indicative levels of overall pricing in regions	MATERIALS			LABOUR				RATES		
	Del to Site	Waste	Material Cost	Craft Optve	Lab	Labour Cost	Sunds	Nett Rate	Unit	Gross Rate (+12.5%)
	£	%	£	Hrs	Hrs	£	£	£		£
R11: FOUL DRAINAGE ABOVE GROUND Cont.										
CAST IRON SOIL AND WASTE PIPES - Pipe supports Cont.										
Aluminium wall hook - type 101										
for 50mm pipes; for M8 fixing bolt (fixing included elsewhere)	3.62	2.00	3.69	-	-	-	-	3.69	nr	4.15
for 75mm pipes; for M8 fixing bolt (fixing included elsewhere)	4.28	2.00	4.37	-	-	-	-	4.37	nr	4.91
for 100mm pipes; for M8 fixing bolt (fixing included elsewhere)	4.95	2.00	5.05	-	-	-	-	5.05	nr	5.68
UNPLASTICIZED MPVC WASTE PIPES (SOLVENT WELDED) - MuPVC pipes and fittings, B.S.5255 Section Three; solvent welded joints										
Pipes; straight										
36mm; in standard plastics pipe brackets fixing to timber with screws	2.00	10.00	2.20	-	-	-	-	2.20	m	2.48
extra; straight expansion couplings	1.35	2.00	1.38	-	-	-	-	1.38	nr	1.55
extra; connections to plastics pipe socket; socket adaptor; solvent welded joint	1.35	2.00	1.38	0.30	-	3.73	-	5.11	nr	5.75
extra; connections to plastics pipe boss; boss adaptor; solvent welded joint	1.38	2.00	1.41	0.30	-	3.73	-	5.14	nr	5.78
extra; fittings; one end	1.42	2.00	1.45	0.15	-	1.87	-	3.31	nr	3.73
extra; fittings; two ends	1.20	2.00	1.22	0.30	-	3.73	-	4.96	nr	5.58
extra; fittings; three ends	1.70	2.00	1.73	0.30	-	3.73	-	5.47	nr	6.15
42mm; in standard plastics pipe brackets fixing to timber with screws	2.15	10.00	2.37	0.40	-	4.98	-	7.34	m	8.26
extra; straight expansion couplings	1.63	2.00	1.66	0.30	-	3.73	-	5.39	nr	6.07
extra; connections to plastics pipe socket; socket adaptor; solvent welded joint	1.38	2.00	1.41	0.30	-	3.73	-	5.14	nr	5.78
extra; connections to plastics pipe boss; boss adaptor; solvent welded joint	1.38	2.00	1.41	0.30	-	3.73	-	5.14	nr	5.78
extra; fittings; one end	1.68	2.00	1.71	0.15	-	1.87	-	3.58	nr	4.03
extra; fittings; two ends	1.34	2.00	1.37	0.30	-	3.73	-	5.10	nr	5.74
extra; fittings; three ends	2.15	2.00	2.19	0.30	-	3.73	-	5.92	nr	6.67
55mm; in standard plastics pipe brackets fixing to timber with screws	3.87	10.00	4.26	0.48	-	5.97	-	10.23	m	11.51
extra; straight expansion couplings	2.21	2.00	2.25	0.40	-	4.98	-	7.23	nr	8.13
extra; connections to plastics pipe socket; socket adaptor; solvent welded joint	1.90	2.00	1.94	0.40	-	4.98	-	6.91	nr	7.78
extra; connections to plastics pipe boss; boss adaptor; solvent welded joint	1.90	2.00	1.94	0.40	-	4.98	-	6.91	nr	7.78
extra; fittings; one end	2.72	2.00	2.77	0.20	-	2.49	-	5.26	nr	5.92
extra; fittings; two ends	2.22	2.00	2.26	0.40	-	4.98	-	7.24	nr	8.15
extra; fittings; three ends	4.24	2.00	4.32	0.40	-	4.98	-	9.30	nr	10.46
36mm; in standard plastics pipe brackets fixing to masonry with screws	2.00	10.00	2.20	0.48	-	5.97	-	8.17	m	9.19
42mm; in standard plastics pipe brackets fixing to masonry with screws	2.15	10.00	2.37	0.48	-	5.97	-	8.34	m	9.38
55mm; in standard plastics pipe brackets fixing to masonry with screws	3.87	10.00	4.26	0.56	-	6.97	-	11.22	m	12.63
POLYPROPYLENE WASTE PIPES ('O' RING) - Polypropylene pipes and fittings; butyl ring joints										
Pipes; straight										
36mm; in standard plastics pipe brackets fixing to timber with screws	0.74	10.00	0.81	0.40	-	4.98	-	5.79	m	6.51
extra; connections to plastics pipe socket; socket adaptor; solvent welded joint	1.38	2.00	1.41	0.30	-	3.73	-	5.14	nr	5.78
extra; connections to plastics pipe boss; boss adaptor; solvent welded joint	1.38	2.00	1.41	0.30	-	3.73	-	5.14	nr	5.78
extra; fittings; one end	0.72	2.00	0.73	0.15	-	1.87	-	2.60	nr	2.93
extra; fittings; two ends	1.25	2.00	1.27	0.30	-	3.73	-	5.01	nr	5.63
extra; fittings; three ends	1.85	2.00	1.89	0.30	-	3.73	-	5.62	nr	6.32
42mm; in standard plastics pipe brackets fixing to timber with screws	0.90	10.00	0.99	0.40	-	4.98	-	5.97	m	6.71
extra; connections to plastics pipe socket; socket adaptor; solvent welded joint	1.38	2.00	1.41	0.40	-	4.98	-	6.38	nr	7.18
extra; connections to plastics pipe boss; boss adaptor; solvent welded joint	1.38	2.00	1.41	0.40	-	4.98	-	6.38	nr	7.18
extra; fittings; one end	0.88	2.00	0.90	0.20	-	2.49	-	3.39	nr	3.81
extra; fittings; two ends	1.30	2.00	1.33	0.40	-	4.98	-	6.30	nr	7.09
extra; fittings; three ends	1.91	2.00	1.95	0.40	-	4.98	-	6.92	nr	7.79
36mm; in standard plastics pipe brackets fixing to masonry with screws	0.74	10.00	0.81	0.48	-	5.97	-	6.79	m	7.63
42mm; in standard plastics pipe brackets fixing to masonry with screws	0.90	10.00	0.99	0.48	-	5.97	-	6.96	m	7.83
UNPLASTICIZED SOIL AND VENTILATING PIPES - PVC-U pipes and fittings, B.S.4514; rubber ring joints; pipework self coloured grey										
Pipes; straight										
82mm; in or on supports (included elsewhere)	5.15	5.00	5.41	0.53	-	6.59	-	12.00	m	13.50
extra; pipes with inspection door bolted and sealed	12.14	2.00	12.38	0.40	-	4.98	-	17.36	nr	19.53
extra; short radius bends	6.81	2.00	6.95	0.40	-	4.98	-	11.92	nr	13.41
extra; single angled branches	10.70	2.00	10.91	0.40	-	4.98	-	15.89	nr	17.88
extra; straight wc connectors	6.27	2.00	6.40	0.40	-	4.98	-	11.37	nr	12.79
pipes with one boss socket	6.38	2.00	6.51	0.40	-	4.98	-	11.48	nr	12.92
110mm; in or on supports (included elsewhere)	4.90	5.00	5.14	0.60	-	7.46	-	12.61	m	14.19

Labour hourly rates: (except Specialists) Craft Operatives £12.44 Labourer £7.02 Rates are national average prices. Refer to REGIONAL VARIATIONS for indicative levels of overall pricing in regions	MATERIALS			LABOUR				RATES		
	Del to Site	Waste	Material Cost	Craft Optve	Lab	Labour Cost	Sunds	Nett Rate	Unit	Gross Rate (+12.5%)
	£	%	£	Hrs	Hrs	£	£	£		£

R11: FOUL DRAINAGE ABOVE GROUND Cont.

UNPLASTICIZED SOIL AND VENTILATING PIPES - PVC-U pipes and fittings, B.S.4514; rubber ring joints; pipework self coloured grey Cont.

	Del to Site	Waste	Material Cost	Craft Optve	Lab	Labour Cost	Sunds	Nett Rate	Unit	Gross Rate
Pipes; straight Cont.										
extra; connections to cast iron pipe socket caulking bush; caulked lead and hempen spun yarn joint	17.73	2.00	18.08	1.50	-	18.66	-	36.74	nr	41.34
extra; pipes with inspection door bolted and sealed	11.97	2.00	12.21	0.60	-	7.46	-	19.67	nr	22.13
extra; short radius bends	8.20	2.00	8.36	0.60	-	7.46	-	15.83	nr	17.81
extra; single angled branches	10.67	2.00	10.88	0.60	-	7.46	-	18.35	nr	20.64
extra; straight wc connectors	7.16	2.00	7.30	0.60	-	7.46	-	14.77	nr	16.61
pipes with one boss socket	6.02	2.00	6.14	0.60	-	7.46	-	13.60	nr	15.30
160mm; in or on supports (included elsewhere)	12.94	5.00	13.59	0.90	-	11.20	-	24.78	m	27.88
extra; pipes with inspection door bolted and sealed	26.46	2.00	26.99	0.90	-	11.20	-	38.19	nr	42.96
extra; short radius bends	21.57	2.00	22.00	0.90	-	11.20	-	33.20	nr	37.35
extra; single angled branches	35.52	2.00	36.23	0.90	-	11.20	-	47.43	nr	53.35
pipes with one boss socket	12.25	2.00	12.49	0.90	-	11.20	-	23.69	nr	26.65

UNPLASTICIZED SOIL AND VENTILATING PIPES - Plastics pipework ancillaries, self coloured grey

	Del to Site	Waste	Material Cost	Craft Optve	Lab	Labour Cost	Sunds	Nett Rate	Unit	Gross Rate
Weathering aprons										
fitted to 82mm pipes	1.85	2.00	1.89	0.40	-	4.98	-	6.86	nr	7.72
fitted to 110mm pipes	1.96	2.00	2.00	0.60	-	7.46	-	9.46	nr	10.65
fitted to 160mm pipes	5.38	2.00	5.49	0.90	-	11.20	-	16.68	nr	18.77

UNPLASTICIZED SOIL AND VENTILATING PIPES - Pipe supports

	Del to Site	Waste	Material Cost	Craft Optve	Lab	Labour Cost	Sunds	Nett Rate	Unit	Gross Rate
Plastics coated metal holderbats										
for 82mm pipes; for building in (fixing included elsewhere)	3.94	2.00	4.02	-	-	-	-	4.02	nr	4.52
for 110mm pipes; for building in (fixing included elsewhere)	3.90	2.00	3.98	-	-	-	-	3.98	nr	4.48
for 82mm pipes; fixing to timber with galvanized screws	3.16	2.00	3.22	0.20	-	2.49	-	5.71	nr	6.43
for 110mm pipes; fixing to timber with galvanized screws	3.12	2.00	3.18	-	-	-	-	3.18	nr	3.58
for 160mm pipes; fixing to timber with galvanized screws	4.82	2.00	4.92	-	-	-	-	4.92	nr	5.53

WIRE BALLOON GRATINGS - Copper pipework ancillaries

	Del to Site	Waste	Material Cost	Craft Optve	Lab	Labour Cost	Sunds	Nett Rate	Unit	Gross Rate
Wire balloon gratings										
fitted to 50mm pipes	2.55	3.00	2.63	0.12	-	1.49	-	4.12	nr	4.63
fitted to 75mm pipes	3.00	3.00	3.09	0.12	-	1.49	-	4.58	nr	5.16
fitted to 100mm pipes	3.45	3.00	3.55	0.12	-	1.49	-	5.05	nr	5.68

WIRE BALLOON GRATINGS - Steel pipework ancillaries, galvanized

	Del to Site	Waste	Material Cost	Craft Optve	Lab	Labour Cost	Sunds	Nett Rate	Unit	Gross Rate
Wire balloon gratings										
fitted to 50mm pipes	2.02	2.00	2.06	0.12	-	1.49	-	3.55	nr	4.00
fitted to 75mm pipes	2.28	2.00	2.33	0.12	-	1.49	-	3.82	nr	4.30
fitted to 100mm pipes	2.60	2.00	2.65	0.12	-	1.49	-	4.14	nr	4.66

PLASTIC TRAPS - Plastics pipework ancillaries self coloured white

	Del to Site	Waste	Material Cost	Craft Optve	Lab	Labour Cost	Sunds	Nett Rate	Unit	Gross Rate
Traps; B.S.3943										
P, two piece, 75mm seal, inlet with coupling nut, outlet with seal ring socket										
36mm outlet	3.08	2.00	3.14	0.40	-	4.98	-	8.12	nr	9.13
42mm outlet	3.55	2.00	3.62	0.40	-	4.98	-	8.60	nr	9.67
S, two piece, 75mm seal, inlet with coupling nut, outlet with seal ring socket										
36mm outlet	3.90	2.00	3.98	0.40	-	4.98	-	8.95	nr	10.07
42mm outlet	4.58	2.00	4.67	0.40	-	4.98	-	9.65	nr	10.85
P, two piece, 75mm seal, with flexible polypropylene pipe for overflow connection, inlet with coupling nut, outlet with seal ring socket										
42mm outlet	6.60	2.00	6.73	0.60	-	7.46	-	14.20	nr	15.97

COPPER TRAPS - Copper pipework ancillaries

	Del to Site	Waste	Material Cost	Craft Optve	Lab	Labour Cost	Sunds	Nett Rate	Unit	Gross Rate
Traps; B.S.1184 Section Two (Solid Drawn)										
P, two piece, 75mm seal, inlet with coupling nut, compression outlet										
35mm outlet	15.34	2.00	15.65	0.50	-	6.22	-	21.87	nr	24.60
42mm outlet	19.53	2.00	19.92	0.50	-	6.22	-	26.14	nr	29.41
54mm outlet	70.04	2.00	71.44	0.75	-	9.33	-	80.77	nr	90.87
S, two piece, 75mm seal, inlet with coupling nut, compression outlet										
35mm outlet	17.58	2.00	17.93	0.50	-	6.22	-	24.15	nr	27.17
42mm outlet	21.11	2.00	21.53	0.50	-	6.22	-	27.75	nr	31.22
54mm outlet	73.98	2.00	75.46	0.75	-	9.33	-	84.79	nr	95.39
bath, two piece, 75mm seal, overflow connection, cleaning eye and plug, inlet with coupling nut, compression outlet										
42mm outlet	28.82	2.00	29.40	0.80	-	9.95	-	39.35	nr	44.27

DISPOSAL SYSTEMS

Labour hourly rates: (except Specialists) Craft Operatives £9.23 Labourer £7.02 Rates are national average prices. Refer to REGIONAL VARIATIONS for indicative levels of overall pricing in regions	PLANT & TRANSPORT			LABOUR				RATES		
	Plant Cost	Trans Cost	P & T Cost	Craft Optve	Lab	Labour Cost	Sunds	Nett Rate	Unit	Gross Rate (+12.5%)
	£	£	£	Hrs	Hrs	£	£	£		£

R12: DRAINAGE BELOW GROUND

EXCAVATION FOR DRAINS - BY MACHINE - Excavating trenches by machine to receive pipes not exceeding 200mm nominal size; disposing of surplus excavated material by removing from site

Excavations commencing from natural ground level
average depth

	Plant Cost	Trans Cost	P & T Cost	Craft Optve	Lab	Labour Cost	Sunds	Nett Rate	Unit	Gross Rate
500mm	1.44	2.52	3.96	-	0.24	1.68	0.94	6.58	m	7.41
750mm	2.35	2.52	4.87	-	0.35	2.46	1.41	8.74	m	9.83
1000mm	3.19	2.52	5.71	-	0.47	3.30	1.88	10.89	m	12.25
1250mm	4.10	2.52	6.62	-	0.64	4.49	2.95	14.06	m	15.82
1500mm	4.93	2.52	7.45	-	0.76	5.34	3.54	16.33	m	18.37
1750mm	5.85	2.52	8.37	-	0.90	6.32	4.13	18.82	m	21.17
2000mm	6.68	2.52	9.20	-	1.02	7.16	4.72	21.08	m	23.72
2250mm	8.62	2.52	11.14	-	1.77	12.43	6.03	29.60	m	33.29
2500mm	9.57	2.52	12.09	-	1.97	13.83	6.70	32.62	m	36.70
2750mm	10.60	2.52	13.12	-	2.16	15.16	7.37	35.65	m	40.11
3000mm	11.54	2.52	14.06	-	2.36	16.57	8.04	38.67	m	43.50
3250mm	12.57	2.52	15.09	-	2.56	17.97	8.71	41.77	m	46.99
3500mm	13.52	2.52	16.04	-	2.76	19.38	9.38	44.80	m	50.39
3750mm	14.55	2.52	17.07	-	2.96	20.78	10.05	47.90	m	53.89
4000mm	15.49	2.52	18.01	-	3.15	22.11	10.72	50.84	m	57.20
4250mm	16.52	2.52	19.04	-	4.46	31.31	12.92	63.27	m	71.18
4500mm	17.47	2.52	19.99	-	4.72	33.13	13.68	66.80	m	75.15
4750mm	18.50	2.52	21.02	-	4.98	34.96	14.44	70.42	m	79.22
5000mm	19.44	2.52	21.96	-	6.56	46.05	15.20	83.21	m	93.61
5250mm	20.47	2.52	22.99	-	6.88	48.30	15.96	87.25	m	98.15
5500mm	21.42	2.52	23.94	-	7.21	50.61	16.72	91.27	m	102.68
5750mm	22.45	2.52	24.97	-	7.53	52.86	17.48	95.31	m	107.22
6000mm	23.39	2.52	25.91	-	7.87	55.25	18.24	99.40	m	111.82

EXCAVATION FOR DRAINS - BY MACHINE - Excavating trenches by machine to receive pipes 225mm nominal size; disposing of surplus excavated material by removing from site

Excavations commencing from natural ground level
average depth

	Plant Cost	Trans Cost	P & T Cost	Craft Optve	Lab	Labour Cost	Sunds	Nett Rate	Unit	Gross Rate
500mm	1.56	2.66	4.22	-	0.25	1.75	0.94	6.92	m	7.78
750mm	2.47	2.66	5.13	-	0.37	2.60	1.41	9.14	m	10.28
1000mm	3.38	2.66	6.04	-	0.49	3.44	1.88	11.36	m	12.78
1250mm	4.28	2.66	6.94	-	0.67	4.70	2.95	14.59	m	16.42
1500mm	5.26	2.66	7.92	-	0.80	5.62	3.54	17.08	m	19.21
1750mm	6.17	2.66	8.83	-	0.94	6.60	4.13	19.56	m	22.00
2000mm	7.08	2.66	9.74	-	1.08	7.58	4.72	22.04	m	24.80
2250mm	9.07	2.66	11.73	-	1.85	12.99	6.03	30.75	m	34.59
2500mm	10.18	2.66	12.84	-	2.06	14.46	6.70	34.00	m	38.25
2750mm	11.20	2.66	13.86	-	2.26	15.87	7.37	37.10	m	41.73
3000mm	12.23	2.66	14.89	-	2.47	17.34	8.04	40.27	m	45.30
3250mm	13.26	2.66	15.92	-	2.68	18.81	8.71	43.44	m	48.87
3500mm	14.36	2.66	17.02	-	2.89	20.29	9.38	46.69	m	52.52
3750mm	15.39	2.66	18.05	-	3.10	21.76	10.05	49.86	m	56.09
4000mm	16.42	2.66	19.08	-	3.30	23.17	10.72	52.97	m	59.59
4250mm	17.45	2.66	20.11	-	4.68	32.85	12.92	65.88	m	74.12
4500mm	18.55	2.66	21.21	-	4.95	34.75	13.68	69.64	m	78.34
4750mm	19.58	2.66	22.24	-	5.22	36.64	14.44	73.32	m	82.49
5000mm	20.61	2.66	23.27	-	6.88	48.30	15.20	86.77	m	97.61
5250mm	21.64	2.66	24.30	-	7.23	50.75	15.96	91.01	m	102.39
5500mm	22.74	2.66	25.40	-	7.66	53.77	16.72	95.89	m	107.88
5750mm	23.77	2.66	26.43	-	7.91	55.53	17.48	99.44	m	111.87
6000mm	24.80	2.66	27.46	-	8.24	57.84	18.24	103.54	m	116.49

EXCAVATION FOR DRAINS - BY MACHINE - Excavating trenches by machine to receive pipes 300mm nominal size; disposing of surplus excavated material by removing from site

Excavations commencing from natural ground level
average depth

	Plant Cost	Trans Cost	P & T Cost	Craft Optve	Lab	Labour Cost	Sunds	Nett Rate	Unit	Gross Rate
500mm	1.73	2.94	4.67	-	0.26	1.83	0.94	7.44	m	8.36
750mm	2.78	2.94	5.72	-	0.37	2.60	1.41	9.73	m	10.94
1000mm	3.82	2.94	6.76	-	0.52	3.65	1.88	12.29	m	13.83
1250mm	4.87	2.94	7.81	-	0.70	4.91	2.95	15.67	m	17.63
1500mm	5.92	2.94	8.86	-	0.84	5.90	3.54	18.30	m	20.58
1750mm	6.97	2.94	9.91	-	0.99	6.95	4.13	20.99	m	23.61
2000mm	8.02	2.94	10.96	-	1.15	8.07	4.72	23.75	m	26.72
2250mm	10.30	2.94	13.24	-	1.95	13.69	6.03	32.96	m	37.08
2500mm	11.48	2.94	14.42	-	2.17	15.23	6.70	36.35	m	40.90
2750mm	12.67	2.94	15.61	-	2.36	16.57	7.37	39.55	m	44.49
3000mm	13.85	2.94	16.79	-	2.60	18.25	8.04	43.08	m	48.47
3250mm	15.04	2.94	17.98	-	2.81	19.73	8.71	46.42	m	52.22
3500mm	16.22	2.94	19.16	-	3.04	21.34	9.38	49.88	m	56.12
3750mm	17.41	2.94	20.35	-	3.25	22.82	10.05	53.22	m	59.87
4000mm	18.59	2.94	21.53	-	3.46	24.29	10.72	56.54	m	63.61
4250mm	19.78	2.94	22.72	-	4.91	34.47	12.92	70.11	m	78.87
4500mm	20.96	2.94	23.90	-	5.19	36.43	13.68	74.01	m	83.27
4750mm	22.15	2.94	25.09	-	5.49	38.54	14.44	78.07	m	87.83
5000mm	23.33	2.94	26.27	-	7.22	50.68	15.20	92.15	m	103.67
5250mm	24.52	2.94	27.46	-	7.58	53.21	15.96	96.63	m	108.71
5500mm	25.70	2.94	28.64	-	7.92	55.60	16.72	100.96	m	113.58
5750mm	26.89	2.94	29.83	-	8.29	58.20	17.48	105.51	m	118.69
6000mm	28.07	2.94	31.01	-	8.66	60.79	18.24	110.04	m	123.80

Labour hourly rates: (except Specialists) Craft Operatives £9.23 Labourer £7.02 Rates are national average prices. Refer to REGIONAL VARIATIONS for indicative levels of overall pricing in regions	PLANT & TRANSPORT			LABOUR				RATES		
	Plant Cost	Trans Cost	P & T Cost	Craft Optve	Lab	Labour Cost	Sunds	Nett Rate	Unit	Gross Rate (+12.5%)
	£	£	£	Hrs	Hrs	£	£	£		£

R12: DRAINAGE BELOW GROUND Cont.

EXCAVATION FOR DRAINS - BY MACHINE - Excavating trenches by machine to receive pipes 400mm nominal size; disposing of surplus excavated material by removing from site

Excavations commencing from natural ground level
average depth

	Plant Cost	Trans Cost	P & T Cost	Craft Optve	Lab	Labour Cost	Sunds	Nett Rate	Unit	Gross Rate
2250mm	12.03	3.50	15.53	-	2.13	14.95	6.03	36.51	m	41.08
2500mm	13.37	3.50	16.87	-	2.37	16.64	6.70	40.21	m	45.23
2750mm	14.79	3.50	18.29	-	2.59	18.18	7.37	43.84	m	49.32
3000mm	16.13	3.50	19.63	-	2.83	19.87	8.04	47.54	m	53.48
3250mm	17.56	3.50	21.06	-	3.06	21.48	8.71	51.25	m	57.66
3500mm	18.90	3.50	22.40	-	3.31	23.24	9.38	55.02	m	61.89
3750mm	20.32	3.50	23.82	-	3.54	24.85	10.05	58.72	m	66.06
4000mm	21.66	3.50	25.16	-	3.78	26.54	10.72	62.42	m	70.22
4250mm	23.09	3.50	26.59	-	5.34	37.49	12.92	77.00	m	86.62
4500mm	24.43	3.50	27.93	-	5.67	39.80	13.68	81.41	m	91.59
4750mm	25.85	3.50	29.35	-	5.98	41.98	14.44	85.77	m	96.49
5000mm	27.19	3.50	30.69	-	7.87	55.25	15.20	101.14	m	113.78
5250mm	28.63	3.50	32.13	-	8.56	60.09	15.96	108.18	m	121.70
5500mm	29.96	3.50	33.46	-	8.66	60.79	16.72	110.97	m	124.84
5750mm	31.38	3.50	34.88	-	9.03	63.39	17.48	115.75	m	130.22
6000mm	32.72	3.50	36.22	-	9.45	66.34	18.24	120.80	m	135.90

EXCAVATION FOR DRAINS - BY MACHINE - Excavating trenches by machine to receive pipes 450mm nominal size; disposing of surplus excavated material by removing from site

Excavations commencing from natural ground level
average depth

	Plant Cost	Trans Cost	P & T Cost	Craft Optve	Lab	Labour Cost	Sunds	Nett Rate	Unit	Gross Rate
2250mm	12.87	3.64	16.51	-	2.21	15.51	6.03	38.05	m	42.81
2500mm	14.37	3.64	18.01	-	2.46	17.27	6.70	41.98	m	47.23
2750mm	15.79	3.64	19.43	-	2.70	18.95	7.37	45.75	m	51.47
3000mm	17.30	3.64	20.94	-	2.94	20.64	8.04	49.62	m	55.82
3250mm	18.80	3.64	22.44	-	3.20	22.46	8.71	53.61	m	60.32
3500mm	20.29	3.64	23.93	-	3.45	24.22	9.38	57.53	m	64.72
3750mm	21.72	3.64	25.36	-	3.69	25.90	10.05	61.31	m	68.98
4000mm	23.22	3.64	26.86	-	3.93	27.59	10.72	65.17	m	73.31
4250mm	24.72	3.64	28.36	-	5.56	39.03	12.92	80.31	m	90.35
4500mm	26.22	3.64	29.86	-	5.91	41.49	13.68	85.03	m	95.66
4750mm	27.64	3.64	31.28	-	6.24	43.80	14.44	89.52	m	100.72
5000mm	29.15	3.64	32.79	-	8.19	57.49	15.20	105.48	m	118.67
5250mm	30.65	3.64	34.29	-	8.61	60.44	15.96	110.69	m	124.53
5500mm	32.14	3.64	35.78	-	9.03	63.39	16.72	115.89	m	130.38
5750mm	33.57	3.64	37.21	-	9.39	65.92	17.48	120.61	m	135.68
6000mm	35.07	3.64	38.71	-	9.81	68.87	18.24	125.82	m	141.54

Excavating trenches by hand to receive pipes not exceeding 200mm nominal size; disposing of surplus excavated material on site

Excavations commencing from natural ground level
average depth

	Plant Cost	Trans Cost	P & T Cost	Craft Optve	Lab	Labour Cost	Sunds	Nett Rate	Unit	Gross Rate
500mm	-	-	-	-	1.20	8.42	3.46	11.88	m	13.37
750mm	-	-	-	-	2.15	15.09	3.93	19.02	m	21.40
1000mm	-	-	-	-	3.10	21.76	4.40	26.16	m	29.43
1250mm	-	-	-	-	3.90	27.38	5.47	32.85	m	36.95
1500mm	-	-	-	-	4.85	34.05	6.06	40.11	m	45.12
1750mm	-	-	-	-	5.80	40.72	6.65	47.37	m	53.29
2000mm	-	-	-	-	6.80	47.74	7.24	54.98	m	61.85
2250mm	-	-	-	-	8.00	56.16	8.55	64.71	m	72.80
2500mm	-	-	-	-	9.20	64.58	9.22	73.80	m	83.03
2750mm	-	-	-	-	10.40	73.01	9.89	82.90	m	93.26
3000mm	-	-	-	-	11.60	81.43	10.56	91.99	m	103.49
3250mm	-	-	-	-	12.80	89.86	11.23	101.09	m	113.72
3500mm	-	-	-	-	14.05	98.63	11.90	110.53	m	124.35
3750mm	-	-	-	-	15.25	107.06	12.57	119.63	m	134.58
4000mm	-	-	-	-	16.50	115.83	13.24	129.07	m	145.20

EXCAVATION FOR DRAINS - BY HAND - Excavating trenches by hand to receive pipes 225mm nominal size; disposing of surplus excavated material on site

Excavations commencing from natural ground level
average depth

	Plant Cost	Trans Cost	P & T Cost	Craft Optve	Lab	Labour Cost	Sunds	Nett Rate	Unit	Gross Rate
500mm	-	-	-	-	1.35	9.48	3.60	13.08	m	14.71
750mm	-	-	-	-	2.35	16.50	4.07	20.57	m	23.14
1000mm	-	-	-	-	3.35	23.52	4.54	28.06	m	31.56
1250mm	-	-	-	-	4.35	30.54	5.61	36.15	m	40.67
1500mm	-	-	-	-	5.35	37.56	6.20	43.76	m	49.23
1750mm	-	-	-	-	6.35	44.58	6.79	51.37	m	57.79
2000mm	-	-	-	-	7.35	51.60	7.38	58.98	m	66.35
2250mm	-	-	-	-	8.80	61.78	8.69	70.47	m	79.27
2500mm	-	-	-	-	10.10	70.90	9.36	00.26	m	90.29
2750mm	-	-	-	-	11.40	80.03	10.03	90.06	m	101.32
3000mm	-	-	-	-	12.70	89.15	10.70	99.85	m	112.34
3250mm	-	-	-	-	14.05	98.63	11.37	110.00	m	123.75
3500mm	-	-	-	-	15.40	108.11	12.04	120.15	m	135.17
3750mm	-	-	-	-	16.70	117.23	12.71	129.94	m	146.19
4000mm	-	-	-	-	18.10	127.06	13.38	140.44	m	158.00

DISPOSAL SYSTEMS

Labour hourly rates: (except Specialists) Craft Operatives £9.23 Labourer £7.02 Rates are national average prices. Refer to REGIONAL VARIATIONS for indicative levels of overall pricing in regions	PLANT & TRANSPORT			LABOUR				RATES		
	Plant Cost £	Trans Cost £	P & T Cost £	Craft Optve Hrs	Lab Hrs	Labour Cost £	Sunds £	Nett Rate £	Unit	Gross Rate (+12.5%) £

R12: DRAINAGE BELOW GROUND Cont.

EXCAVATION FOR DRAINS - BY HAND - Excavating trenches by hand to receive pipes 300mm nominal size; disposing of surplus excavated material on site

Excavations commencing from natural ground level
  average depth

500mm	-	-	-	-	1.50	10.53	3.88	14.41	m	16.21
750mm	-	-	-	-	2.60	18.25	4.35	22.60	m	25.43
1000mm	-	-	-	-	3.65	25.62	4.82	30.44	m	34.25
1250mm	-	-	-	-	4.75	33.34	5.89	39.23	m	44.14
1500mm	-	-	-	-	5.80	40.72	6.48	47.20	m	53.10
1750mm	-	-	-	-	6.90	48.44	7.07	55.51	m	62.45
2000mm	-	-	-	-	7.95	55.81	7.66	63.47	m	71.40
2250mm	-	-	-	-	9.45	66.34	8.97	75.31	m	84.72
2500mm	-	-	-	-	10.85	76.17	9.64	85.81	m	96.53
2750mm	-	-	-	-	12.30	86.35	10.31	96.66	m	108.74
3000mm	-	-	-	-	13.70	96.17	10.98	107.15	m	120.55
3250mm	-	-	-	-	15.15	106.35	11.65	118.00	m	132.75
3500mm	-	-	-	-	16.60	116.53	12.32	128.85	m	144.96
3750mm	-	-	-	-	18.05	126.71	12.99	139.70	m	157.16
4000mm	-	-	-	-	19.50	136.89	13.66	150.55	m	169.37

**This section continues
  on the next page**

Labour hourly rates: (except Specialists) Craft Operatives £9.23 Labourer £7.02 Rates are national average prices. Refer to REGIONAL VARIATIONS for indicative levels of overall pricing in regions	MATERIALS			LABOUR				RATES		
	Del to Site £	Waste %	Material Cost £	Craft Optve Hrs	Lab Hrs	Labour Cost £	Sunds £	Nett Rate £	Unit	Gross Rate (+12.5%) £
R12: DRAINAGE BELOW GROUND Cont.										
BEDS, BENCHINGS AND COVERS - Granular material, 10mm nominal size pea shingle, to be obtained off site										
Beds										
400 x 150mm	2.18	10.00	2.40	–	0.18	1.26	–	3.66	m	4.12
450 x 150mm	2.54	10.00	2.79	–	0.19	1.33	–	4.13	m	4.64
525 x 150mm	2.90	10.00	3.19	–	0.20	1.40	–	4.59	m	5.17
600 x 150mm	3.27	10.00	3.60	–	0.26	1.83	–	5.42	m	6.10
700 x 150mm	3.99	10.00	4.39	–	0.29	2.04	–	6.42	m	7.23
750 x 150mm	3.99	10.00	4.39	–	0.32	2.25	–	6.64	m	7.46
Beds and surrounds										
400 x 150mm bed; 150mm thick surround to 100mm internal diameter pipes -1	5.45	10.00	6.00	–	0.55	3.86	–	9.86	m	11.09
450 x 150mm bed; 150mm thick surround to 150mm internal diameter pipes -1	6.90	10.00	7.59	–	0.62	4.35	–	11.94	m	13.44
525 x 150mm bed; 150mm thick surround to 225mm internal diameter pipes -1	8.71	10.00	9.58	–	0.82	5.76	–	15.34	m	17.25
600 x 150mm bed; 150mm thick surround to 300mm internal diameter pipes -1	10.53	10.00	11.58	–	1.01	7.09	–	18.67	m	21.01
700 x 150mm bed; 150mm thick surround to 400mm internal diameter pipes -1	13.07	10.00	14.38	–	1.20	8.42	–	22.80	m	25.65
750 x 150mm bed; 150mm thick surround to 450mm internal diameter pipes -1	14.52	10.00	15.97	–	1.39	9.76	–	25.73	m	28.95
Beds and filling to half pipe depth										
400 x 150mm overall; to 100mm internal diameter pipes - 1	2.18	10.00	2.40	–	0.14	0.98	–	3.38	m	3.80
450 x 225mm overall; to 150mm internal diameter pipes - 1	3.63	10.00	3.99	–	0.24	1.68	–	5.68	m	6.39
525 x 263mm overall; to 225mm internal diameter pipes - 1	5.08	10.00	5.59	–	0.34	2.39	–	7.97	m	8.97
600 x 300mm overall; to 300mm internal diameter pipes -1	6.53	10.00	7.18	–	0.43	3.02	–	10.20	m	11.48
700 x 350mm overall; to 400mm internal diameter pipes -1	9.08	10.00	9.99	–	0.59	4.14	–	14.13	m	15.90
750 x 375mm overall; to 450mm internal diameter pipes -1	10.16	10.00	11.18	–	0.67	4.70	–	15.88	m	17.86
BEDS, BENCHINGS AND COVERS - Sand; to be obtained off site										
Beds										
400 x 50mm	0.52	5.00	0.55	–	0.05	0.35	–	0.90	m	1.01
400 x 100mm	1.03	5.00	1.08	–	0.10	0.70	–	1.78	m	2.01
450 x 50mm	0.52	5.00	0.55	–	0.06	0.42	–	0.97	m	1.09
450 x 100mm	1.29	5.00	1.35	–	0.11	0.77	–	2.13	m	2.39
525 x 50mm	0.77	5.00	0.81	–	0.07	0.49	–	1.30	m	1.46
525 x 100mm	1.29	5.00	1.35	–	0.13	0.91	–	2.27	m	2.55
600 x 50mm	0.77	5.00	0.81	–	0.07	0.49	–	1.30	m	1.46
600 x 100mm	1.55	5.00	1.63	–	0.14	0.98	–	2.61	m	2.94
700 x 50mm	1.03	5.00	1.08	–	0.08	0.56	–	1.64	m	1.85
700 x 100mm	1.80	5.00	1.89	–	0.17	1.19	–	3.08	m	3.47
750 x 50mm	1.03	5.00	1.08	–	0.10	0.70	–	1.78	m	2.01
750 x 100mm	2.06	5.00	2.16	–	0.18	1.26	–	3.43	m	3.85
BEDS, BENCHINGS AND COVERS - Plain in-situ concrete; B.S.5328, ordinary prescribed mix, ST3, 20mm aggregate										
Beds										
400 x 100mm	2.08	5.00	2.18	–	0.22	1.54	2.70	6.43	m	7.23
400 x 150mm	3.12	5.00	3.28	–	0.32	2.25	4.04	9.56	m	10.76
450 x 100mm	2.60	5.00	2.73	–	0.24	1.68	2.70	7.11	m	8.00
450 x 150mm	3.63	5.00	3.81	–	0.36	2.53	4.04	10.38	m	11.68
525 x 100mm	2.60	5.00	2.73	–	0.29	2.04	2.70	7.47	m	8.40
525 x 150mm	4.15	5.00	4.36	–	0.42	2.95	4.04	11.35	m	12.76
600 x 100mm	3.12	5.00	3.28	–	0.32	2.25	2.70	8.22	m	9.25
600 x 150mm	4.67	5.00	4.90	–	0.46	3.23	4.04	12.17	m	13.69
700 x 100mm	3.63	5.00	3.81	–	0.38	2.67	2.70	9.18	m	10.33
700 x 150mm	5.71	5.00	6.00	–	0.56	3.93	4.04	13.97	m	15.71
750 x 100mm	4.15	5.00	4.36	–	0.41	2.88	2.70	9.94	m	11.18
750 x 150mm	5.71	5.00	6.00	–	0.61	4.28	4.04	14.32	m	16.11
Beds and surrounds										
400 x 100mm bed; 150mm thick surround to 100mm internal diameter pipes -1	6.75	5.00	7.09	–	0.86	6.04	5.38	18.50	m	20.82
400 x 150mm bed; 150mm thick surround to 100mm internal diameter pipes -1	7.79	5.00	8.18	–	0.97	6.81	6.06	21.05	m	23.68
450 x 100mm bed; 150mm thick surround to 150mm internal diameter pipes -1	8.31	5.00	8.73	–	1.09	7.65	6.06	22.44	m	25.24
450 x 150mm bed; 150mm thick surround to 150mm internal diameter pipes -1	9.35	5.00	9.82	–	1.21	8.49	6.74	25.05	m	28.18
525 x 100mm bed; 150mm thick surround to 225mm internal diameter pipes -1	10.90	5.00	11.45	–	1.49	10.46	6.74	28.64	m	32.23
525 x 150mm bed; 150mm thick surround to 225mm internal diameter pipes -1	12.46	5.00	13.08	–	1.63	11.44	7.74	32.27	m	36.30
600 x 100mm bed; 150mm thick surround to 300mm internal diameter pipes -1	13.50	5.00	14.18	–	1.94	13.62	8.09	35.88	m	40.37
600 x 150mm bed; 150mm thick surround to 300mm internal diameter pipes -1	15.06	5.00	15.81	–	2.11	14.81	8.76	39.39	m	44.31
700 x 100mm bed; 150mm thick surround to 400mm internal diameter pipes -1	17.13	5.00	17.99	–	2.65	18.60	9.44	46.03	m	51.78

DISPOSAL SYSTEMS – SMALL WORKS

	MATERIALS			LABOUR				RATES		
Labour hourly rates: (except Specialists) Craft Operatives £9.23 Labourer £7.02 Rates are national average prices. Refer to REGIONAL VARIATIONS for indicative levels of overall pricing in regions	Del to Site £	Waste %	Material Cost £	Craft Optve Hrs	Lab Hrs	Labour Cost £	Sunds £	Nett Rate £	Unit	Gross Rate (+12.5%) £
R12: DRAINAGE BELOW GROUND Cont.										
BEDS, BENCHINGS AND COVERS - Plain in-situ concrete; B.S.5328, ordinary prescribed mix, ST3, 20mm aggregate Cont.										
Beds and surrounds Cont.										
700 x 150mm bed; 150mm thick surround to 400mm internal diameter pipes -1	18.69	5.00	19.62	-	2.83	19.87	10.12	49.61	m	55.81
750 x 100mm bed; 150mm thick surround to 450mm internal diameter pipes -1	19.21	5.00	20.17	-	3.04	21.34	10.12	51.63	m	58.09
750 x 150mm bed; 150mm thick surround to 450mm internal diameter pipes -1	20.77	5.00	21.81	-	3.24	22.74	10.77	55.32	m	62.24
Beds and haunchings										
400 x 150mm; to 100mm internal diameter pipes -1	4.15	5.00	4.36	-	0.48	3.37	4.04	11.77	m	13.24
450 x 150mm; to 150mm internal diameter pipes -1	4.67	5.00	4.90	-	0.55	3.86	4.04	12.80	m	14.41
525 x 150mm; to 225mm internal diameter pipes -1	6.23	5.00	6.54	-	0.74	5.19	4.04	15.78	m	17.75
600 x 150mm; to 300mm internal diameter pipes -1	8.31	5.00	8.73	-	0.97	6.81	4.04	19.57	m	22.02
700 x 150mm; to 400mm internal diameter pipes -1	10.38	5.00	10.90	-	1.32	9.27	4.04	24.21	m	27.23
750 x 150mm; to 450mm internal diameter pipes -1	11.42	5.00	11.99	-	1.62	11.37	4.04	27.40	m	30.83
Vertical casings										
400 x 400mm to 100mm internal diameter pipes - 1	7.79	5.00	8.18	-	0.96	6.74	22.32	37.24	m	41.89
450 x 450mm to 150mm internal diameter pipes - 1	9.35	5.00	9.82	-	1.21	8.49	25.11	43.42	m	48.85
525 x 525mm to 225mm internal diameter pipes - 1	14.54	5.00	15.27	-	1.66	11.65	29.30	56.22	m	63.25
600 x 600mm to 300mm internal diameter pipes - 1	15.06	5.00	15.81	-	2.16	15.16	33.48	64.46	m	72.51
700 x 700mm to 400mm internal diameter pipes - 1	18.69	5.00	19.62	-	2.94	20.64	39.06	79.32	m	89.24
750 x 750mm to 450mm internal diameter pipes - 1	20.77	5.00	21.81	-	3.37	23.66	41.85	87.32	m	98.23
VITRIFIED CLAY FLEXIBLE JOINT DRAINS - Drains; vitrified clay pipes and fittings, B.S.EN295, normal; flexible mechanical joints										
Pipework in trenches										
100mm	8.05	5.00	8.45	0.19	0.07	2.25	0.21	10.91	m	12.27
extra; bends	11.65	5.00	12.23	0.10	-	0.92	-	13.16	nr	14.80
extra; branches	16.16	5.00	16.97	0.19	-	1.75	-	18.72	nr	21.06
150mm	10.46	5.00	10.98	0.22	0.10	2.73	0.24	13.96	m	15.70
extra; bends	19.20	5.00	20.16	0.11	-	1.02	-	21.18	nr	23.82
extra; branches	25.07	5.00	26.32	0.22	-	2.03	-	28.35	nr	31.90
225mm	20.22	5.00	21.23	0.26	0.13	3.31	0.30	24.84	m	27.95
extra; bends	39.17	5.00	41.13	0.13	-	1.20	-	42.33	nr	47.62
extra; branches	58.90	5.00	61.84	0.26	-	2.40	-	64.24	nr	72.28
300mm	31.62	5.00	33.20	0.40	0.17	4.89	0.50	38.59	m	43.41
extra; bends	77.28	5.00	81.14	0.20	-	1.85	-	82.99	nr	93.36
extra; branches	121.42	5.00	127.49	0.40	-	3.69	-	131.18	nr	147.58
400mm	64.90	5.00	68.14	0.66	0.22	7.64	0.76	76.54	m	86.11
extra; bends	243.85	5.00	256.04	0.47	-	4.34	-	260.38	nr	292.93
450mm	84.29	5.00	88.50	0.84	0.24	9.44	0.92	98.86	m	111.22
extra; bends	321.10	5.00	337.15	0.84	-	7.75	-	344.91	nr	388.02
Pipework in trenches; vertical										
100mm	8.05	5.00	8.45	0.22	0.07	2.52	0.21	11.18	m	12.58
150mm	10.46	5.00	10.98	0.24	0.10	2.92	0.24	14.14	m	15.91
225mm	20.22	5.00	21.23	0.28	0.13	3.50	0.30	25.03	m	28.16
300mm	31.62	5.00	33.20	0.43	0.17	5.16	0.50	38.86	m	43.72
400mm	64.90	5.00	68.14	0.72	0.22	8.19	0.76	77.09	m	86.73
450mm	84.29	5.00	88.50	0.92	0.24	10.18	0.92	99.60	m	112.05
VITRIFIED CLAY SLEEVE DRAINS - Drains; vitrified clay pipes and fittings, B.S.EN295; sleeve joints, push-fit polypropylene standard ring flexible couplings										
Pipework in trenches										
100mm	4.63	5.00	4.86	0.19	0.07	2.25	0.25	7.36	m	8.28
extra; bends	8.53	5.00	8.96	0.10	-	0.92	-	9.88	nr	11.11
extra; branches	16.00	5.00	16.80	0.19	-	1.75	-	18.55	nr	20.87
150mm	8.86	5.00	9.30	0.22	0.10	2.73	0.28	12.32	m	13.86
extra; bends	13.04	5.00	13.69	0.11	-	1.02	-	14.71	nr	16.55
extra; branches	17.27	5.00	18.13	0.22	-	2.03	-	20.16	nr	22.68
Pipework in trenches; vertical										
100mm	4.63	5.00	4.86	0.22	0.07	2.52	0.25	7.63	m	8.59
150mm	8.86	5.00	9.30	0.24	0.10	2.92	0.28	12.50	m	14.06
VITRIFIED CLAY SLEEVE DRAINS - Drains; vitrified clay pipes and fittings, B.S.EN295; sleeve joints, push-fit polypropylene neoprene ring flexible couplings										
Pipework in trenches										
100mm	5.35	5.00	5.62	0.19	0.07	2.25	0.25	8.11	m	9.13
150mm	10.20	5.00	10.71	0.22	0.10	2.73	0.34	13.78	m	15.51
Pipework in trenches; vertical										
100mm	5.35	5.00	5.62	0.22	0.07	2.52	0.25	8.39	m	9.44
150mm	10.20	5.00	10.71	0.24	0.10	2.92	0.34	13.97	m	15.71
VITRIFIED CLAY DRAINS - Drains; vitrified clay pipes and fittings, B.S.EN295, normal; cement mortar (1:2) and tarred gasket joints										
Pipework in trenches										
100mm	5.63	5.00	5.91	0.40	0.04	3.97	0.34	10.22	m	11.50
extra; double collars	6.79	5.00	7.13	0.12	-	1.11	-	8.24	nr	9.27
extra; bends	5.63	5.00	5.91	0.24	-	2.22	-	8.13	nr	9.14
extra; rest bends	9.26	5.00	9.72	0.24	-	2.22	-	11.94	nr	13.43

Labour hourly rates: (except Specialists) Craft Operatives £9.23 Labourer £7.02 Rates are national average prices. Refer to REGIONAL VARIATIONS for indicative levels of overall pricing in regions	MATERIALS			LABOUR				RATES		
	Del to Site	Waste	Material Cost	Craft Optve	Lab	Labour Cost	Sunds	Nett Rate	Unit	Gross Rate (+12.5%)
	£	%	£	Hrs	Hrs	£	£	£		£

R12: DRAINAGE BELOW GROUND Cont.

VITRIFIED CLAY DRAINS - Drains; vitrified clay pipes and fittings, B.S.EN295, normal; cement mortar (1:2) and tarred gasket joints Cont.

	Del to Site	Waste	Material Cost	Craft Optve	Lab	Labour Cost	Sunds	Nett Rate	Unit	Gross Rate
Pipework in trenches Cont.										
extra; branches	10.35	5.00	10.87	0.26	-	2.40	-	13.27	nr	14.93
150mm	8.68	5.00	9.11	0.52	0.05	5.15	0.45	14.71	m	16.55
extra; double collars	11.31	5.00	11.88	0.18	-	1.66	-	13.54	nr	15.23
extra; bends	8.68	5.00	9.11	0.32	-	2.95	-	12.07	nr	13.58
extra; rest bends	15.66	5.00	16.44	0.32	-	2.95	-	19.40	nr	21.82
extra; branches	17.14	5.00	18.00	0.34	-	3.14	-	21.14	nr	23.78
225mm	17.18	5.00	18.04	0.66	0.08	6.65	0.56	25.25	m	28.41
extra; double collars	26.49	5.00	27.81	0.28	-	2.58	-	30.40	nr	34.20
extra; bends	27.16	5.00	28.52	0.40	-	3.69	-	32.21	nr	36.24
300mm	28.80	5.00	30.24	0.96	0.12	9.70	0.82	40.76	m	45.86
Pipework in trenches; vertical										
100mm	5.63	5.00	5.91	0.60	0.04	5.82	0.34	12.07	m	13.58
150mm	8.68	5.00	9.11	0.58	0.05	5.70	0.45	15.27	m	17.18
225mm	17.18	5.00	18.04	0.72	0.08	7.21	0.56	25.81	m	29.03
300mm	28.80	5.00	30.24	1.06	0.12	10.63	0.82	41.69	m	46.90

PVC-U DRAINS - Drains; Wavin; PVC-U pipes and fittings, B.S.4660; ring seal joints

	Del to Site	Waste	Material Cost	Craft Optve	Lab	Labour Cost	Sunds	Nett Rate	Unit	Gross Rate
Pipework in trenches										
110mm, ref. 4D.076	3.57	5.00	3.75	0.40	0.04	3.97	-	7.72	m	8.69
extra; connections to cast iron and clay sockets, ref. 4D.107	3.22	5.00	3.38	0.09	-	0.83	-	4.21	nr	4.74
extra; 45 degree bends, ref. 4D.163	7.39	5.00	7.76	0.15	0.02	1.52	-	9.28	nr	10.44
extra; branches, ref. 4D.210	10.89	5.00	11.43	0.15	0.02	1.52	-	12.96	nr	14.58
extra; connections cast iron and clay spigot, ref. 4D.128	7.57	5.00	7.95	0.09	-	0.83	-	8.78	nr	9.88
160mm, ref. 6D.076	7.89	5.00	8.28	0.22	0.03	2.24	-	10.53	m	11.84
extra; connections to cast iron and clay sockets, ref. 6D.107	12.64	5.00	13.27	0.11	-	1.02	-	14.29	nr	16.07
extra; 45 degree bends, ref. 6D.163	17.58	5.00	18.46	0.22	0.03	2.24	-	20.70	nr	23.29
extra; branches, ref. 6D.210	31.39	5.00	32.96	0.22	0.03	2.24	-	35.20	nr	39.60
extra; connections to cast iron and clay spigot, ref. 6D.128	15.31	5.00	16.08	0.11	-	1.02	-	17.09	nr	19.23
Pipework in trenches; vertical										
110mm, ref. 4D.076	3.57	5.00	3.75	0.31	0.02	3.00	-	6.75	m	7.59
160mm, ref. 6D.076	7.89	5.00	8.28	0.26	0.03	2.61	-	10.89	m	12.26

PVC-U DRAINS - Drains; PVC-U solid wall concentric external rib-reinforced Wavin 'Ultra-Rib' pipes and fittings with sealing rings to joints

	Del to Site	Waste	Material Cost	Craft Optve	Lab	Labour Cost	Sunds	Nett Rate	Unit	Gross Rate
Pipework in trenches										
150mm 6UR 046	4.35	5.00	4.57	0.25	0.02	2.45	-	7.02	m	7.89
extra; connectors 6UR 205	5.75	5.00	6.04	0.06	-	0.55	-	6.59	nr	7.42
extra; bends 6UR 563	7.68	5.00	8.06	0.25	-	2.31	-	10.37	nr	11.67
extra; branches 6UR 213	18.68	5.00	19.61	0.25	-	2.31	-	21.92	nr	24.66
extra; connections to clayware pipe ends 6UR 129	14.08	5.00	14.78	0.10	-	0.92	-	15.71	nr	17.67
225mm 9UR 046	9.79	5.00	10.28	0.35	0.03	3.44	-	13.72	m	15.44
extra; connectors 9UR 205	12.01	5.00	12.61	0.10	-	0.92	-	13.53	nr	15.23
extra; bends 9UR 563	31.04	5.00	32.59	0.35	-	3.23	-	35.82	nr	40.30
extra; branches 9UR 213	55.61	5.00	58.39	0.35	-	3.23	-	61.62	nr	69.32
extra; connections to clayware pipe ends 9UR 109	31.10	5.00	32.66	0.12	-	1.11	-	33.76	nr	37.98
300mm 12UR 043	19.74	5.00	20.73	0.45	0.05	4.50	-	25.23	m	28.39
extra; connectors 12UR 205	24.22	5.00	25.43	0.12	-	1.11	-	26.54	nr	29.86
extra; bends 12UR 563	50.08	5.00	52.58	0.45	-	4.15	-	56.74	nr	63.83
extra; branches 12UR 213	118.51	5.00	124.44	0.45	-	4.15	-	128.59	nr	144.66
extra; connections to clayware pipe ends 12UR 112	81.80	5.00	85.89	0.15	-	1.38	-	87.27	nr	98.18
Pipework in trenches; vertical										
150mm	4.35	5.00	4.57	0.50	0.02	4.76	-	9.32	m	10.49
225mm	9.79	5.00	10.28	0.45	0.03	4.36	-	14.64	m	16.47
300mm	19.74	5.00	20.73	0.55	0.05	5.43	-	26.15	m	29.42

CONCRETE DRAINS - Drains; unreinforced concrete pipes and fittings B.S.5911, Class H, flexible joints

	Del to Site	Waste	Material Cost	Craft Optve	Lab	Labour Cost	Sunds	Nett Rate	Unit	Gross Rate
Pipework in trenches										
300mm	12.90	5.00	13.55	0.71	0.29	8.59	0.21	22.34	m	25.14
extra; bends	77.40	5.00	81.27	0.36	0.14	4.31	0.12	85.70	nr	96.41
extra; 100mm branches	42.50	5.00	44.63	0.29	0.07	3.17	0.17	47.96	nr	53.96
extra; 150mm branches	45.00	5.00	47.25	0.48	0.13	5.34	0.28	52.87	nr	59.48
extra; 225mm branches	52.50	5.00	55.13	0.67	0.19	7.52	0.51	63.15	nr	71.05
extra; 300mm branches	57.00	5.00	59.85	0.86	0.26	9.76	0.71	70.32	nr	79.11
375mm	17.70	5.00	18.59	0.86	0.36	10.47	0.24	29.29	m	32.95
extra; bends	106.20	5.00	111.51	0.43	0.18	5.23	0.14	116.88	nr	131.49
extra; 100mm branches	42.50	5.00	44.63	0.29	0.07	3.17	0.17	47.96	nr	53.96
extra; 150mm branches	45.00	5.00	47.25	0.48	0.13	5.34	0.20	52.87	nr	59.48
extra; 225mm branches	52.50	5.00	55.13	0.67	0.19	7.52	0.51	63.15	nr	71.05
extra; 300mm branches	57.00	5.00	59.85	0.86	0.26	9.76	0.71	70.32	nr	79.11
extra; 375mm branches	77.00	5.00	80.85	1.06	0.32	12.03	0.92	93.80	nr	105.53
450mm	21.10	5.00	22.16	1.02	0.44	12.50	0.29	34.95	m	39.32
extra; bends	126.60	5.00	132.93	0.52	0.22	6.34	0.18	139.45	nr	156.89
extra; 100mm branches	42.50	5.00	44.63	0.29	0.07	3.17	0.17	47.96	nr	53.96
extra; 150mm branches	45.00	5.00	47.25	0.48	0.13	5.34	0.29	52.88	nr	59.49
extra; 225mm branches	52.50	5.00	55.13	0.67	0.19	7.52	0.51	63.15	nr	71.05
extra; 300mm branches	57.00	5.00	59.85	0.86	0.26	9.76	0.71	70.32	nr	79.11

DISPOSAL SYSTEMS

Labour hourly rates: (except Specialists) Craft Operatives £9.23 Labourer £7.02 Rates are national average prices. Refer to REGIONAL VARIATIONS for indicative levels of overall pricing in regions	MATERIALS			LABOUR				RATES		
	Del to Site £	Waste %	Material Cost £	Craft Optve Hrs	Lab Hrs	Labour Cost £	Sunds £	Nett Rate £	Unit	Gross Rate (+12.5%) £

R12: DRAINAGE BELOW GROUND Cont.

**CONCRETE DRAINS - Drains; unreinforced concrete pipes
and fittings B.S.5911, Class H, flexible joints Cont.**

Pipework in trenches Cont.

extra; 375mm branches	77.00	5.00	80.85	1.06	0.32	12.03	0.90	93.78	nr	105.50
extra; 450mm branches	95.00	5.00	99.75	1.25	0.40	14.35	1.12	115.22	nr	129.62
525mm	24.30	5.00	25.52	1.18	0.52	14.54	0.30	40.36	m	45.40
extra; bends	145.80	5.00	153.09	0.59	0.26	7.27	0.18	160.54	nr	180.61
extra; 100mm branches	42.50	5.00	44.63	0.29	0.07	3.17	0.17	47.96	nr	53.96
extra; 150mm branches	45.00	5.00	47.25	0.48	0.13	5.34	0.28	52.87	nr	59.48
extra; 225mm branches	52.50	5.00	55.13	0.67	0.19	7.52	0.51	63.15	nr	71.05
extra; 300mm branches	57.00	5.00	59.85	0.86	0.26	9.76	0.71	70.32	nr	79.11
extra; 375mm branches	77.00	5.00	80.85	1.06	0.32	12.03	0.92	93.80	nr	105.53
extra; 450mm branches	95.00	5.00	99.75	1.25	0.40	14.35	1.16	115.26	nr	129.66
600mm	31.00	5.00	32.55	1.32	0.60	16.40	0.40	49.35	m	55.51
extra; bends	186.00	5.00	195.30	0.66	0.30	8.20	0.21	203.71	nr	229.17
extra; 100mm branches	42.50	5.00	44.63	0.29	0.07	3.17	0.17	47.96	nr	53.96
extra; 150mm branches	45.00	5.00	47.25	0.48	0.13	5.34	0.28	52.87	nr	59.48
extra; 225mm branches	52.50	5.00	55.13	0.67	0.19	7.52	0.51	63.15	nr	71.05
extra; 300mm branches	57.00	5.00	59.85	0.86	0.26	9.76	0.71	70.32	nr	79.11
extra; 375mm branches	77.00	5.00	80.85	1.06	0.32	12.03	0.92	93.80	nr	105.53
extra; 450mm branches	95.00	5.00	99.75	1.25	0.40	14.35	1.12	115.22	nr	129.62

**CONCRETE DRAINS - Drains; reinforced concrete pipes
and fittings, B.S.5911, Class H, flexible joints**

Pipework in trenches

450mm	32.00	5.00	33.60	1.02	0.44	12.50	0.28	46.38	m	52.18
extra; bends	256.00	5.00	268.80	0.52	0.22	6.34	0.16	275.30	nr	309.72
extra; 100mm branches	35.00	5.00	36.75	0.29	0.07	3.17	0.16	40.08	nr	45.09
extra; 150mm branches	40.00	5.00	42.00	0.48	0.13	5.34	0.27	47.61	nr	53.56
extra; 225mm branches	45.00	5.00	47.25	0.67	0.19	7.52	0.49	55.26	nr	62.17
extra; 300mm branches	69.00	5.00	72.45	0.80	0.26	9.21	0.69	82.35	nr	92.64
extra; 375mm branches	74.00	5.00	77.70	1.06	0.32	12.03	0.89	90.62	nr	101.95
extra; 450mm branches	96.00	5.00	100.80	1.25	0.40	14.35	1.09	116.24	nr	130.76
525mm	36.50	5.00	38.33	1.18	0.52	14.54	0.31	53.18	m	59.82
extra; bends	292.00	5.00	306.60	0.59	0.26	7.27	0.17	314.04	nr	353.30
extra; 100mm branches	35.00	5.00	36.75	0.29	0.07	3.17	0.16	40.08	nr	45.09
extra; 150mm branches	40.00	5.00	42.00	0.48	0.13	5.34	0.27	47.61	nr	53.56
extra; 225mm branches	45.00	5.00	47.25	0.67	0.19	7.52	0.49	55.26	nr	62.17
extra; 300mm branches	60.00	5.00	63.00	0.86	0.26	9.76	0.69	73.45	nr	82.63
extra; 375mm branches	74.00	5.00	77.70	1.06	0.32	12.03	0.89	90.62	nr	101.95
extra; 450mm branches	96.00	5.00	100.80	1.25	0.40	14.35	1.09	116.24	nr	130.76
600mm	39.75	5.00	41.74	1.32	0.60	16.40	0.39	58.52	m	65.84
extra; bends	318.00	5.00	333.90	0.66	0.30	8.20	0.20	342.30	nr	385.09
extra; 100mm branches	35.00	5.00	36.75	0.29	0.07	3.17	0.16	40.08	nr	45.09
extra; 150mm branches	40.00	5.00	42.00	0.48	0.13	5.34	0.27	47.61	nr	53.56
extra; 225mm branches	45.00	5.00	47.25	0.67	0.19	7.52	0.49	55.26	nr	62.17
extra; 300mm branches	69.00	5.00	72.45	0.86	0.26	9.76	0.69	82.90	nr	93.27
extra; 375mm branches	74.00	5.00	77.70	1.06	0.32	12.03	0.89	90.62	nr	101.95
extra; 450mm branches	96.00	5.00	100.80	1.25	0.40	14.35	1.09	116.24	nr	130.76

This section continues
on the next page

DISPOSAL SYSTEMS – SMALL WORKS

Labour hourly rates: (except Specialists) Craft Operatives £12.44 Labourer £7.02 Rates are national average prices. Refer to REGIONAL VARIATIONS for indicative levels of overall pricing in regions	MATERIALS			LABOUR				RATES		
	Del to Site £	Waste %	Material Cost £	Craft Optve Hrs	Lab Hrs	Labour Cost £	Sunds £	Nett Rate £	Unit	Gross Rate (+12.5%) £

R12: DRAINAGE BELOW GROUND Cont.

CAST IRON DRAINS - Drains; cast iron pipes and cast iron fittings B.S.437, coated; bolted cast iron coupling and synthetic rubber gasket joints

Pipework in trenches
100mm	22.34	5.00	23.46	1.20	-	14.93	-	38.38	m	43.18
extra; connectors, large socket for clayware	31.06	2.00	31.68	1.20	-	14.93	-	46.61	nr	52.44
extra; bends	40.46	2.00	41.27	1.20	-	14.93	-	56.20	nr	63.22
extra; branches	52.79	2.00	53.85	1.20	-	14.93	-	68.77	nr	77.37
150mm	39.45	5.00	41.42	1.80	-	22.39	-	63.81	m	71.79
extra; connectors, large socket for clayware	45.54	2.00	46.45	1.80	-	22.39	-	68.84	nr	77.45
extra; diminishing pieces, reducing to 100mm	42.98	2.00	43.84	1.80	-	22.39	-	66.23	nr	74.51
extra; bends	58.72	2.00	59.89	1.80	-	22.39	-	82.29	nr	92.57
extra; branches	91.64	2.00	93.47	2.70	-	33.59	-	127.06	nr	142.94
in runs not exceeding 3m										
100mm	30.20	5.00	31.71	1.80	-	22.39	-	54.10	m	60.86
150mm	48.96	5.00	51.41	2.70	-	33.59	-	85.00	m	95.62

Pipework in trenches; vertical
100mm	22.34	5.00	23.46	1.44	-	17.91	-	41.37	m	46.54
150mm	39.45	5.00	41.42	2.16	-	26.87	-	68.29	m	76.83

STAINLESS STEEL DRAINS - Drains; stainless steel pipes and fittings, B.M. Stainless Steel Drains Ltd; ring seal push fit joints (AIS 1316)

Pipework in trenches
75mm	22.94	2.00	23.40	0.66	-	8.21	-	31.61	m	35.56
extra; bends	29.29	2.00	29.88	0.66	-	8.21	-	38.09	nr	42.85
extra; branches	31.72	2.00	32.35	0.99	-	12.32	-	44.67	nr	50.25
110mm	29.15	5.00	30.61	0.83	-	10.33	-	40.93	m	46.05
extra; diminishing pieces, reducing to 75mm	30.04	2.00	30.64	0.83	-	10.33	-	40.97	nr	46.09
extra; bends	37.40	2.00	38.15	0.83	-	10.33	-	48.47	nr	54.53
extra; branches	38.24	2.00	39.00	1.23	-	15.30	-	54.31	nr	61.09
in runs not exceeding 3m										
75mm	25.77	5.00	27.06	0.99	-	12.32	-	39.37	m	44.30
110mm	32.54	5.00	34.17	1.23	-	15.30	-	49.47	m	55.65

Pipework in trenches; vertical
75mm	22.94	5.00	24.09	0.83	-	10.33	-	34.41	m	38.71
110mm	29.15	5.00	30.61	1.03	-	12.81	-	43.42	m	48.85

This section continues
on the next page

DISPOSAL SYSTEMS

Labour hourly rates: (except Specialists) Craft Operatives £9.23 Labourer £7.02 Rates are national average prices. Refer to REGIONAL VARIATIONS for indicative levels of overall pricing in regions	MATERIALS			LABOUR				RATES		
	Del to Site £	Waste %	Material Cost £	Craft Optve Hrs	Lab Hrs	Labour Cost £	Sunds £	Nett Rate £	Unit	Gross Rate (+12.5%) £
R12: DRAINAGE BELOW GROUND Cont.										
VITRIFIED CLAY GULLIES - Vitrified clay accessories										
Gullies; Supersleve; joint to pipe; bedding and surrounding in concrete to B.S.5328, ordinary prescribed mix ST3, 20mm aggregate; Hepworth Building Products										
100mm outlet, reference RG1/1, trapped, round; 255mm diameter top	19.14	5.00	20.10	0.66	0.09	6.72	5.44	32.26	nr	36.29
100mm outlet, reference SG2/1, trapped, square; 150 x 150 top	30.11	5.00	31.62	0.66	0.09	6.72	4.45	42.79	nr	48.14
100mm outlet, reference SG1/1, trapped, reversible, round; 100mm trap; hopper reference SH1, 100mm outlet, 150 x 150mm rebated top	22.69	5.00	23.82	1.10	0.11	10.93	7.43	42.18	nr	47.45
100mm outlet, reference SG1/1, trapped, reversible, round; 100mm trap; hopper reference SH2, 100mm outlet, 100mm horizontal inlet, 150 x 150mm rebated top	30.61	5.00	32.14	1.27	0.11	12.49	7.43	52.06	nr	58.57
100mm outlet, reference SG1/1, trapped, reversible, round; 100mm trap; raising pieces - 1, reference RRP2/2, 100mm diameter, 225mm high; hopper reference SH1, 100mm outlet, 150 x 150mm rebated top	30.17	5.00	31.68	1.54	0.17	15.41	9.91	57.00	nr	64.12
100mm outlet, reference SG1/1, trapped, reversible, round; 100mm trap; raising pieces - 1, reference RRP2/2, 100mm diameter, 225mm high; hopper reference SH2, 100mm outlet, 100mm horizontal inlet, 150 x 150mm rebated top	38.09	5.00	39.99	1.60	0.18	16.03	9.91	65.94	nr	74.18
100mm outlet, reference RGP5, trapped, round; 225mm diameter x 600mm deep; perforated galvanised steel bucket reference IBP3	86.06	5.00	90.36	2.75	0.28	27.35	9.41	127.12	nr	143.01
100mm outlet, reference RGR1, trapped, round; 300mm diameter x 600mm deep	59.28	5.00	62.24	2.75	0.36	27.91	10.40	100.55	nr	113.12
150mm outlet, reference RGR2, trapped, round; 300mm diameter x 600mm deep	63.63	5.00	66.81	2.75	0.36	27.91	10.40	105.12	nr	118.26
150mm outlet, reference RGR3, trapped, round; 400mm diameter x 750mm deep	72.77	5.00	76.41	4.40	0.50	44.12	17.33	137.86	nr	155.09
150mm outlet, reference RGR4, trapped, round; 450mm diameter x 900mm deep	96.21	5.00	101.02	6.05	0.55	59.70	23.76	184.48	nr	207.54
100mm outlet, reference RGU2, trapped; internal size 450 x 300 x 525mm deep; perforated tray and galvanised cover and frame	421.21	5.00	442.27	8.80	1.10	88.95	8.91	540.13	nr	607.64
100mm outlet, reference RGU1, trapped; internal size 600 x 450 x 600mm deep; perforated tray and galvanised cover and frame	532.87	5.00	559.51	11.00	1.65	113.11	16.83	689.46	nr	775.64
Rainwater shoes; Supersleve; joint to pipe; bedding and surrounding in concrete to B.S.5328, ordinary prescribed mix ST3, 20mm aggregate										
100mm outlet, reference RRW/S3/1 trapless, round; 100mm vertical inlet, 250 x 150mm rectangular access opening	23.73	5.00	24.92	0.55	0.06	5.50	3.96	34.37	nr	38.67
CONCRETE GULLIES - Unreinforced concrete accessories										
Gullies; B.S.5911; joint to pipe; bedding and surrounding in concrete to B.S.5328 ordinary prescribed mix ST3, 20mm aggregate										
150mm outlet, trapped, round; 375mm diameter x 750mm deep internally, stopper	22.91	5.00	24.06	3.63	0.55	37.37	13.37	74.79	nr	84.14
150mm outlet, trapped, round; 375mm diameter x 900mm deep internally, stopper	24.43	5.00	25.65	3.85	0.55	39.40	16.34	81.39	nr	91.56
150mm outlet, trapped, round; 450mm diameter x 750mm deep internally, stopper	24.25	5.00	25.46	3.85	0.66	40.17	16.83	82.46	nr	92.77
150mm outlet, trapped, round; 450mm diameter x 900mm deep internally, stopper	24.74	5.00	25.98	4.07	0.66	42.20	20.30	88.48	nr	99.54
150mm outlet, trapped, round; 450mm diameter x 1050mm deep internally, stopper	26.19	5.00	27.50	4.40	0.66	45.25	23.76	96.50	nr	108.57
150mm outlet, trapped, round; 450mm diameter x 1200mm deep internally, stopper	34.06	5.00	35.76	4.95	0.66	50.32	26.74	112.82	nr	126.93
PVC-U GULLIES - PVC-U accessories										
Gullies; Osma; joint to pipe; bedding and surrounding in concrete to B.S.5328, ordinary prescribed mix ST3, 20mm aggregate										
110mm outlet trap, base reference 4D.500, outlet bend with access reference 4D.569, inlet raising piece 300mm long, plain hopper reference 4D.503	36.60	5.00	38.43	1.20	0.20	12.48	2.97	53.88	nr	60.62
110mm outlet bottle gulley, reference 4D.900 trapped, round; 200mm diameter top with grating reference 4D.919	24.91	5.00	26.16	1.80	0.25	18.37	4.95	49.47	nr	55.66
Sealed rodding eyes; Osma; joint to pipe; surrounding in concrete to B.S.5328; ordinary prescribed mix ST3, 20mm aggregate										
110mm outlet, reference 4D.316, with cover	26.17	5.00	27.48	2.25	0.45	23.93	2.97	54.38	nr	61.17
GULLY GRATING AND SEALING PLATES - Cast iron accessories, painted black										
Gratings; Hepworth Building Products										
reference IG6C; 140mm diameter	2.55	-	2.55	-	0.03	0.21	-	2.76	nr	3.11
reference IG7C; 197mm diameter	3.89	-	3.89	-	0.03	0.21	-	4.10	nr	4.61
reference IG8C; 284mm diameter	8.47	-	8.47	-	0.03	0.21	-	8.68	nr	9.77

Labour hourly rates: (except Specialists) Craft Operatives £9.23 Labourer £7.02 Rates are national average prices. Refer to REGIONAL VARIATIONS for indicative levels of overall pricing in regions	MATERIALS			LABOUR				RATES		
	Del to Site	Waste	Material Cost	Craft Optve	Lab	Labour Cost	Sunds	Nett Rate	Unit	Gross Rate (+12.5%)
	£	%	£	Hrs	Hrs	£	£	£		£
R12: DRAINAGE BELOW GROUND Cont.										
GULLY GRATING AND SEALING PLATES - Cast iron accessories, painted black Cont.										
Gratings; Hepworth Building Products Cont.										
reference IG2C; 150 x 150mm	2.55	–	2.55	–	0.03	0.21	–	2.76	nr	3.11
reference IG3C; 225 x 225mm	7.57	–	7.57	–	0.03	0.21	–	7.78	nr	8.75
reference IG4C; 300 x 300mm	17.05	–	17.05	–	0.03	0.21	–	17.26	nr	19.42
Gratings and frames; bedding frames in cement mortar (1:3); Hepworth Building Products										
reference IH5C; 100mm diameter	9.13	–	9.13	1.10	–	10.15	0.25	19.53	nr	21.97
reference IH6C; 150mm diameter	15.79	–	15.79	1.38	–	12.74	0.36	28.89	nr	32.50
reference IH7C; 225mm diameter	31.54	–	31.54	1.87	–	17.26	0.49	49.29	nr	55.45
reference IH2C; 150 x 150mm with lock and key	10.15	–	10.15	1.38	–	12.74	0.25	23.14	nr	26.03
reference IH3C; 230 x 230mm with lock and key	18.58	–	18.58	1.93	–	17.81	0.36	36.75	nr	41.35
reference IH4C; 316 x 316mm with lock and key	49.26	–	49.26	2.09	–	19.29	0.42	68.97	nr	77.59
Sealing plates and frames; bedding frames in cement mortar (1:3); Hepworth Building Products										
reference IS5C; 140mm diameter	7.96	–	7.96	1.10	–	10.15	0.25	18.36	nr	20.66
reference IS6C; 197mm diameter	11.45	–	11.45	1.38	–	12.74	0.36	24.55	nr	27.62
reference IS7C; 273mm diameter	18.32	–	18.32	1.87	–	17.26	0.49	36.07	nr	40.58
reference IS2C; 150 x 150mm	9.83	–	9.83	1.38	–	12.74	0.25	22.82	nr	25.67
reference IS3C; 225 x 225mm	17.88	–	17.88	1.93	–	17.81	0.36	36.05	nr	40.56
reference IS4C; 318 x 318mm	44.92	–	44.92	2.09	–	19.29	0.42	64.63	nr	72.71
GULLY GRATING AND SEALING PLATES - Cast iron accessories, galvanised										
Gratings; Hepworth Building Products										
reference IG6G; 140mm diameter	4.66	–	4.66	–	0.03	0.21	–	4.87	nr	5.48
reference IG7G; 197mm diameter	5.94	–	5.94	–	0.03	0.21	–	6.15	nr	6.92
reference IG8G; 4mm diameter	12.71	–	12.71	–	0.03	0.21	–	12.92	nr	14.54
reference IG2G; 150 x 150mm	3.81	–	3.81	–	0.03	0.21	–	4.02	nr	4.52
reference IG3G; 225 x 225mm	11.37	–	11.37	–	0.03	0.21	–	11.58	nr	13.03
reference IG4G; 300 x 300mm	25.60	–	25.60	–	0.03	0.21	–	25.81	nr	29.04
Gratings and frames; bedding frames in cement mortar (1:3); Hepworth Building Products										
reference IH6G; 193mm diameter	23.71	–	23.71	1.38	–	12.74	0.36	36.81	nr	41.41
reference IH7G; 265mm diameter	47.33	–	47.33	1.87	–	17.26	0.49	65.08	nr	73.22
reference IH2G; 150 x 150mm with lock and key	15.24	–	15.24	1.38	–	12.74	0.25	28.23	nr	31.76
reference IH3G; 225 x 225mm with lock and key	27.84	–	27.84	1.93	–	17.81	0.36	46.01	nr	51.77
Sealing plates and frames; bedding frames in cement mortar (1:3); Hepworth Building Products										
reference IS7G; 273mm diameter	27.53	–	27.53	1.87	–	17.26	0.49	45.28	nr	50.94
reference IS2G; 150 x 150mm	14.75	–	14.75	1.38	–	12.74	0.25	27.74	nr	31.20
GULLY GRATING AND SEALING PLATES - Cast iron accessories, painted black										
Gratings and frames; Drainage Systems Ltd.; bedding frames in cement mortar (1:3)										
reference B6064; 300 x 300mm	24.79	–	24.79	1.10	–	10.15	0.64	35.58	nr	40.03
reference B5883; 325 x 325mm, 69kg	75.39	–	75.39	2.75	–	25.38	0.80	101.57	nr	114.27
reference B5885; 400 x 350mm, 87kg	101.76	–	101.76	3.65	–	33.69	0.95	136.40	nr	153.45
reference B5887; 500 x 350mm, 140kg	139.64	–	139.64	4.70	–	43.38	1.06	184.08	nr	207.09
reference B5884; 340 x 340mm, 80kg	111.35	–	111.35	3.05	–	28.15	0.80	140.30	nr	157.84
SURFACE DRAINAGE SYSTEMS - Precast polyester concrete ACO Drain surface drainage systems; Technologies Plc; bedding and haunching in concrete to B.S.5328 ordinary prescribed mix ST5, 20mm aggregate										
K100 Channel Class C system with galvanised steel slotted lockable gratings										
constant depth or complete with 0.6% fall	40.00	5.00	42.00	1.65	2.15	30.32	6.75	79.07	m	88.96
extra; end cap	4.00	5.00	4.20	–	0.22	1.54	–	5.74	nr	6.46
extra; drain union	1.80	5.00	1.89	–	0.17	1.19	–	3.08	nr	3.47
Q100 Channel Class D monolithic system										
constant depth or complete with 0.6% fall	50.00	5.00	52.50	1.87	2.31	33.48	14.33	100.31	m	112.84
extra; end cap	4.00	5.00	4.20	–	0.22	1.54	–	5.74	nr	6.46
extra; drain union	1.80	5.00	1.89	–	0.17	1.19	–	3.08	nr	3.47
Raindrain constant depth system with galvanised steel slotted lockable gratings										
invert depth 115mm	12.00	5.00	12.60	1.54	1.85	27.20	5.11	44.91	m	50.53
extra; end cap	2.00	5.00	2.10	–	0.22	1.54	–	3.64	nr	4.10
extra; drain union	1.80	5.00	1.89	–	0.17	1.19	–	3.08	nr	3.47
extra; sump unit	40.00	5.00	42.00	1.10	1.10	17.88	7.28	67.16	nr	75.55

This section continues
on the next page

DISPOSAL SYSTEMS

Labour hourly rates: (except Specialists) Craft Operatives £9.23 Labourer £7.02 Rates are national average prices. Refer to REGIONAL VARIATIONS for indicative levels of overall pricing in regions	PLANT & TRANSPORT			LABOUR				RATES		
	Plant Cost £	Trans Cost £	P & T Cost £	Craft Optve Hrs	Lab Hrs	Labour Cost £	Sunds £	Nett Rate £	Unit	Gross Rate (+12.5%) £
R12: DRAINAGE BELOW GROUND; MANHOLES AND SOAKAWAYS										
Excavating - by machine										

Excavating pits
  commencing from natural ground level; maximum
  depth not exceeding										

0.25m	4.90	-	4.90	-	0.95	6.67	-	11.57	m3	13.02
1.00m	6.04	-	6.04	-	1.00	7.02	-	13.06	m3	14.69
2.00m	6.09	-	6.09	-	1.15	8.07	-	14.16	m3	15.93
4.00m	6.94	-	6.94	-	1.45	10.18	-	17.12	m3	19.26
Working space allowance to excavation pits	5.70	-	5.70	-	0.70	4.91	-	10.61	m2	11.94
Compacting bottoms of excavations	0.17	-	0.17	-	-	-	-	0.17	m2	0.19

Excavated material
  depositing on site in temporary spoil heaps where
  directed										

25m	1.86	-	1.86	-	-	-	-	1.86	m3	2.09
50m	1.94	-	1.94	-	-	-	-	1.94	m3	2.18
100m	2.12	-	2.12	-	-	-	-	2.12	m3	2.38
Excavating - by hand										

Excavating pits
  commencing from natural ground level; maximum
  depth not exceeding										

0.25m	-	-	-	-	3.70	25.97	-	25.97	m3	29.22
1.00m	-	-	-	-	3.85	27.03	-	27.03	m3	30.41
2.00m	-	-	-	-	4.40	30.89	-	30.89	m3	34.75
4.00m	-	-	-	-	5.60	39.31	-	39.31	m3	44.23
Working space allowance to excavation pits	-	-	-	-	3.45	24.22	-	24.22	m2	27.25
Compacting bottoms of excavations	-	-	-	-	0.13	0.91	-	0.91	m2	1.03

Excavated material
  depositing on site in temporary spoil heaps where
  directed										

25m	-	-	-	-	1.40	9.83	-	9.83	m3	11.06
50m	-	-	-	-	1.65	11.58	-	11.58	m3	13.03
100m	-	-	-	-	2.20	15.44	-	15.44	m3	17.37

This section continues
on the next page

DISPOSAL SYSTEMS – SMALL WORKS

Labour hourly rates: (except Specialists) Craft Operatives £9.23 Labourer £7.02 Rates are national average prices. Refer to REGIONAL VARIATIONS for indicative levels of overall pricing in regions	MATERIALS			LABOUR				RATES		
	Del to Site £	Waste %	Material Cost £	Craft Optve Hrs	Lab Hrs	Labour Cost £	Sunds £	Nett Rate £	Unit	Gross Rate (+12.5%) £
R12: DRAINAGE BELOW GROUND; MANHOLES AND SOAKAWAYS Cont.										
Earthwork support										
Earthwork support distance between opposing faces not exceeding 2.00m; maximum depth not exceeding										
1.00m	0.94	-	0.94	-	0.21	1.47	0.03	2.44	m2	2.75
2.00m	1.18	-	1.18	-	0.26	1.83	0.04	3.05	m2	3.43
4.00m	1.34	-	1.34	-	0.32	2.25	0.05	3.64	m2	4.09
distance between opposing faces 2.00 - 4.00m; maximum not exceeding										
1.00m	1.02	-	1.02	-	0.22	1.54	0.04	2.60	m2	2.93
2.00m	1.25	-	1.25	-	0.27	1.90	0.04	3.19	m2	3.58
4.00m	1.48	-	1.48	-	0.34	2.39	0.05	3.92	m2	4.41
Earthwork support; in unstable ground distance between opposing faces not exceeding 2.00m; maximum depth not exceeding										
1.00m	1.60	-	1.60	-	0.35	2.46	0.06	4.12	m2	4.63
2.00m	1.92	-	1.92	-	0.44	3.09	0.07	5.08	m2	5.71
4.00m	2.22	-	2.22	-	0.53	3.72	0.08	6.02	m2	6.77
distance between opposing faces 2.00 - 4.00m; maximum depth not exceeding										
1.00m	1.76	-	1.76	-	0.37	2.60	0.06	4.42	m2	4.97
2.00m	2.10	-	2.10	-	0.45	3.16	0.07	5.33	m2	6.00
4.00m	2.45	-	2.45	-	0.56	3.93	0.08	6.46	m2	7.27
FILLING - Excavated material arising from excavations										
Filling to excavation average thickness exceeding 0.25m	-	-	-	-	1.44	10.11	-	10.11	m3	11.37
Excavated material obtained from on site spoil heaps 25m distant										
Filling to excavation average thickness not exceeding 0.25m	-	-	-	-	1.85	12.99	-	12.99	m3	14.61
average thickness exceeding 0.25m	-	-	-	-	1.60	11.23	-	11.23	m3	12.64
Excavated material obtained from on site spoil heaps 50m distant										
Filling to excavation average thickness not exceeding 0.25m	-	-	-	-	2.10	14.74	-	14.74	m3	16.58
average thickness exceeding 0.25m	-	-	-	-	1.95	13.69	-	13.69	m3	15.40
Excavated material obtained from on site spoil heaps 100m distant										
Filling to excavation average thickness not exceeding 0.25m	-	-	-	-	4.40	30.89	-	30.89	m3	34.75
average thickness exceeding 0.25m	-	-	-	-	4.00	28.08	-	28.08	m3	31.59
Hard, dry broken brick or stone, 100 - 75mm gauge, to be obtained off site										
Filling to excavation average thickness exceeding 0.25m	13.02	25.00	16.27	-	1.50	10.53	-	26.81	m3	30.16
CONCRETE WORK - Plain in-situ concrete; B.S.5328, ordinary prescribed mix ST3, 20mm aggregate										
Beds; poured on or against earth or unblinded hardcore										
thickness 150 - 450mm	51.92	5.00	54.52	-	4.45	31.24	-	85.75	m3	96.47
thickness not exceeding 150mm	51.92	5.00	54.52	-	4.75	33.34	-	87.86	m3	98.84
Benching in bottoms; rendering with 13mm cement and sand (1:2) before final set										
600 x 450 x average 225mm; trowelling	3.17	5.00	3.33	-	0.62	4.35	3.50	11.18	nr	12.58
675 x 675 x average 225mm; trowelling	5.68	5.00	5.96	-	1.06	7.44	5.73	19.14	nr	21.53
800 x 675 x average 225mm; trowelling	6.24	5.00	6.55	-	1.27	8.92	6.74	22.21	nr	24.98
800 x 800 x average 300mm; trowelling	10.01	5.00	10.51	-	2.00	14.04	7.98	32.53	nr	36.60
1025 x 800 x average 300mm; trowelling	13.34	5.00	14.01	-	2.54	17.83	10.23	42.07	nr	47.33
Reinforced in-situ concrete; B.S.5328, designed mix ST4, 20mm aggregate, minimum cement content 240 kg/m3; vibrated										
Slabs										
thickness 150 - 450mm	53.32	5.00	55.99	-	3.95	27.73	-	83.72	m3	94.18
thickness not exceeding 150mm	53.32	5.00	55.99	-	4.40	30.89	-	86.87	m3	97.73
Formwork and basic finish										
Edges of suspended slabs; plain vertical height not exceeding 250mm	1.13	12.50	1.27	0.80	0.18	8.65	0.27	10.19	m	11.46
Soffits of slabs; horizontal slab thickness not exceeding 200mm; height to soffit not exceeding 1.50m	9.13	10.00	10.04	2.42	0.46	25.57	1.09	36.70	m2	41.29
slab thickness not exceeding 200mm; height to soffit 1.50 - 3.00m	9.13	10.00	10.04	2.20	0.46	23.54	1.09	34.67	m2	39.00

DISPOSAL SYSTEMS

429

Labour hourly rates: (except Specialists) Craft Operatives £9.23 Labourer £7.02 Rates are national average prices. Refer to REGIONAL VARIATIONS for indicative levels of overall pricing in regions	MATERIALS			LABOUR				RATES		
	Del to Site £	Waste %	Material Cost £	Craft Optve Hrs	Lab Hrs	Labour Cost £	Sunds £	Nett Rate £	Unit	Gross Rate (+12.5%) £
R12: DRAINAGE BELOW GROUND; MANHOLES AND SOAKAWAYS Cont.										
Formwork and basic finish Cont.										
Holes; rectangular										
girth 1.00 - 2.00m; depth not exceeding 250mm	2.04	15.00	2.35	1.33	0.26	14.10	-	16.45	nr	18.50
girth 2.00 - 3.00m; depth not exceeding 250mm	2.66	15.00	3.06	1.73	0.34	18.35	-	21.41	nr	24.09
Reinforcement fabric; B.S.4483; hard drawn plain round steel, welded; including laps, tying wire, all cutting and bending and spacers and chairs which are at the discretion of the Contractor										
Reference A142, 2.22kg/m2; 200mm side laps; 200mm end laps										
generally	0.78	15.00	0.90	0.30	0.02	2.91	0.06	3.87	m2	4.35
BRICKWORK - Common bricks, B.S.3921, Category M, 215 x 102.5 x 65mm, compressive strength 20.5 N/mm2; in cement mortar (1:3)										
Walls; vertical										
102mm thick; stretcher bond	9.67	5.00	10.15	1.50	1.65	25.43	1.98	37.56	m2	42.26
215mm thick; English bond	19.50	5.00	20.48	2.60	2.90	44.36	4.75	69.58	m2	78.28
Milton Hall Second hard stock bricks, B.S.3921, Category M, 215 x 102.5 x 65mm; in cement mortar (1:3)										
Walls; vertical										
102mm thick; stretcher bond	18.82	5.00	19.76	1.50	1.65	25.43	1.98	47.17	m2	53.07
215mm thick; English bond	37.96	5.00	39.86	2.60	2.90	44.36	4.75	88.96	m2	100.08
Engineering bricks, Category F, B.S.3921, 215 x 102.5 x 65mm, class B; in cement mortar (1:3)										
Walls; vertical										
102mm thick; stretcher bond	19.28	5.00	20.24	1.50	1.65	25.43	1.98	47.65	m2	53.61
215mm thick; English bond	38.90	5.00	40.84	2.60	2.90	44.36	4.75	89.95	m2	101.19
Common bricks, B.S.3921, Category M, 215 x 102.5 x 65mm, compressive strength 20.5 N/mm2; in cement mortar (1:3); flush smooth pointing as work proceeds										
Walls; vertical										
102mm thick; stretcher bond; facework one side ...	9.67	5.00	10.15	1.73	1.90	29.31	1.98	41.44	m2	46.62
215mm thick; English bond; facework one side	19.50	5.00	20.48	2.99	3.34	51.04	4.75	76.27	m2	85.80
Milton Hall Second hard stock bricks, B.S.3921, Category M, 215 x 102.5 x 65mm; in cement mortar (1:3); flush smooth pointing as work proceeds										
Walls; vertical										
102mm thick; stretcher bond; facework one side ...	18.82	5.00	19.76	1.73	1.90	29.31	1.98	51.05	m2	57.43
215mm thick; English bond; facework one side	37.96	5.00	39.86	2.99	3.34	51.04	4.75	95.65	m2	107.61
Engineering bricks, B.S.3921, Category F, 215 x 102.5 x 65mm, class B; in cement mortar (1:3); flush smooth pointing as work proceeds										
Walls; vertical										
102mm thick; stretcher bond; facework one side ...	19.28	5.00	20.24	1.73	1.90	29.31	2.56	52.11	m2	58.62
215mm thick; English bond; facework one side	38.90	5.00	40.84	2.50	3.34	46.52	5.32	92.69	m2	104.27
Accessories/sundry items for brick/block/stone walling										
Building in ends of pipes; 100mm diameter										
making good fair face one side										
102mm brickwork.............................	-	-	-	0.10	0.10	1.63	0.06	1.69	nr	1.90
215mm brickwork.............................	-	-	-	0.15	0.15	2.44	0.11	2.55	nr	2.87
Building in ends of pipes; 150mm diameter										
making good fair face one side										
102mm brickwork.............................	-	-	-	0.15	0.15	2.44	0.08	2.52	nr	2.83
215mm brickwork.............................	-	-	-	0.24	0.24	3.90	0.15	4.05	nr	4.56
Building in ends of pipes; 225mm diameter										
making good fair face one side										
102mm brickwork.............................	-	-	-	0.24	0.24	3.90	0.11	4.01	nr	4.51
215mm brickwork.............................	-	-	-	0.30	0.30	4.88	0.21	5.09	nr	5.72
Building in ends of pipes; 300mm diameter										
making good fair face one side										
102mm brickwork.............................	-	-	-	0.30	0.30	4.88	0.25	5.13	nr	5.77
215mm brickwork.............................	-	-	-	0.40	0.40	6.50	0.36	6.86	nr	7.72
Mortar, cement and sand (1:3)										
13mm work to walls on brickwork base; one coat; trowelled										
width exceeding 300mm	1.26	-	1.26	0.45	0.45	7.31	0.36	8.93	m2	10.05

DISPOSAL SYSTEMS – SMALL WORKS

Labour hourly rates: (except Specialists) Craft Operatives £9.23 Labourer £7.02 Rates are national average prices. Refer to REGIONAL VARIATIONS for indicative levels of overall pricing in regions	MATERIALS			LABOUR				RATES		
	Del to Site £	Waste %	Material Cost £	Craft Optve Hrs	Lab Hrs	Labour Cost £	Sunds £	Nett Rate £	Unit	Gross Rate (+12.5%) £
R12: DRAINAGE BELOW GROUND; MANHOLES AND SOAKAWAYS Cont.										
Mortar, cement and sand (1:3) Cont.										
13mm work to floors on concrete base; one coat; trowelled										
level and to falls only not exceeding 15 degrees form horizontal	1.26	–	1.26	0.70	0.70	11.38	0.36	12.99	m2	14.62
CHANNELS, VITRIFIED CLAY – Channels in bottoms; vitrified clay, B.S.65, normal, glazed; cement mortar (1:3) joints; bedding in cement mortar (1:3)										
Half section										
straight; 600mm effective length										
100mm	2.02	5.00	2.12	0.35	0.05	3.58	0.23	5.93	nr	6.67
150mm	3.41	5.00	3.58	0.45	0.07	4.64	0.34	8.57	nr	9.64
straight; 1000mm effective length										
100mm	2.83	5.00	2.97	0.60	0.06	5.96	0.23	9.16	nr	10.31
150mm	5.35	5.00	5.62	0.75	0.08	7.48	0.34	13.44	nr	15.12
225mm	12.51	5.00	13.14	1.00	0.12	10.07	0.44	23.65	nr	26.60
300mm	26.33	5.00	27.65	1.50	0.18	15.11	0.53	43.29	nr	48.70
curved; 500mm effective length										
100mm	3.59	5.00	3.77	0.34	0.05	3.49	0.23	7.49	nr	8.42
150mm	5.93	5.00	6.23	0.45	0.07	4.64	0.34	11.21	nr	12.61
225mm	19.79	5.00	20.78	0.60	0.10	6.24	0.44	27.46	nr	30.89
300mm	40.36	5.00	42.38	0.90	0.10	9.01	0.53	51.92	nr	58.41
curved; 900mm effective length										
100mm	3.59	5.00	3.77	0.70	0.06	6.88	0.23	10.88	nr	12.24
150mm	5.93	5.00	6.23	0.90	0.08	8.87	0.34	15.44	nr	17.36
225mm	19.79	5.00	20.78	1.20	0.12	11.92	0.44	33.14	nr	37.28
300mm	40.36	5.00	42.38	1.80	0.20	18.02	0.53	60.93	nr	68.54
Three quarter section										
branch bends										
100mm	8.11	5.00	8.52	0.40	0.04	3.97	0.23	12.72	nr	14.31
150mm	14.09	5.00	14.79	0.60	0.07	6.03	0.34	21.16	nr	23.81
225mm	49.63	5.00	52.11	1.20	0.12	11.92	0.44	64.47	nr	72.53
Channels in bottoms; P.V.C.; 'O' ring joints; bedding in cement mortar (1:3)										
For 610mm clear opening										
straight										
110mm	16.60	5.00	17.43	0.45	0.04	4.43	2.53	24.39	nr	27.44
160mm	31.40	5.00	32.97	0.60	0.05	5.89	5.71	44.57	nr	50.14
short bend										
110mm	7.66	5.00	8.04	0.25	0.01	2.38	2.27	12.69	nr	14.28
160mm	18.19	5.00	19.10	0.35	0.02	3.37	5.50	27.97	nr	31.47
long radius bend										
110mm	27.22	5.00	28.58	0.40	0.02	3.83	2.40	34.81	nr	39.17
160mm	59.28	5.00	62.24	0.60	0.05	5.89	5.50	73.63	nr	82.84
PRECAST CONCRETE INSPECTION CHAMBERS – Precast concrete; standard or stock pattern units; B.S.5911; jointing and pointing in cement mortar (1:3)										
Chamber or shaft sections										
900mm internal diameter										
250mm high	19.74	5.00	20.73	0.58	1.16	13.50	0.09	34.31	nr	38.60
500mm high	19.74	5.00	20.73	0.67	1.34	15.59	0.09	36.41	nr	40.96
750mm high	29.60	5.00	31.08	0.87	1.74	20.24	0.09	51.41	nr	57.84
1000mm high	39.47	5.00	41.44	0.93	1.86	21.64	0.09	63.17	nr	71.07
1050mm internal diameter										
305mm high	27.51	5.00	28.89	0.81	1.62	18.85	0.16	47.89	nr	53.88
458mm high	41.31	5.00	43.38	0.87	1.74	20.24	0.16	63.78	nr	71.75
610mm high	27.51	5.00	28.89	0.93	1.86	21.64	0.16	50.69	nr	57.02
762mm high	34.37	5.00	36.09	0.99	1.98	23.04	0.16	59.29	nr	66.70
914mm high	41.22	5.00	43.28	1.04	2.08	24.20	0.16	67.64	nr	76.10
1200mm internal diameter										
250mm high	29.60	5.00	31.08	0.99	1.98	23.04	0.23	54.35	nr	61.14
500mm high	29.60	5.00	31.08	1.04	2.08	24.20	0.22	55.50	nr	62.44
750mm high	44.39	5.00	46.61	1.22	2.44	28.39	0.22	75.22	nr	84.62
1000mm high	59.19	5.00	62.15	1.28	2.56	29.79	0.22	92.16	nr	103.67
Cutting holes										
cutting holes for pipes; 100mm diameter; making good; pointing	10.72	–	10.72	–	0.09	0.63	0.14	11.49	nr	12.93
cutting holes for pipes; 150mm diameter; making good; pointing	16.08	–	16.08	–	0.11	0.77	0.16	17.01	nr	19.14
cutting holes for pipes; 225mm diameter; making good; pointing	24.12	–	24.12	–	0.13	0.91	0.21	25.24	nr	28.40
cutting holes for pipes; 300mm diameter; making good; pointing	32.16	–	32.16	–	0.18	1.26	0.29	33.71	nr	37.93
Step irons										
galvanised malleable cast iron step irons B.S.1247, cast in	3.35	–	3.35	–	0.17	1.19	–	4.54	nr	5.11
Cover slabs, heavy duty										
for chamber or shaft sections 900mm internal diameter, 125mm thick; access opening 600mm x 600mm	39.87	5.00	41.86	1.70	2.55	33.59	1.43	76.89	nr	86.50

DISPOSAL SYSTEMS

Labour hourly rates: (except Specialists) Craft Operatives £9.23 Labourer £7.02 Rates are national average prices. Refer to REGIONAL VARIATIONS for indicative levels of overall pricing in regions	MATERIALS			LABOUR				RATES		
	Del to Site £	Waste %	Material Cost £	Craft Optve Hrs	Lab Hrs	Labour Cost £	Sunds £	Nett Rate £	Unit	Gross Rate (+12.5%) £
R12: DRAINAGE BELOW GROUND; MANHOLES AND SOAKAWAYS Cont.										
PRECAST CONCRETE INSPECTION CHAMBERS - Precast concrete; standard or stock pattern units; B.S.5911; jointing and pointing in cement mortar (1:3) Cont.										
Cover slabs, heavy duty Cont.										
for chamber or shaft sections 1050mm internal diameter, 125mm thick; access opening 600mm x 600mm	45.77	5.00	48.06	2.25	3.35	44.28	1.70	94.04	nr	105.80
for chamber or shaft sections 1200mm internal diameter, 125mm thick; access opening 600mm x 600mm	60.53	5.00	63.56	2.80	4.20	55.33	1.91	120.79	nr	135.89
for chamber or shaft sections 1350mm internal diameter, 125mm thick; access opening 600mm x 600mm	88.81	5.00	93.25	3.35	5.05	66.37	2.23	161.85	nr	182.08
for chamber or shaft sections 1500mm internal diameter, 150mm thick; access opening 600mm x 600mm	109.67	5.00	115.15	3.95	5.90	77.88	2.44	195.47	nr	219.90
for chamber or shaft sections 1800mm internal diameter, 150mm thick access opening 600mm x 600mm	126.95	5.00	133.30	4.55	6.75	89.38	-	222.68	nr	250.51
for chamber or shaft sections 2100mm internal diameter, 150mm thick access opening 600mm x 600mm	262.47	5.00	275.59	5.15	7.60	100.89	-	376.48	nr	423.54
for chamber or shft sections 2400mm internal diameter, 150mm thick access opening 600mm x 600mm	457.46	5.00	480.33	5.75	8.45	112.39	-	592.72	nr	666.82
for chamber or shaft sections 2700mm internal diameter, 150mm thick access opening 600mm x 600mm	592.22	5.00	621.83	6.35	9.30	123.90	-	745.73	nr	838.94
for chamber or shaft sections 3000mm internal diameter, 150mm thick access opening 600mm x 600mm	697.51	5.00	732.39	6.95	10.15	135.40	-	867.79	nr	976.26
Precast concrete inspection chambers; B.S.5911; jointing and pointing in cement mortar (1:3)										
Rectangular inspection chambers, internal size 457 x 610mm										
100mm high	6.45	5.00	6.77	0.50	0.35	7.07	0.09	13.93	nr	15.68
150mm high	7.10	5.00	7.46	0.65	0.45	9.16	0.09	16.70	nr	18.79
250mm high	9.60	5.00	10.08	0.75	0.53	10.64	0.09	20.81	nr	23.41
extra; galvanised malleable cast iron step irons B.S.1247, built in	2.90	5.00	3.04	-	-	-	-	3.04	nr	3.43
Rectangular inspection chambers, internal size 760 x 610mm										
100mm high	9.05	5.00	9.50	0.81	0.58	11.55	0.17	21.22	nr	23.87
150mm high	9.05	5.00	9.50	1.05	0.75	14.96	0.17	24.63	nr	27.71
225mm high	11.50	5.00	12.07	1.21	0.88	17.35	0.17	29.59	nr	33.29
extra; galvanised malleable cast iron step irons B.S.1247, built in	2.90	5.00	3.04	-	-	-	-	3.04	nr	3.43
Rectangular inspection chambers, internal size 990 x 610mm										
150mm high	10.90	5.00	11.45	1.05	0.75	14.96	0.17	26.57	nr	29.89
Rectangular inspection chambers, internal size 1200 x 750mm										
150mm high	21.05	5.00	22.10	1.20	0.85	17.04	0.17	39.32	nr	44.23
200mm high	25.95	5.00	27.25	1.38	1.00	19.76	0.17	47.17	nr	53.07
250mm high	29.05	5.00	30.50	1.57	1.14	22.49	0.17	53.17	nr	59.81
Base units										
450 x 610mm internal size; 360mm deep	26.05	5.00	27.35	1.05	0.75	14.96	0.09	42.40	nr	47.70
760 x 610mm internal size; 360mm deep	29.90	5.00	31.40	1.05	0.75	14.96	0.17	46.52	nr	52.34
990 x 610mm internal size; 360mm deep	42.10	5.00	44.20	1.05	0.75	14.96	0.17	59.33	nr	66.75
GRP INSPECTION CHAMBERS										
Inspection chambers; Terrain; GRP with preformed benching; 4 nr pvc adaptors; A15 light duty single seal cast iron cover to B.S.EN124 and plastic frame set in position										
475mm internal diameter; depth to invert										
450mm	161.44	5.00	169.51	5.50	0.90	57.08	-	226.60	nr	254.92
585mm	191.59	5.00	201.17	5.50	0.90	57.08	-	258.25	nr	290.53
750mm	221.74	5.00	232.83	5.90	0.95	61.13	-	293.95	nr	330.70
930mm	251.89	5.00	264.48	5.90	0.95	61.13	-	325.61	nr	366.31
POLYPROPYLENE INSPECTION CHAMBERS										
Osma; Shallow inspection chamber Polypropylene with preformed benching. A15 light duty single sealed cover and frame set in position, 250mm diameter; for 100mm diameter pipes ref 4D960, depth to invert										
600mm	58.16	5.00	61.07	2.50	1.44	33.18	-	94.25	nr	106.03
Osma; universal inspection chamber; Polypropylene; 450mm diameter										
preformed chamber base benching, ref. 40.922; for100mm pipes	53.62	5.00	56.30	1.50	0.10	14.55	-	70.85	nr	79.70
chamber shaft 230mm effective length, ref. 40.925	13.66	5.00	14.34	-	0.60	4.21	-	18.56	nr	20.87
cast iron cover and frame; single seal A15 light duty to suit, ref. 40.344	31.27	5.00	32.83	-	0.60	4.21	-	37.05	nr	41.68

Labour hourly rates: (except Specialists) Craft Operatives £12.44 Labourer £7.02 Rates are national average prices. Refer to REGIONAL VARIATIONS for indicative levels of overall pricing in regions	MATERIALS			LABOUR				RATES		
	Del to Site	Waste	Material Cost	Craft Optve	Lab	Labour Cost	Sunds	Nett Rate	Unit	Gross Rate (+12.5%)
	£	%	£	Hrs	Hrs	£	£	£		£
CAST IRON INSPECTION CHAMBERS										
100mm diameter straight coated cast iron bolted inspection chambers; coupling joints; bedding in cement mortar (1:3)										
with one branch one side, fig 110	103.52	2.00	105.59	3.00	-	37.32	-	142.91	nr	160.77
with one branch each side, fig 111	138.35	2.00	141.12	4.00	-	49.76	-	190.88	nr	214.74
with two branches one side, fig 210	195.48	2.00	199.39	4.00	-	49.76	-	249.15	nr	280.29
with two branches each side, fig 212	261.69	2.00	266.92	6.00	-	74.64	-	341.56	nr	384.26
100-150mm diameter straight coated cast iron bolted inspection chambers; coupling joints; bedding in cement mortar (1:3)										
with one branch one side, fig 110	146.23	2.00	149.15	3.75	-	46.65	-	195.80	nr	220.28
150mm diameter straight coated cast iron bolted inspection chambers; coupling joints; bedding in cement mortar (1:3)										
with one branch one side, fig 110	182.09	2.00	185.73	4.00	-	49.76	-	235.49	nr	264.93
with one branch each side, fig 111	212.14	2.00	216.38	5.25	-	65.31	-	281.69	nr	316.90
with two branches one side, fig 210	332.18	2.00	338.82	5.25	-	65.31	-	404.13	nr	454.65
with two branches each side, fig 212	409.50	2.00	417.69	7.75	-	96.41	-	514.10	nr	578.36

This section continues on the next page

DISPOSAL SYSTEMS

Labour hourly rates: (except Specialists) Craft Operatives £9.23 Labourer £7.02 Rates are national average prices. Refer to REGIONAL VARIATIONS for indicative levels of overall pricing in regions	MATERIALS			LABOUR				RATES		
	Del to Site	Waste	Material Cost	Craft Optve	Lab	Labour Cost	Sunds	Nett Rate	Unit	Gross Rate (+12.5%)
	£	%	£	Hrs	Hrs	£	£	£		£
CAST IRON INSPECTION CHAMBERS Cont.										
MANHOLE ACCESSORIES - Manhole step irons										
Step irons; B.S.1247 malleable cast iron; galvanised; building in to joints										
figure 1 general purpose pattern; 110mm tail	2.01	2.50	2.06	0.25	-	2.31	1.48	5.85	nr	6.58
figure 1 general purpose pattern; 225mm tail	2.51	2.50	2.57	0.27	-	2.49	1.48	6.54	nr	7.36
Manhole covers - A15 light duty										
Access covers; B.S.EN124; coated; bedding frame in cement mortar (1:3); bedding cover in grease and sand										
reference MC1-45/45; clear opening 450 x 450mm ...	30.66	2.50	31.43	0.90	0.90	14.63	0.37	46.42	nr	52.22
reference MC1-60/45; clear opening 600 x 450mm ...	30.83	2.50	31.60	1.15	1.15	18.69	0.37	50.66	nr	56.99
reference MC1R-60/60; clear opening 600 x 600mm ...	86.83	2.50	89.00	2.15	2.15	34.94	0.53	124.47	nr	140.03
reference MC2-45/45; clear opening 450 x 450mm ...	47.18	2.50	48.36	1.65	1.65	26.81	0.53	75.70	nr	85.16
reference MC2-60/45; clear opening 600 x 450mm ...	61.79	2.50	63.33	1.90	1.90	30.88	0.53	94.74	nr	106.58
reference MC2-60/60; clear opening 600 x 600mm ...	87.14	2.50	89.32	2.15	2.15	34.94	0.53	124.79	nr	140.38
reference MC2R-45/45; clear opening 450 x 450mm ...	78.36	2.50	80.32	1.95	1.95	31.69	0.53	112.54	nr	126.60
reference MC2R-60/45; clear opening 600 x 450mm ..	72.79	2.50	74.61	2.35	2.35	38.19	0.53	113.33	nr	127.49
reference MC2R-60/60; clear opening 600 x 600mm ..	104.99	2.50	107.61	2.75	2.75	44.69	0.53	152.83	nr	171.94
lifting keys	3.50	2.50	3.59	-	-	-	-	3.59	nr	4.04
Manhole covers - B125 medium duty										
Access covers; B.S.EN124; coated; bedding frame in cement mortar (1:3); bedding cover in grease and sand										
reference MB1-60; clear opening 600mm diameter ...	66.69	2.50	68.36	3.65	3.65	59.31	0.81	128.48	nr	144.54
reference MB2-50; clear opening 500mm diameter ...	74.49	2.50	76.35	2.90	2.90	47.13	0.59	124.07	nr	139.58
reference MB2-55; clear opening 550mm diameter ...	69.36	2.50	71.09	2.40	2.40	39.00	0.59	110.68	nr	124.52
reference MB2-60; clear opening 600mm diameter ...	57.05	2.50	58.48	2.90	2.90	47.13	0.59	106.19	nr	119.47
reference MB2-60/45; clear opening 600 x 450mm ...	51.74	2.50	53.03	2.90	2.90	47.13	0.89	101.05	nr	113.68
reference MB2-60/60; clear opening 600 x 600mm ...	63.16	2.50	64.74	3.15	3.15	51.19	1.11	117.04	nr	131.67
reference MB2R-60/45; clear opening 600 x 450mm ..	92.78	2.50	95.10	2.90	2.90	47.13	0.89	143.11	nr	161.00
reference MB2R-60/60; clear opening 600 x 600mm ..	105.55	2.50	108.19	3.15	3.15	51.19	1.11	160.49	nr	180.55
lifting keys	4.00	2.50	4.10	-	-	-	-	4.10	nr	4.61
Manhole covers - D400 heavy duty										
Access covers; B.S.EN124; coated; bedding frame in cement mortar (1:3); bedding cover in grease and sand										
reference MD-60; clear opening 600mm diameter	103.06	2.50	105.64	1.95	1.95	31.69	1.11	138.43	nr	155.74
reference MA-60; clear opening 600 diameter	69.03	2.50	70.76	1.95	1.95	31.69	1.11	103.55	nr	116.50
reference MA-T; clear opening 550 x 495mm	106.52	2.50	109.18	2.15	2.15	34.94	1.11	145.23	nr	163.38
lifting keys	4.00	2.50	4.10	-	-	-	-	4.10	nr	4.61
INTERCEPTING TRAPS - Vitrified clay intercepting traps										
Intercepting traps; B.S.65; joint to pipe; building into side of manhole; bedding and surrounding in concrete to B.S.5328 ordinary prescribed mix C15P, 20mm aggregate										
cleaning arm and stopper										
100mm outlet, ref. R1 1/1....................	42.65	5.00	44.78	1.45	1.15	21.46	14.40	80.64	nr	90.72
150mm outlet, ref.R1 1/2....................	61.51	5.00	64.59	2.00	1.75	30.75	18.39	113.72	nr	127.94
225mm outlet, ref. R1 1/3...................	191.96	5.00	201.56	2.60	2.30	40.14	26.09	267.79	nr	301.27
FRESH AIR INLETS - Aluminium; fresh air inlets										
Air inlet valve, mice flap; set in cement mortar (1:3)										
100mm diameter................................	19.20	5.00	20.16	1.25	-	11.54	-	31.70	nr	35.66
SETTLEMENT TANKS - Fibreglass settlement tanks; including cover and frame										
Settlement tank; fibreglass; set in position										
1000mm deep to invert of inlet, standard grade										
2700 litres capacity.........................	486.00	-	486.00	1.00	2.00	23.27	-	509.27	nr	572.93
3750 litres capacity.........................	648.00	-	648.00	1.10	2.20	25.60	-	673.60	nr	757.80
4500 litres capacity.........................	767.00	-	767.00	1.20	2.40	27.92	-	794.92	nr	894.29
7500 litres capacity.........................	1435.00	-	1435.00	1.30	2.60	30.25	-	1465.25	nr	1648.41
9000 litres capacity.........................	1563.00	-	1563.00	1.45	2.90	33.74	-	1596.74	nr	1796.33
1500mm deep to invert of inlet, heavy grade										
2700 litres capacity.........................	629.00	-	629.00	1.00	2.00	23.27	-	652.27	nr	733.80
3750 litres capacity.........................	830.00	-	830.00	1.10	2.20	25.60	-	855.60	nr	962.55
4500 litres capacity.........................	984.00	-	984.00	1.20	2.40	27.92	-	1011.92	nr	1138.41
7500 litres capacity.........................	1831.00	-	1831.00	1.30	2.60	30.25	-	1861.25	nr	2093.91
9000 litres capacity.........................	2002.00	-	2002.00	1.45	2.90	33.74	-	2035.74	nr	2290.21
R12: DRAINAGE BELOW GROUND; WORK TO EXISTING DRAINS										
Work to disused drains										
Fly ash filling to disused drain										
100mm diameter............................	0.53	-	0.53	0.40	-	3.69	-	4.22	m	4.75
150mm diameter............................	1.11	-	1.11	0.60	-	5.54	-	6.65	m	7.48
225mm diameter............................	2.13	-	2.13	0.90	-	8.31	-	10.44	m	11.74
Sealing end of disused drain with plain in-situ concrete										
100mm diameter............................	0.94	-	0.94	0.22	-	2.03	-	2.97	nr	3.34
150mm diameter............................	1.87	-	1.87	0.33	-	3.05	-	4.92	nr	5.53
225mm diameter............................	2.82	-	2.82	0.44	-	4.06	-	6.88	nr	7.74

Labour hourly rates: (except Specialists) Craft Operatives £9.23 Labourer £7.02 Rates are national average prices. Refer to REGIONAL VARIATIONS for indicative levels of overall pricing in regions	MATERIALS			LABOUR				RATES		
	Del to Site £	Waste %	Material Cost £	Craft Optve Hrs	Lab Hrs	Labour Cost £	Sunds £	Nett Rate £	Unit	Gross Rate (+12.5%) £
R13: LAND DRAINAGE										
Drains; clayware pipes, B.S.1196; butt joints										
Pipework in trenches										
75mm	1.14	5.00	1.20	-	0.26	1.83	-	3.02	m	3.40
100mm	2.03	5.00	2.13	-	0.29	2.04	-	4.17	m	4.69
150mm	4.35	5.00	4.57	-	0.36	2.53	-	7.09	m	7.98
Drains; porous concrete pipes, B.S.5911; butt joints										
Pipework in trenches										
150mm	5.24	5.00	5.50	-	0.34	2.39	-	7.89	m	8.87
225mm	6.71	5.00	7.05	-	0.42	2.95	-	9.99	m	11.24
300mm	11.03	5.00	11.58	-	0.50	3.51	-	15.09	m	16.98
375mm	15.03	5.00	15.78	-	0.58	4.07	-	19.85	m	22.33
450mm	21.14	5.00	22.20	-	0.66	4.63	-	26.83	m	30.18
525mm	31.32	5.00	32.89	-	0.74	5.19	-	38.08	m	42.84
600mm	38.90	5.00	40.84	-	0.82	5.76	-	46.60	m	52.43
Drains; OGEE pipes, B.S.5911 Part 110; butt joints										
Pipework in trenches										
150mm	4.28	5.00	4.49	-	0.34	2.39	-	6.88	m	7.74
225mm	6.92	5.00	7.27	-	0.42	2.95	-	10.21	m	11.49
300mm	11.36	5.00	11.93	-	0.50	3.51	-	15.44	m	17.37
375mm	15.43	5.00	16.20	-	0.58	4.07	-	20.27	m	22.81
450mm	19.60	5.00	20.58	-	0.66	4.63	-	25.21	m	28.36
525mm	26.48	5.00	27.80	-	0.74	5.19	-	33.00	m	37.12
600mm	35.61	5.00	37.39	-	0.82	5.76	-	43.15	m	48.54
750mm	53.16	5.00	55.82	-	0.90	6.32	-	62.14	m	69.90
Drains; clayware perforated pipes, butt joints; 150mm **hardcore to sides and top**										
Pipework in trenches										
100mm	5.28	5.00	5.54	-	0.44	3.09	1.14	9.77	m	10.99
150mm	8.91	5.00	9.36	-	0.65	4.56	1.51	15.43	m	17.36
225mm	14.88	5.00	15.62	-	0.70	4.91	2.03	22.57	m	25.39

DISPOSAL SYSTEMS

Spence Geddes' Estimating for Building and Civil Engineering Works

This is a fully updated edition of **Spence Geddes'** standard reference publication on estimating. The opportunity has been taken to make certain alterations whilst maintaining the essential features of earlier editions.

It deals in a practical and reasonable way with many of the estimating problems which can arise where building and civil engineering works are carried out and to include comprehensive estimating data within the guidelines of good practice.

£35.00, Hardback, 1996, 448pp, isbn 0 7506 2797 2

Piped supply systems

Labour hourly rates: (except Specialists) Craft Operatives £12.44 Labourer £7.02 Rates are national average prices. Refer to REGIONAL VARIATIONS for indicative levels of overall pricing in regions	MATERIALS			LABOUR				RATES		
	Del to Site £	Waste %	Material Cost £	Craft Optve Hrs	Lab Hrs	Labour Cost £	Sunds £	Nett Rate £	Unit	Gross Rate (+12.5%) £
S115: HOT AND COLD WATER AND GAS										
BLACK AND GALVANISED STEEL TUBES AND FITTINGS - Steel pipes, B.S.1387 Table 2; steel fittings B.S.1740 Part 1; screwed and PTFE tape joints; pipework black										
Pipes, straight										
15mm; in malleable iron pipe brackets fixing to										
timber with screws	1.87	5.00	1.96	0.60	-	7.46	-	9.43	m	10.61
extra; made bends	-	-	-	0.20	-	2.49	-	2.49	nr	2.80
extra; connections to copper pipe ends;										
compression joint	0.64	2.00	0.65	0.12	-	1.49	-	2.15	nr	2.41
extra; fittings; one end	0.40	2.00	0.41	0.15	-	1.87	-	2.27	nr	2.56
extra; fittings; two ends	0.46	2.00	0.47	0.30	-	3.73	-	4.20	nr	4.73
extra; fittings; three ends	0.62	2.00	0.63	0.30	-	3.73	-	4.36	nr	4.91
20mm; in malleable iron pipe brackets fixing to										
timber with screws	2.17	5.00	2.28	0.68	-	8.46	-	10.74	m	12.08
extra; made bends	-	-	-	0.30	-	3.73	-	3.73	nr	4.20
extra; connections to copper pipe ends;										
compression joint	1.04	2.00	1.06	0.20	-	2.49	-	3.55	nr	3.99
extra; fittings; one end	0.46	2.00	0.47	0.20	-	2.49	-	2.96	nr	3.33
extra; fittings; two ends	0.62	2.00	0.63	0.40	-	4.98	-	5.61	nr	6.31
extra; fittings; three ends	0.91	2.00	0.93	0.40	-	4.98	-	5.90	nr	6.64
25mm; in malleable iron pipe brackets fixing to										
timber with screws	2.99	5.00	3.14	0.80	-	9.95	-	13.09	m	14.73
extra; made bends	-	-	-	0.40	-	4.98	-	4.98	nr	5.60
extra; connections to copper pipe ends;										
compression joint	2.03	2.00	2.07	0.30	-	3.73	-	5.80	nr	6.53
extra; fittings; one end	0.57	2.00	0.58	0.30	-	3.73	-	4.31	nr	4.85
extra; fittings; two ends	0.98	2.00	1.00	0.60	-	7.46	-	8.46	nr	9.52
extra; fittings; three ends	1.31	2.00	1.34	0.60	-	7.46	-	8.80	nr	9.90
32mm; in malleable iron pipe brackets fixing to										
timber with screws	3.74	5.00	3.93	0.96	-	11.94	-	15.87	m	17.85
extra; made bends	-	-	-	0.60	-	7.46	-	7.46	nr	8.40
extra; connections to copper pipe ends;										
compression joint	5.95	2.00	6.07	0.40	-	4.98	-	11.05	nr	12.43
extra; fittings; one end	0.83	2.00	0.85	0.40	-	4.98	-	5.82	nr	6.55
extra; fittings; two ends	1.60	2.00	1.63	0.80	-	9.95	-	11.58	nr	13.03
extra; fittings; three ends	2.17	2.00	2.21	0.80	-	9.95	-	12.17	nr	13.69
40mm; in malleable iron pipe brackets fixing to										
timber with screws	4.43	5.00	4.65	1.20	-	14.93	-	19.58	m	22.03
extra; made bends	-	-	-	0.80	-	9.95	-	9.95	nr	11.20
extra; connections to copper pipe ends;										
compression joint	9.57	2.00	9.76	0.60	-	7.46	-	17.23	nr	19.38
extra; fittings; one end	1.05	2.00	1.07	0.50	-	6.22	-	7.29	nr	8.20
extra; fittings; two ends	2.69	2.00	2.74	1.00	-	12.44	-	15.18	nr	17.08
extra; fittings; three ends	2.98	2.00	3.04	1.00	-	12.44	-	15.48	nr	17.41
50mm; in malleable iron pipe brackets fixing to										
timber with screws	6.16	5.00	6.47	1.60	-	19.90	-	26.37	m	29.67
extra; made bends	-	-	-	1.20	-	14.93	-	14.93	nr	16.79
extra; connections to copper pipe ends;										
compression joint	14.51	2.00	14.80	0.80	-	9.95	-	24.75	nr	27.85
extra; fittings; one end	2.05	2.00	2.09	0.60	-	7.46	-	9.55	nr	10.75
extra; fittings; two ends	3.15	2.00	3.21	1.20	-	14.93	-	18.14	nr	20.41
extra; fittings; three ends	4.29	2.00	4.38	1.20	-	14.93	-	19.30	nr	21.72
in malleable iron pipe brackets fixing to masonry with screws										
15mm	1.87	5.00	1.96	0.64	-	7.96	-	9.93	m	11.17
20mm	2.17	5.00	2.28	0.72	-	8.96	-	11.24	m	12.64
25mm	2.99	5.00	3.14	0.84	-	10.45	-	13.59	m	15.29
32mm	3.74	5.00	3.93	1.00	-	12.44	-	16.37	m	18.41
40mm	4.43	5.00	4.65	1.24	-	15.43	-	20.08	m	22.59
50mm	6.16	5.00	6.47	1.64	-	20.40	-	26.87	m	30.23
in malleable iron single rings, 165mm long screwed both ends steel tubes and malleable iron backplates fixing to timber with screws										
15mm	2.48	5.00	2.60	0.72	-	8.96	-	11.56	m	13.01
20mm	2.78	5.00	2.92	0.80	-	9.95	-	12.87	m	14.48
25mm	3.59	5.00	3.77	0.92	-	11.44	-	15.21	m	17.12
32mm	4.20	5.00	4.41	1.08	-	13.44	-	17.85	m	20.08
40mm	4.83	5.00	5.07	1.32	-	16.42	-	21.49	m	24.18

Labour hourly rates: (except Specialists) Craft Operatives £12.44 Labourer £7.02 Rates are national average prices. Refer to REGIONAL VARIATIONS for indicative levels of overall pricing in regions	MATERIALS			LABOUR				RATES		
	Del to Site £	Waste %	Material Cost £	Craft Optve Hrs	Lab Hrs	Labour Cost £	Sunds £	Nett Rate £	Unit	Gross Rate (+12.5%) £
S115: HOT AND COLD WATER AND GAS Cont.										
BLACK AND GALVANISED STEEL TUBES AND FITTINGS - Steel pipes, B.S.1387 Table 2; steel fittings B.S.1740 Part 1; screwed and PTFE tape joints; pipework black Cont.										

Pipes, straight Cont.
    in malleable iron single rings, 165mm long screwed
    both ends steel tubes and malleable iron
    backplates fixing to timber with screws Cont.										

50mm	6.47	5.00	6.79	1.72	-	21.40	-	28.19	m	31.71
BLACK AND GALVANISED STEEL TUBES AND FITTINGS - Steel pipes, B.S.1387 Table 3; steel fittings B.S.1740 Part 1; screwed and PTFE tape joints; pipework black										

Pipes, straight
    15mm; in malleable iron pipe brackets fixing to
    timber with screws ....
	2.16	5.00	2.27	0.60	-	7.46	-	9.73	m	10.95

extra; made bends	-	-	-	0.20	-	2.49	-	2.49	nr	2.80
extra; connections to copper pipe ends;										
compression joint	0.64	2.00	0.65	0.12	-	1.49	-	2.15	nr	2.41
extra; fittings; one end	0.40	2.00	0.41	0.15	-	1.87	-	2.27	nr	2.56
extra; fittings; two ends	0.46	2.00	0.47	0.30	-	3.73	-	4.20	nr	4.73
extra; fittings; three ends	0.62	2.00	0.63	0.30	-	3.73	-	4.36	nr	4.91
20mm; in malleable iron pipe brackets fixing to timber with screws	2.52	5.00	2.65	0.68	-	8.46	-	11.11	m	12.49
extra; made bends	-	-	-	0.30	-	3.73	-	3.73	nr	4.20
extra; connections to copper pipe ends;										
compression joint	1.04	2.00	1.06	0.20	-	2.49	-	3.55	nr	3.99
extra; fittings; one end	0.46	2.00	0.47	0.20	-	2.49	-	2.96	nr	3.33
extra; fittings; two ends	0.62	2.00	0.63	0.40	-	4.98	-	5.61	nr	6.31
extra; fittings; three ends	0.91	2.00	0.93	0.40	-	4.98	-	5.90	nr	6.64
25mm; in malleable iron pipe brackets fixing to timber with screws	3.55	5.00	3.73	0.80	-	9.95	-	13.68	m	15.39
extra; made bends	-	-	-	0.40	-	4.98	-	4.98	nr	5.60
extra; connections to copper pipe ends;										
compression joint	2.03	2.00	2.07	0.30	-	3.73	-	5.80	nr	6.53
extra; fittings; one end	0.57	2.00	0.58	0.30	-	3.73	-	4.31	nr	4.85
extra; fittings; two ends	0.98	2.00	1.00	0.60	-	7.46	-	8.46	nr	9.52
extra; fittings; three ends	1.31	2.00	1.34	0.60	-	7.46	-	8.80	nr	9.90
32mm; in malleable iron pipe brackets fixing to timber with screws	4.45	5.00	4.67	0.96	-	11.94	-	16.61	m	18.69
extra; made bends	-	-	-	0.60	-	7.46	-	7.46	nr	8.40
extra; connections to copper pipe ends;										
compression joint	5.95	2.00	6.07	0.40	-	4.98	-	11.05	nr	12.43
extra; fittings; one end	0.83	2.00	0.85	0.40	-	4.98	-	5.82	nr	6.55
extra; fittings; two ends	1.60	2.00	1.63	0.80	-	9.95	-	11.58	nr	13.03
extra; fittings; three ends	2.17	2.00	2.21	0.80	-	9.95	-	12.17	nr	13.69
40mm; in malleable iron pipe brackets fixing to timber with screws	5.25	5.00	5.51	1.20	-	14.93	-	20.44	m	23.00
extra; made bends	-	-	-	0.80	-	9.95	-	9.95	nr	11.20
extra; connections to copper pipe ends;										
compression joint	9.57	2.00	9.76	0.60	-	7.46	-	17.23	nr	19.38
extra; fittings; one end	1.05	2.00	1.07	0.50	-	6.22	-	7.29	nr	8.20
extra; fittings; two ends	2.69	2.00	2.74	1.00	-	12.44	-	15.18	nr	17.08
extra; fittings; three ends	2.98	2.00	3.04	1.00	-	12.44	-	15.48	nr	17.41
50mm; in malleable iron pipe brackets fixing to timber with screws	7.23	5.00	7.59	1.60	-	19.90	-	27.50	m	30.93
extra; made bends	-	-	-	1.20	-	14.93	-	14.93	nr	16.79
extra; connections to copper pipe ends;										
compression joint	14.51	2.00	14.80	0.80	-	9.95	-	24.75	nr	27.85
extra; fittings; one end	2.05	2.00	2.09	0.60	-	7.46	-	9.55	nr	10.75
extra; fittings; two ends	3.15	2.00	3.21	1.20	-	14.93	-	18.14	nr	20.41
extra; fittings; three ends	4.29	2.00	4.38	1.20	-	14.93	-	19.30	nr	21.72
in malleable iron pipe brackets fixing to masonry with screws										
15mm	2.16	5.00	2.27	0.64	-	7.96	-	10.23	m	11.51
20mm	2.52	5.00	2.65	0.72	-	8.96	-	11.60	m	13.05
25mm	3.55	5.00	3.73	0.84	-	10.45	-	14.18	m	15.95
32mm	4.45	5.00	4.67	1.00	-	12.44	-	17.11	m	19.25
40mm	5.25	5.00	5.51	1.24	-	15.43	-	20.94	m	23.56
50mm	7.23	5.00	7.59	1.64	-	20.40	-	27.99	m	31.49

in malleable iron single rings, 165mm long screwed
    both ends steel tubes and malleable iron
    backplates fixing to timber with screws										

15mm	2.77	5.00	2.91	0.72	-	8.96	-	11.87	m	13.35
20mm	3.13	5.00	3.29	0.80	-	9.95	-	13.24	m	14.89
25mm	4.15	5.00	4.36	0.92	-	11.44	-	15.80	m	17.78
32mm	4.91	5.00	5.16	1.08	-	13.44	-	18.59	m	20.91
40mm	5.65	5.00	5.93	1.32	-	16.42	-	22.35	m	25.15
50mm	7.54	5.00	7.92	1.72	-	21.40	-	29.31	m	32.98
BLACK AND GALVANISED STEEL TUBES AND FITTINGS - Steel pipes, B.S.1387 Table 2; steel fittings B.S.1740 Part 1; screwed and PTFE tape joints; pipework galvanised										

Pipes, straight
    15mm; in malleable iron pipe brackets fixing to
    timber with screws ....
	2.77	5.00	2.91	0.60	-	7.46	-	10.37	m	11.67

extra; made bends	-	-	-	0.20	-	2.49	-	2.49	nr	2.80
extra; connections to copper pipe ends;										
compression joint	0.64	2.00	0.65	0.12	-	1.49	-	2.15	nr	2.41
extra; fittings; one end	0.56	2.00	0.57	0.15	-	1.87	-	2.44	nr	2.74
extra; fittings; two ends	0.65	2.00	0.66	0.30	-	3.73	-	4.39	nr	4.94
extra; fittings; three ends	0.88	2.00	0.90	0.30	-	3.73	-	4.63	nr	5.21

Labour hourly rates: (except Specialists) Craft Operatives £12.44 Labourer £7.02 Rates are national average prices. Refer to REGIONAL VARIATIONS for indicative levels of overall pricing in regions	MATERIALS			LABOUR				RATES		
	Del to Site £	Waste %	Material Cost £	Craft Optve Hrs	Lab Hrs	Labour Cost £	Sunds £	Nett Rate £	Unit	Gross Rate (+12.5%) £
S115: HOT AND COLD WATER AND GAS Cont.										
BLACK AND GALVANISED STEEL TUBES AND FITTINGS – Steel pipes, B.S.1387 Table 2; steel fittings B.S.1740 Part 1; screwed and PTFE tape joints; pipework galvanised Cont.										
Pipes, straight Cont.										
20mm; in malleable iron pipe brackets fixing to timber with screws	3.12	5.00	3.28	0.68	-	8.46	-	11.74	m	13.20
extra; made bends	-	-	-	0.30	-	3.73	-	3.73	nr	4.20
extra; connections to copper pipe ends;										
compression joint	1.04	2.00	1.06	0.20	-	2.49	-	3.55	nr	3.99
extra; fittings; one end	0.65	2.00	0.66	0.20	-	2.49	-	3.15	nr	3.54
extra; fittings; two ends	0.88	2.00	0.90	0.40	-	4.98	-	5.87	nr	6.61
extra; fittings; three ends	1.29	2.00	1.32	0.40	-	4.98	-	6.29	nr	7.08
25mm; in malleable iron pipe brackets fixing to timber with screws	4.21	5.00	4.42	0.80	-	9.95	-	14.37	m	16.17
extra; made bends	-	-	-	0.40	-	4.98	-	4.98	nr	5.60
extra; connections to copper pipe ends;										
compression joint	2.03	2.00	2.07	0.30	-	3.73	-	5.80	nr	6.53
extra; fittings; one end	0.81	2.00	0.83	0.30	-	3.73	-	4.56	nr	5.13
extra; fittings; two ends	1.38	2.00	1.41	0.60	-	7.46	-	8.87	nr	9.98
extra; fittings; three ends	1.86	2.00	1.90	0.60	-	7.46	-	9.36	nr	10.53
32mm; in malleable iron pipe brackets fixing to timber with screws	5.25	5.00	5.51	0.96	-	11.94	-	17.45	m	19.64
extra; made bends	-	-	-	0.60	-	7.46	-	7.46	nr	8.40
extra; connections to copper pipe ends;										
compression joint	5.95	2.00	6.07	0.40	-	4.98	-	11.05	nr	12.43
extra; fittings; one end	1.17	2.00	1.19	0.40	-	4.98	-	6.17	nr	6.94
extra; fittings; two ends	2.26	2.00	2.31	0.80	-	9.95	-	12.26	nr	13.79
extra; fittings; three ends	3.07	2.00	3.13	0.80	-	9.95	-	13.08	nr	14.72
40mm; in malleable iron pipe brackets fixing to timber with screws	6.25	5.00	6.56	1.20	-	14.93	-	21.49	m	24.18
extra; made bends	-	-	-	0.80	-	9.95	-	9.95	nr	11.20
extra; connections to copper pipe ends;										
compression joint	9.57	2.00	9.76	0.60	-	7.46	-	17.23	nr	19.38
extra; fittings; one end	1.50	2.00	1.53	0.50	-	6.22	-	7.75	nr	8.72
extra; fittings; two ends	3.80	2.00	3.88	1.00	-	12.44	-	16.32	nr	18.36
extra; fittings; three ends	4.20	2.00	4.28	1.00	-	12.44	-	16.72	nr	18.81
50mm; in malleable iron pipe brackets fixing to timber with screws	8.66	5.00	9.09	1.60	-	19.90	-	29.00	m	32.62
extra; made bends	-	-	-	1.20	-	14.93	-	14.93	nr	16.79
extra; connections to copper pipe ends;										
compression joint	14.51	2.00	14.80	0.80	-	9.95	-	24.75	nr	27.85
extra; fittings; one end	2.91	2.00	2.97	0.60	-	7.46	-	10.43	nr	11.74
extra; fittings; two ends	4.43	2.00	4.52	1.20	-	14.93	-	19.45	nr	21.88
extra; fittings; three ends	6.06	2.00	6.18	1.20	-	14.93	-	21.11	nr	23.75
in malleable iron pipe brackets fixing to masonry with screws										
15mm	2.77	5.00	2.91	0.64	-	7.96	-	10.87	m	12.23
20mm	3.12	5.00	3.28	0.72	-	8.96	-	12.23	m	13.76
25mm	4.21	5.00	4.42	0.84	-	10.45	-	14.87	m	16.73
32mm	5.25	5.00	5.51	1.00	-	12.44	-	17.95	m	20.20
40mm	6.25	5.00	6.56	1.24	-	15.43	-	21.99	m	24.74
50mm	8.66	5.00	9.09	1.64	-	20.40	-	29.49	m	33.18
in malleable iron single rings, 165mm long screwed both ends steel tubes and malleable iron backplates fixing to timber with screws										
15mm	3.50	5.00	3.67	0.72	-	8.96	-	12.63	m	14.21
20mm	3.85	5.00	4.04	0.80	-	9.95	-	13.99	m	15.74
25mm	4.91	5.00	5.16	0.92	-	11.44	-	16.60	m	18.68
32mm	5.77	5.00	6.06	1.08	-	13.44	-	19.49	m	21.93
40mm	6.68	5.00	7.01	1.32	-	16.42	-	23.43	m	26.36
50mm	8.94	5.00	9.39	1.72	-	21.40	-	30.78	m	34.63
BLACK AND GALVANISED STEEL TUBES AND FITTINGS – Steel pipes, B.S.1387 Table 3; steel fittings B.S.1740 Part 1; screwed and PTFE tape joints; pipework galvanised										
Pipes, straight										
15mm; in malleable iron pipe brackets fixing to timber with screws	3.19	5.00	3.35	0.60	-	7.46	-	10.81	m	12.17
extra; made bends	-	-	-	0.20	-	2.49	-	2.49	nr	2.80
extra; connections to copper pipe ends;										
compression joint	0.64	2.00	0.65	0.12	-	1.49	-	2.15	nr	2.41
extra; fittings; one end	0.56	2.00	0.57	0.15	-	1.87	-	2.44	nr	2.74
extra; fittings; two ends	0.65	2.00	0.66	0.30	-	3.73	-	4.39	nr	4.94
extra; fittings; three ends	0.88	2.00	0.90	0.30	-	3.73	-	4.63	nr	5.21
20mm; in malleable iron pipe brackets fixing to timber with screws	3.60	5.00	3.78	0.68	-	8.46	-	12.24	m	13.77
extra; made bends	-	-	-	0.30	-	3.73	-	3.73	nr	4.20
extra; connections to copper pipe ends;										
compression joint	1.04	2.00	1.06	0.20	-	2.49	-	3.55	nr	3.99
extra; fittings; one end	0.65	2.00	0.66	0.20	-	2.49	-	3.15	nr	3.54
extra; fittings; two ends	0.88	2.00	0.90	0.40	-	4.98	-	5.87	nr	6.61
extra; fittings; three ends	1.29	2.00	1.32	0.40	-	4.98	-	6.29	nr	7.08
25mm; in malleable iron pipe brackets fixing to timber with screws	4.97	5.00	5.22	0.80	-	9.95	-	15.17	m	17.07
extra; made bends	-	-	-	0.40	-	4.98	-	4.98	nr	5.60
extra; connections to copper pipe ends;										
compression joint	2.03	2.00	2.07	0.30	-	3.73	-	5.80	nr	6.53
extra; fittings; one end	0.81	2.00	0.83	0.30	-	3.73	-	4.56	nr	5.13
extra; fittings; two ends	1.38	2.00	1.41	0.60	-	7.46	-	8.87	nr	9.98
extra; fittings; three ends	1.86	2.00	1.90	0.60	-	7.46	-	9.36	nr	10.53

Labour hourly rates: (except Specialists) Craft Operatives £12.44 Labourer £7.02 Rates are national average prices. Refer to REGIONAL VARIATIONS for indicative levels of overall pricing in regions	MATERIALS			LABOUR				RATES		
	Del to Site £	Waste %	Material Cost £	Craft Optve Hrs	Lab Hrs	Labour Cost £	Sunds £	Nett Rate £	Unit	Gross Rate (+12.5%) £

S115: HOT AND COLD WATER AND GAS Cont.

BLACK AND GALVANISED STEEL TUBES AND FITTINGS - Steel pipes, B.S.1387 Table 3; steel fittings B.S.1740 Part 1; screwed and PTFE tape joints; pipework galvanised Cont.

Pipes, straight Cont.

	Del to Site £	Waste %	Material Cost £	Craft Optve Hrs	Lab Hrs	Labour Cost £	Sunds £	Nett Rate £	Unit	Gross Rate (+12.5%) £
32mm; in malleable iron pipe brackets fixing to timber with screws	6.21	5.00	6.52	0.96	-	11.94	-	18.46	m	20.77
extra; made bends	-	-	-	0.60	-	7.46	-	7.46	nr	8.40
extra; connections to copper pipe ends; compression joint	5.95	2.00	6.07	0.40	-	4.98	-	11.05	nr	12.43
extra; fittings; one end	1.17	2.00	1.19	0.40	-	4.98	-	6.17	nr	6.94
extra; fittings; two ends	2.26	2.00	2.31	0.80	-	9.95	-	12.26	nr	13.79
extra; fittings; three ends	3.07	2.00	3.13	0.80	-	9.95	-	13.08	nr	14.72
40mm; in malleable iron pipe brackets fixing to timber with screws	7.38	5.00	7.75	1.20	-	14.93	-	22.68	m	25.51
extra; made bends	-	-	-	0.80	-	9.95	-	9.95	nr	11.20
extra; connections to copper pipe ends; compression joint	9.57	2.00	9.76	0.60	-	7.46	-	17.23	nr	19.38
extra; fittings; one end	1.50	2.00	1.53	0.50	-	6.22	-	7.75	nr	8.72
extra; fittings; two ends	3.80	2.00	3.88	1.00	-	12.44	-	16.32	nr	18.36
extra; fittings; three ends	4.20	2.00	4.28	1.00	-	12.44	-	16.72	nr	18.81
50mm; in malleable iron pipe brackets fixing to timber with screws	10.14	5.00	10.65	1.60	-	19.90	-	30.55	m	34.37
extra; made bends	-	-	-	1.20	-	14.93	-	14.93	nr	16.79
extra; connections to copper pipe ends; compression joint	14.51	2.00	14.80	0.80	-	9.95	-	24.75	nr	27.85
extra; fittings; one end	2.91	2.00	2.97	0.60	-	7.46	-	10.43	nr	11.74
extra; fittings; two ends	4.43	2.00	4.52	1.20	-	14.93	-	19.45	nr	21.88
extra; fittings; three ends	6.06	2.00	6.18	1.20	-	14.93	-	21.11	nr	23.75
in malleable iron pipe brackets fixing to masonry with screws										
15mm	3.19	5.00	3.35	0.64	-	7.96	-	11.31	m	12.72
20mm	3.60	5.00	3.78	0.72	-	8.96	-	12.74	m	14.33
25mm	4.97	5.00	5.22	0.84	-	10.45	-	15.67	m	17.63
32mm	6.21	5.00	6.52	1.00	-	12.44	-	18.96	m	21.33
40mm	7.38	5.00	7.75	1.24	-	15.43	-	23.17	m	26.07
50mm	10.14	5.00	10.65	1.64	-	20.40	-	31.05	m	34.93
in malleable iron single rings, 165mm long screwed both ends steel tubes and malleable iron backplates fixing to timber with screws										
15mm	3.92	5.00	4.12	0.72	-	8.96	-	13.07	m	14.71
20mm	4.33	5.00	4.55	0.80	-	9.95	-	14.50	m	16.31
25mm	5.67	5.00	5.95	0.92	-	11.44	-	17.40	m	19.57
32mm	6.73	5.00	7.07	1.08	-	13.44	-	20.50	m	23.06
40mm	7.81	5.00	8.20	1.32	-	16.42	-	24.62	m	27.70
50mm	10.42	5.00	10.94	1.72	-	21.40	-	32.34	m	36.38
Pipes; straight; in trenches										
15mm	2.67	5.00	2.80	0.26	-	3.23	-	6.04	m	6.79
20mm	3.02	5.00	3.17	0.30	-	3.73	-	6.90	m	7.77
25mm	4.30	5.00	4.51	0.35	-	4.35	-	8.87	m	9.98
32mm	5.34	5.00	5.61	0.40	-	4.98	-	10.58	m	11.91
40mm	6.23	5.00	6.54	0.48	-	5.97	-	12.51	m	14.08
50mm	8.63	5.00	9.06	0.60	-	7.46	-	16.53	m	18.59

LIGHT GAUGE STAINLESS STEEL TUBES AND FITTINGS - Stainless steel pipes, B.S.4127 Part 2; C.P. copper fittings, capillary, B.S.864 Part 2

Pipes, straight

	Del to Site £	Waste %	Material Cost £	Craft Optve Hrs	Lab Hrs	Labour Cost £	Sunds £	Nett Rate £	Unit	Gross Rate (+12.5%) £
15mm; in C.P. pipe brackets fixing to timber with screws	4.36	5.00	4.58	0.40	-	4.98	-	9.55	m	10.75
extra; fittings; one end	3.37	3.00	3.47	0.06	-	0.75	-	4.22	nr	4.74
extra; fittings; two ends	1.03	3.00	1.06	0.12	-	1.49	-	2.55	nr	2.87
extra; fittings; three ends	1.73	3.00	1.78	0.12	-	1.49	-	3.27	nr	3.68
22mm; in C.P. pipe brackets fixing to timber with screws	6.44	5.00	6.76	0.43	-	5.35	-	12.11	m	13.63
extra; fittings; one end	3.94	3.00	4.06	0.10	-	1.24	-	5.30	nr	5.96
extra; fittings; two ends	2.41	3.00	2.48	0.20	-	2.49	-	4.97	nr	5.59
extra; fittings; three ends	4.78	3.00	4.92	0.20	-	2.49	-	7.41	nr	8.34
28mm; in C.P. pipe brackets fixing to timber with screws	8.42	5.00	8.84	0.48	-	5.97	-	14.81	m	16.66
extra; fittings; one end	6.23	3.00	6.42	0.15	-	1.87	-	8.28	nr	9.32
extra; fittings; two ends	5.62	3.00	5.79	0.30	-	3.73	-	9.52	nr	10.71
extra; fittings; three ends	8.71	3.00	8.97	0.30	-	3.73	-	12.70	nr	14.29
35mm; in C.P. pipe brackets fixing to timber with screws	10.35	5.00	10.87	0.60	-	7.46	-	18.33	m	20.62
extra; fittings; one end	13.47	3.00	13.87	0.20	-	2.49	-	16.36	nr	18.41
extra; fittings; two ends	12.68	3.00	13.06	0.40	-	4.98	-	18.04	nr	20.29
extra; fittings; three ends	21.79	3.00	22.44	0.40	-	4.98	-	27.42	nr	30.85
42mm; in C.P. pipe brackets fixing to masonry with screws	14.74	5.00	15.48	0.80	-	9.95	-	25.43	m	28.61
extra; fittings; one end	20.68	3.00	21.30	0.30	-	3.73	-	25.03	nr	28.16
extra; fittings; two ends	22.58	3.00	23.26	0.60	-	7.46	-	30.72	nr	34.56
extra; fittings; three ends	34.13	3.00	35.15	0.60	-	7.46	-	42.62	nr	47.95
in C.P. pipe brackets fixing to masonry with screws										
15mm	4.36	5.00	4.58	0.45	-	5.60	-	10.18	m	11.45
22mm	6.44	5.00	6.76	0.48	-	5.97	-	12.73	m	14.32
28mm	8.42	5.00	8.84	0.53	-	6.59	-	15.43	m	17.36
35mm	10.35	5.00	10.87	0.65	-	8.09	-	18.95	m	21.32
42mm	21.79	5.00	22.88	0.85	-	10.57	-	33.45	m	37.64

Labour hourly rates: (except Specialists) Craft Operatives £12.44 Labourer £7.02 Rates are national average prices. Refer to REGIONAL VARIATIONS for indicative levels of overall pricing in regions	MATERIALS			LABOUR				RATES		
	Del to Site	Waste	Material Cost	Craft Optve	Lab	Labour Cost	Sunds	Nett Rate	Unit	Gross Rate (+12.5%)
	£	%	£	Hrs	Hrs	£	£	£		£
S115: HOT AND COLD WATER AND GAS Cont.										
LIGHT GAUGE STAINLESS STEEL TUBES AND FITTINGS - **Stainless steel pipes, B.S.4127 Part 2; C.P. copper** **alloy fittings, compression, B.S.864 Part 2, Type A**										
Pipes, straight										
15mm; in C.P. pipe brackets fixing to timber with										
screws	4.25	5.00	4.46	0.40	-	4.98	-	9.44	m	10.62
extra; fittings; one end	1.47	3.00	1.51	0.06	-	0.75	-	2.26	nr	2.54
extra; fittings; two ends	1.12	3.00	1.15	0.12	-	1.49	-	2.65	nr	2.98
extra; fittings; three ends	1.65	3.00	1.70	0.12	-	1.49	-	3.19	nr	3.59
22mm; in C.P. pipe brackets fixing to timber with										
screws	6.45	5.00	6.77	0.43	-	5.35	-	12.12	m	13.64
extra; fittings; one end	1.76	3.00	1.81	0.10	-	1.24	-	3.06	nr	3.44
extra; fittings; two ends	1.91	3.00	1.97	0.20	-	2.49	-	4.46	nr	5.01
extra; fittings; three ends	2.69	3.00	2.77	0.20	-	2.49	-	5.26	nr	5.92
28mm; in C.P. pipe brackets fixing to timber with										
screws	8.75	5.00	9.19	0.48	-	5.97	-	15.16	m	17.05
extra; fittings; one end	5.02	3.00	5.17	0.15	-	1.87	-	7.04	nr	7.92
extra; fittings; two ends	6.03	3.00	6.21	0.30	-	3.73	-	9.94	nr	11.19
extra; fittings; three ends	10.69	3.00	11.01	0.30	-	3.73	-	14.74	nr	16.59
35mm; in C.P. pipe brackets fixing to timber with										
screws	11.12	5.00	11.68	0.60	-	7.46	-	19.14	m	21.53
extra; fittings; one end	8.66	3.00	8.92	0.20	-	2.49	-	11.41	nr	12.83
extra; fittings; two ends	14.50	3.00	14.94	0.40	-	4.98	-	19.91	nr	22.40
extra; fittings; three ends	19.14	3.00	19.71	0.40	-	4.98	-	24.69	nr	27.78
42mm; in C.P. pipe brackets fixing to masonry with										
screws	15.70	5.00	16.48	0.80	-	9.95	-	26.44	m	29.74
extra; fittings; one end	14.14	3.00	14.56	0.30	-	3.73	-	18.30	nr	20.58
extra; fittings; two ends	20.25	3.00	20.86	0.60	-	7.46	-	28.32	nr	31.86
extra; fittings; three ends	29.47	3.00	30.35	0.60	-	7.46	-	37.82	nr	42.55
in C.P. pipe brackets fixing to masonry with screws										
15mm	4.25	5.00	4.46	0.45	-	5.60	-	10.06	m	11.32
22mm	6.45	5.00	6.77	0.48	-	5.97	-	12.74	m	14.34
28mm	8.75	5.00	9.19	0.53	-	6.59	-	15.78	m	17.75
35mm	11.12	5.00	11.68	0.65	-	8.09	-	19.76	m	22.23
42mm	15.70	5.00	16.48	0.85	-	10.57	-	27.06	m	30.44
COPPER TUBING AND FITTINGS - Copper pipes, B.S.2871 **Part 1 Table X; copper fittings, capillary, B.S.864** **Part 2**										
Pipes, straight										
15mm; in two piece copper spacing clips to timber										
with screws	1.04	5.00	1.09	0.85	-	10.57	-	11.67	m	13.12
extra; made bends	-	-	-	0.10	-	1.24	-	1.24	nr	1.40
extra; connections to iron pipe ends; screwed										
joint	1.63	3.00	1.68	0.12	-	1.49	-	3.17	nr	3.57
extra; fittings; one end	0.88	3.00	0.91	0.06	-	0.75	-	1.65	nr	1.86
extra; fittings; two ends	0.25	3.00	0.26	0.12	-	1.49	-	1.75	nr	1.97
extra; fittings; three ends	0.42	3.00	0.43	0.12	-	1.49	-	1.93	nr	2.17
22mm; in two piece copper spacing clips to timber										
with screws	1.98	5.00	2.08	0.43	-	5.35	-	7.43	m	8.36
extra; made bends	-	-	-	0.16	-	1.99	-	1.99	nr	2.24
extra; connections to iron pipe ends; screwed										
joint	2.87	3.00	2.96	0.20	-	2.49	-	5.44	nr	6.12
extra; fittings; one end	1.68	3.00	1.73	0.10	-	1.24	-	2.97	nr	3.35
extra; fittings; two ends	0.59	3.00	0.61	0.20	-	2.49	-	3.10	nr	3.48
extra; fittings; three ends	1.20	3.00	1.24	0.20	-	2.49	-	3.72	nr	4.19
28mm; in two piece copper spacing clips to timber										
with screws	2.71	5.00	2.85	0.48	-	5.97	-	8.82	m	9.92
extra; made bends	-	-	-	0.20	-	2.49	-	2.49	nr	2.80
extra; connections to iron pipe ends; screwed										
joint	4.55	3.00	4.69	0.30	-	3.73	-	8.42	nr	9.47
extra; fittings; one end	2.65	3.00	2.73	0.15	-	1.87	-	4.60	nr	5.17
extra; fittings; two ends	1.33	3.00	1.37	0.30	-	3.73	-	5.10	nr	5.74
extra; fittings; three ends	3.71	3.00	3.82	0.30	-	3.73	-	7.55	nr	8.50
35mm; in two piece copper spacing clips to timber										
with screws	6.70	5.00	7.04	0.60	-	7.46	-	14.50	m	16.31
extra; made bends	-	-	-	0.30	-	3.73	-	3.73	nr	4.20
extra; connections to iron pipe ends; screwed										
joint	8.02	3.00	8.26	0.40	-	4.98	-	13.24	nr	14.89
extra; fittings; one end	5.73	3.00	5.90	0.20	-	2.49	-	8.39	nr	9.44
extra; fittings; two ends	5.40	3.00	5.56	0.40	-	4.98	-	10.54	nr	11.86
extra; fittings; three ends	9.27	3.00	9.55	0.40	-	4.98	-	14.52	nr	16.34
42mm; in two piece copper spacing clips to timber										
with screws	8.41	5.00	8.83	0.80	-	9.95	-	18.78	m	21.13
extra; made bends	-	-	-	0.40	-	4.98	-	4.98	nr	5.60
extra; connections to iron pipe ends; screwed										
joint	10.34	3.00	10.65	0.60	-	7.46	-	18.11	nr	20.38
extra; fittings; one end	8.80	3.00	9.06	0.30	-	3.73	-	12.80	nr	14.40
extra; fittings; two ends	9.61	3.00	9.90	0.60	-	7.46	-	17.36	nr	19.53
extra; fittings; three ends	14.52	3.00	14.96	0.60	-	7.46	-	22.42	nr	25.22
54mm; in two piece copper spacing clips to timber										
with screws	11.37	5.00	11.94	1.20	-	14.93	-	26.87	m	30.22
extra; made bends	-	-	-	0.60	-	7.46	-	7.46	nr	8.40
extra; connections to iron pipe ends; screwed										
joint	15.69	3.00	16.16	0.80	-	9.95	-	26.11	nr	29.38
extra; fittings; one end	12.29	3.00	12.66	0.40	-	4.98	-	17.63	nr	19.84
extra; fittings; two ends	19.67	3.00	20.26	0.80	-	9.95	-	30.21	nr	33.99
extra; fittings; three ends	27.44	3.00	28.26	0.80	-	9.95	-	38.22	nr	42.99
in two piece copper spacing clips fixing to masonry with screws										
15mm	1.04	5.00	1.09	0.45	-	5.60	-	6.69	m	7.53

PIPED SUPPLY SYSTEMS

	MATERIALS			LABOUR				RATES		
Labour hourly rates: (except Specialists) Craft Operatives £12.44 Labourer £7.02 Rates are national average prices. Refer to REGIONAL VARIATIONS for indicative levels of overall pricing in regions	Del to Site £	Waste %	Material Cost £	Craft Optve Hrs	Lab Hrs	Labour Cost £	Sunds £	Nett Rate £	Unit	Gross Rate (+12.5%) £
S115: HOT AND COLD WATER AND GAS Cont.										
COPPER TUBING AND FITTINGS - Copper pipes, B.S.2871 Part 1 Table X; copper fittings, capillary, B.S.864 Part 2 Cont.										
Pipes, straight Cont.										
in two piece copper spacing clips fixing to masonry with screws Cont.										
22mm	1.98	5.00	2.08	0.48	-	5.97	-	8.05	m	9.06
28mm	2.71	5.00	2.85	0.53	-	6.59	-	9.44	m	10.62
35mm	6.70	5.00	7.04	0.65	-	8.09	-	15.12	m	17.01
42mm	8.41	5.00	8.83	0.85	-	10.57	-	19.40	m	21.83
54mm	11.37	5.00	11.94	1.25	-	15.55	-	27.49	m	30.92
in pressed brass pipe brackets fixing to timber with screws										
15mm	1.39	5.00	1.46	0.40	-	4.98	-	6.44	m	7.24
22mm	2.39	5.00	2.51	0.43	-	5.35	-	7.86	m	8.84
28mm	3.19	5.00	3.35	0.48	-	5.97	-	9.32	m	10.49
35mm	7.19	5.00	7.55	0.60	-	7.46	-	15.01	m	16.89
42mm	8.87	5.00	9.31	0.80	-	9.95	-	19.27	m	21.67
54mm	11.96	5.00	12.56	1.20	-	14.93	-	27.49	m	30.92
in cast brass pipe brackets fixing to timber with screws										
15mm	2.04	5.00	2.14	0.40	-	4.98	-	7.12	m	8.01
22mm	3.17	5.00	3.33	0.43	-	5.35	-	8.68	m	9.76
28mm	4.14	5.00	4.35	0.48	-	5.97	-	10.32	m	11.61
35mm	8.53	5.00	8.96	0.60	-	7.46	-	16.42	m	18.47
42mm	10.76	5.00	11.30	0.80	-	9.95	-	21.25	m	23.91
54mm	14.33	5.00	15.05	1.20	-	14.93	-	29.97	m	33.72
in cast brass single rings and back plates fixing to timber with screws										
15mm	2.34	5.00	2.46	0.45	-	5.60	-	8.05	m	9.06
22mm	3.32	5.00	3.49	0.48	-	5.97	-	9.46	m	10.64
28mm	4.16	5.00	4.37	0.53	-	6.59	-	10.96	m	12.33
35mm	8.14	5.00	8.55	0.65	-	8.09	-	16.63	m	18.71
42mm	9.83	5.00	10.32	0.85	-	10.57	-	20.90	m	23.51
54mm	12.88	5.00	13.52	1.25	-	15.55	-	29.07	m	32.71
66.7mm	21.24	5.00	22.30	1.80	-	22.39	-	44.69	m	50.28
76.1mm	31.82	5.00	33.41	2.40	-	29.86	-	63.27	m	71.18
108mm	46.10	5.00	48.41	3.60	-	44.78	-	93.19	m	104.84
COPPER TUBING AND FITTINGS - Copper pipes, B.S.2871 Part 1 Table X; copper alloy fittings, compression, B.S.864 Part 2, Type A										
Pipes, straight										
15mm; in two piece copper spacing clips to timber with screws	1.14	5.00	1.20	0.33	-	4.11	-	5.30	m	5.96
extra; made bends	-	-	-	0.08	-	1.00	-	1.00	nr	1.12
extra; connections to iron pipe ends; screwed joint	0.64	3.00	0.66	0.12	-	1.49	-	2.15	nr	2.42
extra; fittings; one end	1.09	3.00	1.12	0.06	-	0.75	-	1.87	nr	2.10
extra; fittings; two ends	0.93	3.00	0.96	0.12	-	1.49	-	2.45	nr	2.76
extra; fittings; three ends	1.22	3.00	1.26	0.12	-	1.49	-	2.75	nr	3.09
22mm; in two piece copper spacing clips to timber with screws	2.12	5.00	2.23	0.43	-	5.35	-	7.58	m	8.52
extra; made bends	-	-	-	0.16	-	1.99	-	1.99	nr	2.24
extra; connections to iron pipe ends; screwed joint	1.04	3.00	1.07	0.20	-	2.49	-	3.56	nr	4.00
extra; fittings; one end	1.30	3.00	1.34	0.10	-	1.24	-	2.58	nr	2.91
extra; fittings; two ends	1.42	3.00	1.46	0.20	-	2.49	-	3.95	nr	4.44
extra; fittings; three ends	1.99	3.00	2.05	0.20	-	2.49	-	4.54	nr	5.10
28mm; in two piece copper spacing clips to timber with screws	3.16	5.00	3.32	0.48	-	5.97	-	9.29	m	10.45
extra; made bends	-	-	-	0.20	-	2.49	-	2.49	nr	2.80
extra; connections to iron pipe ends; screwed joint	2.03	3.00	2.09	0.30	-	3.73	-	5.82	nr	6.55
extra; fittings; one end	3.72	3.00	3.83	0.15	-	1.87	-	5.70	nr	6.41
extra; fittings; two ends	4.47	3.00	4.60	0.30	-	3.73	-	8.34	nr	9.38
extra; fittings; three ends	7.92	3.00	8.16	0.30	-	3.73	-	11.89	nr	13.38
35mm; in two piece copper spacing clips to timber with screws	7.61	5.00	7.99	0.60	-	7.46	-	15.45	m	17.39
extra; made bends	-	-	-	0.30	-	3.73	-	3.73	nr	4.20
extra; connections to iron pipe ends; screwed joint	5.95	3.00	6.13	0.40	-	4.98	-	11.10	nr	12.49
extra; fittings; one end	6.41	3.00	6.60	0.20	-	2.49	-	9.09	nr	10.23
extra; fittings; two ends	10.74	3.00	11.06	0.40	-	4.98	-	16.04	nr	18.04
extra; fittings; three ends	14.17	3.00	14.60	0.40	-	4.98	-	19.57	nr	22.00
42mm; in two piece copper spacing clips to timber with screws	9.62	5.00	10.10	0.80	-	9.95	-	20.05	m	22.56
extra; made bends	-	-	-	0.40	-	4.98	-	4.98	nr	5.60
extra; connections to iron pipe ends; screwed joint	9.57	3.00	9.86	0.60	-	7.46	-	17.32	nr	19.49
extra; fittings; one end	10.48	3.00	10.79	0.30	-	3.73	-	14.53	nr	16.34
extra; fittings; two ends	15.00	3.00	15.45	0.60	-	7.46	-	22.91	nr	25.78
extra; fittings; three ends	21.83	3.00	22.48	0.60	-	7.46	-	29.95	nr	33.69
54mm; in two piece copper spacing clips to timber with screws	12.67	5.00	13.30	1.20	-	14.93	-	28.23	m	31.76
extra; made bends	-	-	-	0.60	-	7.46	-	7.46	nr	8.40
extra; connections to iron pipe ends; screwed joint	14.51	3.00	14.95	0.80	-	9.95	-	24.90	nr	28.01
extra; fittings; one end	14.61	3.00	15.05	0.40	-	4.98	-	20.02	nr	22.53
extra; fittings; two ends	25.51	3.00	26.28	0.80	-	9.95	-	36.23	nr	40.76
extra; fittings; three ends	35.13	3.00	36.18	0.80	-	9.95	-	46.14	nr	51.90

Labour hourly rates: (except Specialists) Craft Operatives £12.44 Labourer £7.02 Rates are national average prices. Refer to REGIONAL VARIATIONS for indicative levels of overall pricing in regions	MATERIALS			LABOUR				RATES		
	Del to Site £	Waste %	Material Cost £	Craft Optve Hrs	Lab Hrs	Labour Cost £	Sunds £	Nett Rate £	Unit	Gross Rate (+12.5%) £
S115: HOT AND COLD WATER AND GAS Cont.										
COPPER TUBING AND FITTINGS – Copper pipes, B.S.2871 **Part 1 Table X; copper alloy fittings, compression,** **B.S.864 Part 2, Type A Cont.**										
Pipes, straight Cont.										
in two piece copper spacing clips fixing to masonry with screws										
15mm	1.14	5.00	1.20	0.45	–	5.60	–	6.79	m	7.64
22mm	2.12	5.00	2.23	0.48	–	5.97	–	8.20	m	9.22
28mm	3.16	5.00	3.32	0.53	–	6.59	–	9.91	m	11.15
35mm	7.61	5.00	7.99	0.65	–	8.09	–	16.08	m	18.09
42mm	9.62	5.00	10.10	0.85	–	10.57	–	20.68	m	23.26
54mm	12.67	5.00	13.30	1.25	–	15.55	–	28.85	m	32.46
in pressed brass pipe brackets fixing to timber with screws										
15mm	1.49	5.00	1.56	0.40	–	4.98	–	6.54	m	7.36
22mm	2.53	5.00	2.66	0.43	–	5.35	–	8.01	m	9.01
28mm	3.64	5.00	3.82	0.48	–	5.97	–	9.79	m	11.02
35mm	8.10	5.00	8.51	0.60	–	7.46	–	15.97	m	17.97
42mm	10.08	5.00	10.58	0.80	–	9.95	–	20.54	m	23.10
54mm	13.26	5.00	13.92	1.20	–	14.93	–	28.85	m	32.46
in cast brass pipe brackets fixing to timber with screws										
15mm	2.14	5.00	2.25	0.40	–	4.98	–	7.22	m	8.13
22mm	3.31	5.00	3.48	0.43	–	5.35	–	8.82	m	9.93
28mm	4.59	5.00	4.82	0.48	–	5.97	–	10.79	m	12.14
35mm	9.44	5.00	9.91	0.60	–	7.46	–	17.38	m	19.55
42mm	11.97	5.00	12.57	0.80	–	9.95	–	22.52	m	25.34
54mm	15.63	5.00	16.41	1.20	–	14.93	–	31.34	m	35.26
in cast brass single rings and back plates fixing to timber with screws										
15mm	2.44	5.00	2.56	0.40	–	4.98	–	7.54	m	8.48
22mm	3.46	5.00	3.63	0.43	–	5.35	–	8.98	m	10.10
28mm	4.61	5.00	4.84	0.48	–	5.97	–	10.81	m	12.16
35mm	9.05	5.00	9.50	0.60	–	7.46	–	16.97	m	19.09
42mm	11.04	5.00	11.59	0.80	–	9.95	–	21.54	m	24.24
54mm	14.18	5.00	14.89	1.20	–	14.93	–	29.82	m	33.54
66.7mm	28.93	5.00	30.38	1.80	–	22.39	–	52.77	m	59.36
76.1mm	41.20	5.00	43.26	2.40	–	29.86	–	73.12	m	82.26
108mm	63.14	5.00	66.30	3.60	–	44.78	–	111.08	m	124.97
COPPER TUBING AND FITTINGS – Copper pipes, B.S.2871 **Part 1 Table X; copper alloy fittings, compression,** **B.S.864 Part 2 Type B**										
Pipes, straight										
15mm; in two piece copper spacing clips to timber with screws	1.64	5.00	1.72	0.40	–	4.98	–	6.70	m	7.54
extra; made bends	–	–	–	0.10	–	1.24	–	1.24	nr	1.40
extra; connections to iron pipe ends; screwed joint	3.38	3.00	3.48	0.16	–	1.99	–	5.47	nr	6.16
extra; fittings; one end	3.69	3.00	3.80	0.08	–	1.00	–	4.80	nr	5.40
extra; fittings; two ends	4.38	3.00	4.51	0.16	–	1.99	–	6.50	nr	7.31
extra; fittings; three ends	6.13	3.00	6.31	0.16	–	1.99	–	8.30	nr	9.34
22mm; in two piece copper spacing clips to timber with screws	2.92	5.00	3.07	0.43	–	5.35	–	8.42	m	9.47
extra; made bends	–	–	–	0.16	–	1.99	–	1.99	nr	2.24
extra; connections to iron pipe ends; screwed joint	4.81	3.00	4.95	0.25	–	3.11	–	8.06	nr	9.07
extra; fittings; one end	4.50	3.00	4.63	0.13	–	1.62	–	6.25	nr	7.03
extra; fittings; two ends	7.13	3.00	7.34	0.25	–	3.11	–	10.45	nr	11.76
extra; fittings; three ends	10.31	3.00	10.62	0.25	–	3.11	–	13.73	nr	15.45
28mm; in two piece copper spacing clips to timber with screws	4.21	5.00	4.42	0.48	–	5.97	–	10.39	m	11.69
extra; made bends	–	–	–	0.20	–	2.49	–	2.49	nr	2.80
extra; connections to iron pipe ends; screwed joint	7.38	3.00	7.60	0.37	–	4.60	–	12.20	nr	13.73
extra; fittings; one end	8.44	3.00	8.69	0.19	–	2.36	–	11.06	nr	12.44
extra; fittings; two ends	12.06	3.00	12.42	0.37	–	4.60	–	17.02	nr	19.15
extra; fittings; three ends	17.38	3.00	17.90	0.37	–	4.60	–	22.50	nr	25.32
35mm; in two piece copper spacing clips to timber with screws	9.25	5.00	9.71	0.60	–	7.46	–	17.18	m	19.32
extra; made bends	–	–	–	0.30	–	3.73	–	3.73	nr	4.20
extra; connections to iron pipe ends; screwed joint	14.56	3.00	15.00	0.48	–	5.97	–	20.97	nr	23.59
extra; fittings; one end	22.38	3.00	23.05	0.24	–	2.99	–	26.04	nr	29.29
extra; fittings; two ends	23.19	3.00	23.89	0.48	–	5.97	–	29.86	nr	33.59
extra; fittings; three ends	31.50	3.00	32.45	0.48	–	5.97	–	38.42	nr	43.22
42mm; in two piece copper spacing clips to timber with screws	11.81	5.00	12.40	0.80	–	9.95	–	22.35	m	25.15
extra; made bends	–	–	–	0.40	–	4.98	–	4.98	nr	5.60
extra; connections to iron pipe ends; screwed joint	21.50	3.00	22.15	0.76	–	9.45	–	31.60	nr	35.55
extra; fittings; one end	31.31	3.00	32.25	0.38	–	4.73	–	36.98	nr	41.60
extra; fittings; two ends	34.75	3.00	35.79	0.76	–	9.45	–	45.25	nr	50.90
extra; fittings; three ends	51.25	3.00	52.79	0.76	–	9.45	–	62.24	nr	70.02
54mm; in two piece copper spacing clips to timber with screws	16.01	5.00	16.81	1.20	–	14.93	£	31.74	m	35.71
extra; made bends	–	–	–	0.60	–	7.46	£	7.46	nr	8.40
extra; connections to iron pipe ends; screwed joint	31.13	3.00	32.06	1.00	–	12.44	–	44.50	nr	50.07
extra; fittings; one end	48.06	3.00	49.50	0.50	–	6.22	–	55.72	nr	62.69
extra; fittings; two ends	55.31	3.00	56.97	1.00	–	12.44	–	69.41	nr	78.09

PIPED SUPPLY SYSTEMS

Labour hourly rates: (except Specialists) Craft Operatives £12.44 Labourer £7.02 Rates are national average prices. Refer to REGIONAL VARIATIONS for indicative levels of overall pricing in regions	MATERIALS			LABOUR				RATES		
	Del to Site £	Waste %	Material Cost £	Craft Optve Hrs	Lab Hrs	Labour Cost £	Sunds £	Nett Rate £	Unit	Gross Rate (+12.5%) £

S115: HOT AND COLD WATER AND GAS Cont.

COPPER TUBING AND FITTINGS - Copper pipes, B.S.2871 Part 1 Table X; copper alloy fittings, compression, B.S.864 Part 2 Type B Cont.

Pipes, straight Cont.

	Del to Site £	Waste %	Material Cost £	Craft Optve Hrs	Lab Hrs	Labour Cost £	Sunds £	Nett Rate £	Unit	Gross Rate £
extra; fittings; three ends	82.19	3.00	84.66	1.00	-	12.44	-	97.10	nr	109.23
in two piece copper spacing clips fixing to masonry with screws										
15mm	1.64	5.00	1.72	0.45	-	5.60	-	7.32	m	8.23
22mm	2.92	5.00	3.07	0.48	-	5.97	-	9.04	m	10.17
28mm	4.21	5.00	4.42	0.53	-	6.59	-	11.01	m	12.39
35mm	9.25	5.00	9.71	0.65	-	8.09	-	17.80	m	20.02
42mm	11.81	5.00	12.40	0.85	-	10.57	-	22.97	m	25.85
54mm	16.01	5.00	16.81	1.25	-	15.55	-	32.36	m	36.41
in pressed brass pipe brackets fixing to timber with screws										
15mm	1.99	5.00	2.09	0.45	-	5.60	-	7.69	m	8.65
22mm	3.33	5.00	3.50	0.48	-	5.97	-	9.47	m	10.65
28mm	4.69	5.00	4.92	0.53	-	6.59	-	11.52	m	12.96
35mm	9.74	5.00	10.23	0.65	-	8.09	-	18.31	m	20.60
42mm	12.27	5.00	12.88	0.85	-	10.57	-	23.46	m	26.39
54mm	16.60	5.00	17.43	1.25	-	15.55	-	32.98	m	37.10
in cast brass pipe brackets fixing to timber with screws										
15mm	2.64	5.00	2.77	0.45	-	5.60	-	8.37	m	9.42
22mm	4.11	5.00	4.32	0.48	-	5.97	-	10.29	m	11.57
28mm	5.64	5.00	5.92	0.53	-	6.59	-	12.52	m	14.08
35mm	11.08	5.00	11.63	0.65	-	8.09	-	19.72	m	22.18
42mm	14.16	5.00	14.87	0.85	-	10.57	-	25.44	m	28.62
54mm	18.97	5.00	19.92	1.25	-	15.55	-	35.47	m	39.90
in cast brass single rings and back plates fixing to timber with screws										
15mm	2.94	5.00	3.09	0.45	-	5.60	-	8.69	m	9.77
22mm	4.26	5.00	4.47	0.48	-	5.97	-	10.44	m	11.75
28mm	4.71	5.00	4.95	0.53	-	6.59	-	11.54	m	12.98
35mm	10.69	5.00	11.22	0.65	-	8.09	-	19.31	m	21.72
42mm	13.23	5.00	13.89	0.85	-	10.57	-	24.47	m	27.52
54mm	17.52	5.00	18.40	1.25	-	15.55	-	33.95	m	38.19

COPPER TUBING AND FITTINGS - Copper pipes, B.S.2871 Part 1 Table Y; copper fittings, capillary, B.S.864 Part 2

Pipes; straight; in trenches

	Del to Site £	Waste %	Material Cost £	Craft Optve Hrs	Lab Hrs	Labour Cost £	Sunds £	Nett Rate £	Unit	Gross Rate £
15mm	2.55	5.00	2.68	0.20	-	2.49	-	5.17	m	5.81
extra; made bends	-	-	-	0.10	-	1.24	-	1.24	nr	1.40
extra; connections to iron pipe ends; screwed joint	1.63	3.00	1.68	0.16	-	1.99	-	3.67	nr	4.13
extra; fittings; one end	0.88	3.00	0.91	0.08	-	1.00	-	1.90	nr	2.14
extra; fittings; two ends	0.25	3.00	0.26	0.16	-	1.99	-	2.25	nr	2.53
extra; fittings; three ends	0.42	3.00	0.43	0.16	-	1.99	-	2.42	nr	2.73
22mm	4.45	5.00	4.67	0.22	-	2.74	-	7.41	m	8.34
extra; made bends	-	-	-	0.16	-	1.99	-	1.99	nr	2.24
extra; connections to iron pipe ends; screwed joint	2.87	3.00	2.96	0.25	-	3.11	-	6.07	nr	6.82
extra; fittings; one end	1.68	3.00	1.73	0.13	-	1.62	-	3.35	nr	3.77
extra; fittings; two ends	0.59	3.00	0.61	0.25	-	3.11	-	3.72	nr	4.18
extra; fittings; three ends	1.20	3.00	1.24	0.25	-	3.11	-	4.35	nr	4.89
28mm	5.89	5.00	6.18	0.24	-	2.99	-	9.17	m	10.32
extra; made bends	-	-	-	0.20	-	2.49	-	2.49	nr	2.80
extra; connections to iron pipe ends; screwed joint	4.55	3.00	4.69	0.37	-	4.60	-	9.29	nr	10.45
extra; fittings; one end	2.65	3.00	2.73	0.19	-	2.36	-	5.09	nr	5.73
extra; fittings; two ends	1.33	3.00	1.37	0.37	-	4.60	-	5.97	nr	6.72
extra; fittings; three ends	3.71	3.00	3.82	0.37	-	4.60	-	8.42	nr	9.48
35mm	8.37	5.00	8.79	0.27	-	3.36	-	12.15	m	13.67
extra; made bends	-	-	-	0.30	-	3.73	-	3.73	nr	4.20
extra; connections to iron pipe ends; screwed joint	8.02	3.00	8.26	0.50	-	6.22	-	14.48	nr	16.29
extra; fittings; one end	5.73	3.00	5.90	0.25	-	3.11	-	9.01	nr	10.14
extra; fittings; two ends	5.40	3.00	5.56	0.50	-	6.22	-	11.78	nr	13.25
extra; fittings; three ends	9.27	3.00	9.55	0.50	-	6.22	-	15.77	nr	17.74
42mm	10.14	5.00	10.65	0.35	-	4.35	-	15.00	m	16.88
extra; made bends	-	-	-	0.40	-	4.98	-	4.98	nr	5.60
extra; connections to iron pipe ends; screwed joint	10.34	3.00	10.65	0.76	-	9.45	-	20.10	nr	22.62
extra; fittings; one end	8.80	3.00	9.06	0.38	-	4.73	-	13.79	nr	15.52
extra; fittings; two ends	9.61	3.00	9.90	0.76	-	9.45	-	19.35	nr	21.77
extra; fittings; three ends	14.52	3.00	14.96	0.76	-	9.45	-	24.41	nr	27.46
54mm	17.56	5.00	18.44	0.39	-	4.85	-	23.29	m	26.20
extra; made bends	-	-	-	0.60	-	7.46	-	7.46	nr	8.40
extra; connections to iron pipe ends; screwed joint	15.69	3.00	16.16	1.00	-	12.44	-	28.60	nr	32.18
extra; fittings; one end	12.29	3.00	12.66	0.50	-	6.22	-	18.88	nr	21.24
extra; fittings; two ends	19.67	3.00	20.26	1.00	-	12.44	-	32.70	nr	36.79
extra; fittings; three ends	27.44	3.00	28.26	1.00	-	12.44	-	40.70	nr	45.79

COPPER TUBING AND FITTINGS - Copper pipes, B.S.2871 Part 1 Table Y; non-dezincifiable fittings; compression, B.S.864 Part 2, Type B

Pipes; straight; in trenches

	Del to Site £	Waste %	Material Cost £	Craft Optve Hrs	Lab Hrs	Labour Cost £	Sunds £	Nett Rate £	Unit	Gross Rate £
15mm	2.79	5.00	2.93	0.20	-	2.49	-	5.42	m	6.09
extra; made bends	-	-	-	0.10	-	1.24	-	1.24	nr	1.40

Labour hourly rates: (except Specialists) Craft Operatives £12.44 Labourer £7.02 Rates are national average prices. Refer to REGIONAL VARIATIONS for indicative levels of overall pricing in regions	MATERIALS			LABOUR				RATES		
	Del to Site £	Waste %	Material Cost £	Craft Optve Hrs	Lab Hrs	Labour Cost £	Sunds £	Nett Rate £	Unit	Gross Rate (+12.5%) £
S115: HOT AND COLD WATER AND GAS Cont.										
COPPER TUBING AND FITTINGS - Copper pipes, B.S.2871 **Part 1 Table Y; non-dezincifiable fittings;** **compression, B.S.864 Part 2, Type B Cont.**										
Pipes; straight; in trenches Cont.										
extra; connections to iron pipe ends; screwed										
joint	3.38	3.00	3.48	0.16	-	1.99	-	5.47	nr	6.16
extra; fittings; one end	3.69	3.00	3.80	0.08	-	1.00	-	4.80	nr	5.40
extra; fittings; two ends	4.38	3.00	4.51	0.16	-	1.99	-	6.50	nr	7.31
extra; fittings; three ends	6.13	3.00	6.31	0.16	-	1.99	-	8.30	nr	9.34
22mm	4.83	5.00	5.07	0.22	-	2.74	-	7.81	m	8.78
extra; made bends	-	-	-	0.16	-	1.99	-	1.99	nr	2.24
extra; connections to iron pipe ends; screwed										
joint	4.81	3.00	4.95	0.25	-	3.11	-	8.06	nr	9.07
extra; fittings; one end	4.50	3.00	4.63	0.13	-	1.62	-	6.25	nr	7.03
extra; fittings; two ends	7.13	3.00	7.34	0.25	-	3.11	-	10.45	nr	11.76
extra; fittings; three ends	10.31	3.00	10.62	0.25	-	3.11	-	13.73	nr	15.45
28mm	6.49	5.00	6.81	0.24	-	2.99	-	9.80	m	11.03
extra; made bends	-	-	-	0.20	-	2.49	-	2.49	nr	2.80
extra; connections to iron pipe ends; screwed										
joint	7.38	3.00	7.60	0.37	-	4.60	-	12.20	nr	13.73
extra; fittings; one end	8.44	3.00	8.69	0.19	-	2.36	-	11.06	nr	12.44
extra; fittings; two ends	12.06	3.00	12.42	0.37	-	4.60	-	17.02	nr	19.15
extra; fittings; three ends	17.38	3.00	17.90	0.37	-	4.60	-	22.50	nr	25.32
35mm	9.39	5.00	9.86	0.27	-	3.36	-	13.22	m	14.87
extra; made bends	-	-	-	0.30	-	3.73	-	3.73	nr	4.20
extra; connections to iron pipe ends; screwed										
joint	14.56	3.00	15.00	0.48	-	5.97	-	20.97	nr	23.59
extra; fittings; one end	22.38	3.00	23.05	0.24	-	2.99	-	26.04	nr	29.29
extra; fittings; two ends	23.19	3.00	23.89	0.48	-	5.97	-	29.86	nr	33.59
extra; fittings; three ends	31.50	3.00	32.45	0.48	-	5.97	-	38.42	nr	43.22
42mm	11.50	5.00	12.07	0.35	-	4.35	-	16.43	m	18.48
extra; made bends	-	-	-	0.40	-	4.98	-	4.98	nr	5.60
extra; connections to iron pipe ends; screwed										
joint	14.56	3.00	15.00	0.76	-	9.45	-	24.45	nr	27.51
extra; fittings; one end	31.29	3.00	32.23	0.38	-	4.73	-	36.96	nr	41.58
extra; fittings; two ends	34.75	3.00	35.79	0.76	-	9.45	-	45.25	nr	50.90
extra; fittings; three ends	51.25	3.00	52.79	0.76	-	9.45	-	62.24	nr	70.02
54mm	19.42	5.00	20.39	0.39	-	4.85	-	25.24	m	28.40
extra; made bends	-	-	-	0.60	-	7.46	-	7.46	nr	8.40
extra; connections to iron pipe ends; screwed										
joint	31.13	3.00	32.06	1.00	-	12.44	-	44.50	nr	50.07
extra; fittings; one end	48.06	3.00	49.50	0.50	-	6.22	-	55.72	nr	62.69
extra; fittings; two ends	55.31	3.00	56.97	1.00	-	12.44	-	69.41	nr	78.09
extra; fittings; three ends	82.19	3.00	84.66	1.00	-	12.44	-	97.10	nr	109.23
POLYBUTYLENE TUBES AND FITTINGS - Polybutylene pipes; **Hepworth Hep20 flexible plumbing system B.S.7291** **Class H; polybutylene slimline fittings**										
Pipes, flexible; in ducts										
15mm; in pipe clips to timber	1.54	10.00	1.69	0.36	-	4.48	-	6.17	m	6.94
extra; made bend	-	-	-	0.01	-	0.12	-	0.12	nr	0.14
extra; connection to copper pipe ends	1.36	2.00	1.39	0.12	-	1.49	-	2.88	nr	3.24
extra; fittings; one end	1.66	2.00	1.69	0.06	-	0.75	-	2.44	nr	2.74
extra; fittings; two ends	1.78	2.00	1.82	0.12	-	1.49	-	3.31	nr	3.72
extra; fittings; three ends	2.55	2.00	2.60	0.12	-	1.49	-	4.09	nr	4.61
22mm; in pipe clips to timber	2.39	10.00	2.63	0.40	-	4.98	-	7.61	m	8.56
extra; made bend	-	-	-	0.01	-	0.12	-	0.12	nr	0.14
extra; connection to copper pipe ends	1.80	2.00	1.84	0.30	-	3.73	-	5.57	nr	6.26
extra; fittings; one end	2.10	2.00	2.14	0.15	-	1.87	-	4.01	nr	4.51
extra; fittings; two ends	2.50	2.00	2.55	0.30	-	3.73	-	6.28	nr	7.07
extra; fittings; three ends	3.24	2.00	3.30	0.30	-	3.73	-	7.04	nr	7.92
in pipe clips fixing to masonry with screws										
15mm	1.54	10.00	1.69	0.36	-	4.48	-	6.17	m	6.94
22mm	2.39	10.00	2.63	0.40	-	4.98	-	7.61	m	8.56
POLYBUTYLENE TUBES AND FITTINGS - Polybutylene pipes; **Hepworth Hep20 flexible plumbing system B.S.7291** **Class H; polybutylene demountable fittings**										
Pipes, flexible										
15mm; in pipe clips to timber with screws	1.53	10.00	1.68	0.45	-	5.60	-	7.28	m	8.19
extra; connection to copper pipe ends	1.36	2.00	1.39	0.12	-	1.49	-	2.88	nr	3.24
extra; fittings; one end	1.63	2.00	1.66	0.06	-	0.75	-	2.41	nr	2.71
extra; fittings; two ends	1.78	2.00	1.82	0.12	-	1.49	-	3.31	nr	3.72
extra; fittings; three ends	2.55	2.00	2.60	0.12	-	1.49	-	4.09	nr	4.61
22mm; in pipe clips to timber with screws	2.38	10.00	2.62	0.48	-	5.97	-	8.59	m	9.66
extra; connection to copper pipe ends	1.80	2.00	1.84	0.30	-	3.73	-	5.57	nr	6.26
extra; fittings; one end	2.05	2.00	2.09	0.15	-	1.87	-	3.96	nr	4.45
extra; fittings; two ends	2.50	2.00	2.55	0.30	-	3.73	-	6.28	nr	7.07
extra; fittings; three ends	3.24	2.00	3.30	0.30	-	3.73	-	7.04	nr	7.92
in pipe clips fixing to masonry with screws										
15mm	1.53	10.00	1.68	0.45	-	5.60	-	7.28	m	8.19
22mm	2.38	10.00	2.62	0.48	-	5.97	-	8.59	m	9.66
Pipes, flexible										
28mm; in pipe clips to timber with screws	3.46	10.00	3.81	0.60	-	7.46	-	11.27	m	12.68
extra; fittings; two ends	5.40	2.00	5.51	0.40	-	4.98	-	10.48	nr	11.79
extra; fittings; three ends	7.53	2.00	7.68	0.40	-	4.98	-	12.66	nr	14.24
in pipe clips fixing to masonry with screws										
28mm	3.46	10.00	3.81	0.60	-	7.46	-	11.27	m	12.68

PIPED SUPPLY SYSTEMS

Labour hourly rates: (except Specialists) Craft Operatives £12.44 Labourer £7.02 Rates are national average prices. Refer to REGIONAL VARIATIONS for indicative levels of overall pricing in regions	MATERIALS			LABOUR				RATES		
	Del to Site £	Waste %	Material Cost £	Craft Optve Hrs	Lab Hrs	Labour Cost £	Sunds £	Nett Rate £	Unit	Gross Rate (+12.5%) £
S115: HOT AND COLD WATER AND GAS Cont.										
POLYETHYLENE TUBES AND FITTINGS - Polythene pipes, B.S.6572 Blue; copper alloy fittings, compression, B.S.864 Part 3, Type A										
Pipes; straight; in trenches										
20mm	0.46	10.00	0.51	0.17	-	2.11	-	2.62	m	2.95
extra; connections to copper pipe ends; compression joint	3.59	3.00	3.70	0.16	-	1.99	-	5.69	nr	6.40
extra; fittings; one end	3.50	3.00	3.61	0.08	-	1.00	-	4.60	nr	5.18
extra; fittings; two ends	4.90	3.00	5.05	0.16	-	1.99	-	7.04	nr	7.92
extra; fittings; three ends	6.70	3.00	6.90	0.16	-	1.99	-	8.89	nr	10.00
25mm	0.61	10.00	0.67	0.20	-	2.49	-	3.16	m	3.55
extra; connections to copper pipe ends; compression joint	5.45	3.00	5.61	0.25	-	3.11	-	8.72	nr	9.81
extra; fittings; one end	5.25	3.00	5.41	0.13	-	1.62	-	7.02	nr	7.90
extra; fittings; two ends	7.14	3.00	7.35	0.25	-	3.11	-	10.46	nr	11.77
extra; fittings; three ends	10.39	3.00	10.70	0.25	-	3.11	-	13.81	nr	15.54
32mm	1.02	10.00	1.12	0.24	-	2.99	-	4.11	m	4.62
extra; connections to copper pipe ends; compression joint (Type A)	9.33	3.00	9.61	0.37	-	4.60	-	14.21	nr	15.99
extra; fittings; one end	11.87	3.00	12.23	0.19	-	2.36	-	14.59	nr	16.41
extra; fittings; two ends	12.56	3.00	12.94	0.37	-	4.60	-	17.54	nr	19.73
extra; fittings; three ends	15.63	3.00	16.10	0.37	-	4.60	-	20.70	nr	23.29
50mm	2.38	10.00	2.62	0.40	-	4.98	-	7.59	m	8.54
extra; fittings; two ends	29.30	3.00	30.18	0.60	-	7.46	-	37.64	nr	42.35
extra; fittings; three ends	38.97	3.00	40.14	0.60	-	7.46	-	47.60	nr	53.55
63mm	3.41	10.00	3.75	0.54	-	6.72	-	10.47	m	11.78
extra; fittings; two ends	34.76	3.00	35.80	1.00	-	12.44	-	48.24	nr	54.27
extra; fittings; three ends	57.21	3.00	58.93	1.00	-	12.44	-	71.37	nr	80.29
UNPLASTICISED PVC TUBES AND FITTINGS - PVC-U pipes, B.S.3505 Class E; fittings, solvent welded joints, B.S.4346										
Pipes, straight										
3/8; in standard plastics pipe brackets fixing to timber with screws	1.45	5.00	1.52	0.40	-	4.98	-	6.50	m	7.31
extra; connections to iron pipe ends; screwed joint	0.81	2.00	0.83	0.12	-	1.49	-	2.32	nr	2.61
extra; fittings; one end	0.51	2.00	0.52	0.06	-	0.75	-	1.27	nr	1.42
extra; fittings; two ends	0.78	2.00	0.80	0.12	-	1.49	-	2.29	nr	2.57
extra; fittings; three ends	0.85	2.00	0.87	0.12	-	1.49	-	2.36	nr	2.65
1/2'; in standard plastics pipe brackets fixing to timber with screws	1.81	5.00	1.90	0.44	-	5.47	-	7.37	m	8.30
extra; connections to iron pipe ends; screwed joint	1.10	2.00	1.12	0.20	-	2.49	-	3.61	nr	4.06
extra; fittings; one end	0.61	2.00	0.62	0.10	-	1.24	-	1.87	nr	2.10
extra; fittings; two ends	0.88	2.00	0.90	0.20	-	2.49	-	3.39	nr	3.81
extra; fittings; three ends	1.01	2.00	1.03	0.20	-	2.49	-	3.52	nr	3.96
3/4'; in standard plastics pipe brackets fixing to timber with screws	2.24	5.00	2.35	0.52	-	6.47	-	8.82	m	9.92
extra; connections to iron pipe ends; screwed joint	1.19	2.00	1.21	0.30	-	3.73	-	4.95	nr	5.56
extra; fittings; one end	0.73	2.00	0.74	0.15	-	1.87	-	2.61	nr	2.94
extra; fittings; two ends	1.05	2.00	1.07	0.30	-	3.73	-	4.80	nr	5.40
extra; fittings; three ends	1.28	2.00	1.31	0.30	-	3.73	-	5.04	nr	5.67
1'; in standard plastics pipe brackets fixing to timber with screws	2.50	5.00	2.63	0.60	-	7.46	-	10.09	m	11.35
extra; connections to iron pipe ends; screwed joint	1.84	2.00	1.88	0.40	-	4.98	-	6.85	nr	7.71
extra; fittings; one end	0.81	2.00	0.83	0.20	-	2.49	-	3.31	nr	3.73
extra; fittings; two ends	1.45	2.00	1.48	0.40	-	4.98	-	6.46	nr	7.26
extra; fittings; three ends	1.93	2.00	1.97	0.40	-	4.98	-	6.94	nr	7.81
1 1/4'; in standard plastics pipe brackets fixing to timber with screws	3.56	5.00	3.74	0.72	-	8.96	-	12.69	m	14.28
extra; connections to iron pipe ends; screwed joint	2.59	2.00	2.64	0.60	-	7.46	-	10.11	nr	11.37
extra; fittings; one end	1.28	2.00	1.31	0.30	-	3.73	-	5.04	nr	5.67
extra; fittings; two ends	2.54	2.00	2.59	0.60	-	7.46	-	10.05	nr	11.31
extra; fittings; three ends	2.71	2.00	2.76	0.60	-	7.46	-	10.23	nr	11.51
1 1/2'; in standard plastics pipe brackets fixing to timber with screws	4.35	5.00	4.57	0.90	-	11.20	-	15.76	m	17.73
extra; connections to iron pipe ends; screwed joint	3.50	2.00	3.57	0.80	-	9.95	-	13.52	nr	15.21
extra; fittings; one end	2.15	2.00	2.19	0.40	-	4.98	-	7.17	nr	8.07
extra; fittings; two ends	3.29	2.00	3.36	0.80	-	9.95	-	13.31	nr	14.97
extra; fittings; three ends	3.94	2.00	4.02	0.80	-	9.95	-	13.97	nr	15.72
in standard plastics pipe brackets fixing to masonry with screws										
3/8'	1.45	5.00	1.52	0.46	-	5.72	-	7.24	m	8.15
1/2'	1.81	5.00	1.90	0.50	-	6.22	-	8.12	m	9.14
3/4'	2.24	5.00	2.35	0.58	-	7.22	-	9.57	m	10.76
1'	2.50	5.00	2.63	0.66	-	8.21	-	10.84	m	12.19
1 1/4'	3.56	5.00	3.74	0.78	-	9.70	-	13.44	m	15.12
1 1/2'	4.35	5.00	4.57	0.96	-	11.94	-	16.51	m	18.57
Pipes; straight; in trenches										
3/8'	0.79	5.00	0.83	0.20	-	2.49	-	3.32	m	3.73
1/2'	1.15	5.00	1.21	0.22	-	2.74	-	3.94	m	4.44
3/4'	1.61	5.00	1.69	0.24	-	2.99	-	4.68	m	5.26
1'	1.88	5.00	1.97	0.30	-	3.73	-	5.71	m	6.42
1 1/4'	2.81	5.00	2.95	0.35	-	4.35	-	7.30	m	8.22
1 1/2'	3.61	5.00	3.79	0.40	-	4.98	-	8.77	m	9.86

Labour hourly rates: (except Specialists) Craft Operatives £12.44 Labourer £7.02 Rates are national average prices. Refer to REGIONAL VARIATIONS for indicative levels of overall pricing in regions	MATERIALS			LABOUR				RATES		
	Del to Site	Waste	Material Cost	Craft Optve	Lab	Labour Cost	Sunds	Nett Rate	Unit	Gross Rate (+12.5%)
	£	%	£	Hrs	Hrs	£	£	£		£
S115: HOT AND COLD WATER AND GAS Cont.										
UNPLASTICISED OVERFLOW PIPES - PVC-U pipes and fittings; solvent welded joints; pipework self coloured white										
Pipes, straight										
19; in standard plastics pipe brackets fixing to										
timber with screws	1.11	5.00	1.17	0.30	-	3.73	-	4.90	m	5.51
extra; fittings; two ends	0.77	2.00	0.79	0.25	-	3.11	-	3.90	nr	4.38
extra; fittings; three ends	0.92	2.00	0.94	0.25	-	3.11	-	4.05	nr	4.55
extra; fittings; tank connector	1.13	2.00	1.15	0.25	-	3.11	-	4.26	nr	4.80
DUCTILE IRON PIPES AND FITTINGS - Ductile iron pipes and fittings, Tyton socketed flexible joints										
Pipes; straight; in trenches										
80mm	18.58	5.00	19.51	0.96	-	11.94	-	31.45	m	35.38
extra; bends, 90 degree	48.59	2.00	49.56	0.96	-	11.94	-	61.50	nr	69.19
extra; duckfoot bends, 90 degree	112.97	2.00	115.23	0.96	-	11.94	-	127.17	nr	143.07
extra; tees	66.57	2.00	67.90	1.44	-	17.91	-	85.81	nr	96.54
extra; hydrant tees	99.38	2.00	101.37	1.44	-	17.91	-	119.28	nr	134.19
extra; branches, 45 degree	190.74	2.00	194.55	1.44	-	17.91	-	212.47	nr	239.03
extra; flanged sockets	38.96	2.00	39.74	0.96	-	11.94	-	51.68	nr	58.14
extra; flanged spigots	36.91	2.00	37.65	0.96	-	11.94	-	49.59	nr	55.79
100mm	18.44	5.00	19.36	1.20	-	14.93	-	34.29	m	38.58
extra; bends, 90 degree	51.82	2.00	52.86	1.20	-	14.93	-	67.78	nr	76.26
extra; duckfoot bends, 90 degree	122.52	2.00	124.97	1.20	-	14.93	-	139.90	nr	157.39
extra; tees	69.74	2.00	71.13	1.80	-	22.39	-	93.53	nr	105.22
extra; hydrant tees	70.80	2.00	72.22	1.80	-	22.39	-	94.61	nr	106.43
extra; branches, 45 degree	291.98	2.00	297.82	1.80	-	22.39	-	320.21	nr	360.24
extra; flanged sockets	42.21	2.00	43.05	1.20	-	14.93	-	57.98	nr	65.23
extra; flanged spigots	39.00	2.00	39.78	1.20	-	14.93	-	54.71	nr	61.55
150mm	24.41	5.00	25.63	1.80	-	22.39	-	48.02	m	54.03
extra; bends, 90 degree	106.37	2.00	108.50	1.80	-	22.39	-	130.89	nr	147.25
extra; duckfoot bends, 90 degree	269.10	2.00	274.48	1.80	-	22.39	-	296.87	nr	333.98
extra; tees	106.01	2.00	108.13	2.70	-	33.59	-	141.72	nr	159.43
extra; hydrant tees	108.09	2.00	110.25	2.70	-	33.59	-	143.84	nr	161.82
extra; branches, 45 degree	344.25	2.00	351.13	2.70	-	33.59	-	384.72	nr	432.81
extra; flanged sockets	64.15	2.00	65.43	1.80	-	22.39	-	87.83	nr	98.80
extra; flanged spigots	45.21	2.00	46.11	1.80	-	22.39	-	68.51	nr	77.07
STOPCOCKS - Brass stopcocks										
Stopcocks; B.S.1010, crutch head; screwed and PTFE joints										
each end threaded internally										
13mm	4.54	2.00	4.63	0.30	-	3.73	-	8.36	nr	9.41
19mm	7.03	2.00	7.17	0.40	-	4.98	-	12.15	nr	13.66
25mm	16.03	2.00	16.35	0.60	-	7.46	-	23.81	nr	26.79
32mm	29.84	2.00	30.44	0.80	-	9.95	-	40.39	nr	45.44
40mm	33.24	2.00	33.90	0.90	-	11.20	-	45.10	nr	50.74
50mm	55.09	2.00	56.19	1.20	-	14.93	-	71.12	nr	80.01
Stopcocks; B.S.1010, crutch head; DZR; joints to polythene B.S.6572										
each end										
20mm	16.89	2.00	17.23	0.40	-	4.98	-	22.20	nr	24.98
25mm	26.85	2.00	27.39	0.60	-	7.46	-	34.85	nr	39.21
32mm	36.04	2.00	36.76	0.80	-	9.95	-	46.71	nr	52.55
50mm	90.10	2.00	91.90	0.90	-	11.20	-	103.10	nr	115.99
Stopcocks B.S.1010, crutch head; compression joints to copper (Type A)										
each end										
15mm	3.55	2.00	3.62	0.30	-	3.73	-	7.35	nr	8.27
22mm	6.18	2.00	6.30	0.40	-	4.98	-	11.28	nr	12.69
28mm	14.83	2.00	15.13	0.60	-	7.46	-	22.59	nr	25.41
35mm	44.56	2.00	45.45	0.80	-	9.95	-	55.40	nr	62.33
42mm	56.24	2.00	57.36	0.90	-	11.20	-	68.56	nr	77.13
54mm	84.89	2.00	86.59	1.20	-	14.93	-	101.52	nr	114.21
Stopcocks B.S.1010, crutch head; DZR; compression joints to copper (Type A)										
each end										
15mm	10.32	2.00	10.53	0.30	-	3.73	-	14.26	nr	16.04
22mm	16.94	2.00	17.28	0.40	-	4.98	-	22.25	nr	25.04
28mm	28.04	2.00	28.60	0.60	-	7.46	-	36.06	nr	40.57
35mm	51.60	2.00	52.63	0.80	-	9.95	-	62.58	nr	70.41
42mm	73.87	2.00	75.35	0.90	-	11.20	-	86.54	nr	97.36
54mm	100.61	2.00	102.62	1.20	-	14.93	-	117.55	nr	132.24
STOPCOCKS - Polybutylene stopcocks										
Stopcocks; fitted with Hep20 ends										
each end										
15mm	4.67	2.00	4.76	0.30	-	3.73	-	8.50	nr	9.56
22mm	5.59	2.00	5.70	0.40	-	4.98	-	10.68	nr	12.01
GATE VALVES - Brass gate valves										
Gate valves; B.S.5154 series B; screwed and PTFE joints										
each end threaded internally										
13mm	5.30	2.00	5.41	0.30	-	3.73	-	9.14	nr	10.28
19mm	6.46	2.00	6.59	0.40	-	4.98	-	11.57	nr	13.01

PIPED SUPPLY SYSTEMS

Labour hourly rates: (except Specialists) Craft Operatives £12.44 Labourer £7.02 Rates are national average prices. Refer to REGIONAL VARIATIONS for indicative levels of overall pricing in regions	MATERIALS			LABOUR				RATES		
	Del to Site £	Waste %	Material Cost £	Craft Optve Hrs	Lab Hrs	Labour Cost £	Sunds £	Nett Rate £	Unit	Gross Rate (+12.5%) £
S115: HOT AND COLD WATER AND GAS Cont.										
GATE VALVES - Brass gate valves Cont.										
Gate valves; B.S.5154 series B; screwed and PTFE joints Cont.										
each end threaded internally Cont.										
25mm	8.96	2.00	9.14	0.60	-	7.46	-	16.60	nr	18.68
32mm	13.09	2.00	13.35	0.80	-	9.95	-	23.30	nr	26.22
38mm	19.18	2.00	19.56	0.90	-	11.20	-	30.76	nr	34.60
51mm	27.50	2.00	28.05	1.20	-	14.93	-	42.98	nr	48.35
Gate valves; B.S.5154 series B; compression joints to copper (Type A)										
each end										
15mm	6.41	2.00	6.54	0.30	-	3.73	-	10.27	nr	11.55
22mm	7.77	2.00	7.93	0.40	-	4.98	-	12.90	nr	14.51
28mm	10.71	2.00	10.92	0.60	-	7.46	-	18.39	nr	20.69
35mm	20.16	2.00	20.56	0.80	-	9.95	-	30.52	nr	34.33
42mm	32.89	2.00	33.55	0.90	-	11.20	-	44.74	nr	50.34
54mm	50.93	2.00	51.95	1.20	-	14.93	-	66.88	nr	75.24
Gate valves; DZR; fitted with Hep20 ends										
each end										
15mm	4.37	2.00	4.46	0.30	-	3.73	-	8.19	nr	9.21
22mm	6.15	2.00	6.27	0.40	-	4.98	-	11.25	nr	12.66
BALL VALVES - Brass ball valves										
Float operated valves for low pressure; B.S.1212; copper float										
inlet threaded externally; fixing to steel										
13mm; part 1	11.78	2.00	12.02	0.40	-	4.98	-	16.99	nr	19.12
13mm; part 2	11.88	2.00	12.12	0.40	-	4.98	-	17.09	nr	19.23
19mm; part 1	21.39	2.00	21.82	0.60	-	7.46	-	29.28	nr	32.94
25mm; part 1	57.15	2.00	58.29	1.20	-	14.93	-	73.22	nr	82.37
Float operated valves for low pressure; B.S.1212; plastics float										
inlet threaded externally; fixing to steel										
13mm; part 1	7.22	2.00	7.36	0.40	-	4.98	-	12.34	nr	13.88
13mm; part 2	7.32	2.00	7.47	0.40	-	4.98	-	12.44	nr	14.00
19mm; part 1	15.18	2.00	15.48	0.60	-	7.46	-	22.95	nr	25.82
25mm; part 1	48.89	2.00	49.87	1.20	-	14.93	-	64.80	nr	72.90
Float operated valves for high pressure; B.S.1212; copper float										
inlet threaded externally; fixing to steel										
13mm; part 1	10.11	2.00	10.31	0.40	-	4.98	-	15.29	nr	17.20
13mm; part 2	11.71	2.00	11.94	0.40	-	4.98	-	16.92	nr	19.04
19mm; part 1	21.18	2.00	21.60	0.60	-	7.46	-	29.07	nr	32.70
25mm; part 1	57.15	2.00	58.29	1.20	-	14.93	-	73.22	nr	82.37
Float operated valves for high pressure; B.S.1212; plastics float										
inlet threaded externally; fixing to steel										
13mm; part 1	5.55	2.00	5.66	0.40	-	4.98	-	10.64	nr	11.97
13mm; part 2	7.16	2.00	7.30	0.40	-	4.98	-	12.28	nr	13.81
19mm; part 1	14.97	2.00	15.27	0.60	-	7.46	-	22.73	nr	25.58
25mm; part 1	48.89	2.00	49.87	1.20	-	14.93	-	64.80	nr	72.90
DRAINING TAPS - Brass draining taps										
Drain cocks; B.S.2879, square head Type2										
13mm	2.11	2.00	2.15	0.12	-	1.49	-	3.65	nr	4.10
19mm	15.21	2.00	15.51	0.16	-	1.99	-	17.50	nr	19.69
STORAGE CISTERNS - Galvanized water storage cisterns										
Cold water storage cisterns; galvanized steel; B.S.417 Part 2 Grade A with cover and byelaw 30 kit										
reference SCM45; 18 litres	97.70	-	97.70	1.80	-	22.39	-	120.09	nr	135.10
reference SCM70; 36 litres	119.60	-	119.60	1.80	-	22.39	-	141.99	nr	159.74
reference SCM90; 54 litres	132.55	-	132.55	1.80	-	22.39	-	154.94	nr	174.31
reference SCM110; 68 litres	140.25	-	140.25	2.10	-	26.12	-	166.37	nr	187.17
reference SCM135; 86 litres	148.95	-	148.95	2.10	-	26.12	-	175.07	nr	196.96
reference SCM180; 114 litres	165.75	-	165.75	2.10	-	26.12	-	191.87	nr	215.86
reference SCM230; 159 litres	207.70	-	207.70	2.40	-	29.86	-	237.56	nr	267.25
reference SCM270; 191 litres	221.70	-	221.70	2.40	-	29.86	-	251.56	nr	283.00
reference SCM320; 227 litres	241.80	-	241.80	3.00	-	37.32	-	279.12	nr	314.01
reference SCM450-1; 327 litres	284.50	-	284.50	3.60	-	44.78	-	329.28	nr	370.44
reference SCM450-2; 336 litres	276.40	-	276.40	3.60	-	44.78	-	321.18	nr	361.33
reference SCM570; 423 litres	372.40	-	372.40	4.80	-	59.71	-	432.11	nr	486.13
reference SCM680; 491 litres	420.60	-	420.60	4.80	-	59.71	-	480.31	nr	540.35
reference SCM910; 709 litres (excluding insulation)	412.40	-	412.40	4.80	-	59.71	-	472.11	nr	531.13
reference SCM1130; 841 litres (excluding insulation)	458.55	-	458.55	7.20	-	89.57	-	548.12	nr	616.63
reference SCM1600; 1227 litres (excluding insulation)	690.90	-	690.90	8.40	-	104.50	-	795.40	nr	894.82
reference SCM2270; 1727 litres (excluding insulation)	813.55	-	813.55	9.60	-	119.42	-	932.97	nr	1049.60
reference SCM2720; 2137 litres (excluding insulation)	875.85	-	875.85	10.80	-	134.35	-	1010.20	nr	1136.48
reference SCM4540; 3364 litres (excluding insulation)	1549.90	-	1549.90	13.20	-	164.21	-	1714.11	nr	1928.37

Labour hourly rates: (except Specialists) Craft Operatives £12.44 Labourer £7.02 Rates are national average prices. Refer to REGIONAL VARIATIONS for indicative levels of overall pricing in regions	MATERIALS			LABOUR				RATES		
	Del to Site £	Waste %	Material Cost £	Craft Optve Hrs	Lab Hrs	Labour Cost £	Sunds £	Nett Rate £	Unit	Gross Rate (+12.5%) £
S115: HOT AND COLD WATER AND GAS Cont.										
STORAGE CISTERNS - Galvanized water storage cisterns Cont.										
Cold water storage cisterns; galvanized steel; B.S.417 Part 2 Grade A with cover and byelaw 30 kit Cont.										
drilled holes for 13mm pipes	-	-	-	0.25	-	3.11	-	3.11	nr	3.50
drilled holes for 19mm pipes	-	-	-	0.25	-	3.11	-	3.11	nr	3.50
drilled holes for 25mm pipes	-	-	-	0.30	-	3.73	-	3.73	nr	4.20
drilled holed for 32mm pipes	-	-	-	0.30	-	3.73	-	3.73	nr	4.20
drilled holes for 38mm pipes	-	-	-	0.40	-	4.98	-	4.98	nr	5.60
drilled holes for 51mm pipes	-	-	-	0.40	-	4.98	-	4.98	nr	5.60
STORAGE CISTERNS - Plastics water storage cisterns										
Rectangular cold water storage cisterns; plastics; B.S.4213 with sealed lid and byelaw 30 kit										
18 litres	12.32	2.00	12.57	1.50	-	18.66	-	31.23	nr	35.13
114 litres	37.36	2.00	38.11	2.10	-	26.12	-	64.23	nr	72.26
182 litres	57.38	2.00	58.53	2.40	-	29.86	-	88.38	nr	99.43
227 litres	56.31	2.00	57.44	3.00	-	37.32	-	94.76	nr	106.60
drilled holes for 13mm pipes	-	-	-	0.20	-	2.49	-	2.49	nr	2.80
drilled holes for 19mm pipes	-	-	-	0.20	-	2.49	-	2.49	nr	2.80
drilled holes for 25mm pipes	-	-	-	0.25	-	3.11	-	3.11	nr	3.50
drilled holed for 32mm pipes	-	-	-	0.25	-	3.11	-	3.11	nr	3.50
drilled holes for 38mm pipes	-	-	-	0.30	-	3.73	-	3.73	nr	4.20
drilled holes for 51mm pipes	-	-	-	0.30	-	3.73	-	3.73	nr	4.20
STORAGE TANKS AND CYLINDERS - Galvanized hot water tanks										
Hot water storage tanks; galvanized steel; B.S.417 Part 2 Grade A										
reference TM114-1; 95 litres	145.40	-	145.40	3.60	-	44.78	-	190.18	nr	213.96
reference TM114-2; 95 litres	153.80	-	153.80	3.60	-	44.78	-	198.58	nr	223.41
reference TM136-1; 114 litres	156.40	-	156.40	3.60	-	44.78	-	201.18	nr	226.33
STORAGE TANKS AND CYLINDERS - Copper direct hot water cylinders; pre-insulated										
Direct hot water cylinders; copper cylinder, B.S.699 Grade 3										
reference 3; 116 litres; four bosses	104.25	2.00	106.33	2.50	-	31.10	-	137.44	nr	154.61
reference 7; 120 litres; four bosses	92.25	2.00	94.09	2.50	-	31.10	-	125.19	nr	140.84
reference 8; 144 litres; four bosses	96.50	2.00	98.43	2.50	-	31.10	-	129.53	nr	145.72
reference 9; 166 litres; four bosses	113.50	2.00	115.77	2.50	-	31.10	-	146.87	nr	165.23
STORAGE TANKS AND CYLINDERS - Copper indirect hot water cylinders; pre-insulated										
Double feed indirect hot water cylinders; copper cylinder, B.S.1566 Part 1 Grade 3										
reference 3; 114 litres; four bosses	96.00	2.00	97.92	2.50	-	31.10	-	129.02	nr	145.15
reference 7; 117 litres; four bosses	81.00	2.00	82.62	2.50	-	31.10	-	113.72	nr	127.94
reference 8; 140 litres; four bosses	104.00	2.00	106.08	2.50	-	31.10	-	137.18	nr	154.33
reference 9; 162 litres; four bosses	136.50	2.00	139.23	2.50	-	31.10	-	170.33	nr	191.62
Single feed indirect hot water cylinders; copper cylinder, B.S.1566 Part 2 Grade 3										
reference 3; 104 litres; four bosses	198.00	2.00	201.96	2.50	-	31.10	-	233.06	nr	262.19
reference 7; 108 litres; four bosses	198.00	2.00	201.96	2.50	-	31.10	-	233.06	nr	262.19
reference 8; 130 litres; four bosses	220.75	2.00	225.16	2.50	-	31.10	-	256.26	nr	288.30
reference 9; 152 litres; four bosses	270.00	2.00	275.40	2.50	-	31.10	-	306.50	nr	344.81
STORAGE TANKS AND CYLINDERS - Copper combination hot water storage units; pre-insulated										
Combination direct hot water storage units; copper unit, B.S.3198 with lid										
450mm diameter x 1200mm high; 115 litres hot water; 45 litres cold water	173.25	2.00	176.72	3.75	-	46.65	-	223.37	nr	251.29
500mm diameter x 1400mm high; 115 litres hot water; 115 litres cold water	210.50	2.00	214.71	3.75	-	46.65	-	261.36	nr	294.03
Combination double feed indirect hot water storage units; copper unit, B.S.3198 with lid										
450mm diameter x 1200mm high; 115 litres hot water; 45 litres cold water	223.00	2.00	227.46	3.75	-	46.65	-	274.11	nr	308.37
500mm diameter x 1400mm high; 115 litres hot water; 115 litres cold water	307.00	2.00	313.14	3.75	-	46.65	-	359.79	nr	404.76
Combination single feed indirect hot water storage units; copper unit, B.S.3198 with lid										
450mm diameter x 1200mm high; 115 litres hot water; 45 litres cold water	235.75	2.00	240.47	3.60	-	44.78	-	285.25	nr	320.91
500mm diameter x 1400mm high; 115 litres hot water; 115 litres cold water	342.50	2.00	349.35	3.60	-	44.78	-	394.13	nr	443.40
PIPE INSULATION - Denso tape wrapping										
Insulation to pipework around one pipe										
15mm	0.61	5.00	0.64	0.20	-	2.49	-	3.13	m	3.52
22mm	0.72	5.00	0.76	0.22	-	2.74	-	3.49	m	3.93

Labour hourly rates: (except Specialists) Craft Operatives £12.44 Labourer £7.02 Rates are national average prices. Refer to REGIONAL VARIATIONS for indicative levels of overall pricing in regions	MATERIALS			LABOUR				RATES		
	Del to Site	Waste	Material Cost	Craft Optve	Lab	Labour Cost	Sunds	Nett Rate	Unit	Gross Rate (+12.5%)
	£	%	£	Hrs	Hrs	£	£	£		£
S115: HOT AND COLD WATER AND GAS Cont.										
PIPE INSULATION - Denso tape wrapping Cont.										
Insulation to pipework Cont.										
around one pipe Cont.										
28mm	0.84	5.00	0.88	0.24	–	2.99	–	3.87	m	4.35
35mm	1.22	5.00	1.28	0.27	–	3.36	–	4.64	m	5.22
42mm	1.67	5.00	1.75	0.30	–	3.73	–	5.49	m	6.17
54mm	2.16	5.00	2.27	0.40	–	4.98	–	7.24	m	8.15
PIPE INSULATION - Thermal insulation; glass fibre preformed lagging, butt joints in the running length; secured with metal bands										
19mm thick insulation to copper pipework										
around one pipe										
15mm	3.08	3.00	3.17	0.20	–	2.49	–	5.66	m	6.37
22mm	3.24	3.00	3.34	0.26	–	3.23	–	6.57	m	7.39
28mm	3.44	3.00	3.54	0.30	–	3.73	–	7.28	m	8.18
35mm	3.66	3.00	3.77	0.36	–	4.48	–	8.25	m	9.28
42mm	4.01	3.00	4.13	0.40	–	4.98	–	9.11	m	10.24
54mm	4.60	3.00	4.74	0.50	–	6.22	–	10.96	m	12.33
25mm thick insulation to copper pipework										
around one pipe										
15mm	3.44	3.00	3.54	0.24	–	2.99	–	6.53	m	7.34
22mm	3.53	3.00	3.64	0.26	–	3.23	–	6.87	m	7.73
28mm	3.78	3.00	3.89	0.30	–	3.73	–	7.63	m	8.58
35mm	4.19	3.00	4.32	0.36	–	4.48	–	8.79	m	9.89
42mm	4.54	3.00	4.68	0.40	–	4.98	–	9.65	m	10.86
54mm	5.33	3.00	5.49	0.50	–	6.22	–	11.71	m	13.17
PIPE INSULATION - Thermal insulation; foamed polyurethane preformed lagging, butt joints in the running length; secured with adhesive bands										
13mm thick insulation to copper pipework										
around one pipe										
15mm	1.99	3.00	2.05	0.20	–	2.49	–	4.54	m	5.10
22mm	2.38	3.00	2.45	0.22	–	2.74	–	5.19	m	5.84
28mm	2.73	3.00	2.81	0.24	–	2.99	–	5.80	m	6.52
35mm	2.98	3.00	3.07	0.27	–	3.36	–	6.43	m	7.23
42mm	3.43	3.00	3.53	0.30	–	3.73	–	7.26	m	8.17
54mm	4.50	3.00	4.63	0.40	–	4.98	–	9.61	m	10.81
INSULATION JACKETS AND LAGGING UNITS FOR CYLINDERS AND TANKS - Thermal insulation; glass fibre filled insulating jacket in strips, white pvc covering both sides; secured with metal straps and wire holder ring at top										
75mm thick insulation to equipment										
sides and tops of cylinders; overall size										
400mm diameter x 1050mm high	10.75	5.00	11.29	1.00	–	12.44	–	23.73	nr	26.69
450mm diameter x 900mm high	8.75	5.00	9.19	1.00	–	12.44	–	21.63	nr	24.33
450mm diameter x 1050mm high	10.75	5.00	11.29	1.00	–	12.44	–	23.73	nr	26.69
450mm diameter x 1200mm high	11.50	5.00	12.07	1.00	–	12.44	–	24.52	nr	27.58
FUEL OIL STORAGE TANKS - Steel fuel oil storage tanks										
Oil storage tanks; carbon steel tank, B.S.799 Part 5 Type III, rectangular; 457mm diameter manhole cover, oil tight washer and screwed socket for fill, vent, sludge and draw off; painted one coat black bituminous paint										
14 gauge; 1520 x 610 x 1220mm; 1130 litre capacity	147.20	–	147.20	4.25	–	52.87	–	200.07	nr	225.08
14 gauge; 1830 x 610 x 1220mm; 1360 litre capacity	148.65	–	148.65	4.50	–	55.98	–	204.63	nr	230.21
12 gauge; 1830 x 1220 x 1220mm; 2730 litre capacity	227.15	–	227.15	5.00	–	62.20	–	289.35	nr	325.52
1/8' plate; 2440 x 1520 x 1220mm; 4550 litre capacity	575.60	–	575.60	5.50	–	68.42	–	644.02	nr	724.52

Labour hourly rates: (except Specialists) Craft Operatives £12.44 Labourer £7.02 Rates are national average prices. Refer to REGIONAL VARIATIONS for indicative levels of overall pricing in regions	MATERIALS			LABOUR				RATES		
	Del to Site £	Waste %	Material Cost £	Craft Optve Hrs	Lab Hrs	Labour Cost £	Sunds £	Nett Rate £	Unit	Gross Rate (+12.5%) £
T10: GAS/OIL FIRED BOILERS										
Gas flue pipes comprising galvanised steel outer skin and aluminium inner skin with air space between, B.S.715; socketed joints										
Pipes; straight										
100mm	13.56	2.00	13.83	0.80	-	9.95	-	23.78	m	26.76
extra; connections to appliance	3.37	2.00	3.44	0.60	-	7.46	-	10.90	nr	12.26
extra; adjustable pipes	7.94	2.00	8.10	0.60	-	7.46	-	15.56	nr	17.51
extra; terminals	10.34	2.00	10.55	0.60	-	7.46	-	18.01	nr	20.26
extra; bends, 90 degrees	9.06	2.00	9.24	0.60	-	7.46	-	16.71	nr	18.79
extra; bends, 45 degrees	9.06	2.00	9.24	0.60	-	7.46	-	16.71	nr	18.79
extra; tees	20.11	2.00	20.51	0.60	-	7.46	-	27.98	nr	31.47
125mm	16.25	2.00	16.57	0.96	-	11.94	-	28.52	m	32.08
extra; connections to appliance	3.88	2.00	3.96	0.80	-	9.95	-	13.91	nr	15.65
extra; adjustable pipes	8.91	2.00	9.09	0.80	-	9.95	-	19.04	nr	21.42
extra; terminals	11.35	2.00	11.58	0.80	-	9.95	-	21.53	nr	24.22
extra; bends, 90 degrees	10.71	2.00	10.92	0.80	-	9.95	-	20.88	nr	23.49
extra; bends, 45 degrees	10.71	2.00	10.92	0.80	-	9.95	-	20.88	nr	23.49
extra; tees	21.69	2.00	22.12	0.80	-	9.95	-	32.08	nr	36.09
150mm	18.14	2.00	18.50	1.20	-	14.93	-	33.43	m	37.61
extra; connections to appliance	4.29	2.00	4.38	1.20	-	14.93	-	19.30	nr	21.72
extra; adjustable pipes	11.25	2.00	11.48	1.20	-	14.93	-	26.40	nr	29.70
extra; terminals	14.57	2.00	14.86	1.20	-	14.93	-	29.79	nr	33.51
extra; bends, 90 degrees	13.41	2.00	13.68	1.20	-	14.93	-	28.61	nr	32.18
extra; bends, 45 degrees	13.41	2.00	13.68	1.20	-	14.93	-	28.61	nr	32.18
extra; tees	22.59	2.00	23.04	1.20	-	14.93	-	37.97	nr	42.72
Flue supports; galvanised steel wall bands fixing to masonry with screws										
for 100mm pipes	4.66	2.00	4.75	0.20	-	2.49	-	7.24	nr	8.15
for 125mm pipes	4.96	2.00	5.06	0.20	-	2.49	-	7.55	nr	8.49
for 150mm pipes	6.22	2.00	6.34	0.20	-	2.49	-	8.83	nr	9.94
Fire stop spacers										
for 100mm pipes	1.93	2.00	1.97	0.30	-	3.73	-	5.70	nr	6.41
for 125mm pipes	1.93	2.00	1.97	0.40	-	4.98	-	6.94	nr	7.81
for 150mm pipes	2.20	2.00	2.24	0.60	-	7.46	-	9.71	nr	10.92
Gas fired boilers										
Gas fired boilers for central heating and hot water; automatically controlled by thermostat with electrical control, gas governor and flame failure device										
free standing boiler; conventional flue; approximate output rating B Th U per hour										
40000	374.75	1.00	378.50	6.00	-	74.64	-	453.14	nr	509.78
50000	392.00	1.00	395.92	6.00	-	74.64	-	470.56	nr	529.38
60000	417.00	1.00	421.17	6.00	-	74.64	-	495.81	nr	557.79
80000	491.00	1.00	495.91	7.00	-	87.08	-	582.99	nr	655.86
100000	693.75	1.00	700.69	9.00	-	111.96	-	812.65	nr	914.23
free standing boiler; balanced flue; approximate output rating B Th U per hour										
40000	463.00	1.00	467.63	7.00	-	87.08	-	554.71	nr	624.05
50000	480.50	1.00	485.31	7.00	-	87.08	-	572.38	nr	643.93
60000	515.75	1.00	520.91	7.00	-	87.08	-	607.99	nr	683.99
80000	707.25	1.00	714.32	7.00	-	87.08	-	801.40	nr	901.58
100000	900.75	1.00	909.76	9.00	-	111.96	-	1021.72	nr	1149.43
wall mounted boiler; conventional flue; approximate output rating B Th U per hour										
40000	383.75	1.00	387.59	6.50	-	80.86	-	468.45	nr	527.00
50000	402.25	1.00	406.27	6.50	-	80.86	-	487.13	nr	548.02
60000	512.00	1.00	517.12	6.50	-	80.86	-	597.98	nr	672.73
wall mounted boiler; balanced flue; approximate output rating B Th U per hour										
40000	398.25	1.00	402.23	7.50	-	93.30	-	495.53	nr	557.47
50000	452.25	1.00	456.77	7.50	-	93.30	-	550.07	nr	618.83
60000	557.00	1.00	562.57	7.50	-	93.30	-	655.87	nr	737.85

Labour hourly rates: (except Specialists) Craft Operatives £12.44 Labourer £7.02 Rates are national average prices. Refer to REGIONAL VARIATIONS for indicative levels of overall pricing in regions	MATERIALS			LABOUR				RATES		
	Del to Site	Waste	Material Cost	Craft Optve	Lab	Labour Cost	Sunds	Nett Rate	Unit	Gross Rate (+12.5%)
	£	%	£	Hrs	Hrs	£	£	£		£

T10: GAS/OIL FIRED BOILERS Cont.

Oil fired boilers

Oil fired boilers for central heating and hot water; automatically controlled by thermostat, with electrical control box
 free standing boiler; conventional flue; approximate output rating BThU per hour

	Del to Site	Waste	Material Cost	Craft Optve	Lab	Labour Cost	Sunds	Nett Rate	Unit	Gross Rate
40000 - 48000	768.00	1.00	775.68	6.00	-	74.64	-	850.32	nr	956.61
50000 - 65000	801.00	1.00	809.01	6.00	-	74.64	-	883.65	nr	994.11
70000 - 85000	913.50	1.00	922.63	6.00	-	74.64	-	997.27	nr	1121.93
88000 - 110000	1001.75	1.00	1011.77	7.50	-	93.30	-	1105.07	nr	1243.20

T11: COAL FIRED BOILERS

Equipment

Solid fuel boilers for central heating and hot water; thermostatically controlled; with tools
 free standing boiler, gravity feed; approximate output rating B Th U per hour

	Del to Site	Waste	Material Cost	Craft Optve	Lab	Labour Cost	Sunds	Nett Rate	Unit	Gross Rate
45000	1210.00	1.00	1222.10	6.00	-	74.64	-	1296.74	nr	1458.83
60000	1410.00	1.00	1424.10	6.00	-	74.64	-	1498.74	nr	1686.08
80000	1757.00	1.00	1774.57	7.50	-	93.30	-	1867.87	nr	2101.35
100000	2026.00	1.00	2046.26	9.00	-	111.96	-	2158.22	nr	2428.00

T31: LOW TEMPERATURE HOT WATER HEATING

Accelerator pumps

Variable head accelerator pumps for forced central heating; small bore indirect systems
 BSP connections; with valves

	Del to Site	Waste	Material Cost	Craft Optve	Lab	Labour Cost	Sunds	Nett Rate	Unit	Gross Rate
20mm	43.50	2.00	44.37	1.00	-	12.44	-	56.81	nr	63.91
25mm	44.75	2.00	45.65	1.00	-	12.44	-	58.09	nr	65.35

Radiators

Pressed steel radiators single panel with convector; air cock, plain plug
 450mm high; fixing brackets to masonry with screws

	Del to Site	Waste	Material Cost	Craft Optve	Lab	Labour Cost	Sunds	Nett Rate	Unit	Gross Rate
600mm long	18.81	1.00	19.00	3.75	-	46.65	-	65.65	nr	73.85
900mm long	28.06	1.00	28.34	3.75	-	46.65	-	74.99	nr	84.36
1200mm long	36.63	1.00	37.00	3.75	-	46.65	-	83.65	nr	94.10
1600mm long	47.44	1.00	47.91	3.75	-	46.65	-	94.56	nr	106.38
2400mm long	76.81	1.00	77.58	3.75	-	46.65	-	124.23	nr	139.76
600mm high; fixing brackets to masonry with screws										
600mm long	24.13	1.00	24.37	3.75	-	46.65	-	71.02	nr	79.90
900mm long	35.50	1.00	35.85	3.75	-	46.65	-	82.50	nr	92.82
1200mm long	46.50	1.00	46.97	3.75	-	46.65	-	93.61	nr	105.32
1600mm long	60.63	1.00	61.24	3.75	-	46.65	-	107.89	nr	121.37
2400mm long	98.75	1.00	99.74	3.75	-	46.65	-	146.39	nr	164.69
700mm high; fixing brackets to masonry with screws										
600mm long	28.75	1.00	29.04	3.75	-	46.65	-	75.69	nr	85.15
900mm long	41.75	1.00	42.17	3.75	-	46.65	-	88.82	nr	99.92
1200mm long	52.75	1.00	53.28	3.75	-	46.65	-	99.93	nr	112.42
1600mm long	77.50	1.00	78.28	3.75	-	46.65	-	124.93	nr	140.54
2400mm long	113.63	1.00	114.77	3.75	-	46.65	-	161.42	nr	181.59

Pressed steel radiators; double panel with single convector; air cock, plain plug
 450mm high; fixing brackets to masonry with screws

	Del to Site	Waste	Material Cost	Craft Optve	Lab	Labour Cost	Sunds	Nett Rate	Unit	Gross Rate
600mm long	31.69	1.00	32.01	3.75	-	46.65	-	78.66	nr	88.49
900mm long	47.25	1.00	47.72	3.75	-	46.65	-	94.37	nr	106.17
1200mm long	61.81	1.00	62.43	3.75	-	46.65	-	109.08	nr	122.71
1600mm long	81.06	1.00	81.87	3.75	-	46.65	-	128.52	nr	144.59
2400mm long	135.56	1.00	136.92	3.75	-	46.65	-	183.57	nr	206.51
600mm high; fixing brackets to masonry with screws										
600mm long	39.94	1.00	40.34	3.75	-	46.65	-	86.99	nr	97.86
900mm long	59.25	1.00	59.84	3.75	-	46.65	-	106.49	nr	119.80
1200mm long	77.50	1.00	78.28	3.75	-	46.65	-	124.93	nr	140.54
1600mm long	115.75	1.00	116.91	3.75	-	46.65	-	163.56	nr	184.00
2400mm long	170.25	1.00	171.95	3.75	-	46.65	-	218.60	nr	245.93
700mm high; fixing brackets to masonry with screws										
600mm long	45.31	1.00	45.76	3.75	-	46.65	-	92.41	nr	103.96
900mm long	66.25	1.00	66.91	3.75	-	46.65	-	113.56	nr	127.76
1200mm long	100.19	1.00	101.19	3.75	-	46.65	-	147.84	nr	166.32
1600mm long	131.31	1.00	132.62	3.75	-	46.65	-	179.27	nr	201.68
2400mm long	192.94	1.00	194.87	3.75	-	46.65	-	241.52	nr	271.71

Brass valves for radiators; self colour

Valves for radiators, B.S.2767; wheel head
 inlet for copper; angle pattern

	Del to Site	Waste	Material Cost	Craft Optve	Lab	Labour Cost	Sunds	Nett Rate	Unit	Gross Rate
15mm	5.31	2.00	5.42	0.30	-	3.73	-	9.15	nr	10.29
22mm	7.22	2.00	7.36	0.40	-	4.98	-	12.34	nr	13.88
inlet for copper; straight pattern										
15mm	5.96	2.00	6.08	0.30	-	3.73	-	9.81	nr	11.04
22mm	8.60	2.00	8.77	0.40	-	4.98	-	13.75	Unr	15.47

Valves for radiators, B.S.2767; lock shield
 inlet for copper; angle pattern

	Del to Site	Waste	Material Cost	Craft Optve	Lab	Labour Cost	Sunds	Nett Rate	Unit	Gross Rate
15mm	5.31	2.00	5.42	0.30	-	3.73	-	9.15	nr	10.29
22mm	7.22	2.00	7.36	0.40	-	4.98	-	12.34	nr	13.88

Labour hourly rates: (except Specialists) Craft Operatives £12.44 Labourer £7.02 Rates are national average prices. Refer to REGIONAL VARIATIONS for indicative levels of overall pricing in regions	MATERIALS			LABOUR				RATES		
	Del to Site £	Waste %	Material Cost £	Craft Optve Hrs	Lab Hrs	Labour Cost £	Sunds £	Nett Rate £	Unit	Gross Rate (+12.5%) £
T31: LOW TEMPERATURE HOT WATER HEATING Cont.										
Brass valves for radiators; self colour Cont.										
Valves for radiators, B.S.2767; lock shield Cont. inlet for copper; straight pattern										
15mm	5.96	2.00	6.08	0.30	-	3.73	-	9.81	nr	11.04
22mm	8.60	2.00	8.77	0.40	-	4.98	-	13.75	nr	15.47
Brass valves for radiators; chromium plated										
Valves for radiators, B.S.2767; wheel head inlet for copper; angle pattern										
15mm	7.22	2.00	7.36	0.30	-	3.73	-	11.10	nr	12.48
22mm	9.67	2.00	9.86	0.40	-	4.98	-	14.84	nr	16.69
inlet for copper; straight pattern										
15mm	8.60	2.00	8.77	0.30	-	3.73	-	12.50	nr	14.07
22mm	11.19	2.00	11.41	0.40	-	4.98	-	16.39	nr	18.44
Valves for radiators, B.S.2767; lock shield inlet for copper; angle pattern										
15mm	7.22	2.00	7.36	0.30	-	3.73	-	11.10	nr	12.48
22mm	9.67	2.00	9.86	0.40	-	4.98	-	14.84	nr	16.69
inlet for copper; straight pattern										
15mm	8.60	2.00	8.77	0.30	-	3.73	-	12.50	nr	14.07
22mm	11.19	2.00	11.41	0.40	-	4.98	-	16.39	nr	18.44
Brass thermostatic valves for radiators; chromium plated										
Thermostatic valves for radiators, one piece; ABS plastics head inlet for copper; angle pattern										
15mm	13.92	2.00	14.20	0.30	-	3.73	-	17.93	nr	20.17
inlet for copper; straight pattern										
15mm	13.92	2.00	14.20	0.30	-	3.73	-	17.93	nr	20.17
inlet for iron; angle pattern										
19mm	17.65	2.00	18.00	0.40	-	4.98	-	22.98	nr	25.85
Skirting heaters										
Pressed metal skirting heaters with fins on copper tube; straight										
900mm long	28.27	2.00	28.84	1.50	-	18.66	-	47.50	nr	53.43
1200mm long	37.43	2.00	38.18	2.00	-	24.88	-	63.06	nr	70.94
1500mm long	41.75	2.00	42.59	2.50	-	31.10	-	73.69	nr	82.90
1800mm long	46.06	2.00	46.98	3.00	-	37.32	-	84.30	nr	94.84
extra; internal corners	5.29	2.00	5.40	0.40	-	4.98	-	10.37	nr	11.67
extra; external corners	5.29	2.00	5.40	0.40	-	4.98	-	10.37	nr	11.67
extra; end stops	4.86	2.00	4.96	0.40	-	4.98	-	9.93	nr	11.17
extra; valve boxes	14.35	2.00	14.64	0.40	-	4.98	-	19.61	nr	22.06
T32: LOW TEMPERATURE HOT WATER HEATING (SMALL SCALE)										
Central heating installations - indicative prices										
Note the following are indicative prices for installation in two storey three bedroomed dwellings with a floor area of approximately 85m2 Installation; boiler, copper piping, pressed steel radiators, heated towel rail; providing hot water to sink , bath, lavatory basin; complete with all necessary pumps, controls etc										
solid fuel fired	-	-	Spclist	-	-	Spclist	-	3325.00	it	3740.63
gas fired	-	-	Spclist	-	-	Spclist	-	3100.00	it	3487.50
oil fired, including oil storage tank	-	-	Spclist	-	-	Spclist	-	3800.00	it	4275.00

MECHANICAL HEATING/COOLING/REFRIGERATION SYSTEMS

Estimating for Builders and Quantity Surveyors

R D Buchan, F W Fleming, J R Kelly

Written for students taking courses in building and surveying at HNC/D and BSc level, this textbook explains the calculation of rates for the items in bills of quantities, based on SMM7. For convenience of use in practice, the arrangement is by individual topic or trade with appropriate reference to subsections of SMM7. Care has been taken, in both discussion and examples; to emphasize the different approaches used when pricing the various types of work. The worked examples reflect both traditional and up-to-date technology.

For all the trades included, a core of the most useful examples is provided, with particular attention paid to those areas that require rather more detailed explanation than has been provided in some more summary texts.

A chapter on computerized estimating is complemented by the Appendix which shows how a spreadsheet program can be developed for the calculation of the cost of mortars etc. Chapters on tendering strategy are included to set estimating in its professional context.

£17.99, Paperback, 1991, 300pp, isbn 0 7506 0041 1

TO ORDER: CREDIT CARD HOTLINE - 01865 888180
BY MAIL: Technical Marketing Dept., Butterworth-Heinemann, FREEPOST, Oxford OX2 8BR
BY FAX: Heinemann Customer Services – 01865 314091
BY EMAIL: Send orders to: bhuk.orders@repp.co.uk

Ventilation/Air conditioning systems

Labour hourly rates: (except Specialists) Craft Operatives £14.38 Labourer £7.02 Rates are national average prices. Refer to REGIONAL VARIATIONS for indicative levels of overall pricing in regions	MATERIALS			LABOUR				RATES		
	Del to Site £	Waste %	Material Cost £	Craft Optve Hrs	Lab Hrs	Labour Cost £	Sunds £	Nett Rate £	Unit	Gross Rate (+12.5%) £
U10: GENERAL VENTILATION										
Ventilating fans										
Note										
the following prices exclude cutting holes in glass										
Equipment										
Extract fans, window mounted; shutter and incorporated switch unit										
152mm diameter	271.51	2.50	278.30	1.50	-	21.57	-	299.87	nr	337.35
229mm diameter	323.02	2.50	331.10	1.80	-	25.88	-	356.98	nr	401.60
305mm diameter	385.42	2.50	395.06	1.80	-	25.88	-	420.94	nr	473.56

Regional Factors

Scotland
0.95

Northern Ireland
0.75

Northern
0.94

Yorks & Humber
0.94

North West
1.00

East Midlands
0.93

West Midlands
0.94

East Anglia
0.96

Wales
0.94

South East
1.06

G.L 1.18

South West
0.99

Laxton's Building Price Book is based upon national average prices (=1.00). Indicative levels of tender pricing in regions as at the second quarter 1999 are given on this map and the factors shown may be applied to adjust overall pricing.

MAJOR WORKS – TENDER VALUES	SMALL WORKS – TENDER VALUES
The measured rates are based on contracts valued in the range of £250,000 to £1,000.00. As a guide to pricing works of larger value and for cost planning or budgetary purposes, the following adjustments may be applied to overall contract values. Contract Value £1,000,000 to £2,000,000.........deduct 2.5% £2,000,000 to £3,000,000.........deduct 5.0% £3,000,000 to £5,000,000.........deduct 7.5%	The measured rates are based on contracts valued in the range of £25,000 to £75,000. As a guide to pricing works of smaller or larger value and for cost planning or budgetary purposes, the following adjustments may be applied to overall contract values. Contract Value £5,000 to £15,000.............................add 20.0% £15,000 to £25,000............................add 10.0% £25,000 to £75,000.......................rate as shown £75,000 to £100,000.......................deduct 5.0% £100,000 to £150,000.....................deduct7.5% £150,000 to £250,000.....................deduct10.%

Electrical supply/power/lighting systems

Labour hourly rates: (except Specialists) Craft Operatives £14.38 Labourer £7.02 Rates are national average prices. Refer to REGIONAL VARIATIONS for indicative levels of overall pricing in regions	MATERIALS			LABOUR				RATES		
	Del to Site	Waste	Material Cost	Craft Optve	Lab	Labour Cost	Sunds	Nett Rate	Unit	Gross Rate (+12.5%)
	£	%	£	Hrs	Hrs	£	£	£		£
V41: STREET/AREA/FLOOD LIGHTING										
Street lighting columns and set in ground; excavate hole, remove surplus spoil, filling with concrete mix ST4, working around base		.								
Aluminium; B.S.5649, nominal mounting height										
4.88m	246.56	–	246.56	5.50	10.95	155.96	22.97	425.49	nr	478.68
8.00m	607.34	–	607.34	6.50	12.65	182.27	35.73	825.34	nr	928.51
10.00m	771.74	–	771.74	7.25	14.50	206.04	77.83	1055.62	nr	1187.57
12.00m	1081.87	–	1081.87	8.00	16.00	227.36	77.83	1387.06	nr	1560.44
Steel; B.S.5649, nominal mounting height										
4.00m	198.00	–	198.00	5.50	10.95	155.96	22.97	376.93	nr	424.05
5.00m	204.03	–	204.03	6.50	12.65	182.27	35.73	422.03	nr	474.79
6.00m	251.44	–	251.44	7.25	14.50	206.04	77.83	535.32	nr	602.23
8.00m	373.14	–	373.14	8.00	16.00	227.36	77.83	678.33	nr	763.12
GENERAL LIGHTING AND POWER - COMMERCIAL										
Steel trunking										
Straight lighting trunking; PVC lid										
50 x 50mm	5.57	5.00	5.85	0.40	–	5.75	0.50	12.10	m	13.61
extra; end cap	2.15	–	2.15	0.08	–	1.15	0.25	3.55	nr	3.99
extra; internal angle	10.55	–	10.55	0.25	–	3.60	0.25	14.40	nr	16.19
extra; external angle	10.55	–	10.55	0.25	–	3.60	0.25	14.40	nr	16.19
extra; tee	11.30	–	11.30	0.25	–	3.60	0.25	15.15	nr	17.04
extra; four way intersection	11.70	–	11.70	0.40	–	5.75	0.25	17.70	nr	19.91
Straight dado trunking, two compartment; steel lid										
150 x 30mm	19.92	5.00	20.92	0.45	–	6.47	0.50	27.89	m	31.37
extra; end cap	4.80	–	4.80	0.08	–	1.15	0.25	6.20	nr	6.98
extra; internal angle	22.25	–	22.25	0.35	–	5.03	0.25	27.53	nr	30.97
extra; external angle	22.25	–	22.25	0.35	–	5.03	0.25	27.53	nr	30.97
extra; single socket plate	8.75	–	8.75	0.16	–	2.30	0.25	11.30	nr	12.71
extra; twin socket plate	9.30	–	9.30	0.16	–	2.30	0.25	11.85	nr	13.33
Straight dado trunking, three compartment; steel lid										
200 x 38mm	26.37	5.00	27.69	0.50	–	7.19	0.75	35.63	m	40.08
extra; end cap	5.25	–	5.25	0.08	–	1.15	0.25	6.65	nr	7.48
extra; internal angle	27.50	–	27.50	0.40	–	5.75	0.25	33.50	nr	37.69
extra; external angle	27.50	–	27.50	0.40	–	5.75	0.25	33.50	nr	37.69
extra; single socket plate	9.05	–	9.05	0.16	–	2.30	0.25	11.60	nr	13.05
extra; twin socket plate	9.70	–	9.70	0.16	–	2.30	0.25	12.25	nr	13.78
Straight skirting trunking; two compartments										
200 x 38mm	26.32	5.00	27.64	0.50	–	7.19	0.75	35.58	m	40.02
extra; internal bend	26.60	–	26.60	0.40	–	5.75	0.25	32.60	nr	36.68
extra; external bend	26.60	–	26.60	0.40	–	5.75	0.25	32.60	nr	36.68
Straight underfloor trunking; two compartments										
150 x 25mm	17.92	5.00	18.82	0.70	–	10.07	–	28.88	m	32.49
extra; floor junction boxes; adjustable frame and trap	79.20	–	79.20	1.85	–	26.60	–	105.80	nr	119.03
extra; horizontal bend	17.50	–	17.50	0.50	–	7.19	–	24.69	nr	27.78
PVC trunking; white										
Straight mini-trunking; clip on lid										
16 x 16mm	1.97	5.00	2.07	0.25	–	3.60	0.50	6.16	m	6.93
extra; end cap	0.54	–	0.54	0.06	–	0.86	0.13	1.53	nr	1.72
extra; internal bend	0.55	–	0.55	0.20	–	2.88	0.13	3.56	nr	4.00
extra; external bend	0.55	–	0.55	0.20	–	2.88	0.13	3.56	nr	4.00
extra; flat bend	0.55	–	0.55	0.20	–	2.88	0.13	3.56	nr	4.00
extra; equal tee	0.92	–	0.92	0.25	–	3.60	0.13	4.64	nr	5.23
25 x 16mm	2.35	5.00	2.47	0.25	–	3.60	0.13	6.19	m	6.97
extra; end cap	0.55	–	0.55	0.06	–	0.86	0.13	1.54	nr	1.74
extra; internal bend	0.57	–	0.57	0.20	–	2.88	0.13	3.58	nr	4.02
extra; external bend	0.57	–	0.57	0.20	–	2.88	0.13	3.58	nr	4.02
extra; flat bend	0.57	–	0.57	0.20	–	2.88	0.13	3.58	nr	4.02

ELECTRICAL

Labour hourly rates: (except Specialists) Craft Operatives £14.38 Labourer £7.02 Rates are national average prices. Refer to REGIONAL VARIATIONS for indicative levels of overall pricing in regions	MATERIALS			LABOUR				RATES		
	Del to Site	Waste	Material Cost	Craft Optve	Lab	Labour Cost	Sunds	Nett Rate	Unit	Gross Rate (+12.5%)
	£	%	£	Hrs	Hrs	£	£	£		£
GENERAL LIGHTING AND POWER – COMMERCIAL Cont.										
PVC trunking; white Cont.										
Straight mini-trunking; clip on lid Cont.										
extra; equal tee	0.92	–	0.92	0.25	–	3.60	0.13	4.64	nr	5.23
40 x 16mm	2.93	5.00	3.08	0.30	–	4.31	0.13	7.52	m	8.46
extra; end cap	0.56	–	0.56	0.06	–	0.86	0.13	1.55	nr	1.75
extra; internal bend	0.67	–	0.67	0.25	–	3.60	0.13	4.39	nr	4.94
extra; external bend	0.67	–	0.67	0.25	–	3.60	0.13	4.39	nr	4.94
extra; flat bend	0.67	–	0.67	0.25	–	3.60	0.13	4.39	nr	4.94
extra; equal tee	1.03	–	1.03	0.30	–	4.31	0.13	5.47	nr	6.16
Straight dado trunking, three compartment; clip on lid										
145 x 40mm	23.20	5.00	24.36	0.40	–	5.75	0.50	30.61	m	34.44
extra; end caps	1.50	–	1.50	0.08	–	1.15	0.25	2.90	nr	3.26
extra; internal bend	5.30	–	5.30	0.30	–	4.31	0.25	9.86	nr	11.10
extra; external bend	5.30	–	5.30	0.30	–	4.31	0.25	9.86	nr	11.10
extra; tee	63.00	–	63.00	0.40	–	5.75	0.25	69.00	nr	77.63
extra; flat angle	46.40	–	46.40	0.30	–	4.31	0.25	50.96	nr	57.33
extra; mini trunking adaptor	1.45	–	1.45	0.20	–	2.88	0.25	4.58	nr	5.15
230 x 40mm	35.84	5.00	37.63	0.45	–	6.47	0.50	44.60	m	50.18
extra; end caps	4.75	–	4.75	0.08	–	1.15	0.25	6.15	nr	6.92
extra; internal bend	9.96	–	9.96	0.30	–	4.31	0.25	14.52	nr	16.34
extra; external bend	9.96	–	9.96	0.30	–	4.31	0.25	14.52	nr	16.34
extra; mini trunking adaptor	1.48	–	1.48	0.20	–	2.88	0.25	4.61	nr	5.18
Straight skirting trunking, three compartment; clip on lid										
100 x 40mm	11.00	5.00	11.55	0.35	–	5.03	0.50	17.08	m	19.22
extra; end cap	2.10	–	2.10	0.06	–	0.86	0.25	3.21	nr	3.61
extra; internal corner	2.70	–	2.70	0.25	–	3.60	0.25	6.54	nr	7.36
extra; external corner	2.70	–	2.70	0.25	–	3.60	0.25	6.54	nr	7.36
extra; mini trunking adaptor	1.90	–	1.90	0.16	–	2.30	0.25	4.45	nr	5.01
extra; socket adaptor, one gang	4.25	–	4.25	0.25	–	3.60	0.25	8.10	nr	9.11
extra; socket adaptor, two gang	6.10	–	6.10	0.25	–	3.60	0.25	9.95	nr	11.19
PVC heavy gauge conduit and fittings; push fit joints; spacer bar saddles at 600mm centres										
Conduits; straight										
20mm diameter; plugging to masonry; to surfaces	3.12	10.00	3.43	0.17	–	2.44	0.42	6.30	m	7.08
extra; small circular terminal boxes	4.55	2.50	4.66	0.10	–	1.44	0.27	6.37	nr	7.17
extra; small circular angle boxes	3.88	2.50	3.98	0.11	–	1.58	0.27	5.83	nr	6.56
extra; small circular three way boxes	4.48	2.50	4.59	0.13	–	1.87	0.27	6.73	nr	7.57
extra; small circular through way boxes	3.88	2.50	3.98	0.11	–	1.58	0.27	5.83	nr	6.56
25mm diameter; plugging to masonry; to surfaces	4.12	10.00	4.53	0.18	–	2.59	0.42	7.54	m	8.48
extra; small circular terminal boxes	5.66	2.50	5.80	0.10	–	1.44	0.27	7.51	nr	8.45
extra; small circular angle boxes	5.14	2.50	5.27	0.11	–	1.58	0.27	7.12	nr	8.01
extra; small circular three way boxes	6.01	2.50	6.16	0.13	–	1.87	0.27	8.30	nr	9.34
extra; small circular through way boxes	5.14	2.50	5.27	0.11	–	1.58	0.27	7.12	nr	8.01
20mm diameter; plugging to masonry; in chases	3.12	10.00	3.43	0.15	–	2.16	0.42	6.01	m	6.76
extra; small circular terminal boxes	4.55	2.50	4.66	0.10	–	1.44	0.27	6.37	nr	7.17
extra; small circular angle boxes	3.88	2.50	3.98	0.11	–	1.58	0.27	5.83	nr	6.56
extra; small circular three way boxes	4.48	2.50	4.59	0.13	–	1.87	0.27	6.73	nr	7.57
extra; small circular through way boxes	3.88	2.50	3.98	0.11	–	1.58	0.27	5.83	nr	6.56
25mm diameter; plugging to masonry; in chases	4.12	10.00	4.53	0.16	–	2.30	0.27	7.10	m	7.99
extra; small circular terminal boxes	5.66	2.50	5.80	0.10	–	1.44	0.27	7.51	nr	8.45
extra; small circular angle boxes	5.14	2.50	5.27	0.11	–	1.58	0.27	7.12	nr	8.01
extra; small circular three way boxes	6.01	2.50	6.16	0.13	–	1.87	0.27	8.30	nr	9.34
extra; small circular through way boxes	5.14	2.50	5.27	0.11	–	1.58	0.27	7.12	nr	8.01
Oval conduits; straight										
20mm (nominal size); plugging to masonry; in chases	2.25	10.00	2.48	0.15	–	2.16	0.42	5.05	m	5.68
25mm (nominal size); plugging to masonry in chases	2.94	10.00	3.23	0.16	–	2.30	0.42	5.95	m	6.70
20mm (nominal size); fixing to timber	2.25	10.00	2.48	0.13	–	1.87	0.21	4.55	m	5.12
25mm (nominal size); fixing to timber	2.94	10.00	3.23	0.15	–	2.16	0.21	5.60	m	6.30
Steel welded heavy gauge conduits and fittings, screwed joints; spacer bar saddles at 1000mm centres; enamelled black inside and outside by manufacturer										
Conduits; straight										
20mm diameter; plugging to masonry; to surfaces	4.50	10.00	4.95	0.33	–	4.75	0.25	9.95	m	11.19
extra; small circular terminal boxes	5.98	2.50	6.13	0.13	–	1.87	0.27	8.27	nr	9.30
extra; small circular angle boxes	7.21	2.50	7.39	0.16	–	2.30	0.27	9.96	nr	11.21
extra; small circular three way boxes	8.19	2.50	8.39	0.20	–	2.88	0.27	11.54	nr	12.98
extra; small circular through way boxes	7.21	2.50	7.39	0.16	–	2.30	0.27	9.96	nr	11.21
25mm diameter; plugging to masonry; to surfaces	6.01	10.00	6.61	0.42	–	6.04	0.27	12.92	m	14.54
extra; small circular terminal boxes	7.49	2.50	7.68	0.13	–	1.87	0.27	9.82	nr	11.04
extra; small circular angle boxes	9.09	2.50	9.32	0.16	–	2.30	0.27	11.89	nr	13.37
extra; small circular three way boxes	10.08	2.50	10.33	0.20	–	2.88	0.27	13.48	nr	15.16
extra; small circular through way boxes	9.09	2.50	9.32	0.16	–	2.30	0.27	11.89	nr	13.37
32mm diameter; plugging to masonry; to surfaces	8.72	10.00	9.59	0.50	–	7.19	0.27	17.05	m	19.18
Steel welded heavy gauge conduits and fittings, screwed joints; spacer bar saddles at 1000mm centres; galvanised inside and outside by manufacturer										
Conduits; straight										
20mm diameter; plugging to masonry; to surfaces	5.85	10.00	6.43	0.33	–	4.75	0.25	11.43	m	12.86
extra; small circular terminal boxes	7.18	2.50	7.36	0.13	–	1.87	0.27	9.50	nr	10.69
extra; small circular angle boxes	8.66	2.50	8.88	0.16	–	2.30	0.27	11.45	nr	12.88
extra; small circular three way boxes	9.82	2.50	10.07	0.20	–	2.88	0.27	13.21	nr	14.86

Labour hourly rates: (except Specialists) Craft Operatives £14.38 Labourer £7.02 Rates are national average prices. Refer to REGIONAL VARIATIONS for indicative levels of overall pricing in regions	MATERIALS			LABOUR				RATES		
	Del to Site £	Waste %	Material Cost £	Craft Optve Hrs	Lab Hrs	Labour Cost £	Sunds £	Nett Rate £	Unit	Gross Rate (+12.5%) £

GENERAL LIGHTING AND POWER – COMMERCIAL Cont.

Steel welded heavy gauge conduits and fittings, screwed joints; spacer bar saddles at 1000mm centres; galvanised inside and outside by manufacturer Cont.

Conduits; straight Cont.

	Del to Site	Waste	Material Cost	Craft Optve	Lab	Labour Cost	Sunds	Nett Rate	Unit	Gross Rate
extra; small circular through way boxes	8.66	2.50	8.88	0.16	-	2.30	0.27	11.45	nr	12.88
25mm diameter; plugging to masonry; to surfaces ..	7.73	10.00	8.50	0.42	-	6.04	0.25	14.79	m	16.64
extra; small circular terminal boxes	8.99	2.50	9.21	0.13	-	1.87	0.27	11.35	nr	12.77
extra; small circular angle boxes	10.91	2.50	11.18	0.16	-	2.30	0.27	13.75	nr	15.47
extra; small circular three way boxes	12.10	2.50	12.40	0.20	-	2.88	0.27	15.55	nr	17.49
extra; small circular through way boxes	10.91	2.50	11.18	0.16	-	2.30	0.27	13.75	nr	15.47
32mm diameter; plugging to masonry; to surfaces ..	10.54	10.00	11.59	0.50	-	7.19	0.25	19.03	m	21.41

Galvanised standard cable tray and fittings

Light duty; straight

	Del to Site	Waste	Material Cost	Craft Optve	Lab	Labour Cost	Sunds	Nett Rate	Unit	Gross Rate
50mm tray..............................	2.71	5.00	2.85	0.25	-	3.60	0.09	6.53	m	7.35
extra; flat bend	6.91	-	6.91	0.20	-	2.88	0.25	10.04	nr	11.29
extra; tee	8.96	-	8.96	0.25	-	3.60	0.25	12.81	nr	14.41
extra; cross piece	15.12	-	15.12	0.40	-	5.75	0.50	21.37	nr	24.04
extra; riser bend	10.78	-	10.78	0.20	-	2.88	0.25	13.91	nr	15.64
100mm tray.....................................	4.05	5.00	4.25	0.33	-	4.75	0.09	9.09	m	10.22
extra; flat bend	7.46	-	7.46	0.23	-	3.31	0.25	11.02	nr	12.39
extra; tee	9.86	-	9.86	0.30	-	4.31	0.25	14.42	nr	16.23
extra; cross piece	16.12	-	16.12	0.41	-	5.90	0.50	22.52	nr	25.33
extra; riser bend	12.43	-	12.43	0.23	-	3.31	0.25	15.99	nr	17.99
extra; reducer 100 to 50mm	9.45	-	9.45	0.23	-	3.31	0.25	13.01	nr	14.63
150mm tray.....................................	5.30	5.00	5.57	0.42	-	6.04	0.09	11.69	m	13.16
extra; flat bend	8.71	-	8.71	0.25	-	3.60	0.25	12.56	nr	14.12
extra; tee	11.96	-	11.96	0.33	-	4.75	0.25	16.96	nr	19.07
extra; cross piece	18.92	-	18.92	0.42	-	6.04	0.50	25.46	nr	28.64
extra; riser bend	16.78	-	16.78	0.25	-	3.60	0.25	20.63	nr	23.20
extra; reducer 150 to 100mm	12.45	-	12.45	0.25	-	3.60	0.25	16.30	nr	18.33
300mm tray.....................................	12.61	5.00	13.24	0.50	-	7.19	0.09	20.52	m	23.09
extra; flat bend	16.06	-	16.06	0.33	-	4.75	0.25	21.06	nr	23.69
extra; tee	22.01	-	22.01	0.50	-	7.19	0.25	29.45	nr	33.13
extra; cross piece	33.82	-	33.82	0.58	-	8.34	0.50	42.66	nr	47.99
extra; riser bend	29.53	-	29.53	0.33	-	4.75	0.25	34.53	nr	38.84
extra; reducer 300 to 150mm	22.65	-	22.65	0.33	-	4.75	0.25	27.65	nr	31.10
450mm tray.....................................	25.48	5.00	26.75	0.58	-	8.34	0.09	35.18	m	39.58
extra; flat bend	25.71	-	25.71	0.50	-	7.19	0.25	33.15	nr	37.29
extra; tee	35.91	-	35.91	0.75	-	10.79	0.25	46.95	nr	52.81
extra; cross piece	55.82	-	55.82	1.00	-	14.38	0.50	70.70	nr	79.54
extra; riser bend	41.83	-	41.83	0.50	-	7.19	0.25	49.27	nr	55.43
extra; reducer 450 to 300mm	37.60	-	37.60	0.50	-	7.19	0.25	45.04	nr	50.67

Medium duty; straight

	Del to Site	Waste	Material Cost	Craft Optve	Lab	Labour Cost	Sunds	Nett Rate	Unit	Gross Rate
75mm tray......................................	5.45	5.00	5.72	0.26	-	3.74	0.09	9.55	m	10.75
extra; flat bend	26.93	-	26.93	0.21	-	3.02	0.25	30.20	nr	33.97
extra; tee	35.53	-	35.53	0.26	-	3.74	0.25	39.52	nr	44.46
extra; cross piece	56.31	-	56.31	0.42	-	6.04	0.50	62.85	nr	70.71
extra; riser bend	14.64	-	14.64	0.21	-	3.02	0.25	17.91	nr	20.15
100mm tray.....................................	5.53	5.00	5.81	0.34	-	4.89	0.09	10.79	m	12.13
extra; flat bend	26.93	-	26.93	0.24	-	3.45	0.25	30.63	nr	34.46
extra; tee	35.53	-	35.53	0.31	-	4.46	0.25	40.24	nr	45.27
extra; cross piece	54.76	-	54.76	0.43	-	6.18	0.50	61.44	nr	69.12
extra; riser bend	13.39	-	13.39	0.24	-	3.45	0.25	17.09	nr	19.23
extra; reducer 100 to 75mm	12.50	-	12.50	0.24	-	3.45	0.25	16.20	nr	18.23
150mm tray.....................................	6.63	5.00	6.96	0.44	-	6.33	0.09	13.38	m	15.05
extra; flat bend	30.38	-	30.38	0.26	-	3.74	0.25	34.37	nr	38.66
extra; tee	40.43	-	40.43	0.34	-	4.89	0.25	45.57	nr	51.27
extra; cross piece	60.76	-	60.76	0.44	-	6.33	0.50	67.59	nr	76.04
extra; riser bend	17.59	-	17.59	0.26	-	3.74	0.25	21.58	nr	24.28
extra; reducer 150 to 100mm	10.85	-	10.85	0.26	-	3.74	0.25	14.84	nr	16.69
300mm tray.....................................	13.35	5.00	14.02	0.52	-	7.48	0.09	21.59	m	24.28
extra; flat bend	41.68	-	41.68	0.34	-	4.89	0.25	46.82	nr	52.67
extra; tee	51.68	-	51.68	0.52	-	7.48	0.25	59.41	nr	66.83
extra; cross piece	79.56	-	79.56	0.61	-	8.77	0.50	88.83	nr	99.94
extra; riser bend	27.94	-	27.94	0.34	-	4.89	0.25	33.08	nr	37.21
extra; reducer 300 to 150mm	22.85	-	22.85	0.34	-	4.89	0.25	27.99	nr	31.49
450mm tray.....................................	19.42	5.00	20.39	0.61	-	8.77	0.09	29.25	m	32.91
extra; flat bend	55.13	-	55.13	0.52	-	7.48	0.25	62.86	nr	70.71
extra; tee	72.23	-	72.23	0.78	-	11.22	0.25	83.70	nr	94.16
extra; cross piece	108.26	-	108.26	1.05	-	15.10	0.50	123.86	nr	139.34
extra; riser bend	42.14	-	42.14	0.52	-	7.48	0.25	49.87	nr	56.10
extra; reducer 450 to 300mm	37.70	-	37.70	0.52	-	7.48	0.25	45.43	nr	51.11

Heavy duty; straight

	Del to Site	Waste	Material Cost	Craft Optve	Lab	Labour Cost	Sunds	Nett Rate	Unit	Gross Rate
75mm tray......................................	9.58	5.00	10.06	0.27	-	3.88	0.09	14.03	m	15.79
extra; flat bend	38.20	-	38.20	0.22	-	3.16	0.25	41.61	nr	46.82
extra; tee	46.95	-	46.95	0.27	-	3.88	0.25	51.08	nr	57.47
extra; cross piece	75.85	-	75.85	0.44	-	6.33	0.50	82.68	nr	93.01
extra; riser bend	26.25	-	26.25	0.22	-	3.16	0.25	29.66	nr	33.37
100mm tray.....................................	9.73	5.00	10.22	0.36	-	5.18	0.09	15.48	m	17.42
extra; flat bend	38.70	-	38.70	0.25	-	3.60	0.25	42.55	nr	47.86
extra; tee	47.35	-	47.35	0.33	-	4.75	0.25	52.35	nr	58.89
extra; cross piece	76.45	-	76.45	0.45	-	6.47	0.50	83.42	nr	93.85
extra; riser bend	26.90	-	26.90	0.25	-	3.60	0.25	30.75	nr	34.59
extra; reducer 100 to 75mm	17.50	-	17.50	0.25	-	3.60	0.25	21.34	nr	24.01
150mm tray.....................................	10.78	5.00	11.32	0.46	-	6.61	0.09	18.02	m	20.28
extra; flat bend	42.00	-	42.00	0.27	-	3.88	0.25	46.13	nr	51.90
extra; tee	52.05	-	52.05	0.36	-	5.18	0.25	57.48	nr	64.66
extra; cross piece	82.80	-	82.80	0.46	-	6.61	0.50	89.91	nr	101.15
extra; riser bend	29.20	-	29.20	0.27	-	3.88	0.25	33.33	nr	37.50

ELECTRICAL

Labour hourly rates: (except Specialists) Craft Operatives £14.38 Labourer £7.02 Rates are national average prices. Refer to REGIONAL VARIATIONS for indicative levels of overall pricing in regions	MATERIALS			LABOUR				RATES		
	Del to Site £	Waste %	Material Cost £	Craft Optve Hrs	Lab Hrs	Labour Cost £	Sunds £	Nett Rate £	Unit	Gross Rate (+12.5%) £
GENERAL LIGHTING AND POWER - COMMERCIAL Cont.										
Galvanised standard cable tray and fittings Cont.										
Heavy duty; straight Cont.										
extra; reducer 150 to 100mm	17.55	-	17.55	0.27	-	3.88	0.25	21.68	nr	24.39
300mm tray	15.69	5.00	16.47	0.55	-	7.91	0.09	24.47	m	27.53
extra; flat bend	48.20	-	48.20	0.36	-	5.18	0.25	53.63	nr	60.33
extra; tee	62.90	-	62.90	0.55	-	7.91	0.25	71.06	nr	79.94
extra; cross piece	99.55	-	99.55	0.64	-	9.20	0.50	109.25	nr	122.91
extra; riser bend	32.10	-	32.10	0.36	-	5.18	0.25	37.53	nr	42.22
extra; reducer 300 to 150mm	28.80	-	28.80	0.36	-	5.18	0.25	34.23	nr	38.51
450mm tray	27.14	5.00	28.50	0.64	-	9.20	0.09	37.79	m	42.51
extra; flat bend	81.75	-	81.75	0.55	-	7.91	0.25	89.91	nr	101.15
extra; tee	98.00	-	98.00	0.83	-	11.94	0.25	110.19	nr	123.96
extra; cross piece	153.80	-	153.80	1.10	-	15.82	0.25	169.87	nr	191.10
extra; riser bend	56.40	-	56.40	0.55	-	7.91	0.25	64.56	nr	72.63
extra; reducer 450 to 300mm	45.05	-	45.05	0.55	-	7.91	0.25	53.21	nr	59.86
Supports for cable tray										
Cantilever arms; mild steel; hot dip galvanised; fixing to masonry										
65mm wide	4.35	2.50	4.46	0.33	-	4.75	0.25	9.45	nr	10.64
90mm wide	4.35	2.50	4.46	0.33	-	4.75	0.25	9.45	nr	10.64
120mm wide	4.35	2.50	4.46	0.33	-	4.75	0.25	9.45	nr	10.64
Stand-off bracket; mild steel; hot dip galvanised; fixing to masonry										
75mm wide	4.35	2.50	4.46	0.33	-	4.75	0.25	9.45	nr	10.64
100mm wide	5.60	2.50	5.74	0.33	-	4.75	0.25	10.74	nr	12.08
150mm wide	5.60	2.50	5.74	0.33	-	4.75	0.25	10.74	nr	12.08
300mm wide	12.35	2.50	12.66	0.33	-	4.75	0.25	17.65	nr	19.86
450mm wide	16.90	2.50	17.32	0.33	-	4.75	0.25	22.32	nr	25.11
PVC insulated cables; single core; reference 6491X; stranded copper conductors; to B.S.6004										
Drawn into conduits or ducts or laid or drawn into trunking										
1.5mm2	0.53	15.00	0.61	0.03	-	0.43	-	1.04	m	1.17
2.5mm2	0.78	15.00	0.90	0.03	-	0.43	-	1.33	m	1.49
4.0mm2	1.30	15.00	1.50	0.04	-	0.58	-	2.07	m	2.33
6.0mm2	1.86	15.00	2.14	0.05	-	0.72	-	2.86	m	3.22
10.0mm2	3.47	15.00	3.99	0.05	-	0.72	-	4.71	m	5.30
PVC insulated and PVC sheathed cables; multicore; copper conductors and bare earth continuity conductor 6242Y; to B.S.6004										
Drawn into conduits or ducts or laid or drawn into trunking										
1.00mm2; twin with bare earth	0.87	15.00	1.00	0.05	-	0.72	-	1.72	m	1.93
1.50mm2; twin with bare earth	1.09	15.00	1.25	0.05	-	0.72	-	1.97	m	2.22
2.5mm2; twin with bare earth	1.49	15.00	1.71	0.05	-	0.72	-	2.43	m	2.74
4.00mm2; twin with bare earth	4.61	15.00	5.30	0.06	-	0.86	-	6.16	m	6.93
6.00mm2; twin with bare earth	5.47	15.00	6.29	0.07	-	1.01	-	7.30	m	8.21
Fixed to timber with clips										
1.00mm2; twin with bare earth	0.87	15.00	1.00	0.07	-	1.01	0.06	2.07	m	2.33
1.50mm2; twin with bare earth	1.09	15.00	1.25	0.07	-	1.01	0.06	2.32	m	2.61
2.5mm2; twin with bare earth	1.49	15.00	1.71	0.07	-	1.01	0.07	2.79	m	3.14
4.00mm2; twin with bare earth	4.61	15.00	5.30	0.08	-	1.15	0.08	6.53	m	7.35
6.00mm2; twin with bare earth	5.47	15.00	6.29	0.09	-	1.29	0.08	7.66	m	8.62
PVC insulated SWA armoured and PVC sheathed cables; multicore; copper conductors to B.S.6346										
Laid and laced on cable tray										
16mm2; three core	13.55	5.00	14.23	0.17	-	2.44	0.11	16.78	m	18.88
Fixed to masonry with screwed clips at average 300mm centres; plugging										
16mm2; three core	13.55	5.00	14.23	0.25	-	3.60	0.94	18.76	m	21.11
Mineral insulated copper sheathed cables; PVC outer sheath; copper conductors; 600V light duty to B.S.6207										
Fixed to masonry with screwed clips at average 300mm centres; plugging										
1.5mm2; two core	2.28	10.00	2.51	0.17	-	2.44	0.94	5.89	m	6.63
extra; termination including gland, seal and shroud	2.52	2.50	2.58	0.42	-	6.04	-	8.62	nr	9.70
2.5mm2; two core	2.89	10.00	3.18	0.17	-	2.44	0.94	6.56	m	7.38
extra; termination including gland, seal and shroud	2.52	2.50	2.58	0.42	-	6.04	-	8.62	nr	9.70
4.0mm2; two core	3.85	10.00	4.24	0.20	-	2.88	0.94	8.05	m	9.06
extra; termination including gland, seal and shroud	2.52	2.50	2.58	0.50	-	7.19	-	9.77	nr	10.99

Labour hourly rates: (except Specialists) Craft Operatives £14.38 Labourer £7.02 Rates are national average prices. Refer to REGIONAL VARIATIONS for indicative levels of overall pricing in regions	MATERIALS			LABOUR				RATES		
	Del to Site	Waste	Material Cost	Craft Optve	Lab	Labour Cost	Sunds	Nett Rate	Unit	Gross Rate (+12.5%)
	£	%	£	Hrs	Hrs	£	£	£		£
GENERAL LIGHTING AND POWER - COMMERCIAL Cont.										
HV switchgear										
Switch fuse 500V, metal cased, fixed to masonry with screws; plugging										
32 amp; single pole and neutral; short circuit rating 6.4Ka	111.06	2.50	113.84	1.60	-	23.01	0.50	137.34	nr	154.51
Consumer control units, 250V, metal cased, fitted with 100 amp D.P. isolator and MCB's , fixed to masonry with screws; plugging										
eight way, single pole and neutral	92.11	2.50	94.41	1.60	-	23.01	0.50	117.92	nr	132.66
Distribution boards 500V, metal cased, MCB pattern, fitted with 100 amp D.P. isolator, fixed to masonry with screws; plugging										
nine way, single pole and neutral, 20 amp	109.96	2.50	112.71	1.80	-	25.88	0.50	139.09	nr	156.48
twelve way, triple pole and neutral, 32 amp ...	488.56	2.50	500.77	8.00	-	115.04	0.50	616.31	nr	693.35
Luminaires										
Note										
for lighting fittings and lamps see General Lighting and Power - Domestic Section										
Accessories; white plastics, with boxes, fixing to masonry with screws; plugging										
Flush plate switches										
5 amp; one gang; one way; single pole	3.69	2.50	3.78	0.42	-	6.04	0.25	10.07	nr	11.33
5 amp; two gang; one way; single pole	5.35	2.50	5.48	0.43	-	6.18	0.25	11.92	nr	13.41
Surface plate switches										
5 amp; one gang; one way; single pole	3.46	2.50	3.55	0.42	-	6.04	0.25	9.84	nr	11.07
5 amp; two gang; one way; single pole	5.10	2.50	5.23	0.43	-	6.18	0.25	11.66	nr	13.12
Ceiling switches										
5 amp; one gang; one way; single pole	4.10	2.50	4.20	0.43	-	6.18	0.25	10.64	nr	11.97
Flush switched socket outlets										
13 amp; one gang	6.63	2.50	6.80	0.43	-	6.18	0.25	13.23	nr	14.88
13 amp; two gang	12.43	2.50	12.74	0.43	-	6.18	0.25	19.17	nr	21.57
Surface switched socket outlets										
13 amp; one gang	6.31	2.50	6.47	0.43	-	6.18	0.25	12.90	nr	14.51
13 amp; two gang	12.06	2.50	12.36	0.43	-	6.18	0.25	18.79	nr	21.14
Flush switched socket outlets with RCCB (residual current device) protected at 30 mAmp										
13 amp; one gang	69.74	2.50	71.48	0.43	-	6.18	0.25	77.92	nr	87.66
Flush fused connection units										
13 amp; one gang; switched; flexible outlet	10.49	2.50	10.75	0.50	-	7.19	0.25	18.19	nr	20.47
13 amp; one gang; switched; flexible outlet; pilot lamp	14.15	2.50	14.50	0.50	-	7.19	0.25	21.94	nr	24.69
Accessories, metalclad, with boxes, fixing to masonry with screws; plugging										
Surface plate switches										
5 amp; one gang; two way; single pole	4.96	2.50	5.08	0.42	-	6.04	0.25	11.37	nr	12.80
5 amp; two gang; two way; single pole	6.03	2.50	6.18	0.43	-	6.18	0.25	12.61	nr	14.19
Surface switched socket outlets										
13 amp; one gang	9.69	2.50	9.93	0.43	-	6.18	0.25	16.37	nr	18.41
13 amp; two gang	19.15	2.50	19.63	0.43	-	6.18	0.25	26.06	nr	29.32
Weatherproof accessories, with boxes, fixing to masonry with screws; plugging										
Surface plate switches										
5 amp; one gang; two way; single pole	6.57	2.50	6.73	0.43	-	6.18	0.25	13.17	nr	14.81
GENERAL LIGHTING AND POWER - DOMESTIC										
Electric wiring										
Note										
the following approximate prices of various types of installations are dependent on the number and disposition of points; lamps and fittings together with cutting and making good are excluded										
Electric wiring in new building										
Installation with PVC insulated and sheathed cables										
lighting points	18.17	7.50	19.53	3.58	-	51.48	1.65	72.66	nr	81.75
socket outlets; 5A	24.41	7.50	26.24	4.84	-	69.60	1.27	97.11	nr	109.25
socket outlets; 13A ring main	23.41	7.50	25.17	4.07	-	58.53	1.15	84.84	nr	95.45
socket outlets; 13A radial circuit	26.59	7.50	28.58	4.24	-	60.97	1.27	90.83	nr	102.18
cooker points; 45A	109.54	7.50	117.76	6.93	-	99.65	1.96	219.37	nr	246.79
immersion heater points	32.44	7.50	34.87	5.06	-	72.76	1.40	109.04	nr	122.67
shaver sockets (transformer)	58.08	7.50	62.44	1.82	-	26.17	1.40	90.01	nr	101.26
Installation with mineral insulated copper sheathed cables										
lighting points	29.37	7.50	31.57	6.82	-	98.07	0.82	130.46	nr	146.77
socket outlets; 5A	39.68	7.50	42.66	5.34	-	76.79	0.75	120.20	nr	135.22

ELECTRICAL

Labour hourly rates: (except Specialists) Craft Operatives £14.38 Labourer £7.02 Rates are national average prices. Refer to REGIONAL VARIATIONS for indicative levels of overall pricing in regions	MATERIALS			LABOUR				RATES		
	Del to Site £	Waste %	Material Cost £	Craft Optve Hrs	Lab Hrs	Labour Cost £	Sunds £	Nett Rate £	Unit	Gross Rate (+12.5%) £
GENERAL LIGHTING AND POWER - DOMESTIC Cont.										
Electric wiring in new building Cont.										
Installation with mineral insulated copper sheathed cables Cont.										
socket outlets; 13A ring main	35.68	7.50	38.36	4.57	–	65.72	0.63	104.70	nr	117.79
socket outlets; 13A radial circuit	41.52	7.50	44.63	6.11	–	87.86	0.75	133.25	nr	149.90
cooker points; 45A	127.40	7.50	136.96	10.12	–	145.53	1.44	283.92	nr	319.41
immersion heater points	54.14	7.50	58.20	7.26	–	104.40	0.88	163.48	nr	183.91
shaver sockets (transformer)	78.86	7.50	84.77	3.30	–	47.45	0.88	133.11	nr	149.75
Installation with black enamel heavy gauge conduit with PVC cables										
lighting points	61.63	7.50	66.25	5.12	–	73.63	2.51	142.39	nr	160.19
socket outlets; 5A	65.45	7.50	70.36	5.17	–	74.34	2.76	147.46	nr	165.90
socket outlets; 13A ring main	60.09	7.50	64.60	4.68	–	67.30	2.26	134.16	nr	150.92
socket outlets; 13A radial circuit	75.29	7.50	80.94	6.60	–	94.91	2.76	178.60	nr	200.93
cooker points; 45A	156.57	7.50	168.31	10.62	–	152.72	3.26	324.29	nr	364.82
immersion heater points	79.66	7.50	85.63	7.87	–	113.17	2.76	201.57	nr	226.76
shaver sockets (transformer)	91.88	7.50	98.77	2.53	–	36.38	2.51	137.66	nr	154.87
Installation with black enamel heavy gauge conduit with coaxial cable										
T.V. sockets	69.57	7.50	74.79	6.33	–	91.03	2.76	168.57	nr	189.64
Installation with black enamel heavy gauge conduit with draw wire										
telephone points	56.17	7.50	60.38	6.00	–	86.28	2.76	149.42	nr	168.10
Electric wiring in extension to existing building										
Installation with PVC insulated and sheathed cables										
lighting points	18.17	7.50	19.53	3.58	–	51.48	1.65	72.66	nr	81.75
socket outlets; 5A	24.41	7.50	26.24	4.84	–	69.60	1.27	97.11	nr	109.25
socket outlets; 13A ring main	23.41	7.50	25.17	4.07	–	58.53	1.15	84.84	nr	95.45
socket outlets; 13A radial circuit	26.59	7.50	28.58	4.24	–	60.97	1.27	90.83	nr	102.18
cooker points; 45A	95.79	7.50	102.97	6.93	–	99.65	1.96	204.59	nr	230.16
immersion heater points	32.44	7.50	34.87	5.06	–	72.76	1.40	109.04	nr	122.67
shaver sockets (transformer)	58.08	7.50	62.44	1.82	–	26.17	1.40	90.01	nr	101.26
Installation with mineral insulated copper sheathed cables										
lighting points	29.37	7.50	31.57	6.82	–	98.07	0.82	130.46	nr	146.77
socket outlets; 5A	39.68	7.50	42.66	5.34	–	76.79	0.75	120.20	nr	135.22
socket outlets; 13A ring main	35.68	7.50	38.36	4.57	–	65.72	0.63	104.70	nr	117.79
socket outlets; 13A radial circuit	41.52	7.50	44.63	6.11	–	87.86	0.75	133.25	nr	149.90
cooker points; 45A	127.40	7.50	136.96	10.12	–	145.53	1.44	283.92	nr	319.41
immersion heater points	54.14	7.50	58.20	7.26	–	104.40	0.88	163.48	nr	183.91
shaver sockets (transformer)	78.86	7.50	84.77	3.30	–	47.45	0.88	133.11	nr	149.75
Installation with black enamel heavy gauge conduit with PVC cables										
lighting points	61.63	7.50	66.25	5.12	–	73.63	2.51	142.39	nr	160.19
socket outlets; 5A	65.45	7.50	70.36	5.17	–	74.34	2.76	147.46	nr	165.90
socket outlets; 13A ring main	60.09	7.50	64.60	4.68	–	67.30	2.26	134.16	nr	150.92
socket outlets; 13A radial circuit	75.29	7.50	80.94	6.60	–	94.91	2.76	178.60	nr	200.93
cooker points; 45A	156.57	7.50	168.31	10.62	–	152.72	3.26	324.29	nr	364.82
immersion heater points	79.66	7.50	85.63	7.87	–	113.17	2.76	201.57	nr	226.76
shaver sockets (transformer)	91.88	7.50	98.77	2.53	–	36.38	2.51	137.66	nr	154.87
Installation with black enamel heavy gauge conduit with coaxial cable										
T.V. sockets	74.52	7.50	80.11	6.33	–	91.03	2.76	173.89	nr	195.63
Installation with black enamel heavy gauge conduit with draw wire										
telephone points	56.17	7.50	60.38	6.00	–	86.28	2.76	149.42	nr	168.10
Electric wiring in extending existing installation in existing building										
Installation with PVC insulated and sheathed cables										
lighting points	18.17	7.50	19.53	4.13	–	59.39	1.65	80.57	nr	90.64
socket outlets; 5A	24.41	7.50	26.24	5.56	–	79.95	1.27	107.46	nr	120.90
socket outlets; 13A ring main	23.41	7.50	25.17	4.68	–	67.30	1.15	93.61	nr	105.32
socket outlets; 13A radial circuit	26.59	7.50	28.58	4.90	–	70.46	1.27	100.32	nr	112.86
cooker points; 45A	109.54	7.50	117.76	7.98	–	114.75	1.96	234.47	nr	263.78
immersion heater points	32.44	7.50	34.87	5.83	–	83.84	1.40	120.11	nr	135.12
shaver sockets (transformer)	58.08	7.50	62.44	2.09	–	30.05	1.40	93.89	nr	105.63
Installation with mineral insulated copper sheathed cables										
lighting points	29.37	7.50	31.57	7.87	–	113.17	0.82	145.56	nr	163.76
socket outlets; 5A	39.68	7.50	42.66	6.16	–	88.58	0.75	131.99	nr	148.49
socket outlets; 13A ring main	35.68	7.50	38.36	5.28	–	75.93	0.63	114.91	nr	129.26
socket outlets; 13A radial circuit	41.52	7.50	44.63	7.04	–	101.24	0.75	146.62	nr	164.95
cooker points; 45A	127.40	7.50	136.96	11.66	–	167.67	1.44	306.07	nr	344.32
immersion heater points	54.14	7.50	58.20	8.36	–	120.22	0.88	179.30	nr	201.71
shaver sockets (transformer)	78.86	7.50	84.77	3.80	–	54.64	0.88	140.30	nr	157.84
Installation with black enamel heavy gauge conduit with PVC cables										
lighting points	61.63	7.50	66.25	5.88	–	84.55	2.51	153.32	nr	172.48
socket outlets; 5A	65.45	7.50	70.36	5.94	–	85.42	2.76	158.54	nr	178.35
socket outlets; 13A ring main	60.09	7.50	64.60	5.39	–	77.51	2.26	144.36	nr	162.41

Labour hourly rates: (except Specialists) Craft Operatives £14.38 Labourer £7.02 Rates are national average prices. Refer to REGIONAL VARIATIONS for indicative levels of overall pricing in regions	MATERIALS			LABOUR				RATES		
	Del to Site £	Waste %	Material Cost £	Craft Optve Hrs	Lab Hrs	Labour Cost £	Sunds £	Nett Rate £	Unit	Gross Rate (+12.5%) £
GENERAL LIGHTING AND POWER - DOMESTIC Cont.										
Electric wiring in extending existing installation in existing building Cont.										
Installation with black enamel heavy gauge conduit with PVC cables Cont.										
socket outlets; 13A radial circuit	75.29	7.50	80.94	7.60	-	109.29	2.76	192.98	nr	217.11
cooker points; 45A	156.57	7.50	168.31	12.21	-	175.58	3.26	347.15	nr	390.55
immersion heater points	79.66	7.50	85.63	9.08	-	130.57	2.76	218.96	nr	246.34
shaver sockets (transformer)	91.88	7.50	98.77	2.92	-	41.99	2.51	143.27	nr	161.18
Installation with black enamel heavy gauge conduit with coaxial cable										
T.V. sockets	69.57	7.50	74.79	7.26	-	104.40	2.76	181.95	nr	204.69
Installation with black enamel heavy gauge conduit with draw wire										
telephone points	56.17	7.50	60.38	6.88	-	98.93	2.76	162.08	nr	182.34
LAMP FITTINGS, FANS, HEATERS etc., ALL INSTALLATIONS										
Fluorescent lamp fittings, mains voltage operations; switch start; including lamps										
Batten type; single tube										
1200mm; 36W	19.90	2.50	20.40	0.83	-	11.94	0.50	32.83	nr	36.94
1500mm; 58W	22.51	2.50	23.07	1.10	-	15.82	0.50	39.39	nr	44.31
1800mm; 70W	27.19	2.50	27.87	1.10	-	15.82	0.50	44.19	nr	49.71
2400mm; 100W	37.20	2.50	38.13	1.38	-	19.84	0.50	58.47	nr	65.78
Batten type; twin tube										
1200mm; 36W	38.16	2.50	39.11	0.83	-	11.94	0.50	51.55	nr	57.99
1500mm; 58W	45.33	2.50	46.46	1.10	-	15.82	0.50	62.78	nr	70.63
1800mm; 70W	49.70	2.50	50.94	1.10	-	15.82	0.50	67.26	nr	75.67
2400mm; 100W	65.11	2.50	66.74	1.38	-	19.84	0.50	87.08	nr	97.97
Metal trough reflector; single tube										
1200mm; 36W	33.41	2.50	34.25	1.10	-	15.82	0.50	50.56	nr	56.88
1500mm; 58W	36.59	2.50	37.50	1.32	-	18.98	0.50	56.99	nr	64.11
1800mm; 70W	43.12	2.50	44.20	1.32	-	18.98	0.50	63.68	nr	71.64
2400mm; 100W	61.37	2.50	62.90	1.60	-	23.01	0.50	86.41	nr	97.21
Metal trough reflector; twin tube										
1200mm; 36W	51.67	2.50	52.96	1.10	-	15.82	0.50	69.28	nr	77.94
1500mm; 58W	59.41	2.50	60.90	1.32	-	18.98	0.50	80.38	nr	90.42
1800mm; 70W	65.63	2.50	67.27	1.32	-	18.98	0.50	86.75	nr	97.60
2400mm; 100W	89.28	2.50	91.51	1.60	-	23.01	0.50	115.02	nr	129.40
Plastics diffused type; single tube										
1200mm; 36W	34.48	2.50	35.34	1.10	-	15.82	0.50	51.66	nr	58.12
1500mm; 58W	38.61	2.50	39.58	1.32	-	18.98	0.50	59.06	nr	66.44
1800mm; 70W	47.25	2.50	48.43	1.32	-	18.98	0.50	67.91	nr	76.40
2400mm; 100W	63.66	2.50	65.25	1.60	-	23.01	0.50	88.76	nr	99.85
Plastics diffused type; twin tube										
1200mm; 36W	64.35	2.50	65.96	1.00	-	14.38	0.50	80.84	nr	90.94
1500mm; 58W	78.17	2.50	80.12	1.32	-	18.98	0.50	99.61	nr	112.06
1800mm; 70W	88.30	2.50	90.51	1.32	-	18.98	0.50	109.99	nr	123.74
2400mm; 100W	112.77	2.50	115.59	1.60	-	23.01	0.50	139.10	nr	156.48
Lighting fittings complete with tungsten lamps										
Bulkhead; alloy										
100W G.L.S. lamp	25.41	2.50	26.05	0.72	-	10.35	0.25	36.65	nr	41.23
Wall glass; alloy; corner bracket										
100W G.L.S. lamp	27.87	2.50	28.57	0.83	-	11.94	0.25	40.75	nr	45.85
Ceiling sphere										
152mm; 60W G.L.S. lamp	14.85	2.50	15.22	0.72	-	10.35	0.25	25.82	nr	29.05
203mm; 100W G.L.S. lamp	21.64	2.50	22.18	0.72	-	10.35	0.25	32.78	nr	36.88
Down light; semi recessed										
175mm; 150W PAR38 lamp	18.54	2.50	19.00	0.72	-	10.35	0.25	29.61	nr	33.31
Lighting track; single circuit; surface mounted										
1250mm; starter pack	18.85	2.50	19.32	1.32	-	18.98	0.50	38.80	nr	43.65
plug-in fitting; 100W G.L.S. lamp	14.46	2.50	14.82	0.22	-	3.16	-	17.99	nr	20.23
Fan heaters										
3kw unit fan heaters										
for commercial application	159.66	2.50	163.65	1.65	-	23.73	0.50	187.88	nr	211.36
Thermostatic switch units comprising ON/OFF switch and selector switch between winter and summer (fan only) operation	54.45	2.50	55.81	1.10	-	15.82	0.25	71.88	nr	80.86
Tubular heaters										
60W per 305mm loading										
600mm long	11.34	2.50	11.62	1.10	-	15.82	0.75	28.19	nr	31.72
910mm long	12.48	2.50	12.79	1.10	-	15.82	0.75	29.36	nr	33.03
1220mm long	14.34	2.50	14.70	1.38	-	19.84	1.00	35.54	nr	39.99
1520mm long	18.74	2.50	19.21	1.38	-	19.84	1.00	40.05	nr	45.06

ELECTRICAL

Labour hourly rates: (except Specialists) Craft Operatives £14.38 Labourer £7.02 Rates are national average prices. Refer to REGIONAL VARIATIONS for indicative levels of overall pricing in regions	MATERIALS			LABOUR				RATES		
	Del to Site £	Waste %	Material Cost £	Craft Optve Hrs	Lab Hrs	Labour Cost £	Sunds £	Nett Rate £	Unit	Gross Rate (+12.5%) £
LAMP FITTINGS, FANS, HEATERS etc., ALL INSTALLATIONS Cont.										
Tubular heaters Cont.										
60W per 305mm loading Cont.										
1830mm long	19.15	2.50	19.63	1.65	-	23.73	1.50	44.86	nr	50.46
Two way mounting brackets										
pair	5.45	2.50	5.59	0.44	-	6.33	0.50	12.41	nr	13.97
Night storage heaters; installed complete in new building including wiring										
Plastics insulated and sheathed cabled										
1.7KW	226.76	2.50	232.43	7.15	-	102.82	6.43	341.68	nr	384.39
2.55KW	303.94	2.50	311.54	7.70	-	110.73	6.43	428.69	nr	482.28
3.4KW	352.01	2.50	360.81	8.25	-	118.64	6.72	486.17	nr	546.94
Mineral insulated copper sheathed cables										
1.7KW	268.90	2.50	275.62	11.00	-	158.18	5.53	439.33	nr	494.25
2.55KW	314.58	2.50	322.44	11.55	-	166.09	5.53	494.06	nr	555.82
3.4KW	362.56	2.50	371.62	12.10	-	174.00	5.77	551.39	nr	620.32
Bell equipment										
Transformers										
3-5-8V	14.77	2.50	15.14	0.44	-	6.33	0.24	21.71	nr	24.42
12V	18.90	2.50	19.37	0.44	-	6.33	0.24	25.94	nr	29.18
Bells for transformer operation										
chime 2-note	10.61	2.50	10.88	0.55	-	7.91	0.24	19.02	nr	21.40
76mm domestic type	5.85	2.50	6.00	0.55	-	7.91	0.24	14.15	nr	15.91
150mm round bell type	21.27	2.50	21.80	0.44	-	6.33	0.24	28.37	nr	31.92
Bell pushes										
domestic	1.48	2.50	1.52	0.28	-	4.03	-	5.54	nr	6.24
industrial	5.84	2.50	5.99	0.33	-	4.75	-	10.73	nr	12.07
Immersion heaters and thermostats - Note - the following prices exclude builders work										
2 or 3KW immersion heaters; without flanges										
305mm; non-withdrawable elements	23.37	2.50	23.95	0.72	-	10.35	4.65	38.96	nr	43.83
457mm; non-withdrawable elements	23.61	2.50	24.20	0.88	-	12.65	4.65	41.50	nr	46.69
686mm; non-withdrawable elements	23.72	2.50	24.31	1.05	-	15.10	4.65	44.06	nr	49.57
Thermostats										
immersion heater	10.24	2.50	10.50	0.55	-	7.91	-	18.41	nr	20.71
15 amp a.c. for air heating; without switch	15.47	2.50	15.86	0.44	-	6.33	0.25	22.43	nr	25.24
Water heaters - Note - the following prices exclude builders work										
Storage type units; free outlet										
7 litre	114.78	-	114.78	1.65	-	23.73	5.15	143.66	nr	161.61
23 litre	506.73	-	506.73	2.48	-	35.66	5.65	548.04	nr	616.55
55 litre	695.18	-	695.18	4.95	-	71.18	7.98	774.34	nr	871.13
Storage type units; multi-point										
50 litre	696.76	-	696.76	4.95	-	71.18	7.98	775.92	nr	872.91
75 litre	828.49	-	828.49	5.78	-	83.12	7.98	919.59	nr	1034.53
125 litre	948.19	-	948.19	6.60	-	94.91	10.30	1053.40	nr	1185.07
Shower units - Note - the following prices exclude Plumbers work										
Instantaneous shower units										
complete with fittings	156.19	2.50	160.09	4.40	-	63.27	7.73	231.10	nr	259.98
TESTING										
Testing; existing installations										
Point										
from	-	-	-	2.25	-	32.35	-	32.35	nr	36.40
to	-	-	-	3.85	-	55.36	-	55.36	nr	62.28
Complete installation; three bedroom house										
from	-	-	-	11.00	-	158.18	-	158.18	nr	177.95
to	-	-	-	17.50	-	251.65	-	251.65	nr	283.11

Communications/Security/Control systems

Labour hourly rates: (except Specialists) Craft Operatives £9.23 Labourer £7.02 Rates are national average prices. Refer to REGIONAL VARIATIONS for indicative levels of overall pricing in regions	MATERIALS			LABOUR				RATES		
	Del to Site	Waste	Material Cost	Craft Optve	Lab	Labour Cost	Sunds	Nett Rate	Unit	Gross Rate (+12.5%)
	£	%	£	Hrs	Hrs	£	£	£		£
W52: LIGHTNING PROTECTION										
Lightning conductors										
Note the following are indicative average prices										
Air termination rods 610mm long; in clips, fixing to masonry	-	-	-	-	-	-	-	39.50	nr	44.44
Air termination tape fixing to masonry	-	-	-	-	-	-	-	9.50	m	10.69

Copper tape down conductors
   19 x 3mm; in clips, fixing to masonry ....
   25 x 3mm; in clips, fixing to masonry ....
	-
	-
		-
		-
			-
			-
				-
				-
					-
					-
						-
						-
							-
							-
								14.35
								15.35
									m
									m
										16.14
										17.27

Test clamps with securing bolts	-	-	-	-	-	-	-	22.00	nr	24.75
Copper earth rods; tape connectors 2438 x 16mm; driving into ground	-	-	-	-	-	-	-	140.00	nr	157.50

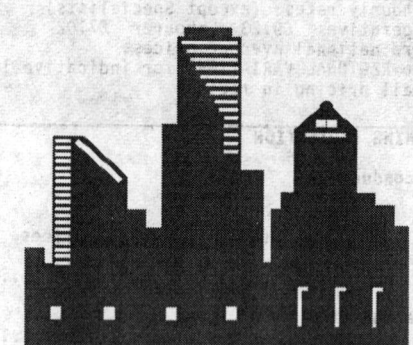

Transport systems

TRANSPORT – SMALL WORKS

Labour hourly rates: (except Specialists) Craft Operatives £9.23 Labourer £7.02 Rates are national average prices. Refer to REGIONAL VARIATIONS for indicative levels of overall pricing in regions	MATERIALS			LABOUR				RATES		
	Del to Site £	Waste %	Material Cost £	Craft Optve Hrs	Lab Hrs	Labour Cost £	Sunds £	Nett Rate £	Unit	Gross Rate (+12.5%) £
X10: LIFTS										
Generally										
Note the following are average indicative prices										
Light passenger lifts; standard range										
Electro Hydraulic drive; 630 kg, 8 person, 0.63m/s, 3 stop										
basic; primed car and entrances	-	-	Spclist	-	-	Spclist	-	15500.00	nr	17437.50
median; laminate walls, standard carpet floor, painted entrances	-	-	Spclist	-	-	Spclist	-	18200.00	nr	20475.00
ACVF drive; 1000 kg, 13 person, 0.63m/s, 4 stop										
basic; primed car and entrances	-	-	Spclist	-	-	Spclist	-	34600.00	nr	38925.00
median; laminate walls, standard carpet floor, painted entrances	-	-	Spclist	-	-	Spclist	-	35900.00	nr	40387.50
Extras on the above lifts										
800/900mm landing doors in cellulose paint finish	-	-	Spclist	-	-	Spclist	-	150.00	nr	168.75
800/900mm landing doors in brushed stainless steel	-	-	Spclist	-	-	Spclist	-	420.00	nr	472.50
800/900mm landing doors in brushed brass	-	-	Spclist	-	-	Spclist	-	1210.00	nr	1361.25
landing position indicators	-	-	Spclist	-	-	Spclist	-	237.00	nr	266.63
landing direction arrows and gongs	-	-	Spclist	-	-	Spclist	-	140.00	nr	157.50
car preference key switch	-	-	Spclist	-	-	Spclist	-	120.00	nr	135.00
car entrance safety (Progard L)	-	-	Spclist	-	-	Spclist	-	400.00	nr	450.00
fluorescent emergency light	-	-	Spclist	-	-	Spclist	-	372.00	nr	418.50
half height mirror to rear wall	-	-	Spclist	-	-	Spclist	-	420.00	nr	472.50
full height mirror to rear wall	-	-	Spclist	-	-	Spclist	-	610.00	nr	686.25
vandal resistant buttons; per car/entrance	-	-	Spclist	-	-	Spclist	-	136.00	nr	153.00
car bumper rail	-	-	Spclist	-	-	Spclist	-	820.00	nr	922.50
trailing cable for Warden Call System	-	-	Spclist	-	-	Spclist	-	349.00	nr	392.63
car telephone	-	-	Spclist	-	-	Spclist	-	274.00	nr	308.25
General purpose passenger lifts										
ACVF drive; 630 kg, 8 person, 1.0m/s, 4 stop										
median; laminate walls, standard carpet floor, painted entrances	-	-	Spclist	-	-	Spclist	-	35600.00	nr	40050.00
Variable speed a/c drive; 1000 kg, 13 person, 1.6m/s, 5 stop										
median; laminate walls, standard carpet floor, painted entrances	-	-	Spclist	-	-	Spclist	-	38200.00	nr	42975.00
Variable speed a/c drive; 1600 kg, 21 person, 1.6m/s, 6 stop										
median; laminate walls, standard carpet floor, painted entrances	-	-	Spclist	-	-	Spclist	-	92000.00	nr	103500.0
If a high quality architectural finish is required, e.g. veneered panelling walls, marble floors, special ceiling, etc, add to the above prices										
from	-	-	Spclist	-	-	Spclist	-	15000.00	nr	16875.00
to	-	%	Spclist	-	-	Spclist	-	42000.00	nr	47250.00
Bed/Passenger lifts										
Electro Hydraulic drive; 1800 kg, 24 person, 0.63m/s, 3 stop										
standard specification	-	-	Spclist	-	-	Spclist	-	48200.00	nr	54225.00
Variable speed a/c drive; 2000 kg, 26 person, 1.6m/s, 4 stop										
standard specification	-	-	Spclist	-	-	Spclist	-	87500.00	nr	98437.50

TRANSPORT SYSTEMS

Labour hourly rates: (except Specialists) Craft Operatives £9.23 Labourer £7.02 Rates are national average prices. Refer to REGIONAL VARIATIONS for indicative levels of overall pricing in regions	MATERIALS			LABOUR				RATES		
	Del to Site £	Waste %	Material Cost £	Craft Optve Hrs	Lab Hrs	Labour Cost £	Sunds £	Nett Rate £	Unit	Gross Rate (+12.5%) £
X10: LIFTS Cont.										
Goods lifts (direct coupled)										
Electro Hydraulic drive; 1500 kg, 0.4m/s, 3 stop stainless steel car lining with chequer plate floor and galvanised shutters	-	-	Spclist	-	-	Spclist	-	38200.00	nr	42975.00
Electro Hydraulic drive; 2000 kg, 0.4m/s, 3 stop stainless steel car lining with chequer plate floor and galvanised shutters	-	-	Spclist	-	-	Spclist	-	39400.00	nr	44325.00
Service hoists										
Single speed a/c drive; 50 kg, 0.63m/s, 2 stop standard specification	-	-	Spclist	-	-	Spclist	-	5400.00	nr	6075.00
Single speed a/c drive; 250 kg, 0.4m/s, 2 stop standard specification	-	-	Spclist	-	-	Spclist	-	6850.00	nr	7706.25
Extra on the above hoists two hour fire resisting shutters; per landing	-	-	Spclist	-	-	Spclist	-	227.00	nr	255.38
X11: ESCALATORS										
Escalators										
30 degree inclination; 3.00m vertical rise department store specification	-	-	Spclist	-	-	Spclist	-	58000.00	nr	65250.00

Basic Prices of Materials

This section gives basic prices of materials delivered to site. Prices are exclusive of Value Added Tax

EXCAVATION AND EARTHWORK

Filling
broken brick or stone	m3	13.02
clinker ashes	m3	13.00
MOT Type 1	m3	15.23
MOT Type 2	m3	14.90

Earthwork support
timber	m3	250.00

CONCRETE WORK

Ready mixed concrete
Prescribed mix
ST1 - 40mm aggregate	m3	48.97
ST2 - 40mm aggregate	m3	49.62
ST3 - 40mm aggregate	m3	51.92
ST3 - 20mm aggregate	m3	52.22
ST4 - 20mm aggregate	m3	53.32
ST5 - 20mm aggregate	m3	54.14
P390 - 20mm aggregate	m3	55.79

Design mix
C 7.5 - 7.5 N/mm2 - 40mm	m3	50.63
C 10 - 10 N/mm2 - 40mm	m3	51.00
C 15 - 15 N/mm2 - 40mm	m3	52.25
C 15 - 15 N/mm2 - 20mm	m3	52.25
C 20 - 20 N/mm2 - 20mm	m3	52.89
C 25 - 25 N/mm2 - 20mm	m3	53.64
C 30 - 30 N/mm2 - 20mm	m3	54.93

Specified mix
1:12 cement concrete 40mm	m3	50.98
1:3:6 cement concrete 40mm	m3	50.47
1:2:4 cement concrete 20mm	m3	59.25
1:1.5:3 cement concrete 20mm	m3	68.02

Cement in bags 5t and over	tonne	84.00

Washed graded shingle
40-5mm	tonne	20.30
20-5mm	tonne	18.90

Washed sharp sand for concreting
	tonne	17.90

"All in" ballast
40mm and down	tonne	16.70
20mm and down	tonne	15.90

Rapid hardening cement	tonne	90.00
Sulphate resisting cement	tonne	90.00

Waterproofing powder	18 kg	22.90
Waterproofing liquid	litre	0.82

Lightweight aggregate medium grade
10-5mm, 919 kg/m3	tonne	40.00
fine grade 5mm to dust, 1200 kg/m3	tonne	50.00

Mild steel rods cut, bent, bundled and labelled
25mm diameter	tonne	320.00
20mm diameter	tonne	320.00
16mm diameter	tonne	320.00
12mm diameter	tonne	320.00
10mm diameter	tonne	320.00
8mm diameter	tonne	330.00
6mm diameter	tonne	365.00

High yield steel rods cut, bent, bundled and labelled
25mm diameter	tonne	330.00
20mm diameter	tonne	330.00
16mm diameter	tonne	330.00
12mm diameter	tonne	330.00
10mm diameter	tonne	330.00
8mm diameter	tonne	340.00
6mm diameter	tonne	375.00

Stainless steel rods cut, bent, bundled and labelled
25mm diameter	tonne	3518.00
20mm diameter	tonne	3518.00
16mm diameter	tonne	3518.00
12mm diameter	tonne	3518.00
10mm diameter	tonne	3518.00
8mm diameter	tonne	3518.00

Fabric reinforcement, B.S. Ref 4483
A393	m2	2.15
A252	m2	1.38
A193	m2	1.06
A142	m2	0.78
A98	m2	0.55
B1131	m2	3.88
B785	m2	2.87
B503	m2	2.09
B385	m2	1.61
B283	m2	1.32
B196	m2	1.09
C785	m2	2.38
C503	m2	1.48
C385	m2	1.20
C283	m2	0.92
D98	m2	0.50
D49	m2	0.56
C636	m2	1.97

Formwork
timber	m3	248.00
19mm plywood	m2	7.50

Claymaster expanded polystyrene permanent formwork
2400 x 1200 x 75mm thick	nr	21.12
2400 x 1200 x 100mm thick	nr	28.16
2400 x 1200 x 150mm thick	nr	42.24

Bitumen/latex solution	litre	2.50
Bituminous emulsion	litre	1.58
Asphaltic bitumen	litre	1.90

Polythene sheeting
125 mu	m2	0.33
250 mu	m2	0.40
300 mu	m2	0.54

Waterproof building paper to BS 1521
Grade B2	m2	0.42
Grade B1F	m2	0.58

BRICKWORK AND BLOCKWORK

Common bricks
plain	1000	163.90
keyed	1000	169.00

Selected regrades	1000	118.90
Second hard stocks	1000	520.30

Engineering bricks
Class A	1000	548.90
Class B	1000	326.70

Cement (7.5 - 10.0 tonnes)
portland	tonne	100.76
sulphate resisting	tonne	143.00

Cement including unloading (7.5 - 10 tonnes)
portland	tonne	104.45
sulphate resisting	tonne	146.69

Hydrated lime	tonne	154.40

Hydrated lime including
unloading	tonne	157.69

Sand	tonne	11.00

203 x 3mm Butterfly pattern wall ties to BS 1243 Table 1
galvanized steel	1000	86.35
stainless steel	1000	93.50

203 x 19 x 3mm Vertical twisted wall ties to BS 1243 Table 3
galvanized steel	1000	265.10
stainless steel	1000	426.80

BRICKWORK AND BLOCKWORK (Cont'd)

Stainless steel furfix wall extension profiles
100mm wide 2338mm high.	nr	27.06
150mm wide 2338mm high.	nr	34.05
215mm wide 2338mm high.	nr	48.95

Staffordshire blue bricks
wirecuts	1000	423.50
pressed	1000	407.00
splayed plinth or bullnose	1000	2013.00

Facing bricks including crane offloading
Claydon Red Multi	1000	146.75
Tudors	1000	204.30
Milton Buff Ridgefaced	1000	188.35
Heathers	1000	204.30
Brecken Grey	1000	175.80
Leicester Red Stock	1000	244.55
West Hoathly Medium multi-stock	1000	329.25
Himley dark brown rustic	1000	328.15
Holbrook smooth red	1000	299.55
Tonbridge hand made multi-coloured	1000	579.00
Waingroves smooth red	1000	222.20
Old English Russet	1000	203.70
First hard stocks	1000	407.00
Second hard stocks	1000	319.00

Thermalite blocks
Shield 2000 blocks
75mm	m2	6.26
90mm	m2	7.50
100mm	m2	8.34
140mm	m2	11.67
150mm	m2	12.51
190mm	m2	15.84
200mm	m2	16.68

Smooth faced blocks
100mm	m2	14.30
140mm	m2	20.02
150mm	m2	21.45
190mm	m2	27.17
200mm	m2	28.60
215mm	m2	30.75

Turbo blocks
100mm	m2	8.65
115mm	m2	9.94
125mm	m2	10.81
130mm	m2	11.24
150mm	m2	12.97
190mm	m2	16.42
200mm	m2	17.29
215mm	m2	18.59

Party Wall 650 blocks
215mm	m2	18.22

Hi-Strength 7 blocks
100mm	m2	10.47
140mm	m2	14.66
150mm	m2	15.71
190mm	m2	19.81
200mm	m2	20.94
215mm	m2	22.52

Trench blocks
255mm	m2	21.60
275mm	m2	23.30
305mm	m2	25.84
355mm	m2	30.07

Lignacite blocks
solid Fair Face blocks
7.0 N/mm
100mm	m2	10.62
140mm	m2	14.75
190mm	m2	18.87
215mm	m2	21.88

Toplite blocks
solid GT1 blocks 2.8 N/mm2
115mm	m2	8.64
125mm	m2	9.39
130mm	m2	9.76
140mm	m2	10.52
150mm	m2	11.26
200mm	m2	15.02
215mm	m2	16.15

solid Standard blocks 3.5 N/mm2
75mm	m2	5.95
90mm	m2	6.55
100mm	m2	7.28
140mm	m2	10.19
150mm	m2	10.91
190mm	m2	13.83
200mm	m2	14.55
215mm	m2	15.65

Topcrete blocks
solid Standard blocks 7.0 N/mm2
75mm	m2	5.95
90mm	m2	6.55
100mm	m2	7.28
140mm	m2	10.19
190mm	m2	13.83
215mm	m2	15.65

hollow Standard blocks 7.0 N/mm2
140mm	m2	8.83
190mm	m2	12.02
215mm	m2	11.61

Hemelite blocks
solid Standard blocks 3.5 N/mm2
75mm	m2	4.95
90mm	m2	6.51
100mm	m2	5.74
140mm	m2	8.39
190mm	m2	12.10
215mm	m2	13.11

7.0 N/mm2
90mm	m2	6.95
100mm	m2	6.18
140mm	m2	8.62
190mm	m2	12.53
215mm	m2	13.64

Damp proof courses
bitumen/latex solution	litre	2.50
bituminous emulsion	litre	1.58
bituminous hessian based weighing 3.90kg/m2	m2	6.69
pitch polymer	m2	6.85
slates size 350 x 225mm	1000	1199.00

Type G general purpose Cavitray without lead
900mm long	nr	7.00
220 x 332mm external angle	nr	7.00
230 x 117mm internal angle.	nr	7.00

Type G general purpose Cavitray with 150mm Code 4 lead attached
900mm long	nr	11.50
220 x 332mm external angle	nr	12.24
230 x 117mm internal angle	nr	12.24

Type X gable abutment standard stepped Cavitray without lead to suit 40 degree pitched roof
intermediate tray	nr	2.09
Ridge tray	nr	4.08
catchment tray	nr	2.09
corner catchment angle tray	nr	4.08

Type X gable abutment standard stepped Caivtray with Code 4 lead for plain tiles to suit 40 degree pitched roof
intermediate tray	nr	4.42
ridge tray	nr	9.33
catchment tray	nr	4.42
corner catchment angle tray	nr	9.76

Type W
Cavity weep ventilator	nr	0.45
Extension duct	nr	0.46

Stainless steel mesh reinforcement
65mm	m	0.59
115mm	m	1.01
175mm	m	1.64
225mm	m	2.22

150 x 25 x 3mm frame cramp
galvanized iron	nr	0.20
stainless steel	nr	0.43

Air bricks
terra cotta, red or buff
225 x 75mm	nr	1.67
225 x 150mm	nr	2.31
225 x 225mm	nr	6.35

iron, light
225 x 75mm	nr	1.96
225 x 150mm	nr	3.72
225 x 225mm	nr	5.56

iron, light, sliding
225 x 75mm	nr	5.18
225 x 150mm	nr	9.46
225 x 225mm	nr	12.55

iron, light, louvre
225 x 75mm	nr	2.68
225 x 150mm	nr	4.42
225 x 225mm	nr	7.52

galvanized iron, light
225 x 75mm	nr	3.55
225 x 150mm	nr	6.51
225 x 225mm	nr	9.48

galvanized iron, light sliding
225 x 75mm	nr	8.06
225 x 150mm	nr	15.43
225 x 225mm	nr	19.55

galvanized iron, light, louvre
225 x 75mm	nr	4.42
225 x 150mm	nr	7.89
225 x 225mm	nr	13.27

64mm Fire bricks	1000	1348.00

230x114x76mm MPK supra
flue bricks	1000	920.00

Fireclay	tonne	585.00

MPK dribrik mortar	50kg	25.00

Cavity wall insulation
glass fibre slabs
50mm	m2	3.41
75mm	m2	4.66
expanded polystyrene pellets	m3	89.00

ROOFING

Welsh Blue/grey slates, 5.5mm thick
610 x 305mm	1000	4703.00
510 x 255mm	1000	2126.00
405 x 205mm	1000	979.00
510 x 305mm	1000	2430.00
460 x 305mm	1000	2059.00
460 x 255mm	1000	1665.00

Fibre cement slates
600 x 300mm	1000	1551.00
500 x 250mm	1000	1167.00
400 x 240mm	1000	747.00

Plain tiles
machine made	1000	400.00
hand made	1000	662.00

Sand-faced pantiles
machine made	1000	493.10

ROOFING (Cont'd)

Concrete tiles

plain.................................. 1000	298.70
interlocking 418 x 330mm 1000	665.20

Treated softwood battens

25 x 19mm......................	m	0.22
38 x 19mm......................	m	0.26
50 x 19mm......................	m	0.38
50 x 25mm......................	m	0.47

Woodwool slabs
standard quality

25mm.............................	m2	6.53
38mm.............................	m2	5.41
50mm.............................	m2	6.95
75mm.............................	m2	9.23
100mm............................	m2	12.88

pre-screeded finish

25mm.............................	m2	7.92
38mm.............................	m2	6.62
50mm.............................	m2	8.16
75mm.............................	m2	10.44
100mm............................	m2	14.09

channel reinforced, standard quality

50mm.............................	m2	17.69
75mm.............................	m2	25.96
100mm............................	m2	30.42

channel reinforced, standard quality with pretextured soffit finish

50mm.............................	m2	20.61
75mm.............................	m2	28.89
100mm............................	m2	33.34

Felt to BS 747
reinforced bitumen underfelt type 1F

15kg/10m2........................	m2	1.26

with 50mm glass fibre insulation

bonded on.........................	m2	5.76

fibre based roofing felt type 1B,

25kg/10m2........................	m2	1.20

mineral surfaced fibre based roofing felt type 1E,

38kg/10m2........................	m2	1.45

glass fibre based roofing felt

36kg/10m2........................	m2	0.77

mineral surfaced glass fibre based roofing felt type 3E 28kg/

10m2..............................	m2	1.71

Type 1F

Reinforced underslating felt.	m2	1.26

Reinforced underslating felt, aluminium foil faced 20kg/

10m2..............................	m2	3.57

Sheet lead in 250 Kg lots...tonne 789.00

Sheet zinc, commercial quality in 150-250kg lots

12 Gauge 0.65mm thick..100kg	216.50
14 Gauge 0.80mm thick.100kg	216.50

Sheet copper in 150-250kg lots

0.70mm thick.................100kg	306.00
0.60mm thick.................100kg	324.00

Sheet aluminium, commercial purity S1B0 in 10 roll lots
0.60mm thick

150mm wide..................8m roll	11.69
225mm wide..................8m roll	16.06
300mm wide..................8m roll	21.85
450mm wide..................8m roll	30.40
600mm wide..................8m roll	40.75
900mm wide..................8m roll	58.05

WOODWORK

Softwood for carcassing
General Structural (GS) Grade1 Class

SC3...............................m3	198.00

Special Structural (SS) Grade1 Class

SC4...............................m3	220.00

Machine General Structural (MGS)

Grade............................ m3	198.00

Machine Special Structural (MSS)

Grade............................ m3	220.00

Timber for First and Second Fixings and Composite items

softwood..................... m3	330.00
British Columbian Pine.... m3	1010.00
English Oak.................. m3	1430.00
Japenese Oak............... m3	3090.00
African Mahogany.......... m3	825.00

South American

Mahogany.................... m3	1270.00
Sapele........................ m3	855.00
Iroko.......................... m3	850.00
Idigbo......................... m3	815.00
Utile........................... m3	1012.00
Afrormosia.................... m3	1188.00
Beech......................... m3	891.00
Teak........................... m3	4675.00
Walnut........................ m3	2156.00

Plywood
Marine, WBP

4mm............................	m2	5.47
6 or 6.5mm...................	m2	5.50
9mm............................	m2	7.25
12mm...........................	m2	9.02
15mm...........................	m2	11.09
18mm...........................	m2	13.30
24 or 25mm...................	m2	17.73

Far Eastern Redwood/Whitewood B/BB Quality, WBP

4mm............................	m2	3.28
6 or 6.5mm...................	m2	4.19
9mm............................	m2	6.00
12mm...........................	m2	7.78
15mm...........................	m2	9.78
18mm...........................	m2	11.61
24 or 25mm...................	m2	16.13

Far Eastern Redwood/Whitewood B/BB Quality, MR

4mm............................	m2	3.11
6 or 6.5mm...................	m2	3.98
9mm............................	m2	5.71
12mm...........................	m2	7.39
15mm...........................	m2	9.30
18/19mm.......................	M2	11.02

Gaboon, B/BB Quality, WBP

4mm............................	m2	3.85
6 or 6.5mm...................	m2	4.84
9mm............................	m2	7.00
12mm...........................	m2	8.82
15mm...........................	m2	10.30
18mm...........................	m2	12.35
24 or 25mm...................	m2	17.52

Decorative plywood with balancing veneer one side, 2438 x 1219mm
figured oak

4mm............................	m2	7.22
6 or 6.5mm...................	m2	9.08
9mm............................	m2	10.31
12mm...........................	m2	11.40

sapele

4mm............................	m2	5.10
6 or 6.5mm...................	m2	6.37
9mm............................	m2	7.63
12mm...........................	m2	8.70

teak

4mm............................	m2	8.00
6 or 6.5mm...................	m2	10.40
9mm............................	m2	12.61
12mm...........................	m2	15.37

Blockboard
Birch faced, Finnish manufacturer

16mm...........................	m2	10.58
18mm...........................	m2	10.58
25mm...........................	m2	16.10

Oak faced

18mm...........................	m2	15.32

Oak faced both sides

18mm...........................	m2	17.47

Afrormosia faced

18mm...........................	m2	19.26

Teak faced

18mm...........................	m2	23.86

Chipboard, wood base

12mm...........................	m2	2.37
15mm...........................	m2	2.73
18mm...........................	m2	3.17
22mm...........................	m2	3.93
25mm...........................	m2	4.44

STRUCTURAL STEELWORK

BASIS PRICES OF ROLLED STEEL

Steel joists and beams in small quantities required for quick delivery can be obtained from local merchants at a basis cost of about £440 per tonne subject to the usual extras for size. For larger quantities obtained direct from the mills see below.

The following prices are extracted, by permission, from the lists issued by British Steel from whom complete details can be obtained.

Prices at December 1997 (list 5)
BS EN 10025 Grade S275JR

Basis prices ex basing point 10 tonnes and over in standard lengths 6000mm to 18500mm of one quality, one serial size of section and one thickness for one delivery to one destination.

Joists

Size mm	Weight kg/m		
76 x 76	12.65	tonne	415.00
89 x 89	19.35	tonne	365.00
102 x 102	23.06	tonne	365.00
114 x 114	26.79	tonne	365.00
127 x 114	26.79	tonne	365.00
127 x 114	29.79	tonne	365.00
152 x 127	37.20	tonne	375.00
203 x 152	52.09	tonne	375.00
254 x 203	81.85	tonne	385.00

Quantity extras

10 tonnes and over.................Basis
under 10 tonnes to 5 tonnes
add per tonne	10.00
under 5 tonnes...... ..add per tonne	25.00

Size mm	Weight kg/m		
Universal Beams			
914 x 419	388]		
	343]	tonne	440.00
914 x 305	289]		
	253]		
	224]		
	201]	tonne	435.00
838 x 292	226]		
	194]		
	176]	tonne	430.00

STRUCTURAL STEELWORK(Cont'd)

Universal Beams (Cont'd)

762 x 267	197]		
	173]		
	147]	tonne	430.00
686 x 254	170]		
	152]		
	140]		
	125]	tonne	430.00
610 x 305	238]		
	179]		
	149]	tonne	420.00
610 x 229	140]		
	125]		
	113]		
	101]	tonne	420.00
533 x 210	122]		
	109]		
	101]		
	92]		
	82]	tonne	405.00
457 x 191	98]		
	89]		
	82]		
	74]		
	67]	tonne	395.00
457 x 152	82]		
	74]		
	67]		
	60]		
	52]	tonne	395.00
406 x 178	74]		
	67]		
	60]		
	54]	tonne	400.00
406 x 140	46]		
	39]	tonne	400.00
356 x 171	67]		
	57]		
	51]		
	45]	tonne	400.00
356 x 127	39]		
	33]	tonne	400.00
305 x 165	54]		
	46]		
	40]	tonne	395.00
305 x 127	48]		
	42]		
	37]	tonne	395.00
305 x 102	33]		
	28]		
	25]	tonne	395.00
254 x 146	43]		
	37]		
	31]	tonne	380.00
254 x 102	28]		
	25]		
	22]	tonne	380.00
203 x 133	30]		
	25]	tonne	340.00
203 x 102	23]	tonne	360.00
178 x 102	19]	tonne	330.00
152 x 89	16]	tonne	370.00
127 x 76	13]	tonne	390.00

Universal columns

356 x 406	634]		
	551]		
	467]		
	393]		
	340]		
	287]		
	235]	tonne	440.00
356 x 368	202]		
	177]		
	153]		
	129]	tonne	440.00
305 x 305	283]		
	240]		
	198]		
	158]		
	137]		
	118]		
	97]	tonne	420.00
254 x 254	167]		
	132]		
	107]		
	89]		
	73]	tonne	400.00
203 x 203	86]		
	71]		
	60]		
	52]		
	46]	tonne	390.00
152 x 152	37]		
	30]		
	23]	tonne	360.00

Quantity extras

10 tonnes and over.......................Basis
under 10 tonnes to 5 tonnes
add per tonne 10.00
under 5 tonnes...... ..add per tonne 25.00

METALWORK

Mild steel sections (BS 4848)
equal angles

25 x 25 x 3mm............	m	0.60
25 x 25 x 4mm............	m	0.74
25 x 25 x 5mm............	m	0.86
30 x 30 x 3mm............	m	0.74
30 x 30 x 4mm............	m	0.91
30 x 30 x 5mm............	m	1.08
40 x 40 x 3mm............	m	1.00
40 x 40 x 4mm............	m	1.22
40 x 40 x 5mm............	m	1.35
40 x 40 x 6mm............	m	1.76
45 x 45 x 3mm............	m	1.13
45 x 45 x 4mm............	m	1.38
45 x 45 x 5mm............	m	1.56
45 x 45 x 6mm............	m	2.03
50 x 50 x 3mm............	m	1.25
50 x 50 x 4mm............	m	1.52
50 x 50 x 5mm............	m	1.80
50 x 50 x 6mm............	m	2.39
50 x 50 x 8mm............	m	2.96

unequal angles

40 x 25 x 4mm	m	0.93
60 x 30 x 5mm	m	1.56
60 x 30 x 6mm	m	2.13
65 x 50 x 5mm	m	2.01
65 x 50 x 6mm............	m	2.67
65 x 50 x 8mm............	m	3.27

circular hollow sections

21.3mm dia. x 3.2mm	m	2.36
26.9mm dia. x 3.2mm	m	2.95
33.7mm dia. x 2.6mm	m	3.15
33.7mm dia. x 3.2mm	m	3.77
33.7mm dia. x 4.0mm	m	4.59
42.4mm dia. x 2.6mm	m	4.00
42.4mm dia. x 3.2mm	m	4.84
42.4mm dia. x 4.0mm	m	5.98
48.3mm dia. x 3.2mm	m	5.61
48.3mm dia. x 4.0mm	m	6.88
48.3mm dia. x 5.0mm	m	8.16

rectangular hollow sections

50 x 30 x 2.6mm.........	m	4.38
50 x 30 x 3.2mm.........	m	5.27
60 x 40 x 3.2mm.........	m	7.24
60 x 40 x 4.0mm.........	m	8.88
80 x 40 x 3.2mm.........	m	8.22
80 x 40 x 4.0mm.........	m	10.09
90 x 50 x 3.6mm.........	m	11.75
90 x 50 x 5.0mm.........	m	15.22
100 x 50 x 3.2mm.........	m	11.13
100 x 50 x 4.0mm.........	m	13.73
100 x 50 x 6.0mm........	m	16.26

METALWORK (Cont'd)

Mild steel sections (BS 4848) (Cont'd.)
square hollow sections
20 x 20 x 2.0mm	m	2.22
20 x 20 x 2.6mm	m	2.78
30 x 30 x 2.6mm	m	4.38
30 x 30 x 3.2mm	m	5.27
40 x 40 x 2.6mm	m	5.11
40 x 40 x 3.2mm	m	5.68
40 x 40 x 4.0mm	m	7.82
50 x 50 x 3.2mm	m	7.22
50 x 50 x 4.0mm	m	8.88
50 x 50 x 5.0mm	m	10.47

Mild steel bars (BS 6722)
round
10mm diameter	m	0.42
12mm diameter	m	0.64
16mm diameter	m	1.10
20mm diameter	m	1.72
25mm diameter	m	2.69
30mm diameter	m	3.88
32mm diameter	m	4.40
35mm diameter	m	5.28
40mm diameter	m	6.90
42mm diameter	m	7.59
45mm diameter	m	8.72
48mm diameter	m	9.93
50mm diameter	m	10.77

square
10 x 10mm	m	0.57
12 x 12mm	m	0.77
14 x 14mm	m	1.10
16 x 16mm	m	1.40
18 x 18mm	m	1.79
20 x 20mm	m	2.20
22 x 22mm	m	2.65
24 x 24mm	m	3.17
25 x 25mm	m	3.42
28 x 28mm	m	4.30
30 x 30mm	m	4.94
32 x 32mm	m	5.61
35 x 35mm	m	6.73
38 x 38mm	m	7.91
40 x 40mm	m	8.77
42 x 42mm	m	9.68
45 x 45mm	m	11.12
50 x 50mm	m	13.59

Mild steel flat sheets
26 gauge	m2	2.99
24 gauge	m2	3.66
22 gauge	m2	4.68
20 gauge	m2	6.00

Galvanized mild steel flat sheets
26 gauge	m2	6.25
24 gauge	m2	12.35
22 gauge	m2	9.23
20 gauge	m2	11.26

Galvanized mild steel strip
18 gauge x 25mm wide	kg	2.49
17 gauge x 29mm wide	kg	2.49
16 gauge x 32mm wide	kg	2.49
16 gauge x 38mm wide	kg	2.49

PLUMBING AND MECHANICAL ENGINEERING INSTALLATIONS

Cast iron eaves gutters (BS 460)
half round
102mm	m	7.91
angle	nr	6.03
nozzle piece	nr	5.87
stop end	nr	2.02
stop end outlet	nr	4.40
114mm	m	8.23
angle	nr	6.20

Cast iron eaves gutters(BS 460) (Cont'd)
half round (Cont'd)
114mm (Cont'd)
nozzle piece	nr	6.39
stop end	nr	2.62
stop end outlet	nr	4.84
127mm	m	9.64
angle	nr	7.33
nozzle piece	nr	7.33
stop end	nr	2.62
stop end outlet	nr	6.51
152mm	m	16.47
angle	nr	13.36
nozzle piece	nr	12.68
stop end	nr	3.50
stop end outlet	nr	12.44

Cast iron eaves gutters (BS 460)
ogee
102mm	m	8.82
angle	nr	6.29
nozzle piece	nr	6.29
stop end	nr	1.83
stop end outlet	nr	4.78
114mm	m	9.71
angle	nr	6.82
nozzle piece	nr	6.82
stop end	nr	2.43
stop end outlet	nr	4.78
127mm	nr	10.18
angle	nr	7.44
nozzle piece	nr	7.44
stop end	nr	2.43
stop end outlet	nr	5.68

galvanized mild steel bracket
for half round gutter
102mm	nr	1.63
114mm	nr	1.63
127mm	nr	1.63
152mm	nr	2.07

Unplasticised PVC gutters (BS 4576)
half round
105mm	m	2.52
angle	nr	3.16
outlet	nr	2.89
stop end	nr	1.38
114mm	m	2.50
angle	nr	3.37
outlet	nr	3.09
stop end	nr	1.65

joint clips
100mm	nr	1.90
114mm	nr	2.76

brackets
100mm	nr	0.72
114mm	nr	0.69

Cast iron round rainwater pipes (BS 460)
pipe without ears
76mm	m	14.55
102mm	m	19.86

pipe with ears
64mm	m	15.55
76mm	m	15.55
102mm	m	20.87

bend
64mm	nr	8.41
76mm	nr	10.21
102mm	nr	14.41

shoe (eared)
64mm	nr	13.72
76mm	nr	13.72
102mm	nr	17.88

offset 76mm projection
64mm	nr	12.87
76mm	nr	12.87
102mm	nr	24.27

offset 152mm projection
64mm	nr	12.87

Cast iron round rainwater pipes

(BS 460) (Cont'd)
offset 152mm projection (Cont'd)
76mm	nr	12.87
102mm	nr	24.27

offset 229mm projection
64mm	nr	14.98
76mm	nr	14.98
102mm	nr	29.38

offset 305mm projection
64mm	nr	17.53
76mm	nr	18.34
102mm	nr	29.38

single branch
64mm	nr	16.20
76mm	nr	17.87
102mm	nr	21.22

mild steel holderbat for screwing on
76mm	nr	5.02
102mm	nr	5.30

mild steel holderbat for building in
76mm	nr	6.25
102mm	nr	6.44

Unplasticised PVC rainwater pipes (BS 4576)
68mm pipe with slip joints	m	2.35
bend	nr	3.04
shoe	nr	2.63
offset bend		
socket	nr	1.82
spigot	nr	1.69
single branch	nr	6.07
pipe clip	nr	1.13

Black steel tubes and fittings (BS 1387)
medium weight tubing
15mm	m	1.48
20mm	m	1.74
25mm	m	2.50
32mm	m	3.09
40mm	m	3.60
50mm	m	5.07

heavy weight tubing
15mm	m	1.77
20mm	m	2.09
25mm	m	3.06
32mm	m	3.80
40mm	m	4.42
50mm	m	6.14

long screw and backnut
15mm	nr	2.33
20mm	nr	2.61
25mm	nr	3.86
32mm	nr	5.11
40mm	nr	6.19
50mm	nr	9.22

Galvanized steel tubes and fittings (BS 1387)
medium weight tubing
15mm	m	2.25
20mm	m	2.54
25mm	m	3.55
32mm	m	4.38
40mm	m	5.10
50mm	m	7.15

heavy weight tubing
15mm	m	2.67
20mm	m	3.02
25mm	m	4.30
32mm	m	5.34
40mm	m	6.23
50mm	m	8.63

long screw and backnut
15mm	nr	2.76
20mm	nr	3.24
25mm	nr	4.58
32mm	nr	5.66
40mm	nr	6.93
50mm	nr	10.41

PLUMBING AND MECHANICAL ENGINEERING INSTALLATIONS (Cont'd)

Black malleable cast iron pipe fittings
90 degree elbow

15mm	nr	0.46
20mm	nr	0.62
25mm	nr	0.98
32mm	nr	1.60
40mm	nr	2.69
50mm	nr	3.15

45 degree elbow

15mm	nr	0.98
20mm	nr	1.20
25mm	nr	1.77
32mm	nr	3.26
40mm	nr	4.00
50mm	nr	5.49

tee

15mm	nr	0.62
20mm	nr	0.91
25mm	nr	1.31
32mm	nr	2.17
40mm	nr	2.98
50mm	nr	4.29

pitcher tee

15mm	nr	1.72
20mm	nr	2.12
25mm	nr	3.17
32mm	nr	4.34
40mm	nr	6.71
50mm	nr	9.43

cross

15mm	nr	1.48
20mm	nr	2.22
25mm	nr	2.83
32mm	nr	3.72
40mm	nr	5.01
50mm	nr	7.77

socket, parallel thread

15mm	nr	0.43
20mm	nr	0.52
25mm	nr	0.69
32mm	nr	1.17
40mm	nr	1.60
50mm	nr	2.41

reducing socket

15mm	nr	0.60
20mm	nr	0.62
25mm	nr	0.83
32mm	nr	1.48
40mm	nr	2.29
50mm	nr	3.22

cap

15mm	nr	0.40
20mm	nr	0.46
25mm	nr	0.57
32mm	nr	0.83
40mm	nr	1.05
50mm	nr	2.05

plug

15mm	nr	0.34
20mm	nr	0.40
25mm	nr	0.52
32mm	nr	0.83
40mm	nr	1.03
50mm	nr	1.77

straight union, female, standard pattern

15mm	nr	1.83
20mm	nr	2.00
25mm	nr	2.34
32mm	nr	3.54
40mm	nr	4.00
50mm	nr	6.63

Galvanized malleable cast iron pipe fittings
90 degree elbow

15mm	nr	0.65
20mm	nr	0.88
25mm	nr	1.38
32mm	nr	2.26
40mm	nr	3.80
50mm	nr	4.43

45 degree elbow

15mm	nr	1.38
20mm	nr	1.69
25mm	nr	2.51
32mm	nr	4.60
40mm	nr	5.66
50mm	nr	7.75

tee

15mm	nr	0.88
20mm	nr	1.29
25mm	nr	1.86
32mm	nr	3.07
40mm	nr	4.20
50mm	nr	6.06

pitcher tee

15mm	nr	2.42
20mm	nr	2.99
25mm	nr	4.49
32mm	nr	6.14
40mm	nr	9.49
50mm	nr	13.31

cross

15mm	nr	2.09
20mm	nr	3.15
25mm	nr	3.99
32mm	nr	5.25
40mm	nr	7.06
50mm	nr	10.97

socket, parallel thread

15mm	nr	0.61
20mm	nr	0.73
25mm	nr	0.98
32mm	nr	1.65
40mm	nr	2.26
50mm	nr	3.39

reducing socket

15mm	nr	0.85
20mm	nr	0.88
25mm	nr	1.17
32mm	nr	2.09
40mm	nr	3.22
50mm	nr	4.56

cap

15mm	nr	0.56
20mm	nr	0.65
25mm	nr	0.81
32mm	nr	1.17
40mm	nr	1.50
50mm	nr	2.91

plug

15mm	nr	0.48
20mm	nr	0.56
25mm	nr	0.73
32mm	nr	1.17
40mm	nr	1.46
50mm	nr	2.51

straight union, female, standard pattern

15mm	nr	2.59
20mm	nr	2.82
25mm	nr	3.30
32mm	nr	5.01
40mm	nr	5.66
50mm	nr	9.36

Hepworth flexible polybutylene pipes and fittings
pipe

15mm	m	0.94
22mm	m	1.83
28mm	m	2.51

straight connector

15mm	nr	1.14
22mm	nr	1.53
28mm	nr	3.91

Hepworth flexible polybutylene pipes and fittings (Cont'd)
elbow 90 degrees

15mm	nr	1.40
22mm	nr	2.04
28mm	nr	4.70

tee (Equal)

15mm	nr	1.98
22mm	nr	2.53
28mm	nr	6.48

straight tap connector

15mm	nr	1.55
15mm x 3/4"	nr	1.84

pipe support sleeves

15mm	nr	0.19
22mm	nr	0.23
28mm	nr	0.35

Copper tubing (BS 2871)
table X

15mm	m	0.87
22mm	m	1.77
28mm	m	2.39
35mm	m	6.02
42mm	m	7.37
54mm	m	9.53
66.7mm	m	13.98
76.1mm	m	19.71
108mm	m	28.94

table Y

15mm	m	2.54
22mm	m	4.43
28mm	m	5.84
35mm	m	8.19
42mm	m	9.87
54mm	m	17.00

Compression pipe fittings for copper tubing (BS 864 Part 2)
Fittings type A - all copper
straight coupling

15mm	nr	0.69
22mm	nr	1.18
28mm	nr	3.56
35mm	nr	8.15
42mm	nr	11.32
54mm	nr	16.20

bent coupling

15mm	nr	0.93
22mm	nr	1.42
28mm	nr	4.47
35mm	nr	10.74
42mm	nr	15.00
54mm	nr	25.51

tee (Equal)

15mm	nr	1.22
22mm	nr	1.99
28mm	nr	7.92
35mm	nr	14.17
42mm	nr	21.83
54mm	nr	35.13

Fittings type A - copper to male iron
straight coupling

15mm	nr	0.64
22mm	nr	1.04
28mm	nr	2.03
35mm	nr	5.95
42mm	nr	9.57
54mm	nr	14.51

bent coupling

15mm	nr	1.22
22mm	nr	1.59
28mm	nr	3.41
35mm	nr	10.26
42mm	nr	14.78
54mm	nr	24.15

tee (female iron branch)

15mm	nr	3.65
22mm	nr	5.71
28mm	nr	8.05
35mm	nr	17.13

PLUMBING AND MECHANICAL ENGINEERING INSTALLATIONS (Cont'd)

Compression pipe fittings for copper tubing (BS 864 Part 2) (Cont'd)
Fittings type A - copper to male iron (Cont'd)
back plate wall elbow

15mm	nr	2.68

tank coupling

15mm	nr	3.46
22mm	nr	3.90
28mm	nr	6.14
35mm	nr	9.37
42mm	nr	13.35
54mm	nr	18.11

Fittings type B - all copper
straight coupling

15mm	nr	3.69
22mm	nr	6.00
28mm	nr	9.81
35mm	nr	18.00
42mm	nr	24.50
54mm	nr	36.25

bent coupling

15mm	nr	4.38
22mm	nr	7.13
28mm	nr	12.06
35mm	nr	23.19
42mm	nr	34.75
54mm	nr	55.31

tee (Equal)

15mm	nr	6.13
22mm	nr	10.31
28mm	nr	17.38
35mm	nr	31.50
42mm	nr	51.25
54mm	nr	82.19

Fittings type B - copper to male iron
straight coupling

15mm	nr	3.38
22mm	nr	4.81
28mm	nr	7.38
35mm	nr	14.56
42mm	nr	21.50
54mm	nr	31.13

bent coupling

15mm	nr	4.75
22mm	nr	6.56
28mm	nr	11.19
35mm	nr	25.00

Capillary fittings for copper tubing (BS 864 Part 2)
all copper
straight coupling

15mm	nr	0.13
22mm	nr	0.36
28mm	nr	0.82
35mm	nr	2.72
42mm	nr	4.06
54mm	nr	8.39

bent coupling

15mm	nr	0.25
22mm	nr	0.59
28mm	nr	1.33
35mm	nr	5.40
42mm	nr	9.61
54mm	nr	19.67

tee (Equal)

15mm	nr	0.42
22mm	nr	1.20
28mm	nr	3.71
35mm	nr	9.27
42mm	nr	14.52
54mm	nr	27.44

Capillary fittings for copper tubing (BS 864 Part 2) (Cont'd)
copper to male iron
straight connector

15mm	nr	1.63
22mm	nr	2.87
28mm	nr	4.55
35mm	nr	8.02
42mm	nr	10.34
54mm	nr	15.69

bent connector

15mm	nr	3.17
22mm	nr	4.51
28mm	nr	6.76
35mm	nr	9.57
42mm	nr	11.38
54mm	nr	20.11

tee (1/2 " female iron branch)

15mm	nr	2.09
22mm	nr	2.90
28mm	nr	8.74
35mm	nr	14.46
42mm	nr	17.25

back plate wall elbow

15mm	nr	3.39
22mm	nr	7.14

tank coupling

15mm	nr	3.86
22mm	nr	5.90
28mm	nr	7.75
35mm	nr	10.17
42mm	nr	13.33
54mm	nr	20.35

black malleable iron screw on brackets

15mm	nr	0.66
20mm	nr	0.73
25mm	nr	0.85
32mm	nr	1.16
40mm	nr	1.55
50mm	nr	2.04

galvanized malleable iron screw on brackets

15mm	nr	0.92
20mm	nr	1.03
25mm	nr	1.20
32mm	nr	1.63
40mm	nr	2.17
50mm	nr	2.87

Pipe clips for copper tube
two piece copper spacing clips

15mm	100	14.35
22mm	100	15.26
28mm	100	20.64
35mm	100	30.51
42mm	100	56.54
54mm	100	72.70

pressed brass screw on brackets

15mm	100	85.36
22mm	100	97.32
28mm	100	116.47
35mm	100	128.44
42mm	100	149.97
54mm	100	190.65

cast brass screw on brackets

15mm	100	213.76
22mm	100	252.07
28mm	100	307.11
35mm	100	396.44
42mm	100	524.86
54mm	100	663.68

Polythene tubing
BS 6572, blue

20mm	m	0.30
25mm	m	0.39
32mm	m	0.63
50mm	m	1.43
63mm	m	2.11

Brass compression fittings for polythene tube
straight coupling

20mm	nr	3.99
25mm	nr	5.66
32mm	nr	9.80
50mm	nr	23.73
63mm	nr	32.41

straight coupling with one end threaded

20mm	nr	3.52
25mm	nr	4.89
32mm	nr	7.24
50mm	nr	18.43
63mm	nr	24.38

tee

20mm	nr	6.70
25mm	nr	10.39
32mm	nr	15.63
50mm	nr	38.97
63mm	nr	57.21

elbow

20mm	nr	4.90
25mm	nr	7.14
32mm	nr	12.56
50mm	nr	29.30
63mm	nr	34.76

tap connector

20mm	nr	5.60
25mm	nr	9.59

Stop cocks (BS 1010)
brass
copper x copper

15mm	nr	3.55
22mm	nr	6.18
28mm	nr	14.83
35mm	nr	44.56
42mm	nr	56.24
54mm	nr	84.89

DZR
copper x copper

15mm	nr	10.32
22mm	nr	16.94
28mm	nr	28.04
35mm	nr	51.60
42mm	nr	73.87
54mm	nr	100.61

DZR
polythene x polythene

20mm	nr	16.89
25mm	nr	26.85
32mm	nr	36.04
50mm	nr	90.10
63mm	nr	126.48

Gate valves (BS 5154)
brass, fullway, Series B ends screwed female BSP thread

13mm	nr	5.30
19mm	nr	6.46
25mm	nr	8.96
32mm	nr	13.09
38mm	nr	19.18
50mm	nr	27.50

copper x copper

15mm	nr	6.41
22mm	nr	7.77
28mm	nr	10.71
35mm	nr	20.16
42mm	nr	32.89
54mm	nr	50.93

bronze, fullway, Series B ends screwed female BSP thread

13mm	nr	13.08
19mm	nr	18.55
25mm	nr	24.01
32mm	nr	34.23
38mm	nr	46.84
50mm	nr	67.05

PLUMBING AND MECHANICAL ENGINEERING INSTALLATIONS (Cont'd)

Ball valves (BS 1212)
piston type with copper ball
low pressure

13mm	nr	11.78
19mm	nr	21.39
25mm	nr	57.15

high pressure

13mm	nr	10.11
19mm	nr	21.18
25mm	nr	57.15
32mm	nr	111.05

diaphragm type (brass body) with copper ball
low pressure

13mm	nr	11.88

high pressure

13mm	nr	11.71

Cast iron soil and waste pipes (BS 416)
for coupling joints
spigoted pipe

50mm	m	12.01
75mm	m	11.57
100mm	m	13.97

couplings with gaskets

50mm	nr	5.24
75mm	nr	5.81
100mm	nr	7.56

bend

51mm	nr	9.25
76mm	nr	9.25
100mm	nr	12.80

single junction

50mm	nr	13.92
75mm	nr	13.92
100mm	nr	19.79

double junction

75mm	nr	23.39
100mm	nr	24.47

anti-syphon junction

100mm	nr	20.73

access pipe

50mm	nr	22.24
75mm	nr	22.24
100mm	nr	23.39

offset 150mm projection

75mm	nr	11.39
100mm	nr	16.05

offset 305mm projection

100mm	nr	20.73

roof connector

50mm	nr	25.33
75mm	nr	25.33
100mm	nr	22.32

bossed pipe 150mm long tapped 32mm

75mm	nr	19.15
100mm	nr	22.87

bossed pipe 150mm long tapped 32mm and 38mm

75mm	nr	25.86
100mm	nr	29.58

bossed pipe 240mm long tapped two 38mm

100mm	nr	29.88

WC connector 305mm long

100mm	nr	13.93

wall fixing bracket

75mm	nr	3.51
100mm	nr	3.94

Cast iron Stanton SMU lightweight soil and waste pipes for coupling joints
spigoted pipe

50mm	m	8.21
75mm	m	9.50
100mm	m	11.42

Cast iron Stanton SMU lightweight soil and waste pipes for coupling joints (Cont'd)
stainless steel couplings with gaskets

50mm	nr	3.46
75mm	nr	3.83
100mm	nr	5.00

short pipe with access door

50mm	nr	15.74
75mm	nr	17.20
100mm	nr	18.85

bends

50mm	nr	5.93
75mm	nr	6.67
100mm	nr	7.90

offset bend 75mm projection

50mm	nr	8.60
75mm	nr	11.47
100mm	nr	14.77

offset bend 150mm projection

50mm	nr	11.06
75mm	nr	15.27
100mm	nr	20.06

single angled branch

50mm	nr	9.55
75mm	nr	10.05
100mm	nr	12.78

Universal plug number of inlets 1

50mm	nr	7.14
75mm	nr	10.18

Universal plug no of inlets 3

100mm	nr	12.81

Traditional joint connector

100mm	nr	16.93

W.C. Connector

100mm	nr	9.57

Roof Connector

100mm	nr	34.59

Unplasticised MPVC waste pipes (solvent welded) (BS 5255)
pipe

32mm	m	1.68
40mm	m	1.77
50mm	m	3.04

bend

32mm	nr	1.20
40mm	nr	1.34
50mm	nr	2.22

tee

32mm	nr	1.70
40mm	nr	2.15
50mm	nr	4.24

coupling

32mm	nr	0.94
40mm	nr	1.13
50mm	nr	1.40

expansion coupling

32mm	nr	1.35
40mm	nr	1.63
50mm	nr	2.21

access plug

32mm	nr	1.42
40mm	nr	1.68
50mm	nr	2.72

plastic clips

32mm	nr	0.25
40mm	nr	0.30
55mm	nr	0.75
solvent cement	litre	17.00

Unplasticised PVC soil and ventilating pipes (BS 4514) with rubber ring "push fit" system
Pipe

82mm	m	5.15
110mm	m	4.90

bend

82mm	nr	6.81
110mm	nr	8.20

single junction

82mm	nr	10.70
110mm	nr	10.67

double junction

110mm	nr	19.67

access pipe

82mm	nr	12.14
110mm	nr	11.97

boss for copper

82mm	nr	6.38
110mm	nr	6.02

boss for PVC

82mm	nr	6.38
110mm	nr	6.02

W.C. connector

82mm	nr	6.27
110mm	nr	7.16

weathering apron

82mm	nr	2.48
110mm	nr	1.96

drain connector

110mm	nr	11.15

plastic coated metal holderbat for screwing on

82mm	nr	3.03
110mm	nr	2.99

plastic coated metal holderbat for building in

82mm	nr	3.94
110mm	nr	3.90

Plastic traps (BS 3943)
traps with 'O' ring joint outlet
P trap

36mm	nr	3.08
42mm	nr	3.55

S trap

36mm	nr	3.90
42mm	nr	4.58

bath trap with overflow connection

42mm	nr	6.60

Copper traps (BS 1184)
two piece trap with compression joint outlet
P trap, 76mm seal

35mm	nr	15.34
42mm	nr	19.53
54mm	nr	70.04

S trap, 76mm seal

35mm	nr	17.58
42mm	nr	21.11
54mm	nr	73.98

bath trap with male iron overflow connection, 76mm seal

42mm	nr	28.82

PLUMBING AND MECHANICAL ENGINEERING INSTALLATIONS (Cont'd)

Wire balloon gratings
galvanized iron

51mm	nr	2.02
76mm	nr	2.28
102mm	nr	2.60

copper

51mm	nr	2.55
76mm	nr	3.00
102mm	nr	3.45

plastic

51mm	nr	1.27
76mm	nr	1.85
102mm	nr	1.96

Galvanized water storage cisterns inc lid (BS 417) and byelaw 30 kit
Grade A open top cistern, BS type

SCM45, size 460 x 310 x 310mm, actual capacity 18 litres.. nr		97.70
SCM70, size 610 x 310 x 380mm, actual capacity 36 litres.. nr		119.60
SCM90, size 610 x 410 x 380mm, actual capacity 55 litres.. nr		132.55
SCM110, size 610 x 430 x 430mm, actual capacity 68 litres... nr		140.25
SCM135, size 610 x 460 x 480mm, actual capacity 86 litres... nr		148.95
SCM180, size 690 x 510 x 510mm, actual capacity 114 litres.. nr		165.75
SCM230, size 740 x 560 x 560mm, actual capacity 159 litres.. nr		207.70
SCM270, size 760 x 580 x 610mm, actual capacity 191 litres.. nr		221.70
SCM320, size 910 x 610 x 580mm, actual capacity 227 litres.. nr		241.80
SCM450/1, size 1220 x 610 x 610mm, actual capacity 327 litres.. nr		284.50
SCM450/2, size 970 x 690 x 690mm, actual capacity 336 litres. nr		276.40
SCM570, size 970 x 760 x 790mm, actual capacity 423 litres.. nr		372.40
SCM680, size 1090 x 860 x 740mm, actual capacity 491 litres.. nr		420.60
SCM910, size 1170 x 890 x 890mm, actual capacity 709 litres.. nr		412.40*
SCM1130, size 1520 x 910 x 910mm, actual capacity 841 litres.. nr		458.35*
SCM1600, size 1520 x 1140 x 910mm, actual capacity 1250 litres..nr		690.90*
SCM2270, size 1830 x 1220 x 1020mm, actual capacity 1728 litres nr		813.55*
SCM2720, size 1830 x 1220 x 1220mm, actual capacity 2137 litres..nr		875.85*
SCM4540, size 2440 x 1520 x 1220mm, actual capacity 3364 litres nr		1549.90*

* Price excludes insulation.

Rectangular plastic water storage cisterns (BS 4213)
open top cistern, BS type

C4, minimum capacity to water line 18 litres.........nr		3.78
C25, minimum capacity to water line 114 litres........ nr		19.60
C40, minimum capacity to water line 182 litres........ nr		34.01
C50, minimum capacity to water line 227 litres.........nr		36.62

sealed lid for cistern, byelaw 30 kit

10 litre	nr	0.54
114 litre	nr	17.76
182 litre	nr	23.37
227 litre	nr	29.69

Pipe insulation
13mm Armaflex/Class O for iron pipes

15mm	m	2.38

Pipe insulation (Cont'd)
13mm Armaflex/Class O for iron pipes

20mm	m	2.73
25mm	m	2.98
32mm	m	3.43
40mm	m	4.09
50mm	m	5.19

13mm Armaflex/Class O for copper pipes

15mm	m	1.99
22mm	m	2.38
28mm	m	2.73
35mm	m	2.98
42mm	m	3.43
54mm	m	4.50

19mm rigid section fibre glass for iron pipes

15mm	m	3.24
20mm	m	3.44
25mm	m	3.66
32mm	m	4.01
40mm	m	4.31
50mm	m	4.89

19mm rigid section fibre glass for copper pipes

15mm	m	3.08
22mm	m	3.24
28mm	m	3.44
35mm	m	3.66
42mm	m	4.01
54mm	m	4.60

25mm rigid section fibre glass for iron pipes

15mm	m	3.53
20mm	m	3.78
25mm	m	4.19
32mm	m	4.54
40mm	m	4.84
50mm	m	5.51

25mm rigid section fibre glass for copper pipes

15mm	m	3.44
22mm	m	3.53
28mm	m	3.78
35mm	m	4.19
42mm	m	4.54
54mm	m	5.33

ELECTRICAL INSTALLATIONS

Steel conduits
black enamelled, heavy gauge, welded, screwed

20mm	100m	257.08
25mm	100m	352.36
32mm	100m	488.19
1.5"	100m	624.90
2"	100m	1009.94

galvanized, heavy gauge, welded, screwed

20mm	100m	388.12
25mm	100m	520.32
32mm	100m	658.68
1.5"	100m	868.00
2"	100m	1242.26

PVC Conduits
heavy gauge super high impact

16mm	100m	192.71
20mm	100m	192.71
25mm	100m	260.34
32mm	100m	418.57
38mm	100m	540.15
50mm	100m	898.27

oval section (nominal sizes)

16mm	100m	70.30
20mm	100m	91.52
25mm	100m	127.18

bending springs, heavy gauge

16mm	nr	11.55
20mm	nr	11.55
25mm	nr	17.14

PVC conduits, heavy gauge super high impact (Cont'd)
bending springs, heavy gauge (Cont'd)

32mm	nr	24.39
38mm	nr	38.83
50mm	nr	53.50

adhesive vinyl cement

1/4 litre	nr	8.27

Steel conduit fittings
bends, black enamel, screwed

20mm	10	23.77
25mm	10	37.86
32mm	10	66.47
1.5"	10	101.61
2"	10	201.78

inspection elbows, black enamel, screwed

20mm	10	46.01
25mm	10	57.10
32mm	10	120.70
1.5"	10	134.34
2"	10	334.32

inspection tees, black enamel, screwed

20mm	10	50.37
25mm	10	79.39
32mm	10	218.37

if galvanized, add 20% to above prices
black enamel standard circular boxes screwed to BS 31, 51mm centres, without covers, with 4mm tapped hole in base, for 20mm conduit

back outlet	10	75.44
terminal	10	36.85
terminal and back outlet	10	90.25
angle	10	43.84
angle tangent	10	92.61
through way	10	43.84
branch "U"	10	62.42
through way and back outlet	10	106.09
three way	10	48.17
three way tangent	10	105.74
branch three way	10	90.64
four way	10	61.15
twin through way	10	128.20
extra for light steel covers	10	5.87
extra for heavy steel covers	10	6.93
extra for rubber gaskets	10	11.76

black enamel standard circular boxes screwed to BS 31, 51mm centres, without covers, with 4mm tapped hole in base for 25mm conduit

back outlet	10	114.30
terminal	10	51.06
terminal and back outlet	10	129.73
angle	10	60.92
angle tangent	10	134.84
through way	10	60.92
branch "U"	10	134.84
through way and back outlet	10	145.73
three way	10	64.39
three way tangent	10	142.24
branch three way	10	150.36
four way	10	89.78
extra for light steel covers	10	5.87
extra for heavy steel covers	10	6.99
extra for rubber gaskets	10	11.76

32mm deep looping in boxes standard 4 hole pattern with 20mm knockouts and 51mm centre fixing lugs.

	10	21.27

saddles bar

20mm	10	5.43
25mm	10	6.80
32mm	10	19.41
1.5"	10	19.68
2"	10	23.89

ELECTRICAL INSTALLATIONS (Cont'd)

Steel conduit fittings (Cont'd)

saddles, distance

20mm	10	17.47
25mm	10	22.68
32mm	10	31.92
1.5"	10	55.80
2"	10	79.16

brass bushes, hexagon smooth bore

20mm	10	7.41
25mm	10	13.29
32mm	10	26.83
1.5"	10	25.43
2"	10	75.43

brass bushes, ring

20mm	10	4.12
25mm	10	5.53
32mm	10	10.24
1.5"	10	17.77
2"	10	35.90

locknuts, circular pattern

20mm	10	2.47
25mm	10	4.40
32mm	10	11.12
1.5"	10	13.96
2"	10	28.52

couplers, solid

20mm	10	5.38
25mm	10	6.22
32mm	10	19.11
1.5"	10	25.54
2"	10	41.68
standard hook plates	10	26.02

standard pendant plates, internal thread

20mm	10	37.28
25mm	10	53.04

adaptable steel boxes, plain sides with flanged lid and screws

75 x 75 x 37mm deep	10	41.38
100 x 100 x 37mm deep	10	42.88
150 x 75 x 37mm deep	10	44.50
75 x 75 x 50mm deep	10	41.88
100 x 100 x 50mm deep	10	50.38
150 x 75 x 50mm deep	10	51.00
150 x 150 x 50mm deep	10	71.75
225 x 150 x 50mm deep	10	108.63
225 x 225 x 50mm deep	10	137.75
75 x 75 x 75mm deep	10	55.00
100 x 100 x 75mm deep	10	63.00
150 x 75 x 75mm deep	10	63.00
150 x 150 x 75mm deep	10	87.63
225 x 150 x 75mm deep	10	123.75
225 x 225 x 75mm deep	10	154.00
300 x 300 x 75mm deep	10	214.63
150 x 150 x 100mm deep	10	130.50
225 x 150 x 100mm deep	10	138.38
225 x 225 x 100mm deep	10	176.63
300 x 300 x 100mm deep	10	343.75

PVC conduit fittings

bends

20mm	10	24.76
25mm	10	33.34
32mm	10	56.35
38mm	10	98.14
50mm	10	172.18

inspection elbows

20mm	10	15.97

inspection tees

20mm	10	22.24
25mm	10	44.56

standard circular boxes, M4 threaded inserts, no earth terminal, for 20mm conduit

back outlet	10	23.65
terminal	10	17.15
terminal and back outlet	10	27.91
angle	10	19.49
through way	10	19.49

PVC conduit fittings (Cont'd)

standard circular boxes, M4 threaded inserts, no earth terminal, for 20mm conduit (Cont'd)

branch "U"	10	35.76
through way and back outlet	10	30.69
three way	10	21.33
branch three way	10	44.19
four way	10	25.04
twin through way	10	45.97
extra for standard cover	10	5.16
extra for overlapping cover	10	9.01
extra for rubber gasket	10	5.71
extra for brass earthing terminal	10	13.24

standard circular boxes, M4 threaded inserts, no earth terminal, for 25mm conduit

back outlet	10	28.32
terminal	10	26.84
terminal and back outlet	10	41.00
angle	10	29.13
through way	10	29.13
through way and back outlet	10	43.53
three way	10	32.05
four way	10	36.46
extra for standard cover	10	5.16
extra for overlapping cover	10	9.01
extra for rubber gasket	10	5.71
extra for brass earthing terminal	10	13.24

32mm deep looping in boxes with four 20mm knockouts, no earth

terminal	10	13.63

couplers, heavy gauge

20mm	10	4.19
25mm	10	5.71
32mm	10	12.96
38mm	10	25.22
50mm	10	44.21

couplers, expansion

20mm	10	10.01
25mm	10	12.96

saddles, spacer bar

20mm	10	6.27
25mm	10	8.00
32mm	10	16.19
38mm	10	28.69
50mm	10	48.70

adaptors, female thread to push-in with male bushes

20mm	10	6.45
25mm	10	9.22
32mm	10	18.04
38mm	10	37.74
50mm	10	65.71

adaptors, male thread to push-in with lock rings

20mm	10	6.45
25mm	10	9.22
32mm	10	18.04
38mm	10	37.74
50mm	10	65.71

Cables for use with conduit systems

single PVC insulated only 300/ 500V Class 6491X

1.0mm2 nominal	100m	32.73
1.5mm2 nominal	100m	53.20
2.5mm2 nominal	100m	77.01
4.0mm2 nominal	100m	126.76
6.0mm2 nominal	100m	184.12
10.0mm2 nominal	100m	344.33
16.0mm2 nominal	100m	531.80

Cables for use where conduit systems are not required

single PVC insulated and sheathed cable 300/500V Class 6181Y

1.0mm2 nominal	100m	56.71
1.5mm2 nominal	100m	75.36
2.5mm2 nominal	100m	125.60
4.0mm2 nominal	100m	204.74
6.0mm2 nominal	100m	272.34
10.0mm2 nominal	100m	433.09
16.0mm2 nominal	100m	563.33
25mm2 nominal	100m	1079.21

twin PVC insulated and sheathed cable 300/500V Class 6192Y

1.0mm2 nominal	100m	100.86
1.5mm2 nominal	100m	129.31
2.5mm2 nominal	100m	182.54
4.0mm2 nominal	100m	270.69
6.0mm2 nominal	100m	369.61
10.0mm2 nominal	100m	600.22
16.0mm2 nominal	100m	934.27

twin and earth PVC insulated and sheathed cable 300/500V Class 6242Y

1.0mm2 nominal	100m	87.55
1.5mm2 nominal	100m	109.33
2.5mm2 nominal	100m	149.51
4.0mm2 nominal	100m	464.97
6.0mm2 nominal	100m	552.09
10.0mm2 nominal	100m	891.29
16.0mm2 nominal	100m	1420.61

triple PVC insulated and sheathed cable 300/500V Class 6193Y

1.0mm2 nominal	100m	172.73
1.5mm2 nominal	100m	239.43
2.5mm2 nominal	100m	391.43
4.0mm2 nominal	100m	447.07
6.0mm2 nominal	100m	607.66
10.0mm2 nominal	100m	940.55
16.0mm2 nominal	100m	1482.79

triple and earth PVC insulated and sheathed cable 300/500V Class 6243Y

1.0mm2 nominal	100m	209.69
1.5mm2 nominal	100m	331.03

twin TRS flexible cords Class 3182

0.75mm2 nominal	100m	135.76
1.00mm2 nominal	100m	167.87
1.50mm2 nominal	100m	233.82
2.5mm2 nominal	100m	342.73

triple TRS flexible cords Class 3183

0.75mm2 nominal	100m	170.58
1.00mm2 nominal	100m	199.82
1.50mm2 nominal	100m	275.12
2.5mm2 nominal	100m	407.42

twin circular PVC flexible cords Class 3182Y

0.75mm2 nominal	100m	79.87
1.00mm2 nominal	100m	104.38
1.50mm2 nominal	100m	148.47
2.5mm2 nominal	100m	314.88

triple circular PVC flexible cords Class 3183Y

0.75mm2 nominal	100m	78.33
1.00mm2 nominal	100m	95.20
1.50mm2 nominal	100m	132.16
2.5mm2 nominal	100m	270.34

bell wire, PVC insulated only, twin (figure 8) 100m 37.92

MICS cable (Two complete terminations are required for each length of cable) 600V grade, 2 core

1.0mm2 conductor	100m	193.00
complete termination with earth tail	nr	4.84
1.5mm2 conductor	100m	227.50
complete termination with earth tail	nr	4.84
2.5mm2 conductor	100m	289.00
complete termination with earth tail	nr	4.84

ELECTRICAL INSTALLATIONS (Cont'd)

MICS cable (Two complete terminations are required for each length of cable)
600V grade, 2 core (Contd)

4.0mm2 conductor........ 100m	384.50	
complete termination with earth tail...................... nr	4.84	

600V grade, 3 core

1.0mm2 conductor........... 100m	203.00	
complete termination with earth tail...................... nr	4.84	
1.5mm2 conductor........... 100m	257.52	
complete termination with earth tail...................... nr	4.84	
2.5mm2 conductor.......... 100m	404.00	
complete termination with earth tail...................... nr	4.84	

600V grade, 4 core

1.0mm2 conductor........ 100m	243.00	
complete termination with earth tail...................... nr	4.84	
1.5mm2 conductor.......... 100m	311.00	
complete termination with earth tail...................... nr	4.84	
2.5mm2 conductor.......... 100m	490.50	
complete termination with earth tail...................... nr	4.84	

1000V grade, 2 core

1.5mm2 conductor.......... 100m	303.00	
complete termination with earth tail...................... nr	4.84	
2.5mm2 conductor......... 100m	372.00	
complete termination with earth tail...................... nr	4.84	
4.0mm2 conductor......... 100m	469.00	
complete termination with earth tail...................... nr	10.27	
6.0mm2 conductor.......... 100m	625.00	
complete termination with earth tail...................... nr	10.27	
10.0mm2 conductor......... 100m	809.00	
complete termination with earth tail...................... nr	17.22	
16.0mm2 conductor....... 100m	1165.00	
complete termination with earth tail...................... nr	30.79	
25.0mm2 conductor....... 100m	1635.00	
complete termination with earth tail...................... nr	30.79	

1000V grade, 3 core

1.5mm2 conductor........... 100m	335.00	
complete termination with earth tail......................nr	4.84	
2.5mm2................. 100m	472.00	
complete termination with earth tail...................... nr	4.84	
4.0mm2 conductor...........100m	535.00	
complete termination with earth tail...................... nr	10.27	
6.0mm2 conductor.......... 100m	690.00	
complete termination with earth tail......................nr	17.22	
10.0mm2 conductor........ 100m	998.00	
complete termination with earth tail...................... nr	30.79	
16.0mm2 conductor........ 100m	1401.00	
complete termination with earth tail...................... nr	26.62	
25.0mm2 conductor........ 100m	2150.00	
complete termination with earth tail...................... nr	30.79	

1000V grade, 4 core

1.5mm2 conductor...........100m	417.00	
complete termination with earth tail...................... nr	4.84	
2.5mm2 conductor...........100m	524.00	
complete termination with earth tail...................... nr	4.84	
4.0mm2 conductor..........100m	655.00	

MICS cable (Two complete terminations are required for each length of cable)
1000V grade, 4 core (Cont'd)
4.0mm2 conductor (Cont'd)

complete termination with earth tail........................ nr	10.27	
6.0mm2 conductor...........100m	874.00	
complete termination with earth tail........................ nr	17.22	
10.0mm2 conductor........100m	1240.00	
complete termination with earth tail...................... nr	30.79	
16.0mm2 conductor........100m	1809.00	
complete termination with earth tail...................... nr	30.79	
25.0mm2 conductor.........100m	2628.00	
complete termination with earth tail...................... nr	30.79	

Switch and fuse gear
metal cased switch fuses, H.R.C. pattern
SP and N, 500V

20A.......................... nr	63.89	
32A.......................... nr	80.22	
63A.......................... nr	125.95	
100A........................ nr	201.31	

TP and N, 500V

20A.......................... nr	83.61	
32A.......................... nr	111.06	
63A.......................... nr	186.60	
100A........................ nr	323.16	

metal cased consumers control units
250V SP and N, 60A D.P. main switch, MCB included

four way.................... nr	54.97	
six way..................... nr	73.17	
eight way.................. nr	92.11	

metal cased consumers control units
250V SP and N, fitted with 80A D.P. R.C.C.B (Residual Current Device), 30m A trip, M.C.B. pattern

four way....................nr	105.16	
eight way..................nr	152.31	
ten way.....................nr	172.34	

Lighting switches
400W Tungsten load dimmer, flush, plastic with plaster depth box, white, one gang..................nr 25.88
5A surface metalclad, aluminium finish, two way switch

one gang..........................nr	4.96	
two gang..........................nr	6.03	
three gang........................nr	9.40	
four gang..........................nr	13.23	
six gang...........................nr	16.00	

5A surface plastic with moulded box, white

one gang......................... nr	3.46	
two gang.........................nr	5.10	
three gang....................... nr	6.87	
four gang.........................nr	12.24	

5A flush plastic with plaster depth box, white

one gang......................... nr	3.69	
two gang......................... nr	5.10	
three gang...................... nr	7.10	
four gang........................ nr	12.59	

5A flush brass with standard depth box, matt chrome or satin brass, two way switch

one gang.......................... nr	10.61	
two gang........................... nr	12.30	
three gang........................ nr	17.96	
intermediate...................... nr	11.14	

surface plastic ceiling switch, ivory

5A.................................nr	4.10	
15A............................... nr	13.18	

Lighting Switches (Cont'd)
surfaceplasticceiling switch, ivory (Cont'd)

45A.................................. nr	13.18	
splashproof all insulated switch, 5A		
one gang, two way.............nr	6.57	
two gang, two way............. nr	14.00	

Socket outlets and plugs
three pin socket outlets with boxes or back plates as required
flush, insulated, white

5A plain............................ nr	9.07	
5A switched...................... nr	10.25	
13A single plain...................nr	5.43	
13A single switched............ nr	6.63	
13A twin switched...............nr	12.43	

surface, insulated, white

5A plain............................ nr	8.82	
5A switched...................... nr	9.97	
13A single plain...................nr	5.26	
13A single switched........... nr	6.31	
13A twin switched...............nr	12.06	

metalclad, aluminium, surface

5A plain............................ nr	10.76	
5A switched...................... nr	11.79	
13A single plain.................nr	7.18	
13A single switched............nr	9.69	
13A twin switched...............nr	19.15	

flush, metalclad, matt, chrome or satin brass

5A switched......................nr	16.52	
13A single plain..................nr	18.77	
13A single switched............nr	12.10	
13A twin switched...............nr	20.66	

plugs, white

5A................................... nr	3.14	
13A................................. nr	2.57	

three pin socket outlets R.C.C.B. (Residual Current Device)
protected at 30mA trip
flush insulated, white 1 gang

13A switched......................nr	69.74	

surface, metalclad, 1 gang

13A switched......................nr	74.87	

Power accessories

water heater switch complete with box, flush, white, 20A........... nr	14.18	
cooker unit, white finish, metal box, with pilot lamps and auxiliary 13A socket, flush or surface... nr	48.59	
20A double, pole switches with boxes		
flush, insulated, white........... nr	8.91	
flush, insulated, white with pilot lamp and flex outlet........ nr	12.98	
surface, metalclad with pilot lamp and flex outlet............ nr	14.02	
flush, metalclad with pilot lamp and flex outlet, BMA or matt chrome............................ nr	19.09	
13A fused connection units (spurs) with switches and boxes		
flush, insulated, white........... nr	10.49	
flush, insulated, white with pilot lamp and flex outlet........ nr	14.15	
surface, metalclad with pilot lamp and flex outlet.............. nr	13.62	
flush, metalclad with pilot lamp and flex outlet, BMA or matt chrome............................. nr	21.13	

Wiring accessories
earthing and bonding clamps for pipes
13mm-32mm, conductor

1 x 10mm2.......................... nr	0.75	

32mm-50mm, conductor

2 x 16mm2.......................... nr	1.26	

50mm-75mm, conductor

2 x 16mm2......................... nr	1.90	

BASIC PRICES OF MATERIALS

ELECTRICAL INSTALLATIONS (Cont'd)

Wiring accessories (Cont'd)
junction boxes, bakelite, white or brown
small	nr	1.81
large	nr	2.10

connectors, block type
single pole, porcelain
5A	100	65.89
15A	100	97.86

earthing and bonding clamps for pipes
double pole, porcelain
5A	100	111.56
15A	100	189.48

cable clips, plastic, push fit, with pins
5mm	100	2.10
7mm	100	2.18
10mm	100	2.86

cable clips, for flat twin and earth cables, plastic, push fit, with pins
1.0 and 1.5mm2	1000	21.54
2.5mm2	1000	24.04
4.0mm2	1000	29.82
6.0mm2	1000	31.00
10.0mm2	1000	40.87
16.0mm2	1000	57.11

ceiling roses
surface, bakelite, white
2 plate and earth	nr	2.07
3 plate and earth	nr	3.74

surface, plug type
3 plate and earth	nr	4.56

Lamp holders
bakelite white
cord grip, standard pattern	nr	2.08
batten holder with HO shield	nr	5.09

Electric lamps
240V electric lamps, bayonet cap GLS
25W	nr	1.30
40W	nr	0.90
60W	nr	0.90
100W	nr	0.90
150W	nr	1.89
200W	nr	3.10
150W PAR 38 E.S. cap	nr	10.25

Fluorescent lamps
fluorescent lamps suitable for operation on either instant start or switch start circuits
450mm long, 15W
white	nr	7.70
warm white	nr	8.83

600mm long, 20W
white	nr	8.18
warm white	nr	7.04

600mm long, 40W
white	nr	7.98

900mm long, 30W
white	nr	7.84
warm white	nr	9.24

1200mm long, 40W
white	nr	7.54
warm white	nr	8.11

1500mm long, 65/80W
white	nr	8.32
warm white	nr	8.90

1800mm long, 75/85W
white	nr	12.00
warm white	nr	12.79

2400mm long, 125W
white	nr	13.95
warm white	nr	14.98

2D lamp, 16W 2 pin
warm polylux 2700	nr	7.05

Fluorescent Lamps (Cont'd)
Powersaver fluorescent lamps suitable for operation on switch start circuits only
2D lamp, 28W 2 pin
warm polylux 2700	nr	10.60

1200mm long, 36W
white	nr	5.49
warm white	nr	6.44

Underground Services Warning Tape
150mm wide x 100 micron thick in 365m rolls
for electric cables (yellow)	roll	58.66
for telephone cables (green)	roll	56.24

FLOOR WALL AND CEILING FINISHINGS

Portland cement, 7.5 tonnes to 10 tonnes	tonne	91.75
Sand	tonne	22.50
Granite chippings	tonne	39.00
Surface hardener	25 litre	44.50
Floor sealer	25 litre	176.05
Bonding fluid	5 litre	26.25

Gypsum plaster to BS 1191 Part 1 Class B
universal one coat plaster	tonne	173.50
Multi finish plaster	tonne	109.00
board finish plaster	tonne	109.00
hardwall plaster	tonne	148.50

Lightweight gypsum plaster to BS 1191 Part 2
browning plaster	tonne	150.50
bonding plaster	tonne	148.00
finish plaster	tonne	116.00

Renovating plaster
Thistle undercoat	tonne	155.00
Thistle finish	tonne	142.50

Tyrolean finish
Cullarend mixture	tonne	362.50
machine	hour	1.25

Plaster beads and stops
Expamet, galvanized
angle bead
reference 550	m	0.61

plaster stop bead for the following thickness plaster
10mm reference 562	m	0.71
13mm reference 563	m	0.71
16mm reference 565	m	0.91
19mm reference 566	m	0.91

depth gauge bead
reference 569	m	0.57

thin coat plaster angle bead for the following thickness plaster
3mm reference 553	m	0.73
6mm reference 554	m	0.96

thin coat plaster stop bead for the following thickness plaster
3mm reference 560	m	0.61
6mm reference 561	m	0.79
render stop reference 570	m	0.74

plasterboard edging bead for the following thickness plasterboard
9.5mm reference 567	m	1.04
12.7mm reference 568	m	1.04

Expamet, stainless steel
angle bead reference 545	m	1.77
render stop reference 547	m	1.58

Clay floor quarries to BS 6431
Terracotta
150 x 150 x 12.5mm	1000	296.00
150 x 150 x 20mm	1000	418.00
225 x 225 x 25mm	1000	1550.00

Blended
150 x 150 x 12.5mm	1000	398.00
150 x 150 x 20mm	1000	449.00
225 x 225 x 25mm	1000	1743.00

Dark
150 x 150 x 12.5mm	1000	515.00

With one rounded edge
Terracotta
150 x 150 x 12.5mm	1000	383.00
150 x 150 x 20mm	1000	486.00
225 x 225 x 25mm	1000	1738.00

Blended
150 x 150 x 12.5mm	1000	455.00
150 x 150 x 20mm	1000	486.00
225 x 225 x 25mm	1000	1945.00

Dark
150 x 150 x 12.5mm	1000	600.00

Rounded edge cove skirting tiles
Terracotta
150 x 100 x 12.5mm	1000	503.00
150 x 138 x 12.5mm	1000	604.00

Blended
150 x 100 x 12.5mm	1000	570.00
150 x 138 x 12.5mm	1000	639.00

Dark
150 x 138 x 12.5mm	1000	832.00

Ceramic floor tiles to BS 6431
plain
red
100 x 100 x 9mm	m2	16.65
152 x 152 x 12mm	m2	18.25
200 x 200 x 12mm	m2	19.70

Cream
100 x 100 x 9mm	m2	17.70
152 x 152 x 12mm	m2	19.15
200 x 200 x 12mm	m2	20.45

black
100 x 100 x 9mm	m2	21.75
152 x 152 x 12mm	m2	22.80
200 x 200 x 12mm	m2	23.70

with one rounded edge
red
100 x 100 x 9mm	m2	22.25
152 x 152 x 12mm	m2	23.90
200 x 200 x 12mm	m2	30.85

Cream
100 x 100 x 9mm	m2	23.25
152 x 152 x 12mm	m2	24.35
200 x 200 x 12mm	m2	30.85

black
100 x 100 x 9mm	m2	27.30
152 x 152 x 12mm	m2	26.05
200 x 200 x 12mm	m2	30.85

rounded edge cove skirting tiles
red
152 x 100 x 9mm	100	63.50
152 x 112 x 12mm	100	84.40
152 x 152 x 12mm	100	100.60

Cream
152 x 100 x 9mm	100	71.60
152 x 112 x 12mm	100	84.40
152 x 152 x 12mm	100	100.65

black
152 x 100 x 9mm	100	79.45
152 x 112 x 12mm	100	89.65
152 x 152 x 12mm	100	104.55

Glazed ceramic tiles to BS 6431
white
108 x 108 x 4mm	m2	20.40
152 x 152 x 5.5mm	m2	14.85

light colours
108 x 108 x 4mm	m2	20.40
152 x 152 x 5.5mm	m2	14.95

FLOOR WALL AND CEILING FINISHINGS (Cont'd)

Glazed ceramic tiles to BS 6431 (Cont'd)
dark colours

108 x 108 x 4mm	m2	25.15
152 x 152 x 5.5mm	m2	18.45

Gypsum plasterboard to BS 1230 Part 2
baseboard

9.5mm	m2	1.71
9.5mm insulating	m2	2.33

lath

9.5mm	m2	2.01
12.7mm	m2	2.36

wallboard

9.5mm	m2	1.71
9.5mm insulating	m2	2.33
12.5mm	m2	1.99
12.5mm insulating	m2	2.61
15mm	m2	2.18
15mm insulating	m2	2.86

plank with square edges

19mm	m2	3.23
joint Tape 50mm wide	150m	3.10

Plastic faced plasterboard
9.5mm with aluminium foil

backing	m2	4.98
12.5mm with aluminium foil backing	m2	5.36
PVC batten trim	m	0.84
plastic coated nail caps	1000	9.50

Hardboard
standard

3.2mm	m2	1.03
4.8mm	m	1.70
6.0mm	m2	2.21

oil tempered

3.2mm	m2	1.70
4.8mm	m2	2.40
6.0mm	m2	3.15

flame retardent

3.2mm	m2	5.51
6.4mm	m2	7.16

Low density medium board
type LME

6mm	m2	6.64
9mm	m2	8.24
12mm	m2	10.92

type LMN

6mm	m2	5.67
9mm	m2	7.00

Medium density fibre board
type MDF

9mm	m2	3.86
12mm	m2	4.89
18mm	m2	5.97
25mm	m2	9.01

Insulating board 13mm thick

plain type SBN	m2	2.10
bitumen impregnated type SBI	m2	2.51

Chipboard, wood base
flooring grade square edge

18mm	m2	5.25
22mm	m2	7.11

flooring grade t & g edge

18mm	m2	5.54
22mm	m2	7.38

Asbestos - free board
Masterboard

6mm	m2	7.70
9mm	m2	15.75
12mm	m2	20.75

Supalux

6mm	m2	11.45
9mm	m2	17.05
12mm	m2	22.50

Vermiculux

20mm	m2	18.25
30mm	m2	29.90
40mm	m2	45.35
50mm	m2	60.35
60mm	m2	71.95

Expanded polystyrene
sheet polystyrene, density 16 kg/m3

13mm	m2	0.76
19mm	m2	1.18
25mm	m2	1.49
50mm	m2	2.95
75mm	m2	4.43

sheet polystyrene, non-flammable

13mm	m2	1.19
19mm	m2	1.41
25mm	m2	1.69
50mm	m2	4.08
75mm	m2	6.47

Cellular partitions

57mm panels	m2	7.14
63mm panels	m2	9.69

Laminated partitions
plasterboard components

50mm	m2	7.35
65mm	m2	9.88

Metal stud partitions
studs

48mm	m	0.97

floor and ceiling channels

50mm	m	0.88
12.7mm, tapered edge plaster-board with one face for direct decoration	m2	2.03

Lightweight aggregate, medium grade

10-5mm, 799kg/m3	m3	40.00

Levelling screeding

compound	kg	0.81

Metal lathing
BB galvanized expanded metal lath, 9mm mesh
0.500mm thick x 1.11kg/

m2	m2	4.21
0.725mm thick x 1.61kg/ m2	m2	4.93

Expamet galvanized expanded metal lath, 9mm mesh
0.950mm thick x 2.50kg/

m2	m2	7.56

Expamet galvanized Rib Lath
0.400mm thick x 1.35kg/

m2	m2	5.00
0.500mm thick x 2.25kg/ m2	m2	5.75

Expamet stainless steel Rib-Lath
0.300mm thick x 1.52kg/

m2	m2	13.43

Expamet galvanized Spray-lath
0.500mm thick x 2.25kg/

m2	m2	7.07

Metal lathing (Cont'd)
BB galvanized expanded metal lath, Expamet galvanized Red-rib lath

0.500mm thick x 2.25kg	m2	6.40

Chicken wire
galvanized

13mm mesh x 22 gauge	m2	2.68
25mm mesh x 22 gauge	m2	1.34
38mm mesh x 19 gauge	m2	1.49
50mm mesh x 19 gauge	m2	1.13

GLAZING

Float glass
GG quality

3mm	m2	19.56
4mm	m2	19.56
5mm	m2	28.60
6mm	m2	28.60
10mm	m2	57.92

obscured ground

3mm	m2	47.18
4mm	m2	51.77

Rough cast glass

6mm	m2	36.58

Georgian wired cast glass

7mm	m2	30.12

Georgian safety wired cast glass

7mm	m2	39.18

Patterned glass
white

4mm	m2	24.30
6mm	m2	36.58

tinted

4mm	m2	35.88
6mm	m2	40.60

Thick float glass
GG quality

12mm	m2	66.71
15mm	m2	111.00
19mm	m2	157.24
25mm	m2	249.60

Polished georgian wired glass

6mm	m2	65.10

Polished georgian safety wired glass

6mm	m2	76.66

Antisun float glass
grey

4mm	m2	37.31
6mm	m2	53.84
10mm	m2	96.26
12mm	m2	132.86

bronze

4mm	m2	37.31
6mm	m2	53.84
10mm	m2	96.26
12mm	m2	132.86

green

6mm	m2	69.02

Toughened safety glass
clear float glasses

4mm	m2	28.86
5mm	m2	43.29
6mm	m2	43.29
10mm	m2	68.68
12mm	m2	134.58

solar control glasses, antisun
bronze

4mm	m2	52.91
6mm	m2	55.21
10mm	m2	109.27
12mm	m2	143.78

BASIC PRICES OF MATERIALS

BASIC PRICES OF MATERIALS - SMALL WORKS

GLAZING (Cont'd)

Toughened safety glass (Contd)
solar control glasses, antisun grey

4mm	m2	52.91
6mm	m2	55.21
10mm	m2	109.23
12mm	m2	143.78

white patterned glasses

4mm	m2	58.18
6mm	m2	60.74

Clear laminated glass
safety glass - float quality

4.4mm	m2	43.96
6.4mm	m2	36.85
8.8mm	m2	66.33
10.8mm	m2	77.58

anti-bandit glass - float quality

7.5mm	m2	74.63
9.5mm	m2	73.12
11.5mm	m2	79.37

Clear polycarbonate sheet
standard sheet with latex paper masked both sides
Standard

2mm	m2	42.41
3mm	m2	63.35
4mm	m2	84.72
5mm	m2	108.99
6mm	m2	126.90
8mm	m2	169.21
9.5mm	m2	201.04
12mm	m2	263.76

Abrasion resistant

3mm	m2	98.24
4mm	m2	117.00
5mm	m2	140.29
6mm	m2	156.79
8mm	m2	189.77
9.5mm	m2	214.40

Sundries

linseed oil putty	25 kg	14.08
metal casement putty	25 kg	14.80
washleather strip	m	0.64
rubber glazing strip	m	0.60

PAINTING AND DECORATING

Creosote	200 ltr	149.00
Wood preservative	5 litre	22.95

Emulsion paint
matt

brilliant white	5 litre	8.25
colours	5 litre	9.05

silk

brilliant white	5 litre	11.10
colours	5 litre	12.40

Priming, paint

wood primer	5 litre	18.50
primer sealer	5 litre	19.30

aluminium sealer and wood

primer	5 litre	21.15
alkali-resisting primer	5 litre	20.95
universal primer	5 litre	22.95
red oxide primer	5 litre	18.65
calcium plumbate primer	5 litre	25.00
zinc phosphate primer	5 litre	38.95
acrylic primer	5 litre	18.55

Undercoat paint

brilliant white	5 litre	15.45
colours	5 litre	15.70

Gloss paint

brilliant white	5 litre	15.45
colours	5 litre	15.70

Eggshell paint

brilliant white	5 litre	16.25
colours	5 litre	18.55

Plastic finish

coating compound	25 kg	16.40

Cement painting
cement paint

cream or white	40 kg	44.50
colours	40 kg	44.50
sealer	5 litre	18.35

textured masonry paint

white	10 litre	23.40
colours	10 litre	26.40
stabilising solution	20 litre	59.25

Mulitcolour painting

primer	5 litre	12.45
basecoat	5 litre	21.00
finish	5 litre	26.30

Chlorinated rubber

Primer	5 litre	48.15
finish	5 litre	48.45

Metallic paint

aluminium	5 litre	26.40
gold or bronze	5 litre	34.90
Bituminous paint	5 litre	12.50

Polyurethane lacquer

clear	5 litre	27.75

Linseed oil

raw	5 litre	15.25
boiled	5 litre	16.20

Fire retardant paints and varnishes

fire retardant paint	5 litre	42.35
fire retardant varnish	5 litre	30.65
overcoat varnish	5 litre	46.80

French polishing

clear polish	1 litre	7.60
button polish	1 litre	7.00
garnet polish	1 litre	7.00
methylated spirit	1 litre	1.85

DRAINAGE

Land drains
clay pipe in 300mm lengths

75mm	m	1.14
100mm	m	2.04
150mm	m	4.35

porous concrete pipe B.S.1194

150mm	m	4.20
225mm	m	5.40
300mm	m	8.80

perforated clay pipe

100mm	m	5.06
150mm	m	9.20
225mm	m	16.90

Vitrified clay flexible joint drains (B.S.EN295)
pipe

100mm	m	8.03
150mm	m	10.46
225mm	m	20.22
300mm	m	31.62
400mm	m	64.90
450mm	m	84.29

Vitrified clay flexible joint drains (B.S. EN295) (Cont'd)
90 degree bend

100mm	nr	11.65
150mm	nr	19.20
225mm	nr	39.17
300mm	nr	77.28
400mm	nr	243.86
450mm	nr	321.10

45 degree bend

100mm	nr	11.65
150mm	nr	19.20
225mm	nr	39.17
300mm	nr	77.28
400mm	nr	174.16
450mm	nr	229.35

30 degree bend

100mm	nr	10.90
150mm	nr	19.20
225mm	nr	39.17
300mm	nr	77.28

15 degree bend

100mm	nr	10.90
150mm	nr	19.20
225mm	nr	39.17
300mm	nr	77.28

rest bend

100mm	nr	13.83
150mm	nr	22.91
225mm	nr	54.57

90 degree curved square junction

100mm	nr	16.16
150mm	nr	25.07
225mm	nr	58.90
300mm	nr	121.42

45 degree oblique junction with supersleve arm

150mm	nr	50.24
225mm	nr	107.67
300mm	nr	228.48

45 degree oblique junction

100mm	nr	16.16
150mm	nr	25.07
225mm	nr	58.90
300mm	nr	121.42

tumbling bay junction

100mm	nr	16.16
150mm	n	25.07
225mm	nr	58.90
300mm	nr	145.73
400mm	nr	321.00

double collar

100mm	nr	10.36
150mm	nr	17.09
225mm	nr	37.35
300mm	nr	60.71
lubricant	kg	8.46

Vitrified clay sleeve drains (B.S. EN295)
pipe (supersleve)

100mm	m	3.23
150mm	m	6.54

couplings with standard rings

100mm	nr	2.24
150mm	nr	4.05

couplings with neoprene rings

100mm	nr	3.39
150mm	nr	6.40

bend (plain ended)

100mm	nr	4.37
150mm	nr	8.99

rest bend

100mm	nr	9.86
150mm	nr	11.57

single junction(plain ended)

100 x 100mm	nr	9.44
150 x 100mm	nr	12.05
150 x 150mm	nr	13.22

DRAINAGE (Cont'd)

Vitrified clay drains (B.S. EN295)
British Standard(unjointed)
pipe

100mm	m	5.63
150mm	m	8.68
225mm	m	17.18
300mm	m	28.80

bend

100mm	nr	5.63
150mm	nr	8.68
225mm	nr	25.26
300mm	nr	45.91

rest bend

100mm	nr	9.26
150mm	nr	15.66

single junction

100mm	nr	10.35
150mm	nr	17.14

taper reducer

150mm	nr	20.46

double collar

100mm	nr	6.80
150mm	nr	11.31
225mm	nr	26.49

Drains PVC-U B.S.4660
pipe

110mm	m	3.06
160mm	m	6.86

coupling with double socket

110mm	nr	3.06
160mm	nr	6.19

45 degree bend

110mm	nr	7.39
160mm	nr	17.58

single junction

110mm	nr	10.44
160mm	nr	31.75

Drains: PVC-U solid wall
concentric external rib
reinforced Wavin "Ultra-Rib"
pipe

150mm	6UR 046	m	3.39
225mm	9UR 046	m	7.78
300mm	12UR 043	m	11.67

connectors

150mm	6UR 205	nr	5.75
225mm	9UR 205	nr	12.01
300mm	12UR 205	nr	24.22

bends

150mm	6UR 563	nr	7.68
225mm	9UR 563	nr	31.04
300mm	12UR 563	nr	50.58

branches

150mm	6UR 213	nr	18.68
225mm	9UR 213	nr	55.61
300mm	12UR 213	nr	118.51

connection to clayware pipe

150mm	6UR 129	nr	14.08
225mm	9UR 109	nr	31.10
300mm	12UR 112	nr	81.80

Concrete drains (BS 5911)
Class H, unreinforced, flexible
joints
pipe

300mm	m	12.09
375mm	m	16.30
450mm	m	18.92
525mm	m	22.78
600mm	m	27.67

bend

300mm	nr	86.02
375mm	nr	129.70

Concrete drains (BS 5911)
Class H, unreinforced, flexible
joints (Cont'd)
bend (Cont'd)

450mm	nr	150.66
525mm	nr	181.54
600mm	nr	220.66

Class H, reinforced, flexible joints
pipe

450mm	m	32.10
525mm	m	36.60
600mm	m	39.85

bend

450mm	nr	256.10
525mm	nr	292.10
600mm	nr	318.10

junction:extra cost on pipe up to
600mm diameter

100mm	nr	35.10
150mm	nr	40.10
225mm	nr	45.10
300mm	nr	69.10
375mm	nr	74.10
450mm	nr	96.10

Spun iron drains (BS 437) for
coupling joints
spigoted pipe

100mm	m	18.42
150mm	m	34.70

couplings with gaskets

100mm	nr	11.78
150mm	nr	14.26

medium radius bend

100mm	nr	22.03
150mm	nr	44.46

equal branch

100mm	nr	29.23
150mm	nr	63.12

connector with large socket for
clayware

100mm	nr	19.29
150mm	nr	31.28

Vitrified clay manhole channels
British Standard Surface Water
Quality
half round, straight, 600mm long

100mm	nr	2.03
150mm	nr	3.41

half round, straight, 1000mm long

100mm	nr	2.84
150mm	nr	5.36
225mm	nr	12.51
300mm	nr	26.33

half round, curved, 500mm girth

100mm	nr	2.87
150mm	nr	4.97
225mm	nr	19.28
300mm	nr	39.29

half round, curved, 900mm girth

100mm	nr	2.87
150mm	nr	4.97
225mm	nr	19.28
300mm	nr	39.29

three-quarter section branch
channel bend

100mm	nr	8.11
150mm	nr	14.09
225mm	nr	49.64

Precast concrete inspection chambers
Unreinforced

Chamber ring 900mm diameter

305mm high	nr	18.27
458mm high	nr	27.04
610mm high	nr	22.93
762mm high	nr	28.45
914mm high	nr	33.96

Precast concrete inspection chambers
Unreinforced (Cont'd)
Chamber ring 1050mm diameter

305mm high	nr	20.78
458mm high	nr	30.81
610mm high	nr	26.13
762mm high	nr	32.44
914mm high	nr	38.75

Chamber ring 1200mm diameter

305mm high	nr	26.96
458mm high	nr	40.08
610mm high	nr	33.90
762mm high	nr	44.14
914mm high	nr	50.40

Chamber ring 1350mm diameter

305mm high	nr	38.62
458mm high	nr	57.58
610mm high	nr	48.52
762mm high	nr	60.40
914mm high	nr	88.47

Chamber ring 1500mm diameter

305mm high	nr	47.22
458mm high	nr	70.49
610mm high	nr	59.31
762mm high	nr	73.90
914mm high	nr	88.47

Cover slab, heavy duty,125mm

900mm diameter	nr	36.54
1050mm diameter	nr	41.89
1200mm diameter	nr	55.15
1350mm diameter	nr	80.29
1500mm diameter	nr	95.44

Cast iron inspection chambers for
coupling joints
straight
with one branch one side

100mm	nr	79.96
100-150mm	nr	120.19
150mm	nr	153.57

with one branch each side,

100mm	nr	103.01
150mm	nr	169.36

with two branches one side,

100mm	nr	160.14
150mm	nr	289.40

with two branches each side,

100mm	nr	202.79
150mm	nr	338.20

Manhole step irons (BS 1247)
galvanized
with 225mm tail, weighing

2.16kg	nr	2.75

Cast iron manhole covers and frames
(B.S. EN 124)
Light duty, single seal, solid top

MC1-45/45, 450 x 450mm	nr	30.66
MC1-60/45, 600 x 450mm	nr	30.83
MC1-60/60, 600 x 600mm	nr	65.53

Light duty, single seal, recessed top

MC1R-60/60, 600 x 600mm	nr	86.83

Light duty, double seal, solid top

MC2-45/45, 450 x 450mm	nr	47.18
MC2-60/45, 600 x 450mm	nr	61.79
MC2-60/60, 600 x 600mm	nr	87.14

Light duty, double seal, recessed top

MC2R-45/45, 450 x 450mm	nr	78.36
MC2R-60/45, 600 x 450mm	nr	72.79
MC2R-60/60, 600 x 600mm	nr	104.99

Medium duty,Class 2, single seal solid
top

MB2-50, 500mm diameter	nr	74.79
MB2-55, 550mm diameter	nr	77.10
MB2-60, 600mm diameter	nr	57.05
MB2-60/45, 600 x 450mm	nr	51.7
MB2-60/60, 600 x 600mm	nr	63.16

DRAINAGE (Cont'd)

Cast iron manhole covers and frames (B.S.EN124) (Cont'd)
Medium duty, Class 2, single seal
recessed top

MB2R-60/45, 600 x 450mm..nr		92.78
MB2R-60/60, 600 x 600mm..nr		105.55
MB1-60, 600mm diameter...nr		66.69

Heavy duty

MD-60, 600mm diameter......nr		103.06
MA-60, 600mm diameter......nr		69.03
MA-T, 490 x 495mm.............nr		106.53
lifting keys............................nr		0.90

FENCING

Post and wire fencing
galvanized line wire

3.15mm diameter.................	m	0.06
4.00mm diameter.................	m	0.10
5.00mm diameter.................	m	0.15

painted angle iron posts
intermediate post

for 900mm high fencing.......	nr	3.04
for 1200mm high fencing.....	nr	3.65
for 1400mm high fencing......	nr	3.96

end straining post with one strut

for 900mm high fencing.......	nr	18.08
for 1200mm high fencing......	nr	23.19
for 1400mm high fencing......	nr	24.71

intermediate straining post with
two struts

for 900mm high fencing.......	nr	28.03
for 1200mm high fencing......	nr	35.86
for 1400mm high fencing......	nr	38.23

corner straining post with two
struts

for 900mm high fencing.......	nr	26.62
for 1200mm high fencing......	nr	33.75
for 1400mm high fencing......	nr	36.13

reinforced concrete posts
intermediate post

for 900mm high fencing.......	nr	4.91
for 1050mm high fencing......	nr	5.57

end straining post with one strut

for 900mm high fencing.......	nr	10.79
for 1050mm high fencing.......	nr	15.49

intermediate straining post with
two struts

for 900mm high fencing.......	nr	15.57
for 1050mm high fencing......	nr	21.00

corner staining post with two
struts

for 900mm high fencing.......	nr	15.57
for 1050mm high fencing.......	nr	21.00

Chain link fencing
galvanized chain link mesh
50mm x 14 G (2.00mm)

900mm.................................	m	0.76
1200mm...............................	m	1.08

50mm x 12 1/2 G (2.50mm)

900mm.................................	m	1.16
1200mm...............................	m	1.57
1400mm...............................	m	1.81
1800mm...............................	m	2.28

50mm x 10 1/2 G (3.00mm)

900mm.................................	m	1.62
1200mm...............................	m	2.04
1400mm...............................	m	2.57
1800mm...............................	m	3.15

plastic coated chain link mesh
50mm x 12 1/2 G

900mm.................................	m	1.98
1200mm...............................	m	2.69
1400mm...............................	m	3.11
1800mm...............................	m	3.62

50mm x 10 1/2 G

900mm.................................	m	2.82
1200mm...............................	m	3.68
1400mm...............................	m	4.12
1800mm...............................	m	4.52

Chain link fencing (Cont'd)
painted angle iron posts
intermediate post

for 900mm high fencing...	nr	3.04
for 1200mm high fencing..	nr	3.65
for 1400mm high fencing..	nr	3.96
for 1800mm high fencing..	nr	5.18

end straining post with one strut

for 900mm high fencing.......	nr	18.08
for 1200mm high fencing......	nr	23.19
for 1400mm high fencing......	nr	24.71
for 1800mm high fencing......	nr	33.30

intermediate straining post with
two struts

for 900mm high fencing.......	nr	28.03
for 1200mm high fencing......	nr	35.86
for 1400mm high fencing......	nr	38.23
for 1800mm high fencing......	nr	50.03

corner straining post with two
struts

for 900mm high fencing.......	nr	26.62
for 1200mm high fencing......	nr	33.75
for 1400mm high fencing......	nr	36.13
for 1800mm high fencing.......	nr	47.93

reinforced concrete posts
intermediate post

for 900mm high fencing.......	nr	4.90
for 1200mm high fencing......	nr	5.81
for 1400mm high fencing......	nr	6.67
for 1800mm high fencing......	nr	8.87

end straining post with one strut

for 900mm high fencing.......	nr	10.79
for 1200mm high fencing......	nr	14.62
for 1400mm high fencing......	nr	16.49
for 1800mm high fencing......	nr	19.93

intermediate straining post with
two struts

for 900mm high fencing.......	nr	15.57
for 1200mm high fencing......	nr	20.13
for 1400mm high fencing......	nr	22.77
for 1800mm high fencing.......	nr	27.57

corner straining post with two
struts

for 900mm high fencing........	nr	15.57
for 1200mm high fencing......	nr	20.13
for 1400mm high fencing......	nr	22.77
for 1800mm high fencing......	nr	27.57

line wire
galvanized

3.15mm diameter.................	m	0.06
4.00mm diameter.................	m	0.10

plastic coated

2.50mm/3.55mm...................	m	0.07
3.00mm/4.00mm...................	m	0.10
3.55mm/4.75mm...................	m	0.23

Chestnut fencing
fencing with three rows of galvanized
steel wire
900mm high, spacing between pales

75mm...................................	m	2.64

1200mm high, spacing between pales

75mm...................................	m	3.09

1500mm high, spacing between pales

75mm...................................	m	4.87

Softwood post 75-100mm girth with
pointed end

1350mm long.......................	nr	1.06
1650mm long.......................	nr	1.31
2100mm long.......................	nr	1.51

Boarded fencing
softwood post 125 x 100mm

for 1200mm high fence.........	nr	8.60
for 1500mm high fence.........	nr	9.98
for 1800mm high fence.........	nr	11.29
two ex 75mm dia. arris rail..	m	0.88
25 x 100mm gravel board.....	m	1.29

75mm dia. centre stump

600mm long.........................	nr	0.77

ex 22 x 100mm feather edge pales

for 1200mm high fence.........	nr	0.48

Boarded fencing (Cont'd)
ex 22 x 100mm feather
edge pales (Cont'd)

for 1500mm high fence.........	nr	0.65
for 1800mm high fence.........	nr	0.75

Panel fencing
preservative treated featherboard
fencing panels

1800 x 1200mm....................	nr	13.79
1800 x 1500mm....................	nr	16.10
1800 x 1800mm....................	nr	18.44

preservative treated lap fencing
panels

1800 x 1200mm....................	nr	12.42
1800 x 1500mm....................	nr	14.46
1800 x 1800mm....................	nr	16.50

75 x 75mm preservative treated fence
post with cap

1800mm long.......................	nr	6.14
2100mm long.......................	nr	7.11
2400mm long.......................	nr	8.06

EXTERNAL WORKS

Fine cold asphalt (BS 4987) with bitumen coated granite, limestone or blast furnace slag aggregate........................tonne		72.03

Bitumen macadam

10mm aggregate.............tonne		34.96
40mm aggregate.............tonne		28.78
bitumen coated grit..........tonne		32.90

Precast concrete kerbs and channels
(BS 7263 Part 1)
edging, figs 10 to 13 inclusive
straight

50 x 150mm.......................	m	1.13
50 x 205mm.......................	m	1.40
50 x 255mm.......................	m	1.51

kerbs, figs 2, 5, 7 and 8
straight

125 x 255mm......................	m	2.19
150 x 305mm......................	m	3.89

curved

125 x 255mm......................	m	3.70
150 x 305mm......................	m	6.11

channel, fig 8
straight

225 x 125mm......................	m	3.01

curved

255 x 125mm......................	m	3.70

quadrant, fig 14

305 x 305 x 150mm..........	nr	5.21
305 x 305 x 225mm..........	nr	5.71
455 x 455 x 150mm..........	nr	6.10
455 x 455 x 225mm..........	nr	6.60

Granite kerbs (BS 435) or equivalent
Edge Kerb
straight

150 x 300mm......................	m	19.54
200 x 300mm......................	m	22.91

curved external radius not exceeding
1000mm

150 x 300mm......................	m	30.02
200 x 300mm......................	m	34.67

curved external radius exceeding
1000mm

150 x 300mm......................	m	25.53
200 x 300mm......................	m	29.63

Flat Kerb
Straight

300 x 150...........................	m	21.63
300 x 200...........................	m	26.28

Curved external radius not exceeding
1000mm

300 x 150...........................	m	33.57
300 x 200mm......................	m	40.31

EXTERNAL WORKS (Cont'd)

Granite kerbs (BS 435)
or equivalent (Cont'd)
Curved external radius exceeding
1000mm
300 x 150mm.................... m
28.46
300 x 200mm.................... m
34.28

Gravel paths
clinker............................. m3
12.60
blinding gravel
10 or 20mm..................... m3
18.00

Precast concrete flag paving (BS 7263)
Part 1
50mm plain flags
600 x 450mm.................... m2   4.43
600 x 600mm.................... m2   4.10
600 x 750mm.................... m2   3.65
600 x 900mm.................... m2   3.63

Granite sett paving
new granite setts to BS 435
100mm.................... m2   23.74
125mm.................... m2   29.69
150mm.................... m2   31.95
reclaimed granite setts
100mm.................... m2   18.90
125mm.................... m2   22.61
150mm.................... m2   24.10

Guard rails
40mm galvanized steel medium weight
tubing to BS 1387 ref
7-2-G........................... m   5.46
Kee Klamp fittings
bend No 15-7................... nr   4.00
three-way intersection
No 20-7........................ nr   6.36
three-way intersection
No 25-7........................ nr   6.11
No 21-7........................ nr   4.97
four-way intersection
No 26-7........................ nr   4.70
five-way intersection
No 35-7........................ nr   7.47
five-way intersection
No 40-7........................ nr   10.53
floor plate No 61-7........... nr   3.83
51 x 51mm welded mesh
infill panel...................... m2   19.96

Estimating and Tendering for Construction Work
Second edition

Estimating and Tendering for Construction Work takes a practical approach to estimating from a contractor's point of view. It explains the estimator's function within the construction team, the techniques and procedures used in building up an estimate, and how to convert an estimate into a tender.

This new edition has been written to reflect recent changes in procurement. These include the recommendations of Sir Michael Latham in his 1994 report 'Constructing the Team,' and new terminology introduced by the 6th edition of the CIOB Code of Estimating Practice 1997. The role of the estimator is covered in detail, from early cost studies through to the preparation of the estimate, and the handover of construction budgets for successful tenders.

The book includes copious examples of an estimator's data sheets and pricing notes and deals with co-ordinated project information. The chapter dealing with computer-aided estimating has been completely re-written to reflect the advantages of electronic exchange of information.

Paperback, 1998, £19.99, 288pp, isbn 0 7506 3404 7

TO ORDER: CREDIT CARD HOTLINE - 01865 888180
BY MAIL: Technical Marketing Dept., Butterworth-Heinemann, FREEPOST, Oxford OX2 8BR
BY FAX: Heinemann Customer Services – 01865 314091
BY EMAIL: Send orders to: bhuk.orders@repp.co.uk

Composite Prices for Approximate Estimating

The purpose of this section is to provide an easy reference to costs of "composite" items of work which would usually be measured for Approximate Estimates. Rates for items which are not included here may be found by reference to the preceding pages of this volume. Allowance has been made in the rates for the cost of works incidental to and forming part of the described item although not described therein, i.e. the cost of formwork to edges of slabs has been allowed in the rate for slabs; the cost of finishings to reveals has been allowed in the rate for finishings to walls.

The rates are based on rates contained in the preceding Small Works section of the book and include the cost of materials, labour fixing or laying and allow 12.5% for overheads and profit: Preliminaries are not included within the rates. The rates are related to a housing or simple commercial building contract in the range of £25,000 to £75,000. For contracts of a higher or lower value, rates should be adjusted by the percentage adjustments given in Essential Information at the beginning of this Volume. Landfill Tax is not included.

The following cost plans have been included in both the Major works and Small works sections, and priced accordingly as examples on how to construct a Cost Plan, for information and to demonstrate how to use the Approximate Estimating section. Both cost plans are above the Small Works section values and should be used for smaller projects.

COST PLAN

COST PLAN FOR AN OFFICE DEVELOPMENT

The Cost Plan that follows gives an example of an Approximate Estimate for an office building as the attached sketch drawings. The estimate can be amended by using the examples of Composite Prices following or the prices included within the preceding pages. Details of the specification used are indicated within the elements. The cost plan is based on a fully measured bill of quantities priced using the detailed rates within this current edition of Laxton's. Preliminaries are shown separately.

Floor area =   932m²

Description	Item	Quantity	Rate	Item Total	Cost per m²	Total
1.0 SUBSTRUCTURE						
Note: Rates allow for excavation by machine.						
1.1 Substructure (271m²)						
Clear site vegetation excavate 150 topsoil and remove spoil.	Site clearance	288m²	0.79	227.52	0.24	
300 x 600 reinforced concrete perimeter ground beams with attached 180 x300 toes; damp proof membrane; 50 concrete blinding; Jablite; 102.5 thick common and facing bricks from toe to top of ground beam; 75 cavity filled with plain concrete; cavity weeps; damp proof courses (excavation included in ground slab excavation)	Ground beam	81m	165.97	13443.57	14.42	
600 x 450 reinforced concrete internal ground beams, mix C30; 50 thick concrete blinding; formwork; Jablite protection (excavation included in ground slab excavation)	Ground beam	34m	128.71	4376.14	4.70	
Excavate to reduce levels; level and compact bottoms; 100 hardcore filling, blinded; damp proof membrane; 50 Jablite insulation; 150 reinforced concrete ground slab 2 layers A252 fabric reinforcement	Ground floor slab	271m²	59.98	16254.58	17.44	

487

OFFICE DEVELOPMENT (Cont'd)

	Item	Quantity	Item Rate	Cost total	per m²	Total
1.0 SUBSTRUCTURE (Cont'd)						
1.1 Substructure (Cont'd)						
1900 x 1900 x 1150 lift pit: excavate; level and compact bottoms; working space; earthwork support; 50 concrete blinding; damp proof membrane; 200 reinforced concrete slab, 200 reinforced concrete walls, formwork; Bituthene tanking and protection board	Lift pit	1nr	1801.39	1801.39	1.93	
1500 x 1500 x 1000 stanchion bases: excavate; level and compact bottoms; earthwork support; reinforced concrete foundations; 102.5 bricks from top of foundation to ground slab level	Stanchion base	13nr	541.74	<u>7042.62</u>	<u>7.56</u>	43145.82
(Element unit rate = £159.21 /m²)				43145.82	46.29	
2.0 SUPERSTRUCTURE						
2.1 Frame (932m²)						
300 x 300 reinforced concrete columns; formwork; two coats RIW to exterior faces	Columns	111m	69.54	7718.94	8.28	
29.5m girth x 1.0m high steel frame to support roof over high level louvres	Frame	1nr	6000.00	6000.00	6.44	
300 x 550 reinforced concrete beams; formwork	Beams	61m	101.00	<u>6161.00</u>	6.61	19879.94
(Element unit rate = £21.33m²)				<u>19879.94</u>	<u>21.23</u>	
2.2 Upper Floors (614m²)						
275 reinforced concrete floor slab; formwork; 300 x 400/675 x 175/300 x 150 reinforced concrete attached beams; formwork	Floor slab	614m²	86.45	53080.30	56.95	
	Attached beams	273m	73.65	<u>20106.45</u>	<u>21.57</u>	73186.75
(Element unit rate = £119.20 /m²)				<u>73186.75</u>	<u>78.52</u>	
2.3 Roof						
2.3.1 Roof Structure (403m² on plan)						
pitched roof trusses, wall plates; 160 thick insulation and Netlon support	Roof structure	403m²	32.53	13109.59	14.07	
100 thick blockwork walls in roof space	Blockwork	27m²	26.27	709.29	0.76	
Lift shaft capping internal size 1900 x 1900;200 thick blockwork walls and 150 reinforced concrete slab	Lift shaft	1nr	535.15	535.15	0.57	
2.3.2 Roof coverings						
Concrete interlocking roof tiles, underlay and battens	Roof finish	465m²	18.28	8500.20	9.12	
Ridge	Ridge	22m	16.87	371.14	0.40	
Hip	Hip	50m	17.43	871.50	0.94	
Valley	Valley	6m	34.07	204.42	0.22	
Abutments to cavity walls, lead flashings and cavity tray damp proof courses	Abutments	9m	59.07	531.63	0.57	

OFFICE DEVELOPMENT (Cont'd)	Item	Quantity	Item Rate	Cost total	per m²	Total

2.0 SUPERSTRUCTURE (Cont'd)

2.3 Roof (Cont'd)

2.3.2 Roof Coverings (Cont'd)

Eaves framing, 25 x 225 softwood fascia 50 wide Resoplan soffit and ventilator, eaves tile ventilator and decoration; 225 x 160 pressed aluminium gutters and fittings — Eaves | 102m | 216.74 | 22107.48 | 23.72 |

2.33 Roof Drainage

62 x 62 aluminium rainwater pipes and rainwater heads; GRP hopper heads — Rain water pipes | 58m | 61.54 | 3569.32 | 3.83 | 50509.72

(Element unit rate = £125.33/m²) — — — — 50509.72 | 54.20

2.4 Stairs

Reinforced in-situ concrete staircase blockwork support walls screed and vinyl floor finishes; plaster paint to soffits; to wall

3nr 1200 wide flights, 1nr half landing and 1nr quarter landing; 3100 rise; handrail one side — Stairs | 1nr | 3524.08 | 3524.08 | 3.78 |

4nr 1200 wide flights, 2nr half landings; 6200 rise; plastics coated steel balustrade one side and handrail one side — Stairs | 1nr | 13483.10 | 13483.10 | 14.47 |

4nr 1200 wide flights, 2nr half landings, mix; 6200 rise; plastics coated steel and glass in-fill panels balustrade one side and handrail one side — Stairs | 1nr | 14247.86 | 14247.86 | 15.29 | 31255.04

— — — — 31255.04 | 33.54

2.5 External Walls (614m²)

Cavity walls, 102.5 thick facing bricks (PC £250 per 1000), 75 wide insulated cavity, 140 thick lightweight concrete blockwork — Facing brick wall | 455m² | 99.83 | 45422.65 | 48.74 |

Cavity walls 102.5 thick facing bricks (PC £250 per 1000), built against concrete beams and columns — Facings | 159m² | 81.01 | 12880.59 | 13.82 |

880 x 725 x 150 cast stonework pier cappings — Stonework | 9nr | 84.98 | 764.82 | 0.82 |

780 x 1300 x 250 cast stonework spandrels — Stonework | 9nr | 169.95 | 1529.55 | 1.64 |

6500 x 2500 attached main entrance canopy — Canopy | 1nr | 7400.83 | 7400.83 | 7.94 |

Milled lead sheet on impregnated softwood boarding and framing, and building paper — Lead sheet | 4m² | 191.11 | 764.44 | 0.82 |

Expansion joint in facing brickwork — Expansion joint | 118m | 14.36 | 1694.48 | 1.82 | 70457.36

(Element unit rate = £114.75/m²) — — — — 70457.36 | 75.60

2.6 Windows and External Doors (214m²)

2.6.1 Windows (190m²)

Polyester powder coated aluminium vertical pivot double glazed windows — Windows | 190m² | 484.73 | 92098.70 | 98.92 |

APPROXIMATE ESTIMATING

OFFICE DEVELOPMENT (Cont'd)

Item	Quantity	Item Rate	Cost total	per m²	Total

2.0 SUPERSTRUCTURE (Cont'd)

2.6 Windows and External Doors (214m²) (Cont'd)

2.6.1 Windows (190m²) (Cont'd)

Item		Quantity	Item Rate	Cost total	per m²	Total
Polyester powder coated aluminium fixed louvre panels	Louvres	35m²	190.19	6656.65	7.14	
Heads 150 x 150 x 10 stainless steel angle bolted to concrete beam, cavity damp proof course, weep vents, plaster and decoration	Heads	98m	68.37	6700.26	7.19	
Reveals closing cavity with facing bricks, damp proof course, plaster and decoration	Reveals	99m	18.80	1861.20	2.00	
Works to reveals of openings comprising closing cavity with facing bricks, damp proof course, 13 thick cement and sand render with angle beads, and ceramic tile finish	Reveals	27m	34.37	927.99	1.00	
Sills closing cavity with blockwork, damp proof course, and 19 x 150 Ash veneered MDF window boards	Sills	8m	29.10	232.80	0.25	
Sills 230 x 150 cast stonework sills and damp proof course	Sills	36m	45.48	1637.28	1.76	
Sills 230 x 150 cast stonework sills, damp proof course, and 19 x 150 Ash veneered MDF window boards	Sills	36m	234.29	8434.44	9.05	
Sills closing cavity with blockwork, damp proof course, and ceramic tile sill finish	Sills	9m	37.14	334.26	0.36	
Surrounds to 900 diameter circular openings stainless steel radius lintels, 102.5 x 102.5 facing brick on edge surround, 13 thick plaster, and eggshell paint decoration	Lintels	8m	130.12	1040.96	1.12	
300 x 100 cast stonework surrounds to openings	Stonework	20m	57.29	1145.80	1.23	

2.6.2 External Doors (24m²)

Item		Quantity	Item Rate	Cost total	per m²	Total
Polyester powder coated aluminium double glazed screens with integral double entrance doors	Entrance screens	24m²	428.16	10275.84	11.03	131346.18
(Element unit rate = £613.77/m²)				131346.18	140.95	

2.7 Internal Walls and Partitions (481m²)

Item		Quantity	Item Rate	Cost total	per m²	Total
100 thick blockwork walls	Block walls	231m²	27.57	6368.67	6.83	
200 thick blockwork walls	Block walls	113m²	47.37	5352.81	5.74	
215 thick common brickwork walls	Brick walls	84m²	76.51	6426.84	6.90	
Demountable partitions	Partitions	42m²	28.44	1194.48	1.28	
Extra over demountable partitions for single doors	Partitions	2nr	27.36	54.72	0.06	
Extra over demountable partitions for double doors	Partitions	1nr	32.83	32.83	0.04	
WC cubicles	Cubicles	9nr	131.76	1185.84	1.27	

OFFICE DEVELOPMENT (Cont'd)

Item	Item	Quantity	Item Rate	Cost total	per m²	Total

2.0 SUPERSTRUCTURE (Cont'd)

2.7 Internal Walls and Partitions (481m²) (Cont'd)

Item	Item	Quantity	Item Rate	Cost total	per m²	Total
Screens in Softwood glazed with 6 thick GWPP glass	Screens	11m²	237.35	2610.85	2.80	
Duct panels in Formica Beautyboard and 50 x 75 softwood framework	Ducts	27m²	58.27	1573.29	1.69	
Duct panels in veneered MDF board and 50 x 75 softwood framework	Ducts	26m²	52.70	<u>1370.20</u>	<u>1.47</u>	26170.53
(Element unit rate = £54.41/m²)				<u>26170.53</u>	<u>28.08</u>	

2.8 Internal Doors (84m²)

Item	Item	Quantity	Item Rate	Cost total	per m²	Total
900 x 2100 half hour fire resistance single doorsets comprising doors, Softwood frames, ironmongery and decoration	Doors	45m²	260.59	11726.55	12.58	
900 x 2100 one hour fire resistance single doorsets comprising doors, Softwood frames, ironmongery and decoration	Doors	9m²	430.55	3874.95	4.16	
900 x 2100 half hour fire resistance single doorsets comprising doors with 2nr vision panels, Softwood frames, ironmongery and decoration	Doors	11m²	302.73	3330.03	3.57	
150 x 2100 half hour fire resistance double doorsets comprising Ash veneered doors with2nr vision panels, Ash frames, ironmongery and decoration	Doors	19m²	294.07	5587.33	5.99	
100 wide lintels	Lintels	32m	9.98	319.36	0.34	
200 wide lintels	Lintels	7m	20.60	144.20	0.15	
215 wide lintels	Lintels	9m	26.25	236.25	<u>0.25</u>	25218.67
(Element unit rate = £300.22 /m²)				<u>25218.67</u>	<u>27.04</u>	

3.0 INTERNAL FINISHES

3.1 Wall Finishes (1086m²)

Item	Item	Quantity	Item Rate	Cost total	per m²	Total
13 thick plaster to blockwork walls with eggshell paint finish	Plaster work	703m²	15.20	10685.60	11.47	
13 thick plaster to blockwork walls with lining paper and vinyl wallpaper finish	Plaster work	124m²	17.88	2217.12	2.38	
10 thick plaster to concrete walls with eggshell paint finish	Plaster work	75m²	21.24	1593.00	1.71	
13 thick cement and sand render to blockwork walls with 150 x 150 coloured ceramic tiles finish	Tiling	184m²	47.84	8802.56	9.44	23298.28
(Element unit rate = £21.45/m²)				<u>23298.28</u>	<u>25.00</u>	

3.2 Floor Finishes (779m²)

Item	Item	Quantity	Item Rate	Cost total	per m²	Total
75 thick sand and cement screed with 300 x 300 x 2.5 vinyl tile finish	Vinyl Tiling	19m²	36.03	684.57	0.73	
75 thick sand and cement screed with 2 thick sheet vinyl finish	Sheet vinyl	108m²	34.05	3677.40	3.95	
75 thick sand and cement screed with 300 x 300 x 6 carpet tile finish	Carpet Tiling	587m²	33.27	19529.49	20.95	

APPROXIMATE ESTIMATING

OFFICE DEVELOPMENT (Cont'd)	Item	Quantity	Item Rate	Cost total	per m²	Total
3.0 INTERNAL FINISHES (Cont'd)						
3.2 Floor Finishes (779m²) (Cont'd)						
75 thick sand and cement screed with fitted carpet finish	Carpeting	17m²	44.54	757.18	0.81	
60 thick sand and cement screed with 150 x 150 x 13 ceramic tile finish	Ceramic Tiling	46m²	54.07	2487.22	2.67	
Entrance mats comprising brass Mat frame, 60 thick cement and sand screed and matting	Matting	2m²	205.01	410.02	0.44	
150 high x 2.5 thick vinyl tile skirting	Skirting	45m	4.84	217.80	0.23	
119 high x 2 thick sheet vinyl skirtings with 20 x 30 lacquered Hardwood cover fillet	Skirting	116m	16.42	1904.72	2.04	
25 x 50 Hardwood skirtings with lacquer finish	Skirting	214m	11.99	2565.86	2.75	
150 x 150 x 13 ceramic tile skirtings on 13 thick sand and cement render	Skirting	74m	15.03	1112.22	1.19	33346.48
(Element unit rate = £42.81 m²)				33346.48	35.76	
3.3 Ceiling Finishes (809m²)						
Suspended ceiling systems to offices	Suspended	587m²	25.36	14886.32	15.97	
Suspended ceiling systems to toilets	Suspended	46m²	32.80	1508.80	1.62	
Suspended ceiling systems to reception area	Suspended	18m²	32.61	586.98	0.63	
Suspended ceiling plain edge trims and 25 x 38 painted softwood shadow battens	Edge Trim	300m	11.30	3390.00	3.64	
Gyproc MF suspended ceiling system with eggshell paint finish	Suspended	127m²	25.98	3299.46	3.54	
12.5 Gyproc Square edge board and skim on softwood supports with eggshell paint finish	Ceiling board	29m²	50.41	1461.89	1.57	
10 thick plaster to concrete with eggshell paint finish	Plastering	2m²	54.59	109.18	0.12	
Plaster coving	Cove	234m	5.66	1324.44	1.42	26567.07
(Element unit rate = £30.32/m²)				26567.07	28.32	
4.0 FITTINGS AND FURNISHINGS						
4.1 Fittings						
2100 x 500 x 900 reception counter in Softwood	Counter	1nr	1061.75	1061.75	1.14	
2700 x 500 x 900 vanity units in Formica faced chipboard	Units	6nr	685.31	4111.86	4.41	
600 x 400 x 6 thick mirrors	Mirror	15nr	80.24	1203.60	1.29	
Vertical fabric blinds	Blinds	161m²	54.05	8702.05	9.34	
900 x 250 x 19 blockboard lipped shelves on Spur support system	Shelves	12nr	29.40	352.80	0.38	
1200 x 600 x 900 Sink base units in melamine finished particle board	Sink base unit	3nr	149.95	449.85	0.48	15881.91
				15881.91	17.04	

OFFICE DEVELOPMENT (Cont'd)	Item	Quantity	Item Rate	Cost total	per m²	Total
5.0 SERVICES						
5.1 Sanitary Appliances						
1200 x 600 stainless steel sink units with pillar taps	Sink	3nr	137.15	411.45	0.44	
White glazed fireclay cleaner's sink units with pillar taps	Cleaners sink	3nr	259.45	778.35	0.84	
Vitreous china vanity basins with pillar taps	Vanity basins	12nr	226.60	2719.20	2.92	
Vitreous china wash hand basin and pedestal with pillar taps	Wash hand basins	1nr	146.65	146.65	0.16	
Vitreous china wash hand basin with pillar taps	Wash hand basins	1nr	101.18	101.18	0.11	
Vitreous china disabled wash hand basin with pillar taps	Wash hand basins	1nr	138.40	138.40	0.15	
Close coupled vitreous china WC suites	WC suite	11nr	200.33	2203.63	2.36	
Showers comprising shower tray, mixer valve and fittings, and corner shower cubicle	Shower	1nr	832.67	832.67	0.89	
Vitreous china urinals with automatic cistern	Urinals	3nr	321.89	965.67	1.04	
Disabled grab rails set	Grab rails	1nr	410.38	410.38	0.44	
Toilet roll holders	Accessories	11nr	15.86	174.46	0.19	
Paper towel dispensers	Accessories	8nr	26.06	208.48	0.22	
Paper towel waste bins	Accessories	8nr	39.68	317.28	0.34	
Sanitary towel disposal units	Accessories	3nr	747.78	2243.34	2.41	
Hand dryers	Accessories	8nr	236.94	1895.52	2.03	13546.66
				13546.66	14.54	
5.2 Services Equipment		1nr	2000.00	2000.00	2.15	2000.00
5.3 Disposal Installations						
Upvc waste pipes and fittings to	Internal Drainage	37nr	116.21	4299.77	4.61	4299.77
5.4 Water Installations						
Hot and cold water supplies		37nr	214.35	7930.95	8.51	7930.95
5.5 Heat Source						
Gas boiler for radiators and hot water from indirect storage tank		1nr	78351.94	78351.94	84.07	78351.94
5.6 Space Heating and Air Treatment						
Low temperature hot water heating by radiators		1nr	2605.90	2605.90	2.80	2605.90
5.7 Ventilation Installations						
Extractor fan to toilet and kitchen areas		9nr	679.80	6118.20	6.56	6118.20

APPROXIMATE ESTIMATING

OFFICE DEVELOPMENT (Cont'd)

	Item	Quantity	Item Rate	Cost total	per m²	Total
5.0 SERVICES (Cont'd)						
5.8 Electrical Installations						
Lighting & Power		1nr	95141.64	95141.64	102.08	95141.64
5.9 Gas Installations						
To boiler room		1nr	2039.40	2039.40	2.19	2039.40
5.10 Lift and Conveyor Installations						
8 person 3 stop hydraulic lift		1nr	26059.80	26059.00	27.96	26059.00
5.11 Protective Installations						
Burglar Alarm System		1nr	6344.80	6344.80	6.81	6344.80
5.12 Communication Installations						
Cabling for telephones		1nr	8497.50	8497.50	9.12	8497.50
5.13 Special Installations						
Door entry to receptions on all floors		1nr	1300.00	1300.00	1.39	1300.00
5.14 Builders Work in Connection With Services						
Builders work in connection with plumbing, mechanical and electrical installations		1nr	5757.32	5757.32	6.18	5757.32
Builders Profit and Attendance on Services		1nr	1000.00	1000.00	1.07	1000.00
6.0 EXTERNAL WORKS						
6.1 Site Works						
6.1.1 Site Preparation						
Clearing site vegetation, filling root voids with excavated material	Site clearance	1716m²	0.69	1184.04	1.27	
Excavate topsoil 150 deep and preserve on site	Clear topsoil	214m³	5.47	1170.58	1.26	
Dispose of surplus topsoil off site	Disposal	70m³	15.75	1102.50	1.18	
6.1.2 Surface Treatments						
Excavate to reduce levels 400 deep; dispose of excavated material off site; level and compact bottoms; herbicide; 250 thick granular material filling, levelled, compacted and blinded; 70 thick two coat coated macadam pavings; white road lines and markings	Pavings	719m²	31.65	22756.35	24.42	
Excavate to reduce levels 400 deep; dispose of excavated material off site; level and compact bottoms; herbicide; 150 thick granular material filling, levelled, compacted and blinded; 200 x 100 x 65 thick clay brick paviors	Pavings	40m²	37.02	1480.80	1.59	

OFFICE DEVELOPMENT (Cont'd)

	Item	Quantity	Item Rate	Cost total	per m²	Total

6.0 EXTERNAL WORKS (Cont'd)

6.1.2 Surface Treatments (Cont'd)

Description	Item	Quantity	Item Rate	Cost total	per m²	Total
Excavate to reduce levels 400 deep; dispose of excavated material off site; level and compact bottoms; herbicide; 150 thick granular material filling, levelled, compacted and blinded; 900 x 600 x 50 thick concrete paving flags	Pavings	149m²	24.00	3576.00	3.84	
125 x 255 precast concrete road kerbs and foundations	Kerbs	265m	20.96	5554.40	5.96	
50 x 150 precast concrete edgings and foundations	Kerbs	48m	13.60	652.80	0.70	
Excavate to reduce levels 400 deep; dispose of excavated material off site; prepare sub-soil; herbicide; 150 thick preserved topsoil filling; rotavate; turf	Landscaping	55m²	18.03	991.65	1.06	
Excavate to reduce levels 400 deep; dispose of excavated material off site; prepare sub-soil; herbicide; 450 thick preserved topsoil filling; cultivate; plant 450 - 600 size herbaceous plants - 3/m2, and 3m high trees - 1nr per 22m2; fertiliser; 25 thick peat mulch	Planting	397m²	28.69	11389.93	12.22	

6.1.3 Site enclosure and division

Description	Item	Quantity	Item Rate	Cost total	per m²	Total
Palisade fencing	Fencing	62m	24.91	1544.42	1.66	
1800 high fencing	Fencing	96m	22.45	2155.20	2.31	
Extra for 1000 x 2000 gate	Gate	1nr	233.87	233.87	0.25	53792.54
				53792.54	57.72	

6.2 Drainage

Description	Item	Quantity	Item Rate	Cost total	per m²	Total
Excavate trench 250 - 500 deep; lay 100 diameter vitrified clay pipe and fittings bedded and surrounded in granular material	Pipes in trenches	44m	36.28	1596.32	1.71	
Excavate trench 500 - 750 deep; lay 100 diameter vitrified clay pipe and fittings bedded and surrounded in granular material	Pipes in trenches	59m	38.92	2296.28	2.46	
Excavate trench 750 - 1000 deep; lay 100 diameter vitrified clay pipe and fittings bedded and surrounded in granular material	Pipes in trenches	49m	41.11	2014.39	2.16	
Excavate trench 1000 -1250 deep; lay 100 diameter vitrified clay pipe and fittings bedded and surrounded in granular material	Pipes in trenches	4m	40.35	172.20	0.18	
Excavate trench 250 - 500 deep; lay 100 diameter vitrified clay pipe and fittings bedded and surrounded in in-situ concrete	Pipes in trenches	8m	47.23	377.84	0.41	
Excavate trench 1000 - 1250 deep; lay 150 diameter vitrified clay pipe and fittings bedded and surrounded in granular material	Pipes in trenches	23m	61.21	1407.83	1.51	
Excavate trench 1500 - 1750 deep; lay 150 diameter vitrified clay pipe and fittings bedded and surrounded in granular material	Pipes in trenches	10m	68.19	681.90	0.73	
100 diameter vertical vitrified clay pipe and fittings cased in in-situ concrete	Pipes vertical	3m	85.37	256.11	0.27	

APPROXIMATE ESTIMATING

OFFICE DEVELOPMENT (Cont'd)

	Item	Quantity	Item Rate	Cost total	per m²	Total

6.0 EXTERNAL WORKS Cont'd

6.2 Drainage (Cont'd)

Description	Item	Quantity	Item Rate	Cost total	per m²	Total
100 outlet vitrified clay yard gully with cast iron grating, surrounded in in-situ concrete	Gully	1nr	67.68	61.68	0.07	
100 outlet vitrified clay road gully with brick kerb and cast iron grating, surrounded in in-situ concrete	Gully	7nr	128.69	900.83	0.97	
Inspection chambers 600 x 900 x 600 deep internally comprising excavation works, in-situ concrete bases, suspended slabs and benchings, engineering brick walls and kerbs, clay channels, galvanised iron step irons and grade A access covers and frames	Inspection chamber	4nr	634.67	2538.68	2.72	
Inspection chambers 600 x 900 x 600 deep internally comprising excavation works, in-situ concrete bases, suspended slabs and benchings, engineering brick walls and kerbs, clay channels, galvanised iron step irons and grade B access covers and frames	Inspection chamber	3nr	635.39	1906.17	2.05	14210.23
				14210.23	15.24	

6.3 External Services

Description	Item	Quantity	Item Rate	Cost total	per m²	Total
Excavate trench for 4nr ducts; lay 4nr 100 diameter uPVC ducts and fittings; lay warning tape; backfill trench		20m	42.29	854.80	0.91	
Sewer connection charges		1nr	5000.00	5000.00	5.36	
Gas main and meter charges		1nr	4000.00	4000.00	4.29	
Electricity main and meter charges		1nr	4000.00	4000.00	4.29	
Water main and meter charges		1nr	3000.00	3000.00	3.22	16845.80
				16845.80	18.07	

PRELIMINARIES

Description	Item	Quantity	Item Rate	Cost total	per m²	Total
Site staff, accommodation, lighting, water safety health and welfare, rubbish disposal, cleaning, drying, protection, security, plant transport, temporary works and scaffolding with a contract period of 38 weeks	Item			65300.00	70.08	65300.00

TOTAL		**£981405.40**
Cost per m2 =		**£1053.01**

OFFICE DEVELOPMENT (Cont'd)

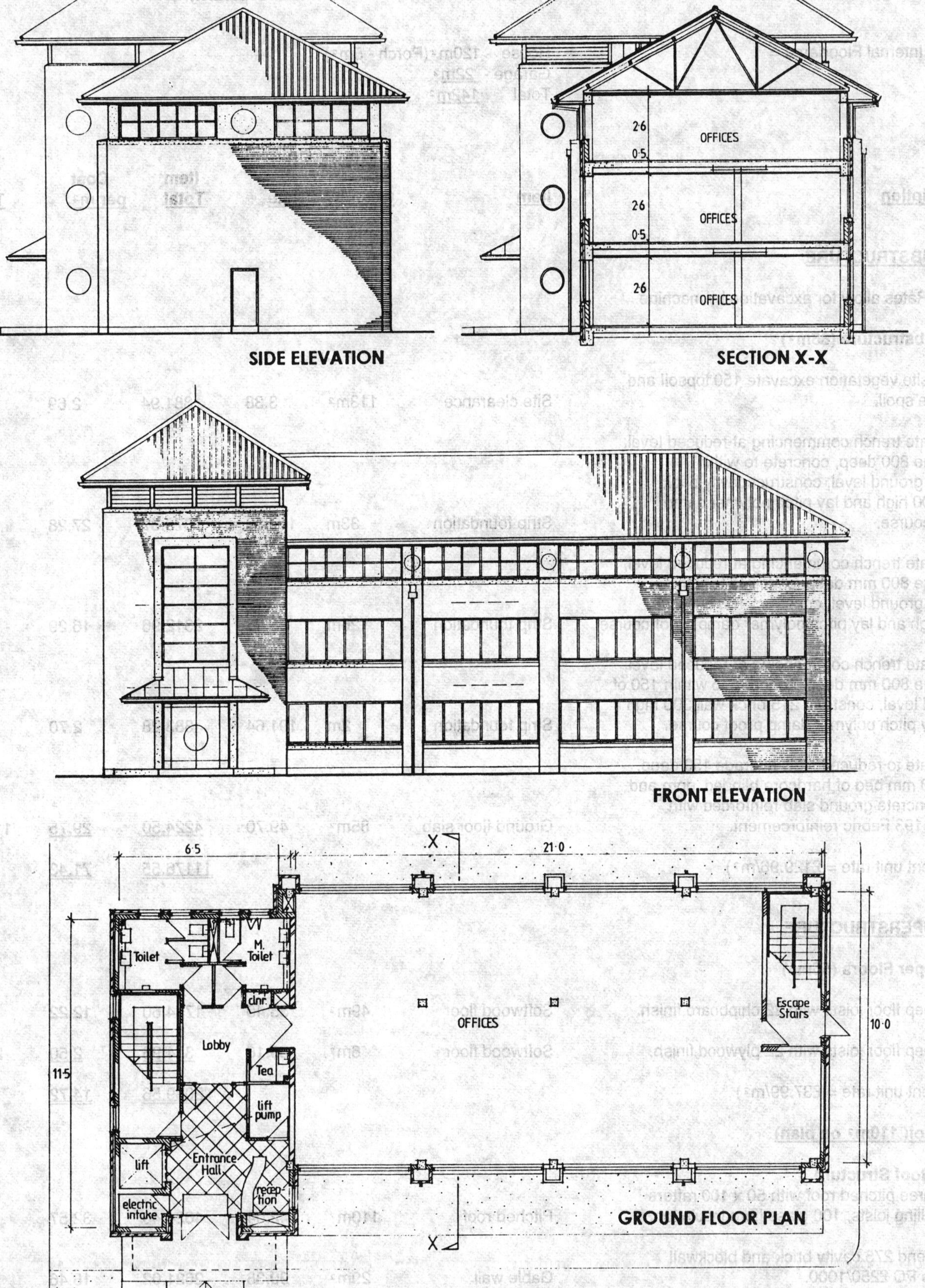

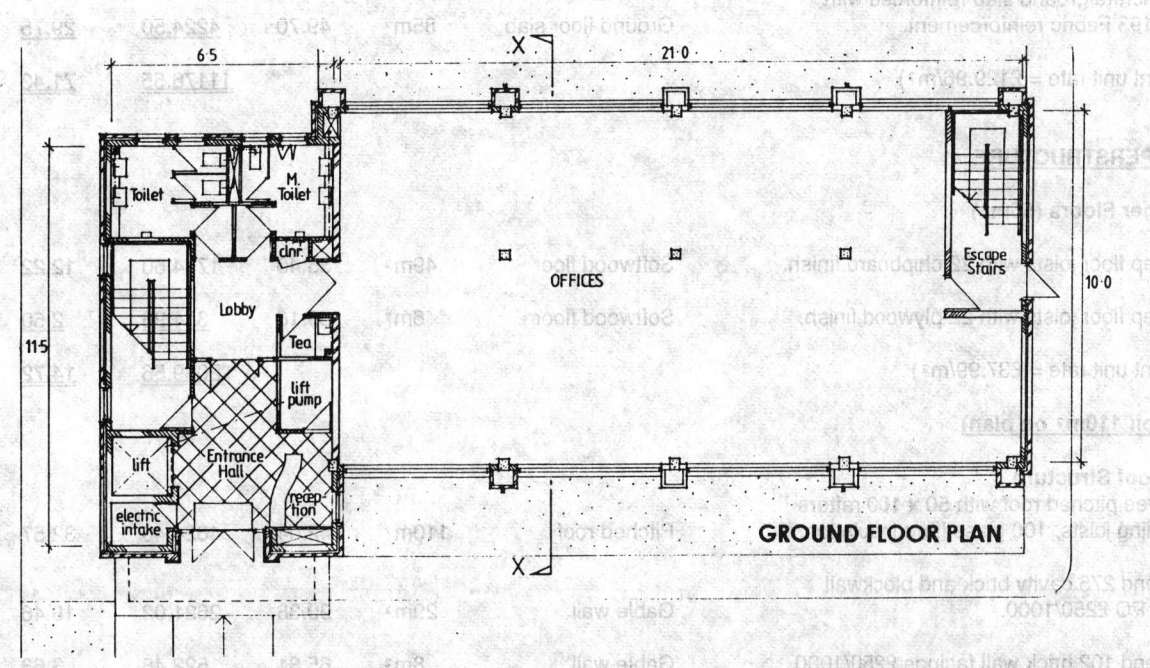

SIDE ELEVATION

SECTION X-X

FRONT ELEVATION

GROUND FLOOR PLAN

OFFICE DEVELOPMENT

© Cotterell Thomas & Thomas Architects

APPROXIMATE ESTIMATING

COST PLAN FOR A DETACHED HOUSE

The Cost Plan that follows gives an example of an Approximate Estimate for a detached house with garage and porch as the attached sketch drawings. The estimate can be amended by using the examples of Composite Prices following or the prices included within the preceding pages. Details of the specification used are indicated within the elements. The cost plan is based on a fully measured bill of quantities priced using the detailed rates within this current edition of Laxton's. Preliminaries are shown separately

Gross Internal Floor area =

House - 120m² (Porch - 5m²)
Garage - 22m²
Total 142m²

Description	Item	Quantity	Rate	Item Total	Cost per m²	Total
1.0 SUBSTRUCTURE						
Note: Rates allow for excavation by machine.						
1.1 Substructure (86m²)						
Clear site vegetation excavate 150 topsoil and remove spoil.	Site clearance	113m²	3.38	381.94	2.69	
Excavate trench commencing at reduced level, average 800 deep, concrete to within 150 of ground level, construct cavity brick wall 300 high and lay pitch polymer damp proof course.	Strip foundation	33m	117.39	3873.87	27.28	
Excavate trench commencing at reduced level, average 800 mm deep, concrete to within 150 of ground level, construct 103 brick wall 300 high and lay pitch polymer damp proof course.	Strip foundation	26m	88.96	2312.96	16.29	
Excavate trench commencing at reduced level, average 800 mm deep, concrete to within 150 of ground level, construct 215 brick wall 300 high and lay pitch polymer damp proof course.	Strip foundation	2m	191.64	383.28	2.70	
Excavate to reduce levels average 150 deep, lay 150 mm bed of hardcore blinded, dpm and 150 concrete ground slab reinforced with Ref. A 193 Fabric reinforcement.	Ground floor slab	85m²	49.70	4224.50	29.75	11176.55
(Element unit rate = £129.96/m²)				11176.55	71.43	
2.0 SUPERSTRUCTURE						
2.2 Upper Floors (55m²)						
175 deep floor joists with 22 chipboard finish.	Softwood floor	49m²	35.40	1734.60	12.22	
175 deep floor joists with 22 plywood finish.	Softwood floor	6m²	59.16	354.96	2.50	2089.56
(Element unit rate = £37.99/m²)				2089.56	14.72	
2.3 Roof(110m² on plan)						
2.3.1 Roof Structure						
35 degree pitched roof with 50 x 100 rafters and ceiling joists, 100 glass fibre insulation.	Pitched roof	110m²	42.04	4624.40	32.57	
Gable end 275 cavity brick and blockwall facings PC £250/1000.	Gable wall	29m²	90.38	2621.02	18.46	
Gable end 102 brick wall facings £250/1000	Gable wall	8m²	65.31	522.48	3.68	

DETACHED HOUSE (Cont'd)

	Item	Quantity	Rate	Item total	Cost per m²	Total
2.0 SUPERSTRUCTURE(Cont'd)						
2.3 Roof (Cont'd)						
2.3.2 Roof Coverings						
Interlocking concrete tiles, felt underlay, battens.	Roof tiling	143m²	18.28	2614.04	18.41	
Verge, 150 under cloak.	Verge	38m	10.53	400.14	2.82	
Ridge, half round.	Ridge	14m	16.87	236.18	1.66	
Valley trough tiles.	Valley	2m	34.07	68.14	0.48	
Abutment, lead flashings cavity trays.	Abutment	11m	58.97	648.67	4.57	
Sheet lead cladding.	Lead roofing	2m²	129.85	259.70	1.83	
Eaves soffit, fascia and 100 PVC gutter.	Eaves	29m	46.02	1334.58	9.40	
2.3.3 Roof Drainage						
68 pvcu rainwater pipe.	Rainwater pipe	24m	19.88	<u>477.12</u>	<u>3.36</u>	13806.47
(Element unit rate = £125.51/m²)				<u>13806.47</u>	<u>97.24</u>	
2.4 Stairs (1 Nr)						
Softwood staircase 2670 rise, balustrade one side, 910 wide two flights with half landing.	Stairs	1nr	1326.49	<u>1326.49</u>	<u>9.34</u>	1326.49
				<u>1326.49</u>	<u>9.34</u>	
2.5 External Walls (152m2)						
Hollow wall, light weight blocks one skin facings PC £250/1000 one skin ties.	Hollow wall	127m²	91.24	11587.48	81.60	
Facing brick wall PC £250/1000 102 thick.	½ B Facing wall	25m²	62.45	1561.25	10.99	
Facing brick wall PC £250/1000 215 thick.	1 B Facing wall	3m²	111.87	<u>335.61</u>	<u>2.36</u>	13484.34
(Element unit rate = £88.71/m²)				<u>13484.34</u>	<u>94.95</u>	
2.6 Windows and External Doors (35m2)						
2.6.1 Windows (16m2)						
Hardwood double glazed casement windows.	Hardwood windows	16m²	421.49	6743.84	47.49	
Galvanised steel lintel, dpc brick soldier course.	Lintels	16m	34.63	554.08	3.90	
Softwood window board, dpc close cavity.	Window board	15m	7.71	115.65	0.81	
Close cavity at jamb, dpc facings to reveal.	Jamb	32m	13.85	443.20	3.12	
2.6.2 External Doors (19m2)						
Hardwood glazed doors, frames, ironmongery PC £50.	Hardwood doors	6m²	648.90	3893.40	27.42	
Hardwood patio doors.	Patio doors	9m²	374.76	3372.84	23.75	
Galvanised steel garage door.	Garage door	4m²	143.05	572.20	4.03	
Galvanised steel lintel, dpc, brick soldier course.	Lintels	12m	35.41	424.92	2.99	

APPROXIMATE ESTIMATING

DETACHED HOUSE (Cont'd)	Item	Quantity	Rate	Item total	Cost per m²	Total

2.0 SUPERSTRUCTURE (Cont'd)

2.6 Windows and External Doors (35m2) (Cont'd)

	Item	Quantity	Rate	Item total	Cost per m²	Total
Close cavity at jamb, dpc facings to reveal.	Jamb	4m	13.85	55.40	0.39	16175.53
(Element unit rate = £462.16/m²)				16175.53	113.90	

2.7 Internal Walls and Partitions (153m2)

Light weight concrete block 100 thick.	Block walls	100m²	27.21	2829.84	19.93	
75 timber stud partition plasterboard and skim both sides.	Stud partition	53m²	54.16	2870.48	20.21	5700.32
(Element unit rate = £37.26/m²)				5700.32	40.14	

2.8 Internal Doors (40m2)

12 hardboard embossed panel door, linings ironmongery PC £20.00.	Flush doors	18m²	190.40	3427.20	24.14	
3 softwood 2GG glazed door, linings, ironmongery PC £20.00	Glazed doors	5m²	229.65	1148.25	8.09	
6 plywood faced door, linings, ironmongery PC £20.00	Flush doors	5m²	224.52	1122.60	7.91	
1 plywood faced half hour fire check flush door, frame, ironmongery PC £60.00.	Fire doors	2m²	216.31	432.62	3.05	
Plywood ceiling hatch.	Hatch	1nr	68.29	68.28	0.48	
Precast concrete lintels.	Lintels	10m	16.67	166.70	1.17	5700.32
(Element unit rate = £159.14/m²)				5700.32	40.14	

3.0 INTERNAL FINISHES

3.1 Wall Finishes (264m2)

13 light weight plaster.	Plaster work	264m²	12.99	3429.36	24.15	
152 x 152 wall tiling coloured PC £20.00 supply.	Tiling	13m²	37.87	492.31	3.47	3921.67
(Element unit rate = £14.85/m²)				3921.67	23.40	

3.2 Floor Finishes (63m2)

22 Chipboard, 50 insulation.	Board flooring	63m²	26.16	1741.95	12.27	
3 PVC sheet flooring.	Sheet vinyl	18m²	19.57	371.34	2.62	
25 x 125 softwood skirting.	Skirting	131m	7.51	1175.07	8.28	3288.36
(Element unit rate = £52.20/m²)				3288.36	23.17	

3.3 Ceiling Finishes (115m2)

9.5 Plasterboard and set.	Plasterboard	109m²	11.95	1302.55	9.17	
Plywood finish	Plywood	6m²	36.20	217.20	1.53	1519.75
(Element unit rate = £13.22/m²)				1519.75	10.70	

DETACHED HOUSE (Cont'd)	Item	Quantity	Rate	Item total	Cost per m²	Total

4.0 FITTINGS AND FURNISHINGS

4.1 Fittings

Kitchen units	600 base unit	2nr	190.13	380.26	2.68	
	1000 base unit	6nr	307.50	1845.00	12.99	
	500 broom cupboard	1nr	261.01	261.01	1.84	
	600 wall unit	3nr	109.31	327.93	2.31	
	1000 wall unit	2nr	165.27	330.54	2.33	
	Work tops	10m	49.89	498.90	3.51	
	Gas fire surround	1nr	577.50	577.50	4.07	4221.14
				4221.14	29.73	

5.0 SERVICES

5.1 Sanitary Appliances

Lavatory basin with pedestal.	3nr	146.65	439.95	3.10	
Bath with side panel.	1nr	290.01	290.01	2.04	
WC. suite.	3nr	193.82	581.46	4.09	
Stainless steel sink and drainer.	1nr	137.15	137.15	0.97	
Shower	1nr	572.12	572.12	4.03	

5.2 Services Equipment

Electric oven built in.	1nr	583.00	583.00	4.11	
Gas hob built in.	1nr	363.00	363.00	2.56	
Kitchen Extract Unit	1nr	385.00	385.00	2.71	
Gas fire with balanced flue.	1nr	297.00	297.00	2.09	

5.3 Disposal Installations

Upvc soil and vent pipe.	15m	65.17	977.55	6.88	
Automatic air admittance valve.	3nr	78.45	235.35	1.66	

5.4 Water Installations

Cold water to 9 No. draw off points.	item	871.19	871.19	6.14	
Hot water to 7 No. draw off points.	item	958.69	958.69	6.75	

5.5 Heat Source

Boiler pump and controls.	item	901.58	901.58	6.35	

5.6 Space Heating

Low temperature hot water radiator system.	item	3823.88	3823.88	26.93	

5.7 Ventilating Systems

Extract fan units.	3nr	358.77	1076.31	7.58	

5.8 Electrical Installations

Electrical Installation comprising 16Nr lighting points, 30Nr outlet points, cooker points, immersion heater point and 9Nr fittings points to include smoke alarms, bell, telephone and aerial installation.	item	5325.79	5325.79	37.51	

5.9 Gas Installation

Installation for boiler, hob and fire.	item	500.00	500.00	3.52	

5.10 Builders Work

Building work for service installation.	item	1272.02	1272.02	8.96	19591.05
				19591.05	137.98	

(Element unit rate = £123.11/m²)

APPROXIMATE ESTIMATING

DETACHED HOUSE (Cont'd)

	Item	Quantity	Rate	Item total	Cost per m²	Total
6.0 EXTERNAL WORKS						
6.1 Site Work						
Clear site vegetation excavate 150 topsoil and remove spoil	Site Clearance	479 m²	3.38	1619.02	11.40	
Excavate and lay 50 precast concrete flag paving on hardcore.	Paving	120m²	40.87	4904.40	34.54	
Brick wall, excavation 600 deep, concrete foundation, 215 brickwork facing. PC £200/1000 damp proof course brick on edge coping 900 high.	Garden wall	13m	95.30	1238.90	8.72	
Plastic coated chain link fencing 900 high on angle iron posts	Fencing	92m	26.56	2443.52	17.21	
Turfed Areas	Landscaping					
6.2 DRAINAGE						
100 diameter clay pipes, excavation, 150 bed and surround, pea shingle, backfill 750 deep.		85m	48.10	4088.50	28.79	
Polypropylene inspection chambers; universal inspection chamber; polypropylene; 475 diameter; preformed chamber base benching; for 100mm pipe; chamber shaft, 175mm effective length; cast iron cover and frame; single seal A15 light duty to suit.		6nr	142.25	853.50	6.01	
6.3 EXTERNAL SERVICES						
100 diameter clay duct, excavation backfill, polythene warning tape 500 deep.		38m	18.82	715.16	5.04	
Provisional sums for External Services - Gas, electric, water, telephone.		item	4000.00	4000.00	28.17	19863.00
				19863.00	139.88	
PRELIMINARIES						
Site staff, accommodation, lighting, water, safety and health and welfare, rubbish disposal cleaning, drying, protection, security, plant transport, temporary works and scaffolding with a contract period of 28 weeks				10000.00	70.42	10000.00

TOTAL = £ 131864.55

Cost per m² = £956.95

DETACHED HOUSE (Cont'd)

REAR ELEVATION

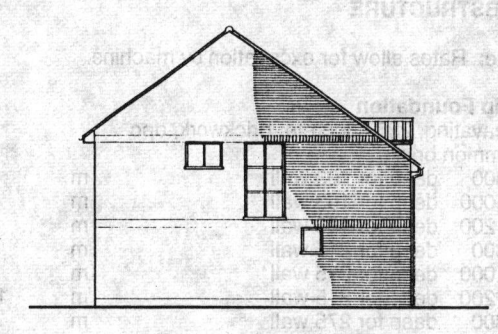

SIDE ELEVATION

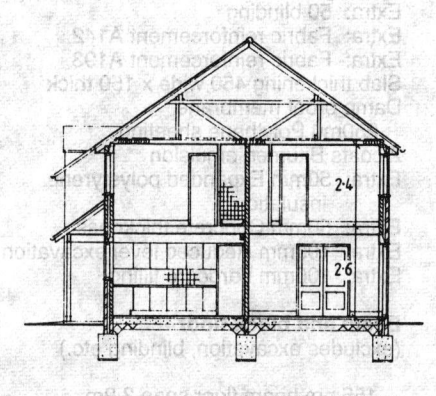

SECTION X-X

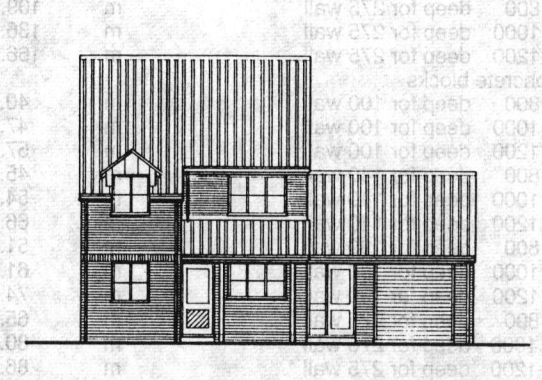

FRONT ELEVATION

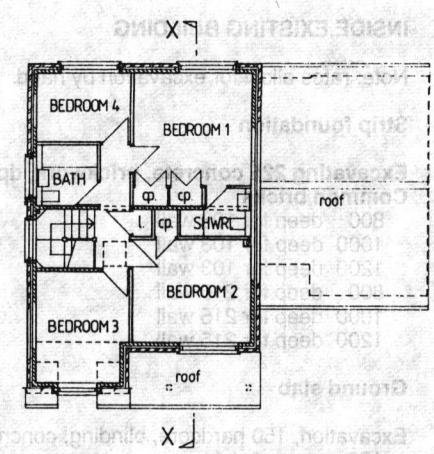

FIRST FLOOR PLAN

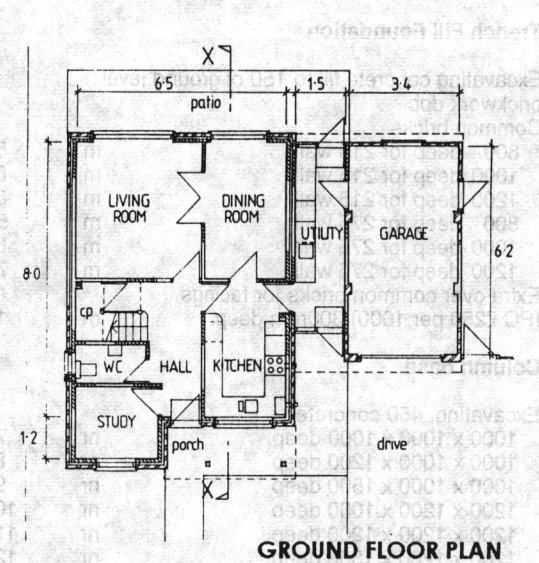

GROUND FLOOR PLAN

APPROXIMATE ESTIMATING

4 – BEDROOM DETACHED HOUSE

© Cotterell Thomas & Thomas Architects

COMPOSITE PRICES

The following composite prices are based on the detailed rates herein and may be used to amend the previous example cost plans or to create new approximate budget estimates.

SUBSTRUCTURE £

Note: Rates allow for excavation by machine.

Strip Foundation
Excavating, 225 concrete, brickwork, dpc
Common bricks

800	deep for 103 wall	m	51.03
1000	deep for 103 wall	m	61.91
1200	deep for 103 wall	m	80.20
800	deep for 215 wall	m	76.40
1000	deep for 215 wall	m	94.14
1200	deep for 215 wall	m	113.77
800	deep for 275 wall	m	89.43
1000	deep for 275 wall	m	110.92
1200	deep for 275 wall	m	134.48

Engineering bricks Class B

800	deep for 215 wall	m	96.32
1000	deep for 215 wall	m	119.55
1200	deep for 215 wall	m	144.67
800	deep for 275 wall	m	109.84
1000	deep for 275 wall	m	136.97
1200	deep for 275 wall	m	166.16

Concrete blocks

800	deep for 100 wall	m	40.07
1000	deep for 100 wall	m	47.92
1200	deep for 100 wall	m	57.84
800	deep for 140 wall	m	45.52
1000	deep for 140 wall	m	54.82
1200	deep for 140 wall	m	66.19
800	deep for 190 wall	m	51.01
1000	deep for 190 wall	m	61.70
1200	deep for 190 wall	m	74.45
800	deep for 275 wall *	m	65.50
1000	deep for 275 wall *	m	80.09
1200	deep for 275 wall *	m	86.44

*Thermalite Trenchblocks

Fabric reinforcement ref A193

350 wide	m	0.86
600 wide	m	1.47
750 wide	m	1.84

Trench Fill Foundation

Excavating concrete fill to 150 of ground level
brickwork dpc
Common bricks

800	deep for 215 wall	m	58.42
1000	deep for 215 wall	m	68.32
1200	deep for 215 wall	m	80.27
800	deep for 275 wall	m	54.87
1000	deep for 275 wall	m	64.76
1200	deep for 275 wall	m	76.06

Extra over common bricks for facings
(PC £250 per 1000) 300mm deep. m 10.70

Column base

Excavating, 450 concrete

1000 x 1000 x 1000 deep	nr	72.35
1000 x 1000 x 1200 deep	nr	82.84
1000 x 1000 x 1500 deep	nr	92.73
1200 x 1200 x 1000 deep	nr	101.55
1200 x 1200 x 1200 deep	nr	115.35
1200 x 1200 x 1500 deep	nr	128.60

Excavating, 750 concrete

1000 x 1000 x 1000 deep	nr	101.09
1000 x 1000 x 1200 deep	nr	111.58
1000 x 1000 x 1500 deep	nr	121.47
1200 x 1200 x 1000 deep	nr	142.94
1200 x 1200 x 1200 deep	nr	156.74
1200 x 1200 x 1500 deep	nr	169.98

Excavating, 450 reinforced concrete

1000 x 1000 x 1000 deep	nr	132.52
1000 x 1000 x 1200 deep	nr	143.46

SUBSTRUCTURE (Cont'd)
Column bases (Cont'd) £

1000 x 1000 x 1500 deep	nr	153.65
1200 x 1200 x 1000 deep	nr	189.28
1200 x 1200 x 1200 deep	nr	203.08
1200 x 1200 x 1500 deep	nr	216.33

Excavating, 750 reinforced concrete

1000 x 1000 x 1000 deep	nr	202.62
1000 x 1000 x 1200 deep	nr	213.11
1000 x 1000 x 1500 deep	nr	223.00
1200 x 1200 x 1000 deep	nr	289.15
1200 x 1200 x 1200 deep	nr	302.95
1200 x 1200 x 1500 deep	nr	316.19

Ground beam

Excavation, blinding, reinforced concrete

300 x 450	m	71.90
300 x 600	m	111.68

Ground slab

Excavation, 150 hardcore, blinding, concrete

150 ground slab	m²	22.05
200 ground slab	m²	26.65
225 ground slab	m²	28.94
Extra: 50 blinding	m²	4.49
Extra: Fabric reinforcement A142	m²	2.66
Extra: Fabric reinforcement A193	m²	2.45
Slab thickening 450 wide x 150 thick	m²	7.26

Damp proof membrane

250mu Polythene sheeting	m²	1.25
2 coats Bitumen emulsion	m²	5.21
Extra: 50mm Expanded polystyrene insulation	m²	6.34
Extra: 50mm Concrete thickness	m²	4.53
Extra: 100mm Reduced level excavation	m²	1.70
Extra: 100mm hardcore filling	m²	2.27

Beam and block floor
(excludes excavation, blinding etc.)

155mm beam floor span 3.9m	m²	24.31

Piling

On/off site charge	nr	4680.00
450 bored cast in-site piles n/e 6m deep	m	27.41
610 bored cast in-site piles n/e 6m deep	m	41.46
Cutting off tops of piles	nr	9.38

INSIDE EXISTING BUILDING

Note: rates allow for excavation by hand.

Strip foundation

Excavating 225 concrete, brickwork, dpc
Common bricks

800	deep for 103 wall	m	63.33
1000	deep for 103 wall	m	77.41
1200	deep for 103 wall	m	101.87
800	deep for 215 wall	m	87.84
1000	deep for 215 wall	m	108.48
1200	deep for 215 wall	m	133.99

Ground slab

Excavation, 150 hardcore, blinding, concrete

150 ground slab	m²	32.80
200 ground slab	m²	38.94
Slab thickening 450 wide x 150 thick	m	9.35

Damp proof membrane

250 mu polythene sheeting	m²	1.25
2 coats bitumen emulsion	m²	5.21

FRAME

Reinforced concrete

225 x 225 column	m	44.83
300 x 300 column	m	65.68
225 x 250 attached beam	m	50.28

FRAME (Cont'd)

		£
300 x 450 attached beam	m	91.51
400 x 650 attached beam	m	146.75
Casing to 152 x 152 steel column	m	32.58
Casing to 203 x 203 steel column	m	42.14
Casing to 146 x 254 steel beam	m	41.86
Casing to 165 x 305 steel beam	m	49.24
Extra: wrought formwork	m²	3.07
Steel		
Columns and beams	tonne	1051.88
Roof trusses	tonne	1687.50
Decorating	tonne	160.00
Purlins and cladding rails	tonne	1080.00
Small sections	tonne	1051.88
Decorating	tonne	160.00

UPPER FLOORS

In-situ reinforced concrete		
150 thick	m²	60.89
200 thick	m²	65.88
225 thick	m²	68.37
Hollow prestressed concrete		
125 thick	m²	34.99
150 thick	m²	36.52
200 thick	m²	40.92
Beam and block		
200 thick	m²	41.42
250 thick	m²	44.18
Softwood joist		
175 thick	m²	17.03
200 thick	m²	18.64
225 thick	m²	20.41
250 thick	m²	22.64
Floor boarding		
19 tongued and grooved softwood	m²	19.57
25 tongued and grooved softwood	m²	20.68
18 tongued and grooved chipboard	m²	10.67
22 tongued and grooved chipboard	m²	13.09
15 tongued and grooved plywood	m²	29.76
18 tongued and grooved plywood	m²	35.33

ROOF

Roof structure

Reinforced in-situ concrete		
150 thick	m²	57.84
200 thick	m²	62.83
50 lightweight screed to falls	m²	17.49
50 cement sand screed to falls	m²	13.57
50 vermiculite screed to falls	m²	16.13
50 flat glass fibre insulation board	m²	8.57
Extra for vapour barrier	m²	0.92
Softwood roof joists		
50 x 175 joists @ 400 centres	m²	18.44
50 x 225 joists @ 400 centres	m²	28.40
50 x 175 joists @ 600 centres	m²	13.16
50 x 225 joists @ 600 centres	m²	19.14
18 chipboard	m²	11.27
25 softwood tongued grooved boarding	m²	20.68
18 plywood decking	m²	16.41
24 plywood decking	m²	21.99
Pitched roof with 50 x 100 rafters and ceiling joist	m²	29.52
Softwood trussed rafters @ 600 centres	m	28.29

Roof coverings

Three layer fibre based felt	m²	14.24
Three layer high performance felt	m²	29.71
20 two coat mastic asphalt	m²	20.25
50 lightweight screed to falls	m²	17.49
50 cement sand screed to falls	m²	13.57
60 Vermiculite screed to falls	m²	19.08
50 flat glass fibre insulation board	m²	8.57
Extra for vapour barrier	m²	1.50
381 x 227 concrete interlocking tiles (on slope)	m²	22.38
268 x 165 plain concrete tiles (on slope)	m²	36.54
268 x 165 machine made clay tiles (on slope)	m²	45.03

ROOF (Cont'd)

Roof Coverings (Cont'd)

		£
265 x 165 hand made tiles (on slope)	m²	63.61
600 x 300 fibre cement slates (on slope)	m²	37.06
610 x 305 natural slates (on slope)	m²	79.85
100 insulation quilt	m²	4.22
Eaves soffit, fascia, gutter and decoration		
150 x 225 with 100 pvc-u gutter	m	34.93
150 x 225 with 114 pvc-u gutter	m	35.27
150 x 225 with 100 half round cast iron gutter	m	47.76
150 x 225 with 114 half round cast iron gutter	m	46.05
150 x 225 with 125 half round cast iron gutter	m	48.93
150 x 225 with 150 half round cast iron gutter	m	60.85
150 x 225 with 100 og cast iron gutter	m	44.47
150 x 225 with 114 og cast iron gutter	m	45.63
150 x 225 with 125 og cast iron gutter	m	47.41
200 x 300 with 100 pvc-u gutter	m	37.49
200 x 300 with 114 pvc-u gutter	m	37.83
200 x 300 with 100 half round cast iron gutter	m	48.11
200 x 300 with 114 half round cast iron gutter	m	48.61
200 x 300 with 125 half round cast iron gutter	m	51.49
200 x 300 with 150 half round cast iron gutter	m	63.41
200 x 300 with 100 og cast iron gutter	m	47.03
200 x 300 with 114 og cast iron gutter	m	48.19
200 x 300 with 125 og cast iron gutter	m	49.97
Verge undercloak, large board and decoration		
200 x 6	m	8.69
Hip rafter and capping		
225 x 50 rafter concrete capping tiles	m	27.99
225 x 50 rafter clay capping tiles	m	32.93
225 x 50 rafter Clay machine made bonnet hip tiles	m	32.93
225 x 50 rafter clay hand made bonnet hip tiles	m	32.93
Ridge board and capping		
200 x 25 ridge board concrete capping tile	m	24.15
200 x 25 ridge board hand made clay tile	m	29.09
Rainwater pipe		
68 pvc-u pipe	m	10.33
63 cast iron pipe and decoration	m	31.20
75 cast iron pipe and decoration	m	31.90
100 cast iron pipe and decoration	m	40.44

STAIRS

Reinforced concrete 2670 rise, mild steel balustrade one side, straight flight		
1000 wide granolithic finish	nr	1739.13
1500 wide granolithic finish	nr	1985.41
1000 wide pvc sheet finish non slip nosings	nr	2276.87
1500 wide pvc sheet finish non slip nosings	nr	2739.25
1000 wide terrazzo finish	nr	3024.61
1500 wide terrazzo finish	nr	3789.59
Reinforced concrete 2670 rise, mild steel balustrade one side in two flights with half landing.		
1000 wide granolithic finish	nr	2107.80
1500 wide granolithic finish	nr	2386.56
1000 wide pvc sheet finish non slip nosings	nr	2675.97
1500 wide pvc sheet finish non slip nosings	nr	3089.56
1000 wide terrazzo finish	nr	3405.68
1500 wide terrazzo finish	nr	4556.68
Softwood staircase 2670 rise, balustrade one side.		
910 wide straight flight	nr	942.00
910 wide one flight with three winders	nr	1153.69
910 wide two flight with half landing	nr	1153.69
Mild steel staircase 2670 rise, balustrade both sides.		
910 wide straight flight	nr	2070.00
910 wide two flights with quarter landing	nr	4241.25
910 wide two flights with half landing	nr	5135.63

EXTERNAL WALLS

Common bricks		
102 wall	m²	41.24
215 wall	m²	73.54
327 wall	m²	98.55
Hollow wall with galvanised ties		
275 common bricks both skins	m²	84.90
275 facings one side P.C. £150 per 1000	m²	89.42
275 facings one side P.C. £200 per 1000	m²	92.90
275 facings one side P.C. £250 per 1000	m²	96.39

APPROXIMATE ESTIMATING

EXTERNAL WALLS (Cont'd)

		£
275 facings one side P.C. £300 per 1000	m²	99.87
Facing bricks		
102 wall P.C. £150 per 1000	m²	45.74
102 wall P.C. £200 per 1000	m²	49.22
102 wall P.C. £250 per 1000	m²	52.70
102 wall P.C. £300 per 1000	m²	56.19
215 wall P.C. £150 per 1000	m²	77.05
215 wall P.C. £200 per 1000	m²	84.08
215 wall P.C. £250 per 1000	m²	91.11
215 wall P.C. £300 per 1000	m²	98.14
Light weight concrete block 4.0n/mm²		
100 wall	m²	26.12
140 wall	m²	32.72
190 wall	m²	39.03
Hollow wall in light weight blocks with galvanised ties.		
275 facings one side P.C. £150 per 1000	m²	74.26
275 facings one side P.C. £200 per 1000	m²	77.74
275 facings one side P.C. £250 per 1000	m²	81.23
275 facings one side P.C. £300 per 1000	m²	84.77
Extra for:		
50 mm glass fibre slab insulation	m²	8.71
13 mm two coat plain render painted	m²	13.28
15 mm pebbled dash render	m²	14.16

WINDOWS AND EXTERNAL DOORS

Windows

		£
Softwood glazed casement window		
type 110v size 631 x 1050	nr	180.49
type 212c size 1220 x 1220	nr	311.73
type 312c size 1769 x 1200	nr	404.49
type 312ww size 1769 x 1200	nr	489.28
average cost	m²	216.68
purpose made in panes 0.10-0.50m²		
(double glazed)	m²	307.35
Hardwood double glazed standard casement windows		
type X 107A size 630 x 750	nr	300.59
type X 112A size 630 x 1200	nr	384.70
type X 210A size 1200 x 1050	nr	554.65
type X 310AE size 1770 x 1050	nr	720.92
average cost	m²	451.08
Afrormosia double glazed casement windows purpose made in pane 0.l0-0.50m²	m²	468.14
European Oak double glazed casement windows purpose made in panes 0.10-0.50m²	m²	503.54
Softwood glazed double hung sash window		
type GS3 size 825 x 1094	nr	506.39
type GS4 size 825 x 1394	nr	579.31
type GSW4 size 1051 x 1394	nr	659.08
average cost	m²	496.00
Galvanised steel windows		
type NC05F size 508 x 1067	nr	146.12
type ND2F size 997 x 1218	nr	304.91
type NC02F size 997 x 1067	nr	316.14
type NC04F size 1486 x 1067	nr	428.42
type ND4 size 1486 x 1218	nr	439.69
Upvc windows tilt / turn factory double glazed		
size 600 x 1050 window	nr	472.12
size 900 x 1200 window	nr	548.61
size 900 x 1500 window	nr	611.10

Doors

		£
44 External plywood flush door, 50 x 100 softwood frame, architrave one side, iron mongery P. C. £50.00 per leaf and decorate		
762 x 1981 door	nr	275.08
838 x 1981 door	nr	283.74
1372 x 1981 pair doors	nr	467.47
44 Softwood external panel door glazed panels, 50 x 100 softwood frame, architrave one side, iron mongery P.C. £50.00 decorated.		
type 2XG size 762 x 1981	nr	319.65
type 2XG size 838 x 1981	nr	333.23
type 2XGG size 762 x 1981	nr	366.48
type 2XGG size 838 x 1981	nr	385.53
type KXT size 762 x 1981	nr	421.24
type KXT size 838 x 1981	nr	437.83

WINDOWS AND EXTERNAL DOORS (Cont'd)

Doors (Cont'd)

		£
44 Hardwood Magnet Alicante door, bevelled glass 50 x 100 hardwood frame architrave one side, iron mongery P.C. £50.00 decorated.		
Size 838 x 1981	nr	620.22
Galvanised steel garage door		
2135 x 1980	nr	494.10
Opening in cavity wall		
Galvanised steel lintel, DPC brick soldier course	m	57.14
Softwood window board DPC close cavity	m	15.51
Close cavity at jamb, DPC, facings to reveal	m	7.96

INTERNAL WALLS

		£
Common bricks		
102 wall	m²	41.34
215 wall	m²	73.71
Light weight concrete block 4.0 n/m²		
75 wall	m²	21.72
90 wall	m²	25.19
100 wall	m²	26.18
140 wall	m²	32.78
190 wall	m²	39.10
Timber stud partition with plasterboard both sides		
75 partition, direct decoration	m²	43.58
100 partition direct decoration	m²	47.49
75 partition with skim coat plaster and decorate	m²	54.20
100 partition with skim coat plaster and decorate	m²	58.11
75 partition with plastic faced plasterboard	m²	56.08
100 partition with plastic faced plasterboard	m²	59.99
Extra for 75 insulation	m²	3.25
Cellular core partition		
57 partition direct decoration	m²	35.46
63 partition direct decoration	m²	42.66
57 partition with skim coat plaster and decorate	m²	46.08
63 partition with skim coat plaster and decorate	m²	53.28

INTERNAL DOORS

		£
34 flush door 38 x 150 lining, stop, architrave both sides, iron mongery PC £20.00 per leaf, decorated.		
762 x 1981 hardboard faced	nr	229.66
838 x 1981 hardboard faced	nr	238.72
1372 x 1981 pair hardboard faced	nr	333.41
762 x 1981 plywood faced	nr	239.79
838 x 1981 plywood faced	nr	249.20
1372 x 1981 pair plywood faced	nr	353.46
44 half hour fire check flush door 50 x 100 frame architrave both sides, iron mongery PC £60.00 per leaf, decorated.		
762 x 1981 plywood faced	nr	276.80
838 x 1981 plywood faced	nr	302.13
1372 x 1981 pair plywood faced	nr	491.90
Extra 6mm georgian polished wired glass		
508 x 508 panel	nr	15.38
584 x 584 panel	nr	19.09
Precast concrete lintels 100 x 145	m	9.78

WALL FINISHES

		£
Hardwall plaster in two coats		
13 thick	m²	10.09
13 thick emulsion painted	m²	12.87
13 thick oil painted	m²	14.18
13 thick textured plastic coating stipple finish	m²	12.96
Lightweight plaster in two coats		
13 thick	m²	8.84
13 thick emulsion painted	m²	11.62
13 thick oil painted	m²	12.93
13 thick textured plastic coating stipple finish	m²	11.71
Plasterboard for direct decoration		
9.5 thick	m²	8.38
9.5 thick emulsion painted	m²	11.16
9.5 thick oil painted	m²	12.47
9.5 thick textured plastic coating stipple finish	m²	11.25
12.5 thick	m²	9.72
12.5 thick emulsion painted	m²	12.50
12.5 thick oil painted	m²	13.81

WALL FINISHES (Cont'd)

		£
12.5 thick textured plastic coating stipple finish	m²	12.59
Plasterboard with 3 plaster skim coat finish		
9.5 thick plasterboard and skim	m²	13.69
9.5 thick plasterboard and skim emulsion painted	m²	16.47
9.5 thick plasterboard and skim oil painted	m²	17.78
9.5 thick textured plastic coating stipple finish	m²	16.56
12.5 thick plasterboard and skim	m²	15.03
12.5 thick plasterboard and skim emulsion painted	m²	17.81
12.5 thick plasterboard and skim oil painted	m²	19.12
12.5 thick textured plastic coating stipple finish	m²	17.90
Extra wall paper PC £4.00 per roll		3.77
Extra vinyl coated wall paper PC £7.00 per roll		4.66
Extra wall tiling 108 x 108 x 6.5 white glazed	m²	49.46
Extra wall tiling 152 x 152 x 5.5 white glazed	m²	37.88
Extra wall tiling 152 x 152 x 5.5 light colour	m²	39.61

FLOOR FINISHES

Cement sand bed		
25 thick	m²	7.53
32 thick	m²	9.51
38 thick	m²	9.78
50 thick	m²	12.01
25 thick with quarry tile paving	m²	43.99
32 thick with quarry tile paving	m²	45.97
38 thick with quarry tile paving	m²	46.24
50 thick with quarry tile paving	m²	48.47
25 thick with 3 pvc sheet flooring	m²	29.34
32 thick with 3 pvc sheet flooring	m²	31.32
38 thick with 3 pvc sheet flooring	m²	31.59
50 thick with 3 pvc sheet flooring	m²	33.82
25 thick with 2.5 pvc tile flooring	m²	30.92
32 thick with 2.5 pvc tile flooring	m²	32.90
38 thick with 2.5 pvc tile flooring	m²	33.17
50 thick with 2.5 pvc tile flooring	m²	35.40
25 thick with 20 oak wood block flooring	m²	84.53
32 thick with 20 oak wood block flooring	m²	86.51
38 thick with 20 oak wood block flooring	m²	86.78
50 thick with 20 oak wood block flooring	m²	89.01
25 thick with 20 maple wood block flooring	m²	78.50
32 thick with 20 maple wood block flooring	m²	80.48
38 thick with 20 maple wood block flooring	m²	80.75
50 thick with 20 maple wood block flooring	m²	82.98
25 thick with 20 merbau wood block flooring	m²	83.93
32 thick with 20 merbau wood block flooring	m²	85.91
38 thick with 20 merbau wood block flooring	m²	86.18
50 thick with 20 merbau wood block flooring	m²	88.41
25 thick with carpet pc £25 / m² and underlay	m²	48.55
32 thick with carpet pc £25 / m² and underlay	m²	50.53
38 thick with carpet pc £25 / m² and underlay	m²	50.80
50 thick with carpet pc £25 / m² and underlay	m²	53.03
3 PVC sheet flooring	m²	21.81
Floor boarding and insulation to concrete		
22 tongued and grooved chipboard and 50 insulation board	m²	26.99
Softwood skirting decorated		
13 x 95 to BS 1186 reference 13RS95	m	7.17
20 x 70 to BS 1186 reference 20CA70	m	7.14
20 x 120 to BS 1186 reference 20CS120	m	8.41
19 x 100 square	m	8.65
19 x 150 square	m	9.05
25 x 100 square	m	9.05
15 x 125 moulded	m	7.94
19 x 125 moulded	m	8.65

CEILING FINISHES

Hardwall plaster in two coats		
13 thick	m²	10.09
13 thick emulsion painted	m²	13.51
13 thick oil painted	m²	15.70
13 thick textured plastic coatings stipple finish	m²	12.96
Lightweight plaster in two coats		
10 thick	m²	9.77
10 thick emulsion painted	m²	13.19
10 thick oil painted	m²	15.38
10 thick textured plastic coatings stipple finish	m²	12.64

CEILING FINISHES(Cont'd)

		£
Plasterboard for direct decoration		
9.5 thick	m²	9.21
9.5 thick emulsion painted	m²	12.63
9.5 thick oil painted	m²	14.82
9.5 thick textured plastic coatings stipple finish	m²	12.68
12.5 thick	m²	10.70
12.5 thick emulsion painted	m²	14.12
12.5 thick oil painted	m²	16.31
12.5 thick textured plastic coating stipple finish	m²	13.57
Plasterboard with 3 plaster skim coat finish		
9.5 thick plasterboard and skim	m²	15.56
9.5 thick plasterboard and skim emulsion painted	m²	18.98
9.5 thick plasterboard and skim oil painted	m²	21.17
9.5 thick textured plastic coating stipple finish	m²	18.43
12.5 thick plasterboard and skim	m²	17.05
12.5 thick plasterboard and skim emulsion painted	m²	26.47
12.5 thick plasterboard and skim oil painted	m²	22.66
12.5 thick textured plastic coating stipple finish	m²	19.92
Suspended ceiling		
300 x 300 x 15.8 plain mineral fibre tile	m²	30.06
300 x 300 x 15.8 textured mineral fibre tile	m²	36.03
300 x 300 patterned tile P.C. £15.00	m²	36.03
600 x 600 x 15.8 plain mineral fibre tile	m²	25.36
600 x 600 x 15.8 drilled mineral fibre tile	m²	32.61
600 x 600 patterned tile P.C. £15.00	m²	32.61

FITTINGS AND FURNISHINGS

Preassembled kitchen units		
500 floor unit	nr	120.50
600 floor unit	nr	125.36
1000 floor unit	nr	189.79
500 floor drawer unit	nr	204.23
1000 sink unit	nr	296.17
500 wall unit	nr	88.30
600 wall unit	nr	121.40
1000 wall unit	nr	137.18
28 work top	m	36.78
40 work top	m	49.89
plinth	m	32.18
cornice	m	37.02
pelmet	m	30.42

SANITARY APPLIANCES

Sanitary appliance with allowance for waste fittings, traps, overflows, taps and builders work.

Domestic building with plastic wastes		
lavatory basin	nr	110.31
bath with panels to front and one end	nr	299.68
low level w.c. suite	nr	193.82
stainless steel sink and drainer	nr	146.82
shower	nr	476.02
Commercial building with copper wastes		
lavatory basin	nr	125.78
low level w.c. suite	nr	193.82
bowl type urinal	nr	351.30
stainless steel sink	nr	166.56
drinking fountain	nr	420.63
shower	nr	505.43

WATER INSTALLATIONS

Soil and vent pipe for ground and first floor		
plastic wastes and SVP	nr	228.94
copper wastes and cast iron SVP	nr	719.72
Hot and cold water services domestic building	nr	1221.02
gas instantaneous sink water heater	nr	691.14
electric instantaneous sink water heater	nr	625.71
undersink water heater	nr	679.24
electric shower unit	nr	307.29

HEATING INSTALLATIONS

		£
Boiler, radiators, pipework, controls and pumps		
100m² domestic building, solid fuel	m²	44.01
100m² domestic building, gas	m²	41.03
100m² domestic building, oil	m²	50.29
150m² domestic building, solid fuel	m²	47.70

APPROXIMATE ESTIMATING

HEATING INSTALLATIONS (Cont'd)

		£
150m² domestic building, gas	m²	41.34
150m² domestic building, oil	m²	45.58
500m² commercial building, gas	m²	64.33
500m² commercial building, oil	m²	74.88
1000m² commercial building, gas	m²	77.50
1000m² commercial building, oil	m²	89.97
1500m² commercial building, gas	m²	88.64
1500m² commercial building, oil	m²	103.38
2000m² commercial building, gas	m²	98.27
2000m² commercial building, oil	m²	113.44
Underfloor heating with polymer pipework and insulation - excludes screed.	m²	29.53

ELECTRICAL INSTALLATIONS

Concealed installation with lighting, power, and ancillary cooker, mersion heater etc. circuit.

100m² domestic building, PVC insulated and sheathed cables	m²	50.22
100m² domestic building, MICS cables	m²	66.42
100m² domestic building, steel conduit	m²	77.54
150m² domestic building, PVC insulated and sheathed cables	m²	44.38
150m² domestic building, MICS cables	m²	59.57
150m² domestic building, steel conuit	m²	68.55
500m² commercial building, light fittings, steel conduit	m²	96.64
1000m² commercial building, light fittings, steel conduit	m²	103.39
1500m² commercial building, light fittings, steel conduit	m²	109.01
2000m² commercial building, light fittings, steel conduit	m²	114.62

LIFT AND ESCALATOR INSTALLATIONS

Light passenger lift, laminate walls, stainless steel doors.

3 stop, 8 person	nr	23811.73
4 stop, 13 person	nr	44463.35

General purpose passenger lift, laminate walls stainless steel doors.

4 stop, 8 person	nr	44125.85
5 stop, 13 person	nr	47789.97
6 stop, 21 person	nr	109054.09

Service hoist

2 stop, 50 kg	nr	6075.00

Escalator 30 degree inclination department store specification

3.00 m rise	nr	65250.00

SITE WORK

Pavings

250 excavation, 150 hardcore

150 concrete road tamped	m²	28.12
200 concrete road tamped	m²	32.57
150 concrete and 70 bituman macadam	m²	41.61
200 concrete and 70 bituman macadam	m²	46.06
Extra for Ref A252 fabric reinforcement	m²	2.99
Extra for Ref C636 fabric reinforcement	m²	3.90

Walls

Brick wall, excavation 600 deep concrete foundation, 215 wall with piers at 3m centres in facings PC £200 / 1000 damp proof course.

900 high brick on edge coping	m	131.26
1800 high brick on edge coping	m	212.49
900 high 300 x 100 splayed precast concrete coping	m	134.33
1800 high 300 x 100 splayed precast concrete coping	m	212.49

SITE WORKS (Cont'd)
£
Fences

Plastic coated chain link fencing, posts at 2743 centres.

900 high 38 x 38 angle iron posts two line wires	m	10.72
1200 high 38 x 38 angle iron posts two line wires	m	13.21
1400 high 38 x 38 angle iron posts three line wires	m	14.30
1800 high 38 x 38 angle iron posts three line wires	m	17.18
900 high concrete posts two line wires	m	14.22
1200 high concrete posts two line wires	m	16.11
1400 high concrete posts three line wires	m	17.95
1800 high concrete posts three line wires	m	21.83

Close boarded fencing, 100 x 100 posts @ 2743 centres 25 x 100 pales, 25 x 150 gravel boards, creosote, post holes filled with concrete.

1200 high two arris rails	m	31.20
1500 high three arris rails	m	39.18
1800 high three arris rails	m	43.43

DRAINAGE

Drains

Vitrified clay flexible joint drain, excavation and backfill.

100 diameter 500 deep	m	16.65
100 diameter 750 deep	m	19.23
100 diameter 1000 deep	m	21.99
100 diameter 1500 deep	m	28.80
150 diameter 500 deep	m	22.45
150 diameter 750 deep	m	25.06
150 diameter 1000 deep	m	27.86
150 diameter 1500 deep	m	34.75

Vitrified clay flexible joint drain, excavation 150 bed and surround pea shingle and backfill.

100 diameter 500 deep	m	29.48
100 diameter 750 deep	m	32.20
100 diameter 1000 deep	m	34.96
100 diameter 1500 deep	m	41.77
150 diameter 500 deep	m	38.00
150 diameter 750 deep	m	40.78
150 diameter 1000 deep	m	43.59
150 diameter 1500 deep	m	50.48

Vitrified clay flexible joint drain, excavation 150 bed and haunch concrete and backfill.

100 diameter 500 deep	m	28.16
100 diameter 750 deep	m	30.81
100 diameter 1000 deep	m	33.57
100 diameter 1500 deep	m	40.63
150 diameter 500 deep	m	37.69
150 diameter 750 deep	m	40.37
150 diameter 1000 deep	m	43.17
150 diameter 1500 deep	m	50.06

Vitrified clay flexible joint drain, excavation 150 bed and surround concrete and back fill.

100 diameter 500 deep	m	39.21
100 diameter 750 deep	m	41.93
100 diameter 1000 deep	m	44.69
100 diameter 1500 deep	m	51.50
150 diameter 500 deep	mm2	50.10
150 diameter 750 deep	m	52.88
150 diameter 1000 deep	m	55.67
150 diameter 1500 deep	m	62.58

Upvc drain, excavation 50 mm bed sand and backfill.

110 diameter 500 deep	m	18.32
110 diameter 750 deep	m	20.92
110 diameter 1000 deep	m	23.68
110 diameter 1500 deep	m	30.49
160 diameter 500 deep	m	21.82
160 diameter 750 deep	m	24.45
160 diameter 1000 deep	m	27.20
160 diameter 1500 deep	m	34.15

Upvc drain, excavation bed, and surround pea shingle and back fill.

110 diameter 500 deep	m	29.89
110 diameter 750 deep	m	32.61
110 diameter 1000 deep	m	35.37
110 diameter 1500 deep	m	42.18
160 diameter 500 deep	m	35.99

DRAINAGE (Cont'd)

Drains (Cont'd)

		£
160 diameter 750 deep	m	38.77
160 diameter 1000 deep	m	41.58
160 diameter 1500 deep	m	48.47

Upvc drain, excavation bed and haunch concrete and back fill.

		£
110 diameter 500 deep	m	28.68
110 diameter 750 deep	m	31.33
110 diameter 1000 deep	m	34.09
110 diameter 1500 deep	m	40.90
160 diameter 500 deep	m	35.99
160 diameter 750 deep	m	38.70
160 diameter 1000 deep	m	41.50
160 diameter 1500 deep	m	48.39

Upvc drain, excavation bed and surround concrete and back fill.

		£
110 diameter 500 deep	m	39.62
110 diameter 750 deep	m	42.34
110 diameter 1000 deep	m	45.10
110 diameter 1500 deep	m	51.91
160 diameter 500 deep	m	48.09
160 diameter 750 deep	m	50.87
160 diameter 1000 deep	m	53.68
160 diameter 1500 deep	m	60.57

Cast iron drain, excavation and back fill.

		£
100 diameter 500 deep	m	51.55
100 diameter 750 deep	m	51.71
100 diameter 1000 deep	m	56.89
100 diameter 1500 deep	m	63.70
150 diameter 500 deep	m	80.38
150 diameter 750 deep	m	82.99
150 diameter 1000 deep	m	85.79
150 diameter 1500 deep	m	92.68

Cast iron drain, excavation, bed and haunch concrete and back fill.

DRAINAGE (Cont'd)

Drains (Cont'd)

		£
100 diameter 500 deep	m	63.14
100 diameter 750 deep	m	63.36
100 diameter 1000 deep	m	68.55
100 diameter 1500 deep	m	75.35
150 diameter 500 deep	m	92.24
150 diameter 750 deep	m	94.93
150 diameter 1000 deep	m	97.74
150 diameter 1500 deep	m	104.63

Cast iron drain, excavation, bed and surround concrete and back fill.

		£
100 diameter 500 deep	m	74.11
100 diameter 750 deep	m	74.41
100 diameter 1000 deep	m	79.59
100 diameter 1500 deep	m	86.40
150 diameter 500 deep	m	108.04
150 diameter 750 deep	m	110.82
150 diameter 1000 deep	m	113.63
150 diameter 1500 deep	m	120.52

Brick manhole

Excavation, 150 concrete bed, 215 class B engineering brick walls, 100 clayware main channel, concrete benching, step irons and cast iron manhole cover.

		£
600 x 450 with two branches 750 deep	nr	454.43
600 x 450 with two branches 1000 deep	nr	573.88
600 x 450 with two branches 1500 deep	nr	808.10
750 x 450 with three branches 750 deep	nr	609.21
750 x 450 with three branches 1000 deep	nr	721.19
750 x 450 with three branches 1500 deep	nr	968.23
900 x 600 with five branches 750 deep	nr	797.26
900 x 600 with five branches 1000 deep	nr	930.09
900 x 600 with five branches 1500 deep	nr	1234.76

APPROXIMATE ESTIMATING

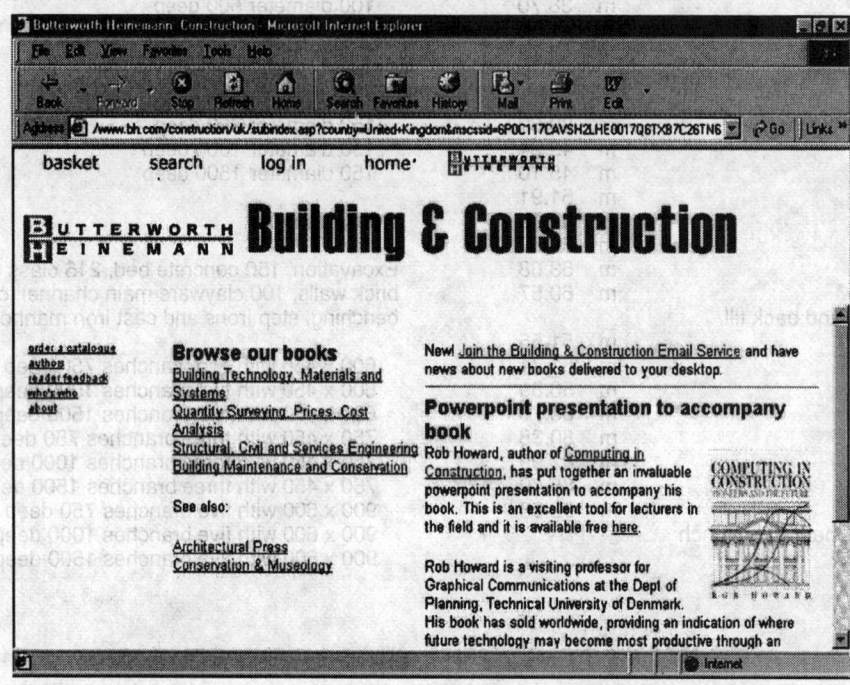

GENERAL INFORMATION
Standard Rates of Wages

ENGLAND, SCOTLAND AND WALES

In the terms of settlement the standard consolidated wages for building craft operatives and labourers due to be operated with effect on and from Monday, 28th June, 1999 will be as indicated below.

Note: The rates shown below are weekly rates. The hourly equivalents are shown in brackets.

1. ENTITLEMENT TO BASIC AND ADDTIONAL RATES-WR.1

		Rate per week £	
Craft operatives		235.95	(6.05p)
Skill Rate	1	224.64	(5.76p)
	2	216.45	(5.55p)
	3	202.80	(5.20p)
	4	191.10	(4.90p)
General Building Operative		177.45	(4.55p)

Additional Payment for Skilled Work

Skilled Operative Additonal Rate:		
1	44.85	(1.15p)
2	27.30	(0.70p)
3	7.80	(0.20p)

2. BONUS - WR.2

It shall be open to employers and employees on any job to agree a bonus scheme based on measured output and productivity for any operation or operations on that particular job.

3. STORAGE OF TOOLS - WR.12

Employers maximum liability shall be £400.

4. LOSS OF CLOTHING - WR.13

Employers maximum liability shall be £30.00

5. SICK PAY - WR.20

Effective on and from 28th June 1999 employers whose terms and conditions of employment incorporate the Working Rule Agreement of the CIJC should make a payment to operatives, who are absent from work due to sickness or injury, of £60.50 per 39 hour week.

Entitlement to addtional sick pay for intermittent skill, responsibility or working in adverse conditions as Schedule 2.

Extra Rate	P/hr	Weekly Rate
A	12p	4.68
B	20p	7.80
C	25p	9.75
D	30p	11.70
E	45p	17.55

RATES OF WAGES SINCE 1982

The following tables give the rates of wages since 1981.

		Grade A and Scotland District		London & Liverpool District	
		Craft Operatives per week £	Labourers per week £	Craft Operatives per week £	Labourers per week £
June	28th, 1982	79.56	68.055	79.755	68.25
June	27th, 1983	84.045	71.76	84.24	71.95
June	25th, 1984	88.335	75.27	88.53	75.465
June	24th, 1985	93.015	79.365	93.21	79.56
June	30th, 1986	98.28	83.855	98.475	84.045
June	29th, 1987	103.35	88.14	103.545	88.335
June	27th, 1988	110.565	94.185	110.76	94.38
June	26th, 1989	120.12	102.375	120.315	102.57
June	25th, 1990	131.82	112.32	132.015	112.515
June	24th, 1991	139.035	118.365	139.23	118.56
June	29th, 1992	159.51	135.915	159.705	136.11
June	27th, 1994	163.41	139.23	163.605	139.425
		All areas			
June	26th, 1995	173.55	143.52		
June	24th, 1996	178.62	147.81		
August	18th, 1997	188.37	156.00		

	Skilled Operative Rate:				General
Craft Operative	1	2	3	4	Building Operative
214.50	209.04	201.24	188.76	177.84	164.97
235.95	224.64	216.45	202.80	191.10	177.45

June 29th, 1999 — 214.50
June 28th, 1999 — 235.95

GUARANTEED MINIMUM BONUS

		Grade A and Scotland District		London and Liverpool District	
		Craft Operatives per week £	Labourers per week £	Craft Operatives per week £	Labourers per week £
November 2nd, 1981		12.09	10.14	12.09	10.14
June	27th, 1983	13.455	11.31	13.455	11.31
June	25th, 1984	14.04	11.895	14.04	11.89
June	24th, 1985	14.82	12.48	14.82	12.48
June	30th, 1986	15.21	12.87	15.21	12.87
June	27th, 1988	15.015	12.675	15.015	12.675
June	26th, 1989	15.015	12.675	15.015	12.675
June	25th, 1990	16.38	14.04	16.38	14.04
June	24th, 1991	16.575	14.235	16.575	14.235
June	29th, 1992	-	-	-	-

HOLIDAYS WITH PAY

Employers Contributions for Annual Holiday Stamps

Date effective from		Adults £	Under 18 £	* Contribution included to Retirement and Death Benefit Scheme £
August	1st, 1983	10.85*	7.65	0.65
July	30th, 1984	11.45*	8.05	0.75
July	29th, 1985	12.15*	8.50	0.90
August	4th, 1986	12.90*	8.95	1.05
August	1st, 1987	13.60*	9.30	1.20
August	1st, 1988	14.45*	9.85	1.30
July	31st, 1989	16.10*	11.05	1.40
July	30th, 1990	17.70*	12.15	1.55
July	29th, 1991	18.70*	12.75	1.75
August	3rd, 1992	19.20*	13.05	1.85
August	1st, 1994	19.30*	13.05	1.95
July	29th, 1996	19.60*	13.05	2.05
August	4th, 1997	20.00*	13.45	2.25
August	3rd, 1998	20.80*	14.25	2.45
August	2nd 1999	21.30	14.75	TBA

NORTHERN IRELAND

Authorised rates of wages in the Building and Civil Engineering Industry in Northern Ireland as agreed by the Joint Council, for the Building and Civil Engineering Industry (Northern Ireland).

With effect on and from Monday, August 2nd 1999, the rates of wages for craftsmen and labourers shall be as follows:

	Craftsmen £	General Building Operative £
Weekly Minimum Earnings....................................	244.92 (6.28p per hour)	214.50 (5.50p per hour)

Overtime, bonus and travelling time shall be calculated on the basic rate only.

With effect on and from Monday 26th July 1993 the Guaranteed Minimum Bonus for Craftsmen, Labourers and Apprenticies will be consolidated into the basic rate of pay.

The Notice of Promulgation dated 7th May 1996 stated that a review of the Working Rules Agreement has resulted in a fundamental restructuring of the wages, conditions and training arrangements in the industry. As part of the review further amendments are being considered.

RATES OF WAGES SINCE 1988

The following tables give the rates of wages since 1988

GUARANTEED MINIMUM BONUS

	Craftsmen Per Week	Labourers Per Week	Craftsmen Per Week	Labourers Per Week
August 1st, 1988.................................	£123.63	£107.25	7.00	7.00
July 31st, 1989..................................	£132.99	£115.83	7.00	7.00
July 30th, 1990..................................	£146.25	£127.14	7.00	7.00
July 29th, 1991..................................	£159.51	£134.84	7.00	7.00
July 27th, 1992..................................	£165.75	£144.30	7.00	7.00
July 26th, 1993..................................	£177.06	£155.22	-	-
July 25th, 1994..................................	£180.96	£158.73	-	-
July 31st, 1995..................................	£188.37	£164.97	-	-

	Craftsmen Per Week	General Building Operative		
August 6th, 1996................................	£194.22	£170.04	-	-
February 3rd, 1997.............................	£200.07	£175.11	-	-
September 1st, 1997...........................	£205.92	£180.18	-	-
February 2nd, 1998.............................	£219.96	£192.27	-	-
August 3rd, 1998................................	£229.71	£200.85	-	-
January 4th, 1999...............................	£236.73	£207.09	-	-
August 2nd, 1999................................	£244.92	£214.50	-	-

HOLIDAYS WITH PAY

Employers Contributions for Operatives:

Date effective	consolidated holiday credit (including benefit contribution)
May 8th, 1989....................................	£19.25
May 7th, 1990....................................	£20.80
May 6th, 1991....................................	£22.90
May 4th, 1992....................................	£24.20
May 3rd, 1993....................................	£25.00
May 2nd, 1994....................................	£25.20
May 1st, 1995....................................	£26.00
April 29th, 1996..................................	£27.15
April 28th, 1997..................................	£29.25
April 27th, 1998..................................	£32.25
April 26th, 1999..................................	£34.35

CONSTRUCTION SKILLS REGISTER

The register was introduced in March 1997 by the Joint Council for the Building and Civil Engineering Industry (Northern Ireland) and will form part of the Working Rules Agreement for the Industry. The Register has been designed to meet the needs of clients, contractors, experienced operatives and trainees across all sectors of the industry.
The Register is based on the achievement of National Vocational Qualifications and Skills Testing and will provide the only route to craft status in the main construction trades.

HEATING, VENTILATING AND DOMESTIC ENGINEERING CONTRACTS

Rates agreed between the Heating and Ventilating Contractors' Association, and the Manufacturing, Science and Finance Union.

RATES OF WAGES

All districts of the United Kingdom
Note : Ductwork Erection Operatives, are entitled to the same rates and allowances
allowances before
as the parallel fitter grades shown.

The Wage Agreement is on the basis that there will be no further increase in: hourly rates and Monday 4th September 2000; or, as already agreed, in weekly Sickness and Accident Benefit or in Weekly Holiday Credit and Welfare Contribution rates before Monday 2nd October 2000. Additionally, there will be no further increase in Daily Travelling Allowance rates before Monday 4th September 2000.

HEATING, VENTILATING AND DOMESTIC ENGINEERING CONTRACTS (Cont'd)

	From 23 August 1999
Main Grades	Hourly Rate

Foreman...	8.73
Senior Craftsman.......................................	7.48
Craftsman..	6.86

	From 23.8.99
Installer..	6.23
Improver..	6.23
Assistant...	5.87
Adult Trainee..	5.25
Mate (over 18)..	5.25
Mate (17-18)..	3.37
Mate (under 17).......................................	2.43

Craft apprentice -

Year 1 -	3.42
Year 2 -	4.05
Year 3 -	5.17
Year 4 -	6.23

Responsibility Allowance

2 Units..	0.62
1 Unit...	0.31

DAILY TRAVELLING ALLOWANCE

C = Craftsmen including improvers.
M and A = Assistants, mates, junior mates and craft apprentices.

Direct distance from centre to job in miles   From 23rd August, 1999

Over	Not exceeding	C p	M & A p
10	20.....................................	349p	300p
20	30.....................................	626p	540p
30	40.....................................	824p	710p
40	50.....................................	1028p	881p

DAILY ABNORMAL CONDITIONS MONEY

From 23rd August, 1999

	p
Per day....................	299p

LODGING ALLOWANCE

From 24th August, 1999

	£
Per day....................	22.05

WEEKLY HOLIDAY CREDIT/WELFARE CONTRIBUTIONS

From 4th October 1999

Credit Value Category (Note 5)	Weekly Holiday Credit £	Combined Weekly Credit and Welfare Contribution £
a	39.64	44.13
b	35.92	40.41
c	34.39	38.88
d	31.47	35.96
e	27.62	32.11
f	23.06	27.55
g	16.57	21.06
h	11.05	15.54

NOTES TO THE APPENDIX ON WAGE RATES, ALLOWANCES AND OTHER PROVISIONS

1. As of 24 August 1998, the grades of Foreman (Pipefitting) and Foreman (Ductwork) will be re-titled as Foreman. From this date there will be only one set of rates and allowances promulgated to cover the new Foreman grade.

2. The Chargehand grade shall be discontinued from 24th August 1998, but existing employees in this grade shall retain the 24 August 1998 rate, on a mark time basis, until this is overtaken by the hourly rate for the Senior Craftsman.

3. From 3 April 2000, the Improver and Assistant grades shall be discontinued, and existing employees in either grade shall be transferred into the Operative grade (New Grade) on that date. This changeover date has been agreed to allow Improvers and Assistants, as at the date of this Agreement, to complete an Improvership to enable them to qualify as a Fitter (with effect from 24 August 1998, Craftsman).

NOTES TO THE APPENDIX ON WAGE RATES, ALLOWANCES AND OTHER PROVISIONS (CONT'D)

4. Responsibility Allowance is payable to Craftsmen and Senior Craftsmen at the lower rate for a second welding skill **or** supervisory responsibility, and at the higher rate for a second welding skill **and** supervisory responsibility. Further details about the Responsibility Allowance system can be found in Clause 8 (i-n) of the National Agreement, published at Appendix 3 ofJCC Letter 67, dated 8 April 1998

5. From 4 October 1999, the grades of H&V Employees covered by the range of credit values and entitled to the different rates of Sickness and Accident Benefit, and weekly Holiday Credit and Welfare Contributions are as follows:

a	b	c
Foreman	Senior Craftsman(+2nd Welding Skill) Senior Craftsman	Craftsmen(+2nd Welding Skill)

d	e	f
Craftsman	Installer Improver Assistant 4th Year Apprentice Senior Modern Apprentice	Adult Trainee(New Grade) Mate(over 18) 3rd Year Apprentice

g	h	
2nd Year Apprentice Intermediate Modern Apprentice Mate (up to 18)	1st Year Apprentice Junior Modern Apprentice	

For explanations, fuller details and general conditions the reader is advised to obtain a copy of the National Working Rule Agreement available from the Heating and Ventilation Contractor's Association Publications Department, Old Mansions House, Eamont Bridge, Penrith, Cumbria, CA10 2BX.

JOINT INDUSTRY BOARD FOR THE ELECTRICAL CONTRACTING INDUSTRY

WAGES

(i) **National Standard Rates**

The JIB hourly rates of wages shall be as set out below.

	From 30th March 1998	From 4th January 1999
Electrician, Instrument Mechanic, Instrument Pipefitter (or equivalent specialist grade)	£6.45	£6.76
Approved Electrician, Approved Instrument Mechanic, Approved Instrument Pipe fitter, (or equivalent specialist grade)	£7.11	£7.42
Technician, (or equivalent specialist grade)	£8.09	£8.48
Labourer	£5.03	£5.24
Adult Trainee (under 21)	£3.77	£3.93
Adult Trainee	£5.03	£5.24
Senior Graded Electrical Trainee	£5.82	£6.08

(ii) **London Weighting**

	From 30th March 1998	From 4th January 1999
Electrician, Instrument Mechanic, Instrument Pipefitter (or equivalent specialist grade)	£6.89	£7.22
Approved Electrician, Approved Instrument Mechanic, Approved Instrument Pipefitter, (or equivalent specialist grade)	£7.55	£7.88
Technician, (or equivalent specialist grade)	£8.53	£8.94
Labourer	£5.47	£5.70
Adult Trainee (under 21)	£4.21	£4.39
Adult Trainee	£5.47	£5.70
Senior Graded Electrical Trainee	£6.26	£6.54

Further details of the wage agreement, allowances and working Rules may be obtained on written application to the Joint Industry Board for the Electrical Contracting Industry, Kingswood House, 47/51 Sidcup Hill, Sidcup, Kent DA14 6HP.

GENERAL INFORMATION

THE NATIONAL JOINT COUNCIL FOR THE LAYING SIDE OF THE MASTIC ASPHALT INDUSTRY

Rates of wages as approved by the National Joint Council for the Laying Side of the Mastic Asphalt Industry.

Rates from Monday 27th June, 1999

	Basic hourly rate (incl. GMPP) £	Guaranteed Basic Weekly Pay £
Chargehands Responsibility Allowance.................	10.00 (per day)	
Spreaders Basic (39 hours)..................................	4.84	188.76 / week
Bonus..	4.00	
Mixermen/Potmen Basic (39 hours).......................	4.48	174.72 / week
Bonus	3.70	
Classified Labourers Basic (39 hours)....................	4.48	174.72 / week
Bonus...........................	3.39	
Unclassified Labourers..	Building Trade Labourer's Rate	

Further information can be obtained from the Employers' Secretary, Mastic Asphalt Council Ltd, Claridge House, 5 Elwick Road, Ashford Kent TN23 1PD (01233 634411)

THE JOINT INDUSTRY BOARD FOR PLUMBING MECHANICAL ENGINEERING SERVICES IN ENGLAND AND WALES

Hourly rates with effect from 24th August 1998

Operatives	£		£
Trained Plumber and Gas Service Fitter...........	5.98	Apprentice Plumbers and Apprentice Service Fitters	
Advanced Plumber and Gas Service Engineer..	6.70	1st year of training.............................	2.62
Technical Plumber and Gas Service Technician	7.48	2nd year of training............................	3.12
		3rd year of training............................	3.52
		3rd year of training with NVQ level 2*..	4.28
		4th year of training............................	4.33
		4th year of training with NVQ level 2*..	4.93
		4th year of training with NVQ level 3*..	5.44

Adult Trainees

1st 6 months of employment	4.69	
2nd 6 " " "	5.01	
3rd 6 " " "	5.23	

* Note
Where apprentices have achieved NVQ's, the appropriate rate is payable from the date of attainment except that it shall not be any earlier than the commencement of the promulgated year of Training in which it applies.

WORKING HOURS AND OVERTIME

a)   The normal working weeks shall be 37.5 hours
b)   Overtime rates shall be payable after 39 hours are worked
c)   After 8.00pm overtime shall be paid at double time

Further information can be obtained from The Joint Industry Board for Plumbing Mechanical Engineering Services in England and Wales, Brook House, Brook Street, St. Neots, Huntingdon, Cambs. PE19 2HW. Tel: 01480 476925 Fax: 01480 403081

NATIONAL INSURANCES

Employers Contribution for Men:

April   1988...............................   5.00% of employee's earnings where £41.00 per week and not exceeding £69.99 per week.
7.00% of employee's earnings where £70.00 per week and not exceeding £104.99 per week.
9.00% of employee's earnings where £105.00 per week and not exceeding £154.99 per week.
10.45% of employee's earnings where £155.00 per week and over.

April   1989...............................   5.00% of employee's earnings where £43.00 per week and not exceeding £74.99 per week.
7.00% of employee's earnings where £75.00 per week and not exceeding £114.99 per week.
9.00% of employee's earnings where £115.00 per week and not exceeding £164.99 per week.
10.45% of employee's earnings where £165.00 per week and over.

April   1990...............................   5.00% of employee's earnings where £46.00 per week and not exceeding £79.99 per week.
7.00% of employee's earnings where £80.00 per week and not exceeding £124.99 per week.
9.00% of employee's earnings where £125.00 per week and not exceeding £174.99 per week.
10.45% of employee's earnings where £175.00 per week and over.

April   1991...............................   4.6% of employee's earnings where £52.00 per week and not exceeding £84.99 per week.
6.6% of employee's earnings where £85.00 per week and not exceeding £129.99 per week.
8.6% of employee's earnings where £130.00 per week and not exceeding £184.99 per week.
10.4% of employee's earnings where £185.00 per week and over.

April   1992...............................   4.6% of employee's earnings where £54.00 per week and not exceeding £89.99 per week.
6.6% of employee's earnings where £90.00 per week and not exceeding £134.99 per week.
8.6% of employee's earnings where £135.00 per week and not exceeding £189.99 per week.
10.4% of employee's earnings where £190.00 per week and over.

NATIONAL INSURANCES

April 1993.............................
4.6% of employee's earnings where £56.00 per week and not exceeding £94.99 per week.
6.6% of employee's earnings where £95.00 per week and not exceeding £139.99 per week.
8.6% of employee's earnings where £140.00 per week and not exceeding £194.99 per week.
10.4% of employee's earnings where £195.00 per week and over.

April 1994.............................
3.6% of employee's earnings where £57.00 per week and not exceeding £99.99 per week.
5.6% of employee's earnings where £100.00 per week and not exceeding £144.99 per week.
7.6% of employee's earnings where £145.00 per week and not exceeding £199.99 per week.
10.2% of employee's earnings where £200.00 per week and over.

April 1995.............................
3% of employee's earnings where £58.00 per week and not exceeding £104.99 per week.
5% of employee's earnings where £105.00 per week and not exceeding £149.99 per week.
7% of employee's earnings where £150.00 per week and not exceeding £204.99 per week.
10.2% of employee's earnings where £205.00 per week and over.

April 1996.............................
3% of employee's earnings where £61.00 per week and not exceeding £109.99 per week.
5% of employee's earnings where £110.00 per week and not exceeding £154.99 per week.
7% of employee's earnings where £155.00 per week and not exceeding £209.99 per week.
10.2% of employee's earnings where £210.00 per week and not exceeding £445.00 per week.
10.2% of employee's earnings where £445.00 per week and over.

April 1997...............................
3% of employee's earnings where £62.00 per week and not exceeding £109.99 per week.
5% of employee's earnings where £110.00 per week and not exceeding £154.99 per week.
7% of employee's earnings where £155.00 per week and not exceeding £209.99 per week.
10% of employee's earnings where £210.00 per week and not exceeding £465.00 per week.
10% of employee's earnings where £465.00 per week and over.

April 1998..............................
3% of employee's earnings where £64.00 per week and not exceeding £109.99 per week.
5% of employee's earnings where £110.00 per week and not exceeding £154.99 per week.
7% of employee's earnings where £155.00 per week and not exceeding £309.99 per week.
10% of employee's earnings where £210.00 per week and not exceeding £485.00 per week.
10% of employee's earnings where £485.00 per week and over.

April 1999..........................
12.2% of employee's earnings above the earnings threshold where £83.01 per Week or more

CITB LEVY

The CITB levies have been increased following The Industrial Training Levy (Construction Board)
order which came into force on 26th January 1999, The increases are as follows:

PAYE rate from 0.29% to 0.38% (plus 31%)

LOSC rate from 2.00% to 2.28% (plus 14%)

Current arrangements whereby 2% of labour-only receipts may be offset against the total levy liability will remain in place.

GENERAL INFORMATION

Estimating for Builders and Quantity Surveyors
R D Buchan, F W Fleming, J R Kelly

Written for students taking courses in building and surveying at HNC/D and BSc level, this textbook explains the calculation of rates for the items in bills of quantities, based on SMM7. For convenience of use in practice, the arrangement is by individual topic or trade with appropriate reference to subsections of SMM7. Care has been taken, in both discussion and examples; to emphasize the different approaches used when pricing the various types of work. The worked examples reflect both traditional and up-to-date technology.

For all the trades included, a core of the most useful examples is provided, with particular attention paid to those areas that require rather more detailed explanation than has been provided in some more summary texts.

A chapter on computerized estimating is complemented by the Appendix which shows how a spreadsheet program can be developed for the calculation of the cost of mortars etc. Chapters on tendering strategy are included to set estimating in its professional context.

£17.99, Paperback, 1991, 300pp, isbn 0 7506 0041 1

Builders' and Contractors' Plant

MECHANICAL PLANT

The following rates are inclusive of operator as noted, fuel, consumables, repairs and maintenance.

Excavators including Driver and Banksman

Hydraulic, full circle slew, crawler mounted with single equipment (shovel)

0.70m3	per hour	24.00
1.00m3	per hour	26.30
1.50m3	per hour	28.00

Hydraulic, offset or centre post, half circle slew, wheeled, dual purpose (back/loader)

0.60m3	per hour	18.00
0.80m3	per hour	19.50
1.00m3	per hour	20.00
1.20m3	per hour	21.00

Mini excavators, full circle slew, crawler mounted with single equipment (back actor) including driver only

1.4 tonne capacity	per hour	12.00
2.4 tonne capacity	per hour	14.00
3.6 tonne capacity	per hour	15.50

Compactors including operator

Vibrating plate

90kg	per hour	7.60
120kg	per hour	7.80
176kg	per hour	7.95
310kg	per hour	8.10

Rollers including Driver

Road deadweight

2.54 tonnes	per hour	11.40
6.10 tonnes	per hour	12.20
8.13 tonnes	per hour	14.70
10.20 tonnes	per hour	15.15

Vibratory pedestrian operated
single roller

550kg	per hour	8.05
680kg	per hour	8.40
twin roller		
650kg	per hour	9.65
950kg	per hour	9.90
1300kg	per hour	10.55

Self-propelled vibratory tandem

1.07 tonne	per hour	11.25
2.03 tonne	per hour	12.20
3.05 tonne	per hour	12.85
6.10 tonne	per hour	14.65

Pumps including attendant part time

Single diaphragm (excluding hoses)

51mm	per hour	2.40
76mm	per hour	2.70
102mm	per hour	3.90

Double diaphragm (excluding hoses)

51mm	per hour	2.50
76mm	per hour	3.00
102mm	per hour	4.20

Suction or delivery hose, flexible, including coupling, valve and strainer, per 7.62m length

£

51mm	per hour	0.16
76mm	per hour	0.16
102mm	per hour	0.32

Lorries including Driver

Ordinary of the following plated gross vehicle weights

3.56 tonnes	per hour	14.00
5.69 tonnes	per hour	16.50
8.74 tonnes	per hour	17.50
10.16 tonnes	per hour	21.00

Tipper of the following plated gross vehicle weights

6.60 tonnes	per hour	15.15
8.74 tonnes	per hour	18.70
9.65 tonnes	per hour	21.00

Vans, including driver carrying capacity

0.4 tonnes	per hour	9.00
1.0 tonnes	per hour	10.00
1.25 tonnes	per hour	10.50

Dumpers including Driver

Manual gravity tipping 2 wheel drive

750 kg	per hour	10.00
1000 kg	per hour	10.25
2 tonne	per hour	10.55
3 tonnes	per hour	11.25

Hydraulic tipping 4 wheel drive

2 tonnes	per hour	10.80
3 tonnes	per hour	11.80
4 tonnes	per hour	13.80
5 tonnes	per hour	14.30

Forklifts, rough terrain including driver

1.0 tonne 2 wheel drive	per hour	13.50
2.0 tonnes 4 wheel drive	per hour	14.50
2.6 tonnes 4 wheel drive	per hour	16.75

Hoists

Scaffold type hoist excluding operator

250 kg	per week	29.00

Ladder type hoist excluding operator

150 kg	per week	65.00

GENERAL INFORMATION

BUILDERS AND CONTRACTORS PLANT

MECHANICAL (Cont'd)

Hoist (Cont'd)

Goods hoist up to 10m height including operator, excluding erecting and dismantling

Mobile type 500 kg.................. per week	300.00
Tied-in type 750 kg.................. per week	320.00

Concrete Mixers including Driver

Closed drum

0.20/0.15m3........................... per hour	7.00
0.30/0.20m3........................... per hour	7.80

Closed drum with swing batch weighing gear

0.40/0.30m3........................... per hour	9.50

Mixer with batch weighing gear and power loading shovel

0.30/0.20m3........................... per hour	9.20
0.40/0.30m3........................... per hour	10.40

Mortar Mixer including Driver

Mortar mixer

0.15/0.10m3........................... per hour	6.90

Concrete Equipment including Driver/Operator

Concrete pump, skid mounted, exclusive of piping

20/26m3.............................. per hour	31.00
30/45m3.............................. per hour	33.00
46/54m3.............................. per hour	35.00

Piping per 3.05m length

102mm diameter..................... per hour	0.15
127mm diameter..................... per hour	0.18
152mm diameter..................... per hour	0.21

Vibrator poker

petrol........................... per hour	7.50
diesel........................... per hour	7.60
electric........................ per hour	7.50
air.............................. per hour	6.75

Vibrator, external type, clamp on

small........................... per hour	6.90
medium........................ per hour	7.25
large........................... per hour	8.30

Power float

petrol........................... per hour	7.70
electric........................ per hour	7.80

Compressors including Driver

Portable, normal deliver of free air at 7kg/cm3

2.0m3/min........................... per hour	9.00
2.6m3/min........................... per hour	8.00
4.0m3/min........................... per hour	8.00

Compressor tools with 15.3m of hose (excluding operator)

breaker including steels........... per hour	2.00
light pick including steels......... per hour	1.80
clay spade including blade........ per hour	1.80

Compressor tool muffler.................... per hour	0.15

CHARGES FOR HIRE

The following Trade Prices have been provided by HSS Hire Shops, 25 Willow Lane, Mitcham, Surrey, CR4 4TS, Tel: 0181 685 9900 to whom acknowledgement is hereby given.

All hire charges are per week, day rates available on application. Guide prices only subject to revision.

Builders' Ladders

Extension Ladders (Aluminium)
2 part

up to 6.0m.............................. each	24.00
up to 9.0m.............................. each	32.00
up to 11.0m.............................. each	58.00

3 part

up to 6.0m.............................. each	24.00
up to 9.1m.............................. each	32.00
up to 13.7m.............................. each	68.00
up to 16.0m.............................. each	80.00

Builders' Timber Trestles

Examples	£
1.8m.......... each	21.00
2.4m.......... each	24.00

Adjustable Steel Trestles, 4 boards wide (fixed or folding leg)

	closed	extended		
no 1	0.5m	0.8m...................... each	4.00
no 2	0.8m	1.2m...................... each	4.00
no 3	1.1m	1.8m...................... each	4.00
no 4	1.4m	2.4m...................... each	4.00

Aluminium Work Towers

Span Towers (Multilevel Work Platforms)

0.85 x 1.8 x 2.2m....................... each	68.00
0.85 x 1.8 x 4.2m....................... each	100.00
1.45 x 1.8 x 6.2m....................... each	132.00
1.45 x 1.8 x 8.2m....................... each	164.00
1.45 x 1.8 x 10.2m....................... each	196.00

Stagings

Lightweight staging 460mm wide

3.0m high.................................. each	23.00
4.2m high.................................. each	29.00
6.0m high.................................. each	36.00
7.2m high.................................. each	50.00

All carriage to and from site charged extra. All prices are subject to alteration without notice.

PURCHASE PRICES

LIGHT ALLOY INDUSTRIAL LADDERS TO DIN 4568/EN 131

The following Trade Prices have been provided by Light Alloy Ltd., Dales Road, Ipswich, Suffolk, IP1 4JR, Tel 01473 740445, to whom acknowledgement is hereby given.

Single part ladders

Alloy ladder with rungs

Ref. 41511 overall length 1.9m..... each	53.32
Ref. 41512 overall length 2.5m..... each	67.01
Ref. 41513 overall length 3.0m..... each	79.97
Ref. 41514 overall length 3.6m..... each	94.69
Ref. 41515 overall length 4.2m..... each	112.15
Ref. 41516 overall length 4.7m..... each	137.88
Ref. 41517 overall length 5.8m..... each	182.00
Ref. 41518 overall length 7.0m..... each	223.38

GRP ladder

Ref. 41251 overall length 1.9m..... each	143.40
Ref. 41253 overall length 3.0m..... each	208.66
Ref. 41255 overall length 4.2m..... each	284.96

GRP ladder with aluminium alloy rungs

Ref. 41151 overall length 1.9m..... each	113.99
Ref. 41153 overall length 3.0m..... each	174.66
Ref. 41155 overall length 4.2m..... each	237.15

PURCHASE PRICES (cont'd)

Single part ladders

Alloy ladder with double depth rungs

Ref. 40451	overall length 1.8m.....	each	82.73
Ref. 40452	overall length 2.4m.....	each	102.04
Ref. 40453	overall length 2.9m.....	each	124.10
Ref. 40454	overall length 3.5m.....	each	146.15
Ref. 40455	overall length 4.0m.....	each	169.13
Ref. 40456	overall length 4.6m.....	each	201.31

Double sided step ladders

Alloy trestle steps

Ref. 40311	vertical height 1.75m...	each	129.61
Ref. 40312	vertical height 2.29m...	each	159.95
Ref. 40313	vertical height 2.83m...	each	196.71
Ref. 40314	vertical height 3.37m...	each	234.40
Ref. 40315	vertical height 3.91m...	each	273.93
Ref. 40316	vertical height 4.45m...	each	330.92
Ref. 40317	vertical height 5.57m...	each	467.89

GRP trestle steps

Ref. 41256	overall height 1.82m....	each	301.51
Ref. 41257	overall height 2.35m....	each	376.89
Ref. 41258	overall height 2.89m....	each	463.29
Ref. 41260	overall height 3.95m....	each	646.21

GRP trestle steps with double depth aluminium alloy rungs

Ref. 41165	vertical height 1.27m....each		145.25
Ref. 41166	vertical height 1.53m....each		193.96
Ref. 41167	vertical height 1.80m....each		244.51

Alloy trestle steps with double depth rungs

Ref. 40461	vertical height 1.71m....each		180.17
Ref. 40462	vertical height 2.23m....each		229.80
Ref. 40463	vertical height 2.76m....each		281.28
Ref. 40464	vertical height 3.28m....each		329.09

Ref. 40465	vertical height 3.81m....	each	401.70
Ref. 40466	vertical height 4.34m....	each	477.08

Push-up and rope operated ladders

Alloy two part push-up ladder

Ref. 40245	extended length 3.0m...	each	102.04
Ref. 40246	extended length 4.1m...	each	128.69
Ref. 40247	extended length 4.9m...	each	166.37
Ref. 40248	extended length 6.1m...	each	210.50
Ref. 40249	extended length 7.2m...	each	250.95
Ref. 40214	extended length 8.3m...	each	296.92
Ref. 40215	extended length 9.2m...	each	366.78
Ref. 40216	extended length 10.3m.	each	448.59
Ref. 40217	extended length 12.2m.	each	560.72

Two part push-up ladder in GRP with aluminium alloy rungs

Ref. 41163	extended length 4.9m...	each	398.94
Ref. 41164	extended length 6.0m...	each	459.61

Alloy two part rope operated ladder

Ref. 40206	extended length 7.2m...	each	331.84
Ref. 40207	extended length 8.3m...	each	379.65
Ref. 40208	extended length 9.2m...	each	454.09
Ref. 40209	extended length 10.3m.	each	492.70
Ref. 40210	extended length 12.2m.	each	598.42

Two part rope operated ladder in GRP with alumimium alloy rungs

Ref. 41162	extended length 8.3m...	each	622.32

Alloy three part rope operated ladder

Ref. 40446	extended length 10.6m.	each	806.16
Ref. 40447	extended length 12.5m.	each	940.37
Ref. 40448	extended length 14.2m.	each	1349.42
Ref. 40449	extended length 15.9m.	each	1473.52

Prices and full catalogue available on application.

RAMSAY INDUSTRIAL CLASS 1 LADDERS: BS 2037 1984

The following Trade Prices have been provided by Ramsay & Sons (Forfar) Ltd., 61, West High Street, Forfar, Angus, DD8 1BH, Tel: 01307 462255, to whom acknowledgement is hereby given.

Aluminium Ladders

Single (parallel sides)	length (m)	rungs	approx weight (kg)		£
SA3. OMK	3.05	12	6	each	61.00
SA3. 5MK	3.55	14	7	each	70.00
SA4. OMK	4.05	16	8	each	80.00
SA4. 5MK	4.55	18	9	each	90.00
SA5. OML	5.05	20	12	each	108.00
SA5. 5MF	5.55	22	16	each	151.00
SA6. OMF	6.05	24	18	each	163.00

Double Extending (push up type)	closed height (m)	extended height (m)	rungs	approx weight (kg)		
DE3. OML	3.05	5.55	24	13	each	154.00
DE3. 5ML	3.55	6.30	28	17	each	172.00
DE4. OMF	4.05	7.55	32	27	each	262.00
DE4. 5MF	4.55	8.55	36	30	each	290.00
DE5. OMF	5.05	9.30	40	33	each	317.00
DE5. 5MF	5.55	10.05	44	36	each	345.00
DE6. OMJ	6.05	11.05	47	52	each	529.00
DE6. 5MJ	6.55	11.75	51	56	each	567.00
DE7. OMJ	7.05	12.75	55	63	each	605.00

GENERAL INFORMATION

PURCHASE PRICES (cont'd)

RAMSAY INDUSTRIAL CLASS 1 LADDERS: BS 2037 1984 (cont'd)

Ramsay Standard Range Ladders

Double Extending (push up type)	closed height (m)	extended height (m)	rungs	approx weight (kg)		
DE4. OML	4.05	7.30	32	19	each	186.30
DE4. 5ML	4.55	8.30	36	21	each	207.40
DE5. OML	5.05	9.30	40	23	each	231.90
DE6. OMF	6.05	11.05	47	39	each	371.30
DE6. 5MF	6.55	11.75	51	42	each	397.80

Non-skid rubber suction grips..	per pair	15.00
Ropes and pulleys...	Per Double Ladder	41.00

Triple Extending (push up type)	closed height (m)	extended height (m)	rungs	approx weight (kg)		£
TE3. OMD	3.05	7.30	35	24	each	275.30
TE3. 5MD	3.55	8.80	41	27	each	307.00
TE4. OMF	4.05	10.30	47	42	each	423.80
TE4. 5MF	4.50	11.80	53	47	each	471.60
TE5. OMF	5.05	13.30	59	52	each	521.20

Triple Extending (push up type)						£
TE6. OMJR	6.05	15.80	69	87	each	921.80
TE7. OMJR	7.05	18.80	81	101	each	1055.70

Non-skid rubber suction grips..	per pair	15.00
Ropes and pulleys...	Per Triple Ladder	61.50

Loft Ladders

AL1	2.29m to 2.54m..	each	179.50
AL2	2.57m to 2.82m..	each	185.70
AL3	2.84m to 3.10m..	each	191.80
AL4	3.12m to 3.38m..	each	211.20
AL5	3.40m to 3.66m..	each	223.40

N.B. The above prices are inclusive of one safety handrail per ladder

Additional safety handrail..	per ladder	18.00

Other sizes quoted for on request.

Aluminium Industrial Platforms

Single Sided	height (m)	approx weight (kg)		
DPS 01	0.240	6	each	80.00
DPS 02	0.475	7	each	100.00
DPS 03	0.710	9	each	120.00
DPS 04	0.945	10	each	140.00
DPS 05	1.180	11	each	160.00
DPS 06	1.415	12	each	180.00

Double Sided				
DPD 02	0.475	8	each	150.00
DPD 03	0.710	10	each	165.00
DPD 04	0.945	12	each	180.00
DPD 05	1.180	15	each	195.00
DPD 06	1.415	17	each	210.00
DPD 07	1.645	19	each	225.00

Standard trade discounts available on all the above items.

Guide Prices to Building Types Per Square Metre

The prices throughout this section are ranged over the most commonly occurring average building prices and are based upon the total floor area of all storeys measured inside external walls and over internal walls and partitions. They exclude contingencies, external works, furniture, fittings and equipment, professional fees and VAT.

Building prices generally are influenced significantly by location, market and local conditions, size and specification. The prices given are intended only as an indicative guide to representative building types and must be used with caution, professional skill and judgement. Prices outside the ranges stated can, of course, be encountered.

New Build

Industrial Buildings	per m2 £
Agricultural Shed	250-480
Factory, light industrial for letting	
incoming mains only	220-440
including lighting and power, toilets and basic office accommodation	280-600
for owner occupation including all services and basic accommodation	300-650
high tech: with controlled environment	400-1600
Garage/Showroom	300-1000
Workshop	300-750
Warehouse/Store	220-600

Administrative Buildings

Council Offices	600-1000
Magistrates Court	730-1370
Offices, low rise (for letting)	
non air conditioned	500-1000
air conditioned	600-1300
Offices, High rise (for owner occupation) air conditioned	1000-1800
Bank (branch)	720-1340
Shops	300-800
Retail Warehouse	200-500
Shopping Mall air-conditioned	1500-1800

Police Station	600-1300
Ambulance Station	400-800
Fire Station	600-1200

Health and Welfare Buildings

General Hospital	600-1300
Private Hospital	800-1350
Day Hospital	700-1300
Health Centre	500-850
Hospice	700-1150
Day Centre	550-1000
Home for the Mentally Ill	500-950
Children's Home	450-1000
Home for the Aged	450-900

Refreshment, Entertainment and Recreation Buildings

Restaurant	700-1200
Public House	500-1200
Kitchen block - with equipment	1000-2100
Community Centre	400-1000
Swimming Pool (covered)	850-1600
Sports Hall	400-800
Squash Courts	500-650
Sports Pavilion	500-1200

Religious Buildings

Church	600-1400
Church Hall	600-1000

Educational, Cultural and Scientific Buildings

Nursery School	600-1200
Primary School	500-900
Middle School	450-900
Secondary School	450-1000
Special School	480-1000

Universities	500-1200
College	500-1000
Training Centre	450-900
Laboratory building	550-1500
Computer Building	750-2000
Libraries	550-1100

Residential Buildings

Local Authority and Housing Association Schemes	
low rise flats	350-600
4/5 person two storey houses	320-600
Private Developments	
low rise flats	350-700
4/5 person two storey houses	350-650
Private Houses, detached	420-1300
Sheltered Housing	
bungalows	400-750
low rise	400-750
flats with lifts and Warden's accommodation, etc	450-800
Student's Residence	500-1200
Hotel	500-1400

Refurbishment

Factories	100-500
Offices (basic refurbishment)	200-800
Offices (high quality including air conditioning)	1250-1700
Banks	350-1000
Shops	300-1000
Hospitals	250-1200
Public Houses	300-850
Churches	150-900
Schools	160-800
Housing & Flats	100-800
Hotels	250-1000

Other available titles

Laxton's General Specification Volumes 1 & 2
(0 7506 3352 0)

Laxton's General Specification – electronic version
(0 7506 3693 9)

Laxton's Guide to Term Maintenance Contracts
(0 7506 2977 0)

Laxton's Measurement Rules for Contractors Quantities
(0 7506 2977 0)

Laxton's Trades Price Book: Small Works Repairs and Maintenance
(0 7506 2978 9)

Laxton's Guide to Budget Estimating
(0 7506 2967 3)

Cost Allowances

HOUSING ASSOCIATION SCHEMES

The following is based on the circular F2 - 29/98 dated August 1998 and the total Cost Indicators 1999/2000 Guidance Notes Effective from 1st April 1999 both reproduced here by kind permission of the Housing Corporation.

CIRCULAR F2 - 29/98 August 1998

Total Cost Indicators, for 1999/2000

Advises associations of the Grant Rates and Total Cost Indicators which will apply from 1 April 1997.

There are two enclosures with this Circular.

Introduction

1.1 The Total Cost Indicators, grant rates, value limits, discount amounts, rent caps and administration allowances which will apply to registered social landlord (RSL) schemes sponsored by the Housing Corporation or by local authorities and receiving grant confirmation on or after 1 April 1999 have now been determined. Details are set out together with guidance notes, in the booklet: 'Total Cost Indicators, grant rates, rent caps and administration allowances 1999/2000: guidance notes effective from 1 April 1999'.

Total cost indicators (TCI)

2.1 TCI are the Housing Corporation's estimate of the norm total cost of providing different parts of the country. TCI are set out in part 1 of the enclosed booklet.

2.2 The TCI system is the Housing Corporation's benchmark for assessing the value for money of estimated scheme costs at the point at which grant is formally confirmed. Projects on which costs exceed their TCI are subject to technical scrutiny prior to grant confirmation to ensure that costs above norm can be justified on value for money grounds. In normal circumstances schemes on which costs are expected to exceed 130% TCI cannot be given grant confirmation. However in recognition of the programme delivery problems that can arise in some areas where costs are at the higher end of the range for their designated TCI group, the Housing Corporation and the Department of the Environment, Transport and the Regions (DETR) have agreed a special framework for setting discretionary higher limits for a very small number of local authorities. These are considered on a case by case basis and will be confirmed separately.

2.3 The changes to 1999/2000 TCI are relatively minor. There are some changes to the level of TCI in certain cost group areas and for certain types of scheme. A number of local authority areas change cost groups, reflecting the latest independent advice on cost relativities.

3. Grant Rates

3.1 The published grant rates represent for each type of scheme in each area the maximum percentage of total qualifying scheme costs that may be funded with public capital subsidy. Grant rates are set out in part 2 of the enclosed booklet.

3.2 The Minister's decision is a national average grant rate of 54% Within this average, the matrix of maximum grant rates for individual types and locations of property have been updated to take account of the TCI changes and of latest evidence on both RSLs' running costs and of the incomes of households being allocated RSL tenancies.

4 Value limits and discounts amounts

4.1 Value limits for Homebuy and discount amounts for Voluntary Purchase Grant and Right to Acquire are also set out in part 2 of the enclosed booklet.

4.2 **The TIS and DIYSO schemes will no longer operate for 1999/2000. These will be replaced by a new initiative 'Homebuy"** for which value limits are published in the enclosed booklet (table 2.7). Further details of Homebuy will be provided in the bidding guidance and under a separate circular.

4.3 Discount amounts for Voluntary Purchase Grant and Right to Acquire schemes are set out in table 2.8 in part 2 of the enclosed booklet.

5 Rent caps

5.1 For 1999/2000, the Housing Corporation is publishing rent caps in line with the TCI and grant rates – these are set out in part 3 of the enclosed booklet. The rent caps are modelled on TCI and grant rates for 1999/2000 and therefore represent cost rents at 100% TCI and published grant rates, assuming that RSLs' operating and borrowing costs are the same as those used in the Housing Corporation's grant rate model.

t
5.2 The rent caps apply to all general needs housing for rent including temporary housing schemes, to category 1 and 2 schemes nor to shared ownership schemes.

5.3 The caps will be used as a control on stating rent levels (i.e. rents plus housing benefit – eligible service charges at first lettings) at the point at which grant is formally confirmed. Projects on which rents (including housing benefit – eligible service charges) exceed the published caps will be subject to close scrutiny and RSLs will be required to provide a convincing justification on overall value for money grounds before grant confirmation is considered. As with TCI, schemes on which proposed rents exceed 130% of published cps will not be given grant confirmation.

5.4 RSLs will continue to be required to certify that average rents on their 1996 Act stock will not increase at more than RPI +1% following first letting.

GENERAL INFORMATION

6 Revised administration allowance

6.1 We have brought forward the annual review of administration allowances for 1999/2000 in order to announce the increase under this one circular rather than produce a separate circular. The revised allowances is shorter than in previous years.

7 Further information

7.1 The Housing Corporation publishes details of the methodology and data sources for TCI reviews, grant rate modelling and rent caps. Copies are available from Investment Division, The Housing Corporation, 149 Tottenham Court Road, London W1P 0BN. Telephone 0171-393 2000

8 Enquiries

8.1 Please direct any enquiries on this circular to the appropriate regional office of the Housing Corporation.
These guidance notes are designed to explain the use of Total Cost Indicators (TCI) in the Social Housing Grant (SHG) funding framework, and help Registered Social Landlords (RSLs) complete a scheme submission. The notes are effective from 1st April 1997, they are issued together with separate grant rate guidance notes.

Enquiries about the contents of this guidance should be directed to the regional offices of the Housing Corporation.

CHANGES FROM THE 1998/99 TCI

The main changes from 1998/1999 Total Cost Indicators(TCI) is that a number of local authority areas have changed TCI cost groups, reflecting latest evidence of relative costs.

EXPLANATION OF TCI

A key objective of the funding system is to achieve value for money in return for grant, and to ensure the correct level of grant is paid. TCI form the basis of this system, and are divided into unit type and cost group area categories.

TCI apply equally to units funded with Social Housing Grant (SHG) by the Corporation, or those sponsored by a local authority.

TCI represent the basis for a cost evaluation of SHG funded units. TCI are also used to calculate the maximum level of grant or other public subsidy payable. Further details on this inter-relationship between TCI and grant levels are given in the part 2 of this guidance.

Key and supplementary multipliers are applied to the base TCI figures to allow for scheme variations as outlined in the multiplier tables. (tables 1.6 to 1.9) Thus, there is a relationship between the base norm cost of a unit and its unit type.

Pro

TYPES OF ACCOMMODATION

Different types of accommodation other than self contained housing for general needs are classified as follows:-

Accommodation for older people

i) Category 1 - self-contained accommodation for the more active older person, which may include an element of support and/or additional communal facilities;

ii) Category 2 - self-contained accommodation for the less active older person, which includes an element of support and the full range of communal facilities.

The term `sheltered' is used generally to describe Category 1 and Category 2 schemes.

iii) Frail Older people – supported extra care accommodation, which may be either shared or self-contained, for the frail older person. Includes the full range of communal facilities, plus additional special features, including wheelchair user environments and supportive management.

Shared Accommodation

Accommodation predominantly for single persons, which includes a degree of sharing between tenants of some facilities (e.g. kitchens, bathrooms, living room) and may include an element of support and/or additional communal facilities.

Supported housing

Accommodation, which may be either shared or self-contained, designed to meet the needs of particular user groups for intensive housing management (see the Housing Corporation's Guide to Supported Housing) Such accommodation may also include additional communal facilities

Accommodation for wheelchair users

Accommodation, which may be either shared or self-contained, designed for independent living by people with physical disabilities and wheelchair users. Where such accommodation is incorporated within schemes containing communal facilities, these facilities should be wheelchair accessible.

Communal Facilities

Ancillary communal accommodation, the range of which comprises:

i) Common room - consisting of common room/s of adequate size to accommodate tenants and occasional visitors, chair store and kitchenette for tea-making.

ii) Associated communal facilities - consisting of warden's office, laundry room and guest room.

TREATMENT OF COMBINED SUPPORTIVE HOUSING AND GENERAL NEEDS SCHEMES

Arrangements exist which allow the combination of special supportive housing and general needs units within a single scheme. Further guidance with regards to supported housing schemes is set out under the heading "Selection of Supplementary Multipliers"

TEMPORARY HOUSING

Temporary Social Housing (TSH) Grant and Temporary Market Rent Housing (TMRH) Grant are terms used to describe SHG paid to registered social landlords (RSLs) to cover the cost of bringing properties into temporary use.

TSH and TMRH are procedurally the same. In practice, there has Been little difference between rent levels on each programme. For 1999/2000 therefore there will be a single temporary housing programme, referred to in this guidance as TSH.

properties are eligible for TSH Grant if they are available for use by the RSL for a period of time covered by a lease or licence for not less than two years and not more than 29 years

A capital grant contribution will be available towards initial acquisition costs or periodic lease charges up to a grant maximum. The maxima are based upon the Housing Corporation's own assessment of what constitutes a reasonable contribution to the capitalised lease value. This is calculated using capitalised lease premium factors.

The TSH *multipliers and capitalised lease premium factor tables for 1999/2000* (tables 1.9 and 1.10) are included in this guidance.

THE COMPOSITION OF TCI

TCI comprise the following elements:-

Acquisition
(i)   purchase price of land/property

HOUSING ASSOCIATION SCHEMES (CONT'D)

Works

(i)   main works contract costs (including where applicable adjustments for additional claims and fluctuations, but excluding any costs defined as on costs below);

(ii)   major site development works (where applicable). These include piling, soil stabilisation, road/sewer construction, major demolition;

(iii)   major pre-works (rehabilitation) where applicable;

(iv)   statutory agreements, associated bonds and party wall agreements (including all fees and charges directly attributable to such works) where applicable;

(v)   additional costs associated with complying with archaeological works and party wall agreement awards (including all fees, charges and claims attributable to such works) where applicable;

(vi)   home loss and associated costs. This applies to new build only;

(vii)   VAT on the above, where applicable.

On-costs

(i)   Legal fees, disbursements and expenses;

(ii)   stamp duty;

(iii)   net gains/losses via interest charges on development period loans;

(iv)   building society or other valuation and administration fees;

(v)   fees for building control and planning permission;

(vi)   fees and charges associated with compliance with European Community directives, and the Housing Corporation's requirements relating to energy rating of dwellings;

(vii)   in house or external consultants' fees, disbursements and expenses (where the development contract is a design and build contract) (see note below);

(viii)   insurance premiums including building warranty and defects/liability insurance (except contract insurance included in works costs);

(ix)   contract performance bond premiums;

(x)   borrowing administration charges (including associated legal and valuation fees);

xi)   an appropriate proportion of the RSL's development and administration costs (formerly Acquisition and Development allowances), excluding Co-operative Promotional Allowance (CPA) and Special Projects Promotion Allowances (SPPA) and including an appropriate proportion of any abortive scheme costs;

(xii)   furniture, loose fittings and furnishings;

(xiii)   home loss and disturbance payments for rehabilitation;

(xiv)   preliminary minor site development works (new build), pre-works (rehabilitation), and minor works (off the shelf) and minor works and repairs in connection with existing satisfactory purchases;

(xv)   marketing costs - for sale schemes only;

(xvi)   post completion interest - for sale schemes only;

(xvii)   legal, administrative and related fees and costs associated with negotiating and arranging leases – for TSH only;

(xviii) VAT on the above, where applicable.

Note:
Where the development contract is design and build, the on-cots are deemed to include the builder's design fee element of the contract sum. Therefore the amount included by the builder for design fees should be deducted from the works cost element submitted by the RSL to the Housing Corporation.

Similarly other non-works costs that may be included by the builder such as fees for building and planning permission, building warranty and defects/liability insurance, contract performance bond and energy rating of dwellings should also be deducted from the works cost element submitted by the RSL to the housing Corporation.

The Housing Corporation will subsequently check compliance through its compliance audit framework.

EXPLANATION OF ON-COSTS

TCI are inclusive of on-costs contained in the relevant on-cost table. The on-costs vary according to TCI cost group and the general purpose of the scheme. TCI levels are set with the assumption that the RSL's development and administrative costs will be contained within the percentages in the relevant on cost table.

In order to allow a proper comparison between the total eligible costs and the relevant TCI it is necessary to add the percentage on-cost (from the relevant table) to the estimated eligible costs of acquisition and works.

Major repair schemes and adaptation schemes are not measured against TCI. However, on costs (from the relevant table) should be added to such schemes. Supplementary on-costs may not be used in any circumstances in connection with major repairs or adaptation schemes.

Table 1.1 KEY ON-COSTS 1999/2000 BY COST GROUP
Note: Only one of the following to be used.

Key On-Costs		A %	B %	C %	D %	E %
a) New Build						
i)	acquisition and Works	12	13	13	14	15
	ii) off the Shelf	7	7	7	7	7
	iii)works only	16	16	16	16	16
b) Rehabilitation						
i)	acquisition and Works					
	- vacant	12	13	13	14	14
	- tenanted	15	15	16	18	18
ii)	existing satisfactory	9	9	9	9	9
iii)	purchase and repair	7	7	7	7	7
iv)	works only					
	- vacant	21	21	21	21	21
	- tenanted	26	26	26	26	26
v)	reimprovements	21	21	21	21	21
vi)	TSH					
	-unimproved vacant	21	21	21	21	21
	-improved vacant	9	9	9	9	9
c) Major repairs and miscellaneous works		25	25	25	25	25
d) Adaptation works		13	13	13	13	13

Table 1.2 SUPPLEMENTARY ON-COSTS 1999/2000

Supplementary On-Costs (All Cost Groups)	Purchase & repair and aquistion Works %	Works Only and Reimprovements %	Off-the-Shelf & Existing Satisfactory %
a) Sheltered with common room or communal facilities	+2	+3	+1
b) Frail older persons	+4	+3	
c) Supported housing:			
(i) supported housing	+4	+5	+3
(ii) with common room or communal facilities	+5	+7	+4
d) Shared	+5	+7	+5
e) Standard House Types	-1	-2	0
f) Housing for Sale	+8	+11	+5
g) TSH			
i) shared unimproved	+7		
ii) shared improved	+5		

HOUSING ASSOCIATION SCHEMES (CONT'D)

SELECTION OF ON-COSTS

One key on-cost will apply per scheme. To this should be added any appropriate supplementary on-cost.

Supplementary on-costs may be used when the accommodation is designed to meet the relevant standards set out in the Housing Corporation publication 'Scheme Development Standards'.

The appropriate key or supplementary on-cost is determined by the predominant dwelling type in a scheme. Predominance is established, where necessary, by the largest number of persons in total.

Where two supplementary on-costs are equally applicable (e.g. special needs and shared), the higher should be used exclusively

TRANCHES

For units developed under the Housing Act 1988, a set percentage of approved grant can be paid once a unit reaches certain key development stages. These grant payments are known as tranches.

The key stages are:

(i)   exchange of purchase contracts (ACQ);

(ii)   start on site of main contract works (SOS); this is deemed to be the date when the contractor took possession of the site/property in accordance with the main building contract;

(iii)   practical completion of the scheme (PC).

Where a *public subsidy* is given by way of discounted land, (acquisition public subsidy), the whole subsidy is deducted from the first tranche; any excess balance should be deducted from the second tranche. This is to ensure that grant is not paid in advance of need.

In other circumstances any other public subsidy will be deducted from each tranche on a pro-rata basis.

In the case of package deals (including land) the first and second tranches are paid together at start on site stage. Tranche details for Special Projects Promotion Allowances (SPPA) and Co-op Promotion Allowances (CPA) can be found in the Capital Funding System Procedure Guide.

The grant on outstanding mortgages for reimprovement schemes will be paid in accordance with the tranche percentage for the scheme.

All tranche payments may be paid directly to RSL's rather than via solicitors.

Table 1.3 GRANT PAYMENTS FOR TRANCHES 1999/2000

TRANCHE PERCENTAGES Scheme type Groups	Cost Groups	Acquisition & Works			Works Only & Reimprovements		Existing Satisfactory off-the-shelf	Purchase & Repair	
		(i) %	(ii) %	(iii) %	(ii) %	(iii) %	(i) %	(i) %	(ii) %
Housing for rent									
Mixed funded: New Build	ALL	40	40	20	65	35	100	80	20
Mixed funded Rehabilitation	ALL	50	30	20	60	40	100		
100% SHG: New Build	A,B and C	30	35	35	50	50	50		100
100% SHG: New Build	D and E	20	35	45	45	55	100		
100% SHG: Rehabilitation	A,B, and C	50	20	30	40	60	100	80	20
100% SHG: Rehabilitation	D and E	45	20	35	35	65	100	80	20
TSH unimproved					65	35			
TSH improved							100		
Housing for sale									
Mixed funded: New Build & Rehabilitation	ALL	50	45	5	95	5	100	95	5

THE USE OF THE TCI BASE TABLE

The unit size in square metres shown on the *TCI base table 1999/2000; self-contained accommodation* (table 1.4) relates to the total floor area of the unit. The probable occupancy figure in the tables is only a guideline figure. The number of occupants is derived from the total number of bedspaces provided.

The TCI for a unit where the total floor area exceeds 120m2 will be the cost of a unit of 115-120m2 plus for each additional 5m2 or part thereof, the difference between the cost of a 110-115m2 unit and the cost of a 115-120m2 unit.

In calculating the appropriate TCI floor area band the relevant floor area should be rounded to the nearest whole number.

For Self-Contained Accommodation

Self contained units provide each household, defined as a tenancy, with all their basic facilities behind their own lockable front door.

For self-contained units the base TCI is determined by its total floor area and the cost group in which it is located. The dwelling

GENERAL INFORMATION

floor area is determined by the area of each unit for the private use of a single household. Communal areas or any facilities shared by two or more households should be excluded.

The total floor area of self-contained accommodation is measured to the internal faces of the main containing walls on each floor of the accommodation and includes the space, on plan, taken up by private staircases, partitions, internal walls (but not 'party' or similar walls), chimney breasts, flues and heating appliances. It includes the area of tenant's internal and/or external essential storage space.

It excludes:-

i)   any space where the height to the ceiling is less than 1.5m (eg. areas in rooms with sloping ceilings, external dustbin enclosures);

ii)   any porch, covered way, etc, open to air;

iii) all balconies (private, escape and access) and decks

iv)   non-habitable basements, attics or sheds;

v)   external storage space in excess of 2.5m2;

vi)   all space for purposes other than housing (e.g. garages, commercial premises, etc).

The TCI for Frail older persons dwellings should always be calculated as *self-contained* units even if they have some characteristics which are more typical of shared accommodation.

For Shared Accommodation

Shared accommodation is defined as one household (i.e. one tenancy or licence) which shares facilities (i.e. bathroom, kitchen) with other householders. Each household sharing such accommodation may comprise more than one person. The base TCI for shared accommodation should be calculated separately to any self-contained accommodation in the scheme. The base TCI for shared accommodation is calculated on a per bed space basis and may include any staff with a residential tenancy for shared accommodation. Staff sleepover accommodation which is not subject to a tenancy is not regarded as a bedspace for TCI purposes. The relevant base TCI cost for the cost group from the TCI *base table* 1999/2000; *shared accommodation* (table 1.5) is used.

For TSH shared accommodation

The relevant floor area, for TCI and capitalised lease premium calculation purposes only, should be taken as the overall building gross floor area measured to the finished internal faces of the main containing walls all as otherwise described for self-contained units above, divided by the number of people sharing. This calculation will result in the correct band size per person sharing which will then be used to select the base figure from the TCI base table: self contained (table 1.4. This figure should then be multiplied by the number of people sharing prior to applying the relevant TSH key and supplementary multipliers or capitalised lease premium factor.

Table 1.4 TCI BASE TABLE 1997/98
SELF-CONTAINED ACCOMMODATION

Total unit costs	Probable occupancy	ALL SELF-CONTAINED ACCOMMODATION (including all frail older persons and TSH) £ per unit Cost Group				
Unit floor area m2 Persons		A	B	C	D	E
Up to 25	1	51,400	43,000	38,300	35,300	32,900
Exceeding/not exceeding						
25/30	1	56,500	47,100	41,700	38,100	35,400
30/35	1 and 2	61,600	51,200	45,100	40,900	37,800
35/40	1 and 2	66,700	55,300	48,400	43,700	40,300
40/45	2	71,800	59,400	51,800	46,500	42,800
45/50	2	76,900	63,500	55,100	49,300	45,200
50/55	2 and 3	82,000	67,500	58,500	52,100	47,700
55/60	2 and 3	87,100	71,600	61,900	54,900	50,200
60/65	3 and 4	92,200	75,700	65,200	57,700	52,700
65/70	3 and 4	97,300	79,800	68,600	60,500	55,100
70/75	3, 4 and 5	102,500	83,900	71,900	63,300	57,600
75/80	3, 4 and 5	107,000	87,500	74,900	65,700	59,700
80/85	4, 5 and 6	111,600	91,200	77,900	68,200	61,900
85/90	4, 5 and 6	116,100	94,800	80,900	70,600	64,000
90/95	5 and 6	120,700	98,400	83,800	73,100	66,200
95/100	5 and 6	125,300	102,000	86,800	75,300	68,300
100/105	6 and 7	129,800	105,600	89.800	78,000	70,500
105/110	6 and 7	134,400	109,300	92,700	80,400	72,600
110/115	6, 7 and 8	138,900	112,900	95,700	82,900	74,800
115/120	6, 7 and 8	143,500	116,500	98,700	85,300	76,900

HOUSING ASSOCIATION SCHEMES (CONT'D)

Table 1.5 TCI BASE TABLE 1999/2000: SHARED ACCOMMODATION

Total costs per person	ALL SHARED ACCOMMODATION £ per PERSON sharing Cost Group				
	A	**B**	**C**	**D**	**E**
Each Person bedspace	60,600	49,500	42,400	37,100	33,700

SELECTION OF KEY MULTIPLIERS

Only one key multiplier can be used per unit.

New build acquisition and works is the basic key multiplier, hence its neutral value.

The *off-the-shelf* multiplier is used where new dwellings to a standard suitable for social housing letting are purchased, following inspection, from contractors/developers or their agents. The cost of any minor works required should be set against the on cost allowance.

The *existing satisfactory* multiplier is used where existing dwellings to a standard and a condition suitable for social housing letting are purchased, following inspection, from the second hand property market. The cost of any minor works or repairs required should be set against the on cost allowance.

The *purchase and repair* multiplier is used where existing dwellings are purchased, following inspection, from the property market which necessitate a degree of repair to bring them to a standard and a condition suitable for social housing but not full rehabilitation. Purchase and repair classification will apply where the estimated repair/improvement costs of each dwelling exceed £1,500 but are less than £10,000.

The *works only* multiplier is used for accommodation which involves the development of land or property already in the RSL's ownership and for which no acquisition costs (other than legal charges) apply.

The *reimprovement* multiplier is used for dwellings which have already received some form of public/grant subsidy and are now being rehabilitated. Reimprovement schemes are generally expected to be submitted no less than 15 years after a rehabilitation or 30 years after construction. It should be noted that the outstanding mortgage of the original works can be considered an eligible cost and will attract the grant rate applicable to the new scheme as a whole. The reimprovement key multiplier is now linked to that of the Rehabilitation works only (vacant) multiplier rather than the Rehabilitation works only (tenanted) multiplier in order to reflect the average cost of provision.

The TSH *improved and unimproved vacant* multiplier is used for accommodation with a lease of between two and 29 years.

Table 1.6 KEY MULTIPLIERS BY COST GROUP 19999/2000

Note: Only one of the following to be used.

Scheme type		TCI calc. form Line No.	GROUP				
			A	**B**	**C**	**D**	**E**
a) New Build							
i)	acquisition and works	0010	1.00	1.00	1.00	1.00	1.00
ii)	off-the-shelf	0020	0.96	0.95	0.95	0.94	0.93
iii)	works only	0090	0.65	0.68	0.75	0.83	0.87
b) Rehabilitation							
i)	acquisition and works						
	- vacant	0100	1.23	1.13	1.07	1.03	1.01
	- tenanted	0105	1.16	1.06	1.01	0.97	0.96
ii)	existing satisfactory	0120	0.96	0.95	0.95	0.94	0.93
iii)	purchase and repair	0110	0.95	0.93	0.93	0.92	0.91
iv)	works only						
	- vacant	0125	0.61	0.57	0.57	0.61	0.63
	- tenanted	0130	0.64	0.59	0.60	0.63	0.66
v)	reimprovements	0135	0.61	0.57	0.57	0.61	0.63
vi)	TSH						
	-vacant improved	0030	1.23	1.13	1.07	1.03	1.01
	-vavant unimproved	0035	1.23	1.13	1.07	1.03	1.01

SELECTION OF SUPPLEMENTARY MULTIPLIERS

Table 1.7 is used for the acquisition and works, off-the-shelf, existing satisfactory and purchase and repair schemes. For other schemes types use the table 1.8. None of these supplementary multipliers are applicable to TSH schemes.

Supplementary multipliers can be applied to new build and rehabilitation units when the accommodation is designed to meet the relevant standards set out in the Housing Corporation publication 'Scheme Development Standards'

More than one supplementary multiplier can be used per unit. However certain combinations of multipliers are invalid.

GENERAL INFORMATION

HOUSING ASSOCIATION SCHEMES (CONT'D)

SELECTION OF SUPPLEMENTARY MULTIPLIERS (CONT'D)

Multipliers for sheltered, frail older people, supported housing and extended families cannot be combined. The matrix (table 1.11) gives a comprehensive list of valid combinations of multipliers.

The main reason for combinations of multipliers being invalid is that the combination of those multipliers would lead to a duplication of the financial provision for the facilities accounted for in the multipliers. e.g. category 2 includes allowance new lifts or single storey. This means that in these circumstances the other relevant multiplier does not apply i.e. a new lift or single storey multiplier cannot be used with a category 2 multiplier.

The *supported housing* multiplier and on-costs will only apply to schemes approved within the new funding framework introduced in 1995. A scheme which is developed within this framework must be eligible to receive SHMG whether or not SHMG is actually being claimed for the scheme. Applications for approval of capital only supported housing schemes which utilise the supported housing multiplier must be accompanied by form TS1 (revenue budget). This multiplier should only be used where it is the RSL's plan to use the accommodation to provide supported housing in the long term.

Supported housing schemes with shared facilities cannot be combined with either of the three common room multipliers in *part (h)* of the supplementary multiplier table.

The supported housing multiplier cannot be used for staff units.

The *extended families* multiplier is used when a self contained dwelling is to cater for eight or more persons and additional or duplicate facilities ie kitchen and/or sanitary fittings/equipment are provided. The additional space requirement is accounted for in the size band selection.

The appropriate shared supplementary multiplier is determined by the total number of bedspaces provided within all of the households sharing facilities in the scheme. Where a cluster (i.e more than one) of shared accommodation is provided the appropriate shared supplementary multiplier is determined by the total number of bedspaces contained within the households comprising each independent and self sufficient shared accommodation arrangement. The *TCI base table: shared accommodation* (table 1.5) is deemed to include all communal and ancillary facilities.

The shared multiplier does not apply to either sheltered or frail old people schemes.

The *served by new lifts* multiplier is used when new vertical passenger lift provision is incorporated for access to dwelling entrances and communal accommodation. The multiplier does not apply to dwellings with entrances at ground floor level unless, exceptionally, the scheme includes essential communal accommodation provided at a level other than at ground floor level.

The *wheelchair with individual carport* multiplier should be used only with individual self-contained dwellings designed in accordance with the relevant standards set out in the Housing Corporation publication Scheme Development Standards. Where non-individual self-contained wheelchair user dwellings are provided, e.g. on some category 2 schemes, or where a waiver for non-compliance with the carport provisions has been obtained from the regional office, the wheelchair without individual carport multiplier should be used.

The *wheelchair* multipliers do not allow for any fixed additional equipment for people with disabilities; these needs should be met via the adaptations funding framework.

The *Standard House type* multiplier is used for new build when the design of the dwelling internal shell is derived, without significant amendment, from generic house plan designs or pattern books produced by contractors, developers, architects, surveyors, RSL's or other similar organisations or persons, resulting in a significant reduction in the detailed design input commissioned by the client RSL. In assessing whether application of the multipliers is appropriate the Housing Corporation will have regard to:

(i) the extent to which detailed design is to be specifically commissioned by the RSL; and

(ii) the scope which exists for reductions in design fees due to repetition of design

The *parent with children refuges* multiplier applies to shared accommodation specifically designed to meet the needs of parents with children, including vulnerable women with babies and women and children at risk of domestic violence. The TCI is calculated on a per bed space basis with children counted as full bedspaces. The provision of additional bunk-bedspaces should be disregarded for TCI calculation purposes.

The *housing for sale* multiplier is used for all sale schemes. It cannot be used with the supported housing or shared multipliers.

The *rehabilitation to pre-1919 properties* multiplier is used where the scope of the refurbishment or conversion work is carried out on a property or properties originally constructed prior to 1919. It cannot be used in connection with existing satisfactory or purchase and repair multiplier.

The *No VAT rated rehabilitation* multiplier is used when the scope of the refurbishment or conversion work is such that the relevant local Customs & Excise office determines that VAT is not chargeable on the Works.

The *rural housing* multiplier is used for schemes identified by rural investment codes R, G and F. Rural areas mainly comprise those with 1,000 or less inhabitants and a minimum of 60% of the rural programme is targeted to relieve housing need in these settlements. Exceptionally, the limit may be extended to 3,000 inhabitants on a case by case basis by the relevant regional office of the Housing Corporation. (See the Housing Corporation's Guide to the Allocations Process and, for details of population settlements, the Housing Corporation's Rural Settlements Gazetteer, last published in 1998. In addition the rural housing multiplier may be applied to any scheme within a formally designated National Park.

The TSH supplementary multiplier is used for all schemes with a lease length of between two and 29 years requiring works, (see table 1.9) and it can only be used with the TSH shared supplementary multiplier. It does not apply to improved TSH schemes.

Note: Where a RSL is in any doubt concerning the appropriateness of the application of any supplementary multiplier, it is advised to seek advice from the relevant regional office of the Housing Corporation. RSL's should also ensure that relevant information relating to the appropriateness of the application of supplementary multipliers is maintained on file for Scheme Audit Purposes.

TABLE 1.7

SUPPLEMENTARY MULTIPLIERS FOR ACQUISITION AND WORKS, OFF-THE-SHELF AND EXISTING SATISFACTORY SCHEMES ONLY.1999/2000

Scheme type	TCI Calc Form Line No.	GROUP A	B	C	D	E
a) Sheltered:						
i) Category 1	0140	1.01	1.01	1.01	1.01	1.01
ii) Category 2 (includes New lifts /single storey)	0170	1.26	1.27	1.29	1.31	1.32
b) Frail elderly (use alone only)	0180	1.35	1.37	1.39	1.42	1.44
c) Supported housing	0186	1.09	1.09	1.10	1.27	1.11
d) Extended general families	0195	1.13	1.13	1.15	1.17	1.18
e) Shared (not used with Frail elderly) Bedspaces per unit						
i) 2 to 3	0200	1.00	1.00	1.00	1.00	1.00
ii) 4 to 6	0210	0.90	0.90	0.90	0.90	0.90
iii) 7 to 10	0220	0.81	0.81	0.81	0.81	0.81
iv) 11 and over	0235	0.76	0.76	0.76	0.76	0.76
f) Served by new lifts (not used with Frail older person and sheltered Category 2)	0245	1.08	1.08	1.09	1.10	1.10
g) Single storey (New build only not used with Frail elderly and sheltered Category 2 or Wheelchair with carport)	0095	1.19	1.18	1.16	1.13	1.12
h) Common room etc (Sheltered Category 1, and Self-contained supported housing)						
i) Common room only	0250	1.08	1.08	1.08	1.09	1.09
ii) Associated communal facilities only	0260	1.06	1.06	1.06	1.07	1.07
iii) Common room and communal facilities	0270	1.14	1.14	1.15	1.16	1.16
i) Standard house-types (not used with Off-the-shelf, existing satisfactory and purchase and repair)	0300	0.99	0.99	0.99	0.99	0.99
j) Rehabilitation to pre-1919 properties (not used with existing satisfactory or purchase and repair)	0310	1.08	1.07	1.07	1.08	1.08
k) No VAT rehabilitation	0320	0.92	0.92	0.92	0.91	0.91
l) Wheelchair (except Single storey and where included as above)						
i) With individual carport (not used with older people or Single storey)	0380	1.30	1.30	1.27	1.24	1.22
ii) Without individual carport (not used with older person) (Note: Provision (ii) requires a waiver from the regional office)	0385	1.08	1.08	1.09	1.10	1.10
m) Parent with children refuges (Shared general needs and Shared supported housing only)	0420	0.48	0.48	0.48	0.48	0.48
n) Housing for sale	5100	1.05	1.05	1.05	1.04	1.04
o) Rural housing	0430	----	1.10	1.11	1.12	1.13

GENERAL INFORMATION

Table 1.8 SUPPLEMENTARY MULTIPLIERS FOR WORKS ONLY AND REIMPROVEMENT SCHEMES 1999/2000

SCHEME TYPE	TCI Calc Form Line No.	GROUP				
		A	B	C	D	E
a) Sheltered:						
i) Category 1	0140	1.01	1.01	1.01	1.01	1.01
ii) Category 2 (includes New lift/Single storey)	0170	1.40	1.39	1.38	1.38	1.37
b) Frail older person (use alone only)	0180	1.54	1.53	1.52	1.51	1.50
c) Supported housing	0186	1.14	1.14	1.13	1.13	1.12
d) Extended general families	0195	1.19	1.20	1.20	1.20	1.21
e) Shared (not used with Frail older person) Bedspaces per unit						
i) 2 to 3	0200	1.00	1.00	1.00	1.00	1.00
ii) 4 to 6	0210	0.85	0.85	0.87	0.88	0.88
iii) 7 to 10	0220	0.71	0.72	0.75	0.77	0.78
iv) 11 and over	0235	0.63	0.65	0.68	0.71	0.72
f) Served by new lifts (not used with Frail older person and Sheltered Category 2)	0245	1.12	1.12	1.12	1.12	1.12
g) Single storey (New build only not used with Frail older person and Sheltered Category 2 or Wheelchair with carport)	0095	1.07	1.07	1.07	1.07	1.07
h) Common room etc (Sheltered category 1, and Self-contained supported housing only)						
i) Common room only	0250	1.07	1.07	1.07	1.07	1.07
ii) Associated communal facilities only	0260	1.08	1.08	1.08	1.08	1.08
iii) Common room and communal facilities	0270	1.16	1.16	1.16	1.15	1.15
i) Standard house-types	0300	0.98	0.98	0.98	0.99	0.99
j) Rehabilitation to pre-1919 properties	0310	1.13	1.13	1.13	1.13	1.13
k) No VAT rehabilitation	0320	0.86	0.86	0.86	0.86	0.86
l) Wheelchair (except Single storey and where included as above)						
i) With individual carport (not used with Frail older person or Single storey)	0380	1.15	1.16	1.16	1.16	1.16
ii) Without individual carport (not used with Frail older person) (Note: provision (ii) requires a waiver from the regional office)	0385	1.12	1.12	1.12	1.12	1.12
m) Parent with children refuges (Shared general needs and Shared supported housing only)	0420	0.20	0.24	0.31	0.37	0.40
n) Housing for sale	5100	1.08	1.07	1.06	1.05	1.05
o) Rural housing	0430	----	1.15	1.15	1.15	1.15

Table 1.9 SUPPLEMENTARY MULTIPLIERS (INCLUSIVE OF ON-COSTS) FOR TEMPORARY HOUSING (TSH)

	TCI calc. Form Line no.	Cost group A	B	C	D	E	
TSH	4000						
Term: 29 years		0.53	0.51	0.53	0.57	0.59	
28 years		0.53	0.51	0.52		0.56	0.58
27 years		0.52	0.50	0.52	0.55	0.58	
26 years		0.51	0.49	0.51	0.55	0.57	
25 years		0.51	0.49	0.50	0.54	0.56	
24 years		0.49	0.47	0.49	0.52	0.55	
23 years		0.48	0.46	0.48	0.51	0.53	
22 years		0.46	0.45	0.46	0.49	0.51	
21 years		0.45	0.43	0.45	0.48	0.50	
20 years		0.44	0.42	0.44	0.47	0.48	
19 years		0.42	0.41	0.42	0.42	0.47	
18 years		0.41	0.40	0.41	0.44	0.46	
17 years		0.40	0.38	0.40	0.43	0.44	
16 years		0.39	0.37	0.39	0.41	0.43	
15 years		0.38	0.36	0.38	0.40	0.42	
14 years		0.33	0.32	0.33	0.35	0.36	
13 years		0.29	0.28	0.29	0.30	0.32	
12 years		0.25	0.24	0.25	0.27	0.28	
11 years		0.22	0.21	0.22	0.23	0.24	
10 years		0.19	0.18	0.19	0.20	0.21	
9 years		0.19	0.18	0.19	0.20	0.21	
8 years		0.18	0.18	0.18	0.19	0.20	
7 years		0.18	0.17	0.18	0.19	0.20	
6 years		0.18	0.17	0.18	0.19	0.20	
5 years		0.15	0.15	0.15	0.16	0.17	
4 years		0.13	0.12	0.13	0.14	0.14	
3 years		0.10	0.10	0.10	0.11	0.11	
2 years		0.08	0.08	0.08	0.09	0.09	
TSH (shared) Term 2-29 years		1	1	1	1	1	

Table 1.10 CAPITALISED LEASE PREMIUM FACTORS FOR TEMPORARY HOUSING (TSH)

	TCI calc. Form Line no.	Cost group A	B	C	D	E
TSH						
Term: 29 years		0.20	0.20	0.19	0.18	0.17
28 years		0.19	0.20	0.19	0.17	0.17
27 years		0.19	0.19	0.18	0.17	0.16
26 years		0.18	0.19	0.18	0.16	0.16
25 years		0.18	0.18	0.18	0.16	0.15
24 years		0.17	0.18	0.17	0.16	0.15
23 years		0.17	0.17	0.17	0.15	0.15
22 years		0.16	0.17	0.16	0.15	0.14
21 years		0.16	0.16	0.16	0.14	0.14
20 years		0.15	0.16	0.15	0.14	0.13
19 years		0.15	0.15	0.15	0.14	0.13
18 years		0.15	0.15	0.14	0.13	0.13
17 years		0.14	0.14	0.14	0.13	0.12
16 years		0.14	0.14	0.14	0.12	0.12
15 years		0.13	0.14	0.13	0.12	0.11
14 years		0.13	0.13	0.13	0.12	0.11
13 years		0.12	0.13	0.12	0.11	0.11
12 years		0.12	0.12	0.12	0.11	0.10
11 years		0.12	0.12	0.11	0.10	0.10
10 years		0.11	0.11	0.11	0.10	0.10
9 years		0.11	0.11	0.11	0.10	0.09
8 years		0.10	0.10	0.10	0.09	0.09
7 years		0.10	0.10	0.10	0.09	0.08
6 years		0.09	0.10	0.09	0.08	0.08
5 years		0.07	0.08	0.07	0.07	0.06
4 years		0.06	0.06	0.06	0.05	0.05
3 years		0.05	0.05	0.05	0.04	0.04
2 years		0.04	0.04	0.04	0.03	0.03

GENERAL INFORMATION

Table 1.11 VALID COMBINATION OF SUPPLEMENTARY MULTIPLIERS 1999/2000

	CAT1 0140	CAT2 0170	FE 0180	SN 0186	ExFm 0195	SH23 0200	SH46 0210	SH710 0220	SH11+ 0235	LIFT 0245	SS 0095	CR 0250	FAC 0260	CR&FC 0270	SHT 0300	1919 0310	VAT 0320	WC&C 0380	WC 0385	PWCR 0420	HFS 5100	RH 0430
CAT1 0140	■									●	●	●	●	●	●	●	●	●	●		●	●
CAT2 0170		■													●	●	●	●	●		●	●
FE 0180			■												●	●	●				●	●
SN 0186				■		●	●	●	●	●	●	●	●	●	●	●	●	●	●	●		●
ExFm 0195					■					●	●				●	●	●	●	●		●	●
SH23 0200				●		■				●	●				●	●	●	●	●	●		●
SH46 0210				●			■			●	●				●	●	●	●	●	●		●
SH710 0220				●				■		●	●				●	●	●	●	●	●		●
SH11+ 0235				●					■	●	●				●	●	●	●	●	●		●
LIFT 0245	●			●	●	●	●	●	●	■		●	●	●	●	●	●	●	●	●	●	●
SS 0095	●			●	●	●	●	●	●		■	●	●	●	●	●	●	●	●	●	●	●
CR 0250	●			●						●	●	■			●	●	●	●	●		●	●
FAC 0260	●			●						●	●		■		●	●	●	●	●		●	●
CR&FC 0270	●			●						●	●			■	●	●	●	●	●		●	●
SHT 0300	●	●	●	●	●	●	●	●	●	●	●	●	●	●	■		●	●	●	●	●	●
1919 0310	●	●	●	●	●	●	●	●	●	●	●	●	●	●	●	■	●	●	●	●	●	●
VAT 0320	●	●	●	●	●	●	●	●	●	●	●	●	●	●	●	●	■	●	●	●	●	●
WC&C 0380	●	●		●	●	●	●	●	●	●	●	●	●	●	●	●	●	■		●	●	●
WC 0385	●	●		●	●	●	●	●	●	●	●	●	●	●	●	●	●		■		●	●
PWCR 0420				●		●	●	●	●		●				●	●	●	●	●	■		●
HFS 5100	●	●	●		●					●	●	●	●	●	●	●	●	●	●		■	
RH 0430	●	●	●	●	●	●	●	●	●	●	●	●	●	●	●	●	●	●	●	●	●	■

Key

CAT1	0140	Category 1
CAT2	0170	Category 2
FE	0180	Frail Old person
SN	0186	Supported housing
ExFm	0195	Extended Families
SH23	0200	Shared 2 to 3 bedspaces per unit
SH46	0210	Shared 4 to 6 bedspaces per unit
SH710	0220	Shared 7 to 10 bedspaces per unit
SH11+	0235	Shared 11+ bedspaces per unit
LIFT	0245	Served by new lifts
SS	0095	Single Storey
CR	0250	Common room
FAC	0260	Common room associated communal facilities only
CR&FC	0270	Common room and communal facilities
SHT	0300	Standard house types
1919	0310	Rehab pre - 1919 properties
VAT	0320	No VAT rehabilitation
WC&C	0380	Wheelchair and individual carport
WC	0385	Wheelchair without individual carport
PWCR	0420	Parent with children refuges
HFS	5100	Housing for sale
RH	0430	Rural housing

● indicates a valid combination

It is valid to combine each supplementary multiplier singularly with each key multiplier with the following exceptions:
*The Housing for Sale multiplier cannot be combined with the Rehabilitation reimprovement key multiplier.

The Single storey multiplier can only be combined with the New Build acquisition and works, New build off-the-shelf and New build works only key multipliers.

The Standard House Type multiplier cannot be combined with the New build off-the-shelf, Rehabilitation existing satisfactory or Purchase and repair key multipliers.

The Rehabilitation pre-1919 multiplier cannot be combined with the Existing satisfactory or Purchase and repair key multipliers.

CALCULATION OF MAXIMUM GRANT CONTRIBUTION TO LEASE COSTS

The factors in table 1.10 are used to calculate maximum grant contributions for lease premiums outside the normal TCI framework. Whilst the factors are applied to the appropriate figures in the TCI base table and the relevant TSH key multiplier to determine the maximum contribution, any grant paid is additional to that paid in respect of works costs, which is subject to separate value for money assessment using the TCI multipliers
The grant payment will be calculated according to the actual cost of the lease premium.

The above factors should be used only in conjunction with the *TCI Base table for 1999/2000: self contained accommodation* (table 1.4) and the relevant TSH key multiplier. The resultant value reflects the maximum grant contribution towards any capitalised lease premium payable to acquire the lease. No other factors or supplementary multipliers apply. The on-costs associated with setting up the lease are included within the TCI element of eligible costs. The different principles underlying the calculation of TCI for self-contained accommodation and for shared accommodation as outlined apply equally to the calculation of capitalised lease premiums

TCI COST GROUP TABLES

TCI shall apply to schemes in cost groups as follows:

GROUP A

Comprises:

The following Inner London Boroughs:

Brent	Kensington and Chelsea
Camden	Lambeth
City of London	Southwark
Hackney	Tower Hamlets
Hammersmith and Fulham	Wandsworth
Islington	Westminster

The following Outer London Boroughs:

Barnet	Ealing
Haringey	Harrow
Hillingdon	Hounslow
Kingston upon Thames	Merton
Richmond upon Thames	

The following unitary authority:

Windsor and Maidenhead

The following Local Authorities in the counties of

Buckinghamshire
South Buckinghamshire

Cornwall and Isles of Scilly
Isles of Scilly

Hertfordshire
Three Rivers

Surrey
Guildford

GROUP B:

Comprises:

The following Inner London Boroughs:

Greenwich	Lewisham

The following Outer London Boroughs:

Barking and Dagenham	Bexley
Bromley	Croydon
Enfield	Newham
Redbridge	Sutton

Waltham Forest

The following unitary authorities:

Bracknell Forest	Reading
Slough	Wokingham

The following local authorities in the Counties of:

Buckinghamshire

Chiltern	Wycombe

Cambridgeshire

Cambridge

Essex

Epping Forest	Harlow

Hampshire

Basingstoke and Deane	East Hampshire
Hart	Rushmoor
Winchester	

Hertfordshire

Broxbourne	Dacorum
East Hertforshire	Hertsmere
St. Albans	Watford
Welwyn Hatfield	

Kent

Sevenoaks	Tonbridge and Malling

Oxfordshire

Oxford

Surrey

Elmbridge	Epsom and Ewell
Mole Valley	Reigate and Banstead
Runnymede	Spelthorne
Surrey Heath	Tandridge
Waverley	Woking

West Sussex

Crawley	Horsham
Mid Sussex	

GROUP C

Comprises:

The following outer London borough:

Havering

The following unitary authorities:

Bath and North East Somerset	Bournemouth
Brighton and Hove	Bristol
Luton	Milton Keynes
Plymouth	Poole
Portsmouth	Southampton
Southend-on-sea	Swindon
The Meadway Towns	Thurrock
West Berkshire	York

The following local authorities in the counties of:

Bedfordshire

Mid Bedfordshire	North Bedfordshire
South Bedfordshire	

Buckinghamshire

Aylesbury Vale

GENERAL INFORMATION

Cambridgeshire

South Cambridgeshire

Cheshire

Macclesfield

Devon

South Hams

Dorset

Christchurch	East Dorset
North Dorset	Purbeck
West Dorset	Weymouth and Portland

East Sussex

Eastbourne	Hastings
Lewes	Wealden

Essex

Basildon	Braintree
Brentwood	Castle Point
Chelmsford	Colchester
Malden	Rochford
Uttlesford	

Gloucestershire

Cheltenham	Cotswold

Greater Manchester

Manchester	Stockport
Trafford	

Hampshire

Eastleigh	Fareham
Gosport	Havant
New Forest	Test Valley

Hertfordshire

North Hertfordshire	Stevenage

Kent

Ashford	Canterbury
Dartford	Dover
Gravesham	Maidstone
Swale	Thanet
Tunbridge Wells	

Lancashire

West Lancashire

North Yorkshire

Harrogate

Oxfordshire

Cherwell	South Oxfordshire
Vale of White Horse	West Oxfordshire

Warwickshire

Stratford-upon-Avon	Warwick

West Midlands

Birmingham	Solihull

West Sussex

Adur	Arun
Chichester	Worthing

Wiltshire

Salisbury	Kennet
North Wiltshire	

Worcestershire

Bromsgrove	Wyre Forest

GROUP D

Comprises:

The following unitary authorities:

Blackburn	Blackpool
Darlington	East Riding
Harlton	Hereford
Isle of Wight	Kingston upon Hull
Leicester	Middlesborough
North East Lincolnshire	North West Somerset
Nottingham City	Peterborough
Rutland	Stoke-on-Trent
The Wrekin	Torbay
Warrington	

The following local authorities in the counties of:

Cambridgeshire

East Cambridgeshire	Huntingdonshire

Cheshire

Chester	Congleton
Crewe and Nantwich	Ellesmere Port and Neston
Vale Royal	

Cornwall and Isles of Scilly

Carrick	Kerrier
Penwith	

Cumbria

Allerdale	Barrow-in-Furness
Carlisle	Copeland
Eden	South Lakeland

Derbyshire

Chesterfield	Derbyshire Dales
High Peak	North East Derbyshire

Devon

East Devon	Exeter
Mid Devon	North Devon
Teignbridge	Torridge
West Devon	

Durham

Durham

East Sussex

Essex

Tendering

Gloucestershire

Forest of Dean	Gloucester
Stroud	Tewkesbury

Greater Manchester

Bolton	Bury
Oldham	Rochdale
Salford	Tameside
Wigan	

Kent

Shepway

Lancashire

Burnley	Chorley
Fylde	Hyndburn
Lancaster	Pendle
Preston	Ribble Valley
Rossendale	South Ribble
Wyre	

Leicestershire

Blaby	Charnwood
Harborough	Hinckley and Boxsworth
Melton	North West Leicestershire

Oadby and Wigston

Lincolnshire

Lincoln

Merseyside

Knowsley	Liverpool
Sefton	St. Helens
Wirral	

Norfolk

Broadland	King's Lynn and West Norfolk
Norwich	

North Yorkshire

Craven	Hambleton
Richmondshire	Ryedale
Scarborough	Selby

Northamptonshire

Corby	Daventry
East Northamptonshire	Kettering
Northampton	South Northamptonshire
Wellingborough	

Northumberland

Alnwick	Berwick-upon-Tweed
Castle Morpeth	Tynedale

Nottinghamshire

Broxtowe	Gedling
Rushcliffe	

Shropshire

Bridgnorth	Oswestry
Shrewsbury and Atcham	South Shropshire

Somerset

Mendip	South Somerset
Taunton Deane	West Somerset

South Yorkshire

Doncaster	Sheffield

Staffordshire

Cannock Chase	Lichfield
South Staffordshire	Stafford
Staffordshire Moorlands	Tamworth

Suffolk

Babergh	Forest Heath
Ipswich	Mid Suffolk
St Edmundsbury	Suffolk Coastal

Tyne and Wear

Gateshead	Newcastle upon Tyne
North Tyneside	South Tyneside
Sunderland	

Warwickshire

Noth Warwickshire	Nuneaton and Bedworth
Rugby	

West Midlands

Coventry	Dudley
Sandwell	Walsall
Wolverhampton	

West Yorkshire

Bradford	Calderdale
Kirklees	Leeds

Wiltshire

West Wiltshire

Worcestershire

Malvern Hills	Redditch
Worcester City	Wychav

GROUP E:

Comprises:

The following unitary authorities:

City of Derby
Hartlepool
North Lincolnshire
Redcar and Cleveland
Stockton-on-Tees

The following local authorities in the Counties of:

Cambridgeshire
Fenland

Cornwall and Isles of Scilly

Caradon	North Cornwall
Restormel	

Durham

Chester-le-Street	Derwentside
Easington	Sedgefield
Teesdale	Waer Valley

Lincolnshire

Boston	East Lindsey
North Kesteven	South Holland
South Kesteven	West Lindey

Norfolk

Breckland	Great Yarmouth
North Norfolk	South Norfolk

Northumberland

Blyth Valley   Wansbeck

Nottinghamshire

Ashfield   Bassetlaw
Mansfield   Newark and Sherwood

Shropshire

North Shropshire

Somerset

Sedgemoor

South Yorkshire

Barnsley   Rotherham

Staffordshire

East Staffordshire   Newcastle-under-Lyme

Suffolk

Waveney

West Yorkshire

Wakefield

Fax. (01902) 795001

North Eastern
St Paul's House
23 Park Square South
Leeds
LS1 2ND
Tel. (0113) 233 7100
Fax. (0113) 233 7101

North West
Elisabeth House
16 St Peter's Square
Manchester
M2 3DF
Tel. (0161) 242 2000
Fax. (0161) 242 2001

Merseyside
6th Floor
Corn Exchange Building
Fenwick Street
Liverpool
L2 7RB
Tel. (0151) 242 1200
Fax. (0151) 242 1201

The Housing Corporation's Offices

Headquarters
149 Tottenham Court Road
London
W1P 0BN
Tel. (0171) 393 2000
Fax. (0171) 393 2111

Regions:
London
Waverley House
7-12 Noel street
London
W1V 4BA
Tel. (0171) 292 4400
Fax. (0171) 292 4401

South East
Leon House
High Street
Croydon
Surrey
CR9 1UH
Tel. (0181) 253 1400
Fax. (0181) 253 1444

South West
Beaufort House
51 New North Road
Exeter
EX4 4EP
Tel. (01392) 428200

East
Attenborough House
109/119 Charles Street
Leicester
LE1 1FQ
Tel. (0116) 242 4800
Fax. (0116) 242 4801

West Midlands
31 Waterloo Road
Wolverhampton
WV1 4BP
Tel. (01902) 795000

National Working Rule Agreement

WR.1 ENTITLEMENT TO BASIC AND ADDITIONAL RATES OF PAY

Operatives employed to carry outwork in the Building and Civil Engineering Industry are entitled to basic pay in accordance with this working Rule (W. R. 1) Rates of pay are set out in a separate schedule, published periodically by the Council.

Classification of basic and additional pay rates of operatives:

General operative
Skilled Operative Rate:- 4
3
2
1

Craft Operative.

1.1 General Operatives

1.1.1 General Operatives employed to carry out general building and/or civil engineering work are entitled to receive the "General Operatives Basic Rate of Pay".

Payment for Occasional Skilled Work

1.1.2 General Operatives, employed as such, who are required to carry out building and/or civil engineering work defined in Schedule 1, on an occasional basis, are entitled to receive the "General Operative Basic Rate of Pay" increased to the rate of pay specified in Schedule 1 for the hours they are engaged to carry out the defined work.

1.2 Skilled Operatives

1.2.1 Skilled Operatives engaged and employed whole time as such, who are required to carry out skilled building and/or civil engineering work defined in Schedule 1. on a continuous basis, are entitled to the "Basic Rate of Pay" specified in Schedule 1.

Additional payment for Skilled Work

1.2.2 In addition to the "Basic Rate of Pay" Skilled Operatives who are required to carry out skilled work defined in Schedule 1 are entitled to an additional payment in accordance with the following scale:

Skilled Operative Additional Rate iii
ii
i

1.3 Craft operatives

Craft operatives employed to carry out craft building and/or civil engineering work are entitled to receive the "Craft Operative Basic Rate of Pay".

1.4 Additional payments for Intermittent Skill, Responsibility or Working in Adverse Conditions.

1.4.1. While carrying out duties specified in Schedule 2 of the Working Rule Agreement, an operative shall be entitled, during the hour he is so engaged, to an additional payment as specified in Schedule 1.

Note: Normal hourly rate
The expression 'normal hourly rate' in this agreement means the craft, skilled or general operatives' weekly basic rate of pay as above, divided by the hours defined in Working Rule. 3 "Working Hours". Additional payments for intermittent skill or responsibility, occasional skilled work, work at heights, work in adverse conditions or bonus payments are not taken into account for calculating the "normal hourly rate".

WR.2 BONUS

It shall be open to employers and employees on any job to agree a bonus scheme based on measured output and productivity for any operation or operations on that particular job.

WR.3 WORKING HOURS

Working Hours
The normal working hours shall be:
Monday to Thursday   8 hours per day
Friday   7 hours per day
Total   39 hours per week

except for operatives working shifts whose working hours shall continue to be 8 hours per weekday.   40 hours per week.

The expression "normal working hours" means the number of hours prescribed above for any day (or night) when work is actually undertaken reckoned from the starting time fixed by the employer.

3.1 Rest/Meal Breaks

At each site or job there shall be a break or breaks for rest and/or refreshment at times to be fixed by the employer. The breaks, which shall not exceed one hour per day in aggregate, shall include a meal break of not less than a half an hour.

3.2. Average Weekly Working Hours

3.2.1 Where there are objective or technical reasons concerning the organisation of work average weekly working hours will be calculated by reference to a twelve month period subject to the employer complying with the general principles relating to the protection of health and safety of workers and

3.2. Average Weekly Working Hours (Cont'd)

3.2.1 (Cont'd)

providing equivalent compensatory rest periods or, in exceptional cases where it is not possible for objective reasons to grant such periods, ensure appropriate protection for the operatives concerned.

3.2.2 The calculation of average weekly working hours set out in WR 3.2.3 only applies where the operatives concerned are employed under the Working Rule Agreement of the Construction Industry Joint Council.

3.2.3 The total of hours worked over the relevant period ending with the date on which the calculation is made will be divided by the number of weeks in the period where work is carried out (complete weeks of absence where work is not carried out are excluded) to give the weekly average working hours.

3.2.4 Disputes regarding the application of WR. 3.2 shall be referred to the National Conciliation Panel of the Construction Industry Joint Council and a hearing shall be convened within ten working days of the date the application is received by the Joint Secretaries.

WR.4 OVERTIME RATES

The employer may require overtime to be worked and the operative may not unreasonably refuse to work overtime.

Overtime will be calculated on a daily basis, but overtime premium rates will not be payable until the normal hours (39 hours-WR3) have been worked in the pay week unless the short time is authorised by the employer on compassionate or other grounds or is a certified absence due to sickness or injury.

Note: The number of hours worked in excess of normal hours will be reduced by the number of hours at unauthorised absence before the overtime premium is calculated.

Overtime shall be calculated as follows:

Monday to Friday
For the first four hours after completion of the normal working hours of the day time and a half; thereafter at the rate of double time until starting time the following day.

Saturday
Time and a half, until completion of the first four hours, and thereafter at double time.

Sunday
At the rate of double time, until starting time on Monday morning.

When an operative is called out after completing the normal working hours of the day or night, he shall be paid at overtime rates for the additional time worked as if he had worked continuously. Any intervening period shall count to determine the rate, but shall not be paid.

Overtime shall be calculated on the normal hourly rate. Additional payments for intermittent skill or responsibility or adverse conditions and bonus shall not be included when calculating overtime payments.

In no case shall payment exceed double time.

WR.5 DAILY FARE AND TRAVEL ALLOWANCES

5.1 Extent of Payment

Operatives are entitled to a daily fare and travel allowance, measured one way from their home to the job/site. The allowances will be paid in accordance with the table published periodically by the Council. There is no entitlement to allowances under this Rule for operatives

5.1 Extent of Payment (Cont'd)

who report, at normal starting time, to a fixed establishment, such as the employers yard, shop or permanent depot. The distance travelled will be calculated by reference to Working Rule 5.2. There is no entitlement under this Rule to allowances for distances less than 15 kilometres or in excess of 75 kilometres. The Employer will, on production of a receipt, reimburse the cost of ferries and tolls incurred by the operative.

Notes:Part of a kilometre counts as a full kilometre. Time spent in daily travelling (except in transfer during working hours) is not to be reckoned as part of the normal working hours.

5.2 Measurement of Distance

All distances shall be measured on an Ordnance Survey map in the 1: 50,000 series in a straight line from the centre of the kilometre square in which his home lies to the centre of the kilometre square where his place of work is located (job centre).

Where special physical conditions such as river, mountain range or motorway make straight-line measurement inequitable, the distance shall be measured from the job centre to the nearest crossing place or exit/entry point, then in a straight line across the crossing place where applicable and then in a straight line to the home centre.

An operative's home is the address at which he is living while employed on the work to which he travels daily.

5.3 Transport provided free by the employer

Where the employer provides free transport, the operative shall not he entitled to fare allowance. However, operatives who travel to the pick up point for transport provided free by the employer are entitled to fare allowance for that part of the journey, in accordance with the table.

5.4 Transfer during Working Day

An operative transferred to another place of work during work hours shall, on the day of the transfer only, be paid any fares actually incurred:

(i) in the transfer, and
(ii) in travelling home from the place where he finishes work if this differs from the place where he reported for work at starting time, subject to deduction of half the daily fare allowance.

5.5 Emergency Work

An operative called from the home (or temporary place of residence) outside normal working hours to carry out emergency work shall be paid travelling expenses and his normal hourly rate for the time spent travelling to and from the job.

WR.6 ROTARY SHIFT WORKING

Rotary shift working means a situation in which more than one shift of not less than eight hours is worked on a job in a 24 hour period and an operative employed on that job rotates between the different shifts (whether in the same or different pay weeks).

On all work which is carried out on two or more shifts in a 24 hour period the following provisions shall apply:

The first shift in the week shall be the first shift that ends after midnight on Sunday

The normal hours of a shift shall be eight hours, excluding meal breaks, notwithstanding which, the hours to be worked on any particular shift shall be established by the employer. The rate payable for the normal hours of the shift shall be the operative's normal hourly rate plus, in the

WR.6 ROTARY SHIFT WORKING (Cont'd)

case of an operative completing a shift, a shift allowance of 14% of the normal rate.

An operative required to work continuously for over eight hours on a shift or shifts shall be paid at the rate of time and a half, plus a shift allowance of 14% of his normal hourly rate, for the first four hours beyond eight and thereafter at double time but such double time payment shall not be enhanced by the 14% shift allowance (i.e. the maximum rate in any circumstance shall be double the normal hourly rate).

After having worked five complete scheduled shifts in a week (i.e. discounting any additional overtime shifts worked) an operative shall on the first shift thereafter be paid at the rate of time and a half of normal rate plus 12.5% shift allowance for the first eight hours of the shift and thereafter and on any subsequent shift in that week at the rate of double time but with no shift allowance.

Where the work so requires, an operative shall be responsible for taking over from and handing over to his counterpart at commencement and on completion of duty unless otherwise instructed by his employer.

Where the nature of the work is such as to require an operative to remain at his workplace and remain available for work during mealtimes a shift allowance of 20% shall apply instead of the 14% or 12.5% otherwise referred to in this rule.

The shift allowance shall be regarded as a conditions payment and shall not be included when calculating overtime payments.

Under this rule the first five complete shifts worked by an operative shall count as meeting the requirements of the Guaranteed Minimum Rule.

This Rule does not apply to operatives employed on night work, continuous working or tunnel work.

Employers and operatives may agree alternative shift working arrangements and rates of pay where, at any job or site, flexibility is essential to effect completion of the work.

WR.7 NIGHT WORK

Where work is carried out at night by a separate gang of operatives from those working during daytime, operatives so working shall **be** paid at their normal hourly rate plus an allowance of 25% of the normal hourly rate.

Overtime shall be calculated on the normal hourly rate provided that in no case shall the total rate exceed double the normal hourly rate. Overtime shall therefore be paid as follows:

(a)   Monday to Friday:
      after completion of the normal working hours at the rate of time and a half plus the night work allowance (i.e. time and a half plus 25% of normal hourly rate) for the next four working hours and thereafter at double time.

(b)   Weekends:
      All hours worked on Saturday or Sunday night at double time until the start of working hours on Monday.

This rule does not apply to operatives employed on shift work, tunnel work or continuous working.

WR.8 CONTINUOUS WORKING

An operative whose normal duties are such as to require him to be available for work during mealtimes and consequently

WR.8 CONTINUOUS WORKING (Cont'd)

has no regular mealtime, shall be deemed a "continuous worker" and shall be responsible for taking over from and handing over to his counterpart at commencement and completion of duty unless otherwise instructed by his employer. He will be paid for the number of hours on duty on the job at normal hourly rate plus 20% as a continuous working allowance. If in the normal cycle of operations for the particular job he is required to be on duty between 10 p.m. Saturday and 10 p.m. Sunday, he will during these hours be paid the rate of time and a half, plus 20% of his normal hourly rate as a continuous working allowance, provided that the continuous working allowance of 20% of normal hourly rate shall be deemed to be a conditions payment and shall not be reckoned for the purpose of calculating overtime payments. If work between 10 p.m. Saturday to 10 p.m. Sunday is not within the normal cycle of operations for the particular job, then no continuous working allowance shall be paid but the rate of payment shall be double the normal hourly rate.

This rule does not apply to operatives on night-work, shift work or tunnel work.

WR.9 TIDE WORK

9.1   Where work governed by tidal conditions is carried out during part only of the normal working hours, and an operative is employed on other work for the remainder of the normal working hours, the normal hourly rate (with the addition of any additional rate payable in respect of the conditions under which work is done), shall be paid during the normal working hours a thereafter shall be in accordance with the rule on Overtime Rates.

9.2   Where work governed by tidal conditions necessitates operatives turning out for each tide and they are not employed on other work, they shall be paid a minimum for each tide of six hours pay at ordinary rates, provided they do not work more than eight hours in the two tides. Payment for hours worked beyond a total of eight on two tides shall be calculated proportionately i.e. those hours worked in excess of eight multiplied by the total hours worked and the result divided by eight, to give the number of hours to be paid in addition to the twelve paid hours (six for each tide) provided for above. Work done on Saturday after 4 p.m. and all Sunday shall be paid at the rate of double time. Operatives shall be guaranteed eight hours at ordinary rate for time work between 4 p.m. and midnight on Saturday and 16 hours at ordinary rates for two tides work on Sunday. Payment under the rule on Overtime Rates does not apply to this sub-paragraph.

WR.10 TUNNEL WORK

The long-standing custom of the industry that tunnel work is normally carried out by day and by night is reaffirmed. Where shifts are being worked within and in connection with the driving tunnels the first period of a shift equivalent to the normal working hours specified in the Working Hours rule for that day shall be deemed to be the ordinary working day. Thereafter the next four working hours shall be paid at time and a half and thereafter at double time provided that

(a) In the case of shifts worked on Saturday the first four hours shall be paid at time and a half and thereafter at double time.

(b) In the case of shifts worked wholly on Sunday payment shall be made double time.

(c) In the case of shifts commencing on Saturday but continuing into Sunday, payment shall be made for all hours worked at double time.

(d) In case of shifts commencing on Sunday but continuing into Monday, hours worked before midnight shall be paid for at double time and thereafter four working hours calculated from midnight shall be at time and a half and thereafter at double time.

GENERAL INFORMATION

WR.10 TUNNEL WORK (Cont'd)

This rule does not apply to an operative employed under the Continuous Working rule.

WR.11 REFUELLING, SERVICING, MAINTENANCE AND REPAIRS

Operators of mechanical plant, such as excavators, cranes, derricks, rollers, locomotives, compressors or concrete mixers and boiler attendants shall, if required, work and be paid at their normal hourly rate for half an hour before and half an hour after the working hours prescribed by the employer for preparatory or finishing work such as refuelling firing up, oiling, greasing, getting out, starting up, banking down, checking over, cleaning out and parking and securing the machine or equipment.

Refuelling, Servicing, Maintenance and Repair work carried out on Saturday and Sundays shall be paid in accordance with the rule on Overtime Rates.

WR.12 STORAGE OF TOOLS

When practicable and reasonable on a site, job or in a shop the employer shall provide an adequate lock-up or lock-up boxes, where tools can be left at the owner's risk, provided always that the employer shall accept liability up to a maximum amount specified by the Council for any loss caused by fire or theft of tools properly secured by an operative in such lock-up or lock-up boxes, on the understanding that the employee shall comply with all the employer's requirements as regards the storage of tools. At all times an operative shall take good care of his tools and personal property and act in a responsible manner to ensure their reasonable safety.

WR.13 LOSS OF CLOTHING

Where an operative leaves clothing in accommodation provided by the employer the Employer shall be liable up to a maximum amount specified by the Council for loss of such clothing through fire.

WR.14 TRANSFER ARRANGEMENTS

14.1 At any time during the period of employment, the operative may, at the discretion of the employer be transferred from one job to another.

14.2 The employer shall have the right to transfer an operative to any site within daily travelling distance of where the operative is living. A site is within daily travelling distance if:

(a) when transport is provided free by the employer the operative can normally get from where he is living to the pickup point designated by the employer within one hour, using public transport if necessary or

(b) in any other case the operative, by using the available public transport on the most direct surface route, can normally get to the site within two hours.

14.3 Transfer to a job which requires the operative to live away from home shall be by mutual consent. The consent shall not be necessary where the operative has been in receipt of subsistence allowance in accordance with the Subsistence Allowance Rule from the same employer at any time within the preceding 12 months.

WR.15 SUBSISTENCE ALLOWANCE

When an operative is recruited on the job or site and employment commences on arrival at job or site he shall not be entitled to payment of subsistence allowance. An operative necessarily living away from the place in which

WR.15 SUBSISTENCE ALLOWANCE (Cont'd)

he normally resides shall be entitled to a subsistence allowance of an amount specified by the Council.

Subsistence allowance shall not be paid in respect of any day on which an operative is absent from work except when that absence is due to sickness or industrial injury and he continues to live in the temporary accommodation and meets the industry sick pay requirements.

Alternatively, the employer may make suitable arrangements for a sick or injured operative to return home, the cost of which shall be met in full by the employer

An operative in receipt of subsistence allowance shall only be entitled to daily fare allowance under the Daily Fare Allowance rule between his accommodation and the job if he satisfies the employer that he is living as near to the job as there is a accommodation available.

WR.16 PERIODIC LEAVE

16.1 When an operative is recruited on the job or site and employment commences on arrival at the job or site he shall not be entitled to the periodic leave allowances in this rule. In other cases when an operative is recruited or a sent to a job which necessitates his living away from the place in which he normally resides, he shall he entitled to:

Payment of his fares or conveyance in transport provided by the employer as follows:-

(a) from a convenient centre to the job at commencement

(b) to the convenient centre and back to the job at the following periodic leave intervals

(i) for jobs up to 128 kilometres from the convenient centre (measured in a straight line), every four weeks

(ii) for jobs over 128 kilometres from the convenient centre (measured in a straight line) at an interval fixed by mutual arrangement between the employer and operative before he goes to the job.

(c) from the job to the convenient centre at completion.

16.2 **Payment of fares**

Where an employer does not exercise the option to provide free transport, the obligation to pay fares may at the employer's option be discharged by the provision of a free railway or bus ticket or travel voucher or the rail fare.

Payment for the time spent travelling between the convenient centre and the job as follows:

(a) On commencement of his employment at the job, the time travelling from the convenient centre to the job, provided that an operative shall not be entitled to such payment if within one month from the date of commencement of his employment on the job he discharges himself voluntarily or is discharged for misconduct.

(b) When returning to the job (i.e. one way only) after periodic leave, provided that he returns to the job at the time specified by the employer and provided also that an operative shall not be entitled to such payment if, within one month from the date of his return to the jobs, he discharges himself voluntarily or is discharged for misconduct.

(c) On termination of his employment on the job by his employer, the time spent travelling from the job to the convenient centre, provided that he is not discharged for misconduct.

16.2 Payment of fares (Cont'd)
(d) Time spent in periodic travelling is not to be reckoned as part of the normal working hours; periodic travelling time payments shall in all cases be at the operative's normal hourly rate to the nearest quarter of an hour and shall not exceed payment for eight hours per journey.

16.3 Convenient centre
The convenient centre shall be a railway station, bus station or other similar suitable place in the area in which the operative normally resides.

WR.17 GUARANTEED MINIMUM WEEKLY EARNINGS

Where an operative, who is employed under the Working Rules of the Council, has been available for work throughout the normal working hours for the week whether or not work has been provided by the employer. He shall be entitled to guaranteed minimum weekly earnings as defined in Rule 1.

17.1 Loss of Guarantee

There shall be no entitlement to guaranteed minimum weekly earnings where the employer is unable to provide continuity of work due to industrial action or for Tide Work Rule 9 or work paid by shift Rule,6.

17.2 Proportional Reduction

Where an operative is absent for part of normal working hours due to certified sickness or injury or for one or more days of annual or recognised public holiday the requirement for the operative to be available for work will be deemed to be met and the payment of Guaranteed Minimum Weekly Earnings will be proportionately reduced. The proportionate reduction will not apply where the employer authorises the absence on compassionate or other grounds.

17.3 Availability for Work

An operative has satisfied the requirements to keep himself available for work during normal working hours if he complies with the following conditions:

That, unless otherwise instructed by the employer, he has presented himself for work at the starting time and location prescribed by the employer and has remained available for work during normal working hours.

That he carries out satisfactorily the work for which he was engaged or suitable alternative work if instructed by the employer and

He complies with the instructions of the employer as to when, during normal working hours, work is to be carried out, interrupted or resumed.

17.4 Temporary Lay-off

17.4.1 Where work is temporarily stopped or is not provided by the employer the operative may be temporarily laid off. The operative shall, subject to the provisions of Rule 17.4.2, paid one fifth of his "Guaranteed Minimum Weekly Earnings" as defined in Rule 17 for the day on which he is notified of the lay-off and for each of the first five days of temporary lay-off. While the stoppage of work continues and the operative is prevented from actually working, he will be required by the employer to register as available for work at the operatives local job centre.

17.4.2 The payment described in Rule. 174.1 will be made, provided that in the three months prior to any lay-off there has not been a previous period or periods of lay-off in respect of which a guaranteed payment was made for 5 consecutive days or 5 days cumulative, excluding the day or days of

17.4 Temporary Lay-off (Cont'd)

notification of lay-off. In any such case the operative will not be entitled to a further guaranteed payment until a total of three months has elapsed from the last day of the period covered by the previous payment. Thereafter and for so long as the stoppage lasts, he shall be entitled to a further guaranteed payment of up to five days.

17.5 Disputes

A dispute arising under this agreement concerning guaranteed minimum payment due may, at the option of the claimant, be referred to ACAS and/or an industrial tribunal in the event of decision by the Council.

WR.18 ANNUAL HOLIDAYS

Holidays with Pay shall be as provided for in the Agreement for Holidays with Pay Scheme dated 28 October 1942 as subsequently amended from time to time.

The Winter Holiday shall be seven working days taken in conjunction with Christmas Day, Boxing Day and New Year's Day, to give a Winter Holiday of two calendar weeks. The dates each Winter Holiday shall be published by the Council.

The Easter (Spring) Holiday shall be the four working days immediately following Easter Monday, to give an Easter Holiday of one calendar week.

It shall be open to employers and operatives to agree that all or some of the 'Winter Holiday' and/or the 'Easter (Spring) Holiday' will be taken on alternative dates.
The Summer Holiday shall be two calendar weeks, by mutual agreement not necessarily consecutive, to be granted in the "summer period" - i.e. between 1st April and 30th September. Except by mutual agreement, neither week is to be taken in conjunction with the Easter (Spring) Holiday.

Notwithstanding the conditions prescribed in that Agreement in regard to the provision of holiday credits, the employer shall not be required to provide holiday credits for the calendar week in question if an operative terminates his employment without the required notice (see Termination of Employment rule) or when the operative without good cause is available for work for less than a minimum period of four normal working days in a pay week.

NOTE: The Building and Civil Engineering Annual Holidays and Benefits Schemes (see also the Benefit Schemes rule) are designed to provide holiday, accidental injury and retirement benefits for all building and civil engineering operatives and death benefit for their dependants. The benefits are paid for by employer through a weekly surcharge on the holiday credits and it is an essential feature of the Scheme that the operative is normally available for work throughout normal working hours. Absence without good cause cannot be condoned and the Parties to this agreement support this principle. Jury Service shall be deemed to be a good cause for the purpose of this Rule.

The Agreement on benefits for retirement and death do not apply to an operative aged 65 or over (see Benefits Scheme rule). In the case of such an operative the employer shall either:

(a) continue operation of the Annual Holidays Scheme on the understanding that, with the exception of the provisions relating to accidental injury and accidental death (see Benefits Schemes rule), the Benefits Schemes shall not apply; or

(b) honour the outstanding holiday credits in accordance with the Agreement and, thereafter make alternative arrangements for holiday pay notifying the operative in writing of this change in his terms of employment

GENERAL INFORMATION

WR.19 PUBLIC HOLIDAYS

19.1 The following are recognised as public holiday for the purpose of this agreement:

(a) **England and Wales**

Christmas Day, Boxing Day, New Year's Day, Good Friday, Easter Monday, the May Day, Bank Holiday, the Spring Bank Holiday, the Summer Bank Holiday shall be recognised as public holidays in England and Wales, provided that such days are generally recognised as holidays in the locality in which the work is being done.

(b) **Scotland**

Christmas Day, Boxing Day, New Year's Day, 2nd January, Easter Monday, the first Monday in May, the Friday immediately preceding the Annual Summer Local Trades Holiday and the Friday and Monday at the Autumn Holiday, as fixed by the competent Local Authority.

(c) **Local Variations**

Where, in any locality, any of the above public holidays is generally worked and another day is recognised instead as a general holiday, such other day shall be recognised as the alternative holiday.

(d) **Alternative Days**

When Christmas Day, Boxing Day or New Year's Day falls on a Saturday or Sunday an alternative day or days of public holiday will be promulgated. Any reference in this Rule to Christmas Day, Boxing Day or New Year's Day shall be taken to apply to the alternative day so fixed.

19.2 **Payment in respect of Public Holidays**

Payment for days of public holiday recognised under this rule shall be made by the employer to an operative in his employment at the time of each such holiday on the pay day in respect of the payweek in which such holiday occurs, except that payment for Christmas Day, Boxing Day and New Year's Day shall be made on the last pay day before the Winter Holiday. The amount shall be the equivalent of payment for the number of normal working hours specified in the Working Hours rule for that day at basic rate, provided that:

(i) an operative shall be entitled for each day of recognised public holiday to be pal appropriate proportion of his basic rate of pay, provided that on the last normal we day before and the first normal working day after the day or days of recognised public holiday he is:

a) available for work within the meaning of the Rule on Guaranteed Mini Earnings throughout normal working hours; or

b) incapacitated for work and produces satisfactory evidence of the reason.

(ii) The conditions in sub-paras (a) and (b) do not apply to Christmas Day, Boxing Day and New Year's Day or where a day or days of agreed Easter or Summer holiday fall between the recognised public holiday(s) and the first working day thereafter.

(iii) An operative who has completed 4 weeks in employment, but who is no longer in employment of employer on such a public holiday shall, nevertheless, be entitled to payment in respect of such holiday if he was in the employer's employment on the last working day preceding the day or days of public holiday but the

19.2 **Payment in respect of Public Holidays(Cont'd)**

employment terminated by the employer (otherwise than for misconduct) before the holiday occurred. Where the holiday is Christmas Day, payment shall also be made for Boxing Day and Year's Day. Where the holiday is Good Friday payment shall also be made for Easter Monday. Payment shall be made on termination of employment.

(iv) An operative who is required to work on Christmas Day, Boxing Day, Good Friday or Easter Monday has the option by arrangement with the employer of an alternative day of holiday as soon thereafter as is mutually convenient, in which case the payment prescribed by this rule shall be made in respect of such alternative day instead of the public holiday. When the employment is terminated before such alternative day occurs, the operative shall receive such payment on the termination of employment.

(v) There is no entitlement to payment for public holidays for an operative who has been unavailable for work, due to sickness or injury for a continuous period of ten weeks or more and who remains unavailable for work on the last working day preceding the or days of public holiday.

19.3 **Payment for Work on a Public Holiday**

All hours worked on a day designated as a public holiday shall be paid for at double time.

WR.20 SICK PAY

Absence from Work due to Sickness or Injury

20.1 **Relationship of Industry Sick Pay with Statutory Sick Pay**

Under existing legislation there is an entitlement to statutory sick pay. Any payment due under this rule shall be increased by an amount equivalent to any statutory sick pay that may be payable in respect of the same day of incapacity for work under the Regulations made under that Act. These are referred to elsewhere in this Rule as "SSP Regulations".

20.2 **Scope**

This Rule applies to all adult operatives i.e. all operatives aged 18 years and over.

20.3 **Qualifying Days**

For the purpose of both this Rule and the SSP Regulations, the "qualifying days" that shall generally apply in the industry are Monday to Friday in each week.

While the qualifying days referred to above shall generally be the same five days as those which form the normal week of guaranteed employment under this agreement, it is accepted that there might be certain exceptions, e.g. where the particular circumstances of the workplace require continuous six or seven day working. In these situations it is in order, where there is mutual agreement, for other days to be regarded as 'qualifying days' for the purpose of this Rule and SSP.

20.4 **Amount of Payment**

An operative who during employment with an employer is absent from work on account of sickness or injury shall, subject to satisfying all the conditions set out in this Rule be paid the appropriate proportion of a weekly amount specified by the Council for each qualifying day of incapacity for work. For this purpose, the appropriate proportion due for a day shall be the weekly rate divided by the number of qualifying days specified under Rule 20.3 above.

20.5 Notification of Incapacity for Work

An operative shall not be entitled to payment under this Rule unless, during the first qualifying day in the period of incapacity, his employer is notified that he is unable to work due to sickness or injury and when the incapacity for work started. Thereafter, the operative shall, at intervals not exceeding one week throughout the whole period of absence, keep the employer informed of his continuing incapacity for work. Where the employer is notified later than this rule requires, he may nevertheless make payment under the rule if satisfied that there was good cause for the delay.

20.6 Certification of Incapacity for Work

The whole period of absence from work shall be covered by a certificate or certificates of incapacity for work to the satisfaction of the employer. For the first seven consecutive days of sickness absence, including weekends and public holidays, a self certificate will normally suffice for this purpose. Any additional days of the same period of absence must be covered by a certificate or certificates given by a registered medical practitioner.

NOTE: For the purpose of this paragraph a self certificate means a signed statement made by the operative in a form that is approved by the employer, that he has been unable to work due to sickness/injury for the whole period specified in the statement.

20.7 Qualifying Conditions for Payment

An operative shall not be entitled to the payment prescribed in this rule unless the following conditions are satisfied:

20.7.1 That incapacity has been notified to the employer in accordance with Rule 20.5 above

20.7.2 That the requirements of Rule 20.6 above to supply certificate(s) of incapacity for work have been complied with.

20.7.3 That the first three qualifying days (for which no payment shall be due) have elapsed each period of absence.

20.7.4 That none of the qualifying days concerned is a day of annual or public holiday granted in accordance with the respective provisions of the Annual Holidays with Pay Agreement and this Working Rule Agreement.

20.7.5 That the incapacity does not arise directly or indirectly from insurrection or war, attempted suicide or self-inflicted injury, the operative's own misconduct, any gainful occupation outside working hours or participation as a professional in sports or games.

20.7.6 That the operative has had four or more weekly holiday credits in the period of eight calendar weeks ending with the calendar week immediately prior to that in which absence from work starts. For this purpose, weekly holiday credits in respect of sickness absence under the Annual Holidays Agreement are not to be counted.

20.7.7 That the limit of payment has not been reached in the Accounting Period. The limit is to be determined as follows:

20.7.7.1 For an operative who has holiday credits current from the beginning of an Accounting Period for the Annual Holidays Scheme, the maximum number qualifying days for which there shall be entitlement to payment during that Accounting Period shall be the number occurring in a total of 10 weeks including days of absence for which payment has been made by previous employers. In no case shall an operative receive payment for more than 10 weeks in respect of any one period of absence due to sickness.

20.7 Qualifying Conditions for Payment (Cont'd)

20.7.7.2 For a new entrant to the industry, including an operative who on entering employment has no holiday credits, the foregoing maximum during the current Accounting Period shall be reduced on the following basis. The time that has elapsed between the beginning of that Accounting Period and the date of entry into employment shall be assessed as a proportion of the full Accounting Period and the number of qualifying days for which there shall be entitlement to payment shall be reduced by that proportion of the above maximum.

NOTE: For the purpose of this paragraph:

(i) The maximum number of qualifying days in Rule 20.7.7.1 above will be 50: and

(ii) The reduction under Rule 20.7.7.2 above will generally be 4 1/6 days for each calendar month that has elapsed (rounded down to a full day).

20.8 Record of Absence

A record of days of absence for which payment has been made under this Rule in any one Accounting Period shall be kept by each employer.

WR.21 BENEFIT SCHEMES

21.1 An operative is entitled to be provided by his employer with cover for:

(a) retirement benefit, of an amount dependent on length of service in the industry, unless the operative has exercised his right under legislation to forgo this retirement benefit (in which case he must advise the Administrator of the Building and Civil Engineering Benefits Scheme accordingly) and

(b) death benefit, of an amount promulgated from time to time by the Trustees of the Benefits Scheme, with an additional discretionary amount payable if the cause of death is an accident at the place of work or while travelling to or from work, all such amounts payable at the discretion of the Trustees, and

(c) accidental injury benefit for a specified injury (or injuries) sustained as a result of an accident at the place of work or an accident while travelling to or from work, the amount being dependent on the nature of the qualifying injury or injuries, all such payments being payable at the discretion of the Trustees of the Building and Civil Engineering Accident Benefit Scheme.

In each case in accordance with the Trust Deed and Rules of the relevant Scheme, which are published separately.

21.2 Cover for the above benefits is to be provided by payment by the employer of a surcharge on the value of the weekly holiday credits under the Annual Holidays Scheme.

21.3 There is no entitlement to cover in the case of an operative under the age of 18 years or over the age of 65 years, except for (a) accidental injury benefit as set out in 1(c) above, and (b) death benefit (of up to the sum of the amounts set out in 1(b) above) if the cause of death is an accident at the place of work or an accident while travelling to or from work when the operative is under the age of 18 years, or on or after his 65th birthday.

21.4 Additional Voluntary Contributions

Where an operative wishes to make additional voluntary contributions in respect of retirement cover, and requests his employer to do so, the employer shall deduct from the operative's weekly pay an amount duly authorised by the

GENERAL INFORMATION

21.4 Additional Voluntary Contributions (Cont'd)

operative, and shall arrange for that amount to be transmitted to the Trustees of the Building and Civil Engineering Benefits Scheme in accordance with the Rules of that Scheme.

WR.22 GRIEVANCE PROCEDURE

Procedure for dealing with grievances

Any issue which may give rise to or has given rise to a grievance (including issues relating to discipline) affecting the employer's workplace and operatives employed by that employer at that workplace shall be dealt with in accordance with the following procedure.

There shall be no stoppage of work, either partial or general, including a 'go-slow', strike, lock out or any other kind of disruption or restriction in output or departure from normal working, in relation to any grievance unless the grievance procedure has been fully used and exhausted at all levels.

Every effort should be made by all concerned to resolve any issue at the earliest stage.

A written record shall be kept of meetings held and conclusions reached or decisions taken. The appropriate management or union representative should indicate at each stage of the procedure when an answer to questions arising is likely to be given, which should be as quickly as practicable.

Stage 1

Any operative who has a grievance concerning his employment shall raise the matter orally or in writing, with his immediate supervisor.

Stage 2

If the issue remains unresolved, the operative or his immediate supervisor shall bring it to the attention of the next level of supervision. The operative may involve an appointed steward of a recognised trade union from this stage. A steward may also be involved if a grievance affects a number of operatives.

Stage 3

Failing resolution of the issue at stage 2, the appropriate supervisor or the operative(s) concerned or the steward (if appointed) shall, if there are good grounds for so doing, refer the matter to the agent or his nominee. If the matter then remains unresolved, the steward shall report the matter to the appropriate full time union official who shall, if he considers it appropriate, pursue any outstanding issue with the agent or his nominee after advising him in writing of the issue(s) he wishes to pursue.

Stage 4

Failing resolution of the issue at stage 3, the full time local union official shall report the matter up to the appropriate full time union official and the agent (or nominee) shall report the matter to an appropriate representative of the employer. Such union official, if there are good grounds for so doing shall pursue the issue with the appropriate representative of the employer.

Stage 5

Failing resolution of the issue at stage 4, the union official concerned shall, if it is decided to pursue the matter further, put the issue in writing to the employer and it is the duty of such official and/or the employer to submit the matter, as quickly as practicable, to the Construction Industry Joint Council for settlement.

The decisions of the Construction Industry Joint Council shall be accepted and implemented by all concerned.

WR.23 DISCIPLINARY PROCEDURE

It is recognised that, in order to maintain good morale, the employer has the right to discipline those:

- who fail to fulfil competently and to instructions of the company the duties and responsibilities called for by the position they hold; and/or

- whose behaviour is unsatisfactory; and/or

- who fail to make appropriate use of the disputes procedure for the resolution of questions arising without recourse to strike or other industrial action.

It is equally recognised that the employer must exercise this right consistently and with justice and care.

Discipline shall be applied in accordance with the following procedure:

23.1 An oral warning shall be given to the operative of the employer's dissatisfaction and the improvement called for within a stated period. The steward, if appointed, may be present and the warning shall be entered in the operative's record.

23.2 If the required improvement does not occur, or a new complaint arises, a written warning shall be issued stating that it is a final warning failure to improve within a stated period will result in dismissal.

The steward, if appointed, shall be given a copy of the written warning.

23.3 On continued failure, the operative shall be dismissed with appropriate notice in writing, stating, if the operative so requests, the reason for dismissal. A copy of the written notice shall be given to the steward, if appointed.

23.4 If, at any stage, alleged gross misconduct arises, the case shall be investigated promptly and a decision taken by the employer, after a fair hearing of the operative and his steward, if appointed. Such decision may be summary dismissal; return to normal working; or transfer to another workplace. Where the decision is dismissal, it shall be in writing stating, if the operative so requests, the reason for dismissal. A copy of the written decision will be given to the steward, if appointed.

At the discretion of the employer, the operative(s) may be warned of the investigation and/or suspended from the place of work on pay. Such pay will be forfeited if the investigation confirms that dismissal is appropriate.

23.5 At any stage in the procedure, an appeal may be made by or on behalf of the operative in accordance with the grievance procedure.

WR.24 TERMINATION OF EMPLOYMENT

24.1 The minimum period of notice of termination of employment that an employer shall give to an employee is:-

(i) During the first four weeks — One day's notice

(ii) After four weeks' continuous employment but less than two years — One week's notice

(iii) After two years' continuous each employment but less than twelve years — One weeks notice for full year of continuous employment.

(iv) Twelve years' continuous employment or more — Twelve weeks' notice

WR.24 TERMINATION OF EMPLOYMENT (Cont'd)

24.2 The minimum period of notice of termination of employment that an employee shall give an employer is:

    (i) During the first four weeks -   one day's notice

    (ii) After four weeks'
       continuous employment -   one   week's
       notice.

24.3 The employment may be terminated at any time by mutual consent which should preferably be expressed in writing.

24.4 All outstanding wages are to be paid at the expiration of the period of notice and the employee advised of his entitlement to holiday credits and PAYE certificates or, in lieu thereof, a written statement that they will be forwarded as soon as possible.

24.5 In the event of gross misconduct, an operative may be summarily discharged at any time subject to a sufficient investigation into the circumstances being carried out in accordance with the recommended procedures.

Where gross misconduct is alleged which may result in dismissal without previous warning, the case shall be investigated, a hearing shall be arranged and a decision taken by the employer as quickly as practicable.

At the discretion of the employer; the operative(s) may be suspended on full pay whilst the case is investigated. Such pay will be forfeited if the investigation confirms that dismissal is appropriate.

24.6 If an operative (who is not under notice of termination by his employer) voluntarily terminates his employment without giving the required notice, then the employee shall not be entitled to a weekly holiday credit for the calendar week in which the employment is terminated.

WR.25 TRADE UNIONS

25.1 Since it is essential to good industrial relations that both employers and operatives comply with the Working Rules, the Employers are encouraged to:

    25.1.1 recognise the trade unions who are signatories to the Agreement;

    25.1.2 ensure that all operatives are in the direct employment of the company or its sub contractors and are engaged under the terms and conditions of the Working Rule Agreement.

25.2 Deduction of Union Subscriptions

When requested by a signatory union, employers should not unreasonably refuse facilities for the deduction of union subscriptions (check-oft) from the pay of union members.

25.3 Full Time Trade Union Officials

A full time official of a trade union which is party to the Agreement shall be entitled, by prior arrangement with the employer's agent or other senior representative in charge and on presenting his credentials, to visit a workplace to carry out trade union duties and to see that the Working Rule Agreement is being properly observed.

25.4 Trade Union Stewards

An operative is eligible for appointment as a steward on completion of not less than four weeks continuous work in the employment of the employer. Where an operative has been properly appointed as a steward in accordance with the rules of his trade union (being a trade union signatory to the Agreement) and issued with written credentials by the trade union concerned, the trade union shall notify the employer's agent or representative of the appointment, for formal recognition by the employer of the steward. On completion of

WR.25 TRADE UNIONS (Cont'd)

25.4 Trade Union Stewards (Cont'd)
this procedure the employer will recognise the steward, unless the employer has any objection to granting recognition, in which case he shall immediately notify the trade union with a view to resolving the question.

An employer shall not be required to recognise for the purposes of representation more than one officially accredited steward for each trade or union at any one site or workplace.

25.5 Convenor Stewards

Where it is jointly agreed by the employer and the unions, having regard to the number of operatives employed by the employer at the workplace and/or the size of the workplace the recognised trade unions may appoint a convenor steward, who should normally be in the employment of the main contractor, from among the stewards and such appointment shall be confirmed in writing by the Operatives' side. On completion of this procedure the employer will recognise the Convenor Steward, unless the employer has any objection to granting recognition in which case he shall immediately notify the trade union with a view to resolving the question.

25.6 Duties and Responsibilities of Stewards and Convenor Stewards.

    25.6.1 The duties of the Stewards shall be:

    i)   To represent the members of their union engaged on the site/factory/depot,
    ii)   to ensure that the Working rule is observed,
    iii)   to recruit members on the site/factory/depot into a signatory union,
    iv)   to participate in the Grievance Procedure at the appropriate stage under WR. 22,
    v)   to assist in the resolution of disputes in accordance with the Working Rule Agreement.

    25.6.2 The duties of the Convenor Steward, in addition to those set in WR 25.6.1 shall be:

    i)   to represent the operatives on matters concerning members of more than one union,
    ii)   to co-operate with Management and to assist individual Shop Stewards
    iii)   to ensure that disputes are resolved by negotiation in accordance with the Working rule Agreement.

    25.6.3 No Steward or Convenor Steward shall leave their place of work to conduct union business without prior permission of their immediate supervisor. Such permission should not be unreasonably withheld but should only be given where such business is urgent and relevant to the site, job or shop.

25.7 **Steward Training**

To assist them in carrying out their functions, Shop Stewards will be allowed reasonable release to attend training courses approved by their union.

25.8 **Stewards Facilities**

Management shall give recognised union officials and/or convenor stewards reasonable facilities for exercising their proper duties and functions. These facilities, which must not be unreasonably withheld, must not be abused. The facilities should include (a) use of a meeting room, (b) access to a telephone and (c) the use of a noticeboard on site. If the convenor steward so requests, the employer shall provide him regularly with the names of operatives newly engaged by that contractor for work on that site or job or in that shop.

GENERAL INFORMATION

WR.25 TRADE UNIONS (Cont'd)

25.9 Meetings

Meetings of operatives, stewards or convenor stewards may be held during working hours only with the prior consent of Management.

WR.26 SCAFFOLDERS

The following provisions are to be read in conjunction with the provisions of the Construction Industry Scaffolders' Record Scheme published separately by the Council and the Construction Industry Training Board. Special attention is drawn to the provision in the Scheme that if a scaffolder or trainee scaffolder is not in possession of a CITB training record card at the time of engagement, application should be made immediately by the new employer on the prescribed form. Any difficulties should be referred to the Joint Secretaries for action in accordance with the Scheme.

26.1 Scaffolders employed whole time as such are to be in one of three categories, as defined below:

Trainee Scaffolder: an operative who can produce one of the following types of CITB training record card:

(i) a valid Trainee Scaffolder's card, or

(ii) a Basic Scaffolder's card with postdated validity.

Basic Scaffolder: an operative who has at least one year's whole-time experience of scaffolding and who can produce one of the following types of CITB training record cards:

(i) a valid Basic Scaffolder's card, or

(ii) an Advanced Scaffolder's card with Postdated validity.

Advanced Scaffolder: an operative who has at least two years' wholetime experience of scaffolding and at least one year's experience as a Basic Scaffolder and who can produce an Advanced Scaffolder's CITB training record card.

26.2 No operative other than a Basic or Advanced Scaffolder, as defined above, may be employed on basic scaffolding operations and no operative other than an Advanced Scaffolder may be employed on advanced scaffolding operations (see schedule Rule 26.5) unless

(i) under adequate supervision, or

(ii) working together with a Basic or Advanced Scaffolder, or

(iii) erecting, altering or dismantling simple access scaffolding with a working platform no higher than 5 metres, or

(iv) a craft operative who is normally required to erect, alter or dismantle scaffolding in the course of his work and who has received adequate training in the scaffolding operation(s) concerned.

These exceptions do not affect the application to such scaffolding work of the provisions of safety legislation and, in particular, the Construction (Health, Safety and Welfare) Regulations 1996.

26.3 The onus of proof of training and experience required under this Working Rule is on the operative concerned and the onus of checking the proof submitted is on the employer.

26.4 Scaffolders covered by Rule 26.1 are entitled to the "Skilled Operative Basic Rate of Pay" 4,3,2,1 or Craft Rate as follows:

26.4 (Cont'd)	Rate
Trainee Scaffolder	4.
Basic Scaffolder with less than $2^1/_2$ years whole time experience in scaffolding and less than $1^1/_2$ years whole time experience as a Basic Scaffolder	2.
Basic Scaffolder with at least $2^1/_2$ years whole time experience in scaffolding and at least $1^1/_2$ years whole time experience as a Basic Scaffolder and Advanced Scaffolder	Craft Rate

26.5 Approved list of scaffolding operations,

Basic:
Erecting, adapting and dismantling:
Independent, putlog and birdcage scaffolds and static and mobile towers.
Beams to form gantries and openings, correctly braced.
Hoist frameworks and guides
Protective fans.
Stack scaffolds.
Roof scaffolds
Proprietary systems.
Fixing wire netting to hoist framework or scaffold framework.
Fixing sheeting to scaffold framework.
Interpreting simple design layout drawing for scaffolding detailed above.
Applying knowledge of Construction Regulations to operations listed above.

Advanced:
All work in list of Basic Operations:
Erecting, adapting and dismantling:
Scaffolding to form truss-out scaffolds.
Slung scaffolds including use of lifting equipment, wire ropes, chains and shackling.
Raking and flying shores.
Other forms of designed structures, e.g. larger truss-outs, cantilevers, lifting structures, ramps, footbridges and temporary roofs.
Scaffolding or standard props (including all bracing) to form a dead shore, including adjustable bases and forkheads.
Scaffolding and proprietary systems (including levelling to within reasonable tolerances) to support formwork as laid out in engineering scaffold drawings.
Interpreting scaffold design drawings.
Applying knowledge of Construction Regulations to operations listed above.

WR.27 HEALTH SAFETY AND WELFARE

27.1 The nature of work in the building and civil engineering industry presents particular hazards to those engaged in it. The employers and operatives Organisation who are signatories to the Working Rule Agreement shall take all reasonable steps to ensure that employers and employees, whether operatives or management, comply with the requirement of legislation dealing with health, safety and welfare.

Safety Representatives

27.2 Legislation provides that recognised trade unions may appoint safety representatives to represent operatives. Provision is also made for the establishment of safety committees where a formal request, in writing, is made to an employer by at least two safety representatives who have been appointed in accordance with legislation.

WR. 28 REFERENCE PERIODS AND DEFINITIONS

For the purpose of compliance with the Working Rule Agreement, statutory definitions, entitlements and calculations the reference period shall, subject to the requirements of WR 3.2 - "AVERAGE WEEKLY WORKING HOURS", be twelve months.

WR.29 SUPPLEMENTARY AGREEMENTS

When it is agreed between the employer(s) and operative(s) that at any particular workplace it would be appropriate to enter into an agreement specifically for that workplace, any such agreement shall be supplementary to and not in conflict with this Working Rule Agreement. Where any dispute arises in this respect the National Working Rule Agreement takes precedence.

WR.30 DURATION OF AGREEMENT

This Agreement shall continue in force and the parties to it agree to honour its terms until the expiration of three calendar month's notice to withdraw from it, given by either the Employers' side or the Operatives' side.

SCHEDULE 1

Specified Work Establishing Entitlement to the Skilled Operative Pay Rate 4,3,2 1. or Craft Rate

Rates of Pay
Basic Additional

QUALIFIED BAR BENDERS AND REINFORCEMENT FIXERS
Bender and fixer of Concrete Reinforcement capable of reading and understanding drawings and bending schedules and able to set out work — Craft Rate

CONCRETE
Concrete Leveller or Vibrator Operator} Screeder and Concrete Surface Finisher man working off forms or other datum (e.g. road-form or concrete haunch, or edge beam or wire) man required by his employer to use trowel or hand or powered float to produce high quality finished concrete} — 4.

DRILLING AND BLASTING
Drills, rotary or percussive: mobile rigs, operator of — 3.
Operative attending drill rig — 4.
Shotfirer, operative in control of and responsible for explosives, including placing, connecting and detonating charges — 3.
Operatives attending on shotfirer including stemming — 4.

FORMWORK CARPENTERS
1st year trainee — 4.
2nd year trainee — 3.
Formwork Carpenters — Craft Rate

GANGERS AND TRADE CHARGEHANDS
(Higher Grade payments may be made at the employer's discretion) — 4.

GAS DISTRIBUTION
Operatives who have successfully completed approved training to the standard of:
Gas Distribution Operative
Category 1 and 2 — 4.
Category 3 — 3.
Team Leader - Service layer — 2.
Team Leader-Main Layer — 1.
Team Leader - Main and Service Layer — 1.

LINESMEN - ERECTORS
1st Grade — 2.
(Skilled in all works associated with the erection of O.H. Transmission Lines including Assembly and Erection of Steel Towers; Concrete and Wood Poles of all types; Insulators and O.H. Line Switchgear; Stringing, tensioning, sagging, jointing, clamping and making off all types of O.H. conductors; making off stay wires; also similar structures and otherworks appertaining thereto.)
2nd Grade — 3.

(As above but lesser degree of skill - or competent and fully skilled either in assembly and erection of all types of supports or assembly and erection of the insulators, conductors, stay wires - all as specified above.)

LINESMEN - ERECTORS' MATE — 4.
(Semi-skilled in works specified above and a general helper) — 4.

SKILLED MECHANICS
Maintenance Mechanic on site, i.e, man capable of carrying out field service, maintenance, and minor repairs ancillary to civil engineering works — 2.

Contractors' Plant Mechanic in shop, depot or on site, i.e, man capable of carrying out major repairs and overhauls including welding work, operating turning lathe or similar machine and/or carrying out highly skilled work ancillary to civil engineering works — 1.

Contractors' Plant Mechanics' Mate on site or in depot — 4.
Greaser on site, servicing machines — 4.
Tyre Fitter, heavy earth mover tyres — 2.

MECHANICAL PLANT
Compressors and Generators
Air compressors diesel or petrol or generators over 10 KW operator or Driver of — 4.

Concrete Placing Equipment
Portable or static concrete pumps; pneumatic concrete placers; concrete placing booms; operator of — 4.

Self-propelled Mobile Concrete Pump, with or without boom, mounted on lorry or lorry chassis; driver/operator of — 3.

Mobile Cranes, Hoists or Fork-Lift Trucks
Self-propelled Mobile Crane on road wheels or caterpillar tracks including lorry mounted:

Capacity up to and including 5 tons (or metric tonnes); driver of — 4.

Capacity up to and including 10 tons (or metric tonnes); driver of — 3.

Capacity over 10 tons (or metric tonnes) and up to and including 120 tons (or metric tonnes); driver of — Craft Rate

Capacity over 120 tons (or metric tonnes); driver of Craft Rate

Where grabs are attached to Cranes the next higher skilled basic rate of pay applies except over 120 tons (or metric tonnes) where the rate is at the employer's discretion.

Tower Cranes (including travelling or climbing)
Up to and including 2 tons (or metric tonnes) maximum lifting capacity at minimum radius; driver of — 4.

Over 2 tons (or metric tonnes) up to and including 10 tons (or metric tonnes) lifting capacity at minimum radius; driver of — 3.

Over 10 tons (or metric tonnes) up to and including 20 tons (or metric tonnes) lifting capacity at minimum radius; driver of — 2

Over 20 tons (or metric tonnes) lifting capacity at minimum radius; driver of — 1.

Miscellaneous Cranes, etc.

Overhead traveller or gantry up to and including 10 tons (or metric tonnes) capacity; driver of — 3.

Loco traveller up to and including 10 tons (or metric tonnes) capacity; driver of — 3.

Power driven Hoist/crane; driver of — 4.

Where grabs are attached to Cranes — 3.

Derricks (Stationary or Travelling)
Power-driven derrick: capacity up to and including 20 tons (or metric tonnes); driver of — 3.

Capacity over 20 tons (or metric tonnes); driver of — 2.

Where grabs are attached to Derricks the next higher skilled Operative Basic Rate of pay applies.

Banksmen appointed to attend Crane, hoist or Derrick and be responsible for fastening or slinging loads and generally direct Crane hoist or Derrick Driver — 4.

Smooth or Rough Terrain fork lift trucks or side loader up to and including 3 tons (or metric tonnes) capacity driver of. — 4.

Over 3 tons (or metric tonnes) capacity driver of — 3.

Dumpers and Dump Trucks
Up to and including 7 tons (or metric tonnes) carrying capacity; driver of — 4.

Over 7 tons (or metric tonnes) and up to and including 16 tons (or metric tonnes) carrying capacity; driver of — 3.

Over 16 tons (or metric tonnes) and up to and including 60 tons (or metric tonnes) carrying capacity; driver of — 2.

Over 60 tons (or metric tonnes) and up to and including 125 tons (or metric tonnes) carrying capacity; driver of — 1.

Over 125 tons (or metric tonnes) carrying capacity; driver of — Craft Rate

Excavators (360 degree slewing, rope or hydraulic)
Excavators with rated bucket capacity up to and including cu. yd. (0.6 cu. metres); driver of — 3.

Excavator with rated bucket capacity over - cu. yd. (0.6 cu. metres) and up to and including S cu. Yds (0.7 (3.85 cu. metres); driver of — 2.

Excavator with rated bucket capacity over 5 cu. yds. (3.85 cu. metres) and up to and including 10 cu. yds. (7.65 cu. metres);driver of — 1.

Excavator with rated bucket capacity over 10 cu. yds. (765 cu. metres); and up to 20 cu. yds (15.28 cu metres) driver of — Craft Rate (iii)

Excavator with rated bucket capacity over 20 cu. Yds (15.28 cu. metres) and up to and including 40 cu. Yds (30.57 cu. metres) driver of — Craft Rate (ii)

Excavator with rated bucket capacity over 40 cu. Yds (30.57cu. metres) driver of — Craft Rate (i)

Operative attending excavator or responsible for positioning vehicles during loading of tipping — 4.

Shovel Loaders, Wheeled or Tracked
See Tractors and Equipments.

Locos
Loco all; driver of — 4.

Mixers
Mortar Pan or Concrete Mixer up to but not including 21/4 (or 400 litres) wet capacity, to apply to one man only per machine; man employed on, and actually responsible for, operating. — 4.

Mixer Concrete 21/14 and up to and including 2 cu. yds. (or 400 litres and up to but not including 1,500 litres) wet capacity to apply to one man only per machine; man employed on, and actually responsible for, operating — 3.

Mixer Concrete of Over 2 cu, yds. (or 1,500 litres) wet capacity, to apply to one man only per machine; man employed on, and actually responsible for, operating — 3

Mixer Concrete, mobile self-loading and batching up to 2,500 litres wet capacity, to apply to one man only per machine; man employed on, and actually responsible for, operating — 2.

Drag Shovel: operative of — 4.

Motor Graders
Motor Grader; driver of — 2.

Motor Vehicles (Road Licensed Vehicles within the Construction and Use Regulations)
Vehicle up to and including 10 tons (or metric tonnes) Gross Vehicle Weight: driver of — 4.

Vehicle over 10 tons (or metric tonnes); driver of — 3.

Operations requiring a LGV Licence of either or both Classes C & E — 1.

Lorry drivers employed whole time as such required to hold Class C & E licence — 1.

Motorised Scrapers
Motorised Scraper; driver of — 2.

Power Driven Tools
Operatives using power-driven tools such as breakers, percussive drills, picks and spades, rammers and tamping machines. — 4.

Pumps
Power-driven pump(s); attendant of — 4.

Power Rollers
Roller up to but not including 4 tons (or metric tonnes);driver of — 4.

Roller, 4 tons (or metric tonnes) and upwards; driver of — 3.

Tractors (Agricultural)

Tractor, rubber-tyred agricultural type, when used to tow trailer and/or with mounted compressor; driver of. — 4.

Tractors (Wheeled or Tracked - with or without equipment such as buckets, blades, loaders, backhoes)

Tractor, up to and including 100 h.p.; driver of — 3.

Tractor, over 100 h.p. up to and including 400 h.p.; driver of — 2.

Tractor, over 400 h.p. up to and including 650 h.p. driver of — 1.

Tractor, over 650 h.p.; driver of — Craft Rate

Trenching Machine (Multi Bucket)

Trenching Machine, up to and including 30 h.p, driver of — 4.

Trenching Machine, over 30 h.p. and up to and including 70 h.p.; driver of   3.

Trenching Machine, over 70 h.p., driver of   2.

Winches
Power driven winch, driver of   4.

PAVIORS, MASONS, ETC.
Paviors Rammerman   4.
Kerb & Paving Jointer   4.
Operative engaged in face pitching or dry walling   3.

PILING
General Skilled Piling Operative   4.
Piling Ganger/Chargehand   3.
Pile Frame Winch Driver   3.
Rotary or Specialist Mobile Piling Rig Driver   2.

PIPE JOINTERS
Jointers, Stoneware or Concrete Pipes   4.

Jointers working on flexible joints or setting up or caulking lead joints under 12 in. (or 300 mm.) dia   4.

Jointers working on flexible joints or setting up or caulking lead joints of 12in. (or 300 mm.) dia and over   4.

Jointers, Cast Iron or Steel Pipes using lead   4.

EXCEPT
(a)   in mains of 12 in. dia. up to and including 21in. dia. (or 300 mm. dia, up to and including 535 mm. dia.) when experienced in all or any of the following operations: 3.

     i) Jointing with run lead
     ii) Jointing with cold strip lead, wool or similar type

(b)   on mains of over 21 in. (or 535 mm.) dia. when so experienced   2.

PIPELAYERS
Men preparing the bed and laying pipes under 12 in (or 300 mm.) dia   4.

PLATE LAYER not labourer in gang   3.

POST TENSIONING AND PRE-STRESSING CONCRETE
Operative in control of and responsible for hydraulic jack and other tensioning devices engaged in post-tensioning and/or pre-stressing whilst so employed (not labourer in gang)   3.

SCAFFOLDERS
See WR. 26

STEELWORK CONSTRUCTION
Men fully skilled in and engaged in the assembly erection and fixing of steel-framed construction of a permanent nature ancillary to civil engineering works   1.

Operative capable of and engaged in fixing simple steelwork and structures ancillary to civil engineering works   3.

TAR SPRAYING, MANUAL OR MECHANICAL
Chargehand   4.
Spraybar Operators   4.
Gritter Operators   4.

TIMBERMAN
Timberman   3.

Except in cases where the timbering required calls for a special degree of skill, as, for example, in running sand and/or cofferdams, or where walings, struts and/or rakers are of 10 in. by 5 in, (or 250 mm, by 125 mm.) nominal or large timbers, in which cases the rates shall be   2.

Operative attending   4.

TUNNELS
Operative working below ground on the construction or reconstruction of Tunnels or sinking shafts for Tunnels in hard rock or soft ground:

Face Tunnelling machine; operator of   2.

Tunnel Miner (the skilled operative working at the face, or machineman working a drifter type of machine).   3.

Tunnel Miner's mates (the operative who assists the miner at the face or break-up, including such men who work pneumatic tools breakers, and/or in a soft tunnel, prepare the timber).   4.

(The above tunnel rates are exclusive of plus rate payments prescribed for work in active foul or surface water sewers but inclusive of any other plus rate payment for conditions which might otherwise apply)

Other operatives engaged in driving headings over 2 metres in length from the entrance, in connection with drain cable and main laying   4.

WELDERS
Gas or electric arc welder capable of welding mild steel and building up to normal welds, ancillary to civil engineering work, when so required   2.

Electric arc welder able and required to weld to highest standards for structural fabrication and simple pressure vessels (air receivers), including CO_2 processes, ancillary to civil engineering works   1.

Electric arc welder able and required to undertake all welding processes on all weldable materials, ancillary to civil engineering works, including working on his own initiative from drawings   Craft Rate

Opencast Coal Washeries and Screening Plants (where work is ancillary to civil engineering works).
Screening Plant Machine Operators, Operatives employed and responsible for controlling:

Mechanical equipment in a dry coal plant, or a washing plant machine   4

ROLLED ASPHALT ROAD SURFACING WORK, TAR AND/OR BITUMEN MACADAM SURFACING WORK

(1) The Working Rule Agreement of the Civil Engineering Construction Conciliation Board so far as if is applicable will apply to operatives engaged in this class of work, subject as hereinafter provided.
(2) Operatives employed on this class of work to be paid as follows:

Raker   3.

Tamperman   4.

Chipper   4.

Mixing Platform Chargehand   4.

Power Roller, 4 tons (or metric tonnes) and upwards driver of   3.

Operator of Mechanical Spreader of the Barber-Oreene or similar type   3.

GENERAL INFORMATION

Leveller on Mechanical Spreader of the Barber
Oreene or similar type   3.

(3) The hours of work as set in this Working Rule Agreement shall apply to this class of work, subject to the proviso that due to the exceptional circumstances of the work and the desirability of keeping work going continuously, it is necessary to limit intervals for meals to a period of half-hour each.

*(4) In the event of a breakdown of plant (including haulage plant) or of conditions arising which make it necessary or desirable in the interests of the job to cease work on the job for any period exceeding one hour, the following arrangements shall apply:

(a) If the operatives who have to cease work are given alternative work, hours worked either on that work, or on Rolled Asphalt Road SurfacingWork after the end of the normal working day they shall be paid at the overtime rates in accordance with WR.4 Overtime Rates.

(b) If the operatives who have to cease work are not given alternative work but are notwithstanding paid for the period of such stoppage, work may be continued beyond what would otherwise be the normal working day for a period equal to the total of the periods of such stoppages, but not exceeding three hours at ordinary hourly rates, and after a period of three hours, at the overtime rate prescribed by WR.4 Overtime Rates as it the day had ended at the time at which ordinary hourly rates cease to be payable.

* NOTE Para 4 does not apply to Tar Macadam Surfacing Work,

Operatives below 18 years of age will receive payment of a percentage of the General Operative Basic Rate:
At 16 years of age 50% of the relevant rate
At 17 years of age 75% of the relevant rate
At 18 years of age or over 100% of the relevant rate.

Dry-liners
Operatives undergoing approved training in dry-lining operations   4.

Operatives who can produce a certificate of training achievement indicating satisfactory completion of approved dry-lining training module (iii) (iv) or (v)   3.

SCHEDULE 2

Entitlement to additional rates of pay for intermittent skill, responsibility or working in adverse conditions.

	Extra Rate

1.4.1 STONE CLEANING

Operatives other than craft operatives employed on dry-cleaning stone-work by mechanical process for the removal of protective material and/or discoloration   E

1.4.2 TUNNELS

Operatives (other than Tunnel Machine Operators, Tunnel Miners and Tunnel Miners' Mates) wholly or mainly engaged in work of actual construction including the removal and dumping of mined materials but excluding operatives whose employment in the tunnel is occasional and temporary:

In unlined tunnels   A

(The above tunnels are exclusive of extra payments prescribed for work in active foul or surface water sewers, but inclusive of any other extra payment for conditions which might otherwise apply).

1.4.3 SEWER WORK

Operatives working within a totally enclosed active foul sewer of any nature or condition   D

Operatives working within a totally enclosed active surface water sewer of any nature or condition .   B

Operatives working outside existing sewers excavating or removing foul materials emanating from existing sewers.   C

1.4.4 WORKING AT HEIGHT

Operatives (including drivers of tower cranes, but excluding the drivers of power driven derricks on high stages and linesmen-erectors and their mates and scaffolders) employed on "detached work" shall receive in respect of conditions extra payments on the following scale calculated from the "point of departure".

Above 45m and up to 90m   C

Thereafter a further extra payment at the same rate for each additional 45 metres

For the purpose of this rule "detached work" shall be deemed to comprise work on detached "tower like structures" whether completely detached or rising above and being connected to a main structure. In the latter case the "point of departure" shall be the level at which the detached structure leaves the top of the main structure.

CONSTRUCTION INDUSTRY JOINT COUNCIL

Joint Secretaries Guidance Notes on the Working Rule Agreement of the Construction Industry Joint Council.

Introduction

These Notes of Guidance, whilst not forming part of the Working Rules, are intended to assist employers and operatives to understand and implement the Working Rule Agreement.

It is the intent of all parties to this Agreement that operatives employed in the building and civil engineering industry are engaged under the terms and conditions of the CIJC Working Rule Agreement.

Requests for definitions, clarification or resolution of disputes in relation to this Agreement should be addressed to the Construction Industry Joint Council at the address set out on the front cover of the Working Rule Agreement.

WR. 1 Entitlement to Basic Pay and Additional Rates of Pay.

WR. 1 sets out the entitlement to the basic rate of pay, additional payments for skilled work and intermittent skill, responsibility or for working in adverse conditions.

There are six basic rates of pay under this Agreement; General Operative rate, four rates for Skilled Operatives and a rate for a Craft Operative.

Payment for Occasional Skilled Work

WR. 1.1.2 deals with the payment for occasional skilled work and provides that general operatives who are required to carry out work defined in Schedule 1 on an occasional basis should receive an increased rate of pay commensurate with the work they are carrying out for the period such work is undertaken. This sets out the flexibility to enable enhanced payment to be made to general operatives undertaking skilled work for a limited amount of time but should not be used where the operative is engaged whole time on skilled work.

Skilled Operatives

WR.1.2.1 sets out a permanent rate of pay for skilled operatives who are engaged whole time on the skilled activity and does not permit the operative engaged on whole time skilled work to have his pay reduced to the General Operative basic rate when occasional alternative work is undertaken.

Additional Payment for Skilled Work.
WR.1.2.2 provides additional payment for skilled work for those operatives engaged in the work defined in Schedule 1. These payments are permanent payments for skilled operatives carrying out the specified work.

WR. 1.4. Additional Payments for Intermittent Skill, Responsibility or Working in Adverse Conditions.

WR 1.4.1 Provides additional payments for intermittent skill, responsibility or working in adverse conditions. These payments are made for the duration of the requirement for the skilled work or responsibility or to the extent the work is undertaken in adverse conditions and are not permanent payments.

N.B. For definition of normal hourly rate see the notes contained in the Working Rule Agreement.

WR.3.2 Average Weekly Working Hours.

Working Rules 3.2.1 to 3.2.4 provide, where there are objective or technical reasons, for the calculation of average weekly working hours by reference to a period of twelve months.

Whilst it is open to employers and employees to agree to work additional hours over the "normal working hours," Rule 3.2 does not give the employer the unilateral right to introduce excessive hours on a job or site. The 12 month averaging period referred to may only be applied where, for objective reasons, it is necessary to ensure completion of the work efficiently and effectively or where there are technical reasons that require additional hours to be worked.

Examples of objective and/or technical reasons which may require average weekly working hours to be calculated using a twelve month reference period are set out below. The list is not exhaustive other objective or technical reasons may apply.

Objective Reasons:

Work on infrastructure, roads, bridges, tunnels and tide work etc.

Client requirements for work to be completed within a tightly defined period, work undertaken for exhibitions, schools, retail outlets, shopfitting and banks etc.

Emergency work, glazing and structural safety etc.

Technical Reasons;

Work requiring a continuous concrete pour surfacing and coating work, tunnelling etc.

N.B. Any disputes regarding the validity of objective and/or technical reasons may be referred to the National Conciliation Panel of the Construction Industry Joint Council.

W.R. 4 Overtime Rates.

Where an operative who has worked overtime tails without authorisation to be available for work during normal weekly working hours he may suffer a reduction in or may not be entitled to premium payments in respect of overtime worked.

To calculate the number of hours paid at premium rate (overtime) you subtract the number of hours of unauthorised absence from the total number of hours overtime worked. This is in effect using a part of the hours of overtime worked to make the hours paid at the normal hourly rate up to 39 hours per week, the balance of overtime is paid at the appropriate premium rate.

EXAMPLES

Example 1.

An operative works three hours overtime on Monday. Tuesday and Wednesday works normal hours on, Thursday and is unavailable for work on Friday due to unauthorised absence. In these circumstances overtime premia will be calculated as follows:

	Normal Hours	Overtime Hours
Monday	8	3
Tuesday	8	3
Wednesday	8	3
Thursday	8	-
Friday	-	-
Total hours worked	(A) 32	(B) 9

Normal weekly working hours	(C)	39
Less total normal hours worked	(B)	32
Hours required to make up to 39	(D)	7
Total overtime hours	(B)	9
Less	(D)	7
Hours to be paid at premium rate	(E)	2

The operative is therefore, entitled to be paid:

(A+D)	39	Hours at "Normal Hourly Rate"
(E)	2	Hours at premium rate (time and a half)
Total	41	Hours

Example 2.

An operative works four hours overtime on a Monday, three hours on Tuesday five hours (one hour double time) on the Wednesday, no overtime on either Thursday or Friday and absents himself from work on Friday at 12 noon without authorisation and then works six hours on Saturday. This calculation would be as follows:

	Normal hours	Overtime Hours
Monday	8	4
Tuesday	8	3
Wednesday	8	5 (1hour double time)
Thursday	8	-
Friday	3	-
Saturday	-	6 (2hours double time)
Total Hours worked	(A) 35$\frac{1}{2}$	(B) 18

Normal weekly working hours	(C)	39
Less total normal hours worked	(B)	35$\frac{1}{2}$
Hours required to make up to 39	(D)	3$\frac{1}{2}$
Total overtime hours	(B)	18
Less	(D)	3$\frac{1}{2}$
Hours to be paid at premium rate	(E)	14$\frac{1}{2}$

The operative is, therefore, entitled to be paid:

(A+D)	39	Hours at "Normal Hourly Rate"
(E)	14$\frac{1}{2}$	Hours at premium rate (time and a half)
Total	53$\frac{1}{2}$	Hours

The entitlement to 3 hours pay at double time is extinguished in this example by the hours of unauthorised absence. If the number of hours worked at double time exceeds the number of hours of unauthorised absence the balance must be paid at the rate of double time.

NB. There shall be no reduction in overtime premium payments for operatives who are absent from work with the permission of the employer or who are absent due to sickness or injury.

W.R. 17.4 Temporary Lay-off.

The temporary lay-off provisions may only be used when the employer has a reasonable expectation of being able to provide work within a reasonable time.

In this context an example of an employer who has a reasonable expectation to be able to provide work may be where a tender has been accepted but commencement delayed, where work is temporarily stopped due to weather conditions or for some other reason outside the employer's control. Reasonable time is not legally defined, however, an employee who has been temporarily laid off for four or more consecutive weeks or six weeks cumulative in any thirteen week period may claim a redundancy payment.

In no circumstances may the temporary lay-off rule be used where a genuine redundancy situation exists or to evade statutory obligations.

An employee who is temporarily laid off is entitled to payment of one fifth of his guaranteed minimum earnings for the day of notification of lay-off and for each of the first five days of the layoff subject to the limitations in WR 17.4.2. limitations in WR 17.4.2.

WR 22. Grievance Procedure.

The Grievance Procedure allows companies and operatives to resolve problems at a local level and in the event of a failure to agree to refer the dispute to a CIJC National Conciliation Panel.

The Working Rule Agreement sets out a five stage procedure for dealing with grievances; Stage 4 permits employer representatives and full time trade union officials to hold a local meeting to discuss resolution of the grievance or dispute. In the rare event that such a meeting does not resolve the problem a reference may be made to the National Conciliation Panel. Both operatives and employers have the right of appeal to the Construction Industry Joint Council from the decisions made by the National Conciliation Panel.

N.B.   **For full details of the conciliation dispute machinery please see the Constitution and Rules of the CIJC. Any question relating to Working Rules not covered by the notes of guidance should be referred to the Construction Industry Joint Council at the address on the front cover of this booklet.**

Daywork Charges

CONTRACT WORK

DEFINITION OF PRIME COST OF DAYWORK CARRIED OUT UNDER A BUILDING CONTRACT

Reproduced by permission of the Royal Institution of Chartered Surveyors.

This Definition of Prime Cost is published by The Royal Institution of Chartered Surveyors and the National Federation of Building Trades Employers, for convenience and for use by people who choose to use it. Members of the National Federation of Building Trades Employers are not in any way debarred from defining Prime Cost and rendering their accounts for work carried out on that basis in any way they choose. Building owners are advised to reach agreement with contractors on the Definition of Prime Cost to be used prior to issuing instructions.

SECTION 1

Application

1.1 This definition provides a basis for the valuation of daywork executed under such building contracts as provide for its use (e.g. contracts embodying the Standard Forms issued by the Joint Contracts Tribunal).

1.2 It is not applicable in any other circumstances, such as jobbing or other work carried out as a separate or main contract nor in the case of daywork executed during the Defects Liability Period of contracts embodying the above mentioned Standard Forms.

SECTION 2

Composition of total charges

2.1 The prime cost of daywork comprises the sum of the following costs:

(a) Labour as defined in Section 3.

(b) Materials and goods as defined in Section 4.

(c) Plant as defined in Section 5.

2.2 Incidental costs, overheads and profit as defined in Section 6, as provided in the building contract and expressed therein as percentage adjustments, are applicable to each of 2.1 (a)-(c).

SECTION 3

Labour

3.1 The standard wage rates, emoluments and expenses referred to below and the standard working hours referred to in 3.2 are those laid down for the time being in the rules or decisions of the National Joint Council for the Building Industry and the terms of the Building and Civil Engineering Annual and Public Holiday Agreements applicable to the works, or the rules or decisions or agreements of such body, other than the National Joint Council for the Building Industry, as may be applicable relating to the class of labour concerned at the time when and in the area where the daywork is executed.

3.2 Hourly base rates for labour are computed by dividing the annual prime cost of labour, based upon standard working hours and as defined in 3.4 (a)-(i), by the number of standard working hours per annum (see examples)

3.3 The hourly rates computed in accordance with 3.2 shall be applied in respect of the time spent by operatives directly engaged on daywork, including those operating mechanical plant and transport and erecting and dismantling other plant (unless otherwise expressly provided in the building contract).

3.4 The annual prime cost of labour comprises the following:

(a) Guaranteed minimum weekly earnings (e.g. Standard Basic Rate of Wages, Joint Board Supplement and Guaranteed Minimum Bonus Payment in the case of NJCBI rules).

(b) All other guaranteed minimum payments (unless included in Section 6).

(c) Differentials or extra payments in respect of skill, responsibility, discomfort, inconvenience or risk (excluding those in respect of supervisory responsibility - see 3.5).

(d) Payments in respect of public holidays.

(e) Any amounts which may become payable by the Contractor to or in respect of operatives arising from the operation of the rules referred to in 3.1 which are not provided for in 3.4 (a)-(d) or in Section 6.

(f) Employer's National Insurance contributions applicable to 3.4 (a)-(e).

(g) Employer's contributions to annual holiday credits.

(h) Employer's contributions to death benefit scheme.

(i) Any contribution, levy or tax imposed by statute, payable by the Contractor in his capacity as an employer.

3.5 Note

Differentials or extra payments in respect of supervisory responsibility are excluded from the annual prime cost (see Section 6). The time of principals, foremen, gangers, leading hands and similar categories, when working manually, is admissible under this Section at the appropriate rates for the trades concerned.

SECTION 4

Materials and Goods

4.1 The prime cost of materials and goods obtained from stockists or manufacturers is the invoice cost after deduction of all trade discounts but including cash discounts not exceeding 5 per cent and includes the cost of delivery to site.

CONTRACT WORK (Cont'd)
SECTION 4 (cont'd).
Materials and Goods (cont'd).

4.2 The prime cost of materials and goods supplied from the Contractor's stock is based upon the current market prices plus any appropriate handling charges.

4.3 Any Value Added Tax which is treated, or is capable of being treated, as input tax (as defined in the Finance Act, 1972) by the Contractor is excluded.

SECTION 5

Plant

5.1 The rates for plant shall be as provided in the building contract.

5.2 The costs included in this Section comprise the following:

(a) Use of mechanical plant and transport for the time employed on daywork.

(b) Use of non-mechanical plant (excluding non-mechanical

hand tools) for the time employed on daywork.

5.3 Note: The use of non-mechanical hand tools and of erected scaffolding, staging, trestles or the like is excluded (see Section 6).

SECTION 6

Incidental Costs, Overheads and Profit

6.1 The percentage adjustments provided in the building contract, which are applicable to each of the totals of Sections 3, 4 and 5, comprise the following:

(a) Head Office charges

(b) Site staff including site supervision.

(c) The additional cost of overtime (other than that referred to in 6.2).

(d) Time lost due to inclement weather.

e) The additional cost of bonuses and all other incentive payments in excess of any guaranteed minimum included in 3.4 (a).

(f) Apprentices study time.

(g) Subsistence and periodic allowances.

(h) `Fares and travelling allowances.

(i) Sick pay or insurance in respect thereof.

(j) Third party and employer's liability insurance.

(k) Liability in respect of redundancy payments to employees.

(l) Employer's National Insurance contributions not included in Section 3.4

(m) Tool allowances.

(n) Use, repair and sharpening of non-mechanical hand tools.

(o) Use of erected scaffolding, staging, trestles or the like.

(p) Use of tarpaulins, protective clothing, artificial lighting, safety and welfare facilities, storage and the like that may be available on the site.

(q) Any variation to basic rates required by the Contractor in cases where the building contract provides for the use of a specified schedule of basic plant charges (to the extent that no other provision is made for such variation).

(r) All other liabilities and obligations whatsoever not specifically referred to in this Section nor chargeable under any other section.

(s) Profit.

6.2 Note: The additional cost of overtime, where specifically ordered by the Architect/Supervising Officer, shall only be chargeable in the terms of prior written agreement between the parties to the building contract

Example of calculation of typical standard hourly base rate (as defined in Section 3) for NJCBI building craft operative and labour in Grade A areas **based upon rates ruling at 1st July 1975.**

		Rate £	Craft Operative £	Rate £	Labourer £
Guaranteed minimum weekly earnings					
Standard Basic Rate	49 wks	37.00	1813.00	31.40	1538.60
Joint Board Supplement	49 wks	5.00	245.00	4.20	205.80
Guaranteed Minimum Bonus	49 wks	4.00	196.00	3.60	176.40
			2254.00		1920.80
Employer's National Insurance Contribution at 8.5%			191.59		63.27
			2445.59		2084.07
Employer's Contributions to:					
CITB annual levy			15.00		3.00
Annual holiday credits	49 wks	2.80	137.20	2.80	137.20
Public holidays (included in guaranteed minimum weekly earnings above)			-		-
Death benefit scheme	49 wks	0.10	4.90	0.10	4.90
Annual labour cost as defined in Section 3		£	2602.69	£	2229.17

Hourly base rates as defined
in Section 3, clause 3.2

Craft operative
£ $\dfrac{2602.69}{1904}$ = £ 1.37

Labourer
£$\dfrac{2229.17}{1904}$ = £ 1.17

Note: (1) Standard working hours per annum calculated as follows:

	52 weeks at 40 hours	=	2080
Less	3 weeks holiday at 40hours	= 120	
	7 days public holidays at 8 hours	= 56	176
			1904

CONTRACT WORK (Cont'd).
Example of Calculation (Cont'd)

Note:
(2)   It should be noted that all labour costs incurred by the Contractor in his capacity as an employer, other than those contained in the hourly base rate, are to be taken into account under Section 6.

(3)   The above example is for the convenience of users only and does not form part of the Definition; all the basic costs are subject to re-examination according to the time when and in the area where the daywork is executed.

BUILD UP OF STANDARD HOURLY BASE RATES - RATES APPLICABLE AT JUNE 29th, 1998

Under the JCT Standard Form, dayworks are calculated in accordance with the Definition of the Prime Cost of Dayworks carried out under a Building Contract published by the RICS and the NFBTE. The following build-up has been calculated by the Building Cost Information Service (BCIS) in liaison with the NFBTE. The example is for the calculation of the standard hourly base rates for craftsmen and labourers in Grade A and is for convenience only and does not form part of the Definition; all basic rates are subject to re-examination according to when and where the dayworks are executed.

Standard working hours per annum
52 weeks at 39 hours   2028   hours

Less -   21 days annual holiday
           16 days at 8 hours   128
           5 days at 7 hours   35
            8 days public holiday
           7 days at 8 hours   56
           1 day at 7 hours   7   226   hours

                                                                1802   hours

	Craftsman £	Labourer £
Guaranteed Minimum weekly earnings		
Standard basic rate	214.50	164.97
Guaranteed Minimum Earnings	-	-
Joint Board Supplement	-	-
	£ 214.50	£ 164.97

Hourly Base Rate

	Rate £	Craftsman £	Rate £	Labourer £
Guaranteed minimum weekly earnings 47.8 weeks at	214.50	10253.10	164.97	7885.57
Extra payments for skill, responsibility, discomfort, inconvenience or risk* - 1802 hours	-	-	-	-
Employer's National Insurance Contribution**	10%	1025.31	7%	551.99
*** CITB Annual Levy 0.29%		29.73		22.87
Annual holiday credits and death benefit schemes 47 weeks at 20.10		944.70		944.70
Public holidays (included in guaranteed minimum weekly earnings above).		-		-
		£12252.84		£9405.13

Hourly Base Rate

Craftsman
$$\frac{£12252.84}{1802} = £6.80$$

Labourer
$$\frac{£9438.03}{1802} = £5.22$$

*   Only included in hourly base rate for operatives receiving such payments.
**   National Insurance Contribution based on weekly earnings bracket of employee.
        21 days annual holiday taken in conjunction with public holidays as 2 calendar weeks holidays at Christmas, 1 calendar week at Easter and 2 weeks summer holiday.
***   From 13th March 1998, the CITB levy is calculated at 0.29% of the PAYE payroll of each building contractor having a payroll of £45,000 or more. The levy is included in the daywork rate of each working operative, the remainder being a constituent part of the overheads percentage.
****   From 29th June 1992, the Guaranteed Minimum Bonus ceased and is consolidated in the Guaranteed Minimum Earnings.

Note The holiday/benefits stamp value is discounted by a contribution of 70p by the Management Company.

DAYWORK, OVERHEADS AND PROFIT

Fixed percentage additions to cover overheads and profit no longer form part of the published Daywork schedules issued by the Royal Institution of Chartered Surveyors and Contractors are usually asked to state the percentages they require as part of their tender. These will vary from firm to firm and before adding a percentage, the list of items included in Section 6 of the above Schedule should be studied and the cost of each item assessed and the percentage overall additions worked out.

As a guide, the following percentages are extracted from recent tenders for Contract Work:
On Labour costs 80-200%
On Material costs 10-20%
On Plant costs 15-30%

Much higher percentages may occur on small projects or where the amount of daywork envisaged is low.

GENERAL INFORMATION

JOBBING WORK

DEFINITION OF PRIME COST OF BUILDING WORKS

OF A JOBBING OR MAINTENANCE CHARACTER

Reproduced by permission of the Royal Institution of Chartered Surveyors.

This Definition of Prime Cost is published by the Royal Institution of Chartered Surveyors and the National Federation of Building Trades Employers for convenience and for use by people who choose to use it. Members of the National Federation of Building Trades Employers are not in any way debarred from defining prime cost and rendering their accounts for work carried out on that basis in any way they choose.

Building owners are advised to reach agreement with contractors on the Definition of Prime Cost to be used prior to issuing instructions.

SECTION 1

Application

1.1   This definition provides a basis for the valuation of work of a jobbing or maintenance character executed under such building contracts as provide for its use.

1.2   It is not applicable in any other circumstances such as daywork executed under or incidental to a building contract.

SECTION 2

Composition of Total Charges

2.1   The prime cost of jobbing work comprises the sum of the following costs:

(a)   Labour as defined in Section 3.

(b)   Materials and goods as defined in Section 4.

(c)   Plant, consumable stores and services as defined in Section 5.

(d)   Sub-contracts as defined in Section 6.

2.2   Incidental costs, overhead and profit as defined in Section 7 and expressed as percentage adjustments are applicable to each of 2.1 (a)-(d).

SECTION 3

Labour

3.1   Labour costs comprise all payments made to or in respect of all persons directly engaged upon the work, whether on or off the site, except those included in Section 7.

3.2   Such payments are based upon the standard wage rates, emoluments and expenses as laid down for the time being in the rules or decisions of the National Joint Council for the Building Industry and the terms of the Building and Civil Engineering Annual and Public Holiday Agreements applying to the works, or the rules or decisions or agreements of such other body as may relate to the class of

labour concerned, at the time when and in the area where the work is executed, together with the Contractor's statutory obligations, including:

(a)   Guaranteed minimum weekly earnings (e.g. Standard Basic Rate of Wages and Guaranteed Minimum Bonus Payment in the case of NJCBI rules).

(b)   All other guaranteed minimum payments (unless included in Section 7).

(c)   Payments in respect of incentive schemes or productivity agreements applicable to the works.

(d)   Payments in respect of overtime normally worked; or

necessitated by the particular circumstances of the work; or as otherwise agreed between parties.

(e)   Differential or extra payments in respect of skill responsibility, discomfort, inconvenience or risk.

(f)   Tool allowance.

(g)   Subsistence and periodic allowances.

(h)   Fares, travelling and lodging allowances.

(j)   Employer's contributions to annual holiday credits.

(k)   Employer's contributions to death benefit schemes.

(l)   Any amounts which may become payable by the Contractor to or in respect of operatives arising from the operation of the rules referred to in 3.2 which are not provided for in 3.2 (a)-(k) or in Section 7.

(m)   Employer's National Insurance contributions and any contributions, levy or tax imposed by statute, payable by the Contractor in his capacity as employer.

Note: Any payments normally made by the Contractor which are of a similar character to those described in 3.2 (a)-(c) but which are not within the terms of the rules and decisions referred to above are applicable subject to the prior agreement of the parties, as an alternative to 3.2 (a)-(c).

3.3   The wages or salaries of supervisory staff, timekeepers, storekeepers, and the like, employed on or regularly visiting site, where the standard wage rates etc. are not applicable, are those normally paid by the Contractor together with any incidental payments of a similar character to 3.2 (c)-(k).

3.4   Where principals are working manually their time is chargeable, in respect of the trades practised, in accordance with 3.2.

SECTION 4

Materials and Goods

4.1   The prime cost of materials and goods obtained by the Contractor from stockists or manufacturers is the invoice cost after deduction of all trade discounts but including cash

discounts not exceeding 5 per cent, and includes the cost of

delivery to site.

4.2   The prime cost of materials and goods supplied from the Contractor's stock is based upon the current market prices plus any appropriate handling charges.

4.3   The prime cost under 4.1 and 4.2 also includes any costs of:

(a)   non-returnable crates or other packaging.

(b)   returning crates and other packaging less any credit obtainable.

4.4   Any Value Added Tax which is treated, or is capable of being treated, as input tax (as defined in the Finance Act, 1972 or any re-enactment thereof) by the Contractor is excluded.

SECTION 5

Plant, Consumable Stores and Services

5.1   The prime cost of plant and consumable stores as listed below is the cost at hire rates agreed between the parties or in the absence of prior agreement at rates not exceeding those normally applied in the locality at the time when the works are carried out, or on a use and waste basis where applicable:

(a)   Machinery in workshops.

b)   Mechanical plant and power-operated tools.

JOBBING WORK (Cont'd).

5.1 (c) Scaffolding and scaffold boards.

     (d) Non-Mechanical plant excluding hand tools.

     (e) Transport including collection and disposal of rubbish.

     (f) Tarpaulins and dust sheets.

     (g) Temporary roadways, shoring, planking and strutting, hoarding, centering, formwork, temporary fans, partitions or the like.

     (h) Fuel and consumable stores for plant and power-operated tools unless included in 5.1 (a), (b), (d) or (e) above.

     (j) Fuel and equipment for drying out the works and fuel for testing mechanical services.

5.2 The prime cost also includes the net cost incurred by the Contractor of the following services, excluding any such cost included under Sections 3, 4 or 7:

     (a) Charges for temporary water supply including the use of temporary plumbing and storage

     (b) Charges for temporary electricity or other power and lighting including the use of temporary installations.

     (c) Charges arising from work carried out by local authorities or public undertakings.

     (d) Fee, royalties and similar charges.

     (e) Testing of materials

     (f) The use of temporary buildings including rates and telephone and including heating and lighting not charged under (b) above.

     (g) The use of canteens, sanitary accommodation, protective clothing and other provision for the welfare of persons engaged in the work in accordance with the

         current Working Rule Agreement and any Act of Parliament, statutory instrument, rule, order, regulation or bye-law.

     (h) The provision of safety measures necessary to comply with any Act of Parliament.

     (j) Premiums or charges for any performance bonds or insurances which are required by the Building Owner and which are not referred to elsewhere in this Definition.

SPECIMEN ACCOUNT FORMAT

If this Definition of Prime Cost is followed the Contractor's account could be in the following format:-   £

Labour (as defined in Section 3)
Add % (see Section 7)
Materials and goods (as defined in Section 4)
Add % (see Section 7)
Plant, consumable stores and services (as defined in Section 5)
Add % (see Section 7)
Sub-contracts (as defined in Section 6)
Add % (see Section 7)

VAT to be added if applicable

SECTION 6

Sub-Contracts

6.1 The prime cost of work executed by sub-contractors, whether nominated by the Building Owner or appointed by the Contractor is the amount which is due from the Contractor to the sub-contractors in accordance with the terms of the sub-contracts after deduction of all discounts except any cash discount offered by any sub-contractor to the Contractor not exceeding 2 1/2 per cent.

SECTION 7

Incidental Costs, Overheads and Profit

7.1 The percentages adjustments provided in the building contract, which are applicable to each of the totals of Sections 3-6, provide for the following:-

     (a) Head Office Charges.

     (b) Off-site staff including supervisory and other administrative staff in the Contractor's workshops and yard.

     (c) Payments in respect of public holidays.

     (d) Payments in respect of apprentices' study time.

     (e) Sick pay or insurance in respect thereof.

     (f) Third Party and employer's liability insurance.

     (g) Liability in respect of redundancy payments made to employees.

     (h) Use, repair and sharpening of non-mechanical hand tools.

     (j) Any variation to basic rates required by the Contractor in cases where the building contract provides for the use of a specified schedule of basic plant charges (to the extent that no other provision is made for such variation).

     (k) All other liabilities and obligations whatsoever not specifically referred to in this Section nor chargeable under any other section.

     (l) Profit.

£

GENERAL INFORMATION

SCHEDULE OF BASIC PLANT CHARGES

For use in connection with Dayworks under a Building Contract (Fourth revision - 1st January 1990)
This schedule is reproduced by permission of the Royal Institution of Chartered Surveyors.

Explanatory Notes

1. The rates in the Schedule are intended to apply solely to daywork carried out under and incidental to a Building Contract. They are NOT intended to apply to:-
   (i) Jobbing or any other work carried out as a main or separate contract; or
   (ii) Work carried out after the date of commencement of the Defects Liability Period.

2. The rates in the Schedule are basic and may be subject to an overall adjustment to be quoted by the Contractor prior to the placing of the Contract.

3. The rates apply to plant and machinery already on site, whether hired or owned by the Contractor.

4. The rates, unless otherwise stated, include the cost of fuel and power of every description, lubricating oils, grease, maintenance, sharpening of tools, replacement of spare parts, all consumable stores and for licences and insurances applicable to items of plant. They do not include the costs of drivers and attendants (unless otherwise stated)

5. The rates should be applied to the time during which the plant is actually engaged in daywork.

6. Whether or not plant is chargeable on daywork depends on the daywork agreement in use and the inclusion of an item of plant in this schedule does not necessarily indicate that that item is chargeable.

7. Rates for plant not included in the Schedule or which is not on site and is specifically hired for daywork shall be settled at prices which are reasonably related to the rates in the Schedule having regard to any overall adjustment quoted by the Contractor in the Conditions of Contract.

Note:-
All rates are expressed per hour and were calculated during the first quarter of 1989.

MECHANICAL PLANT AND TOOLS

Item of Plant	Description	Unit	Rate per hour
Bar Bending and Shearing Machines			
Bar bending machine	Power driven - up to 2" (51mm) dia rods	Each	2.01
Bar Shearing machine	Power driven - up to 1 1/2" (38mm) dia rods	Each	2.00
	Power driven - up to 2" (51mm) dia rods	Each	2.00
Bar cropper machine	Power driven - up to 2" (51mm) dia rods	Each	1.51
Block and Stone Splitter	Hydraulic	Each	0.83
Brick Saws			
Brick saw (use of abrasive disc to be charged net and credited)	Power driven (bench type, clipper or similar or portable)	Each	1.83
Compressors			
Portable compressor (machine only)	Nominal delivery of free air per min at 100lb/sq. inch (7kg/sq.m) pressure		
	(cfm) (m3/min)		
	80/85 2.41	Each	1.30
	125-140 3.50-3.92	Each	1.58
	160-175 4.50-4.95	Each	2.37
	250 7.08	Each	3.25
	380 10.75	Each	4.80
	600-630 16.98-17.84	Each	6.00
Lorry Mounted Compressors			
(Machine plus lorry only)	Nominal delivery of free air per min at 100 PSI (7kg/cm2)		
	101-150 2.86-4.24	Each	4.12
Tractor Mounted Compressors			
(Machine plus rubber tyred tractor only)	Nominal delivery of free air per min at 100 PSI (7kg/cm2)		
	101-120 2.86-3.40	Each	3.63
Compressed Air Equipment			
(With and including up to 50ft (15.24m) of air hose)			
Breakers (with six steels)	Light	Each	0.47
	Medium (65lbs)	Each	0.47
	Heavy (85lbs)	Each	0.47

SCHEDULE OF BASIC PLANT CHARGES (cont'd)
MECHANICAL PLANT AND TOOLS (cont'd)

Item of Plant	Description	Unit	Rate per hour
Light pneumatic pick (with six steels)		Each	0.37
Pneumatic clay spade (one blade)		Each	0.22
Chipping hammer (plus 6 steels)		Each	0.65
Hand held rock drill with rod (bits to be paid for at net cost and credited)	7Kg (15lbs) class (light)	Each	0.33
	16Kg (35lbs) class (medium)	Each	0.37
	20Kg (45lbs) class (medium)	Each	0.37
	25Kg (55lbs) class (heavy)	Each	0.39
Rotary drill (bits to be paid for at net cost and credited)	Up to 3/4" (19mm)	Each	0.37
	1 1/4" (32mm)	Each	0.73
Sander/Grinder		Each	0.39
Scabbler (heads extra)	Single head	Each	0.48
	Triple head	Each	0.65
Compressor Equipment Accessories			£
Additional hoses	Per 50ft (15m) length	Each	0.09
Muffler, tool silencer		Each	0.08
Concrete Breaker			
Concrete breaker, portable hydraulic complete with power pack		Each	1.33
Concrete/Mortar Mixers			
Concrete mixer	Diesel, electric or petrol		
	cu ft m3		
Open drum without hopper	3/2 0.09/0.06	Each	0.40
	4/3 0.12/0.09	Each	0.44
	5/3 1/2 0.15/0.10	Each	0.54
Open drum with hopper	7/5 0.20/0.15	Each	0.67
Closed drum	8/6 1/2 0.25/0.18	Each	0.90
Reversing drum with hopper weigher and feed shovel	10/7 0.28/0.20	Each	2.44
	21/14 0.60/0.40	Each	4.17
Concrete Pump			
Lorry mounted concrete pump (metreage charge to be added net)		Each	13.70
Concrete Equipment			
Vibrator, poker type	Petrol, diesel or electric		
	Up to 3" diameter	Each	0.87
	Air, excluding compressor and hose		
	Up to 3" diameter	Each	0.65
	Extra heads	Each	0.50
Vibrator, tamper	With tamping board	Each	0.87
	Double beam screeder	Each	1.19
Power float	29"-36"	Each	1.13
Conveyor Belts	Power operated up to 25' (7.62m) long, 16" (400mm) wide	Each	3.15
Cranes			
Mobile rubber tyred	Maximum capacity up to 15cwt (762kg)	Each	5.15
Lorry mounted, telescopic jib, 2 wheel (All rates inclusive of driver)	6 tons (tonnes)	Each	10.47
	7 tons (tonnes)	Each	11.22
	8 tons (tonnes)	Each	13.67
	10 tons (tonnes)	Each	14.99

GENERAL INFORMATION

SCHEDULE OF BASIC PLANT CHARGES (cont'd)
MECHANICAL PLANT AND TOOLS (cont'd)

Item of Plant	Description		Unit	Rate per hour
Lorry mounted, telescopic jib, 2 wheel (All rates inclusive of driver) (con'td)	12 tons (tonnes)		Each	16.91
	15 tons (tonnes)		Each	19.21
	18 tons (tonnes)		Each	20.34
	20 tons (tonnes)		Each	22.04
	25 tons (tonnes)		Each	24.86
Lorry mounted, telescopic jib, 4 wheel (All rates inclusive of driver)	10 tons (tonnes)		Each	15.30
	12 tons (tonnes)		Each	17.26
	15 tons (tonnes)		Each	19.59
	20 tons (tonnes)		Each	22.48
	25 tons (tonnes)		Each	25.36
	30 tons (tonnes)		Each	28.82
	45 tons (tonnes)		Each	31.12
	50 tons (tonnes)		Each	33.64
Track-mounted tower crane (electric) (Capacity = max lift in tons x max radius at which can be lifted) (All rates inclusive of driver)	Capacity (metre/tonnes)	Height under hook above ground (m)		
	up to 10	17	Each	7.40
	15	17	Each	7.95
	20	18	Each	8.50
	25	20	Each	10.70
	30	22	Each	12.76
	40	22	Each	16.75
	50	22	Each	21.70
	60	22	Each	22.52
	70	22	Each	23.07
	80	22	Each	23.99
	110	22	Each	24.49
	125	30	Each	27.20
	150	30	Each	29.95
Static tower cranes	To be charged at a percentage of the above rates		Each of the above	90%

Crane Equipment

Item of Plant	Description	Unit	Rate per hour
Tipping bucket	Circular		
	0.19m3 (1/4 cu.yd.)	Each	0.25
	0.57m3 (3/4 cu.yd.)	Each	0.28
Skip, muck	Up to 0.38m3 (1/2 cu.yd.)	Each	0.25
	0.57m3 (3/4 cu.yd.)	Each	0.25
	0.76m3 (1 1/2 cu.yd.)	Each	0.30
Skip, concrete	Up to 0.38m3 (1/2 cu.yd.)	Each	0.44
	0.57m3 (3/4 cu.yd.)	Each	0.53
	0.76m3 (1 cu.yd.)	Each	0.76
	1.15m3 (1 1/2 cu.yd.)	Each	0.61
Skip, concrete lay down or roll over	0.38m3 (1/2 cu.yd.)	Each	0.44

Dehumidifiers

Item of Plant	Description	Unit	Rate per hour
110/240v Water extraction per 24 hours	68 litres (15 gallons)	Each	0.84
	90 litres (20 gallons)	Each	0.97

Diamond Drilling and Chasing

Item of Plant	Description	Unit	Rate per hour
Chasing machine drilling (Diamond core)	6" (152mm)	Each	1.35
	Mini rig up to 35mm	Each	3.16
	3"-8" (76mm-203mm) electric 110v	Each	3.33
	3"-8" (76mm-203mm) Air	Each	4.23

Dumpers

Dumper (site use only, excluding Tax, Insurance and extra cost of DERV etc. when operating on highway

Item of Plant	Description	Unit	Rate per hour
2-wheel drive	Makers capacity		
Gravity tip	15 cwt (762 kg)	Each	1.04
Hydraulic tip	20 cwt (1016 kg)	Each	1.25
Hydraulic tip	23 cwt (1168 kg)	Each	1.31
4-wheel drive			
Gravity tip	23 cwt (762 kg)	Each	1.42
Hydraulic tip	25 cwt (1270 kg)	Each	1.61

SCHEDULE OF BASIC PLANT CHARGES (cont'd)
MECHANICAL PLANT AND TOOLS (cont'd)

Item of Plant	Description		Unit	Rate per hour
Dumpers (Cont'd)				
Hydraulic tip	30 cwt (1524 kg)		Each	1.83
Hydraulic tip	35 cwt (1778 kg)		Each	2.08
Hydraulic tip	40 cwt (2023 kg)		Each	2.33
Hydraulic tip	50 cwt (2540 kg)		Each	2.96
Hydraulic tip	60 cwt (3048 kg)		Each	3.68
Hydraulic tip	80 cwt (4064 kg)		Each	4.72
Hydraulic tip	100 cwt (5080 kg)		Each	6.38
Hydraulic tip	120 cwt (6096 kg)		Each	7.54
Electric Hand Tools				
Breakers	Heavy: Kango 2500		Each	0.90
	Kango 1800		Each	0.87
	Medium: Hilti TP800		Each	0.73
Rotary hammers	Heavy: Kango 950		Each	0.55
	Medium: Kango 627/637		Each	0.47
	Light: Hilti TE12/TE17		Each	0.39
Pipe drilling tackle	Ordinary type		Set	0.51
	Under pressure type		Set	0.83
Excavators				
Hydraulic full circle slew	Crawler mounted, backactor			
	5/8 cu.yd. (0.50m3)		Each	5.01
	3/4 cu.yd. (0.60m3)		Each	5.92
	7/8 cu.yd. (0.70m3)		Each	8.24
	1 cu. yd. (0.75m3)		Each	13.25
Wheeled tractor type	Hydraulic excavator, JCB type 3C or similar		Each	9.23
Mini Excavators				
Kubota mini excavators	360 tracked - 1 tonne		Each	5.38
	360 tracked - 3.5 tonnes		Each	7.23
Forklifts				
2-wheel drive "Rough terrain"	Payload	Max. lift		
	20cwt (1016kg)	21'4" (6.50m)	Each	4.38
	20cwt (1016kg)	26'0" (7.92m)	Each	4.38
	30cwt (1524kg)	20'0" (6.09m)	Each	4.50
	36cwt (1829kg)	18'0" (5.48m)	Each	4.50
	50cwt (2540kg)	12'0" (3.66m)	Each	4.73
4-wheel drive "Rough terrain"	30cwt (1524kg)	20'0" (6.09m)	Each	5.60
	40cwt (2032kg)	20'0" (6.09m)	Each	5.78
	50cwt (2540kg)	12'0" (3.66m)	Each	7.79
Hammers				
Cartridge (excluding cartridges and studs)	Hammer DX450		Each	0.40
	Hammer DX600		Each	0.48
	Hammer spit TS		Each	0.48
Heaters, Space	Paraffin/electric Btu/hr			
	50,000 - 75,000		Each	0.73
	80,000 - 100,000		Each	0.77
	150,000		Each	0.88
	320,000		Each	1.20
Hoists				
Scaffold	Up to 5cwt (254kg)		Each	0.90
Mobile (goods only)	Up to 10cwt (508kg)		Each	1.43
Static (goods only)	10 to 15cwt (508-762kg)		Each	1.67
Rack and pinion (goods only)	16cwt (813kg)		Each	2.75
	20cwt (1016kg)		Each	3.00
	25cwt (1270kg)		Each	3.30

GENERAL INFORMATION

SCHEDULE OF BASIC PLANT CHARGES - contd
MECHANICAL PLANT AND TOOLS - contd

Item of Plant	Description		Unit	Rate per hour	
Hoists (Cont'd)					
Rack and pinion	8 person, 1433lbs (650kg)		Each	4.50	
(goods and passenger)	12 person, 2205lbs (1000kg)		Each	4.95	
Lorries	Plated gross vehicle weight (ton/tonnes)				
Fixed body	Up to 5.50		Each	9.60	
	Up to 7.50		Each	10.20	
Tipper	Up to 16.00		Each	11.73	
	Up to 24.00		Each	14.98	
	Up to 30.00		Each	17.73	
Pipe Work Equipment					
Pipe bender	Power driven, 50-150mm dia		Each	0.75	
	Hydraulic, 2" capacity		Each	0.96	
Pipe cutter	Hydraulic		Each	1.02	
Pipe defrosting equipment	Electrical		Set	1.83	
Pipe testing equipment	Compressed air		Set	1.04	
	Hydraulic		Set	0.68	
Pipe threading equipment	Up to 4"		Set	1.65	
Pumps					
Including 20'(6m) length of suction/delivery hose, couplings, valves and strainers					
Diaphragm:"Simplite" 2" (50mm)	Petrol or electric		Each	0.79	
"Wickham" 3" (76mm)	Diesel		Each	1.22	
Submersible: 2" (50mm)	Electric		Each	0.85	
Induced flow: "Spate" 3" (76mm)	Diesel		Each	1.42	
"Spate" 4" (102mm)	Diesel		Each	1.88	
Centrifugal, self priming: "Univac" 2" (50mm)	Diesel		Each	1.87	
"Univac" 4" (102mm)	Diesel		Each	2.40	
"Univac" 6" (152mm)	Diesel		Each	3.69	
Pumping Equipment					
Pump hoses (per 20ft (6m) flexible	in Diameter	mm			
section/delivery including coupling valve	2	51	Each	0.15	
and strainer)	3	76	Each	0.17	
	4	102	Each	0.19	
	6	152	Each	0.32	
Rammers and Compactors					
Power rammer, Pegson or similar			Each	0.88	
Soil compactor, plate type					
12 x 13" (305 x 330mm)	172lb (78kg)		Each	0.78	
20 x 18" (508 x 457mm)	264lb (120kg)		Each	0.93	
Rollers	Cwt	Kg			
Vibrating rollers	7 1/4 - 8 1/4	368 - 420	Each	1.00	
Single roller	10 1/2	533	Each	1.53	
Twin roller	13 3/4	698	Each	1.75	
	16 3/4	851	Each	2.01	
Twin roller with seat and steering wheel	21	1067		Each	2.75
	27 1/2	1397		Each	2.88

SCHEDULE OF BASIC PLANT CHARGES - contd
MECHANICAL PLANT AND TOOLS - contd

Item of Plant	Description		Unit	Rate per hour
Rollers Cont'd)	ton/tonne			£
Pavement rollers	3-4		Each	2.89
dead weight	over 4-6		Each	3.75
	over 6-10		Each	4.40
Saws, Mechanical				
Chain saw	21	0.53	Each	0.78
	30	0.76	Each	1.12
Bench saw	in	m		
	Up to 20	0.51 blade	Each	0.80
	Up to 24	0.61 blade	Each	1.20
Screed Pump				
Working volume 7 cu. ft (200/litres)	Maximum delivery Vertical 300ft (91m) Horizontal 600ft (182m)		Each	8.33
Screed pump hose	50/65mm diameter 13.3m long		Each	1.83
Screwing Machine	13-50mm dia		Each	0.43
	25-100mm dia		Each	0.87
Tractors				
Shovel, tractor (crawler), any type of bucket	cu.yd	cu.m		
	Up to 3/4	0.57	Each	4.70
	Up to 1	0.76	Each	5.70
	Up to 1 1/4	0.96	Each	6.62
	Up to 1 1/2	1.15	Each	8.15
	Up to 1 3/4	1.34	Each	8.98
	Up to 2	1.53	Each	10.69
	2 3/4 - 3 1/2	2.10-2.70	Each	17.27
Shovel, tractor (wheeled)	Up to 1/2	0.38	Each	3.36
	Up to 3/4	0.57	Each	4.14
	Up to 1	0.76	Each	5.02
	Up to 2 1/2	1.91	Each	7.30
Tractor (crawler) with dozer	Makers rated flywheel horsepower			
	75		Each	9.31
	140		Each	12.52
Tractor, wheeled (rubber-tyred) agricultural type	Light 48 h.p.		Each	4.23
	Heavy 65 h.p.		Each	4.68
Traffic Lights	Mains/generator 2-way		Set	2.23
	Mains/generator 3-way		Set	4.40
	Mains/generator 4-way		Set	5.45
	Trailer mounted 2-way		Set	2.21
Welding and Cutting and Burning Sets				
Welding and cutting set (including oxygen and acetylene, excluding underwater equipment and thermic boring)			Each	4.33
Welding set, diesel (excluding electrodes)	300 amp single operator		Each	1.84
	480 amp double operator		Each	1.89
	600 amp double operator		Each	2.35

GENERAL INFORMATION

SCHEDULE OF BASIC PLANT CHARGES - Cont'd　　　　NON-MECHANICAL PLANT

Item of Plant	Description	Unit	Rate per hour _
Bar Bending and Shearing Machines			£
Bar bending machine, manual	Up to 1 in. (25mm) dia. rods	Each	0.26
Shearing machine, manual	Up to 5/8 in. (16mm) dia. rods	Each	0.15
Brother or Sling Chain	Not exceeding 2 tons/tonnes	Set	0.28
	Exceeding 2 tons/tonnes, not exceeding 5 tons/tonnes	Set	0.40
	Exceeding 5 tons/tonnes, not exceeding 10 tons/tonnes	Set	0.48
Drain Testing Equipment		Set	0.42
Lifting and Jacking Gear	ton/tonne		
Pipe winch including shear legs	1/2	Sets	0.35
Pipe winch including gantry	2	Sets	0.72
	3	Sets	0.84
Chain blocks up to 20ft (6.10m) lift	ton/tonne		
	1	Each	0.35
	2	Each	0.44
	3	Each	0.51
	4	Each	0.66
Pull lift (Tirfor type)	3/4 tons/tonnes	Each	0.35
	1 1/2 tons/tonnes	Each	0.47
	3 tons/tonnes	Each	0.61
Pipe Benders	13-75mm dia	Each	0.32
	50-100mm dia	Each	0.52
Plumbers Furnace	Calor gas or similar	Each	1.20
Road Works - Equipment			
Barrier trestles or similar		10	0.50
Crossing plates (steel sheets)		Each	0.32
Danger lamp (including oil)		Each	0.03
Warning signs		Each	0.07
Road Cones		10	0.11
Flasher unit (battery to be charged at cost)		Each	0.06
Flashing bollard (battery to be charged at cost)		Each	0.10
Rubbish Chutes			
Length of sections 154cm giving working length of 130cm			
Standard section		Each	0.15
Brand section		Each	0.20
Hopper		Each	0.16
Fixing frame		Each	0.20
Winch Type A	20m for erection	Each	0.29
Winch Type B	40m for erection	Each	0.67
Scaffolding			
Boards		100ft (30.48m)	0.06
Castor wheels	Steel or rubber tyred	100	0.87
Fall ropes	Up to 200ft (61m)	Each	0.44
Fittings (including couplers and base plates)	Steel or alloy	100	0.07
Ladders, pole	20 rung	Each	0.11
	30 rung	Each	0.15
	40 rung	Each	0.26

SCHEDULE OF BASIC PLANT CHARGES (cont'd)
NON-MECHANICAL PLANT (Cont'd)

Item of Plant	Description		Unit	Rate per hour
Ladder, extension	Extended length			
	ft	(m)		
	20	(6.10)	Each	0.31
	26	(7.92)	Each	0.43
	35	(10.67)	Each	0.64
Putlogs	Steel or alloy		100	0.06
Splitheads	Small		10	0.12
	Medium		10	0.15
	Large		10	0.16
Staging, lightweight			100ft (30.48m)	1.00
Tube	Steel		100ft	0.02
	Alloy		(30.48m)	0.04
Wheeled tower				
7ft x 7ft (2.13 x 2.13m)	Working platform up to 20ft (6.00m) high		Each	0.58
10ft x 10ft (3.05 x 3.05m)	including castors and boards		Each	0.59
Tarpaulins			10 m2	0.04
Trench Struts and Sheets				
Adjustable steel trench strut	All sizes from 1ft (305mm) to 5ft 6in (1.68m) extended		10	0.06
Steel trench sheet	5-14ft (1.52-4.27m) lengths		100ft (30.48m)	0.16

ELECTRICAL CONTRACTORS

A "Definition of Prime Cost of Daywork carried out under an electrical contract" dated March 1st 1981, agreed between the Royal Institution of Chartered Surveyors and the Electrical Contractors Associations can be obtained from the RICS, 12 Great George Street, London SW1, or the ECA, 32-34 Palace Court, Bayswater, London W2 4HY.

TERRAZZO-MOSAIC SPECIALISTS

Daywork rates approved by the National Federation of Terrazzo Marble & Mosaic Specialists, P.O. Box 50, Banstead, Surrey SM7 2RD.

Minimum Rates for Dayworks

		£
Terrazzo or mosaic craftsman	per hour	18.50
Labourers	per hour	17.35

All travelling and waiting time to be paid at above rates.

The above rates cover Wages, National Insurance, Graduated Pension Contribution, Workmen's Compensation and Third Pary Insurance, Annual and Public Holidays Contributions, Head Office supervision, overheads and profit. Value added Tax is not included. All known wage increases up to 30th June, 1997 are covered.

Non-productive overtime or overtime premiums should be charged at £11.95 per hour for all classes of labour.

Increased labour costs should be plus 20% on basis of clause 31 (e) of R.I.B.A. Standard Form of Building Contract.

Floor polishing machines (single head) £31.20 per day or part day. Dry sanding machines £31.20 per day or part day. Multi-head floor polishing machines, £99.30 per day or part day. Exclusive of cost of fuel and abrasives.

Fares to and from the site, lodging allowances and materials will be charged at cost plus 20%.
Delivery and collection of materials/plant not included in any of the above rates.

HEATING, VENTILATING, AIR CONDITIONING, REFRIGERATION, PIPEWORK AND/OR DOMESTIC ENGINEERING CONTRACTS

DEFINITION OF PRIME COST OF DAYWORK CARRIED OUT UNDER A HEATING, VENTILATING, AIR CONDITIONING, REFRIGERATION, PIPEWORK AND/OR DOMESTIC ENGINEERING CONTRACT

Reproduced by permission of the Royal Institution of Chartered Surveyors and the Heating and Ventilating Contractors' Association.

This definition of Prime Cost is published by the Royal Institution of Chartered Surveyors and the Heating and Ventilating Contractors' Association for convenience and for use by people who choose to use it. Members of the Heating and Ventilating Contractors' Association are not in any way debarred from defining Prime Cost and rendering their accounts for work carried out on that basis in any way they choose. Building owners are advised to reach agreement with contractors on the Definition of Prime Cost to be used prior to entering into a contract or sub-contract.

SECTION 1

Application

1.1 This definition provides a basis for the valuation of daywork executed under such heating, ventilating, air conditioning, refrigeration, pipework, and/or domestic engineering contracts as provide for its use.

1.2 It is not applicable in any other circumstances, such as jobbing or other work carried out as a separate or main contract nor in the case of daywork executed after a date of practical completion.

1.3 The terms "contract" and "contractor" herein shall be read as "sub-contract" and "sub-contractor" as applicable.

GENERAL INFORMATION

HEATING, VENTILATING, ETC (contd)

SECTION 2

Composition of Total Charges

2.1 The Prime Cost of daywork comprises the sum of the following costs:

(a) Labour as defined in Section 3.

(b) Materials and goods as defined in Section 4.

(c) Plant as defined in Section 5.

2.2 Incidental costs, overheads and profit as defined in Section 6, as provided in the contract and expressed therein as percentage adjustments, are applicable to each of 2.1 (a)-(c).

SECTION 3

Labour

3.1 The standard wage rates, emoluments and expenses referred to below and the standard working hours referred to in 3.2 are those laid down for the time being in the rules or decisions or agreements of the Joint Conciliation Committee of the Heating, Ventilating and Domestic Engineering Industry applicable to the Works (or those of such other body as may be appropriate) and to the grade of operative concerned at the time when and the area where the daywork is executed.

3.2 Hourly base rates for labour are computed by dividing the annual prime cost of labour, based upon the standard working hours and as defined in 3.4, by the number of standard working hours per annum. See example at end of Section 6.

3.3 The hourly rates computed in accordance with 3.2 shall be applied in respect of the time spent by operatives directly engaged on daywork, including those operating mechanical plant and transport and erecting and dismantling other plant (unless otherwise expressly provided in the contract) and handling and distributing the materials and goods used in the daywork.

3.4 The annual prime cost of labour comprises the following:

(a) Standard weekly earnings (i.e. the standard working
week as determined at the appropriate rate for the operative concerned).

(b) Any supplemental payments.

(c) Any guaranteed minimum payments (unless included in Section 6.1 (a)-(p)).

(d) Merit money.

(e) Differentials or extra payments in respect of skill, responsibility, discomfort, inconvenience or risk (excluding those in respect of supervisory responsibility - see 3.5).

(f) Payments in respect of public holidays.

(g) Any amounts which may become payable by the contractor to or in respect of operatives arising from the rules etc. referred to in 3.1 which are not provided for in 3.4 (a)-(f) nor in Section 6.1 (a)-(p).

(h) Employer's contributions to the Annual Holiday with
Pay and Welfare Benefits Scheme or payments in lieu thereof.

(i) Employer's National Insurance contributions as applicable to 3.4 (a)-(h).

(j) Any contribution, levy or tax imposed by Statute, payable by the contractor in his capacity as an employer.

3.5 Differentials or extra payments in respect of supervisory responsibility are excluded from the annual prime cost (see Section 6). The time of principals, staff, foremen, chargehands and the like when working manually is admissible under this Section at the rates for the appropriate grades.

SECTION 4

Materials and Goods

4.1 The prime cost of materials and goods obtained specifically for the daywork is the invoice cost after deducting all trade discounts and any portion of cash discounts in excess of 5 per cent.

4.2 The prime cost of all other materials and goods used in the daywork is based upon the current market prices plus any appropriate handling charges.

4.3 The prime cost referred to in 4.1 and 4.2 includes the cost of delivery to site.

4.4 Any Value Added Tax which is treated, or is capable of being treated, as input tax (as defined by the Finance Act 1972, or any re-actment or amendment thereof or substitution therefor) by the contractor is excluded.

SECTION 5

Plant

5.1 Unless otherwise stated in the contract, the prime cost of plant comprises the cost of the following:

(a) use or hire of mechanically-operated plant and transport for the time employed on and/or provided

or

retained for the daywork;

(b) use of non-mechanical plant (excluding non-mechanical hand tools) for the time employed on and/or provided or retained for the daywork;

(c) transport to and from site and erection and dismantling where applicable.

5.2 The use of non-mechanical hand tools and of erected scaffolding, staging, trestles or the like is excluded (see Section 6), unless specifically retained for the daywork.

SECTION 6

Incidental Costs, Overheads and Profit

6.1 The percentage adjustments provided in the contract which are applicable to each of the totals of Sections 3, 4 and 5, comprise the following:

(a) Head office charges.

(b) Site staff including site supervision.

(c) The additional cost of overtime (other than that referred to in 6.2).

(d) Time lost due to inclement weather.

(e) The additional cost of bonuses and all other incentive payments in excess of any included in 3.4.

(f) Apprentices' study time.

(g) Fares and travelling allowances.

(h) Country, lodging and periodic allowances.

(i) Sick pay or insurances in respect thereof, other than as included in 3.4.

HEATING, VENTILATING, ETC - contd
Section 6 contd.

(j)   Third party and employers' liability insurance.

(k)   Liability in respect of redundancy payments to employees.

(l)   Employer's National Insurance contributions not included in 3.4.

(m)   Use and maintenance of non-mechanical hand tools.

(n)   Use of erected scaffolding, staging, trestles or the like (but see 5.2).

(o)   Use of tarpaulins, protective clothing, artificial lighting, safety and welfare facilities, storage and the like that may be available on site.

(p)   Any variation to basic rates required by the contractor in cases where the contract provides for the use of a specified schedule of basic plant charges (to the extent that no other provision is made for such variation - see 5.1).

(q)   In the case of a sub-contract which provides that the sub-contractor shall allow a cash discount, such provision as is necessary for the allowance of the prescribed rate of discount.

(r)   All other liabilities and obligations whatsoever not specifically referred to in this Section nor chargeable under any other Section.

(s)   Profit.

6.2   The additional cost of overtime where specificaly ordered by the Architect/Supervising Officer shall only be chargeable in the terms of a prior written agreement between the parties.

Example of calculation of typical standard hourly base rate (as defined in section 3) for Advanced Fitter (qualified gas and arc) and mate employed in the Heating, Ventilating, Air Conditioning, Piping and Domestic Engineering Industry under NJIC Rules based upon rates ruling at 4th February 1980.

	Rate £	Advanced Fitter (qual. gas & arc) £	Rate £	Mate £
Standard Weekly Earnings: 48 weeks x 38 hours	2.26	4122.24	1.64	2991.36
Welding Supplement (gas and arc): 48 weeks x 38 hours	.20	364.80	-	-
Merit Money and other variables as applicable		*		*
		4487.04		2991.36
Employer's National Insurance Contribution at 13.5%		605.75		403.83
		5092.79		3395.19
CITB Annual Levy		60.00		6.00
Weekly Holiday Credit and Welfare Stamp: 48 weeks	9.83	471.84	8.08	387.84
Annual Labour Cost as Defined in Section 3		£ 5624.63		£ 3789.03
Hourly base rates as defined in Section 3 clause 3.2	$\frac{5624.63}{1763.2}$ =	£3.19	$\frac{3789.03}{1763.2}$ =	£2.15

Note: (1)   Standard working hours per annum calculated as follows:

52 weeks at 38 hours	=		1976
Less			
4 weeks holiday at 38 hours	=	152	
8 days Public Holidays at (average) 7.6 hours per day	=	60.8	212.8
			1763.2

(2)   Where applicable, Merit Money and other variables (e.g. Daily Abnormal Conditions Money), which attract Employer's National Insurance Contribution, should be included at*.

It should be noted that all labour costs incurred by the Contractor in his capacity as an Employer, other than those contained in the hourly base rate, as defined under Section 3, are to be taken into account under Section 6.

(3)   The above example is for the convenience of users only and does not form part of the Definition; all the basic costs are subject to re-examination according to the time when and the area where the daywork is executed.

GENERAL INFORMATION

THE ASSOCIATION OF CIVIL ENGINEERING CONTRACTORS
I.C.E. SCHEDULES of DAYWORKS CARRIED OUT INCIDENTAL
TO CONTRACT WORK

(For operation thoroughout ENGLAND, SCOTLAND and WALES)

1. **LABOUR**

2. **MATERIALS**

3. **SUPPLEMENTARY CHARGES**

4. **PLANT**

These Schedules are the Schedules referred to in the I.C.E. Conditions of Contract and have been prepared for use in connection with Dayworks carried out incidental to contract work where no other rates have been agreed. They are not intended to be applicable for Dayworks ordered to be carried out after the contract works have been substantially completed or to a contract to be carried out wholly on a daywork basis. The circumstances of such works vary so widely that the rates applicable call for special consideration and agreement between contractor and employing authority.

1st July 1998

1. LABOUR

Add to the amount of wages paid to operatives.....................148%

1. "Amount of wages" means:-
   Actual wages and bonus paid, daily travelling allowances (fare and/or time) including those in respect of time lost due to inclement weather paid to operatives at plain time rates and/or at overtime rates.

2. The percentage addition provides for all statutory charges at the date of publication and other charges including:-

   National Insurance & Surcharge.
   Normal Contract Works, Third Party, & Employer's Liability Insurances.
   Annual and Public Holidays with Pay and Benefit Scheme.
   Non-contributory Sick Pay Scheme.
   Industrial Training Levy.
   Redundancy Payments.
   The Employment Rights Act 1996
   Site Supervision and Staff including foremen and walking gangers, but the time of the gangers or charge hands working their gangs is to be paid for as for operatives.
   Small Tools - such as Picks, Shovels, Barrows, Trowels, Hand Saws, Buckets, Trestles, Hammers, Chisels and all items of a like nature.
   Protective Clothing.
   Head Office charges and Profit.

3. The time spent in training, mobilisation, demobilisation etc for the Dayworks operation is chargeable.

4. All hired plant drivers and labour sub-contractors' accounts to be charged in full (without deduction of any cash discounts not exceeding 2.5%) plus 88%.

5. Subsistence or lodging allowances and periodic travel allowances (fare and/or time) paid to or incurred on behalf of operatives are chargeable at cost plus 12.5%.

2. MATERIALS

Add to cost of materials delivered to site...............................12.5%

1. The percentage addition provides for Head Offices charges and Profit.

2. The cost of materials means the invoiced price of materials including delivery to site without deduction of any cash discounts not exceeding 2.5%.

3. Unloading of materials:-

   The percentage added to the cost of materials excludes the cost of handling which shall be charged in addition. An Allowance for unloading into site stock or storage including wastage should be added where materials are taken from existing stock.

3. SUPPLEMENTARY CHARGES

1. Transport provided by contractors for operatives to, from in and around the site to be charged at the appropriate Schedule Rates.

2. Any other charges incurred in respect of any Dayworks operation including, tipping charges, professional fees, sub-contractors's accounts and the like shall be paid for in full plus 12.5% (without deduction of cash discounts not exceeding 2.5%). Labour subcontractor's being dealt with in Section 1.

3. The cost of operating welfare facilities to be charged by the contractor and all plus 12.5% payments

4. The cost of additional insurance premiums for abnormal contract work or special site conditions to be charged at cost plus 12.5%.

5. The cost of watching and lighting specially necessitated by Dayworks is to be paid for seperately at Schedule Rates.

4. PLANT

1. These rates apply only to plant already on site, exclusive of drivers and attendants, but inclusive of fuel and consumable stores unless stated to be charged in addition, repairs and maintenace, insurance of plant but excluding time spent in general servicing.

2. Where plant is hired in specifically for dayworks: plant hire (exclusive of drivers and attendants), fuel, oil and grease, insurance, transport etc., to be charged at full amount of invoice (without deduction of any cash discount not exceeding 2.5%) to which should be added consumables where supplied by the Contractor, all plus 12.5%.

3. Fuel distribution, mobilisation and demoblisation are not included in the rates quoted which shall be an additional charge.

4. Metric capacities are adopted and these are not necessarily exact conversions from their imperial equivalents, but cater for the variations arising from comparison of plant manufacturing firms' ratings.

5. Minimum hire charge will be for the period quoted.

6. Hire rates for plant not included below, shall be settled at prices reasonably related to the rates quoted.

7. The rates provide for Head Office charges and Profit.

Section & Item No.	Description	Nominal Size or Capacity		Hire Rate	Remarks
		Capacity Kg	Platform Height m		
	1. ACCESS PLATFORMS				
(1)	Scissor lift self-propelled - yard travel..................	Up to 400	Up to 5.0	5.82	
(2)	Ditto..	Up to 400	Up to 10.0	6.02	
(3)	Ditto..	Up to 800	Up to 9.0	8.91	

CIVIL ENGINEERING - PLANT (Cont'd)

Section & Item No.	Description	Nominal Size or Capacity		Hire Rate	Remarks
		Capacity Kg	Platform Height m		
	1. ACCESS PLATFORMS Contd.				
(4)	Ditto..	Up to 800	Up to 15.0	12.67	
(5)	Ditto..	Over 800	any height	15.47	
(6)	Scissor lift self-propelled - rough terrain travel.......	Up to 800	Up to 9.0	7.74	
(7)	Ditto..	Over 800	any height	16.57	
(8)	Scissor lift - hand-propelled or towed....................	Any Capacity	Up to 7.0	1.94	
(9)	Ditto..	-	Up to 10.0	7.73	
(10)	Telescopic self-propelled - yard travel....................	Up to 400	Up to 10.0	7.74	
(11)	Ditto..	Up to 400	Up to 15.0	18.17	
(12)	Telescopic self-propelled - rough terrain travel........			+ 85% on above	
(13)	Vehicle mounted platform - telescopic/ articulated (Excluding Vehicle)................................	Any Capacity	Up to 10.0	6.64	
(14)	Ditto..	"	Up to 20.0	8.98	
(15)	Ditto..	"	Up to 30.00	48.57	
(16)	Ditto..	"	Up to 40	167.10	
	2. ASPHALT EQUIPMENT				
(1)	Asphalt and/or coated macadam spreader crawler or wheeled................................	Up to 37kW		20.21	
(2)	Ditto..	Up to 56kW		46.19	
(3)	Ditto..	Up to 82kW		66.76	
(4)	Ditto..	Up to 150kW		120.88	
(5)	Extra for above m/c's used with Dry Lean Concrete................................			+ 23%	
(6)	Joint matcher. Separately or in conjunction with longitudial beams (extra over machine)...................			2.81	
(7)	Joint matcher with addition of transverse control devices or fully automatic equipment capable of control from reference guides (extra over machine)..	Up to 150kW		4.11	
(8)	Self-propelled, metered coated chipping applicator for asphalt work (including trailer).........................	Up to 3.5m width		38.05	
(9)	Ditto..	Up to 4.0m width		38.87	
(10)	Ditto..	Up to 4.5m width		42.30	
(11)	Heater planer..	Up to 2.5m width		309.39	
(12)	Cold planer..	Up to 0.5m width incl. picks		71.30	
(13)	Ditto..	Up to 1.00m incl. picks		104.98	
(14)	Ditto..	Up to 2.25m incl. picks		247.32	
(15)	Asphalt road burner (portable)............................	-		7.96	
(16)	Ditto................(self propelled)	-		117.78	
	3. BAR BENDING AND BAR SHEARING MACHINES	For mild steel rods Up to mm. diameter			
(1)	Bar bending machine,......................hand operated	Up to 25		3.61	Per day
(2)	Ditto.....................................power driven	Up to 40		3.22	Per day
(3)	Ditto.....................................Ditto	Up to 50		3.66	
(4)	Bar shearing machine,....................hand operated	Up to 25		1.71	
(5)	Ditto.....................................Ditto	Up to 40		2.07	
(6)	Ditto.....................................Ditto	Up to 50		3.26	
		For high tensile steel rods			
(7)	Bar bending machine,.......................power driven	Up to 55		6.52	
(8)	Bar shearing machine,....................................Ditto	Up to 55		10.06	
	4. COMPRESSORS	Normal delivery of free air per minute at 7kg per cm2 (for compressors with higher working pressure, rates to be negotiated)			
(1)	Compressor (silenced) (machine only)	Up to 2.5m3/min		4.13	
(2)	Ditto..	Up to 3.0m3/min		4.46	
(3)	Ditto..	Up to 4.0m3/min		5.49	
(4)	Ditto..	Up to 5.0m3/min		6.24	
(5)	Ditto..	Up to 6.0m3/min		7.11	
(6)	Ditto..	Up to 8.0m3/min		10.87	
(7)	Ditto..	Up to 11.0m3/min		13.52	
(8)	Ditto..	Up to 14.0m3/min		17.77	
(9)	Ditto..	Up to 17.0m3/min		18.92	
(10)	Ditto..	Up to 20.0m3/min		20.71	

The rates shown above are per unit machine per hour unless otherwise stated.

GENERAL INFORMATION

CIVIL ENGINEERING - PLANT (Cont'd)

Section & Item No.	Description	Nominal Size or Capacity	Hire Rate	Remarks
	4. COMPRESSORS (Cont'd)			
(11)	Ditto..	Up to 26.0m3/min	25.43	
(12)	Lorry mounted silenced compressor (machine + lorry only) (site and public highway use - including Tax, Insurance and extra cost of D.E.R.V.)	Up to 3.0m3/min	12.16	
(13)	Ditto..	Up to 4.5m3/min	16.53	
(14)	Ditto..	Up to 6.0m3/min	18.54	
(15)	Tractor mounted compressor (machine plus rubber tyred tractor with or without front loading bucket) (site use - excluding Tax, Insurance and extra cost D.E.R.V.)..	Up to 4.0m3/min	13.25	
(16)	Air receiver.......................................	Up to 2.0m3	2.14	Per day
(17)	Ditto..	Up to 4.0m3	5.27	Per day
(18)	Ditto..	Up to 8.0m3	11.31	Per day

Note　(i)　"Super Silenced" compressors (less than 81 dBA at 7m in any direction) would carry a premium of 50% at the above rates.

Section & Item No.	Description	Nominal Size or Capacity	Hire Rate	Remarks
	5. CONCRETE MIXERS	Wet Capacity		
(1)	Tilting Drum mixer...............................	Up to 90 litres	0.98	
(2)	Ditto..	Up to 100 litres	1.11	
(3)	Ditto..	Up to 150 litres	1.89	
(4)	Ditto..	Up to 185 litres	2.48	
(5)	Ditto..	Up to 200 litres	2.75	
(6)	Non-Tilting mixer with integral batch weighing......	Up to 300 litres	8.04	
(7)	Ditto..	Up to 400 litres	8.42	
(8)	Ditto..	Up to 500 litres	9.83	
(9)	Extra for power driven loading shovel (items (1)-(8) inclusive).................................	-	1.27	
(10)	Extra for aggregate feed apron..........................	2 compartment	0.25	
(11)	Ditto..	3 compartment	0.29	
(12)	Tilting or non-tilting drum mixers or pan mixers incorporated into a structure with cement, etc., silos at high or low level and built in weighing and batching equipment	-	Rates to be negotiated	
(13)	Mortar or Roller pan mixer....................	Up to 200 litres	6.59	
(14)	Power driven batch loader for mortar pan.............	-	2.61	
		Capacity		
(15)	Cement Silo - low level...........................	Up to 13 tonnes	1.26	
(16)	Ditto..	Up to 21 tonnes	1.50	
(17)	Ditto..	Up to 31 tonnes	1.79	
(18)	Ditto..	Up to 41 tonnes	2.05	
(19)	Ditto..	Up to 51 tonnes	2.94	
(20)	Extra for Silo aeration...........................	-	0.22	
(21)	Extra for autofeed, hydraulic cement weighgear and screw discharge...........................	-	3.62	
(22)	Ditto, with electronic weighgear...........................	-	4.26	

Section & Item No.	Description	Maximum Output litre/min	Batch Capacity litre	Hire Rate
(23)	Grout mixer(Single drum).....................	Up to 60	Up to 120	6.73
(24)	Ditto(Double drum).............................	Up to 110	Up to 230	9.28
(25)	Ditto(ditto)..	Up to 170	Up to 230	14.01
(26)	Loading hopper c/w weighgear and loading shovel..	Up to 250kg		8.33

Section & Item No.	Description	Maximum Output		Hire Rate
(27)	Grout mixer, Roller type............................	Up to 10 litre/min		2.55
(28)	Ditto..	Up to 22 litre/min		4.99
(29)	Ditto..	Up to 43 litre/min		6.97

Section & Item No.	Description	Maximum Output litre/min	Batch Capacity litre	Hire Rate
(30)	Grout mixer/pump c/w loading hopper and weighgear and double drum.....................	Up to 22	Up to 480	24.50
(31)	Grouting machine including pump and hopper..	Up to 40	Up to 40	6.08
(32)	Ditto..	Up to 70	Up to 40	7.37
(33)	Ditto..	Up to 110	Up to 40	8.47
(34)	Grout pump (electric)............................	Up to 310	-	5.41
(35)	Ditto..	Up to 480	-	5.93
(36)	Agitating tank for use with grout pumps.....			2.53
(37)	Grout pump hand operated - 38m.............	Up to 25	Up to 40	1.58

The rates shown above are per unit machine per hour unless otherwise stated.

CIVIL ENGINEERING - PLANT (Cont'd)

Section & Item No.	Description	Nominal Size or Capacity		Hire Rate	Remarks
5. CONCRETE MIXERS (Cont'd)					
(38)	Concrete truck/mixer/agitator (including separate engine with mechanical or hydraulic drive to drum)	Mixer	Agitator		
		Up to 4.5	Up to 5.0m3	35.63	
(39)	Ditto	Up to 5.0	Up to 6.0m3	37.60	
(40)	Ditto	Up to 6.0	Up to 7.0m3	48.28	
6. CONCRETE EQUIPMENT		Maximum Output per hour		Per hour	
(1)	Concrete pump, skid or trailer mounted including pipe cleaning equipment (excluding piping)	Up to 26m3		25.31	
(2)	Ditto	Up to 40m3		27.85	
(3)	Ditto	Up to 55m3		34.69	
(4)	Ditto	Up to 85m3		49.79	
(5)	Ditto	Up to 110m3		81.72	
(6)	Concrete pump (lorry mounted) (exclusive of piping) (including boom up to 30 metre radius and pipe cleaning equipment)	Up to 50m3		64.30	
(7)	Ditto	Up to 80m3		89.05	
(8)	Ditto	Up to 120m3		154.69	
		Diameter			
(9)	Piping per 3m. straight length	Up to 102mm/per 3m		0.60	Per day
(10)	Ditto	Up to 127mm/per 3m		0.75	Per day
(11)	Ditto	Up to 152mm/per 3m		1.26	Per day
(12)	Ditto	Up to 102mm/per 3m		0.78	Per day
(13)	Ditto	Up to 127mm/per 3m		1.24	Per day
(14)	Ditto	Up to 152mm/per 3m		1.73	Per day
(15)	Flexible distributor hose per 4m length	Up to 102mm/per 3m		4.74	Per day
(16)	Ditto	Up to 127mm/per 3m		7.96	Per day
(17)	Ditto	Up to 152mm/per 3m		21.77	Per day
(18)	Shut off pipe section	Up to 102mm/per 3m		2.11	Per day
(19)	Ditto	Up to 127mm/per 3m		2.28	Per day
(20)	Ditto	Up to 152mm/per 3m		5.41	Per day
(21)	Vibrator poker(petrol driven)	-		1.19	
(22)	Ditto(diesel driven)	-		2.24	
(23)	Ditto(electric)	-		1.19	
(24)	Ditto...............air (excluding compressor)	-		0.54	
(25)	Alternator/frequency converter	Up to 2.5 kVA		1.63	
(26)	Ditto	Up to 4.5 kVA		2.48	
(27)	Vibrator poker (H/F motor in head type)	-		1.31	
(28)	Vibrator external type clamp-on electric	small		0.84	
(29)	Ditto	medium		0.95	
(30)	Ditto	large		1.70	
(31)	Ditto air (excluding compressor)	-		0.79	
(32)	Vibrator tamper type including single timber or metal screed board	screed length Up to 3.25m		1.51	
(33)	Extra for additional single screed length	additional per 300mm		0.05	
(34)	Vibrator tamper type including double screed board	screed length Up to 3.25m		2.22	
(35)	Extra for additional double screed length	additional per 300mm		0.08	
(36)	Power float(petrol driven)	-		2.13	
(37)	Ditto(electric driven)	-		2.09	
(38)	Concrete saw (exclusive of blades and water supply to be charged in addition) manually propelled	blade diameter Up to 350mm		2.29	
(39)	Ditto	Up to 450mm		2.91	
(40)	Self-propelled	Up to 500mm		3.43	
(41)	Ditto(with hydraulic system for blade movement)	Up to 600mm		7.84	
		Up to 900mm		9.73	
(42)	Ditto	width			
(43)	Joint former hand-propelled with vibratory blade	Up to 2.4m		6.80	
(44)	Ditto	Up to 8.2m		9.51	
(45)	Grinder pedestrian operated	-		2.73	
(46)	Concrete pram dobbin rubber tyred tipping.	Struck Capacity Up to 120 litre		1.60	per day
(47)	Ditto	Up to 200 litre		1.86	per day
(48)	Ditto	Up to 280 litre		2.74	per day
(49)	Power driven barrow	Up to 170 litre		0.87	

The rates shown above are per unit machine per hour unless otherwise stated

CIVIL ENGINEERING - PLANT (Cont'd)

Section & Item No.	Description	Nominal Size or Capacity	Hire Rate	Remarks
7. CRANES		Maximum S.W.L. to BS 1757 (1964)		
	Moble rubber tyred (site use only)	Clause II - Stability		
(1)	Full circle slew	Up to 6.5 tonnes	20.97	
(2)	Ditto	Up to 8.5 tonnes	22.12	
(3)	Ditto	Up to 12.5 tonnes	26.67	
(4)	Ditto	Up to 16.0 tonnes	32.18	
(5)	Ditto	Up to 19.0 tonnes	38.63	
(6)	Mobile rubber tyred rough terrain type (site use only excluding Tax, Insurance and extra cost of D.E.R.V.)	Up to 13 tonnes	35.35	
(7)	Ditto	Up to 18 tonnes	45.86	
(8)	Ditto	Up to 25 tonnes	66.54	
(9)	Ditto	Up to 40 tonnes	82.59	
(10)	Ditto	Up to 55 tonnes	104.03	
(11)	Ditto	Up to 70 tonnes	137.70	
(12)	Crawler mounted cranes (site use only)	Up to 28 tonnes	30.30	
(13)	Ditto	Up to 36 tonnes	35.41	
(14)	Ditto	Up to 56 tonnes	65.23	
(15)	Ditto	Up to 75 tonnes	89.76	
(16)	Ditto	Up to 85 tonnes	103.49	
(17)	Ditto	Up to 115 tonnes	116.95	
(18)	Ditto	Up to 150 tonnes	141.80	
(19)	Lorry mounted or all terrain Crane (including) Tax, Insurance and extra cost of D.E.R.V.)	Up to 12 tonnes	38.73	
(20)	Ditto	Up to 20 tonnes	52.63	
(21)	Ditto	Up to 26 tonnes	59.89	
(22)	Ditto	Up to 30 tonnes	65.56	
(23)	Ditto	Up to 45 tonnes	74.74	
(24)	Ditto	Up to 60 tonnes	121.03	
(25)	Ditto	Up to 90 tonnes	150.06	
(26)	Ditto	Up to 110 tonnes	193.97	
(27)	Ditto	Up to 135 tonnes	251.81	
(28)	Ditto	Up to 200 tonnes	301.14	
(29)	Ditto	Up to 300 tonnes	511.13	
(30)	Tower Crane - electric - (erected complete with ballast). Standard trolley or luffing jib, rail mounted (static or travelling) or fixed on anchor shoes or fixing angles. Capacity (in metre tonnes) is the maximum lift in tonnes x maximum radius at which that load can be lifted when the crane is fitted with its maximum jib length. Metre tonnes Up to 30m.t.	Height under hook above ground for trolley jib cranes (Jib pivot height for luffing jib cranes) Up to 25 m	13.79	
(31)	Up to 40m.t.	Up to 30 m	20.13	
(32)	Up to 50m.t.	Up to 30 m	21.86	
(33)	Up to 65m.t.	Up to 30 m	31.15	
(34)	As above at (30) but excluding luffing jib cranes 65 - 80m.t.	Up to 36 m	33.72	
(35)	Up to 110m.t.	Up to 36 m	38.29	
(36)	Up to 140m.t.	Up to 36 m	49.21	
(37)	Up to 170m.t.	Up to 36 m	56.63	
(38)	Up to 200m.t.	Up to 36 m	68.22	
(39)	Up to 250m.t.	Up to 36 m	83.30	
(40)	Luffing jib cranes exceeding 65m.t. rating to be charged at the following percentage of the above rates		150%	
(41)	Extra for extend height of tower (including accessories) for trolley jib cranes. Up to 65m.t.	per 1m of height	0.44	
(42)	Up to 140m.t.	per 1m of height	0.46	
(43)	Up to 170m.t.	per 1m of height	0.58	
(44)	Up to 250m.t.	per 1m of height	0.63	
(45)	Extra for extended height of tower (including accessories) for luffing jib cranes. Up to 60m.t.	per 1m of height	0.46	
(46)	Up to 100m.t.	per 1m of height	0.58	
(47)	Up to 170m.t.	per 1m of height	0.63	
(48)	Tower crane track complete with Fish plates, bolts, concr. sleeper pads, and gauge holders	Up to 61m.t./per m	0.76	per day
(49)	Ditto	Up to 300m.t./per m	1.05	per day

The rates shown above are per unit machine per hour unless otherwise stated.

CIVIL ENGINEERING - PLANT (Cont'd)

ection & Item No.	Description	Nominal Size or Capacity	Hire Rate	Remarks
	8. CRANE EQUIPMENT	struck capacity		
(1)	Crane Grab, all types, excavating (normal weight)..	Up to .55m3	3.29	
(2)	Ditto..	Up to .80m3	4.21	
(3)	Ditto..	Up to 1.25m3	5.18	
(4)	Ditto(heavy duty)...............................	Up to .55m3	4.83	
(5)	Ditto(Ditto).................................	Up to .80m3	6.04	
(6)	Ditto(Ditto).................................	Up to 1.25m3	7.88	
(7)	Crane Grab, all types, rehandling (normal weight)..	Up to .55m3	2.11	
(8)	Ditto..	Up to .80m3	2.18	
(9)	Ditto..	Up to 1.25m3	3.29	
(10)	Ditto(Heavy duty)............................	Up to .55m3	3.66	
(11)	Ditto(Ditto).................................	Up to .80m3	4.69	
(12)	Ditto(Ditto).................................	Up to 1.25m3	5.75	
(13)	Skip, muck tipping circular.......................	Up to .08m3	0.12	
(14)	Ditto..	Up to .15m3	0.15	
(15)	Ditto..	Up to .20m3	0.15	
(16)	Ditto..	Up to .40m3	0.17	
(17)	Ditto..	Up to .60m3	0.22	
(18)	Ditto..	Up to .80m3	0.26	
(19)	Ditto..	Up to 1.20m3	0.39	
(20)	Ditto..	Up to 1.60m3	0.56	
(21)	Skip concrete lay down or roll over standard front or bottom discharge.........	Up to .20m3	0.44	
(22)	Ditto..	Up to .40m3	0.56	
(23)	Ditto..	Up to .60m3	0.65	
(24)	Ditto..	Up to .80m3	0.79	
(25)	Ditto..	Up to 1.20m3	1.06	
(26)	Skip concrete roll over geared or hydraulic hand operated clamshell.......................	Up to .40m3	0.86	
(27)	Ditto..	Up to .60m3	0.97	
(28)	Ditto..	Up to .80m3	1.22	
(29)	Ditto..	Up to 1.20m3	1.85	
(30)	Ditto..	Up to 1.60m3	2.47	

ection & Item No.	Description	chain diameter mm.	safe working load tonnes	Hire Rate	Remarks
(31)	Chain slings or brothers Single...............	6	1.27	0.18	per day
(32)	Double..(at 90 degrees).......................	6	1.78	0.34	per day
(33)	Single..	10	2.19	0.29	per day
(34)	Double(Ditto)................................	10	3.24	0.54	per day
(35)	Single..	13	5.08	0.46	per day
(36)	Double(Ditto)................................	13	7.36	0.79	per day
(37)	Single..	16	7.27	0.55	per day
(38)	Double(Ditto)................................	16	11.18	1.24	per day
(39)	Single..	19	11.68	0.91	per day
(40)	Double(Ditto)................................	19	15.42	2.02	per day
(41)	Single..	22	15.85	1.31	per day
(42)	Double(Ditto)................................	22	21.46	3.03	per day
(43)	Grab tag lines................................	Up to 1.20m3 Grab		6.80	per day
(44)	Ditto..	Over 1.20m3 Grab		10.78	per day
(45)	Demolition Ball...............................	Dead Weight Up to 260Kg		1.73	per day
(46)	Ditto..	Up to 510Kg		2.11	per day
(47)	Ditto..	Up to 770Kg		2.61	per day
(48)	Ditto..	Up to 1020Kg		3.11	per day
(49)	Ditto..	Up to 1300Kg		3.60	per day
(50)	Ditto..	Up to 1550Kg		4.35	per day
(51)	Ditto..	Up to 1800Kg		4.84	per day
(52)	Ditto..	Up to 2050Kg		5.21	per day

ection & Item No.	Description	Nominal Size or Capacity	Hire Rate	Remarks
	9. DUMPERS	Maker's Rated payload		
(1)	Small Dumper (site use only - excluding Tax, Insurance and extra cost of D.E.R.V.) (Manual gravity tipping) (2 wheel drive).....	Up to 1200Kg	3.00	
(2)	Ditto..	Up to 1500Kg	4.04	
(3)	Ditto - Hydraulic Tipping........................	Up to 1500Kg	4.04	
(4)	Ditto..	Up to 2150Kg	4.38	
(5)	Ditto - 4 wheel drive, Hydraulic Tipping......	Up to 2100Kg	6.88	
(6)	Ditto..	Up to 2550Kg	7.93	
(7)	Ditto..	Up to 3100Kg	9.22	
(8)	Ditto..	Up to 4100Kg	13.40	
(9)	Ditto..	Up to 5100Kg	15.21	
(10)	Ditto..	Up to 6100Kg	16.99	
(11)	Extra for high discharge on equivalent standard dumper rate............................		+ 15%	

The rates shown above are per unit machine per hour unless otherwise stated.

GENERAL INFORMATION

CIVIL ENGINEERING - PLANT (Cont'd)

Section & Item No.	Description	Nominal Size or Capacity	Hire Rate	Remarks
	9. DUMPERS - contd	Maker's Rated payload		
(12)	Extra for Turntable Skip on equivalent standard dumper rate............................		+ 22%	
(13)	Rear Dump Truck..................................	Up to 15500Kg	31.30	
(14)	Ditto..	Up to 17000Kg	32.99	
(15)	Ditto..	Up to 22000Kg	41.60	
(16)	Ditto..	Up to 25000Kg	47.75	
(17)	Ditto..	Up to 28000Kg	57.86	
(18)	Ditto..	Up to 32000Kg	78.54	
(19)	Ditto..	Up to 38000Kg	81.30	
(20)	Ditto..	Up to 45000Kg	108.22	
(21)	Ditto..	Up to 50000Kg	110.57	
(22)	Articulated Dump Truck......................	Up to 12200Kg	24.66	
(23)	Ditto..	Up to 18500Kg	32.66	
(24)	Ditto..	Up to 23000Kg	44.80	
(25)	Ditto..	Up to 30000Kg	56.32	
(26)	Ditto..	Up to 35000Kg	90.05	
(27)	Ditto..	Up to 40000Kg	102.91	
	10. EXCAVATORS			
(1)	Rope operated, full circle slew, crawler mounted, with single equipment (dragline)...	Maker's Rated Dragline Capacity Up to .70m3	27.64	
(2)	Ditto..	Up to .80m3	33.46	
(3)	Ditto..	Up to 1.00m3	36.75	
(4)	Ditto..	Up to 1.20m3	43.34	
(5)	Ditto..	Up to 1.40m3	49.10	
(6)	Ditto..	Up to 1.50m3	52.15	
(7)	Ditto..	Up to 2.00m3	67.08	
(8)	Ditto..	Up to 2.30m3	76.66	
(9)	Ditto..	Up to 2.70m3	90.68	
(10)	Roper operated, full circle slew, crawler Mounted, with single equipment (face shovel)......................................	Maker's Rated Face Shovel Capacity Up to .70m3	32.45	
(11)	Ditto..	Up to 1.00m3	43.71	
(12)	Ditto..	Up to 1.40m3	58.74	
(13)	Ditto..	Up to 1.50m3	61.78	
(14)	Ditto..	Up to 2.50m3	100.62	
(15)	Hydraulic, full circle slew, crawler or wheel mounted, with single equipment..	Maker's Rated Nominal weight of machine Up to 2.0 tonnes	14.83	
(16)	Ditto..	Up to 3.0 tonnes	15.60	
(17)	Ditto..	Up to 4.0 tonnes	15.89	
(18)	Ditto..	Up to 6.0 tonnes	16.87	
(19)	Ditto..	Up to 11.0 tonnes	19.57	
(20)	Ditto..	Up to 14.0 tonnes	22.85	
(21)	Ditto..	Up to 17.0 tonnes	29.30	
(22)	Ditto..	Up to 21.0 tonnes	34.99	
(23)	Ditto..	Up to 25.0 tonnes	43.49	
(24)	Ditto..	Up to 30.0 tonnes	50.85	
(25)	Ditto..	Up to 38.0 tonnes	66.36	
(26)	Ditto..	Up to 55.0 tonnes	105.06	
(27)	Ditto..	Up to 75.0 tonnes	122.58	
(28)	Extra for m/c mounted percussion breaker..	Unit weight less cradle Up to 100Kg	4.50	
(29)	Ditto..	Up to 500Kg	8.03	
(30)	Ditto..	Up to 1000Kg	14.10	
(31)	Ditto..	Up to 1500Kg	17.21	
(32)	Ditto..	Up to 2000Kg	21.80	
(33)	Ditto..	Up to 3000Kg	30.89	
(34)	Hydraulic, offset or centre post, half circle slew, wheeled, dual purpose (back hoe/loader)..	Maker's Rated Loader Bucket Capacity Up to .60m3	11.54	
(35)	Ditto..	Up to .80m3	17.03	
(36)	Ditto..	Up to 1.00m3	18.14	
(37)	Excavator mats, Light, thickness 150mm	per sq.m	0.39	
(38)	Ditto...............Heavy, thickness 300mm	per sq.m -	0.46	

The rates shown above are per unit machine per hour unless otherwise stated.

CIVIL ENGINEERING - PLANT (Cont'd)

Section & Item No.	Description	Nominal Size or Capacity	Hire Rate	Remarks
	11. GENERATING SETS AND TRANSFORMERS	Nominal rating		
(1)	Generating set..	Up to 2.0kVa	2.46	
(2)	Ditto..	Up to 4.0kVa	3.17	
(3)	Ditto..	Up to 10.0kVa	4.14	
(4)	Ditto..	Up to 16.0kVa	7.11	
(5)	Ditto..	Up to 30.0kVa	9.31	
(6)	Ditto..	Up to 50.0kVa	11.80	
(7)	Ditto..	Up to 70.0kVa	13.06	
(8)	Ditto..	Up to 90.0kVa	15.60	
(9)	Ditto..	Up to 100.0kVa	18.61	
(10)	Ditto..	Up to 120.0kVa	20.59	
(11)	Ditto..	Up to 160.0kVa	24.50	
(12)	Ditto..	Up to 210.0kVa	29.56	
(13)	Ditto..	Up to 280.0kVa	33.69	
(14)	Ditto..	Up to 330.0kVa	33.46	
(15)	Transformer, stationary...(air cooled)	Up to 1.0kVva	0.22	
(16)	Ditto(Ditto)...	Up to 2.5kVa	0.32	
(17)	Ditto(Ditto)...	Up to 5.0kVa	0.56	
(18)	Ditto(Ditto)...	Up to 7.5kVa	0.90	
(19)	Ditto(Ditto)...	Up to 10.0kVa	1.16	
(20)	Ditto(Ditto)...	Up to 12.5kVa	1.39	
(21)	Ditto(Ditto)...	Up to 15.0kVa	1.66	
(22)	Ditto(Ditto)...	Up to 35.0kVa	1.49	
(23)	Ditto(Ditto)...	Up to 50.0kVa	2.01	
(24)	Ditto(Ditto)...	Up to 60.0kVa	2.18	
(25)	Ditto(Ditto)...	Up to 100.0kVa	2.39	
(26)	Ditto(Ditto)...	Up to 200.0kVa	3.59	
(27)	Ditto(Ditto)...	Up to 300.0kVa	4.47	
(28)	Ditto(Ditto)...	Up to 500.0kVa	5.42	
(29)	Mobile lighting unit, 2 light tungsten halogen including 1 1/2kVa generator.......	Tower/mast height Up to 8.0m	4.17	
(30)	Ditto, 4 light mercury vapour including 6 1/4kVa generator.................................	Up to 16.0m	11.13	
(31)	Ditto, 4 light tungsten halogen including 7 1/2kVa generator.................................	Up to 20.0m	14.11	
	12. HOISTS			
(1)	Goods hoist, cantilever or rope suspended c/w safety devices and limits, bottom gate and one landing gate, interlocked. Erected to 9m top landing height...............................	Up to 500Kg	3.62	
(2)	Extra for mast extension including ties.......	per 1m	0.05	
(3)	Extra for additional landing gate.................		0.06	
(4)	Goods hoists, rack and pinion complete with safety devices and limits, bottom gate and one landing gate interlocked. Erected to 50m landing height.....................................	Up to 400Kg	4.88	
(5)	Extra for mast extension for item (4) including cable, ties etc...........................	per 1m	0.06	
(6)	Extra for additional interlocking landing gate for item (4)............................		0.16	
(7)	As item (4) - Single cage..........................	Up to 800Kg	8.67	
(8)	Ditto...	Up to 1500Kg	8.98	
(9)	Item (7) or (8) when fitted with twin cage....		+ 66% on above rates	
(10)	Extra for mast extension including cable, ties, etc. for items (7) or (8)	per 1m	0.13	
(11)	Ditto (for item (9))....................................	per 1m	0.20	
(12)	Extra for additional interlocking landing gate for item (7) or (8)........................		0.13	
(13)	Ditto (for item (9))...................................		0.22	
(14)	Passenger/Goods hoist, rack and pinion complete with base unit, bottom gate and one set interlocking landing gate, cable and mast to 60m top landing height(Single Cage)	Up to 1000Kg or 12 men	11.91	
(15)	Ditto...	Up to 1600Kg or 20 men	13.43	
(16)	Ditto(Twin Cage)	Up to 1000Kg or 12 men	18.36	
(17)	Ditto...	Up to 1600Kg or 20 men	20.29	

The rates shown above are per unit machine per hour unless otherwise stated.

GENERAL INFORMATION

CIVIL ENGINEERING - PLANT (Cont'd)

Section & Item No.	Description	Nominal Size or Capacity	Hire Rate	Remarks
	12. HOISTS - cont			
(18)	Extra for mast extension for single cage Pass/Goods hoist (including cable, ties etc.)......................................	per 1m	0.15	
(19)	Extra for additional interlocking landing gates for single cage pass/goods hoist......		0.24	
(20)	Extra for mast extension for Twin cage Pass/Goods hoist (including cable, ties etc.)......................................	per 1m	0.19	
(21)	Extra for additional interlocking landing gates for Twin cage Pass/Goods hoist......		0.44	
(22)	Scaffold Hoist..............................	Up to 75Kg	0.67	
(23)	Ditto..	Up to 200Kg	1.24	
	13. LIFTING AND JACKING GEAR			
(1)	Shear legs, steel, 5 metres......................	Up to 1 tonne	2.61	per day
(2)	Pipe gantry..	Up to 3 tonnes	6.03	per day
(3)	Chain blocks....................................	Up to 1 tonne	3.60	per day
(4)	Ditto..	Up to 2 tonnes	3.98	per day
(5)	Ditto..	Up to 3.5 tonnes	4.66	per day
(6)	Ditto..	Up to 4.5 tonnes	5.46	per day
(7)	Ditto..	Up to 5.5 tonnes	6.02	per day
(8)	Ditto..	Up to 6.5 tonnes	6.83	per day
(9)	Ditto..	Up to 8 tonnes	8.10	per day
(10)	Ditto..	Up to 10 tonnes	10.12	per day
(11)	Ditto..	Up to 15 tonnes	19.90	per day
(12)	Ditto..	Up to 20 tonnes	25.57	per day
(13)	Hydraulic Jack (Hand operated)................	Up to 10.00 tonnes	2.06	per day
(14)	Ditto..	Up to 15.00 tonnes	2.33	per day
(15)	Ditto..	Up to 20.00 tonnes	2.72	per day
(16)	Ditto..	Up to 30.00 tonnes	3.15	per day
(17)	Ditto..	Up to 60.00 tonnes	4.78	per day
(18)	Ditto..	Up to 100.00 tonnes	6.99	per day
(19)	Ratchet Jack......................................	Up to 5.50 tonnes	1.82	per day
(20)	Ditto..	Up to 10.50 tonnes	2.94	per day
(21)	Ditto..	Up to 16.00 tonnes	3.22	per day
(22)	Ditto..	Up to 21.00 tonnes	4.23	per day
		Safe working load		
		Lifting · Pulling		
(23)	Lifting and pulling machine......................	800Kg · 1200Kg	1.52	per day
(24)	Ditto..	1600Kg · 2500Kg	2.05	per day
(25)	Ditto..	3000Kg · 5000Kg	3.19	per day
	14. LORRIES, VANS, ETC.			
(1)	Lorry, ordinary (site use and public highway use - including Tax, Insurance and extra cost of petrol or D.E.R.V.).	Plated Gross Vehicle Weight Up to 6.5 tonnes	9.48	
(2)	Ditto..	Up to 7.5 tonnes	10.63	
(3)	Ditto..	Up to 12.0 tonnes	12.59	
(4)	Ditto..	Up to 14.0 tonnes	14.74	
(5)	Ditto..	Up to 17.0 tonnes	18.91	
(6)	Ditto..	Up to 24.5 tonnes	26.02	
(7)	Ditto..	Up to 30.0 tonnes	30.73	
(8)	Ditto (Articulated).............................	Up to 33.0 tonnes	27.32	
(9)	Ditto (Articulated).............................	Up to 39.0 tonnes	32.03	
(10)	Extra for lorry fitted with crane attachment (whether used or not).............	Maximum load Up to 1.5 tonnes	1.02	
(11)	Ditto..	Up to 2.5 tonnes	1.30	
(12)	Ditto..	Up to 3.5 tonnes	2.17	
(13)	Ditto..	Up to 5.0 tonnes	2.49	
(14)	Ditto..	Over 5.1 tonnes	3.10	
(15)	Lorry, tipper (site use and public highway use - including Tax, Insurance and extra cost of D.E.R.V.).................................	Plated Gross Vehicle Weight Up to 11.0 tonnes	11.64	

The rates shown above are per unit machine per hour unless otherwise stated.

CIVIL ENGINEERING - PLANT (Cont'd)

Section & Item No.	Description	Nominal Size or Capacity	Hire Rate	Remarks
	14. LORRIES, VANS, ETC. (cont'd)			
(16)	Ditto..	Up to 17.0 tonnes	18.62	
(17)	Ditto..	Up to 25.0 tonnes	27.41	
(18)	Ditto..	Up to 31.0 tonnes	33.77	
(19)	Extra for Side Tipping............................	-	+ 20%	
(20)	Van, pick up or similar utility vehicle........	Carrying Capacity		
		Up to .60 tonnes	6.07	
(21)	Ditto..	Up to 1.10 tonnes	6.84	
(22)	Ditto..	Up to 1.30 tonnes	8.22	
(23)	Ditto..	Up to 1.60 tonnes	8.73	
(24)	Ditto..	Up to 2.10 tonnes	10.75	
(25)	Ditto..	Up to 2.60 tonnes	12.06	
(26)	Passenger/goods, cross country.............	Wheelbase up to 2.4m	9.48	
(27)	Ditto..	Wheelbase 2.40m and over	10.05	
(28)	Ditto, Station wagon..........................	Up to 7 seater	10.40	
(29)	Ditto..	Up to 12 seater	11.39	
(30)	Personnel carrier/coach/bus....................	Up to 9/13 seater	10.56	
(31)	Ditto..	Up to 14/17 seater	11.94	
		Hopper Capacity		
(32)	Road sweeper/cleaner self-propelled........	Up to 1.0m3	10.63	
(33)	Ditto..	Up to 2.0m3	13.35	
(34)	Road sweeper complete with gulley emptier (site and public highway use including Tax, Insurance and use of D.E.R.V.).......................................	Up to 3.0m3	14.86	
(35)	Ditto..	Up to 6.0m3	20.05	
(36)	Tractor mounted brush (front or rear) non-collector (excluding tractor)................	-	3.09	
	15. MATERIALS HANDLING			
(1)	Trailer flat (towed with knock on brakes) (site use only excluding Tax and Insurance)...	Carrying Capacity Up to 2.0 tonnes	6.69	per day
(2)	Ditto..	Up to 3.0 tonnes	10.09	per day
(3)	Ditto..	Up to 4.0 tonnes	11.50	per day
(4)	2 wheeled tipping trailer with overrun brakes, powered tipping.......................	Up to 4 tonnes	13.23	per day
(5)	4 wheeled flat trailer with Hydraulic or air brakes...	Up to 7 tonnes	15.11	per day
(6)	Ditto..	Up to 10 tonnes	25.43	per day
(7)	Ditto..	Up to 12 tonnes	31.25	per day
		Maximum Lifting Capacity (forward loading)		
(8)	Fork lift truck (yard type)...........................	Up to 1000Kg	5.34	
(9)	Ditto..	Up to 1250Kg	5.92	
(10)	Ditto..	Up to 1500Kg	6.68	
(11)	Ditto..	Up to 2000Kg	7.91	
(12)	Ditto..	Up to 2500Kg	8.77	
(13)	Fork lift truck (rough terrain type) (2 wheel drive)...	Up to 1000Kg	7.19	
(14)	Ditto..	Up to 1250Kg	7.81	
(15)	Ditto..	Up to 1500Kg	8.56	
(16)	Ditto..	Up to 2000Kg	10.11	
(17)	Ditto..	Up to 2500Kg	11.41	
(18)	Ditto..	Up to 3000Kg	12.90	
(19)	Ditto..	Up to 4000Kg	15.88	
(20)	Ditto..	Up to 5000Kg	18.84	
(21)	Fork lift truck (rough terrain) (4 wheel drive)...	Up to 1500Kg	8.81	
(22)	Ditto..	Up to 2000Kg	9.92	

The rates shown above are per unit machine per hour unless otherwise stated.

GENERAL INFORMATION

CIVIL ENGINEERING - PLANT (Cont'd)

Section & Item No.	Description	Nominal Size or Capacity	Hire Rate	Remarks
	15. MATERIALS HANDLING (Cont'd)			
(23)	Ditto....................................	Up to 2500Kg	11.91	
(24)	Ditto....................................	Up to 3000Kg	13.08	
(25)	Ditto....................................	Up to 3500Kg	15.34	
(26)	Ditto....................................	Up to 4000Kg	18.84	
(27)	Ditto....................................	Up to 5000Kg	24.75	
(28)	Ditto....................................	Up to 6000Kg	25.53	
(29)	Ditto....................................	Up to 7000Kg	26.76	
(30)	Ditto....................................	Up to 8000Kg	29.02	
(31)	Telescopic Site Handling Machines (2 wheel drive)............................	Up to 2500Kg	10.67	
(32)	Ditto....................................	Up to 4000Kg	12.94	
(33)	Telescopic Site Handing Machines (4 wheel drive)............................	Up to 2500Kg	13.15	
(34)	Ditto....................................	Up to 4000kg	15.78	
	16. OFFICES, STORES, ETC.			
(1)	Offices on site with usual fittings ie, heating and lighting, desks, tables, chairs, plan chests, etc........................	Per 10m² floor area	29.48	per week
(2)	Mess room on site with usual fittings, ie, heating and lighting, tables, forms etc. but excluding kitchen equipment......................	"	26.34	per week
(3)	Stores on site with usual fittings ie, counter, desk, chairs, racks, shelves or bins, heating and lighting..............................	"	24.44	per week
(4)	Toilet unit, single chemical type, (excluding servicing and consumables).........................	"	8.67	per week
(5)	Toilet unit, single mains flushing type, with wash basin, excluding consumables............. Note:- Rate for multiple toilet units: Add 70% of single unit rate for each additional unit.	"	15.21	per week
(6)	Mobile office including heating and lighting (mains) and usual fittings,	Up to 5m long	51.16	per week
(7)	Ditto....................................	Up to 7m long	64.03	per week
(8)	I.S.O. Containers used as Tool Stores, Fitting shop or storage........................	Up to 13m long	6.84	per day
	17. PILING PLANT (Excluding Compressor)	Hammer weight		
(1)	Piling Hammer double-acting (air)................	Up to 155Kg	2.90	
(2)	Ditto....................................	Up to 305Kg	3.15	
(3)	Ditto....................................	Up to 1143Kg	4.27	
(4)	Ditto....................................	Up to 2132Kg	5.75	
(5)	Ditto....................................	Up to 3006Kg	7.26	
(6)	Ditto....................................	Up to 4334Kg	11.11	
(7)	Ditto....................................	Up to 6350Kg	20.34	
(8)	Hydraulically operated drop hammer c/w hydraulic power pack....................................	Up to 3000Kg	95.23	
(9)	Ditto....................................	Up to 5000kg	109.88	
(10)	Ditto....................................	Up to 7000Kg	124.53	
(11)	Air driven impulse hammer..........................	Total weight Up to 5 tonnes	36.63	
(12)	Drop Hammer (bare) (suitable for use with channel or tubular leaders fitted with leather guides and rubber inserts)...........................	Hammer weight Up to 1016Kg	2.73	
(13)	Ditto....................................	Up to 2032Kg	3.40	
(14)	Ditto....................................	Up to 3048Kg	4.07	
(15)	Internal drop hammer (for use with cased piles): Internal diameter of cased pile..............................Up to 305mm	Up to 1270Kg	2.05	
(16)	Ditto.........................Up to 406mm	Up to 3556Kg	2.58	
(17)	Ditto.........................Up to 508mm	Up to 5588Kg	3.27	
(18)	Air operated Extractor (excluding compressor)..............................	Unit weight Up to 764Kg	6.93	
(19)	Ditto....................................	Up to 1705Kg	9.95	

The rates shown above are per unit machine per hour unless otherwise stated.

CIVIL ENGINEERING - PLANT (Cont'd)

Section & Item No.	Description	Nominal Size or Capacity	Hire Rate	Remarks
	17. PILING PLANT (Cont'd)			
(20)	Ditto..	Up to 3000Kg	12.22	
(21)	Ditto..	Up to 4590Kg	14.94	
(22)	Flexible Reinforced Rubber Hose (for compressed air)......................................	Diameter		
		Up to 25mm/per m	0.31	per day
(23)	Ditto..	Up to 32mm/per m	0.38	per day
(24)	Ditto..	Up to 38mm/per m	0.45	per day
(25)	Ditto..	Up to 51mm/per m	0.55	per day
(26)	Ditto..	Up to 64mm/per m	0.68	per day
(27)	Diesel Hammer (including complete set of guide equipment) (single acting)................	Piston weight		
		Up to 500Kg	14.64	
(28)	Ditto..	Up to 800Kg	16.49	
(29)	Ditto..	Up to 1300Kg	19.49	
(30)	Ditto..	Up to 1500Kg	22.08	
(31)	Ditto..	Up to 1900Kg	23.28	
(32)	Ditto..	Up to 2500Kg	34.18	
(33)	Ditto..	Up to 3000Kg	36.32	
(34)	Ditto..	Up to 3500Kg	48.83	
(35)	Ditto..	Up to 4600Kg	56.49	
(36)	Ditto..	Up to 6200Kg	77.33	
(37)	Ditto..	Up to 8000Kg	89.94	
(38)	Ditto..	Up to 10000Kg	100.80	
		Centrifugal force Tonnes		
(39)	Hydraulic vibrating hammer/extractor including power pack..............................	Up to 21	49.81	
(40)	Ditto..	Up to 38	72.51	
(41)	Ditto..	Up to 62	107.68	
(42)	Ditto..	Up to 143	190.76	
(43)	Ditto..	Up to 207	366.28	
(44)	Pile Helmet for Pile.............................	203mm x 203mm	0.84	
(45)	Ditto..	254mm x 254mm	1.16	
(46)	Ditto..	305mm x 305mm	1.62	
(47)	Ditto..	356mm x 356mm	1.69	
(48)	Ditto..	406mm x 406mm	1.90	
(49)	Ditto..	457mm x 457mm	2.02	
(50)	Ditto..	508mm x 508mm	2.15	
	(Plastic Dollies or equivalent to be paid for in addition)			
(51)	Hanging Leaders Channel type (for use with drop hammer...............................	Length of jib 12.2m	7.52	
(52)	Ditto..	15.3m	8.58	
(53)	Hanging Leaders, rectangular section .61m x .61m (for use with drop or diesel hammers)..	Crane Boom 9.2 — Nominal Length 15.3	13.39	
(54)	Ditto..	15.3 — 24.4	15.44	
(55)	Hanging Leaders, rectangular section .84 x .84m...................................	12.2 — 21.4	19.28	
(56)	Ditto..	18.3 — 29.9	23.47	
(57)	Ditto..	24.4 — 36.6	26.27	
(58)	Auger Piling and diaphragm Walling machines and other specialist equipment			To be Negotiated

18. PILING

Temporary steel piling and steel trench sheeting to be charged as a material. The residual value of recovered steel piling and steel trench sheeting to be the subject of special agreement.

The rates shown above are per unit machine per hour unless otherwise stated.

GENERAL INFORMATION

CIVIL ENGINEERING - PLANT (Cont'd)

Section & Item No.	Description	Nominal Size or Capacity	Hire Rate	Remarks
	19. PIPE BENDING EQUIPMENT			
(1)	Special purpose pipe winches (for standard winches - See Section 36).......................	Nominal pipe diameter	To be negotiated	
(2)	Pipe bending machine (Hand operated).....	Up to 50mm	0.60	
(3)	Ditto..	Up to 75mm	0.72	
(4)	Ditto..	Up to 150mm	3.69	
(5)	Ditto..	Up to 50mm	1.70	
(6)	Ditto..	Up to 75mm	1.90	
(7)	Ditto..	Up to 100mm	3.33	
(8)	Ditto..	Up to 150mm	4.57	

Section & Item No.	Description	Nominal Size or Capacity	Basic Rate per Day	Extra per Working Hour	Remarks
	20. PUMPS, PORTABLE				
	(exclusive of all hoses)				
(1)	Semi rotary (hand).....................................	Up to 19mm	0.53	-	
(2)	Ditto..	Up to 25mm	0.66	-	
(3)	Single diaphragm.......................................	Up to 51mm	5.21	0.37	
(4)	Ditto..	Up to 76mm	7.44	0.56	
(5)	Ditto..	Up to 102mm	7.91	0.59	
(6)	Double diaphragm......................................	Up to 51mm	7.25	0.50	
(7)	Ditto..	Up to 76mm	8.27	0.62	
(8)	Ditto..	Up to 102mm	15.40	1.11	
(9)	Self priming centrifugal.............................	Up to 38mm	3.95	0.32	
(10)	Ditto..	Up to 51mm	10.00	0.75	
(11)	Ditto..	Up to 76mm	14.18	1.41	
(12)	Ditto..	Up to 102mm	17.13	1.59	
(13)	Ditto..	Up to 152mm	23.20	1.95	
(14)	Sludge and Sewage...................................	Up to 76mm	16.50	1.32	
(15)	Ditto..	Up to 102mm	18.22	1.43	
(16)	Ditto..	Up to 152mm	29.57	2.73	
(17)	Ditto..	Up to 204mm	36.73	3.37	
(18)	Sump pneumatic (excluding compressor)..	Up to 51mm	4.13	0.21	
(19)	Ditto..	Up to 64mm	5.10	0.26	
(20)	Ditto..	Up to 76mm	6.86	0.36	
(21)	Electric submersible..................................	Up to 38mm	2.34	0.17	
(22)	Ditto..	Up to 51mm	2.96	0.30	
(23)	Ditto..	Up to 64mm	4.61	0.40	
(24)	Ditto..	Up to 76mm	6.55	0.67	
(25)	Ditto..	Up to 102mm	10.83	1.71	

Section & Item No.	Description	Nominal Size or Capacity	Hire Rate	Remarks
	21. PUMPING EQUIPMENT			
	Pump hoses, flexible, suction or delivery, including coupling, valve and strainer	Diameter		per metre/day
(1)	Suction..	Up to 38mm	0.20	per metre/day
(2)	Ditto..	Up to 51mm	0.21	per metre/day
(3)	Ditto..	Up to 76mm	0.29	per metre/ day
(4)	Ditto..	Up to 102mm	0.39	per metre/day
(5)	Ditto..	Up to 152mm	0.74	per metre/ day
(6)	Ditto..	Up to 204mm	1.22	per metre/ day
(7)	Delivery..	Up to 38mm	0.13	per metre/ day
(8)	Ditto..	Up to 51mm	0.15	per metre/ day
(9)	Ditto..	Up to 76mm	0.26	per metre/ day
(10)	Ditto..	Up to 102mm	0.39	per metre/ day
(11)	Ditto..	Up to 152mm	0.69	per metre/ day
(12)	Ditto..	Up to 204mm	1.22	per metre/ day

The rates shown above are per unit machine per hour unless otherwise stated.

CIVIL ENGINEERING - PLANT (Cont'd)

Section & Item No.	Description	Nominal Size or Capacity	Hire Rate	Remarks
	21. PUMPING EQUIPMENT (Cont'd)			
(13)	Steel suction or delivery, including flanges, bolts and joint rings (excluding valve and strainer)...............................	Up to 76mm	0.15	per metre/ day
(14)	Ditto..	Up to 152mm	0.22	per metre/ day
(15)	Ditto..	Up to 204mm	0.49	per metre/ day
(16)	Bend to be charged as 9.0m length..........	-		
(17)	Valve to be charged as 9.0m length..........	-		
(18)	Steel pipe suction or delivery, screwed and socketed joints (excluding valve and strainer.................................	Up to 51mm	0.05	per metre/ day
(19)	Ditto..	Up to 102mm	0.12	per metre/ day
(20)	Ditto..	Up to 152mm	0.19	per metre/day
(21)	Bend to be charged as 3.50m length........	-		
(22)	Valve to be charged as 3.50m length........	-		
	22. RAMMERS AND COMPACTORS			
(1)	Vibro or vibration rammer........................	Up to 60Kg	1.16	
(2)	Ditto..	Up to 70Kg	1.35	
(3)	Ditto..	Up to 90Kg	1.60	
(4)	Ditto..	Up to 115Kg	1.95	
(5)	Ditto..	Up to 200Kg	2.01	
(6)	Vibrating plate compactor........................	Up to 80Kg	1.11	
(7)	Ditto..	Up to 110Kg	1.51	
(8)	Ditto..	Up to 150Kg	1.94	
(9)	Ditto..	Up to 200Kg	2.52	
(10)	Ditto..	Up to 300Kg	3.26	
(11)	Ditto..	Up to 400Kg	4.70	
(12)	Ditto..	Up to 500Kg	5.92	
(13)	Ditto..	Up to 600Kg	7.21	
(14)	Jumping rammer including trolley......	Up to 100Kg	1.77	
	23. ROLLERS			
(1)	Road Deadweight (steel 3 wheel/3 roll, Diesel)..	Unballasted weight Up to 6.10 tonnes	13.53	
(2)	Ditto..	Up to 8.50 tonnes	13.88	
(3)	Ditto..	Up to 10.50 tonnes	15.25	
(4)	Ditto..	Up to 13.00 tonnes	17.21	
(5)	Ditto(tandem) Diesel........................	Up to 8.50 tonnes	14.24	
(6)	Rubber tyred self propelled....................	Average Weight Up to 1.0 tonne/wheel	21.04	
(7)	Ditto(Ditto).................................	Up to 2.0 tonnes/wheel	32.75	
(8)	Ditto(Ditto).................................	Up to 3.0 tonnes/wheel	44.29	
(9)	Vibratory pedestrian operated............................(single roller)	Makers weight Up to 550Kg	2.72	
(10)	Ditto..	Up to 700Kg	3.01	
(11)	Vibratory pedestrian operated................................(twin roller)	Up to 650Kg	3.72	
(12)	Ditto..	Up to 950Kg	4.52	
(13)	Ditto..	Up to 1300Kg	5.88	
(14)	Ditto..	Up to 1750Kg	10.17	
(15)	Towed, Vibratory trailer...........................	Up to 1500Kg	5.21	
(16)	Ditto..	Up to 6500Kg	9.88	
(17)	Ditto..	Up to 8800Kg	11.66	
(18)	Ditto..	Up to 11700Kg	14.46	
(19)	Ditto..	Up to 13300Kg	15.51	
(20)	Self propelled vibratory tandem (seated control)......................................	Up to 1.4 tonnes	7.21	
(21)	Ditto..	Up to 2.5 tonnes	9.72	
(22)	Ditto..	Up to 5.0 tonnes	16.37	
(23)	Ditto..	Up to 9.0 tonnes	27.44	
(24)	Ditto..	Up to 12.0 tonnes	34.37	
(25)	Ditto, single roll, rubber tyred driving wheels..	Up to 8.2 tonnes	29.03	
(26)	Ditto..	Up to 10.5 tonnes	32.84	
(27)	Scarifier (working time) extra over Roller including sharpening tines)......................	1 Tine	4.19	
(28)	Ditto..	2 Tine Hydraulic	6.70	
(29)	Ditto..	3 Tine Hydraulic	9.41	

The rates shown above are per unit machine per hour unless otherwise stated.

GENERAL INFORMATION

CIVIL ENGINEERING - PLANT (Cont'd)

Section & Item No.	Description	Nominal Size or Capacity	Hire Rate	Remarks
	23. ROLLERS - cont'd	Maker's rated flywheel		
(30)	Compactor sheeps foot self-propelled Tamping foot heavy duty wheeled...........	Up to 130Kw	52.25	
(31)	Ditto..	Up to 200Kw	66.55	
(32)	Ditto..	Up to 300Kw	97.52	
	24. SAWS MECHANICAL			
(1)	Chain Saw including Heavy Duty protective clothing.....................	Guide bar length Up to 310mm	1.31	
(2)	Ditto..	Up to 410mm	1.49	
(3)	Ditto..	Up to 510mm	1.87	
(4)	Ditto..	Up to 650mm	2.11	
(5)	Ditto..	Up to 800mm	2.30	
(6)	Ditto..	Up to 950mm	2.47	
(7)	Ditto..	Up to 1300mm	2.56	
		Saw diametre		
(8)	Protable Saw Bench........................	Up to 260mm	1.14	
(9)	Ditto..	Up to 310mm	1.26	
(10)	Ditto..	Up to 410mm	1.53	
(11)	Ditto..	Up to 460mm	1.69	
(12)	Ditto..	Up to 510mm	1.89	
(13)	Ditto..	Up to 610mm	2.29	
(14)	Ditto..	Up to 660mm	2.53	
(15)	Ditto..	Up to 760mm	3.03	
(16)	Band Saw.......................................	-	2.81	
	25. SCAFFOLDING			
(1)	Tubular steel..................................	.51mm dia. Nominal per m	.02	per week
(2)	Ditto, alloy....................................	.51mm dia. Nominal per m	0.05	per week
(3)	Putlog steel...................................	1.52m long Nominal	0.05	per week
(4)	Ditto, alloy....................................	1.52m long Nominal	0.07	per week
(5)	Ditto, steel....................................	1.83m long Nominal	0.05	per week
(6)	Putlog alloy...................................	1.83m long Nominal	0.08	per week
(7)	Fitting steel, single double or swivel coupler, joint pin, fixed base plate..........	-	.02	per week
(8)	Ditto, Adjustable Base Plate....................	-	0.12	per week
(9)	Ditto, Hop up Bracket.....................	-	0.18	per week
(10)	Ditto, Split Head trestle folding type/adjustable.....................	.48m - .84m	0.57	per week
(11)	Ditto..	.76m - 1.37m	0.74	per week
(12)	Ditto..	1.07m - 1.83m	0.85	per week
(13)	Ditto..	1.37m - 2.44m	1.05	per week
(14)	Castor...	-	0.60	per week
(15)	Ditto, rubber tyred.........................		0.66	per week
(16)	Ditto, nylon tyred...........................	-	1.26	per week
(17)	Jenny Wheel including 30m rope..........	Up to 254mm diameter	2.33	per week
(18)	Ditto..	Up to 305mm diameter	2.57	per week
(19)	Board..	Up to 4.0m long	0.29	per week
	Rates for special items of prefabricated scaffold units owned by the main contractor to be similar to those charge by Proprietory Firms, plus 12.5%			
	26. SHORING, PLANKING AND STRUTTING			
(1)	Baulk timber, use and waste (excluding nails, dogs, wedges, etc., to be charged in addition as consumables)................. (Minimum charge 18 days hire)	Per m^3	3.08	per day
(2)	Timber for planking and strutting use and waste (excluding nails, dogs, wedges, etc., to be charged in addition as consumables)................................... (Minimum charge 12 days hire)	Per m^3	4.37	per day
(3)	Adjustable Steel Strut, extending..........	457mm - 711mm	0.50	per week
(4)	Ditto..	686mm - 1.09m	0.58	per week
(5)	Ditto..	1.04m - 1.70m	0.66	per week

The rates shown above are per unit machine per hour unless otherwise stated.

CIVIL ENGINEERING - PLANT (Cont'd)

Section & Item No.	Description	Nominal Size or Capacity	Hire Rate	Remarks
	27. SHUTTERING			
(1)	Steel Shutters, all types. Rates for items of steel shuttering to be based on those chargeable by Proprietary Firms. (Wedges, keys and consumables to be added) plus 12.5%...............................	per 10 metres		per day
(2)	Steel road forms...................................	102mm per 10 metres	0.57	per day
(3)	Ditto...	152mm per 10 metres	0.62	per day
(4)	Ditto...	203mm per 10 metres	0.72	per day
(5)	Ditto...	203mm (heavy section) per 10 metres	0.93	per day
(6)	Ditto...	203mm (heavy section with rail) per 10 metres	1.19	per day
(7)	Ditto...	254mm (heavy section with rail) per 10 metres	1.49	per day
(8)	Ditto...	305mm (heavy section with rail) per 10 metres	1.97	per day
(9)	Telescopic steel floor centre, extending lattice girder type................................	Span up to 1.59m	1.09	per week
(10)	Ditto...	Span up to 2.76m	1.35	per week
(11)	Ditto...	Span up to 4.16m	1.80	per week
(12)	Ditto...	Span up to 5.56m	2.13	per week
(13)	Telescopic steel prop extending............	1.04m - 1.83m	0.71	per week
(14)	Ditto...	1.75m - 3.12m	0.88	per week
(15)	Ditto...	1.98m - 3.35m	0.93	per week
(16)	Ditto...	2.44m - 3.96m	1.02	per week
(17)	Column Shutter Clamps, extending........	254mm - 508mm per set	0.69	per week
(18)	Ditto...	406mm - 813mm per set	0.87	per week
(19)	Ditto...	.61m - 1.22m per set	1.22	per week
(20)	Beam Shutter Clamps, clamping width 114mm to 850mm..............................	305mm arm per set	0.88	per week
(21)	Ditto...	457mm arm per set	0.92	per week
(22)	Ditto...	610mm arm per set	1.40	per week
(23)	Wall Shutter Clamps, concrete thickness	102mm to 305mm per set	0.75	per week
(24)	Ditto...	102mm to 610mm per set	0.90	per week
(25)	Ditto...	102mm to 915mm per set	1.13	per week
(26)	Timber used for shutter. Rough (minimum charge 10 days hire) (excluding wedges, nails, screws, bolts, etc., to be charged in addition as consumables + 12.5%).........................	per m^3	6.73	per day
(27)	Ditto, Wrot (minimum charge 10 days hire)..	per m^3	8.86	per day
(28)	Douglas Fir Plywood Sheeting (Excluding timber and steel or timber supports to be charged in addition to + 12.5%) (Thickness 13mm (minimum charge 12 days Hire).....................................	Good 1 side per 3m^2	0.59	per day
(29)	Ditto...	Good 2 sides per 3m^2	0.82	per day
(30)	Thickness 19mm (minimum charge 12 days hire)....................................	Good 1 sides per 3m^2	0.79	per day

The rates shown above per unit machine per hour unless otherwise stated.

GENERAL INFORMATION

CIVIL ENGINEERING - PLANT (Cont'd)

Section & Item No.	Description	Nominal Size or Capacity	Hire Rate	Remarks
	27. SHUTTERING (cont'd)			
(31)	Ditto..	Good 2 sides per 3m^2	1.09	per day
(32)	Thickness 25mm (minimum charge 12 days hire)...	Good 1 sides per 3m^2	1.10	per day
(33)	Ditto..	Good 2 sides per 3m^2	1.26	per day
	Plastic Faced			
(34)	Thickness 13mm (minimum charge 12 days hire)...	per 3m^2	1.22	per day
(35)	Thickness 19mm (minimum charge 12 days hire)...	per 3rn^2	1.75	per day

Item (28) to (35) exclude wedges, nails screws, bolts etc., to be charged in addition as consumables at cost plus 12 1/2%

Section & Item No.	Description	Nominal Size or Capacity	Hire Rate	Remarks
	28. SURVEYING INSTRUMENTS			
(1)	Dumpy level c/w Tripod and Staff......	-	2.44	per day
		Standard Deviation		
(2)	Auto level c/w Tripod and Staff.........	2.5mm/km	3.90	per day
(3)	Engineers Precise Auto level c/w Tripod and Staff..................................	2.0mm/km	4.41	per day
(4)	Engineers Precise Auto level with parallel plate c/w Tripod and Staff......	1.0mm/km	8.31	per day
(5)	Digital Theodolite................................	20 seconds	10.87	per day
(6)	Ditto..	6 seconds	13.38	per day
(7)	Ditto..	1 second	21.72	per day
(8)	Ranging Rod or Pole in timber..............	-	0.06	per day
(9)	Ranging Rod or Pole in alloy...............	-	0.12	per day
	Rates for laser levelling or distance measuring to be based on those charged by proprietory firms + 12 1/2%.			
	29. TAR SPRAYING AND COLD EMULSION PLANT			
(1)	Tar Boiler and Sprayer (including firing) hand operated.....................................	Up to 700 litres	4.07	
(2)	Ditto, power operated...........................	Up to 1200 litres	6.63	
(3)	Gritter (attached to lorry) extra over lorry...	-	1.29	
(4)	Gritter (towed hopper type) ditto..........	-	2.90	
(5)	Cold Emulsion Sprayer hand operated..	Up to 250 litres	0.84	
(6)	Ditto, power operated...........................	Over 700 litres	1.77	
	30 TOOLS PNEUMATIC (excluding compressors but including 15m of hose)			
(1)	Breaker including steels.........................	-	1.90	
(2)	Light pneumatic pick including steels.....	-	1.62	
(3)	Pneumatic clay Spade including blade...	-	1.72	
(4)	Chipping/Scaling/Caulking hammer.........	-	0.22	
	The rates for the above items includes the sharpening of the steel or blade.			
(5)	Hand-held Rock Drill without Drill Rods or Detachable Bits................light weight	16-20Kg	0.72	
(6)	Ditto..................................middle weight	Up to 24Kg	1.00	
(7)	Ditto......................................heavy weight	Up to 32Kg	1.07	
	Drill Rods and Bits for Hand-held Rock Drill to be paid for in addition + 12 1/2%			
(8)	Addition for Silencer Tool or Muffler........	-	0.15	
(9)	Additional hoses.....................................	Per 15m	0.17	
(10)	Drill(excluding consumables).............	Up to 13mm chuck	0.30	
(11)	Reversible drill(Ditto).....................	Up to 13mm chuck	0.48	
(12)	Grinder(lightweight)(Ditto).............	76mm diameter	0.26	
(13)	Ditto(heavy weight)(Ditto)..............	117mm diameter	0.35	
(14)	Sander (excluding pad and disc)(Ditto)...	-	0.48	
(15)	Riveting Hammer(Ditto)......................	-	0.28	
(16)	Pneumatic paint scraper tool....................	-	0.26	
(17)	Polisher..	-	0.20	

Consumables to be charged in addition + 12 1/2% except where otherwise stated.

The rates shown above are per unit machine per hour unless otherwise stated.

CIVIL ENGINEERING - PLANT (Cont'd)

Section & Item No.	Description	Nominal Size or Capacity	Hire Rate	Remarks
	31. TOOLS, PORTABLE ELECTRIC (excluding generator or power) (consumables, to be charged in addition + 12 1/2%)			
(1)	Drill..	Up to 10mm diameter	0.16	
(2)	Ditto...	Up to 13mm diameter	0.18	
(3)	Ditto...	Up to 19mm diameter	0.45	
(4)	Ditto...	Up to 32mm diameter	0.60	
(5)	Extra for Stand.................................	-	0.15	
(6)	Ditto, magnetic Stand........................	-	0.71	
(7)	Bench Grinder and pedestal..............	Up to 152mm diameter	0.28	
(8)	Ditto...	Up to 204mm diameter	0.36	
(9)	Ditto...	Up to 250mm diameter	1.26	
(10)	Angle Grinder....................................	Up to 230mm diameter	0.32	
(11)	Straight Grinder................................	Up to 152mm diameter	0.59	
(12)	Sander..	Up to 180mm diameter	0.40	
(13)	Ditto...	Up to 230mm diameter	0.50	
(14)	Polisher/Sander................................	-	0.36	
(15)	Electric Demolition Hammer/Drill Light weight...	Up to 25mm diameter	0.75	
(16)	Ditto(heavy weight)...........................	Up to 51mm diameter	0.92	
(17)	Portable electric Saw (circular)...........	Up to 250mm diameter	0.39	
(18)	Extra for power to be added to above items where applicable.......................		+ 50%	
	32. TRACTORS, SCRAPERS, ETC.	Maker's rated fly wheel Kw		
(1)	Tractor (crawler) with bull or angle dozer (hydraulically or winch operated)............	Up to 70.0Kw	26.76	
(2)	Ditto...	Up to 85.0Kw	33.50	
(3)	Ditto...	Up to 100.0Kw	43.56	
(4)	Ditto...	Up to 115.0Kw	44.22	
(5)	Ditto...	Up to 135.0Kw	51.56	
(6)	Ditto...	Up to 185.0Kw	69.09	
(7)	Ditto...	Up to 215.0Kw	101.88	
(8)	Ditto...	Up to 250.0Kw	115.32	
(9)	Ditto...	Up to 350.0Kw	142.72	
(10)	Ditto...	Up to 450.0Kw	213.81	
	Ripper attachements to items (1) to (10) subject to negotiation	S.A.E. rated capacity		
(11)	Tractor Loading Shovel (Crawler)...........	Up to .80m3	24.17	
(12)	Ditto...	Up to 1.00m3	27.63	
(13)	Ditto...	Up to 1.20m3	31.91	
(14)	Ditto...	Up to 1.40m3	35.55	
(15)	Ditto...	Up to 1.80m3	45.24	
(16)	Ditto...	Up to 2.00m3	55.38	
(17)	Ditto...	Up to 2.10m3	64.39	
(18)	Ditto...	Up to 3.50m3	84.88	
(19)	Tractor Loading Shovel (Crawler) with Back Hoe Equipment..........................	Add to above rates	+ 17%	
(20)	Tractor Loading Shovel (Crawler) with 4 in 1 attachement...............................	Add to above rates	+ 10%	
(21)	Tractor Loading Shovel (Crawler) with additional hydraulic mounted ripper.........	S.A.E. rated capacity Up to .60m3	18.73	
(22)	Ditto...	Up to .80m3	27.29	
(23)	Ditto...	Up to 1.00m3	34.34	
(24)	Ditto...	Up to 1.20m3	42.72	
(25)	Ditto...	Up to 1.40m3	47.51	
(26)	Ditto...	Up to 1.60m3	54.50	
(27)	Ditto...	Up to 1.80m3	62.90	
(28)	Tractor Loading Shovel (wheeled) with 4 wheel drive - Articulated.....................	S.A.E. rated capacity Up to 1.00m3	20.66	
(29)	Ditto...	Up to 1.20m3	23.46	
(30)	Ditto...	Up to 1.40m3	29.56	
(31)	Ditto...	Up to 1.50m3	33.25	
(32)	Ditto...	Up to 1.60m3	35.30	
(33)	Ditto...	Up to 1.80m3	36.11	
(34)	Ditto...	Up to 2.00m3	38.06	
(35)	Ditto...	Up to 2.30m3	41.84	

The rates shown above are per unit machine per hour unless otherwise stated.

GENERAL INFORMATION

CIVIL ENGINEERING - PLANT (Cont'd)

Section & Item No.	Description	Nominal Size or Capacity	Hire Rate	Remarks
	32. TRACTORS, SCRAPERS, ETC. - contd			
(36)	Ditto..	Up to 2.70m3	46.98	
(37)	Ditto..	Up to 3.10m3	63.65	
(38)	Ditto..	Up to 3.50m3	66.74	
(39)	Ditto..	Up to 3.85m3	85.48	
(40)	Ditto..	Up to 5.00m3	101.01	
(41)	Ditto..	Up to 6.00m3	181.61	
(42)	Tractor Loading Shovel (wheeled) with 4 wheel drive - Articulated with 4 in 1 attachement....................................	S.A.E. rated capacity Up to 1.00m3	23.85	
(43)	Ditto..	Up to 1.40m3	30.98	
(44)	Ditto..	Up to 1.60m3	39.36	
(45)	Ditto..	Up to 2.10m3	41.96	
(46)	Ditto..	Up to 2.70m3	54.72	
(47)	Ditto..	Up to 3.50m3	93.53	
(48)	Tractor Loading Shovel (wheeled, skid steer loader).......................	S.A.E. rated capacity Up to 0.15m3	8.05	
(49)	Ditto..	Up to 0.30m3	9.98	
(50)	Ditto..	Up to 0.40m3	13.89	
(51)	Tractor Loading Shovel (wheeled) with 2 wheel drive....................................	S.A.E. rated capacity Up to 1.00m3	17.82	
(52)	Ditto..	Up to 1.10m3	20.19	
(53)	Ditto..	Up to 1.20m3	22.95	
		Variable Blade/Flywheel Kw		
(54)	Motor Grader................................	Up to 80Kw	27.85	
(55)	Ditto..	Up to 110Kw	35.28	
(56)	Ditto..	Up to 120Kw	40.74	
(57)	Ditto..	Up to 160Kw	47.42	
(58)	Ditto..	Up to 200Kw	71.80	
(59)	Wheeled Tractor (rubber tyred).............	Up to 40Kw	6.23	
(60)	Ditto..	Up to 50Kw	9.06	
(61)	Ditto..	Up to 60Kw	11.92	
(62)	Ditto..	Up to 70Kw	13.42	
(63)	Ditto..	Up to 75Kw	15.58	
(64)	Ditto..	Up to 90Kw	21.46	
(65)	Ditto..	Up to 95Kw	22.95	
(66)	Ditto..	Up to 105Kw	25.25	
(67)	Ditto..	Up to 120Kw	29.95	
(68)	Ditto..	Up to 140Kw	36.38	
(69)	Motorised Scraper (rubber tyred) (single engine)..	Heaped Capacity (S.A.E.) Up to 16.0m3	99.42	
(70)	Ditto..	Up to 23.0m3	130.37	
(71)	Motorised Scraper (rubber tyred) (twin engine)..	Up to 16.0m3	113.57	
(72)	Ditto..	Up to 25.0m3	162.54	
(73)	Ditto..	Up to 35.0m3	233.89	
(73)	Motorised elevating Scraper (rubber tyred)...	Up to 9.0m3	67.79	
(75)	Ditto..	Up to 15.0m3	92.63	
(76)	Ditto..	Up to 20.0m3	127.28	
	33. TRENCHERS	Flywheel Kw		
(1)	Chain bucket or wheel type....................	Up to 10 Kw	5.51	
(2)	Ditto..	Up to 30 Kw	12.94	
(3)	Ditto..	Up to 50 Kw	23.56	
(4)	Ditto..	Up to 70 Kw	32.58	
(5)	Ditto..	Up to 100 Kw	39.19	
(6)	Ditto..	Up to 200 Kw	76.85	
(7)	Ditto..	Up to 300 Kw	119.42	
	34. WATER AND FUEL SUPPLY	(Exclusive of supporting structure)		
(1)	Water storage tank............................	Up to 1140 litres	1.56	per day
(2)	Ditto..	Up to 2280 litres	2.06	per day
(3)	Ditto..	Up to 4550 litres	3.52	per day
(4)	Ditto..	Up to 9100 litres	6.25	per day
(5)	Fuel storage tank............................	Up to 2280 litres	1.43	per day
(6)	Ditto..	Up to 4450 litres	1.91	per day
(7)	Ditto..	Up to 9100 litres	5.69	per day

The rates shown above are per unit machine per hour unless otherwise stated.

CIVIL ENGINEERING - PLANT (Cont'd)

Section & Item No.	Description	Nominal Size or Capacity	Hire Rate	Remarks
	34. WATER AND FUEL SUPPLY - contd	(Exclusive of supporting structure)		
(8)	Water or fuel storage tank trailer (rubber tyred)...	Up to 1140 litres	6.05	per day
(9)	Ditto..	Up to 2280 litres	10.30	per day
(10)	Water or fuel tanker mobile - self propelled...	Up to 4450 litres	11.50	
(11)	Ditto..	Up to 6820 litres	12.36	
	35. WELDING AND CUTTING SETS			
(1)	Oxy-acetylene cutting and welding set (inclusive of oxygen and acetylene) (excluding under-water equipment)........	-	4.37	
(2)	Welding set, diesel..................................	150 amp single operator	2.39	
	(exclusive of electrodes to be charged in addition +12.5%)			
(3)	Ditto........................Ditto..........................	300 amp single operator	5.21	
(4)	Ditto........................Ditto.........................	480 amp double operator	8.91	
(5)	Ditto..........................transformer electric	150 amp single operator	0.98	
(6)	Ditto........................Ditto.......................	300 amp single operator	2.52	
(7)	Head screen..	-	0.49	per day
(8)	Helmet..	-	0.63	per day
(9)	Standard kit 300 amp. set........................	-	4.88	per day

The rates shown above are per unit machine per hour unless otherwise stated.

CIVIL ENGINEERING - PLANT (Con'td)

Section & Item No.	Description	Nominal Size or Capacity	Hire Rate	Remarks
	36. WINCHES	Line pull tonnes		
(1)	Double drum diesel friction winch...........	Up to 1.00 tonnes	6.90	
(2)	Ditto.....................	Up to 3.00 tonnes	10.71	
(3)	Ditto.....................	Up to 4.00 tonnes	13.68	
(4)	Single drum winch, diesel hydraulic........	Up to 10.00 tonnes	40.97	
(5)	Ditto.....................	Up to 15.00 tonnes	58.79	
(6)	Single drum winch, electric.....................	Up to 1.00 tonnes	2.94	
(7)	Ditto.....................	Up to 2.00 tonnes	7.14	
(8)	Ditto.....................	Up to 3.00 tonnes	9.43	
(9)	Ditto.....................	Up to 5.00 tonnes	12.79	
(10)	Ditto.....................	Up to 10.00 tonnes	20.95	
(11)	Ditto.....................	Up to 15.00 tonnes	28.02	
(12)	Hand operated winch...........................	Up to 1.00 tonnes	0.63	
(13)	Ditto.....................	Up to 2.00 tonnes	0.74	
(14)	Ditto.....................	Up to 3.00 tonnes	0.79	
(15)	Ditto.....................	Up to 5.00 tonnes	0.90	
(16)	Ditto, mounted on lorry (power ex lorry battery)......................................	Up to 2.00 tonnes	1.86	
(17)	Air powered winches (excluding supply of compressed air)................................	Up to 2.00 tonnes	2.96	
(18)	Ditto.....................	Up to 4.00 tonnes	7.68	
(19)	Ditto.....................	Up to 5.00 tonnes	12.78	
	N.B. Special rope requirements to be charged extra.			
	37. MISCELLANEOUS PLANT AND CONSUMABLE STORES			
(1)	Air testing machine for drains..................	-	1.19	per day
(2)	Fencing Chestnut.....................................	per metre	0.30	per day
(3)	Fencing, post and rail or similar..............	per metre	0.24	per day
(4)	L.P.G. for tar boiler.................................	-	2.72	
(5)	Watchmans heater including LPG...........	-	0.20	
(6)	Road barrier...	per metre	0.11	per day
(7)	Tarpaulin..	per m^2	0.04	per day
(8)	Tilley Lamp (flood) including LPG............	-	1.67	per night
(9)	Traffic signals, two way (excluding generator and including mains power)....	per set	120.12	per week
(10)	Watchman's lamp including paraffin and attendance..	-	1.44	per night
		BTU's hour		
(11)	Space Heater (paraffin/electric)..............	Up to 30000	0.90	
(12)	Ditto.....................Ditto......................	Up to 63000	1.40	
(13)	Ditto.....................Ditto......................	Up to 100000	2.00	
(14)	Ditto.....................Ditto......................	Up to 150000	2.61	
(15)	Ditto.....................Ditto......................	Up to 240000	3.91	
(16)	Ditto.....................Ditto......................	Up to 325000	4.72	
(17)	Extra for space heater using LPG in lieu of paraffin..		add 10%	
(18)	Brickwork and masonry saw (excluding abrasive discs to be charged at cost +12.5%..	Blade diameter Up to 356mm	1.72	
(19)	Ditto.....................	Up to 457mm	1.87	
(20)	Crossing plates (steel sheets)...............	Thickness 20mm per m^2	0.21	per day
(21)	Flashing traffic warning lamp unit including batteries.....................................	-	1.44	per day
(22)	Flashing traffic warning lamp unit including batteries, bollard or tripod	-	1.79	per day
(23)	Pendent barrier marker including fencing pins..	per 26m	1.20	per day
(24)	Drainage rods including accessories......	102 to 152mm diameter per 10m	0.87	per day
(25)	Ditto, extra per rod...............................	.91m length each	0.07	per day
(26)	Drain stopper (expanding)......................	Up to 102mm diameter each	0.08	per day
(27)	Ditto.....................	Up to 152mm diameter each	0.12	per day
(28)	Ditto.....................	Up to 305mm diameter each	0.35	per day
(29)	Ditto.....................	Up to 457mm diameter each	1.67	per day
(30)	Ditto.....................	Up to 610mm diameter each	1.88	per day
(31)	Ditto.....................	Up to 762mm diameter each	2.68	per day
(32)	Ditto.....................	Up to 914mm diameter each	4.85	per day

The rates shown above are per unit machine per hour unless otherwise stated.

CIVIL ENGINEERING - PLANT (Cont'd)

Section & Item No.	Description	Nominal Size or Capacity	Hire Rate	Remarks
	37. MISCELLANEOUS PLANT AND CONSUMABLE STORES (Cont'd)			
(33)	Ditto..	Up to 1220mm diameter each	8.62	per day
		Water extraction rate per 24 hours		
(34)	Dehumidifier.................................	Up to 68 litres	5.72	per day
(35)	Ditto...	Up to 91 litres	8.21	per day
(36)	Ditto...	Up to 160 litres	16.43	per day
(37)	Ditto, with temperature control...............	Up to 160 litres	19.81	per day
(38)	Ladders - all types and lengths.............	-	to be negotiated	
		Height		
(39)	Traffic warning cone (plastic)..............	Up to 500mm	0.15	per day
(40)	Ditto...	Up to 600mm	0.24	per day
(41)	Ditto...	Up to 800mm	0.28	per day
(42)	Ditto...	Up to 1000mm	0.56	per day
(43)	Traffic warning signs (circular)............	Up to 600mm diameter	0.55	per day
(44)	Ditto...	Up to 750mm diameter	0.84	per day
(45)	Ditto...	Up to 900mm diameter	1.10	per day
(46)	Ditto...	Up to 1200mm diameter	1.72	per day
(47)	Traffic warning sign (triangular)...........	Up to 600mm height	0.43	per day
(48)	Ditto...	Up to 700mm height	0.60	per day
(49)	Ditto...	Up to 900mm height	0.82	per day
(50)	Ditto...	Up to 1200mm height	1.30	per day
(51)	Traffic warning board (rectangular).....	1050 x 750mm	1.10	per day
(52)	Paint spraying machine, 1 gun type with 7.5m lengths of air and fluid hose and 10.0 litre pressure tank...................	(Excluding compressor)	0.62	
(53)	Ditto (power driven)............................	(Electric or petrol)	3.30	
(54)	Paint spraying machine, 2 gun type with 15.0m lengths of air and fluid hose and 25 litre pressure tank.....................	(Excluding compressor)	1.11	
(55)	Ditto (power driven)............................	(Electric, petrol or diesel)	4.31	

The rates shown above are per unit machine per hour unless otherwise stated .

the contractor.

CONVERSION TABLE

Plant items whose specifications and data are related to now discarded imperial capacities may still be in use on contracts. To allocate them to appropriate current daywork items the following conversions should be used:-

LINEAR
0.03937 in. - 1 - mm 25.4
3.28084 ft. - 1 - metre 0.3048
1.0936 yd. - 1 - metre 0.9144

WEIGHT
0.0196 cwt. - 1 - kg 50.820
0.9842 ton - 1 - ton 1.016
2.20463 lb. - 1 - kg .45359

AREA
0.00155 in2 - 1 - mm2 645.16
1.1960 yd2 - 1 - m2 0.83613
2.4711 acre - 1 - ha 0.40469
0.3861 sq. mile - 1 - km2 2.59

CAPACITY
1.7598 pint - 1 - litre 0.56826
0.21997 gallon - 1 - litre 4.54609

VOLUME
0.0610 in3 - 1 - cm3 16.387
35.315 ft3 - 1 - m3 0.0283
1.3079 yd3 - 1 - m3 0.7645

POWER
1.310 HP - 1 - kW 0.7457

Regional Factors

Laxton's Building Price Book is based upon national average prices (=1.00). Indicative levels of tender pricing in regions as at the second quarter 1999 are given on this map and the factors shown may be applied to adjust overall pricing.

MAJOR WORKS – TENDER VALUES	SMALL WORKS – TENDER VALUES
The measured rates are based on contracts valued in the range of £250,000 to £1,000.00. As a guide to pricing works of larger value and for cost planning or budgetary purposes, the following adjustments may be applied to overall contract values.	The measured rates are based on contracts valued in the range of £25,000 to £75,000. As a guide to pricing works of smaller or larger value and for cost planning or budgetary purposes, the following adjustments may be applied to overall contract values.
Contract Value £1,000,000 to £2,000,000.........deduct 2.5% £2,000,000 to £3,000,000.........deduct 5.0% £3,000,000 to £5,000,000.........deduct 7.5%	Contract Value £5,000 to £15,000............................add 20.0% £15,000 to £25,000...........................add 10.0% £25,000 to £75,000.......................rate as shown £75,000 to £100,000.......................deduct 5.0% £100,000 to £150,000.....................deduct7.5% £150,000 to £250,000.....................deduct10.%

Fees

ROYAL INSTITUTE OF BRITISH ARCHITECTS

STANDARD FORM OF AGREEMENT FOR THE

APPOINTMENT OF AN ARCHITECT (SFA/92)

The Standard Form of Agreement for the appointment of an Architect (July 1992) has been issued to replace the previous Architects Appointment (July 1982). Extracts from the Standard Form of Agreement (July 1992) and A guide to the Standard Form of Agreement for the Appointment of an Architect printed by permission of the Royal Institute of British Architects, 66 Portland Place, London, W1N 4AD from whom full copies can be obtained.

The Standard Form of Agreement for the Appointment of an Architect (SFA/92) consists of:-

Memorandum of Agreement

Definitions

Schedule One - Information to be supplied by client

Schedule Two - Conditions of Appointment
(Reproduced in outline below)

> Historic Buildings (alternative Schedule Two)
>
> Design and Build (Employer)
> (alternative Schedule Two and associated Conditions)
>
> Design and Build (Contractor)
> (alternative Schedule Two and associated Conditions)
>
> Community Architecture
> (supplmentary Schedule of Services and additional Conditions)

Schedule Three - Fees and Expenses
Schedule Four - Appointment of Consultants, Specialists and Site Staff.

Conditions of Appointment

PART ONE   CONDITIONS COMMON TO ALL COMMISSIONS

1.1   Governing law/interpretation

1.1.1   The application of the Appointment shall be governed by the laws of (England and Wales) (Northern Ireland) (Scotland). Delete those parts not applicable.

1.1.2   The conditions headings and side notes are for the convenience of the parties to this Agreement only and do not effect its interpretation.

1.1.3   Words denoting the masculine gender include the feminine gender and words denoting natural persons include corporations and firms and shall be construed interchangeably in that manner.

1.2   Architect's obligations

Duty of care

1.2.1   The Architect shall in providing the Services exercise reasonable skill and care in conformity with the normal standards of the Architect's profession.

Architect's authority

1.2.2   The Architect shall act on behalf of the Client in the matters set out or necessarily implied in the Appointment.

1.2.3   The Architect shall at those points and/or dates referred to in the Timetable obtain the authority of the Client before proceeding with the Services.

No alteration to services

1.2.4   The Architect shall make no material alteration to or addition to or omission from the Services without the knowledge and consent of the Client except in case of emergency when the Architect shall inform the Client without delay.

Variations

1.2.5   The Architect shall inform the Client upon its becoming apparent that there is any incompatibility between any of the Client's Requirements; or between the Client's Requirements, the Budget and the Timetable; or any need to vary any part of them.

1.2.6   The Architect shall inform the Client on its becoming apparent that the Services and/or the fees and/or any other part of the Appointment and/or any information or approvals need to be varied. The Architect shall confirm in writing any agreement reached.

1.3   Client's obligations

Client's representative

1.3.1   The Client shall name the person who shall exercise the powers of the Client under the Appointment and through whom all instructions to the Architect shall be given.

Information

1.3.2   The Client shall provide to the Architect the information specified in Schedule 1.

1.3.3   The Client shall provide to the Architect such further information as the Architect shall reasonably and necessarily request for the performance of the Services: all such information to be provided free of charge and at such times as shall permit the Architect to comply with the Timetable.

Conditions of Appointment

PART ONE   CONDITIONS COMMON TO ALL COMMISSIONS (Cont'd)

1.3.4 The Client accepts that the Architect will rely on the accuracy, sufficiency and consistency of the information supplied by the Client.

1.3.5 The Client shall advise the Architect of the relative priorities of the Client's Requirements, the Budget and the Timetable and shall inform the Architect of any variations to any of them.

Decisions and approvals
1.3.6 The Client shall give such decisions and decisions and approvals as are necessary for the performance of the Services and at such times as to enable the Architect to comply with the Timetable.

Architect does not warrant
1.3.7 The Client acknowledges that the Architect does not warrant the work or products of others nor warrants that the Services will or can be completed in accordance with the Timetable.

1.4   Assignment and sub-contracting

Assignment
1.4.1 Neither the Architect nor the Client shall assign the whole or any part of the benefit or in anyway transfer the obligation of the Appointment without the consent in writing of the other.

Sub-contracting
1.4.2 The Architect shall not sub-contract any of the Services without the consent in writing of the Client, which consent shall not be unreasonably withheld.

1.5   Payment

Payment
1.5.1 Payment for the Services shall be calculated, charged and paid as set out in Schedule Three.

Percentage fees
1.5.2 Where it is stated in Schedule Three that fees and/or expenses are payable on a percentage basis, then, unless any other basis has been agreed between the Architect and the Client and confirmed by the Architect to the Client in writing, the fees and/or expenses shall be based on the Total Construction Cost of the Works. On the issue of the final certificate under the building contract the fees and/or expenses shall be recalculated on the actual Total Construction Cost.

1.5.3 The following bases shall be used for the calculation of percentage fees based on the Total Construction Cost until that cost has been ascertained:
. until tenders are obtained - the cost estimate;
. after tenders have been obtained - the lowest acceptable tender;
. after the contract is let - the contract sum.

Revise rates
1.5.4 Unless otherwise stated in Schedule Three, time rates and mileage rates for vehicles shall be revised every twelve months from the date of the appointment.

Fee variation
1.5.5 Where any change is made to the Architect's Services, the Procurement Method, the Client's Requirements, the Budget, or the Timetable, or where the Architect consents to enter into any Collateral Agreement the form or beneficiary of which had not been agreed by the Architect at the date of the Appointment, the fees specified in Schedule Three shall be varied.

Vary lump sum
1.5.6 Where fees and/or expenses are specified in Schedule Three to be a lump sum, that lump sum shall also be varied in accordance with the provisions of Schedule Three.

Additional fees
1.5.7 Where the Architect is involved in extra work and/or expense for which the Architect is not otherwise remunerated caused by:
. the Clients variations to completed work or services
. the examination and/or negotiation of notices, applications or claims under a building contract;
. delay or for any other reason beyond the Architect's control; the Architect shall be entitled to additional fees calculated on a time basis.

1.5.8 Where fees and/or expenses are varied under conditions 1.2.6, 1.5.4, 1.5.5 and/or 1.5.6 or where additional fees are payable under condition 1.5.7, the additional or varied fees and/or expenses shall be stated by the Architect in writing.

Incomplete Services
1.5.9 Where the Architect carries out only part of the Services specified in Schedule Two, fees shall be calculated as described in Schedule Three for:
. completed Work Stage (Schedule Two)
. completed Service (Schedule Two)
. completed part (Timetable, Schedule One) and for the balance of any of the above fee shall be on the basis of the Architect's estimate of the percentage of completion.

Expenses and disbursements
1.5.10 The Client shall pay the expenses specified in Schedule Three. Expenses other than those specified shall only be charged with the prior authorisation of the Client.

1.5.11 The Client shall reimburse the Architect as specified in Schedule Three for any disbursements made on the Client's behalf.

Maintain records
1.5.12 The Architect shall maintain records of expenses and of disbursements and shall make these available to the Client on reasonable request.

Instalments
1.5.13 All payments due under the Appointment shall be made by instalments specified in Schedule Three. Where no such basis is specified, payments shall be made monthly on the basis of the Architect's estimate of percentage of completion of the Services.

Conditions of Appointment

PART ONE	CONDITIONS COMMON TO ALL COMMISSIONS (Cont'd)

Payment 1.5.14 Payment shall become due to the Architect on submission of the Architect's account.

No setoff 1.5.15 The Client may not withhold or reduce any sum payable to the Architect under the Appointment by reason of claims or alleged claims against the Architect. All rights of setoff which the Client may otherwise exercise in common law are hereby expressly excluded.

Disputed accounts 1.5.16 If any item or part of an item of any account is disputed or subject to question by the Client, the payment by the Client of the remainder of that account shall not be withheld on those grounds.

Interest on outstanding accounts 1.5.17 Any sums remaining unpaid at the expiry of twenty-eight days from the date of submission of an account shall bear interest thereafter, such interest to accrue from day to day at the rate specified in Schedule Three.

Payment on suspension or termination 1.5.18 On suspension or termination of the Appointment the Architect shall be entitled to, and shall be paid, fees for all Services provided to that time calculated as incomplete Services, and to expenses and disbursements reasonably incurred to that time.

1.5.19 During any period of suspension the Architect shall be reimbursed by the Client for expenses, disbursements and other costs reasonably incurred as a result of the suspension.

1.5.20 On the resumption of a suspended Service within six months, fees paid prior to resumption shall be regarded solely as payments on account of the total fee.

1.5.21 Where the Appointment is suspended or terminated by the Client or suspended or terminated by the Architect on account of a breach of the Appointment by the Client, the Architect shall be paid by the Client for all expenses and other costs necessarily incurred as a result of any suspension and any resumption or termination.

VAT 1.5.22 All fees, expenses and disbursements under the Appointment are exclusive of Value Added Tax. Any Value Added Tax on the Architect's services shall be paid by the Client.

1.6 Suspension, resumption and termination

Services impracticable 1.6.1 The Architect shall give reasonable notice in writing to the Client of any circumstances which make it impracticable for the Architect to carry out any of the Services in accordance with the Timetable.

Suspension 1.6.2 The Client may suspend the performance of any or all of the Services by giving reasonable notice in writing to the Architect.

1.6.3 In the event of the Client's being in default of payment of any fees, expenses and/or disbursements, the Architect may suspend the performance of any or all of the Services on giving notice in writing to the Client.

Resumption 1.6.4 If the Architect has not been given instructions to resume any suspended Service within six months from the date of suspension, the Architect shall request in writing such instructions. If written instructions have not been received within twenty-eight days of the date of such request the Architect shall have the right to treat the Appointment as terminated.

Termination 1.6.5 The Appointment may be terminated by either party on the expiry of reasonable notice in writing.

Architect's death or incapacity 1.6.6 Should the Architect through death or incapacity be unable to provide the Services, the Appointment shall thereby be terminated.

Accrued rights 1.6.7 Termination of the Appointment shall be without prejudice to the accrued rights and remedies of either party.

1.7 Copyright

Copyright 1.7.1 Copyright in all documents and drawings prepared by the Architect and in any work executed from those documents and drawings shall remain the property of the Architect.

1.8 Dispute resolution

Arbitration 1.8.1 In England and Wales, and subject to the provisions of conditions 1.8.2 and 1.8.3 in Northern Ireland, any difference or dispute arising out of the Appointment shall be referred by either of the parties to arbitration by a person to be agreed between the parties or, failing agreement within fourteen days after either party has given the other a written request to concur in the appointment of an arbitrator, a person to be nominated at the request of either party by the President of the Chartered Institute of Arbitrators provided that in a difference or dispute arising out of the conditions relating to copyright the arbitrator shall, unless otherwise agreed, be an architect.

Scotland 1.8.1S In Scotland, subject to the provisions of conditions 1.8.2 and 1.8.3, any difference or dispute arising out of the Appointment shall be referred to arbitration by a person to be agreed between the parties or, failing agreement within fourteen days after either party has given the other a written request to concur in the appointment of an arbiter, a person to be nominated at the request of either party by the Dean of the Faculty of Advocates, provided that in a difference or dispute arising out of the conditions relating to copyright the arbiter shall, unless otherwise agreed, be an architect.

GENERAL INFORMATION

Conditions of Appointment

PART ONE	CONDITIONS COMMON TO ALL COMMISSIONS (Cont'd)

Opinion   1.8.2   In Northern Ireland or Scotland, any difference or dispute arising from the Appointment may be referred respectively to the RSUA or the RIAS for an opinion provided that:
. the opinion is sought on a joint statement of undisputed facts;
. the parties agree to be bound by the opinion.

Negotiation   1.8.3   In Northern Ireland or Scotland, the parties shall attempt to settle any dispute by negotiation and no procedure shall be commenced under condition 1.8.1 or 1.8.1S until the expiry of twenty-eight days after notification has been given in writing by one to the other of a difference or dispute.

1.8.4   Nothing herein shall prevent the parties agreeing to settle any difference or dispute arising out of the Appointment without recourse to arbitration.

PART TWO	CONDITIONS SPECIFIC TO DESIGN OF BUILDING PROJECTS, STAGES A-H

2.1   **Architect's obligations**

Architect's authority   2.1.1   The Architect shall, where specified in the Timetable, obtain the authority of the Client before initiating any Work Stage and shall confirm that authority in writing.

Procurement Method   2.1.2   The Architect shall advise on the options for the Procurement Method for the Project.

No alteration to design   2.1.3   The Architect shall make no material alteration, addition to or omission from the approved design without the knowledge and consent of the Client and shall confirm such consent in writing.

2.2   **Client's obligations**

Statutory requirements   2.2.1   The Client shall instruct the making of applications for planning permission and approval under Building Acts, Regulations and other statutory requirements and applications for consents by freeholders and all others having an interest in the Project and shall pay any statutory charges and any fees, expenses and disbursements in respect of such applications.

2.2.2   The Client shall have informed the Architect prior to the date of the Appointment whether any third party will acquire or is likely to acquire an interest in the whole or any part of the Project.

Collateral Agreements   2.2.3   The Client shall not require the Architect to enter into any Collateral Agreement with a third party which imposes greater obligations or liabilities on the Architect than does the Appointment.

Procurement Method   2.2.4   The Client shall confirm the Procurement Method for the Project.

2.3   **Copyright**

2.3.1   Notwithstanding the provisions of condition 1.7.1, the Client shall be entitled to reproduce the Architect's design by proceeding to execute the Project provided that:
. the entitlement applies only to the Site or part of the Site to which the design relates, and
. the Architect has completed a scheme design or
. has provided detail design and production information, and
. any fees, expenses and disbursements due to the Architect have been paid.
This entitlement shall also apply to the maintenance repair and/or renewal of the Works.

2.3.2   Where the Architect has not completed a scheme design, the Client shall not reproduce the design by proceeding to execute the Project without the consent of the Architect.

2.3.3   Where the Services are limited to making and negotiating planning applications, the Client may not reproduce the Architect's design without the Architect's consent, which consent shall not be unreasonably withheld, and payment of any additional fees.

2.3.4   The Architect shall not be liable for the consequences of any use of any information or designs prepared by the Architect except for the purposes for which they were provided.

PART THREE	CONDITIONS SPECIFIC TO CONTRACT ADMINISTRATION AND INSPECTION OF THE WORKSSTAGES J-L

3.1   **Architect's obligations**

Visits to   3.1.1   The Architect shall in providing the Works the Services specified in stages K and L of Schedule Two make such visits to the Works as the Architect at the date of the Appointment reasonably expected to be necessary. The Architect shall confirm such expectation in writing.

Variations to visits to the works   3.1.2   The Architect shall, on its becoming apparent that the expectation of the visits to the Works needs to be varied, inform the Client in writing of his recommendations and any consequential variation in fees.

More frequent visits to the Works   3.1.3   The Architect shall, where the Client requires more frequent visits to the Works than that specified by the Architect in condition 3.1.1 inform the Client of any consequential variation in fees. The Architect shall confirm in writing any agreement reached.

Alteration to design only in emergency   3.1.4   The Architect may in an emergency make an alteration, addition or omission without the Client's knowledge and consent but shall inform the Client without delay and shall confirm that in writing. Otherwise the Architect shall make no material alteration or addition to or omission from the approved design during construction without the knowledge and consent of the Client and the Architect shall confirm such consent in writing.

Conditions of Appointment

PART THREE CONDITIONS SPECIFIC TO CONTRACT ADMINISTRATION AND INSPECTION OF THE WORKS STAGES J-L (Cont'd)

3.2 Client's obligations

Contractor 3.2.1 The Client shall employ a contractor under a separate agreement to undertake construction or other works relating to the Project.

Responsibilities of contractor 3.2.2 The Client shall hold the contractor and not the Architect responsible for the contractor's management and operational methods and for the proper carrying out and completion of the Works and for health and safety provisions on the Site.

Products and materials 3.2.3 The Client shall hold the contractor and not the Architect responsible for the proper installation and incorporation of all products and materials into the Works.

Collateral Agreements 3.2.4 The Client shall, where the Architect consents to enter into a Collateral Agreement with a third party in respect of the Project, procure that the contractor is equally bound.

Instructions 3.2.5 The Client shall only issue instructions to the contractor through the Architect, and the Client shall not hold the Architect responsible for any instructions issued other than through the Architect.

3.3 Site Staff

3.3.1 The Architect shall recommend the appointment of Site Staff to the Client if in his opinion such appointments are necessary to provide the Services specified in K-L 04-08 of Schedule Two.

3.3.2 The Architect shall confirm in writing to the Client the Site Staff to be appointed, their disciplines, the expected duration of their employment, the party to appoint them and the party to pay, and the method of recovery of payment to them.

3.3.3 All Site Staff shall be under the direction and control of the Architect.

PART THREE CONDITIONS SPECIFIC TO APPOINTMENT OF CONSULTANTS AND SPECIALISTS WHERE ARCHITECT IS LEAD CONSULTANT

4.1 Consultants

Nomination 4.1.1 The Architect shall identify professional services which require the appointment of consultants. Such consultants may be nominated at any time by either the Client or the Architect subject to acceptance by each party.

Appointment 4.1.2 The Client shall appoint and pay the nominated consultants.

4.1.3 The consultants to be appointed at the date of the Appointment and the services to be provided by them shall be confirmed in writing by the Architect to the Client.

Collateral Agreements 4.1.4 The Client shall, where the Architect consents to enter into a Collateral Agreement with a third party in respect of the Project, procure that all consultants are equally bound.

Lead Consultant 4.1.5 The Client shall appoint and give authority to the Architect as Lead Consultant in relation to all consultants however employed. The Architect shall be the medium of all communication and instruction between the Client and the consultants, co-ordinate and integrate into the overall design the services of the consultants, require reports from the consultants.

4.1.6 The Client shall procure that the provisions of condition 4.1.5 above are incorporated into the conditions of appointment of all consultants however employed and shall provide a copy of such conditions of appointment to the Architect.

Responsibilities of Consultants 4.1.7 The Client shall hold each consultant however appointed and not the Architect responsible for the competence and performance of the services to be performed by the consultant and for the general inspection of the execution of the work designed by the consultant.

Responsibilities of Architect 4.1.8 Nothing in this Part shall affect any responsibility of the Architect for issuing instructions under the building contract or for other functions ascribed to the Architect under the building contract in relation to work designed by a consultant.

4.2 Specialists

Nomination 4.2.1 A Specialist who is to be employed directly by the Client or indirectly through the contractor to design any part of the Works may be nominated by either the Architect or the Client subject to acceptance by each party.

Appointment 4.2.2 The Specialists to be appointed at the date of the Appointment and the services to be provided by them shall be those confirmed in writing by the Architect to the Client.

Collateral Agreements 4.2.3 The Client shall, where the Architect consents to enter into a Collateral Agreement with a third party in respect of the Project, procure that all Specialists are equally bound.

Co-ordination and integration 4.2.4 The Client shall give the authority to the Architect to co-ordinate and integrate the services of all Specialists into the overall design and the Architect shall be responsible for such co-ordination and integration.

Responsibilities of Specialists 4.2.5 The Client shall hold any Specialist and not the Architect responsible for the products and materials supplied by the Specialist and for the competence, proper execution and performance of the work with which such Specialists are entrusted.

GENERAL INFORMATION

ROYAL INSTITUTE OF BRITISH ARCHITECTS

STANDARD FORM OF AGREEMENT FOR THE APPOINTMENT OF AN ARCHITECT (SFA/92)

Schedule Two   Services to be provided by Architect

1 Design Skills

1.01 Provide interior design services

1.02 Advise on the selection of furniture and fittings

1.03 Design furniture and fittings

1.04 Inspect the making up of furnishings

1.05 Advise on works of special quality, e.g. shopfittings

1.06 Prepare information for installation of works of special quality

1.07 Inspect installation of works of special quality

1.08 Advise on commissioning or selection of works of art

1.09 Prepare information for installation of works of art

1.10 Inspect installation of works of art

1.11 Provide industrial design services

1.12 Develop a building system or components for mass production

1.13 Examine and advise on existing building systems

1.14 Monitor testing of prototypes, mock-ups or models of building systems

1.15 Provide town planning and urban design services

1.16 Provide landscape design services

1.17 Provide graphic design services

1.18 Provide exhibition design services

1.19 Provide presentation material design services

1.20 Provide perspective and other illustrations

1 Design Skills Contd

1.21 Provide model-making services

1.22 Provide photographic record services

2 Consultancy Services

2.01 Provide services as a consultant Architect on a regular or intermittent basis

2.2 Consult statutory authorities

2.03 Provide information in connection with local authority, government and other grants

2.04 Make applications for local authority, governments and other grants

2.05 Conduct negotiations for local authority, government and other grants

2.06 Make submissions to RFAC, UK heritage bodies and/or non-statutory bodies

2.07 Provide information to advisory bodies

2.08 Negotiate with advisory bodies

2.09 Advise on rights including easements and responsibilities of owners and lessees

2.10 Provide information on rights including easements and responsibilities of owners and lessees

2.11 Negotiate rights including easements

2.12 Provide services in connection with party wall negotiations

2.13 Provide services in connection with planning appeals and/or inquiries

2.14 Advise on the use of energy in new or existing buildings

2.15 Carry out life cycle analyses of proposed or existing buildings to determine their likely cost in use

2 Consultancy Services Contd

2.16 Provide services in connection with environmental studies

2.17 Act as coordinator in health and safety matters

2.18 Prepare, settle proofs, attend conferences and give evidence

2.19 Act as witness as to fact

2.20 Act as expert witness

2.21 Act as arbitrator

2.22 Provide project management services

3 Buildings/Sites

3.01 Advise on the suitability and selection of sites

3.02 Make measured surveys, take levels and prepare plans of sites

3.03 Arrange for investigations of soil conditions of sites

3.04 Advise of the suitability and selection of buildings

3.05 Make measured surveys and prepare drawings of existing buildings

3.06 Inspect and prepare report and schedule of existing buildings

3.07 Inspect and prepare report and schedule of dilapidations

3.08 Prepare estimates for the replacement and reinstatement of buildings and plant

3.09 Prepare, submit, negotiate claims following damage by fire and other causes

3.10 Investigate and advise on means of escape in existing buildings

3.11 Investigate and advise on change of use in existing buildings

3 Buildings/Sites Contd

3.12 Investigate and report on building failures

3.13 Arrange for and inspect exploratory work by contractors and specialists in connection with building failures

3.14 Prepare a layout for the development of a site

3.15 Prepare a layout for a greater area than that which is to be developed immediately

3.16 Prepare development plans for a site or a large building or a complex of buildings

3.17 Prepare drawings and specifications of materials for the construction of estate roads and sewers

3.18 Make structural surveys and report on the structural elements of buildings

3.19 Investigate and advise on floor loadings in existing buildings

3.20 Investigate and advise on sound insulation in existing buildings

3.21 Investigate and advise on fire protection and alarms in existing buildings

3.22 Inspect and advise on security systems in existing buildings

3.23 Inspect and prepare a valuation report for mortgage or other purpose

4 All Commissions

4.01 Obtain the Client's Requirements, Budget and Timetable

4.02 Advise on the need for and the scope of consultants' services and the conditions of their appointment

4.03 Arrange for and assist in the selection of other consultants

ROYAL INSTITUTE OF BRITISH ARCHITECTS

STANDARD FORM OF AGREEMENT FOR THE APPOINTMENT OF AN ARCHITECT (SFA/92)

Schedule Two   Services specific to Building Projects - Stages

A-B Inception and Feasibility

01 Obtain information about the Site from the Client

02 Visit the Site and carry out an initial appraisal

03 Assist the Client in preparation of Client's Requirements

04 Advise the Client on methods of procuring construction

05 Advise on the need for specialist contractors, sub-contractors and suppliers to design and execute parts of the Works

06 Prepare proposals and make application for outline planning permission

07 Carry out such studies as may be necessary to determine the feasibility of the Client's Requirements

08 Review with the Client alternative design and construction approaches and cost implications

09 Advise on the need to obtain planning permission, approvals under Building Acts and/or Regulations and other statutory requirements

10 Develop the Client's Requirements

11 Advise on environmental impact and prepare report

C Outline Proposals

01 Analyse the Client's Requirements; prepare outline proposals

02 Provide information to discuss proposals with and incorporate input of other consultants

03 Provide information to other consultants for their preparation of an approximation of construction cost

C Outline Proposals Contd

03A Prepare an approximation of construction cost

04 Submit outline proposals and approximation of construction cost for the Client's preliminary approval

05 Propose a procedure forcost planning and control

06 Provide information to others for cost planning and control throughout the Project

06A Operate the procedure for cost planning and control throughout the Project

07 Prepare and keep updated a Client's running expenditure plan for the Project

08 Prepare special presentation drawings, brochures, models or technical information for use of the Clients or others

09 Carry out negotiations with tenants or others identified by the Client

D Scheme Design

01 Develop scheme design from approved outline proposals

02 Provide information to, discuss proposals with and incorporate input of other consultants into scheme design

03 Provide information to other consultants for their preparation of cost estimate

03A Prepare cost estimate

04 Prepare preliminary timetable for construction

05 Consult with planning authorities

06 Consult with building control authorities

07 Consult with fire authorities

D Scheme Design Contd

8 Consult with environmental authorities

09 Consult with licensing authorities

10 Consult with statutory undertakers

11 Prepare an application for full planning permission

12 Submit scheme design showing spatial arrangements, materials and appearance, together with cost estimate, for the Client's approval

13 Consult with tenants or others identified by the Client

14 Conduct exceptional negotiations with planning authorities

15 Submit an application for full planning permission

16 Prepare multiple applications for full planning permission

17 Submit multiple applications for full planning permission

18 Make revisions to scheme design to deal with requirements of planning authorities

19 Revise planning application

20 Resubmit planning application

21 Carry out special constructional research for the Project including design of prototypes, mock-ups or models

22 Monitor testing of prototypes, mock-ups or models, etc.

E Detail Design

01 Develop detail design from approved scheme design

02 Provide information to, discuss proposals with and incorporate input of other consultants into detail design

03 Provide information to other consultants for their revision of cost estimate

03A Revise cost estimate

04 Prepare applications for approvals under Building Acts and/or Regulations and other statutory requirements

04A Prepare building notice under Building Acts and/or Regulations*

05 Agree form of building contract and explain the Client's obligations thereunder

06 Obtain the Client's approval of the type of construction, quality of materials and standard of workmanship

07 Apply for approvals under Building Acts and/or Regulations and other statutory requirements

07A Give building notice under Building Acts and/or Regulations*

08 Negotiate if necessary over Building Acts and/or Regulations and other statutory requirements and revise production information

09 Conduct exceptional negotiations for approvals by statutory authorities

10 Negotiate waivers or relaxations under Building Acts and/or Regulations and other statutory requirements

* Not applicable in Scotland

GENERAL INFORMATION

ROYAL INSTITUTE OF BRITISH ARCHITECTS

STANDARD FORM OF AGREEMENT FOR THE APPOINTMENT OF AN ARCHITECT (SFA/92)

Schedule Two   Services Specific to Building Projects - stages cont'd

F-GProduction Information

01 Prepare production drawings

02 Prepare specification

03 Provide information for the preparation of bills of quantities and/or schedules of works

03A Prepare schedule of rates and/or quantities and/or schedules of works for tendering purposes

04 Provide information to, discuss proposals with and incorporate input of other consultants into production information

05 Co-ordinate production information

06 Provide information to other consultants for their revision of cost estimate

06A Revise cost estimate

07 Review timetable for construction

08 Prepare other production information

09 Submit plans for proposed building works for approval of landlords, funders, free-holders, tenants or others as requested by the Client

H   Tender Action

01 Advise on and obtain the Client's approval to a list of tenderers for the building contract

02 Invite tenders

03 Appraise and report on tenders with other consultants

03A Appraise and report on tenders

04 Assist other consultants in negotiating with a tenderer

04A Negotiate with a tenderer

05 Assist other consultants in negotiating a price with a contractor

H   Tender Action Contd

05A Negotiate a price with a contractor

06 Select a contractor by other means

07 Revise production information to adjust tender sum

08 Arrange for other contracts to be let prior to the main building contract

J   Project Planning

01 Advise the Client on the appointment of the contractor and on the responsibilities of the parties and the Architect under the building contract

02 Prepare the building contract and arrange for it to be signed

03 Provide production information as required by the building contract

04 Provide services in connection with demolitions

05 Arrange for other contracts to be let subsequent to the commencement of the building contract

K-L Operations on Site and Completion

01 Administer the terms of the building contract

02 Conduct meetings with the contractor to review progress

03 Provide information to other consultants for the preparation of financial reports to the Client

03A Prepare financial reports for the Client

04 Generally inspect materials delivered to the site

K-L Operations on Site and Completion Contd

05 As appropriate instruct sample taking and carrying out tests of materials, components, techniques and work manship and examine the conduct and results of such tests whether on or off site

06 As appropriate instruct the opening up of completed work to determine that it is generally in accordance with the Contract Documents

07 As appropriate visit the sites of the extraction and fabrication and assembly of materials and components to inspect such materials and workmanship before delivery to site

08 At intervals appropriate to the stage of construction visit the Works to inspect the progress and quality of the Works and to determine that they are being executed generally in accordance with the Contract Documents

09 Direct and control the activities of Site Staff

10 Provide drawings showing the building and the main lines of drainage

11 Arrange for drawings of building services installations to be provided

12 Give general advice on maintenance

13 Administer the terms of other contracts

14 Monitor the progress of the Works against the contractor's programmne and report to the Client

15 Prepare valuations of work carried out and completed

16 Provide specially prepared drawings of a building as built

K-L Operations on Site

17 Prepare drawings for conveyancing purposes

18 Compile maintenance

19 Incorporate information prepared by others in maintenance manuals

20 Prepare a programme for the maintenance of a building

21 Arrange maintenance contracts

Work Stages
are specified
by circling
the stage letters

Basic Services

C01-C04
D01-D12
E01-E06
FG01-FG07
H01-H03A
J01-J03
KL01-12
are specified unless
struck out

Additional Services
are specified by
circling the relevant
numbered items

SFA GUIDE

A guide to the Standard Form of Agreement for the Appointment of an Architect.

The SFA Guide introduces the SFA, describes its component parts and gives guidance on their use, function and completion together with:-

Notes on Use and Completion

The Memorandum of Agreement
The Conditions of Appointment
Schedule One
Schedule Two
Schedule Three
Schedule Four

Example Letters

Preliminary Appointment
Confirmation of Agreement to proceed
Confirmation of Amendment
Activating an Appointment by Stages

Methods of Charging

Percentage Fees
Time Charges
Lump Sum Fees
Unit Price Fees
Betterment Fees
Equity Shares
Incentive Fees
Expenses
Disbursements
Other Costs

Payment of Fees by Programmed Instalments

Appendices

The SFA/92 Documents: Worked Examples
Alternative/Supplementary Schedules: Specimen Forms
Summary of Actions arising from the Conditions of Appointment
Negotiating Appointment: Notes for Architects
Survey of Fees

Recommended fees are no longer published by the professional bodies. The most recent information on levels of fees charged by Architects is that obtained from an historical survey of ACA and RIBA members on jobs undertaken between 1985 and 1990.

A survey is by definition a descriptive rather than an analytical exercise. Members were asked to include projects which in their opinion showed a reasonable return, although this was not defined. A `reasonable return' is a matter for individual judgement in the particular circumstances. The Tables show the middle ground of fees received. These fee levels will not necessarily be appropriate for architectural commissions in future years.

Fees agreed for particular jobs may be above or below the figures in Tables 1 and 2, according to the size of practice, the complement of services to be provided and the quality of design, particularly detailed design.

The classification of building types used for the survey is set out in Table 3.

The survey was constructed on the following bases.

. Appointments were concluded at the outset of the job and were for the Basic Services C-L as described in Architect's Appointment (1982). It is thought likely that where appointments were made stage by stage, were postponed or determined or otherwise disrupted, fees would have been higher.

. Works to buildings of architectural or historical interest or to buildings within conservation areas were not surveyed. It is thought likely that fees for such jobs would be significantly higher.

. For every job, the Architect was appointed as lead consultant, and other consultants were appointed directly by the Client.

. The construction contracts used were nationally accepted forms with the contractors in traditional roles. Collateral agreements, where used, were only provided to a funder and not more than one tenant, and imposed no greater obligations on the Architect than the principal Agreement.

. Jobs did not include significant repetition.

. Fees were based on the total construction cost as defined in the SFA Conditions. Fees were not charged on the VAT element of construction costs, nor do these costs include expenses, disbursements or VAT.

A SURVEY OF FEES

Table 1   New Works - Percentage fee bands for classes of buildings (see Table 3)

Construction costs bands*	Class 1	Class 2	Class 3	Class 4	Class 5
	%	%	%	%	%
£50,000 to £75,000	7.20-7.45	7.75-8.60	-	-	-
£75,000 to £100,000	6.85-7.20	7.15-7.75	7.45-8.10	-	-
£100,000 to £250,000	6.25-6.85	6.00-7.15	6.70-7.45	7.10-7.60	7.75-8.00
£250,000 to £500,000	5.90-6.25	5.60-6.00	6.25-6.70	6.75-7.10	7.50-7.75
£500,000 to £750,000	5.75-5.90	5.45-5.60	6.00-6.25	6.50-6.75	7.35-7.50
£750,000 to £1 million	5.65-5.75	5.35-5.45	5.90-6.00	6.35-6.50	7.20-7.35
£1m to £2.5 million	5.50-5.65	5.05-5.35	5.70-5.90	5.95-6.35	6.85-7.20
£2.5m to £5 million	-		5.60-5.70	5.70-5.95	6.30-6.85
£5m to £7.5 million	-	-	5.35-5.50	5.65-5.70	6.20-6.30
£7.5m to £10 million	-	-	5.25-5.35	5.60-5.65	5.90-6.20
£10m to £50 million	-	-	5.00-5.25	5.25-5.60	5.30-5.90

•   Where no figures are given, insufficient data was available.

GENERAL INFORMATION

- **SFA GUIDE - cont'd**

A Survey of Fees - cont'd

Table 2   **Works to Existing Buildings - Percentage fee bands for classes of buildings**
(see Table 3)

Construction costs bands*	Class 1	Class 2	Class 3	Class 4	Class 5
	%	%	%	%	%
£50,000 to £75,000	12.30-14.00	11.40-12.40	-	-	-
£75,000 to £100,000	11.75-12.30	10.80-11.40	10.65-11.15	-	-
£100,000 to £250,000	10.20-11.75	9.60-10.80	9.60-10.65	10.60-11.40	11.40-11.90
£250,000 to £500,000	9.70-10.20	9.35- 9.60	8.95- 9.60	9.80-10.60	11.10-11.40
£500,000 to £750,000	9.50- 9.70	9.30- 9.35	8.60- 8.95	9.40- 9.80	10.70-11.10
£750,000 to £1 million	9.30- 9.50	9.20- 9.30	8.40- 8.60	9.10- 9.40	10.30-10.70
£1m to £2.5 million	-	-	7.70- 8.40	8.70- 9.10	9.90-10.30
£2.5m to £5 million	-	-	7.25- 7.70	8.40- 8.70	9.60- 9.90
£5m to £7.5 million	-	-	7.10- 7.25	8.20- 8.40	9.30- 9.60
£7.5m to £10 million	-	-	6.90- 7.10	8.10- 8.20	8.90- 9.30
£10m to £50 million	-	-	6.20- 6.90	7.40- 8.10	7.40- 8.90

* Where no figures are given, insufficient data was available.

Table 3   **Classification of Building Types**

Type	Class 1	Class 2	Class 3	Class 4	Class 5
Industrial	. Storage sheds	. Speculative factories and warehouses . Assembly and machine workshops . Transport garages	. Purpose-built factories and warehouses		
Agricultural	. Barns and sheds	. Stables	. Animal breeding units		
Commercial	. Speculative shops . Surface car parks	. Multi-storey and underground car parks	. Supermarkets . Banks . Purpose-built shops . Office developments . Retail warehouses . Garages/showrooms	. Department stores . Shopping centres . Food processing units . Breweries . Telecommuni-cations and com-puter buildings . Restaurants . Public Houses	. High risk research and production buildings . Research and development labs . Radio, TV and recording studios

SFA GUIDE - cont'd
A Survey of Fees - cont'd

Table 3   Classification of Building Types (Cont'd)

Type	Class 1	Class 2	Class 3	Class 4	Class 5

Community		. Community halls	. Community centres
			. Branch libraries
			. Ambulance and
			  fire stations
			. Bus stations
			. Railway stations
			. Airports
			. Police stations
			. Prisons
			. Postal buildings
			. Broadcasting
				. Civic centres
				. Churches and
				  crematoria
				. Specialist
				  libraries
				. Museums and art
				  galleries
				. Magistrates/
				  County Courts
					. Theatres
					. Opera houses
					. Concert halls
					. Cinemas
					. Crown Courts

Residential		. Dormitory
		  hostels
			. Estates housing
			  and flats
			. Barracks
			. Sheltered housing
			. Housing for
			  single people
			. Student housing
				. Parsonages/manses
				. Apartment blocks
				. Hotels
				. Housing for the
				  handicapped
				. Housing for the
				  frail elderly
					. Houses and flats
					  for individual
					  clients

Education			. Primary/nursery/
			  first schools
				. Other schools
				  including middle
				  and secondary
				. University
				  complexes
					. University
					  laboratories

Recreation			. Sports halls
			. Squash courts
				. Swimming pools
				. Leisure
				  complexes
					. Leisure pools
					. Specialised
					  complexes

Medical/Social Services			. Clinics	. Health Centres . General hospitals . Nursing homes . Surgeries	. Teaching hospitals . Hospital laboratories . Dental surgeries

CONDITIONS OF ENGAGEMENT FOR THE APPOINTMENT OF AN ARCHITECT (CE/95)

The RIBA *Conditions of Engagement for the Appointment of an Architect (CE/95)* is compatible with the *Standard Form of Agreement for the Appointment of an Architect (SFA/92)*.

CE/95 will be suitable for a wide range of projects, including those of a relatively simple content for which the JCT standard forms of building contract, IFC84 or MW 80, might be appropriate.

CE/95 can be used with the printed Memorandum of Agreement provided or with a covering Letter Appoinntment. A model text for a letter of appointment from architect to client is included within the document.

CE/95 takes account of the Construction (Design and Management) Regulations 1994, effective 31 March 1995. Architects as well as clients may find it helpful to read *Engaging and Architect: Guidance for Clients on the Health and Safety Regulations*, published by the RIBA, which alerts clients to their duties under the CDM Regulations.

It should be noted that the appointment of an architect as Planning Supervisor is a separate appointment from the provision of architectural services under CE/95, and should be made using the RIBA's *Form of Appointment as Planning Supervisor (PS/95)*.

Architects' appointing documents and guides can be obtained from RIBA Publications by mail order (tel: 0171-251 0791) or from RIBA bookshops.

GENERAL INFORMATION

THE ROYAL INSTITUTION OF CHARTERED SURVEYORS

The Scales of Professional Charges for Quantity Surveying Services have been withdrawn and replaced by the RICS information document "Appointing a Quantity Surveyor: A guide for Clients and Surveyors". The document gives no indication of the level of fees to be anticipated although a Form of Enquiry, Schedule of Services, Fee Offer, Form of Agreement and Terms of Appointment are included. The view being that consultants are now appointed less often by reputation and recommendation and more often after a process of competition. To assist employers, quantity surveyors and those procuring professional services and where consultants are being appointed by reputation or recommendation the following will assist in assessing the reasonableness of any fee proposal.

Extracts from Appointing a Quantity Surveyor: A guide for Clients and Surveyors
Scales of Professional Charges
Example of calculation of fees for lump sum and % basis as clause 5.6 of the guide
Anticipated Percentage fees by project and value based on calculations as example.

Appointing a Quantity Surveyor

The following extracts are for information and the full document should be refered to for further advice.

Contents

Selection and Appointment Advice

This Section outlines for Clients and Surveyors when and how to select and appoint a Quantity Surveyor

1. **Introduction**

1.1 The Royal Institution of Chartered Surveyors ('the RICS') has published this Guide to assist Clients when they wish to appoint a chartered Quantity Surveyor. The contents of the Guide should also assist Quantity Surveyors when concluding an agreement with their Client.

1.2 Over the last few years changes have taken place in the market place for professional services. Consultants are now appointed less often by reputation and recommendation and more often after a process of competition. Competition takes place in the private and public sectors and is supported by European Union requirements and UK Compulsory Competitive Tendering legislation.

1.3 This Guide is applicable to a range of situations - whether details of the services required are proposed to or by the Quantity Surveyor; or whether appointment is of a single Quantity Surveyor or following selection from a limited list. The Guide sets out principles to be considered and applied rather than a detailed step-by-step procedure. Details often vary from project to project depending on the type of Client and the nature of the project. A detailed procedure that attempts to cover all projects is not therefore considered appropriate.

1.4 Quantity Surveyors may be appointed directly by a Client or as part of a multidisciplinary team. The RICS recommends that the quantity surveying appointment is made directly by a Client so as to ensure that independent financial advice is made directly to that Client by the Quantity Surveyor

1.5 The Guide provides a basis for the appointment of a Quantity Surveyor so that those concerned are clear about:
• the services being requested and/or offered
• the terms and conditions of the contract
• the fee payable and the method of payment.

A Client and his Quantity Surveyor should set out clearly their requirements in their agreement so that the services to be provided and the conditions of engagement are certain. It is necessary for a Client to set out his requirements with the same clarity as when seeking tenders from contractors on construction projects. It is now compulsory for Members of the RICS to provide written notification to their Client of the terms and conditions of the appointment. This documentation is intended to form a good basis upon which this certainty can be achieved and is recommended for use between the Client and his Quantity Surveyor.

2. Notes on use and completion of documents

2.1 Form of Enquiry
The Form of Enquiry sets out the details of the services a Client wishes a Quantity Surveyor to provide and should be attached to the Form of Agreement and identified as belonging to it.

2.2 Form of Agreement and Terms of Appointment
The Form of Agreement should be completed and signed only when the services, fees and expenses have been agreed between a Client and his Quantity Surveyor. The Form of Agreement states the names of the parties, their intentions and their arrangement.

The names and addresses of the parties should be inserted, a brief description of the project given and any site identified. If the Form of Agreement is to made as a simple contract it should be signed by both parties and the date entered. If it is to be executed as a Deed the alternative version should be used. See notes *1 and *2 on page 40.

3. When to appoint a Quantity Surveyor

In order that maximum benefit can be gained from his skills a Quantity Surveyor should be appointed by a Client as soon as possible in the life of a project, preferably at the inception of a scheme, so that advice can be provided on:
- the costs of the project so that a realistic budget can be set at inception and cost management can be applied throughout
- the procurement method best suited to the requirements of the Client
- the implications of the appointment of other consultants and contractors.

It is recommended that a Client and his prospective Quantity Surveyor should meet and discuss the appointment before an agreement is reached, unless the services of the Quantity Surveyor are to be restricted only to some of those from the range available and shown in the Form of Enquiry.

4. How to select and appoint a Quantity Surveyor

Detailed guidance on the selection and appointment of chartered surveyors is given in the RICS publication A Guide to Securing the Services of a Chartered Surveyor. Chapters 2, 3 and 4 also provide useful information on preparing a brief for a Client's requirements, producing tender and contract documentation, organising competitions and evaluating offers.

Methods of selecting a Quantity Surveyor include:
Selection based on existing knowledge
A Client may select and then appoint a Quantity Surveyor using existing knowledge of that Quantity Surveyor's performance and reputation. This knowledge may arise from a previous working relationship or be based on the recommendation of others
Selection from a panel maintained by a Client
A Client may maintain a panel of Quantity Surveyors. He will have records of their experience which will enable him to make his selection and appointment
Selection from an ad hoc list produced by a Client
If a Client is unable to make an appointment based on knowledge or reputation or by selection from a standing panel it may be more appropriate for an ad hoc list to be prepared.

4.3 Whichever of the above methods of selection is used it is important for the selection criteria to include the following:
- the financial standing of the Quantity Surveyor under consideration
- the experience, competence and reputation of each Quantity Surveyor in the area of the project/skill being considered so that selection is made using comparable standards between firms.
- the ability of each Quantity Surveyor to provide the required service at the relevant time.

There is no one correct way to use competitive tendering as a basis for appointment and no single way of complying with legislation. It is important, however, that the procedure is always fair and open. Where legislation means that competition is compulsory, reference should be made to Clause 4.8 below and any other guidance available.

4.4 If competition is to be used in the selection process the following should be borne in mind:
- a Client should choose between the various methods of competition from either open, selective or restrictive lists, or single- or two-stage tenders, according to the complexity of the project which is to be the subject of competition
- the basis of selection can be on quality, competence or price, or a combination of all three. Whichever basis is chosen it must be clearly stated in the documentation
inviting interest in the commission
- it is recommended that price is never regarded in isolation, as will be clear from the detailed guidance set out below.

4.5 To ensure that any competition is well organised the key ingredients required are that:
- specification of the service required is clear
- definition of the skills and competencies required to deliver the services is clear
- criteria on which offers will be evaluated will be made available with the tender documents
- any weighting to be applied to skills, competencies and price is made available with the tender documents
- each stage of the tender process is documented and the procedures for selection and invitation to tender match the requirements of the commission.

4.6 Every effort should be made to ensure that contract documentation is complete. If circumstances change to the extent that criteria are altered all tenderers should immediately be advised and given the opportunity to respond and confirm a willingness to continue.

4.7 **Private contracts**
In the case of private contracts there is no upper or lower limit to the number of private sector firms that may be invited to tender, nor formal rules as to their selection. Nevertheless, the RICS recommends that not less than 3 and not more than 6 firms are invited to submit proposals.

4.8 **Public contracts**
In the case of public sector contracts Compulsory Competitive Tendering rules and European Union rules, together with relevant Treasury regulations and DETR guidance, need to be followed. More information is available in the RICS publication A Guide to Securing the Services of a Chartered Surveyor. Chapter 2, pages 3 to 6, provides details on regulations but it is emphasised

GENERAL INFORMATION

that current requirements should always be verified before seeking public sector offers.

5. Specifying the quantity surveying service and determining the fee options and expense costs

5.1 It is emphasized that the fee for a project can either be negotiated, sought as a sole offer or sought in competition. The following considerations may apply to any of the methods of selecting a Quantity Surveyor and determining his fee in relation to the services to be provided.

5.2 When fee quotations are to be sought from more than one Quantity Surveyor it is of paramount importance that the enquiry and/or the submissions relate precisely to the same details of service(s) required. Proper use of the Form of Enquiry should ensure that this is achieved. In particular, the services required should be specified in detail, by using the Schedule of Services provided and/or by adding or deleting any special services not listed or required in that Schedule.

5.3 Information provided by a Client in an offer document should include, as a minimum:
- the complete scope of the project
- a full and precise description of the quantity surveying service(s5 to be provided
- the terms and conditions which will apply, preferably the Form of Agreement and Terms of Appointment contained in this Guide, together with any further requirements that will apply
- the anticipated time scales that will apply, both for the quantity surveying service and for the project
- information on the inclusion or exclusion of expense costs
- the basis of the fee upon which the offer is being invited or offered including:
  - where a fee submission is to deal separately with different components of the service, the way in which any components will be evaluated in order to give comparable totals for selection purposes; and
  - wherever a percentage is to be quoted by a Quantity Surveyor; the definition of the total construction cost to which that percentage is to be applied
- provisions for stage or instalment payments
- where the nature, size, scope, time or value of the project or of the quantity surveying service(s) are likely to vary, the method(s) of appropriately adjusting the fee.

5.4 In determining the basis of the services and the fees required the following should be considered:
- where quantity surveying services are incapable of precise definition at the time of appointment, or where they could change substantially, enough information should be contained in an agreement to enable possible variations in those services to be negotiated
- where services are likely to be executed over a particularly long period, or where the service and/or the project might be delayed or postponed, the method and timing of any increase(s) in the fees should be stated
- the basis of the fees to be offered might be one of, or a combination of:
  - a single percentage encompassing all the described services
  - separate percentages for individually defined components of services
  - a single lump sum encompassing all the described services
  - separate lump sums for individually defined components of services
  - a time charge with either hourly rates quoted or a multiplier or adjustment factor
  to be applied to a specified definition of hourly cost.

5.5 Where the nature or scale of the project warrants it Quantity Surveyors may be invited to:
- explain in outline their proposed method of operation for the project, and/or
- demonstrate their ability to provide the services required from their own resources, and/or
- state the extent of partner/director involvement, and/or
- name the key personnel to be allocated to the project with details of relevant experience.

5.6 Fee options
Before offering any fee proposal a Quantity Surveyor will need to evaluate the costs of carrying out the service being offered. Options for charging fees are given below and may be used alone or in combination.

5.6.1 Percentage fees
This method of charging fees is appropriate when a construction project is reasonably straightforward and the quantity surveying service can be clearly defined but the exact amount of the total construction cost cannot be determined with much certainty

In determining a fee, judgements need to be made by tthe Quantity Surveyor on the size of the project, its complexity, degree of repetition, method of procurement and contract arrangements. This method may be seen as a 'broad brush' way of assessment which sometimes may be vulnerable to market forces and their influence on contractors' tenders.

Fees are expressed as a percentage of a sum, which is invariably the total construction cost, for the provision of a defined service. This sum will generally be calculated from the value of all construction contracts and other items of work carried out directly by a Client on a project.

It is advisable to have a definition of total construction cost agreed between a Client and his Quantity Surveyor Exclusions from the total construction cost should also be defined.

5.6.2 Lump sum fee
This method of charging fees is appropriate when a construction project is reasonably certain in programme time, project size and construction cost and the quantity surveying service can be clearly defined. A total fee is agreed for providing a defined service or amount of work. If appropriate, percentage-based or time-based fees may be converted to a lump sum fee once a project has become sufficiently defined.

If time, size, cost or circumstances of the project are significantly varied by more than specified amounts the lump sum fee may be varied, perhaps according to a formula contained in the Agreement. Otherwise the fee remains fixed. It should be borne in mind that a lump sum fee, whilst giving certainty to both parties at the start of a commission, may not always be appropriate unless the circumstances of the project remain significantly as stated at the time that the Agreement was signed.

5.6.3 Time charge fees
This method of charging fees is appropriate where the scope and/or the extent of the services to be provided cannot reasonably be foreseen and/or cannot reasonably

be related to the cost of construction. It is often appropriate where open-ended services are required on feasibility studies, negotiation, claims consultancy etc. It is also a method that allows for additional or varied services to be provided, in addition to the provision of basic services quoted for in a fee agreement.

Fees are calculated from all the time expended on a project by partners/directors and staff and charged at rates previously agreed. The rates may be increased periodically to allow for the effects of inflation.

It is advisable to have previously agreed rates for grades of staff and partners/directors and methods of revising those rates periodically to reflect subsequent changes in salaries and costs. The inclusion of overhead allowances, generally including secretarial and administrative staff, within the rates needs careful calculation by the Quantity Surveyor Principles for the reimbursement of the time involved also need to have been agreed in advance. Sometimes this method may appear to be open ended, perhaps with little incentive for working efficiently. Periodic review of the charges is important so that progress is monitored against budgets.

5.6.4 Target cost fees

This method is appropriate where time charge fees are the basis for the Agreement but a Client wishes to have a guaranteed maximum target for fees.

Fees are recovered on a time charge basis but a 'capped' or guaranteed target cost is agreed between the parties before the work is carried out by the Quantity Surveyor If the target cost is exceeded then the Quantity Surveyor may not receive all of his costs, for instance part of his overhead allowances. If alternatively the total amount of the time charges falls below the budget cost then the Quantity Surveyor may receive an additional payment, a 'bonus' to reflect or share in the 'saving' to his Client. A formula for this should be agreed at the outset.

5.7 Expense costs

Expense costs are the costs to a firm such as the provision of cars, mileage-based payments to staff for travel, other travel costs, hotels, subsistence, meals, reproduction of documents, photocopying, postage and phone/fax charges, purchase costs of items bought specifically for a project, expenses in connection with appointing and engaging site-based staff, exceptional time spent by staff in travelling outside of usual hours and/or beyond usual distances etc.

In any agreement on fees it is advisable to be clear about the inclusion or exclusion of any category of expense costs within the fee charges, to identify which expenses are chargeable and to have a machinery for adjustment where necessary.

5.7.1 Recovering expense costs in fee arrangements Expense costs in fee bids and fee arrangements may be dealt with in a number of ways. Tender enquiries may prescribe the way required or there may be flexibility in how expense costs are included within the fee figures or shown separately from them. Three options exist for the recovery of expense costs in fee arrangements:

- **Expense costs may be included within a lump sum fee**
  Expense costs of any variety can be assessed and included within a fee and not shown separately. The fee offer thus includes a lump sum for all expense costs. This is often referred to as a 'rolled-up' offer in that the costs have not been shown separately in the bid. Although expense costs have been so included it is often necessary to consider how any adjustment to those costs may be made if changed circumstances require. For instance, if travel costs have been included in a 'rolled-up' fee and much more travel becomes necessary, it may then be required to demonstrate the basis of travel expenses that had been included in the fee offer in order to agree an adjustment

- **Expense costs may be converted to a percentage of the total construction cost of the project**
  An assessment is made of the likely expense costs that may be incurred and the amount is then converted into a percentage of the total construction cost. This percentage may be added to the percentage shown as a fee offer for the quantity surveying service or it may be shown separately. The relationship between construction costs, to which a percentage fee would be applied, and the amount of expense costs that will be incurred is often tenuous and this method should only be entered into after careful consideration by a Client and his Quantity Surveyor

- **Expense costs may be paid as a separate lump sum**
  A lump sum can be calculated for expense costs and shown separately from any we lump sum, percentage-based or time charge fee for the project. Whilst this method will give a Client certainty at the outset, provision should be made for adjustment by the parties if circumstances change.

5.7.2 Constituents of expense costs and payment considerations

If expense costs are not to be included in a lump sum, percentage-based or time charge fee they may be recovered by the Quantity Surveyor on the submission of the cost records. If this method is used it is important to establish how expense costs will be calculated and reimbursed to a Quantity Surveyor by his Client. There are three methods for this:

- **Recovery of actual expense costs of disbursements incurred by a Quantity Surveyor with the authority of his Client**
  This category of cost would normally include items such as public transport costs, hotels and subsistence, and sundry expenses incurred, for instance, in obtaining financial reports on companies when considering prospective tender lists, or in establishing a site office

- **Recovery of expense costs for in-house resources provided by an organisation**
  This would include, for example, photocopying and the reproduction of tender and contract documentation. The amounts due under this category of costs are not always as clearly demonstrable as the costs referred to in the previous category. It may be helpful to set up an agreement between the parties that allows the Quantity Surveyor to recover an amount in addition to the actual material costs incurred so that some staff and overhead costs can be recovered in addition to the basic costs

- **Recovery of expense costs at market rates**
  This category is for resources provided by an organisation from its in-house facilities, for example photocopying or drawing reproduction, travel carried out by personnel using company cars etc. These costs would be recovered at what is agreed to be the market rate for the service provided.

GENERAL INFORMATION

5.8 Implication of termination or suspension

If a project is terminated or suspended it may become necessary to consider the implications on fees and on expenses, for instance for staff, offices, cars, rented accommodation or leased equipment taken on especially for a project. It may be that some of these costs cannot readily be avoided if termination or suspension occurs and a fee agreement should contain provisions to allow adjustments to be made, particularly to lump sum, 'rolled-up' or percentage fees.

6. Submission and comparison of fee offers and selection and notification of results

6.1 Submission

6.1.1 Clients are advised to invite fee offers only from firms of comparable capability and to make a selection taking into account value as well as the fee bid.

6.1.2 Where competitive offers have been sought, instructions on the submission of offers should be clear; giving the date, time and place for their delivery. Offers should all be opened at the same time and treated as confidential until notification of results is possible.

6.2 Comparison and selection

6.2.1 Competitive offers should be analysed and compared (including where there are different components of a fee evaluation, as referred to above). Comparison of offers should be on a basis which incorporates all the component parts of an offer in order to indicate the lowest offer and its relationship with other offers.

6.2.2 If two or more submissions give identical or very close results then it may be appropriate to apply a sensitivity test, namely to check the impact of possible or probable changes in the scope, size or value of the project on the fee bids.

6.2.3 In comparing offers it is advisable to weigh quality criteria against the offer price. Further detailed guidance is given in Appendix 10 of the RICS publication A Guide to Securing the Services of a Chartered Surveyor. It is usual to weigh quality of service criteria at a minimum of 60 per cent relative to price criteria at a maximum of 40 per cent.

6.3 Notification

6.3.1 Once a decision has been taken to appoint a Quantity Surveyor; the successful firm should be notified. Unsuccessful firms should also be notified of the decision and given information on the bids received (where appropriate this can be by the use of 'indices' which do not link the names of each firm to its offer). All notifications should be made in writing and be sent as soon as possible after the decision to appoint has been made.

7. Confirmation of the Agreement between the Client and the Quantity Surveyor

7.1 The Agreement between a Client and his Quantity Surveyor should be effected by either:
- using the Form of Agreement and Terms of Appointment contained in this Guide, with cross-reference to the Form of Enquiry, Schedule of Services and Fee Offer
- using a separate form of agreement, terms and conditions, with cross reference to the services to be provided and to the fee offer

- a simple exchange of letters incorporating the information given above, making reference to the fee offer.

The 1996 Housing Grants, Construction and Regeneration Act ('the HGCR Act') introduced a right to refer any dispute to adjudication and certain provisions relating to payment. This documentation complies with the HGCR Act and includes the necessary adjudication and payment provisions. Other forms of agreement will need to comply with the HGCR Act or the adjudication and payment provisions of the Statutory Scheme for Construction Contracts will apply. An exchange of letters is unlikely to comply and the Scheme will therefore also apply to these.

8. Complaints handling and disputes resolution

8.1 Quantity Surveyors who are partners or directors in firms providing surveying services must operate an internal complaints handling procedure, which applies to disputes less than £50,000, under the RICS Bye-Laws. The RICS also sets a minimum standard of complaints handling, as laid out in its Professional Conduct – Rules of Conduct and Disciplinary Procedures. If the complaint cannot be resolved internally by the firm then the matter must go to final resolution by a third party. A reference to adjudication under the HGCR Act would be sufficient for the purposes of satisfying RICS regulations, pending any final resolution of the dispute at the end of the contract period. See also Clause 11 of the Terms of Appointment.

8.2 The HGCR Act introduced a compulsory scheme of third party neutral dispute resolution – which can occur at any time during the construction process – called adjudication. Any party may refer any dispute to an adjudicator at any time.

1. Category One: General services

The following services may be provided on any project, whatever its nature and whatever the method of procurement adopted.

☐ The following services shall, where relevant, also apply to environmental engineering service (mechanical and electrical engineering) if indicated by placing a cross in this box.

1.1 Inception and feasibility

1.1.1 ☐ Liaise with Client and other consultants to determine Client's initial requirements and subsequent development of the full brief

1.1.2 ☐ Advise on selection of other consultants if not already appointed

1.1.3 ☐ Advise on implications of proposed project and liaise with other experts in developing such advice

1.1.4 ☐ Advise on feasibility of procurement options

1.1.5 ☐ Establish Client's order of priorities for quality, time and cost

1.1.6 ☐ Prepare initial budget estimate from feasibility proposals

1.1.7 ☐ Prepare overall project cost calculation and cash flow projections

1.2 Pre-contract cost control

1.2.1. ☐ Prepare and develop preliminary cost plan

1.2.2 ☐ Advise on cost of design team's proposals, including effects of site usage, shape of buildings, alternative forms of design and construction as design develops

1.2.3 ☐ Monitor cost implications during detailed design stage

1.2.4 ☐ Maintain and develop cost plan, and prepare periodic reports and updated cash flow forecasts

1.3 Tender and contractual documentation

1.3.1 ☐ Advise on tendering and contractual arrangements taking into account the Client's priorities and information available from designers

1.3.2 ☐ Advise on insurance responsibilities and liaise with Client's insurance adviser

1.3.3 ☐ Advise on warranties

1.3.4 ☐ Advise on bonds for performance and other purposes

1.3.5 ☐ Prepare tender and contract documentation in conjunction with the Client and members of the design team

1.3.6 ☐ Provide copies of documentation as agreed

1.3.7 ☐ Advise on use and/or amendment of standard forms of contract or contribute to drafting of particular requirements in association with Client's legal advisers

1.3.8 ☐ Draw up forms of contract, obtain contract drawings from members of design team and prepare and deliver to both parties contract copies of all documents

1.4 Tender selection and appraisal

1.4.1 ☐ Advise on shortlisting prospective tenderers

1.4.2 ☐ Investigate prospective tenderers and advise Client on financial status and experience

1.4.3 ☐ Attend interviews with tenderers

1.4.4 ☐ Arrange delivery of documents to selected tenderers

1.4.5 ☐ Check tender submissions for accuracy, level of pricing, pricing policy etc.

1.4.6 ☐ Advise on errors and qualifications and, if necessary, negotiate thereon

1.4.7 ☐ Advise on submission of programme of work and method statement

1.4.8 ☐ Prepare appropriate documentation,

if required, to adjust the tender received to an acceptable contract sum

1.4.9 ☐ Review financial budget in view of tenders received and prepare revised cash flow

1.4.10 ☐ Prepare report on tenders with appropriate recommendations

1.4.11 ☐ Advise on letters of intent and issue in conjunction with Client's advisers

1.5 Interim valuations

1.5.1 ☐ Prepare recommendations for interim payments to contractors, subcontractors and suppliers in accordance with contract requirements

1.6 Post-contract cost control

1.6.1 ☐ Value designers' draft instructions for varying the project before issue

1.6.2 ☐ Prepare periodic cost reports in agreed format at specified intervals including any allocations of cost and/or copies as requested by third parties

1.7 Final account

1.7.1 ☐ Prepare the final account

1.8 Attendance at meetings

1.8.1 ☐ Attend meetings as provided for under this Agreement

19 Provision of printing/reproduction/copying of documents and the like

1.9.1 ☐ Provide copies of documentation as provided for under this Agreement

2. Category Two: Services particular to non-traditional methods of procurement

These services relate to particular methods of procurement and contract arrangement and should be incorporated into the Agreement as required in conjunction with services from Categories One and Three of the Schedule.

2.1 Services particular to prime cost and management or construction management contracts

2.1.1 ☐ Obtain agreement of a contractor to the amount of the approximate estimate and confirm the amount of the fee for the contract

2.1.2 ☐ Prepare recommendations for interim payments to contractor based on contractor's prime costs

2.1.3 ☐ Adjust the approximate estimate to take account of variations and price fluctuations

2.1.4 ☐ Check the final amounts due to contractors, subcontractors and suppliers

GENERAL INFORMATION

2.2 Services particular to management and construction management contracts

The terms 'management contracting' and 'construction management' mean contractual arrangements where a firm is employed for a fee to manage, organise, supervise and secure the carrying out of the work by other contractors.

2.2.1 ☐ If required, assist in drafting special forms of contract

2.2.2 ☐ Prepare tender documents for the appointment of a management contractor or construction manager

2.2.3 ☐ Attend interviews of prospective contractors or managers

2.2.4 ☐ Obtain manager's agreement to contract cost plan and confirm amount of manager's fee

2.2.5 ☐ Assist in allocation of cost plan into work packages

2.2.6 ☐ Assist in preparation of tender and contract documents

2.2.7 ☐ Price tender documents to provide an estimate comparable with tenders

2.2.8 ☐ Review cost plan as tenders are obtained and prepare revised forecasts of cash flow

2.2.9 ☐ Prepare periodic cost reports to show effect of variations, tenders let and prime costs

2.2.10 ☐ Check the final amounts due to managers, contractors, subcontractors or works contractors and suppliers

2.3 Services particular to design and build contracts - services available to a Client

2.3.1 ☐ Draft the Client's brief, in association with the Client and his designers

2.3.2 ☐ Prepare tender documents incorporating the Client's requirements

2.3.3 ☐ Prepare contract documentation, taking into account any changes arising from the contractor's proposals

2.3.4 ☐ Prepare recommendations for interim and final payments to the contractor; including compliance with statutory requirements of the 1996 Housing Grants, Construction and Regeneration Act

2.3.5 ☐ Assist in agreement of settlement of the contractor's final account

2.4 Services particular to design and build contracts - services available to a contractor

2.4.1 ☐ Prepare bills of quantities to assist in the preparation of a contractor's tender

2.4.2 ☐ Prepare alternative cost studies to assist in determining the optimum scheme for a contractor's submission

2.4.3 ☐ Draft specifications forming the contractor's proposals

2.4.4 ☐ Assist with specialist enquiries in compiling the contractor's tender

2.4.5 ☐ Measure and price variations for submission to the Client's representative

2.4.6 ☐ Prepare applications for interim payments

2.4.7 ☐ Agree final account with Client's representative

2.5 Services particular to measured term contracts

2.5.1 ☐ Take measurements, price from agreed schedule of rates and agree totals with contractor

2.5.2 ☐ Check final amounts due to contractor(s)

3. Category Three: Services not always required in Categories One and Two

These services should be incorporated into the Agreement in conjunction with services from Categories One and Two of the Schedule as required. Where points are left blank, the Client and/or the Quantity Surveyor should specify their own service as required.

3.1 Bill of quantities

3.1.1 ☐ Provide bills of quantities for mechanical and engineering services

3.1.2 ☐ Price bills of quantities to provide an estimate comparable with tenders

3.2 Cost analysis

3.2.1 ☐ Prepare cost analysis based on agreed format or special requirement

3.3 Advise on financial implications as follows:

3.3.1 ☐ Cost options of developing different sites

3.3.2 ☐ Preparation of development appraisals

3.3.3 ☐ Cost implications of alternative development programmes

3.3.4 ☐ Effect of capital and revenue expenditure

3.3.5 ☐ Life cycle cost studies and estimate of annual running costs

3.3.6 ☐ Availability of grants

3.3.7 ☐ Assist in applications for grants and documentation for these

3.3.8 ☐ Evaluation of items for capital allowances, grant payments or other such matters

3.4 Advice on use of areas and provide:

3.4.1 ☐ Measurement of gross floor areas

3.4.2 ☐ Measurement of net lettable floor areas

3.5 **Provide advice on contractual matters affecting the following:**

3.5.1 ☐ Entitlement to liquidated and ascertained damages

3.5.2 ☐ Final assessment of VAT

3.5.3 ☐ Opinion on delays and/or disruptions and requests for extensions of time

3.5.4 ☐ Consequences of acceleration

3.5.5 ☐ Assessment of the amount of loss and expense or other such matters and if instructed carrying out negotiations with contractors to reach a settlement

3.6 **Provide value management and value engineering services as follows:**

3.6.1 ☐ (state)_____

3.6.2 ☐ (state)_____

3.7 **Provide risk assessment and management services as follows:**

3.7.1 ☐ (state)_____

3.7.2 ☐ (state)_____

3.8 **Adjudication services**

3.8.1 ☐ Provide services acting as an adjudicator in construction disputes

3.8.2 ☐ Provide services in connection with advising the Client in relation to active or threatened adjudication proceedings

3.9 **Provide services in connection with arbitration and/or litigation as follows:**

3.9.1 ☐ (state)_____

3.9.2 ☐ (state)_____

3.10 **Provide services arising from fire or other damage to buildings including preparing and negotiating claims with loss adjusters as follows:**

3.10.1 ☐ (state)_____

3.10.2 ☐ (state)_____

3.11 **Provide services to a contractor in connection with negotiation of claims as follows:**

3.11.1 ☐ (state)_____

3.11.2 ☐ (state)_____

3.12 **Project Management**

Project management services are available from Quantity Surveyors and are set out in detail (together with a Form of Agreement and Guidance Notes) in the RICS publication Project Management Agreement and Conditions of Engagement.

Clients are refered to the Project Management Agreement if the service is mainly for project management. This section is intended to be used if ancillary project management is required to a mainly quantity surveying service.

Provide project management services as follows:

3.12.1 ☐ (state)_____

3.12.2 ☐ (state)_____

3.13 **Provide programme co-ordination and monitoring services as follows:**

3.13.1 ☐ (state)_____

3.13.2 ☐ (state)_____

3.14 **Provide planning supervisor services as follows:**

3.14.1 ☐ (state)_____

3.14.2 ☐ (state)_____

3.15 **Provide information for use in future management and / or maintenance of the building as follows:**

3.15.1 ☐ (state)_____

3.15.2 ☐ (state)_____

3.16 **Any other services not listed elsewhere in the Form of Enquiry:**

3.16.1 ☐ (state)_____

3.16.2 ☐ (state)_____

3.16.3 ☐ (state)_____

Other services required or provided by quantity surveyors include:-
Planning Supervisor duties
Contractors Services – Measurement
               Pricing
               Procurement advice -
                    Subcontractors
                    Suppliers
Post Contract services

Adjudication
Mediation
Arbitration
Expert Witness

GENERAL INFORMATION

Example Quantity Surveying Fee Calculation (see clause 5.6 of Guide)

Example Complex Project Resource Schedule

Project Value £1,500,000 (Including alteration works value £150,000)

	Hours	Rate	Cost
1. COST PLANNING			
Client Liaison	20	25.00	500.00
Attendance at meetings	20	25.00	500.00
Pre tender cost planning	72	20.00	1440.00
			2440.00
2. BILLS OF QUANTITIES			
Client Liaison	12	25.00	300.00
Attendance at meetings	20	25.00	500.00
Demolitions & Alterations	32	16.00	512.00
Substructures	24	16.00	384.00
Frame	16	16.00	256.00
Upper floors	12	16.00	192.00
Roof	16	16.00	256.00
Stairs	8	16.00	128.00
External walls	16	16.00	256.00
Windows & Doors	32	16.00	512.00
Internal Doors	16	16.00	256.00
Wall Finishes	16	16.00	256.00
Floor Finishes	12	16.00	192.00
Ceiling Finishes	8	16.00	128.00
Fittings & Furnishings	8	16.00	128.00
Sanitary appliances	8	16.00	128.00
Services equipment	8	16.00	128.00
Disposal installations	12	16.00	192.00
Water installations	8	16.00	128.00
Heat source	8	16.00	128.00
Space heating & air treatment	12	16.00	192.00
Ventilation installations	8	16.00	128.00
Electrical installations	24	16.00	384.00
Gas Installations	8	16.00	128.00
Lift installations	8	16.00	128.00
Protective installations	8	16.00	128.00
Communication installations	8	16.00	128.00
Special installations	8	16.00	128.00
Builders work in connection	24	16.00	384.00
External works	42	16.00	672.00
Drainage	16	16.00	256.00

	Hours	Rate	Cost
External services	8	16.00	128.00
Preliminaries & Preambles	24	16.00	384.00
Pre-tender estimates	24	16.00	384.00
Tender report & Cost Analysis	16	16.00	256.00
Contract documentation	8	20.00	160.00
			9120.00
3. POST CONTRACT			
Client liaison	24	25.00	600.00
Attendance at meetings	100	17.00	1700.00
Valuations	80	17.00	1360.00
Cost Reports	80	17.00	1360.00
Pre Estimate Instructions	40	17.00	680.00
Final Account	80	17.00	1360.00
			7060.00

Total estimated cost		8620.00
Add overheads	50 %	9310.00
		27930.00
Add profit	25 %	6982.00
		£34913.00

Cost Planning only

Estimated Cost	2440.00
Add Overhead @ 50 %	1220.00
	3660.00
Add Profit @ 25 %	915.00
DATE 20-Jul-99	4575.00

Bills of Quantities only

Estimated Cost		9120.00
Add Overheads	@ 50%	4560.00
		13680.00
Add Profit	@ 25%	3420.00
		17100.00

Post Contract only

Estimated Cost		7060.00
Add Overheads	@ 50%	3530.00
		10590.00
Add Profit	@ 25%	2847.00
		13437.00
Total		£34913.00

The above example excludes all disbursements – travel, reproduction of documents, forms of contracts and the like.

The rates indicated equate to:

Partner	cost rate £25.00	Charge Rate £46.87
Senior Surveyor	cost rate £20.00	Charge Rate £37.50
Taker off	cost rate £16.00	Charge Rate £30.00
Post Contract Surveyor	cost rate £17.00	Charge Rate £31.87

The hours indicated and rates used should be adjusted as appropriate to suit each individual project and the costs, overheads and profit requirements of the practice bidding for work.

Other factors that should be considered when calculating fees include timescale, workload, relationship with client and other consultants, prestige, likely ongoing programme of work, availability of labour

Typical fees calculated as above for provision of basic services

	Cost Planning	Bills of Quantities	Post Contract services
Complex Works with Little repetition Work Value			
50,000	250	1100	700
150,000	650	2850	1900
300,000	1250	4900	3400
600,000	2500	8100	6300
1,500,000	4600	16200	13300
3,000,000	8200	27000	24000
6,000,000	14500	46800	43800
10,000,000	21500	70800	67800
Less Complex work with some repetition Work Value			
50,000	250	1000	700
150,000	650	2700	1900
300,000	1250	4450	3400
600,000	2500	7100	6100
1,500,000	4600	13000	12000
3,000,000	8200	22000	21000
6,000,000	14500	38200	37200
10,000,000	21500	57400	56400

Simple works or works
with a substantial amount
of repetition
Work Value

50,000	200	900	600
150,000	500	2400	1600
300,000	1000	4000	2900
600,000	1800	6100	5400
1,500,000	3400	10900	11300
3,000,000	6100	18100	19400
6,000,000	10600	30700	33800
10,000,000	15400	45000	50600

Alteration Works – Additional Fee on value of alteration works
Work Value

50,000	300
150,000	900

Decoration Works – Additional Fee on value of alteration works
Work Value

50,000	450
150,000	1350

Where Provisional Sums, Prime Cost Sums, Subcontracts, Dayworks contingencies form a substantial part of the project the above figures may require adjustment accordingly.

Mechanical and Electrical Services – Additional Fee for provision of Bills of Quantities

	Cost Planning	Bills of Quantities	Post Contract services
Value of M & E work			
120,000	1800		1500
240,000	3420		2600
600,000	7500		5300
1,000,000	11000		8250
4,000,000	31700		26200
6,000,000	45500		44500

Negotiating Tenders – Additional Fee
Work Value

150,000	450
600,000	1250
1,200,000	1950
3,000,000	3000

The following Scales of Professional Charges printed by permission of The Royal Institution of Chartered Surveyors, 12 Great George Street, Westminster, London SW1P 3AD, are now obsolete and are reproduced for information only.

SCALE NO. 36

INCLUSIVE SCALE OF PROFESSIONAL CHARGES FOR QUANTITY SURVEYING SERVICES FOR BUILDING WORKS

(Effective from 29th July, 1988)

This Scale has been determined by the Fees Committee appointed by the Quantity Surveyors Divisional Council of The Royal Institution of Chartered Surveyors. The Scale is for guidance and is not mandatory.

1.0 GENERALLY

1.1 This scale is for use when an inclusive scale of professional charges is considered to be appropriate by mutual agreement between the employer and the quantity surveyor.

1.2 This scale does not apply to civil engineering works, housing schemes financed by local authorities and the Housing Corporation and housing improvement work for which separate scales of fees have been published.

1.3 The fees cover quantity surveying services as may be required in connection with a building project irrespective of the type of contract from initial appointment to final certification of the contractor's account such as:

(a) Budget estimating; cost planning and advice on tendering procedures and contract arrangements.

(b) Preparing tendering documents for main contract and specialist sub-contracts; examining tenders received and reporting thereon or negotiating

tenders and pricing with a selected contractor and/or sub-contractors.

(c) Preparing recommendations for interim payments on account to the contractor; preparing periodic assessments of anticipated final cost and reporting thereon; measuring work and adjusting variations in accordance with the terms of the contract and preparing final account, pricing same and agreeing totals with the contractor.

(d) Providing a reasonable number of copies of bills of quantities and other documents; normal travelling and other expenses. Additional copies of documents, abnormal travelling and other expenses (e.g. in remote areas or overseas) and the provision of checkers on site shall be charged in addition by prior arrangement with the employer.

1.4 If any of the materials used in the works are supplied by the employer or charged at a preferential rate, then the actual or estimated market value thereof shall be included in the amounts upon which fees are to be calculated.

1.5 If the quantity surveyor incurs additional costs due to exceptional delays in building operations or any other cause beyond the control of the quantity surveyor then the fees shall be adjusted by agreement between the employer and the quantity surveyor to cover the reimbursement of these additional costs.

1.6 The fees and charges are in all cases exclusive of value added tax which will be applied in accordance with legislation.

1.7 Copyright in bills of quantities and other documents prepared by the quantity surveyor is reserved to the quantity surveyor.

2.0 INCLUSIVE SCALE

2.1 The fees for the services outlined in paragraph 1.3, subject to the provision of paragraph 2.2, shall be as follows:

GENERAL INFORMATION

(a) CATEGORY A. Relatively complex works and/or works with little or no repetition.

Examples:

Ambulance and fire stations; banks; cinemas; clubs; computer buildings; council offices; crematoria; fitting out of existing buildings; homes for the elderly; hospitals and nursing homes; laboratories; law courts; libraries; 'one off' houses; petrol stations; places of religious worship; police stations; public houses, licensed premises; restaurants; sheltered housing; sports pavilions; theatres; town halls; universities, polytechnics and colleges of further education (other than halls of residence and hostels); and the like.

Value of Work		Category A Fee	
	£	£	£
Up to	150000	380 + 6.0% (Minimum Fee £3380)	
150000 -	300000	9380 + 5.0% on balance over	150000
300000 -	600000	16880 + 4.3% on balance over	300000
600000 -	1500000	29780 + 3.4% on balance over	600000
1500000 -	3000000	60380 + 3.0% on balance over	1500000
3000000 -	6000000	105380 + 2.8% on balance over	3000000
Over	6000000	189380 + 2.4% on balance over	6000000

(b) CATEGORY B. Less complex works and/or works with some element of repetition.

Examples:

Adult education facilities; canteens; church halls; community centres; departmental stores; enclosed sports stadia and swimming baths; halls of residence; hostels; motels; offices other than those included in Categories A and C; railway stations; recreation and leisure centres; residential hotels; schools; self-contained flats and maisonettes; shops and shopping centres; supermarkets and hypermarkets; telephone exchanges; and the like.

Value of Work		Category B Fee	
	£	£	£
Up to	150000	360 + 5.8% (Minimum Fee £3260)	
150000 -	300000	9060 + 4.7% on balance over	150000
300000 -	600000	16110 + 3.9% on balance over	300000
600000 -	1500000	27810 + 2.8% on balance over	600000
1500000 -	3000000	53010 + 2.6% on balance over	1500000
3000000 -	6000000	92010 + 2.4% on balance over	3000000
Over	6000000	164010 + 2.0% on balance over	6000000

(c) CATEGORY C. Simple works and/or works with a substantial element of repetition.

Examples:

Factories; garages; multi-storey car parks; open-air sports stadia; structural shell offices not fitted out; warehouses; workshops; and the like.

Value of Work		Category C Fee	
	£	£	£
Up to	150000	300 + 4.9% (Minimum Fee £2750)	
150000 -	300000	7650 + 4.1% on balance over	150000
300000 -	600000	13800 + 3.3% on balance over	300000
600000 -	1500000	23700 + 2.5% on balance over	600000
1500000 -	3000000	46200 + 2.2% on balance over	1500000
3000000 -	6000000	79200 + 2.0% on balance over	3000000
Over	6000000	139200 + 1.6% on balance over	6000000

THE ROYAL INSTITUTION OF CHARTERED SURVEYORS

SCALE NO. 36 (cont'd)

2.1 (d) Fees shall be calculated upon the total of the final account for the whole of the work including all nominated sub-contractors' and nominated suppliers' accounts. When work normally included in a building contract is the subject of a separate contract for which the quantity surveyor has not been paid fees under any other clause hereof, the value of such work shall be included in the amount upon which fees are charged.

(e) When a contract comprises buildings which fall into more than one category, the fee shall be calculated as follows:

(i) The amount upon which fees are chargeable shall be allocated to the categories of work applicable and the amounts so allocated expressed as percentages of the total amount upon which fees are chargeable.

(ii) Fees shall then be calculated for each category on the total amount upon which fees are chargeable.

(iii) The fee chargeable shall then be calculated by applying the percentages of work in each category to the appropriate total fee and adding the resultant amounts.

(iv) A consolidated percentage fee applicable to the total value of the work may be charged by prior agreement between the employer and the quantity surveyor. Such a percentage shall be based on this scale and on the estimated cost of the various categories of work and calculated in accordance with the principles stated above.

(f) When a project is the subject of a number of contracts then, for the purpose of calculating fees, the values of such contracts shall not be aggregated but each contract shall be taken separately and the scale of charges (paragraphs 2.1 (a) to (e)) applied as appropriate.

2.2 Air Conditioning, Heating, Ventilating and Electrical Services

(a) When the services outlined in paragraph 1.3 are provided by the quantity surveyor for the air conditioning, heating, ventilating and electrical services there shall be a fee for these services in addition to the fee calculated in accordance with paragraph 2.1 as follows:

Value of Work			Additional Fee	
	£		£	£
Up to	120000		5.0%	
120000	-	240000	6000 + 4.7% on balance over	120000
240000	-	480000	11640 + 4.0% on balance over	240000
480000	-	750000	21240 + 3.6% on balance over	480000
750000	-	1000000	30960 + 3.0% on balance over	750000
1000000	-	4000000	38460 + 2.7% on balance over	1000000
Over	4000000		119460 + 2.4% on balance over	4000000

(b) The value of such services, whether the subject of separate tenders or not, shall be aggregated and the total value of work so obtained used for the purpose of calculating the additional fee chargeable in accordance with paragraph (a). (Except that when more than one firm of consulting engineers is engaged on the design of these services, the separate values for which each such firm is responsible shall be aggregated and the additional fees charged shall be calculated independently on each such total value so obtained.)

(c) Fees shall be calculated upon the basis of the account for the whole of the air conditioning, heating, ventilating and electrical services for which bills of quantities and final accounts have been prepared by the quantity surveyor.

2.3 Works of Alteration

On works of alteration or repair, or on those sections of the work which are mainly works of alteration or repair, there shall be a fee of 1.0% in addition to the fee calculated in accordance with paragraphs 2.1 and 2.2.

2.4 Works of Redecoration and Associated Minor Repairs

On works of redecoration and associated minor repairs, there shall be a fee of 1.5% in addition to the fee calculated in accordance with paragraphs 2.1 and 2.2.

2.5 Generally

If the works are substantially varied at any stage or if the quantity surveyor is involved in an excessive amount of abortive work, then the fees shall be adjusted by agreement between the employer and the quantity surveyor.

3.0 ADDITIONAL SERVICES

3.1 For additional services not normally necessary, such as those arising as a result of the termination of a contract before completion, liquidation, fire damage to the buildings, services in connection with arbitration, litigation and investigation of the validity of contractors' claims, services in connection with taxation matters and all similar services where the employer specifically instructs the quantity surveyor, the charges shall be in accordance with paragraph 4.0 below.

GENERAL INFORMATION

THE ROYAL INSTITUTION OF CHARTERED SURVEYORS

SCALE NO. 36 - (cont'd)

4.0 TIME CHARGES

4.1 (a) For consultancy and other services performed by a principal, a fee by arrangement according to the circumstances including the professional status and qualifications of the quantity surveyor.

(b) When a principal does work which would normally be done by a member of staff, the charge shall be calculated as paragraph 4.2 below.

4.2 (a) For services by a member of staff, the charges for which are to be based on the time involved, such charges shall be calculated on the hourly cost of the individual involved plus 145%.

4.2 (b) A member of staff shall include a principal doing work normally done by an employee (as paragraph 4.1 (b) above), technical and supporting staff, but shall exclude secretarial staff or staff engaged upon general administration.

(c) For the purpose of paragraph 4.2 (b) above, a principal's time shall be taken at the rate applicable to a senior assistant in the firm.

(d) The supervisory duties of a principal shall be deemed to be included in the addition of 145% as paragraph 4.2 (a) above and shall not be charged separately.

(e) The hourly cost to the employer shall be calculated by taking the sum of the annual cost of the member of staff of:

(i) Salary and bonus but excluding expenses;
(ii) Employer's contributions payable under any Pension and Life Assurance Schemes;
(iii) Employer's contributions made under the National Insurance Acts, the Redundancy Payments Act and any other payments made in respect of the employee by virtue of any statutory requirements; and
(iv) Any other payments or benefits made or granted by the employer in pursuance of the terms of employment of the member of staff; and dividing by 1650.

5.0 INSTALMENT PAYMENTS

5.1 In the absence of agreement to the contrary, fees shall be paid by instalments as follows:

(a) Upon acceptance by the employer of a tender for the works, one half of the fee calculated on the amount of the accepted tender.

(b) The balance by instalments at intervals to be agreed between the date of the first certificate and one month after final certification of the contractor's account.

5.2 (a) In the event of no tender being accepted, one half of the fee shall be paid within three months of completion of the tender documents. The fee shall be calculated upon the basis of the lowest original bona fide tender received. In the event of no tender being received, the fee shall be calculated upon a reasonable valuation of the works based upon the tender documents.

Note: In the foregoing context "bona fide tender" shall be deemed to mean a tender submitted in good faith without major errors of computation and not subsequently withdrawn by the tenderer.

(b) In the event of the project being abandoned at any stage other than those covered by the foregoing, the proportion of the fee payable shall be by agreement between the employer and the quantity surveyor.

SCALE NO. 37

ITEMISED SCALE OF PROFESSIONAL CHARGES FOR QUANTITY SURVEYING SERVICES FOR BUILDING WORKS
(29th July, 1988)

This Scale has been determined by the Fees Committee appointed by the Quantity Surveyors Divisional Council of The Royal Institution of Chartered Surveyors. The Scale is for guidance and is not mandatory.

1.0 GENERALLY

1.1 The fees are in all cases exclusive of travelling and other expenses (for which the actual disbursement is recoverable unless there is some prior arrangement for such charges) and of the cost of reproduction of bills of quantities and other documents, which are chargeable in addition at net cost.

1.2 The fees are in all cases exclusive of services in connection with the allocation of the cost of the works for purposes of calculating value added tax for which there shall be an additional fee based on the time involved (see paragraphs 19.1 and 19.2)

1.3 If any of the materials used in the works are supplied by the employer or charged at a preferential rate, then the actual or estimated market value thereof shall be included in the amounts upon which fees are to be calculated.

1.4 The fees are in all cases exclusive of preparing a specification of the materials to be used and the works to be done, but the fees for preparing bills of quantities and similar documents do include for incorporating preamble clauses describing the materials and workmanship (from instructions given by the architect and/or consulting engineer).

1.5 If the quantity surveyor incurs additional costs due to exceptional delays in building operations or any other cause beyond the control of the quantity surveyor then the fees may be adjusted by agreement between the employer and the quantity surveyor to cover the reimbursement of these additional costs.

1.6 The fees and charges are in all cases exclusive of value added tax which will be applied in accordance with legislation.

1.7 Copyright in bills of quantities and other documents prepared by the quantity surveyor is reserved to the quantity surveyor.

CONTRACTS BASED ON BILLS OF QUANTITIES: PRE-CONTRACT SERVICES

2.0 BILLS OF QUANTITIES

2.1 Basic Scale: For preparing bills of quantities and examining tenders received and reporting thereon.

THE ROYAL INSTITUTION OF CHARTERED SURVEYORS

SCALE NO. 37 (cont'd)

2.1 (a)   CATEGORY A: Relatively complex works and/or works with little or no repetition.

Examples:
Ambulance and fire stations; banks; cinemas; clubs; computer buildings; council offices; crematoria; fitting out of existing buildings; homes for the elderly; hospitals and nursing homes; laboratories; law courts; libraries; 'one off' houses; petrol stations; places of religious worship; police stations; public houses, licensed premises; restaurants; sheltered housing; sports pavilions; theatres; town halls; universities, polytechnics and colleges of further education (other than halls of residence and hostels); and the like.

Value of Work		Category A Fee	
£	£	£
Up to	150000	230 + 3.0% (minimum fee £1730)	
150000 -	300000	4730 + 2.3% on balance over	150000
300000 -	600000	8180 + 1.8% on balance over	300000
600000 -	1500000	13580 + 1.5% on balance over	600000
1500000 -	3000000	27080 + 1.2% on balance over	1500000
3000000 -	6000000	45080 + 1.1% on balance over	3000000
Over	6000000	78080 + 1.0% on balance over	6000000

(b)   CATEGORY B: Less complex works and/or works with some element of repetition.

Examples:
Adult education facilities; canteens; church halls; community centres; departmental stores; enclosed sports stadia and swimming baths; halls of residence; hostels; motels; offices other than those included in Categories A and C; railway stations; recreation and leisure centres; residential hotels; schools; self-contained flats and maisonettes; shops and shopping centres; supermarkets and hypermarkets; telephone exchanges; and the like.

Value of Work		Category B Fee	
£	£	£
Up to	150000	210 + 2.8% (minimum fee £1610)	
150000 -	300000	4410 + 2.0% on balance over	150000
300000 -	600000	7410 + 1.5% on balance over	300000
600000 -	1500000	11910 + 1.1% on balance over	600000
1500000 -	3000000	21810 + 1.0% on balance over	1500000
3000000 -	6000000	36810 + 0.9% on balance over	3000000
Over	6000000	63810 + 0.8% on balance over	6000000

(c)   CATEGORY C: Simple works and/or works with a substantial element of repetition.

Examples:
Factories; garages; multi-storey car parks; open-air sports stadia; structural shell offices not fitted out; warehouses; workshops; and the like.

Value of Work		Category C Fee	
£	£	£
Up to	150000	180 + 2.5% (minimum fee £1430)	
150000 -	300000	3930 + 1.8% on balance over	150000
300000 -	600000	6630 + 1.2% on balance over	300000
600000 -	1500000	10230 + 0.9% on balance over	600000
1500000 -	3000000	18330 + 0.8% on balance over	1500000
3000000 -	6000000	30330 + 0.7% on balance over	3000000
Over	6000000	51330 + 0.6% on balance over	6000000

THE ROYAL INSTITUTION OF CHARTERED SURVEYORS

SCALE NO. 37 (cont'd)

2.1   (d) The scales of fees for preparing bills of quantities (paragraphs 2.1 (a) to (c)) are overall scales based upon the inclusion of all provisional and prime cost items, subject to the provision of paragraph 2.1 (g). When work normally included in a building contract is the subject of a separate contract for which the quantity surveyor has not been paid fees under any other clause hereof, the value of such work shall be included in the amount upon which fees are charged.

    (e) Fees shall be calculated upon the accepted tender for the whole of the work subject to the provisions of paragraph 2.6. In the event of no tender being accepted, fees shall be calculated upon the basis of the lowest original bona fide tender received. In the even of no such tender being received, the fees shall be calculated upon a reasonable valuation of the works based upon the original bills of quantities.

        Note: In the foregoing context "bona fide tender" shall be deemed to mean a tender submitted in good faith without major errors of computation and not subsequently withdrawn by the tenderer.

    (f) In calculating the amount upon which fees are charged the total of any credits and the totals of any alternative bills shall be aggregated and added to the amount described above. The value of any omission or addition forming part of an alternative bill shall not be added unless measurement or abstraction from the original dimension sheets was necessary.

    (g) Where the value of the air conditioning, heating, ventilating and electrical services included in the tender documents together exceeds 25% of the amount calculated as described in paragraphs 2.1 (d) and (e), then, subject to the provisions of paragraph 2.2, no fee is chargeable on the amount by which the value of these services exceeds the said 25% in this context the term "value" excludes general contractor's profit, attendance, builder's work in connection with the services, preliminaries and any similar additions.

    (h) When a contract comprises buildings which fall into more than one category, the fee shall be calculated as follows:

        (i) The amount upon which fees are chargeable shall be allocated to the categories of work applicable and the amounts so allocated expressed as percentages of the total amount upon which fees are chargeable.

        (ii) Fees shall then be calculated for each category on the total amount upon which fees are chargeable.

        (iii) The fee chargeable shall then be calculated by applying the percentages of work in each category to the appropriate total fee and adding the resultant amounts.

    (j) When a project is the subject of a number of contracts then, for the purpose of calculating fees, the values of such contracts shall not be aggregated but each contract shall be taken separately and the scale of charges (paragraphs 2.1 (a) to (h)) applied as appropriate.

    (k) Where the quantity surveyor is specifically instructed to provide cost planning services the fee calculated in accordance with paragraphs 2.1 (a) to (j) shall be increased by a sum calculated in accordance with the following table and based upon the same value of work as that upon which the aforementioned fee has been calculated:

2.1 (k)   CATEGORIES A & B: (as defined in paragraphs 2.1 (a) and (b)).

Value of Work			Fee
	£	£	£
Up to	600000	0.7%	
600000 -	3000000	4200 + 0.4% on balance over	600000
3000000 -	6000000	13800 + 0.35% on balance over	3000000
Over	6000000	24300 + 0.3% on balance over	6000000

CATEGORY C: (as defined in paragraph 2.1 (c)).

Value of Work			Fee
	£	£	£
Up to	600000	0.5%	
600000 -	3000000	3000 + 0.3% on balance over	600000
3000000 -	6000000	10200 + 0.25% on balance over	3000000
Over	6000000	17700 + 0.2% on balance over	6000000

THE ROYAL INSTITUTION OF CHARTERED SURVEYORS

SCALE NO. 37 (cont'd)

2.2 Air Conditioning, Heating, Ventilating and Electrical Services

(a) Where bills of quantities are prepared by the quantity surveyor for the air conditioning, heating, ventilating and electrical services there shall be a fee for these services (which shall include examining tenders received and reporting thereon), in addition to the fee calculated in accordance with paragraph 2.1 as follows:

Value of Work				Additional Fee
	£		£	£
Up to	120000		2.5%	
120000	-	240000	3000 + 2.25% on balance over	120000
240000	-	480000	5700 + 2.0% on balance over	240000
480000	-	750000	10500 + 1.75% on balance over	480000
750000	-	1000000	15225 + 1.25% on balance over	750000
Over	1000000		18350 + 1.15% on balance over	1000000

(b) The values of such services, whether the subject of separate tenders or not, shall be aggregated and the total value of work so obtained used for the purpose of calculating the additional fee chargeable in accordance with paragraph (a).

(Except that when more than one firm of consulting engineers is engaged on the design of these services, the separate values for which each such firm is responsible shall be aggregated and the additional fees charged shall be calculated independently on each such total value so obtained).

(c) Fees shall be calculated upon the accepted tender for the whole of the air conditioning, heating, ventilating and electrical services for which bills of quantities have been prepared by the quantity surveyor. In the event of no tender being accepted, fees shall be calculated upon the basis of the lowest original bona fide tender received. In the event of no such tender being received, the fees shall be calculated upon a reasonable valuation of the services based upon the original bills of quantities.

Note: In the foregoing context "bona fide tender" shall be deemed to mean a tender submitted in good faith without major errors of computation and not subsequently withdrawn by the tenderer.

(d) When cost planning services are provided by the quantity surveyor for air conditioning, heating, ventilating and electrical services (or for any part of such services) there shall be an additional fee based on the time involved (see paragraphs 19.1 and 19.2). Alternatively the fee may be on a lump sum or percentage basis agreed between the employer and the quantity surveyor.

Note: The incorporation of figures for air conditioning, heating, ventilating and electrical services provided by the consulting engineer is deemed to be included in the quantity surveyor's services under paragraph 2.1.

2.3 Works of Alteration

On works of alteration or repair, or on those sections of the works which are mainly works of alteration or repair, there shall be a fee of 1.0% in addition to the fee calculated in accordance with paragraphs 2.1 and 2.2.

2.4 Works of Redecoration and Associated Minor Repairs

On works of redecoration and associated minor repairs, there shall be a fee of 1.5% in addition to the fee calculated in accordance with paragraphs 2.1 and 2.2.

2.5 Bills of Quantities Prepared in Special Forms

Fees calculated in accordance with paragraphs 2.1, 2.2, 2.3 and 2.4 include for the preparation of bills of quantities on a normal trade basis. If the employer requires additional information to be provided in the bills of quantities or the bills to be prepared in an elemental, operational or similar form, then the fee may be adjusted by agreement between the employer and the quantity surveyor.

2.6 action of Tenders

(a) When cost planning services have been provided by the quantity surveyor and a tender, when received, is reduced before acceptance and if the reductions are not necessitated by amended instructions of the employer or by the inclusion in the bills of quantities of items which the quantity surveyor has indicated could not be contained within the approved estimate, then in such a case no charge shall be made by the quantity surveyor for the preparation of bills of reductions and the fee for the preparation of the bills of quantities shall be based on the amount of the reduced tender.

(b) When cost planning services have not been provided by the quantity surveyor and if a tender, when received, is reduced before acceptance, fees are to be calculated upon the amount of the unreduced tender. When the preparation of bills of reductions is required, a fee is chargeable for preparing such bills of reductions as follows:

(i) 2.0% upon the gross amount of all omissions requiring measurement or abstraction from original dimension sheets.

(ii) 3.0% upon the gross amount of all additions requiring measurement.

(iii) 0.5% upon the gross amount of all remaining additions.

Note: The above scale for the preparation of bills of reductions applies to work in all categories.

2.7 Generally

If the works are substantially varied at any stage or if the quantity surveyor is involved in an excessive amount of abortive work, then the fees shall be adjusted by agreement between the employer and the quantity surveyor.

THE ROYAL INSTITUTION OF CHARTERED SURVEYORS

SCALE NO. 37 (cont'd)

3.0 NEGOTIATING TENDERS

3.1 (a) For negotiating and agreeing prices with a contractor:

Value of Work			Fee	
	£		£	£
Up to	–	150000	0.5%	
150000	-	600000	750 + 0.3% on balance over	150000
600000	-	1200000	2100 + 0.2% on balance over	600000
Over		1200000	3300 + 0.1% on balance over	1200000

3.1 (b) The fee shall be calculated on the total value of the works as defined in paragraphs 2.1 (d), (e), (f), (g) and (j).

(c) For negotiating and agreeing prices with a contractor for air conditioning, heating, ventilating and electrical services there shall be an additional fee as paragraph 3.1 (a) calculated on the total value of such services as defined in paragraph 2.2 (b).

4.0 CONSULTATIVE SERVICES AND PRICING BILLS OF QUANTITIES

4.1 Consultative Services

Where the quantity surveyor is appointed to prepare approximate estimates, feasibility studies or submissions for the approval of financial grants or similar services, then the fee shall be based on the time involved (see paragraphs 19.1 and 19.2) or, alternatively, on a lump sum or percentage basis agreed between the employer and the quantity surveyor.

4.2 Pricing Bills of Quantities

(a) For pricing bills of quantities, if instructed, to provide an estimate comparable with tenders, the fee shall be one-third (33 1/3%) of the fee for negotiating and agreeing prices with a contractor, calculated in accordance with paragraphs 3.1 (a) and (b).

(b) For pricing bills of quantities, if instructed, to provide an estimate comparable with tenders for air conditioning, heating, ventilating and electrical services the fee shall be one-third (33 1/3%) of the fee calculated in accordance with paragraph 3.1 (c).

CONTRACTS BASED ON BILLS OF QUANTITIES: POST-CONTRACT SERVICES

Alternative Scales (1 and 11) for post-contract services are set out below to be used at the quantity surveyor's discretion by prior agreement with the employer.

ALTERNATIVE 1

5.0 OVERALL SCALE OF CHARGES FOR POST-CONTRACT SERVICES

5.1 If the quantity surveyor appointed to carry out the post-contract services did not prepare the bills of quantities then the fees in paragraphs 5.2 and 5.3 shall be increased to cover the additional services undertaken by the quantity surveyor.

5.2 Basic Scale: For taking particulars and reporting valuations for interim certificates for payments on account to the contractor, preparing periodic assessments of anticipated final cost and reporting thereon, measuring and making up bills of variations including pricing and agreeing totals with the contractor, and adjusting fluctuations in the cost of labour and materials if required by the contract.

(a)   CATEGORY A: Relatively complex works and/or works with little or no repetition.

Examples:
Ambulance and fire stations; banks; cinemas; clubs; computer buildings; council offices; crematoria; fitting out of existing buildings; homes for the elderly; hospitals and nursing homes; laboratories; law courts; libraries; 'one-off' houses; petrol stations; places of religious worship; police stations; public houses, licensed premises; restaurants; sheltered housing; sports pavilions; theatres; town halls; universities, polytechnics and colleges of further education (other than halls of residence and hostels); and the like.

Value of Work			Category A Fee	
	£		£	£
Up to		150000	150 + 2.0% (minimum fee £1150)	
150000	-	300000	3150 + 1.7% on balance over	150000
300000	-	600000	5700 + 1.6% on balance over	300000
600000	-	1500000	10500 + 1.3% on balance over	600000
500000	-	3000000	22200 + 1.2% on balance over	1500000
3000000	-	6000000	40200 + 1.1% on balance over	3000000
Over		6000000	73200 + 1.0% on balance over	6000000

SCALE NO. 37 (cont'd)

(b)   CATEGORY B: Less complex works and/or works with some element of repetition.

Examples:
Adult education facilities; canteens; church halls; community centres; departmental stores; enclosed sports stadia and swimming baths; halls of residence; hostels; motels; offices other than those included in Categories A and C; railway stations; recreation and leisure centres; residential hotels; schools; self-contained flats and maisonettes; shops and shopping centres; supermarkets and hypermarkets; telephone exchanges; and the like.

Value of Work			Category B Fee	
	£		£	£
Up to	150000		150 + 2.0% (minimum fee £1150)	
150000	-	300000	3150 + 1.7% on balance over	150000
300000	-	600000	5700 + 1.5% on balance over	300000
600000	-	1500000	10200 + 1.1% on balance over	600000
1500000	-	3000000	20100 + 1.0% on balance over	1500000
3000000	-	6000000	35100 + 0.9% on balance over	3000000
Over	6000000		62100 + 0.8% on balance over	6000000

(c)   CATEGORY C: Simple works and/or works with a substantial element of repetition.

Examples:
Factories; garages; multi-storey car parks; open-air sports stadia; structural shell offices not fitted out; warehouses; workshops; and the like.

Value of Work			Category C Fee	
	£		£	£
Up to	150000		120 + 1.6% (minimum fee £920)	
150000	-	300000	2520 + 1.5% on balance over	150000
300000	-	600000	4770 + 1.4% on balance over	300000
600000	-	1500000	8970 + 1.1% on balance over	600000
1500000	-	3000000	18870 + 0.9% on balance over	1500000
3000000	-	6000000	32370 + 0.8% on balance over	3000000
Over	6000000		56370 + 0.7% on balance over	6000000

(d)   The scales of fees for post-contract services (paragraphs 5.2 (a) and (c) are overall scales, based upon the inclusion of all nominated sub-contractors' and nominated suppliers' accounts, subject to the provision of paragraph 5.2 (g). When work normally included in a building contract is the subject of a separate contract for which the quantity surveyor has not been paid fees under any other clause hereof, the value of such work shall be included in the amount on which fees are charged.

(e)   Fees shall be calculated upon the basis of the account for the whole of the work, subject to the provisions of paragraph 5.3.

(f)   In calculating the amount on which fees are charged the total of any credits is to be added to the amount described above.

(g)   Where the value of air conditioning, heating, ventilating and electrical services included in the tender documents together exceeds 25% of the amount calculated as described in paragraphs 5.2 (d) and (e) above, then, subject to the provisions of paragraph 5.3, no fee is chargeable on the amount by which the value of these services exceeds the said 25%. In this context the term "value" excludes general contractor's profit, attendance, builders work in connection with the services, preliminaries and any other similar additions.

(h)   When a contract comprises buildings which fall into more than one category, the fee shall be calculated as follows:

(i)   The amount upon which fees are chargeable shall be allocated to the categories of work applicable and the amounts so allocated expressed as percentages of the total amount upon which fees are chargeable.

(ii)   Fees shall then be calculated for each category on the total amount upon which fees are chargeable.

(iii) The fee chargeable shall then be calculated by applying the percentages of work in each category to the appropriate total fee and adding the resultant amounts.

(j)   When a project is the subject of a number of contracts then, for the purposes of calculating fees, the values of such contracts shall not be aggregated but each contract shall be taken separately and the scale of charges (paragraphs 5.2 (a) to (h)) applied as appropriate.

GENERAL INFORMATION

SCALE NO. 37 (cont'd)

(k) When the quantity surveyor is required to prepare valuations of materials or goods off site, an additional fee shall be charged based on the time involved (see paragraphs 19.1 and 19.2).

(l) The basic scale for post-contract services includes for a simple routine of periodically estimating final costs. When the employer specifically requests a cost monitoring service which involves the quantity surveyor in additional or abortive measurement an additional fee shall be charged based on the time involved (see paragraphs 19.1 and 19.2), or alternatively on a lump sum or percentage basis agreed between the employer and the quantity surveyor.

(m) The above overall scales of charges for post contract services assume normal conditions when the bills of quantities are based on drawings accurately depicting the building work the employer requires. If the works are materially varied to the extent that substantial remeasurement is necessary then the fee for post-contract services shall be adjusted by agreement between the employer and the quantity surveyor.

5.3 Air Conditioning, Heating, Ventilating and Electrical Services

(a) Where final accounts are prepared by the quantity surveyor for the air conditioning, heating, ventilating and electrical services there shall be a fee for these services, in addition to the fee calculated in accordance with paragraph 5.2, as follows:

Value of Work			Additional Fee	
	£		£	3
Up to	120000		2.0%	
120000	-	240000	2400 + 1.6% on balance over	120000
240000	-	1000000	4320 + 1.25% on balance over	240000
1000000	-	4000000	13820 + 1.0% on balance over	1000000
Over	4000000		43820 + 0.9% on balance over	4000000

(b) The values of such services, whether the subject of separate tenders or not, shall be aggregated and the total value of work so obtained used for the purpose of calculating the additional fee chargeable in accordance with paragraph (a).

(Except that when more than one firm of consulting engineers is engaged on the design of these services the separate values for which each such firm is responsible shall be aggregated and the additional fee charged shall be calculated independently on each such total value so obtained).

(c) The scope of the services to be provided by the quantity surveyor under paragraph (a) above shall be deemed to be equivalent to those described for the basic scale for post-contract services.

(d) When the quantity surveyor is required to prepare periodic valuations of materials or goods off site, an additional fee shall be charged based on the time involved (see paragraphs 19.1 and 19.2).

(e) The basic scale for post-contract services includes for a simple routine of periodically estimating final costs. When the employer specifically requests a cost monitoring service which involves the quantity surveyor in additional or abortive measurement an additional fee shall be charged based on the time involved (see paragraph 19.1 and 19.2), or alternatively on a lump sum or percentage basis agreed between the employer and the quantity surveyor.

(f) Fees shall be calculated upon the basis of the account for the whole of the air conditioning, heating, ventilating and electrical services for which final accounts have been prepared by the quantity surveyor.

Scale 37   Alternative 11
Separate stages of post contract services and Scale 44 are not represented here.

Building Act 1984
The Building Regulations 1991 (as amended)
The Building (Local Authority Charges) Regulations 1998

FULL PLAN AND INSPECTION CHARGES

Description of work		Plan charge £	VAT @ 17.5% £	Total plan charge £	Inspection charge £	VAT @ 17.5% £	Total inspection charge payable £
Domestic extensions *(not exceeding 3 storeys md. basement)* Total areas to be aggregated	10m2 & under	200	35	235	-	-	-
	40m2 & under	75	13.13	88.13	225	39.38	264.38
	60m2 & under	100	17.50	117.50	300	52.50	352.50
	over 60m2	Estimate required see table below for other work					
Detached garages/ Carports *(not otherwise exempt)*	40m & under	25	4.38	29.38	75	13.13	88.13
	Over 40m	Estimate required see table below for other work					
New houses and flats *(Small domestic buildings not exceeding 300m2 floor area and no more than 3 storeys incl. Basement*	**No. of dwellings**						
	1.	140	24.50	164.50	160	28.00	188.00
	2.	205	35.88	240.88	300	52.50	352.50
	3.	270	47.25	317.25	435	76.13	511.13
	4.	335	58.63	393.63	575	100.63	675.63
	5.	405	70.88	475.88	710	124.25	834.25
	6.	475	83.13	558.13	785	137.38	922.38
	7.	495	86.63	581.63	945	165.38	1,110.38
	8.	515	90.13	605.13	1,105	193.38	1,298.38
	9.	535	93.63	628.63	1,265	221.38	1,486.38
	10.	540	94.50	634.50	1,440	252.00	1,692.00
	11.	545	95.38	640.38	1,580	276.50	1,856.50
	12.	550	96.25	646.25	1,720	301.00	2,021.00
	13.	555	97.13	652.13	1,860	325.50	2,185.50
	14.	560	98.00	658.00	2,000	350.00	2,350.00
	15.	565	98.88	663.88	2,140	374.50	2,514.50
	16.	570	99.75	669.75	2,280	399.00	2,679.00
	17.	575	100.63	675.63	2,420	423.50	2,843.50
	18.	580	101.50	681.50	2,560	448.00	3,008.00
	19.	585	102.38	687.38	2,700	472.50	3,172.50
	20.	590	103.25	693.25	2,840	497.00	3,337.00
	21to30	600	105	705.00	2,940	514.50	3,454.50
		Plus £10 +£1.75 VAT for each additional dwelling above 21			**Plus £100 +£17.50 VAT for each additional dwelling above 21**		
	31 &over	700	122.50	822.50	3,940	689.50	4,629.50
		Plus £5 +£0.88 VAT for each additional dwelling above 31.			**Plus £75 +£13.13 VAT for each additional dwelling above 31.**		

OTHER WORK

Estimated cost of building work (excluding VAT) (£.)	Plan charge £.	VAT @ 17.5% £.	Total plan charge payable £.	Inspection charge £.	VAT @ 17.5% £.	Total inspection charge payable £
2,000 or less	100	17.50	117.50	-	-	-
Exceeding 2,000 but not exceeding 5,000	165	28.88	193.88	-	-	-
Exceeding 5,000 but not exceeding 20,000	41.25	7.22	48.47	123.75	21.66	145.41
Plus for every £1,000 (or part thereof) by which the cost exceeds £5,000:	2.25	0.39	2.64	6.75	1.18	7.93
Exceeding 20,000 but not exceeding 100,000	75	13.13	88.13	225	39.38	264.38
Plus for every £1,000 (or part thereof) by which the cost exceeds £20,000:	2	0.35	235	6	1.05	7.05
Exceeding 100,000 but not exceeding 1,000,000	235	41.13	276.13	705	123.38	828.38
Plus for every £1,000 (or part thereof) by which the cost exceeds £100,000:	0.88	0.15	1.03	2.62	0.46	3.08
Exceeding 1,000,000 but not exceeding 10,000,000	1,022.50	178.94	1,201.44	3,067.50	536.81	3,604.31
Plus for every £1,000 (or part thereof) by which the cost exceeds £1,000,000:	0.69	0.12	0.81	2.06	0.36	2.42
Exceeding 10,000,000	7,210	1,261.75	8,471.75	21,630	3,785.25	25,415.25
Plus for every £1,000 (or part thereof) by which the cost exceeds £10,000,000:	0.50	0.09	0.59	1.50	0.26	1.76

GENERAL INFORMATION

WATFORD COUNCIL
The Building (Local Authority Charges) Regulations 1998

BUILDING NOTICE & REVERSION CHARGES   REGULARISATION CHARGES

Description of Work		Charge £.	VAT @ 17.5% £.	Total payable £.	Charge £.	VAT @ 0% £	Total payable £
Domestic extensions							
(not exceeding 3 storeys	10m² & under	200	35.00	235.00	240		240
incl. basement). Total	40m² & under	300	52.50	352.50	360		360
floor areas to be	60m² & under	400	70.00	470.00	480		480
aggregated	Over 60m²	Estimate required see table below			Estimate required see table below		
Detached garages/carports							
(not otherwise exempt)	40m2 & under	100	17.50	117.50	120	-	120
	Over 40m	Estimate required see table below			Estimate required see table below		
New houses and flats	**No. of dwellings**						
(small domestic buildings)	1.	300	52.50	352.50	360	-	360
not exceeding 300m² floor	2.	505	88.38	593.38	606	-	606
area and no more than	3.	705	123.38	828.38	846	-	846
3 storeys incl. basment	4.	910	159.25	1,069.25	1.092	-	1.092
	5.	1,115	195.13	1,310.13	1,338	-	1,338
	6.	1,260	220.50	1,480.50	1,512	-	1,512
	7.	1,440	252.00	1,692.00	1,728	-	1,728
	8.	1,620	283.50	1,903.50	1,944	-	1,944
	9.	1,800	315.00	2,115.00	2,160	-	2,160
	10.	1,980	346.50	2,326.50	2,376	-	2,376
	11.	2,125	371.88	2,496.88	2,550	-	2,550
	12.	2,270	397.25	2,667.25	2,724	-	2,724
	13.	2,415	422.63	2,837.63	2,898	-	2,898
	14.	2,560	448.00	3,008.00	3,072	-	3,072
	15.	2,705	473.38	3,178.38	3,246	-	3,246
	16.	2,850	498.75	3,348,.75	3,420	-	3,420
	17.	2,995	524.13	3,519.13	3,594	-	3,594
	18.	3,140	549.50	3,689.50	3,768	-	3,768
	19.	3,285	574.88	3,859.88	3,942	-	3,942
	20.	3,430	600.25	4,030.25	4,116	-	4,116
	21 to 30	3,540	619.50	4,159.50	4,248	-	4,248
	Plus £110 + £19.25 VAT for each additional dwelling above 21				**Plus £132 for each additional dwelling above 21**		
	31 & over	4,640	812.00	5,452.00	5,568	-	5,568
	Plus £80 + £14.00 VAT for each additional dwelling over 31				**Plus £96 for each additional dwelling over 31.**		

OTHER WORK

Estimated cost of building work (excluding VAT) (£)	Charge £.	VAT @ 17.5% £	Total payable £	Charge £	VAT @ 0% £	Total payable £
2,000 or less	100	17.50	117.50	120	-	120
Exceeding 2,000 but not exceeding 5,000	165	28.88	193.88	198	-	198
Exceeding 5,000 but not exceeding 20,000	165	28.88	193.88	198		198
Plus for every £1,000 (or part thereof) by which the cost exceeds £5,000.						
	9	1.58	10.58	10.80		10.80
Exceeding 20,000 but not exceeding 100,000	300	52.50	352.50	360	-	360
Plus for every £1,000 (or part thereof) by which the cost exceeds £20,000.						
	8	1.40	9.40	9.60		9.60
Exceeding 100,00 but not exceeding 1,000,000	940	164.50	1,104.50	1,128	-	1,128
Plus for every £1,000 (or part thereof) by which the cost exceeds £100,000						
	3.50	0.61	4.11	4.20		4.20
Exceeding 1,000,000 but not exceeding 10,000,000	4,090	715.75	4,805.75	4,908	-	4,908
Plus for every £1,000 (or part thereof) by which the cost exceeds £1,000,000						
	2.75	0.48	3.23	3.30		3.30
Exceeding 10,000,000	28,840	5,047	33,887	34,608	-	34,608
Plus for every £1,000 (or part thereof) by which the cost exceeds £10,000,000						
	2.00	0.35	2.35	2.4		2.4

Construction (Design and Management) Regulations 1994

CONSTRUCTION (DESIGN AND MANAGEMENT) REGULATIONS 1994

The Construction (Design and Management) Regulations 1994 came into force on 31st March 1995.

The "Approved Code of Practice - Managing Construction for Health and Safety" gives advice on how to comply with the Law and has legal status.

If you have not followed the relevant provisions of the code you may be prosecuted for breach of health and safety law.

The regulations place duties upon clients, client's agents, designers and contractors so that health and safety is taken into account and then co-ordinated and managed effectively throughout all stages of a construction project; from conception to execution of works on site and subsequent maintenance and repair and demolitions.

The following is an outline of the requirements of the code including excerpts together with extracts from the HSE leaflet "CDM Regulations - How the regulations affect you". No attempt has been made to cover all the detail and readers are advised to study the code in full.

The Client

Clients in particular should note the obligations put on them, Key duties are:-

a) To appoint a health and safety planning supervisor and principal contractor.

b) Be satisfied that the planning supervisor and principal contractor are competent and will allocate adequate resources for health and safety.

c) Be satisfied that designers and contractors are also competent and will allocate adequate resources.

d) Provide the planning supervisor with information relevant to health and safety on the project.

e) Ensure construction work does not start until the principal contractor has prepared a satisfactory health and safety plan.

f) Ensure the health and safety file is available for inspection after the project is completed.

Duties on clients do not apply to domestic households having construction projects carried out or very minor works carried out in occupied premises.

The CDM Regulations will generally apply to construction work which is notifiable i.e. lasts for more than 30 days or will involve more than 500 person days of work. CDM will apply to any design which involves five people or more on site at any one time or if the work includes demolition.

The Planning Supervisor

The planning supervisor has to co-ordinate the health and safety aspects project design and the initial planning to ensure as much as they can that:

a) designers comply with their duties - in particular, the avoidance and reduction of risk;

b) designers co-operate with each other for the purposes of health and safety;

c) a health and safety plan is prepared before arrangements are made for a principal contractor to be appointed;

d) they are able to give advice, if requested, to the client on the competence and allocation of resources by designers and all contractors; advise contractors appointing designers; and also advise the client on the health and safety plan before the construction phase starts;

e) the project is notified to the Health and Safety Executive;

f) the health and safety file is prepared and deliver it to the client at the end of a project.

The Designer

Designers must design in a way which avoids, reduces, or controls risks to health and safety as far as is reasonably practicable so that projects they design can be constructed and maintained safely. Where risks remain, they have to be stated to the extent necessary to enable reliable performance by a competent contractor.

The designer's key duties are, as far as reasonably practicable, to:

a) alert clients to their duties;

b) consider during the development of designs the hazards and risks which may arise to those constructing and maintaining the structure;

c) design to avoid risks to health and safety so far as is reasonable practicable;

d) reduce risks at source if avoidance is not possible;

e) consider measures which will protect all workers if neither avoidance nor reduction to a safe level is possible;

The Designer (Cont'd)

(f) ensure that the design includes adequate information on health and safety.

g) pass this information on to the planning supervisor so that it can be included in the health and safety plan; and ensure that it is given on drawings or in specifications etc.;

h) co-operate with the planning supervisor and, where necessary, other designers involved in the project.

Note:

Designer means any person who prepares a design.

Design includes drawings, design details, specifications and bills of quantities (Quantity Surveyors Note!)

The Principal Contractor

The principal contractor should take account of the specific requirements of a project when preparing and presenting tenders or similar documents, take over and develop the health and safety plan and co-ordinate the activities of all contractors so that they comply with health and safety law.

The principal contractor's key duties are to:

a) develop and implement the health and safety plan;

b) arrange for competent and adequately resourced contractors to carry out the work where it is subcontracted;

c) ensure the co-ordination and co-operation of contractors;

d) obtain from contractors the main findings of their risk assessments and details of how they intend to carry out high risk operations;

e) ensure that contractors have information about risks on site;

f) ensure that workers on site have been given adequate training;

g) ensure that contractors and workers comply with any site rules which may have been set out in the health and safety plan;

h) monitor health and safety performance;

I) ensure that all workers are properly informed and consulted;

j) make sure only authorised people are allowed onto the site;

k) display the notification of the project to HSE;

l) pass information to the planning supervisor for the Health and Safety file;

Contractors and the Self-employed

Contractors in general have duties to play their part in the successful management of health and safety during construction work. The key duties are to;

a) provide information for the health and safety plan about risks to health and safety arising from their work and the steps they will take to control and manage the risks;

b) manage their work so that they comply with rules in the health and safety plan and directions from the principal contractor;

c) provide information for the health and safety file, and about injuries, dangerous occurrences and ill health;

d) provide information to their employees.

The self-employed also have these duties when they act as contractors.

Employees

Under CDM employees will benefit by being better informed and more able to play an active part in health and safety.

Employees are:

a) entitled to information about health and safety during the construction phase;

b) able to express their views about health and safety to the principal contractor.

The principal contractor has to check that employees have been provided with adequate information and training.

The pre-tender health and safety plan

The pre-tender plan which the planning supervisor has to ensure is prepared, should include:

a) a general description of the work;

b) details of timings within the project;

c) details of risks to workers as far as possible at that stage;

d) information required by potential principal contractors to demonstrate competence or adequacy of resources;

e) information for preparing a health and safety plan for the construction phase and information for welfare provision.

The health and safety plan for the construction phase

The plan developed by the principal contractor is the foundation on which health and safety management of construction work is based. It should include:

a) arrangements for ensuring the health and safety of all who may be affected by the construction work;

b) arrangements for the management of health and safety of construction work and monitoring of compliance with health and safety law;

c) information about welfare arrangements.

Health and safety file

The health and safety file amounts to a normal maintenance manual enlarged to alert those who will be responsible for a structure after handover to risks that must be managed when the structure and associated plant is maintained, repaired, renovated or demolished. It is a record of information to inform future decisions on the management of health and safety.

Competence and provision for health and safety

Checking on competence and on the allocation of adequate resources for health and safety may be done in a way that fits in with established procedures, with the degree of detail suited for the project.

The Approved Code of Practice

The Construction (Design and Management) Regulations 1994 are reproduced in italics in the Code with each part of the Code printed after the relevant regulation. The Code covers the following topics.

The following is an outline of the Appendices for information

Outline of Appendix 1   Summary of the Management of Health and Safety at Work Regulations 1992

1   The Regulations aim to improve health and safety management and make more explicit what is required of employers under the Health and Safety at Work etc. Act 1974 with a more systematic and better organised approach to dealing with health and safety.

2   The Regulations place duties on employers (and in some cases, the self-employed) whether they are clients, designers, planning supervisors, principal contractors or other contractors.

3   Employers have to:

(a) assess the risks to the health and safety of their employees and others who may be affected by the work activity. This is to identify the preventive and protective measures necessary. Employers with five or more employees must record the significant findings of the assessment (regulation 3);

(b) ensure that arrangements are made for implementing the measures following the risk assessment. This will cover planning, organisation, control, monitoring and review, i.e. the management of health and safety (regulation 4);

(c) provide health surveillance for employees whenever the risk assessment determines it to be needed (regulation 5);

(d) appoint competent people to help devise and apply the measures (regulation 6);

(e) set up emergency procedures (regulation 7);

(f) provide employees with relevant information on health and safety (regulation 8);

(g) co-operate with other employers sharing the same work site and co-ordinate the preventive and protective measures needed (regulations 9 and 10);

(h) make sure employees have adequate health and safety training (regulation 11);

(I) provide temporary workers with health and safety information to meet special needs (regulation 13);

4   Employees also have duties (regulation 12) to:

(a) use equipment in accordance with training and instruction;

(b) report dangerous situations;

(c) report any shortcomings in health and safety arrangements.

Outline of Appendix 2   The principles of prevention and protection

The principles of Prevention and Protection are:

(a) **If possible avoid the risk completely.**

(b) **Combat risks at source.**

(c) **Wherever possible, adapt work to the individual,** particularly in the choice of work equipment and methods of work.

(d) **Take advantage of technological progress.**

(e) **Incorporate the prevention measures into a coherent plan** to reduce progressively those risks which cannot altogether be avoided and which takes into account working conditions, organisational factors, the working environment and social factors. On individual projects, the health and safety plan (regulation 15) will act as the focus for bringing together and co-ordinating the individual policies of everyone involved. Where an employer is required under section 2(3) of the Health and Safety at Work etc. Act 1974 to have a health and safety policy, this should be prepared and applied by reference to these principles.

(f) **Give priority to those measures which protect the whole workforce or activity,** i.e. give collective protective measures, such as suitable working platforms with edge protection, priority over individual measures, such as safety harnesses.

GENERAL INFORMATION

| Outline of
Appendix 2 | The principles of prevention and
protection (Contd) |

(g) **Employees and the self-employed need to understand what they need to do,** e.g. by training, instruction, and communication of plans and risk assessments.

(h) **The existence of an active safety culture affecting the organisations responsible for developing and executing the project needs to be assured.**

| Appendix 3 | Where the local authority is the enforcing authority under the Health and Safety (Enforcing Authority) Regulations 1989 |

(Not reproduced here)

Note:

For construction work, HSE is the enforcing authority in all premises for which it is responsible and in all premises which would be inspected by the local authority except in respect of minor construction works as noted in the regulations.

| Appendix 4 | The Health and Safety plan prepared under Regulation 15(1)-(3) |

1 Nature of the project

(a) Name of client.

(b) Location.

(c) Nature of construction work to be carried out.

(d) Timescale for completion of the construction work.

2 The existing environment

(a) Surrounding land uses and related restrictions, e.g. premises (schools, shops or factories) adjacent to proposed construction site, planning restrictions which might affect health and safety.

(b) Existing services, e.g. underground and overheads lines.

(c) Existing traffic systems and restrictions, e.g. access for fire appliances, times of delivery, ease of delivery and parking.

(d) Existing structures, e.g. special health problems from materials in existing structures which are being demolished or refurbished, any fragile materials which require special safety precautions or instability problems.

(e) Ground conditions, e.g. contamination, gross instability, possible subsidence, old mine workings or underground obstructions.

3 Existing drawings

(a) Available drawings of structure(s) to be demolished or incorporated in the proposed structure(s) (this may include a health and safety file prepared for the structure(s) and held by the client).

4 The design

(a) Significant hazards or work sequences identified by designers which cannot be avoided or designed out and, where appropriate, a broad indication of the precautions assumed for dealing with them.

(b) The principles of the structural design and any precautions that might be needed or sequences of assembly that need to be followed during construction.

(c) Detailed reference to specific problems where contractors will be required to explain their proposals for managing these problems.

5 Construction materials

(a) Health hazards arising from construction materials where particular precautions are required either because of their nature or the manner of their intended use. These will have been identified by designers as hazards which cannot be avoided or designed out. They should be specified as far as is necessary to ensure reliable performance by a competent contractor who may be assumed to know the precautionary information that suppliers are, by law, required to provide.

6 Site-wide elements

(a) Positioning of site access and egress points (e.g. for deliveries and emergencies).

(b) Location of temporary site accommodation.

(c) Location of unloading, layout and storage areas.

(d) Traffic/pedestrian routes.

7 Overlap with client's undertaking

(a) Consideration of the health and safety issues which arise when the project is to be located in premises occupied or partially occupied by the client.

8 Site rules

(a) Specific site rules which the client or the planning supervisor may wish to lay down as a result of points 2 to 7 above or for other reasons, e.g. specific permit-to-work rules, emergency procedures.

9 Continuing liaison

(a) Procedures for considering the health and safety implications of design elements of the principal contractor's and other contractor's packages.

(b) Procedures for dealing with unforeseen eventualities during project execution resulting in substantial design change and which might affect resources.

| Appendix 5 | The health and safety file prepared under regulation 14(d)-(f) |

Information contained in the file needs to include that which will assist persons carrying out construction work on the structure at any time after completion of the current project and may include:

(a) record or 'as built' drawings and plans used and produced throughout the construction process along with the design criteria;

(b) general details of the construction methods and materials used;

(c) details of the structure's equipment and maintenance facilities;

(d) maintenance procedures and requirements for the structure;

Appendix 5   The health and safety file prepared under regulation 14(d)-(f) (Contd)

(e) manuals produced by specialist contractors and suppliers which outline operating and maintenance procedures and schedules for plant and equipment installed as part of the structure; and

(f) details on the location and nature of utilities and services, including emergency and fire-fighting systems.

ARRANGEMENT OF REGULATIONS

1. Citation and commencement

2. Interpretation

3. Application of regulations

4. Clients and agents of clients

5. Requirements on developer

6. Appointments of planning supervisor and principal contractor

7. Notification of project

8. Competence of planning supervisor, designer and contractors

9. Provision for health and safety

10. Start of construction phase

11. Client to ensure information is available

12. Client to ensure health and safety file is available for inspection

13. Requirements on designer

14. Requirements on planning supervisor

15. Requirements relating to the health and safety plan

16. Requirements on and powers of principal contractor

17. Information and training

18. Advice from, and views of, persons at work

19. Requirements and prohibitations on contractors

20. Extension outside Great Britain

21. Exclusion of civil liability

22. Enforcement

23. Transitional provisions

24. Repeals, revocations and modifications

Schedule 1 Particulars to be notified to the Executive

Schedule 2 Transitional provisions

Schedule 1   Particulars to be notified to the Executive Regulation 7

1 Date of forwarding
2 Exact address of the construction site
3 Name and address of the client or clients (see note)
4 Type of project
5 Name and address of the planning supervisor
6 A declaration signed by or on behalf of the planning supervisor that he has been appointed as such.
7 Name and address of the principal contractor
8 A declaration signed by or on behalf of the principal that he has been appointed as such
9 Date planned for start of the construction phase
10 Planned duration of the construction phase
11 Estimated maximum number of people at work on the construction site
12 Planned number of contractors on the construction site
13 Name and address of any contractor or contractors already chosen

Note: Where a declaration has been made in accordance with regulation 4(4), item 3 above refers to the client or clients on the basis that that declaration has not yet taken effect.

Acknowledgement

Excerpts reproduced from HSE publications -

(a) CDM Regulations - How the regulations affect you!

(b) Managing Construction for Health and Safety, Construction (Design and Management) Regulation 1994, Approved Code of Practice.

(c) Statutory Instruments 1994 No.3140 HEALTH AND SAFETY the Construction (Design and Management) Regulations 1994.

GENERAL INFORMATION

NOTES

Arbitration

ADJUDICATION

A more 'immediate' and statutory method of resolving disputes has been introduced by The Housing Grants, Construction and Regeneration Act 1996 (commonly known as 'The Construction Act'), which became law on 1st May 1998. It provides for the Act's "Scheme for Construction Contracts" to be read into all defined construction contracts, entered into after 1st May 1998, which do not already contain similar provisions.

Included in the provisions is a statutory right for the parties to a construction contract to refer all disputes to adjudication by a third party - the adjudicator.

The aim of the provisions is to resolve disputes speedily and effectively as they occur. The adjudicator is to be appointed either by agreement or by appointment by an 'adjudicator nominating body'.

The object is to ensure the appointment of the adjudicator and the referral of the dispute to him within 7 days of a notice being given by either party that he intends to refer a dispute to adjudication.

The adjudicator has then to reach a decision within 28 days of referral or within such longer period as is agreed by the parties. If the parties fail to agree on an extension of time, requested by the adjudicator, an extension of 14 days may be granted with the sole consent of the party by whom the dispute was referred.

The adjudicator has to act impartially but is not bound to follow any set procedure. He is expected to take the initiative in ascertaining the facts and the law.

The parties may agree to accept the decision of the adjudicator as final. However, the contract may provide that his decision is binding until the dispute is eventually determined by arbitration, by legal proceedings or by agreement.

Until now, the standard forms of sub-contract have provided adjudication only as a means of establishing the allocation of money in set-off disputes. The new provisions, which are statutory and apply to all defined construction contracts (including consultancy agreements), can be invoked for all types of dispute at any time.

An adjudicator can be named in the contract. If no name appears in the contract documents, application can be made by a party to an 'adjudicator nominating body' for an appointment to be made. A variety of organisations will offer the service of selecting a suitable adjudicator and will run training programmes for 'would be' adjudicators. The 'adjudicator nominating bodies' include the Chartered Institute of Building, Confederation of Construction Specialists, Construction Confederation, Construction Industry Council, ICE, RIBA, RICS and the Chartered Institute of Arbitrators.

ARBITRATION

Most standard forms of building contract and subcontract conditions contain an arbitration clause. This means that the parties agree to refer any dispute, arising under the contract, to be agreed between the parties, to act as arbitrator.

If the parties are unable to agree on a person then the president or vice-president of one of the professional institutes will appoint a person from a panel of experienced arbitrators. Depending upon the wording in the conditions of contract, the request for an appointment by either party may be made to the president of the Royal Institute of British Architects, The Royal Institution of Chartered Surveyors, the Institution of Civil Engineers or the Chartered Institute of Arbitrators.

Compared with High Court procedures, arbitration has the benefit of being private.

It is recommended that a negotiated settlement of a dispute should be pursued as far as possible before the attempt at an agreement is abandoned and arbitration commenced. Once arbitration is started it may be necessary to search files for all relevant documents and witnesses may have to appear to give evidence. The arbitrator will need to see or hear all the necessary evidence upon which to base his award. The process can be very demanding in management and staff time. Sometimes even the "winner" gets little financial reward for the amount of time and trouble spent on preparing for the arbitration, particularly where small amounts of money are involved.

However, a recent Act of Parliament now gives the parties, in an arbitration, the opportunity to avoid unnecessary delay and expense.

The substantive provisions of the latest Arbitration Act - The Arbitration Act 1996 - came into force on 31 January 1997. The Act is printed in full on the following pages (except Schedules 1-4) Sections 85 to 87, which make special provisions in relation to domestic arbitration agreements, are not yet in force.

The underlying philosophy of the Act is that the parties should be free to agree between themselves how their dispute should be resolved. The Act is founded on three principles- a) the object of arbitration is to obtain the fair resolution of disputes by an impartial tribunal without unnecessary delay and expense; b) the parties should be free to agree how their disputes are resolved, subject only to safeguards as are necessary in the public interest; c) the Court should not intervene except in the very limited circumstances provided by the Act.

The arbitrator now has the power to decide all procedural and evidential matters, subject to the right of the parties to agree any matter if they so wish. The strict rules of evidence, adhered to by the Courts, may be ignored. The arbitrator may take the initiative in ascertaining the facts and law instead of relying upon the parties or their representatives to ascertain them by examination and cross examination. The absolute right, of either party, to an oral hearing has been swept away.

GENERAL INFORMATION

The arbitrator will seek to establish the wishes of the parties at a Preliminary Meeting. He will be made aware of the nature of the dispute and assist the parties in agreeing the most suitable procedure for the resolution of the dispute. Subsequently he, will issue an order giving the timetable for the arbitration process.

The 1996 Act gives arbitrators a completely new power. Unless otherwise agreed between the parties, the arbitrator may direct that the recoverable costs of the arbitration or any part of the arbitral proceedings, shall be limited to a specified amount. The direction has to be given sufficiently in advance of the costs being incurred for the limit to be taken into account.

The Act also gives the arbitrator immunity from liability for anything he may do or omit to do in the discharge of his duties except acting in bad faith.

The right to challenge an award for the technical misconduct of the arbitrator is limited and the Court will only intervene in those cases of serious irregularity where substantial injustice has occurred. The kind of irregularity is narrowly defined by the Act.

The Act also contains guidelines concerning those circumstances in which the Court will give leave to appeal on a point of law. It generally needs to be on a point of general public importance for the Court to give leave. Thus it will be very difficult to overturn the award of the arbitrator.

The arbitration process, prior to the Act coming into force, had become almost a copy of Court procedures. Arbitration had become expensive and time consuming. It is hoped that the new Act will enable parties to make agreements on procedure and costs and arbitrators to use their discretion for the sake of the fair and swift resolution of damaging disputes.

Following the enactment of the Arbitration Act, the Society of Construction Arbitrators initiated the production of Model Arbitration Rules for adoption by all construction institutions and other bodies having interests in construction arbitration.

The outcome was The Construction Industry Model Arbitration Rules (CIMAR). These rules together with supplementary and advisory procedures were published by the Joint Contracts Tribunal in 1998 and are incorporated in all current editions of the JCT standard forms of building contract.

The rules provide procedures and time limits for document only, sort hearing or full procedure arbitrations. Sections of the Act, which need to be read with the Rules, are printed in the published text and other sections are reproduced in the Appendix.

Where arbitration agreements are incorporated in non-standard contract conditions, without any rules being stated, it might be helpful to the parties, in the interest of speed and cost effectiveness, to adopt the Model Rules, by agreement, when the arbitration commences.

ARBITRATION ACT 1996

CHAPTER 23

ARRANGEMENT OF SECTIONS

PART I

ARBITRATION PURSUANT TO AN ARBITRATION
AGREEMENT

Introductory

Section
1.   General principles.
2.   Scope of application of provisions.
3.   The seat of the arbitration.
4.   Mandatory and non-mandatory provisions.
5.   Agreements to be in writing.

The arbitration agreement

6.   Definition of arbitration agreement.
7.   Separability of arbitration agreement.
8.   Whether agreement discharged by death of a party.

Stay of legal proceedings

Section
9.   Stay of legal proceedings.
10.   Reference of interpleader issue to arbitration.
11.   Retention of security where Admiralty proceeding stayed.

Commencement of arbitral proceedings

12.   Power of court to extend time for beginning arbitral proceedings, &c.
13.   Application of Limitation Acts.
14.   Commencement of arbitral proceedings.

The arbitral tribunal

15.   The arbitral tribunal.
16.   Procedure for appointment of arbitrators.
17.   Power in case of default to appoint sole arbitrator.
18.   Failure of appointment procedure.
19.   Court to have regard to agreed qualifications.
20.   Chairman
21.   Umpire.
22.   Decision-making where no chairman or umpire.
23.   Revocation of arbitrator's authority.
24.   Power of court to remove arbitrator.
25.   Resignation of arbitrator.
26.   Death of arbitrator or person appointing him.
27.   Filling of vacancy, &c.
28.   Joint and several liability of parties to arbitrators for fees and expenses.
29.   Immunity of arbitrator.

Jurisdiction of the arbitral tribunal

30.   Competence of tribunal to rule on its own jurisdiction.
31.   Objection to substantive jurisdiction of tribunal.
32.   Determination of preliminary point of jurisdiction.

The arbitral proceedings

33.   General duty of the tribunal.
34.   Procedural and evidential matters.
35.   Consolidation of proceedings and concurrent hearings.
36.   Legal or other representation.
37.   Power to appoint experts, legal advisers or assessors.
38.   General powers exercisable by the tribunal.
39.   Power to make provisional awards.
40.   General duty of parties.
41.   Powers of tribunal in case of party's default.

Powers of court in relation to arbitral proceedings

42.   Enforcement of peremptory orders of tribunal.
43.   Securing the attendance of witnesses.
44.   Court powers exercisable in support of arbitral proceedings.
45.   Determination of preliminary point of law.

The award

46.   Rules applicable to substance of dispute.
47.   Awards on different issues, &c.
48.   Remedies.
49.   Interest.
50.   Extension of time for making award.
51.   Settlement.
52.   Form of award.
53.   Place where award treated as made.
54.   Date of award.
55.   Notification of award.
56.   Power to withhold award in case of non-payment.
57.   Correction of award or additional award.
58.   Effect of award.

Costs of the arbitration

59.   Costs of the arbitration.
60.   Agreement to pay costs in any event.
61.   Award of costs.
62.   Effect of agreement or award about costs.
63.   The recoverable costs of the arbitration.
64.   Recoverable fees and expenses of arbitrators.
65.   Power to limit recoverable costs.

PART II

OTHER PROVISIONS RELATING TO ARBITRATION

Domestic arbitration agreements

Consumer arbitration agreements

Small claims arbitration in the county court

Appointment of judges as arbitrators

Statutory arbitrations

PART III

RECOGNITION AND ENFORCEMENT OF CERTAIN FOREIGN AWARDS

Enforcement of Geneva Convention awards

Recognition and enforcement of New York Convention awards

PART IV

GENERAL PROVISIONS

GENERAL INFORMATION

Arbitration Act 1996

ELIZABETH II

Arbitration Act 1996

1996 CHAPTER 23

An Act to restate and improve the law relating to arbitration pursuant to an arbitration agreement; to make other provision relating to arbitration and arbitration awards; and for connected purposes. [17th June 1996]

BE IT ENACTED by the Queen's most Excellent Majesty, by and with the advice and consent of the Lords Spiritual and Temporal, and Commons, in this present Parliament assembled, and by the authority of the same, as follows:-

PART I

ARBITRATION PURSUANT TO AN ARBITRATION AGREEMENT

Introductory

General principles

1. The provisions of this Part are founded on the following principles, and shall be construed accordingly-

(a) the object of arbitration is to obtain the fair resolution of disputes by an impartial tribunal without unnecessary delay or expense;

(b) the parties should be free to agree how their disputes are resolved, subject only to such safeguards as are necessary in the public interest;

(c) in matters governed by this Part the court should not intervene except as provided by this Part.

Scope of application of provisions

2. (l) The provisions of this Part apply where the seat of the arbitration is in England and Wales or Northern Ireland.

(2) The following sections apply even if the seat of the arbitration is outside England and Wales or Northern Ireland or no seat has been designated or determined-

(a) sections 9 to 11 (stay of legal proceedings, &c.), and

(b) section 66 (enforcement of arbitral awards).

(3) The powers conferred by the following sections apply even if the seat of the arbitration is outside England and Wales or Northern Ireland or no seat has been designated or determined-

(a) section 43 (securing the attendance of witness), and

(b) section 44 (court powers exercisable in support of arbitral proceedings);

Part I

but the court may refuse to exercise any such power, if in the opinion of the court, the fact that the seat of the arbitration is outside England and Wales or Northern Ireland, or that when designated or determined the seat is likely to be outside England and Wales or Northern Ireland, makes it inappropriate to do so.

(4) The court may exercise a power conferred by any provision of this Part not mentioned in subsections (2) or (3) for the purpose of supporting the arbitral process where-

(a) no seat of the arbitration has been designated or determined, and

(b) by reason of a connection with England and Wales or Northern Ireland the court is satisfied that it is appropriate to do so.

(5) Section 7 (separability of arbitration agreement) and section 8 (death of a party) apply where the law of England and Wales or Northern Ireland even if the seat of the arbitration is outside England and Wales or Northern Ireland or has not been designated or determined.

The seat of the arbitration

3. In this Part "the seat of the arbitration" means the juridical seat of the arbitration designated-

(a) by the parties to the arbitration agreement, or

(b) by any arbitral or other institution or person vested by the parties with power in that regard, or

(c) by the arbitral tribunal if so authorised by the parties,

or determined, in the absence of any such designation, having regard to the parties' agreement and all the relevant circumstances.

Mandatory and non-mandatory provisions

4.(1) The mandatory provisions of this Part are listed in Schedule 1 and have effect notwithstanding any agreement to the contrary,

(2) The other provisions of this Part (the "non-mandatory provisions") allow the parties to make their own arrangements by agreement but provide rules which apply in the absence of such agreement

(3) The parties may make such arrangements by agreeing to the application of institutional rules or providing any other means by which a matter may be decided.

(4) It is immaterial whether or not the law applicable to the parties' agreement is the law of England and Wales or, as the case may be, Northern Ireland.

(5) The choice of a law other than the law of England and Wales or Northern Ireland as the applicable law in respect of a matter provided for by a non-mandatory provision of this Part is equivalent to an agreement making provision

Arbitration Act 1996

Part I

about that matter.

For this purpose an applicable law determined in accordance with the parties' agreement, or which is objectively determined in the absence of any express or implied choice, shall be treated as chosen by the parties.

Agreements to be in writing

5.(1) The provisions of this Part apply only where the arbitration agreement is in writing, and any other agreement between the parties as to any matter is effective for the purposes of this Part only if in writing.

The expressions "agreement", "agree" and "agreed" shall be construed accordingly.

(2) There is an agreement in writing-

(a) if the agreement is made in writing (whether or not it is signed by the parties),

(b) if the agreement is made by exchange of communications in writing, or

(c) if the agreement is evidenced in writing.

(3) Where parties agree otherwise than in writing by reference to terms which are in writing, they make an agreement in writing.

(4) An agreement is evidenced in writing if an agreement made otherwise than in writing is recorded by one of the parties, or by a third party, with the authority of the parties to the agreement.

(5) An exchange of written submissions in arbitral or legal proceedings in which the existence of an agreement otherwise than in writing is alleged by one party against another party and not denied by the other party in his response constitutes as between those parties an agreement in writing to the effect alleged.

(6) References in this Part to anything being written or in writing include its being recorded by any means.

The arbitration agreement

Definition of arbitration agreement

6.(1) In this Part an "arbitration agreement" means an agreement to submit to arbitration present or future disputes (whether they are contractual or not).

(2) The reference in an agreement to a written form of arbitration clause or to a document containing an arbitration clause constitutes an arbitration agreement if the reference is such as to make the clause part of the agreement.

Separability or arbitration agreement

7. Unless otherwise agreed by the parties, an an arbitration agreement which forms or was intended to form part of another agreement (whether or not in writing) shall not be regarded as invalid, non-existent or ineffective because that other agreement is invalid, or did not come into existence or has become ineffective, and it shall for that purpose be treated as a distinct

Arbitration Act 1996

Part I

agreement.

Whether agreement discharged by death of a party

8.(1) Unless otherwise agreed by the parties, an arbitration agreement is not discharged by the death of a party and may be enforced by or against the personal representatives of that party.

(2) Subsection (1) does not affect the operation of any enactment or rule of law by virtue of which a substantive right or obligation is extinguished by death.

Stay of legal proceedings

Stay of Legal proceedings

9.(1) A party to an arbitration agreement against whom legal proceedings are brought (whether by way of claim or counterclaim in respect of a matter which under the agreement is to be referred to arbitration may (upon notice to the other parties to the proceedings) apply to the court in which the proceedings have been brought to stay the proceedings so far as they concern that matter.

(2) An application may be made notwithstanding that the matter is to be referred to arbitration only after the exhaustion of other dispute resolution procedures.

(3) An application may not be made by a person before taking the appropriate procedural step (if any) to acknowledge the legal proceedings against him or after he has taken any step in those proceedings to answer the substantive claim.

(4) On an application under this section the court shall grant a stay unless satisfied that the arbitration agreement is null and void, inoperative, or incapable of being performed.

(5) If the court refuses to stay the legal proceedings, any provision that an award is a condition precedent to the bringing of legal proceedings in respect of any matter is of no effect in relation to those proceedings.

Reference of interpleader issue to arbitration

10.(1)Where in legal proceedings relief by way of interpleader is granted and any issue between the claimants is one in respect of which there is an arbitration agreement between them, the court granting the relief shall direct that the issue be determined in accordance with the agreement unless the circumstances are such that proceedings brought by a claimant in respect of the matter would not be stayed.

(2) Where subsection (1) applies but the court does not direct that the issue be determined in accordance with the arbitration agreement, any provision that an award is a condition precedent to the bringing of legal proceedings in respect of any matter shall not affect the determination of that issue by the court.

Retention of security where

11.(1)Where Admiralty proceedings are stayed on the ground that the dispute in question should

GENERAL INFORMATION

Arbitration Act 1996

Part I

Admiralty proceedings stayed

be submitted to arbitration, the court granting the stay may, if in those proceedings property has been arrested or bail or other security has been given to prevent or obtain release from arrest.

(a) order that the property arrested be retained as security for the satisfaction of any award given in the arbitration in respect of that dispute, or

(b) order that the stay of those proceedings be conditional on the provision of equivalent security for the satisfaction of any such award.

(2) Subject to any provision made by rules of court and to any necessary modifications, the same law and practice shall apply in relation to property retained in pursuance of an order as would apply if it were held for the purposes of proceedings in the court making the order.

Commencement of arbitral proceedings

Power of court to extend time for beginning arbitral proceedings, &c.

12.(1)Where an arbitration agreement to refer future disputes to arbitration provides that a claim shall be barred, or the claimant's right extinguished, unless the claimant takes within a time fixed by the agreement some step.

(a) to begin arbitral proceedings, or

(b) to begin other dispute resolution procedures which must be exhausted before arbitral proceedings can be begun.

the court may by order extend the time for taking that step.

(2) Any party to the arbitration agreement may apply for such an order (upon notice to the other parties), but only after a claim has arisen and after exhausting any available arbitral process for obtaining an extension of time.

(3) The court shall make an order only if satisfied.

(a) that the circumstances are such as were outside the reasonable contemplation of the parties when they agreed the provision in question, and that it would be just to extend the time, or

(b) that the conduct of one party makes it unjust to hold the other party to the strict terms of the provision in question.

(4) The court may extend the time for such period and on such terms as it thinks fit, and may do so whether or not the time previously fixed (by agreement or by a previous order) has expired.

(5) An order under this section does not affect the operation of the Limitation Acts (see section 13).

(6) The leave of the court is required for any

Arbitration Act 1996

Part I

appeal from a decision of the court under this section.

Application of Limitation Acts

13.(1)The Limitation Acts apply to arbitral proceedings as they apply to legal proceedings.

(2) The court may order that in computing the time prescribed by the Limitation Acts for the commencement of proceedings (including arbitral proceedings) in respect of a dispute which was the subject matter.

(a) of an award which the court orders to be set aside or declares to be of no effect, or

(b) of the affected part of an award which the court orders to be set aside in part, or declares to be in part of no effect,

the period between the commencement of the arbitration and the date of the order referred to in paragraph (a) or (b) shall be excluded.

(3) In determining for the purposes of the Limitation Acts when a cause of action accrued, any provision that an award is a condition precedent to the bringing of legal proceedings in respect of a matter to which an arbitration agreement applies shall be disregarded.

(4) In this Part "the Limitation Acts" means-

1980 c.58
1984 c.16

(a) in England and Wales, the Limitation Act 1980, the Foreign Limitation Periods Act 1984 and any other enactment (whenever passed) relating to the limitation of actions;

S.I. 1989/1339 (N.I.11)
S.I.1985/754 (N.I.5)

(b) in Northern Ireland, the Limitation (Northern Ireland) Order 1989, the Foreign Limitation Periods (Northern Ireland) Order 1985 and any other enactment (whenever passed) relating to the limitation of actions.

Commencement of arbitral for proceedings

14.(1)The parties are free to agree when arbitral proceedings are to be regarded as commenced for the purposes of this Part and for the purposes of the Limitations Acts.

(2) If there is no such agreement the following provisions apply.

(3) Where the arbitrator is named or designated in the arbitration agreement, arbitral proceedings are commenced in respect of a matter when one party serves on the other party or parties a notice in writing requiring him or them to submit that matter to the person so named or designated.

(4) Where the arbitrator or arbitrators are to be appointed by the parties, arbitral proceedings are commenced in respect of a matter when one party serves on the other party or parties notice in writing requiring him or them to appoint an arbitrator or to agree to the appointment of an arbitrator in respect of that matter.

Arbitration Act 1996

Arbitration Act 1996

Part I

(5) Where the arbitrator or arbitrators are to be appointed by a person other than a party to the proceedings, arbitral proceedings are commenced in respect of a matter when one party gives notice in writing to that person requesting him to make the appointment in respect of that matter.

The arbitral tribunal

The arbitral tribunal

15.(1)The parties are free to agree on the number of arbitrators to form the tribunal and whether there is to be a chairman or umpire.

(2) Unless otherwise agreed by the parties, an agreement that the number of arbitrators shall be two or any other even number shall be understood as requiring the appointment of an additional arbitrator as chairman of the tribunal.

(3) If there is no agreement as to the number of arbitrators, the tribunal shall consist of a sole arbitrator.

Procedure for appointment of arbitrators

16.(1)The parties are free to agree on the procedure for appointing the arbitrator or arbitrators including the procedure for appointing any chairman or umpire.

(2) If or to the extent that there is no such agreement, the following provisions apply.

(3) If the tribunal is to consist of a sole arbitrator, the parties shall jointly appoint the arbitrator not later than 28 days after service of a request in writing by either party to do so.

(4) If the tribunal is to consist of two arbitrators, each party shall appoint one arbitrator not later than 14 days after service of a request in writing by either party to do so.

(5) If the tribunal is to consist of three arbitrators-

(a) each party shall appoint one arbitrator not later than 14 days after service of a request in writing by either party to do so, and

(b) the two so appointed shall forthwith appoint a third arbitrator as the chairman of the tribunal.

(6) If the tribunal is to consist of two arbitrators and an umpire-

(a) each party shall appoint one arbitrator not later than 14 days after service of a request in writing by either party to do so, and

(b) the two so appointed may appoint an umpire at any time after they themselves are appointed and shall do so before any substantive hearing or forthwith if they cannot agree on a matter relating to the arbitration.

(7) In any other case (in particular, if there are more than two parties) section 18 applies as in the case of a failure of the agreed appointment procedure.

Part I

Power in case of default to appoint sole arbitrator

17.(1)Unless the parties otherwise agree, where each of two parties to an arbitration agreement is to appoint an arbitrator and one party ("the party in default") refuses to do so, or fails to do so within the time specified, the other party, having duly appointed his arbitrator, may give notice in writing to the party in default that he proposes to appoint his arbitrator to act as sole arbitrator.

(2) If the party in default does not within 7 clear days of that notice being given-

(a) make the required appointment, and

(b) notify the other party that he has done so,

the other party may appoint his arbitrator as sole arbitrator whose award shall be binding on both parties as if he had been so appointed by agreement.

(3) Where a sole arbitrator has been appointed under subsection (2), the party in default may (upon notice to the appointing party) apply to the court which may set aside the appointment.

(4) The leave of the court is required for any appeal from a decision of the court under this section.

Failure of appointment procedure

18.(1)The parties are free to agree what is to happen in the event of a failure of the procedure for the appointment of the arbitral tribunal.

There is no failure if an appointment is duly made under section 17 (power in case of default to appoint sole arbitrator), unless that appointment is set aside.

(2) If or to the extent that there is no such agreement any party to the arbitration agreement may (upon notice to the other parties) apply to the court to exercise its powers under this section.

(3) Those powers are-

(a) to give directions as to the making of any necessary appointments;

(b) to direct that the tribunal shall be constituted by such appointments (or any one or more of them) as have been made;

(c) to revoke any appointments already made;

(d) to make any necessary appointments itself.

(4) An appointment made by the court under this section has effect as if made with the agreement of the parties.

(5) The leave of the court is required for any appeal from a decision of the court under this section.

Court to have regard to

19. In deciding whether to exercise, and in considering how to exercise, any of its powers

GENERAL INFORMATION

Arbitration Act 1996

Part I

agreed
qualifications

under section 16 (procedure for appointment of arbitrators) or section 18 (failure of appointment procedure), the court shall have due regard to any agreement of the parties as to the qualifications required of the arbitrators.

Chairman

20.(1)Where the parties have agreed that there is to be a chairman, they are free to agree what the functions of the chairman are to be in relation to the making of decisions, orders and awards.

(2) If or to the extent that there is no such agreement, the following provisions apply.

(3) Decisions, orders and awards shall be made by all or a majority of the arbitrators (including the chairman).

(4) The view of the chairman shall prevail in relation to a decision, order or award in respect of which there is neither unanimity nor a majority under subsection (3).

Umpire

21.(1)Where the parties have agreed that there is to be an umpire, they are free to agree what the functions of the umpire are to be, and in particular-

(a) whether he is to attend the proceedings, and

(b) when he is to replace the other arbitrators as the tribunal with power to make decisions, orders and awards.

(2) If or to the extent that there is no such agreement, the following provisions apply.

(3) The umpire shall attend the proceedings and be supplied with the same documents and other materials as are supplied to the other arbitrators.

(4) Decisions, orders and awards shall be made by the other arbitrators unless and until they cannot agree on a matter relating to the arbitration.

In that event they shall forthwith give notice in writing to the parties and the umpire, whereupon the umpire shall replace them as the tribunal with power to make decisions, orders and awards as if he were sole arbitrator.

(5) If the arbitrators cannot agree but fail to give notice of that fact, or if any of them fails to join in the giving of notice, any party to the arbitral proceedings may (upon notice to the other parties and to the tribunal) apply to the court which may order that the umpire shall replace the other arbitrators as the tribunal with power to make decisions, orders and awards as if he were sole arbitrator.

(6) The leave of the court is required for any appeal from a decision of the court under this section.

Part I

Decision-
making where
no chairman
or umpire

22.(1)Where the parties agreed that there shall be two or more arbitrators with no chairman or umpire, the parties are free to agree how the tribunal is to make decisions, orders and awards.

(2) If there is no such agreement, decisions, orders and awards shall be made by all or a majority of the arbitrators.

Revocation
arbitrator's
authority

23.(1)The parties are free to agree in what circumstances the authority of an arbitrator may be revoked.

(2) If or to the extent that there is no such agreement the following provisions apply.

(3) The authority of an arbitrator may not be revoked except-

(a) by the parties acting jointly, or

(b) by an arbitral or other institution or person vested by the parties with powers in that regard.

(4) Revocation of the authority of an arbitrator by the parties acting jointly must be agreed in writing unless the parties also agree (whether or not in writing) to terminate the arbitration agreement.

(5) Nothing in this section affects the power of the court-

(a) to revoke an appointment under section 18 (powers exercisable in case of failure of appointment procedure), or

(b) to remove an arbitrator on the grounds specified in section 24.

Power of court
to remove
arbitrator

24.(1)A party to arbitral proceedings may (upon notice to the other parties, to the arbitrator concerned and to any other arbitrator) apply to the court to remove an arbitrator on any of the following grounds-

(a) that circumstances exist that give rise to justifiable doubts as to his impartiality;

(b) that he does not possess the qualifications required by the arbitration agreement;

(c) that he physically or mentally incapable of conducting the proceedings or there are justifiable doubts as to his capacity to do so;

(d) that he has refused or failed-

(i) properly to conduct the proceedings, or

(ii) to use all reasonable despatch in conducting the proceedings or making an award,and that substantial injustice has been or will caused to the applicant.

Arbitration Act 1996

Part 1

(2) If there is arbitral or other institution or person vested by the parties with power to remove an arbitrator, the court shall not exercise its power of removal unless satisfied that the applicant has first exhausted any available recourse to that institution or person.

(3) The arbitral tribunal may continue the arbitral proceedings and make an award while an application to the court under this section is pending.

(4) Where the court removes an arbitrator, it may make such order as it thinks fit with respect to his entitlement (if any) to fees or expenses, or the repayment of any fees or expenses already paid.

(5) The arbitrator concerned is entitled to appear and be heard by the court before it makes any order under this section.

(6) The leave of the court is required for any appeal from a decision of the court under this section.

Resignation of arbitrator
25.(1)The parties are free to agree with an arbitrator as to the consequences of his resignation as regards-

  (a) his entitlement (if any) to fees or expenses, and
  (b) any liability thereby incurred by him.

(2) If or to the extent that there is no such agreement the following provisions apply.

(3) An arbitrator who resigns his appointment may (upon notice to the parties) apply to the court-

  (a) to grant him relief from any liability thereby incurred by him, and

  (b) to make such order as it thinks fit with respect to his entitlement (if any) to fees or expenses or the repayment of any fees or expenses already paid.

(4) If the court is satisfied that in all the circumstances it was reasonable for the arbitrator to resign, it may grant such relief as it mentioned in subsection (3)(a) on such terms as it thinks fit.

(5) The leave of the court is required for any appeal from a decision of the court under this section.

Death of arbitrator or person appointing him
26.(1)The authority of an arbitrator is personal and ceases on his death.

(2) Unless otherwise agreed by the parties, the death of the person by whom an arbitrator was appointed does not revoke the arbitrator's authority.

Arbitration Act 1996

Part I

Filling of Vacancy &c
27.(1)Where an arbitrator ceases to hold office, the parties are free to agree-

  (a) whether and if so how the vacancy is to be filled,

  (b) whether and if so to what extent the previous proceedings should stand, and

  (c) what effect (if any) his ceasing to hold office has on any appointment made by him (alone or jointly).

(2) If or to the extent that there is no such agreement, the following provisions apply.

(3) The provisions of sections 16 (procedure for appointment of arbitrators) and 18 (failure of appointment procedure) apply in relation to the filling of the vacancy as in relation to an original appointment.

(4) The tribunal (when reconstituted) shall determine whether and if so to what extent the previous proceedings should stand.

This does not affect any right of a party to challenge those proceedings on any ground which had arisen before the arbitrator ceased to hold office.

(5) His ceasing to hold office does not affect any appointment by him (alone or jointly) of another arbitrator, in particular any appointment of a chairman or umpire.

Joint and several liability of parties to arbitrators for fees and expenses
28.(1)The parties are jointly and severally liable to pay to the arbitrators such reasonable fees and expenses (if any) as are appropriate in the circumstances.

(2) Any party may apply to the court (upon notice to the other parties and to the arbitrators) which may order that the amount of the arbitrators' fees and expenses shall be considered and adjusted by such means and upon such terms as it may direct.

(3) If the application is made after any amount has been paid to the arbitrators by way of fees or expenses, the court may order the repayment of such amount (if any) as is shown to be excessive, but shall not do so unless it is shown that it is reasonable in the circumstances to order repayment.

(4) The above provisions have effect subject to any order of the court under section 24(4) or 25(3)(b) (order as to entitlement to fees or expenses in case of removal or resignation of arbitrator).

(5) Nothing in this section affects any liability of a party to any other party to pay all or any of the costs of arbitration (see sections 59 to 65) or any contractual right of an arbitrator to payment of his fees and expenses.

Arbitration Act 1996

Part I

(6) In this section references to arbitrators include an arbitrator who has ceased to act and an umpire who has not replaced the other arbitrators.

Immunity of arbitrator

29.(1)An arbitrator is not liable for anything done or omitted in the discharge or purported discharge of his functions as arbitrator unless the act or omission is shown to have been in bad faith.

(2) Subsection (1) applies to an employee or agent of an arbitrator as it applies to the arbitrator himself.

(3) This section does not affect any liability incurred by an arbitrator by reason of his resigning (but see section 25).

Jurisdiction of the arbitral tribunal

Competence of tribunal to rule on its own jurisdiction

30. (1)Unless otherwise agreed by the parties, the arbitral tribunal may rule on its own substantive jurisdiction, that is, as to-

(a) whether there is a valid arbitration agreement

(b) whether the tribunal is properly constituted, and

(c) what matters have been submitted to arbitration in accordance with the arbitration agreement.

(2) Any such ruling may be challenged by any available arbitral process of appeal or review or in accordance with the provisions of this Part.

Objection to substantive jurisdiction of tribunal

31.(1)An objection that the arbitral tribunal lacks substantive jurisdiction at the outset of the proceedings must be raised by a party not later than the time he takes the first step in the proceedings to contest the merits of any matter in relation to which he challenges the tribunal's jurisdiction.

A party is not precluded from raising such an objection by the fact that he has appointed or participated in the appointment of an arbitrator.

(2) Any objection during the course of the arbitral proceedings that the arbitral tribunal is exceeding its substantive jurisdiction must be made as soon as possible after the matter alleged to be beyond its jurisdiction is raised.

(3) The arbitral tribunal may admit an objection later than the time specified in subsection (1) or (2) if it considers the delay justified.

(4) Where an objection is duly taken to the tribunal's substantive jurisdiction and the tribunal has power to rule on its own jurisdiction, it may-

(a) rule on the matter in an award as to jurisdiction, or

Arbitration Act 1996

Part I

(b) deal with the objection in its award on the merits.

If the parties agree which of these courses the tribunal should take, the tribunal shall proceed accordingly.

(5) The tribunal may in any case, and shall if the parties so agree, stay proceedings whilst an application is made to the court under section 32 (determination of preliminary point of jurisdiction).

Determination of preliminary point of jurisdiction

32.(1)The court may, on the application of a party to arbitral proceedings (upon notice to the other parties), determine any question as to the substantive jurisdiction of the tribunal.

A party may lose the right to object (see section 73).

(2) An application under this section shall not be considered unless-

(a) it is made with the agreement in writing of all the other parties to the proceedings, or

(b) it is made with the permission of the tribunal and the court is satisfied-

(i) that the determination of the question is likely to produce substantial savings in costs,

(ii) that the application was made without delay, and

(iii) that there is good reason why the matter should be decided by the court.

(3) An application under this section, unless made with the agreement of all the other parties to the proceedings, shall state the grounds on which it is said that the matter should be decided by the court.

(4) Unless otherwise agreed by the parties, the arbitral tribunal may continue the arbitral proceedings and make an award while an application to the court under this section is pending.

(5) Unless the court gives leave, no appeal lies from a decision of the court whether the conditions specified in subsection (2) are met.

(6) The decision of the court on the question of jurisdiction shall be treated as a judgement of the court for the purposes of an appeal.

But no appeal lies without the leave of the court which shall not be given unless the court considers that the question involves a point of law which is one of general importance or is one which for some other special reason should be considered by the Court of Appeal.

The arbitral proceedings

General duty of the tribunal

33.(1)The tribunal shall-

Arbitration Act 1996

Arbitration Act 1996

Part I

(a) act fairly and impartially as between the parties, giving each party a reasonable opportunity of putting his case and dealing with that of his opponent, and

(b) adopt procedures suitable to the circumstances of the particular case, avoiding unnecessary delay or expense, so as to provide a fair means for the resolution of the matters falling to be determined.

(2) The tribunal shall comply with that general duty in conducting the arbitral proceedings, in its decisions on matters of procedure and evidence and in the exercise of all other powers conferred on it.

Procedural and evidential matters

34.(1)It shall be for the tribunal to decide all procedural and evidential matters, subject to the right of the parties to agree any matter.

(2) Procedural and evidential matters include-

(a) when and where any part of the proceedings is to be held;

(b) the language or languages to be used in the proceedings and whether translations of any relevant documents are to be supplied;

(c) whether any and if so what form of written statements of claim and defence are to be used, when these should be supplied and the extent to which such statements can be later amended;

(d) whether any and if so which documents or classes of documents should be disclosed between and produced by the parties and at what stage;

(e) whether any and if so what questions should be put to and answered by the respective parties and when and in what form this should be done;

(f) whether to apply strict rules of evidence (or any other rules) as to the admissibility, relevance or weight of any material (oral, written or other) sought to be tendered on any matters of fact or opinion, and the time, manner and form in which such material should be exchanged and presented;

(g) whether and to what extent the tribunal should itself take the initiative in ascertaining the facts and the law;

(h) whether and to what extent there should be oral or written evidence or submissions.

(3) The tribunal may fix the time within which any directions given by it are to be complied with, and may if it thinks fit extend the time so fixed (whether or not it has expired).

Consolidation of proceedings and concurrent hearings

35.(1)The parties are free to agree-

(a) that the arbitral proceedings shall be consolidated with other arbitral

Part I

proceedings, or

(b) that concurrent hearings shall be held, on such terms as may be agreed.

(2) Unless the parties agree to confer such power on the tribunal, the tribunal has no power to order consolidation of proceedings or concurrent hearings.

Legal or other representation

36. Unless otherwise agreed by the parties, a party to arbitral proceedings may be represented in the proceedings by a lawyer or other person chosen by him.

Power to appoint experts, legal advisers or assessors

37.(1)Unless otherwise agreed by the parties-

(a) the tribunal may-

(i) appoint experts or legal advisers to report to it and the parties, or

(ii) appoint assessors to assist it on technical matters,

and may allow any such expert, legal adviser or assessor to attend the proceedings; and

(b) the parties shall be given a reasonable opportunity to comment on any information, opinion or advice offered by any such person.

(2) The fees and expenses of an expert, legal adviser or assessor appointed by the tribunal for which the arbitrators are liable are expenses of the arbitrators for the purposes of this Part.

General powers exercisable by the tribunal

38.(1)The parties are free to agree on the powers exercisable by the arbitral tribunal for the purposes of and in relation to the proceedings.

(2) Unless otherwise agreed by the parties the tribunal has the following powers.

(3) The tribunal may order a claimant to provide security for the costs of the arbitration.

This power shall not be exercised on the ground that the claimant is-

(a) an individual ordinarily resident outside the United Kingdom, or

(b) a corporation or association incorporated or formed under the law of a country outside the United Kingdom, or whose central management and control is exercised outside the United Kingdom.

(4) The tribunal may give directions in relation to any property which is the subject of the proceedings or as to which any question arises in the proceedings, and which is owned by or is in the possession of a party to the proceedings-

(a) for the inspection, photographing, preservation, custody or detention of the property by the tribunal, an expert or a party, or

GENERAL INFORMATION

Arbitration Act 1996

Part I

(b) ordering that samples be taken from, or any observation be made of or experiment conducted upon, the property.

(5)

The tribunal made direct that a party or witness shall be examined on oath or affirmation, and may for that purpose administer any necessary oath or take any necessary affirmation.

(6) The tribunal may give directions to a party for the preservation for the purposes of the proceedings of any evidence in his custody or control.

Power to make provisional awards

39.(1)The parties are free to agree that the tribunal shall have power to order on a provisional basis any relief which it would have power to grant in a final award.

(2) This includes, for instance, making-

(a) a provisional order for the payment of money or the disposition of property as between the parties, or

(b) an order to make an interim payment on account of the costs of the arbitration.

(3) Any such order shall be subject to the tribunal's final adjudication; and the tribunal's final award, on the merits or as to costs, shall take account of any such order.

(4) Unless the parties agree to confer such power on the tribunal, the tribunal has no such power.

This does not affect its powers under section 47 (awards on different issues, &c).

General duty of parties

40.(1)The parties shall do all things necessary for the proper and expeditious conduct of the arbitral proceedings.

(2) This includes-

(a) complying without delay with any determination of the tribunal as to procedural or evidential matters, or with any order or directions of the tribunal, and

(b) where appropriate, taking without delay any necessary steps to obtain a decision of the court on a preliminary question of jurisdiction or law (see sections 32 and 45).

Powers of tribunal in case of party's default

41.(1)The parties are free to agree on the powers of the tribunal in case of a party's failure to do something necessary for the proper and expeditious conduct of the arbitration.

(2) Unless otherwise agreed by the parties, the following provisions apply.

(3) If the tribunal is satisfied that there has been inordinate and inexcusable delay on the part of the claimant in pursuing his claim and that the delay-

Arbitration Act 1996

Part I

(a) gives rise, or is likely to give rise, to a substantial risk that it is not possible to have a fair resolution of the issues in that claim, or

(b) has caused, or is likely to cause, serious prejudice to the respondent,

the tribunal may make an award dismissing the claim.

(4) If without showing sufficient cause a party-

(a) fails to attend or be represented at an oral hearing of which due notice was given, or

(b) where matters are to be dealt with in writing, fails after due notice to submit written evidence or make written submissions,

the tribunal may continue the proceedings in the absence of that party of, as the case may be, without any written evidence or submissions on his behalf, and may make an award on the basis of the evidence before it.

(5) If without showing sufficient cause a party fails to comply with any order or directions of the tribunal, the tribunal may make a peremptory order to the same effect, prescribing such time for compliance with it as the tribunal considers appropriate.

(6) If a claimant fails to comply with a peremptory offer of the tribunal to provide security for costs, the tribunal may make an award dismissing his claim.

(7) If a party fails to comply with any other kind of peremptory order, then, without prejudice to section 42 (enforcement by court of tribunal's peremptory orders), the tribunal may do any of the following-

(a) direct that the party in default shall not be entitled to rely upon any allegation or material which was the subject matter of the order;

(b) draw such adverse inferences from the act of non-compliance as the circumstances justify;

(c) proceed to an award on the basis of such materials as have been properly provided to it;

(d) make such order as it thinks fit as to the payment of costs of the arbitration incurred in consequence of the non-compliance.

Powers of court in relation to arbitral proceedings

Enforcement of peremptory orders of tribunal

42.(1)Unless otherwise agreed by the parties, the court may make an order requiring a party to comply with a peremptory order made by the tribunal.

Arbitration Act 1996

Part I

(2) An application for an order under this section may be made-

(a) by the tribunal (upon notice to the parties),

(b) by a party to the arbitral proceedings with the permission of the tribunal (and upon notice to the other parties), or

(c) where the parties have agreed that the powers of the court under this section shall be available.

(3) The court shall not act unless it is satisfied that the applicant has exhausted any available arbitral process in respect of failure to comply with the tribunal's order.

(4) No order shall be made under this section unless the court is satisfied that the person to whom the tribunal's order was directed has failed to comply with it within the time prescribed in the order, or, if no time was prescribed, within a reasonable time.

(5) The leave of the court is required for any appeal from a decision of the court under this section.

Securing the attendance of witnesses

43.(1)A party to arbitral proceedings may use the same court procedures as are available in relation to legal proceedings to secure the attendance before the tribunal of a witness in order to give oral testimony or to produce documents or other material evidence.

(2) This may only be done with the permission of the tribunal or the agreement of the other parties.

(3) The court procedures may only be used if-

(a) the witness is in the United Kingdom, and

(b) the arbitral proceedings are being conducted in England and Wales or, as the case may be, Northern Ireland.

(4) A person shall not be compelled by virtue of this section to produce any document or other material evidence which he could not be compelled to produce in legal proceedings.

Court powers exercisable in support of arbitral proceedings

44.(1)Unless otherwise agreed by the parties, the court has for the purposes of and in relation to arbitral proceedings the same power of making orders about the matters listed below as it has for the purposes of and in relation to legal proceedings.

(2) Those matters are-

(a) the taking of the evidence of witnesses;

(b) the preservation of evidence;

(c) making orders relating to property which is the subject of the proceedings or as to which any question arises in the proceedings-

Arbitration Act 1996

Part I

(i) for the inspection, photographing, preservation, custody or detention of the property, or

(ii) ordering that samples to be taken from, or any observation be made of or experiment conducted upon, the property;

and for that purpose authorising any person to enter any premises in the possession or control of a party to the arbitration;

(d) the sale of any goods the subject of the proceedings;

(e) the granting of an interim injunction or the appointment of a receiver.

(3) If the case is one of urgency, the court may, on the application of a party or proposed party to the arbitral proceedings, make such orders as it thinks necessary for the purpose of preserving evidence or assets.

(4) If the case is not one of urgency, the court shall act only on the application of a party to the arbitral proceedings (upon notice to the other parties and to the tribunal) made with the permission of the tribunal or the agreement in writing of the other parties.

(5) In any case the court shall act only if or to the extent that the arbitral tribunal, and any arbitral or other institution or person vested by the parties with power in that regard, has no power or is unable for the time being to act effectively.

(6) If the court so orders, an order made by it under this section shall cease to have effect in whole or in part on the order of the tribunal or of any such arbitral or other institution or person having power to act in relation to the subject-matter of the order.

(7) The leave of the court is required for any appeal from a decision of the court under this section.

Determination of preliminary point of law

45.(1)Unless otherwise agreed by the parties, the court may on the application of a party to arbitral proceedings (upon notice to the other parties) determine any question of law arising in the course of the proceedings which the court is satisfied substantially affects the rights of one or more of the parties.

An agreement to dispense with reasons for the tribunal's award shall be considered an agreement to exclude the court's jurisdiction under this section.

(2) An application under this section shall not be considered unless-

(a) it is made with the agreement of all the other parties to the proceedings, or

(b) it is made with the permission of the

Arbitration Act 1996

Arbitration Act 1996

Part I

tribunal and the court is satisfied-

(i) that the determination of the question is likely to produce substantial savings in costs, and

(ii) that the application was made without delay.

(3) The application shall identify the question of law to be determined and, unless made with the agreement of all the other parties to the proceedings, shall state the grounds on which it is said that the question should be decided by the court.

(4) Unless otherwise agreed by the parties, the arbitral tribunal may continue the arbitral proceedings and made an award while an application to the court under this section is pending.

(5) Unless the court gives leave, no appeal lies from a decision of the court whether the conditions specified in subsection (2) are met.

(6) The decision of the court on the question of law shall be treated as a judgement of the court for the purposes of an appeal.

But no appeal lies without the leave of the court which shall not be given unless the court considers that the question is one of general importance, or is one which for some other special reason should be considered by the Court of Appeal.

The award

Rules applicable to substance of dispute

46.(1)The arbitral tribunal shall decide the dispute-

(a) in accordance with the law chosen by the parties as applicable to the substance of the dispute, or

(b) if the parties so agree, in accordance with such other considerations as are agreed by them or determined by the tribunal.

(2) For this purpose the choice of the laws of a country shall be understood to refer to the substantive laws of that country and not its conflict of laws rules.

(3) If or to the extent that there is no such choice or agreement, the tribunal shall apply the law determined by the conflict of laws rules which it considers applicable.

Awards on different issues, &c.

47.(1)Unless otherwise agreed by the parties, the tribunal may make more than one award at different times on different aspects of the matters to be determined.

(2) The tribunal may, in particular, make an award relating-

(a) to an issue affecting the whole claim, or

Part I

(b) to a part only of the claims or cross-claims submitted to it for decision.

(3) If the tribunal does so, it shall specify in its award the issue, or the claim or part of a claim, which is the subject matter of the award.

Remedies

48.(1)The parties are free to agree on the powers exercisable by the arbitral tribunal as regards remedies.

(2) Unless otherwise agreed by the parties, the tribunal has the following powers.

(3) The tribunal may make a declaration as to any matter to be determined in the proceedings.

(4) The tribunal may order the payment of a sum of money, in any currency.

(5) The tribunal has the same powers as the court-

(a) to order a party to do or refrain from doing anything;

(b) to order specific performance of a contract (other than a contract relating to land);

(c) to order the rectification, setting aside or cancellation of a deed or other document.

Interest

49.(1)The parties are free to agree on the powers of the tribunal as regards the award of interest.

(2) Unless otherwise agreed by the parties the following provisions apply.

(3) The tribunal may award simple or compound interest from such dates at such rates and with such rests as it considers meets the justice of the case-

(a) on the whole or part of any amount awarded by the tribunal, in respect of any period up to the date of the award;

(b) on the whole or part of any amount claimed in the arbitration and outstanding at the commencement of the arbitral proceedings but paid before the award was made, in respect of any period up to the date of payment.

(4) The tribunal may award simple or compound interest from the date of the award (or any later date) until payment, at such rates and with such rests as it considers meets the justice of the case, on the outstanding amount of any award (including any award of interest under subsection (3) and any award as to costs).

(5) References in this section to an amount awarded by the tribunal include an amount payable in consequence of a declaratory award by the tribunal.

(6) The above provisions do not affect any other power of the tribunal to award interest.

Arbitration Act 1996

Arbitration Act 1996

Part I

Part I

Extension of time for making award

50.(1) Where the time for making an award is limited by or in pursuance of the arbitration agreement, then, unless otherwise agreed by the parties, the court may in accordance with the following provisions by order extend that time.

(2) An application for an order under this section may be made-

(a) by the tribunal (upon notice to the parties), or

(b) by any party to the proceedings upon notice to the tribunal and the other parties),

but only after exhausting any available arbitral process for obtaining an extension of time.

(3) The court shall only make an order if satisfied that a substantial injustice would otherwise be done.

(4) The court may extend the time for such period and on such terms as it thinks fit, and may do so whether or not the time previously fixed (by or under the agreement or by a previous order) has expired.

(5) The leave of the court is required for any appeal from a decision of the court under this section.

Settlement

51.(1) If during arbitral proceedings the parties settle the dispute, the following provisions apply unless otherwise agreed by the parties.

(2) The tribunal shall terminate the substantive proceedings and, if so requested by the parties and not objected to by the tribunal, shall record the settlement in the form of an agreed award.

(3) An agreed award shall state that it is an award of the tribunal and shall have the same status and effect as any other award on the merits of the case.

(4) The following provisions of this Part relating to awards (sections 52 to 58) apply to an agreed award.

(5) Unless the parties have also settled the matter of the payment of the costs of the arbitration, the provisions of this Part relating to costs (sections 59 to 65) continue to apply.

Form of award

52.(1) The parties are free to agree on the form of an award.

(2) If or to the extent that there is no such agreement, the following provisions apply.

(3) The award shall be in writing signed by all the arbitrators or all those assenting to the award.

(4) The award shall contain the reasons for the award unless it is an agreed award or the parties have agreed to dispense with reasons.

(5) The award shall state the seat of the arbitration and the date when the award is made.

Place where award treated as made

53. Unless otherwise agreed by the parties, where the seat of the arbitration is in England and Wales or Northern Ireland, any award in the proceedings shall be treated as made there, regardless of where it was signed, despatched or delivered to any of the parties.

Date of award

54.(1) Unless otherwise agreed by the parties, the tribunal may decide what is to be taken to be the date on which the award was made.

(2) In the absence of any such decision, the date of the award shall be taken to be the date on which it is signed by the arbitrator or, where more than one arbitrator signs the award, by the last of them.

Notification of award

55.(1) The parties are free to agree on the requirements as to notification of the award to the parties.

(2) If there is no such agreement, the award shall be notified to the parties by service on them of copies of the award, which shall be done without delay after the award is made.

(3) Nothing in this section affects section 56 (power to withhold award in case of non-payment).

Power to withhold award in case of Non-payment

56.(1) The tribunal may refuse to deliver an award to the parties except upon full payment of fees and expenses of the arbitrators.

(2) If the tribunal refuses on that ground to deliver an award, a party to the arbitral proceedings may (upon notice to the other parties and the tribunal) apply to the court, which may order that-

(a) the tribunal shall deliver the award on the payment into court by the applicant of the fees and expenses demanded, or such lesser amount as the court may specify,

(b) the amount of the fees and expenses properly payable shall be determined by such means and upon such terms as the court may direct, and

(c) out of the money paid into court there shall be paid out such fees and expenses as may be found to be properly payable and the balance of the money (if any) shall be paid out to the applicant.

(3) For this purpose the amount of fees and expenses properly payable is the amount the applicant is liable to pay under section 28 or any agreement relating to the payment of the arbitrators.

(4) No application to the court may be made where there is any available arbitral process for appeal or review of the amount of the fees or

GENERAL INFORMATION

Arbitration Act 1996

Part I

expenses demanded.

(5) References in this section to arbitrators include an arbitrator who has ceased to act and an umpire who has not replaced the other arbitrators.

(6) The above provisions of this section also apply in relation to any arbitral or other institution or person vested by the parties with powers in relation to the delivery of the tribunal's award.

As they so apply, the references to the fees and expenses of the arbitrators shall be construed as including the fees and expenses of that institution or person.

(7) The leave of the court is required for any appeal from a decision of the court under this section.

(8) Nothing in this section shall be construed as excluding an application under section 28 where payment has been made to the arbitrators in order to obtain the award.

Correction of award or additional award

57.(1)The parties are free to agree on the powers of the tribunal to correct an award or make an additional award.

(2) If or to the extent there is no such agreement, the following provisions apply.

(3) The tribunal may on its own initiative or on the application of a party-

(a) correct an award so as to remove any clerical mistake or error arising from an accidental slip or omission or clarify or remove any ambiguity in the award, or

(b) make an additional award in respect of any claim (including a claim for interest or costs) which was presented to the tribunal but was not dealt with in the award.

These powers shall not be exercised without first affording the other parties a reasonable opportunity to make representations to the tribunal.

(4) Any application for the exercise of those powers must be made within 28 days of the date of the award or such longer period as the parties may agree.

(5) Any correction of an award shall be made within 28 days of the date the application was received by the tribunal or, where the correction is made by the tribunal on its own initiative, within 28 days of the date of the award or, in either case, such longer period as the parties may agree.

(6) Any additional award shall be made within 56 days of the date of the original award or such longer period as the parties may agree.

(7) Any correction of an award shall form part of the award.

Arbitration Act 1996

Part I

Effect of award

58.(1)Unless otherwise agreed by the parties, an award made by the tribunal pursuant to an arbitration agreement is final and binding both on the parties and on any persons claiming through or under them.

(2) This does not affect the right of a person the challenge the award by any available arbitral process of appeal or review or in accordance with the provisions of this Part.

Costs of the arbitration

Costs of the arbitration

59.(1)References in this Part to the costs of the arbitration are to-

(a) the arbitrators' fees and expenses,

(b) the fees and expenses of any arbitral institution concerned, and

(c) the legal or other costs of the parties.

(2) Any such reference includes the costs of or incidental to any proceedings to determine the amount of the recoverable costs of the arbitration (see section 63).

Agreement to pay costs in any event

60. An agreement which has the effect that a party is to pay the whole or part of the costs of the arbitration in any event is only valid if made after the dispute in question has arisen.

Award of costs

61.(1)The tribunal may make an award allocating the costs of the arbitration as between the parties, subject to any agreement of the parties.

(2) Unless the parties otherwise agree, the tribunal shall award costs on the general principle that costs should follow the event except where it appears to the tribunal that in the circumstances this is not appropriate in relation to the whole or part of the costs.

Effect of agreement or award about costs

62. Unless the parties otherwise agree, any obligation under an agreement between them as to how the costs of the arbitration are to be borne, or under an award allocating the costs of the arbitration, extends only to such costs as are recoverable.

The recoverable costs of the Arbitration

63.(1)The parties are free to agree what costs of arbitration are recoverable.

(2) If or to the extent there is no such agreement, the following provisions apply.

(3) The tribunal may determine by award the recoverable costs of the arbitration on such basis as it thinks fit.

If it does so, it shall specify-

(a) the basis on which it has acted, and

(b) the items of recoverable costs and the amount referable to each.

Arbitration Act 1996

Arbitration Act 1996

Part I

(4) If the tribunal does not determine the recoverable costs of the arbitration, any party to the arbitral proceedings may apply to the court (upon notice to the other parties) which may-

(a) determine the recoverable costs of the arbitration on such basis as it thinks fit, or

(b) order that they shall be determined by such means and upon such terms as it may specify.

(5) Unless the tribunal or the court determines otherwise-

(a) the recoverable costs of the arbitration shall be determined on the basis that there shall be allowed a reasonable amount in respect of all costs reasonably incurred, and

(b) any doubt as to whether costs were reasonably incurred or were reasonable in amount shall be resolved in favour of the paying party.

(6) The above provisions have effect subjection to section 64 (recoverable fees and expenses of arbitrators).

(7) Nothing in this section affects any right of the arbitrators, any expert, legal adviser or assessor appointed by the tribunal, or any arbitral institution, to payment of their fees and expenses.

Recoverable fees and expenses of arbitrators

64.(1)Unless otherwise agreed by the parties, the recoverable costs of the arbitration shall include in respect of the fees and expenses of the arbitrators only such reasonable fees and expenses as are appropriate in the circumstances.

(2) If there is any question as to what reasonable fees and expenses are appropriate in the circumstances, and the matter is not already before the court on an application under section 63(4), the court may on the application of any party (upon notice to other parties)-

(a) determine the matter, or

(b) order that it be determined by such means and upon such terms as the court may specify.

(3) Subsection (1) has effect subject to any order of the court under section 24(4) or 25(3)(b) (order as to entitlement to fees or expenses in case of removal or resignation of arbitrator).

(4) Nothing in this section affects any right of the arbitrator to payment of his fees and expenses.

Power to limit recoverable costs

65.(1)Unless otherwise agreed by the parties, the tribunal may direct that the recoverable costs of the arbitration, or of any part of the arbitral proceedings, shall be limited to a specified amount.

Part I

(2) Any direction may be made or varied at any stage, but this must be done sufficiently in advance of the incurring of costs to which is relates, or the taking of any steps in the proceedings which may be affected by it, for the limit to be taken into account.

Powers of the court in relation to award

Enforcement of the award

66.(1)An award made by the tribunal pursuant to an arbitration agreement may, by leave of the court, be enforced in the same manner as a judgement or order of the court to the same effect.

(2) Where leave is so given, judgement may be entered in terms of the award.

(3) Leave to enforce an award shall not be given where, or to the extent that, the person against whom it is sought to be enforced shows that the tribunal lacked substantive jurisdiction to make the award.

The right to raise such an objection may have been lost (see section 73).

1950 c.27

(4) Nothing in this section affects the recognition or enforcement of an award under any other enactment or rule of law, in particular under Part II of the Arbitration Act 1950 (enforcement of awards under Geneva Convention) or the provisions of Part III of this Act relating to the recognition and enforcement of awards under the New York Convention or by an action on the award.

Challenging the award: substantive jurisdiction

67.(1)A party to arbitral proceedings may (upon notice to the other parties and to the tribunal) apply to the court-

(a) challenging any award of the arbitral tribunal as to its substantive jurisdiction; or

(b) for an order declaring an award made by the tribunal on the merits to be of no effect, in whole or in part, because the tribunal did not have substantive jurisdiction.

A party may lose the right to object (see section 73) and the right to apply is subject to the restrictions in section 70(2) and (3).

(2) The arbitral tribunal may continue the arbitral proceedings and make a further award while an application to the court under this section is pending in relation to an award as to jurisdiction.

(3) On an application under this section challenging an award of the arbitral tribunal as to its substantive jurisdiction, the court may by order-

(a) confirm the award,

(b) vary the award, or

(c) set aside the award in whole or in part.

GENERAL INFORMATION

Arbitration Act 1996

Part I

(4) The leave of the court is required for any appeal from a decision of the court under this section.

Challenging the award: serious irregularity

68.(1) A party to arbitral proceedings may (upon notice to the other parties and to the tribunal) apply to the court challenging an award in the proceedings on the ground of serious irregularity affecting the tribunal, the proceedings or the award.

A party may lose the right to object (see section 73) and the right to apply is subject to the restrictions in section 70(2) and (3).

(2) Serious irregularity means an irregularity of one or more of the following kinds which the court considers has caused or will cause substantial injustice to the applicant-

(a) failure by the tribunal to comply with section 33 (general duty of tribunal);

(b) the tribunal exceeding its powers (otherwise then by exceeding its substantive jurisdiction: see section 67);

(c) failure by the tribunal to conduct the proceedings in accordance with the procedure agreed by the parties;

(d) failure by the tribunal to deal with all the issues that were put to it;

(e) any arbitral or other institution or person vested by the parties with powers in relation to the proceedings or the award exceeding its powers;

(f) uncertainty or ambiguity as to the effect of the award;

(g) the award being obtained by fraud or the award or the way in which it was procured being contrary to public policy;

(h) failure to comply with the requirements as to the form of the award; or

(i) any irregularity in the conduct of the proceedings or in the award which is admitted by the tribunal or by any arbitral or other institution or person vested by the parties with powers in relation to the proceedings or the award.

(3) If there is shown to be serious irregularity affecting the tribunal, the proceedings or the award, the court may-

(a) remit the award to the tribunal, in whole or in part, for reconsideration,

(b) set the award aside in whole or in part, or

(c) declare the award to be of no effect, in whole or in part.

The court shall not exercise its power to set

Arbitration Act 1996

Part I

aside or to declare an award to be of no effect, in whole or in part, unless it is satisfied that it would be inappropriate to remit the matters in question to the tribunal for reconsideration.

(4) The leave of the court is required for any appeal from a decision of the court under this section.

Appeal on point of law

69.(1) Unless otherwise agreed by the parties, a party to arbitral proceedings may (upon notice to the other parties and to the tribunal) appeal to the court on a question of law arising out of an award made in the proceedings.

An agreement to dispense with reasons for the tribunal's award shall be considered an agreement to exclude the court's jurisdiction under this section.

(2) An appeal shall not be brought under this section except-

(a) with the agreement of all the other parties to the proceedings, or

(b) with the leave of the court.

The right to appeal is also subject to the restrictions in section 70(2) and (3).

(3) Leave to appeal shall be given only if the court is satisfied-

(a) that the determination of the question will substantially affect the rights of one or more of the parties,

(b) that the question is one which the tribunal was asked to determine,

(c) that, on the basis of the findings of fact in the award-

(i) the decision of the tribunal on the question is obviously wrong, or

(ii) the question is one of general public importance and the decision of the tribunal is at least open to serious doubt, and

(d) that, despite the agreement of the parties to resolve the matter by arbitration, it is just and proper in all the circumstances for the court to determine the question.

(4) An application for leave to appeal under this section shall identify the question of law to be determined and state the grounds on which it is alleged that leave to appeal should be granted.

(5) The court shall determine an application for leave to appeal under this section without a hearing unless it appears to the court that a hearing is required.

(6) The leave of the court is required for any appeal from a decision of the court under this section to grant or refuse leave to appeal.

Arbitration Act 1996

Part I

(7) On an appeal under this section the court may by order-

(a) confirm the award,

(b) vary the award,

(c) remit the award to the tribunal, in whole or in part, for reconsideration in the light of the court's determination, or

(d) set aside the award in whole or in party.

The court shall not exercise its power to set aside an award, in whole or in part, unless it is satisfied that it would be inappropriate to remit the matters in question to the tribunal for reconsideration.

(8) The decision of the court on an appeal under this section shall be treated as a judgement of the court for the purposes of a further appeal.

But no such appeal lies without the leave of the court which shall not be given unless the court considers that the question is one of general importance or is one which for some other special reason should be considered by the Court of Appeal.

Challenge or appeal: supplementary provisions

70.(1) The following provisions apply to an application or appeal under section 67, 68 or 69.

(2) An application or appeal may not be brought if the applicant or appellant has not first exhausted-

(a) any available arbitral process of appeal or review, and

(b) any available recourse under section 57 (correction of award or additional award).

(3) Any application or appeal must be brought within 28 days of the date of the award or, if there has been any arbitral process of appeal or review, of the date when the applicant or appellant was notified of the result of that process.

(4) If on an application or appeal it appears to the court that the award-

(a) does not contain the tribunal's reasons, or

(b) does not set out the tribunal's reasons in sufficient detail to enable the court properly to consider the application or appeal,

the court may order the tribunal to state the reasons for its award in sufficient detail for that purpose.

(5) Where the court makes an order under subsection (4), it may make such further order as it thinks fit with respect to any additional costs of the arbitration resulting from its order.

Arbitration Act 1996

Part I

(6) The court may order the applicant or appellant to provide security for the costs of the application or appeal, and may direct that the application or appeal be dismissed if the order is not complied with.

The power to order security for costs shall not be exercised on the ground that the applicant or appellant is-

(a) an individual ordinarily resident outside the United Kingdom, or

(b) a corporation or association incorporated or formed under the law of a country outside the United Kingdom, or whose central management and control is exercised outside the United Kingdom.

(7) The court may order that any money payable under the award shall be brought into court or otherwise secured pending the determination of the application or appeal, and may direct that the application or appeal be dismissed if the order is not complied with.

(8) The court may grant leave to appeal subject to conditions to the same or similar effect as an order under subsection (6) or (7).

This does not affect the general discretion of the court to grant leave subject to conditions.

Challenge or appeal: effect of order of court

71.(1) The following provisions have effect where the court makes an order under section 67, 68 or 69 with respect to an award.

(2) Where the award is varied, the variation has effect as part of the tribunal's award.

(3) Where the award is remitted to the tribunal, in whole or in part, for reconsideration, the tribunal shall make a fresh award in respect of the matters remitted within three months of the date of the order for remission or such longer or shorter period as the court may direct.

(4) Where the award is set aside or declared to be of no effect, in whole or in part, the court may also order that any provision that an award is a condition precedent to the bringing of legal proceedings in respect of a matter to which the arbitration agreement applies, is of no effect as regards the subject matter of the award or, as the case may be, the relevant part of the award.

Miscellaneous

Saving for rights of person who takes no part in proceedings

72.(1) A person alleged to be a party to arbitral proceedings but who takes no part in the proceedings may question-

(a) whether there is a valid arbitration agreement,

(b) whether the tribunal is properly constituted, or

(c) what matters have been submitted to arbitration in accordance with the

GENERAL INFORMATION

Arbitration Act 1996

Part I

arbitration agreement,

by proceedings in the court for a declaration or injunction or other appropriate relief.

(2) He also has the same right as a party to the arbitral proceedings to challenge an award-

(a) by an application under section 67 on the ground of lack of substantive jurisdiction in relation to him, or

(b) by an application under section 68 on the ground of serious irregularity (within the meaning of that section) affecting him;

and section 70(2) (duty to exhaust arbitral procedures) does not apply in his case.

Loss of right to object

73.(1)If a party to arbitral proceedings takes part, or continues to take part, in the proceedings without making, either forthwith or within such time as is allowed by the arbitration agreement or the tribunal or by any provision of this Part, any objection-

(a) that the tribunal lacks substantive jurisdiction,

(b) that the proceedings have been improperly conducted,

(c) that there has been a failure to comply with the arbitration agreement or with any provision of this Part, or

(d) that there has been any other irregularity affecting the tribunal or the proceedings.

He may not raise that objection later, before the tribunal or the court, unless he shows that, at the time he took part or continued to take part in the proceedings, he did not know and could not with reasonable diligence have discovered the grounds for the objection.

(2) Where the arbitral tribunal rules that it has substantive jurisdiction and a party to arbitral proceedings who could have questions that ruling-

(a) by any available arbitral process of appeal or review, or

(b) by challenging the award,

does not do so, or does not do so within the time allowed by the arbitration agreement or any provision of this Part, he may not object later to the tribunal's substantive jurisdiction on any ground which was the subject of that ruling.

Immunity of arbitral institutions, &c

74.(1) An arbitral or other institution or person designated or requested by the parties to appoint or nominate an arbitrator is not liable for anything done or omitted in the discharge or purported discharge of that function unless the act or omission is shown to have been in bad faith.

(2) An arbitral or other institution or person by

Arbitration Act 1996

Part I

whom an arbitrator is appointed or nominated is not liable, by reason of having appointed or nominated him, for anything done or omitted by the arbitrator (or his employees or agents) in the discharge or purported discharge of his functions as arbitrator).

(3) The above provisions apply to an employee or agent of an arbitral or other institution or person as they apply to the institution or person himself.

Charge to secure payment of solicitors' costs
1974 c.47
S.I. 1976/582
(N.I.12)

75. The powers of the court to make declarations and orders under section 73 of the Solicitors Act 1974 or Article 71H of the Solicitors (Northern Ireland) Order 1976 (power to charge property recovered in the proceedings with the payment of solicitors' costs) may be exercised in relation to arbitral proceedings as if those proceedings were proceedings in the court.

Supplementary

Service of notices, &c

76.(1) The parties are free to agree on the mannder of service of any notice or other document required or authorised to be given or served in pursuance of the arbitration agreement or for the purposes of the arbitral proceedings.

(2) If or to the extent that there is no such agreement the following provisions apply.

(3) A notice or other document may be served on a person by any effective means.

(4) If a notice or other document is addressed, pre-paid and delivered by post-

(a) to the addressee's last known principal residence or, if he is or has been carrying on a trade, profession or business, his last known principal business address, or

(b) where the addressee is a body corporate, to the body's registered or principal office,

it shall be treated as effectively served.

(5) This section does not apply to the service of documents for the purposes of legal proceedings, for which provision is made by rules of court.

(6) References in this Part to a notice or other document include any form of communication in writing and references to giving or serving a notice or other document shall be construed accordingly.

Powers of court in relation to service of documents

77.(1) this section applies where service of a document on a person in the manner agreed by the parties, or in accordance with provisions of section 76 having effect in default of agreement, is not reasonably practicable.

(2) Unless otherwise agreed by the parties, the court may make such order as it thinks fit-

(a) for service in such manner as the court may direct, or

Arbitration Act 1996

Arbitration Act 1996

Part I

(b) dispensing with service of the document.

(3) Any party to the arbitration agreement may apply for an order, but only after exhausting any available arbitral process for resolving the matter.

(4) The leave of the court is required for any appeal from a decision of the court under this section.

Reckoning periods of time

78.(1) The parties are free to agree on the method of reckoning periods of time for the purposes of any provision agreed by them or any provision of this Part having effect in default of such agreement.

(2) If or to the extent there is no such agreement, periods of time shall be reckoned in accordance with the following provisions.

(3) Where the act is required to be done within a specified period after or from a specified date, the period begins immediately after that date.

(4) Where the act is required to be done a specified number of clear days after a specified date, at least that number of days must intervene between the day on which the act is done and that date.

(5) Where the period is a period of seven days or less which would include a Saturday, Sunday or a public holiday in the place where anything which has to be done within the period falls to be done, that day shall be excluded.

1971 c.80

In relation to England and Wales or Northern Ireland, a "public holiday" means Christmas Day, Good Friday or a day which under the Banking and Financial Dealings Act 1971 is a bank holiday.

Power of court to extend time limits relating to arbitral proceedings

79.(1) Unless the parties otherwise agree, the court may by order extend any time limit agreed by them in relation to any matter relating to the arbitral proceedings or specified in any provision of this Part having effect in default of such agreement.

This section does not apply to a time limit to which section 12 applies (power of court to extend time for beginning arbitral proceedings, &c).

(2) An application for an order may be made-

(a) by any party to the arbitral proceedings (upon notice to the other parties and to the tribunal), or

(b) by the arbitral tribunal (upon notice to the parties).

(3) The court shall not exercise its power to extend a time limit unless it is satisfied-

(a) that any available recourse to the tribunal, or to any arbitral or other institution or person vested by the parties with power in that regard, has first been exhausted, and

Part I

(b) that a substantial injustice would otherwise be done.

(4) The court's power under this section may be exercised whether or not the time has already expired.

(5) An order under this section may be made on such terms as the court thinks fit.

(6) The leave of the court is required for any appeal from a decision of the court under this section.

Notice and other requirements in connection with legal proceedings

80.(1) References in this Part to an application, appeal or other step in relation to legal proceedings being taken "upon notice" to the other parties to the arbitral proceedings, or to the tribunal, are to such notice of the originating process as is required by rules of court and do not impose any separate requirement.

(2) Rules of court shall be made-

(a) requiring such notice to be given as indicated by any provision of this Part, and

(b) as to the manner, form and content of any such notice.

(3) Subject to any provision made by rules of court, a requirement to give notice to the tribunal of legal proceedings shall be construed-

(a) if there is more than one arbitrator, as a requirement to give notice to each of them; and

(b) if the tribunal is not fully constituted, as a requirement to give notice to any arbitrator who has been appointed.

(4) References in this Part to making an application or appeal to the court within a specified period are to the issue within that period of the appropriate originating process in accordance with rules of court.

(5) Where any provision of this Part requires an application or appeal to be made to the court within a specified time, the rules of the court relating to the reckoning of periods, the extending or abridging of periods, and the consequences of not taking a step within the period prescribed by the rules, apply in relation to that requirement.

(6) Provision may be made by rules of court amending the provisions of this Part-

(a) with respect to the time within which any application or appeal to the court must be made,

(b) so as to keep any provision made by this Part in relation to arbitral proceedings in step with the corresponding provision of rules of court applying in relation to proceedings in the court, or

GENERAL INFORMATION

Arbitration Act 1996

Part I

(c) so as to keep any provision made by this Part in relation to legal proceedings in step with the corresponding provision of rules of court applying generally in relation to proceedings in the court.

(7) Nothing in this section affects the generality of the power to make rules of court.

Saving for certain matters governed by common law

81.(1) Nothing in this Part shall be construed as excluding the operation of any rule of law consistent with the provisions of this Part, in particular, any rule of law as to-

(a) matters which are not capable of settlement by arbitration;

(b) the effect of an oral arbitration agreement; or

(c) the refusal of recognition or enforcement of an arbitral award on grounds of public policy.

(2) Nothing in this Act shall be construed as reviving any jurisdiction of the court to set aside or remit an award on the ground of errors of fact or law on the face of the award.

Minor definitions 82.(1) In this Part-

"arbitrator", unless the context otherwise requires, includes an umpire;

"available arbitral process", in relation to any matter, includes any process of appeal to or review by an arbitral or other institution or person vested by the parties with powers in relation to that matter;

"claimant", unless the context otherwise requires, includes a counter claimant, and related expressions shall be construed accordingly;

"dispute" includes any difference;

"enactment" includes an enactment contained in Northern Ireland legislation;

"legal proceedings" means civil proceedings in the High Court or a county court;

"peremptory order" means an order made under section 41(5) or made in exercise of any corresponding power conferred by the parties;

"premises" includes land, buildings, moveable structures, vehicles, vessels, aircraft and hovercraft;

"question of law" means-

(a) for a court in England and Wales, a question of the law of England and Wales, and

(b) for a court in Northern Ireland, a question of the law of Northern Ireland;

"substantive jurisdiction", in relation to an arbitral tribunal, refers to the matters specified in

Arbitration Act 1996

Part I

section 30(1)(a) to (c), and references to the tribunal exceeding its substantive jurisdiction shall be construed accordingly.

(2) References in this Part to a party to an arbitration agreement include any person claiming under or through a party to the agreement.

Index of defined expressions: Part I

83. In this Part the expressions listed below are defined or otherwise explained by the provisions indicated-

agreement, agree and agreed	section 5(1)
agreement in writing	section 5(2) to (5)
arbitration agreement	sections 6 and 5(1)
arbitrator	section 82(1)
available arbitral process	section 82(1)
claimant	section 82(1)
commencement (in relation to arbitral proceedings)	section 14
costs of the arbitration	section 59
the court	section 105
dispute	section 82(1)
enactment	section 82(1)
legal proceedings	section 82(1)
Limitation Acts	section 13(4)
notice (or other document)	section 76(6)
party-	
- in relation to an arbitration agreement	section 82(2)
- where section 106(2) or (3) applies	section 106(4)
peremptory order	section 82(1) (and see section 41(5))
premises	section 82(1)
question of law	section 82(1)
recoverable costs	sections 63 and 64
seat of the arbitration	section 3
serve and service (of notice or other document)	section 76(6)
substantive jurisdiction (in relation to an arbitral tribunal)	section 82(1) (and see section 30(1)(a) to (c))
upon notice (to the parties or the tribunal)	section 80
written and in writing	section 5(6)

Transitional provisions

84.(1) The provisions of this Part do not apply to arbitral proceedings commenced before the date on which this Part comes into force.

(2) They apply to arbitral proceedings commenced on or after that date under an arbitration agreement whenever made.

(3) The above provisions have effect subject to any transitional provision made by an order under section 109(2) (power to include transitional provisions in commencement order).

Arbitration Act 1996

PART II

OTHER PROVISIONS RELATING TO ARBITRATION

Domestic arbitration agreements
(Note Sections 85 to 87 are not yet inforced)

Part II

Modification of Part I in relation to domestic arbitration agreement

85.(1)In the case of a domestic arbitration agreement the provisions of Part I are modified in accordance with the following sections.

(2) For this purpose a "domestic arbitration agreement" means an arbitration agreement to which none of the parties is-

(a) an individual who is a national of, or habitually resident in, a state other than the United Kingdom, or

(b) a body corporate which is incorporated in, or whose central control and management is exercised in, a state other than the United Kingdom,

and under which the seat of the arbitration (if the seat has been designated or determined) is in the United Kingdom.

(3) In subsection (2) "arbitration agreement" and "seat of the arbitration" have the same meaning as in Part I (see sections 3, 5(1) and 6).

Staying of legal proceedings

86.(1)In section 9 (stay of legal proceedings), subsection (4) (stay unless the arbitration agreement is null and void, inoperative, or incapable of being performed) does not apply to a domestic arbitration agreement.

(2) On an application under that section in relation to a domestic arbitration agreement the court shall grant a stay unless satisfied-

(a) that the arbitration agreement is null and void, inoperative, or incapable of being performed, or

(b) that there are other sufficient grounds for not requiring the parties to abide by the arbitration agreement.

(3) The court may treat as a sufficient ground under subsection (2)(b) the fact that the applicant is or was at any material time not ready and willing to do all things necessary for the proper conduct of the arbitration or of any other dispute resolution procedures required to be exhausted before resorting to arbitration.

(4) For the purposes of this section the question whether an arbitration agreement is a domestic arbitration agreement shall be determined by reference to the facts at the time the legal proceedings are commenced.

Effectiveness of agreement to exclude court's jurisdiction

87.(1)In the case of a domestic arbitration agreement any agreement to exclude the jurisdiction of the court under-

(a) section 45 (determination of preliminary point of law), or

(b) section 69 (challenging the award: appeal on point of law),

Arbitration Act 1996

Part II

is not effective unless entered into after the commencement of the arbitral proceedings in which the question arises or the award is made.

(2) For this purpose the commencement of the arbitral proceedings has the same meaning as in Part I (see section 14).

(3) For the purposes of this section the question whether an arbitration agreement is a domestic arbitration agreement shall be determined by reference to the facts at the time the agreement is entered into.

Power to repeal or amend sections 85 to 87

88.(1)The Secretary of State may by order repeal or amend the provisions of sections 85 to 87.

(2) An order under this section may contain such supplementary, incidental and transitional provisions as appear to the Secretary of State to be appropriate.

(3) An order under this section shall be made by statutory instrument and no such order shall be made unless a draft of it has been laid before and approved by a resolution of each House of Parliament.

Consumer arbitration agreements

Application of unfair terms regulations to consumer arbitration agreements
S.I.1994/3159

89.(1)The following sections extend the application of the Unfair Terms in Consumer Contracts Regulations 1994 in relation to a term which constitutes an arbitration agreement

For this purpose "arbitration agreement" means an agreement to submit to arbitration present or future disputes or differences (whether or not contractual).

(2) In those sections "the Regulations" means those regulations and includes any regulations amending or replacing those regulations.

(3) Those sections apply whatever the law applicable to the arbitration agreement.

Regulations apply where consumer is a legal person

90. The Regulations apply where the consumer is a legal person as they apply where the consumer is a natural person.

Arbitration agreement unfair where modest amount sought

91.(1)A term which constitutes an arbitration agreement is unfair for the purposes of the Regulations so far as it relates to a claim for a pecuniary remedy which does not exceed the amount specified by order for the purposes of this section.

(2) Orders under this section may make different provision for different cases and for different purposes.

(3) The power to make orders under this section is exercisable-

(a) for England and Wales, by the Secretary of State with the concurrence of the Lord

Arbitration Act 1996

Part II   Chancellor

(b) for Scotland, by the Secretary of State with the concurrence of the Lord Advocate, and

(c) for Northern Ireland, by the Department of Economic Development for Northern Ireland with the concurrence of the Lord Chancellor.

(4) Any such order for England and Wales or Scotland shall be made by statutory instrument which shall be subject to annulment in pursuance of a resolution of either House of Parliament.

S.I.1979/1573 (N.I.12)
1954 c.33 (N.I.)
(5) Any such order for Northern Ireland shall be a statutory rule for the purposes of the Statutory Rules (Northern Ireland) Order 1979 and shall be subject to negative resolution, within the meaning of section 41(6) of the Interpretation Act (Northern Ireland) 1954.

Small claims arbitration in the county court

Exclusion of Part I in relation to small claims arbitration in the county court
1984 c.28
92. Nothing in Part I of this Act applies to arbitration under section 64 of the County Courts Act 1984.

Appointment of judges as arbitrators

Appointment of judges as arbitrators
93.(1)A judge of the Commercial Court or an official referee may, if in all the circumstances he thinks fit, accept appointment as a sole arbitrator or as umpire by or by virtue of an arbitration agreement.

(2) A judge of the Commercial Court shall not do so unless the Lord Chief Justice has informed him that, having regard to the state of business in the High Court and the Crown Court, he can be made available.

(3) An official referee shall not do so unless the Lord Chief Justice has informed him that, having regard to the state of official referees' business, he can be made available.

(4) The fees payable for the services of a judge of the Commercial Court or official referee as arbitrator or umpire shall be taken in the High Court.

(5) In this section-

"arbitration agreement" has the same meaning as in Part I; and

1981 c.54
"official referee" means a person nominated under section 68(1)(a) of the Supreme Court Act 1981 to deal with official referees' business.

(6) The provisions of Part I of this Act apply to arbitration before a person appointed under this section with the modifications specified in Schedule 2.

Arbitration Act 1996

Part II   *Statutory arbitrations*

Application of Part I to statutory arbitrations
94.(1)The provisions of Part I apply to every arbitration under an enactment (a "statutory arbitration"), whether the enactment was passed or made before or after the commencement of this Act, subject to the adaptations and exclusions specified in sections 95 to 98.

(2) The provisions of Part I do not apply to a statutory arbitration if or to the extent that their application-

(a) is inconsistent with the provisions of the enactment concerned, with any rules or procedure authorised or recognised by it, or

(b) is excluded by any other enactment.

(3) In this section and the following provisions of this Part "enactment"-

1978 c.30
(a) in England and Wales, includes an enactment contained in subordinate legislation within the meaning of the Interpretation Act 1978;

1954 c.33 (N.I.)
(b) in Northern Ireland, means a statutory provision within the meaning of section 1(f) of the Interpretation Act (Northern Ireland) 1954.

General adaptations of provisions in relation to statutory arbitrations
95.(1)The provisions of Part I apply to a statutory arbitration-

(a) as if the arbitration were pursuant to an arbitration agreement and as if the enactment were that agreement, and

(b) as if the persons by and against whom a claim subject to arbitration in pursuance of the enactment may be or has been made were parties to that agreement.

(2) Every statutory arbitration shall be taken to have its seat in England and Wales or, as the case may be, Northern Ireland.

Specific adaptations of provisions in relation to statutory arbitrations
96.(1)The following provisions of Part I apply to a statutory arbitration with the following adaptations.

(2) In section 30(1) (competence of tribunal to rule on its own jurisdiction), the reference in paragraph (a) to whether there is a valid arbitration agreement shall be construed as a reference to whether the enactment applies to the dispute or difference in question.

(3) Section 35 (consolidation of proceedings and concurrent hearings) applies only so as to authorise the consolidation of proceedings, or concurrent hearings in proceedings, under the same enactment.

(4) Section 46 (rules applicable to substance of dispute) applies with the omission of subsection (1)(b) (determination in accordance with considerations agreed by parties).

Arbitration Act 1996

Part II

Provisions excluded from applying to statutory arbitrations

97. The following provisions of Part I do not apply in relation to a statutory arbitration-

   (a) section 8 (whether agreement discharged by death of a party);

   (b) section 12 (power of court to extend agreed time limits);

   (c) sections 9(5), 10(2) and 71(4) (restrictions on effect of provision that award condition precedent to right to bring legal proceedings).

Power to make further provision by regulations

98.(1)The Secretary of State may make provision by regulations for adapting or excluding any provision of Part I in relation to statutory arbitrations in general or statutory arbitrations of any particular description.

   (2) The power is exercisable whether the enactment concerned is passed or made before or after the commencement of this Act.

   (3) Regulations under this section shall be made by statutory instrument which shall be subject to annulment in pursuance of a resolution of either House of Parliament.

Part III

PART III

RECOGNITION AND ENFORCEMENT OF CERTAIN FOREIGN AWARDS

Enforcement of Geneva Convention awards

Continuation of Part II of the Arbitration Act 1950

1950 c.27

99. Part II of the Arbitration Act 1950 (enforcement of certain foreign awards) continues to apply in relation to foreign awards within the meaning of that Part which are not also New York Convention awards

Recognition and enforcement of New York Convention awards

New York Convention awards

100.(1)In this Part a "New York Convention award" means an award made, in pursuance of an arbitration agreement, in the territory of a state (other than the United Kingdom) which is a party to the New York Convention.

   (2) For the purposes of subsection (1) and of the provisions of this Part relating to such awards-

   (a) "arbitration agreement" means an arbitration agreement in writing, and

   (b) an award shall be treated as made at the seat of arbitration, regardless of where it was signed, despatched or delivered to any of the parties.

In this subsection "agreement in writing" and "seat of the arbitration" have the same meaning

Part III

as in Part I.

   (3) If Her Majesty by Order in Council declares that a state specified in the Order is a party to the New York Convention, or is a party in respect of any territory so specified, the Order shall, while in force, be conclusive evidence of that fact.

   (4) In this section, "the New York Convention" means the Convention on the Recognition and Enforcement of Foreign Arbitral Awards adopted by the United Nations Conference on International Commercial Arbitration on 10th June 1958.

Recognition and enforcement of awards

101.(1)A New York Convention award shall be recognised as binding on the persons as between whom it was made, and may accordingly be relied on by those persons by way of defence, set-off or otherwise in any legal proceedings in England and Wales or Northern Ireland.

   (2) A New York Convention award may, by leave of the court, be enforced in the same manner as a judgement or order of the court to the same effect.

As to the meaning of "the court" see section 105.

   (3) Where leave is so given, judgement may be entered in terms of the award.

Evidence to be produced by party seeking recognition or enforcement

102.(1)A party seeking the recognition or enforcement of a New York Convention award must produce-

   (a) the duly authenticated original award or a duly certified copy of it, and

   (b) the original arbitration agreement or a duly certified copy of it.

   (2) If the award or agreement is in a foreign language, the party must also produce a translation of it certified by an official or sworn translator or by a diplomatic or consular agent.

Refusal of recognition or enforcement

103.(1)Recognition or enforcement of a New York Convention award shall not be refused except in the following cases.

   (2) Recognition or enforcement of the award may be refused if the person against whom it invoked proves-

   (a) that a party to the arbitration agreement was (under the law applicable to him) under some incapacity;

   (b) that the arbitration agreement was not valid under the law to which the parties subjected it or, failing any indication thereon, under the law of the country where the award was made;

   (c) that he was not given proper notice of the appointment of the arbitrator or of the arbitration proceedings or was otherwise unable to present his case;

Arbitration Act 1996

Part III

(d) that the award deals with a difference not contemplated by or not falling within the terms of the submission to arbitration or contains decisions on matters beyond the scope of the submission to arbitration (but see subsection (4));

(e) that the composition of the arbitral tribunal or the arbitral procedure was not in accordance with the agreement of the parties or, failing such agreement, with the law of the country in which the arbitration took place;

(f) that the award has not yet become binding on the parties, or has been set aside or suspended by a competent authority of the country in which, or under the law of which, it was made.

(3) Recognition or enforcement of the award may also be refused if the award is in respect of a matter which is not capable of settlement by arbitration, or if it would be contrary to public policy to recognise or enforce the award.

(4) An award which contains decisions on matters not submitted to arbitration may be recognised or enforced to the extent that it contains decisions on matters submitted to arbitration which can be separated from those on matters not so submitted.

(5) Where an application for the setting aside or suspension of the award has been made to such a competent authority as is mentioned in subsection (2)(f), the court before which the award is sought to be relied upon, may if it considers it proper, adjourn the decision on the recognition or enforcement of the award.

It may also on the application of the party claiming recognition or enforcement of the award order the other party to give suitable security.

Saving for other bases of recognition or enforcement **104** Nothing in the preceding provisions of this Part affects any right to rely upon or enforce a New York Convention award at common law or under section 66.

PART IV

GENERAL PROVISIONS

Meaning of "the court": jurisdiction of High Court and county court **105**.(1)In this Act "the court" means the High Court or a county court, subject to the following provisions.

(2) The Lord Chancellor may be order make provision-

(a) allocating proceedings under this Act to the High Court or to county courts; or

(b) specifying proceedings under this Act which may be commenced or taken only in the High Court or in a county court.

Arbitration Act 1996

Part IV

(3) The Lord Chancellor may be order make provision requiring proceedings of any specified description under this Act in relation to which a county court has jurisdiction to be commenced or taken in one or more specified county courts.

Any jurisdiction so exercisable by a specified county court is exercisable throughout England and Wales or, as the case may be, Northern Ireland.

(4) An order under this section-

(a) may differentiate between categories of proceedings by reference to such criteria as the Lord Chancellor sees fit to specify, and

(b) may make such incidental or transitional provision as the Lord Chancellor considers necessary or expedient.

(5) An order under this section for England and Wales shall be made by statutory instrument which shall be subject to annulment in pursuance of a resolution of either House of Parliament.

.S.I. 1979/1573 (N.I. 12)

1946 c.36

(6) An order under this section for Northern Ireland shall be a statutory rule for the purposes of the Statutory Rules (Northern Ireland) Order 1979 which shall be subject to annulment in pursuance of a resolution of either House of Parliament in like manner as a statutory instrument and section 5 of the Statutory Instruments Act 1946 shall apply accordingly.

Crown application **106**.(1)Part I of this Act applies to any arbitration agreement to which Her Majesty, either in right of the Crown or of the Duchy of Lancaster or otherwise, or the Duke of Cornwall is a party.

(2) Where Her Majesty is party to an arbitration agreement otherwise than in right of the Crown, Her Majesty shall be represented for the purposes of any arbitral proceedings-

(a) where the agreement was entered into by Her Majesty in right of the Duchy of Lancaster, by the Chancellor of the Duchy or such person as he may appoint, and

(b) in any other case, by such person as Her Majesty may appoint in writing under the Royal Sign Manual.

(3) Where the Duke of Cornwall is party to an arbitration agreement, he shall be represented for the purposes of any arbitral proceedings by such person as he may appoint.

(4) References in Part I to a party or the parties to the arbitration agreement or to arbitral proceedings shall be construed, where subsection (2) or (3) applies, as references to the person representing Her Majesty or the Duke of Cornwall.

Arbitration Act 1996

Arbitration Act 1996

Part IV

Consequential amendments and repeals

107.(1)The enactments specified in Schedule 3 are amended in accordance with that Schedule, the amendments being consequential on the provisions of this Act.

(2) The enactments specified in Schedule 4 are repealed to the extent specified.

Extent

108.(1)The provisions of this Act extend to England and Wales and, except as mentioned below, to Northern Ireland.

(2) The following provisions of Part II do not extend to Northern Ireland-

section 92 (exclusion of Part I in relation to small claims arbitration in the county court), and

section 93 and Schedule 2 (appointment of judges as arbitrators).

(3) Sections 89, 90 and 91 (consumer arbitration agreements) extend to Scotland and the provisions of Schedules 3 and 4 (consequential amendments and repeals) extend to Scotland so far as they relate to enactments which so extend, subject as follows.

1975 c.3.

(4) The repeal of the Arbitration Act 1975 extends only to England and Wales and Northern Ireland.

Commencement **109**.(1)The provisions of this Act come into force on such day as the Secretary of State may appoint by order made by statutory instrument, and different days may be appointed for different purposes.

(2) An order under subsection (1) may contain such transitional provisions as appear to the Secretary of State to be appropriate.

Short title

110. This Act may be cited as the Arbitration Act 1996.

Please note: **THE SCHEDULES TO THE ARBITRATION ACT ARE NOT REPRODUCED HERE.**

GENERAL INFORMATION

The Housing Grants, Construction and Regeneration Act 1996

This act, provides a new right to fast impartial ADJUDICATION usually within 28 days.

The Act requires relevant contracts to provide for-

Adjudication for disputes which may be binding until the matter is resolved by litigation, arbitration or agreement.

Periodic payments for contracts greater than 45 days.

Notice to be given of intention to withhold payments.

A right to suspend performance for non payment.

The "Scheme for Construction Contracts" will provide for the matters raised by the act including supplementing contracts that do not provide for the Act.

The following extracts from the Act should be noted-

PART II

CONSTRUCTION CONTRACTS

Introductory provisions

104. Construction contracts.
105. Meaning of "construction operations".
106. Provisions not applicable to contract with residential occupier.
107. Provisions applicable only to agreements in writing.

Adjudication

108. Right to refer disputes to adjudication.

Payment

109. Entitlement to stage payments.
110. Dates for payment.
111. Notice of intention to withhold payment.
112. Right to suspend performance for non-payment.
113. Prohibition of conditional payment provisions.

Supplementary provisions

114. The Scheme for Construction Contracts.
115. Service of notices, etc.
116. Reckoning periods of time.
117. Crown application.

Adjudication

Right to refer disputes to adjudication

108.(1) A party to a construction contract has the right to refer a dispute arising under the contract for adjudication under a procedure complying with this section.

For this purpose "dispute" includes any difference.

(2) The contract shall-

(a) enable a party to give notice at any time of his intention to refer a dispute to adjudication;

(b) provide a timetable with the object of securing the appointment of the adjudicator and referral of the dispute to him within 7 days of such notice;

(c) require the adjudicator to reach a decision within 28 days of referral or such longer period as is agreed by the parties after the dispute has been referred;

(d) allow the adjudicator to extend the period of 28 days by up to 14 days, with the consent of the party by whom the dispute was referred;

(e) impose a duty on the adjudicator to act impartially; and

(f) enable the adjudicator to take the initiative in ascertaining the facts and the law.

(3) The contract shall provide that the decision of the adjudicator is binding until the dispute is finally determined by legal proceedings, by arbitration (if the contract provides for arbitration or the parties otherwise agree to arbitration) or by agreement.

The parties may agree to accept the decision of the adjudicator as finally determining the dispute.

(4) The contract shall also provide that the adjudicator is not liable for anything done or omitted in the discharge or purported discharge of his functions as adjudicator unless the act or omission is in bad faith, and that any employee or agent of the adjudicator is similarly protected from liability.

(5) If the contract does not comply with the requirements of subsections (1) to (4), the adjudication provisions of the Scheme for Construction Contracts apply.

(6) For England and Wales, the Scheme may apply the provisions of the Arbitration Act 1996 with such adaptations and modifications as appear to the Minister making the scheme to be appropriate.

For Scotland, the Scheme may include provision conferring powers on courts in relation to adjudication and provision relating to the enforcement of the adjudicator's decision.

1996 c.23.

Tables and Memoranda

AVERAGE COVERAGE OF PAINTS

The following information has been provided by the British Decorators Association, whose permission to publish is hereby acknowledged.

In this revision a range of spreading capacities is given. Figures are in square metres per litre, except for oil-bound water paint and cement-based paint which are in square metres per kilogram.

For comparative purposes figures are given for a single coat, but users are recommended to follow manufactures' recommendations as to when to use single or multicoat systems.

It is emphasised that the figures quoted in the schedule are practical figures for brush application, achieved in big scale painting work and take into account losses and wastage. They are not optima figures based upon ideal conditions of surface, nor minima figures reflecting the reverses of these conditions.

There will be instances when the figures indicated by paint manufacturers in their literature will be higher than those shown in the schedule. The Committee realize that under ideal conditions of application, and depending on such factors as the skill of the applicator, type and quality of the product, substantially better covering figures can be achieved.

The figures given below are for application by brush and to appropriate systems on each surface. They are given for guidance and are qualified to allow for variation depending on certain factors.

SCHEDULE OF AVERAGE COVERAGE OF PAINTS IN SQUARE METRES PER LITRE

Coating per litre	Finishing plaster	Wood floated ren-dering	Smooth concrete/ cement	Fair faced brick-work	block-work	* Roughcast pebble-dash	Hardboard	Surfaces Soft fibre insulating board
Water thinned primer/ undercoat								
as primer........................	13-15	-	-	-	-	-	10-12	7-10
as undercoat....................	-	-	-	-	-	-	-	10-12
Plaster primer (including building board).................	9-11	8-12	9-11	7-9	5-7	2-4	8-10	7-9
Alkali resistant primer.......	7-11	6-8	7-11	6-8	4-6	2-4	-	-
External wall primer sealer.	6-8	6-7	6-8	5-7	4-6	2-4	-	-
Undercoat........................	11-14	7-9	7-9	6-8	6-8	3-4	11-14	10-12
Oil based thixotropic finish................................	Figures should be obtained from individual manufacturers							

AVERAGE COVERAGE OF PAINTS

SCHEDULE OF AVERAGE COVERAGE OF PAINTS IN SQUARE METRES PER LITRE (cont'd)   Surfaces

Coating per litre	Finishing plaster	Wood floated ren-dering	Smooth concrete/ cement	Fair faced brick-work	block-work	Roughcast pebble-dash *	Hardboard	Soft fibre insulating board
Eggshell/semi-gloss finish (oil based)	11-14	9-11	11-14	8-10	7-9	-	10-13	10-12
Acrylic eggshell	11-14	10-12	11-14	8-11	7-10	-	11-14	10-12
Emulsion paint standard	12-15	8-12	11-14	8-12	6-10	2-4	12-15	8-10
contract	10-12	7-11	10-12	7-10	5-9	2-4	10-12	7-9
Glossy emulsion.	Figures should be obtained from individual manufacturers							
Heavy textured coating	2-4	2-4	2-4	2-4	2-4	-	2-4	2-4
Masonry paint	5-7	4-6	5-7	4-6	3-5	2-4	-	-
per kilogram Oil bound water paint	7-9	6-8	7-9	6-8	5-7	-	-	-
Cement based paint	-	4-6	6-7	3-6	3-6	2-3	-	-

Surfaces

Coating per litre	Fire retardent fibre insulating board	Smooth paper faced board	Hard asbestos sheet **	Struc-tural steel-work	Metal sheeting	Joinery	Smooth primed surfaces	Smooth under-coated surfaces
Woodprimer (oil based)	-	-	-	-	-	8-11	-	-
Water thinned primer/ undercoat as primer	-	8-11	* *	-	-	10-14	-	-
as undercoat	-	10-12	-	-	-	12-15	12-15	-
Aluminium sealer + spirit based	-	-	-	-	-	7-9	-	-
oil based	-	-	-	-	9-13	9-13	-	-
Metal primer conventional	-	-	-	7-10	10-13	-	-	-
specialised	Figures should be obtained from individual manufacturers							
Plaster primer (including building board)	8-10	10-12	10-12	-	-	-	-	-
Alkali resistant primer	-	-	8-10	-	-	-	-	-
External wall primer sealer.	-	-	6-8	-	-	-	-	-
Undercoat	10-12	11-14	10-12	10-12	10-12	10-12	11-14	-
Gloss finish	10-12	11-14	10-12	10-12	10-12	10-12	11-14	11-14
Eggshell/semi-gloss finish (oil based)	10-12	11-14	10-12	10-12	10-12	10-12	11-14	11-14
Acrylic eggshell	10-12	11-14	-	-	11-14	11-13	11-14	-
Emulsion paint standard	8-10	12-15	10-12	-	-	10-12	12-15	12-15
contract	-	10-12	8-10	-	-	10-12	10-12	10-12
Heavy textured coating	2-4	2-4	2-4	2-4	2-4	2-4	2-4	2-4
Masonry paint	-	-	5-7	-	-	-	6-8	6-8
per kilogram Oil bound water paint	-	-	7-9	-	-	-	-	-
Cement based paint	-	-	4-6	-	-	-	-	-
Glossy emulsion.	Figures should be obtained from individual manufacturers							

AVERAGE COVERAGE OF PAINTS (cont'd)

SCHEDULE OF AVERAGE COVERAGE OF PAINTS IN SQUARE METRES PER LITRE (cont'd)
Surfaces

+ Aluminium primer/sealer is normally used over "bitumen" painted surfaces.

* On some roughcast/pebbledash surfaces appreciably lower coverage may be obtained.

In many instances the coverages achieved will be affected by the suction and texture to the backing; for example the suction and texture of brickwork can vary to such an extent that coverages outside those quoted may on occasions be obtained.

It is necessary to take these factors into account when using this table.

** Owing to new legislation (COSHH) further advice regarding the encapsulation of asbestos should be sought.

APPROXIMATE WEIGHT OF SUNDRY MATERIALS

	lb per ft2	lb per ft3		lb per ft2	lb per ft3
Asbestos-cement sheeting			Partition slabs (hollow)		
1/4 in corrugated	3 1/4	-	2 in Coke breeze or pumice	9 1/4	-
1/4 in flat	2 1/4	-	2 in Clinker	11 1/2	-
Asbestos-cement slating			2 in Moler	2 1/2	-
Diamond	3	-	Plastering		
Rectangular	4	-	3/4 in Lime or gypsum	7 1/2	-
Asphalt 1 in thick	12	-	on lathing	8 3/4	-
Ballast, river	-	120	Roof boarding, 1 in	2 1/2	-
Bituminous felt roofing	1	-	Reconstructed stone	-	145
Brickwork, Commons	-	113			
			Sand, pit	-	90
Cement, Portland	-	90	river	-	120
Cement screeding (1:3) 1/2 in	6	-	Shingle	-	90
Concrete, cement			Slag wool	-	17
Ballast	-	140	Slate, 1 in slab	15	-
Brick	-	115	Slating, 3 in lap		
Clinker	-	95	Cornish (medium)	7 1/2	-
Pumice	-	70	Welsh (medium)	8 1/2	-
Reinforced (about 2% steel)		150	Westmorland (medium)	11 1/2	-
Cork slabs 1 in thick	1	-	Westmorland (thin)	9	-
Fibre boards 1/2 in thick	3/4	-	Terrazzo pavings 5/8 in	7	-
Compressed 1/4 in thick	2/3	-			
			Tiling, roof		
Fibrous plaster 5/8 in thick	3	-	Machine made 4 in gauge	13	-
Flooring			Hand made 4 in gauge	14	-
1 in Magnesium Oxychloride					
(sawdust filler)	7 1/2	-	Tiling floor (1/2 in)	5 3/4	-
			wall (3/8 in)	4	-
1 in Magnesium Oxychloride					
(mineral filler)	11 1/2	-	Timber, seasoned		
			Elm	-	39
1/4 in Rubber	2 3/4	-	Baltic Fir	-	35-38
1 in (Nominal) softwood	2 1/4	-	Red Pine	-	40
1 in (Nominal) pitchpine	3	-	Yellow Pine	-	28
1 in (Nominal) hardwood	3 1/4	-	Douglas Fir	-	33
			Canadian Spruce	-	29
			White Pine	-	27
Glass, 1/4 in plate	3 1/2	-	Yellow Birch	-	44
Gravel	-	115	Canadian Maple	-	47
			Honduras Mahogany	-	34
Lime, chalk	-	44	Spanish Mahogany	-	44
Partition slabs (solid)			English Oak	-	45
2 in Coke breeze or pumice	11 1/2	-	American Oak	-	47
2 in Clinker	15	-	Baltic Oak	-	46 2 2
in Terrazzo	25		Indian Teak	-	41
			African Teak	-	60
			Blackbean	-	40-47
			Water	-	62 1/2
			rain	-	62 1/2
			sea	-	64

GENERAL INFORMATION

WEIGHTS OF VARIOUS METALS

Weight in 1lb per Square Foot

Inches	1/16	1/8	1/4	3/8	1/2	5/8	3/4	7/8	1
Steel	2.55	5.1	10.2	15.3	20.4	25.5	30.6	35.7	40.8
Wrought Iron	2.50	5.0	10.0	15.0	20.0	25.0	30.0	35.0	40.0
Cast iron	2.35	4.69	9.37	14.06	18.75	23.44	28.12	32.81	37.5
Brass	2.75	5.48	10.94	16.42	21.88	27.34	32.81	38.29	43.76
Copper	2.89	5.78	11.56	17.34	23.12	28.99	36.68	40.46	46.24
Cast lead	3.70	7.39	14.78	22.17	29.56	36.95	44.34	51.73	59.12

Weights of Steel Flat Bar in lb per Foot Lineal

Width Inches	Thickness 1/8	1/4	5/16	3/8	7/16	1/2	5/8	3/4	7/8	1
1	0.43	0.85	1.06	1.28	1.49	1.70	2.13	2.55	2.98	3.40
1 1/8	0.48	0.96	1.20	1.43	1.67	1.91	2.39	2.87	3.35	3.83
1 1/4	0.53	1.06	1.33	1.59	1.86	2.13	2.66	3.19	3.72	4.25
1 3/8	0.58	1.17	1.46	1.75	2.05	2.34	2.92	3.51	4.09	4.68
1 1/2	0.64	1.28	1.59	1.91	2.23	2.55	3.19	3.83	4.46	5.10
1 5/8	0.69	1.38	1.73	2.07	2.42	2.76	3.45	4.14	4.83	5.53
1 3/4	0.74	1.49	1.86	2.23	2.60	2.98	3.72	4.46	5.21	5.95
1 7/8	0.80	1.59	1.99	2.39	2.79	3.19	3.98	4.78	5.58	6.38
2	0.85	1.70	2.13	2.55	2.98	3.40	4.25	5.10	5.95	6.80
2 1/4	0.96	1.91	2.39	2.87	3.35	3.83	4.78	5.74	6.69	7.65
2 1/2	1.06	2.13	2.66	3.19	3.72	4.25	5.31	6.38	7.44	8.50
2 3/4	1.17	2.34	2.92	3.51	4.09	4.68	5.84	7.01	8.18	9.35
3	1.28	2.55	3.19	3.83	4.46	5.10	6.38	7.65	8.93	10.20
3 1/4	1.38	2.76	3.45	4.14	4.83	5.53	6.91	8.29	9.67	11.05
3 1/2	1.49	2.98	3.72	4.46	5.21	5.95	7.44	8.93	10.41	11.90
3 3/4	1.59	3.19	3.98	4.78	5.58	6.38	7.97	9.56	11.16	12.75
4	1.70	3.40	4.25	5.10	5.95	6.80	8.50	10.20	11.90	13.60
4 1/4	1.81	3.61	4.52	5.42	6.32	7.23	9.03	10.84	12.64	14.45
4 1/2	1.91	3.83	4.78	5.74	6.69	7.65	9.56	11.48	13.39	15.30
4 3/4	2.02	4.04	5.05	6.06	7.07	8.08	10.09	12.11	14.13	16.15
5	2.13	4.25	5.31	6.38	7.44	8.50	10.63	12.75	14.88	17.00
5 1/2	2.34	4.68	5.84	7.01	8.18	9.35	11.69	14.03	16.36	18.70
6	2.55	5.10	6.38	7.56	8.93	10.20	12.75	15.30	17.85	20.40
6 1/2	2.76	5.53	6.91	8.29	9.67	11.05	13.81	16.58	19.34	22.10
7	2.98	5.95	7.44	8.93	10.41	11.90	14.88	17.85	20.83	23.80
7 1/2	3.19	6.38	7.97	9.56	11.16	12.75	15.94	19.13	22.31	25.50
8	3.40	6.80	8.50	10.20	11.90	13.60	17.00	20.40	23.80	27.20
9	3.83	7.65	9.56	11.48	13.39	15.30	19.13	22.95	26.78	30.60
10	4.25	8.50	10.63	12.75	14.88	17.00	21.25	25.50	29.75	34.00
11	4.68	9.35	11.69	14.03	16.36	18.70	23.38	28.05	32.73	37.40
12	5.10	10.20	12.75	15.30	17.85	20.40	25.50	30.60	35.70	40.80

Weights of Steel Round and Square Bars in lb per Foot Lineal

Inches	Round	Square	Inches	Round	Square	Inches	Round	Square
1/8	0.042	0.053	1 3/8	5.049	6.428	4	42.73	54.40
3/16	0.094	0.120	1 1/2	6.008	7.650	4 1/4	48.23	61.41
1/4	0.167	0.213	1 5/8	7.051	8.978	4 1/2	54.07	68.85
5/16	0.261	0.332	1 3/4	8.178	10.412	4 3/4	60.25	76.71
3/8	0.376	0.478	1 7/8	9.388	11.953	5	66.76	85.00
7/16	0.511	0.651	2	10.681	13.600	5 1/4	73.60	93.71
1/2	0.668	0.849	2 1/8	12.06	15.35	5 1/2	80.78	102.85
9/16	0.845	1.076	2 1/4	13.52	17.21	5 3/4	88.29	112.41
5/8	1.043	1.328	2 3/8	15.06	19.18	6	96.13	122.40
11/16	1.262	1.607	2 1/2	16.69	21.25	6 1/4	104.31	138.81
3/4	1.502	1.912	2 5/8	18.40	23.43	6 1/2	112.82	143.65
13/16	1.763	2.245	2 3/4	20.19	25.71	6 3/4	121.67	154.88
7/8	2.044	2.603	2 7/8	22.07	28.10	7	130.85	-
15/16	2.347	2.988	3	24.03	30.60	7 1/2	150.21	-
1	2.670	3.400	3 1/4	28.21	35.91	8	170.90	-
1 1/8	3.380	4.303	3 1/2	32.71	41.65	9	216.00	-
1 1/4	4.172	5.312	3 3/4	37.55	47.81	10	267.00	-

WEIGHTS OF STEEL JOISTS TO BS 4: PART 1: 1980

inches	Size millimetres	Weight lb/ft	kg/m	inches	Size millimetres	Weight lb/ft	kg/m
10 x 8	254 x 203	55.0	81.85	5 x 4.5	127 x 114	18.0	26.79
10 x 4.5	254 x 114	25.0	37.20	4.5 x 4.5	114 x 114	18.0	26.79
8 x 6	203 x 152	35.0	52.09	4 x 4	102 x 102	15.5	23.07
6 x 5	152 x 127	25.0	37.20	3.5 x 3.5	89 x 89	13.0	19.35
5 x 4.5	127 x 114	20.0	29.76	3 x 3	76 x 76	8.5	12.65

WEIGHTS OF STEEL CHANNELS TO BS 4: PART 1: 1980

Size of channel in	mm	Weight lb/ft	kg/m	Size of channel in	mm	Weight lb/ft	kg/m
17 x 4	432 x 102	44	65.54	8 x 3.5	203 x 89	20	29.78
15 x 4	381 x 102	37	55.10	8 x 3	203 x 76	16	23.82
12 x 4	305 x 102	31	46.18	7 x 3.5	178 x 89	18	26.81
12 x 3.5	305 x 89	28	41.69	7 x 3	178 x 76	14	20.84
10 x 3.5	254 x 89	24	35.74	6 x 3.5	152 x 89	16	23.84
10 x 3	254 x 76	19	28.29	6 x 3	152 x 76	12	17.88
9 x 3.5	229 x 89	22	32.76	5 x 2.5	127 x 64	10	14.90
9 x 3	229 x 76	17.5	26.06				

WEIGHTS OF STRUCTURAL STEEL TEE BARS SPLIT FROM UNIVERSAL BEAMS

Size of Tee in	mm	lb/ft	Weight kg/m	Size of Tee in	mm	lb/ft	Weight kg/m
12 x 12	305 x 305	60.0	90	7 x 8	178 x 203	18.0	27
10 x 13.5	254 x 343	42.0	63	5.5 x 8	140 x 203	15.5	23
9 x 12	229 x 305	47.0	70	6.75 x 7	171 x 178	22.5	34
9 x 12	229 x 305	34.0	51	6.75 x 7	171 x 178	15.0	23
8.25 x 10.5	210 x 267	41.0	61	5 x 7	127 x 178	13.0	20
8.25 x 10.5	210 x 267	27.5	41	6.5 x 6	165 x 152	18.0	27
7 x 9	191 x 229	33.0	49	5 x 6	127 x 152	16.0	24
7 x 9	191 x 229	22.5	34	4 x 6	102 x 152	11.0	17
6 x 9	152 x 229	27.5	41	5.75 x 5	146 x 127	14.5	22
6 x 9	152 x 229	17.5	26	4 x 5	102 x 127	9.5	14
7 x 8	178 x 203	25.0	37	5.25 x 4	133 x 102	10.0	15

WEIGHTS OF STRUCTURAL STEEL TEE BARS SPLIT FROM UNIVERSAL COLUMNS

Size of Tee in	mm	Weight lb/ft	kg/m	Size of Tee in	mm	lb/ft	kg/m	Weight	
12 x 6	305 x 152	53.0	79	8 x 4	203 x 102			29.0	43
12 x 6	305 x 152	32.5	49	8 x 4	203 x 102			15.5	23
10 x 5	254 x 127	44.5	66	6 x 3	152 x 76	12.5		19	
10 x 5	254 x 127	24.5	37	6 x 3	152 x 76	7.85		12	

WEIGHTS OF STEEL EQUAL ANGLES TO B.S. 4848 PART 4: 1972

Size mm	Thickness mm	Weight kg/m	Size mm	Thickness mm	Weight kg/m
250 x 250	35	128.0	120 x 120	15	26.6
	32	118.0		12	21.6
	28	104.0		10	18.2
	25	93.6		8	14.7
200 x 200	24	71.1	110 x 110	16	25.8
	20	59.9		12	19.8
	18	54.2		10	16.7
	16	48.5		8	13.5
150 x 150	18	40.1	100 x 100	15	21.9
	15	33.8		12	17.8
	12	27.3		10	15.0
	10	23.0		8	12.2
130 x 130	16	30.8	90 x 90	12	15.9
	12	23.5		10	13.4
	10	19.8		8	10.9
	8	16.0		6	8.3

WEIGHTS OF STEEL UNEQUAL ANGLES TO B.S. 4848 PART 4: 1972
(except where marked *)

Size mm	Thickness mm	Weight kg/m	Size mm	Thickness mm	Weight kg/m
200 x 150	18	47.1	*137 x 102	9.5	17.3
	15	39.6		7.9	14.5
	12	32.0		6.4	11.7

GENERAL INFORMATION

WEIGHTS OF STEEL UNEQUAL ANGLES TO B.S. 484 PART 4: 1972 (cont'd)
(except where marked *)

Size mm	Thickness mm	Weight kg/m	Size mm	Thickness mm	Weight kg/m
200 x 100	15	33.7	125 x 75	12	17.8
	12	27.3		10	15.0
	10	23.0		8	12.2
				6.5	9.98
150 x 90	15	26.6	100 x 75	12	15.4
	12	21.6		10	13.0
	10	18.2		8	10.6
150 x 75	15	24.8	100 x 65	10	12.3
	12	20.2		8	9.94
	10	17.0		7	8.77

WEIGHTS OF UNIVERSAL BEAMS TO B.S. 4 PART 1: 1980

Size in	mm	Weight lb/ft	kg/m	Size in	mm	Weight lb/ft	kg/m
36 x 16.5	914 x 419	260	388	18 x 6	457 x 152	55	82
		230	343			50	74
						45	67
						40	60
						35	52
36 x 12	914 x 305	194	289	16 x 7	406 x 178	50	74
		170	253			45	67
		150	224			40	60
		135	201			36	54
33 x 11.5	838 x 292	152	226	16 x 5.5	406 x 140	31	46
		130	194			26	39
		118	176				
30 x 10.5	762 x 267	132	197	14 x 6.75	356 x 171	45	67
		116	173			38	57
		99	147			34	51
						30	45
27 x 10	686 x 254	114	170	14 x 5	356 x 127	26	39
		102	152			22	33
		94	140				
		84	125				
24 x 12	610 x 305	160	238	12 x 6.5	305 x 165	35	54
		120	179			31	46
		100	149			27	40
24 x 9	610 x 229	94	140	12 x 5	305 x 127	32	48
		84	125			28	42
		76	113			25	37
		68	101				
21 x 8.25	533 x 210	82	122	12 x 4	305 x 102	22	33
		73	109			19	28
		68	101			16.5	25
		62	92				
		55	82				
18 x 7.5	457 x 191	66	98	10 x 5.75	254 x 146	29	43
		60	89			25	37
		55	82			21	31
		50	74				
		45	67				
				10 x 4	254 x 102	19	28
						17	25
						15	22
				8 x 5.25	203 x 133	20	30
						17	25

WEIGHTS OF UNIVERSAL COLUMNS TO B.S. 4 PART 1: 1980

Size in	mm	Weight lb/ft	kg/m	Size in	mm	Weight lb/ft	kg/m
14 x 16	356 x 406	426	634	14 x 14.5	356 x 368	136	202
		370	551			119	177
		314	467			103	153
		264	393			87	129
		228	340				
		193	287				
		158	235				

Size in	mm	Weight lb/ft	kg/m	Size in	mm	Weight lb/ft	kg/m
12 x 12	305 x 305	109	283	10 x 10	254 x 254	112	167
		161	240			89	132
		133	198			72	107
		106	158			60	89
		92	137			49	73
		79	118				
		65	97				

Size	mm	lb/ft	kg/m	Size	mm	lb/ft	kg/m
8 x 8	203 x 203	58	86	6 x 6	152 x 152	25	37
		48	71			20	30
		40	60			15.7	23
		35	52				
		31	46				

BRITISH WEIGHTS AND MEASURES

2.25 in..............................	= 1 nail		**Lineal Measure**	
4 in..................................	= 1 hand		3 in.................................	= 1 palm
12 in................................	= 1 ft		9 in or 4 nails....................	= span, or 1/4 yd
5 quarters of yd...................	= 1 ell		3 ft or 4 quarters................	= 1 yd
16 ft 6 in or 5.5 yd...............	= 1 rod, pole or perch		6 ft or 2 yd.......................	= 1 fathom
220 yd or 40 poles.................	= 1 furlong		4 poles or 22 yd..................	= 1 chain
7.92 in.............................	= 1 link		1760 yd or 8 furlongs.............	= 1 mile = 5280 ft
10 chains...........................	= 1 furlong		100 links or 66 ft.................	= 1 chain
3 miles.............................	= 1 league		80 chains..........................	= 1 mile
6075.5 ft..........................	= 1 nautical mile		2027 yd............................	= 1 Admiralty knot
69.16 miles.........................	= 1 degree of longitude at Equator		69.04 miles........................	= 1 degree of latitude
The Cheshire pole..................	= 8 yd		43.08 miles........................	= 1 degree of longitude at London

144 in 2.............................	= 1 ft2		**Square Measure**	
9 ft2................................	= 1 yd2		10 square chains..................	= 1 acre
100 ft2..............................	= 1 square		30 acres..........................	= 1 yard land
272.25 ft2..........................	= 1 rod, pole or perch		100 acres..........................	= 1 hide
30.25 yd2...........................	= 1 rod, pole or perch		460 acres..........................	= 1 square mile
40 square rods.....................	= 1 rood		1 mile long by 80 chains........	= 1 square mile
4 roods or 4840 yd2.............	= 1 acre		7840 yd2..........................	= 1 Irish acre
43560 ft2...........................	= 1 acre		6084 yd2..........................	= 1 Scotch acre
67.7264 in2........................	= 1 square link		1 mile long x 1 chain wide.......	= 8 acres
10000 square links..............	= 1 square chain			

1728 in3............................	= 1 ft		**Solid or Cubic Measure**	
42 ft3 of timber..................	= 1 shipping ton		327 ft3.............................	= 1 yd3
165 ft3.............................	= 1 standard of wood		108 ft3.............................	= 1 stack of wood
			128 ft3.............................	= 1 cord of wood

16 drachms........................	= 1 ounce		**Avoirdupois Weight**	
16 ounces..........................	= 1 lb		20 cwt.............................	= 1 ton
28 lb...............................	= 1 qr cwt		7000 grains........................	= 1 lb avoirdupois
112 lb..............................	= 1 cwt		437.5 grains.......................	= 1 oz avoirdupois

24 grains..........................	= 1 dwt		**Troy Weight**	
20 dwts............................	= 1 ounce		5760 grains........................	= 1 lb troy
12 ounces..........................	= 1 lb		480 grains.........................	= 1 oz troy

Dry Measure

2 gal...............................	= 1 peck		6.232 gal..........................	= 1 ft3
8 gal...............................	= 1 bushel		168.264 gal........................	= 1 yd3
64 gal..............................	= 1 quarter		4.893 gal..........................	= 1 cyl ft

Liquid Measure

8.665 in2...........................	= 1 gill		4 gills.............................	= 1 pint
2 pints.............................	= 1 quart		4 quarts...........................	= 1 gal
1 gal...............................	= 277.25 in3 or 0.16 ft3		1 gal..............................	= 10 lb of distilled water
1 ft3...............................	= 6.232 gal		1 cwt water........................	= 1.8 ft3
1 bushel...........................	= 1.28 ft		38 gal.............................	= 1 bushel

Imperial gallons = 277.275 in3, and 10 lb of distilled water at 62 degrees F

GENERAL INFORMATION

HVAC
Engineer's Handbook 10ed

- **Concise, definitive reference for day-to-day work of the engineer or technician**
- **Covers the design of HVAC systems, domestic hot and cold water services, gas supply and steam services.**
- **Fifty years history – previously the *Handbook of Heating, Ventilating and Air Conditioning***

Changes made for this edition include additional material in the air conditioning section to provide data for recently introduced systems, and a further increase in the range of U-values to cover modern wall constructions and internal partitions. There is more information on expansion of gases and information has been added on replacement refrigerants for CFCs and HCFCs.

Rather than having to work through pages of theory to extract the required data in a usable form; the reader is given brief summaries of the various types of system which can be considered, together with the formulae, physical constants and typical design parameters needed for actual design and estimating. This information is backed up by a comprehensive bibliography and list of British, European and International Standards.

Hardback, 272pp, £50.00, 1995, isbn 0 7506 2594 5

Tables and Memoranda

SYMBOLS USED

Length	m = linear metre	Area	m2 = square metre
	mm = linear millimetre		mm2 = square millimetre
Volume	m3 = cubic metre	Mass	kg = kilogramme
	mm3 = cubic millimetre		g = gramme
Pressure	N/m2 = Newton per square metre	Density	kg/m = kilogrammes per linear metre
			kg/m3 = kilogrammes per cubic metre

CONVERSION FACTORS

To convert Imperial weights and measures into Metric or vice versa, multiply by the following conversion factors.

Length	Factor	Length	Factor
Inches to millimetres (mm)............................	25.4000	Millimetres to inches.............................	0.0394
Feet to metres (m)...	0.3048	Metres to feet.....................................	3.281
Yards to metres (m)......................................	0.9144	Metres to yards...................................	1.094
Area			
Square feet to square metres (m2)...............	0.0929	Square metres to square feet................	10.764
Square yards to ditto....................................	0.8361	Ditto to square yards...........................	1.196
Volume			
Cubic feet to cubic metres (m3)...................	0.0283	Cubic metres to cubic feet....................	35.315
Cubic yards to ditto......................................	0.7645	Ditto to cubic yards.............................	1.308
Mass			
Tons (2240 lbs) to tonnes............................	1.0161	Tonnes (1000 kg) to tons......................	0.9842
Cwts to kilogrammes....................................	50.802	Kilogrammes to cwts............................	0.0196
Pounds to kilogrammes (kg)........................	0.454	Kilogrammes to pounds.........................	2.205
Ounces to grammes (g)...............................	28.350	Grammes to ounces.............................	0.0353
Pressure			
Pounds per square inch to Newtons per square metre (N/mm2)..............................	6894.800	Newtons per square metre to pounds per square inch......................................	0.000145
Density			
Pounds per linear foot to kilogrammes per linear metre (kg/m).............................	1.488	Kilogrammes per linear metre to pounds per linear foot.............................	0.6720
Pounds per square foot to kilogrammes per metre (k/m2)...	4.882	Kilogrammes per square metre to pounds per square foot..........................	0.2048
Pounds per square yard to ditto..................	0.542	Ditto to pounds per square yard..............	1.845
Pounds per cubic foot to kilogrammes per cubic metre (kg/m3)....................................	16.019	Kilogrammes per cubic metre to pounds per cubic foot............................	0.0624
Capacity			
Gallons to litres..	4.546	Litres to gallons....................................	0.220

To convert items priced in Imperial units into Metric or vice versa, mulitply by the following conversion factors.

	£ Factor		£ Factor
Linear feet to linear metres........................	3.281	Linear metres to linear feet.......................	0.3048
Linear yards to linear metres......................	1.094	Linear metres to linear yards...................	0.9144
Square feet to square metres.....................	10.764	Square metres to square feet...................	0.0929
Square yards to square metres..................	1.196	Square metres to square yards...............	0.8361
Cubic feet to cubic metres.........................	35.315	Cubic metres to cubic feet.......................	0.028
Cubic yards to cubic metres	1.308	Cubic metres to cubic yards..................	0.7045
Pounds to kilogrammes..............................	2.2046	Kilogrammes to pounds...........................	0.454
Cwts to kilogrammes..................................	0.0196	Kilogrammes to cwts...............................	50.802
Cwts to tonnes..	19.684	Tonnes to cwts.......................................	0.051
Tons to tonnes..	0.9842	Tonnes to tons..	1.0161
Gallons to litres...	0.220	Litres to gallons......................................	4.546

METRIC EQUIVALENTS

Linear inches to linear millimetres   12 in = 304.80 millimetres

Inches	0	1	2	3	4	5	6	7	8	9	10	11
0		25.40	50.80	76.20	101.60	127.00	152.40	177.80	203.20	228.60	154.00	279.40
1/32	0.79	26.19	51.59	76.99	102.39	127.79	153.19	178.59	203.99	229.39	254.79	280.19
1/16	1.59	26.99	52.39	77.79	103.19	128.59	153.99	179.39	204.79	230.19	255.59	280.99
3/32	2.38	27.78	53.18	78.58	103.98	129.38	154.78	180.18	205.58	230.98	256.38	281.78
1/8	3.18	28.58	53.98	79.38	104.78	130.18	155.58	180.98	206.38	231.78	257.18	282.58
5/32	3.97	29.37	54.77	80.17	105.57	130.97	156.37	181.77	207.17	232.57	257.97	283.37
6/16	4.76	30.16	55.56	80.96	106.36	131.76	157.16	182.56	207.96	233.36	258.76	284.16
7/32	5.56	30.96	56.36	81.76	107.16	132.56	157.96	183.36	208.76	234.16	259.56	284.96
1/4	6.35	31.75	57.15	82.55	107.95	133.35	158.75	184.15	209.55	234.95	260.35	285.75
3/8	9.53	34.93	60.33	85.73	111.13	136.53	161.93	187.33	212.73	238.13	263.53	288.93
1/2	12.70	38.10	63.50	88.90	114.30	139.70	165.10	190.50	215.90	241.30	266.70	292.10
5/8	15.88	41.28	66.68	92.08	117.48	142.88	168.28	193.68	219.08	244.48	269.88	295.28
3/4	19.05	44.45	69.85	95.25	120.65	146.05	171.45	196.85	222.25	247.65	273.05	298.45
7/8	22.23	47.63	98.43	98.43	123.83	149.23	174.63	200.03	225.43	150.83	276.23	301.63

Linear feet and inches to linear millimetres

Feet	1	2	3	4	5	6	7	8	9	10	11	12
0 in	304.8	609.6	914.4	1219.2	1524.0	1828.8	2133.6	2438.4	2743.2	3048.0	3352.8	3657.6
1 in	330.2	635.0	939.8	1244.6	1549.4	1854.2	2159.0	2463.8	2768.6	3073.4	3378.2	2683.0
2 in	355.6	660.4	965.2	1270.0	1574.8	1879.6	2184.4	2489.2	2794.0	3098.8	3403.6	3708.4
3 in	381.0	385.8	990.6	1295.4	1600.2	1905.0	2209.8	2514.6	2819.4	3124.2	3429.0	3733.8
4 in	406.4	711.2	1016.0	1320.8	1625.6	1930.4	2235.2	2540.0	2844.8	3149.6	3454.4	3759.2
5 in	431.8	736.6	1041.4	1346.4	1651.0	1955.8	2260.6	2565.4	2870.2	3175.0	3479.8	3784.6
6 in	457.2	762.0	1066.8	1371.6	1676.4	1981.2	2286.0	2590.8	2895.6	3200.4	3505.2	3810.0
7 in	482.6	787.4	1092.2	1397.0	1701.8	2006.6	2311.4	2616.2	2921.0	3225.8	3530.6	3835.4
8 in	508.0	812.8	1117.6	1422.4	1727.2	2032.0	2336.8	2641.6	2946.4	3251.2	3556.0	3860.8
9 in	533.4	838.2	1143.0	1447.8	1752.6	2057.4	2362.2	2667.0	2971.8	3276.6	3581.4	3886.2
10 in	558.8	863.6	1168.4	1473.2	1778.0	2082.8	2387.6	2692.4	2997.2	3302.0	3606.8	3911.6
11 in	584.2	889.0	1193.8	1498.6	1803.4	2108.2	2413.0	2717.8	3022.6	3327.4	3632.2	3937.0

METRIC WEIGHTS OF MATERIALS

Material	kg/m3	Material	kg/m3
Aerated concrete..............................	800-960	Lime	
Aluminium		grey chalk, lump.................	700
cast.............................	2550	grey stone, lump..........................	800
rolled...........................	2700	hydrate, bags.......................	510
Asphalt		Macadam.....................	2100
natural..............................	1000	Mahogany	
paving..............................	2080	African.......................	560
Ballast, loose, graded........................	600	Honduras......................	540
Beech.........................	700	Spanish.......................	690
Birch		Marble........................	2590-2830
American..................................	640	Mortar	
yellow..............................	700	cement, set......................	1920-2080
Brass		lime, set......................	1600-1760
cast...........................	8330	Oak	
rolled...........................	8570	African..........................	960
Bricks (common burnt clay)		American red.................	720
stacked........................	1600-1920	English.........................	800-880
sand cement........................	1840	Padouk...........................	780
sand lime............................	2080	Paint	
ballast.........................	1200	aluminium....................	1200
brickwork.....................	1920	bituminous emulsion............	1120
Cement		red lead.........................	3120
bags...........................	1280	white lead......................	2800
bulk...........................	1281-1442	zinc.........................	2400
casks..........................	960	Pine	
Concrete (cement, plain)		American red.....................	530
brick aggregate........................	1840	British Columbian.................	530
clinker.........................	1440	Christina......................	690
stone ballast.........................	2240	Oregon.......................	530
Concrete (cement, reinforced)		Pitch..........................	650
1% steel..............................	2370	Plywood......................	480-640
2% steel..............................	2420	Polyvinyl chloride acetate.............	1200-1350
5% steel..............................	2580	Poplar........................	450
Copper		Portland cement	
cast...........................	8760	loose......................	1200-1360
drawn or sheet........................	8940	bags......................	1120-1280
Cork...........................	130-240	drums......................	1200
Deal, yellow........................	430	Redwood	
Ebony........................	1180-1330	American......................	530
Elm		Baltic......................	500
American.......................	670	non-graded...................	430
Dutch.......................	580	Rhodesian....................	910
English.......................	580	Slate	
Wych.......................	690	Welsh......................	2800
Fir		Westmorland....................	3000
Douglas.......................	530	Steel...........................	7830
Silver.......................	480	Stone	
Flint...........................	2560	Ancaster.....................	2500
Foam slag.........................	700	Bath.....................	2080
Freestone.........................	2243-4280	Darley Dale....................	2370
masonry, dressed......................	2400	Forest of Dean..................	2430
rubble.........................	2240	Granite.....................	2640
Glass		Hopton Wood...................	2530
bottle...........................	2720	Kentish rag...................	2670
flint, best...........................	3080	Mansfield....................	2260
flint, heavy...........................	5000-6000	Portland....................	2240
plate...........................	2800	Purbeck....................	2700
Granolithic...........................	2240	York....................	2240
Hardcore...........................	1920	Tarmacadam....................	2080
Hoggin...........................	1760	Teak	
Iroko...........................	600	Burma, African........................	650
Iron...........................		Walnut.......................	660
cast...........................	7200	Water	
malleable cast........................	7370-7500	fresh......................	1001
wrought........................	7690	salt......................	1009-1201
Lead		Whitewood	460
cast or rolled.........................	11300	Zinc	
Lime		cast...........................	6840
acetate or, bags......................	1280	rolled...........................	7190
blue lias, ground......................	850	sheets, packed.............	900
carbonate of, barrels.............	1280		
chloride of, lead-lined cases......	450		

GENERAL INFORMATION

WEIGHTS OF VARIOUS METALS

in kilogrammes per square metre

mm	in	Steel kg	Wt. iron kg	Cast iron kg	Brass kg	Copper kg	Cast Lead kg
1.59	1/16	12.45	12.21	11.47	13.43	14.11	18.07
3.18	1/8	24.90	24.41	22.90	26.76	28.22	36.08
6.35	1/4	49.80	48.82	45.75	53.41	56.44	72.16
9.53	3/8	74.70	73.24	68.65	80.17	84.66	108.24
12.70	1/2	99.60	97.65	91.55	106.83	112.88	144.32
15.88	5/8	124.50	122.06	144.44	133.49	141.54	180.41
19.05	3/4	149.40	146.47	137.29	160.19	169.32	216.49
22.23	7/8	174.30	170.88	160.19	186.95	197.54	252.57
25.40	1	199.20	195.30	183.09	213.65	225.76	288.65

WEIGHTS OF ROUND AND SQUARE STEEL BARS

Round mm	kg/m	lb/lin ft	Square kg/m	lb/lin ft	Round mm	kg/m	lb/lin ft	Square kg/m	lb/lin ft
6	0.222	0.149	0.283	0.190	20	2.466	1.657	3.139	2.110
8	0.395	0.265	0.503	0.338	25	3.854	2.590	4.905	3.296
10	0.616	0.414	0.784	0.527	32	6.313	4.243	8.035	5.400
12	0.888	0.597	1.130	0.759	40	9.864	6.629	12.554	8.437
16	1.579	1.061	2.010	1.351	50	15.413	10.358	19.617	13.183

WEIGHTS OF FLAT STEEL BARS

in kilogrammes per linear metre

Thickness in millimetres and inches

Width mm	in	3.18 (1/8 in)	6.35 (1/4 in)	7.94 (5/16 in)	9.53 (3/8 in)	11.11 (7/16 in)	12.70 (1/2 in)	15.88 (5/8 in)	19.05 (3/4 in)	22.23 (7/8 in)	25.40 (1 in)
25.40	1	0.64	1.27	1.58	1.91	2.22	2.53	3.17	3.80	4.44	5.06
28.58	1 1/8	0.71	1.43	1.79	2.13	2.49	2.84	3.56	4.27	4.99	5.70
31.75	1 1/4	0.79	1.58	1.98	2.37	2.77	3.17	3.96	4.75	5.54	6.33
34.93	1 3/8	0.86	1.74	2.17	2.60	3.05	3.48	4.35	5.22	6.09	6.97
38.10	1 1/2	0.95	1.91	2.37	2.84	3.32	3.80	4.75	5.70	6.64	7.59
41.28	1 5/8	1.03	2.05	2.58	3.08	3.60	4.11	5.13	6.16	7.19	8.23
44.45	1 3/4	1.10	2.22	2.77	3.32	3.87	4.44	5.54	6.64	7.75	8.86
50.80	2	1.27	2.53	3.17	3.80	4.44	5.06	6.33	7.59	8.86	10.12
57.15	2 1/4	1.44	2.84	3.56	4.27	4.99	5.70	7.11	8.54	9.96	11.39
63.50	2 1/2	1.58	3.17	3.96	4.75	5.54	6.33	7.90	9.50	11.07	12.68
69.85	2 3/4	1.74	3.48	4.35	5.22	6.09	6.97	8.69	10.43	12.17	13.92
76.20	3	1.91	3.80	4.75	5.70	6.64	7.59	9.50	11.39	13.29	15.18
82.55	3 1/4	2.05	4.11	5.13	6.16	7.19	8.23	10.28	12.34	14.39	16.44
88.90	3 1/2	2.22	4.44	5.54	6.64	7.75	8.86	11.07	13.29	15.49	17.71
95.25	3 3/8	2.37	4.75	5.92	7.11	8.30	9.50	11.86	14.23	16.61	18.97
101.60	4	2.53	5.06	6.33	7.59	8.86	10.12	12.65	15.18	17.71	20.24
107.95	4 1/4	2.69	5.37	6.73	8.07	9.41	10.76	13.44	16.13	18.81	21.50
114.30	4 1/2	2.84	5.70	7.11	8.54	9.96	11.39	14.23	17.08	19.93	22.77
120.65	4 3/4	3.01	6.01	7.52	9.02	10.52	12.02	15.02	18.02	21.03	24.03
127.00	5	3.17	6.33	7.90	9.50	11.07	12.65	15.82	18.97	22.14	25.30
139.70	5 1/2	3.48	6.97	8.69	10.43	12.17	13.92	17.40	20.88	24.35	27.83
152.40	6	3.80	7.59	9.50	11.39	13.29	15.18	18.97	22.77	26.56	30.36
165.10	6 1.2	4.11	8.23	10.28	12.34	14.39	16.44	20.55	24.67	28.78	32.89
177.80	7	4.44	8.86	11.07	13.29	15.49	17.71	22.14	26.56	31.00	35.42
190.50	7 1/2	4.75	9.50	11.86	14.23	16.61	18.97	23.72	28.47	33.20	37.95
203.20	8	5.06	10.12	12.65	15.18	17.71	20.24	25.30	30.36	35.42	40.48
228.60	9	5.70	11.39	14.23	17.08	19.93	22.77	28.47	34.15	39.85	45.54
254.00	10	6.73	12.65	15.82	18.97	22.14	25.30	31.62	37.95	44.27	50.50
279.40	11	6.97	13.92	17.40	20.88	24.35	27.83	34.79	41.74	48.71	55.66
304.80	12	7.59	15.18	18.97	22.77	26.56	30.36	37.95	45.54	53.13	60.72

Brands and Trade Names

+K - Anders + Kern UK Ltd
-Line - Desking Systems Ltd
AF Ltd - McQuay UK Ltd
ardee Spring & Lock Co Ltd - Aardee Locks and Shutters Ltd
aztec - Aaztec Cubicles
bacus - DA Systems Ltd
bacus Lighting Ltd - Abacus Lighting Ltd
banaki - Furmanite International Ltd
bbcol - The Abbseal Group incorporating Everseal (Thermovitrine Ltd)
bbey - Abbey Building Supplies Co
bbotsford - Brintons Ltd
bco Metal Refiner - ABCO Products Ltd.
bet - Viking Laminates Ltd
BG - Ingersoll-Rand European Sales Ltd
bingdon - Carpets International (UK) PLC
bloy - Abloy Security Ltd
bmags - Abloy Security Ltd
bratract - Carborundum Abrasives G.B Ltd
brobility - Gabriel & Co Ltd
broclamps - Gabriel & Co Ltd
brotube - Gabriel & Co Ltd
BS Floors - Optiroc Ltd
ABS Pumps - ABS Pumps Ltd
bSence/ PreSense - ECS Lighting Controls Ltd
btex - ABG Ltd
ABZ - ABB Flakt Products
AC - Howden Buffalo
cc-q-line - Metalrax Ltd
ccent - Caradon M K Electric Ltd
ccess - Ecophon Ltd
ccess - Gradus Ltd
ccess 1000 - CEM Systems Ltd
ccess 2000 - CEM Systems Ltd
ccessories - Ingersoll - Rand Architectural Hardware
cclaim - Dermide Ltd
cclaim - Dudley Thomas Ltd
ccoflex - Armstrong World Industries Ltd
ccolade - Gledhill Water Storage Ltd
ccolade - Jaga Heating Poducts (UK) Ltd
ccommodation - Checkmate Industries Ltd
ccrodal - Smithbrook Building Products Ltd
ccuro - Draeger Ltd
ACE - Masterbill Micro Systems Ltd
Acme - Eternit Clay Tiles Ltd
ACO - ACO Technologies Plc
Aco Drain - ACO Technologies Plc
ACO Fulbora - ACO Technologies Plc
ACO vinyl seal - ACO Technologies Plc
Acolade - Armstrong World Industries Ltd
Acorn - Oakdale (Contracts) Ltd
Acoustic Panel - Illbruck Ltd
Acoustic Vents - Simon R W Ltd
Acousticel M20AD - Sound Service (Oxford) Ltd
Acousticel R10 - Sound Service (Oxford) Ltd
Acousticurtain - Acousticabs Industrial Noise Control Ltd
Acoustifoam - Acousticabs Industrial Noise Control Ltd
Acoustilouvre - Acousticabs Industrial Noise Control Ltd
Acoustislab - Acousticabs Industrial Noise Control Ltd
Acrylacote - Dacrylate Paints Ltd
Actfast - Lycetts (Burslem) Ltd
Actimatic - Wade International (UK) Ltd
Activ-Ox - Feedwater Ltd
Activa - Sylvania Lighting International INC.
Acumina - Orsogril Sarl UK Department
Adams Rite - Permclose Group PLC
Adanced counters & Serveries - Viscount Catering Ltd
Adaptalok - Adaptaflex Ltd
Adaptaring - Adaptaflex Ltd
Adaptaseal - Adaptaflex Ltd
Adaptasteel - Adaptaflex Ltd
Addacrete - Addagrip Surface Treatments Ltd
Addaflor - Addagrip Surface Treatments Ltd
Addagrip1000 System - Addagrip Surface Treatments Ltd
Addalevel - Addagrip Surface Treatments Ltd
Addaprime - Addagrip Surface Treatments Ltd
Addastone - Addagrip Surface Treatments Ltd
Adesilex Range - Mapei UK Ltd
Adjustabars - Badderley Rose Ltd
Adjustagate - Badderley Rose Ltd
Adjustex - Exitex Ltd

Admiral - Guest & Chrimes Ltd
Admix - Sheardown Engineering Ltd
Admont - McLoughlin Wood Ltd
Adorn - Valor Heating
ADPRD - Vision Sytems (Europe) Ltd
Adria - Armstrong World Industries Ltd
ADT Modern - ADT Modern Security Systems Ltd
Advance - Nilfisk Advance Ltd
Advance Gaffa - Advance Tapes International Ltd
Advance Sectional Buildings - UK Cabin Co
Advantage - Dauphin PLC
Advantage - Dixon Turner Wallcoverings Ltd
Advantage - Ecophon Ltd
Advantage Respirators - MSA (Britain) Ltd
Advisor & Infoboard - Glasdon Designs Ltd
AEG - Electrolux Domestic Appliances
AEG - Parkinson Cowan and Tricity Bendix Co, The
AEL Video - ADT Modern Security Systems Ltd
Aelrad - AEL
AEM - ABB Flakt Products
Aercon - Aercon Wiring Systems Ltd
Aero - Blakey John H Ltd
Aero hardboard - Wood International Agency
Aero-flor - Dex-o-Tex International Ltd
Aerocleve - Clearwater Plc
Aerodyne cowl - Brewer Metalcraft
Aerodyne range - Guest & Chrimes Ltd
Aeroguard - Manton Industrial Seals Ltd
Affinity - Tektura Plc
Affinity Disposable Nespinctod - MSA (Britain) Ltd
Afghan - Carpets of Worth Ltd
AFP - C-Tec (Computionics Ltd)
Aga - Aga-Rayburn, Glynwed Consumer & Construction Products
Aga-Cookware - Aga-Rayburn, Glynwed Consumer & Construction Products
Aga-Rayburn - Aga-Rayburn, Glynwed Consumer & Construction Products
Agastat - Thomas & Betts Ltd
Agc 200 - Stewart Wales, Somerville Ltd
Agglio Conglomerate Marble - Reed Harris - Reed Harris
Agrement certificates - British Board of Agrement
Agriseal - Kleeneze Sealtech Ltd
Agrob - Buchtal (U.K.) Limited
Agrob Buchtal - Buchtal (U.K.) Limited
Agru - Capper Pipe Systems Ltd
Aidalarm - Hoyles Fire & Safety Ltd
Ainsworth OSB - Wood International Agency
Ainsworth Pourform 107 - Wood International Agency
Ainsworth Pourform H00 - Wood International Agency
Air Marshall - Rapaway Energy Ltd
Air Miser - Rapaway Energy Ltd
Aircore-Core - Ryton's Building Products Ltd
Airdor - S & P Coil Products Ltd
AirDuct - Airflow (Nicoll Ventilators) Ltd
Airfix - Flamco UK Ltd
Airguard - Aquastat Ltd
Airline - Ves Andover Ltd
Airliner - Ryton's Building Products Ltd
Airspeed - Crosbie Coatings
Airspeed - Warner Howard Ltd
Airsprung Beds Ltd - Airsprung Beds Ltd
Airstar - Rapaway Energy Ltd
Airstrip - Simon R W Ltd
Airworld - Warner Howard Ltd
Akademy Classroom System - Portakabin Ltd
AL 50 - SG System Products Ltd
Alabastine - Polycell Products Ltd
Alag Aggregate - Lafarge Aluminates Ltd
Alan Butcher - Reinforced Plastic Products Limited
Alarmcomm - Alarmcomm Ltd
Alarming UK - UK Fire International Ltd
Alarmsense - Apollo Fire Detectors Ltd
Alba De-Fence Systems - Albion Fencing Ltd
Albany - Caradon M K Electric Ltd
Albany - Firth Carpets Ltd
Albany - Gloster Furniture Ltd
Albany Paints - Brewer C & Sons Ltd
Albany Range - Caradon Bathrooms Ltd
Albany Wallcoverings - Brewer C & Sons Ltd
Albi-Clear - Rentokil Initial U.K. Ltd.
Albi-guard - Rentokil Initial U.K. Ltd.
Albi-Max - Rentokil Initial U.K. Ltd.
Albi-Pruf - Rentokil Initial U.K. Ltd.
Albi-Steel 90 - Rentokil Initial U.K. Ltd.

Albion - Swedecor Ltd
Alcione - Orsogril Sarl UK Department
Alclad - Solair Ltd
Alcoclad - Alcover UK Ltd
Alcoglass - Alcover UK Ltd
Alcon - Baresford Pumps
Aldyl - Uponor Ltd
Aleonard - Smithbrook Building Products Ltd
Alertcall - C-Tec (Computionics Ltd)
Alexandra - BRB Industrial Services Ltd
Alexio Roofing & Building Co. Ltd - Alexio Roofing & Building Co Ltd
Alfab - The Safety Tread Ltd
Alfacryl - Alfas Industries Ltd
Alfacryl FR - Alfas Industries Ltd
Alfapower - MJ Electronics Services (International) Ltd
Alfas 'C' seal - Alfas Industries Ltd
Alfas Bond - Alfas Industries Ltd
Alfas Bond FR - Alfas Industries Ltd
Alfas Bono FR - Compriband Ltd
Alfas Flash & Seal - Alfas Industries Ltd
Alfas GBT - Alfas Industries Ltd
Alfas SGT - Alfas Industries Ltd
Alfas TMT - Alfas Industries Ltd
Alfas UVA - Alfas Industries Ltd
Alfasil - Alfas Industries Ltd
Alfatherm - AEL
Alfed - Aluminium Federation (Alfed) Ltd
Algopilot - Siemens Building Technologies Ltd, Cerberus Division
Algorex - Siemens Building Technologies Ltd, Cerberus Division
Aliflex - Flexible Ducting Ltd
Aliflex - Oldham Signs Ltd
Aligator ® - Marley Alutec Ltd
Alimaster Shelving - Beford & Soar Ltd
Alitherm - Smart Systems Ltd
Alkorflex - Alkor Draka Ltd
Alkorplan - Alkor Draka Ltd
All Abrasive Strip - The Safety Tread Ltd
Allart Deco - Allart, Frank, & Co Ltd
Allegra - Hansgrohe
Allegro - Erlau AG
Allgood Hardware - Allgood plc
Allgood Secure - Allgood plc
Allibert - Allibert (UK) Ltd
Allite - The Safety Tread Ltd
Allmat - Allmat (East Surrey) Ltd
Allweiler - New Haden Pumps Ltd
Alno - Alno (United Kingdom) Ltd
Alno 2000 - Alno (United Kingdom) Ltd
Alnova - Alno (United Kingdom) Ltd
Aloxite - Carborundum Abrasives G.B Ltd
Alpha - Premdor
Alphaline - ABG Ltd
Alpine Automation - Alpine Autimation Ltd.
Alpine Finish - Snowcem
Althon - Allpipe Ltd
Althon-Lite - Allpipe Ltd
Altis - Legrand Electric Ltd
Alto - Alto Cleaning Systems (UK) Ltd
Alton Green Houses - Compton Building Ltd
Alton Greenhouses - Phillip Ader Ltd
ALU - Osram Ltd
Alu-lift - Metreel Ltd
Alu-Steps - Quantum Profile Systems
Alucobond - Booth Muirie Ltd
Alucobond ACM - Technology Telford Ltd
Aluglaze - Panel Systems Ltd
Alulux - Brown F plc
Alumasc - Alumasc Exterior Building Products Ltd
Alumasc - Drainage Systems
Alumatit - Caldwell Hardware (UK) Ltd
Aluminium A Vent - Areco Roofing Supplies Company
Alutrim - Areco Roofing Supplies Company
Aluzink - SSAB Dobol Coated Steel Ltd
Alwitra - ICB (International Construction Bureau) Ltd
Ama - Kwikform UK Ltd
Ama Drainer - KSB Ltd
Ama Rex - KSB Ltd
Ama Rex - Kwikform UK Ltd
Ambassador - Metlex
Amber Booth - Designed for Sound Ltd
Amberclems - Ambersil Ltd
Amberklene - Ambersil Ltd

Ambi-Rad - Ambi-Rad Ltd
Ambiance - Gerland Ltd
Ambience - Helvar Electrosonic
Amcor - Amcor (Appliances) Ltd
Amdega - Amdega Ltd
Amercoat - Kenyon Industrial Paints & Adhesives
American Lincoln - Alto Cleaning Systems (UK) Ltd
American Sanders - Alto Cleaning Systems (UK) Ltd
Amerlock - Kenyon Industrial Paints & Adhesives
Ameron - Andrews J S (Coatings) Ltd
Amie - Tunstal Telecom Ltd
Amiran Anti-Reflective Glass - Schott Glass Ltd, Arcitectural Glass
Amphistoma - Adams- Hydraulics Ltd
AN2000 - McMillan Fire Alarm Systems Ltd
Analok Revetment - Ruthin Precast Concrete Ltd
Anaplast Sheeting - Anaplast Ltd
Ancastel Hard White - The Rare Stone Group Ltd
Ancaster - Realstone Ltd
Ancaster Hard White - Chilmark Quarries Ltd
Ancaster Hard White Limestone - Gregory Quarries Ltd, The
Ancaster Limestone - The Rare Stone Group Ltd
Ancaster Weatherbed - The Rare Stone Group Ltd
Ancaster Weatherbed Limestone - Gregory Quarries Ltd, The
Anchorlite slate - Forticrete Ltd
Anchorlite slate - Forticrete Roofing Products
Ancon - Ancon CCL Ltd
Ancon CCL - Ancon CCL Ltd
Anda-Crib - PHI Group Ltd
Andante - Kinsley-Cooke Furniture Ltd
Anders & Kern - Anders + Kern UK Ltd
Andersen ® - Andersen/ Black Millwork
Andersons - Andersons Ltd
Andrews - Andrews Sykes Hire Ltd
Andrews Air Conditioning - Andrews Sykes Hire Ltd
Andrews Heat for Hire - Andrews Sykes Hire Ltd
Andura - Andura Textured Masonry Coatings Ltd
Anglo - Caradon Bathrooms Ltd
Anglo Abrasives - Carborundum Abrasives G.B Ltd
Angolar - Provex Products Ltd
Animo - Provex Products Ltd
Anki - Sparkes K & L
Annabelle Classique - Brintons Ltd
Ansell - BRB Industrial Services Ltd
Anstone - Ibstock Building Products Ltd
Antel - Andrews J S (Coatings) Ltd
Anti Bird Line - Birdscarer Products
Anti Bird Matting - Birdscarer Products
Anti Bird Mesh - Birdscarer Products
Anti Bird Net - Birdscarer Products
Anti Nesting Cage - Birdscarer Products
Anti- Bird-Curtain - Birdscarer Products
Anti-Bird- Spices - Birdscarer Products
Antifreeze - Jimi-Heat Ltd
Antilla - Dryad Architectural Ltd
Antique Coloured Glass - Schott Glass Ltd, Arcitectural Glass
Antium - Decra Cubicles Ltd
Apa plywood - Wood International Agency
Apex - Vulcanite Limited
Apex Piling - Structural Bonding Cambridge Ltd
Apollo - Brierley B (Garstang) Ltd
Apollo - Guardall
Apollo - Iles Waste Systems
Apollo Architectural Facing Masonry - Lignacite (Brandon) Ltd
Apollo incinerators - Combustion Linings Ltd
Apollo Windows - Heywood Williams Architectural
Applause - Pegler Ltd
Appro - Robot (UK) Ltd
APT Barriers - APT Controls Ltd
APT Bollards - APT Controls Ltd
APTcard - APT Controls Ltd
Aption Partitions - Adex Storage Equipment Ltd
APTkey - APT Controls Ltd
Apton - Adex Storage Equipment Ltd
Aqua Blue Designs - Aqua-Blue Ltd
Aqua Drolics - Aqua Design Ltd
Aqua-epoxeel - IFT Supplies Ltd
Aqua-ifthane - IFT Supplies Ltd
Aqua-Lite - Naylor Drainage Ltd
Aquachill - J & E Hall Ltd
Aquadish - Jones of Oswestry Ltd
Aquaflex - Richard Hose Ltd

Aquaflow - Armorex Ltd
Aquagard - Marley Waterproofing
Aquagrip - Viking Johnson
Aquaguard - Aquastat Ltd
Aquahib - Aquastat Ltd
Aqualife - Tor Coatings Group
Aqualine - Leisure, Glynwed Consumer Foodservice Products Ltd
Aqualock Surface DPM - Isocrete Group Sales Ltd
Aquameter - Channel Electronics
Aquamix - Martin Orgee & Associates
Aquamixa - Aqualisa Products
Aquapac - HRS Hevac Ltd
Aquarian - Aqualisa Products
Aquarius - Guardall
Aquasave - HRS Hevac Ltd
Aquaseal - FEB Ltd
Aquaslique - Midland Stom Ltd
Aquaslot - Jones of Oswestry Ltd
Aquaspeed - Crosbie Coatings
Aquastone - Bollom J W & Co Ltd
Aquastream - Aqualisa Products
Aquastyle - Aqualisa Products
Aquatec - Loblite Ltd
Aquatech - ICI Dulux Trade Paints
Aquatique - Aqualisa Products
Aquatread - Tor Coatings Group
Aquavalve - Aqualisa Products
AR2000 - McMillan Fire Alarm Systems Ltd
Arabesque - Hille Ltd
Arbo - Adshead Ratcliffe & Co Ltd
Arbo Wessel Servais - Buchtal (U.K.) Limited
Arbocrylic - Adshead Ratcliffe & Co Ltd
Arbofoam - Adshead Ratcliffe & Co Ltd
Arbokol - Adshead Ratcliffe & Co Ltd
Arbomast - Adshead Ratcliffe & Co Ltd
Arborslot - Jones of Oswestry Ltd
Arboseal - Adshead Ratcliffe & Co Ltd
Arbosil - Adshead Ratcliffe & Co Ltd
Arbothane - Adshead Ratcliffe & Co Ltd
Arcade - Shackerley (Holdings) Group Ltd incorporating Designer ceramics
Arcal - Hanson Aggregates
Arch-Line - Angle Ring Co Ltd
Archco-Rigidon - Winn & Coales
Architecton - Gerland Ltd
Architectural - Protim Solignum Ltd
Architectural Concrete Ltd - Marble Mosaic Co Ltd, The
Architectural coping - Alifabs (Woking) Ltd
Architectural Masonry - Ibstock Building Products Ltd
Archliner - Lafarge Plasterboard Ltd
Archmaster® - Simpson Strong-Tie®
Archtec - Cavity Lock Systems Ltd
Arclime - Hanson Aggregates
Arclion - Warner H & Son Ltd
Arcus - Erlau AG
Arcus - Laidlaw Architectural Hardware
Arden lighting - Arden Manufacturing Birmingham Ltd
Ardenbrite - Tor Coatings Group
Ardicol - Ardex UK Ltd
Ardion - Ardex UK Ltd
Ardipox - Ardex UK Ltd
Ardit - Ardex UK Ltd
Arditex - Ardex UK Ltd
Ardu-flex - Ardex UK Ltd
Arducem - Ardex UK Ltd
Ardurapid - Ardex UK Ltd
Ardurit - Ardex UK Ltd
Areana Job Costing - Arena Software Ltd
Areco Polycarbonate - Areco Roofing Supplies Company
Arecotrim - Areco Roofing Supplies Company
Arena Accounts - Arena Software Ltd
Argenta - Imperial Machine Co Ltd
Argo - TSE Brownson
Argo-Braze - Johnson Matthey PLC - Metal Joining
Argosteel - NT Partition Systems
Argyll - Richard Hose Ltd
Arioso - Kinsley-Cooke Furniture Ltd
Arioso Joccata - Kinsley-Cooke Furniture Ltd
Ark - Vulcanite Limited
Arlon - Armstrong World Industries Ltd
Armacast baths - Armitage Shanks Ltd
Armacolour - Yonder Hill Ltd
Armada - Legrand Electric Ltd
Armadillo Products - Yonder Hill Ltd
Armaglaze - Yonder Hill Ltd
Armamesh - Yonder Hill Ltd
Armapod - Chubb Security Installations Ltd
Armatherm - Crosbie Coatings
Armbond - Armstrong (Concrete Blocks), Thomas, Ltd
Armitage Shanks - Barflow-LCA
Armor Gard - MSC Speciality Films (UK) Ltd
Armorcote - Armorex Ltd
Armordon - Attwater & Sons Ltd
Armorsol - Armorex Ltd
Armspan - Armfibre Ltd
Armstrong - Kitsons Insulation Products Ltd
Armstrong - Nevill Long Limited
Armstrong - New Forest Ceilings Ltd
Arnold Montrose - Denmans Montrose Ltd
Aro Pumps - Ingersoll-Rand European Sales Ltd
Arresta Rail - Unistrut Limited
Arrone - HOPPE (UK) Ltd
Arrow - Walls & Ceiling International Ltd
Arstyl - nmc (UK) Ltd
Artek - Astron Building Systems Commercial Intertech Ltd
Artemide - Quip Lighting, Div of Bruce & Macintyre Ltd
Artemis - Desking Systems Ltd
Artesit - Erlau AG
Artizan - Sovereign Brush Co Ltd, The
Artoleum - Forbo - Nairn Ltd
Asco - Asco Extinguishers Co Ltd
Ascott - Pouliot Designs
Ascott Designs - Desney Products
Ashbury - Paragon by Heckmondwike F B Ltd
Ashdown - Ibstock
Ashfab - Ash & Lacy Building Products Ltd
Ashfix - Ash & Lacy Building Products Ltd
Ashflow - Ash & Lacy Building Products Ltd
Ashgrid - Ash & Lacy Building Products Ltd
Ashjack - Ash & Lacy Building Products Ltd
Ashpark Old London Yellow - Cranleigh Brick & Tile Co Ltd
Ashtherm - Ash & Lacy Building Products Ltd
Ashworth Weyrite - Salter Weigh-Tronix
ASL-S - Stannah Lifts Ltd
Aslux - Yorian Garden Products
Aspiromatic - Docherty H Ltd
Aspirotor - Marflex Chimney Systems
ASSA - ASSA Ltd
Associated Akam - Associated Holdings Ltd
Associated Asphalt - Associated Holdings Ltd
Associated Construction - Associated Holdings Ltd
Associated facilities services - Associated Holdings Ltd
Asthetics - ARC Partitioning
Astoria - Muraspec Ltd
Astra - Guardall
Astra - Hometex Trading Ltd
Astra Miera - Guardall
Astracast - Astracast PLC
Astraclad - Stoakes Systems Ltd
Astralite Rooflights - Stoakes Systems Ltd
Astraroof - Stoakes Systems Ltd
Astrawall - Stoakes Systems Ltd
Astrawall SSG - Stoakes Systems Ltd
Astro - Zellweger Analytic Ltd, Sieger Division
Astron - Astron Building Systems Commercial Intertech Ltd
Astronet - Astron Building Systems Commercial Intertech Ltd
Astrotec - Astron Building Systems Commercial Intertech Ltd
AT ™ - Maestro International Ltd
Atelier Sedap - Optelma Lighting Ltd
Athena - James Fairley Athena
Athens - Metlex
Athlaprene - En-tout-cas plc
Athlon - Trespa UK Ltd
Athmer - Strand Hardware Ltd
Atlas - Cego Frameware Ltd
Atlas - T & D Plastech
Atlas (Smart & Brown) - Bernlite Ltd
Atlas Buildings - Atlas Ward Structures Ltd
Atlas Concorde - Swedecor Ltd
Atlas Copco - F. W. Harris & Co Ltd
Atlaz - Woolliscroft Tiles Ltd
Audiodor - Leaderflush & Shapland
Audiopath - PEL Services Limited
Audioscreen - Pilkington Plyglass
Audioscren - Pilkington Birmingham
Aura worksurfaces - Bushboard Parker Ltd
Auro - Top Office Equipment Ltd
Ausmark - Ramset Fasteners Ltd
Auto Flow - Hattersley Newman Hender Ltd
Auto M - Airmaster Engineering Ltd
Auto Seal - Exitex Ltd
Auto-Freway - Tuke & Bell Ltd
Autodor - Garador Ltd
Autoflow - Pillinger G C & Co (Engineers) Ltd
Autogas - Calor Gas Ltd
Autoglide - Cardale Doors Ltd
Automat - Deakin Davenset Rectifiers Ltd
Automet - Metalrax Ltd
Auton - Harrison Drape, Divsion of McKenie (UK)
Autosec - Gunnebo Mayor Ltd
Autowalk - O & K Escalators
Autowaste - Fordwater Pumping Supplies
Autowaste - Fordwater Pumping Supplies Ltd
Avalon - Legrand Electric Ltd
Avancia - Blyde-Barton Ltd
Avanti Partitions - Davies John Interiors Ltd
Avenue - Neptune Outdoor Furniture Ltd
Avenue - President Office Furniture Ltd
AVL-P - Stannah Lifts Ltd
AVL-S - Stannah Lifts Ltd
Avon Gutter - Dales Fabrications Ltd
AWS - Buchtal (U.K.) Limited
Axcess - Rackline Systems Storage Ltd
Axedstone - Hampton Stone Ltd
Axim - Permclose Group PLC
Axim-Parkside Group Ltd - Parkside Group Ltd, The
Axis - Ledu UK Ltd
Axjet - Howden Buffalo
Axor - Hansgrohe
B - Line - Glasdon Designs Ltd
B 3000 - IBP Conex
B&R - Power Breaker
B-Line - Bartlett Catering Equipment Ltd
B75 - Birchwood Concrete Products Ltd
Babcock Transformers - Babcock Transformers
Baby Ben - General Time Europe
Backarod - Fillcrete Ltd
Baco Alcan, now A Line Systems - Alframes Holdings Ltd
Badge Master - Cardkey Systems
Badger - Baggeridge Brick PLC
Baggeridge - Baggeridge Brick PLC
baggio - Zon International
Baka - Industrial Devices Ltd
Bal - Admiix AD1/GT1 - Norcros Adhesives Ltd
Bal - CEM Gold Star - Norcros Adhesives Ltd
Bal - CFT3 - Norcros Adhesives Ltd
Bal - Easypoxy - Norcros Adhesives Ltd
Bal - Epoxy LV - Norcros Adhesives Ltd
Bal - Floor Epoxy - Norcros Adhesives Ltd
Bal - Grout - Norcros Adhesives Ltd
Bal - Pourable Thik Bed - Norcros Adhesives Ltd
Bal - Rapidset - Norcros Adhesives Ltd
Bal - Supergrout - Norcros Adhesives Ltd
Bal - Wall Green/Blue/White Star - Norcros Adhesives Ltd
Bal- Wide Joint Grout - Norcros Adhesives Ltd
Bald worksurfaces - Bushboard Parker Ltd
Ball - Millar Dennis (1998) Ltd
Ball-O-Star Ball Valves - Klinger Fluid Instrumentation Ltd
Ball-O-Top - Klinger Fluid Instrumentation Ltd
Balmoral - Davant Products Ltd
Balmoral - Muraspec Ltd
Balmoral Porcelain - Balmoral Porcelain Limited/Longmead Ceramics
Baltic - Neptune Outdoor Furniture Ltd
Baltimore - Townscape Products Ltd
Banbro Classic Residential Doors - Birtley Building Products Ltd
Banbro Loft Access Doors - Birtley Building Products Ltd
Band IT - Kem Edwards Ltd
Band-seal - Naylor Drainage Ltd
Bandit - Brissco Equipment Ltd
Bandit - Surmak Products Ltd
Bandseal - Ashmead Buidling Supplies Ltd
Bandseal - Ashmead Building Supplies Ltd
Baraflor - Carpet Specifier Services Ltd
Barbican - Carpets of Worth Ltd
Barbican ® - Jackson H S & Son (Fencing) Ltd
Barclay Kellett - Stanhope Barclay Kellet Crown Division
Barco Pantile - Blyth William
Barcooler - Imperial Machine Co Ltd
Bardeau - Smithbrook Building Products Ltd
Bardoline - Onduline Building Products Ltd
Barduct BL - Barduct Ltd
Barduct CL - Barduct Ltd
Barduct ML - Barduct Ltd
Barduct UL - Barduct Ltd
Barduct XL - Barduct Ltd
Barkeller - Imperial Machine Co Ltd
Barlo - Diamond Merchants
Barnwood - Barnwood Shopfitting Ltd
Baron - Bartlett Catering Equipment Ltd
Barrier Floor Seal SB/30 - Tank Storage & Services Ltd
Barrier Foil - Thatching Advisory Services Ltd
Barrow Bold Roman - Sandtoft Roof Tiles Ltd
Bartangle - Barton Storage Systems Ltd
Bartender - Imperial Machine Co Ltd
Bartex - Bartoline Ltd
Bartisan - Clenaware Systems Ltd
Bartoline - Bartoline Ltd
Barton & Merton - Anthony de Grey Trellises
Bartrak - Barton Storage Systems Ltd
Bartspan - Barton Storage Systems Ltd
Barwil - Barber Wilsons & Co Ltd
Bas 2800+ - Satchwell Control Systems Ltd
Basepak - Flamco UK Ltd
Basic - Kermi (UK) Ltd
Bat - Expamet Building Products
Batchmatic - BSA Tools Ltd
Bathscreens by Coram - Coram (UK) Ltd
Batterywatch - Chloride Industrial Batteries Ltd
Bax- net - bespoke network - Baxall Ltd
Baxi Air Management - Actionair
Bayferrox - Bayer Plc
Baysilone - GE Bayer Silicones
BB1 Fastrack Cubicles - Bushboard Parker Ltd
BB2 Cubicles - Bushboard Parker Ltd
BB3 Cubicles - Bushboard Parker Ltd
BB4 Cubicles - Bushboard Parker Ltd
BBA - British Board of Agrement
Bc 8000 - IBP Conex
BCMA - British Carpet Manufacturers Association
BCT - Ferham Products
Beamform - Rom Ltd
Beany Block - Marshalls Mono Ltd
Beasalton Revetment - Ruthin Precast Concrete Ltd
Beaufort - Bassett & Findley Ltd
Beaufort - Neptune Outdoor Furniture Ltd
Beaumont Chimneys - Beaumont Ltd F E
Beauty of marble collection - Emsworth Fireplaces Ltd
Beaver - Hipkiss, H, & Co Ltd
Beaver Paints - Philip Johnstone Group Ltd
Becker - F. W. Harris & Co Ltd
Becker Acroma - Anderson Gibb & Wilson
Beckers Paints - Hamron Group
Beeline - Ward Bekker Sales Ltd
Beemul - Emusol Products (LEICS) Ltd
Belgravia - S & P Coil Products Ltd
BELL - British Electric Lamps Ltd
Bell Fireplaces - Bell & Co Ltd
Bell fires - Bell & Co Ltd
Bell Twist - Brintons Ltd
Bella - Kermi (UK) Ltd
Belle - Chippindale Plant Ltd
Bendix - Electrolux Domestic Appliances
Bendix - Parkinson Cowan and Tricity Bendix Co, The
Benga W - Jotun Henry Clark Ltd
Berber Twist - Brockway Carpets Ltd
Beresford - Baresford Pumps
Bergen - Loft Centre Products
Berkeley - Beeston Heating Group Ltd
Berkeley - Carpets of Worth Ltd
Berkeley - Crittall Steel Windows Ltd
Berkeley Variations - Carpets of Worth Ltd
Berkshire Multi - Charnwood Brick Group Ltd
Berry - Seaque Technologies Ltd
BESA - British Electrical Systems Association
Best In the World - Stockten Heath Forge (Caldwells) Ltd
Beta - Beta Lighting Ltd
Beta Naco - Air Diffusion Ltd
Betec - Beton Construction Ltd
Betec - Forma Lighting Ltd
Betoatlas - Grass Concrete Ltd
Betoatlas - Landscape Grass (Concrete) Ltd
Betoflor - Grass Concrete Ltd
Betoflor - Landscape Grass (Concrete) Ltd
Betojard - Grass Concrete Ltd
Betojard - Landscape Grass (Concrete) Ltd
Betokem - Optiroc Ltd
Betonap - Grass Concrete Ltd
Betonap - Landscape Grass (Concrete) Ltd
Betotitan - Landscape Grass (Concrete) Ltd
Beverley - Beeston Heating Group Ltd
Bexor - BRB Industrial Services Ltd
BFT - Crimegard Ltd
Biarritz - Pegler Ltd
Biasi - Diamond Merchants
BICC - Kem Edwards Ltd
Bicester Court Floors - Bicester Products Ltd
Bicester Play Wall System - Bicester Products Ltd
Bieri - Biral Pumps Ltd
Bifjet - Howden Buffalo
Biflo - Grohe Ltd
Big Ben - General Time Europe
Bighead ® - Bighead Bonding Fasteners Ltd
Bigwood Stokers - Hodgkinson Bennis Ltd
Bilco - Bilco UK Ltd
Binnlock Mesh Fencing - Binns Fencing Ltd
Binwall - Grass Concrete Ltd
Binwall - Landscape Grass (Concrete) Ltd
Bio-Rolls - PHI Group Ltd
Biocleve - Clearwater Plc
BioDisc ® - Klargester Environmental Engineering Ltd
Biomass Grease Traps - Progressive Product Developments Ltd
Biospiral - Clearwater Plc
Biostat - Tuke & Bell Ltd
Biotec - Entec (Polution Control) Ltd
Biral - Beeston Heating Group Ltd
Biral - Biral Pumps Ltd
Birco - Marshalls Mono Ltd
Birkdale Bathroom Suite - Ideal-Standard Ltd
Birthday - Hoskins Medical Equipment Ltd
Birtley Collection Garage Doors - Birtley Building Products Ltd
Birtley Galvanizing - Birtley Building Products
Bishore - Euroroof Ltd
Bisley - Beeston Heating Group Ltd
Bisley - Bisley Office Equipment
Bison Beams - Armstrong Concrete Products Ltd
Bitite - Callenders Ltd
Bitubond Cork - Schlegel Engineering
Bitufor - Bekaert Building Products
Bituline - Onduline Building Products Ltd
Bitusheet - Schlegel Engineering
Bituthene - Grace Services
Biwater - Biwater Industries Ltd
Biz - Builder's Iron & Zincwork Ltd
BK - Imperial Machine Co Ltd
Black Beauty - Valor Heating
Black Box Wipes - Deb Ltd
Black Ebony - Manders Paint Ltd
Blackfriar - Blackfriar Paints and Varnishes
Blackfriar - Parsons E. & Sons Ltd
Blackjack Square Spiral - Loft Centre Products
Blake - Tuke & Bell Ltd
Blakely Electrics - Blakley Electrics Ltd
Blanc de Bierges - Blanc de Bierges
Blanc de Bierges - Milner Delvaux Ltd
Blaw-Knox - Ingersoll-Rand European Sales Ltd
Blenheim - Adam Carpets Ltd
Blenheim - Bassett & Findley Ltd
Blockmaster - CAEC Howard (Holdings) Ltd
Blockmaster - ECC Building Products Ltd
Blockspan - Lees Richard Ltd
Blocnail - Glasgow Steel Nail Co Ltd
Blok-N-Mesh - Wade Building Services Ltd
BLP - BLP (Hamble) Ltd
BLP - BLP UK Ltd
BluBat - Junckers Limited
BLW - BLW Associates Ltd
BM 310 - Monarflex Ltd
Bn - Wilo Samson Pumps Ltd
Board) - Vencel Resil Ltd
Boardwalk - Gradus Ltd
BOB Stevensons - BOB Stevenson Ltd
Boda - Industrial Devices Ltd
Boge - F. W. Harris & Co Ltd
Boiler Crown Valves - Millar Dennis (1998) Ltd
Boilermate - Gledhill Water Storage Ltd
Bold roll - Forticrete Roofing Products
Bolderaja - Boardcraft Ltd
Bolivar - IBP Conex
Boltawall - Tektura Plc
Bomac Level Crossings - Tarmac Precast Concrete
Bomag - Chippindale Plant Ltd
Bond It - Avocet Hardware Ltd
Bonda - Bondaglass Voss Ltd
Bonda Adhesives - Bondaglass Voss Ltd
Bonda G4 Damp Seal - Bondaglass Voss Ltd

Bonda PU Adhesive - Bondaglass Voss Ltd
Bondax - Carpets of Worth Ltd
Bondcrete - Industrial Adhesives Ltd
Bondel - Boardcraft Ltd
Bondite - Trade Sealants
Bondwave - Flexible Reinforcements Ltd
Bonsack Doors - Regency Garage Door Services Ltd
Bonus by Coram - Coram (UK) Ltd
Bonus Plus by Coram - Coram (UK) Ltd
Booths - Booth Samuel & Co Ltd
Borems - Luxo UK Ltd
Bosch - Arrow Supply Company Ltd
Bosch - Crimegard Ltd
Bosch - Kem Edwards Ltd
Bosch - Robert Bosch Ltd
Boss - SGB Youngman
Boss White - BSS (UK) Ltd
Boulevard - Marshalls Mono Ltd
Boulton & Paul - Rugby Joinery
BowTie - Helifix Ltd
BQ the delph - Realstone Ltd
BQ+ - Masterbill Micro Systems Ltd
BRAD - NBS Services
Bradley - Relcross Ltd
Bradstone - ECC Building Products Ltd
Braemar suite - Armitage Shanks Ltd
Brahams - Bitmen Products Ltd
Bramah - Bramah Security Equipment Ltd
Brampton - Anthony de Grey Trellises
Brandy Crag - Burlington Slate Ltd
Brathay Blue/Black Slate - Kirkstone
Brava - Cego Frameware Ltd
Brazilian Pinc plywood - Wood International Agency
BRC Crack Control - BRC Building Products
Breathalyzer - Draeger Ltd
Brecon - Sashless Window Co Ltd
Breeam - BRE
Breeza - London Fan Company Ltd
Breezax - London Fan Company Ltd
Bretona - Tegral Building Materials Ltd
Bretshaw - Brettell & Shaw Ltd
Brewer - Brewer Metalcraft
Brick & Stone - Optiroc Ltd
Brickfill - Fillcrete Ltd
Brickforce - BRC Building Products
Brickhouse - Drainage Systems
Bricklifter - Benton Co Ltd Edward
Brickslot - ACO Technologies Plc
Bricktie - BRC Building Products
Bricktor - BRC Building Products
Bridge-Coat - Andura Textured Masonry Coatings Ltd
Bridgestone - Braithaite Engineers Ltd
Bridgman Doors - Bridgman Doors Ltd
Brierkrete - Brierley B (Garstang) Ltd
Briflo - Brickhouse
Briflor - Brickhouse
Brig - Applied Acoustics (a part of Henry Venables Limited)
Brig - Henry Venables Ltd
Brigadier - Richard Hose Ltd
Brillite - Witham Oil And Paint Ltd
Brimar Tanks - Brimar Plastic Fabrications Ltd
Brimax - Brickhouse
Bripave - Brickhouse
Briseal - Brickhouse
Brisol - British Solvent Oils, A Division of Careless
Bristlex - Exitex Ltd
Bristol Armour - Meggitt Lavmour Systems
Bristol Maid - Hospital Metalcraft Ltd
Bristowes - Bitmen Products Ltd
Brit Clips - Kem Edwards Ltd
Britannia - Diamond Merchants
Britannia - Marflow Eng Ltd
Britannia Extinguishers - UK Fire International Ltd
Britflux - Metal Walde Lamps - Sylvania Lighting International INC.
British Electric Lamps Ltd - British Electric Lamps Ltd
British Gypsum - Brikenden (Builders Merchants) Ltd
British Gypsum - Hatmet Ltd
British Gypsum - Kitsons Insulation Products Ltd
British Gypsum - Nevill Long Limited
British Monorail - Morris Mechanical Handling Ltd
Britlock - Sandtoft Roof Tiles Ltd
Briton - Ingersoll - Rand Architectural Hardware
Briton - Laidlaw Architectural Hardware
Briton - Laidlaw Architectural Hardware Ltd
Briton - Stanton plc
Britosterope - Orsogril Sarl UK Department
Britslate - Sandtoft Roof Tiles Ltd
Briwax - Bollom J W & Co Ltd
BRIX - Federation of Master Builders
Broadland Conservatories - Phillip Ader Ltd
Broadmead - Broadmead Products
Broadstairs Batten - Durabella Ltd
Broadstel - Brickhouse
Broadstone - Hamworthy Heating Ltd
Broadway - Atlas Stone Products
Broag-Remeha - Broag Ltd
Brodclad - Broderick Structures Limited
Brodeck - Broderick Structures Limited
Brodford - Broderick Structures Limited
Brodform - Broderick Structures Limited
Brodsteam - Broderick Structures Limited
Broflame - Bollom Fire Protection Ltd
Broflor - Bollom J W & Co Ltd
Brolac - Akzo Nobel Decorating Coatings Ltd
Brolac - Crown Trade Paints
Bromatex - Bollom J W & Co Ltd
Bromel - Bollom J W & Co Ltd
Bromgard - Feedwater Ltd
Bron-Lock - Booth Muirie Ltd
Bronz-Steps - Quantum Profile Systems
Brook Air Changer - Brook Airchanger
Brook vent - Brook Airchanger
Brookes - Deakin Davenset Rectifiers Ltd
Brooking - Clement Steel Windows
Brooks Roof Units - Matthews & Yates Ltd
Brosteel - Bollom Fire Protection Ltd
Broughton Moor - Burlington Slate Ltd
Broxap ® - Broxap & Corby Ltd
Broxley - Beeston Heating Group Ltd
Brummer - Clam-Brummer Ltd
Brunel (External bin) - Glasdon UK Ltd
Brunner Mould - Anderson Gibb & Wilson
Bruynzeel - Dexion Ltd
BSA - BSA Tools Ltd
Buccaneer - Brickhouse
Buchanan - Thomas & Betts Ltd
Buchtal - Buchtal (U.K.) Limited
Buckden - Anthony de Grey Trellises
Buckingham - Cape Boards Ltd
Buckingham - Firth Carpets Ltd
Buckingham - Metlex
Budget - Spur Shelving Ltd
Buflon - Hamron Group
Buggie - Manitou (Site Lift) Ltd
Buggiescopic - Manitou (Site Lift) Ltd
Buildaid - Armorex Ltd
Builders Plant - Kwikform UK Ltd
Building Bookshop - The Building Centre Group Ltd
Building Centre - The Building Centre Group Ltd
Bulb-Tite Rivet - SFS Stadler Ltd
Bulletin Board - Forbo - Nairn Ltd
Bulpack - Bullock & Driffill Ltd
Bunnie Dryers - Wandsworth Elecrtrical Ltd
Burley - Beeston Heating Group Ltd
Buroflex - Metalliform Products Plc
Bursting Stone - Burlington Slate Ltd
Bush Circulators - Turney Turbines (pumps) Ltd
Business Alert - Contract Journal Business Alert
Butterley - Hanson Brick Ltd
Butterley - Hanson Brick Ltd, Northern Regonal Sales Office
Button diamond - Lionweld Kennedy Ltd
Butzbach - Envirodoor Markus Ltd
Bvent Universal Gas Vent - Rite-Vent Ltd
BX2000 - Horstmann Timers & Controls Ltd
C & R - Accomodex
C - Bus - Clips Partitions Ltd
C Series - Tenby Industries Ltd
C+/C plywood - Wood International Agency
C.D.S. Car Deck Systems - Ferguson G A & Co Ltd
C.F.P - Westpile Ltd
C.R. window systems - BLW Associates Ltd
CAB-RAD - Property Mechanical Products Ltd
Cabaret - Qualitas Bathrooms
Cabeline Ambassador - Mita (UK) Ltd
Cabeline Classic - Mita (UK) Ltd
Cabeline Powertrack - Mita (UK) Ltd
Cable - Cable Lift Installations
Cable-Con - Aercon Wiring Systems Ltd
Cableline Ambassador Duo - Mita (UK) Ltd
Cablofil - Mita (UK) Ltd
Cabsys - Mita (UK) Ltd
Cadeby natural stone - Ibstock Building Products Ltd
Cadet - Blyde-Barton Ltd
Cadet cabinets - Glasdon Designs Ltd
Cadet Kerb - Redland Precast
Cadevza - Cego Frameware Ltd
Cadweld ® - Erico Europa (GB) Ltd
Cafco - Firebarrier Services Ltd
Caimi - Lesco Products Ltd
Cal Master - ABB Kent Taylor Ltd.
Calder - Calder Industrial Materials Ltd
Calderdale - Sandtoft Roof Tiles Ltd
Caldwells - Stockten Heath Forge (Caldwells) Ltd
Caleffi - Altecnic Ltd
Calibre - BRE
Callendrite - Callenders Ltd
Calypso - Eltron Chromalox
Calypso - LazyLawn
Cam-Jack Technology - Structural Bonding Cambridge Ltd
Camargue - Carpets of Worth Ltd
Camargue - Qualitas Bathrooms
Camborne - Eternit UK Ltd
Cambourne Fabrics - Interface Europe Ltd
Cambrian - Brickhouse
Cambridge Gazebo - Anthony de Grey Trellises
Camelot - Brintons Ltd
Camer - Anthony de Grey Trellises
Camer Rose Arch - Anthony de Grey Trellises
Camer Rose Walk - Anthony de Grey Trellises
Camfine Thermsaver - Daylight Centre
Camlok - Camlok Lifting Clamps Ltd
Campden Pitched Stone - Atlas Stone Products
Campden Stone - Atlas Stone Products
Camray - Boulter Boilers Ltd
Can-Corp - Bush Nelson PLC
Canal - Optelma Lighting Ltd
Canonbury - Canonbury Asphalte Co. Ltd.
Canply Cofi plywood - Wood International Agency
Canterbury - Atlas Stone Products
Canti-Bolt - Hi-Store Ltd
Canti-Clad - Hi-Store Ltd
Canti-Frame - Hi-Store Ltd
Canti-Guide - Hi-Store Ltd
Canti-Lec - Hi-Store Ltd
Canti-Lock - Hi-Store Ltd
Canti-Track - Hi-Store Ltd
Canti-Weld - Hi-Store Ltd
Canvas - Tektura Plc
Capaphenk - Cape Insulation Products Ltd
Capco - Capco Test Equipment Ltd
Cape - Firebarrier Services Ltd
Cape Acousticfloor - Cape Boards Ltd
Cape Boards - Kitsons Insulation Products Ltd
Capella - Ward Building Components Ltd
Capex - Exitex Ltd
Caplock - Ibstock Building Products Ltd
Capoplastic GP - Thermica Ltd
Cappit - Swish Building Products Ltd
Capricorn Range - Caradon Bathrooms Ltd
Capstat - Heat Trace Ltd
Capyt SS100 composition - Thermica Ltd
Carborundum Abrasives - Carborundum Abrasives G.B Ltd
Cardale - Cardale Doors Ltd
Cardale Doors - Regency Garage Door Services Ltd
Cardax - Vidionics Security Systems Ltd
Cardoc - Kent & Co (Twines) Ltd
Care LST - Hudevad Britain
Carefree Buffable - S. C. Johnson Professional
Carefree Emulsion - S. C. Johnson Professional
Carefree Eternum - S. C. Johnson Professional
Carefree Gloss Restorer - S. C. Johnson Professional
Carefree Maintainer - S. C. Johnson Professional
Carefree Satin - S. C. Johnson Professional
Carefree Speed Stripper - S. C. Johnson Professional
Carefree Stride 1000 - S. C. Johnson Professional
Carefree Stride 2000 - S. C. Johnson Professional
Carefree Stride 3000 - S. C. Johnson Professional
Carefree Stripper - S. C. Johnson Professional
Carefree Undercoat - S. C. Johnson Professional
Caremix - Altecnic Ltd
Carerscreen - Contour Showers Ltd
Caretaker III - Rapaway Energy Ltd
Cargo 2000 - Otis plc
Caribe - Holophane Europe Ltd
Carlite - British Gypsum Ltd
Carlton - Metlex
Carlton - Mita (UK) Ltd
Carlton - Tuke & Bell Ltd
Carlton Brick - Carlton Main Brickworks Ltd
Carnival - Gaskell Textiles Ltd
Carpet Industry Training Council - British Carpet Manufacturers Association
Carr - Carr Gymnasium Equipment Ltd
Carron Phoenix - Carron Phoenix Ltd
Carson Futuka - Paragon Business Furniture
Carson Motif - Paragon Business Furniture
Cas zips - Power Plastics Ltd
Casana - Air Improvement Centre Ltd
Cascade - Calomax (Engineers) Ltd
Cascamite - Humbrol Ltd
Cascamite - Wessex Resins & Adhesives Ltd
Casco-Tape - Wessex Resins & Adhesives Ltd
Cascofil - Humbrol Ltd
Cascophen - Humbrol Ltd
Cascophen - Wessex Resins & Adhesives Ltd
Cascorez - Humbrol Ltd
Cashel Rose Limestone - Kirkstone
Cashflow - Masterbill Micro Systems Ltd
Casing & Eenclosures - Alumasc Interior Building Product Limited
Casscom - Telelarmcare Ltd
CAST Rooflight - Clement Steel Windows
Castalia Filter Clear - Opella Ltd
Castalia Tap Range - Opella Ltd
Castell - Castell Safety International
Castell 150 - Lok - Castell Safety International
Castelli - Haworth UK Ltd
Castle AV1 - Leaderflush & Shapland
Castle Bricks - Castle Brick (Wales) Ltd
Castle Multicem - Castle Cement Ltd
Cathedral - Redland Precast
Cathedral Axminster - Mackay Carpets, Hugh
Catnic - Caradon Catnic Ltd
Catnic Classic - Caradon Catnic Ltd
Catnic Unique - Garador Ltd
Catseye - Reflecting Roadstuds Ltd
CATV & Telephony Cabinets - Climet Ltd
Caviarch - Cavity Trays Ltd
Cavicloser - Cavity Trays Ltd
Caviflash - Cavity Trays Ltd
Cavilintel - Cavity Trays Ltd
Caviroll - Cavity Trays Ltd
Cavitray - Cavity Trays Ltd
Cavity Trays of Yeovil - Cavity Trays Ltd
Cavity Wall Batts - Rockwool Ltd
Cavity Wallboard - Callenders Ltd
Cavivent - Cavity Trays Ltd
Caviweep - Cavity Trays Ltd
CCL - Contract Components Ltd
CD 500 - Pegler Ltd
CD 5517 - Pegler Ltd
CD 9000 Cameras - Baxall Ltd
CDI - USF Permutit/Ionpure
CDSP 9000 cameras - Baxall Ltd
CE SI Wall Tiles - Stokes R J & Co Ltd
Cedar Green Houses - Compton Building Ltd
Cedec footpath gravel - Civil Engineering Development Ltd
Cefndy - Cefndy Enterprises
Celadon - Euroroof Ltd
Celafelt - Davant Products Ltd
Celbor - Rentokil Ltd
Celbrite - Rentokil Ltd
Celbronze - Rentokil Ltd
Celco - Helvar Electrosonic
Celcure - Rentokil Ltd
Celeno - Orsogril Sarl UK Department
Celeste - Pegler Ltd
Celeste - Ubbink (UK) Ltd
Cellarguard - Trade Sealants
Celoflex - Celotex Ltd
Celpruf - Rentokil Ltd
Celtic Pantile - Blyth William
Celuform - Caradon Duraflex Systems Ltd
Celuform - Celuform
Cem-Grout - Sealocrete Ltd
Cemboard - Torvale Building Products, A Division of Stadium Group PLC
Cemglaze - Cement Glaze Decorators Ltd
Cemlevel - Armorex Ltd
Cemprover - Snowcem
Cemrend - Snowcem
CemTie - Helifix Ltd
Centafoam - The Modelshop
Centaur Plus - Horstmann Timers & Controls Ltd
Centaurstat - Horstmann Timers & Controls Ltd
Centemmial - Midland Stom Ltd
CentralScotland Plumbers - Ogilvie Construction Ltd
Centre - Cool - Johnson & Starley
Centrel - Emergi-Lite Safety Systems Ltd
Centro - Kermi (UK) Ltd
Centurian - Laidlaw Architectural Hardware Ltd
Centurion - Boulter Boilers Ltd
Centurion - Heat Trace Ltd
Centurion - Laidlaw Architectural Hardware
Centurion - Shackerley (Holdings) Group Ltd incorporating Designer ceramics
Centurion 12.5 - Forticrete Roofing Products
Century - Yale Security Products Ltd
Cepac - Emusol Products (LEICS) Ltd
Ceramica Alhambra Ceramics - Kirkstone
Ceramicsteel - Alliance Europe
Ceramitz Sprayed Stone - Textured European Finishes Ltd
Cerdisastone - Stokes R J & Co Ltd
Cermamica Impex - Stokes R J & Co Ltd
Cerofil - Firebarrier Services Ltd
Certikin - Certikin International Ltd
Certite - Webrr & Broutin
CF - Davis International Ltd
Chainex - BRB Industrial Services Ltd
Challenge - ECC Building Products Ltd
Challenger - Dunhams of Norwich
Challenger - Flamco UK Ltd
Champion - Forson Design & Engineering Ltd
Chamry Surface Range - Lumitron Ltd
Chancellor - Firth Carpets Ltd
Chancery - Firth Carpets Ltd
Channel Plus - Horstmann Timers & Controls Ltd
Chantal Range - Caradon Bathrooms Ltd
Character stone - Dockra Concrete Co Ltd, The
Charcon Tunnels - Tarmac Precast Concrete
Charlotte - Loft Centre Products
Charlotte - Qualitas Bathrooms
Check - Millar Dennis (1998) Ltd
Checkstat Plus - Checkmate Industries Ltd
Chelsea - Gloster Furniture Ltd
Chelsea Setts - RMC Concrete Products (UK) Ltd
Cheltenham - Chase James & Son (Furnishing) Ltd
Cheltenham - Ellis J T & Co Ltd
Chelwood - Chelwood Brick Ltd
Chemi-epoxeel - IFT Supplies Ltd
Cheminert epoxy system - Dex-o-Tex International Ltd
Cheminert terrazzo system - Dex-o-Tex International Ltd
Chemset - Ramset Fasteners Ltd
Cherwell PV - Portacel
Cheshire Heritage - Optiroc Ltd
Chess - Flexiform Business Furniture Ltd
Chevin - Crompton Lighting Ltd
Chevlok Walling - Ruthin Precast Concrete Ltd
Chieftain - Falcon Catering Equipment
Childrey seat - Anthony de Grey Trellises
Chilled ceilings - Clestra Hauserman Ltd
Chilmalic Limestone - Gregory Quarries Ltd, The
Chilmaric Limestone - The Rare Stone Group Ltd
Chilmark Limestone - Chilmark Quarries Ltd
Chilstone Architectural Stonework - Chilstone Garden Ornaments
Chiltern - LazyLawn
Chiltern Plus - LazyLawn
Chimaster - Docherty H Ltd
Chimflex LW/SB - Rite-Vent Ltd
Chimliner - Marflex Chimney Systems
Chimney Capper - Brewer Metalcraft
Chimney Cowls - Birdscarer Products
Chocflex - Bonar Floors Ltd
Chocflex - Bonar Floors Ltd
Chroma - Buchtal (U.K.) Limited
Chroma - Caradon M K Electric Ltd
Chromaclad - Cape Boards Ltd
Churchill - BSA Tools Ltd
Churchill - Northcot Brick Ltd
Churchouse Bollard - Crompton Lighting Ltd
CIJC - Construction Industry Joint Council
Cilplan - CIL International Ltd
Cimberio - SAV United Kingdom Ltd
Ciment Fondu Lafarge - Lafarge Aluminates Ltd
Cimm - SAV United Kingdom Ltd
CINCA Mosaic & Tiles - Reed Harris
Cintec - Cavity Lock Systems Ltd
Circulux - Poselco Lighting
Cirrus - Dermide Ltd
Cissell Dryers - Warner Howard Ltd
Cistermiser - Cistermiser Ltd
Citadel - Chubb Security Installations Ltd
Citadel - Fitzpatrick Door Frames Ltd
CITC - British Carpet Manufacturers Association

Citizen - Townscape Products Ltd
Citycabin - Wernick S & Sons Ltd
Cityspace - Kone Lifts Ltd
Civa - Optelma Lighting Ltd
CL200 - Davis International Ltd
Clarendon - Tenby Industries Ltd
Clarke - Alto Cleaning Systems (UK) Ltd
Classic - Dunhams of Norwich
Classic - Elkington Gatic Ltd
Classic - Hill Leigh (Joinery) Ltd
Classic - Kermi (UK) Ltd
Classic - Movitex Signs Ltd
Classic - Reinforced Plastic Products Limited
Classic 62 - Spectus UPVC Systems Ltd
Classic Apex - Piper Windows Doors & Conservatories
Classic Defence - Piper Windows Doors & Conservatories
Classic Gutter - Dales Fabrications Ltd
Classic Imperial - Gerland Ltd
Classic Mouldings - Regency Garage Door Services Ltd
Classica - Matki plc
Classicair louvres - Grille Diffuser & Louvre Co Ltd The
Classics - Muraspec Ltd
Classidur - Blackfriar Paints and Varnishes
Classidur - Parsons E. & Sons Ltd
Classique - Balmoral Porcelain Limited/Longmead Ceramics
Claudgen - Consort Equipment Products Ltd
Claylite - Kay Metzeler Ltd
Clean Guard - Collie Carpets Ltd
Clean steam - Fulton Boiler Works (Great Britain) Ltd
Clean Step - Collie Carpets Ltd
Clean Tread Custom - Collie Carpets Ltd
Clean Tread Tradition - Collie Carpets Ltd
Cleanflow - Johnson & Starley
Cleanguard - National Britannia Ltd
Cleanline - Neslo Partitioning & Interiors
Cleanline - Welconstruct Co
Cleanroom - Neslo Partitioning & Interiors
Cleanshield ™ - Accent Hansen
Clearline - Dyer Environmental Controls
Clearline - Morse Controls Ltd
Clearline - Neslo Partitioning & Interiors
Clearsec - Gunnebo Mayor Ltd
Clearstor Shelving - Moresecure Ltd
Clenaglass - Clenaware Systems Ltd
Clenewall - Brown F plc
Clereflo - Conder Products Ltd
Cleveland - Sashless Window Co Ltd
Clic - Signs & Labels Ltd
Clickfix coping - Alifabs (Woking) Ltd
Cliffhanger - Cliffhanger Shelving
Clifton - Gloster Furniture Ltd
Climaflex - Davant Products Ltd
Climaflex - nmc (UK) Ltd
Climatronic - Satchwell Control Systems Ltd
Climet 100 Series - Climet Ltd
Clinicall - Tunstal Telecom Ltd
Clintinade limestone - The Rare Stone Group Ltd
Clip-in - PFP Electrical Products Ltd
Clip-Top - Gradus Ltd
Clipfix 750 - European Profiles
Clipfold - Lycetts (Burslem) Ltd
Cliplan 10 - CIL International Ltd
Clipper - Calomax (Engineers) Ltd
Clipsal - Clips Partitions Ltd
Clipsexecutive - ClipsPartitions Ltd
Clipsharmonie - ClipsPartitions Ltd
Clipsoclips - ClipsPartitions Ltd
Clipsolab - ClipsPartitions Ltd
Clipstrip - Ryton's Building Products Ltd
Clorocel - Portacel
Clorocote - Witham Oil And Paint Ltd
Cloudburst - T & D Plastech
Club - Lappset UK Ltd
Clyde - IBP Conex
Clyde - Portacel
Clyde Rail Spikes - Glasgow Steel Nail Co Ltd
Co-Seal Liquid Plastic Dressing - Stewart Wales, Somerville Ltd
Coalbrookdale - Aga-Rayburn, Glynwed Consumer & Construction Products
Cobra - Allmat (East Surrey) Ltd
Cobra - MSA (Britain) Ltd
Coburn - Hillaldam Coburn Ltd
Coda - Kinsley-Cooke Furniture Ltd
Coex - Elkington Gatic Ltd
Cofast adhesive - Combustion Linings Ltd
Cofax castables - Combustion Linings Ltd
Cofax cements - Combustion Linings Ltd
Cog - Intralux UK Ltd
Coir - Land Wood & Water Co. Ltd
Cokeglow - Interoven Ltd
Colaquex - Emusol Products (LEICS) Ltd
Coldblocker - Space - Ray UK
Colman - Senior Air Systems
Colonade - Matki plc
Colorail - Rothley Burn Ltd
Colorduct 2 - Hotchkiss Air Supply
Colorette - DLW Flooring Ltd
Colorex - Forbo - Nairn Ltd
Colorfects - Crown Trade Paints
Colorflek - Crown Trade Paints
Colorslat - Barker W H Ltd
Colosseum - Metalliform Products Plc
Colour dimensions - ICI Dulux Trade Paints
Colour Index - Muraspec Ltd
Colour Palette - ICI Dulux Trade Paints
Colourcast - Procter Bros Limited
Colourfast - Bollom J W & Co Ltd
Colourmark - Armorex Ltd
Colourstep - Forbo - Nairn Ltd
Colourtex - Bardon (England) Ltd
Colourworld - Bollom J W & Co Ltd
Colring - Legrand Electric Ltd
Colson - Avocet Hardware Ltd
Colson - Legrand Electric Ltd
Colt Houses - Colt W H & Co Ltd
Columbus - President Office Furniture Ltd
Columbus timers - Elkay Electrical Manufacturing Co Ltd
Comar - Parkside Group Ltd, The
Comax - RFA Group Ltd
Combimate - Cistermiser Ltd
Cometec - CGL Cometec
Comforto - Haworth UK Ltd
Commander - Bawn W. B. & Co Ltd
Commander Mass Meter - ABB Kent Taylor Ltd.
Commando - Paragon by Heckmondwike F B Ltd
Commando - Richard Hose Ltd
Common (SP3) - Beacon Hill Brick Co Ltd
Communicall - Tunstal Telecom Ltd
Community - Checkmate Industries Ltd
Compac - Rycroft Ltd
Compact - Calomax (Engineers) Ltd
Compact Activ - Alno (United Kingdom) Ltd
Compact Sack - Assi Doman Sacks (UK)
Compactdoor - Applied Acoustics (a part of Henry Venables Limited)
Compactdoor - Henry Venables Ltd
Companion 90 - Telelarmcare Ltd
Compleat Kit - Protimeter plc
Complete Bathrooms - Tropravit (UK) Ltd
Compli - Pump Technical Services Ltd
Compo-Seal - Keeling Oliver
Compriband - Compriband Ltd
Comprisil - Compriband Ltd
Compton - Dawson McDonald & Dawson Ltd
Compton Buildings Ltd - Phillip Ader Ltd
Compton Garages - Compton Building Ltd
Computaquant Ver. 2.OB - EITE & Associates
Conbex - Fosroc Expandite Ltd
Conbextra - Fosroc Expandite Ltd
Concept - Kermi (UK) Ltd
Concept Suite - Armitage Shanks Ltd
Concierge - Movitex Signs Ltd
Concord - Concord Lighting
Concord - Dunhams of Norwich
Concrete Dustproofing and Sealing - Ferguson G A & Co Ltd
Concrete Master - Protimeter plc
Concrete surfaces - Ferguson G A & Co Ltd
Concrex - Watco UK Ltd
Concryl - IFT Supplies Ltd
Concute - Sempol Surfaces Ltd
Condensator - Protimeter plc
Condensor - Kedddy (Poujoulat) (UK) Ltd
Conderbrake - Conder Products Ltd
Conductite - Laybond Products Ltd
Cone Optic™ - Windsor Limited D W
Conex - Hipkin F W Ltd
Conex - IBP Conex
Congress - Project Office Furniture PLC
Connect grid - Ecophon Ltd
Connoisseur - Dunhams of Norwich
Consens - Girsberger London
Consolite - Cheshire Concrete Products
Consort - Clenaware Systems Ltd
Consort - Consort Equipment Products Ltd
Consort - Marflow Eng Ltd
Consort - Mita (UK) Ltd
Constructa - Taylor & Portway Ltd
Construction and building abstracts - NBS Services
Construction Information Service - NBS Services
Constructor - Dexion Ltd
Conta-clip terminals - Elkay Electrical Manufacturing Co Ltd
Contact - Girsberger London
Contact - President Office Furniture Ltd
Contain-It - George Fischer Sales Ltd
Contetch - IFT Supplies Ltd
Contigym - Continental Sports Ltd
Contimat - Continental Sports Ltd
Continental - Matki plc
Continental Range - Pedley Furniture international Ltd
Contitramp - Continental Sports Ltd
Contour - City Hardware and Security
Contour - Hambleside Danelaw Ltd
Contour 2 - Armitage Shanks Ltd
Contour 2000 - Bolton Gate Co Ltd
Contour Cisterns - Dudley Thomas Ltd
Contour Counters - Balmforth Engineering Ltd
Contour Factor 20 - Hambleside Danelaw Ltd
Contour OTT - Hambleside Danelaw Ltd
Contour Stepsafe - Hambleside Danelaw Ltd
Contour Vinyl Wallcoverings - Forbo (Lancaster) Ltd
Contourail - Jendico
Contract 'E' - Eltron Chromalox
Contract Journal - Contract Journal Business Alert
Contracters - GE Bayer Silicones
Contravent - Contravent Regal Ltd
Convectastream - Falcon Catering Equipment
Coo-Var ® - Coo-Var Ltd
Coo-Var ® - Teal & Mackrill Ltd
Coolfit - Sylvania Lighting International INC.
Coolflow - Biddle Air Systems Ltd
Coolguard - Feedwater Ltd
Coolkote - Bonwyke Ltd
Coolplex - Feedwater Ltd
CoolView - Brandon Medical
Copal - Alpha Fry Ltd
Copal - Fernox Fry Technology UK
Copeland - Copeland Corporation Ltd
Copenhagen - Loft Centre Products
Copley Decor - Copeley Decor Ltd
Copon - Wood E Ltd
Copper-Flo - Johnson Matthey PLC - Metal Joining
Copydex - Henkel Home Improvement & Adhesive Products
Coradoor Preliminary Zone - Tufton Ltd
Coragrip Preliminary Zone - Tufton Ltd
Coral Brush - Bonar Floors Ltd
Coral Classic - Bonar Floors Ltd
Coral Clean Off Zone - Tufton Ltd
Coral Door - Bonar Floors Ltd
Coral Grip - Bonar Floors Ltd
Coral Luxe - Bonar Floors Ltd
Coral Plus Clean Off Zone - Tufton Ltd
Corex - Cordek Ltd
Corflam - British Cork Mills Ltd
Corian - Du - pont - Decra Cubicles Ltd
Corkfast - Western Cork Ltd
Corkmaster - Wicanders (GB) Ltd
Cormet systems - Lafarge Plasterboard Ltd
Cornice - Southern Sanitary Specialists Ltd
Corofil - Alumasc Exterior Building Products Ltd
Corofix - Corofil Woodall Ltd
Coroflow - Corofil Woodall Ltd
Coroline - Ariel Plastics Ltd
Corolux - Ariel Plastics Ltd
Corolux 2000 - Ariel Plastics Ltd
Corona - SCHUCO International KG
Coroseal - Corofil Woodall Ltd
Corotherm - Ariel Plastics Ltd
Corovent - Corofil Woodall Ltd
Corovin - Willan Building Services Ltd
Corporate 2000 - Crittall Steel Windows Ltd
Corporate W20 - Crittall Steel Windows Ltd
Corralux - Rada Lighting Ltd
Corridor - Ecophon Ltd
Corroban - Feedwater Ltd
Corrocure - Corroless International Association
Corrogiene - Corroless International
Corrogiene - Corroless International Association
Corroguard - Corroless International
Corroguard - Corroless International Association
Corroless - Corroless International
Corroless - Corroless International Association
Corroshield - Corroless International Association
Corrosperse - Feedwater Ltd
Corruspan - Compriband Ltd
Cortal - Haworth UK Ltd
Cortega - Armstrong World Industries Ltd
Cosmofin - Beton Construction Ltd
Coss - IBP Conex
Cosybug - Harton Heating Appliances
Cottage suite - Armitage Shanks Ltd
Cottarstone - Camas Building Materials
Cottarstone - Fyfe John Ltd
Cougar - Caradon Catnic Ltd
Country collection - Emsworth Fireplaces Ltd
Country- CLAD - Hadley Industries Plc
Countryside - ECC Building Products Ltd
Countrystone - ECC Building Products Ltd
County Kerb - Redland Precast
County pantile - Sandtoft Roof Tiles Ltd
County Walling - Atlas Stone Products
Courtaulds performance films - Banafix Limited
Courtrai - Smithbrook Building Products Ltd
Covadbreak - Square D, Electrical Distribution Division
Cove red - Realstone Ltd
Coverlife - EBC UK Ltd
Coverseal - Armorex Ltd
Coverwalk Paving System - Coverite Ltd
Coxclad vinyl cladding - Cox Building Products Ltd
Coxdome T.P.X - Cox Building Products Ltd
Coxdome Trade - Cox Building Products Ltd
Coxdone 2000 - Cox Building Products Ltd
Coxdone Mark 5 - Cox Building Products Ltd
Coxlite Jetstream - Cox Building Products Ltd
Coxspan Baselock - Cox Building Products Ltd
Coxspan GlassPlank - Cox Building Products Ltd
Coxspan Modular roof glazing - Cox Building Products Ltd
CoxwindowsScape - Cox Building Products Ltd
CPL Fineliner - Composite Panels Ltd
CPL Fireclad - Composite Panels Ltd
CPL Micro - V - Composite Panels Ltd
CPL Super 4 - Composite Panels Ltd
Cpv - df - CPV Ltd
Crack Bond - Helifix Ltd
Craftsman Trio - Brockway Carpets Ltd
Craftsman Twist - Brockway Carpets Ltd
Craftwall - Neslo Partitioning & Interiors
Cranleigh Light Multi - Cranleigh Brick & Tile Co Ltd
Cranleigh Red Multi - Cranleigh Brick & Tile Co Ltd
Craven - Morris Mechanical Handling Ltd
Creation Stone - Gerland Ltd
Creation Wood - Gerland Ltd
Credo - Kermi (UK) Ltd
Creighton Thomson Associates - Structural Bonding Cambridge Ltd
Creosote - Coalite Chemicals
Cresco - Creteco Sales Ltd
Cresfinex - Exitex Ltd
Cresset - Adams- Hydraulics Ltd
Cresta - MSA (Britain) Ltd
Creteangle - Benton Co Ltd Edward
Cricket Gas Detectors - MSA (Britain) Ltd
Crimegard - Crimegard Ltd
Criterion - Dermide Ltd
Crittall Composite - Crittall Steel Windows Ltd
Crocodile - Simpson Strong-Tie®
Croda - Andrews J S (Coatings) Ltd
Cromleigh - Allan Harris & Sons Ltd
Cromleigh S.V.K - Allan Harris & Sons Ltd
Crompack - Crompton Lighting Ltd
Crompton - Reynolds UK Ltd
Cromwell Canopy - James Smellie Fabrications Ltd
Crossgrip - Plastic Extruders Ltd
Crossguard - Bradbury Security Grilles Ltd
Crossley - Carpets International (UK) PLC
Crossvent - Howden Buffalo
Crown Trade - Akzo Nobel Decorating Coatings Ltd
Crown Trade - Crown Trade Paints
Crown Wool - Owens Corning Building Products (UK) Ltd
Cryclad - Cryotherm Insulation Ltd
Cryosil - Cryotherm Insulation Ltd
Cryostop - Cryotherm Insulation Ltd
Cryotherme - Firebarrier Services Ltd
Crystal skylights - Formwood
CS - Davis International Ltd
CU Lights - Phosco Ltd
Cubes Range - Pedley Furniture international Ltd
Cubic - Elkington Gatic Ltd
CubicleTrack - Integra Products
Cullamix Tyrolean - Snowcem
Cuplok - SGB Youngman
Cupola Brighton - Goods Directions Ltd
Cupola Cambridge - Goods Directions Ltd
Cupola Canterbury - Goods Directions Ltd
Cupola Chester - Goods Directions Ltd
Cupola Cranleigh - Goods Directions Ltd
Cupola Edinburgh - Goods Directions Ltd
Cupola Ellesmere - Goods Directions Ltd
Cupola Newmarket - Goods Directions Ltd
Cupola Oxford - Goods Directions Ltd
Cupola Salisbury - Goods Directions Ltd
Cupola Sarum - Goods Directions Ltd
Cupola Winchester - Goods Directions Ltd
Cuprinol - ICI Paints
Cuproright 70 - Righton Ltd
Cuproright 90 - Righton Ltd
Curvatura - USG (UK) Ltd
Curve - Ecophon Ltd
Cutan Hospital Range - Deb Ltd
CWP Waterproofer - David Ball Group Plc
Cwt y Bugail - McAlpine Slate Ltd, Alfred
Cypherlok - Abloy Security Ltd
d Line - Allgood plc
D & H - Dyer Environmental Controls
D Lock Access - Durabella Ltd
D S 90 - Salex Interiors Ltd
D Shield - Bollom J W & Co Ltd
D-Bus - Home Automation Ltd
D.A.C. - Hepworth Building Products Ltd
D.F.D - Dudley Factory Doors Ltd - Dudley Factory Doors Limited
D.F.D - Dudley Factory Doors Ltd - Neway Doors Ltd
D.F.D - Dudley Factory Doors Ltd - Priory Shutter & Door Co Ltd
D.F.D - Dudley Factory Doors Ltd - Shutter Door Repair & Maintenance Ltd
Dac Chlor - Dacrylate Paints Ltd
Dac Crete - Dacrylate Paints Ltd
Dac flex - Dacrylate Paints Ltd
Dac Roc - Dacrylate Paints Ltd
Dac Shield - Dacrylate Paints Ltd
Daikin - HV Air Conditioning Ltd
Daikin - Purified Air Ltd
Dales Decor - Copeley Decor Ltd
Dalesauna - Dalesauna Ltd
Damcor - TDI (UK) Ltd
Damixa - Berglen Group Ltd
Dampa 10 - Dampa (UK) Ltd
Dampa 100 - Dampa (UK) Ltd
Dampa 200 - Dampa (UK) Ltd
Dampa 300 - Dampa (UK) Ltd
Dampa Chess - Dampa (UK) Ltd
Dampa Interval - Dampa (UK) Ltd
Dampa Lamel - Dampa (UK) Ltd
Dampa Sport - Dampa (UK) Ltd
Dampa Tile - Dampa (UK) Ltd
Dampa X3 - Dampa (UK) Ltd
Dampcoursing 132 - Kiltox Damp Free Solutions
Dampseal - Marley Waterproofing
Danelaw - Drainage Systems
Danelaw - Hambleside Danelaw Ltd
Danfoss - Danfoss Randall Ltd
Daniel Platt - Stokes R J & Co Ltd
Danpalon - EBC UK Ltd
Danum - Pegler Ltd
Dark Victorian Red - Charnwood Brick Group Ltd
Darkroom safelights - Encapsulite International Ltd
Darksky - Crompton Lighting Ltd
Daryl - Daryl Industries Ltd
Dashing Render - Snowcem
Dasphalte - IFT Supplies Ltd
Datacall Aquarius - Blick Communication Systems Ltd
Datacall Gemini - Blick Communication Systems Ltd
Datacall Minder - Blick Communication Systems Ltd
Datasafe - Chloride Industrial Batteries Ltd
Dauphin Trendline - Dauphin PLC
Davenset - Deakin Davenset Rectifiers Ltd
David Ball Test Sands - David Ball Group Plc
David Mclean Contractors Ltd - Mclean Group, David
Day star - Full Spectrum Lighting Ltd
Day-Lite - Emergi-Lite Safety Systems Ltd
DCE - DCE - BTR Environmental Ltd
Deamant - Novellini Bathroom Products
Deanlite - Deans Blinds & Awnings (UK) Ltd

Deanmaster - Deans Blinds & Awnings (UK) Ltd
Deanox - Elementis Pigments
Deans - Deans Blinds & Awnings (UK) Ltd
Deans signs - Deans Blinds & Awnings (UK) Ltd
Deb Natural - Deb Ltd
Deb Protect - Deb Ltd
Deb Restore - Deb Ltd
Debline - Deb Ltd
Debut - Ness Furniture Ltd
Debut Range - Caradon Bathrooms Ltd
Decadex - Liquid Plastics Ltd
Decamel - Formica Ltd
Decathlon - Hodkin Jones (Sheffield) Ltd
Deck drain - ABG Ltd
Deckdrain - Hauration Kaskade Ltd
Deckshield - Flowcrete Plc
Deco Gard - MSC Speciality Films (UK) Ltd
Decoflair - nmc (UK) Ltd
Decoline - Deceuninck Ltd
Deconyl - Plascoat Systems Ltd
Decor - Deceuninck Ltd
Décor Pro File - Samafil Ltd
Decor-flor - Dex-o-Tex International Ltd
Decorative - Redland Precast
Decorators Choice - Akzo Nobel Decorating Coatings Ltd
Decorfix - Decorfix
Decoroc - Deceuninck Ltd
Decostar - Osram Ltd
Decostyle - Cape Boards Ltd
Decothane - Liquid Plastics Ltd
Decra Stratos - Decra Roofing Systems
Decra tiles - Decra Roofing Systems
Decra Vent - Decra Roofing Systems
Deedlock - Dryad Architectural Ltd
Deeplas - Deceuninck Ltd
Deepster - Dexion Ltd
Defender - Guest & Chrimes Ltd
Defender - Securistyle Ltd
Defendor - Henderson-Bostwick
Defensor - Geoquip Ltd
Degreasol - IFT Supplies Ltd
Degussa - Anderson Gibb & Wilson
Dekor - Rockfon Ltd
Delabole - Delabole Slate
Delbraze - IBP Conex
Delcop B - IBP Conex
Delflo - Hattersley Newman Hender Ltd
Deliplan - DLW Flooring Ltd
Dellchem - Opella Ltd
Delmag - Burlington Engineers Ltd
Delstar - Bolivar Stamping Ltd
Delta - Henderson-Bostwick
Delta downlights - LB Lighting Ltd
Delta Fire Chests - ISOKERN (UK) Ltd
Deltascreen - Quartet-GBC
Deltos - Lightfoot Charles Ltd
Delvo - Feb MBT
DEM Tip-Up Seating System - Race Furniture Ltd
Demidekk - Jotun Henry Clark Ltd
Deminpac - Feedwater Ltd
Denchem - Naylor Drainage Ltd
Denduct - Ashmead Buidling Supplies Ltd
Denduct - Ashmead Building Supplies Ltd
Denduct - Naylor Drainage Ltd
Denka Lifts - Denka International
Denline - Ashmead Buidling Supplies Ltd
Denline - Ashmead Building Supplies Ltd
Denline - Naylor Drainage Ltd
Denlok - Naylor Drainage Ltd
Dennis Ruabon - Stokes R J & Co Ltd
Denrod - Naylor Drainage Ltd
Denseal - Ashmead Buidling Supplies Ltd
Denseal - Ashmead Building Supplies Ltd
Denseal - Naylor Drainage Ltd
Densleeve - Ashmead Buidling Supplies Ltd
Densleeve - Ashmead Building Supplies Ltd
Densleeve - Naylor Drainage Ltd
Denso - Winn & Coales
Derbigum - Euroroof Ltd
Dernier & Hamilyn - Denmans Montrose Ltd
Derwent Gutter - Dales Fabrications Ltd
Designer - Clark Door Ltd
Designer - Pilkington's Tiles Ltd
Designer 2000 - Bassett & Findley Ltd
Designer Collection - Brintons Ltd
Desimpel - Hanson Brick Ltd
Desimpel - Hanson Brick Ltd, Northern Regonal Sales Office
DeskTop - Forbo - Nairn Ltd
Desney - Pouliot Designs
Desney Products Ltd - Desney Products
Destiny - Baxall Ltd
Detan - Ancon CCL Ltd
Detectamesh - Boddingtons Ltd
Detectatape - Boddingtons Ltd
Developer - Checkmate Industries Ltd
Devweld structural - Kenyon Industrial Paints & Adhesives
Dialock - Haffele UK Ltd
Diamard - Emusol Products (LEICS) Ltd
Diamond - Iles Waste Systems
Diamond - Paragon by Heckmondwike F B Ltd
Diamond 600 - Durabella Ltd
Diamond décor safety glass - Chelsea Artisans Ltd
Diamond Glaze - ICI Dulux Trade Paints
Diamond Granite - Stonecraft & Chelsea Artisans Ltd
Diamond granite & marble lightweight panels - Chelsea Artisans Ltd
Diamond Marble - Stonecraft & Chelsea Artisans Ltd
Diamond Mirror - Stonecraft & Chelsea Artisans Ltd
Diamond Optic™ - Windsor Limited D W
Diamond Sign System - Drakard & Humble Ltd
Diamond system mirrors - Chelsea Artisans Ltd
Dibond ® - Righton Ltd
Digico - Masterbill Micro Systems Ltd
Digital Mini - Protimeter plc
Diklon Sunshade fabric - Dallas & Forrest Ltd
Dimplex - Dimplex (UK) Ltd
Dinella - Provex Products Ltd
Diplomat - Dudley Thomas Ltd
Diplomat Range - Lumitron Ltd
Direct Worktops - Viking Laminates Ltd
Directions - Desking Systems Ltd
Direx - Universal Components Ltd
Diricall - Telelarmcare Ltd
Disbocrete - PermaRock Products Ltd
Discovery - Apollo Fire Detectors Ltd
Discreet - Integra Products
Disklock Pro - Abloy Security Ltd
Disposapad - Sissons W & G Ltd
Disque - Ledu UK Ltd
DMP Bonder - Isocrete Group Sales Ltd
DNA - Bernlite Ltd
Dobelshield stucco Aluzink - SSAB Dobel Coated Steel Ltd
Dolphin - Dolphin Showers Ltd
Domestic Gutter - Dales Fabrications Ltd
Domino - OWA UK Ltd
Domo - Dalesauna Ltd
Domostyl - nmc (UK) Ltd
DON - Record & Parkes Ltd
Don - The Safety Tread Ltd
Don Brown - Davidson C & Sons
Don Fileboard - Davidson C & Sons
Donaldson filter components - Encon Air Systems Ltd
Donio - Dalesauna Ltd
Donn - USG (UK) Ltd
Donn - Brown - Brown F plc
Donn DX Grid - USG (UK) Ltd
Doorfit - Doorfit Products Ltd
Dorchester - Hamworthy Heating Ltd
Dorma - City Hardware and Security
Dorma - Permclose Group PLC
Dorma BST - DORMA Entrance Systems Ltd
Dorma Compact Slide - DORMA Entrance Systems Ltd
Dorma ED200 - DORMA Entrance Systems Ltd
Dorma ED200I - DORMA Entrance Systems Ltd
Dorma ED800 - DORMA Entrance Systems Ltd
Dorma ES - DORMA Entrance Systems Ltd
Dorma FFT - DORMA Entrance Systems Ltd
Dorma HSW - DORMA Entrance Systems Ltd
Dorma KTC - DORMA Entrance Systems Ltd
Dorma KTV - DORMA Entrance Systems Ltd
Dorma RST - DORMA Entrance Systems Ltd
Dorman Long - Wade Building Services Ltd
Dorset - Pilkington's Tiles Ltd
Dorset Range - Beacon Hill Brick Co Ltd
Double Duty - Integra Products
Dove - Contour Showers Ltd
Dovedale - Chiltern Invadex
DowCorning - Dow Corning Hansil Ltd
Dowty Spats - Corofil Woodall Ltd
DP5 - Daylight Centre
DP91 - Daylight Centre
DPn - Wilo Samson Pumps Ltd
Draegertubes - Draeger Ltd
Dragon - Fulton Boiler Works (Great Britain) Ltd
Dragon HLO - Elan-Dragonair
Dragon HNG - Elan-Dragonair
Dragonair Hlo - Dragonair Ltd
Dragonair Hng - Dragonair Ltd
Drainmaster - Pump Technical Services Ltd
Dramix- Steelwire Fibres - Bekaert Building Products
Drape - Harrison Drape, Divsion of McKenie (UK)
Draper - Kem Edwards Ltd
Draught Buster - Airflow (Nicoll Ventilators) Ltd
Draught Dodger - Map Hardware Ltd
Draughtbuster - Timloc Expamet Building Products
Dravo - Dravo Environmental Services
Dravo AE unit heater - Dravo Environmental Services
Dravo DF - Dravo Environmental Services
Dravo Directflow - Dravo Environmental Services
Dravo Euroflow - Dravo Environmental Services
Dravo FS Directflow - Dravo Environmental Services
Dravo Heatmiser - Dravo Environmental Services
Dravo NDS - Dravo Environmental Services
Dravo ventilation - Dravo Environmental Services
Drawmet - Drawn Metal Ltd
Dreadnought - Baydale Architectural Systems Ltd
Dreadnought Tiles - Hinton Perry & Davenhill Ltd
Dream - Orsogril Sarl UK Department
Drehmax - Hyde Brian Ltd
Dremel - Robert Bosch Ltd
Dricon - Hickson Timber Products Ltd
Dritherm - Owens Corning Building Products (UK) Ltd
Driveway - Hedland Precast
Drum Form - Scapa Tapes UK Ltd.
Drury Lane collection - Firth Carpets Ltd
Dry Fix - Helifix Ltd
Dry Shake Armorshield - Armorex Ltd
Dryangles - Atlas Stone Products
Dryflow - Johnson & Starley
Dryseal - Hambleside Danelaw Ltd
Drystone - Atlas Stone Products
DS - 3 - Fernox Fry Technology UK
DS - 40 - Fernox Fry Technology UK
DS System - Lewes Design Contracts Ltd
DS-40 - Alpha Fry Ltd
Dual - Airflow (Nicoll Ventilators) Ltd
Dualcase - Universal Components Ltd
Dubix - Electrolux Laundry Systems
Ducale - Polybau Ltd
Duck Tapes - Henkel Home Improvement & Adhesive Products
Ductex - Rega Metal Products Ltd
Duets - Forticrete Ltd
Dukes Whatstand well - Realstone Ltd
Dulux - Osram Ltd
Dulux Trade - ICI Dulux Trade Paints
Dulux Trade Emulsions - ICI Dulux Trade Paints
Dulux Trade Gloss - ICI Dulux Trade Paints
Dulux Trade Natural Wood finishes - ICI Dulux Trade Paints
DUNA - Reiner Fixing Devices
Dune Plus - Armstrong World Industries Ltd
Dunham Strip - Dunham Bush Ltd
Dunsfold Selected Dark - Cranleigh Brick & Tile Co Ltd
Dunvent - Dunbrik (Yorks) Ltd
Duo-Fast - Duo-Fast (UK) Ltd
Duotherm - Smart Systems Ltd
Duplex - Airsprung Beds Ltd
Duplex Building System - Portakabin Ltd
Duplix - Legrand Electric Ltd
Duquesa Slate - Blunn Slates Ltd
Duracem - Eternit UK Ltd
Duraflex - Caradon Duraflex Systems Ltd
Durafort - Dixon Turner Wallcoverings Ltd.
Duraglas - Amari Plastics Plc
Duralite - Pryorsign
Duralock range post and rails - Duralock UK Ltd
Durama X - Quartet-GBC
Durapipe - Capper Pipe Systems Ltd
Duraplug - Caradon M K Electric Ltd
Durastat - Heat Trace Ltd
Durastep - Alpha Therm United House Ltd
DVS Switchers - Baxall Ltd
DWM Copeland - Copeland Corporation Ltd
Dycell Revetment - Ruthin Precast Concrete Ltd
Dycem Clean-Zone - Dycem Ltd
Dycem Protectamat - Dycem Ltd
Dycem Work - Zone - Dycem Ltd
Dyke-Aluminium - Dyke Chemicals Ltd
Dyke-Flashing - Dyke Chemicals Ltd
Dyke-Glass - Dyke Chemicals Ltd
Dyke-Mastic - Dyke Chemicals Ltd
Dyke-Roof - Dyke Chemicals Ltd
Dyke-Roof colour finish - Dyke Chemicals Ltd
Dyke-Roof universal primer - Dyke Chemicals Ltd
Dyke-Seal - Dyke Chemicals Ltd
Dyke-Sil - Dyke Chemicals Ltd
Dyke-Silver - Dyke Chemicals Ltd
Dynabolt - Ramset Fasteners Ltd
Dynaset - Ramset Fasteners Ltd
Dyno-locks - Dyno Rod PLC
Dyno-rod - Dyno Rod PLC
Dytap Revetment - Ruthin Precast Concrete Ltd
E Range - James Gibbons Format Ltd
E-fix - Arthur Fisher (UK) Ltd
E-Pack - MTE Ltd
E. D. L - EDL Cable Supports Ltd
E. F.S - Fulton Boiler Works (Great Britain) Ltd
E2000 - Eternit UK Ltd
E240 - Bolton Gate Co Ltd
E30 - Bolton Gate Co Ltd
E7BX - Horstmann Timers & Controls Ltd
EAC - Unicorn Abrasives (UK) Ltd
Easi-away - Easi-Fall International Ltd
Easi-Fall - Easi-Fall International Ltd
Easi-fill - Easi-Fall International Ltd
Easi-flex - Easi-Fall International Ltd
Easi-mend - Easi-Fall International Ltd
Easi-rect - Harkness Hall Ltd
Easiclamp - Viking Johnson
Easicollar - Viking Johnson
Easipipe 100, 125, 150 - Domus Ventilation Ltd.
Easitap - Viking Johnson
Easitee - Viking Johnson
Easitrax - Oldham Signs Ltd
Easiwork - Metreel Ltd
Easy Change - Movitex Signs Ltd
Easy Finish - RMC Readymix Limited
Easy-Flo - Johnson Matthey PLC - Metal Joining
Easyclad - Advanced Hygienic Contracting Ltd
Easyfix - Tinsley & Co Ltd, Eliza
Easyhang - Spur Shelving Ltd
Eavesguard - Ryton's Building Products Ltd
Eavesguard - Vulcanite Limited
EB - Clement Steel Windows
EBC UK Ltd - EBC UK Ltd
Ebco - Barber Edward & Co Ltd
Ebm - Ziehl-Ebm (UK) Ltd
EC 2000 - ABB Flakt Products
Ecamatic - Kwikform UK Ltd
Echomaster - Tarmac Topblock Ltd
Ecletec Lighting - Woodhouse UK Plc
Eclipse - Eclipse Blind Systems Ltd
Eclipse - Marley Floors Ltd
Eclipse - MTE Ltd
Eclipse - Neill Tools Ltd
Eclipse - Northern Incinerators (GB) Ltd
Eclipse - Safetell Ltd
Eclipse - Southern Sanitary Specialists Ltd
Eclipse Spiralux - Neill Tools Ltd
Eco - Townscape Products Ltd
Ecodisc - Kone Lifts Ltd
Ecofibre - Gilmour Ecometal
Ecohome - Gilmour Ecometal
Ecomax Reduc - Trim Acoustics
Ecomax Soundslab - Trim Acoustics
Ecometal - George Gilmour (Metals) Ltd
Ecometal - Gilmour Ecometal
Econocoat - Stewart Wales, Somerville Ltd
Econoflame - Stokvis R S & Sons Ltd
Econoloc - Camloc (UK) Ltd
Economy 7 QTZ - Horstmann Timers & Controls Ltd
Econoplas Voge - Stewart Wales, Somerville Ltd
Econospan - Gilmour Ecometal
Econotex - Stewart Wales, Somerville Ltd
Ecopanel - Gilmour Ecometal
Ecophon - Nevill Long Limited
Ecophon - New Forest Ceilings Ltd
Ecospan - Gilmour Ecometal
Ecotherm - Prestoplan
Ecotray - Gilmour Ecometal
Ecovent - Ves Andover Ltd
Ecowall - Gilmour Ecometal
Ecowarm-Cavity - Springvale E P S
Ecowarm-Flat Roofing - Springvale E P S
Ecowarm-Fulfil - Springvale E P S
Ecowarm-Warmfloor - Springvale E P S
Ecowarm-Warmlath - Springvale E P S
Ecowarm-Warmsark - Springvale E P S
Ecowarm-Warmsqueez - Springvale E P S
Ecozip - Gilmour Ecometal
ECS Reset - ECS Lighting Controls Ltd
Ecurse TS - Interlux Ltd
Edale - Chiltern Invadex
Edge Slot - Spur Shelving Ltd
Edilkamin - Emsworth Fireplaces Ltd
EDL - Luxo UK Ltd
EEKO - Electric Elements Co, The
Eeto - Davant Products Ltd
Eetofoam - Davant Products Ltd
Effets scale - Higginson Staircases Ltd
EFP1 Single Zone Panel - C-Tec (Computionics Ltd)
Egoluce - Forma Lighting Ltd
8 French Limestones - Kirkstone
800 series - C-Tec (Computionics Ltd)
Ekom-50 - Komfort Systems Ltd
Elastaseal - Tor Coatings Group
Elastotex 100 - Dex-o-Tex International Ltd
Elastotex 500 - Dex-o-Tex International Ltd
Elcock - Purdie Elcock Ltd
Electrace - Jimi-Heat Ltd
Electrak - Electrak International Ltd
Electrisaver - Horstmann Timers & Controls Ltd
Electro-Fence - Advanced Perimeter Systems
Electro-flor - Dex-o-Tex International Ltd
Electroline - Jebron Ltd
Electrolux - Electrolux Domestic Appliances
Electrolux - Electrolux Laundry Systems
Electrolux - Parkinson Cowan and Tricity Bendix Co, The
Electrolux commercial service - Electrolux food service
Electromatic - Tuke & Bell Ltd
Electronic 7 - Horstmann Timers & Controls Ltd
Electropack - Fulton Boiler Works (Great Britain) Ltd
Electrosonic - Helvar Electrosonic
Electrospot - Illuma Lighting Ltd
Electroway - Viscount Catering Ltd
Elegance - Gerland Ltd
Elegance - Provex Products Ltd
Elegance - Unilock Ltd
Eleganza - Matki plc
Elementis Pigments - Elementis Pigments
Elica Handrail - Lewes Design Contracts Ltd
Elite - Cardale Doors Ltd
Elite - Dudley Thomas Ltd
Elite - Gradus Ltd
Elite - Nevill Long Limited
Elite - Shackerley (Holdings) Group Ltd incorporating Designer ceramics
Elite Xtra - Marley Floors Ltd
Elith Slate - Blunn Slates Ltd
Eliturbo - S & P Coil Products Ltd
Eliza Tinsley - Challenge Fencing & Ltd
ElJan - Johnson & Starley
Elkington Gatic - Elkington Gatic Ltd
Elkosta - Elkosta; Rodney Coate & Partners Ltd
Elkosta - Fernden Construction (South Western) Limited
Ellacrete - Resdev Ltd
Ellacure - Resdev Ltd
Ellamod - Resdev Ltd
Elliott Lucas - Neill Tools Ltd
Elson - Elsy & Gibbons Ltd
Elterwater - Burlington Slate Ltd
Eltex - Brettell & Shaw Ltd
Elvaco - Elvaco UK Ltd
EMA Model Supplies - The Modelshop
Emaco - Feb MBT
EmaW - FEB Ltd
Emerald - En-tout-cas plc
Emerald - En-tout-cas plc (Scotland)
Emerald - En-tout-cas South West
Emergi - Man - Emergi-Lite Safety Systems Ltd
Emil Ceramica - Stokes R J & Co Ltd
Emir - Emmerich (Berlon) Ltd
Emit - Allgood plc
Empressa - Eternit UK Ltd
Energy Beta - Beta Lighting Ltd
Energymaster - Lumitron Ltd
Engerseal - Callenders Ltd
Ensign - Drainage Systems
Ensudisc - Ensor Metal Products
Enterprise - Accomodex
Enterprise - Dunsley Heat Ltd
Envirograf - Firebarrier Services Ltd
Enviroguard - Leigh's Paints

Envirolite - Envirodoor Markus Ltd
Enviromat - Linatex Ltd
Enviropave - Linatex Ltd
Enviroplast - Envirodoor Markus Ltd
Envirotector - Conder Products Ltd
Enyiroseal - Mann McGowan Fabrication Ltd
Epandite - Fosroc Expandite Ltd
Epidox - Witham Oil And Paint Ltd
Epigrip - Leigh's Paints
Epo-dur - ICI Dulux Trade Paints
Epoch Soventless Plastic - Stewart Wales, Somerville Ltd
Epoxeel - IFT Supplies Ltd
Epoxy Plus - Webrr & Broutin
Epoxy Wall Shield - Anglo Building Products Ltd
ERA - Reynolds UK Ltd
Ercall - Powys Hire Ltd/ P.I.B
Ergoflex - BRB Industrial Services Ltd
Erico - Kem Edwards Ltd
Erico-caddy ® - Erico Europa (GB) Ltd
Erosamat - ABG Ltd
Erskines - Newton John & Co Ltd
Esavian - 1200 - Jewers Doors Ltd
Esavian - 126 - Jewers Doors Ltd
Esavian - 127 - Jewers Doors Ltd
Esavian - 128 - Jewers Doors Ltd
Escofet - Woodhouse UK Plc
Escol - Escol Panels (S & G) Ltd
ESP - President Office Furniture Ltd
Espero - Dividers Ltd
Esplanade - Gradus Ltd
Essar Aquacoat SP - Smith & Rodger Ltd
Essar Aquacoat Xtra - Smith & Rodger Ltd
Essar Bar Top Lacquer - Smith & Rodger Ltd
Essar French Polish - Smith & Rodger Ltd
Essar Precatalysed Lacquers - Smith & Rodger Ltd
Essar Woodshield - Smith & Rodger Ltd
Estoryl Sunshade fabric - Dallas & Forrest Ltd
Eternit-Promat - Firebarrier Services Ltd
EU 2000 - ABB Flakt Products
Eureka - Eureka Products Ltd
Euro Shelving - Moresecure Ltd
Euro-Guard ® - Jackson H S & Son (Fencing) Ltd
Eurobar - Imperial Machine Co Ltd
Eurobench - Quigley Metal Products Ltd
Eurodellor - Egger (UK) Ltd
Eurodoor 2000 - James & Bloom Ltd
Eurofix - Lastite Building Products Ltd
Eurofloor - Atkinson & Kirby Ltd
Eurofoil - Roof Units Group
Eurofold - Bolton Gate Co Ltd
Euroglas - Clow Group Ltd
Eurogullies - ACO Technologies Plc
Eurohose - Flexible Ducting Ltd
Euroline - Leisure, Glynwed Consumer Foodservice Products Ltd
Euromesh - Boddingtons Ltd
Euromond - Cego Frameware Ltd
Euronox - Wellman Robey Ltd
Europa - Marley Floors Ltd
Europa - Sandtoft Roof Tiles Ltd
Europak - Roof Units Group
Europave - Bardon (England) Ltd
Europclip - Bolivar Stamping Ltd
European - Wellman Robey Ltd
Europipe Drainage Pipework System - BM Stainless Steel Drains Ltd
Europitch - Roof Units Group
Euroroof - Alumasc Exterior Building Products Ltd
Euroseries - Roof Units Group
Euroslate - Euroroof Ltd
Euroslide - Quigley Metal Products Ltd
Eurospan - Egger (UK) Ltd
Eurotape - Boddingtons Ltd
Eurotec - Astron Building Systems Commercial Intertech Ltd
Eurotherm - Martin Orgee & Associates
Eurotile - Euroroof Ltd
Evalastic - ICB (International Construction Bureau) Ltd
Evalon - ICB (International Construction Bureau) Ltd
Evamatic - KSB Ltd
Evans - Evans Vanodine International PLC
Evans Pourform - Wood International Agency
Evelite - Amari Plastics Plc
Evenceil - Polycell Products Ltd
Evenlode - Gloster Furniture Ltd
Everclad - The Abbseal Group incorporating Everseal (Thermovitrine Ltd)
Everdry - Timloc Expamet Building Products
Everest - Caradon Everest Ltd
Everflex - Everlac (GB) Ltd
Evergold - Bollom J W & Co Ltd
Everseal - Everlac (GB) Ltd
Evertaut - Evertaut
Evertex - Everlac (GB) Ltd
Everywear - Firth Carpets Ltd
Evo-Stik - Evode Ltd
Evolution radiant strip - Dunham Bush Ltd
Evolve desking and seating - Flexiform Business Furniture Ltd
Excalibur - Bartlett Catering Equipment Ltd
Excel - Axter Ltd
Excel - James Gibbons Format Ltd
Excel 5000 - Honeywell Control Systems Ltd
Excel Life Safety - Honeywell Control Systems Ltd
Excel Security Manager - Honeywell Control Systems Ltd
Excelastic - Geosynthetic Technology Ltd
Excelflex - Axter Ltd
Excell - OWA UK Ltd
Excludoor - Sunray Engineering Ltd
Exec - Abloy Security Ltd
Executive - Midland Alloy Ltd
Executive - Rollalong
Exedra - BPT Security Systems (UK) Ltd
Exhausterl blowers - BVC-BIVAC, Division of DD Lamson
Exide - MJ Electronics Services (International) Ltd
Exitguard - Hoyles Fire & Safety Ltd
Exmet - Expamet Building Products
Exodus - Orsogril Sarl UK Department
Exos 4000 - Kaba UK Ltd
Expamet - Advanced Fencing Systems Ltd
Expamet - Expamet Building Products
Expamet - M & M Fencing Limited
Expandafoam - Fosroc Expandite Ltd
Expelex - Exitex Ltd
Explorer and Harlequin Range - Lumitron Ltd
Explosit - Erlau AG
Expowall - Deceuninck Ltd
Express Evans - Express Evans Lift
Express Evans Lifts - Express Evans Lifts
ExteriorGuard - Everlac (GB) Ltd
Extra Aqua Vent - Areco Roofing Supplies Company
Extral - Rada Lighting Ltd
Extraproof - Briggs Roofing & Cladding Ltd, Roofing Contract
Extrusions Direct - Extrusions Direct
Ezelift - Maswell Engineering Ltd
Ezi - lift - Didsbury Engineering Co Ltd
F Sereis - Swedecor Ltd
F Shield - Bollom Fire Protection Ltd
F&p Supertube Clip-In - PFP Electrical Products Ltd
F100 - Sempol Surfaces Ltd
F300 - Sempol Surfaces Ltd
F600 - Sempol Surfaces Ltd
FAAC - Crimegard Ltd
Faber - Faber Blinds Ltd
Faber 1800 - Faber Blinds Ltd
Faber 2000 - Faber Blinds Ltd
Faber Autostop - Faber Blinds Ltd
Faber Maximatic - Faber Blinds Ltd
Faber Metalet - Faber Blinds Ltd
Faber Metamatic - Faber Blinds Ltd
Faber Midimatic - Faber Blinds Ltd
Faber Minimatic - Faber Blinds Ltd
Faber Multistop - Faber Blinds Ltd
Faber Softline - Faber Blinds Ltd
Fablok - SFS Stadler Ltd
Fabresa - Stokes R J & Co Ltd
Fadini - Automatic Doors & Gates
Fairley - James Fairley Athena
Fairmile - Advanced Fencing Systems Ltd
Fairmile - Fairmile Fencing Ltd
Fairmile A - Pale - Fairmile Fencing Ltd
Fairmile Crownguard - Fairmile Fencing Ltd
Fairmile Palisade - Fairmile Fencing Ltd
Fairmile Roadguard - Fairmile Fencing Ltd
Fairmile Safeguard - Fairmile Fencing Ltd
Fairmile Siteguard - Fairmile Fencing Ltd
Fairmile Sovereign Grille - Fairmile Fencing Ltd
Fairmile Vertical Bar - Fairmile Fencing Ltd
Fairmitre - Fairmitre Ltd
Fairoaks Red Multi - Cranleigh Brick & Tile Co Ltd
Fairway - Metalrax Ltd
Faiveley - Automatic Doors & Gates
Falcon - James Gibbons Format Ltd
Falcon - Tenby Industries Ltd
Falcon 3.6.9 - Falcon Catering Equipment
Falcon 350 - Falcon Catering Equipment
Falcon warewash - Falcon Catering Equipment
Fanaire - Fantasia Distribution Ltd
Fantasia - Lappset UK Ltd
Fantasia ceiling fans - Fantasia Distribution Ltd
Fantasy - Novellini Bathroom Products
FAR - SAV United Kingdom Ltd
Faral Etal - Faral Radiators
Faral Green - Faral Radiators
Faral Lampo - Faral Radiators
Faral Lineal - Faral Radiators
Faral Longo - Faral Radiators
Faral Magic - Faral Radiators
Faral Mera - Faral Radiators
Faral Piuma - Faral Radiators
Faral Riva - Faral Radiators
Faral Tondo - Faral Radiators
Faral Tropical - Faral Radiators
Faral Vega - Faral Radiators
Farm Pave - RMC Readymix Limited
Farmington - Farmington Natural Stone Ltd
Farr 3D/3D - Farr Filtration Ltd
Fascinating finishes - Kingstonian Point Ltd
Fascinations - Carpets of Worth Ltd
Fascut Thin Wheels - Unicorn Abrasives (UK) Ltd
Fasset - Petrarch Claddings Ltd
Fast brolly - Unifix Ltd
Fast Finish - RMC Readymix Limited
Fast-Track - Tektura Plc
Fastcall - Tunstal Telecom Ltd
Fastel - P4 Limited
Fastline - IFT Supplies Ltd
Fastnet NT - Tegral Building Materials Ltd
Fastwall - Fastwall Ltd
Favourite - Valor Heating
FavxFinsh - Dixon Turner Wallcoverings Ltd.
FCGR - Ferham Products
FCR - Ferham Products
FD - Davis International Ltd
Featherfinish - Ardex UK Ltd
FEB - FEB Ltd
FEB - Strand Hardware Ltd
Fecon Baffle Filtors - Vianen Kitchen Ventilation (UK) Ltd.
Federal electric - Schneider Ltd
Feedmaster - CAEC Howard (Holdings) Ltd
Fence Decor - Jotun Henry Clark Ltd
Fenceguard - Guardall
Fencemaker - Fleetwood Fengate Co Ltd
Fencetone - Protim Solignum Ltd
Fencing Division - May Gurney (Technical Services) Ltd - Fencing Division
Fep 'O' rings - Ashton Seals Ltd
Fermacell dry flooring elements - Fels UK
Fermacell wallboards - Fels UK
Feron - Clayton-Munroe Ltd
Ferrocrete - Blue Circle Cement
FF - Davis International Ltd
Ffestiniog - McAlpine Slate Ltd, Alfred
FG Wilson - Wilson (Engineering) Ltd, F G
Fibaform - FIBAFORM PRODUCTS
Fibarack - Mita (UK) Ltd
Fibastrut - Mita (UK) Ltd
Fibatray - Mita (UK) Ltd
Fibercill - Fibercill
Fibermat - Vandex (UK) Ltd
Fiberscript - Fibercill
Fibershield - Bolton Gate Co Ltd
Fibertrane - Fibercill
Fibo - Lignacite (North London) Ltd
Fibral - Rockfon Ltd
Fibrespan - Armfibre Ltd
Fibrotrace ® - Maestro International Ltd
Fido bin - Glasdon UK Ltd
Fiesta - Gaskell Textiles Ltd
Fiesta - Shackerley (Holdings) Group Ltd incorporating Designer ceramics
Fife Alarms - Fife Fire Engineers & Consultants Ltd
Fife Fire - Fife Fire Engineers & Consultants Ltd
Fiji 2 - Leisure, Glynwed Consumer Foodservice Products Ltd
Filatrap - Conder Products Ltd
Fildek - GRAB Industrial Flooring Ltd
Fillaboard - Fillcrete Ltd
Fillcrete - Fillcrete Ltd
Fillter&Poulten - PFP Electrical Products Ltd
Filmatic - Martin Orgee & Associates
Filon - Filion Products Ltd
Filon DR - Filion Products Ltd
Filon Multiclad - Filion Products Ltd
Filon Overproof with Profix - Filion Products Ltd
Filon Plate - Filion Products Ltd
Filtacleve - Clearwater Plc
Filterball - Marflow Eng Ltd
Filtermate - Marflow Eng Ltd
Filtrasol ® - Hunter Douglas Ltd (Contracts Division) Luxaflex Blinds ®
Fine Textured Renovation Blend - Charnwood Brick Group Ltd
Fineline - Jebron Ltd
Fineline - Matki plc
Fineline - Neslo Partitioning & Interiors
Finesse - ABG Ltd
Finesse - Nevill Long Limited
Finno - Lappset UK Ltd
Finsa - Boardcraft Ltd
Finvector - Dunham Bush Ltd
Firaqua - Firwood Paint & Varnish Co Ltd
Fire Protection Prducts - Smith & Rodger Ltd
Fire Safe - Kingspan Building Products Ltd
Fireater Ltd - Macron Fireater Ltd
FireBar - Sealmaster
Firebeam - Photain Controls Plc
Firebrand - Hart Door Systems Ltd
Firecel - Fitzpatrick Door Frames Ltd
Firecheck - Liquid Plastics Ltd
Firecheck 3000 Series - Tann Synchronome
Firecheck Beta - Tann Synchronome
Firecheck Delta - Tann Synchronome
Firecheck midi - Tann Synchronome
Firechef - Profilex Ltd
Fireclamps® - Abesco Ltd
Firecom - PEL Services Limited
Firedata - McMillan Fire Alarm Systems Ltd
Firedex - JSB Electrical PLC
Firedex 3302 - JSB Electrical PLC
Firedex Biwire - JSB Electrical PLC
FireFace 2000 - Sealmaster
FireFace Extra - Sealmaster
FireFace Plus - Sealmaster
FireFace Standard - Sealmaster
FireFoam - Sealmaster
FireGlaze - Sealmaster
Fireguard - TCW Services (Control) Ltd
Firelite - Valor Heating
Firelock - Sunray Engineering Ltd
Fireman - Clark Door Ltd
Firemaster - Docherty H Ltd
Firemaster - Premdor
Fireplan - Internal Partitions Systems
FirePlugs - Sealmaster
Firesheild - Premdor
Fireshield - Bollom Fire Protection Ltd
Fireshield ™ - Accent Hansen
Firesine - McMillan Fire Alarm Systems Ltd
Firestile - Mann McGowan Fabrication Ltd
Firestop - Bollom Fire Protection Ltd
Firestop - Dow Corning Hansil Ltd
Firestop - Yonder Hill Ltd
Firestore - Franklin Hodge Industries Ltd
Fireswiss - C. G. I. International Ltd, trading as Colebrand Glass
Firetainer - Franklin Hodge Industries Ltd
Firetech - Owens Corning Building Products (UK) Ltd
Fireter - AEI Cables Limited
Firetex - Henderson-Bostwick
Firetex - Leigh's Paints
Firewall - Neslo Partitioning & Interiors
Firewarn - PEL Services Limited
Firewatch - TCW Services (Control) Ltd
Firex - James Smellie Fabrications Ltd
Firglo - Firwood Paint & Varnish Co Ltd
Firlene - Firwood Paint & Varnish Co Ltd
Firlex - Firwood Paint & Varnish Co Ltd
Firpavar - Firwood Paint & Varnish Co Ltd
First Alert - BRK Brands Europe Ltd
First class - Lappset UK Ltd
Firsyn - Firwood Paint & Varnish Co Ltd
Firth - Interface Europe Ltd
Firths Carpets - Interface Europe Ltd
Firwood - Firwood Paint & Varnish Co Ltd
Fischer - Arrow Supply Company Ltd
Fischer - Arthur Fisher (UK) Ltd
Fischer - Kem Edwards Ltd
Fitex - Airsprung Beds Ltd
Fitter & Paulton - PFP Electrical Products Ltd
5 Star signs - Glasdon Designs Ltd
500 Series - Opella Ltd
513 NPE - Otis plc
5 Star - Webrr & Broutin
506 NCE - Otis plc
Fix - Grorud Industries Ltd
Fix - Sampson T F Ltd
Flair - Taylor & Portway Ltd
Flairline - Hadley Industries Plc
Flame Bar - Fire Protection Ltd
Flameguard - Bollom Fire Protection Ltd
Flameguard - Bonar Floors Ltd
Flamenco - MMP International
Flamenco Super - Valor Heating
Flamestat - Jimi-Heat Ltd
Flametex - Airsprung Beds Ltd
FlamPac - Bollom Fire Protection Ltd
Flanders - Forticrete Roofing Products
Flapmaster - Paddock Fabrications Ltd
Flapstile - Gunnebo Mayor Ltd
Flare - Intralux UK Ltd
Flash - Realstone Ltd
Flash - Schlumberger
Flashpoint infra red - Emergi-Lite Safety Systems Ltd
Flavel - Flavel - Leisure Glynwed Consumer & Construction Products Ltd
Flaviker Tiles - Reed Harris
Fleck Coat - Stewart Wales, Somerville Ltd
Flectoline - Rockfon Ltd
Fleet - Ingersoll - Rand Architectural Hardware
Flemish tile - Sandtoft Roof Tiles Ltd
Flex - seal - Flex-Seal Couplings Ltd
Flexcase - Universal Components Ltd
Flexcell - Fosroc Expandite Ltd
Flexcon - Flamco UK Ltd
Flexconpak - Flamco UK Ltd
Flexfiller - Flamco UK Ltd
Flexflyte - Flexible Ducting Ltd
Flexi - Ledu UK Ltd
Flexibar 2000 - Manton Industrial Seals Ltd
Flexiboard - Salex Interiors Ltd
Flexiburo - Flexiform Business Furniture Ltd
Flexicon - Polypipe Civils Ltd
Flexidoor - EAP International Ltd
Flexidoor - Telford Doors and Barriers Ltd
Flexiframe ® - Andersen/ Black Millwork
Flexiglide - Flexiform Business Furniture Ltd
Flexiguard - Advanced Perimeter Systems
Fleximetric - Flexiform Business Furniture Ltd
Flexion - Envirodoor Markus Ltd
Flexirend Highbuild - Snowcem
Flexiseal - RIW Ltd
Flexisound - Salex Interiors Ltd
Flexistor - Flexiform Business Furniture Ltd
Flexithane - Tor Coatings Group
Flexitrack - Beta Lighting Ltd
Flexitrack - Metlex
Flexlite - Ventlane Ltd
Flexlock - Viking Johnson
Flexmaster - Docherty H Ltd
Flextight seccoral systems - P.C.I Construction Systems Ltd
Flextract 2000 - Flexible Ducting Ltd
Flexvent - Flamco UK Ltd
Flint Range - Beacon Hill Brick Co Ltd
Flipflap - Envirodoor Markus Ltd
Flo-jo ® - Maestro International Ltd
Floclad - Cladding & Decking UK Ltd
Flocoat - Hub Tubes
Flolift - Conder Products Ltd
Floodline - Illuma Lighting Ltd
Floor to Floor - Supreme Lifts Ltd
Floor-fast - Lindapter International
Floorcote - Coo-Var Ltd
Floorfast - P.C.I Construction Systems Ltd
Floorline - Egger (UK) Ltd
Florad - FloRad Heating Systems
Floralis - Sanderson & Sons Ltd, Arthur
FlorAwall - PHI Group Ltd
Florentine - Forticrete Ltd
Florette - Carpets of Worth Ltd
Florex - Tor Coatings Group
Florida - Leisure, Glynwed Consumer Foodservice Products Ltd
Flos/Arteluce - Quip Lighting, Div of Bruce & Macintyre Ltd
Flotex - Bonar Floors Ltd
Flotex - Bonar Floors Ltd
Floway - CPV Ltd
Flowcrete - Flowcrete Plc
Flowforge® - Redman Fisher Engineering
Flowgrid - Redman Fisher Engineering
FlowGRip TM - Redman Fisher Engineering
Flowguard - Isopad Ltd

Flowlok - Redman Fisher Engineering
Flowpac - Barr & Wray Ltd
Flowtop - Watco UK Ltd
FLT - Ferham Products
Flueflex - Ventlane Ltd
Fluidair - F. W. Harris & Co Ltd
Flush glazing - Clestra Hauserman Ltd
Flushplan - Internal Partitions Systems
Flushwall - Neslo Partitioning & Interiors
FMB - Federation of Master Builders
Foamalux ® - Righton Ltd
Foamglas - Pittsburg Corning (UK) Ltd
Focal Point - Brintons Ltd
Focal point - nmc (UK) Ltd
Focus - Ecophon Ltd
Focus - James Gibbons Format Ltd
Fogo Montanha - Emsworth Fireplaces Ltd
Fogtec Water Mist - Macron Fireater Ltd
Foilboard - Callenders Ltd
Foldaway - Tilley International Ltd
Fondaline - Onduline Building Products Ltd
Fondu - Lafarge Aluminates Ltd
Footnotes - Woolliscroft Tiles Ltd
Footprint - Footprint Tools Ltd
Force - Axter Ltd
Forceflow - Biddle Air Systems Ltd
Fordham Pland - Astracast PLC
Fordham Pland - Fordham Pland
Forest - Lappset UK Ltd
Forest-Saver - Earth Anchors Ltd
Forgecrete - Cheshire Concrete Products
Forgecrete plus - Cheshire Concrete Products
Formagrid - Formwood
Formalux - Formwood
Format - James Gibbons Format Ltd
Format 50 - James Gibbons Format Ltd
Formawall - European Profiles
Formica - Decra Cubicles Ltd
Formica - Formica Ltd
Formica - Viking Laminates Ltd
Formica Spawall - Formica Ltd
Formica Cubewall - Formica Ltd
Formica Firewall - Formica Ltd
Formica Lifeseal - Formica Ltd
Formica Prima - Formica Ltd
Formica Unipanel - Formica Ltd
Forpave - Bardon (England) Ltd
Fortic - IMI Range Ltd
Foscarini - Quip Lighting, Div of Bruce & Macintyre Ltd
Fospro - Hellerman Electric, Member of Bopwthorpe Holdings PLC
Foulmaster - Pump Technical Services Ltd
Foundation 2000 - Computer Foundations Ltd
4D Castings - The Modelshop
4D Scenics - The Modelshop
4200 Sidewalk fibre bonded carpet - Burmatex Ltd
4000 AC - GE Bayer Silicones
4400 Broadway fibre bonded carpet - Burmatex Ltd
425 Range - Horstmann Timers & Controls Ltd
4-hour fire door - Rolflex Doors Ltd
Fox Tools - Hyde Brian Ltd
Foxxx swimming pool kits - Fox Pool (UK) Ltd
FP 2-14 Zone Panel - C-Tec (Computionics Ltd)
Frame Fast - HW Systems Ltd
Frameseal - Fosroc Expandite Ltd
Frametherm - Davant Products Ltd
Frank - Creteco Sales Ltd
Freedom 5 Vertical Lift - Brooks Stairlifts Ltd
Freeflex - William Freeman Ltd
FreeFlush - Cistermiser Ltd
Freestanding Systems - Spur Shelving Ltd
Freestyle - Swedecor Ltd
Freeway - Vantrunk Engineering Ltd
Freeway 11 - Cistermiser Ltd
Freezstop - Heat Trace Ltd
Frenchwood ® - Andersen/ Black Millwork
Fridgewatch - J & E Hall Ltd
Friedland - Caradon Friedland Ltd
Frimeda Rapid Lift - Halfen Ltd
Froggo (Frog bin) - Glasdon UK Ltd
Frontrunner - Plastic Extruders Ltd
Frostar - Imperial Machine Co Ltd
Frostguard - Isopad Ltd
FRT 80 - Thatching Advisory Services Ltd
FSB - Allgood plc
FSB Door Furniture - JB Architectural Ironmongery Ltd
FST - Binns Fencing Ltd
Fugue - Kinsley-Cooke Furniture Ltd
Fujitsu - HV Air Conditioning Ltd
Fujitsu - Purified Air Ltd
Fulbora - ACO Technologies Plc
Fulfil - Vencel Resil Ltd
Fulscope Aquamag Aquaprobe - ABB Kent Taylor Ltd.
Fulton-Series E - Fulton Boiler Works (Great Britain) Ltd
Fungrass - LazyLawn
Futech - Arthur Fisher (UK) Ltd
FW 50 - SCHUCO International KG
Fyfestone - Camas Building Materials
Fyfestone - Fyfe John Ltd
Fyrebag 240 - Mann McGowan Fabrication Ltd
Fyrex - Iles Waste Systems
G series tanks - Glasdon Designs Ltd
G-Flex - Johnston Pipes Ltd
G.B. Wendland - Caradon Duraflex Systems Ltd
G146 Gutter - Alifabs (Woking) Ltd
G3 Gates - Albion Fencing Ltd
G55 Gutter - Alifabs (Woking) Ltd
Gainsborough Hardware - Harwood & Welpac Hardware Ltd
Galante - Tegral Building Materials Ltd
Galaxy Ultra - Brandon Medical
Galena Spencer Ltd - Fire Protection Services PLC
Galerie Range - Caradon Bathrooms Ltd
Gallery - Symphony Group PLC, The
Galletti Fan Coil Units - Therminal Technology (sales) Ltd
Galon - Muraspec Ltd
Galtres - York Handmade Brick Co Ltd
Gameflor - Halstead James Ltd
Garador - Garador Ltd
Gardex - Palgrave Brown Timber & Fireproofing Co
Gardwall - Brown F plc
Garmex - Metalrax Ltd
Gascogne Limestone - Kirkstone
Gascool - Birdsall Services Ltd
Gasgard - Marley Waterproofing
Gasgard - MSA (Britain) Ltd
Gasgard gas detector - MSA (Britain) Ltd
Gass - SGB Youngman
GasSeal - Marley Waterproofing
Gasstyle 2 - Dunbrik (Yorks) Ltd
Gastech - Birdsall Services Ltd
Gate - Millar Dennis (1998) Ltd
Gates - APT Controls Ltd
Gatic 2000 - Elkington Gatic Ltd
Gatic Easylift - Elkington Gatic Ltd
Gatic Slotdrain - Elkington Gatic Ltd
GEC Anderson - Anderson GEC Ltd
Gedina - Ecophon Ltd
Geepee - Swish Building Products Ltd
Gelva - Atkinson & Kirby Ltd
Gem - Tunstal Telecom Ltd
Gemini - Forticrete Roofing Products
Gemini - Hoyles Fire & Safety Ltd
Gemini - Leisure, Glynwed Consumer Foodservice Products Ltd
Gemini - Roof Units Group
Gemini Shelving - Balmforth Engineering Ltd
Gemini tile - Forticrete Ltd
Genesis - Dixon Turner Wallcoverings Ltd
Genesis - Dixon Turner Wallcoverings Ltd.
Genesis - Eltron Chromalox
Genie - Genie UK Ltd
Genovent - Gebhardt Kiloheat
Genpac - Sandhurst MFG Co Ltd
Gent - Caradon Gent Ltd
Geo-Coir - PHI Group Ltd
Geofix - FEB Ltd
Geojute - Land Wood & Water Co. Ltd
Geojute - PHI Group Ltd
Geomat - Land Wood & Water Co. Ltd
George Fischer - Capper Pipe Systems Ltd
Geotextiles - Land Wood & Water Co. Ltd
Gerard - Decra Roofing Systems
Gerbergraphix - Green Brothers Ltd
Gerni - Nilfisk Advance Ltd
Geze - City Hardware and Security
Gifcrete - GRAB Industrial Flooring Ltd
Gildeloc - HCL Verta
Gillair smoke vent products - Matthews & Yates Ltd
GL1000 - Gleno Industries Ltd
GL24 - Gleno Industries Ltd
GL24W - Gleno Industries Ltd
GL252 - Gleno Industries Ltd
GL3 - Gleno Industries Ltd
GL32 - Gleno Industries Ltd
GL32W - Gleno Industries Ltd
GL38 - Gleno Industries Ltd
GL38W - Gleno Industries Ltd
GL6 - Gleno Industries Ltd
GL914 - Gleno Industries Ltd
Glacier - Tenby Industries Ltd
Glada - Novellini Bathroom Products
Glamox - Glamox Electric (UK) Ltd
Glamur - Jotun Henry Clark Ltd
Glasflex - Flexible Ducting Ltd
Glasgrid - Associated Holdings Ltd
Glasroc - British Gypsum Ltd
Glasroc - Nevill Long Limited
Glass Blocks - Swedecor Ltd
Glass chalkboards - Lightfoot Charles Ltd
Glassgard - Bonwyke Ltd
Glasstile - Gunnebo Mayor Ltd
Glasstrim GRP - Areco Roofing Supplies Company
Glasswhite - Cardale Doors Ltd
Glasswood - Cardale Doors Ltd
Glasurit - Marcel Guest Paints
Glazing Accessories - Simon R W Ltd
Glehmoor - Tegral Building Materials Ltd
Glendale - Lycetts (Burslem) Ltd
Glendyne Slate - Blunn Slates Ltd
Glenfield - Bennett Windows Ltd
Glenium - Feb MBT
Gleno - Therm - Gleno Industries Ltd
Glidden - ICI Dulux Trade Paints
Glidebolt - Paddock Fabrications Ltd
Glidemaster - Hillaldam Coburn Ltd
Glidetrak - Rackline Systems Storage Ltd
Glidevale - Willan Building Services Ltd
Glo-dac - Dacrylate Paints Ltd
Global Sign Sales - Signs International
Globe - Millar Dennis (1998) Ltd
Globelite - LB Lighting Ltd
Glocote - Coo-Var Ltd
Gloria - Kermi (UK) Ltd
Glostal - Glostal - Monarch
Gloster - Gloster Furniture Ltd
Gloucester - Atlas Stone Products
Glover Series 2000 Extinguishers - Fife Fire Engineers & Consultants Ltd
Glow-Worm - Hepworth Heating Ltd
Gloy - Henkel Home Improvement & Adhesive Products
Glulam - Structural Timbers Ltd
Glynn-Johnson - Relcross Ltd
Go Green - ABG Ltd
Godwin Pumps - Goodwin HJ Ltd
Gold - Novellini Bathroom Products
Gold - Wilo Samson Pumps Ltd
Goldbach - Mauser Interiors (UK) Ltd
Goldseal - Marley Waterproofing
Goldstar - Wilo Samson Pumps Ltd
Goods-rol - Lycetts (Burslem) Ltd
Goodsmaster - Stannah Lifts Ltd
Gordon Russell - Harvey Furniture
Gordon Russell - Steelcase Strafor Plc
Goslo - Fyfe John Ltd
Gothic - Oswestry Reinforced Plastics Ltd
GR 1 Graffiti Remover - Stewart Wales, Somerville Ltd
Grabber screws - Lafarge Plasterboard Ltd
Gradelux - Laidlaw Architectural Hardware Ltd
Gradus - Gradus Ltd
Graffiti-Gard - Andura Textured Masonry Coatings Ltd
Gragreen - En-tout-cas South West
Graham Avis Details - The Modelshop
Grampian - Sashless Window Co Ltd
Grando UK - Buckingham Swimming Pools Ltd
Granette - DLW Flooring Ltd
Graniplast Trowelled Marble - Textured European Finishes Ltd
Granirapid - Mapei UK Ltd
Gransprung - Granwood Flooring Ltd
Grantile - Granflex Roofing Ltd
Granwood - Granwood Flooring Ltd
Grassblock - Grass Concrete Ltd
Grassblock - Landscape Grass (Concrete) Ltd
Grasscel Paving - Ruthin Precast Concrete Ltd
Grasscrete - Grass Concrete Ltd
Grasscrete - Landscape Grass (Concrete) Ltd
Grasspave - ABG Ltd
Grassroad - Grass Concrete Ltd
Grassroad - Landscape Grass (Concrete) Ltd
Grate-fast - Lindapter International
Grateglow - Robinson Willey Ltd
Grease Bugs - Progressive Product Developments Ltd
Greaves Portmadoc Slate - Greaves Welsh Slate Co Ltd
Greenfix - PHI Group Ltd
Greenmoor sandstone - Lindley George & Sons Ltd
Greenwood - Sandtoft Roof Tiles Ltd
Gres de Valles - Stokes R J & Co Ltd
Grid Plus - Caradon M K Electric Ltd
Grid-lok - Vantrunk Engineering Ltd
Gridplex - Cameron & Moore (Gridplex) Ltd
Griffin - Tinsley & Co Ltd, Eliza
Griffin Fire - Griffin and General Fire Services Ltd
Gripfalt SMA - Johnston Roadstone Ltd
Gripfast - Vallance
Gripfill - Laybond Products Ltd
Grippa - Talbot
Griptop - GRAB Industrial Flooring Ltd
Grisco Bauder - Phoenix Roofing
Gro-Lux - Sylvania Lighting International INC.
Groheart - Grohe Ltd
Grohedal - Grohe Ltd
Grohetec - Grohe Ltd
Grohetec special fittings - Grohe Ltd
Groutex - Sealocrete Ltd
Groutfast - P.C.I Construction Systems Ltd
Grovesnor - ciptek - Grosvenor Pumps Ltd
GRP 4 - Group Four Glassfibre Ltd
Grundy Kuppersbusch - Grundy Catering
Grundy Olner Toms - Grundy Catering
Grundy Shopfitting - Grundy Catering
Grundy Survice - Grundy Catering
Grundy Systems - Grundy Catering
Guard - Andrews J S (Coatings) Ltd
Guard - Dixon Turner Wallcoverings Ltd.
Guard Dog - Poselco Lighting
Guardian - Intersolar Group Ltd
Guardian - Portasilo Ltd
Guardian - Sissons W & G Ltd
Guardstation - Guardall
Guardwire - Geoquip Ltd
Guideline - The Building Centre Group Ltd
Guildford Fabrics - Interface Europe Ltd
Guiraud Freres - Smithbrook Building Products Ltd
Gulfstream - Gledhill Water Storage Ltd
Gullwing See-Saw - Record Playground Equipment Ltd
Gun-point - Gun-Point Ltd
Gunk - Unicorn Abrasives (UK) Ltd
GW Axial - Matthews & Yates Ltd
Gwerthiannau Arwyddion Bydeang - Signs International
Gym-flor - Dex-o-Tex International Ltd
Gymnasium Side Mounting Range - Lumitron Ltd
Gypframe - Mestec PLC
Gypglas - Nevill Long Limited
Gyproc - British Gypsum Ltd
Gyproc - Nevill Long Limited
Gyvlon - Flowcrete Plc
Gyvlon - Isocrete Group Sales Ltd
H & R Johnson - Stokes R J & Co Ltd
H.S.M - Tony Team Ltd
H20 - E grp pipes - Johnston Pipes Ltd
Haddon - Tecstone - Haddonstone Ltd
Haddonstone - Haddonstone Ltd
Hadfields - Kalon Decorative Products
Hadrian - Relcross Ltd
Hadron slates - Forticrete Roofing Products
Haffele UK Ltd - Haffele UK Ltd
Hagley - Welconstruct Co
Hairfelt - Davant Products Ltd
Hairy snake - Manton Industrial Seals Ltd
Halcyon - Rixonway Kitchens Ltd
Hale Hamilton Ltd - Hale Hamilton Ltd
Halex - AEI Cables Limited
Halfen Channels - Halfen Ltd
Halfen fixing system - Halfen Ltd
Hallmark - Forticrete Ltd
Hallscrew - J & E Hall Ltd
Halo - Bowater Containers Ltd
Halolux - JSB Electrical PLC
Halostar - Osram Ltd
Halstead - Diamond Merchants
Hambleton - Sashless Window Co Ltd
Hambleton - York Handmade Brick Co Ltd
Hamilton Litestat - Hamilton R & Co Ltd
Hamron products - Hamron Group
Hamron Steel 6 Intumescent Coating - Hamron Group
Hamron Steel 7 Intumescent Coating - Hamron Group
Hamron WD-05 Intumescent Paint - Hamron Group
Handi-Access - Profilex Ltd
Handi-Lift 4 Wheelchair - Bison Bede Ltd
Handiclamp - Viking Johnson
Handifoot - Anderton Concrete Products Ltd
Handitap - Viking Johnson
Handmade Haven Red - Cranleigh Brick & Tile Co Ltd
Handmade Light Multi - Cranleigh Brick & Tile Co Ltd
Handmade Red Multi - Cranleigh Brick & Tile Co Ltd
Handmade Victorian Buff - Cranleigh Brick & Tile Co Ltd
Handy Angle - Link 51 (Storage Products) Ltd
Handy Dri - Heatrae Sadia Heating Ltd
Handy Tube - Link 51 (Storage Products) Ltd
Hansgrohe - Hansgrohe
Hanson Bricks - Brikenden (Builders Merchants) Ltd
Happy Shower - Hansgrohe
Har-Tru - En-tout-cas plc
Harbex - Harris & Bailey Ltd
Hard Deck - Ferguson G A & Co Ltd
Hardac - Webrr & Broutin
Hardrive - Westpile Ltd
Hardrow - Forticrete Ltd
Hardside - Salex Interiors Ltd
Harewood - Ellis J T & Co Ltd
Harlequin - Ferham Products
Harmer - Alumasc Exterior Building Products Ltd
Harmer - Drainage Systems
Harmonise - Paragon by Heckmondwike F B Ltd
Harmony - Bartlett Catering Equipment Ltd
Harmony - Novellini Bathroom Products
Harmony Shower Canopies - Jendico
Harmony Shower Curtains - Jendico
Harmony Shower Panels - Jendico
Harmony Thermostatic - Jendico
Harpoon - BRB Industrial Services Ltd
Hart - Hart Door Systems Ltd
Hartam Park (Box Ground) Stone - Hanson Bath & Portland Stone
Harton and Harco Pak. - Harton Heating Appliances
Harton Boost - Harton Heating Appliances
Harvey - Harvey Furniture
Harwood Hardware - Harwood & Welpac Hardware Ltd
HAScheck - National Britannia Ltd
Hasflex - Hotchkiss Air Supply
Hatcrete - Hatfield Ltd, Roy
Hathernware architectural faience - Ibstock Hathernware Ltd
Hathernware architectural terracotta - Ibstock Hathernware Ltd
Hathernware faience - Ibstock Hathernware Ltd
Hathernware terracotta - Ibstock Hathernware Ltd
Haucphalt - Bardon (England) Ltd
Haunchwood - Ibstock
Haven Red - Cranleigh Brick & Tile Co Ltd
Hawkins - Marley Building Materials Ltd
Haworth - Haworth UK Ltd
Hayward - Fox Pool (UK) Ltd
Hazguard - Thurston Building Systems
HBS flagpoles - Harrision Flagpoles
HCS (HOPPE Compact System) - HOPPE (UK) Ltd
HD Vinyl - Marley Floors Ltd
HDS High Density Stacking Chair - Race Furniture Ltd
Healthguard® - Earth Anchors Ltd
Heartbeat - Valor Heating
Heat Tracer - Heat Trace Ltd
Heat-Rad - Horizon International Ltd
Heat-Saver - Horizon International Ltd
Heatec - Loblite Ltd
Heather Barber - Brintons Ltd
Heatherbrown - Ruabon Dennis Ltd
Heathfield Slate - Blunn Slates Ltd
Heathland - Checkmate Industries Ltd
Heatkeeper Homes - Walker Timber Ltd
Heatovent - Seaque Technologies Ltd
Heavyside - Oakdale (Contracts) Ltd
Hebefix - Pump Technical Services Ltd
Helagrip - Hellerman Electric, Member of Bopwthorpe Holdings PLC
Helatemp - Hellerman Electric, Member of Bopwthorpe Holdings PLC
Heldite - Heldite Ltd
Heli Bar - Helifix Ltd

HeliBond - Helifix Ltd
Helit - Lesco Products Ltd
Helix - ABB Kent Meters
Hellerman - Hellerman Insuloid
Hellermann - Kem Edwards Ltd
Hellermark - Hellerman Electric, Member of Bopwthorpe Holdings PLC
Helmsman - Bawn W. B. & Co Ltd
Helo - Finnish Sauna Baths (Sales) Ltd
Helsyn - Hellerman Electric, Member of Bopwthorpe Holdings PLC
Helvar - Bemlite Ltd
Helvar - Helvar Electrosonic
Helvin - Hellerman Electric, Member of Bopwthorpe Holdings PLC
Hemelite - Tarmac Topblock Ltd
Hemelite Blocks - CAEC Howard (Holdings) Ltd
Hempe - Hyde Brian Ltd
Hempstead systems - Parkwood Mellows Ltd
Hemsec - Hemsec Manufacturing Ltd
Henlow - Hill Leigh (Joinery) Ltd
Henry Nuttall - Viscount Catering Ltd
Henslowe & Fox - Commodore Kitchens Ltd
Hep v0 - Hepworth Building Products Ltd
Hep20 - Hepworth Building Products Ltd
Hep30 - Hepworth Building Products Ltd
Heplock - Hepworth Building Products Ltd
Hepseal - Hepworth Building Products Ltd
Hepsleve plus - Hepworth Building Products Ltd
Hepworth Building Products - Hepworth Building Products Ltd
Hepworth Concrete - Hepworth Building Products Ltd
Hepworth Drainage - Hepworth Building Products Ltd
Hepworth Plumbing - Hepworth Building Products Ltd
Herald - Iles Waste Systems
Heraperm - Isocrete Group Sales Ltd
Heras - Advanced Fencing Systems Ltd
Herbert - BSA Tools Ltd
Herbol - Marcel Guest Paints
Heritage - Bennett Windows Ltd
Heritage - Marshalls Mono Ltd
Heritage Brass - Marcus M Ltd
Heritage Range - Pedley Furniture international Ltd
Heritage System - Hyperion Wall Furniture Ltd
Herz - Ellis Miller Ltd
Hessian 51 - Muraspec Ltd
Heuga - Interface Europe Ltd
Hewi - City Hardware and Security
HEWI - HEWI (UK) Ltd
Hexadeck - Turner Access Ltd
Hexcell - Interlux Ltd
Hexslot - Elkington Gatic Ltd
Heydal - OSC Process Engineering Ltd
Heyes - PFP Electrical Products Ltd
Hi Spot - Mains Voltage Halogen Lamps - Sylvania Lighting International INC.
Hi Watt Batteries - J.F.Poynter Ltd
Hi-Deck - Hi-Store Ltd
Hi-Frame - Hi-Store Ltd
Hi-Slim - Oldham Lighting Ltd
Hi-Style 2000 - HW Systems Ltd
Hi-Ten Elastomeric - Euroroof Ltd
Hi-Trolley - Hi-Store Ltd
Hi-way - Vantrunk Engineering Ltd
Hi-Wide - Welconstruct Co
HiBuild - Everlac (GB) Ltd
Hickson Decor - Hickson Timber Products Ltd
Hicom Communication Server - Siemens Communication Ltd
Hideaway - Hillaldam Coburn Ltd
Highlander - Richard Hose Ltd
Highlight - Leisure, Glynwed Consumer Foodservice Products Ltd
Highlight - Valor Heating
Highspot - Illuma Lighting Ltd
Higlide - Hadley Industries Plc
HiJan - Johnson & Starley
Hilastic 44 - Fosroc Expandite Ltd
Hiline - CPV Ltd
Hille - Evertaut
Hilltop - McAlpine Slate Ltd, Alfred
Hilmor - Hilmor
Hilo - MSA (Britain) Ltd
Hiltons Italian Plaster Coving - Hilton Banks Ltd
Hinton Pavilion - Anthony de Grey Trellises
Hiper Bar - Exitex Ltd
Hiperlan - Alcatel Data Cable
Hiperlan-E - Alcatel Data Cable
Hiperlink - Alcatel Data Cable
Hipernet - Alcatel Data Cable
Hiperway - Alcatel Data Cable
Hirecabin - UK Cabin Co
Hit - Ledu UK Ltd
Hitachi - Duo-Fast (UK) Ltd
Hitachi - HV Air Conditioning Ltd
Hitec Stair - Lewes Design Contracts Ltd
HL - Davis International Ltd
Hocoplast Profile - Juno Roplasto Ltd
Hodgkinson Bennis Stokers - Hodgkinson Bennis Ltd
Hodgkisson - TSE Brownson
Hollo-bolt - Lindapter International
Holodeck - Lees Richard Ltd
Home 'n' dry - Johnson & Starley
Home Automation - Home Automation Ltd
Home-Loc 6000 - Timloc Expamet Building Products
Homefit - Doorfit Products Ltd
Homegard - Marley Waterproofing
HomeGuard - Securikey Ltd
Homelift - Stannah Lifts Ltd
Homelight - Crittall Steel Windows Ltd
Homelux - Homelux
Homer - Interoven Ltd
Homerette - Interoven Ltd
Hometrim - LB Plastics Ltd
Horizon - Hudevad Britain
Horizon - Telelarmcare Ltd
Hormann - Hormann (UK) Ltd
Hoskins Medical Equipment - Hoskins Medical Equipment Ltd
Hot wall - TWS Presentation Systems (Mcmillan UK Ltd T/A)
Hotfoil - Isopad Ltd
Hotrace - Jimi-Heat Ltd
Hotwat - Heat Trace Ltd
Housesafe - Securikey Ltd
Howden - Howden H D Ltd
HQI - Osram Ltd
HR - Davis International Ltd
HR nails, tacks & pins - John Reynolds & Sons (Birmingham) Ltd
HRV - Kiltox Damp Free Solutions
HS/FR Armouring Compound - Thermica Ltd
HT Elastomeric - Anderson D & Son
HT Polyflex - Anderson D & Son
HTA Rain - Kwikform UK Ltd
Huflor - Henderson-Bostwick
Humber Plain - Sandtoft Roof Tiles Ltd
Humbrol - The Modelshop
Humidity Vents - Simon R W Ltd
Humidvent - Airflow Developments Ltd
Hunsingore - York Handmade Brick Co Ltd
Hunter - Capper Pipe Systems Ltd
Huntonit - Norske Interiors (UK) Ltd
Hurdcott Green Sandstone - Chilmark Quarries Ltd
Hurdcott Green sandstone - Gregory Quarries Ltd, The
Hurdcott Green sandstone - The Rare Stone Group Ltd
Hurricane 1000 Gutter - Dales Fabrications Ltd
Hurricane 2000 Gutter - Dales Fabrications Ltd
Hy-rib - Expamet Building Products
Hya-ran - KSB Ltd
Hyclean - Potter & Soar Ltd
Hyclene - Welconstruct Co
Hyde - Hyde Brian Ltd
Hydome - Axter Ltd
Hydrasenta - Daryl Industries Ltd
Hydroban - TRC Service Group
Hydroban ® - TRC (Midlands) Ltd
Hydrocell - Fosroc Expandite Ltd
Hydrocol - Hawley W & Son Ltd
Hydrodeck - Golden Coast Ltd
Hydroduct - Grace Services
Hydroferrox - Hawley W & Son Ltd
Hydroflow - Hydropath (UK) Ltd
Hydronic Air Handling Units - Therminal Technology (sales) Ltd
Hydrotech - Euroroof Ltd
Hydrotite - Laybond Products Ltd
Hyflek - Flowcrete Plc
Hyflex 15 - Hyflex Roofing
Hyflex 5 - Hyflex Roofing
Hyflex BW - Hyflex Roofing
Hyflex Corricouer - Hyflex Roofing
Hyflex CP - Hyflex Roofing
Hyflex Flexstep - Hyflex Roofing
Hyflex10 - Hyflex Roofing
Hyflex10 C.A - Hyflex Roofing
Hygiene - Ecophon Ltd
Hygiene Safe - Kingspan Building Products Ltd
Hygienic - Rockfon Ltd
Hyperscan - Robot (UK) Ltd
Hypra - Legrand Electric Ltd
Hyranger - Axter Ltd
Hyslide - Welconstruct Co
Hytech - Welconstruct Co
Hytex - Potter & Soar Ltd
Hytherm - Axter Ltd
Hyvodex - Hyflex Roofing
Hyweld - Potter & Soar Ltd
I Tre - Forma Lighting Ltd
I-Line - Square D, Electrical Distribution Division
I.C.P - Rockfon Ltd
Ibiza - Kermi (UK) Ltd
IBM - INMAC (a division of microwarehouse Ltd)
Iboflor - Checkmate Industries Ltd
Ibstock Hathernware - Ibstock Building Products Ltd
Ibstock Scottish brick - Ibstock Building Products Ltd
Ibstock Scottish Stone - Ibstock Scottish Brick
Icelert - Findlay Irvine Ltd
Icemate - CSSP (Computer Software Services Partnership)
Iceni - Blyde-Barton Ltd
Iceni Extinguishers - UK Fire International Ltd
Icepac - CSSP (Computer Software Services Partnership)
ICI - Anderson Gibb & Wilson
Icon - Valor Heating
ICS chimney system - Rite-Vent Ltd
ICS5000 chimney system - Rite-Vent Ltd
Ideal - Caradon Heating Ltd
Ideal - Ideal Furniture Enterprises Ltd
Ideal - Tuke & Bell Ltd
Ideal Caradon - Brikenden (Builders Merchants) Ltd
IDM - Kone Lifts Ltd
Ifthane - IFT Supplies Ltd
IG - IG Limited
IG Sectional Doors - Regency Garage Door Services Ltd
Ikos Encaustic Plaster - Textured European Finishes Ltd
Illac resin - Illbruck Ltd
Illmod - Illbruck Ltd
Illroc - Illbruck Ltd
Illsonic - Salex Interiors Ltd
Illtec - Illbruck Ltd
Illtec FM - Illbruck Ltd
IMA Pro spec and insolvency guarantees - Anderson D & Son
Image - Interface Europe Ltd
Imagine - Helvar Electrosonic
Imit - Altecnic Ltd
Impactor - Imperial Machine Co Ltd
Impel - Marflow Eng Ltd
Imper Italia - Intergrated Polymer Systems (UK) Ltd
Imperial - Miele Co Ltd
Imperial - Smart Systems Ltd
Imperial Access Lift - Brooks Stairlifts Ltd
Imperial locks - Guardian Lock and Engineering Co Ltd
Impermia - Callenders Ltd
Impisx - Dexion Ltd
Impression - Komfort Systems Ltd
Imprint - Prismo Ltd
Impro - BPT Security Systems (UK) Ltd
In stock - Dixon Turner Wallcoverings Ltd
In-Sit - Unifix Ltd
Indeline - Bollom J W & Co Ltd
Index cable glands - Elkay Electrical Manufacturing Co Ltd
Index of Quality Names - British Carpet Manufacturers Association
Indicator Mesh - Associated Holdings Ltd
Industrial premier strip - Manton Industrial Seals Ltd
Infil - Infil Ltd
Infinity - Shackerley (Holdings) Group Ltd incorporating Designer ceramics
Inforail ® - Broxap & Corby Ltd
Ingersol-Rand - Ingersoll-Rand European Sales Ltd
Ingersoll - Yale Security Products Ltd
Ingersoll Rand - Duo-Fast (UK) Ltd
Ingersoll rand - F. W. Harris & Co Ltd
Initial - Deborah Services
Initial Sherrock Monitoring - Initial Electronic Security Systems Ltd
Initial Sherrork Surveilance - Initial Electronic Security Systems Ltd
Initial Shorrock Fire - Initial Electronic Security Systems Ltd
Initial Shorrock Monitoring - Ingersoll-Rand European Sales Ltd
Initial Shorrock Security - Initial Electronic Security Systems Ltd
Initial Shorrock Surveillance - Ingersoll-Rand European Sales Ltd
Initial Shorrock Systems - Initial Electronic Security Systems Ltd
Injection Plastics - Hines & Sons, P E, Ltd
Inlay Range - Pedley Furniture international Ltd
Innerspace Storage Wall - ARC Partitioning
Innovair - Biddle Air Systems Ltd
Inova - Guardall
Inperim - TDI (UK) Ltd
Inpro by Gradus - Gradus Ltd
Insitex - Creteco Sales Ltd
InSkew - Helifix Ltd
Insta - Ball David Group plc
Instacem concrete repair system - David Ball Group Plc
Instacem polymer floor system - David Ball Group Plc
Instacem Underwater mortar - David Ball Group Plc
Instacem Wall Render - David Ball Group Plc
Instacom - Rollalong
Instacoustic - InstaGroup Ltd
Instafibre - InstaGroup Ltd
Instaflex - George Fischer Sales Ltd
Instaflex flexible jointing mastic - David Ball Group Plc
Instalastic flexible waterproof coating - David Ball Group Plc
Instamac instant repair material - David Ball Group Plc
Instant Adhesive - Loctite UK Ltd
Instapruf Polymerflooring topping - David Ball Group Plc
Instaset rapid repair mortar - David Ball Group Plc
Insudoor - Stokvis R S & Sons Ltd
Insufloor - RMC Readymix Limited
Insugard - Henderson-Bostwick
Insuglaze - Panel Systems Ltd
Insul-Sheet - nmc (UK) Ltd
Insul-Tube - nmc (UK) Ltd
Insulath - BRC Building Products
Insulderm - Crosbie Coatings
Insulight - Pilkington Plyglass
Insulink - Vencel Resil Ltd
Insulite - European Profiles
Insulite FR - European Profiles
Insuloid - Hellermann Insuloid
Insulok - Hellermann Insuloid
Insultube - Davant Products Ltd
Insuroll - Stokvis R S & Sons Ltd
Intec - Creteco Sales Ltd
Integra - Dauphin PLC
integra - Hawley W & Son Ltd
Integra - Integra Products
Intelekt - Eclipse Blind Systems Ltd
Intellect - Crompton Lighting Ltd
Intellect Media Walls - TWS Presentation Systems (Mcmillan UK Ltd T/A)
Intelock - Vantrunk Engineering Ltd
Interbond FP - International Paint, Protective Coatings,
Intercell - Interface Europe Ltd
Intercure rapid curing epoxies - International Paint, Protective Coatings,
Interface - Interface Europe Ltd
Interfine b29 iso cyonate free finishes - International Paint, Protective Coatings,
Interlock - Howarth Timber (Elland) Ltd
Interplan - Internal Partitions Systems
Interpritations - Paragon by Heckmondwike F B Ltd
Interscreen - Safetell Ltd
Interspan - Compriband Ltd
Interspan All Metal - Compriband Ltd
Interstrip - Shackerley (Holdings) Group Ltd incorporating Designer ceramics
Interzinc 72 epoxy zinc rich - International Paint, Protective Coatings,
Interzone high solid expoxies - International Paint, Protective Coatings,
Intrad - Fixatrad Ltd
Intralux - Intralux UK Ltd
Intravent - Gebhardt Kiloheat
Intuclear - Bollom Fire Protection Ltd
Intufoam - Quelfire
Intumescent Compound - Sealmaster
IntuPac - Bollom Fire Protection Ltd
Intuvar - Bollom Fire Protection Ltd
Intuwrap - Quelfire
Invadex - Chiltern Invadex
Invaflo Shower Drain - Chiltern Invadex
Invalux - Chiltern Invadex
Inversa - Ecophon Ltd
Invicta - Invicita Plastics Ltd
Invisidor - Biddle Air Systems Ltd
Ioniq - Trespa UK Ltd
Ionpure - USF Permutit/Ionpure
IPA - Sampson T F Ltd
IPAF Training - Turner Access Ltd
Ipe - Wilo Samson Pumps Ltd
Ipn - Wilo Samson Pumps Ltd
IPS washroom systems - Armitage Shanks Ltd
IRT - Isopad Ltd
Isis - Stuart Turner Ltd
Island - Neptune Outdoor Furniture Ltd
Isocrete - Flowcrete Plc
Isocrete Self Level - Isocrete Group Sales Ltd
Isofast - SFS Stadler Ltd
Isoflex - Isolated Systems Ltd
Isofoam - Baxenden Chemicals Ltd
Isogran - Isocrete Group Sales Ltd
Isokern - ISOKERN (UK) Ltd
Isokoat - ISOKERN (UK) Ltd
Isola Marble & Granite - Reed Harris
Isolgomma - Sound Service (Oxford) Ltd
IsoSystems Original - IsoSystems
IsoSystems Traditional - IsoSystems
Isotape - Isopad Ltd
Isotherm - Biddle Air Systems Ltd
Isothin - Isocrete Group Sales Ltd
Isouosl - Nevill Long Limited
Isowool - British Gypsum Ltd
ISPO Products - Snowcem
Italiana - Phoenix Roofing
Italiana Camini - Emsworth Fireplaces Ltd
Itconi - Lewes Design Contracts Ltd
ITS (Interior Transformation Services) - ITS Ceilings & Partitions
Ivanhoe - Ferham Products
J & S Pumps - ABS Pumps Ltd
J R Pearson - Drawn Metal Ltd
J R Pearson Revolvers - Pearsons J R (Birmingham) Ltd
J/S Warm Flow - Johnson & Starley
Jabcore - Vencel Resil Ltd
Jabdec - Vencel Resil Ltd
Jablite - Vencel Resil Ltd
Jabroll - Vencel Resil Ltd
Jabsqueeze - Vencel Resil Ltd
Jabtherm - Vencel Resil Ltd
Jackson - Jebron Ltd
Jackson - Viscount Catering Ltd
Jacksons fencing ® - Jackson H S & Son (Fencing) Ltd
Jacksons finefencing ® - Jackson H S & Son (Fencing) Ltd
Jackstone - J & E Hall Ltd
Jacuzzi - Jacuzzi UK Ltd, Division of Jacuzzi Inc
Jakcure ® - Jackson H S & Son (Fencing) Ltd
Jaktop ® - Jackson H S & Son (Fencing) Ltd
James Athena - James Fairley Athena
James Fairley - James Fairley Athena
Janssen-Dings - Smithbrook Building Products Ltd
Janus - Johnson & Starley
Jasper - Project Office Furniture PLC
Javelin 2 - CIL International Ltd
Jazz - Gaskell Textiles Ltd
JCB - Bamford J C Excavators Ltd
Jebron - Jebron Ltd
Jeka - Smithbrook Building Products Ltd
Jendispa - Jendico
Jet-rite - Flamco UK Ltd
Jetfloor - Marshalls Flooring Ltd
Jetmaster - Dimplex (UK) Ltd
Jetslab - Marshalls Flooring Ltd
Jimi-Heat - Jimi-Heat Ltd
Jizer - Deb Ltd
Jobomulch - ABG Ltd
John Carr - Rugby Joinery
John Tann - Rosengerns Tann Ltd
Johnstone's Paints - Kalon Decorative Products
Joltec - Ronacrete Ltd

Jorgian Wire - Oswestry Reinforced Plastics Ltd
Joseph Bramah - Bramah Security Equipment Ltd
Jotalakk - Jotun Henry Clark Ltd
Jotaplast - Jotun Henry Clark Ltd
Jotapro - Jotun Henry Clark Ltd
Jotawall - Jotun Henry Clark Ltd
JP - Ward Bekker Sales Ltd
JPA - Pulsford Associates Ltd, John
Jubilee - JSB Electrical PLC
Jubilee - Wade International (UK) Ltd
Juggenaut Poles - Harrision Flagpoles
Juliette - Loft Centre Products
Jumbo - Swish Building Products Ltd
Jumbotec - Swish Building Products Ltd
Jumbovent - Swish Building Products Ltd
Jung - New Haden Pumps Ltd
Jung - Pump Technical Services Ltd
Junior Kwiktower - Kwikform UK Ltd
Jupiter - Fabrikat (Nottingham) Ltd
Jupiter - Guardall
Jupiter Range - Caradon Bathrooms Ltd
Jutlandia Doors - Jutlandia Doors
JVC - Vidionics Security Systems Ltd
JW Green Swimming Pools Ltd - Regency Swimming Pools Ltd
K - fill - Kay Metzeler Ltd
K -Rend - Kilwaughter Chemical Co Ltd
K Range - Kooltherm Insulation Production Ltd
K Screed - Isocrete Group Sales Ltd
K&S Metals - The Modelshop
K-Flex - Isopad Ltd
K-Seal range - Kingfisher Chemicals Ltd
K-System - Hauration Kaskade Ltd
K4 - Komfort Systems Ltd
K40 - Henderson-Bostwick
K500 / K700 Delivery hose - UK Fire International Ltd
K7 - Komfort Systems Ltd
Kaba Benzine - Kaba UK Ltd
Kaba Legic - Kaba UK Ltd
Kaba Vario - Kaba UK Ltd
Kadett - Dermide Ltd
Kair - Kair - The Ventilation Division of Kiltox Chemicals Limited
Kair - Kiltox Damp Free Solutions
Kalahari - Kiltox Damp Free Solutions
Kaloric - Kaloric Heater Co Ltd
Kalwall - Stoakes Systems Ltd
Kalzip - E. H. S. Roofing Ltd
Kameo - Komfort Systems Ltd
Kanaline - Flexible Ducting Ltd
Kappa - Rockwell Sheet Sales Ltd
Kappa Curvo - Rockwell Sheet Sales Ltd
Kara - Neptune Outdoor Furniture Ltd
Karbo - Stanton-Thompson (Agencies) Ltd
Karibia - Kompan (UK) Ltd
Kassel Kerb - Redland Precast
Kassette Roofing System - Kwikform UK Ltd
Kawneer - EBS Services Ltd
Kawneer - Lakeside Group
Kay -fill - Kay Metzeler Ltd
Kaycel - Kay Metzeler Ltd
KD 100 - Hauration Kaskade Ltd
Kee Guard ® - Kee Klamps Ltd
Kee Klamp ® - Kee Klamps Ltd
Kee Line ® - Kee Klamps Ltd
Kee nect ™ - Kee Klamps Ltd
Keepheat - Yule & Son Ltd, A C
Kemnay Ashlar - Camas Building Materials
Kemnay Ashlar - Fyfe John Ltd
Kemperol - Granflex Roofing Ltd
Kempston - Hanson Brick Ltd
Kempston - Hanson Brick Ltd, Northern Regonal Sales Office
Kenco - The Kenco Coffee Company
Kenco In-Cup - The Kenco Coffee Company
Kenco Singles - The Kenco Coffee Company
Kenrick - Kenrick Archibald & Sons Ltd
Kenyon floor paint - Kenyon Industrial Paints & Adhesives
Keps - Springvale E P S
Kerabond - Mapei UK Ltd
Keracolor - Mapei UK Ltd
Kerafloor - Mapei UK Ltd
Keraion - Buchtal (U.K.) Limited
Keralastic - Mapei UK Ltd
Keraquick - Mapei UK Ltd
Kerbdrain - ACO Technologies Plc
Kerbmaster - Benton Co Ltd Edward
Kermira - Anderson Gibb & Wilson
Kerridge - Kerridge Stone
Kerridge - Macc Stone Ltd
Kestrel series - Falcon Catering Equipment
Kew - Alto Cleaning Systems (UK) Ltd
Keybak - Securikey Ltd
Keyblock - Marshalls Mono Ltd
Keyboard - Kronospan Ltd
Keyguard - Hoyles Fire & Safety Ltd
Keylex - Relcross Ltd
Keyline 100 Master - NT Partition Systems
Keymer - Keymer Hand Made Clay Tiles
Keymer Inline Ventilation System - Keymer Hand Made Clay Tiles
Keypac - Wernick S & Sons Ltd
Keystone - PHI Group Ltd
Kiddi-Ride - Record Playground Equipment Ltd
Killaspray - Hozelock Ltd
Kiltox - Kair - The Ventilation Division of Kiltox Chemicals Limited
Kiltox - Kiltox Damp Free Solutions
King - Novellini Bathroom Products
King Door Products - Beaver Industrial Doors Ltd
King Hoists - PCT Group Sales Ltd
Kingfisher - MacBee Manufacturing Ltd
Kingley - Flowflex Components Ltd
Kingshield - Power Breaker
Kingsley Doors - JRM Doors Ltd
Kingsley Trademaster - JRM Doors Ltd
Kingsley XL™ - JRM Doors Ltd
Kingsley-Cooke Wallstore - Kinsley-Cooke Furniture Ltd
Kingspan - E. H. S. Roofing Ltd
Kingston - Balmoral Porcelain Limited/Longmead Ceramics
Kingston - Gloster Furniture Ltd
Kingstonian - Kingstonian Point Ltd
Kingswood - Allied Manufacturing Co (London) Ltd
Kinmar ® - Kinnings Marlow Ltd
Kirkby - Burlington Slate Ltd
Kirkstone Volcanic Stone - Kirkstone
Kitchen Confidence - Symphony Group PLC, The
Kito Bitumen Emultions - South Western Tar Distilleries
Kiz - Interlux Ltd
Kizfireguard - Interlux Ltd
Klassic - Komfort Systems Ltd
Kleertred - Exitex Ltd
Klemm - Ingersoll-Rand European Sales Ltd
Klik - Ashley & Rock Ltd
KM - Ferham Products
KM 3 - Komfort Systems Ltd
KMC - Shapland & Petter Ltd
Knauf - Firebarrier Services Ltd
Knauf - Kitsons Insulation Products Ltd
Knight - Iles Waste Systems
Knobs Range - Pedley Furniture international Ltd
Knorr - The Kenco Coffee Company
Kohlangaz - Flavel - Leisure Glynwed Consumer & Construction Products Ltd
Koi-sen - Lotus Water Garden Products Ltd
Kolorbond - Amari Plastics Plc
Kolorcourt - Emusol Products (LEICS) Ltd
Komfire - Komfort Systems Ltd
Komfort - HEM Interiors Group Ltd
Komfort - New Forest Ceilings Ltd
Komfort Partitions - Davies John Interiors Ltd
Komplan - Kompan (UK) Ltd
Konexion - Kone Lifts Ltd
Konfigure - Komfort Systems Ltd
Kontrakt - Pedley Furniture international Ltd
Koolshade - Cooper Group Ltd
Kooltherm - Kingspan Insulation Ltd
Kooltherm - Kitsons Insulation Products Ltd
Kopex - Kem Edwards Ltd
Koral - Rockfon Ltd
Korean plywood - Wood International Agency
Korifit - Adaptaflex Ltd
Koroseal - Tektura Plc
Kosset - Carpets International (UK) PLC
Kove - Intralux UK Ltd
Koverflor - Witham Oil And Paint Ltd
Kraftex - Package Products Ltd
Krantz - Designed for Sound Ltd
Kreicor - Shapland & Petter Ltd
Kriblok Crb Walling - Ruthin Precast Concrete Ltd
Kromax - Booth Samuel & Co Ltd
Kromital Trowelled Stone - Textured European Finishes Ltd
Kromofin Multicolour Spray - Textured European Finishes Ltd
Kronofloor - Kronospan Ltd
Kronolam - Kronospan Ltd
Kronolaq - Kronospan Ltd
Kronoplus - Kronospan Ltd
Kronospan - Boardcraft Ltd
KS1000 - Kingspan Building Products Ltd
Kue System 918 - Cleanstone Co (Bradford) Ltd
Kungsater - Scandanavian Window Systems Ltd
Kuterlex - IMI Yorkshire Copper Tube Ltd
Kuterlex plus - IMI Yorkshire Copper Tube Ltd
Kuterlon - IMI Yorkshire Copper Tube Ltd
KV/2 - BPT Security Systems (UK) Ltd
Kvent - Rite-Vent Ltd
KVM - ABB Kent Meters
Kwickroll - Henderson-Bostwick
Kwikbeam - Kwikform UK Ltd
Kwikfence - Kwikform UK Ltd
Kwikfence - Kwikform UK Ltd
Kwikguard - Kwikform UK Ltd
Kwikhire - Kwikform UK Ltd
Kwikpoint - Kingfisher Chemicals Ltd
Kwikshor - Kwikform UK Ltd
Kwikstage - Kwikform UK Ltd
Kwikstage System Scaffold - Kwikform UK Ltd
Kwikstair - Kwikform UK Ltd
Kwikstair - Kwikform UK Ltd
Kwikstor - Kwikform UK Ltd
KYO - President Office Furniture Ltd
Kyomi Bathroom Suite - Ideal-Standard Ltd
L F Knight - Amdega Ltd
L.A Range - Balmoral Porcelain Limited/Longmead Ceramics
L2 Highflow - Armorex Ltd
Labline - NT Partition Systems
Labline 2 - Norwood Partition Systems Ltd
Lachat - Zellweger Analytic Ltd, Sieger Division
Lacuna - Opto International Ltd
Ladder Up Safety Post - Dilco UK Ltd
Ladderlatch - Latchways plc
Lafarge Homespan - Lafarge Plasterboard Ltd
Lafarge - Lafarge Aluminates Ltd
Lafarge firecheck - Lafarge Plasterboard Ltd
Lafarge fresco - Lafarge Plasterboard Ltd
Lafarge Moisturecheck - Lafarge Plasterboard Ltd
Lafarge profiles - Lafarge Plasterboard Ltd
Lafarge thermalcheck - Lafarge Plasterboard Ltd
Lafarge Toughcheck - Lafarge Plasterboard Ltd
Lakeshore - Ward Bekker Sales Ltd
Lakeside Group - Lakeside Group
Lakshield - Degussa Ltd
Lamax - Booth Muirie Ltd
Lamcor - Shapland & Petter Ltd
Lamella - McLoughlin Wood Ltd
Lami - Lami Doors UK
Lamina Filtertrap - Progressive Product Developments Ltd
Lamode - Gradus Ltd
Lampro - Laminated Profiles Ltd
LamproSystem - Laminated Profiles Ltd
Lamwood - Laminated Wood Ltd
Lancaster - Bassett & Findley Ltd
Lancaster - Carpets International (UK) PLC
Lancer - Forson Design & Engineering Ltd
Langford - Langford Joinery Ltd
Langley By Shavrin - Shavrin Levatap Co Ltd
Lanka - Swedecor Ltd
LANmark - Alcatel Data Cable
Lap - Polycell Products Ltd
Laser - Flexiform Business Furniture Ltd
Laser - Ingersoll - Rand Architectural Hardware
Lastogum - P.C.I Construction Systems Ltd
Latham Clad - Latham James plc
Laticrete UK - CH Agencies
Laumans - Smithbrook Building Products Ltd
Lavanda - Ecophon Ltd
Laybond adhesives - Laybond Products Ltd
Lbofall - Accomodex
LC-35 - APT Controls Ltd
LCN - Relcross Ltd
LCS - Daylight Centre
Le Farge - Brikenden (Builders Merchants) Ltd
Leakguard - Isopad Ltd
Lean-to - CIL International Ltd
Leca - Hanson Quarry Products Europe Ltd
Leca Energysavers - ARC Conbloc
Ledcor - Callenders Ltd
Ledumite - Callenders Ltd
Lee Bishop - Viscount Catering Ltd
Legge - Ingersoll - Rand Architectural Hardware
Legge - Laidlaw Architectural Hardware
Leicester Square - Brintons Ltd
Leigh's - Leigh's Paints
Leisure - Flavel - Leisure Glynwed Consumer & Construction Products Ltd
Leisuretex - Bardon (England) Ltd
Lentex - Dixon Turner Wallcoverings Ltd.
Lenton ® - Erico Europa (GB) Ltd
Leromur - Grass Concrete Ltd
Les Actuels de Lucien Gau - Myddleton Hall Lighting, Division of Pearless Design Ltd
Lesco - Lesco Products Ltd
Letter Box Company (Tebworth) Ltd - Signs Of The Times
Levolux - Western Avery Ltd
Lexic - Legrand Electric Ltd
Leyland - Kalon Decorative Products
LFH - Hellerman Electric, Member of Bopwthorpe Holdings PLC
Li-Flat - Aremco Products
Liberator - Allart, Frank, & Co Ltd
Liberty - Southern Sanitary Specialists Ltd
Libra - Balmforth Engineering Ltd
Liebig - Arrow Supply Company Ltd
Life-Line ® - Absolute Action Ltd
Lift 'n' Lock - Airsprung Beds Ltd
Lift-Slab - Hevilifts Ltd
Liftboy Remote Control - Regency Garage Door Services Ltd
Light Fantastic - Intersolar Group Ltd
Light Russet - Charnwood Brick Group Ltd
Light Victorian Red - Charnwood Brick Group Ltd
Lighting - litetronics - Formwood
Lighting Systems UK - Phosco Ltd
Lightmanager - ECS Lighting Controls Ltd
Lightmaster - Crompton Lighting Ltd
Lightmaster - ECS Lighting Controls Ltd
Lightmatic - Optex (Europe) Ltd
Lightpack - Fitzgerald Lighting Ltd
Lightrak - Electrak International Ltd
Lightseal - Illuma Lighting Ltd
Lightspan - Duplus Domes Ltd
Lignacite - Lignacite (Brandon) Ltd
Lignacite - Lignacite (North London) Ltd
Lignacite - Tarmac Topblock Ltd
Lignacrete - Lignacite (Brandon) Ltd
Lignacrete - Lignacite (North London) Ltd
Lignastone Architectural Dressings - Lignacite (Brandon) Ltd
Lignoform - Salex Interiors Ltd
Lignum - Opto International Ltd
Lilliput - Hamworthy Heating Ltd
Lilliput Nursery - Portakabin Ltd
Lime quick - Legge Thompson F & Co Ltd
Limelight - Tilcon Ltd
Limoges - Qualitas Bathrooms
Limpet BD6 self setting composition - Thermica Ltd
Lina 25 - Legrand Electric Ltd
Lincoln Model Stairlift - Brooks Stairlifts Ltd
Lincoln Pantile - Blyth William
Lincoln Stairlifts - Brooks Stairlifts Ltd
Lindapter - Kem Edwards Ltd
Lindapters - Arrow Supply Company Ltd
Lindiclip - Lindapter International
Lindum - York Handmade Brick Co Ltd
Linea - Jaga Heating Poducts (UK) Ltd
Lineadiamanti - Polybau Ltd
Lineagraniti - Polybau Ltd
lineaquarzo - Polybau Ltd
Linear - Leisure, Glynwed Consumer Foodservice Products Ltd
Linelight - Universal Components Ltd
Liner Major Concrete Mixers - Multi Marque Production Engineering Ltd
Liner Portasaw - Multi Marque Production Engineering Ltd
Liner Rolpanit Pan Mixers - Multi Marque Production Engineering Ltd
Liner Roughrider Dumpers - Multi Marque Production Engineering Ltd
Linergrip - Viking Johnson
Linesman - MSA (Britain) Ltd
Linflex - Polypipe Civils Ltd
Linido - Southern Sanitary Specialists Ltd
Linkdeco - Signs & Labels Ltd
Linkex - AEI Cables Limited
Linkguard - Bonar Floors Ltd
Linklight - Beta Lighting Ltd
Linolite - Concord Lighting
Linton - Anthony de Grey Trellises
Linurette - DLW Flooring Ltd
Lion - Boardcraft Ltd
Lion Abrasives - Acton & Borman
Lion Brand - Davis Burrow & Sons
Liqued Trent - Portacel
Liquid 99 - British Nova Works Ltd
Liquid Nails - Vallance
Liquistore - Franklin Hodge Industries Ltd
Liquitainer - Franklin Hodge Industries Ltd
Liquitite - RMC Readymix Limited
Lisbon - Swedecor Ltd
Lisson ® - Cooper H W & Co Ltd
Lister-Petter - Martin (Marine) Ltd Alec
Liteglaze - Ariel Plastics Ltd
Litelink LSC - MEM
Lithos Venetian Stucco - Textured European Finishes Ltd
Litta Pikka - Glasdon UK Ltd
Little crown - Kompan (UK) Ltd
LL - Davis International Ltd
LM Series - Baxall Ltd
LMS Modular - Siemens Building Technologies Ltd, Cerberus Division
Loadbank - Dorman Smith Switchgear Ltd
Loadcentre - Square D, Electrical Distribution Division
Loadframe - Dorman Smith Switchgear Ltd
Loadlimiter - Dorman Smith Switchgear Ltd
Loadline - Dorman Smith Switchgear Ltd
Loadswitch - Dorman Smith Switchgear Ltd
Loblite - Loblite Ltd
Locate - Universal Components Ltd
Lochrin ® - Bain WM & Co Fencing (1990) Ltd
Lockblock - Camas Building Materials
Lockblock - Fyfe John Ltd
Lockhart Project services - Lockhart Catering Equipment
Lockmaster - Paddock Fabrications Ltd
Lockstone - PHI Group Ltd
Loctite - Henkel Home Improvement & Adhesive Products
Loctite - Loctite UK Ltd
Lodlifta - Supreme Lifts Ltd
Logic Plus - Caradon M K Electric Ltd
Logsensation Ultra - Valor Heating
Lokmesh - Potter & Soar Ltd
Lokset - Fosroc Expandite Ltd
Lomond toilet and shower cubicles - Combat Polystyrene Group Ltd
London - Hanson Brick Ltd, Northern Regonal Sales Office
London Brick - Hanson Brick Ltd
Longden - Leaderflush & Shapland
Longglow Photoluminescent PVC - Signs International
Longley - Longley James & Co Ltd
Longlife - Nicholson Plastics Ltd
Longline - Nevill Long Limited
Longline 2000 - Nevill Long Limited
Longline 3000 - Nevill Long Limited
Longline 4000 - Nevill Long Limited
Longline Elite - Nevill Long Limited
Longline Finesse - Nevill Long Limited
Longmead Ceramics - Balmoral Porcelain Limited/Longmead Ceramics
Longspan Shelving - Moresecure Ltd
Lonox Pulse - Fulton Boiler Works (Great Britain) Ltd
Lonsdale - EBS Services Ltd
Loovent - Airflow Developments Ltd
Lotamax - Lotus Water Garden Products Ltd
Lotus - Lotus Water Garden Products Ltd
Lotus - Qualitas Bathrooms
Louvolite - Dallas & Forrest Ltd
Louvolite - Louvre-Lite Ltd
Louvrelite - Durable Berkeley Co Ltd
Louvrestyle - Hallmark Blinds Ltd
Lowatt - Vent-Axia Ltd
Lowline - Thermal Radiators Ltd
Loxwood Multi - Cranleigh Brick & Tile Co Ltd
Lucas Furniture Systems - Carleton Furniture Group Ltd
Luce Plan - Quip Lighting, Div of Bruce & Macintyre Ltd
Lulworth - Hamworthy Heating Ltd
Lumaseal - Illuma Lighting Ltd
Lumaspot - Illuma Lighting Ltd
Lumiere Polychromatic Coating - Textured European Finishes Ltd
Lumilux - Osram Ltd
Lumina - Forma Lighting Ltd
Luminex - Interlux Ltd
Luna - Rada Lighting Ltd
Luraflex ® - Hunter Douglas Ltd (Contracts Division) Luxaflex Blinds ®
Lux Lift - Park Products

Luxaclair - Pilkington Plyglass
Luxblocks - Luxcrete Ltd
Luxcrete - Luxcrete Ltd
Luxe - Leisure, Glynwed Consumer Foodservice Products Ltd
Luxfibre - Luxcrete Ltd
Luxfix - Luxcrete Ltd
Luxfloor - Kronospan Ltd
Luxfloor Plus - Kronospan Ltd
Luxline Plus - Sylvania Lighting International INC.
Luxmix - Luxcrete Ltd
Luxo - Luxo UK Ltd
Lynx - Sylvania Lighting International INC.
Lynx aluminium towers - Kwikform UK Ltd
Lynx Aluminium Towers - Kwikform UK Ltd
Lyric - President Office Furniture Ltd
Lyssand - Lyssand UK Ltd
M-Bond - Isocrete Group Sales Ltd
M-Bond Xtra - Isocrete Group Sales Ltd
M-E flooring - Dex-o-Tex International Ltd
M100 - Sempol Surfaces Ltd
M11 - Maby Hire Ltd
M11 - Matbro Ltd
M7 - Maby Hire Ltd
M7 - Matbro Ltd
Mab Door Closers - JB Architectural Inronmongery Ltd
MAC - Maestro International Ltd
Macallay - McCalls Special Products
Macaw - Paragon by Heckmondwike F B Ltd
Maccast - Macc Stone Ltd
Macclex - Exitex Ltd
Maccstone - Kerridge Stone
Machin - Amdega Ltd
MacLarenline - Johnson Controls Systems Ltd
Macmount - MacLellan Rubber Ltd
Macpherson - Akzo Coatings Plc
Macphersons - Akzo Nobel Decorating Coatings Ltd
Macphersons - Crown Trade Paints
Macvent - MacLellan Rubber Ltd
Maddalena - SAV United Kingdom Ltd
Madico/ M.S.C. Specialty Films - Hi-Tec Window Films
Madison - Gloster Furniture Ltd
Maestro ® - Maestro International Ltd
Maestroflo ® - Maestro International Ltd
Maestropac ® - Maestro International Ltd
Mag Master - ABB Kent Taylor Ltd.
Magiboards - Magiboards Ltd
Magma - Thatching Advisory Services Ltd
Magnapleat - Farr Filtration Ltd
Magnaseal - Magnet Ltd
Magnastar - Magnet Ltd
Magnaview - Magnet Ltd
Magnetek - Bernlite Ltd
Magnetek - Magnatek Ltd
Magnetic accessories - Clestra Hauserman Ltd
Magsi - Elliott John H (Monostamp) Ltd
Main - Potterton Myson Ltd
Maine Reception System - Hyperion Wall Furniture Ltd
Maintain - MMP International
Majestic - Brintons Ltd
Majestic - Majestic Shower Co Ltd
Majestic - Metlex
Makita - Arrow Supply Company Ltd
Makroswing - Envirodoor Markus Ltd
Maldives - Novellini Bathroom Products
Maltistore - RMC Readymix Limited
Manade - Lesco Products Ltd
Manby - Mandor Engineering Ltd
Mancuna - Mandor Engineering Ltd
Manders - Kalon Decorative Products
Mandolite - Cafco Construction Products
Mandor - Mandor Engineering Ltd
Mangers - Davant Products Ltd
Mangers - Vallance
Manhattan - JSB Electrical PLC
Manhattan Furniture - Dennis & Robinson Ltd
Manifix - Manitou (Site Lift) Ltd
Manireach - Manitou (Site Lift) Ltd
Maniscopic - Manitou (Site Lift) Ltd
Mannesmann SS Suply Pipework - BM Stainless Steel Drains Ltd
Manor - Metlex
Manor Range Of Firechests - ISOKERN (UK) Ltd
Mansafe - Latchways plc
Mansafe - Unistrut Limited
Mansfield White - Chilmark Quarries Ltd
Mansfield White - Gregory Quarries Ltd, The
Mansfield White - The Rare Stone Group Ltd
Mansfield White limestone - The Rare Stone Group Ltd
Mansion - Metlex
Mantal - Latham James plc
Manton - Manton Industrial Seals Ltd
Map Vents - Map Hardware Ltd
Mapecem - Mapei UK Ltd
Marcaddy - WF Electrical Distributors
Margard - Dacrylate Paints Ltd
Marine - Richard Hose Ltd
Marineflex - Trade Sealants
Marineseal - Trade Sealants
Mark 1 - Channel Electronics
Markar - Relcross Ltd
Market - Buchtal (U.K.) Limited
Marktrack - Illuma Lighting Ltd
Markus - Envirodoor Markus Ltd
Marlborough - Ellis J T & Co Ltd
Marlborough - Gloster Furniture Ltd
Marlborough - Metlex
Marley - Capper Pipe Systems Ltd
Marley Paving - Marley Building Materials Ltd
Marley Roof Tiles - Marley Building Materials Ltd
Marleyflex - Marley Floors Ltd
Marleyflor Plus - Marley Floors Ltd
Marleyseal - Marley Waterproofing
Marleytorch - Marley Waterproofing
Marlin 3F - Marlin Lighting Ltd
Marlin Matrix - Marlin Lighting Ltd
Marlin Optics - Marlin Lighting Ltd
Marlow - Bartlett Catering Equipment Ltd
Marlux ® - Marshall Contracts
Marmarette - DLW Flooring Ltd
Marmerina - Francesca Di Blasi Co. Ltd, The
Marmoleum - Forbo - Nairn Ltd
Marmora Mosaics - Kirkstone
Marples - Record Tools Ltd
Marquis - Brintons Ltd
Marrakesh - Brintons Ltd
Marshalite - Marshalls Mono Ltd
Martin Lunel - WF Electrical Distributors
Masonite - Boardcraft Ltd
Masonite Beams - Fillcrete Ltd
Masons Timber Products - Mason FW & Sons Ltd
Masonstone - Hampton Stone Ltd
Master - ABB Kent Meters
Master - Norwood Partition Systems Ltd
Master - Securikey Ltd
Master Brass - Home Automation Ltd
Master Craftsman - Brockway Carpets Ltd
Master Flash - Ventlane Ltd
Master Plug - Perma Industries Ltd
Masterbill '97 - Masterbill Micro Systems Ltd
Masterboard - Cape Boards Ltd
Masterclad - Cape Boards Ltd
Masterfloat stone raised access floors - Chelsea Artisans Ltd
Masterpiece - Checkmate Industries Ltd
Masters Collection - Firth Carpets Ltd
Masterseal - Caradon M K Electric Ltd
Masterseal - FEB Ltd
Masterseal - Sealmaster
Mastertop - FEB Ltd
Mastertop - Feb MBT
Matador - Cimex International Ltd
Matador - Crosbie Coatings
Match-up Range - Beacon Hill Brick Co Ltd
Matchplay - En-tout-cas plc
Matchplay - En-tout-cas plc (Scotland)
Matchplay Supercushion - En-tout-cas plc
Matki - Matki plc
Matrex - Terrapin Ltd
Matterson Cranes - PCT Group Sales Ltd
Matthews & Yates Centrifugal Fans - Matthews & Yates Ltd
Mattseal - Armorex Ltd
Mauser - Mauser Interiors (UK) Ltd
Mawrob Incinerators - Mawrob Co (Engineers) Ltd
Max - Jotun Henry Clark Ltd
Max - Ves Andover Ltd
Maxi - Dexion Ltd
Maxi - Isolated Systems Ltd
Maxi-Mixer - Cardale Doors Ltd
Maxifit - Viking Johnson
Maxifrost - Shearflow Phoenix
Maxigard - Lycetts (Burslem) Ltd
Maxilift - Stannah Lifts Ltd
Maxim - Cego Frameware Ltd
Maxim - J.F.Poynter Ltd
Maxim - MTE Ltd
Maxima - Norwood Partition Systems Ltd
Maxima - NT Partition Systems
Maximiser - Rycroft Ltd
Maximixam - O'Brian Manufacturing Ltd
Maxivent - Airflow Developments Ltd
Maxmatic - Max Appliances
Maxmatic - Sussex Brassware Ltd
Maxol 12D - Burco Maxol
Maxol M10D - Burco Maxol
Maxol M15 - Burco Maxol
Maxol Microturbo - Burco Maxol
Maxol Mirage - Burco Maxol
Maxol Montana - Burco Maxol
Maxol Morocco - Burco Maxol
Maxpax - Craft Jacobs Suchard Ltd
Maxum R ™ - Dezurik International Ltd
Maxum V ™ - Dezurik International Ltd
Maxweld - Twin County Fan Company Ltd
Maxwell House - The Kenco Coffee Company
May & Christe - Magnatek Ltd
Mayfair - Gower Furniture Ltd
Mayfair - Kaloric Heater Co Ltd
Mayfair Chrome Accessories - Allibert (UK) Ltd
Mayfield Brol - Kingstonian Point Ltd
MB - 1 - Fernox Fry Technology UK
Mb1 - Alpha Fry Ltd
Mbrace - FEB Ltd
Mbrace - Feb MBT
MBT - FEB Ltd
MBT-Feb - Ashmead Buidling Supplies Ltd
MBT-Feb - Ashmead Building Supplies Ltd
MC - Howden Buffalo
Mc-Wall - Smart Systems Ltd
McAlpine - Hipkin F W Ltd
MCR - APT Controls Ltd
MCS - APT Controls Ltd
MDA Scientific - Zellweger Analytic Ltd, Sieger Division
Meadrain - Mea UK Ltd
Meagard - Mea UK Ltd
Meagard S - Mea UK Ltd
Mealine - Mea UK Ltd
Mec - AEL
Mechbond - Vulcanite Limited
Meckel - HCL Verta
Mectherm - AEL
Mecury - Fabrikat (Nottingham) Ltd
Medallion - Ashley & Rock Ltd
Media line - TWS Presentation Systems (Mcmillan UK Ltd T/A)
Medici - Forticrete Ltd
Medite - Boardcraft Ltd
Medite 313 MR MDF - Nillamette Europe Ltd
Medite Exterior - Nillamette Europe Ltd
Medite FQ - Nillamette Europe Ltd
Medite FR Class 0 - Nillamette Europe Ltd
Medite FR Class 1 - Nillamette Europe Ltd
Medite HD - Nillamette Europe Ltd
Medite MDF - Nillamette Europe Ltd
Medite ZF - Nillamette Europe Ltd
Medway - Portacel
Meg - Abet Ltd
Mega Vent - Areco Roofing Supplies Company
Megafit - Viking Johnson
Melbourn - Eternit UK Ltd
Meleto - Panel Systems Ltd
meltaway - Wirsbo
Meltone - Redland Precast
Memera 2000 - MEM
Memera 2000 AD - MEM
Mempower - MEM
Memshield 2 - MEM
Memstyle - MEM
Menveir Security - Cooper Security
Menvier - Crimegard Ltd
Merchant Hardware - Avocet Hardware Ltd
Mercury Litestat - Hamilton R & Co Ltd
Meridian - Chubb Security Installations Ltd
Merino - Gaskell Textiles Ltd
Merlin - T & D Plastech
Merlin Gerin - Schneider Ltd
Mermaid - Norske Interiors (UK) Ltd
Mervene - Firwood Paint & Varnish Co Ltd
Meshlite - Oswestry Reinforced Plastics Ltd
Messagetrax - Oldham Signs Ltd
Met-track - Metreel Ltd
Metabin - Metalrax Ltd
Metabolt - Metalrax Ltd
Metaclip - Metalrax Ltd
Metalarc - Sylvania Lighting International INC.
Metalclad Plus - Caradon M K Electric Ltd
Metalgrid - Tenby Industries Ltd
Metalliform - Metalliform Products Plc
Metalloxyd Ano-Coil - Metalloxyd Ano-Coil Ltd
Metalphoto - Pryorsign
Metalset - Loctite UK Ltd
Metasys - Johnson Controls Systems Ltd
Metavent - Blunn Slates Ltd
Metbar - Mid Essex Trading Co Ltd
Meteon - Trespa UK Ltd
Meterheat - Jimi-Heat Ltd
Metframe - Mestec PLC
Metiflash - Metra Non Ferrous Metals Ltd
Metizinc - Metra Non Ferrous Metals Ltd
Metlex - Triton PLC
Metpost - Challenge Fencing & Ltd
Metreboard - Callenders Ltd
Metreisk - Callenders Ltd
Metrex - Hattersley Newman Hender Ltd
Metro - Harton Heating Appliances
Metro Therm - Harton Heating Appliances
Metro Walling - Chubb Security Installations Ltd
MetroPous - Hille Ltd
Metrowel - Mid Essex Trading Co Ltd
Metsec - Metsec Building Products Ltd
Metsec floor beams - Mestec PLC
Metsec lattice joists - Mestec PLC
Metsec Zed purlins - Mestec PLC
Meva screens - Adams- Hydraulics Ltd
Meynell - Caradon Plumbing Solutions
Mezane - ClipsPartitions Ltd
MFP 4-28 Zone Microprocessor - C-Tec (Computionics Ltd)
Mi - Warner Howard Ltd
Mic - Ledu UK Ltd
Micafil - Dupre Vermiculite
MICE Kaymar Ltd - MICE Kaymar Limited
Michelmersh - Michelmersh Brick & Tile Co Ltd
Mico - ASSA Ltd
MicraFlo ™ - Johnson Matthey Pigments & Dispersions
Micro-Air - Feb MBT
Microbore - Wednesbury Tube
Microcard Readers - Time and Data Systems International Ltd (TDSI)
Microcem - Blue Circle Cement
Microfloor 600 - Quiligotti Access Flooring Limited
Microfoam X205 Flooring - Springvale E P S
Microfoam X32 Cavity - Springvale E P S
Microlift - Stannah Lifts Ltd
Microlock Systems - Time and Data Systems International Ltd (TDSI)
Microlook - Armstrong World Industries Ltd
MicroPac - Draeger Ltd
Microtracer - Heat Trace Ltd
Midland Portable Buildings Ltd - Phillip Ader Ltd
Miele - Miele Co Ltd
Miko - Optelma Lighting Ltd
Mildsil - Notcutt W P Ltd
Milestone - Forticrete Ltd
Millars - Bitmen Products Ltd
Millenium - Tektura Plc
Millenium - Troika Architectural Mouldings Ltd
Millenium Bed Range - Sidhil Care
Millenium Freedom Bed Range - Sidhil Care
Millennium Aluplast Aluminium - Signs International
Milliken - Hattersley Newman Hender Ltd
Millwood - Project Office Furniture PLC
Mimo - Helvar Electrosonic
Minaboard - Armstrong World Industries Ltd
Minatone - Armstrong World Industries Ltd
Minerac - Treetex Acoustics Ltd
Mineralite - Snowcem
Mingardi - Dyer Environmental Controls
Mini Clearflow - Dunbrik (Yorks) Ltd
Mini Compack - Fulton Boiler Works (Great Britain) Ltd
Mini Loovent - Airflow Developments Ltd
Mini slate - Forticrete Ltd
Mini-compacta - KSB Ltd
Mini-Compacta - Kwikform UK Ltd
Mini-Lynx - Sylvania Lighting International INC.
Minibore - IMI Yorkshire Copper Tube Ltd
Minifrost - Shearflow Phoenix
Minilift - Didsbury Engineering Co Ltd
Minipack - Ramset Fasteners Ltd
Miniseal - Sheardown Engineering Ltd
Minislats - Forticrete Roofing Products
Minispace - Kone Lifts Ltd
Minitec - Entec (Polution Control) Ltd
Minitracer - Heat Trace Ltd
Miniwarn - Draeger Ltd
Minized - Hadley Industries Plc
Minor - Norwood Partition Systems Ltd
Minor - NT Partition Systems
Minster - Minsterstone Ltd
Minster Range - Beacon Hill Brick Co Ltd
Miraflex - Owens Corning Building Products (UK) Ltd
Mirage - Dudley Thomas Ltd
Mirage - Firth Carpets Ltd
Mirage - Komfort Systems Ltd
Mirage - Mode Lighting (UK) Ltd
Mirage - Muraspec Ltd
Mirage - Shackerley (Holdings) Group Ltd incorporating Designer ceramics
Miralyte - Harkness Hall Ltd
Mirazzo - Camas Building Materials
Mirazzo - Fyfe John Ltd
Mirogard Anti-Reflective Glass - Schott Glass Ltd, Architectural Glass
Mission - Flex-Seal Couplings Ltd
Mistral - HEM Interiors Group Ltd
Mistral - Imperial Machine Co Ltd
Mistral - Unilock Ltd
Mitlite - Oswestry Reinforced Plastics Ltd
Mitraclad - Oswestry Reinforced Plastics Ltd
Mitsubishi - HV Air Conditioning Ltd
Mitsubishi - Purified Air Ltd
Mitx - Oswestry Reinforced Plastics Ltd
Mixcal - Altecnic Ltd
Mixed Golden Russet - Charnwood Brick Group Ltd
Mixed Hampshire Red - Charnwood Brick Group Ltd
Mixed Regency Red - Charnwood Brick Group Ltd
Mixmaster - CAEC Howard (Holdings) Ltd
MJC - Airmaster Engineering Ltd
MJX - Airmaster Engineering Ltd
MK - Caradon M K Electric Ltd
ML - Davis International Ltd
MM2 - Helifix Ltd
MMS - Kone Lifts Ltd
Mobilver International - Haworth UK Ltd
Mock Birds of Prey - Birdscarer Products
Mockridge labels - Mockridge Labels & Nameplates Ltd
Mod 4 - Accomodex
Mode - Allgood plc
Mode - Bernlite Ltd
Model S - Hudevad Britain
Modelmark - Elliott John H (Monostamp) Ltd
Modlag - Cape Insulation Products Ltd
Modric - Allgood plc
Moducel - Senior Air Systems
Moducell - Crompton Lighting Ltd
Moduclad - Ward Building Components Ltd
Modula - Ashley & Rock Ltd
Modulair - Biddle Air Systems Ltd
Modular 2000 - ACO Technologies Plc
Modulay - Crompton Lighting Ltd
Module 4 - Harrison Drape, Divsion of McKenie (UK)
Modulite - Cheshire Concrete Products
Modulite plus - Cheshire Concrete Products
Modulor - The Modelshop
Moduspec - Crompton Lighting Ltd
Moelven - Moelven Laminated Timber Structures Ltd
Moffett Thallon - Moffett Thallon & Co Ltd
Mogaspaan - Metra Non Ferrous Metals Ltd
Mogat - Metra Non Ferrous Metals Ltd
Molar Major - Manton Industrial Seals Ltd
Molar Minor - Manton Industrial Seals Ltd
Molyneux concealed weepholes - Molyneux G (Products) Ltd
Moment - Tektura Plc
Monafloor - Monarflex Ltd
Monaframe - Glostal - Monarch
Monarch - Glostal - Monarch
Monarch aluminium, now GrangiuosSystems - Alframes Holdings Ltd
Monarch Brass - MEM
Monarfol 250 - Monarflex Ltd
Monarperm 450 - Monarflex Ltd
Mondial - Uponor Ltd
Mondo Sportflek Athletic Surface - Bernhards Landscapes Ltd
Mondo Sportflex Athletic Surface - Bernhard's Sports Surfaces Ltd
Monitohm - Heat Trace Ltd
Monks Park Bath/Cotswald Stone - Hanson Bath & Portland Stone

Monmouth Canopy - James Smellie Fabrications Ltd
Mono - Optelma Lighting Ltd
Monobloc - ClipsPartitions Ltd
Monoceram - Stokes R J & Co Ltd
Monocouche - Webrr & Broutin
Monofilament 250 - Monarflex Ltd
Monojet - Clearwater Plc
Monolastex smooth - Liquid Plastics Ltd
Monolastex textured - Liquid Plastics Ltd
Monolux - Cape Boards Ltd
Monopoly - Checkmate Industries Ltd
Monoscape - Marshalls Mono Ltd
Monoset - Ronacrete Ltd
Monospace - Kone Lifts Ltd
Monotrak - Rackline Systems Storage Ltd
Monotrex 2000 - Kingspan Building Products Ltd
Monsoon - Stuart Turner Ltd
Montage - Satchwell Control Systems Ltd
Monument - Monument Tools Ltd
Monza - Cummins Power Generation Ltd
Moonlight Marker - Intersolar Group Ltd
Moorwood Vulcan - Viscount Catering Ltd
Morris - Morris Mechanical Handling Ltd
Mosaiq - Kompan (UK) Ltd
Motastile - Gunnebo Mayor Ltd
Motex® Dry - Webrr & Broutin
Mouldguard - Trade Sealants
Mouldshield - ICI Dulux Trade Paints
Movitex - Movitex Signs Ltd
MR - Davis International Ltd
MR Swisslab - Alumasc Exterior Building Products Ltd
MRS 7 - MR Limited
MSM - ABB Kent Meters
MT70 IP67 waterproof fitting - Encapsulite International Ltd
Mufti-Lag - Salex Interiors Ltd
Muhlhauser - Burlington Engineers Ltd
Mulchmat 2000 - Land Wood & Water Co. Ltd
Mulsicoat - Webrr & Broutin
Mulsifix - Webrr & Broutin
Multi Brindle - Charnwood Brick Group Ltd
Multi Red Brown - Charnwood Brick Group Ltd
Multi Versatile - Woolliscroft Tiles Ltd
Multi wall panelling - Combat Polystyrene Group Ltd
Multi-Tex collection - Firth Carpets Ltd
Multibeam - Ward Building Components Ltd
Multibond - Loctite UK Ltd
Multibond Gold - Laybond Products Ltd
Multichannel - Ward Building Components Ltd
Multicryl - Laybond Products Ltd
Multideck - Ward Building Components Ltd
Multiduct - Ward Building Components Ltd
Multifix Airbrick - Ryton's Building Products Ltd
Multiflex - Marflex Chimney Systems
Multiflex Slab Formwork - PERI Ltd
Multigrid - Tenby Industries Ltd
Multiguard - Hoyles Fire & Safety Ltd
Multiguard - Manton Industrial Seals Ltd
Multikwik - Hipkin F W Ltd
Multikwik - Phetco (England) Ltd
Multiline - Opto International Ltd
Multiplate - Asset International Ltd
Multipot - TSE Brownson
Multishield ™ - Accent Hansen
Multistep - Forbo - Nairn Ltd
Multisystem disabled product - Armitage Shanks Ltd
Multitack - Laybond Products Ltd
Multitex - Multitex GRP & GRG Products
Multitrak - Rackline Systems Storage Ltd
Multivent - Domus Ventilation Ltd.
Multivent - Ubbink (UK) Ltd
Multiwarn - Draeger Ltd
Mupro Pipe Supports - Industrial Hangers Ltd
Muralon - Muraspec Ltd
Murek - Muraspec Ltd
Murfor - Bekaert Building Products
Mvi - Wilo Samson Pumps Ltd
MWS35 - Manton Industrial Seals Ltd
MX - Carborundum Abrasives G.B Ltd
Myflo - Ves Andover Ltd
Myotex hp - Eclipse Blind Systems Ltd
Myotex sp - Eclipse Blind Systems Ltd
Myrtha - Aqua Design Ltd
Myson - Potterton Myson Ltd
N T Access - Permclose Group PLC
N.T.Master - Nevill Long Limited
N.T.Minor - Nevill Long Limited
NAD - Project Office Furniture PLC
Nailweb - Alpine Autimation Ltd.
Nappigon - Imperial Machine Co Ltd
Nappigon - Sissons W & G Ltd
Narcis Mirror Radiator - Jaga Heating Poducts (UK) Ltd
Narrowline ® - Andersen/ Black Millwork
Narvik - Smithbrook Building Products Ltd
National bondmaster - Kenyon Industrial Paints & Adhesives
National Britannia Training - National Britannia Ltd
National Range - Lumitron Ltd
National Register of Warranted Builders - Federation of Master Builders
Naturefloor - Pilkington's Tiles Ltd
Nautilus - IBP Conex
Navada Western - Thomerson & Brett Ltd
Navigator - Bawn W. B. & Co Ltd
Navigator - Project Office Furniture PLC
Navivia - Tegral Building Materials Ltd
Naylor - Armstrong Concrete Products Ltd
Naylor - Ashmead Buidling Supplies Ltd
Naylor - Ashmead Building Supplies Ltd
Naylor Drainage - Brikenden (Builders Merchants) Ltd
NBS - NBS Services
NBS Landscape - NBS Services
Neaco - Norton Engineering Alloys Co Ltd
Neaco Support Systems - Norton Engineering Alloys Co Ltd
Neatalever - Sheardown Engineering Ltd
Neataseal - Sheardown Engineering Ltd
Neatatap - Sheardown Engineering Ltd
Neatdek 2 - Norton Engineering Alloys Co Ltd
Neatgrille - Norton Engineering Alloys Co Ltd
Neatmat - Norton Engineering Alloys Co Ltd
NebulaZT - Burmatex Ltd
Nedair - Ellis Miller Ltd
Nederman Ltd - Encon Air Systems Ltd
Neff - Diamond Merchants
Nelson - Marlow Ropes Ltd
Nelton - The Nelton Group Ltd
Nene 1 Gutter - Dales Fabrications Ltd
Nene 2 Gutter - Dales Fabrications Ltd
Neotechnik - DCE - BTR Environmental Ltd
Neotex - Eclipse Blind Systems Ltd
Neotran - Mode Lighting (UK) Ltd
Neotronics - Zellweger Analytic Ltd, Sieger Division
Neptune - Neptune Outdoor Furniture Ltd
Neptune - Neptune Showers Ltd
Nera Design - Gerland Ltd
Nestor - Glasdon UK Ltd
Neta - Harrison Drape, Divsion of McKenie (UK)
New c3m/p5. Moisture Resistant Carcase - Rixonway Kitchens Ltd
New Century - Hoskins Medical Equipment Ltd
New Era - Bennett Windows Ltd
New Generation steel - Parkwood Mellows Ltd
New Terrier - Collie Carpets Ltd
Neway - Dudley Factory Doors Limited
Neway - Neway Doors Ltd
Neway Doors Ltd. - Priory Shutter & Door Co Ltd
Neway Doors Ltd. - Shutter Door Repair & Maintenance Ltd
Newdawn - Newdawn & Sun Ltd
Newlath - Newton John & Co Ltd
Newline - Tunstal Telecom Ltd
Newmwn Tonks - Ingersoll - Rand Architectural Hardware
Newspack - Accomodex
Newtherm - Cape Insulation Products Ltd
Niagara rainwater system - FloPlast Ltd
Nic-cool - Nicholson Plastics Ltd
Nico - Nico Manufacturing Ltd
Nico-O-Grout - Nicholls & Clarke Ltd
Nicobond - Nicholls & Clarke Ltd
Night 'n' Day - Eclipse Blind Systems Ltd
Nika3 - Industrial Devices Ltd
9000 Series - Jebron Ltd
900SX series - Falcon Catering Equipment
Niroflex - BRB Industrial Services Ltd
Nitobond - Fosroc Expandite Ltd
Nitofill - Fosroc Expandite Ltd
Nitoflor - Fosroc Expandite Ltd
Nitromatric - Pillinger G C & Co (Engineers) Ltd
Nitromors - Henkel Home Improvement & Adhesive Products
Nocturne Range - Caradon Bathrooms Ltd
Nomafoam - nmc (UK) Ltd
Nomapack - nmc (UK) Ltd
Nomastyl - nmc (UK) Ltd
Non Com X - Hickson Timber Products Ltd
Norbo - Booth Muirie Ltd
Nordal Gutter - Dales Fabrications Ltd
Nordic Shelving - Balmforth Engineering Ltd
Nordica - Deceuninck Ltd
Nordstar - Atkinson & Kirby Ltd
Nordyl - Dixon Turner Wallcoverings Ltd
Nordyl - Dixon Turner Wallcoverings Ltd.
Norges Vinduet - Jutlandia Doors
Nori - Marshalls Mono Ltd
Norit - Mauser Interiors (UK) Ltd
Norlyn - Smart F & G (Shopfittings)Ltd
Normban - Laidlaw Architectural Hardware Ltd
Normbau - City Hardware and Security
Normbau - Ingersoll - Rand Architectural Hardware
Normet - Burlington Engineers Ltd
Norslo - Power Breaker
North - BRB Industrial Services Ltd
Norusto - Protim Services Ltd
Norwegian 'H' - Scandanavian Window Systems Ltd
NOS - Strand Hardware Ltd
Nostalgia - Scholes Windows Ltd
Notifier Fire Detection - Fife Fire Engineers & Consultants Ltd
Nouvelle - Qualitas Bathrooms
Nova - Vicon Industries (UK) Ltd
Nova / Novaplex - British Nova Works Ltd
Novacare - British Nova Works Ltd
Novacryl - British Nova Works Ltd
Novalift - British Nova Works Ltd
Novalin - Dixon Turner Wallcoverings Ltd.
Novalite - British Nova Works Ltd
Novaract - British Nova Works Ltd
Novashield - British Nova Works Ltd
Novastet - British Nova Works Ltd
Novastone - Forticrete Ltd
Novatone - USG (UK) Ltd
Novatreet - British Nova Works Ltd
Novatron - British Nova Works Ltd
Novaways - British Nova Works Ltd
Novellini - Novellini Bathroom Products
Noviasol - Dalesauna Ltd
Novoment - P.C.I Construction Systems Ltd
Novus - Harris T A Ltd
Noyant - Eternit UK Ltd
NP100 Slurry - ECC International Ltd
NRWB - Federation of Master Builders
Nu-klad - Kenyon Industrial Paints & Adhesives
Nuance worksurfaces - Bushboard Parker Ltd
Nuastyle taps - Armitage Shanks Ltd
Nucana - Ward's Flexible Rod Company Ltd
Nufins - Hines & Sons, P E, Ltd
Nuflex - Ward's Flexible Rod Company Ltd
Nuflex 95 - Lastite Building Products Ltd
Nullifire - Firebarrier Services Ltd
Nuralite - Nuralite (UK) Ltd
Nuraply - Nuralite (UK) Ltd
Nursecall 800 Panel - C-Tec (Computionics Ltd)
Nutex - Webrr & Broutin
Nutrim - Areco Roofing Supplies Company
Nuway - Bonar Floors Ltd
Oakey - Unicorn Abrasives (UK) Ltd
Oasis - Rovacabin
Oatmeal Range - Beacon Hill Brick Co Ltd
Oberon - Decra Roofing Systems
Ocr - Snowcem
Octadoor - Righton Ltd
OCTO - Turner Access Ltd
Odyssey worksurfaces - Bushboard Parker Ltd
OEP contract - OEP Furniture Group PLC
OEP Engineering - OEP Furniture Group PLC
OEP matrix - OEP Furniture Group PLC
Offsigns - Movitex Signs Ltd
Ogilive Sealants - Ogilvie Construction Ltd
Ogilvie Builders Ltd - Ogilvie Construction Ltd
Oikos - Hamron Group
Oil slurps - IFT Supplies Ltd
Old Clamp - York Handmade Brick Co Ltd
Old colonial brick - Fyfe John Ltd
Old English pantile - Sandtoft Roof Tiles Ltd
Olsen - Scandanavian Window Systems Ltd
Olympia - Erlau AG
Olympian - Wilson (Engineering) Ltd, F G
Olympus - Carpets of Worth Ltd
Omega - Bennett Windows Ltd
Omega - Dryad Architectural Ltd
Omega - Viking Laminates Ltd
Omega 4 - JSB Electrical PLC
Omega worksurfaces - Bushboard Parker Ltd
Omni-Pro - En-tout-cas South West
Omnia Range of Products - Birchwood Concrete Products Ltd
Omnicourt Pro - En-tout-cas plc
Omnirack - Drainage Systems
Onduclair - Onduline Building Products Ltd
Onduline - Onduline Building Products Ltd
Ondusteel - Onduline Building Products Ltd
Ondutile - Onduline Building Products Ltd
132 - Kair - The Ventilation Division of Kiltox Chemicals Limited
100 High - Hudevad Britain
£1 coin locks for coats - Locking Systems Ltd
1200 - GE Bayer Silicones
1900 Series Cutting & Grinding Discs - Acton & Borman
Onyx - Pland
Onyx - Project Office Furniture PLC
Open Lok - Unifix Ltd
Open Options - Project Office Furniture PLC
Opiocolour Mosaics - Reed Harris
Optelma - Optelma Lighting Ltd
Optiflame - Dimplex (UK) Ltd
Optima - Cardale Doors Ltd
Optima - FDB Electrical Ltd
Optimum Cut - Gaskell Textiles Ltd
Optimum Loop - Gaskell Textiles Ltd
Optimum Pattern - Gaskell Textiles Ltd
Optimum Plain - Gaskell Textiles Ltd
Options - Bartlett Catering Equipment Ltd
Options - Bathrooms of Wycombe
Options - Sanderson & Sons Ltd, Arthur
Optix - Intralux UK Ltd
Opto Contour Showcases - Opto International Ltd
Optoma - Magnatek Ltd
Optorex - Siemens Building Technologies Ltd, Cerberus Division
Oracle - TWS Presentation Systems (Mcmillan UK Ltd T/A)
Orator - Caradon Gent Ltd
Oratorio carpet tiles - Burmatex Ltd
Orbik - Bernlite Ltd
Orbis - Laidlaw Architectural Hardware
Orbis - Laidlaw Architectural Hardware Ltd
Orcal - Armstrong World Industries Ltd
Orglas - RIW Ltd
Origlia Spa - Origlia Spa
Orion - Dryad Architectural Ltd
Orion - Rada Lighting Ltd
Orion - Ubbink (UK) Ltd
Orkla Chipboard - Woodmark Ltd
Orkla Elite Chipboard - Woodmark Ltd
Orona - Cable Lift Installations
Orsogril - Orsogril Sarl UK Department
Orsogril - Squires Metal Fabrications Ltd
Orsopanel - Orsogril Sarl UK Department
Osma - Wavin Building Products Ltd
Osma Amazon - Wavin Building Products Ltd
Osma DeepLine - Wavin Building Products Ltd
Osma Drain - Wavin Building Products Ltd
Osma Gold - Wavin Building Products Ltd
Osma RoofLine - Wavin Building Products Ltd
Osma Round Line - Wavin Building Products Ltd
Osma Soil - Wavin Building Products Ltd
Osma Square Line - Wavin Building Products Ltd
Osma SuperLine - Wavin Building Products Ltd
Osma UltraRib - Wavin Building Products Ltd
Osma Weld - Wavin Building Products Ltd
Osram - Bernlite Ltd
Otis 2000 - Otis plc
Otter - Baresford Pumps
Ovalgrip - Hellerman Electric, Member of Bopwthorpe Holdings PLC
Overlord - Boddingtons Ltd
OWAcontruct - OWA UK Ltd
OWAcoustic - OWA UK Ltd
OWAlux - OWA UK Ltd
OWAsigna - OWA UK Ltd
OWAspectra - OWA UK Ltd
OWAtecta - OWA UK Ltd
Owens Corning - Kitsons Insulation Products Ltd
Oxan WP - Jotun Henry Clark Ltd
Oxylene - Palgrave Brown Timber & Fireproofing Co
P. J. Paints - Philip Johnstone Group Ltd
P. S. Fan units - Elan-Dragonair
P.P.P - Wade International (UK) Ltd
P1 - Sempol Surfaces Ltd
P5 - Hudevad Britain
P6 Sheeting - Eternit UK Ltd
PA90+Range - Draeger Ltd
PAC - BPT Security Systems (UK) Ltd
PAC - Vidionics Security Systems Ltd
PAC Access Control - APT Controls Ltd
Pacemaker - Portakabin Ltd
Pacesottca - Fyfe John Ltd
Pacific - Deceuninck Ltd
Pacifyre Fire Stops - Industrial Hangers Ltd
Pahlens - Golden Coast Ltd
Pakyderm - Crosbie Coatings
Palace Switches - Wandsworth Elecrtrical Ltd
Palletstor - Moresecure Ltd
Palmyco - Palmyra Ltd
Pan Diut - Brissco Equipment Ltd
Panasonic - Connought Communication Systems Ltd
Panasonic - Vidionics Security Systems Ltd
Panatrim - Universal Components Ltd
Panda - Horizon International Ltd
Panelmaster - Timeguard Ltd
Panorama - Muraspec Ltd
Panther - Eclipse Blind Systems Ltd
Panther - Hilladam Coburn Ltd
Papst - Ziehl-Ebm (UK) Ltd
Paptrim Aluminium - Pitchmastic PLC, Building Products Division
Paptrim Roof Edge Trim - Pitchmastic PLC, Building Products Division
Parabolt - Allman Fasteners Ltd
Parabolt - Tucker Fasteners Ltd
Parade - Iles Waste Systems
Parade - Welconstruct Co
Parafon - Armstrong World Industries Ltd
Paralock - Locking Systems Ltd
Paralon membranes - Intergrated Polymer Systems (UK) Ltd
Parasquare - Interlux Ltd
Parat C - Draeger Ltd
Parcon ® - Composite Structures
Parctile - Decra Roofing Systems
Parigi - Altecnic Ltd
Parinox - Altecnic Ltd
Park-k-Ply - Lydney Products Ltd
Parkiflex - Western Cork Ltd
Parkray - Hepworth Heating Ltd
Paroc - Hemsec Manufacturing Ltd
Parol - Owens Corning Building Products (UK) Ltd
Parqcolor - Abet Ltd
Parquet Countryhouse - DLW Flooring Ltd
Parquet Professional - DLW Flooring Ltd
Parquet Shipsdeck - DLW Flooring Ltd
Parquet Virginia - DLW Flooring Ltd
Paslite Aggregates - CAEC Howard (Holdings) Ltd
PASMA Training - Turner Access Ltd
Passivent - Willan Building Services Ltd
Pastoral - Alno (United Kingdom) Ltd
Patay Pumps - Pump International Ltd
Pathfinder - Boulter Boilers Ltd
Pathfinder - Gradus Ltd
Pathfinder - Intersolar Group Ltd
Patina - Shadbolt F R & Sons Ltd
Patio Magic - Resiblock Ltd
PAVAC - Park Products
Pavia - Pilkington's Tiles Ltd
PC Lamp - Ledu UK Ltd
PDA - C-Tec (Computionics Ltd)
PE100 - Uponor Ltd
PE80 - Uponor Ltd
Peakmoor - Realstone Ltd
Peakstone - RMC Concrete Products
Pearce Homes - Pearce Construction(Barnstaple)Ltd
Pearce Property Services - Pearce Construction(Barnstaple)Ltd
Pedalo - Erlau AG
Pedestal Tip-Up Seating System - Race Furniture Ltd
Pedmon - CEM Systems Ltd
Pegasus - Bison Bede Ltd
Pegasus - Powys Hire Ltd/ P.I.B
Pegasys - Cardkey Systems
Pehdennis - Tegral Building Materials Ltd
Pel-Job - Chippindale Plant Ltd
Pelican - Iles Waste Systems
Pelican - Powys Hire Ltd/ P.I.B
Pelican - Tuke & Bell Ltd
Pella - Pellfold Parthos Ltd.
Pemberley - CPV Ltd
Pemko - Relcross Ltd
Penn - Johnson Controls Systems Ltd
Pennine - Laidlaw Architectural Hardware
Pennine - Laidlaw Architectural Hardware Ltd
Penrhyn - McAlpine Slate Ltd, Alfred
Pepex Pipes - Wirsbo

Perfecta Circulators - Turney Turbines (pumps) Ltd
Performa - Pegler Ltd
Performance Plus steel - Parkwood Mellows Ltd
PERI UP System scaffolding - PERI Ltd
Periplan - P.C.I Construction Systems Ltd
Periscope - Ryton's Building Products Ltd
Periwarm - Unilock Ltd
Perko - Perkins & Powell
Perkomatic - Perkins & Powell
Perlux - Harkness Hall Ltd
Perma Plug - Perma Industries Ltd
Permaclick - Brohome Ltd
Permacrib - PHI Group Ltd
Permafoam - Isocrete Group Sales Ltd
Permagrain - Wicanders (GB) Ltd
Permaguard - PermaRock Products Ltd
Permalux Sealed Range - Lumitron Ltd
Permaprene - En-tout-cas plc
Permarend - PermaRock Products Ltd
PermaRock - PermaRock Products Ltd
Permashield ® - Andersen/ Black Millwork
Permatite - RMC Readymix Limited
Permatred - Permali Gloucester Ltd
Permoglaze - Akzo Coatings Plc
Permoglaze - Akzo Nobel Decorating Coatings Ltd
Permoglaze - Crown Trade Paints
Permutit - USF Permutit/Ionpure
Perspectives - Gradus Ltd
Perstorp - Viking Laminates Ltd
Petraflex - Grass Concrete Ltd
Petraflex - Landscape Grass (Concrete) Ltd
Petrarch - Petrarch Claddings Ltd
Petrel - PFP Electrical Products Ltd
Petrelux - PFP Electrical Products Ltd
Pexapipe - IPPEC Sytsems
Pexatherm - IPPEC Sytsems
PH Products - PH Products Ltd
Phantom - Dudley Thomas Ltd
Pharo - Hansgrohe
Phenblox Pipe Support Inserts - Industrial Hangers Ltd
Phenomenon - Paragon by Heckmondwike F B Ltd
Phillips - Bernlite Ltd
Phlexicare - Nicholls & Clarke Ltd
Phoenix - Cooke Brothers Ltd
Phoenix - Intersolar Group Ltd
Phoenix Kingfisher - Jewers Doors Ltd
Phoenix Osprey - Jewers Doors Ltd
Phoenix Swift - Jewers Doors Ltd
Phosco - Phosco Ltd
Piazza - Erlau AG
Piazza - Muraspec Ltd
Piccolo lift - Stannah Lifts Ltd
Pickering Europe Ltd - Pickerings Ltd
Pikes - Pike Signals Ltd
Pilkington 'K' Glass - Pilkington Birmingham
Pilkington Antisun - Pilkington UK Ltd
Pilkington Audioscreen - Pilkington UK Ltd
Pilkington Insulating Units - Pilkington UK Ltd
Pilkington K Glass - Pilkington UK Ltd
Pilkington Planar Structural Glazing - Pilkington UK Ltd
Pilkington Planar Systems - Stoakes Systems Ltd
Pilkington Planarclad - Pilkington UK Ltd
Pilkington Pyrodur - Pilkington UK Ltd
Pilkington Pyroshield - Pilkington UK Ltd
Pilkington Pyroshield Safety - Pilkington UK Ltd
Pilkington Pyrostop - Pilkington UK Ltd
Pilkington Suncool - Pilkington UK Ltd
Pilkington Tiles - Stokes R J & Co Ltd
Pilkington UV Screen - Pilkington UK Ltd
Pinescan - Scandanavian Window Systems Ltd
Pinnacle Envioromental Plastic - Signs International
Pinpiontplus - Pergola Products Ltd
Piper Lifeline - Tunstal Telecom Ltd
Pirouette - Rackline Systems Storage Ltd
Pivette - Leaderflush & Shapland
Placon - Plastic Construction Fabrications Ltd
Plain Tile - Decra Roofing Systems
Plan - Hudevad Britain
PLAN-RAD - Property Mechanical Products Ltd
Planal - Axter Ltd
Planar - Pilkington Birmingham
Planer - Ide T & W Ltd
Planet - Pland
Planet Projects - Planet Projects Ltd
Plano - Rockfon Ltd
Planolit - Mapei UK Ltd
Planta - Erlau AG
Plasflow - Fullflow Ltd
Plasform Original - Leaderflush & Shapland
Plastel - Armstrong World Industries Ltd
Plastene - Webrr & Broutin
Plaster Arris Protector - Areco Roofing Supplies Company
Plastex - Plastic Extruders Ltd
Plastic Padding - Henkel Home Improvement & Adhesive Products
Plastic padding - Loctite UK Ltd
Plastilac - Trimite Ltd
Plastiment - Sika Ltd
Plastimetal - EBC UK Ltd
Plastocrete - Sika Ltd
Plastruct - The Modelshop
Playdek - En-tout-cas South West
Playtime - Record Playground Equipment Ltd
Plaza Range - Pedley Furniture international Ltd
Plazgard - Lonsdale Metal Company
Pleatex - Sampson T F Ltd
Pleione - Orsogril Sarl UK Department
Plexiglas - Amari Plastics Plc
Plugless - Millar Dennis (1998) Ltd
Plumbers bits - Hunter Plastics Ltd
Plus - Osram Ltd
Plus Eight Universal System - Turner Plus Eight Ltd
Plus Gas - Unicorn Abrasives (UK) Ltd
Plycorapid - Laybond Products Ltd
PmB : Waterproofing - Pitchmastic PLC,
PMF - E. H. S. Roofing Ltd
PMP-RAD - Property Mechanical Products Ltd
PNC 3 Vision - Tunstal Telecom Ltd
Polagard - Manders Paint Ltd
Polar - Komfort Systems Ltd
Polar - Rockfon Ltd
Polcarb 40 S - ECC International Ltd
Polcarb 45 Slurry - ECC International Ltd
Polcarb 60 (S) - ECC International Ltd
Polcarb 60 Slurry - ECC International Ltd
Polcarb 90 - ECC International Ltd
Polcarb SB - ECC International Ltd
Policor C.C. - TDI (UK) Ltd
Poligras - En-tout-cas plc
Polished Plaster - Francesca Di Blasi Co. Ltd, The
Politarp - Visqueen Agri
Polo - Pegler Ltd
Polo - Swish Building Products Ltd
Polow - Troika Architectural Mouldings Ltd
Polow - Tropravit (UK) Ltd
Poly Buytl - Alfas Industries Ltd
Polycell - ICI Paints
Polycell - Polycell Products Ltd
Polyclens - Polycell Products Ltd
Polycolor - Shapland & Petter Ltd
Polycup - Rockwell Sheet Sales Ltd
Polydor - Shapland & Petter Ltd
Polydrain - Polybau Ltd
Polydry - Polypipe Civils Ltd
Polyduct - Polypipe Civils Ltd
Polyfilla - Polycell Products Ltd
Polyflex - Halstead James Ltd
Polyflor - Halstead James Ltd
Polyfoam - Owens Corning Building Products (UK) Ltd
Polyforme - Shapland & Petter Ltd
Polylath - BRC Building Products
PolyLay - Junckers Limited
Polymatic - CPV Ltd
Polymer Modified Bitumen Emultions and cutbacks. - South Western Tar Distilleries
Polymetron - Zellweger Analytic Ltd, Sieger Division
Polypipe - Ashmead Buidling Supplies Ltd
Polypipe - Brikenden (Builders Merchants) Ltd
Polypipe - Drainage Systems
Polypipe - Hipkin F W Ltd
PolyPlus - Helifix Ltd
Polyrex - Neural Smoke Detector - Siemens Building Technologies Ltd, Cerberus Division
Polyrey - Decra Cubicles Ltd
Polyrey - Viking Laminates Ltd
Polyripple - Polycell Products Ltd
Polysafe - Halstead James Ltd
Polyself - Polybau Ltd
Polyshield - Feedwater Ltd
Polystrippa - Polycell Products Ltd
Polytan - Feedwater Ltd
Polytek - Notcutt W P Ltd
Polytex - Polycell Products Ltd
Polytherm - Smart Systems Ltd
Polytred - Halstead James Ltd
Polytron - Draeger Ltd
Polyu - Rockwell Sheet Sales Ltd
Polyvent 225, 300 - Domus Ventilation Ltd.
Polyvision - Shapland & Petter Ltd
Polywrak - Shapland & Petter Ltd
Pool Plus - LazyLawn
Pop - Tucker Fasteners Ltd
POP Rivets - Allman Fasteners Ltd
Porcupipe - Polypipe Civils Ltd
Pordorfilm - RIW Ltd
Porta Master - Pergola Products Ltd
Portaflex - Hughes Safety Showers
Portaloo - Portasilo Ltd
Portasilo - Portasilo Ltd
Portaspray - Dawson McDonald & Dawson Ltd
Portastor - Portasilo Ltd
Portcullis - Chubb Security Installations Ltd
Portland - Hamworthy Heating Ltd
Portway - Taylor & Portway Ltd
Poselco - Poselco Lighting
Posi-Strut - MiTek Industries Ltd
Posi-Web - MiTek Industries Ltd
Posilok - Exitex Ltd
Positred - Dex-o-Tex International Ltd
Potissimum - Orsogril Sarl UK Department
Pottelberg - Smithbrook Building Products Ltd
Poulica - Pouliot Designs
Poulot Designs - Desney Products
Power breaker - Power Breaker
Power-Style - Square D, Electrical Distribution Division
Powerbond - Gilmour Ecometal
Powerbrace - Maby Hire Ltd
Powerbrace - Matbro Ltd
Powercyl - Gledhill Water Storage Ltd
Powerdrain - ACO Technologies Plc
Powerflow Flux - Alpha Fry Ltd
Powerflow Plux - Fernox Fry Technology UK
Powerflow Solder - Alpha Fry Ltd
Powerflow Solder - Fernox Fry Technology UK
Powerheat - Heat Trace Ltd
Powerline - Morse Controls Ltd
Powerline - Roof Units Group
Powerlink Plus - Caradon M K Electric Ltd
Powermatic - Sussex Brassware Ltd
PowerPlas 519 - Power Plastics Ltd
PowerPlas 541 - Power Plastics Ltd
Powerpole - Mita (UK) Ltd
Powerrac - Dezurik International Ltd
Powersafe - Chloride Industrial Batteries Ltd
Powerscape - Game Time UK Ltd
Powershade - Eclipse Blind Systems Ltd
Powershare - Maby Hire Ltd
Powershore - Matbro Ltd
Powersoft - MTE Ltd
Powerstar - Osram Ltd
Powerstart - MTE Ltd
Powerstation - Ves Andover Ltd
Powerstock - Hamworthy Heating Ltd
Powerswitch - MTE Ltd
Powertrak - Rackline Systems Storage Ltd
Powertrim - Heat Trace Ltd
Powervent - Kiltox Damp Free Solutions
Powervent - Martin Orgee & Associates
Pozidrain - ABG Ltd
Pozzolith - Feb MBT
PPA 571 - Plascoat Systems Ltd
Prandina - Forma Lighting Ltd
PREcheck - National Britannia Ltd
Precious Gems - Carpets of Worth Ltd
Prefixx - Tektura Plc
Preiser - The Modelshop
Premier - Cego Frameware Ltd
Premier - Integra Products
Premier - Manton Industrial Seals Ltd
Premier - Securikey Ltd
Premier - Xpelair Ltd
Premier by Coram - Coram (UK) Ltd
Premier Collection - Marley Floors Ltd
Premier Service - Flexiform Business Furniture Ltd
Premier tools - Lafarge Plasterboard Ltd
Premix - Hanson Aggregates
Premix Concretes - Hanson Aggregates
Prepakt - Westpile Ltd
Prescor - Flamco UK Ltd
PreSense/ AbSence - ECS Lighting Controls Ltd
President - Deans Blinds & Awnings (UK) Ltd
President - Securistyle Ltd
Presspak - Flamco UK Ltd
Pressure Pak - Fordwater Pumping Supplies
Pressurepak - Fordwater Pumping Supplies Ltd
Prestcold - Copeland Corporation Ltd
Prestex - Pegler Ltd
Prestige - Balmoral Porcelain Limited/Longmead Ceramics
Prestige - Caradon Gent Ltd
Prestige Plus - Caradon M K Electric Ltd
Prestressed Beam Construction P.B.C - Birchwood Concrete Products Ltd
Presweb - Prestoplan
Prima - Armstrong World Industries Ltd
Prima - Cardale Doors Ltd
Prima - Cego Frameware Ltd
Prima - Dermide Ltd
Prima - Prima Security & Fencing Products
Prima - Sampson & Partners Fencing
Prima 2 Gas Vent - Rite-Vent Ltd
Prima Plus Single Wall System - Rite-Vent Ltd
Prima SW Single Wall System - Rite-Vent Ltd
Prima worksurfaces - Bushboard Parker Ltd
Primacalc - Fullflow Ltd
Primaflow - Fullflow Ltd
Primar - Girsberger London
Primary Options - Project Office Furniture PLC
Primatic - IMI Range Ltd
Primera - Gaskell Textiles Ltd
Primetime - Game Time UK Ltd
Primo - Industrial Devices Ltd
Primus Washers/ Ironers - Warner Howard Ltd
Prinmuls Lite LMP90K Polymer Emultion - South Western Tar Distilleries
Prinmuls Mac H.A.U.C. Binder - South Western Tar Distilleries
Prinmuls MP90X Polymer Emultion - South Western Tar Distilleries
Print HPL - Abet Ltd
Priory - Dudley Factory Doors Limited
Priory - Priory Shutter & Door Co Ltd
Priory - RMC Concrete Products
Priory The Priory Shutter & Door Co. Ltd. - Neway Doors Ltd
Priory The Priory Shutter & Door Co. Ltd. - Shutter Door Repair & Maintenance Ltd
Prisma 4000 - Project Office Furniture PLC
Pritt - Henkel Home Improvement & Adhesive Products
Pro Bounce - En-tout-cas plc (Scotland)
Pro Surv Kit - Protimeter plc
Pro-Bounce - En-tout-cas South West
Pro-Sport - Hepworth Minerals & Chemicals Ltd
Proarc - PCT Group Sales Ltd
Probea - Ferham Products
ProBounce - En-tout-cas plc
Probuild - Vallance
Procast - Procter Bros Limited
PROcheck - National Britannia Ltd
Proclad - Cladding & Decking UK Ltd
ProCon 250 - Caradon Catnic Ltd
Procor - Grace Services
Procut - PCT Group Sales Ltd
Prodec coatings - Intergrated Polymer Systems (UK) Ltd
Prodorbond - RIW Ltd
Prodorbond Terrazzo - RIW Ltd
Prodorcrete - RIW Ltd
Prodorflor - RIW Ltd
Prodorglaze - RIW Ltd
Prodorguard - RIW Ltd
Prodorite - RIW Ltd
Prodorshield - RIW Ltd
Profile - Ashley & Rock Ltd
Profile - Kermi (UK) Ltd
PROfile - National Britannia Ltd
Profile - Rackline Systems Storage Ltd
Profile 22 - Plastal Commercial Ltd
Profile Paving - RMC Readymix Limited
Profiles - Anderson D & Son
Profiles Cubicles - Bushboard Parker Ltd
Profiles solo - Anderson D & Son
Profilia - Applied Acoustics (a part of Henry Venables Limited)
Profilia - Henry Venables Ltd
Profilptin - DLW Flooring Ltd
Proform - Rackline Systems Storage Ltd
Profuse - Uponor Ltd
Proglas - Marley Waterproofing
Programastat - Timeguard Ltd
Progressive - Paragon Business Furniture
Prohire - PCT Group Sales Ltd
Prolift - PCT Group Sales Ltd
Proline - Leisure, Glynwed Consumer Foodservice Products Ltd
Promaseal - Promat Fire Protection Ltd
Promaster - En-tout-cas plc
Promat - Kitsons Insulation Products Ltd
Promatect - Promat Fire Protection Ltd
Promenade - Lappset UK Ltd
Promisol Bi-Modular Flat Panel - TAC Metal Forming Ltd
Pronto - Girsberger London
Propadek - Propaflor Ltd
Proplene - CPV Ltd
Propower - PCT Group Sales Ltd
Proseal - Armorex Ltd
Proseal - Illuma Lighting Ltd
Prospex - Terrapin Ltd
Protal - Winn & Coales
Protectacap - Interlux Ltd
Protector - Amber Doors Ltd
Protector - Securikey Ltd
Protector Frames - Climet Ltd
Protectosie 300 E - Degussa Ltd
Protectosie 800 E - Degussa Ltd
Protectosie Antigraffiti - Degussa Ltd
Protectosie WS 405 - Degussa Ltd
Protim - Protim Solignum Ltd
Protim- Prevac - Protim Solignum Ltd
ProTime - Cardkey Systems
Protimeter - Protimeter plc
Proton - Paddock Fabrications Ltd
Provincial - Sandtoft Roof Tiles Ltd
Ps-3 - Alpha Fry Ltd
Psicon - Geoquip Ltd
PSM - ABB Kent Meters
Pudlo - Ball David Group plc
Puldo - David Ball Group Plc
Pullman - Portakabin Ltd
Pulsacoil - Gledhill Water Storage Ltd
Pulsar - Rada Lighting Ltd
Puma - Mode Lighting (UK) Ltd
Puma - Resdev Ltd
Pumadur 97 - Resdev Ltd
Pumaflor - Resdev Ltd
Pumalite - Fyfe John Ltd
Pumpkn - Pump Technical Services Ltd
Pure - Isolated Systems Ltd
Purewell - Hamworthy Heating Ltd
Purilan - Ubbink (UK) Ltd
Purogene - Hertel Services
Purogene - Verna Ltd
Push and Lock - Ryton's Building Products Ltd
Push Flush - Sheardown Engineering Ltd
Pushfit - Talbot
Pushlock - Latchways plc
PVA FLAKE - Sempol Surfaces Ltd
PVCU-Steps - Quantum Profile Systems
Pygme - Grosvenor Pumps Ltd
Pyra - Webrr & Broutin
Pyramid - Baxall Ltd
Pyramid - Vulcanite Limited
Pyranova Insulated Fire-resisting Glass - Schott Glass Ltd, Arcitectural Glass
Pyrans Fire-Resisting Glass - Schott Glass Ltd, Arcitectural Glass
Pyro-Guard - Timber Treatments Ltd
Pyrocoil - Mann McGowan Fabrication Ltd
Pyrocryl - Vallance
Pyrodur - Pilkington Birmingham
Pyroglaze - Mann McGowan Fabrication Ltd
Pyrogrille - Mann McGowan Fabrication Ltd
Pyroguard - C. G. I. International Ltd, trading as Colebrand Glass
Pyrok - Cape Boards Ltd
Pyrolieth - Hickson Timber Products Ltd
Pyromas - Mann McGowan Fabrication Ltd
Pyropol - Vallance
Pyroputty - Mann McGowan Fabrication Ltd
Pyroshield - Pilkington Birmingham
Pyrosleeve - Mann McGowan Fabrication Ltd
Pyrospali - Mann McGowan Fabrication Ltd
Pyrosteel - Cape Boards Ltd
Pyrostem - C. G. I. International Ltd, trading as Colebrand Glass
Pyrostop - Ide T & W Ltd
Pyrostop - Pilkington Birmingham
Pyrostrip - Mann McGowan Fabrication Ltd
Pyrotenax - BICC plc
Pyrotenax - BICC Thermoheat Ltd
Pyroyista - Mann McGowan Fabrication Ltd
Pyruma - Purimachos Ltd
Pyxel - Grosvenor Pumps Ltd
Q-Rail 2000 - Quartet-GBC
Q.S. Elite - Masterbill Micro Systems Ltd

QB - Welconstruct Co
Qbs - Metalrax Ltd
QD 90 - Blackfriar Paints and Varnishes
QD 90 - Parsons E. & Sons Ltd
QD2 - Dacrylate Paints Ltd
QD3 - Dacrylate Paints Ltd
QT eQTEL - Pegler Ltd
Quadrascott - Fyfe John Ltd
Quadrasett - Camas Building Materials
Quadro 4 - Rockfon Ltd
Quadroline - Illbruck Ltd
Quaker deck drain - Fox Pool (UK) Ltd
Quality Water Group - British Water
Quantec Call System - C-Tec (Computionics Ltd)
Quantum - Buchtal (U.K.) Limited
Quarrycast - Troika Architectural Mouldings Ltd
Quartermaster Shelving - Beford & Soar Ltd
Quartz 750 - Durabella Ltd
Queensfil 240 - ECC International Ltd
Queensfil 25 - ECC International Ltd
Queensfil 300 - ECC International Ltd
Quelcote - Quelfire
Quelfire - Firebarrier Services Ltd
Quelfire - Quelfire
Quelfire QF1 Mortar - Quelfire
Quelfire QF2 Mortar - Quelfire
Quelfire QF4 Mortar - Quelfire
Quicbuilt - Brierley B (Garstang) Ltd
Quick Clip - Junckers Limited
Quickcool - IPPEC Sytsems
Quickfit - Vantrunk Engineering Ltd
Quickfit - Viking Johnson
Quickframe - IPPEC Sytsems
Quickline - Dyer Environmental Controls
Quickpipe - IPPEC Sytsems
Quickstrip - Laybond Products Ltd
Quicktronic - Osram Ltd
Quiet Mat - Salex Interiors Ltd
Quietpack - Roof Units Group
Quietzone - Owens Corning Building Products (UK) Ltd
Quikaboard - Quinton & Kaines (Holdings) Ltd
Quikpost - Purdie Elcock Ltd
Quintesse - Matki plc
Quip - Quip Lighting, Div of Bruce & Macintyre Ltd
Quorndon - Redland Precast
Qwickline - Square D, Electrical Distribution Division
Raak - Electrak International Ltd
Racklight - IMP Lighting Ltd
Rada - Rada Lighting Ltd
Rada Controls - Caradon Plumbing Solutions
Radcliffe King of Diamonds - Radcliffe & Sons (London)
Radcliffe London Warrented - Radcliffe & Sons (London)
Radcrete - Radflex Contract Services Limetid
Radflex 125 - Radflex Contract Services Limetid
Radflex Expantion Joints - Radflex Contract Services Limetid
Radflex S150 - Radflex Contract Services Limetid
Radflex S200 - Radflex Contract Services Limetid
Radiance - Matki plc
Radipex - Wirsbo
Radius - Hudevad Britain
Radjoint - Radflex Contract Services Limetid
Radroof - Radflex Contract Services Limetid
Rafid - Geoquip Ltd
Ragdale - Redland Precast
Rail, Roll & Ride - Record Playground Equipment Ltd
Raincoat - Tor Coatings Group
Raindrain - ACO Technologies Plc
Rainstopper - Andura Textured Masonry Coatings Ltd
Rainwater Goods - Eternit UK Ltd
Rallibondite - Astor-Stag Ltd
Rallikol - Astor-Stag Ltd
Rallisil - Astor-Stag Ltd
Rallisil - Trade Sealants
Rallithane - Astor-Stag Ltd
Rambolt - Ramset Fasteners Ltd
Ramplug - Ramset Fasteners Ltd
Ramsay Ladders - Ramsay & Sons (Forfar) Ltd
Ramset - Arrow Supply Company Ltd
Ramset - Ramset Fasteners Ltd
Randi - Laidlaw Architectural Hardware Ltd
Randi-line - Laidlaw Architectural Hardware
Range Cooker Co - Diamond Merchants
Rangemaster - Leisure, Glynwed Consumer Foodservice Products Ltd
Rapid - Clark Door Ltd
Rapid Clamp - Tenby Industries Ltd
Rapid Jit delivery service - Seco Aluminium Ltd
Rapid-Roll - Stokvis R S & Sons Ltd
Rapidac - Dacrylate Paints Ltd
Rapidair - Draeger Ltd
Rapide - Project Office Furniture PLC
Rapide - Ves Andover Ltd
Rapide Shelving - Balmforth Engineering Ltd
Rapide solo - Anderson D & Son
Rapidplan - Wernick S & Sons Ltd
Rapier - Kenrick Archibald & Sons Ltd
Rapier 2 - CIL International Ltd
Rapier Marine - CIL International Ltd
Rapitech - Stokvis R S & Sons Ltd
Ravenna - Paragon by Heckmondwike F B Ltd
Ravenna Bathroom Suite - Ideal-Standard Ltd
Rawlplug - Arrow Supply Company Ltd
Rawlplug - Kem Edwards Ltd
Rayburn - Aga-Rayburn, Glynwed Consumer & Construction Products
Raypak - AEL
RB - R-B International Ltd
RB Horizontal - Fulton Boiler Works (Great Britain) Ltd
RCC - RCC, Division of Tarmac Precast Concrete Ltd
RCC-retaining Walls - Tarmac Precast Concretre
RD. - Hille Ltd
RDM - Apollo Fire Detectors Ltd
Readyblock - RMC Concrete Products (UK) Ltd
Readypave - RMC Concrete Products (UK) Ltd
Readypave 50 - RMC Concrete Products (UK) Ltd
Readyscreed - RMC Readymix Limited
Readyspan - RMC Concrete Products (UK) Ltd
ReadySpread - RMC Readymix Limited
Realitis DX Communication Server - Siemens Communication Ltd
Realstone Ltd - Realstone Ltd
Recollections - Pegler Ltd
Record - Polybau Ltd
Rectaleen - Flexible Reinforcements Ltd
Recupovent - Therminal Technology (sales) Ltd
Recycling - Lappset UK Ltd
Recyfix - Hauration Kaskade Ltd
Recyfix Channels - Hauration Kaskade Ltd
Red Box Wipes - Deb Ltd
Reddicord - Reddiglaze Ltd
Reddifoam - Reddiglaze Ltd
Reddihinges - Reddiglaze Ltd
Reddilock - Reddiglaze Ltd
Reddipile - Reddiglaze Ltd
Reddiprene - Reddiglaze Ltd
Redgra - Hanson Aggregates
Rediguard - Hoyles Fire & Safety Ltd
Refatex - Astron Building Systems Commercial Intertech Ltd
Reflections Bathroom Suite - Ideal-Standard Ltd
Reflex 275 - Monarflex Ltd
Reflex Flameproof - Monarflex Ltd
ReflexSuper - Monarflex Ltd
Reflux - Lumitron Ltd
Regaflex S - Rega Metal Products Ltd
Regaflex W - Rega Metal Products Ltd
Regal - Contravent Regal Ltd
Regency - Forticrete Ltd
Regency Buff - Cranleigh Brick & Tile Co Ltd
Regency Divestone - Fyfe John Ltd
Regency Drestone - Camas Building Materials
Regency Swimming Pools - Regency Swimming Pools Ltd
Regent - Clenaware Systems Ltd
Regent - JSB Electrical PLC
Regent Mastersraft 1890's - Barber Wilsons & Co Ltd
Regular - Integra Products
Rehau - Architectural Plastics (Handrail) Ltd
Rehau - Ideal Williams Ltd
Rehau - Masterframe Windows Ltd
Reheat - Heat Transfer Pressurisation
Reid Span - Reid John & Sons (Strucsteel) Ltd
Relax - Provex Products Ltd
Relite - Fantasia Distribution Ltd
Relrofix - Pitchmastic PLC,
Remclamp - LPA Industries PLC
Remstrap - LPA Industries PLC
Renaissance - Tenby Industries Ltd
Rend Aid - Snowcem
Rendabrick - MR Limited
Rendalath - BRC Building Products
Renfor - Super Cement Ltd
Renovaction - Aqua Design Ltd
Renovisions - Interface Europe Ltd
Rentokil LFB30/60 - Rentokil Initial U.K. Ltd.
Renz Mail Boxes - JB Architectural Inronmongery Ltd
Repairite - MMP International
Resdev - Resdev Ltd
Resiblock 22 - Resiblock Ltd
Resiblock ER - Resiblock Ltd
Resiblock OR - Resiblock Ltd
Resiblock Superior - Resiblock Ltd
Resiclean - Resiblock Ltd
Resifix 3P Anchor Grout - Exchem Mining and Construction
Resiflex - Resiblock Ltd
Resimac Blacktop - Resiblock Ltd
Resistite - Dex-o-Tex International Ltd
Resistoid - Harrison Drape, Divsion of McKenie (UK)
ResiTie - Helifix Ltd
Responseline - National Britannia Ltd
Retriever 55 bing - Glasdon UK Ltd
Retro-seal - Granflex Roofing Ltd
Retroflame - Timber Treatments Ltd
RetroTie - Helifix Ltd
Revelation - Spectus UPVC Systems Ltd
Revo Actuators - Alfa Laval Saunders Ltd
Revosec - Gunnebo Mayor Ltd
Revue Bathroom Suite - Ideal-Standard Ltd
Rex - Legrand Electric Ltd
Rex - Ves Andover Ltd
Rex - Ward Bekker Sales Ltd
Reynobond - Booth Muirie Ltd
Reynobond ACM - Technology Telford Ltd
RH 'U' Tube Heater - Radiant Services Ltd
Rhapsody Range - Caradon Bathrooms Ltd
RHD Double Linear Heater - Radiant Services Ltd
Rheobuild - Feb MBT
Rheocell - Feb MBT
Rheofinish - Feb MBT
Rheomix - Feb MBT
Rhino Corlon - Armstrong World Industries Ltd
Rhino Doors - European Profiles
Rhino Flex - Armstrong World Industries Ltd
Rhino Floor - Armstrong World Industries Ltd
Rhino Geneous - Armstrong World Industries Ltd
Rhino Quartz - Armstrong World Industries Ltd
Rhino Safe - Armstrong World Industries Ltd
Rhino Tex - Armstrong World Industries Ltd
Rhino Tile - Armstrong World Industries Ltd
RHL Linear Heater - Radiant Services Ltd
Rhodius - Hyde Brian Ltd
Ribblelite - Richard Hose Ltd
Ribloc - Forticrete Ltd
Richardson - Amdega Ltd
Richter spielgerate - Kompan (UK) Ltd
Ridgicoil - Polypipe Civils Ltd
Ridgidrain Twin Wall - Polypipe Civils Ltd
Ridgiduct - Polypipe Civils Ltd
Ridgigully - Polypipe Civils Ltd
Rigaflo - Farr Filtration Ltd
Righton PQ - Righton Ltd
Rigibeam - Southern Sanitary Specialists Ltd
Rigid-Cor - TDI (UK) Ltd
Rigidal - Alcan Building Products
Rigidline - Phoenix Metal Products Ltd
Rigidlock 15 - Phoenix Metal Products Ltd
Rigidlock 24 - Phoenix Metal Products Ltd
Rigifix - Huntley & Sparks
Rilass V - Farr Filtration Ltd
Rinaldi - Atkinson & Kirby Ltd
Rio Ferrada Brazilian Slate - Kirkstone
Rio Neblina Brazilian State - Kirkstone
Rio Verde Brazilian Slate - Kirkstone
Rior - Ward's Flexible Rod Company Ltd
Ripac - CSSP (Computer Software Services Partnership)
Riposeal - Shackerley (Holdings) Group Ltd incorporating Designer ceramics
Ritz - New Haden Pumps Ltd
Ritz-atro - New Haden Pumps Ltd
Rivendale - Eternit UK Ltd
RMB 300 - Monarflex Ltd
RMC Polybead - RMC Panel Products Ltd
Robar - JCL Engineering Ltd
Robec - Tuke & Bell Ltd
Robert Lynam collection - Emsworth Fireplaces Ltd
Robey - Wellman Robey Ltd
Robinhood - Beeston Heating Group Ltd
Robinlight - Torvale Building Products, A Division of Stadium Group PLC
Robinson Willey - Robinson Willey Ltd
Robinsons Greenhouses - Compton Building Ltd
Robocal - Altecnic Ltd
Robofil - Altecnic Ltd
Robokit - Altecnic Ltd
Robolink - Altecnic Ltd
Robot - Robot (UK) Ltd
Rockbloc Acoustic Barriers - Ruthin Precast Concrete Ltd
Rockcem - Cryctherm Insulation Ltd
Rockfon - Nevill Long Limited
Rockfon - New Forest Ceilings Ltd
Rockliner - Cryotherm Insulation Ltd
Rockranger Crushing outfit - Parker Plant Ltd
Rocksil - Owens Corning Building Products (UK) Ltd
Rockwall Walling/Cladding - Ruthin Precast Concrete Ltd
Rockwell - Rockwell Sheet Sales Ltd
Rockwool - Firebarrier Services Ltd
Rockwool - Rockwool Ltd
Rofatop Membrane - Blunn Slates Ltd
Rola - Bramah Security Equipment Ltd
Roll-fix ridge - Klober Ltd
Rollajack - Rollalong
Rollaskid - Rollalong
Rollatape - Advance Tapes International Ltd
Rollawheel - Rollalong
Rollercash - Safetell Ltd
Rollertrax - Oldham Signs Ltd
Rolls Rotary - Tuke & Bell Ltd
Rollstick - Axter Ltd
Rolux - Ubbink (UK) Ltd
Rolyat - Elsy & Gibbons Ltd
Romil - Anderson Gibb & Wilson
Ronabond - Ronacrete Ltd
Ronafix - Ronacrete Ltd
Ronascreed - Ronacrete Ltd
Ronascreed DB Fast Floor - Ronacrete Ltd
Ronaset - Ronacrete Ltd
Rondelux - Poselco Lighting
Roofite - Watco UK Ltd
Roofmate - Dow Construction Products
Roofmax - Owens Corning Building Products (UK) Ltd
Roomstat - Horstmann Timers & Controls Ltd
Roomvent - Airflow Developments Ltd
Rootfast® - Earth Anchors Ltd
Rope and Bow - Balmoral Porcelain Limited/Longmead Ceramics
Ropetwist Bathroom Suite - Ideal-Standard Ltd
Rosamon - Heating Equipment Ltd
Roscal - Higginson Staircases Ltd
Rosebury - Qualitas Bathrooms
Rosengerns - Rosengerns Tann Ltd
Rossella - Dermide Ltd
Rossetti - Symphony Group PLC, The
Rossflex - Dermide Ltd
Rota spike ® - Jackson H S & Son (Fencing) Ltd
Rotaflo - Fluidair Compressors Ltd
Rotaflow - Gebhardt Kiloheat
Rotapak - Fluidair Compressors Ltd
Rotasec - Gunnebo Mayor Ltd
Rotastar - Fluidair Compressors Ltd
Rotatune - Sabroe Refigeration Ltd
Rotavent - Gebhardt Kiloheat
Rotaworm - Ward's Flexible Rod Company Ltd
Rothersky - Paragon by Heckmondwike F B Ltd
Rotocleve - Clearwater Plc
Rougeite - Civil Engineering Development Ltd
Rough at the edges - Clayton-Munroe Ltd
Roulcabin - UK Cabin Co
Roulston - UK Cabin Co
Rovacabin - Rovacabin
Rovalink - Rovacabin
Rovaspan - Rovacabin
Royair - Royair Ltd
Royal Granite - Pilkington's Tiles Ltd
Royale - Firth Carpets Ltd
Royale - Hill Leigh (Joinery) Ltd
Royale - Midland Stom Ltd
Royale Doors - Kind, J B, Ltd
Royals - SCHUCO International KG
Royalux - JSB Electrical PLC
Royce - Morris Mechanical Handling Ltd
Royde & Tucker - City Hardware and Security
RSS - Rouse Security Service
Ruabon Red - Ruabon Dennis Ltd
Rubberfuse - E. H. S. Roofing Ltd
Rubberfuse - Intergrated Polymer Systems (UK) Ltd
Rubbergard - Firestone Building Products Ltd
Rubberline - Dudley Thomas Ltd
Rubberwell - Dudley Thomas Ltd
Rubboseal - Trade Sealants
Ruberoid - Ashmead Buidling Supplies Ltd
Ruberoid - Ashmead Building Supplies Ltd
Rubino - Novellini Bathroom Products
Rudgwick - Baggeridge Brick PLC
Rugby cement - Brikenden (Builders Merchants) Ltd
Rugby Hydrated Lime - Rugby Cement
Rugby Masonry - Rugby Cement
Rugby Oilwell - Rugby Cement
Rugby Portland - Rugby Cement
Rugby Rapid - Rugby Cement
Rugby Special - Rugby Cement
Rugby Sulfate - Rugby Cement
Rugby White - Rugby Cement
Ruko - ASSA Ltd
Rundflex Circular wall formwork - PERI Ltd
Russell Hobbs - Pifco Ltd
Rustins - Rustins Ltd
Rustoleum - Andrews J S (Coatings) Ltd
Rutex - Rockfon Ltd
Rutland - Dryad Architectural Ltd
Rytweep - Ryton's Building Products Ltd
S Shield - Bollom J W & Co Ltd
S.F.S - Mestec PLC
S.R. Epo Resin Floor Screed - Ferguson G A & Co Ltd
S.R. Resin Floor Sealer - Ferguson G A & Co Ltd
S.R. Tanking System - Ferguson G A & Co Ltd
S.V.K. Cromleigh - Allan Harris & Sons Ltd
Sabre - CIL International Ltd
Sadia Refrigeration - Viscount Catering Ltd
Sadolin Classic - Sadolin UK Ltd
Sadolin Extra - Sadolin UK Ltd
Sadolin Focus - Sadolin UK Ltd
Sadolin Harmoni - Sadolin UK Ltd
Sadolin Holdex - Sadolin UK Ltd
Sadolin New Base - Sadolin UK Ltd
Sadolin Prestige - Sadolin UK Ltd
Sadolin PV67 - Sadolin UK Ltd
Sadolin Sadolac - Sadolin UK Ltd
Sadolin Superdec - Sadolin UK Ltd
Saf 2 - The Safety Tread Ltd
Safe 'n' Sound - Premdor
Safe Call - Wandsworth Elecrtrical Ltd
Safe Glas - Cliffhanger Shelving
Safe T Epoxy - Anglo Building Products Ltd
Safe-T-net - Granflex Roofing Ltd
Safe-x-Scape - Samways P A & Co
Safedress - Associated Holdings Ltd
Safegard - James & Bloom Ltd
Safegrate - Lionweld Kennedy Ltd
Safegrid - Lionweld Kennedy Ltd
Safeguard - Securikey Ltd
Safeguard - Unicorn Abrasives (UK) Ltd
SafeGuard - Wandsworth Elecrtrical Ltd
Safelock - Lionweld Kennedy Ltd
Saferail - Lionweld Kennedy Ltd
Saferglaze - C. G. I. International Ltd, trading as Colebrand Glass
SafeSeal - Granflex Roofing Ltd
Safestep - Forbo - Nairn Ltd
Safeticurb - ECC Building Products Ltd
Safetread - Lionweld Kennedy Ltd
Safetred Range - Marley Floors Ltd
Saffire - Baydale Architectural Systems Ltd
Saga - Gerland Ltd
Salex Acoustic Products - The Salex Group Ltd.
Salient - Paragon Business Furniture
Salinas - Tektura Plc
Salmen's - Fine H & Son Ltd
Salter - Salter Weigh-Tronix
Sam - Ves Andover Ltd
Samontec - Arthur Fisher (UK) Ltd
Samson - Clow Group Ltd
Samson - Rockfon Ltd
Samsung - Connought Communication Systems Ltd
Samtor - Axter Ltd
Samuel Booth collection - Booth Samuel & Co Ltd
Sanador - EDS Shutter Door Systems Ltd
Sand Drain - ABG Ltd
Sanderson - Sanderson & Sons Ltd, Arthur
Sandringham - Loft Centre Products
Sandringham suite - Armitage Shanks Ltd
Sandtex - Akzo Coatings Plc
Sandtex - Crown Trade Paints

Sandtex Trade - Akzo Nobel Decorating Coatings Ltd
Sandwell - Chiltern Invadex
Sangamo - Schlumberger
Sani Panel - Jaga Heating Poducts (UK) Ltd
Sani Ronda Towel Rail - Jaga Heating Poducts (UK) Ltd
Sanigas Guillot - HRS Hevac Ltd
Sanimacerator - Max Appliances
Sanimacerator - Sussex Brassware Ltd
Sanistrel - Imperial Machine Co Ltd
Sanistrel - Sissons W & G Ltd
Sanitas - OWA UK Ltd
Santane - Webrr & Broutin
Sapodrain - Ashmead Buidling Supplies Ltd
Sapodrain - Ashmead Building Supplies Ltd
Sarnafast - Sarnafil Ltd
Sarnafil - Sarnafil Ltd
Sarnatherm - Sarnafil Ltd
Sarnatred - Sarnafil Ltd
Sartec - Anderson D & Son
Sartec vaperm breather membrane - Anderson D & Son
Sasmox - McLoughlin Wood Ltd
Satchnet - Satchwell Control Systems Ltd
Satellite - Paragon Business Furniture
Saturn - Fabrikat (Nottingham) Ltd
Saunier Duval - Diamond Merchants
Saunier Duval - Hepworth Heating Ltd
Savanna - En-tout-cas plc (Scotland)
Savanna - En-tout-cas South West
Savanna County - En-tout-cas plc
Save Pave - Associated Holdings Ltd
Saver - Draeger Ltd
Saver MK2 - Sunray Engineering Ltd
Sawco - The Saw Centre Group
Sawcut - The Saw Centre Group
Sawflex - nmc (UK) Ltd
Saxon - Legrand Electric Ltd
SB - Wilo Samson Pumps Ltd
SBS - Burlington Engineers Ltd
SBS - Chloride Industrial Batteries Ltd
Scala - DLW Flooring Ltd
Scan Pumps - ABS Pumps Ltd
Scantronic - Cooper Security
Scantronic - Crimegard Ltd
Sceneset - Helvar Electrosonic
Sceneview - Helvar Electrosonic
Sceptre - Paragon Business Furniture
Schlage - Relcross Ltd
Schoema - Burlington Engineers Ltd
Scholar - Ellis J T & Co Ltd
Schuco - EBS Services Ltd
Schwarzwald - Erlau AG
Schwing - Burlington Engineers Ltd
Scion - USF Permutit/Ionpure
Scratch Plaster - Snowcem
Screedmaster - Laybond Products Ltd
Screen Three - Pergola Products Ltd
Sculpture - Marley Floors Ltd
SDR - Stoves PLC
Seagull - Gledhill Water Storage Ltd
Sealanstone - Sealocrete Ltd
Sealdeck - Laybond Products Ltd
Sealex - Klinger Fluid Instrumentation Ltd
Sealmaster - Sealmaster
Sealobond - Sealocrete Ltd
Sealocalk - Sealocrete Ltd
Sealoclean - Sealocrete Ltd
Sealocote - Sealocrete Ltd
Sealocure - Sealocrete Ltd
Sealoflash - Sealocrete Ltd
Sealoflor - Sealocrete Ltd
Sealofoam - Sealocrete Ltd
Sealoform - Sealocrete Ltd
Seamaker - Barr & Wray Ltd
Secar Cements - Lafarge Aluminates Ltd
Securesheild ™ - Accent Hansen
Securicel - Fitzpatrick Door Frames Ltd
Securigard - Solaglas Laminated
Securus - Hill Leigh (Joinery) Ltd
Sedia - Erlau AG
Seedmat 2000 - Land Wood & Water Co. Ltd
Seekure - Cordek Ltd
Sefac - Elkington Gatic Ltd
Seip Remote Control - Regency Garage Door Services Ltd
Sela - SFS Stadler Ltd
Selandia Dan Parket - Jutlandia Doors
Seldex - Haworth UK Ltd
Select Range - Pedley Furniture international Ltd
Selectarail - Rothley Burn Ltd
Selectos - Nu-Way Ltd
Selfix Capsules - Exchem Mining and Construction
Selfix Style 3 Bolts - Exchem Mining and Construction
Selflock - Kwikform UK Ltd
Selkirk - Docherty H Ltd
Sellite - Sellite Blocks Ltd
Sellite Blocks - CAEC Howard (Holdings) Ltd
Sellotape - Amari Plastics Plc
Selux Lighting - Woodhouse UK Plc
Sembla - Allgood plc
Senator - Deans Blinds & Awnings (UK) Ltd
Senator - Forticrete Roofing Products
Senator - Securistyle Ltd
Senior Aluminium systems - EBS Services Ltd
Sensor - Tenby Industries Ltd
Sensor Coil 600 - Geoquip Ltd
Sensor flow & electronic taps - Armitage Shanks Ltd
Sentinal - Advanced Fencing Systems Ltd
Sentinel - Chubb Security Installations Ltd
Sentinel - Shackerley (Holdings) Group Ltd incorporating Designer ceramics
Sentrilock - Kenrick Archibald & Sons Ltd
Sentry - Caradon M K Electric Ltd
Sentry - Spectus UPVC Systems Ltd
Sentry ® - Jackson H S & Son (Fencing) Ltd
Sequel - Pegler Ltd
Series 200 low assistants range - Parker Bath Division
Series 2100 - James Gibbons Format Ltd
Series 300 Mid/high assistance range - Parker Bath Division
Series 400 High assistance range - Parker Bath Division
Series 500 Hospital range - Parker Bath Division
Series 55 Hose Reel - Macron Fireater Ltd
Series 60 - Apollo Fire Detectors Ltd
Series 60 Low Voltage - Apollo Fire Detectors Ltd
Series 600 Hygiene system range - Parker Bath Division
Series 700 Showering systems range - Parker Bath Division
Series 800 Transfer systems range - Parker Bath Division
Series F - Dunham Bush Ltd
Series UH - Dunham Bush Ltd
Serpo RG - Optiroc Ltd
Serpo Rock - Optiroc Ltd
Serv-o-Slide - Store Development Ltd
Serviflex - Uponor Ltd
Servisol - Ambersil Ltd
Servitherm - Servitherm
Servowarm - Servowarm
SES - Shepherd Engineering Services Ltd
Sesam - Dividers Ltd
Sesame - Profilex Ltd
Seth Thomas - General Time Europe
Sewerdrain - Polypipe Civils Ltd
Sewtec - Guest & Chrimes Ltd
SFS-Ashgrid - SFS Stadler Ltd
SFS-Masterflash - SFS Stadler Ltd
SG Autelio - Solaglas, Technical Advisory Service
SG Baldwin - Tarmac Precast Concrete
SG Cool-Lite - Solaglas, Technical Advisory Service
SG Curved - Solaglas, Technical Advisory Service
SG Diamont - Solaglas, Technical Advisory Service
SG Elco Plus - Solaglas, Technical Advisory Service
SG Planilux - Solaglas, Technical Advisory Service
SG Priva-Lite - Solaglas, Technical Advisory Service
SG Screen-Lite - Solaglas, Technical Advisory Service
SG Shockgard - Solaglas, Technical Advisory Service
SG Shotguard - Solaglas, Technical Advisory Service
SG Stadip Silence - Solaglas, Technical Advisory Service
Shadacrete - Hawley W & Son Ltd
Shades of light - Eclipse Blind Systems Ltd
Shadflam - Shadbolt F R & Sons Ltd
Shadmaster - Shadbolt F R & Sons Ltd
Shaftbrace - Maby Hire Ltd
Shaftbrace - Matbro Ltd
Shaftesbury - Hamworthy Heating Ltd
Shallovent - Therminal Technology (sales) Ltd
Sharazaar - Dixon Turner Wallcoverings Ltd
Shavrin - Shavrin Levatap Co Ltd
Shavrin Bijoux - Shavrin Levatap Co Ltd
Shavrin Levamixa - Shavrin Levatap Co Ltd
Shavrin Levatap - Shavrin Levatap Co Ltd
Shavrin Modesta - Shavrin Levatap Co Ltd
Shaw TTL - Shaw Arthur Manufacturing Ltd
Shawflow Ventilators - Shaw Arthur Manufacturing Ltd
Shaws glazed bricks - Ibstock Building Products Ltd
Shearflame - Shearflow Phoenix
Shearflow - Shearflow Phoenix
Shearfrost - Shearflow Phoenix
Shearstone - Ibstock Building Products Ltd
Shearveil - Shearflow Phoenix
Sheerframe - LB Plastics Ltd
Sheerline - LB Plastics Ltd
Sheffield - Diamond Merchants
Sheffield - Henderson-Bostwick
Shell - Balmoral Porcelain Limited/Longmead Ceramics
Shepherd - Kenrick Archibald & Sons Ltd
Sherwood Shelving - Balmforth Engineering Ltd
Shev - Dunsley Heat Ltd
Shieldoor - Norwood Partition Systems Ltd
Shirehill - Realstone Ltd
Shires - Shires Bathrooms
Shopline - Smart Systems Ltd
Shoptrack - CIL International Ltd
Shoreseal - Shackerley (Holdings) Group Ltd incorporating Designer ceramics
Shotax Cement Accelerator - Lafarge Aluminates Ltd
Shower valves and kits - Jendico
Showercare - Contour Showers Ltd
Showercyl - Gledhill Water Storage Ltd
Showerforce - NewTeam Ltd
Showermate - Stuart Turner Ltd
ShowerSport - ShowerSport Ltd
Showertime - Neptune Showers Ltd
Showertrays by Coram - Coram (UK) Ltd
Showerworld products - Armitage Shanks Ltd
Shutter Door Repair & Maintenance Ltd. S.D.R.M. - Dudley Factory Doors Limited
Shutter Door Repair & Maintenance Ltd. S.D.R.M. - Neway Doors Ltd
Shutter Door Repair & Maintenance Ltd. S.D.R.M. - Priory Shutter & Door Co Ltd
Shutter Door Repair & Maintenance Ltd. S.D.R.M. - Shutter Door Repair & Maintenance Ltd
Sibe Scale - Higginson Staircases Ltd
Sidetrak - Rackline Systems Storage Ltd
Sidewinder - Sidewinder Concrete Pumps Ltd
Sieger - Zellweger Analytic Ltd, Sieger Division
Siemens - F. W. Harris & Co Ltd
Siesta - Erlau AG
Siforms - Burlington Engineers Ltd
Sightline 70 - Spectus UPVC Systems Ltd
Sigma - Andrews J S (Coatings) Ltd
Sigma Paints - Philip Johnstone Group Ltd
Sigmaclad - Sigma Coatings Ltd
Sigmacover - Sigma Coatings Ltd
Sigmadur - Sigma Coatings Ltd
Sigmafast - Sigma Coatings Ltd
Sigmaferro - Sigma Coatings Ltd
Sigmaguard - Sigma Coatings Ltd
Sigmarite - Sigma Coatings Ltd
Sigmatherm - Sigma Coatings Ltd
Sigmaweld - Sigma Coatings Ltd
Sigmetal - Sigma Coatings Ltd
Signal - Shadbolt F R & Sons Ltd
Signature - Unilock Ltd
Signet - George Fischer Sales Ltd
Signtrax 2 - Oldham Signs Ltd
Sikador - Sika Ltd
Sikaflex - Sika Ltd
Sikkens - Akzo Coatings Plc
Silabe - ClipsPartitions Ltd
Silbond - Vallance
Silbralloy - Johnson Matthey PLC - Metal Joining
Silcabond sp - Liquid Plastics Ltd
Silent Gliss - Silent Gliss Ltd
Silfos - Johnson Matthey PLC - Metal Joining
Silglaze - GE Bayer Silicones
Silica Paint Potassium Silicate - Textured European Finishes Ltd
Silicone Inc - Notcutt W P Ltd
Silpruf - GE Bayer Silicones
SilvaSport Premium - Junckers Limited
Silvent - Designed for Sound Ltd
Silver-Flo - Johnson Matthey PLC - Metal Joining
Silverline - Aluminium RW Supplies Ltd
Simon Fiesta - Simon R W Ltd
Simplan - Dryad Architectural Ltd
Simplan-3 - Booth Muirie Ltd
Simplay Felts - Gaskell Textiles Ltd
Simply Elegant - Hodkin Jones (Sheffield) Ltd
Simpson Strong-Tie® - Simpson Strong-Tie®
Single Touch Comm System - BPT Security Systems (UK) Ltd
Sinzig - Buchtal (U.K.) Limited
Sirius - Bison Bede Ltd
Sirius - Poselco Lighting
Sirocco - Paragon by Heckmondwike F B Ltd
Sirplus - GE Bayer Silicones
Sirrus - Gummers A & J Ltd
Sisalite - British Sisalkraft Ltd
Sisalkraft - British Sisalkraft Ltd
SissonsCF - Sissons W & G Ltd
Sitecop - Rediweld Rubber & Plastics Ltd
Siteguard - GE Capital Modular Space
Sitesealer Membrane - Cavity Trays Ltd
600 Series - Komfort Systems Ltd
606 NCT - Otis plc
SK 60 - SCHUCO International KG
Skaala - Jutlandia Doors
Skanform - Norske Interiors (UK) Ltd
Skarsten - Skarsten Tools
Skarsten - Skarsten Tools Ltd
Skil - Robert Bosch Ltd
Skilmatic - Skil Controls Ltd
Skinneer for Fencing - WA Skinner & Co Limited
Skydeck - PERI Ltd
Skygard - Lonsdale Metal Company
Skyline - Ves Andover Ltd
Skyway - Scholes Windows Ltd
SL500 Beam Mounted Seating System - Race Furniture Ltd
Slab Stress - Freyssinet Ltd
Slatevent II - Blunn Slates Ltd
Sleepsafe - PH Products Ltd
Slim Vent - Ryton's Building Products Ltd
Slimglide - Integra Products
Slimline - ACO Technologies Plc
Slimline - Cooke Brothers Ltd
Slimline - Dudley Thomas Ltd
Slimpack - Roof Units Group
Slimstar - AEL
Slimstile - Gunnebo Mayor Ltd
Slimstyle - Hallmark Blinds Ltd
Slimstyle Groundfloor Treatments - Heywood Williams Architectural
Slite - Intralux UK Ltd
Sloan & Davidson - Longbottom J & J W Ltd
Slotline - Royair Ltd
Slotz - Browne Winther & Co Ltd
Slotz - Winther Browne & Co Ltd
Smaragd - Forbo - Nairn Ltd
Smart systems - Alframes Holdings Ltd
Smartscreen - Smart Systems Ltd
Smartstar - Roof Units Group
Smatex - Bardon (England) Ltd
SMC - Potterton Myson Ltd
Smiths - Timeguard Ltd
Smog-Eater - Horizon International Ltd
Smog-Mobile - Horizon International Ltd
Smog-Rambler - Horizon International Ltd
Smoke Ventilation Louvres - Matthews & Yates Ltd
Smokex - Kedddy (Poujoulat) (UK) Ltd
Smometa - Photain Controls Plc
Smooth - On - Notcutt W P Ltd
Smoothline - Rega Metal Products Ltd
Smoothline - WP Metals Ltd
SMU - Stanton plc
Snaptite - Newdawn & Sun Ltd
Snowcem Paint - Snowcem
Sofco ® - Broxap & Corby Ltd
Soil Panel - PHI Group Ltd
Soilcrete - Keller Ground Engineering
Soilfrac - Keller Ground Engineering
Solair - Solair Ltd
Solar 4000 - Radway, Door and Windows Ltd.
Solar AV 4000 Commercial Doors - Radway, Door and Windows Ltd.
Solar Eclipse - Gaskell Textiles Ltd
Solar Gard - MSC Speciality Films (UK) Ltd
Solar Knight - Intersolar Group Ltd
Solar Moler - Intersolar Group Ltd
Solar Protect - Eclipse Blind Systems Ltd
Solar Pumpkit - Intersolar Group Ltd
Solargard - Hi-Tec Window Films
Solaris - Shackerley (Holdings) Group Ltd incorporating Designer ceramics
Solarvent - Intersolar Group Ltd
Solarvent Turbo - Intersolar Group Ltd
Solid - ASSA Ltd
Solifer - Industrial Devices Ltd
Solignum - Protim Solignum Ltd
Solitude - Cep Ceilings Ltd
Soll - HCL Verta
Solo - Ellis J T & Co Ltd
Solo - Helvar Electrosonic
Solo - Paragon Business Furniture
Solo - Vent-Axia Ltd
Solocate - Telelarmcare Ltd
Solomat - Zellweger Analytic Ltd, Sieger Division
Solos - Forticrete Roofing Products
Solstar - Crompton Lighting Ltd
Solutions - Hands of Wycombe
Solutions 2000 SHR - Beeston Heating Group Ltd
Solvite - Henkel Home Improvement & Adhesive Products
Somerset Doorset Range - Hill Leigh (Joinery) Ltd
Somerton - Top Office Equipment Ltd
Sonae - Boardcraft Ltd
Sonar - Rockfon Ltd
Sonata - Cego Frameware Ltd
Sonata - Quartet-GBC
Sonex - Illbruck Ltd
Sonex - Salex Interiors Ltd
Sonneborn - FEB Ltd
Sonneborn - Feb MBT
Sonpac - Williamson Floor & Wall Specialists Ltd
Sony - Vidionics Security Systems Ltd
Sorrus - Lumitron Ltd
Sorviseal - Grace Services
SOS - Maestro International Ltd
Soss Invisible Hinges - Notcutt W P Ltd
Sottini bathrooms - Ideal-Standard Ltd
Soundcel - Fitzpatrick Door Frames Ltd
Soundcote - Cafco Construction Products
Soundpac - Hotchkiss Air Supply
Soundshield ™ - Accent Hansen
Southampton - Neptune Outdoor Furniture Ltd
Sovereign - Clenaware Systems Ltd
Sovereign - Sovereign Brush Co Ltd, The
Sovereign Doors - Boulton & Paul Ltd
Sovereign Range - Eternit Clay Tiles Ltd
Sovereign Windows - Boulton & Paul Ltd
SP 5 - SCHUCO International KG
SPAC - Brighton (Handrails), W
Space -Ray - Space - Ray UK
Space Bathroom Suite - Ideal-Standard Ltd
Spacebuild - Wernick S & Sons Ltd
Spacedeck - Turner Access Ltd
Spacepac - Wernick S & Sons Ltd
Spacewhirl - Record Playground Equipment Ltd
Spandex - Dexion Ltd
Spandex Modular System - Green Brothers Ltd
Spanfast - Carter Concrete Ltd
Spangard - Lonsdale Metal Company
Spania - Marshalls Mono Ltd
Spanseal - Compriband Ltd
Spantile Callette Ventilated Ceilling System - Vianen Kitchen Ventilation (UK) Ltd.
Spanwell - Welconstruct Co
Spartacus - Metalliform Products Plc
Spartan - Spartan Tiles Ltd
Spartus - General Time Europe
SPC - Furmanite International Ltd
Spear & Jackson - Neill Tools Ltd
Specification Manager - NBS Services
Specification Writer - NBS Services
Specifier Choice - Checkmate Industries Ltd
Spectra Tile System - Collie Carpets Ltd
Spectral 2000 - Harkness Hall Ltd
Spectralux® - Absolute Action Ltd
Spectrum - Norton Engineering Alloys Co Ltd
Spectrum Paint - Sanderson & Sons Ltd, Arthur
Specula - Shadbolt F R & Sons Ltd
Spedec - SFS Stadler Ltd
Speed - Speed Saws Ltd.
Speed Patch - Anglo Building Products Ltd
Speed Screed - Anglo Building Products Ltd
Speedal ® - Righton Ltd
Speeddeck - E. H. S. Roofing Ltd
Speedfit - Guest (Speedfit) Ltd, John,
Speedframe - Dexion Ltd
Speedlock - Dexion Ltd
Speedor - Hart Door Systems Ltd

Speedplug - Ramset Fasteners Ltd
Speedstile - Gunnebo Mayor Ltd
Speedturn - BSA Tools Ltd
Speedwall - Bell & Webster Concrete Ltd
Speedway - Vantrunk Engineering Ltd
Speedy - Protimeter plc
Spelsburg enclosures - Elkay Electrical Manufacturing Co Ltd
Spencers - Anderson Gibb & Wilson
Sphinx - Royal Dutch Sphinx Ltd
Spicing - LazyLawn
Spider Glass - Solaglas, Technical Advisory Service
Spike - SFS Stadler Ltd
Spilguard - Portasilo Ltd
Spiral construction - Consculpt Spiral Stairs Ltd
Spiralift - Caldwell Hardware (UK) Ltd
Spiralite - Cryotherm Insulation Ltd
Spiratube - Flexible Ducting Ltd
Spiratube HRS Heat Exchangers Ltd - HRS Hevac Ltd
Spirex - Caldwell Hardware (UK) Ltd
Spirotred - Pedley Furniture international Ltd
Splash (dolphin bin) - Glasdon UK Ltd
Sportswall - Record Playground Equipment Ltd
Sporturf - En-tout-cas plc
Sporturf - En-tout-cas plc (Scotland)
Spraydon - Cafco Construction Products
Sprayed Limpet Mineral Wool-GP - Thermica Ltd
Sprayed Limpet Mineral Wool-HT - Thermica Ltd
Sprayed Limpet Mineral Wool-TI - Thermica Ltd
Sprayed Limpet Vermiculite-external - Thermica Ltd
Sprayed Limpet Vermiculite-Internal - Thermica Ltd
Springvin - Flexible Ducting Ltd
Sprint - Paddock Fabrications Ltd
Sprint - SGB Youngman
Sproughton Arbour Seat - Anthony de Grey Trellises
Spyhawk - Hunting Engineering Ltd
Spyscan - Hunting Engineering Ltd
Square D - Schneider Ltd
Squareline - Illbruck Ltd
Squire - Squire Henry & Sons
SRS acoustic floating floor - Christie & Grey Ltd
SS System - Quelfire
SS200 - Sempol Surfaces Ltd
SS300 - Sempol Surfaces Ltd
St James - Marflow Eng Ltd
Sta - Tie - Fix Fast (Fasteners) Wales Ltd
Sta-Lok Rigging - Scotia Rigging Services (Industrial Stainless Steel Rigging Specialists)
Stabil Binders - Huntley & Sparks
Stabila - Hyde Brian Ltd
Stabilo - Kompan (UK) Ltd
Staffline - Chubb Security Installations Ltd
Staifix Range - Ancon CCL Ltd
Stainless Steel - Swift & Sure - Camloc (UK) Ltd
Stal - Sabroe Refigeration Ltd
Stalectronic - Sabroe Refigeration Ltd
Stalwart - Chubb Security Installations Ltd
Stamfordstone - Williamson Cliff Ltd
Standard Patent Glazing - Standard Patent Glazing Co Ltd
Stang Monitors - UK Fire International Ltd
Stanhope - Stanhope Barclay Kellet Crown Division
Stanley - Relcross Ltd
Stanley - Stanley Access Technologies
Stanley Hardware - Harwood & Welpac Hardware Ltd
Stanton - Drainage Systems
Stantons - Ashmead Buidling Supplies Ltd
Stantons - Ashmead Building Supplies Ltd
Stanweld - Lionweld Kennedy Ltd
Stanwin - Stanhope Barclay Kellet Crown Division
Stanza - Cego Frameware Ltd
Stapoco - Automatic Safety Lighting Ltd (T/as 'Stapoco')
Stapoco - Standby Power Company
Star - Novellini Bathroom Products
Star - Rada Lighting Ltd
Star Premium PVC - Signs International
Stargard - SG System Products Ltd
Starlifter - Sandhurst MFG Co Ltd
Starlight - IMP Lighting Ltd
Starline - Deceuninck Ltd
Starsystem - Watco UK Ltd
Statesman - Project Office Furniture PLC
Stayflex - Adaptaflex Ltd
Steden - Ingersoll - Rand Architectural Hardware
Steden - Laidlaw Architectural Hardware
Steel -Lok - Spur Shelving Ltd
Steel Line - Cardale Doors Ltd
Steel monoblock - Clestra Hauserman Ltd
Steelcase Strafor - Steelcase Strafor Plc
Steelcote - Quelfire
Steelex - Lee Steel Strip Ltd
Steeline Roller Shutter - Regency Garage Door Services Ltd
Steelkane - Ward's Flexible Rod Company Ltd
Steeltone - Applied Acoustics (a part of Henry Venables Limited)
Steeltone - Henry Venables Ltd
Steelwall - Brown F plc
Steelwall - Neslo Partitioning & Interiors
Stelduct - Brickhouse
Stelrad - Caradon Heating Ltd
Steriboard - Cape Boards Ltd
Steridex - Liquid Plastics Ltd
Sterisept - Liquid Plastics Ltd
Sterisheen - Liquid Plastics Ltd
Steristeel - Associated Metal (Stainless) Ltd
Sterling - Marlow Ropes Ltd
Sterling - Securistyle Ltd
Sterling range of cable management systems - Marshall-Tufflex Ltd
Sterope - Orsogril Sarl UK Department
Sterreberg - Smithbrook Building Products Ltd
Stetter - Burlington Engineers Ltd
Stewart Film Screen - Anders + Kern UK Ltd
Stick-Lite - Encapsulite International Ltd
Stik - Iles Waste Systems
Stilpro - Allpipe Ltd
STO products - CCS Scotseal Ltd
Stockfloor - RMC Readymix Limited
Stockmaster Shelving - Beford & Soar Ltd
Stocks Blocks - Brikenden (Builders Merchants) Ltd
Stomflow - Midland Stom Ltd
Stomstar - Midland Stom Ltd
Stone slate - Forticrete Roofing Products
Stonecor - TDI (UK) Ltd
Stoneraise red - Realstone Ltd
Stontex - Atlas Stone Products
Stopdust - GRAB Industrial Flooring Ltd
Stopgap - Ball F & Co Ltd
Stoplite - Cryotherm Insulation Ltd
Storage Wall - Davies John Interiors Ltd
Storbox - Moresecure Ltd
Storemaster - T & D Plastech
Storeshield - Lycetts (Burslem) Ltd
Storm - Blakey John H Ltd
Storm Tile - Eternit Clay Tiles Ltd
Stormflo Rainwater Sytems - Hunter Plastics Ltd
Stormgard - Corroless International
Stormor Clearstor - Link 51 (Storage Products) Ltd
Stormor Drawerstor - Link 51 (Storage Products) Ltd
Stormor Euro - Link 51 (Storage Products) Ltd
Stormor Longspan - Link 51 (Storage Products) Ltd
Stormor Roller Edge - Link 51 (Storage Products) Ltd
Stormor XL Pallet - Link 51 (Storage Products) Ltd
Stormproof - Premdor
Stortech - Rothley Burn Ltd
Stoves - Stoves PLC
Stoves Newhome - Stoves PLC
Stowell - Ashmead Buidling Supplies Ltd
Stowell - Ashmead Building Supplies Ltd
Stra - Optiroc Ltd
Strading - SG System Products Ltd
Strand FEB - Strand Hardware Ltd
Strata - Stokvis R S & Sons Ltd
Stratabord - European Profiles
Stratacem NT - Tegral Building Materials Ltd
Stratalite - LB Lighting Ltd
Stratogrid - HT Martingale Ltd
Stratos - Rada Lighting Ltd
Stratotrack - HT Martingale Ltd
Stratton Canopy - James Smellie Fabrications Ltd
Streamline - Clenaware Systems Ltd
Streamline - Wade International (UK) Ltd
Streetcrete ™ - Broxap & Corby Ltd
Streetiron ™ - Broxap & Corby Ltd
Streetscene ® - Broxap & Corby Ltd
Strip Form - Scapa Tapes UK Ltd.
Stroke - Bernhard's Sports Surfaces Ltd
Stroke - Bernhards Landscapes Ltd
Strongbeam - Spur Shelving Ltd
Strongcast concrete - Challenge Fencing & Ltd
Stronghold - Caradon Catnic Ltd
Stronghold - SG System Products Ltd
Strongoat - GRAB Industrial Flooring Ltd
Strypit - Rustins Ltd
Stuart - Stuart Turner Ltd
Stucco Lustro - Francesca Di Blasi Co. Ltd, The
Studflex vibration isolation mat - Christie & Grey Ltd
Studio 3 - Home Automation Ltd
Studio Bathroom Suite - Ideal-Standard Ltd
Studio System - Hyperion Wall Furniture Ltd
Studrail system - Christie & Grey Ltd
Sturdee - Marlow Ropes Ltd
Styccobond - Ball F & Co Ltd
Stylemaster - Checkmate Industries Ltd
Styrene - Ariel Plastics Ltd
Styroclad - Panel Systems Ltd
Styrodur - Kingspan Insulation Ltd
Styrofloor - Panel Systems Ltd
Styroglaze - Panel Systems Ltd
Styroliner - Panel Systems Ltd
Suba Care - William Freeman Ltd
Suba Seal - William Freeman Ltd
Suffolk - Dixon Turner Wallcoverings Ltd
Suffolk - Dixon Turner Wallcoverings Ltd.
Sumlock - Sumlock J.H.P.
Summit - Rada Lighting Ltd
Sumo - Pump Technical Services Ltd
Sumpmaster - Pump Technical Services Ltd
Suncell - CPV Ltd
Suncool - Pilkington Birmingham
Sundeala A Grade - Sundeala Division of Celotex Ltd
Sundeala K Grade - Sundeala Division of Celotex Ltd
Sundeala superspec FR Class 0 - Sundeala Division of Celotex Ltd
Sundeala Surespec FR Class 1 - Sundeala Division of Celotex Ltd
Sunflex ® - Hunter Douglas Ltd (Contracts Division) Luxaflex Blinds ®
Sungard - Bonwyke Ltd
Sunstar - Space - Ray UK
Suntech - Deans Blinds & Awnings (UK) Ltd
Suntech systems - Deans Blinds & Awnings (UK) Ltd
Suntime - Faber Blinds Ltd
Sunway ® - Hunter Douglas Ltd (Contracts Division) Luxaflex Blinds ®
Supabore - Avesta Sheffield Ltd
Supachute - Watkins
Supafil - Owens Corning Building Products (UK) Ltd
Supaflo - RMC Readymix Limited
Supakey - BPT Security Systems (UK) Ltd
Supalux - Cape Boards Ltd
Supamate - Watkins
Supapac - Rycroft Ltd
Supasafe - Ashley & Rock Ltd
Supastrike box striking plate - Guardian Lock and Engineering Co Ltd
Supatube - Avesta Sheffield Ltd
Supawrap Pink - Owens Corning Building Products (UK) Ltd
Super 100NC - Hauration Kaskade Ltd
Super 150 - 500 - Hauration Kaskade Ltd
Super Clean - Dunsley Heat Ltd
Super Seven Class 1 - Dunbrik (Yorks) Ltd
Super Shaftbrace - Maby Hire Ltd
Super Slot - Versatile Fittings Ltd
Super Top - Armorex Ltd
Super Twosome - Integra Products
Supercoil - Airsprung Beds Ltd
Supercomfort - Dunham Bush Ltd
Superconcentrate Boiler Noise Silencer - Alpha Fry Ltd
Superconcentrate Boiler Noise Silencer - Fernox Fry Technology UK
Superconcentrate Leaksealer - Alpha Fry Ltd
Superconcentrate Leaksealer - Fernox Fry Technology UK
Superconcentrate Limescale Preventer - Fernox Fry Technology UK
Superconcentrate Limestone preventer - Alpha Fry Ltd
Superconcentrate Protector - Alpha Fry Ltd
Superconcentrate Protector - Fernox Fry Technology UK
Superconcentrate Restorer - Alpha Fry Ltd
Superconcentrate Restorer - Fernox Fry Technology UK
Supercote - Witham Oil And Paint Ltd
Superfine - Integra Products
Superfloc - Fernox Fry Technology UK
Superflux - Alpha Fry Ltd
Superfold - Bolton Gate Co Ltd
Superfoot - Bernhard's Sports Surfaces Ltd
Superfoot - Bernhards Landscapes Ltd
SuperG - Ecophon Ltd
Supergalv Lintels - Birtley Building Products
Supergalv Lintels - Birtley Building Products Ltd
Supergrid 8000 - MEM
Superior - Beacon Machine Tools Ltd
Superior - Bolton Gate Co Ltd
SuperJan - Johnson & Starley
Superlintel - Jones of Oswestry Ltd
Supersafe - Chloride Industrial Batteries Ltd
Supersave - Hemsec Manufacturing Ltd
Superseal - Kleeneze Sealtech Ltd
Superset AC1 - Cape Insulation Products Ltd
Supershaftbrace - Matbro Ltd
Supersleve House Drainage - Hepworth Building Products Ltd
Superspan - Compriband Ltd
Superstrong - Boddingtons Ltd
Superswing - Record Playground Equipment Ltd
Superswitch - Caradon Friedland Ltd
Supertube - PFP Electrical Products Ltd
Supertube 125 - Domus Ventilation Ltd.
Supervent - Airflow Developments Ltd
Supervisor - Caradon Gent Ltd
Supplyline - National Britannia Ltd
Supplymaster - Timeguard Ltd
Supporto - Evertaut
Supporto - Hille Ltd
Supra-Arch - Jones of Oswestry Ltd
Suprabloc - Jones of Oswestry Ltd
Supragate - Jones of Oswestry Ltd
Supragrid - Jones of Oswestry Ltd
Suprasteel - Jones of Oswestry Ltd
Supratrim - Jones of Oswestry Ltd
Surefire - Thurston Building Systems
Surefit Rainwater Systems - Hunter Plastics Ltd
Suregrip - Wincro Metal Industires Ltd
Suregrip ® - Coo-Var Ltd
Suregrip ® - Teal & Mackrill Ltd
Sureguard - Thurston Building Systems
Surell - Decra Cubicles Ltd
Surell - Formica Ltd
Surelock - Cego Frameware Ltd
Sureloo - Thurston Building Systems
Sureplay - En-tout-cas plc (Scotland)
Surespace - Thurston Building Systems
Surespan - Thurston Building Systems
Suresport - Thurston Building Systems
Surestep - Forbo - Nairn Ltd
Suretex - Ward Bekker Sales Ltd
Suretred - BRB Industrial Services Ltd
Suretwin - Polypipe Civils Ltd
Surfking - Barr & Wray Ltd
Surmabond - Surmak Products Ltd
Surmaglaze - Surmak Products Ltd
Surmaseal - Surmak Products Ltd
Surrey-Sac - Surrey Sackholders Ltd.
Surveymaster SM - Protimeter plc
Surveyor - Vicon Industries (UK) Ltd
Swakdale - Paragon by Heckmondwike F B Ltd
Swaledale - Oakdale (Contracts) Ltd
Swan - Boardcraft Ltd
Swanglide - Integra Products
Swarfega - Deb Ltd
Swarfega Orange - Deb Ltd
Swarfega Power - Deb Ltd
Swash - Clenaware Systems Ltd
Swedeglaze - Swedecor Ltd
Swedhouse - Swedhouse Ltd
Swedish Plain roller blind fabric - Dallas & Forrest Ltd
Swedsign - Signs & Labels Ltd
Swift-and-Sure - Camloc (UK) Ltd
Swiftlay - Swifts Of Scarborough Ltd
Swiftplan - Wernick S & Sons Ltd
Swiftrack - Swifts Of Scarborough Ltd
Swimmer - Golden Coast Ltd
Swimmer - Sanua UK Ltd
Swimtile - Pilkington's Tiles Ltd
Swisslab - MR Limited
Swissplan - MR Limited
Swisswall - C. G. I. International Ltd, trading as Colebrand Glass
Sws Envirocoat - Stewart Wales, Somerville Ltd
Sygnette - Wellman Robey Ltd
Sykes Pumps - Andrews Sykes Hire Ltd
SylvaFoam - Junckers Limited
SylvaKet - Junckers Limited
Sylvania Lamps - Concord Lighting
SylvaRed - Junckers Limited
SylvaSport Club - Junckers Limited
SylvaSquash - Junckers Limited
SylvaThene - Junckers Limited
Sylvatone - Applied Acoustics (a part of Henry Venables Limited)
Sylvatone - Henry Venables Ltd
Symphony - Hudevad Britain
Symphony - Symphony Group PLC, The
Syncrolux - Rada Lighting Ltd
Synergy - H. B. Fuller Coatings Ltd
Syntha Pulvin - H. B. Fuller Coatings Ltd
Syntha Pulvin Plus - H. B. Fuller Coatings Ltd
Synthatel - HB Fuller Coatings Ltd
Synthatel Plus - H. B. Fuller Coatings Ltd
Syntropal - ICB (International Construction Bureau) Ltd
System 10 - Masterframe Windows Ltd
System 100 - BPT Security Systems (UK) Ltd
System 100 - Domus Ventilation Ltd.
System 120 Gutter - Dales Fabrications Ltd
System 160 - Maby Hire Ltd
System 200 - BPT Security Systems (UK) Ltd
System 2000 - nmc (UK) Ltd
System 25 - Metalrax Ltd
System 3000 - TDI (UK) Ltd
System 500 - Newton John & Co Ltd
System 75 Gutter - Dales Fabrications Ltd
System 9000 - Timloc Expamet Building Products
System Airtech - Phoenix Roofing
System160 - Matbro Ltd
Systemate - Gledhill Water Storage Ltd
Systemfor plant room floors - Ferguson G A & Co Ltd
Systemform - Leaderflush & Shapland
Systemglas - Promat Fire Protection Ltd
Systems 600 - Marley Waterproofing
T Grade Seal - IFT Supplies Ltd
T Vent - Areco Roofing Supplies Company
T-Pren - Matthew Hebden Ltd
T-series - Vent-Axia Ltd
T-T controls - T-T Pumps & T-T controls
T.D.I. Firestop - TDI (UK) Ltd
T.I.P.S. - Hellerman Electric, Member of Bopwthorpe Holdings PLC
T100 - Davis International Ltd
T14 storage wall - Gifford Grant Ltd
T5 Lighting - Encapsulite International Ltd
Tablo - Dermide Ltd
Tacdeck - TAC Metal Forming Ltd
Tactray 90 - TAC Metal Forming Ltd
Tadelakt - Francesca Di Blasi Co. Ltd, The
Tahiti - Qualitas Bathrooms
Takpave - Rediweld Rubber & Plastics Ltd
Talia - Orsogril Sarl UK Department
Talisman - Tilley International Ltd
Taller Uno - Optelma Lighting Ltd
Tamiya - The Modelshop
Tanalith - Hickson Timber Products Ltd
Tanda - Drawn Metal Ltd
Tanking 134 - Kiltox Damp Free Solutions
Tantofex Bathroom taps and mixers - Ideal-Standard Ltd
Tantofex Kitchen taps and mixers - Ideal-Standard Ltd
Tanums - Scandanavian Window Systems Ltd
Tapes4 Builders - Advance Tapes International Ltd
Tapetex - Architectural Textiles Ltd
Tappex - Wirsbo
Taralay Comfort - Gerland Ltd
Taranis - Girsberger London
Tarasafe Standard - Gerland Ltd
Tarasafe Ultra - Gerland Ltd
Tarkett - Atkinson & Kirby Ltd
Tarrant - Hamworthy Heating Ltd
Taskmaster - Doors & Hardware Ltd
Taskmaster Select - Lumitron Ltd
Tasso - Property Mechanical Products Ltd
Tate access floor system - Alumasc Interior Building Product Limited
Tate Floors - Hatmet Ltd
Tatra - Armstrong World Industries Ltd
Tauatex - Sempol Surfaces Ltd
Taura - Sempol Surfaces Ltd
Tayler tools - Hyde Brian Ltd
TD Board - Cafco Construction Products
Teaching Wall Systems - TWS Presentation Systems (Mcmillan UK Ltd T/A)
Teamac - Teal & Mackrill Ltd

Tecfix - Contract Components Ltd
Tecflex / Tectherm - Contract Components Ltd
Techdek - Norton Engineering Alloys Co Ltd
Technal - Cantifix of London Limited
Techniflam - Eclipse Blind Systems Ltd
Technostone Ltd - Technostone Ltd
Tecroc - Drainage Systems
Tecseal - Contract Components Ltd
Tecstrut - Contract Components Ltd
Tectape - Contract Components Ltd
Tectite - Contract Components Ltd
Tedlar - Tektura Plc
Teegee - Swish Building Products Ltd
Tegola - Matthew Hebden Ltd
Tegometall - Smart F & G (Shopfittings)Ltd
Tegracem NT - Tegral Building Materials Ltd
Tegral Classic Natural Slates - Tegral Building Materials Ltd
Tekfix - Tektura Plc
Teksafe - Portasilo Ltd
Tektaroot - Watkins
Telabon ACM - Technology Telford Ltd
Telebloc - Jones of Oswestry Ltd
Telecom Security Ltd - ADT Modern Security Systems Ltd
Teleflex - Dyer Environmental Controls
Teleflex - Morse Controls Ltd
Telelight - Sandhurst MFG Co Ltd
Tempaclaclad - Yule & Son Ltd, A C
Tempaflam - Yule & Son Ltd, A C
Tempaglas - Yule & Son Ltd, A C
Tempest - Rada Lighting Ltd
Tempo - Jaga Heating Poducts (UK) Ltd
10 Plus - Kompan (UK) Ltd
Tenon - HEM Interiors Group Ltd
Tensabarrier - Tensator Ltd, Sales Department
Tensaguide - Tensator Ltd, Sales Department
Tensar - The Nelton Group Ltd
Tequla - Marshalls Mono Ltd
Terasafe Trend - Gerland Ltd
Terminator - Marflow Eng Ltd
Terraglass - Diespeker Marble & Terrazzo Ltd
Terrain - Capper Pipe Systems Ltd
Terrain - Drainage Systems
Terranova - Pilkington's Tiles Ltd
Terrapin - Terrapin Ltd
Terrex - ABG Ltd
Terrier 2 - Collie Carpets Ltd
Terrings - Compriband Ltd
Terry's - The Kenco Coffee Company
Tesa - Beiersdorf (UK) Ltd
Tetraduct - Polypipe Civils Ltd
Teviot - Gaskell Textiles Ltd
Tex - Ves Andover Ltd
Texitone - Redland Precast
Texsol - Trade Sealants
Texspra - En-tout-cas plc (Scotland)
Textura - Gradus Ltd
Texture - Pilkington Birmingham
Texwood - McLoughlin Wood Ltd
Thames - Portacel
Thatchbatt - Thatching Advisory Services Ltd
The 127 Range - Architectural Textiles Ltd
The All Seeing Eye - ISOKERN (UK) Ltd
The Invisible Lightswitch - Forbes & Lomax Ltd
The Lead Workshop Ltd - Alexio Roofing & Building Co Ltd
The Ultimate Collection - Ball William Ltd
The White Group - ADT Modern Security Systems Ltd
The window centre - BLW Associates Ltd
Theben - Timeguard Ltd
Themerend - Thermica Ltd
Thermabate - RMC Panel Products Ltd
Thermaclip - Flexible Ducting Ltd
Thermadoor - Stokvis R S & Sons Ltd
Thermafibre - Davant Products Ltd
Thermafit - Brohome Ltd
Thermaflex - Flexible Ducting Ltd
Thermafloor - Kingspan Insulation Ltd
Thermafold - Bolton Gate Co Ltd
Thermal Line - Thermal Radiators Ltd
Thermal Mitx - Oswestry Reinforced Plastics Ltd
Thermal Panel - Thermal Radiators Ltd
Thermal Rad - Thermal Radiators Ltd
Thermal Safe - Kingspan Building Products Ltd
Thermalath - BRC Building Products
Thermaline - Harton Heating Appliances
Thermaliner - Kingspan Insulation Ltd
Thermalite - Marley Building Materials Ltd
Thermalux - Hill Leigh (Joinery) Ltd
Thermapipe 100, 125, 150 - Domus Ventilation Ltd.
Thermapitch - Kingspan Insulation Ltd
Thermaroof - Kingspan Insulation Ltd
Thermataper - Kingspan Insulation Ltd
Thermatic - Martin Orgee & Associates
Thermawall - Kingspan Insulation Ltd
Thermaweld - Anderson D & Son
Thermax - Birdsall Services Ltd
Thermgard - Lonsdale Metal Company
Therminox - Kedddy (Poujoulat) (UK) Ltd
Thermo-Brite - Fillcrete Ltd
Thermoclear - Daylight Centre
Thermocool - Thermoscreens Ltd
Thermocor C.C. - TDI (UK) Ltd
Thermodek (Pre-felted warm flat roof insulation - Vencel Resil Ltd
Thermoflue - Marflex Chimney Systems
Thermoframe - HIS Ltd
Thermoheat - BICC Thermoheat Ltd
Thermolan - Owens Corning Building Products (UK) Ltd
Thermolier unit heaters - Turnbull & Scott (Engineers) Ltd
Thermonda - Rockwell Sheet Sales Ltd
Thermorex - Heat Detector - Siemens Building Technologies Ltd, Cerberus Division
Thermoscreens - Thermoscreens Ltd
Thermoson - Axter Ltd
Thermotapes - Jimi-Heat Ltd
Thirkleby - York Handmade Brick Co Ltd
Thistle - British Gypsum Ltd
Thomas Dudley - Hipkin F W Ltd
Thor - Thor Hammer Co Ltd
Thorace - Thor Hammer Co Ltd
Thorex - Thor Hammer Co Ltd
Thorlite - Thor Hammer Co Ltd
Thoroseal - Thoro Systems Products Ltd
Thortex - Wood E Ltd
3 in 1 Mould killer - Polycell Products Ltd
316 Range - James Gibbons Format Ltd
3C Tricontrol - Martin Orgee & Associates
3G Vulcavent - Vulcanite Limited
3K - Boulter Boilers Ltd
3M - Clean Step Ltd
3M Nomad Matting - 3M United Kingdom plc
3M Optical Lighting Film - 3M United Kingdom plc
3M Safety Walk - 3M United Kingdom plc
3M Scotch-Clad - 3M United Kingdom plc
3M Scotch-Gard - 3M United Kingdom plc
3M Scotch-Lane - 3M United Kingdom plc
3M Scotch-Seal - 3M United Kingdom plc
3M Scotchcal - 3M United Kingdom plc
3M Scotchcal Graphics Vinyl - 3M United Kingdom plc
3M Scotchlite - 3M United Kingdom plc
3M Scotchshield - Durable Berkeley Co Ltd
3M Scotchshield - 3M United Kingdom plc
3M Scotchtint - Durable Berkeley Co Ltd
3M Scotchtint - 3M United Kingdom plc
Threshex - Exitex Ltd
Thrutone N/T - Eternit UK Ltd
Thwaites - Chippindale Plant Ltd
Thyssen Stairlifts - Thyssen Stairlifts Ltd
Tiara - Hometex Trading Ltd
Tiberon Tileworks - Kirkstone
Tico - Tiflex Ltd
Tiger Diamond Blades - Unicorn Abrasives (UK) Ltd
Tile-A-Door - Howe-Green Ltd
Tilefast - P.C.I Construction Systems Ltd
Tileflex - Compriband Ltd
Tilene - Visqueen Agri
Tilevent PT - Blunn Slates Ltd
Tilley Lamp - Tilley International Ltd
Tilt Wash - Andersen/ Black Millwork
Timber - Cardale Doors Ltd
Timbercare - Manders Paint Ltd
Timberflex - P.C.I Construction Systems Ltd
Timberlogger - Protimeter plc
Timbersound noise barrier - Chris Wheeler Construction Limited
Timbertone - Protim Solignum Ltd
Timberwrap - Monarflex Ltd
Time-Piece - Arena Software Ltd
Timesaver - Drainage Systems
Timonox - Crown Trade Paints
Tisca - Hometex Trading Ltd
Titan - Fabrikat (Nottingham) Ltd
Titan - Ingersoll-Rand European Sales Ltd
Titan - Osram Ltd
Titan - Titan Ladder and Case Co Ltd
Titan Garden Buildings - Phillip Ader Ltd
Titan maga blind - Hallmark Blinds Ltd
Titazel - The Safety Tread Ltd
Tivoli 21 carpet tiles loop pile - Burmatex Ltd
TMC - Andura Textured Masonry Coatings Ltd
Tokstrip - Winn & Coales
Toledo - Townscape Products Ltd
Tonewood - Applied Acoustics (a part of Henry Venables Limited)
Tonewood - Henry Venables Ltd
Tony dustbins - Iles Waste Systems
Tony Team - Tony Team Ltd
Top Fixing Solution for Built Up Roofs - Latchways plc
Top Fixing Solution for Composite Roofs - Latchways plc
Top office - OEP Furniture Group PLC
Top Plan - Top Office Equipment Ltd
Top Star Rooflights - Klober Ltd
Top Tech - Top Office Equipment Ltd
Top Vent - Klober Ltd
Topcrete - Tarmac Topblock Ltd
TopForm - SFS Stadler Ltd
Toplab - Trespa UK Ltd
Toplite - Tarmac Topblock Ltd
Topper Range - Lumitron Ltd
Toprak Tiles - Tropravit (UK) Ltd
Toprax - Barton Storage Systems Ltd
Topshelf - Barton Storage Systems Ltd
Topspot - Illuma Lighting Ltd
Topstore - Barton Storage Systems Ltd
Topsy - Glasdon UK Ltd
Topsy Jubilee (bins) - Glasdon UK Ltd
Torbeck - Opella Ltd
Toreador carpet tiles loop pile - Burmatex Ltd
Toreboda - Moelven Laminated Timber Structures Ltd
Torlife - Tor Coatings Group
Torlife WB - Tor Coatings Group
Torlite - Torvale Building Products, A Division of Stadium Group PLC
Toro - Hille Ltd
Torprufe CRC - Tor Coatings Group
Torres - Eternit UK Ltd
Torshield - Tor Coatings Group
Torso - Caldwell Hardware (UK) Ltd
Tortread - Tor Coatings Group
Toscana - Kermi (UK) Ltd
Toscana - Marshalls Mono Ltd
Toscana - Paragon by Heckmondwike F B Ltd
Toshiba - HV Air Conditioning Ltd
Toshiba - Purified Air Ltd
Total Roof Systems - Phoenix Roofing
Tottime - Game Time UK Ltd
Toughliner - Lotus Water Garden Products Ltd
Toughness - BRB Industrial Services Ltd
Toughseal - RIW Ltd
Toupret - Toupret Fillers Agency, Hill & Rednall Ltd
Tower - Dixon Turner Wallcoverings Ltd.
Tower - Pifco Ltd
Towerpak - Tower Manufacturing
Tplus - Monarflex Ltd
Trackelast - Tiflex Ltd
Trackranger crushing outfit - Parker Plant Ltd
Tradesman - Turner Access Ltd
Tradgalv - Direct Wire Ties Ltd
Traditional - Northcot Brick Ltd
Traffic 2 - Woolliscroft Tiles Ltd
Traficop - Rediweld Rubber & Plastics Ltd
Trane - HV Air Conditioning Ltd
Trans-Oxide ™ - Johnson Matthey Pigments & Dispersions
Transcab - AEI Cables Limited
Translite - Harkness Hall Ltd
Transmit - Tunstal Telecom Ltd
Transplex - EDS Shutter Door Systems Ltd
Transtor - Desking Systems Ltd
Transvario - O & K Escalators
Trav-O-lators ® - Otis plc
Travertine - Marley Floors Ltd
Travertines - Kirkstone
Treacle Taps - Millar Dennis (1998) Ltd
Tread Safe - Anglo Building Products Ltd
Treadmaster - Tiflex Ltd
Treadspire - Consculpt Spiral Stairs Ltd
Trebitt - Jotun Henry Clark Ltd
Trebitt Opaque - Jotun Henry Clark Ltd
Tredaire - Gates Rubber Co Ltd The
Tredaire contract - Gates Rubber Co Ltd The
Tredaire Safe 'N' Sound - Gates Rubber Co Ltd The
Tredaire Silentfloor - Gates Rubber Co Ltd The
Tredaire Silentwood - Gates Rubber Co Ltd The
Tredaire Technics - Gates Rubber Co Ltd The
Tredaire Tredsafe FR - Gates Rubber Co Ltd The
Trelaze - Eternit UK Ltd
Trellidor - Telford Doors and Barriers Ltd
Trent - Portacel
Trent - Woolliscroft Tiles Ltd
Trent Gutter - Dales Fabrications Ltd
Trentobond - Tremco Ltd
Trentocrete - Tremco Ltd
Trentodel - Tremco Ltd
Trentoflex - Tremco Ltd
Trentoshield - Tremco Ltd
Trentothane - Tremco Ltd
Trespa - Amari Plastics Plc
Trespa - Decra Cubicles Ltd
Trespa - Trespa UK Ltd
Tretol - Viking Laminates Ltd
Trevi showers - Ideal-Standard Ltd
Tri-Guard ® - Jackson H S & Son (Fencing) Ltd
Tri-Plugs - Tower Manufacturing
Tri-Shell - Dudley Thomas Ltd
Tribol Limited - MMP International
Tribond - Vulcanite Limited
Tribond"E" - Vulcanite Limited
Tricity - Electrolux Domestic Appliances
Tricity - Parkinson Cowan and Tricity Bendix Co, The
Triclamp - Opto International Ltd
Tricosal - Beton Construction Ltd
Tridgent - Baresford Pumps
Tridonic - Bemlite Ltd
Trief Kerb - Redland Precast
Triflow - IBP Conex
Triglide - Brickhouse
Trilax - Girsberger London
Triline - Smart Systems Ltd
Trilock - Grass Concrete Ltd
Trilock - Landscape Grass (Concrete) Ltd
Trim Defender Door Bars - Trim Acoustics
Trim Defender Wall - Trim Acoustics
Trimalac - Trimite Ltd
Trimapanel - European Profiles
Trimawall - European Profiles
Trimguard - Manton Industrial Seals Ltd
Trimite - Andrews J S (Coatings) Ltd
Trimite - Trimite Ltd
Trimloc Interchange directories - Drakard & Humble Ltd
Trimwall - Neslo Partitioning & Interiors
Trio Panel wall formwork - PERI Ltd
Trio-ving - Sampson T F Ltd
Triple Seven - Dunbrik (Yorks) Ltd
Tripower - Blakley Electrics Ltd
Trisomet - European Profiles
Tristile - Gunnebo Mayor Ltd
Tritainer - Franklin Hodge Industries Ltd
Triton - Stanhope Barclay Kellet Crown Division
Triton - Triton PLC
Tritorch - Vulcanite Limited
Trizone - Airsprung Beds Ltd
Troikafix - Troika Architectural Mouldings Ltd
Trojan - Bitmen Products Ltd
Trojan - Legrand Electric Ltd
Trojan - Neptune Showers Ltd
Trojan - Shackerley (Holdings) Group Ltd incorporating Designer ceramics
Trolleylift - Stannah Lifts Ltd
Tropic Trend - Valor Heating
Tropicano - Xpelair Ltd
Trubolt - Ramset Fasteners Ltd
Trucrete - Tilcon Ltd
True-lite - Full Spectrum Lighting Ltd
Trueflue - Marflex Chimney Systems
Trufrule - Rabone Chesterman Ltd
Truline - Simpson Strong-Tie®
Truss Tray - European Profiles
Truswal Twinaplate - Alpine Autimation Ltd.
TS60P - Ramset Fasteners Ltd
TS750P - Ramset Fasteners Ltd
TTF - Simon R W Ltd
Tub-box - Creteco Sales Ltd
TUBA-RAD - Property Mechanical Products Ltd
Tubeclamps - Tubeclamps Ltd
Tubelite - LB Lighting Ltd
Tubesec - Gunnebo Mayor Ltd
Tubinox - Kedddy (Poujoulat) (UK) Ltd
Tucker - Arrow Supply Company Ltd
Tuf Light - Wilsons Fibreglass Ltd
Tuf Light roof sheets - Alpha Glass Fibres Ltd
Tufanega - Deb Ltd
Tuffblocks - RFA Group Ltd
Tuffgrip - Hanson Aggregates
Tuffwall - Brown F plc
Tuflok - Rabone Chesterman Ltd
Tuftiguard - Bonar Floors Ltd
Tunnel Mortar - Vandex (UK) Ltd
Tuppeck-Bridge Decking - Tarmac Precast Concrete
Turbo - Roof Units Group
Turbo cast 300 - Glasdon UK Ltd
Turbo-Dry - Cistermiser Ltd
Turbo-Flo - MSA (Britain) Ltd
TurboFast - Helifix Ltd
TurboSyphous - Dudley Thomas Ltd
Turicor - EBC UK Ltd
Turkish room - Dalesauna Ltd
Turlok - Armstrong World Industries Ltd
Turmix - Air Improvement Centre Ltd
Turnerised - TRC Service Group
Tusc - Unistrut Limited
Tuscany - Muraspec Ltd
Tweeny - Supreme Lifts Ltd
Tweeny - The Haigh Tweeny Co Ltd
20/20 Interlocking clay plain - Sandtoft Roof Tiles Ltd
Twergrip - Tower Manufacturing
Twiclad - Ide T & W Ltd
Twill - Advanced Fencing Systems Ltd
Twill Wire products - Challenge Fencing & Ltd
Twilweld - United Wire Ltd
Twin 22 - Ledu UK Ltd
Twin Traps - Midland Stom Ltd
Twin-Drain - Naylor Drainage Ltd
Twinaplate - Alpine Autimation Ltd.
Twinarc - Sylvania Lighting International INC.
Twinbreak - Square D, Electrical Distribution Division
Twinfix - Amari Plastics Plc
Twinimum - Ves Andover Ltd
Twinlok Self Sealing Test Plugs - Industrial Hangers Ltd
Twinloop - Checkmate Industries Ltd
Twinlux - Illuma Lighting Ltd
2000 Range safety wall tiles - Ensor Metal Products
2200 Antistat carpet - Burmatex Ltd
Twosome - Integra Products
TWS - TWS Presentation Systems (Mcmillan UK Ltd T/A)
Twyfords - Caradon Bathrooms Ltd
Tyco - Golden Coast Ltd
Tyfords - Barflow-LCA
Tyglas - Fothergill Engineered Fabrics Ltd
Tylo Sauna - Sanua UK Ltd
Tylo Sauna - Tylo UK Ltd
Tylo Showers - Sanua UK Ltd
Tylo Showers - Tylo UK Ltd
Tylo Steam - Sanua UK Ltd
Tylo Steam - Tylo UK Ltd
Type E - Cavity Trays Ltd
Type X - Cavity Trays Ltd
Typhoo - The Kenco Coffee Company
Tyrespan - Metalrax Ltd
Tyton - Hellermann Insuloid
Tyveks - Klober Ltd
U.S.G Donn - New Forest Ceilings Ltd
UbiFreash - Ubbink (UK) Ltd
Ubigas - Ubbink (UK) Ltd
UbiSoil - Ubbink (UK) Ltd
Ubivent - Ubbink (UK) Ltd
Ucrete - Thoro Systems Products Ltd
UF - Davis International Ltd
UK Crane Service - Morris Mechanical Handling Ltd
Ultima - Cego Frameware Ltd
Ultima - Marley Waterproofing
Ultima - Midland Stom Ltd
Ultima - Rixonway Kitchens Ltd
Ultimate - Valor Heating
Ultimate flue outlet - Brewer Metalcraft
Ultra - Monarflex Ltd
Ultra 80 - Norwood Partition Systems Ltd
Ultra 80 - NT Partition Systems
Ultra-Cool - Beta Lighting Ltd
Ultra-Rib - Uponor Ltd
Ultra-Safe - Uponor Ltd
Ultrabeam - Hadley Industries Plc
Ultracolor - Mapei UK Ltd
Ultraflex - Dyer Environmental Controls
Ultraflex - P.C.I Construction Systems Ltd
Ultraframe - Ultraframe PLC

Ultragard Software - Time and Data Systems International Ltd (TDSI)
Ultragard Vision-ID - Time and Data Systems International Ltd (TDSI)
Ultraglaze - GE Bayer Silicones
Ultragrain ® - Righton Ltd
Ultralift - Caldwell Hardware (UK) Ltd
Ultralite 500 - Ultraframe PLC
Ultramastic - Mapei UK Ltd
Ultraplan - Mapei UK Ltd
Ultrasheet - Hadley Industries Plc
Ultraspec reversible windows - Dale Joinery Ltd
Ultrasteel - Hadley Industries Plc
Ultrastor - Norwood Partition Systems Ltd
Ultrastrong - Boddingtons Ltd
Ultraturn - Righton Ltd
Ultrazed - Hadley Industries Plc
Ultrum - Game Time UK Ltd
Ulverston Limestone - Gregory Quarries Ltd, The
Ulverston Limestone - The Rare Stone Group Ltd
Ulverston Marble - Chilmark Quarries Ltd
Ulveston Limestone - The Rare Stone Group Ltd
Uni -Walten - DLW Flooring Ltd
Uni- Ecoloc - RMC Concrete Products
Uni-Ecoloc - RMC Concrete Products (UK) Ltd
Uni-Fix Hammer-Set - Unifix Ltd
Uni-Fix Hi-Load - Unifix Ltd
Uni-Fix Klip-Lok - Unifix Ltd
Uni-Junction - Johnston Pipes Ltd
Uni-Kliplock - Unifix Ltd
Uni-Trex - Terrapin Ltd
Unibond - Henkel Home Improvement & Adhesive Products
Unibond No More Nails - Henkel Home Improvement & Adhesive Products
Unibond No More Sealent Guns - Henkel Home Improvement & Adhesive Products
Unibond Unifilla - Henkel Home Improvement & Adhesive Products
Unicell - Ronacrete Ltd
Unicryl - Witham Oil And Paint Ltd
Unidare - Seaque Technologies Ltd
Unidox - Witham Oil And Paint Ltd
Unifact - Satchwell Control Systems Ltd
Uniflair - Quiligotti Access Flooring Limited
Uniflow - FC Precast Concrete Ltd
Uniframe - FC Precast Concrete Ltd
Unigas - GP Burners (CIB) Ltd
Uniglow - Witham Oil And Paint Ltd
Uniline - Leisure, Glynwed Consumer Foodservice Products Ltd
Unimate - Quartet-GBC
Unimotor - Dyer Environmental Controls
Union - Parkes Josiah & Sons Ltd
Union Jack Label Fittings - Spegelstein S & Son
Uniperf - United Wire Ltd
Uniplant - Witham Oil And Paint Ltd
Unipleat - Eclipse Blind Systems Ltd
Uniply - Marley Waterproofing
Unique - Wilson & Garden Ltd
Unique Adio & Visual Systems. Ltd - Wilson & Garden Ltd
Unistrut - Kem Edwards Ltd
Unistrut - Unistrut Limited
Unit Swimming Pools - Regency Swimming Pools Ltd
Unitas - Witham Oil And Paint Ltd
Universal - EBS Services Ltd
Universal - Matki plc
Universal - Unicorn Abrasives (UK) Ltd
Universal Gulley Adaptor - Johnston Pipes Ltd
Universal Metricove - Pilkington's Tiles Ltd
Universal range - Marshall-Tufflex Ltd
Universal roofboard - Callenders Ltd
Upat - Arthur Fisher (UK) Ltd
UPE - Talbot
Upex - Uponor Ltd
Upline - Illuma Lighting Ltd
Uponal - Uponor Ltd
USP Intruder Guard - United Safety Products UK Ltd
UV Dermagard - Bonwyke Ltd
UV Free Lighting - Encapsulite International Ltd
Uyeda - Samways P A & Co
V Trim GRP - Areco Roofing Supplies Company
V-Gard - MSA (Britain) Ltd
V/ Slider - Hill Leigh (Joinery) Ltd
Vac Vac - Hickson Timber Products Ltd
Vaccari - Swedecor Ltd
Vacsol - Hickson Timber Products Ltd
Valiant - Cego Frameware Ltd
Vallance - Vallance
Valley - Iles Waste Systems
Valley Range - Dale Joinery Ltd
Valspar - Akzo Coatings Plc
Valves - Horstmann Timers & Controls Ltd
Van Besouw - E. H. S. Roofing Ltd
Van Erven - McLoughlin Wood Ltd
Vandalene ® - Coo-Var Ltd
Vandex BB75 - Vandex (UK) Ltd
Vandex BB75E - Vandex (UK) Ltd
Vandex Injection Mortar - Vandex (UK) Ltd
Vandex Plug - Vandex (UK) Ltd
Vandex Premix - Vandex (UK) Ltd
Vandex Super - Vandex (UK) Ltd
Vandex Unimortar - Vandex (UK) Ltd
Vandgard - Vandgard Ltd
Vanguard - Forticrete Roofing Products
Vanguard - Top Office Equipment Ltd
Vanity flair - Ellis J T & Co Ltd
Vanodine - Evans Vanodine International PLC
Vapourfoil - Vulcanite Limited
Vapourguard - Trade Sealants
Varedplan - Edmonds A & Co Ltd
Vari-Cleat - EDL Cable Supports Ltd
Vari-Level - Wade International (UK) Ltd
Variflow - Grosvenor Pumps Ltd
Varilift - Camloc (UK) Ltd
Vario - Kermi (UK) Ltd
Vario - Kompan (UK) Ltd
Vario Wall formwork - PERI Ltd
Variplan - Walsall Cable Management Ltd
Varispace - Harkness Hall Ltd
Varsity - Crompton Lighting Ltd
Varsity - Metalrax Ltd
Vasura - Erlau AG
VCD - Air Diffusion Ltd
Vector - Securistyle Ltd
Vector - Vantrunk Engineering Ltd
Vector Exluder - Securistyle Ltd
Vee - Project Office Furniture PLC
Vega - Guardall
Veha Radiators - Brikenden (Builders Merchants) Ltd
VEKA - ABB Flakt Products
Velour - Burmatex Ltd
Velux - VELUX Co Ltd, The
Vemo - RFA Group Ltd
Venetian - Symphony Group PLC, The
Venice - Metlex
Vent A Matic - Simon R W Ltd
Ventform - Cordek Ltd
Ventlite - Ventlane Ltd
Ventmaster - Docherty H Ltd
Ventraflow - Jones of Oswestry Ltd
Ventrust - Callenders Ltd
Venture Parry - Bemlite Ltd
Venturer - Record Playground Equipment Ltd
Venwall - Applied Acoustics (a part of Henry Venables Limited)
Verco - Verco Office Furniture Ltd
Vermeer - B-Trac Equipment Ltd
Vermiculite - Dupre Vermiculite
Vermiculux - Cape Boards Ltd
Vermont - Deceuninck Ltd
Vernacare - Verna Ltd
Vernagene - Verna Ltd
Verona Range - Caradon Bathrooms Ltd
Versailles - Qualitas Bathrooms
Versalite - Righton Ltd
Versalux - JSB Electrical PLC
Vertalux - Rada Lighting Ltd
Verti-Frame - Hi-Store Ltd
Vesda - McMillan Fire Alarm Systems Ltd
Vesda - Vision Sytems (Europe) Ltd
Vesselpak - Flamco UK Ltd
Vestos - Hickson Timber Products Ltd
Vetisol - Axter Ltd
Vetonit Self Level - Isocrete Group Sales Ltd
Vibraseal - Williamson Floor & Wall Specialists Ltd
Vibro - Keller Ground Engineering
Vicrtex - Tektura Plc
Victaulic - Midland Tube and Fabrications
Victaulic - Victaulic Systems
Victorglos - Bollom J W & Co Ltd
Victorian - Property Mechanical Products Ltd
Victory 2000 - Thermal Radiators Ltd
Victory STD - Thermal Radiators Ltd
Vicucase - Promat Fire Protection Ltd
Viero Decorative Coatings - Textured European Finishes Ltd
Vieroquartz Masonry Paint - Textured European Finishes Ltd
Viewguard - Hadley Industries Plc
Viewpoint - Pryorsign
Vigilante - Blick Communication Systems Ltd
Vigilon - Caradon Gent Ltd
Viking - Anglowest Distributors Ltd
Viking - Macron Fireater Ltd
Viking - Roof Units Group
Village Stone - RMC Concrete Products
Village Stone - RMC Concrete Products (UK) Ltd
Vinadac - Dacrylate Paints Ltd
Vinyllon - Dixon Turner Wallcoverings Ltd
Vinyllon - Dixon Turner Wallcoverings Ltd.
Vinylock - HW Systems Ltd
Viper spike ® - Jackson H S & Son (Fencing) Ltd
Vipps - CEM Systems Ltd
Virage - Securistyle Ltd
Virgil - Interoven Ltd
Visage - Valor Heating
Viscacid - Kenyon Industrial Paints & Adhesives
Viscount - Forticrete Roofing Products
Viscount - Securistyle Ltd
Visedge - Howe-Green Ltd
Visions - Shires Bathrooms
Visofold - Smart Systems Ltd
Visoglide - Smart Systems Ltd
Visoline - Smart Systems Ltd
Visolplast textured coatings - Textured European Finishes Ltd
Visqueen - Visqueen Agri
Vista - Dallas & Forrest Ltd
Vista - Fabrikat (Nottingham) Ltd
Vista - Welconstruct Co
Vista Range - Caradon Bathrooms Ltd
Vista-Fix - Vista Engineering Ltd
Vista-Plas - Vista Engineering Ltd
Vistalux - Ariel Plastics Ltd
Visual Builder - Opto International Ltd
Viternus - Crosbie Coatings
Vitra - Vitra (UK) Ltd
Vitraclad - Panel Systems Ltd
Vitraglaze - Panel Systems Ltd
Vitral - EBS Services Ltd
Vitros - Crosbie Coatings
Vixalit Lime-based paint - Textured European Finishes Ltd
VLT ® - Danfoss Ltd
Vodafone - Vodafone Paging Ltd
Vogue - Dermide Ltd
Voidak - Dunham Bush Ltd
Vokera - Diamond Merchants
Vokes - BTR Environmental Ltd Vokes
Volacel - Volcrepe Ltd
Voltan - Townscape Products Ltd
Volumeter - Gebhardt Kiloheat
Von Duprin - Relcross Ltd
Vortax - Axial - Howden Buffalo
Vossloh Schwabe - Bemlite Ltd
Vtech - Vantrunk Engineering Ltd
Vulcadome - Vulcan Plastics Ltd
Vulcaflex - Vulcanite Limited
Vulcalap - Vulcan Plastics Ltd
Vulcalon Spg - Vulcan Plastics Ltd
Vulcan - Leaderflush & Shapland
Vulcaplast Spg - Vulcan Plastics Ltd
Vulcathene - Capper Pipe Systems Ltd
Vulcatorch - Vulcanite Limited
Vulcatuf - Vulcan Plastics Ltd
Vylon Plus - Marley Floors Ltd
Vynarac - Hispack Limited
W Schneider - Schneider (UK)
W/A formwork plywood - Wood International Agency
W/A pine doors - Wood International Agency
W20 - Clement Steel Windows
W40 - Clement Steel Windows
Wacker Silicones - Notcutt W P Ltd
Wade - Drainage Systems
Wade - Wade International (UK) Ltd
Walk On Ceiling - Norwood Partition Systems Ltd
Walker Timber Frame - Walker Timber Ltd
Wall - glas - Dex-o-Tex International Ltd
Wall panels - Ecophon Ltd
Wallflex - Lastite Building Products Ltd
Wallforce - BRC Building Products
Wallis - Bitmen Products Ltd
WallLok - Vencel Resil Ltd
Wallmaster - Hillaldam Coburn Ltd
Wallmate - Dow Construction Products
Wallstor - Desking Systems Ltd
Walsall - Walsall Cable Management Ltd
Walton Sheds - Challenge Fencing & Ltd
Wanit Prima Slate - Blunn Slates Ltd
Wanit Repro Slate - Blunn Slates Ltd
Ward - E. H. S. Roofing Ltd
Warmcel - Fillcrete Ltd
Warmsafe LST - Dunham Bush Ltd
Warmwell - Hamworthy Heating Ltd
Wascator - Electrolux Laundry Systems
Washroom Control - Cistermiser Ltd
Waste King - Berglen Group Ltd
Wastematic - Max Appliances
Wastematic - Sussex Brassware Ltd
Watasaver - Sheardown Engineering Ltd
Watchdog - Heat Trace Ltd
Watchkeepers - Rapaway Energy Ltd
Water Biotreatment Club - British Water
WaterFlow - Airflow (Nicoll Ventilators) Ltd
Waterless - Relcross Ltd
Waterloo - Marston A & Co Ltd
Watermatic - Rexam Harcostar
Waterproof Surface Range - Lumitron Ltd
Watertimer - Sheardown Engineering Ltd
Watts cliff - Realstone Ltd
Wavelay - Boddingtons Ltd
Waveney - Blyde-Barton Ltd
Wavespan - Flexible Reinforcements Ltd
Wayland - Advanced Fencing Systems Ltd
Waylite 800 - JSB Electrical PLC
Ways with Doorways - Kern (Design Products) J T Ltd
Weather Master - Ves Andover Ltd
Weather Ten - Exitex Ltd
Weatherbeater doors - IG Limited
Weatherbeater garage doors - IG Limited
WeatherGuard - Everlac (GB) Ltd
Weathershield - ICI Dulux Trade Paints
Weatherwhite - Cardale Doors Ltd
Weatherwood - Cardale Doors Ltd
Weaverstone - Anderton Concrete Products Ltd
Weavespread - Flexible Reinforcements Ltd
Weavetop - Flexible Reinforcements Ltd
Web wall - ABG Ltd
Weep - Airflow (Nicoll Ventilators) Ltd
Weholite - Asset International Ltd
Welex - Welconstruct Co
Wellington - Bassett & Findley Ltd
Wells Trinidad Lake Asphalt - Associated Holdings Ltd
Welmade - Quigley Metal Products Ltd
Welpac - Harwood & Welpac Hardware Ltd
Wensleydale - Carpets of Worth Ltd
Wensum - Blyde-Barton Ltd
Wernick - Wernick S & Sons Ltd
Wescol Glosford - Wescol Glosford
Wessex - Hamworthy Heating Ltd
Wessex - Multitex GRP & GRG Products
Wessex Doors - Regency Garage Door Services Ltd
Wessex Range - Beacon Hill Brick Co Ltd
Westbond - Hometex Trading Ltd
Westbourne Acoustic - Durabella Ltd
Westbury Canopy - James Smellie Fabrications Ltd
Westclox - General Time Europe
Westdale - Chiltern Invadex
Westman Systems - MJ Electronics Services (International) Ltd
Westminster - Atlas Stone Products
Westminster - Branson Leisure
Westminster - Firth Carpets Ltd
Weston - Gloster Furniture Ltd
Westwood Ground Bath. Cotswald Stone - Hanson Bath & Portland Stone
Weyroc - Egger (UK) Ltd
Weyroc HDX - Egger (UK) Ltd
Weyroc PFB - Egger (UK) Ltd
Weyroc V313 - Egger (UK) Ltd
WF - Westbury Filters Limited
WF1 - Westbury Filters Limited
WFB - Westbury Filters Limited
WFC - Westbury Filters Limited
WFF - Westbury Filters Limited
WFG - Westbury Filters Limited
WFH HEPA - Westbury Filters Limited
WFK - Westbury Filters Limited
WFV - Westbury Filters Limited
Whirl-A-Waste - Sissons W & G Ltd
Whirline - Hipkin F W Ltd
Whirlwind - Envirodoor Markus Ltd
Whispair - Xpelair Ltd
Whisper Door Heater - Shearflow Phoenix
Whisperply - Pilkington Plyglass
Whitehill - Whitehill Spindle Tools Ltd
Whiting & Davis safety - BRB Industrial Services Ltd
Whizard - BRB Industrial Services Ltd
Wicanders cork Parquet - Wicanders (GB) Ltd
Wicanders cork-o-floor - Wicanders (GB) Ltd
Wicanders Wood Parquet - Wicanders (GB) Ltd
Wicanders Woodline - Wicanders (GB) Ltd
Widespan - Metalrax Ltd
Wilclo - Clow Group Ltd
Wildgoose - Derby Timber Supplies
Wilka Locks & Cylinders - JB Architectural Ironmongery Ltd
Wilkinson's Furniture - Carleton Furniture Group Ltd
Willseal Firestop - Illbruck Ltd
Wilo SE - Wilo Samson Pumps Ltd
Wilo Top E - Wilo Samson Pumps Ltd
Wilo Top S - Wilo Samson Pumps Ltd
Wilo Top SD - Wilo Samson Pumps Ltd
Wilsonart® - Wilsonart Ltd
Wilsons Fibreglass ltd - Alpha Glass Fibres Ltd
Wilton Royal - Carpets International (UK) PLC
Wimborne - Hamworthy Heating Ltd
Winchester - BSA Tools Ltd
Windowgard - Henderson-Bostwick
Windsor - Guardall
Wings - Bisley Office Equipment
Wingstile - Gunnebo Mayor Ltd
Winterfast - Leigh's Paints
Winther Browne & Co Ltd - Browne Winther & Co Ltd
Wintun - Caradon Duraflex Systems Ltd
Wiremaster - ECS Lighting Controls Ltd
WirralRange - AS Newbould Ltd
Wisa-Deck - Schauman (UK) Ltd
Wispa Hoist - Chiltern Invadex
Wizard - T & D Plastech
WMS - Avocet Hardware Ltd
Wolfin - Beton Construction Ltd
Wonderex - Optex (Europe) Ltd
Wondertrack - Optex (Europe) Ltd
Wood-n-Play - Record Playground Equipment Ltd
Wood-o-Cork - Wicanders (GB) Ltd
Wood-o-Floor - Wicanders (GB) Ltd
Woodacoustic - Applied Acoustics (a part of Henry Venables Limited)
Woodacoustic - Henry Venables Ltd
Woodcelid - Torvale Building Products, A Division of Stadium Group PLC
Woodcemair - Torvale Building Products, A Division of Stadium Group PLC
Woodcemax - Torvale Building Products, A Division of Stadium Group PLC
Woodcraft - Dallas & Forrest Ltd
Woodcraft - Design Line
Woodflex - Polycell Products Ltd
Woodgrip - Pedley Furniture international Ltd
Woodland - Neptune Outdoor Furniture Ltd
Woodland System - Hyperion Wall Furniture Ltd
Woodro - Hille Ltd
Woodscape - Woodscape Ltd
Woodstock Accent - Adam Carpets Ltd
Woodstock Classic - Adam Carpets Ltd
Woodstyle - Hallmark Blinds Ltd
Wooliscroft tiles - Stokes R J & Co Ltd
World - Warner Howard Ltd
Worldspan - Clow Group Ltd
Write Easy Chalkboards - Signs International
WSM - Welconstruct Co
Wyrem - Flexible Ducting Ltd
Wyvern Fireplaces - Wyvern Marlborough Ltd
X-Plas PVC - Signs International
Xen - Optelma Lighting Ltd
Xenex - Caradon Gent Ltd
Xenex - RMC Concrete Products
Xenex - RMC Concrete Products (UK) Ltd
Xenflex - Intralux UK Ltd
Xerra - Girsberger London
Xodus - Xpelair Ltd
XP95 - Apollo Fire Detectors Ltd
XP95 Intrinsically safe - Apollo Fire Detectors Ltd
Xtra load DPC - Anderson D & Son
Xtra load elite - Anderson D & Son
Xtra load preformed cavity trays - Anderson D & Son
Xtraflex - Adaptaflex Ltd
Xtraflex Sanding Pads - Acton & Borman
Xypex - Fullstop Technology Ltd
Yackham Wilton - Mackay Carpets, Hugh

Yale - Yale Security Products Ltd
Yanmar - Martin (Marine) Ltd Alec
Yeoman - Bartlett Catering Equipment Ltd
Yeoman Custom mouldings - Harrison Thompson & Co Ltd
Yeoman Formula One seating - Harrison Thompson & Co Ltd
Yeoman Rainguard - Harrison Thompson & Co Ltd
Yeoman Shield total surface protection - Harrison Thompson & Co Ltd
Ygnette - Wellman Robey Ltd
Ygnis - Wellman Robey Ltd
York Terracotta - York Handmade Brick Co Ltd

Yorkdale - Atlas Stone Products
Yorkex - IMI Yorkshire Copper Tube Ltd
Yorkshire - IMI Yorkshire Copper Tube Ltd
Yorkshire Imperial Metals - Brikenden (Builders Merchants) Ltd
Yorktone - Redland Precast
Youngman - SGB Youngman
Yuma - Optelma Lighting Ltd
Z purlins - Brohome Ltd
Zampano - Girsberger London
Zannussi - Electrolux Domestic Appliances
Zanussi - Parkinson Cowan and Tricity Bendix Co, The

Zanussi Professional - Zanussi Professional
Zap - The Modelshop
Zap! - Vodafone Paging Ltd
Zedex - Zedcor Ltd - Zedcor Ltd
Zehnder - Diamond Merchants
Zemdrain - Creteco Sales Ltd
Zenith Club Class - Brintons Ltd
Zerostat 2000 - Checkmate Industries Ltd
Zest - Contract Components Ltd
Zestseal - Contract Components Ltd
Zeta - Provex Products Ltd
Zeta II - JSB Electrical PLC
Ziehl-Abegg - Ziehl-Ebm (UK) Ltd

Zig Zag - Light Alloy Ltd
Ziggurat - Poselco Lighting
Zilmet - Altecnic Ltd
Ziptube - Hadley Industries Plc
ZMX multiplexers - Baxall Ltd
Zolpacryl - Ronacrete Ltd
ZR Receivers - Baxall Ltd
ZS Stacking Linking Chair - Race Furniture Ltd
ZTX3 to 6 Telemetry Transmitters - Baxall Ltd
Zykon - Arthur Fisher (UK) Ltd
ZZ System – Quelfire

Company Information Section

Aardee Locks and Shutters Ltd Hillington Ind. Estate 453, Hillington Road Glasgow G52 4BL tel:(0141) 810 3444 fax:(0141) 553 2080 E_Mail:101771.2655@compuserve.com

Aaztec Cubicles Unit 189 Thorp Arch Trading Estate Wetherby West Yorkshire LS23 7BJ tel:(01937) 844633 fax:(01937) 842756

Abacus Integrated Systems Ltd Unit 3, Redland Centre 5 Oaks Way Coulsdon Surrey CR5 2UT tel:(0181) 763 8800 fax:(0181) 763 9996

Abacus Lighting Ltd Sutton-in-Ashfield Nottinghamshire NG17 5FT tel:(01623) 511111 fax:(01623) 552133 E_Mail:sales@abacus-lighting.com Web_Site:http://www.abacus-lighting.com

Abacus Signs Fairways Wenvoe Castle, Wenvoe Cardiff Vale of Glamorgan CF5 6BE tel:(01222) 593214 fax:(01222) 593897

ABB Flakt Products Grovelands House Longford Road Exhall Coventry CV7 9ND tel:(01203) 368500 fax:(01203) 364499

ABB Industrial Systems Ltd District Heating Division Stafford Park 12 Telford Shropshire TF3 3BJ tel:(01952) 205450 fax:(01852) 205451

ABB Kent Meters Lea Works Ponwicks Road Luton Beds. LU1 3LJ tel:(01582) 402020 fax:(01582) 438052 E_Mail:sales@gbkem_mail_abb_com Web_Site:http://www.ab.co.uk/kentmeters

ABB Kent Taylor Ltd. Howard Road Eaton Socon St Neots Huntingdon PE19 3EU tel:(01480) 475321 fax:(01480) 217948

Abbex Fire Protection Ltd The Fine House 140 Taremadile St. London EC2 84SD tel:(0171) 336 0776739 6251

Abbey Building Supplies Co 213 Stourbridge Road Halesowen W. Midlands B63 3QY tel:(0121) 550 7674 fax:(0121) 585 5031

Abbey Craftsmen Ltd 56 High Street Haslemere Surrey GU27 2LA tel:(01428) 652666 fax:(01428) 642125

Abbott Bros (Southall) Ltd Abbess House 39-47 High Street Southall Middx. UB1 3HE tel:(0181) 574 6961 fax:(0181) 571 4735

ABC Studios (Plaster Mouldings) Ltd. Oxford Lane City Road Routh Cardiff CF2 3DU tel:(01222) 482886

ABCO Products Ltd. Bldg. 45, Membury Airfield Lambourne Newbery Berks. RG16 7TJ tel:(01488) 72414

Abel Alarm Co Ltd Detection House 4 Vaughan Way Leicester LE1 4ST tel:(0116) 265 4200 fax:(0116) 251 5341 E_Mail:sales@abelalarm.co.uk Web_Site:http://www.abelalarm.co.uk

Abesco Ltd Abesco House Laurence Kirk Business Park Laurence Kirk AB30 1EY tel:(01561) 377766 fax:(01561) 378887 Web_Site:http://www.abesco.co.uk

Abet Ltd 70 Roding Road London Industrial Park London E6 4LS tel:(0171) 473 6910 fax:(0171) 476 6935 E_Mail:sales@abet.ltd.uk Web_Site:http://www.abet-laminati.it/

ABG Ltd Unit E7, Meltham Mills Meltham Mills Road Meltham W. Yorks. HD7 3AR tel:(01484) 852096 fax:(01484) 851562 E_Mail:sales@abg_georgthetics.com Web_Site:http://www.abg_georgthetics.com

Ableson L & D Ltd Thwaite Gate Low Road Leeds LS10 2SL tel:(01532) 713527 fax:(01532) 711352

Abloy Security Ltd 2-3 Hatters Lane Croxley Business Park Watford Herts. WD1 8YY tel:(01923) 255066 fax:(01923) 655001 E_Mail:sales@abloysecurity.co.uk Web_Site:http://www.abloysecurity.co.uk

Abrafract Ltd Beulah Road Sheffield S6 2AR tel:(01742) 348971

ABS Pumps Ltd Station Road Horley Surrey RH6 9HN tel:(01293) 821975 fax:(01293) 821976

Absolute Action Ltd Mantle House Broomhill Road Wandsworth London SW18 4JQ tel:(0181) 871 5005 fax:(0181) 877 9498 telex:917003 LPC G E_Mail:enquiries@absolute-action.com Web_Site:http://www.absolute-action.com

Abstracta Construction Ltd 187 Brent Crescent London NW10 7RX tel:(0181) 965 8845 fax:(0181) 965 5120 telex:262284 Ref:3737

ABT Gibbons Ashmore Lake Way Willenhall W. Midlands WV12 4LL tel:(01902) 368080 fax:(01902) 602431

Accent Hansen Greengate Industrial Park Greengate Middleton Manchester M24 1SW tel:(0161) 284 4100 fax:(0161) 655 3119 E_Mail:accent-han@aol.com

Access Industries Ltd 16 Acacia Close Cherrycourt Way Leighton Buzzard Beds. LU7 8UH tel:(01525) 383101 fax:(01525) 384381

Access Industries U.K. 72 Jay Avenue Teeside Industrial Estate Stockton Cleveland TS17 9LZ tel:(01642) 750707 fax:(01642) 750709

Access Piling Ltd Lyngarth Sutton Field Road Sutton Doncaster DN6 9JX tel:(01302) 707506 fax:(01302) 707506

Accomodex Leofric House Ryton-on-Dunsmore Coventry CV8 3ED tel:(01203) 301301 fax:(01203) 301148 E_Mail:100577.2713@compuserve.com

Accrington Brick Tile Co Ltd Acrrington Nori Factory Whinney Hill Accrington Lancs. BB5 6NR tel:(01254) 232684 fax:(01254) 399128

Acmeflooring Ltd, Marketing Department St Peters Road Huntington Cambs. PE18 7DN tel:(01480) 52101

Acmexd Doors Ltd Unit 2 Stone Close West Drayton Middx. UB7 8JU tel:(01895) 444044

ACO Technologies Plc Hitchin Road Shefford Beds. SG17 5TE tel:(01462) 816666 fax:(01462) 851490 E_Mail:drainsales@aco.co.uk Web_Site:http://www.aco.co.uk

Acorn Mill Co Ltd Mellor Street Lees, Oldham Lancs. OL4 3DA tel:(0161) 624 4259 fax:(0161) 624 4259

Acousticabs Industrial Noise Control Ltd Unit 52 Pocklington Industrial Estate Pocklington Yorks Y042 1NR tel:(01759) 305266 fax:(01759) 305268

Acousticabs Industrial Noise Control ltd Unit 52 Pocklington Ind Estate Pocklington Yorks. YO42 1NR tel:(01759) 305266 fax:(01759) 305268

ACP Concrete (Contracts) Limited Unit 11 Westside Industrial Estate Jackson Street St Helens Merseyside WA9 3AT tel:(01744) 24600 fax:(01744) 451815

ACR Heat Transfer Manufacturing Ltd Rollesby Road, Hardwick Industrial Estate Kings Lynn Norfolk PE30 4CN tel:(01553) 763371 fax:(01553) 771322

Actionair South Street Whistable Kent CT5 3DU tel:(01227) 276100 fax:(01227) 264262

Acton & Borman Cavendish Road Stevenage Herts. SG1 2EG tel:(01438) 312243 fax:(01438) 741335

Acustic + Environmental Technolgy 23 Cranford Drive Acton Hants GU34 4HJ tel:(01420) 85478 fax:(01420) 85478

Adam Carpets Ltd Greenhill Works Birmingham Road Kidderminster DY10 2SH tel:(01562) 822247 fax:(01562) 751471

Adam Furniture Group PLC Fairfield Road Croylsden Manchester M35 6AR tel:(0161) 370 8317 fax:(0161) 370 4307

Adamant Engineering Co Ltd Headley Road East Woodley Reading RG5 4SN tel:(01734) 690980 fax:(01734) 442093

Adams James & Sons Ltd 26 Blackfriers Road London SE1 8NY tel:(0181) 928 7375

Adams- Hydraulics Ltd PO Box 15 York YO30 4TA tel:(01904) 695695 fax:(01904) 695600

Adaptaflex Ltd Station Road Coleshill Birmingham B46 1HT tel:(01675) 468200 fax:(01675) 464930 E_Mail:mturner@adaptaflex.co.uk

Adda Systems RB Ltd Stockholm Road Sutton Fields industrial Estate Hull HU8 0XW tel:(01482) 831555 fax:(01482) 837698 telex:507418

Addacabin Ltd Southend Thornton Kircaldy Fife KY1 4ED tel:(01592) 774387

Addagrip Surface Treatments Ltd Bird-in-Eye Hill Uckfield E. Sussex TN22 5HA tel:(01825) 761333 fax:(01825) 768566 E_Mail:roger@addagrip.co.uk Web_Site:http://www.addagrip.co.uk

Adex Storage Equipment Ltd 5 Avbury Court Mark Road Hemel Hempstead Herts. HP2 7TA tel:(01442) 232327 fax:(01442) 62713

Adhesive Solutions Ltd Moor Road Chesham Bucks. HP5 1SB tel:(01494) 784444

Adshead Ratcliffe & Co Ltd Derby Road Belper Derbys. DE56 1WJ tel:(01773) 826661 fax:(01773) 821215 E_Mail:arbo@arbo.co.uk Web_Site:http://www.arbo.co.uk

ADT Modern Security Systems Ltd The Clock House The Campus Hemel Hempstead Herts. HP2 7TL tel:(01442) 234123 fax:(01442) 252815

ADT Security Systems Security House Trafalger Way Camberley Surrey GU15 3BN tel:(01276) 692737 fax:(01276) 681378

Advance Tapes International Ltd P.O Box 122 Abbey Meadows Leics. LE4 5RA tel:(0116) 251 0191 fax:(0116) 2653070 Web_Site:http://www.advancetapes.com

Advance Technical Panels Ltd Longland Industrial Estate Milner Way Ossett W. Yorks. WF5 9JE tel:(01924) 263655 fax:(01924) 280193

Advanced Air (UK) Ltd 3-4 Cavendish Road Bury St Edmunds Suffolk IP33 3TE tel:(01284) 701356 fax:(01284) 701357 E_Mail:sales@advancedair.co.uk Web_Site:http://www.advancedair.co.uk

Advanced Fencing Systems Ltd 104-120 Blackburn Road Wincobank Sheffield S. Yorks. S61 2DN tel:(0114) 2891891 fax:(0114) 2891892

Advanced Hygienic Contracting Ltd Hammerain House Hookstone Avenue Harrogate N. Yorks. HG2 8ER tel:(01423) 870049 fax:(01423) 870051 E_Mail:advanced@cladding.co.uk Web site:http://www.cladding.co.uk

Advanced Industries Ltd West Richardson Street High Wycombe Bucks. HP11 2SB tel:(01494) 450722 fax:(01494) 448998 E_Mail:info@advanced-industries.co.uk

Advanced Panels & Products Grosvenor Road, Gillingham Business Park Gillingham Kent ME8 0SA tel:(01634) 378880 fax:(01634) 378381

Advanced Perimeter Systems 16 Cunningham Road, Springkerse Ind Est Stirling FK7 7TP tel:(01786) 479862 fax:(01786) 470331 E_Mail:admin@aps-stirling.co.uk

AEI Cables Limited Birtley Chester-Le-Street Durham DH3 2RA tel:(0191) 4103111 fax:(0191) 4108312 telex:25829 E_Mail:salen@aeic.co.uk

AEL 6 Berkley Court Manor Park Buncorn Cheshire WA7 1TQ tel:(01928) 579068 fax:(01928) 579523

AEL Furniture Group A27 Hastingwood Trading Estate Harbet Road Edmonton London N18 3LP tel:(0181) 807 7476 fax:(0181) 807 7579 Web_Site:http://www.yell.co.uk

Aercon Ltd Woodside Road Eastleigh Hampshire SO5 4DZ tel:(01703) 614322 fax:(01703) 620649

Aercon Wiring Systems Ltd Aercon Works Alfred Road Gravesend Kent DA11 7QF tel:(01268) 418822 fax:(01268) 418822

Aga-Rayburn, Glynwed Consumer & Construction Products PO BOX 30 Ketley Telford Shropshire TF1 4DD tel:(01952) 642000 fax:(01952) 641961 telex:35196 Web_Site:http://www.aga.rayburn.co.uk

AGN Fencing Limited Plumtree Farm Industrial Estate Bawtry Road, Harworth, Doncaster S. Yorks. DN11 8EW tel:(01302) 710247 fax:(01302) 711751

Ainsworth B R (Southern) Ltd Scarisbrick House Brunnel Way Fareham Hampshire PO15 5TX tel:(01489) 885565 fax:(01489) 885258

Air Diffusion Ltd Stourbridge Road Bridgnorth Shropshire WV15 5BB tel:(01746) 761921 fax:(01746) 766450

Air Heating Ltd Murray Street Paisley Renfrewshire tel:(0141) 889 4802 telex:776626 PAIEN G

Air Improvement Centre Ltd 23 Denbigh Street London SW1V 2HF tel:(0171) 834 2834 fax:(0171) 630 8485

Airdri Ltd Oakfield Industrial Est Eynsham Oxford OX8 1TH tel:(01865) 882330 fax:(01865) 881647 E_Mail:airdri-sales@msn.com Web_Site:http://www.airdri.com

Airedale International Air Conditioning Ltd Leeds Road Rawdon Leeds LS19 6JY tel:(0113) 239 1000 fax:(0113) 250 7219 E_Mail:marketing@airedale.co.uk Web_Site:http://www.Airedale.co.uk

Airflow (Nicoll Ventilators) Ltd Unit 2, Queensway Stem Lane Industrial Estate New Milton Hampshire BH25 5NN tel:(01425) 6112547 fax:(01425) 638912 E_Mail:sales@airflow.vent.co.uk Web_Site:http://www.airflow.vent.co.uk

Airflow Air Bricks Ltd Oxhey Lane Watford WD1 4RQ tel:(01923) 232736

Airflow Developments Ltd Lancaster Road High Wycombe Bucks. HP12 3QP tel:(01494) 525252 fax:(01494) 461073 E_Mail:info@airflow.co.uk Web_Site:http://www.airflow.co.uk

Airmaster Engineering Ltd Limewood Approach Seacroft Leeds LS14 1NG tel:(0113) 273 3333 fax:(0113) 265 0735

Airsprung Beds Ltd Canal Road Trowbridge Wilts. BA14 8RQ tel:(01225) 754411 fax:(01225) 779123

Airstream Products Ltd Airstream House Brook Street Cheadle Cheshire SK8 2BN tel:(0161) 428 7544 fax:(0161) 428 7135

Akzo Coatings Plc 135 Milton Park Abingdon Oxon. OX14 4SB tel:(01235) 862226 fax:(01235) 862236 telex:833376

Akzo Nobel Decorating Coatings Ltd P.O. Box 37 Hollins Road Darwen Lancashire BB3 OBG tel:(01254) 704951 fax:(01254) 774414

Alarmcomm Ltd Baddow Park West Hangingfield Road Great Baddow, Chelmsford Essex CM2 7SY tel:(01245) 478585 fax:(01245) 478530

Albany Engineering Co Ltd Church Road Lydney Glos. GL15 5EQ tel:(01594) 842275 fax:(01594) 842574 E_Mail:sales@albany.pumps.co.uk Web_Site:http://www.albany.pumps.co.uk

Albion Concrete Products Pipehouse Wharf Morfa Road Swansea SA1 1TD tel:(01792) 655968 fax:(01792) 644461

Albion Fencing Ltd Albion House 2239 London Road Glasgow G32 8XN tel:(0141) 778 1672 fax:(0141) 778 6688

Albion Manufacturer's The Granary Silfields Road Windon Norfolk NR18 NAU tel:(01953) 605983 fax:(01953) 606764

Alcan Building Products Blackpole Trading Estate Worcester WR3 8TJ tel:(01905) 754030 fax:(01905) 754037

Alcatel Data Cable Felixstowe Road Abbey Wood London SE2 9AA tel:(0181) 557 3456 fax:(0181) 557 3535

Alcover UK Ltd 110 Gloucester Avenue Primrose Hill London NW1 8JA tel:(0171)483 2681 fax:(0171)209 5095 E_Mail:cmi@cmiltd.demon.co.uk Web_Site:http://www.cmiltd.co.uk

Aldous & Stamp Services Ltd 86-90 Avenue Road Beckenham Kent BR3 4SA tel:(0181) 659 1833 fax:(0181) 676 9676 E_Mail:sales@aldous-stamp.co.uk Web_Site:http://www.aldous-stamp.co.uk

Alexio Roofing & Building Co Ltd Alexio House 3 Blondin Street Bow London E3 2TR tel:(0181) 981 6080 fax:(0181) 981 4614 Web_Site:http://www.alexio.co.uk

Alfa Laval Saunders Ltd Grange Road Cwmbran Gwent NP44 3XX tel:(01633) 486666 fax:(01633) 486777

Alfas Industries Ltd Bentall Business Park Glover, District 11 Washington Tyne & Wear NE37 3JD tel:(0191) 419 0505 fax:(0191) 419 2200 E_Mail:alfas.ind@aol.com.uk Web_Site:http://www.alfas.com

Alframes Holdings Ltd 1A Arnold Road Tooting London SW17 9HU tel:(0181) 648 9394 fax:(0181) 648 4985

Alifabs (Woking) Ltd Forsyth Road Sheerwater Woking Surrey GU21 5SB tel:(01483) 755744 fax:(01483) 769529

Alkor Draka Ltd Odhams Trading Estate St Albans Road Watford Herts. WD2 5DG tel:(01923) 249511 fax:(01923) 227427

Allan Blunn Ltd Surrey Commercial Wharf 165 Rotherhithe Street London SE16 1QU tel:(0171) 232 2926 fax:(0171) 237 0154 telex:884176 BLUNN

Allan Harris & Sons Ltd Station Road St Georges Weston-super-Mare N. Somerset BS22 7XN tel:(01934) 511166 fax:(01934) 513066

Allart, Frank, & Co Ltd 15-35 Great Tindal Street Ladywood Birmingham B16 8DR tel:(0121) 454 2977 fax:(0121) 456 2234 E_Mail:sales@allart.co.uk Web_Site:http://www.allart.co.uk

Allaway Acoustics Ltd 1 Queens Road Hertford SG14 1EN tel:(01992) 550825 fax:(01992) 554982

Allen (Concrete) Ltd Govett Avenue Shepperton Middx. TW17 8AH tel:(01932) 224051

Allen (Fencing) Ltd Birch Walk West Byfleet Surrey KT14 6EJ tel:(01932) 349607 fax:(01932) 354868 E_Mail:allenfencingltd@btinternet.com

Allen P W & Co Ltd 25 Swan Lane Evesham Worcs. WR11 4PE tel:(01386) 40148 fax:(01386) 765351 E_Mail:sales@pwallen.co.uk Web_Site:http://www.inspection.pwallen.co.uk

Allermuir Contract Furniture Ltd Branch Road Lower Darwen Lancs. BB3 0PR tel:(01254) 682421 fax:(01254) 673793 E_Mail:sales@allermuir.co.uk.

Allgood plc 297 Euston Road London NW1 3AQ tel:(0171) 387 9951 fax:(0171) 380 1232 E_Mail:info@allgood.co.uk

Alliance Europe 2nd Floor Suite 21 Lower Street Kettering Northamptonshire NN16 8HE tel:(01536) 522473 fax:(01536) 410805

Allibert (UK) Ltd PO Box 66 St Andrews Square Droitwich Spa Worcs. WR9 8XE tel:(01905) 770469 fax:(01905) 771959

Allied Guilds Unit 19, Reddicap Trading Estate Coleshill Road Sutton Coalfield W. Midlands B75 7 BU tel:(0121) 329 2874 fax:(0121) 311 1883

Allied Manufacturing Co (London) Ltd Sarena House Grove Park London NW9 0EB tel:(0181) 205 8844 fax:(0181) 200 9510 telex:23719

Allied Reinforcements PO Box 41 Meadowhall Road Sheffield S9 1ED tel:(01742) 560152 fax:(01742) 821736

Allis Mineral Systems (UK) Ltd Netherton Road Wishaw Lanarkshire ML2 0EJ tel:(01698) 355921 fax:(01698) 351376 telex:777281

Allman Fasteners Ltd Unit 42, Heaton Mersey Industrial Estate Station Road Stockport Cheshire SK4 3QT tel:(0161) 442 6366 fax:(0161) 443 1755

Allmat (East Surrey) Ltd Kenley Treatment Works Godstone Road Kenley Surrey CR8 5AE tel:(0181) 668 6666 fax:(0181) 763 2110 E_Mail:allmat.demon.co.uk

Allpipe Ltd Vulcan Road South Norwich NR6 6AF tel:(01603) 488700 fax:(01603) 488598

Alltons (Structural Steel) Ltd PO Box 4 Ure Bank Rippon Yorks. HG4 1JE tel:(01765) 604351

Alltype Fencing Specialists Ye Wentes Wayes, High Road Langdon Hills Basildon Essex SS16 6HY tel:(01268) 545192 fax:(01268) 545192

Alno (United Kingdom) Ltd Unit 10, Hampton Farm Industrial Estate Hampton Road West Hanworth Middx. TW13 6DB tel:(0181) 898 4781 fax:(0181) 898 0268

Alpha Fry Ltd Tandem House, Marlowe Way Beddington Farm Road Croydon Surrey CR0 4XS tel:(0181) 665 6666 fax:(0181) 665 4695 E_Mail:sales@frytechnology.cooleson.com Web_Site:http://www.frytechnology.com

Alpha Glass Fibres Ltd Fitzherbert Road Farlington Industrial Estate Portsmouth PO6 1RU tel:(01705) 379990 fax:(01705) 210716

Alpha M & T Ltd Unit 2, Munro Drive Cline Road London N11 2LZ tel:(0181) 368 2230 fax:(0181) 368 2301

Alpha Therm United House Ltd United House Goldsel Road Swanley Kent BR8 8EX tel:(01322) 613924 fax:(01322) 662313 E_Mail:marketing@unitedhouse.co.uk

Alpine Autimation Ltd. Threemilestone Industrial Estate Threemilestone Truro Cornwall TR4 9LD tel:(01872) 279525 fax:(01872) 222150

Altecnic Ltd Airfield Industrial Estate Hixon Staffs. ST18 0PF tel:(01889) 271371 fax:(01889) 270577

Alto Cleaning Systems (UK) Ltd Bowerbank Way Gilwilly Industrial Estate Penrith Cumbria CA11 9BN tel:(01768) 868995 fax:(01768) 864713 E_Mail:sales@alto.uk.com

Altro Floors Works Road Letchworth Herts. SG6 1NW tel:(01462) 480480 fax:(01462) 480010

Alumasc Exterior Building Products Ltd White House Works Bold Road Sutton, St Helens Merseyside WA9 4JG tel:(01744) 820103 fax:(01744) 818997 E_Mail:info@abp-drainage.co.uk Web_Site:http://www.abp-drainage.co.uk

Alumasc Interior Building Product Limited Halesfield 19 Telford Shropshire TF7 4QT tel:(01952) 580590 fax:(01952) 587805 E_Mail:alumascinterior@compuserve.com

Aluminium Federation (Alfed) Ltd Broadway House Calthorpe Road Five Ways Birmingham B15 1TN tel:(0121) 456 1103 fax:(0121) 456 2274 E_Mail:alfed@alfed.org.uk Web_Site:http://www.alfed.org.uk

Aluminium RW Supplies Ltd Ryan House, Unit 6 Dumballs Road Cardiff CF1 6JE tel:(01222) 390576 fax:(01222) 238410

Amari Plastics Plc Holmes House 24-30 Baker Street Weybridge Surrey KT13 8AU tel:(01932) 835000 fax:(01932) 835001

Ambec Fencing Doctors Lane Eccleston Chorley Lancs. PR7 5QZ tel:(01257) 451412 fax:(01257) 450094

Amber Doors Ltd Mason Way, Platts Common Industrial Estate Hoyland, Barnsley S. Yorks. S74 9TG tel:(01226) 351135 fax:(01226) 350176

Amberol Ltd The Plantation 80 Spencer road Belper Derbys. D56 1JW tel:(01773) 823907 fax:(01773) 829445

Ambersil Ltd Wylds Road, Castlefield Industrial Estate Bridgwater Somerset TA6 4DD tel:(01278) 424200 fax:(01278) 425644 E_Mail:ambersil@btinternet.com Web_Site:http://www.ukindustry.co.uk/ambersil

Ambi-Rad Ltd Penspool Avenue Wallows Industrial Estate Brierley Hill W. Midlands DY5 1QA tel:(01384) 489700 fax:(01384) 489707 E_Mail:sales@ambirad.co.uk Web_Site:http://www.ambirad.co.uk.

Amcor (Appliances) Ltd Unit 2, Canal Court 152-156 High Street Brentford Middx. TW8 8JA tel:(0181) 560 4141 fax:(0181) 232 8814

Amdega Ltd Faverdale Darlington Co. Durham DL3 0PW tel:(01325) 489209 fax:(01325) 381708 E_Mail:info@amdega.co.uk Web_Site:http://www.amegda.co.uk

Amey Grilles & Doors Ltd Westminster Road Vauxhall Industrial Estate Canterbury Kent CT1 1TX tel:(01227) 456081 fax:(01227) 450977

Amitico 18 Hanover Square London W1R tel:(0171) 629 6258 telex:28788

AMK Fence-In Limited Wallace Road Parkwood Springs Sheffield Yorks. S3 9SR tel:(0114) 2739372 fax:(0114) 2739372

Anaplast Ltd 96 Port Glazgow Road Greenock Strathclyde PA15 2RP tel:(01475) 501100 fax:(01550) 00180

Ancon CCL Ltd President Way President Park Sheffield S. Yorks. S4 7UR tel:(0114) 275 5224 fax:(0114) 276 8543 E_Mail:info@anconccl.com Web_Site:http://www.anconccl.com

Anda Products Ltd Terminus Road Chichester W. Sussex PO19 2TR tel:(01243) 787943 fax:(01243) 780384

Anders + Kern UK Ltd Norderstedt House James Carter Road Mildenhall Suffolk IP28 7RQ tel:(01638) 510900 fax:(01638) 510901 E_Mail:sales@anders-kern.co.uk Web_Site:http://www.anders-kern.co.uk

Andersen/ Black Millwork Andersen House Dallow Street Burton-on-Trent Staffs. DE14 2PQ tel:(01283) 511122 fax:(01283) 510863 Web_Site:http://www.andersenwindows.com/international

Anderson C F & Son Ltd 228 London Road, Marks Tey Colchester CO6 1HD tel:(01206) 211666 fax:(01206) 212450 telex:987101

Anderson D & Son Barton Dock Road Stretford Manchester M32 0YL tel:(0161) 865 4444 fax:(0161) 864 1178 E_Mail:marketing@anderson-roofing.co.uk Web_Site:http://www.anderson-roofing.co.uk

Anderson GEC Ltd Oakengrove Shire Lane Hastoe Herts HP23 6LY tel:(01442) 826999 fax:(01442) 825999 E_Mail:gec@ndirect.co.uk Web_Site:http://www.ascwebindex.com/geca

Anderson Gibb & Wilson A Division of Charles Tennant & Co Ltd 543 Gorgie Road Edinburgh E11 3AR tel:(0131) 443 4556 fax:(0131) 455 7608

Andersons Ltd Hillview Enerprise Park Belfast BT14 7BZ tel:(01232) 741222 fax:(01232) 351440 E_Mail:anderson.doors.co.ltd@itnet

Anderton Concrete Products Ltd Anderton Wharf Soot Hill Anderton, Northwich Cheshire CW9 6AA tel:(01606) 79436 fax:(01606) 871590

Andrews J S (Coatings) Ltd 9a Walsall Street West Bromwich W. Midlands B70 7NX tel:(0121) 525 1080 fax:(0121) 500 5453

Andrews Sykes Hire Ltd Premier House Darlington Street Wolverhampton W. Midlands WV1 4JJ tel:(01902) 328700 fax:(01902) 422466 E_Mail:info@andrews-sykes.com Web_Site:http://www.andrews-sykes.com

Andura Textured Masonry Coatings Ltd 20 Murdock Road Bicester Oxon. OX6 7PP tel:(01869) 240374 fax:(01869) 240375

Andy Thornton Architectural Antiques Ltd Ainleys Industrial Estate Elland W. Yorks. HX5 9JP tel:(01422) 375595 fax:(01422) 377455 E_Mail:email@ataa.co.uk Web_Site:http://www.ataa.co.uk

Angle Ring Co Ltd Bloomfield Road Tipton W. Midlands DY4 9EH tel:(0121) 557 7241 fax:(0121) 522 4555 E_Mail:sales@anglering.co.uk Web_Site:http://www.anglering.co.uk

Anglepoise Lighting Ltd Unit 51 Enfield Industrial Area Redditch Worcs. B97 6DR tel:(01527) 63771 fax:(01527) 61232 telex:336918

Anglia Lead Ltd 49 Barker Street Norwich Norfolk NR2 4TN tel:(01603) 630979 fax:(01603) 619171

Anglian Daneshill Estate Whitney Road Basingstoke Hampshire RG24 8NS tel:(01256) 472247 fax:(01256) 346697

Anglo Building Products Ltd Branksome House Filmer Grove Godalming Surrey GU7 3AB tel:(01483) 427777 fax:(01483) 428888

Anglowest Distributors Ltd 1 Carlisle Road London NW9 0HZ tel:(0181) 205 7285 fax:(0181) 200 3741

Angus Fire Thame Park Road Thame Oxon. OX9 3RT tel:(01844) 214545 fax:(01844) 213511

Anixter (UK) Ltd 1 York Road Uxbridge Middx. UB8 1RN tel:(01895) 818181 fax:(01895) 818182

Anthony de Grey Trellises Broadhinton Yard 77a North Street London SW4 0HQ tel:(0171) 738 8866 fax:(0171) 498 9075

Anti Corrosion Services Ltd Carrington Business Park Carrington Urmston Manchester M31 4QW tel:(0161) 776 4019 fax:(0161) 775 8995 telex:667678

Antocks Lairn Ltd Lancaster Road, Cressex Industrial Estate High Wycombe Bucks. HP12 3HZ tel:(01494) 465454 fax:(01494) 465901

APA Youlditch Barns Peter Tavy Tavistock Devon PL19 9LY tel:(01822) 810187/8 fax:(01822) 810189 E_Mail:sales@alanpow.co.uk Web_Site:http://www.alanpow.co.uk

APE Crossley Ltd PO Box 1 Manchester M11 2DP tel:(01616) 223 1353 telex:688975

Apex Roofing Contractors Victor Lane Heavitree Exeter Devon EX1 3BJ tel:(01392) 466297 fax:(01392) 432202

Apollo Fire Detectors Ltd 36 Brookside Road Havant Hampshire PO9 1JR tel:(01705) 492412 fax:(01705) 492754 Web_Site:http://www.apollo-fire.co.uk

Apollo Space Systems Ltd Appolo House Wharf Road Pinxton Notts. NG16 6LF tel:(01773) 812800 fax:(01773) 861607

Applied Acoustics (a part of Henry Venables Limited) Castletown Stafford ST16 2EN tel:(01785) 259131 fax:(01785) 215087 E_Mail:enquiries@henryvenables.co.uk Web_Site:http://www.henryvenables.co.uk

Applied Chemicals Ltd Applied House Wilsons Lane Coventry CV6 6JA tel:(01203) 363575 fax:(01203) 366639

APT Controls Ltd The Power House, Chantry Place Headstone Lane Harrow Middx. HA3 6NY tel:(0181) 421 2411 fax:(0181) 421 2411 E_Mail:rachel@aptcontrols.co.uk Web_Site:http://www.aptcontrols.co.uk

APV Howard Pumps Ltd Fort Road Eastbourne E. Sussex BN22 7SE tel:(01323) 722804 fax:(01323) 648955 telex:87672

Aqua Design Ltd Gratton Way Roundswell Industrial Estate Barnstaple Devon EX31 3NL tel:(01271) 325825 fax:(01271) 371699 E_Mail:aquadesign@goldenc.com

Aqua-Blue Ltd 24 Hazlemere Gardens Worcester Park Surrey KT4 8AH tel:(0181) 337 5401 fax:(0181) 330 3964 Web_Site:http://www.teelframgroup@btinternet

Aqualisa Products The Flyer's Way Westerham Kent TN16 1DE tel:(01959) 560000 fax:(01959) 560030 E_Mail:marketing@aqualisa.co.uk Web_Site:http://www.aqualisa.co.uk

Aquastat Ltd Aquastat House, Tudor Works Beaconsfield Road Hayes Middx. UB4 0SL tel:(0181) 848 8811 fax:(0181) 756 0841 E_Mail:david_a.garland@msn.com

Aquatanks Unit 3, Westcombe Trading Estate Station Road Illminster Somerset TA19 9DW tel:(01460) 55664 fax:(01460) 53338

Arbion manufacturers The Granary Seelfield Road Windon Norfolk NR18 NAV tel:(01953) 605983 fax:(01953) 606764

ARC Partitioning 212-214 Great Portland Street London W1N 5HG tel:(0171) 637 8156 fax:(0171) 631 3721

ARC Conbloc PO Box 14 Appleford Road Sutton Courtenay, Abbingdon Oxon. OX14 4UB tel:(01235) 848877 fax:(01235) 848767

ARC Northern Clifford House York Road Wetherby W. Yorks. LS22 4NS tel:(01937) 581977 fax:(01937) 581610

ARC Pipes Mells Road Mells Frome Somerset BA11 3PD tel:(01179) 812791 fax:(01179) 814516 E_Mail:sales@arc-pipes.co.uk

Architectural & Building Products Ltd Unit 4, Ponders End Indstrial Estate Ducks Lees Lane Enfield Middx. EN3 7SP tel:(0181) 805 4444 fax:(0181) 805 0022

Architectural AssociationThe 34 Bedford Square London WCIB 3ES tel:(0171)636 0974

Architectural Plastics (Handrail) Ltd Unit One 2 Robert Street Harrogate N. Yorks. HG1 1HP tel:(01423) 561852 fax:(01423) 520728

Architectural Textiles Ltd Units 13 & 14, Heckford Business Centre Heckford Street London E1 9HS tel:(0171) 790 2902 fax:(0171) 790 3699

Arden Manufacturing Birmingham Ltd 47-50 Tenby Street North Birmingam B1 3EG tel:(0121) 693 1818 fax:(0121) 693 1819 E_Mail:info@arden-lighting.co.uk Web_Site:http://www.arden-lighting.co.uk

Ardex UK Ltd Homefield Road Haverhill Suffolk CB9 8QP tel:(01440) 714939 fax:(01440) 703424

Areco Roofing Supplies Company Unit 3, Weston Works Weston Lane Tyseley Birmingham B11 3RP tel:(0121) 706 4909 fax:(0121) 707 3031

Aremco Products Foxoak Street Cradley Heath W. Midlands B64 5DQ tel:(01384) 68566 fax:(01384) 638919

Arena Software Ltd 3 Rayleigh Close Cambridge CB2 2AZ tel:(01223) 302220 fax:(01223) 460920 E_Mail:arena@arenasoftware.co.uk. Web_Site:http//www.arenasoftware.co.uk.

Ariel Plastics Ltd Speedwell Ind. Est. Staveley Derbys. S34 3JP tel:(01246) 561122 fax:(01246) 561111

Arjo Ltd St Catherine Street Gloucester GL1 2SL tel:(08702) 430430 fax:(01452) 525207

Arjo Ltd St Cathrine Street Gloucester GL1 2SL tel:(01452) 500200 fax:(01452) 525207

Arkinstall Galvanizing Ltd 38 Coventry Street Birmingham W. Midlands B5 5NQ tel:(0121) 643 6455 fax:(0121) 643 0192 E_Mail:arkinstall@msn.com Web_Site:http://www.galvanizing.com

Armes, Williams, Ltd Armes Trading Estate Cronard Road Sudbury Suffolk CO10 6XB tel:(01787) 372988 fax:(01787) 379383

Armfibre Ltd Drove Road Everton Sandy Beds. SG19 2HX tel:(01767) 651811 fax:(01767) 651901 E_Mail:sales@armfibre.u-net.com Web_Site:http://www.armfibre.u-net.com

Armitage Shanks Ltd Armitage Rugeley Staffs. WS15 4BT tel:(01543) 490253 fax:(01543) 491677

Armorex Ltd Riverside House Bury Road lavenham Suffolk CO10 9QD tel:(01787) 248482 fax:(01787) 248277 E_Mail:sales@armorex.co.uk

Armstrong (Concrete Blocks), Thomas, Ltd Whinfield Industrial Estate Rowlands Gill Tyne & Wear NA39 1EH tel:(01207) 544214 fax:(01207) 542761

Armstrong (Timber), Thomas, Ltd Workington Road Flimby, Maryport Cumbria CA15 8RY tel:(01900) 68226 fax:(01900) 870800

Armstrong Concrete Products Ltd Risehow Industrial Estate Firmby, Maryprot Cumbria CA15 8PD tel:(01900) 814659 fax:(01900) 816200

Armstrong Pumps Ltd Peartree Road Stanway, Colchester Essex CO3 5JX tel:(01206) 579491 fax:(01206) 760532

Armstrong World Industries Ltd Armstrong House 38 Market Square Uxbridge Middx. UB8 1NG tel:(0800) 371849 fax:(01895) 274287

Armstrong World Industries Ltd Fleck Way Teeside Industrial Estate Thornaby on Tees Cleveland TS17 9JT tel:(01642) 763224 fax:(01642) 750213

Arnold Wragg Ltd. Bradley Street Sandiacre Nottingham NG10 5AJ tel:(0115) 939 4646 fax:(0115) 939 8321

Arnull Bernard J & Co Ltd 17-21 Sunbeam Road Park Royal London NW10 6JP tel:(0181) 965 6094 fax:(0181) 961 1585 E_Mail:bernard.arnill@easynet.co.uk.

Arrow Supply Company Ltd Sunbeam Road Woburn Industrial Estate Kempston Beds. MK42 7BZ tel:(01234) 840404 fax:(01234) 840374 E_Mail:information@arrow-supply.co.uk Web_Site:http://www.arrow-supply.co.uk

Artex Ltd Artex Avenue Newhaven E. Sussex BN9 9DD tel:(01273) 513100 fax:(01273) 513100

Arthur Fisher (UK) Ltd Hithercroft Trading Estate Wallingford Oxon OX10 9AT tel:(01491) 833000 fax:(01491) 827953 E_Mail:info@fischer.co.uk Web_Site:http://www.fisher.co.uk

Artistic Plastercraft Lyndhurst Studios 16-18 Lyndhurst Road, Oldfield Park Bath Avon BA2 3JH tel:(01225) 315404 fax:(01225) 315404

Arts Council 14 Great Peter Street London SW1P 3NQ tel:(0171) 333 0100 fax:(0171) 973 6590 Web_Site:http://www.arts council.org.uk

AS Newbould Ltd 19 Tarran Way West Tarran Way Industrial Estate Moreton, Wirral Merseyside L46 4TT tel:(0151) 677 6906 fax:(0151) 678 0680 E_Mail:newbould@btinternet.com

Asbestos Removal & Demolition Services 20 Brown Street Dundee DD1 5ED tel:(01382) 225767 fax:(01382) 203659

Asco Extinguishers Co Ltd Melisa House Unit 3, Festival Court, Brand Street Glasgow G51 1DR tel:(0141) 427 1144 fax:(0141) 427 6644

Ascot Industrial Doors Ltd Unit 16, Bleak Hill Way Hermitage Lane Mansfield Notts. NG18 5EZ tel:(01623) 422966 fax:(01623) 424346

Ascot Industrial Doors Ltd Britannia Way Industrial Park Union Road Bolton Lancs. BL2 2HE tel:(0990) 556644 fax:(01204) 545800 E_Mail:100613.3133@compuserv.com

Ascot Lamps & Lighting Ltd Unit 4, Pedham Place Estate Wested Lane Swanley Kent BR8 8TE tel:(01322) 667334 fax:(01322) 614418

Ash & Lacy Building Products Ltd Unit 5, Shaw Street Hill Top Ind Estate West Bromwich W.Mids B70 OTX tel:(0121) 556 1444 fax:(0121) 556 0444 E_Mail:name@ash-and-lacy.demon.co.uk

Ash Joseph Galvanizing PO Box 16 Charles Henry Street Birmingham B12 0SP tel:(0121) 622 4661 fax:(0121) 622 4317

Ashby & Horner Joinery Ltd 795 London Road West Thurrock Grays Essex RM16 1LH tel:(01708) 866841

Ashford Marine & Industrial Ltd Ashford Works 9 Ashford Road Fordingbridge Hampshire SP6 1DA tel:(019425) 52422

Ashley & Rock Ltd Morecambe Road Ulverston Cumbria LA12 9BN tel:(01229) 583333 fax:(01229) 587659

Ashmead Buidling Supplies Ltd Portview Road Avonmouth Bristol BS11 9LD tel:(0117) 982 8281 fax:(0117) 982 0135 Web_Site:http://www.ashmead.co.uk

Ashmead Building Supplies Ltd Devon Distribution Centre Willand Cullompton Devon EX15 2QW tel:(01884) 820078 fax:(01884) 820040 Web_Site:http://www.ashmead.co.uk

Ashton Seals Ltd PO Box 133, Speedwell Works Sidney Street Sheffield S. Yorks. S1 3QB tel:(0114) 2766770 fax:(0114) 2723748

Ashworth Diecasting Sycamore Avenue Burnley Lancs. BB12 6QR tel:(01282) 439911 fax:(01282) 453293 E_Mail:castings@ashworth-diecasting.co.uk Web_Site:http://www.ashworth-dieasting.co.uk

ASSA Ltd 75 Sumner Road Croydon Surrey CR0 3LN tel:(0181) 688 5191 fax:(0181) 688 0285 E_Mail:sales@assa.co.uk Web_Site:http://www.assa.co.uk

Asset International Ltd Stephenson Street Newport Gwent NP9 0XH tel:(01633) 271906 fax:(01633) 290519 E_Mail:postbox@assetint.co.uk Web_Site:http://www.assetint.co.uk

Assi Doman Sacks (UK) Medway House New Hythe Lane Aylesford Kent ME20 6SH tel:(01622) 717855 fax:(01622) 716360 E_Mail:sales.sacksuk@asdo.com Web_Site:http://www.asdo.com

Associated Holdings Ltd Highlands Lane Henley-on-Thames Oxon. RG9 4PS tel:(01491) 575921 fax:(01491) 579713 E_Mail:marketing@associatedholdings.com Web_Site:http://www.associatedholdings.com

Associated Metal (Stainless) Ltd 101 Brook Street Glasgow G40 3AP tel:(0141) 551 0707 fax:(0141) 551 0690 E_Mail:info@assoc-metal.co.uk

Associated Perforators & Weavers Ltd PO Box 75 Church Street Warrington WA1 2SN tel:(01925) 632402 fax:(01925) 413810 E_Mail:sales@apw.co.uk

Association of Cost Engineers, The Administrative Office Lea House, 5 Middlewich Road Sandbach Cheshire CW11 1XL tel:(01270) 764798 fax:(01270) 766180 E_Mail:a.coste.@btinternet.com

Association of Professional Foresters of Great Britain 7-9 West Street Belford Northumberland NE70 7QA tel:(01668) 213937 fax:(01668) 213555 E_Mail:jane@apfs.demon.co.uk Web_Site:http://www.apfs.demon.co.uk/whyjoin/homepage.htm

Astolat Co Ltd Portsmouth Road Peasmarsh Guildford Surrey GU3 1NE tel:(01483) 575211 fax:(01483) 578094

Astor-Stag Ltd Tavistock Road West Drayton Middx. UB7 7RA tel:(01895) 445511 fax:(01895) 449199 telex:28559 E_Mail:astorwd@aol.com Web_Site:http://www.astorcorp.com

Astracast PLC PO Box 20 Spring Ram Business Park Birstall W. Yorks. WF17 9XD tel:(01924) 477466 fax:(01924) 475801 E_Mail:marketing@astra.co.uk

Astrofade Ltd Kyle Road Gateshead Tyne & Wear NE8 2YE tel:(0191) 420 0515 fax:(0191) 460 4185 E_Mail:astrofade@aol.com

Astron Building Systems Commercial Intertech Ltd Tachbrook Park Drive Tachbrook Park Warwick CV34 6TU tel:(01926) 888080 fax:(01926) 885088

Atcost Building Ltd Spa House 18 Upper Grosvenor Road Tunbridge Wells Kent TN1 2EP tel:(01892) 526288 fax:(01892) 515348

Atkinson & Kirby Ltd 81 Wigan Road Ormskirk Lancs. L39 2AR tel:(01695) 573234 fax:(01695) 573859

Atlas Fire Engineering Ltd 67A Boston Manor Rd. Middx. TW8 9JQ tel:(0181) 570 8805 fax:(0181) 577 2692

Atlas Stone Co Ltd Harbour Road Rye E. Sussex TN31 7TE tel:(01797) 223955

Atlas Stone Products Westington Quarry Chipping Campden Glos. GL55 6EG tel:(01386) 841104 fax:(01386) 841356

Atlas Ward Structures Ltd Sherburn, Malton N. Yorks. YO17 8PZ tel:(01944) 710421 fax:(01944) 710759 E_Mail:enquiries@atlasward.com Web_Site:http://www.atlasward.com

Attwater & Sons Ltd Hopwood Street Mills Preston Lancs. PR1 1TH tel:(01772) 258245 fax:(01772) 203361 telex:67105 E_Mail:info@attwater.co.uk

Auchard Development Co Ltd Old Road Southam, Leamington Spa Warwks. CV33 0HP tel:(01926) 812419 fax:(01926) 817425

Auckland Construction Ltd Mill Lane Arlesey Beds SG15 6RF

Auld Valves Ltd Cowlairs Industrial Estate Finlas Street Glasgow Scotland G22 5DQ tel:(0141) 557 0515 fax:(0141) 558 1059 telex:77262

Auto-Klean Filtration Ltd Lascar Works Hounslow Middx. TW3 3JL tel:(0181) 570 7722 fax:(0181) 570 4438 telex:24789

Automatic Doors & Gates 18a St Johns Road Isleworth Middx. TW7 6NW tel:(0181) 568 6781 fax:(0181) 847 2682

Automatic Safety Lighting Ltd (T/as 'Stapoco') 311 Lidgett Lane Leeds 17 Yorks. LS17 6PD tel:(01132) 682682 fax:(01132) 682682 Web_Site:http://www.yorkshirenet.co.uk/businfo/stapco

Aveling-Barford (Machines) plc Houghton Road Grantham Lincs. NG31 6JE tel:(01476) 5551 fax:(01476) 79947 telex:377861

Avesta Sheffield Ltd Stelco Hardy Blaenrhondda Treorchy M. Glam. CF42 5BY tel:(01443) 771774 fax:(01443) 776009 E_Mail:lyn.curtis@avestasheffield.com

Avocet Hardware Ltd Brookfoot Mills Elland Road Brighouse W. Yorks. HD6 2RW tel:(01484) 711700 fax:(01484) 720124

Axter Ltd Cliff Road Ipswich Suffolk IP3 0AY tel:(01473) 217154 fax:(01473) 232118 E_Mail:mail@axtor.u-net.c0m

Azimex Fabrications Ltd Cartwright House Cartwright Road Kingsthorpe Northampton NN2 6HF tel:(01604) 717712 fax:(01604) 791087

B & D Clays & Chemicals Ltd 10 Wandle Way Willow Lane Trading Estate Mitcham Surrey CR4 4TE tel:(0181) 640 9221 fax:(0181) 648 5033 telex:945938

B & M Fencing Limited Kingsbridge Copse Newnham Road Hook Hampshire RG27 9AE tel:(01256) 762739 fax:(01256) 763895

B & M Laminaites PO Box 9 Alfreton Derbys. DE55 5ZZ tel:(01773) 812608 fax:(01773) 812608

B-Trac Equipment Ltd 45-51 Rixon Road Wellingborough Northamptonshire NN8 4BA tel:(01933) 274400 fax:(01933) 274403

Babcock Joinery Rosyth Royal Dockyard Bell Road Rosyth Fife KY11 2YD tel:(01383) 423049 fax:(01383) 423727

Babcock Transformers Oxford Street Bilston W. Midlands WV14 7DL tel:(01902) 492681 fax:(01902) 491116

BAC Ltd Edingburghs Drive Eastern Avenue (West) Romford Essex RM7 7PX tel:(01708) 724824 fax:(01708) 728114

Bacon WM & RW Ltd Walter House, Wickford Business Park Hodgson Way Wickford Essex SS11 8YG tel:(01268) 561035 fax:(01268) 561036 E_Mail:sales@wmbacon.co.uk

Badderley Rose Ltd Marshmoor Works Great North Road North Mymms, Hatfield Herts. AL9 5SD tel:(01707) 257689 fax:(01707) 257690

Baggeridge Brick PLC Fir Street Sedgley Dudley W. Midlands DY3 4AA tel:(01902) 880555 fax:(01902) 880432 Web_Site:http://www.Baggeridge.co.uk

Bain WM & Co Fencing (1990) Ltd Lochrin Works Waverley Street Coatbridge Lanarkshire ML5 2BB tel:(01236) 423471 fax:(01236) 435097

Bainbridge Engineering Ltd Woodhill Road Bury Lancs. BL8 1BW tel:(0161) 764 5034 fax:(0161) 764 5020

Baker Fencing Limited Hillside Grange, Warren Road Trellech Monmouth Gwent NP5 4PQ tel:(01600) 860600 fax:(01600) 860888 E_Mail:baker_fencing@msn.com

Ball David Group plc Huntington Road Bar Hill Cambridge CB3 8HN tel:(01954) 780687 fax:(01954) 782912 telex:817213 BALLCO G

Ball F & Co Ltd Churnetside Business Park Station Road Cheddleton, Leek Staffs. ST13 7RS tel:(01538) 361633 fax:(01538) 361622 E_Mail:webmaster@f-ball.co.uk. Web_Site:http://www.f-ball.co.uk

Ball William Ltd Haydock Industrial Estate Bahama Road Haydock Merseyside WA11 9XB tel:(01942) 270000 fax:(01942) 270225

Ball William Ltd Gumley Road Grays Essex RM20 4WB tel:(01375) 375151 fax:(01375) 393355 E_Mail:marketing@wball.co.uk Web_Site:http://www.wballl.co.uk

Ballast Wiltshier PLC, Construction Specialist Prospect House 19-21 Holmesdale Road Bromley Kent BR2 9LY tel:(0181) 464 4111 fax:(0181) 466 5911

Ballentine, Bo'Ness Iron Co Ltd Links Road Bo'ness Scotland EH51 9PW tel:(01506) 822721 fax:(01506) 827326 E_Mail:ballantine@sol.co.uk

Ballofix Valves Ltd Bishops Gate Works 68 Lower City Road Tividale, Warley W. Midlands B69 2HF tel:(0121) 552 5281 fax:(0121) 552 6895

Balmforth Engineering Ltd Unit 5-7, Finway Dallow Road Luton Beds. LU1 1TR tel:(01582) 455115 fax:(01582) 453569

Balmoral Group Ltd Balmoral Park Loirston Aberdeen AB12 3GY tel:(01224) 859000 fax:(01224) 859059 E_Mail:group@balmoral.co.uk Web_Site:http://www.balmoral-group.com

Balmoral Porcelain Limited/Longmead Ceramics Millwey ind. Est. Axminster Devon EX13 5HU tel:(01297) 32578 fax:(01297) 32710

Baltcon Ltd Tor Hill Concrete Works Tor Hill Wells Somerset BA5 3NT tel:(01749) 675757 fax:(01749) 677750

Bamford J C Excavators Ltd Rocester Staffs. ST14 5JP tel:(01889) 590312 fax:(01889) 590588 telex:36372JCB ROC G

Banafix Limited Banafix House Hillbottom Road High Wycombe Bucks. HP12 4HJ tel:(01494) 539898 fax:(01494) 539191

Banbury Windows Ltd Long Bank Bewdley Worcs. DY12 2UJ tel:(01299) 266332

Barber Edward & Co Ltd Paxton Road Tottenham London N17 0BS tel:(0181) 808 5161 fax:(0181) 801 5718

Barber Wilsons & Co Ltd Crawley Road London N22 6AH tel:(0181) 888 3461 fax:(0181) 8882041 E_Mail:barwil@compuserve.com

Barbour Enquiry Service New Lodge Drift Road Windsor SL4 4RQ tel:(01344) 884999 fax:(01344) 899377

Bardon (England) Ltd Bardon Hill Coalville Leics. LE67 1TL tel:(01530) 510066 fax:(01530) 510123

Bardsleys-Colchester Ltd 196 Bergholt Road Colchester Essex CO4 5AL tel:(01206) 853670 fax:(01206) 845829

Barduct Ltd Gatehouse Close Aylesbury Bucks HP19 3DJ tel:(01296) 339388 fax:(01296) 339969 E_Mail:elainec@barduct.co.uk. Web_Site:http://www.barduct.co.uk

Baresford Pumps Carlton Road Foleshill Coventry CV6 7FL tel:(01203) 638484 fax:(01203) 637891

Barflow-LCA 116 London Road Hailsham E. Sussex BN27 3AL tel:(01765) 690690

Barker & Co (Leeds) Ltd George Highfield Works Highfield Road Bradford Yorks. BD10 8RF tel:(01274) 611141 telex:517326 GEOBAR G

Barker & Geary Limited The Yard, Romsey Road, Kings Somborne Nr Stockbridge Hampshire SO20 6PW tel:(01794) 388205 fax:(01794) 388205

Barker Bros Aggregates Ltd The Green Downham Market Norfolk PE38 9DY tel:(01366) 382525 fax:(01366) 383002

Barker W H Ltd Etna Works Duke Street Fenton Stoke-on-Trent ST4 3NS tel:(01782) 319264 fax:(01782) 599724

Barlow Architectural & Security Ltd Broadfield Road Sheffield S8 0XU tel:(01742) 556222 fax:(01742) 589220

Barlow Shopfitting Ltd 136 London Road Sheffield S. Yorks. S2 4NX tel:(0114) 255 6331 fax:(0114) 258 9627 Web_Site:http://www.barlowgroup.co.uk

Barlow Tyrie Ltd Braintree Essex CM7 7RN tel:(01376) 322505 fax:(01376) 347052 Web_Site:http://www.teak.com

Barna Buildings Ltd Dublin Road Enniscorthy Co Wexford Ireland

Barnwood Shopfitting Ltd 203 Barnwood Road Gloucester GL4 3HT tel:(01452) 614124 fax:(01452) 372933 E_Mail:barnwood@shopfitting.demon.co.uk

Barr & Wray Ltd 324 Drumoyne Road Glasgow G51 4DY tel:(0141) 882 9991 fax:(0141) 882 3690 E_Mail:carey@barlandwray.com

Barrow Hepburn Sala Ltd 4 Old Mill Road Portishead Avon BS20 9BX tel:(01275) 846119 fax:(01275) 8409114

Bartlett Catering Equipment Ltd Maylands Avenue Hemel Hempsted Herts. HP2 7EN tel:(01442) 284284 fax:(01442) 231265 telex:825093

Bartoline Ltd Barmston Close Beverley E. Yorks. HU17 0GL tel:(01482) 882185 fax:(01482) 872606 E_Mail:info@bartoline.co.uk Web_Site:http://www.bartoline.co.uk

Barton B C & Son Ltd PO Box 67 No 1 Hainge Road Tividale, Oldbury W. Midlands B69 2NJ tel:(0121) 557 2272 fax:(0121) 557 2276 E_Mail:sales@b-c-b.co.uk Web_Site:http://www.b-c-b.co.uk

Barton Engineering Birchills Walsall WS2 8QE tel:(01922) 626581 fax:(01922) 646675

Barton Storage Systems Ltd Barton Industrial Park Mount Pleasant, Bilston W. Midlands WV14 7NG tel:(01902) 499500 fax:(01902) 353098 E_Mail:email@barton - industrial. Co.uk Web_Site:http://www.barton-indusrial.co.uk

Bassaire Ltd Duncan Road Park Gate Southampton SO31 1ZS tel:(01489) 885111 fax:(01489) 885211 E_Mail:bassaire@compuserve.com

Bassett & Findley Ltd Talbot Road North Wellingborough Northamptonshire NN8 1QS tel:(01933) 224898 fax:(01933) 227731 E_Mail:info@basseltandfinleyltd.ok

Bauer Foundations Limited Bauer House Woodrow Way, Fairhill Industrial Estate Irlam Manchester M44 6ZQ tel:(0161) 777 4400 fax:(0161) 776 2446 E_Mail:bauerltd@acol.com

Bawn W. B. & Co Ltd Northern Way Bury St Edmunds Suffolk IP32 6HN tel:(01284) 752812 fax:(01284) 752844

Baxall Ltd Unit 1, Castlehill Horsefield Way, Bredbury Park Industrial Estate Stockport Cheshire SK6 2SU tel:(0161) 406 6611 fax:(0161) 406 8988 E_Mail:info@baxall.com Web_Site:http://www.baxall.com

Baxenden Chemicals Ltd Paragon Works Baxenden Nr Accrington Lancs. BB5 2SL tel:(01254) 872278 fax:(01284) 871 247 E_Mail:mail@baxchem.co.uk Web_Site:http://www.baxchem.co.uk.

Baydale Architectural Systems Ltd 14 Northfield Way, Aycliffe Industrial Estate Newton Aycliffe Co. Durham DL5 6EJ tel:(01325) 307030 fax:(01325) 308030 E_Mail:baydale@onyxnet.co.uk

Bayer Plc Bayer House Strawberry Hill Newbure Berks. RG14 1JA tel:(01635) 563000 fax:(01635) 563135

BB & EA Ltd 63-65 London Road Sandy Beds SG19 1DJ tel:(01767) 680291 fax:(01767) 691288

Be-Modern Ltd Western Approach South Shields Tyne & Wear NE33 5QZ tel:(0191) 455 3571 fax:(0191) 456 5556

Beacon Hill Brick Co Ltd Beacon Hill Lane Wareham Road Corfe Mullen,Wimborne Dorset BH21 3RX tel:(01202) 697633 fax:(01202) 605141

Beacon Machine Tools Ltd Mission Works Purdy Road Bilston W. Midlands WV4 8UB tel:(01902) 493331 fax:(01902) 493241

Beaker Acroma Ltd Landywood Lane Cheslyn Hay Walsall W. Midlands WS6 7AL tel:(01922) 410101 fax:(01992) 419563

Beama Ltd Westminster Tower 3 Albert Embankment London SE1 7SL tel:(0171) 793 3000 fax:(0171) 793 3003

Beasley Joiners Ltd Famley Low Mills Whitehall Road Leeds LS12 5PS tel:(0113) 263 0524 fax:(0113) 279 2389

Beaumont Ltd F E Woodlands Road Mere Wilts. BA12 6BT tel:(01747) 860481 fax:(01747) 861076 E_Mail:sales@beaumont-chimneys.co.uk Web_Site:http://www.beaumont-chimneys.co.uk

Beaver Industrial Doors Ltd PO Box 1772 Yate Bristol BS17 5FW tel:(01454) 325632 fax:(01454) 323099

Becker (SLIDING PARTITIONS) Ltd Wemco House 477 Whippendell Road Watford Herts WD1 7PS tel:(01923) 236906 fax:(01923) 230149

Bedford Fencing Co. Limited High Street Works Lye Stourbridge W. Midlands DY9 8NF tel:(01384) 422668 fax:(01384) 422688

Beeston Heating Group Ltd Derwentside Industrial Park Derby Road Belper Derbys. DE5 1UX tel:(01773) 828383 fax:(01773) 829091 E_Mail:infoatbeestonheating.co.uk

Beford & Soar Ltd David Road Poyle Trading Estate Colnbrook, Slough Berks. SL3 0DB tel:(01753) 680666 fax:(01753) 680520 E_Mail:bedfords@btinternet.com Web_Site:http://www.bedfordshelf.co.uk

Beiersdorf (UK) Ltd Tesa Division Yeomans Drive Blakelands, Milton Keynes Bucks. MK14 5LS tel:(01908) 211333 fax:(01908) 211555 telex:825598

Bekaert Building Products PO Box 119 Shepcote Lane Sheffield S9 1TY tel:(0114) 256 1561 fax:(0114) 261 1529

Beldam Lascar Seals Ltd Lascar Works Hounslow Middx. TW3 3JL tel:(0181) 570 7722 fax:(0181) 570 4438

Bell & Co Ltd Kindsthorpe Road Northampton NN2 6LT tel:(01604) 712505 fax:(01604) 721028

Bell & Webster Concrete Ltd Alma Park Road Grantham Lincs. NG31 9SE tel:(01476) 562277 fax:(01476) 562944 E_Mail:bellandwebster.co.uk Web_Site:http://www.m.scott@bellandwebster.co.uk

Belzona Ltd Claro Road Harrogate N. Yorks. tel:(01423) 567641 fax:(01423) 505967

Benfield Fencing Limited Unit 22 Philadelphia Workshops Houghton-Le-Spring Tyne & Wear DH4 4TG tel:(0191) 584 7272 fax:(0191) 584 9222

Bennett Windows Ltd Park Road Ratby Leics. LE6 0JL tel:(0116) 239 5353 fax:(0116) 238 7295

Bennett Jr Ltd Ben Lisle Road Rotherham Yorks. S60 2RL tel:(01709) 382251 fax:(01706) 369206

Benson Industries Ltd Unit 5, Norcroft Industrial Estate Norcroft Street Bradford W. Yorks. BD7 1JA tel:(01274) 722204 fax:(01274) 306319

Benton Co Ltd Edward Creteangle Works Brook Lane Ferring, Worthing W. Sussex BN12 5LP tel:(01903) 241349 fax:(01903) 700213 E_Mail:benton@creteangle.co.uk Web_Site:http://www.creteangle.co.uk

Berglen Group Ltd Unit 1, Kingsbury Trading Estate Barningham Way Kingsbury London NW9 8AU tel:(0181) 205 1133 fax:(0181) 200 0074

Berkeley Co Ltd 10-13 Southview Park Marsack Street Caversham Reading RG4 5AF tel:(0118) 9483500 fax:(01734) 462114 telex:849021 FRAN G

Bernhard's Sports Surfaces Ltd Bilton Road Rugby Warwks CV22 7DT tel:(01788) 811500 fax:(01788) 816803

Bernhards Landscapes Ltd Bilton Road Rugby Warwks. CV22 7DT tel:(01788) 811500 fax:(01788) 816803

Bernlite Ltd 3 Brookside Colne Way Watford Herts. WD2 4QJ tel:(01923) 200160 fax:(01923) 246057 E_Mail:sales@bernlite.co.uk

Bernstein Group PLC Silburn House Great Bank Road Westhoughton Bolton BL5 3XU tel:(01942) 840840 fax:(01942) 840084

Besam Ltd Washington House Brooklands House Sunbury on Thames Middx. TW16 7EQ tel:(01932) 765888 fax:(01932) 812235 telex:946231

Best & Lloyd Ltd Cambray Works William Street West Smethwick, Warley W. Midlands B66 2NX tel:(0121) 558 1191 fax:(0121) 565 3547

Beta Lighting Ltd 383 /387 Leeds Road Bradford W. Yorks. BD3 9LZ tel:(01274) 721129 fax:(01274) 305007 E_Mail:beta@betalighting.com Web_Site:http://www.betalighting.com

Beton Construction Ltd PO Box 11 Basingstoke Hampshire RG21 8EL tel:(01256) 353146 fax:(01256) 840621

Betta Fencing Co Green Gates Yard Lundy Lane off Wilson Road Reading Berks. RG30 2RR tel:(01189) 502282 fax:(01189) 588259

BetzDearborn Ltd Foundry Lane Widnes Cheshire WA8 8UD tel:(0151) 424 5351 fax:(0151) 423 2722

Bevan Funnell Ltd Beach Road Newhaven E. Sussex BN9 0BZ tel:(01273) 513762 fax:(01273) 516735

BFRC Services Ltd 177 Bagnall Road Basford Nottingham NG6 8SJ tel:(0115) 942 4200 fax:(0115) 942 4488 E_Mail:jim@bfrc.demon.co.uk Web_Site:http://www.roofinguk.com

BICC plc Devonshire House Mayfair London W1X 5FH tel:(0171) 629 6622 fax:(0171) 409 0070

BICC Thermoheat Ltd Hedgeley Road Hebburn Tyne & Wear NE31 1XR tel:(0191) 483 2244 fax:(0191) 483 4127 E_Mail:sales@biccthermoheat.co.uk Web_Site:http://www.biccthermoheat.co.uk

Bicester Products Ltd 55 West End Witney Oxon. OX8 6NJ tel:(01993) 774426 fax:(01993) 779569

Biddle Air Systems Ltd St Mary's Road Nuneaton Warwks. CV11 5AU tel:(01203) 384233 fax:(01203) 373621 E_Mail:sales@biddle-air.co.uk Web_Site:http://www.biddle.air.co.uk

Bighead Bonding Fasteners Ltd Unit 15-16 Elliott Road West Howe Ind Estate Bournemouth Dorset BH11 8LZ tel:(01202) 574601 fax:(01202)

578300 E_Mail:anybody@bighead.co.uk Web_Site:http://www.bighead.co.uk

Bilco UK Ltd 3 Park Farm Business Centre Fornham St Genevieve Bury St Edmunds Suffolk IP28 6TS tel:(01284) 701696 fax:(01284) 702531 Web_Site:http://www.bilco.com

Bill Circuit Protection and Control Aston Lane Perry Barr Birmingham B20 3BT tel:(0121) 332 3000 fax:(0121) 356 1962 E_Mail:sales@billcpc.com Web_Site:http://www.billcpc.com

Billericay Fencing Ltd Morbec Farm Arterial Road Wickford Essex SS12 9JF tel:(01268) 727712 fax:(01268) 590225

Billington Structures Limited Barnsley Road Wombwell, Barnsley S Yorks. S73 8DS tel:(01226) 340666 fax:(01266) 755947 E_Mail:postroom@billington-structures.co.uk Web_Site:http://www.billington-structures

Binder Old Ipswich Road Claydon, Ipswich IP6 0AG tel:(01473) 830582 fax:(01473) 832175

Binks Bullows Ltd Pensall Road Brownhills Wallsall WS9 7HW tel:(01543) 372571 fax:(01543) 360702

Binns Fencing Ltd Harvest House Cranborne Road Potters Bar Hertfordshire EN6 3JF tel:(01707) 855555 fax:(01707) 857565

Biral Pumps Ltd Derwentside Industrial Park Derby Road Belper Derbys. DE56 1UX tel:(01773) 821020 fax:(01773) 829091 E_Mail:info@beeston-heating.co.uk

Birchwood Concrete Products Ltd Birchwood Way Cotes Park Industrial Estate Somercotes Derbys. DE55 4NH tel:(01773) 602432 fax:(01773) 603134 telex:377106 CHAMCOM E_Mail:admin@birchwood.demon.co.uk

Bird Howard & Co Ltd Manor Works 168 Worcester Road Bromsgrove Worcs. B61 7AZ tel:(01527) 575832 fax:(01527) 833466

Birdsall Services Ltd 6 Frogmore Road Hemel Hempstead Herts. HP3 9RW tel:(01442) 212501 fax:(01442) 248989

Birdscarer Products Oak House 24 Lovedean Lane Horndean Hampshire P08 8HJ tel:(01705) 363208 fax:(01705) 363208 E_Mail:stevemarston@birdscarer.freeserve.co.uk Web_Site:http://www.timed.co.uk

Birkinshaw L. Ltd Park House Ings Road, Osmondthorpe Lane Leeds W. Yorks. LS9 9HG tel:(0113) 249 6641 fax:(0113) 248 8968

Birmingham Guild Ltd Guild House Cradley Road Netherton, Dudley W. Midlands DY2 9TH tel:(01384) 411511 fax:(01384) 411234 telex:338024 BIRCOM G (PREFIX GUILD)

Birtley Building Products Mary Avenue Birtley Co. Durham DH3 1JF tel:(0191) 410 6631 fax:(0191) 410 0650

Birtley Building Products Ltd Mary Avenue Birtley Co. Durham DH3 1JF tel:(0191) 410 6631 fax:(0191) 410 0650 E_Mail:info@birtley-building.co.uk Web_Site:http://www.birtley-building.co.uk

Bisley Office Equipment Queens Road Bisley, Woking Surrey GU24 9BJ tel:(01483) 474577 fax:(01483) 489962

Bison Bede Ltd Castleside Industrial Estate Consett Co. Durham DH8 8JB tel:(01207) 585000 fax:(01207) 585085 E_Mail:marketing@bisonbede.co.uk

Bison Concrete Products Ltd Amington House Silica Road Tamworth Staffs. B77 4AZ tel:(01827) 64141 fax:(01827) 69009

Bison Structures Ltd London Road Tetbury Glos. GL8 8HH tel:(01666) 502792 fax:(01666) 504246 E_Mail:enquiry@bison.co.uk Web_Site:http://www.bison.co.uk

Bitmen Products Ltd PO Box 339 Over Cambridge Cambs. CB4 5TU tel:(01954) 231315 fax:(01954) 231315

Biwater Industries Ltd Clay Cross Chesterfield Derbys. S45 9NG tel:(01246) 250740 fax:(01246) 250741

Biwater Valves (Glenfield & Kennedy Ltd) PO Box 3 Queens Drive Kilmarnock Ayrshire KA1 3XH tel:(01563) 21150 fax:(01563) 41013 telex:77258 BWATER G

Blackfriar Paints and Varnishes Blackfriars Road Nailsea Nr. Bristol BS48 4DJ tel:(01275) 854911 fax:(01275) 858108

Blagg & Johnson Ltd Massey Street Newark Notts. NG24 1PF tel:(01636) 703137 fax:(01636) 701914

Blair Joinery 9 Baker Street Greenock Strathclyde PA15 4TU

Blake John Clayton le Moors Accrington BB5 5LP tel:(01254) 235441 fax:(01254) 382899

Blakell Europlacer Ltd Blanford Heights Blandford Dorset DT11 7TE tel:(01258) 451353 fax:(01258) 480183 telex:51324 BLAKEL G

Blakey John H Ltd New Bridge Mill Lord Street Industrial Estate Radcliffe Manchester M26 3BZ tel:(0161) 723 2414 fax:(0161) 725 9148

Blakley Electrics Ltd Conington Road London SE13 7LJ tel:(0181) 852 4383 fax:(0181) 318 5284

Blanc de Bierges Eastrea Road Whittlesey, Peterborough Cambs. PE7 2AG tel:(01733) 202566 fax:(01733) 205405

Blaze Neon Ltd Patricia Way Pysons Road Broadstairs Kent CT10 2XZ tel:(01843)601075 fax:(01843) 867924 E_Mail:blazeneon@compuserve.com

Blick Communication Systems Ltd Blick House Bramble Road Swindon Wilts. SN2 6ER tel:(01793) 692401 fax:(01793) 615848 telex:44332 BLKINT G E_Mail:mtweedie@blick.co.uk Web_Site:http://www.blick.co.uk

Blindcraft Industries 12 Edgefauld Avenue Glasgow G21 4BB tel:(0141) 287 0800 fax:(0141) 287 0802

Blockleys Brick Ltd Sommerfield Road Trench Lock, Telford Shropshire TF1 4RY tel:(01952) 251933 fax:(01952) 641900 E_Mail:info@blockleys.com

Blount Shutters Ltd Unit B 734 London Road West Thurrock Essex RM16 1NL tel:(01708) 860000 fax:(01708) 861272

BLP (Hamble) Ltd Race Shop Hamble Point Quay, School Lane Hamble Southampton SO31 5NB tel:(01703) 455537 fax:(01703) 456466

BLP UK Ltd Sandallstones Road Kirk Sandall, Doncaster S. Yorks. DN3 1QR tel:(01302) 890555 fax:(01302) 886724 E_Mail:marketing@blpuk.com Web_Site:http://www.blpuk.com

Blue Circle Cement Portland House, Aldermaston Park Church Road Aldermaston Berks. RG7 4HP tel:(01734) 818000 fax:(01734) 81399

Blunn Slates Ltd 57 Kellner Road London SE28 0AX tel:(0181) 301 8900 fax:(0181) 301 8901 E_Mail:sales@blunn.co.uk Web_Site:http://www.blunn.co.uk

BLV Licht-und Vakuumtecnik Units 4 & 5 Rabans Close, Rabans Lane Industrial Estate Aylesbury Bucks. HP19 3RS tel:(01296) 399334 fax:(01296) 393422

BLW Associates Ltd Head Office, Caledonian Works Alexandra Drive Lockerbie DG11 2PD tel:(01576) 203595 fax:(01576) 202276

Blyde-Barton Ltd Unit D, Frenbury Estate Drayton High Road Norwich Norfolk NR6 5DP tel:(01603) 789000 fax:(01603) 405476

Blyth William Hoe Hill Barton-on-Humber North Lincolnshire DN18 5RB tel:(01652) 632175 fax:(01652) 660966

BM Stainless Steel Drains Ltd Station Road Industrial Estate Tadcaster N. Yorks. LS24 9SG tel:(01937) 838000 fax:(01937) 832454 E_Mail:bm-stainless@compuserve.com Web_Site:http://www.blucher.com

BMK Ltd Burnside Street Kilmarnock Ayrshire KA1 1SX tel:(01563) 578000 fax:(01563) 578011

Boardcraft Ltd Howard Road Eaton Socon St Neots, Huntingdon Cambs. PE19 3ET tel:(01480) 213266 fax:(01480) 219095 E_Mail:info@boardcraft.co.uk Web_Site:http://www.boardcraft.co.uk

Boatman Plastics Ltd GT Hales Street Market Drayton Shropshire TF9 2AA tel:(01630) 657286 fax:(01630) 655545

BOB Stevenson Ltd Coleman Street Derby DE24 8NN tel:(01332) 574112 fax:(01332) 757286

Boddingtons Ltd Unit 10 Chelmsford Road Industrial Estate Gt Dunmow Essex CM6 1HF tel:(01371) 875101 fax:(01371) 874906 E_Mail:sales@boddingtons-ltd.com Web_Site:http://www.boddingtons-ltd.com

Boilden MKM Ltd Middlemore Lane Aldridge Walsall WS9 8ND tel:(01922) 743321 fax:(01922) 51566

Bolivar Stamping Ltd Crown Works Luton Street Keighley W. Yorks. BD21 2LB tel:(01535) 605766 fax:(01535) 609915

Bollom Fire Protection Ltd Croydon Road Elmers End, Beckenham Kent BR3 4BL tel:(0181) 658 2299 fax:(0181) 658 8672 E_Mail:fire@bollom.com Web_Site:http://www.bollom.com

Bollom J W & Co Ltd Croydon Road Elmers End, Beckenham Kent BR3 4BL tel:(0181) 658 2299 fax:(0181) 658 8672 telex:946804 E_Mail:sales@bollom.com Web_Site:http://www.bollom.com

Bolton Brady Ltd Unit J3, Colchester Avenue Factory Estate Colchester Avenue Cardiff CF3 7XG tel:(01222) 494771 fax:(01222) 494191

Bolton Brady Repair & Service Ltd 405 Hillington Road Glasgow G52 4BL tel:(0141) 883 2131 fax:(0141) 883 4502

Bolton Gate Co Ltd Waterloo Street Bolton Gtr. Manchester BL1 2SP tel:(01204) 871000 fax:(01204) 871049 E_Mail:boltongate@dial.pipex.com

Bolton, Thomas, Ltd PO Box 1 Froghall nr Cheadle Stock-on-Trent ST10 2HF tel:(01538) 752241 fax:(01538) 756715

Bonar Floors Ltd High Holborn Road Ripley Derby DE5 3NT tel:(01773) 744121 fax:(01773) 744142

Bonar Floors Ltd High Holborn Road Ripley Derbys. DE5 3NT tel:(01773) 744121 fax:(01773) 744142 E_Mail:bonarfloors.uk@lowandbonar.com Web_Site:http://www.nuwaymats.com

Bondaglass Voss Ltd 158-164 Ravenscroft Road Beckenham Kent BR3 4TW tel:(0181) 778 0071 fax:(0181) 659 5297

Bonwyke Ltd Bonwyke House 41-43 Redlands Lane Fareham Hampshire PO14 1HL tel:(01329) 289621 fax:(01329) 822768 E_Mail:sungard@bonwyke.demon.co.uk. Web_Site:http://www.bonwyke.demon.co.uk.

Booth Industries Ltd PO Box 50, Hulton Steel Works St Helens Road Bolton BL3 3ST tel:(01204) 61191 fax:(01204) 64646

Booth Muirie Ltd 870 South Street Glasgow G14 0SY tel:(0141) 959 1183 fax:(0141) 958 1173 E_Mail:acm@booth-muirie-ltd.demon.co.uk

Booth Samuel & Co Ltd Cheapside Works Cheapside, Birmingham B12 0PS tel:(0121) 772 2717 fax:(0121) 766 6962 Web_Site:http://www.samuelbooth.co.uk

Border Concrete Products Ltd Jedburgh Road Kelso Roxburghshire TD5 8 JG tel:(01573) 224393 fax:(01573) 226360

Bostik Ltd Ulverscroft Road Leicester LE4 6BW tel:(01533) 510015 fax:(01533) 513943 telex:34625

Boston Chemical Co Ltd Thorpe Arch Trading Estate Avenue 4 Wetherby W. Yorks. LS23 7BZ tel:(01937) 843413 fax:(01937) 841458 telex:557894

Bostwick Doors (UK) Ltd Grove Works, Mersey Industrial Estate Heaton Mersey Stcokport Gtr. Manchester SK4 3ED tel:(0161) 442 7227 fax:(0161) 431 6406

Boulter Boilers Ltd Magnet House White House Road Ipswich Suffolk IP1 5JA tel:(01473) 241555 fax:(01473) 241321

Boulton & Paul Ltd Woodland Place Pinetrees Road Norwich Norfolk NR7 9BB tel:(01603) 706000 fax:(01603) 706050

Bourne Steel Ltd St Clements House St Clements Road Poole Dorset BH12 4GP tel:(01202) 746666 fax:(01202) 732002 E_Mail:sales@bourne-steel.demon.co.uk

Bovingdon Brickworks Ltd Leyhill Road Hemel Hempstead Herts. HP3 0NW tel:(01442) 833176 fax:(01442) 834539

Bowater Containers Ltd Monmouthshire Broad Mills Alexandra Docks Newport Gwent NPT 2WE tel:(01633) 246666

Bowden Fencing Leicester Lane Great Bowden Market Harborough Leics. LE16 7HA tel:(01858) 410660

Bowthorpe EMP Ltd Stevenson Road Brighton Sussex BN2 2DF tel:(01273) 692591 fax:(01273) 676637

BPT Automation Ltd Unit 16, Sovereign Park Cleveland Way Hemel Hempstead Herts. HP2 7DA tel:(01442) 235355 fax:(01442) 244729

BPT Security Systems (UK) Ltd Unit 16, Sovereign Park Cleveland Way Hemel Hempstead Herts. HP2 7DA tel:(01442) 230800 fax:(01442) 244729 E_Mail:sales@bpt.co.uk Web_Site:http://www.bpt.co.uk

Bradbury Security Grilles Ltd Dunlop Way, Queensway Enterprise Estate Scunthorpe N. Lincolnshire DN16 3RN tel:(01724) 271999 fax:(01724) 271888 E_Mail:sales@bradburyuk.com Web_Site:http://www.bradburyuk.com

Brads Fencing Co. Limited 22 Hare Lane Farncombe Godalming Surrey GU7 3EE tel:(01483) 414745 fax:(01483) 419394

Braithaite Engineers Ltd Neptune Works Usk Way Newport S. Wales NP9 2UY tel:(01633) 262141 fax:(01633) 250631 telex:498237

Bramah Security Equipment Ltd 31 Oldbury Place London W1M 3AP tel:(0171) 486 1739 fax:(0171) 935 2779

Bran & Luebbe Ltd Scaldwell Road Brixworth Northamptonshire NN6 9UD tel:(01604) 880751 fax:(01604) 880145 E_Mail:branluebbe@compuserve.com

Brandon Medical Leathley Road Leeds West Yorkshire LS10 1BG tel:(0113) 245 7311 fax:(0113) 242 6708 E_Mail:enquiries@brandon-medical.com Web_Site:http://www.brandon-medical.com

Brannan S & Sons Ltd Leconfield Ind Est Cleator Moor Cumbria CA25 5QE tel:(01946) 816600 fax:(01946) 816625 E_Mail:sales@brannan.co.uk

Brannan Thermometers & Gauges Leconfield Industrial Estate Cleator Cumbria CA25 5QE tel:(01946) 816624 fax:(01946) 816625 telex:64248

Branson Leisure Roman House, Temple Bank River Way Harlow Essex CM20 2DY tel:(01279) 432151 fax:(01279) 450542 E_Mail:sales@bransonleisure.co.uk Web_Site:http://www.bransonleisure.co.uk

Brash John & Co Ltd The Old ShipYard Gainsborough Lincs. DN21 1NG tel:(01427) 613858 fax:(01427) 810218

Brass Tracks Hardware Ltd 177 Bilton Road Perivale, Greenford Middx. UB6 7HG tel:(0181) 566 9669 fax:(0181) 566 9339

BRB Industrial Services Ltd Douro Place Off Edge Lane Liverpool Merseyside L13 1AG tel:(0151) 259 6161 fax:(0151) 220 4410 (0151) 259 7097 - Zetafax

BRC Building Products Carver Road, Astonfields Industrial Estate Stafford ST16 3BP tel:(01785)222288 fax:(01785)240029 E_Mail:email@brc-building-products.co.uk

BRC Square Grip 79 - 81 Station Road Mansfield Nottinghamshire ng17 5fk tel:(01924) 369551 fax:(01924) 290050

BRE Bucknalls Lane Garston Watford Herts WD2 7JR tel:(01923) 664000 fax:(01923) 664010 E_Mail:enquiries@bre.co.uk Web_Site:http://www.bre.co.uk

Brett Martin Ltd Speedwell Industrial Estate Staveley Chesterfield Derbys. S43 3JP tel:(01246) 280000 fax:(01246) 280001

Brettell & Shaw Ltd West Street Quarry Bank Brierley Hill W. Midlands DY5 2DS tel:(01384) 566838 fax:(01384) 569123

Brewer C & Sons Ltd Albany House Ashford Road Eastbourne E. Sussex BN21 3TR tel:(01323) 411080 fax:(01323) 721435

Brewer Metalcraft Quarry Lane Chichester W. Sussex PO19 2NY tel:(01243) 539639 fax:(01243) 533184

Brewer T. & Co Ltd Old Station Yard Springbank Road Hither Green London SE13 6SS tel:(0181) 461 2471 fax:(0181) 461 4822

Brick Development Association Woodside House Winkfield Windsor Berks. SL4 2DX tel:(01344) 885651 fax:(01344) 890129 E_Mail:brick@brick.org.uk Web_Site:http://www.brick.org.uk

Brickhouse Pontymister Industrial Estate Risca Gwent NP1 6YL tel:(01633) 612833 fax:(01633) 601593 E_Mail:brickhse@celtic.co.uk

Bridgman Doors Ltd High Street Harrold Bedford MK43 7DA tel:(01234) 720561 fax:(01234) 270844 E_Mail:sales@bridgmandoors.co.uk Web_Site:http://www.bridgmandoors.co.uk

Bridon International Ltd Carr Hill Doncaster S. Yorks. DN4 8DG tel:(01302) 344010 fax:(01302) 382593

Brierley B (Garstang) Ltd Catterall Gates Lane Garstang, nr Preston Lancs. PR3 1XP tel:(01995) 603431 fax:(01995) 604234

Briggs Industrial Footwear Ltd Cornwall Road South Wigston Leicester LE8 2YU tel:(01533) 786353 fax:(01533) 477649

Briggs Roofing & Cladding Ltd, Roofing Contract Forsyth Road Sheerwater, Woking Surrey GU21 5RR tel:(01483) 756055 fax:(01483) 756077

Bright's of London Ltd Westgate Business Park Pickering North Yorkshire YO18 8LX tel:(01751) 474333

Brighton (Handrails), W 55 Quarry Hill Tamworth Staffs. B77 5BW tel:(01827) 284488 fax:(01827) 250907 E_Mail:wbrighton1@aol.com Web_Site:http://www.members.aol.com/wbrighton1

Brightside Firesnow Ltd Brightside House 122 Waterloo Raod Manchester MA8 8AS tel:(0161) 832 9761 fax:(0161) 835 2554

Brikenden (Builders Merchants) Ltd Grange Lane North Scunthorpe North Lincolnshire DN16 1DP tel:(01724) 282829 fax:(01274) 281375

Brimar Plastic Fabrications Ltd North Road Yate Bristol BS17 5PR tel:(01454) 322111 fax:(01454) 316955 E_Mail:brimar@brimar.co.uk Web_Site:http://www.brimar.co.uk

Brintons Ltd PO Box 16 Exchange Street Kidderminster Worcs. DY10 1AG tel:(01562) 820000 fax:(01562) 515597 Web_Site:http://www.brintons.co.uk

Brissco Equipment Ltd Carter Road Bishopsworth Bristol BS13 7TX tel:(0117) 964 3700 fax:(0117) 964 7906

Briticent International Ltd Crow Arch Lane Ringwood Hants. BH24 1NZ tel:(01425) 474617 fax:(01425) 471595

British Aluminium Extrusions Ltd Southam Road Banbury Oxon. OX16 7SN tel:(01295) 454545 fax:(01295) 454454 E_Mail:bae.sales@british-aluminium.ltd.uk

British Board of Agrement PO Box 195 Bucknalls Lane Garston , Watford Herts WD2 7NG tel:(01923) 665300 tel: (01923) 665301 E_Mail:bba@btinternet.com Web_Site:http://www.bbacerts.co.uk

British Building & Engineering Appliances plc 63-65 London Road Sandy Beds SG19 1DJ tel:(01767) 680291 fax:(01767) 691288

British Carpet Manufacturers Association P. O. Box 1155 Kidderminster Worcs. DY10 2ZH tel:(01562) 747351 fax:(01562) 747359 E_Mail:bcma@clara.net Web_Site:http://www.home.clara.net/bcma

British Cement Association Century House Telford Avenue Crowthorne RG45 6YS tel:(01344) 762676 tel: (01344) 727202 E_Mail:cement@bca.org.uk Web_Site:http://www.bca.org.uk

British Constructional Steelwork Association Ltd 4 Whitehall Court Westminster London SW1A 2ES tel:(0171) 839 8566 fax:(0171) 976 1634 E_Mail:postroom@bcsa.org.uk Web_Site:http://www.bcsa.org.uk

British Cork Mills Ltd Gerona House 17 Merton Road Bootle Merseyside L20 2PG tel:(0151) 922 1917 fax:(0151) 922 5843

British Decorators Association 32 Coton Road Nuneaton Warwks. CV11 5TW tel:(01203) 353776 fax:(01203) 354513

British Electric Lamps Ltd Spencer Hill Road London SW19 4EN tel:(0181) 946 5035 fax:(0181)8790602 telex:261507

British Electrical Systems Association Granville Chambers 2 Radford Street Stone Staffs. ST15 8DA tel:(01785) 812426 fax:(01785) 818157 E_Mail:besaeabman@webfactory.co.uk

British Fire Protection Systems Asssociation Neville House 55 Eden Street Kingston upon Thames Surrey KT1 1BW tel:(0181) 549 5855 fax:(0181) 547 1564 E_Mail:bfpsa@abft.org.uk Web_Site:http://www.bfpsa.org.uk

British Gypsum Ltd Technical Services Dept East Leake Loughborough Leics. LE12 6JT tel:(0990) 456123 fax:(0990) 456356

British Lead Mills Peartree Lane Welwyn Garden City Herts AL7 3UB tel:(01707) 324595 fax:(01707) 328941 E_Mail:sales@britishlead.co.uk Web_Site:http://www.britishlead.co.uk

British Library Business Information Service The British Library 96 Euston Road London NW1 2DB tel:(0171) 412 7454 fax:(0171) 412

7453 E_Mail:business-information@bl.uk Web_Site:http://www.bl.uk

British Nova Works Ltd 57-61 Lea Road Southall Middx. UB2 5QB tel:(0181) 574 6531 fax:(0181) 571 7572

British Sisalkraft Ltd Commissioners Road Strood Kent ME2 4ED tel:(01634) 290505 fax:(01634) 291029 telex:96330 SISAL G

British Solvent Oils, A Division of Careless PO Box 6 Metcalf Drive, Altham Industrial Estate Altham, nr Accrington Lancs. BB5 5TU tel:(01282) 79112 fax:(01282) 78660 telex:63300 BURCOM BRISTOL G

British Standards Institution 389 Chiswick High Road London W4 4AL tel:(0181) 996 9000

British Steel Sections Plates & Commercial Steels, Commercial Office PO Box 1 Frodingham House Scunthorpe S. Humberside DN16 1BP tel:(01724) 280280 fax:(01724) 282040 telex:52601

British Steel Special Sections PO Box 9, Carr House Whessoe Road Darlington DL3 0RG tel:(01325) 382382 fax:(01325) 380038

British Wall Tie Association PO Box 22 Goring Reading RG8 9YX tel:(01189) 842674 fax:(01189) 845396

British Water 1 Queen Anne's Gate London SW1H 9BT tel:(0171) 957 4554 fax:(0171) 957 4665 E_Mail:britishwater@onyxnet.co.uk Web_Site:http://www.britishwater.org

BRK Brands Europe Ltd Fountain House Canal View Road Newbury Berks. RG14 5XF tel:(01635) 528080 fax:(01635) 521635 E_Mail:sales@brk-firstalert.co.uk

Broadbent Machine Tools Ltd Mytholmroyd Yorks. HX7 5EH tel:(01422) 882816

Broadcrown Projects Ltd Alliance Works Airfield Industrial Estate Hixon Stafford ST18 0PF tel:(01889) 272200 fax:(01889) 272220

Broadmead Products Broadmead Works Hart Street Maidstone Kent ME16 8RE tel:(01622) 690960 fax:(01622) 765484 E_Mail:sales@broadmead.co.uk Web_Site:http://www.broadmead.co.uk

Broag Ltd Remeha House Molly Millars Lane Wokingham Berks. RG41 2QP tel:(0118) 978 3434 fax:(0118) 978 6977 E_Mail:boiler@broagltd.uk

Brockway Carpets Ltd Hoobrook Kidderminster Worcs. DY10 1XW tel:(01562) 824737 fax:(01562) 752010 E_Mail:sales@brockway.co.uk

Broderick Structures Limited Forsyth Road Sheerwater Woking Surrey GU21 5RR tel:(01483) 750207 fax:(01483) 750209

Broen Valves Ltd 7 Clecton Street Business Park Clecton Street Tipton W. Midlands DY4 7TR tel:(0121) 522 4515 fax:(0121) 552 4535

Brohome Ltd Duffryn Industrial Estate Ystrad Mynach Hengoed, South Wales CF82 TRJ tel:(01443) 813814 fax:(01443) 815813

Bromag Structures Ltd Monarch Court House 2 Mill Lane Benson Wallingford, Oxon. OX10 6SA tel:(01491) 838808 fax:(01491) 34183

Bromford Iron & Steel Co Ltd Bromford Lane West Bromwich W. Midlands B70 7JJ tel:(0121) 525 3110 fax:(0121) 525 4673

Brook Airchanger Brook House Dummarry Industrial Estate Dummarry Belfast BT17 9HU tel:(01232) 616505 fax:(01232) 616518

Brooks & Walker Ltd Century House 82 Tanner Street London SE1 3JP tel:(0171) 231 4030 fax:(0171) 231 5369 telex:264691

Brooks Stairlifts Ltd Westminster Industrial Esate Station Raod North Hykenham Lincoln LN6 3QY tel:(01522) 500288 fax:(01522) 500448 E_Mail:help@stairlifts.co.uk Web_Site:http://www.stairlifts.co.uk

Broome Bros (Doncaster) Ltd Line Side York Road Doncaster DN5 8AR tel:(01302) 361733 fax:(01302) 328536

Broughton Controls Ltd Shaw Road Oldham OL1 4AW tel:(0161) 627 0060 fax:(0161) 627 1362

Brown & Tawse Ltd Crown Works Mucklow Hill Halesowen W. Midlands B62 8DB tel:(0121) 550 9921 fax:(0121) 501 1013 telex:336966

Brown F plc Moor Lane Preston Lancs. PR1 1JQ tel:(01772) 824141 fax:(01772) 203383 E_Mail:fbrownplc@btconnect.com

Brown F plc Unit 17, Block 1 Vale of Leven Industrial Estate Dumbarton Dumbartonshire G82 3PD tel:(01389) 59911 fax:(01389) 58986

Brown N C (Storage Equipment) Ltd Hutchinson House, Firwood Industrial Estate Thicketford Road Bolton BL2 3TR tel:(01204) 596777 fax:(01204) 596555

Browne Winther & Co Ltd Nobel Road, Eley Estate Edmonton London N18 3DX tel:(0181) 803 3434 fax:(0181) 807 0544 E_Mail:sales@wintherbrowne.co.uk

Broxap & Corby Ltd Rowhurst Industrial Estate Chesteron Newcastle-under-Lyme Staffs. ST5 6BD tel:(01782) 564411 fax:(01782) 565357 E_Mail:e-mail@broxap.co.uk Web_Site:http://www.broxap.co.uk

Bruntons (Musselburgh) Ltd Musselburgh Scotland EH21 7UG tel:(0131) 665 3888 fax:(0131) 653 2236 telex:72212

Brymor Ltd Ytonbridge Road East Peckham Kent TN12 5JX tel:(01622) 871384 fax:(01622) 871011

BSA Tools Ltd Mackadown Lane Kitts Green Birmingham B33 0LE tel:(0121) 783 4071 fax:(0121) 784 5921 telex:338838 E_Mail:ch&ceo@bsatools.co.uk

BSS (UK) Ltd Fleet house Lee Circle Leicester LE1 3QQ tel:(01533) 623232 fax:(01533) 531343 telex:342761 BSSG

BSS Zenith 981 Scotswood Road Newcastle-upon-Tyne NE99 1CS tel:(0191) 273 5938 fax:(0191) 272 3487

BTR Environmental Ltd Vokes Henley Park Guildford Surrey GU3 2AF tel:(01483) 569971 fax:(01483) 235384 telex:859235 VOKES G E_Mail:vokes@btrinc.com Web_Site:http://www.vokes.com

Buchtal (U.K.) Limited 18 Smithbrook Kilns Cranleigh Surrey GU6 8JJ tel:(01483) 268980 fax:(01483) 274754 E_Mail:buchtaluk@compuserve.com

Buckingham Swimming Pools Ltd Dalehouse Lane Kenilworth Warwks. CV32 5UX tel:(01926) 852351 fax:(01926) 512387 E_Mail:buckswimpools.freeserve.co.uk Web_Site:http://www.scoot

BUFCA Ltd PO Box 23 Kenilworth Warwks. CV8 2YZ tel:(01926) 513187 fax:(01926) 513187

Builder's Equipment (Norwich) Limited City Road Norwich Norfolk NR1 3AN tel:(01603) 612211 fax:(01603) 630408

Builder's Iron & Zincwork Ltd Millmarsh Lane Brimsdown Enfield Middx. EN3 7QA tel:(0181) 443 3300 fax:(0171) 804 6672

Building Additions Ltd Unit 1 Vallis Road Frome Somerset BA11 3EQ tel:(01373) 454577 fax:(01373) 454578 E_Mail:buildadd@globalnet.co.uk

Building Bookshop 26 Store Street London WC1E 7BT tel:(0171) 692 4040

Building Cost Information Service BCIS 12 Great George Street Parliament Square London SW1P 3AD tel:(0171) 222 7000 fax:(0171) 695 1501 E_Mail:bcis@bcis.co.uk

Building Maintenance Information BCIS 12 Great George Street Parliament Square London SW1P 3AD tel:(0171) 222 7000 fax:(0171) 695 1501 E_Mail:bcis@bcis.co.uk

Bull Group of Companies Ltd Masonry Works Priors Haw Road, North Weldon Industrial Estate Corby Northants NN17 1JG tel:(01536) 69579 fax:(01536) 240265

Bullock & Driffill Ltd Staunton Works Newark Road Staunton in the Vale, Nottingham Nottinghamshire NG13 9PF tel:(01400) 280000 fax:(01400) 280010

Bundy Plastic Products Ltd Unit 2, Block C Capel Hendre Industrial Estate Ammanford Dyfed SA18 3SJ tel:(01269) 845800 fax:(01269) 845785

Burco Maxol Rosegrove Burnley Lancs. BB12 6AL tel:(01282) 427241 fax:(01282) 831206

Burgess Architectural Products Ltd Brookfield Road Hinkley Leics. LE10 2LN tel:(01455) 618787 fax:(01455) 251061

Burlington Engineers Ltd Unit 11, Perivale Industrial Estate Horseden Lane South Greenford Middx. UB6 7RL tel:(0181) 997 1515 fax:(0181) 998 3517 E_Mail:info@burlington-eng.com

Burlington Slate Ltd Cavendish House Kirkby in Furness Cumbria LA17 7UN tel:(01229) 889661 fax:(01229) 889466 E_Mail:sales@burstone.demon.co.uk

Burmatex Ltd Victoria Mills The Green Ossett W. Yorks. WF5 0AN tel:(01924) 276333 fax:(01924) 280033 E_Mail:info@burmatex.co.uk Web_Site:http://www.burmatex.co.uk.burmatex.htm.

Burn Fencing Limited West End Farm, West End Lane Balne Goole E. Riding DN14 0EH tel:(01302) 708706 fax:(01302) 707377

Burn Tubes Ltd Radway Road Shirley, Soilhull W. Midlands B90 4NS tel:(0121) 704 2211 fax:(0121) 704 2217 telex:339758

Burndept Electronics Ltd Tom Cribb Road Thamesmead London SE28 0BM tel:(0181) 316 4477 telex:896299 BURNDT G

Burnett N R Ltd Union Building Clarence Street Hull HU9 1DH tel:(01482) 20648 fax:(01482) 219600

Burnham Signs Burnham Way Kangley Bridge Road London SE26 5AE tel:(0181) 659 1525 fax:(0181) 659 4707

Bush Nelson PLC Stephenson Way Three Bridges Crawley W. Sussex RH10 1TN tel:(01293) 547361 fax:(01293) 531432 E_Mail:sales@bush-nelson.com Web_Site:http://www.bush-nelson.com

Bushboard Parker Ltd Rixon Road Finedon Road Industrial Estate Wellingborough Northants NN8 4BA tel:(01933) 224983 fax:(01933) 223553 Web_Site:http://www.bushboard.co.uk

Butler Building Systems Mitchelston Industrial Estate Kirkclady Fife KY1 3LZ tel:(01592) 52300 fax:(01592) 53135

Butler Cladding Systems No 1 The Parade Lodge Drive Culcheth, nr Warrington Cheshire WA3 4ES tel:(01925) 764335 fax:(01925) 763445

Butler Davies (Fencing) Limited 24 Mount Pleasant Road Northam Southampton Hampshire SO14 0QA tel:(01703) 222686 fax:(01703) 231201

Butterworth-Heinemann Linacre House Jordan Hill Oxford OX2 8DP tel:(01865) 314091 Web_Site:http://www.bh.com

BVC-BIVAC, Division of DD Lamson Harbour Road Gosport Hampshire PO12 1BG tel:(01705) 584281 fax:(01705) 504648 E_Mail:marketing@bvc.co.uk Web_Site:http://www.bvc.co.uk

C & K Systems Unit 24, Walkers Road, North Moons Moat Industrial Estate Reditch Worcs. B98 9HE tel:(01527) 68111 fax:(01527) 68222

C & P Developments Co (Lndon) Ltd Birch Road Eastbourne E. Sussex BN23 6PB tel:(01323) 22195 fax:(01323) 643764

C & R Building Systems Ltd Shay Lane Ovenden Halifax W. Yorks. HX3 6SF tel:(01422) 359311 fax:(01422) 340272 telex:517538

C & W Fencing Ltd Causeway End, Station Road, Lawford Manningtree Essex CO11 2LN tel:(01206) 395595 fax:(01206) 395583

C-Tec (Computionics Ltd) Stephens Way Goose Green, Wigan Lancs. WN3 6PH tel:(01942) 322744 fax:(01942) 829867 E_Mail:sales@c-tec.co.uk Web_Site:http://www.c-tec.co.uk

C. G. I. International Ltd, trading as Colebrand Glass Dallam Lane Warrington Cheshire WA2 7NT tel:(01925) 629111 fax:(01925) 245865 E_Mail:glass@colebrand.com Web_Site:http://www.colebrand.com

Cable Lift Installations 50 Brooksbys Walk Hackney London E9 6DA tel:(0181) 986 9416 fax:(0181) 986 0229

Cablok Ltd Ryeford Stonehouse Gloucester GL10 3HE tel:(01453) 824341 fax:(01453) 828908

Cadisch Precision Meshes Ltd Unit 1, Finchley Industrial Centre 879 High Road Finchley London N12 8QA tel:(0181) 492 0444 fax:(0181) 492 0333

CAEC Howard (Holdings) Ltd Grove Crescent House 18 Grove Place Bedford Beds. MK40 3JJ tel:(01234) 363171 fax:(01234) 357160

Cafco Construction Products Gainsborough Business Park Forbes Close, Fields Farm Long Eaton Nottingham NG10 1PX tel:(0115) 9464454 fax:(0115) 9464405

Caird Environmental Ltd, Industrial Services Division Unit 21, Freshwharf Highbridge Road Barking Essex IG11 7BW tel:(0181) 594 5099 fax:(0181) 594 8499

Calder Industrial Materials Ltd Leadworks Lane Chester CH1 3BS tel:(01244) 321022 fax:(01244) 315041

Calder Industrial Materials Ltd Cresent House Redheugh Bridge Road Newcastle-upon-Tyne NE99 1GE tel:(0191) 261 0161 fax:(0191) 261 1001 E_Mail:info@caldergroup.co.uk

Caldwell Hardware (UK) Ltd Berrington Road, Sydenham Industrial Estate Lemington Spa Warwks. CV31 1NB tel:(01926) 451767 fax:(01926) 315370

Callenders Ltd Harvey Road Basildon Essex SS13 1QJ tel:(01268) 591155 fax:(01268) 591165 E_Mail:technical@callendars.co.uk Web_Site:http://www.callenders.co.uk

Calomax (Engineers) Ltd Lupton Avenue Leeds W. Yorks. LS9 7DD tel:(0113) 249 6681 fax:(0113) 235 0358

Calor Gas Ltd Athena Drive Tachbrook Park Warwick CV34 6RL tel:(01926) 330088 fax:(01926) 420609

Calpeda Limited Wedgwood Road Industrial Estate Bicester Oxon. OX6 7UL tel:(01869) 241441 fax:(01869) 240681 E_Mail:pumps@calpeda.co.uk

Camas Aggregates Ltd Huncote Road Croft Leics. LE9 3GT tel:(01455) 285200 fax:(01455) 283837

Camas Building Materials Kemnay Quarry Aquithie Road Kemnay Aberdeenshire AB51 9PD tel:(01467) 643861 fax:(01467) 642342

Camas Building Materials Ltd Hulland Ward Ashbourne Derbys. DE6 3ET tel:(01335) 372222 fax:(01335) 370074 E_Mail:camas.demon.co.uk Web_Site:http://www.camas.com

Cambridge Asphalte Co Ltd Ely Road Waterbeach Cambridge CB5 9PG tel:(01223) 863000 fax:(01223) 440006

Cambridge Structures Ltd Huntington Road Bar Hill Cambridge CB 3 8HN tel:(01954) 781077 fax:(01954) 782912 telex:817213 BALLCO G

Cameron & Moore (Gridplex) Ltd Rannock Industrial Estate Geddington Road Corby Northamptonshire NN18 8AA tel:(01536) 260261 fax:(01536) 400208

Camloc (UK) Ltd 15 New Star Road Leicester LE4 9JD tel:(0116) 274 3600 fax:(0116) 274 3620

Camlok Lifting Clamps Ltd Knutsford Way Sealand Industrial Estate Chester Cheshire CH14NZ tel:(01244) 375375 fax:(01244) 377403 E_Mail:camlok@deeweld.u-net.com

Campbell Marson & Co Unit 34, Wimbledon Business Centre Riverside Road London SW17 0BA tel:(0181)8791909 fax:(0181) 946 9395

Camtwix Engineering (Division of Terrapan Ltd) Unit 1, Bond Trading Estate Bond Avenue Bletchley Milton Keynes MK1 1RE tel:(01908) 372238 fax:(01908) 373277

Candy & Co Ltd Great Western Potteries Healthfield Newton Abbot Devon TQ12 6RF tel:(01626) 831333 fax:(01626) 831318

Cannon Hygiene Ltd Middlegate White Lund Industrial Estate Morecambe Lancs. LA3 3BJ tel:(01524) 60894 fax:(01524) 64393 E_Mail:100676.36@compuserv.com Web_Site:http://www.ocs-group.comm

Canonbury Asphalte Co. Ltd. 453 Wick Lane Bow London E3 2TB tel:(0181) 980 6501 fax:(0181) 981 2956

Cantifix of London Limited Unit 9, Garrick Industrial Centre 22 Irving Way London NW9 6AQ tel:(0181) 203 6203 fax:(0181) 203 6454 E_Mail:e-mail@cantifix.co.uk Web_Site:http://www.cantifix.co.uk

Capco Test Equipment Ltd Riverside View Wickham Market Woodbridge Suffolk IP13 0TA tel:(01728) 747407 fax:(01728) 747599 E_Mail:sales@capco.co.uk Web_Site:http://www.capco.co.uk

Cape Boards Ltd Iver Lane Uxbridge UB8 2JQ tel:(01895) 463000 fax:(01895) 259262 telex:23471

Cape Ceilings Ltd Veralum Road Stafford ST16 3EA tel:(01785) 223435 fax:(01785) 51309 telex:367100

Cape Durasteel Ltd Bradfield Road Finedon Industrial Estate Wellingborough NN8 4HB tel:(01933) 440055 fax:(01933) 440066

Cape Insulation Products Ltd Patterson Industrial Estate Washington Tyne & Wear NE38 8JL tel:(0191) 416 1111 fax:(0191) 416 9409 telex:53413

Capel Fencing Contractors Ltd. 22 Sychem Lane Five Oak Green Tonbridge Kent TNl2 6TR

Capper Pipe Systems Ltd Chantry House High Street Coleshill Birmingham B46 3BP tel:(01675) 467557 fax:(01675) 465972

Caradon Bathrooms Ltd Lawton Road Alsager Stoke ST7 2DF tel:(01270) 879777 fax:(01270) 873864

Caradon Catnic Ltd Pontygwindy Industrial Estate Caerphilly M. Glam. CF83 2WJ tel:(01222) 337900 fax:(01222) 863178

Caradon Duraflex Systems Ltd Customer Centre Arle Road Cheltenham Glos. GL51 8LX tel:(0990) 351351 fax:(01242) 239604

Caradon Everest Ltd Everest House Sopers Road Cuffley, Potters Bar Herts. EN6 4SG tel:(01707) 875700 fax:(01707) 875621

Caradon Friedland Ltd Houldsworth Street Reddish, Stockport Gtr. Manchester SK5 6BP tel:(0161) 432 0277 fax:(0161) 431 4385 Web_Site:http://www.friedland.co.uk

Caradon Gent Ltd 140 Waterside Road Hamilton Industrial Park Leicester LE5 1TN tel:(0116) 246 2000 fax:(0116) 246 2300 E_Mail:gent_enquiry@caradon.com Web_Site:http://www.gent.co.uk

Caradon Heating Ltd PO Box 103 National Avenue Hull HU5 4JN tel:(01482) 492251 fax:(01482) 448858

Caradon Ideal Ltd PO Box 103 National Avenue Hull HU5 4JN tel:(01482) 492251 fax:(01482) 448859

Caradon Jones Ltd Whittington Road Oswestry Shropshire SY11 1HZ tel:(01691) 653251 fax:(01691) 658222

Caradon M K Electric Ltd The Arnold Centre Paycocke Road Basildon Essex SS14 3EA tel:(01268) 563000 fax:(01268) 563563 Web_Site:http://www.mkelectric.co.uk

Caradon Mira Ltd Cromwell Road Cheltenham Glos. GL52 5EP tel:(01242) 221221 fax:(01242) 221925 telex:43242

Caradon Plumbing Solutions Lawton Road Alsager Stoke on Trent ST7 2DF tel:(01270) 871598 fax:(01270) 871302

Caradon Terrain Ltd Aylesford,Maidstone Kent ME20 7PJ tel:(01622) 717811 fax:(01622) 716920

Caradon Trend Ltd PO Box 34 Horsham W. Sussex RH12 2YF tel:(01403) 211888 fax:(01403) 241608

Carborundum Abrasives G.B Ltd Trafford Park Road Trafford Park Manchester M17 1HP tel:(0161) 872 2381 fax:(0161) 953 2982 E_Mail:marketing@carboagb.co.uk Web_Site:http://www.carbogb.co.uk

Cardale Doors Ltd Farm Road Brackley Northants NN13 7EA tel:(01280) 703022 fax:(01280) 701138 telex:832171 E_Mail:marketing@cardale.co.uk Web_Site:http://www.cardale.uk.inter.net

Cardale Engineering Ltd Selbourne Road Luton Beds. LU4 8NT tel:(01582) 572546 fax:(01582) 505503

Cardinal Fencing (Northern) Limited Unit 3A, Royce Industrial Estate Ashburton Road West Trafford Park Manchester M17 1RY tel:(0161) 872 0725 fax:(0161) 872 0107

Cardkey Systems 23 Stadium Way Portman Road Reading RG30 6ER tel:(0118) 965 1200 fax:(0118) 941 7676 E_Mail:info@cardkey.co.uk Web_Site:http://www.cardkey.co.uk

Carless Refining & Marketing Ltd St James House Eastern Road Romford Essex RM1 3NL tel:(01708) 716600 fax:(01708) 716716

Carleton Furniture Group Ltd Mill Dam Lane Monkill Pontefract W. Yorks. WF8 2NS tel:(01977) 700770 fax:(01977) 708740 E_Mail:cfgit@compuserve.com

Carlton Benbow Bradley Mill Newton Abbott Devon TQ12 1NF tel:(01626) 367861 fax:(01626) 355591 E_Mail:sales@carltonbenbow.co.uk Web_Site:http://www.carltonbenbow.co.uk

Carlton Main Brickworks Ltd Grimethorpe Barnsley S. Yorks. S72 7BG tel:(01226) 711521 fax:(01226) 780417

Carnock Building Services 24 Norwood Drive Giffnock Glasgow G46 8JE tel:(0141) 638 1830 fax:(0141) 620 3006

Carpet Specifier Services Ltd Netherfield Park Netherfield Road Ravensthorpe W. Yorks WF13 3JY tel:(01924) 454566 fax:(01924) 468742 E_Mail:mglcarpets@compuserve.com

Carpets International (UK) PLC PO Box 255 Toftshaw Lane Bradford W. Yorks. BD4 6QW tel:(01274) 651155 fax:(01274) 652491

Carpets of Worth Ltd Severn Valley Mills Stourport on Severn Worcs. DY13 9HA tel:(01299) 827222 fax:(01299) 827049 telex:338324 CWORTH G
Carr Gymnasium Equipment Ltd Ronald Street Radford Nottingham NG7 3GY tel:(0115) 942 2252 fax:(0115) 942 2276 E_Mail:carrofnottm@btconnect.com Web_Site:http://www.carrofnottm.co.uk
Carrier Air Conditioning Units 13 & 14 Airport Trading Estate Biggin Hill Kent TN16 3BW tel:(0870) 6001100 fax:(01959) 571009
Carrier Holdings Ltd Cameron Court Cameron Street Hillington Industrial Estate Glasgow G52 4JH tel:(0141) 882 9692 fax:(0141) 882 9291
Carrington Tom & Co Ltd EGA-KUT Works Willenhall Lane Bloxwich Walsall WS3 2XN tel:(01922) 406611 fax:(01922) 493493
Carron Phoenix Ltd Stenhouse Road Carron Falkirk FK2 8DW tel:(01324) 638321 fax:(01324) 620978
Carrs Paper Ltd Shirley Solihull B90 4LJ tel:(0121) 733 3030 fax:(0121) 733 3811 telex:337836
Carter Concrete Ltd Briton's lane Beeston Regis Sheringham Norfolk NR26 8TP tel:(01263) 823434 fax:(01263) 825678
Carter Industrial Products Ltd Bedford Road Birmingham B11 1AY tel:(0121) 772 4300 fax:(0121) 772 3672 telex:339219
Carter Refrigeration Display Ltd Redhill Road Birmingham B25 8EY tel:(0121) 772 4300 fax:(0121) 772 4300
Carved Pine Mantelpieces Ltd High Street Dorchester on Thames Oxon OX10 7HL tel:(01865) 340028 fax:(01865) 341149
Casaire Ltd Raebarn House Northolt Road Harrow Middlesex HA2 0DY tel:(0181) 423 2323 fax:(0181) 864 2952
Castelco (Great Britain) Ltd Castle Works High Street Old Woking Surrey GU22 9LE tel:(01483) 714172 fax:(01483) 715317
Castell Safety International Kingsbury Works Kingsbury Road Kingsbury London NW9 8UR tel:(0181) 200 1200 fax:(0181) 205 0055 E_Mail:sales@castell.co.uk Web_Site:http://www.castell.co.uk
Castle Brick (Wales) Ltd Waterloo Road Penygroes Camarthenshire SA147NR tel:(01269) 831003 fax:(01269) 831003
Castle Care-Tech North Street Winfield, nr Windsor Berks. SL4 4SY tel:(01344) 886446 fax:(01344) 890024 E_Mail:sales@castle-caretech.com Web_Site:http://www.castle-caretech.com
Castle Cement Ltd Park Square, 3160 Solihull Parkway Brimingham Business Park Birmingham West Midlands B37 7YN tel:(0121) 779 7771 fax:(0121) 779 7609 Web_Site:http://www.castlecement.co.uk
Castrol (UK) Ltd Burmah House Pipers Way Swindon SB3 1RE tel:(01793) 512712 fax:(01793) 513506 telex:449221
Caswell & Co Ltd Chelsea Works St Michael's Road Kettering Northants NN15 6AU tel:(01536) 518340 fax:(01536) 310059
Catomance Ltd 96 Bridge Road East Welwyn Garden City Herts. AL7 1JW tel:(01707) 324373 fax:(01707) 372191 telex:267418 CANTIC G
Caunton Engineering Ltd Moorgreen Industrial Estate Moorgreen Nottingham NG16 3QU tel:(01773) 531111 fax:(01773) 5322020 E_Mail:sales@caunton.co.uk
Cavity Lock Systems Ltd Factory Road Newport Gwent S. Wales NP9 5FA tel:(01633) 246614 fax:(01633) 246110 E_Mail:cintec@aol.com Web_Site:http://www.cintec.com
Cavity Trays Ltd Administration Centre Boundary Road Yeovil Somerset BA22 8HU tel:(01935) 474769 fax:(01935) 428223 E_Mail:cavitytrays.co.uk
Cavtie Ltd Conway Tolhurst Works Termorfa Cardiff
Caxton Name Plate Manufacturing Co Ltd Kew Green Richmond Gtr. London TW9 3AR tel:(0181) 940 0041 fax:(0181) 940 0642
CCS Scotseal Ltd Unit 3 Lyon Road Lynwood Industrial Estate Paisley Scotland PA3 3BQ tel:(01505) 324262 fax:(01505) 323618
CCTCO Europe Ltd Scots Quays East Street Birkenhead Merseyside L41 1FB tel:(0151) 606 5900
Cefndy Enterprises Cefndy Road Rhyl Clwyd LL18 2HG tel:(01745) 343877 fax:(01745) 355806
Cegelec Industrial Control Ltd Kidsgrove Stoke-on-Trent ST7 1TW tel:(01784) 163551 telex:36293
Cego Frameware Ltd Western Road Silver End, Witham Essex CM8 3QB tel:(01376) 507507 fax:(01376) 584687 E_Mail:sales@cego.co.uk Web_Site:http://www.cego.co.uk
Ceiling Partitioning Systems Co Domum House Domum Road Winchester Hants. SO23 9NN tel:(01962) 844666 fax:(01962) 840351
CEL Instruments Ltd 35-37 Bury Mead Road Hitchin Herts. SG5 1RT tel:(01462) 422411 fax:(01462) 422511
Celcon Blocks Ltd Celcon House 289-293 High Holborn London WC1V 7HU tel:(0171) 242 9766 fax:(0171) 430 0038
Celmac Ltd Saxon Street Denton Manchester M34 3AB tel:(0161) 336 4401 fax:(0161) 335 0026 telex:667641
Celotex Ltd Warwick House 27-31 St Mary's Road Ealing London W5 5PR tel:(0181) 579 0811 telex:263863
Celuform Customer Centre Arle Road Cheltenham Glos. GL51 8LX tel:(0990) 920930 fax:(01242) 241983
CEM Fencing Contractors The Old Cattle Market Watling Street East Towcester Northants NN12 6HN tel:(01327) 351234 fax:(01327) 350410
CEM Systems Ltd Unit 4, Ravenhill Business Park Ravenhill Road Belfast BT6 8AW tel:(01232) 456767 fax:(01232) 454535 E_Mail:sales@devel.cemsys.com Web_Site:http://www.cemsys.com
Cement Admixtures Association(CAA) 4 The Knowl Churton Cheshire C43 6NE tel:(01829) 270230 fax:(01829) 270230
Cement Glaze Decorators Ltd 5 Barry Parade Barry Road London SE22 OJA tel:(0181) 299 2553 fax:(0181) 299 2346
Centro Engineering Ltd High Green House Highwood Chelmsford Essex CM1 3QH tel:(01245) 248509 fax:(01245) 248739
Cep Ceilings Ltd Veralum Road Common Road Industrial Estate Stafford ST16 3EA tel:(01785) 223435 fax:(01785) 251309 E_Mail:cep@cep-ms.demon.co.uk Web_Site:http://www.cep-ms.demon.co.uk
Ceram Research Queens Road Penkhull Stoke-on-Trent ST4 7LQ tel:(01782) 845431 fax:(01782) 412331
Cerro (Maganese Bronze) Ltd PO Box 22, Hanford Works Hadleigh Road Ipswich Suffolk IP2 0EG tel:(01473) 252127 fax:(01473) 218229
Cerro Extruded Metals Ltd Greets Green Road West Bromwich W. Midlands B70 9ER tel:(0121) 500 6188 fax:(0121) 553 6505
Certifield Laboratories (Div or CPS Industries Ltd) Landchard House Victoria Street West Bromwich W. Midlands B70 8ER tel:(0121) 525 6678 fax:(0121) 500 5386
Certikin International Ltd Unit 9, Witan Park, Avenue 2 Station Lane Industrial Estate Witney Oxon. OX8 6FH tel:(01993) 778855 fax:(01993) 778620 E_Mail:info@certikin.co.uk Web_Site:http://www.certikin.co.uk
CFS Carpets Unit 1 Goldsworth Trading Estate Woking Surrey GU21 3AZ tel:(01483) 770661 fax:(01483) 724494 E_Mail:cfsfloors@btinternet.com
CGL Cometec 2 Lithgow Place College Milton North East Kilbridge Glasgow G74 1PL tel:(01355) 235561 fax:(01355) 247189 E_Mail:101606.1730@compuserve. com
CH Agencies 133 Newhaven Road Edinburgh EH6 4AP tel:(0131) 554 2283 fax:(0131) 553 4533 E_Mail:laticreteuk@ibm.net Web_Site:http://www.laticrete.com
Chaffoteaux et Maury Ltd Trench Lock Trench Telford Shropshire TF1 4SZ tel:(01852) 222727 fax:(01952) 243493
Challenge Fencing & Ltd Downside Road, Cobham Surrey KT11 3LY tel:(01932) 866555 fax:(01932) 866445 E_Mail:challengefencing@clara.net
Chalmers & Mitchell Ltd 388 Hillington Road Glasgow G52 4BL tel:(0141) 882 5555 fax:(0141) 883 3704 telex:779378 CHALMT G
Channel Electronics 23a Cradle Hill Industrial Estate Seaford E. Sussex BN25 3JE tel:(01323) 894961 fax:(01323) 893149
Channel Safety Systems Ltd 9 Petersfield Business Park Bedford Road Petersfield Hants. GU32 3QA tel:(01730) 268231 fax:(01730) 265552
Chapel Studio Bridge Road Hunton Bridge Kings Langley Herts. WD4 8RE tel:(01923) 226386 fax:(01923) 269707
Charcon Tunnels (Division of Tarmac plc) PO Box 1 1 Southwell Lane Kirby-in-Ashfield Notts. G17 8GQ tel:(01623) 754493 fax:(01623) 759825
Charles Collinge (Architectural Ironmongery) Ltd 44 Loman Street London SE1 0EH tel:(0171) 928 3541 fax:(0171) 620 1326
Charles Homes (Halesowen) Ltd Burrell Way Thetford Norfolk IP24 3QT
Charnwood Brick Group Ltd Old Station Close Shepshed nr Loughborough Leics. LE12 9RJ tel:(01509) 503203 fax:(01509) 507566
Charnwood Fencing Limited Beveridge Lane Bardon Hill Leicester Leics. LE67 1TB tel:(01530) 835835 fax:(01530) 814545
Chartered Building Company Scheme Englemere Kings Ride Ascot Berks. SL5 7TB tel:(01344) 23355 fax:(01344) 23467
Chartered Institute of Building Englemere Kings Ride Ascot Berks. SL5 7TB tel:(01344) 630700 fax:(01344) 630777 E_Mail:irc@ciob.org.uk Web_Site:http://www.ciob.org.uk
Chartered Institution of Building Services Engineers (CIBSE) Delta House 222 Balham High Road London SW12 9BS tel:(0181) 675 5211 fax:(0181) 675 5449 E_Mail:info@cibse.org Web_Site:http://www.cibse.org
Chase Equipment Ltd Hoebridge Farm Old Woking Road Old Woking Surrey GU22 8JH tel:(01483) 722922
Chase James & Son (Furnishing) Ltd 191 Thornton Rd. Bradford BD1 2JT tel:(01274) 738 282
Checkmate Industries Ltd Bridge House Bridge Street Halstead Essex CO9 1HT tel:(01787) 477272 fax:(01787) 476334
Chelsea Artisans Ltd Unit C2, Sandown Industrial Park Mill Road Esher Surrey KT10 8BL tel:(01372) 469301 fax:(01372) 470590 E_Mail:info@chelsea-artisans.co.uk Web_Site:http://www.chelsea-artisans.co.uk
Chelwood Brick Ltd Adswood Road Cheadle Hulme Cheadle Cheshire SK8 5QY tel:(0161) 485 8211 fax:(0161) 488 4351
Chemsearch (Div of NCH UK Ltd) Lanchard House Victoria Street West Bromwich W. Midlands B70 8ER tel:(0121) 525 1666 fax:(0121) 500 5386
Cheshire Concrete Products Brooks Lane Middlewich Cheshire CW10 0JQ tel:(01606) 837364 fax:(01606) 837365
Chesterfelt Ltd Foxwood Way Sheepbridge Chesterfield Derbys. S41 9RX tel:(01246) 268000 fax:(01246) 268001
Chestnut Products Limited Unit No 8, Gaza Trading Estate Hildenborough Tonbridge Kent TN11 8PL
Child Bros Town Street Earlsheaton Dewsbury W. Yorks. WF12 8JA tel:(01924) 464901
Chilmark Quarries Ltd c/o The Rare Stone Group Ltd 184 Nottingham Road Mansfield Notts. NG18 5AP tel:(01623) 623092 fax:(01623) 622509 E_Mail:enquiries@rarestonegroup.co.uk Web_Site:http://www.rarestonegroup.co.uk
Chilstone Garden Ornaments Victoria Park Fordcombe Road, Langton Green Tunbridge Wells Kent TN3 0 RE tel:(01892) 740866 fax:(01892) 740249 E_Mail:chilstone@hndl.demon.co.uk Web_Site:http://www.greatbritain.co.uk/chilstone
Chiltern Concrete Products Ltd Faldo Road Barton - Le - Clay Beds. MK45 4RH tel:(01582) 881414 fax:(01582) 881855 E_Mail:chilternprecast.co.uk
Chiltern Invadex 66-68 Manchester Road Chapel-en-Le-Firth High Peak SK23 9TP tel:(01298) 816366 fax:(01298) 816106
Chindwell Co Ltd Hyde House Edgware Road The Hyde London NW9 6JT tel:(0181) 205 6171 fax:(0181) 205 8800
Chiorino UK Ltd Highlands Road Shirley, Solihull W. Midlands B90 4HN tel:(0121) 705 8271 fax:(0121) 711 3741
Chippindale Plant Ltd Butterbowl Works Lower Wortley Ring Road Leeds LS12 5AJ tel:(0113) 263 2344 fax:(0113) 279 1710
Chloride Industrial Batteries Ltd Rake Lake Clifton Junction Swinton Manchester M27 8LR tel:(0161) 794 4611 fax:(0161) 793 6606 E_Mail:sales.cibl@btrinc.com Web_Site:http://www.hawker.co.uk
Chris Wheeler Construction Limited Church Farm Burbage Nr Marlborough Wilts. SN8 3AT tel:(01672) 810315 fax:(01672) 810309
Christie & Grey Ltd Sovereign Way Tonbridge Kent TN9 1RH tel:(01732) 371100 fax:(01732) 359666 E_Mail:sales@christiegrey.com Web_Site:http://www.christiegrey.com
Chubb Fire Ltd Chubb House Sunbury-on-Thames Surrey TW16 7AR tel:(0800) 321666 fax:(01932) 765630 telex:263169 CHBFIR G
Chubb Safe Equipment Company PO Box 61 Wednesfield Road Wolverhampton WV10 0EW tel:(01902) 455111 fax:(01902) 450949 E_Mail:info@chubb-safes.com Web_Site:http://www.chubb-safes.com
Chubb Security Installations Ltd Ronald Close Kempston Bedford MK42 7SH tel:(01234) 840840 fax:(01234) 840333
Church & Bramhall Shearing Pedmore Road Woodside Dudley DY2 0RF tel:(01384) 357357 fax:(01384) 357350
Ciba-Geigy Polymers Duxford Cambridge CB2 1QA tel:(01223) 832121 telex:81101
Cico Chimney Linings Westleton Saxmundham Suffolk IP17 3BS tel:(01728) 73608 fax:(01728) 73428
CIL International Ltd CIL Trading Estate Fonthill Road London N4 3HN tel:(0171) 272 0222 fax:(0171) 272 6402 E_Mail:info@cil_international. com Web_Site:http://www.cil_international. com
Cila Landscapes 15 Cave Street Cwmdu, Swansea Swansea SA5 8JY tel:(017920 585545 fax:(01792) 585545
Cimex International Ltd Somerford Road Christchurch Dorset BH23 3PS tel:(01202) 499699 fax:(01202) 499465
Cinderford Enginering Co Ltd Valley Road Works Cinderford Glos. GL14 2NZ tel:(01594) 822226/823531
Cistermiser Ltd Unit 1, Woodley Park Estate 59-69 Reading Raod Woodley Reading RG5 3AN tel:(0118) 969 1611 fax:(0118) 944 1426 E_Mail:cistermiser@btinternet.com Web_Site:http://www.cistermiser.co.uk
City Hardware and Security Avon House Queens Drive Kings Norton Business Centre Birmingham B30 3HH tel:(0121) 459 5599 fax:(0121) 459 4101
Civic Trust 17 Carlton House Terrace London SW1Y 5AW tel:(0171) 930 0914
Civil & Marine Slag Cement Ltd London Road Grays Essex RM20 3NL tel:(01708) 864813 fax:(01708) 865907
Civil Engineering Contractors Association Construction House 56-64 Leonard Street London EC2A 4JX tel:(0171) 608 5060 fax:(0171) 608 5061
Civil Engineering Development Ltd 728 London Road West Thurrock, Grays Essex RM20 3LU tel:(01708) 867237 fax:(01708) 867230
Cladcolour Profiling Whitehead Estate Docks Way Newport Gwent tel:(01633) 252191
Cladding & Decking UK Ltd Shaw Street Hilltop, West Bromwich W. Midlands B70 0TX tel:(0121) 556 4211 fax:(0121) 502 4385
Cladding & Decking UK Ltd William Nadin Way Swadlincote Derbys. DE11 0BB tel:(01283) 211700 fax:(01283) 229259
Cladding Consultants (UK) Ltd Burton End Stanstead Essex CM24 8UE tel:(01279) 812224 fax:(01279) 812693
Clam-Brummer Ltd London Road Spellbook, Nr. Bishops Stortford Herts. CM23 4BA tel:(0171) 476 3171 fax:(0171) 474 0098 E_Mail:sales@brummer.ltd.com Web_Site:http://www.brummer.ltd.com
Clare R S & Co Ltd Stanhope Street Liverpool L8 5RQ tel:(0151) 709 2902 fax:(0151) 709 0518
Clark & Fenn Ltd Unit 19, Mitcham Industrial Estate Streatham Road Mitcham Surrey CR4 2AJ tel:(0181) 648 4343 fax:(0181) 640 1986
Clark Door Ltd Willowholme Carlisle CA2 5RR tel:(01228) 522321 fax:(01228) 401854 E_Mail:mail@clarkdoor.com Web_Site:http://www.clarkdoor.com
Clarke Instruments Ltd Distloc House, Old Sarum Airfield The Portway Salisbury Wilts. SP4 6DZ tel:(01722) 323451 fax:(01722) 335154 E_Mail:sales@clarke-inst.com Web_Site:http://www.clarke-inst.com
Clarke Stairways Edwin Coultham Street Lincoln LN5 8HQ tel:(01522) 530912/520376 fax:(01522) 544100
Clarksteel Ltd Station Works Station Road Yaxley Peterborough PE7 3EG tel:(01733) 240811 fax:(01733) 240201 E_Mail:sales@clarksteel.com Web_Site:http://www.clarksteel.com
Clay Pipe Development Association Ltd Copsham House 53 Broad Street Chesham Bucks. HP5 3EA tel:(01494) 791456 fax:(01494) 792378 E_Mail:cpda@aol.com Web_Site:http://www.cpda.co.uk
Claydon Architectural Metalwork Ltd Units 11/12 Claydon Industrial Park Great Blakenham Suffolk IP6 0NL tel:(01473) 831000 fax:(01473) 832154
Clayton-Munroe Ltd West Drive Kingston Staverton, Totnes Devon TQ9 6AR tel:(01803) 762626 fax:(01803) 762584 E_Mail:mail@claytonmunroe.co.uk Web_Site:http://www.claytonmunroe.co.uk
Clean Step Ltd Unit 2, Ravenseft Park Cheney Manor Industrial Estate Swindon SN2 2QJ tel:(01793) 420088 fax:(01793) 431113
Cleaner Surfaces (Repair & Restoration) Ltd Basford Works Egypt Road Basford Nottingham NG7 7GL tel:(01602)783177 fax:(01602) 422798
Cleanstone Co (Bradford) Ltd Dick Lane Bradford BD4 8JW tel:(01274) 669516 fax:(01274) 665356
Clear Span Ltd Wellington Road Greenfield Oldham Lancs. OL3 7AG tel:(01457) 873244 fax:(01457) 870151
Clearwater Environmental Engineering Ltd Little London Spalding Lincs. PE11 2UE tel:(01775) 768694 fax:(01775) 710294
Clearwater Plc Clearwater House Clearwater Industrial Estate Bristol Road Bridgwater, Somerset TA6 4AW tel:(01278) 433443 fax:(01278) 433455
Clement Steel Windows Clement House Haslemere Surrey GU27 1HR tel:(01428) 647700 fax:(01428) 661369 E_Mail:csw@clement-windows.demon.co.uk Web_Site:http://www.clement-windows.demon.co.uk
Clenaware Systems Ltd The Trading Estate Farnham Surrey GU9 9PQ tel:(01252) 712789 fax:(01252) 723719 E_Mail:cleanaware@aol.com
Clestra Hauserman Ltd Hamilton House 3 North Street Carshalton Surrey SM5 2HZ tel:(0181) 773 2121 fax:(0181) 773 4793
Cliffhanger Shelving 8 Fletchers Square Temple Farm Industrial Estate Southend-on-Sea Essex SS2 5RN tel:(01702) 613135 fax:(01702) 602997 E_Mail:sales@cliffhgr.demon.uk Web_Site:http://www.cliffhanger.co.uk
Clifford Partitioning Co Ltd Champion House Burlington Road New Malden KT3 4NB tel:(0181) 942 6646 fax:(0181) 336 2369
Clifton Nurseries 5a Clifton Villas Little Venice London W9 2PH tel:(0171) 289 6851 fax:(0171) 286 4215 E_Mail:e-mail@clifton.co.uk
Climet Ltd Port Marsh Industrial Estate Port Marsh Road Calne Wilts. SN11 9BW tel:(01249) 813600 fax:(01249) 814134 E_Mail:sales@climet.ltd.uk Web_Site:http://www.climet.ltd.uk
Clinicon Fire Charwell House Chestnut Avenue Haslemere Surrey GU27 2AT tel:(01428) 654044 fax:(01428) 656881 telex:859872 CLINIC
Clips Partitions Ltd 24 Dalston Gardens Stanmore Middx. HA7 1BU tel:(0181) 204 9494 fax:(0181) 204 6505
ClipsPartitions Ltd Unit B5, Haslemere Industrial Estate Pig Lane Bishop's Stortford Herts. CM23 3HG tel:(01279) 506587 fax:(01279) 503402
Clough (Croydon) Ltd Suprete House 85 Manor Road Wallington Surrey SM6 0DH tel:(0181) 395 8787 fax:(0181) 647 9235
Clow Group Ltd Diamond Ladder Factory 562-584 Lea Bridge Road Leyton London E10 7DW tel:(0181) 558 0300 fax:(0181) 558 0301
Clyde Canvas Fabric Structures Ltd 50 Lindsay Road Leith Edinburgh EH6 6NW tel:(0131) 554 1331 fax:(0131) 553 7655
Clyde Combustion Ltd Cox Lane Chesington Surrey KT9 1SL tel:(0141) 882 3291/8 fax:(0141) 883 3846

Clyde Combustions Ltd Cox Lane Chessington Surrey tel:(0181) 391 2020 fax:(0181) 397 4598
Co Steel Sheerness PLC Sheerness Kent ME12 1TH tel:(01795) 663333 fax:(01795) 660233
Coalite Chemicals Buttermilk Lane Chesterfield Derbys. S44 6 AZ tel:(01246) 826816 fax:(01246) 240309 telex:547624
Coastal Ltd D'Oriel House Holton Heath Trading Park Poole Dorset BH16 6LE tel:(01202) 624011 fax:(01202) 622465
Coates Fencing Unit 3, Barhams Close Wylds Road Bridgwater Somerset TA6 4DP tel:(01278) 423577 fax:(01278) 427760
Cobham Composites Ltd Davey House Gelders Hall Road Shepshed, Loughborough Leics. LE12 9NH tel:(01509) 504541 fax:(01509) 507563
Coblands Landscapes Ltd South Farm Barn South Farm Lane, Langton Green Tunbridge Wells Kent TN3 9JN tel:(01892) 863535 fax:(01892) 863778
Colas Roofing Harvey Road Basildon Essex SS13 1ES tel:(01268) 591155 fax:(01268) 591165 telex:99364
Cold Rolled Sections Association Robson Rhodes, Centre City Tower 7 Hill Street Birmingham B5 4UU tel:(0121) 697 6000 fax:(0121) 697 6113
Colebrand Ltd 20 Warwick Street London W1R 6BE tel:(0171) 549 9191 telex:261495
Coleford Brick & Tile Co Ltd Hawkwell Green Cinderford Glos. GL14 3JJ tel:(01594) 822160 fax:(01594) 826655
Collie Carpets Ltd Causeway Mill, Express Trading Estsate Stone Hill Road Farnworth Nr. Bolton BL4 9TP tel:(01204) 571108 fax:(01204) 795176 E_Mail:colliecarpets@dial.pipex.com
Collier W H Ltd Church Lane Marks Tey Colchester Essex CO6 1LN tel:(01206) 230301 fax:(01206) 212540
Collins Walker Ltd, Elec Steam & Hot Water Boilers Unit 7a, Nottingham S & Wilford Ind. Est. Ruddington Lane Nottingham NG11 3EP tel:(01602) 818044 fax:(0160) 455376 E_Mail:malcolm@collinswalker.co.uk Web_Site:http://www.collins/walker.co.uk
Colman Greeves Unit 3 Rutland Street Ashton under Lyne OL6 6TX tel:(0161) 330 9316 fax:(0161) 339 5016
Colortype Ltd Poughill Bude Cornwall EX23 9EZ tel:(01288) 352879
Colt International New Lane Havant Hants. PO9 2LY tel:(01705) 451111 fax:(01705) 454220
Colt W H & Co Ltd Colt Works, Pluckley Road Bethersden Nr Ashford Kent TN26 3DD tel:(01233) 820456 fax:(01233) 820991
Coltman Precast Concrete Ltd London Road Canwell, Sutton Coldfield W. Midlands B75 5SX tel:(01543) 480482 fax:(01543) 481587
Colton Electrical Equipment Co 329 Front Lane Cranham, Upminster Essex RM14 1LN tel:(014022) 24454 fax:(014022) 21191
Combat Engineering Ltd Oxford Street Bilston W. Midlands WV14 7EG tel:(01902) 494425 fax:(01902) 403200 E_Mail:enquiry@combat.co.uk Web_Site:http://www.combat.co.uk
Combat Polystyrene Group Ltd Unit 5c, Grange Road Houston Industrial Estate Livingston EH54 5DE tel:(01506) 441188 fax:(01506) 431133
Comber Models 17 London Lane London E8 3PR tel:(0181) 533 6592 fax:(0181) 533 5333
Combustion Linings Ltd Unit 18, The Railway Enterprise Centre Shelton New Road Stoke-on-Trent ST4 7SH tel:(01782) 266766 fax:(01782) 266890
Comet Pump & Engineering Co Ltd 2A Selhurst Road South Norwood London SE25 5QF tel:(0181) 684 3816 fax:(0181) 689 4307
Commercial Floorings Exhibition 19 Woodland Rise Ascot Berks. SL5 9HP tel:(01344) 876123 fax:(01344) 28838
Commodore Kitchens Ltd Acorn House Gumley Road Grays Essex RM20 4XP tel:(01375) 382323 fax:(01375) 394955 E_Mail:info@commodorekitchens.co.uk Web_Site:http://www.commodorekitchens.co.uk
CompAir Holman Ltd Camborne Cornwall TR14 8DS tel:(01209) 712750 fax:(01209) 713955 telex:45501 COMAIR G
Component Developments Park Avenue Madeley Telford Shropshire TF7 5 LG tel:(01952) 588488 fax:(01952) 581901 E_Mail:ww_ltd@compuserve.com
Composite Panels Ltd Newmains Avenue, Inchinnan Business Park Renfrew Strathclyde PA 4 9RR tel:(0141) 812 6866 fax:(0141) 812 7721 E_Mail:info@compositepanels.co.uk Web_Site:http://www.compositepanels.co.uk
Composite Structures Eastleigh House Upper Market Street Eastleigh Hants. SO50 9FD tel:(01703) 616712 fax:(01703) 643665 E_Mail:email@compstruct.co.uk
Compriband Ltd Bentall Business Park Glover. District II Washington Tyne & Wear NE37 3JD tel:(0191) 417 8700 fax:(0191) 417 8707 E_Mail:compriband@aol.com
Compton Building Ltd Station Works Fenny Crompton Leamington Spa Warwks. CV33 0XB tel:(01295) 770291 fax:(01295) 770748
Computer Foundations Ltd Foundation House 75- 77 London Road Lexden, Colchester Essex CO3 5AL tel:(01206) 541455 fax:(01206) 761930 E_Mail:info@foundation.ltd.uk
Computertel Ltd CTL House, The Maltings, Unit 53 Bath Street Gravesend Kent DA11 0DF tel:(01474) 561111 fax:(01474) 561122 E_Mail:computertel_ltd@compuserve.com
Comyn Ching & Co Lrd (SOLRAY) 7 Skylines Village Lime Harbour London E14 9TS tel:(0171)987 8787
Concord Lighting Avis Way Newhaven E. Sussex BN9 0ED tel:(01273) 515811 fax:(01273) 611101
Concord Lighting Ltd 174 High Holborn London WC1V 7AA tel:(0171) 497 1400 fax:(0171) 497 1404 telex:263084
Concordia Electric Wire & Cable Co. Ltd Derwent Street Long Eaton Nottingham NG10 3LP tel:(0115) 946 7400 fax:(0115) 946 1026 E_Mail:sales@concordia.ltd.uk Web_Site:http://www.concordia.ltd.uk
Concrete Products (Lincoln) 1980 Ltd Riverside Industrial Estate Skellingthorpe Road Saxilby Lincoln LN1 2LR tel:(01522) 704158 fax:(01522) 714233
Concrete Repairs Ltd Cathite House 23a Willow Lane Mitcham Surrey CR4 4TU tel:(0181) 288 4848 fax:(0181) 288 4847
Conder Cladding Shaw Street Hilltop, West Bromwich W. Midlands B70 0TX tel:(0121) 556 4211 fax:(0121) 502 5385
Conder Group Plc Moorside Road Winnall, Winchester Hampshire SO23 7SL tel:(01962) 863555 fax:(01962) 870658
Conder Products Ltd Abbotts Barton House Worthy Road Winchester Hants. SO23 7SH tel:(01962) 841313 fax:(01962) 841759 E_Mail:sales@conder.co.uk Web_Site:http://www.conder.co.uk
Conex- Sanbra Ltd Whithall Road Tipton W. Midlands DY4 7JU tel:(0121) 5772831 fax:(0121) 520 8778 telex:339351
Connect Lighting Systems (UK) Ltd Murrils Estate Portchester Hants. PO16 9RD tel:(01705) 375140 fax:(01705) 210475
Connoleys (Blackley) Ltd Delaunays Road Manchester M9 3FP tel:(01928) 722727 fax:(01928) 762486
Connought Communication Systems Ltd Systems House Reddicap Estate Sutton Coalfield Birmingham B75 7BU tel:(0121) 311 1010 fax:(0121) 311 1890
Conren Ltd Redwither Works Redwither Road Wrexham Clwyd LL13 9RD tel:(01978) 611991 fax:(01978) 661120
Consculpt Spiral Stairs Ltd The Grange Rectory Road Camborne Cornwall TR14 7DN tel:(01209) 714761 fax:(01209) 612464 E_Mail:info@spiralstairs.uk.com Web_Site:http://www.spiralstairs.uk.com
Conservatory Roof Systems Ltd Bedwas Road Caerphilly M. Glam. CF83 1WG tel:(01222) 862900 fax:(01222) 884552 E_Mail:conservatoryroofsystems@ukbusiness.co m Web_Site:http://www.ukbusiness.com/conservat oryroofsystems
Consort Equipment Products Ltd Thornton Industrial Estate Milford Haven Pembrokeshire SA73 2RT tel:(01646) 692172 fax:(01646) 695195 E_Mail:consortepl @ msn. com Web_Site:http://www.consortepl.co.uk
Construction Confederation Construction House 56 - 64 Leonard Street London EC2A 4JX tel:(0171) 608 5000 fax:(0171) 608 5001 E_Mail:enquiries@constructionconfederation.co. uk Web_Site:http://www.constructionconfederation. co.uk
Construction Employers Federation 143 Malone Road Belfast BT9 6SU tel:(01232) 877143 fax:(01232) 877155 E_Mail:email@cefni.co.uk Web_Site:http://www.cefni.co.uk
Construction Industry Joint Council Construction House 56-64 Leonard Street London EC2A 4JX tel:(0171) 608 5027 fax:(0171) 608 5031
Construction Specialist (UK) Ltd Conspec House Bicester Road Aylesbury Bucks. HP19 3AF tel:(01296) 399 700
Contactum Ltd Victoria Works Edgware Road Cricklewood London NW2 6LF tel:(0181) 452 6366 fax:(0181) 208 3340 E_Mail:general@contactum.co.uk Web_Site:http://www.contactum.co.uk
Contano Aluminium Ltd Horseshoe Road Spalding Lincs. PE11 3JB tel:(01775) 724351 fax:(01775) 760414 telex:32196
Continental Shutters Ltd Unit 1 Heatway Industrial Estate Manchester Way, Wantz Road Dagenham Essex RH10 8PN tel:(0181) 517 8877 fax:(0181) 593 5721
Continental Sports Ltd Hill Top Road Paddock Huddersfield HD1 4SD tel:(01484) 542051 fax:(01484) 539148 E_Mail:sales@contisports.co.uk Web_Site:http://www.contisports.co.uk
Contour Showers Ltd Siddorn Street Winsford Cheshire CW7 2BA tel:(01606) 592586 fax:(01606) 861260 E_Mail:sales@contour-shower.co.uk Web_Site:http://www.contour-showers.co.uk
Contract Components Ltd 10 Woodall Road, Redburn Industrial Estate Enfield Middx. EN3 4LE tel:(0181) 805 3656 fax:(0181) 805 0558
Contract Flooring Association 4c St Marys Place The Lace Market Nottingham NG1 1PH tel:(0115) 941 1126 fax:(0115) 941 2238 E_Mail:info@cfa.org.uk Web_Site:http://www.cfa.org.uk
Contract Journal Business Alert Subscriptions Department, Quadrant House The Quadrant Sutton Surrey SM5 5AS tel:(0181) 652 4642 fax:(0181) 652 8958
Contract Maintenance Ltd Newgrow Avenue Dublin Ireland
Contravent Regal Ltd 5 Cowley Mill Trading Estate Longbridge Way Uxbridge Middx. UB8 2YG tel:(01895) 257766 fax:(01895) 257860 E_Mail:contravent@compuserve.com Web_Site:http://www.ourworld.compuserve.com/ home pages/ contravent
Coo-Var Ltd Elenshaw Works Lockwood Street Hull HU2 0HN tel:(01482) 328053 fax:(01482) 219266
Cooke Brothers Ltd Northgate Aldridge Walsall W. Midlands WS9 8TL tel:(01922) 53141 fax:(01922) 56227 E_Mail:sales@cookebrothers.co.uk
Cookson and Zinn Ltd Station Road Works Hadleigh Ipswich IP7 5PN tel:(01473) 825200 fax:(01473) 824164
Cooper & Turner Sheffield Road Sheffield S9 1RS tel:(0114) 256 0057 fax:(0114) 244 5529
Cooper Dryad Ltd Omega House Blackbird Road Leicester LE4 0AJ tel:(0116) 253 8844 fax:(0116) 251 3623
Cooper Group Ltd Unit 18, The Tanneries Brockhampton Lane Havent Hants. PO9 1JB tel:(01705) 454405 fax:(01705) 492732 E_Mail:coopergroup@msn.com
Cooper H W & Co Ltd Page House Page Walk London SE1 4SF tel:(0171) 237 1767 fax:(0171) 237 6480
Cooper Security Security House Xerox Business Park Mitcheldean Glos. GL17 0SZ tel:(01594) 545400 fax:(01594) 545400 E_Mail:enquiries@coopersecurity.co.uk Web_Site:http://www.coopersecurity.co.uk
Cope & Trimmins Ltd Angel Road Works Edmonton London N18 3AY tel:(0181) 803 6481 fax:(0181) 884 232
Copeland Corporation Ltd Colthrop Way Colthrop, Thatcham Berks. RG19 4NQ tel:(01635) 876161 fax:(01635) 877111
Copeley Decor Ltd Unit 1 Leyburn Business Park Leyburn N. Yorks. DL8 5QA tel:(01696) 623410 fax:(01696) 624398
Copper Development Association Verulam Industrial Estate 224 London Road St Albans Herts. AL1 1AQ tel:(01727) 731200 fax:(01727) 731216 E_Mail:copperdev@compuserve.com Web_Site:http://www.cda.org.uk
Coram (UK) Ltd Stanmore Industrial Estate Bridgnorth Shropshire WV15 5HP tel:(01746) 766466 fax:(01746) 764140 E_Mail:sales@coram.co.uk Web_Site:http://www.coram.co.uk
Corbett & Co (Galvanising) W Ltd Alexandra Works Hollies Road Wellington,Telford Shropshire TF1 1QJ tel:(01952) 412777 fax:(01952) 412888
Cordek Ltd Spring Copse Business Park Slinford W. Sussex RH13 7RD tel:(01403) 791717 fax:(01403) 791718 E_Mail:sales@cordek.com Web_Site:http://www.cordek.com
Cornwall Parker Ltd The Courtyard Frogmore Hygh Wycombe Bucks. HP13 5DD tel:(01494) 521144 fax:(01494) 461028
Corofil Woodall Ltd Unit 2 Nauigation Drive Hurst Business park Brierley Hill, West Midlands DY5 1UT tel:(01384) 489400 fax:(01384) 489401 telex:337121 E_Mail:sales@corofilwoodall.co.uk
Corroless International Nottingham Road Derby DE21 6AR tel:(01332) 295959 fax:(01332) 259252
Corroless International Association Regent House Regent Street Oldham Gtr. Manchester OL1 3TZ tel:(0161) 624 4941 fax:(0161) 627 5072
Corroless International Association Suite 25, IMEX Business Park Upper Villiers Street Wolverhampton WV2 4NU tel:(01902) 426990 fax:(07070) 716464
Corrugated Sheets & Profiles Ltd Ridgecare Road Black Lake West Bromwich W. Midlands B71 1BB tel:(0121) 553 6671 fax:(0121) 500 6133
Cottage Craft Spirals The Barn, Gorsty Low Farm The Wash Chapel-en-le-Firth High Peak SK23 OQL tel:(01663) 750716 fax:(01663) 751093 E_Mail:sales@cottspirals.u-net.com Web_Site:http://www.personal.u-net.com/cottspirals/home.htm
Cottam & Preedy Ltd 68 Lower City Road Tividale W. Midlands B69 2HF tel:(0121) 552 5281 fax:(0121) 454 9613 telex:4556363
Cottam Bros Ltd Wilson Street North Sheepfolds Industrial Estate Sunderland SR5 1BB tel:(0191) 567 1091 fax:(0191) 510 8187 E_Mail:cottam.bros@dial.pipex.com
Council for Aluminium in Building 191 Cirencester Road Charlton Kings Cheltenham Glos. GL53 8DF tel:(01242) 578278 fax:(01242) 578283
Country Manor Bricks The Brickyard East Wall Road Dublin 1 tel:(00 353) 1836 6901 fax:(00 353) 1855 4743 E_Mail:sales@cmb.ie Web_Site:http://www.cmb.ie
Courtaulds/Celanese Foleshill Road Coventry tel:(01203) 688771
Coverite Ltd Bridge Road Wood Green London N22 7SP tel:(0181) 888 7821 fax:(0181) 889 0731 E_Mail:frankc@lineone.net
Covers Timber Structures Ltd Sussex House Quarry Lane Industrial Estate Chichester W. Sussex PO19 2PE tel:(01243) 531818 fax:(01243) 537373
Cowley Structural Timberwork Ltd The Quarry Grantham Road Waddington Lincoln LN5 9NT tel:(01522) 720913 fax:(01522) 722778 E_Mail:cowleytimberwork@compuserve.com
Cox Building Products Ltd Ickfield Way Tring Herts. HP23 4RF tel:(0144) 282 4222 fax:(0144) 282 3351 telex:825389 E_Mail:coxdome@btinternet.com Web_Site:http://www.ascwebindex.com
Cox Peter Products Oakhurst Drive Cheadle Heath Stockport Cheshire SK3 0XT tel:(0161) 428 0622 fax:(0161) 428 1239
CPV Ltd Woodington Mill East Wellow Romsey Hants SO51 6DQ tel:(01794) 322884 fax:(01794) 322885 E_Mail:sales@cpv.co.uk Web_Site:http://www.cpv.co.uk
Cradley Castings Ltd Mill Street Cradley Halesowen W. Midlands B63 2UB tel:(01384) 560601 fax:(01384) 639181
Craft Jacobs Suchard Ltd Banbury Oxon. OX16 7QU tel:(01295) 264433 fax:(01295) 259018
Crane Fluid Systems Nacton Road Ipswich Suffolk IP3 9QH tel:(01473) 277300 fax:(01473) 277301 E_Mail:enquiries@crane-ltd.co.uk Web_Site:http://www.crane-ltd.co.uk
Cranleigh Brick & Tile Co Ltd Baynards Rudgwick Nr Horsham W. Sussex RH12 3AG tel:(01403) 823251 fax:(01403) 823351
Crawford Door Limited Wittle road Meir, Stoke-on -Trent Staffs. ST3 7QA tel:(01782) 599899 fax:(01782) 599989
Creative Systems Ltd Brooks Road E. Sussex BN7 2BY tel:(01273) 478881 telex:877354 CFC G
Crescent lighting 8 Rivermead, Pipers Lane Thatcham Berks. RG19 4EP tel:(01635) 878888 fax:(01635) 873888
Crescent of Cambridge Ltd Edison Road St. Ives Cambs. PE17 4LF tel:(01480) 301522 fax:(01480) 494001 E_Mail:info@crescentstairs.co.uk Web_Site:http://www.crescentstairs.co.uk
Creteco Sales Ltd 17 St Martin's Street Wallingford Oxon. OX10 0EA tel:(01491) 839488 fax:(01491) 833879
Crimegard Ltd Unit 14-19, Acton Business Centre School Road London NW10 6TD tel:(0181) 961 6630 fax:(0181) 961 6654
Crittall Steel Windows Ltd Springwood Drive Braintree Essex CM7 2YN tel:(0141) 427 4931 fax:(0141) 427 1463 E_Mail:hq@crittall-windows.co.uk Web_Site:http://www.crittall-windows.co.uk
Crittall Steel Windows Ltd 39 Durham Street Glasgow G41 1BS tel:(0141) 427 4931 fax:(0141) 427 1463 E_Mail:hq@crittall-windows.co.uk Web_Site:http://www.crittall-windows.co.uk
Croft Stone Services (Scotland) Ltd 379 Blythswood Court Anderston Centre Glasgow G2 7PH tel:(0141) 221 2003 fax:(0141) 221 4710
Crompton Lighting Ltd Wheatley Hall Road Doncaster S. Yorks. DN2 4NB tel:(01302) 344555 fax:(01302) 367155 E_Mail:sales@crompton.lighting.co.uk Web_Site:http://www.crompton-lighting.co.uk
Crompton Stud Welding Lathkill Street Market Harborough Leicester LE16 9EZ tel:(01858) 410600 fax:(01858) 466536
Crosbie Coatings Walsall Street Wolverhampton WV1 3LP tel:(01902) 352020 fax:(01902) 456392
Crown Nail Co Ltd Commercial Road Wolverhampton WV1 3QS tel:(01902) 351806 fax:(01902) 871212
Crown Trade Paints Akzo Nobel Decorative Coatings Crown House, P O Box 37, Hollins Road Darwen Lancs BB3 OBG tel:(01254) 704951 fax:(01254) 870155
Crowthorne Fencing Englemere Sawmill London Road Ascot Berks. SL5 8DG tel:(01344) 885451 fax:(01344) 893101
Croxton + Garry Ltd Curtis Road Dorking Surrey RH4 1XA tel:(01306) 886688 telex:859567
Crwford Door Ltd Whittle Road Meir,Stoke on Trent Staffs. ST3 7QA tel:(01782) 599899 fax:(01782) 599989
Cryotherm Insulation Ltd Hirst Wood Works Hirst Wood Road Shipley W. Yorks. BD18 4BU tel:(01274) 589175 fax:(01274) 593315
Cryselco Ltd Cryselco House 274 Ampthill Road Bedford MK42 9QL tel:(01234) 273355 fax:(01234) 210867
CSL (Services) Limited Malt Cottages Basford Nottingham NG7 7DX tel:(0115) 978 3177 fax:(0115) 942 2798
CSR Humes (UK) Ltd Oakley House London Road Hartley Witney, Basingstoke Hants. RG27 8PE tel:(01252) 844901 fax:(01252) 845157
CSR Humes Ltd Oakley House London Road Hartley Wintney, Basingstoke Hants. RG27 8PE tel:(01252) 844901 fax:(01252) 845157
CSSP (Computer Software Services Partnership) One West Street Bromley Kent BR1 1RE tel:(0181) 460 0022 fax:(0181) 460 1196 E_Mail:enq@cssp.demon.co.uk Web_Site:http://www.cssp.demon.co.uk
CTS Ltd Abbey Road Shepley, Huddersfield W. Yorks. HD8 8BX tel:(01484) 606416 fax:(01484) 608763
CU Lighting Ltd Great Amwell Ware Herts. SG12 9AT tel:(01920) 462272 fax:(01920) 461370 telex:81398

Cumberland Construction Units 6c-6f Albright Ind Estate Ferry Lane Rainham Essex RM13 9YH tel:(01708) 555226 fax:(01708) 520231 E_Mail:contracts@cumberland-construction.co.uk Web_Site:http://www.cumberlandconstruction.co.uk

Cummins Power Generation Ltd Manston Park Columbus Avenue, Manston Ramsgate Kent CT12 5BF tel:(01843) 255000 fax:(01843) 255902 E_Mail:cpg.uk@cummins.com

Cuprinol Ltd Adderswell Frome Somerset BA11 1NL tel:(01373) 465151 fax:(01373) 474124 telex:44269

Curtis Steel Ltd Brookside Works Mill Road Radstock Bath BA3 5TX tel:(01761) 432841 fax:(01761) 433919

Cutler Hammer UK Mill Street Ottery St. Mary Devon EX11 1AG tel:(01404) 812131 fax:(01404) 815471

Cutting R C & Co Arcadia Avenue London N3 2JU tel:(0181) 371 0001 fax:(0181) 371 0003

Cyclone Stripper Ltd Phoenix Works 8 Avery Hill Road New Eltham London SE9 2BD tel:(0181) 850 1458

D. D Lamson plc Harbour Road Gosport Hampshire PO12 1BG tel:(01705) 584271 fax:(01705) 504648 E_Mail:marketing@lamson.co.uk Web_Site:http://www.lamson.co.uk

DA Systems Ltd Cranleigh Gardens Southall Middx. UB1 2BZ tel:(0181) 843 2282 fax:(0181) 843 9637 E_Mail:email@dasystems.co.uk Web_Site:http://www.dasystems.co.uk

Dacrylate Paints Ltd Lime Street Kirkby in Ashfield Nottingham NG17 8AL tel:(01623) 753845 fax:(01623) 757151 E_Mail:sales@dacrylate.co.uk Web_Site:http://www.dacrylate.co.uk

Dalair Ltd Southern Way Wednesbury W. Midlands WS10 7BU tel:(0121) 556 9944 fax:(0121) 502 3124

Dale Joinery Ltd Queensway Rochdale Lancs. OL11 2PR tel:(01706) 350350 fax:(01706) 350351

Dale Power Systems Electricity Buildings Filey N. Yorks. YO14 9PJ tel:(01723) 514141 fax:(01723) 515723

Dales Fabrications Ltd Crompton Road Industrial Estate Ilkeston Derbys. DE7 4BG tel:(0115) 930 1521 fax:(0115) 930 7625 E_Mail:sales@dales-eaves.co.uk

Dalesauna Ltd Chatsworth Road Harrogate N. Yorks. NG1 5HS tel:(01423) 522241 fax:(01423) 509450

Dallas & Forrest Ltd 21 Old Castle Road Cathcart Glasgow G44 4BM tel:(0141) 637 3073 fax:(0141) 637 1582

Dampa (UK) Ltd Berinsfield Wallingford Oxford OX10 7LZ tel:(01865) 342200 fax:(01865) 341482

Dampcoursing Limited 10-12 Dorset Road Tottenham London N15 5AJ tel:(0181) 802 2233 fax:(0181) 809 1839

Danbury Fencing Limited Olivers Farm Maldon Road Witham Essex CM8 3HY tel:(01376) 502020 fax:(01376) 520500

Danfoss Ltd Perivale Industrial Park Horsenden Lane South Greenford Middx. UB6 7QE tel:(0181) 991 7000 fax:(0181) 991 7171 Web_Site:http://www.danfoss.co.uk

Danfoss Randall Ltd Ampthill Road Bedford Beds. MK42 9ER tel:(01234) 364621 fax:(01234) 219705 E_Mail:danfossrandall@danfoss.com Web_Site:http://www.danfoss-randall.co.uk

Dantherm Ltd Hither Green Clevedon Avon BS21 6XT tel:(01275) 876851 fax:(01275) 343086 E_Mail:ikf.ltd.uk@dantherm.com

Darfen Durafencing, Central Region 15 - 21 Speedwell Road Birmingham W. Midlands B25 8HU tel:(0121) 7728666 fax:(0121) 7728648

Darfen Durafencing, Eastern Region Clearview Works Norwich Road Lenwade Norwich NR9 5SL tel:(01603) 872060 fax:(01603) 871093

Darfen Durafencing, North East Region 3 Swan Road, South West Industrial Estate Peterlee Co. Durham SR8 2HS tel:(0191) 5862200 fax:(0191) 5865533

Darfen Durafencing, North West Region Bradman Road Knowsley Industrial Park North Liverpool L33 7UR tel:(0151) 547 3626 fax:(0151) 549 1205

Darfen Durafencing, Northern Region Carr Hill Doncaster S. Yorks. DN4 8DQ tel:(01302) 360242 fax:(01302) 364359 telex:547981

Darfen Durafencing, Scottish Region Units 7-8 Rochsolloch Road Industrial Estate Airdrie Scotland ML6 9BA tel:(01236) 755001 fax:(01236) 747012

Dartford Portable Buildings & Supplies Ltd 389-397 Princes Road Dartford Kent DA1 1JU tel:(01322) 229521 fax:(01322) 221948

Daryl Industries Ltd Alfred Road Wallasey Wirral L44 7HY tel:(0151) 606 5000 fax:(0151) 638 0303 E_Mail:daryl@daryl-showers.co.uk

Dauphin PLC Peter Street Blackburn Lancs. BB1 5HL tel:(01254) 52220 fax:(01254) 680401 E_Mail:sb@dauphin.plc.uk Web_Site:http://www.dauphin.plc.uk

Davant Products Ltd Davant House, Jugs Green Business Park Jugs Green Staplow, nr Ledbury Herefordshire HR8 1NR tel:(01531) 640880 fax:(01531) 640827

David Ball Group Plc David Ball Buildings Huntington Road Bar Hill Cambridge CB3 8HN tel:(01954) 780687 fax:(01954) 782912 E_Mail:sales@davidballgroup.com Web_Site:http://www.davidballgroup.com

Davidson C & Sons Mugiemoss Mills Bucksburn Aberdeen AB2 9AA tel:(01224) 712821 fax:(01224) 712453 telex:73185

Davies John Interiors Ltd Century House Century Road Oldbury W. Midlands B69 4AD tel:(0121) 511 1000 fax:(0121) 511 1116

Davis Burrow & Sons Wilson Street North Sunderland Tyne & Wear SR5 1BB tel:(0191) 567 5889 fax:(0191) 510 8187 E_Mail:cottam.bros@dial.pipex.com

Davis International Ltd 4-6 Rookwood Way Haverhill Suffolk CB9 8PB tel:(01440) 704411 fax:(01440) 702822

Davis Ridley & Co Ltd Beaconsfield Road Hatfield Herts. AL10 8BA tel:(01707) 264237 fax:(01707) 268448

Davis Security Communications Ltd Davis House Barrow Road Sheffield S9 1JQ tel:(01742) 431577 fax:(01742) 430108

Dawson McDonald & Dawson Ltd Compton Works Ashbourne Derbys. DE6 1DB tel:(01335) 343184 fax:(01335) 342540 E_Mail:dmd@ndirect.co.uk

Dawson-Keith Ltd Dekay House Brookhampton Lane Havant Hants. PO9 1HQ tel:(01705) 474122 fax:(01705) 470921 telex:86491 DEEKAY G

Day Marketing Ltd 64A Guildford Street Chertsey Surrey KT16 9BD tel:(01932) 567341 fax:(01932) 568858

Daylight Centre Princes Drive Industrial Estate Kenilworth Warwks. CV8 2FD tel:(01926) 511411 fax:(01926) 854155 E_Mail:daylightcentre@btinternet.com

DCE - BTR Environmental Ltd Humberstone Lane Thurnaston Leicester LE4 8HP tel:+44 (0)116 269 6161 fax:+44 (0)116 269 3028 E_Mail:environment@btrinc.com Web_Site:http://www.btrenvironmental.co.uk

De Longhi Radiators srl 19 Intake Lane Dunnington York YO1 5NX tel:(01904) 488212

Deakin Davenset Rectifiers Ltd Hunters Lane Rugby Warwks. CV21 1EA tel:(01788) 541326 fax:(01788) 540937 E_Mail:davenset@aol.com.uk

Deanes Furniture Ltd Wycombe Lane Wooburn Green, nr High Wycombe Bucks. HP10 0HH tel:(01628) 525011

Deans Blinds & Awnings (UK) Ltd Unit 4, Haslemere Industrial Estate Ravensbury Terrace London SW18 4SE tel:(0181) 947 8931 fax:(0181) 947 8336 E_Mail:info@deansblinds.com Web_Site:http://www.deansblinds.com

Deb Ltd Spencer Road Belper Derbys. DE56 1JX tel:(01773) 596700 fax:(01773) 822548 E_Mail:enquiry@deb.co.uk Web_Site:http://www.deb.co.uk

Deborah Services Initial Plant Services 10 South Parade Wakefield W. Yorks. WF1 1LS tel:(01924) 378222 fax:(01924) 366250

Deceuninck Ltd Unit 2 Stanier Road, Port Marsh Industrial Estate Calne Wilts. SN11 9PX tel:(01249) 816969 fax:(01249) 815234 E_Mail:marketing@deceuninck.co.uk Web_Site:http://www.deceuninck.com

Decorfix Chapel Works Chapel Lane Lower Halstow, Sittingbourne Kent ME9 7AB tel:(01795) 843124 fax:(01795) 842465

Decra Cubicles Ltd 34 Forest Business Park South Access Road Walthamstow London E17 8BA tel:(0181) 520 4371 fax:(0181) 521 0605

Decra Roofing Systems Unit 7, Church Road Industrial Estate Lowfield Heath Crawley W. Sussex RH11 0YA tel:(01293) 545058 fax:(01293) 562709 E_Mail:technical@decra.co.uk Web_Site:http://www.decra.co.uk/

Dee-Organ Ltd 5 Sandyford Road Paisley Renfrewshire PA3 4HP tel:(0141) 889 7000 fax:(0141) 889 7764

Deepdale Engineering Co Ltd Pedmore Road Dudley W. Midlands DY2 0RD tel:(01384) 480022 fax:(01384) 480489 E_Mail:sales@deepdale-eng.sbx.co.uk Web_Site:http://www.sbx.co.uk/deepdale-eng

Degussa Ltd Winterton House Winterton Way Macclesfield Cheshire SK11 0LP tel:(01625) 503050 fax:(01625) 502096

DEL Piling Contractors Park Road Holmewood Industrial Park Chesterfield Derbys. S42 5UY tel:(01246) 855855 fax:(01246) 854909

Delabole Slate Pengelly Delabole Cornwall PL33 9AZ tel:(01840) 212242 fax:(01840) 212948 E_Mail:info@delaboleslate.com Web_Site:http://www.delaboleslate.com

Delbridge Lifts Ltd Steetley Industrial Estate Sangwin Road Corseley, Bilston W. Midlands WV14 9EE tel:(01902) 662998 fax:(01902) 674998

Deloro Satellite Ltd Stratton St Margaret Swindon Wilts. SN3 4QA tel:(01793) 822451 fax:(01793) 823814 telex:44215

Delta Crompton Cables Ltd Milmarsh Lane Brimsdown Enfield Middx. EN3 7QD tel:(0181) 804 2468 fax:(0181) 443 2281 telex:261749

Denco Ltd Holmer Road Hareford HR4 9SJ tel:(01432) 277277 fax:(01432) 268005 telex:35144

Denka International Red Roofs Chinnor Road Thame Oxon. OX9 3RF tel:(01844) 216754 fax:(01844) 216141 E_Mail:101336.1667@compuserve.com

Denmans Montrose Ltd 28-30 South Bank Business Centre Ponton Road Vauxhall London SW8 5BL tel:(0171) 622 1221 fax:(0171) 622 5152

Dennis & Robinson Ltd Blenhiem Road Churchill Industrial Estate Lancing W. Sussex BN15 8HU tel:(01903) 524300 fax:(01903) 750679

Department of Environment Romney House 43 Marsham Street London SW1B 3EB tel:(0171) 276 3000

Derby Group PLC The Derby House Sunningdale Road Scunthorpe S. Humberside DN17 2SS tel:(01724) 280044 fax:(01724) 868295

Derby Timber Supplies 3 John Street Derby DE1 2LU tel:(01332) 348340 fax:(01332) 385573

Dermide Ltd Westfield Mill Carr Road Barnoldswick Lancs. BB18 5UU tel:(01282) 812581 fax:(01282) 812366

Derwent Macdee Ltd Warmsworth Halt Industrial Estate Warmsworth, Doncaster S. Yorks. DN4 9SL tel:(01302) 310666 fax:(01302) 856421

Design Line 45 Lawley Middleway Aston Birmingham B4 7XH tel:(0121) 359 6941 fax:(0121) 359 5758

Design Windows (South West) Ltd Unit 5 Aldermoore Way Longwell Green Bristol BS15 7DA tel:(0117) 960 5717 fax:(0117) 867 2346

Designed for Sound Ltd 61-67 Rectory Road Wivenhoe Essex CO7 9ES tel:(01206) 827171 fax:(01206) 826936

Designplan Lighting Ltd Wealdstone Road Kimpton Industrial Estate Sutton Surrey SM3 9RW tel:(0181) 254 2020 fax:(0181) 644 4253

Desking Systems Ltd Warpsgrove Lane Chalgrove Oxfordshire OX44 7TH tel:(01865) 891444 fax:(01865) 891427 E_Mail:sales@desking.co.uk Web_Site:http://www.desking.co.uk

Desney Products Units 5-6 Marley Industrial Estate Southam Road Banbury Oxon. OX16 7RL tel:(01295) 270848 fax:(01295) 270965

Deva Tap Co Ltd Brooklands Mill English Street Leigh Lancs. WN7 3EH tel:(01942) 680177 fax:(01942) 680190

Dew Construction Limited 1st Floor, 10 Lake End Court Taplow Road Taplow, Maidenhead Berks. SL9 0JQ tel:(01628) 669888 fax:(01628) 669800

Dewey Waters Ltd Cox Green Wrington Bristol BS40 5QS tel:(01934) 862601 fax:(01934) 862604 E_Mail:sales@deweywaters.co.uk Web_Site:http://www.deweywaters.co.uk

Dewhurst plc Inverness Road Hounslow Middx. tel:(0181) 570 7791 fax:(0181) 572 5986

Dewplan Ltd Maple Court 17-21 Queens Road High Wycombe Bucks. HP13 6AQ tel:(01494) 557300 fax:(01494) 465489 E_Mail:post@dewplan.co.uk

Dex-o-Tex International Ltd Unit 15, Blenheim Road Cressex Business Park High Wycombe Bucks. HP12 3RS tel:(01494) 452515 fax:(01494) 465480 E_Mail:crossfield.uk@btinternet.com Web_Site:http://www.dexotex.com

Dexine Rubber Co Ltd Spotland Bridge Works Rochdale Lancs. OL12 6AU tel:(01706) 40011 fax:(01706) 527714

Dexion Ltd Maylands Avenue Hemel Hempstead Herts. HP2 7DE tel:(01442) 242261 fax:(01442) 217145 E_Mail:marketing@racking.co.uk Web_Site:http://www.dexion.com

Dexter Paints Ltd Dexter House Trafalgar Street Burnley Lancs. B11 1RE tel:(01282) 23361 fax:(01282) 414573

Dezurik International Ltd Nelson Way Nelson Industrial Estate Cramlington Northumberland NE23 9BJ tel:(01670) 714111 fax:(01670) 737813

Diamond Merchants 43 Acre Lane Brixton London SW2 5TN tel:(0171) 274 6624 fax:(0171) 978 8370

Dicon Safety Products (UK) Ltd Unit 7, Manchester Park Tewkesbury Cheltenham Glos. GL51 9EJ tel:(01242) 516241 fax:(01242) 222935

Didsbury Engineering Co Ltd Clifton works Manor Road Levenshulme Manchester M19 3EJ tel:(0161) 224 6224 fax:(0161) 224 2098 E_Mail:sales@didsbury.com

Diespeker Marble & Terrazzo Ltd 132-136 Ormside Street London SE15 1TF tel:(0171) 358 0160 fax:(0171) 639 1695 E_Mail:sales@diespeker.demon.co.uk

Dimplex (UK) Ltd Millbrook Southampton SO15 0AW tel:(01703) 777117 fax:(01703) 771096

Dinnington Fencing Co. Limited North Hill Dinnington Newcastle-upon-Tyne NE13 7LG tel:(01661) 824046 fax:(01661) 872234

Direct Wire Ties Ltd Kingston International Business Park Somerden Road Hull East Yorkshire HU9 5PE tel:(01482) 712630 fax:(01482) 707426

Direction Group Ltd 4 Riverside Business Centre Walnut Tree Close Guildford Surrey GU1 4UG tel:(01483) 455555 fax:(01483) 451590

Dividers Ltd Great Gutter Lane Willerby Hull East Yorkshire HU10 6BS tel:(01482) 651331 fax:(01482) 651497

Dixon Turner Wallcoverings Ltd 29 Store Street London WC1E 7BT tel:(0171) 636 4032 fax:(0171) 636 4044

Dixon Turner Wallcoverings Ltd. 29 Store Street London WC1E 7BT tel:(0171) 636 4032 fax:(0171) 636 4044

DJ Profiles Stafford Park 15 Telford Shropshire TF3 3BB tel:(01952) 290303 fax:(01952) 290226 E_Mail:sales@djprofiles.co.uk Web_Site:http://www.djprofiles.co.uk

DLW Flooring Ltd Centurian Court Milton Park Abingdon Oxon. OX14 4RY tel:(01235) 831296 fax:(01235) 861016 E_Mail:sales@dlw.co.uk Web_Site:http://www.dlw.co.uk

Docherty H Ltd Unit 3 Sawmill Road, Redshute Hill Indusrial Estate Hermitage, Newbury Berks. RG18 9QL tel:(01635) 200145 fax:(01635) 201737

Dockra Concrete Co Ltd, The Old Mill Quarry Beith North Ayrshire Scotland KA15 1HY tel:(01505) 502721 fax:(01505) 506211

Doherty Medical Ltd Doherty House 278 Alma Road Enfield Middx. EN3 7BH tel:(0181) 804 1244 fax:(0181) 804 9314

Dolphin Showers Ltd Bromwich Road Worcester WR2 4DB tel:(01905) 748500 fax:(01905) 429034

Dom-Nemef-Corbin Unit 1, Harolds Close Harolds Road Harlow Essex CM19 5TH tel:(01279) 454709 fax:(01279) 454711

Domus Tiles Ltd 33 Parkgate Road Battersea London SW11 4NP tel:(0171) 223 5555 fax:(0171) 924 2556 E_Mail:info@domustiles.co.uk Web_Site:http://www.domustiles.co.uk

Domus Ventilation Ltd. Bearwalden House, Bearwalden Business Park Royston Road Wendens Ambo, Saffron Waldon Essex CB11 4JX tel:(01799) 540602 fax:(01799) 541143 E_Mail:info@domusducting.co.uk Web_Site:http://www.domusducting.co.uk

Don & Low Ltd Newfordpark House Glamis Road Forfar Angus DD8 IFR tel:(01307) 452200 fax:(01307) 452201

Don Engineering (South West) Ltd Wellington Trading Estate Wellington Somerset TA21 8SS tel:(01823) 663181 fax:(01823) 665889

Donkin (Bryan Donkin Co Ltd) Derby Road Chesterfield S40 2EB tel:(01246) 273153 fax:(01246) 235273

Door Panels PLC Rectory Road Upton-upon-Severn Worcs. WR8 0LX tel:(01684) 594561 fax:(01684) 593431

Doorfit Products Ltd Icknield House Heaton Street, Hockley Birmingham W. Midlands B18 5BA tel:(0121) 523 4171 fax:(0121) 554 3859

Doors & Hardware Ltd Taskmaster Works Maybrook Road Minworth, Sutton Coldfield W. Midlands B76 1AL tel:(0121) 351 5276 fax:(0121) 313 1228

Dorma Door Controls Ltd Dorma Trading Park Staffa Road London E10 7QX tel:(0181) 558 8411 fax:(0181) 558 6122

DORMA Entrance Systems Ltd Woodside Industrial Park Works Road Letchworth Herts. SG6 1LA tel:(01462) 480544 fax:(01462)480588 E_Mail:dormaentrance@msn.com Web_Site:http://www.dorma.co.uk

Dorman Smith Switchgear Ltd Blackpool Road Preston Lancs. PR2 2DQ tel:(01772) 728271 fax:(01772) 726276

Douglas Industrial Ltd Wards Farm Industrial Estate Grennmoore, Woodcote Reading RG8 0RB tel:(01491) 682100 fax:(01491) 682037

Dover Engineering Works Ltd Dour Iron Foundrey Dover Kent CT16 2LE tel:(01304) 203545 telex:963171

Dover Trussed Roof Co Ltd Shelvin Manor Shelvin, Canterbury Kent CT4 6RL tel:(01303) 844303 fax:(01303) 844342

Dow Construction Products Lakeside House Stockley Park Uxbridge Middx. UB11 1BE tel:(0181) 848 5050 fax:(0181) 848 5413 telex:934626

Dow Corning Hansil Ltd Meridian Business Park Copse Drive, Allesley Coventry CV5 9RG tel:(01676) 528000 fax:(01676) 528001

Down & Francis Ltd, Division of Metalrax PLC Ardath Road Kings Norton Birmingham B38 9PN tel:(0121) 433 3300 fax:(0121) 459 9222 E_Mail:down_francis@csi.com

DPC Screeding Ltd Brunwick Industrial Estate Brunswick Village Newcastle-upon-Tyne NE13 7BA tel:(0191) 236 4226 fax:(0191) 236 2242

DPS Security Systems Ltd Security House 62A Spital Street Dartford Kent DA1 2DT tel:(01322) 278178 fax:(01322) 284315 E_Mail:dpsecurity@btinternet.com Web_Site:http://www.dpsecurity.co.uk

Draeger Ltd Ullswater Close Kitty Brewster Industrial Estate Blyth Northumberland NE24 4RG tel:(01670) 352891 fax:(01670) 356266 E_Mail:richard.beckwith@dreagar.ltd.uk

Dragonair Ltd Fitzherbert Road Units 7-8, Dragon Industrial Centre Fallington, Portsmouth Hants. PO6 1SQ tel:(01705) 376451 fax:(01705) 370411

Drainage Systems Brickhouse Lane West Bromwich West Midlands B70 0DX tel:(0121) 521 2900 fax:(0121) 522 2197 E_Mail:drainage.system@dial.pipex.com

Drakard & Humble Ltd Brunleys Kiln Farm Milton Keynes MK11 3EP tel:(01908) 567909 fax:(01908) 563527 E_Mail:sales@drakardsigns.co.uk

Draught Proofing Advisory Association PO Box 12 Haslemere Surrey GU27 3AH tel:(01428) 654011 fax:(01428) 651401 E_Mail:113052.2106@compuserve.com Web_Site:http://www.nationline.co.uk/ceed

Dravo Environmental Services Sir Henry Parkes Road Canley Coventry CV5 6BN tel:(01203) 717170 fax:(01203) 714978

Drawn Metal Ltd Swinnow Lane Bramley Leeds LS13 4NE tel:(0113) 256 5661 fax:(0113) 239 3194 E_Mail:drawmet@btinternet.com Web_Site:http://www.drawmet.com

Drayton Controls (Engineering) Ltd Chantry Close West Drayton Middx. UB7 7SP tel:(01895) 444012 fax:(01895) 441288

Drugasar Ltd Deans Road Swinton Manchester M27 3JH tel:(0161) 793 8700 fax:(0161) 727 8057

Dry Stone Walling Association of Great Britain c /o YFC Centre National Agricultural Centre Stoneleigh Park Warwks. CV8 2LG tel:(0121) 378 0493 fax:(0121) 378 0493 E_Mail:j.simkins@dswagb.ndirect.co.uk Web_Site:http://www.dswagb.ndirect.co.uk/dswa

Dryad Architectural Ltd Orion House, Queens Drive Kings Norton Business Centre Birmingham B30 3HH tel:(0121) 458 6387 fax:(0121) 458 6152 E_Mail:dryad@btconnect.com

Dryad Deelock Ltd 27 Howard Business Park Howard Close Waltham Abbey Essex EN9 1XE tel:(01992) 653344 fax:(01992) 652375

DSM Engineering Plastic Products Ltd 83 Bridge Road East Welwyn Garden City Herts. AL7 1LA tel:(01707) 361800 fax:(01707) 361801

DSM UK Ltd DSM House Papermill Drive Redditch Worcs. B98 8QJ tel:(01527) 590590 fax:(01527) 590555

Du Pont (UK) Ltd Maylands Avenue Hemel Hempstead Herts. HP2 7DP tel:(01442) 61251 fax:(01442) 249463

Ducal Ltd Andover Hants. SP10 5AZ tel:(01264) 332720 fax:(01264) 334046

Ducana Surrey Ltd Elmsleigh School House 3 The Fairfield Farnham Surrey GU9 8AH tel:(01252) 715768 fax:(01252) 737152

Duct Access Covers Ltd Newtown Industrial Estate Crosskeys Newport Gwent NP1 7PZ tel:(01495) 272727 fax:(01495) 270978

Ductform Engineering Ltd 141 Barkby Road Leicester LE4 7LW tel:(0116) 276 6636 fax:(0116) 240 0426

Ductile Sections Planetary Road Willenhall W. Midlands WV13 3SW tel:(01902) 739739 fax:(01902) 862841

Dudley Factory Doors Limited Unit G6 Grice Street West Bromwich West Midlands B70 7EZ tel:(0121) 558 8989 fax:(0121) 558 4616

Dudley Thomas Ltd PO Box 28 Birmingham New Road Dudley DY1 4SN tel:(0121) 557 5411 fax:(0121) 557 5345 E_Mail:info@thomasdudley.co.uk

Dufaylite Developments Ltd Cromwell Road St Neots Huntingdon Cambs. PE19 1QW tel:(01480) 215000 fax:(01480) 405526

Dunbrik (Yorks) Ltd Ferry Lane Stanley Ferry Wakefield West Yorkshire WF3 4LT tel:(01924) 373694 fax:(01924) 383459

Dunham Bush Ltd European Headquarters Downley Road Havant Hants. PO9 2JD tel:(01705) 477700 fax:(01705) 450396

Dunhams of Norwich Hellesdon Park Road Norwich NR6 5DR tel:(01603) 424855

Dunlop Adhesives Chester Road Birmingham B35 7AL tel:(0121) 373 8101 fax:(0121) 384 2826

Dunphy Combustion Ltd Queensway Rochdale Lancs. OL11 2SL tel:(01706) 649217 fax:(01706) 655512 Web_Site:http://www.dunphy.co.uk

Dunsley Heat Ltd Fearnough Huddersfield Road Holmfirth W. Yorks. HD7 2TU tel:(01484) 682635 fax:(01484) 688428 Web_Site:http://www.i-inetdunsleyht

Duo-Fast (UK) Ltd Northfield Drive, Northfield Milton Keynes Bucks MK15 0DR tel:(01908) 667788 fax:(01908) 672689 E_Mail:info@duo-fast.co.uk

Duplus Domes Ltd 370 Melton Road Leicester Leicestershire LE4 7SL tel:(0116) 261 0710 fax:(0116) 261 0539 E_Mail:sales@duplus.co.uk Web_Site:http://www.duplus.co.uk

Duplus Domes Ltd 370 Melton Road Leicester LE4 7SL tel:(01533) 610710 fax:(01533) 610539

Dupox (Epoxy Resins) Ltd Sadler Street Industrial Estate Church, Accrington Lancs. BB5 0HP tel:(01254) 396062 fax:(01254) 872783

Dupre Vermiculite Tamworth Road Hertford Herts SG13 7DL tel:(01992) 582541 fax:(01992) 553436 E_Mail:dupre@microfine.co.uk Web_Site:http://www.dupre.microfine.co.uk

Durabella Ltd Talisman Square Kenilworth Warwickshire CV8 1JB tel:(0870) 7894000 fax:(0870) 7894100

Durable Berkeley Co Ltd 10-13 Southview Park Marsack Street Caversham Reading RG4 5AF tel:(0118) 948 3500 fax:(0118) 946 2114 telex:849021 FRANG E_Mail:durable_co@msn.com Web_Site:http://www.durable.co.uk

Durable Contracts Ltd Durable House Crabtree Manorway Belvedere Kent DA17 6AB tel:(0181) 311 1211 fax:(0181) 310 7893

Duralock UK Ltd 36 Springfield Road Elburton Plymouth Devon PL9 8EN tel:(01752) 484085 fax:(01752) 484086 Web_Site:http://www.business.virgin.net/duraloc k.ltd

Duratan Ltd West Street Buckingham MK18 1HE tel:(01280) 814048 fax:(01280) 817842 telex:837005 DURTAN G

Durey Casting Ltd Hawley Road Dartford Kent DA1 1PU tel:(01322) 272424 fax:(01322) 288073

Durox Building Products Ltd Linford Stanford le Hope Essex SS17 0PY tel:(01375) 673344 fax:(01735) 642460

Duroy Fibreglass Mouldings Unit 1 Mercury Yacht Harbour Satchell Lane Hamble Southampton SO31 4HQ tel:(01703) 453781 fax:(01703) 455538

Dycem Ltd Ashley Trading Estate Bristol BS2 9BB tel:(0117) 955 9921 fax:(0117) 954 1194 E_Mail:uk@dycem.com Web_Site:http://www.dycem.com

Dyer Environmental Controls 1 Brooklyn Road Cheadle Cheshire SK8 1BS tel:(0161) 4914840 fax:(0161) 4914841

Dyke Chemicals Ltd PO Box 158 Cardiff CF1 9RU tel:(01932) 866096 fax:(01932) 866097

Dynaspan (UK) Ltd Parkhouse High Street Colnbrook, Slough Berks. SL3 0LX tel:(01753) 689977 fax:(01753) 689970

Dyno Rod PLC Zockoll House 143 Maple Road Surbiton Surrey KT6 4BJ tel:(0181) 481 2200 fax:(0181) 481 2288 E_Mail:postmaster@dyno.com

E. H. S. Roofing Ltd 42 Flint Green Road Acocks Green Birmingham B27 6QA tel:(0121) 764 4500 fax:(0121) 706 2650

Eagles, William, Ltd Liverpool Street Salford Manchester M5 4LP tel:(0161) 736 1661 fax:(0161) 745 7765

EAP International Ltd Flexion Division, Manchester Industrial Centre Water Street Manchester M3 4JU tel:(0161) 832 6784 fax:(0161) 835 2619

Earth Anchors Ltd 15 Campbell Road Croydon Surrey CR0 2SQ tel:(0181) 684 9601 fax:(0181) 684 2230 Web_Site:http://www.earth-anchors.com

Ease-E-Load Trolleys Ltd Crown Works Baltimore Road Perry Barr Birmingham B42 1DP tel:(0121) 356 7411 fax:(0121) 344 3358

Easi-Fall International Ltd Brindley House Brindley Avenue Sale Cheshire M33 7BE tel:(0161) 973 0304 fax:(0161) 969 5009 E_Mail:e.fallint@zen.co.uk

Eastdale R M & Co Ltd 67 Washington Street Glasgow G3 8BB tel:(0141) 204 2708 fax:(0141) 204 3159 telex:77464

Easy Arches Ltd Lodgefield Road Halesowen W. Midlands B62 8RT tel:(0121) 559 5613 fax:(0121) 561 2895

Eaton Elizabeth 85 Bourne Street London SW1W 8HF tel:(0171) 730 2262 fax:(0171) 730 7294

Eaton Ltd Plymbridge Road Estover Plymouth PL6 7PN tel:(01752) 731155 fax:(01752) 737372

Eaton-Williams Group Ltd Station Road Edenbridge Kent TN8 6EG tel:(01732) 866055 fax:(01732) 863461

EBC UK Ltd Unit 16, Old brewery Yard Kilton Road Worksop Notts. S80 2DE tel:(01909) 479276 fax:(01909) 479278 E_Mail:rob@ebc.force9.co.uk

Eberle GmbH 25 Gosforth Close Sandy Business Park Sandy Beds. SG19 1RB tel:(01767) 692323 fax:(01767) 692333

Ebor Concretes Ltd PO Box 4 Ure Bank Ripon N. Yorks. HG4 1JE tel:(01765) 604351 fax:(01765) 690065 E_Mail:sales@eborconcrete.co.uk

EBS Services Ltd Thames House Longreach Road Barking Essex IG11 0JR tel:(0181) 594 5255 fax:(0181) 594 5404

ECC Building Products Ltd Head Office Hulland Ward Ashbourne Derbys. DE6 3ET tel:(01335) 372222 fax:(01335) 370074

ECC International Ltd Westwood Beverley N. Humberside HU17 8RQ tel:(01482) 881234 fax:(01482) 872301

ECC Spares & Services Ltd Bushbury Engineering Works Showell Road Wolverhampton WV10 9LN tel:(01902) 27831 fax:(01902) 865237 telex:339618

Eclipse Blind Systems Ltd Inchinnan Business Park !0 Founain Cress Renfrew Renfrewshire PA4 9RE tel:(0141) 812 3322 fax:(0141) 812 5253 telex:113747,2612 E_Mail:email@compuserv.com Web_Site:http://www.eclipse-blinds.com

Eclipse Sprayers Ltd 120 Beakes Road Smethwick W. Midlands B67 5AB tel:(0121) 420 2494 fax:(0121) 429 1668

Ecolec Dale Street Bilston W. Midlands WV14 7JU tel:(01902) 353752 fax:(01902) 353144 E_Mail:ecolec@thama.co.uk Web_Site:http://www.thama.co.uk

Ecophon Ltd Old Brick Kiln Ramsdell Tadley Hants. RG26 5PP tel:(01256) 850977 fax:(01256) 850600 E_Mail:info@ecophon.co.uk Web_Site:http://www.ecophon.co.uk

Ecowater Systems Ltd Unit 1, The Independant Business Park Mill Road Stockenchurch Bucks. HP14 3TP tel:(01494) 484000 fax:(01494) 484396

ECS Lighting Controls Ltd Enterprise House Central Way Feltham Middx. TW14 0RX tel:(0181) 751 6514 fax:(0181) 890 438 E_Mail:mail@ecs-control.co.uk Web_Site:http://www.ecs-control.co.uk

Edenaire Ltd Station Road Edenbridge Kent TN8 6EG tel:(01732) 866066 fax:(01732) 866653

EDL Cable Supports Ltd Redbrook Lane Brereton Rugeley Staffs. WS15 1QY tel:(01889) 582112 fax:(01889) 584012

Edmonds A & Co Ltd 91 Constitution Hill Birmingham B19 3JY tel:(0121) 236 8351 fax:(0121) 236 4793 telex:337781

EDS Shutter Door Systems Ltd Sava Works Barlow Way, Fairview Industrial Park Rainham Essex RM13 8UE tel:(01708) 551964 fax:(01708) 551964

EFG Business Furniture (UK) Ltd 3 Finway Road Hemel Hempstead Herts. HP2 7PT tel:(01442) 216666 fax:(01442) 217023

Ega Ltd St Asaph Clwyd N. Wales Ll17 0ER tel:(01745) 584927

Egetaepper (UK) Ltd Ege House Drumhead Road, Chorley North Business Park Chorley Lancs. PR6 7BZ tel:(01257) 239000 fax:(01257) 239001

Egger (UK) Ltd Anick Range Road Hexham Northumberland NE46 4JS tel:(01434) 602191 fax:(01434) 605103 E_Mail:enquiries@egger.com

EITE & Associates 58 Carsington Cresent Allestree Derby DE22 2QZ tel:(01332) 559929 fax:(01332) 559929

Ejot Ecofast Ltd Ejot House Fox Way Hunslett Leeds LS10 IPS tel:(0113) 2470880 fax:(0113) 2470882

Elan-Dragonair Units 7-8, Dragon Industrial Centre Fitzherbert Road Farlington Portsmouth PO6 1SQ tel:(01705) 376451 fax:(01705) 370411

Eldyke Ltd PO Box 194 Stourbridge W. Midlands DY9 7YT tel:(01384) 892378 fax:(01384) 891369

Electrak International Ltd 1 Industrial Estate Medomsley Road Consett Co. Durham DH8 6SR tel:(01207) 503400 fax:(01207) 501799 E_Mail:sales@electrak.co.uk Web_Site:http://www.electrak.co.uk

Electric Elements Co, The Tokenhouse Yard Bridlesmith Gate Nottingham NG1 2HH tel:(0115) 9505253 fax:(0115) 9588283

Electrical Contractors Association, The Esca House 34 Palace Court London W2 4HY tel:(0171) 313 4800 fax:(0171) 221 7344 E_Mail:electricalcontractors@eca.co.uk Web_Site:http://www.eca.co.uk

Electrical Review Quadrant House The Quadrant Sutton SM2 tel:(0181) 652 3492

Electrical Times Quadrant House The Quadrant Sutton SM2 tel:(0181) 652 8735

Electrix (Northern) Ltd 1a-1b Dovecot Hill South Church Enterprise Park Bishop Auckland Co. Durham DL14 6XP tel:(01388) 774455 fax:(01388) 777359

Electro Mechanical Systems Ltd Eros House Calleva Industrial Park Aldermaston Reading RG7 8LN tel:(0118) 981 7391 fax:(0118) 981 613 E_Mail:barriers@ems-ltd.com Web_Site:http://www.ems-limited.demon.co.uk

Electrolux Domestic Appliances PO Box 545 55-57 High Street Slough Berks. SL1 9BG tel:(0990) 146146 fax:(01753) 872381

Electrolux food service P. O. Box 7377 Birmingham B9 4UG tel:(0121)7666655 fax:(0121) 6231801 E_Mail:fse.info@electrolux.co.uk Web_Site:http://www.electrolux.se

Electrolux Laundry Systems Unit 3A Humphrys Road, Woodside Estate Dunstable Bedfordshire LU5 4TP tel:(01582) 593273 fax:(01582) 588871

Electrosonic Ltd Hawley Mill Hawley Road Dartford Kent DA2 7SY tel:(01322) 222211 fax:(01322) 282282

Elementis Pigments Liliput Road Bracksmills Industrial Estate Northampton NN4 7DT tel:(01604) 827403 fax:(01604) 827400 E_Mail:pigmentsinfo.eu@elementis.com Web_Site:http://www.elementis.com

Elf Atochem UK Ltd Colthorp Way Thatcham, Newbury Berks. RG19 4NR tel:(01635) 870000 fax:(01635) 861212

Elga Ltd High Street Lane End High Wycombe Bucks. HP14 3JH tel:(01494) 887700 fax:(01494) 881007 E_Mail:elga@elga.demon.co.uk Web_Site:http://www.elga.co.uk

Elkay Electrical Manufacturing Co Ltd Unit 18, Mochdre Industrial Estate Newtown Powys SY16 4LF tel:(01686) 627000 fax:(01686) 628276 E_Mail:elkay@co.uk

Elkington Gatic Ltd Hammond House Holmestone Road, Poulton Close Dover Kent CT17 0UF tel:(01304) 203545 fax:(01304) 215001

Elkosta; Rodney Coate & Partners Ltd Bishops Hull, Taunton Somerset TA1 5EA tel:(01823) 271911 fax:(01823) 335763

Elliott Group Construction Division Ltd Braemar House Snelsins Road Cleckheaton W. Yorks. BD19 3UE tel:(01274) 861221 fax:(01274) 861582

Elliott Group Ltd Delta Way Cannock Staffordshire WS11 3BE tel:(01543) 404040 fax:(01543) 572710 E_Mail:sales@elliot-group.co.uk Web_Site:http://www.elliot-group.co.uk

Elliott John H (Monostamp) Ltd 4 Evans Street Off Milton Street Sheffield S3 7WE tel:(0114) 272 6141 fax:(0114) 272 3218

Elliott Medway FineLine Commisioners Road Strood, Rochester Kent ME2 4ET tel:(01634) 719701 fax:(01634) 716394 E_Mail:fineline.windows@virgin.net Web_Site:http://www.fineline-windows.co.uk

Elliott-Medway Construction Ltd Glebe Court Peterborough PE2 8EE tel:(01733) 52151 fax:(01733) 313002

Ellis & McDougall Lifts Ltd Argill Works 86 Broad Street Bridgestone Glasgow G40 2PX tel:(0141) 554 7604 fax:(0141) 554 6762

Ellis J T & Co Ltd Kilner Bank Industrial Esatae Silver Street Huddersfield W. Yorks. HD5 9BA tel:(01484) 514212 fax:(01484) 456433/533454

Ellis Miller Ltd Market House 62 St Judes Road Englefield Green Surrey TW20 0BU tel:(01784) 435302 fax:(01784) 471846 telex:21792 E_Mail:Sales@ellismiller.com

Ellis Patents High Street Rillington, Malton N. Yorks. YO17 8LA tel:(01944) 758395 fax:(01944) 758808

Elmstone 135 Allport Street Cannock Staffs. WS11 1JZ tel:(01543) 573050 fax:(01543) 573040

Elsy & Gibbons Ltd Simonside South Shields Tyne & Wear N34 9PE tel:(0191) 427 0777 fax:(0191) 427 0888

Elta Fans Ltd 15 Barnes Wallis Road Segensworth East Industrial Estate Fareham Hants. PO15 5ST tel:(01489) 583044 fax:(01489) 584699

Eltron Chromalox Eltron House 28 Whitehorse Road Croydon Surrey CR9 2NA tel:(0181) 665 8900 fax:(0181) 689 0571 telex:946649 E_Mail:eltronchromalox@btinternet.com Web_Site:http://www.chromaloxeurope.com

Elvaco UK Ltd Cambridge House Gogmore Lane Chertsey Surrey KT16 9AP tel:(01932) 564444 fax:(01932) 570707

Elvet Structures Low Willington Industrial Estate Willington Crook Co. Durham DL5 0UH tel:(01388) 747120

Elwell Buildings Ltd 204 Oldbury Road West Bromwich W. Midlands B70 9DE tel:(0121) 553 5723 fax:(0121) 553 5723

Elwell Sections Ltd Phoenix Street West Bromwich W. Midlands B70 0AQ tel:(0121) 553 4274 fax:(0121) 553 4272

Emcol International Ltd Westfordmill Wellington Somerset TA21 0DT tel:(01823) 660769 fax:(01823) 665719

Emergi-Lite Safety Systems Ltd Bruntcliffe Lane Morley, Leeds W. Yorks. LS27 9LL tel:(0113) 2810600 fax:(0113) 2810601 E_Mail:mailbox@emergi-lite.co.uk

Emlux Ltd Industrial Estate Black Bourton Road Carterton Oxford OX8 3EX tel:(01993) 841574 fax:(01993) 843186

Emmerich (Berlon) Ltd Kingsnorth Industrial Estate Wotton Road Ashford Kent TN23 6JY tel:(01233) 622684 fax:(01233) 645801 E_Mail:sales@emir.co.uk Web_Site:http://www.emiv.co.uk

Emmott George (Pawsons) Ltd Wadsworth Mill Oxlenhope Keighley W. Yorks. BD22 9NE tel:(01535) 643733 fax:(01535) 642108

Emsworth Fireplaces Ltd Unit 3 Station Approach North Street Emsworth Hampshire PO10 7PW tel:(01243) 373431 fax:(01243) 371023 E_Mail:info@emsworth.co.uk Web_Site:http://www.emsworth.co.uk

Emusol Products (LEICS) Ltd 10 Mandervell Road Oadby Leicester LE2 5LQ tel:(0116) 271 3763 fax:(0116) 271 2133 E_Mail:emusol@aol.com

En-tout-cas plc PO Box 36 Melton Mowbray Leics. LE13 0QT tel:(01664) 411616 fax:(01664) 411552

En-tout-cas plc Unit 2 Bartlett Park Gazelle Rd. Pontefract Yeourlle BA20 2PJ tel:(01977) 701885 fax:(01977) 600158

En-tout-cas plc 10 Digby Drive Melton Mowbray Leics. LE13 0RQ tel:(01664) 411711 fax:(01664) 481290

En-tout-cas plc (Scotland) Axwel House East Mains Broxburn W. Lothian EH52 5AU tel:(01506) 857044 fax:(01506) 857250

En-tout-cas South West Unit11, Bartlett Park Gazelle Road, Lynx Trading Estate Yoevil Somerset BA20 2PJ tel:(01935) 425372 fax:(01935) 432413

Enamelled Signs (UK) Ltd 23 Cranford Street Smethwick, Warley W. Midlands B66 2RT tel:(0121) 558 6060 fax:(0121) 555 5532

Encapsulite International Ltd Youngs Trading Estate Stanbridge Road Leighton Buzzard Beds. LU7 8QF tel:(01525) 376974 fax:(01525) 850306 E_Mail:info@encapsulite.co.uk Web_Site:http://www.encapsulite.co.uk

Encon Air Systems Ltd 31 Quarry Park Close Charter Gate, Moulton Park Industrial Estate Northampton NN3 6QB tel:(01604) 494187 fax:(01604) 645848

Enfield Speciality Doors Alexandra Road The Ride Enfield Middx. EN3 7HE tel:(0181) 805 6662 fax:(0181) 443 1290

English Abrasives & Chemicals Co Ltd Doxey Road Stafford ST16 1EA tel:(01785) 251288 fax:(01785) 259428 E_Mail:info@english-abrasives.co.uk Web_Site:http://www.unicom-abrasives.co.uk

English Architectural Glazing Ltd Chiskwick Avenue Midenhall Suffolk IP28 7AY tel:(01638) 510000 fax:(01638) 510400

English Heritage 429 Oxford Street London W1R 2HD tel:(0171) 973 3000

Ensor (Sandbeach) Ltd Shepley Lane Industrial Estate Hawk Green Stockport SK6 7JW tel:(01270) 762623 fax:(01270) 768461

Ensor Metal Products Shepley Lane Industrial Estate Hawks Green Marple Stockport SK6 7JW tel:(0161) 427 2746 fax:(0161) 427 3074 E_Mail:sales@ensormetal.co.uk Web_Site:http://www.ensormetal.co.uk

Entec (Polution Control) Ltd West Portway Andover Hants. SP10 3LF tel:(01264) 357666 fax:(01264) 366446

Envair Ltd York Avenue Haslingden Rossendale Lancs. BB4 4HX tel:(01706) 228416 fax:(01706) 83957 E_Mail:envair@envair.co.uk Web_Site:http://www.envair.co.uk

Envirodoor Markus Ltd Great Gutter Lane East Willerby Hull Humberside HU10 6BS tel:(01482)

659375 fax:(01482) 655131 E_Mail:sales@envirodoor.com
Environmental Acoustics Ltd Goose Lane Barwell Leics. LE9 8DH tel:(01455) 843678 fax:(01455) 841740 E_Mail:riklewis@clara.net
Envirtec 11 Sherwood Close Whitstable Kent CT5 4PE tel:(01227) 275015 fax:(01227) 275015
Envopak Group Ltd Envopak House Edgington Way Sidcup Kent DA14 5EF tel:(0181) 308 8000 fax:(0181) 300 3832 E_Mail:sales@envopak.co.uk
Erico Europa (GB) Ltd 52 Milford Road Reading RG1 8LJ tel:(01189) 588386 fax:(01189) 594856
Erlau AG UK Sales Office 42 Macclesfield Road Hazel Grove Stockport SK7 6BE tel:(01625) 877277 fax:(01625) 850242 E_Mail:erlau@zuppinger.u-net.com Web_Site:http://www.erlau.de
Escol Panels (S & G) Ltd Paisley Works Bevan Close Wellingborough Northants NN8 4BL tel:(01933) 276136 fax:(01933) 440121
Esk Manufacturing Co Ltd Calsil Brickworks Brisco Carlisle Cumbria tel:(01228) 22438 fax:(01228) 514019
Estec Environmental Ltd Old Pump House, Elmer Works Hawks Hill Leatherhead Surrey KT22 9DA tel:(01372) 361451 fax:(01372) 361453 E_Mail:melperkins@esteceenv.demon.co.uk
Eternit Clay Tiles Ltd Ridgehill Drive Madeley Heath Crewe Cheshire CW3 9LY tel:(01782) 750243 fax:(01782) 750901 E_Mail:marketing@eternitclaytiles.co.uk Web_Site:http://www.eternitclaytiles.co.uk
Eternit UK Ltd Meldreth Nr Royston Herts. SG8 5RL tel:(01763) 260421 fax:(01763) 262338 E_Mail:marketing@eternit.co.uk Web_Site:http://www.eternit.co.uk
Eureka Products Ltd Norfolk Street Nelson Lancs. BB9 7SY tel:(01282) 615661 fax:(01282) 699542
Euro Materials Handling Ltd Emec Buildings Forge Lane, Mucklow Hill Halesowen W. Midlands B62 8EA tel:(0121) 585 5818 fax:(0121) 585 6127
Eurocom Enterprise Ltd Index House St Georges Lane Ascot Berks. SL5 7EU tel:(01344) 23404 fax:(01344) 874696
Eurogrid Ltd Halesfield 19 Telford Shropshire TF7 4QT tel:(01952) 581988 fax:(01952) 586285
European Profiles Llandybie Ammanford Carmarthenshire SA18 3JG tel:(01269) 850691 fax:(01269) 851081 E_Mail:epl@dial.pipex.com Web_Site:http://www.europeanprofiles.co.uk
Euroroof Ltd Denton Drive Northwich Cheshire CW9 7LU tel:(01606) 48222 fax:(01606) 49940 E_Mail:em@euroroof.u-net.com Web_Site:http://www.euroroof.u-net.com
Evans Concrete Products Ltd Pease Hill Road Ripley Derbys. DE5 3HZ tel:(01773) 748026 fax:(01773) 570354
Evans Howard Roofing Ltd Tyburn Road Erdington Birmingham B24 8NB tel:(0121) 327 1336 fax:(0121) 327 3423 E_Mail:roofherl@ukbusiness.com
Evans Lifts Unit 46 Oak Hill Trading Estate Devonshire Road Worsley Manchester M28 5PT tel:(01204) 861234 fax:(01204) 861242
Evans Vanodine International PLC Brierley Road Walton Summit Centre Bamber Bridge Preston PR5 8AH tel:(01772) 322200 fax:(01772) 626000
Everlac (GB) Ltd The Maltings Fordham Road Newmarket Suffolk CB8 7AA tel:(01638) 664241 fax:(01638) 560015
Evertaut Cross Street Darwen Lancs. BB3 2PW tel:(01254) 778800 fax:(0845) 6066334 E_Mail:sales@evertaut.co.uk Web_Site:http://www.evertaut.co.uk
Everybright Fasteners Ltd Stainless House 4 Edwin Road Twickenham CH6 5BD tel:(0181) 891 0111 fax:(0181) 891 4236 telex:933506
Evode Ltd Common Road Stafford ST16 3EH tel:(01785) 257755 fax:(01785) 252337 E_Mail:evo-stik@evode.co.uk Web_Site:http://www.evode.co.uk
Evode Spaciality Adhesives Ltd Anglo House Scudamore Road Leicester LE3 1UQ tel:(01533) 322922 fax:(01533) 322922 telex:34485
Exchem Mining and Construction PO Box 7 Venture Cresent Alfreton Derbys. DE55 7RE tel:(01773) 540440 fax:(01773) 607638 E_Mail:exchem-emc@compuserve.com
Exitex Ltd Dundalk Ireland tel:(00 353) 4293 71244 fax:(00 353) 4293 71221
Expamet Building Products PO Box 52 Longhill Industrial Estate (North) Hartlepool TS25 1PR tel:(01429) 866611 fax:(01429) 866633 E_Mail:expamet@compuserve.com
Expanded Piling Co Ltd, The Cheapside Works Waltham Grimsby N. E. Lincolnshire DN37 0JD tel:(01472) 822552 fax:(01472) 220675 E_Mail:expanded@tarmac.co.uk
Express Evans Lift 123 Abbey Lane Abbey Lane Leicester LE4 5QX tel:(0116) 201 1200 fax:(0116) 266 8592
Express Evans Lifts Unit 1E Wavetree Boulevard South Wavetree Technology Park Liverpool L7 9PF tel:(0151) 472 1500 fax:(0151) 472 1520
ExpressLift Co Ltd, The PO Box 19, Abbey Works Weedon Road Northampton NN5 5BT tel:(01604) 751221 fax:(01604) 756231 telex:3311314 EXLIFT G
External Wall Insulation Association PO Box 12 Haslemere Surrey GU27 3AH tel:(01428) 654011 fax:(01428) 651401 E_Mail:113052.2106@compuserve.com Web_Site:http://www.nationline.co.uk/ceed
Extrusions Direct Crittall Road Witham Essex CM8 3AW tel:(01376) 503504 fax:(0800) 526005 E_Mail:extrusions.direct@btinternet.com
Eyre & Baxter (Stampcraft) Ltd 229 Derbyshire Lane Sheffield S8 8SD tel:(01742) 2500153 fax:(01742) 580856
F. W. Harris & Co Ltd Moorland Road Burslem Stoke-on-Trent ST6 1DR tel:(01782) 575181 fax:(01782) 839439
Faber Blinds Ltd Kilvey Road Brackmills Northampton NN4 7PB tel:(01604) 766251 fax:(01604) 705209 E_Mail:contracts@faberblinds.co.uk Web_Site:http://www.faberblinds.co.uk
Fabricated Aluminium Services Ltd 97-99 Beddington Lane Croydon Surrey CR0 4TD tel:(0181) 664 3333 fax:(0181) 664 3300
Fabrikat (Nottingham) Ltd Hamilton Road Sutton-in-Ashfield Nottingham NG17 5LN tel:(01623) 442200 fax:(01623) 442233
Fairmile Fencing Ltd Gower Street St Georges, Telford Shropshire TF2 9DB tel:(01952) 620062 fax:(01952) 610011
Fairmitre Ltd Ashburton Road East Trafford Park Manchester M17 1BB tel:(0161) 877 7200 fax:(0161) 877 7300
Falcon Catering Equipment PO Box 37 Foundry Loan Larbert Stirlingshire FK5 4PL tel:(01324) 554221 fax:(01324) 552211
Fantasia Distribution Ltd Unit B The Flyers Way Westerham Kent TN16 1DE tel:(01959) 564440 fax:(01959) 564829
Faral Radiators Tropical House Charlwoods Road East Grinstead W. Sussex RH19 2HJ tel:(01342) 305420 / 317171 fax:(01342) 315362
Farefence Limited Pinfold House Pinfold Road Worsley Manchester M28 5DZ tel:(0161) 799 4925 fax:(0161) 703 8542
Farmington Natural Stone Ltd Northleach Cheltenham Glos. GL54 3NZ tel:(01451) 860280 fax:(01451) 860115 E_Mail:cotswold.stone@farmington.co.uk
Farr Filtration Ltd 272 Kings Road Tyseley Birmingham B11 2AB tel:(0121) 707 8211 fax:(0121) 706 9986
Fastclean Scotland Ltd Burngrange Lodge Dalziel Estate Motherwell Strathclyde ML1 2BL tel:(01689) 263963 fax:(01698) 263963
Fastwall Ltd PO Box 500 Epsom Surrey KT18 6YB tel:(01372) 278111 fax:(01372) 279222 E_Mail:sales@fastwall.com Web_Site:http://www.fastwall.com
Faversham Fencing Ltd 62-66 Leatherhead Road Chessington Surrey KT9 2HZ tel:(0181) 397 6164 fax:(0181) 397 5115
FC Precast Concrete Ltd Alfreton Road Derby DE21 4BN tel:(01332) 364314 fax:(01332) 372208 E_Mail:fcprecast.co.uk
FDB Electrical Ltd Reynard Mills Trading Estate Windmill Road Brentford Middx. TW8 9NZ tel:(0181) 568 4621 fax:(0181) 569 7899 E_Mail:fdbelectricalltd@btinternet.com
FDI- Samborn Ltd 820 Yeoville Road Slough Berks. tel:(01753) 6923557 fax:(01753) 692358
FEB Ltd Albany House Swinton Hall Road Swinton Manchester M27 4DT tel:(0161) 794 7411 fax:(0161) 727 8547
Feb MBT Albany House Swinton Hall Road Swinton Manchester M27 4DT tel:(0161) 794 7411 fax:(0161) 727 8547
Federation of Master Builders Gordon Fisher House 14-15 Great James Street London WC1N 3DP tel:(0171) 242 7583 fax:(0171) 404 0296 Web_Site:http://www.fmb.org.uk
Feedwater Ltd Tarran Way Moreton Wirral L46 4TP tel:(0151) 606 0808 fax:(0151) 678 5459 E_Mail:enquiries@feedwater.co.uk Web_Site:http://www.feedwater.co.uk
Fels UK Trinity Place Midland Drive Sutton Coalfield W. Midlands B72 1TX tel:(0121) 321 1155 fax:(0121) 321 1018 E_Mail:uksales@fels.de Web_Site:http://www.fels.de
Fence Hire Limited Stubs Industrial Estate Holly Bush Lane Aldershot Hants. GU11 2PX
Fencelines Ltd Unit 16 Westbrook Road Trafford Park Manchester M17 1AY tel:(0161) 848 8311 fax:(0161) 872 9643
Fencing Construction (Yorkshire) Limited Bankwood Lane Industrial Estate Rossington Doncaster S. Yorks. DN11 0PS tel:(01302) 868567 fax:(01302) 864249 E_Mail:fencingconstruct@compuserve.com
Fencing Contractors Association Warren Road Monmouth Trellech, Gwent NP5 4PQ tel:(07000) 560722 fax:(01600) 860888
Fencing Design & Development Co Limited Harold House 73 High Street Waltham Cross Herts. EN8 7AF tel:(01992) 767235 fax:(01992) 767260
Fenlock-Hansen Ltd Heworth House William Street Felling, Gateshead Tyne & Wear NE10 0JP tel:(0191) 438 3222 fax:(0191) 438 1686
Ferguson G A & Co Ltd Western Road Oldbury Warley W. Midlands B69 4LY tel:(0121) 552 3674 Fax: (0121) 552 3676
Ferham Products PO Box 164 Greasbro Road Tinsley, Sheffield S. Yorks. S9 1TJ tel:(01142) 446451 fax:(01142) 560071
Fermec Holdings Ltd Barton Dock Road Stretford Manchester M32 0YH tel:(0161) 865 4400 fax:(0161) 875 5316 telex:667145
Fernden Construction (South Western) Limited Bishops Hull Taunton Somerset TA1 5EA tel:(01823) 271911 fax:(01823) 335763 E_Mail:elkosta.uk@btinternet.com Web_Site:http://www.btinternet.com/~elkosta.uk
Fernden Construction (Winchester) Limited Barfield Close Winchester Hampshire SO23 9SQ tel:(01962) 866400 fax:(01962) 864139
Fernden Fencing & Construction Co Ltd Bemnhall Mill Road Tunbridge Wells Kent TN2 5JW tel:(01892) 514120
Fernox Fry Technology UK Tandem House, Marlowe Way Beddington Farm Road Croydon Surrey CRO 4XS tel:(0181) 665 6666 fax:(0181) 665 4695 E_Mail:sales@frytechnology.cookson.com Web_Site:http://www.frytechnology.com
Ferrograph Limited New York Way New York Industrial Park Newcastle-upon-Tyne NE27 OQF tel:(0191) 280 8800 fax:(0191) 280 8810 E_Mail:ferrograph@btinternet.com Web_Site:http://www.ferrograph.com
Ferrous Gate Co Albion Works, Orbital One Green Street Road Dartford Kent DA1 1QQ tel:(01322) 272119 fax:(01322) 272339
Fersina Windows Ltd Industry Road Carlton Barnsley S. Yorks. S79 3NH tel:(01226) 728310 fax:(01226) 722090
FGF (Ashton) Ltd Shadwell House Shadwell Street Birmingham B4 6LJ tel:(0121) 233 1144 fax:(0121) 212 2539
Fibaflo Reinforced Plastics Ltd Industrial Estate Holton Heath Poole Dorset BH16 6LP tel:(01202) 624141 fax:(01202) 631589
FIBAFORM PRODUCTS Unit 22a Lansil Industrial Estate Caton Road Lancaster Lancs LA1 3PQ tel:(01524) 60182 fax:(01524) 389289/60182
Fibercill Unit 8 Navigation Drive Brierley Hill W. Midlands DY5 1UT tel:(01384) 482221 fax:(01384) 482212
Fibre Reinforced Products Ltd 2 Whitehouse Way South West Industrial Estate Paterlee Co. Durham SR8 2HZ tel:(0191) 586 5311 fax:(0191) 586 1274
Fibremesh Synthetic Industries Europe Ltd Fibremesh House Smeckley Wood Close Sheepsbridge, Chesterfield Derbys. S41 9PZ tel:(01246) 453102 fax:(01246) 455841 E_Mail:fibremesh@aol.com
Fiche (UK) Ltd Security House Acrewood Way St Albans Herts. AL4 0JL tel:(01727) 863863 fax:(01727) 41079
Fieldmount (Terrazzo) Ltd 152 West End Lane West Hampstead London NW6 1SD tel:(0171) 624 8866 fax:(0171) 328 1836
Fife Fire Engineers & Consultants Ltd Waverley Road, Mitchelston Industrial Estate Kirkcaldy Fife KY1 3NH tel:(01592) 653661 fax:(01592) 653990
Filion Products Ltd Aldridge Road Streetly Sutton Coalfield W. Midlands B74 2DZ tel:(0121) 353 0814 fax:(0121) 352 0886
Fillcrete Ltd Grindell Street Hull HU9 1RT tel:(01482) 223405 fax:(01482) 327957
Findlay Irvine Ltd Bog Road Penicuik Midlothian EH26 9BU tel:(01968) 671200 fax:(01968) 671237 E_Mail:sales@findlayirvine.com Web_Site:http://www.findlayirvine.com
Fine H & Son Ltd Victoria House 93 Manor Farm Road Wembley Middx. HA0 1XB tel:(0181) 997 5055 fax:(0181) 997 8410 Web_Site:http://www.hfine.co.uk
Finecraft Joinery Ltd Arundel House Arundel Road Uxbridge Middx. UB8 2RX tel:(01895) 233101 fax:(01895) 231933
Finish Plywood International PO Box 99 Welwyn Garden City Herts. AL6 0HS tel:(01438) 798746 fax:(01438) 798305
Finlock Gutters Leacroft Works London Road East Grinstead W. Sussex RH19 1PQ tel:(01342) 321433 fax:(01342) 317276
Finnish Sauna Baths (Sales) Ltd 4 Shelbourne Close Pinner Middx. HA5 3AF tel:(0181) 868 7170
Finwood Leisure Ltd 41 Feus Auchterarder Tayside PH3 1EP tel:(01764) 664920 fax:(01764) 664923
Fire Escapes & Fabrications (UK) Ltd Foldhead Mills Newgate Mirfield W. Yorks. WF14 8DD tel:(01924) 498787 fax:(01924) 497424
Fire Protection Ltd Millars 3 Southmill Road Bishops Stortford Herts. CM23 3DH tel:(01279) 467077 fax:(01279) 466994 E_Mail:fire.protection@btinternet.com
Fire Protection Services PLC 400 Dallows Road Luton Beds. LU1 1UR tel:(01582) 413694 fax:(01582) 402339
Fire Security (Sprinkler Installations) Ltd Homefield Road Haverhill Suffolk CB9 8QP tel:(01440) 705815 fax:(01440) 704352
Firebarrier Services Ltd 10 A Braddons Hill Road West Torquay Devon TQ1 1BG tel:(01803) 291185 fax:(01803) 290026 E_Mail:firebarrier.co.uk
Fired Earth Twyford Mill Oxford Road Adderbury Oxon. OX17 3HP tel:(01295) 812088 fax:(01295) 810832
Fired Earth Twyford Mill Oxford Road Adderbury Oxon. OX17 3HP tel:(01295) 812088 fax:(01295) 812189
Firemaster Extinguisher Ltd Firex House 174-176 Hither Green Lane London SE13 6QB tel:(0181) 852 8585 fax:(0181) 297 8020 E_Mail:ffirex@aol.com Web_Site:http://www.member.aol.com/fifirex
Firestone Building Products Ltd Meridian House Road One Winsford Cheshire CW7 3QG tel:(01606) 552026 fax:(01606) 592666
Firetecnics Systems Ltd Southbank House Black Prince Road London SE1 7SJ tel:(0171) 587 1493 fax:(0171) 582 3496
First Stop Builders Merchants Ltd Queens Drive Kilmarnock Ayrshire KA1 3XA tel:(01563) 534818 fax:(01563) 537848
Firsteel Metal Products Ltd Hirwaun Industrial Estate Aberdere M. Glam. CF44 9YG tel:(01685) 811919 fax:(01685) 814301
Firth Carpets Ltd PO Box 17 Clifton Mills Brighouse W. Yorks. HD6 4EJ tel:(01484) 713371 fax:(01484) 711128
Firth-Vickers Special Steels Ltd Staybrite Works Weedon Street Sheffield S9 2FU tel:(01742) 449955 fax:(01742) 445390
Firwood Paint & Varnish Co Ltd Victoria Works Oakenbottom Road Bolton Lancs. BL2 6DP tel:(01204) 525231 fax:(01204) 362522 E_Mail:sales@firwood.co.uk
Fisher Scientific (UK) Ltd Bishops Meadow Road Loughborough Leics. LE11 5RG tel:(01509) 231166 fax:(01509) 231893
Fitchett & Woolacott Ltd Willow Road Lenton Lane Nottingham NG7 2PR tel:(01602) 700691 fax:(01602) 420912
Fitzgerald Lighting Ltd Normandy Way Bodmin Cornwall PL31 1HH tel:(01208) 262200 fax:(01208) 74893 E_Mail:sales@flg.co.uk Web_Site:http://www.flg.co.uk
Fitzpatrick Door Frames Ltd Rushey Lane Tyseley Birmingham B11 2BL tel:(0121) 706 6363 fax:(0121) 708 2250
Fix Fast (Fasteners) Wales Ltd Penallta Industrial Estate Ystrad Mynach Hengoed M. Glam. CF82 7QZ tel:(01443) 815441 fax:(01443) 815021
Fixatrad Ltd Unit 29, Robert Cort Industrial Estate Britten Road Reading RG2 OAD tel:(0118) 921 2100 fax:(0118) 921 0634
Fixing Delivery 6 Three Arch Road Redhill RH1 5SS tel:(01737) 767656 fax:(01737) 772867
Flamco UK Ltd PO Box 16, Brookhouse Peel Green Eccles Manchester M30 7QA tel:(0161) 789 8111 fax:(0161) 789 1189
Flavel - Leisure Glynwed Consumer & Construction Products Ltd Clarence Street Leamington Spa Warwks. CV31 2AD tel:(01926) 427027 fax:(01926) 450526 E_Mail:flavel-leisure.market@dial.pipex.com
Fleetwood Fengate Co Ltd 62-66 Leatherhead Road Chessington Surrey KT9 2HZ tel:(0181) 397 6161 fax:(0181) 397 5115
Fleetwood Paints (Sales) Ltd 117 Wood Lane Isleworth Middx. TW7 5EG tel:(0181) 560 7634
Fleming Buildings Ltd 23 Auchinloch Road Lenzie Glasgow G66 5ET tel:(0141) 776 1181 fax:(0141) 775 1394
Fleming Homes Ltd Coldstream Road Duns Berwickshire TD11 3BR tel:(01361) 883785 fax:(01361) 883898
Flex-Seal Couplings Ltd Endeavour Works Newlands Way, Valley Park Wombwell, Barnsley South Yorkshire S73 OUW tel:(01226) 340888 fax:(01226) 340999 E_Mail:marketing@flexseal.co.uk Web_Site:http://www.flexseal.co.uk
Flexcrete Ltd PO Box 7 London Road Preston Lancs. PR1 4AJ tel:(01772) 259477 fax:(01772) 202902
Flexel International Ltd Queensway Industrial Estate Glenrothes Fife KY7 5QF tel:(01592) 757313 fax:(01592) 75435
Flexello Limited 136 Edinburgh Avenue Slough Berks. SL1 4SS tel:(01753) 775200 fax:(01753) 775300 E_Mail:sales@flexello.co.uk
Flexi-Plan Partitions Ltd Unit J1 Halesfield 19 Telford Shropshire TF7 4QT tel:(01952) 586126 fax:(01952) 581174
Flexible Ducting Ltd Cloberfield Milngavie Glasgow G62 7LW tel:(0141) 956 4551 fax:(0141) 956 4847 E_Mail:info@flexibleducting.co.uk Web_Site:http://www.flexibleducting.co.uk
Flexible Reinforcements Ltd Queensway House Queensway Clitheroe Lancs. BB7 1AU tel:(01200) 442266 fax:(01200) 452010
Flexicote Ltd Grinstead House Partridge Green Horsham W. Sussex RH13 8EJ tel:(01306) 741599 fax:(01306) 885901
Flexiform Business Furniture Ltd The Business Furniture Centre 1392 Leeds Road Bradford W. Yorks. BD3 7AE tel:(01274) 656013 fax:(01274) 660867 E_Mail:flexiform@aol.com Web_Site:http://www.flexiform.co.uk
Flexitallic Gaskets Ltd Marsh Wood Dewsbury Rd. Cleckheaton W. Yorks. BD19 5BT tel:(01274) 851273 fax:(01274) 876211
Flexwood Veneers 9 Egsign Drive Palmers Green London N13 5AF tel:(0181) 447 8479 fax:(0181) 886 2483
Floor Protection Services Ltd P. O. Box 359 Kempston Beds. MK42 7FP tel:(01234) 303090 fax:(01234) 303080 E_Mail:sales@fpsmatting.co.uk
FloPlast Ltd Sheppey Way Howt Green, Sittingbourne Kent ME9 8QT tel:(01795) 431731 fax:(01795) 431188
FloRad Heating Systems Unit 1, Horseshoe Business Park Lye Lane Bricket Wood, St. Albans Herts. AL2 3TA tel:(01923) 893025 fax:(01923) 670723
Flowcrete Plc Flowcrete Business Park Booth Lane Moston, Sandbach Cheshire CW11 3QF tel:(01270) 753000 fax:(01270) 753333 E_Mail:marketing@flowcrete.co.uk Web_Site:http://www.flowcrete.co.uk

Flowflex Componants Ltd Samuel Blaser Works Tounge Lane Industrial Estate Buxton Derbys. SK17 7LR tel:(01298) 77211 fax:(01298) 72362

Flowline Grilles Block 3 Narrow Boat Way Dudley W. Midlands DY2 0XQ tel:(01384) 459260 fax:(01384) 458381

Fluidair Compressors Ltd Miller Street Radcliffe Manchester M26 4JB tel:(0161) 723 2421 fax:(0161) 724 8727

Fluorel Ltd 312 Broadmead Road Woodford Green Essex IG8 8PG tel:(0181) 504 9691 fax:(0181) 506 1792

Focal Signs Ltd 12 Wandle Way Mitcham Surrey CR4 4NB tel:(0181) 687 5300 fax:(0181) 687 5301 E_Mail:sales@focalsigns.co.uk Web_Site:http://www.focalsigns.co.uk

Footprint Tools Ltd PO Box 19 Hollis Croft Sheffield S. Yorks. S1 3HY tel:(01142) 753200 fax:(01142) 759613 E_Mail:sales@dfootprint-tools.co.uk Web_Site:http://www.footprint-tools.co.uk

Forbes & Lomax Ltd 205a St John's Hill London SW11 1TH tel:(0171) 738 0202 fax:(0171) 738 9224

Forbo (Lancaster) Ltd Lune Mills Lancaster LA1 5QN tel:(01524) 65222 fax:(01524) 61638

Forbo - Nairn Ltd PO Box 1 Den Road Kirkcaldy Fife KY1 2SB tel:(01592) 643777 fax:(01592) 643999 E_Mail:headoffice@forbo-nairn.co.uk Web_Site:http://www.forbo-nairn.co.uk

Fordham Pland PO Box 20 Spring Ram Business Park Birstall W. Yorks. WF17 9XD tel:(01924) 351351 fax:(01924) 351333 E_Mail:marketing@astracast.co.uk

Fordwater Pumping Supplies 11 Lea Road Abingdon Northampton NN1 4PE tel:(01604) 39805 fax:(01604) 20143

Fordwater Pumping Supplies Ltd Unit 32 Forestvale Ind Estate Cinderford Glos. GL14 2PH tel:(01594) 826780 fax:(01594) 826780

Fordwater Pumping Supplies Ltd 49-53 Stratford Road Sparkbrook Birmingham W. Midlands B11 1RQ tel:(0121) 7728336 fax:(0121) 7710530

Forest City Signs Ltd 50 Park Road Timperley Altrincham WA14 5QX tel:(0161) 9720245 fax:(0161) 972 0255 E_Mail:forestcitysigns@compuserve.com

Forest Dean Stone Firms Ltd Bixslade Stone Works Parkend Nr Lydney Glos. GL15 4JS tel:(01594) 562304 fax:(01594) 564184

Forma Lighting Ltd Unit 3, Mitcham Industrial Estate 85 Streatham Road Mitcham Surrey CR4 2AP tel:(0181) 640 6811 fax:(0171) 640 6910

Formica Ltd Coast Road North Shields Tyne & Wear NE29 8RE tel:(0191) 259 3000 fax:(0191) 258 2719

Formline Ltd Formak House St Johns Road Caversham Reading RG4 0BD tel:(01734) 483377 fax:(01734) 461304

Formpave Ltd Tufthorne Avenue Coleford Glos. GL16 8PR tel:(01594) 836999 fax:(01594) 810577

Formwood Tufthorn Industries Ltd Coleford Gloucester GL16 8PR tel:(01594) 833305 fax:(01594) 835318 E_Mail:formwood@tufthorn-industries.uk Web_Site:http://www.tufthorn-industries.ltd.uk

Forson Design & Engineering Ltd Commerce Way Lancing W. Sussex BN15 8TQ tel:(01903) 752835 fax:(01903) 756677 E_Mail:forsons@mistral.co.uk

Fortafix Ltd First Drove Fengate Peterborough Cambs. PE1 5BJ tel:(01733) 566136 fax:(01733) 315393 Web_Site:http://www.ourworld.compuserve.com/homepages/fortafix

Fortec International 6 Sandfield Close Lowton Nr Warrington WA3 2TT tel:(01942) 722896 fax:(01942) 719250

Forticrete Ltd Bridle Way Bootle Merseyside L30 4UA tel:(0151) 521 3545 fax:(0151) 521 5696 E_Mail:forticretetechnical@compuserve.com Web_Site:http://www.forticrete.co.uk

Forticrete Roofing Products Heath Road Leighton Buzzard Beds. LU7 8ER tel:(01525) 851100 fax:(01525) 850432

Fosroc Expandite Ltd Coleshill Road Tamworth Staffs. B78 3TZ tel:(01827) 262222 fax:(01827) 281696

Foster C G & Sons, (Contract Soft Furnishers) Sherbourne House 247 Humber Avenue Coventry CV1 2AQ tel:(01203) 224229 fax:(01203) 222786

Foster W H & Sons Ltd 3 Cardale Street Rowley Regis Warley W. Midlands B65 0LX tel:(0121) 561 1103 fax:(0121) 559 2620

Fosters Upholstery Services More Hall Ibstock Road Ellistown Leics. LE67 1ED tel:(01530) 260211

Fothergill Engineered Fabrics Ltd PO Box 1 Summit Littleborough Lancs. OL15 OLU tel:(01706) 372414 fax:(01706) 376422 E_Mail:sales@fothergill.co.uk Web_Site:http://www.fothergill.co.uk

Fox Pool (UK) Ltd Mere House Stow Lincoln Lincs. LN1 2BZ tel:(01427) 788662 fax:(01427) 788526

Framford Kitchens Ltd Middlefield Industrial Estate Sunderland Road Sandy Beds. tel:(01767) 680831

Francesca Di Blasi Co. Ltd, The 14 Holland Street London W8 4LT tel:(0171) 938 2244 fax:(0171) 938 1920 E_Mail:info@thecube.co.uk Web_Site:http://www.thecube.co.uk

Francis Firth Collection The Old Rectory Bimport Shaftesbury Dorset SP7 8AT tel:(01747) 55669 fax:(01747) 55065

Franklin Hodge Industries Ltd P. O. Box 11, Holmer Road Hereford Herefordshire HR4 9SJ tel:(01432) 277277 fax:(01432) 277454 E_Mail:fhisales@denco.co.uk

Fray Design Ltd The Bradley Centre Bradley, Keighley W. Yorks. BD20 9LE tel:(01535) 636433 fax:(01535) 634642

Frazer Safety & Construction Station Road Hebburn Tyne & Wear NE31 1BD tel:(0191) 428 0242 fax:(0191) 483 3628

Freeman T R Beadle Trading Estate Ditton Walk Cambridge CB5 8PD tel:(01223) 410600 fax:(01223) 411061 E_Mail:t.r.freeman@dial.pipex.com

Freshfield Lane Brickworks Ltd Freshfield Lane Dane Hill, nr Haywards Heath W. Sussex RH17 7HH tel:(01825) 790350 fax:(01825) 790779

Freyssinet Ltd The Ridgeway Trading Estate Iver Bucks. SL0 9JE tel:(01753) 652844 fax:(01753) 655479 E_Mail:psc@freyssinet.co.uk

Friends of the Earth 26 Underwood Street London N1 7JQ tel:(0171) 490 1555

Fritztile (UK) Ltd 110 Ashley Down Road Bristol BS7 9JR tel:(01272) 420221 fax:(01272) 241696 telex:449700 ZEDBEE G

Frosts Landscape Construction Ltd Newport Road Woburn Sands Milton Keynes MK17 8UE tel:(01908) 583511 fax:(01908) 585238

Full Spectrum Lighting Ltd 19 Lincoln Road Cressex Business Park High Wycombe Bucks. HP12 3FX tel:(01494) 526051 fax:(01494) 527005

Fullflow Ltd Fullflow House Holbrook Avenue Holbrook, Sheffield S. Yorks. S20 3FF tel:(0114) 247 3655 fax:(0114) 247 7805 E_Mail:info@fullflow.com

Fullstop Technology Ltd Caidan House Canal Road Timperley Cheshire WA14 1TE tel:(0161) 962 8093 fax:(0161) 905 1436

Fulton Boiler Works (Great Britain) Ltd Broomhill Road Brislington Bristol BS4 4TU tel:(0117) 972 3322 fax:(0117) 972 3358 E_Mail:uk-info@fulton.com

Furmanite International Ltd Furman House Shap Road Kendal Cumbria LA9 6RU tel:(01539) 729009 fax:(01539) 729359 E_Mail:uk.enquiry@furmanite.com Web_Site:http://www.furmanite.com

Furmanite International Ltd Owens Road Skipper Lane Industrial Estate South Bank, Middlesborough Cleveland TS6 6HE tel:(01642) 455111 fax:(01642) 465692

Furse W J & Co Ltd Wilford Road Nottingham NG2 1EB tel:(0115) 986 3471 fax:(0115) 986 0538 E_Mail:eip@wjfurse.com

Futmis Ltd Futimis House 11 Mead Park, River Way Harlow Essex CM20 2SE tel:(01279) 411131 fax:(01279) 453107

Fyfe John Ltd Westhill Industrial Estate Westhill Skene Aberdeen AB32 6TQ tel:(01224) 744144 fax:(01224) 744500 telex:739529 FYFE G

Gabriel & Co Ltd Abro Works 10 Hay Hall Road Tyseley Birmingham B11 2AU tel:(0121) 248 3333 fax:(0121) 248 3330 E_Mail:john@gabrielco.com Web_Site:http://www.gabrielco.com

Galvanizers Association Wrens Court 56 Victoria Road Sutton Coldfield W. Midlands B72 1SY tel:(0121) 355 8838 fax:(0121) 355 8727 E_Mail:ga@hdg.org.uk Web_Site:http://www.hdg.org.uk

Game Time UK Ltd Harbury Road Deppers Bridge Bishops Itchington, Leamington Spa Warwks. CV33 0SZ tel:(01926) 612105 fax:(01926) 613609

Gang-Nail Systems Ltd Christy Estate Ivy Road Aldershot Hants. GU12 4XG tel:(01252) 334691 fax:(01252) 334562

Garador Ltd Bunford Lane Yeovil Somerset BA20 2YA tel:(01935) 443700 fax:(01935) 443744

Garlick E & Sons Ltd 78 Hoyle Street Sheffield Yorks. S3 7EX tel:(01742) 725412 fax:(01742) 725412

Garog 14 Leacroft Road Birchwood Warrington WA3 6GG tel:(01925) 825225 fax:(01925) 822239

Garran Lockers Ltd PO Box 4 Nantgarw Road Caerphilly M. Glam. CF83 1WW tel:(01222) 869924 fax:(01222) 882841 E_Mail:peter@garran-lockers.com Web_Site:http://www.garran-lockers.com

Garrett & Campbell Ltd 19/21 Nile Street London N1 tel:(0171) 490 0656

Gas Measurement Instruments Ltd Inchinnan Business Park Renfrewshire PA4 9RG tel:(0141) 812 3211 fax:(0141) 812 7820 E_Mail:sales@gmiuk.com Web_Site:http://www.gmiuk.com

Gaskell Textiles Ltd Clayton Park Clayton - Le - Moors Accrington Lancs. BB5 5GT tel:(01254) 724000 fax:(01254) 724200 Web_Site:http://www.gaskill.co.uk

Gates Rubber Co Ltd The Edinburgh Road Heathhall Dumfries DG1 1QA tel:(01387) 253111 fax:(01387) 250542 telex:778785

Gazco Ltd Osprey Road Sowton Industrial Estate Exeter EX2 7JG tel:(01392) 444030 fax:(01392) 444148

GB Metals Ltd Grange Works, Cromwell Centre Sellina's Lane Dagenham Essex RM8 1QH tel:(0181) 984 7470 fax:(0181) 984 7970

GBG Fences Limited 25 Barns Lane Rushall Walsall Staffs. WS4 1HQ tel:(01922) 623207 fax:(01922)722110 E_Mail:113105.1311@compuserve.com

GE Bayer Silicones Old Hall Road Sale Gtr. Manchester M33 2HG tel:(01257) 473867 fax:(01257) 473868 E_Mail:alan.lacey@gepex.ge.com

GE Capital Modular Space Sandtoft Ind . Estate Sandtoft Road Belton, Doncaster S. Yorks. DN9 1PN tel:(01427) 871111 fax:(01427) 874543

GE Lighting Ltd Conquest House 42-44 Wood Streets Kingston Upon Thames Surrey KT1 1UZ tel:(0181) 626 8500 fax:(0181) 626 8501

GE Plastics Ltd Old Hall Road Sale Gtr. Manchester M33 2HG tel:(0161) 905 5000 fax:(0161) 905 5160

Gearing F T Landscape Services Ltd Crompton Road Depot Stevenage Herts. SG1 2EE tel:(01438) 369321 fax:(01483) 353039

Geberit UK Metcalf House 25 Kirkgate Rippon N. Yorks. HG4 1PB tel:(01765) 602082 fax:(01765) 640765

Gebhardt Kiloheat Kiloheat House Enterprise Way Edenbridge Kent TN8 6HF tel:(01732) 866000 fax:(01732) 866370

GEC Alsthom Paxman Diesels Ltd Paxman Works Hythe Hill Colchester CO1 2HW tel:(01206) 795151 fax:(01206) 797869 telex:98152

GEC Anderson Ltd 89 Herkomer Road Bushey Watford WD2 3LS tel:(0181) 950 1826 fax:(0181) 950 3626

Gehlmax Limited Kingston International Bus Park Hedon Road Hull HU9 5PA tel:(01482) 705705 fax:(01482) 701920

General Combustion Ltd Beverley Division Billinghurst W. Sussex RH14 9SA tel:(01403) 782091 fax:(01403) 782087

General Security Systems Ltd Unit 5, 1st Floor Chelwood House 61-65 Perrymount Road Haywards Heath W. Sussex RH16 1DN tel:(01444) 410199 fax:(01444) 410188

General Time Europe 8 Heathcote Way Warwick CV34 6TE tel:(01926) 885400 fax:(01926) 885723 E_Mail:nestcloxeurope@generaltime.com Web_Site:http://www.westclox.com

Genie UK Ltd Brunel Drive Newark Notts. NG24 2EG tel:(01636) 605030 fax:(01636) 611090 Web_Site:http://www.genielift.com

Geoquip Ltd Unit C, Kingsfield Industrial Estate Derby Road Wirksworth, Matlock Derbys. DE4 4BG tel:(01629) 824891 fax:(01429) 824896

George Fischer Sales Ltd Coventry Walsgrave Triangle, Paradise Way Hinckley Road Coventry CV2 2ST tel:(02476) 535535 fax:(02476) 530450 E_Mail:info@georgefischer.co.uk Web_Site:http://www.georgefischer.co.uk

George Gilmour (Metals) Ltd 245 Govan Road Glasgow G51 2SQ tel:(0141) 427 1264 fax:(0141) 427 2205

George James Partnership Tincklers House Tincklers Lane Eccleston, Chorley Lancs. PR7 5QX tel:(01257) 452151 fax:(01257) 450094

Geosynthetic Technology Ltd Nags Corner Wiston Road Nayland, Colchester Essex CO6 4LT tel:(01206) 262676 fax:(01206) 262998 Web_Site:http://www.geosynthetic.co.uk

Gerflor Ltd 43 Crawford Street London W1H 2AP tel:(0171) 723 6601 fax:(0171) 723 9557

Gerland Ltd 43 Crawford Street London W1H 2AP tel:(0171) 723 6601 fax:(0171) 723 9557 telex:262555 GERFLEX G

Geze UK Chelmsford Business Park Colchester Road Chelmsford Essex CM2 5LA tel:(01245) 451093 fax:(01245) 451108

Giehews Building Technologies Ltd. Landis & Staefa Division Hawthorne Road Staines Middx. TW18 3AY tel:(01784) 461616 fax:(01784) 464646 Web_Site:http://www.landisstaefa.co.uk

Gifford Grant Ltd 144 Clarence Road Fleet Hants. GU13 9RS tel:(01252) 816188 fax:(01252) 625980 E_Mail:giffgrant@aol.com

Gilmer Limited Southdown Works Laughton Road Ringmer, Nr Lewes E. Sussex BN8 5PN tel:(01273) 813370 fax:(01273) 813963

Gilmour Ecometal 245 Govan Road Glasgow G51 2SQ tel:(0141) 427 7000 fax:(0141) 427 5345 Web_Site:http://www.gilmour.ecometal.co.uk

Giraffe Products Ltd Units 1-3, Fowler Industrial Park Chorley New Road Horwich Bolton BL6 5LU tel:(01204) 690052 fax:(01204) 695636

Girdlestone Pumps Ltd Station Road Melton, Woodbridge Suffolk IP12 1ER tel:(01394) 383777 fax:(01394) 386733 E_Mail:girdlestoneenquiries@north.com.co.uk Web_Site:http://www.girdlestone.co.uk

Girsberger London 104 - 110 Goswell Road London EC1V 7DH tel:(0171) 490 3223 fax:(0171) 490 5665

Giscol Bricks 20 Clifton Street Glasgow G3 7XS tel:(0141) 332 3786 fax:(0141) 332 7588

GJ Durafencing Limited Silverlands Park Nursery Holloway Hill Chertsey Surrey KT16 0AE tel:(01932) 568727 fax:(01932) 567799 E_Mail:gavinjones@msn.com Web_Site:http://www.gavinjones.co.uk

Glamox Electric (UK) Ltd Unit 3E Howden Green Industrial Estate Wallsend Tyne & Wear NE28 6SX tel:(0191) 262 7126 fax:(0191) 262 4118

Glasdon Designs Ltd Clitheroe Road Brierfield Lancs. BB9 5PT tel:(01282) 616221 fax:(01282) 603049

Glasdon UK Ltd Preston New Road Blackpool FY4 4UL tel:(01253) 694811 fax:(01253) 792558 telex:677288

Glasgow Steel Nail Co Ltd Lowmoss Bishopbriggs Glasgow G64 2HX tel:(0141) 762 3355 fax:(0141) 762 0914 E_Mail:101453.3140@compuserve.com Web_Site:http://www.ourworld.compuserve.com/homepages/glasgowsteelnail

Glass & Glazing Federation The 44-48 Borough High Street London SE1 1XB tel:(0171) 403 7177 fax:(0171) 357 7458 E_Mail:info@ggf.org.uk Web_Site:http://www.ggf.org.uk

Glaverbel (UK) Ltd Chesnut Field Regents Place Rugby Warwks. CV21 2XH tel:(01788) 535353 fax:(01788) 560853

Glazzard (Dudley) Ltd The Washington Centre Netherton, Dudley W. Midlands DY2 9RE tel:(01384) 233151 fax:(01384) 250224 E_Mail:glazzard@btinternet.com Web_Site:http://www.glazzard.co.uk

Gledhill Water Storage Ltd Sycamore Trading Estate Squires Gate Lane Blackpool Lancs. FY4 3RL tel:(01253) 401494 fax:(01253) 349657

Glenigan Ltd 41-47 Seabourne Road Bournemouth Dorset BH5 2HU tel:(0800) 373771 fax:(01202) 423441 E_Mail:info@glenigan.emap.co.uk

Gleno Industries Ltd Grange Road Houstoun Industrial Estate Livingston W. Lothian EH54 5DH tel:(01506) 32551 fax:(01506) 34386 E_Mail:glenosales@aol.com

Glidevale Building & Products Ltd Plymouth Avenue Brookhill Industrial Estate Pinxton Notts. NG16 6NS tel:(01773) 814123 fax:(01773) 814101

Glixtone Ltd Westminster Works Alvechurch Road West Heath Birmingham B31 3PG tel:(0121) 683 6349 fax:(0121) 683 6348

Gloddfa Ganol Slate Mine Blaenau Ffestiniog Gwynedd LL41 3NB tel:(01766) 830664 fax:(01766) 831527

Glostal - Monarch Ashchurch Tewksbury Glos. GL20 8NB tel:(01684) 853500 fax:(01684) 851850 E_Mail:enquire@granges.co.uk Web_Site:http://www.granges.co.uk

Glostal Systems Ltd Ashchurch Tewkesbury Glos. GL20 8NB tel:(01684) 297073 fax:(01684) 293904

Gloster Furniture Ltd Concorde Road Patchway Bristol BS34 5TB tel:(0117) 931 5335 fax:(0117) 931 5334

Glymwood Ltd Amari House 52 High Street Kingston-upon-Thames Surrey KT1 1HN tel:(0181) 549 6122 fax:(0181) 481 3081

Gold & Wassall (Hinges) Ltd Castle Works Tamworth Staffs. tel:(01827) 63391 fax:(01827) 310819 telex:341225

Golden Coast Ltd Gratton Way Roundswell Industrial Estate Barnstaple Devon EX31 3NL tel:(01271) 378100 fax:(01271) 371699 E_Mail:swimmer@goldenc.com

Goldschmidt Th Ltd Yego House Chippenham Drive Kingston Milton Keynes HA4 0YL tel:(01908) 582250

Goodman Croggon Ltd 6-8 Clothier Road Brislington Trading Estate Bristol BS4 5PE tel:(01272) 770721 fax:(01272) 723676

Goodman Croggon Ltd PO Box 40 Alma Street Smethwick, Warley W. Midlands B66 2SN tel:(0121) 558 3657 fax:(0121) 555 5618

Goods Directions Ltd 11 - 15 Tailsman Business Centre Duncan Road Park Gate Hants. SO31 7GA tel:(01489) 577828 fax:(01489) 577858 E_Mail:office@good-directions.co.uk Web_Site:http://www.good-directions.co.uk

Goodwin HJ Ltd Quenington, nr Cirencester Glos. GL7 5BX tel:(01285) 750271 fax:(01285) 750352

Goodwin Tanks Ltd Pontefract Street Derby DE24 8JD tel:(01332) 363112

Gower Furniture Ltd Holmfield Industrial Estate Holmefield Halifax W. Yorks. tel:(01422) 246201 fax:(01422) 249932 telex:517396 GOWER G

GP Burners (CIB) Ltd 2d Hargreaves Road Groundwell Industrial Estate Swindon Wilts. SN2 5AZ tel:(01793) 709050 fax:(01793) 709060 E_Mail:gpburnerd.co.uk Web_Site:http://www.gpburnerd.co.uk

GRAB Industrial Flooring Ltd Park Road South Wigston Leicester LE18 4DQ tel:(0116) 277 0024 fax:(0116) 277 4038

Grace Services Ajax Avenue Slough Berkshire SL1 4BH tel:(01753) 692929 fax:(01753) 691623 telex:846966 SERVCD G Web_Site:http://www.Grace.com

Gradient Insulations (UK) Ltd Station Road Four Ashes Wolverhampton WV10 7DB tel:(01902) 791888 fax:(01902) 791886 E_Mail:sales@gradient.infotrade.co.uk

Gradus Ltd Park Green Macclesfield Cheshire SK11 7LZ tel:(01625) 428922 fax:(01625) 433949 E_Mail:sales@gradusworld.com Web_Site:http://www.gradusword.com

Gradwood Ltd Landsdowne House 85 BuxtonRoad Stockport Cheshire SK2 6LR tel:(0161) 480 9629 fax:(0161) 474 7433

Grafton Magna Ltd 10 Lyall Court, Commerce Way Maulden Road Industrial Estate Flitwick Beds. MK45 1UE tel:(01525) 718811 fax:(01525) 718920

Grando Ltd Cotton Drive Kenilworth Warwks. CV8 2EB tel:(01926) 854977 fax:(01926) 856772

Granflex Roofing Ltd Brick Kiln Lane Basford Stoke-on-Trent ST4 7BT tel:(01782) 202208 fax:(01782) 273601
Grange Fencing Erectors Limited Halesfield 21 Telford Shropshire TF7 4PA tel:(01952) 587892 fax:(01952) 684461
Grant & Livingston Ltd Kings Road Charfleets Industrial Estate Canvey Island Essex SS8 0RA tel:(01268) 696855 fax:(01268) 697018
Granwood Flooring Ltd PO Box 60 Alfreton DE55 4ZX tel:(01773) 606060 fax:(01773) 606030
Grass Concrete Ltd Walker House 22 Bond Street Wakefield WF1 2QP tel:(01924) 374818 fax:(01924) 290289
Gravesend Fencing Limited The Fencing Works Codrington Crescent Gravesend Kent DA12 5DD tel:(01474) 326016 fax:(01474) 324562
Gray Hill Trading Ltd PO Box 521 Great Dunmow Essex CM6 1UY tel:(01371) 874662 fax:(01371) 876483
Gray J. W. & Son Ld 163 Wantz Road Maldon Essex tel:(01621) 859950
Great Metropolitan Flooring Co Ltd 83 Kinnerton Street Knightsbridge London SW1X 8ER tel:(0171) 235 1161 fax:(0171) 235 9027
Greaves Welsh Slate Co Ltd Llechwedd Slate Mines Blaenau Ffestiniog Gwynedd LL41 3NB tel:(01766) 830522 fax:(01766) 830711 E_Mail:llechwedd@aol.com Web_Site:http://www.llechwedd.co.uk
Green A P Refractories Ltd Dock Road South Bromborough Wirral Merseyside L62 4SP tel:(0151) 645 0701 fax:(0151) 645 8261 telex:627549
Green Alexander Lough Road Lurgan Co. Armagh BT66 6LY tel:(01762) 322086 fax:(01762) 325502
Green Bros (Geebro) Ltd Hailsham E. Sussex BN27 3DT tel:(01323) 840771 fax:(01323) 440109
Green Brothers Ltd Condor Works Rusholme Place Manchester M14 5GB tel:(0161) 224 2831 fax:(0161) 257 2713 E_Mail:sales@greensigns.co.uk Web_Site:http://www.greensigns.co.uk
Green Gerald 211 Hinckley Road Nuneaton Warwks. CV11 6LL tel:(01203) 325059 fax:(01203) 325059 E_Mail:gerald@ggarts.co.uk
Greenham Consruction Materials Ltd Salt Lane Rochester Kent ME3 7SZ tel:(01634) 221801
Greenwood Air Management Brookside Industrial Estate Rustington W. Sussex BN16 3LH tel:(01903) 771021 fax:(01903) 782398 telex:87297 E_Mail:info@greenwood.co.uk Web_Site:http://www.greenwood.btinternet.com
Gregory Quarries Ltd, The 184 Nottingham Road Mansfield Notts. NG18 5AP tel:(01623) 623092 fax:(01623) 622509 E_Mail:enquiries@rarestonegroup.co.uk Web_Site:http://www.rarestonegroup.co.uk
Griff Chains Ltd Quarry Road Dudley Wood Dudley DY2 0ED tel:(01384) 69415 fax:(01384) 410580
Griffin and General Fire Services Ltd Unit F 7 Willow Street London EC2A 4BH tel:(0171) 251 9379 fax:(0171) 729 5652 E_Mail:headoffice@griffinfire.demon.co.uk
Grille Diffuser & Louvre Co Ltd The Air Diffusion Works Woolley Bridge Road Hollingworth, Hyde Cheshire SK14 7BW tel:(01457) 861538 fax:(01457) 866010
Gripperrods Ltd Wyrley Brook Park Walkmill Lane Bridgtown, Cannock Staffs. WS11 3RX tel:(01922) 417777 fax:(01922) 419411
Grohe Ltd 1 River Road Barking Essex IG11 0HD tel:(0181) 594 7292 fax:(0181) 594 8898
Grorud Industries Ltd Castleside Industrial Estate Consett Co. Durham DH8 8HG tel:(01207) 581485 fax:(01207) 580036
Grosvenor Pumps Ltd 8 Shaftesbury Industrial Estate The Runnings Cheltenham Glos. GL51 9NH tel:(01242) 227400 fax:(01242) 227404 E_Mail:grovesnor.pumps@virgin.net
Grothkarst UK Ltd Unit 28, Balfour Business Centre Balfour Road Southall Middx. UB2 5BD tel:(0181) 574 6464 fax:(0181) 573 1189
Group Four Glassfibre Ltd Church Road Business Centre Murston Sittingbourne Kent ME10 3RS tel:(01795) 429424 fax:(01795) 476248
Groupco Ltd 18 Thresham Road Orton Southgate Peterborough PE2 6SG tel:(01733) 234750 fax:(01733) 235246
Grundfos Pumps Ltd Groveberry Road Leighton Buzzard Beds. LU7 8TL tel:(01525) 850000 fax:(01525) 850011
Grundman UK 45-46 Riverside Medway City Estate Rochester Kent ME2 4DP tel:(01634) 290585 fax:(01634) 290594
Grundy Catering Grundy House Ascot Road Bedfont, Feltham Middx. TW14 8QH tel:(01784) 251244 fax:(01784) 246137
Guardall Lochend Industrial Estate Newbridge Midlothian EH28 8PL tel:(0131) 333 2900 fax:(0131) 333 4919
Guardian Lock and Engineering Co Ltd Imperial Works Wednesfield Road Willenhall W. Midlands WV13 1AL tel:(01902) 635964 fax:(01902) 630675
Guardian Wire Ltd Guardian Works Stock Lane Chadderton Lancs. OL9 9EY tel:(0161) 624 6020 fax:(0161) 620 2880
Guest & Chrimes Ltd P O Box 9 Don Street Rotherham Yorks. S60 1AH tel:(01709) 828001 fax:(01709) 828012 E_Mail:info@guest-chrimes.demon.co.uk Web_Site:http://www.guest-chrimes.com
Guest (Speedfit) Ltd, John, Horton Road West Drayton Middx. UB7 8JL tel:(01895) 449233 fax:(01895) 420321
Gummers A & J Ltd Unit H, Redfern Park Way Tilsley Birmingham B11 2DN tel:(0121) 706 2241 fax:(0121) 706 2960
Gun-Point Ltd Thavies Inn House 3-4 Holborn Circus London EC1N 2PL tel:(0171) 353 1759 fax:(0171) 583 7259
Gunnebo Ltd Bell Brook Business Park Uckfield E. Sussex TN22 1QQ tel:(01252) 373700 fax:(01252) 373707
Gunnebo Mayor Ltd Bellbrook Business Park Uckfield E. Sussex TN22 1QQ tel:(01825) 761022 fax:(01825) 763835 E_Mail:mayormarketing@mayor.co.uk Web_Site:http://www.gunnebomayor.com
Gunton Leisure Ltd David Longley House East Park Crawley W. Sussex RH10 6AP tel:(01293) 564777 fax:(01293) 564789
Guthrie Allesbrook & Co Ltd The Old Hall Works Arborfield Road Shinfield Reading RG2 9DP tel:(01734) 882022 fax:(01734) 884678
Guthrie Douglas Ltd 12 Heathcote Way Heathcote Industrial Estate Warwick CV34 6TE tel:(01926) 452452 fax:(01926) 336417
H Pickup Mechanical & Electrical Services Ltd Durham House Lower Clark Street Scarborough N. Yorks. YO12 7PW tel:(01723) 369191 fax:(01723) 362044 E_Mail:pickup@hpickup.demon.co.uk
H. B. Fuller Coatings Ltd 95 Ashton Church Road Birmingham B7 5RQ tel:(0121) 322 6900 fax:(0121) 322 6901 Web_Site:http://www.hbfuller.com
Haddonstone Ltd The Forge House East Haddon Northants NN6 8DB tel:(01604) 770711 fax:(01604) 770027 E_Mail:info@haddonstone.co.uk Web_Site:http://www.businessconnections.com/haddonstone
Hadida Ltd, M R Old Foley Pottery King Street Fenton Stoke-on-Trent ST4 3DH tel:(01782) 597700 fax:(01782) 597702 E_Mail:mrh@hadida.demon.co.uk Web_Site:http://www.hadida.co.uk
Hadley Industries Plc PO Box 92 Downing Street Smethwick W. Midlands B66 2PA tel:(0121) 558 3222 fax:(0121) 558 1795 E_Mail:hi@hadleyindustries.plc.uk Web_Site:http://www.hadleyindustries.plc.uk
Haffele UK Ltd Swift Valley Industrial Estate Rugby Warwks. CV21 1RD tel:(01788) 542020 fax:(01788) 544440 E_Mail:email@hafele.co.uk Web_Site:http://www.hafele.co.uk
Hager Powertech Ltd Hortonwood 50 Telford Shropshire TF1 4FT tel:(01952) 677899 fax:(01952) 675581
Hago Products Ltd Durban Road, South Berstead Industrial Estate Bogner Regis W. Sussex PO22 9RD tel:(01243) 863131 fax:(01243) 827687
Hale Hamilton Ltd Frays Mills Works Cowley Road Uxbridge Middx. UB8 2AF tel:(01895) 236525 fax:(01895) 231407 E_Mail:gpearson@halehamilton.com
Halfen Ltd Interstates House Horsecroft Road Pinnacles, Harlow Essex CM19 5BZ tel:(01279) 641600 fax:(01279) 451650
Halfen Ltd Unit 31 Humpfreys Road, Woodside Estate Dunstable Beds. LU5 4TP tel:(08705) 316300 fax:(08705) 316304 E_Mail:info@halfen.co.uk
Hall & Watts Ltd 266 Hatfield Road St Albans Herts. AL1 4UN tel:(01727) 859288 fax:(01727) 835683 telex:267001
Hall Stage Products The Gate Studios Station Road Borehamwood Herts. WD6 1DQ tel:(0181) 953 9371 fax:(0181) 207 3657 telex:8955602 PERLUX G
Hallmark Blinds Ltd Hallmark House 173 Caladonian Road Islington London N1 0SL tel:(0171) 837 0964 fax:(0171) 833 1693
Halmark Panels Ltd Valletta House, Valletta Street Hedon Road Hull E. Yorks. HU9 5NP tel:(01482) 781111 fax:(01482) 701185
Halmatic Ltd Saxon Wharf Lower York Street Northam Southampton SO14 5QF tel:(01703) 337477 fax:(01703) 337478
Halo, W H S Division Of Bowater Industries 1 Cranfield Road, Lostock Industrial Estate Lostock Bolton BL6 4SB tel:(01204) 699211 fax:(01204) 699559
Halstead Boilers Ltd 4 First Avenue, Bluebridge Industrial Estate Halstead Essex CO9 2EX tel:(01787) 475557 fax:(01787) 474585
Halstead James Ltd Polyflor Division Radcliffe New Road Whitefield Manchester M45 7NR tel:(0161) 767 1111 fax:(0161) 767 1128
Hambleside Danelaw Ltd 2-8 Bentley Way Royal Oak Industrial Estate Daventry Northants NN11 5QH tel:(01327) 871737 fax:(01327) 310886 Web_Site:http://www.hambleside.internet2.co.uk
Hambo Contrywide Security PLC Securite House, The Loddon Business Centre Roentgen Road Basingstoke Hants. RG24 8NG tel:(01256) 322299 fax:(01256) 841100
Hamilton Acorn Ltd Halford Road Attleborough Norfolk N17 2HZ tel:(01953) 453201 fax:(01953) 454943 telex:97328 BRITON G
Hamilton Building Products Unit 38, Churchill Way Flecney Industrial Estate Fleckney Leics. LE8 0UD tel:(01533) 403981 fax:(01533) 402899
Hamilton R & Co Ltd Unit G Quarry Industrial Estate Mere Wilts. BA12 6LA tel:(01747) 860088 fax:(01747) 861032 E_Mail:hamilton_litestat@msn.com Web_Site:http://www.rhamilton.co.uk
Hammond & Champness Ltd 4-5 St Johns Square London EC1M 4HS tel:(0171) 490 2316 fax:(0171) 250 3012
Hampton Stone Ltd The Hampton Centre Cirencester Road Chalford Glos. GL6 8PE tel:(01453) 882180 fax:(01453) 886975
Hamron Group 58 Trench Road Mallusk Newtown Abbey BT36 4TY tel:(01232) 834910
Hamworthy Heating Ltd Fleets Corner Poole Dorset BH17 0HH tel:(01202) 662500 fax:(01202) 665111
Hand Tools Ltd Wreakes Lane Dronfeild Sheffield S18 6PN tel:(01246) 413139 fax:(01246) 415208
Hands of Wycombe 36 Dashwood Avenue High Wycombe Bucks. HP12 3DX tel:(01494) 524222 fax:(01494) 526508 E_Mail:info@hands.co.uk Web_Site:http://www.hands.co.uk
Hanovia Ltd 145 Farnham Road Slough Berks. SL1 4XB tel:(01753) 515300 fax:(01753) 534277 E_Mail:sales@hanovia.co.uk Web_Site:http://www.hanovia.co.uk
Hansen Windows Ltd 6 and 7 The Rickyard Clifton Reynes Olney Bucks MK46 5LQ tel:(01234) 713151 fax:(01234) 713094 E_Mail:jrmayhew@btinternet.com
Hansgrohe Units D1 & D2 Sandown Park Trading Estate Royal Mills, Esher Surrey KT10 8BL tel:(01372) 465655 fax:(01372) 470670 E_Mail:sales@hansgrohe.co.uk Web_Site:http://www.hansgrohe.co.uk
Hanson Aggregates Asby Road East Shepshed, Loughborough Leics. LE12 9BU tel:(01509) 503161 fax:(01509) 504120
Hanson Aggregates Stoneleigh House Regional Office Frome Somerset BA11 2HB tel:(01373) 463211 fax:(01373) 465843 Web_Site:http://.www.hanson-aggregates.com
Hanson Bath & Portland Stone Bumpers Lane Wakeham Portland Dorset DT5 1HY tel:(01305) 820207 fax:(01305) 860275
Hanson Brick Ltd Stewartby Bedford MK43 9LZ tel:(0990) 258258 fax:(01234) 762040 E_Mail:marketing@hanson-brick.co.uk Web_Site:http://www.hanson-bricksevrope.com
Hanson Brick Ltd, Northern Regonal Sales Office Unicorn House Wellington Street Ripley Derbys. DE5 3DZ tel:(0990) 258258 fax:(01733) 206040
Hanson Brick Ltd, Southern Regional Sales Office 222 Peterborough Road Whittlesey Peterborough PE7 1PD tel:(0990) 258258 fax:(01733) 206040
Hanson Quarry Products Europe Ltd The Ridge Chipping Sodbury Bristol BS37 6AY tel:(01454) 316000 fax:(01454) 325161 Web_Site:http://www.hanson-quarryproducts.com
Harbro Supplies 62 Princes Street Bishops Auckland Co. Durham DL14 7BA tel:(01388) 605363 fax:(01388) 603263
Hardall International Ltd 34 Clarke Road Mount Farm Milton Keynes MK1 1LG tel:(01908) 274441 fax:(01908) 367265
Hardman John Studios Lightwood House, Lightwood Park Hagley Road West Bearwood, Warley W. Midlands B67 5DP tel:(0121) 429 7609
Hargreaves Quarries Ltd Moota Qyarry Cockermouth Cumbria CA13 0QE tel:(016973) 20234 fax:(016973) 20179
Harkness Hall Ltd The Gate Studios Station Road Borehamwood Herts. WD6 1DQ tel:(0181) 953 3611 fax:(0181) 207 3657 E_Mail:info@harknesshall.com Web_Site:http://www.harknesshall.com
Harris & Bailey Ltd 50 Hastings Road Croydon Surrey CR9 6BR tel:(0181) 654 3181 fax:(0181) 656 9369 E_Mail:mail@harris-bailey.co.uk
Harris Road Roofing & Flooring Ltd Skates Lane Pamber Green Tadley, Basingstoke Hants. RG26 3AB tel:(01189) 811466 fax:(0118) 981 9021
Harris T A Ltd Hargreen House 134 New Kent Road London SE1 6TY tel:(0171) 703 3842 fax:(0171) 708 2596
Harrision Flagpoles Borough Road Darlington Co. Durham DL1 1SW tel:(01325) 355433 fax:(01325) 461726 E_Mail:sales@flagpoles.co.uk Web_Site:http://www.flagpoles.co.uk
Harrison Drape, Divsion of McKenie (UK) PO Box 233 Bradford Street Birmingham B12 0PE tel:(0121) 766 6111 fax:(0121) 772 0696 E_Mail:mlaydon@harrisondrape.com
Harrison Thompson & Co Ltd Yeoman House, Whitehall Estate Whitehall Road Leeds LS12 5JB tel:(0113) 279 5854 fax:(0113) 231 0406 E_Mail:info@yeomanshield.com Web_Site:http://www.yeomanshield.com
Hart Door Systems Ltd Redburn Road Westerhope Industrial Estate Newcastle-upon-Tyne Tyne & Wear NE5 1PJ tel:(0191) 214 0404 fax:(0191) 271 1611 E_Mail:response@hart-speedor.co.uk Web_Site:http://www.hart-speedor.co.uk
Hartley & Sugden Atlas Works Gibbet Street Halifax W. Yorks. HX1 4DB tel:(01422) 355651 fax:(01422) 359636
Hartley Clear Span Ltd Wellington Road Greenfield, Oldham Lancs. OL3 7AG tel:(01457) 873244 fax:(01457) 870151
Hartley John W Ltd Lunds Fields Carnforth Lancs. LA5 9AB tel:(01455) 272457 fax:(01455) 274564
Harton Heating Appliances Unit 6, Thistlebrook Industrial Est. Eynsham Drive Abbey Wood SE2 9RB tel:(0181) 310 0421 fax:(0181) 310 6785
Harty Holdings Ltd Crossbeg Industrial Estate Upper Ballymount Road Tallaght Dublin 24 Ireland tel:(00 353) 1 4566482 fax:(00 353) 1 4501214
Harvey Fabrication Ltd Hancock Road London E3 3DA tel:(0181) 981 7811 fax:(0181) 981 7815
Harvey Furniture Newlands Drive Poyle Berks. SL3 0DX tel:(01753) 680200 fax:(01753) 683673
Harwood & Welpac Hardware Ltd Harwood House Harrison Steet Briercliffe, Burnley Lancs. BB10 2HP tel:(01282) 451110 fax:(01282) 451160 E_Mail:harwood.welpac.@virgin.net
Hastings & Folkstone Glass Works Ltd Mill Bay Folkestone Kent CT20 1JS tel:(01303) 851277 fax:(01303) 851278
Hatcham Rubber Ltd Mead Lane Hertford SG13 7AU tel:(01992) 515600 fax:(01992) 515601
Hatema (UK) Ltd St Pancras Commercial Centre 63 Prat Street London NW1 0BY tel:(0171) 482 4466 fax:(0171) 267 3239 telex:262619
Hatfield Ltd, Roy Fullerton Road Rotherham S. Yorks. S60 1DL tel:(01709) 820855 fax:(01709) 374062
Hatmet Ltd Interiors House Lynton Road Crouch End London N8 8SL tel:(0181) 348 9262 fax:(0181) 341 9878
Hatt Waste Management Ltd PO Box 21 Monmouth Gwent NP5 4YN tel:(01600) 860900 fax:(01600) 860913
Hattersley & Davidson Ltd 71 Greystock Street Shefield S4 7WA tel:(01742) 728957 fax:(01742) 729330
Hattersley Newman Hender Ltd Burscough Road Ormskirk Lancs. L39 2XG tel:(01695)577199 fax:(01695)578775 telex:267571 E_Mail:uksales@hatterley-values.co.uk Web_Site:http://www.hatterley-values.co.uk
Hauration Kaskade Ltd Forge Works The Swillett Chorleywood Herts. WD3 5BN tel:(01923) 285601 fax:(01923) 283386 E_Mail:haur.kask@btinternet.com
Havering Fencing Co. 237 Chase Cross Road Collier Row Romford Essex RM5 3XS tel:(01708) 747855 fax:(01708) 721010
Hawkins Insulation Ltd Oakdale Court Downend Bristol Avon BS16 6DY tel:(0117) 957 0887 fax:(0117) 970 1136
Hawkins Tiles, Tarmac Bricks & Tiles Ltd Watling Street Cannock Staffs. WS11 3BJ tel:(01543) 503744 fax:(01534) 466434
Hawley W & Son Ltd Colour Works Duffield Derbys. DE56 4FG tel:(01332) 840294 fax:(01332) 842570 E_Mail: info@hawley.co.uk Web_Site:http://www.hawley.co.uk
Haworth UK Ltd 10 New Oxford Street London WC1A 1EE tel:(0171) 404 1617 fax:(0171) 404 1607
Hay Joinery Lingwood Norwich NR13 4NY tel:(01603) 712392 fax:(01603) 714248
Haydon & Hagan Construction Ltd Palfrey House 71A Palfrey Place London SW8 1AR tel:(0171) 587 1150 fax:(0171) 820 9080
Haymills (Contractors) Ltd Empire House Hanger Green London W5 3BD tel:(0181) 991 4200 fax:(0181) 991 4201
Hazard Safety Products Ltd 55-57 Bristol Road Edgbaston Birmingham B5 7TU tel:(0121) 446 4433 fax:(0121) 446 4230
HB Fuller Coatings Ltd 95 Aston Church Road Nechells Birmingham B7 5RQ tel:(0121) 322 6900 fax:(0121) 322 6902
HCL Contracts Bridge House Commerce Road Brentford Middx. TW8 8LQ tel:(0800) 212867 fax:(0181) 568 7066 E_Mail:hclgroup@compuserve.com
HCL Verta Unit 6 Wharfeilde Business Centre Trafford W Wharf Road Manchester M17 1EX tel:(0161) 877 6619 fax:(0161) 877 7197 E_Mail:hclverta@compuserve.com
HE Services (Plant Hire) Ltd West Midlands Depot Berry Hill Industrial Estate Droitwich Worcs. WR9 9AL tel:(01905) 774466 fax:(01905) 774466 E_Mail:mail@heservices.co.uk Web_Site:http://www.heservices.co.uk
HE Services (Plant Hire) Ltd Head Office & Accounts Whitewall Road Strood Kent ME2 4DZ tel:(01634) 291491 fax:(01634) 295626 E_Mail:mail@heservices.co.uk Web_Site:http://www.heservices.co.uk
HE Services (Plant Hire) Ltd East Midlands Depot Long March Industrial Estate Daventry Northants NN11 4HB tel:(01327) 301111 fax:(01327) 301111 E_Mail:mail@heservices.co.uk Web_Site:http://www.heservices.co.uk
HE Services (Plant Hire) Ltd Hungerford Depot Lambourn Woodlands Hungerford Berks. RG17 7TJ tel:(01488) 73444 fax:(01488) 73444 E_Mail:mail@heservices.co.uk Web_Site:http://www.heservices.co.uk
HE Services (Plant Hire) Ltd Strood Depot Whitewall Road Strood Kent ME2 4DZ tel:(01634) 291290 fax:(01634) 295355 E_Mail:mail@heservices.co.uk Web_Site:http://www.heservices.co.uk
HE Services (Plant Hire) Ltd Ashford Depot Ellingham Industrial Estate Ashford Kent TN22 2LZ tel:(01233) 639831 fax:(01233) 639831

E_Mail:mail@heservices.co.uk Web_Site:http://www.heservices.co.uk

HE Services (Plant Hire) Ltd North London Depot Eley Industrial Estate, Kynoch Road Edmonton London N18 3BH tel:(0181) 803 1924 fax:(0181) 807 3165 E_Mail:mail@heservices.co.uk Web_Site:http://www.heservices.co.uk

HE Services (Plant Hire) Ltd Liverpool Depot Merton Bank Road St Helens Lancs. WA9 1NH tel:(01744) 453446 fax:(01744) 453446 E_Mail:mail@heservices.co.uk Web_Site:http://www.heservices.co.uk

Heap Joshua & Co Ltd Boodle Street Ashton-under-Lyne Lancs. OL6 8NG tel:(0161) 330 1868 fax:(0161) 339 2040

Heat Trace Ltd Tracer House Cromwell Road Bredbury, Stockport Gtr. Manchester SK6 2RF tel:(0161) 430 8333 fax:(0161) 430 8654 E_Mail:sales@heat-trace.ltd.uk Web_Site:http://www.heat-trace.ltd.uk

Heat Transfer Pressurisation PO Box 473 Bradford W. Yorks. BD8 9YF tel:(01422) 349560 fax:(01274) 483495

Heath Samuel & Sons PLC Cobden Works Leopold Street Birmingham B12 0UJ tel:(0121) 772 2303 fax:(0121) 772 3334

Heating & Ventilating Contractors Associstion Esca House 34 Palace Court London W2 4JG tel:(0171) 313 4900 fax:(0171) 727 9268 E_Mail:contact@hvca.org.uk Web_Site:http://www.hvca.org.uk

Heating Equipment Ltd Hughenden Manor Petersfinger Salisbury Wilts. SP5 3DF tel:(01722) 335139 fax:(01722) 323592

Heatrae Sadia Heating Ltd Hurrican Way Norwich NR6 6EA tel:(01603) 42144 fax:(01603) 409409 telex:97331

Heldite Ltd Heldite Centre Bristow Road Hounslow Middx. TW3 1UP tel:(0181) 577 9157/9257 fax:(0181) 577 9057

Helifix Ltd Second Floor Suite 21 Warple Way London W3 0RX tel:(0181) 735 5200 fax:(0181) 735 5201 E_Mail:sales@helifix.co.uk Web_Site:http://www.helifix.co.uk

Hellerman Electric, Member of Bopwthorpe Holdings PLC Pennycross Close Plymouth Devon PL2 3NX tel:(01752) 761222 fax:(01752) 790058 E_Mail:cad.marketing@hellermanelectric.co.uk

Hellermann Insuloid Leestone Road Wythenshawe Manchester M22 4RH tel:(0161) 998 8551 fax:(0161) 954 3708

Helvar Electrosonic Hawley Mill Hawley Road Dartford Kent DA2 7SY tel:(01322) 222211 fax:(01322) 282282 Web_Site:http://www.helvar.com

HEM Interiors Group Ltd HEM House Kirkstall Road Leeds LS4 2BT tel:(0113) 263 2222 fax:(0113) 231 0237

Hemsec Manufacturing Ltd Stoney Lane Rainhill Prescot Merseyside L35 9LL tel:(0151) 426 7171 fax:(0151) 493 1331

Henderson-Bostwick Grange Close, Clover Nook Industrial Park Somercotes Derbyshire DE55 4QT tel:(01773) 523300 fax:(01773) 523301 E_Mail:sales@hbid.co.uk Web_Site:http://www.hbid.co.uk

Henkel Home Improvement & Adhesive Products Henkel House 292-308 Southbury Road Enfield Middx. EN1 1TS tel:(0181) 804 3343 fax:(0181) 443 4321 Web_Site:http://www.unibond.co.uk

Henkel Industrial Adhesives (AIV) Marketing and Sales Department Watchmead Welwyn Garden City Hertfordshire AL7 1JB tel:(01707) 355303 fax:(01707) 355302

Henkel Industrial Adhesives (AIV) Cromwell Road St. Neots Cambs. PE19 2JW tel:(01480) 404505 fax:(01480) 406111

Henry Venables Ltd Castletown Stafford Staffordshire ST16 2EN tel:(01785) 259131 fax:(01785) 215087 E_Mail:enquiries@henryvenables.co.uk Web_Site:http://www.henryvenables.co.uk

Hepworth Building Products Ltd Hazlehead Crow Edge Sheffield S. Yorks. S36 4HG tel:(01226) 763561 fax:(01226) 764827 E_Mail:info@hepworthbp.co.uk Web_Site:http://www.hepworthbp.co.uk/hepwort h

Hepworth Heating Ltd Nottingham Road Belper Derbys. DE56 1JT tel:(01773) 824141 fax:(01773) 820569 telex:37586 Web_Site:http://www.hep-heat.co.uk

Hepworth Minerals & Chemicals Ltd Brookside Hall Sandbach Cheshire CW11 4TF tel:(01270) 752 752 fax:(01270) 752753 E_Mail:sales@h.m.c.co.uk Web_Site:http://www.h.m.c.co.uk

Herberts, Division of HPG Freshwater Road Dagenham Essex RM8 1RU tel:(0181) 590 6030 fax:(0181) 599 3527

Hercules Piling Ltd Birchwood Way Cotes Park West Somercotes Derbys. DE55 4PY tel:(01770) 603091 fax:(01773) 540406

Herga Electric Ltd Northern Way Bury St Edmunds Suffolk IP32 6NN tel:(01284) 701422 fax:(01284) 753112

Hertel Services Hertel (UK) Ltd Poolspringe Llanwarne Hereford HR2 8JJ tel:(01981) 540020 fax:(01981) 540853

Hettich UK Ltd Unit 200, Metroplex Business Park Broadway Salford Manchester M5 2UW tel:(0161) 872 9552 fax:(0161) 848 7605

Hevilifts Ltd Brickyard Road Aldridge W. Midlands WS9 8TA tel:(01922) 458333 fax:(01922) 743043

Hewetson Floors Ltd Marfleet Hull HU9 5SG tel:(01482) 781701 fax:(01482) 799276 E_Mail:enquiries@hewetson.co.uk Web_Site:http://www.hewetson.co.uk

HEWI (UK) Ltd Scimitar Close Gillingham Business Park Gillingham Kent ME8 0RN tel:(01634) 377688 fax:(01634) 370612

Heywood Williams Architectural Birds Royd Lane Brighouse W. Yorks. HD6 1NG tel:(01484) 717677 fax:(01484) 400148

HF (GB) Ltd York House Vicarage Lane Bowdon, Altrincham Cheshire WA14 3BA tel:(0161) 928 4572 fax:(0161) 929 8513

Hi-Store Ltd Station Approach Four Marks Alton Hants. GU34 5HN tel:(01420) 562522 fax:(01420) 564420 E_Mail:hi-store@msn.com Web_Site:http://www.hi-store.com

Hi-Tec Window Films The Stone House 79 Mount Road Tettenhall Wood Wolverhampton WV6 8HQ tel:(01902) 761303 fax:(01902) 766225

Hi-Teck Insulation Unit 1 Beck's Green Common St Andrews Beccles Suffolk NR34 8NB tel:(01986) 781342

Hi-Vee Ltd Southdown Western Road Crowborough E. Sussex TN6 3EW tel:(01892) 662177 fax:(01892) 667225

Hickson Timber Products Ltd Sowgate Lane Knottingley W. Yorks. WF11 0BS tel:(01977) 671771 fax:(01977) 671701 E_Mail:jones@htp.co.uk

Hickton & Co Ltd, John PO Box 10 Stourbridge Road Holesowen W. Midlands B63 3QY tel:(0121) 550 1671 fax:(0121) 585 5031

Higginson Staircases Ltd Unit 1 Carlisle Road London NW9 0HD tel:(0181) 200 4848 fax:(0181) 200 8249 E_Mail:sales@higginson.co.uk Web_Site:http://www.higginson.co.uk

Hill & Noyes Office Furniture Abess House 39-47 High Street Southall Middx. UB1 3HE tel:(0181) 867 9999 fax:(0181) 867 9988

Hill Leigh (Joinery) Ltd Brue Way Walrow Industrial Estate Highbridge Somerset TA9 4AW tel:(01278) 789156 fax:(01278) 781768

Hill Top Sections Ltd Ridgacre Road West Bromwich W. Midlands B71 1BB tel:(0121) 553 5271 fax:(0121) 500 5611 E_Mail:hts@hadleyindustries.plc.uk Web_Site:http://www.hadleyindustries.plc.uk

Hillaldam Coburn Ltd Unit 6 Wyvern Estate Beverley Way New Malden Surrey KT3 4PH tel:(0181) 336 1515 fax:(0181) 336 1414 E_Mail:sales@hillaldam.co.uk Web_Site:http://www.coburn.co.uk

Hillday 1 Haverscroft Industrial Estate Attleborough Norfolk NR17 1YE tel:(01953) 454014 fax:(01953) 454014

Hille Ltd Cross Street Darwen Lancashire Cheshire BB3 2PW tel:(01254) 778856 fax:(01254) 778870

Hills Industries Ltd Pontgwindy Industrial Estate Caerphilly M. Glam. CF8 3HU tel:(01222) 883951 fax:(01222) 886102

Hilmor Caxton Way Stevenage Herts. SG1 2DQ tel:(01438) 312466 fax:(01438) 728827

Hilti (Great Britain) Ltd 1 Trafford Wharf Road Manchester M17 1BY tel:(0800) 886100 fax:(0800) 886200 Web_Site:http://www.hilti.com

Hilton Banks Ltd 48-52 Burhill Road Hersham Walton-on-Thames Surrey KT12 4JE tel:(01932) 221385 fax:(01932) 244936 E_Mail:hilton banks@btinternet.com

Hines & Sons, P E, Ltd Whitbridge Lane Stone Staffs. ST15 8LU tel:(01785) 814921 fax:(01785) 818808

Hinton Perry & Davenhill Ltd Dreadnought Works Pensnett Brierley Hill Staffs. DY5 4TH tel:(01384) 77405 fax:(01384) 74553

Hipkin F W Ltd Victory Works Maitland Road Stratford London E14 4HF tel:(0181) 555 8201 fax:(0181) 519 6606

Hipkiss, H, & Co Ltd 40-45 George Street Birmingham B3 1QA tel:(0121) 236 5342 fax:(0121) 236 4174

Hird Hastie Paints Ltd 73 Milnpark Street Glasgow G41 1EH tel:(0141) 429 2775 fax:(0141) 429 1407

HIS Ltd Huntworth Business Park Bridgewater Somerset TA6 6PS tel:(01278) 450600 fax:(01278) 452160

Hispack Limited 8 School Road Downham, Billericay Essex CM11 1QU tel:(01268) 711499 fax:(01268) 711068

HMSO Publications Office Parliamentary Press Mandela Way London SE1 tel:(0171) 394 4200

Hochiki Europe (UK) Ltd Grosvenor Road, Gillingham Business Park Gillingham Kent ME8 0SA tel:(01634) 260133 fax:(01634) 260132

Hodgkinson Bennis Ltd Unit 7A Highfield Road Little Hulton, Worsley Manchester M38 9SS tel:(0161) 790 4411 fax:(0161) 703 8505

Hodkin Jones (Sheffield) Ltd Callywhite Lane Dronfield Sheffield S18 2XP tel:(01246) 290890 fax:(01246) 290292 E_Mail:info@hodkin-jones.co.uk Web_Site:http://www.hodkin-jones.co.uk

Holden Brooke Ltd Haveford Street Manchester M12 5JL tel:(0161) 274 4242 fax:(0161) 274 3068

Hollaender Rainer PO Box 52 Walsall W. Midlands WS2 7PL tel:(01922) 495777 fax:(01922) 497943

Holophane Europe Ltd Bond Avenue Bletchley Milton Keynes MK1 1JG tel:(01908) 649292 fax:(01908) 270006

Holts Shutters, Bolton Bradey Turton Street Bolton BL1 2SP tel:(01204) 32111 fax:(01204) 396936

Homa Castors (Cosby) Ltd Cambridge Road Cosby Leicester LE9 5SH tel:(01533) 848266 fax:(01533) 867692

Home Automation Ltd Bumpers Way Chippenham Wilts. SN14 6LF tel:(01249) 443422 fax:(01249) 443315

Homelux Airfield Industrial Airfield Industrial Estate Ashbourne Derbys. DE6 1HA tel:(01335) 347300 fax:(01335) 340333 Web_Site:http://www.homelux.co.uk

Hometex Trading Ltd 13 Wheatley Road Ikley W. Yorks. LS29 8TS tel:(01943) 608197 fax:(01943) 816370

Honeywell Control Systems Ltd Honeywell House Arlington Business Park Bracknell Berks. RG12 1EB tel:(01344) 656000 fax:(01344) 656240

Hoover European Appliances Group Pentrbach, Merthyr Tydfil M. Glam. CF48 4TU tel:(01685) 721222 fax:(01685) 382946

Hopkinsons Ltd PO Box B27 Britania Works Huddersfield HD2 2UR tel:(01484) 422171 fax:(01484) 518092

HOPPE (UK) Ltd Gailey Park Gravelly Way Standeford Wolverhampton WV10 7GW tel:(01902) 484400 fax:(01902) 484406 E_Mail:info.uk@hoppe.com Web_Site:http://www.hoppe.com

Horizon International Ltd Willment Way Avonmouth Bristol BS11 8DJ tel:(0117) 982 1415 fax:(0117) 982 0630 E_Mail:horizon.int@lineone.net

Hormann (UK) Ltd Whiteacres Whetstone Leicester LE8 6ZG tel:(01533) 861404 fax:(01533) 867440

Horne Engineering Co Ltd, The PO Box 7 Rankine Street Johnstone Renfrewshire PA5 8BD tel:(01505) 321455 fax:(01505) 336287

Horstmann Timers & Controls Ltd Newbridge Road Bath Somerset BA1 3EF tel:(01225) 421141 fax:(01225) 423070 E_Mail:sales@horstmann.co.uk Web_Site:http://www.horstmann.co.uk

Hoskins Medical Equipment Ltd Admail 1001 Birmingham B1 1HJ tel:(0121) 707 6600 fax:(0121) 707 6688

Hospital Metalcraft Ltd Blandford Heights Blandford Forum Dorset DT11 7TG tel:(01258) 451338 fax:(01258) 455056

Hostess Furniture Ltd Vulcan Road Bilston Staffs. WV14 7JR tel:(01902) 493681 fax:(01902) 353185 telex:335261

Hotchkiss Air Supply Heath Mill Road Wombourne Wolverhampton WV5 8AP tel:(01902) 895161 fax:(01903) 892045 E_Mail:hotchkissairsupply@btinternet.com

Housing Corporation 149 Tottenham Court Road London W1P 0BN tel:(0171) 393 2000 fax:(0171) 393 2111

Hovair Systems Ltd - A Division of Air-Log Ltd North Lane Aldershot Hampshire GU12 4QH tel:(01252) 319922 fax:(01252) 341872 E_Mail:sales@airlog.co.uk Web_Site:http://www.airlog.co.uk

Hoval Ltd Northgate Newark-on-Trent Notts. NG24 1JN tel:(01636) 72711 fax:(01636) 73532

Howard Bros Joinery Ltd Station Approach Battle E. Sussex TN33 0DE tel:(01424) 773272 fax:(01424) 773836 E_Mail:sales@howard-bros-joinery.com Web_Site:http://www.howard-bros-Joinery.com

Howard Evans Roofing Limited 75 Tyburn Road Erdington Birmingham B24 8NB tel:(0121) 327 1336 fax:(0121) 327 3423 E_Mail:roofherl@ukbusiness.com

Howard Home Improvements Ltd 45 Mawney Road Romford Essex RM7 7HL tel:(01708) 722245 / 720048 fax:(01708) 736090

Howarth Timber (Elland) Ltd Elland Lane Elland Halifax Yorks. HX5 9DZ tel:(01422) 376012 fax:(01422) 310107

Howden Buffalo Old Govan Road Renfrew Renfrewshire PA4 8XJ tel:(0141) 885 7500 fax:(0141) 886 1961 E_Mail:buffalo.marketing@howden.com Web_Site:http://www.howden.com

Howden H D Ltd 73 Coronation Road New Stevenston Motherwell Lanarkshire ML1 4JF tel:(01698) 732303 fax:(01698) 832229

Howe Fencing & Sectional Buildings Horse Cross, Standon Road Standon Nr Ware Herts. SG11 2PU tel:(01920) 822055 fax:(01920) 822871

Howe-Green Ltd Merchant Drive Mead Lane Industrial Estate Hertford Herts. SG13 7BH tel:(01992) 554388 fax:(01992) 584612 E_Mail:info@howegreen.co.uk Web_Site:http://www.howegreen.co.uk

Howse Thomas Ltd Cakemore Road Rowley Regis Waroloy Worcs. tel.(0121) 559 1451

Hoyles Fire & Safety Ltd Sandwash Close Rainford Industrial Estate Rainford, St. Helens Merseyside WA11 8LY tel:(01744) 885161 fax:(01744) 882410

Hozelock Ltd Haddenham Aylesbury Bucks. HP17 8JD tel:(01844) 291881 fax:(01844) 290344 telex:837901 (HOZLOK)

HRS Hevac Ltd PO Box 230 Watford Herts. WD1 8TX tel:(01923) 232335 fax:(01923) 230266 E_Mail:mail@hrs.co.uk Web_Site:http://www.hrs.co.uk

HSS Hire Shops 25 Willow Lane Mitcham Surrey CR4 4TS tel:(0181) 260 3100 fax:(0181) 687 5005 E_Mail:hire@hss.co.uk Web_Site:http://www.hss.co.uk

HT Martingale Ltd Ridgeway Industrial Estate Iver Bucks. SL0 9HU tel:(01753) 654411 fax:(01753) 630002

Hub Tubes 14-16 McDowall Street Paisley Strathclyde PA3 2NB tel:(0141) 848 6767 fax:(0141) 848 6991

Hubbard Architectural Metalwork Ltd 3 Hurricane Way Norwich Norfolk NR6 6HS tel:(01603) 424817 fax:(01603) 487158 E_Mail:tony@hubbardsfreeserve.co.uk

Hubbell Ltd Ronald Close, Woburn Road Industrial Estate Kempston Bedford MK42 7SH tel:(01234) 855444 fax:(01234) 854008

Hudevad Britain Bridge House Bridge Street Walton-on-Thames Surrey KT12 1AL tel:(01932) 247835 fax:(01932) 247694

Hufcor (Partitions) Ltd Newbold Drive Castle Donington Derby DE74 2QX tel:(01332) 810576 fax:(01332) 811059

Hughes Safety Showers Whitefield Road Bredbury, Stockport Cheshire SK6 2SS tel:(0161) 430 6618 fax:(0161) 4307928 E_Mail:sales@hughes-safety-showers.co.uk Web_Site:http://www.hughes-safety-showers.co.uk

Humbrol Ltd Marfleet Hull HU9 5NE tel:(01482) 701191 fax:(01482) 712908

Humphrey & Stretton Ltd Pindar Road Hoddesdon Herts. tel:(01992) 460359 fax:(01992) 463996

Hunt Europe Ltd Chester Hall Lane Basildon Essex SS14 3BG tel:(01268) 530331 fax:(01268) 527211

Hunter Douglas Ltd (Contracts Division) Luxaflex Blinds ® Unit 8A Business Centre 17 London Road Swanscombe Kent DA10 0LH tel:(01322) 624580 fax:(01322) 624558

Hunter Flemming Associates Ltd 3 Tythe Barn Brumstead Road Stalham Norwich NR12 9DH tel:(01692) 581545 fax:(01692) 582893

Hunter Plastics Ltd Nathan Way London SE28 0AE tel:(0181) 855 9851 fax:(0181) 317 7764 telex:25770 E_Mail:hurplas@hunplas.co.uk. Web_Site:http://www.hunterplastics.co.uk

Hunter Timber Ireland Ltd 2-4 Milewater Road Belfast BT3 9AB tel:(01232) 744201 fax:(01232) 748952

Hunter Timber Wholesale Distribution East Side Tyne Dock South Shields Tyne & Wear NE33 5SP tel:(0191) 455 4311 fax:(0191) 454 7017

Hunting Engineering Ltd Reddings Wood Ampthill Bedford MK45 2HD tel:(01525) 840162 fax:(01525) 843730

Hunting Industrial Coatings Ltd Derby Road Widnes Cheshire WA8 9ND tel:(0151) 495 3505 fax:(0151) 495 3522

Huntley & Sparks Sterling House Blacknell Lane Crewkerne Somerset TA18 8LL tel:(01460) 72222 fax:(01460) 76402 E_Mail:mail@sterling-hydraulics.co.uk

Huntree Fencing Ltd 104 Green End Road Great Barford Beds. MK44 3HD tel:(01234) 870864 fax:(01480) 471082

HV Air Conditioning Ltd Coach Fold Works Coach Fold Road Haley Hill, Halifax W. Yorks. HX3 6ED tel:(01422) 366631 fax:(01422) 366061

HW Systems Ltd Olympus Park Quedgeley Gloucester GL2 4NF tel:(01452) 722227 fax:(01452) 727600

Hy-Ten Reinforcement Co Ltd Whitewall Road Medway City Industrial Estate Strood, Rochester Kent ME2 4DZ tel:(01634) 290102 fax:(01634) 290574 E_Mail:sales@hy-ten.co.uk

Hyde Brian Ltd Stirling Road Shirley Solihull W. Midlands B90 4LZ tel:(0121) 705 7988 fax:(0121) 711 2465 E_Mail:directtools@roh.globalnet.co.uk Web_Site:http://www.directtools.co.uk

Hydraseeders Ltd Coxbench Derby Derbys. DE21 5BH tel:(01332) 880364 fax:(01332) 883241

Hydro Polymers Ltd Newton Acliffe Co. Durham DL5 6EA tel:(01325) 300555 fax:(01325) 300215

Hydropath (UK) Ltd Linpac House Otterspool Way Watford Herts. WD2 8HL tel:(01923) 210028 fax:(01923) 800243

Hydropure Group Ltd Hydropure House Alington Road St Neots, Huntingdon Cambs. PE19 2RB tel:(01480) 407900 fax:(01480) 407911

Hydrovane Compressor Co Ltd Claybrook Drive Washford Industrial Estate Redditch B98 0DS tel:(01527) 525522 fax:(01527) 510862 telex:339843

Hyflex Roofing Halfords Lane Smethwick W. Midlands B66 1BJ tel:(0121) 555 6464 fax:(0121) 555 5862

Hygood Ltd Woodlands Place Woodlands Road Guilford Surrey GU1 1RN tel:(01483) 572222 fax:(01483) 302180

Hymo Ltd Ferro Fields Brixworth Northampton NN6 9AU tel:(01604) 880724 fax:(01604) 881314

Hynes Aden K Sculpture Studios 3 Hornsby Square Southfield Industrial Park Laindon, Basildon Essex SS15 6DS tel:(01268) 418837 fax:(01268) 414118

Hyperion Wall Furniture Ltd Business Park 7 Brook Way Leatherhead Surrey KT22 7NA tel:(01372) 378279 fax:(01372) 362004

IBL Lighting Unit C60, Barwell Business Park Leatherhead Road Chessington Surrey KT9 2NY tel:(0181) 391 7500 fax:(0181) 974 1629

E_Mail:info@ibl.co.uk Web_Site:http://www.ibl.co.uk

IBP Conex Whitehall Road Tipton W. Midlands DY4 7JU tel:(0121) 557 2831 fax:(0121) 520 8778 Web_Site:http://www.ibp.co.uk

Ibstock Greylands Horsham W. Sussex RH12 4QG tel:(01403) 211222 fax:(01403) 21077 telex:87442

Ibstock Building Products Ltd Anstone Office Kiveton Park Station, Kiveton Park Sheffield S. Yorks. S26 6NP tel:(01909) 771122 fax:(01909) 515281

Ibstock Building Products Ltd Leicester Road Ibstock Leics. LE67 6HS tel:(01530) 261999 fax:(01530) 261888 E_Mail:marketing@ibstock.co.uk Web_Site:http://www.ibstock.co.uk

Ibstock Hathernware Ltd Station Works, Rempstone Road Normanton on Soar Loughborough Leics. LE12 5EW tel:(01509) 842273 fax:(01509) 843629 E_Mail:hathernware@compuserve.com Web_Site:http://www.hathernware.co.uk

Ibstock Scotish Brick Tannochside Works Old Edingburgh Road Tanockside, Uddingston Strathclyde G71 6HL tel:(01698) 810686 fax:(01698) 812364 Web_Site:http://www.ibstock.co.uk

ICB (International Construction Bureau) Ltd Unit 9, Elliott Road West Howe Industrial Estate Bournemouth Dorset BH11 8JX tel:(01202) 579208 fax:(01202) 581748 Web_Site:http://www.alwitra.co.uk

ICI Chemicals & Polymers Ltd PO Box 14 The Heath Runcorn Cheshire WA7 4QF tel:(01928) 514444 fax:(01928) 572296

ICI Dulux Trade Paints Wexham Road Slough Berks. SL1 5HU tel:(01753) 691690 fax:(01753) 530336 Web_Site:http://www.dulux.com

ICI Paints Wexham Road Slough Berks SL1 5HU tel:(01753) 550000 fax:(01753) 530336 Web_Site:http://www.dulux.com

ICI Polyurethanes Hitchen Lane Shepton Mallet BA4 5TZ tel:(01749) 343061 fax:(01749) 346283

Ide T & W Ltd Glasshouse Fields London E1 9JA tel:(0171) 790 2333 fax:(0171) 790 0201 E_Mail:sales@hetleys.co.uk

Ideal Building Systems Ltd Lancaster Road Carnaby Industrial Estate Bridlington E. Yorks. YO15 3QY tel:(01262) 606750 fax:(01262) 671960 E_Mail:ideal.build@fsbdial.co.uk

Ideal Furniture Enterprises Ltd Broadmeadow Industrial Estate Dumbarton G82 2RG tel:(01389) 42588 fax:(01389) 61819

Ideal Williams Ltd High Holburn Road Condor Gate Industrial Estate Ripley Derbys. DE5 3NW tel:(01773) 570408 fax:(01773) 743478

Ideal-Standard Ltd The Bathroom Works National Avenue Kingston Upon Hull HU5 4HS tel:(01482) 346461 fax:(01482) 445886 telex:592113 E_Mail:brochures@ideal-standard.co.uk Web_Site:http://www.ideal-standard.co.uk

IDIA International Ltd 50-54 Farm Road Hove E. Sussex BN3 1GR tel:(01273) 220090 fax:(01273) 207993

IFT Supplies Ltd Unit 39, Cromwell Estate Staffa Road Leyton London E10 7QZ tel:(0181) 556 0678 fax:(0181) 556 0678

IG Limited Avondale Road Cwmbran Gwent NP44 1XY tel:(01633) 486486 fax:(01633) 486465 E_Mail:info@igltd.demon.co.uk Web_Site:http://www.igltd.co.uk

Iles Waste Systems Valley Mills Valley Road Bradford W. Yorks. BD1 4RU tel:(01274) 728837 fax:(01274) 734351

Illbruck Ltd Croesfield Industrial Park Rhostyllen, Wrexham LL14 4BJ tel:(01978) 294932 fax:(01978) 316005

Illuma Lighting Ltd Illuma House Gelders Hall Road Shepshed Leicestershire LE12 9NH tel:(01509) 601050 fax:(01509) 601036 E_Mail:info@illuma.co.uk Web_Site:http://www.illuma.co.uk

Imagetrim (UK) Ltd Unit 3, Teakcroft Centre, Fairview Industrial Park Marsh Way Rainham Essex RM13 8UH tel:(01708) 630472 fax:(01708) 630470

IMI Air Conditioning Ltd Armytage Road Brighouse W. Yorks. HD6 1QF tel:(01484) 714361 fax:(01484) 400543 telex:517487

IMI Bailey Birkett Ltd Sharp Street Worsley Manchester M28 3NA tel:(0161) 790 7741 fax:(0161) 703 8451 Web_Site:http://www.imibb.co.uk

IMI Range Ltd PO Box 1 Stalybridge Cheshire SK15 1PQ tel:(0161) 338 3353 fax:(0161) 303 2634

IMI Stanton Ltd Somerton Works Newport Gwent NP9 0XU tel:(01633) 277711 fax:(01633) 276004

IMI Yorkshire Copper Tube Ltd East Lancashire Road Kirkby Liverpool L33 7TU tel:(0151) 546 2700 fax:(0151) 549 2139 E_Mail:sales@yorkshirecoppertube.co.uk Web_Site:http://www.yorkshirecoppertube.co.uk

IMP Lighting Ltd Spring Lane Malven Link Worcester WR14 1AT tel:(01684) 891211 fax:(01684) 891046 E_Mail:enquiries@implighting.co.uk Web_Site:http://www.implighting.co.uk

Imperial Electrical Wholesalers Unit 19 Capitol Park Capitol Way Colindale London NW9 0EQ tel:(0181) 205 0333 fax:(0181) 205 3330

Imperial Machine Co Ltd Harvey Road Croxley Green Herts. WD3 3AX tel:(01923) 718000 fax:(01923) 777273 E_Mail:mail@imc.co.uk Web_Site:http://www.imco.co.uk

Incinerator The Co Ltd Howard Road Eaton Socon St Neots Huntington PE19 3ER tel:(01480) 213171 telex:32337

Indalex Ltd Kingsditch Road Cheltenham Glos. GL51 9PD tel:(01242) 521641 fax:(01242) 513304

Industrial Acoustics Ltd IEC House Moorside Road Winchester Hamp. SO23 7UJ tel:(01962) 873 000

Industrial Adhesives Ltd Moor Road Chesham Bucks. HP5 1SB tel:(01494) 784444 fax:(01494) 791903 telex:837356

Industrial Brushware Ltd 77 Malt Mill Lane Halesowen West Midlands B62 8JJ tel:(0121) 559 3862 fax:(0121) 559 9404 E_Mail:brushware@compuserve.com

Industrial Building Componants Ltd PO Box 36 Longhill Industrial Estate (North) Hartlepool Cleveland TS25 1QR tel:(01429) 221111 fax:(01429) 274035

Industrial Devices Ltd 309 West End Lane London NW6 1RG tel:(0171) 431 1118 fax:(0171) 794 3913

Industrial Hangers Ltd IHL House Thorpe Close, Thorpe Way Industrial Estate Banbury Oxon. OX16 8UU tel:(01295) 753400 fax:(01295) 753428 E_Mail:sales@ihl.co.uk Web_Site:http://www.ihl.co.uk

Industrial Lighting Ltd Pinnacle House Stanley Road Bootle Merseyside L20 7JF tel:(0151) 922 0052 fax:(0151) 922 3959

Infil Ltd Infil House Hadnall Shrewsbury Shropshire SY4 4AG tel:(01939) 210320 fax:(01939) 210690

Infraglo (Sheffield) Ltd Dannemora Drive Greenland Road Industrial Park Sheffield S. Yorks. S9 5DF tel:(0114) 2495445 fax:(0114) 2495066

Ingersol Locks Ltd Wood Street Willenhall W. Midlands WV13 1LA tel:(01902) 366911 fax:(01902) 363535 telex:338251

Ingersoll - Rand Architectural Hardware Wallows Lane Walsall W. Midlands WS1 4NA tel:(01922) 707400 fax:(01922) 612855 E_Mail:kparmar@compuserve.com Web_Site:http://www.ingersoll-rand-ahg.co.uk

Ingersoll-Rand European Sales Ltd PO Box 2 Chorley New Road Horwich Bolton BL6 6JN tel:(01204) 690690 fax:(01204) 690388

Initial Electronic Security Systems Ltd Shadsworth Road Blackburn BB1 2PR tel:(01254) 688688 fax:(01254) 662571

Initial Plant Services Ltd, Redispace and Johnson Valletta Street Hedon Road Hull HU9 5NP tel:(01482) 781202 fax:(01482) 712157

INMAC (a division of microwarehouse Ltd) Corporative Sales Office Unit 6, Wolsey Business Park Tolpits Lane, Watford Herts. WD1 8QP tel:(01923) 471288 fax:(01923) 227215 Web_Site:http://www.inmac.co.uk

Insect-O-Cutor Oakhurst Drive Cheadle Heath, Stockport Cheshire SK3 0XT tel:(0161) 428 0622 fax:(0161) 428 1239

Insituform Permaline Ltd Roundwood Industrial Estate Ossett W. Yorks. WF5 9SQ tel:(01924) 277076 fax:(01924) 265107

InstaGroup Ltd Insta House Ivanhoe Road, Hogwood Business Park Finchampstead, Wokingham Berks. RG40 4PZ tel:(0118) 932 8811 fax:(0118) 932 8314 E_Mail:101470.3501@compuserve.com Web_Site:http://www.instagroup.co.uk

Instant Zip-Up Ltd Audley Avenue Newport Shropshire TF10 7DP tel:(01952) 825815 fax:(01952) 825635

Institute of Concrete Technology The PO Box 7827 Crowthorne Berks RG45 6FR tel:(01344) 752096 fax:(01344) 752096

Institute of Plumbing The 64 Station Lane Hornchurch Essex RM12 6NB tel:(01708) 472791 fax:(01708) 448987 E_Mail:info@plumbers.org.uk Web_Site:http://www.plumbers.org.uk

Institute of Wastes Management 9 Saxon Court St Peters Gardens Northampton Northamptonshire NN1 1SX tel:(01604) 620426 fax:(01604) 621339 E_Mail:iwm.technical@iwm.co.uk Web_Site:http://www.iwm.co.uk

Institution of Civil Engineers 1 Great George Street London SW1P 3AA tel:(0171) 222 7722

Institution of Electrical Engineers Savoy Place London WC2R OBL tel:(0171) 240 1871

Institution of Incorporated Engineers in electronic, electrical and mechanical engineering (IIE) Savoy Hill House Savoy Hill London WC2R 0BS tel:(0171) 836 3357 fax:(0171) 497 9006 E_Mail:iie@dial.pipex.com Web_Site:http://www.iie.org.uk/iie

Institution of Mechanical Engineers 1 Birdcage Walk London SW1H 9JJ tel:(0171) 235 4535

Institution of Structural Engineers 11 Upper Belgrave Street London SW1X 8BH tel:(0171) 235 4535 fax:(0171) 235 4294 Web_Site:http://www.istructe.org.uk

Integra Products Eastern Avenue Lichfield Staffs. WS13 7SB tel:(01543) 267100 fax:(01543) 267104 E_Mail:cuscare@integra-products.co.uk Web_Site:http://www.integra-proucts.co.uk

Intercraft Designs Ltd The Arenson Centre Arenson Way Dunstable Beds LU5 5UL tel:(01582) 678300 fax:(01582) 678333 E_Mail:info@intercraft.co.uk Web_Site:http://www.intercraft.co.uk

Interface Europe Ltd The Don E. Russell Plant Shelf Mills Halifax W. Yorks. HX3 7PA tel:(01274) 690690 fax:(01274) 694095

Interface Europe Ltd Ashlyns Hall Chesham Road Berkhamsted Herts. HP4 2ST tel:(01442) 285000 fax:(01442) 876053 Web_Site:http://www.uksite.tbc

Interframe Ltd Aspen Way Yalberton Industrial Estate Paignton Devon TQ4 7QR tel:(01803) 666633 fax:(01803) 663030

Interglass Ltd Greenfields Industrial Estate Tindale Crescent Bishop Auckland Co. Durham DL1 9TF tel:(01388) 603667 fax:(01388) 609744 telex:587126

Intergrated Polymer Systems (UK) Ltd Swinton Meadows Mexborough S. Yorks. S64 8AB tel:(01969) 663377 fax:(01969) 6633000

Interlink Group Interlink House Commerse Way, Lancing Industrial Estate Lancing W. Sussex BN15 8TA tel:(01903) 763663 fax:(01903) 762621

Interlux Ltd 204 Oldbury Road West Bromwich W. Midlands B70 9DE tel:(0121) 5537551 fax:(0121) 580 0238 E_Mail:sales@interlux.co.uk Web_Site:http://www.interlux.co.uk

Internal Partitions Systems Interplan House, Dunmow Industrial Estate Chelmsford Road Great Dunmow Essex CM6 1HD tel:(01371) 874241 fax:(01371) 873848 E_Mail:contact@ipsinteriors.co.uk Web_Site:http://www.ips-interiors.co.uk

International Paint, Protective Coatings, Stoneygate Lane Felling on Tyne Tyne & Wear NE10 0JY tel:(0191) 469 6111 fax:(0191) 495 0676

Interoven Ltd 9 Shaftesbury Parade South Harrow Middx. HA2 0AJ tel:(0181) 422 2541 fax:(0181) 864 7514

Interphone Security Group PLC Interphone House, 5 Oxgate Centre Oxgate lane London NW2 7JE tel:(0181) 208 2311 fax:(0181) 208 0747

Intersolar Group Ltd Cock Lane High Wycombe Bucks. HP13 7DE tel:(01494) 452941 fax:(01494) 437045 E_Mail:intersolar@intersolar.com Web_Site:http://www.intersolar.com

Intralux UK Ltd Unit7A, The Seedbed Business Centre Vanguard Way Shoeburyness, Southend-on-Sea Essex SS3 9QX tel:(01702) 299929 fax:(01702) 296961

Invicita Plastics Ltd Harborough Road Oadby Leics. LE2 4LB tel:(0116) 272 0555 fax:(0116) 272 0626

IPPEC Sytsems 66 Rea Street South Digbeth Birmingham B5 6LB tel:(0121) 622 4333 fax:(0121) 622 5768 Web_Site:http://www.co.ippec.co.uk

Ironcrafts (Stotfold) Ltd Baldock Road Stotfold Hitchin Herts. SG5 4PA tel:(01462) 730671 fax:(01462) 835045

Island Cottages Ltd 30 Carrisbrooke Road Newport Isle of Wight PO30 1BW tel:(01983) 525985 fax:(0181) 525985

Isocrete Group Sales Ltd The Flooring Technology Centre Booth Lane Moston Sandbach, Cheshire CW11 9QK tel:(01270) 753000 fax:(01270) 753333

ISOKERN (UK) Ltd 14 Haviland Road Ferndown Industrial Estate Wimborne Dorset BH21 7RF tel:(01202) 861650 fax:(01202) 861632 E_Mail:sales@isokern.co.uk Web_Site:http://www.isokern.co.uk

Isolated Systems Ltd Adams Close, Heanor Gate Industrial Estate Heanor Derbys. DE75 7SW tel:(01773) 761226 fax:(01773) 760408 E_Mail:sales@isolatedsystems.com Web_Site:http://www.isolatedsystems.com

Isopad Ltd Oldmixon Industrial Estate Oldmixon Crescent Weston super Mare N. Somerset BS24 9AY tel:(01934) 629273 fax:(01934) 626125 E_Mail:rob.gunning@isopad.trimail.com Web_Site:http://www.isopad.co.uk

IsoSystems Unit 145 Hartlebury Trading Estate Hartlebury, Kidderminster Worcs. DY10 4JB tel:(01299) 251404 fax:(01299) 251468

Isowall UK) Ltd Winchester Hill, Romsey Hants. SO51 7UT tel:(01794) 522910 fax:(01794) 523243

ITM 2 Springlakes Estate Aldershot Hants. GU12 4UH tel:(01252) 28512 fax:(01252) 343217

ITS Ceilings & Partitions 44 Portman Road Reading Berks. RG30 1EA tel:(0118) 9500225 fax:(0118) 9503267

ITT Flygt Ltd Colwick Nottingham NG4 2AN tel:(0115) 940 0111 fax:(0115) 940 0444 telex:37316 Web_Site:http://www.flygt.com

ITW Buldex 37 Suttons Business Park Reading RG6 1HF tel:(01734) 261044 fax:(01189) 268568

ITW Redhead Queenslie Industrial Estate Glasgow G33 4JD tel:(0141) 774 2267 fax:(0141) 774 5802

ITW Spit Crompton Way Crawley W. Sussex RH10 2QR tel:(01293) 523372 fax:(01293) 515186

J & E Hall Ltd 6 Prospect Place Dartford Kent DA1 1BU tel:(01332) 223456 fax:(01322) 291458 telex:25594

J & M Fencing Services Unit 3, Wrexham Road, Laindon Basildon Essex SS15 6PX tel:(01268) 415233 fax:(01268) 417357

J Pugh-Lewis Limited Bushypark Farm Pilsley Chesterfield Derbys. S45 8HW

J. B. Corrie & Co Limited Frenchmans Road Petersfield Hampshire GU32 3AP tel:(01730) 262552 fax:(01730) 264915 E_Mail:sales@jbcorrie.co.uk

J. D. Goodacre & Son Saxelby Lodge Saxelby Pastures,Saxelby Melton Mowbray Leics. LE14 3NA tel:(01664) 823718 fax:(01664) 823719

J. Scott (Thrapston) Ltd Bridge Street Thrapston Northants NN14 4LR tel:(01832) 732366 fax:(01832) 733703

J.F.Poynter Ltd Maxim Lamp Works Cuckfield Hurstpierpoint W. Sussex BN6 9RW tel:(01273) 834890 fax:(01273) 834979

Jackson & Sons, G, ITD Unit 19, Mitcham Ind Estate Streatham Road Surrey CR4 2AP tel:(0181) 648 4343

Jackson H S & Son (Fencing) Ltd Stowting Common Ashford Kent TN25 6BN tel:(01233) 750393 fax:(01233) 750403 Web_Site:http://www.jacksons-fencing.co.uk

Jackson James & Co (London) Ltd 76-89 Alscot Road Bermondsey London SE1 3AW tel:(0171) 237 2862 fax:(0171) 232 2362

Jacuzzi UK Ltd, Division of Jacuzzi Inc 17 Mount Street Mayfair London W1Y 5RA tel:(0171) 409 1776 fax:(0171) 495 2353

Jaga Heating Poducts (UK) Ltd Jaga House, Orchard Business Park Bromyard Road Ledbury Herefordshire HR8 1LG tel:(01531) 631533 fax:(01531) 631534 E_Mail:jaga_uk@msm.com Web_Site:http://www.jaga.be

Jakem Timbers Ltd The Old Malt House 125 High Street Uckfield E. Sussex TN22 1EG tel:(01825) 768555 fax:(01825) 768483 E_Mail:richard@jakem.demon.co.uk

Jamak Fabrications (Europe) Ltd Unit H1 & H2, Europa Trading Estate Stoneclough Road Radcliffe Manchester M26 1GG tel:(01204) 794554 fax:(01204) 574521 E_Mail:jamak.fabrication@btinternet.com

James & Bloom Ltd Crossley Park Crossley Road Stockport Cheshire SK4 5DF tel:(0161) 432 5555 fax:(0161) 432 5312

James D Cocks (Enterprises) St Michaels, Brightlingsea Road Thorrington Colchester Essex CO7 8JJ tel:(01206) 302285 fax:(01206) 302285

James Fairley Athena Arley Road Saltley Birmingham B8 1BB tel:(0121) 326 5000 fax:(0121) 326 5005 E_Mail:sales.bs@j-f-a.co.uk

James Gibbons Format Ltd Colliery Road Wolverhampton West Midlands WV1 2QW tel:(01902) 458585 fax:(01902) 351336 E_Mail:deptname@jgf.co.uk Web_Site:http://www.jgf.co.uk

James Smellie Fabrications Ltd Unit F Leona Industrial Estate Nimmings Road Halesowen W. Midlands B62 9JQ tel:(0121) 561 1167 fax:(0121) 559 0336

Jandor Architectural Ltd Unit 26, Perryvale Industrial Park Horsenden Lane South Perryvale Middx. UB6 7RJ tel:(0181) 998 4321 fax:(0181) 991 1374

Jarvis Construction (UK) Ltd Suite 7/6, Clydeway Skypark 8 Elliott Place Glasgow G3 8EP tel:(0141) 248 7733 fax:(0141) 248 8779

Jaton Opal Way Stone Business Park Stone Staffs. S15 0SS tel:(01785) 811300 fax:(01785) 813331

Jaymart Rubber & Plastics Ltd Woodland Trading Estate Edenvale Road Westbury Wilts. BA13 3QS tel:(01373) 864926 telex:449776 J.BUNNY G

JB Architectural Inronmongery Ltd Avis Way Newhaven E. Sussex BN9 0DU tel:(01273) 514961 fax:(01273) 516764

JCL Engineering Ltd Bank Mill Huxley Street Oldham OL4 5JX tel:(0161) 665 2721 fax:(0161) 627 3773 E_Mail:thurst.jdengineering@zen.co.uk

Jebron Ltd Bright Street Wednesbury W. Midlands WS10 9HY tel:(0121) 526 2212 fax:(0121) 526 5322

Jendico Jendico House 1 Cork Lane Glen Parva Leicester LE2 9JR tel:(0116) 277 0474 fax:(0116) 277 2941

Jenkins Newell Dunford Ltd Thrumpton Lane Retford Notts. DN22 7AN tel:(01777) 706777 fax:(01777) 708141

Jerrards Commercial Lighting Division Arcadia House Cairo New Road Croydon London CR0 1XP tel:(0181) 251 5555 fax:(0181) 251 5500 E_Mail:jerrards@compuserve.com

Jewers Doors Ltd Stratton Business Park Normandy Lane Biggleswade Beds. SG18 8QB tel:(01767) 317090 fax:(01767) 312305 E_Mail:postroom@jewersdoors.co.uk Web_Site:http://www.jewersdoors.co.uk

Jewson Ltd High Street Brasted Westerham Kent TN16 1NG tel:(01959) 563856 fax:(01959) 564666

JI Case Europe Ltd PO Box 121 Wheatley Hall Road Doncaster S. Yorks. DN2 4PN tel:(01302) 366631 fax:(01302) 738581 telex:547242

Jimi-Heat Ltd Jimi-Heat House 200 Rickmansworth Road Watford Herts. WD1 7JS tel:(01923) 234477 fax:(01923) 240264 E_Mail:trace@jimiheat.demon.co.uk.

JKO Ltd Hughenden Avenue High Wycombe Bucks. HP13 5SQ tel:(01494) 521051 fax:(01494) 461176

JLC Pumps & Engineering Co Ltd PO Box 225 Barton-le-Clay Beds. MK45 4PN tel:(01582) 881946 fax:(01582) 881951

Jobling Perser Ltd Paradise Works Scotswood Road Newcastle-upon-Tyne NE4 8NY tel:(0191) 273 2331 fax:(0191) 226 0129

John Davidson (Pipes) Ltd Townfoot Industrial Estate Longtown Carlisle Cumbria CA6 5LY tel:(01228) 791503 fax:(01228) 792051

John Flower Fencing Limited Britannia House Kingsditch Lane Cheltenham Glos. GL51 9NF

John Reynolds & Sons (Birmingham) Ltd Church Lane West Bromwich W. Midlands B71 1DJ tel:(0121) 553 1287 / 2754 fax:(0121) 500 5460

Johnathon James Ltd 15-17 New Road Rainham Essex RM13 8DJ tel:(01708) 556921 fax:(01708) 520751

Johnson & Johnson PLC Units 12-19, Guiness Road Trading Estate Trafford Park Manchester M17 1SB tel:(0161) 872 7041 fax:(0161) 872 7351

Johnson & Starley Rhosili Road Brackmills Northampton NN4 7LZ tel:(01604) 762881 fax:(01604) 767408

Johnson Controls Systems Ltd Johnson House Randalls Way Leatherhead Surrey KT22 7TS tel:(01372) 376111 fax:(01372) 376823 E_Mail:products.uk@jci.com Web_Site:http://www.johnson-controls

Johnson H & R Tiles Ltd Highgate Tile Works Tunstall Stoke-on-Trent ST6 4JX tel:(01782) 575575 fax:(01782) 577377 telex:36146 HRJSTK G

Johnson Matthey Pigments & Dispersions Liverpool Road East Kidsgrove Stoke-on-Trent ST7 3AA tel:(01782) 794400 fax:(01782) 787338 E_Mail:pd@matthey.com Web_Site:http://www.cmd.matthey.com

Johnson Matthey PLC - Metal Joining York Way Royston Herts. SG8 5HJ tel:(01763) 253200 fax:(01763) 253168 telex:817351 E_Mail:metal-joining@matthey.com Web_Site:http://www.matthey.com

Johnson V B & Partners St John's House 304-310 St Albans Road Watford Herts. WD2 5PE tel:(01923) 227236 fax:(01923) 231134 E_Mail:office@vbjw.demon.co.uk Web_Site:http://www.vbjw.demon.co.uk

Johnston Pipes Ltd Doseley Telford Shropshire TF4 3BX tel:(01952) 630300 fax:(01952) 501537 telex:35179JONPIPG

Johnston Roadstone Ltd Leinthall Quarry Leinthall Earls Wigmore Nr Leominster Herefordshire HR6 9TR tel:(01568) 770521 fax:(01568) 770700

Johnston Roadstone Ltd Leaton Quarry Leaton Nr Wellington Telford TF6 5HA tel:(01952)740351 fax:(01952) 740413

Joint Council For The Building & Civil Engineering Industry (N. Ireland) 143 Marlone Road Belfast BT9 6SU tel:(01232) 661711 fax:(01232) 666323

Jolec Electrical Supplies Unit 2, The Empire Centre Imperial Way Watford Herts. WD2 4YH tel:(01923) 243656 fax:(01923) 230301

Jones & Attwood Ltd Titan Works Stourbridge DY8 4LR tel:(01384) 392181 fax:(01384) 371937 telex:338120

Jones & Thomas (Wireworks) Ltd Unit 5, Burtonwood Industrial Centre Burtonwood Warrington Cheshire WA5 4HX tel:(01925) 229795 fax:(01925) 228069

Jones of Oswestry Ltd Whittington Road Oswestry Shropshire SY11 1HZ tel:(01691) 653251 fax:(01691) 658222 telex:35517 Web_Site:http://www.jonesofoswestry.co.uk

Joseph Ash Galvanizing PO Box 16 Charles Henry Street Birmingham B12 OSP tel:(0121) 622 4661 fax:(0121) 622 4317

Jotun Henry Clark Ltd Unit 16 Alston Drive Bradwell Abbey Milton Keynes MK13 9HA tel:(01908) 321818 fax:(01908) 315073 telex:885421

JRM Doors Ltd Kingsley House, Bordon Trading Estate Oakhanger Road Bordon Hampshire GU35 9HH tel:(01420) 487681 fax:(01420) 489432 E_Mail:jrmdoors@compuserve.com

JSB Electrical PLC Manor Lane Holmes Chapel, Crewe Cheshire CW4 8AF tel:(01477) 537773 fax:(01477) 535722 E_Mail:marketng@jsb-electrical.com Web_Site:http://www.jsb-electrical.com

JT Contract Marketing Ltd 29 Store Street London WC1E 7BS tel:(0171) 580 8535 fax:(0171) 436 2620

Junckers Limited Wheaton Court Wheaton Road Witham Essex CM8 3UJ tel:(01376) 517512 fax:(01376) 514401

Juno Roplasto Ltd Wellington House Wellington Crescent New Malden Surrey KT3 3NE tel:(0181) 942 9323 fax:(0181) 949 5661

Jutlandia Doors Unit 14 & 18 Hazel Road Four Marks, Alton Hants. GU34 5EY tel:(01420) 561720 fax:(01420) 561730 E_Mail:info@norlandia.co.uk

KAB Seating Ltd Round Spinney Northampton NN3 8RS tel:(01604) 790500 fax:(01604) 790155 E_Mail:marketing@kabseating.com Web_Site:http://www.kabseating.com

Kaba UK Ltd Lower Moorway Tiverton Devon EX16 6SS tel:(01884) 256464 fax:(01884) 255818

KAC Alarm Company Ltd KAC House Tything Road Alcester Warwks. B49 6EP tel:(01789) 763338 fax:(01789) 400027 E_Mail:marketing@kac.co.uk Web_Site:http://www.kac.co.uk

Kair - The Ventilation Division of Kiltox Chemicals Limited Kiltox House Park Row Greenwich London SE10 9NL tel:(0181) 858 6277 fax:(0181) 853 3572 E_Mail:info@kiltox.co.uk Web_Site:http://www.kiltox.co.uk

Kalon Decorative Products Huddersfield Road Birstall, Batley W. Yorks. WF17 9XA tel:(01924) 354000 fax:(01924) 354001

Kalon Group PLC Ploughland House 62 Goerge Street Wakefield Yorks. WF1 1DL tel:(01924) 330100 fax:(01924) 330102

Kaloric Heater Co Ltd 31-33 Beethoven Street London W10 4LJ tel:(0181) 969 1367 fax:(0181) 968 8913

Kasenit Furnaces Ltd Bourne Works Collingbourne, Ducis Marlborough Wilts. SN8 3EH tel:(01264) 850105 fax:(01264) 850445

Kawneer UK Ltd Astmoor Road Astmoor Runcorn Cheshire WA7 1QQ tel:(01928) 563732 fax:(01928) 569297

Kay Metzeler Ltd The Hemmels Laindon North Essex SS15 6ED tel:(01268) 540054 fax:(01268) 540106

Kayanson Engineers Ltd Unit 34, Lydney Industrial Estate Harbour Road Lydney Glos. GL15 4EJ tel:(01594) 843494 fax:(01594) 843501

Kaye Aluminium plc Shaw Lane Industrial Estate Ogden Road Wheatley Hills, Doncaster S. Yorks. DN2 4SG tel:(01302) 762500 fax:(01302) 360307 E_Mail:sales@kayealuminium.co.uk Web_Site:http://www.kayealuminium.co.uk

KCW Windows & Conservatories Ltd 25A Shuttleworth Road Goldington Bedford Bedfordshire MK41 0HS tel:(01234) 269911 fax:(01234) 325034

Kedddy (Poujoulat) (UK) Ltd 6B Ascot Road Feltham Middlesex TW14 8QH tel:(01784) 256585 fax:(01784) 256586 E_Mail:sales@keddy.co.uk

Kee Klamps Ltd 10 Worton Drive Worton Grange Reading Berks. RG2 0TQ tel:(0118) 931 1022 fax:(0118) 931 1146 telex:849335 E_Mail:sales@keeklamp.comk Web_Site:http://www.keeklamp.com

Keeling Oliver Wellsway Keynsham Bristol BS18 1HU tel:(0117) 986 3177 fax:(0117) 986 3412

Keller Ground Engineering Oxford Road Ryton-on-Dunsmore Coventry CV8 3EG tel:(01203) 511266 fax:(01203) 305230

Kelsey Roofing Industries Ltd Kelsey House, Paper Mill Drive Church Hill South Redditch Worcestershire B98 8QJ tel:(01527) 594400 fax:(01527) 594444

Kem Edwards Ltd Edson House 143 Hersham Road Walton-on-Thames Surrey KT12 1RR tel:(01932) 227700 fax:(01932) 244095 E_Mail:sales@kemedwards.co.uk

Kemira Coatings Warth Road, off Radcliffe Road Bury Gtr. Manchester BL9 9NB tel:(0161) 764 6030 fax:(0161) 764 6102

Kenrick Archibald & Sons Ltd PO Box 9 Union Street West Bromwich W. Midlands B70 6BD tel:(0121) 553 2741 fax:(0121) 500 6332

Kensal CMS Kensal House President Way Luton Beds LU2 9NR tel:(01582) 425777 fax:(01582) 425776

Kent & Co (Twines) Ltd Hartley Trading Estate Long Lane Liverpool L9 7DE tel:(0151) 525 1601 fax:(0151) 523 1410

Kenyon Industrial Paints & Adhesives Regent House Regent Street Oldham OL1 3TZ tel:(0161) 633 6328 fax:(0161) 927 5027 E_Mail:ian@kenyon-group.co.uk Web_Site:http://www.kenyon-group.co.uk

Kenyon T H & Sons PLC Chancellors House Brampton Lane Hendon London NW4 2DX tel:(0181) 202 3816 fax:(0181) 202 9768

Kermi (UK) Ltd Unit A, Marconi Courtyard Brunel Road, Earlstrees Industrial Estate Corby Northants NN17 4LT tel:(01536) 400004 fax:(01536) 203774

Kern (Design Products) J T Ltd The Grange Budworth Road Noctorum, Birkenhead Merseyside L43 9TL tel:(0151) 647 9323 fax:(0151) 647 9324

Kerridge Stone Bridge Quarry Windmill Lane Kerridge, Macclesfield Cheshire SK10 5AZ tel:(01782) 514353 fax:(01782) 516783

Kershaw Mechanical Services Ltd The Beadle Trading Estate Ditton Walk Cambridge CB5 8PD tel:(01223) 410600 fax:(01223) 411061 E_Mail:k.m.s@dial.pipex.com

Kestner Building Products Ltd Station Road Greenhithe Kent DA9 9NG tel:(01322) 383281 fax:(01322) 386606 telex:896356

Ketley Brick Co Ltd , The Dreadnought Road Pensnett Brierley Hill W. Midlands DY5 4TH tel:(01384) 78361 fax:(01384) 74553

Keymer Hand Made Clay Tiles Nye Road Burgess Hill W. Sussex RH15 0LZ tel:(01444) 232931 fax:(01444) 871852 E_Mail:info@keymer.co.uk. Web_Site:http://www.keymer.co.uk

Keys W H Ltd Wall End Works Church Lane West Bromwich W. Midlands B71 1BN tel:(0121) 553 0206 fax:(0121) 500 5820

Kidde-Graviner Ltd Belvue Road Northolt Middx. UB5 5QW tel:(0181) 845 7711 fax:(0181) 845 4304

Kiltox Damp Free Solutions Kiltox House Park Row Greenwich London SE10 9NL tel:(0181) 858 6277 fax:(0181) 853 3572 E_Mail:info@kiltox.co.uk Web_Site:http://www.kiltox.co.uk

Kilwaughter Chemical Co Ltd Kilwaughter Lime Works 9 Starbog Road Larne Co. Antrim BT40 2JT tel:(01574) 260766 fax:(01574) 260136

Kimberly-Clark Ltd Cobdown House Larkfield Aylesford Kent ME20 9JS tel:(01622) 615000 fax:(01622) 615070 telex:95356

Kind, J B, Ltd Shobnall Street Burton-on-Trent Staffs. DE14 2HP tel:(01283) 564631 fax:(01283) 511132 Web_Site:http://www.jbkind.internet2.co.uk

Kinder-Janes Engineering Ltd Porters Wood St Albans Herts. AL3 6HU tel:(01727) 844441 fax:(01727) 844247

Kinetico Water Softeners 11 The Maltings Thorney Peterborough Cambs. PE6 0QF tel:(01733) 270463

Kingfisher Chemicals Ltd Cooper Lane Bardsea, Ulverston Cumbria LA12 9RA tel:(01229) 869100 fax:(01229) 869101

Kingsforth Security Fencing Limited Mangham Way Barbot Hall Industrial Estate Rotherham S. Yorks. S61 4RL tel:(01709) 378977 fax:(01709) 838992

Kingsnorth Bitumen Products Ltd Kingsnorth, Hoo, Rochester Kent ME3 9ND tel:(01634) 250722 fax:(01634) 253601

Kingspan Building Products Ltd Greenfield Business Park 2 Greenfield Holleywell Flintshire CH8 7HU tel:(01352) 716100 fax:(01352) 712444 E_Mail:bestprice@kbp.kingspan.co.uk Web_Site:http://www.kbp.kinspan.com

Kingspan Insulation Ltd Pembridge Leominster Herefordshire HR6 9LA tel:(01544) 388601 fax:(01544) 388888

Kingston Craftsmen Structural Timber Engineers Ltd Cannon Street Hull HU2 0AD tel:(01482) 225171 fax:(01482) 217032

Kingstonian Point Ltd Mayfield House Sculcoates Lane Hull E. Yorks. HU5 1DR tel:(01482) 342216 fax:(01482) 493096

Kinnings Marlow Ltd Franchise Street Wednesbury W. Midlands WS10 9RF tel:(0121) 526 6692 fax:(0121) 526 7166

Kinsley-Cooke Furniture Ltd Unit B2, Newton Business Park Talbot Road Newton, Hyde Gtr. Manchester SK14 4UQ tel:(0161) 367 9150 fax:(0161) 367 9157

Kirkpatrick Ltd Frederick Street Walsall Staffs. WS2 9NF tel:(01922) 620026 fax:(01922) 722525

Kirkstone 128 Walham Green Court Moore Park Road Fulham London SW6 4DG tel:(0171) 381 0424 fax:(0171) 381 0434 Web_Site:http://www.kirkstone.com

Kitsons Insulation Products Ltd Kitsons House Centurion Way Meridian Business Park Leicester LE3 2WH tel:(0116) 201 4499 fax:(0116) 201 4498

Klargester Environmental Engineering Ltd College Road Aston Clinton nr Aylesbury Bucks. HP22 5EW tel:(01296) 633000 fax:(01296) 633001 E_Mail:uksales@klargester.co.uk Web_Site:http://www.klargester.co.uk.

Klaxon Signal Ltd Warwick Road Tyseley Birmingham B11 2HB tel:(0121) 706 1654 fax:(0121) 708 1220 telex:339585

Klaxon Signals Ltd 502 Honey Pot Lane Stanmore Middx. HA71BE tel:(0181) 952 5566 fax:(0181) 952 6983

Kleeneze Sealtech Ltd Superseal Division Ansteys Road Hanham Bristol BS15 3SS tel:(0117) 947 5149 fax:(0117) 960 0141 E_Mail:enq@ksl.uk.com Web_Site:http://www.ksltd.com

Klick Technology Ltd Claverton Road Wythenshawe Manchester M23 9FT tel:(0161) 998 9726 fax:(0161) 946 0419

Klinger Fluid Instrumentation Ltd Edgington Way Sidcup Kent DA14 5AG tel:(0181) 300 7777 fax:(0181) 302 8145

Klober Ltd Pear Tree Industrial Estate Upper Langford Bristol N. Somerset BS18 7DJ tel:(01934) 853224 fax:(01934) 853221 E_Mail:support@klober.co.uk Web_Site:http://www.klober.co.uk

Klockner-Moeller Ltd PO Box 35 Gatehouse Close Aylesbury Bucks. HP19 3DH tel:(01296) 393322 fax:(01296) 21854

Knauf PO Box 133 Sittingbourne Kent ME10 3HW tel:(01795) 424499 fax:(01795) 428651 E_Mail:knauf@knauf.co.uk Web_Site:http://www.knauf.co.uk

Knight & Smith Ltd, Electrical Contractors 3 High Elms Cottages Woodside Road Watford Herts. WD2 7LA tel:(01923) 894744 fax:(01923) 679073

Knightbridge Furniture Productions Ltd 191 Thornton Road Bradford W. Yorks. BD1 2JT tel:(01274) 731442 fax:(01274) 736641 E_Mail:email@knightsbridge-furniture.co.uk Web_Site:http://www.knightsbridge-furniture.co.uk

Knowles Ltd R E Knowles Industrial Estate Furness Vale High Peak SK23 7PJ tel:(01663) 44127 fax:(01663) 741 562

Knowles W T & Sons Ltd Elland W. Yorks. HX5 9JA tel:(01422) 372833 fax:(01422) 370900

Knurr (UK) Ltd Burrel Road St. Ives Cambs. PE17 4LE tel:(01480) 496125 fax:(01480) 496373

Kobi Ltd Unit 25, St. Marks Industrial Estate Oriental Road Silvertown London E16 2BS tel:(0171) 474 3464 fax:(0171) 511 2306 E_Mail:cradles@kobi.co.uk Web_Site:http://www.kobi.co.uk

Komfort Systems Ltd Unit 1-10 Whittle Way Crawley W. Sussex RH10 2RW tel:(01293) 592500 fax:(01293) 549357 E_Mail:general@komfort.co.uk Web_Site:http://www.komfort.co.uk

Kompan (UK) Ltd 5 Holdom Avenue Bletchley Milton Keynes MK1 1QU tel:(01908) 642466 fax:(01908) 379081

Konaflex Ltd Unit 2 Northcote Road Stechford Birmingham B33 9EB tel:(0121) 783 9778

Kone Lifts Ltd 168-170 Wellington Road South Hounslow Middx. TW4 5JN tel:(0181) 570 7799 fax:(0181) 572 8389 Web_Site:http//www.kone.com

Kooltherm Insulation Production Ltd PO Box 3 Charlestown Glossop Derbys. SK13 8LE tel:(01457) 861611 fax:(01457) 852319 telex:669867

KPMC Ltd 64 Moyser Road Streatham London SW16 6SQ tel:(0181) 677 1059 fax:(0181) 357 6982 E_Mail:ppmc@compuserve.com

Kronospan Ltd Chirk Wrexham LL14 5NT tel:(01691) 773361 fax:(01691) 773292 E_Mail:kronospan@kronospan.btinternet.com Web_Site:http://www.kronospan.co.uk

KSB Ltd 2 Cotton Way Loughborough Leics. LE11 5TF tel:(01509) 231872 fax:(01509) 215228

Kufa Plastics Ltd 2 Lyon Close Chantry Woburn Road Estate Kempston Beds. MK42 7SB tel:(01234) 854464

Kush UK Ltd Unit 3B, Northway Trading Estate Tewkesbury Glos. GL20 8JH tel:(01684) 850787 fax:(01684) 850758

Kvaerner Cementation Foundation Ltd Denham Way Maple Cross Rickmansworth Herts. WD3 2SW tel:(01923) 423100 fax:(01923) 777834

Kwikform UK Ltd Waterloo Road Hay Mills, Birmingham B25 8LE tel:(0121) 275 0200 fax:(0121) 275 0300 E_Mail:enquiries@kwikform.co.uk Web_Site:http://www.kwikform.co.uk

Kwikform UK Ltd 192 Waterloo Road Hay Mills Birmingham B25 8LE tel:(0121) 275 0200 fax:(0121) 275 0300

L & H Polymers Ltd Crow Lane Gt Billing Northampton NN3 9BY tel:(01604) 410202 fax:(01604) 406439

Lab Furnishings Ltd Catwick Lane Brandesburton, Driffield E. Yorks. YO25 8RW tel:(01964) 542131 fax:(01964) 544121 E_Mail:labfurnishings@aaf.co.uk Web_Site:http//www.labfurn.co.uk

Lab-Craft Ltd Church Road Harold Wood Romford Essex RM3 OHT tel:(01708) 349320 fax:(01708) 376394 E_Mail:sales@labcraft.co.uk

Lafarge Aluminates Ltd 730 London Road Grays Essex RM20 3NJ tel:(01708) 863333 fax:(01708) 861033

Lafarge Plasterboard Ltd Easton-in-Gordano Marsh Lane Bristol BS20 ONF tel:(01275) 377773 fax:(01275) 377737

Laidlaw Architectural Hardware 124-126 Denmark Hill London SE5 8RX tel:(0171) 733 2101 fax:(0171) 737 2743 E_Mail:london.sales@laidlaw.net Web_Site:http//www.laidlaw.net

Laidlaw Architectural Hardware Ltd Pennine House Dakota avenue Ssauford Manchester MS 2PU tel:(0161) 848 1700 fax:(0161) 848 1700

Lakeside Group Bruce Grove Forest Fach Industrial Estate Swansea SA5 4US tel:(01792) 561117 fax:(01792) 587046

Lamb Macintosh 415-416 Montrose Avenue Slough Trading Estate Slough Berks. SL1 4TP tel:(01753) 522369 fax:(01753) 517216

Lami Doors UK Lower Wade House 2A Church Street Clitheroe Lancs. BB7 2AA tel:(01200) 444707 fax:(01200) 444717

Laminated Profiles Ltd 25 Caker Stream Road Alton Hants. GU34 2BR tel:(01420) 86512 fax:(01420) 541124 Web_Site:http://www.lampro.co.uk

Laminated Wood Ltd Unit 7, PHC Trading Estate Stibb Cross Torrington N. Devon EX38 8LH tel:(01805) 601596 fax:(01805) 601313 E_Mail:lamwood@clara.net Web_Site:http://www.ascwebindex.com/lamwood

Lampost Construction Co Ltd Greenore Co. Louth tel:(00 353) 42 73554 fax:(00 353) 42 73378

Lampways Ltd Allenby House Knowles Lane, Wakfield Road Bradford W. Yorks. BD4 9AB tel:(01274) 686666 fax:(01274) 680157

Lamson, D D, PLC Harbour Road Gosport Hants. PO12 1BG tel:(01705) 584271 fax:(01705) 504648

Lanarkshire Bolt Bretram Street Burnbank Hamilton Lanarkshire ML3 0QU tel:(01698) 284444 fax:(01698) 284842

Lancashire Fittings Ltd The Science Village Claro Road Harrogate N. Yorks. HG1 4AF tel:(01423) 522355 fax:(01423) 506111 E_Mail:kenidle@lancs-fittings.co.uk Web_Site:http://www.lancs-fittings.co.uk

Lancashire Steel Fabrication Co Ltd 5 Gardeners Place West Gillibrands Skelmersdale WN8 9SP tel:(01695) 727011 fax:(01695) 727611

Lancaster & Co (Bow) Ltd Hancock Road Bromley by Bow London E3 3DA tel:(0181) 980 2827 fax:(0181) 981 7815

Land Wood & Water Co. Ltd The Barn High Street Hartfield East Sussex TN7 4AE tel:(01892) 770470 fax:(01892) 770760

Landscape Grass (Concrete) Ltd Walker House 22 Bond Street Wakefield W. Yorks. WF1 2QP tel:(01924) 374818/375997 fax:(01924) 290289

Landscape Institute The 6 Barnard Mews London SW11 1QU tel:(0171) 738 9166

fax:(0171) 7389134 E_Mail:mail@l-i.org.uk Web_Site:http//www.l-i.org.uk

Langford Joinery Ltd 44 Largy Road Langford Lodge Crumlin Co. Antrim BT29 4RN tel:(01849) 452787 fax:(01849) 422415

Langley Systems Ltd 12-14 Magdelen Street London SE1 2EW tel:(0171) 407 6271 fax:(0171) 378 1754

Lappset UK Ltd Lappset House, Henson Way Telford Industrial Estate Kettering Northants NN16 8PX tel:(01536) 412612 fax:(01536) 521703 E_Mail:lappset@intac.co.uk Web_Site:http://www.intac.co.uklappset

Lastite Building Products Ltd Tenax Road Trafford Park Manchester M17 1JT tel:(0161) 876 7559 fax:(0161) 877 0117

Latchways plc Redman Road Porte Marsh Calne Wilts. SN11 9PL tel:(01249) 816326 fax:(01249) 813283 E_Mail:info@latchways.com Web_Site:http://www.latchways.com

Latham James plc Leeside Wharf, Mount Pleasant Hill Clapton London E5 9NG tel:(0181) 806 3333 fax:(0181) 806 7249 E_Mail:marketing@lathams.co.uk Web_Site:http//www.lathamtimber.co.uk

Laurier M & Sons 18 Marshgate Lane London E15 2NH tel:(0181) 534 7211 fax:(0181) 555 8991

Lawrence Surfaces Plc, Charles Newbridge Industrial Estate Newbridge Mid Lothian EH28 8PJ tel:(0131) 333 3030 fax:(0131) 333 4154 E_Mail:surfaces@clgplc.co.uk

Lawton Tube Co Ltd, The Torrington Avenue Coventry CV4 9AB tel:(01203) 466203 fax:(01203) 694183

Laybond Products Ltd Riverside Saltney Chester CH4 8RS tel:(01244) 674774 fax:(01244) 680215 E_Mail:sales_info@laybond.co.uk Web_Site:http//www.laybond.co.uk

LazyLawn The Dog House 45a Church Street, Langham Oakham Rutland LE15 7JE tel:(01572) 722923 fax:(01572) 724386 E_Mail:lazylawn.uk@btinternet.com

LB Lighting Ltd 14 West Burrowfield Burrowfield Industrial Estate Welwyn Garden City Herts. AL7 4TW tel:(01707) 880066 fax:(01707) 880077 E_Mail:lblighting.co.uk Web_Site:http//www.lblighting.co.uk

LB Plastics Ltd Firs Works Nether Heage Belper Derbys. DE56 2JJ tel:(01773) 852311 fax:(01773) 857080

LB Structures Ltd Spring Road Ettingshall Wolverhampton WV4 6JT tel:(01902) 353311 fax:(01902) 353049

Leach Dandridge Ltd Henwood Industrial Estate Ashford Kent TN24 8DH tel:(01233) 634471 fax:(01233) 62226

Lead Development Association 42 Weymouth Street London W1N 3LQ tel:(0171) 499 8422

Leaderflush & Shapland Po Box 5404 Nottingham NG16 4BU tel:(0870) 240 0666 fax:(0870) 240 0777 E_Mail:marketing@leaderflushshapland.co.uk Web_Site:http://www.leaderflushshapland.co.uk

Leakale Taurus Visual Aids Bargate Manchester Road Linthwaite Huddersfield HD7 5QX tel:(01484) 844016 fax:(01484) 847254

Leander Architectural Fletcher Foundry, Hallsteads Close Dove Holes Buxton Derbys. SK17 8BP tel:(01298) 814941 fax:(01298) 814970 E_Mail:sales@leanderarch.demon.co.uk Web_Site:http://www.leanderarch.demon.co.uk

Leavlite Electropaint Ltd Linchfield Industrial Estate Tamworth Staffs. B79 7TA tel:(01827) 52083 fax:(01827) 66938

Lectros International Ltd Ringstead Nr Kettering Northants NN14 4AT tel:(01933) 624340

Ledite Glass (Southend) Ltd 168 London Road Southend on Sea Essex SS1 1PH tel:(01702) 345893 fax:(01702) 435099

Ledu UK Ltd 12 Barmeston Road London SE6 3BF tel:(0181) 265 1681 fax:(0181) 265 1829 E_Mail:leduuklimited@compuserve.com

Lee Smith Wires ltd Station Road Ecclesfield South Yorkshire S35 9YR tel:(01142) 467161 fax:(01142) 462952

Lee Steel Strip Ltd PO Box 54 Meadow Hall Sheffield S. Yorks. S9 1HU tel:(0114) 243 7272 fax:(0114) 243 1277 E_Mail:sales@leesteelstrip.com Web_Site:http//www.leesteelstrip.com

Lees Richard Ltd Weston Underwood Derby DE6 4PH tel:(01335) 60601 fax:(01335) 60014 telex:377493 RDLLEE G

Legge N T Ltd Willenhall W. Midlands WV13 1TD tel:(01902) 366332 fax:(01902) 603935

Legge Thompson F & Co Ltd 1 Norfolk Street Liverpool L1 0BE tel:(0151) 709 7494 fax:(0151) 709 3774 telex:628761 (LIGROL) A/B BULTEL G

Legrand Electric Ltd 7-8 Foster Avenue Woodside Park Dunstable Beds. LU5 5TA tel:(01582) 676767 fax:(01582) 676771 E_Mail:legrand.sales@legrand.co.uk

Legrand Power Centre Brookside Wednesbury W. Midlands WS10 0QF tel:(0121) 506 4506 fax:(0121) 506 4507

Leicester Barfitting Co Ltd West Avenue Wigston Leics. LE18 2FB tel:(0116) 288 4897 fax:(0116) 281 3122

Leicester Fencing Contracts Limited Station Road Great Glen Leics. LE8 0FP tel:(0116) 2592112 fax:(0116) 2592090

Leigh's Paints Tower Works Kestor Street Bolton Lancs. BL2 2AL tel:(01204) 21771 fax:(01204) 382115 telex:63489

Leisure, Glynwed Consumer Foodservice Products Ltd Meadow Lane Long Eaton Notts. NG10 2AT tel:(0115) 946 4000 fax:(0115) 946 0374 Web_Site:http://www.leisure-sinks.co.uk

Lenesco Ltd Grove Road Northfleet Kent DA11 9AX tel:(01474) 564692 fax:(01474) 327329

Leofric Broadspan Building Ltd Leofric Works Ryton on Dunsmore Coventry CV8 3ED tel:(01203) 301301 fax:(01203) 301148

Lesco Products Ltd Wincheap Industrial Estate Canterbury Kent CT1 3RH tel:(01227) 763637 fax:(01227) 762239 E_Mail:sales@lesco.co.uk

Lever James & Sons Ltd Orient Mill Brandwood Street Bolton Lancs. BL3 4BH tel:(01204) 61121 fax:(01204) 658154

Lewden Electrical Industries Argall Avenue Leyton London E10 7QD tel:(0181) 539 0237 fax:(0181) 558 2718

Lewes Design Contracts Ltd The Mill Glynde Lewes E. Sussex BN8 6SS tel:(01273) 858341 fax:(01273) 858200 E_Mail:spiral@pavillion.co.uk Web_Site:http://www.freepages.pavillion.net/users/spiral/

Lewis Spring Products Ltd Studley Road Redditch Worcs. B98 7HJ tel:(01527) 510535 fax:(01527) 500868 E_Mail:sales@lewis-spring.com Web_Site:http://www.lewis-spring.com

Leyland Paint Company, The Huddersfield Road Birstall Batley W. Yorks. WF17 9XA tel:(01924) 420202 fax:(01924) 420403

Liebert Swindon Ltd Elgin Drive Swindon SN2 6DX tel:(01793) 553355 fax:(01793) 553400

Lift Boy Ltd Bessemer Close Ebblake Industrial Estate Verwood Dorset tel:(01202) 827388 fax:(01202) 823242

Light (Multiforms), H A Woods Lane Cradley Heath Warley W. Midlands B64 7AL tel:(01384) 569283 fax:(01384) 633712

Light Alloy Ltd Dales Road Ipswich Suffolk IP1 4JR tel:(01473) 740445 fax:(01473) 240002

Lightfoot Charles Ltd 42 Higher Ardwick Manchester M12 6DA tel:(0161) 273 1134 fax:(0161) 273 6417

Lighting Association, The Stafford Park 7 Telford Shropshire TF3 3BQ tel:(01952) 290905 fax:(01952) 290906 Web_Site:http://www.lightingassociation.com

Lighting Systems UK, Division of Phosco Ltd Charles House Furlong Way Gt Amwell, Ware Herts. SG12 9TA tel:(01920) 462272 fax:(01920) 485915

Lignacite (Brandon) Ltd Norfolk House High Street Brandon Suffolk IP27 0AX tel:(01842) 810678 fax:(01842) 814602 E_Mail:info@lignacite.co.uk Web_Site:http://www.lignicite.co.uk

Lignacite (North London) Ltd Meadgate Works Nazeing Waltham Abbey Essex EN9 2PD tel:(01992) 464661 fax:(01992) 445713

Linatex Ltd Wilkinson House Galway Road, Blackbush Business Park Yateley Hants. GU46 6GE tel:(01252) 743000 fax:(01252) 743030 E_Mail:linatex.sales@linatex.co.uk Web_Site:http//www.linatex.net

Lindapter International Lindsay House Brackenbeck Road Bradford W. Yorks. BD7 2NF tel:(01274) 521444 fax:(01274) 521130 E_Mail:lindapter@dial.pipex.com Web_Site:http://www.lindapter.com

Lindley George & Sons Ltd Sovereign Quarries Carr Lane Shepley Nr. Huddersfield HD8 8BH tel:(01484) 606203 fax:(01484) 605274

Linear Ltd Coatham Avenue Newton Aycliffe Industrial Estate Aycliffe Co. Durham DL5 6DB tel:(01325) 310151 fax:(01325) 310801

Link 51 (Storage Products) Ltd Link House Halesfield 6 Telford Shropshire TF7 4LN tel:(01952) 682200 fax:(01952) 680352 E_Mail:sales@link51.co.uk Web_Site:http://www.link51.co.uk

Lionweld Kennedy Ltd Marsh Road Middlesbrough Cleveland T51 5JS tel:(01642) 245151 fax:(01642) 224710 E_Mail:sales@lionweldkennedy.co.uk

Liquid Plastics Ltd PO Box 7 London Road Preston Lancs. PR1 4AJ tel:(01772) 259781 fax:(01772) 882016 E_Mail:info@liquidplastics.co.uk Web_Site:http://www.liquidplastics.co.uk

Liver Grease Oil & Chemical Co Ltd 11 Norfolk Street Liverpool L1 0BE tel:(0151) 709 7494 fax:(0151) 709 3774 telex:628761 (LIGROL) A/B BULTEL G Web_Site:http://www.merseyworld.com/

Llewellyn (Stonecare) Limited Bleak Hall Milton Keynes Bucks. MK6 1LA tel:(01908) 679222 fax:(01908) 235250

Llewellyn Homes Ltd Courtlands Road Eastbourne E. Sussex BN22 8TS tel:(01323) 735271 fax:(01323) 410246

Lloyd John & Sons (Engineers) Ltd 12 Frederick Street Birmingham B1 3HE tel:(0121) 236 6592

Loblite Ltd Third Avenue Team Valley Gateahead Tyne & Wear NE11 0QQ tel:(0191) 487 8103 fax:(0191) 482 0270 E_Mail:loblite@btinternet.com

Lockhart Catering Equipment Lockhart House Arrowhead Road Theale, Reading Berks. RG7 4AH tel:(0118) 9303900 fax:(0118) 9303931 E_Mail:lockhart@gardnermerchant.com Web_Site:http://www.gardnermerchant.com

Locking Systems Ltd Hopwood Lane / New Bond Street Halifax Yorks. HX1 5EZ tel:(01422) 330312 fax:(01422) 349987

Loctite UK Ltd Watchmead Welwyn Garden City Herts. AL7 1JB tel:(01707) 358800 fax:(01707) 358900

Loft Centre Products Quarry Lane Chichester W. Sussex PO19 2NY tel:(01243) 785246 fax:(01243) 533184

Logic Office Interiors 748 London Road Hounslow Middx. TW3 1SE tel:(0181) 572 7474 fax:(0181) 570 9643

London Fan Company Ltd 75-81 Stirling Road Acton London W3 8DJ tel:(0181) 992 6923 fax:(0181) 992 6928

London Lamplighting Co 21 Old Newtown Road Newbury Berks. RG14 7DP tel:(01635) 48131 fax:(01635) 35675

Longbottom & Co (Keighley) Ltd Dalton Mill Dalton Lane Keighley Yorks. BD21 4JL tel:(01535) 604007 fax:(01535) 609947

Longbottom J & J W Ltd Bridge Foundry Holmfirth Huddersfield W. Yorks. HD7 1AW tel:(01484) 682141 fax:(01484) 681513

Longden Doors Ltd Parkwood Road Sheffield S3 8AH tel: fax:

Longley C R & Co Ltd Ravensthorpe Road Thornhill Lees Dewsbury W. Yorks. WF12 9EF tel:(01924) 464283 fax:(01924) 459183

Longley James & Co Ltd East Park Crawley W. Sussex RH10 6AP tel:(01293) 561212 fax:(01293) 564564 E_Mail:mailroom@longleyco.uk

Lonsdale Metal Company Millmead Industrial Centre Mill Mead Road London N17 9QU tel:(0181) 801 4221 fax:(0181) 801 1287

Lord Isaac Ltd Desborough Road High Wycombe Bucks. HP11 2QN tel:(01494) 459191 fax:(01494) 461376 E_Mail:info@isaaclord.co.uk Web_Site:http://www.isaaclord.co.uk

Lotus Water Garden Products Ltd Junction Street Burnley Lancs. BB12 0NA tel:(01282) 420771 fax:(01282) 412719

Louvre-Lite Ltd Newton Mill Ashton Road Hyde Cheshire SK14 4BG tel:(0161) 366 6872 fax:(0161) 368 1133

Lowe & Fletcher Ltd Moorcroft Drive Wednesbury W. Midlands WS10 7DR tel:(0121) 5050400 fax:(0121) 5050420

LPA Industries PLC PO Box 15, Tudor Works Debden Road Saffron Walden Essex CB11 4AN tel:(01779) 512802 fax:(01779) 512828 E_Mail:sales@lpa-ind.demon.co.uk

LSA Projects Ltd The Barn White Horse Lane Witham Essex CM8 2UB tel:(01376) 501199 fax:(01376) 502027 E_Mail:sales@lsaprojects.demon.co.uk

Lumitron Ltd Chandos Road London NW10 6PA tel:(0181) 965 0211 fax:(0181) 965 8629 telex:291828 Web_Site:http://www.lumitron.co.uk

Luxcrete Ltd Premier House Disraeli Road Park Royal London NW10 7BT tel:(0181) 965 7292 fax:(0181) 961 6337

Luxo UK Ltd 4 Barmeston Road London SE6 3BN tel:(0181) 698 7238 fax:(0181) 698 6134

Lycetts (Burslem) Ltd Glendale Street Burslem Stock-on-Trent Staffs. ST6 2EP tel:(01782) 575236 fax:(01782) 577841 E_Mail:sales@lycettdoors.co.uk Web_Site:http//www.lycettdoors.co.uk

Lydney Access Floor Ltd Pine End Works Lydney Glos. GL15 5EW tel:(01594) 844855 fax:(01594) 843727 telex:43252 LYDPLY G

Lydney Containers Ltd Harbour Road Lydney Glos. GL15 5EW tel:(01594) 842378 fax:(01594) 843213

Lydney Products Ltd Pine End Works Harbour Road Lydney Glos. GL15 5EW tel:(01594) 842213 fax:(01594) 843727

Lyktan Lighting (UK) Ltd 31 a Bruton Place Mayfair London W1X 7AB tel:(0171) 409 3454 fax:(0171) 409 3414 E_Mail:info@lyktan-lighting.co.uk Web_Site:http//www.lyktan-lighting.co.uk

Lyncrete Burkle Services Ltd 20 Burns Avenue Church Crookham Fleet Hants. GU13 0BN tel:(01252) 816227 fax:(01252) 816227

Lynman Windows Unit 10, Parkgate Business Centre Chandlers Way Swanwick, Southampton Hants. SO31 1FQ tel:(01489) 577599 fax:(01489) 579472

Lyssand UK Ltd 11 Acorn Business Centre Arran Road Perth PH1 3DZ tel:(01738) 444456 fax:(01738) 444452

M & G Brickcutters Hockley Works Hooley Lane Redhill RH1 6JE tel:(01737) 771171

M & M Fencing Limited 8 Lightborne Road Sale Cheshire M33 5EA tel:(0161) 973 0816 fax:(0161) 962 7322

Maby Hire Ltd 1 Railway Street Scout Hill Ravensthorpe, Dewsbury W. Yorks. WF13 3EJ tel:(01924) 460601 fax:(01924) 457932 E_Mail:generalemabyhire.freeserve.co.uk

MacBee Manufacturing Ltd Unit 1A Building 11 Stanmore Industrial Estate Bridgnorth Shropshire WV15 5HR tel:(01746) 763103/763123 fax:(01746) 764659

Macc Stone Ltd 2 Robin Hill Biddulph Moor Stoke-on-Trent ST8 7NN tel:(01782) 514353 fax:(01782) 516783 E_Mail:macclesfield@stone.comm

Macclesfield Stone Quarries Ltd 194/198 New Street Buddulph Moor Stoke-on-Trent ST8 7NW tel:(01782) 514353 fax:(01782) 516783 E_Mail:sales@macclesfieldstone.com

Mackay Carpets, Hugh PO Box 1 Dragon lane Durhum DH1 2RX tel:(0191) 386 4444 fax:(0191) 384 0530

Mackwell Electronics Ltd Virgo Place Aldridge W. Midlands WS9 8UG tel:(01922) 58255 fax:(01922) 51263

MacLellan Rubber Ltd Shuna Street Maryhill Glasgow G20 9QA tel:(0141) 946 5111 fax:(0151) 946 5222 E_Mail:macrubber@compuserve.com

Maco Door & Window Hardware (UK) Ltd New Marlborough House Arnolde Close Medway City East, Rochester Kent ME2 4QW tel:(01634) 294555 fax:(01634) 294345 E_Mail:craig.maco@btinternet.com Web_Site:http//www.maco-europe.com

Macron Fireater Ltd Fireater House South Denes Road Great Yarmouth Norfolk NR30 3PJ tel:(01493) 859822 fax:(01493) 858374 telex:97409MACRON G E_Mail:sales@fireater.co.uk Web_Site:http://www.fireater.co.uk

Madewel Electronics Systems Ltd 220 Higher Road Urmston Manchester M41 1BH tel:(0161) 748 1115 fax:(0161) 746 8415

Maestro International Ltd Powerscroft Road Sidcup Kent DA14 5NH tel:(0181) 302 4035 fax:(0181) 302 8933

Magiboards Ltd Unit B3, Stafford Park 11 Telford Shropshire TF3 3AY tel:(01952) 292111 fax:(01952) 292280 E_Mail:sales@magiboards.co.uk Web_Site:http://www.magiboards.co.uk

Magnatek Ltd Unit 22, Mead Park Riverway Harlow Essex CM20 2SE tel:(01279) 422540 fax:(01279) 421570 E_Mail:magnetec.ltd@virgin.net Web_Site:http://www.magnetic.com

Magnet Applications Ltd Northbridge Road Berkhamptsead Herts. HP4 1EH tel:(01442) 875081 fax:(01442) 875009

Magnet Ltd Royd Ings Avenue Keighley W. Yorks. BD21 4BY tel:(01535) 661133 fax:(01535) 610363

Magnum Scaffolding (Bristol) Ltd Yard Brook Estate Stockwood Vale Keynshame Bristol BS31 2AL tel:(0117) 986 0123

Majestic Shower Co Ltd 1 North Place Edinburgh Way Harlow Essex CM20 2SL tel:(01279) 443644 fax:(01279) 635074 E_Mail:enquiries@majesticshowers.com Web_Site:http//www.majesticshowers.com

Makay's of Cambridge Ltd 85 East Road Cambridge CB1 1BY tel:(01223) 363132 fax:(01223) 366350

Maldon Fencing Co. Burnham Road Latchingdon Nr Chelmsford Essex CM3 6HA tel:(01621) 740415 fax:(01621) 740769

Manders Paint Ltd PO Box 9 Old Heath Road Wolverhampton WV1 2XG tel:(01902) 871028 fax:(01902) 452435 telex:338354

Mandor Engineering Ltd Turner Street Works Turner Street Ashton Under Lyne Gtr. Manchester OL6 8LU tel:(0161) 330 6837 fax:(0161) 308 3336 E_Mail:sales@mandor.co.uk Web_Site:http//www.asewetindex.com/mandor

Mandrain Slate Ltd Grange Mill Raglan Gwent WP5 2AA tel:(01291) 691048 fax:(01291) 691049

Manitou (Site Lift) Ltd Ebblake Industrial Estate Verwood Winborne Dorset BH31 6BB tel:(01202) 825331 fax:(01202) 813027

Mann McGowan Fabrication Ltd Intumescent House, 9 Springlakes Estate Deadbrook Lane Aldershot Hants. GU12 9UH tel:(01252) 333601 fax:(01252) 322724

Mansfield Brick Co Ltd Sanhurst Avenue Mansfield Notts. NG18 4BE tel:(01623) 622441 fax:(01623) 420904 telex:377778 MANSTD

Mansfield Keith Illustration 3 Johns Place Romiley Stockport Cheshire SK6 4BP tel:(0161) 494 9912

Manton Industrial Seals Ltd Unit 5B Bydand Industrial Estate Little Paxton Cambs. PE19 4ES tel:(01480) 214300 fax:(01480) 218987 E_Mail:gculley251@aol.com

Map Hardware Ltd King Alfred Way Cheltenham Glos. GL52 6QP tel:(01242) 512503 fax:(01242) 525767 telex:43589

Mapei UK Ltd 17-18 Drake Court, Britannia Park Riverside Road Middlesbrough Cleveland TS2 1RS tel:(01642) 245166 fax:(01642) 240993

Mar.Com Systems Ltd 1 MarCom House Heath Gardens Twickenham Middx. TW1 4BP tel:(0181) 891 5061 fax:(0181) 892 9028 E_Mail:kmccalney@mar-com.co.uk Web_Site:http//www.mar-com.co.uk

Marble Mosaic Co Ltd, The Winterstoke Road Weston Super Mare N. Somerset BS23 3YE tel:(01934) 419941 fax:(01934) 625479 E_Mail:marblemosa@aol.com

Marcel Guest Paints Riverside Works Collyhurst Road Collyhurst Manchester M40 7RU tel:(0161) 205 7631 fax:(0161) 205 4829 E_Mail:sales@hmguest.u-net.com Web_Site:http://www.h.marcel.guest.com

Marcrist Industries Ltd Kirk Sandell Industrial Estate Doncaster DN3 1QR tel:(01302) 890888 fax:(01302)883864

Marcus M Ltd Unit 7, Blackbrook Industrial Estate Peartree Lane Dudley W. Midlands DY2 0XW tel:(01384) 457900 fax:(01384) 457903 telex:335032 E_Mail:enquiries@m-marcus.com Web_Site:http//www.m-marcus.com

Marflex Chimney Systems Unit 40 Vale Business Park Cowbridge S. Glam. CF71 7PF tel:(01446)

775551 fax:(01446) 772468 Web_Site:http://www.marflexchimneys.internet2.co.uk.
Marflow Eng Ltd Britannia House Austin Way Hamstead Industrial Estate Birmingham B42 1DU tel:(0121) 358 1555 fax:(0121) 358 1444 E_Mail:sales@marflow.co.uk Web_Site:http//www.marflow.co.uk
Marine & Commercial Rigging Systems 31 Chapel Road Burnham-on -Crouch Essex CM0 8JB tel:(01621) 782284 fax:(01621) 782263
Marleton Cross Ltd Unit 2 Alpha Close Delta Drive, Tewkesbury Industrial Estate Tewklesbury Glos. GL20 8JF tel:(01684) 293311 fax:(01684) 293900
Marley Alutec Ltd Dickley Lane Lenham, Maidstone Kent ME17 2DE tel:(01622) 858888 fax:(01622) 858725 Web_Site:http://www.demon.co.uk/mel
Marley Building Materials Ltd Station Road Coleshill Birmingham B46 1HP tel:(01675) 468400 fax:(01675) 468485 E_Mail:mbmmark@mbpurrh.demon.co.uk Web_Site:http://www.marley.co.uk
Marley Floors Ltd Lenham Maidstone Kent ME17 2DE tel:(01622) 854000 fax:(01622) 854299 E_Mail:info@marfaw.u-net.co Web_Site:http://www.marley.co.uk
Marley Waterproofing Dickley Lane Lanham Maidstone Kent ME17 2DE tel:(01622) 854000 fax:(01622) 854229 E_Mail:info@marfaw.u-net.com Web_Site:http://www.marley.co.uk
Marlin Lighting Ltd Hampton Road West Hanworth Trading Estate Feltham Middx. TW13 6DR tel:(0181) 894 5522 fax:(0181) 755 1215 telex:935256 MARLING
Marlow Ropes Ltd Diplocks Way Hailsham E. Sussex BN27 3JS tel:(01323) 847234 fax:(01323) 440093 telex:87676 E_Mail:industrial@marlowropes.com Web_Site:http://www.marlowropes.com
Marlows Timber Engineering Marlow House Hollow Road Bury St Edmonds Suffolk IP32 7AP tel:(01284) 772700 fax:(01284) 755567 E_Mail:ted@marlowsteng.co.uk Web_Site:http//www.marlowsteng.co.uk
Marples William & Sons Ltd Oscar Works Meadow Street Sheffield S. Yorks. S3 7BQ tel:(0114) 272 6662 fax:(0114) 273 0257
Marriott & Price Ltd Station House Banstead Road Banstead Surrey SM7 1PZ tel:(01737) 352735 fax:(01737) 359192
Marshall Contracts 1704 Coventry Road Yardley Birmingham B25 8DT tel:(0121) 772 8485 fax:(0121) 772 5433 E_Mail:info@marshallcontract.co.uk Web_Site:http://www.marshallcontracts.co.uk.
Marshall J A & Co (Brushes) Ltd Parkside House Bent Lane Prestwich Manchester M25 1DL tel:(0161) 773 2526 fax:(0161) 773 1545
Marshall-Tufflex Ltd 55-65 Castleham Road Castleham Industrial Estate Hastings E. Sussex TN38 9NU tel:(01424) 427691 fax:(01424) 720670 E_Mail:general@marshall-tufflex.ltd.uk Web_Site:http//www.marshall-tufflew.ltd.uk
Marshalls Clay Products Ltd Robin Hood Wakefield W. Yorks. WF3 3BH tel:(01532) 822141 fax:(01532) 829213
Marshalls Flooring Ltd Hoveringham Nottingham NG14 7JX tel:(01636) 832000 fax:(01636) 832020
Marshalls Mono Ltd Southowram Halifax Yorks. HX3 9SY tel:(01422) 306000 fax:(01422) 330185 telex:517087 MARSHAL G Web_Site:http://www.marshalls.co.uk
Marsland & Co Ltd Station Road Edenbridge Kent TN8 6EE tel:(01732) 862501 fax:(01732) 866737
Marston A & Co Ltd Wellington Works Planetary Road Willenhall W. Midlands WV13 3ST tel:(01902) 305511 fax:(01902) 865290
Martella PLC President Drive Rooksley Milton Keynes MK13 8PD tel:(01908) 667418 fax:(01908) 604799
Martello Plastics Ltd Unit 11, Ross Way Shorncliffe Industrial Estate Folkstone Kent CT20 3UJ tel:(01303) 256848 fax:(01303) 246301
Martin & Co Ltd 119 Camden Street Birmingham B1 3DJ tel:(0121) 233 2111 fax:(0121) 236 0488
Martin (Marine) Ltd Alec 59 Norman Street Birkenhead Merseyside L41 0AS tel:(0151) 652 1663 fax:(0151) 653 3835
Martin Orgee & Associates Cherry Cottage Overwood Cleobury Mortimer Kidderminster, Worcs. DY14 8LG tel:(01299) 271177 fax:(01299) 271188
Martin Roberts Grimrod Place East Gillibrands Skelmersdale Lancs. WN8 9UU tel:(01695) 733068 fax:(01695) 50227 E_Mail:rascroft@martin-roberts.co.uk Web_Site:http://www.martin-roberts.co.uk
Masfield Pollard & Co Ltd Crown Works Parry Lane Bradford BD4 8TL tel:(01274) 724466 fax:(01274) 390175
Mason FW & Sons Ltd Colwick Industrial Estate Colwick Nottingham NG4 2EQ tel:(0115) 961 1555 fax:(0115) 940 0106 telex:377833 MASONS G
Masonite CP Ltd Jason House Kerry Hill Horsforth Leeds LS18 4JR tel:(01532) 587689 fax:(01532) 590015
Massey Furguson Tractors Ltd PO Box 62 Banner Lane Coventry CV4 9GF tel:(01203) 694400 fax:(01203) 852495
Masterbill Micro Systems Ltd Woodland Court Soothouse Spring, Valley Road St. Albans Herts. AL3 6NR tel:(01727) 855563 fax:(01727) 854626 E_Mail:sales@masterbill.com Web_Site:http://www.masterbill.com
Masterframe Windows Ltd 4 Perry Road Witham Essex CM8 3YZ tel:(01376) 510410 fax:(01376) 510400 E_Mail:sales@masterframe.co.uk
Mastex Coatings Ltd St Peters House Cambridge Road Kingston Upon Thames Surrey KT1 3JY tel:(0181) 546 9445 fax:(0181) 547 1836
Mastic Ashphalt Council 8 North Strret Ashford Kent TN24 8JN tel:(01233) 634411 fax:(01233) 634466
Maswell Engineering Ltd 33 Burlington Road Isleworth Middx. TW7 4LU tel:(0181) 568 3630 fax:(0870) 1641217 E_Mail:mxwell@maswell.demon.co.uk
Matbro Ltd London Road Tetbury Glos. GL8 8JD tel:(01666) 502502 fax:(01666) 504483 telex:43361
Mathews Maclay & Manson Ltd 136/138 Hydepark Street Glasgow G3 8BW tel:(0141) 221 0544 fax:(0141) 221 9807
Matki plc Churchward Road Yate Bristol BS37 5PL tel:(01454) 322888 fax:(01454) 315284 E_Mail:matki.co.uk
Mattersons Cranes PO Box 31, Healey Works Shawclough Road Shawclough, Rochdale Lancs. OL12 6LP tel:(01706) 49321 fax:(01706) 57452
Matthew Hebden Ltd 123 Lonsdale Drive Oakwood Enfield Middx. EN2 7LS tel:(0181) 367 6463 fax:(0181) 367 0166
Matthews & Yates Ltd Peartree Road Stanway Colchester Essex CO3 5LD tel:(01206) 543311 fax:(01206) 760497 telex:98231 GAXIL G E_Mail:sales@matthewyates.co.uk Web_Site:http//www.matthewyates.co.uk
Matthews Office Furniture Ltd PO Box 81 61-63 Dale Street Liverpool Merseyside L69 2DN tel:(0151) 236 9851 fax:(0151) 236 8096
Mauser Interiors (UK) Ltd 1 Canterbury Court, Kennington Park 1-3 Brixton Road London SW9 6DE tel:(0171) 735 4565 fax:(0171) 735 4989 E_Mail:info@mauser.co.uk Web_Site:http//www.mauser.co.uk
Mawdsley's Ltd Uley Road Dursley Glos. GL11 5ES tel:(01453) 544131 fax:(01453) 546571
Mawrob Co (Engineers) Ltd 121a/125a Sefton Street Southport Merseyside PR8 5DR tel:(01704) 501011 fax:(01704) 541403
Max Appliances Napier Road Castleham Industrial Estate St Leonard-on-Sea E. Sussex TN38 9NY tel:(01424) 854444 fax:(01424) 853862 E_Mail:sales@max-appliances.co.uk Web_Site:http//www.max-appliances.co.uk
Max Appliances Ltd Castleham Industrial Estate Napier Rd. St. Leonards on Sea East sussex PN3 89NY tel:(01424) 854 444
Maxblast UK Ltd Unit 11, Vanguard Trading Estate, Storforth Lane Chesterfield S40 2T2 tel:(01246) 209926 fax:(01246) 221620 E_Mail:maxblastuk@btinternet.com
Maxwell Hart Ltd Winnersh Berks. RG11 5HF tel:(01734) 785655 fax:(01734) 785805
May Gurney (Technical Services) Ltd Fencing Division Ringland Lane, Costessey Norwich Norfolk NR8 5BG tel:(01603) fax:(01603)
May Gurney (Technical Services) Ltd - Fencing Division Trowse Norwich Norfolk NR14 8SZ tel:(01603) 744440 fax:(01603) 747310 E_Mail:piling@maygurney.co.uk Web_Site:http://www.maygurney.co.uk
May William (Ashton) Ltd Cavendish Street Ashton Under Lyne Lancs. OL6 7BR tel:(0161) 330 3838/4879 fax:(0161) 339 1097
Mayplas Ltd Butterworth Street Little Borough Lancs. OL15 8JS tel:(01706) 374321 fax:(01706) 374473
McAlpine & Co Ltd Hillington Industrial Estate Glasgow G52 4LF tel:(0141) 882 3213 telex:778846
McAlpine Slate Ltd, Alfred Penrhyn Quarry Bethesda Bangor Gwynedd LL57 4YG tel:(01248) 600656 fax:(01248) 601171 E_Mail:pensales@alfred_mcalpine.co.uk Web_Site:http://www.amslete.com
McArthur Group Limited Foundry Lane Bristol BS5 7UE tel:(0117) 965 6242 fax:(0117) 958 3536
McCalls Special Products PO Box 71 Hawke Street Sheffield S9 2LN tel:(0114) 242 6704 fax:(0114) 243 1324 E_Mail:sales@mccalls.demon.co.uk Web_Site:http://www.msp.co.uk
McCarthy & Stone Construction Services 3 Queensway New Milton Hants. BH25 5PB tel:(01425) 638855 fax:(01425) 638343
McKenzie-Martin Ltd Eton Hill Works Eton Hill Road Radcliffe Manchester M26 2US tel:(0161) 723 2234 fax:(0161) 725 9531
Mclean Group, David Enterprise House Aber Road Flint Flintshire CH6 5EX tel:(01352) 762388 fax:(01352) 761112 E_Mail:kud@davidmclean.demon.co.uk Web_Site:http://www.davidmclean.co.uk
McLoughlin Wood Ltd 5 Upper Street South New Ash Green Longfield Kent DA3 8JJ tel:(01474) 872019 fax:(01474) 873343 E_Mail:mcwood@ashgreen.telme.com
McMillan Fire Alarm Systems Ltd Detection House, Brooklands Approach North Street Romford Essex RM1 1DX tel:(01708) 769314 fax:(01708) 725868 telex:893534 MFASJS
McQuay UK Ltd Bassington lane Cramlington Northumberland NE23 8AF tel:(01670) 566159 fax:(01670) 566206
Mea UK Ltd Rectors Lane Pentre Deeside Flintshire CH5 2DH tel:(01244) 534455 fax:(01244) 534477
Medina Gimson 60 Churchill Square Kingshill West Malling Kent ME19 4YU tel:(01732) 849440 fax:(01732) 847099
Meech Static Eliminators Ltd Burford House 15 Thorney Leys Business Park Whitney Oxon. OX8 7GE tel:(01993) 706700 fax:(01993) 776977 E_Mail:sales@meech.co.uk
Meggitt Lavmour Systems New Road Netherton, Dudley W. Midlands DY2 9AF tel:(01384) 357799 fax:(01384) 357700 Web_Site:http//www.meggit.com
Meldrum (Thermoplastics) Ltd Upper Basildon Pangbourne Berks. RG8 8ST tel:(01491) 671333 fax:(01491) 671504
Mells Roofing Ltd The Barns, Wickham Hall Wickham Bishops Witham Essex CM8 3JQ tel:(01621) 893388 fax:(01621) 893444
MEM Whitegate Broadway Chadderton Oldham Lancs. OL9 9QG tel:(0161) 652 1111 fax:(0161) 626 1709 Web_Site:http//www.memonline.com
Mentha & Hessal (Shopfitters) Ltd 95a Linaker Street Southport Merseyside PR8 5BU tel:(01704) 530800 fax:(01704) 500601
Menvier Ltd Southam Road Banbury Oxon. OX16 7RX tel:(01295) 755200 fax:(01295) 270102 E_Mail:enquiries@menvier.co.uk Web_Site:http://www.menvier.co.uk
Mercian Industrial Doors Pearsall Drive Oldbury W. Midlands B69 2RA tel:(0121) 544 6124 fax:(0121) 552 6793
Merlin Gerin Stafford Park 5 Telford Shropshire TF3 3BL tel:(01952) 290029 fax:(01952) 290534 telex:35433
Merronbrook Ltd Hazeley Bottom Hartley Wintney Basingstoke Hampshire RG27 8LX tel:(01252) 844747 fax:(01252) 845304
Mestec PLC Broadwell Road Oldbury Wareley W. Midlands B69 4HE tel:(0121) 601 6000 fax:(0121) 554 5520
Metalcraft (Tottenham) Ltd 6-40 Durnford Street Seven Sisters Road Tottenham London N15 5NQ tel:(0181) 802 1715 fax:(0181) 802 1258
Metalliform Products Plc Chambers Road Hoyand, Barnsley S. Yorks. S74 0EZ tel:(01226) 350555 fax:(01226) 350112 E_Mail:sales@metalliform-plc.demon.co.uk
Metalline Signs Ltd Unit 18 Barton Hill Trading Estate Bristol BS5 9TE tel:(01179) 555291 fax:(01179) 557518
Metalloxyd Ano-Coil Ltd Chippenham Drive Kingston Milton Keynes MK10 0AN tel:(01908) 282044 fax:(01908) 282032
Metalrax Ltd Bordesley Green Road Birmingham W Midlands B9 4TP tel:(0121) 772 8151 fax:(0121) 772 6135 E_Mail:sales@metalrax-storage.co.uk Web_Site:http//www.metalrax-storage-co.uk
Metalwood Fencing (Contracts) Limited Soothouse Spring Valley Road Industrial Estate St. Albans Herts. AL3 6PF
Metcraft Ltd Harwood Industrial Estate Harwood Road Littlehampton E. Sussex BN17 7BB tel:(01903) 714226 fax:(01903) 723206
Metlex Shepperton Park Caldwell Road Nuneaton Warwks. CV11 4NR tel:(01203) 324502 fax:(01203) 352974
Metra Non Ferrous Metals Ltd Pindar Road Hoddesdon Herts. EN11 0DE tel:(01992) 460455 fax:(01992) 451207
Metreel Ltd Cossall Industrial Estate Coronation Road Ilkeston Derby DE7 5UA tel:(0115) 932 7010 fax:(0115) 930 6263 E_Mail:sales@metreel.co.uk Web_Site:http://www.metreel.co.uk
Metsec Building Products Ltd Broadwell Road Oldbury, Warley W. Midlands B69 4HE tel:(0121) 552 1541 fax:(0121) 544 5520
Metway Electriacal Industries Ltd Barrie House 18 North Street Portslade Brighton BN41 1DE tel:(01273) 439266 fax:(01273) 439288 telex:877166
Meyer Drenham Ltd Westfield Road Toftwood, Drenham Norfolk NR19 1JA tel:(01362) 697951 fax:(01362) 698980
MGB Randalls Hoyle Mill Road Kinsley Pontefract W. Yorks. WF9 5JB tel:(01977) 615132 fax:(01977) 610059
MICE Kaymar Limited Brookhill Road, Brookhill Industrial Estate Pinxton Nottingham NG16 6NS tel:(01773) 810107 fax:(01773) 580286 E_Mail:micekaymar@compuserve.com
Michelmersh Brick & Tile Co Ltd Hillview Road Michelmarsh, Romsey Hants. SO51 0NN tel:(01794) 368506 fax:(01794) 368845
Mid Essex Trading Co Ltd Montrose Road Dukes Park Industrial Estate Springfield, Chelmsford Essex CM2 6TH tel:(01245) 469922 fax:(01245) 450755
Midgley C T & Co Ltd Scotts Road Hawkhead, Paisley Strathclyde PA2 7AN tel:(0141) 889 3118 fax:(0141) 889 2823
Midland Advertising Products Ltd 19 Melchett Road Kings Norton Birmingham B30 3HG tel:(0121) 451 1336 fax:(0121) 433 3278
Midland Alloy Ltd Stafford Park 17 Telford Shropshire TF3 3DG tel:(01952) 290961 fax:(01952) 290441
Midland Installation Ltd Heanor Street Leicester LE1 4DB tel:(01533) 625464 fax:(01533) 512435
Midland Stom Ltd Chem Road Millfields Wolverhampton W. Midlands WV4 6JJ tel:(01902) 353021 fax:(01902) 353635
Midland Tube and Fabrications Corngreave Works, Unit 4 Corngreave Road Cradley Heath, Warley W. Midlands B64 7DA tel:(01384) 566364 fax:(01384) 566365
Midland Wire Cordage Co Ltd Orchard Works Arthur Street Lakeside, Redditch Worcs. B98 8LJ tel:(01527) 520262 fax:(01527) 510052
Midven Ltd Hayseech Road Halesowen W. Midlands B63 3PE tel:(0121) 550 6441 fax:(0121) 585 5045
Miele Co Ltd Fairacres Marcham Road Abingdon Oxon. OX14 1TW tel:(01235) 554455 fax:(01235) 554477
Miles-Stone, Natural Stone Merchants Quarry Yard Toynbee Road Eastleigh Hants. SO50 9DN tel:(01703) 613178
Milesahead Graphic Design 11 Aylsham Close, The Willows Widnes Cheshire WA8 9FF tel:(0151) 422 0004 fax:(0151) 422 0004
Millar Dennis (1998) Ltd Victoria Works 142 Thornton Road Bradford West Yorkshire BD1 2EB tel:(01274) 722118 fax:(01274) 370977
Millenium Commission, The 2 Little Smith Street London SW1P 3DH tel:(0171) 340 2001
Miller Construction Ltd Metic House, Ripley Drive Normanton Industrial Estate Normanton West Yorkshire WF6 1QT tel:(01924) 224370 fax:(01924) 224380 Web_Site:http//www.miller.construction.co.uk
Miller Construction Ltd Parsons House Parsons Green Washington Tyne & Wear NE37 1EZ tel:(0191) 419 0461 fax:(0191) 419 3515 Web_Site:http://www.miller.construction.co.uk
Miller Construction Ltd 35 Baird Street Glasgow G4 0EE tel:(0141) 303 5678 fax:(0141) 303 5758 Web_Site:http://www.miller.construction.co.uk
Miller Construction Ltd. Miller House Corporation Street Rugby CV21 2DW Tel. (01788) 534 500 Fax: (01788) 534 510 Web_Site:http://www.miller.construction.co.uk
Miller Construction Ltd. Winnall Moorside Road Winchester Hampshire SO23 7SJ tel:(01962) 874 000 fax:(01962) 874 077 Web_Site:http://www.miller.construction.co.uk
Milliken Carpets Beech Hill Plant Gidlow Lane Wigan Lancs. WN6 8RN tel:(01942) 826073 fax:(01942) 826569
Mills John Ltd 72 Albert Street London NW1 7NR tel:(0171) 388 7212 fax:(0171) 388 3454
Milner Delvaux Ltd Eastrea Road Whittlesey Peterborough PE7 2AG tel:(01733) 202566 fax:(01733) 205405
Miltek 11 James Watt Place College Milton North E. Kilbride G74 5 HG tel:(013552) 42199 fax:(013552) 64334
Milton Pipes Ltd Milton Regis Sittingbourne Kent ME6 2QF tel:(01795) 425191 fax:(01795) 420360 E_Mail:sales@miltonpipes.karoo.co.uk
Minsterstone Ltd Pondhayes Farm Dinnington Hinton St. George Somerset TA17 8SU tel:(01460) 52277 fax:(01460) 57865 E_Mail:varyl@minsterstoneltduk Web_Site:http//www.minsterstoneltduk
Mita (UK) Ltd Bodelwyddan Business Park Bodelwyddan Denbighshire LL18 5SX tel:(01745) 586010 fax:(01475) 586015
MiTek Industries Ltd MiTek House, Grazebrook Industrial Park Pear Tree Lane Dudley W. Midlands DY2 OXW tel:(01384) 451400 fax:(01384) 451411 Web_Site:http//www.mitek.co.uk
Mitel Telecom Ltd Mitel Business Park Portskewett Monmouthshire NP6 4YR tel:(01291) 430000 fax:(01291) 430400
Mitsui Babcock Energy Services Ltd Porterfield Road Renfrew PA4 8DJ tel:(0141) 886 4141 fax:(0141) 885 3359 telex:778731 BABCON G Web_Site:http://www.mitsuibabcock.com
MJ Electronics Services (International) Ltd Unit 8, Axe Road Colley Lane Industrial Estate Bridgewater Somerset TA6 5LJ tel:(01278) 422882 fax:(01278) 453331 E_Mail:101756.501@compuserve.com
MMP International Bilton Court Wetherby Road Harrogate N. Yorks. HG3 1LN tel:(01423) 889441 fax:(01423) 889866 E_Mail:info@mmp-international.co.uk Web_Site:http://www.mmp-international.co.uk
Mockridge Labels & Nameplates Ltd Cavendish Street Ashton-under-Lyne Lancs. OL6 7QL tel:(0161) 308 2331 fax:(0161) 343 1958 E_Mail:sales@mockridge.com Web_Site:http://www.mockridge.com
Mode Lighting (UK) Ltd Chelsing Lodge Tonwell Ware Herts. SG12 0LB tel:(01920) 462121 fax:(01920) 466881
Modelscape Adams House Dickerage New Malden Surrey KT3 3SF tel:(0181) 949 9286 fax:(0181) 949 7418
Modern Engineering (Bristol) Ltd 456 Badminton Road Yate Bristol BS17 5HX tel:(01454) 318181 fax:(01454) 318231
Moelven Laminated Timber Structures Ltd Unit 10, Vicarage Farm Winchester Road Fair Oak, Eastleigh Hants. SO50 7HD tel:(01703) 695566 fax:(01703) 695577
Moffat Ltd, E & R Seabegs Road Bonnybridge Stirlingshire FK4 2BS tel:(01324) 812272 fax:(01324) 814107
Moffett Thallon & Co Ltd 143 Northumberland Street Belfast BT13 2JF tel:(01232) 322802 fax:(01232) 241428

Molyneux G (Products) Ltd 27 Blue Cap Road Stratford-upon-Avon Warwickshire CV37 6TQ tel:(01789) 299601 fax:(01789) 267867

Monarflex Ltd Lyon Way St Albans Herts. AL4 0LB tel:(01727) 830116 fax:(01727) 868045

Monk Metal Windows Ltd Hansons Bridge Road Erdington Birmingham B24 0QP tel:(0121) 351 4411 fax:(0121) 351 3673

Mono Pumps Ltd Martin Street Audenshaw Manchester M34 5DQ tel:(0161) 339 9000 fax:(0161) 344 0727

Monodraught Flues Ltd Lancaster Court Cressex Business Park High Wycombe Bucks. HP12 3TD tel:(01494) 464858 fax:(01494) 532465

Montrose Safety Fixings 6 Bowers Road Shoreham, Sevenoaks Kent TN14 7SS tel:(0181) 304 6615 fax:(0181) 854 1680

Monument Tools Ltd Restmor Way Hackbridge Road Hackbridge, Wallington Surrey SM6 7AH tel:(0181) 288 1100 fax:(0181) 288 1100 E_Mail:info@monument-tools.com Web_Site:http//www.monument-tools.com

Moorlite Electrical Ltd Burlington Street Ashton Under Lyne Lancs. OL7 0AX tel:(0161) 330 6811 fax:(0161) 330 2815

Moray Timber Ltd The Woodshop, Springfield Yard The Wards Elgin Morayshire IV30 3AA tel:(01343) 545151 fax:(01343) 549518

Moresecure Ltd Haldane House Halesfield 1 Telford TF7 4EH tel:(01952) 683943 fax:(01952) 683982 E_Mail:sales@moresecure.co.uk Web_Site:http://www.moresecure.co.uk

Morris Mechanical Handling Ltd PO Box 7 Loughborough Leics. LE11 1RL tel:(01509) 643200 fax:(01509) 610666 telex:34408 E_Mail:mail@p-h.co.uk

Morrison Chemicals Ltd 331-337 Derby Road Liverpool L20 8LQ tel:(0151) 933 0044

Morse Controls Ltd Christopher Martin Road Basildon Essex SS14 3ES tel:(01268) 522861 fax:(01268) 282994 telex:99237

Moseley GRP Product (Division of Moseley Rubber Co Ltd) Hoyle Street Mancunian Way Manchester M12 6HL tel:(0161) 273 3341 fax:(0161) 274 3743

Moss WM & Sons (Stove Anamellers Ripon) Ltd PO Box 4 Ure Bank Ripon N. Yorks. HG4 1JE tel:(01765) 604351 fax:(01765) 690065

Mould J H & Co Ltd Bearmore Road Cradley Heath W. Midlands B60 6DW tel:(01384) 568241 fax:(01384) 411104

Mountford Rubber & Plastics Ltd 44 Bracebridge Street Aston Birmingham B6 4PE tel:(0121) 359 0135/6 fax:(0121) 333 3204

Movitex Signs Ltd Unit 6, Abbey Mead Industrial Park Brooker Road Waltham Abbey Essex EN9 1HU tel:(01992) 719662 fax:(01992) 710101

MR Limited White House Works Bold Road Sutton, St Helens Merseyside WA9 4JG tel:(01744) 820103 fax:(01744) 818997

MSA (Britain) Ltd East Shawhead Coatbridge N. Lanarkshire ML5 4TD tel:(01236) 424966 fax:(01236) 440881 E_Mail:info@msabritain.co.uk Web_Site:http://www.msabritain.co.uk/msa

MSC Speciality Films (UK) Ltd 4-6 De Salis Drive Hampton Lovett Industrial Estate Droitwich Spa Worcs. WR9 0QE tel:(01905) 797797 fax:(01905) 794436

MTE Ltd Stephenson Road Eastwood Leith-on-Sea Essex SS9 5LS tel:(01702) 421124 fax:(01702) 420365 E_Mail:info@mte_co Web_Site:http//www.mte.co.uk

Multi Marque Production Engineering Ltd 32-33 Monkton Road Industrial Estate Wakefield W. Yorks. WF2 7AL tel:(01924) 290231 fax:(01924) 382241

Multi Mesh Ltd Eurolink House Lea Green Industrial Estate St Helens Merseyside WA9 4QU tel:(01744) 820666 fax:(01744) 821417

Multibeton Ltd 15 Oban Court Wickford Business Park Wickford Essex SS11 8YB tel:(01268) 561688 fax:(01268) 561690

Multitex GRP & GRG Products Unit 5, Dolphin Industrial Estate Southampton Road Salisbury Wilts. SP1 2NB tel:(01722) 332139 fax:(01722) 338458 Web_Site:http://www.multitex.co.uk

Muraspec Ltd 74-78 Wood Lane End Hemel Hempstead Herts. HP2 4RF tel:(08705) 329020 fax:(08705) 329020 Web_Site:http//www.muraspec.com

Muswell Manufacturing Co Ltd Unit D1, Lower Park Road New Southgate Industrial Estate London N11 1QD tel:(0181) 368 8738 fax:(0181) 368 4726

Myddleton Hall Lighting, Division of Pearless Design Ltd Myddleton Hall Almeida Street London N1 1TD tel:(0171) 226 3443 fax:(0171) 354 1397

National Association of Loft Insulation Contractors PO Box 12 Haslemere Surrey GU27 3AH tel:(01428) 654011 fax:(01428) 651401 E_Mail:113052.2106@compuserve.com Web_Site:http://www.nationline.co.uk/ceed

National Britannia Ltd Caerphilly Business Park Van Road Caerphilly CF8 3ED tel:(01222) 852000 fax:(01222) 867738 E_Mail:enquiries@national-britannia.co.uk

National Cavity Insulation Association Ltd PO Box 12 Haslemere Surrey GU27 3AH tel:(01428) 654011 fax:(01428) 651401 E_Mail:113052.2106@compuserve.com Web_Site:http://www.nationline.co.uk/ceed

National Federation of Builders Contstruction House 56-64 Leonard Street London EC2A4JX Tel. (0171)608 5150 Fax: (0171)608 5151 E_Mail:nfbnat@builders.org.uk Web_Site:http://www.builders.org.uk

National Federation of Terrazzo Marble & Mosaic Specialists PO.Box 2843 London WIA SPG tel:(0845) 609 0050 Fax (0845) 607 8610 E_Mail :slade@nttmms.demon.co.uk

National Starch & Chemical Ltd Galvin Road Slough Berks. SL1 4DF tel:(01753) 533494 fax:(01753) 539338 telex:848386

National Trust The 36 Queens Anee's Gate London SW1H 9AS tel:(0171) 222 9251

Nationwide Filter Co Ltd Unit 5, Rufus Business Centre Ravensbury Terrace London SW18 4RL tel:(0181) 944 8877 fax:(0181) 944 8833

Naylor Drainage Ltd Clough Green Cawthorne Nr Barnsley S. Yorks. S75 4AD tel:(01226) 790591 fax:(01226) 790531 E_Mail:info@naylor.co.uk Web_Site:http://www.naylor.co.uk

NBS Services Mansion House Chambers The Close Newcastle-upon-Tyne NE1 3RE tel:(0191) 232 9594 fax:(0191)232 5714 E_Mail:info@nbsservices.co.uk Web_Site:http://www.nbsservices.org.uk

NCMP Ltd Falcon Way Feltham Middx. TW14 0XJ tel:(0181) 751 0986 fax:(0181) 751 2446

Nederman Ltd PO Box 503 91 Walton Summit Bamber Bridge Preston PR5 8AF tel:(01772) 315293 fax:(01772) 315273 E_Mail:sales@nederman.co.uk Web_Site:http://www.nederman.se

NEI Reyrolle Ltd Hebburn Tyne & Wear NE31 1UP tel:(0191) 483 2451 fax:(0191) 483 2446 telex:537866

Neill Tools Ltd Atlas Way Atlas North Sheffield S. Yorks. S4 7QQ tel:(0114) 281 4242 fax:(0114) 281 4343 telex:54278 J NEILL G E_Mail:sales@neill.tools.co.uk Web_Site:http://www.spear-and-jackson.com

Nendle Acoustic Company Ltd Tacitus House 153a High Street Aldershot Hants. GU11 1TT tel:(01252) 344222 fax:(01252) 333782

Nenplas Airfield Industrial Estate Ashbourne Derbys. DE6 1HA tel:(01335) 343821 fax:(01335) 340271 E_Mail:enquiries@nenplas.co.uk Web_Site:http://www.nenplas.co.uk

Neocrylic Signs Ltd Unit 7 Mather Road Eccles Manchester M30 0WQ tel:(0161) 707 8933 fax:(0161) 707 8934

Neopost Ltd Neopost House South Street Romford Essex RM1 2AR tel:(01708) 746000 fax:(01708) 727192 E_Mail:mktg@neopost.co.uk Web_Site:http://www.neopost-group.com

Neptune Outdoor Furniture Ltd Thompsons Lane Marwell Nr Winchester Hampshire SO21 1JH tel:(01962) 777799 fax:(01962) 777723 E_Mail:sales@nofi.co.uk Web_Site:http://www.nofi.co.uk

Neptune Showers Ltd Neptune House 45 Pember Road London NW10 5LT tel:(0181) 960 3185 fax:(0181) 968 5895

Neslo Partitioning & Interiors Construction House Birch Street Wolverhampton WV1 4HY tel:(01902) 316842 fax:(01902) 428774

Ness Furniture Ltd Croxdale Durham DH6 5HT tel:(01388) 816109 fax:(01388) 812416 E_Mail:sales@nessfurniture.co.uk

Nettlefolds Ltd Heath Street Smethwick W. Midlands B66 2SA tel:(0121) 626 0600 fax:(0121) 555 5640

Network Flooring Cardiff Units B7 and B8 Springmeadow Road, Springmeadow Industrial Park Rumney Cardiff CF3 8GA tel:(01222) 360800 fax:(01222) 360007

Nevill Long Limited Centre House, Victory Way Southall Lane Heston Middx. TW5 9NS tel:(0181) 5739898 fax:(0181) 8135127

Nevill Long Limited Southfork Ind Est Dartmouth Way off Garnet Road Leeds LS11 5JL tel:(0113) 2718000 fax:(0113) 271 8111

Nevill Long Limited Badminton Road Trading Estate Badminton Road Yate Bristol BS17 5JX tel:(01454) 326622 fax:(01454)325800

Nevill Long Limited Chartwell Drive West Avenue Wigston Leicester LE18 2FL tel:(0116) 2570670 fax:(0116) 2570044 Web_Site:http://www.nevilllong.co.uk

Nevill Long Limited Unit 5, Sterling Industrial Estate Rainham Road South Dagenham Essex RM10 8TX tel:(0181) 593 7121 fax:(0181) 593 0406

Neville, Scales & Co Ltd 34 Priests Bridge London SW14 8TA tel:(0171) 254 1033 fax:(0181) 392 2026

New Centurian Industry Services Ltd Sovereign House, 2 Sovereign Park Coronation Road Park Royal London NW10 7 QP tel:(0181) 965 7300 fax:(0181) 965 2615

New Forest Ceilings Ltd 61-65 High Street Totton Southampton SO40 9HL tel:(01703) 869510 fax:(01703) 862244

New Haden Pumps Ltd New Haden Works Cheadle, Stoke on Trent Staffs. ST10 2NW tel:(01538) 752351 fax:(01538) 756569

Neway Doors Ltd 89-91 Rolfe Street Smethwick Warley W.Midlands B66 2AY tel:(0121) 558 6406 fax:(0121) 558 7140

Newdawn & Sun Ltd Tything Road Arden Forest Inustrial Estate Alcester Warwks. B49 6EP tel:(01789) 764444 fax:(01789) 400164 E_Mail:sales@newdawn-sun.co.uk Web_Site:http://www.newdawn-sun.co.uk

Newdome Ltd PO Box 45 Hall Street St Helens Lancs. tel:(01744) 22189/28661 fax:(01744) 26934 telex:627441 PB ST

Newey Ceilings 68 Florence Road Wylde Green Sutton Coldfield W. Midlands B73 5NG tel:(0121) 382 2098 fax:(0121) 382 1764

Newington Bricks Ltd London Road Newington Nr Sittingbourne Kent ME9 7NU tel:(01795) 842338

Newlay Concrete Ltd Thornhill Works Calder Road Dewsbury W. Yorks. WF12 9HY tel:(01924) 456416 fax:(01924) 430697

NewTeam Ltd Brunel Road Earlstrees Industrial Estate Corby Northants NN17 2LS tel:(01536) 409222 fax:(01536) 400144

Newton & Frost Downsview Yard North Corner, Horam Heathfield E. Sussex TN21 9HJ tel:(01435) 813535 fax:(01435) 813687

Newton John & Co Ltd 12 Verney Road London SE16 3DH tel:(0171) 237 1217 fax:(0171) 252 2769 E_Mail:enquiries@newtonandco.co.uk Web_Site:http://www.newtonandco.uk

Newtons Computer Services Ltd Wandle House Riverside Drive Mitcham Surrey CR4 4BU tel:(0181) 648 8707 fax:(0181) 648 2240

Nicholls & Clarke Ltd Niclar House Shoreditch High Street London E1 6PE tel:(0171) 247 5432 fax:(0171) 247 7738

Nicholson Plastics Ltd Riverside Road Kirkfieldbank Lanarkshire ML11 9JS tel:(01555) 664316 fax:(01555) 663056

Nico Manufacturing Ltd 109 Oxford Road Clacton on Sea Essex CO15 3TJ tel:(01255) 422333 fax:(01255) 432909 E_Mail:sales@nico.co.uk Web_Site:http://www.nico.co.uk

Nilfisk Advance Ltd Newmarket Road Bury St Edmunds Suffolk IP33 3SR tel:(01284) 763163 fax:(01284) 750562

Nillamette Europe Ltd 10th Floor, Maitland House Warrior Square Southend-on-Sea Essex SS1 2JY tel:(01702) 619044 fax:(01702) 617162 Web_Site:http://www.meditemdf.com

Nimlok Ltd Booth Drive Park Farm Wellingborough Northants NN8 6NL Tel: (01933) 409409 Fax: (01933) 409451 E_Mail:info@nimlok.co.uk Web_Site:http://www.nimlok.co.uk

Nittan (UK) Ltd Hipley Street Old Woking Surrey GU22 9LQ tel:(01483) 769555 fax:(01483) 756686

Niven Thomas Ltd Murrell Hill Sawmills Dalsten Road Carlisle CA2 5NS tel:(01228) 26277/8 fax:(01228) 593854

nmc (UK) Ltd Prospect Road Beechburn Industrial Estate Crook Co. Durham DL15 8JN tel:(01388) 761300 fax:(01388) 761400 E_Mail:enquiries@nmc.uk.wm Web_Site:http://www.svrgb2/nmc-uk/

Norcros Adhesives Ltd Longton Road Trentham Stoke-on-Trent ST4 8JB tel:(01782) 59110 fax:(01782) 591101 telex:36574 BALAD G

Nordson (UK) Ltd Wenman Road Thame Industrial Estate Thame Oxon. OX9 3XB tel:(01844) 264500 fax:(01844) 245358

Norske Interiors (UK) Ltd Estate Road One South Humberside Industrial Estate Grimsby Humberside DN31 2TA tel:(01472) 240832 fax:(01472) 240304

North Enfield Precast Concrete Co Ltd BR Goods Depot Whitehart Lane London N17 8DP tel:(0181) 801 7217 fax:(0181) 808 0854

North Herts Asphalte Ltd Unit 15, The Cuttings Station Approach Hitchin Herts. SG4 9UW tel:(01462) 434877/454119 fax:(01462) 421539

North Wales Slate Roofing Co, The PO Box 2 Pormadoc Gwynedd tel:(01766) 831532 fax:(01766) 830711

North Yorkshire Artstone Ltd Showfield Lane Malton Yorks. YO17 6BT tel:(01653) 697714 fax:(01653) 692427

Northcot Brick Ltd Blockley, Nr Moreton in Marsh Glos. GL56 9LH tel:(01386) 700551 fax:(01386) 700852

Northern Fencing Contractors Limited Sandbeck House Sandbeck Way Wetherby W. Yorks. LS22 4DL tel:(01937) 583131 fax:(01937) 580034

Northern Incinerators (GB) Ltd Old Grammar School House School Gardens Shrewsbury Shropshire SY1 2AJ tel:(01743) 368134

Northern Joinery Ltd Daniel Street Whitworth, Rochdale Lancs. OL12 8DA tel:(01706) 852345 fax:(01706) 853114

Northgate Solar Controls PO Box 200 Barnet Herts. EN4 9EW tel:(0181) 441 4545

Northwest Pre-cast Ltd Holmefield Works Garstang Road Pilling Nr. Preston PR3 6AN tel:(01253) 790444 fax:(01253) 790085

Norton Abrasives Ltd Bridge Road East Welwyn Garden City Herts. AL7 1HZ tel:(01707) 323484 fax:(01707) 325231

Norton Construction Products Unit 2, Meridian West Meridian Business Park Leicester LE3 2WX tel:(0116) 263 0606 fax:(0116) 282 7292

Norton Engineering Alloys Co Ltd Norton Grove Industrial Estate Norton Malton N. Yorks. YO17 9HQ tel:(01653) 695721 fax:(01653) 697276 E_Mail:106063,1222@compuserve.com

Norwood Partition Systems Ltd Oxleasow Road East Moons Moat Redditch Worcs B98 ORE tel:(01527) 510015 fax:(01527) 517222 E_Mail:marketing@norwood.co.uk Web_Site:http://www.norwood.co.uk

Notcutt W P Ltd 25 Church Road Teddington Middx. TW11 8PF tel:(0181) 977 2252 fax:(0181) 977 6423

Nova Garden Furniture Ltd The Faversham Group Graveney Road Faversham Kent ME13 8UN tel:(01795) 535511 fax:(01795) 539215

Nova Group Ltd Norman Road Board Heath Altrincham Cheshire WA14 4EN tel:(0161) 941 5174 fax:(0161) 926 8405

Novellini Bathroom Products Unit 5G Tewkesbury Industrial Centre Delta Drive Tewkesbury Glos. GL20 8HD tel:(01684) 290088 fax:(01684) 850912 E_Mail:sales@novellini.freeserve.co.uk

NSA Environmental Products Belasis Business Centre Belasis Hall Technology Park Billingham Cleveland TS23 4EA tel:(01642) 345679 fax:(01642) 345680

NT Architectural Hardware Straight Road Short Heath Willenhall W. Midlands WS1 4NA tel:(01922) 401606 fax:(01922) 495344

NT Brass Art Ltd Regent Works Attwood Street Lye, Stourbridge W. Midlands DY9 8Ry tel:(01384) 894814 fax:(01384) 423824

NT Partition Systems Oxleasow Road East Moons Moat Redditch Worcs. B98 0RE tel:(01527) 510015 fax:(01527) 517222

NT Worcester Parsons Ltd Lifford Lane Kings Norton Birmingham B30 3RJ tel:(0121) 486 2211 fax:(0121) 459 6405

Nu-Swift International Ltd Elland W. Yorks. HX5 9DS tel:(01422) 372852 fax:(01422) 379569 Web_Site:http://www.nu-swift.co.uk

Nu-Way Ltd PO Box 1 Vines lane Droitwich Worcs. WR9 8AN tel:(01905) 794331 fax:(01905) 794017

NuAir Ltd Western Industrial Estate Caerphilly M. Glam. CF83 1XH tel:(01222) 885911 fax:(01222) 887033

Nullifire Ltd Torrington Avenue Coventry CV4 9TJ tel:(01203) 470022 fax:(01203) 469547

Nuralite (UK) Ltd Nuralite House, Canal Road Higham Rochester Kent ME3 7JA tel:(01474) 823451 fax:(01474) 823961

Nuttall, Henry (Viscount Catering Ltd) Green Lane Ecclesfield Sheffield S. Yorks. S35 9ZY tel:(0114) 257 0100 fax:(0114) 257 0251 E_Mail:sales@viscountcatering.demon.co.uk Web_Site:http://www.popltd.co.uk/vbpyork/visco unt

O & K Escalators Worth Bridge Road Keighley W. Yorks. BD21 4YA tel:(01535) 662841 fax:(01535) 680498

O'Brian Manufacturing Ltd Robian Way Swadlincote Derbys. DE11 9DH tel:(01238) 215613 fax:(01238) 215613

O'Connor Fencing Limited Winston House Haile Nr Egremont Cumbria CA22 2PD tel:(019467) 29202 fax:(019467) 29212

Oakdale (Contracts) Ltd Walkerville Industrial Estate Catterick Garrison N. Yorks. DL9 4SA tel:(01748) 834184 fax:(01748) 833003

Oakden A & Sons Ltd 86-90 Curtain Road London EC2A 3AB tel:(0171) 739 7611 fax:(0171) 739 0785

Oakland Elevators Ltd Mandervell Road Oadby Leics. LE2 5LL tel:(0116) 272 0800 fax:(0116) 272 0904 E_Mail:oakland@elevator.co.uk

Oakleaf Reproductions Ltd Ling Bob Wilsden Bradford W. Yorks. BD15 0JP tel:(01535) 272878 fax:(01535) 275748 E_Mail:sales@oakleaf.co.uk Web_Site:http://www.oakleaf.co.uk

Oce (UK) Ltd Langston Road Loughton Essex IG10 3SL tel:(0181) 508 5544 fax:(0181) 508 6689

OEP Furniture Group PLC 7-9 Cartersfield Road Waltham Abbey Essex EN9 1JD tel:(01992) 767014 fax:(01992) 762968

Ogilvie Construction Ltd Pirnhall Works 200 Glasgow Road Whins of Milton Stirling FK7 8ES tel:(01786) 812273 fax:(01786) 816287 E_Mail:enq@ogilvie.co.uk

Ogley Bros Ltd Allen Street Smithfield Sheffield S3 7AS tel:(0114) 276 8948 fax:(0114) 275 8948

Olby H E & Co Ltd 229-313 Lewisham High Street London SE13 6NW tel:(0181) 690 3401 fax:(0181) 690 1408

Oldham Lighting Ltd 4 Rowan Court 56 High Street Wimbledon Village London SW19 5EE tel:(0181) 946 5555 fax:(0181) 946 5522

Oldham Signs Ltd Peter Road Lancing Pusness Park Lancing W. Sussex BN15 8TH tel:(01903) 753333 fax:(01903) 766677 E_Mail:email @oldhamsigns.co.uk Web_Site:http://www.odhamsigns.co.uk

Oldham Signs Ltd PO Box YR15 Cross Green Approach Leeds LS9 0TQ tel:(0113) 240 4142 fax:(0113) 249 6249

Olivand Metal Windows Ltd Chesley House 43a Chesley Gardens London E6 3LN tel:(0181) 471 8111 fax:(0181) 552 7015

Olley & Sons Ltd, C Iberia House Finchley Avenue Mildenhall Suffolk IP28 7BJ tel:(01638) 712076 fax:(01638) 717304

Onduline Building Products Ltd Eardley House 182-184 Campden Hill Road Kensington London W8 7AS tel:(0171) 727 0533 fax:(0171) 792 1390

Opals (Mirror-Flex) Co Ltd 14 Herbert Road Clacton-on-Sea Essex CO15 3BE tel:(01225) 423927 fax:(01225) 221117

Opella Ltd Twyford Road Rotherwas Industrial Estate Hereford Herefordshire HR2 6JR tel:(01432) 357331 fax:(01432) 264014

Optelma Lighting Ltd 14 Napier Court The Science Park Abingdon Oxon. OX14 3NB tel:(01235) 553769 fax:(01235) 523005

Optex (Europe) Ltd Clivemont Road Cordwallis Park Maidenhead Berks. SL6 7BU tel:(01628) 631000 fax:(01628) 636311

Optiroc Ltd Adamson House Pomona Strand Manchester M16 0BA tel:(0161) 873 7701 fax:(0161) 872 4385

Opto International Ltd Tower Mill Park Road Dukinfield Cheshire SK16 5LN tel:(0161) 330 9136 fax (0161) 343 7332 E_Mail:p.wrigley@optoint.demon.co.uk Web_Site:http://www.boxes.demon.co.uk

Opus 4 Ltd 6c Chapman Way Haslemere Estate Tunbridge Wells Kent TN2 3EF tel:(01892) 515157 fax:(01892) 515417

Orbik Electrics Ltd Orbik House Northgate Way Aldridge Walsall WS9 8TX tel:(01922) 743515 fax:(01922) 743173

Orchard Street Furniture, Division of Camas Building Materials Orchard House Moreton Avenue Wallingford Oxon. OX10 9RH tel:(01491) 826100 fax:(01491) 825073

Ordnance Survey Romney Road Maybush Southhampton SO9 4DH tel:(01703) 792 792

Origlia Spa 41 Sydney Grove Hendon London NW4 2EJ tel:(0181) 203 3248 fax:(0181) 203 3248

Orsogril Sarl UK Department Geddings Road Hoddesdon Herts. EN11 0NZ tel:(0870) 606 2070 fax:((07070) 646 200

OSC Process Engineering Ltd Europa 67 PO Box 57 Stockport Cheshire SK3 0JT tel:(0161) 428 0747 fax:(0161) 491 1565 E_Mail:106023.E_Mail:1061@compuserve.com

OSF Ltd Unit 6, The Four Ashes Industrial Estate Station Road Four Ashes Wolverhampton WV10 7DB tel:(01902) 798080 fax:(01902) 798226 E_Mail:osf@woverhampton.co.uk Web_Site:http://www.wolverhampton.co.uk/osf

Osram Ltd PO Box 17 East Lane Wembley Middx. HA9 7PG tel:(0181) 904 4321 fax:(0181) 901 1222 Web_Site:http://www.osram.co.uk.

Oswestry Reinforced Plastics Ltd The Old Hall Cheswardine Market Drayton Shropshire TF9 2RN tel:(01630) 661410 fax:(01630) 661410

Otis plc 43-59 Clapham Road London SW9 0JZ tel:(0171) 735 9131 fax:(0171) 735 4639 telex:915348 Web_Site:http://www.otis.com

Outdoor Power Products Dolmer House Clare Street Denton Manchester M34 3LQ tel:(0161) 320 8100 fax:(0161) 335 0114

Ova Bargellini UK Ltd Unit 3 Severn Link Distibution Centre Chepstow Gwent NP6 6UN tel:(01291) 626252 fax:(01291) 626272

OWA UK Ltd Elm House 23-25 Elmshott Lane Cippenham Slough SL1 5QS tel:(01628) 663797 fax:(01628) 662167

Owens Corning Building Products (UK) Ltd PO Box 10 St Helens Merseyside WA10 3NS tel:(01744) 24022 fax:(01744) 612007 Web_Site:http://www.owenscorning.com/ owens/ uk

Owens Corning Polyfoam UK Ltd Hunter House Industrial Estate Brenda Road Hartlepool Cleveland TS25 2BE tel:(01429) 855100 fax:(01429) 855138

Oxford Double Glazing Ltd Ferry Hinksey Road Oxford OX2 0BY tel:(01865) 248287 fax:(01865) 251070

P & L Systems Ltd Pals House Halfpenny Road Knaresborough N. Yorks. HG5 0PS tel:(01423) 881664 fax:(01423) 863497

P & R Electrical (London) Ltd 11 Hanger Lane Ealing Common London W5 3HH tel:(0181) 992 0174

P & R Laboratory Group Ltd Unit 5 Brindley Road St Helens Merseyside WA9 4HY tel:(01744) 831800 fax:(01744) 831888 E_Mail:sales@pandr.co.uk Web_Site:http://www.pandr.co.uk

P.C.I Construction Systems Ltd Bedewell Industrial Park Adair Way Hebburn Tyne & Wear NE31 2HG tel:(0191) 428 2266 fax:(0191) 428 1717 E_Mail:pciconstruction@compuse

P4 Limited Unit 4A Sratton Park Dunton Lane Biggleswade Befordshire SG18 8QS tel:(01767) 600024 fax:(01767) 600808

Pac International Ltd 1 Park Gate Close Bredbury, Stockport Gtr. Manchester SK6 2SZ tel:(0161) 494 1331 fax:(0161) 430 9658 Web_Site:http://www.pac.co.uk

Package Products Ltd PO Box 2 Collyhurst Road Manchester M10 7RX tel:(0161) 205 4181 fax:(0161) 203 4678

Paddock Fabrications Ltd Fryer Road Walsall, Bloxwich W. Midlands WS2 7NF tel:(01922) 711722 fax:(01922) 476021

Palgrave Brown Timber & Fireproofing Co Station Road Market Bosworth Nuneaton Warwks. CV13 0PQ tel:(01455) 290917 fax:(01455) 290918

Palmers Scaffolding Ltd Penhall Road Charlton SE7 tel:(0181) 858 6511

Palmyra Ltd Tavistock Works Glasson Estate Maryport Cumbria CA15 8NX tel:(01900) 812796 fax:(01900) 815509

Panasonic UK Ltd Panasonic House Willoughby Road Bracknell Berks. RG12 8FP tel:(01344) 862444 fax:(01344) 857034

Panel Systems Ltd 3-9 Welland Close Parkwood Industrial Estate Rutland Road, Sheffield S. Yorks. S3 9QY tel:(0114) 275 2881 fax:(0114) 276 8807

Panelite Green Lane Garforth Leeds LS25 2AE tel:(0113) 2875111 fax:(0113) 286 3377

Pannell Signs Ltd Duke Street New Basford Nottingham NG7 7JN tel:(0115) 970 0371 fax:(0115) 942 2452

Paragon Business Furniture 100 Queensbury Rd. Wembley Middy HA0 1WP tel:(0181) 991 9999 fax:(0181) 991 9933 E_Mail:sale@parafurn.co.uk Web_Site:http://www.paragon-businessfurn.com

Paragon by Heckmondwike F B Ltd Farfield Park Manvers Wath on Dearne Rotherham, South Yorkshire S63 5DB tel:(01924) 406161 fax:(01924) 413613

Park Products Greenbank Technology park Challenge way Blackburn Lancs. BB1 5QJ tel:(01254) 614000 fax:(01254) 614001 E_Mail:sales@parkproducts.co.uk Web_Site:http://www.parkproducts.co.uk

Parker AG Engineering Ltd Burrowfield Welwyn Garden City Herts. AL7 4SS tel:(01707) 327066 fax:(01707) 372441

Parker Bath Division Queensway Stem Lane New Milton Hants. BH25 5NN tel:(01425) 624000 fax:(01425) 624008

Parker Kislingbury Ltd Wotton Road Brill Bucks. HP18 9UB tel:(0800) 220142 fax:(01844) 238016

Parker Plant Ltd PO Box 146 Canon Street Leicester LE4 6HD tel:(0116) 266 5999 fax:(0116) 268 1254 E_Mail:sales@parkerplant.demon.co.uk Web_Site:http://www.parkerplant.com

Parkes Josiah & Sons Ltd Union Works Willenhall W. Midlands WV13 1JX tel:(01902) 366931 fax:(01902) 366888 telex:337642

Parking Technology Calder Wearing Scott Road W. Yorks. HX2 6JB tel:(01422) 882382

Parkinson Cowan and Tricity Bendix Co, The Cornwall House 55-57 High Street Slough Berks. SL1 1BG tel:(0990) 146146 fax:(01953) 872381

Parklines Building Ltd Gala House 3 Raglan Road Edgbaston Birmingham B5 7RA tel:(0121) 446 6030 fax:(0121) 446 5991

Parkside Group Ltd, The Unit 5, The Willows Business Centre 17 Willow Lane Mitcham Surrey CR4 4YD tel:(0181) 685 9685 fax:(0181) 646 5096 E_Mail:technical@parksidegroup.co.uk

Parkwood Mellows Ltd Ridgeacre Road West Bromwich W. Midlands B71 1BB tel:(0121) 553 4011 fax:(0121) 553 4019 telex:338948

Parsons E. & Sons Ltd Blackfriars Road Nailsea Nr. Bristol BS48 4DJ tel:(01275) 854911 fax:(01275) 858108

Parsons Brothers Steelwork Griffiths Road West Bromwich W. Midlands B71 2EL tel:(0121) 588 4871 fax:(0121) 588 7513

Parsons C B Co Ltd, The Ashby de la Zouch Leics. LE6 5DR tel:(01530) 412885 fax:(01530) 417315 telex:451313 PARSONS

Partitioning & Interiors Association Jago House 692 Warwick Road Solihull W. Midlands B91 3DX tel:(0121) 705 9270 fax:(0121) 711 2892

Parton Fibreglass Ltd PFG House Claymore, Tame Valley Industrial Estate Tamworth Staffs. B77 5DQ tel:(01827) 261771 fax:(01827) 261390

PCT Group Sales Ltd Milton Chain Works Shirland Lane Sheffield S. Yorks. S9 3FG tel:(0114 2449761 fax:(0114)2441282

Peal furniture (Durham) Ltd Littleburn Industrial Estate Langley Moor Co. Durham DH7 8HE tel:(0191) 378 0232 fax:(0191) 378 1660

Pearce Construction(Barnstaple)Ltd Great Western House Old Station Road Barnstaple N. Devon EX32 8PB tel:(01271) 45261 fax:(01271) 22164

Pearce Signs Ltd Murray Road Orpington Kent BR5 3QY tel:(01689) 892500 fax:(01689) 892518 E_Mail:signs@pearcegroup.com Web_Site:http://www.//pearcegroup.com

Pearl Paints Ltd Severn Road Teforest Industrial Estate Pontypridd M. Glam. CF37 5SR tel:(01443) 843231 fax:(01443) 843705

Pearman Fencing Greystone Yard, Notting Hill Way Lower Weare Axbridge Somerset BS26 2JU tel:(01934) 733380 fax:(01934) 733398

Pearsons J R (Birmingham) Ltd 212/213 New John Street West, Hockley Birmingham West Midlands B19 3UD tel:(0121) 551 9411 fax:(0121) 551 9412 E_Mail:jrpearson@btinternet.com Web_Site:http://www.drawmet.com

Pebblecote Products Sussex Ltd Woods Way Goring by Sea Worthing W. Sussex BN12 4RF tel:(01903) 244835 fax:(01903) 506587

Pedley Furniture international Ltd Shirehill Works Saffron Walden Essex CB11 3AL tel:(01799) 522461 fax:(01799) 513403 E_Mail:sales@pedley.com Web_Site:http://www.pedley.com

Pegler Ltd St Cathrines Avenue Doncaster S Yorks. DN4 8DF tel:(01302) 560560 fax:(01302) 367661

Pegson Ltd Coalville Leics. LE67 3GN tel:(01530) 510051 fax:(01530) 510041

Pel Plc Oldbury Warley W. Midlands B69 4HN tel:(0121) 552 3377 fax:(0121) 552 6067

PEL Services Limited Belvue Business Center Belvue Road Northolt Middx. UB5 5QQ tel:(0181) 841 6251/ (0181) 839 2100 fax:(0181) 841 1948 E_Mail:sales@pel.co.uk

Pellfold Parthos Ltd. 1 The Quadrant Howarth Road Maidenhead Berks. SL6 1AP tel:(01628) 773353 fax:(01628) 773363

Pembury Fencing Limited Leys Industrial Park Maidstone Road Paddock Wood Kent TN12 6PX tel:(01892) 838833 fax:(01892) 838834

Penrose A & Co Ltd 123 packington Street London N1 7EA tel:(0171) 226 6467

Pentas Office Furniture Asher Lane Ripley Derbys. DE5 3RE tel:(01773) 570 700 fax:(01773) 570 760

Performance Roofing Ltd 80 Sunleigh Road Wembley Middx. HA0 4LR tel:(0181) 902 7823

Pergola Products Ltd Leigh Court Leigh Street High Wycombe Bucks. HP11 2QU tel:(01494) 520495 fax:(01494) 449145

PERI Ltd Market Harborough Road Clifton Upon Dunsmore Rugby Warwickshire CV23 0AN tel:(01788) 861600 fax:(01788) 861610 Web_Site:http://www.peri.de

Perkins & Powell Cobden Works Leopold Street Birmingham B12 0UJ tel:(0121) 772 2303 fax:(0121) 772 3334 E_Mail:mail@samvel-heath.com Web_Site:http://www. samvel-heath.com

Perma Industries Ltd Winchelsea Works Winchelsea Road Dover Kent CT17 9SS tel:(01304) 213255 fax:(01304) 206258

Permali Gloucester Ltd Bristol Road Gloucester Glos. GL1 5TT tel:(01452) 528528 fax:(01452) 507409

Perman Briggs Ltd 224 Cheltenham Road Longlevens Gloucester GL2 0JW tel:(01452) 524192 fax:(01452) 309879

Permanoid Ltd Hulme Hall Road Manchester M10 8HH tel:(0161) 205 6161 fax:(0161) 205 9325

PermaRock Products Ltd Jubilee Drive Loughborough Leics. LE11 5TW tel:(01509) 262924 fax:(01509) 230063 E_Mail:permarock@permarock.com Web_Site:http://www.permarock.com

Permaseal Roofing 4 Fivehouse Tonedale Wallington TA21 0AH tel:(01823) 662262 fax:(01823) 662262

Permclose Group PLC Permclose House 505a Kingsland Road London E8 4AU tel:(0171) 254 9767 fax:(0171) 923 1447

Permic Emergency Lighting Ltd PO Box 3 Chesterfield Derbys. S40 1EX tel:(01246) 277776 fax:(01246) 275879

Peta Construction Co Ltd 55-69 Willenhall Road Darlaston W. Midlands WS10 8JG tel:(0121) 526 2983 fax:(0121) 526 3531

Petit Roque Ltd 5a New Road Croxley Green Rickmansworth Herts. WD3 3EJ tel:(01923) 799291 fax:(01923) 896728

Petrarch Claddings Ltd Wainwright Close Churchfields Hastings E. Sussex TN38 9PP tel:(01424) 852641 fax:(01424) 852797 E_Mail:petrarch@btinternet.com

PFP Electrical Products Ltd Fortnum Close Mackadown Lane Kits Green Birmingham B33 0LB tel:(0121) 783 7161 fax:(0121) 783 5717 telex:339519 FAP-G E_Mail:mail@pfp-elec.co.uk Web_Site:http://www.pfp-elec.co.uk/pfp

PFP Fabrications Ltd Quarry Mills Quarry Road Godstone Surrey RH9 8DQ tel:(01883) 742797 fax:(01883) 742943

PH Products Ltd C7 Baird Court Park Farm Industrial Estate Wellingborough Northants NN8 6QJ tel:(01933) 402212 fax:(01933) 679478 E_Mail:sales@ph-products.com Web_Site:http://www.ph-products.com

Phetco (England) Ltd 37 High Street Totton Hampshire SO40 9HL tel:(011703) 663777 fax:(011703) 869996

PHI Group Ltd Harcourt House 13 Royal Crescent Cheltenham Gloucester GL50 3DA tel:(0870) 333 4120 fax:(0870) 333 4121 E_Mail:info@phigroup.co.uk Web_Site:http://www.phigroup.co.uk

Philip Johnstone Group Ltd Tingewick Road Buckingham Bucks. MK18 1AN tel:(01208) 823823 fax:(01208) 813910 telex:83529 BVRCEM G

Philips Lighting The Philips Centre 420-430 London Road Croydon CR9 3QR tel:(0181) 665 6655 fax:(0181) 781 8929

Phillip Ader Ltd 9 Green Lane Shepperton Middx. TW17 8DP tel:(01932) 221474 fax:(01932) 224789

Phoenix Engineering Co Ltd, The Combe Street Chard Somerset TA20 1JE tel:(01460) 63531 fax:(01460) 67388 E_Mail:sales@phoenixeng.co.uk

Phoenix Metal Products Ltd Stanmore Industrial Estate Bridgnorth Shropshire WV15 5JA tel:(01746) 765444 fax:(01746) 765440 Web_Site:http://www.phoenix-metal.com

Phoenix Roofing Unit 21 Dukes Close Thurmaston Leicester LE4 8EY tel:(0116) 269 5854 fax:(0116) 269 5854

Phoenix Timber Engineering Ltd 44 Ferry Lane Rainham Essex RM13 9DD tel:(01402) 755311 fax:(01402) 750122

Phosco Ltd Furlong Way, Lower Road Great Amwell Ware Herts. SG12 9TA tel:(01920) 462272 fax:(01920) 485915

Photain Controls Plc Rudford Estate Ford Aerodrome Arundel W. Sussex BN18 0BE tel:(01903) 721531 fax:(01903) 726795 E_Mail:enquiry@photaincontrols.plc.uk Web_Site:http://www.photaincontrols.plc.uk

Photo Fabrications Services Ltd 14 Cromwell Road St Neots, Huntingdon Cambs. PE19 2HP tel:(01480) 475831 fax:(01480) 475801

PhotoScan Ltd Dolphin Estate Windmill Road Sunbury on Thames Middx. TW16 7HG tel:(01932) 789741 fax:(01932) 787067 telex:8952449

Pickerings Ltd Globe Elevator Works PO Box 19 Stoke-on-Tees Cleveland TS20 2AD tel:(01642) 607161 fax:(01642) 677638 E_Mail:info@pickering.co.uk Web_Site:http://www.pickerings.co.uk

Pickersgill Kaye Ltd Pepper Road Leeds LS10 2PP tel:(0113) 277 5531 fax:(0113) 276 0221

Pifco Ltd Failsworth Manchester M35 0HS tel:(0161) 681 8321 fax:(0161) 682 1708 telex:669972

Pike Ltd, Geoffrey Garnet Close Greycaines Industrial Estate Watford Herts. WD2 4JL tel:(01923) 224884 fax:(01923) 238124

Pike Signals Ltd Equipment Works Alma Street Birmingham West Midlands B19 2RS tel:(0121) 359 4034 fax:(0121) 333 3167 E_Mail:enquiries@pikesignals.co.uk Web_Site:http://www.pikesignals.co.uk

Pilkington Birmingham Nechells Park Road Nechells Birmingham B7 5NQ tel:(0121) 326 5300 fax:(0121) 328 4277 Web_Site:http://www.Pilkington.co.uk

Pilkington Plyglass Cotes Park Somercotes Derbys. DE55 4PL tel:(01773) 620000 fax:(01773) 6200052 E_Mail:nr50@dial.pipex.com

Pilkington UK Ltd Prescot Road St Helens Merseyside WA10 3TT tel:(01744) 692000 fax:(01744) 613049 E_Mail:lipscombes@pilkington.com Web_Site:http://www.pilkington.co.uk

Pilkington's Tiles Ltd PO Box 4 Clifton Junction Manchester M27 8LP tel:(0161) 727 1111 fax:(0161) 727 1122

Pillinger G C & Co (Engineers) Ltd Pillinger House 4 Wells Place Merstham,Redhill Surrey RH1 3DR tel:(01737) 733456 fax:(01737) 645036 E_Mail:sales@pillingerengineers.co.uk

Pims Pumps Ltd 22 Invincible Rd Ind Est Farnborough Hampshire GU14 7QU tel:(01252) 513366 fax:(01252) 516404

Pioneer Concrete Holdings PLC Pioneer House 56-60 Northolt Road South Harrow HA2 0EY tel:(0181) 423 3066 fax:(0181) 864 1975

Piper Windows Doors & Conservatories Hamelin Court 140 Newington Road Ramsgate Kent CT12 6PP tel:(01843) 850500 fax:(01843) 852626 E_Mail:piperwindows@btinternet.com Web_Site:http://www.piperwindows.com

Pitchmastic PLC, Royds Works Attercliffe Road Sheffield S4 7WZ tel:(0114) 2700100 fax:(0114) 2768782 E_Mail:admin@dew-pitch-shefield.co.uk

Pitchmastic PLC, Building Products Division Royds Works Attercliffe Road Sheffield South Yorks tel:(0114) 270 0100 fax:(0114) 276 8782 E_Mail:admin@dew-pitch-shefield.co.uk

Pittsburg Corning (UK) Ltd Southcourt 29 South Street Reading Berks. RG1 4QU tel:(01189) 500655 fax:(01189) 509019

PJP plc Dixon Hill Road Wellham Green Hatfield Herts. AL9 7JE tel:(01707) 266726 fax:(01707) 263614

PK Hardwood Melsa Prima Brill Sawmills Woutton Road Brill Bckinhamshire HP18 9W tel:(0151) 546 9321 fax:(0151) 5491423

Pland Lower Wortley Ring Road Leeds W. Yorks. LS12 6AA tel:(0113) 263 4184 fax:(0113) 231 0560 E_Mail:plandstainless@compuserve.com

Planet Architectural Glazing Ltd Stansdale House Cliftonville Road Northampton NN1 5BU tel:(01604) 29721 fax:(01604) 29726

Planet Projects Ltd Aegis House, Ground Floor Castle Hill Maidenhead Berks. SL6 4JL tel:(01628) 549249 fax:(01628) 549292 Web_Site:http://www.planet.co.uk

Plansee Tizit (UK) Ltd Grappenhall Warrington Cheshire WA4 3JX tel:(01925) 261161 fax:(01925) 267933

Plascoat Systems Ltd Trading Estate Farnham Surrey GU90 0NY tel:(01252) 733777 fax:(01252) 721250 E_Mail:sales@plascoat.com Web_Site:http://www.plascoat.com

Plastal Commercial Ltd 7 Sopwith Way Drayton Fields Ind Est Daventry Northants NN11 5PB tel:(01327) 867627 fax:(01327) 871344

Plastic Coatings Ltd Woodbridge Industrial Estate Guildford Surrey GU1 1BG tel:(01483) 31155 fax:(01483) 33534 telex:859237

Plastic Construction Fabrications Ltd Evelyn Road Sparkhill Birmingham B11 3JJ tel:(0121) 773 4951 fax:(0121) 772 3588

Plastic Extruders Ltd Russell Gardens Wickford Essex SS11 8DN tel:(01268) 735231 fax:(01268) 560027 telex:995101 E_Mail:plastic@compuserve.com Web_Site:http://www.plastex.co.uk

Plasticable Ltd 112 Hawley Lane Farnborough Hampshire GU14 8JE tel:(01252) 541385 fax:(01252) 373816

PLC Hunwick Ltd Harrision Works Kings Road Halstead Essex CO9 1HD tel:(01787) 474547 fax:(01787) 475741

PMD Technologies Ltd Cromwell House Elland Road Brighouse W. Yorks. HD6 2RG tel:(01422) 370938 fax:(01422) 377730

Polybau Ltd Trafford Park Road Newbridge Trafford Park Manchester M17 1JD tel:(0161) 872 1472 fax:(0161) 877 6592 E_Mail:gml@poly.co.uk Web_Site:http://www.poly.co.uk

Polycell Products Ltd Wexham Road Slough Berkshire SL2 5DS tel:(01753) 550000 fax:(01753) 578218 Web_Site:http://www.polycell.co.uk

Polypipe Civils Ltd Union Works Bishop Meadow Road Loughborough Leics. LE11 5RE tel:(01509) 615100 fax:(01509) 610215 E_Mail:polypipe@polypipecivils.co.uk

Porn Dunwoody ERS (Lifts & Escalators) Ltd Union Works Bear Gardens London SE1 9EB tel:(0171) 261 1162 fax:(0171) 261 1434

Portacel Winnall Valley Road Winchester SO23 0LL tel:(01962) 705200 fax:(01962) 705279 E_Mail:portacel_sales@tmproducts.com Web_Site:http://www. Portacel.co.uk
Portakabin Ltd Huntington York YO32 9PT tel:(01904) 611655 fax:(01904) 611644 E_Mail:solutions@portakabin.com Web_Site:http://www.portakabin.com
Portasilo Ltd New Lane Huntington York YO32 9PR tel:(01904) 624872 fax:(01904) 611760 E_Mail:action@portasilo.co.uk Web_Site:http://www.portasilo.co.uk
Portia Engineering Ltd PO Box 9 Star Lane Ipswich IP4 1JJ tel:(01473) 252334 fax:(01473) 233863 E_Mail:portia@btinternet.com
Poselco Lighting 1 The Metropolitan Centre Bristol Road Greenford Middx. UB6 8UW tel:(0181) 813 0101 fax:(0181) 813 0099
Potter & Soar Ltd Beaumont Road Banbury Oxon. OX16 7SD tel:(01295) 253344 fax:(01295) 272132 E_Mail:potter.soar@btinternet.com Web_Site:http://www.wiremesh.co.uk
Potters Insulations (Sales) Ltd Tameside Mills Park Road Dukinfield Cheshire SK16 5LS tel:(0161) 308 3535 fax:(0161) 343 1706 E_Mail:sales@potters-insulations.co.uk
Potterton Myson Ltd Eastern Avenue Team Valley Trading Estate Gateshead Tyne & Wear NE11 0PG tel:(0191) 491 4466 fax:(0191) 491 7568 telex:53265
Pouliot Designs Unit 5, Marley Industrial Estate Southam Road Banbury Oxon. OX16 7RL tel:(01295) 270848 fax:(01295) 270965
Power Breaker Temple Fields, South Rd Harlow Essex CM20 2BG tel:(01279) 434561 fax:(01279) 635285 E_Mail:gbe@greenbrook.co.uk
Power Plastics Ltd Station Road Thirsk Yorks. YO7 1PZ tel:(01845) 525503 fax:(01845) 525485 E_Mail:info@powerplastics.co.uk
Powys Hire Ltd/ P.I.B PO Box 238 Ennerdale Road, Harlescott Industrial Estate Shrewsbury Shropshire SY1 1YB tel:(01743) 442624 fax:(01743) 462293
Precolor Sales Ltd Newport Road Market Drayton Shropshire TF9 2AA tel:(01630) 657281 fax:(01630) 655545 E_Mail:precolor@aol.com Web_Site:http://www.members.aol.com?precolor
Preformed Components Ltd Davis Road Chessington Surrey KT9 1TU tel:(0181) 391 0533 fax:(0181) 391 2723
Premdor Groundwell Industrial Estate Hargreaves Road Swindon Wilts. SN25AZ tel:(01793) 708200 fax:(01793) 706179 E_Mail:ukmarketing@premdor.com
President Blinds Ltd 211-215 High Street Penge London SE20 7PF tel:(0181) 776 7766 fax:(0181) 676 0916
President Office Furniture Ltd The Arenson Centre Arenson Way Dunstable Beds. LU5 5UL tel:(01582) 678200 fax:(01582) 678222 E_Mail:enquires@president.co.uk
Pressalit Ltd Riverside Business Park Dansk Way, Leeds Road Ilkley W. Yorks. LS29 8JT tel:(01943) 607651 fax:(01943) 607214 E_Mail:ltd@pressalit.com
Pressurisation Ltd, Member of Hoval Group PO Box 473 Bradford W. Yorks. BD8 9YF tel:(01422) 349560 fax:(01274) 483459
Presto Engineers Cutting Tools Ltd Penistone Road Sheffield S6 2FN tel:(01742) 349361 fax:(01742) 347446 telex:54223
Preston & Thomas Ltd Woodville Engineering Works Heron Road, off Greenway Road Rumney Cardiff CF3 8YF tel:(01222) 793331 fax:(01222) 779195
Prestoplan Four Oaks Road Walton Summit Centre Preston PR5 8AS tel:(01772) 627373 fax:(01772) 627575
Preussag Fire Protection Ltd Field Way Greenford Middx. UB6 8UZ tel:(0181) 832 2000 fax:(0181) 832 2200
Prima Security & Fencing Products Aubrey Works 15 Aubrey Avenue London Colney Herts. AL2 1NE tel:(01727) 822222 fax:(01727) 826307 E_Mail:101625.2443@compuserve.com
Prima Systems (South East) Ltd The Old Malt House Easole Street Nonington, Dover Kent CT15 4HF tel:(01304) 842888 fax:(01304) 842840
Priory Castor & Engineering Co Ltd Aston Hall Road Aston Birmingham B6 7LA tel:(0121) 327 0832 fax:(0121) 322 2123
Priory Shutter & Door Co Ltd 89-91 Rolfe Street Smethwick Warley W. Midlands B66 2AY tel:(0121) 558 6406 fax:(0121) 558 7140
Prismo Ltd Gleneagles Court Brighton Road Crawley W. Sussex RH10 6UE tel:(01293) 530356 fax:(01293) 530362 E_Mail:roadsafe@prismo.co.uk Web_Site:http://prismo.co.uk
Pro-Fence (Midlands) 3 Barley Croft Perton Wolverhampton W. Midlands WV6 7XX tel:(01902) 894747 fax:(01902) 897007
Procter Bros Limited New Victoria Works Bowling Green Terrace Leeds W. Yorks. LS11 9SZ tel:(0113) 2430531 fax:(0113) 2422649
Procter Bros. Limited Pantglas Industrial Estate Bedwas Newport Gwent NP1 8XD tel:(01222) 882111 fax:(01222) 887005
Procter Concrete Products Isabella Works East Garforth Leeds LS25 2HQ tel:(01132) 862586 fax:(01132) 867376
Procter Johnson & Co Ltd Excelsior Works Castle Park, Evans Street Flint Clwyd CH6 5NT tel:(01352) 732157 fax:(01352) 735530 telex:61469 PROJO G
Producta Ltd Mount Pleasant Stillingfleet Yorks. YO4 6HS tel:(01904) 728453 fax:(01904) 728140
Profile 22 Systems Ltd Stafford Park 6 Telford Shropshire TF3 3AT tel:(01952) 290910 fax:(01952) 290460
Profile Lighting Services Ltd Links Business Centre Raynham Road Bishops Stortford Herts. CM23 5NZ tel:(01279) 757595 fax:(01279) 755599
Profilex Ltd Dodwells Road Dodwells Bridge Industrial Estate Hinckley Leics. LE10 3BU tel:(01455) 612009 fax:(01455) 614002 E_Mail:enquiries@profilex.com Web_Site:http://www.profilex.com
Profolia Ltd Unit 3, Plymouth Avenue Brookhill Industrial Estate Pinxton Notts. NG16 6NS tel:(01773) 510505 fax:(01773) 510534
Progress Furnishings Systems Ltd 14 Sextant Park Neptune Close, Medway City Estate Rochester Kent ME2 4LU tel:(01634) 290988 fax:(01634) 291028
Progressive Product Developments Ltd 24 Beacon Bottom Swanwick South Hampton Hampshire SO31 7GQ tel:(01489) 576787 fax:(01489) 578463 E_Mail:ppd.limited@btinternet.com Web_Site:http://www.btinternet.com/~ppd.limited
Project Aluminium Ltd 418-420 Limpsfield Road Warlington Surrey CR6 9LA tel:(01883) 624004 fax:(01883) 627201
Project Office Furniture PLC Hamlet Green Haverhill Suffolk CB9 8QJ tel:(01440) 705411 fax:(01440) 703376 E_Mail:enquiries@project.co.uk Web_Site:http://www.project.co.uk
Promat Fire Protection Ltd Whaddon Road Meldreth Nr Royston Herts SG8 5RL tel:(01763) 262310 fax:(01763) 262342 telex:817521 ETERNT G
Propaflor Ltd 2-6 Bilton Way Luton Bedfordshire LU1 1UU tel:(01582) 734161 fax:(01582) 400946 E_Mail:propaflor@propaflor.co.uk. Web_Site:http://www.propaflor.co.uk.
Property Mechanical Products Ltd Stanton Square Stanton Way Sydenham London SE26 5AB tel:(0181) 676 0911 fax:(0181) 659 1017 E_Mail:info@pmp-walney.co.uk
Protecter Lamp & Lighting Co Ltd Lansdowne Road Eccles Nr. Manchester M30 9PH tel:(0161) 789 5680 fax:(0161) 787 8257
Protective Rubber Coatings (Bristol) Ltd Paynes Shipyard Coronation Road Bristol BS3 1RP tel:(01272) 661155 fax:(01272) 661158
Protim Services Ltd Aquis House 27-33 Station Road Hayes Middx. UB3 4XD tel:(0181) 606 1000 fax:(0181) 569 1743
Protim Solignum Ltd Fieldhouse Lane Marlow Bucks. SL7 1LS tel:(01628) 486644 fax:(01628) 476757 telex:847057
Protimeter plc Meter House Fieldhouse Lane Marlow Bucks. SL7 1LW tel:(01628) 472722 fax:(01628) 474312 E_Mail:moisture@protimeter.com Web_Site:http://www.protimeter.com
Provend Services Ltd 19 Aintree Road Perivale, Greenford Middx. UB6 7LG tel:(0181) 998 2828 fax:(0181) 998 0704
Provex Products Ltd Unit 5G Tewkesbury Ind Centre Delta Drive Tewkesbury Glos. GL20 8HD tel:(01684) 297570 fax:(01684) 850912
Pryorsign Field View Brinsworth Lane Brinsworth, Rotherham S. Yorks. S60 5DG tel:(01709) 839559 fax:(01709) 837659 Web_Site:http://www.buyersguide.co.uk/docume nt/pryor-sign/
Pulsford Associates Ltd, John Unit 4 Sphere Industrial Estate Campfield Road St Albans Herts. AL1 5HT tel:(01727) 840800 fax:(01727) 840083 E_Mail:jpafurniture.demon.co.uk
Pump International Ltd Trevoole Praze Camborne Cornwall TR14 0PJ tel:(01209) 831937 fax:(01209) 831939 E_Mail:pump.int@tesco.net Web_Site:http://www.housepages.tesco.net/-pump.int
Pump Technical Services Ltd Beco Works, Cricket Lane off Kent House Lane Beckenham Kent BR3 1LA tel:(020) 8778 4271 fax:(020) 8659 3576 E_Mail:sales@ptsjung.co.uk Web_Site:http://www.pts-jung.co.uk
Purdie Elcock Ltd Frankfort Street Works Birmingham B19 2YL tel:(0121) 359 5898 fax:(0121) 359 5510
Purewell Timber Works Stony Lane Christchurch Dorset BH23 1EY tel:(01202) 484422 fax:(01202) 490151
Purified Air Ltd Lyon House Lyon Road Romford Essex RM1 2BG tel:(01708) 755414 fax:(01708) 721488 E_Mail:enq@purifieldair.co.uk Web_Site:http://www.purifield.co.uk
Purimachos Ltd 14 Waterloo Road St Phillips Bristol BS2 0PG tel:(0117) 995 4361 fax:(0117) 935 0844
Purpose Built Ltd Spring Lane South Malvern Link Worcester WR14 1AQ tel:(01684) 892602 fax:(01684) 892801
Pyramid Plastics Ltd Britannia Works Beaumont Street Gainsborough Lincs. DN21 2EN tel:(01427) 677990 fax:(01427) 612204
Quail Paul Boundary Farm House Swanton Road Gunthorpe,Melton Constable Norfolk NR24 2NS tel:(01263) 860826 fax:(01263) 860 826
Qualbatch Minimix Mattersey Road Ranskill Retford Notts. DN22 8NH tel:(01777) 818155 fax:(01777) 817031
Qualitas Bathrooms Hartshorne Road Woodville Swadlincote Derbys. DE11 7JD tel:(01283) 550550 fax:(01283) 550314
Quantum Profile Systems Salmon Fields Royton, Oldham Lancs. OL2 6JG tel:(0161) 627 4222 fax:(0161) 627 4333 E_Mail:qps@quantum-ps.co.uk Web_Site:http://www.quantum-ps.co.uk
Quartet-GBC 2 Springlakes Estate Deadbrook Lane Aldershot Hampshire GU12 4UH tel:(01252) 28512 fax:(01252) 343217 E_Mail:sales@quartetmfg.co.uk
Quelfire PO Box 35 Altrincham Cheshire WA14 5QA tel:(0161) 928 7308
Quentplass Ltd Thorp Arch Trading Estate Wetherby W. Yorks. LS23 7BZ tel:(01937) 843388 fax:(01937) 541458
Quigley Metal Products Ltd Timmis Road Lye, Stourbridge W. Midlands DY9 7BQ tel:(01384) 424181 fax:(01384) 892042 E_Mail:qmp@qmpgroup.com
Quiligotti Access Flooring Limited PO Box 4 Clifton Junction Manchester M27 8LP tel:(0161) 727 1070 fax:(0161) 727 1080 E_Mail:quiligotti.access@pilkingtons.com Web_Site:http://www.quiligotti.com
Quinton & Kaines (Holdings) Ltd Creeting Road Stowmarket Suffolk IP14 5AS tel:(01449) 612145 fax:(01449) 677604 Web_Site:http://www.bbaduiven.demon.nl
Quip Lighting, Div of Bruce & Macintyre Ltd 71 Tenison Road Cambridge Cambs. CB1 2DG tel:(01223) 321277 fax:(01223) 321277
R & C Plant Sales (Scotland) Carseview Road Forfar Tayside DD8 3BT tel:(01307) 468200 fax:(01307) 467770
R A International Kingshill Business Park Darlaston Road Wednesbury W. Midlands WS10 7SH tel:(0121) 556 4271 fax:(0121) 502 2562
R-B International Ltd Beevor Street Lincoln Lincs. LN6 7DJ tel:(01522) 525261 fax:(01522) 512230 telex:56191 RBLIN G E_Mail:sales@rbcranes.com Web_Site:http://www.rberanes.com
Raab Karcher Energy Services Ltd, Clorius Division Unit 5, Weighbridge Row Cardiff Road Reading RG1 8LX tel:(0118) 956 0060 fax:(0118) 956 7610
Rabone Chesterman Ltd Woodside Sheffield S3 9PD tel:(01742) 768888 fax:(01742) 739038 telex:54150 STANLEY G
Racal Guardall Ltd Lochend Industrial Estate Newbridge Edinburgh EH28 8PL tel:(0131) 333 2900 fax:(0131) 333 4919
Race Furniture Ltd Bourton Industrial Park Bourton-on-the-Water Glos. GL54 2HQ tel:(01451) 821446 fax:(01451) 821686
Rackline Systems Storage Ltd Oaktree Lane Talke Newcastle-under-Lyme Staffs. ST7 1RX tel:(01782) 777666 fax:(01782) 777444 E_Mail:now@rackline.co.uk Web_Site:http://www.rackline.co.uk
Rada Lighting Ltd Hollies Way High Street Potters Bar Herts. EN6 5BH tel:(01707) 643401 fax:(01707) 645548
Radcliffe & Sons (London) Ventura House 5-7 St Leonards Road Horsham W. Sussex RH13 6EH tel:(01403) 251444 fax:(01403) 217174
Radflex Contract Services Limetid Unit 35, Wilks Avenue Dartford Trade Park Dartford Kent DA1 1JS tel:(01322) 276363 fax:(01322) 270606 E_Mail:expjiont@radflex.co.uk Web_Site:http://www.radflex.co.uk
Radial & Axial Fans Radial House Coldharbour Lane Harpenden Herts. AL5 4UN tel:(01582) 460624 fax:(01582) 460629
Radiant Services Ltd Barrett House 111 Millfields Road Ettingshall Wolverhampton WV4 6JQ tel:(01902) 494266 fax:(01902) 494153 E_Mail:radser@globainet.co.uk Web_Site:http://www.radiantservices.co.uk
Radiodetection Ltd Western Drive Bristol BS14 0AZ tel:(0117) 976 7776 fax:(0117) 976 7775 Web_Site:http://www.radiodetection.co.uk/
Radius Lighting Ltd 15 South Street Havant Hampshire PO9 1BU tel:(01705) 455525 fax:(01705) 451130
Radway, Door and Windows Ltd. Merse Road North Moons Moat Redditch Worcs. B98 9HL tel:(01527) 585558 fax:(01527) 63965
RAF Quedgley Unit 5 Site 1 Quedgley Gloucester GL2 5ZZ tel:(01452) 417575 fax:(01452) 524446
Railex Systems Ltd The Wilson Building 1 Curtain Road London EC2A 3JX tel:(0171) 377 1777 fax:(0171) 247 6181
Railex Systems Ltd, Elite Division Elite Works Station Road Manningtree Essex CO11 1DZ tel:(01206) 392171 fax:(01206) 391465 telex:987675
Rainham Steel Co Ltd Kathryn House Manor Way Rainham Essex RM13 8RE tel:(01708) 556034 fax:(01708) 559024
Ramsay & Sons (Forfar) Ltd 61 West High Street Forfar Angus DD8 1BH tel:(01307) 462255 fax:(01307) 466956
Ramset Fasteners Ltd Ramset House Galleymead Road Colnbrook Slough SL3 0EN tel:(01753) 682277 fax:(01753) 680187 E_Mail:pwalsh@ramsetfastenersltd.com Web_Site:http://www.ramsetuk.com
Ranalah Gates Ltd Gloucester Road Industrial Estate Malmesbury Wilts. SN16 9JT tel:(01666) 823001 fax:(01666) 824175
Rapaway Energy Ltd 35 Park Avenue Solihull W. Midlands B91 3EJ tel:(0121) 246 0441 fax:(0121) 246 0442 E_Mail:rapaway@btinternet
Rapid Metal Developments Stubbers Green Road Aldridge Walsall Staffs. WS9 8BW tel:(01922) 743743 fax:(01922) 743400
Ratcliffe J H & Co (Paints) Ltd 135 A Linaker Street Southport Merseyside PR8 5DF tel:(01704) 537999 fax:(01704) 544138
Rawlings(SE) Ltd 8 Faversham Road Maidstone Kent ME17 2PN tel:(01622) 859841 fax:(01622) 858851
Rawson W E Ltd Castle Bank Mills Portobello Road Wakefield W. Yorks. WF1 5PS tel:(01924) 73421 telex:55438 RAWSON G
RB Farquhar Manufacturing Ltd Deveronside Works Huntly Aberdeenshire AB54 4PS tel:(01466) 793231 fax:(01466) 793098 E_Mail:sales@rbfarquhar.co.uk Web_Site:http://www.rbfarquhar.co.uk
RB Shelving Systems (UK) Ltd Element House Napier Road Bedford MK41 0QX tel:(01234) 272717 fax:(01234) 270202
RCC, Division of Tarmac Precast Concrete Ltd Barholm Road Tallington Stamford Lincolnshire PE9 4RL tel:(01778) 344460 fax:(01778) 345949
RDP-Howden Ltd Althorpe Street Leamington Spa Warwks. CV31 2BA tel:(01926) 427782 fax:(01926) 429873 telex:335430 RDPG
Rea Metal Windows Ltd 126-136 Green Lane Liverpool L13 7ED tel:(0151) 228 6373 fax:(0151) 254 1828 E_Mail:rea.metal@cwcom.net
Real Door Company The Unit 5 Cadwell Lane Hitchin Herts. SG4 OSA tel:(01462) 768324 fax:(01462) 768537
Realstone Ltd Wingerworth, Chesterfield Derbys. S42 6RG tel:(01246) 270244 fax:(01246) 220095 E_Mail:john_c@realstone.co.uk
Record & Parkes Ltd Crown Wharf Iron Works Dace Road Old Ford London E3 2NL tel:(0181) 985 5383 fax:(0181) 986 8212
Record Electrical Atlantic Street Altrincham Cheshire WA14 5DB tel:(0161) 928 6211 fax:(0161) 926 9750
Record Playground Equipment Ltd Waterfront Complex, Shipyard Industrial Estate Carr Street Selby N. Yorks. YO8 8AP tel:(01757) 703620 fax:(01757) 705158
Record Power Ltd Parkway Works Sheffield S9 3BL tel:(01742) 434370 fax:(01742) 617141
Record Tools Ltd Parkway Works Sheffield S9 3BL tel:(01742) 449066 fax:(01742) 434302 telex:547139
Red Bank Maufacturing Co Ltd Atherstone Road Measham, Swandlincote Derbys. DE12 7EL tel:(01530) 270333 fax:(01530) 273667
Red Head, Division of ITW Ltd 118 Coltness Street Queenslie Industrial Estate Glasgow G33 4JD tel:(0141) 774 2267 fax:(0141) 774 5802
Reddiglaze Ltd The Furlong Droitwich Worcs. WR9 9BG tel:(01905) 795432 fax:(01905) 795757
Redditch Plastic Products Pipers Road Park Farm Industrial Estate Redditch Worcs. B98 0HU tel:(01527) 528024 fax:(01527) 520236
Rediweld Rubber & Plastics Ltd 6-8 Newman Lane Alton Hampshire GU34 2QR tel:(01420) 543007 fax:(01420) 544090 E_Mail:info@rediwell.co.uk Web_Site:http://www.rediwell.co.uk
Redland Precast Six Hills Melton Mowbray Leics. LE14 3PD tel:(01509) 882121 fax:(01509) 880799
Redman Fisher Engineering Birmingham New Road Tipton W. Midlands DY4 9AA tel:(01902) 880880 fax:(01902) 880446 telex:337380 E_Mail:flooring@redmanfisher.co.uk Web_Site:http://www.redmanfisher.co.uk
Reed Harris Riverside House 27 Carnwath Road Fulham/London SW6 3HR tel:(0171) 736 7511 fax:(0171) 736 2988
Reflecting Roadstuds Ltd Boothtown Halifax W. Yorks. HX3 6TR tel:(01422) 360208 fax:(01422) 349075
Reflex-Rol (UK) Ryeford House Ryeford Nr Ross-on-Wye Worcs. HR9 7PU tel:(01989) 750704 fax:(01989) 750768
Rega Metal Products Ltd Station Road Sandy Beds. SG19 1BH tel:(01767) 691291 fax:(01767) 692451 E_Mail:sales@rega-uk.com
Regency Garage Door Services Ltd Unit 8, Central Trading Estate Signal Way Swindon Wilts. SN3 1PD tel:(01793) 611688 fax:(01793) 495144
Regency Swimming Pools Ltd Regency House 88a Brick Kiln Street Graisley Wolverhampton WV3 0PU tel:(01902) 427709 fax:(01902) 422632
Regon Ltd Station Lane Trading Estate Witney Oxon. OX8 6YE tel:(01993) 771441 fax:(01993) 774105
Rehau Ltd Waterside Drive Langley Berks. SL3 6EZ tel:(01753) 549974 fax:(01753) 544461
Reiber & Son PLC California Drive Castleford W. Yorks. WF10 5QZ tel:(01977) 512121 fax:(01977) 550996 telex:55160
Reid John & Sons (Strucsteel) Ltd Strucsteel House 264-266 Reid Street Christchurch Dorset BH23 2BT tel:(01202) 483333 fax:(01202) 470103 E_Mail:sales@reidsteel.co.uk Web_Site:http://www.reidsteel.co.uk
Reiner Fixing Devices Hall Farm Church Lane North Ockendon, Upminster Essex RM14 3QH tel:(01708) 856601 fax:(01708) 852293

Reinforced Plastic Products Limited Slough House Slough Lane Horton, Wimborne Dorset BH21 7JL tel:(01202) 828744 fax:(01202) 822055

Relcross Ltd Hambleton Avenue Devizes Wilts. SN10 2RT tel:(01380) 729600 fax:(013080) 729888 E_Mail:sales@relcross.freeserve.co.uk Web_Site:http://www.relcross.co.uk

Remploy Ltd Unit 2, Brocastle Avenue Waterton Industrial Estate Bridgend M. Glam. CF31 3YN tel:(01656) 653982 fax:(01656) 767499

Remtox Silexine 14 Spring Road Smethwick Warley W. Midlands B66 1PE tel:(0121) 525 2299 fax:(0121) 525 1740

Rentokil Healthcare Service Felcourt East Grinstead W. Sussex RH19 2JY tel:(01342) 833022 fax:(01342) 326229

Rentokil Initial U.K. Ltd. Rentokil Fire Protection Products Felcourt East Grimstend West Sussex RH19 2JY tel:(01342) 833022 fax:(01342) 326229

Rentokil Ltd Timber Preserving Products Division Felcourt East Grinstead W. Sussex RH19 2JY tel:(01342) 833022 fax:(01342) 326229 telex:95456 RNTKIL G

Resdev Ltd Pumaflor House Ainleys Industrial Estate Elland W. Yorks. HX5 9JP tel:(01422) 379131 fax:(01422) 370943 E_Mail:resdevltd@compuserve.com Web_Site:http://www.resdev.co.uk

Resiblock Ltd Resiblock house Archers Field Close Basildon Essex SS13 1DW tel:(01268) 273344 fax:(01268) 273355

Rettig Heating Hartlebury Trading Estate Nr Kidderminster Worcs. DY10 4JB tel:(01299) 250700 fax:(01299) 251192

Reventa Products (UK) Ltd 12 Wheatley Way Chalfont St Peter Bucks. SL9 0JE tel:(012407) 5315 fax:(012407) 71394 telex:837456

Revol Products Ltd Samson Close Killingworth Newcastle-upon-Tyne NE12 0DZ tel:(0191) 268 4555 fax:(0191) 216 0004

Rexam Harcostar Windover Road Huntingdon Cambs. PE18 7EE tel:(01480) 445113 fax:(01480) 413203

Reynolds John & Son (Birmingham) Ltd Church Lane West Bromwich W. Midlands B71 1DJ tel:(0121) 553 2754/1287 fax:(0121) 500 5460

Reynolds UK Ltd Straight Road Short Heath, Willenhall W. Midlands WV12 5RA tel:(01922) 710222 fax:(01922) 710066

Reyven (Sportsfields) Ltd Greenhill Farm Tilsworth Leighton Buzzard Beds. LU7 9PU tel:(01525) 210714 fax:(01525) 211130

Reznor UK Ltd Park Farm Road Park Farm Industrial Estate Folkestone Kent CT19 5DR tel:(01303) 259141 fax:(01303) 850002 E_Mail:sales@reznor.co.uk Web_Site:http://www.reznor.co.uk

RFA Group Ltd Bullhouse Works Bullhouse Bridge Millhouse Green, Sheffield S. Yorks. S36 9NF tel:(01226) 764425 fax:(01226) 767562

Rheinzink UK Cedar House Cedar Lane Frimley, Camberley Surrey GU16 5HY tel:(01276) 686725 fax:(01276) 64480

Rhodar Limited Beza Road Hunslet Leeds LS10 2BR tel:(0113) 270 0775 fax:(0113) 270 4124 E_Mail:rhodar@legend.co.uk Web_Site:http://www.Rhodar.co.uk

Rhone-Poulenc Chemicals Ltd Liverpool Road Barton Moss Site Eccles Manchester M30 7RT tel:(0161) 707 9779 fax:(0161) 910 8812

Rib (UK) Ltd Ellard House Baxter Road, Foxhill Industrial Estate Wadesley Bridge, Sheffield S. Yorks. S6 1JF tel:(0114) 285 2030 fax:(0114) 285 2060

Richard Hose Ltd Unit 7, Roman Way Centre Longridge Road Ribbleton, Preston Lancs. PR2 5BB tel:(01772) 651550 fax:(01772) 651325 E_Mail:richards.fire@btinternet.com

Richards H S Ltd King Street Smethwick Warley W. Midlands B66 2JW tel:(0121) 558 2261 fax:(01785) 840 603

Righton Ltd Righton House Brookvale Road Witton Birmingham B6 7EY tel:(0121) 356 1141 fax:(0121) 331 1347 E_Mail:righton@righton.co.uk Web_Site:http://www.righton.co.uk

Rigidal Industries Ltd. Blackpole Trading Estate West Worcester WR3 8ZJ tel:(01905) 750500 fax:(01905) 750555 E_Mail:rigidal@compuserve.com Web_Site:http://www.rigidal.co.uk

Rigized Metals Ltd Aden Road Ponders End Enfield Middx. EN3 7SU tel:(0181) 804 0633 fax:(0181) 804 7275

Ring-Gard (UK) Ltd Unit 18, Pedmore Road Industrial Estate Pedmore Road Brierley Hill W. Midlands DY5 1TJ tel:(01384) 74849 fax:(01384) 74348

Rite-Vent Ltd Crowrther Estate Washington Tyne & Wear NE38 0AQ tel:(0191) 416 1150 fax:(0191) 415 1263 E_Mail:sales@rite-vent.co.uk Web_Site:http://www.write-vent.co.uk

Ritherdon & Co Ltd Lorne Street Darwen Lancs. BB3 1QW tel:(01254) 819100 fax:(01254) 819101 E_Mail:sales@ritherdon.co.uk Web_Site:http://www.ritherdon.co.uk

RIW Ltd Arc House Terrace Road South Binfield. Bracknell Berks RG42 4PZ tel:(01344) 861988 fax:(01344) 862010

RIW Ltd Arc House Terrace Road South Binfield, Bracknell Berks. RG42 4PZ tel:(01344) 861988 fax:(01344)862010

Rixonway Kitchens Ltd Shaw Cross Busness Park Dewsbury West Yorkshire WF12 7RD tel:(01924) 431300 fax:(01924) 431301 E_Mail:info@rixonway.co.uk.ltd. Web_Site:http://www.@rixonway.co.uk.ltd.

RMC Concrete Products 82-87 Feeder Road Bristol B52 0UE tel:(0117) 9373371 fax:(0117) 9373002

RMC Concrete Products (UK) Ltd Dale Road Dove Holes Buxton Derbys. SK17 8BH tel:(01298) 22244 fax:(01298) 815221

RMC Concrete Products (UK) Ltd Yaxham Road Dereham Norfolk NR19 1HF tel:(01362) 692261 fax:(01362) 694120

RMC Panel Products Ltd Waldorf Way Denby Dale Road Wakefield W. Yorks. WF2 8DH tel:(01924) 362081 fax:(01924) 290126 E_Mail:thermabate@dial.pipex.com Web_Site:http://www.thermabate.co.uk

RMC Readymix Limited RMC House High Street Feltham Middx. TW13 4HA tel:(0800) 667827 fax:(0181) 7510006 E_Mail:parker_n@rmc-readymix.co.uk

RNB Industrial Doors Services Unit 6, Davenport Centre Renwick Road Barking Essex IG11 0SP tel:(0181) 595 1242 fax:(0181) 595 3849

Roadways & Car Parks Ltd 174 Twickenham Road Isleworth Middx. TW7 7DW tel:(0181) 560 7211 fax:(0181) 560 1894

Robert Bosch Ltd PO Box 98, Broadwater Park North Orbital Road Denham, Uxbridge Middx. UB9 5HJ tel:(01895) 834466 fax:(01895) 838388 telex:935244 S

Robertson H H (UK) Ltd Ellesmere Port South Wirral L65 4DS tel:(0151) 355 3622 fax:(0151) 355 1276 telex:629639

Robinson J J & Son Ltd Eden Lane Gainford Darlington DL2 3BN tel:(01325) 730241 fax:(01325) 730596

Robinson Willey Ltd Mill Lane Old Swan Liverpool L134AJ tel:(0151) 228 9111 fax:(0151) 228 6661

Robot (UK) Ltd Wincanton House Wincanton Close Derby Derbys. DE24 8NB tel:(01332) 758000 fax:(01332) 758202

Rock Electrical Accessories Ltd Commerce Road Brentford Middx. TW8 8LN tel:(0181) 560 8151 fax:(0181) 560 0857

Rockfon Ltd Pencoed Bridgend County Borough CF35 6NY tel:(01656) 864696 fax:(01656) 864549 Web_Site:http://www.rockfon.co.uk

Rockwell Sheet Sales Ltd Main Road Meriden W. Midlands CV7 7NH tel:(01676) 523386 fax:(01676) 523630 E_Mail:info@rockwell-sheet.co.uk Web_Site:http://www.rockwell-sheet.co.uk

Rockwool Ltd Pencoed/Bridgend M. Glam. CF35 6NY tel:(01656) 862621 fax:(01656) 862302

Rogers Concrete Ltd Puddletown Road Wareham Dorset BH20 6AU tel:(01929) 462373 fax:(01929) 405326

Rohm & Hass (UK) Ltd Lennig House 2 Mason's Avenue Croydon Surrey CR9 3NB tel:(0181) 686 8844 fax:(0181) 686 8329

Rok-Crete Units Co Ltd 112 Oxford Road Clacton on Sea Essex CO15 3TN tel:(01255) 424884 fax:(01255) 221155

Rolflex Doors Ltd Bldg 63 The Pensnett Estate Kingswinford W. Midlands DY6 7PP tel:(01384) 401555 fax:(01384) 401556

Rollalong Woolsbridge Industrial Estate Three Legged Cross Wimborne Dorset BH21 6SF tel:(01202) 824541 fax:(01202) 826525 telex:41483

Rollins & Sons (London) Ltd Rollins House Murray Road Hertford Herts. SG14 1NR tel:(01992) 587555 fax:(01992) 500159

Rom Ltd Eastern Avenue Trent Valley Lichfield Staffs. WS13 6RN tel:(01543) 414111 fax:(01543) 268221 E_Mail:sales@rom.co.uk

Roman Mosaic Contracts Ltd Bloomfield Road Tipton W. Midlands DY4 9ES tel:(0121) 557 2267 fax:(0121) 557 0975

Ronacrete Ltd Ronac House Selinas Lane Dagenham Essex RM8 1QL tel:(0181) 593 7621 fax:(0181) 595 6969 E_Mail:sales@ronacrete.co.uk Web_Site:http://www.ronacrete.co.uk

Roof Units Group Peartree House Pear Tree Lane Dudley W. Midlands DY2 0QU tel:(01384) 74062 fax:(01384) 70435 E_Mail:ru@airmovement.co.uk

Roofing Contractors (Cambridge) Ltd 5 Winship Road Milton Cambridge CB4 6BQ tel:(01223) 423059 fax:(01223) 420737

Roofrite (East Anglia) Ltd The Street Sheering Bishop's Stortford Herts. CM22 7LY tel:(01279) 734515 fax:(01279) 734568

Rosa Paula Kitchens Water Lane Trading Estate Storrington W. Sussex RH20 3DS tel:(01903) 746666 fax:(01903) 742410 E_Mail:info@paularose.com Web_Site:http://www.paularose.com

Rose Emergency Equipment Co Unit 1A, North Orbital Trading Estate Napsbury Lane St Albans Herts. AL1 1XB tel:(01727) 840311 fax:(01727) 844143

Rosengerns Tann Ltd Rosengerns House Whillbury Way Hitchen Herts. SG4 0TY tel:(01462) 471 200

Rotafix Resins Rotafix House Abercraf Swansea SA9 1UX tel:(01639) 730481 fax:(01639) 730858

Rother Boiler Co Ltd Meadow Bank Works Rotherham S61 2ND tel:(01709) 555416 fax:(01709) 564341

Rothley Burn Ltd Macrome Road Wolverhampton West Midlands WV6 9HG tel:(01902) 756461 fax:(01902) 745554 E_Mail:sales@rothley.co.uk Web_Site:http://www.rothley.co.uk

Roto Frank Ltd Swift Point Rugby Warwks. CV21 1QH tel:(01788) 541840 fax:(01788) 541835

Rouse Security Service 345-347 Harrow Road London W9 3RA tel:(0171) 286 1184 fax:(0171) 266 4146

Rovacabin Priest End Works Rycote Lane Thame Oxon OX9 2HD tel:(01844)267200 fax:(01844)267201

Rowberry Technical Services 15 Copse Cross Street Ross-on-Wye Herefordshire HR9 5PD tel:(01989) 563850 fax:(01989) 563854

Royair Ltd Heathpark Honiton Devon EX14 8SE tel:(01404) 41651 fax:(01404) 46227 E_Mail:royairltd@aol.com Web_Site:http://www.royair.co.uk

Royal Dutch Sphinx Ltd Units 1-2 Pipers Court Thatcham Berks. RG19 4ER tel:(01635) 865475 fax:(01635) 861189

Royal Fine Art Commission 7 St James's Square London SW1Y 4JU tel:(0171) 839 6537

Royal Incorporation of Architects in Scotland The 15 Rutland Square Edinburgh EH1 2BE tel:(0131) 229 7545 fax:(0131) 228 2188 E_Mail:rias@rias.org.uk Web_Site:http://www.rias.org.uk

Royal Institute of British Architects (RIBA) 66 Portland Place London W1N 4AD tel:(0171) 580 5533

Royal Institution of Chartered Surveyors 12 Great George Street London SW1P 3AD tel:(0171) 222 7000 fax:(0171) 222 9430 E_Mail:infocentre@rics.org.uk Web_Site:http://www.rics.org

Royal Label Factory, The Unit 9 Goldicote Business Park Goldicote Bridge, Banbury Road Stratford Upon Avon Warwickshire CV37 7N5 tel:(01789) 740557 fax:(01789) 740876

Royal Town Planning Institute 26 Portland Place London W1N 9BE tel:(0171) 636 9107

Royde & Tucker Unit 15-16 High Cross Centre Fountayne Road Tottenham London N15 4QN tel:(0181) 801 7717 fax:(0181) 801 5747

Royston Lead Ltd Pogmoor Works Stocks Lane Barnsley S. Yorks. S75 2DS tel:(01226) 770110 fax:(01226) 730359 E_Mail:info@roystonlead.co.uk Web_Site:http://www.roystonlead.co.uk

Royston Steel Fencing Limited Tadlow Road Tadlow Royston Herts. SG8 0EP tel:(01767) 631721 fax:(01767) 631721

RT Display Systems Ltd 212 New Kings Road London SW6 4NZ tel:(0171) 731 4181 fax:(0171) 731 1907

Ruabon Dennis Ltd Hafod Tileries Ruabon, Wrexham Clwyd LL14 6ET tel:(01978) 843484 fax:(01978) 843276

Ruabon Dennis Ltd Hafod Tileries Ruabon Wrexham LL14 6ET tel:(01978) 843484 fax:(01978) 843276 E_Mail:sales@dennisruabon.co.uk

Ruberoid Architectural Patent Glazing/Cutain Wall Contractor St Mungo Street Bishopbriggs Glasgow G64 1QX tel:(0141) 772 1117 fax:(0141) 762 2217

Ruberoid Building Products Ltd 14 Tewin Road Welwyn Garden City Herts. AL7 1BP tel:(01707) 822222 fax:(01707) 375060

Rubert & Co Ltd Acru Works Demmings Road Cheadle Cheshire SK8 2PG tel:(0161) 428 6058 fax:(0161) 428 1146

Rudgwick Brickworks Co Ltd Lynwick Street Rudgwick Horsham W. Sussex RH12 3DH tel:(01403) 822212 fax:(01403) 823352

Rugby Cement Crown House Rugby Warwks. CV21 2DT tel:(01788) 564345 fax:(01788) 564452 E_Mail:tsd@rugbycement.co.uk Web_Site:http://www.rugbycement.co.uk

Rugby Joinery Watch House Lane Doncaster S. Yorks. DN5 9LR tel:(01302) 394000 fax:(01302) 787383 telex:547160 E_Mail:rugby-joinery@co.uk Web_Site:http://www.rugby-joinery.co.uk

Runtalrad (1970) Ltd The Ridgeway Iver Bucks. SL0 9JQ tel:(01753) 655215 fax:(01753) 630073

Rustin Allen Ltd, Belbien Divsion Mill House and Works Columbia Avenue Edgware Middx. HA8 5DQ tel:(0181) 951 4177 fax:(0181) 951 0801 E_Mail:rustin.allen@dail.pipex.com

Rustins Ltd Waterloo Road Cricklewood London NW2 7TX tel:(0181) 450 4666 fax:(0181) 452 2008 E_Mail:rustins@rustins.co.uk Web_Site:http://www.rustins.co.uk

Ruthin Precast Concrete Ltd 1st. Floor 11 Paul Street Taunton Sumerset TA1 3PF tel:(01824) 274232 fax:(01824) 274231 E_Mail:enquiries@rpcltd.co.uk Web_Site:http://www.rpcltd.co.uk

Ruthin Precast Concrete Ltd Thornfalcon Works Henlade Taunton Somerset TA3 5DN tel:(01823) 444606 fax:(01823) 444616

Ryall & Edwards Limited Green Lane Sawmills Outwood Redhill Surrey RH1 5QP tel:(01342) 842288 fax:(01342) 843312

Rycroft Ltd Dunlombe Road Bradford West Yorkshire BD8 9TB tel:(01274) 490911 fax:(01274) 498580

Ryton's Building Products Ltd Design House, Orion Way Kettering Business Park Kettering Northants NN15 6NL tel:(01536) 511874 fax:(01536) 310455 E_Mail:vents@rytons.com Web_Site:http://www.rytons.com

S & B Ltd Labtec Street Swinton Manchester M27 8SE tel:(0161) 793 9333 fax:(0161) 728 2233

S & P Coil Products Ltd SPC House Evington Valley Road Leicester LE5 5LU tel:(0116) 2490044 fax:(0116) 2490033 E_Mail:spc@spcoils.co.uk

S. C. Johnson Professional Frimley Green Camberley Surrey GU16 5AJ tel:(01276) 852852 fax:(01276) 852800

Saacke Ltd Marshland Spur Farlington Portsmouth PO6 1RX tel:(01705) 383111 fax:(01705) 327120

Sabre Access Control Systems Ltd Unit 4, Lismirrane Industrial Park Elstree Road Elstree Herts. WD6 3EE tel:(0181) 953 6724 fax:(0181) 207 4575

Sabroe Refigeration Ltd Unit 5, The Grand Union Office Park Packet Boat Lane Uxbridge Middx. UB8 2GH tel:(01895) 445661 fax:(01895) 420781 E_Mail:sabroeuxb.aol.com

Sadolin UK Ltd Sadolin House Meadow Lane St Ives Cambs. PE17 4UY tel:(01480) 496868 fax:(01480) 496801 E_Mail:sadolin@akzonobel.co.uk

Safetell Ltd Unit 46, Fawkes Avenue Dartford Trade Park Dartford Kent DA1 1JQ tel:(01322) 223233 fax:(01322) 277751

Safety Stairways Ltd 45 Owen Road Industrial Estate Owen Road Willenhall W. Midlands WV13 2PX tel:(0121) 526 3133 fax:(0121) 526 2833

Sage Brent Ltd Alperton Lane Wembley Middx. HA0 1JX tel:(0181) 998 4791 fax:(0181) 991 0673

Saint-Gobain Glass UK Ltd Waterside Drive Langley Business Park Langley, Slough Berks. SL3 6EZ tel:(01753) 731400 fax:(01753) 731501

Sala Ltd, B H Old Mill Road Portishead Bristol BS20 9BX tel:(01275) 846119 fax:(01275) 849114

Salamandre Davis Hunts Rise South Marston Park South Marston Swindon SN3 4RE tel:(01793) 828000 fax:(01793) 828597

Salex Interiors Ltd Crown Gate Newcomen Way, Severalls Industrial Park Colchester Essex CO4 4YR tel:(01206) 508111 fax:(01206) 852795

Salter Weigh-Tronix George Street West Bromwich W. Midlands B70 6AD tel:(0121) 553 1855 fax:(0121) 553 0494 E_Mail:sales@salter-wtx.co.uk Web_Site:http://www.salter-wtx.co.uk

Samas Roneo Ltd PO Box 10 Hawley Road Dartford Kent DA1 1NY tel:(01322) 223477 fax:(01322) 222890

Sampson & Partners Fencing Aubrey Works, 15 Aubrey Avenue London Colney St.Albans Herts. AL2 1NE tel:(01727) 822222 fax:(01727) 826307 E_Mail:101625.2443@compuserve.com

Sampson T F Ltd Creeting Road Stowmarket Suffolk IP14 5BA tel:(01449) 613535 fax:(01449) 678381 E_Mail:sales@t-f-sampson.co.uk

Sampson Windows Ltd Maitland Road, Lion Barn Business Park Needham Market Ipswich Suffolk IP6 8NZ tel:(01449) 722922 fax:(01449) 722911

Samuel Heath & Sons Cobden Works Leopod Street Birmingham B12 0UJ tel:(0121) 772 2303 fax:(0121) 772 3334 E_Mail:mail@samuel-heath.com Web_Site:http://www.sauel-heath.com

Samways P A & Co The Lodge 1 Pytchley Road Kettering Northants NN15 6NE tel:(01536) 522824 fax:(01536) 416150

Sanderson & Sons Ltd, Arthur 100 Acres Sanderson Road Uxbridge Middx. UB8 1DH tel:(01895) 238244 fax:(01895) 231450

Sandersons Building (Milnrow) Ltd Uncouth Road Works Firgrove Rochdale Lancs. OL16 3DD tel:(01706) 41022 fax:(01706) 59989

Sandhurst MFG Co Ltd Roman Bank Cherry Holt Road Bourne Lincs. PE10 9LQ tel:(01778) 426426 fax:(01778) 393007 E_Mail:info@sandhurst.co.uk

Sandler Seating Ltd 58-64 Three Colts Lane London E2 6JR tel:(0171) 729 4777 fax:(0171) 729 2843 Web_Site:http://www.Sandlerseating.com

Sandoms (Batley) Ltd 491 Bradford Road Batley Yorks. WF17 8LQ tel:(01924) 474108 fax:(01924) 475409

Sandtoft Roof Tiles Ltd Sandtoft Doncaster South Yorkshire DN8 5SY tel:(01427) 871200 fax:(01427) 871222 E_Mail:support@sandtoft.co.uk Web_Site:http://www.sandtoft.co.uk

Sangwin Concrete Products Dansom Lane Hull HU8 7LN tel:(01964) 622339 fax:(01964) 624787

Sanhall Limited Victoria Stables South Road Bourne Lincs. PE10 9JZ tel:(01778) 426655 fax:(01778) 426899

Sanitary Appliances Ltd 3 Sandiford Road Kimpton Road Industrial Estate Sutton Surrey SM3 9RN tel:(0181) 641 0310 fax:(0181) 641 6426 telex:2422REF 2962

Sanua UK Ltd Gratton Way Roundswell Industrial Estate Barnstaple Devon EX31 3NL tel:(01271) 371676 fax:(01271) 371699 E_Mail:swimmer@goldenc.com

Sarena Plastics Ltd Beeching Way Gillingham Kent ME8 6PT tel:(01634) 370887 fax:(01634) 370915 telex:23719 G

Sarnafil Ltd Robberds Way Bowthorpe Industrial Estate Norwich Norfolk NR5 9JF tel:(01603) 748985 fax:(01603) 743054 E_Mail:roofing@sarnafil.co.uk

COMPANY INFORMATION

Sashless Window Co Ltd Standard Way Northallerton N. Yorks. DL6 2XA tel:(01609) 780202 fax:(01609) 779820

Satchwell Control Systems Ltd PO Box 57 Farnham Road Slough Berkshire SL1 4UH tel:(01753) 550550 fax:(01753) 824078 telex:848186 Web_Site:http://www.satchwell.com

Satec (Holdings) Ltd, Public Health Engineering PO Box 12 Weston Road Crewe CW1 1DE tel:(01270) 584012 fax:(01270) 589936 telex:36421 SATEC G

SAV United Kingdom Ltd Scandia House 131 Armfield Close Molesey Surrey KT8 2RJ tel:(0181) 941 4153 fax:(0181) 783 1132

Sawyer & Fisher 30-32 High Street Epsom Surrey KT19 8AH tel:(01372) 742815 fax:(01372) 729710 E_Mail:sawyerfish@aol.com

SB Woodworking Ltd Satinacre Works Stainacre Lane Whitby N. Yorks. YO22 4NN tel:(01947) 603312 fax:(01947) 820667

SC Johnson Wax Frimley Green Camberley Surrey GU16 5AJ tel:(01276) 852000 fax:(01276) 852308

Scandanavian Timber 30a Roxborough Park Harrow on the Hill Middx. HA1 3AY tel:(0181) 864 0131 fax:(0181) 426 9151 E_Mail:scantimber@compuserve.com

Scandanavian Window Systems Ltd Holmelea House Top Street East Drayton Retford Notts. DN22 0LG tel:(01777) 248112 fax:(01777) 248689 E_Mail:scanwin@lineone.net

Scapa Tapes UK Ltd. Gordleton Industrial Park Hannah Way Lymington Hampshire SO41 8JD tel:(01590) 684400 fax:(01590) 683728 E_Mail:jon.kitcher@Scapatapes.com Web_Site:http://www.sellotape.co.uk

Schauman (UK) Ltd Stags End House Gaddesden Row Hemel Hempstead Herts. HP2 6HN tel:(01582) 794661 fax:(01582) 794661

Schindler Ltd Schindler House 3 Plane Tree Crescent Feltham Middx. TW13 7HX tel:(0181) 818 7900 fax:(0181) 818 7999

Schlegel Engineering Henlow Industrial Estate Henlow Camp Bedford Beds. SG16 6DS tel:(01462) 815500 fax:(01462) 811963

Schlumberger Industrial Estate Port Glasgow Renfrewshire PA14 5XG tel:(01475) 745131 fax:(01475) 744567

Schmidlin (UK) Ltd White Lion Court Swan Street Old Isleworth Middx. TW7 6RN tel:(0181) 560 9944 fax:(0181) 568 7081

Schneider (UK) Ten Acres Chobham Road Ottershaw Surrey KT16 0QB tel:(01932) 872137 fax:(01932) 875366

Schneider Ltd Fordhouse Road Wolverhampton Staffs. WV10 9ED tel:(01952) 290029 fax:(01902)303853

Scholes Windows Ltd St Thomas Road Longroyd Bridge Huddersfield W. Yorks. HD1 3LF tel:(01484) 435116 fax:(01484) 435360

Schott Glass Ltd, Arcitectural Glass Drummond Road Stafford ST16 3EL tel:(01785) 223166 fax:(01785) 223522

SCHUCO International KG Whitehall Avenue Kingston Milton Keynes MK10 0AL tel:(01908) 282111 fax:(01908) 282124 E_Mail:schuecosupport@compuserve.com Web_Site:http://www.schueco.de

Scotia Rigging Services (Industrial Stainless Steel Rigging Specialists) 68 Bridge Street Linwood Paisley Renfrewshire PA3 3DR tel:(01505) 321127 fax:(01505) 321333 telex:776766 SCOTIA G

Scott Appleton Furniture Ltd Unit 2 New Rock Industrial Estate Chilcompton Somerset BA3 4JE tel:(01761) 232997 fax:(01761) 233526

Scott Fencing Limited Brunswick Industrial Estate Newcastle-upon-Tyne Tyne & Wear NE13 7BA tel:(0191) 236 5314 fax:(0191) 217 0193

Screen Systems Ltd PO Box 237, The West Site Britannia Works Bewsey Road Warrington WA5 5JZ tel:(01925) 653306 fax:(01925) 571060

Sealmaster Brewery Road Pampisford Cambridge CB2 4HG tel:(01223) 832851 fax:(01223) 837215 E_Mail:sales@sealmaster.co.uk Web_Site:http://www.sealmaster.co.uk

Sealocrete Ltd Queensway Rockdale OL11 2LD tel:(01706) 352 259

Seaque Technologies Ltd Church Road Portadown Co. Armagh BT63 5HU tel:(01762) 333131 fax:(01762) 333042

Sebry Staircase & Ironworks Ltd The Granary Silfield Road Norfolk NR18 9AW tel:(01953) 605983 fax:(01953) 606764

Seco Aluminium Ltd Crittall Road Witham Essex CM8 3AW tel:(01376) 515141 fax:(01376) 500542 E_Mail:seco.aluminium@btinternet.com Web_Site:http://www.btinternet.com/~seco.aluminium/

Securikey Ltd PO Box 18 Aldershot Hampshire GU12 4SL tel:(01252) 311888 fax:(01252) 343950 Web_Site:http://www.securikey.co.uk

Securistyle Ltd Units D & E, Kingsmead Industrial Estate Princess Elizabeth Way Cheltenham Glos. GL51 7RE tel:(01242) 221200 fax:(01242) 520828 E_Mail:info@securistyle.co.uk Web_Site:http://www.securistyle.co.uk

Security Lighting Unit 1A, Southgate Way Orton Southgate Peterborough Cambs. PE2 6YG tel:(01733) 371500 fax:(01733) 371560

Selkin W E Ltd Selray House Ludlow Hill Road West Bridgford Nottingham NG2 6HF tel:(01159) 232286 fax:(01159) 233816

Sellite Blocks Ltd Old Quarry Long Lane Great Heck N. Yorks. DN14 0BT tel:(01977) 661631 fax:(01977) 662155 E_Mail:sales@sellite.co.uk Web_Site:http://www.sellite.co.uk

Seltek Instruments Ltd Leeside Works Lawrence Avenue Stanstead Abbots Herts SG12 8DL tel:(01920) 871094 fax:(01920) 871853 E_Mail:sales@seltekltd.co.uk Web_Site:http://www.seltekltd.co.uk

Sempol Surfaces Ltd Taura House Coppice Side Brownhills WS8 7EY tel:(01543) 370041 fax:(01543) 361041 telex:338212CHACOM G

Senior Air Systems Oldfields Business Park Birrell Street Fenton Stoke-on-Trent ST4 3ES tel:(01782) 599995 fax:(01782) 559220

Servitherm Vale House 100 Vale Road Windsor Berks. SL4 5JL tel:(01753) 829449 fax:(01753) 829448

Servowarm Stuart House Coronation Road High Wycombe Bucks. HP12 3TA tel:(01494) 474474 fax:(01494) 472906 E_Mail:sales@servowarm.co.uk Web_Site:http://www.servowarm.co.uk

SFC (Midlands) Ltd Brooklands 1655 Melton Road East Goscote Leicester LE7 6YQ tel:(0116) 260 1688 fax:(0116) 264 0209

SFS Stadler Ltd Idsall House High Street Cheltenham Glos. GL52 3AX tel:(01242) 585400 fax:(01242) 520682 E_Mail:phamby4824@aol.com Web_Site:http://www.sfs-online.com

SG System Products Ltd Unit 22 Wharfedale Road Ipswich Suffolk IP1 4JP tel:(01473) 240055 fax:(01473) 461616

SGB plc White Lion Court Swan Street Isleworth Middx. tel:(0181) 568 3600 fax:(0181) 568 2929

SGB Youngman Stane Street Slinfold W. Sussex RH13 7RD tel:(01403) 790456 fax:(01403) 790511

Shackerley (Holdings) Group Ltd incorporating Designer ceramics PO Box 20 139 Wigan Road Euxton, Chorley Lancs. PR7 6JJ tel:(01257) 273114 fax:(01257) 262386 E_Mail:enquiries@shackerley.co.uk

Shadbolt F R & Sons Ltd North Circular Road South Chingford London E4 8PZ tel:(0181) 527 6441 fax:(0181) 523 2774 E_Mail:sales@shadbolt.co.uk Web_Site:http://www.shadbolt.co.uk

Shap Concrete Products Ltd Shap Penrith Cumbria CA10 3QQ tel:(01931) 716444 fax:(01931) 716617

Shapland & Petter Ltd Raleigh Works Barnstable N. Devon EX31 2AA tel:(01271) 322501 fax:(01271) 322111 E_Mail:shapland-letter@compuserve.com

Sharman Fencing & Gate Supplies Robinswood Farm Bere Regis Wareham Dorset BH20 7JJ tel:(01929) 472181 fax:(01929) 472182

Sharpe & Fisher Kelston Road Southmead Bristol Avon BS10 5ER tel:(0117) 950 4700 fax:(0117) 950 4500

Shavrin Levatap Co Ltd 32 Waterside Kings Langley Herts. WD4 8HH tel:(01923) 267678 fax:(01923) 265050 E_Mail:shavrin@levatap.freeserve.co.uk

Shaw Arthur Manufacturing Ltd 1 Rose Hill Willenhall W. Midlands WV13 2AS tel:(01902) 368368 fax:(01902) 366766

Shaw Carpets plc PO Box 4 Darton Barnsley S. Yorks. S75 5NH tel:(01226) 390390 fax:(01226) 390549

Shaws of Darwen Waterside Darwen Lancs. BB3 3NX tel:(01254) 775111 fax:(01254) 873462

Sheardown Engineering Ltd 15-17 South Road Templefields Harlow Essex CM20 2AP tel:(01279) 421788 fax:(01279) 435642 E_Mail:sales@sheardown.co.uk Web_Site:http://www.sheardown.co.uk

Shearflow Phoenix Shearflow Works 61 Markfield Road Tottenham London N15 4RE tel:(0181) 808 0571 fax:(0181) 801 7984 E_Mail:shearflow@divcon.co.uk Web_Site:http://www.shearflow.co.uk

Shell Chemical Ireland Ltd Dublin Ireland tel:(0110) 3531 785177 fax:(0110) 3531 767489

Shepherd Engineering Services Ltd Mill Mount York N. Yorks. YO24 1GH tel:(01904) 629151 fax:(01904) 610175 E_Mail:rhpearson@ses-ltd.co.uk Web_Site:http://www.shepherd-group.co.uk

Shipley Fan Co Ltd The Propeller Works 106 Dockfield Road Shipley Yorks. BD17 7BA tel:(01247) 581337 fax:(01274) 531337

Shires Bathrooms Beckside Road Bradford W. Yorks. BD7 2JE tel:(01274) 521199 fax:(01274) 521583 E_Mail:marketing@shires-bathrooms.co.uk Web_Site:http://www.shires-bathrooms.co.uk

Shower Products Ltd Unit 1, High Cross Centre Fountayne Road Tottenham London N15 4QW tel:(0181) 801 4465 fax:(0181) 801 9936 telex:261744

Showerlux UK Ltd Sibree Road Coventry W. Midlands CV3 4EL tel:(01203) 639400 fax:(01203) 305457

ShowerSport Ltd 1 North Place Edinburgh Way Harlow Essex CM20 2SL tel:(01279) 451450 fax:(01279) 635074

Shutter Door Repair & Maintenance Ltd 89-91 Rolfe Street Smethwick Warley W.Midlands B66 2AY tel:(0121) 558 6406 fax:(0121) 558 7140

Sidewinder Concrete Pumps Ltd Unit 1, Grange Works Lomond Road Coatbridge MJ5 2NN tel:(01236) 422075 fax:(01236) 427274

Sidhil Care Boothtown Halifax W. Yorks. HX3 6NT tel:(01422) 363447 fax:(01422) 344270 E_Mail:info@sidhil-care.ltd.uk Web_Site:http://www.sidhil-care.ltd.uk

Siemens Communication Ltd Brickhill Street Willen Lake Milton Keynes MK15 0DJ tel:(01908) 855000 fax:(01908) 855001 Web_Site:http://siemenscomms.co.uk

Siemens Building Technologies Ltd, Cerberus Division Trinity Court Woosehill Wokingham Berks. RG41 3EA tel:(0118) 989 4488 fax:(0118) 979 5750

Sierra Windows Yaberton Industrial Estate Alders Way Paignton Devon TQ4 7QE tel:(01803) 697000 fax:(01803) 697071

Sigma Coatings Ltd Tingewick Road Buckingham Bucks. MK18 1ED tel:(01280) 812081 fax:(01280) 815656 telex:83139 SIGMA G E_Mail:pcuk@sigmacoatings.com Web_Site:http://www.sigmacoatings.com

Signs & Labels Ltd Latham Close, Bredbury Industrial Park Stockport Cheshire SK6 2SD tel:(0800) 132323 fax:(0161) 430 8514 E_Mail:sales@signsandlabel.co.uk Web_Site:http://www.signsandlabels.co.uk

Signs International PO Box 25 Kingswood Bristol BS30 6LU tel:(0117) 932 4049 fax:(0117) 932 4924 telex:+51 449752SIGINT G E_Mail:safetysigns@signsinternational.co.uk Web_Site:http://www.signsinternational.co.uk/signs/

Signs Of The Times Tebworth Leighton Buzzard Beds. LU7 9QG tel:(01525) 874185 fax:(01525) 875746

Sika Ltd Watchmead Welwyn Garden City Herts. AL7 1BQ tel:(01707) 394444 fax:(01707) 393296 E_Mail:technical@uk.sika.com Web_Site:http://www.sika.com

Silavent Ltd 32 Blyth Road Hayes Middx. UB3 1DG tel:(0181) 573 2822 fax:(0181) 573 6621 telex:849041 SHARET G

Silent Gliss Ltd Star Lane Margate Kent CT9 4EF tel:(01843) 863571 fax:(01843) 864503 E_Mail:info@silent.gliss.co.uk

Silkolene Lubricants PLC Silkolene Oil Refinery Belper Derby DE5 1WF tel:(01773) 824151 fax:(01773) 823659

Silverwood Fencing Co. Limited 56 Leeds Road Mirfield W. Yorks. WF14 0DE tel:(01924) 498689 fax:(01924) 498689

Simon R W Ltd System Works Hatchmoor Industrial Estate Torrington Devon EX38 7HP tel:(01805) 623721 fax:(01805) 624578

Simon-Hartley Garner Street Stoke-on-Trent Staffs. ST4 7BH tel:(01782) 202300 fax:(01782) 260534 Web_Site:http://www.hnashe.com/shl.htm

Simons Construction Ltd 401 Monks Road Lincoln Lincs. LN3 4NU tel:(01522) 510000 fax:(01522) 521812

Simplex Lighting Ltd Groveland Road Tipton W. Midlands DY4 7XB tel:(0121) 557 2828 fax:(0121) 557 8900

Simpson Strong-Tie® Winchester Road Cardinal Point Tamworth Staffs. B78 3HG tel:(01827) 255600 fax:(01827) 255616 Web_Site:http://www.strongtie.com

Sinclair Foundry Products Sinclair Works, PO Box 3 Ketley Telford Shropshire TF1 4AD tel:(01952) 641414 fax:(01952) 243760 E_Mail:sales@sinclair-foundry.co.uk Web_Site:http://www.sinclair-foundry.co.uk

Singleton Flint Newland Works Deacon Road Lincoln LN2 4LE tel:(01522) 524542 fax:(01522) 514426

Sissons W & G Ltd Carrwood Road Sheepbridge Chesterfield Derbys. S41 9QB tel:(01246) 450 255 fax:(01246) 451 276 telex:547465 E_Mail:wgs@sissons.co.uk Web_Site:http://www.sissons.co.uk

Skarsten Tools Trafalgar Business Park 1 Baird Road Corby Northants NN17 5ZA tel:(01536) 200976 fax:(01536) 400656

Skarsten Tools Ltd 1 Baird Road Trafalgar Park Corby Northants NN17 5ZA tel:(01536) 200976 fax:(01536) 400656

Skil Controls Ltd Greenhay Place East Gillibrands Skelmersdale Lancs. WN8 9SB tel:(01695) 714600 fax:(01695) 714611 E_Mail:sales@skilcontrols.com Web_Site:http://www.skillcontrols.u-net.com

Slingsby H C plc Preston Street Bradford W. Yorks. BD7 1JF tel:(01274) 721591 fax:(01274) 723044

Slottseal Plc Flemming Road Earlstrees Industrial Estate Corby Northants NN17 2TY tel:(01536) 200555 fax:(01536) 204217

Smart F & G (Shopfittings)Ltd The Shopfitting Centre, Tyseley Industrial Estate Seeleys Road Greet Birmingham B11 2LA tel:(0121) 772 5634 fax:(0121) 766 8995

Smart Systems Ltd North End Road Yatton, nr Bristol BS49 4AW tel:(01934) 876100 fax:(01934) 835169 E_Mail:sales@smartsystems.co.uk Web_Site:http://www.smartsystems.co.uk

Smith & Rodger Ltd 24-36 Elliot Street Glasgow G3 8EA tel:(0141) 248 6341 fax:(0141) 248 6475 E_Mail:postmater@imcaiansmitror.demon.co.uk

Smithbrook Building Products Ltd 18 Smithbrook Kilns Cranleigh Surrey GU6 8JJ tel:(01483) 268808 fax:(01483) 278258

Smoke Control Services Units 24/25 Valley Enterprise Bedwas House Industrial Estate Caerphilly NP1 8DW tel:(01222) 881848 fax:(01222) 881948 E_Mail:enquiry2smokcontrol.co.uk Web_Site:http://www.smokecontrol.co.uk

Snowcem Brookside Off Station Street Longport, Stoke on Trent Staffs ST6 4NF tel:(01782) 813028 fax:(01782) 839028 E_Mail:snowcem@wallmaster.co.uk

Society for the Protection of Ancient Buildings (SPAB) 37 Spital Square London E1 6DY tel:(0171) 377 1644 fax:(0171) 247 5296

Solaglas Laminated Saffron Way Sittingbourne Kent ME10 2PD tel:(01795) 421534 fax:(01795) 473651 telex:965935

Solaglas, Technical Advisory Service Herald Way Binley Coventry CV3 2ND tel:(01203) 458844 fax:(01203) 636473 Web_Site:http://www.glassfact.co.uk

Solair Ltd Pennington Close Albion Road West Bromwich W. Midlands B70 8BA tel:(0121) 525 2722 fax:(0121) 525 6786

Solair GRP Architectural Products Smeaton Road Churchfields Industrial Estate Sailsbury Wilts. SP2 7NQ tel:(01722) 323036 fax:(01722) 337546

Sommerfeld Flexboard Ltd Frame Lane Doseley Telford Shropshire TF4 3BY tel:(01952) 503737 fax:(01952) 630132

Sonneborn & Rieck Ltd Jaxa Works 91-95 Peregrine Road Hainault Ilford IG6 3XH tel:(0181) 500 0251 fax:(0181) 500 3696

Sound Service (Oxford) Ltd 55 West End Witney Oxon. OX8 6NJ tel:(01993) 704981 fax:(01993) 779569 E_Mail:stephen@ssol.freeserve.co.uk

South Durham Structures Limited Dovecot Hill South Church Enterprise Park Bishop Auckland Co. Durham DL14 6XR tel:(01388) 777350 fax:(01388) 775225

South Western Tar Distilleries Totton Works High Street Totton Southampton SO40 9TN tel:(01703) 861169 fax:(01703) 864274

Southern Sanitary Specialists Ltd Cerdic House West Portway Industrial Estate Andover Hampshire SP10 3LF tel:(01264) 324131 fax:(01264) 333476

Southeron James Ltd Martsmith Works Sutton Coldfield W. Midlands B75 7AY tel:(0121) 378 0194 fax:(0121) 378 3438

Sovereign Brush Co Ltd, The 29-43 Sydney Road Watford Herts. WD1 7PZ tel:(01923) 227301 fax:(01923) 817121

Sovereign Fencing Limited Sovereign House 261 Monton Road, Monton Eccles Manchester M30 9LF tel:(0161) 789 5479

Space - Ray UK Chapel Lane Claydon Ipswich Suffolk IP6 0JL tel:(01473) 830551 fax:(01473) 832055 Web_Site:http://www.spaceray.com

Sparkes K & L The Forge Claverdon Nr Warwick Warwks. CV35 8P tel:(01926) 842545 fax:(01926) 842559

Spartan Tiles Ltd Slough Lane Ardleigh Colchester Essex CO7 7RU tel:(01206) 230553 fax:(01206) 230516

Spear & Jackson Interntional Ltd. Atlas Way, Atlasnorth. Sheffield S4 7QQ tel:(0114) 256 1133 fax:(0114) 243 1360

Spectra Glaze Services Unit 7 Rosehill Court St Helier Avenue Morden Surrey SM4 6JT tel:(0181) 640 6737 Fax: (1081) 648 4692 E_Mail:sales@spectraglaze.co.uk Web_Site:http://www.spectraglaze.co.uk

Spectus UPVC Systems Ltd Charter Way Macclesfield Cheshire SK10 2NG tel:(01625) 420400 fax:(01625) 433946

Speed Saws Ltd. Speed Works Lancaster Street Sheffield S3 8AQ tel:(0114) 275 3333 fax:(0114) 279 5511

Spegelstein S & Son 80-84 Wallis Road London E9 5LW tel:(0181) 986 0114 fax:(0181) 985 0706 E_Mail:df@dwo.co.uk Web_Site:http://www.dwo.co.uk

Spencer & Halstead Ltd Ossett W. Yorks. WF5 9AW tel:(01924) 276303 fax:(01924) 277829

Sperrin Metal Products Ltd Cahore Road Draperstown Co. Derry BT45 7AP tel:(01648) 28362 fax:(01648) 28972 E_Mail:raymond@sperrin-metal.com

Spital Tile Co Spital Mills Spital Lane Chesterfield Derbys. S41 0EX tel:(01246) 273657 fax:(01246) 220313

Spontex Ltd St Nicholas Quay Maritime Quarter Swansea SA1 1UT tel:(01792) 482100 fax:(01792) 654009

Sportmark (Leisure Products) , Part of Sportsmark Group Ltd Sportsmark House Ealing Road Brentford Middx. TW8 0LH tel:(0181) 560 2010 fax:(0181) 568 2177 E_Mail:sportsmark@compuserve.com

Springvale E P S 75 Springvale Road Ballyclare Co. Antrim BT39 0SS tel:(01960) 340203 fax:(01960) 341159 E_Mail:sales@springvale.com Web_Site:http://www.springvale.com

Spur Shelving Ltd Spur House Otterspool Way Watford Herts WD2 8HT tel:(01923) 226071 fax:(01923) 241040 telex:22410 Web_Site:http://www.spurshelving.com

Square D, Electrical Distribution Division Fordhouse Road Wolverhampton WV10 9ED tel:(01902) 301150 fax:(01902) 303353

Squire Henry & Sons Linchfield Road New Invention Willenhall W. Midlands WV12 5BD tel:(01922) 476711 fax:(01922) 493490 E_Mail:info@henry-squire.co.uk

Squires Metal Fabrications Ltd Burgess Road Ivyhouse Lane industrial Estate Hastings Sussex

TN35 4NR tel:(01424) 428794 fax:(01424) 431567
SSAB Dobel Coated Steel Ltd Narrow Boat Way Hurst Business Park Brierley Hill W. Midlands DY5 1UF tel:(01384) 74660 fax:(01384) 77575 E_Mail:sales@dobel.co.uk Web_Site:http://www.dobel.co.uk
Stadium Ltd Unit 10, Riverside Industrial Estate Morson Road Ponders End, Enfield Middx. EN3 4TU tel:(0181) 804 4343 fax:(0181) 805 7336
Staines Steel Gate Co 20 Ruskin Road Staines Middx. TW18 2PX tel:(01784) 454456 fax:(01784) 466668
Stainton Metal Co Ltd Dukesway Teeside Industrial Estate Thornaby Cleveland TS17 9LT tel:(01642) 766242 fax:(01642) 765509 E_Mail:genenq@stainton-metal.co.uk
Standard Flat Roofing Co Ltd The Street, Sheering Bishops Stortford Herts. CM22 7LY tel:(01279) 734521 fax:(01279) 734568 E_Mail:standardflatroofing@btinternet.com
Standard Patent Glazing Co Ltd Forge Lane Dewsbury W. Yorks. WF12 9EL tel:(01924) 461213 fax:(01924) 458083
Standby Power Company 311 Lidgett Lane Leeds 17 W. Yorks. LS17 6PD tel:(0113) 2682682 fax:(0113) 2682682
Stanhope Barclay Kellet Crown Division Richter Works Garnett Street Bradford W. Yorks. BD3 9HB tel:(01274) 725351 fax:(01274) 742467
Stanhope Chemical Products Ltd 96 Bridge Road East Welwyn Garden City Herts. AL7 1JW tel:(01707) 324373 fax:(01707) 372191 telex:267418 CATC G
Stanley Access Technologies Halesfield 4 Telford Shropshire TF7 4AP tel:(01952) 682100 fax:(01952) 682101
Stannah Lifts Ltd Anton Mill Andover Hampshire SP10 2NX tel:(01264) 339090 fax:(01264) 337942
Stanton plc Lows Lane Stanton- By- Dale Ilkeston Derbyshire DE7 4QU tel:(0115) 930 5000 fax:(0115) 932 9513 E_Mail:sales@stanton.plc.uk Web_Site:http://www.stanton.plc.uk
Stanton-Thompson (Agencies) Ltd Lys Mill Watlington Oxon. OX9 5EP tel:(01491) 613515 fax:(01491) 613516
Stantons Timber Services Old Golf Course Fishtoft Road Boston Lincs. PE21 OBJ tel:(01205) 362461 fax:(01205) 354488
Staytite Ltd Crown Works Goerge Street High Wycombe Bucks. HP11 2RZ tel:(01494) 462322 fax:(01494) 464747
Steel Line Ltd 2 Long Acre Close Holbrook Sheffield S20 3FR tel:(0114) 2477555 fax:(0114) 2484100
Steel-Crete Ltd Stuart House Church Hill Coleshill West Midlands B46 3AF tel:(01675) 463555 fax:(01675) 467265
Steelcase Strafor Plc Newlands Drive Poyle Berks. SL3 0DX tel:(01753) 680200 fax:(01753) 680345 Web_Site:http://www.steelcase.com
Steelway Fensecure Glynwed Steels & Engineering Ltd Queensgate Works Bilston Road Wolverhampton WV2 2NJ tel:(01902) 452256 fax:(01902) 452256
Stenoak Fencing Stenoak House New Town Uckfield Sussex TN22 5DL tel:(01825) 762266 fax:(01825) 765432
Stent Foundations Ltd Osborn Way Hook Hampshire RG27 9HX tel:(01256) 763161 fax:(01256) 768614
Steralic Filters Ltd 34-36 Oak End Way Gerrards Cross Bucks. SL9 8BR tel:(01753) 884646 fax:(01753) 887163
Stevenson & Cheyne Ltd 60 New Haven Road Leigh Edinburgh EH6 5QT tel:(0131) 554 5211 fax:(0131) 554 1272
Stewart Wales, Somerville Ltd Glenburn Road College Milton East Kilbride Glasgow G74 5BA tel:(013552) 22101 fax:(013552) 33847 E_Mail:info@swsinternational.com
Stiebel Eltron Ltd Lyveden Road Brackmills Northampton NN4 7ED tel:(01604) 766421 fax:(01604) 765283 E_Mail:info@stiebel-eltron.co.uk Web_Site:http://www. Stiebel-eltron.co.uk
Stoakes Systems Ltd 1 Banstead Road Purley Surrey CR8 3EB tel:(0181) 6607667 fax:(0181) 660 5707 E_Mail:mailbox@stoakes.co.uk Web_Site:http://www.stoakes.co.uk
Stockten Heath Forge (Caldwells) Ltd Dallam Lane Warrington Cheshire WA2 7PZ tel:(01925) 636387 fax:(01925) 234158
Stoke Hall Quarry(Stone Sales)Ltd Eyam Road Grindleford Hope Valley S32 2HW tel:(01433) 630313 fax:(01433) 631353
Stokes R J & Co Ltd 12 Moore Street Sheffield S3 7UQ tel:(01142) 737211 fax:(01142) 755656
Stokvis R S & Sons Ltd Pool Road West Molesey Surrey KT8 2HN tel:(0181) 941 1212 fax:(0181) 941 4136 E_Mail:info@stokvis.co.uk Web_Site:http://www.stokvis.co.uk
Stone Fasterners Ltd Woolwich Road London SE7 8SL tel:(0181) 293 5080 fax:(0181) 293 4935
Stonecraft & Chelsea Artisans Ltd Unit C2 Sandown Industrial Park Mill Road Esher Surrey KT10 8BL tel:(01372) 469301 fax:(01372) 470590
Stonewest Ltd Lambert 's Place St James's Road Croydon Surrey CR9 2HX tel:(0181) 684 6646 fax:(0181) 684 9323 E_Mail:stonewest@cwcom.net
Store Development Ltd Upton Road Rugby Warwks. CV22 7DL tel:(01788) 338833 fax:(01788 338832 E_Mail:sdg@storedevelopment.co.uk
Stoves PLC Stoney Lane Prescot Liverpool Merseyside L35 2XW tel:(0151) 426 6551 fax:(0151) 426 3261 Web_Site:http://www.stoves.co.uk
Stowell Concrete Ltd Arnolds Way Yatton, Bristol North Sumerset BS49 4QN tel:(01934) 834000 fax:(01934) 835474
Stramit Industries Ltd Yaxley Eye Suffolk IP23 8BW tel:(01379) 783465 fax:(01379) 783659
Strand Hardware Ltd Strand House, Premier Business Park Long Street Wallsall W. Midlands WS2 9DY tel:(01922) 639111 fax:(01922) 626025 E_Mail:strandhw@aol.com
Stratford Joists Ltd London Industrial Estate 1 Whiting Way E6 6LR tel:(0171) 474 0550
Strebor Diecasting Co Ltd Hutchingson Way Radcliffe Manchester M26 3AR tel:(0161) 723 2661 fax:(0161) 725 9086
Stringer's Charles Sons & Co Ltd Charterhouse Works Northfield Road Coventry CV1 2BL tel:(01203) 226363 fax:(01203) 227080
Structural Bonding Cambridge Ltd Orchard House 22 Nethergate Street Clare Suffolk CO10 8NP tel:(01787) 277888 fax:(01787) 278900 E_Mail:sbcl@cwcom.net
Structural Soils Ltd Chevet House A1 Great North Road Knottenley W. Yorks. WF11 0BS tel:(01977) 674461 fax:(01977) 674461
Structural Timbers Ltd Whitchurch Lane Whitchurch Bristol BS14 0BH tel:(01275) 832724 fax:(01275) 892593
Stuart Turner Ltd Market Place Henley on Thames Oxon. RG9 2AD tel:(01491) 572655 fax:(01491) 573704
Stuarts Industrial Flooring Ltd Stuarts House Church Hill Coleshill Birmingham B46 3AF tel:(01675) 463555 fax:(01675) 467265
Sulzer Infra (UK) Ltd Westmead Farnborough Hampshire GU14 7LP tel:(01252) 544400 fax:(01252)378988
Sumlock J.H.P. Sutherland House Matlock Road Foleshill Coventry CV1 4JQ tel:(01203) 667891 fax:(01203) 668075
Sundeala Division of Celotex Ltd Warwick House 27 St Marys Road Ealing London W5 5PR tel:(0181) 579 0811 fax:(0181) 579 0106 E_Mail:sundeala@celotex.co.uk Web_Site:http://www.celotex.co.uk
Sundwel Solar Ltd 7 Tower Road Glover Washington NE37 2SH tel:(0191) 416 3001 fax:(0191) 416 3001 E_Mail:solar@sundwel.com Web_Site:http://www.sundwel.com
Sunray Engineering Ltd Kingsnorth Industrial Estate Wotton Road Ashford Kent TN23 6LL tel:(01233) 639039 fax:(01233) 625137 E_Mail:sunrayengineering@ukbusiness.com Web_Site:http://www.leonarddo.com/sunray
Sunrite Blinds Ltd Blackchapel Road New Craig Hall Edinburgh EH15 3HL tel:(0131) 669 2345 fax:(0131) 657 3595
Sunway UK Ltd Mersey Industrial Estate Heaton Mersey Stockport Cheshire SK4 3EQ tel:(0161) 432 5303 fax:(0161) 431 5087
Super Cement Ltd 108 High Street Uckfield E. Sussex TN22 1DX tel:(01825) 767125 fax:(01825) 767125
Supra Acoustics Ltd Hainge Road Tividale, Warley W. Midlands B69 2NF tel:(0121) 557 6884 fax:(0121) 557 6884
Supreme Lifts Ltd Finchwell Close Handsworth Sheffield S13 9DF tel:(0114) 243 4241 fax:(0114) 256 0991
Surmak Products Ltd 99 Mabgate Leeds LS9 7DR tel:(01532) 450371 fax:(01532) 428701
Surrey Sackholders Ltd. Trinity Road Richmond Surrey TW9 2LG tel:(0181) 878 7878 fax:(0181) 948 1998
Sussex Brassware Ltd Napier Road Castleham Industrial Estate St Leonards-on-Sea E. Sussex TN38 9NY tel:(01424) 440734 fax:(01424) 853862 E_Mail:sales@sussex-brasework.co.uk Web_Site:http://www.sussex-brassware.co.uk
Sutton Castings Ltd 57 Roe Street Macclesfield Cheshire SK11 6XD tel:(01625) 425911 fax:(01625) 613633
Swanlux Horticultural Services Ltd Unit 10 Courtyard Aurelia Croydon CR0 3BF tel:(0181) 764 6217 fax:(0181) 764 6227
Swedecor Ltd Manchester Street Hull E. Yorks. HU3 4TX tel:(01482) 329691 fax:(01482) 212988 E_Mail:info@swedecor.com Web_Site:http://www.swedecor.com
Swedhouse Ltd PO Box 10 East Horsley Surrey KT24 6TU tel:(01483) 284004 fax:(01483) 285315
Swift Hitch Ltd Trinity Street Kingsmand Trading Estate St Phillips Bristol BS2 0NT tel:(0117) 941 2121 fax:(0117) 941 3093
Swifts Of Scarborough Ltd Cayton Low Road Eastfield Scarborough N. Yorks. YO11 3BY tel:(01723) 583131 fax:(01723) 584625
Swintex Ltd Derby Works Manchester Road Bury Lancs. BL9 9NX tel:(0161) 761 4933 fax:(0161) 797 1146
Swish Building Products Ltd Lichfield Rd Industrial Estate Tamworth Staffs. B79 7TF tel:(01827) 317200 fax:(01827) 317201
SWJ Group Ltd Goodridge Industrial Estate Goodridge Avenue Gloucester Glos. GL2 5EB tel:(01452) 306181 fax:(01452) 300806
Sylvania Lighting International INC. Otley Road Charlestown, Shipley W. Yorks. BD17 7SN tel:(01274) 537777 fax:(01274) 51268 E_Mail:silishipley@mistral.co.uk
Symphony Group PLC, The Gelderd Lane Leeds LS12 6AL tel:(0113) 230 8000 fax:(0113) 230 8134 E_Mail:marketing@symphonygroup.co.uk Web_Site:http://www.symphonygroup.co.uk
Symplex Systems Ltd 418-420 Limpsfield Road Warlingham Surrey CR6 9LA tel:(01883) 624004 fax:(01883) 627201
Syston Rolling Shutters Ltd 33 Albert Street Syston Leicester LE7 2JB tel:(01533) 608841 fax:(01533) 640846
Szerelmey Ltd 369 Kennington Lane Vauxhall London SE11 5QY tel:(0171) 735 9995 fax:(0171) 793 9800
T & D Plastech Bowling House Bowling Iron Works Bradford W. Yorks. BD4 8SX tel:(01274) 707050 fax:(01274) 707060
T-T Pumps & T-T controls Onneley Works Newcastle Road Woore CW3 9RU tel:(01630) 647200 fax:(01630) 647167 E_Mail:response@ttpumps.com Web_Site:http://www.ttpumps.com
TAC Metal Forming Ltd Abbotsfield Road Abbotsfield Industrial Park St Helens Merseyside WA9 4HU tel:(01744) 818181 fax:(01744) 851555 E_Mail:106703.475@compuserve.com Web_Site:http://www.tacmetal.co.uk
Talbot Winnal Valley Road Winchester Hampshire SO23 0LL tel:(01962) 705200 fax:(01962) 841344 E_Mail:talbot_sales@tmproducts.com Web_Site:http://www.talbot.co.uk
Tank Storage & Services Ltd Lilley Farm Thurlow Road Withersfield Suffolk CB9 7SA tel:(01440) 712614 fax:(01440) 712615
Tankard Carpets Ltd York Mills York Street, Fairweather Green Bradford W. Yorks. BD8 0HR tel:(01274) 495646 fax:(01274) 544820
Tann Synchronome Wildmere Road Banbury Oxon Ox16 7JU tel:(01295) 755287 fax:(01295) 755288 E_Mail:enquiries@tannfire.co.uk Web_Site:http://www.tannfire.co.uk
Tarkett Ltd Poyle House, PO Box 173 Blackthorne Road Colnbrook Slough SL3 0AZ tel:(01753) 684533 fax:(01753) 684334
Tarmac Heavy Building Hays Road Ham Hill Snodland Kent ME6 5LA tel:(01634) 242514 fax:(01634) 243927
Tarmac Pellite Ltd Tarmac Wharf Teesport Grangetown Middlesborough TS6 6UG tel:(01642) 452691 fax:(01642) 457793
Tarmac Precast Concretre Tallington Factory Tallington Stamford Lincs. PE9 4RL tel:(01778)381000 fax:(01778)348041
Tarmac Topblock Ltd Wergs Hall Wergs Hall Road Wolverhampton WV8 2HZ tel:(01902) 754131 fax:(01902) 754171
Taskworthy Llantellen Cross Ash Abergavenny Gwent NP7 8LU tel:(01873) 821430 fax:(01873) 821316 E_Mail:taskworthy.co.uk Web_Site:http://www.taskworthy.co.uk
Tate Fencing Chase Wood Works Frant Road, Frant Tunbridge Wells Kent TN3 9HG tel:(01892) 750230 fax:(01892) 750130 E_Mail:sales@tate-fencing.co.uk
Tavac Insulation Ltd Mitchell Hey Mills College Road Rochdale Lancs. OL12 6AE tel:(01706) 715500 fax:(01706) 715511
Taylor & Portway Ltd 52 Broton Drive Trading Estate off Rosemary Lane Halstead Essex CO9 1HB tel:(01787) 472551 fax:(01787) 476589 E_Mail:sales@portway.demon.co.uk
TBA Sealing Materials Ltd PO Box 21 Rochdale Lancs. OL12 7EQ tel:(01706) 715000 fax:(01706) 42284
TBR Joinery Station Road Lenwade Norwich NR9 5LY tel:(01603) 872485 fax:(01603) 871762
TCW Services (Control) Ltd Bradshaw Works Honley Huddersfield Yorks. HD7 2DT tel:(01484) 662865 fax:(01484) 667574 Web_Site:http://www.tcw-services.com
TDI (UK) Ltd Unit 3, Unity Complex Dale Road North Darley Dale, Matlock Derbys. DE4 2HX tel:(01629) 733177 fax:(01629) 732779 Web_Site:http://www.polypipe.plc.uk
Teal & Mackrill Ltd Ellenshaw Works Lockwood Street Hull E. Yorkshire HU2 0HN tel:(01482) 320194 fax:(01482) 219266
Technical Control Systems Ltd Treefield Industrial Estate Gildersome Leeds LS27 7JU tel:(0113) 252 5977 fax:(0113) 238 0095
Technical Indexes Ltd Willoughby Road Bracknell Berks. RG12 8DW tel:(01344) 426311 fax:(01344) 424971 E_Mail:systems@techindex.co.uk Web_Site:http://www.techindex.co.uk
Technology Telford Ltd Halesfield 19 Telford Shropshire TF7 4QT tel:(01952) 585580 fax:(01952) 680240
Technostone Ltd Littleton Lane Shepperton Middx. TW17 0NF tel:(01932) 566999 fax:(01932) 564365
Techrete (UK) Ltd Station Road Scawby Brigg N. Lincolnshire DN20 9AA tel:(01652) 659454 fax:(01652) 659458 E_Mail:ok41@dial.pipex.com
Techspan Systems PLC Church Lane Chalfont St Peter Bucks. SL9 9RF tel:(01753) 889911 fax:(01753) 887496 E_Mail:techspan@dial.pipex.com
Teddington Controls Ltd Daniels Lane Holmbush St Austell Cornwall PL25 3HS tel:(01726) 74400 fax:(01726) 67953 E_Mail:teddingtoncontrols@btinternet.com Web_Site:http://www.teddington-controls.ltd.uk
Tefcote Surface Systems Central House 4 Christchurch Road Bournemouth BH1 3NE tel:(01202) 551212 fax:(01202) 559090 E_Mail:sales@telcote.co.uk Web_Site:http://www.telfcote.co.uk
Tegral Building Materials Ltd The Common Aylesbeare Exeter Devon EX5 2DG tel:(01395) 233225 fax:(01395) 232339
Tektura Plc 1 Heron Quay Docklands London E14 4JA tel:(0171) 536 3300 fax:(0171) 536 3322
Tektura Plc 2 Lapwing Centre Ordsall Lane Salford Quays Manchester M5 3EY tel:(0161) 877 7754 fax:(0161) 877 8626
Telelarmcare Ltd Latour House, Chertsey Boulevard Hanworth Lane Chertsey Surrey KT16 9JR tel:(01932) 577700 fax:(01932) 577744 E_Mail:carlosgomezz@telelarmcare.co.uk
Telford Doors and Barriers Ltd Unit 11, The Bridges Business park Bridge Road Horsham , Telford Shropshire TF4 3EE tel:(01952) 503050 fax:(01952) 5050 E_Mail:telfordoors@demon.co.uk Web_Site:http://www.telfordoors.demon.co.uk
Temperature Ltd Dewar Close Segensworth West, Fareham Hampshire PO15 5UB tel:(01489) 572238 fax:(01489) 573033
Templer C G & Co Ltd 41-9 Stirling Road Acton London W3 8DJ tel:(0181) 992 7777 fax:(0181) 993 7050
Tenby Industries Ltd Great King Street North Birmingham B19 2LF tel:(0121) 515 0515 fax:(0121) 515 0516 E_Mail:tenby.sales@legrand.co.uk
Tensator Ltd, Sales Department Continental House Sherbourne Tilbrook Milton Keynes MK7 8BW tel:(01908) 271153 fax:(01908) 274572 E_Mail:dtuppin@tensator.demon.co.uk Web_Site:http://www.tensator.co.uk
Terminix Peter Cox Ltd Heritage House 234 High Street Sutton Surrey SM1 1NX tel:(0181) 661 6600 fax:(0181) 642 0677
Termstall Limited 50 Burman Street Droylsden Manchester M43 6TE tel:(0161) 370 2835 fax:(0161) 370 9264
Terrapin Ltd Bond Avenue Bletchley Milton Keynes MK1 1JJ tel:(01908) 270900 fax:(01908) 270052 E_Mail:sales@terrapin-ltd.co.uk Web_Site:http://www.terrapin-ltd.co.uk
Terry of Redditch Ltd Millsbro Road Redditch Worcs. B98 7BU tel:(01527) 64261 fax:(01527) 62268
Tex Steel Tubes Ltd Unit 35, Claydon Ind Park Gipping Road Great Blakenham, Ipswich Suffolk IP6 0NL tel:(01473) 830030 fax:(01473) 831664
Textile Bonding Ltd Midland Road Higham Ferrers Northants NN10 8ER tel:(01933) 410100 fax:(01933) 410200
Textron Fastening Systems Ltd Mundells Welwyn Garden City Herts. AL7 1EZ tel:(01707) 668668 fax:(01707) 338828 Web_Site:http://www.advel.textron.com
Textured European Finishes Ltd 1 Dunedin Road Ilford Essex IG1 4LW tel:(0181) 533 1091 fax:(0181) 478 7870 E_Mail:tef.uk@virgin.net Web_Site:http://www.ascwebindex.com/tef
Thakeham Tiles Ltd Rock Road Heath Common Storrington W. Sussex RH20 3AD tel:(01903) 742381 fax:(01903) 746341
Thanet Ware Ltd Ellington Works Princes Road Ramsgate Kent CT11 7RZ tel:(01843) 591076) fax:(01843) 586198
Thatching Advisory Services Ltd Faircross Offices Stratfield Saye Reading Berks. RG7 2BT tel:(01256) 880828 fax:(01256) 880866
The Abbseal Group incorporating Everseal (Thermovitrine Ltd) P O Box 7 Broadway Hyde Cheshire SK14 4QW tel:(0161) 368 5711 fax:(0161) 366 8155
The Building Centre Group Ltd 26 Store Street Camden London WC1E 7BT tel:(0171) 692 4000 fax:(0171) 580 9641 E_Mail:information@buildingcentre.co.uk Web_Site:http://www.buildingcentre.co.uk
The Cotswold Casement Company Fosse Way Industrial Estate Stratford Road Moreton in Marsh Glos. GL56 9NQ tel:(01608) 650568 fax:(01608) 651699
The Cube 14 Holland Street London W8 4LT tel:(0171) 938 2244 fax:(0171) 938 1920 E_Mail:thefrancesdiblasi@ukbusiness.com Web_Site:http://www.ukbusiness.com/thefrances diblasi
The Haigh Tweeny Co Ltd Haigh Industrial Estate Alton Road Ross-on-Wye Herefordshire HR9 5LA tel:(01989) 566222 fax:(01989) 767498 E_Mail:sales@tweeny.co.uk Web site:http://www.tweeny.co.uk
The Kenco Coffee Company Cheltenham Glos GL50 3AE tel:(0800) 242000 fax:(01295) 264602
The Modelshop 151 City Road London EC1V 1JH tel:(0171) 253 1996 fax:(0171) 253 1998 E_Mail:info@modelshop.demon.co.uk Web_Site:http://www.modelshop.co.uk
The National Roof Truss Company 1 Nelson Way Boston Lincs. PE21 8TS tel:(01205) 362468 fax:(01205) 350982
The Nelton Group Ltd New Wellington Street Mill Hill Blackburn Lancs. BB2 4PJ tel:(01254) 262431 fax:(01254) 680008 E_Mail:sales@nelton.co.uk
The Rare Stone Group Ltd 184 Nottingham Road Mansfield Notts. NG18 5AP tel:(01623) 623092 fax:(01623) 622509 E_Mail:enquiries@rarestonegroup.co.uk Web_Site:http://www.rarestonegroup.co.uk
The Safety Tread Ltd Crown Wharf Iron Works Dace Road Old Ford London E3 2NL tel:(0181) 985 4407 fax:(0181) 986 8212

The Salex Group Ltd. Eastgates Colchester Essex CO1 2TW tel:(01206) 866911 fax:(01206) 865987 E_Mail:sales@salexgroup.co.uk Web_Site:http://www.salexgroup.co.uk/salex
The Saw Centre Group 650 Eglinton Street Glasgow G5 9RP tel:(0141) 429 4444 fax:(0141) 429 5609 Web_Site:http://www.thesawcentre.co.uk
Thermal Radiators Ltd Airport Service Road Portsmouth PO3 5PD tel:(01705) 671821 fax:(01705) 664103
Thermalite (Marley Building Materials) Station Road Coleshill Birmingham B46 1HP tel:(01675) 468400 fax:(01675) 468455 telex:335969 E_Mail:Thermalite@mbm-marley.co.uk Web_Site:http://www.marley.co.uk
Thermax Interglass Ltd Greenfield Industrial Estate Tindale Cresent Bishop Auckland Co. Durham DL14 9TF tel:(01388) 603667 fax:(01388) 609744
Thermica Ltd Stoneford House Chamberlain Road Stoneferry Hull HU8 8HH tel:(01482) 329618 fax:(01482) 227723 E_Mail:sales@thermica.co.uk Web_Site:http://www.thermica.co.uk
Therminal Technology (sales) Ltd Bridge House, Station Road Westbury Wilts. BA13 4HS tel:(01373) 865454 fax:(01373)864425
Thermoscreens Ltd Chandlers Ford Industrial Estate Avenger Close Eastleigh Hampshire SO53 4DQ tel:(01703) 254731 fax:(01703) 266563 E_Mail:info@thermoscreens.ltd.co.uk
Thomas & Betts Ltd White Lion Road Amersham Bucks HP7 9JQ tel:(01494) 764661 (X 230) fax:(01494) 766678
Thomerson & Brett Ltd Foster Avenue Woodside Park Dunstable Beds. LU5 5TA tel:(01582) 677000 fax:(01582) 608816
Thompson Kennicott (Incorporating Howl) PO Box 100 Ettingshall Wolverhampton WV4 6JY tel:(01902) 353353 fax:(01902) 354619
Thor Hammer Co Ltd Highlands Road Shirley Birmingham B90 4NJ tel:(0121) 705 4695 fax:(0121) 705 4727 E_Mail:thorhammer@btinternet.com
Thorlux Lighting Merse Road North Moons Moat Redditch Worcs. B98 9HH tel:(01527) 583200 fax:(01527) 584177 E_Mail:thorlux@thorlux.co.uk
Thorn Lighting Ltd 3 King George Close Eastern Avenue West Romford Essex RM7 7PP tel:(01708) 766033 fax:(01708) 776303
Thorn Security Ltd Security House, The Summit Hanworth Road Sunbury on Thames Middx. TW16 5DB tel:(01932) 743333 fax:(01932) 743155 telex:8814916
Thoro Systems Products Ltd 19 Broad Ground Road Lakeside, Redditch Worcs. B98 8YP tel:(01527) 517989 fax:(01527) 510299
ThreeM United Kingdom plc 3M House, PO Box 1 Market Place Bracknell Berks. RG12 1JU tel:(01344) 858000 fax:(01344) 858306 telex:849371 MMMUKA G
Thurston Building Systems Quarry Hill Industrial Estate Hawkingcroft Road Horbury, Wakefield W. Yorks. WF4 6AJ tel:(01924) 265461 fax:(01924) 280246
Thwaites Engineering Co Ltd Welsh Road Works Cubbington Leamington Spa Warwks. tel:(01926) 422471 fax:(01926) 337155 telex:31667
Thyssen Stairlifts Ltd 62 Boston Road Leicester LE4 1AW tel:(0116) 234 4300 fax:(0116) 236 5460 E_Mail:thyssen.stairlifts@btinternet.com Web_Site:http://www.thyssen-stairlifts.co.uk
Tidmarsh & Sons 32 Hyde Way Wewyn Garden City Hertfordshire AL7 3AW tel:(01707) 886226 fax:(01707) 886227 E_Mail:blinds@tidmarsh.co.uk
Tiflex Ltd Tico Works Hipley Street Old Woking Surrey GU22 9LL tel:(01483) 757757 fax:(01483) 757715/755374/771474 E_Mail:hipleystreet@tiflex.co.uk Web_Site:http://www.tiflex.co.uk
Tilcon Ltd Lingerfield Scooton Knaresborough N. Yorks. HG5 9JN tel:(01423) 864041 fax:(01423) 864049
Tilcon South Ltd Buxton Derbys. SK17 8TG tel:(01298) 768555 fax:(01298) 768454
Tileform Ltd, Sales Office Spital Farm Grimsbury Banbury Oxon. OX16 8RZ tel:(01295) 250998 fax:(01295) 271068
Tilley International Ltd 30-32 High Street Frimley Surrey GU16 5JD tel:(01276) 691996 fax:(01276) 27282
Tillicoutry Quarries Ltd Tulliallan Tulliallan, Kincardine By Alloa FK10 4DT tel:(01259) 730481 fax:(01259) 731201
Timber Supplies 60-62 Pretoria Road Patchway Bristol Avon BS12 5PX tel:(0117) 969 1356 fax:(0117) 969 4519
Timber Treatments Ltd Old Mixon Cresent Winterstroke Road Weston-Super-Mare N. Somerset BS24 9AY tel:(01934) 632651 fax:(01934) 623849 E_Mail:retroframe@aol.com Web_Site:http://www.timberwise
Timberwise (UK) Plc Caladonian House Tatton Street Knutsford Cheshire WA16 6AG tel:(01565) 621100 fax:(01565) 621000 E_Mail:hq@timberwise.co.uk Web_Site:http://www.timberwise.co.uk
Timbmet Ltd PO Box 39 Chawley Works Cumnor Hill Oxford OX2 9PP tel:(01865) 862223 fax:(01865) 864367 E_Mail:sales@timbmet.demon.co.uk Web_Site:http://www.gabriel.co.uk/timbmet
Time and Data Systems International Ltd (TDSI) Sentinel House Nuffield Road Poole Dorset BH17 0RE tel:(01202) 666222 fax:(01202) 679730 E_Mail:info@tdsi.co.uk Web_Site:http://www.tdsi.co.uk
Timeguard Ltd Apsley Way London NW2 7UR tel:(0181) 450 8944 fax:(0181) 452 5143 E_Mail:timeguard.com Web_Site:http://www.timeguard.com
Timloc Expamet Building Products Rawcliffe Road Goole E. Yorks. DN14 6UQ tel:(01405) 765567 fax:(01405) 720479
Timtec International Ltd Cotes Park Lane Somercotes, Alfreton Derby DE55 4NJ tel:(01773) 836262 fax:(01773) 835526
Tinsley & Co Ltd, Eliza Reddal Hill Road Cradley Heath W. Midlands B64 5JF tel:(01384) 566066 fax:(01384) 639156 Web_Site:http://www.exportcourier.co.uk/etinsley
Tipper (Hardware) Ltd, Joseph Century Works Moat Street Willenhall W. Midlands WV13 1FZ tel:(01902) 608444 fax:(01902) 608445
Titan Ladder and Case Co Ltd 191-201 Mendip Road Yatton Bristol BS19 4ET tel:(01934) 832161 fax:(01934) 876180
TLC Southern Ltd Unit 10, Chelsea Fields Industrial Estate 278 Western Road Merton London SW19 2QA tel:(0181) 646 6866 fax:(0181) 646 6750
Today Interiors Ltd Hollis Road Grantham Lincs. NG31 7QH tel:(01476) 574401 fax:(01476) 590208 E_Mail:info@today-interiors.co.uk Web_Site:http://www.todayinteriors.internet2.co.uk
Toffolo Jackson(UK) Ltd Burnfield Road Thornliebank Glasgow G46 7TQ tel:(0141) 649 5601 fax:(0141) 632 9314
Tonbridge Fencing Limited Mount Pleasant Road Hildenborough Nr Tonbridge Kent TN11 9JQ tel:(01732) 833191 fax:(01732) 833856
Tonglen Fencing Ltd. West Park Lodge Farm, Broxhill Road Havering-Atte-Bower Romford Essex RM4 1QH tel:(01708) 379722 fax:(01708) 379723
Tony Team Ltd Unit 5 Station Road Bakewell Derbys. DE45 1GE tel:(01629) 813859 fax:(01629) 814334 E_Mail:sales@tonyteam.co.uk Web_Site:http://www.tonyteam.co.uk
Top Coatings Group Portobello Industrial Estate Birtley Co. Durham DH3 2RE tel:(0191) 410 6611 fax:(0191) 492 0125
Top Office Equipment Ltd 7 Heron Industrial Estate Cooks Road London E15 2PW tel:(0181) 519 3330 fax:(0181) 519 5142
Topcon Corporation Boundary Road Newbury RG1 45RR
Tor Coatings Group Portobello Industrial Estate Birtley Chester-le-Street Co. Durham DH3 2RE tel:(0191) 410 6611 fax:(0191) 492 0125
Tormo Ltd Units 4-5, Hurricane Trading Estate Grahame Park Way London NW9 5QY tel:(0181) 205 5533 fax:(0181) 201 3656
Torvale Building Products, A Division of Stadium Group PLC Pembridge Leominster Herefordshire HR6 9LA tel:(01544) 388262 fax:(01544) 388568
Toupret Fillers Agency, Hill & Rednall Ltd 99 Revelstoke Road London SW18 5NL tel:(0181) 946 2701 fax:(0181) 946 2862
Tower Flue Componants Ltd Tower House Vale Rise Tonbridge Kent TN9 1TB tel:(01732) 351555 fax:(01732) 354445
Tower Manufacturing Navigation Road Worcester WR5 3DE tel:(01905) 763012 fax:(01905) 763610 E_Mail:clips@towerman.co.uk Web_Site:http://www.towerman.co.uk
Town & Country Landscapes Ltd Burnsall House 6 Birk Dale Bexhill-On-Sea E. Sussex TN39 3TR tel:(01424) 773834 fax:(01424) 774633
Town and Country Planning Association 17 Carlton House Terrace London SW1Y 5AS tel:(0171) 930 8903 Fax: (0171) 930 3280 E_Mail:tcpa@tcpa.org.uk Web_Site:http://www.tcpa.org.uk
Townscape Products Ltd Fulwood Road South Sutton in Ashfield Notts. NG17 2JZ tel:(01623) 513355 fax:(01623) 440267 E_Mail:townscape-products.co.uk Web_Site:http://www.townscape-products.co.uk
Trada Technology Ltd Stocking Lane Hughenden Valley High Wycombe HP14 4ND tel:(01494) 563091
Trade Fireseal Systems Ltd Unit 4 127a Reading Road Wokingham Berks. RG11 1HD tel:(01734) 894614 fax:(01734) 890107
Trade Sealants 34 Aston Road Waterlooville Hampshire PO7 7XQ tel:(01705) 251321 fax:(01705) 264307 E_Mail:tsl@newnet.co.uk Web_Site:http://www.brit-net.com/sealents
Trans World Enterprises Ltd 101 Stephenson Street Canning Town London E16 4SA tel:(0171) 511 2288 fax:(0171) 511 1466 telex:884227 SLADN G E_Mail:trt@sealandair.com
Transline Ltd Brandsburton Nr Driffield Nr. Humberside YO25 8RW tel:(01964) 542131 fax:(01964) 542971 telex:527068
Transtar Units 1-10, Victoria Industrial Estate Victoria Road West Hebburn Tyne & Wear NE31 1UB tel:(0191) 483 2797 fax:(0191) 428 0262
Travis Perkins Cobtree House Forstal Road Aylesford, Maidstone Kent ME20 7AG tel:(01622) 710111 fax:(01622) 719800
Travis Perkins (Southern) 149 Harrow Road London W2 6NA tel:(0171) 262 6602 fax:(0171) 724 7485
TRC (Midlands) Ltd 1 Mount Pleasant Street West Bromwich W. Midlands B70 7DL tel:(0121) 5006181 fax:(0121) 5005075
TRC Service Group Unit 36, Acorn Industrial Park Crayford Road Crayford Kent DA1 4AL tel:(01322) 555979 fax:(01322) 524804 E_Mail:tnc@btconect.com
Treetex Acoustics Ltd 23 Kelvin Way Trading Estate West Bromich W. Midlands B70 7TW tel:(0121) 525 1111 fax:(0121) 500 5406
Trelleborg Ltd 9 Somers Road Rugby Warwks. CV22 7DB tel:(01788) 532400 fax:(01788) 579959
Tremco Ltd 86-88 Bestobell Road Slough Berks. SL1 4SZ tel:(01753) 691696 fax:(01753) 822640
Trent Bathrooms PO Box 290 Hanley Stoke-on-Trent ST1 3RR tel:(01782) 202334 fax:(01782) 285474
Trespa UK Ltd Grosvenor House Hollinswood, Central Park Telford Shropshire TF2 9TW tel:(01952) 290707 fax:(01952) 290101 Web_Site:http://www.trespa.internet2.co.uk
Trianco Redfyre Ltd Thorncliffe, Chappeltown Sheffield S. Yorks. S30 4PZ tel:(0114) 257 2300 fax:(0114) 245 3021
Trim Acoustics Unit 4, Leaside Industrial Estate Stockingswater Lane Enfield London EN3 7PH tel:(0181) 443 0099 fax:(0181) 443 1919
Trimite Ltd Arundel Road Uxbridge Middx. UB8 2SD tel:(01895) 951234 fax:(01895) 256789 telex:234444 TRIMIT G
Trimite Ltd, Special Coating Division Bush Road Cuxton, Rochester Kent ME2 1HD tel:(01634) 724422 fax:(01634) 711289
Trimite Scotland Ltd 38 Welbeck Road Darnley Industrial Estate Glasgow G53 7RG tel:(0141) 881 9595 fax:(0141) 881 9333
Triton PLC Shepperton Park Caldwell Road Nuneaton Warwks. CV11 4NR tel:(01203) 344441 fax:(01203) 324424
Troika Architectural Mouldings Ltd Troika House Clun Street Sheffield S. Yorks. S4 7JS tel:(01142) 753222 fax:(01142) 753781
Tropravit (UK) Ltd Unit B2, Hubert Road Hubert Road Industrial Estate Brentwood Essex CM14 4JY tel:(01277) 218282 fax:(01277) 218284
Trox (UK) Ltd Caxton Way Thetford Norfolk IP24 3SQ tel:(01842) 754545 fax:(01842) 763057
Truesigns Limited 204 Oldbury Road West Bromwich W. Midlands B70 9DE tel:(0121) 553 2695 fax:(0121) 580 0238
TSE Brownson Valmar Trading Estate Valmar Road Camberwell, London SE5 9NP tel:(0171) 274 4577 fax:(0171) 978 8141 telex:262433
Tube Products (A Division of Senior Tube) PO Box 13 Popes Lane Oldbury Warley B69 4PF tel:(0121) 552 1511 fax:(0121) 544 5735
Tubeclamps Ltd PO Box 41 Petford Street Cradley Heath, Warley W. Midlands B64 6EJ tel:(01384) 565241 fax:(01384) 410490 telex:337462
Tubela Engineering Co Ltd 2-6 Fowler Road Hainault, Ilford Essex IG6 3UP tel:(0181) 500 1253 fax:(0181) 5001259
Tucker Fasteners Ltd Emhart Fastening Teknologies Walsall Road Birmingham W. Midlands B42 1BP tel:(0121) 356 4811 fax:(0121) 356 1598 telex:337680 Web_Site:http://www.emhart.com
Tudor RooF Tile CO Ltd Denge Marsh Road Lydd Kent TN29 9JH tel:(01797) 320202 fax:(01797) 320700
Tufton Ltd High Holborn Road Ripley Derbys. DE5 3NT tel:(01773) 740620 fax:(01773) 740629
Tuke & Bell Ltd Lomard house No 1 Cross Keys Lichfield Staffs. WS1 6DN tel:(01543) 414161 fax:(01543) 250462
Tungstone Batteries Ltd Lathkill Street Market Harborough Leics. LE16 9EZ tel:(01858) 410900 fax:(01858) 410505
Tunnel Building Products Ltd PO Box 1 Westerham Kent TN16 2BY tel:(01959) 561776 fax:(01959) 561776
Tunstal Telecom Ltd Whitley Lodge Whitley Bridge N. Yorks. DN14 0HR tel:(01977) 661234 fax:(01977) 661993 E_Mail:sales @ tunstall.co.uk Web_Site:http://www.tunstallgroup.com
Turnbull & Scott (Engineers) Ltd 20 Commercial Road Hawick Roxburghshire TD9 7AQ tel:(01450) 372053 fax:(01450) 377800
Turner Access Ltd 65 Craigton Road Glasgow G51 3EQ tel:(0141) 309 5555 fax:(0141) 309 5436
Turner Plus Eight Ltd 65 Craigton Road Glasgow G51 3EQ tel:(0141) 440 0663 fax:(0141) 425 1073
Turney Turbines (pumps) Ltd High Mead Works 73 Station Road Harrow Middx. HA1 2TZ tel:(0181) 427 1355 fax:(0181) 863 4215
TWI Abington Hall Abington Cambridge CB1 6AL tel:(01223) 891162 fax:(10223) 892588
Twide-Paragon Ltd Glasshouse Fields Stepney London E1 9JA tel:(0171) 790 2333 fax:(0171) 790 0201
Twiflex Ltd The Green Twickenham Middx. tel:(0181) 894 1161 fax:(0181) 894 6056 telex:261704 TWIFLX G
Twil Ltd PO Box 119 Shepcote Lane Sheffield S9 1TY tel:(0114) 256 1561 fax:(0114) 261 9351
TWIL Wire Products Ltd PO Box 119 Sheffield S. Yorks. S9 1TY tel:(01742) 561561 fax:(01742) 425947 telex:54136
Twin County Fan Company Ltd Hirwaun Industrial Estate Hirwaun Aberdare M. Glam. CF44 9YA tel:(01685) 813454 fax:(01685) 813325
Twin Technology Lighting PO Box 207 Brixworth Northampton NN6 9HU tel:(01604) 881234 fax:(01604) 882217
TWS Presentation Systems (Mcmillan UK Ltd T/A) 185 Walton Summit Centre Bamber Bridge Preston Lancs. PR5 8AJ tel:(01772) 337249 fax:(01772) 626830
Tylo UK Ltd Gratton Way Roundswell Industrial Estate Barnstaple Devon EX31 3NL tel:(01271) 371676 fax:(01271) 371699 E_Mail:swimmer@goldenc.com
Tyrrell Tanks Ltd 37 Seagoe Industrial Estate Portadown Co. Armagh BT63 5QD tel:(01762) 330668 fax:(01762) 350171
Ubbink (UK) Ltd Borough Road Brackley Northants NN13 7TB tel:(01280) 700211 fax:(01280) 705332 E_Mail:admin@ubbink.demon.co.uk
UK Cabin Co Kings Court Kingsway TVTE Gateshead NE11 0SH tel:(0191) 461 1000 fax:(0191) 461 1001
UK Fasteners Ltd C1 Liddington Trading Estate Leckhampton Road Cheltenham Glos. GL53 0DL tel:(01242) 577077 fax:(01242) 577078
UK Fire International Ltd The Safety Centre Mountergate Norwich Norfolk NR1 1PY tel:(01603) 727000 fax:(01603) 727072 E_Mail:admin@ukfire.co.uk
Ulster Carpet Mills Ltd Castleisland Factory Portadown Co. Armagh BT62 1EE tel:(01762) 334433 fax:(01762) 333142 Web_Site:http://www.ulstercarpets.com
Ultraframe PLC Salthill Road Clitheroe Lancs. BB7 1PE tel:(01200) 443311 fax:(01200) 425455 E_Mail:marketing@ultraframe.co.uk Web_Site:http://www.ultraframe.co.uk/ultra
Unicorn Abrasives (UK) Ltd Doxey road Stafford ST16 1EA tel:(01785) 223281 fax:(01785) 213487 E_Mail:info@unicorn-abrasives.co.uk Web_Site:http://www.unicorn-abrasives.co.uk
Unicorn Containers Ltd Banbridge Road Waringstown Co. Armagh BT66 7QB tel:(01762) 881346 fax:(01762) 882254
Unifix Ltd Bridge House Grove Lane Smethwick, Warley W. Midlands B66 2SA tel:(0121) 609 0099 fax:(0121) 626 0586 E_Mail:unifix@dial.pipex.com
Unilock Ltd Napier Road, Castleham Industrial Estate St Leonards-on-Sea E. Sussex TN38 9NY tel:(01424) 853362 fax:(01424) 852390 E_Mail:unilock_hastings@btinternet.com
Unistrut Limited Edison Road Elms Industrial Estate Bedford Beds. MK41 0HU tel:(01234) 211331 fax:(01234) 270819
United Flexibles (A Division of Senior Flex) Abercanaid Merthyr Tydfil M. Glam. CF48 1UX tel:(01685) 385641 fax:(01685) 389683 telex:497533
United Safety Products UK Ltd Unit 15 Bramley Business Centre Station Road Bramley Surrey GU6 0AZ tel:(01483) 898 666 fax:(01483) 894 666 E_Mail:uspuk@btinternet.com
United Wire Ltd Granton Edinburgh EH5 1HT tel:(0131) 552 6241 fax:(0131) 552 8462 telex:72116 UNITED G E_Mail:info@unitedwire.com Web_Site:http://www.unitedwire.com
Universal Components Ltd Universal House Pennywell Road Bristol Avon BS5 0ER tel:(0117) 955 9091 fax:(0117) 955 6091 E_Mail:sales@universal-aluminiaum.co.uk
Universal Filters Ltd Unfil House, Unit 12, The Woodford Trading Estate Southend Road Woodford Green Essex IG8 8HF tel:(0181) 551 4447 fax:(0181) 550 9101
Uniwall-Lockwall 34 Westside Watford Herts. WD1 8SB tel:(01923) 245612 fax:(01923) 245408
Uponor Ltd Hillcote Plant, PO Box 1 Blackwell, Nr Alfreton Derbys. DE55 5JD tel:(01773) 811112 fax:(01773) 812343 Web_Site:http://www.uponor.co.uk
Urban Planters 202 Pasture Lane Bradford W. Yorks. BD7 2SE tel:(01274) 579331 fax:(01274) 521150 E_Mail:sales@urbanplanters.co.uk
Urbis Lighting Ltd Telford Road Houndsmill, Basingstoke Hampshire RG21 6YW tel:(01256) 354446 fax:(01256) 841314
USF Permutit/lonpure Harforde Court John Tate Road Hertford Herts. SG13 7NW tel:(01992) 823 300 fax:(01992) 501528
USG (UK) Ltd 1 Swan Road South West Industrial Estate Peterlee Co. Durham SR8 2HS tel:(0191) 586 1121 fax:(0191) 586 0097
V B Johnson & Partners St John's House 304-310 St Albans Road Watford Herts. WD2 5PE tel:(01923) 227236 fax:(01923) 231134 E_Mail:office@vbjw.demon.co.uk Web_Site:http://www.vbjw.demon.co.uk
Vallance Leeds 27, Trading Estate Moreley Leeds LS27 0LL tel:(0113) 253 7211 fax:(0113) 252 5829 E_Mail:richard.beene@kalon.co.uk
Valor Heating Wood Lane Erdington Birmingham B24 9QP tel:(0121) 373 8111 fax:(0121) 373 8181
Valspar UK Corporation Ltd, The Avenue 1 Station Lane Witeney Oxon. OX8 6XZ tel:(01993) 707400 fax:(01993) 775579

Valtti Specialist Coatings Ltd Unit 3B South Gyle Cresent Lane Edinburgh EH12 9EG tel:(0131) 334 4999 fax:(0131) 334 3987

Van Leer (UK) Ltd Meadow Lane Works Ellesmere Port South Wirral L65 4EU tel:(0151) 355 3341 fax:(0151) 356 4219

Vandex (UK) Ltd PO Box 88 Leatherhead Surrey KT22 7YF tel:(01372) 363040 fax:(01372) 363373 E_Mail:tech@vandex.demon.co.uk

Vandgard Ltd PO Box 51 Edenbridge Kent TN8 7QH tel:(01732) 865901 fax:(01732) 867567 E_Mail:vandgard@easynet.co.uk Web_Site:http://www./hi-media.co.uk/vandgard

Vanesco Ltd 165 Garth Road Morden Surrey SM4 4LH tel:(0181) 330 0101 fax:(0181) 337 5532

Vantrunk Engineering Ltd Goodard Road Astmoor Runcorn Cheshire WA7 1QF tel:(01928) 564211 fax:(01928) 580157 E_Mail:sales@vantrunk.co.uk

Varley & Gulliver Ltd 57-70 Alfred Street Sparkbrook Birmingham B12 8JR tel:(0121) 773 2441 fax:(0121) 776 6875

Veha Radiators Ltd Spinning Jenny Way Leigh Lancs. WN7 4PE tel:(01942) 261777 fax:(01942) 262200

Veka PLC Farrington Road Rossendale Road Indstrial Estate Burnley Lancs. BB11 5DA tel:(01282) 416611 fax:(01282) 439260 telex:635571 VEKAVK

Velfac Ltd Merlin Place Milton Road Cambridge CB4 4DP tel:(01223) 426606 fax:(01223) 426607

VELUX Co Ltd, The Woodside Way Glenrothes E. Fife KY7 4ND tel:(01592) 772211 fax:(01592) 771839 E_Mail:velux-gb@compuserve.com Web_Site:http://www.velux.co.uk

Vencel Resil Ltd Arndale House 18-20 Spital Street Dartford Kent DA1 2HT tel:(01322) 626600 fax:(01322) 626610

Vent-Axia Flemming Way Crawley W. Sussex RH10 2NN tel:(01293) 526062 fax:(01293) 551188 E_Mail:info@vent-axia.com Web_Site:http://www.vent-axia.com

Vent-Axia Ltd Fleming Way Crawley W. Sussex RH10 2NN tel:(01293) 526062 fax:(01293) 551188 E_Mail:info@vent-axia.com Web_Site:http://www.vent-axia.com

Ventlane Ltd Betha Street Bolton Gtr. Manchester BL1 8AH tel:(01204) 365858 fax:(01204) 364440

Verco Office Furniture Ltd Chapel lane Sands High Wycombe Bucks. HP12 4BG tel:(01494) 448000 fax:(01494) 464216 E_Mail:info@verco.co.uk Web_Site:http://www.verco.co.uk

Verine Ltd Folly Faunts House Goldhanger Maldon Essex CM9 8AP tel:(01621) 788611 fax:(01621) 788754 E_Mail:verine@dial.pipex.com Web_Site:http://www.verine.co.uk/verine/

Verna Ltd Folds Road Bolton Lancs. BL1 2TX tel:(01204) 529494 fax:(01204) 521862

Versatile Fittings Ltd Bicester Road Aylesbury Bucks. HP19 3AU tel:(01296) 483481 fax:(01296) 437596 E_Mail:general@versatile-fittings.ibmail.com

Vertika International Ltd PO Box 66 Macclesfield Cheshire SK11 7PR tel:(01625) 611622 fax:(01625) 501510

Ves Andover Ltd 10-12 Crown Way Walworth Industrial Estate Andover Hampshire SP10 5LU tel:(01264) 366325 fax:(01264) 333115 E_Mail:vesltd@interalpha.co.uk Web_Site:http://www.wes.co.uk

Vescom (UK) Ltd Witan Court 274 Witan Gate West Central Milton Keynes MK9 1EJ tel:(01908) 696222 fax:(01908) 696111

Vianen Kitchen Ventilation (UK) Ltd. Welton Road Warwick CV34 5PZ tel:(01926) 496644 fax:(01926) 493977 E_Mail:vainen@ao.com

Vicaima Ltd Marlowe Avenue Greenbridge Industrial Estate Swindon Wilts. SN3 3JF tel:(01793) 532333 fax:(01793) 530193

Vicon Industries (UK) Ltd Brunel Way Fareham Hampshire PO15 5TX tel:(01489) 566300 fax:(01489) 566322 Web_Site:http://www.vicon-cctv.com

Victaulic Systems PO Box 13 46-48 Wibury Way Hitchin Herts. SG4 0UD tel:(01462) 422622 fax:(01462) 431670 E_Mail:info@victaulic.co.uk Web_Site:http://www.victaulic.co.uk

Vidionics Security Systems Ltd Systems House, Desborough Industrial Park Desborough Park Road High Wycombe Bucks. HP12 3BG tel:(01494) 459606 fax:(01494) 461936 E_Mail:vidionics@pixelgroup.co.uk

Viking Johnson PO Box 13 46-48 Wilbury Way Hitchin Herts. SG4 0UD tel:(01462) 443322 fax:(01462) 443311 E_Mail:info@vikingjohnson.co.uk Web_Site:http://www.vikingjohnson.co.uk

Viking Laminates Ltd Spanbourne Avenue Chippenham Wilts. SN15 1LQ tel:(01249) 656050 fax:(01249) 659633 Web_Site:http://www.viklam.co.uk

Viking Security Systems Ltd Security House, 4 Lismirrane Industrial Park Elstree Road Elstree Herts. WD6 3EE tel:(0181) 207 3838 fax:(0181) 2074575 E_Mail:100140.1364@compuserve.com Web_Site:http://www.vikingsecurity.co.uk

Visco Ltd Stafford Road Croydon Surrey CR9 4DT tel:(0181) 686 3861 fax:(0181) 681 5930 E_Mail:info@visco.co.uk

Viscount Catering Ltd PO Box 16 Green Lane Eccelsfield, Sheffield S. Yorks. S35 9ZY tel:(0114) 257 0100 fax:(0114) 257 0257 E_Mail:sales@viscount.demon.co.uk Web_Site:http://www.viscount-catering.co.uk

Vision Sytems (Europe) Ltd Vision House, Focus 31 Mark Road Hemel Hempstead Herts. HP2 7BW tel:(01442) 242330 fax:(01442) 252619

Visqueen Agri PO Box 343 Yarm Road Stockton-on-Tees Cleveland TS18 3GE tel:(01642) 672288 fax:(01642) 664325

Vista Engineering Ltd Carr Brook Works Shallcross Mill Road Whaley Bridge High Peak SK23 7JL tel:(01663) 734524 fax:(01663) 732403 E_Mail:106656.1263@compuserve.com

Vista Plan International PLC High March Daventry Northants NN11 4QE tel:(01327) 704767 fax:(01327) 300243

Visua Ltd Chalfont Grove, Narcot Lane Chalfont St Peter Gerrards Cross Bucks. SL9 8TN tel:(01494) 878384 fax:(01494) 878006 E_Mail:peterw@visua.com Web_Site:http://www.visua.com

Vita Liquid Polymers Ltd Harling Road Wythenshaw Manchester M22 4TP tel:(0161) 998 3226

Vitalighting Ltd Unit 4, Sutherland Court Moor Park Industrial Estate, Tolpits Lane Watford Herts. WD1 8SP tel:(01923) 896476 fax:(018923) 897741

Vitra (UK) Ltd 121 Milton Park Milton, Abingdon Oxon. OX14 4SA tel:(01235) 820400 fax:(01235) 820404

Vitra UK Ltd 13 Grosvenor Street London W1X 9FB tel:(0171) 408 1122 fax:(0171) 499 1967

Vodafone Paging Ltd 21-22 Park Way Newbury Berks. RG14 1EE tel:(01635) 521800 fax:(01635) 523016 Web_Site:http://www.vodafone.co.uk

Volcrepe Ltd V.C. Works High Street East Glossop Derbys. SK13 8QB tel:(01457) 853254 fax:(01457) 869470 E_Mail:info@volcrepe.com Web_Site:http://www.volcrepe.com

Volex Group PLC Dornoch House Kelvin Close Birchwood Warrington WA3 7JX tel:(01925) 830101 fax:(01925) 830141

Vortice Ltd Milley Lane Hare Hatch Reading Berks. RG10 9TH tel:(0118) 9404211 fax:(0118) 9403787

VT Plastics Ltd Shoreham House Shoreham Street Sheffield S. Yorks. S1 4SR tel:(01742) 757161 fax:(01742) 758920

Vulcan Plastics Ltd Hosey Hill Westerham Kent TN16 1TZ tel:(01959) 562304

Vulcan Tanks Ltd Cates Park Lane Coates Park Industrial Estate Alfreton Derbys. DE55 4NJ tel:(01773) 835321 fax:(01773) 836578

Vulcanite Limited High Street Crigglestone Wakefield WF4 3HT tel:(01924) 240404/257716 fax:(01924) 252006/240435 telex:55114

Vulcascot Ltd Gatwick Gate Estate Lowfield Heath Crawley W. Sussex RH11 0TG tel:(01293) 560130 fax:(01293) 537743

W Ward Fencing Contractors K & M Hauliers Indust. Estate The Aerodrome, Watnall Road Hucknall Notts. NG15 6EQ tel:(0115) 963 6948 fax:(0115) 963 0864

WA Skinner & Co Limited Dorset Way, off Abbot Close Byfleet Surrey KT14 7LB tel:(01932) 344228 fax:(01932) 348517

Wade Building Services Ltd Groveland Road Tipton W. Midlands DY4 7TN tel:(0121) 520 8121 fax:(0121) 557 7061

Wade Ceramics Ltd Westport Road Burslem Stoke-on-Trent ST6 4AP tel:(01782) 577321 fax:(01782) 575195

Wade International (UK) Ltd 20 Broton Drive Halstead Essex CO9 1HE tel:(01787) 475151 fax:(01787) 475579

Wagner (GB) Ltd VBH House Bailey Drive, Gillingham Business Park Gillingham Kent ME8 0WG tel:(01634) 263300 fax:(01634) 263504

Walker James & Co Ltd Lion House Woking Surrey GU22 8AP tel:(01483) 757575 fax:(01483) 755711

Walker Management Services PO Box 525 Leigh on Sea Essex SS9 5RN tel:(01702) 523628 fax:(01702) 77161

Walker Timber Ltd Carriden Sawmills Bo'ness W. Lothian EH51 9SQ tel:(01506) 823331 fax:(01506) 822590 E_Mail:walker.timber.ltd@dial.pipex.com

Wallbank Wallbank Farm, Wallbank Road, Bramhall, Cheshire SK7 3AP tel:(0161) 439 0908 fax:(0161) 439 0908

Wallis Office Furniture Ltd 8-18 Fowler Road Hainault Industrial Estate Ilford Essex IG6 3UP tel:(0181) 500 9991 fax:(0181) 500 1949

Walls & Ceiling International Ltd Tything Road Arden Forest Industrial Estate Alcester Warwks. B49 6EP tel:(01789) 763727 fax:(01789) 400312 Web_Site:http://www.walls-and-ceilings.co.uk

Walsall Cable Management Ltd Dial Lane Hill Top Bromwich W. Midlands B70 0EB tel:(0121) 5222222 fax:(0121) 5222251 E_Mail:sales@walsall-cable-management.co.uk

Wandsworth Elecrtrical Ltd Albert Drive Sheerwater Woking Surrey GU21 5SE tel:(01483) 740740 fax:(01483) 740384 E_Mail:info@wandsworth-electrical.com Web_Site:http://www.wandsworth-electrical.com

Wanson & Co Ltd 7 Elstree Way Borehamwood Herts. WD6 1SA tel:(0181) 953 7111 fax:(0181) 207 5177

Ward Bekker Sales Ltd Northgate White Lund Estate Morecambe Lancs. LA3 3PA tel:(01524) 63233 fax:(01524) 65792 E_Mail:info@beeline-rex.co.uk Web_Site:http://www.beeline-rex.co.uk

Ward Building Components Ltd Sherburn Malton N. Yorks. YO17 8PQ tel:(01944) 712000 fax:(01944) 710555 E_Mail:wbc@wards.co.uk Web_Site:http://www.wards.co.uk

Ward's Flexible Rod Company Ltd 22 James Carter Road Mildenhall Suffolk IP28 7DE tel:(01638) 713800 fax:(01638) 716863 E_Mail:sales@wardflex.co..uk

Wardle Storeys Grove Mill Earby via Colne Lancs. BB8 6UT tel:(01282) 842511 fax:(01282) 843170

Warefence Limited Clare Terrace Carterton South Industrial Estate Carterton Oxon. OX18 3ES tel:(01993)845279 fax:(01993)8400551

Wareing Bros & Co Ltd Carlton Street Works Bolton Lancs. BL2 1DJ tel:(01204) 521566 fax:(01204) 361491

Warner H & Son Ltd Arclion House Hadleigh Road Ipswich Suffolk IP2 0EQ tel:(01473) 253702 fax:(01473) 233330 telex:987703 HWS Web_Site:http://www.warners-engineers.co.uk

Warner Howard Ltd 170-172 Honeypot Lane Stanmore Middx. HA7 1EE tel:(0181) 206 2900 fax:(0181) 206 1313 E_Mail:info@warnerhoward.co.uk Web_Site:http://www.warnerhoward.w.uk

Warsop-Metrix Hever Road Edenbridge Kent TN8 5DL tel:(01732) 863081 fax:(01732) 865680

Watco UK Ltd Watco House Filmer Grove Godalming Surrey GU7 3AL tel:(01483) 425000 fax:(01483) 428888 E_Mail:sales@watco.co.uk

Waterhouse Denbigh Ltd Bolton Wood Quarries Bolton Hall Road Bradford BD2 1BQ tel:(01274) 593433 fax:(01274) 582267

Watkins Litchfield Auction Centre Mradley Staffs. WS13 89F tel:(01543) 250088 fax:(01543) 415254

Wavin Building Products Ltd Parsonage Way Chippenham Wilts. SN15 5PN tel:(01249) 654121 fax:(01249) 443286 E_Mail:wbpgeneqy@wavin.com Web_Site:http://www.wavin.co.uk

Wearparts Ltd Oaks Industrial Estate Gilmorton Road Lutterworth Leics. LE17 4HA tel:(01455) 558107 fax:(01455) 554838

Weatherley Fencing Contractors Limited The Orchard 135 North Cray Road Sidcup Kent DA14 5HE tel:(0181) 308 6421 fax:(0181) 308 1317

Webrr & Broutin Dickens House Maulden Road Flitwick Bedford MK45 5BY tel:(01525) 718877 fax:(01525) 718988 E_Mail:mail@sbd.ltd.uk

Wednesbury Fencing Co Limited Hollybush Farm Warstone Road Shareshill Wolverhampton WV10 7LX tel:(01922) 417648 fax:(01922) 413420

Wednesbury Tube Oxford Street Bilston W. Midlands WV14 7DS tel:(01902) 491133 fax:(01902) 405838

Welconstruct Co Woodgate Business Park Kettles Wood Drive Birmingham B32 3GH tel:(0121) 421 9000 fax:(0121) 421 9888 E_Mail:mail@welconstruct.co.uk Web_Site:http://www.welconstruc.co.uk

Wellman Robey Ltd Newfield Road off Dudley Road East Oldbury, Warley W. Midlands B69 3ET tel:(0121) 552 3311 fax:(0121) 552 4571 telex:338711 E_Mail:sales@wellmanrobey.com Web_Site:http://www.wellmanrobey.com

Wendland GB Ltd, Division of Caradon Doors & Windows Ltd Central Way Swindon Road Cheltenham Glos. GL5 9LY tel:(0990) 388377 fax:(01242) 239604

Wenmore Rooflights Ltd Hatters Lane Chipping Sodbury Bristol BS37 6AA tel:(01454) 318995 fax:(01454) 311569

Wentworth Sawmills Ltd, Barrowfield Lane, Wentworth, Nr. Rotherham S. Yorks. S62 7TP tel:(01226) 742206 fax:(01226) 742484

Wernick S & Sons Ltd Molineux House Russell Gardens Wickford Essex SS11 8BL tel:(01268) 735544 fax:(01268) 560026

Wescol Glosford Westercroft Lane Northowram Halifax HX3 7TY tel:(01422) 203522 fax:(01422) 205365

Wessenden Products Ltd Prospect Road Alresford Hampshire SO24 9QQ tel:(01962) 732867 fax:(01962) 735331

Wessex Medical Equipment Ltd Budds Lane Romsey Hampshire SO51 0HA tel:(01794) 830303 fax:(01794) 512621

Wessex Resins & Adhesives Ltd Cuperham House Cuperham Lane Romsey Hants. SO51 7LF tel:(01794) 521111

West London Security Security House 19 Stannary Street London SE11 4AA tel:(0171) 823 2077 fax:(0171) 582 6266

Westbrick Tarmac Bricks & Tiles Ltd Pinhoe Exeter Devon EX4 8JT tel:(01392) 66561 fax:(01392) 68188

Westbury Filters Limited The Studio New Chapel Road Lingfield Surrey RH7 6LE tel:(0181) 665 5525 fax:(0181) 665 5572 E_Mail:sales@westburnfilters.co.uk Web_Site:http://www.westburnfilter.co.uk

Westco Group Ltd Penarth Road Cardiff CF1 7YN tel:(01222) 233926 fax:(01222) 383573 E_Mail:westcodly@aol.com Web_Site:http://www.estcofloors.co.uk

Western Avery Ltd Levolux House 24 Easville Close Eastern Avenue Gloucester GL2 5EU tel:(01452) 500007 fax:(01452) 527496 E_Mail:113675.2405@compuserve.com

Western Cork Ltd Penarth Road Cardiff CF1 7YN tel:(01222) 233926 fax:(01222) 383573 E_Mail:westcodiy@aol.com Web_Site:http://www.westcofloors.co.uk

Western landscapes Ltd t/a Wyevale Landscapes Upper Buckover Farm Buckover Wotton-under-Edge Glos. GL12 8DZ tel:(01454) 419175 fax:(01454) 412901 E_Mail:sale@westlondensec.co.uk Web_Site:http://www.westlondonsec.

Westminster Scaffolding Ltd Westminster House Pensbury Place London SW8 4TP tel:(0171) 720 3404 fax:(0171) 622 5846

Westpile Ltd Dolphin Bridge House Rockingham Road Uxbridge Middx. UB8 2UB tel:(01895) 258266 fax:(01895) 271805 E_Mail:estimating@westpile.co.uk Web_Site:http://www.westpile.co.uk

WF Electrical Distributors 313-333 Rainham Road South Dagenham Essex RM10 8SX tel:(0181) 515 7000 fax:(0181) 595 0519 telex:897289 Web_Site:http://www.wf-online.com

WH Wesson Fencing Limited 126 Connaught Road Brookwood Woking Surrey GU24 0AS tel:(01483) 472124 fax:(01483) 472115

Whipp & Bourne Switchgear Works Manchester Road Castleton, Rochdale Gtr. Manchester OL11 2SS tel:(01706) 632051 fax:(01706) 345896 E_Mail:whipp+bourne@dial.pipex.com

White Gates Westcoe Hill Lane Weeton Nr. Leeds LS17 0AS tel:(01423) 734466

White Seal Stairways Ltd 88 Hopewel Drive Chatham Kent tel:(01634) 812371 fax:(01634) 812663

Whitecroft Lighting Limited Burlington Street Ashton-Under-Lyne Lancs. OL7 0AX tel:(0990) 087087 fax:(0990) 084210 E_Mail:email@whitecroftlight.com

Whitehill Spindle Tools Ltd Unit C1 Hollystreet Trading Estate Holly Street Luton Beds. LU1 3XG tel:(01582) 736881 fax:(01582) 488987

Whitmore's Timber Co Ltd Main Road Claybrooke Magna, Nr Lutterworth Leics. LE17 5AQ tel:(01455) 209121 fax:(01455) 209041

Wicanders (GB) Ltd Amorim House Star Road, Partridge Green Horsham W. Sussex RH13 8RA tel:(01403) 710001 fax:(01403) 710003 Web_Site:http://www.wicanders-amorim.co.uk

Wilec Ltd Cumnor House 11 Cumnor Road Bournemouth BH1 1JR tel:(01202) 290370

Wilkey Block (Colcuester) Ltd Martells Pit Ardleight Colchester Essex CO7 7RU tel:(01206) 230694 fax(01206) 231785

Willan Building Services Ltd 2 Brooklands Road Sale Cheshire M33 3SS tel:(0161) 962 7113 fax:(0161) 905 2085 E_Mail:postmaster@willan.co.uk Web_Site:http://www.willan.co.uk

William Freeman Ltd Wakefield Road Staincross Barnsley S. Yorks. S75 6DH tel:(01226) 284081 fax:(01226) 731832 telex:547186

Williams & Farmer Monk Meadow Hempstead Lane Gloucester GL2 5JJ tel:(01452) 300300 fax:(01452) 300791

Williamson Cliff Ltd Little Casterton Road Stamford Lincs. PE9 4DA tel:(01780) 764383 fax:(01780) 753295

Williamson Floor & Wall Specialists Ltd Neil House Twining Road Ashburton, Trafford Park Manchester M17 1AT tel:(0161) 872 5147

Williamson T & R Ltd 36 Stonebridge Gate Rippon N. Yorks. HG4 1TP tel:(01765) 607711 fax:(01765) 60798

Wilo Samson Pumps Ltd 11 Ashlyn Road West Meadows Ind Estate Derby DE21 6XE tel:(01332)385181 fax:(01332)344423 E_Mail:sales@wilo.co.uk Web_Site:http://www.wilo.co.uk

Wilson & Garden Ltd 17-21 Newtown Street Kilsyth Glasgow G65 0JX tel:(01236) 823291 fax:(01236) 825683 telex:777967 CHAMCOM G E_Mail:wilsongar@aol.com Web_Site:http://www.besaret.org.uk/wilsong

Wilson (Engineering) Ltd, F G Old Glenarm Road Larne Co. Antrim BT40 1EJ tel:(01574) 261000 fax:(01574) 261111 Web_Site:http://www.fgwilson.com

Wilsonart Ltd 2 Norfolk Court Norfolk Road Rickmansworth Herts WD3 1LA tel:(01923) 712846 fax:(01923) 774642 Web_Site:http://www.wilsonart.com

Wilsons Fibreglass Ltd Fitzherbert Road Farlington Industrial; Estate Portsmouth PO6 1RU tel:(01705) 384921 fax:(01705) 210716

Wiltshire G R & Co Main Road Claybrooke Magna, Lutterworth Leics. LE17 5AQ tel:(01455) 202666 fax:(01455) 209041

Wincro Metal Industires Ltd Fife Street Wincobank Sheffield S. Yorks. S9 1NJ tel:(0114) 242 2171 fax:(0114) 243 4306 E_Mail:sales@wincro.com Web_Site:http://www.wincro.com

Window Fabrication & Fixing Supplies Whittle House Leicester Road Lutterworth Leics. LE17 4HE tel:(01455) 553754 fax:(01455) 557949

Windowbuild 55-56 Lewis Road East Moors Cardiff CF1 5EB tel:(01222) 307200 fax:(01222) 480030

Windsor Limited D W Marsh Lane Ware Herts. SG12 9QL tel:(01920) 466499 fax:(01920) 460327 E_Mail:info@dwwindsor.co.uk

Winn & Coales Denso House Chapel Road London SE27 0TR tel:(0181) 670 7511 fax:(0181) 761 2456 E_Mail:mail@denso.net Web_Site:http://www.denso.net

Winther Browne & Co Ltd Nobel Road Eleys Estate, Angel Road London N18 3DX tel:(0181) 803 3434 fax:(0181) 807 0544 E_Mail:sales@wintherbrowne.co.uk

Wirsbo Space House Satellite Business Village Crawley W. Sussex RH102NE tel:(01293) 548512 fax:(01293) 548552 E_Mail:Enquireiswiuk@uponor.com Web_Site:http://www.wirsbo.co.uk

Witham Oil And Paint Ltd Stanley Road, Oulton Broad Lowestoft Suffolk NR33 9ND tel:(01502) 563434 fax:(01502) 500010

WL West & Son Limited Selham Petworth W. Sussex GU28 0PJ tel:(01798) 861611 fax:(01798) 861635

Wood Bros (Warsop) Ltd Mansfield Road Warsop, Nr Mansfield Notts. NG20 0EH tel:(01623) 842533 fax:(01623) 846415

Wood E Ltd Standard Way North Allerton N. Yorks. DL6 2XA tel:(01609) 780170 fax:(01609) 780438 telex:587577 EWOOD G E_Mail:grant@ewood.co.uk

Wood International Agency Wood House 16 King Edward Road Brentwood Essex CM4 4HL tel:(01277) 232991 fax:(01277) 222108 Web_Site:http//www.ourworld.compuserve.com/homepages/wood_international_agency

Woodburn Engineering Ltd Rosganna Works Trailcock Road Carrickfergus Co. Antrim BT38 7NU tel:(01960) 366404 fax:(01960) 367539

Woodhouse UK Plc Spartan Close Tachbrook Park Warwick CV34 6RR tel:(01926) 314313 fax:(01926) 883778 E_Mail:sales@woodhouse.co.uk

Woodmark Ltd 109 Fairfax Road Teddington Middx. TW11 9DA tel:(0181) 977 1069 fax:(0181) 943 1722

Woodscape Ltd Upfield Pike lowe Brinscall,Nr .chorley Lancs. PR6 8SP tel:(01254) 830886 fax:(01254) 831846 telex:67674 AWTLX G E_Mail:wscape@lancs.co.uk

Woolliscroft Tiles Ltd Melville Street Hanley Stoke-on-Trent ST1 3ND tel:(01782) 208082 fax:(01782) 202631 E_Mail:wtiles@btinternet.com Web_Site:http://www.wtiles.com

Worcester Controls Burrell Road Haywards Heath W Sussex RH16 1TL tel:(01444) 414133 fax:(01444) 459468 telex:87189

Worcester Heat Systems Ltd Cotswold Way Warndon Worcester WR4 9SW tel:(01905) 754624 fax:(01905) 754619

Wormald Ansul UK Ltd Wormald Park Grimshaw Lane Newton Heath Manchester M40 2WL tel:(0161) 205 2321 fax:(0161) 455 4459

WP Metals Ltd Westgate Aldridge W. Midlands WS9 8DJ tel:(01922) 743111 fax:(01922) 743344 E_Mail:info@wpmetals.co.uk

Wragg Bros (Aluminium Equipment) Ltd Robert Way Wickford Business Park Wickford Essex S11 8DQ tel:(01268) 732607 fax:(01268) 768499

Wright & Offland Ltd Floats Road Roundthorn, Wythenshawe Manchester M23 9WF tel:(0161) 946 8000 fax:(0161) 946 8092

WT Henley Ltd Cretehall Road Gravesend Kent DA11 9DA tel:(01474)564466 fax:(01474)566703 telex:965959

WT Henley Ltd Crete Hall Road Gravesend Kent DA11 9DA tel:(01474) 564466 fax:(01474) 566703 Web_Site:http://www.wt-henley.com

Wybone Ltd Mason Way, Platts Common Industrial Estate Hoyland Nr Barnsley S. Yorks. S74 9TF tel:(01226) 744010 fax:(01226) 350105 E_Mail:admin@wybone.co.uk / sales@wybone.co.uk Web_Site:http://www.wybone.co.uk

Wyckham Blackwell (Joinery) Ltd Old Station Road Hampton in Arden, Solihull W. Midlands B92 0HB tel:(01675) 442233 fax:(01675) 442227

Wylex Ltd Wylex Works Wythenshaw Manchester M22 4RA tel:(0161) 998 5454 fax:(0161) 945 1587

Wymark Ltd Runnings Road Industrial Estate Cheltenham Glos. G51 9NQ tel:(01242) 520966 fax:(01242) 519925

Wyndawaye Systems Ltd 9/10 Kelvin Way Trading Estate Kelvin Way West Bromwich W.

Midlands B70 9LL tel:(0121) 553 0989 fax:(0121) 553 3280

Wyvern Marlborough Ltd Grove Trading Estate Dorchester Dorset DT1 1SU tel:(01305) 264716 fax:(01305) 264717

Xpelair Ltd PO Box 279 Morley Way Peterbourough PE2 9JJ tel:(01733) 456789 fax:(01733) 310606 E_Mail:info@xpelair.co.uk Web_Site:http://www.xpelair.co.uk

XuXu Joinery Ltd Block 51 Deeside Industrial Estate Queensferry, Deeside Flintshire CH5 2LR tel:(01244) 288743 fax:(01244) 281881

Yaght Parts (London) 99 Fulham Palace Road London W6 8JA tel:(0181) 741 9803 fax:(0181) 741 5727

Yale Security Products Ltd Wood Street Willenhall W. Midlands WV13 1LA tel:(01902) 366911 fax:(01902) 368535

Yarco Fencing Limited Riverside Road Gorleston Great Yarmouth Norfolk NR31 6PX tel:(01493) 656672 fax:(01493) 440225

Yonder Hill Ltd Yonder Hill Chard Junction Chard Somerset TA20 4QR tel:(01460) 220751 fax:(01460) 221392

Yorian Garden Products 311 Lidgett Lane Leeds 17 LS17 6PD tel:(0113) 268 2682 fax:(0113) 268 2682

York Handmade Brick Co Ltd Winchester House Forest Lane Alne, York W. Yorks. Y061 1TU tel:(01347) 838881 fax:(01347) 838885 E_Mail:sales@yorkhandmade.co.uk Web_Site:http://www.yorkhandmade.co.uk

Yorkon Ltd New Lane Huntington Yorks. YO3 9PT tel:(01904) 610990 fax:(01904) 610880 E_Mail:contact@yorkon.com Web_Site:http://www.yorkon.com

Yorkstone Products Ltd Britannia Quarries Morley Leeds LS7 0SW tel:(0113) 253 0464 fax:(0113) 252 7520

Yougman S G B Ltd Stane Street Slinford, Horsesham W. Sussex RH13 7RD tel:(01403) 790456 fax:(01403) 790881

Youngs Doors Ltd City Road Works Norwich Norfolk NR1 3AN tel:(01603) 629889 fax:(01603) 764650 Web_Site:http://www.youngsdoors.internet2.co.uk

Yule & Son Ltd, A C Craigshaw Road West Tullos Industrial Estate Aberdeen Aberdeen AB12 3ZG tel:(01224) 230000 fax:(01224) 230011 E_Mail:info@acyule.com Web_Site:http://www.acyule.com

Zanussi Professional Yeoman House, Parsonage Square Station Road Dorking Surrey RH4 1UP tel:(01306) 741239 fax:(01306) 741243

Zedcor Ltd Zedcor Business Park Bridge Street Whitney Oxon. OX8 6LJ tel:(01993) 776346 fax:(01995) 776233

Zehnder Ltd Unit 6 Invincible Road Farnborough Hampshire GU14 7QU tel:(01252) 515151 fax:(01252) 522528

Zellweger Analytic Ltd, Sieger Division Hatch Pond House 4 Stinsford Road Poole Dorset BH17 0RZ tel:(01202) 676161 fax:(01202) 678011 E_Mail:sales@zellweger-analytics.co.uk Web_Site:http://www.zelana.co.uk

Zephyr Flags & Banners Midland Road Thrapston Northants NN14 4LX tel:(01832) 734484 fax:(01832) 733064

Ziehl-Ebm (UK) Ltd 17-19 Richmond Road Dukes Park Industrial Estate Chelmsford Essex CM2 6TL tel:(01245) 468555 fax:(01245) 466336 E_Mail:sales@ziehl-ebm.co.uk

Zinc Alloy Ltd Shakespeare Street Wolverhampton WV1 3LR tel:(01902) 452915 fax:(01902) 352917

Zinc Development Association (ZDA) 46 Weymouth Street London W1N 3CQ tel:(0171) 499 6636

Zon International PO Box 329 Edgware Middx HA8 6NH tel:(0181) 381 1222 fax:(0181) 381 1333

Products and Services

A01 Institutions/ Associations

Accrediting body and learned society - Chartered Institution of Building Services Engineers (CIBSE)
Advice given to Contractors - National Federation of Builders
Aluminium trade association - Aluminium Federation (Alfed) Ltd
Aluminium window association - Council for Aluminium in Building
Architectural aluminium association - Council for Aluminium in Building
Association of concrete technologists - Institute of Concrete Technology The
Brick Information Service - Brick Development Association
Builders federation - Federation of Master Builders
Builders federation - National Federation of Builders
Building cost information service - Building Cost Information Service
Building industry joint council - Construction Industry Joint Council
Building maintenance information service - Building Maintenance Information
Business research service - British Library Business Information Service
Cable management products trade association - British Electrical Systems Association
Carpet manufacturers trade association - British Carpet Manufacturers Association
Cavity wall insulation association - National Cavity Insulation Association Ltd
Civil engineering contractors association - Civil Engineering Contractors Association
Clay pipe trade asssociation - Clay Pipe Development Association Ltd
Construction confederation - Construction Confederation
Construction employers federation - Construction Employers Federation
Copper development association - Copper Development Association
Cost Engineers' association - Association of Cost Engineers, The
Decorators trade association - British Decorators Association
Draught proofing trade association - Draught Proofing Advisory Association
Dry stone walling association - Dry Stone Walling Association of Great Britain
Electrical association - Electrical Contractors Association, The
Estimators - V B Johnson & Partners
Fire protection association - British Fire Protection Systems Asssociation
Galvinisers trade association - Galvanizers Association
Glazing trade association - Glass & Glazing Federation The
Heating and ventilating trade association - Heating & Ventilating Contractors Assocition
Housing association - Island Cottages Ltd
Housing associations / registered social landlords - Housing Corporation
Information service to businesses - British Library Business Information Service
Institution of Incorporated Engineers in electronic, electrical and mechanical engineering (IIE) - Institution of Incorporated Engineers in electronic, electrical and mechanical engineering (IIE)
Institutions and association - Architectural AssociationThe
Institutions and association - British Standards Institution
Institutions and association - Building Bookshop
Institutions and association - Civic Trust
Institutions and association - Department of Environment
Institutions and association - English Heritage
Institutions and association - Friends of the Earth
Institutions and association - HMSO Publications Office
Institutions and association - Institution of Civil Engineers
Institutions and association - Institution of Electrical Engineers
Institutions and association - Institution of Mechanical Engineers
Institutions and association - Institution of Structural Engineers
Institutions and association - Lead Development Association
Institutions and association - Millenium Commission, The
Institutions and association - National Trust The
Institutions and association - Ordnance Survey
Institutions and association - Royal Fine Art Commission
Institutions and association - Royal Institute of British Architects (RIBA)
Institutions and association - Royal Town Planning Institute
Institutions and association - Society for the Protection of Ancient Buildings (SPAB)
Institutions and association - Zinc Development Association (ZDA)
Journals - Contract Journal Business Alert
Lighting association - Lighting Association, The
Loft insulation rade association - National Association of Loft Insulation Contractors
Manages water biotreatment club - British Water
National funding body for arts in England - Arts Council
Partitioning and interiors association - Partitioning & Interiors Association
Patent glazing contractors association - Council for Aluminium in Building
Perspective artist - Green Gerald
Product information service, technical publications, library service. - British Cement Association
Professional & trade association - Association of Professional Foresters of Great Britain
Professional body - Chartered Institute of Building
Professional Body - Royal Institution of Chartered Surveyors
Professional body for wastes managment - Institute of Wastes Management
Professional body registered as an educational charity - Institute of Plumbing The
Professional recognition and development for Incorporated Engineers and Engineering Technicians in electronic, electrical and mechanical engineering - Institution of Incorporated Engineers in electronic, electrical and mechanical engineering (IIE)
Publications: Contract Flooring Association Members Handbook - Contract Flooring Association
Scotish architectural trade association - Royal Incorporation of Architects in Scotland The
TCPA Publications - Town and Country Planning Association
Telephone help line for building professionals - Barbour Enquiry Service
Terrazzo, marble and mosaic federation - National Federation of Terrazzo Marble & Mosaic Specialists
Testing, certification and and technical approval - British Board of Agrement
Thatched roof advice, training and insurance - Thatching Advisory Services Ltd
The welding institute - TWI
Town and Country Planing Journal - Town and Country Planning Association
Trade association - Brick Development Association
Trade Association - British Constructional Steelwork Association Ltd
Trade association for external wall insulation - External Wall Insulation Association
Trade Association Service - Mastic Ashphalt Council
Trade associations of electrical and electronics sector - Beama Ltd
Urethane foam industry trade association - BUFCA Ltd
Wall Ties Trade Association - British Wall Tie Association
Water and wastewater association - British Water

A02 Building contractors

Asbestos removal - Rhodar Limited
Builders - Ogilvie Construction Ltd
Building and civil enginering contractors - Dew Construction Limited
Building Contractor - Kenyon T H & Sons PLC
Building contractors - Jarvis Construction (UK) Ltd
Building Contractors-Associated Holdingd Ltd - Associated Holdings Ltd
Civil engineering - Bardon (England) Ltd
Civil engineering - Freyssinet Ltd
Civil Engineering - Mclean Group, David
Civil engineering and coastal protection works - Wiltshire G R & Co
Construction - Miller Construction Ltd
Construction company - Rawlings(SE) Ltd
Design & Build - Mclean Group, David
Design and build - MGB Randalls
Design and build contractor - Longley James & Co Ltd
Design and build fencing contractors - Baker Fencing Limited
Design and build office fitting out works - Advanced Industries Ltd
Design and build surfacings contractors - Easi-Fall International Ltd
Environmental health, pest control - National Britannia Ltd
Fencing contractors - Burn Fencing Limited
Interior contractors - Gifford Grant Ltd
Landscape contracting - Land Wood & Water Co. Ltd
Mechcanical services Contractor - Kershaw Mechanical Services Ltd
New build and refurbishment contractor - Longley James & Co Ltd
Partnering - Mclean Group, David
Pest control - Insect-O-Cutor
Plumbers - Ogilvie Construction Ltd
Roofing contractor - Evans Howard Roofing Ltd
Roofing contractor - Roofrite (East Anglia) Ltd
Sports field contractor - Bernhard's Sports Surfaces Ltd
Waterproofing - Pitchmastic PLC,

A10 Project particulars

Chartered quantity surveyors - Sawyer & Fisher
Chartered quantity surveyors - V B Johnson & Partners
Construction project management, planning and cost consultants, civil, structural and electrical engineers - KPMC Ltd
Electrical services - H Pickup Mechanical & Electrical Services Ltd
Estimators - Johnson V B & Partners
General engineering - Turnbull & Scott (Engineers) Ltd
Mechanical services - H Pickup Mechanical & Electrical Services Ltd
Mechanical, electrical and process services engineers - Shepherd Engineering Services Ltd
Quantity Surveyors - Johnson V B & Partners

A11 Tender and contract documents

Models - Comber Models

A12 The site/ Existing buildings

Building Regulations Approved Documents England & Wales - NBS Services
Civil Engineering - Mclean Group, David
Design and Build - Miller Construction Ltd
Design, Manage and Construct - Miller Construction Ltd
Management Contracting - Miller Construction Ltd
Management Insurance - Abbott Bros (Southall) Ltd
Partnering - Mclean Group, David
Partnering - Miller Construction Ltd
Publishers of The National Building Spesification. - NBS Services
Windows based Q.S, civils and estimating software - CSSP (Computer Software Services Partnership)

A33 Quality standards/ control

Civil engineering test equipment - Capco Test Equipment Ltd

A34 Security/ Safety/ Protection

Asbestos Removal - Kitsons Insulation Products Ltd
Breathing apparatus - Draeger Ltd
Floor protection - Floor Protection Services Ltd
Hand and electric tools - BRB Industrial Services Ltd
High security locks - Chubb Safe Equipment Company
Installation and testing of chemical anchors - Montrose Safety Fixings
Protection rubber products - Gates Rubber Co Ltd The
Protection sheets - Cordek Ltd
Respiratory protection and safety helmets - MSA (Britain) Ltd
Rolled Protection - Cordek Ltd
Safes - Chubb Safe Equipment Company
Safety access systems - Kobi Ltd
Safety systems - Unistrut Limited
Safety/ Security/ Protection - Pilkington UK Ltd
Security / Safety Protections - UK Fire International Ltd
Strongrooms - Chubb Safe Equipment Company
Temporary electric surface heating for maintenance of pipework - Isopad Ltd
Time & attendance management - Cardkey Systems
Vaults - Chubb Safe Equipment Company

A36 Facilities/ Temporary works/ services

Cabins, site accommodation and event toilets - Rollalong
Graphic design services - Milesahead Graphic Design
Temporary fencing and site guard panels - Guardian Wire Ltd

A37 Operation/ Maintenance of the finished building

Operational and maintenance contracts and technical advice - Mitsui Babcock Energy Services Ltd

A41 Site accommodation

Modular buildings - RB Farquhar Manufacturing Ltd
Modular buildings - Wernick S & Sons Ltd
Portable accommodation - Rovacabin
Portacabins, steel cabins and portaloos - UK Cabin Co
Site Accomadation - Thurston Building Systems
Site cabins - RB Farquhar Manufacturing Ltd
Steel and timber system buildings - Elliott Group Ltd
Timber sectional buildings - Ableson L & D Ltd

A42 Services and facilities

24 hour service - Servowarm
24 Hour Service Call Out - Chubb Safe Equipment Company
Abrasives tungsten carbide burs - PCT Group Sales Ltd
Access Equipment hire - HSS Hire Shops
Advisory service - Birdscarer Products
air qulity monitoring - Zellweger Analytic Ltd, Sieger Division
Alloy tower and aerial platform training - Turner Access Ltd
Aluminium towers, aerial platforms, scissors and booms - Turner Access Ltd
Annual Services - Servowarm
Asbestos abatement and consultancy services - Rhodar Limited
Bookshop - The Building Centre Group Ltd
Breakdown repairs - Servowarm
Brooms and paint, industrial, specialist brushes - Davis Burrow & Sons

Builders cleans and one offs - CSL (Services) Limited
Building Products - Quantum Profile Systems
Bulk handling systems - Portasilo Ltd
Coal tar B.P - Legge Thompson F & Co Ltd
Conferance & Seminar Facilities - The Building Centre Group Ltd
Conferences and seminars - Town and Country Planning Association
Construction equipment - Burlington Engineers Ltd
Construction products - GE Bayer Silicones
Construction research and development - BRE
Construction tools - Ingersoll-Rand European Sales Ltd
Creative materials, plastic, metal, timber, adhesives, tools, scenery and paints - The Modelshop
Decontamination units - Rollalong
Decorators Tools - Ward Bekker Sales Ltd
Drainage industry and plumbers tools - Monument Tools Ltd
Engineering Stockist - Thanet Ware Ltd
Exhibitions, conferences, and meeting rooms - The Building Centre Group Ltd
Formliners - Gray Hill Trading Ltd
Generators - Standby Power Company
Global after-sales support network - Wilson (Engineering) Ltd, F G
Guardian safety buildings and products - Portasilo Ltd
Guideline - The Building Centre Group Ltd
Hand & power tools - Kem Edwards Ltd
Hand and garden tools - Neill Tools Ltd
Hand tools - Footprint Tools Ltd
Hand tools, power tool accessories and abrasives - Hyde Brian Ltd
Handtools, shovels, spades and hammers - Stockten Heath Forge (Caldwells) Ltd
High security units - Portasilo Ltd
Highway maintenance contractors - Tarmac Heavy Building
Hot dip galvanising - Jones of Oswestry Ltd
Information - The Building Centre Group Ltd
Inspection an testing - Pitchmastic PLC,
Installation - Servowarm
Ladders - Kwikform UK Ltd
Lifting Equipment hire - HSS Hire Shops
Local engineers (corgi regestered) - Servowarm
Market & Technical Information - The Building Centre Group Ltd
Masking tape - Ward Bekker Sales Ltd
Open and enclosed skips, cargo containers and custom built units - MGB Randalls
Paint brushes and rollers - Sovereign Brush Co Ltd, The
Paper sacks for cementious products and building materials - Assi Doman Sacks (UK)
Pipe tools - Hilmor
Plant - Linatex Ltd
Portable toilets and amenity products - Portasilo Ltd
Power tools-pneumatic/electric/hydraulic - PCT Group Sales Ltd
Props and struts - Kwikform UK Ltd
Publications and Technical Information on Timber - Trada Technology Ltd
Rail Track works and regeneration - Pitchmastic PLC,
Recoverable anchor screws - Gray Hill Trading Ltd
Scaffold hire - HSS Hire Shops
Services & Facilities - Certifield Laboratories (Div or CPS Industries Ltd)
Servocare economic insurance - Servowarm
Servocare priority insurance - Servowarm
Skin care products - Deb Ltd
Soft faced hammers and mallets - Thor Hammer Co Ltd
Soft soap - Liver Grease Oil & Chemical Co Ltd
Special Works / Maintenance - Durabella Ltd
Spindle tooling - Whitehill Spindle Tools Ltd
Supplier of contract wall coverings to the commercial, hospitality and healthcare markets - Muraspec Ltd
Suppliers of Bituminous Binders Road Surfface Dressing Contractors - South Western Tar Distilleries
Supply of Electrical products to the trade & industry - Imperial Electrical Wholesalers
Tool hire - HSS Hire Shops
Tools - ITW Spit
Tools - Skarsten Tools
Tools - Tinsley & Co Ltd, Eliza
Tools & Ironmongery - Spital Tile Co
Trade in non mchanical plant - Gray Hill Trading Ltd
Transportation of Brituminous Products, Liquid and Oils - South Western Tar Distilleries
Upgrade to existing system - Servowarm
Wall protection - Moffett Thallon & Co Ltd
Wall tie condition surveys - HCL Contracts
Wallpaper pasting machines - Ward Bekker Sales Ltd
Water analysis - Zellweger Analytic Ltd, Sieger Division

A43 Mechanical plant

Access platforms - Manitou (Site Lift) Ltd
Air Conditioning - Stiebel Eltron Ltd
Bandsaw blades, hacksaw blades, mini blades and frames - Speed Saws Ltd.
Boring machines, boring tools and trenchers - B-Trac Equipment Ltd
Builders tools - Benson Industries Ltd
Burnishers and sanders - Alto Cleaning Systems (UK) Ltd
Compact, telescopic crawler cranes and mobile site lighting - Sandhurst MFG Co Ltd
Compressor crankcase heaters - Jimi-Heat Ltd
Construction equipment - Watkins
Construction equipment, compressors, rock drills and accessories, compaction and paving equipment - Ingersoll-Rand European Sales Ltd
Contractors plant - Leach Dandridge Ltd
Design and manufacture of portable industrial vacuum cleaners - BVC-BIVAC, Division of DD Lamson
Diamond tipped glasscutters and emery wheel dressers - Radcliffe & Sons (London)
Diesel power generators - Sandhurst MFG Co Ltd
Double vacuum impregnation plant - Protim Solignum Ltd
Drill bits - Arthur Fisher (UK) Ltd
Dumper - HE Services (Plant Hire) Ltd
Dumptruck hire - HE Services (Plant Hire) Ltd
Excavating and earth moving plant - Bamford J C Excavators Ltd
Excavator hire - HE Services (Plant Hire) Ltd
Excavators, crawler mounted cranes and draglines - R-B International Ltd
Fillers and surfacers - Toupret Fillers Agency, Hill & Rednall Ltd
Grinding machines, surface and double ended grindes - Beacon Machine Tools Ltd
Grinding wheels and abrasive products - Carborundum Abrasives G.B Ltd
Hand and electric tools - BRB Industrial Services Ltd
Hand engraving tools - Elliott John H (Monostamp) Ltd
Hight Safety, Controlled Access & Retrieval Systems - Sala Ltd, B H
Hirepoint - Jewson Ltd
Horticultural and agricultural tools - Neill Tools Ltd
Inspection & measurement equipment - Allen P W & Co Ltd
Lifting clamps and grabs - Camlok Lifting Clamps Ltd
Machine tools for the manufacture of building products - BSA Tools Ltd
Machinery - The Saw Centre Group
Manual and powered hoists, winches and jacks - PCT Group Sales Ltd
Mechanical plant spares - O'Brian Manufacturing Ltd
Mechcanical services Contractor - Kershaw Mechanical Services Ltd
Mini Excavators - Sandhurst MFG Co Ltd
Mini excavators, dumper trucks, rollers and cement mixers - Chippindale Plant Ltd
Pan type mixing machinery - Benton Co Ltd Edward
Plant - Certifield Laboratories (Div or CPS Industries Ltd)
Plant hire - HE Services (Plant Hire) Ltd
Plant hire - HSS Hire Shops
Pocket tapes and folding rules - Rabone Chesterman Ltd
Power tools and accessories - Robert Bosch Ltd
Road Maintenance Equipment - Phoenix Engineering Co Ltd, The
Rollers - HE Services (Plant Hire) Ltd
Rough terrain forklifts - Manitou (Site Lift) Ltd
Site dumpers, concrete and mortar mixers, sawbenches and telescopic handlers - Multi Marque Production Engineering Ltd
Space Heating - Stiebel Eltron Ltd
Special rigging structures - Scotia Rigging Services (Industrial Stainless Steel Rigging Specialists)
Telehandeler Hire - HE Services (Plant Hire) Ltd
Telehandler - HE Services (Plant Hire) Ltd
Telescopic loaders - Manitou (Site Lift) Ltd
Tools - Ashmead Building Supplies Ltd
Tools - Ashmead Building Supplies Ltd
Tools - Skarsten Tools Ltd
Tools and plant - Vulcanite Limited
Tree equipment, woodchippers, stump cutters and tub grinders - B-Trac Equipment Ltd
Truck mounted forklift - Manitou (Site Lift) Ltd
Vehicle Turntables - Hovair Systems Ltd - A Division of Air-Log Ltd
Water Heating - Stiebel Eltron Ltd
Welding equipment - PCT Group Sales Ltd
Wheeled loader hire - HE Services (Plant Hire) Ltd
Wheeled loader hire - HE Services (Plant Hire) Ltd
Wood working tools - Record Tools Ltd

A44 Temporary works

Aluminium towers - Kwikform UK Ltd
Aluminium towers - SGB Youngman
Ladders - Titan Ladder and Case Co Ltd
Ladders & steps - SGB Youngman
Plaited and braided ropes, sash and builders lines - Blakey John H Ltd
Scaffold security services - Alpine Autimation Ltd.
Scaffold sheeting - Monarflex Ltd
Scaffolding - Deborah Services
Scaffolding - Kwikform UK Ltd
Scaffolding - Schauman (UK) Ltd
Scaffolding - Wade Building Services Ltd
Scaffolding and access equipment - Kwikform UK Ltd
Scaffolding and ladders - SGB Youngman
Scaffolding restraints - HCL Contracts
Scaffolding service - Turner Plus Eight Ltd
Scaffolding systems - SGB Youngman
Temporary fencing - Rom Ltd
Temporary roadway - Sommerfeld Flexboard Ltd
Temporary security fencing and control barriers - Kwikform UK Ltd
Tempory protection - British Sisalkraft Ltd

B10 Prefabricated buildings/ structures

Aluminium green houses, chalets, garages, workshops and sheds - Compton Building Ltd
Aluminium special structures - Alifabs (Woking) Ltd
Bridges - Structural Timbers Ltd
Bridges in timber or steel, boardwalks, decking, structures pergolas, screens and planters - CTS Ltd
Builders merchant - First Stop Builders Merchants Ltd
Builders merchant - Jewson Ltd
Builders merchant - Nicholls & Clarke Ltd
Building and development - Pearce Construction(Barnstaple)Ltd
Building systems to buy or hire, self contained buildings and modular buildings - Portakabin Ltd
Cantilever and pigeon hole racking - Hi-Store Ltd
Cedar green houses, chalets, garages, workshops and sheds - Compton Building Ltd
Columbariums - Milner Delvaux Ltd
Complete buildings - Elliott Group Construction Division Ltd
Composite housing and containers and portable buildings - Cobham Composites Ltd
Concrete Workshops & Sheds - Phillip Ader Ltd
Concrete garages - Phillip Ader Ltd
Conservatories - Caradon Duraflex Systems Ltd
Conservatories - Conservatory Roof Systems Ltd
Conservatories - Hill Leigh (Joinery) Ltd
Conservatories - Marshall-Tufflex Ltd
Conservatories - Phillip Ader Ltd
Conservatories - Solair Ltd
Conservatories, garden buildings - Amdega Ltd
Design and build fast track buildings - Elliott Group Ltd
Design and build of timber frame components and floor joist system - Prestoplan
Electrical engineering - Park Products
Environmental trade products - Furmanite International Ltd
Erection and design service - Llewellyn Homes Ltd
Export Buildings - Wernick S & Sons Ltd
Extruded conservatories - Smart Systems Ltd
Fabric tensile structures - Clyde Canvas Fabric Structures Ltd
Flat Pack Buildings - Wernick S & Sons Ltd
Garden buildings - Anthony de Grey Trellises
Greenhouses - Phillip Ader Ltd
Grp modular buildings - Dewey Waters Ltd
Hire of buildings - Wernick S & Sons Ltd
Industrial and gate houses - FIBAFORM PRODUCTS
Jack Leg Cabins - Ableson L & D Ltd
Maintenance - Oldham Signs Ltd
Mechcanical services - Kershaw Mechanical Services Ltd
Modular and volumetric buildings - Rollalong
Modular buildings - Ableson L & D Ltd
Modular buildings - Bell & Webster Concrete Ltd
Modular buildings - Rovacabin
Modular buildings - Terrapin Ltd
Modular buildings - Wernick S & Sons Ltd
Modular buildings amd relocatable buildings - GE Capital Modular Space
Modular buildings and garages - Panelite
Modular portable and prefabricated buildings - Ideal Building Systems Ltd
Modular solutions - Composite Structures
Module based buildings - Thurston Building Systems
Multi block garages and storage units - Panelite
Overhead monorail systems - Down & Francis Ltd, Division of Metalrax PLC
Portable sectional buildings - UK Cabin Co
Portable, mobile, demountable, sectional, industrial, system, prefabricated buildings, mobile homes and chalets - Powys Hire Ltd/ P.I.B
Pre engineered buildings - Terrapin Ltd
Pre engineered metal buildings - Atlas Ward Structures Ltd
Prefabricated buildings - Parklines Building Ltd
Proprietary buildings - Leander Architectural
PVCu & Aluminium Conservatories - KCW Windows & Conservatories Ltd
Sale and hire of relocatable buildings - Initial Plant Services Ltd, Redispace and Johnson
Squash and racquet ball courts - Bicester Products Ltd
Steel bus shelters - Planet Architectural Glazing Ltd
Steel frame buildings - Terrapin Ltd
Steel portal frame buildings - Accomodex
System Buildings - Wernick S & Sons Ltd
Tailor made steel buildings - Astron Building Systems Commercial Intertech Ltd
Technical consultancy - BFRC Services Ltd
Timber Building Systems - Bullock & Driffill Ltd
Timber framed houses and structures - Colt W H & Co Ltd
Timber roof design consultancy - Alpine Autimation Ltd.
Timber sheds/ Workshops/ Gazebos - Phillip Ader Ltd
Timber site offices - Ableson L & D Ltd

B11 Prefabricated building units

Conservatories - Elliott Group Ltd
Design and build fast track buildings - Elliott Group Ltd
Horizontal panel concrete buildings - Accomodex
Modular buildings - Rovacabin
Portastor equipment housing - Portasilo Ltd
Prefabricated building units - Ideal Building Systems Ltd
Prefabricated building units - Initial Plant Services Ltd, Redispace and Johnson
Steel container toilets - Ableson L & D Ltd
Vertical panel concrete buildings – Accomodex

C10 Site survey

Geotechnical consultancy and contamination assessment - Structural Soils Ltd
Ground investigation - May Gurney (Technical Services) Ltd - Fencing Division
Site Investigation - Haydon & Hagan Construction Ltd
Site investigation,soils and material testing - Structural Soils Ltd
Surveying instruments - Topcon Corporation

C14 Building services survey

Energy efficiency in lighting advisory service - Poselco Lighting

C20 Demolition

Alterations - Haydon & Hagan Construction Ltd
Demolition - HE Services (Plant Hire) Ltd
Demolition - Parker Plant Ltd
Hydrodemolition - Llewellyn (Stonecare) Limited
Office refurbishment - Cumberland Construction
Structural alterations - Structural Bonding Cambridge Ltd

C21 Toxic/ Hazardous material removal

Asbestos removal - Hawkins Insulation Ltd
Toxic / Hazardous material removal - Hertel Services

C30 Shoring/ Facade retention

Proping and needling - Maby Hire Ltd
Propping and needling - Matbro Ltd
Shoring - Kwikform UK Ltd
Shoring - SGB Youngman
Shoring and propping - Structural Bonding Cambridge Ltd
Trench strutting equipment - Maby Hire Ltd

C40 Cleaning masonry/ concrete

Brick and stone cleaning - Gun-Point Ltd
Building, cleaning and restoration - Cleanstone Co (Bradford) Ltd

C41 Repairing/ Renovating/ Conserving masonry

Brick replacement - Gun-Point Ltd
Brickwork repair materials - PermaRock Products Ltd
Faience Blocks - Shaws of Darwen
High strength, quick setting repair materials - David Ball Group Plc
Masonry and brickwork repair materials - PermaRock Products Ltd
Pressure pointing - Gun-Point Ltd
Remedial pointing - Kingfisher Chemicals Ltd
Remedial Wall Ties - Terminix Peter Cox Ltd
Repairing concrete and masonry structures - Vandex (UK) Ltd
Reparing and renovating - Canonbury Asphalte Co. Ltd.
Repointing brickwall and stonework - Gun-Point Ltd
Restoration of facades in stone, brick & terracotta - Stonewest Ltd
Stainless steel remedial wall ties - Helifix Ltd
Stainless steel replacement - Fix Fast (Fasteners) Wales Ltd
Stone and brickwork repair and restoration - CSL (Services) Limited
Technical publications - Brick Development Association
Terracotta Blocks - Shaws of Darwen
Wall tie replacement - Timberwise (UK) Plc

C42 Repairing/ Renovating/ Conserving concrete

Concrete maintenance and repair products - Marcel Guest Paints
Concrete repair - CCS Scotseal Ltd
Concrete repair - FEB Ltd
Concrete Repair - Feb MBT
Concrete repair - Fullstop Technology Ltd
Concrete repair - Rok-Crete Units Co Ltd
Concrete repair mortars - Hines & Sons, P E, Ltd
Concrete repair mortars, crack injection resins - Ronacrete Ltd
Concrete repair products - Flexcrete Ltd
Concrete repair systems - Beton Construction Ltd
Concrete repair systems - Exchem Mining and Construction
Concrete repair systems - PermaRock Products Ltd
Concrete repairs - Llewellyn (Stonecare) Limited
Concrete Repairs - Pitchmastic PLC,
Epoxy resin repairs - Gun-Point Ltd
Floor seals and concrete repair kits - Tank Storage & Services Ltd
Repairing / renovation / conserving concrete - Freyssinet Ltd
Repairing and conserving concrete - Anti Corrosion Services Ltd
Repairing concrete - Concrete Repairs Ltd
Repairs and restorations - Alpha M & T Ltd
Resin and precision grout - Armorex Ltd
Stonework restoration/cleaning - PermaRock Products Ltd

C45 Damp proof course renewal/ insertion

Chemical DPC and remedial treatments - Protim Solignum Ltd
Chemical DPC Installers - Dampcoursing Limited
Chemical dpcs to existing walls - Terminix Peter Cox Ltd
Condensation control and DPC - Kair - The Ventilation Division of Kiltox Chemicals Limited
Damp Coursing - Kiltox Damp Free Solutions
Damp-profing - Timberwise (UK) Plc
Physical DPC Insertion - Dampcoursing Limited

C50 Repairing/ Renovating/ Conserving metal

Corrosion protection for metal fastenings - Interlux Ltd
Repairing/ renovating metal - Certifield Laboratories (Div or CPS Industries Ltd)
Repairs and maintenance to doors and shutters - Andersons Ltd
Restoration work in metal - Ballentine, Bo'Ness Iron Co Ltd
Site welding service - C & R Building Systems Ltd

C51 Repairing/ Renovating/ Conserving timber

Solvent based and water based preservatives - Protim Solignum Ltd
Timber resin repairs - Timberwise (UK) Plc
Timber treatment - Rentokil Ltd

C52 Fungus/ Beetle eradication

Dry rot and woodworm treatment - Kair - The Ventilation Division of Kiltox Chemicals Limited
Fungus and beetle eradication - Terminix Peter Cox Ltd
Timber Preservation - Kiltox Damp Free Solutions
Timber preservatives - Celotex Ltd
Timber preservatives - Rentokil Ltd
Timber treatment - Timberwise (UK) Plc
Timber treatment and dry rot eradication - Dampcoursing Limited
Woodworm - Kiltox Damp Free Solutions

D11 Soil stabilisation

Chemicals - Baxenden Chemicals Ltd
Compressable fill - Kay Metzeler Ltd
Erosion control mats - ABG Ltd
Green roof systems - ABG Ltd
Ground improvement - Keller Ground Engineering
Ground improvement - Keller Ground Engineering
Retaining walls - ABG Ltd
Retaining Walls - Landscape Grass (Concrete) Ltd
Retaining walls, soil stabilisation and erosion control - PHI Group Ltd
Soil Stabilisation - Associated Holdings Ltd
Soil stabilisation - Kvaerner Cementation Foundation Ltd
Soil Stabilisation - The Nelton Group Ltd
Soil stabilisation and errosion control - Land Wood & Water Co. Ltd
Soil stabilization - Grass Concrete Ltd
Soil stabilization - May Gurney (Technical Services) Ltd - Fencing Division
Soil stabilization - Vencel Resil Ltd
Vibro and dynamic compaction - Bauer Foundations Limited

D12 Site dewatering

Dewatering - Keller Ground Engineering
Dewatering - Simon-Hartley
Pumping systems for ground water control - Andrews Sykes Hire Ltd

D20 Excavating and filling

Aggregates - Barker Bros Aggregates Ltd
Aggregates - Brikenden (Builders Merchants) Ltd
Construction materials - Hanson Aggregates
Crushed rock - Hanson Aggregates
Crushed rock, sands and gravels - Hanson Aggregates
Dumptruck hire - HE Services (Plant Hire) Ltd
Excavating and filling - HE Services (Plant Hire) Ltd
Excavating and filling - Vencel Resil Ltd
Excavator hire - HE Services (Plant Hire) Ltd
Filling rock - Bardon (England) Ltd
Land reclamation - Bernhards Landscapes Ltd
Landfill blocks - Kay Metzeler Ltd
Sand and gravels - Hanson Aggregates
Wheeled loader hire - HE Services (Plant Hire) Ltd

D21 Ground gas venting

Gas Venting - Cordek Ltd
Ground gas venting - Keller Ground Engineering
Methane gas venting - ABG Ltd

D30 Piling

Ateel piling - Stent Foundations Ltd
Cased flight auger piling - Westpile Ltd
Cast in place concrete piling - Expanded Piling Co Ltd, The
Cast in place concrete piling - Kvaerner Cementation Foundation Ltd
Cast in place concrete piling - May Gurney (Technical Services) Ltd - Fencing Division
Cast in place concrete piling - Stent Foundations Ltd
Concrete Piling - Access Piling Ltd
Continuous flight auger bored piling - Westpile Ltd
Driven precast concrete piling, steel sheet piling and steel bearing - DEL Piling Contractors
Driven precast - Westpile Ltd
Driven Tube Piling - Westpile Ltd
Ground engineering, piling - Keller Ground Engineering
Ground engineering, Piling - Keller Ground Engineering
Ground engineering, piling - Keller Ground Engineering
Large diameter bored piling - Westpile Ltd
Preformed concrete piling - Expanded Piling Co Ltd, The
Preformed concrete piling - Kvaerner Cementation Foundation Ltd
Preformed concrete piling - May Gurney (Technical Services) Ltd - Fencing Division
Preformed concrete piling - Stent Foundations Ltd
Restricted access Piling - Structural Bonding Cambridge Ltd
Rotary bored piling - Westpile Ltd
Steel Piling - Access Piling Ltd
Steel Piling - Deepdale Engineering Co Ltd
Trench shoring equipment - Matbro Ltd
Trench strutting equipment - Maby Hire Ltd
Tripod bored piling - Westpile Ltd
Tubular steel piles - DEL Piling Contractors

D40 Embedded retaining walls

Diaphragm walling - Keller Ground Engineering
Diaphragm walling - Kvaerner Cementation Foundation Ltd
Diaphragm Walling - Stent Foundations Ltd

D41 Crib walls/ Gabions/ Reinforced earth

Crib walls/Gabions/Reinforced earth - Keller Ground Engineering
Crib walls/Gabions/Reinforced earth - Kvaerner Cementation Foundation Ltd
Cribs walls - Associated Holdings Ltd
Decfin walls - Bell & Webster Concrete Ltd
Retaining walls - ABG Ltd
Tee walls - Bell & Webster Concrete Ltd

D50 Underpinning

Piled raft underpinning - Structural Bonding Cambridge Ltd
Underpinning - Haydon & Hagan Construction Ltd
Underpinning - Rok-Crete Units Co Ltd
Underpinning including jacking - Structural Bonding Cambridge Ltd

E05 In situ concrete construction generally

Cement - Brikenden (Builders Merchants) Ltd
Concretes - Hanson Aggregates
Construction accessories - Creteco Sales Ltd
Spacer blocks for reinforcement - Creteco Sales Ltd

E10 Mixing/ Casting/ Curing in situ concrete

Additives - Mapei UK Ltd
Aggregates - Pioneer Concrete Holdings PLC
Cavity fill - Owens Corning Building Products (UK) Ltd
Cement - Blue Circle Cement
Cement - Rugby Cement
Cement - Super Cement Ltd
Cement, high alumina cement, calcium aluminate aggregate, synthetic cement accelerator - Lafarge Aluminates Ltd
Coated road stone - Hanson Aggregates
Concrete admixtures - Beton Construction Ltd
Concrete plasticizers - Sika Ltd
Concrete repair materials - Optiroc Ltd
Concretes - Hanson Aggregates
Controlled permeability formliner - Creteco Sales Ltd
Fire cement - Purimachos Ltd
Lightweight and dense aggregates - CAEC Howard (Holdings) Ltd
Ready mixed concrete - Fyfe John Ltd
Ready mixed concrete - Pioneer Concrete Holdings PLC
Readymix - Tillicoutry Quarries Ltd
Sand and aggregate - Ashmead Buidling Supplies Ltd
Sand and aggregate - Ashmead Building Supplies Ltd
Slag cement - Civil & Marine Slag Cement Ltd
Underwater concrete - Armorex Ltd

E11 Sprayed in situ concrete

Concrete admixtures - Beton Construction Ltd
Process pumps - Sidewinder Concrete Pumps Ltd
Tunnel concrete pumps - Sidewinder Concrete Pumps Ltd

E20 Formwork for in situ concrete

Column formers - circular, square, rectangular - Creteco Sales Ltd
Formwork - Kwikform UK Ltd
Formwork - SGB Youngman
Formwork & falsework equipment - Matbro Ltd
Formwork and Scaffolding - PERI Ltd
Formwork coatings - Hines & Sons, P E, Ltd
Formwork equipment - Maby Hire Ltd
Hy-rib permanent formwork - Expamet Building Products
Metal deck flooring - Caunton Engineering Ltd
Specialist post forming - P & R Laboratory Group Ltd

E30 Reinforcement for in situ concrete

Reinforcement - Expamet Building Products
Reinforcement accessories and spacers - Hines & Sons, P E, Ltd
Reinforcement for in situ concrete - Rother Boiler Co Ltd
Reinforcement for insitu concrete - Freyssinet Ltd
Reinforcement mesh and bar - BRC Square Grip
Reinforcing bar and mesh - Hy-Ten Reinforcement Co Ltd
Reinforcing fabric, bars and accessories - Bromford Iron & Steel Co Ltd
Stainless steel reinforcement - Helifix Ltd
Stainless steel reinforcing rods - Helifix Ltd
Steel reinforcement, fibre, mesh and industrial wire - Rom Ltd
Steel bars and rods for concrete reinforcing applications - Co Steel Sheerness PLC
Steel wire fibres for concrete reinforcement - Bekaert Building Products

E40 Designed joints in in situ concrete

Designed grout anchoring system - Cavity Lock Systems Ltd
Designed joints - Associated Holdings Ltd
Grouting - Gun-Point Ltd
Structural expansion joints - Radflex Contract Services Limetid
Waterstops - Beton Construction Ltd
Waterstops - RFA Group Ltd

E41 Worked finishes/ Cutting to in situ concrete

Anti slip surfacing - Addagrip Surface Treatments Ltd
Concrete airfield pavement protection - Addagrip Surface Treatments Ltd
Dustproofing / Anti-Skidding Concrete Floors - Ferguson G A & Co Ltd
Floor ventilators - Airflow (Nicoll Ventilators) Ltd
Formliner for textured concrete - Creteco Sales Ltd
Grouts - Beton Construction Ltd
Resin bonded surface dressing - Addagrip Surface Treatments Ltd
Surface hardeners - Armorex Ltd
Synthetic ironoxide colouring agents - Bayer Plc

E42 Accessories cast into in situ concrete

Accessories cast into in situ concrete - Rother Boiler Co Ltd
Cast in dovetail slots and ties - Abbey Building Supplies Co
Concrete related accessories - BRC Square Grip
Corner protection - Fixatrad Ltd
Grout Equipment - Parker AG Engineering Ltd
Reinforcement couplers - Halfen Ltd
Ties to bind reinforcing rods together - Huntley & Sparks
Tying wire, spacers and chemicals for concrete - Hy-Ten Reinforcement Co Ltd
Water stops - Creteco Sales Ltd

E50 Precast concrete frame structures

Bespoke concrete products - Bell & Webster Concrete Ltd
Concrete products - Bell & Webster Concrete Ltd
Engineering castings - Ballentine, Bo'Ness Iron Co Ltd
Factory Engineered Precast concrete components - Tarmac Precast Concrete
Platform walls - Bell & Webster Concrete Ltd
Precast concrete and cast stone products - Sangwin Concrete Products
Precast concrete fixings - Halfen Ltd
Precast concrete floor beams, staircases, sills and lintols - Armstrong Concrete Products Ltd
Precast concrete framing - Composite Structures
Precast concrete joists - Longley C R & Co Ltd
Precast concrete large units - Benton Co Ltd Edward
Precast concrete lifting - Halfen Ltd
Precast concrete manholes - Albion Concrete Products
Precast concrete products - Redland Precast
Precast concrete products - Ruthin Precast Concrete Ltd
Precast concrete retaining wall units and safety barriers - RCC, Division of Tarmac Precast Concrete Ltd
Precast concrete retaining wall. Precast concrete staircases - Ebor Concretes Ltd
Precast concrete structural components - FC Precast Concrete Ltd
Precast concrete tunnels - Charcon Tunnels (Division of Tarmac plc)
Precast concrete units - Evans Concrete Products Ltd
Precast concrete walling - Rogers Concrete Ltd
Precast staircases - Bison Concrete Products Ltd
Precast Terrace units/ Stadia elements - Bison Concrete Products Ltd
Retaining panels - Blanc de Bierges
Retaining walls, tee walls and raft floors - Bell & Webster Concrete Ltd
Spiral stairways - Blanc de Bierges
Stadia components - Bell & Webster Concrete Ltd
Staircases, bridge beams, stadia terracing and retaining wall units - Birchwood Concrete Products Ltd
Steps - Blanc de Bierges
Structural frames - Baltcon Ltd
Structural frames - Bison Structures Ltd
Synthetic iron oxide pigments and coloured pigments - Elementis Pigments
Timberframed multi storey modular buildings – Accomodex

E60 Precast/ Composite concrete decking

Concrete Beam/ Block floors - Bison Concrete Products Ltd
Concrete Floors - RMC Concrete Products
Floors - Baltcon Ltd
Ground Beams - Bell & Webster Concrete Ltd
Hollow core and solid composite floors - Bison Concrete Products Ltd
Interior flooring - Blanc de Bierges
Precast concrete decking - Freyssinet Ltd
Precast concrete flooring - Marshalls Mono Ltd
Precast concrete units - Caunton Engineering Ltd
Prestressed flooring slabs - Coltman Precast Concrete Ltd
Staircases, staduim terraces, columns, wall and ground beams - Coltman Precast Concrete Ltd
Various prestressed precast concrete floorings - Birchwood Concrete Products Ltd

F10 Brick/ Block walling

Aircrete, lightweight and dense concrete building blocks - Tarmac Topblock Ltd
Aireared Concrete Building Blocks - Marley Building Materials Ltd
Architectural facing masonry, medium and dense concrete blocks - Lignacite (Brandon) Ltd
Autoclaved Aerated Concrete Building Blocks - Thermalite (Marley Building Materials)
Block walling - Cheshire Concrete Products
Blocks - Brikenden (Builders Merchants) Ltd
Blocks - Hanson Quarry Products Europe Ltd
Blocks - RMC Concrete Products (UK) Ltd
Brick and block walling, cut and bond specials - Castle Brick (Wales) Ltd
Brick specialists - M & G Brickcutters
Brick trade association - Brick Development Association
Brick/ Block Walling - Collier W H Ltd
Brick/ Blockwork - Roadways & Car Parks Ltd
Brick/Block Walling - Mansfield Brick Co Ltd
Bricks - Baggeridge Brick PLC

Bricks - Bovingdon Brickworks Ltd
Bricks - Brikenden (Builders Merchants) Ltd
Bricks - Chelwood Brick Ltd
Bricks - Country Manor Bricks
Bricks - Cranleigh Brick & Tile Co Ltd
Bricks - Freshfield Lane Brickworks Ltd
Bricks - Giscol Bricks
Bricks - Ibstock
Bricks - Ibstock Scotish Brick
Bricks - Rudgwick Brickworks Co Ltd
Bricks - York Handmade Brick Co Ltd
Bricks and blocks - Camas Building Materials
Bricks and blocks - Ashmead Buidling Supplies Ltd
Bricks and blocks - Ashmead Building Supplies Ltd
Bricks and blocks - Blockleys Brick Ltd
Building blocks - ECC Building Products Ltd
Building blocks - RMC Concrete Products (UK) Ltd
Calcium silicate facing and loadbearing bricks - Beacon Hill Brick Co Ltd
Chimney systems - Kedddy (Poujoulat) (UK) Ltd
Chimney systems - Marflex Chimney Systems
Clay commons, fencing bricks and engineering bricks - Hanson Brick Ltd
Clay facing bricks - Ibstock Building Products Ltd
Clay Handmade Facing Bricks - Charnwood Brick Group Ltd
Concrete blocks - Forticrete Ltd
Concrete blocks - RMC Concrete Products
Concrete blocks - Sellite Blocks Ltd
Concrete building blocks - Lignacite (North London) Ltd
Concrete Drainage System - Hepworth Building Products Ltd
Coping stones - Chiltern Concrete Products Ltd
Dense and lightweight aggregate building blocks - CAEC Howard (Holdings) Ltd
Dense, lightweight, insulating solid, cellular and hollow concrete Blocks - Armstrong (Concrete Blocks), Thomas, Ltd
Double module block chimney systems - ISOKERN (UK) Ltd
Energy saving blocks - ARC Conbloc
Engineering and facing bricks - Carlton Main Brickworks Ltd
Facing Bricks - Hanson Brick Ltd, Northern Regonal Sales Office
Faience - Ibstock Building Products Ltd
Faience Blocks - Shaws of Darwen
Fair face blocks - Thakeham Tiles Ltd
Glazed bricks - Smithbrook Building Products Ltd
Handmade facing bricks - Michelmersh Brick & Tile Co Ltd
Handmade facing bricks - Williamson Cliff Ltd
Insulation - TDI (UK) Ltd
Iron cement - Legge Thompson F & Co Ltd
Library and matching service - Mid Essex Trading Co Ltd
Lintels - RMC Concrete Products (UK) Ltd
Partitions - ClipsPartitions Ltd
Reclaimed bricks - Optiroc Ltd
Slab and block - Lindley George & Sons Ltd
Staffordshire blue bricks and pavers - Ketley Brick Co Ltd , The
Terracotta - Ibstock Building Products Ltd
Terracotta Blocks - Shaws of Darwen
Terracotta wall facings - Smithbrook Building Products Ltd
Wirecut rustic facing bricks, handmade facing bricks, common, engineering, reclaimed and special bricks - Northcot Brick Ltd

F11 Glass block walling

Bullet restraint/ proof glass block panels - Luxcrete Ltd
Glass block systems - Swedecor Ltd
Glass Blocks - Cooper H W & Co Ltd
Glass blocks - Luxcrete Ltd
Glass blocks - Shackerley (Holdings) Group Ltd incorporating Designer ceramics

F20 Natural stone rubble walling

Architectural masonry - Farmington Natural Stone Ltd
Granite and natural stone - Bardon (England) Ltd
Natural limestone quarried blocks - Hanson Bath & Portland Stone
Natural stone - Elmstone
Natural stone - Hanson Aggregates
Natural stone - Miles-Stone, Natural Stone Merchants
Natural stone - Realstone Ltd
Natural stone rubble walling - Chilmark Quarries Ltd
Natural stone rubble walling - Gregory Quarries Ltd, The
Natural stone rubble walling - The Rare Stone Group Ltd
Paving and walling stone - Delabole Slate
Portland Stone aggregates - Hanson Bath & Portland Stone
Rock face walling - Thakeham Tiles Ltd
Stone masonry specialists - Macc Stone Ltd
Stone walling - Macclesfield Stone Quarries Ltd
Traditional stone - Chelsea Artisans Ltd

F21 Natural stone ashlar walling/ dressings

Architectural terracotta and faience - Ibstock Hathernware Ltd
Ashlar masonry - Macc Stone Ltd
Cleft stone and screen walling - Thakeham Tiles Ltd
English limestones and sandstones - Chilmark Quarries Ltd
English limestones and sandstones - Gregory Quarries Ltd, The
English limestones and sandstones - The Rare Stone Group Ltd
Facades in stone, brick & terracotta - Stonewest Ltd
Granite and natural stone - Bardon (England) Ltd
Natural derbyshire sandstone - Stoke Hall Quarry(Stone Sales)Ltd
Natural stone - Hanson Aggregates
Natural stone - Miles-Stone, Natural Stone Merchants
Natural stone - Realstone Ltd
Natural stone walling - Macclesfield Stone Quarries Ltd
Sawn ashlar - Lindley George & Sons Ltd

F22 Cast stone walling/ dressings

Architectural cast stone - Hampton Stone Ltd
Architectural masonry - Forticrete Ltd
Architectural Stonework - Haddonstone Ltd
Artificial stone - Procter Concrete Products
Artificial stone units - Northwest Pre-cast Ltd
Cast stone - Border Concrete Products Ltd
Cast Stone - Dockra Concrete Co Ltd, The
Cast stone - Ibstock Building Products Ltd
Cast stone walling - Macclesfield Stone Quarries Ltd
Cast stone walling - Technostone Ltd
Cast stone walling/ dressings - Chilstone Garden Ornaments
Fencing - Procter Bros Limited
Precast walling - Oakdale (Contracts) Ltd
Reconstructed cast stone architectural units - Broadmead Products
Reconstructed stone fireplaces - Taylor & Portway Ltd
Reconstructed stone walling - Atlas Stone Products
Stone - Ibstock Scotish Brick

F30 Accessories/ Sundry items for brick/ block/ stone walling

Accessories/ sundry items for brick/ block/ stone - Neill Tools Ltd
Accessories/ sundry items for brick/ block/ stone - Rother Boiler Co Ltd
Accessories/Sundry items for brick/ block/stone - Mansfield Brick Co Ltd
Accessories/sundry items for brick/block/stone - Haddonstone Ltd
Accessoriesfor brick and block walling - Ensor Metal Products
Aluminium cappings & flashings - Aluminium RW Supplies Ltd
Aluminium copings - Alifabs (Woking) Ltd
Aluminium fascias and soffits - Aluminium RW Supplies Ltd
Arches - Cavity Trays Ltd
Brick and stone support systems - Halfen Ltd
Brick reinforcement - BRC Square Grip
Brick reinforcement - RFA Group Ltd
Brick ties - Halfen Ltd
Brickties and windposts - Halfen Ltd
Brickwork reinforcement - BRC Building Products
Brickwork support - Halfen Ltd
Cast in channel - Halfen Ltd
Cast in dovetail slots and ties - Abbey Building Supplies Co
Cavity and through wall ventilators - Ryton's Building Products Ltd
Cavity fixings - Arthur Fisher (UK) Ltd
Cavity floor boxes - Salamandre Davis
Cavity tanking tiles - Atlas Stone Products
Cavity trays - Cavity Trays Ltd
Cavity trays - Glidevale Building & Products Ltd
Cavity trays - IG Limited
Cavity trays, hight performance dpc - Timloc Expamet Building Products
Cavity wall ties - Fix Fast (Fasteners) Wales Ltd
Cavity wall ties - Fix Fast (Fasteners) Wales Ltd
Chimney flues and linings - Rite-Vent Ltd
Chimney systems - Marflex Chimney Systems
Chimney terminals - Dunbrik (Yorks) Ltd
Closers - Cavity Trays Ltd
Columns, canopies, cornices, porticos, bay canopies - Solair GRP Architectural Products
Connectors steel - Allmat (East Surrey) Ltd
Copings and cills - Burlington Slate Ltd
Damp proof courses - Anderson D & Son
Damp proof courses - Marley Waterproofing
Dense and Lightweight blocks - Newlay Concrete Ltd
DPC - Visqueen Agri
External wall insulation and cladding - Axter Ltd
External wall insulation systems - PermaRock Products Ltd
Faience Blocks - Shaws of Darwen
Fireplaces - Farmington Natural Stone Ltd
Floor ventilators - Airflow (Nicoll Ventilators) Ltd
Flue linings - Dunbrik (Yorks) Ltd
Flues and chimneys - Docherty H Ltd
Foundation Blocks - Newlay Concrete Ltd
Gas Flue blocks - Dunbrik (Yorks) Ltd
Granite and natural stone - Bardon (England) Ltd
GRP Dummy Chimneys - Wilsons Fibreglass Ltd
GRP Roof turrets - Alpha Glass Fibres Ltd
High performance DPC and cavity trays - Zedcor Ltd
Insulated cavity closers - RMC Panel Products Ltd
Insulating cavity closers and insulated DPC - TDI (UK) Ltd
Insulation Blocks - Newlay Concrete Ltd
Insulation fixings - Arthur Fisher (UK) Ltd
Lateral restraint - HCL Contracts
Light weight insulating concrete flue system - Taylor & Portway Ltd
Lintels - Ancon CCL Ltd
Lintels - Cavity Trays Ltd
Lintels - RMC Concrete Products
Lintels and other hand made shapes - Lignacite (Brandon) Ltd
Lintels, cils and padstones - Chiltern Concrete Products Ltd
Manufacturers of Dense and Lightweight Concrete Blocks - Wilkey Block (Colcuester) Ltd
Masonry coatings - Jotun Henry Clark Ltd
Masonry Design Development - Hanson Bath & Portland Stone
Masonry reinforcement - Bekaert Building Products
Masonry support systems - Ancon CCL Ltd
Masonry support systems - Mestec PLC
Masonry to masonry connectors - Simpson Strong-Tie®
Movement joint - Shackerley (Holdings) Group Ltd incorporating Designer ceramics
Movement joints - Allmat (East Surrey) Ltd
Natural stone fixing - Halfen Ltd
Natural ventilation systems - Willan Building Services Ltd
Polythene damp-proof courses - Zedcor Ltd
Prestressed concrete lintels - RMC Concrete Products (UK) Ltd
Pumice stone chimney systems - Sparkes K & L
Remedial Wall Ties - Terminix Peter Cox Ltd
Repair Mortars - Armorex Ltd
Roof ventilation, underfloor ventilation, through the wall ventilation - Timloc Expamet Building Products
Specialist waterproofing materials, admixtures & grouts - David Ball Group Plc
Stainless steel remedial wall ties - Helifix Ltd
Steel Lintels - Birtley Building Products
Steel Lintels - Expamet Building Products
Steel lintels - IG Limited
Steel lintels - Wade Building Services Ltd
Steel lintols - Caradon Catnic Ltd
Steel wall connectors - Allmat (East Surrey) Ltd
Stone and brickwork, cleaning - CSL (Services) Limited
Straps - Expamet Building Products
Terracotta Blocks - Shaws of Darwen
Timber to masonry connectors - Simpson Strong-Tie®
Under floor and cavity wall vents - Willan Building Services Ltd
Ventilators - Cavity Trays Ltd
Wall connectors - Caradon Catnic Ltd
Wall connectors, steel - Allmat (East Surrey) Ltd
Wall insulation - Callenders Ltd
Wall insulation - Dow Construction Products
Wall starters - Expamet Building Products
Wall tie replacement - HCL Contracts
Wall ties - Caradon Catnic Ltd
Wall ties - Direct Wire Ties Ltd
Wall ties - Ensor Metal Products
Wall ties and reinforcements - Ancon CCL Ltd
Wall tiles - Vista Engineering Ltd
Water repellents, antigraffiti and bridge deck membrane system - Degussa Ltd
Weepholes for cavity walls - Molyneux G (Products) Ltd
Wire tray - Salamandre Davis

F31 Precast concrete sills/ lintels/ copings/ features

Chimney systems - Marflex Chimney Systems
Coping and caps - Rogers Concrete Ltd
General precast Items - Brierley B (Garstang) Ltd
Hydrant boxes for gas and water - Brierley B (Garstang) Ltd
Lintels - Procter Concrete Products
Plastic cavity closers - LB Plastics Ltd
Precast concrete - Border Concrete Products Ltd
Precast concrete and reconstructed stone manufacturers - Rok-Crete Units Co Ltd
Precast concrete lintels, steps and copings - Dockra Concrete Co Ltd, The
Precast concrete sills/ lintels/ copings/ features - Benton Co Ltd Edward
Precast concrete sills/ lintels/ copings/ features - Chilstone Garden Ornaments
Precast concrete sills/ lintels/ copings/ features - Evans Concrete Products Ltd
Precast concrete sills/lintels/copings/features - Broadmead Products
Precast concrete sills/lintels/copings/features - Haddonstone Ltd
Precast concrete sills/lintels/copings/features - Technostone Ltd
Precast concrete units - Northwest Pre-cast Ltd
Precast lintels - Carter Concrete Ltd
Precast lintels and manhole covers - Caradon Jones Ltd
Precast lintels and prestressed lintels - Procter Bros Limited
PVCu products - Howarth Timber (Elland) Ltd
Reconstituted Artstone and Precast Concrete Products - North Yorkshire Artstone Ltd
Steel cavity fixings - Lindapter International
Supergalv Steel Lintels - Birtley Building Products Ltd

G10 Structural steel framing

Aluminium and steel fabricators - Ramsay & Sons (Forfar) Ltd
Aluminium fabrications - Midland Alloy Ltd
Architectural & General Metalwork - Thanet Ware Ltd
Brazing alloys, solder and fluxes - Johnson Matthey PLC - Metal Joining
Bridges in steel, boardwalks, decking, lake edging and bespoke landscape structures - CTS Ltd
CEG section Mezzanine floor beams - Mestec PLC
Coated steel - SSAB Dobel Coated Steel Ltd
Cold rolled purlins and rails - Ward Building Components Ltd
Cold rolled steel sections - Hadley Industries Plc
Collled rolled sections - Walls & Ceiling International Ltd
Copper wire and rods - United Wire Ltd
Domestic and industrial tongue and grooved flooring - Egger (UK) Ltd
Fabric tensile structures - Clyde Canvas Fabric Structures Ltd
Gypframe (steel framed houses) - Mestec PLC
Heavy duty protective coatings - International Paint, Protective Coatings,
Hot and cold pressings - Barton B C & Son Ltd
Hot dip galvanisers - Lancaster & Co (Bow) Ltd
Intumescent coatings for steel - Quelfire
Meshes - Cadisch Precision Meshes Ltd
Metal fabrication - OEP Furniture Group PLC
Metal fabricators to the construction industry - Builder's Iron & Zincwork Ltd
Metal framing - Unistrut Limited
Metsec framing - panelised light steel buildings - Mestec PLC
Open grill flooring - Norton Engineering Alloys Co Ltd
Rail fixings - Lindapter International
Roof and soffit ventilators - Ryton's Building Products Ltd
Roof truss system - MiTek Industries Ltd
Roof trusses - Llewellyn Homes Ltd
Special steel sections - British Steel Special Sections
Stainless steel fabrications - Midland Alloy Ltd
Steel fabrication - WA Skinner & Co Limited
Steel fabrication - Woodburn Engineering Ltd
Steel framed buildings - Yorkon Ltd
Steel framed buildings complete with fixtures, fittings and cladding - Reid John & Sons (Strucsteel) Ltd
Steel framing - Vantrunk Engineering Ltd
Steel tube stockholders, fabrications, welding and fittings - Midland Tube and Fabrications
Steel wire rope - Bridon International Ltd
Steel wire rope rigging - Marine & Commercial Rigging Systems
Steel wire, cables, rods and bars - McCalls Special Products
Steel wires, cables, rods and bars - Light (Multiforms), H A
Steelwork fabrication - Gilmer Limited
Structural frames - Bison Structures Ltd
Structural steel - Hy-Ten Reinforcement Co Ltd
Structural steel - Woodburn Engineering Ltd
Structural steel framing - Caunton Engineering Ltd
Structural steel framing - Hubbard Architectural Metalwork Ltd
Structural steel framing and general fabrication - Camtwix Engineering (Division of Terrapan Ltd)
Structural steel towers - Franklin Hodge Industries Ltd
Structural steelwork - Atlas Ward Structures Ltd
Structural steelwork - Billington Structures Limited
Structural steelwork - CAEC Howard (Holdings) Ltd
Structural steelwork - South Durham Structures Limited
Structural steelwork - Wescol Glosford
Structural steelwork, portal frames and general steelwork - Bromag Structures Ltd
Structural, mezzanine floors, conveyor support steelwork and architectural structures - Down & Francis Ltd, Division of Metalrax PLC
Support systems - Lindapter International
Timbers - Power Breaker
Wire - Twil Ltd
Wire rope assemblies and fittings - Midland Wire Cordage Co Ltd

G11 Structural aluminium framing

Aluminium extrusions/ sections - Seco Aluminium Ltd
Aluminium Structural Framing - Hartley Clear Span Ltd
Curved canopies, barrel vaults and fabrications - Midland Alloy Ltd
Ground floor framing - Parkside Group Ltd, The

G12 Isolated structural metal members

Cold roll forming - Mestec PLC
Cold roll sections - Metsec Building Products Ltd
Curved steel sections, tubes, roof and cambered beams - Angle Ring Co Ltd
Isolated structural members - Hubbard Architectural Metalwork Ltd
Isolated structural members - Hy-Ten Reinforcement Co Ltd
Steel chimneys and associated structural steelwork - Beaumont Ltd F E
Structural secondary steelwork - Down & Francis Ltd, Division of Metalrax PLC
Zed purlin and CEG purlins - Mestec PLC
Zed purlins - Brohome Ltd

G20 Carpentry/ Timber framing/ First fixing

Associated joinery - ARC Partitioning
Bespoke joinery - Howard Bros Joinery Ltd
Bridges in timber boardwalks, decking, lake edging and bespoke landscape structures - CTS Ltd
British and European hardwoods - Henry Venables Ltd
Carpentry/ Timber framing/ First fixing - Medina Gimson
Cellulose fibre insulation - Fillcrete Ltd
Composite timber 'I' beams - Fillcrete Ltd
Design, manufacture and supply only of timber engineered floor structures - Marlows Timber Engineering
Design, manufacture and supply only of trussed rafter and associated roof structures - Marlows Timber Engineering
Eaves fillers - Corofil Woodall Ltd
Fibreglass products - Yonder Hill Ltd
Fire resisting joinery - Decorfix
Fixings to trusses - MiTek Industries Ltd
Glulam beams - Laminated Wood Ltd
Hardwood - Latham James plc
Hardwood timber - Jakem Timbers Ltd
Industrial pre-treatment preservatives for timber - Protim Solignum Ltd
Intumescent coatings for timber - Quelfire
Joinery - XuXu Joinery Ltd
Joist hangers - Caradon Catnic Ltd
Joist hangers - Expamet Building Products
Laminated timber beams and structures - Moelven Laminated Timber Structures Ltd
Lightweight lattice joists and trusses - Mestec PLC
Load bearing and non LB site fixed panels - Mestec PLC
Nail plates - Gang-Nail Systems Ltd
Pneumatic nailers, staples and fastenings - Duo-Fast (UK) Ltd
Portal frames and cranked beams - Laminated Wood Ltd
Radiant heat barrier - Fillcrete Ltd
Radon barrier - Fillcrete Ltd
Restraint straps - Caradon Catnic Ltd
Roof space ventilation products - Willan Building Services Ltd
Roof trusses - Armstrong (Timber), Thomas, Ltd
Roof trusses - J. Scott (Thrapston) Ltd
Roof trusses - Merronbrook Ltd
Roof trusses - Meyer Drenham Ltd
Roofing contractor - Evans Howard Roofing Ltd
Roofing ventilators - Airflow (Nicoll Ventilators) Ltd
Rooftrusses - Sandersons Building (Milnrow) Ltd
Roofvents - Caradon Catnic Ltd
Saw millers and timber merchants - Whitmore's Timber Co Ltd
Stainless steel timber to masonry fixings - Helifix Ltd
Stainless steel warm roof fixings - Helifix Ltd
Steel decking - Lees Richard Ltd
Structural timber engineers - Kingston Craftsmen Structural Timber Engineers Ltd
Timber - Ashmead Buidling Supplies Ltd
Timber - Ashmead Building Supplies Ltd
Timber - Hunter Timber Wholesale Distribution
Timber - Moray Timber Ltd
Timber and log building specialists - Finwood Leisure Ltd
Timber for structural applications - Wiltshire G R & Co
Timber frame buildings - Fleming Buildings Ltd
Timber frame housing - Fleming Homes Ltd
Timber frame sheathing - Fillcrete Ltd
Timber frames and roof trusses - Walker Timber Ltd
Timber framing design and manufacture - Purpose Built Ltd
Timber merchants - Brewer T. & Co Ltd
Timber merchants - Derby Timber Supplies
Timber merchants - Timber Supplies
Timber merchants - Timbmet Ltd
Timber pallets - Yonder Hill Ltd
Timber panels - Challenge Fencing & Ltd
Timber portal frames engineered products, laminated beams, trussed rafters and bolted trusses - Structural Timbers Ltd
Timber preservative treatments - Kingfisher Chemicals Ltd
Timber roof trusses - Dover Trussed Roof Co Ltd
Timber Supplies - Bullock & Driffill Ltd
Timber to masonry connectors - Simpson Strong-Tie®
Timber to timber connectors - Simpson Strong-Tie®
Timber Trussed Rafters - Bullock & Driffill Ltd
Truss shoes - Caradon Catnic Ltd
Trussed rafter nailplate system, builders fixings and nailweb beams - Alpine Autimation Ltd.
Trussed rafters - The National Roof Truss Company
Trussed rafters - Wyckham Blackwell (Joinery) Ltd
Trussed rafters and timber frame buildings - Covers Timber Structures Ltd
Trusses, beams, frames, panels , timber engineering - Cowley Structural Timberwork Ltd
Wood finishes - Bollom J W & Co Ltd

G30 Metal profiled sheet decking

Composite decking - Broderick Structures Limited
Open steel industrial flooring - Eurogrid Ltd
Profiled metal roof decking - Ward Building Components Ltd
PVC Decking - LB Plastics Ltd
Sheet metal work - Barton B C & Son Ltd

G31 Prefabricated timber unit decking

Flat and pitched decking products - Torvale Building Products, A Division of Stadium Group PLC
Structural tongue and groove decking - Structural Timbers Ltd
Timber decking - McLoughlin Wood Ltd
Timber frame kits - Merronbrook Ltd
Timber Trussed Rafters - Bullock & Driffill Ltd

H10 Patent glazing

Aluminium and lead clothed steel patent glazing systems - Standard Patent Glazing Co Ltd
Aluminium extrusions/ sections - Seco Aluminium Ltd
Cladding/ covering - Pilkington UK Ltd
Commercial aluminium window doors - Barlow Architectural & Security Ltd
Curtain walling - Astrofade Ltd
Curtain Walling - Freeman T R
Curtain Walling - Heywood Williams Architectural
Curtain Walling - Ide T & W Ltd
Curtain walling - Parkwood Mellows Ltd
Curtain Walling - Yule & Son Ltd, A C
Firerated screens - Baydale Architectural Systems Ltd
Glazed steel frame, covered walkways and canopies - Planet Architectural Glazing Ltd
Installation - Oldham Signs Ltd
Patent glazing - Cantifix of London Limited
Patent glazing - Duplus Domes Ltd
Patent glazing - EBS Services Ltd
Patent glazing - Lonsdale Metal Company
Patent glazing, prorietary glazing systems - Kelsey Roofing Industries Ltd
Rooflights, lantern lights and pyramid lights - Standard Patent Glazing Co Ltd

H11 Curtain walling

Aluminium curtain walling - Glostal - Monarch
Aluminium extrusions/ sections - Seco Aluminium Ltd
Commercial curtain walling - Barlow Architectural & Security Ltd
Curtain wall systems - Parkside Group Ltd, The
Curtain walling - Alcover UK Ltd
Curtain walling - BLW Associates Ltd
Curtain walling - Duplus Domes Ltd
Curtain Walling - Duroy Fibreglass Mouldings
Curtain walling - English Architectural Glazing Ltd
Curtain Walling - KCW Windows & Conservatories Ltd
Curtain Walling - SCHUCO International KG
Curtain walling - Stoakes Systems Ltd
Curtain walling and cladding - Schmidlin (UK) Ltd
Extruded curtain walling - Smart Systems Ltd
Non- combustible roofing and cladding systems - Gilmour Ecometal
Patent glazing - Astrofade Ltd
Patent Glazing - Heywood Williams Architectural
Patent glazing - Ide T & W Ltd
Patent glazing - Parkwood Mellows Ltd
Patent glazing - Solaglas, Technical Advisory Service
Patent Glazing - Yule & Son Ltd, A C
Rainscreen - Gilmour Ecometal

H12 Plastics glazed vaulting/ walling

Plastic glazed vaulting - Duplus Domes Ltd
Plastic glazed vaulting/ walling - Astrofade Ltd
Polycarbonate glazing for flat and industrial roofs - Cox Building Products Ltd
Profiled plastic sheets, conservatory roofs and flat plastic sheets - Daylight Centre

H13 Structural glass assemblies

Barrelvaults, ridgelights, pyramids in glass or polycarbonate - Cox Building Products Ltd
Glass arches - Luxcrete Ltd
Project Management - Oldham Signs Ltd
Structural glass assemblers - Astrofade Ltd
Structural Glass Assemblies - Pilkington UK Ltd
Structural Glass Assemblies - SCHUCO International KG
Structural Glass Assemblies - Yule & Son Ltd, A C
Structural glazing - Alcover UK Ltd
Structural silicone glazing - Stoakes Systems Ltd

H20 Rigid sheet cladding

Advanced hygienic walls and ceilings - Advanced Hygienic Contracting Ltd
Armour sheeting - Meggitt Lavmour Systems
Cedar shakes for roofing and cladding - Brash John & Co Ltd
Cladding - Burlington Slate Ltd
Cladding - Coverite Ltd
Cladding panels - Blaze Neon Ltd
Curtain Walling - Solaglas, Technical Advisory Service
Exterior grade high pressure laminate - Abet Ltd
External cladding boards - Cape Boards Ltd
Fire boards - Cape Boards Ltd
Metal Roofing and cladding - Ash & Lacy Building Products Ltd
PVC claddings and fascias - Rockwell Sheet Sales Ltd
Pvc fascias, soffits, barge boards and claddings - Swish Building Products Ltd
Sheet materials - Mason FW & Sons Ltd
Vitreous enamelled steel panels - Alliance Europe
Wall Cladding, roof cladding, insulated wall and roof cladding - Kingspan Building Products Ltd
Wall lining systems - Formica Ltd

H30 Fibre cement profiled sheet cladding/ covering/ siding

Cladding - Eternit UK Ltd
Extruded plastics - Rustin Allen Ltd, Belbien Divsion
Fastenings - Cooper & Turner
Fibre cement profiled sheet cladding / covering - TRC (Midlands) Ltd
Fibre cement sheeting - Gleno Industries Ltd
Resin bound mineral coatings - Certifield Laboratories (Div or CPS Industries Ltd)

H31 Metal profiled sheet cladding/ covering/ siding

Aluminium cladding and roofing - Rigidal Industries Ltd.
Cladding - Caunton Engineering Ltd
Cladding - Orsogril Sarl UK Department
Cladding - Unilock Ltd
Cladding and siding - Bromag Structures Ltd
Composite aluminium and steel flat cladding systems - Booth Muirie Ltd
Composite aluminium and steel rain screens - Booth Muirie Ltd
Composite panels - Composite Panels Ltd
External Refurbishment & Repair - Pitchmastic PLC,
Fastenings - Cooper & Turner
Flat cladding panels - Technology Telford Ltd
Industrial roofing - Brown F plc
Insulated wall and roof panels and cladding sheets - Ward Building Components Ltd
Metal cladding - Decra Roofing Systems
Metal fabrication, ancillary cladding components - Interlink Group
Metal profiled / flat sheet cladding / covering - TRC (Midlands) Ltd
Metal profiled and flat sheet cladding, covering and siding - Brohome Ltd
Metal profiled/ flat sheet cladding/ covering - Alexio Roofing & Building Co Ltd
Metal profiled/ flat sheet cladding/ covering - Hadley Industries Plc
Metal profiled/ flat sheet cladding/ covering - Reid John & Sons (Strucsteel) Ltd
Metal roof, wall and decking - TAC Metal Forming Ltd
Metal Roofing - Coverite Ltd
Metal Roofing and cladding - Ash & Lacy Building Products Ltd
Metal roofing and cladding - Briggs Roofing & Cladding Ltd, Roofing Contract
Metal tiles - EBC UK Ltd
Prebonded claddding - Broderick Structures Limited
Prefabricated bathroom structures - IPPEC Sytsems
Profiled aluminium - Alcan Building Products
Profiled cladding - European Profiles
Profiled metal cladding, decking and roofing systems - Cladding & Decking UK Ltd
Profiled metal floor decking - Ward Building Components Ltd
Profiled metal sheeting - Gleno Industries Ltd
Profiled sheet stockholders - Builder's Iron & Zincwork Ltd
Rainscreen cladding - E. H. S. Roofing Ltd
Sheet Metal Work - Thanet Ware Ltd
Sheeting and cladding - TRC Service Group
Single ply flat roof system - Intergrated Polymer Systems (UK) Ltd
Stainless steel traditional longstrip roofing and cladding material - Lee Steel Strip Ltd
Steel and aluminium profiled roofing - Kelsey Roofing Industries Ltd
Traditional flat roofing systems - Ruberoid Building Products Ltd
Vitreous enamelled steel cladding panels - Escol Panels (S & G) Ltd
Wall cladding - Drawn Metal Ltd
Waterproofing - Euroroof Ltd

H32 Plastics profiled sheet cladding/ covering/ siding

Cellular PVC Fascias and Soffits - Celuform
Cladding and roof glazing - Vulcan Plastics Ltd
Corrugated or flat PVC and polycarbonate clear or coloured plastic sheets - Rockwell Sheet Sales Ltd
Covers for new and existing pools of almost any shape - Grando Ltd
Fascia, soffit and cladding systems - FloPlast Ltd
Fastenings - Cooper & Turner
Glass reinforced polyesther profiled and flat sheeting - Filion Products Ltd
GRP Cladding - Wilsons Fibreglass Ltd
Multiwall polycarbonate sheets - Rockwell Sheet Sales Ltd
Plastic fabrication - Colman Greeves
Plastic mouldings - Invicita Plastics Ltd
Plastic profile extrusion - HIS Ltd
Plastic profiled sheet cladding - Brett Martin Ltd
Plastic profiled sheet cladding /covering - TRC (Midlands) Ltd
Plastic profiled sheet cladding and roofing - Ariel Plastics Ltd
Plastic profiled sheet cladding, facias, soffits, interior wall and ceiling panelling - Deceuninck Ltd
Plastic profiled sheets - Leicester Barfitting Co Ltd
Plastic sheets and laminates - Vulcascot Ltd
PVC Cladding - LB Plastics Ltd
PVC single-ply roofing - Granflex Roofing Ltd
PVCu cladding - Hastings & Folkstone Glass Works Ltd
PVCu sills, trims, cladding and roof line products - Caradon Duraflex Systems Ltd
Thermoplastic roofing and cladding - EBC UK Ltd
Vinyl cladding system for fascias soffits and full elevations - Cox Building Products Ltd
Wall cladding - E. H. S. Roofing Ltd
Wall cladding - Opto International Ltd
Woodgrain weather boarding - Vulcan Plastics Ltd

H33 Bitumen and fibre profiled sheet cladding/ covering

Bitumious felt roofing systems - Ruberoid Building Products Ltd
Foundation protection - Onduline Building Products Ltd
Plastic profiled sheet cladding and roofing - Ariel Plastics Ltd

H40 Glassfibre reinforced cement panel cladding/ features

Glass reinforced cement cladding - Techrete (UK) Ltd

H41 Glassfibre reinforced plastics panel cladding/ features

Architectural GRP - Dewey Waters Ltd
Architectural mouldings - Hodkin Jones (Sheffield) Ltd
Columns, canopies, dormer windows and cladding - Reinforced Plastic Products Limited
G. R. P . Porches - Solair Ltd
Glass reinforced plastic cladding - Laminated Profiles Ltd
Glass reinforced resins and plastics - BLP (Hamble) Ltd
Glassfibre reinforced plastics cladding/features - Duroy Fibreglass Mouldings
GRG internal cladding - Multitex GRP & GRG Products
GRP building products - B & M Laminaites
GRP Canopies, Roofs, Mouldings - Birtley Building Products Ltd
GRP cladding - Adams- Hydraulics Ltd
GRP cladding and mouldings - Armfibre Ltd
GRP cladding, specialist moulding, door canopies and columns - Multitex GRP & GRG Products
GRP Coloumns - Alpha Glass Fibres Ltd
GRP door canopies and features - MR Limited
GRP flashings - Hambleside Danelaw Ltd
GRP housings - Precolor Sales Ltd
GRP Materials - Wilsons Fibreglass Ltd
GRP Pilasters - Alpha Glass Fibres Ltd
GRP roof lights and cladding products - Hambleside Danelaw Ltd
GRP Roof Sheets - Wilsons Fibreglass Ltd
GRP roofing systems for flat, low pitched and barrelled applications - Hambleside Danelaw Ltd
GRP structures & plant rooms - APA
Interior and exterior laminated glassfibre components - Pyramid Plastics Ltd
Plastic profiled sheet cladding and roofing - Ariel Plastics Ltd

Polycarbonate sheeting - Areco Roofing Supplies Company

H50 Precast concrete slab cladding/ features

Concrete products - Fyfe John Ltd
Concrete slab cladding - Marble Mosaic Co Ltd, The
Precast concrete cladding - Techrete (UK) Ltd
Precast concrete panels - Anderton Concrete Products Ltd
Precast concrete slab cladding/ features - Benton Co Ltd Edward
Precast concrete slab cladding/ features - Evans Concrete Products Ltd
Precast concrete slab cladding/features - Technostone Ltd
Retaining walls - Grass Concrete Ltd
Retaining Walls - Landscape Grass (Concrete) Ltd

H51 Natural stone slab cladding/ features

Architectural granite - Fyfe John Ltd
Lightweight stone, traditional marble and granite - Stonecraft & Chelsea Artisans Ltd
Masonry - Kerridge Stone
Natural stone - Macclesfield Stone Quarries Ltd
Natrual Stone Slab Cladding / Features - Toffolo Jackson(UK) Ltd
Natural granite & marble lightweight panels - Chelsea Artisans Ltd
Natural Limestone Masonry - Hanson Bath & Portland Stone
Natural stone cladding - Kirkstone
Natural stone slab cladding - Marble Mosaic Co Ltd, The
Natural stone slab cladding and features - Chilmark Quarries Ltd
Natural stone slab cladding/ features - Gregory Quarries Ltd, The
Natural stone slab cladding/ features - Ibstock Building Products Ltd
Natural stone slab cladding/ features - The Rare Stone Group Ltd
Natural stone slab cladding/features - Marriott & Price Ltd
Stonework restoration/cleaning - PermaRock Products Ltd
Terracotta rain screen cladding - Smithbrook Building Products Ltd

H52 Cast stone slab cladding/ features

Architectural cast stone - Hampton Stone Ltd
Cast Stone Slab Cladding / Features - Toffolo Jackson(UK) Ltd
Cast stone slab cladding /features - Haddonstone Ltd
Cast stone slab cladding/ features - Benton Co Ltd Edward
Cast stone slab cladding/ features - Chilstone Garden Ornaments
Cast stone slab cladding/ features - Ibstock Building Products Ltd
Cast stone slab cladding/features - Technostone Ltd
Cast stone slab/ cladding/ features - Evans Concrete Products Ltd
Cladding - Milner Delvaux Ltd
Mullion windows, quoins, cornices and doorways - Minsterstone Ltd
Reconstructed stone slab cladding - Marble Mosaic Co Ltd, The

H60 Plain roof tiling

Acrylic glass tiles - Klober Ltd
Clay and concrete tiles - Sandtoft Roof Tiles Ltd
Clay and concrete tiling - Standard Flat Roofing Co Ltd
Clay roof tile manufacturer - Blyth William
Clay roof tiles - Hinton Perry & Davenhill Ltd
Clay roof tiles and glazed clay roof tiles - Smithbrook Building Products Ltd
Clay roof tiling - HF (GB) Ltd
Clay roofing tiles - Michelmersh Brick & Tile Co Ltd
Clay, roof tiles and under tile vent - Tudor RooF Tile CO Ltd
Concrete and Clay Roof tiles - Marley Building Materials Ltd
Concrete roof tiles - Forticrete Ltd
Concrete roof tiles - Forticrete Roofing Products
Flat roofing - Mells Roofing Ltd
Handmade clay tiles - Keymer Hand Made Clay Tiles
Machine and hand crafted clay plain, roof tiles, creasing and interlocking clay tiles - Eternit Clay Tiles Ltd
Ornamental Tiles and Fittings - Hinton Perry & Davenhill Ltd
Pitched roofs and accessories - Anderson D & Son
Plain Clay Roof Tiles - Hinton Perry & Davenhill Ltd
Roof Coverings - Howard Evans Roofing Limited
Roof space ventilating tiles - Klober Ltd
Roof ventilation - Hambleside Danelaw Ltd
Roof ventilation - HV Air Conditioning Ltd
Roofing claywork - Red Bank Maufacturing Co Ltd
Roofing underlay felt - Newton John & Co Ltd

Ventilated ridge system - Klober Ltd

H61 Fibre cement slating

Fibre cement slates - Tegral Building Materials Ltd
Fibre cement slating - Eternit UK Ltd
Roof space ventilating slates - Klober Ltd
Roofing slate - Allan Harris & Sons Ltd
Roofing slates, ventilation products and breather membranes - Blunn Slates Ltd

H62 Natural slating

Copings and sills - Kirkstone
Natural slate cladding - McAlpine Slate Ltd, Alfred
Natural slate copings - McAlpine Slate Ltd, Alfred
Natural slate counters - McAlpine Slate Ltd, Alfred
Natural slate flooring - McAlpine Slate Ltd, Alfred
Natural slate landscaping - McAlpine Slate Ltd, Alfred
Natural slate roofing - Burlington Slate Ltd
Natural slate roofing - McAlpine Slate Ltd, Alfred
Natural slate sills - McAlpine Slate Ltd, Alfred
Natural slate walling - McAlpine Slate Ltd, Alfred
Natural slate worktops - McAlpine Slate Ltd, Alfred
Natural slating - Alexio Roofing & Building Co Ltd
Natural slating - Eternit UK Ltd
Natural Stone - Kerridge Stone
Polythene under slating sheet - Zedcor Ltd
Roofing - E. H. S. Roofing Ltd
Roofing - Kirkstone
Roofing slate - Allan Harris & Sons Ltd
Roofing slate - Delabole Slate
Slating - Standard Flat Roofing Co Ltd
Tegral Classic Natural slates - Tegral Building Materials Ltd
Welsh roofing slates - Greaves Welsh Slate Co Ltd
Welsh, Spanish and Canadian slates - Blunn Slates Ltd

H63 Reconstructed stone slating/ tiling

Green roofs - Euroroof Ltd
Manmade slating - Eternit UK Ltd
Recon: stone - Chiltern Concrete Products Ltd
Reconstituted slates - Sandtoft Roof Tiles Ltd
Reconstructed stone roofing tiles - Atlas Stone Products
Reconstructed stone slating - Delabole Slate

H64 Timber shingling

Cedar shingles for roofing and cladding - Brash John & Co Ltd

H65 Single lap roof tiling

Finials - Hinton Perry & Davenhill Ltd
Thatching, thatching materials and fire retardants - Thatching Advisory Services Ltd
Torch-on promenade tiles - Granflex Roofing Ltd

H66 Bituminous felt shingling

Foundation protection - Onduline Building Products Ltd
Tegola asphalt shingles - Matthew Hebden Ltd

H70 Malleable metal sheet prebonded coverings/ cladding

Corrugated sheeting - Eternit UK Ltd
Fully supported metal roofing - Kelsey Roofing Industries Ltd
Malleable metal sheet prebonded covering/ cladding - Reid John & Sons (Strucsteel) Ltd
Metal cladding coatings - Jotun Henry Clark Ltd

H71 Lead sheet coverings/ flashings

Aluminium flashing - George Gilmour (Metals) Ltd
Cloaks - Cavity Trays Ltd
Flashings - Cavity Trays Ltd
Lead and sheet flashings - Royston Lead Ltd
Lead clad steel and lead clad stainless steel cladding - Calder Industrial Materials Ltd
Lead roofing, sand cast lead sheet - Anglia Lead Ltd
Lead sheet and flashing - Calder Industrial Materials Ltd
Lead sheet coverings/ flashings - Alexio Roofing & Building Co Ltd
Lead sheet coverings/ flashings - Freeman T R
Lead sheet coverings/ flashings - Harris & Bailey Ltd
Leadwork - TRC Service Group
White and red lead - Liver Grease Oil & Chemical Co Ltd

H72 Aluminium strip/ sheet coverings/ flashings

Alucobond and aluminium cladding - Alcover UK Ltd
Aluminium covers - Franklin Hodge Industries Ltd
Aluminium & Steel Composite Roof & Wall Cladding - Rigidal Industries Ltd.
Aluminium and GRP roof edge trim - Areco Roofing Supplies Company
Aluminium coping - WP Metals Ltd
Aluminium Coping Systems - Dales Fabrications Ltd
Aluminium flashings - WP Metals Ltd
Aluminium rain screens cladding - Axter Ltd
Aluminium roof outlets - Marley Alutec Ltd
Aluminium roofing and cladding sheets - Gilmour Ecometal
Aluminium sheet coverings/ flashings - Alexio Roofing & Building Co Ltd
Aluminium sheet coverings/ flashings - Freeman T R
Aluminium sheet coverings/ flashings - Heywood Williams Architectural
Aluminium sheet coverings/ flashings - Reid John & Sons (Strucsteel) Ltd
Anodised Aluminium sheet - Metalloxyd Ano-Coil Ltd
Flashings - Corofil Woodall Ltd
Metal fabricators to the construction industry - Builder's Iron & Zincwork Ltd
Metal Roofing and cladding - Ash & Lacy Building Products Ltd
Perforated panels in aluminium - Browne Winther & Co Ltd
Powder coated aluminium - Dales Fabrications Ltd
Powdercoat aluminium - Leavlite Electropaint Ltd
Roofing material suppliers - Builder's Iron & Zincwork Ltd
Standing seam aluminium roofing - Kelsey Roofing Industries Ltd

H73 Copper strip/ sheet coverings/ flashings

Copper sheet coverings/ flashings - Alexio Roofing & Building Co Ltd
Copper sheet coverings/ flashings - Freeman T R
Copper strip - Metra Non Ferrous Metals Ltd
Traditional metal roofing contractors - Builder's Iron & Zincwork Ltd

H74 Zinc strip/ sheet coverings/ flashings

Traditional metal roofing contractors - Builder's Iron & Zincwork Ltd
Zinc and titanium alloy sheets - Metra Non Ferrous Metals Ltd
Zinc roofing - Rheinzink UK
Zinc sheet coverings/ flashings - Alexio Roofing & Building Co Ltd
Zinc sheet coverings/ flashings - Freeman T R

H75 Stainless steel strip/ sheet coverings/ flashings

Stainless steel sheet coverings/ flashings - Alexio Roofing & Building Co Ltd
Stainless steel sheet coverings/ flashings - Freeman T R
Traditional metal roofing contractors - Builder's Iron & Zincwork Ltd

H92 Rainscreen cladding

Cellular PVC Cladding - Celuform
Cladding & Rainscreening Panels Systems - Technology Telford Ltd
Facade systems/specialist wallcladding - Kelsey Roofing Industries Ltd
Faience Blocks - Shaws of Darwen
Metal Roofing and cladding - Ash & Lacy Building Products Ltd
Overcladding - Pitchmastic PLC,
Rainscreen cladding - Eternit UK Ltd
Rainscreen cladding - Miller Construction Ltd
Rainscreen Cladding - SCHUCO International KG
Rainscreen Panels - Rheinzink UK
Terracotta Blocks - Shaws of Darwen

J10 Specialist waterproof rendering

Cementitious and dry line membrane tanking - Dampcoursing Limited
Cementitious waterproof systems - Kingfisher Chemicals Ltd
Conductive coatings - PermaRock Products Ltd
Damp Proofing - Kiltox Damp Free Solutions
External rendering systems - PermaRock Products Ltd
Lliquid applied DPM - P.C.I Construction Systems Ltd
Renders & Castings - Snowcem - Webrr & Broutin
Rooflighting - Euroroof Ltd
Sika 1 - Sika Ltd - Sika Ltd
Specialist waterproof rendering - Concrete Repairs Ltd
Specialist waterproof rendering - Terminix Peter Cox Ltd
Water proofing systems - Beton Construction Ltd
Waterproofing - FEB Ltd
Waterproofing - Feb MBT
Waterproofing - Fullstop Technology Ltd
Waterproofing - Kiltox Damp Free Solutions
Waterproofing products - Ruberoid Building Products Ltd

J20 Mastic asphalt tanking/ damp proofing

Asphalt - Tillicoutry Quarries Ltd
Asphalt reinforcement - Bekaert Building Products
Basement tanking - Vandex (UK) Ltd
Damp proof membranes - Anderson D & Son
Damp proof membranes - Monarflex Ltd
Mastic asphalt tanking / damp proof membranes - Durable Contracts Ltd
Mastic asphalt tanking/ damp proof membranes - Alexio Roofing & Building Co Ltd
Mastic tanking and damp proof membranes - Canonbury Asphalte Co. Ltd.
Tanking - Dyke Chemicals Ltd
Tanking - Roofing Contractors (Cambridge) Ltd

J21 Mastic asphalt roofing/ insulation/ finishes

Asphalt work - TRC Service Group
Damp-proofing systems - Callenders Ltd
Flat Roofing - Pitchmastic PLC,
GRP and aluminium roof edge trims and flashings - Pitchmastic PLC, Building Products Division
Mastic Asphalt - Coverite Ltd
Mastic asphalt roofing - Standard Flat Roofing Co Ltd
Mastic asphalt roofing and car parks - Briggs Roofing & Cladding Ltd, Roofing Contract
Mastic asphalt roofing/ insulation/ finishes - Alexio Roofing & Building Co Ltd
Mastic asphalt roofing/ insulation/ finishes - BFRC Services Ltd
Mastic roofing - Canonbury Asphalte Co. Ltd.
Porous concrete promenade tiles - Spartan Tiles Ltd
Roof breather vents - Areco Roofing Supplies Company
Roof waterproofing systems, colour solar reflective roof finishes - Dyke Chemicals Ltd
Roofing - North Herts Asphalte Ltd
Roofing - Roofing Contractors (Cambridge) Ltd
Roofing contractors - Durable Contracts Ltd
Roofing systems - Callenders Ltd
Tarpauling - Monarflex Ltd
Tarpauling - Power Plastics Ltd

J22 Proprietary roof decking with asphalt finish

Bitumen boilers - Bitmen Products Ltd
Flat roof decking - Isocrete Group Sales Ltd
Flat Roofing - Coverite Ltd
Mastic asphalt mixers - Bitmen Products Ltd
Pre coated chipspreaders - Bitmen Products Ltd
Proprietary roof decking with asphalt finish - Durable Contracts Ltd
Roof tile underlays - Monarflex Ltd
Roofing contractor - Evans Howard Roofing Ltd
Roofing membranes - Onduline Building Products Ltd
Torch on roofing - Ruberoid Building Products Ltd

J30 Liquid applied tanking/ damp proofing

Basement tanking - Vandex (UK) Ltd
Basement water proofing - Timberwise (UK) Plc
Cement Waterproofing - Keeling Oliver
Damp proof coatings - RIW Ltd
DPC tanking systems - Ruberoid Building Products Ltd
Flashings - Dyke Chemicals Ltd
Grouts and sealers for tiles and surface of floors - P.C.I Construction Systems Ltd
Heldite Joining Compound - Heldite Ltd
High performance floor coatings - Intergrated Polymer Systems (UK) Ltd
Hot air and torch applied membranes - Phoenix Roofing
Internal Tanking of Plant/ Tank Rooms & Suspended Floors - Ferguson G A & Co Ltd
Liquid applied damp proof membranes - Kingfisher Chemicals Ltd
Liquid applied tanking/ damp proof membranes - Dex-o-Tex International Ltd
Liquid applied tanking/ damp proof membranes - GRAB Industrial Flooring Ltd
Liquid applied tanking/damp proof membranes - Trade Sealants
Liquid applied waterproof membranes - Kelsey Roofing Industries Ltd
Structural waterproofing - Euroroof Ltd
Wall coatings - Intergrated Polymer Systems (UK) Ltd
Wall coatings - Lastite Building Products Ltd
Waterproofing membranes - Kenyon Industrial Paints & Adhesives
Waterproofing products - Kingsnorth Bitumen Products Ltd

J31 Liquid applied waterproof roof coatings

Bitumen based flat roofing membranes - Kelsey Roofing Industries Ltd
Bituminous roofing - TRC (Midlands) Ltd
Bitumious coatings - Protim Services Ltd

Civil Engineering Repair - Pitchmastic PLC,
Damp-proof membranes - Dyke Chemicals Ltd
Elastometric seamless roofing membrane - Resiblock Ltd
Flat roof waterproofing - Alumasc Exterior Building Products Ltd
Fluid applied roof coatings - Liquid Plastics Ltd
High performance roof coatings - Intergrated Polymer Systems (UK) Ltd
Liquid applied waterproof coating - Bondaglass Voss Ltd
Liquid applied waterproof coatings - BFRC Services Ltd
Liquid applied waterproof coatings - Certifield Laboratories (Div or CPS Industries Ltd)
Liquid applied waterproof coatings - Concrete Repairs Ltd
Liquid applied waterproof coatings - Dex-o-Tex International Ltd
Liquid applied waterproof coatings - GRAB Industrial Flooring Ltd
Liquid applied waterproof coatings - Harris Road Roofing & Flooring Ltd
Liquid applied waterproof coatings - Tor Coatings Group
liquid applied waterproof coatings - Trade Sealants
Liquid applied waterproof roof coatings - Hyflex Roofing
Liquid membranes - TRC Service Group
Liquid-applied roofing and waterproofing - Granflex Roofing Ltd
Masonry paints - MR Limited
Roof coatings - Crosbie Coatings
Roof coatings - Lastite Building Products Ltd
Roof repair membranes - Kingfisher Chemicals Ltd
Roof weather proofing - Infil Ltd
Sealing/ Dust and Oil Proofing Multi-Storey Car Park Decks - Ferguson G A & Co Ltd
Structural Waterproofing - Coverite Ltd
Waterproof roof coatings - RIW Ltd
Waterproofing products - Kingsnorth Bitumen Products Ltd
Waterproofing Roofs of Multi-Storey Car Parks - Ferguson G A & Co Ltd

J32 Sprayed vapour control layers

Radon and methane gas barriers - Monarflex Ltd
Sprayer vapour barriers - Trade Sealants
Vapour barriers and breather membranes - Monarflex Ltd

J33 In situ glassfibre reinforced plastics

Fibertex geotextiles - Tex Steel Tubes Ltd
In Situ GRP - Precolor Sales Ltd

J40 Flexible sheet tanking/ damp proofing

Basement tanking - Vandex (UK) Ltd
Bitumen membranes - Vulcanite Limited
Damp proof courses - Cavity Trays Ltd
Damp proof membranes - Isocrete Group Sales Ltd
Damp proof membranes - Marley Waterproofing
Damp proof tanking - Callenders Ltd
Damproof membranes - Newton John & Co Ltd
DPM - Visqueen Agri
Flexible sheet damp proof membranes - Kingfisher Chemicals Ltd
Flexible Sheet Membrane - Anaplast Ltd
Flexible sheet membranes - RIW Ltd
Flexible sheet tanking / damp proof membranes - Durable Contracts Ltd
Flexible sheet tanking and damp proof membranes - Harris Road Roofing & Flooring Ltd
Flexible sheet tanking/ damp proof membranes - BFRC Services Ltd
Flexible sheet tanking/ damp proof membranes - John Davidson (Pipes) Ltd
Floor coatings - Lastite Building Products Ltd
Gas barrier membranes and pond liners - Geosynthetic Technology Ltd
Membranes - Cavity Trays Ltd
Polythene dampproof membranes - Zedcor Ltd - Zedcor Ltd
Single Ply Roofing Systems - Sarnafil Ltd
Vapour permeable membranes - Willan Building Services Ltd
Waterproof expantion joints - Radflex Contract Services Limetid
Waterproof mebranes - Grace Services
Waterproofing membranes - Schlegel Engineering
Waterproofing sheets - ABG Ltd

J41 Built up felt roof coverings

Apex roofing - Apex Roofing Contractors
Bituminous roofing felts - Marley Waterproofing
Bituminous roofing system - Vulcanite Limited
Built up felt roof coverings - Alexio Roofing & Building Co Ltd
Built up felt roof coverings - Canonbury Asphalte Co. Ltd.
Built up felt roof coverings - Durable Contracts Ltd
Built up felt roof coverings - Harris Road Roofing & Flooring Ltd
Built up felt roof coverings - Roofrite (East Anglia) Ltd
Built up felt roof coverings - Standard Flat Roofing Co Ltd
Built up felt roof coverings - TRC (Midlands) Ltd
Built up felt roofing - Briggs Roofing & Cladding Ltd, Roofing Contract
Built Up Felt Roofing - North Herts Asphalte Ltd
Built up felt roofing - Roofing Contractors (Cambridge) Ltd
Built up roofing - Anderson D & Son
Built-up felt roofing - Granflex Roofing Ltd
Felt roofing - TRC Service Group
Flat roofing - Anderson D & Son
Flat roofing built up felt roofing - Chesterfelt Ltd
High Performance Felt - Coverite Ltd
Membrane waterproofing - BFRC Services Ltd
Pitched roofing - Mells Roofing Ltd
Porous concrete promenade tiles - Spartan Tiles Ltd
Roof coverings - Howard Evans Roofing Limited
Roof membranes - Axter Ltd
Roofing compound - Bartoline Ltd
Roofing contractor - Evans Howard Roofing Ltd
Roofing membrane - Nuralite (UK) Ltd
Roofing products - Cavity Trays Ltd
Roofing products - Klober Ltd
Structural water proofing - Alumasc Exterior Building Products Ltd
Vapour membranes - Klober Ltd
Weather protection sheeting and sheeting - Power Plastics Ltd

J42 Single layer polymeric roof coverings

Conservatory roof systems, vents and glazing bars - Newdawn & Sun Ltd
Flat roofing - E. H. S. Roofing Ltd
Polymeric roof coatings - MMP International
Single layer plastic roof coverings - BFRC Services Ltd
Single layer plastic roof coverings - Marley Waterproofing
Single layer plastic roof coverings - Standard Flat Roofing Co Ltd
Single layer polymeric roof coverings - Durable Contracts Ltd
Single layer PVC and CPE membrane roofing systems - Alkor Draka Ltd
Single layer roofing - TRC Service Group
Single ply polymetric membranes - Roofing Contractors (Cambridge) Ltd
Single ply roofing - Coverite Ltd
Single ply roofing - ICB (International Construction Bureau) Ltd
Single ply roofing and roof accessories - Anderson D & Son
Single ply roofing membrane - Kelsey Roofing Industries Ltd
Single ply systems - Briggs Roofing & Cladding Ltd, Roofing Contract
Water repellent solutions - Dyke Chemicals Ltd
Waterproofing systems - Beton Construction Ltd

J43 Proprietary roof decking with felt finish

Proprietary roof decking with felt finish - Durable Contracts Ltd
Roofing ventilators - Airflow (Nicoll Ventilators) Ltd
Single ply roofing - Firestone Building Products Ltd
Standing seam roofing - E. H. S. Roofing Ltd
Waterproofing - Alumasc Exterior Building Products Ltd

J44 Sheet linings for pools/ lakes/ waterways

Pond / landfill lining systems - ABG Ltd
Single Ply Roofing Systems - Sarnafil Ltd
Waterproofing systems - Beton Construction Ltd

K10 Plasterboard dry linings/ partitions/ ceilings

Custom Rolled Sections to Client Specification - Hill Top Sections Ltd
Dry lining systems - British Gypsum Ltd
Drylining - Nevill Long Limited
Drywalling collated screws, nails and tools - Duo-Fast (UK) Ltd
Fire Barrier Angle & Strap - Hill Top Sections Ltd
Fire stopping and barriers - Firebarrier Services Ltd
Gysum fibreboards - Fels UK
Moveable Walls - Becker (SLIDING PARTITIONS) Ltd
Partitions - Unilock Ltd
Plasterboard dry lining - Harris & Bailey Ltd
Plasterboard dry lining/partitions/linings - Fastwall Ltd
Plasterboards and drywall Accessories - Knauf
Relocatable aluminium trimmed system - Neslo Partitioning & Interiors
Sliding Partitions - Becker (SLIDING PARTITIONS) Ltd
Stud & Track Partitioning Sections - Hill Top Sections Ltd
Supplier of Ceilings, Partitioning, Dry lining and Interior Building products. - Nevill Long Limited
Wall Liner System - Hill Top Sections Ltd

K11 Rigid sheet flooring/ sheathing/ decking/ sarking/ linings/ casings

Acoustic insulation boards - Fels UK
Chipboard - Woodmark Ltd
Column and beam encasement systems - Knauf
Cutting, drilling and machining - Boardcraft Ltd
Dry Flooring elements - Fels UK
Fire protection boards - Fels UK
Floating floors - Designed for Sound Ltd
Flooring accessories - Atkinson & Kirby Ltd
Flooring panels - Cape Boards Ltd
Hardwood and softwood - Boardcraft Ltd
Impact resistant boards - Fels UK
Insulating laminates and plasters - Knauf
Laminate joinery - Decra Cubicles Ltd
Laminates and MDF's - Kronospan Ltd
Lightweight composite boards and panels - Quinton & Kaines (Holdings) Ltd
MDF sheets - Nillamette Europe Ltd
Moisture resistant boards - Fels UK
OSB - Kronospan Ltd
Panel products - Latham James plc
Plain and melamine faced clipsboard - Kronospan Ltd
Plain and melamine faced flooring - Kronospan Ltd
Plain, melamine faced and lacquered MDF - Kronospan Ltd
Plywood, hard, fibre, chip and particle boards - Boardcraft Ltd
Roof deck panels - Cape Boards Ltd
Semi finished plastics and building products - Amari Plastics Plc
Soffit lining boards - Cape Boards Ltd
Sprays fire proof - Firebarrier Services Ltd
Thermal insulated panels - Panel Systems Ltd
Timber floors - Jutlandia Doors
Wet wall and showerwall panels, wet area wall lining and ceiling panels - Norske Interiors (UK) Ltd
Woodbased panels - Egger (UK) Ltd

K13 Rigid sheet fine linings/ panelling

Acoustic flooring and acoustic wall products - Trim Acoustics
Bespoke joinery - Howard Bros Joinery Ltd
Carved cornice, frieze and embellishments - Oakleaf Reproductions Ltd
Cladding panel - Panel Systems Ltd
Cladding panels - Petrarch Claddings Ltd
Cubicle systems - Formica Ltd
Decorative lining boards - Cape Boards Ltd
Facing laminates - Wilsonart Ltd
Faux book pines, mirrors and frames - Oakleaf Reproductions Ltd
High performance building panel - Trespa UK Ltd
High pressure decorative laminates - Abet Ltd
Hygienic and railway arch lining materials - Rockwell Sheet Sales Ltd
Hygienic finish lining boards - Cape Boards Ltd
Hygienic GRP Cladding - Laminated Profiles Ltd
Laminated Amiran - Pilkington Plyglass
Linenfold wall panelling, riven oak wall panelling - Oakleaf Reproductions Ltd
Partitions - Unilock Ltd
Plastic laminate fabricators - PFP Fabrications Ltd
Plywood, hard, fibre, chip and particle boards - Boardcraft Ltd
Rigid sheet panelling - Advanced Panels & Products
Rigid sheet wall panelling high performance cladding - Combat Polystyrene Group Ltd
Softwood - Latham James plc
Structural wall panels - Marshalls Mono Ltd
Toilet cubicles - Flexi-Plan Partitions Ltd
Veneered boards - Atkinson & Kirby Ltd
Wall lining systems - Knauf
Wall paneling - Viking Laminates Ltd
Wall systems - Neslo Partitioning & Interiors

K20 Timber board flooring/ decking/ sarking/ linings/ casings

Cane woven panels - Browne Winther & Co Ltd
Decorative laminated flooring - Egger (UK) Ltd
Densified wood and industrial flooring - Permali Gloucester Ltd
Fire protection boards - Knauf
Glazed aluminium system with internal blinds - Neslo Partitioning & Interiors
Hardwood Timber Importers - Parker Kislingbury Ltd
Plywood, hard, fibre, chip and particle boards - Boardcraft Ltd
Plywood, hard, fibre, chip and particle boards - Boardcraft Ltd
Sanding and Sealing wood floors - Sportmark (Leisure Products) , Part of Sportsmark Group Ltd
Specialised timber panels - McLoughlin Wood Ltd
Timber cladding, mouldings and flooring - Latham James plc

K21 Timber strip/ board fine flooring/ linings

Hardwood flooring - Atkinson & Kirby Ltd
Hardwood flooring - Gerland Ltd
Laminates - Egger (UK) Ltd
Perforated panels in hardboard and MDF - Browne Winther & Co Ltd
Solid hardwood flooring - Junckers Limited
Strip floor - McLoughlin Wood Ltd
Teak Specialist - Parker Kislingbury Ltd
Wooden strip flooring - Lydney Products Ltd

K30 Panel partitions

Acoustic folding partitions - Pellfold Parthos Ltd.
Acoustic movable walls - Hillaldam Coburn Ltd
Aluminium extrusions/ sections - Seco Aluminium Ltd
Associated drylining - ARC Partitioning
Demountable partitioning - Internal Partitions Systems
Demountable Partitioning Systems - Brown F plc
Demountable partitions - Flexi-Plan Partitions Ltd
Demountable partitions - HEM Interiors Group Ltd
Demountable partitions - Komfort Systems Ltd
Demountable partitions - Norwood Partition Systems Ltd
Demountable partitions and associated electrical works - Opus 4 Ltd
Demountable suspended ceilings - Clestra Hauserman Ltd
Dry lining systems - British Gypsum Ltd
Drywall systems - Lafarge Plasterboard Ltd
Drywalling collated screws, nails and tools - Duo-Fast (UK) Ltd
Industrial and cleanroom partition systems - NT Partition Systems
Interior contractors - Gifford Grant Ltd
Mesh partitioning - Barton Storage Systems Ltd
Mineral fibre and vermiculite board - Cryotherm Insulation Ltd
Mobile partitions - Dividers Ltd
Movable walls - Pellfold Parthos Ltd.
Partitioning - Hatmet Ltd
Partitioning - Hatmet Ltd
Partitioning - New Forest Ceilings Ltd
Partitioning - Newey Ceilings
Partitioning - Welconstruct Co
Partitioning and dry lining - Nevill Long Limited
Partitioning and wall storage systems - Fastwall Ltd
Partitioning systems - Knauf
Partitioning systems - Nevill Long Limited
Partitions - Adex Storage Equipment Ltd
Partitions - Brown F plc
Partitions - Davies John Interiors Ltd
Partitions - Grafton Magna Ltd
Partitions - ITS Ceilings & Partitions
Plaster boards - Ashmead Building Supplies Ltd
Plaster boards - Ashmead Building Supplies Ltd
Plywood, hard, fibre, chip and particle boards - Boardcraft Ltd
Relocatable partitioning and ceilings - Clestra Hauserman Ltd
Relocatable timber and glazed partition systems - Neslo Partitioning & Interiors
Relocateable and demountable partitions - ARC Partitioning
Sliding and folding partitions - Applied Acoustics (a part of Henry Venables Limited)
Solid, decorated plasterboard partitions - Neslo Partitioning & Interiors
Trimless decorated partition systems - Neslo Partitioning & Interiors

K32 Panel cubicles

Changing cubicles - Bawn W. B. & Co Ltd
Cubical partitions - NT Partition Systems
Cubicle and washroom systems - Nevill Long Limited
Cubicle partitions - Architectural & Building Products Ltd
Cubicle systems - Panel Systems Ltd
Cubicles - Viking Laminates Ltd
Framed cubicle partitions - Harris & Bailey Ltd
Framed panel cubical partitions - HEM Interiors Group Ltd
Office partition systems - NT Partition Systems
Office partitions - Origlia Spa
Plywood, hard, fibre, chip and particle boards - Boardcraft Ltd
Purpose made doors, toilet cubicles and washroom systems - Moffett Thallon & Co Ltd
Toilet and shower cubicles - Decra Cubicles Ltd
Toilet cubicles - Combat Polystyrene Group Ltd
Toilet, shower & changing cubicles - Aaztec Cubicles
Trimless double glazed partition systems - Neslo Partitioning & Interiors
Washroom cubicle systems - Bushboard Parker Ltd
WC cubicles and washroom systems - Industrial Building Componants Ltd

K33 Concrete/ Terrazzo partitions

Mosaic and terrazzo - Alpha M & T Ltd

K40 Demountable suspended ceilings

Acoustic ceilings - CCS Scotseal Ltd
Ceiling systems - Formwood
Ceiling systems - OWA UK Ltd

Ceiling systems - USG (UK) Ltd
Ceiling tiles - Cep Ceilings Ltd
Ceilings - Adex Storage Equipment Ltd
Celling furring System - Hill Top Sections Ltd
Celling Suspention Components - Hill Top Sections Ltd
Demountable suspended ceilings - Clestra Hauserman Ltd
Dry lining systems - British Gypsum Ltd
Drylining - Neslo Partitioning & Interiors
Fire rated ceiling systems - Fire Protection Ltd
Fire retardent and acoustic suspended ceilings - Howard Home Improvements Ltd
Grid, plaster, metal and soft fibre suspended ceilings - Armstrong World Industries Ltd
Metal and 3D ceiling systems - USG (UK) Ltd
Relocatable partitioning and ceilings - Clestra Hauserman Ltd
Specialist in Cold Formed Sections - Hill Top Sections Ltd
Supply and installation of suspended ceilings - ITS Ceilings & Partitions
Suspended ceiling brackets and accessories - Lewis Spring Products Ltd
Suspended ceiling panels - Cape Boards Ltd
Suspended ceiling systems - Rockfon Ltd
Suspended ceilings - Brown F plc
Suspended ceilings - Clark & Fenn Ltd
Suspended ceilings - Dampa (UK) Ltd
Suspended ceilings - Davies John Interiors Ltd
Suspended ceilings - Flexi-Plan Partitions Ltd
Suspended ceilings - Grafton Magna Ltd
Suspended ceilings - Hatmet Ltd
Suspended ceilings - HEM Interiors Group Ltd
Suspended ceilings - HT Martingale Ltd
Suspended Ceilings - Illbruck Ltd
Suspended ceilings - Internal Partitions Systems
Suspended ceilings - Nevill Long Limited
Suspended ceilings - New Forest Ceilings Ltd
Suspended ceilings - Newey Ceilings
Suspended ceilings - NT Partition Systems
Suspended ceilings - Opus 4 Ltd
Suspended ceilings - Walls & Ceiling International Ltd
Suspended ceilings, shaftwall Systems - Knauf

K41 Raised access floors

Access flooring - Moseley GRP Product (Division of Moseley Rubber Co Ltd)
Cable management - Interface Europe Ltd
Flooring and access - Wincro Metal Industires Ltd
Full fit out interiors service - Neslo Partitioning & Interiors
Partial Access Timber flooring - Durabella Ltd
Plywood, hard, fibre, chip and particle boards - Boardcraft Ltd
Raised & Access floors. - Nevill Long Limited
Raised access flooring - Brown F plc
Raised access flooring - Mauser Interiors (UK) Ltd
Raised Access flooring - USG (UK) Ltd
Raised Access Floors - Hatmet Ltd
Raised access floors - HEM Interiors Group Ltd
Raised access floors - Hewetson Floors Ltd
Raised access floors - New Forest Ceilings Ltd
Raised access floors - Redman Fisher Engineering
Rased Access Floors - Tate - Alumasc Interior Building Product Limited
Steel Raised Access flooring - Durabella Ltd
Structural refurbishment, fit-out and furnishing of buildings - Brown F plc
Timber Raised Access flooring - Durabella Ltd

L10 Windows/ Rooflights/ Screens/ Louvres

Aluminium conservatories - Ducana Surrey Ltd
Aluminium extrusions/ sections - Seco Aluminium Ltd
Aluminium louvres - WP Metals Ltd
Aluminium windows - Alcover UK Ltd
Aluminium windows - Fabricated Aluminium Services Ltd
Aluminium windows - Glostal - Monarch
Aluminium windows - Hastings & Folkstone Glass Works Ltd
Aluminium windows - KCW Windows & Conservatories Ltd
Aluminium windows - Lakeside Group
Aluminium windows - Midland Alloy Ltd
Aluminium windows - Parkside Group Ltd, The
Aluminium windows - Planet Architectural Glazing Ltd
Aluminium windows - Prima Systems (South East) Ltd
Aluminium/ Wood Composit Windows - Scandanavian Window Systems Ltd
Anti-Reflective Glass - Schott Glass Ltd, Arcitectural Glass
Armour plate door assemblies - Ide T & W Ltd
Atria - Lonsdale Metal Company
Atrium glazing - EBS Services Ltd
Bespoke joinery - AS Newbould Ltd
Bespoke joinery - Howard Bros Joinery Ltd
Bird screen - Timberwise (UK) Plc
Building hinges - Nico Manufacturing Ltd
Bullet resistant window doors and counter screen - Barlow Architectural & Security Ltd
Composite aluminium windows - Scandanavian Timber
Composite windows - Crittall Steel Windows Ltd
Composite windows - Crittall Steel Windows Ltd
Composite windows - Velfac Ltd
Conservatory roofing, atria, canopies and rooflights - Ultraframe PLC
Double glazed roof windows - Klober Ltd
Double glazed units - Ledite Glass (Southend) Ltd
Double glazed windows - Alframes Holdings Ltd
Double glazed windows - Alframes Holdings Ltd
Entrance screens, shopfronts - Pearsons J R (Birmingham) Ltd
Extruded aluminium composite windows - Smart Systems Ltd
Extruded shopfronts and entrance screens - Smart Systems Ltd
Extruded UPVC windows - Halo, W H S Division Of Bowater Industries
Extruded UPVC windows - Smart Systems Ltd
Firerated windows - Baydale Architectural Systems Ltd
French opening door bar - Badderley Rose Ltd
G. R. P. Canopies - Solair Ltd
Glass & Window Supplies - Spectra Glaze Services
Glazed malls - Lonsdale Metal Company
Glazing Maintenance - Spectra Glaze Services
GRP Canopies - Alpha Glass Fibres Ltd
GRP Dormer Windows - Wilsons Fibreglass Ltd
GRP louvres - APA
High performance timber windows - Applied Acoustics (a part of Henry Venables Limited)
High performance timber windows - Henry Venables Ltd
High performance windows - Sashless Window Co Ltd
Industrial grade plastic framed roof windows for slate/tile roofs - Cox Building Products Ltd
Insect screens - President Blinds Ltd
Internal and external doors - Rugby Joinery
Lead Light Glazing - Yule & Son Ltd, A C
Loft access traps - Willan Building Services Ltd
Louvres - McKenzie-Martin Ltd
Manual control window opening systems - Morse Controls Ltd
Metal and glass louvres grilles, diffusers and dampers - Air Diffusion Ltd
Metal framed double glazed windows - Nova Group Ltd
Metal Windows - Anglian
Metal windows - Olivand Metal Windows Ltd
Metal Windows - Thanet Ware Ltd
Metal windows - The Cotswold Casement Company
Metal windows/ rooflights/ screens/ louvres - Drawn Metal Ltd
Metal windows/ rooflights/ screens/ louvres - Heywood Williams Architectural
Metal windows/ rooflights/ screens/ louvres - Reid John & Sons (Strucsteel) Ltd
Metal windows/Rooflights/screens - Yule & Son Ltd, A C
Modular and continuous rooflights - Willan Building Services Ltd
Plastic framed double glazed windows - Nova Group Ltd
Plastic framed double glazed windows - Radway, Door and Windows Ltd.
Plastic profiled sheet cladding and roofing - Ariel Plastics Ltd
Plastic rooflights - Brett Martin Ltd
Plastic verical sliding windows - Masterframe Windows Ltd
Plastic windows - Andersen/ Black Millwork
Plastic windows - Anglian
Plastic windows - Caradon Everest Ltd
Plastic windows - Ideal Williams Ltd
Plastic windows - Lynman Windows
PVC extrusions - Nenplas
PVC Window and door systems - LB Plastics Ltd
PVC window profiles - Profile 22 Systems Ltd
PVC windows - Prima Systems (South East) Ltd
PVC windows - Scholes Windows Ltd
PVC windows - Swish Building Products Ltd
PVC-u windows - IsoSystems
PVC-U windows, doors and conservatories fabricated and installed - Elliott Medway FineLine
PVCU extrusions for window systems - Spectus UPVC Systems Ltd
Pvcu Plastic Windows/roof lights/screens/louvres - Yule & Son Ltd, A C
PVCU window fabrication - HIS Ltd
PVCu window systems - HW Systems Ltd
PVCu windows - Dale Joinery Ltd
PVCu windows - Hastings & Folkstone Glass Works Ltd
PVCu windows - Interframe Ltd
PVCu windows - KCW Windows & Conservatories Ltd
PVCu windows - Plastal Commercial Ltd
Replacement Window service - Spectra Glaze Services
Roof lights - Lonsdale Metal Company
Roof lights - Luxcrete Ltd
Roof vents - Lonsdale Metal Company
Roof windows - VELUX Co Ltd, The
Roofglazing - EBS Services Ltd
Rooflights - Astrofade Ltd
Rooflights - Astrofade Ltd
Rooflights - Astrofade Ltd
Rooflights - Axter Ltd
Rooflights - Daylight Centre
Rooflights - McKenzie-Martin Ltd
Rooflights - Stoakes Systems Ltd
Rooflights - Torvale Building Products, A Division of Stadium Group PLC
Rooflights - Ubbink (UK) Ltd
Rooflights and skylights - Wenmore Rooflights Ltd
Rooflights, skylights, Atria, covered walkways, barrel vaults, deadlights, aluminium windows - Duplus Domes Ltd
Ropes and sashcords - Marlow Ropes Ltd
Sash balences - Garador Ltd
Sash Windows - J. Scott (Thrapston) Ltd
Screens - Komfort Systems Ltd
Screens/Louvres - Allaway Acoustics Ltd
Security screens and equipment - Pearce Signs Ltd
Security screens, counters and partitioning - Chubb Security Installations Ltd
Shopfront systems - Aardee Locks and Shutters Ltd
Shopfronts - Bassett & Findley Ltd
Sliding patio door bar - Badderley Rose Ltd
Stained glass windows - Ide T & W Ltd
Steel and aluminium windows - Parkwood Mellows Ltd
Steel doors - Badderley Rose Ltd
Steel screens - Fitzpatrick Door Frames Ltd
Steel windows - Crittall Steel Windows Ltd
Steel windows - Monk Metal Windows Ltd
Steel windows and doors - Clement Steel Windows
Steel Windows Installation - Rea Metal Windows Ltd
Steel Windows Manufacture - Rea Metal Windows Ltd
Structured Steel framing - Phoenix Metal Products Ltd
Supply & Install PVCu Windows, Doors & Conservatories - Piper Windows Doors & Conservatories
Timber conservatories - Fairmitre Ltd
Timber framed double glazed windows - Nova Group Ltd
Timber porches - Fairmitre Ltd
Timber windows - Andersen/ Black Millwork
Timber windows - Anglian
Timber windows - Bennett Windows Ltd
Timber windows - Boulton & Paul Ltd
Timber windows - Fairmitre Ltd
Timber windows - Hill Leigh (Joinery) Ltd
Timber windows - Jutlandia Doors
Timber windows - Langford Joinery Ltd
Timber windows - Lyssand UK Ltd
Timber Windows - Premdor
Timber windows - Sampson Windows Ltd
Timber windows - Scandanavian Timber
Timber windows - Scandanavian Window Systems Ltd
Timber windows - Swedhouse Ltd
Timber windows - Walker Timber Ltd
Timber windows and screens - Hansen Windows Ltd
Translucent rooflights - Filion Products Ltd
Translucent wall and roof systems - Stoakes Systems Ltd
UPVC conservatories - Ducana Surrey Ltd
UPVC double glazed windows - Sierra Windows
UPVC window casements - Windowbuild
UPVC window fabrication and fixing supplies - Window Fabrication & Fixing Supplies
UPVC window systems - Profolia Ltd
UPVC window systems - Rehau Ltd
UPVC windows - BLW Associates Ltd
UPVC windows - Fabricated Aluminium Services Ltd
UPVC windows - Juno Roplasto Ltd
Vertical sliding Windows - Radway, Door and Windows Ltd.
Window control gear - Cooper H W & Co Ltd
Window hardware - Shaw Arthur Manufacturing Ltd
Window hardware - Shaw Arthur Manufacturing Ltd
Window hardware - Shaw Arthur Manufacturing Ltd
Window hinges - Nico Manufacturing Ltd
Window insulation - Llewellyn (Stonecare) Limited
Window insulation - Llewellyn (Stonecare) Limited
Window insulation - Llewellyn (Stonecare) Limited
Window systems - Deceuninck Ltd
Window ventilator manufactures - Simon R W Ltd
Windows - Bowater Containers Ltd
Windows - English Architectural Glazing Ltd
Windows - Magnet Ltd
Windows - Magnet Ltd
Windows - Magnet Ltd
Windows - Marshall-Tufflex Ltd
Windows - Solair Ltd
Windows (Alum) rooflights - Duplus Domes Ltd
Windows and conservatories - Elliott Group Ltd
Windows/ Rooflights/ Screens/ Louvres - SCHUCO International KG
Windows/Rooflights/Screens/Louvres - Duroy Fibreglass Mouldings

L20 Doors/ Shutters/ Hatches

Acoustic doors - Jewers Doors Ltd
Acoustic seals - Sealmaster
Aircraft hanger doors - Jewers Doors Ltd
All types of fencing & gates - WA Skinner & Co Limited
Aluminium Entrances and patio doors - KCW Windows & Conservatories Ltd
Aluminium door - Parkside Group Ltd, The
Aluminium doors - Ducana Surrey Ltd
Aluminium doors - Duplus Domes Ltd
Aluminium doors - Fabricated Aluminium Services Ltd
Aluminium doors - Glostal - Monarch
Aluminium doors - Hastings & Folkstone Glass Works Ltd
Aluminium doors - Lakeside Group
Aluminium doors - Planet Architectural Glazing Ltd
Aluminium doors - Prima Systems (South East) Ltd
Automatic and revolving doors - DORMA Entrance Systems Ltd
Automatic door operating equipment - Garog
Ballistic doors - Jewers Doors Ltd
Bespoke joinery - AS Newbould Ltd
Bespoke joinery - Howard Bros Joinery Ltd
Blast doors - Jewers Doors Ltd
Blast resistant windows/ curtain walling - Barlow Architectural & Security Ltd
Cedarwood doors - Regency Garage Door Services Ltd
Cold store doors - Ascot Industrial Doors Ltd
Conservatories - Hill Leigh (Joinery) Ltd
Conservatories - Solair Ltd
Contract Galvanizing - Birtley Building Products
Door hardware - Shaw Arthur Manufacturing Ltd
Door hardware - Shaw Arthur Manufacturing Ltd
Door hardware - Shaw Arthur Manufacturing Ltd
Door manufactures - Edmonds A & Co Ltd
Door panels - Door Panels PLC
Door protection - Moffett Thallon & Co Ltd
Door sets - Hill Leigh (Joinery) Ltd
Door systems - Deceuninck Ltd
Doors - BLP UK Ltd
Doors - BLP UK Ltd
Doors - J. Scott (Thrapston) Ltd
Doors - Magnet Ltd
Doors - Magnet Ltd
Doors - Magnet Ltd
Doors - Marshall-Tufflex Ltd
Doors - Scandanavian Window Systems Ltd
Doors (Alum) - Duplus Domes Ltd
Doors and conservatories - Elliott Group Ltd
Doors and entrances - Bassett & Findley Ltd
Doors and revolving doors - Pearsons J R (Birmingham) Ltd
Doors architectural - Timbmet Ltd
Doors panelled - Kern (Design Products) J T Ltd
Doors/ Shutters/ Hatches - SCHUCO International KG
Doors/Hatches - Allaway Acoustics Ltd
Doorsets - JRM Doors Ltd
Double glazed doors - Alframes Holdings Ltd
Double glazed doors - Alframes Holdings Ltd
Electric openers/ spare parts - Cardale Doors Ltd
Electric openers/ spare parts - Cardale Doors Ltd
Electric openers/ spare parts - Cardale Doors Ltd
Espagnolette locking systems - Nico Manufacturing Ltd
Extruded aluminium composite doors - Smart Systems Ltd
Extruded UPVC doors - Halo, W H S Division Of Bowater Industries
Extruded UPVC doors - Smart Systems Ltd
Fd30 and Fd60 fire doors - Real Door Company The
Fibre Composite Residential Doors - Birtley Building Products Ltd
Fire and security doors - Andersons Ltd
Fire and security doors - Bridgman Doors Ltd
Fire doors - Enfield Speciality Doors
Fire doors - J. Scott (Thrapston) Ltd
Fire doors - Rolflex Doors Ltd
Fire doors - Vicaima Ltd
Fire doors and door sets - Shadbolt F R & Sons Ltd
Fire doors and roller shutters - Wormald Ansul UK Ltd
Fire rated door sets - Lami Doors UK
Fire rated door sets - Lami Doors UK
Fire resisting doors and joinery - Decorfix
Fire Shutters - Ascot Industrial Doors Ltd
Fire shutters - Lakeside Group
Fire shutters - Lycetts (Burslem) Ltd
Fire Shutters/ Fire Doors - Stokvis R S & Sons Ltd
Fire, folding, roller shutters and grilles - Bolton Brady Ltd
Fire, folding, roller shutters, grilles and overhead doors - Bolton Brady Repair & Service Ltd
Fire, sound, security, smoke, hospital and disabled doors - Shapland & Petter Ltd
Fire-Resisting Glass - Schott Glass Ltd, Arcitectural Glass
Firegass Doors - Premdor
Firerated doors - Baydale Architectural Systems Ltd
Flat composite panel systems - European Profiles
Flexible door - EAP International Ltd
Flexible doors - Stokvis R S & Sons Ltd
Flush and feature doors - Vicaima Ltd
Folding doors - Dividers Ltd
Folding doors - Hillaldam Coburn Ltd
Folding doors and partitions - Pellfold Parthos Ltd.
Folding doors, partitions and screens - Kern (Design Products) J T Ltd
Food industry doors - Ascot Industrial Doors Ltd
Frames and covers - Bilco UK Ltd
G. R. P . Porches - Solair Ltd
G.R.P Doors - Regency Garage Door Services Ltd
Garage and industrial doors - Hormann (UK) Ltd
Garage Doors - Birtley Building Products
Garage doors - Doorfit Products Ltd
Garage doors - Solair GRP Architectural Products
Gas/ Oil Fired Boilers - Stokvis R S & Sons Ltd
Gates and Barriers - Hart Door Systems Ltd
Glazing Maintenance - Spectra Glaze Services
Grilles - Bradbury Security Grilles Ltd
GRP Door surrounds - Alpha Glass Fibres Ltd
GRP hatches - Group Four Glassfibre Ltd
Hardwood doors - Chindwell Co Ltd
Hardwood doors - Wood International Agency

Hatches - metal - McKenzie-Martin Ltd
High performance doors - Sashless Window Co Ltd
High Sped Rapid Roll Doors - Stokvis R S & Sons Ltd
High Speed Overhead Doors - Stokvis R S & Sons Ltd
High speed vertical folding and roll -up doors - Mandor Engineering Ltd
Hinged access panels for ceilings and walls - Profilex Ltd
Industrial and cold storage strip curtain systems - Flexible Reinforcements Ltd
Industrial & commercial doors, shutters, grilles and partitions - Henderson-Bostwick
Industrial & commercial doors, shutters, grilles and partitions - Henderson-Bostwick
Industrial doors - CPV Ltd
Industrial doors - RNB Industrial Doors Services
Industrial doors - Telford Doors and Barriers Ltd
Industrial doors and loading bay equipment - Envirodoor Markus Ltd
Industrial doors rolling, folding and sliding - Lycetts (Burslem) Ltd
Industrial doors, roller, folding and insulated - Beaver Industrial Doors Ltd
Industrial fire resistant steel doors and shutters - Bolton Gate Co Ltd
Industrial sliding and folding doors - European Profiles
Insulated bifolding doors - Jewers Doors Ltd
Insulated doors - Hemsec Manufacturing Ltd
Insulated Overhead Doors - Stokvis R S & Sons Ltd
Insulated Side Folding/ Sliding Doors - Stokvis R S & Sons Ltd
Insulated sliding doors - Jewers Doors Ltd
Insulated sliding folding doors - Jewers Doors Ltd
Internal and external moulded timber panel doors - JRM Doors Ltd
Loading bay equipment - Lycetts (Burslem) Ltd
Loft Access Doors - Birtley Building Products Ltd
Loft access doors - Timloc Expamet Building Products
Maintenance of industrial doors, shutters and shopfronts - Shutter Door Repair & Maintenance Ltd
Manufaturers Of framed and panelled doors, frames, skirtings and architraves. - Real Door Company The
Matching Front and Side Doors - Regency Garage Door Services Ltd
Metal doors - EDS Shutter Door Systems Ltd
Metal doors - Leaderflush & Shapland
Metal doors - Premdor
Metal doors - The Cotswold Casement Company
Metal doors, shutters and hatches - Amber Doors Ltd
Metal doors/ shutters/ hatches - Heywood Williams Architectural
Metal doors/ shutters/ hatches - Nico Manufacturing Ltd
Metal doors/ shutters/ hatches - Reid John & Sons (Strucsteel) Ltd
Metal meter boxes and replacement architrave units - Ritherdon & Co Ltd
Metal roller shutters - Lakeside Group
P.V.C. Strip curtains - Dudley Factory Doors Limited
Patio doors - Andersen/ Black Millwork
Patio doors - Andersen/ Black Millwork
Patio Doors - Solair Ltd
Patio Doors - Solair Ltd
Performance Doors and Doorsets - Leaderflush & Shapland
Pine doors - Wood International Agency
Pine furniture doors - Mason FW & Sons Ltd
Plastic door sets - Lami Doors UK
Plastic doors - Advanced Panels & Products
Plastic doors - Leaderflush & Shapland
Plastic doors - Lynman Windows
Profiled composite panel systems - European Profiles
Purpose made flush timber doors - Youngs Doors Ltd
PVC crash and strip doors - Beaver Industrial Doors Ltd
PVC door profiles - Profile 22 Systems Ltd
PVC doors - Prima Systems (South East) Ltd
PVC doors and conservatories - Scholes Windows Ltd
PVC doors and strip curtains - Bolton Brady Ltd
PVC doors and strip curtains - Bolton Brady Repair & Service Ltd
PVC Strip Curtains - Stokvis R S & Sons Ltd
PVCU door fabrication - HIS Ltd
PVCu doors - Dale Joinery Ltd
PVCu doors - Hastings & Folkstone Glass Works Ltd
PVCu doors - Interframe Ltd
PVCu doors - Plastal Commercial Ltd
PVCu Entrances and patio doors - KCW Windows & Conservatories Ltd
PVCU extrusions for door systems - Spectus UPVC Systems Ltd
Rapid roll doors - Ascot Industrial Doors Ltd
Rapid roll doors - Lycetts (Burslem) Ltd
Remote control garage doors - Regency Garage Door Services Ltd
Residential garage doors - Garador Ltd
Residential garage doors - Garador Ltd
Residential garage doors - Garador Ltd
Revolving doors - Grothkarst UK Ltd
Roll - A - Door - Birtley Building Products
Roll-A-Door - Birtley Building Products Ltd
Roller Doors (including insulation) - Stokvis R S & Sons Ltd
Roller doors/ side-hinged doors - Cardale Doors Ltd
Roller doors/ side-hinged doors - Cardale Doors Ltd
Roller doors/ side-hinged doors - Cardale Doors Ltd
Roller shutter and grilles - James & Bloom Ltd
Roller shutter and grilles - James & Bloom Ltd
Roller shutter and grilles - James & Bloom Ltd
Roller shutter doors - Ring-Gard (UK) Ltd
Roller shutter doors, Brickbond grilles - Dudley Factory Doors Limited
Roller shutter doors, fast action doors - Priory Shutter & Door Co Ltd
Roller shutters - Ascot Industrial Doors Ltd
Roller shutters - Dallas & Forrest Ltd
Roller shutters and industrial doors - Regency Garage Door Services Ltd
Ropes and sashcords - Marlow Ropes Ltd
Rubber and PVC flexible crash doors, PVC strip curtains - Neway Doors Ltd
Rubber and PVC flexible doors - Mandor Engineering Ltd
Sectional doors - Lakeside Group
Sectional Doors - Regency Garage Door Services Ltd
Sectional garage doors - IG Limited
Sectional overhead doors - Ascot Industrial Doors Ltd
Security doors - Chubb Security Installations Ltd
Security doors - Jewers Doors Ltd
Security Doors (GRP) - Oswestry Reinforced Plastics Ltd
Security doors, grilles with built in fly screens - United Safety Products UK Ltd
Security Grilles - Crimegard Ltd
Security metal doors and shutters - Ascot Industrial Doors Ltd
Security roller shutters - Lakeside Group
Security shutters - Aardee Locks and Shutters Ltd
Shutters and grilles - Permclose Group PLC
Side hung and double action door sets - Lami Doors UK
Side hung and double action door sets - Lami Doors UK
Side hung and double action door sets - Lami Doors UK
Sliding and folding doors - Hillaldam Coburn Ltd
Sliding and folding partitions - Building Additions Ltd
Sliding closures - Hillaldam Coburn Ltd
Sliding door sets - Lami Doors UK
Sliding door sets - Lami Doors UK
Sliding doors, partitions and screens - Kern (Design Products) J T Ltd
Sliding folding doors - Ascot Industrial Doors Ltd
Special timber flush doors and door sets - Industrial Building Componants Ltd
Specialised timber doors - AS Newbould Ltd
Specialist steel doorsets - Accent Hansen
Stainless steel, acoustic and insulated freezer doors - Clark Door Ltd
Stainless steel, bronze and aluminium panel doors, entrance screens, shopfronts and window frames - Drawn Metal Ltd
Steel and aluminium security shutters - Lycetts (Burslem) Ltd
Steel Door Installation - Rea Metal Windows Ltd
Steel Door Manufacture - Rea Metal Windows Ltd
Steel door sets - Ascot Industrial Doors Ltd
Steel doors - Crittall Steel Windows Ltd
Steel doors - IG Limited
Steel doors - Monk Metal Windows Ltd
Steel Doors - Regency Garage Door Services Ltd
Steel doors and frames - Fitzpatrick Door Frames Ltd
Steel doors, Ceilings - Norwood Partition Systems Ltd
Steel doors, frames and entrance systems - Doors & Hardware Ltd
Steel faced residential doors - BLW Associates Ltd
Steel Firescreen Installation - Rea Metal Windows Ltd
Steel Firescreen Manufacture - Rea Metal Windows Ltd
Steel hinged doors - Lycetts (Burslem) Ltd
Steel industrial doors - Mercian Industrial Doors
Steel louvred doorsets - Sunray Engineering Ltd
Steel personal doors - Lakeside Group
Steel Residential Doors - Birtley Building Products Ltd
Steel security doorsets - Sunray Engineering Ltd
Steel windows and doors - Clement Steel Windows
Supply only Casements Tilt & Turn, Top Swing, Box Sash, Pivots, Residential Doors, Ptio Doors. - Piper Windows Doors & Conservatories
Synthetic rubber moulding components and ironmongery - Notcutt W P Ltd
Teak Specialist - Parker Kislinghury Ltd
Timber door - Leaderflush & Shapland
Timber door panels - Halmark Panels Ltd
Timber door panels - Wendland GB Ltd, Division of Caradon Doors & Windows Ltd
Timber doors - Applied Acoustics (a part of Henry Venables Limited)
Timber doors - Boulton & Paul Ltd
Timber doors - Hansen Windows Ltd
Timber doors - Jutlandia Doors
Timber doors - Kind, J B, Ltd
Timber Doors - Langford Joinery Ltd
Timber doors - Lyssand UK Ltd
Timber doors - Premdor
Timber doors - Swedhouse Ltd
Timber patio doors - Fairmitre Ltd
Transformer chamber doorsets - Sunray Engineering Ltd
Up & Over Garage Doors - Birtley Building Products Ltd
Up and over doors/ sectional doors - Cardale Doors Ltd
Up and over doors/ sectional doors - Cardale Doors Ltd
Up and over doors/ sectional doors - Cardale Doors Ltd
UPVC door systems - Profolia Ltd
UPVC doors - BLW Associates Ltd
UPVC Doors - Ducana Surrey Ltd
UPVC doors - Fabricated Aluminium Services Ltd
UPVC Doors - Regency Garage Door Services Ltd
UPVC Doors and conservatories - Juno Roplasto Ltd
Wall access hatches - Howe-Green Ltd
Wall panelling - Kern (Design Products) J T Ltd
Wedge wire screens - Screen Systems Ltd
Windows - Rugby Joinery

L30 Stairs/ Walkways/ Balustrades

Aluminium extrusions/ sections - Seco Aluminium Ltd
Aluminium floor outlets - Marley Alutec Ltd
Aluminium handrails/balustrades - Norton Engineering Alloys Co Ltd
Aluminium ladders and step ladders - Ramsay & Sons (Forfar) Ltd
Aluminium stairways - Light Alloy Ltd
Architectural metal workers - Builder's Iron & Zincwork Ltd
Balustrade - City Hardware and Security
Balustrades - Steel Line Ltd
Balustrades and Handrails - Gabriel & Co Ltd
Balustrades and wallrails - HEWI (UK) Ltd
Balustrades, staircases and fire escapes - Staines Steel Gate Co
Balustrading - Marine & Commercial Rigging Systems
Bespoke joinery - AS Newbould Ltd
Commercial grilles and shutters - Andersons Ltd
Double ball industrial handrail systems - Eurogrid Ltd
Energy saving doors - Andersons Ltd
Escape and electric stairway ladders, roof exit systems, spiral, spacesaver and traditional stairs - Loft Centre Products
Fabricators of open metal flooring, handrails and standards - OSF Ltd
Fibreglass ladders and step ladders - Ramsay & Sons (Forfar) Ltd
Fire Escapes - Consculpt Spiral Stairs Ltd
Fire escapes, gates, railings, balustrades, guardrails and spiral staircases - Metalcraft (Tottenham) Ltd
Fire shutters - Andersons Ltd
Flooring - Orsogril Sarl UK Department
Grab rail and disabled equipment - Dryad Architectural Ltd
GRP String course - Alpha Glass Fibres Ltd
Guardrail - Kee Klamps Ltd
Handrail and balustarde - Dryad Architectural Ltd
Handrail systems in steel, stainless steel, aluminium and GRP - Lionweld Kennedy Ltd
Handrailing - Steel Line Ltd
Handrailing & Balustrading - Laidlaw Architectural Hardware Ltd
Handrailing, balustrading and fire escapes - Cameron & Moore (Gridplex) Ltd
Handrails and balustrades - Drawn Metal Ltd
Handrails and balustrades - Pearsons J R (Birmingham) Ltd
Industrial doors - Andersons Ltd
Industrial flooring, handrailing systems and stairtreads - Redman Fisher Engineering
Insulated Door panels - Ward Building Components Ltd
Internal External Stairs - Camtwix Engineering (Division of Terrapan Ltd)
Ladders, walkways and hand railing - Moseley GRP Product (Division of Moseley Rubber Co Ltd)
Metal railings - Varley & Gulliver Ltd
Metal staircases - Sebry Staircase & Ironworks Ltd
Metal stairs - Cottage Craft Spirals
Metal stairs/ walkways/ balustrades - Hubbard Architectural Metalwork Ltd
Metal, spiral, straight, and fire starcases - Crescent of Cambridge Ltd
Open steel grating and stairtreads - Lionweld Kennedy Ltd
Perforated sheets and balustrading - Associated Perforators & Weavers Ltd
Plywood, hard, fibre, chip and particle boards - Boardcraft Ltd
PVC Handrails - Architectural Plastics (Handrail) Ltd
Rails for special needs - Gabriel & Co Ltd
Ropes and sashcords - Marlow Ropes Ltd
Spiral Staircase Systems - Lewes Design Contracts Ltd
Spiral staircases and special precast stairs. - Consculpt Spiral Stairs Ltd
Spiral stairs - Gabriel & Co Ltd
Spiral stairways - Milner Delvaux Ltd
Stainless steel handrails - Aqua-Blue Ltd
Stainless steel, aluminium, PVC covered, galvenized and brass balustrades - SG System Products Ltd
Stair and ladder treads - Eurogrid Ltd
Stair parts - Winther Browne & Co Ltd
Staircases - Hill Leigh (Joinery) Ltd
Staircases, fire escapes, ladders, handrails and balustrades - Down & Francis Ltd, Division of Metalrax PLC
Stairs - Kirkstone
Stairs - Magnet Ltd
Stairs, walkways and ladders - Makay's of Cambridge Ltd
Stairs/ walkways/ balustrades - Portia Engineering Ltd
Steel fabrication railings, gates, fire escapes - SFC (Midlands) Ltd
Steel spiral stairs - Higginson Staircases Ltd
Steel staircases - Gabriel & Co Ltd
Steel stairs - Fire Escapes & Fabrications (UK) Ltd
Teak Specialist - Parker Kislingbury Ltd
Timber loft and escape ladders, roof exit systems, spiral, spacesaver and traditional stairs - Loft Centre Products
Timber spiral stairs - Higginson Staircases Ltd
Timber staircases - Boulton & Paul Ltd
Timber staircases - Northern Joinery Ltd
Timber stairs - Cottage Craft Spirals
Timber stairs - Rugby Joinery
Towers - Clow Group Ltd
Walkways - Clow Group Ltd
Walkways - Gabriel & Co Ltd
Walkways and platforms - Woodburn Engineering Ltd
Wallrails and balustrades - HEWI (UK) Ltd
Wooden ladders and step ladders - Ramsay & Sons (Forfar) Ltd

L40 General glazing

Antique Coluoered Glass - Schott Glass Ltd, Arcitectural Glass
Architectural aluminium systems for glazing - Universal Components Ltd
Atrium glazing - English Architectural Glazing Ltd
Bomb blast - Hi-Tec Window Films
Cell windows and observation units - Luxcrete Ltd
Curved and flat toughened glass - Thermax Interglass Ltd
Decorated and stained glass - Ide T & W Ltd
Decorative glass service - Andy Thornton Architectural Antiques Ltd
Double glazed units - Ledite Glass (Southend) Ltd
Double glazing - Coastal Ltd
Double glazing - Elliott Medway FineLine
Dry glazing systems and ventilated glazing beads etc - Exitex Ltd
Etched and sand blasted glass - Lightfoot Charles Ltd
Fire rated glazing - English Architectural Glazing Ltd
Fire resistant glass and glazing beads - C. G. I. International Ltd, trading as Colebrand Glass
Fire resistant glazing - Ide T & W Ltd
Fire resistant glazing - Mann McGowan Fabrication Ltd
Ganeral glazing - Saint-Gobain Glass UK Ltd
General glazing - Glaverbel (UK) Ltd
General Glazing - Pilkington UK Ltd
General glazing - Solaglas, Technical Advisory Service
General glazing - The Cotswold Casement Company
General glazing - Yule & Son Ltd, A C
Glass - Fenlock-Hansen Ltd
Glass - Pilkington Birmingham
Glass - Solaglas, Technical Advisory Service
Glass - Spital Tile Co
Glass and sheet plastic processing - Cooper H W & Co Ltd
Glass bending - Ide T & W Ltd
Glass cutters - Stanton-Thompson (Agencies) Ltd
Glaziers and glassmerchants - Ledite Glass (Southend) Ltd
Glazing - Ide T & W Ltd
Glazing - The Abbseal Group incorporating Everseal (Thermovitrine Ltd)
Glazing Accessories - Simon R W Ltd
Glazing and factory finishing - Bennett Windows Ltd
Glazing bars for glass sealed units - Newdawn & Sun Ltd
Glazing Bars for Timber roofs - Newdawn & Sun Ltd
Glazing gaskets to windows and doors - DJ Profiles

L41 Lead light glazing

Glazing Maintenance - Spectra Glaze Services
Glazing Maintenance - Spectra Glaze Services
High performance window films - Banafix Limited
Insulated Fire Resisting Glass - Schott Glass Ltd, Arcitectural Glass
Laminated glass - Solaglas Laminated
Mirror - Hi-Tec Window Films
Opaque bullet resistant glass - Attwater & Sons Ltd
Patent glazing - Ide T & W Ltd
Plastic profiled sheet cladding and roofing - Ariel Plastics Ltd
Polycarbonate glazing - Cox Building Products Ltd
Polycarbonate glazing systems - EBC UK Ltd

Privacy - Hi-Tec Window Films
Putty - Adshead Ratcliffe & Co Ltd
Safety mirror and glass - Stonecraft & Chelsea Artisans Ltd
Safety/ Security - Hi-Tec Window Films
Secondary glazing - Alframes Holdings Ltd
Secondary glazing manufactures - Simon R W Ltd
Self support glazing beams - Newdawn & Sun Ltd
Shopfronts - Parkside Group Ltd, The
Silicone jointed glazing - ARC Partitioning
Solar, anti shatter glass films - Cooper H W & Co Ltd
Tinting - Hi-Tec Window Films
Toughened and stock glass - Wright & Offland Ltd
Toughened, laminate, ceramic, acoustic glass and screen printing on glass - Pilkington Plyglass
Window films - Bonwyke Ltd
Window films - Durable Berkeley Co Ltd
Window films - MSC Speciality Films (UK) Ltd
Window films - ThreeM United Kingdom plc
Window films to existing glass - Hi-Tec Window Films
Glazing Maintenance - Spectra Glaze Services
Impact resistant safety mirrors and glass - Chelsea Artisans Ltd
Lead Light Glazing - Yule & Son Ltd, A C
Leaded lights - Ledite Glass (Southend) Ltd
Leaded lights - Lightfoot Charles Ltd
Restoration/ Conservation/ Design and production of stained glass - Chapel Studio
Specialist glazing - Lightfoot Charles Ltd
Stained glass - Lightfoot Charles Ltd
Stained glass windows - Ide T & W Ltd
Stained glass, new and restoration - Quail Paul

L42 Infill panels/ sheets

Anti vandal cladding and glazing products - Oswestry Reinforced Plastics Ltd
Cane woven panels - Winther Browne & Co Ltd
Perforated panels in hardboard, mdf and aluminium - Winther Browne & Co Ltd
Wall & Ceiling Panels - McLoughlin Wood Ltd

M10 Cement: sand/ Concrete screeds/ toppings

Calcium Carbonate - ECC International Ltd
Cement - Rugby Cement
Cement based screeds - Flowcrete Plc
Cementitious screed - IFT Supplies Ltd
Chemical resistant screed - Addagrip Surface Treatments Ltd
Concrete screeds and toppings - Anti Corrosion Services Ltd
Concretes - RMC Readymix Limited
Cristobalile Flour - Hepworth Minerals & Chemicals Ltd
Cristobalile Sand - Hepworth Minerals & Chemicals Ltd
Floor coatings - RIW Ltd
Floor finishes - RIW Ltd
Floor insulation - Kay Metzeler Ltd
Floor screeds and floor repair mortars - Ronacrete Ltd
Floor screeds and roof screeds - Isocrete Group Sales Ltd
Floor seals - Evans Vanodine International PLC
Flooring - FEB Ltd
Flooring - Feb MBT
Flooring - Marshalls Flooring Ltd
Flooring - Thoro Systems Products Ltd
Industrial flooring - Exchem Mining and Construction
Industrial Flooring - Sempol Surfaces Ltd
Industrial flooring - Stuarts Industrial Flooring Ltd
Mezzanine floors and allied equipment - Bromag Structures Ltd
Norina Screed Cavity flooring - Durabella Ltd
Pavement Repair - Webrr & Broutin
Quick drying cementicious screeds - P.C.I Construction Systems Ltd
Sand cement/ Concrete/ Granolithic screeds/ flooring - Dex-o-Tex International Ltd
Sand cement/ Concrete/ Granolithic screeds/ flooring - GRAB Industrial Flooring Ltd
Screeds - RMC Readymix Limited
Screeds and floor coatings - Addagrip Surface Treatments Ltd
Self levelling cementicious screeds - P.C.I Construction Systems Ltd
Silica Flour - Hepworth Minerals & Chemicals Ltd
Silica Sand - Hepworth Minerals & Chemicals Ltd
Steel framed multi storey volumetric buildings - Accomodex
Surface hardeners - Armorex Ltd
Tanking - Kiltox Damp Free Solutions
Underwater grouts and specialist flooring materials - David Ball Group Plc
Waterproofing - Webrr & Broutin
Wilsonart solid surfacing - Wilsonart Ltd

M12 Trowelled bitumen/ resin/ rubber-latex flooring

Anti slip surfacing - Addagrip Surface Treatments Ltd
Epoxy floor coatings - Ronacrete Ltd
Epoxy resin coatings - IFT Supplies Ltd
Epoxy resin coatings for floors - Watco UK Ltd
Epoxy resin repair materials - Anglo Building Products Ltd
Finishing equipment - Electrolux Laundry Systems
Floor coatings - Jotun Henry Clark Ltd
Flooring systems - Dex-o-Tex International Ltd
Resin based floor & wall finishes - Flowcrete Plc
Resin coatings - Armorex Ltd
Trowelled bitumen/ resins/ rubber latex flooring - GRAB Industrial Flooring Ltd

M13 Calcium sulphate based screeds

Calcium sulphate based screeds - Flowcrete Plc

M20 Plastered/ Rendered/ Roughcast coatings

Angle beads and metal arch frames - Easy Arches Ltd
Arch formers - Expamet Building Products
Building plasters - British Gypsum Ltd
Cement based floor screeds, external render systems - Optiroc Ltd
Coil mesh - Caradon Catnic Ltd
Coloured limestone, dolomite and white sand internal and external renders - Kilwaughter Chemical Co Ltd
Decorative coatings - Textured European Finishes Ltd
Epoxy resin coatings for walls - Watco UK Ltd
Expanded metal products - Ensor Metal Products
Exterior renders - Alumasc Exterior Building Products Ltd
Grout pumps - Sidewinder Concrete Pumps Ltd
High performance building boards - Knauf
Lightwieght plaster - Tilcon Ltd
Metal lathing - Expamet Building Products
Metal beads - Expamet Building Products
Metal lath - Caradon Catnic Ltd
Monocouche renders - Webrr & Broutin
Plaster arris protector - Areco Roofing Supplies Company
Plaster beads - Walls & Ceiling International Ltd
Plasterboards - Brikenden (Builders Merchants) Ltd
Plasterers profiles - Caradon Catnic Ltd
Plasters - Kair - The Ventilation Division of Kiltox Chemicals Limited
Renders - Snowcem
Specialist plaster and wall finishes - Francesca Di Blasi Co. Ltd, The
Synthetic resin renders - CCS Scotseal Ltd
Textured coatings & grout - Ward Bekker Sales Ltd

M21 Insulation with rendered finish

Expanded polystyrene insulation - Vencel Resil Ltd
Exterior render on insluation - Alumasc Exterior Building Products Ltd
External wall insulation - CCS Scotseal Ltd
Floor insulation - Callenders Ltd
Floor insulation - Kay Metzeler Ltd
Industrial flooring - Webrr & Broutin
Insulated external render - Optiroc Ltd
Insulation - Rockwool Ltd
Insulation with rendered finish - Concrete Repairs Ltd
Insulation with rendered finish - Llewellyn (Stonecare) Limited
Polymer cement renders - MR Limited
Stonework restoration/cleaning - PermaRock Products Ltd

M22 Sprayed monolithic coatings

External wall insulation -Eglinton - Webrr & Broutin
Intimescent paint - Firebarrier Services Ltd
Specialist coatings - Valtti Specialist Coatings Ltd
Specialist wall coatings - MR Limited
Spray coating and fire protection - Cafco Construction Products

M23 Resin bound mineral coatings

Manufacture of Industrial Resin Flooring in Polyurethane or Epoxy materials - Resdev Ltd
Mineral Coating T.E.F. - Textured European Finishes Ltd
Rainscreen cladding - Eglinton - Webrr & Broutin
Resin bound mineral coatings - GRAB Industrial Flooring Ltd
Resin sealing coats - Armorex Ltd
Resins - DSM UK Ltd
Specialist coatings - Valtti Specialist Coatings Ltd

M30 Metal mesh lathing/ Anchored reinforcement for plastered coatings

Cladding systems - BRC Building Products
Stainless Steel Surface Protection - Component Developments

M31 Fibrous plaster

Decorative mouldings, cornices, dados, ceiling roses and coving - Copeley Decor Ltd
Decorative plaster coving and ceiling roses - Hilton Banks Ltd
Fibrous plaster - Clark & Fenn Ltd
Fibrous plaster mouldings - Troika Architectural Mouldings Ltd
Fibrous plasterware, cornices and ceiling roses - ABC Studios (Plaster Mouldings) Ltd.
Ornamental fibrous plasterwork - Allied Guilds
Plaster cornices and archways - Artistic Plastercraft

M40 Stone/ Concrete/ Quarry/ Ceramic tiling/ Mosaic

Ceramic fibre products - Combustion Linings Ltd
Ceramic floor and wall tiles - Stokes R J & Co Ltd
Ceramic floor, wall ties and terracotta floor tiles - Smithbrook Building Products Ltd
Ceramic marble and granite tiles - Reed Harris
Ceramic tiles - Bell & Co Ltd
Ceramic tiles - Candy & Co Ltd
Ceramic tiles - Swedecor Ltd
Ceramic wall and floor tiles - Buchtal (U.K.) Limited
Ceramic wall and floor tiles - Vitra (UK) Ltd
Ceramic, marble and glass mosaics, swimming pool tiles and porcelain stoneware - Domus Tiles Ltd
Ceramic, marble, granite tiles - Reed Harris
Ceramic, porcelain and mosaic tiles - Arnull Bernard J & Co Ltd
Ceramics - Wade Ceramics Ltd
Ceramics and mosaics - Shackerley (Holdings) Group Ltd incorporating Designer ceramics
Concrete topping - Ardex UK Ltd
Faience Blocks - Shaws of Darwen
Floor slabs and floor tiles - Delabole Slate
Floor systems - MiTek Industries Ltd
Floor tile separators - Compriband Ltd
Floor tiles - Fired Earth
Floor Tiling - Fired Earth
Flooring - Burlington Slate Ltd
Flooring - Farmington Natural Stone Ltd
Flooring - Kirkstone
Glass mosaic -sandstone - Reed Harris
Glazed wall tiles, unglazed floor tiles,swimming pool tiles and mosaics - Pilkington's Tiles Ltd
Industrial floor tiles - Swedecor Ltd
Industrial tiling - Anti Corrosion Services Ltd
Interior flooring - Milner Delvaux Ltd
Keope Tiles - Italian Porecellian - Ruabon Dennis Ltd
Limestone quarry tiles - Reed Harris
Marble and ceramic tiling - Marriott & Price Ltd
Marble and granite flooring - Diespeker Marble & Terrazzo Ltd
Marble tiles - Emsworth Fireplaces Ltd
Marble, granite, limestone, precast and insitu terrazzo mosaic - Fieldmount (Terrazzo) Ltd
Marble, granite, slate and limestone - Toffolo Jackson(UK) Ltd
Marble, granite, slate, limestone & terrazzo - Diespeker Marble & Terrazzo Ltd
Mosaic - Roman Mosaic Contracts Ltd
Natural floor covering - Fired Earth
Quality ceramic tile adhesives - Ardex UK Ltd
Quarry tiles - Marley Building Materials Ltd
Quarry tiles - Ruabon Dennis Ltd
Quarry tiling - Ruabon Dennis Ltd
Rapid set high strength bedding mortars - Ronacrete Ltd
Slate - Reed Harris
Specialist contractors in marble and granite mosaics - Alpha M & T Ltd
Sphinx Ttiles - Glazed Wall and Floor - Ruabon Dennis Ltd
Stone faced raised access flooring - Chelsea Artisans Ltd
Stone quarry - Chilmark Quarries Ltd
Stone/ Ceramic lining - The Rare Stone Group Ltd
Stone/ Concrete/ Quarry/ Ceramic tiling/ Mosaic - Network Flooring Cardiff
Stone/ Quarry/ - Gregory Quarries Ltd, The
Stone/Concrete/Quarry/Ceramic tiling/Mosaic - Spartan Tiles Ltd
Subfloor preparation products - Ardex UK Ltd
Swimming pool tiles - Swedecor Ltd
Terracotta - Reed Harris
Terracotta Blocks - Shaws of Darwen
Terrazzo - Reed Harris
Tiles - Spital Tile Co
Tiling adhesives - Axter Ltd
Tiling Cement For Metal Substrates - P.C.I Construction Systems Ltd
Tiling cements (adhesive) for natural stone, marble and ceramic tiles - P.C.I Construction Systems Ltd
Tiling Trims - Imagetrim (UK) Ltd
Unglazed ceramic floor tiles - Woolliscroft Tiles Ltd
Wall and floor tiles - Royal Dutch Sphinx Ltd
Wall tiles - Fired Earth

M41 Terrazzo tiling/ In situ terrazzo

Insitu Terrazzo - RIW Ltd
Marble, granite, limestone, precast and insitu terrazzo mosaic - Fieldmount (Terrazzo) Ltd
Specialist contractors in terrazzo - Alpha M & T Ltd
Terracotta - York Handmade Brick Co Ltd
Terrazzo - Marriott & Price Ltd
Terrazzo flooring - Diespeker Marble & Terrazzo Ltd
Terrazzo tile manufacturers - Toffolo Jackson(UK) Ltd
Terrazzo tiles - Quiligotti Access Flooring Limited
Terrazzo tiling/ In situ terrazzo - Dex-o-Tex International Ltd
Terrazzo tiling/ In situ terrazzo - GRAB Industrial Flooring Ltd
Terrazzo tiling/ In situ terrazzo - Network Flooring Cardiff

M42 Wood block/ Composition block/ Parquet flooring

Floor sanding machines - Eureka Products Ltd
Flooring - Lees Richard Ltd
HPL flooring - Abet Ltd
Laminated flooring - Kronospan Ltd
Parquet and laminate flooring - Western Cork Ltd
Sports flooring systems and sprung flooring - Granwood Flooring Ltd
Wood block/ Composition block/ Parquet flooring - Network Flooring Cardiff
Wood block/ Composition block/ Parquet Flooring - Westco Group Ltd
Wood flooring - DLW Flooring Ltd
Wooden flooring - Wicanders (GB) Ltd

M50 Rubber/ Plastics/ Cork/ Lino/ Carpet tiling/ sheeting

Aluminium edge trim for vinyl flooring - Howe-Green Ltd
Anti-slip surfaces - Redman Fisher Engineering
Authentic tribal rugs - Fired Earth
Bitumen backed carpet tiles - Halstead James Ltd
Bonded tiles - Carpets of Worth Ltd
Carpet and tiles - Adam Carpets Ltd
Carpet tiles - Brintons Ltd
Carpet tiles - Firth Carpets Ltd
Carpet tiles - Gaskell Textiles Ltd
Carpet tiles - Mackay Carpets, Hugh
Carpet tiles - Milliken Carpets
Carpet, cork and vinyl tiles and laminate flooring - Western Cork Ltd
Carpets - CFS Carpets
Carpets - Collie Carpets Ltd
Carpets - HEM Interiors Group Ltd
Carpets - LazyLawn
Carpets - Opus 4 Ltd
Cellular rubber products - Volcrepe Ltd
Commercial carpet tiles, broadloom and entrance clean off zones - Checkmate Industries Ltd
Commercial, barrier, hospital carpet and tiles - Carpet Specifier Services Ltd
Contamination control flooring - Dycem Ltd
Contract carpet and carpet tile manufacturer - Burmatex Ltd
Cork - Olley & Sons Ltd, C
Cork flooring - Wicanders (GB) Ltd
Cork, rubber and non slip surfaces - Tiflex Ltd
Data installations - HEM Interiors Group Ltd
Decoration and home improvements - Cement Glaze Decorators Ltd
Edging - Egger (UK) Ltd
Entrance matting systems and fire retardant flooring - Tufton Ltd
Entrance matting, flooring and roof walkway matting - Plastic Extruders Ltd
Flameproof sheeting - Monarflex Ltd
Floor coverings - Bonar Floors Ltd
Floor coverings - Hometex Trading Ltd
Floor smoothing underlay - Ball F & Co Ltd
Floorcoverings - Interface Europe Ltd
Flooring - HEM Interiors Group Ltd
Flooring accessories, stairnosings and PVC accessories - Quantum Profile Systems
Flooring Adhesives - Ball F & Co Ltd
Heavy duty carpet tiles - Paragon by Heckmondwike F B Ltd
High pressure decorative laminates - Abet Ltd
Laminate flooring - Wicanders (GB) Ltd
Lino and PVC flooring - DLW Flooring Ltd
Linoleum and vinyl floorcoverings - Forbo - Nairn Ltd
Matting and marking tape - ThreeM United Kingdom plc
Mezane floors & partitions - ClipsPartitions Ltd
Natural floor coverings - Fired Earth
Nylon reinfoced PVC sheeting - Flexible Reinforcements Ltd
Publications: Contract Flooring Association Guide to Contract Flooring - Contract Flooring Association
PVC floorcoverings and accessories - Halstead James Ltd
PVC safety floor coverings - Halstead James Ltd
PVC sheeting - Flexible Reinforcements Ltd
Room Improvement Products - nmc (UK) Ltd
Rubber flooring - DLW Flooring Ltd
Rubber flooring - Gates Rubber Co Ltd The
Rubber sheeting and expansion joints & bellows - MacLellan Rubber Ltd
Rubber sheeting and flooring - Linatex Ltd
Rubber sheets - Manton Industrial Seals Ltd
Rubber stud tiles - Halstead James Ltd

Rubber/ Plastics/ Cork/ Lino/ Carpet tiling/ sheeting - Bonar Floors Ltd
Rubber/ Plastics/ Cork/ Lino/ Carpet tiling/ sheeting - Interface Europe Ltd
Rubber/ Plastics/ Cork/ Lino/ Carpet tiling/ sheeting - Network Flooring Cardiff
Rubber/ Plastics/ Cork/ Lino/ Carpet tilling/ Sheeting - Westco Group Ltd
Self-adhesive surface coverings - Rustin Allen Ltd, Belbien Divsion
Specialised flloring materials, nosing and adhesives - Tiflex Ltd
Specialist acid resisting tiling and coatings - Williamson Floor & Wall Specialists Ltd
Sponge rubber carpet - Gates Rubber Co Ltd The
Stair edgings, floor trims, and entrance barrier matting - Gradus Ltd
Stair treads, nosing and matting - The Safety Tread Ltd
Underlay - Gaskell Textiles Ltd
Vinyl floor coverings - Dermide Ltd
Vinyl floorcovering sheets, tiles and accessories - Marley Floors Ltd
Vinyl flooring - Gerflor Ltd
Vinyl flooring - Tarkett Ltd
Vinyl sheet and tile flooring - Armstrong World Industries Ltd
Vinyl, safety and rubber floor coverings - Gerland Ltd
Wall furnishings - Vescom (UK) Ltd
Wood and laminate underlay - Gates Rubber Co Ltd The

M51 Edge fixed carpeting

Bespoke carpets - Acorn Mill Co Ltd
Carpet - Egetaepper (UK) Ltd
Carpet - Hometex Trading Ltd
Carpet for domestic and contract use - Carpets of Worth Ltd
Carpeting - Davies John Interiors Ltd
Carpets - CFS Carpets
Carpets - Foster C G & Sons, (Contract Soft Furnishers)
Carpets - Shaw Carpets plc
Carpets - Ulster Carpet Mills Ltd
Commercial carpet tiles, broadloom and entrance clean off zones - Checkmate Industries Ltd
Commercial, barrier and hospital carpets and tiles - Carpet Specifier Services Ltd
Decoration and home improvements - Cement Glaze Decorators Ltd
Edge fixed carpeting - Bonar Floors Ltd
Edge fixed carpeting - Brockway Carpets Ltd
Edge fixed carpeting - Carpets International (UK) PLC
Edge fixed carpeting - Firth Carpets Ltd
Edge fixed carpeting - Interface Europe Ltd
Edge fixed carpeting - Mackay Carpets, Hugh
Edge fixed carpeting - Network Flooring Cardiff
Floorcovering accessories - Gripperrods Ltd
Floorcoverings - Interface Europe Ltd
Heavy duty carpets - Paragon by Heckmondwike F B Ltd
Impervious backed carpet - Halstead James Ltd
Surface membranes and release systems for carpet - Laybond Products Ltd
Woven and tufted carpet - Brintons Ltd

M52 Decorative papers/ fabrics

135cm wide fabric backed vinyls - Dixon Turner Wallcoverings Ltd.
53cm x 10m paper backed vinyls - Dixon Turner Wallcoverings Ltd.
Abrasive Papers & Decorating Sundry Products - Acton & Borman
Authentic tribal rugs - Fired Earth
Blown vinyl wallcoverings - Dermide Ltd
Contract wallcoverings - Dixon Turner Wallcoverings Ltd
Contract Wallcoverings - Dixon Turner Wallcoverings Ltd.
Decoration and home improvements - Cement Glaze Decorators Ltd
Decorative finishes - Glixtone Ltd
Decorative Finishes - Snowcem
Decorative Laminates - Viking Laminates Ltd
Decorative papers/fabrics - Hamron Group
Decorative wallcoverings, fabrics and borders - Dixon Turner Wallcoverings Ltd
Duplex wallcoverings - Dermide Ltd
Fabric - Fired Earth
Fabrics - Interface Europe Ltd
Fabrics and wall papers - Today Interiors Ltd
Fabrics and wallpapers - Sanderson & Sons Ltd, Arthur
Linolem finish material for desks - Forbo - Nairn Ltd
Notice board material - Forbo - Nairn Ltd
Paint stripper, wall paper stripping machines - Ward Bekker Sales Ltd
Paints and wallcoverings - Kalon Decorative Products
PVC coated fabrics - Flexible Reinforcements Ltd
Simplex wallcoverings - Dermide Ltd
Textile wallcoverings - Dixon Turner Wallcoverings Ltd.
Vinyl Wallcoverings - Forbo (Lancaster) Ltd
Wallcoverings - Tektura Plc
Wallcoverings and adhesives - Architectural Textiles Ltd

M60 Painting/ Clear finishing

Air drying enamels - Crosbie Coatings
Alkyd, epoxy, acrylated rubber, vinyl, polyurethane and silicate high performance coatings - Leigh's Paints
Anti - static coatings - Kenyon Industrial Paints & Adhesives
Anti carbonation coatings - Ronacrete Ltd
Anti corrosion coatings - Corroless International Association
Anti corrosion coatings - Kenyon Industrial Paints & Adhesives
Anti corrosive primers - Crosbie Coatings
Anti Graffiti Coatings - Bollom J W & Co Ltd
Bitumen and aluminium bit paint - Liver Grease Oil & Chemical Co Ltd
Black varnish - Liver Grease Oil & Chemical Co Ltd
Brushing tar, creosote, tallow and linseed oil - Legge Thompson F & Co Ltd
Cadmium Pigments - Johnson Matthey Pigments & Dispersions
Cement paint - Hunting Industrial Coatings Ltd
Cleaners for paving - Resiblock Ltd
Coal tar creosote & pitch - Liver Grease Oil & Chemical Co Ltd
Coal, barn, wood and stockholm tar - Liver Grease Oil & Chemical Co Ltd
Coatings for concrete floors - Everlac (GB) Ltd
Cold emulsion sprayers - Bitmen Products Ltd
Complex inorganic pigments - Johnson Matthey Pigments & Dispersions
Creosote - Coalite Chemicals
Decorating - Davies John Interiors Ltd
Decorating materials - Brewer C & Sons Ltd
Decoration - HEM Interiors Group Ltd
Decoration, painting and home improvements - Cement Glaze Decorators Ltd
Decorative ancillary products - Bartoline Ltd
Decorative coatings for timber - Protim Solignum Ltd
Decorative finishes - ICI Paints
Decorative paints - Akzo Coatings Plc
Decorative Paints - Philip Johnstone Group Ltd
Decorative paints, opaque and translucent wood stains - Jotun Henry Clark Ltd
Decorative Sundries - Kalon Decorative Products
Decorative wall coatings - RIW Ltd
Electroplating, powder coating and painting - Ritherdon & Co Ltd
Epoxy paints - Trade Sealants
Epoxy resin coatings - Anglo Building Products Ltd
Epoxy resins - Addagrip Surface Treatments Ltd
Fillers - ICI Paints
Fire retardant treatment - Palgrave Brown Timber & Fireproofing Co
Flame Retardant Lacquers - Bollom Fire Protection Ltd
Flame retardant paints - Bollom Fire Protection Ltd
Flame retardant paints - Bollom J W & Co Ltd
Flame Retardant Varnishes - Bollom Fire Protection Ltd
Flame retardent coatings - Rentokil Initial U.K. Ltd.
Floor coating systems - Corroless International Association
Floor coatings - Crosbie Coatings
Floor coatings, self levellers - Kenyon Industrial Paints & Adhesives
Floor paints - Kingstonian Point Ltd
Fungicidal Coatings - Bollom J W & Co Ltd
Glass flake reinforced coatings - Corroless International Association
High performance paints and specialist coatings - Stewart Wales, Somerville Ltd
Hygiene coatings - Kenyon Industrial Paints & Adhesives
Hygienic coatings - Corroless International Association
Industrial Coatings - Bollom J W & Co Ltd
Industrial coatings - Crosbie Coatings
Industrial Paints - Firwood Paint & Varnish Co Ltd
Interior coatings - Jotun Henry Clark Ltd
Intumescent coatings - Bollom Fire Protection Ltd
Lime putty - Legge Thompson F & Co Ltd
Linseed oils - Liver Grease Oil & Chemical Co Ltd
Long life exterior wall coating - Everlac (GB) Ltd
Lump and french chalk - Legge Thompson F & Co Ltd
Marking equipment - Elliott John H (Monostamp) Ltd
Masonry paint - CCS Scotseal Ltd
Masonry Paints - Bollom J W & Co Ltd
Metal Primers - Bollom J W & Co Ltd
Oakum - Liver Grease Oil & Chemical Co Ltd
Paint - Fired Earth
Paint - Sanderson & Sons Ltd, Arthur
Paint effects - Kingstonian Point Ltd
Paint spraying equipment - Dawson McDonald & Dawson Ltd
Painting - Deborah Services
Painting aids - Polycell Products Ltd
Painting/ Clear finishes - RIW Ltd
Painting/ Clear finishing - Concrete Repairs Ltd
Painting/ Clear finishing - Coo-Var Ltd
Painting/ Clear finishing - GRAB Industrial Flooring Ltd
Painting/clear finishing - Hamron Group
Painting/Clear finishing - Tefcote Surface Systems
Paints - Akzo Nobel Decorating Coatings Ltd
Paints - Bollom J W & Co Ltd
Paints - Crown Trade Paints
Paints - Dexter Paints Ltd
Paints - Emusol Products (LEICS) Ltd
Paints - Herberts, Division of HPG
Paints - Hird Hastie Paints Ltd
Paints - ICI Dulux Trade Paints
Paints - Kalon Group PLC
Paints - Manders Paint Ltd
Paints - Marcel Guest Paints
Paints - Snowcem
Paints - Surmak Products Ltd
Paints - Tor Coatings Group
Paints - Wareing Bros & Co Ltd
Paints & coatings - Sigma Coatings Ltd
Paints and coatings for tennis courts and other sports surfaces - Everlac (GB) Ltd
Paints and high temperature resistant coatings - Fortafix Ltd
Paints and surface coatings - Dacrylate Paints Ltd
Paints Primers and specialised coatings - Coo-Var Ltd
Paints, wood finishes, slovents and chemicals - Anderson Gibb & Wilson
Pigments for the surface coatings - Johnson Matthey Pigments & Dispersions
Polish and sealers - British Nova Works Ltd
Polyurethane floor paint - IFT Supplies Ltd
Powder and ready mixed fillers - Ward Bekker Sales Ltd
Protection products - Sadolin UK Ltd
Protective and decorative masonry coatings - Andura Textured Masonry Coatings Ltd
Protective coatings - Andrews J S (Coatings) Ltd
Protective epoxy resin coatings to concrete - Anti Corrosion Services Ltd
Protective maintenance coatings - Corroless International
Resin coatings - Armorex Ltd
Rust stabilising primers - Corroless International Association
Sandpaper - Atkinson & Kirby Ltd
Seals and coatings for wood floors - Everlac (GB) Ltd
Slurry granules - Bayer Plc
Specialised paints - Teal & Mackrill Ltd
Specialist industrial paints - Witham Oil And Paint Ltd
Specialist paints and varnishes - Blackfriar Paints and Varnishes
Specialist paints and varnishes - Parsons E. & Sons Ltd
Spray equipment - Bollom J W & Co Ltd
Standard, Paint Quality and Fair Face Finishes - Tarmac Topblock Ltd
Stensils, branding tools and ink marking - Elliott John H (Monostamp) Ltd
Structural painting and spraying - Maxblast UK Ltd
Surface dressings - Vallance
Tapes - ThreeM United Kingdom plc
Tennis court spraying - Sportmark (Leisure Products) , Part of Sportsmark Group Ltd
Tile transfers - Homelux
Timber preservatives, fire retardants and decorative finishes - Hickson Timber Products Ltd
Transparent ironoxide pigments - Johnson Matthey Pigments & Dispersions
Vapour corrosion inhibitors - Corroless International Association
Weatherproof decorative wall coatings and hygiene finishes - Liquid Plastics Ltd
Wood finishes, speciality paints, paint removers and decorating sundries - Rustins Ltd
Woodfinishes - Smith & Rodger Ltd
Intumescent coatings for fire protection - Hamron Group

N10 General fixtures/ furnishings/ equipment

Accommodation furniture and electronic workbenches - Blyde-Barton Ltd
Aluminium frame seating - Origlia Spa
Architectural Antiques - Andy Thornton Architectural Antiques Ltd
Architectural exibitions and model making - Modelscape
Awnings - Tidmarsh & Sons
Basket units - Smart F & G (Shopfittings)Ltd
Bathroom cabinets and tube fittings - MacBee Manufacturing Ltd
Bathroom mirrored cabinets and mirrors with lighting - Schneider (UK)
Bathrooms - Rosa Paula Kitchens
Beam setting - Origlia Spa
Bedroom fittings - Bernstein Group PLC
Bedroom furniture - OEP Furniture Group PLC
Bespoke museum display cases - Carlton Benbow
Blackout/ sunscreen blinds - Tidmarsh & Sons
Blind and awning fabrics - Flexible Reinforcements Ltd
Blinds - Design Line
Blinds - Durable Berkeley Co Ltd
Blinds - Eclipse Blind Systems Ltd
Blinds - Foster C G & Sons, (Contract Soft Furnishers)
Blinds - Hallmark Blinds Ltd
Blinds - Opus 4 Ltd
Blinds - VELUX Co Ltd, The
Blinds and curtains - Contravent Regal Ltd
Blinds and shutters - Andersons Ltd
Blinds in sealed units - Pilkington Plyglass
Bomb curtains and blankets - Meggitt Lavmour Systems
Boot and shoe cleaning equipment - Mawrob Co (Engineers) Ltd
Bracketry - Vantrunk Engineering Ltd
Brushes - Industrial Brushware Ltd
Canteen and office furniture for cabins - Ableson L & D Ltd
Caravan Security Boxes - Ritherdon & Co Ltd
Carpets - Gradus Ltd
Cash and security boxes - Securikey Ltd
Castors and wheels - Flexello Limited
Changing room furniture - Bawn W. B. & Co Ltd
Chimney sweeping equipment - Wessenden Products Ltd
Climate Dials - Goods Directions Ltd
Coal and log effect gas fires - Verine Ltd
Conservatory blinds - Tidmarsh & Sons
Contract furniture - Metalliform Products Plc
Contract furniture - Ness Furniture Ltd
Contract hardwood seating - Branson Leisure
Contract seating & tables - Knightbridge Furniture Productions Ltd
Contract/Letail Beds - Blindcraft Industries
Cupboards - ClipsPartitions Ltd
Curtain Fabrics - Eclipse Blind Systems Ltd
Curtain poles - Rothley Burn Ltd
Curtain tracks, cubicle rails and window blinds - Silent Gliss Ltd
Curtain tracks, curtains and blinds - Marshall Contracts
Curtains - Foster C G & Sons, (Contract Soft Furnishers)
Curtains - Sanderson & Sons Ltd, Arthur
Custom mouldings - Harrison Thompson & Co Ltd
Customer guidence and queue management systems - Tensator Ltd, Sales Department
Customised Panel Work - Ritherdon & Co Ltd
Decorative melamine faced chipboard - Egger (UK) Ltd
Decorative profits - PVC - Rustin Allen Ltd, Belbien Divsion
Desk and door plates - Movitex Signs Ltd
Desks and furniture in metalwork, glass, marble and granite - Howard Bros Joinery Ltd
Desks and seating - Mauser Interiors (UK) Ltd
Domestic bedroom furniture - Ball William Ltd
Doors - BLP UK Ltd
Electric blankets - Parking Technology
Electronic fly killers - P & L Systems Ltd
Entrance matting - Bonar Floors Ltd
Entrance matting - Cimex International Ltd
Entrance matting - Norton Engineering Alloys Co Ltd
Entrance matting - Shackerley (Holdings) Group Ltd incorporating Designer ceramics
Enviropol products - Glasdon UK Ltd
Equestrian arenas and products pvc-u - Duralock UK Ltd
Equipment cabinets - Glasdon Designs Ltd
External blinds - Tidmarsh & Sons
Filing and storage cabinets - Railex Systems Ltd, Elite Division
Fine surfaced furniture chip board - Egger (UK) Ltd
Fire blankets - Rentokil Initial U.K. Ltd.
Fire extinguishers - Asco Extinguishers Co Ltd
Fire extinguishers - Fife Fire Engineers & Consultants Ltd
Fire extinguishers - Fire Protection Services PLC
Fire extinguishers - Firetecnics Systems Ltd
Fire extinguishers - Griffin and General Fire Services Ltd
Fire extinguishers - Hoyles Fire & Safety Ltd
Fire extinguishers - Richard Hose Ltd
Fire fighting equipment - Eagles, William, Ltd
Fire surrounds, radiator covers, moulding and carvings - Winther Browne & Co Ltd
Fireplace frames and canopies - James Smellie Fabrications Ltd
Fireplaces - Emsworth Fireplaces Ltd
Fireplaces - Wyvern Marlborough Ltd
Fireplaces and fires - Bell & Co Ltd
Fireplaces, surrounds and hearths - Browne Winther & Co Ltd
Fitted Bedrooms - Symphony Group PLC, The
Fittings - Kee Klamps Ltd
Flags, flagstaffs, banners and bunting - Zephyr Flags & Banners
Flame retardant fitting, wrought iron furniture - Pouliot Designs
Free standing notice boards - Movitex Signs Ltd
Furniture - Andy Thornton Architectural Antiques Ltd
Furniture - Barlow Tyrie Ltd
Furniture - Bevan Funnell Ltd
Furniture - Chase James & Son (Furnishing) Ltd
Furniture - Cornwall Parker Ltd
Furniture - Ducal Ltd
Furniture - Welconstruct Co
Furniture and equipment - TWS Presentation Systems (Mcmillan UK Ltd T/A)
Furniture fittings - Nico Manufacturing Ltd
Furniture fittings and accessories - Haffele UK Ltd
Furniture for restuarants and catering facilities - Sandler Seating Ltd
Furniture for schools, public buildings, hotels and restaurants - AEL Furniture Group
Furniture lifting device - Interface Europe Ltd
Furniture manufacturers and contract furnishers - Pulsford Associates Ltd, John
Galvanized dustbins - Brettell & Shaw Ltd
Garment rails/shoe fittings - Smart F & G (Shopfittings)Ltd
Gas fires - Flavel - Leisure Glynwed Consumer & Construction Products Ltd
Geeral fixture/ furnishings and equipment - Hipkiss, H, & Co Ltd
General fixtures - Anders + Kern UK Ltd

General fixtures and furnishings - Be-Modern Ltd
General fixtures/ furnishings/ Equipment - Laidlaw Architectural Hardware
General furnishings - Hostess Furniture Ltd
General Furniture - Ellis J T & Co Ltd
General furniture and perspex furniture - The Cube
General purpose fittings and furnishings - Hettich UK Ltd
GRP Customised Mouldings - Wilsons Fibreglass Ltd
GRP enclosures and kiosks - Dewey Waters Ltd
Handling equipment - Welconstruct Co
Hinges - Cooke Brothers Ltd
Home furniture - Vitra UK Ltd
Hotel bedroom furniture and contract furniture - Pedley Furniture international Ltd
Hotel Security Boxes - Ritherdon & Co Ltd
Import furniture - Kush UK Ltd
Industrial textiles - Don & Low Ltd
Instrument cabinet - Ruthin Precast Concrete Ltd
Interchangable notice boards - Movitex Signs Ltd
Interior and exterior blinds and awnings - Deans Blinds & Awnings (UK) Ltd
Interior fit outs - OEP Furniture Group PLC
Internal litter bins - Glasdon UK Ltd
Iron and aluminium sand castings - Cradley Castings Ltd
Iso-design venetian blinds - Eclipse Blind Systems Ltd
Key filing, security cabinets and safes - Securikey Ltd
Kiosks - FIBAFORM PRODUCTS
Kitchen furniture - OEP Furniture Group PLC
Laminate wood table tops - Origlia Spa
Letter Boxes - Signs Of The Times
Lightweight fire barriers - Rentokil Initial U.K. Ltd.
Lockers - Decra Cubicles Ltd
Lockers and cubicles - Shackerley (Holdings) Group Ltd incorporating Designer ceramics
Loft hatches - Glidevale Building & Products Ltd
Manufacturers of Display & Exhibition Systems - Nimlok Ltd
Marble vanity tops - Emsworth Fireplaces Ltd
Material handeling equipment - Moresecure Ltd
Mats and matting - Floor Protection Services Ltd
Matwell frames and entrance matting - Clean Step Ltd
Mechanical extract units - Greenwood Air Management
Mirrors - Hastings & Folkstone Glass Works Ltd
Mirrors - Ledite Glass (Southend) Ltd
Mirrors - Lightfoot Charles Ltd
Modular shelving systems - Screen Systems Ltd
Modular shelving systems - Versatile Fittings Ltd
Nameplates and fireplaces - Delabole Slate
Natural ventilation products - Greenwood Air Management
Notice boards - Glasdon Designs Ltd
P.O.S support - Eclipse Blind Systems Ltd
Pallet racking, shelving systems and lockers - Sperrin Metal Products Ltd
Parafin pressure lamp - Tilley International Ltd
Personnel equipment - Welconstruct Co
Pinboards - Movitex Signs Ltd
Pine beds, bunk beds and headboards - Airsprung Beds Ltd
Pine room panelling, bookcases and display cupboards - Carved Pine Mantelpieces Ltd
Plastic wheeled bins - Brettell & Shaw Ltd
Portable auditorium seating - Sandler Seating Ltd
Portable gas cookers - Tilley International Ltd
Portable units - Lab Furnishings Ltd
Poster frames - Movitex Signs Ltd
Presentation wall systems - TWS Presentation Systems (Mcmillan UK Ltd T/A)
Press Tooling - Cooke Brothers Ltd
Pressings - Cooke Brothers Ltd
Product Finishing - Cooke Brothers Ltd
Public area bins, ashtrays and fire safe bins - Lesco Products Ltd
PVC strip curtains - Mandor Engineering Ltd
Radiator covers - Browne Winther & Co Ltd
Radiator covers - Winther Browne & Co Ltd
Ready made curtain - Eclipse Blind Systems Ltd
Reconstructed stone fireplaces - Minsterstone Ltd
Refuse sack holders and fire retardant bins - Unicorn Containers Ltd
Roller blinds - Tidmarsh & Sons
Ropes and twines - Kent & Co (Twines) Ltd
sackholders - Surrey Sackholders Ltd.
Sanibins - Surrey Sackholders Ltd.
Sculptures - Hynes Aden K Sculpture Studios
Sealer units - Pilkington Plyglass
Seating and desking - Davies John Interiors Ltd
Seating systems - Harrison Thompson & Co Ltd
Security mirrors - Securikey Ltd
Self retracting key reels - Securikey Ltd
Sheet metal enclosures - Ritherdon & Co Ltd
Shelves, panels - Anderson GEC Ltd
Shelving - RB Shelving Systems (UK) Ltd
Shelving - Smart F & G (Shopfittings)Ltd
Shelving Brackets-decorative - Winther Browne & Co Ltd
Shelving systems - Moresecure Ltd
Showroom design - Eclipse Blind Systems Ltd
Slat board - Smart F & G (Shopfittings)Ltd
Slotted angle plastic containers and boxes - Moresecure Ltd
Slotvents - Greenwood Air Management
Soft Fabrics - Eclipse Blind Systems Ltd
Soft furnishings fabrics - Gradus Ltd
Solar blinds and tension systems - EBS Services Ltd
Solar shading - Western Avery Ltd
Solid surfaces - Viking Laminates Ltd
Special Stainless Fabrications - Pland
Sprung mattresses, divan bases, pillows and accessories - Airsprung Beds Ltd
Square tube construction systems - Moresecure Ltd
Stainless steel general fixings - Associated Metal (Stainless) Ltd
Stainless steel, aluminium, PVC covered, galvenized and brass handrails - SG System Products Ltd
Stainless steel, cabinets, worktops, mirrors - Anderson GEC Ltd
Steel cabinets - Badderley Rose Ltd
Steel cabinets - Climet Ltd
Steel tube and fittings - Rothley Burn Ltd
Storage - Rothley Burn Ltd
Storage cabinets - Richard Hose Ltd
Storage wall equipment - ARC Partitioning
Study, lounge, bedroom, office, reception fitted furniture - Hyperion Wall Furniture Ltd
Sunbeds, lockers and cubicles - Dalesauna Ltd
Sunblinds - Sampson T F Ltd
Support rails and systems - Southern Sanitary Specialists Ltd
Tracking systems - Moresecure Ltd
Training Rooms Installations - Planet Projects Ltd
Trolley - Unicorn Containers Ltd
Upholstery fabrics - Gradus Ltd
Vanity bathroom units - Tropravit (UK) Ltd
Vanity units - Architectural & Building Products Ltd
Vertical and horizontal carousels - Dexion Ltd
Waiting area seating for train stations, airports, ferry terminals, hospitals and surgeries - Hille Ltd
Wall rails - Fixatrad Ltd
Waste bins - Surrey Sackholders Ltd.
Weathervanes - Goods Directions Ltd
Weighing scales and weigh bridges - Salter Weigh-Tronix
Window blinds - Hunter Douglas Ltd (Contracts Division) Luxaflex Blinds ®
Window blinds and sunscreening systems - Faber Blinds Ltd
Window blinds and sunshades - Dallas & Forrest Ltd
Window furnishings, curtain tracks, poles and valences - Integra Products
Window vertical, roller, venetian and pleated blinds - Louvre-Lite Ltd
Wood and mirror sliding wardrobe systems - Howard Home Improvements Ltd
Wooden Shutters - Design Line
Wooden Venetian blinds - Design Line
Wooden venetian blinds - Eclipse Blind Systems Ltd
Wooden venetian blinds, Timbershades - Tidmarsh & Sons
Work benches - Fine H & Son Ltd
Work benches and cupboards - Emmerich (Berlon) Ltd
Work surfaces - McLoughlin Wood Ltd
Worktops - Delabole Slate
Worktops - Viking Laminates Ltd
Worktops and vanity units - Domus Tiles Ltd

N11 Domestic kitchen fittings

Bath and kitchen seals - Homelux
Built -in cookers - Stoves PLC
Built-in dishwashers - Stoves PLC
Built-in refrigeerators - Stoves PLC
Cookers - Aga-Rayburn, Glynwed Consumer & Construction Products
Cookers - Flavel - Leisure Glynwed Consumer & Construction Products Ltd
Cookers - Stoves PLC
Cookers and electrical appliances - Electrolux Domestic Appliances
Cookers and electrical appliances - Parkinson Cowan and Tricity Bendix Co, The
Domestic kitchen fitting - Johnson & Johnson PLC
Domestic kitchen fittings - Be-Modern Ltd
Domestic kitchen fittings - Commodore Kitchens Ltd
Domestic kitchen fittings - Hipkiss, H, & Co Ltd
Domestic kitchen fittings - Jenkins Newell Dunford Ltd
Domestic kitchen fittings - Sissons W & G Ltd
Domestic kitchen fittings, Lever action - Shavrin Levatap Co Ltd
Domestic kitchen furniture - Alno (United Kingdom) Ltd
Domestic Kitchen furniture - Ball William Ltd
Domestic kitchen sinks, taps and accessories - Carron Phoenix Ltd
Domestic waste disposal units - Max Appliances
Electrical appliances - Pifco Ltd
Fitted kitchens - Miele Co Ltd
Free standing cookers - Stoves PLC
Granite kitchen work top - Emsworth Fireplaces Ltd
Kitchen and bathroom furniture - Dennis & Robinson Ltd
Kitchen brassware - Midland Stom Ltd
Kitchen fittings - Bernstein Group PLC
Kitchen fittings - Magnet Ltd
Kitchen fittings - Symphony Group PLC, The
Kitchen furniture - Allied Manufacturing Co (London) Ltd
Kitchen furniture - Gower Furniture Ltd
Kitchen showrooms - Jewson Ltd
Kitchen worksurfaces - Bushboard Parker Ltd
Kitchen worktops - Kronospan Ltd
Kitchens - Bell & Co Ltd
sackholders - Surrey Sackholders Ltd.
Sanibins - Surrey Sackholders Ltd.
Self assembley kitchens - Ideal Furniture Enterprises Ltd
Sinks - Astracast PLC
Sinks, taps, waste disposal and water filters - Berglen Group Ltd
Stoves inset and free standing - Emsworth Fireplaces Ltd
Traditional kitchen taps and Mixers - Shavrin Levatap Co Ltd
Undersink pumping units - Fordwater Pumping Supplies Ltd
Vanitories worktops - Burlington Slate Ltd
Waste bins - Surrey Sackholders Ltd.
Work surfaces - Kirkstone
Worktops - Formica Ltd

N12 Catering Equipment

Airport design and service unit systems - Store Development Ltd
Bar Tender Sinks - Pland
Catering Equipment - BRB Industrial Services Ltd
Catering Equipment - Calpeda Limited
Catering equipment - Electrolux food service
Catering Equipment - Falcon Catering Equipment
Catering equipment - Grundy Catering
Catering Equipment - Lockhart Catering Equipment
Catering equipment - Sissons W & G Ltd
Catering grease and oil traps - Conder Products Ltd
Catering sinks and tables - Pland
Catering storage - Hispack Limited
Catering water boilers - Calomax (Engineers) Ltd
Coffee percolators - TSE Brownson
Commercial catering and bar equipment - Imperial Machine Co Ltd
Commercial scale kitchen equipment - Zanussi Professional - Zanussi Professional
Commercial waste disposal units - Max Appliances
Counters and Serveries - Viscount Catering Ltd
Dishwashers, glasswashers and bar refrigeration - Clenaware Systems Ltd
Display carts - TSE Brownson
Domestic kitchen foodwaste disposers - The Haigh Tweeny Co Ltd
Drinks vending machines - Craft Jacobs Suchard Ltd
Elements for catering trade - Howden H D Ltd
Fish and chip frying ranges - Preston & Thomas Ltd
Fish and Chip Ranges & Catering equipment - Nuttall, Henry (Viscount Catering Ltd)
Food and drink vending machines - Hillday
Food technology furniture (education) - Klick Technology Ltd
Furniture for the quick service restaurant industry - Store Development Ltd
Heater service counter tops - TSE Brownson
Heavy duty catering equipment - Viscount Catering Ltd
High performance kitchen furniture - Rixonway Kitchens Ltd
Industrial humidifiers - Calomax (Engineers) Ltd
Insulated Urns - TSE Brownson
Kitchen equipment - Magnet Ltd
Light duty catering equipment - Viscount Catering Ltd
Medium duty cateing equipment - Viscount Catering Ltd
Milk heaters - TSE Brownson
Rifrgeration - Viscount Catering Ltd
sackholders - Surrey Sackholders Ltd.
Sanibins - Surrey Sackholders Ltd.
Show cases - TSE Brownson
Stainless steel catering equipment - Associated Metal (Stainless) Ltd
Stainless steel catering equipment - Moffat Ltd, E & R
Stainless steel commercial catering equipment - Bartlett Catering Equipment Ltd
Stainless steel shelving systems for kitchens and coldrooms - Beford & Soar Ltd
Tray/plate dispensers - TSE Brownson
Trolleys - TSE Brownson
Vending machines - Climet Ltd
Vending machines - Provend Services Ltd
Vending systems - The Kenco Coffee Company
Waste bins - Surrey Sackholders Ltd.
Waste disposal equipment, balers, shredders, drinks can crushers - Tony Team Ltd
Waste disposal units and catering cupboards - Pland

N13 Sanitary appliances/ fittings

Accessories - Caradon Plumbing Solutions
Accessories - Ideal-Standard Ltd
Aluminium shower outlets - Marley Alutec Ltd
Anti-vandal showers - Shavrin Levatap Co Ltd
Bath and kitchen seals - Homelux
Bath and shower mats - William Freeman Ltd
Bath screens - Majestic Shower Co Ltd
Bathing equipment for elderly and disabled - Chiltern Invadex
Bathroom accessories - HEWI (UK) Ltd
Bathroom Accessories - Metlex
Bathroom accessories - Neptune Showers Ltd
Bathroom accessories, shower enclosures, overbath screens, bath seats and grab rails, shower seats, safety rails, mirrors - Provex Products Ltd
Bathroom and shower accessories - Marleton Cross Ltd
Bathroom and showering equipment - Southern Sanitary Specialists Ltd
Bathroom cabinets, accessories, toilet seats and vanityware - Allibert (UK) Ltd
Bathroom fittings - MacBee Manufacturing Ltd
Bathroom fittings - Rothley Burn Ltd
Bathroom Pods - RB Farquhar Manufacturing Ltd
Bathroom products - Qualitas Bathrooms
Bathroom showrooms - Jewson Ltd
Bathrooms - Spital Tile Co
Bathrooms - Vitra (UK) Ltd
Baths - Ideal-Standard Ltd
Baths and taps - Midland Stom Ltd
Bedroom furniture - Bell & Co Ltd
Brassware, bath and toilet seats - Tropravit (UK) Ltd
Brassware, showers and accessories - Hansgrohe
Cabinet furniture - HEWI (UK) Ltd
Ceramic and acrylic sanitaryware - Armitage Shanks Ltd
Ceramic bathroom accessories - Balmoral Porcelain Limited/Longmead Ceramics
Cubicle rails - Harrison Drape, Divsion of McKenie (UK)
Cubicle tracks - Marshall Contracts
Domestic kitchen sinks and taps - Fordham Pland
Drinking Fountains - Harris T A Ltd
Drinking fountains - Hughes Safety Showers
Drinking fountains and water coolers - Maestro International Ltd
Duct panels - Dunhams of Norwich
Electric mixer and power showers, hand wash units and water heaters - Triton PLC
Electric showers - Dolphin Showers Ltd
Emergency showers - Rose Emergency Equipment Co
Enclosures - Ideal-Standard Ltd
Ensuite shower pods - RB Farquhar Manufacturing Ltd
Fitted Bathrooms - Symphony Group PLC, The
Fittings for disabled and aged - HEWI (UK) Ltd
Frameless shower screens - ShowerSport Ltd
Grab rail and disabled equipment - Dryad Architectural Ltd
Grab rails - Rothley Burn Ltd
Hand dryers - Heatrae Sadia Heating Ltd
Hand dryers - Warner Howard Ltd
Handryers and hygiene products - Anda Products Ltd
Hygiene equipment - Imperial Machine Co Ltd
Incinerators - Combustion Linings Ltd
Intumescent Mastics - Bollom Fire Protection Ltd
Jacuzzi, baths and showers - Jacuzzi UK Ltd, Division of Jacuzzi Inc
Kitchens - Bell & Co Ltd
Kits - Ideal-Standard Ltd
Lever taps for disabled people, actuators for concealed cisterns - Sheardown Engineering Ltd
Mixing valves - Douglas Industrial Ltd
Modern bathroom taps and mixers - Shavrin Levatap Co Ltd
Pan connectors - Derwent Macdee Ltd
Pan connectors - Phetco (England) Ltd
Pillar taps - Booth Samuel & Co Ltd
Plumbing fittings - Flowflex Componants Ltd
Power, mixer valve and electric showers - NewTeam Ltd
Push taps - Caradon Plumbing Solutions
Sanitary appliances - Arjo Ltd
Sanitary appliances and fittings - Barflow-LCA
Sanitary appliances and fittings - Cefndy Enterprises
Sanitary appliances and fittings - Guest (Speedfit) Ltd, John,
Sanitary appliances and fittings - Gummers A & J Ltd
Sanitary appliances and fittings - Hadida Ltd, M R
Sanitary appliances/ fittings - Harris & Bailey Ltd
Sanitary appliances/fittings - Allgood plc
Sanitary appliances/fittings - Broadmead Products
Sanitary Appliances/Fittings - Wybone Ltd
Sanitary disposal units, washroom and baby room products - Cannon Hygiene Ltd
Sanitary equipment and accessories - Ideal-Standard Ltd
Sanitary fittings - Sissons W & G Ltd
Sanitary fixings - Arthur Fisher (UK) Ltd
Sanitary manifolds - SAV United Kingdom Ltd
Sanitary towel incinerator and towel rails - Consort Equipment Products Ltd
Sanitary ware - SAV United Kingdom Ltd
Sanitary ware and appliances - Diamond Merchants
Sanitary ware and fittings - Caradon Bathrooms Ltd
Sanitaryware, associated fittings and accessories - Shires Bathrooms
Sauna - Golden Coast Ltd
Sauna cabins and heaters, steam rooms and steam showers - Sanua UK Ltd
Sauna cabins and heaters, steam rooms and steam showers - Tylo UK Ltd
Saunas - Finnish Sauna Baths (Sales) Ltd
Saunas, steam rooms and spa baths - Dalesauna Ltd
Self closing non concussive press and lever taps - Sheardown Engineering Ltd
Shower accessories - Gummers A & J Ltd

Shower accessories - NewTeam Ltd
Shower booster pumps - Baresford Pumps
Shower controls - Douglas Industrial Ltd
Shower curtains - Neptune Showers Ltd
Shower doors - Majestic Shower Co Ltd
Shower doors and carer screens - Norton Engineering Alloys Co Ltd
Shower doors, enclosures, bath screen and shower trays - Daryl Industries Ltd
Shower doors, trays and bath screens - Matki plc
Shower enclosers - Majestic Shower Co Ltd
Shower enclosures - Contour Showers Ltd
Shower enclosures and towel rails - Kermi (UK) Ltd
Shower enclosures, bathscreens, shower trays, baths and bathroom equipment - Novellini Bathroom Products
Shower grilles - Norton Engineering Alloys Co Ltd
Shower products - Aqualisa Products
Shower pumps - NewTeam Ltd
Shower rails - Harrison Drape, Divsion of McKenie (UK)
Shower rails - Neptune Showers Ltd
Shower screens - Majestic Shower Co Ltd
Shower trays, cubicles and bath screens - Coram (UK) Ltd
Shower valves - Ideal-Standard Ltd
Shower valves - Sheardown Engineering Ltd
Showering systems - Daryl Industries Ltd
Showers - Caradon Plumbing Solutions
Showers - Jendico
Spa - Golden Coast Ltd
Specialist bathing, transfer, hygiene and showering systems - Parker Bath Division
Spray mixing valves - Sheardown Engineering Ltd
Stainless steel sanitary appliances and fittings - BM Stainless Steel Drains Ltd
Stainless steel sanitaryware - Pland
Stainless steel sanitaryware and washroom equipment - Associated Metal (Stainless) Ltd
Stainless steel sinks, wc's, urinals, sluices, baths and foot baths - Anderson GEC Ltd
Stainless steel, synthetic sinks and shower cubicles - Leisure, Glynwed Consumer Foodservice Products Ltd
Taps - Astracast PLC
Taps - Deva Tap Co Ltd
Taps - Shavrin Levatap Co Ltd
Taps and mixers - Ideal-Standard Ltd
Taps and valves - Opella Ltd
Taps for the disabled and elderly - Shavrin Levatap Co Ltd
Taps, mixers, and showers - Pegler Ltd
Taps, showers and thermostatic mixers - Grohe Ltd
Thermostatic mixing valves - Caradon Plumbing Solutions
Thermostatic shower valves - Shavrin Levatap Co Ltd
Toilet Cubicales - Relcross Ltd
Toilet cubicles and washroom systems - Moffett Thallon & Co Ltd
Toilet seats - Derwent Macdee Ltd
Toilet Seats - Pressalit Ltd
Toilet, shower and changing cubicles - Dunhams of Norwich
Towel and soap dispensers - Kimberly-Clark Ltd
Towel rails - Faral Radiators
Trays - Ideal-Standard Ltd
Urinal flush controls - Cistermiser Ltd
Valves - Home Engineering Co Ltd, The
Vandal resistant taps and fittings - Sheardown Engineering Ltd
Vanity units - Dunhams of Norwich
Victoriana brass shower roses - Purdie Elcock Ltd
Vitreous china ceramic sanitaryware and brasswear - Vitra (UK) Ltd
W. C.'s - Ideal-Standard Ltd
Warm air hand and face dryers - Airdri Ltd
Warm air hand dryers - Cistermiser Ltd
Warm air hand dryers - Wandsworth Elecrtrical Ltd
Wash basins - Ideal-Standard Ltd
Wash room hygiene products - Airstream Products Ltd
Washroom Equipment - Relcross Ltd
Washroom system - Bushboard Parker Ltd
Water taps and mixers - Barber Wilsons & Co Ltd
Waterless Urinals - Relcross Ltd
WC cisterns - Derwent Macdee Ltd

N15 Signs/ Notices

Architectural aluminium systems for signs - Universal Components Ltd
Architectural signage - Signs & Labels Ltd
Bus Stops - Burnham Signs
Cast signs - Signs Of The Times
Chalk boards, pin boards and display boards - Leakale Taurus Visual Aids
Commemorative plaques, coats of arms and architectural lettering - Metalline Signs Ltd
Contract enamelling - Enamelled Signs (UK) Ltd
Electric, illuminated and neon signs - Pearce Signs Ltd
Enamelling cladding panels - Enamelled Signs (UK) Ltd
Engravers - Abbey Craftsmen Ltd
Fibre Optic Signs - Burnham Signs
Hand engraving tools and metal etching pens - Elliott John H (Monostamp) Ltd
Illuminated signs - Burnham Signs
Instrumentation panels - Pryorsign
Internal and external illuminated signs - Interlux Ltd
Large letter signs - Pannell Signs Ltd
Letter plates - Map Hardware Ltd
Litter Bins - Burnham Signs
Manufacture - Oldham Signs Ltd
Marking inks and products - Eyre & Baxter (Stampcraft) Ltd
Nameplates - Eyre & Baxter (Stampcraft) Ltd
Nameplates - Mockridge Labels & Nameplates Ltd
Nameplates and signs - Caxton Name Plate Manufacturing Co Ltd
Notice board material - Forbo - Nairn Ltd
Notice panels - Opto International Ltd
Pinboards - Sundeala Division of Celotex Ltd
Professional site signs - Abacus Signs
Safety signs - Asco Extinguishers Co Ltd
Safety Signs - Burnham Signs
Safety signs - Eyre & Baxter (Stampcraft) Ltd
Safety signs - Fife Fire Engineers & Consultants Ltd
Safety signs - Movitex Signs Ltd
Sign manufacturers - Blaze Neon Ltd
Sign posts - Fabrikat (Nottingham) Ltd
Signs - Allgood plc
Signs - Dee-Organ Ltd
Signs - Drakard & Humble Ltd
Signs - Forest City Signs Ltd
Signs - Green Brothers Ltd
Signs - Movitex Signs Ltd
Signs - Neocrylic Signs Ltd
Signs - Oldham Signs Ltd
Signs and graphics - Alliance Europe
Signs and notices - Enamelled Signs (UK) Ltd
Signs and notices - Focal Signs Ltd
Signs and notices - Glasdon Designs Ltd
Signs and notices - Hoyles Fire & Safety Ltd
Signs and notices - Oldham Signs Ltd
Signs/ notices - Laidlaw Architectural Hardware
Signs/ Notices - Leander Architectural
Signs/Notices - Duroy Fibreglass Mouldings
Specialist signs - Signs International
Stencils - Eyre & Baxter (Stampcraft) Ltd
Teaching walls, column boards and white boards - TWS Presentation Systems (Mcmillan UK Ltd T/A)
Traffic and street names signs - Royal Label Factory, The
Traffic management systems - Marshalls Mono Ltd
Transport signage and wayfinding - Burnham Signs
Visual communication equipment - Wilson & Garden Ltd
Vitreous enamel signs - Burnham Signs
Vitreous enamelled steel cladding signs - Escol Panels (S & G) Ltd
Wall plaques - Rogers Concrete Ltd
Waterproof, vandal resistant illuminated, health and safety and shop signs, nameplates and fascias labels - Pryorsign
Whiteboards - Magiboards Ltd

N20 Safety Equipment

Anti-slip sighting bars - Redman Fisher Engineering
Body armour - Meggitt Lavmour Systems
Confined space equipment - MSA (Britain) Ltd
Emergency safety showers and eye and face wash units - Maestro International Ltd
Explosion containment tubes - Attwater & Sons Ltd
Fire extinguishers - Firemaster Extinguisher Ltd
Fire extinguishers - Macron Fireater Ltd
Fire hose and fittings - Macron Fireater Ltd
Fire hose reels - Fire Protection Services PLC
Gas detection systems - Blakell Europlacer Ltd
Harnesses - HCL Verta
Hazardous area equipment - Legrand Electric Ltd
Height Safety - Barrow Hepburn Sala Ltd
Industrial safety clothing and equipment - BRB Industrial Services Ltd
Ladder safety post - Bilco UK Ltd
Ladders and safety cages - Redman Fisher Engineering
Ladders, lowering line and vertical inclined chute systems - Samways P A & Co
Personal Protective equipment - MSA (Britain) Ltd
Ropes and sashcords - Marlow Ropes Ltd
Safety and security equipment - Jones & Thomas (Wireworks) Ltd
Safety equipment - Barrow Hepburn Sala Ltd
Safety equipment - MSA (Britain) Ltd
Safety Equipment - Radiodetection Ltd
Safety harnesses etc - Sala Ltd, B H
Safety interlocking systems - Castell Safety International
Safety signs - Movitex Signs Ltd
Safety treads - Record & Parkes Ltd
Safety, drench, decontamination, emergency showers and eyebaths - Hughes Safety Showers
Security grilles and collapsible gates - Badderley Rose Ltd
Special purpose safety equipment - Hazard Safety Products Ltd
Steel security doorsets - Sunray Engineering Ltd
Vertical fall arrest safe access laders - HCL Verta
Water safety products - Glasdon Designs Ltd
Welding curtains and hangers - Nederman Ltd

N21 Storage Equipment

Adjustable pallet racking - Link 51 (Storage Products) Ltd
Adjustable Shelving - Spur Shelving Ltd
Archive storage - Dexion Ltd
Construction products - Link 51 (Storage Products) Ltd
Cupboards - Brown N C (Storage Equipment) Ltd
Document storage shelving systems - Link 51 (Storage Products) Ltd
Drawer Units - Brown N C (Storage Equipment) Ltd
Filing systems - Dexion Ltd
Freestanding Shelf Systems - Spur Shelving Ltd
Garment Collectors - Brown N C (Storage Equipment) Ltd
Garment Dispensers - Brown N C (Storage Equipment) Ltd
Heavy duty shelving systems - Link 51 (Storage Products) Ltd
Library shelving - Sperrin Metal Products Ltd
Lockers - Brown N C (Storage Equipment) Ltd
Lockers and cubicles - LSA Projects Ltd
Megyanine Floors - Hi-Store Ltd
Mobile and static shelving equipment - Railex Systems Ltd, Elite Division
Mobile shelving - Dexion Ltd
Mobile shelving - Link 51 (Storage Products) Ltd
Pallet raking - Dexion Ltd
Personal storage lockers - Bawn W. B. & Co Ltd
Picking systems - Dexion Ltd
Shelving - Lesco Products Ltd
Shelving and furniture for libraries and archives - Balmforth Engineering Ltd
Shelving and storage racking - Dexion Ltd
Shelving systems and safes - Rosengems Tann Ltd
Single Shelf Brackets - Spur Shelving Ltd
Small parts storage - Link 51 (Storage Products) Ltd
Small storage - Moresecure Ltd
Steel storage and office shelving systems - Balmforth Engineering Ltd
Storage cabinets - Unicorn Containers Ltd
Storage equipment - Barton Storage Systems Ltd
Storage Equipment - Envopak Group Ltd
Storage equipment - Garran Lockers Ltd
Storage Equipment - Hi-Store Ltd
Storage equipment - Hispack Limited
Storage Equipment - Rackline Systems Storage Ltd
Storage equipment - Welconstruct Co
Storage Equipment - Wybone Ltd
Storage racks and shelving - Metalrax Ltd
Storage systems - Komfort Systems Ltd
Storage systems - Moresecure Ltd
Storage walls - ClipsPartitions Ltd
Toughened shelves - Cliffhanger Shelving
Waste containers and systems - Iles Waste Systems
Waste containment - MGB Randalls

N22 Office Equipment

Acoustic Screens - Pergola Products Ltd
Binders - Quartet-GBC
Boardroom installations - Planet Projects Ltd
Chairs - Kinsley-Cooke Furniture Ltd
Church seating & accessories - Knightbridge Furniture Productions Ltd
Conference Room Installations - Planet Projects Ltd
Contract School/ Office furniture - Blindcraft Industries
Cupboards - Brown N C (Storage Equipment) Ltd
Display Stands - Pergola Products Ltd
Domestic and commercial office furniture - Ball William Ltd
Ergonomic products - Lesco Products Ltd
Fire resistant cabinets - Chubb Safe Equipment Company
Fire safes - INMAC (a division of microwarehouse Ltd)
Flipcharts, hanging rail systems and visual aids - Leakale Taurus Visual Aids
Floor, fire vents, ceiling and roof hatches - Bilco UK Ltd
Furniture - Deanes Furniture Ltd
Furniture - HEM Interiors Group Ltd
Hand stamps - Elliott John H (Monostamp) Ltd
Interactive whiteboards - Quartet-GBC
IT Furniture - MICE Kaymar Limited
Laminators - Quartet-GBC
Notice Boards - Pergola Products Ltd
Office and industrial cleaning - CSL (Services) Limited
Office and reception fitted furniture - Hyperion Wall Furniture Ltd
Office chairs - KAB Seating Ltd
Office equipment - Antocks Lairn Ltd
Office equipment - Carleton Furniture Group Ltd
Office Equipment - Envopak Group Ltd
Office equipment - Hago Products Ltd
Office equipment - Hill & Noyes Office Furniture
Office equipment - Intercraft Designs Ltd
Office equipment - Lamb Macintosh
Office equipment - Logic Office Interiors
Office equipment - Martela PLC
Office equipment - Matthews Office Furniture Ltd
Office equipment - Vista Plan International PLC
Office equipment - Vitra UK Ltd
Office filing systems - Railex Systems Ltd
Office flooring - Propaflor Ltd
Office furniture - EFG Business Furniture (UK) Ltd
Office furniture - Fray Design Ltd
Office furniture - Girsberger London
Office furniture - Haworth UK Ltd
Office furniture - INMAC (a division of microwarehouse Ltd)
Office furniture - Kinsley-Cooke Furniture Ltd
Office furniture - Komfort Systems Ltd
Office furniture - OEP Furniture Group PLC
Office furniture - Origlia Spa
Office furniture - President Office Furniture Ltd
Office furniture - Progress Furnishings Systems Ltd
Office furniture - Project Office Furniture PLC
Office furniture - Samas Roneo Ltd
Office furniture - Scott Appleton Furniture Ltd
Office furniture - Top Office Equipment Ltd
Office furniture and accessories - Lesco Products Ltd
Office furniture and seating - Blyde-Barton Ltd
Office furniture including storage cabinets and desks - Flexiform Business Furniture Ltd
Office furniture, desking, storage, boardroom and conference furniture - Hands of Wycombe
Office furniture, screens and seating - Intercraft Designs Ltd
Office furniture, seating and board room tables - Steelcase Strafor Plc
Office furniture, seating and storage - Desking Systems Ltd
Office partitions - ClipsPartitions Ltd
Office racking, storage cabinets, mobile shelving and filing systems - Rackline Systems Storage Ltd
Office seating - Dauphin PLC
Office seating - Formline Ltd
Office seating and desking - Metalliform Products Plc
Office seating, steel filing, storage products, tables and chairs - Harvey Furniture
Office storage equipment and desking - Bisley Office Equipment
Paragon Business Furniture - Paragon Business Furniture
Photocopiers - Oce (UK) Ltd
Post room furniture - Lesco Products Ltd
Rail presentation systems - Quartet-GBC
Residential furniture - Carleton Furniture Group Ltd
Rubber stamps - Eyre & Baxter (Stampcraft) Ltd
Screens - Kinsley-Cooke Furniture Ltd
Seating and executive furniture - Verco Office Furniture Ltd
Seating Equipment - Evertaut
Shredders - Quartet-GBC
Storage - Kinsley-Cooke Furniture Ltd
Task lights - Lesco Products Ltd
Type - Elliott John H (Monostamp) Ltd
Visual display products - Lesco Products Ltd
Waiting area seating for train stations, airports, ferry terminals, hospitals and surgeries - Hille Ltd
Work stations - Moresecure Ltd
Workstation furniture and mobile filing systems - Mauser Interiors (UK) Ltd
Writing Boards - Pergola Products Ltd

N23 Educational Equipment

Acoustic Screens - Pergola Products Ltd
Auditorium, theatre, multi purpose, public area and courtroom seating - Race Furniture Ltd
Binders - Quartet-GBC
Computer furniture (education) - Klick Technology Ltd
Corporate seating & tables - Knightbridge Furniture Productions Ltd
Display Stands - Pergola Products Ltd
Education Equipment - Evertaut
Education storage furniture - Blyde-Barton Ltd
Educational equipment - Invicita Plastics Ltd
Educational furniture - AEL Furniture Group
Educational furniture - MICE Kaymar Limited
Educational furniture for science and technology - S & B Ltd
Educational seating and desking - Metalliform Products Plc
Educational workshop equipment - Emmerich (Berlon) Ltd
Interactive whiteboards - Quartet-GBC
Laminators - Quartet-GBC
Notice Boards - Pergola Products Ltd
Rail presentation systems - Quartet-GBC
Technology furniture (education) - Klick Technology Ltd
Waiting area seating for train stations, airports, ferry terminals, hospitals and surgeries - Hille Ltd
White boards - Leakale Taurus Visual Aids
Writing Boards - Pergola Products Ltd

N24 Shopfitting

Automatic electric drive through window - Store Development Ltd
Commercial shelving - Dexion Ltd
Contour showcases - Opto International Ltd
Contract fit outs - J. Scott (Thrapston) Ltd
Entrances - Parkside Group Ltd, The
Extrusions - Opto International Ltd
Fast security screen counters - Safetell Ltd
Illuminated and non illuminated show cases - Movitex Signs Ltd
Interior design - Eaton Elizabeth
Interior fit outs - Welconstruct Co
Interior fitting out - Timtec International Ltd
Maintenance of industrial doors, shutters and shopfronts - Shutter Door Repair & Maintenance Ltd

Management systems and lobby shutters - Safetell Ltd
Panel Connectors - Opto International Ltd
Purpose made joinery - Opto International Ltd
Reception counters - Sandoms (Batley) Ltd
Reception desks - Opto International Ltd
Retail furniture - Blyde-Barton Ltd
Retail merchandising equipment - Store Development Ltd
Retractable glazed screen counters and secure access systems - Safetell Ltd
Shop and office fitters - Grafton Magna Ltd
Shop, office, bank and bar fitters - Beasley Joiners Ltd
Shopfit Shelf Systems - Spur Shelving Ltd
Shopfitters - Cumberland Construction
Shopfitters, display cabinet manufacturers - Sandoms (Batley) Ltd
Shopfitting - Andy Thornton Architectural Antiques Ltd
Shopfitting - Blaze Neon Ltd
Shopfitting - Carlton Benbow
Shopfitting - CIL International Ltd
Shopfitting - Smart F & G (Shopfittings)Ltd
Shopfitting - Steel Line Ltd
Shopfitting and specialist joinery - Barlow Shopfitting Ltd
Shopfitting and specialist joinery - Barnwood Shopfitting Ltd
Shopfitting products - Interlux Ltd
Shopfronts - Parkside Group Ltd, The
Static filing cabinets - Rackline Systems Storage Ltd
Tube connectors - Opto International Ltd

N25 Hospital/ Health equipment

Beds, mattresses, accessories, cauches, commodes, overbed tables, living aids - Sidhil Care
Clean room partitions - ClipsPartitions Ltd
Disabled and elderly equipment - Nicholls & Clarke Ltd
Disposable hospital products - Verna Ltd
Eye and face wash equipment - Rose Emergency Equipment Co
Health care products - William Freeman Ltd
Hospital / Health Equipment - Evertaut
Hospital and health equipment - Arjo Ltd
Hospital and health equipment - Brandon Medical
Hospital and laboratory equipment - Associated Metal (Stainless) Ltd
Hospital Curtains - Marshall Contracts
Hospital Equipment, Elbow control - Shavrin Levatap Co Ltd
Hospital furniture - Sidhil Care
Hospital Sinks - Pland
Hospital/ Health Equipment - LSA Projects Ltd
Hospital/ Health Equipment - Sissons W & G Ltd
Hospital/ Health Equipment - The Kenco Coffee Company
Hospital/Health Equipment - Wybone Ltd
Hotel & Leisure seating & tables - Knightbridge Furniture Productions Ltd
Laboratory furniture - Decra Cubicles Ltd
Lifting, moving and handling products for disabled - Chiltern Invadex
Medical and first aid furniture - Hoskins Medical Equipment Ltd
Medical and first aid furniture - Hospital Metalcraft Ltd
Medical examination lamps - Wandsworth Elecrtrical Ltd
Medical waste containers - MGB Randalls
Shower enclosures - Contour Showers Ltd
Slophers - Pland
Specialist bathing, transfer, hygiene and showering systems - Parker Bath Division
Sprung mattresses, divan bases, pillows and accessories - Airsprung Beds Ltd
Stainless steel shelving systems for sterile areas - Beford & Soar Ltd
Support systems for elderly/disabled - Norton Engineering Alloys Co Ltd
Supportive Bathroom Products for Disabled People - Pressalit Ltd
Surgeons scrub up trough - Pland
Waiting area seating for train stations, airports, ferry terminals, hospitals and surgeries - Hille Ltd
X-ray door sets - Lami Doors UK

N26 Gymnastic/ Sport/ Play Equipment

Activity Nets - Record Playground Equipment Ltd
Aluminium extrusions/ sections - Seco Aluminium Ltd
Automatic slatted pvc swimming pool - Grando Ltd
Children's play areas and playground equipment - Roadways & Car Parks Ltd
Claybody - Hepworth Minerals & Chemicals Ltd
Cloakroom equipment - Locking Systems Ltd
Cloakroom equipment including security locks - Locking Systems Ltd
Furniture for the leisure industry - Store Development Ltd
Furniture, Leisure - Allermuir Contract Furniture Ltd
Gymnastic matting - Continental Sports Ltd
Gymnastic, sport and play equipment - Carr Gymnasium Equipment Ltd
Medical and first aid furniture - Doherty Medical Ltd
Nappy changers - LSA Projects Ltd
Physical education equipment - Continental Sports Ltd
Play equipment - En-tout-cas plc (Scotland)
Play equipment - Game Time UK Ltd
Playground equipment - Lappset UK Ltd
Playground equipment - Record Playground Equipment Ltd
Playground equipment and associated works - Kompan (UK) Ltd
Pool surrounds - LazyLawn
Safety surfacings - En-tout-cas plc (Scotland)
Skateboard Equipment - Record Playground Equipment Ltd
Sports arenas products - pvc-u - Duralock UK Ltd
Sports benches/Play equipment - Hispack Limited
Sports equipment - Continental Sports Ltd
Sports equipment - En-tout-cas plc (Scotland)
Sports Sands and Rootzones - Hepworth Minerals & Chemicals Ltd
Sports surfaces - LazyLawn
Stadia and auditorium seating - Metalliform Products Plc
Track systems and rigging equipment for theatre applications - Harkness Hall Ltd

N27 Cleaning Equipment

Brooms - Cottam Bros Ltd
Cleaning equipment - Arjo Ltd
Cleaning maintenance equipment - Centro Engineering Ltd
Cleaning materials suppliers - Spontex Ltd
Cleaning Products - SC Johnson Wax
Fixed and portable cleaning systems - Lamson, D D, PLC
Hand cleaner - Ward Bekker Sales Ltd
Hygiene products - Applied Chemicals Ltd
Image-Maintenance - Interface Europe Ltd
Industrial and commercial cleaning equipment - Cimex International Ltd
Industrial and commercial scrubbers, vacuum cleaners and floor maintenance equipment - Nilfisk Advance Ltd
Maintenance, polish and cleaning chemicals - British Nova Works Ltd
Power washers, scrubber driers, sweepers, polishers and vacuums - Alto Cleaning Systems (UK) Ltd
Specialist brushes - Cottam Bros Ltd
Specialist cleaning - New Centurian Industry Services Ltd

N28 Laboratory furniture/ equipment

Aluminium extrusions/ sections - Seco Aluminium Ltd
Ceramic and acrylic sanitaryware - Armitage Shanks Ltd
Decontamination booths - Rose Emergency Equipment Co
Industrial laboratory furniture - S & B Ltd
Laboratory Eqyiptment - USF Permutit/Ionpure
Laboratory furniture - ClipsPartitions Ltd
Laboratory furniture - Fisher Scientific (UK) Ltd
Laboratory furniture - MICE Kaymar Limited
Laboratory furniture (education) - Klick Technology Ltd
Laboratory furniture / equipment - Evertaut
Laboratory furniture and fume cupboards - P & R Laboratory Group Ltd
Laboratory furniture/ equipment - Benton Co Ltd Edward
Laboratory furniture/equipment - Sissons W & G Ltd
Laboratory Sinks - Pland

N29 Industrial Equipment

Bowl Pressings - Pland
Commercial and industrial equipment - Slingsby H C plc
Electric water heaters - Industrial - Howden H D Ltd
Hopper feeder systems - Benton Co Ltd Edward
Industrial batteries for standby applications - Chloride Industrial Batteries Ltd
Industrial doors - Telford Doors and Barriers Ltd
Industrial Equipment - Benton Co Ltd Edward
Industrial Equipment - BRB Industrial Services Ltd
Industrial Flooring - Sempol Surfaces Ltd
Industrial Hoses & Flexibles - Capper Pipe Systems Ltd
Industrial materials handling equipment - Barton B C & Son Ltd
Industrial shelving - Dexion Ltd
Industrial steel workbenches and workstations - Eldyke Ltd
Industrial workbenches and storage drawer cabinets - Quigley Metal Products Ltd
Maintenance of industrial doors, shutters and shopfronts - Shutter Door Repair & Maintenance Ltd
Office and industrial cleaning - CSL (Services) Limited
Paint brushes, rollers and brooms - Cottam Bros Ltd
Specialist brushes - Cottam Bros Ltd
Ultra violet chambers, lamps and control panels - Hanovia Ltd
Vibratory equipment and moulds - Benton Co Ltd Edward
Warehouse partitions - ClipsPartitions Ltd
X-ray door sets - Lami Doors UK

N30 Laundry Equipment

Clothes lines - Lever James & Sons Ltd
Domestic appliances - Miele Co Ltd
Garment Collectors - Brown N C (Storage Equipment) Ltd
Garment Dispensers - Brown N C (Storage Equipment) Ltd
Ironing tables - Electrolux Laundry Systems
Laundry equipment - Anglowest Distributors Ltd
Laundry Equipment - Calpeda Limited
Lockers - Brown N C (Storage Equipment) Ltd
Roller ironers - Electrolux Laundry Systems
Tumble dryers - Electrolux Laundry Systems
Washers, dryers and ironers - Warner Howard Ltd
Washing machines - Electrolux Laundry Systems
Washroom control systems - Cistermiser Ltd

N31 Entertainment/ Auditorium Equipment

Auditorium & retractable seating - AEL Furniture Group

P10 Sundry insulation/ proofing work/ fire stops

Acoustic Door Seals - Acustic + Environmental Technolgy
Acoustic insulation - InstaGroup Ltd
Acoustic Panels - Acustic + Environmental Technolgy
Bitumen building papers, film/paper laminated papers - Package Products Ltd
Building film - Visqueen Agri
Cavity wall insulation - Kay Metzeler Ltd
Cellular glass insulation, tapered roof systems - Pittsburg Corning (UK) Ltd
Coating of Chimneys - ISOKERN (UK) Ltd
Composite panel - Gradient Insulations (UK) Ltd
Draught Excluders - Homelux
Expanded polystyrene insulation - Springvale E P S
Expanded polystyrene insulation - Vencel Resil Ltd
Expanding foam filler - Davant Products Ltd
Exterior insulation - Alumasc Exterior Building Products Ltd
External wall insulation systems - MR Limited
Factory insulated composite panels - Cladding & Decking UK Ltd
Fibre glass fabrics - Fothergill Engineered Fabrics Ltd
Fibreglass loft insulation - Davant Products Ltd
Fire and smoke seals - Mann McGowan Fabrication Ltd
Fire Barriers - Bollom Fire Protection Ltd
Fire collars - FloPlast Ltd
Fire proof cladding - Firebarrier Services Ltd
Fire proof panels - Hemsec Manufacturing Ltd
Fire protection - Nevill Long Limited
Fire protection and acoustic insulation - Ainsworth B R (Southern) Ltd
Fire protection casings - Cryotherm Insulation Ltd
Fire protection insulation spray - Cryotherm Insulation Ltd
Fire Resistant Pillows - Bollom Fire Protection Ltd
Fire resisting doors - Mann McGowan Fabrication Ltd
Fire sound thermal insulation - Lafarge Plasterboard Ltd
Fire stopping products - Bollom Fire Protection Ltd
Fire stopping systems - Cryotherm Insulation Ltd
Fire stops - FGF (Ashton) Ltd
Fire stops - Fire Protection Ltd
Fire stops - Hamron Group
Fire stops - Quelfire
Fire stops around pipes - Industrial Hangers Ltd
Fire stops for cables pipes and ductwork - Quelfire
Fire stops for joints - Quelfire
Fire walls - Gilmour Ecometal
Fire walls and barriers - Cryotherm Insulation Ltd
Firechests - ISOKERN (UK) Ltd
Fireproofing and insulation - Thermica Ltd
Firestopping materials - Preformed Components Ltd
Flat roof insulation - Isocrete Group Sales Ltd
Floor and wall insulation - Kooltherm Insulation Production Ltd
Floor insulation - Springvale E P S
Foam and rubber products - R A International
Hot and cold water cylinder jackets - Davant Products Ltd
Insuation panels - Cryotherm Insulation Ltd
Insulation - British Gypsum Ltd
Insulation - Callenders Ltd
Insulation - Cape Insulation Products Ltd
Insulation - CPV Ltd
Insulation - Deborah Services
Insulation - Rockwool Ltd
Insulation - Rockwool Ltd
Insulation accessories - Astron Building Systems Commercial Intertech Ltd
Insulation, fire protection and fire stopping - Kitsons Insulation Products Ltd
Intumecsent mastics - Promat Fire Protection Ltd
Intumecsent seals - Mann McGowan Fabrication Ltd
Intumescent seals for PVC pipes - Hamron Group
Passive fire prevention - Potters Insulations (Sales) Ltd
Passive fire protection seals - Abesco Ltd
Pipe Collars - Bollom Fire Protection Ltd
Plasterboard Dry Lining - Phoenix Metal Products Ltd
Plastic extrusions - Manton Industrial Seals Ltd
Polyethylene and PVC pipe insulation - Davant Products Ltd
Polythene vapour barrier - Zedcor Ltd
Polyurethane and polystyrene insulation panels - Hemsec Manufacturing Ltd
Public Address System - Acustic + Environmental Technolgy
Rigid urathane insulation boards - Kingspan Insulation Ltd
Roof insulation - Dow Construction Products
Roof insulation - Springvale E P S
Roof tile underlay - Visqueen Agri
Roofing insulation - Kooltherm Insulation Production Ltd
Rubber extrusions and strips - Manton Industrial Seals Ltd
Rubber waterstops - Fosroc Expandite Ltd
Seals and draught strips - Manton Industrial Seals Ltd
Seals and intumescents - Laidlaw Architectural Hardware Ltd
Self sealing test plugs - Industrial Hangers Ltd
Sound insulation - Sound Service (Oxford) Ltd
Standard corkboard - Gradient Insulations (UK) Ltd
Sundry Insulation - Strand Hardware Ltd
Sundry/ Insulation/ proofing work/ fire stops - Owens Corning Building Products (UK) Ltd
Tapered, corkboard, cork/PUR composite, polystyrene and PUR - Gradient Insulations (UK) Ltd
Thermal insulation - Axter Ltd
Underfloor Cable Management - Brown F plc
Vessel,Tank and Pipe insulation - nmc (UK) Ltd
Wall insulation - Springvale E P S
Washers and spacers - Manton Industrial Seals Ltd

P11 Foamed/ Fibre/ Bead cavity wall insulation

Cavity wall bead - Kay Metzeler Ltd
Cavity wall insulation - Callenders Ltd
Cavity wall insulation and loft insulation - InstaGroup Ltd
Expanded polystyrene insulation - Vencel Resil Ltd
Extruded polystyrene - Owens Corning Polyfoam UK Ltd
Foamed/ Fibre/ Bead cavity wall insulation - Owens Corning Building Products (UK) Ltd
Insulation - FGF (Ashton) Ltd
Polystyrene bead cavity wall insulation - RMC Panel Products Ltd

P20 Unframed isolated trims/ skirtings/ sundry items

Architectural and decorative mouldings - Mason FW & Sons Ltd
Bath and Kitchen Trims - Homelux
Decorative ridges and finials - Eternit Clay Tiles Ltd
GRP Dentil cornice - Alpha Glass Fibres Ltd
Hardwood mouldings - Atkinson & Kirby Ltd
MDF mouldings - Fibercill
Metal Trims - Homelux
Plastic and Foam Components - nmc (UK) Ltd
Plastic handrails - Brighton (Handrails), W
Protection rails and corner angles - Harrison Thompson & Co Ltd
PVC trims, angles & edges - Homelux
Ropes and sashcords - Marlow Ropes Ltd
Sashcords, clothes lines and building - Kent & Co (Twines) Ltd
Simulated wood mouldings, reproduction oak beams - Oakleaf Reproductions Ltd
Timber and associated products - Brewer T. & Co Ltd
UPVC Plaster Beading & Trims - Homelux
Window films to existing glass - Hi-Tec Window Films

P21 Door/ Window ironmongery

Architectural door furniture and security hardware - Allgood plc
Architectural door furniture, hangers, screws and hinges - Harwood & Welpac Hardware Ltd
Architectural hardware - Laidlaw Architectural Hardware
Architectural hardware including hinges, locks, levers, flush bolts and other ancilliary items including ceiling fans - James Gibbons Format Ltd
Architectural hinges, door closers and thresholds - Relcross Ltd
Architectural ironmongerers - Moffett Thallon & Co Ltd
Architectural ironmongery - City Hardware and Security
Architectural Ironmongery - Cooper Dryad Ltd
Architectural Ironmongery - Doorfit Products Ltd

Architectural Ironmongery - Haffele UK Ltd
Architectural ironmongery - Industrial Devices Ltd
Architectural Ironmongery - JCL Engineering Ltd
Architectural ironmongery, cabinet hardware and fittings - Lord Isaac Ltd
Architectural ironmongery, security locks and panic equipment - Birkinshaw L. Ltd
Architecturl ironmongery - Dryad Architectural Ltd
Black Antique Ironmongery - Kirkpatrick Ltd
Brass furniture fittings - Martin & Co Ltd
Brass hardware - Marcus M Ltd
Brass, bronze door and window hardware - Allart, Frank, & Co Ltd
Brush strip door and window seals - Kleeneze Sealtech Ltd
Butt hinges - Caldwell Hardware (UK) Ltd
Cash transfer units - Chubb Security Installations Ltd
Castors and wheels - Flexello Limited
Catches & bolts - Caldwell Hardware (UK) Ltd
Computer lockdown units - Badderley Rose Ltd
Concealed door closers, solid brass architectural ironmongery - Perkins & Powell
Cotton twine - Lever James & Sons Ltd
Door and Window Locks - Securikey Ltd
Door Closers - Forson Design & Engineering Ltd
Door closers - Permclose Group PLC
Door fixings - Fairmitre Ltd
Door furniture - HEWI (UK) Ltd
Door furniture and hinges - Laidlaw Architectural Hardware Ltd
Door furniture, catches, hook and castors - Map Hardware Ltd
Door hardware - ASSA Ltd
Door Hardware - Paddock Fabrications Ltd
Dor plates - Fixatrad Ltd
Draught Excluders - Homelux
Draught, weather, intumescent, fire and smoke seals - Sealmaster
Electric Door Operator - Relcross Ltd
Emergency Exit hardware - Jebron Ltd
Emergency exit locks, break glass locks - Pickersgill Kaye Ltd
Espagnolette locking systems - Nico Manufacturing Ltd
Floor springs - Forson Design & Engineering Ltd
Floor springs - Jebron Ltd
Folding openers - Caldwell Hardware (UK) Ltd
Friction stays - Securistyle Ltd
Furniture castors and hardware - Kenrick Archibald & Sons Ltd
Galvanised water bar - Mid Essex Trading Co Ltd
Gas springs - Camloc (UK) Ltd
Gate Locks - Clarke Instruments Ltd
GRP Decorative brackets - Alpha Glass Fibres Ltd
Handles - Securistyle Ltd
Hardware - Tipper (Hardware) Ltd, Joseph
Hardware for windows and doors - Avocet Hardware Ltd
Hardware for windows and doors - Roto Frank Ltd
Hardware for windows and doors - Wagner (GB) Ltd
Heavy Duty Hinges - Securistyle Ltd
High security locks - Chubb Safe Equipment Company
High security locks and locking systems - Abloy Security Ltd
Hinged steel security door sets - Bolton Brady Repair & Service Ltd
Hinges - Cooke Brothers Ltd
Intumescent fire stopping products - Sealmaster
Ironmongery - Ashmead Buidling Supplies Ltd
Ironmongery - Ashmead Building Supplies Ltd
Ironmongery - Caldwell Hardware (UK) Ltd
Ironmongery - Challenge Fencing & Ltd
Ironmongery - Clayton-Munroe Ltd
Ironmongery - Grorud Industries Ltd
Ironmongery - Groupco Ltd
Ironmongery - Harris & Bailey Ltd
Ironmongery - Hipkiss, H, & Co Ltd
Ironmongery - HOPPE (UK) Ltd
Ironmongery - Ingersoll - Rand Architectural Hardware
Ironmongery - Maco Door & Window Hardware (UK) Ltd
Ironmongery - Sampson T F Ltd
Ironmongery - Strand Hardware Ltd
Ironmongery - Tinsley & Co Ltd, Eliza
Locks - Aardee Locks and Shutters Ltd
Locks - Bramah Security Equipment Ltd
Locks - Grundman UK
Locks - Guardian Lock and Engineering Co Ltd
Locks - Ingersoll Locks Ltd
Locks - Lowe & Fletcher Ltd
Locks - Parkes Josiah & Sons Ltd
Locks and latches - Marston A & Co Ltd
Locksmiths - Doorfit Products Ltd
Mail boxes, door closers, door furniture, door locks and cylinders - JB Architectural Ironmongery Ltd
Manual and electrical systems for opening windows - Dyer Environmental Controls
Multi point locking systems - Paddock Fabrications Ltd
Overhead door closers - Jebron Ltd
Padlockable bars - Badderley Rose Ltd
Padlocks - Securikey Ltd
Padlocks and locks - Dom-Nemef-Corbin
Padlocks security products, cyclelocks and motorcycle locks - Squire Henry & Sons
Pivot hinges - Caldwell Hardware (UK) Ltd
Porcelain door furniture - Balmoral Porcelain Limited/Longmead Ceramics
Porcelain finger plates - Balmoral Porcelain Limited/Longmead Ceramics
Porcelain furniture knobs - Balmoral Porcelain Limited/Longmead Ceramics
Porcelain mortice sets - Balmoral Porcelain Limited/Longmead Ceramics
Push biutton and cobination locks - Relcross Ltd
Restrictors - Caldwell Hardware (UK) Ltd
Rim, mortice, padlocks, cylinders and panic escape locks - Yale Security Products Ltd
Rim-lock boxes - Badderley Rose Ltd
Ropes and sashcords - Marlow Ropes Ltd
Sash balances - Caldwell Hardware (UK) Ltd
Sash cord - Lever James & Sons Ltd
Security ironmongery - Chubb Security Installations Ltd
Security products, black ironmongery and hardware - Reynolds UK Ltd
Shelving brackets - decorative - Browne Winther & Co Ltd
Shootbolt - Securistyle Ltd
Sliding door gear - Beaver Industrial Doors Ltd
Smoke Ventilation systems - Dyer Environmental Controls
Solenoid Locks - Clarke Instruments Ltd
Tills, Drawers and paystations - Chubb Security Installations Ltd
Transom closers - Jebron Ltd
Wall, corner, door protection - Gradus Ltd
Weatherstriping joinery seal, roof cappings, glazing bars and ridge systems - Exitex Ltd
Weatherstripping - Reddiglaze Ltd
Window and door hardware - Shaw Arthur Manufacturing Ltd
Window and door ironmongery - Cego Frameware Ltd
Window hardware - Paddock Fabrications Ltd
Window ironmongery - Securistyle Ltd
Window ventilators - Glidevale Building & Products Ltd

P22 Sealant joints

Acoustic seals - Mann McGowan Fabrication Ltd
Bath and kitchen seals - Homelux
Expansion joints - Fillcrete Ltd
Fire protection seals - Abesco Ltd
Fire resistant expansion joint seals - Mann McGowan Fabrication Ltd
Jointing compounds - BSS (UK) Ltd
Mechanical movement joints and fire rated sealants - Compriband Ltd
Sealant joints - Firebarrier Services Ltd
Sealants, solvents, wood preservatives and creosote - Bartoline Ltd
Sealent joints - Canonbury Asphalte Co. Ltd.
Silicones for industry - Ambersil Ltd
Weather draft seals - Mann McGowan Fabrication Ltd

P30 Trenches/ Pipeways/ Pits for buried engineering services

Cable and pipeline locators - Boddingtons Ltd
Cable warning - Boddingtons Ltd
Chambers - Talbot
Compressible fill - Kay Metzeler Ltd
Detection and protection products - Boddingtons Ltd
Leak Location Service - Boddingtons Ltd
Plastic Drainage System - Hepworth Building Products Ltd
Underground warning tape - Boddingtons Ltd
Window ventilators, condensation and drainage channels - Willan Building Services Ltd

P31 Holes/ Chases/ Covers/ Supports for services

Access covers - Bilco UK Ltd
Aluminium and brass recessed floor access covers - Howe-Green Ltd
Builders cast iron products - Dudley Thomas Ltd
Drainage covers & pressed grating - Barton B C & Son Ltd
Pipe clip manufacturers - Bolivar Stamping Ltd
Pipe support componants - Industrial Hangers Ltd

Q10 Kerbs/ Edgings/ Channels/ Paving accessories

Concrete gutter blocks - Finlock Gutters
Concrete paving, blocks and kerbs - Camas Building Materials
Edging - Rogers Concrete Ltd
Edgings and copings - Oakdale (Contracts) Ltd
Edgings for paving - Camas Building Materials Ltd
Granite Kerbs - Civil Engineering Development Ltd
Grease traps - ACO Technologies Plc
Kerb and channels - Bardon (England) Ltd
Kerbs / Edgings / Channels / Paving Accessories - Formpave Ltd
Kerbs ans edging - Macclesfield Stone Quarries Ltd
Path edging - Atlas Stone Products
Paving bricks - Baggeridge Brick PLC
Paving slab supports - Compriband Ltd
Precast concrete and GRC surface water channels - Allpipe Ltd
Protective skirting - ACO Technologies Plc
Stone kerbs, edgings and channels - Chilmark Quarries Ltd
Stone kerbs/ edgings/ channels - The Rare Stone Group Ltd
Stone/ edgings/ channels / kerbs - Gregory Quarries Ltd, The

Q20 Granular sub-bases to roads/ pavings

Clay Pavers - Hanson Brick Ltd, Northern Regonal Sales Office
Granular sub-bases - Associated Holdings Ltd
Paving - Camas Building Materials Ltd
Paving supports - Euroroof Ltd
Suspended ceilings - Phoenix Metal Products Ltd

Q22 Coated macadam/ Asphalt roads/ pavings

Asphalt - Pioneer Concrete Holdings PLC
Asphalts - Hanson Aggregates
Bitumen boilers - Bitmen Products Ltd
Bitumen macadam, hot rolled asphalt, dense tar surfacing and sub base aggregates - Johnston Roadstone Ltd
Bituminous coated stone products - Fyfe John Ltd
Coated macadam and asphalt road surfacing - Jobling Perser Ltd
Coated macadam/ Asphalt roads/ pavings - Associated Holdings Ltd
Cold emulsion sprayers - Bitmen Products Ltd
Infra ray heaters - Bitmen Products Ltd
Macadam - Bardon (England) Ltd
Macadam and asphalt - Canonbury Asphalte Co. Ltd.
Mastic asphalt road contractors - Tarmac Heavy Building
Paver Hire - Johnston Roadstone Ltd
Pre coated chipspreaders - Bitmen Products Ltd
Red Macadam - Johnston Roadstone Ltd
Reflective tarffic markings - Reflecting Roadstuds Ltd
Road surfacing - Fyfe John Ltd

Q23 Gravel/ Hoggin/ Woodchip roads/ pavings

Aggregates - Bardon (England) Ltd
Decorative aggregates - Civil Engineering Development Ltd
Footpath Gravels & Hoggin - Civil Engineering Development Ltd
Gravel/ Hoggin roads/ pavings - Benton Co Ltd Edward

Q24 Interlocking brick/ block roads/ pavings

Block paving - Williamson Cliff Ltd
Block paving - Thakeham Tiles Ltd
Blockpaving contractors - Gearing F T Landscape Services Ltd
Clay pavers - Marshalls Mono Ltd
Concrete block paving - RMC Concrete Products
Concrete block paving - RMC Concrete Products (UK) Ltd
Decorative Paving - RMC Concrete Products (UK) Ltd
Grass reinforcement systems - Landscape Grass (Concrete) Ltd
Interlocking block pavings - Formpave Ltd
Interlocking brick/ block roads/ pavings - Associated Holdings Ltd
Interlocking brick/ block roads/ pavings - Grass Concrete Ltd
Interlocking brick/ block roads/ pavings - Resiblock Ltd
Luxury paving systems - Polybau Ltd
Pavers - Cranleigh Brick & Tile Co Ltd
Pavers - Freshfield Lane Brickworks Ltd
Pavers - Ibstock Building Products Ltd
Pavers - York Handmade Brick Co Ltd
Paving - Milner Delvaux Ltd
Paving - Townscape Products Ltd
Paving bricks - Baggeridge Brick PLC
Paving hard landscaping - Blanc de Bierges
Pavings - Prismo Ltd
Precast paving - Oakdale (Contracts) Ltd

Q25 Slab/ Brick/ Sett/ Cobble pavings

Boulders, Cobbles and Pebbles - Civil Engineering Development Ltd
Clay pavers - Castle Brick (Wales) Ltd
Clay pavers - Hanson Brick Ltd
Clay Pavers - Ruabon Dennis Ltd
Concrete slab, block paving - Marshalls Mono Ltd
Concrete, clay block and decorative paving - Marley Building Materials Ltd
Granite Paving - Civil Engineering Development Ltd
Granite setts - Civil Engineering Development Ltd
Grass reinforcement systems - Grass Concrete Ltd
Grass reinforcement systems - Landscape Grass (Concrete) Ltd
Limestone paving - Civil Engineering Development Ltd
Natural stone paving - Marshalls Mono Ltd
Pavac - Park Products
Paving - Atlas Stone Products
Paving - Bardon (England) Ltd
Paving - Burlington Slate Ltd
Paving - Macclesfield Stone Quarries Ltd
Paving and flooring - Roofing Contractors (Cambridge) Ltd
Paving bricks - Baggeridge Brick PLC
Pavings - Prismo Ltd
Pavings - Ruthin Precast Concrete Ltd
Porphyry setts and edging - Civil Engineering Development Ltd
Precast concrete paving - Rogers Concrete Ltd
Precast concrete paving slabs - Anderton Concrete Products Ltd
Regency paving and flagstone paving - Thakeham Tiles Ltd
Roofing and paving - Kerridge Stone
Slab / Brick / Sett / Cobble pavings - Formpave Ltd
Slab / Brick/ Sett/ Cobble pavings - Harris & Bailey Ltd
Slab/ Sett/ Cobble pavings - Gregory Quarries Ltd, The
Slab/ Brick/ Sett/ Cobble pavings - Associated Holdings Ltd
Slab/ Brick/ Sett/ Cobble pavings - Chilstone Garden Ornaments
Slab/ Sett cobble pavings - Chilmark Quarries Ltd
Slab/ Sett/ Cobble pavings - The Rare Stone Group Ltd
Slabs - RMC Concrete Products (UK) Ltd
Stone paving - Macc Stone Ltd
Swimming pool surrounds, hard landscaping and steps - Milner Delvaux Ltd
Yorkstone paving - Civil Engineering Development Ltd

Q26 Special surfacings/ pavings for sport/ general amenity

Anti slip surfacing - Addagrip Surface Treatments Ltd
Artificial sports surfaces and safety surfacing - En-tout-cas plc (Scotland)
Athletic tracks and synthetic grass sports pitches - Bernhard's Sports Surfaces Ltd
Bowling green equipment and channels - Sportmark (Leisure Products) , Part of Sportsmark Group Ltd
Clay pavers - Michelmersh Brick & Tile Co Ltd
Construction, referbishing and al types of tennis courts - En-tout-cas South West
Gauged arches - Michelmersh Brick & Tile Co Ltd
Indoor sports surfaces - Gerland Ltd
Permeable paving - RMC Concrete Products
Playground surfacing - Record Playground Equipment Ltd
Road lining paints - Bollom J W & Co Ltd
Road markings - Prismo Ltd
Rubber paving - Linatex Ltd
Rubber surfaces for play areas, paddling and swimming pools surrounds, "kick about" play areas and lockers rooms in golf clubs - Easi-Fall International Ltd
Sports court marking - Sportmark (Leisure Products) , Part of Sportsmark Group Ltd
Sports paving - Bardon (England) Ltd
Sports surfacing and pavings - Lawrence Surfaces Plc, Charles
Squash court play surfaces - Bicester Products Ltd
Synthetic grass - Palmyra Ltd
Synthetic sports surfaces - En-tout-cas plc
Tactile surfaces for visually impaired - Rediweld Rubber & Plastics Ltd
Tennis courts - Roadways & Car Parks Ltd

Q30 Seeding/ Turfing

Aquatic landscaping services - Land Wood & Water Co. Ltd
Artificial grass - LazyLawn
Grass seed and turf - Sportmark (Leisure Products) , Part of Sportsmark Group Ltd
Hydraulic seedings - Hydraseeders Ltd
Landscape contracting - Land Wood & Water Co. Ltd
Landscapes - Ogilvie Construction Ltd
Landscaping - Bernhards Landscapes Ltd
Landscaping - Cila Landscapes
Recycled landscape products - Deceuninck Ltd
Seeding/ Turfing - Clifton Nurseries
Seeding/ Turfing - Coblands Landscapes Ltd
Seeding/ Turfing - Frosts Landscape Construction Ltd
Seeding/ Turfing - Reyven (Sportsfields) Ltd
Turfing - Town & Country Landscapes Ltd

Q31 External planting

Aquatic landscaping services - Land Wood & Water Co. Ltd
Artificial plants - Desney Products
Artificial trees - Desney Products
Bespoke landscape - CTS Ltd
External Planting - Landscape Institute The
External planting - Urban Planters
Garden landscaping products - RMC Concrete Products (UK) Ltd

General exterior works contractors - Gearing F T Landscape Services Ltd
Landscape contracting - Land Wood & Water Co. Ltd
Landscape contractors - Gearing F T Landscape Services Ltd
Landscape mulcher - ABG Ltd
Landscape ornaments - Haddonstone Ltd
Landscapes - Ogilvie Construction Ltd
Landscaping - Cila Landscapes
Landscaping and grounds maintenance - Reyven (Sportsfields) Ltd
Landscaping products - ECC Building Products Ltd
Planters - Blanc de Bierges
Planting - Clifton Nurseries
Planting - Coblands Landscapes Ltd
Planting - Frosts Landscape Construction Ltd
Soft and hard landscaping - Western landscapes Ltd t/a Wyevale Landscapes
Tree and shrub planting - Town & Country Landscapes Ltd

Q32 Internal planting

Artificial plants - Desney Products
Artificial plants and trees - Pouliot Designs
Artificial trees - Desney Products
Interior landscape - Chilmark Quarries Ltd
Interior landscape - Gregory Quarries Ltd, The
Interior landscpe - The Rare Stone Group Ltd
Interior planting - Urban Planters
Internal planting - Frosts Landscape Construction Ltd
Internal planting - Haddonstone Ltd
Internal Planting - Landscape Institute The

Q35 Landscape maintenance

Landscape Maintenance - Coblands Landscapes Ltd
Landscape maintenance - Frosts Landscape Construction Ltd
Landscape Maintenance - Landscape Institute The
Light landscaping installation and maintenance - Sportmark (Leisure Products) , Part of Sportsmark Group Ltd
Soft/ Hard Landscaping - Roadways & Car Parks Ltd
Summer houses - Ableson L & D Ltd
Tree surrounds - Jones of Oswestry Ltd

Q40 Fencing

Acoustic barriers - PHI Group Ltd
Alarm fencing - Allen (Fencing) Ltd
All types of fencing - Dinnington Fencing Co. Limited
Barbican steel fencing - Jackson H S & Son (Fencing) Ltd
Bowtop fencing - Jackson H S & Son (Fencing) Ltd
BS1722 Parts 1-16 Inclusive - Burn Fencing Limited
Car parking barriers - Clarke Instruments Ltd
Chain link fencing, concrete fence posts - Procter Concrete Products
Chainlink fencing - Hy-Ten Reinforcement Co Ltd
Chestnut steel palisade, chain link, railings, weld mesh, barbed and razor wire - Guardian Wire Ltd
Concrete Fencing - Ebor Concretes Ltd
Concrete, wire fencing and gates - Challenge Fencing & Ltd
Crash barriers and motorway type railings - Fencing Design & Development Co Limited
Entrance barrier systems - Collie Carpets Ltd
Environmental barrier/ acoustic fencing - Newton & Frost
Euroguard mesh fencing - Jackson H S & Son (Fencing) Ltd
Fence panel manufacturers - Derby Timber Supplies
Fencing - Advanced Fencing Systems Ltd
Fencing - AGN Fencing Limited
Fencing - Alltype Fencing Specialists
Fencing - Ambec Fencing
Fencing - AMK Fence-In Limited
Fencing - Associated Holdings Ltd
Fencing - Auckland Construction Ltd
Fencing - B & M Fencing Limited
Fencing - Barker & Geary Limited
Fencing - Bedford Fencing Co. Limited
Fencing - Benfield Fencing Limited
Fencing - Betta Fencing Co
Fencing - Bowden Fencing
Fencing - Brads Fencing Co. Limited
Fencing - Burn Fencing Limited
Fencing - Butler Davies (Fencing) Limited
Fencing - Capel Fencing Contractors Ltd.
Fencing - Cardinal Fencing (Northern) Limited
Fencing - CEM Fencing Contractors
Fencing - Challenge Fencing & Ltd
Fencing - Chestnut Products Limited
Fencing - Cila Landscapes
Fencing - Coates Fencing
Fencing - Danbury Fencing Limited
Fencing - Darfen Durafencing, Northern Region
Fencing - Dinnington Fencing Co. Limited
Fencing - Fabrikat (Nottingham) Ltd
Fencing - Fairmile Fencing Ltd
Fencing - Faversham Fencing Ltd
Fencing - Fence Hire Limited
Fencing - Fencing Construction (Yorkshire) Limited
Fencing - Fencing Contractors Association
Fencing - Fernden Construction (South Western) Limited
Fencing - Fernden Construction (Winchester) Limited
Fencing - Fleetwood Fengate Co Ltd
Fencing - Frosts Landscape Construction Ltd
Fencing - GBG Fences Limited
Fencing - Gilmer Limited
Fencing - GJ Durafencing Limited
Fencing - Gravesend Fencing Limited
Fencing - Hadley Industries Plc
Fencing - Havering Fencing Co.
Fencing - Howe Fencing & Sectional Buildings
Fencing - Huntree Fencing Ltd
Fencing - J Pugh-Lewis Limited
Fencing - J. B. Corrie & Co Limited
Fencing - J. D. Goodacre & Son
Fencing - John Flower Fencing Limited
Fencing - Kingsforth Security Fencing Limited
Fencing - Kwikform UK Ltd
Fencing - Leicester Fencing Contracts Limited
Fencing - Maldon Fencing Co.
Fencing - May Gurney (Technical Services) Ltd - Fencing Division
Fencing - Metalwood Fencing (Contracts) Limited
Fencing - Newton & Frost
Fencing - Northern Fencing Contractors Limited
Fencing - Pearman Fencing
Fencing - Pembury Fencing Limited
Fencing - Portia Engineering Ltd
Fencing - Pro-Fence (Midlands)
Fencing - Procter Bros. Limited
Fencing - Record Playground Equipment Ltd
Fencing - Reyven (Sportsfields) Ltd
Fencing - Royston Steel Fencing Limited
Fencing - Ryall & Edwards Limited
Fencing - Sampson & Partners Fencing
Fencing - Sanhall Limited
Fencing - Scott Fencing Limited
Fencing - SFC (Midlands) Ltd
Fencing - Sharman Fencing & Gate Supplies
Fencing - Silverwood Fencing Co. Limited
Fencing - Sovereign Fencing Limited
Fencing - Termstall Limited
Fencing - The Nelton Group Ltd
Fencing - Tonglen Fencing Ltd.
Fencing - Town & Country Landscapes Ltd
Fencing - W Ward Fencing Contractors
Fencing - Wallbank
Fencing - Warefence Limited
Fencing - Weatherley Fencing Contractors Limited
Fencing - Wednesbury Fencing Co Limited
Fencing - Wentworth Sawmills Ltd,
Fencing - WH Wesson Fencing Limited
Fencing - White Gates
Fencing - WL West & Son Limited
Fencing - Yarco Fencing Limited
Fencing and barriers - Baker Fencing Limited
Fencing and barriers - George James Partnership
Fencing and gates - Farefence Limited
Fencing and gates - May Gurney (Technical Services) Ltd
Fencing and gates - Tonbridge Fencing Limited
Fencing and gates, security fencing - Binns Fencing Ltd
Fencing and security fencing - Jackson H S & Son (Fencing) Ltd
Fencing contractors - Gearing F T Landscape Services Ltd
Fencing contractors - Grange Fencing Erectors Limited
Fencing contractors - James D Cocks (Enterprises)
Fencing including steel palisade, power operated gates and metal railings - Charnwood Fencing Limited
Fencing security - Geoquip Ltd
Fencing suppliers - Crowthorne Fencing
Fencing, gates and barriers - Albion Fencing Ltd
Fencing, gates, automatic gates and barriers - Tate Fencing
Fencing, gates, railings, pedestrian guard rail and crash barrier - Billericay Fencing Ltd
Fencing, gates, railings, traffic barriers and turnstiles - Elkosta; Rodney Coate & Partners Ltd
Gates - Alltype Fencing Specialists
Gates - Fernden Construction (South Western) Limited
Gates - WA Skinner & Co Limited
Gates & Railings - Crimegard Ltd
Gates, barriers, shutters, fencing and parking post - DPS Security Systems Ltd
Gates, coloured concrete posts, ornamental, vertical bar and bow top railings - Procter Bros Limited
General fencing, safety and noise barrier - Chris Wheeler Construction Limited
Grating fencing - Orsogril Sarl UK Department
Guard rail and barrier fencing - Stainton Metal Co Ltd
High Security - Alltype Fencing Specialists
Industrial and commercial fencing and gates - M & M Fencing Limited
Industrial and domestic fencing - ACP Concrete (Contracts) Limited
Industrial and security fencing - Darfen Durafencing, North East Region
Industrial and security fencing - Darfen Durafencing, North West Region
Jakcure treated timber fencing - Jackson H S & Son (Fencing) Ltd
Local authority suppliers of fencing - Duralock UK Ltd
Metal fencing, railings and barriers - Barker W H Ltd
On site welding - C & W Fencing Ltd
Paladin - Alltype Fencing Specialists
Palisade - Alltype Fencing Specialists
Palisade fencing and gates - Bain WM & Co Fencing (1990) Ltd
Pedestrian barrier - Wade Building Services Ltd
Perimeter protection - Guardall
Power fencing - Allen (Fencing) Ltd
Precast concrete fencing - Anderton Concrete Products Ltd
PVC Fencing - LB Plastics Ltd
Railings - Alltype Fencing Specialists
Railings - Fernden Construction (South Western) Limited
Railings, gates - Sebry Staircase & Ironworks Ltd
Rotating anti climb guards - Vandgard Ltd
Safety fencing - Newton & Frost
Security fencing - Allen (Fencing) Ltd
Security fencing - C & R Building Systems Ltd
Security fencing - C & W Fencing Ltd
Security Fencing - Dinnington Fencing Co. Limited
Security fencing - WA Skinner & Co Limited
Security fencing and gates - Stenoak Fencing
Security fencing, gates and access systems - O'Connor Fencing Limited
Security fencing, gates, barriers and turnstiles - Darfen Durafencing, Central Region
Security grilles & guards - WA Skinner & Co Limited
Security steel palisade fencing and gates - Prima Security & Fencing Products
Security, industrial fencing, railings, crash barriers, gates, palisade, posts and rails - J & M Fencing Services
Security, Industrial, commercial new housing and highways - Ambec Fencing
Sentry bar steel fencing - Jackson H S & Son (Fencing) Ltd
Site welding service - C & R Building Systems Ltd
Steel aluminium grating - Orsogril Sarl UK Department
Steel fencing and gates - C & W Fencing Ltd
Steel gates - Badderley Rose Ltd
Steel gates - Staines Steel Gate Co
Steel palisading, welded mesh, chainlink fencing - Fencelines Ltd
Temporary security fences - Wade Building Services Ltd
Timber, chainlink and weldmesh fencing systems - Prima Security & Fencing Products
Traffic control barriers - Fernden Construction (South Western) Limited
Tubular construction, barriers and rials - Tubeclamps Ltd
Turnstiles - Fernden Construction (South Western) Limited
Vehicle fencing, envirnomental barriers,pedestrian guardrail - Stenoak Fencing
Vertical tube fencing - Orsogril Sarl UK Department
Weld mesh - Alltype Fencing Specialists
Welded mesh fencing - Orsogril Sarl UK Department
Windposts - Ancon CCL Ltd
Wood - Alltype Fencing Specialists

Q41 Barriers/ Guard-rails

Barriers - May Gurney (Technical Services) Ltd - Fencing Division
Barriers and gaurdrails - Portia Engineering Ltd
Barriers and guard rails - Associated Holdings Ltd
Cantilever gates - Jackson H S & Son (Fencing) Ltd
Control barriers - Kwikform UK Ltd
Endless cords - Lever James & Sons Ltd
Garden twines - Lever James & Sons Ltd
Gates - May Gurney (Technical Services) Ltd - Fencing Division
Gates and Barriers - Hart Door Systems Ltd
Handrail - Kee Klamps Ltd
Housing projects post and rail pvc-u - Duralock UK Ltd
Pedguard rail - Jackson H S & Son (Fencing) Ltd
Perimeter protection - APT Controls Ltd
Post and rail products -schools pvc-u - Duralock UK Ltd
Racecourse rails and products pvc-u - Duralock UK Ltd
Ropes and sashcords - Marlow Ropes Ltd
Security barriers - APT Controls Ltd
Stagings - Kwikform UK Ltd
Swing gates - Jackson H S & Son (Fencing) Ltd
Traffic Barriers - Bell & Webster Concrete Ltd
Traffic barriers - Bolton Brady Repair & Service Ltd

Q50 Site/ Street furniture/ equipment

Access covers and frames - Jones of Oswestry Ltd
Architectural trellises and planters - Anthony de Grey Trellises
Arm Chairs - Gloster Furniture Ltd
Automatic gates and barriers - BPT Automation Ltd
Balustrade - Kee Klamps Ltd
Barrier posts - Aremco Products
Barriers - Telford Doors and Barriers Ltd
Benches - Erlau AG
Benches - Gloster Furniture Ltd
Bird deterent measures - Birdscarer Products
Bollards - Erlau AG
Bollards - Glasdon UK Ltd
Bollards, cable covers, troughs and bespoke products - Anderton Concrete Products Ltd
Bridges in steel, boardwalks and decking - CTS Ltd
Bus & Rail Shelters - Abacus Lighting Ltd
Bus and glazed shelters - Midland Alloy Ltd
Car parks in pvc-u - Duralock UK Ltd
Cast iron table components - Origlia Spa
Concrete Street Furniture - Ebor Concretes Ltd
Cycle shelters, bollards, perking posts, benches and outdoor seating - Mawrob Co (Engineers) Ltd
Cycle stands - Erlau AG
Dining chairs - Gloster Furniture Ltd
Dog waste bins - Glasdon UK Ltd
Drainage gratings - Jones of Oswestry Ltd
Electrical Termination/ Service Pillar - Ritherdon & Co Ltd
Engraving service - Gloster Furniture Ltd
Exterior Lighting - Abacus Lighting Ltd
External litter bins - Glasdon UK Ltd
Folding chairs - Gloster Furniture Ltd
Garden furniture - Nova Garden Furniture Ltd
Garden sheds - Ableson L & D Ltd
Gates and barriers - Rib (UK) Ltd
Glass fibre and aluminium flag poles - Harrision Flagpoles
Granite table tops - Origlia Spa
Grit and salt bins - MGB Randalls
Ground and Wall Mounted Banner Poles - Harrision Flagpoles
Interior Lighting - Abacus Lighting Ltd
landscape furniture, seatings, shelters, bridges, recycling and litter bins - Lappset UK Ltd
Lighting control systems - Abacus Lighting Ltd
Litter bins - Erlau AG
Loungers - Erlau AG
Motorway crash barriers and pedestrian guard rails - Lionweld Kennedy Ltd
Occasional furniture - Gloster Furniture Ltd
Ornamental statues, pots and urns - Rogers Concrete Ltd
Outdoor furniture - Andy Thornton Architectural Antiques Ltd
Outdoor metal tables and chairs - Origlia Spa
Pavior infill covers and frames - Jones of Oswestry Ltd
Pedestrian guard rails - Fabrikat (Nottingham) Ltd
Pergolas, screens and planters - CTS Ltd
Planters - Erlau AG
Plastic and metal static and portable wheeled bins - MGB Randalls
Public seats, litter bins, benches, bollards, picnic tables and planters - Neptune Outdoor Furniture Ltd
Reconstructed stone balustrading, paving, garden ornaments and architectural dressings - Minsterstone Ltd
Recycling containers - MGB Randalls
Retaining panels, planters and street furniture - Milner Delvaux Ltd
Road signs and road markers - Glasdon UK Ltd
Seating - Erlau AG
Seating - Glasdon UK Ltd
Seats - Record Playground Equipment Ltd
Security gates, barriers, road blockers and turnstiles - Broughton Controls Ltd
Sign posts - Stainton Metal Co Ltd
Site and street furniture equipment - Woodscape Ltd
Site furniture, car park furniture and equipment - APT Controls Ltd
Site/ Street furniture/ equipment - Chilstone Garden Ornaments
Site/ Street furniture/ equipment - Haddonstone Ltd
Site/ Street furniture/ equipment - Leander Architectural
Site/Street Furniture/Equipment - Wybone Ltd
Sleeping policemen and traffic calming products - Rediweld Rubber & Plastics Ltd
Smoking disposable units - Glasdon UK Ltd
Sports equipment - En-tout-cas plc
Stone Street Furniture - Civil Engineering Development Ltd
Street Furniture - Abacus Lighting Ltd
Street furniture - Ballentine, Bo'Ness Iron Co Ltd
Street furniture - Blanc de Bierges
Street Furniture - Broxap & Corby Ltd
Street furniture - Dee-Organ Ltd
Street Furniture - Earth Anchors Ltd
Street furniture - Interlux Ltd
Street furniture - Marshalls Mono Ltd
Street furniture - Orsogril Sarl UK Department
Street Furniture - Polybau Ltd
Street Furniture - Townscape Products Ltd
Street furniture - Urbis Lighting Ltd
Street furniture - Woodhouse UK Plc
Street furniture and equipment - Game Time UK Ltd
Summer houses - Ableson L & D Ltd
Tables - Erlau AG
Tables - Gloster Furniture Ltd
Timber, steel, cast iron street and garden furniture - Orchard Street Furniture, Division of Camas Building Materials
Traffic barriers - Electro Mechanical Systems Ltd
Waiting area seating for train stations, airports, ferry terminals, hospitals and surgeries - Hille Ltd

Water gardens and water features - Lotus Water Garden Products Ltd
Winter grit spreaders - Glasdon UK Ltd
Winter grit storage bins - Glasdon UK Ltd
Wrought iron furniture - Desney Products

R10 Rainwater pipework/ gutters

Accessories - Corofil Woodall Ltd
Aluminium flashings - Gilmour Ecometal
Aluminium gutter and rainwater pipe products - Marley Alutec Ltd
Aluminium gutters - George Gilmour (Metals) Ltd
Aluminium gutters - Gilmour Ecometal
Aluminium pipework and gutters - Aluminium RW Supplies Ltd
Aluminium rainwater good - Alifabs (Woking) Ltd
Aluminium rainwater goods - WP Metals Ltd
Architectural rainwater systems - Alumasc Exterior Building Products Ltd
Cast iron rainwater and soil goods - Longbottom J & J W Ltd
Clay, Plastic and concrete drainage systems - Hepworth Building Products Ltd
Copper guttering - Goods Directions Ltd
Copper rainwater system - Klober Ltd
Fascia and soffit systems in powder coated aluminium and steel - Dales Fabrications Ltd
Fire protection - Deborah Services
GRP Gutters - Wilsons Fibreglass Ltd
GRP valley troughs - Timloc Expamet Building Products
Gutter heating - Findlay Irvine Ltd
Gutters - Corofil Woodall Ltd
Industrial valley gutters - Alumasc Exterior Building Products Ltd
Insulated gutters and rainscreen facade systems - CGL Cometec
Insulation boards - Vulcanite Limited
Metal fabricators to the construction industry - Builder's Iron & Zincwork Ltd
Pitched roofing - Euroroof Ltd
Plastic deck drainage - Fox Pool (UK) Ltd
Plastic drainage systems - Hunter Plastics Ltd
Plumbing & Heating - Spital Tile Co
Polypropylene pipe systems - Capper Pipe Systems Ltd
Rainwater and plumbing systems - Marshall-Tufflex Ltd
Rainwater containers and accessories - T & D Plastech
Rainwater drainage products - Drainage Systems
Rainwater goods - Broderick Structures Limited
Rainwater goods - Eternit UK Ltd
Rainwater Goods - Rheinzink UK
Rainwater goods - Rockwell Sheet Sales Ltd
Rainwater pipework and gutters - Geberit UK
Rainwater pipework/ gutters - Harris & Bailey Ltd
Rainwater pipework/ gutters - Interlink Group
Rainwater pipework/ gutters - John Davidson (Pipes) Ltd
Rainwater systems - FloPlast Ltd
Rainwater systems - Harrison Thompson & Co Ltd
Rainwater systems in powder coated aluminium and steel - Dales Fabrications Ltd
Relining of concrete gutters - Finlock Gutters
Roof outlets - ACO Technologies Plc
Siphonic Rainwater drainage systems - Dales Fabrications Ltd
Stainless steel drainage - BM Stainless Steel Drains Ltd
Surface water drainage - Hodkin Jones (Sheffield) Ltd
Syphonic drainage systems - Fullflow Ltd
Vapour barrier - Visqueen Agri
Water storage containers and accessories - Rexam Harcostar
Zinc rainwater goods - Metra Non Ferrous Metals Ltd

R11 Foul drainage above ground

Above ground drainage - Ballentine, Bo'Ness Iron Co Ltd
Above ground drainage - FloPlast Ltd
Air admittance vales - FloPlast Ltd
Bacteria Grease traps - Progressive Product Developments Ltd
Bricks - Hepworth Building Products Ltd
Engineered drainage systems - Alumasc Exterior Building Products Ltd
Filter beds - Milton Pipes Ltd
Foul drainage above ground - Drainage Systems
Grease traps, oil and grease water separators - Progressive Product Developments Ltd
Grease treatment plants - Progressive Product Developments Ltd
Gutters, drainage pipes and fittings - Longbottom J & J W Ltd
Plastic drainage - Capper Pipe Systems Ltd
Plastic drainage systems - Hunter Plastics Ltd
Plastic soil pipes, urinal pipework and traps - Midland Stom Ltd
Rainwater pipework/gutters - BM Stainless Steel Drains Ltd
Seamless cold drawn copper tube - Lawton Tube Co Ltd, The
Soil and drain down systems - Stanton plc
Soil and vent terminals - Klober Ltd
Stainless Steel drainage channels, gratings and gullies - Component Developments
Tanks (retention) - Milton Pipes Ltd
Traps - FloPlast Ltd
Traps and waste fittings - Opella Ltd

R12 Drainage below ground

Access covers and drainage products - Elkington Gatic Ltd
Access covers and frames linear surface water drainage system and special fabrication - Clarksteel Ltd
Bacteria Grease traps - Progressive Product Developments Ltd
Below ground drainage - FloPlast Ltd
Below ground drainage and maholes - Ballentine, Bo'Ness Iron Co Ltd
Box culverts - Brierley B (Garstang) Ltd
Box culverts - Milton Pipes Ltd
Cesspits and septic tanks - Klargester Environmental Engineering Ltd
Channel grating - Shackerley (Holdings) Group Ltd incorporating Designer ceramics
Coissons - Milton Pipes Ltd
Concrete inspection chambers - Brierley B (Garstang) Ltd
Concrete inspection chambers and pipes - Milton Pipes Ltd
Cover slabs - Chiltern Concrete Products Ltd
Design, installation and maintenance of building services - Sulzer Infra (UK) Ltd
Drain and pipework cleaning - Dyno Rod PLC
Drain and pipework inspection - Dyno Rod PLC
Drain and pipework installation - Dyno Rod PLC
Drain and pipework repairs - Dyno Rod PLC
Drainage - Marshalls Mono Ltd
Drainage - Polypipe Civils Ltd
Drainage & rainwater sysytems - Alumasc Exterior Building Products Ltd
Drainage below ground - ARC Pipes
Drainage below ground - Certifield Laboratories (Div or CPS Industries Ltd)
Drainage below ground - Formpave Ltd
Drainage below ground - Geberit UK
Drainage below ground - Harris & Bailey Ltd
Drainage below ground - John Davidson (Pipes) Ltd
Drainage Clay & plastic - Brikenden (Builders Merchants) Ltd
Drainage products - Ashmead Buidling Supplies Ltd
Drainage products - Ashmead Building Supplies Ltd
Drainage products - Brett Martin Ltd
Drainage products - Drainage Systems
Flexible couplings - Flex-Seal Couplings Ltd
Floor gratings - Screen Systems Ltd
Floor gullies, roof outlets, access covers, grease converters and linear drainage - Wade International (UK) Ltd
Glass reinforced concrete and HDPE drainage channels - Hauration Kaskade Ltd
Gravity drain systems - Uponor Ltd
Grease traps, oil and grease water separators - Progressive Product Developments Ltd
Grease treatment plants - Progressive Product Developments Ltd
GRP Covers - Armfibre Ltd
GRP pipes, concrete pipes and fittings - Johnston Pipes Ltd
Gullies, gratings, channels and traps - ACO Technologies Plc
Gully grating and access covers - Brickhouse
Inspection/ drainage concrete chambers - Ruthin Precast Concrete Ltd
Jetters cold and hot, winches, rods and accessories - Ward's Flexible Rod Company Ltd
Large tanks and sewage treatment plant - Conder Products Ltd
Manhole Chamberes - Uponor Ltd
Manhole covers - Duct Access Covers Ltd
Manhole covers - Stanton plc
Manhole covers and gratings - Durey Casting Ltd
Manhole covers and reservoir lid - Ruthin Precast Concrete Ltd
Petrol/ Oil Interceptors - NSA Environmental Products
Plastic building products - Haymills (Contractors) Ltd
Plastic drainage - Capper Pipe Systems Ltd
Plastic drainage systems - Hunter Plastics Ltd
Pumping stations for storm water/Foul Water - Pims Pumps Ltd
PVC & ABS Pipe systems - Capper Pipe Systems Ltd
Septic tanks - Brierley B (Garstang) Ltd
Settelment tanks and componants - Armfibre Ltd
Sewage treatment products - Albion Concrete Products
Silt traps, septic tanks, cesspools, silage effluent tanks, grease traps and grease filters - NSA Environmental Products
Soil and drain down systems - Stanton plc
Stainless steel channels, gratings, gullies, access covers and grease separators - BM Stainless Steel Drains Ltd
Stainless Steel Manhole Covers - Component Developments
Surface drainage products - Airflow (Nicoll Ventilators) Ltd
Surface water drainage systems - Mea UK Ltd
Surface water drainage systems - Polybau Ltd
Systems (Rio) - KSB Ltd
Terracotta - Hepworth Building Products Ltd
Ultra rib sewer systems - Uponor Ltd
Underground drainage - Naylor Drainage Ltd

R13 Land drainage

Clay Drainage System - Hepworth Building Products Ltd
GRP Soakers - Wilsons Fibreglass Ltd
Land drainage - Drainage Systems
Land drainage - John Davidson (Pipes) Ltd
Land Drainage - Naylor Drainage Ltd
Land Drainage - Polypipe Civils Ltd
Land Drainage - T-T Pumps & T-T controls
Petrol/oil inceptors, catering grease/oil traps, balancing tanks - Conder Products Ltd
Plumbing and Drainning - Wavin Building Products Ltd
Preformed sheets, cavity and channel drainage - ABG Ltd

R14 Laboratory/ Industrial waste drainage

Bacteria Grease traps - Progressive Product Developments Ltd
Grease traps, oil and grease water separators - Progressive Product Developments Ltd
Grease treatment plants - Progressive Product Developments Ltd
Laboratory and industrial waste - Drainage Systems
Laboratory/ Industrial Waste - Naylor Drainage Ltd
Stainless steel channels, gratings, gullies, access covers and grease separators - BM Stainless Steel Drains Ltd
Stainless steel drainage systems - ACO Technologies Plc
Thermoplastic pipes - CPV Ltd
Waste water treatment - NSA Environmental Products

R20 Sewage pumping

Centrifugal pumps - Calpeda Limited
Drainage and sewage package systems - Pump Technical Services Ltd
Interceptors - Klargester Environmental Engineering Ltd
Maintenance and repair of pumping equipment - Pims Pumps Ltd
Pumping stations - NSA Environmental Products
Sewage pumping - Certifield Laboratories (Div or CPS Industries Ltd)
Sewage pumping - Conder Products Ltd
Sewage Pumping - T-T Pumps & T-T controls
Sewage pumping sets - Fordwater Pumping Supplies Ltd
Sewage pumps and pumping systems - New Haden Pumps Ltd
Sewage, drainage, pumps and systems - KSB Ltd

R21 Sewage treatment/ sterilisation

Cesspools - Entec (Polution Control) Ltd
Drain clearing rods - Wessenden Products Ltd
Effluent treatment - Armfibre Ltd
Grease traps - Entec (Polution Control) Ltd
Interceptors - Entec (Polution Control) Ltd
Package sewage treatment - NSA Environmental Products
Package sewage treatment plants - Entec (Polution Control) Ltd
Pump stations - Entec (Polution Control) Ltd
Sample chambers - Entec (Polution Control) Ltd
Septic tanks - Entec (Polution Control) Ltd
Sewage purification equipment - Tuke & Bell Ltd
Sewage storage tanks - Franklin Hodge Industries Ltd
Sewage Treatment - Balmoral Group Ltd
Sewage treatment - Klargester Environmental Engineering Ltd
Sewage Treatment - Simon-Hartley
Sewage treatment equipment - Adams- Hydraulics Ltd
Sewage treatment plant - Clearwater Plc
Sewage treatment/sterilisation - Conder Products Ltd
Ultra violet chambers, lamps and control panels - Hanovia Ltd

R30 Centralised vacuum cleaning

Centralised vacuum cleaning - DCE - BTR Environmental Ltd
Centralised vacuum cleaning systems - BVC-BIVAC, Division of DD Lamson

R31 Refuse chutes

Refuse chutes - Kwikform UK Ltd

R32 Compactors/ Macerators

Compactors - Tony Team Ltd
Compactors/ Macerators - Sissons W & G Ltd
Compactors/Macerators - The Haigh Tweeny Co Ltd
Demountable compactors - MGB Randalls
Sanitary towel disposal units - Max Appliances
Sluiceroom macerators - Verna Ltd
Waste management equipment, compactors/ macerators - Imperial Machine Co Ltd

R33 Incineration plant

Incinerators - Mawrob Co (Engineers) Ltd
Incinerators - Northern Incinerators (GB) Ltd

S10 Cold Water

Cold water - John Davidson (Pipes) Ltd
Cold Water - Polypipe Civils Ltd
Design, installation and maintenance of building services - Sulzer Infra (UK) Ltd
Ductwork clearing - Aquastat Ltd
Level gauges and transmitters - Klinger Fluid Instrumentation Ltd
MDPE pipe and fittings - FloPlast Ltd
Packaged Cold Water Storage/Pump - Harton Heating Appliances
Piping systems - George Fischer Sales Ltd
Plumbing products - Drainage Systems
Plumbing systems - Hepworth Building Products Ltd
Polyethylene pipes for water and gas - Capper Pipe Systems Ltd
Polyethylene systems - Uponor Ltd
Potable water pipe fittings - Uponor Ltd
Radiators - Brikenden (Builders Merchants) Ltd
Seamless cold drawn copper tube - Lawton Tube Co Ltd, The
Submersible pumps - Pillinger G C & Co (Engineers) Ltd
Water boundary control boxes - Barber Edward & Co Ltd
Water fittings and controls - Flamco UK Ltd
Water meters - Raab Karcher Energy Services Ltd, Clorius Division
Water meters - SAV United Kingdom Ltd
Water meters and oil Meters - ABB Kent Meters

S11 Hot Water

Control panels and controls - Pillinger G C & Co (Engineers) Ltd
Copper tube and fittings - Brikenden (Builders Merchants) Ltd
Copper tube for water - IMI Yorkshire Copper Tube Ltd
Design, installation and maintenance of building services - Sulzer Infra (UK) Ltd
Plumbing products - Drainage Systems
Pre-commisson clearing of heating & chilled water systems - Aquastat Ltd
Switches, valves and manifolds - Klinger Fluid Instrumentation Ltd
Thermoplastic pipes - CPV Ltd
Unvented Combination Units with Unvented Hot Water Storage & Boilers - Harton Heating Appliances

S12 Hot and cold water (self-contained specification)

Cable Duct - Hepworth Building Products Ltd
Copper Unvented Units - Harton Heating Appliances
Cylinder Packs With Integral Controls - Harton Heating Appliances
Enamelled Steel Unvented Units - Harton Heating Appliances
Fittings - Pegler Ltd
Flexible plastic plumbing - IPPEC Sytsems
Hot & cold water - Hertel Services
Hot and cold water cylinder jackets - Davant Products Ltd
Plumbing - Marflow Eng Ltd
Plumbing accessories - BSS Zenith
Plumbing products - Drainage Systems
Plumbing systems - Rehau Ltd
Prefabricated Plumbing Units - Harton Heating Appliances
PTFE sealant - Klinger Fluid Instrumentation Ltd
Rigid plastic plumbing - IPPEC Sytsems
Sealed system equipment - Flamco UK Ltd
Stainless steel water pipework - BM Stainless Steel Drains Ltd
Swivel ferrule straps, stop taps, leadpacks and boundery boxes - Booth Samuel & Co Ltd
Water heaters - Beeston Heating Group Ltd

S13 Pressurised water

Design, installation and maintenance of building services - Sulzer Infra (UK) Ltd
Expansion vessels for portable water - Flamco UK Ltd
High pressure washdown units - Dewey Waters Ltd
Pressure sprayers - Hozelock Ltd
Pressure washers - Fastclean Scotland Ltd
Pressurisation - Beeston Heating Group Ltd
Pressurised water - Calpeda Limited
Pressurised water - John Davidson (Pipes) Ltd
Pressurised Water - New Haden Pumps Ltd
Pressurised water - Polypipe Civils Ltd
PVC pressure pipes - Uponor Ltd
Rainwater utilization pump - KSB Ltd
Valves for high pressure pneumatic and hydraulic appliances - Hale Hamilton Ltd
Water boosters and pressurisation units - Pillinger G C & Co (Engineers) Ltd
Water management - Verna Ltd

Water Suplly Pipe - Hepworth Building Products Ltd

S14 Irrigation

Irrigation - Calpeda Limited
Irrigation - Polypipe Civils Ltd

S15 Fountains/ Water features

Fountains/water features - Calpeda Limited

S20 Treated/ Deionised/ Distilled water

Cooling tower water treatment, cleaning and refurb - Aquastat Ltd
Demineralised, deionised, softened water and filters - Dewplan Ltd
Disinfection systems - Portacel
Evaporative water cooling tower - Visco Ltd
Flexible tubes - United Flexibles (A Division of Senior Flex)
Limescale inhibitors - Hydropath (UK) Ltd
Treated/Deionised/Distilled water - Hertel Services
Treated/Deionised/Distilled Water - USF Permutit/Ionpure
Ultra violet chambers, lamps and control panels - Hanovia Ltd
Waste water treatment - Clearwater Environmental Engineering Ltd
Water chlorination - Portacel
Water conditiners - Cistermiser Ltd
Water filtration and treatment - Barr & Wray Ltd
Water filtration equipment - Hydropure Group Ltd
Water purification - USF Permutit/Ionpure
Water purification systems - Elga Ltd
Water Treatment - BetzDearbom Ltd
Water treatment - Estec Environmental Ltd
Water treatment - Feedwater Ltd

S21 Swimming pool water treatment

Calorifiors - Beeston Heating Group Ltd
Hydrotherapy pools - Aqua-Blue Ltd
Modular precoated stainless steel swimming pools - Aqua Design Ltd
Pool covers - Amdega Ltd
Saunas, steam rooms and jacazzi spa pools - Aqua-Blue Ltd
Swimming pool surrounds - Blanc de Bierges
Swimming pool accessories - Airflow (Nicoll Ventilators) Ltd
Swimming pool equipment - Golden Coast Ltd
Swimming pool installations - Barr & Wray Ltd
Swimming pool kits and accessories-domestic - Fox Pool (UK) Ltd
Swimming pool water treatment - Calpeda Limited
Swimming pools and accessories - Aqua-Blue Ltd
Swimming pools and equipment, spas, heat pumps, water features and chemicals. - Certikin International Ltd
Swimming pools, spa baths, sauna and steam rooms - Buckingham Swimming Pools Ltd
Swimming pools, spas, saunas, steam rooms, baptisteries and mosaic murals - Regency Swimming Pools Ltd
Ultra violet chambers, lamps and control panels - Hanovia Ltd
Water features - Aqua-Blue Ltd
Watertank refurbishment - Aquastat Ltd

S30 Compressed air

Air compressors - F. W. Harris & Co Ltd
Compressed air - Hertel Services
Compressed air nozzles - Designed for Sound Ltd
Compressed air supply systems - Fluidair Compressors Ltd
Hand and actuated valves - George Fischer Sales Ltd
Oil free air compressors, vacuum pumps and gas circulators. - Dawson McDonald & Dawson Ltd

S31 Instrument air

Industrial instrumentation - ABB Kent Taylor Ltd.

S32 Natural gas

Decorative gas fires - Taylor & Portway Ltd
Gas - Sulzer Infra (UK) Ltd
Gas and sanitation applications - IMI Yorkshire Copper Tube Ltd
Gas detection equipment - Draeger Ltd
Gas Detection Equipment - TCW Services (Control) Ltd
Gas detection systems - Blakell Europlacer Ltd
Gas fired equipment - Valor Heating
Gas isolation devices - TCW Services (Control) Ltd
Gas Isolation Equipment - TCW Services (Control) Ltd
Gas meters - Raab Karcher Energy Services Ltd, Clorius Division
Gas pipe fittings - Uponor Ltd
Natural gas - John Davidson (Pipes) Ltd
Respiratory filter masks - Draeger Ltd

Seamless cold drawn copper tube - Lawton Tube Co Ltd, The

S33 Liquefied petroleum gas

Bulk tank bottled gas - Calor Gas Ltd
Calor gas - Bell & Co Ltd
Cylinder bottled gas - Calor Gas Ltd
Decorative gas fires - Taylor & Portway Ltd
Gas Detection Equipment - TCW Services (Control) Ltd
Gas Isolation Equipment - TCW Services (Control) Ltd
Gas storage tanks - NSA Environmental Products
Heating Appliences - Bell & Co Ltd
LDG vessels - Flamco UK Ltd
Portable gas heaters - Valor Heating

S40 Petrol/ Diesel storage/ distribution

Fill Point cabinets - Metcraft Ltd
Lubricants - MMP International

S41 Fuel oil storage/ distribution

Fuel Oil storage/ distribution - John Davidson (Pipes) Ltd
Fuel oil storage/ distribution - Metcraft Ltd
Fuel storage tanks - T & D Plastech
Liquid storage tanks - Franklin Hodge Industries Ltd
Oil refinery - Carless Refining & Marketing Ltd
Oil storage tanks - Davant Products Ltd
Oil tanks and storage bunkers - Ferham Products

S50 Vacuum

Vacuum pumps - F. W. Harris & Co Ltd

S60 Fire hose reels

Fire fighting hoses, nozzles, adaptors and standpipes - Richard Hose Ltd
Fire fighting sets - Fordwater Pumping Supplies Ltd
Fire hose reels - Nu-Swift International Ltd
Fire hose reels - UK Fire International Ltd
Hosereels - Fixed, automatic and swinging - Macron Fireater Ltd

S61 Dry risers

Dry riser - Eagles, William, Ltd
Dry riser equipment - Fire Protection Services PLC
Dry risers - Firetecnics Systems Ltd
Dry risers - Macron Fireater Ltd
Hose reels for fire fighting - Richard Hose Ltd

S62 Wet risers

Hydro extractors - Electrolux Laundry Systems

S63 Sprinklers

Fire sprinkler tanks - Franklin Hodge Industries Ltd
GRP sprinkler and pump suction tanks - Dewey Waters Ltd
Sprinkler systems - Firetecnics Systems Ltd
Sprinklers - Fire Security (Sprinkler Installations) Ltd
Sprinklers - Richard Hose Ltd
Sprinklers - UK Fire International Ltd
Sprinklers - Wormald Ansul UK Ltd

S64 Deluge

Deluge - UK Fire International Ltd
Deluge - Wormald Ansul UK Ltd

S65 Fire hydrants

Fire Extinguishers Supply and Maintinance - Nu-Swift International Ltd
Fire hydrant valves including dry riser equipment - Richard Hose Ltd

S70 Gas fire fighting

Fire Extinguishers Supply and Maintinance - Nu-Swift International Ltd
Fire suppression systems - Macron Fireater Ltd
Gas detection equipment - MSA (Britain) Ltd
Gas detection equipment - Zellweger Analytic Ltd, Sieger Division
Gas detection systems - Blakell Europlacer Ltd
Gas fire fighting - UK Fire International Ltd
Gaseous fire extinguishing system - Wormald Ansul UK Ltd
Polyethylene pipes for water and gas - Capper Pipe Systems Ltd

S71 Foam fire fighting

Extinguishers - Channel Safety Systems Ltd
Fire extinguisher systems - Fife Fire Engineers & Consultants Ltd
Fire extinguishers - Richard Hose Ltd
Fire Extinguishers Supply and Maintinance - Nu-Swift International Ltd
Foam equipment - Eagles, William, Ltd
Foam equipment and concentrate - Macron Fireater Ltd
Foam fire fighting - UK Fire International Ltd
Foam systems - Wormald Ansul UK Ltd
Hose assemblies for fire fighting, industrial and drinking purposes - Richard Hose Ltd

T10 Gas/ Oil fired boilers

Boiler maintenance - H Pickup Mechanical & Electrical Services Ltd
Boilers - Alpha Therm United House Ltd
Boilers - Caradon Heating Ltd
Boilers Oil, Gas & solid fuel - Brikenden (Builders Merchants) Ltd
Central heating equipment - Hudevad Britain
Combustion equipment - Nu-Way Ltd
Flexible liners - Ventlane Ltd
Flue and chimney systems - Flamco UK Ltd
Flue systems - Marflex Chimney Systems
Flue terminals - Tower Flue Componants Ltd
Gas and solid fuel heating appliances - Hepworth Heating Ltd
Gas boilers - AEL
Gas boilers - Halstead Boilers Ltd
Gas central heating boilers - Burco Maxol
Gas fired boilers - Fulton Boiler Works (Great Britain) Ltd
Gas Fired Hot water heaters - HRS Hevac Ltd
Gas fired radiant heating equipment - Horizon International Ltd
Gas flue systems - Ubbink (UK) Ltd
Gas vents - Ventlane Ltd
Gas, oil fired boilers - Broag Ltd
Gas, oil fired boilers - Don Engineering (South West) Ltd
Gas, oil fired boilers - Hartley & Sugden
Gas, oil fired boilers - Kayanson Engineers Ltd
Gas, Oil, Dual fuel - GP Burners (CIB) Ltd
Gas/Oil Fired Boilers - service maintenance - Hodgkinson Bennis Ltd
Greenhouse heaters - Brettell & Shaw Ltd
Heating - Dantherm Ltd
Heating - HV Air Conditioning Ltd
Heating - Potterton Myson Ltd
Heating manifolds - SAV United Kingdom Ltd
Instantaneous hot water systems - HRS Hevac Ltd
Oil and gas fired boilers and ancillaries - Boulter Boilers Ltd
Plumbing & Heating - Spital Tile Co
Refractory boiler linings, bricks, castables, ceramic fibre products & incinerators - Combustion Linings Ltd
Sectional boilers - Beeston Heating Group Ltd
Service Terminals - Ubbink (UK) Ltd
Steam and hot water boilers and pressure vessels - Wellman Robey Ltd
Steel boilers - Beeston Heating Group Ltd
Water boilers - TSE Brownson

T11 Coal fired boilers

Back boilers for solid fuel open fires - Interoven Ltd
Combstion equipment - Saacke Ltd
Combustion units - IMI Range Ltd
Flexible liners and solid fuel - Ventlane Ltd
Flue and chimney systems - Kedddy (Poujoulat) (UK) Ltd
Flue liners - ISOKERN (UK) Ltd
Flue systems - Marflex Chimney Systems
Plastic coal bunkers - Tyrrell Tanks Ltd
Refractory boiler linings, bricks, castables, ceramic fibre products & incinerators - Combustion Linings Ltd
Solid fuel central HT boiler - Dunsley Heat Ltd
Steam and hot water boilers and pressure vessels - Wellman Robey Ltd
Stokers for coal fired boilers and associated equipment - service and maintenance - Hodgkinson Bennis Ltd

T12 Electrode/ Direct electric boilers

Electric boilers - Eltron Chromalox
Electric boilers - Fulton Boiler Works (Great Britain) Ltd
Electric boilers - Technical Control Systems Ltd
Electric calorifiers - Eltron Chromalox
Electrode/Direct electric boilers - Collins Walker Ltd, Elec Steam & Hot Water Boilers
Flue systems - Marflex Chimney Systems
Refractory boiler linings, bricks, castables, ceramic fibre products & incinerators - Combustion Linings Ltd

T13 Packaged steam generators

Packaged steam generators - Rother Boiler Co Ltd
Packaged steam generators - Rycroft Ltd
Steam and hot water boilers and pressure vessels - Wellman Robey Ltd
Steam boilers - Fulton Boiler Works (Great Britain) Ltd

T14 Heat pumps

Heat pumps - Rycroft Ltd
Water Chillers - Therminal Technology (sales) Ltd

T15 Solar collectors

Solar - Stiebel Eltron Ltd
Solar collectors, solar electric powered products - Intersolar Group Ltd
Solar panels - CPV Ltd
Solar water heating systems - Sundwel Solar Ltd

T16 Alternative fuel boilers

Alternative fuel boilers - Hartley & Sugden
Alternative fuel boilers - Rother Boiler Co Ltd
Combustion Burners - GP Burners (CIB) Ltd
Energy saving equipment - Saacke Ltd
Flue and chimney systems - Kedddy (Poujoulat) (UK) Ltd
Flue systems - Marflex Chimney Systems
Refractory boiler linings, bricks, castables, ceramic fibre products & incinerators - Combustion Linings Ltd

T20 Primary heat distribution

Design, installation and maintenance of building services - Sulzer Infra (UK) Ltd
District heating - ABB Industrial Systems Ltd
Drum heaters - Jimi-Heat Ltd
Flue systems - Marflex Chimney Systems
Heater cables - Jimi-Heat Ltd
Heater mats - Jimi-Heat Ltd
Heating - Birdsall Services Ltd
Heating - BSS Zenith
Heating Accessories - Faral Radiators
Industrial and commercial heating systems - Ambi-Rad Ltd
Industrial heating equipment - Turnbull & Scott (Engineers) Ltd
Radiators - Faral Radiators
Temperature controls - Jimi-Heat Ltd

T30 Medium temperature hot water heating

Heating - Dunham Bush Ltd
MTHW Heating - Rycroft Ltd
Steam and hot water boilers and pressure vessels - Wellman Robey Ltd

T31 Low temperature hot water heating

Boiler and calorifier descaling - Aquastat Ltd
Boilers - HRS Hevac Ltd
Buffer and storage vessels - Flamco UK Ltd
Central heating - Property Mechanical Projects Ltd
Central heating equipment - Hudevad Britain
Central heating programmers, water heating controllers, motorised valves, room stats and thermostats - Horstmann Timers & Controls Ltd
Domestic and commercial central heating controls - Danfoss Randall Ltd
Electric heating system - Elvaco UK Ltd
Gas convector heaters - Burco Maxol
Heating - BSS Zenith
Heating - Dunham Bush Ltd
Heating - Encon Air Systems Ltd
Heating and plumbing - Diamond Merchants
Heating to pool areas - Buckingham Swimming Pools Ltd
Hot water heating - Jaga Heating Poducts (UK) Ltd
Low temperature hot water - Hamworthy Heating Ltd
Low temperature hot water - Rother Boiler Co Ltd
LTHW Heating - Rycroft Ltd
Manifolds - Martin Orgee & Associates
Perimeter heating - Unilock Ltd
Pressure reducing Valves - Martin Orgee & Associates
Radiator Valves - Martin Orgee & Associates
Radiators - De Longhi Radiators srl
Steam and hot water boilers and pressure vessels - Wellman Robey Ltd
Thermostatic Mixinet Valuves - Martin Orgee & Associates
Underfloor heating - FloRad Heating Systems
Warm water underfloor heating - IPPEC Sytsems

T32 Low temperature hot water heating (self-contained specification)

Central heating and radiators - Thermal Radiators Ltd
DHW generators and aluminium radiators - AEL
Domestic Central Heating - Boulter Boilers Ltd
Domestic radiators - Veha Radiators Ltd
Electric radiators - Loblite Ltd
Electric water heating - Kaloric Heater Co Ltd
Gas Central Heating - Servowarm
Heating - ACR Heat Transfer Manufacturing Ltd
Heating - Servitherm
Heating and boiler engineers - Caradon Ideal Ltd
Heating and plumbing - Diamond Merchants
Heating systems - Marflow Eng Ltd
Low temperature hot water - Rother Boiler Co Ltd

Low temperature hot water heating - Multibeton Ltd
Low temperature hot water(small scale) - Hamworthy Heating Ltd
LTHW Heating (self contained) - Rycroft Ltd
Radiator valves - Pegler Ltd
Radiators - Caradon Heating Ltd
Radiators - Kermi (UK) Ltd
Radiators - Novellini Bathroom Products
Storage water heating - Johnson & Starley
Underfloor heating - Multibeton Ltd
Water heaters - P & R Electrical (London) Ltd
Water treatment of heating and chilled water systems - Aquastat Ltd

T33 Steam heating

Air conditioning - Dunham Bush Ltd
Steam heating - Rycroft Ltd

T40 Warm air heating

Air conditioning - Dunham Bush Ltd
Air warmers - Seaque Technologies Ltd
Fan convecters - S & P Coil Products Ltd
Gas and oil fired indirect warm air heaters and systems - Dragonair Ltd
Gas fired air heaters - Reznor UK Ltd
Heat exchangers gasketted plate - HRS Hevac Ltd
Heating equipment - NuAir Ltd
Industrial & commercial, gas or oil indirect fired warm air heating systems - Elan-Dragonair
Industrial and commercial heating systems - Ambi-Rad Ltd
Industrial heating - Hoval Ltd
Solid fuel fires - Dimplex (UK) Ltd
Unit heater - Dravo Environmental Services
Warm air heating - Combat Engineering Ltd
Warm air heating - Consort Equipment Products Ltd
Warm Air heating - Nu-Way Ltd
Warm air heating products - Jaga Heating Poducts (UK) Ltd

T41 Warm air heating (self-contained specification)

Electric warm air heating - Kaloric Heater Co Ltd
Heat exchangers - Heat Transfer Pressurisation
Natural and fan assisted floor convectors - Multibeton Ltd
Warm Air Heating - Vortice Ltd

T42 Local heating units

Anti condensation heaters - Eltron Chromalox
Electric fires - Valor Heating
Flameproof heaters, radiant panel heaters - Seaque Technologies Ltd
Gas fired air heaters - Reznor UK Ltd
Gas fired space heating - domestic/commercial - Johnson & Starley
Gas fires - Dimplex (UK) Ltd
Gas Fires - Gazco Ltd
Gas fires/ Gas conductors - Robinson Willey Ltd
Gas Stoves - Gazco Ltd
Heaters - Aga-Rayburn, Glynwed Consumer & Construction Products
Heating - Andrews Sykes Hire Ltd
Heating products - Pressurisation Ltd, Member of Hoval Group
Immersion heaters - Heating Equipment Ltd
Infra ray heaters - Bitmen Products Ltd
Local heating units - Consort Equipment Products Ltd
Local heating units - Harvey Fabrication Ltd
Products for commercial and industrial applications - Bush Nelson PLC
Radiant gas heaters - Infraglo (Sheffield) Ltd
Radiant infra-red plaque gas heaters - Space - Ray UK
Radiant infra-red poultry gas brooders - Space - Ray UK
Radiant infra-red tube gas heaters - Space - Ray UK
Radiant tube and radiant heaters - Radiant Services Ltd
Refractory boiler linings, bricks, castables, ceramic fibre products & incinerators - Combustion Linings Ltd
Removable cove heaters - Howden H D Ltd
Storage and panel heaters - Seaque Technologies Ltd
Storage gas water heaters, focal point fan convector - Burco Maxol
Unit heaters - Biddle Air Systems Ltd
Unit Heaters - Therminal Technology (sales) Ltd
VAT heaters - Howden H D Ltd

T50 Heat recovery

Brazed plate heat exchangers - HRS Hevac Ltd
Coil heat exchangers - S & P Coil Products Ltd
Condensation control - Kiltox Damp Free Solutions
Heat recovery - Brook Airchanger
Heat recovery - Dantherm Ltd
Heat recovery - Kair - The Ventilation Division of Kiltox Chemicals Limited
Heat recovery - Rycroft Ltd
Heat Recovery - S & P Coil Products Ltd
Heat Recovery - Vent-Axia
Heat recovery unit - Dunsley Heat Ltd
Heat recovery units - Greenwood Air Management
Heat recovery units - Therminal Technology (sales) Ltd
Heat recovery ventilation units - Ves Andover Ltd
Heating and heat recovery equipment thermal wheels - AEL
Steam and hot water boilers and pressure vessels - Wellman Robey Ltd

T60 Central refrigeration plant

Chillers, coolers and freezers - Sabroe Refigeration Ltd
Design, installation and maintenance of building services - Sulzer Infra (UK) Ltd
Industrial refridgeration equipment and systems - J & E Hall Ltd
Refrigeration - Birdsall Services Ltd
Refrigeration (McQuay UK Ltd) - McQuay UK Ltd
Refrigeration plant - Copeland Corporation Ltd

T61 Chilled water

Chilled water - J & E Hall Ltd
Chilled water - Rycroft Ltd

T70 Local cooling units

Absorption chillers - Birdsall Services Ltd
Chilled beams - ABB Flakt Products
Portable coolers - London Fan Company Ltd
Portable Heating - Andrews Sykes Hire Ltd

T71 Cold rooms

Cold room insulation panels - Kay Metzeler Ltd
Cold rooms - Isowall UK) Ltd
Cold rooms - J & E Hall Ltd
Industrial refrigeration systems - Sabroe Refigeration Ltd
Modular coldrooms - Hemsec Manufacturing Ltd

T72 Ice pads

Ice pads - J & E Hall Ltd

U10 General ventilation

Access Doors - Hotchkiss Air Supply
Air conditioners, dampers, air ductline - Advanced Air (UK) Ltd
Air conditioning - Birdsall Services Ltd
Air conditioning - Purified Air Ltd
Air conditioning - Trox (UK) Ltd
Air conditioning products, systems, parts and maintenance services - Carrier Air Conditioning
Air conditioning systems - Elvaco UK Ltd
Air distribution equipment - Designed for Sound Ltd
Air seperators - Martin Orgee & Associates
Air transfer grilles - Mann McGowan Fabrication Ltd
Air-conditioning - Andrews Sykes Hire Ltd
Airflow instrumentation - Airflow Developments Ltd
Anti-condensation for motors heater tapes - Jimi-Heat Ltd
Design, installation and maintenance of building services - Sulzer Infra (UK) Ltd
Domestic and commercial exract fans - Greenwood Air Management
Domestic ventilation systems - Airflow (Nicoll Vontilators) Ltd
Domestic, commercial and industrial ventilating systems - Vent-Axia Ltd
Exhausterl blowers - BVC-BIVAC, Division of DD Lamson
Extract fan units - Ves Andover Ltd
Fan convectors - Biddle Air Systems Ltd
Fire resistant air transfer grilles - Hamron Group
Flanging systems - Hotchkiss Air Supply
Flatoval Tube - Hotchkiss Air Supply
Flexible Ducting - Hotchkiss Air Supply
Flue systems - Marflex Chimney Systems
Full freshair systems - Dantherm Ltd
General supply/ extract - DCE - BTR Environmental Ltd
General supply/ extract - Emsworth Fireplaces Ltd
General Ventilation - Dyer Environmental Controls
Housing ventilation - ABB Flakt Products
Humidity control - Kiltox Damp Free Solutions
Mechanical extract terminals - Klober Ltd
Modular ventilation ducting - Domus Ventilation Ltd.
Roof extract units - Ves Andover Ltd
Roof ventilator - Eternit UK Ltd
Roof Ventilators - (Ridge & Slope mounted) - McKenzie-Martin Ltd
Sealed system equipment and air separation and dirt removal equipment - Flamco UK Ltd
Spiral Tube - Hotchkiss Air Supply
Support systems - Hotchkiss Air Supply
Trade suppliers of commercial and decorative ceiling fans - Fantasia Distribution Ltd
Underfloor ventilators - Ryton's Building Products Ltd
Ventelated Ceilings - Vianen Kitchen Ventilation (UK) Ltd.
Ventilation - Dravo Environmental Services
Ventilation - Encon Air Systems Ltd
Ventilation - HV Air Conditioning Ltd
Ventilation - NuAir Ltd
Ventilation - Shearflow Phoenix
Ventilation - Therminal Technology (sales) Ltd
Ventilation - Ubbink (UK) Ltd
Ventilation - domestic, commercial and industrial - Vent-Axia
Ventilation and ductwork components - Contract Components Ltd
Ventilation equipment - Johnson & Starley
Ventilation equipment - Roof Units Group
Ventilation of pool areas - Buckingham Swimming Pools Ltd
Ventilation Products - Hinton Perry & Davenhill Ltd
Ventilation products - Map Hardware Ltd
Ventilation systems - Eternit Clay Tiles Ltd
Ventilators - Kair - The Ventilation Division of Kiltox Chemicals Limited
Window ventilator manufactures - Simon R W Ltd

U11 Toilet ventilation

Domestic extractor fans - Domus Ventilation Ltd.
Toilet ventilation - Kair - The Ventilation Division of Kiltox Chemicals Limited

U12 Kitchen ventilation

Design, Supply and Installation of Commercial Kitchen Ventelation - Vianen Kitchen Ventilation (UK) Ltd.
Domestic extractor fans - Domus Ventilation Ltd.
Extraction systems and odour control - Purified Air Ltd
Grease filters - Farr Filtration Ltd
Grease filters - Westbury Filters Limited
Kitchen Canopies - Vianen Kitchen Ventilation (UK) Ltd.
Kitchen extract - Fire Protection Ltd
Kitchen ventilation - Kair - The Ventilation Division of Kiltox Chemicals Limited
Kitchen ventilation ceiling - OSC Process Engineering Ltd
Ventilation - Kiltox Damp Free Solutions
Ventilation - domestic, commercial and industrial - Vent-Axia

U13 Car parking ventilation

Car park extract - Fire Protection Ltd
Vehicle exhaust extraction - Nederman Ltd

U14 Smoke extract/ Smoke control

Ancillaries - Advanced Air (UK) Ltd
Automatic smoke blinds/ sunshields - Cooper Group Ltd
Birdguards - Brewer Metalcraft
Chimney cappers - Brewer Metalcraft
Chimney cowls - Brewer Metalcraft
Chimneys and flues - Metcraft Ltd
Commercial and domestic electrostatic air filters - Robinson Willey Ltd
Powered cowls - Brewer Metalcraft
Rain caps - Brewer Metalcraft
Smoke control systems - Smoke Control Services
Smoke Extract - McKenzie-Martin Ltd
Smoke extract fans - London Fan Company Ltd
Smoke extract/ Smoke control - DCE - BTR Environmental Ltd
Smoke extract/ smoke control - Fire Protection Ltd
Smoke Extract/ Smoke control - Photain Controls Plc
smoke lights - Luxcrete Ltd
Spinning cowls - Brewer Metalcraft
Terminals - Brewer Metalcraft
Ventilation - domestic, commercial and industrial - Vent-Axia
Ventilation and smoke extract equipment - Twin County Fan Company Ltd

U15 Safety cabinet/ Fume cupboard extract

Domestic and commercial flue and ventilation products - Rega Metal Products Ltd
Fume cupboards - S & B Ltd
Fume extract equipment - HV Air Conditioning Ltd
Maintenance/ servicing/ testing/ inspection - Cooper Group Ltd
Safety Cabinet / fume cupboards - Envair Ltd
Safety extract/ Smoke control - DCE - BTR Environmental Ltd
Safety extract/ smoke control - Fire Protection Ltd
Safety extract/ Smoke Control - Redland Precast
Smoke management systems - Advanced Air (UK) Ltd

U16 Fume extract

Air pollution control plant - Plastic Construction Fabrications Ltd
Dust and fume extraction - Encon Air Systems Ltd
Flue systems - Marflex Chimney Systems
Fume cupboards - ClipsPartitions Ltd
Fume extract - DCE - BTR Environmental Ltd
Fume extract - Emsworth Fireplaces Ltd
Fume Extract - Redland Precast
Fume extraction - Nederman Ltd
Fume Extraction - PCT Group Sales Ltd
Nox reduction combustion equipment and technical advice - Mitsui Babcock Energy Services Ltd
Portable Heating - Andrews Sykes Hire Ltd

U20 Dust collection

Dust Collection - BVC-BIVAC, Division of DD Lamson
Dust collection - DCE - BTR Environmental Ltd
Dust collection - Farr Filtration Ltd
Dust control equipment - Airmaster Engineering Ltd
Dust extraction - Nederman Ltd

U30 Low velocity air conditioning

Air conditioning - Airedale International Air Conditioning Ltd
Air conditioning - Carrier Air Conditioning
Air conditioning - Dalair Ltd
Air conditioning - Eaton-Williams Group Ltd
Air conditioning - Eberle GmbH
Air conditioning - Edenaire Ltd
Air conditioning - HV Air Conditioning Ltd
Air handling units - ABB Flakt Products
Design, installation and maintenance of building services - Sulzer Infra (UK) Ltd
Domestic and commercial flue and ventilation products - Rega Metal Products Ltd
Portable Heating - Andrews Sykes Hire Ltd

U31 VAV air conditioning

VAV air conditioning - Senior Air Systems
Vav boxes - ABB Flakt Products
VAV terminal units - Advanced Air (UK) Ltd

U41 Fan-coil air conditioning

Air conditioning - Airedale International Air Conditioning Ltd
Air conditioning units - IPPEC Sytsems
Fan coil units - Therminal Technology (sales) Ltd
Fan coil Valves - Martin Orgee & Associates
Fan coils - ABB Flakt Products
Fan coils (McQuay UK Ltd) - McQuay UK Ltd
Fan coils units - Biddle Air Systems Ltd
Fan-coil air conditioning - Senior Air Systems
Legionella prevention and control - Aquastat Ltd
Refrigeration equipment - Dunham Bush Ltd

U42 Terminal re-heat air conditioning

Air terminal devices - ABB Flakt Products
Coil heat exchangers - S & P Coil Products Ltd
Heat exchangers gasketted plate - HRS Hevac Ltd
Terminal re-heat air - S & P Coil Products Ltd
Terminal re-heat air - Senior Air Systems

U60 Air conditioning units

Air conditioning - Airedale International Air Conditioning Ltd
Air conditioning - Johnson & Starley
Air conditioning units - Amcor (Appliances) Ltd
Design, installation and maintenance of building services - Sulzer Infra (UK) Ltd
Free standing air conditioning - Senior Air Systems
Hygrometers - Air Improvement Centre Ltd
Mobile air conditioners, humidifiers, dehumidifiers and air purifiers - Air Improvement Centre Ltd
Portable air conditioners - Thermoscreens Ltd
Portable Heating - Andrews Sykes Hire Ltd
Refrigeration equipment - Dunham Bush Ltd
Thermometers - Air Improvement Centre Ltd

U70 Air curtains

Air curtains - Biddle Air Systems Ltd
Air curtains - Shearflow Phoenix
Air curtains - Thermoscreens Ltd
Fires - Dimplex (UK) Ltd
Warm air curtains - S & P Coil Products Ltd

V10 Electricity generation plant

Aluminium extrusions/ sections - Seco Aluminium Ltd
Automatic stant generators - Cummins Power Generation Ltd
Automatic synchronising generators - Cummins Power Generation Ltd
Baseload Diesel Generator - Broadcrown Projects Ltd
Bio Gas Generators - Broadcrown Projects Ltd
Central battery systems - Channel Safety Systems Ltd
Continuous and emergency standby power generator sets from 7.5kVA to 6500kVA power range - Wilson (Engineering) Ltd, F G
Diesel engined generation plant - Dawson-Keith Ltd
Electrical components - Thomerson & Brett Ltd
General electrical products - WF Electrical Distributors

General electrical products - wholesale - Jolec Electrical Supplies
Generation plant - Broadcrown Projects Ltd
Generation plant, diesel, and gas turbine - Dale Power Systems
Generator sets manufactured to suit specific requirements - Wilson (Engineering) Ltd, F G
Industrial batteries - Tungstone Batteries Ltd
Landfill Generators - Broadcrown Projects Ltd
Mobile generator sets for on-set power requirements - Wilson (Engineering) Ltd, F G
Mobile generators - Cummins Power Generation Ltd
Mobile generators and electricity distribution equipment - Automatic Safety Lighting Ltd (T/as 'Stapoco')
Natural Gas Generators - Broadcrown Projects Ltd
Peak Lopping Diesel Generators - Broadcrown Projects Ltd
Power generators - Cummins Power Generation Ltd
Range of generator canopies including sound attenuated - Wilson (Engineering) Ltd, F G
Silenced generators - Cummins Power Generation Ltd
Standby Diesel Generators - Broadcrown Projects Ltd
Super silenced generators - Cummins Power Generation Ltd

V11 HV supply/ distribution/ public utility supply

Electrical cables - AEI Cables Limited
Industrial plugs, sockets, switched sockets - LPA Industries PLC
UPVC cable tray and fixings - LPA Industries PLC

V12 LV supply/ public utility supply

Control products - Legrand Electric Ltd
Electrical cables - AEI Cables Limited
Electricity meters - Raab Karcher Energy Services Ltd, Clorius Division

V20 LV distribution

Accessories for electrical services - Lewden Electrical Industries
Circuit protection and control - Bill Circuit Protection and Control
Consumer units - Square D, Electrical Distribution Division
Electrical cables - AEI Cables Limited
Electrical components - Thomerson & Brett Ltd
Electrical distribution - Clips Partitions Ltd
Electrical distribution equipment - Square D, Electrical Distribution Division
Electrical installations - HEM Interiors Group Ltd
Lighting and distribution equipment - Blakley Electrics Ltd
Low voltage spots - Lumitron Ltd

V21 General lighting

Access panels to ceiling void - Ecophon Ltd
Accessories for electrical services - Lewden Electrical Industries
Architectural lighting - Electrak International Ltd
Cold cathode - Oldham Lighting Ltd
Commercial lighting - Radius Lighting Ltd
Compact Fluorecent Downlights - Profile Lighting Services Ltd
Diffusers louvres and glare control products - Interlux Ltd
Display lighting - Illuma Lighting Ltd
Downlights - Illuma Lighting Ltd
Electrical Accessories - Sussex Brassware Ltd
Electrical cables - AEI Cables Limited
Electrical components - Thomerson & Brett Ltd
Energy efficiency in lighting - Poselco Lighting
Energy efficient lighting for commercial, retail and industrial markets - IMP Lighting Ltd
Exterior lighting - Windsor Limited D W
Fibre optic lighting systems - Absolute Action Ltd
Fibre optics - Oldham Lighting Ltd
Fluorescent lighting and spotlights - Howard Home Improvements Ltd
General and bespoke lighting systems - Concord Lighting
General lighting - Arden Manufacturing Birmingham Ltd
General lighting - Brandon Medical
General lighting - Crompton Lighting Ltd
General lighting - Glamox Electric (UK) Ltd
General lighting - IBL Lighting
General lighting - Intralux UK Ltd
General lighting - Jerrards Commercial Lighting Division
General lighting - Lampways Ltd
General lighting - Oldham Lighting Ltd
General Lighting - Ova Bargellini UK Ltd
General Lighting - Sylvania Lighting International INC.
General lighting - The Cube
General lighting and bulbs - GE Lighting Ltd
General lighting and purpose built lighting - Quip Lighting, Div of Bruce & Macintyre Ltd
General lighting products - Whitecroft Lighting Limited
General lighting, lighting controls and dimmer switches - Home Automation Ltd
General lighting, magnetic and electronic ballasts - Helvar Electrosonic
General, modular, compact Fluorescent - Lumitron Ltd
Hazardous area lighting - Legrand Electric Ltd
Hazardous Area Lighting - PFP Electrical Products Ltd
Highbay lighting - Lighting Systems UK, Division of Phosco Ltd
Highbay lighting - Phosco Ltd
Interior and exterior lighting - Andy Thornton Architectural Antiques Ltd
Light bulbs, tungsten halogen and metal halide - BLV Licht-und Vakuumtecnik
Light fittings - Profile Lighting Services Ltd
Light fittings - Rada Lighting Ltd
Light fittings - Thorn Lighting Ltd
Light fittings - Vitalighting Ltd
Lighting - Davies John Interiors Ltd
Lighting - Designplan Lighting Ltd
Lighting - Forma Lighting Ltd
Lighting - Full Spectrum Lighting Ltd
Lighting - HT Martingale Ltd
Lighting - Optelma Lighting Ltd
Lighting - Power Breaker
Lighting - Twin Technology Lighting
Lighting busbar - Salamandre Davis
Lighting control equipment - ECS Lighting Controls Ltd
Lighting control gear and compact fluorescent lamps - Bemlite Ltd
lighting equipment - Thorlux Lighting
Lighting equipment - Transtar
Lighting products - Caradon Friedland Ltd
Lighting support systems - Vantrunk Engineering Ltd
Lighting systems - Absolute Action Ltd
Lighting trunking - Salamandre Davis
Lighting, switches, accessories and weatherproof sockets - Loblite Ltd
Low energy compact fluorescent lighting - Beta Lighting Ltd
Office, desk and task lighting - Ledu UK Ltd
Office, medical, architectural, industrial and workstation lighting - Luxo UK Ltd
Period and decorative lighting - Denmans Montrose Ltd
Purpose made luminaires - LB Lighting Ltd
Recessed and Surface LG3 Luminaires - Profile Lighting Services Ltd
Retail Display Luminaires - Profile Lighting Services Ltd
Retrofit lighting - ECS Lighting Controls Ltd
Security PIR lighting - Timeguard Ltd
Spotlights - Illuma Lighting Ltd
Suspended acoustic ceiling systems - Ecophon Ltd
Track - Illuma Lighting Ltd
Transformers, lighting and distribution equipment - Blakley Electrics Ltd
Uplights - Illuma Lighting Ltd

V22 General LV power

Cable tray - Mita (UK) Ltd
Electrical cables - AEI Cables Limited
Floor boxes - Mita (UK) Ltd
Power - Davies John Interiors Ltd
Powerpoles and powerposts - Mita (UK) Ltd
Support channel - Mita (UK) Ltd

V30 Extra low voltage supply

Accessories for electrical services - Lewden Electrical Industries
Electrical components - Thomerson & Brett Ltd
Entry sound communoication products - Caradon Friedland Ltd
Exrta low voltage supply - Brandon Medical
General electrical products - WF Electrical Distributors
General electrical products - wholesale - Jolec Electrical Supplies
Low voltage fittings - Bemlite Ltd
Miniature precision switches, push button and cord pulls etc - Castelco (Great Britain) Ltd
Transformers - Blakley Electrics Ltd
Transformers for low voltage lighting - Mode Lighting (UK) Ltd

V31 DC supply

Batteries - MJ Electronics Services (International) Ltd
Battery chargers - MJ Electronics Services (International) Ltd
Direct current supply - Brandon Medical

V32 Uninterruptible power supply

Central battery systems - Emergi-Lite Safety Systems Ltd
Continuous and emergency standby power generator sets from 7.5 kVA to 6500 kVA power range - Wilson (Engineering) Ltd, F G
Diesel UPS - Broadcrown Projects Ltd
Generator sets manufactured to suit specific requirements - Wilson (Engineering) Ltd, F G
Industrial batteries for standby applications - Chloride Industrial Batteries Ltd
Power Electronic repair and service - MJ Electronics Services (International) Ltd
Uninterrupted power supply - Liebert Swindon Ltd
Uniterupted power supply - Channel Safety Systems Ltd
UPS Services - MJ Electronics Services (International) Ltd
UPS systems and inverters - MJ Electronics Services (International) Ltd

V40 Emergency lighting

Dimming controls - Jerrards Commercial Lighting Division
Electronic starter switches - Jerrards Commercial Lighting Division
Emergency lighting - Caradon Gent Ltd
Emergency lighting - Channel Safety Systems Ltd
Emergency lighting - Crompton Lighting Ltd
Emergency lighting - Designplan Lighting Ltd
Emergency lighting - Emergi-Lite Safety Systems Ltd
Emergency lighting - Glamox Electric (UK) Ltd
Emergency lighting - Gradus Ltd
Emergency lighting - Lab-Craft Ltd
Emergency lighting - Loblite Ltd
Emergency lighting - Madewel Electronics Systems Ltd
Emergency lighting - Menvier Ltd
Emergency Lighting - MJ Electronics Services (International) Ltd
Emergency lighting - Orbik Electrics Ltd
Emergency lighting - PFP Electrical Products Ltd
Emergency lighting - Poselco Lighting
Emergency Lighting - Rada Lighting Ltd
Emergency lighting - Security Lighting
Emergency lighting equipment - Ova Bargellini UK Ltd
Emergency lighting systems - JSB Electrical PLC
Emergency lighting systems - Mackwell Electronics Ltd
Emergency lighting testing systems - Emergi-Lite Safety Systems Ltd
Emergency lighting, linear lighting systems, up and downlighters - Beta Lighting Ltd
Emergency lighting, switch tripping and battery chargers - Deakin Davenset Rectifiers Ltd
Emergency lights - Firetecnics Systems Ltd
Fibre optic lighting - Absolute Action Ltd
Fire protective covers for recessed light fittings - Hamron Group
General electrical products - WF Electrical Distributors
General electrical products - wholesale - Jolec Electrical Supplies
High frequency ballast - Jerrards Commercial Lighting Division
Industrial batteries for standby applications - Chloride Industrial Batteries Ltd
Light fittings - Thorn Lighting Ltd
Lighting components - Jerrards Commercial Lighting Division
lighting equipment - Thorlux Lighting
Lighting systems - Formwood
Magnetic ballast - Jerrards Commercial Lighting Division
Security lighting - Asco Extinguishers Co Ltd

V41 Street/ Area/ Flood lighting

Accessories for electrical services - Lewden Electrical Industries
Architectural interior and exterior lighting - Lyktan Lighting (UK) Ltd
Exterior uplighters - Lumitron Ltd
Lighting - Abacus Lighting Ltd
lighting equipment - Thorlux Lighting
Outdoor lighting - Holophane Europe Ltd
Pavement lights - Cooper H W & Co Ltd
Pavement lights - Luxcrete Ltd
Security lighting systems - Yorian Garden Products
Street lighting - Urbis Lighting Ltd
Street lighting - Windsor Limited D W
Street lighting - Woodhouse UK Plc
Street lighting and floodlighting - Lighting Systems UK, Division of Phosco Ltd
Street lighting columns and accessories - Lampost Construction Co Ltd
Street lighting columns and high masts - Stainton Metal Co Ltd
Street lighting columns and sign posts - Fabrikat (Nottingham) Ltd
Street lighting systems - Industrial Lighting Ltd
Street lighting, floodlights - Phosco Ltd
Street/ Area & Flood Lighting - Sylvania Lighting International INC.
Street/ Area/ Flood lighting - Crompton Lighting Ltd
Street/ Area/ Flood lighting - Designplan Lighting Ltd
Street/ Area/ Flood lighting - Rada Lighting Ltd

V42 Studio/ Auditorium/ Arena lighting

Decorative and display lighting - Myddleton Hall Lighting, Division of Pearless Design Ltd
Fibre optics lighting - Absolute Action Ltd
Lux Lift - Park Products
Specialist industrial and commercial lighting - Martin Roberts
Studio / Auditorium / Arena lighting - Formwood
Studio/ Auditorium/ Arena - Sylvania Lighting International INC.
Theatrical lighting - Multitex GRP & GRG Products

V50 Electric underfloor/ ceiling heating

Electric matting - MacLellan Rubber Ltd
Electric underfloor heating - Bush Nelson PLC
Electric underfloor heating - Heat Trace Ltd
Electric underfloor, road and ramp Heating - BICC Thermoheat Ltd
Ramp Heating - Findlay Irvine Ltd
Underfloor heating - H Pickup Mechanical & Electrical Services Ltd
Underfloor heating - Wirsbo
Underfloor heating systems - Rehau Ltd

V51 Local electric heating units

Electric air duct heaters - Eltron Chromalox
Electric convector heaters - Eltron Chromalox
Electric fan heaters - Eltron Chromalox
Electric fire manufacturer - Dimplex (UK) Ltd
Electric heater batteries - Ves Andover Ltd
Electric Heaters - HRS Hevac Ltd
Electric heaters - oil - Howden H D Ltd
Electric Heating - Kaloric Heater Co Ltd
Electric heating system controls - Heat Trace Ltd
Electric water heaters - domestic - Howden H D Ltd
Electrical heating units - Consort Equipment Products Ltd
General electrical products - WF Electrical Distributors
General electrical products - wholesale - Jolec Electrical Supplies
Local electric heating units - Bush Nelson PLC
Low energy electric panel heaters - Ecolec

V90 Electrical installation (self-contained specification)

Accessories for electrical services - Lewden Electrical Industries
Design of electrical services - Aercon Wiring Systems Ltd
Domestic wiring accessories and circuit protection equipment - Contactum Ltd
Electrical accessories - MEM
Electrical cables - AEI Cables Limited
Electrical distribution - Clips Partitions Ltd
Electrical installation - Walsall Cable Management Ltd
Electrical installation equipment - Legrand Power Centre
Electrical installation products - Barton Engineering
Electrical Installations - Knight & Smith Ltd, Electrical Contractors
Electrical works - Davies John Interiors Ltd
Electronic door control systems - Dryad Deelock Ltd
F & P Super Tube Clip-in Instalation System - PFP Electrical Products Ltd
General electrical products - Jolec Electrical Supplies
General electrical products - WF Electrical Distributors
Mains lighting - JSB Electrical PLC
Power pole.posts - Salamandre Davis
Ropes and sashcords - Marlow Ropes Ltd
Wire accessories - Hellerman Electric, Member of Bopwthorpe Holdings PLC

W10 Telecommunications

Audio Visual Hire - Planet Projects Ltd
Audio visual solutions & Sales - Planet Projects Ltd
Cables - Alcatel Data Cable
Communication systems - General Security Systems Ltd
Communications - Ogilvie Construction Ltd
Hospital communications systems - Tunstal Telecom Ltd
Industrial batteries for standby applications - Chloride Industrial Batteries Ltd
Multimedia Creation - Planet Projects Ltd
Telecommunication - Illbruck Ltd
Telecommunication Systems - Siemens Communication Ltd
Telecommunications - Connought Communication Systems Ltd
Telecommunications - Mar.Com Systems Ltd
Telecommunications - Mitel Telecom Ltd
Telecommunications - Vodafone Paging Ltd
Telemetry control - Baxall Ltd
Telephone enclosures - Midland Alloy Ltd
TV/ Telephone systems for hospitals - Wandsworth Elecrtical Ltd
Video conferencing - Planet Projects Ltd
Voice recording, processing and training enhancement software - Computertel Ltd

W11 Paging/ Emergency call

Call systems - C-Tec (Computionics Ltd)
Data Paging systems - Channel Safety Systems Ltd
Dispersed alarms - Telealarmcare Ltd
Doctor Call systems - Channel Safety Systems Ltd

Intercoms - PEL Services Limited
Nurse call and on-site paging systems - Blick Communication Systems Ltd
Nurse call systems - Channel Safety Systems Ltd
Nurse call systems - Madewel Electronics Systems Ltd
Paging - Wandsworth Elecrtrical Ltd
Staff paging/ location - Johnson Controls Systems Ltd
Warden call systems - Telelarmcare Ltd

W12 Public address/ Conference audio facilities

Background music - PEL Services Limited
Conference equipment - Anders + Kern UK Ltd
Events & Confrencing Mgmt. - Planet Projects Ltd
Intruder alarm and CCTV - PEL Services Limited
Public address - Blick Communication Systems Ltd
Public address - PEL Services Limited
Public address/conference audio - Mar.Com Systems Ltd
Sound Masking Systems - Acustic + Environmental Technolgy

W20 Radio/ TV/ CCTV

Aaudio visual system design and installation - Visua Ltd
Audio Visual - Anders + Kern UK Ltd
Audio visual equipment - Visua Ltd
Audio visual equipment systems - TWS Presentation Systems (Mcmillan UK Ltd T/A)
C.C.T.V - Sampson & Partners Fencing
CCTV - Abel Alarm Co Ltd
CCTV - Allen (Fencing) Ltd
CCTV - Allgood plc
CCTV - General Security Systems Ltd
CCTV - Hambo Contrywide Security PLC
CCTV - Prima Security & Fencing Products
CCTV - UK Fire International Ltd
CCTV - Ward's Flexible Rod Company Ltd
CCTV - White Gates
CCTV cameras - Baxall Ltd
CCTV cameras, switches, multiplexers and telemetry control video transmission systems - Baxall Ltd
CCTV equipment - Hunting Engineering Ltd
CCTV systems - Interphone Security Group PLC
CCTV systems - Viking Security Systems Ltd
CCTV Video Systems - ISOKERN (UK) Ltd
CCTV's - Initial Electronic Security Systems Ltd
CCTV's - Madewel Electronics Systems Ltd
CCTVs, fastscan video and audio telephone transmission system - Vision Sytems (Europe) Ltd
Closed circuit television - Crimegard Ltd
Closed circuit television - Vicon Industries (UK) Ltd
Closed circuit television - Vidionics Security Systems Ltd
Design, manufacture and installation of audio visual systems - Wilson & Garden Ltd
Induction loop amplifiers - C-Tec (Computionics Ltd)
Monitoring - Mar.Com Systems Ltd
Presentation and conference equipment - Anders + Kern UK Ltd
Radio/ TV/ CCTV - Johnson Controls Systems Ltd
Radio/CCTV - Mar.Com Systems Ltd
Transmission poles - Stainton Metal Co Ltd
TV distribution - Blick Communication Systems Ltd
Video transmission systems - Baxall Ltd
Visual communications systems - Quartet-GBC

W21 Projection

Projection screen systems - Harkness Hall Ltd
Projections - Mar.Com Systems Ltd

W22 Information/ Advertising display

Banner installations - Marine & Commercial Rigging Systems
Construction information on CD-Rom. - Technical Indexes Ltd

W23 Clocks

Clock towers, cupolas and turrets - Goods Directions Ltd
Clocks - General Time Europe
Clocks - Lesco Products Ltd
Clocks - Internal & External - Goods Directions Ltd
Space management systems for power on manual, vertical or horizontal movement - Harkness Hall Ltd
Timer switches - Schlumberger
Timer switches for central heating and controls for central heating - Tower Flue Componants Ltd
Timers - Elkay Electrical Manufacturing Co Ltd

W30 Data transmission

Cables - Alcatel Data Cable
Construction software - Computer Foundations Ltd
Data cabling - Davies John Interiors Ltd
Data entry and calculating of bills of quantities or other volume information - Sumlock J.H.P.
Data Transmission - Mar.Com Systems Ltd
Electric information systems - Ferrograph Limited
Electronic information display systems - Techspan Systems PLC
Job costing, accounts computer software - Arena Software Ltd
Software for the construction industry - Masterbill Micro Systems Ltd
Take off, bills of quantities and estimating for the PQS or builder - EITE & Associates

W40 Access control

Access Control - Abel Alarm Co Ltd
Access control - Allen (Fencing) Ltd
Access control - Allgood plc
Access control - City Hardware and Security
Access Control - Crimegard Ltd
Access control - Garog
Access control - General Security Systems Ltd
Access control - Guardall
Access control - Hambo Contrywide Security PLC
Access control - Honeywell Control Systems Ltd
Access control - Initial Electronic Security Systems Ltd
Access control - Johnson Controls Systems Ltd
Access control - Laidlaw Architectural Hardware
Access control - Madewel Electronics Systems Ltd
Access control - PEL Services Limited
Access control - Permclose Group PLC
Access Control - Pike Signals Ltd
Access control - Sampson & Partners Fencing
Access control - UK Fire International Ltd
Access control - Vidionics Security Systems Ltd
Access control - Viking Security Systems Ltd
Access control & security products - Clarke Instruments Ltd
Access control and security - Dryad Architectural Ltd
Access control and visual imaging pass production systems - CEM Systems Ltd
Access control equipment - Haffele UK Ltd
Access control personal barriers - Electro Mechanical Systems Ltd
Access control products - Abloy Security Ltd
Access Control Security Solutions - Laidlaw Architectural Hardware Ltd
Access Control System Design - Clarke Instruments Ltd
Access control systems - Cardkey Systems
Access control systems - Interphone Security Group PLC
Access control systems - Pac International Ltd
Access control systems - Sabre Access Control Systems Ltd
Access control systems - Time and Data Systems International Ltd (TDSI)
Access control systems - Wandsworth Elecrtrical Ltd
Access control systems - Yale Security Products Ltd
Access control systems / equipment - APT Controls Ltd
Access control systems, ID and access card production and registered key systems - Kaba UK Ltd
Access control, exit devices, electromagnetic locks and electric strikes - Relcross Ltd
Alarm monitoring - Cardkey Systems
Auto sliding gates - Allen (Fencing) Ltd
Automatic doors - Permclose Group PLC
Automatic doors - Stanley Access Technologies
Automatic gates, barriers and railings - Parsons Brothers Steelwork
Automatic traffic barriers, gates and doors - Crimegard Ltd
Barriers and fully automated equipment - Prima Security & Fencing Products
Bird control - Timberwise (UK) Plc
Call systems - Wandsworth Elecrtrical Ltd
Car park barriers - Maswell Engineering Ltd
Controlled access - Barrow Hepburn Sala Ltd
Door & window control equipment - Geze UK
Door and gate automation - Automatic Doors & Gates
Door control / Accessories - Abloy Security Ltd
Door controls and locking devices - Laidlaw Architectural Hardware Ltd
Door entry systems - Aardee Locks and Shutters Ltd
Door entry systems and enclosures - Legrand Electric Ltd
Electric openers/ spare parts - Cardale Doors Ltd
Electric releases - Clarke Instruments Ltd
Electrical garage door accessories - Garador Ltd
Electrical remote control window opening systems - Morse Controls Ltd
Electromagnetic closers - Jebron Ltd
Electronic security equipment - Racal Guardall Ltd
Entrance control solutions - Gunnebo Mayor Ltd
Entrance systems - Doors & Hardware Ltd
Gate Automation - Clarke Instruments Ltd
High speed vertical folding and roll -up doors - Mandor Engineering Ltd
Installation Service - Clarke Instruments Ltd
Intercom - Wandsworth Elecrtrical Ltd
Maintenance service - Clarke Instruments Ltd
Mechanical and electrical locks - ASSA Ltd
Photo ID systems - Cardkey Systems
Proximity access control systems - BPT Security Systems (UK) Ltd
Security access and control - Hart Door Systems Ltd
Security roller shutters - Lakeside Group
Sliding Door gear - Haffele UK Ltd
Strongrooms and strongroom doors - Rosengerns Tann Ltd
Turnstiles - Allen (Fencing) Ltd
Turnstiles - Clarke Instruments Ltd
Video and audio entry access control systems - BPT Security Systems (UK) Ltd

W41 Security detection and alarm

Access control and security - Dryad Architectural Ltd
Alarm systems - Cooper Security
Burglar alarms - Initial Electronic Security Systems Ltd
Cable warning - Boddingtons Ltd
Door security locks - Haffele UK Ltd
Electronic security services - West London Security
Fall arrest systems - Latchways plc
Gates and Barriers - Hart Door Systems Ltd
Installation and maintenance of intruder alarms - Rouse Security Service
Integrated security systems - Vidionics Security Systems Ltd
Intruder Alarms - Crimegard Ltd
Intruder alarms - UK Fire International Ltd
Intruder detection - Guardall
Intruder detection and alarm panels - DA Systems Ltd
Lone worker protection - Telelarmcare Ltd
Passive infra red detectors and photoelectric beams - Optex (Europe) Ltd
Personal security system - Wandsworth Elecrtrical Ltd
Remote verifications alarms and passive infra detection - Vision Sytems (Europe) Ltd
Security systems - Pac International Ltd
Security ans surveillance - Telford Doors and Barriers Ltd
Security detection - Caradon Friedland Ltd
Security detection - Honeywell Control Systems Ltd
Security detection - Madewel Electronics Systems Ltd
Security detection and alarm - Castle Care-Tech
Security detection and alarm - Johnson Controls Systems Ltd
Security detection and alarm systems - Abel Alarm Co Ltd
Security detection and alarm systems - ADT Modern Security Systems Ltd
Security detection and alarm systems - Alarmcomm Ltd
Security detection and alarm systems - C & K Systems
Security detection and alarm systems - C-Tec (Computionics Ltd)
Security detection and alarm systems - Hambo Contrywide Security PLC
Security detection and alarms - Geoquip Ltd
Security detection and intruder alarms - General Security Systems Ltd
Security doors - Jewers Doors Ltd
Security equipment and alarms - ADT Security Systems
Security equipment and alarms - Advanced Perimeter Systems
Security equipment, cameras, multiplexes and phoneline video transmission systems - Robot (UK) Ltd
Security fencing - C & R Building Systems Ltd
Security fencing - C & W Fencing Ltd
Security grilles - Ring-Gard (UK) Ltd
Security management systems and high security locks - Relcross Ltd
Security markers - Elliott John H (Monostamp) Ltd
Security mirrors - Smart F & G (Shopfittings)Ltd
Security products - Viking Security Systems Ltd
Security roller shutters - Lakeside Group
Security Systems - JSB Electrical PLC
Voice alarms - PEL Services Limited

W50 Fire detection and alarm

Break glass and call points - KAC Alarm Company Ltd
Emergency lighting - P4 Limited
Fire alarm equipment - Lab-Craft Ltd
Fire alarms - Fife Fire Engineers & Consultants Ltd
Fire alarms - Griffin and General Fire Services Ltd
Fire alarms - Initial Electronic Security Systems Ltd
Fire alarms and smoke detectors - Asco Extinguishers Co Ltd
Fire and gas control systems - Zellweger Analytic Ltd, Sieger Division
Fire detection - Abel Alarm Co Ltd
Fire detection - Honeywell Control Systems Ltd
Fire detection and alarm - Johnson Controls Systems Ltd
Fire detection and alarm - PEL Services Limited
Fire detection and alarm - Photain Controls Plc
Fire detection and alarm systems - Blick Communication Systems Ltd
Fire detection and alarm systems - Caradon Gent Ltd
Fire detection and alarm systems - Channel Safety Systems Ltd
Fire detection and alarm systems - Dicon Safety Products (UK) Ltd
Fire detection and alarm systems - Fire Protection Services PLC
Fire detection and alarm systems - Firetecnics Systems Ltd
Fire detection and alarm systems - Hochiki Europe (UK) Ltd
Fire detection and alarm systems - Hygood Ltd
Fire detection and alarm systems - Siemens Building Technologies Ltd, Cerberus Division
Fire detection and alarm systems - Tann Synchronome
Fire detection and alarm systems - Wormald Ansul UK Ltd
Fire detection and alarms - C-Tec (Computionics Ltd)
Fire detection equipment - Nittan (UK) Ltd
Fire detection products - McMillan Fire Alarm Systems Ltd
Fire detection systems - Emergi-Lite Safety Systems Ltd
Fire detection Systems - JSB Electrical PLC
Fire detection systems - Menvier Ltd
Fire protection - Madewel Electronics Systems Ltd
Fire protection equipment - Preussag Fire Protection Ltd
Fire protection products for doors and building openings - Moffett Thallon & Co Ltd
Fire safety products and alarms - UK Fire International Ltd
Fire screens - Parkwood Mellows Ltd
Gates and Barriers - Hart Door Systems Ltd
Radio Fire Alarm systems - Channel Safety Systems Ltd
Smoke and heat detectors - Apollo Fire Detectors Ltd
Smoke detection, carbon monoxide and heat alarms - BRK Brands Europe Ltd
Very early smoke detecting apparatus - Vision Sytems (Europe) Ltd

W51 Earthing and bonding

Earth rods and earthing systems - Erico Europa (GB) Ltd
Earthing and bonding - Cutting R C & Co
Earthing protection - Furse W J & Co Ltd

W52 Lightning protection

Lightning protection - Cutting R C & Co
Lightning protection - Furse W J & Co Ltd
Lightning protection, earthing and electronic protection - Bacon WM & RW Ltd

W53 Electromagnetic screening

Anti static equipment - Meech Static Eliminators Ltd
Patient tagging system - Wandsworth Elecrtrical Ltd

W60 Central control/ Building management

Building automation - Guthrie Douglas Ltd
Building Automation - Johnson Controls Systems Ltd
Building control and management systems - Giehews Building Technologies Ltd. Landis & Staefa Division
Building management systems - Honeywell Control Systems Ltd
Central control - Johnson Controls Systems Ltd
Central control software - Telelarmcare Ltd
Central control systems - Honeywell Control Systems Ltd
Community care alarm and monitering equipment - Tunstal Telecom Ltd
Control equipment - George Fischer Sales Ltd
Control systems - Technical Control Systems Ltd
Electric radiator controls - Loblite Ltd
Energy saving controls - Caradon Plumbing Solutions
Home control products - Caradon Friedland Ltd
Industrial controls, air compressor controls and building management systems - Rapaway Energy Ltd
Lighting control systems - Electrak International Ltd
Meter Boxes - Talbot
Monitoring - Initial Electronic Security Systems Ltd
Portable moisture, temperature and noise level meters - Channel Electronics
Remote management and monitoring - Guardall
Urinal filters - Cistermiser Ltd
Voice recording, processing and training enhancement software - Computertel Ltd
Wireless Internal system - Telelarmcare Ltd

X10 Lifts

Access, hydraulic and towable lifts - Denka International
Goods, passenger, car, freight lifts, hydraulic or electric traction - Porn Dunwoody ERS (Lifts & Escalators) Ltd
Hydraulic lifting equipment - Hymo Ltd
Hydraulic passenger lifts, access wheelchair platforms lifts - Stannah Lifts Ltd
Installation, reburbishment and repair of lifts - Cable Lift Installations
Lift Maintenance - Express Evans Lift
Lift maintenance, lift modernisation - Express Evans Lifts

Lift manufacturers, installers service and maintenance - Pickerings Ltd
Lifts - Ellis & McDougall Lifts Ltd
Lifts - Express Evans Lift
Lifts - Kone Lifts Ltd
Lifts - Otis plc
Lifts - Schindler Ltd
Mechanical screws, electro hydralic goods and lift platforms - Supreme Lifts Ltd
New lifts 5kg service two ton goods lifts - Express Evans Lifts
Passenger, scenic, goods and vehicle lifts - Oakland Elevators Ltd
Puplic access and domestic stairlifts - Thyssen Stairlifts Ltd
Service lifts, maintenance - Oakland Elevators Ltd
Through floor home lift and service lifts - Stannah Lifts Ltd
Utility Lift - Bison Bede Ltd
Vertical, stair, and public access lifts - Brooks Stairlifts Ltd
Wheelchair platform stairlifts - Bison Bede Ltd

X11 Escalators

Escalators - Kone Lifts Ltd
Escalators - Otis plc
Escalators - Schindler Ltd
Escalators and passenger conveyers, service and maintenance - O & K Escalators

X12 Moving pavements

Moving pavements - Schindler Ltd
Passenger conveyors - Kone Lifts Ltd
Transit Systems - Otis plc
Stair Lifts - Bison Bede Ltd

X20 Hoists

Access lifting equipment - Genie UK Ltd
Access platforms - Kobi Ltd
Hoists - Metreel Ltd
Hoists - Morris Mechanical Handling Ltd
Hoists - PCT Group Sales Ltd
Hoists, davits, tripods, confined space access equipment and manhole cover lifter - Didsbury Engineering Co Ltd
Hydraulic movement systems - Hevilifts Ltd
Lifting and moving whole buildings - Structural Bonding Cambridge Ltd
Lifting equipment, systems for stage and studio applications - Harkness Hall Ltd
Permanent access cradles - Kobi Ltd
Ropes and sashcords - Marlow Ropes Ltd
Vertical lifting platform - Bison Bede Ltd

X21 Cranes

Cranes - Metreel Ltd
Cranes - Morris Mechanical Handling Ltd
Cranes - PCT Group Sales Ltd
Mini Crane Hire - HE Services (Plant Hire) Ltd

X22 Travelling cradles/ gantries/ ladders

Access ladders - Clow Group Ltd
Gantries - Clow Group Ltd
Travelling gantries - Kobi Ltd
Travelling ladders - Kobi Ltd

X23 Goods distribution/ Mechanised warehousing

Automated warehouse equipment - Morris Mechanical Handling Ltd
Conveyors - Metalrax Ltd
Mobile food conveyors - TSE Brownson

X31 Pneumatic document conveying

Air tube systems - Lamson, D D, PLC
Conveying systems - D. D Lamson plc
Mailroom equipment - Envopak Group Ltd
Mailroom equipment - Neopost Ltd

X32 Automatic document filing and retrieval

Automated systems - Dexion Ltd
Retrieval Systems - Barrow Hepburn Sala Ltd

Y10 Pipelines

Corrugated steel buried structures and pipes - Asset International Ltd
GRE GRP pipes, tubes and fabrications - Fibaflo Reinforced Plastics Ltd
Main Fittings - Talbot
Mechanical joints - Victaulic Systems
Pipe clips - Unifix Ltd
Pipe couplings - Viking Johnson
Pipelines - ARC Pipes
Pipelines - Deepdale Engineering Co Ltd
Pipelines - Grant & Livingston Ltd
Pipelines - IMI Yorkshire Copper Tube Ltd
Pipelines - Insituform Permaline Ltd
Pipelines - John Davidson (Pipes) Ltd

Pipes ductile iron - Biwater Industries Ltd
Pipework - Woodburn Engineering Ltd
Plastic pipelines - IPPEC Sytsems
Plumbing accessories - BSS Zenith
Polyethylene plastic pipes - Wirsbo
Service fittings - Talbot
Spiral weld steel tube - Tex Steel Tubes Ltd
Stainless steel fittings, flanges and fabrications - Lancashire Fittings Ltd
Stainless steel tubes and sections - Avesta Sheffield Ltd
Stainless steel valves - Lancashire Fittings Ltd
Structured wall HDPE pipes - Asset International Ltd

Y11 Pipeline ancillaries

Accessories - Talbot
Automatic Flow Balancing valves - Hattersley Newman Hender Ltd
Balencing valves - Crane Fluid Systems
Ball valves - Crane Fluid Systems
Ballvalves - Warner H & Son Ltd
Bronze & Cast Iron double regulating valves and commissioning sets - Hattersley Newman Hender Ltd
Bronze thermostatic, wheel head & lockshield radiator valves - Hattersley Newman Hender Ltd
Bronze, cast iron and ductile gate globe, check and butterfly valves - Hattersley Newman Hender Ltd
Butterfly valves - Crane Fluid Systems
Check valves - Crane Fluid Systems
DZR & Bronze ball valves - Hattersley Newman Hender Ltd
Fire and gas safety valves - Teddington Controls Ltd
Fitttings, valves and tanks - CPV Ltd
Gate valves - Crane Fluid Systems
Globe valves - Crane Fluid Systems
Hydrant valves and pressure valves - Eagles, William, Ltd
Isolation valves and expansion vessels - SAV United Kingdom Ltd
Malleable Iron pipe fittings - Crane Fluid Systems
Mixer Valves - Gummers A & J Ltd
Pipe clips and brackets - Regon Ltd
Pipeline ancillaries - Deepdale Engineering Co Ltd
Pipeline ancillaries - Flex-Seal Couplings Ltd
Pipeline ancillaries - Hamworthy Heating Ltd
Pipeline ancillaries - Isolated Systems Ltd
Pipeline ancillaries - John Davidson (Pipes) Ltd
Plug valves, control valves, gate valves, butterfly valves, check valves and consistency transmitters - Dezurik International Ltd
Plumbing accessories - Ashmead Buidling Supplies Ltd
Plumbing accessories - Ashmead Building Supplies Ltd
Plumbing accessories - BSS Zenith
Plumbing fittings, pipes, tools and brassware - Hipkin F W Ltd
Plumbing products and pipe fixings - IBP Conex
Radiator valves - Crane Fluid Systems
Relief valves - Crane Fluid Systems
Sealed system equipment, air vents, pressure reducers, mixing valves, flow switches etc. - Altecnic Ltd
Strainers - Crane Fluid Systems
Thermostatic and manual shower controls - Gummers A & J Ltd
Valves - Faral Radiators
Valves - Guest & Chrimes Ltd
Valves - Pegler Ltd
Waste fittings - Midland Stom Ltd
Water valves and plumbing equipment - Barber Edward & Co Ltd
White petroleum jelly - Legge Thompson F & Co Ltd

Y20 Pumps

Chemical resistant pumps - Plastic Construction Fabrications Ltd
Circulating, heating and ventilating pumps - Biral Pumps Ltd
Diaphragm hand pumps - Pump International Ltd
Fire Hose reel and wet riser boosters - Pillinger G C & Co (Engineers) Ltd
Fixing systems for building services - Erico Europa (GB) Ltd
Gear and centrifugal pumps - Stanhope Barclay Kellet Crown Division
Grout pumps - Sidewinder Concrete Pumps Ltd
GRP packaged pumping stations - Pims Pumps Ltd
Pressure sets - Fordwater Pumping Supplies Ltd
Process pumps - Sidewinder Concrete Pumps Ltd
Pump repair - Turney Turbines (pumps) Ltd
Pumps - Albany Engineering Co Ltd
Pumps - Andrews Sykes Hire Ltd
Pumps - Barr & Wray Ltd
Pumps - Beeston Heating Group Ltd
Pumps - Bran & Luebbe Ltd
Pumps - Calpeda Limited
Pumps - Fordwater Pumping Supplies Ltd
Pumps - Girdlestone Pumps Ltd
Pumps - Goodwin HJ Ltd
Pumps - Grosvenor Pumps Ltd
Pumps - Ingersoll-Rand European Sales Ltd
Pumps - Mono Pumps Ltd
Pumps - New Haden Pumps Ltd
Pumps - T-T Pumps & T-T controls
Pumps - Turney Turbines (pumps) Ltd

Pumps - Wilo Samson Pumps Ltd
Pumps, fountains, pressure sets and waste water pumping sets - Fordwater Pumping Supplies
Pumps, fountains, pressure sets and waste water pumping sets - Fordwater Pumping Supplies Ltd
Pumps, mixers and aerators - ABS Pumps Ltd
Submersible electric pumps and mixers - ITT Flygt Ltd
Submersilble pumps, servicing, maintenance and technical back up - Pump Technical Services Ltd
Tunnel Concrete Pumps - Sidewinder Concrete Pumps Ltd
Water pumps - Stuart Turner Ltd
Water, chemical, submersible, magnetic, portable and heavy duty pumps - Baresford Pumps

Y21 Water tanks/ cisterns

Carbon and stainless steel storage tanks - Cookson and Zinn Ltd
Cold water storage - Derwent Macdee Ltd
Domestic hot water tanks - Elsy & Gibbons Ltd
Domestic hot water tanks - Howden H D Ltd
Fibreglass water storage tanks - Parton Fibreglass Ltd
General GRP moulding - Boatman Plastics Ltd
GRP conical settlement tanks, covers and effluent treatment - Armfibre Ltd
GRP housings - Boatman Plastics Ltd
GRP one piece and sectional water storage tanks - Dewey Waters Ltd
GRP one piece, semi sectional and sectional water tanks. - Boatman Plastics Ltd
GRP storage cisterns - Precolor Sales Ltd
GRP storage tanks - APA
GRP tank housings - Group Four Glassfibre Ltd
GRP tanks - Fibre Reinforced Products Ltd
GRP water storage tanks - Group Four Glassfibre Ltd
Hot water cylinders and tanks - Gledhill Water Storage Ltd
Hot water storage tanks - Quigley Metal Products Ltd
One piece/ semi sectional/ sectional GRP water storage tanks - Aquatanks
Plastic cisterns - Dudley Thomas Ltd
Plastic water storage tanks - Ferham Products
Plastic water tanks, plastic oil tanks and plastic septic tanks - Tyrrell Tanks Ltd
Sectional and one piece GRP cold water tanks - Nicholson Plastics Ltd
Storage tanks - Franklin Hodge Industries Ltd
Storage tanks and vessels - Metcraft Ltd
Tanks - Balmoral Group Ltd
Tanks - Braithaite Engineers Ltd
Tanks - FIBAFORM PRODUCTS
Tanks - Glasdon Designs Ltd
Water storage containers and accessories - Rexam Harcostar
Water storage systems and cylinders - T & D Plastech
Water storage tanks - Brimar Plastic Fabrications Ltd
Water storage tanks - Vulcan Tanks Ltd
Water tanks - Harris & Bailey Ltd
Water tanks/ cisterns - Grant & Livingston Ltd
Water tanks/ cisterns - Harvey Fabrication Ltd

Y22 Heat exchangers

Corrugated shell and tube heat exchanger - HRS Hevac Ltd
Heat exchangers - Grant & Livingston Ltd
Heat exchangers - Rycroft Ltd
Heat recovery - Kiltox Damp Free Solutions
Manufacturer of heating elements - Electric Elements Co, The
Plate heat exchangers - Beeston Heating Group Ltd

Y23 Storage cylinders/ Calorifiers

Calorifiers and pressurising equipment - Heat Transfer Pressurisation
Carbon and stainless steel pressure vessels - Cookson and Zinn Ltd
Cold water storage cisterns - Davant Products Ltd
Corrugated shell and tube heat exchanger - HRS Hevac Ltd
Storage cylinders/ calorifiers - Grant & Livingston Ltd
Storage cylinders/ calorifiers - Harris & Bailey Ltd
Storage cylinders/ calorifiers - Harvey Fabrication Ltd
Storage cylinders/ Calorifiers - Rycroft Ltd
Storage cylinders/calorifires - Hamworthy Heating Ltd
Unvented cylinders - Ferham Products
Water heaters/calorfirers - Flamco UK Ltd

Y24 Trace heating

Electric space heating - Bush Nelson PLC
Self- regulating cut-to-length heater tapes - Jimi-Heat Ltd
Sone parallel constant wattage heater tapes - Jimi-Heat Ltd
Trace Heating - Heat Trace Ltd

Y25 Cleaning and chemical treatment

Building, cleaning and restoration - Cleanstone Co (Bradford) Ltd
Central heating system corrosion protectors, cleansers + leak sealers - Fernox Fry Technology UK
Central heating system corrosion protectors, Cleansers and leak sealers - Alpha Fry Ltd
Chemical photo etchings - Photo Fabrications Services Ltd
Chemical, Solvents, Greases and Lubricants - Ambersil Ltd
Chemicals - Baxenden Chemicals Ltd
Chemicals - Baxenden Chemicals Ltd
Chemicals - DSM UK Ltd
Chemicals - P & R Laboratory Group Ltd
Chemicals - Rhone-Poulenc Chemicals Ltd
Chlorination of water services - Aquastat Ltd
Cleaning and chemical treatment - Hertel Services
Cleaning and chemical treatments - Deb Ltd
Cleaning chemicals - Certifield Laboratories (Div or CPS Industries Ltd)
Cleaning materials - Evans Vanodine International PLC
Construction chemicals - Arthur Fisher (UK) Ltd
Detergents - Evans Vanodine International PLC
Disinfectants - Evans Vanodine International PLC
Disinfection systems - Portacel
Grit blasting, chemical cleaning, high pressure water jetting - Maxblast UK Ltd
Industrial Aerosols - Ambersil Ltd
Industrial chemical supplies - IFT Supplies Ltd
Limescale remedies, jointing compounds + quick repair products - Fernox Fry Technology UK
Limescale remedies, jointing compounds, quick repair products and solder & flux - Alpha Fry Ltd
Office and industrial cleaning - CSL (Services) Limited
Oil water separators - Furmanite International Ltd
Polishes - Evans Vanodine International PLC
Polishes and floor cleaners - S. C. Johnson Professional
Specialist Pigments : Artist colours - Hawley W & Son Ltd
Specialist Pigments : Asphalt - Hawley W & Son Ltd
Specialist Pigments : Concrete paving - Hawley W & Son Ltd
Specialist Pigments : Concrete roofing - Hawley W & Son Ltd
Specialist Pigments : Concrete walling - Hawley W & Son Ltd
Specialist Pigments : Mortar - Hawley W & Son Ltd
Specialist Pigments : Paint - Hawley W & Son Ltd
Specialist Pigments : Plastic - Hawley W & Son Ltd
Stone and brickwork cleaning - CSL (Services) Limited
U.V treatment - Entec (Polution Control) Ltd
Water treatment chemicals - Tower Flue Components Ltd
Water treatment products and services - Aldous & Stamp Services Ltd

Y30 Air ductlines/ ancillaries

Air ductlines/Ancillaries - Isolated Systems Ltd
Aluminium air ductwork - Flowline Grilles
Connection and sleves for air ducts - Westbury Filters Limited
Ducting - Alumasc Interior Building Product Limited
Ducting - MacLellan Rubber Ltd
Ducting and precision sheet metalwork - Ductform Engineering Ltd
Fire rated ductwork systems - Fire Protection Ltd
Flexible ducting - Flexible Ducting Ltd
Rubber mouldings and elastomeric bearings - MacLellan Rubber Ltd
Sheet metal ductwork - Hi-Vee Ltd

Y40 Air handling units

Air conditioning - Dantherm Ltd
Air Handling Unit (AAF Ltd) - McQuay UK Ltd
Air handling units - Dunham Bush Ltd
Air handling units - Ellis Miller Ltd
Air handling units - Hi-Vee Ltd
Air Handling Units - Senior Air Systems
Air handling units - Therminal Technology (sales) Ltd
Air handling units - Ves Andover Ltd
Air Management - Actionair
Connect grid and air handling - Ecophon Ltd
Curve and coffers - Ecophon Ltd
Displacement Ventelation - Senior Air Systems
Integral emergency lighting - Lumitron Ltd

Y41 Fans

Axial, centrifugal, mixed flow fans and dust extraction units - Howden Buffalo
Axial, roof extract and centrifugal fans - Matthews & Yates Ltd
Chimney extractor fans - Kedddy (Poujoulat) (UK) Ltd
Crossflow fans - Consort Equipment Products Ltd
Destratification fans - S & P Coil Products Ltd
Domestic and commercial and industrial fans - Xpelair Ltd

Domestic and industrial fans - Airflow Developments Ltd
Fans - BOB Stevenson Ltd
Fans - Elta Fans Ltd
Fans - Gebhardt Kiloheat
Fans - Klaxon Signals Ltd
Fans - Mawdsley's Ltd
Fans - McKenzie-Martin Ltd
Fans - Power Breaker
Fans - Roof Units Group
Fans - Shipley Fan Co Ltd The
Fans - Vortice Ltd
Fans and motors - Ziehl-Ebm (UK) Ltd
Industrial axial fans and impellers - London Fan Company Ltd
Industrial fans - ABB Flakt Products
Industrial fans units - Radial & Axial Fans
Industrial, bifurcated, centrifugal and axial fans - Twin County Fan Company Ltd
Jet tunnel fans - Matthews & Yates Ltd
Plastic fans - Plastic Construction Fabrications Ltd
Road fans - Matthews & Yates Ltd
Road tunnel fans - Matthews & Yates Ltd
Twin fans - Ves Andover Ltd

Y42 Air filtration

Air filtration - Air Improvement Centre Ltd
Air ailtration - Bassaire Ltd
Air cleaners - Warner Howard Ltd
Air cleaning equipment - Horizon International Ltd
Air filters - Auchard Development Co Ltd
Air filters - BTR Environmental Ltd Vokes
Air filters - Nationwide Filter Co Ltd
Air filters - Universal Filters Ltd
Air filtration - Estec Environmental Ltd
Air filtration - Farr Filtration Ltd
Air Filtration - Vortice Ltd
Air filtration (McQuay UK Ltd) - McQuay UK Ltd
Air purifiers - Amcor (Appliances) Ltd
Air purifiers - Purified Air Ltd
Air quality testing - Westbury Filters Limited
Carbon filters - Westbury Filters Limited
Filter loss gauges - Westbury Filters Limited
Panel filters - Westbury Filters Limited
Pleateed filters - Westbury Filters Limited
Roll filters - Westbury Filters Limited
Slimline filters - Westbury Filters Limited
Washable filters - Westbury Filters Limited

Y43 Heating/ Cooling coils

Coil heat exchangers - S & P Coil Products Ltd
Heat exchangers gasketted plate - HRS Hevac Ltd
Heaters - Klaxon Signals Ltd

Y44 Air treatment

Air treament - Air Improvement Centre Ltd
Air treatment - Hertel Services
Dehumidifier and air treatment - Amcor (Appliances) Ltd
Dehumification - Dantherm Ltd
Humidifiers - Eaton-Williams Group Ltd
Industrial humidifiers - Calomax (Engineers) Ltd
Moisture eliminators - Plastic Construction Fabrications Ltd
Ultra violet chambers, lamps and control panels - Hanovia Ltd

Y45 Silencers/ Acoustic treatment

Accoustic flooring - Monarflex Ltd
Acoustic cladding and lining products - Gilmour Ecometal
Acoustic Correction & Noise Control - Brown F plc
Acoustic curtains, attenuators, enclosures, panels and louvres - Acousticabs Industrial Noise Control Ltd
Acoustic doors and frames - Fitzpatrick Door Frames Ltd
Acoustic enclosures - Designed for Sound Ltd
Acoustic Flooring Products - Hillaldam Coburn Ltd
Acoustic flooring solutions - Durabella Ltd
Acoustic Inserts - Hotchkiss Air Supply
Acoustic insulation - Potters Insulations (Sales) Ltd
Acoustic insulation,wall ceiling panels,stretchwall systems, - Salex Interiors Ltd
Acoustic linings - Applied Acoustics (a part of Henry Venables Limited)
Acoustic panels and blocks in terracotta - Smithbrook Building Products Ltd
Acoustic products - Nendle Acoustic Company Ltd
Acoustic roofing & cladding - Axter Ltd
Acoustic solutions - Nevill Long Limited
Acoustic tiles - Treetex Acoustics Ltd
Acoustic Walling Products - Hillaldam Coburn Ltd
Attenuators/ silencers - Allaway Acoustics Ltd
Ductwork, silencers, acoustic products - Isolated Systems Ltd
Floating Floors - The Salex Group Ltd.
Hinged acoustic doorsets - Scandanavian Window Systems Ltd
Industrial Sound Protection solutions - Illbruck Ltd
Noise absorbing panels - The Salex Group Ltd.
Noise control systems - Acousticabs Industrial Noise Control Ltd
Noise resistant doors - The Salex Group Ltd.
Room Acoustic solutions - Illbruck Ltd
Silencers - Ves Andover Ltd
Silencers - Acoustic - Sound Service (Oxford) Ltd
Silencers, acoustic enclosures and telephone hoods - Environmental Acoustics Ltd
Timber acoustic panels and doors - Salex Interiors Ltd

Y46 Grilles/ Diffusers/ Louvres

Acoustic products - Nendle Acoustic Company Ltd
Air diffusers - HT Martingale Ltd
Anodised Aluminium coils - Metalloxyd Ano-Coil Ltd
Design and CAD support - Ecophon Ltd
Grilles and Diffusers and Louvres - Senior Air Systems
Grilles, diffusers and louvres - Grille Diffuser & Louvre Co Ltd The
Grilles, diffusers and louvres - Royair Ltd
Grilles/Diffusers/Louvres - Isolated Systems Ltd
Grills and diffusers - ABB Flakt Products
Integrated lighting systems - Ecophon Ltd
Louvre, hit and miss ventilators - Ryton's Building Products Ltd
Louvres - McKenzie-Martin Ltd
Smoke ventilation louvres - Matthews & Yates Ltd
Solar shading systems - Dales Fabrications Ltd

Y50 Thermal insulation

Aluzink Insulated Jacketing - SSAB Dobel Coated Steel Ltd
Bobelshield - SSAB Dobel Coated Steel Ltd
Cork/ Rubber - Olley & Sons Ltd, C
Expanded polystyrene insulation - Vencel Resil Ltd
Floor insulation - Springvale E P S
Garden Products - nmc (UK) Ltd
Insulation - Deborah Services
Insulation - Nevill Long Limited
Insulation - Rockwool Ltd
Insulation Cork - Olley & Sons Ltd, C
Insulation materials - Gradient Insulations (UK) Ltd
Roof insulation - Springvale E P S
Thermal insulation - Ainsworth B R (Southern) Ltd
Thermal insulation - Firebarrier Services Ltd
Thermal Insulation - Hawkins Insulation Ltd
Thermal Insulation - Hertel Services
Thermal Insulation - Isolated Systems Ltd
Thermal Insulation - Owens Corning Building Products (UK) Ltd
Thermal insulation - Potters Insulations (Sales) Ltd
Thermal insulation - Tavac Insulation Ltd
Wall insulation - Springvale E P S

Y51 Testing and commissioning of mechanical services

Cable and pipeline locators - Boddingtons Ltd
Carbon Monoxide - Detectors - Dicon Safety Products (UK) Ltd
Carbon Monoxide Detectors - Schlumberger
Contract Smoke Pellets, Matches & Detector Sprays - PH Products Ltd
Corbon Monoxide Detectors - PH Products Ltd
Gas Engineers Record Pads, Lables & Tape - PH Products Ltd
Gas Engineers Tools & Accessories - PH Products Ltd
Gas Leak Etector Spray & Fluids - PH Products Ltd
Leak Location Service - Boddingtons Ltd
Portable moisture, temperature and noise level meters - Channel Electronics
Silicone grease & spray WRI Approved - PH Products Ltd
Sleep Safe home Products - PH Products Ltd
Smoke Matches - PH Products Ltd
Smoke Pellets - PH Products Ltd
Testing facilities for GRP tanks - APA
Underground warning tape - Boddingtons Ltd

Y52 Vibration isolation mountings

Acoustic products - Nendle Acoustic Company Ltd
Anti - vibration mountings - MacLellan Rubber Ltd
Anti vibration mounting - Designed for Sound Ltd
Anti-vibration mounts - Isolated Systems Ltd
Anti-vibration pads - Olley & Sons Ltd, C
Machine mounting, anti vibration materials, structural and sliding bearings and resilient seatings - Tiflex Ltd
Noise and vibration control products for mechanical service engineers - The Salex Group Ltd.
Vibration isolation - Allaway Acoustics Ltd
Vibration Isolation - Sound Service (Oxford) Ltd
Vibration isolation mountings - Christie & Grey Ltd
Vibration sensors - Optex (Europe) Ltd

Y53 Control components - mechanical

Ancilleries - Dawson-Keith Ltd
Ancilliary Security Device - Securistyle Ltd
Building Controls - Satchwell Control Systems Ltd
Building Management Systems - Satchwell Control Systems Ltd
Control - T-T Pumps & T-T controls
Control components - Johnson Controls Systems Ltd
Control components - mechanical - Guthrie Douglas Ltd
Control equipment, timers and relays - Thomas & Betts Ltd
Controls - Ves Andover Ltd
Customer Training - Satchwell Control Systems Ltd
Diaphragm, butterfly, ball valves and actuators - Alfa Laval Saunders Ltd
Embeded Control Systems - MJ Electronics Services (International) Ltd
Enclosures - Knurr (UK) Ltd
Energy Control Products - Schlumberger
Energy Management Systems - Satchwell Control Systems Ltd
Fail safe actuators - Skil Controls Ltd
Hazardous Area Control Gear - PFP Electrical Products Ltd
Heating programmers - Timeguard Ltd
Millennium Management - Satchwell Control Systems Ltd
Moisture meters, hygrometers - Protimeter plc
Nexus Bureau - Satchwell Control Systems Ltd
Performance Contracting - Satchwell Control Systems Ltd
Process systems and analytical equipment - Bran & Luebbe Ltd
Service BMS Systems - Satchwell Control Systems Ltd
Thermometers, hygrometers, hydrometers, pressure and altitude gauges - Brannan S & Sons Ltd
Thermometers, pressure gauges and associated instruments - Brannan Thermometers & Gauges
Thermostats - Teddington Controls Ltd

Y59 Sundry common mechanical items

Air receivers - Flamco UK Ltd
Ancillary components - Llewellyn Homes Ltd
Balancing valves - Ellis Miller Ltd
Diesel engines - Martin (Marine) Ltd Alec
Exterior surface heating - Wirsbo
Flexible plumbing - Wirsbo
Fry's solder and plux - Fernox Fry Technology UK
Gas Fires - Spital Tile Co
Gaskets - Olley & Sons Ltd, C
GRP Kiosks & modular enclosures - Group Four Glassfibre Ltd
Lockshield radiator valves - Ellis Miller Ltd
Manual radiator valves - Ellis Miller Ltd
Manufacturerers of: Metal Cutting Circular Saws - Spear & Jackson Interntional Ltd.
O Rings, bonded seals, grease guns, grease nipples and bucket pumps - Ashton Seals Ltd
Parallel slide valves - IMI Bailey Birkett Ltd
Pressure reducing valves - IMI Bailey Birkett Ltd
Pressure regulating valves - IMI Bailey Birkett Ltd
Pulley lines - Lever James & Sons Ltd
Resharpners of Circular saw blades - Spear & Jackson Interntional Ltd.
Safety relief valves - IMI Bailey Birkett Ltd
Seals - Atkinson & Kirby Ltd
System 21 - the latest in enrgy efficiency - Servowarm
Thermostatic radiator valves - Ellis Miller Ltd
Valve Manufacturer - Millar Dennis (1998) Ltd
Wear, heat and impact resistant parts - Wearparts Ltd

Y60 Conduit and cable trunking

Cable management - Kem Edwards Ltd
Cable management - Walsall Cable Management Ltd
Cable management systems - Caradon M K Electric Ltd
Cable management systems - Rehau Ltd
Cable routing - Hellerman Electric, Member of Bopwthorpe Holdings PLC
Cable tray - Davis International Ltd
Cable trunking - Davis International Ltd
Conduit and trunking - Mita (UK) Ltd
Dado trunking - Salamandre Davis
Electrical cable management systems - Marshall-Tufflex Ltd
Flexible conduit systems - Adaptaflex Ltd
Floor trunking - Salamandre Davis
Flooring systems - Davis International Ltd
Ladder rack - Davis International Ltd
Lighting trunking - Davis International Ltd
Metal cable trunking - Legrand Power Centre
Perimeter trunking systems - Mita (UK) Ltd
Pipe supports - Unistrut Limited
Pre wired PVC conduit - Aercon Wiring Systems Ltd
Skirting trunking - Salamandre Davis
stainless steel conduit - Kensal CMS
Stainless steel trunking and conduit - Electrix (Northern) Ltd
Underfloor Cable Management - Brown F plc

Y61 HV/ LV cables and wiring

Cable glands and connectors - Elkay Electrical Manufacturing Co Ltd
Cable protection - Hellerman Electric, Member of Bopwthorpe Holdings PLC
Cables - Alcatel Data Cable
Cables and wiring accessories - Ellis Patents
Clips and brackets for pipes - Longbottom & Co (Keighley) Ltd
HV and LV cables and wiring - Concordia Electric Wire & Cable Co. Ltd
Insulated cables - BICC plc
Mineral insulated cables - BICC Thermoheat Ltd
Wiring accessories - Caradon M K Electric Ltd

Y62 Busbar trunking

Bus bar power distribution systems - Electrak International Ltd
Busbar trunking - MEM
Busbar Trunking - Walsall Cable Management Ltd
Busbar trunking systems - Barduct Ltd
Busbar trunking, low power - Salamandre Davis
Busbar, cable management systems for floors - Legrand Power Centre
HV and LV wiring product - Legrand Electric Ltd
stainless steel enclosures 1P65 - Kensal CMS
stainless steel trunking - Kensal CMS

Y63 Support components - cables

Cable ladders - Vantrunk Engineering Ltd
Cable routing - Hellerman Electric, Member of Bopwthorpe Holdings PLC
Cable support systems - Corofil Woodall Ltd
Cable support systems - Elwell Sections Ltd
Cable supports - EDL Cable Supports Ltd
Cable supports - Unistrut Limited
Cable ties and cable clips - Unifix Ltd
Cable tray - Ductile Sections
Cable tray - Electrix (Northern) Ltd
Cable tray - Vantrunk Engineering Ltd
Cable trays, cable ladder, channel support systems - Swifts Of Scarborough Ltd
Cleats and cable ties - Vantrunk Engineering Ltd
Clips and brackets for tubes and cables - Longbottom & Co (Keighley) Ltd
Control systems - Eltron Chromalox
GRP ladders and trays - Mita (UK) Ltd
Industrial cable glands, cleats and clamps - LPA Industries PLC
Plastic cable ties, accessories and security seals - Hellermann Insuloid
stainless steel cable tray - Kensal CMS
Support components-cables - Hadley Industries Plc
Support systems - Davis International Ltd
Wire basket cable tray - Vantrunk Engineering Ltd

Y70 HV switchgear

Circuit breakers - Schneider Ltd
Electrical services circuit breakers - Power Breaker
HV switchgear - Hager Powertech Ltd
HV switchgear - Hubbell Ltd
Industrial switch and fusegear - MEM
Residual current devices - Power Breaker
Switchboards - Technical Control Systems Ltd
Switchgear - Power Breaker

Y71 LV switchgear and distribution boards

Circuit protection - Caradon M K Electric Ltd
Consumer units - Schneider Ltd
Control panels - Skil Controls Ltd
Control products - Schneider Ltd
Distribution boards - Schneider Ltd
Domestic switch gear - MEM
Earth monitoring - FDB Electrical Ltd
Electrical control panels and systems - Warner H & Son Ltd
Electrical switchgear - Dorman Smith Switchgear Ltd
Enclosures - Elkay Electrical Manufacturing Co Ltd
LV electronic switchgear - MTE Ltd
LV switchgear - Klockner-Moeller Ltd
LV switchgear - T-T Pumps & T-T controls
LV switchgear and distribution boards - Hager Powertech Ltd
MV and LV cubicle switchboards - MEM
Panel boards - Schneider Ltd
Panel boards - Square D, Electrical Distribution Division
Pre-wired distribution units - Mita (UK) Ltd
RCDs - Power Breaker
Residual current circuit breakers - FDB Electrical Ltd
Residual current monitors - FDB Electrical Ltd
Residual current protested skt outlets - FDB Electrical Ltd
Switch and fusegear - Square D, Electrical Distribution Division
Switch boards - Square D, Electrical Distribution Division
Switchers - Baxall Ltd
Switches - Home Automation Ltd
Switchgear - Power Breaker
Switchgear - Whipp & Bourne
Three phase distribution board - Square D, Electrical Distribution Division
Transformers - Babcock Transformers

Y72 Contactors and starters

Electronic motor protection - MTE Ltd
LV motor control equipment - MTE Ltd
Safety guards photoelectric - MTE Ltd
Softstarters - MTE Ltd

Y73 Luminaires and lamps

Associated accessories - Bernlite Ltd
Cable management systems - Vantrunk Engineering Ltd
Coloured and clear covers for fluorescent lamps and fittings - Encapsulite International Ltd
Columns and brackets - Windsor Limited D W
Converters for cold cathode and ballasts - Mode Lighting (UK) Ltd
Decorative lighting systems - Gradus Ltd
Direct and indirect luminaires - Lumitron Ltd
Equipment and luminaires, control gear and accessories - Poselco Lighting
Flameproof EExd Fluorecents - PFP Electrical Products Ltd
Fluorescent and discharge lighting - Fitzgerald Lighting Ltd
Fluorescent lighting - Nevill Long Limited
General fluorescent lighting - Fluorel Ltd
Incandesant light bulbs and other domestic lighting - J.F.Poynter Ltd
Increased Safety Fluorecents - PFP Electrical Products Ltd
Lamps and electronic control gear - Osram Ltd
Light bulbs and fittings - British Electric Lamps Ltd
Light fittings - Marlin Lighting Ltd
Light sensors - Timeguard Ltd
Lighting components - Magnatek Ltd
Luminaires and lamps - Brandon Medical
Luminares & Lamps - Glamox Electric (UK) Ltd
Power distribution - Legrand Electric Ltd
Project management services - Ecophon Ltd
Refurbishment and upgrading service for existing luminaires - Poselco Lighting
Traffic signals, radar detectors, wigwag signals - Pike Signals Ltd
Zone 1 Floodlight - PFP Electrical Products Ltd

Y74 Accessories for electrical services

Accessories for electrical - Barton Engineering
Dimmers, switches, sockets, modern and traditional styles on brass some with concealed fixing, 'specials made' - Hamilton R & Co Ltd
Dimming switches - Mode Lighting (UK) Ltd
Din-rail time switches and surge protectors - Timeguard Ltd
Domestic electrical accessories and dry cell batteries and torches - J.F.Poynter Ltd
Electrical Accessories - Walsall Cable Management Ltd
Electrical products - Panasonic UK Ltd
Electrical swithces - Wandsworth Elecrtrical Ltd
Electrical wiring accessories - Tenby Industries Ltd
General electrical accessories - Ashley & Rock Ltd
General Electrical Products - WF Electrical Distributors
General electrical products - wholesale - Jolec Electrical Supplies
Invisible light switches and sockets - Forbes & Lomax Ltd
Light fittings - Marlin Lighting Ltd
Light switches - Redditch Plastic Products
Painted, nickel silver, brass and stainless steel switches and sockets - Forbes & Lomax Ltd
Plugs and socket outlets - Perma Industries Ltd
Switch and socket boxes - Legrand Power Centre
Switches - Teddington Controls Ltd
Thermocouples & RTDs - BICC Thermoheat Ltd
Time controllers, security light switches and RCD switched sockets - Timeguard Ltd
Track support Materials - Tiflex Ltd
Ultrasonic sensors - Thomas & Betts Ltd
Wire accessories - Hellerman Electric, Member of Bopwthorpe Holdings PLC

Y80 Earthing and bonding components

Cable identification - Hellerman Electric, Member of Bopwthorpe Holdings PLC
Detection and protection products - Boddingtons Ltd
Identification Labels - Mockridge Labels & Nameplates Ltd

Y89 Sundry common electrical items

Air circuit breakers - MEM
Automatic Pool covers - Golden Coast Ltd
Electrical Distribution Equipment - WT Henley Ltd
Electronic accessories - VELUX Co Ltd, The
Emergency lighting and accessories - Legrand Electric Ltd
Flat Wire in Mild, Carbon & Stainless - Lee Smith Wires ltd
Installation accessories - Mita (UK) Ltd
Memshield 2 circuit breaker systems - MEM
Rotary packet switches - MEM

Round Wire in Corbon & Spring Grades, Plain or galvanised - Lee Smith Wires ltd
stainless steel isolators - Kensal CMS
Trunking - Vantrunk Engineering Ltd
Wiring accessories - Mita (UK) Ltd

Y90 Fixing to building fabric

Fixings and fastenings - Tower Manufacturing
Frame cramps, restraint straps, sliding anchor stems - Vista Engineering Ltd
High Load Anchors - Exchem Mining and Construction
Insulation fixings and sainless steel helexial nails - Reiner Fixing Devices
Mild steel and non-ferrous nails, tacks and pins - John Reynolds & Sons (Birmingham) Ltd
Nails - Glasgow Steel Nail Co Ltd
Nails - Stone Fastemers Ltd
Nuts and bolts - Jaton
Rotary impact and masonry drills - Unifix Ltd
Screws, nuts and bolts - Nettlefolds Ltd
Stainless steel building components - Wincro Metal Industires Ltd
Through bolt fixings - Tucker Fasteners Ltd
Wood, coach, self tapping, fastbrolley screws, frame anchors, hammer and expansion plugs - Unifix Ltd

Y91 Off-site painting/ Anti-corrosion treatments

Anti-corrosion and sealing systems - Winn & Coales
Coatings - Thoro Systems Products Ltd
Corrosion protection for metal fastenings - Interlux Ltd
Corrosion protection, building maintenance coatings and composition - Wood E Ltd
Flame retardant treatment - Timber Treatments Ltd
Galvanised buckets and cans - Brettell & Shaw Ltd
Galvanisers - Joseph Ash Galvanizing
Hot dip galvanizing - Arkinstall Galvanizing Ltd
Hot dip galvinising - Ash Joseph Galvanizing
Inks and metal coatings - Valspar UK Corporation Ltd, The
Paint and powder coatings - Trimite Ltd
Powder coatings - HB Fuller Coatings Ltd
Printing solutions - Hellerman Electric, Member of Bopwthorpe Holdings Ltd
Protective coatings - MMP International
Thermoplastic coating powders - Plascoat Systems Ltd

Y92 Motor drives – electric

Automatic gates - Jackson H S & Son (Fencing) Ltd
Electric motor repairs - Turney Turbines (pumps) Ltd
Electrical motors - Klockner-Moeller Ltd
Motor control products - MEM
Motor drives (electric) - T-T Pumps & T-T controls
Motor Drives-electric - Danfoss Ltd
Motors - Electro Mechanical Systems Ltd

Z10 Purpose made joinery

Aluminium extrusions, standard (Angle, Bars, Tubes, Channels etc) and custom profiles available - Extrusions Direct
Architectural joinery - Timtec International Ltd
Architectural joinery manufacturers - Edmonds A & Co Ltd
Architectural joinery restoration oak, European hardwoods - Applied Acoustics (a part of Henry Venables Limited)
Architectural joinery - Henry Venables Ltd
Bespoke joinery - Howard Bros Joinery Ltd
Bespoke Joinery - Taskworthy
Bespoke museum display cases - Carlton Benbow
Blockboard - Wood International Agency
Concrete formwork plywood - Wood International Agency
Decorative wood moulding and carvings, cornice, picture rail, dado rail, skirting, architrave and pediments, radiator cabinet grilles - Browne Winther & Co Ltd
Film Faced plywood - Wood International Agency
Fire resisting joinery - Decorfix
Fixings- nuts and bolts - Lanarkshire Bolt
Hardboard - Wood International Agency
Hardwood plywood - Wood International Agency
Joinery - Grafton Magna Ltd
Machined softwood and machined MDF mouldings - Stantons Timber Services
Made to measure joinery - Dale Joinery Ltd
MDF - Wood International Agency
OSB - Wood International Agency
Pine room panelling, bookcases and display cupboards - Carved Pine Mantelpieces Ltd
Purpose made joinery - Carlton Benbow
Purpose made joinery - Babcock Joinery
Purpose made joinery - Boulton & Paul Ltd
Purpose made joinery - Decorfix
Purpose made joinery - SWJ Group Ltd
Softwood plywood - Wood International Agency
Specialist Bespoke Joinery - J. Scott (Thrapston) Ltd
Specialist Joinery - Andy Thornton Architectural Antiques Ltd

Specialist joinery - Barlow Shopfitting Ltd
Specialist joinery - Barnwood Shopfitting Ltd
Specialist joinery - Cumberland Construction
Specialist joinery manufacturers - Sandoms (Batley) Ltd
Specialist laminate fabrication - P & R Laboratory Group Ltd
Study, lounge, bedroom, office and reception fitted furniture - Hyperion Wall Furniture Ltd
Teak Specialist - Parker Kislingbury Ltd
Timber engineering services - Wyckham Blackwell (Joinery) Ltd
Timber specialist - Challenge Fencing & Ltd
Wooden moulding and carvings - Winther Browne & Co Ltd

Z11 Purpose made metalwork

Aluminium and steel fabricators - Ramsay & Sons (Forfar) Ltd
Aluminium extrusions, powder coating, anodising and fabrication - Kaye Aluminium plc
Aluminium fabrication - Panel Systems Ltd
Aluminium special structures - Alifabs (Woking) Ltd
Aluminium, bronzes, resistance welding alloys and cupro nickels - Cerro (Maganese Bronze) Ltd
Architectural metalwork - Grafton Magna Ltd
Architectural metalwork - Multitex GRP & GRG Products
Architectural metalwork - Pearsons J R (Birmingham) Ltd
Architectural metalwork - Sebry Staircase & Ironworks Ltd
Architectural metalwork - Woodburn Engineering Ltd
Architectural metalworkers - Edmonds A & Co Ltd
Architectural metalworks - Glazzard (Dudley) Ltd
Bespoke metalwork - Andy Thornton Architectural Antiques Ltd
Bespoke museum display cases - Carlton Benbow
Brass Fabrications - Andy Thornton Architectural Antiques Ltd
Brass handrails - Cerro Extruded Metals Ltd
Brick bond grilles - Priory Shutter & Door Co Ltd
Cast stanchions - Gabriel & Co Ltd
Cast steelwork - Gabriel & Co Ltd
Cattle grids, duct covers - Cameron & Moore (Gridplex) Ltd
Copper tubing - Wednesbury Tube
Custom sheet metal - Azimex Fabrications Ltd
Decorative Castings - Andy Thornton Architectural Antiques Ltd
Design, manufacture and painting/ anodising - Seco Aluminium Ltd
Engineering castings - Ballentine, Bo'Ness Iron Co Ltd
Fast action doors, rubber & pvc crash doors - Dudley Factory Doors Limited
GRP, Aluminium, mild and stainless steel - Redman Fisher Engineering
Guards for corner protection on walls and columns - Huntley & Sparks
High pressure & gravity die casting - Ashworth Diecasting
Hot dip galvanising - Corbett & Co (Galvanising) W Ltd
Iron and aluminium sand castings - Cradley Castings Ltd
Marine Craft - Thanet Ware Ltd
Metal fabrications - Barton B C & Son Ltd
Metal fabrications - Squires Metal Fabrications Ltd
Metal finishers - Rigized Metals Ltd
Metal framing - Unistrut Limited
Metal Roofing and cladding - Ash & Lacy Building Products Ltd
Metal Spraying - Thanet Ware Ltd
Metalwork - Glymwood Ltd
Metalwork - WA Skinner & Co Limited
On site welding - C & W Fencing Ltd
Precision sheet metal workers - Builder's Iron & Zincwork Ltd
Purpose made metalwork - Carlton Benbow
Purpose made metalwork - Claydon Architectural Metalwork Ltd
Purpose made metalwork - Grant & Livingston Ltd
Purpose made metalwork - Hickton & Co Ltd, John
Purpose made metalwork - Hipkiss, H, & Co Ltd
Purpose made metalwork - Hubbard Architectural Metalwork Ltd
Purpose made metalwork - Leander Architectural
Seamless cold drawn copper tube - Lawton Tube Co Ltd, The
Special arichitectural metalwork - Bassett & Findley Ltd
Stainless steel and sheet metal - Kensal CMS
Stainless Steel Fabrications - Component Developments
Stainless steel wire mesh - Multi Mesh Ltd
Stainless steel, angles, beams, tees, and channels - James Fairley Athena
Steel canopies - Gabriel & Co Ltd
Steel fabrication - Lancashire Steel Fabrication Co Ltd
Steel fabrications - Modern Engineering (Bristol) Ltd
Steel lintels - Jones of Oswestry Ltd
Steel tube and hollow section - Hub Tubes
Structures - Kee Klamps Ltd
Welding engineers - Woodbourn Engineering Ltd
Wirecloth, screens, filters, vent mesh and welded mesh - United Wire Ltd

Woven wirecloth, welded wiremesh, perforated metal, quarry screening, wire products - Potter & Soar Ltd

Z12 Preservative/ Fire retardant treatments for timber

Fire retardant treatments for timber - Hamron Group
Preservative and fire retardant treatments for timber - Dover Trussed Roof Co Ltd

Z20 Fixings/ Adhesives

Adhesive - Loctite UK Ltd
Adhesive - Tremco Ltd
Adhesive products - Henkel Home Improvement & Adhesive Products
Adhesive Tapes - Advance Tapes International Ltd
Adhesives - Atkinson & Kirby Ltd
Adhesives - Brissco Equipment Ltd
Adhesives - CH Agencies
Adhesives - Dunlop Adhesives
Adhesives - Evode Ltd
Adhesives - Fortafix Ltd
Adhesives - Henkel Industrial Adhesives (AIV)
Adhesives - Industrial Adhesives Ltd
Adhesives - Kenyon Industrial Paints & Adhesives
Adhesives - Mapei UK Ltd
Adhesives - Marley Floors Ltd
Adhesives - Norcros Adhesives Ltd
Adhesives - Viking Laminates Ltd
Adhesives and fillers - Polycell Products Ltd
Adhesives and grouts - Shackerley (Holdings) Group Ltd incorporating Designer ceramics
Adhesives and pastes - Clam-Brummer Ltd
Adhesives for wall covering and wood - Ward Bekker Sales Ltd
Adhesives, wood filler and glue - Humbrol Ltd

Z21 Mortars

Admixtures - FEB Ltd
Ahesives - Combustion Linings Ltd
Bevel bonding UV adhesive - Alfas Industries Ltd
Blind fastening systems - Textron Fastening Systems Ltd
Blind fastening systems - Tucker Fasteners Ltd
Bracketry - Vantrunk Engineering Ltd
Building adhesives - Vallance
Cable clips - Arthur Fisher (UK) Ltd
Cast-in fixing channel - Halfen Ltd
Cavity Fixings - Ramset Fasteners Ltd
Chemical Anchors - Ramset Fasteners Ltd
Cold formed components - Textron Fastening Systems Ltd
Concrete repair and protection - Webrr & Broutin
Connectors - Simpson Strong-Tie®
Corrosion resistant alloys fasteners - Cerro (Maganese Bronze) Ltd
Cut steel and leak-proof nails, tacks, pins and screws - Reynolds John & Son (Birmingham) Ltd
Drill Bits - Ramset Fasteners Ltd
Drop In Anchors - Ramset Fasteners Ltd
Engineered assemblies - Textron Fastening Systems Ltd
Engineering and construction fasteners - Allman Fasteners Ltd
Epoxy resins - Dupox (Epoxy Resins) Ltd
Expanding Fre Rated Foam - Ramset Fasteners Ltd
Fasteners - UK Fasteners Ltd
Fasteners and fastening systems - Duo-Fast (UK) Ltd
Fasteners fixings and tools - Arrow Supply Company Ltd
Fixing and adhesives - FEB Ltd
Fixings - Corofil Woodall Ltd
Fixings - Exchem Mining and Construction
Fixings - Hipkiss, H, & Co Ltd
Fixings - ITW Buldex
Fixings - ITW Spit
Fixings - Power Plastics Ltd
Fixings - Red Head, Division of ITW Ltd
Fixings - RFA Group Ltd
Fixings and Adhesives - Caswell & Co Ltd
Fixings and adhesives - Feb MBT
Fixings and fastenings - Kem Edwards Ltd
Fixings/ Adhesives - Bondaglass Voss Ltd
Flat roofing fasteners - SFS Stadler Ltd
Flooring Adhesives - Ball F & Co Ltd
Flooring adhesives - Lastite Building Products Ltd
Flooring adhesives, smoothing compounds - Laybond Products Ltd
Frame fixings - Arthur Fisher (UK) Ltd
General fixings - Arthur Fisher (UK) Ltd
Heavy Duty Anchors - Ramset Fasteners Ltd
Heavy Duty fixings - Arthur Fisher (UK) Ltd
Hot melt adhesives - Kenyon Industrial Paints & Adhesives
Industrial resins - Carborundum Abrasives G.B Ltd
Large perforated headed, steel and stainless steel anchorages - Bighead Bonding Fasteners Ltd
Lightweight fixings - Arthur Fisher (UK) Ltd
Manilla rope - Marine & Commercial Rigging Systems
Metal repair compounds - MMP International
Metal Roofing and cladding - Ash & Lacy Building Products Ltd
Plastic components - Textron Fastening Systems Ltd

VISIT THE LAXTON'S WEBSITE TODAY!

www.laxtonsprices.co.uk

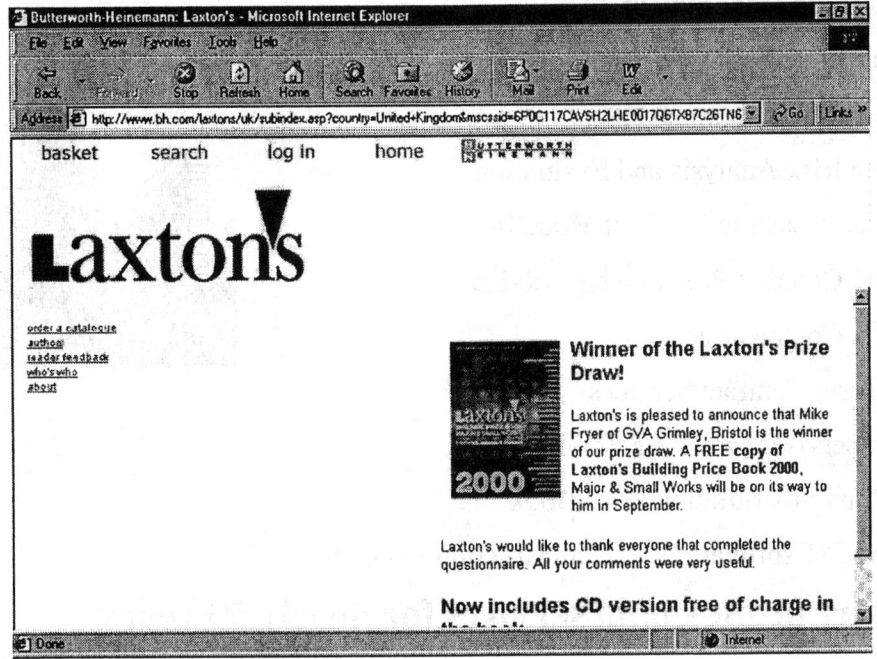

- FIND OUT MORE ABOUT BUTTERWORTH-HEINEMANN'S RANGE OF TITLES

- ORDER ON-LINE

- BENEFIT FROM SPECIAL OFFERS

- REGULAR EMAIL UPDATES

BOOKMARK THIS SITE TODAY!

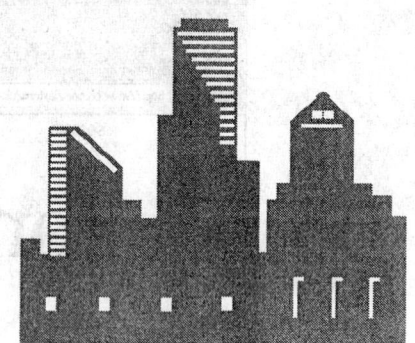

Other available titles

Laxton's General Specification Volumes 1 & 2
(0 7506 3352 0)

Laxton's General Specification – electronic version
(0 7506 3693 9)

Laxton's Guide to Term Maintenance Contracts
(0 7506 2977 0)

Laxton's Measurement Rules for Contractors Quantities
(0 7506 2977 0)

Laxton's Trades Price Book: Small Works Repairs and Maintenance
(0 7506 2978 9)

Laxton's Guide to Budget Estimating
(0 7506 2967 3)

VISIT THE LAXTON'S WEBSITE TODAY!

www.laxtonsprices.co.uk

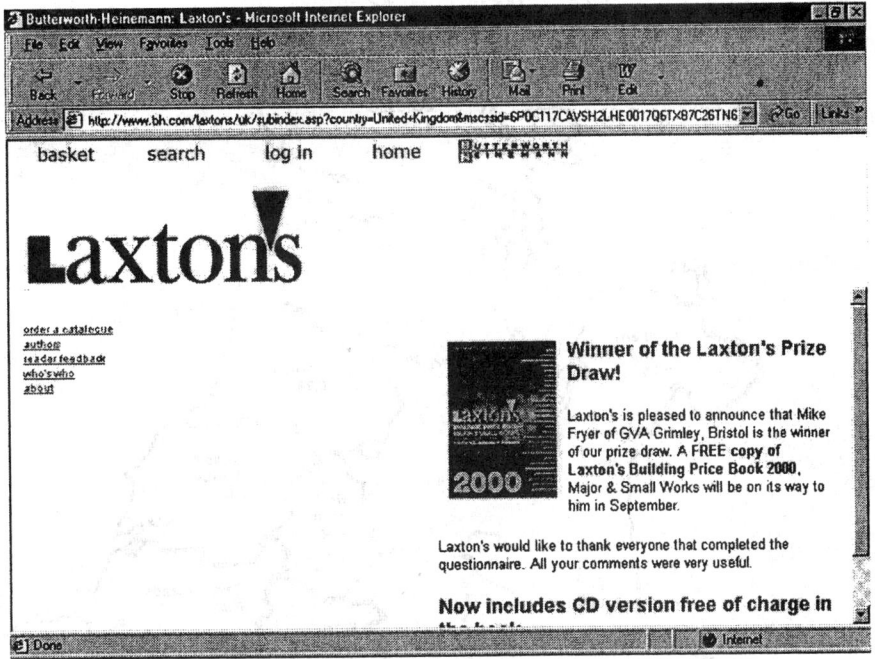

- FIND OUT MORE ABOUT BUTTERWORTH-HEINEMANN'S RANGE OF TITLES

- ORDER ON-LINE

- BENEFIT FROM SPECIAL OFFERS

- REGULAR EMAIL UPDATES

BOOKMARK THIS SITE TODAY!

Regional Factors

Laxton's Building Price Book is based upon national average prices (=1.00). Indicative levels of tender pricing in regions as at the second quarter 1999 are given on this map and the factors shown may be applied to adjust overall pricing.

MAJOR WORKS – TENDER VALUES	SMALL WORKS – TENDER VALUES
The measured rates are based on contracts valued in the range of £250,000 to £1,000.00. As a guide to pricing works of larger value and for cost planning or budgetary purposes, the following adjustments may be applied to overall contract values. Contract Value £1,000,000 to £2,000,000.........deduct 2.5% £2,000,000 to £3,000,000.........deduct 5.0% £3,000,000 to £5,000,000.........deduct 7.5%	The measured rates are based on contracts valued in the range of £25,000 to £75,000. As a guide to pricing works of smaller or larger value and for cost planning or budgetary purposes, the following adjustments may be applied to overall contract values. Contract Value £5,000 to £15,000..............................add 20.0% £15,000 to £25,000............................add 10.0% £25,000 to £75,000.......................rate as shown £75,000 to £100,000.......................deduct 5.0% £100,000 to £150,000.......................deduct7.5% £150,000 to £250,000.......................deduct10.%

Estimating for Builders and Quantity Surveyors
R D Buchan, F W Fleming, J R Kelly

Written for students taking courses in building and surveying at HNC/D and BSc level, this textbook explains the calculation of rates for the items in bills of quantities, based on SMM7. For convenience of use in practice, the arrangement is by individual topic or trade with appropriate reference to subsections of SMM7. Care has been taken, in both discussion and examples; to emphasize the different approaches used when pricing the various types of work. The worked examples reflect both traditional and up-to-date technology.

For all the trades included, a core of the most useful examples is provided, with particular attention paid to those areas that require rather more detailed explanation than has been provided in some more summary texts.

A chapter on computerized estimating is complemented by the Appendix which shows how a spreadsheet program can be developed for the calculation of the cost of mortars etc. Chapters on tendering strategy are included to set estimating in its professional context.

£17.99, Paperback, 1991, 300pp, isbn 0 7506 0041 1

Estimating for Builders and Quantity Surveyors
R D Buchan, F W Fleming, J R Kelly

Written for students taking courses in building and surveying at HNC/D and BSc level, this textbook explains the calculation of rates for the items in bills of quantities, based on SMM7. For convenience of use in practice, the arrangement is by individual topic or trade with appropriate reference to subsections of SMM7. Care has been taken, in both discussion and examples; to emphasize the different approaches used when pricing the various types of work. The worked examples reflect both traditional and up-to-date technology.

For all the trades included, a core of the most useful examples is provided, with particular attention paid to those areas that require rather more detailed explanation than has been provided in some more summary texts.

A chapter on computerized estimating is complemented by the Appendix which shows how a spreadsheet program can be developed for the calculation of the cost of mortars etc. Chapters on tendering strategy are included to set estimating in its professional context.

£17.99, Paperback, 1991, 300pp, isbn 0 7506 0041 1

Masterbill's ESTIMATOR

Incorporating **Laxton's** small and major works electronic price book.

ESTIMATOR With Laxtons provides an easy to use tool enabling you to quickly search for any of the Laxtons priced items and identify their price build-ups.

INSTALLATION

To install the program, simply load the CD and follow the auto set-up instructions on your screen.

If the CD does not auto-run, view the 'readme.txt' file in the route directory.

System requirements – IBM compatible PC, Pentium 166 or above, Win95 or above

ESTIMATOR Features	Trial version	Full version
• Extensive Search facilities to find the Laxtons rates in seconds	✓	✓
• Select Bill items from Laxtons Small or Major Works library	✓	✓
• Create new Bill items of your own	✓	✓
• Import BQs using CITE or .txt	✓	✓
• Measure your own quantities using our 'on-screen' dimsheet	✓	✓
• Create and manage your own resource library	✓	✓
• Produce a priced, unpriced or resourced Bill of Quantities	x	✓
• Print a priced, unpriced or resourced Bill of Quantities	x	✓
• Produce a Buying list of Materials required	x	✓
• Print a Buying list of Materials required	x	✓
• Analyse the Labour and Plant requirements	x	✓
• Adjust for Overheads, Profit and Waste	✓	✓
• Print presentable Reports and Estimates	x	✓
• Maintain your own pricing library	✓	✓
• On screen step-by-step help assistant	✓	✓
• Automatic Bill sorting	✓	✓

This trial version can also be enable to provide a full estimating tool in order to produce fully priced and resourced BQs, created from standard library items, your own entries or electronically imported BQs.

For just **£295** (ex VAT) you can enable **ESTIMATOR** to the fully working program.

Simply contact Masterbill on 01727 855563 with the Unique code (when the program is installed) and your credit card details and you will be supplied with an enabling Pass Code.